PETERSON'S

GRADUATE PROGRAMS IN ENGINEERING & APPLIED SCIENCES

2000

THIRTY-FOURTH EDITION

BOOK 5

Peterson's
Thomson Learning™

Australia • Canada • Denmark • Japan • Mexico • New Zealand • Philippines
Puerto Rico • Singapore • Spain • United Kingdom • United States

About Peterson's

Peterson's is the country's largest educational information/communications company, providing the academic, consumer, and professional communities with books, software, and online services in support of lifelong education access and career choice. Well-known references include Peterson's annual guides to private schools, summer programs, colleges and universities, graduate and professional programs, financial aid, international study, adult learning, and career guidance. Peterson's Web site at petersons.com is the only comprehensive—and most heavily traveled—education resource on the Internet. The site carries all of Peterson's fully searchable major databases and includes financial aid sources, test-prep help, job postings, direct inquiry and application features, and specially created Virtual Campuses for every accredited academic institution and summer program in the U.S. and Canada that offers in-depth narratives, announcements, and multimedia features.

The colleges and universities represented in this book recognize that federal laws, where applicable, require compliance with Title IX (Education Amendments of 1972), Title VII (Civil Rights Act of 1964), and Section 504 of the Rehabilitation Act of 1973 as amended, prohibiting discrimination on the basis of sex, race, color, handicap, or national or ethnic origin in their educational programs and activities, including admissions and employment.

Editorial inquiries concerning this book should be addressed to:
Editor, Peterson's, P.O. Box 2123, Princeton, New Jersey 08543-2123

ISSN 1097-1068
ISBN 0-7689-0268-1

Composition and design by Peterson's

Printed in the United States of America

10 9 8 7 6 5 4 3 2 1

Contents

Introduction

How to Use These Guides

OVERVIEW

The six volumes of Peterson's Annual Guides to Graduate Study, the only annually updated reference work of its kind, provide wide-ranging information on the graduate and professional programs offered by accredited colleges and universities in the United States and U.S. territories and by those institutions in Canada, Mexico, Europe, and Africa that are accredited by U.S. accrediting bodies. More than 37,000 individual academic and professional programs at more than 1,700 institutions are listed. Peterson's Annual Guides to Graduate Study have been used for more than thirty years by prospective graduate and professional students, placement counselors, faculty advisers, and all others interested in postbaccalaureate education.

- Book 1, *Graduate & Professional Programs: An Overview*, contains information on institutions as a whole, while Books 2 through 6 are devoted to specific academic and professional fields.
- Book 2—*Graduate Programs in the Humanities, Arts & Social Sciences*
- Book 3—*Graduate Programs in the Biological Sciences*
- Book 4—*Graduate Programs in the Physical Sciences, Mathematics, Agricultural Sciences, the Environment & Natural Resources*
- Book 5—*Graduate Programs in Engineering & Applied Sciences*
- Book 6—*Graduate Programs in Business, Education, Health, Information Studies, Law & Social Work*

The books may be used individually or as a set. For example, if you have chosen a field of study but do not know what institution you want to attend or if you have a college or university in mind but have not chosen an academic field of study, the best place to begin is Book 1.

Book 1 presents several directories to help you identify programs of study that might interest you; you can then research those programs further in Books 2 through 6. The Directory of Graduate and Professional Programs by Field lists the 397 fields for which there are program directories in Books 2 through 6 and gives the names of those institutions that offer graduate degree programs in each.

For geographical or financial reasons, you may be interested in attending a particular institution and will want to know what it has to offer. You should turn to the Directory of Institutions and Their Offerings, which lists the degree programs available at each institution, again, in the 397 academic and professional fields for which Books 2 through 6 have program directories. As in the Graduate and Professional Programs by Field directory, the level of degrees offered is also indicated.

Finally, the Directory of Combined-Degree Programs lists the areas in which two graduate degrees may be earned concurrently and the schools that offer them.

CLASSIFICATION OF PROGRAMS

After you identify the particular programs and institutions that interest you, use both Book 1 and the specialized volumes to obtain detailed information—Book 1 for information on the institutions overall and Books 2 through 6 for details about the individual graduate units and their degree programs.

Books 2 through 6 are divided into sections that contain one or more directories devoted to programs in a particular field. If you do not find a directory devoted to your field of interest in a specific book, consult the Index of Directories and Subject Areas in Books 2–6; this index appears at the end of each book. After you have identified the correct book, consult the Index of Directories and Subject Areas in This Book, which shows (as does the more general directory) what directories cover subjects not specifically named in a directory or section title. This index in Book 2, for example, will tell you that if you are interested in sculpture, you should see the directory entitled Art/Fine Arts. The Art/Fine Arts entry will direct you to the proper page.

Books 2 through 6 have a number of general directories. These directories have entries for the largest unit at an institution granting graduate degrees in that field. For example, the general Engineering and Applied Sciences directory in Book 5 consists of profiles for colleges, schools, and departments of engineering and applied sciences.

General directories are followed by other directories, or sections, in Books 2, 3, 5, and 6 that give more detailed information about programs in particular areas of the general field that has been covered. The general Engineering and Applied Sciences directory, in the example above, is followed by nineteen sections in specific areas of engineering, such as Chemical Engineering, Industrial Engineering, and Mechanical Engineering.

Because of the broad nature of many fields, any system of organization is bound to involve a certain amount of overlap. Environmental studies, for example, is a field whose various aspects are studied in several types of departments and schools. Readers interested in such studies will find information on relevant programs in Book 3 under Ecology and Environmental Biology; in Book 4 under Environmental Policy and Resource Management and Natural Resources; in Book 5 under Energy Management and Policy and Environmental Engineering; and in Book 6 under Environmental and Occupational Health. To help you find all of the programs of interest to you, the introduction to each section of Books 2 through 6 includes, if applicable, a paragraph suggesting other sections and directories with information on related areas of study to consult.

In addition, this book contains more than 385 listings of academic centers and institutes, including information about the graduate students served, affiliated faculty members, and research budgets. This information can be found in the Research and Training Opportunities section in this volume.

SCHOOL AND PROGRAM INFORMATION

In all of the books, information is presented in three forms: profiles—capsule summaries of basic information—and the short announcements and in-depth descriptions written by graduate school and program administrators. The format of the profiles is constant, making it easy to compare one institution with another and one program with another. A description of the information in the profiles in Books 2 through 6 may be found below; the Book 1 profile description is found immediately preceding the profiles in Book 1. A number of graduate school and program administrators have attached brief announcements to the end of their profile listings. In them you will find information that an institution or program wants to emphasize. The in-depth descriptions are by their very nature more expansive and flexible than the profiles, and the administrators who have written them may emphasize different aspects of their programs. All of these in-depth descriptions are organized in the same way, and in each one you will find information on the same basic topics, such as programs of study, research facilities, tuition and fees, financial aid, and application procedures. If an institution or program has submitted an in-depth description, a boldface cross-reference appears below its profile. As with the profile announcements, all of the in-depth descriptions in the guides have been submitted by choice of administrators; the absence of an announcement or in-depth description does not reflect any type of editorial judgment on the part of Peterson's.

Interdisciplinary Programs

In addition to the regular directories that present profiles of programs in each field of study, many sections in Books 2 through 6 contain special notices under the heading Cross-Discipline Announcements. Appearing at the end of the profiles in many sections, these Cross-Discipline Announcements inform you about programs that you may find of interest described in a different section. A biochemistry department, for example, may place a notice under Cross-Discipline Announcements in the Chemistry section (Book 4) to alert chemistry students to their current description in the Biochemistry section of Book 3. Cross-discipline announcements, also written by administrators to highlight their programs, will be helpful to you not only in finding out about programs in fields related to your own but also in locating departments that are actively recruiting students with a specific undergraduate major.

Profiles of Graduate Units (Books 2–6)

The profiles found in the 397 directories in Books 2 through 6 provide basic data about the graduate units in capsule form for quick reference. To make these directories as useful as possible, profiles are generally listed for an institution's smallest academic unit within a subject area. In other words, if an institution has a College of Liberal Arts that administers many related programs, the profile for the individual program (e.g., Program in History), not the entire College, appears in the directory.

There are some programs that do not fit into any current directory and are not given individual profiles. The directory structure is reviewed annually in order to keep this number to a minimum and to accommodate major trends in graduate education.

The following outline describes the profile information found in the guides and explains how best to use that information. Any item that does not apply to or was not provided by a graduate unit is omitted from its listing.

Identifying Information. The institution's name, in boldface type, is followed by a complete listing of the administrative structure for that field of study. (For example, **University of Akron,** Buchtel College of Arts and Sciences, Department of Mathematical Sciences and Statistics, Program in Mathematics.) The last unit listed is the one to which all information in the profile pertains. The institution's address follows.

Offerings. Each field of study offered by the unit is listed with all postbaccalaureate degrees awarded. Degrees that are not preceded by a specific concentration are awarded in the general field listed in the unit name. Frequently, fields of study are broken down into subspecializations, and those appear following the degrees awarded; for example, "Offerings in secondary education (M Ed), including English education, mathematics education, science education." Students enrolled in the M.Ed. program would be able to specialize in any of the three fields mentioned.

Professional Accreditation. Profiles indicate whether a program is professionally accredited. Specific information on the accreditation status of a unit is obtained directly from the accreditation agency's most current listing at the time of publication. However, because it is possible for a program to receive or lose professional accreditation at any time, students entering fields in which accreditation is important to a career should verify the status of programs by contacting either the chairperson or the appropriate accrediting association (see Accreditation and Accrediting Agencies in each book).

Jointly Offered Degrees. Explanatory statements concerning programs that are offered in cooperation with other institutions are included in the list of degrees offered. This occurs most commonly on a regional basis (for example, two state universities offering a cooperative Ph.D. in special education) or where the specialized nature of the institutions encourages joint efforts (a J.D./M.B.A. offered by a law school at an institution with no formal business programs and an institution with a business school but lacking a law school). Only programs that are truly cooperative are listed; those involving only limited course work at another institution are not. Interested students should contact the heads of such units for further information.

Part-Time and Evening/Weekend Programs. When information regarding the availability of part-time or evening/weekend study appears in the profile, it means that students are able to earn a degree exclusively through such study.

Postbaccalaureate Distance Learning Degrees. A postbaccalaureate distance learning degree program signifies that course requirements can be fulfilled with minimal or no on-campus study. If these programs require minimal on-campus study or no on-campus study it may be indicated here.

Faculty. Figures on the number of faculty members actively involved with graduate students through teaching or research are separated into full- and part-time as well as men and women whenever the information has been supplied.

Matriculated Students. Figures for the number of students enrolled in graduate and professional programs pertain to the semester of highest enrollment from the 1998–99 academic year. These figures are broken down into full- and part-time and men and women whenever the data have been supplied. Information on the number of matriculated students enrolled in the unit who are members of a minority group or are international students appears here. The average age of the matriculated students is followed by the number of applicants and the percentage accepted for fall 1998.

Degrees Awarded. In addition to the number of degrees awarded in the 1998 calendar year, this section contains information on the percentages of students who have gone on to continue full-time study, entered university research or teaching, or chosen other work related to their field and information on the average amount of time required to earn the degree for full-time and part-time students. Many doctoral programs offer a terminal master's degree if students leave the program after completing only part of the requirements for a doctoral degree; that is indicated here. All degrees are classified into one of four types: master's, doctoral, first professional, and other advanced degrees. A unit may award one or several degrees at a given level; however, the data are only collected by type and may therefore represent several different degree programs.

Degree Requirements. The information in this section is also broken down by type of degree, and all information for a degree level pertains to all degrees of that type unless otherwise specified. Degree requirements are collected in a simplified form to provide some very basic information on the nature of the program and on foreign language, computer language, and thesis or dissertation requirements. Many units also provide a short list of additional requirements, such as fieldwork or an internship. No information is listed on the number of courses or credits required for completion or whether a minimum or maximum number of years or semesters is needed. For complete information on graduation requirements, contact the graduate school or program directly.

Entrance Requirements. Entrance requirements are broken down into the four degree levels of master's, doctoral, first professional, and other advanced degrees. Within each level, information may be provided in two basic categories, entrance exams and other requirements. The entrance exams use the standard acronyms used by the testing agencies, unless they are not well known. Additional information on each of the common tests is provided in the section Tests Required of Applicants. The usual format in this part of the profile is a test name followed by a minimum score. When a minimum or average combined score is given for the GRE General Test, it is for the verbal and quantitative sections combined (without the analytical section) unless otherwise specified. More information on the scale and other aspects of the test may be obtained directly from the testing agency. Other entrance requirements are quite varied, but they often contain an undergraduate or graduate grade point average (GPA). Unless otherwise stated, the GPA is calculated on a 4.0 scale and is listed as a minimum required for admission. The standard application deadlines, any nonrefundable application fee, and whether electronic applications are accepted may be listed here. Note that the deadline should be used for reference only; these dates are subject to change, and students interested in applying should contact the graduate unit directly about application procedures and deadlines.

Expenses. The typical cost of study for the 1999–2000 academic year is given in two basic categories, tuition and fees. It is not possible to represent the complete tuition and fees schedule for each graduate unit, so a simplified version of the cost of studying in that unit is provided. In general, the costs of both full- and part-time study are listed if the unit allows for both types of programs and lists separate costs. For public institutions, the tuition and fees are listed for both state residents and nonresidents. Cost of study may be quite complex at a graduate institution. There are often sliding scales for part-time study, a different cost for first-year students, and other variables that make it impossible to completely cover the cost of study for each graduate program. To provide the most usable information, figures are given for full-time study for a full year where available and for part-time study in terms of a per-unit rate (per credit, per semester hour, etc.). Occasionally, variances may be noted in tuition and fees for reasons such as the type of program, whether courses are taken during the day or evening, whether courses are at the master's or doctoral level, or other institution-specific reasons. Expenses are usually subject to change; for exact costs at any given time, contact your chosen schools and programs directly. Keep in mind that the tuition of Canadian institutions is usually given in Canadian dollars.

Financial Aid. This section contains data on the number of awards that are administered by the institution and were given to graduate students during the 1998–99 academic year. The first figure given represents the total number of students receiving financial aid enrolled in that unit. If the unit has provided information on graduate appointments, these are broken down into three major categories: *fellowships* give money to graduate students to cover the cost of study and living expenses and are not based on a work obligation or research commitment, *research assistantships* provide stipends to graduate students for assistance in a formal research project with a faculty member, and *teaching assistantships* provide stipends to graduate students for teaching or for assisting faculty members in teaching undergraduate classes. Within each category, figures are given for the total number of awards, the average yearly amount per award, and whether full or partial tuition reimbursements are awarded.

In addition to graduate appointments, the availability of several other financial aid sources is covered in this section. *Tuition waivers* are routinely part of a graduate appointment, but units sometimes waive part or all of a student's tuition even if a graduate appointment is not available. *Federal Work-Study* is made available to students who demonstrate need and meet the federal guidelines; this form of aid normally includes 10 or more hours of work per week in an office of the institution. *Institutionally sponsored loans* are low-interest loans available to graduate students to cover both educational and living expenses. *Career-related internships* or *fieldwork* offer money to students who are participating in a formal off-campus research project or practicum. Grants, scholarships, traineeships, unspecified assistantships, and other awards may also be noted. The availability of financial aid to part-time students is also indicated here. Some programs list the financial aid application deadline and the forms that need to be completed for students to be eligible for financial aid. There are two forms: FAFSA, the Free Application for Federal Student Aid, which is required for federal aid; and the CSS Financial Aid PROFILE, if required.

Faculty Research. Each unit has the opportunity to list several keyword phrases describing the current research involving faculty members and graduate students. Space limitations prevent the unit from listing complete information on all research programs. The total expenditure for funded research from the previous academic year may also be included.

Unit Head and Application Contact. The head of the graduate program for each unit is listed with the academic title and telephone, fax, and e-mail numbers if available. In addition to the unit head, many graduate programs list separate contacts for application and admis-

sion information, which follows the listing for the unit head. If no unit head or application contact is given, you should contact the overall institution for information on graduate admissions.

Data Collection and Editorial Procedures

DIRECTORIES AND PROFILES

The information published in the directories and profiles of all the books is collected through Peterson's Annual Survey of Graduate Institutions. The survey is sent each spring to more than 1,700 institutions offering postbaccalaureate degree programs, including accredited institutions in the United States and U.S. territories and those institutions in Canada, Mexico, Europe, and Africa that are accredited by U.S. accrediting bodies. Deans and other administrators complete these surveys, providing information on programs in the 397 academic and professional fields covered in the guides as well as overall institutional information. Peterson's staff then goes over each returned survey carefully and verifies or revises responses after further research and discussion with administrators at the institutions. Extensive files on past responses are kept from year to year.

While every effort has been made to ensure the accuracy and completeness of the data, information is sometimes unavailable or changes occur after publication deadlines. All usable information received in time for publication has been included. The omission of any particular item from a directory or profile signifies either that the item is not applicable to the institution or program or that information was not available. Profiles of programs scheduled to begin during the 1999–2000 academic year cannot, obviously, include statistics on enrollment or, in many cases, the number of faculty members. If no usable data were submitted by an institution, its name, address, and program name where appropriate nonetheless appear in order to indicate the existence of graduate work.

ANNOUNCEMENTS AND IN-DEPTH DESCRIPTIONS

The announcements and in-depth descriptions are supplementary insertions submitted by deans, chairs, and other administrators who wish to make an additional, more individualized statement to readers. Those who have chosen to write these insertions are responsible for the accuracy of the content, but Peterson's editors have reserved the right to delete irrelevant material or questionable self-appraisals and to edit for style. Statements regarding a university's objectives and accomplishments are a reflection of its own beliefs and are not the opinions of the editors. Since inclusion of announcements and descriptions is by choice, their presence or absence in the guides should not be taken as an indication of status, quality, or approval.

The Graduate Adviser

This section consists of two essays and information about admissions tests and accreditation. The first essay, Applying to Graduate and Professional Schools, was written by Jane E. Levy of Cornell University and Elinor R. Workman. It covers topics of interest to students considering post-baccalaureate work, including types of degrees, choosing a specialization, researching programs, applying, and some issues for returning, part-time, and international students. The second essay is Financing Your Graduate and Professional Education, by Patricia McWade of Georgetown University. It covers determining your need for financial aid, the types of aid available, and how and when to apply for aid as it relates to degree programs in engineering and applied sciences. Both essays appear in each of the six Graduate Guides. Tests Required of Applicants lists all standardized admissions tests relevant to programs in engineering and applied sciences. Accreditation and Accrediting Agencies gives information on accreditation and its purpose. Institutional accrediting agencies are listed first; specialized accrediting agencies relevant to engineering and applied sciences are listed after that. This section is filled with crucial information for all students; it is addressed to the reader who is still in college but also contains information specifically for returning, part-time, and international students.

Applying to Graduate and Professional Schools

The decision to attend graduate school and the choice of an institution and degree program require serious consideration. The time, money, and energy you will expend doing graduate work are significant, and you will want to analyze your options carefully. Before you begin filing applications, you should evaluate your interests and goals, know what programs are available, and be clear about your reasons for pursuing a particular degree.

There are two excellent reasons for attending graduate school, and if your decision is based on one of these, you probably have made the right choice. There are careers such as medicine, law, and college and university teaching that require specialized training and, therefore, necessitate advanced education. Another motivation is to specialize in a subject that you have decided is of great importance, either for career goals or for personal satisfaction.

Degrees

Traditionally, graduate education has involved acquiring and communicating knowledge gained through original research in a particular academic field. The highest earned academic degree, which requires the pursuit of original research, is the Doctor of Philosophy (Ph.D.). In contrast, professional training stresses the practical application of knowledge and skills; this is true, for example, in the fields of business, law, and medicine. At the doctoral level, degrees in these areas include the Doctor of Business Administration (D.B.A.), Juris Doctor (J.D.), and the Doctor of Medicine (M.D.).

Master's degrees are offered in most fields and may also be academic or professional in orientation. In many fields, the master's degree may be the only professional degree needed for employment. This is the case, for example, in fine arts (M.F.A.), library science (M.L.S.), and social work (M.S.W.). (For a list of the graduate and professional degrees currently being offered in the United States and Canada, readers may refer to the appendix of degree abbreviations.)

Some people decide to earn a master's degree at one institution and then select a different university or a somewhat different program of study for doctoral work. This can be a way of acquiring a broad background: you can choose a master's program with one emphasis or orientation and a doctoral program with another. The total period of graduate study may be somewhat lengthened by proceeding this way, but probably not by much.

In recent years, the distinctions between traditional academic programs and professional programs have become blurred. The course of graduate education has changed direction in the last thirty years, and many programs have redefined their shape and focus. There are centers and institutes for research, many graduate programs are now interdepartmental and interdisciplinary, off-campus graduate programs have multiplied, and part-time graduate programs have increased. Colleges and universities have also established combined-degree programs, in many cases in order to enable students to combine academic and professional studies. As a result of such changes, you now have considerable freedom in determining the program best suited to your current needs as well as your long-term goals.

Choosing a Specialization and Researching Programs

There are several sources of information you should make use of in choosing a specialization and a program. A good way to begin is to consult the appropriate directories in these guides, which will tell you what programs exist in the field or fields you are interested in and, for each one, will give you information on degrees, research facilities, the faculty, financial aid resources, tuition and other costs, application requirements, and so on.

Talk with your college adviser and professors about your areas of interest and ask for their advice about the best programs to research. Besides being very well informed themselves, these faculty members may have colleagues at institutions you are investigating, and they can give you inside information about individual programs and the kind of background they seek in candidates for admission.

The valuable perspective of educators should not be overlooked. If the faculty members you know through your courses are not involved in your field of interest, do not hesitate to contact other appropriate professors at your institution or neighboring institutions to ask for advice on programs that might suit your goals. In addition, talk to graduate students studying in your field of interest; their advice can be valuable also.

Your decision about a field of study may be determined by your research interests or, if you choose to enter a professional school, by the appeal of a particular career. In either case, as you attempt to limit the number of institutions you will apply to, you will want to familiarize yourself with publications describing current research in your discipline. Find related professional journals and note who is publishing in the areas of specialization that interest you, as well as where they are teaching. Take note of the institutions represented on the publications' editorial boards (they are usually listed on the inside cover); such representation usually reflects strength in the discipline.

Being aware of who the top people are and where they are will pay off in a number of ways. A graduate department's reputation rests heavily on the reputation of its faculty, and in some disciplines it is more important to study under someone of note than it is to study at a college or university with a prestigious name. In addition, in certain fields graduate funds are often tied to a particular research project and, as a result, to the faculty member directing that project. Finally, most Ph.D. candidates (and nonprofessional master's degree candidates) must pick an adviser and one or more other faculty members who form a committee that directs and approves their work. Many times this choice must be made during the first semester, so it is important to learn as much as you can about faculty members before you begin your studies. As you research the faculties of various departments, keep in mind the following questions: What is their academic training? What are their research activities? What kind of concern do they have for teaching and student development?

There are other important factors to consider in judging the educational quality of a program. First, what kind of students enroll in the program? What are their academic abilities, achievements, skills, geographic representation, and level of professional success upon completion of the program? Second, what are the program's resources? What kind of financial support does it have? How complete is the library? What laboratory equipment and computer facilities are available? And third, what does the program have to offer in terms of both curriculum and services? What are its purposes, its course offerings, and its job placement and student advisement services? What is the student-faculty ratio, and what kind of interaction is there between students and professors? What internships, assistantships, and other experiential education opportunities are available?

When evaluating a particular institution's reputation in a given field, you may also want to look at published graduate program ratings. There is no single rating that is universally accepted, so you would be well advised to read several and not place too much importance on any one. Most consist of what are known as "peer ratings"; that is, they are the results of polls of respected scholars who are asked to rate graduate departments in their field of expertise. Many academicians feel that these ratings are too heavily based upon traditional concepts of what constitutes quality—such as the publications of the faculty—and that they perpetuate the notion of a research-oriented department as the only model of excellence in graduate education. Depending on whether your own goals are research-oriented, you may want to attribute more or less importance to this type of rating.

If possible, visit the institutions that interest you and talk with faculty members and currently enrolled students. Be sure, however, to write or call the admissions office a week in advance to give the person in charge a chance to set up appointments for you with faculty members and students.

The Application Process

TIMETABLE

It is important to start gathering information early to be able to complete your applications on time. Most people should start the process a full year and a half before their anticipated date of matriculation. There are, however, some exceptions to this rule. The time frame will be different if you are applying for national scholarships or if your undergraduate institution has an evaluation committee through which you are applying, for example, to a health-care program. In such a situation, you may have to begin the process two years before your date of matriculation in order to take your graduate admission test and arrange for letters of recommendation early enough to meet deadlines.

Application deadlines may range from August (a year prior to matriculation) for early decision programs at medical schools using the American Medical College Application Service (AMCAS) to late spring or summer (when beginning graduate school in the fall) for a few programs with rolling admissions. Most deadlines for entry in the fall are between January and March. You should in all cases plan to meet formal deadlines; beyond this, you should be aware of the fact that many schools with rolling admissions encourage and act upon early applications. Applying early to a school with rolling admissions is usually advantageous, as it shows your enthusiasm for the program and gives admissions committees more time to evaluate the subjective components of your application, rather than just the "numbers." Applicants are not rejected early unless they are clearly below an institution's standards.

The timetable that appears below represents the ideal for most applicants.

Six months prior to applying
- Research areas of interest, institutions, and programs.
- Talk to advisers about application requirements.
- Register and prepare for appropriate graduate admission tests.
- Investigate national scholarships.
- If appropriate, obtain letters of recommendation.

Three months prior to applying
- Take required graduate admission tests.
- Write for application materials or request them online.
- Write your application essay.
- Check on application deadlines and rolling admissions policies.
- For medical, dental, osteopathy, podiatry, or law school, you may need to register for the national application or data assembly service most programs use.

Fall, a year before matriculating
- Obtain letters of recommendation.
- Take graduate admission tests if you haven't already.
- Send in completed applications.

Winter, before matriculating in the fall
- Complete the Free Application for Federal Student Aid (FAFSA) and Financial Aid PROFILE, if required.

Spring, before matriculating in the fall
- Check with all institutions before their deadlines to make sure your file is complete.
- Visit institutions that accept you.
- Send a deposit to your institution of choice.
- Notify other colleges and universities that accepted you of your decision so that they can admit students on their waiting list.
- Send thank-you notes to people who wrote your recommendation letters, informing them of your success.

You may not be able to adhere to this timetable if your application deadlines are very early, as is the case with medical schools, or if you decide to attend graduate school at the last minute. In any case, keep in mind the various application requirements and be sure to meet all deadlines. If deadlines are impossible to meet, call the institution to see if a late application will be considered.

OBTAINING AAPPLICATION FORMS AND INFORMATION

To obtain the materials you need, send a neatly typed or handwritten postcard requesting an application, a bulletin, and financial aid information to the address provided in this Guide. However, you may want to request an application by writing a formal letter directly to the department chair in which you briefly describe your training, experience, and specialized research interests. If you want to write to a particular faculty member about your background and interests in order to explore the possibility of an assistantship, you should also feel free to do so. However, do not ask a faculty member for an application, as this may cause a significant delay in your receipt of the forms.

NATIONAL APPLICATION SERVICES

In a few professional fields, there are national services that provide assistance with some part of the application process. These services are the Law School Data Assembly Service (LSDAS), American Medical College Application Service (AMCAS), American Association of Colleges of Osteopathic Medicine Application Service (AACOMAS), American Association of Colleges of Podiatric Medicine Application Service (AACPMAS), and American Association of Dental Schools Application Service (AADSAS). Many programs require applicants to use these services because they simplify the application process for both the professional programs' admissions committees and the applicant. The role these services play varies from one field to another. The LSDAS, for example, analyzes your transcript(s) and submits the analysis to the law schools to which you are applying, while the other services provide a more complete application service. More information and applications for these services can be obtained from your undergraduate institution.

MEETING APPLICATION REQUIREMENTS

Requirements vary from one field to another and from one institution to another. Read each program's requirements carefully; the importance of this cannot be overemphasized.

Graduate Admission Tests

Colleges and universities usually require a specific graduate admission test, and departments sometimes have their own requirements as well. Scores are used in evaluating the likelihood of your success in a particular program (based upon the success rate of past students with similar scores). Most programs will not accept scores more than three to five years old. The various tests are described a little later in this book.

Transcripts

Admissions committees require official transcripts of your grades to evaluate your academic preparation for graduate study. Grade point averages are important but are not examined in isolation; the rigor of the courses you have taken, your course load, and the reputation of the undergraduate institution you have attended are also scrutinized. To have your college transcript sent to graduate institutions, contact your college registrar.

Letters of Recommendation

Choosing people to write recommendations can be difficult, and most graduate schools require two or three letters. While recommendations from faculty members are essential for academically oriented programs, professional programs may seriously consider nonacademic recommendations from professionals in the field. Indeed, often these nonacademic recommendations are as respected as those from faculty members.

To begin the process of choosing references, identify likely candidates from among those you know through your classes, extracurricular activities, and jobs. A good reference will meet several of the following criteria: he or she has a high opinion of you, knows you well in more than one area of your life, is familiar with the institutions to which you are applying as well as the kind of study you are pursuing, has taught or worked with a large number of students and can make a favorable comparison of you with your peers, is known by the admissions committee and is regarded as someone whose judgment should be given weight, and has good written communication skills. No one person is likely to satisfy all these criteria, so choose those people who come closest to the ideal.

Once you have decided whom to ask for letters, you may wonder how to approach them. Ask them if they think they know you well enough to write a meaningful letter. Be aware that the later in the semester you ask, the more likely they are to hesitate because of time constraints; ask early in the fall semester of your senior year. Once those you ask to write letters agree in a suitably enthusiastic manner, make an appointment to talk with them. Go to the appointment with recommendation forms in hand, being sure to include addressed, stamped envelopes for their convenience. In addition, give them other supporting materials that will assist them in writing a good, detailed letter on your behalf. Such documents as transcripts, a résumé, a copy of your application essay, and a copy of a research paper can help them write a thorough recommendation.

On the recommendation form, you will be asked to indicate whether you wish to waive or retain the right to see the recommendation. Before you decide, discuss the confidentiality of the letter with each writer. Many faculty members will not write a letter unless it is confidential. This does not necessarily mean that they will write a negative letter but, rather, that they believe it will carry more weight as part of your application if it is confidential. Waiving the right to see a letter does, in fact, usually increase its validity.

If you will not be applying to graduate school as a senior but you plan to pursue further education in the future, open a credentials file if your college or university offers this service. Letters of recommendation can be kept on file for you until you begin the application process. If you are returning to school after working for several years and did not establish a credentials file, it may be difficult to obtain letters of recommendation from professors at your undergraduate institution. In this case, contact the graduate schools you are applying to and ask what their policies are regarding your situation. They may waive the requirement of recommendation letters, allow you to substitute letters from employment supervisors, or suggest you enroll in relevant courses at a nearby institution and obtain letters from professors upon completion of the course work. Program policies vary considerably, so it is best to check with each school.

Application Essays

Writing an essay, or personal statement, is often the most difficult part of the application process. Requirements vary widely in this regard. Some programs request only one or two paragraphs about why you want to pursue graduate study, while others require five or six separate essays in which you are expected to write at length about your motivation for graduate study, your strengths and weaknesses, your greatest achievements, and solutions to hypothetical problems. Business schools are notorious for requiring several time-consuming essays.

An essay or personal statement for an application should be essentially a statement of your ideas and goals. Usually it includes a certain amount of personal history, but, unless an institution specifically requests autobiographical information, you do not have to supply any. Even when the requirement is a "personal statement," the possibilities are almost unlimited. There is no set formula to follow, and, if you do write an autobiographical piece, it does not have to be arranged chronologically. Your aim should be a clear, succinct statement showing that you have a definite sense of what you want to do and enthusiasm for the field of study you have chosen. Your essay should reflect your writing abilities; more important, it should reveal the clarity, the focus, and the depth of your thinking.

Before writing anything, stop and consider what your reader might be looking for; the general directions or other parts of the application may give you an indication of this. Admissions committees may be trying to evaluate a number of things from your statement, including the following things about you:

- Motivation and commitment to a field of study
- Expectations with regard to the program and career opportunities
- Writing ability
- Major areas of interest
- Research or work experience
- Educational background
- Immediate and long-term goals
- Reasons for deciding to pursue graduate education in a particular field and at a particular institution
- Maturity
- Personal uniqueness—what you would add to the diversity of the entering class

There are two main approaches to organizing an essay. You can outline the points you want to cover and then expand on them, or you can put your ideas down on paper as they come to you, going over them, eliminating certain sentences, and moving others around until you achieve a logical sequence. Making an outline will probably lead to a well-organized essay, whereas writing spontaneously may yield a more inspired piece of writing. Use the approach you feel most comfortable with. Whichever approach you use, you will want someone to critique your essay. Your adviser and those who write your letters of recommendation may be very helpful to you in this regard. If they are in the field you plan to pursue, they will be able to tell you what things to stress and what things to keep brief. Do not be surprised, however, if you get differing opinions on the content of your essay. In the end, only you can decide on the best way of presenting yourself.

If there is information in your application that might reflect badly on you, such as poor grades or a low admission test score, it is better not to deal with it in your essay unless you are asked to. Keep your essay positive. You will need to explain anything that could be construed as negative in your application, however, as failure to do so may eliminate you from consideration. You can do this on a separate sheet entitled "Addendum," which you attach to the application, or in a cover letter that you enclose. In either form, your explanation should be short and to the point, avoiding long, tedious excuses. In addition to supplying your own explanation, you may find it appropriate to ask one or more of your recommenders to address the issue in their recommendation letter. Ask them to do this only if they are already familiar with your problem and could talk about it from a positive perspective.

In every case, essays should be word processed or typed. It is usually acceptable to attach pages to your application if the space provided is insufficient. Neatness, spelling, and grammar are important.

Interviews, Portfolios, and Auditions

Some graduate programs will require you to appear for an interview. In certain fields, you will have to submit a portfolio of your work or schedule an audition.

Interviews. Interviews are usually required by medical schools and are often required or suggested by business schools and other programs. An interview can be a very important opportunity for you to persuade an institution's admissions officer or committee that you would be an excellent doctor, dentist, manager, etc. Interviewers will be interested in the way you think and approach problems and will probably concentrate on questions that enable them to assess your thinking skills, rather than questions that call upon your grasp of technical knowledge. Some interviewers will ask controversial questions, such as "What is your viewpoint on abortion?" or give you a hypothetical situation and ask how you would handle it. Bear in mind that the interviewer is more interested in how you think than in what you think. As in your essay, you may be asked to address such topics as your motivation for graduate study, personal philosophy, career goals, related research and work experience, and areas of interest.

You should prepare for a graduate school interview as you would for a job interview. Think about the questions you are likely to be asked and practice verbalizing your answers. Think too about what you want interviewers to know about you so that you can present this information when the opportunity is given. Dress as you would for an employment interview.

Portfolios. Many graduate programs in art, architecture, journalism, environmental design, and other fields involving visual creativity may require a portfolio as part of the application. The function of the portfolio is to show your skills and ability to do further work in a particular field, and it should reflect the scope of your cumulative training and experience. If you are applying to a program in graphic design, you may be required to submit a portfolio showing advertisements, posters, pamphlets, and illustrations you have prepared. In fine arts, applicants must submit a portfolio with pieces related to their proposed major. Individual programs have very specific requirements regarding what your portfolio should contain and how it should be arranged and labeled. Many programs request an interview and ask you to present your portfolio at that time. They may not want you to send the portfolio in advance or leave it with them after the interview, as they are not insured against its loss. If you do send it, you usually do so at your own risk, and you should label all pieces with your name and address.

Auditions. Like a portfolio, the audition is a demonstration of your skills and talent, and it is often required by programs in music, theater, and dance. Although all programs require a reasonable level of proficiency, standards vary according to the field of study. In a nonperformance area like music education, you need only show that you have attained the level of proficiency normally acquired through an undergraduate program in that field. For a performance major, however, the audition is the most important element of the graduate application. Programs set specific requirements as to what material is appropriate, how long the performance should be, whether it should be memorized, and so on. The audition may be live or taped, but a live performance is usually preferred. In the case of performance students, a committee of professional musicians will view the audition and evaluate it according to prescribed standards.

SUBMITTING COMPLETED APPLICATIONS

Graduate schools have established a wide variety of procedures for filing applications, so read each institution's instructions carefully. Some may request that you send all application materials in one package (including letters of recommendation). Others—medical schools, for example—may have a two-step application process. This system requires the applicant to file a preliminary application; if this is reviewed favorably, he or she submits a second set of documents and a second application fee. Pay close attention to each school's instructions.

Graduate schools generally require an application fee. Sometimes this fee may be waived if you meet certain financial criteria. Check with your undergraduate financial aid office and the graduate schools to which you are applying to see if you qualify.

ADMISSION DECISIONS

At most institutions, once the graduate school office has received all of your application materials, your file is sent directly to the academic department. A faculty committee (or the department chairperson) then makes a recommendation to the chief graduate school officer (usually a graduate dean or vice president), who is responsible for the final admission decision. Professional schools at most institutions act independently of the graduate school office; applications are submitted to them directly, and they make their own admission decisions.

Usually a student's grade point average, letters of recommendation, and graduate admission test scores are the primary factors considered by admissions committees. The appropriateness of the undergraduate degree, an interview, and evidence of creative talent may also be taken into account. Normally the student's total record is examined closely, and the weight assigned to specific factors fluctuates from program to program. Few, if any, institutions base their decisions purely on numbers, that is, admission test scores and grade point average. A study by the Graduate Record Examinations Board found that grades and recommendations by known faculty members were considered to be somewhat more important than GRE General Test scores and that GRE Subject Test scores were rated as relatively unimportant (Oltman and Hartnett, 1984). This indicates that some graduate admission test scores may be of less importance than is commonly believed, but this will of course differ from program to program.

Some of the common reasons applicants are rejected for admission to graduate schools are inappropriate undergraduate curriculum; poor grades or lack of academic prerequisites; low admission test scores; weak or ineffective recommendation letters; a poor interview, portfolio, or audition; and lack of extracurricular activities, volunteer experience, or research activities. To give yourself the best chances of being admitted where you apply, try to make a realistic assessment of an institution's admission standards and your own qualifications. Remember, too, that missing deadlines and filing an incomplete application can also be a cause for rejection; be sure that your transcripts and recommendation letters are received on time.

Returning Students

Many graduate programs not only accept the older, returning student but actually prefer these "seasoned" candidates. Programs in business administration, social work, and other professional fields value mature applicants with work experience, for they have found that these students often show a higher level of motivation and commitment and work harder than 21-year-olds. Many programs also seek the diversity older students bring to the student body, as differences in perspective and experience make for interesting—and often intense—class discussions. Nonprofessional programs also view older students favorably if their academic and experiential preparation is recent enough and sufficient for the proposed fields of study.

Many institutions have programs designed to make the transition to academic life easier for the returning student. Such programs include low-cost child-care centers, emotional support programs for both the returning student and his or her spouse, and review courses of various kinds.

Other than making the necessary changes in their life-style, older students report that the most difficult aspect of returning to school is recovering, or developing, appropriate study habits. Initially, older students often feel at a disadvantage compared to students fresh out of an undergraduate program who are accustomed to preparing research papers and taking tests. This feeling can be overcome by taking advantage of noncredit courses in study skills and time management and review courses in math and writing, as well as by taking a tour of the library and becoming thoroughly familiar with it. By the end of the graduate program, most returning students feel that their life experience gave them an edge, because they could use concrete experiences to help them understand academic theory.

If you choose to go back to school, you are not alone. A significant number of adults are currently enrolled in some kind of educational program in order to make their lives or careers more rewarding.

Part-Time Students

As graduate education has changed over the past thirty years, the number of part-time graduate programs has increased. Traditionally, graduate programs were completed by full-time students. Graduate schools instituted residence requirements, demanding that students take a full course load for a certain number of consecutive semesters. It was felt that total immersion in the field of study and extensive interaction with the faculty were necessary to achieve mastery of an academic area.

In most academic Ph.D. programs as well as many health-care fields, this is still the only approach. However, many other programs now admit part-time students or allow a portion of the requirements to be completed on a part-time basis. Professional schools are more likely to allow part-time study because many students work full-time in the field and pursue their degree in order to enhance their career credentials. Other applicants choose part-time study because of financial considerations. By continuing to work full-time while attending school, they take fewer economic risks.

Part-time programs vary considerably in quality and admissions standards. When evaluating a part-time program, use the same criteria you would use in judging the reputation of any graduate program. Some schools use more adjunct faculty members with weaker academic training for their night and weekend courses, and this could lower the quality of the program; however, adjunct lecturers often have excellent experiential knowledge. Admissions standards may be lower for a part-time program than for an equivalent full-time program at the same school, but, again, your fellow students in the part-time program may be practicing in the field and may have much to add to class discussions. Another concern is placement opportunities upon completion of the program. Some schools may not offer placement services to part-time students, and many employers do not value part-time training as highly as a full-time education. However, if a part-time program is the best option for you, do not hesitate to enroll after carefully researching available programs.

International Students

If you are an international student, you will follow the same application procedures as other graduate school applicants. However, you will have to meet additional requirements.

Since your success as a graduate student will depend on your ability to understand, write, read, and speak English, if English is not your native language, you will be required to take the Test of English as a Foreign Language (TOEFL), or a similar test. Some schools will waive the language test requirement, however, if you have a degree from a college or university in a country where the native language is English or if you have studied two or more years in an undergraduate or graduate program in a country where the native language is English. As for all other tests, score requirements vary, but some schools admit students with lower scores on the condition that they enroll in an intensive English program before or during their graduate study. You should ask each school or department about its policies.

In addition to scores on your English test, or proof of competence in English, your formal application must be accompanied by a certified English translation of your academic transcripts. You may also be required to submit records of immunization and certain health certificates as well as documented evidence of financial support at the time of application. However, since you may apply for financial assistance from graduate schools as well as other sources, some institutions require evidence of financial support only as the last step in your formal admittance and may grant you conditional acceptance first.

Once you have been formally admitted into a graduate program and have submitted evidence of your source or sources of financial support, the school will send you Form I-20 or Form IAP-66, Certificate of Eligibility for Non-Immigrant Status. You must present this document, along with a passport from your own government, and evidence of financial support (some schools will require evidence of support for the entire course of study, while others require evidence of support only for the first year of study, if there is also documentation to show reasonable expectation of continued support) to a U.S. embassy or consulate to obtain an international student visa (F-1 with the Form I-20 or J-1 with the Form IAP-66).

Your own government may have other requirements you must meet to study in the United States. Be sure to investigate those requirements as well.

Once all the paperwork has been completed and approved, you are ready to make your travel arrangements. If your port of entry into the United States will be New York's Kennedy Airport, you can arrange, for a fee, to be met and assisted by a representative of the YMCA Arrivals Program. This person will help you through customs and assist you in making travel connections. He or she can also help you find temporary overnight accommodations, if needed. To inquire about fees for this service, contact the Arrivals Program by phone (212-727-8800 Ext. 130), fax (212-727-8814), or e-mail (jholt@ymcanyc.org). If you decide to take advantage of this assistance, you must provide the Arrivals Program with the following information: your name, age, sex, date and time of arrival, airline and flight number, college or university you will be attending, sponsoring agency (if any), and connecting flight information. Include a photo to help identify you, and note if you need overnight accommodations in New York. This information should be sent well in advance to YMCA Arrivals Program, 71 West 23rd Street, Suite 1904, New York, New York 10010.

When you arrive on your American college campus, you will want to contact the international student adviser. This person's job is to help international students in their academic and social adjustment. The adviser often coordinates special orientation programs for new students, which may consist of lectures on American culture, intensive language instruction, campus tours, academic placement examinations, and visits to places of cultural interest in the community. This adviser will also help you with travel and employment questions as well as financial concerns and will keep copies of your visa documents on file, which is required by U.S. immigration law.

A number of nonprofit educational organizations are available throughout the world to assist international students in planning graduate study in the United States. To learn how to get in touch with these organizations for detailed information, contact the U.S. embassy in your country.

Jane E. Levy
Senior Associate Director
University Career Center
Cornell University
and
Elinor R. Workman

Financing Your Graduate and Professional Education

If you're considering attending graduate school but fear you don't have enough money, don't despair. Financial support for graduate study does exist, although, admittedly, the information about support sources can be difficult to find.

Support for graduate study can take many forms, depending upon the field of study and program you pursue. For example, some 60 percent of doctoral students receive support in the form of either grants/fellowships or assistantships, whereas most students in master's programs rely on loans to pay for their graduate study. In addition, doctoral candidates are more likely to receive grants/fellowships and assistantships than master's degree students, and students in the sciences are more likely to receive aid than those in the arts and humanities.

For those of you who have experience with financial aid as an undergraduate, there are some differences for graduate students you'll notice right away. For one, aid to undergraduates is based primarily on need (although the number of colleges that now offer undergraduate merit-based aid is increasing). But graduate aid is often based on academic merit, especially in the arts and sciences. Second, as a graduate student, you are automatically "independent" for federal financial aid purposes, meaning your parents' income and asset information is not required in assessing your need for federal aid. And third, at some graduate schools, the awarding of aid may be administered by the academic departments or the graduate school itself, not the financial aid office. This means that at some schools, you may be involved with as many as three offices: a central financial aid office, the graduate school, *and* your academic department.

FINANCIAL AID MYTHS

- **Financial aid is just for poor people.**
- **Financial aid is just for smart people.**
- **Financial aid is mainly for minority students.**
- **I have a job, so I must not be eligible for aid.**
- **If I apply for aid, it will affect whether or not I'm admitted.**
- **Loans are not financial aid.**

Be Prepared

Being prepared for graduate school means you should put together a financial plan. So, before you enter graduate school, you should have answers to these questions:

- What should I be doing now to prepare for the cost of my graduate education?
- What can I do to minimize my costs once I arrive on campus?
- What financial aid programs are available at each of the schools to which I am applying?
- What financial aid programs are available outside the university, at the federal, state, or private level?
- What financing options do I have if I cannot pay the full cost from my own resources and those of my family?
- What should I know about the loans I am being offered?
- What impact will these loans have on me when I complete my program?

You'll find your answers in three guiding principles: think ahead, live within your means, and keep your head above water.

Think Ahead

The first step to putting together your financial plan comes from thinking about the future: the loss of your income while you're attending school, your projected income after you graduate, the annual rate of inflation, additional expenses you will incur as a student and after you graduate, and any loss of income you may experience later on from unintentional periods of unemployment, pregnancy, or disability. The cornerstone of thinking ahead is following a step-by-step process.

1. *Set your goals.* Decide what and where you want to study, whether you will attend full- or part-time, whether you'll work while attending, and what an appropriate level of debt would be. Consider whether you would attend full-time if you had enough financial aid or whether keeping your full-time job is an important priority in your life. Keep in mind that some employers have tuition reimbursement plans for full-time employees.
2. *Take inventory.* Collect your financial information and add up your assets—bank accounts, stocks, bonds, real estate, business and personal property. Then subtract your liabilities—money owed on your assets including credit card debt and car loans—to yield your net worth.
3. *Calculate your need.* Compare your net worth with the costs at the schools you are considering to get a rough estimate of how much of your assets you will need to use for your schooling.
4. *Create an action plan.* Determine how much you'll earn while in school, how much you think you will receive in grants and scholarships, and how much you plan to borrow. Don't forget to consider inflation and possible life changes that could affect your overall financial plan.
5. *Review your plan regularly.* Measure the progress of your plan every year and make adjustments for such things as increases in salary or other changes in your goals or circumstances.

Live Within Your Means

The second step in being prepared is knowing how much you spend now so you can determine how much you'll spend when you're in school. Use the standard cost of attendance budget published by your school as a guide. But don't be surprised if your estimated budget is higher than the one the school provides, especially if you've been out of school for a while. Once you've figured out your budget, see if you can pare down your current costs and financial obligations so the lean years of graduate school don't come as too large a shock.

Keep Your Head Above Water

Finally, the third step is managing the debt you'll accrue as a graduate student. Debt is manageable only when considered in terms of five things:

1. Your future income
2. The amount of time it takes to repay the loan
3. The interest rate you are being charged
4. Your personal lifestyle and expenses after graduation
5. Unexpected circumstances that change your income or your ability to repay what you owe

To make sure your educational debt is manageable, you should borrow an amount that requires payments of between 8 and 15 percent of your starting salary.

The approximate monthly installments for repaying borrowed principal at 5, 8–10, and 12 percent are indicated on page 10.

Estimated Loan Repayment Schedule
Monthly Payments for Every $1000 Borrowed

Rate	5 years	10 years	15 years	20 years	25 years
5%	$18.87	$10.61	$ 7.91	$ 6.60	$ 5.85
8%	20.28	12.13	9.56	8.36	7.72
9%	20.76	12.67	10.14	9.00	8.39
10%	21.74	13.77	10.75	9.65	9.09
12%	22.24	14.35	12.00	11.01	10.53

You can use this table to estimate your monthly payments on a loan for any of the five repayment periods (5, 10, 15, 20, and 25 years). The amounts listed are the monthly payments for a $1000 loan for each of the interest rates. To estimate your monthly payment, choose the closest interest rate and multiply the amount of the payment listed by the total amount of your loan and then divide by 1,000. For example, for a total loan of $15,000 at 9 percent to be paid back over ten years, multiply $12.67 times 15,000 (190,050) divided by 1,000. This yields $190.05 per month.

If you're wondering just how much of a loan payment you can afford monthly without running into payment problems, consult the chart below.

HOW MUCH CAN YOU AFFORD TO REPAY?

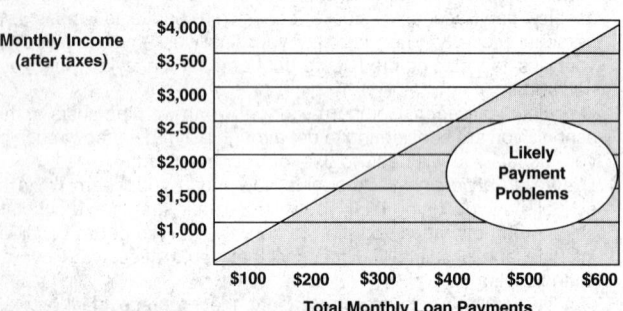

This graph shows the monthly cash-flow outlook based on your total monthly loan payments in comparison with your monthly income earned after taxes. Ideally, to eliminate likely payment problems, your monthly loan payment should be less than 15 percent of your monthly income.

Of course, the best way to manage your debt is to borrow less. While cutting your personal budget may be one option, there are a few others you may want to consider:

- *Ask Your Family for Help:* Although the federal government considers you "independent," your parents and family may still be willing and able to help pay for your graduate education. If your family is not open to just giving you money, they may be open to making a low-interest (or deferred-interest) loan. Family loans usually have more attractive interest rates and repayment terms than commercial loans. They may also have tax consequences, so you may want to check with a tax adviser.
- *Push to Graduate Early:* It's possible to reduce your total indebtedness by completing your program ahead of schedule. You can either take more courses per semester or during the summer. Keep in mind, though, that these options reduce the time you have available to work.
- *Work More, Attend Less:* Another alternative is to enroll part-time, leaving more time to work. Remember, though, to qualify for aid, you must be enrolled at least half time, which is usually considered six credits per term. And if you're enrolled less than half time, you'll have to start repaying your loans once the grace period has expired.

Roll Your Loans into One

There's a good chance that as a graduate student you will have two or more loans included in your aid package, plus any money you borrowed as an undergraduate. That means when you start repaying, you could be making loan payments to several different lenders. Not only can the recordkeeping be a nightmare, but with each loan having a minimum payment, your total monthly payments may be more than you can handle. If that is the case, you may want to consider consolidating your federal loans.

There is no minimum or maximum on the amount of loans you must have in order to consolidate. Also, there is no consolidation fee. The interest rate varies annually, is adjusted every July 1, and is capped at 8.25 percent. Your repayment can also be extended to up to thirty years, depending on the total amount you borrow, which will make your monthly payments lower (of course, you'll also be paying more total interest). With a consolidated loan, some lenders offer graduated or income-sensitive repayment options. Consult with your lender or the U.S. Department of Education about the types of consolidation provisions offered.

Plastic Mania

Any section on managing debt would be incomplete if it didn't mention the responsible use of credit cards. Most graduate students hold one or more credit cards, and many students find themselves in financial difficulties because of them. Here are two suggestions: use credit cards only for convenience, never for extended credit; and, if you have more than one credit card, keep only the one that has the lowest finance charge and the lowest limit.

Credit: Don't Let Your Past Haunt You

Many schools will check your credit history before they process any private educational loans for you. To make sure your credit rating is accurate, you may want to request a copy of your credit report before you start graduate school. You can get a copy of your report by sending a signed, written request to one of the four national credit reporting agencies at the address listed below. Include your full name, social security number, current address (and proof of current address such as from a driver's license or utility bill), any previous addresses for the past five years, date of birth, and daytime phone number. Call the agency before you request your report so you know whether there is a fee for this report. Note that you are entitled to a free copy of your credit report if you have been denied credit within the last sixty days. In addition, Experian currently provides complimentary credit reports once every twelve months.

Credit criteria used to review and approve student loans can include the following:
- Absence of negative credit
- No bankruptcies, foreclosures, repossessions, charge-offs, or open judgments
- No prior educational loan defaults, unless paid in full or making satisfactory repayments
- Absence of excessive past due accounts; that is, no 30-, 60-, or 90-day delinquencies on consumer loans or revolving charge accounts within the past two years

CREDIT REPORTING AGENCIES

Experian
P.O. Box 2104
Allen, Texas 75013-2104
888-397-3742

Equifax
P.O. Box 105873
Atlanta, Georgia 30348
800-685-1111

CSC Credit Services
Consumer Assistance Center
P.O. Box 674402
Houston, Texas 77267-4402
800-759-5979

Trans Union Corporation
P.O. Box 390
Springfield, Pennsylvania 19064-0390
800-888-4213

Types of Aid Available

There are three types of aid: money given to you (grants, scholarships, and fellowships), money you earn through work, and loans.

GRANTS, SCHOLARSHIPS, AND FELLOWSHIPS

Most grants, scholarships, and fellowships are outright awards that require no service in return. Often they provide the cost of tuition and fees plus a stipend to cover living expenses. Some are based exclusively on financial need, some exclusively on academic merit, and some on a combination of need and merit. As a rule, grants are awarded to those with financial need, although they may require the recipient to have expertise in a certain field. Fellowships and scholarships often connote selectivity based on ability—financial need is usually not a factor.

Federal Support

Several federal agencies fund fellowship and trainee programs for graduate and professional students. The amounts and types of assistance offered vary considerably by field of study. The following are programs available for those studying engineering or applied sciences:

Graduate Assistantships in Areas of National Need. This program is designed to offer fellowships to outstanding doctoral candidates of superior ability. It is designed to offer financial assistance to students enrolled in specific programs for which there is both a national need and lack of qualified personnel. The definition of national need is determined by the Secretary of Education. Current areas include chemistry, engineering, mathematics, physics, and area studies. Funds are awarded to schools who then select their recipients, based on academic merit. Awardees must also demonstrate financial need. Awards include tuition plus a living expense stipend. Awards are not to exceed four years. Contact the graduate dean's office or academic department to see whether the school participates in this program.

National Science Foundation. Graduate Research Program Fellowships include tuition and fees plus a $15,000 stipend for three years of graduate study in engineering, mathematics, the natural sciences, the social sciences, and the history and philosophy of science. The application deadline is in early November. For more information, write to the National Science Foundation at Oak Ridge Associated Universities, P.O. Box 3010, Oak Ridge, Tennessee 37831-3010, call 423-241-4300, or visit their Web site at http://www.orau.org/nsf/nsffel.htm.

National Institutes of Health (NIH). NIH sponsors many different fellowship opportunities. For example, it offers training grants administered through schools' research departments. Training grants provide tuition plus a twelve-month stipend. For more information, call 301-435-0714 or e-mail grantsinfo@nih.gov.

Veterans' Benefits. Veterans may use their educational benefits for training at the graduate and professional levels. Contact your regional office of the Veterans Administration for more details.

State Support

Some states offer grants for graduate study, with California, Michigan, New York, North Carolina, Texas, and Virginia offering the largest programs. States grant approximately $2.9 billion per year to graduate students. Due to fiscal constraints, however, some states have had to reduce or eliminate their financial aid programs for graduate study. To qualify for a particular state's aid you must be a resident of that state. Residency is established in most states after you have lived there for at least twelve consecutive months prior to enrolling in school. Many states provide funds for in-state students only; that is, funds are not transferable out of state. Contact your state scholarship office to determine what aid it offers.

Institutional Aid

Educational institutions using their own funds provide more than $3 billion in graduate assistance in the form of fellowships, tuition waivers, and assistantships. Consult each school's catalog for information about aid programs.

Corporate Aid

Some corporations provide graduate student support as part of the employee benefits package. Most employees who receive aid study at the master's level or take courses without enrolling in a particular degree program.

Aid from Foundations

Most foundations provide support in areas of interest to them. For example, for those studying for the Ph.D., the Howard Hughes Institute funds students in the biomedical sciences, while the Spencer Foundation funds dissertation research in the field of education.

The Foundation Center of New York City publishes several reference books on foundation support for graduate study. For more information, call 212-620-4230, or access their Web site at http://www.fdncenter.org.

Financial Aid for Minorities and Women

Bureau of Indian Affairs. The Bureau of Indian Affairs (BIA) offers aid to students who are at least one quarter American Indian or native Alaskan and from a federally recognized tribe. Contact your tribal education officer, BIA area office, or call the Bureau of Indian Affairs at 202-208-3710.

The Ford Foundation Doctoral Fellowship for Minorities. Provides three-year doctoral fellowships and one-year dissertation fellowships. Predoctoral fellowships include an annual stipend of $14,000 to the fellow and an annual institutional grant of $7500 to the fellowship institution in lieu of tuition and fees. Dissertation fellows receive a stipend of $21,500 for a twelve-month period. Applications are due in early November. For more information, contact the Fellowship Office, National Research Council at 202-334-2872 or visit their Web site at http://www2.nas.edu/fo/.

National Consortium for Graduate Degrees in Engineering and Science (GEM). GEM was founded in 1976 to help minority men and women pursue graduate study in engineering by helping them obtain practical experience through summer internships at consortium worksites and finance graduate study toward a master's or Ph.D. degree. For more information, contact GEM, Box 537, Notre Dame, Indiana 46556, call 219-631-7771, or visit their Web site at http://www.nd.edu/~gem/.

National Physical Sciences Consortium. Graduate fellowships are available in astronomy, chemistry, computer science, geology, materials science, mathematics, and physics for women and Black, Hispanic, and Native American students. These fellowships are available only at member universities. Awards may vary by year in school and the application deadline is November 5. Fellows receive tuition plus a stipend between $12,500 and $15,000. For more information, contact National Physical Sciences Consortium, MSC 3NPS, New Mexico State University, P.O. Box 30001, Las Cruces, New Mexico 88033-8001, call 800-952-4118, or visit their Web site at http://www.npsc.org.

In addition, the following are books that describe financial aid opportunities for women and minorities. *The Directory of Financial Aids for Women* by Gail Ann Schlachter (Reference Service Press,

1998) lists sources of support and identifies foundations and other organizations interested in helping women secure funding for graduate study.

The Association for Women in Science publishes *Grants-at-a-Glance,* a booklet highlighting fellowships for women in science. It can be ordered by calling 202-326-8940 or by visiting their Web site at http://www.awis.org.

Books such as *Financial Aid for Minorities* (Garrett Park, MD: Garrett Park Press, 1998) describe financial aid opportunities for minority students. For more information, call 301-946-2553.

Reference Service Press also publishes four directories specifically for minorities: *Financial Aid for African Americans, Financial Aid for Asian Americans, Financial Aid for Hispanic Americans,* and *Financial Aid for Native Americans.*

For more information on financial aid for minorities, see the Minority On-Line Information Service (MOLIS) Web site at http://web.fie.com/web/mol/.

Disabled students are eligible to receive aid from a number of organizations. *Financial Aid for the Disabled and Their Families, 1998–2000* by Gail Ann Schlachter and David R. Weber (Reference Service Press) lists aid opportunities for disabled students. The Vocational Rehabilitation Services in your home state can also provide information.

Researching Grants and Fellowships

The books listed below are good sources of information on grant and fellowship support for graduate education and should be consulted before you resort to borrowing. Keep in mind that grant support varies dramatically from field to field.

Annual Register of Grant Support: A Directory of Funding Sources, Wilmette, Illinois: National Register Publishing Co. This is a comprehensive guide to grants and awards from government agencies, foundations, and business and professional organizations.

Corporate Foundation Profiles, 10th ed. New York: Foundation Center, 1998. This is an in-depth, analytical profile of 250 of the largest company-sponsored foundations in the United States. Brief descriptions of all 700 company-sponsored foundations are also included. There is an index of subjects, types of support, and geographical locations.

The Foundation Directory, edited by Stan Olsen. New York: Foundation Center, 1999. This directory, with a supplement, gives detailed information on U.S. foundations, with brief descriptions of the purpose and activities of each.

The Grants Register 1999, 17th ed. Edited by Lisa Williams. New York: St. Martin's, 1999. This lists grant agencies alphabetically and gives information on awards available to graduate students, young professionals, and scholars for study and research.

Peterson's Grants for Graduate & Postdoctoral Study, 5th ed. Princeton: Peterson's, 1998. This book includes information on 1,400 grants, scholarships, awards, fellowships, and prizes. Originally compiled by the Office of Research Affairs at the Graduate School of the University of Massachusetts at Amherst, this guide is updated periodically by Peterson's.

Graduate schools sometimes publish listings of support sources in their catalogs, and some provide separate publications, such as the *Graduate Guide to Grants,* compiled by the Harvard Graduate School of Arts and Sciences. For more information, call 617-495-1814.

THE INTERNET AS A SOURCE OF FUNDING INFORMATION

If you have not explored the financial resources on the World Wide Web (the Web, for short), your research is not complete. Now available on the Web is a wealth of information ranging from loan and entrance applications to minority grants and scholarships.

University-Specific Information on the Web

Most university financial aid offices have Web sites. Florida, Virginia Tech, Massachusetts, Emory, and Georgetown are just a few. Applications of admission can now be downloaded from the Web to start the graduate process. After that, detailed information can be obtained on financial aid processes, forms, and deadlines. University-specific grant and scholarship information can also be found, and more may be learned about financing information by using the Web than by an actual visit. Questions can be answered on line.

Scholarships on the Web

Many benefactors and other scholarship donors have Web sites. You can reach this information through a variety of methods. For example, you can find a directory listing minority scholarships, quickly look at the information on line, decide if it applies to you, and then move on. New scholarship pages are being added to the Web daily. Library and Web resources are productive and--free.

The Web also lists many services that will look for scholarships for you. Some of these services cost money and advertise more scholarships per dollar than any other service. While some of these might be helpful, beware. Check references to make sure a bona fide service is being offered. Your best bet initially is to surf the Web and use the traditional library resources on available scholarships.

Bank and Loan Information on the Web

Banks and loan servicing centers have pages on the Web, making it easier to access loan information on the Web. Having the information on screen in front of you instantaneously is more convenient than being put on hold on the phone. Any loan information such as interest rate variations, descriptions of loans, loan consolidation programs, and repayment charts can all be found on the Web. Also, many lenders now allow you to fill out loan applications online.

WORK PROGRAMS

Certain types of support, such as teaching, research, and administrative assistantships, require recipients to provide service to the university in exchange for a salary or stipend; sometimes tuition is also provided or waived.

Teaching Assistantships

If your field of study is taught at the undergraduate level, you stand a good chance of securing a teaching assistantship. These positions usually involve conducting small classes, delivering lectures, correcting class work, grading papers, counseling students, and supervising laboratory groups. Usually about 20 hours of work is required each week.

Teaching assistantships provide excellent educational experience as well as financial support. TAs generally receive a salary (now considered taxable income). Sometimes tuition is provided or waived as well. In addition, at some schools, TAs can be declared state residents, qualifying them for the in-state tuition rates. Appointments are based on academic qualifications and are subject to the availability of funds within a department. If you are interested in a teaching assistantship, contact the academic department. Ordinarily you are not considered for such positions until you have been admitted to the graduate school.

Research Assistantships

Research Assistantships usually require that you assist in the research activities of a faculty member. Appointments are ordinarily made for the academic year. They are rarely offered to first-year students. Contact the academic department, describing your particular research interests. As is the case with teaching assistantships, research assistantships provide excellent academic training as well as practical experience and financial support.

Administrative Assistantships

These positions usually require 10 to 20 hours of work each week in an administrative office of the university. For example, those seeking a graduate degree in education may work in the admissions, financial aid, student affairs, or placement office of the school they are attending. Some administrative assistantships provide a tuition waiver, others a salary. Details concerning these positions can be found in the school catalog or by contacting the academic department directly.

Federal Work-Study Program (FWS)

This federally funded program provides eligible students with employment opportunities, usually in public and private nonprofit organizations. Federal funds pay up to 75 percent of the wages, with the remainder paid by the employing agency. FWS is available to graduate students who demonstrate financial need. Not all schools have these funds, and some only award undergraduates. Each school sets its application deadline and work-study earnings limits. Wages vary and are related to the type of work done.

Additional Employment Opportunities

Many schools provide on-campus employment opportunities that do not require demonstrated financial need. The student employment office on most campuses assists students in securing jobs both on and off the campus.

LOANS

Most needy graduate students, except those pursuing Ph.D.'s in certain fields, borrow to finance their graduate programs. There are basically two sources of student loans—the federal government and private loan programs. You should read and understand the terms of these loan programs before submitting your loan application.

Federal Loans

Federal Stafford Loans. The Federal Stafford Loan Program offers government-sponsored, low-interest loans to students through a private lender such as a bank, credit union, or savings and loan association.

There are two components of the Federal Stafford Loan program. Under the *subsidized* component of the program, the federal government pays the interest accruing on the loan while you are enrolled in graduate school on at least a half-time basis. Under the *unsubsidized* component of the program, you pay the interest on the loan from the day proceeds are issued. Eligibility for the federal subsidy is based on demonstrated financial need as determined by the financial aid office from the information you provide on the Free Application for Federal Student Aid (FAFSA). A cosigner is not required, since the loan is not based on creditworthiness. Although Unsubsidized Federal Stafford Loans may not be as desirable as Subsidized Federal Stafford Loans from the consumer's perspective, they are a useful source of support for those who may not qualify for the subsidized loans or who need additional financial assistance.

Graduate students may borrow up to $18,500 per year through the Stafford Loan Program, up to a maximum of $138,500, including undergraduate borrowing. This may include up to $8500 in Subsidized Stafford Loans, depending on eligibility, up to a maximum of $65,000, including undergraduate borrowing. The amount of the loan borrowed through the Unsubsidized Stafford Program equals the total amount of the loan (as much $18,500) minus your eligibility for a Subsidized Stafford Loan (as much as $8500). You may borrow up to the cost of the school in which you are enrolled or will attend, minus estimated financial assistance from other federal, state, and private sources, up to a maximum of $18,500.

The interest rate for the Federal Stafford Loans varies annually and is set every July. The rate during in-school, grace, and deferment periods is based on the 91-Day U.S. Treasury Bill rate plus 2.5 percent, capped at 8.25 percent. The rate in repayment is based on the 91-Day U.S. Treasury Bill rate plus 3.1 percent, capped at 8.25 percent.

Two fees are deducted from the loan proceeds upon disbursement: a guarantee fee of up to 1 percent, which is deposited in an insurance pool to ensure repayment to the lender if the borrower defaults, and a federally mandated 3 percent origination fee, which is used to offset the administrative cost of the Federal Stafford Loan Program.

Under the *subsidized* Federal Stafford Loan Program, repayment begins six months after your last enrollment on at least a half-time basis. Under the *unsubsidized* program, repayment of interest begins within thirty days from disbursement of the loan proceeds, and repayment of the principal begins six months after your last enrollment on at least a half-time basis. Some lenders may require that some payments be made even while you are in school, although most lenders will allow you to defer payments and will add the accrued interest to the loan balance. Under both components of the program repayment may extend over a maximum of ten years with no prepayment penalty.

Federal Direct Loans. Some schools are participating in the Department of Education's Direct Lending Program instead of offering Federal Stafford Loans. The two programs are essentially the same except that with the Direct Loans, schools themselves originate the loans with funds provided from the federal government. Terms and interest rates are virtually the same except that there are a few more repayment options with Federal Direct Loans.

Federal Perkins Loans. The Federal Perkins Loan is a long-term loan available to students demonstrating financial need and is administered directly by the school. Not all schools have these funds, and some may award them to undergraduates only. Eligibility is determined from the information you provide on the FAFSA. The school will notify you of your eligibility. Eligible graduate students may borrow up to $5000 per year, up to a maximum of $30,000, including undergraduate borrowing (even if your previous Perkins Loans have been repaid). The interest rate for Federal Perkins Loans is 5 percent, and no interest accrues while you remain in school at least half-time. There are no guarantee, loan, or disbursement fees. Repayment begins nine months after your last enrollment on at least a half-time basis and may extend over a maximum of ten years with no prepayment penalty.

Deferring Your Federal Loan Repayments. If you borrowed under the Federal Stafford Loan Program or the Federal Perkins Loan Program for previous undergraduate or graduate study, some of your repayments may be deferred (i.e., suspended) when you return to graduate school, depending on when you borrowed and under which program. There are other deferment options available if you are temporarily unable to repay your loan. Information about these deferments is provided at your entrance and exit interviews. If you believe you are eligible for a deferment of your loan repayments, you must contact your lender to complete a deferment form. The deferment must be filed prior to the time your repayment is due, and it must be refiled when it expires if you remain eligible for deferment at that time.

Supplemental Loans

Many lending institutions offer supplemental loan programs and other financing plans, such as the ones described below, to students seeking assistance in meeting their expected contribution toward educational expenses.

If you are considering borrowing through a supplemental loan program, you should carefully consider the terms of the program and be sure to "read the fine print." Check with the program sponsor for the most current terms that will be applicable to the amounts you intend to borrow for graduate study. Most supplemental loan programs for graduate study offer unsubsidized, credit-based loans. In general, a credit-ready borrower is one who has a satisfactory credit history or no credit history at all. A creditworthy borrower generally must pass a credit test to be eligible to borrow or act as a cosigner for the loan funds.

Many supplemental loan programs have a minimum annual loan limit and a maximum annual loan limit. Some offer amounts equal to the cost of attendance minus any other aid you will receive for graduate study. If you are planning to borrow for several years of graduate study, consider whether there is a cumulative or aggregate limit on the amount you may borrow. Often this cumulative or aggregate limit will include any amounts you borrowed and have not repaid for undergraduate or previous graduate study.

The combination of the annual interest rate, loan fees, and the repayment terms you choose will determine how much the amount is that you will repay over time. Compare these features in combination before you decide which loan program to use. Some loans offer interest rates that are adjusted monthly, some quarterly, some annually. Some offer interest rates that are lower during the in-school, grace, and deferment periods, and then increase when you begin repayment. Most programs include a loan "origination" fee, which is usually deducted from the principal amount you receive when the loan is disbursed, and must be repaid along with the interest and other principal when you graduate, withdraw from school, or drop below half-time study. Sometimes the loan fees are reduced if you borrow with a qualified cosigner. Some programs allow you to defer interest and/or principal payments while you are enrolled in graduate school. Many programs allow you to capitalize your interest payments; the interest due on your loan is added to the outstanding balance of your loan, so you don't have to repay immediately, but this increases the amount you owe. Other programs allow you to pay the interest as you go, which will reduce the amount you later have to repay.

For more information about supplemental loan programs, visit http://www.estudentloans.com. This Web site has the most up-to-date information about supplemental loans.

International Education and Study Abroad

A variety of funding sources are offered for study abroad and for foreign nationals studying in the United States. The Institute of International Education in New York assists students in locating such aid. It publishes *Funding for U.S. Study—A Guide for International Students and Professionals* and *Financial Resources for International Study,* a guide to organizations offering awards for overseas study. To learn more, visit the institute's Web site at http://www.iie.org.

The Council on International Educational Exchange in New York publishes the *Student Travel Catalogue,* which lists fellowship sources and explains the council's services both for United States students traveling abroad and for foreign students coming to the United States. For more information, see the council's Web site at http://www.ciee.org.

The U.S. Department of Education administers programs that support fellowships related to international education. Foreign Language and Area Studies Fellowships and Fulbright-Hays Doctoral Dissertation Awards were established to promote knowledge and understanding of other countries and cultures. They offer support to graduate students interested in foreign languages and international relations. Discuss these and other foreign study opportunities with the financial aid officer or someone in the graduate school dean's office at the school you will attend.

How to Apply

All applicants for federal aid must complete the Free Application for Federal Student Aid (FAFSA). This application must be submitted *after* January 1 preceding enrollment in the fall. It is a good idea to submit the FAFSA as soon as possible after this date. You can fill out the paper form, or you can apply online at http://www.fafsa.ed.gov. On this form you report your income and asset information for the preceding calendar year and specify which schools will receive the data. Two to four weeks later you'll receive an acknowledgment, the Student Aid Report (SAR), on which you can make any corrections. The schools you've designated will also receive the information and may begin asking you to send them documents, usually your U.S. income tax return, verifying what you reported.

In addition to the FAFSA, some graduate schools want additional information and will ask you to complete the CSS Financial Aid PROFILE. If your school requires this form, it will be listed in the PROFILE registration form available in college financial aid offices. Other schools use their own supplemental application. Check with your financial aid office to confirm which forms they require.

If you have already filed your federal income tax for the year, it will be much easier for you to complete these forms. If not, use estimates, but be certain to notify the financial aid office if your estimated figures differ from the actual ones once you have calculated them.

APPLICATION DEADLINES

Application deadlines vary. Some schools require you to apply for aid when applying for admission; others require that you be admitted before applying for aid. Aid application instructions and deadlines should be clearly stated in each school's application material. The FAFSA must be filed after January 1 of the year you are applying for aid but the Financial Aid PROFILE should be completed earlier, in October or November.

Determining Financial Need

Eligibility for need-based financial aid is based on your income during the calendar year prior to the academic year in which you apply for aid. Prior-year income is used because it is a good predictor of current-year income and is verifiable. If you have a significant reduction in income or assets after your aid application is completed, consult a financial aid counselor. If, for example, you are returning to school after working, you should let the financial aid counselor know your projected income for the year you will be in school. Aid counselors may use their "professional judgment" to revise your financial need, based on the actual income you will earn while you are in graduate school.

Need is determined by examining the difference between the cost of attendance at a given institution and the financial resources you bring to the table. Eligibility for aid is calculated by subtracting your resources from the total cost of attendance budget. These standard student budgets are generally on the low side of the norm. So if your expenses are higher because of medical bills, higher research travel, or more costly books, for example, a financial aid counselor can make an adjustment. Of course, you'll have to document any unusual expenses. Also, keep in mind that with limited grant and scholarship aid, a higher budget will probably mean either more loan or more working hours for you.

Tax Issues

Since the passage of the Tax Reform Act of 1986, grants, scholarships, and fellowships may be considered taxable income. That portion of the grant used for payment of tuition and course-required fees, books, supplies, and equipment is excludable from taxable income. Grant support for living expenses is taxable. A good rule of thumb for determining the tax liability for grants and scholarships is to view anything that exceeds the actual cost of tuition, required fees, books, supplies related to courses, and required equipment as taxable.

- If you are employed by an educational institution or other organization that gives tuition reimbursement, you must pay tax on the value that exceeds $5250.
- If your tuition is waived in exchange for working at the institution, the tuition waiver is taxable. This includes waivers that come with teaching or research assistantships.
- Other student support, such as stipends and wages paid to research assistants and teaching assistants, is also taxable income. Student loans, however, are not taxable.
- If you are an international student you may or may not owe taxes depending upon the agreement the U.S. has negotiated with your home country. The United States has tax treaties with more than forty countries. You are responsible for making sure that the school you attend follows the terms of the tax treaty. If your country does not have a tax treaty with the U.S., you may have as much as 30 percent withheld from your paycheck.

A Final Note

While amounts and eligibility criteria vary from field to field as well as from year to year, with thorough research you can uncover many opportunities for graduate financial assistance. If you are interested in graduate study, discuss your plans with faculty members and advisers. Explore all options. Plan ahead, complete forms on time, and be tenacious in your search for support. No matter what your financial situation, if you are academically qualified and knowledgeable about the different sources of aid, you should be able to attend the graduate school of your choice.

Patricia McWade
Dean of Student Financial Services
Georgetown University

Tests Required of Applicants

Many graduate schools require that applicants submit scores on one or more standardized tests, often the Graduate Record Examinations (GRE) or the Miller Analogies Test (MAT). Professional schools usually require that applicants take a specific admission test. Virtually all graduate and professional schools ask students whose native language is not English to take the Test of English as a Foreign Language (TOEFL), and some also ask for TOEFL's Test of Written English (TWE) or the Test of Spoken English (TSE).

Brief descriptions of these tests and the addresses to write to for additional information are given below.

GRADUATE RECORD EXAMINATIONS

The GRE General Test and Subject Tests are designed to assess academic knowledge and skills relevant to graduate study. The General Test measures verbal, quantitative, and analytical reasoning skills, and the Subject Tests measure achievement in particular fields of study. The GRE tests are administered worldwide by Educational Testing Service (ETS) of Princeton, New Jersey, under policies established by the Graduate Record Examinations Board, an independent board affiliated with the Association of Graduate Schools and the Council of Graduate Schools.

Subject Tests, offered only as paper-based tests, are available in fourteen areas: biochemistry, cell and molecular biology; biology; chemistry; computer science; economics; engineering; geology; history; literature in English; mathematics; music; physics; psychology; and sociology.

The CAT (Computer Adaptive Test) General Test is offered year-round at more than 600 test centers around the world. The CAT offers convenient scheduling, immediate viewing of unofficial scores, and faster score reporting. To schedule an appointment in the U.S., U.S. Territories, or Canada, call 800-GRE-CALL. For international testing, refer to the 1999–2000 *GRE Information and Registration Bulletin* or the GRE Web site (http://www.gre.org) for a list of the regional registration centers. The *GRE Bulletin* contains registration and program services information.

The 1999–2000 GRE Subject Test dates are November 6, December 11, and April 8; the economics, geology, history, music, and sociology tests will not be offered on the November test date. The General Test is offered on the November and April test dates only.

Fees for the General Test and Subject Tests are $96 for testing in the U.S. and U.S. Territories and $120 in all other locations. Fees are subject to change.

Nonstandard testing accommodations are available for test takers with disabilities through the testing programs. Students who cannot test on Saturdays for religious reasons may request a Monday paper-based administration immediately following a regular Saturday test date. Refer to the *GRE Bulletin* for more information.

Test takers can register by phone, fax, mail, or online. Test takers should consider admission deadlines and register early to get their preferred test dates.

Further information on registration is available from the GRE Web site or by writing to GRE-ETS, P.O. Box 6000, Princeton, New Jersey 08541-6000 or by calling 609-771-7670.

Peterson's offers *GRE Success*, a complete guide to the GRE. Visit your local bookstore or contact Peterson's at 800-225-0261.

MILLER ANALOGIES TEST

The MAT is published and administered by The Psychological Corporation, a division of Harcourt Brace & Company. The MAT is a high-level mental ability test that requires the solution of 100 problems stated in the form of analogies. The MAT is accepted by more than 2,300 graduate school programs as part of their admission process. The test items use different types of analogies to sample general information and a variety of fields, such as fine arts, literature, mathematics, natural science, and social science. Examinees are allowed 50 minutes to complete the test.

The MAT is offered at more than 600 test centers in the United States and Canada. For examinee convenience, the test is given on an as-needed basis at most test centers. Fees are also determined by each test center.

Additional information about the MAT, including preparatory materials and test center locations, is available from The Psychological Corporation, 555 Academic Court, San Antonio, Texas 78204. Telephone: 210-299-1061 or 800-622-3231 (7 a.m. to 7 p.m., Monday through Friday, Central time).

TEST OF ENGLISH AS A FOREIGN LANGUAGE

The purpose of the TOEFL is to evaluate the English proficiency of people whose native language is not English.

The TOEFL is administered as a computer-based test throughout most of the world. The computer-based TOEFL is available year-round by appointment only. It is not necessary to have previous computer experience to take the test. Examinees will be given all the instructions and practice needed to perform the necessary computer tasks before the actual test begins. The test consists of four sections—listening, reading, structure, and writing. Total testing time is approximately 4 hours. The fee for the computer-based TOEFL is $100, which must be paid in U.S. dollars. The *Information Bulletin for Computer-Based Testing* contains information about the new testing format, registration procedures, and testing sites.

TOEFL will remain paper-based in Bangladesh, Bhutan, Cambodia, Hong Kong, India, Japan, Korea, Laos, Macau, Pakistan, People's Republic of China, Taiwan, Thailand, and Vietnam. For 1999–2000, in these countries tests will be offered on December 17, March 17, and June 9.

The paper-based TOEFL consists of three sections—listening comprehension, structure and written expression, and reading comprehension. Testing time is approximately 3 hours. In December, February, and May, the Test of Written English (TWE) will also be given. TWE is a 30-minute essay that measures the examinee's ability to compose in English. Examinees receive a TWE score separate from their TOEFL score. The fee for the paper-based TOEFL is $75, which must be paid in U.S. dollars. There is no additional charge for TWE. The *Information Bulletin* contains information on local fees and registration procedures.

The TOEFL is given at many test centers throughout the world and is administered by Educational Testing Service (ETS) under the general direction of a policy council established by the College Board and the Graduate Record Examinations Board.

Additional information and registration material is available from the TOEFL Program Office, P.O. Box 6151, Princeton, New Jersey 08541-6151. Telephone: 609-771-7100. E-mail: toefl@ets.org. World Wide Web: http://www.toefl.org.

Peterson's offers *TOEFL Success*, a complete guide to the TOEFL. Visit your local bookstore or contact Peterson's at 800-225-0261.

TEST OF SPOKEN ENGLISH

The major purpose of the TSE is to evaluate the spoken English proficiency of people whose native language is not English. The test, which takes about 30 minutes, requires examinees to demonstrate their ability to speak English by answering a variety of questions presented in printed and recorded form. All the answers to test questions are recorded on tape; no writing is required. TSE is given at selected TOEFL test centers worldwide. The test is administered by Educational Testing Service (ETS) under the general direction of a policy council established by the College Board and the Graduate Record Examinations Board.

The 1999–2000 test dates are November 20, January 15, February 26, April 15, and May 13. The registration fee is $125, which must be paid in U.S. dollars.

Additional information and registration material can be found in the *Information Bulletin for the Test of Spoken English,* available from the TOEFL Program Office, P.O. Box 6151, Princeton, New Jersey 08541-6151, U.S.A. Telephone: 609-771-7100.

Accreditation and Accrediting Agencies

Colleges and universities in the United States, and their individual academic and professional programs, are accredited by nongovernmental agencies concerned with monitoring the quality of education in this country. Agencies with both regional and national jurisdictions grant accreditation to institutions as a whole, while specialized bodies acting on a nationwide basis—often national professional associations—grant accreditation to departments and programs in specific fields.

Institutional and specialized accrediting agencies share the same basic concerns: the purpose an academic unit—whether university or program—has set for itself and how well it fulfills that purpose, the adequacy of its financial and other resources, the quality of its academic offerings, and the level of services it provides. Agencies that grant institutional accreditation take a broader view, of course, and examine universitywide or collegewide services that a specialized agency may not concern itself with.

Both types of agencies follow the same general procedures when considering an application for accreditation. The academic unit prepares a self-evaluation, focusing on the concerns mentioned above and usually including an assessment of both its strengths and weaknesses; a team of representatives of the accrediting body reviews this evaluation, visits the campus, and makes its own report; and finally, the accrediting body makes a decision on the application. Often, even when accreditation is granted, the agency makes a recommendation regarding how the institution or program can improve. All institutions and programs are also reviewed every few years to determine whether they continue to meet established standards; if they do not, they may lose their accreditation.

Accrediting agencies themselves are reviewed and evaluated periodically by the U.S. Department of Education and the Council for Higher Education Accreditation (CHEA). Agencies recognized adhere to certain standards and practices, and their authority in matters of accreditation is widely accepted in the educational community.

This does not mean, however, that accreditation is a simple matter, either for schools wishing to become accredited or for students deciding where to apply. Indeed, in certain fields the very meaning and methods of accreditation are the subject of a good deal of debate. For their part, those applying to graduate school should be aware of the safeguards provided by regional accreditation, especially in terms of degree acceptance and institutional longevity. Beyond this, applicants should understand the role that specialized accreditation plays in their field, as this varies considerably from one discipline to another. In certain professional fields, it is necessary to have graduated from a program that is accredited in order to be eligible for a license to practice, and in some fields the federal government also makes this a hiring requirement. In other disciplines, however, accreditation is not as essential, and there can be excellent programs that are not accredited. In fact, some programs choose not to seek accreditation, although most do.

Institutions and programs that present themselves for accreditation are sometimes granted the status of candidate for accreditation, or what is known as "preaccreditation." This may happen, for example, when an academic unit is too new to have met all the requirements for accreditation. Such status signifies initial recognition and indicates that the school or program in question is working to fulfill all requirements; it does not, however, guarantee that accreditation will be granted.

Readers are advised to contact agencies directly for answers to their questions about accreditation. The names and addresses of all agencies recognized by the U.S. Department of Education and the Council for Higher Education Accreditation are listed below.

Institutional Accrediting Agencies—Regional

MIDDLE STATES ASSOCIATION OF COLLEGES AND SCHOOLS
Accredits institutions in Delaware, District of Columbia, Maryland, New Jersey, New York, Pennsylvania, Puerto Rico, and the Virgin Islands.
Jean Avnet Morse, Executive Director
Commission on Higher Education
3624 Market Street
Philadelphia, Pennsylvania 19104-2680
Telephone: 215-662-5606
Fax: 215-662-5950
E-mail: jamorse@msache.org

NEW ENGLAND ASSOCIATION OF SCHOOLS AND COLLEGES
Accredits institutions in Connecticut, Maine, Massachusetts, New Hampshire, Rhode Island, and Vermont.
Charles M. Cook, Director
Commission on Institutions of Higher Education
209 Burlington Road
Bedford, Massachusetts 01730-1433
Telephone: 781-271-0022
Fax: 781-271-0950
E-mail: ccook@neasc.org

NORTH CENTRAL ASSOCIATION OF COLLEGES AND SCHOOLS
Accredits institutions in Arizona, Arkansas, Colorado, Illinois, Indiana, Iowa, Kansas, Michigan, Minnesota, Missouri, Nebraska, New Mexico, North Dakota, Ohio, Oklahoma, South Dakota, West Virginia, Wisconsin, and Wyoming.
Steve Crow, Executive Director
Commission on Institutions of Higher Education
30 North LaSalle, Suite 2400
Chicago, Illinois 60602-2504
Telephone: 312-263-0456
Fax: 312-263-7462
E-mail: crow@ncacihe.org

NORTHWEST ASSOCIATION OF SCHOOLS AND COLLEGES
Accredits institutions in Alaska, Idaho, Montana, Nevada, Oregon, Utah, and Washington.
Sandra E. Elman, Executive Director
Commission on Colleges
11130 Northeast 33rd Place, Suite 120
Seattle, Washington 98004
Telephone: 425-827-2005
Fax: 425-827-3395
E-mail: selman@u.washington.edu

SOUTHERN ASSOCIATION OF COLLEGES AND SCHOOLS
Accredits institutions in Alabama, Florida, Georgia, Kentucky, Louisiana, Mississippi, North Carolina, South Carolina, Tennessee, Texas, and Virginia.
James T. Rogers, Executive Director
Commission on Colleges
1866 Southern Lane
Decatur, Georgia 30033-4097
Telephone: 404-679-4500
Fax: 404-679-4558
E-mail: jrogers@sacscoc.org

WESTERN ASSOCIATION OF SCHOOLS AND COLLEGES
Accredits institutions in California, Guam, and Hawaii.
Ralph A. Wolff, Executive Director
Accrediting Commission for Senior Colleges and Universities
Mills College
P.O. Box 9990
Oakland, California 94613-0990
Telephone: 510-632-5000
Fax: 510-632-8361
E-mail: rwolff@wasc.mills.edu

Institutional Accrediting Agencies—Other

ACCREDITING COUNCIL FOR INDEPENDENT COLLEGES AND SCHOOLS
Stephen D. Parker, Executive Director
750 First Street, NE, Suite 980
Washington, D.C. 20002-4241
Telephone: 202-336-6780
Fax: 202-842-2593
E-mail: acics@digex.net
World Wide Web: www.acics.org

DISTANCE EDUCATION AND TRAINING COUNCIL
Michael P. Lambert, Executive Secretary
1601 Eighteenth Street, NW
Washington, D.C. 20009-2529
Telephone: 202-234-5100
Fax: 202-332-1386
E-mail: detc@detc.org
World Wide Web: www.detc.org

Specialized Accrediting Agencies
[Only Book 1 of Peterson's Annual Guides to Graduate Study includes the complete list of specialized accrediting groups recognized by the U.S. Department of Education and the Council on Higher Education Accreditation (CHEA). The lists in Books 2, 4, 5, and 6 are abridged, and there are no such recognized specialized accrediting bodies for the programs in Book 3.]

ENGINEERING
George D. Peterson, Executive Director
Accreditation Board for Engineering and Technology, Inc.
111 Market Place, Suite 1050
Baltimore, Maryland 21202
Telephone: 410-347-7700
Fax: 410-625-2238
E-mail: kaberle@abet.ba.md.us

Directory of Institutions with Programs in Engineering and Applied Sciences

This directory lists institutions in alphabetical order and includes beneath each name the academic fields in engineering and applied sciences in which each institution offers graduate programs. The degree level in each field is also indicated, provided that the institution has supplied that information in response to Peterson's Annual Survey of Graduate Institutions. An *M* indicates that a master's degree program is offered; a *D* indicates that a doctoral degree program is offered; a *P* indicates that the first-professional degree is offered; an *O* signifies that other advanced degrees (e.g., certificates or specialist degrees) are offered; and an * (asterisk) indicates that an In-Depth Description and/or Announcement is located in this volume. See the index for the page number of the In-Depth Description and/or Announcement.

ACADIA UNIVERSITY
Computer Science — M

AIR FORCE INSTITUTE OF TECHNOLOGY
Aerospace/Aeronautical Engineering — M,D
Computer Engineering — M,D
Computer Science — M,D
Electrical Engineering — M,D
Engineering Physics — M,D
Engineering and Applied Sciences—General — M,D
Environmental Engineering — M
Nuclear Engineering — M,D
Operations Research — M,D
Systems Engineering — M,D

ALABAMA AGRICULTURAL AND MECHANICAL UNIVERSITY
Computer Science — M
Engineering and Applied Sciences—General — M
Materials Sciences — M,D

ALASKA PACIFIC UNIVERSITY
Telecommunications Management — M

ALCORN STATE UNIVERSITY
Computer Science — M
Information Science — M

ALFRED UNIVERSITY
Ceramic Sciences and Engineering — M,D*
Electrical Engineering — M
Materials Sciences — M,D
Mechanical Engineering — M

ALLEGHENY UNIVERSITY OF THE HEALTH SCIENCES (SEE MCP HAHNEMANN UNIVERSITY)

ALLENTOWN COLLEGE OF ST. FRANCIS DE SALES
Information Science — M

AMERICAN UNIVERSITY
Computer Science — M*
Information Science — M,O

AMERICAN UNIVERSITY IN CAIRO
Computer Science — M
Engineering and Applied Sciences—General — M,O

ANGELO STATE UNIVERSITY
Computer Science — M

APPALACHIAN STATE UNIVERSITY
Computer Science — M

ARIZONA STATE UNIVERSITY
Aerospace/Aeronautical Engineering — M,D*
Bioengineering — M,D*
Biomedical Engineering — M,D
Chemical Engineering — M,D*
Civil Engineering — M,D
Computer Science — M,D*
Construction Engineering and Management — M
Electrical Engineering — M,D*
Engineering and Applied Sciences—General — M,D*
Geological Engineering — M,D
Industrial/Management Engineering — M,D*
Materials Engineering — M,D
Materials Sciences — M,D*
Mechanical Engineering — M,D

ARIZONA STATE UNIVERSITY EAST
Aerospace/Aeronautical Engineering — M
Computer Engineering — M
Electrical Engineering — M
Engineering and Applied Sciences—General — M*
Information Science — M
Manufacturing Engineering — M

ARKANSAS STATE UNIVERSITY
Computer Science — M

ATHABASCA UNIVERSITY
Management of Technology — M,O

AUBURN UNIVERSITY
Aerospace/Aeronautical Engineering — M,D*
Agricultural Engineering — M,D
Chemical Engineering — M,D
Civil Engineering — M,D*

Computer Engineering — M,D
Computer Science — M,D
Construction Engineering and Management — M,D
Electrical Engineering — M,D
Engineering and Applied Sciences—General — M,D
Environmental Engineering — M,D
Geotechnical Engineering — M,D
Hydraulics — M,D
Industrial/Management Engineering — M,D
Materials Engineering — M,D
Mechanical Engineering — M,D
Structural Engineering — M,D
Systems Engineering — M,D
Textile Sciences and Engineering — M
Transportation and Highway Engineering — M,D

AZUSA PACIFIC UNIVERSITY
Computer Science — M,O*
Software Engineering — M,O
Telecommunications — M,O

BALL STATE UNIVERSITY
Computer Science — M
Information Science — M

BARRY UNIVERSITY
Information Science — M

BARUCH COLLEGE OF THE CITY UNIVERSITY OF NEW YORK
Operations Research — M

BAYLOR COLLEGE OF MEDICINE
Biomedical Engineering

BAYLOR UNIVERSITY
Computer Science — M

BENTLEY COLLEGE
Management of Technology — M,O

BOISE STATE UNIVERSITY
Computer Science — M

BOSTON UNIVERSITY
Aerospace/Aeronautical Engineering — M,D*
Biomedical Engineering — M,D*
Computer Engineering — M,D
Computer Science — M,D
Electrical Engineering — M,D*
Energy Management and Policy — M
Engineering and Applied Sciences—General — M,D*
Management of Technology — M
Manufacturing Engineering — M,D*
Mechanical Engineering — M,D
Systems Engineering — M,D
Telecommunications — M

BOWIE STATE UNIVERSITY
Computer Science — M

BOWLING GREEN STATE UNIVERSITY
Computer Science — M
Manufacturing Engineering — M

BRADLEY UNIVERSITY
Civil Engineering — M
Computer Science — M
Construction Engineering and Management — M
Electrical Engineering — M
Engineering and Applied Sciences—General — M
Industrial/Management Engineering — M
Information Science — M
Manufacturing Engineering — M
Mechanical Engineering — M

BRANDEIS UNIVERSITY
Computer Science — M,D*

BRIDGEWATER STATE COLLEGE
Computer Science — M

BRIGHAM YOUNG UNIVERSITY
Chemical Engineering — M,D
Civil Engineering — M,D
Computer Science — M,D
Electrical Engineering — M,D
Engineering and Applied Sciences—General — M,D*
Management of Technology — M
Manufacturing Engineering — M
Mechanical Engineering — M,D

BROOKLYN COLLEGE OF THE CITY UNIVERSITY OF NEW YORK
Computer Science — M,D
Information Science — M,D

BROWN UNIVERSITY
Aerospace/Aeronautical Engineering — M,D
Biomedical Engineering — M,D
Biotechnology — M,D
Chemical Engineering — M,D
Computer Science — M,D
Electrical Engineering — M,D
Engineering and Applied Sciences—General — M,D*
Materials Sciences — M,D
Mechanical Engineering — M,D
Mechanics — M,D

BUCKNELL UNIVERSITY
Chemical Engineering — M
Civil Engineering — M
Electrical Engineering — M
Engineering and Applied Sciences—General — M
Mechanical Engineering — M

CALIFORNIA INSTITUTE OF TECHNOLOGY
Aerospace/Aeronautical Engineering — M,D,O
Chemical Engineering — M,D
Civil Engineering — M,D,O
Computer Science — M,D*
Electrical Engineering — M,D,O
Engineering and Applied Sciences—General — M,D,O*
Environmental Engineering — M,D*
Materials Sciences — M,D
Mechanical Engineering — M,D,O
Mechanics — M,D
Systems Engineering — D

CALIFORNIA NATIONAL UNIVERSITY FOR ADVANCED STUDIES
Engineering and Applied Sciences—General — M

CALIFORNIA POLYTECHNIC STATE UNIVERSITY, SAN LUIS OBISPO
Aerospace/Aeronautical Engineering — M*
Biochemical Engineering — M
Civil Engineering — M
Computer Science — M
Electrical Engineering — M
Engineering and Applied Sciences—General — M*
Engineering Management — M
Environmental Engineering — M
Industrial/Management Engineering — M
Management of Technology — M
Materials Sciences — M
Mechanical Engineering — M
Water Resources Engineering — M

CALIFORNIA STATE POLYTECHNIC UNIVERSITY, POMONA
Computer Science — M
Electrical Engineering — M
Engineering and Applied Sciences—General — M

CALIFORNIA STATE UNIVERSITY, CHICO
Computer Science — M
Electrical Engineering — M
Engineering and Applied Sciences—General — M
Mechanical Engineering — M

CALIFORNIA STATE UNIVERSITY, FRESNO
Civil Engineering — M
Computer Science — M
Electrical Engineering — M
Engineering and Applied Sciences—General — M
Industrial/Management Engineering — M
Mechanical Engineering — M

CALIFORNIA STATE UNIVERSITY, FULLERTON
Civil Engineering — M
Computer Science — M
Electrical Engineering — M
Engineering and Applied Sciences—General — M
Information Science — M
Mechanical Engineering — M
Mechanics — M
Operations Research — M
Systems Engineering — M

CALIFORNIA STATE UNIVERSITY, HAYWARD
Computer Science — M
Operations Research — M

CALIFORNIA STATE UNIVERSITY, LONG BEACH
Aerospace/Aeronautical Engineering — M
Civil Engineering — M,O
Computer Engineering — M
Computer Science — M
Electrical Engineering — M
Engineering and Applied Sciences—General — M,D,O*
Mechanical Engineering — M

CALIFORNIA STATE UNIVERSITY, LOS ANGELES
Civil Engineering — M
Electrical Engineering — M
Engineering and Applied Sciences—General — M
Mechanical Engineering — M
Technology and Public Policy — M

CALIFORNIA STATE UNIVERSITY, NORTHRIDGE
Aerospace/Aeronautical Engineering — M
Biomedical Engineering — M
Civil Engineering — M
Computer Engineering — M
Computer Science — M
Electrical Engineering — M
Engineering Management — M
Engineering and Applied Sciences—General — M
Industrial/Management Engineering — M
Materials Engineering — M
Mechanical Engineering — M
Mechanics — M
Structural Engineering — M

CALIFORNIA STATE UNIVERSITY, SACRAMENTO
Civil Engineering — M
Computer Science — M
Electrical Engineering — M
Engineering and Applied Sciences—General — M
Mechanical Engineering — M
Software Engineering — M

CALIFORNIA STATE UNIVERSITY, SAN BERNARDINO
Computer Science — M

CALIFORNIA STATE UNIVERSITY, SAN MARCOS
Computer Science — M

CAPITOL COLLEGE
Information Science — M
Systems Engineering — M
Telecommunications Management — M

CARLETON UNIVERSITY
Aerospace/Aeronautical Engineering — M,D
Civil Engineering — M,D
Computer Science — M,D
Electrical Engineering — M,D
Engineering and Applied Sciences—General — M,D
Environmental Engineering — M,D
Information Science — M,D
Materials Engineering — M,D
Mechanical Engineering — M,D*
Systems Engineering — M,D
Systems Science — M
Telecommunications Management — M

CARLOW COLLEGE
Management of Technology — M

CARNEGIE MELLON UNIVERSITY
Artificial Intelligence/Robotics — M,D*
Bioengineering — M,D
Biomedical Engineering — M,D*
Chemical Engineering — M,D
Civil Engineering — M,D*
Computer Engineering — M,D*
Computer Science — M,D*
Electrical Engineering — M,D
Engineering and Applied Sciences—General — M,D
Environmental Engineering — M,D
Information Science — M,D*
Materials Engineering — M,D
Materials Sciences — M,D*
Mechanical Engineering — M,D*
Operations Research — D

Carnegie Mellon University (continued)

Polymer Science and Engineering M,D
Software Engineering M,D*
Technology and Public Policy M,D*

CASE WESTERN RESERVE UNIVERSITY

Bioengineering D
Biomedical Engineering M,D*
Chemical Engineering M,D
Civil Engineering M,D
Computer Engineering M,D
Computer Science M,D
Electrical Engineering M,D
Engineering and Applied Sciences—General M,D*
Information Science M,D
Materials Sciences M,D
Mechanical Engineering M,D
Medical Informatics M,D
Operations Research M,D*
Polymer Science and Engineering M,D*
Systems Engineering M,D

THE CATHOLIC UNIVERSITY OF AMERICA

Artificial Intelligence/Robotics M,D
Biomedical Engineering M,D
Civil Engineering M,D*
Computer Science M,D
Construction Engineering and Management M
Electrical Engineering M,D
Engineering Design D
Engineering Management M
Engineering and Applied Sciences—General M,D*
Environmental Engineering M
Geotechnical Engineering M
Mechanical Engineering M
Mechanics M,D
Structural Engineering M,D

CENTRAL MICHIGAN UNIVERSITY

Computer Science M
Software Engineering M,O

CENTRAL MISSOURI STATE UNIVERSITY

Engineering and Applied Sciences—General M,D,O
Industrial/Management Engineering M,D
Management of Technology M,D
Transportation and Highway Engineering M

CHRISTIAN BROTHERS UNIVERSITY

Engineering and Applied Sciences—General M

CHRISTOPHER NEWPORT UNIVERSITY

Computer Science M

CITY COLLEGE OF THE CITY UNIVERSITY OF NEW YORK

Chemical Engineering M,D
Civil Engineering M,D
Computer Science M,D
Electrical Engineering M,D
Engineering and Applied Sciences—General M,D*
Mechanical Engineering M,D

CLAREMONT GRADUATE UNIVERSITY

Information Science M,D*
Operations Research M,D

CLARK ATLANTA UNIVERSITY

Computer Science M
Information Science M

CLARKSON UNIVERSITY

Chemical Engineering M,D
Civil Engineering M,D
Computer Engineering M,D
Computer Science M,D*
Electrical Engineering M,D
Engineering Management M
Engineering and Applied Sciences—General M,D*
Environmental Engineering M,D
Mechanical Engineering M,D

CLEMSON UNIVERSITY

Agricultural Engineering M,D
Bioengineering M,D*
Biomedical Engineering M,D
Ceramic Sciences and Engineering M,D
Chemical Engineering M,D
Civil Engineering M,D
Computer Engineering M,D
Computer Science M,D*
Construction Engineering and Management M
Electrical Engineering M,D
Engineering and Applied Sciences—General M,D*
Environmental Engineering M,D*
Industrial/Management Engineering M,D
Materials Engineering M,D*
Materials Sciences M,D*
Mechanical Engineering M,D
Mechanics M,D
Operations Research M,D
Polymer Science and Engineering M,D
Textile Sciences and Engineering M,D

CLEVELAND STATE UNIVERSITY

Biomedical Engineering D*
Chemical Engineering M,D
Civil Engineering M,D
Computer Engineering M,D
Electrical Engineering M,D
Engineering and Applied Sciences—General M,D*
Industrial/Management Engineering M,D
Mechanical Engineering M,D

COLEMAN COLLEGE

Information Science M

THE COLLEGE OF SAINT ROSE

Computer Science M
Information Science M

COLLEGE OF ST. SCHOLASTICA

Medical Informatics M

COLLEGE OF STATEN ISLAND OF THE CITY UNIVERSITY OF NEW YORK

Computer Science M,D

COLLEGE OF WILLIAM AND MARY

Computer Science M,D*
Engineering and Applied Sciences—General M,D*
Operations Research M*

COLORADO SCHOOL OF MINES

Chemical Engineering M,D
Computer Science M,D
Energy Management and Policy M,O
Engineering and Applied Sciences—General M,D,O
Environmental Engineering M,D
Geological Engineering M,D,O
Materials Engineering M,D
Materials Sciences M,D
Metallurgical Engineering and Metallurgy M,D
Mineral/Mining Engineering M,D
Petroleum Engineering M,D
Systems Engineering M,D

COLORADO STATE UNIVERSITY

Agricultural Engineering M,D
Bioengineering M,D
Biomedical Engineering M,D
Chemical Engineering M,D
Computer Engineering M,D
Computer Science M,D*
Construction Engineering and Management M,D
Electrical Engineering M,D
Energy and Power Engineering M,D
Engineering Management M
Engineering and Applied Sciences—General M,D
Environmental Engineering M,D
Geotechnical Engineering M,D
Hydraulics M,D
Industrial/Management Engineering M,D
Management of Technology M,D
Manufacturing Engineering M,D
Materials Engineering M,D
Mechanical Engineering M,D
Mechanics M,D
Structural Engineering M,D
Technology and Public Policy M,D
Waste Management M,D

COLORADO TECHNICAL UNIVERSITY

Computer Engineering M
Computer Science M,D
Electrical Engineering M

COLORADO TECHNICAL UNIVERSITY DENVER CAMPUS

Computer Engineering M
Computer Science M,D
Electrical Engineering M

COLUMBIA UNIVERSITY

Biomedical Engineering M,D
Chemical Engineering M,D,O
Civil Engineering M,D,O
Computer Science M,D,O*
Construction Engineering and Management M
Electrical Engineering M,D,O*
Engineering and Applied Sciences—General M,D,O*
Environmental Engineering M,D
Geological Engineering M,D*
Industrial/Management Engineering M,D,O*
Materials Engineering M,D,O
Materials Sciences M,D,O
Mechanical Engineering M,D
Mechanics M,D,O
Medical Informatics M,D*
Metallurgical Engineering and Metallurgy M,D,O
Mineral/Mining Engineering D,O
Operations Research M,D,O
Telecommunications M,D,O

COLUMBUS STATE UNIVERSITY

Computer Science M

CONCORDIA UNIVERSITY (CANADA)

Aerospace/Aeronautical Engineering M
Civil Engineering M,D,O
Computer Engineering M,D
Computer Science M,D,O
Construction Engineering and Management M,D,O
Electrical Engineering M,D
Engineering and Applied Sciences—General M,D,O*
Mechanical Engineering M,D,O
Software Engineering M,D,O

CORNELL UNIVERSITY

Aerospace/Aeronautical Engineering M,D
Agricultural Engineering M,D*
Artificial Intelligence/Robotics M,D
Biochemical Engineering M,D
Bioengineering M,D
Biomedical Engineering M,D
Chemical Engineering M,D
Civil Engineering M,D*
Computer Engineering M,D
Computer Science M,D*
Electrical Engineering M,D*
Engineering Physics M,D
Engineering and Applied Sciences—General M,D
Environmental Engineering M,D
Ergonomics and Human Factors M
Geotechnical Engineering M,D
Industrial/Management Engineering M,D
Manufacturing Engineering M,D
Materials Engineering M,D
Materials Sciences M,D
Mechanical Engineering M,D
Mechanics M,D
Nuclear Engineering M,D*
Operations Research M,D*
Polymer Science and Engineering M,D
Structural Engineering M,D
Textile Sciences and Engineering M,D
Transportation and Highway Engineering M,D
Water Resources Engineering M,D

CREIGHTON UNIVERSITY

Computer Science M

DALHOUSIE UNIVERSITY

Agricultural Engineering M,D
Bioengineering M,D
Chemical Engineering M,D
Civil Engineering M,D
Computer Engineering M,D
Computer Science M,D
Electrical Engineering M,D
Engineering and Applied Sciences—General M,D*
Industrial/Management Engineering M,D
Mechanical Engineering M,D
Metallurgical Engineering and Metallurgy M,D
Mineral/Mining Engineering M,D

DARTMOUTH COLLEGE

Biochemical Engineering M,D
Biomedical Engineering M,D
Biotechnology M,D
Computer Engineering M,D
Computer Science M,D*
Electrical Engineering M,D
Engineering Management M
Engineering and Applied Sciences—General M,D*
Environmental Engineering M,D
Materials Engineering M,D
Materials Sciences M,D
Mechanical Engineering M,D

DEPAUL UNIVERSITY

Computer Science M,D*
Information Science M
Software Engineering M
Telecommunications M*

DREXEL UNIVERSITY

Biochemical Engineering M
Biomedical Engineering M,D*
Chemical Engineering M,D*
Civil Engineering M,D*
Computer Engineering M,D
Computer Science M*
Electrical Engineering M,D*
Engineering Management M,D*
Engineering and Applied Sciences—General M,D*
Environmental Engineering M,D*
Geological Engineering M*
Information Science M,D,O*
Manufacturing Engineering M,D*
Materials Engineering M,D*
Mechanical Engineering M,D
Mechanics M,D*
Software Engineering M
Telecommunications M*

DUKE UNIVERSITY

Biomedical Engineering M,D*
Civil Engineering M,D*
Computer Engineering M,D
Computer Science M,D*
Electrical Engineering M,D*
Engineering Management M*
Engineering and Applied Sciences—General M,D
Environmental Engineering M,D
Materials Sciences M,D
Mechanical Engineering M,D*
Medical Informatics M,O

EAST CAROLINA UNIVERSITY

Biotechnology M
Computer Science M
Engineering and Applied Sciences—General M

EASTERN ILLINOIS UNIVERSITY

Engineering and Applied Sciences—General M

EASTERN KENTUCKY UNIVERSITY

Manufacturing Engineering M

EASTERN MICHIGAN UNIVERSITY

Computer Science M
Industrial/Management Engineering M
Polymer Science and Engineering M
Technology and Public Policy M

EASTERN WASHINGTON UNIVERSITY

Computer Science M

EAST STROUDSBURG UNIVERSITY OF PENNSYLVANIA

Computer Science M

EAST TENNESSEE STATE UNIVERSITY

Computer Science M
Engineering and Applied Sciences—General M
Information Science M
Manufacturing Engineering M

ÉCOLE DES HAUTES ÉTUDES COMMERCIALES

Energy Management and Policy O

ÉCOLE POLYTECHNIQUE DE MONTRÉAL

Aerospace/Aeronautical Engineering M,D,O
Biomedical Engineering M,D,O
Chemical Engineering M,D,O
Civil Engineering M,D,O
Computer Engineering M,D,O
Computer Science M,D,O
Electrical Engineering M,D,O
Engineering Physics M,D,O
Engineering and Applied Sciences—General M,D,O
Environmental Engineering M,D,O
Geological Engineering M,D,O
Geotechnical Engineering M,D,O
Hydraulics M,D,O
Industrial/Management Engineering M,O
Management of Technology M,O
Materials Engineering M,D,O
Materials Sciences M,D,O
Mechanical Engineering M,D,O
Mechanics M,D,O
Metallurgical Engineering and Metallurgy M,D,O
Mineral/Mining Engineering M,D,O

Nuclear Engineering	M,D,O
Operations Research	M,D,O
Structural Engineering	M,D,O
Transportation and Highway Engineering	M,D,O

EMBRY-RIDDLE AERONAUTICAL UNIVERSITY

Aerospace/Aeronautical Engineering	M*
Ergonomics and Human Factors	M*
Operations Research	M*
Software Engineering	M*
Systems Engineering	M

EMBRY-RIDDLE AERONAUTICAL UNIVERSITY, EXTENDED CAMPUS

Aerospace/Aeronautical Engineering	M
Management of Technology	M

EMORY UNIVERSITY

Computer Science	M,D*

FAIRFIELD UNIVERSITY

Engineering and Applied Sciences—General	M
Management of Technology	M
Software Engineering	M

FAIRLEIGH DICKINSON UNIVERSITY, FLORHAM-MADISON CAMPUS

Computer Science	M

FAIRLEIGH DICKINSON UNIVERSITY, TEANECK–HACKENSACK CAMPUS

Computer Engineering	M
Computer Science	M
Electrical Engineering	M
Engineering and Applied Sciences—General	M
Systems Science	M

FITCHBURG STATE COLLEGE

Computer Science	M

FLORIDA AGRICULTURAL AND MECHANICAL UNIVERSITY

Chemical Engineering	M,D
Civil Engineering	M,D*
Electrical Engineering	M,D
Engineering and Applied Sciences—General	M,D*
Environmental Engineering	M,D
Industrial/Management Engineering	M
Mechanical Engineering	M,D*

FLORIDA ATLANTIC UNIVERSITY

Civil Engineering	M
Computer Engineering	M,D
Computer Science	M,D
Electrical Engineering	M,D
Engineering and Applied Sciences—General	M,D
Manufacturing Engineering	M
Mechanical Engineering	M,D
Ocean Engineering	M,D
Systems Engineering	M

FLORIDA INSTITUTE OF TECHNOLOGY

Aerospace/Aeronautical Engineering	M,D*
Biotechnology	M
Chemical Engineering	M,D
Civil Engineering	M,D*
Computer Engineering	M,D
Computer Science	M,D*
Electrical Engineering	M,D*
Engineering Management	M
Engineering and Applied Sciences—General	M,D
Environmental Engineering	M
Ergonomics and Human Factors	M
Geotechnical Engineering	M,D
Information Science	M
Mechanical Engineering	M,D*
Ocean Engineering	M,D*
Operations Research	M,D
Structural Engineering	M,D
Water Resources Engineering	M,D

FLORIDA INTERNATIONAL UNIVERSITY

Civil Engineering	M,D
Computer Engineering	M
Computer Science	M,D*
Construction Engineering and Management	
Electrical Engineering	M,D
Engineering and Applied Sciences—General	M,D
Environmental Engineering	M
Industrial/Management Engineering	M

Mechanical Engineering	M,D

FLORIDA STATE UNIVERSITY

Chemical Engineering	M,D
Civil Engineering	M,D*
Computer Science	M,D
Electrical Engineering	M,D
Engineering and Applied Sciences—General	M,D*
Environmental Engineering	M,D
Industrial/Management Engineering	M,D
Mechanical Engineering	M,D*
Software Engineering	M,D

FORDHAM UNIVERSITY

Computer Science	M

GANNON UNIVERSITY

Electrical Engineering	M
Engineering and Applied Sciences—General	M
Mechanical Engineering	M
Software Engineering	M

GEORGE MASON UNIVERSITY

Computer Science	M
Electrical Engineering	M
Engineering Physics	M
Engineering and Applied Sciences—General	M,D*
Information Science	M,D
Medical Informatics	M
Operations Research	M
Software Engineering	M
Systems Engineering	M
Telecommunications	M

THE GEORGE WASHINGTON UNIVERSITY

Aerospace/Aeronautical Engineering	M,D,O
Civil Engineering	M,D,O*
Computer Engineering	M,D,O
Computer Science	M,D,O*
Electrical Engineering	M,D,O
Engineering Management	M,D,O*
Engineering and Applied Sciences—General	M,D,O*
Environmental Engineering	M,D,O
Management of Technology	M,D
Materials Sciences	M,D
Mechanical Engineering	M,D,O*
Operations Research	M,D,O
Systems Engineering	M,D,O
Technology and Public Policy	M*
Telecommunications	M

GEORGIA INSTITUTE OF TECHNOLOGY

Aerospace/Aeronautical Engineering	M,D*
Bioengineering	M,D,O*
Biomedical Engineering	M,D,O
Ceramic Sciences and Engineering	M,D
Chemical Engineering	M,D,O
Civil Engineering	M,D
Computer Engineering	M,D
Computer Science	M,D*
Construction Engineering and Management	M,D
Electrical Engineering	M,D
Engineering and Applied Sciences—General	M,D,O
Environmental Engineering	M,D
Industrial/Management Engineering	M,D*
Management of Technology	M
Materials Engineering	M,D
Mechanical Engineering	M,D*
Mechanics	M,D
Metallurgical Engineering and Metallurgy	M,D
Nuclear Engineering	M,D*
Operations Research	M
Paper and Pulp Engineering	O
Polymer Science and Engineering	M
Systems Engineering	M,D
Textile Sciences and Engineering	M,D

GEORGIA SOUTHERN UNIVERSITY

Engineering and Applied Sciences—General	M

GEORGIA SOUTHWESTERN STATE UNIVERSITY

Computer Science	M
Information Science	M

GEORGIA STATE UNIVERSITY

Computer Science	M*
Operations Research	M,D

GOLDEN GATE UNIVERSITY	
Engineering and Applied Sciences—General	M,O
Telecommunications Management	M,D,O

GONZAGA UNIVERSITY

Electrical Engineering	M
Engineering and Applied Sciences—General	M

GOVERNORS STATE UNIVERSITY

Computer Science	M

GRADUATE SCHOOL AND UNIVERSITY CENTER OF THE CITY UNIVERSITY OF NEW YORK

Chemical Engineering	D
Civil Engineering	D
Computer Science	D
Electrical Engineering	D
Engineering and Applied Sciences—General	D
Mechanical Engineering	D

GRAND VALLEY STATE UNIVERSITY

Information Science	M
Software Engineering	M

HAMPTON UNIVERSITY

Computer Science	M

HARVARD UNIVERSITY

Biomedical Engineering	M,D*
Computer Science	D
Engineering and Applied Sciences—General	M,D,O*
Environmental Engineering	M,D
Information Science	M,O

HARVEY MUDD COLLEGE

Engineering and Applied Sciences—General	M

HOFSTRA UNIVERSITY

Computer Science	M

HOLLINS UNIVERSITY

Computer Science	M,O

HOOD COLLEGE

Computer Science	M
Information Science	M

HOWARD UNIVERSITY

Aerospace/Aeronautical Engineering	M,D
Artificial Intelligence/Robotics	M,D
Biotechnology	M,D
Chemical Engineering	M
Civil Engineering	M*
Computer Science	M
Electrical Engineering	M,D*
Engineering and Applied Sciences—General	M,D*
Materials Engineering	M,D
Materials Sciences	M,D
Mechanical Engineering	M,D
Mechanics	M,D

HUNTER COLLEGE OF THE CITY UNIVERSITY OF NEW YORK

Computer Science	D

IDAHO STATE UNIVERSITY

Engineering and Applied Sciences—General	M,D
Environmental Engineering	M,D
Nuclear Engineering	M,D
Operations Research	M,D
Waste Management	M,D

ILLINOIS INSTITUTE OF TECHNOLOGY

Aerospace/Aeronautical Engineering	M,D
Architectural Engineering	M,D
Biotechnology	M,D
Chemical Engineering	M,D*
Civil Engineering	M,D*
Computer Engineering	M,D
Computer Science	M,D*
Electrical Engineering	M,D*
Engineering and Applied Sciences—General	M,D
Environmental Engineering	M,D
Management of Technology	M,D
Manufacturing Engineering	M,D
Materials Engineering	M,D
Mechanical Engineering	M,D*
Metallurgical Engineering and Metallurgy	M,D
Software Engineering	M,D
Telecommunications	M,D

ILLINOIS STATE UNIVERSITY	
Computer Science	M
Industrial/Management Engineering	M

INDIANA STATE UNIVERSITY

Computer Engineering	M
Engineering and Applied Sciences—General	M,D
Industrial/Management Engineering	M

INDIANA UNIVERSITY BLOOMINGTON

Computer Science	M,D*

INDIANA UNIVERSITY–PURDUE UNIVERSITY FORT WAYNE

Computer Science	M
Engineering and Applied Sciences—General	M
Operations Research	M

INDIANA UNIVERSITY–PURDUE UNIVERSITY INDIANAPOLIS

Biomedical Engineering	M,D
Computer Science	M
Electrical Engineering	M,D
Engineering and Applied Sciences—General	M,D
Mechanical Engineering	M

INSTITUTE OF PAPER SCIENCE AND TECHNOLOGY

Chemical Engineering	M,D
Mechanical Engineering	M,D
Paper and Pulp Engineering	M,D

INSTITUTE OF TEXTILE TECHNOLOGY

Textile Sciences and Engineering	M

INSTITUTO TECNOLÓGICO Y DE ESTUDIOS SUPERIORES DE MONTERREY, CAMPUS CHIHUAHUA

Computer Engineering	M,O
Electrical Engineering	M,O
Engineering Management	M,O
Industrial/Management Engineering	M,O
Mechanical Engineering	M,O
Systems Engineering	M,O

INSTITUTO TECNOLÓGICO Y DE ESTUDIOS SUPERIORES DE MONTERREY, CAMPUS ESTADO DE MÉXICO

Computer Science	M,D
Engineering and Applied Sciences—General	M,D
Environmental Engineering	M,D
Industrial/Management Engineering	M,D
Manufacturing Engineering	M,D
Materials Engineering	M,D

INSTITUTO TECNOLÓGICO Y DE ESTUDIOS SUPERIORES DE MONTERREY, CAMPUS LAGUNA

Industrial/Management Engineering	M

INSTITUTO TECNOLÓGICO Y DE ESTUDIOS SUPERIORES DE MONTERREY, CAMPUS MONTERREY

Agricultural Engineering	M,D
Artificial Intelligence/Robotics	M,D
Biotechnology	M,D
Chemical Engineering	M,D
Civil Engineering	M,D
Computer Science	M,D
Electrical Engineering	M,D
Engineering and Applied Sciences—General	M,D
Environmental Engineering	M,D
Industrial/Management Engineering	M,D
Information Science	M,D
Manufacturing Engineering	M,D
Mechanical Engineering	M,D
Systems Engineering	M,D

INSTITUTO TECNOLÓGICO Y DE ESTUDIOS SUPERIORES DE MONTERREY, CAMPUS MORELOS

Computer Science	M,D
Information Science	M,D
Management of Technology	M,D

INSTITUTO TECNOLÓGICO Y DE ESTUDIOS SUPERIORES DE MONTERREY, CAMPUS SONORA NORTE

Information Science	M

IONA COLLEGE

Computer Science	M

Iona College (continued)
Telecommunications M,O

IOWA STATE UNIVERSITY OF SCIENCE AND TECHNOLOGY
Aerospace/Aeronautical
  Engineering M,D
Agricultural Engineering M,D
Chemical Engineering M,D
Civil Engineering M,D
Computer Engineering M,D
Computer Science M,D*
Construction Engineering and
  Management M,D
Electrical Engineering M,D
Engineering and Applied
  Sciences—General M,D
Environmental Engineering M,D
Geotechnical Engineering M,D
Industrial/Management
  Engineering M,D
Materials Engineering M,D
Materials Sciences M,D
Mechanical Engineering M,D
Mechanics M,D
Operations Research M,D
Structural Engineering M,D
Systems Engineering M
Transportation and Highway
  Engineering M,D

JACKSON STATE UNIVERSITY
Computer Science M
Materials Sciences M

JAMES MADISON UNIVERSITY
Computer Science M

JOHNS HOPKINS UNIVERSITY
Biomedical Engineering M,D*
Chemical Engineering M,D
Civil Engineering M,D
Computer Engineering M,D
Computer Science M,D*
Electrical Engineering M,D*
Engineering and Applied
  Sciences—General M,D
Environmental Engineering M,D
Materials Engineering M,D
Materials Sciences M,D*
Mechanical Engineering M,D*
Mechanics M,D

KANSAS STATE UNIVERSITY
Agricultural Engineering M,D
Architectural Engineering M
Bioengineering M,D
Chemical Engineering M,D
Civil Engineering M,D
Computer Engineering M,D
Computer Science M,D*
Electrical Engineering M,D
Engineering Management M,D
Engineering and Applied
  Sciences—General M,D*
Industrial/Management
  Engineering M,D
Information Science M,D
Manufacturing Engineering M,D
Mechanical Engineering M,D
Nuclear Engineering M,D
Operations Research M,D
Software Engineering M

KELLER GRADUATE SCHOOL OF MANAGEMENT
Telecommunications
  Management M

KENNESAW STATE UNIVERSITY
Information Science M

KENT STATE UNIVERSITY
Computer Science M,D
Engineering and Applied
  Sciences—General M

KETTERING UNIVERSITY
Engineering Design M
Engineering and Applied
  Sciences—General M*
Manufacturing Engineering M*
Mechanical Engineering M

KIRKSVILLE COLLEGE OF OSTEOPATHIC MEDICINE
Medical Informatics M

KNOWLEDGE SYSTEMS INSTITUTE
Computer Science M*
Information Science M

KUTZTOWN UNIVERSITY OF PENNSYLVANIA
Computer Science M
Information Science M

LAKEHEAD UNIVERSITY
Computer Science M
Engineering and Applied
  Sciences—General M

LAMAR UNIVERSITY
Chemical Engineering M,D
Civil Engineering M,D
Computer Science M*
Electrical Engineering M,D
Engineering Management M
Engineering and Applied
  Sciences—General M,D
Environmental Engineering M
Industrial/Management
  Engineering M
Mechanical Engineering M,D

LA SALLE UNIVERSITY
Computer Science M

LAURENTIAN UNIVERSITY
Engineering and Applied
  Sciences—General M
Metallurgical Engineering and
  Metallurgy M
Mineral/Mining Engineering M

LAWRENCE TECHNOLOGICAL UNIVERSITY
Civil Engineering M
Engineering and Applied
  Sciences—General M
Manufacturing Engineering M
Transportation and Highway
  Engineering M

LEHIGH UNIVERSITY
Chemical Engineering M,D
Civil Engineering M,D
Computer Engineering M,D
Computer Science M,D*
Electrical Engineering M,D
Engineering and Applied
  Sciences—General M,D*
Environmental Engineering M,D
Industrial/Management
  Engineering M,D
Management of Technology M,D
Manufacturing Engineering M
Materials Engineering M,D
Materials Sciences M,D
Mechanical Engineering M,D
Mechanics M,D
Polymer Science and
  Engineering M,D
Systems Engineering M

LEHMAN COLLEGE OF THE CITY UNIVERSITY OF NEW YORK
Computer Science M

LONG ISLAND UNIVERSITY, BROOKLYN CAMPUS
Computer Science M

LONG ISLAND UNIVERSITY, C.W. POST CAMPUS
Engineering Management M
Information Science M

LONG ISLAND UNIVERSITY, ROCKLAND GRADUATE CAMPUS
Computer Science M

LOUISIANA STATE UNIVERSITY AND AGRICULTURAL AND MECHANICAL COLLEGE
Agricultural Engineering M,D
Bioengineering M,D
Chemical Engineering M,D*
Civil Engineering M,D*
Computer Engineering M,D
Computer Science M,D
Electrical Engineering M,D*
Engineering and Applied
  Sciences—General M,D
Environmental Engineering M,D
Geotechnical Engineering M,D
Industrial/Management
  Engineering M,D
Mechanical Engineering M,D
Mechanics M,D
Nuclear Engineering M
Petroleum Engineering M,D
Structural Engineering M,D
Systems Science M
Transportation and Highway
  Engineering M,D
Water Resources Engineering M,D

LOUISIANA STATE UNIVERSITY IN SHREVEPORT
Systems Engineering M
Systems Science M

LOUISIANA TECH UNIVERSITY
Biomedical Engineering M,D

Chemical Engineering M,D
Civil Engineering M,D
Computer Science M
Electrical Engineering M,D
Engineering and Applied
  Sciences—General M,D
Industrial/Management
  Engineering M,D
Manufacturing Engineering M,D
Mechanical Engineering M,D
Operations Research M,D

LOYOLA COLLEGE IN MARYLAND
Engineering and Applied
  Sciences—General M

LOYOLA MARYMOUNT UNIVERSITY
Civil Engineering M
Computer Science M
Electrical Engineering M
Engineering Management M
Engineering and Applied
  Sciences—General M
Industrial/Management
  Engineering M
Mechanical Engineering M

LOYOLA UNIVERSITY CHICAGO
Computer Science M*

LOYOLA UNIVERSITY NEW ORLEANS
Computer Science M

MAHARISHI UNIVERSITY OF MANAGEMENT
Computer Science M

MANHATTAN COLLEGE
Biotechnology M
Chemical Engineering M
Civil Engineering M
Computer Engineering M
Electrical Engineering M
Engineering and Applied
  Sciences—General M*
Environmental Engineering M
Mechanical Engineering M

MANKATO STATE UNIVERSITY (SEE MINNESOTA STATE UNIVERSITY, MANKATO)

MARIST COLLEGE
Computer Science M
Information Science M

MARQUETTE UNIVERSITY
Biomedical Engineering M,D
Civil Engineering M,D
Computer Engineering M,D
Construction Engineering and
  Management M,D
Electrical Engineering M,D
Engineering Management M,D
Engineering and Applied
  Sciences—General M,D*
Environmental Engineering M,D
Geotechnical Engineering M,D
Management of Technology M,D
Manufacturing Engineering M,D
Materials Engineering M,D
Materials Sciences M,D
Mechanical Engineering M,D
Structural Engineering M,D
Transportation and Highway
  Engineering M,D
Water Resources Engineering M,D

MARSHALL UNIVERSITY
Engineering and Applied
  Sciences—General M
Information Science M
Management of Technology M

MARYCREST INTERNATIONAL UNIVERSITY
Computer Science M

MARYMOUNT UNIVERSITY
Computer Science M

MASSACHUSETTS INSTITUTE OF TECHNOLOGY
Aerospace/Aeronautical
  Engineering M,D,O*
Bioengineering M,D*
Biomedical Engineering D*
Chemical Engineering M,D,O
Civil Engineering M,D,O
Computer Science M,D,O
Electrical Engineering M,D,O
Engineering Management M,O
Engineering and Applied
  Sciences—General M,D,O*
Environmental Engineering M,D,O
Manufacturing Engineering M*
Materials Engineering M,D,O
Materials Sciences M,D,O

Chemical Engineering M,D
Civil Engineering M,D
Computer Science M
Electrical Engineering M,D
Engineering and Applied
  Sciences—General M,D
Industrial/Management
  Engineering M,D
Manufacturing Engineering M,D
Mechanical Engineering M,D
Operations Research M,D

LOYOLA COLLEGE IN MARYLAND
Engineering and Applied
  Sciences—General M

Mechanical Engineering M,D,O
Medical Informatics M
Metallurgical Engineering and
  Metallurgy M,D,O
Nuclear Engineering M,D,O*
Ocean Engineering M,D,O*
Operations Research M,D*
Systems Engineering M,O
Technology and Public Policy M,D*
Transportation and Highway
  Engineering M,D*

MAYO GRADUATE SCHOOL
Biomedical Engineering D*

MCGILL UNIVERSITY
Agricultural Engineering M,D
Biomedical Engineering M,D
Biotechnology M,D,O
Chemical Engineering M,D
Civil Engineering M,D
Computer Science M,D
Construction Engineering and
  Management M,D
Electrical Engineering M,D*
Engineering and Applied
  Sciences—General M,D,O
Environmental Engineering M,D
Geotechnical Engineering M,D
Hydraulics M,D
Mechanical Engineering M,D
Mechanics M,D
Metallurgical Engineering and
  Metallurgy M,D
Mineral/Mining Engineering M,D,O
Structural Engineering M,D
Water Resources Engineering M,D

MCMASTER UNIVERSITY
Bioengineering M,D
Chemical Engineering M,D
Civil Engineering M,D
Computer Science M
Electrical Engineering M,D
Engineering Physics M,D
Engineering and Applied
  Sciences—General M,D
Materials Engineering M,D
Materials Sciences M,D
Mechanical Engineering M,D
Nuclear Engineering M,D

MCNEESE STATE UNIVERSITY
Chemical Engineering M
Civil Engineering M
Computer Science M
Electrical Engineering M
Engineering and Applied
  Sciences—General M
Mechanical Engineering M

MCP HAHNEMANN UNIVERSITY
Bioengineering D
Biomedical Engineering D

MEDICAL COLLEGE OF GEORGIA
Medical Informatics M

MEDICAL COLLEGE OF WISCONSIN
Medical Informatics M

MEDICAL UNIVERSITY OF SOUTH CAROLINA
Medical Informatics M

MEMORIAL UNIVERSITY OF NEWFOUNDLAND
Civil Engineering M,D
Computer Science M,D
Electrical Engineering M,D
Engineering and Applied
  Sciences—General M,D
Environmental Engineering M
Mechanical Engineering M,D
Ocean Engineering M,D

MERCER UNIVERSITY
Biomedical Engineering M
Electrical Engineering M
Engineering Management M
Engineering and Applied
  Sciences—General M
Management of Technology M
Mechanical Engineering M
Software Engineering M

MERCER UNIVERSITY, CECIL B. DAY CAMPUS
Electrical Engineering M
Engineering Management M
Engineering and Applied
  Sciences—General M
Management of Technology M
Software Engineering M

MIAMI UNIVERSITY
Computer Science M

Engineering and Applied
   Sciences—General — M
Operations Research — M
Paper and Pulp Engineering — M

MICHIGAN STATE UNIVERSITY

Agricultural Engineering — M,D
Chemical Engineering — M,D
Civil Engineering — M,D*
Computer Engineering — M,D
Computer Science — M,D*
Construction Engineering and
   Management — M
Electrical Engineering — M,D*
Engineering and Applied
   Sciences—General — M,D*
Environmental Engineering — M,D
Manufacturing Engineering — M,D
Materials Sciences — M,D*
Mechanical Engineering — M,D*
Mechanics — M,D
Operations Research — M,D
Telecommunications — M

MICHIGAN TECHNOLOGICAL UNIVERSITY

Chemical Engineering — M,D
Civil Engineering — M,D
Computer Science — M,D*
Electrical Engineering — M,D
Engineering and Applied
   Sciences—General — M,D*
Environmental Engineering — M,D
Geological Engineering — M
Geotechnical Engineering — M,D
Materials Engineering — M,D
Mechanical Engineering — M,D*
Mechanics — M
Metallurgical Engineering and
   Metallurgy — M,D
Mineral/Mining Engineering — M,D

MIDDLE TENNESSEE STATE UNIVERSITY

Aerospace/Aeronautical
   Engineering — M
Computer Science — M

MIDWESTERN STATE UNIVERSITY

Computer Science — M

MILLS COLLEGE

Computer Science — M,O

MILWAUKEE SCHOOL OF ENGINEERING

Architectural Engineering — M*
Biomedical Engineering — M*
Engineering Management — M
Engineering and Applied
   Sciences—General — M*
Environmental Engineering — M*
Medical Informatics — M*

MINNESOTA STATE UNIVERSITY, MANKATO

Computer Science — M
Electrical Engineering — M
Manufacturing Engineering — M

MISSISSIPPI COLLEGE

Computer Science — M

MISSISSIPPI STATE UNIVERSITY

Aerospace/Aeronautical
   Engineering — M
Bioengineering — M
Chemical Engineering — M
Civil Engineering — M
Computer Engineering — M,D*
Computer Science — M,D*
Electrical Engineering — M,D
Engineering and Applied
   Sciences—General — M,D
Industrial/Management
   Engineering — M
Mechanical Engineering — M
Mechanics — M

MONMOUTH UNIVERSITY

Computer Science — M
Electrical Engineering — M
Software Engineering — M*

MONTANA STATE UNIVERSITY–BOZEMAN

Chemical Engineering — M,D
Civil Engineering — M,D
Computer Science — M,D
Construction Engineering and
   Management — M,D
Electrical Engineering — M,D
Engineering and Applied
   Sciences—General — M,D
Environmental Engineering — M,D
Industrial/Management
   Engineering — M,D

Mechanical Engineering — M,D

MONTANA TECH THE UNIVERSITY OF MONTANA

Engineering and Applied
   Sciences—General — M*
Environmental Engineering — M
Geological Engineering — M
Industrial/Management
   Engineering — M
Metallurgical Engineering and
   Metallurgy — M
Mineral/Mining Engineering — M
Petroleum Engineering — M

MONTCLAIR STATE UNIVERSITY

Computer Science — M

MORGAN STATE UNIVERSITY

Engineering and Applied
   Sciences—General — M,D

MURRAY STATE UNIVERSITY

Management of Technology — M
Safety Engineering — M

NATIONAL TECHNOLOGICAL UNIVERSITY

Chemical Engineering — M
Computer Engineering — M
Computer Science — M
Electrical Engineering — M
Engineering Management — M
Engineering and Applied
   Sciences—General — M
Management of Technology — M
Manufacturing Engineering — M
Materials Engineering — M
Materials Sciences — M
Software Engineering — M
Systems Engineering — M
Transportation and Highway
   Engineering — M
Waste Management — M

NATIONAL UNIVERSITY

Electrical Engineering — M
Engineering Management — M
Engineering and Applied
   Sciences—General — M
Industrial/Management
   Engineering — M
Management of Technology — M
Software Engineering — M
Telecommunications
   Management — M

NAVAL POSTGRADUATE SCHOOL

Aerospace/Aeronautical
   Engineering — M,D,O
Computer Engineering — M,D,O
Computer Science — M,D
Electrical Engineering — M,D,O
Information Science — M
Mechanical Engineering — M,D,O
Operations Research — M,D

NEW HAMPSHIRE COLLEGE

Artificial Intelligence/Robotics — M,O

NEW JERSEY INSTITUTE OF TECHNOLOGY

Biomedical Engineering — M
Chemical Engineering — M,D,O
Civil Engineering — M,D,O
Computer Engineering — M,D,O
Computer Science — M,D
Electrical Engineering — M,D,O
Energy and Power Engineering — M,D,O
Engineering Management — M
Engineering and Applied
   Sciences—General — M,D,O*
Environmental Engineering — M,D
Industrial/Management
   Engineering — M,D
Information Science — M,D
Manufacturing Engineering — M
Materials Engineering — M,D
Materials Sciences — M
Mechanical Engineering — M,D,O
Medical Informatics — M,D
Safety Engineering — M
Telecommunications — M,D,O
Transportation and Highway
   Engineering — M,D

NEW MEXICO HIGHLANDS UNIVERSITY

Computer Science — M

NEW MEXICO INSTITUTE OF MINING AND TECHNOLOGY

Computer Science — M,D
Environmental Engineering — M
Materials Engineering — M,D
Mechanics — M
Mineral/Mining Engineering — M

Operations Research — M
Petroleum Engineering — M,D
Waste Management — M
Water Resources Engineering — M

NEW MEXICO STATE UNIVERSITY

Chemical Engineering — M,D
Civil Engineering — M,D
Computer Engineering — M,D
Computer Science — M,D*
Electrical Engineering — M,D
Engineering and Applied
   Sciences—General — M,D
Environmental Engineering — M,D
Industrial/Management
   Engineering — M,D
Mechanical Engineering — M,D

NEW YORK INSTITUTE OF TECHNOLOGY

Computer Science — M
Electrical Engineering — M
Energy Management and
   Policy — M,O
Energy and Power Engineering — M,O
Engineering and Applied
   Sciences—General — M,O
Environmental Engineering — M,O

NEW YORK MEDICAL COLLEGE

Medical Informatics — M

NEW YORK UNIVERSITY

Computer Science — M,D*
Information Science — M
Operations Research — M,D,O

NORFOLK STATE UNIVERSITY

Materials Sciences — M

NORTH CAROLINA AGRICULTURAL AND TECHNICAL STATE UNIVERSITY

Agricultural Engineering — M
Architectural Engineering — M
Chemical Engineering — M
Civil Engineering — M
Computer Science — M
Electrical Engineering — M,D
Engineering and Applied
   Sciences—General — M,D*
Industrial/Management
   Engineering — M,D
Management of Technology — M
Mechanical Engineering — M,D

NORTH CAROLINA STATE UNIVERSITY

Aerospace/Aeronautical
   Engineering — M,D
Agricultural Engineering — M,D
Bioengineering — M,D
Biotechnology — M
Chemical Engineering — M,D
Civil Engineering — M,D
Computer Engineering — M,D
Computer Science — M,D
Electrical Engineering — M,D*
Engineering and Applied
   Sciences—General — M,D,O*
Industrial/Management
   Engineering — M,D*
Manufacturing Engineering — M*
Materials Engineering — M,D*
Materials Sciences — M,D
Mechanical Engineering — M,D*
Nuclear Engineering — M,D
Operations Research — M,D
Paper and Pulp Engineering — M,D
Telecommunications — M
Textile Sciences and
   Engineering — M,D

NORTH CENTRAL COLLEGE

Computer Science — M

NORTH DAKOTA STATE UNIVERSITY

Agricultural Engineering — M,D
Civil Engineering — M,D
Computer Science — M,D
Electrical Engineering — M
Engineering and Applied
   Sciences—General — M,D
Environmental Engineering — M,D
Industrial/Management
   Engineering — M
Manufacturing Engineering — M
Mechanical Engineering — M
Mechanics — M
Operations Research — M,D
Polymer Science and
   Engineering — M,D*

NORTHEASTERN ILLINOIS UNIVERSITY

Computer Science — M

NORTHEASTERN UNIVERSITY

Chemical Engineering — M,D
Civil Engineering — M,D

Computer Engineering — M,D
Computer Science — M,D*
Electrical Engineering — M,D*
Engineering Management — M,D
Engineering and Applied
   Sciences—General — M,D*
Environmental Engineering — M,D
Industrial/Management
   Engineering — M,D
Information Science — M
Manufacturing Engineering — M,D
Mechanical Engineering — M,D*
Operations Research — M,D
Systems Engineering — M

NORTHERN ARIZONA UNIVERSITY

Engineering and Applied
   Sciences—General — M

NORTHERN ILLINOIS UNIVERSITY

Computer Science — M
Electrical Engineering — M
Engineering and Applied
   Sciences—General — M*
Industrial/Management
   Engineering — M
Mechanical Engineering — M

NORTHERN KENTUCKY UNIVERSITY

Management of Technology — M

NORTHWESTERN POLYTECHNIC UNIVERSITY

Computer Engineering — M
Computer Science — M
Electrical Engineering — M
Engineering and Applied
   Sciences—General — M

NORTHWESTERN UNIVERSITY

Biomedical Engineering — M,D*
Biotechnology — M,D
Chemical Engineering — M,D
Civil Engineering — M,D
Computer Engineering — M,D
Computer Science — D*
Electrical Engineering — M,D*
Engineering Management — M
Engineering and Applied
   Sciences—General — M,D
Environmental Engineering — M,D
Geotechnical Engineering — M,D
Industrial/Management
   Engineering — M,D*
Information Science — M
Manufacturing Engineering — M
Materials Engineering — M,D
Materials Sciences — M,D
Mechanical Engineering — M,D*
Mechanics — M,D
Operations Research — M,D
Structural Engineering — M,D
Technology and Public Policy — M,O
Telecommunications
   Management — M,O
Telecommunications — M,O*
Transportation and Highway
   Engineering — M,D

NORTHWEST MISSOURI STATE UNIVERSITY

Computer Science — M

NOVA SOUTHEASTERN UNIVERSITY

Computer Science — M,D*
Information Science — M,D

OAKLAND UNIVERSITY

Computer Engineering — M
Computer Science — M
Electrical Engineering — M
Engineering Management — M
Engineering and Applied
   Sciences—General — M,D
Mechanical Engineering — M
Software Engineering — M
Systems Engineering — M,D

THE OHIO STATE UNIVERSITY

Aerospace/Aeronautical
   Engineering — M,D
Agricultural Engineering — M,D
Bioengineering — M,D
Biomedical Engineering — M,D*
Ceramic Sciences and
   Engineering — M,D
Chemical Engineering — M,D*
Civil Engineering — M,D
Computer Science — M,D*
Electrical Engineering — M,D*
Engineering and Applied
   Sciences—General — M,D
Industrial/Management
   Engineering — M,D
Information Science — M,D
Mechanical Engineering — M,D
Mechanics — M,D

The Ohio State University (continued)

Metallurgical Engineering and Metallurgy	M,D
Nuclear Engineering	M,D
Surveying Science and Engineering	M,D
Systems Engineering	M,D

OHIO UNIVERSITY

Artificial Intelligence/Robotics	D
Chemical Engineering	M,D
Civil Engineering	M
Electrical Engineering	M,D*
Engineering Management	M,D
Engineering and Applied Sciences—General	M,D*
Environmental Engineering	M,D
Geotechnical Engineering	M,D
Industrial/Management Engineering	M
Manufacturing Engineering	M
Materials Sciences	D
Mechanical Engineering	M,D
Structural Engineering	M
Systems Engineering	M
Water Resources Engineering	M

OKLAHOMA CITY UNIVERSITY

Computer Science	M

OKLAHOMA STATE UNIVERSITY

Agricultural Engineering	M,D
Architectural Engineering	M
Bioengineering	M,D
Chemical Engineering	M,D
Civil Engineering	M,D
Computer Engineering	M,D
Computer Science	M,D*
Electrical Engineering	M,D*
Engineering and Applied Sciences—General	M,D*
Environmental Engineering	M,D
Industrial/Management Engineering	M,D
Manufacturing Engineering	M
Mechanical Engineering	M,D
Systems Engineering	M
Telecommunications Management	M

OLD DOMINION UNIVERSITY

Aerospace/Aeronautical Engineering	M,D
Civil Engineering	M,D
Computer Engineering	M,D
Computer Science	M,D
Electrical Engineering	M,D
Engineering Management	M,D*
Engineering and Applied Sciences—General	M,D*
Environmental Engineering	M,D
Manufacturing Engineering	M,D
Mechanical Engineering	M,D
Mechanics	M,D
Operations Research	M,D

OREGON GRADUATE INSTITUTE OF SCIENCE AND TECHNOLOGY

Computer Engineering	M,D,O
Computer Science	M,D,O*
Electrical Engineering	M,D,O*
Environmental Engineering	M,D*
Management of Technology	M,O
Materials Engineering	M,D*
Materials Sciences	M,D

OREGON HEALTH SCIENCES UNIVERSITY

Medical Informatics	M*

OREGON INSTITUTE OF TECHNOLOGY

Computer Engineering	M

OREGON STATE UNIVERSITY

Bioengineering	M,D*
Chemical Engineering	M,D
Civil Engineering	M,D
Computer Engineering	M,D
Computer Science	M,D*
Electrical Engineering	M,D
Engineering and Applied Sciences—General	M,D*
Environmental Engineering	M,D
Industrial/Management Engineering	M,D
Manufacturing Engineering	M
Materials Sciences	M
Mechanical Engineering	M,D
Nuclear Engineering	M,D
Ocean Engineering	M
Operations Research	M,D
Paper and Pulp Engineering	M,D
Water Resources Engineering	M,D

PACE UNIVERSITY

Computer Science	M,D,O*
Information Science	M,D,O

Telecommunications	M,D,O

PACIFIC LUTHERAN UNIVERSITY

Management of Technology	M

PACIFIC STATES UNIVERSITY

Computer Science	M
Management of Technology	M

PENNSYLVANIA STATE UNIVERSITY AT ERIE, THE BEHREND COLLEGE

Engineering and Applied Sciences—General	M

PENNSYLVANIA STATE UNIVERSITY GREAT VALLEY SCHOOL OF GRADUATE PROFESSIONAL STUDIES

Electrical Engineering	M
Engineering and Applied Sciences—General	M,D*
Environmental Engineering	M
Industrial/Management Engineering	M
Information Science	M*
Software Engineering	M
Systems Engineering	M

PENNSYLVANIA STATE UNIVERSITY HARRISBURG CAMPUS OF THE CAPITAL COLLEGE

Computer Science	M
Electrical Engineering	M
Engineering and Applied Sciences—General	M
Environmental Engineering	M

PENNSYLVANIA STATE UNIVERSITY UNIVERSITY PARK CAMPUS

Aerospace/Aeronautical Engineering	M,D*
Agricultural Engineering	M,D
Architectural Engineering	M,D*
Bioengineering	M,D
Biomedical Engineering	M,D
Ceramic Sciences and Engineering	M,D
Chemical Engineering	M,D*
Civil Engineering	M,D
Computer Engineering	M,D
Computer Science	M,D
Electrical Engineering	M,D
Engineering Management	M,D
Engineering and Applied Sciences—General	M,D*
Environmental Engineering	M,D
Industrial/Management Engineering	M,D
Manufacturing Engineering	M
Materials Engineering	M,D
Materials Sciences	M,D
Mechanical Engineering	M,D
Mechanics	M,D
Metallurgical Engineering and Metallurgy	M,D
Mineral/Mining Engineering	M,D
Nuclear Engineering	M,D*
Petroleum Engineering	M,D
Polymer Science and Engineering	M,D
Structural Engineering	M,D
Telecommunications	M
Transportation and Highway Engineering	M,D
Water Resources Engineering	M,D

PEPPERDINE UNIVERSITY

Management of Technology	M

PHILADELPHIA COLLEGE OF TEXTILES AND SCIENCE (SEE PHILADELPHIA UNIVERSITY)

PHILADELPHIA UNIVERSITY

Textile Sciences and Engineering	M

PITTSBURG STATE UNIVERSITY

Engineering and Applied Sciences—General	M

POLYTECHNIC UNIVERSITY, BROOKLYN CAMPUS

Aerospace/Aeronautical Engineering	M
Chemical Engineering	M,D
Civil Engineering	M,D
Computer Science	M,D*
Electrical Engineering	M,D*
Environmental Engineering	M
Industrial/Management Engineering	M
Information Science	M
Management of Technology	M
Manufacturing Engineering	M*
Materials Sciences	M
Mechanical Engineering	M,D*
Polymer Science and Engineering	M
Systems Engineering	M

Telecommunications Management	M
Telecommunications	M
Transportation and Highway Engineering	M

POLYTECHNIC UNIVERSITY, FARMINGDALE CAMPUS

Aerospace/Aeronautical Engineering	M
Chemical Engineering	M,D
Civil Engineering	M,D
Computer Science	M,D
Electrical Engineering	M,D
Engineering Physics	M
Environmental Engineering	M
Industrial/Management Engineering	M
Information Science	M
Manufacturing Engineering	M
Mechanical Engineering	M
Systems Engineering	M
Telecommunications Management	M
Telecommunications	M
Transportation and Highway Engineering	M

POLYTECHNIC UNIVERSITY, WESTCHESTER GRADUATE CENTER

Chemical Engineering	M
Civil Engineering	M,D
Computer Science	M,D
Electrical Engineering	M,D
Engineering Physics	M
Environmental Engineering	M
Industrial/Management Engineering	M
Information Science	M
Management of Technology	M
Manufacturing Engineering	M
Materials Sciences	M
Systems Engineering	M
Telecommunications Management	M
Telecommunications	M

PORTLAND STATE UNIVERSITY

Civil Engineering	M,D
Computer Engineering	M,D
Computer Science	M,D
Electrical Engineering	M,D
Engineering Management	M,D
Engineering and Applied Sciences—General	M,D*
Manufacturing Engineering	M
Mechanical Engineering	M,D
Systems Science	D

PRAIRIE VIEW A&M UNIVERSITY

Engineering and Applied Sciences—General	M

PRINCETON UNIVERSITY

Aerospace/Aeronautical Engineering	M,D*
Chemical Engineering	M,D*
Civil Engineering	M,D
Computer Engineering	M,D
Computer Science	M,D*
Electrical Engineering	M,D
Electronic Materials	M,D
Engineering and Applied Sciences—General	M,D
Environmental Engineering	D
Information Science	M,D
Mechanical Engineering	M,D
Operations Research	M,D
Polymer Science and Engineering	M,D
Structural Engineering	M,D
Transportation and Highway Engineering	M,D
Water Resources Engineering	D

PURDUE UNIVERSITY

Aerospace/Aeronautical Engineering	M,D*
Agricultural Engineering	M,D
Bioengineering	M,D
Biomedical Engineering	M,D
Chemical Engineering	M,D*
Civil Engineering	M,D*
Computer Engineering	M,D
Computer Science	M,D*
Electrical Engineering	M,D*
Engineering and Applied Sciences—General	M,D*
Ergonomics and Human Factors	M,D
Industrial/Management Engineering	M,D
Manufacturing Engineering	M,D
Materials Sciences	M,D
Mechanical Engineering	M,D*
Metallurgical Engineering and Metallurgy	M,D
Nuclear Engineering	M,D*
Operations Research	M,D
Systems Engineering	M,D

PURDUE UNIVERSITY CALUMET

Engineering and Applied Sciences—General	M

QUEENS COLLEGE OF THE CITY UNIVERSITY OF NEW YORK

Computer Science	M

QUEEN'S UNIVERSITY AT KINGSTON

Chemical Engineering	M,D
Civil Engineering	M,D
Computer Science	M,D
Electrical Engineering	M,D
Engineering and Applied Sciences—General	M,D
Information Science	M,D
Materials Engineering	M,D
Mechanical Engineering	M,D
Metallurgical Engineering and Metallurgy	M,D
Mineral/Mining Engineering	M,D

REGIS UNIVERSITY

Information Science	M,O
Management of Technology	M,O

RENSSELAER AT HARTFORD

Computer Science	M
Electrical Engineering	M
Information Science	M
Mechanical Engineering	M

RENSSELAER POLYTECHNIC INSTITUTE

Aerospace/Aeronautical Engineering	M,D*
Architectural Engineering	M
Biomedical Engineering	M,D*
Ceramic Sciences and Engineering	M,D
Chemical Engineering	M,D*
Civil Engineering	M,D*
Computer Engineering	M,D*
Computer Science	M,D*
Electrical Engineering	M,D*
Energy and Power Engineering	M,D*
Engineering Management	M,D*
Engineering Physics	M,D
Engineering and Applied Sciences—General	M,D*
Environmental Engineering	M,D*
Ergonomics and Human Factors	M
Geotechnical Engineering	M,D
Industrial/Management Engineering	M*
Information Science	M*
Management of Technology	M,D
Manufacturing Engineering	M
Materials Engineering	M,D
Materials Sciences	M,D*
Mechanical Engineering	M,D
Mechanics	M,D
Medical Informatics	M*
Metallurgical Engineering and Metallurgy	M,D
Nuclear Engineering	M,D*
Operations Research	M,D
Polymer Science and Engineering	M,D
Structural Engineering	M,D
Systems Engineering	M,D
Technology and Public Policy	M,D*
Transportation and Highway Engineering	M,D

RHODE ISLAND COLLEGE

Management of Technology	M

RICE UNIVERSITY

Bioengineering	M,D
Biomedical Engineering	M,D
Chemical Engineering	M,D*
Civil Engineering	M,D
Computer Engineering	M,D
Computer Science	M,D*
Electrical Engineering	M,D
Engineering and Applied Sciences—General	M,D
Environmental Engineering	M,D
Materials Sciences	M,D
Mechanical Engineering	M,D
Structural Engineering	M,D

RICHMOND, THE AMERICAN INTERNATIONAL UNIVERSITY IN LONDON

Systems Engineering	M

RIVIER COLLEGE

Computer Science	M
Information Science	M

ROCHESTER INSTITUTE OF TECHNOLOGY

Computer Engineering	M
Computer Science	M,O*
Electrical Engineering	M
Engineering Management	M

Engineering and Applied
  Sciences—General M,O*
Industrial/Management
  Engineering M
Information Science M
Manufacturing Engineering M
Materials Engineering M
Materials Sciences M
Mechanical Engineering M
Software Engineering M
Systems Engineering M
Telecommunications M

ROOSEVELT UNIVERSITY

Computer Science M
Telecommunications M

ROSE-HULMAN INSTITUTE OF TECHNOLOGY

Biomedical Engineering M
Chemical Engineering M
Civil Engineering M
Electrical Engineering M
Engineering Management M
Engineering and Applied
  Sciences—General M*
Environmental Engineering M
Mechanical Engineering M

ROWAN UNIVERSITY

Engineering and Applied
  Sciences—General M

RUTGERS, THE STATE UNIVERSITY OF NEW JERSEY, NEW BRUNSWICK

Aerospace/Aeronautical
  Engineering M,D
Agricultural Engineering M
Biochemical Engineering M,D
Bioengineering M
Biomedical Engineering M,D*
Chemical Engineering M,D
Civil Engineering M,D
Computer Engineering M,D
Computer Science M,D*
Electrical Engineering M,D*
Engineering and Applied
  Sciences—General M,D*
Environmental Engineering M,D
Industrial/Management
  Engineering M,D
Materials Sciences M,D
Mechanical Engineering M,D
Mechanics M,D
Metallurgical Engineering and
  Metallurgy M,D
Operations Research D
Polymer Science and
  Engineering M,D
Systems Engineering M,D
Waste Management M,D

SACRED HEART UNIVERSITY

Computer Science M*
Information Science M

SAGINAW VALLEY STATE UNIVERSITY

Engineering and Applied
  Sciences—General M
Management of Technology M

ST. AMBROSE UNIVERSITY

Management of Technology M

ST. CLOUD STATE UNIVERSITY

Computer Science M
Engineering and Applied
  Sciences—General M
Technology and Public Policy M

ST. JOHN'S UNIVERSITY (NY)

Computer Science M

SAINT JOSEPH'S UNIVERSITY

Computer Science M

SAINT LOUIS UNIVERSITY

Aerospace/Aeronautical
  Engineering M*
Mechanical Engineering M

SAINT MARTIN'S COLLEGE

Engineering Management M

SAINT MARY'S UNIVERSITY OF MINNESOTA

Telecommunications M

ST. MARY'S UNIVERSITY OF SAN ANTONIO

Computer Engineering M
Computer Science M
Electrical Engineering M
Engineering Management M

Engineering and Applied
  Sciences—General M
Industrial/Management
  Engineering M
Information Science M
Operations Research M

SALEM-TEIKYO UNIVERSITY

Biotechnology M

SALVE REGINA UNIVERSITY

Systems Science M

SAM HOUSTON STATE UNIVERSITY

Computer Science M

SAN DIEGO STATE UNIVERSITY

Aerospace/Aeronautical
  Engineering M,D
Civil Engineering M
Computer Science M
Electrical Engineering M
Engineering and Applied
  Sciences—General M,D
Mechanical Engineering M,D
Mechanics M,D
Telecommunications
  Management M

SAN FRANCISCO STATE UNIVERSITY

Computer Science M
Engineering and Applied
  Sciences—General M

SAN JOSE STATE UNIVERSITY

Aerospace/Aeronautical
  Engineering M
Artificial Intelligence/Robotics M
Chemical Engineering M
Civil Engineering M
Computer Engineering M
Computer Science M
Electrical Engineering M
Engineering and Applied
  Sciences—General M*
Ergonomics and Human
  Factors M
Information Science M
Materials Engineering M
Mechanical Engineering M
Mechanics M
Polymer Science and
  Engineering M
Software Engineering M
Systems Engineering M

SANTA CLARA UNIVERSITY

Civil Engineering M
Computer Engineering M,D,O
Computer Science M,D,O
Electrical Engineering M,D,O
Engineering Management M
Engineering and Applied
  Sciences—General M,D,O*
Mechanical Engineering M,D,O
Software Engineering M,D,O

SEATTLE UNIVERSITY

Engineering and Applied
  Sciences—General M
Software Engineering M

SETON HALL UNIVERSITY

Operations Research M

SHIPPENSBURG UNIVERSITY OF PENNSYLVANIA

Computer Science M
Information Science M

SIMON FRASER UNIVERSITY

Computer Science M,D
Engineering and Applied
  Sciences—General M,D

SOUTH DAKOTA SCHOOL OF MINES AND TECHNOLOGY

Chemical Engineering M,D
Civil Engineering M,D
Computer Science M
Electrical Engineering M,D
Geological Engineering M,D
Management of Technology M
Materials Engineering M,D*
Materials Sciences M,D
Mechanical Engineering M,D
Metallurgical Engineering and
  Metallurgy M,D

SOUTH DAKOTA STATE UNIVERSITY

Agricultural Engineering M,D
Civil Engineering M
Computer Science M
Electrical Engineering M
Engineering and Applied
  Sciences—General M,D

Environmental Engineering M
Industrial/Management
  Engineering M
Mechanical Engineering M

SOUTHEASTERN UNIVERSITY

Computer Science M

SOUTHERN ADVENTIST UNIVERSITY

Software Engineering M

SOUTHERN ILLINOIS UNIVERSITY CARBONDALE

Civil Engineering M
Computer Science M
Electrical Engineering M,D
Energy and Power Engineering D
Engineering and Applied
  Sciences—General M,D*
Manufacturing Engineering M
Mechanical Engineering M
Mechanics M,D
Mineral/Mining Engineering M

SOUTHERN ILLINOIS UNIVERSITY EDWARDSVILLE

Civil Engineering M
Computer Science M
Electrical Engineering M
Engineering and Applied
  Sciences—General M
Mechanical Engineering M

SOUTHERN METHODIST UNIVERSITY

Computer Engineering M,D
Computer Science M,D*
Electrical Engineering M,D*
Engineering Management M,D
Engineering and Applied
  Sciences—General M,D
Manufacturing Engineering M,D
Materials Engineering M,D
Materials Sciences M,D
Mechanical Engineering M,D*
Operations Research M,D
Software Engineering M,D
Systems Engineering M,D
Telecommunications M,D
Waste Management M,D

SOUTHERN OREGON UNIVERSITY

Computer Science M

SOUTHERN POLYTECHNIC STATE UNIVERSITY

Computer Engineering M
Computer Science M
Construction Engineering and
  Management M
Electrical Engineering M
Engineering and Applied
  Sciences—General M
Industrial/Management
  Engineering M
Management of Technology M
Software Engineering M

SOUTHERN UNIVERSITY AND AGRICULTURAL AND MECHANICAL COLLEGE

Computer Science M

SOUTHWEST MISSOURI STATE UNIVERSITY

Materials Sciences M

SOUTHWEST TEXAS STATE UNIVERSITY

Computer Science M
Industrial/Management
  Engineering M
Management of Technology M

STANFORD UNIVERSITY

Aerospace/Aeronautical
  Engineering M,D,O*
Biomedical Engineering M
Chemical Engineering M,D,O
Civil Engineering M,D,O*
Computer Science M,D*
Electrical Engineering M,D,O
Engineering Design M
Engineering Management M,D,O
Engineering and Applied
  Sciences—General M,D,O
Environmental Engineering M,D,O
Industrial/Management
  Engineering M,D,O*
Manufacturing Engineering M*
Materials Engineering M,D,O
Materials Sciences M,D,O
Mechanical Engineering M,D,O*
Medical Informatics M,D*
Operations Research M,D,O
Petroleum Engineering M,D,O*
Systems Engineering M,D,O*
Technology and Public Policy M,D,O

STATE UNIVERSITY OF NEW YORK AT ALBANY

Computer Science M,D*
Information Science M,D*

STATE UNIVERSITY OF NEW YORK AT BINGHAMTON

Computer Science M,D*
Electrical Engineering M,D*
Engineering and Applied
  Sciences—General M,D*
Industrial/Management
  Engineering M,D*
Mechanical Engineering M,D*
Systems Science M,D

STATE UNIVERSITY OF NEW YORK AT BUFFALO

Aerospace/Aeronautical
  Engineering M,D
Chemical Engineering M,D*
Civil Engineering M,D*
Computer Engineering M,D
Computer Science M,D*
Construction Engineering and
  Management M,D
Electrical Engineering M,D*
Engineering and Applied
  Sciences—General M,D*
Environmental Engineering M,D
Geotechnical Engineering M,D
Industrial/Management
  Engineering M,D*
Materials Sciences M
Mechanical Engineering M,D*
Mechanics M,D
Structural Engineering M,D
Water Resources Engineering M,D

STATE UNIVERSITY OF NEW YORK AT NEW PALTZ

Computer Science M
Engineering and Applied
  Sciences—General M

STATE UNIVERSITY OF NEW YORK AT STONY BROOK

Biomedical Engineering M,D,O
Computer Engineering M,D
Computer Science M,D,O
Electrical Engineering M,D*
Engineering and Applied
  Sciences—General M,D,O
Management of Technology M
Materials Engineering M,D
Materials Sciences M,D
Mechanical Engineering M,D
Software Engineering M,D,O
Waste Management M,O

STATE UNIVERSITY OF NEW YORK COLLEGE AT BUFFALO

Industrial/Management
  Engineering M

STATE UNIVERSITY OF NEW YORK COLLEGE OF ENVIRONMENTAL SCIENCE AND FORESTRY

Environmental Engineering M,D

STATE UNIVERSITY OF NEW YORK INSTITUTE OF TECHNOLOGY AT UTICA/ROME

Computer Science M
Engineering and Applied
  Sciences—General M
Information Science M
Telecommunications M

STEPHEN F. AUSTIN STATE UNIVERSITY

Biotechnology M
Computer Science M

STEVENS INSTITUTE OF TECHNOLOGY

Chemical Engineering M,D,O
Civil Engineering M,D,O
Computer Engineering M,D,O
Computer Science M,D,O*
Construction Engineering and
  Management M
Electrical Engineering M,D,O
Engineering Design M
Engineering Physics M,D,O
Engineering and Applied
  Sciences—General M,D,O*
Environmental Engineering M,D,O
Information Science M,O
Management of Technology M,D,O
Materials Engineering M,D,O
Materials Sciences M,D,O
Mechanical Engineering M,D,O
Ocean Engineering M,D
Telecommunications
  Management M,D,O

SUFFOLK UNIVERSITY
Computer Science — M*

SYRACUSE UNIVERSITY
Aerospace/Aeronautical
Engineering — M,D
Bioengineering — M
Biomedical Engineering — M*
Chemical Engineering — M,D
Civil Engineering — M,D
Computer Engineering — M,D,O
Computer Science — M,D
Electrical Engineering — M,D,O*
Engineering Management — M
Engineering and Applied
Sciences—General — M,D,O*
Environmental Engineering — M,D
Information Science — M,D
Manufacturing Engineering — M
Materials Sciences — M,D
Mechanical Engineering — M,D
Systems Science — M,D
Telecommunications
Management — M
Telecommunications — M*

TEMPLE UNIVERSITY
Civil Engineering — M
Computer Engineering — M
Computer Science — M,D
Electrical Engineering — M
Engineering and Applied
Sciences—General — M,D
Environmental Engineering — M
Information Science — M,D
Mechanical Engineering — M

TENNESSEE STATE UNIVERSITY
Engineering and Applied
Sciences—General — M

TENNESSEE TECHNOLOGICAL UNIVERSITY
Chemical Engineering — M,D
Civil Engineering — M,D
Electrical Engineering — M,D
Engineering and Applied
Sciences—General — M,D*
Industrial/Management
Engineering — M,D
Mechanical Engineering — M,D

TEXAS A&M UNIVERSITY
Aerospace/Aeronautical
Engineering — M,D
Agricultural Engineering — M,D
Bioengineering — M,D
Biomedical Engineering — M,D
Chemical Engineering — M,D
Civil Engineering — M,D*
Computer Engineering — M,D
Computer Science — M,D*
Construction Engineering and
Management — M,D
Electrical Engineering — M,D*
Engineering and Applied
Sciences—General — M,D*
Environmental Engineering — M,D
Geotechnical Engineering — M,D
Hydraulics — M,D
Industrial/Management
Engineering — M,D
Materials Engineering — M,D
Mechanical Engineering — M,D*
Nuclear Engineering — M,D*
Ocean Engineering — M,D*
Petroleum Engineering — M,D*
Safety Engineering — M
Structural Engineering — M,D
Transportation and Highway
Engineering — M,D
Water Resources Engineering — M,D

TEXAS A&M UNIVERSITY–COMMERCE
Computer Science — M
Management of Technology — M

TEXAS A&M UNIVERSITY–CORPUS CHRISTI
Computer Science — M

TEXAS A&M UNIVERSITY–KINGSVILLE
Chemical Engineering — M
Civil Engineering — M
Computer Science — M
Electrical Engineering — M
Engineering and Applied
Sciences—General — M*
Environmental Engineering — M*
Industrial/Management
Engineering — M
Mechanical Engineering — M
Petroleum Engineering — M

TEXAS CHRISTIAN UNIVERSITY
Software Engineering — M

TEXAS SOUTHERN UNIVERSITY
Transportation and Highway
Engineering — M

TEXAS TECH UNIVERSITY
Chemical Engineering — M,D
Civil Engineering — M,D*
Computer Science — M,D
Electrical Engineering — M,D
Engineering and Applied
Sciences—General — M,D
Environmental Engineering — M,D
Industrial/Management
Engineering — M,D
Mechanical Engineering — M,D*
Petroleum Engineering — M

THOMAS JEFFERSON UNIVERSITY
Biomedical Engineering — D
Biotechnology — D*

THUNDERBIRD, THE AMERICAN GRADUATE SCHOOL OF INTERNATIONAL MANAGEMENT
Management of Technology — M

TOWSON UNIVERSITY
Computer Science — M

TRENT UNIVERSITY
Computer Science — M

TUFTS UNIVERSITY
Biotechnology — O
Chemical Engineering — M,D
Civil Engineering — M,D
Computer Science — M,D,O
Electrical Engineering — M,D,O
Engineering Management — M
Engineering and Applied
Sciences—General — M,D*
Environmental Engineering — M,D
Ergonomics and Human
Factors — M,D
Geotechnical Engineering — M,D
Manufacturing Engineering — O
Mechanical Engineering — M,D
Structural Engineering — M,D
Water Resources Engineering — M,D

TULANE UNIVERSITY
Biomedical Engineering — M,D*
Chemical Engineering — M,D*
Civil Engineering — M,D
Computer Science — M,D
Electrical Engineering — M,D
Engineering and Applied
Sciences—General — M,D
Environmental Engineering — M,D
Mechanical Engineering — M,D*

TUSKEGEE UNIVERSITY
Electrical Engineering — M
Engineering and Applied
Sciences—General — M,D
Materials Engineering — D
Mechanical Engineering — M

UNION COLLEGE (NY)
Computer Science — M
Electrical Engineering — M
Engineering and Applied
Sciences—General — M
Mechanical Engineering — M

UNITED STATES INTERNATIONAL UNIVERSITY
Management of Technology — M,D

UNIVERSIDAD DE LAS AMÉRICAS–PUEBLA
Chemical Engineering — M
Computer Science — M
Construction Engineering and
Management — M
Electrical Engineering — M
Engineering and Applied
Sciences—General — M
Industrial/Management
Engineering — M

UNIVERSITÉ DE MONCTON
Civil Engineering — M
Electrical Engineering — M
Engineering and Applied
Sciences—General — M
Industrial/Management
Engineering — M
Mechanical Engineering — M

UNIVERSITÉ DE MONTRÉAL
Biomedical Engineering — M,D
Computer Science — M,D
Ergonomics and Human
Factors — O

UNIVERSITÉ DE SHERBROOKE
Chemical Engineering — M,D
Civil Engineering — M,D
Computer Engineering — M,D
Electrical Engineering — M,D
Engineering and Applied
Sciences—General — M,D
Environmental Engineering — M
Mechanical Engineering — M,D

UNIVERSITÉ DU QUÉBEC À CHICOUTIMI
Engineering and Applied
Sciences—General — M,D
Mineral/Mining Engineering — D

UNIVERSITÉ DU QUÉBEC À MONTRÉAL
Ergonomics and Human
Factors — O
Mineral/Mining Engineering — D

UNIVERSITÉ DU QUÉBEC À TROIS-RIVIÈRES
Computer Science — M
Electrical Engineering — M,D
Energy Management and
Policy — M,D
Industrial/Management
Engineering — O
Paper and Pulp Engineering — M,D

UNIVERSITÉ DU QUÉBEC, ÉCOLE DE TECHNOLOGIE SUPÉRIEURE
Structural Engineering — M,O
Systems Engineering — M,O

UNIVERSITÉ DU QUÉBEC, INSTITUT NATIONAL DE LA RECHERCHE SCIENTIFIQUE
Energy Management and
Policy — M,D
Information Science — M,D,O
Materials Sciences — M,D
Software Engineering — M,D,O
Telecommunications — M,D,O

UNIVERSITÉ LAVAL
Aerospace/Aeronautical
Engineering — M
Agricultural Engineering — M
Chemical Engineering — M,D
Civil Engineering — M,D,O
Computer Science — M,D
Electrical Engineering — M,D
Engineering and Applied
Sciences—General — M,D,O
Industrial/Management
Engineering — O
Management of Technology — O
Mechanical Engineering — M,D
Metallurgical Engineering and
Metallurgy — M,D
Mineral/Mining Engineering — M,D

THE UNIVERSITY OF AKRON
Biomedical Engineering — M,D*
Chemical Engineering — M,D
Civil Engineering — M,D
Computer Science — M
Electrical Engineering — M,D
Engineering Management — M
Engineering and Applied
Sciences—General — M
Geological Engineering — M
Mechanical Engineering — M,D*
Polymer Science and
Engineering — M,D*

THE UNIVERSITY OF ALABAMA
Aerospace/Aeronautical
Engineering — M,D
Chemical Engineering — M,D
Civil Engineering — M,D
Computer Science — M,D*
Electrical Engineering — M,D
Engineering and Applied
Sciences—General — M,D
Environmental Engineering — M
Industrial/Management
Engineering — M
Materials Engineering — M,D
Materials Sciences — D
Mechanical Engineering — M,D
Mechanics — M,D
Metallurgical Engineering and
Metallurgy — M,D

THE UNIVERSITY OF ALABAMA AT BIRMINGHAM
Biomedical Engineering — M,D*
Civil Engineering — M,D
Computer Engineering — M,D
Computer Science — M,D*
Electrical Engineering — M,D
Engineering and Applied
Sciences—General — M,D
Environmental Engineering — M,D
Information Science — M,D
Materials Engineering — M,D
Materials Sciences — D

THE UNIVERSITY OF ALABAMA IN HUNTSVILLE
Chemical Engineering — M
Civil Engineering — M
Computer Engineering — M,D
Computer Science — M,D
Electrical Engineering — M,D*
Engineering and Applied
Sciences—General — M,D
Environmental Engineering — M
Industrial/Management
Engineering — M,D
Materials Engineering — M
Materials Sciences — M,D
Mechanical Engineering — M,D
Operations Research — M

Mechanical Engineering — M,D
Medical Informatics — M
Metallurgical Engineering and
Metallurgy — M,D

UNIVERSITY OF ALASKA ANCHORAGE
Civil Engineering — M
Engineering Management — M
Engineering and Applied
Sciences—General — M

UNIVERSITY OF ALASKA FAIRBANKS
Civil Engineering — M
Computer Science — M,D
Electrical Engineering — M
Engineering Management — M
Environmental Engineering — M
Geological Engineering — M,O
Mechanical Engineering — M
Mineral/Mining Engineering — M,O
Petroleum Engineering — M

UNIVERSITY OF ALBERTA
Biomedical Engineering — M,D
Biotechnology — M,D
Chemical Engineering — M,D
Civil Engineering — M,D
Computer Engineering — M,D
Computer Science — M,D*
Construction Engineering and
Management — M,D
Electrical Engineering — M,D
Energy and Power Engineering — M,D
Engineering Management — M,D
Environmental Engineering — M,D
Geotechnical Engineering — M,D
Materials Engineering — M,D
Mechanical Engineering — M,D
Mineral/Mining Engineering — M,D
Petroleum Engineering — M,D
Structural Engineering — M,D
Systems Engineering — M,D
Telecommunications — M,D
Water Resources Engineering — M,D

THE UNIVERSITY OF ARIZONA
Aerospace/Aeronautical
Engineering — M,D*
Agricultural Engineering — M,D
Chemical Engineering — M,D
Civil Engineering — M,D
Computer Engineering — M,D
Computer Science — M,D
Electrical Engineering — M,D*
Engineering and Applied
Sciences—General — M,D
Environmental Engineering — M,D
Geological Engineering — M,D
Industrial/Management
Engineering — M,D
Materials Engineering — M,D
Materials Sciences — M,D
Mechanical Engineering — M,D
Mechanics — M,D
Mineral/Mining Engineering — M,D*
Nuclear Engineering — M,D
Reliability Engineering — M
Systems Engineering — M,D

UNIVERSITY OF ARKANSAS
Agricultural Engineering — M,D
Bioengineering — M,D
Chemical Engineering — M,D
Civil Engineering — M,D
Computer Engineering — M,D
Computer Science — M,D
Electrical Engineering — M,D
Engineering and Applied
Sciences—General — M,D
Environmental Engineering — M
Industrial/Management
Engineering — M,D
Mechanical Engineering — M,D
Operations Research — M
Transportation and Highway
Engineering — M

UNIVERSITY OF ARKANSAS AT LITTLE ROCK
Computer Science — M
Engineering and Applied
Sciences—General — M,D
Information Science — M

UNIVERSITY OF BRIDGEPORT
Computer Engineering	M
Computer Science	M*
Electrical Engineering	M
Engineering and Applied Sciences—General	M*
Industrial/Management Engineering	M
Mechanical Engineering	M

UNIVERSITY OF BRITISH COLUMBIA
Agricultural Engineering	M,D
Bioengineering	M,D
Chemical Engineering	M,D
Civil Engineering	M,D
Computer Science	M,D
Electrical Engineering	M,D
Engineering Physics	M
Engineering and Applied Sciences—General	M,D
Fire Protection Engineering	M,D
Geological Engineering	M,D
Management of Technology	M
Materials Engineering	M,D
Materials Sciences	M,D
Mechanical Engineering	M,D
Metallurgical Engineering and Metallurgy	M,D
Mineral/Mining Engineering	M,D
Paper and Pulp Engineering	M

UNIVERSITY OF CALGARY
Chemical Engineering	M,D
Civil Engineering	M,D
Computer Engineering	M,D
Computer Science	M,D
Electrical Engineering	M,D
Engineering and Applied Sciences—General	M,D
Geotechnical Engineering	M,D
Mechanical Engineering	M,D
Petroleum Engineering	M,D

UNIVERSITY OF CALIFORNIA, BERKELEY
Bioengineering	D*
Biomedical Engineering	D
Ceramic Sciences and Engineering	M,D
Chemical Engineering	M,D
Civil Engineering	M,D*
Computer Science	M,D
Construction Engineering and Management	M,D
Electrical Engineering	M,D
Energy Management and Policy	M,D
Engineering and Applied Sciences—General	M,D*
Environmental Engineering	M,D
Geological Engineering	M,D
Geotechnical Engineering	M,D
Industrial/Management Engineering	M,D*
Materials Engineering	M,D
Materials Sciences	M,D
Mechanical Engineering	M,D*
Mechanics	M,D
Metallurgical Engineering and Metallurgy	M,D
Mineral/Mining Engineering	M,D
Nuclear Engineering	M,D*
Ocean Engineering	M,D*
Operations Research	M,D
Petroleum Engineering	M,D
Structural Engineering	M,D
Transportation and Highway Engineering	M,D
Water Resources Engineering	M,D

UNIVERSITY OF CALIFORNIA, DAVIS
Aerospace/Aeronautical Engineering	M,D,O
Agricultural Engineering	M,D
Bioengineering	M,D
Biomedical Engineering	M,D
Chemical Engineering	M,D,O
Civil Engineering	M,D,O
Computer Engineering	M,D
Computer Science	M,D
Electrical Engineering	M,D
Engineering and Applied Sciences—General	M,D,O
Environmental Engineering	M,D,O
Materials Sciences	M,D,O
Mechanical Engineering	M,D,O
Transportation and Highway Engineering	M,D*

UNIVERSITY OF CALIFORNIA, IRVINE
Aerospace/Aeronautical Engineering	M,D
Biochemical Engineering	M,D
Biomedical Engineering	M,D
Chemical Engineering	M,D*
Civil Engineering	M,D*
Computer Engineering	M,D
Computer Science	M,D*

Electrical Engineering	M,D*
Engineering and Applied Sciences—General	M,D*
Environmental Engineering	M,D
Information Science	M,D
Materials Engineering	M,D
Materials Sciences	M,D
Mechanical Engineering	M,D*

UNIVERSITY OF CALIFORNIA, LOS ANGELES
Aerospace/Aeronautical Engineering	M,D
Biomedical Engineering	M,D
Ceramic Sciences and Engineering	M,D
Chemical Engineering	M,D
Civil Engineering	M,D
Computer Science	M,D
Electrical Engineering	M,D*
Engineering and Applied Sciences—General	M,D*
Environmental Engineering	M,D*
Geotechnical Engineering	M,D
Manufacturing Engineering	M
Materials Engineering	M,D
Materials Sciences	M,D
Mechanical Engineering	M,D*
Metallurgical Engineering and Metallurgy	M,D
Operations Research	M,D
Structural Engineering	M,D
Water Resources Engineering	M,D

UNIVERSITY OF CALIFORNIA, RIVERSIDE
Chemical Engineering	M,D*
Computer Science	M,D*
Electrical Engineering	M,D*
Engineering and Applied Sciences—General	M,D
Environmental Engineering	M,D

UNIVERSITY OF CALIFORNIA, SAN DIEGO
Aerospace/Aeronautical Engineering	M,D
Artificial Intelligence/Robotics	M,D
Bioengineering	M,D*
Biomedical Engineering	M,D
Chemical Engineering	M,D*
Computer Engineering	M,D
Computer Science	M,D*
Electrical Engineering	M,D*
Engineering Physics	M,D
Engineering and Applied Sciences—General	M,D
Management of Technology	M
Materials Sciences	M,D*
Mechanical Engineering	M,D*
Mechanics	M,D
Ocean Engineering	M,D
Structural Engineering	M,D*
Telecommunications	M,D

UNIVERSITY OF CALIFORNIA, SAN FRANCISCO
Bioengineering	D*
Biomedical Engineering	D
Medical Informatics	M,D*

UNIVERSITY OF CALIFORNIA, SANTA BARBARA
Chemical Engineering	M,D
Computer Engineering	M,D*
Computer Science	M,D
Electrical Engineering	M,D
Engineering and Applied Sciences—General	M,D
Environmental Engineering	M,D
Materials Engineering	M,D*
Materials Sciences	M,D
Mechanical Engineering	M,D

UNIVERSITY OF CALIFORNIA, SANTA CRUZ
Computer Engineering	M,D*
Computer Science	M,D*

UNIVERSITY OF CENTRAL FLORIDA
Aerospace/Aeronautical Engineering	M,D,O
Civil Engineering	M,D,O
Computer Engineering	M,D,O
Computer Science	M,D
Electrical Engineering	M,D,O*
Engineering Management	M,D
Engineering and Applied Sciences—General	M,D,O*
Environmental Engineering	M,D,O
Industrial/Management Engineering	M,D*
Manufacturing Engineering	M,D
Materials Engineering	M,D,O
Materials Sciences	M,D,O
Mechanical Engineering	M,D,O*
Operations Research	M,D

UNIVERSITY OF CENTRAL OKLAHOMA
Computer Science	M

UNIVERSITY OF CHICAGO
Computer Science	M,D

UNIVERSITY OF CINCINNATI
Aerospace/Aeronautical Engineering	M,D
Ceramic Sciences and Engineering	M,D
Chemical Engineering	M,D
Civil Engineering	M,D
Computer Engineering	M,D
Computer Science	M,D
Electrical Engineering	M,D*
Engineering and Applied Sciences—General	M,D*
Environmental Engineering	M,D
Industrial/Management Engineering	M,D
Materials Engineering	M,D
Materials Sciences	M,D
Mechanical Engineering	M,D
Mechanics	M,D
Metallurgical Engineering and Metallurgy	M,D
Nuclear Engineering	M,D
Polymer Science and Engineering	M,D

UNIVERSITY OF COLORADO AT BOULDER
Aerospace/Aeronautical Engineering	M,D
Architectural Engineering	M,D
Chemical Engineering	M,D
Civil Engineering	M,D*
Computer Engineering	M,D
Computer Science	M,D*
Construction Engineering and Management	M,D
Electrical Engineering	M,D
Engineering Management	M
Engineering and Applied Sciences—General	M,D*
Environmental Engineering	M,D
Geotechnical Engineering	M,D
Management of Technology	M,D
Mechanical Engineering	M,D
Structural Engineering	M,D
Telecommunications Management	M
Telecommunications	M*
Water Resources Engineering	M,D

UNIVERSITY OF COLORADO AT COLORADO SPRINGS
Aerospace/Aeronautical Engineering	M
Computer Engineering	M,D
Computer Science	M
Electrical Engineering	M,D
Engineering and Applied Sciences—General	M,D

UNIVERSITY OF COLORADO AT DENVER
Civil Engineering	M,D
Computer Science	M
Electrical Engineering	M
Engineering and Applied Sciences—General	M,D*
Mechanical Engineering	M

UNIVERSITY OF CONNECTICUT
Aerospace/Aeronautical Engineering	M,D
Bioengineering	M,D
Biomedical Engineering	M,D
Biotechnology	M,D*
Chemical Engineering	M,D
Civil Engineering	M,D
Computer Science	M,D
Electrical Engineering	M,D*
Engineering and Applied Sciences—General	M,D*
Environmental Engineering	M,D
Materials Sciences	M,D
Mechanical Engineering	M,D
Mechanics	D
Metallurgical Engineering and Metallurgy	M,D
Ocean Engineering	M,D
Polymer Science and Engineering	M,D*
Software Engineering	M,D
Systems Engineering	M,D

UNIVERSITY OF DALLAS
Engineering Management	M
Telecommunications Management	M

UNIVERSITY OF DAYTON
Aerospace/Aeronautical Engineering	M,D

Agricultural Engineering	M
Chemical Engineering	M
Civil Engineering	M
Computer Engineering	M,D
Computer Science	M
Electrical Engineering	M,D
Engineering Management	M
Engineering and Applied Sciences—General	M,D*
Environmental Engineering	M
Industrial/Management Engineering	M
Materials Engineering	M,D
Mechanical Engineering	M,D
Mechanics	M
Structural Engineering	M
Transportation and Highway Engineering	M

UNIVERSITY OF DELAWARE
Biotechnology	D
Chemical Engineering	M,D
Civil Engineering	M,D*
Computer Science	M,D*
Electrical Engineering	M,D
Engineering and Applied Sciences—General	M,D*
Environmental Engineering	M,D
Geotechnical Engineering	M,D
Information Science	M,D
Materials Engineering	M,D
Materials Sciences	M,D
Mechanical Engineering	M,D*
Ocean Engineering	M,D
Operations Research	M,D*
Structural Engineering	M,D
Transportation and Highway Engineering	M,D
Water Resources Engineering	M,D

UNIVERSITY OF DENVER
Computer Engineering	M,D
Computer Science	M,D
Construction Engineering and Management	M
Electrical Engineering	M,D
Engineering Management	M
Engineering and Applied Sciences—General	M,D*
Management of Technology	M
Materials Sciences	M,D
Mechanical Engineering	M,D
Telecommunications Management	M
Telecommunications	M

UNIVERSITY OF DETROIT MERCY
Chemical Engineering	M,D*
Civil Engineering	M,D*
Computer Science	M*
Electrical Engineering	M,D*
Engineering Management	M
Engineering and Applied Sciences—General	M,D*
Environmental Engineering	M,D
Manufacturing Engineering	M,D
Mechanical Engineering	M,D*
Polymer Science and Engineering	M

UNIVERSITY OF FLORIDA
Aerospace/Aeronautical Engineering	M,D,O
Agricultural Engineering	M,D,O
Biomedical Engineering	M,D*
Ceramic Sciences and Engineering	M,D,O
Chemical Engineering	M,D,O*
Civil Engineering	M,D,O
Computer Engineering	M,D,O
Computer Science	M,D,O*
Construction Engineering and Management	M,D
Electrical Engineering	M,D,O*
Engineering Management	M,D,O
Engineering Physics	M,D,O
Engineering and Applied Sciences—General	M,D,O*
Environmental Engineering	M,D,O*
Industrial/Management Engineering	M,D,O*
Information Science	M,D,O
Manufacturing Engineering	M,D,O
Materials Engineering	M,D,O*
Materials Sciences	M,D,O*
Mechanical Engineering	M,D,O
Mechanics	M,D,O
Metallurgical Engineering and Metallurgy	M,D,O
Nuclear Engineering	M,D,O
Ocean Engineering	M,D,O
Operations Research	M,D,O
Polymer Science and Engineering	M,D,O
Systems Engineering	M,D,O

UNIVERSITY OF GEORGIA
Agricultural Engineering	M,D

University of Georgia (continued)

Artificial Intelligence/Robotics	M
Bioengineering	M,D*
Computer Science	M,D

UNIVERSITY OF GREAT FALLS

Information Science	M

UNIVERSITY OF GUELPH

Bioengineering	M,D
Computer Science	M
Engineering and Applied Sciences—General	M,D
Environmental Engineering	M,D
Water Resources Engineering	M,D

UNIVERSITY OF HARTFORD

Engineering and Applied Sciences—General	M

UNIVERSITY OF HAWAII AT MANOA

Agricultural Engineering	M
Bioengineering	M
Civil Engineering	M,D
Computer Science	M,D,O
Electrical Engineering	M,D
Engineering and Applied Sciences—General	M,D
Information Science	D
Mechanical Engineering	M,D
Ocean Engineering	M,D

UNIVERSITY OF HOUSTON

Aerospace/Aeronautical Engineering	M,D
Biomedical Engineering	M
Chemical Engineering	M,D*
Civil Engineering	M,D*
Computer Engineering	M,D
Computer Science	M,D*
Construction Engineering and Management	M
Electrical Engineering	M,D*
Engineering and Applied Sciences—General	M,D*
Environmental Engineering	M,D
Industrial/Management Engineering	M,D
Manufacturing Engineering	M
Materials Engineering	M,D*
Mechanical Engineering	M,D*
Operations Research	M,D
Petroleum Engineering	M
Systems Engineering	M,D

UNIVERSITY OF HOUSTON–CLEAR LAKE

Computer Engineering	M
Computer Science	M
Information Science	M
Software Engineering	M

UNIVERSITY OF IDAHO

Agricultural Engineering	M,D
Chemical Engineering	M,D
Civil Engineering	M,D
Computer Engineering	M
Computer Science	M,D
Electrical Engineering	M,D
Engineering and Applied Sciences—General	M,D
Geological Engineering	M
Mechanical Engineering	M,D
Metallurgical Engineering and Metallurgy	M,D
Mineral/Mining Engineering	M,D
Nuclear Engineering	M,D
Waste Management	M

UNIVERSITY OF ILLINOIS AT CHICAGO

Bioengineering	M,D*
Biomedical Engineering	M,D
Chemical Engineering	M,D*
Civil Engineering	M,D*
Computer Engineering	M,D
Computer Science	M,D*
Electrical Engineering	M,D*
Engineering and Applied Sciences—General	M,D*
Geotechnical Engineering	D
Industrial/Management Engineering	M,D
Materials Engineering	M,D
Mechanical Engineering	M,D*
Operations Research	D

UNIVERSITY OF ILLINOIS AT SPRINGFIELD

Computer Science	M

UNIVERSITY OF ILLINOIS AT URBANA–CHAMPAIGN

Aerospace/Aeronautical Engineering	M,D*
Agricultural Engineering	M,D
Bioengineering*	
Biomedical Engineering	
Chemical Engineering	M,D
Civil Engineering	M,D*
Computer Engineering	M,D
Computer Science	M,D*
Electrical Engineering	M,D*
Engineering Design	M
Environmental Engineering	M,D
Industrial/Management Engineering	M,D
Materials Engineering	M,D
Materials Sciences	M,D
Mechanical Engineering	M,D*
Mechanics	M,D*
Nuclear Engineering	M,D*
Systems Engineering	M*

THE UNIVERSITY OF IOWA

Biochemical Engineering	M,D
Biomedical Engineering	M,D
Chemical Engineering	M,D
Civil Engineering	M,D
Computer Engineering	M,D
Computer Science	M,D*
Electrical Engineering	M,D
Engineering and Applied Sciences—General	M,D*
Environmental Engineering	M,D
Ergonomics and Human Factors	M,D
Industrial/Management Engineering	M,D
Manufacturing Engineering	M,D
Mechanical Engineering	M,D*
Operations Research	M,D

UNIVERSITY OF KANSAS

Aerospace/Aeronautical Engineering	M,D
Architectural Engineering	M
Chemical Engineering	M,D
Civil Engineering	M,D*
Computer Science	M,D*
Electrical Engineering	M,D*
Engineering Management	M
Engineering and Applied Sciences—General	M,D*
Environmental Engineering	M,D
Mechanical Engineering	M,D
Petroleum Engineering	M,D
Water Resources Engineering	M,D

UNIVERSITY OF KENTUCKY

Agricultural Engineering	M,D
Biomedical Engineering	M,D
Chemical Engineering	M,D
Civil Engineering	M,D
Computer Science	M,D*
Electrical Engineering	M,D
Engineering and Applied Sciences—General	M,D
Manufacturing Engineering	M
Materials Sciences	M,D
Mechanical Engineering	M,D
Mechanics	M,D
Mineral/Mining Engineering	M,D

UNIVERSITY OF LOUISVILLE

Chemical Engineering	M,D
Civil Engineering	M,D
Computer Engineering	D
Computer Science	M,D
Electrical Engineering	M
Engineering and Applied Sciences—General	M,D*
Environmental Engineering	M,D
Industrial/Management Engineering	M,D
Mechanical Engineering	M,D

UNIVERSITY OF MAINE

Agricultural Engineering	M
Chemical Engineering	M,D*
Civil Engineering	M,D
Computer Engineering	M,D
Computer Science	M,D
Electrical Engineering	M,D
Engineering Physics	M
Engineering and Applied Sciences—General	M,D
Environmental Engineering	M,D
Geotechnical Engineering	M,D
Mechanical Engineering	M
Structural Engineering	M,D

UNIVERSITY OF MANITOBA

Agricultural Engineering	M,D
Civil Engineering	M,D
Computer Engineering	M,D
Computer Science	M,D
Electrical Engineering	M,D
Engineering and Applied Sciences—General	M,D
Industrial/Management Engineering	M,D
Mechanical Engineering	M,D

UNIVERSITY OF MARYLAND, BALTIMORE COUNTY

Biochemical Engineering	M,D
Chemical Engineering	M,D
Computer Engineering	M,D
Computer Science	M,D*
Electrical Engineering	M,D
Energy Management and Policy	M,D
Engineering Management	M
Engineering and Applied Sciences—General	M,D*
Information Science	M,D*
Mechanical Engineering	M,D
Mechanics	M,D

UNIVERSITY OF MARYLAND, COLLEGE PARK

Aerospace/Aeronautical Engineering	M,D
Agricultural Engineering	M,D*
Bioengineering	M,D*
Chemical Engineering	M,D
Civil Engineering	M,D*
Computer Engineering	M,D
Computer Science	M,D
Electrical Engineering	M,D*
Engineering and Applied Sciences—General	M,D
Environmental Engineering	M,D*
Fire Protection Engineering	M*
Manufacturing Engineering	M,D
Materials Engineering	M,D*
Materials Sciences	M,D*
Mechanical Engineering	M,D
Mechanics	M,D
Nuclear Engineering	M,D
Reliability Engineering	M,D*
Software Engineering	M
Systems Engineering	M
Telecommunications	M*
Water Resources Engineering	M,D

UNIVERSITY OF MARYLAND UNIVERSITY COLLEGE

Engineering Management	M
Management of Technology	M
Software Engineering	M
Telecommunications Management	M

UNIVERSITY OF MASSACHUSETTS AMHERST

Chemical Engineering	M,D
Civil Engineering	M,D
Computer Engineering	M,D
Computer Science	M,D*
Electrical Engineering	M,D
Engineering Management	M
Engineering and Applied Sciences—General	M,D*
Environmental Engineering	M
Industrial/Management Engineering	M,D
Manufacturing Engineering	M
Mechanical Engineering	M,D
Operations Research	M,D
Polymer Science and Engineering	M,D*

UNIVERSITY OF MASSACHUSETTS BOSTON

Biotechnology	M
Computer Science	M,D

UNIVERSITY OF MASSACHUSETTS DARTMOUTH

Computer Science	M
Electrical Engineering	M,D
Engineering and Applied Sciences—General	M,D
Mechanical Engineering	M
Textile Sciences and Engineering	M

UNIVERSITY OF MASSACHUSETTS LOWELL

Biotechnology	M,D
Chemical Engineering	M
Civil Engineering	M
Computer Engineering	M,D
Computer Science	M,D
Electrical Engineering	M,D
Energy and Power Engineering	M,D
Engineering and Applied Sciences—General	M,D*
Environmental Engineering	M
Industrial/Management Engineering	M,D
Mechanical Engineering	M,D
Mechanics	M,D
Polymer Science and Engineering	M,D
Systems Engineering	M,D

UNIVERSITY OF MASSACHUSETTS MEDICAL CENTER AT WORCESTER (SEE UNIVERSITY OF MASSACHUSETTS WORCESTER)

UNIVERSITY OF MASSACHUSETTS WORCESTER

Biomedical Engineering	D*

UNIVERSITY OF MEDICINE AND DENTISTRY OF NEW JERSEY

Biomedical Engineering	M,D
Medical Informatics	M,D,O

THE UNIVERSITY OF MEMPHIS

Architectural Engineering	M
Biomedical Engineering	M,D*
Civil Engineering	M,D
Computer Engineering	M,D
Computer Science	M,D
Electrical Engineering	M,D
Energy and Power Engineering	M,D
Engineering and Applied Sciences—General	M,D
Environmental Engineering	M,D
Industrial/Management Engineering	M
Manufacturing Engineering	M
Mechanical Engineering	M,D
Structural Engineering	M,D
Systems Engineering	M
Transportation and Highway Engineering	M,D
Water Resources Engineering	M,D

UNIVERSITY OF MIAMI

Architectural Engineering	M,D
Biomedical Engineering	M,D
Civil Engineering	M,D
Computer Engineering	M,D
Computer Science	M,D
Electrical Engineering	M,D
Engineering and Applied Sciences—General	M,D*
Ergonomics and Human Factors	M,D
Industrial/Management Engineering	M,D
Management of Technology	M,D
Mechanical Engineering	M,D*
Ocean Engineering	M,D
Telecommunications Management	M,O

UNIVERSITY OF MICHIGAN

Aerospace/Aeronautical Engineering	M,D,O*
Biomedical Engineering	M,D*
Chemical Engineering	M,D,O*
Civil Engineering	M,D,O*
Computer Engineering	M,D
Computer Science	M,D
Construction Engineering and Management	M,D,O
Electrical Engineering	M,D,O*
Engineering and Applied Sciences—General	M,D,O*
Environmental Engineering	M,D,O
Industrial/Management Engineering	M,D,O
Manufacturing Engineering	M,D
Materials Engineering	M,D
Materials Sciences	M,D*
Mechanical Engineering	M,D*
Mechanics	M,D
Nuclear Engineering	M,D,O*
Ocean Engineering	M,D,O
Operations Research	M,D,O
Systems Engineering	M,D
Transportation and Highway Engineering	M,D,O

UNIVERSITY OF MICHIGAN–DEARBORN

Computer Engineering	M
Computer Science	M
Electrical Engineering	M
Engineering Management	M
Engineering and Applied Sciences—General	M,D*
Industrial/Management Engineering	M
Information Science	M
Manufacturing Engineering	M,D
Mechanical Engineering	M
Software Engineering	M
Systems Engineering	M
Transportation and Highway Engineering	M

UNIVERSITY OF MINNESOTA, DULUTH

Computer Science	M

UNIVERSITY OF MINNESOTA, TWIN CITIES CAMPUS

Aerospace/Aeronautical Engineering	M,D*
Agricultural Engineering	M,D
Biomedical Engineering	M,D*
Biotechnology	M*
Chemical Engineering	M,D
Civil Engineering	M,D
Computer Engineering	M,D*
Computer Science	M,D*
Electrical Engineering	M,D*
Engineering and Applied Sciences—General	M,D
Geological Engineering	M,D
Industrial/Management Engineering	M,D*

Information Science	M,D
Management of Technology	M
Manufacturing Engineering	M
Materials Engineering	M,D
Materials Sciences	M,D
Mechanical Engineering	M,D*
Mechanics	M,D
Medical Informatics	M,D*
Software Engineering	M
Technology and Public Policy	M*

UNIVERSITY OF MISSISSIPPI

| Engineering and Applied Sciences—General | M,D |

UNIVERSITY OF MISSOURI–COLUMBIA

Aerospace/Aeronautical Engineering	M,D
Agricultural Engineering	M,D
Bioengineering	M,D
Chemical Engineering	M,D*
Civil Engineering	M,D*
Computer Engineering	M,D*
Computer Science	M,D
Electrical Engineering	M,D*
Engineering and Applied Sciences—General	M,D*
Environmental Engineering	M,D
Geotechnical Engineering	M,D
Industrial/Management Engineering	M,D*
Manufacturing Engineering	M,D
Mechanical Engineering	M,D*
Nuclear Engineering	M,D*
Structural Engineering	M,D
Transportation and Highway Engineering	M,D
Water Resources Engineering	M,D

UNIVERSITY OF MISSOURI–KANSAS CITY

Computer Science	M,D
Polymer Science and Engineering	M,D
Software Engineering	M,D
Telecommunications	M,D*

UNIVERSITY OF MISSOURI–ROLLA

Aerospace/Aeronautical Engineering	M,D
Ceramic Sciences and Engineering	M,D
Chemical Engineering	M,D
Civil Engineering	M,D
Computer Engineering	M,D
Computer Science	M,D
Construction Engineering and Management	M,D
Electrical Engineering	M,D
Engineering Management	M,D
Engineering and Applied Sciences—General	M,D*
Environmental Engineering	M,D
Geological Engineering	M,D
Geotechnical Engineering	M,D
Hydraulics	M,D
Mechanical Engineering	M,D
Mechanics	M,D
Metallurgical Engineering and Metallurgy	M,D
Mineral/Mining Engineering	M,D
Nuclear Engineering	M,D
Petroleum Engineering	M,D
Structural Engineering	M,D

UNIVERSITY OF MISSOURI–ST. LOUIS

| Biotechnology | M,D,O |
| Computer Science | M,D |

THE UNIVERSITY OF MONTANA–MISSOULA

| Computer Science | M |
| Operations Research | M,D |

UNIVERSITY OF NEBRASKA AT OMAHA

| Computer Science | M |

UNIVERSITY OF NEBRASKA–LINCOLN

Agricultural Engineering	M,D
Bioengineering	M,D
Chemical Engineering	M,D
Civil Engineering	M,D
Computer Engineering	M,D*
Computer Science	M,D*
Electrical Engineering	M,D
Engineering and Applied Sciences—General	M,D*
Environmental Engineering	M,D
Industrial/Management Engineering	M,D
Manufacturing Engineering	M,D
Materials Engineering	M,D
Mechanical Engineering	M,D*
Mechanics	M,D

UNIVERSITY OF NEVADA, LAS VEGAS

| Civil Engineering | M,D |

Computer Engineering	M,D
Computer Science	M,D
Electrical Engineering	M,D
Engineering and Applied Sciences—General	M,D*
Environmental Engineering	M,D
Mechanical Engineering	M,D

UNIVERSITY OF NEVADA, RENO

Biomedical Engineering	M,D
Chemical Engineering	M,D
Civil Engineering	M,D
Computer Science	M
Electrical Engineering	M,D
Engineering and Applied Sciences—General	M,D
Environmental Engineering	M,D
Geological Engineering	M,D,O
Mechanical Engineering	M,D
Metallurgical Engineering and Metallurgy	M,D,O
Mineral/Mining Engineering	M,O

UNIVERSITY OF NEW BRUNSWICK

Chemical Engineering	M,D
Civil Engineering	M,D
Computer Engineering	M,D
Computer Science	M,D
Electrical Engineering	M,D
Engineering and Applied Sciences—General	M,D,O
Environmental Engineering	M,D
Geotechnical Engineering	M,D
Mechanical Engineering	M,D
Mechanics	M,D
Structural Engineering	M,D
Surveying Science and Engineering	M,D,O
Transportation and Highway Engineering	M,D

UNIVERSITY OF NEW HAMPSHIRE

Chemical Engineering	M,D
Civil Engineering	M,D
Computer Science	M,D
Electrical Engineering	M,D
Engineering and Applied Sciences—General	M,D
Mechanical Engineering	M,D
Ocean Engineering	M
Systems Engineering	D

UNIVERSITY OF NEW HAVEN

Aerospace/Aeronautical Engineering	M
Computer Science	M
Electrical Engineering	M
Engineering Design	M,O
Engineering and Applied Sciences—General	M,O*
Environmental Engineering	M,O
Fire Protection Engineering	M
Industrial/Management Engineering	M,O
Information Science	M
Mechanical Engineering	M
Operations Research	M
Software Engineering	M
Telecommunications	M

UNIVERSITY OF NEW MEXICO

Chemical Engineering	M,D*
Civil Engineering	M,D
Computer Engineering	M,D
Computer Science	M,D
Electrical Engineering	M,D*
Engineering and Applied Sciences—General	M,D*
Management of Technology	M
Manufacturing Engineering	M,D*
Mechanical Engineering	M,D*
Nuclear Engineering	M
Waste Management	M

UNIVERSITY OF NEW ORLEANS

Civil Engineering	M
Computer Science	M
Electrical Engineering	M
Engineering Management	M,O
Engineering and Applied Sciences—General	M,D,O*
Mechanical Engineering	M

THE UNIVERSITY OF NORTH CAROLINA AT CHAPEL HILL

Biomedical Engineering	M,D*
Computer Science	M,D*
Environmental Engineering	M,D
Materials Sciences	M,D*
Operations Research	M,D

UNIVERSITY OF NORTH CAROLINA AT CHARLOTTE

Civil Engineering	M
Computer Engineering	M,D
Computer Science	M
Electrical Engineering	M,D

Engineering and Applied Sciences—General	M,D*
Information Science	D*
Mechanical Engineering	M,D

UNIVERSITY OF NORTH DAKOTA

Chemical Engineering	M
Civil Engineering	M
Computer Science	M
Electrical Engineering	M
Energy and Power Engineering	D
Engineering and Applied Sciences—General	M,D
Mechanical Engineering	M
Mineral/Mining Engineering	M
Structural Engineering	M

UNIVERSITY OF NORTHERN IOWA

| Computer Science | M |

UNIVERSITY OF NORTH FLORIDA

| Computer Science | M |
| Information Science | M |

UNIVERSITY OF NORTH TEXAS

Computer Science	M,D*
Engineering and Applied Sciences—General	M
Information Science	M,D
Materials Sciences	M,D

UNIVERSITY OF NOTRE DAME

Aerospace/Aeronautical Engineering	M,D*
Bioengineering	M,D
Chemical Engineering	M,D
Civil Engineering	M,D
Computer Engineering	M,D
Computer Science	M,D*
Electrical Engineering	M,D*
Engineering and Applied Sciences—General	M,D
Environmental Engineering	M,D
Mechanical Engineering	M,D

UNIVERSITY OF OKLAHOMA

Aerospace/Aeronautical Engineering	M,D*
Chemical Engineering	M,D
Civil Engineering	M,D*
Computer Engineering	M,D
Computer Science	M,D
Electrical Engineering	M,D*
Engineering Physics	M,D
Engineering and Applied Sciences—General	M,D
Environmental Engineering	M,D
Geological Engineering	M,D
Geotechnical Engineering	M,D
Industrial/Management Engineering	M,D
Mechanical Engineering	M,D
Petroleum Engineering	M,D
Structural Engineering	M,D
Transportation and Highway Engineering	M,D
Waste Management	M,D

UNIVERSITY OF OREGON

| Computer Science | M,D* |
| Information Science | M,D |

UNIVERSITY OF OTTAWA

Aerospace/Aeronautical Engineering	M,D
Chemical Engineering	M,D
Civil Engineering	M,D
Computer Engineering	M,D
Computer Science	M,D
Electrical Engineering	M,D
Engineering Management	M
Engineering and Applied Sciences—General	M,D
Mechanical Engineering	M,D*
Systems Science	M

UNIVERSITY OF PENNSYLVANIA

Bioengineering	M,D*
Biomedical Engineering	M,D
Biotechnology	M*
Chemical Engineering	M,D
Computer Science	M,D*
Electrical Engineering	M,D*
Engineering and Applied Sciences—General	M,D*
Environmental Engineering	M,D
Information Science	M,D
Management of Technology	M*
Materials Engineering	M,D
Materials Sciences	M,D*
Mechanical Engineering	M,D*
Mechanics	M,D
Systems Engineering	M,D*
Technology and Public Policy	M,D
Telecommunications Management	M
Telecommunications	M*

| Transportation and Highway Engineering | M,D |

UNIVERSITY OF PHOENIX

Computer Science	M
Information Science	M
Management of Technology	M

UNIVERSITY OF PITTSBURGH

Bioengineering	M,D*
Biomedical Engineering	M,D
Chemical Engineering	M,D
Civil Engineering	M,D
Computer Science	M,D
Electrical Engineering	M,D*
Engineering and Applied Sciences—General	M,D,O*
Environmental Engineering	M,D
Industrial/Management Engineering	M,D,O*
Information Science	M
Manufacturing Engineering	M,D
Materials Engineering	M,D
Materials Sciences	M,D
Mechanical Engineering	M,D
Metallurgical Engineering and Metallurgy	M,D
Petroleum Engineering	M,D
Systems Engineering	M
Telecommunications	M,O*

UNIVERSITY OF PORTLAND

Civil Engineering	M
Electrical Engineering	M
Engineering and Applied Sciences—General	M
Mechanical Engineering	M

UNIVERSITY OF PUERTO RICO, MAYAGÜEZ CAMPUS

Chemical Engineering	M
Civil Engineering	M,D
Computer Engineering	M
Electrical Engineering	M
Engineering and Applied Sciences—General	M,D
Industrial/Management Engineering	M
Mechanical Engineering	M

UNIVERSITY OF PUERTO RICO, MEDICAL SCIENCES CAMPUS

| Medical Informatics | M |

UNIVERSITY OF REGINA

Computer Engineering	M,D
Computer Science	M,D
Engineering and Applied Sciences—General	M,D
Environmental Engineering	M,D
Industrial/Management Engineering	M,D
Systems Engineering	M,D

UNIVERSITY OF RHODE ISLAND

Chemical Engineering	M,D
Civil Engineering	M,D
Computer Engineering	M,D
Computer Science	M,D
Electrical Engineering	M,D*
Engineering and Applied Sciences—General	M,D
Environmental Engineering	M,D
Geotechnical Engineering	M,D
Industrial/Management Engineering	M
Manufacturing Engineering	M
Mechanical Engineering	M,D
Mechanics	M,D
Ocean Engineering	M,D
Structural Engineering	M,D
Systems Engineering	M,D
Transportation and Highway Engineering	M,D

UNIVERSITY OF ROCHESTER

Biomedical Engineering	M,D*
Chemical Engineering	M,D
Computer Engineering	M,D
Computer Science	M,D
Electrical Engineering	M,D*
Engineering and Applied Sciences—General	M,D*
Materials Sciences	M,D
Mechanical Engineering	M,D

UNIVERSITY OF ST. THOMAS (MN)

Engineering and Applied Sciences—General	M,O
Manufacturing Engineering	M,O
Software Engineering	M,O*
Systems Engineering	M,O

UNIVERSITY OF SAN FRANCISCO

| Computer Science | M |
| Telecommunications Management | M |

UNIVERSITY OF SASKATCHEWAN

Agricultural Engineering	M,D
Biomedical Engineering	M,D
Chemical Engineering	M,D
Civil Engineering	M,D
Electrical Engineering	M,D
Engineering and Applied Sciences—General	M,D,O
Environmental Engineering	M,D,O
Mechanical Engineering	M,D

THE UNIVERSITY OF SCRANTON

Software Engineering	M

UNIVERSITY OF SOUTH ALABAMA

Chemical Engineering	M
Computer Science	M
Electrical Engineering	M
Engineering and Applied Sciences—General	M*
Information Science	M
Mechanical Engineering	M

UNIVERSITY OF SOUTH CAROLINA

Chemical Engineering	M,D
Civil Engineering	M,D
Computer Engineering	M,D
Computer Science	M,D*
Electrical Engineering	M,D
Engineering and Applied Sciences—General	M,D*
Mechanical Engineering	M,D
Waste Management	M,D

UNIVERSITY OF SOUTH DAKOTA

Computer Science	M

UNIVERSITY OF SOUTHERN CALIFORNIA

Aerospace/Aeronautical Engineering	M,D,O
Artificial Intelligence/Robotics	M
Biomedical Engineering	M,D
Chemical Engineering	M,D,O
Civil Engineering	M,D,O*
Computer Engineering	M,D
Computer Science	M,D*
Construction Engineering and Management	M*
Electrical Engineering	M,D,O*
Engineering Management	M
Engineering and Applied Sciences—General	M,D,O*
Environmental Engineering	M,D*
Geotechnical Engineering	M
Industrial/Management Engineering	M,D,O
Manufacturing Engineering	M
Materials Engineering	M
Materials Sciences	M,D,O
Mechanical Engineering	M,D,O*
Mechanics	M
Ocean Engineering	M
Operations Research	M
Petroleum Engineering	M,D,O
Software Engineering	M
Structural Engineering	M
Systems Engineering	M,D,O
Transportation and Highway Engineering	M
Water Resources Engineering	M

UNIVERSITY OF SOUTHERN COLORADO

Engineering and Applied Sciences—General	M
Industrial/Management Engineering	M
Systems Engineering	M

UNIVERSITY OF SOUTHERN INDIANA

Engineering and Applied Sciences—General	M

UNIVERSITY OF SOUTHERN MAINE

Computer Science	M

UNIVERSITY OF SOUTHERN MISSISSIPPI

Computer Science	M,D*
Engineering and Applied Sciences—General	M
Polymer Science and Engineering	M,D

UNIVERSITY OF SOUTH FLORIDA

Chemical Engineering	M,D
Civil Engineering	M,D
Computer Engineering	M,D
Computer Science	M,D*
Electrical Engineering	M,D*
Engineering Management	M,D
Engineering Physics	M,D
Engineering and Applied Sciences—General	M,D*
Environmental Engineering	M,D
Industrial/Management Engineering	M,D
Mechanical Engineering	M,D

UNIVERSITY OF SOUTHWESTERN LOUISIANA

Chemical Engineering	M
Civil Engineering	M
Computer Engineering	M,D
Computer Science	M,D*
Engineering Management	M
Engineering and Applied Sciences—General	M,D
Mechanical Engineering	M
Petroleum Engineering	M
Telecommunications	M

UNIVERSITY OF TENNESSEE AT CHATTANOOGA

Computer Science	M
Engineering Management	M
Engineering and Applied Sciences—General	M

UNIVERSITY OF TENNESSEE, KNOXVILLE

Aerospace/Aeronautical Engineering	M,D
Agricultural Engineering	M,D
Artificial Intelligence/Robotics	M,D
Biomedical Engineering	M,D
Biotechnology	M,D
Chemical Engineering	M,D
Civil Engineering	M,D
Computer Science	M,D*
Electrical Engineering	M,D
Engineering Management	M
Engineering and Applied Sciences—General	M,D*
Environmental Engineering	M
Industrial/Management Engineering	M,D
Information Science	M,D
Manufacturing Engineering	M
Materials Sciences	M,D
Mechanical Engineering	M,D
Mechanics	M,D
Metallurgical Engineering and Metallurgy	M,D
Nuclear Engineering	M,D
Polymer Science and Engineering	M,D

UNIVERSITY OF TENNESSEE, MEMPHIS

Biomedical Engineering	M,D*

UNIVERSITY OF TENNESSEE SPACE INSTITUTE

Aerospace/Aeronautical Engineering	M,D
Chemical Engineering	M
Computer Science	M
Electrical Engineering	M,D
Engineering Management	M
Engineering and Applied Sciences—General	M,D*
Mechanical Engineering	M,D
Mechanics	M,D
Metallurgical Engineering and Metallurgy	M,D

THE UNIVERSITY OF TEXAS AT ARLINGTON

Aerospace/Aeronautical Engineering	M,D
Biomedical Engineering	M,D
Civil Engineering	M,D
Computer Engineering	M,D
Computer Science	M,D*
Electrical Engineering	M,D*
Engineering and Applied Sciences—General	M,D*
Environmental Engineering	M,D
Industrial/Management Engineering	M,D
Materials Engineering	M,D
Materials Sciences	M,D*
Mechanical Engineering	M,D
Software Engineering	M

THE UNIVERSITY OF TEXAS AT AUSTIN

Aerospace/Aeronautical Engineering	M,D*
Architectural Engineering	M
Biomedical Engineering	M,D*
Chemical Engineering	M,D
Civil Engineering	M,D*
Computer Engineering	M,D
Computer Science	M,D*
Electrical Engineering	M,D*
Engineering and Applied Sciences—General	M,D*
Environmental Engineering	M
Geotechnical Engineering	M,D
Industrial/Management Engineering	M,D
Manufacturing Engineering	M,D
Materials Engineering	M,D
Materials Sciences	M,D
Mechanical Engineering	M,D
Mechanics	M,D
Operations Research	M,D
Petroleum Engineering	M,D
Technology and Public Policy	M

Water Resources Engineering	M

THE UNIVERSITY OF TEXAS AT DALLAS

Computer Science	M,D*
Electrical Engineering	M,D
Engineering and Applied Sciences—General	M,D
Telecommunications	M,D

THE UNIVERSITY OF TEXAS AT EL PASO

Civil Engineering	M
Computer Engineering	M,D
Computer Science	M
Electrical Engineering	M,D
Engineering and Applied Sciences—General	M,D
Environmental Engineering	D
Industrial/Management Engineering	M
Manufacturing Engineering	M
Materials Engineering	D
Materials Sciences	D
Mechanical Engineering	M
Metallurgical Engineering and Metallurgy	M

THE UNIVERSITY OF TEXAS AT SAN ANTONIO

Biotechnology	M
Civil Engineering	M
Computer Science	M,D
Electrical Engineering	M
Engineering and Applied Sciences—General	M
Management of Technology	M
Mechanical Engineering	M

THE UNIVERSITY OF TEXAS AT TYLER

Computer Science	M
Engineering and Applied Sciences—General	M

THE UNIVERSITY OF TEXAS–PAN AMERICAN

Computer Science	M

THE UNIVERSITY OF TEXAS SOUTHWESTERN MEDICAL CENTER AT DALLAS

Biomedical Engineering	M,D*

UNIVERSITY OF THE SCIENCES IN PHILADELPHIA

Biotechnology	M

UNIVERSITY OF TOLEDO

Bioengineering	M,D
Chemical Engineering	M,D
Civil Engineering	M,D
Computer Science	M,D
Electrical Engineering	M,D
Engineering and Applied Sciences—General	M,D*
Industrial/Management Engineering	M,D
Mechanical Engineering	M,D

UNIVERSITY OF TORONTO

Aerospace/Aeronautical Engineering	M,D
Biomedical Engineering	M,D
Chemical Engineering	M,D
Civil Engineering	M,D
Computer Engineering	M,D
Computer Science	M,D
Electrical Engineering	M,D
Engineering and Applied Sciences—General	M,D*
Environmental Engineering	M,D
Industrial/Management Engineering	M,D
Manufacturing Engineering	M
Materials Sciences	M,D
Mechanical Engineering	M,D
Metallurgical Engineering and Metallurgy	M,D
Software Engineering	M

UNIVERSITY OF TULSA

Chemical Engineering	M,D
Computer Science	M,D
Electrical Engineering	M
Engineering Management	M
Engineering and Applied Sciences—General	M,D
Management of Technology	M
Mechanical Engineering	M
Petroleum Engineering	M,D

UNIVERSITY OF UTAH

Bioengineering	M,D*
Biomedical Engineering	M,D
Chemical Engineering	M,D
Civil Engineering	M,D
Computer Science	M,D*
Electrical Engineering	M,D,O
Engineering Management	M

THE UNIVERSITY OF TEXAS AT DALLAS

Engineering and Applied Sciences—General	M,D,O
Geological Engineering	M,D
Materials Engineering	M,D
Materials Sciences	M,D
Mechanical Engineering	M,D*
Mechanics	M
Medical Informatics	M,D
Metallurgical Engineering and Metallurgy	M,D
Mineral/Mining Engineering	M,D
Nuclear Engineering	M,D
Petroleum Engineering	M,D

UNIVERSITY OF VERMONT

Biomedical Engineering	M
Civil Engineering	M,D
Computer Science	M
Electrical Engineering	M,D
Engineering Physics	M
Engineering and Applied Sciences—General	M,D*
Materials Sciences	M,D
Mechanical Engineering	M,D

UNIVERSITY OF VICTORIA

Computer Science	M,D
Electrical Engineering	M,D
Engineering and Applied Sciences—General	M,D
Mechanical Engineering	M,D

UNIVERSITY OF VIRGINIA

Aerospace/Aeronautical Engineering	M,D
Biomedical Engineering	M,D
Chemical Engineering	M,D
Civil Engineering	M,D
Computer Science	M,D*
Electrical Engineering	M,D*
Engineering Physics	M,D
Engineering and Applied Sciences—General	M,D*
Environmental Engineering	M,D
Materials Sciences	M,D
Mechanical Engineering	M,D*
Mechanics	M
Medical Informatics	M
Nuclear Engineering	M,D
Structural Engineering	M,D
Systems Engineering	M,D*
Transportation and Highway Engineering	M,D
Water Resources Engineering	M,D

UNIVERSITY OF WASHINGTON

Aerospace/Aeronautical Engineering	M,D
Bioengineering	M,D*
Biomedical Engineering	M,D
Biotechnology	D
Ceramic Sciences and Engineering	M
Chemical Engineering	M,D*
Civil Engineering	M,D
Computer Science	M,D*
Construction Engineering and Management	M,D
Electrical Engineering	M,D*
Engineering and Applied Sciences—General	M,D
Environmental Engineering	M,D
Geotechnical Engineering	M,D
Hydraulics	M,D
Materials Engineering	M,D
Materials Sciences	M,D
Mechanical Engineering	M,D
Metallurgical Engineering and Metallurgy	M
Paper and Pulp Engineering	M,D
Structural Engineering	M,D
Transportation and Highway Engineering	M,D

UNIVERSITY OF WATERLOO

Chemical Engineering	M,D
Civil Engineering	M,D
Computer Science	M,D
Electrical Engineering	M,D
Engineering Management	M,D*
Engineering and Applied Sciences—General	M,D
Management of Technology	M,D
Mechanical Engineering	M,D
Software Engineering	M,D
Systems Engineering	M,D

THE UNIVERSITY OF WESTERN ONTARIO

Computer Science	M,D
Engineering and Applied Sciences—General	M,D

UNIVERSITY OF WEST FLORIDA

Computer Science	M
Systems Engineering	M

UNIVERSITY OF WINDSOR

Civil Engineering	M,D

Computer Science	M
Electrical Engineering	M,D
Engineering and Applied	
Sciences—General	M,D
Environmental Engineering	M,D
Geological Engineering	M
Industrial/Management	
Engineering	M
Materials Engineering	M,D
Mechanical Engineering	M,D

UNIVERSITY OF WISCONSIN–MADISON

Agricultural Engineering	M,D
Chemical Engineering	M,D
Civil Engineering	M,D
Computer Science	M,D*
Electrical Engineering	M,D
Engineering Physics	M,D
Engineering and Applied	
Sciences—General	M,D,O*
Environmental Engineering	M,D
Geological Engineering	M,D
Industrial/Management	
Engineering	M,D
Manufacturing Engineering	M*
Materials Sciences	M,D
Mechanical Engineering	M,D*
Mechanics	M,D
Metallurgical Engineering and	
Metallurgy	M,D
Nuclear Engineering	M,D*
Polymer Science and	
Engineering	M,D

UNIVERSITY OF WISCONSIN–MILWAUKEE

Computer Science	M,D
Engineering and Applied	
Sciences—General	M,D*

UNIVERSITY OF WISCONSIN–STOUT

Management of Technology	M
Safety Engineering	M

UNIVERSITY OF WYOMING

Chemical Engineering	M,D
Civil Engineering	M,D
Computer Science	M,D
Electrical Engineering	M,D
Engineering and Applied	
Sciences—General	M,D
Environmental Engineering	M,D
Mechanical Engineering	M,D
Petroleum Engineering	M,D

UTAH STATE UNIVERSITY

Aerospace/Aeronautical	
Engineering	M,D
Agricultural Engineering	M,D
Civil Engineering	M,D,O
Computer Science	M
Electrical Engineering	M,D,O
Engineering and Applied	
Sciences—General	M,D,O*
Environmental Engineering	M,D,O
Mechanical Engineering	M,D
Water Resources Engineering	M,D

VANDERBILT UNIVERSITY

Biomedical Engineering	M,D*
Chemical Engineering	M,D
Civil Engineering	M,D
Computer Science	M,D
Electrical Engineering	M,D
Engineering and Applied	
Sciences—General	M,D*
Environmental Engineering	M,D
Management of Technology	M,D
Materials Engineering	M,D
Materials Sciences	M,D
Mechanical Engineering	M,D

VILLANOVA UNIVERSITY

Chemical Engineering	M
Civil Engineering	M
Computer Engineering	M
Computer Science	M
Electrical Engineering	M*
Engineering and Applied	
Sciences—General	M,O
Environmental Engineering	M
Manufacturing Engineering	M,O
Mechanical Engineering	M,O*
Transportation and Highway	
Engineering	M

Water Resources Engineering	M

VIRGINIA COMMONWEALTH UNIVERSITY

Biomedical Engineering	M,D*
Computer Science	M
Engineering and Applied	
Sciences—General	M,D
Operations Research	M

VIRGINIA POLYTECHNIC INSTITUTE AND STATE UNIVERSITY

Aerospace/Aeronautical	
Engineering	M,D
Agricultural Engineering	M,D
Bioengineering	M,D
Chemical Engineering	M,D
Civil Engineering	M,D*
Computer Engineering	M,D
Computer Science	M,D*
Electrical Engineering	M,D*
Engineering Management	M
Engineering and Applied	
Sciences—General	M,D*
Environmental Engineering	M,D
Industrial/Management	
Engineering	M,D
Information Science	M
Materials Engineering	M,D
Materials Sciences	M,D
Mechanical Engineering	M,D
Mechanics	M,D*
Mineral/Mining Engineering	M,D
Ocean Engineering	M
Operations Research	M,D
Systems Engineering	M

WAKE FOREST UNIVERSITY

Biomedical Engineering	D
Computer Science	M

WASHINGTON STATE UNIVERSITY

Chemical Engineering	M,D
Civil Engineering	M,D
Computer Science	M,D
Electrical Engineering	M,D
Engineering and Applied	
Sciences—General	M,D
Environmental Engineering	M
Materials Engineering	M,D
Materials Sciences	M,D
Mechanical Engineering	M,D

WASHINGTON UNIVERSITY IN ST. LOUIS

Biomedical Engineering	M,D*
Chemical Engineering	M,D
Civil Engineering	M,D
Computer Science	M,D*
Construction Engineering and	
Management	M
Electrical Engineering	M,D*
Engineering and Applied	
Sciences—General	M,D
Environmental Engineering	M,D
Materials Engineering	M,D
Materials Sciences	M,D
Mechanical Engineering	M,D*
Structural Engineering	M,D
Systems Engineering	D
Systems Science	M,D*
Technology and Public Policy	M,D
Transportation and Highway	
Engineering	D

WAYNE STATE UNIVERSITY

Biomedical Engineering	M,D
Chemical Engineering	M,D
Civil Engineering	M,D*
Computer Engineering	M,D
Computer Science	M,D*
Electrical Engineering	M,D*
Engineering Management	M
Engineering and Applied	
Sciences—General	M,D,O*
Environmental Engineering	M,D
Industrial/Management	
Engineering	M,D
Manufacturing Engineering	M
Materials Engineering	M,D,O
Materials Sciences	M,D,O
Mechanical Engineering	M,D
Operations Research	M

Polymer Science and	
Engineering	M,D,O
Software Engineering	M,D
Waste Management	M,O

WEBB INSTITUTE

Ocean Engineering	M

WEBSTER UNIVERSITY

Aerospace/Aeronautical	
Engineering	M
Computer Science	M,O
Telecommunications	
Management	M

WEST CHESTER UNIVERSITY OF PENNSYLVANIA

Computer Science	M

WESTERN CAROLINA UNIVERSITY

Computer Science	M
Industrial/Management	
Engineering	M

WESTERN CONNECTICUT STATE UNIVERSITY

Computer Science	M

WESTERN ILLINOIS UNIVERSITY

Computer Science	M
Technology and Public Policy	M
Telecommunications	M

WESTERN KENTUCKY UNIVERSITY

Computer Science	M

WESTERN MICHIGAN UNIVERSITY

Chemical Engineering	M
Computer Engineering	M
Computer Science	M,D
Construction Engineering and	
Management	M
Electrical Engineering	M
Engineering Management	M
Engineering and Applied	
Sciences—General	M,D*
Industrial/Management	
Engineering	M
Manufacturing Engineering	M
Materials Engineering	M
Materials Sciences	M
Mechanical Engineering	M,D
Operations Research	M
Paper and Pulp Engineering	M

WESTERN NEW ENGLAND COLLEGE

Electrical Engineering	M
Engineering and Applied	
Sciences—General	M
Industrial/Management	
Engineering	M
Manufacturing Engineering	M
Mechanical Engineering	M

WESTERN WASHINGTON UNIVERSITY

Computer Science	M

WEST TEXAS A&M UNIVERSITY

Engineering and Applied	
Sciences—General	M

WEST VIRGINIA UNIVERSITY

Aerospace/Aeronautical	
Engineering	M,D
Chemical Engineering	M,D
Civil Engineering	M,D
Computer Engineering	D
Computer Science	M,D
Electrical Engineering	M,D
Engineering and Applied	
Sciences—General	M,D*
Environmental Engineering	M,D
Industrial/Management	
Engineering	M,D
Mechanical Engineering	M,D
Mineral/Mining Engineering	M,D
Petroleum Engineering	M
Safety Engineering	M
Software Engineering	M

WEST VIRGINIA UNIVERSITY INSTITUTE OF TECHNOLOGY

Engineering and Applied	
Sciences—General	M

Systems Engineering	M

WICHITA STATE UNIVERSITY

Aerospace/Aeronautical	
Engineering	M,D
Computer Science	M
Electrical Engineering	M,D
Engineering and Applied	
Sciences—General	M,D*
Industrial/Management	
Engineering	M,D*
Manufacturing Engineering	M,D
Mechanical Engineering	M,D*

WIDENER UNIVERSITY

Chemical Engineering	M
Civil Engineering	M
Computer Engineering	M
Electrical Engineering	M
Engineering Management	M
Engineering and Applied	
Sciences—General	M*
Mechanical Engineering	M
Software Engineering	M
Telecommunications	M

WILKES UNIVERSITY

Electrical Engineering	M
Engineering and Applied	
Sciences—General	M

WILLIAM PATERSON UNIVERSITY OF NEW JERSEY

Biotechnology	M

WORCESTER POLYTECHNIC INSTITUTE

Biomedical Engineering	M,D,O*
Biotechnology	M,D
Chemical Engineering	M,D*
Civil Engineering	M,D,O
Computer Engineering	M,D,O
Computer Science	M,D,O*
Electrical Engineering	M,D,O
Energy and Power Engineering	M,D,O
Engineering and Applied	
Sciences—General	M,D,O*
Environmental Engineering	M,D,O
Fire Protection Engineering	M,D,O
Manufacturing Engineering	M,D,O*
Materials Engineering	M,D,O
Materials Sciences	M,D,O
Mechanical Engineering	M,D,O*

WORCESTER STATE COLLEGE

Biotechnology	M

WRIGHT STATE UNIVERSITY

Biomedical Engineering	M
Computer Engineering	M,D*
Computer Science	M,D*
Electrical Engineering	M
Engineering and Applied	
Sciences—General	M,D*
Ergonomics and Human	
Factors	M
Materials Engineering	M
Materials Sciences	M
Mechanical Engineering	M

YALE UNIVERSITY

Chemical Engineering	M,D*
Computer Science	D*
Electrical Engineering	M,D
Engineering Physics	M,D
Engineering and Applied	
Sciences—General	M,D*
Mechanical Engineering	M,D
Mechanics	M,D

YORK UNIVERSITY

Computer Science	M

YOUNGSTOWN STATE UNIVERSITY

Chemical Engineering	M
Civil Engineering	M
Electrical Engineering	M
Engineering and Applied	
Sciences—General	M
Environmental Engineering	M
Industrial/Management	
Engineering	M
Mechanical Engineering	M

Academic and Professional Programs in Engineering and Applied Sciences

This part of Book 5 consists of twenty sections covering engineering and applied sciences. Each section has a table of contents; program directories, which consist of brief profiles of programs in the relevant subject areas followed by 50-word and 100-word Announcements, if programs have chosen to include them; Cross-Discipline Announcements, if any programs have chosen to submit such entries; and In-Depth Descriptions, which are more individualized statements, if programs have chosen to submit them. The first section's directory, Engineering and Applied Sciences, is general for all of the following sections in the book except Section 8, Computer Science and Information Technology.

Section 1
Engineering and Applied Sciences

This section contains a directory of institutions offering graduate work in engineering and applied sciences, followed by in-depth entries submitted by institutions that chose to prepare detailed program descriptions. Additional information about programs listed in the directory but not augmented by an in-depth entry may be obtained by writing directly to the dean of a graduate school or chair of a department at the address given in the directory.

For programs in specific areas of engineering, see all other sections in this book. For work in related areas, see also Applied Arts and Design (Industrial Design) and Architecture (Environmental Design) in Book 2; Ecology, Environmental Biology, and Evolutionary Biology in Book 3; and Agricultural and Food Sciences and Natural Resources in Book 4.

CONTENTS

CONTENTS

Engineering and Applied Sciences—General

Air Force Institute of Technology, School of Engineering, Wright-Patterson AFB, OH 45433-7765. Offers MS, PhD. *Accreditation:* ABET (one or more programs are accredited). Part-time programs available. *Faculty:* 82 full-time, 2 part-time. *Students:* 370 full-time, 35 part-time. In 1998, 50 master's awarded (100% found work related to degree); 18 doctorates awarded (100% found work related to degree). *Degree requirements:* For master's and doctorate, thesis/dissertation required, foreign language not required. *Entrance requirements:* For master's, GRE General Test (minimum score of 500 on verbal section, 600 on quantitative required), minimum GPA of 3.0, must be military officer or U.S. citizen; for doctorate, GRE General Test (minimum score of 550 on verbal section, 650 on quantitative required), minimum GPA of 3.0, must be military officer or U.S. citizen. Application fee: $0. *Financial aid:* Fellowships, research assistantships available. Aid available to part-time students. *Faculty research:* Speech processing and synthesis, battle damage repair analysis, digital flight control laboratory. *Unit head:* Dr. Robert Calico, Dean, 937-255-3025, Fax: 937-255-6569.

Alabama Agricultural and Mechanical University, School of Graduate Studies, School of Engineering and Technology, Normal, AL 35762-1357. Offers M Ed, MS. *Accreditation:* NCATE (one or more programs are accredited). Part-time and evening/weekend programs available. *Faculty:* 5 full-time (1 woman). *Students:* 2 full-time (0 women), 17 part-time (4 women); includes 8 minority (all African Americans), 2 international. In 1998, 4 degrees awarded (100% found work related to degree). *Degree requirements:* For master's, comprehensive exam required, thesis optional, foreign language not required. *Entrance requirements:* For master's, GRE General Test. *Application deadline:* For fall admission, 5/1. Application fee: $15 ($20 for international students). Tuition, state resident: full-time $1,932. Tuition, nonresident: full-time $3,864. Tuition and fees vary according to course load. *Financial aid:* Research assistantships with tuition reimbursements, career-related internships or fieldwork available. Financial aid application deadline: 4/1. *Faculty research:* Ionized gases, hypersonic flow phenomenology, robotics systems development. *Unit head:* Dr. Arthur Bond, Dean, 256-851-5560.

American University in Cairo, Graduate Studies and Research, School of Sciences and Engineering, Cairo, 11511, Egypt. Offers MS, Diploma. *Faculty:* 15 full-time (1 woman). *Students:* 44. *Degree requirements:* For master's, thesis required. *Entrance requirements:* For master's, English entrance exam and/or TOEFL. *Application deadline:* For fall admission, 3/31 (priority date); for spring admission, 1/10 (priority date). Applications are processed on a rolling basis. Application fee: $45. Electronic applications accepted. *Financial aid:* Fellowships, teaching assistantships, scholarships and unspecified assistantships available. *Faculty research:* Construction management and technology, structural engineering, public works engineering. *Unit head:* Dr. Fadel Assabghy, Dean, 202-357-5412, Fax: 202-355-7565, E-mail: assabghy@aucegypt.edu. *Application contact:* Mary Davidson, Coordinator of Student Affairs, 212-730-8800, Fax: 212-730-1600, E-mail: davidson@aucnyo.edu.

Arizona State University, Graduate College, College of Engineering and Applied Sciences, Tempe, AZ 85287. Offers M Eng, MCS, MS, MSE, PhD, MSE/MIMOT. Part-time programs available. *Faculty:* 205 full-time (24 women), 18 part-time (1 woman). *Students:* 876 full-time (168 women), 521 part-time (123 women); includes 135 minority (16 African Americans, 68 Asian Americans or Pacific Islanders, 46 Hispanic Americans, 5 Native Americans), 721 international. Average age 28. 2626 applicants, 51% accepted. In 1998, 371 master's, 63 doctorates awarded. *Degree requirements:* For doctorate, dissertation required. Application fee: $45. *Financial aid:* Fellowships, research assistantships, teaching assistantships, career-related internships or fieldwork available. *Faculty research:* Aerodynamics, computer design, environmental fluid dynamics, solar energy, thermosciences. *Unit head:* Dr. Peter E. Crouch, Dean, 480-965-1722.

See in-depth description on page 61.

Arizona State University East, College of Technology and Applied Sciences, Mesa, AZ 85212. Offers MS. Part-time and evening/weekend programs available. *Faculty:* 19 full-time (2 women), 2 part-time (0 women). *Students:* 53 full-time (23 women), 104 part-time (32 women); includes 17 minority (2 African Americans, 9 Asian Americans or Pacific Islanders, 6 Hispanic Americans), 41 international. Average age 34. In 1998, 43 degrees awarded. *Degree requirements:* For master's, thesis or applied project and oral defense required. *Entrance requirements:* For master's, minimum GPA of 3.0. *Average time to degree:* Master's–2.7 years full-time, 5 years part-time. *Application deadline:* Applications are processed on a rolling basis. Application fee: $45. *Financial aid:* In 1998–99, 31 students received aid, including 21 research assistantships with partial tuition reimbursements available (averaging $10,505 per year); teaching assistantships with partial tuition reimbursements available, career-related internships or fieldwork, Federal Work-Study, grants, scholarships, tuition waivers (full and partial), and unspecified assistantships also available. Aid available to part-time students. Financial aid application deadline: 3/1; financial aid applicants required to submit FAFSA. Total annual research expenditures: $2 million. *Unit head:* Dr. Albert L. McHenry, Dean, 602-727-1093, Fax: 602-727-1089, E-mail: iacaxm@asuvm.inre.asu.edu. *Application contact:* Dr. Albert L. McHenry, Dean, 602-727-1093, Fax: 602-727-1089, E-mail: iacaxm@asuvm.inre.asu.edu.

See in-depth description on page 63.

Auburn University, Graduate School, College of Engineering, Auburn, Auburn University, AL 36849-0002. Offers M Ch E, M Mtl E, MAE, MCE, MCSE, MEE, MIE, MME, MS, PhD. Part-time programs available. *Faculty:* 129 full-time (8 women). *Students:* 219 full-time (37 women), 181 part-time (32 women); includes 32 minority (20 African Americans, 7 Asian Americans or Pacific Islanders, 5 Hispanic Americans), 143 international. 415 applicants, 39% accepted. In 1998, 131 master's, 46 doctorates awarded. *Degree requirements:* For master's, thesis (MS) required; for doctorate, dissertation required. *Entrance requirements:* For master's, GRE General Test; for doctorate, GRE General Test (minimum score of 400 on each section required). *Application deadline:* For fall admission, 9/1; for spring admission, 3/1. Applications are processed on a rolling basis. Application fee: $25 ($50 for international students). Tuition, state resident: full-time $2,760; part-time $76 per credit hour. Tuition, nonresident: full-time $8,280; part-time $228 per credit hour. *Financial aid:* Fellowships, research assistantships, teaching assistantships, Federal Work-Study available. Aid available to part-time students. Financial aid application deadline: 3/15. *Unit head:* Dr. Larry Benefield, Interim Dean, 334-844-2308. *Application contact:* Dr. John F. Pritchett, Dean of the Graduate School, 334-844-4700.

Boston University, College of Engineering, Boston, MA 02215. Offers MS, PhD, MBA/MS, MD/PhD. Part-time programs available. Postbaccalaureate distance learning degree programs offered (no on-campus study). *Faculty:* 109 full-time (13 women), 17 part-time (2 women). *Students:* 296 full-time (70 women), 69 part-time (8 women); includes 24 minority (4 African Americans, 19 Asian Americans or Pacific Islanders, 1 Hispanic American), 151 international. Average age 27. 1063 applicants, 23% accepted. In 1998, 92 master's, 29 doctorates awarded. Terminal master's awarded for partial completion of doctoral program. *Degree requirements:* For doctorate, dissertation required, foreign language not required, foreign language not required. *Entrance requirements:* For master's, GRE General Test, TOEFL (minimum score of 500 required; 213 for computer-based); for doctorate, GRE General Test, TOEFL. *Application deadline:* For fall admission, 4/1; for spring admission, 10/1. Applications are processed on a rolling basis. Application fee: $50. Tuition: Full-time $23,770; part-time $743 per credit. Required fees: $220. Tuition and fees vary according to class time, course level, campus/location and program. *Financial aid:* In 1998–99, 309 students received aid, including 37 fellowships with full tuition reimbursements available (averaging $13,000 per year), 146 research assistantships with full tuition reimbursements available (averaging $11,500 per year), 65 teaching assistantships with full tuition reimbursements available (averaging $11,500 per year); career-related internships or fieldwork, Federal Work-Study, institutionally-sponsored loans, scholarships, and tuition waivers (full and partial) also available. Financial aid application deadline: 12/15; financial aid applicants required to submit FAFSA. *Faculty research:* Biomolecular engineering, optoelectronic devices, materials, speech and signal processing, waves and acoustics, dynamics and controls. Total annual research expenditures: $15.9 million. *Unit head:* Dr. Charles DeLisi, Dean, 617-353-2800, Fax: 617-353-6322. *Application contact:* Cheryl Kelley, Graduate Programs Director, 617-353-9760, Fax: 617-353-0259, E-mail: enggrad@bu.edu.

See in-depth description on page 65.

Bradley University, Graduate School, College of Engineering and Technology, Peoria, IL 61625-0002. Offers MSCE, MSEE, MSIE, MSME, MSMFE. Part-time and evening/weekend programs available. *Degree requirements:* For master's, comprehensive exam required. *Entrance requirements:* For master's, TOEFL (minimum score of 525 required), minimum GPA of 3.0.

Brigham Young University, Graduate Studies, College of Engineering and Technology, Provo, UT 84602-1001. Offers MS, PhD, MBA/MS. Part-time programs available. *Faculty:* 94 full-time (0 women), 7 part-time (1 woman). *Students:* 222 full-time (15 women), 42 part-time (3 women); includes 7 minority (1 African American, 2 Asian Americans or Pacific Islanders, 4 Hispanic Americans), 36 international. Average age 25. 221 applicants, 42% accepted. In 1998, 110 master's, 14 doctorates awarded. *Degree requirements:* For doctorate, dissertation required, foreign language not required. *Average time to degree:* Master's–2 years full-time; doctorate–4 years full-time. *Application deadline:* Applications are processed on a rolling basis. Application fee: $30. Electronic applications accepted. Tuition: Full-time $3,330; part-time $185 per credit hour. Tuition and fees vary according to program and student's religious affiliation. *Financial aid:* In 1998–99, 226 students received aid, including 26 fellowships with partial tuition reimbursements available (averaging $9,000 per year), 73 research assistantships with partial tuition reimbursements available (averaging $4,200 per year), 93 teaching assistantships with partial tuition reimbursements available (averaging $6,600 per year); career-related internships or fieldwork, institutionally-sponsored loans, and scholarships also available. Aid available to part-time students. Financial aid application deadline: 3/15; financial aid applicants required to submit FAFSA. *Faculty research:* Combustion, design, optimization, catalysis, computer modeling. Total annual research expenditures: $6 million. *Unit head:* Dr. Douglas M. Chabries, Dean, 801-378-4327, Fax: 801-378-5705, E-mail: college@et.byu.edu.

See in-depth description on page 67.

Brown University, Graduate School, Division of Engineering, Providence, RI 02912. Offers aerospace engineering (Sc M, PhD); biomedical engineering (Sc M); electrical sciences (Sc M, PhD); fluid mechanics, thermodynamics, and chemical processes (Sc M, PhD); materials science (Sc M, PhD); mechanics of solids and structures (Sc M, PhD). *Degree requirements:* For doctorate, dissertation, preliminary exam required, foreign language not required, foreign language not required.

See in-depth description on page 69.

Bucknell University, Graduate Studies, College of Engineering, Lewisburg, PA 17837. Offers MS, MS Ch E, MSCE, MSEE, MSME. Part-time programs available. *Faculty:* 26 full-time, 3 part-time. *Students:* 25 (6 women). *Degree requirements:* For master's, thesis required, foreign language not required. *Entrance requirements:* For master's, GRE General Test (minimum combined score of 1000 required), GRE Subject Test, TOEFL (minimum score of 550 required), minimum GPA of 2.8. *Application deadline:* For fall admission, 6/1 (priority date); for spring admission, 12/1 (priority date). Applications are processed on a rolling basis. Application fee: $25. Tuition: Part-time $2,600 per course. Tuition and fees vary according to course load. *Financial aid:* Fellowships, research assistantships, teaching assistantships, unspecified assistantships available. Financial aid application deadline: 3/1. *Unit head:* Dr. Joseph Humphrey, Dean, 570-577-3711.

California Institute of Technology, Division of Engineering and Applied Science, Pasadena, CA 91125-0001. Offers aeronautics (MS, PhD, Engr); applied mathematics (PhD); applied mechanics (MS, PhD); applied physics (MS, PhD), including applied physics, plasma physics; civil engineering (MS, PhD, Engr); computation and neural systems (MS, PhD); computer science (MS, PhD); control and dynamical systems (PhD); electrical engineering (MS, PhD, Engr); environmental engineering science (MS, PhD); materials science (MS, PhD); mechanical engineering (MS, PhD, Engr). *Faculty:* 76 full-time (5 women). *Students:* 422 full-time (69 women), 213 international. 1832 applicants, 4% accepted. In 1998, 65 master's, 69 doctorates, 6 other advanced degrees awarded. Terminal master's awarded for partial completion of doctoral program. *Degree requirements:* For doctorate, dissertation required, foreign language not required, foreign language not required. *Application deadline:* For fall admission, 1/15. Application fee: $0. *Financial aid:* Fellowships, research assistantships, teaching assistantships, Federal Work-Study and institutionally-sponsored loans available. Aid available to part-time students. *Unit head:* Dr. John H. Seinfeld, Chairman, 626-395-4101.

See in-depth description on page 71.

California National University for Advanced Studies, College of Engineering, North Hills, CA 91343. Offers MS Eng. Part-time programs available. Postbaccalaureate distance learning degree programs offered (no on-campus study). *Degree requirements:* For master's, computer language, thesis or alternative, project required. *Application deadline:* Applications are processed on a rolling basis. Application fee: $50 ($100 for international students). *Unit head:* Dr. Robert Ryan, Dean, 800-782-2422, Fax: 818-830-2418, E-mail: cnuadmin@mail.cnuas.edu. *Application contact:* Jeanne Cunneff, Admissions Representative, 800-782-2422, Fax: 818-830-2418, E-mail: jeanne@mail.cnuas.edu.

California Polytechnic State University, San Luis Obispo, College of Engineering, San Luis Obispo, CA 93407. Offers MS, MSAE, MSCS, MBA/MS, MCRP/MS. Part-time programs available. *Faculty:* 98 full-time (8 women), 82 part-time (14 women). *Students:* 75 full-time (12 women), 40 part-time (6 women); includes 21 Asian Americans or Pacific Islanders, 9 Hispanic Americans, 11 international. 156 applicants, 44% accepted. In 1998, 62 degrees awarded. *Degree requirements:* Foreign language not required. *Entrance requirements:* For master's, GRE General Test. *Average time to degree:* Master's–2 years full-time, 4 years part-time. *Application deadline:* For fall admission, 5/31 (priority date). Applications are processed on a rolling basis. Application fee: $55. Electronic applications accepted. Tuition, nonresident: part-time $164 per unit. Required fees: $531 per quarter. *Financial aid:* Fellowships, research assistantships, teaching assistantships, career-related internships or fieldwork, Federal Work-Study, and institutionally-sponsored loans available. Financial aid application deadline: 3/2; financial aid applicants required to submit FAFSA. *Faculty research:* Artificial intelligence, fuel systems, solar power, advanced materials, traffic systems. *Unit head:* Dr. Peter Y. Lee, Dean, 805-756-2131, Fax: 805-756-6503, E-mail: plee@calpoly.edu. *Application contact:* Dr. Daniel W. Walsh, Associate Dean, 805-756-2131, Fax: 805-756-6503, E-mail: dwalsh@calpoly.edu.

See in-depth description on page 73.

California State Polytechnic University, Pomona, Graduate Studies, College of Engineering, Pomona, CA 91768-2557. Offers electrical engineering (MSEE); engineering (MSE). Part-time programs available. *Faculty:* 60. *Students:* 23 full-time (6 women), 153 part-time (26 women); includes 83 minority (1 African American, 75 Asian Americans or Pacific Islanders, 6 Hispanic Americans, 1 Native American), 27 international. Average age 30. 91 applicants, 70% accepted. In 1998, 41 degrees awarded. *Degree requirements:* For master's, computer language, thesis or comprehensive exam required. *Entrance requirements:* For master's, TOEFL, GRE General Test or minimum GPA of 3.0 in upper-level course work. *Average time to degree:* Master's–2 years full-time, 4 years part-time. *Application deadline:* Applications are processed on a rolling basis. Application fee: $55. Tuition, nonresident: part-time $164 per unit. *Financial aid:* In 1998–99, 27 students received aid, including 1 fellowship, 6 research assistantships, 5 teaching assistantships; career-related internships or fieldwork, Federal Work-Study,

Engineering and Applied Sciences—General

California State Polytechnic University, Pomona (continued)
institutionally-sponsored loans, and unspecified assistantships also available. Aid available to part-time students. Financial aid application deadline: 3/2; financial aid applicants required to submit FAFSA. *Faculty research:* Aerospace; alternative vehicles; communications, computers, and controls; engineering management. Total annual research expenditures: $650,000. *Unit head:* Dr. Carl E. Rathmann, Interim Dean, 909-869-2600, Fax: 909-869-4370, E-mail: cerathmann@csupomona.edu. *Application contact:* Dr. Elhami T. Ibrahim, Graduate Director, 909-869-2476, Fax: 909-869-4370, E-mail: etibrahim@csupomona.edu.

California State University, Chico, Graduate School, College of Engineering, Computer Science, and Technology, Chico, CA 95929-0722. Offers MS. Part-time programs available. *Faculty:* 41 full-time (4 women), 8 part-time (1 woman). *Students:* 52 full-time (12 women), 24 part-time (7 women); includes 15 minority (11 Asian Americans or Pacific Islanders, 4 Hispanic Americans), 31 international. Average age 31. In 1998, 72 degrees awarded. *Degree requirements:* For master's, thesis or alternative, oral exam required, foreign language not required. *Application deadline:* For fall admission, 4/1. Applications are processed on a rolling basis. Application fee: $55. *Financial aid:* Fellowships, research assistantships, teaching assistantships, career-related internships or fieldwork and Federal Work-Study available. Aid available to part-time students. *Unit head:* Dr. Kenneth Derucher, Dean, 530-898-5963.

California State University, Fresno, Division of Graduate Studies, School of Engineering, Fresno, CA 93740-0057. Offers MS. Part-time and evening/weekend programs available. *Faculty:* 18 full-time (1 woman), 4 part-time (0 women). *Students:* 24 full-time (4 women), 59 part-time (20 women); includes 13 minority (2 African Americans, 9 Asian Americans or Pacific Islanders, 2 Hispanic Americans), 37 international. Average age 31. 90 applicants, 81% accepted. In 1998, 19 degrees awarded. *Degree requirements:* For master's, thesis or alternative required, foreign language not required. *Entrance requirements:* For master's, GRE General Test, TOEFL (minimum score of 550 required). *Average time to degree:* Master's–3.5 years full-time. *Application deadline:* For fall admission, 8/1 (priority date); for spring admission, 12/1. Applications are processed on a rolling basis. Application fee: $55. Electronic applications accepted. Tuition, nonresident: part-time $246 per unit. Required fees: $1,906; $620 per semester. *Financial aid:* In 1998–99, 8 teaching assistantships were awarded.; fellowships, research assistantships, career-related internships or fieldwork, Federal Work-Study, scholarships, and unspecified assistantships also available. Financial aid application deadline: 3/1; financial aid applicants required to submit FAFSA. *Faculty research:* Exhaust emission, blended fuel testing,waste management. *Unit head:* Dr. Karl Longley, Dean, 559-278-2500, Fax: 559-278-7071, E-mail: karl_longley@csufresno.edu. *Application contact:* Dr. Jesus Larralde-Muro, Graduate Program Coordinator, 559-278-2566, E-mail: jesus_larralde-muro@csufresno.edu.

California State University, Fullerton, Graduate Studies, School of Engineering and Computer Science, Fullerton, CA 92834-9480. Offers MS. Part-time programs available. *Faculty:* 39 full-time (6 women), 44 part-time. *Students:* 25 full-time (7 women), 215 part-time (43 women); includes 104 minority (5 African Americans, 88 Asian Americans or Pacific Islanders, 11 Hispanic Americans), 46 international. Average age 31. 206 applicants, 48% accepted. In 1998, 84 degrees awarded. *Degree requirements:* For master's, computer language, comprehensive exam, project or thesis required. *Entrance requirements:* For master's, minimum undergraduate GPA of 2.5. Application fee: $55. Tuition, nonresident: part-time $264 per unit. Required fees: $1,947; $1,281 per year. *Financial aid:* Career-related internships or fieldwork, Federal Work-Study, grants, and institutionally-sponsored loans available. Aid available to part-time students. Financial aid application deadline: 3/1. *Unit head:* Dr. Richard Rolke, Acting Dean, 714-278-3362. *Application contact:* Dr. David Falconer, Associate Dean, 714-278-3362.

California State University, Long Beach, Graduate Studies, College of Engineering, Long Beach, CA 90840. Offers MS, MSAE, MSCE, MSE, MSEE, MSME, PhD, CE. Part-time and evening/weekend programs available. *Faculty:* 83 full-time (7 women), 44 part-time (3 women). *Students:* 180 full-time (40 women), 485 part-time (92 women); includes 292 minority (17 African Americans, 227 Asian Americans or Pacific Islanders, 46 Hispanic Americans, 2 Native Americans), 158 international. Average age 32. 679 applicants, 55% accepted. In 1998, 128 degrees awarded. Terminal master's awarded for partial completion of doctoral program. *Degree requirements:* For doctorate, one foreign language, computer language, dissertation required, foreign language not required. *Entrance requirements:* For master's, TOEFL (minimum score of 550 required). *Application deadline:* For fall admission, 8/1; for spring admission, 12/1. Application fee: $55. Electronic applications accepted. Tuition, nonresident: part-time $246 per unit. Required fees: $569 per semester. Tuition and fees vary according to course load. *Financial aid:* Research assistantships, teaching assistantships, career-related internships or fieldwork, Federal Work-Study, grants, institutionally-sponsored loans, and unspecified assistantships available. Financial aid application deadline: 3/2. *Faculty research:* Aerodynamics, CAD/CAM, structural engineering, artificial intelligence communications. Total annual research expenditures: $13.1 million. *Unit head:* Dr. J. Richard Williams, Dean, 562-985-5190, Fax: 562-985-7561, E-mail: jrw@engr.csulb.edu. *Application contact:* Dr. Mihir K. Das, Associate Dean for Instruction, 562-985-5257, Fax: 562-985-7561, E-mail: mdas@engr.csulb.edu.

See in-depth description on page 75.

California State University, Los Angeles, Graduate Studies, School of Engineering and Technology, Los Angeles, CA 90032-8530. Offers MA, MS. Part-time and evening/weekend programs available. *Faculty:* 33 full-time, 37 part-time. *Students:* 31 full-time (9 women), 131 part-time (22 women); includes 94 minority (11 African Americans, 55 Asian Americans or Pacific Islanders, 28 Hispanic Americans), 30 international. In 1998, 36 degrees awarded. *Degree requirements:* For master's, computer language required, foreign language not required. *Entrance requirements:* For master's, TOEFL (minimum score of 550 required). *Application deadline:* For fall admission, 6/30; for spring admission, 2/1. Applications are processed on a rolling basis. Application fee: $55. *Financial aid:* In 1998–99, 46 students received aid. Federal Work-Study available. Aid available to part-time students. Financial aid application deadline: 3/1. *Unit head:* Dr. Raymond Landis, Dean, 323-343-4500.

California State University, Northridge, Graduate Studies, College of Engineering and Computer Science, Northridge, CA 91330. Offers MS, MSE. Part-time and evening/weekend programs available. *Faculty:* 53 full-time, 35 part-time. *Students:* 63 full-time (20 women), 240 part-time (44 women); includes 93 minority (7 African Americans, 62 Asian Americans or Pacific Islanders, 22 Hispanic Americans, 2 Native Americans), 43 international. Average age 33. 221 applicants, 62% accepted. In 1998, 78 degrees awarded. *Degree requirements:* Foreign language not required. *Entrance requirements:* For master's, GRE General Test, TOEFL, minimum GPA of 2.5. *Application deadline:* For fall admission, 11/30. Application fee: $55. Tuition, nonresident: part-time $246 per unit. International tuition: $7,874 full-time. Required fees: $1,970. Tuition and fees vary according to course load. *Financial aid:* Teaching assistantships, career-related internships or fieldwork and Federal Work-Study available. Aid available to part-time students. Financial aid application deadline: 3/1. *Unit head:* Dr. Laurence Caretto, Interim Dean, 818-677-4501. *Application contact:* Dr. Laurence Caretto, Interim Dean, 818-677-4501.

California State University, Sacramento, Graduate Studies, School of Engineering and Computer Science, Sacramento, CA 95819-6048. Offers MS. Part-time and evening/weekend programs available. *Degree requirements:* For master's, writing proficiency exam required. *Entrance requirements:* For master's, TOEFL (minimum score of 550 required). *Application deadline:* For fall admission, 4/15; for spring admission, 11/1. Application fee: $55. *Financial aid:* Research assistantships, teaching assistantships, career-related internships or fieldwork and Federal Work-Study available. Aid available to part-time students. Financial aid application deadline: 3/1. *Unit head:* Dr. Braja Das, Dean, 916-278-6366.

Carleton University, Faculty of Graduate Studies, Faculty of Engineering and Design, Ottawa, ON K1S 5B6, Canada. Offers M Arch, M Eng, M Sc, PhD. *Faculty:* 97 full-time (7 women). *Students:* 263 full-time (57 women), 133 part-time (25 women). In 1998, 77 master's, 8 doctorates awarded. *Degree requirements:* For doctorate, dissertation required. *Entrance requirements:* For master's, TOEFL (minimum score of 550 required), honors degree; for doctorate, TOEFL (minimum score of 550 required), MA Sc or M Eng. *Average time to degree:* Master's–2.1 years full-time, 3.4 years part-time; doctorate–2.7 years full-time, 6.2 years part-time. *Application deadline:* For fall admission, 3/1 (priority date). Applications are processed on a rolling basis. Application fee: $35. *Financial aid:* Fellowships, research assistantships, teaching assistantships, institutionally-sponsored loans available. Financial aid application deadline: 3/1. Total annual research expenditures: $7.2 million. *Unit head:* Dr. Samy A. Mahmoud, Dean, 613-520-2600 Ext. 5743, Fax: 613-520-7481, E-mail: mahmoud@sce.carleton.ca. *Application contact:* Cate Palmer, Graduate Studies Administrator, 613-520-5659 Ext. 5659, Fax: 613-520-5682, E-mail: cate_palmer@pigeon.carleton.ca.

Carnegie Mellon University, Carnegie Institute of Technology, Pittsburgh, PA 15213-3891. Offers M Ch E, ME, MOM, MS, PhD. Part-time and evening/weekend programs available. *Faculty:* 185 full-time (25 women), 9 part-time (1 woman). *Students:* 538 full-time (99 women), 68 part-time (12 women); includes 46 minority (9 African Americans, 28 Asian Americans or Pacific Islanders, 7 Hispanic Americans, 2 Native Americans), 321 international. Average age 27. In 1998, 164 master's, 97 doctorates awarded. *Degree requirements:* For doctorate, qualifying exam required. *Entrance requirements:* For master's and doctorate, GRE General Test, TOEFL. *Financial aid:* Fellowships, research assistantships, teaching assistantships, career-related internships or fieldwork, Federal Work-Study, institutionally-sponsored loans, and tuition waivers (full and partial) available. Total annual research expenditures: $31. 3 million. *Unit head:* John Anderson, Dean, 412-268-2537. *Application contact:* Chris Hendrickson, Associate Dean, 412-268-2478, Fax: 412-268-7813.

Case Western Reserve University, School of Graduate Studies, The Case School of Engineering, Cleveland, OH 44106. Offers ME, MS, PhD, MD/PhD. Part-time and evening/weekend programs available. Postbaccalaureate distance learning degree programs offered. *Faculty:* 106 full-time (6 women), 58 part-time (4 women). *Students:* 221 full-time (51 women), 475 part-time (93 women). Average age 24. 1370 applicants, 31% accepted. In 1998, 178 master's, 59 doctorates awarded. Terminal master's awarded for partial completion of doctoral program. *Degree requirements:* For doctorate, dissertation required, foreign language not required, foreign language not required. *Entrance requirements:* For master's and doctorate, TOEFL. Application fee: $25. Electronic applications accepted. *Financial aid:* In 1998–99, 65 fellowships, 258 research assistantships, 122 teaching assistantships were awarded.; career-related internships or fieldwork, Federal Work-Study, institutionally-sponsored loans, and tuition waivers (full and partial) also available. Aid available to part-time students. Total annual research expenditures: $27.3 million. *Unit head:* James W. Wagner, Dean, 216-368-4436, Fax: 216-368-6939, E-mail: jww2@po.cwru.edu.

See in-depth description on page 77.

The Catholic University of America, School of Engineering, Washington, DC 20064. Offers MBE, MCE, MEE, MME, MS Engr, D Engr, PhD. Part-time and evening/weekend programs available. *Faculty:* 25 full-time (1 woman), 30 part-time (1 woman). *Students:* 33 full-time (6 women), 131 part-time (15 women); includes 32 minority (19 African Americans, 10 Asian Americans or Pacific Islanders, 3 Hispanic Americans), 45 international. Average age 33. 168 applicants, 75% accepted. In 1998, 70 master's, 12 doctorates awarded. *Degree requirements:* For master's, thesis optional, foreign language not required; for doctorate, dissertation, comprehensive and oral exams required, foreign language not required. *Entrance requirements:* For master's, minimum GPA of 3.0. *Application deadline:* For fall admission, 7/31; for spring admission, 12/10. Applications are processed on a rolling basis. Application fee: $50. *Financial aid:* Fellowships, research assistantships, teaching assistantships, career-related internships or fieldwork, Federal Work-Study, institutionally-sponsored loans, tuition waivers (full and partial), and unspecified assistantships available. Aid available to part-time students. Financial aid application deadline: 2/1. *Faculty research:* Controls, signal processing, multiphaseflow, robotics, composite structures. Total annual research expenditures: $783,133. *Unit head:* Dr. William E. Kelly, Dean, 202-319-5160, Fax: 202-319-4499. *Application contact:* Mary Teresa Stilwell, Administrative Assistant, 202-319-5177, Fax: 202-319-4499, E-mail: stilwell@cua.edu.

See in-depth description on page 79.

Central Missouri State University, School of Graduate Studies, College of Applied Sciences and Technology, Warrensburg, MO 64093. Offers MS, MSE, PhD, Ed S. Part-time programs available. *Faculty:* 75 full-time. *Students:* 88 full-time (25 women), 193 part-time (40 women); includes 24 minority (15 African Americans, 4 Asian Americans or Pacific Islanders, 4 Hispanic Americans, 1 Native American), 29 international. In 1998, 139 degrees awarded. *Degree requirements:* For Ed S, thesis required. *Application deadline:* Applications are processed on a rolling basis. Application fee: $25 ($50 for international students). Tuition, state resident: full-time $3,576; part-time $149 per credit hour. Tuition, nonresident: full-time $7,152; part-time $298 per credit hour. Tuition and fees vary according to course load and campus/location. *Financial aid:* In 1998–99, 9 research assistantships with tuition reimbursements (averaging $3,750 per year), 19 teaching assistantships with tuition reimbursements (averaging $3,750 per year) were awarded.; fellowships with tuition reimbursements, Federal Work-Study, grants, scholarships, unspecified assistantships, and administrative and laboratory assistantships also available. Aid available to part-time students. Financial aid application deadline: 3/1; financial aid applicants required to submit FAFSA. *Unit head:* Dr. Art Rosser, Dean, 660-543-4450, Fax: 660-543-8031, E-mail: rosser@cmsu1.cmsu.edu.

Christian Brothers University, Graduate Programs, School of Engineering, Memphis, TN 38104-5581. Offers MEM. Part-time and evening/weekend programs available. *Faculty:* 1 full-time (0 women), 2 part-time (0 women). *Students:* 10 full-time (6 women), 28 part-time (4 women); includes 6 minority (4 African Americans, 1 Asian American or Pacific Islander, 1 Hispanic American) Average age 32. 16 applicants, 75% accepted. In 1998, 4 degrees awarded. *Degree requirements:* For master's, engineering management project required, foreign language and thesis not required. *Entrance requirements:* For master's, GRE. Application fee: $25. *Financial aid:* Institutionally-sponsored loans available. *Unit head:* Dr. Ray Brown, Dean, 901-321-3408, Fax: 901-321-3494. *Application contact:* C. Blackman, Director, 901-321-3283, Fax: 901-321-3494.

City College of the City University of New York, Graduate School, School of Engineering, New York, NY 10031-9198. Offers ME, MS, PhD. Part-time programs available. *Students:* 104 full-time (39 women), 351 part-time (95 women). In 1998, 148 degrees awarded. *Degree requirements:* For master's, computer language required, thesis optional, foreign language not required; for doctorate, dissertation, comprehensive exams required. *Entrance requirements:* For master's, TOEFL (minimum score of 500 required); for doctorate, GRE General Test, TOEFL. *Application deadline:* Applications are processed on a rolling basis. Application fee: $40. *Financial aid:* Fellowships, research assistantships, teaching assistantships, Federal Work-Study, institutionally-sponsored loans, and tuition waivers (full and partial) available. Aid available to part-time students. *Unit head:* Dr. Muntaz G. Kassir, Associate Dean for Graduate Studies, 212-650-8030. *Application contact:* Graduate Admissions Office, 212-650-6977.

See in-depth description on page 81.

Clarkson University, Graduate School, School of Engineering, Potsdam, NY 13699. Offers ME, MS, PhD. Part-time programs available. *Faculty:* 58 full-time (5 women), 2 part-time (0 women). *Students:* 125 full time (32 women), 2 part-time (1 woman); includes 5 minority (1 African American, 4 Native Americans), 85 international. Average age 27. 494 applicants, 49% accepted. In 1998, 57 master's, 10 doctorates awarded. *Degree requirements:* For master's, thesis required, foreign language not required; for doctorate, dissertation, departmental qualifying exam required, foreign language not required. *Entrance requirements:* For master's, GRE, TOEFL. *Application deadline:* For fall admission, 5/15 (priority date); for spring admission, 10/15 (priority date). Applications are processed on a rolling basis. Application fee: $25 ($35 for international students). Tuition: Part-time $661 per credit hour. Required fees: $215 per semester. *Financial aid:* In 1998–99, 7 fellowships, 61 research assistantships, 34 teaching assistantships were awarded. *Faculty research:* Turbulent flow, structural dynamics, fluid dynamics, soldynamics, electronic manufacturing. Total annual research expenditures: $3.5 million. *Unit head:* Dr. Gregory A. Campbell, Dean, 315-268-6446, Fax: 315-268-3841, E-mail: gac@clarkson.edu. *Application contact:* Dr. Philip K. Hopke, Dean of the Graduate School, 315-268-6447, Fax: 315-268-7994, E-mail: hopkepk@clarkson.edu.

See in-depth description on page 83.

Clemson University, Graduate School, College of Engineering and Science, Clemson, SC 29634. Offers M Engr, MS, PhD. *Accreditation:* ABET (one or more programs are accredited). Part-time programs available. *Faculty:* 263 full-time (20 women), 46 part-time (7 women). *Students:* 777 full-time (191 women), 127 part-time (33 women); includes 32 minority (18 African Americans, 7 Asian Americans or Pacific Islanders, 7 Hispanic Americans), 405 international. 1357 applicants, 56% accepted. In 1998, 226 master's, 44 doctorates awarded. *Degree requirements:* For doctorate, dissertation required. *Entrance requirements:* For master's, TOEFL; for doctorate, GRE General Test, TOEFL. Application fee: $35. Electronic applications accepted. *Financial aid:* Fellowships, research assistantships, teaching assistantships, career-related internships or fieldwork, institutionally-sponsored loans, and unspecified assistantships available. Aid available to part-time students. Financial aid applicants required to submit FAFSA. *Unit head:* Dr. Thomas M. Keinath, Dean, 864-656-3202, Fax: 864-656-0859, E-mail: keinath@clemson.edu. *Application contact:* Dr. Christian Przirembel, Associate Dean, 864-656-3200, Fax: 864-656-0859, E-mail: rutgers@clemson.edu.

See in-depth description on page 85.

Cleveland State University, College of Graduate Studies, Fenn College of Engineering, Cleveland, OH 44115-2440. Offers MS, D Eng. Part-time programs available. *Faculty:* 44 full-time (1 woman). *Students:* 33 full-time (3 women), 169 part-time (17 women); includes 15 minority (3 African Americans, 4 Asian Americans or Pacific Islanders, 6 Hispanic Americans, 2 Native Americans), 78 international. Average age 30. 422 applicants, 69% accepted. In 1998, 115 master's, 6 doctorates awarded. Terminal master's awarded for partial completion of doctoral program. *Degree requirements:* For master's, thesis or alternative required, foreign language not required; for doctorate, dissertation, candidacy and qualifying exams required, foreign language not required. *Entrance requirements:* For master's and doctorate, GRE General Test, GRE Subject Test, TOEFL. *Application deadline:* For fall admission, 7/15 (priority date). Applications are processed on a rolling basis. Application fee: $25. *Financial aid:* In 1998–99, 1 fellowship, 17 research assistantships, 16 teaching assistantships were awarded.; career-related internships or fieldwork, Federal Work-Study, institutionally-sponsored loans, tuition waivers (full), and unspecified assistantships also available. Aid available to part-time students. *Faculty research:* Structural mechanics, environmental manufacturing systems, fluid dynamics, control engineering. Total annual research expenditures: $4.9 million. *Unit head:* Dr. Kenneth Lloyd Keys, Dean, 216-687-2555. *Application contact:* Dr. Chet Jain, Associate Dean, 216-687-2555.

See in-depth description on page 87.

College of William and Mary, Faculty of Arts and Sciences, Department of Applied Science, Williamsburg, VA 23187-8795. Offers MS, PhD. *Faculty:* 5 full-time (0 women), 3 part-time (1 woman). *Students:* 36 full-time (9 women), 3 part-time; includes 4 minority (1 African American, 2 Asian Americans or Pacific Islanders, 1 Hispanic American), 11 international. Average age 31. 31 applicants, 61% accepted. In 1998, 11 master's, 6 doctorates awarded. Terminal master's awarded for partial completion of doctoral program. *Degree requirements:* For master's, thesis required (for some programs), foreign language not required; for doctorate, dissertation required, foreign language not required. *Entrance requirements:* For master's and doctorate, GRE General Test, GRE Subject Test, minimum GPA of 3.0. *Average time to degree:* Master's–2 years full-time; doctorate–5 years full-time. *Application deadline:* For fall admission, 2/1 (priority date); for spring admission, 10/1 (priority date). Application fee: $30. *Financial aid:* In 1998–99, 22 students received aid, including 22 research assistantships (averaging $14,000 per year); teaching assistantships, career-related internships or fieldwork, Federal Work-Study, and grants also available. Financial aid application deadline: 4/1; financial aid applicants required to submit FAFSA. *Faculty research:* Interface/surface science, nondestructive evaluation, applied mathematics and modeling, polymer and composite materials, patent practice. Total annual research expenditures: $1.7 million. *Unit head:* Dr. Dennis Manos, Chair, 757-221-2563. *Application contact:* Kathee Card, Business Manager, 757-221-2563, Fax: 757-221-2050.

Announcement: The Department of Applied Science offers an interdisciplinary program leading to the MS and PhD degrees. Research specializations of core and affiliated faculty members include applied mathematics and modeling, composite and polymer materials science, computational materials physics, interface and surface science, materials processing and characterization, nondestructive evaluation, nonlinear dynamics, and patent practice. Students perform research in laboratories on campus and at nearby NASA Langley Research Center and DoE's Jefferson Lab. Admission is competitive; GRE is required. Chair of Applied Science, College of William and Mary, Williamsburg, VA 23187-8795. Telephone: 757-221-2563. E-mail: info@as.wm.edu. See WWW at http://as.wm.edu

Colorado School of Mines, Graduate School, Golden, CO 80401-1887. Offers ME, MS, PhD, Diplôme d'ingénieur, Diploma. Part-time and evening/weekend programs available. *Faculty:* 274 full-time (83 women), 34 part-time (15 women). *Students:* 480 full-time (102 women), 282 part-time (61 women); includes 47 minority (5 African Americans, 21 Asian Americans or Pacific Islanders, 17 Hispanic Americans, 4 Native Americans), 257 international. 721 applicants, 61% accepted. In 1998, 169 master's, 38 doctorates awarded (95% found work related to degree); 5 other advanced degrees awarded (100% found work related to degree). *Degree requirements:* For doctorate, dissertation, comprehensive exam required, foreign language not required. *Entrance requirements:* For master's, doctorate, and other advanced degree, GRE General Test, minimum GPA of 3.0. *Application deadline:* Applications are processed on a rolling basis. Application fee: $40. Electronic applications accepted. *Financial aid:* In 1998–99, 435 students received aid, including 44 fellowships, 196 research assistantships, 122 teaching assistantships; career-related internships or fieldwork, Federal Work-Study, institutionally-sponsored loans, and unspecified assistantships also available. Aid available to part-time students. Financial aid applicants required to submit FAFSA. *Faculty research:* Energy, environment, materials, minerals, engineering systems. Total annual research expenditures: $20.6 million. *Unit head:* Dr. Phillip R. Romig, Dean of Graduate Studies and Research, 303-273-3247, Fax: 303-273-3244, E-mail: grad-school@mines.edu. *Application contact:* Linda Powell, Graduate Admissions Officer, 303-273-3032, Fax: 303-273-3244, E-mail: lpowell@mines.edu.

Colorado State University, Graduate School, College of Engineering, Fort Collins, CO 80523-0015. Offers MS, PhD. Part-time programs available. *Faculty:* 91 full-time (5 women), 4 part-time (0 women). *Students:* 229 full-time (49 women), 171 part-time (33 women); includes 19 minority (4 African Americans, 6 Asian Americans or Pacific Islanders, 8 Hispanic Americans, 1 Native American), 129 international. Average age 30. 1089 applicants, 49% accepted. In 1998, 135 master's, 46 doctorates awarded. Terminal master's awarded for partial completion of doctoral program. *Degree requirements:* For doctorate, dissertation required, foreign language not required, foreign language not required. *Entrance requirements:* For master's, GRE General Test, TOEFL, minimum GPA of 3.0; for doctorate, GRE General Test, TOEFL. *Application deadline:* For fall admission, 2/1 (priority date). Applications are processed on a rolling basis. Application fee: $30. Electronic applications accepted. *Financial aid:* In 1998–99, 5 fellowships, 140 research assistantships, 44 teaching assistantships were awarded.; career-related internships or fieldwork, Federal Work-Study, institutionally-sponsored loans, and traineeships also available. *Faculty research:* Atmospheric science, optoelectronics, water resources, chemical and bioresource engineering, manufacturing, engine and energy conversion. Total annual research expenditures: $29.9 million. *Unit head:* Dr. Neal Gallagher, Dean, 970-491-3366, Fax: 970-491-8462, E-mail: abt@engr.colostate.edu. *Application contact:* Dr. Steve Abt, Interim Associate Dean, 970-491-8657, Fax: 970-491-8462, E-mail: abt@engr.colostate.edu.

Columbia University, Fu Foundation School of Engineering and Applied Science, New York, NY 10027. Offers ME, MS, Eng Sc D, PhD, CSE, EE, EM, Engr, Met E, MBA/MS. Part-time programs available. Postbaccalaureate distance learning degree programs offered (no on-campus study). *Faculty:* 127 full-time (11 women), 63 part-time (4 women). *Students:* 487 full-time (83 women), 450 part-time (77 women). 2236 applicants, 38% accepted. In 1998, 253 master's, 59 doctorates, 11 other advanced degrees awarded. Terminal master's awarded for partial completion of doctoral program. *Degree requirements:* For doctorate, dissertation, qualifying exam required, foreign language not required. *Entrance requirements:* For master's, doctorate, and other advanced degree, GRE General Test, TOEFL. *Application deadline:* Applications are processed on a rolling basis. Application fee: $55. *Financial aid:* In 1998–99, 10 fellowships, 189 research assistantships, 58 teaching assistantships were awarded.; career-related internships or fieldwork, Federal Work-Study, institutionally-sponsored loans, scholarships, unspecified assistantships, and outside fellowships also available. Aid available to part-time students. Financial aid application deadline: 1/5; financial aid applicants required to submit FAFSA. *Unit head:* Zvi Galil, Dean, 212-854-7996, Fax: 212-854-5900, E-mail: seasgradmit@columbia.edu. *Application contact:* Joan C. Zarodkiewicz, Director of Graduate Admission/Financial Aid, 212-854-6438, Fax: 212-854-5900, E-mail: seasgradmit@columbia.edu.

See in-depth description on page 89.

Concordia University, School of Graduate Studies, Faculty of Engineering and Computer Science, Montréal, PQ H3G 1M8, Canada. Offers M Eng, MA Sc, MCS, PhD, Certificate, Diploma. *Students:* 730 full-time (186 women), 179 part-time (27 women). *Degree requirements:* For master's, computer language required; for doctorate, computer language, dissertation, comprehensive exam required. *Application deadline:* For spring admission, 10/1. Application fee: $50. *Unit head:* Dr. N. Esmail, Dean, 514-848-3062, Fax: 514-848-4509.

See in-depth description on page 91.

Cornell University, Graduate School, Graduate Fields of Engineering, Ithaca, NY 14853-0001. Offers M Eng, MPS, MS, PhD, M Eng/MBA. *Faculty:* 208 full-time. *Students:* 970 full-time (195 women); includes 147 minority (12 African Americans, 105 Asian Americans or Pacific Islanders, 29 Hispanic Americans, 1 Native American), 461 international. Terminal master's awarded for partial completion of doctoral program. *Degree requirements:* For doctorate, dissertation required, foreign language not required. *Entrance requirements:* For master's and doctorate, TOEFL. Application fee: $65. Electronic applications accepted. *Financial aid:* In 1998–99, 638 students received aid, including 174 fellowships with full tuition reimbursements available, 280 research assistantships with full tuition reimbursements available, 184 teaching assistantships with full tuition reimbursements available; career-related internships or fieldwork, institutionally-sponsored loans, scholarships, tuition waivers (full and partial), and unspecified assistantships also available. Financial aid applicants required to submit FAFSA. *Unit head:* Dr. John Hopcroft, Dean. *Application contact:* Graduate School Application Requests, Caldwell Hall, 607-255-4884.

Dalhousie University, Faculty of Graduate Studies, DalTech, Faculty of Engineering, Halifax, NS B3H 3J5, Canada. Offers M Eng, M Sc, MA Sc, PhD, M Eng/MURP, MA Sc/MURP. *Faculty:* 71 full-time (2 women). 262 applicants, 44% accepted. In 1998, 49 master's, 18 doctorates awarded. *Degree requirements:* For master's and doctorate, thesis/dissertation required, foreign language not required. *Entrance requirements:* For master's and doctorate, TOEFL (minimum score of 580 required). *Application deadline:* For fall admission, 6/1 (priority date); for winter admission, 10/1 (priority date); for spring admission, 2/1 (priority date). Applications are processed on a rolling basis. Application fee: $55. *Financial aid:* Fellowships, research assistantships, teaching assistantships available. *Unit head:* Dr. A. Bell, Dean, 902-494-6055, Fax: 902-429-3011, E-mail: dean.engineering@dal.ca. *Application contact:* Shelley Parker, Admissions Coordinator, Graduate Studies and Research, 902-494-1288, Fax: 902-494-3149, E-mail: shelley.parker@dal.ca.

See in-depth description on page 93.

Dartmouth College, Thayer School of Engineering, Hanover, NH 03755. Offers MEM, MS, PhD, MBA/MEM, MD/PhD. *Faculty:* 31 full-time (4 women), 24 part-time (3 women). *Students:* 117 full-time (32 women), 7 part-time (1 woman); includes 10 minority (2 African Americans, 6 Asian Americans or Pacific Islanders, 2 Hispanic Americans), 37 international. Average age 24. 482 applicants, 14% accepted. In 1998, 37 master's, 6 doctorates awarded. *Degree requirements:* For doctorate, dissertation, candidacy oral exam required. *Entrance requirements:* For master's and doctorate, GRE General Test. *Average time to degree:* Master's–2 years full-time; doctorate–4 years full-time. *Application deadline:* For fall admission, 1/15 (priority date). Application fee: $20 ($40 for international students). *Financial aid:* In 1998–99, 12 fellowships with tuition reimbursements (averaging $15,120 per year), 42 research assistantships with tuition reimbursements (averaging $16,320 per year) were awarded.; teaching assistantships, career-related internships or fieldwork, Federal Work-Study, institutionally-sponsored loans, and tuition waivers (full and partial) also available. Financial aid application deadline: 1/15; financial aid applicants required to submit CSS PROFILE. *Faculty research:* Chemical and biochemical engineering, biomedical engineering, electrical and computer engineering, materials and mechanical engineering, environmental science and engineering. Total annual research expenditures: $6.3 million. *Unit head:* Dr. Lewis M. Duncan, Dean, 603-646-2238, Fax: 603-646-3856, E-mail: lewis.m.duncan@dartmouth.edu. *Application contact:* Candace S. Potter, Admissions Coordinator, 603-646-3844, Fax: 603-646-3856, E-mail: candace.potter@dartmouth.edu.

See in-depth description on page 95.

Drexel University, Graduate School, College of Engineering, Philadelphia, PA 19104-2875. Offers MS, MSEE, PhD. Part-time and evening/weekend programs available. *Faculty:* 71 full-time (7 women), 41 part-time (3 women). *Students:* 165 full-time (29 women), 426 part-time (56 women). Average age 32. 1519 applicants, 58% accepted. In 1998, 187 master's, 30 doctorates awarded. *Degree requirements:* For doctorate, dissertation required. *Entrance requirements:* For master's, TOEFL (minimum score of 570 required), minimum GPA of 3.0; for doctorate, TOEFL (minimum score of 570 required). *Application deadline:* For fall admission, 8/21. Applications are processed on a rolling basis. Application fee: $35. Tuition: Full-time $15,795; part-time $585 per credit. Required fees: $375; $67 per term. Tuition and fees vary according to program. *Financial aid:* In 1998–99, 60 research assistantships, 102 teaching assistantships were awarded.; career-related internships or fieldwork, Federal Work-Study, institutionally-sponsored loans, tuition waivers (full and partial), and unspecified assistantships also available. Aid available to part-time students. Financial aid application deadline: 2/1. *Unit head:* Dr. Raj Mutharasan, Dean, 215-895-2210. *Application contact:* Kelli Kennedy, Director of Admissions, 215-895-6706, Fax: 215-895-5939, E-mail: crowlka@duvm.ocs.drexel.edu.

See in-depth description on page 97.

Duke University, Graduate School, School of Engineering, Durham, NC 27708-0586. Offers MEM, MS, PhD, JD/MS, MBA/MS. Part-time programs available. *Faculty:* 105 full-time, 28 part-time. *Students:* 252 full-time, 2 part-time; includes 16 minority (3 African Americans, 10 Asian Americans or Pacific Islanders, 3 Hispanic Americans), 108 international. 596 applicants, 27% accepted. In 1998, 40 master's, 28 doctorates awarded. *Degree requirements:* For doctorate, dissertation required. *Entrance requirements:* For master's and doctorate, GRE General Test. *Application deadline:* For fall admission, 12/31; for spring admission, 11/1. Application fee: $75. *Financial aid:* Fellowships, research assistantships, teaching assistantships, Federal Work-Study available. Financial aid application deadline: 12/31. *Unit head:* Dr. Earl H. Dowell, Dean, 919-660-5389, Fax: 919-684-4860.

East Carolina University, Graduate School, School of Industry and Technology, Greenville, NC 27858-4353. Offers MS. Part-time programs available. *Faculty:* 9 full-time (1 woman), 1 part-time (0 women). *Students:* 18 full-time (4 women), 66 part-time (13 women); includes 9 minority (7 African Americans, 1 Asian American or Pacific Islander, 1 Native American), 1

Engineering and Applied Sciences—General

East Carolina University (continued)
international. Average age 34. 42 applicants, 79% accepted. In 1998, 21 degrees awarded. *Degree requirements:* For master's, comprehensive exams required. *Application deadline:* For fall admission, 6/1 (priority date). Applications are processed on a rolling basis. Application fee: $40. Tuition, state resident: full-time $1,012. Tuition, nonresident: full-time $8,578. Required fees: $1,006. Part-time tuition and fees vary according to course load. *Financial aid:* Fellowships, research assistantships, teaching assistantships, Federal Work-Study available. Aid available to part-time students. Financial aid application deadline: 6/1. *Unit head:* Dr. Elmer Poe, Director of Graduate Studies, 252-328-6103, Fax: 252-328-4250, E-mail: poee@mail.ecu.edu. *Application contact:* Dr. Paul D. Tschetter, Senior Associate Dean, 252-328-6012, Fax: 252-328-6071, E-mail: grad@mail.ecu.edu.

Eastern Illinois University, Graduate School, Lumpkin College of Business and Applied Sciences, School of Technology, Charleston, IL 61920-3099. Offers MS. Part-time and evening/weekend programs available. *Degree requirements:* For master's, foreign language and thesis not required.

East Tennessee State University, School of Graduate Studies, College of Applied Science and Technology, Johnson City, TN 37614-0734. Offers MA, MS. Part-time and evening/weekend programs available. Postbaccalaureate distance learning degree programs offered (no on-campus study). *Degree requirements:* For master's, computer language, exam required. *Entrance requirements:* For master's, TOEFL (minimum score of 550 required). *Faculty research:* Total quality management, software engineering.

École Polytechnique de Montréal, Graduate Programs, Montréal, PQ H3C 3A7, Canada. Offers M Eng, M Sc A, PhD, DESS. Part-time and evening/weekend programs available. *Degree requirements:* For master's and doctorate, one foreign language, computer language, thesis/dissertation required. *Entrance requirements:* For master's, minimum GPA of 2.75; for doctorate, minimum GPA of 3.0. *Faculty research:* Material science, environmental engineering, microelectronics and communications, biomedical engineering.

Fairfield University, School of Engineering, Fairfield, CT 06430-5195. Offers management of technology (MS); software engineering (MS). Part-time and evening/weekend programs available. *Faculty:* 44 part-time (4 women). *Students:* 1 full-time (0 women), 39 part-time (4 women); includes 2 minority (1 African American, 1 Hispanic American), 2 international. *Degree requirements:* For master's, thesis, final exam required. *Entrance requirements:* For master's, interview, minimum GPA of 2.8. *Application deadline:* For fall admission, 6/30 (priority date). Applications are processed on a rolling basis. Application fee: $40. *Financial aid:* Tuition waivers (partial) available. Financial aid applicants required to submit FAFSA. *Unit head:* Dr. Evangelos Hadjimichael, Dean, 203-254-4000 Ext. 4147, Fax: 203-254-4013, E-mail: hadjm@fair1.fairfield.edu.

Fairleigh Dickinson University, Teaneck–Hackensack Campus, University College: Arts, Sciences, and Professional Studies, School of Engineering and Engineering Technology, Teaneck, NJ 07666-1914. Offers electrical engineering (MSEE). *Degree requirements:* For master's, thesis optional, foreign language not required. *Entrance requirements:* For master's, GRE General Test.

Florida Agricultural and Mechanical University, Division of Graduate Studies, Research, and Continuing Education, FAMU-FSU College of Engineering, Tallahassee, FL 32307-3200. Offers MS, PhD. College administered jointly by Florida State University. *Students:* 50 (16 women); includes 40 minority (36 African Americans, 3 Asian Americans or Pacific Islanders, 1 Hispanic American) 9 international. *Entrance requirements:* For master's, GRE General Test (minimum combined score of 1000 required), minimum GPA of 3.0. *Application deadline:* For fall admission, 7/1. Application fee: $20. *Financial aid:* Fellowships, research assistantships, teaching assistantships, tuition waivers (full) available. *Unit head:* Dr. C. J. Chen, Dean, 850-487-6100, Fax: 850-487-6486.

See in-depth description on page 99.

Florida Atlantic University, College of Engineering, Boca Raton, FL 33431-0991. Offers MS, PhD. Part-time and evening/weekend programs available. Postbaccalaureate distance learning degree programs offered (minimal on-campus study). *Faculty:* 75 full-time (5 women), 5 part-time (0 women). *Students:* 144 full-time (42 women), 160 part-time (36 women); includes 66 minority (10 African Americans, 29 Asian Americans or Pacific Islanders, 27 Hispanic Americans), 151 international. Average age 31. In 1998, 112 master's, 15 doctorates awarded. Terminal master's awarded for partial completion of doctoral program. *Degree requirements:* For master's, thesis optional, foreign language not required; for doctorate, dissertation, qualifying exam required, foreign language not required. *Entrance requirements:* For master's, GRE General Test (minimum combined score of 1000 required), TOEFL (minimum score of 550 required), minimum GPA of 3.0; for doctorate, GRE General Test, TOEFL (minimum score of 550 required). *Application deadline:* Applications are processed on a rolling basis. Application fee: $20. Tuition, state resident: part-time $148 per credit hour. Tuition, nonresident: part-time $509 per credit hour. *Financial aid:* Fellowships, research assistantships, teaching assistantships, career-related internships or fieldwork, Federal Work-Study, and unspecified assistantships available. Aid available to part-time students. Financial aid applicants required to submit FAFSA. *Faculty research:* Automated underwater vehicles, communication systems, computer networks, materials, neural networks. Total annual research expenditures: $8.5 million. *Unit head:* Dr. John Jurewicz, Dean, 561-297-3400, Fax: 561-297-2659, E-mail: jurewicz@fau.edu. *Application contact:* Patricia Capozziello, Graduate Admissions Coordinator, 561-297-2694, Fax: 561-297-2659, E-mail: capozzie@fau.edu.

Florida Institute of Technology, Graduate School, College of Engineering, Melbourne, FL 32901-6975. Offers MS, PhD. Part-time and evening/weekend programs available. *Faculty:* 59 full-time (3 women), 22 part-time (5 women). *Students:* 143 full-time (29 women), 313 part-time (65 women); includes 44 minority (9 African Americans, 14 Asian Americans or Pacific Islanders, 20 Hispanic Americans, 1 Native American), 203 international. Average age 31. 775 applicants, 66% accepted. In 1998, 110 master's, 16 doctorates awarded. Terminal master's awarded for partial completion of doctoral program. *Degree requirements:* For doctorate, dissertation required, foreign language not required. *Entrance requirements:* For master's, minimum GPA of 3.0; for doctorate, minimum GPA of 3.2. *Application deadline:* Applications are processed on a rolling basis. Application fee: $50. Electronic applications accepted. Tuition: Part-time $575 per credit hour. Required fees: $100. Tuition and fees vary according to campus/location and program. *Financial aid:* In 1998–99, 112 students received aid, including 52 research assistantships with full and partial tuition reimbursements available (averaging $3,977 per year), 49 teaching assistantships with full and partial tuition reimbursements available (averaging $3,875 per year); career-related internships or fieldwork, institutionally-sponsored loans, and tuition remissions also available. Financial aid application deadline: 3/1; financial aid applicants required to submit FAFSA. *Faculty research:* Electrical and computer science and engineering; aerospace, chemical, civil, mechanical, and ocean engineering; environmental science and oceanography. Total annual research expenditures: $3.5 million. *Unit head:* Dr. Robert L. Sullivan, Dean, 407-674-7318, Fax: 407-676-0883, E-mail: sullivan@zach.fit.edu. *Application contact:* Carolyn P. Farrior, Associate Dean of Graduate Admissions, 407-674-7118, Fax: 407-723-9468, E-mail: cfarrior@fit.edu.

Florida International University, College of Engineering, Miami, FL 33199. Offers MS, PhD. Part-time and evening/weekend programs available. *Faculty:* 54 full-time (5 women), 1 (woman) part-time. *Students:* 98 full-time (22 women), 222 part-time (57 women); includes 158 minority (17 African Americans, 13 Asian Americans or Pacific Islanders, 128 Hispanic Americans), 95 international. Average age 32. 361 applicants, 46% accepted. In 1998, 99 master's, 3 doctorates awarded. *Degree requirements:* For doctorate, dissertation required. *Entrance requirements:* For master's and doctorate, GRE General Test (minimum combined score of 1000), TOEFL. *Application deadline:* For fall admission, 4/1 (priority date); for spring admission, 10/1. Applications are processed on a rolling basis. Application fee: $20. Tuition, state resident: part-time $145 per credit hour. Tuition, nonresident: part-time $506 per credit hour. Required fees: $158; $158 per year. *Financial aid:* Fellowships, research assistantships, teaching assistantships, career-related internships or fieldwork, Federal Work-Study, and institutionally-sponsored loans available. *Unit head:* Dr. Gordon R. Hopkins, Dean, 305-348-2522, Fax: 305-348-1401, E-mail: gordon@eng.fiu.edu.

Florida State University, Graduate Studies, FAMU/FSU College of Engineering, Tallahassee, FL 32306. Offers MS, PhD. Part-time programs available. Postbaccalaureate distance learning degree programs offered (minimal on-campus study). *Faculty:* 74 full-time (7 women), 14 part-time (1 woman). *Students:* 151 full-time (37 women), 43 part-time (8 women). Average age 26. 285 applicants, 35% accepted. In 1998, 59 master's, 4 doctorates awarded (100% found work related to degree). *Degree requirements:* For doctorate, dissertation, preliminary exam, qualifying exam required, foreign language not required, foreign language not required. *Entrance requirements:* For master's and doctorate, GRE General Test. *Average time to degree:* Master's–2 years full-time, 3.5 years part-time; doctorate–3.5 years full-time, 4.5 years part-time. *Application deadline:* Applications are processed on a rolling basis. Application fee: $20. Electronic applications accepted. Tuition, state resident: part-time $139 per credit hour. Tuition, nonresident: part-time $482 per credit hour. Tuition and fees vary according to program. *Financial aid:* In 1998–99, 4 fellowships with full tuition reimbursements, 60 research assistantships with full tuition reimbursements, 36 teaching assistantships with full tuition reimbursements were awarded.; career-related internships or fieldwork, institutionally-sponsored loans, scholarships, tuition waivers (full), and unspecified assistantships also available. *Faculty research:* Fluid mechanics, aerodynamics, electromagnetics, digital signal processing, polymer processing. Total annual research expenditures: $7 million. *Unit head:* Dr. Timothy Beard, Director, Graduate Student Services and Special Programs, 850-410-6120, Fax: 850-410-6344, E-mail: beard@eng.fsu.edu. *Application contact:* Dianne Robertson, Graduate Student Assistant, 850-410-6369, Fax: 850-410-6344, E-mail: graduate_studies@eng.fsu.edu.

See in-depth description on page 99.

Gannon University, School of Graduate Studies, College of Sciences, Engineering, and Health Sciences, School of Sciences and Engineering, Program in Engineering, Erie, PA 16541-0001. Offers electrical engineering (MS); embedded software engineering (MS); mechanical engineering (MS). Part-time and evening/weekend programs available. *Students:* 23 full-time (4 women), 29 part-time (5 women); includes 1 minority (Asian American or Pacific Islander), 13 international. Average age 26. 24 applicants, 100% accepted. In 1998, 12 degrees awarded. *Degree requirements:* For master's, thesis or alternative, comprehensive exam required. *Entrance requirements:* For master's, GRE Subject Test, bachelor's degree in engineering, minimum QPA of 2.5. *Application deadline:* Applications are processed on a rolling basis. Application fee: $25. *Financial aid:* Career-related internships or fieldwork and traineeships available. Aid available to part-time students. Financial aid application deadline: 7/1; financial aid applicants required to submit FAFSA. *Unit head:* Dr. Mehmet Cultu, Co-Director, 814-871-7624. *Application contact:* Beth Nemenz, Director of Admissions, 814-871-7240, Fax: 814-871-5803, E-mail: admissions@gannon.edu.

George Mason University, School of Information Technology and Engineering, Fairfax, VA 22030-4444. Offers MS, PhD. Part-time and evening/weekend programs available. *Faculty:* 95 full-time (14 women), 41 part-time (5 women). *Students:* 189 full-time (68 women), 1,588 part-time (434 women); includes 447 minority (105 African Americans, 303 Asian Americans or Pacific Islanders, 34 Hispanic Americans, 5 Native Americans), 333 international. Average age 33. 1069 applicants, 72% accepted. In 1998, 369 master's, 29 doctorates awarded. *Degree requirements:* For master's, thesis optional, foreign language not required; for doctorate, dissertation, comprehensive oral and written exams required, foreign language not required. *Entrance requirements:* For master's, TOEFL (minimum score of 575 required), minimum GPA of 3.0 in last 60 hours; for doctorate, GRE General Test, TOEFL (minimum score of 575 required), minimum graduate GPA of 3.5. *Application deadline:* For fall admission, 5/1; for spring admission, 11/1. Application fee: $30. Electronic applications accepted. Tuition, state resident: full-time $4,416; part-time $184 per credit hour. Tuition, nonresident: full-time $12,516; part-time $522 per credit hour. Tuition and fees vary according to program. *Financial aid:* Fellowships, research assistantships, teaching assistantships, career-related internships or fieldwork, Federal Work-Study, institutionally-sponsored loans, and unspecified assistantships available. Aid available to part-time students. Financial aid application deadline: 3/1; financial aid applicants required to submit FAFSA. *Faculty research:* Systems management, quality assurance, decision support systems, cognitive ergonomics. Total annual research expenditures: $8.9 million. *Unit head:* Lloyd Griffiths, Dean, 703-993-1500, Fax: 703-993-1734, E-mail: lgriffiths@gmu.edu. *Application contact:* Student Services, 703-993-1511, Fax: 703-993-1734, E-mail: sitegrad@gmu.edu.

See in-depth description on page 101.

The George Washington University, School of Engineering and Applied Science, Washington, DC 20052. Offers MEM, MS, D Sc, App Sc, Engr, MEM/MS. Part-time and evening/weekend programs available. *Faculty:* 76 full-time (7 women), 100 part-time (4 women). *Students:* 331 full-time (81 women), 944 part-time (211 women); includes 222 minority (87 African Americans, 100 Asian Americans or Pacific Islanders, 32 Hispanic Americans, 3 Native Americans), 440 international. Average age 34. 1064 applicants, 92% accepted. In 1998, 446 master's, 42 doctorates, 13 other advanced degrees awarded. *Degree requirements:* For master's, thesis optional, foreign language not required; for doctorate, dissertation, qualifying exam required; for other advanced degree, foreign language and thesis not required. *Entrance requirements:* For master's, TOEFL (minimum score of 550 required; average 580) or George Washington University English as a Foreign Language Test, appropriate bachelor's degree; for doctorate, TOEFL (minimum score of 550 required; average 580) or George Washington University English as a Foreign Language Test, appropriate bachelor's or master's degree, GRE required if highest earned degree is BS; for other advanced degree, TOEFL (minimum score of 550 required; average 580) or George Washington University English as a Foreign Language Test, appropriate master's degree. *Application deadline:* For fall admission, 3/1; for spring admission, 10/1. Applications are processed on a rolling basis. Application fee: $55. Tuition: Full-time $17,328; part-time $722 per credit hour. Required fees: $828; $35 per credit hour. Tuition and fees vary according to campus/location and program. *Financial aid:* In 1998–99, 26 fellowships with full and partial tuition reimbursements, 133 research assistantships with full and partial tuition reimbursements, 75 teaching assistantships with full and partial tuition reimbursements were awarded.; career-related internships or fieldwork, Federal Work-Study, institutionally-sponsored loans, and tuition waivers (full and partial) also available. Financial aid application deadline: 3/1; financial aid applicants required to submit FAFSA. *Faculty research:* Fatigue fracture and structural reliability, computer-integrated manufacturing, materials engineering, artificial intelligence and expert systems, quality assurance. Total annual research expenditures: $6.3 million. *Unit head:* Dr. Thomas Mazzuchi, Interim Dean, 202-994-6080, Fax: 202-994-4522. *Application contact:* Howard M. Davis, Manager, Office of Admissions and Student Records, 202-994-6158, Fax: 202-994-0909, E-mail: data:adms@seas.gwu.edu.

See in-depth description on page 103.

Georgia Institute of Technology, Graduate Studies and Research, College of Engineering, Atlanta, GA 30332-0001. Offers MS, MS Bio E, MS Ch E, MS Env E, MS Poly, MS Stat, MS Text, MSAE, MSCE, MSEE, MSESM, MSHP, MSHS, MSIE, MSME, MSMSE, MSNE, MSOR, MST Ch, MSTE, PhD, Certificate, MCP/MSCE, MD/PhD. *Accreditation:* ABET (one or more programs are accredited). Part-time programs available. Postbaccalaureate distance learning degree programs offered. Terminal master's awarded for partial completion of doctoral program. *Degree requirements:* For doctorate, dissertation required, foreign language not required, foreign language not required. *Entrance requirements:* For master's and doctorate, TOEFL. Electronic applications accepted.

Georgia Southern University, Jack N. Averitt College of Graduate Studies, Allen E. Paulson College of Science and Technology, School of Technology, Statesboro, GA 30460. Offers M Tech.

Part-time and evening/weekend programs available. *Faculty:* 12 full-time (1 woman). *Students:* 2 full-time (1 woman), 13 part-time (4 women); includes 2 minority (both African Americans) Average age 31. 6 applicants, 33% accepted. In 1998, 3 degrees awarded. *Degree requirements:* For master's, thesis, terminal exam required. *Entrance requirements:* For master's, GRE General Test (minimum score of 450 on each section required), minimum GPA of 2.5. *Average time to degree:* Master's–5 years full-time, 4 years part-time. *Application deadline:* For fall admission, 7/1 (priority date); for spring admission, 11/15 (priority date). Applications are processed on a rolling basis. Application fee: $0. Electronic applications accepted. *Financial aid:* In 1998–99, 5 students received aid. Career-related internships or fieldwork and unspecified assistantships available. Financial aid application deadline: 4/15; financial aid applicants required to submit FAFSA. *Faculty research:* Ergonomics, imaging science, printability, productivity, manufacturing technology, assistive technology, water and wastewater treatment, transportation, computer modeling, fiber optic communications. *Unit head:* Dr. John Wallace, Director, 912-681-5761, Fax: 912-871-1455, E-mail: sotecjw@gasoue.edu. *Application contact:* Dr. John R. Diebolt, Associate Graduate Dean, 912-681-5384, Fax: 912-681-0740, E-mail: gradschool@gasou.edu.

Golden Gate University, School of Technology and Industry, San Francisco, CA 94105-2968. Offers hospitality administration and tourism (MS); information systems (MS, Certificate); telecommunications management (MS, Certificate). Part-time and evening/weekend programs available. *Students:* 167 full-time (71 women), 398 part-time (144 women); includes 171 minority (43 African Americans, 105 Asian Americans or Pacific Islanders, 21 Hispanic Americans, 2 Native Americans), 161 international. Average age 34. 299 applicants, 75% accepted. In 1998, 162 degrees awarded. *Degree requirements:* For master's and Certificate, computer language required, foreign language and thesis not required. *Entrance requirements:* For master's, GMAT (MBA), TOEFL (minimum score of 550 required), minimum GPA of 2.5. *Average time to degree:* Master's–2.5 years full-time. *Application deadline:* For fall admission, 7/1 (priority date). Applications are processed on a rolling basis. Application fee: $55 ($70 for international students). *Financial aid:* Career-related internships or fieldwork, Federal Work-Study, and institutionally-sponsored loans available. Aid available to part-time students. *Unit head:* James Koerlin, Dean, 415-442-6540, Fax: 415-442-7049. *Application contact:* Enrollment Services, 415-442-7800, Fax: 415-442-7807, E-mail: info@ggu.edu.

Gonzaga University, Graduate School, School of Engineering, Spokane, WA 99258. Offers MSEE. *Degree requirements:* For master's, project required, foreign language and thesis not required. *Entrance requirements:* For master's, GRE General Test, TOEFL (minimum score of 550 required), minimum GPA of 3.0 during previous 2 years.

Graduate School and University Center of the City University of New York, Graduate Studies, Program in Engineering, New York, NY 10036-8099. Offers chemical engineering (PhD); civil engineering (PhD); electrical engineering (PhD); mechanical engineering (PhD). *Faculty:* 68 full-time (1 woman). *Students:* 105 full-time (16 women), 11 part-time (2 women); includes 12 African Americans, 5 Asian Americans or Pacific Islanders, 4 Hispanic Americans Average age 32. 132 applicants, 55% accepted. In 1998, 22 degrees awarded. *Degree requirements:* For doctorate, dissertation required, dissertation required. *Entrance requirements:* For doctorate, GRE General Test. *Application deadline:* For fall admission, 4/15. Application fee: $40. *Financial aid:* In 1998–99, 67 students received aid, including 64 fellowships; research assistantships, teaching assistantships, Federal Work-Study, institutionally-sponsored loans, and tuition waivers (full and partial) also available. Financial aid application deadline: 2/1; financial aid applicants required to submit FAFSA. *Unit head:* Dr. Mumtaz Kassir, Acting Executive Officer, 212-650-8030.

Harvard University, Extension School, Cambridge, MA 02138-3722. Offers applied sciences (CAS); English for graduate and professional studies (DGP); information technology (ALM); liberal arts (ALM); museum studies (CMS); premedical studies (Diploma); public health (CPH); publication and communication (CPC); special studies in administration and management (CSS). Part-time and evening/weekend programs available. *Faculty:* 450 part-time. *Degree requirements:* For master's, thesis required, foreign language not required; for other advanced degree, computer language required, foreign language and thesis not required. *Entrance requirements:* For master's and other advanced degree, TOEFL (minimum score of 600 required), TWE (minimum score of 5 required). *Application deadline:* Applications are processed on a rolling basis. Application fee: $75. *Unit head:* Michael Shinagel, Dean. *Application contact:* Program Director, 617-495-4024, Fax: 617-495-9176.

Harvard University, Graduate School of Arts and Sciences, Division of Engineering and Applied Sciences, Cambridge, MA 02138. Offers applied mathematics (ME, SM, PhD); applied physics (ME, SM, PhD); computer science (ME, SM, PhD); computing technology (PhD); engineering science (ME); engineering sciences (SM, PhD); medical engineering/medical physics (PhD, Sc D), including applied physics (PhD), engineering sciences (PhD), medical engineering/medical physics (Sc D), physics (PhD). *Students:* 143 full-time (31 women); includes 10 minority (1 African American, 9 Asian Americans or Pacific Islanders), 57 international. 437 applicants, 19% accepted. In 1998, 42 master's, 24 doctorates awarded. Terminal master's awarded for partial completion of doctoral program. *Degree requirements:* For master's, foreign language and thesis not required; for doctorate, dissertation required, foreign language not required. *Entrance requirements:* For master's and doctorate, GRE General Test, GRE Subject Test, TOEFL (minimum score of 550 required). Application fee: $60. *Financial aid:* Fellowships, research assistantships, teaching assistantships, career-related internships or fieldwork, Federal Work-Study, and institutionally-sponsored loans available. Financial aid application deadline: 12/30. *Unit head:* Dr. Paul C. Martin, Dean, 617-495-2833. *Application contact:* Office of Admissions and Financial Aid, 617-495-5315.

See in-depth description on page 105.

Harvey Mudd College, Program in Engineering, Claremont, CA 91711-5994. Offers MS. MS available to Harvey Mudd undergraduates only. *Faculty:* 18 full-time (1 woman). *Students:* 6 full-time (1 woman); includes 1 minority (Asian American or Pacific Islander) Average age 22. 12 applicants, 67% accepted. In 1998, 5 degrees awarded. *Degree requirements:* For master's, clinic, final report required. *Average time to degree:* Master's–1 year full-time. Application fee: $0. *Financial aid:* In 1998–99, 6 fellowships with full tuition reimbursements (averaging $7,500 per year) were awarded.; stipends also available. *Faculty research:* Robotics, finite element analysis and electromagnetics, magnetic recording. Total annual research expenditures: $918,000. *Unit head:* Clive Dym, Chairman, 909-621-8019, Fax: 909-621-8967, E-mail: clive_dym@hmc.edu.

Howard University, College of Engineering, Architecture, and Computer Sciences, School of Engineering and Computer Science, Washington, DC 20059-0002. Offers M Eng, MCS, MS, PhD. Part-time programs available. *Faculty:* 51 full-time (5 women), 4 part-time (0 women). *Students:* 84 full-time (20 women), 12 part-time (4 women); includes 82 minority (80 African Americans, 2 Hispanic Americans), 14 international. 146 applicants, 28% accepted. In 1998, 27 master's, 2 doctorates awarded. Terminal master's awarded for partial completion of doctoral program. *Degree requirements:* For doctorate, dissertation, preliminary exam required, foreign language not required. *Entrance requirements:* For master's and doctorate, GRE General Test, TOEFL, minimum GPA of 3.0. *Application deadline:* For fall admission, 4/1; for spring admission, 11/1. Applications are processed on a rolling basis. Application fee: $45. Electronic applications accepted. *Financial aid:* In 1998–99, 20 research assistantships with full tuition reimbursements (averaging $8,000 per year), 14 teaching assistantships with full and partial tuition reimbursements (averaging $8,000 per year) were awarded.; fellowships with full tuition reimbursements, career-related internships or fieldwork, grants, institutionally-sponsored loans, scholarships, and unspecified assistantships also available. Financial aid application deadline: 4/1; financial aid applicants required to submit FAFSA. *Faculty research:* Environmental engineering, solid-state electronics, dynamics and control of large flexible space structures, power systems, reaction kinetics. Total annual research expenditures: $9.3 million. *Unit head:* Student Services, 703-993-1511, Fax: 703-993-1734, E-mail: sitegrad@gmu.edu.

See in-depth description on page 107.

Idaho State University, Graduate School, College of Engineering, Pocatello, ID 83209. Offers engineering and applied science (PhD); environmental engineering (MS); hazardous waste management (MS); measurement and control engineering (MS); nuclear science and engineering (MS). MS (hazardous waste management), PhD offered jointly with the University of Idaho. Part-time programs available. *Degree requirements:* For master's, thesis required, foreign language not required; for doctorate, dissertation required. *Entrance requirements:* For master's and doctorate, GRE General Test, TOEFL. *Faculty research:* Isotope separation, control technology,two-phase flow, photosonolysis, criticality calculations.

Illinois Institute of Technology, Graduate College, Armour College of Engineering and Sciences, Chicago, IL 60616-3793. Offers M Ch E, M Chem, M Env E, M Geoenv E, M Trans E, MAC, MCEM, MECE, MGE, MHP, MMAE, MME, MMME, MPA, MPW, MS, MSE, MST, MTSE, PhD, JD/MPA, MBA/MPA. Part-time and evening/weekend programs available. *Faculty:* 150 full-time (16 women), 86 part-time (13 women). *Students:* 479 full-time (96 women), 1,435 part-time (354 women); includes 496 minority (146 African Americans, 292 Asian Americans or Pacific Islanders, 56 Hispanic Americans, 2 Native Americans), 632 international. 2323 applicants, 50% accepted. In 1998, 352 master's, 57 doctorates awarded. Terminal master's awarded for partial completion of doctoral program. *Degree requirements:* For master's, thesis (for some programs), comprehensive exam required, foreign language not required; for doctorate, dissertation, comprehensive exam required, foreign language not required. *Entrance requirements:* For master's, TOEFL; for doctorate, GRE (minimum score of 1200 required), TOEFL (minimum score of 550 required), undergraduate GPA of 3.0 required. *Application deadline:* Applications are processed on a rolling basis. Application fee: $30. Electronic applications accepted. *Financial aid:* In 1998–99, 12 fellowships, 77 research assistantships, 129 teaching assistantships were awarded.; career-related internships or fieldwork, Federal Work-Study, institutionally-sponsored loans, scholarships, and graduate assistantships also available. Financial aid application deadline: 3/1. *Faculty research:* Polymers, wastewater control, soil-structure interaction, digital and computer systems, fluid dynamics. Total annual research expenditures: $14.8 million. *Unit head:* Dean, 312-567-3009, Fax: 312-567-5205. *Application contact:* Dr. S. Mohammad Shahidehpour, Dean of Graduate College, 312-567-3024, Fax: 312-567-7517, E-mail: grad@minna.cns.iit.edu.

Indiana State University, School of Graduate Studies, School of Technology, Terre Haute, IN 47809-1401. Offers MA, MS, PhD. *Faculty:* 33 full-time (1 woman). *Students:* 48 full-time (14 women), 79 part-time (42 women); includes 19 minority (16 African Americans, 3 Asian Americans or Pacific Islanders), 25 international. Average age 35. 123 applicants, 49% accepted. In 1998, 38 master's awarded. *Degree requirements:* For doctorate, computer language, dissertation required, foreign language not required, foreign language not required. *Entrance requirements:* For master's, TOEFL (minimum score of 550 required), bachelor's degree in industrial technology or related field, minimum undergraduate GPA of 2.5; for doctorate, GRE General Test. *Average time to degree:* Master's–2 years full-time, 5 years part-time; doctorate–4 years full-time, 7 years part-time. *Application deadline:* For fall admission, 7/1 (priority date); for spring admission, 11/1 (priority date). Applications are processed on a rolling basis. Application fee: $20. Electronic applications accepted. *Financial aid:* In 1998–99, 2 fellowships with partial tuition reimbursements, 16 research assistantships with partial tuition reimbursements, 7 teaching assistantships with partial tuition reimbursements were awarded.; Federal Work-Study and institutionally-sponsored loans also available. Financial aid application deadline: 3/1; financial aid applicants required to submit FAFSA. *Unit head:* Dr. W. Tad Foster, Dean, 812-237-3166.

Indiana University–Purdue University Fort Wayne, School of Engineering, Technology, and Computer Science, Fort Wayne, IN 46805-1499. Offers MS. Part-time and evening/weekend programs available. *Faculty:* 6 full-time (1 woman). *Students:* 1 full-time (0 women), 15 part-time (2 women). Average age 35. *Degree requirements:* For master's, computer language required, foreign language and thesis not required. *Entrance requirements:* For master's, GRE, TOEFL, minimum GPA of 3.0. *Application deadline:* For fall admission, 2/15 (priority date); for spring admission, 9/1. Applications are processed on a rolling basis. Application fee: $30. *Financial aid:* Career-related internships or fieldwork available. Financial aid application deadline: 3/1; financial aid applicants required to submit FAFSA. *Faculty research:* Antenna theory, two-phase heat transfer, experimental stress analysis, power electronics and motor drives. *Unit head:* G. Allen Pugh, Dean, 219-481-6839.

Indiana University–Purdue University Indianapolis, School of Engineering and Technology, Indianapolis, IN 46202-2896. Offers MS, MS Bm E, MSE, MSEE, MSME, PhD. Part-time and evening/weekend programs available. *Faculty:* 24 full-time (2 women). *Students:* 22 full-time (9 women), 41 part-time (10 women); includes 6 minority (2 African Americans, 4 Asian Americans or Pacific Islanders), 23 international. Average age 29. In 1998, 16 degrees awarded. Terminal master's awarded for partial completion of doctoral program. *Degree requirements:* For master's, thesis optional, foreign language not required. *Entrance requirements:* For master's, GRE, TOEFL (minimum score of 550 required), minimum B average. Application fee: $50 for international students. Tuition, state resident: part-time $171 per credit hour. Tuition, nonresident: part-time $490 per credit hour. Required fees: $121 per year. *Financial aid:* In 1998–99, 16 students received aid, including fellowships with tuition reimbursements available (averaging $15,000 per year), research assistantships with full and partial tuition reimbursements available (averaging $13,000 per year); teaching assistantships, Federal Work-Study, institutionally-sponsored loans, and tuition waivers (full and partial) also available. Aid available to part-time students. Financial aid application deadline: 3/1. *Faculty research:* Computational fluid dynamics, heat and mass transfer, robotics and automation, signal processing, biomechanics. Total annual research expenditures: $2.6 million. *Unit head:* Dr. H. Oner Yurtseven, Dean, 317-274-0802, Fax: 317-274-4567. *Application contact:* Vickie Lawrence, Graduate Program Secretary, 317-274-9740, Fax: 317-274-4567, E-mail: grad@engr.iupui.edu.

Instituto Tecnológico y de Estudios Superiores de Monterrey, Campus Estado de México, Graduate Division, Division of Engineering and Architecture, Atizapán de Zaragoza, 52500, Mexico. Offers computer science (MCS); environmental engineering (MEE); industrial engineering (MIE); manufacturing systems (MMS); materials engineering (PhD). *Degree requirements:* For master's; for doctorate, one foreign language, dissertation required. *Entrance requirements:* For master's, interview; for doctorate, research proposal. *Application deadline:* For fall admission, 1/13 (priority date); for spring admission, 4/4. Applications are processed on a rolling basis. Application fee: 750 Mexican pesos. *Unit head:* Juan López Díaz, Headmaster, 5-326-5530, Fax: 5-326-5531, E-mail: jlopez@campus.cem.itesm.mx. *Application contact:* Lourdes Turrubiates, Admissions Officer, 5-326-5776, Fax: 5-326-5788, E-mail: lturrubi@campus.cem.itesm.mx.

Instituto Tecnológico y de Estudios Superiores de Monterrey, Campus Monterrey, Graduate and Research Division, Programs in Engineering, Monterrey, 64849, Mexico. Offers applied statistics (M Eng); artificial intelligence (PhD); automation engineering (M Eng); chemical engineering (M Eng); civil engineering (M Eng); electrical engineering (M Eng); electronic engineering (M Eng); environmental engineering (M Eng); industrial engineering (M Eng, PhD); manufacturing engineering (M Eng); mechanical engineering (M Eng); systems and quality engineering (M Eng). M Eng offered jointly with the University of Waterloo; PhD (industrial engineering) offered jointly with Texas A&M University. Part-time and evening/weekend programs available. Terminal master's awarded for partial completion of doctoral program. *Degree requirements:* For master's and doctorate, one foreign language, computer language, thesis/dissertation required. *Entrance requirements:* For master's, PAEG, TOEFL; for doctorate, GRE, TOEFL, master's in related field. *Faculty research:* Flexible manufacturing cells, materials, statistical methods, environmental prevention, control and evaluation.

Iowa State University of Science and Technology, Graduate College, College of Engineering, Ames, IA 50011. Offers M Eng, MS, PhD. Part-time programs available. *Faculty:* 230 full-time, 17 part-time. *Students:* 465 full-time (92 women), 212 part-time (41 women); includes 22 minority (8 African Americans, 8 Asian Americans or Pacific Islanders, 5 Hispanic Americans,

Engineering and Applied Sciences—General

Iowa State University of Science and Technology (continued)
1 Native American), 423 international. 1534 applicants, 21% accepted. In 1998, 191 master's, 64 doctorates awarded. *Degree requirements:* For doctorate, dissertation required. *Entrance requirements:* For master's and doctorate, TOEFL. Application fee: $20 ($50 for international students). Electronic applications accepted. Tuition, state resident: full-time $3,308. Tuition, nonresident: full-time $9,744. Part-time tuition and fees vary according to course load, campus/location and program. *Financial aid:* In 1998–99, 368 research assistantships with partial tuition reimbursements (averaging $10,964 per year), 115 teaching assistantships with partial tuition reimbursements (averaging $10,857 per year) were awarded.; fellowships, Federal Work-Study and scholarships also available. Aid available to part-time students. *Unit head:* Dr. James L. Melsa, Dean, 515-294-5933, E-mail: melsa@iastate.edu.

Johns Hopkins University, G. W. C. Whiting School of Engineering, Baltimore, MD 21218-2699. Offers MA, MCE, MS, MSE, PhD, MD/PhD. Part-time and evening/weekend programs available. *Faculty:* 115 full-time (8 women), 56 part-time (3 women). *Students:* 498 full-time (141 women), 31 part-time (7 women); includes 46 minority (9 African Americans, 34 Asian Americans or Pacific Islanders, 3 Hispanic Americans), 205 international. Average age 25. 3658 applicants, 11% accepted. In 1998, 92 master's, 51 doctorates awarded. Terminal master's awarded for partial completion of doctoral program. *Degree requirements:* For doctorate, dissertation required, foreign language not required, foreign language not required. *Entrance requirements:* For master's and doctorate, GRE General Test, TOEFL (minimum score of 560 required). Application fee: $50. Tuition: Full-time $23,660. Tuition and fees vary according to program. *Financial aid:* In 1998–99, 160 fellowships (averaging $12,554 per year), 158 research assistantships (averaging $12,606 per year), 118 teaching assistantships (averaging $12,664 per year) were awarded.; Federal Work-Study, grants, and institutionally-sponsored loans also available. Aid available to part-time students. Financial aid application deadline: 4/15. *Faculty research:* Biomedical engineering, environmental systems and engineering, materials science and engineering, signal and image processing, structural dynamics and geomechanics. Total annual research expenditures: $21.8 million. *Unit head:* Dr. Ilene J. Busch-Vishniac, Dean, 410-516-8350 Ext. 3, Fax: 410-516-8627, E-mail: mia@jhu.edu. *Application contact:* Pamela Carey, Senior Academic Advisor, 410-516-7084, Fax: 410-516-8627, E-mail: pcarey@hun14.hcf.jhu.edu.

Kansas State University, Graduate School, College of Engineering, Manhattan, KS 66506. Offers MEM, MS, MSE, PhD. Part-time programs available. Postbaccalaureate distance learning degree programs offered (minimal on-campus study). Electronic applications accepted.

See in-depth description on page 109.

Kent State University, School of Technology, Kent, OH 44242-0001. Offers MA. *Faculty:* 13 full-time. *Students:* 5 full-time (1 woman), 6 part-time; includes 1 minority (Native American), 2 international. 7 applicants, 100% accepted. In 1998, 3 degrees awarded. *Degree requirements:* For master's, thesis optional, foreign language not required. *Entrance requirements:* For master's, minimum GPA of 2.75. *Application deadline:* For fall admission, 7/12; for spring admission, 11/29. Applications are processed on a rolling basis. Application fee: $30. *Financial aid:* Research assistantships, teaching assistantships, Federal Work-Study and tuition waivers (full) available. Financial aid application deadline: 2/1. *Unit head:* Dr. A. Raj Chowdhury, Dean, 330-672-2892, Fax: 330-672-2894.

Kettering University, Graduate School, Flint, MI 48504-4898. Offers MS Eng, MSMM, MSOM. Part-time and evening/weekend programs available. *Faculty:* 42 full-time (4 women), 2 part-time (0 women). In 1998, 164 degrees awarded (100% found work related to degree). *Degree requirements:* For master's, foreign language and thesis not required. *Average time to degree:* Master's–3 years full-time, 6 years part-time. *Application deadline:* For fall admission, 7/15. Applications are processed on a rolling basis. Application fee: $0. *Financial aid:* In 1998–99, 24 students received aid, including 3 fellowships with full tuition reimbursements available, 12 research assistantships with full tuition reimbursements available, 9 teaching assistantships with full tuition reimbursements available; Federal Work-Study, institutionally-sponsored loans, and tuition waivers (partial) also available. Aid available to part-time students. Financial aid application deadline: 7/15. *Unit head:* Dr. C. David Hurt, Associate Dean, Graduate Studies and Extension Services, 810-762-7953, Fax: 810-762-9935, E-mail: dhurt@kettering.edu. *Application contact:* Betty L. Bedore, Coordinator of Publicity, 810-762-7494, Fax: 810-762-9935, E-mail: bbedore@kettering.edu.

Announcement: Kettering's MS in engineering degree is designed for the working professional. The video-based program allows distance learning opportunities at established learning centers throughout the United States. All programs have a strong orientation toward manufacturing, supported by Kettering's history of educating engineers and its long association with industry.

Lakehead University, Graduate Studies and Research, Faculty of Engineering, Thunder Bay, ON P7B 5E1, Canada. Offers control engineering (M Sc Engr). *Entrance requirements:* For master's, TOEFL (minimum score of 550 required).

Lamar University, College of Graduate Studies, College of Engineering, Beaumont, TX 77710. Offers ME, MEM, MES, MS, DE. Part-time and evening/weekend programs available. *Faculty:* 33. *Students:* 143 full-time (28 women), 63 part-time (10 women); includes 16 minority (2 African Americans, 12 Asian Americans or Pacific Islanders, 2 Hispanic Americans), 164 international. Average age 26. In 1998, 65 master's, 2 doctorates awarded. Terminal master's awarded for partial completion of doctoral program. *Degree requirements:* For doctorate, computer language, dissertation required, foreign language not required, foreign language not required. *Entrance requirements:* For master's and doctorate, GRE General Test, TOEFL. *Application deadline:* For fall admission, 5/15 (priority date); for spring admission, 10/1 (priority date). Applications are processed on a rolling basis. Application fee: $0. *Financial aid:* Fellowships with partial tuition reimbursements, research assistantships with partial tuition reimbursements, teaching assistantships with partial tuition reimbursements, career-related internships or fieldwork, Federal Work-Study, institutionally-sponsored loans, scholarships, tuition waivers (full and partial), and laboratory assistantships, graders available. Aid available to part-time students. Financial aid application deadline: 4/1. *Faculty research:* Energy alternatives; process analysis, design, and control; pollution prevention. Total annual research expenditures: $1.6 million. *Unit head:* Dr. Fred M. Young, Dean, 409-880-8741, Fax: 409-880-8121, E-mail: fred@hal.lamar.edu.

Laurentian University, School of Graduate Studies and Research, School of Engineering, Sudbury, ON P3E 2C6, Canada. Offers metallurgy (MA Sc); mining (M Eng). Part-time programs available. *Faculty:* 15 full-time (1 woman), 5 part-time (0 women). *Students:* 10 full-time (0 women), 7 part-time. 11 applicants, 9% accepted. In 1998, 2 degrees awarded. *Degree requirements:* Foreign language not required. *Application deadline:* For fall admission, 9/1. Application fee: $50. *Financial aid:* In 1998–99, 5 fellowships (averaging $2,000 per year), 6 teaching assistantships (averaging $6,500 per year) were awarded.; scholarships also available. *Faculty research:* Mining engineering, rock mechanics (tunneling, rockbursts, rock support), metallurgy (mineral processing, hydro and pyrometallurgy), simulations and remote mining. Total annual research expenditures: $1.7 million. *Unit head:* Dr. Paul Lindon, Director, 705-675-1151 Ext. 2244, Fax: 705-675-4862. *Application contact:* 705-675-1151 Ext. 3909, Fax: 705-675-4843.

Lawrence Technological University, College of Engineering, Southfield, MI 48075-1058. Offers automotive engineering (MAE); civil engineering (MCE); manufacturing systems (MEMS). Part-time and evening/weekend programs available. *Faculty:* 8 full-time (1 woman), 6 part-time (0 women). Average age 31. 29 applicants, 72% accepted. In 1998, 39 degrees awarded. *Degree requirements:* For master's, foreign language and thesis not required. *Average time to degree:* Master's–3 years part-time. *Application deadline:* For fall admission, 8/1 (priority date); for spring admission, 1/1. Applications are processed on a rolling basis. Application fee: $50. Electronic applications accepted. Tuition: Full-time $5,128; part-time $419 per credit hour. Required fees: $100; $100 per year. $50 per semester. Tuition and fees vary according to course level. *Financial aid:* Institutionally-sponsored loans available. Aid available to part-time students. Financial aid application deadline: 6/1. *Faculty research:* Advanced composite materials in bridges, strengthening existing bridges with carbon and glass fiber sheets, development of drive shafts using composite materials. Total annual research expenditures: $150,000. *Unit head:* Dr. George Kartsounes, Dean, 248-204-2500, Fax: 248-204-2509, E-mail: kartsounes@ltu.edu. *Application contact:* Lisa Kujawa, Director of Admissions, 248-204-3160, Fax: 248-204-3188, E-mail: admission@hu.edu.

Lehigh University, College of Engineering and Applied Science, Bethlehem, PA 18015-3094. Offers M Eng, MS, PhD. Part-time and evening/weekend programs available. Postbaccalaureate distance learning degree programs offered (no on-campus study). *Faculty:* 115 full-time (8 women), 7 part-time (0 women). *Students:* 390 full-time (59 women), 182 part-time (37 women); includes 57 minority (15 African Americans, 24 Asian Americans or Pacific Islanders, 18 Hispanic Americans), 238 international. Average age 24. 1180 applicants, 16% accepted. In 1998, 122 master's, 36 doctorates awarded. *Degree requirements:* For doctorate, dissertation required. *Entrance requirements:* For master's and doctorate, TOEFL. *Application deadline:* For fall admission, 7/15; for spring admission, 12/1. Applications are processed on a rolling basis. Application fee: $40. Electronic applications accepted. *Financial aid:* In 1998–99, fellowships with full and partial tuition reimbursements (averaging $11,000 per year), research assistantships with full and partial tuition reimbursements (averaging $15,000 per year), teaching assistantships with full and partial tuition reimbursements (averaging $11,000 per year) were awarded.; career-related internships or fieldwork, Federal Work-Study, institutionally-sponsored loans, scholarships, and tuition waivers (full and partial) also available. Aid available to part-time students. Financial aid application deadline: 1/15. Total annual research expenditures: $20.2 million. *Unit head:* Dr. Harvey Stenger, Dean, 610-758-5308. *Application contact:* Gail Andrews, Administrative Coordinator, 610-758-6310, Fax: 610-758-5623.

See in-depth description on page 111.

Louisiana State University and Agricultural and Mechanical College, Graduate School, College of Engineering, Baton Rouge, LA 70803. Offers MS Ch E, MS Pet E, MSCE, MSEE, MSES, MSIE, MSME, PhD. Part-time and evening/weekend programs available. *Faculty:* 104 full-time (4 women), 3 part-time (0 women). *Students:* 324 full-time (52 women), 96 part-time (8 women); includes 27 minority (12 African Americans, 11 Asian Americans or Pacific Islanders, 3 Hispanic Americans, 1 Native American), 288 international. Average age 29. 954 applicants, 37% accepted. In 1998, 84 master's, 30 doctorates awarded. Terminal master's awarded for partial completion of doctoral program. *Degree requirements:* For doctorate, dissertation required, foreign language not required. *Entrance requirements:* For master's, GRE General Test, minimum GPA of 3.0; for doctorate, GRE General Test. *Application deadline:* For fall admission, 1/25 (priority date). Applications are processed on a rolling basis. Application fee: $25. *Financial aid:* In 1998–99, 22 fellowships, 196 research assistantships with partial tuition reimbursements, 41 teaching assistantships with partial tuition reimbursements were awarded.; career-related internships or fieldwork, Federal Work-Study, institutionally-sponsored loans, scholarships, tuition waivers (full and partial), and unspecified assistantships also available. *Faculty research:* Agricultural, industrial, environmental, mechanical, and nuclear engineering. Total annual research expenditures: $17.8 million. *Unit head:* Dr. Adam T. Bourgoyne, Interim Dean, 225-388-5731, Fax: 225-334-1559, E-mail: ted_bourgoyne@eng.lsu.edu.

Louisiana Tech University, Graduate School, College of Engineering and Science, Ruston, LA 71272. Offers MS, D Eng, PhD. Part-time programs available. Terminal master's awarded for partial completion of doctoral program. *Degree requirements:* For doctorate, dissertation required, foreign language not required, foreign language not required. *Entrance requirements:* For master's, GRE General Test (minimum combined score of 1070 required; average 1245), TOEFL (minimum score of 550 required), minimum GPA of 3.0 in last 60 hours; for doctorate, TOEFL (minimum score of 550 required). *Faculty research:* Trenchless technology, micromanufacturing, radionuclide transport, microbial liquefaction, hazardous waste treatment.

Loyola College in Maryland, Graduate Programs, College of Arts and Sciences, Engineering Science Program, Baltimore, MD 21210-2699. Offers MES, MS. Part-time and evening/weekend programs available. *Faculty:* 5 full-time (1 woman), 16 part-time (0 women). *Students:* 16 full-time (6 women), 167 part-time (26 women); includes 33 minority (22 African Americans, 7 Asian Americans or Pacific Islanders, 4 Hispanic Americans), 8 international. 51 applicants, 86% accepted. In 1998, 50 degrees awarded. *Application deadline:* For fall admission, 8/1 (priority date); for spring admission, 12/1 (priority date). Applications are processed on a rolling basis. Application fee: $35. Tuition: Part-time $385 per credit. Required fees: $25 per semester. *Financial aid:* Research assistantships available. *Unit head:* Dr. Bernard Weigman, Director, 410-617-2260, E-mail: bjw@loyola.edu. *Application contact:* Scott Greatorex, Director, Graduate Admissions, 410-617-5020 Ext. 2407, Fax: 410-617-2002, E-mail: sgreatorex@loyola.edu.

Loyola Marymount University, Graduate Division, College of Science and Engineering, Los Angeles, CA 90045-8350. Offers MS, MSE. Part-time and evening/weekend programs available. *Faculty:* 55 full-time (9 women), 38 part-time (8 women). *Students:* 38 full-time (6 women), 81 part-time (14 women); includes 52 minority (8 African Americans, 32 Asian Americans or Pacific Islanders, 11 Hispanic Americans, 1 Native American), 4 international. 133 applicants, 50% accepted. In 1998, 30 degrees awarded. *Entrance requirements:* For master's, TOEFL. Application fee: $35. Electronic applications accepted. Tuition: Part-time $525 per unit. Required fees: $143; $14 per semester. Tuition and fees vary according to program. *Financial aid:* In 1998–99, 23 students received aid. Federal Work-Study, grants, scholarships, and instructorships available. Aid available to part-time students. Financial aid application deadline: 3/2; financial aid applicants required to submit FAFSA. *Unit head:* Dr. Gerald S. Jakubowski, Dean, 310-338-2834. *Application contact:* Dr. Paul A. Rude, Graduate Director, 310-338-5101.

Manhattan College, Graduate Division, School of Engineering, Riverdale, NY 10471. Offers biotechnology (MS); chemical engineering (MS); civil engineering (MS); computer engineering (MS); electrical engineering (MS); environmental engineering (ME, MS); mechanical engineering (MS). Part-time and evening/weekend programs available. *Degree requirements:* Foreign language not required. *Entrance requirements:* For master's, GRE, TOEFL, minimum GPA of 3.0.

See in-depth description on page 113.

Marquette University, Graduate School, College of Engineering, Milwaukee, WI 53201-1881. Offers MS, PhD. Part-time and evening/weekend programs available. *Students:* 118 full-time (29 women), 182 part-time (22 women); includes 24 minority (3 African Americans, 13 Asian Americans or Pacific Islanders, 7 Hispanic Americans, 1 Native American), 86 international. Terminal master's awarded for partial completion of doctoral program. *Degree requirements:* For doctorate, dissertation required, foreign language not required, foreign language not required. *Entrance requirements:* For master's, TOEFL; for doctorate, GRE General Test, TOEFL. *Average time to degree:* Master's–2 years full-time, 5 years part-time; doctorate–5 years full-time. *Application deadline:* Applications are processed on a rolling basis. Application fee: $40. Tuition: Part-time $510 per credit hour. Tuition and fees vary according to program. *Financial aid:* Fellowships, research assistantships, teaching assistantships, Federal Work-Study, grants, institutionally-sponsored loans, scholarships, and tuition waivers (full and partial) available. Aid available to part-time students. Financial aid application deadline: 2/15. *Faculty research:* Pulmonary hypertension, urban watershed management, pavement texture, microsensors for environmental pollutants, surface mount technology. Total annual research expenditures: $4.7 million. *Unit head:* Dr. G. E. O. Widera, Interim Dean, 414-288-7259, Fax: 414-288-1647,

E-mail: geo.widera@marquette.edu. *Application contact:* Director of Admissions, 414-288-7137, Fax: 414-288-1902, E-mail: mugs@vms.csd.mu.edu.

See in-depth description on page 115.

Marshall University, Graduate College, Graduate School of Information, Technology and Engineering, Huntington, WV 25755-2020. Offers MS, MSE. Part-time and evening/weekend programs available. *Faculty:* 12 full-time (1 woman), 7 part-time (0 women). *Students:* 24 full-time (10 women), 182 part-time (35 women); includes 8 minority (1 African American, 6 Asian Americans or Pacific Islanders, 1 Hispanic American), 2 international. Average age 36. In 1998, 27 degrees awarded. *Degree requirements:* For master's, final project, oral exam required. *Financial aid:* Fellowships, tuition waivers (full) available. Aid available to part-time students. Financial aid application deadline: 8/1; financial aid applicants required to submit FAFSA. *Unit head:* Dr. James Hooper, Dean, 304-696-4748, E-mail: hooper@marshall.edu. *Application contact:* Ken O'Neal, Assistant Vice President, Adult Student Services, 304-746-2500 Ext. 1907, Fax: 304-746-1902, E-mail: oneal@marshall.edu.

Massachusetts Institute of Technology, School of Engineering, Cambridge, MA 02139-4307. Offers M Eng, MBA, MST, SM, PhD, Sc D, CAS, CE, EAA, EE, EE, Mat E, Mech E, Met E, NE, Naval E, Ocean E. *Faculty:* 325 full-time (31 women), 2 part-time (0 women). *Students:* 2,390 full-time (474 women), 41 part-time (10 women); includes 335 minority (37 African Americans, 249 Asian Americans or Pacific Islanders, 48 Hispanic Americans, 1 Native American), 824 international. Average age 27. 4693 applicants, 32% accepted. In 1998, 808 master's, 239 doctorates awarded. *Degree requirements:* For doctorate and other advanced degree, dissertation required. Application fee: $55. *Financial aid:* In 1998–99, 389 fellowships, 1,573 research assistantships, 282 teaching assistantships were awarded.; career-related internships or fieldwork, Federal Work-Study, grants, institutionally-sponsored loans, scholarships, traineeships, and tuition waivers (partial) also available. Financial aid applicants required to submit FAFSA. Total annual research expenditures: $171.3 million. *Unit head:* Thomas L. Magnanti, Dean, 617-253-6604, Fax: 617-253-8549, E-mail: magnanti@mit.edu.

See in-depth description on page 117.

McGill University, Faculty of Graduate Studies and Research, Faculty of Engineering, Montréal, PQ H3A 2T5, Canada. Offers M Arch, M Eng, M Sc, MMM, MUP, PhD, Diploma. Part-time and evening/weekend programs available. *Faculty:* 147 full-time (7 women). *Students:* 681. In 1998, 167 master's, 63 doctorates, 1 other advanced degree awarded. *Degree requirements:* For doctorate, dissertation required. *Entrance requirements:* For master's, TOEFL, minimum GPA of 3.0; for doctorate, TOEFL. *Application deadline:* Applications are processed on a rolling basis. Application fee: $60. *Financial aid:* Fellowships, research assistantships, teaching assistantships, institutionally-sponsored loans, tuition waivers (full and partial), and teaching and research assistantships available. *Faculty research:* Robotics, pulp and paper, metals processing, microelectronics, polymer science and engineering. *Unit head:* John M. Dealy, Dean.

McMaster University, School of Graduate Studies, Faculty of Engineering, Hamilton, ON L8S 4M2, Canada. Offers M Eng, M Sc, PhD. Part-time and evening/weekend programs available. *Faculty:* 115 full-time, 2 part-time. *Students:* 267 full-time, 81 part-time, 47 international. In 1998, 74 master's, 33 doctorates awarded. *Degree requirements:* For doctorate, dissertation, comprehensive exam required, foreign language not required. *Application deadline:* For fall admission, 3/1 (priority date). Applications are processed on a rolling basis. Application fee: $50. *Financial aid:* In 1998–99, teaching assistantships (averaging $7,722 per year); fellowships, research assistantships, career-related internships or fieldwork also available. *Unit head:* Dr. M. Shoukri, Dean.

McNeese State University, Graduate School, College of Engineering and Technology, Lake Charles, LA 70609-2495. Offers chemical engineering (M Eng); civil engineering (M Eng); electrical engineering (M Eng); mechanical engineering (M Eng). Part-time and evening/weekend programs available. *Faculty:* 13 full-time (1 woman). *Students:* 5 full-time (0 women), 3 part-time. In 1998, 3 degrees awarded. *Degree requirements:* For master's, computer language, thesis or alternative required, foreign language not required. *Entrance requirements:* For master's, GRE General Test, TOEFL, minimum undergraduate GPA of 3.0. *Application deadline:* For fall admission, 7/15 (priority date). Applications are processed on a rolling basis. Application fee: $10 ($25 for international students). *Financial aid:* Federal Work-Study available. Aid available to part-time students. Financial aid application deadline: 5/1. *Unit head:* Dr. O. C. Karkalits, Dean, 318-475-5875.

Memorial University of Newfoundland, School of Graduate Studies, Faculty of Engineering and Applied Science, St. John's, NF A1C 5S7, Canada. Offers civil engineering (M Eng, PhD); electrical engineering (M Eng, PhD); mechanical engineering (M Eng, PhD); ocean engineering (M Eng, PhD). Part-time programs available. *Students:* 75 full-time (11 women), 28 part-time (2 women), 31 international. 202 applicants, 8% accepted. In 1998, 27 master's, 7 doctorates awarded. *Degree requirements:* For master's, thesis optional; for doctorate, dissertation, comprehensive exam required. *Application deadline:* For fall admission, 3/1. Application fee: $40. *Financial aid:* Fellowships, research assistantships, teaching assistantships available. *Unit head:* Dr. Rangaswany Seshadri, Dean, 709-737-8810, Fax: 709-737-8975, E-mail: sesh@engr.mun.ca. *Application contact:* Dr. J. J. Sharp, Associate Dean, 709-737-8901, Fax: 709-737-3480, E-mail: jsharp@engr.mun.ca.

Memorial University of Newfoundland, School of Graduate Studies, Interdisciplinary Program in Environmental Engineering and Applied Science, St. John's, NF A1C 5S7, Canada. Offers MA Sc. *Students:* 2 full-time (1 woman), 8 part-time. 6 applicants, 0% accepted. In 1998, 20 degrees awarded. *Degree requirements:* For master's, project required, thesis not required. *Entrance requirements:* For master's, honors B Sc or 2nd class B Eng. *Application deadline:* For fall admission, 5/30. Application fee: $40. *Unit head:* Dr. Tahir Husain, Chair, 709-737-8900, E-mail: thusain@engr.mun.ca. *Application contact:* Dr. M. Haddara, Associate Dean, Faculty of Engineering, 709-737-8900, E-mail: mhaddara@engr.mun.ca.

Mercer University, School of Engineering, Macon, GA 31207-0003. Offers biomedical engineering (MSE); electrical engineering (MSE); engineering management (MSE); mechanical engineering (MSE); software engineering (MSE); software systems (MS); technical management (MS). Part-time and evening/weekend programs available. *Faculty:* 23 full-time (1 woman), 6 part-time (0 women). Average age 35. 10 applicants, 80% accepted. In 1998, 18 degrees awarded. *Degree requirements:* For master's, computer language, thesis or alternative required, foreign language not required. *Entrance requirements:* For master's, GRE, minimum undergraduate GPA of 3.0. *Application deadline:* For fall admission, 7/1; for spring admission, 11/15. Applications are processed on a rolling basis. Application fee: $35 ($50 for international students). *Financial aid:* Federal Work-Study available. *Unit head:* Dr. Benjamin S. Kelley, Dean, 912-752-2459, Fax: 912-752-5593, E-mail: kelley_bs@mercer.edu. *Application contact:* Kathy Olivier, Coordinator, Special Programs, 912-752-2196, E-mail: oliver_kh@mercer.edu.

Mercer University, Cecil B. Day Campus, School of Engineering, Atlanta, GA 30341-4155. Offers electrical engineering (MSE); engineering management (MSE); software engineering (MSE); software systems (MS); technical communication management (MS). Part-time and evening/weekend programs available. Postbaccalaureate distance learning degree programs offered (no on-campus study). *Faculty:* 5 full-time (1 woman), 1 part-time (0 women). Average age 35. 22 applicants, 95% accepted. In 1998, 8 degrees awarded. *Degree requirements:* For master's, computer language, thesis or alternative required, foreign language not required. *Entrance requirements:* For master's, GRE, minimum GPA of 3.0 in major. *Application deadline:* For fall admission, 7/1; for spring admission, 11/15. Applications are processed on a rolling basis. Application fee: $35 ($50 for international students). *Unit head:* Dr. Benjamin S. Kelley, Acting Dean, 912-752-2459, E-mail: kelley_bs@mercer.edu. *Application contact:* Dr. David Leonard, Director of Admissions, 770-986-3203.

Miami University, Graduate School, School of Applied Science, Oxford, OH 45056. Offers MS. *Faculty:* 24. *Students:* 33 full-time (9 women), 8 part-time (3 women); includes 3 minority (1 African American, 2 Asian Americans or Pacific Islanders), 17 international. 54 applicants, 78% accepted. In 1998, 15 degrees awarded. *Degree requirements:* For master's, thesis, final exam required, foreign language not required. *Entrance requirements:* For master's, GRE, minimum undergraduate GPA of 3.0 during previous 2 years or 2.75 overall. *Application deadline:* For fall admission, 3/1 (priority date). Applications are processed on a rolling basis. Application fee: $35. *Financial aid:* Fellowships, research assistantships, teaching assistantships, Federal Work-Study and tuition waivers (full) available. Financial aid application deadline: 3/1. *Unit head:* David C. Haddad, Dean, 513-529-4036.

Michigan State University, Graduate School, College of Engineering, East Lansing, MI 48824-1020. Offers MS, PhD. Part-time programs available. Postbaccalaureate distance learning degree programs offered (minimal on-campus study). *Faculty:* 131. *Students:* 285 full-time (51 women), 307 part-time (48 women); includes 62 minority (22 African Americans, 29 Asian Americans or Pacific Islanders, 11 Hispanic Americans), 351 international. Average age 28. In 1998, 177 master's, 50 doctorates awarded. Terminal master's awarded for partial completion of doctoral program. *Degree requirements:* For doctorate, dissertation required, foreign language not required. *Entrance requirements:* For master's and doctorate, GRE. *Application deadline:* Applications are processed on a rolling basis. Application fee: $30 ($40 for international students). *Financial aid:* In 1998–99, 201 research assistantships with tuition reimbursements (averaging $12,617 per year), 161 teaching assistantships with tuition reimbursements (averaging $12,183 per year) were awarded.; fellowships, Federal Work-Study also available. Aid available to part-time students. Financial aid applicants required to submit FAFSA. *Faculty research:* Materials, environment and energy, information and computation, biotechnology, transportation and automotive engineering, manufacturing and processing. Total annual research expenditures: $15.8 million. *Unit head:* Dr. George VanDusen, Acting Dean, 517-355-5113, Fax: 517-355-2288, E-mail: vandusen@egr.msu.edu. *Application contact:* Dr. Anthony Wojcik, Associate Dean for Graduate Studies and Research, 517-355-3522, Fax: 517-353-7782, E-mail: wojcik@egr.msu.edu.

See in-depth description on page 119.

Michigan Technological University, Graduate School, College of Engineering, Houghton, MI 49931-1295. Offers MS, PhD. Part-time programs available. *Faculty:* 136 full-time (13 women), 8 part-time (2 women). *Students:* 350 full-time (79 women), 4 part-time (2 women); includes 10 minority (7 African Americans, 3 Asian Americans or Pacific Islanders), 157 international. Average age 27. 601 applicants, 57% accepted. In 1998, 82 master's, 24 doctorates awarded. *Degree requirements:* For doctorate, dissertation required, foreign language not required, foreign language not required. *Entrance requirements:* For master's and doctorate, TOEFL. *Average time to degree:* Master's–2.8 years full-time; doctorate–4.8 years full-time. *Application deadline:* For fall admission, 3/15 (priority date). Applications are processed on a rolling basis. Application fee: $30 ($35 for international students). Tuition, state resident: full-time $4,377. Tuition, nonresident: full-time $9,108. Required fees: $126. Tuition and fees vary according to course load. *Financial aid:* In 1998–99, 45 fellowships (averaging $3,376 per year), 116 research assistantships (averaging $8,181 per year), 76 teaching assistantships (averaging $6,833 per year) were awarded.; career-related internships or fieldwork, Federal Work-Study, institutionally-sponsored loans, and unspecified assistantships also available. Aid available to part-time students. Financial aid applicants required to submit FAFSA. Total annual research expenditures: $12.4 million. *Unit head:* Dr. Robert Warrington, Dean, 906-487-2005, Fax: 906-487-2782, E-mail: row@mtu.edu.

See in-depth description on page 121.

Milwaukee School of Engineering, Department of Electrical Engineering and Computer Science, Program in Engineering, Milwaukee, WI 53202-3109. Offers MS. Part-time and evening/weekend programs available. *Faculty:* 6 part-time (0 women). *Students:* 13 full-time (1 woman), 64 part-time (5 women); includes 4 minority (2 African Americans, 1 Asian American or Pacific Islander, 1 Native American), 2 international. Average age 25. *Degree requirements:* For master's, computer language, thesis or alternative, design project required, foreign language not required. *Entrance requirements:* For master's, GRE General Test (combined average 1300), BS in engineering or related field, interview. *Average time to degree:* Master's–5 years part-time. *Application deadline:* For fall admission, 8/15 (priority date); for spring admission, 2/1. Applications are processed on a rolling basis. Application fee: $30. Electronic applications accepted. *Financial aid:* Research assistantships, career-related internships or fieldwork available. Aid available to part-time students. *Faculty research:* Microprocessors, materials, thermodynamics, artificial intelligence, fluid power/hydaulics. Total annual research expenditures: $1.3 million. *Unit head:* Dr. Edward Chandler, Director, 414-277-7337, Fax: 414-277-7465, E-mail: chandler@msoe.edu. *Application contact:* Helen Boomsma, Director, Lifelong Learning Institute, 800-321-6763, Fax: 414-277-7475, E-mail: boomsma@msoe.edu.

Announcement: MSOE offers 6 master's degree programs. The master's programs in architectural engineering, engineering, engineering management, and environmental engineering are designed for working technical and management professionals. Classes meet in the evenings. The master's in perfusion is a full-time program, and the master's in medical informatics is offered full- or part-time. MSOE's applications-oriented curriculum is emphasized in each program.

See in-depth description on page 123.

Mississippi State University, College of Engineering, Mississippi State, MS 39762. Offers MS, PhD. Part-time programs available. Postbaccalaureate distance learning degree programs offered (no on-campus study). *Students:* 284 full-time (60 women), 114 part-time (22 women); includes 153 minority (19 African Americans, 130 Asian Americans or Pacific Islanders, 4 Hispanic Americans), 54 international. Average age 29. 481 applicants, 30% accepted. In 1998, 76 master's, 10 doctorates awarded. *Entrance requirements:* For master's, GRE General Test, TOEFL, minimum GPA of 2.75. *Application deadline:* For fall admission, 7/1; for spring admission, 11/1. Applications are processed on a rolling basis. Application fee: $25 for international students. *Financial aid:* Fellowships with full tuition reimbursements, research assistantships with full tuition reimbursements, teaching assistantships with full tuition reimbursements, Federal Work-Study, institutionally-sponsored loans, and unspecified assistantships available. Financial aid applicants required to submit FAFSA. *Faculty research:* Fluid dynamics, combustion, composite materials, computer design, high-voltage phenomena. Total annual research expenditures: $21 million. *Unit head:* Dr. A. Wayne Bennett, Dean, 662-325-2270, Fax: 662-325-8573, E-mail: bennet@engr.msstate.edu. *Application contact:* Jerry B. Inmon, Director of Admissions, 662-325-2224, Fax: 662-325-7360, E-mail: admit@admissions.msstate.edu.

Montana State University–Bozeman, College of Graduate Studies, College of Engineering, Bozeman, MT 59717. Offers MCEM, MPEM, MS, PhD. Part-time programs available. *Students:* 112 full-time (26 women), 56 part-time (14 women); includes 6 minority (4 Asian Americans or Pacific Islanders, 1 Hispanic American, 1 Native American) Average age 28. 115 applicants, 70% accepted. In 1998, 50 master's, 4 doctorates awarded. *Degree requirements:* For master's, thesis or alternative required, foreign language not required; for doctorate, dissertation required, foreign language not required. *Entrance requirements:* For master's and doctorate, GRE General Test, TOEFL. *Application deadline:* For fall admission, 6/1 (priority date); for spring admission, 11/1. Applications are processed on a rolling basis. Application fee: $50. *Financial aid:* Fellowships, research assistantships with full tuition reimbursements, teaching assistantships with full tuition reimbursements, career-related internships or fieldwork, Federal Work-Study, scholarships, and tuition waivers (full and partial) available. Financial aid application deadline: 3/1; financial aid applicants required to submit FAFSA. Total annual research expenditures: $3.7 million. *Unit head:* Dr. David F. Gibson, Dean, 406-994-2272, Fax: 406-994-6665, E-mail: carolynh@coe.montana.edu.

Montana Tech of The University of Montana, Graduate School, Butte, MT 59701-8997. Offers MPEM, MS, MTC. Part-time and evening/weekend programs available. Postbaccalaureate distance learning degree programs offered (minimal on-campus study). *Faculty:* 62 full-time (8 women). *Students:* 65 full-time (28 women), 36 part-time (7 women); includes 3 minority (1

Engineering and Applied Sciences—General

Montana Tech of The University of Montana (continued)
Asian American or Pacific Islander, 2 Native Americans) 94 applicants, 63% accepted. In 1998, 35 degrees awarded. *Degree requirements:* Foreign language not required. *Entrance requirements:* For master's, GRE General Test, TOEFL (minimum score of 525 required). *Application deadline:* For fall admission, 4/1 (priority date); for spring admission, 10/1 (priority date). Applications are processed on a rolling basis. Application fee: $30. Tuition, state resident: full-time $3,211; part-time $162 per credit hour. Tuition, nonresident: full-time $9,883; part-time $440 per credit hour. International tuition: $15,500 full-time. *Financial aid:* In 1998–99, 56 students received aid, including 26 research assistantships with partial tuition reimbursements available (averaging $4,190 per year), 24 teaching assistantships with partial tuition reimbursements available (averaging $3,400 per year); career-related internships or fieldwork, Federal Work-Study, institutionally-sponsored loans, and tuition waivers (full and partial) also available. Aid available to part-time students. Financial aid application deadline: 4/1; financial aid applicants required to submit FAFSA. *Faculty research:* Mineral processing, environmental restoration, endophytic fungi, renewable and nonrenewable energy. Total annual research expenditures: $347,000. *Unit head:* John Brower, Director, 406-496-4128, Fax: 406-496-4334, E-mail: jbrower@mtech.edu. *Application contact:* Cindy Dunstan, Administrative Assistant, 406-496-4128, Fax: 406-496-4334, E-mail: cdunstan@mtech.edu.

See in-depth description on page 125.

Morgan State University, School of Graduate Studies, School of Engineering, Baltimore, MD 21251. Offers MS, D Eng. Part-time and evening/weekend programs available.

National Technological University, Programs in Engineering, Fort Collins, CO 80526-1842. Offers chemical engineering (MS); computer engineering (MS); computer science (MS); electrical engineering (MS); engineering management (MS); hazardous waste management (MS); health physics (MS); management of technology (MS); manufacturing systems engineering (MS); materials science and engineering (MS); software engineering (MS); special majors (MS); transportation engineering (MS); transportation systems engineering (MS). Part-time programs available. *Faculty:* 600 part-time (20 women). In 1998, 176 degrees awarded. *Entrance requirements:* For master's, BS in engineering or related field. *Application deadline:* Applications are processed on a rolling basis. Application fee: $50. *Unit head:* Lionel V. Baldwin, President, 970-495-6400, Fax: 970-484-0668, E-mail: baldwin@mail.ntu.edu.

National University, Graduate Studies, School of Business and Technology, Department of Technology, La Jolla, CA 92037-1011. Offers e-commerce (MBA, MS); electronic engineering (MS); engineering management (MS); environmental management (MBA, MS); industrial engineering management (MS); software engineering (MS); technology management (MBA, MS); telecommunication systems management (MS). Part-time and evening/weekend programs available. Postbaccalaureate distance learning degree programs offered (minimal on-campus study). *Faculty:* 12 full-time, 125 part-time. *Students:* 305 (79 women); includes 122 minority (34 African Americans, 69 Asian Americans or Pacific Islanders, 17 Hispanic Americans, 2 Native Americans) 53 international. 79 applicants, 100% accepted. In 1998, 125 degrees awarded. *Degree requirements:* For master's, foreign language and thesis not required. *Entrance requirements:* For master's, interview, minimum GPA of 2.5. *Application deadline:* Applications are processed on a rolling basis. Application fee: $60 ($100 for international students). Tuition: Full-time $7,830; part-time $870 per course. One-time fee: $60. Tuition and fees vary according to campus/location. *Financial aid:* Grants, institutionally-sponsored loans, scholarships, and tuition waivers (full and partial) available. Aid available to part-time students. Financial aid application deadline: 5/1; financial aid applicants required to submit FAFSA. *Unit head:* Dr. Leonid Preiser, Chair, 858-642-8425, Fax: 858-642-8716, E-mail: lpreiser@nu.edu. *Application contact:* Nancy Rohland, Director of Enrollment Management, 858-642-8180, Fax: 858-642-8709, E-mail: nrohland@nu.edu.

New Jersey Institute of Technology, Office of Graduate Studies, Newark, NJ 07102-1982. Offers M Arch, MA, MAT, MIP, MS, PhD, Engineer, M Arch/MIP, M Arch/MS. Part-time and evening/weekend programs available. Terminal master's awarded for partial completion of doctoral program. *Degree requirements:* For doctorate, residency required, foreign language not required. *Entrance requirements:* For master's, GRE General Test (minimum score of 450 on verbal section, 600 on quantitative, 550 on analytical required); for doctorate, GRE General Test (minimum score of 450 on verbal section, 600 on quantitative, 550 on analytical required), minimum graduate GPA of 3.5. Electronic applications accepted. *Faculty research:* Toxic and hazardous waste management, transportation, biomedical engineering, computer-integrated manufacturing, management of technology.

See in-depth description on page 127.

New Mexico State University, Graduate School, College of Engineering, Las Cruces, NM 88003-8001. Offers MS Ch E, MS Env E, MSCE, MSEE, MSIE, MSME, PhD. *Accreditation:* ABET (one or more programs are accredited). Part-time programs available. *Faculty:* 62 full-time (4 women), 3 part-time (0 women). *Students:* 157 full-time (32 women), 112 part-time (19 women); includes 47 minority (2 African Americans, 5 Asian Americans or Pacific Islanders, 38 Hispanic Americans, 2 Native Americans), 85 international. Average age 32. 338 applicants, 62% accepted. In 1998, 103 master's, 11 doctorates awarded. *Degree requirements:* For doctorate, dissertation required. *Application deadline:* For fall admission, 7/1 (priority date); for spring admission, 11/1. Applications are processed on a rolling basis. Application fee: $15 ($35 for international students). Electronic applications accepted. Tuition, state resident: full-time $2,682; part-time $112 per credit. Tuition, nonresident: full-time $8,376; part-time $349 per credit. Tuition and fees vary according to course load. *Financial aid:* Fellowships, research assistantships, teaching assistantships, career-related internships or fieldwork and Federal Work-Study available. Aid available to part-time students. Financial aid application deadline: 3/1. *Faculty research:* Structures and nondestructive testing, environmental science and engineering, telecommunication theory and systems, manufacturing methods and systems, high performance computing and software engineering. *Unit head:* Dr. Jay B. Jordan, Interim Dean, 505-646-2914, Fax: 505-646-3549, E-mail: jjordan@nmsu.edu.

New York Institute of Technology, Graduate Division, School of Engineering and Technology, Old Westbury, NY 11568-8000. Offers MS, Certificate. Part-time and evening/weekend programs available. Postbaccalaureate distance learning degree programs offered. *Students:* 192 full-time (56 women), 433 part-time (98 women); includes 186 minority (67 African Americans, 92 Asian Americans or Pacific Islanders, 25 Hispanic Americans, 2 Native Americans), 218 international. Average age 34. 367 applicants, 62% accepted. In 1998, 134 degrees awarded. *Entrance requirements:* For master's, minimum QPA of 2.85. *Application deadline:* For fall admission, 8/1. Applications are processed on a rolling basis. Application fee: $50. Electronic applications accepted. *Financial aid:* Fellowships, research assistantships, teaching assistantships, career-related internships or fieldwork, institutionally-sponsored loans, tuition waivers (full and partial), and unspecified assistantships available. Aid available to part-time students. *Faculty research:* Develop hybrid vehicle, video dial tone network, algorithm development for unsteady viscious flow. *Unit head:* Dr. Heskia Heskiaoff, Dean, 516-686-7931. *Application contact:* Glenn Berman, Executive Director of Admissions, 516-686-7519, Fax: 516-626-0419, E-mail: gberman@iris.nyit.edu.

North Carolina Agricultural and Technical State University, Graduate School, College of Engineering, Greensboro, NC 27411. Offers MSAE, MSCS, MSE, MSEE, MSIE, MSME, PhD. Part-time programs available. *Faculty:* 73 full-time (4 women), 19 part-time (1 woman). *Students:* 210 full-time (70 women), 74 part-time (33 women); includes 202 minority (165 African Americans, 36 Asian Americans or Pacific Islanders, 1 Hispanic American), 48 international. Average age 25. 253 applicants, 62% accepted. In 1998, 62 master's awarded. *Degree requirements:* Foreign language not required. *Application deadline:* For fall admission, 7/1 (priority date); for spring admission, 1/9. Applications are processed on a rolling basis. Application fee: $35. *Financial aid:* Fellowships, research assistantships, teaching assistantships, career-related internships or fieldwork and unspecified assistantships available. Aid available to part-time students. Total annual research expenditures: $9 million. *Unit head:* Dr. Lonnie Sharpe, Dean, 336-334-7589, Fax: 336-334-7540, E-mail: lsharpe@ncat.edu. *Application contact:* Dr. Thoyd Melton, Dean of the Graduate School, 336-334-7920, Fax: 336-334-7282, E-mail: meltont@ncat.edu.

See in-depth description on page 129.

North Carolina State University, Graduate School, College of Engineering, Raleigh, NC 27695. Offers M Ch E, M Eng, MBAE, MC Sc, MCE, MIE, MIMS, MME, MMSE, MNE, MOR, MS, MSIE, PhD, PD. Part-time programs available. *Faculty:* 246 full-time (13 women), 166 part-time (7 women). *Students:* 892 full-time (180 women), 482 part-time (76 women); includes 196 minority (71 African Americans, 101 Asian Americans or Pacific Islanders, 21 Hispanic Americans, 3 Native Americans), 417 international. Average age 30. 1640 applicants, 33% accepted. In 1998, 295 master's, 116 doctorates awarded. Terminal master's awarded for partial completion of doctoral program. *Degree requirements:* For doctorate, dissertation required, foreign language not required, foreign language not required. *Application deadline:* Applications are processed on a rolling basis. Application fee: $45. *Financial aid:* In 1998–99, 117 fellowships (averaging $2,260 per year), 525 research assistantships (averaging $4,755 per year), 188 teaching assistantships (averaging $5,016 per year) were awarded.; career-related internships or fieldwork, Federal Work-Study, institutionally-sponsored loans, scholarships, and traineeships also available. Total annual research expenditures: $50.9 million. *Unit head:* Dr. Nino A. Masnari, Dean, 919-515-2311, Fax: 919-515-7951, E-mail: nino_masnari@ncsu.edu. *Application contact:* Fran Coats, Administrative Assistant, 919-515-2311, Fax: 919-515-7951, E-mail: fran_coats@ncsu.edu.

See in-depth description on page 131.

North Dakota State University, Graduate Studies and Research, College of Engineering and Architecture, Fargo, ND 58105. Offers MS, PhD. Part-time and evening/weekend programs available. Postbaccalaureate distance learning degree programs offered (minimal on-campus study). *Faculty:* 56 full-time (1 woman), 5 part-time (0 women). *Students:* 37 full-time (9 women), 40 part-time (37 women); includes 30 minority (2 African Americans, 26 Asian Americans or Pacific Islanders, 1 Hispanic American, 1 Native American) Average age 25. 286 applicants, 51% accepted. In 1998, 15 master's awarded (100% found work related to degree); 2 doctorates awarded. Terminal master's awarded for partial completion of doctoral program. *Degree requirements:* For doctorate, computer language, dissertation required, foreign language not required, foreign language not required. *Entrance requirements:* For master's, TOEFL; for doctorate, TOEFL (minimum score of 525 required). *Average time to degree:* Master's–3 years full-time, 4 years part-time. *Application deadline:* Applications are processed on a rolling basis. Application fee: $25. *Financial aid:* In 1998–99, 41 students received aid, including 3 fellowships with full tuition reimbursements available (averaging $25,120 per year), 18 research assistantships with full tuition reimbursements available (averaging $11,255 per year), 20 teaching assistantships with full tuition reimbursements available (averaging $13,902 per year); career-related internships or fieldwork, Federal Work-Study, institutionally-sponsored loans, scholarships, and tuition waivers (full) also available. Aid available to part-time students. *Faculty research:* Theoretical mechanics, robotics, automation, CAD/CAM, environmental engineering. Total annual research expenditures: $690,513. *Unit head:* Dr. Otto J. Helweg, Dean, 701-231-7525, Fax: 701-231-8957, E-mail: helweg@badlands.nodak.edu. *Application contact:* Dr. William D. Slanger, Interim Dean of Graduate Studies and Research, 701-231-7033, Fax: 701-231-8098, E-mail: wslanger@gwmail.nodak.edu.

Northeastern University, College of Engineering, Boston, MA 02115-5096. Offers MS, PhD. Part-time programs available. *Faculty:* 87 full-time (7 women), 25 part-time (1 woman). *Students:* 449 full-time (152 women), 562 part-time (123 women); includes 56 minority (14 African Americans, 30 Asian Americans or Pacific Islanders, 11 Hispanic Americans, 1 Native American), 381 international. Average age 26. 1511 applicants, 49% accepted. In 1998, 255 master's, 19 doctorates awarded. Terminal master's awarded for partial completion of doctoral program. *Degree requirements:* For doctorate, dissertation, departmental qualifying exam required, foreign language not required, foreign language not required. *Entrance requirements:* For master's and doctorate, GRE General Test. *Average time to degree:* Master's–2.65 years full-time, 4.39 years part-time; doctorate–5.13 years full-time, 4 years part-time. *Application deadline:* For fall admission, 4/15. Applications are processed on a rolling basis. Application fee: $50. *Financial aid:* In 1998–99, 244 students received aid, including 1 fellowship with tuition reimbursement available (averaging $12,450 per year), 99 research assistantships with full tuition reimbursements available (averaging $12,450 per year), 73 teaching assistantships with full tuition reimbursements available (averaging $12,450 per year); career-related internships or fieldwork, Federal Work-Study, tuition waivers (full), and unspecified assistantships also available. Aid available to part-time students. Financial aid application deadline: 2/15; financial aid applicants required to submit FAFSA. Total annual research expenditures: $13.6 million. *Unit head:* Dr. Yaman Yener, Associate Dean for Research and Graduate Education, 617-373-2711, Fax: 617-373-2501. *Application contact:* Stephen L. Gibson, Associate Director, 617-373-2711, Fax: 617-373-2501, E-mail: grad-eng.@coe.neu.edu.

Announcement: College of Engineering offers various full- and part-time doctoral degrees as well as Master of Science degrees in chemical, civil, computer systems, electrical, mechanical, and industrial engineering; engineering management; information systems; and operations research. Special features: interdisciplinary program (MS or PhD), the Co-operative Education Plan, Women in Information Systems program.

See in-depth description on page 133.

Northern Arizona University, Graduate College, College of Engineering, Flagstaff, AZ 86011. Offers M Eng. *Entrance requirements:* For master's, minimum GPA of 3.0 in final 60 hours of undergraduate work. *Unit head:* Dr. Pamela Eibeck, Coordinator, 520-523-5252, E-mail: m.eng@nau.edu.

Northern Illinois University, Graduate School, College of Engineering and Engineering Technology, De Kalb, IL 60115-2854. Offers MS. Part-time and evening/weekend programs available. *Faculty:* 35 full-time (2 women), 3 part-time (0 women). *Students:* 47 full-time (6 women), 97 part-time (13 women); includes 7 minority (2 African Americans, 3 Asian Americans or Pacific Islanders, 1 Hispanic American, 1 Native American), 81 international. Average age 30. 214 applicants, 47% accepted. In 1998, 60 degrees awarded. *Degree requirements:* For master's, comprehensive exam required. *Entrance requirements:* For master's, GRE General Test, TOEFL (minimum score of 550 required; 213 for computer-based), minimum GPA of 2.75. *Application deadline:* For fall admission, 6/1; for spring admission, 11/1. Applications are processed on a rolling basis. Application fee: $30. *Financial aid:* Fellowships, research assistantships, teaching assistantships, career-related internships or fieldwork, Federal Work-Study, tuition waivers (full), and unspecified assistantships available. Aid available to part-time students. *Unit head:* Dr. Romualdas Kasuba, Dean, 815-753-1281.

See in-depth description on page 135.

Northwestern Polytechnic University, School of Engineering, Fremont, CA 94539-7482. Offers computer science (MS); computer systems engineering (MS); electrical engineering (MS). Part-time and evening/weekend programs available. *Faculty:* 8 full-time, 52 part-time. *Students:* 262. 140 applicants, 94% accepted. In 1998, 43 degrees awarded (100% found work related to degree). *Degree requirements:* For master's, computer language, thesis required, foreign language not required. *Entrance requirements:* For master's, TOEFL. *Average time to degree:* Master's–2 years full-time, 3 years part-time. *Application deadline:* For fall admission, 8/15; for winter admission, 12/15; for spring admission, 7/15. Applications are processed on a rolling basis. Application fee: $50 ($75 for international students). Tuition: Full-time $6,750; part-time $375 per unit. Required fees: $135 per term. Tuition and fees vary according to course load and program. *Unit head:* Dr. Pochang Hsu, Dean, 510-657-5911, Fax: 510-657-8975, E-mail: npuadm@npu0.npu.edu. *Application contact:* Dr. Fred Kuttner, Dean of Academic Affairs and Admissions, 510-657-5911, Fax: 510-657-8975, E-mail: npuadm@npu0.npu.edu.

Northwestern University, The Graduate School, Robert R. McCormick School of Engineering and Applied Science, Evanston, IL 60208. Offers MEM, MIT, MME, MMM, MPM, MS, PhD. MS and PhD admissions and degrees offered through The Graduate School. Part-time programs available. *Faculty:* 161 full-time. *Students:* 846 full-time (199 women), 173 part-time (33 women); includes 164 minority (36 African Americans, 97 Asian Americans or Pacific Islanders, 30 Hispanic Americans, 1 Native American), 329 international. 2305 applicants, 30% accepted. In 1998, 104 master's, 108 doctorates awarded. *Degree requirements:* For doctorate, dissertation required. *Entrance requirements:* For master's and doctorate, GRE General Test, TOEFL (minimum score of 560 required). *Application deadline:* For fall admission, 8/30. Applications are processed on a rolling basis. Application fee: $50 ($55 for international students). *Financial aid:* In 1998–99, 88 fellowships with full tuition reimbursements (averaging $11,673 per year), 268 research assistantships with partial tuition reimbursements (averaging $16,285 per year), 85 teaching assistantships with full tuition reimbursements (averaging $12,042 per year) were awarded.; career-related internships or fieldwork, Federal Work-Study, and institutionally-sponsored loans also available. Financial aid application deadline: 1/15; financial aid applicants required to submit FAFSA. Total annual research expenditures: $45 million. *Unit head:* Jerome Cohen, Dean, 847-491-5220.

See in-depth description on page 137.

Oakland University, Graduate Studies, School of Engineering and Computer Science, Rochester, MI 48309-4401. Offers MS, PhD. Part-time and evening/weekend programs available. *Faculty:* 34 full-time, 9 part-time. *Students:* 200 full-time (59 women), 405 part-time (66 women); includes 58 minority (13 African Americans, 43 Asian Americans or Pacific Islanders, 2 Hispanic Americans), 124 international. 291 applicants, 75% accepted. In 1998, 178 master's, 11 doctorates awarded. *Degree requirements:* For master's, foreign language and thesis not required; for doctorate, dissertation required, foreign language not required. *Entrance requirements:* For master's and doctorate, minimum GPA of 3.0 for unconditional admission. *Application deadline:* For fall admission, 7/15; for spring admission, 3/15. Application fee: $30. Tuition, state resident: part-time $221 per credit hour. Tuition, nonresident: part-time $488 per credit hour. Required fees: $214 per semester. Part-time tuition and fees vary according to program. *Financial aid:* Federal Work-Study, institutionally-sponsored loans, and tuition waivers (full) available. Financial aid application deadline: 3/1; financial aid applicants required to submit FAFSA. *Unit head:* Dr. Michael P. Polis, Dean, 248-370-2217. *Application contact:* Dr. Bhushan Bhatt, Associate Dean, 248-370-2233.

The Ohio State University, Graduate School, College of Engineering, Columbus, OH 43210. Offers M Arch, M Land Arch, MCRP, MS, PhD. Part-time and evening/weekend programs available. *Faculty:* 295 full-time, 158 part-time. *Students:* 1,272 full-time (273 women), 195 part-time (43 women); includes 109 minority (30 African Americans, 56 Asian Americans or Pacific Islanders, 23 Hispanic Americans), 828 international. 3719 applicants, 23% accepted. In 1998, 384 master's, 115 doctorates awarded. *Degree requirements:* For master's, computer language required, foreign language not required; for doctorate, computer language, dissertation required, foreign language not required. *Application deadline:* For fall admission, 8/15. Applications are processed on a rolling basis. Application fee: $30 ($40 for international students). *Financial aid:* Fellowships, research assistantships, teaching assistantships, career-related internships or fieldwork, Federal Work-Study, institutionally-sponsored loans, and unspecified assistantships available. Aid available to part-time students. *Unit head:* David B. Ashley, Dean, 614-292-2836, Fax: 614-292-9615.

Ohio University, Graduate Studies, College of Engineering and Technology, Athens, OH 45701-2979. Offers MS, PhD. Part-time programs available. *Faculty:* 61 full-time (3 women), 11 part-time (3 women). *Students:* 213 full-time (41 women), 77 part-time (9 women); includes 6 minority (1 African American, 5 Asian Americans or Pacific Islanders), 225 international. 811 applicants, 65% accepted. In 1998, 47 master's, 11 doctorates awarded. Terminal master's awarded for partial completion of doctoral program. *Degree requirements:* For doctorate, dissertation required, foreign language not required. *Entrance requirements:* For doctorate, GRE. Application fee: $30. Tuition, state resident: full-time $5,754; part-time $238 per credit hour. Tuition, nonresident: full-time $11,055; part-time $457 per credit hour. Tuition and fees vary according to course load, campus/location and program. *Financial aid:* In 1998–99, 17 fellowships, 42 research assistantships, 33 teaching assistantships were awarded.; career-related internships or fieldwork, Federal Work-Study, institutionally-sponsored loans, tuition waivers (full and partial), and unspecified assistantships also available. Total annual research expenditures: $7.4 million. *Unit head:* Dr. Warren K. Wray, Dean, 740-593-1474.

See in-depth description on page 139.

Oklahoma State University, Graduate College, College of Engineering, Architecture and Technology, Stillwater, OK 74078. Offers M Arch, M Arch E, M Bio E, M En, M Gen E, MIE Mgmt, MS, PhD. *Faculty:* 89 full-time (4 women), 5 part-time (0 women). *Students:* 289 full-time (25 women), 328 part-time (65 women); includes 42 minority (8 African Americans, 17 Asian Americans or Pacific Islanders, 9 Hispanic Americans, 8 Native Americans), 306 international. Average age 30. In 1998, 135 master's, 18 doctorates awarded. *Degree requirements:* For doctorate, dissertation required, foreign language not required, foreign language not required. *Entrance requirements:* For master's and doctorate, TOEFL. *Application deadline:* For fall admission, 7/1 (priority date). Application fee: $25. *Financial aid:* In 1998–99, 260 students received aid, including 140 research assistantships (averaging $9,503 per year), 120 teaching assistantships (averaging $7,798 per year); fellowships, career-related internships or fieldwork, Federal Work-Study, and tuition waivers (partial) also available. Aid available to part-time students. Financial aid application deadline: 3/1. *Unit head:* Dr. Karl N. Reid, Dean, 405-744-5140.

See in-depth description on page 141.

Old Dominion University, College of Engineering and Technology, Norfolk, VA 23529. Offers ME, MEM, MS, PhD. Part-time and evening/weekend programs available. Postbaccalaureate distance learning degree programs offered. *Faculty:* 75 full-time (5 women). *Students:* 140 full-time (19 women), 190 part-time (28 women); includes 30 minority (16 African Americans, 9 Asian Americans or Pacific Islanders, 4 Hispanic Americans, 1 Native American), 131 international. Average age 32. In 1998, 110 master's, 18 doctorates awarded. *Degree requirements:* For master's, comprehensive exam required; for doctorate, dissertation, candidacy exam required, foreign language not required. *Entrance requirements:* For master's and doctorate, TOEFL (minimum score of 550 required). *Application deadline:* For fall admission, 7/1; for spring admission, 10/1. Applications are processed on a rolling basis. Application fee: $30. Electronic applications accepted. *Financial aid:* In 1998–99, 186 students received aid, including 89 research assistantships with tuition reimbursements available (averaging $12,447 per year), 2 teaching assistantships with tuition reimbursements available (averaging $6,813 per year); fellowships, career-related internships or fieldwork, grants, institutionally-sponsored loans, scholarships, and tuition waivers (partial) also available. Aid available to part-time students. Financial aid application deadline: 2/15; financial aid applicants required to submit FAFSA. *Faculty research:* Physical electronics, computational applied mechanics, structural dynamics, computational fluid dynamics, coastal engineering of water resources. Total annual research expenditures: $5.6 million. *Unit head:* Dr. William Swart, Dean, 757-683-5897, Fax: 757-683-4898, E-mail: wswart@coet.odu.edu. *Application contact:* Dr. Griffith McRee, Associate Dean, 757-683-3790, Fax: 757-683-4898, E-mail: gjm200u@coet.odu.edu.

See in-depth description on page 143.

Oregon State University, Graduate School, College of Engineering, Corvallis, OR 97331. Offers M Agr, M Eng, M Oc E, MA, MAIS, MS, PhD. Part-time programs available. *Faculty:* 103 full-time (13 women), 42 part-time (2 women). *Students:* 341 full-time (76 women), 95 part-time (8 women); includes 26 minority (1 African American, 23 Asian Americans or Pacific Islanders, 1 Hispanic American, 1 Native American), 226 international. Average age 26. In 1998, 141 master's, 27 doctorates awarded. Terminal master's awarded for partial completion of doctoral program. *Degree requirements:* For doctorate, dissertation required, foreign language not required. *Entrance requirements:* For master's and doctorate, TOEFL, minimum GPA of 3.0 in last 90 hours. *Application deadline:* Applications are processed on a rolling basis. Application fee: $50. *Financial aid:* Fellowships, research assistantships, teaching assistantships, career-related internships or fieldwork, Federal Work-Study, institutionally-sponsored loans, and instructorships available. Aid available to part-time students. Financial aid application deadline: 3/1. *Faculty research:* Molecular beam epitaxy, wave-structure interaction, pavement materials, toxic wastes, mechanical design methodology. *Unit head:* Ronald L. Adams, Dean, 541-737-3101, Fax: 541-737-1805, E-mail: adamsrl@engr.orst.edu. *Application contact:* Roy C. Rathja, Assistant Dean, 541-737-5236, Fax: 541-737-3124, E-mail: rathja@engr.orst.edu.

See in-depth description on page 145.

Pennsylvania State University at Erie, The Behrend College, Graduate Center, School of Engineering and Engineering Technology, Erie, PA 16563. Offers M Eng. *Students:* 1 (woman) full-time, 5 part-time (3 women). *Entrance requirements:* For master's, GRE General Test. *Application deadline:* For fall admission, 7/26. Application fee: $50. *Unit head:* Richard C. Progelhof, Director, 814-898-6047.

Pennsylvania State University Great Valley School of Graduate Professional Studies, Graduate Studies and Continuing Education, College of Engineering, Malvern, PA 19355-1488. Offers M Eng, PhD. PhD offered jointly with University of Wales. Part-time programs available. *Students:* 3 full-time (1 woman), 213 part-time (41 women). In 1998, 53 degrees awarded. Application fee: $50. *Unit head:* Dr. David Russell, Academic Division Head, 610-648-3335. *Application contact:* 610-648-3242, Fax: 610-889-1334, E-mail: gvengr@psugv.psu.edu.

See in-depth description on page 149.

Pennsylvania State University Harrisburg Campus of the Capital College, Graduate Center, School of Science, Engineering and Technology, Program in Engineering Science, Middletown, PA 17057-4898. Offers M Eng. Evening/weekend programs available. *Students:* 2 full-time (0 women), 38 part-time (5 women). Average age 33. In 1998, 10 degrees awarded. *Degree requirements:* For master's, thesis required, foreign language not required. *Entrance requirements:* For master's, GRE General Test, TOEFL, minimum GPA of 2.5. *Application deadline:* For fall admission, 7/26. Application fee: $50. *Unit head:* Dr. Seroj Mackertich, Chair, 717-948-6131.

Pennsylvania State University University Park Campus, Graduate School, College of Engineering, State College, University Park, PA 16802-1503. Offers M Eng, MAE, MS, PhD. *Students:* 816 full-time (129 women), 319 part-time (37 women). In 1998, 271 master's, 113 doctorates awarded. *Degree requirements:* For doctorate, dissertation required, foreign language not required. *Entrance requirements:* For master's and doctorate, GRE General Test. Application fee: $50. *Financial aid:* Fellowships, research assistantships, teaching assistantships available. *Unit head:* Dr. David N. Wormley, Dean, 814-865-7537.

See in-depth description on page 147.

Pittsburg State University, Graduate School, School of Technology, Department of Technology Studies, Pittsburg, KS 66762-5880. Offers technology education (MS). *Students:* 24 full-time, 16 part-time, 8 international. In 1998, 22 degrees awarded. *Degree requirements:* For master's, thesis or alternative required, foreign language not required. Application fee: $0 ($40 for international students). Tuition, state resident: full-time $2,466; part-time $105 per credit hour. Tuition, nonresident: full-time $6,268; part-time $264 per credit hour. *Financial aid:* Teaching assistantships, career-related internships or fieldwork and Federal Work-Study available. *Unit head:* Dr. John Iley, Chairperson, 316-235-4371. *Application contact:* Marvene Darraugh, Administrative Officer, 316-235-4220, Fax: 316-235-4219, E-mail: mdarraug@pittstate.edu.

Portland State University, Graduate Studies, School of Engineering and Applied Science, Portland, OR 97207-0751. Offers ME, MS, PhD. Part-time and evening/weekend programs available. *Faculty:* 45 full-time (4 women), 35 part-time (2 women). *Students:* 114 full-time (26 women), 144 part-time (31 women); includes 24 minority (18 Asian Americans or Pacific Islanders, 4 Hispanic Americans, 2 Native Americans), 133 international. Average age 30. 263 applicants, 65% accepted. In 1998, 78 master's, 2 doctorates awarded. *Degree requirements:* For doctorate, one foreign language, computer language, dissertation, oral and written exams required. *Entrance requirements:* For master's, TOEFL (minimum score of 550 required), minimum GPA of 3.0 in upper-division course work or 2.75 overall; for doctorate, GRE General Test, GRE Subject Test, minimum GPA of 3.0 in upper-division course work. *Application deadline:* Applications are processed on a rolling basis. Application fee: $50. *Financial aid:* In 1998–99, 27 research assistantships, 56 teaching assistantships were awarded.; career-related internships or fieldwork, Federal Work-Study, and institutionally-sponsored loans also available. Aid available to part-time students. Financial aid application deadline: 3/1; financial aid applicants required to submit FAFSA. *Unit head:* Dr. Robert D. Dryden, Dean, 503-725-4631, Fax: 503-725-4298, E-mail: drydenr@eas.pdx.edu. *Application contact:* Alisia Walton, Administrative Assistant, 503-725-4631, Fax: 503-725-4298, E-mail: waltona@eas.pdx.edu.

See in-depth description on page 151.

Prairie View A&M University, Graduate School, College of Engineering, Prairie View, TX 77446-0188. Offers MS Engr. Part-time and evening/weekend programs available. *Faculty:* 10 full-time (0 women). *Students:* 28 full-time (8 women), 8 part-time (1 woman); includes 17 minority (14 African Americans, 3 Asian Americans or Pacific Islanders), 16 international. Average age 28. In 1998, 22 degrees awarded. *Degree requirements:* For master's, computer language, thesis required, foreign language not required. *Entrance requirements:* For master's, GRE General Test. *Average time to degree:* Master's–2.5 years full-time, 4 years part-time. *Application deadline:* For fall admission, 7/1 (priority date); for spring admission, 11/1. Applications are processed on a rolling basis. Application fee: $10. *Financial aid:* Career-related internships or fieldwork, Federal Work-Study, and institutionally-sponsored loans available. Aid available to part-time students. Financial aid application deadline: 5/31. *Faculty research:* Radiation effects on solid-state devices, circuits, robotic vision, environmental studies. *Unit head:* Dr. Milton R. Bryant, Dean, 409-857-2211, Fax: 409-857-2222. *Application contact:* Dr. James O. Morgan, Graduate Director, 409-857-4200.

Princeton University, Graduate School, School of Engineering and Applied Science, Princeton, NJ 08544-1019. Offers M Eng, MSE, PhD. *Degree requirements:* For doctorate, dissertation required. *Entrance requirements:* For master's and doctorate, GRE General Test.

Purdue University, Graduate School, Schools of Engineering, West Lafayette, IN 47907. Offers MS, MS Bm E, MS Ch E, MS Met E, MSAAE, MSABE, MSCE, MSE, MSIE, MSME, MSNE, PhD. Part-time and evening/weekend programs available. Postbaccalaureate distance learning degree programs offered (no on-campus study). *Faculty:* 309. *Students:* 1,212 full-time (194 women), 423 part-time (62 women); includes 170 minority (30 African Americans, 93 Asian Americans or Pacific Islanders, 42 Hispanic Americans, 5 Native Americans), 834 international. 3285 applicants, 46% accepted. In 1998, 392 master's, 126 doctorates awarded. *Degree requirements:* For doctorate, dissertation required. *Entrance requirements:* For master's and doctorate, TOEFL. Application fee: $30. Electronic applications accepted. *Financial aid:* Fellowships, research assistantships, teaching assistantships, career-related internships or fieldwork available. Aid available to part-time students. Financial aid applicants required to submit FAFSA. *Unit head:* Dr. Richard J. Schwartz, Dean, 765-494-5346.

See in-depth description on page 153.

Purdue University Calumet, Graduate School, School of Professional Studies, Department of Engineering, Hammond, IN 46323-2094. Offers MSE. Evening/weekend programs available. *Entrance requirements:* For master's, TOEFL.

Engineering and Applied Sciences—General

Queen's University at Kingston, School of Graduate Studies and Research, Faculty of Applied Science, Kingston, ON K7L 3N6, Canada. Offers M Sc, M Sc Eng, PhD. Part-time programs available. *Students:* 232 full-time (41 women), 49 part-time (6 women). In 1998, 71 master's, 25 doctorates awarded. *Degree requirements:* For doctorate, dissertation, comprehensive exam required, foreign language not required. *Entrance requirements:* For master's and doctorate, TOEFL. *Application deadline:* For fall admission, 2/28 (priority date). Applications are processed on a rolling basis. Application fee: $60. Electronic applications accepted. *Financial aid:* Fellowships, research assistantships, teaching assistantships, institutionally-sponsored loans available. Financial aid application deadline: 3/1. *Unit head:* T. J. Harris, Dean, 613-533-2055. *Application contact:* Jane Kalin, Registrar, 613-533-6100, Fax: 613-533-6015.

Rensselaer Polytechnic Institute, Graduate School, School of Engineering, Troy, NY 12180-3590. Offers M Eng, MS, D Eng, PhD, MBA/M Eng. Part-time and evening/weekend programs available. Postbaccalaureate distance learning degree programs offered (no on-campus study). *Faculty:* 146 full-time (7 women), 57 part-time (2 women). *Students:* 646 full-time (106 women), 198 part-time (26 women); includes 103 minority (9 African Americans, 62 Asian Americans or Pacific Islanders, 32 Hispanic Americans), 373 international. 1993 applicants, 35% accepted. In 1998, 255 master's, 76 doctorates awarded. Terminal master's awarded for partial completion of doctoral program. *Degree requirements:* For doctorate, dissertation required, foreign language not required, foreign language not required. *Entrance requirements:* For master's and doctorate, GRE, TOEFL (minimum score of 550 required). *Application deadline:* For fall admission, 2/1 (priority date). Applications are processed on a rolling basis. Application fee: $35. *Financial aid:* Fellowships with full tuition reimbursements, research assistantships with full and partial tuition reimbursements, teaching assistantships with full and partial tuition reimbursements, career-related internships or fieldwork, institutionally-sponsored loans, scholarships, and tuition waivers (full and partial) available. Financial aid application deadline: 2/1. *Faculty research:* Data mining, manufacturing, materials, computational mechanics and modeling, infrastructure engineering, electronics technology, composites, multiphase research. Total annual research expenditures: $44.1 million. *Unit head:* Dr. James M. Tien, Acting Dean, 518-276-6298, Fax: 518-276-8788, E-mail: federp@rpi.edu. *Application contact:* Gail Gere, Director, Graduate Academic and Enrollment Services, 518-276-6789, Fax: 518-276-8433, E-mail: grad-services@rpi.edu.

See in-depth description on page 155.

Rice University, Graduate Programs, George R. Brown School of Engineering, Houston, TX 77251-1892. Offers M Ch E, M Stat, MA, MAM Sc, MCE, MCS, MEE, MEE, MES, MME, MMS, MS, PhD, MBA/M Eng, MD/PhD. MD/PhD offered jointly with Baylor College of Medicine or The University of Texas–Houston Health Science Center. Part-time programs available. Terminal master's awarded for partial completion of doctoral program. *Degree requirements:* For doctorate, dissertation required, foreign language not required. *Entrance requirements:* For master's and doctorate, GRE General Test, GRE Subject Test, TOEFL (minimum score of 550 required), minimum GPA of 3.0.

Rochester Institute of Technology, Part-time and Graduate Admissions, College of Engineering, Rochester, NY 14623-5604. Offers ME, MS, MSEE, MSME, AC. Part-time and evening/weekend programs available. *Students:* 43 full-time (10 women), 147 part-time (24 women); includes 23 minority (10 African Americans, 7 Asian Americans or Pacific Islanders, 6 Hispanic Americans), 31 international. 329 applicants, 64% accepted. In 1998, 107 master's, 2 other advanced degrees awarded. *Degree requirements:* Foreign language not required. *Entrance requirements:* For master's, TOEFL, minimum GPA of 3.0. *Application deadline:* For fall admission, 3/1 (priority date). Applications are processed on a rolling basis. Application fee: $40. *Financial aid:* Fellowships, research assistantships, teaching assistantships, career-related internships or fieldwork, Federal Work-Study, institutionally-sponsored loans, and tuition waivers (partial) available. Aid available to part-time students. *Faculty research:* Microprocessors, energy, communication systems. *Unit head:* Dr. Paul Petersen, Dean, 716-475-2146. *Application contact:* Dr. Richard Reeve, Associate Dean, 716-475-7048, E-mail: nrreie@rit.edu.

See in-depth description on page 157.

Rose-Hulman Institute of Technology, Faculty of Engineering and Applied Sciences, Terre Haute, IN 47803-3920. Offers MS, MD/MS. Part-time programs available. Postbaccalaureate distance learning degree programs offered (minimal on-campus study). *Faculty:* 73 full-time (6 women). *Students:* 74 full-time (21 women), 58 part-time (2 women); includes 1 minority (African American), 51 international. Average age 28. 86 applicants, 45% accepted. In 1998, 32 degrees awarded. *Degree requirements:* Foreign language not required. *Entrance requirements:* For master's, GRE, TOEFL (minimum score of 580 required), minimum GPA of 3.0. *Average time to degree:* Master's–2 years full-time, 5 years part-time. *Application deadline:* For fall admission, 2/1 (priority date). Applications are processed on a rolling basis. Application fee: $0. Tuition: Full-time $19,305; part-time $540 per credit hour. Required fees: $800. *Financial aid:* In 1998–99, 50 students received aid, including 43 fellowships with full and partial tuition reimbursements available (averaging $6,000 per year); research assistantships, teaching assistantships, grants, institutionally-sponsored loans, and tuition waivers (full and partial) also available. Financial aid application deadline: 2/1. *Faculty research:* Optics, electromagnetics, robotics and controls, digital electronics, polymers. Total annual research expenditures: $3.5 million. *Unit head:* Dr. Buck F. Brown, Dean for Research and Graduate Studies, 812-877-8403, Fax: 812-877-8102, E-mail: buck.brown@rose-hulman.edu.

Announcement: Graduate programs are available leading to the MS degree in applied optics and engineering management, as well as biomedical, chemical, civil, electrical, environmental, and mechanical engineering. These programs have areas of emphasis in both design and research and offer the graduate excellent preparation either for industry or for further graduate study.

See in-depth description on page 159.

Rowan University, Graduate Studies, School of Engineering, Glassboro, NJ 08028-1701. Offers MS. 6 applicants, 50% accepted. In 1998, 1 degree awarded. *Application deadline:* For fall admission, 11/1 (priority date); for spring admission, 4/1. Applications are processed on a rolling basis. Application fee: $50. Tuition, state resident: full-time $5,051; part-time $281 per semester hour. Tuition, nonresident: full-time $7,715; part-time $429 per semester hour. Tuition and fees vary according to degree level. *Unit head:* Dr. James Tracey, Dean, 609-256-4670. *Application contact:* Dr. T. R. Chandrupata, Program Adviser, 609-256-4632.

Rutgers, The State University of New Jersey, New Brunswick, Programs in Engineering, New Brunswick, NJ 08903. Offers MS, PhD. Degrees offered through the Graduate School.

Announcement: Graduate programs are offered in biomedical engineering, bioresource engineering, ceramic and materials science and engineering, chemical and biochemical engineering, civil and environmental engineering, electrical and computer engineering, industrial and systems engineering, mechanical and aerospace engineering, and mechanics.

Saginaw Valley State University, College of Science, Engineering, and Technology, University Center, MI 48710. Offers technological processes (MS). Part-time and evening/weekend programs available. *Faculty:* 7 full-time (0 women), 5 part-time (1 woman). *Students:* 3 full-time (0 women), 29 part-time (5 women); includes 4 minority (3 African Americans, 1 Asian American or Pacific Islander), 1 international. 55 applicants, 87% accepted. *Entrance requirements:* For master's, TOEFL (minimum score of 525 required), minimum GPA of 3.0. *Application deadline:* Applications are processed on a rolling basis. Application fee: $25. *Financial aid:* In 1998–99, 1 fellowship with partial tuition reimbursement, 1 research assistantship with full tuition reimbursement (averaging $2,500 per year) were awarded.; Federal Work-Study also available. Aid available to part-time students. Financial aid application deadline: 4/1; financial aid applicants required to submit FAFSA. *Unit head:* Dr. Thomas Kullgren, Dean, 517-790-4144, Fax: 517-790-2717, E-mail: kullgren@svsu.edu. *Application contact:* Wynn P. McDonald, Director, Graduate Admissions, 517-249-1696, Fax: 517-790-0180, E-mail: gradadm@svsu.edu.

St. Cloud State University, School of Graduate Studies, College of Science and Engineering, St. Cloud, MN 56301-4498. Offers MA, MS. *Faculty:* 59 full-time (19 women), 2 part-time (0 women). *Students:* 13 full-time (8 women), 22 part-time (8 women). In 1998, 18 degrees awarded. *Degree requirements:* For master's, thesis or alternative required, foreign language not required. *Entrance requirements:* For master's, GRE General Test, minimum GPA of 2.75. Application fee: $20. *Financial aid:* Federal Work-Study and unspecified assistantships available. Financial aid application deadline: 3/1. *Unit head:* Dr. A. I. Musah, Dean, 320-255-3909, Fax: 320-255-4262, E-mail: cose@stcloudstate.edu. *Application contact:* Ann Anderson, Graduate Studies Office, 320-255-2113, Fax: 320-654-5371, E-mail: aeanderson@stcloudstate.edu.

St. Mary's University of San Antonio, Graduate School, Department of Engineering, San Antonio, TX 78228-8507. Offers electrical engineering (MS); electrical/computer engineering (MS); engineering administration (MS); engineering computer application (MS); industrial engineering (MS); operations research (MS). Part-time and evening/weekend programs available. *Students:* 3 full-time (2 women), 49 part-time (6 women); includes 10 minority (2 African Americans, 8 Hispanic Americans), 8 international. Average age 25. In 1998, 15 degrees awarded. *Degree requirements:* For master's, computer language, thesis required, foreign language not required. *Entrance requirements:* For master's, GRE General Test. *Application deadline:* For fall admission, 8/1. Application fee: $15. *Financial aid:* Teaching assistantships, Federal Work-Study available. *Faculty research:* Image processing, control, communication, artificial intelligence, robotics. *Unit head:* Dr. Abe Yazdani, Adviser, 210-436-3305.

San Diego State University, Graduate and Research Affairs, College of Engineering, San Diego, CA 92182. Offers MS, PhD. Part-time and evening/weekend programs available. *Students:* 56 full-time (5 women), 143 part-time (20 women); includes 70 minority (9 African Americans, 32 Asian Americans or Pacific Islanders, 29 Hispanic Americans), 27 international. Average age 29. In 1998, 38 master's, 2 doctorates awarded. Terminal master's awarded for partial completion of doctoral program. *Degree requirements:* For doctorate, dissertation required, foreign language not required. *Entrance requirements:* For master's, GRE General Test (minimum combined score of 950 required), TOEFL (minimum score of 550 required). *Application deadline:* Applications are processed on a rolling basis. Application fee: $55. *Financial aid:* Fellowships, research assistantships, teaching assistantships, career-related internships or fieldwork and Federal Work-Study available. Aid available to part-time students. Total annual research expenditures: $1.3 million. *Unit head:* Pieter A. Frick, Dean, 619-594-6061, Fax: 619-594-6005, E-mail: pieter.frick@sdsu.edu.

San Francisco State University, Graduate Division, College of Science and Engineering, School of Engineering, San Francisco, CA 94132-1722. Offers MS. *Faculty:* 19 full-time (2 women), 6 part-time (0 women). *Students:* 60 (11 women). *Entrance requirements:* For master's, minimum GPA of 2.5 in last 60 units. *Application deadline:* For fall admission, 11/30 (priority date). Applications are processed on a rolling basis. Application fee: $55. *Financial aid:* Federal Work-Study available. Financial aid application deadline: 3/1. *Faculty research:* Signal processing, control systems, rehabilitation technology, computer engineering, power systems. Total annual research expenditures: $400,000. *Unit head:* Dr. Zorica Pantic-Tanner, Director, 415-338-1228, E-mail: zpt@sfsu.edu. *Application contact:* Dr. Hamid Shahnasser, Graduate Coordinator, 415-338-2124, E-mail: hamid@sfsu.edu.

San Jose State University, Graduate Studies, College of Engineering, San Jose, CA 95192-0001. Offers MS. Part-time programs available. *Faculty:* 60 full-time (5 women), 82 part-time (7 women). *Students:* 245 full-time (101 women), 569 part-time (110 women); includes 475 minority (20 African Americans, 427 Asian Americans or Pacific Islanders, 28 Hispanic Americans), 131 international. Average age 30. 722 applicants, 60% accepted. In 1998, 243 degrees awarded. *Application deadline:* For fall admission, 6/1. Applications are processed on a rolling basis. Application fee: $59. Tuition, nonresident: part-time $246 per unit. Required fees: $1,939; $1,309 per year. *Financial aid:* Teaching assistantships, career-related internships or fieldwork, Federal Work-Study, and institutionally-sponsored loans available. Aid available to part-time students. *Unit head:* Dr. Don Kirk, Dean, 408-924-3800, Fax: 408-924-3818.

See in-depth description on page 161.

Santa Clara University, School of Engineering, Santa Clara, CA 95053-0001. Offers MSAM, MSCE, MSCSE, MSE, MSE Mgt, MSEE, MSME, PhD, Certificate, Engineer. Part-time and evening/weekend programs available. *Faculty:* 36 full-time (8 women), 52 part-time (7 women). *Students:* 192 full-time (87 women), 630 part-time (160 women); includes 264 minority (7 African Americans, 235 Asian Americans or Pacific Islanders, 21 Hispanic Americans, 1 Native American), 312 international. Average age 35. 596 applicants, 73% accepted. In 1998, 272 master's, 2 doctorates, 3 other advanced degrees awarded. *Degree requirements:* For master's, thesis or alternative required, foreign language not required; for doctorate and other advanced degree, dissertation required, foreign language not required. *Entrance requirements:* For master's, GRE General Test (combined average 1600 on three sections), TOEFL (minimum score of 550 required), minimum GPA of 2.75; for doctorate, GRE General Test (combined average 1600 on three sections), TOEFL (minimum score of 550 required), master's degree or equivalent; for other advanced degree, master's degree, published paper. *Application deadline:* For fall admission, 6/1; for spring admission, 1/1. Applications are processed on a rolling basis. Application fee: $40. *Financial aid:* Fellowships, research assistantships, teaching assistantships, career-related internships or fieldwork, Federal Work-Study, institutionally-sponsored loans, and scholarships available. Aid available to part-time students. Financial aid application deadline: 2/1; financial aid applicants required to submit CSS PROFILE or FAFSA. Total annual research expenditures: $406,047. *Unit head:* Dr. Terry E. Shoup, Dean, 408-554-4600. *Application contact:* Tina Samms, Assistant Director of Graduate Admissions, 408-554-4313, Fax: 408-554-5474, E-mail: engr-grad@scu.edu.

See in-depth description on page 163.

Seattle University, School of Science and Engineering, Seattle, WA 98122. Offers MSE. Part-time and evening/weekend programs available. *Faculty:* 8 full-time (2 women), 4 part-time (1 woman). *Students:* 9 full-time (4 women), 65 part-time (13 women); includes 19 minority (5 African Americans, 11 Asian Americans or Pacific Islanders, 1 Hispanic American, 2 Native Americans), 10 international. Average age 34. 37 applicants, 95% accepted. In 1998, 28 degrees awarded. *Degree requirements:* For master's, computer language, thesis or alternative required, foreign language not required. *Entrance requirements:* For master's, GRE General Test, 2 years of related work experience. *Average time to degree:* Master's–2 years full-time, 3 years part-time. *Application deadline:* For fall admission, 5/1 (priority date). Application fee: $55. *Financial aid:* Career-related internships or fieldwork and Federal Work-Study available. Aid available to part-time students. Financial aid applicants required to submit FAFSA. *Unit head:* Dr. George Simmons, Dean, 206-296-5500, Fax: 206-296-2071.

Simon Fraser University, Graduate Studies, Faculty of Applied Science, School of Engineering Science, Burnaby, BC V5A 1S6, Canada. Offers M Eng, MA Sc, PhD. *Faculty:* 21 full-time (1 woman). *Students:* 88 full-time (12 women), 32 part-time (5 women). Average age 28. In 1998, 27 master's, 4 doctorates awarded. *Degree requirements:* For master's, thesis required (for some programs); for doctorate, dissertation, qualifying exam required. *Entrance requirements:* For master's, TOEFL (minimum score of 570 required), TWE (minimum score of 5 required), or International English Language Test (minimum score of 7.5 required), minimum GPA of 3.0; for doctorate, TOEFL (minimum score of 570 required), TWE (minimum score of 5 required), or International English Language Test (minimum score of 7.5 required), minimum GPA of 3.5. Application fee: $55. *Financial aid:* In 1998–99, 33 fellowships were awarded.; research assistantships, teaching assistantships *Faculty research:* Signal processing, electronics, communications. *Unit head:* A. Leung, Director, 604-291-4371, Fax: 604-291-4951. *Application contact:* Graduate Secretary, 604-291-4923, Fax: 604-291-4951, E-mail: rabold@sfu.ca.

South Dakota State University, Graduate School, College of Engineering, Brookings, SD 57007. Offers MS, PhD. Part-time programs available. *Degree requirements:* For master's, thesis, oral exam required, foreign language not required; for doctorate, dissertation, preliminary oral and written exams required. *Entrance requirements:* For master's and doctorate, TOEFL. *Faculty research:* Process control and management, ground source heat pumps, water quality, heat transfer, power systems.

Southern Illinois University Carbondale, Graduate School, College of Engineering, Carbondale, IL 62901-6806. Offers MS, PhD. *Faculty:* 64 full-time (3 women), 2 part-time (0 women). *Students:* 149 full-time (15 women), 38 part-time (8 women). 257 applicants, 51% accepted. In 1998, 43 master's, 3 doctorates awarded. *Degree requirements:* For master's, comprehensive exam required; for doctorate, dissertation required. *Entrance requirements:* For master's, TOEFL (minimum score of 550 required), minimum GPA of 2.7; for doctorate, GRE General Test, TOEFL (minimum score of 600 required), minimum GPA of 3.5. *Application deadline:* Applications are processed on a rolling basis. Application fee: $20. *Financial aid:* In 1998–99, 112 students received aid, including 1 fellowship, 58 research assistantships, 95 teaching assistantships; Federal Work-Study, institutionally-sponsored loans, and tuition waivers (full) also available. Aid available to part-time students. *Faculty research:* Electrical systems, all facets of fossil energy, mechanics. *Unit head:* Dr. Sedat Sami, Chair, 618-536-2368. *Application contact:* Dr. James Evers, Associate Dean of Research and Graduate Programs, 618-453-4321, Fax: 618-453-4235, E-mail: evers@sysa.c_engri.siu.edu.

See in-depth description on page 165.

Southern Illinois University Edwardsville, Graduate Studies and Research, School of Engineering, Edwardsville, IL 62026-0001. Offers MS. Part-time programs available. *Faculty:* 33 full-time (1 woman), 9 part-time (1 woman). *Students:* 98 full-time (26 women), 104 part-time (21 women); includes 20 minority (5 African Americans, 14 Asian Americans or Pacific Islanders, 1 Hispanic American), 107 international. 315 applicants, 40% accepted. In 1998, 44 degrees awarded. *Degree requirements:* For master's, thesis or research paper, final exam required. *Entrance requirements:* For master's, TOEFL (minimum score of 550 required). *Application deadline:* For fall admission, 7/24. Application fee: $25. *Financial aid:* In 1998–99, 11 research assistantships with full tuition reimbursements, 12 teaching assistantships with full tuition reimbursements were awarded.; fellowships with full tuition reimbursements, career-related internships or fieldwork, Federal Work-Study, institutionally-sponsored loans, scholarships, traineeships, and unspecified assistantships also available. Aid available to part-time students. *Unit head:* Dr. Harlan Bengtson, Dean, 618-650-2861, E-mail: hbengts@siue.edu. *Application contact:* Dr. Harlan Bengtson, Dean, 618-650-2861, E-mail: hbengts@siue.edu.

Southern Methodist University, School of Engineering and Applied Science, Dallas, TX 75275. Offers MS, MS Cp E, MSEE, MSEM, MSME, DE, PhD. Part-time programs available. Postbaccalaureate distance learning degree programs offered (no on-campus study). *Faculty:* 44 full-time (3 women), 32 part-time (2 women). *Students:* 132 full-time (36 women), 806 part-time (148 women); includes 230 minority (52 African Americans, 126 Asian Americans or Pacific Islanders, 50 Hispanic Americans, 2 Native Americans), 164 international. Average age 34. 590 applicants, 46% accepted. In 1998, 238 master's, 18 doctorates awarded. Terminal master's awarded for partial completion of doctoral program. *Degree requirements:* For master's, thesis optional, foreign language not required; for doctorate, dissertation, oral and written qualifying exams required. *Entrance requirements:* For master's, GRE General Test (minimum score of 650 on quantitative section required), TOEFL (minimum score of 550 required), minimum GPA of 3.0 in last 2 years; bachelor's degree in engineering, mathematics, or sciences; for doctorate, bachelor's degree in related field. *Application deadline:* For fall admission, 8/1; for spring admission, 12/15. Applications are processed on a rolling basis. Application fee: $25. Tuition: Full-time $9,216; part-time $512 per credit hour. Required fees: $88 per credit hour. Part-time tuition and fees vary according to course load and campus/location. *Financial aid:* Fellowships, research assistantships, teaching assistantships, career-related internships or fieldwork, Federal Work-Study, institutionally-sponsored loans, scholarships, and tuition waivers (full and partial) available. Financial aid applicants required to submit FAFSA. *Faculty research:* Mobile and fault-tolerant computing, manufacturing systems, telecommunications, solid state devices and materials, fluid and thermal sciences. Total annual research expenditures: $3 million. *Unit head:* Dr. Andre G. Vacroux, Dean, 214-768-3051, Fax: 214-768-3845, E-mail: vacroux@seas.smu.edu. *Application contact:* Dr. Zeynep Celik-Butler, Assistant Dean for Graduate Studies and Research, 214-768-3979, Fax: 214-768-3845, E-mail: zcb@seas.smu.edu.

Southern Polytechnic State University, College of Technology, Marietta, GA 30060-2896. Offers MS. Part-time and evening/weekend programs available. Postbaccalaureate distance learning degree programs offered. In 1998, 47 degrees awarded. *Degree requirements:* Foreign language not required. *Application deadline:* For fall admission, 7/15 (priority date); for spring admission, 12/1. Applications are processed on a rolling basis. Tuition, state resident: full-time $2,146; part-time $119 per credit hour. Tuition, nonresident: full-time $7,586; part-time $421 per credit hour. *Financial aid:* Teaching assistantships, career-related internships or fieldwork and Federal Work-Study available. Aid available to part-time students. Financial aid application deadline: 5/1; financial aid applicants required to submit FAFSA. *Unit head:* Graduate Admissions, 608-262-2433, Fax: 608-262-5134, E-mail: gradadmiss@mail.bascom.wisc.edu. *Application contact:* Graduate Admissions, 608-262-2433, Fax: 608-262-5134, E-mail: gradadmiss@mail.bascom.wisc.edu.

Stanford University, School of Engineering, Stanford, CA 94305-9991. Offers MS, PhD, Eng, MBA/MS. *Faculty:* 209 full-time (13 women). *Students:* 2,283 full-time (447 women), 645 part-time (117 women); includes 610 minority (54 African Americans, 444 Asian Americans or Pacific Islanders, 105 Hispanic Americans, 7 Native Americans), 1,195 international. Average age 27. 3907 applicants, 46% accepted. In 1998, 937 master's, 205 doctorates awarded. *Degree requirements:* For doctorate and Eng, dissertation required. *Entrance requirements:* For master's, doctorate, and Eng, GRE General Test, TOEFL. *Average time to degree:* Doctorate–6 years full-time. Application fee: $65 ($80 for international students). Electronic applications accepted. Tuition: Full-time $24,588. Required fees: $152. Part-time tuition and fees vary according to course load. *Financial aid:* In 1998–99, 1,365 students received aid, including 436 fellowships, 1,113 research assistantships, 389 teaching assistantships; career-related internships or fieldwork, Federal Work-Study, and institutionally-sponsored loans also available. Financial aid applicants required to submit FAFSA. *Unit head:* John L. Hennessy, Dean, 650-723-3938, Fax: 650-723-8545, E-mail: jlh@vsop.stanford.edu. *Application contact:* Graduate Admissions Office, 650-723-4291.

State University of New York at Binghamton, Graduate School, Thomas J. Watson School of Engineering and Applied Science, Binghamton, NY 13902-6000. Offers M Eng, MS, MSAT, PhD. Part-time and evening/weekend programs available. *Faculty:* 47 full-time, 25 part-time. *Students:* 209 full-time (38 women), 159 part-time (33 women); includes 43 minority (13 African Americans, 23 Asian Americans or Pacific Islanders, 6 Hispanic Americans, 1 Native American), 169 international. Average age 30. 532 applicants, 56% accepted. In 1998, 104 master's, 15 doctorates awarded. Terminal master's awarded for partial completion of doctoral program. *Degree requirements:* For doctorate, dissertation required, foreign language not required, foreign language not required. *Entrance requirements:* For master's and doctorate, GRE General Test, GRE Subject Test, TOEFL (minimum score of 550 required). *Application deadline:* For fall admission, 4/15 (priority date); for spring admission, 11/1. Applications are processed on a rolling basis. Application fee: $50. Electronic applications accepted. Tuition, state resident: full-time $5,100; part-time $213 per credit. Tuition, nonresident: full-time $8,416; part-time $351 per credit. Required fees: $77 per credit. Part-time tuition and fees vary according to course load. *Financial aid:* In 1998–99, 168 students received aid, including 3 fellowships with full tuition reimbursements available (averaging $10,767 per year), 81 research assistantships with full tuition reimbursements available (averaging $8,925 per year), 72 teaching assistantships with full tuition reimbursements available (averaging $8,400 per year); career-related internships or fieldwork, Federal Work-Study, institutionally-sponsored loans, and unspecified assistantships also available. Aid available to part-time students. Financial aid application deadline: 2/15. *Unit head:* Dr. Lyle D. Feisel, Dean, 607-777-2871.

See in-depth description on page 167.

State University of New York at Buffalo, Graduate School, School of Engineering and Applied Sciences, Buffalo, NY 14260. Offers M Eng, MS, PhD. Part-time programs available. Postbaccalaureate distance learning degree programs offered (minimal on-campus study). *Faculty:* 88 full-time (4 women), 22 part-time (1 woman). *Students:* 312 full-time (49 women), 355 part-time (36 women); includes 46 minority (11 African Americans, 33 Asian Americans or Pacific Islanders, 1 Hispanic American, 1 Native American), 361 international. Average age 25. 1464 applicants, 59% accepted. In 1998, 176 master's, 41 doctorates awarded. Terminal master's awarded for partial completion of doctoral program. *Degree requirements:* For doctorate, dissertation required, foreign language not required, foreign language not required. *Entrance requirements:* For master's and doctorate, GRE General Test, TOEFL (minimum score of 550 required). *Application deadline:* Applications are processed on a rolling basis. Application fee: $35. Electronic applications accepted. Tuition, state resident: full-time $5,100; part-time $213 per credit hour. Tuition, nonresident: full-time $8,416; part-time $351 per credit hour. Required fees: $870; $75 per semester. Tuition and fees vary according to course load and program. *Financial aid:* In 1998–99, 22 fellowships with full tuition reimbursements (averaging $16,000 per year), 160 research assistantships with full and partial tuition reimbursements (averaging $12,000 per year), 109 teaching assistantships with full tuition reimbursements (averaging $10,500 per year) were awarded.; career-related internships or fieldwork, Federal Work-Study, grants, institutionally-sponsored loans, scholarships, tuition waivers (full and partial), and unspecified assistantships also available. Aid available to part-time students. Financial aid applicants required to submit FAFSA. *Faculty research:* Biochemical engineering, earthquake engineering, semiconductors, ergonomics, systems design. Total annual research expenditures: $5.4 million. *Unit head:* Dr. Mark Karwan, Dean, 716-645-2771, Fax: 716-645-2495. *Application contact:* Dr. Andres Soom, Associate Dean, 716-645-2772, Fax: 716-645-2495, E-mail: soom@acsu.buffalo.edu.

See in-depth description on page 169.

State University of New York at New Paltz, Graduate School, Faculty of Engineering and Business, Department of Engineering, New Paltz, NY 12561. Offers MS. *Students:* 3 full-time (0 women), 3 part-time; includes 2 minority (both Asian Americans or Pacific Islanders), 1 international. *Entrance requirements:* For master's, GRE General Test, minimum GPA of 3.0. *Application deadline:* For fall admission, 3/15 (priority date). Applications are processed on a rolling basis. Application fee: $50. *Unit head:* Dr. Owen Hill, Director, 914-257-3720.

State University of New York at Stony Brook, Graduate School, College of Engineering and Applied Sciences, Stony Brook, NY 11794. Offers MS, PhD, Certificate. Part-time and evening/weekend programs available. *Faculty:* 125 full-time (10 women), 39 part-time (10 women). *Students:* 429 full-time (138 women), 246 part-time (57 women); includes 103 minority (23 African Americans, 71 Asian Americans or Pacific Islanders, 9 Hispanic Americans), 359 international. 1278 applicants, 52% accepted. In 1998, 197 master's, 47 doctorates, 24 other advanced degrees awarded. *Degree requirements:* For doctorate, dissertation, comprehensive exams required, foreign language not required. *Entrance requirements:* For master's, TOEFL; for doctorate, GRE General Test, TOEFL. *Application deadline:* For fall admission, 1/15. Application fee: $50. *Financial aid:* In 1998–99, 16 fellowships, 122 research assistantships, 147 teaching assistantships were awarded.; career-related internships or fieldwork also available. Total annual research expenditures: $14.4 million. *Unit head:* Dr. Yacov Shamash, Dean, 516-632-8380.

State University of New York Institute of Technology at Utica/Rome, School of Information Systems and Engineering Technology, Utica, NY 13504-3050. Offers advanced technology (MS); computer and information science (MS); telecommunications (MS). Part-time and evening/weekend programs available. *Faculty:* 20 full-time (2 women). *Students:* 19 full-time (4 women), 84 part-time (15 women); includes 9 minority (2 African Americans, 4 Asian Americans or Pacific Islanders, 3 Hispanic Americans), 4 international. Average age 35. 52 applicants, 81% accepted. In 1998, 11 degrees awarded (100% found work related to degree). *Degree requirements:* For master's, computer language required, foreign language not required. *Entrance requirements:* For master's, GRE General Test, TOEFL (minimum score of 550 required), minimum GPA of 3.0. *Average time to degree:* Master's–2 years full-time, 4 years part-time. *Application deadline:* For fall admission, 6/15 (priority date). Applications are processed on a rolling basis. Application fee: $50. *Financial aid:* In 1998–99, 47 students received aid, including 3 fellowships with full tuition reimbursements available (averaging $4,900 per year), 5 research assistantships with full tuition reimbursements available; career-related internships or fieldwork, Federal Work-Study, and unspecified assistantships also available. Aid available to part-time students. Financial aid applicants required to submit FAFSA. *Faculty research:* Artificial intelligence, multimedia, parallel processing, databases and data fusion, networks. *Unit head:* Dr. Rosemary Mullick, Acting Dean, 315-792-7234, Fax: 315-792-7800, E-mail: rosemary@sunyit.edu. *Application contact:* Marybeth Lyons, Director of Admissions, 315-792-7500, Fax: 315-792-7837, E-mail: smbl@sunyit.edu.

Stevens Institute of Technology, Graduate School, Charles V. Schaefer Jr. School of Engineering, Hoboken, NJ 07030. Offers M Eng, MS, PhD, Certificate, Engr. Part-time and evening/weekend programs available. Postbaccalaureate distance learning degree programs offered. Terminal master's awarded for partial completion of doctoral program. *Degree requirements:* For master's and other advanced degree, computer language required, foreign language not required; for doctorate, computer language, dissertation required. *Entrance requirements:* For master's and doctorate, TOEFL. Electronic applications accepted.

See in-depth description on page 171.

Stevens Institute of Technology, Graduate School, Program in Interdisciplinary Sciences and Engineering, Hoboken, NJ 07030. Offers M Eng, MS, PhD. *Degree requirements:* For doctorate, dissertation required, foreign language not required. *Entrance requirements:* For master's and doctorate, TOEFL. Electronic applications accepted.

Stevens Institute of Technology, Graduate School, School of Applied Sciences and Liberal Arts, Hoboken, NJ 07030. Offers M Eng, MS, PhD, Certificate. Part-time and evening/weekend programs available. Terminal master's awarded for partial completion of doctoral program. *Degree requirements:* For doctorate, dissertation required, foreign language not required. *Entrance requirements:* For master's and doctorate, TOEFL. Electronic applications accepted.

See in-depth description on page 173.

Syracuse University, Graduate School, L. C. Smith College of Engineering and Computer Science, Syracuse, NY 13244-0003. Offers MS, PhD, CE, EE, JD/MS. Part-time and evening/weekend programs available. *Faculty:* 128 full-time, 20 part-time. *Students:* 300 full-time (57 women), 306 part-time (53 women); includes 32 minority (7 African Americans, 20 Asian Americans or Pacific Islanders, 5 Hispanic Americans), 354 international. Average age 30. 1331 applicants, 73% accepted. In 1998, 208 master's, 25 doctorates awarded. *Degree requirements:* For doctorate, computer language, dissertation required, foreign language not required, foreign language not required. *Entrance requirements:* For master's and doctorate, GRE General Test, GRE Subject Test. *Application deadline:* Applications are processed on a rolling basis. Application fee: $40. Tuition: Full-time $13,992; part-time $583 per credit hour. *Financial aid:* Fellowships, research assistantships, teaching assistantships, Federal Work-Study and tuition waivers (partial) available. Financial aid application deadline: 3/1. *Unit head:* Dr. Edward Bogucz, Dean, 315-443-4341.

See in-depth description on page 175.

Engineering and Applied Sciences—General

Temple University, Graduate School, College of Science and Technology, College of Engineering, Philadelphia, PA 19122-6096. Offers civil and environmental engineering (MSE); electrical and computer engineering (MSE); engineering (PhD); environmental health sciences (MS), including environmental health; mechanical engineering (MSE). Part-time programs available. *Faculty:* 25 full-time (1 woman). *Students:* 84 (16 women); includes 29 minority (5 African Americans, 20 Asian Americans or Pacific Islanders, 4 Hispanic Americans) 10 international. 178 applicants, 32% accepted. In 1998, 35 master's, 4 doctorates awarded. *Degree requirements:* For doctorate, dissertation, 2 published papers, comprehensive exam required, foreign language not required, foreign language not required. *Entrance requirements:* For master's, GRE General Test, TOEFL (minimum score of 575 required); for doctorate, GRE General Test, TOEFL, minimum graduate GPA of 3.5, MS. Application fee: $40. *Financial aid:* In 1998–99, 30 students received aid, including 24 research assistantships, 6 teaching assistantships; fellowships, career-related internships or fieldwork, Federal Work-Study, and institutionally-sponsored loans also available. Financial aid application deadline: 2/15. *Faculty research:* Artificial intelligence, adaptive control, transportation, materials, vacuum microelectronics. Total annual research expenditures: $500,000. *Unit head:* Dr. Saroj Biswas, Director of Graduate Studies, 215-204-8403, Fax: 215-204-6936.

Tennessee State University, Graduate School, College of Engineering and Technology, Nashville, TN 37209-1561. Offers ME. Part-time and evening/weekend programs available. *Faculty:* 20 full-time (0 women), 5 part-time (0 women). *Students:* 43 full-time (9 women), 31 part-time (10 women); includes 24 minority (17 African Americans, 7 Asian Americans or Pacific Islanders), 36 international. Average age 27. 101 applicants, 83% accepted. In 1998, 22 degrees awarded (90% found work related to degree, 10% continued full-time study). *Degree requirements:* For master's, project required. *Application deadline:* Applications are processed on a rolling basis. Application fee: $15. Electronic applications accepted. Tuition, state resident: full-time $2,962; part-time $182 per credit hour. Tuition, nonresident: full-time $7,788; part-time $393 per credit hour. *Financial aid:* In 1998–99, 7 research assistantships (averaging $13,646 per year), 6 teaching assistantships (averaging $16,665 per year) were awarded.; fellowships Financial aid application deadline: 4/30. *Faculty research:* Intelligence/robotics, computational fluid dynamics, design methodologies, intelligent manufacturing, neural networks. Total annual research expenditures: $1.2 million. *Unit head:* Dr. Decatur B. Rogers, Dean, 615-963-5409, Fax: 615-963-5397. *Application contact:* Dr. Mohan J. Malkani, Associate Dean, 615-963-5400, Fax: 615-963-5397, E-mail: malkani@harpo.tnstate.edu.

Tennessee Technological University, Graduate School, College of Engineering, Cookeville, TN 38505. Offers MS, PhD. Part-time programs available. *Faculty:* 76 full-time (2 women). *Students:* 135 full-time (16 women), 25 part-time (4 women); includes 109 minority (1 African American, 108 Asian Americans or Pacific Islanders) Average age 28. 730 applicants, 62% accepted. In 1998, 48 master's, 3 doctorates awarded. *Degree requirements:* For master's, thesis required, foreign language not required; for doctorate, one foreign language (computer language can substitute), dissertation required. *Entrance requirements:* For master's, GRE General Test, TOEFL (minimum score of 525 required); for doctorate, GRE Subject Test, TOEFL (minimum score of 525 required), minimum GPA of 3.5. *Application deadline:* For fall admission, 3/1 (priority date); for spring admission, 8/1. Application fee: $25 ($30 for international students). Tuition, state resident: part-time $137 per hour. Tuition, nonresident: part-time $361 per hour. Required fees: $17 per hour. Tuition and fees vary according to course load. *Financial aid:* In 1998–99, 1 fellowship, 83 research assistantships (averaging $7,417 per year), 32 teaching assistantships (averaging $7,417 per year) were awarded.; career-related internships or fieldwork also available. Aid available to part-time students. Financial aid application deadline: 4/1. *Unit head:* Dr. Glen Johnson, Dean, 931-372-3172, Fax: 931-372-6172. *Application contact:* Dr. Rebecca F. Quattlebaum, Dean of the Graduate School, 931-372-3233, Fax: 931-372-3497, E-mail: rquattlebaum@tntech.edu.

See in-depth description on page 177.

Texas A&M University, College of Engineering, College Station, TX 77843. Offers M Agr, M Eng, MCE, MCS, MS, D Eng, DE, PhD. Part-time programs available. *Faculty:* 314 full-time (18 women), 35 part-time (6 women). *Students:* 1,648 (262 women); includes 118 minority (24 African Americans, 46 Asian Americans or Pacific Islanders, 48 Hispanic Americans) 1,005 international. 2413 applicants, 51% accepted. In 1998, 249 master's, 103 doctorates awarded. Terminal master's awarded for partial completion of doctoral program. *Entrance requirements:* For master's and doctorate, GRE General Test, TOEFL. Application fee: $50 ($75 for international students). *Financial aid:* Fellowships, research assistantships, teaching assistantships, career-related internships or fieldwork, institutionally-sponsored loans, and scholarships available. Financial aid applicants required to submit FAFSA. Total annual research expenditures: $51.3 million. *Unit head:* C. Roland Haden, Dean, 409-845-7203.

See in-depth description on page 179.

Texas A&M University–Kingsville, College of Graduate Studies, College of Engineering, Kingsville, TX 78363. Offers ME, MS. Part-time and evening/weekend programs available. *Faculty:* 22 full-time, 5 part-time. *Students:* 80 full-time (11 women), 96 part-time (7 women). In 1998, 56 degrees awarded. *Degree requirements:* For master's, computer language, comprehensive exam required. *Entrance requirements:* For master's, GRE General Test, TOEFL (minimum score of 525 required). *Application deadline:* For fall admission, 6/1; for spring admission, 11/15. Applications are processed on a rolling basis. Application fee: $15 ($25 for international students). Tuition, state resident: full-time $2,062. Tuition, nonresident: full-time $7,246. *Financial aid:* Fellowships, research assistantships, teaching assistantships, career-related internships or fieldwork, Federal Work-Study, institutionally-sponsored loans, tuition waivers (partial), and unspecified assistantships available. Aid available to part-time students. Financial aid application deadline: 5/15. *Unit head:* Dr. Phil V. Compton, Dean, 361-593-2001.

See in-depth description on page 181.

Texas Tech University, Graduate School, College of Engineering, Lubbock, TX 79409. Offers M Engr, MENVEGR, MS, MS Ch E, MS Pet E, MSCE, MSEE, MSETM, MSIE, MSME, PhD. Part-time programs available. *Faculty:* 87 full-time (5 women), 3 part-time (0 women). *Students:* 308 full-time (53 women), 105 part-time (14 women); includes 20 minority (2 African Americans, 7 Asian Americans or Pacific Islanders, 10 Hispanic Americans, 1 Native American), 246 international. Average age 29. 830 applicants, 42% accepted. In 1998, 130 master's, 28 doctorates awarded. *Degree requirements:* For master's, computer language, thesis required (for some programs), foreign language not required; for doctorate, computer language, dissertation required, foreign language not required. *Entrance requirements:* For master's, GRE General Test (minimum combined score of 1000 required; average 1182), minimum GPA of 3.0; for doctorate, GRE General Test (minimum combined score of 1000 required), minimum GPA of 3.0. *Application deadline:* For fall admission, 4/15 (priority date); for spring admission, 11/1 (priority date). Applications are processed on a rolling basis. Application fee: $25 ($50 for international students). Electronic applications accepted. *Financial aid:* In 1998–99, 215 students received aid, including 197 research assistantships (averaging $9,401 per year), 18 teaching assistantships (averaging $7,757 per year); fellowships, career-related internships or fieldwork, Federal Work-Study, and institutionally-sponsored loans also available. Aid available to part-time students. Financial aid application deadline: 5/15; financial aid applicants required to submit FAFSA. *Faculty research:* Fluid mechanics/wind engineering, alternative fuels, pulsed power for space, environmental engineering, human factors/performance in the workplace. Total annual research expenditures: $11.6 million. *Unit head:* Dr. William M. Marcy, Interim Dean, 806-742-2011, Fax: 806-742-3493.

Tufts University, Division of Graduate and Continuing Studies and Research, Graduate School of Arts and Sciences, College of Engineering, Medford, MA 02155. Offers ME, MS, MSEM, MSEM, PhD. Part-time programs available. *Faculty:* 64 full-time, 24 part-time. *Students:* 343 (114 women); includes 38 minority (9 African Americans, 19 Asian Americans or Pacific Islanders, 10 Hispanic Americans) 102 international. 363 applicants, 53% accepted. In 1998, 96 master's, 7 doctorates awarded. Terminal master's awarded for partial completion of doctoral program. *Degree requirements:* For doctorate, dissertation required, foreign language not required, foreign language not required. *Entrance requirements:* For master's, TOEFL (minimum score of 550 required); for doctorate, GRE General Test, TOEFL (minimum score of 550 required). *Application deadline:* Applications are processed on a rolling basis. Application fee: $50. *Financial aid:* Research assistantships with full and partial tuition reimbursements, teaching assistantships with full and partial tuition reimbursements, Federal Work-Study, scholarships, and tuition waivers (partial) available. Aid available to part-time students. Financial aid application deadline: 2/15; financial aid applicants required to submit FAFSA. *Unit head:* Dr. Ioannis Miaoulis, Dean, 617-627-3237.

See in-depth description on page 183.

Tulane University, School of Engineering, New Orleans, LA 70118-5669. Offers MS, MSCS, MSE, PhD, Sc D. MS and PhD offered through the Graduate School. Part-time programs available. *Faculty:* 55 full-time, 1 part-time. *Students:* 144 full-time (40 women), 16 part-time (7 women). 533 applicants, 22% accepted. In 1998, 38 master's, 13 doctorates awarded. *Degree requirements:* For doctorate, dissertation required. *Entrance requirements:* For master's and doctorate, GRE General Test, TOEFL, minimum B average in undergraduate course work. Application fee: $35. *Financial aid:* Fellowships, research assistantships, teaching assistantships, career-related internships or fieldwork, Federal Work-Study, institutionally-sponsored loans, and tuition waivers (full and partial) available. Financial aid application deadline: 2/1. *Unit head:* Dr. Paul Michael Lynch, Acting Dean, 504-865-5766. *Application contact:* Dr. E. Michaelides, Associate Dean, 504-865-5764.

Tuskegee University, Graduate Programs, College of Engineering, Architecture and Physical Sciences, Tuskegee, AL 36088. Offers MSEE, MSME, PhD. *Faculty:* 19 full-time (0 women). *Students:* 38 full-time (7 women), 19 part-time (2 women); includes 33 African Americans, 3 Asian Americans or Pacific Islanders, 14 international. Average age 24. 104 applicants, 59% accepted. In 1998, 14 degrees awarded. *Degree requirements:* For master's, computer language, thesis or alternative required, foreign language not required. *Entrance requirements:* For master's, GRE General Test, GRE Subject Test. *Application deadline:* For fall admission, 7/15. Applications are processed on a rolling basis. Application fee: $25 ($35 for international students). *Financial aid:* Fellowships, research assistantships, teaching assistantships, career-related internships or fieldwork, Federal Work-Study, and institutionally-sponsored loans available. Aid available to part-time students. Financial aid application deadline: 4/15. *Unit head:* Dr. Ben Oni, Acting Dean, 334-727-8356.

Union College, Graduate and Continuing Studies, Division of Engineering and Computer Science, Schenectady, NY 12308-2311. Offers MS. Part-time and evening/weekend programs available. *Students:* 8 full-time (3 women), 36 part-time (4 women); includes 6 minority (5 Asian Americans or Pacific Islanders, 1 Hispanic American), 4 international. 29 applicants, 97% accepted. In 1998, 20 degrees awarded. *Degree requirements:* For master's, computer language required. *Entrance requirements:* For master's, minimum GPA of 3.0. *Application deadline:* Applications are processed on a rolling basis. Application fee: $50. Tuition: Part-time $1,786 per course. *Financial aid:* Research assistantships available. Aid available to part-time students. *Unit head:* Dr. Robert T. Balmer, Dean, 518-388-6530.

Universidad de las Américas–Puebla, Division of Graduate Studies, School of Engineering, Cholula, 72820, Mexico. Offers M Adm, MS. Part-time and evening/weekend programs available. *Faculty:* 43 full-time (3 women), 9 part-time (1 woman). *Students:* 118 full-time (48 women), 52 part-time (21 women); all minorities (all Hispanic Americans) Average age 26. 80 applicants, 69% accepted. In 1998, 41 degrees awarded. *Degree requirements:* For master's, one foreign language, computer language, thesis required. *Average time to degree:* Master's–2.5 years full-time, 3.5 years part-time. *Application deadline:* For fall admission, 7/16. Applications are processed on a rolling basis. Application fee: $0. *Financial aid:* In 1998–99, 103 students received aid, including 22 research assistantships, 4 teaching assistantships Aid available to part-time students. Financial aid application deadline: 5/15. *Faculty research:* Artificial intelligence, food technology, construction, telecommunications, computers in education, operations research. Total annual research expenditures: $200,000. *Unit head:* Dr. José F. Tamborero, Dean, 22-29-20-32, Fax: 22-29-20-32, E-mail: jtambore@mail.udlap.mx. *Application contact:* Mauricio Villegas, Chair of Admissions Office, 22-29-20-17, Fax: 22-29-20-18, E-mail: admision@mail.udlap.mx.

Université de Moncton, School of Engineering, Moncton, NB E1A 3E9, Canada. Offers civil engineering (M Sc A); electrical engineering (M Sc A); industrial engineering (M Sc A); mechanical engineering (M Sc A). *Faculty:* 11 full-time (1 woman). *Students:* 9 full-time (2 women), 2 part-time, 2 international. Average age 25. 14 applicants, 86% accepted. In 1998, 4 degrees awarded (25% entered university research/teaching, 50% found other work related to degree, 25% continued full-time study). *Degree requirements:* For master's, thesis, proficiency in French required. *Average time to degree:* Master's–3.5 years full-time. *Application deadline:* For fall admission, 6/1 (priority date); for winter admission, 11/15 (priority date). Application fee: $30. *Financial aid:* In 1998–99, 8 students received aid, including fellowships (averaging $17,200 per year), research assistantships (averaging $650 per year), teaching assistantships (averaging $1,080 per year) Financial aid application deadline: 5/31. *Faculty research:* Structures, energy, composite materials, quality control, geo-environment, telecommunications, instrumentation, analog and digital electronics. Total annual research expenditures: $524,500. *Unit head:* Dr. Soumaya Yacout, Director, 506-858-4300, Fax: 506-858-4082, E-mail: yacouts@umoncton.ca.

Université de Sherbrooke, Faculty of Applied Sciences, Sherbrooke, PQ J1K 2R1, Canada. Offers M Env, M Sc A, PhD. *Degree requirements:* For master's and doctorate, thesis/dissertation required.

Université du Québec à Chicoutimi, Graduate Programs, Program in Engineering, Chicoutimi, PQ G7H 2B1, Canada. Offers M Sc A, PhD. Part-time programs available. *Degree requirements:* For master's and doctorate, thesis/dissertation required. *Entrance requirements:* For master's, appropriate bachelor's degree, proficiency in French.

Université Laval, Faculty of Graduate Studies, Faculty of Sciences and Engineering, Sainte-Foy, PQ G1K 7P4, Canada. Offers M Sc, PhD, Diploma. *Students:* 709 full-time (201 women), 179 part-time (38 women), 192 international. Average age 28. 580 applicants, 54% accepted. In 1998, 151 master's, 69 doctorates, 11 other advanced degrees awarded. *Application deadline:* For fall admission, 3/1. Application fee: $30. *Unit head:* Pierre Moreau, Dean, 418-656-2131 Ext. 2354, Fax: 418-656-5902, E-mail: pierre.moreau@fsg.ulaval.ca.

The University of Akron, Graduate School, College of Engineering, Akron, OH 44325-0001. Offers MS, MS Ch E, MSCE, MSE, MSEE, PhD, MD/PhD. Part-time and evening/weekend programs available. *Faculty:* 61 full-time, 17 part-time. *Students:* 275 (57 women). Average age 29. 327 applicants, 54% accepted. In 1998, 65 master's, 18 doctorates awarded. Terminal master's awarded for partial completion of doctoral program. *Degree requirements:* For doctorate, variable foreign language requirement (computer language can substitute for one), dissertation, candidacy exam, qualifying exam required, foreign language not required. *Entrance requirements:* For master's, TOEFL; for doctorate, GRE, TOEFL. *Average time to degree:* Master's–2 years full-time, 4 years part-time; doctorate–4 years full-time. *Application deadline:* Applications are processed on a rolling basis. Application fee: $25 ($50 for international students). Tuition, state resident: part-time $189 per credit. Tuition, nonresident: part-time $353 per credit. Required fees: $7.3 per credit. *Financial aid:* In 1998–99, 183 students received aid, including 1 fellowship with full tuition reimbursement available, 68 research assistantships with full tuition reimbursements available, 80 teaching assistantships with full tuition reimbursements available; career-related internships or fieldwork, Federal Work-Study, and tuition waivers (full) also available. Financial aid application deadline: 3/1. *Faculty research:* Computational mechanics, signal processing, reaction engineering, control engineering, thermodynamic heat transfer. *Unit head:* Dr. S. Graham Kelly, Interim Dean, 330-972-6978, E-mail: sgraham@uakron.edu.

The University of Alabama, Graduate School, College of Engineering, Tuscaloosa, AL 35487. Offers MS Ch E, MS Met E, MSAE, MSCE, MSCS, MSE, MSEE, MSESM, MSIE, MSME, PhD. Part-time and evening/weekend programs available. Postbaccalaureate distance learning degree programs offered. *Faculty:* 108 full-time (9 women), 12 part-time (1 woman). *Students:* 220 full-time (30 women), 114 part-time (17 women). Average age 28. In 1998, 83 master's, 20 doctorates awarded. Terminal master's awarded for partial completion of doctoral program. *Degree requirements:* For doctorate, dissertation required, foreign language not required. *Application deadline:* Applications are processed on a rolling basis. Application fee: $25. Electronic applications accepted. *Financial aid:* In 1998–99, 13 fellowships with full tuition reimbursements, 82 research assistantships with full tuition reimbursements, 79 teaching assistantships with full tuition reimbursements were awarded.; career-related internships or fieldwork, Federal Work-Study, and institutionally-sponsored loans also available. *Faculty research:* Energy, global environmental change, magnetic information technology, solidification modeling. Total annual research expenditures: $8 million. *Unit head:* Verle N. Schrodt, Dean, 205-348-6405, Fax: 205-348-8573, E-mail: vschrodt@coe.eng.ua.edu. *Application contact:* Ronald Rogers, Assistant Vice President for Academic Affairs, 205-348-8280, Fax: 205-348-0400, E-mail: rrogers@aalan.ua.edu.

The University of Alabama at Birmingham, Graduate School, School of Engineering, Birmingham, AL 35294. Offers MS Mt E, MSBE, MSCE, MSEE, MSME, PhD, DMD/PhD, MD/PhD. Evening/weekend programs available. *Students:* 105 full-time (32 women), 46 part-time (8 women); includes 17 minority (13 African Americans, 3 Asian Americans or Pacific Islanders, 1 Hispanic American), 34 international. 105 applicants, 76% accepted. In 1998, 39 master's, 10 doctorates awarded. *Degree requirements:* For doctorate, dissertation required, foreign language not required. *Entrance requirements:* For master's, GRE General Test. *Application deadline:* Applications are processed on a rolling basis. Application fee: $30 ($60 for international students). Electronic applications accepted. *Financial aid:* Fellowships with full tuition reimbursements, research assistantships with full tuition reimbursements, career-related internships or fieldwork, Federal Work-Study, institutionally-sponsored loans, and tuition waivers (full and partial) available. Aid available to part-time students. *Unit head:* Dr. Stephen Szygenda, Dean, 205-934-8400.

The University of Alabama in Huntsville, School of Graduate Studies, College of Engineering, Huntsville, AL 35899. Offers MSE, MSOR, PhD. Part-time and evening/weekend programs available. Postbaccalaureate distance learning degree programs offered (no on-campus study). *Faculty:* 58 full-time (4 women), 5 part-time (0 women). *Students:* 141 full-time (31 women), 279 part-time (53 women); includes 45 minority (24 African Americans, 12 Asian Americans or Pacific Islanders, 7 Hispanic Americans, 2 Native Americans), 67 international. Average age 34. 270 applicants, 80% accepted. In 1998, 100 master's, 25 doctorates awarded. *Degree requirements:* For master's, oral and written exams required, thesis optional, foreign language not required; for doctorate, dissertation, oral and written exams required, foreign language not required. *Entrance requirements:* For master's and doctorate, GRE General Test (minimum combined score of 1500 on three sections required), minimum GPA of 3.0. *Application deadline:* For fall admission, 7/24 (priority date); for spring admission, 11/15 (priority date). Applications are processed on a rolling basis. Application fee: $20. Tuition and fees vary according to course load. *Financial aid:* In 1998–99, 105 students received aid, including 1 fellowship with full and partial tuition reimbursement available (averaging $14,400 per year), 44 research assistantships with full and partial tuition reimbursements available (averaging $9,827 per year), 53 teaching assistantships with full and partial tuition reimbursements available (averaging $8,233 per year); career-related internships or fieldwork, Federal Work-Study, grants, institutionally-sponsored loans, scholarships, and tuition waivers (full and partial) also available. Aid available to part-time students. Financial aid application deadline: 4/1; financial aid applicants required to submit FAFSA. *Faculty research:* Propulsion, missile systems, automation, robotics, plasma. Total annual research expenditures: $1.7 million. *Unit head:* Dr. Jorge Aunon, Dean, 256-890-6474, Fax: 256-890-6843, E-mail: aunon@eb.uah.edu.

University of Alaska Anchorage, School of Engineering, Anchorage, AK 99508-8060. Offers MCE, MS. Part-time and evening/weekend programs available. *Students:* 13 full-time (2 women), 53 part-time (22 women); includes 10 minority (5 Asian Americans or Pacific Islanders, 1 Hispanic American, 4 Native Americans) 34 applicants, 76% accepted. In 1998, 33 degrees awarded. *Degree requirements:* For master's, computer language required, foreign language not required. *Entrance requirements:* For master's, GRE General Test. *Application deadline:* For fall admission, 5/1 (priority date). Applications are processed on a rolling basis. Application fee: $45. *Financial aid:* In 1998–99, 1 research assistantship was awarded.; Federal Work-Study and traineeships also available. Aid available to part-time students. Financial aid application deadline: 4/1; financial aid applicants required to submit FAFSA. *Unit head:* Dr. Robert Miller, Director, 907-786-1859, Fax: 907-786-1079. *Application contact:* Cecile Mitchell, Director for Enrollment Services, 907-786-1558.

The University of Arizona, Graduate College, College of Engineering and Mines, Tucson, AZ 85721. Offers MS, PhD. Part-time programs available. *Faculty:* 243. *Students:* 468 full-time (81 women), 212 part-time (44 women); includes 50 minority (7 African Americans, 22 Asian Americans or Pacific Islanders, 19 Hispanic Americans, 2 Native Americans), 312 international. Average age 31. 796 applicants, 58% accepted. In 1998, 152 master's, 65 doctorates awarded. *Degree requirements:* For doctorate, dissertation required, foreign language not required. *Entrance requirements:* For master's and doctorate, TOEFL. Application fee: $35. *Financial aid:* Fellowships, research assistantships, teaching assistantships, institutionally-sponsored loans and scholarships available. *Unit head:* Dr. Ernest Smerdon, Dean, 520-621-6601, Fax: 520-621-2232.

University of Arkansas, Graduate School, College of Engineering, Fayetteville, AR 72701-1201. Offers MS, MS Ch E, MS En E, MSBAE, MSCE, MSCSE, MSE, MSEE, MSIE, MSME, MSOR, MSTE, PhD. *Faculty:* 71 full-time (0 women). *Students:* 227 full-time (49 women), 68 part-time (10 women); includes 44 minority (20 African Americans, 14 Asian Americans or Pacific Islanders, 7 Hispanic Americans, 3 Native Americans), 90 international. 331 applicants, 66% accepted. In 1998, 146 master's, 11 doctorates awarded. *Degree requirements:* For doctorate, one foreign language, dissertation required, foreign language not required. Application fee: $40 ($50 for international students). Tuition, state resident: full-time $3,186. Tuition, nonresident: full-time $7,560. Required fees: $378. *Financial aid:* In 1998–99, 55 research assistantships, 59 teaching assistantships were awarded.; fellowships, career-related internships or fieldwork and Federal Work-Study also available. Aid available to part-time students. Financial aid application deadline: 4/1; financial aid applicants required to submit FAFSA. *Unit head:* Dr. Otto Loewer, Dean, 501-575-3054.

University of Arkansas at Little Rock, Graduate School, College of Sciences and Engineering Technology, Department of Applied Science, Little Rock, AR 72204-1099. Offers instrumental sciences (MS, PhD). Part-time programs available. *Degree requirements:* For master's, comprehensive and oral exams required, thesis optional, foreign language not required; for doctorate, computer language, dissertation, 2 semesters of residency, candidacy exams required, foreign language not required. *Entrance requirements:* For master's, GRE General Test (minimum combined score of 1500 on three sections required), TOEFL (minimum score of 550 required), interview, minimum GPA of 3.0; for doctorate, GRE General Test (minimum combined score of 1500 on three sections required), TOEFL (minimum score of 550 required), interview, minimum graduate GPA of 3.5. *Faculty research:* Particle and powder science and technology, optical sensors, process control and automation, signal and image processing, biomedical measurement systems.

University of Bridgeport, College of Graduate and Undergraduate Studies, School of Science, Engineering, and Technology, Bridgeport, CT 06601. Offers MS. Part-time and evening/weekend programs available. *Faculty:* 13 full-time (0 women), 15 part-time (0 women). *Students:* 107 full-time (18 women), 174 part-time (60 women); includes 20 minority (19 Asian Americans or Pacific Islanders, 1 Hispanic American), 245 international. Average age 31. 459 applicants, 75% accepted. In 1998, 83 degrees awarded. *Degree requirements:* For master's, thesis optional, foreign language not required. *Entrance requirements:* For master's, TOEFL. *Application deadline:* Applications are processed on a rolling basis. Application fee: $35 ($50 for international students). *Financial aid:* In 1998–99, 76 students received aid; fellowships, research assistantships, teaching assistantships, career-related internships or fieldwork, Federal Work-Study, institutionally-sponsored loans, and tuition waivers (partial) available. Aid available to part-time students. Financial aid application deadline: 6/1; financial aid applicants required to submit FAFSA. *Faculty research:* Atmospheric chemistry, minicomputers, heat transfer. *Unit head:* Dr. Steven F. Malary, Director, 203-576-4111.

See in-depth description on page 185.

University of British Columbia, Faculty of Graduate Studies, Faculty of Applied Science, Vancouver, BC V6T 1Z2, Canada. Offers M Arch, M Eng, M Sc, MA Sc, MASA, MSN, PhD. Part-time and evening/weekend programs available. *Entrance requirements:* For master's and doctorate, TOEFL.

University of Calgary, Faculty of Graduate Studies, Faculty of Engineering, Calgary, AB T2N 1N4, Canada. Offers M Engr, M Sc, PhD. Part-time and evening/weekend programs available. *Faculty:* 102 full-time (7 women), 26 part-time (0 women). *Students:* 321 full-time (57 women), 115 part-time (19 women). Average age 25. 1326 applicants, 8% accepted. In 1998, 36 master's, 26 doctorates awarded. *Degree requirements:* For doctorate, dissertation, candidacy exam required, foreign language not required, foreign language not required. *Average time to degree:* Master's–4 years full-time, 4 years part-time; doctorate–7.2 years full-time. *Application deadline:* Applications are processed on a rolling basis. *Financial aid:* In 1998–99, 242 students received aid, including 17 fellowships, 54 research assistantships, 54 teaching assistantships; career-related internships or fieldwork and grants also available. *Faculty research:* Chemical and petroleum engineering, civil engineering, electrical and computer engineering, geomatics engineering, mechanical engineering and computer-integrated manufacturing, software engineering, oil and gas engineering. Total annual research expenditures: $5 million. *Unit head:* Dr. S. C. Wirasinghe, Dean, 403-220-5731, Fax: 403-284-3697, E-mail: wirasing@acs.ucalgary.ca.

University of California, Berkeley, Graduate Division, College of Engineering, Berkeley, CA 94720-1500. Offers M Eng, MS, D Eng, PhD, M Arch/MS, MCP/MS, MPP/MS. *Students:* 1,323 full-time (271 women); includes 297 minority (34 African Americans, 211 Asian Americans or Pacific Islanders, 45 Hispanic Americans, 7 Native Americans), 490 international. 3303 applicants, 30% accepted. In 1998, 340 master's, 108 doctorates awarded. *Degree requirements:* For doctorate, dissertation, exam required. *Entrance requirements:* For master's and doctorate, GRE General Test, minimum GPA of 3.0. Application fee: $40. *Financial aid:* Fellowships, research assistantships, teaching assistantships, career-related internships or fieldwork, Federal Work-Study, institutionally-sponsored loans, scholarships, and tuition waivers (full and partial) available. *Unit head:* Dr. Paul R. Gray, Dean, 510-642-5771.

University of California, Berkeley, Graduate Division, Group in Applied Science and Technology, Berkeley, CA 94720-1500. Offers PhD. *Students:* 15 full-time (3 women); includes 4 minority (2 African Americans, 2 Asian Americans or Pacific Islanders), 4 international. 30 applicants, 23% accepted. In 1998, 1 degree awarded. *Degree requirements:* For doctorate, dissertation, preliminary exam, qualifying exam required. *Entrance requirements:* For doctorate, GRE General Test, minimum GPA of 3.0, BA or BS in engineering, physics, mathematics, chemistry, or related field. *Application deadline:* For fall admission, 1/3. Application fee: $40. *Financial aid:* Fellowships, research assistantships, teaching assistantships, career-related internships or fieldwork and minority fellowships available. Financial aid application deadline: 1/3; financial aid applicants required to submit FAFSA. *Unit head:* Timothy D. Sands, Chair. *Application contact:* Patricia M. Berumen, Graduate Assistant for Admission, 510-642-8790, Fax: 510-643-6103, E-mail: ast-program@coe.berkeley.edu.

See in-depth description on page 187.

University of California, Davis, Graduate Studies, College of Engineering, Davis, CA 95616. Offers M Engr, MS, D Engr, PhD, Certificate, M Engr/MBA. Part-time programs available. *Faculty:* 127 full-time (8 women), 63 part-time (4 women). *Students:* 580 full-time (136 women), 9 part-time; includes 111 minority (8 African Americans, 79 Asian Americans or Pacific Islanders, 21 Hispanic Americans, 3 Native Americans), 158 international. In 1998, 97 master's, 48 doctorates, 3 other advanced degrees awarded. Terminal master's awarded for partial completion of doctoral program. *Degree requirements:* For doctorate, dissertation required, foreign language not required, foreign language not required. *Entrance requirements:* For doctorate, GRE. *Application deadline:* Applications are processed on a rolling basis. Application fee: $40. Electronic applications accepted. *Financial aid:* In 1998–99, 398 students received aid; fellowships with full and partial tuition reimbursements available, research assistantships with full and partial tuition reimbursements available, teaching assistantships with full and partial tuition reimbursements available, career-related internships or fieldwork, Federal Work-Study, institutionally-sponsored loans, and tuition waivers (full and partial) available. Aid available to part-time students. Financial aid application deadline: 1/15; financial aid applicants required to submit FAFSA. *Unit head:* Dr. Alan Laub, Dean, 530-752-0554.

University of California, Irvine, Office of Research and Graduate Studies, School of Engineering, Irvine, CA 92697. Offers MS, PhD. Part-time programs available. *Faculty:* 70 full-time (7 women), 1 part-time (0 women). *Students:* 260 full-time (49 women), 49 part-time (11 women); includes 84 minority (3 African Americans, 71 Asian Americans or Pacific Islanders, 10 Hispanic Americans), 130 international. Average age 29. 667 applicants, 50% accepted. In 1998, 81 master's, 33 doctorates awarded. Terminal master's awarded for partial completion of doctoral program. *Degree requirements:* For doctorate, dissertation required, foreign language not required, foreign language not required. *Entrance requirements:* For master's, GRE General Test, minimum GPA of 3.0; for doctorate, GRE General Test. *Application deadline:* For fall admission, 1/15 (priority date). Applications are processed on a rolling basis. Application fee: $40. Electronic applications accepted. *Financial aid:* Fellowships, research assistantships, teaching assistantships, institutionally-sponsored loans and tuition waivers (full and partial) available. Financial aid application deadline: 3/2; financial aid applicants required to submit FAFSA. *Faculty research:* Chemical and biochemical engineering, civil and environmental engineering, electrical and computer engineering, mechanical and aerospace engineering, materials science. *Unit head:* Dr. John C. LaRue, Associate Dean, 949-824-6737, Fax: 949-824-3440, E-mail: jclarue@uci.edu. *Application contact:* John Sommerhauser, Graduate Counselor, 949-824-6475, Fax: 949-824-3440, E-mail: jdsommer@uci.edu.

See in-depth description on page 189.

University of California, Los Angeles, Graduate Division, School of Engineering and Applied Science, Los Angeles, CA 90095. Offers MS, PhD, MBA/MS. *Faculty:* 126 full-time, 70 part-time. *Students:* 972 full-time (149 women); includes 292 minority (15 African Americans, 246 Asian Americans or Pacific Islanders, 27 Hispanic Americans, 4 Native Americans), 359 international. 2037 applicants, 34% accepted. In 1998, 237 master's, 81 doctorates awarded. *Degree requirements:* For master's, comprehensive exam or thesis required; for doctorate, dissertation, qualifying exams required, foreign language not required. *Entrance requirements:* For master's, GRE General Test, minimum GPA of 3.0; for doctorate, GRE General Test, minimum GPA of 3.25. Application fee: $40. Electronic applications accepted. *Financial aid:* In 1998–99, 142 fellowships, 536 research assistantships, 167 teaching assistantships were awarded.; career-related internships or fieldwork, Federal Work-Study, institutionally-sponsored loans, and tuition waivers (full and partial) also available. Financial aid applicants required to submit FAFSA. *Unit head:* Dr. Stephen E. Jacobsen, Associate Dean, Student Affairs, 310-825-1704. *Application contact:* Diane Golomb, Student Affairs Officer, 310-825-1704, Fax: 310-825-2473, E-mail: diane@ea.ucla.edu.

See in-depth description on page 191.

Engineering and Applied Sciences—General

University of California, Riverside, Graduate Division, College of Engineering, Riverside, CA 92521-0102. Offers MS, PhD. Part-time programs available. *Faculty:* 18 full-time (1 woman). *Students:* 50 full-time (14 women), 1 part-time; includes 8 minority (6 Asian Americans or Pacific Islanders, 2 Hispanic Americans), 33 international. Average age 27. 183 applicants, 39% accepted. In 1998, 12 master's, 2 doctorates awarded. Terminal master's awarded for partial completion of doctoral program. *Degree requirements:* For doctorate, dissertation, qualifying exams required, foreign language not required, foreign language not required. *Entrance requirements:* For master's and doctorate, GRE General Test (minimum combined score of 1100 required), TOEFL (minimum score of 550 required), minimum GPA of 3.2. *Average time to degree:* Master's–2 years full-time; doctorate–5 years full-time. *Application deadline:* For fall admission, 5/1; for spring admission, 12/1. Applications are processed on a rolling basis. Application fee: $40. Electronic applications accepted. *Financial aid:* Fellowships, research assistantships, teaching assistantships, career-related internships or fieldwork, Federal Work-Study, institutionally-sponsored loans, and tuition waivers (full and partial) available. Financial aid application deadline: 2/1; financial aid applicants required to submit FAFSA. *Unit head:* Dr. Satish K. Tripathi, Dean, 909-787-2942.

University of California, San Diego, Graduate Studies and Research, Department of Applied Mechanics and Engineering Sciences, La Jolla, CA 92093-5003. Offers aerospace engineering (MS, PhD); applied mechanics (MS, PhD); applied ocean science (MS, PhD); chemical engineering (MS, PhD); engineering physics (MS, PhD); mechanical engineering (MS, PhD); structural engineering (MS, PhD). Part-time programs available. *Faculty:* 36 full-time (5 women), 4 part-time (0 women). *Students:* 116 full-time (17 women), 5 part-time (2 women); includes 27 minority (1 African American, 15 Asian Americans or Pacific Islanders, 8 Hispanic Americans, 3 Native Americans), 28 international. 308 applicants, 33% accepted. In 1998, 10 master's, 19 doctorates awarded. *Degree requirements:* For master's, comprehensive exam or thesis required; for doctorate, dissertation, qualifying exam required. *Entrance requirements:* For master's and doctorate, GRE General Test, TOEFL (minimum score of 550 required), minimum GPA of 3.0. *Application deadline:* For fall admission, 5/31. Application fee: $40. *Financial aid:* In 1998–99, 20 fellowships with full tuition reimbursements (averaging $15,000 per year), 51 research assistantships with full tuition reimbursements (averaging $15,000 per year), 17 teaching assistantships with partial tuition reimbursements (averaging $13,000 per year) were awarded.; career-related internships or fieldwork and scholarships also available. Financial aid application deadline: 1/31; financial aid applicants required to submit FAFSA. *Faculty research:* Solid and structural mechanics, micromechanics, heterogenous composite materials, physically-based virtual reality. *Unit head:* Forman A. Williams, Chair. *Application contact:* AMES Graduate Student Affairs, 619-534-4387, Fax: 619-534-1730, E-mail: bwalton@ames.ucsd.edu.

University of California, Santa Barbara, Graduate Division, College of Engineering, Santa Barbara, CA 93106. Offers MS, PhD. *Faculty:* 108 full-time (7 women), 3 part-time (1 woman). *Students:* 481 full-time (81 women); includes 70 minority (3 African Americans, 57 Asian Americans or Pacific Islanders, 9 Hispanic Americans, 1 Native American), 203 international. 1342 applicants, 39% accepted. In 1998, 102 master's, 59 doctorates awarded. Terminal master's awarded for partial completion of doctoral program. *Degree requirements:* For doctorate, dissertation required, foreign language not required, foreign language not required. *Entrance requirements:* For master's and doctorate, GRE, TOEFL. Application fee: $40. Electronic applications accepted. *Financial aid:* Fellowships, research assistantships, teaching assistantships, career-related internships or fieldwork, Federal Work-Study, institutionally-sponsored loans, and tuition waivers (full and partial) available. Financial aid applicants required to submit FAFSA. *Unit head:* Venkatesh Narayanamurti, Dean, 805-893-3141. *Application contact:* 805-893-3207, E-mail: engrdean@engineering.ucsb.edu.

University of Central Florida, College of Engineering, Orlando, FL 32816. Offers MS, MS Cp E, MS Env E, MS Mfg E, MSCE, MSEE, MSIE, MSME, PhD, Certificate. Part-time and evening/weekend programs available. *Faculty:* 130. *Students:* 400 full-time (79 women), 309 part-time (60 women); includes 124 minority (20 African Americans, 49 Asian Americans or Pacific Islanders, 55 Hispanic Americans), 225 international. Average age 32. 384 applicants, 43% accepted. In 1998, 201 master's, 26 doctorates awarded. *Degree requirements:* For master's, thesis or alternative required, foreign language not required; for doctorate, dissertation, departmental qualifying exam required, foreign language not required. *Entrance requirements:* For master's, GRE General Test (minimum combined score of 1000 required), TOEFL (minimum score of 550 required; 213 computer-based), minimum GPA of 3.0 in last 60 hours; for doctorate, TOEFL (minimum score of 550 required; 213 computer-based), minimum GPA of 3.5 in last 60 hours. *Application deadline:* For fall admission, 7/15; for spring admission, 12/15. Application fee: $20. Tuition, state resident: full-time $2,054; part-time $137 per credit. Tuition, nonresident: full-time $7,207; part-time $480 per credit. Required fees: $47 per term. *Financial aid:* In 1998–99, 419 students received aid, including 93 fellowships with partial tuition reimbursements available (averaging $2,379 per year), 170 teaching assistantships with partial tuition reimbursements available (averaging $2,345 per year); research assistantships with partial tuition reimbursements available, career-related internships or fieldwork, Federal Work-Study, institutionally-sponsored loans, tuition waivers (partial), and unspecified assistantships also available. Financial aid application deadline: 3/1; financial aid applicants required to submit FAFSA. *Faculty research:* Electrooptics, lasers, materials, simulation, microelectronics. *Unit head:* Dr. Martin Wanielista, Dean, 407-823-2156. *Application contact:* Dr. Issa Batarseh, Graduate Coordinator, 407-823-0185.

See in-depth description on page 193.

University of Cincinnati, Division of Research and Advanced Studies, College of Engineering, Cincinnati, OH 45221-0091. Offers MS, PhD, MBA/MS. *Accreditation:* ABET (one or more programs are accredited). Part-time and evening/weekend programs available. *Faculty:* 110 full-time. *Students:* 775 full-time (127 women), 251 part-time (46 women); includes 71 minority (23 African Americans, 38 Asian Americans or Pacific Islanders, 10 Hispanic Americans), 684 international. 2085 applicants, 33% accepted. In 1998, 252 master's, 59 doctorates awarded. Terminal master's awarded for partial completion of doctoral program. *Degree requirements:* For doctorate, dissertation required, foreign language not required. *Entrance requirements:* For master's and doctorate, GRE General Test, TOEFL. *Average time to degree:* Master's–3.1 years full-time; doctorate–6.4 years full-time. *Application deadline:* For fall admission, 2/1 (priority date). Application fee: $40. *Financial aid:* Fellowships, research assistantships, teaching assistantships, career-related internships or fieldwork, tuition waivers (full), and unspecified assistantships available. Aid available to part-time students. Financial aid application deadline: 2/1. Total annual research expenditures: $18.4 million. *Unit head:* Dr. Stephen T. Kowel, Dean, 513-556-2933, Fax: 513-556-3626.

Announcement: UC is designated a Research 1 university by the Carnegie Commission and is ranked in the top 3% of leading research universities by the National Science Foundation. Master of Science and Doctor of Philosophy degrees are offered in 16 majors. Approximately 800 tuition scholarships, 150 teaching assistantships, and 350 research assistantships are available. A $36-million Engineering Research Center (ERC) opened in 1995. The ERC provides state-of-the-art research facilities for students in all of the graduate programs. There are 750 full-time graduate students conducting basic and applied research, leading to innovative technologies and world-class scholarship. Contact Dr. Roy Eckart, Associate Dean for Academic and Administrative Affairs, 513-556-2739.

See in-depth description on page 195.

University of Colorado at Boulder, Graduate School, College of Engineering and Applied Science, Boulder, CO 80309. Offers ME, MS, PhD, MBA/MS. Part-time programs available. Terminal master's awarded for partial completion of doctoral program. *Degree requirements:* For master's, comprehensive exam required; for doctorate, dissertation required.

See in-depth description on page 197.

University of Colorado at Colorado Springs, Graduate School, College of Engineering and Applied Science, Colorado Springs, CO 80933-7150. Offers ME, MS, PhD. Part-time and evening/weekend programs available. *Faculty:* 36 full-time (2 women). *Students:* 123 full-time (24 women), 89 part-time (16 women); includes 29 minority (3 African Americans, 18 Asian Americans or Pacific Islanders, 7 Hispanic Americans, 1 Native American), 24 international. Average age 29. 96 applicants, 90% accepted. *Degree requirements:* For doctorate, dissertation, comprehensive exams required, foreign language not required, foreign language not required. *Entrance requirements:* For master's, GRE General Test (minimum combined score of 1200 required), TOEFL (minimum score of 550 required), minimum GPA of 3.0; for doctorate, GRE General Test (minimum combined score of 1200 required), TOEFL (minimum score of 550 required), minimum GPA of 3.3. Application fee: $40 ($50 for international students). *Financial aid:* Fellowships, research assistantships, teaching assistantships, career-related internships or fieldwork and Federal Work-Study available. *Faculty research:* Ferroelectronics, electronics communication, computer-aided design, electromagnetics. Total annual research expenditures: $750,000. *Unit head:* Dr. Ronald Sega, Dean, 719-262-3246, Fax: 719-262-3542, E-mail: rsega@mail.uccs.edu.

University of Colorado at Denver, Graduate School, College of Engineering and Applied Science, Denver, CO 80217-3364. Offers ME, MS, PhD. Part-time and evening/weekend programs available. *Faculty:* 41. *Students:* 49 full-time (12 women), 234 part-time (44 women); includes 46 minority (5 African Americans, 31 Asian Americans or Pacific Islanders, 9 Hispanic Americans, 1 Native American), 45 international. Average age 33. 154 applicants, 76% accepted. In 1998, 72 master's, 1 doctorate awarded. *Degree requirements:* For doctorate, dissertation required. *Entrance requirements:* For master's and doctorate, GRE. *Application deadline:* Applications are processed on a rolling basis. Application fee: $50 ($60 for international students). Electronic applications accepted. Tuition, state resident: part-time $217 per credit hour. Tuition, nonresident: part-time $783 per credit hour. Required fees: $3 per credit hour. $130 per year. One-time fee: $25 part-time. *Financial aid:* Research assistantships, teaching assistantships, career-related internships or fieldwork and Federal Work-Study available. Financial aid application deadline: 3/1; financial aid applicants required to submit FAFSA. Total annual research expenditures: $593,062. *Unit head:* Peter Jenkins, Dean, 303-556-2870, Fax: 303-556-2511, E-mail: pjenkins@cse.cudenver.edu. *Application contact:* Oren Strom, Associate Dean, 303-556-2870, Fax: 303-556-2511, E-mail: ostrom@castle.cudenver.edu.

See in-depth description on page 199.

University of Connecticut, Graduate School, School of Engineering, Storrs, CT 06269. Offers MS, PhD. *Degree requirements:* For doctorate, dissertation required.

See in-depth description on page 201.

University of Dayton, Graduate School, School of Engineering, Dayton, OH 45469-1300. Offers MS Ch E, MS Mat E, MSAE, MSCE, MSE, MSEE, MSEM, MSEM, MSEO, MSME, MSMS, DE, PhD. Part-time and evening/weekend programs available. *Faculty:* 61 full-time (1 woman), 44 part-time (3 women). *Students:* 172 full-time (33 women), 176 part-time (26 women); includes 52 minority (22 African Americans, 19 Asian Americans or Pacific Islanders, 11 Hispanic Americans), 50 international. Average age 24. In 1998, 110 master's, 12 doctorates awarded. *Degree requirements:* For doctorate, dissertation, departmental qualifying exam required, foreign language not required. *Entrance requirements:* For master's, TOEFL. *Application deadline:* For fall admission, 8/1 (priority date). Applications are processed on a rolling basis. Application fee: $30. *Financial aid:* In 1998–99, 3 fellowships with full tuition reimbursements (averaging $18,000 per year), 68 research assistantships with full tuition reimbursements (averaging $13,500 per year), 7 teaching assistantships with full tuition reimbursements (averaging $9,000 per year) were awarded.; career-related internships or fieldwork, institutionally-sponsored loans, and tuition waivers (full and partial) also available. *Faculty research:* Aerodynamics, energy systems, composite materials, rare-earth magnetics, artificial intelligence. *Unit head:* Dr. Blake Cherrington, Dean, 937-229-2736, Fax: 937-229-2756. *Application contact:* Dr. Donald L. Moon, Associate Dean, 937-229-2241, Fax: 937-229-2471, E-mail: dmoon@engr.udayton.edu.

See in-depth description on page 203.

University of Delaware, College of Engineering, Newark, DE 19716. Offers M Ch E, MAS, MCE, MEE, MEM, MMSE, MSME, PhD. Part-time and evening/weekend programs available. Postbaccalaureate distance learning degree programs offered (minimal on-campus study). *Faculty:* 86. *Students:* 356 full-time (54 women), 61 part-time (23 women); includes 24 minority (3 African Americans, 16 Asian Americans or Pacific Islanders, 5 Hispanic Americans), 219 international. 979 applicants, 38% accepted. In 1998, 56 master's, 31 doctorates awarded. Terminal master's awarded for partial completion of doctoral program. *Degree requirements:* For doctorate, dissertation required, foreign language not required, foreign language not required. *Entrance requirements:* For master's and doctorate, GRE General Test. Application fee: $45. Electronic applications accepted. *Financial aid:* In 1998–99, 21 fellowships, 216 research assistantships, 47 teaching assistantships were awarded.; career-related internships or fieldwork, Federal Work-Study, and institutionally-sponsored loans also available. Aid available to part-time students. Financial aid applicants required to submit FAFSA. Total annual research expenditures: $20 million. *Unit head:* Andras Z. Szeri, Interim Dean, 302-831-8017, Fax: 302-831-8179, E-mail: szeri@me.udel.edu.

See in-depth description on page 205.

University of Denver, Graduate Studies, Faculty of Natural Sciences, Mathematics and Engineering, Department of Engineering, Denver, CO 80208. Offers computer science and engineering (MS); electrical engineering (MS); management and general engineering (MSMGEN); materials science (PhD); mechanical engineering (MS). Part-time and evening/weekend programs available. *Faculty:* 15. *Students:* 23 (9 women) 8 international. 19 applicants, 84% accepted. In 1998, 4 master's, 1 doctorate awarded. Terminal master's awarded for partial completion of doctoral program. *Degree requirements:* For master's, thesis required (for some programs), foreign language not required; for doctorate, dissertation required, foreign language not required. *Entrance requirements:* For master's and doctorate, GRE General Test, TOEFL (minimum score of 570 required), TSE (minimum score of 230 required). *Application deadline:* Applications are processed on a rolling basis. Application fee: $40 ($45 for international students). *Financial aid:* In 1998–99, 12 students received aid, including 7 research assistantships with full and partial tuition reimbursements available (averaging $12,228 per year), 7 teaching assistantships with full and partial tuition reimbursements available (averaging $12,636 per year); fellowships with full and partial tuition reimbursements available, career-related internships or fieldwork, Federal Work-Study, institutionally-sponsored loans, and scholarships also available. Financial aid application deadline: 3/1; financial aid applicants required to submit FAFSA. *Faculty research:* Microelectrics, digital signal processing, robotics, speech recognition, microwaves, aerosols, x-ray analysis, acoustic emissions. Total annual research expenditures: $1 million. *Unit head:* Dr. Albert J. Rosa, Chair, 303-871-2102. *Application contact:* Louise Carlson, Assistant to Chair, 303-871-2107.

See in-depth description on page 207.

University of Detroit Mercy, College of Engineering and Science, Detroit, MI 48219-0900. Offers M Eng Mgt, MA, MATM, ME, MS, MSCS, MSEC, DE, PhD. Part-time and evening/weekend programs available. *Degree requirements:* For doctorate, dissertation required, foreign language not required.

See in-depth description on page 209.

University of Florida, Graduate School, College of Engineering, Gainesville, FL 32611. Offers MCE, ME, MS, PhD, Certificate, Engr. *Accreditation:* ABET (one or more programs are accredited). Part-time programs available. *Faculty:* 424. *Students:* 1,144 full-time (202 women), 351 part-time (73 women); includes 198 minority (37 African Americans, 73 Asian Americans or Pacific Islanders, 82 Hispanic Americans, 6 Native Americans), 580 international. 3009 applicants, 53% accepted. In 1998, 375 master's, 103 doctorates awarded. *Degree requirements:* For doctorate, dissertation required. *Entrance requirements:* For master's, GRE General Test, minimum GPA of 3.0; for doctorate and other advanced degree, GRE General Test. *Application deadline:* For fall admission, 6/1 (priority date). Applications are processed on a rolling basis. Application fee: $20. Electronic applications accepted. *Financial aid:* In 1998–99, 778 students received aid, including 112 fellowships, 715 research assistantships, 72 teaching assistantships; career-related internships or fieldwork, Federal Work-Study, institutionally-sponsored loans, and unspecified assistantships also available. Aid available to part-time students. *Unit head:* Dr. Winifred M. Phillips, Dean, 352-392-6000, Fax: 352-392-9673, E-mail: wphil@eng.ufl.edu. *Application contact:* Dr. Warren Viessman, Associate Dean of Academic Programs, 352-392-0943, Fax: 352-392-9673, E-mail: wvies@eng.ufl.edu.

See in-depth description on page 213.

University of Florida, Graduate School, Graduate Engineering and Research Center (GERC), Gainesville, FL 32611. Offers aerospace engineering (ME, MS, PhD, Engr); electrical and computer engineering (ME, MS, PhD, Engr); engineering mechanics (ME, MS, PhD, Engr); industrial and systems engineering (ME, MS, PhD, Engr). Part-time programs available. Postbaccalaureate distance learning degree programs offered. *Faculty:* 6 full-time (0 women), 16 part-time (1 woman). *Students:* 13 full-time (6 women), 159 part-time (29 women); includes 21 minority (8 African Americans, 8 Asian Americans or Pacific Islanders, 4 Hispanic Americans, 1 Native American) Average age 36. In 1998, 14 master's awarded (100% found work related to degree); 1 doctorate awarded (100% found work related to degree). Terminal master's awarded for partial completion of doctoral program. *Degree requirements:* For master's, computer language required (for some programs), thesis optional, foreign language not required; for doctorate, computer language (for some programs), dissertation required; for Engr, computer language (for some programs), thesis required, foreign language not required. *Entrance requirements:* For master's, GRE General Test (minimum combined score of 1000 required), TOEFL, minimum GPA of 3.0; for doctorate, GRE General Test (minimum combined score of 1200 required), written and oral qualifying exams, TOEFL, minimum GPA of 3.0, master's degree in engineering; for Engr, GRE General Test (minimum combined score of 1000 required), TOEFL, minimum GPA of 3.0, master's degree in engineering. *Average time to degree:* Master's–1 year full-time, 2.5 years part-time; doctorate–3 years full-time, 7 years part-time; Engr–1 year full-time, 2.5 years part-time. *Application deadline:* For fall admission, 6/1; for spring admission, 10/1. Applications are processed on a rolling basis. Application fee: $20. Electronic applications accepted. *Financial aid:* Fellowships, research assistantships, institutionally-sponsored loans available. Aid available to part-time students. Financial aid applicants required to submit FAFSA. *Faculty research:* Aerodynamics, terradynamics, and propulsion; composite materials and stress anaylsis; optical processing of microwave signals and photonics; holography, radar, and communications; system and signal theory; digital signal processing. *Unit head:* Dr. Pasquale M. Sforza, Director, 850-833-9355, Fax: 850-833-9366, E-mail: sforza@gerc.eng.ufl.edu. *Application contact:* Judi Shivers, Program Assistant, 850-833-9350, Fax: 850-833-9366, E-mail: reginfo@gerc.eng.ufl.edu.

See in-depth description on page 211.

University of Guelph, Faculty of Graduate Studies, College of Physical and Engineering Science, School of Engineering, Guelph, ON N1G 2W1, Canada. Offers biological engineering (M Sc, PhD); environmental engineering (M Eng, M Sc, PhD); water resources engineering (M Eng, M Sc, PhD). Part-time programs available. *Faculty:* 18 full-time (1 woman), 29 part-time (4 women). *Students:* 54 full-time (15 women), 13 part-time (2 women); includes 22 minority (2 African Americans, 18 Asian Americans or Pacific Islanders, 2 Hispanic Americans), 7 international. 40 applicants, 33% accepted. In 1998, 15 master's awarded (20% entered university research/teaching, 73% found other work related to degree, 7% continued full-time study); 4 doctorates awarded (100% entered university research/teaching). *Degree requirements:* For master's, thesis required (for some programs); for doctorate, dissertation required. *Entrance requirements:* For master's, minimum B- average during previous 2 years; for doctorate, minimum B average. *Average time to degree:* Master's–2.4 years full-time, 3 years part-time; doctorate–5 years full-time. *Application deadline:* For fall admission, 8/1 (priority date); for winter admission, 11/1 (priority date); for spring admission, 4/1 (priority date). Applications are processed on a rolling basis. Application fee: $60. *Expenses:* Tuition and fees charges are reported in Canadian dollars. Tuition, area resident: Full-time $4,725 Canadian dollars; part-time $1,055 Canadian dollars per term. International tuition: $6,999 Canadian dollars full-time. Required fees: $295 Canadian dollars per term. *Financial aid:* In 1998–99, 20 research assistantships (averaging $13,000 per year), 46 teaching assistantships (averaging $3,781 per year) were awarded.; fellowships, career-related internships or fieldwork also available. Aid available to part-time students. *Faculty research:* Food, systems, and computing engineering. Total annual research expenditures: $1.3 million. *Unit head:* Dr. Lambert Otten, Director, 519-824-4120 Ext. 2043, Fax: 519-836-0227, E-mail: lotten@uoguelph.ca. *Application contact:* Dr. Ramesh P. Rudra, Graduate Coordinator, 519-824-4120 Ext. 2110, Fax: 519-836-0227, E-mail: rrudra@uoguelph.ca.

University of Hartford, College of Engineering, West Hartford, CT 06117-1599. Offers M Eng. Part-time and evening/weekend programs available. *Faculty:* 9 full-time (1 woman), 3 part-time (1 woman). *Students:* 8 full-time (0 women), 21 part-time (3 women); includes 2 minority (1 Asian American or Pacific Islander, 1 Hispanic American), 9 international. Average age 32. 80 applicants, 60% accepted. In 1998, 11 degrees awarded (80% found work related to degree, 20% continued full-time study). *Degree requirements:* For master's, thesis required, foreign language not required. *Entrance requirements:* For master's, GRE General Test, TOEFL, minimum GPA of 3.0. *Application deadline:* Applications are processed on a rolling basis. Application fee: $40 ($55 for international students). Electronic applications accepted. *Financial aid:* In 1998–99, 20 students received aid, including research assistantships (averaging $6,000 per year); Federal Work-Study and unspecified assistantships also available. Aid available to part-time students. Financial aid application deadline: 6/1; financial aid applicants required to submit FAFSA. *Faculty research:* Real time fault diagnostics of electrical power transformers using wavelet transforms and supersonic sensors. *Unit head:* Carolyn P. Lahey, Coordinator, Graduate Student Services, 860-768-4368, Fax: 860-768-5073, E-mail: lahey@mail.hartford.edu. *Application contact:* Nancy Clubb-Lazzerini, Coordinator of Graduate Applications, 860-768-4373, Fax: 860-768-5160, E-mail: gettoknow@mail.hartford.edu.

University of Hawaii at Manoa, Graduate Division, College of Engineering, Honolulu, HI 96822. Offers MS, PhD. *Accreditation:* ABET (one or more programs are accredited). Part-time programs available. *Faculty:* 85 full-time (3 women). *Students:* 81 full-time (23 women), 75 part-time (11 women); includes 44 minority (1 African American, 41 Asian Americans or Pacific Islanders, 2 Hispanic Americans), 94 international. 201 applicants, 60% accepted. In 1998, 39 master's, 3 doctorates awarded. *Degree requirements:* For doctorate, dissertation, exams required, foreign language not required. *Application deadline:* Applications are processed on a rolling basis. Application fee: $25 ($50 for international students). *Financial aid:* In 1998–99, 65 research assistantships (averaging $15,312 per year), 17 teaching assistantships (averaging $12,922 per year) were awarded.; fellowships, career-related internships or fieldwork, Federal Work-Study, and tuition waivers (full and partial) also available. Financial aid applicants required to submit FAFSA. *Unit head:* Dr. Paul Yuen, Dean, 808-956-7727, Fax: 808-956-2291, E-mail: pyuen@wiliki.eng.hawaii.edu.

University of Houston, College of Technology, Houston, TX 77004. Offers construction management (MT); manufacturing systems (MT); microcomputer systems (MT); occupational technology (MSOT). Part-time and evening/weekend programs available. *Faculty:* 23 full-time (7 women), 3 part-time (0 women). *Students:* 17 full-time (11 women), 75 part-time (41 women); includes 27 minority (15 African Americans, 4 Asian Americans or Pacific Islanders, 8 Hispanic Americans), 6 international. Average age 37. 33 applicants, 82% accepted. In 1998, 40 degrees awarded. *Degree requirements:* Foreign language not required. *Entrance requirements:* For master's, GMAT, GRE, or MAT (MSOT); GRE (MT), minimum GPA of 3.0 in last 60 hours. *Application deadline:* For fall admission, 7/1; for spring admission, 11/1. Application fee: $35 ($110 for international students). *Financial aid:* Fellowships, research assistantships, teaching assistantships, career-related internships or fieldwork, Federal Work-Study, and institutionally-sponsored loans available. Aid available to part-time students. *Faculty research:* Educational delivery systems, technical curriculum development, computer-integrated manufacturing, neural networks. Total annual research expenditures: $1.3 million. *Unit head:* Bernard McIntyre, Dean, 713-743-4028, Fax: 713-743-4032, E-mail: bmcintyre@uh.edu. *Application contact:* Holly Rosenthal, Graduate Academic Adviser, 713-743-4098, Fax: 713-743-4032, E-mail: hrosenthal@uh.edu.

University of Houston, Cullen College of Engineering, Houston, TX 77004. Offers M Ch E, MCE, MEE, MIE, MME, MS, MS Ch E, MS Env E, MS Mat, MSAER, MSCE, MSCSE, MSEE, MSIE, MSME, MSPE, PhD, MBA/MIE. Part-time and evening/weekend programs available. *Faculty:* 91 full-time (5 women), 39 part-time (3 women). *Students:* 373 full-time (83 women), 273 part-time (60 women); includes 90 minority (14 African Americans, 57 Asian Americans or Pacific Islanders, 18 Hispanic Americans, 1 Native American), 310 international. Average age 30. 828 applicants, 34% accepted. In 1998, 160 master's, 43 doctorates awarded. Terminal master's awarded for partial completion of doctoral program. *Degree requirements:* For doctorate, dissertation, departmental qualifying exam required, foreign language not required, foreign language not required. *Entrance requirements:* For master's and doctorate, GRE General Test, TOEFL. *Application deadline:* Applications are processed on a rolling basis. Application fee: $25 ($75 for international students). *Financial aid:* Fellowships, research assistantships, teaching assistantships, career-related internships or fieldwork, Federal Work-Study, institutionally-sponsored loans, scholarships, and tuition waivers (partial) available. *Faculty research:* Superconducting materials, microantennas for space packs, direct numerical simulation of pairing vortices. Total annual research expenditures: $9.4 million. *Unit head:* Dr. John Wolfe, Dean, 713-743-4200, Fax: 713-743-4214. *Application contact:* Charles Dalton, Associate Dean, Graduate Programs, 713-743-4200, Fax: 713-743-4214, E-mail: dalton@uh.edu.

See in-depth description on page 215.

University of Idaho, College of Graduate Studies, College of Engineering, Moscow, ID 83844-4140. Offers M Engr, MS, PhD. *Faculty:* 71 full-time (4 women), 3 part-time (1 woman). *Students:* 83 full-time (16 women), 228 part-time (21 women); includes 14 minority (6 African Americans, 7 Asian Americans or Pacific Islanders, 1 Hispanic American), 40 international. In 1998, 65 master's, 5 doctorates awarded. *Degree requirements:* For doctorate, dissertation required. *Entrance requirements:* For doctorate, minimum undergraduate GPA of 2.8, 3.0 graduate. *Application deadline:* For fall admission, 8/1; for spring admission, 12/15. Application fee: $35 ($45 for international students). *Financial aid:* Fellowships, research assistantships, teaching assistantships, career-related internships or fieldwork and Federal Work-Study available. Aid available to part-time students. Financial aid application deadline: 2/15. *Unit head:* Dr. David E. Thompson, Dean, 208-885-6479.

University of Illinois at Chicago, Graduate College, College of Engineering, Chicago, IL 60607-7128. Offers MS, PhD, MD/PhD. Part-time and evening/weekend programs available. *Faculty:* 85 full-time (1 woman). *Students:* 512 full-time (96 women), 344 part-time (64 women); includes 145 minority (18 African Americans, 112 Asian Americans or Pacific Islanders, 14 Hispanic Americans, 1 Native American), 498 international. Average age 25. 1804 applicants, 33% accepted. In 1998, 153 master's, 51 doctorates awarded. Terminal master's awarded for partial completion of doctoral program. *Degree requirements:* For doctorate, dissertation required, foreign language not required, foreign language not required. *Entrance requirements:* For master's, TOEFL (minimum score of 550 required); for doctorate, GRE, TOEFL (minimum score of 550 required). Application fee: $40 ($50 for international students). *Financial aid:* In 1998–99, 167 students received aid; fellowships, research assistantships, teaching assistantships, career-related internships or fieldwork and tuition waivers (full) available. *Unit head:* Lawrence A. Kennedy, Dean, 312-996-2400.

See in-depth description on page 217.

University of Illinois at Urbana-Champaign, Graduate College, College of Engineering, Urbana, IL 61801. Offers MCS, MS, MST, PhD, M Arch/MS, MBA/MS. *Faculty:* 425 (21 women), 4 part-time (0 women). *Students:* 2031 full-time (325 women); includes 226 minority (15 African Americans, 159 Asian Americans, 51 Hispanic Americans, 1 Native American). 3688 applicants, 14% accepted. *Unit Head:* Dr. William R. Schowalter, Dean, 217-333-2150.

See in-depth description on page 219.

The University of Iowa, Graduate College, College of Engineering, Iowa City, IA 52242-1316. Offers MS, PhD. *Faculty:* 78 full-time, 11 part-time. *Students:* 174 full-time (36 women), 126 part-time (22 women); includes 20 minority (4 African Americans, 12 Asian Americans or Pacific Islanders, 3 Hispanic Americans, 1 Native American), 155 international. 1057 applicants, 41% accepted. In 1998, 70 master's, 36 doctorates awarded. *Degree requirements:* For master's, thesis optional; for doctorate, dissertation, comprehensive exam required. *Entrance requirements:* For master's and doctorate, GRE, TOEFL. Application fee: $30 ($50 for international students). *Financial aid:* In 1998–99, 25 fellowships, 177 research assistantships, 79 teaching assistantships were awarded. Financial aid applicants required to submit FAFSA. *Unit head:* P. Barry Butler, Interim Dean, 319-335-5766, Fax: 319-335-6086.

See in-depth description on page 221.

University of Kansas, Graduate School, School of Engineering, Lawrence, KS 66045. Offers ME, MS, DE, PhD. Part-time and evening/weekend programs available. Postbaccalaureate distance learning degree programs offered (no on-campus study). *Faculty:* 84. *Students:* 122 full-time (26 women), 350 part-time (55 women); includes 28 minority (5 African Americans, 14 Asian Americans or Pacific Islanders, 7 Hispanic Americans, 2 Native Americans), 176 international. In 1998, 141 master's, 18 doctorates awarded. Terminal master's awarded for partial completion of doctoral program. *Degree requirements:* For doctorate, dissertation, comprehensive exam required, foreign language not required. *Entrance requirements:* For master's and doctorate, TOEFL, minimum GPA of 3.0. Application fee: $30. *Financial aid:* In 1998–99, 12 fellowships, 131 research assistantships, 62 teaching assistantships were awarded.; career-related internships or fieldwork and Federal Work-Study also available. *Faculty research:* Telecommunications, oil recovery, airplane design, structured materials, robotics. Total annual research expenditures: $7 million. *Unit head:* Carl E. Locke, Dean, 785-864-3881. *Application contact:* Tom Mulinazzi, Associate Dean, 785-864-2928, Fax: 785-864-5445, E-mail: tomm@engr.ukans.edu.

See in-depth description on page 223.

University of Kentucky, Graduate School, Graduate School Programs from the College of Engineering, Lexington, KY 40506-0032. Offers M Eng, MCE, MME, MS, MS Ch E, MS Min, MSAE, MSCE, MSEE, MSEM, MSMAE, MSME, MSMSE, PhD. Part-time programs available. *Degree requirements:* For master's, comprehensive exam required; for doctorate, dissertation, comprehensive exam required. *Entrance requirements:* For master's, GRE General Test; for doctorate, GRE General Test, minimum graduate GPA of 3.0.

University of Louisville, Graduate School, Speed Scientific School, Louisville, KY 40292-0001. Offers M Eng, MS, PhD, M Eng/MBA. *Accreditation:* ABET (one or more programs are accredited). Part-time programs available. *Faculty:* 83 full-time (8 women), 8 part-time (1 woman). *Students:* 203 full-time (40 women), 365 part-time (71 women); includes 58 minority (19 African Americans, 31 Asian Americans or Pacific Islanders, 7 Hispanic Americans, 1 Native American), 158 international. Average age 29. 486 applicants, 56% accepted. In 1998, 121 master's, 11 doctorates awarded. Terminal master's awarded for partial completion of doctoral program. *Degree requirements:* For master's and doctorate, thesis/dissertation required, foreign language not required. *Entrance requirements:* For master's and doctorate, GRE General Test (minimum combined score of 1200 required). *Application deadline:* Applications are processed on a rolling basis. Application fee: $25. Electronic applications accepted. *Financial aid:* In 1998–99, 10 fellowships with full tuition reimbursements (averaging $4,000 per year), 9 research assistantships with full tuition reimbursements (averaging $1,200 per year), 34 teaching assistantships with full tuition reimbursements (averaging $12,036 per year) were awarded.; Federal Work-Study, grants, and scholarships also available. *Faculty research:*

Engineering and Applied Sciences—General

University of Louisville (continued)
Computer vision, image processing, mems, nano technology, bioengineering, materials manufacturing, logistics, computer science, electro-optics. Total annual research expenditures: $3.1 million. *Unit head:* Dr. Thomas R. Hanley, Dean, 502-852-6281, Fax: 502-852-7033, E-mail: trhanl01@gwise.louisville.edu. *Application contact:* Dr. Mickey R. Wilhelm, Associate Dean, 502-852-08002, Fax: 502-852-1577, E-mail: wilhelm@louisville.edu.

See in-depth description on page 225.

University of Maine, Graduate School, College of Engineering, Orono, ME 04469. Offers MS, PhD. Part-time programs available. *Faculty:* 55. *Students:* 102 full-time (28 women), 27 part-time (5 women). In 1998, 29 master's, 6 doctorates awarded. Terminal master's awarded for partial completion of doctoral program. *Degree requirements:* For doctorate, dissertation required, foreign language not required, foreign language not required. *Entrance requirements:* For master's and doctorate, GRE General Test, TOEFL (minimum score of 550 required). *Application deadline:* For fall admission, 2/1 (priority date); for spring admission, 10/15. Applications are processed on a rolling basis. Application fee: $50. *Financial aid:* Fellowships, research assistantships, teaching assistantships, Federal Work-Study, institutionally-sponsored loans, scholarships, and tuition waivers (full and partial) available. Financial aid application deadline: 3/1. *Unit head:* Dr. Chet Rock, Interim Dean, 207-581-2219, Fax: 207-581-2220. *Application contact:* Scott G. Delcourt, Director of the Graduate School, 207-581-3218, Fax: 207-581-3232, E-mail: graduate@maine.edu.

University of Manitoba, Faculty of Graduate Studies, Faculty of Engineering, Winnipeg, MB R3T 2N2, Canada. Offers M Eng, M Sc, PhD. *Unit head:* A. C. Trupp, Chair, Graduate Committee.

University of Maryland, Baltimore County, Graduate School, College of Engineering, Baltimore, MD 21250-5398. Offers MS, PhD. *Students:* 163 full-time (49 women), 105 part-time (23 women); includes 34 minority (8 African Americans, 23 Asian Americans or Pacific Islanders, 3 Hispanic Americans), 123 international. In 1998, 40 master's, 22 doctorates awarded. *Entrance requirements:* For master's and doctorate, GRE General Test. *Application deadline:* For fall admission, 7/1. Applications are processed on a rolling basis. Application fee: $45. *Financial aid:* Fellowships, research assistantships, teaching assistantships available. *Unit head:* Dr. Shlomo Carmi, Dean, 410-455-3270.

See in-depth description on page 227.

University of Maryland, College Park, Graduate School, A. James Clark School of Engineering, College Park, MD 20742-5045. Offers M Eng, MS, PhD. Part-time and evening/weekend programs available. Postbaccalaureate distance learning degree programs offered. *Faculty:* 314 full-time (31 women), 91 part-time (7 women). *Students:* 723 full-time (123 women), 638 part-time (117 women); includes 223 minority (83 African Americans, 108 Asian Americans or Pacific Islanders, 29 Hispanic Americans, 3 Native Americans), 627 international. 2331 applicants, 25% accepted. In 1998, 235 master's, 84 doctorates awarded. *Degree requirements:* For doctorate, dissertation required, foreign language not required. *Application deadline:* Applications are processed on a rolling basis. Application fee: $50 ($70 for international students). Electronic applications accepted. Tuition, state resident: part-time $272 per credit hour. Tuition, nonresident: part-time $475 per credit hour. Required fees: $632; $379 per year. *Financial aid:* In 1998–99, 51 fellowships (averaging $13,462 per year), 478 research assistantships (averaging $11,902 per year), 192 teaching assistantships (averaging $9,977 per year) were awarded.; career-related internships or fieldwork, Federal Work-Study, grants, institutionally-sponsored loans, and scholarships also available. Aid available to part-time students. Financial aid applicants required to submit FAFSA. Total annual research expenditures: $45.4 million. *Unit head:* Dr. Herbert Rabin, Interim Dean, 301-405-3868, Fax: 301-314-9867. *Application contact:* Trudy Lindsey, Director, Graduate Admission and Records, 301-405-4198, Fax: 301-314-9305, E-mail: grschool@deans.umd.edu.

University of Massachusetts Amherst, Graduate School, College of Engineering, Amherst, MA 01003. Offers MS, PhD. *Accreditation:* ABET (one or more programs are accredited). Part-time and evening/weekend programs available. *Faculty:* 106 full-time (7 women). *Students:* 252 full-time (64 women), 163 part-time (25 women); includes 17 minority (2 African Americans, 2 Asian Americans or Pacific Islanders, 12 Hispanic Americans, 1 Native American), 236 international. Average age 27. 1397 applicants, 35% accepted. In 1998, 97 master's, 45 doctorates awarded. Terminal master's awarded for partial completion of doctoral program. *Degree requirements:* For doctorate, dissertation required, foreign language not required. *Entrance requirements:* For master's and doctorate, GRE General Test. *Application deadline:* For fall admission, 2/1 (priority date). Applications are processed on a rolling basis. Application fee: $40. *Financial aid:* In 1998–99, 17 fellowships with full tuition reimbursements (averaging $5,646 per year), 288 research assistantships with full tuition reimbursements (averaging $10,491 per year), 39 teaching assistantships with full tuition reimbursements (averaging $6,510 per year) were awarded.; career-related internships or fieldwork, Federal Work-Study, grants, scholarships, traineeships, and unspecified assistantships also available. Aid available to part-time students. *Unit head:* Dr. Joseph I. Goldstein, Dean, 413-545-0300, Fax: 413-545-0724, E-mail: jigo@ecs.umass.edu.

See in-depth description on page 229.

University of Massachusetts Dartmouth, Graduate School, College of Engineering, North Dartmouth, MA 02747-2300. Offers MS, PhD. Part-time programs available. *Faculty:* 60 full-time (4 women), 3 part-time (2 women). *Students:* 103 full-time (26 women), 89 part-time (10 women); includes 6 minority (3 African Americans, 2 Asian Americans or Pacific Islanders, 1 Hispanic American), 115 international. Average age 30. 268 applicants, 93% accepted. In 1998, 32 master's, 1 doctorate awarded. *Degree requirements:* For doctorate, dissertation, comprehensive exam required, foreign language not required, foreign language not required. *Entrance requirements:* For master's, GRE General Test, TOEFL. *Application deadline:* For fall admission, 4/20 (priority date); for spring admission, 11/15 (priority date). Applications are processed on a rolling basis. Application fee: $40 for international students. Tuition, area resident: Full-time $3,107; part-time $129 per credit. Tuition, state resident: full-time $2,071; part-time $86 per credit. Tuition, nonresident: full-time $7,845; part-time $327 per credit. Required fees: $2,888. Full-time tuition and fees vary according to program and reciprocity agreements. Part-time tuition and fees vary according to course load and reciprocity agreements. *Financial aid:* In 1998–99, 28 research assistantships with full tuition reimbursements (averaging $8,613 per year), 36 teaching assistantships with full tuition reimbursements (averaging $6,717 per year) were awarded.; Federal Work-Study and unspecified assistantships also available. Aid available to part-time students. Financial aid applicants required to submit FAFSA. *Faculty research:* Treatment with multi beam laser, marine sciences, cranberry bog research, acoustics, fiber studies. Total annual research expenditures: $512,000. *Unit head:* Dr. Thomas J. Curry, Dean, 508-999-8539, Fax: 508-999-9137, E-mail: tcurry@umassd.edu. *Application contact:* Carol A. Novo, Graduate Admissions Office, 508-999-8026, Fax: 508-999-8183, E-mail: graduate@umassd.edu.

University of Massachusetts Lowell, Graduate School, James B. Francis College of Engineering, Lowell, MA 01854-2881. Offers MS, MS Eng, D Eng, PhD, Sc D. Part-time and evening/weekend programs available. *Faculty:* 127 full-time (4 women), 20 part-time (0 women). *Students:* 230 full-time (51 women), 604 part-time (106 women); includes 101 minority (30 African Americans, 59 Asian Americans or Pacific Islanders, 7 Hispanic Americans, 5 Native Americans), 137 international. 715 applicants, 47% accepted. In 1998, 143 master's, 20 doctorates awarded. Terminal master's awarded for partial completion of doctoral program. *Degree requirements:* For doctorate, dissertation required, foreign language not required. *Entrance requirements:* For master's and doctorate, GRE General Test. *Application deadline:* For fall admission, 4/1 (priority date); for spring admission, 10/1. Applications are processed on a rolling basis. Application fee: $20 ($35 for international students). *Financial aid:* In 1998–99, 3 fellowships, 40 research assistantships, 125 teaching assistantships were awarded.; career-related internships or fieldwork, Federal Work-Study, and institutionally-sponsored loans also available. Aid available to part-time students. Financial aid application deadline: 4/1. *Unit head:* Dr. Krishna Vedula, Dean, 978-934-2575, E-mail: krishna_vedula@woods.uml.edu.

See in-depth description on page 231.

The University of Memphis, Graduate School, Herff College of Engineering, Memphis, TN 38152. Offers MS, PhD. Part-time programs available. *Faculty:* 41 full-time (1 woman), 20 part-time (2 women). *Students:* 94 full-time (18 women), 71 part-time (4 women); includes 9 minority (3 African Americans, 6 Asian Americans or Pacific Islanders), 81 international. Average age 31. 228 applicants, 41% accepted. In 1998, 46 master's, 7 doctorates awarded. *Degree requirements:* For doctorate, dissertation required. *Entrance requirements:* For master's, GRE General Test. *Application deadline:* For fall admission, 8/1; for spring admission, 12/1. Application fee: $25 ($50 for international students). Electronic applications accepted. Tuition, state resident: full-time $3,410; part-time $178 per credit hour. Tuition, nonresident: full-time $8,670; part-time $408 per credit hour. Tuition and fees vary according to program. *Financial aid:* In 1998–99, 66 research assistantships with full tuition reimbursements, 12 teaching assistantships with full tuition reimbursements were awarded.; fellowships with full tuition reimbursements, career-related internships or fieldwork, tuition waivers (full and partial), and unspecified assistantships also available. *Faculty research:* Soil structure, expert systems development, automatic computer troubleshooting, surface tension experiments. *Unit head:* Dr. Richard C. Warder, Dean, 901-678-2171. *Application contact:* Dr. Frank Claydon, Director of Graduate Studies, 901-678-2171.

University of Miami, Graduate School, College of Engineering, Coral Gables, FL 33124. Offers MS, MSAE, MSBE, MSCE, MSECE, MSEH, MSIE, MSME, DA, PhD, MBA/MSIE. Part-time and evening/weekend programs available. *Faculty:* 37 full-time (2 women), 39 part-time (2 women). *Students:* 103 full-time (27 women), 33 part-time (8 women). Average age 26. 347 applicants, 72% accepted. In 1998, 24 master's, 11 doctorates awarded. *Degree requirements:* For doctorate, dissertation required, foreign language not required, foreign language not required. *Entrance requirements:* For master's, GRE General Test, TOEFL (minimum score of 550 required), minimum GPA of 3.0; for doctorate, GRE General Test, TOEFL (minimum score of 550 required). *Average time to degree:* Master's–2 years full-time, 3 years part-time; doctorate–4 years full-time, 5 years part-time. *Application deadline:* Applications are processed on a rolling basis. Application fee: $35. Tuition: Full-time $15,336; part-time $852 per credit. Required fees: $174. Tuition and fees vary according to program. *Financial aid:* In 1998–99, 83 students received aid, including 6 fellowships with tuition reimbursements available, 33 research assistantships with tuition reimbursements available (averaging $965 per year), 37 teaching assistantships with tuition reimbursements available (averaging $965 per year); career-related internships or fieldwork, Federal Work-Study, institutionally-sponsored loans, scholarships, tuition waivers (partial), and unspecified assistantships also available. Aid available to part-time students. Financial aid application deadline: 2/1; financial aid applicants required to submit FAFSA. *Faculty research:* Ergonomics, biomedical engineering, alternative energy, image processing, environmental engineering. Total annual research expenditures: $4.8 million. *Unit head:* Dr. M. Lewis Temares, Dean, 305-284-2404, Fax: 305-284-4792, E-mail: mtemares@eng.miami.edu. *Application contact:* Thomas D. Waite, Associate Dean, 305-284-2408, Fax: 305-284-2885, E-mail: twaite@miami.edu.

See in-depth description on page 233.

University of Michigan, Horace H. Rackham School of Graduate Studies, College of Engineering, Ann Arbor, MI 48109. Offers M Eng, MS, MSE, D Eng, PhD, Aerospace E, CE, Certificate, EE, IOE, Mar Eng, Nav Arch, Nuc E, M Arch/M Eng, M Arch/MSE, MBA/M Eng, MBA/MS, MBA/MSE, MHSA/MS. Part-time programs available. Postbaccalaureate distance learning degree programs offered. *Faculty:* 303 full-time. *Students:* 1,786 full-time (308 women), 299 part-time (48 women); includes 270 minority (51 African Americans, 184 Asian Americans or Pacific Islanders, 34 Hispanic Americans, 1 Native American), 871 international. 3595 applicants, 42% accepted. In 1998, 552 master's, 192 doctorates awarded. *Degree requirements:* For doctorate, dissertation required. *Application deadline:* Applications are processed on a rolling basis. Electronic applications accepted. *Financial aid:* In 1998–99, 179 fellowships with full tuition reimbursements, 780 research assistantships with full tuition reimbursements, 223 teaching assistantships with full tuition reimbursements were awarded.; career-related internships or fieldwork, Federal Work-Study, institutionally-sponsored loans, scholarships, traineeships, and tuition waivers (full and partial) also available. Aid available to part-time students. Financial aid applicants required to submit FAFSA. Total annual research expenditures: $99.5 million. *Unit head:* Stephen W. Director, Dean, 734-647-7010, Fax: 734-647-7009.

See in-depth description on page 235.

University of Michigan–Dearborn, College of Engineering and Computer Science, Dearborn, MI 48128-1491. Offers MS, MSE, D Eng, MBA/MSE. Part-time and evening/weekend programs available. *Faculty:* 49 full-time (2 women), 38 part-time (4 women). *Students:* 30 full-time (7 women), 678 part-time (140 women). Average age 28. 200 applicants, 75% accepted. In 1998, 173 degrees awarded. *Degree requirements:* For master's, computer language required, thesis optional, foreign language not required. *Application deadline:* Applications are processed on a rolling basis. Application fee: $55. Electronic applications accepted. Tuition, state resident: part-time $259 per credit hour. Tuition, nonresident: part-time $748 per credit hour. Required fees: $80 per course. Tuition and fees vary according to course level, course load and program. *Financial aid:* Fellowships, research assistantships, teaching assistantships, Federal Work-Study available. *Faculty research:* CAD/CAM, expert systems, acoustics, vehicle electronics, engines and fuels. *Unit head:* Dr. S. Sengupta, Dean, 313-593-5290, Fax: 313-593-9967, E-mail: razal@umich.edu. *Application contact:* J. L. Linn, Director, Graduate Student Services, 313-593-0897, Fax: 313-593-9967, E-mail: johnlinn@umich.edu.

See in-depth description on page 237.

University of Minnesota, Twin Cities Campus, Graduate School, Institute of Technology, Minneapolis, MN 55455-0213. Offers M Aero E, M Ch E, M Comp E, M Geo E, M Mat SE, MA, MCE, MCIS, MEE, MIE, MME, MS, MS Ch E, MS Mat SE, MSEE, MSIE, MSME, MSMOT, PhD, MD/PhD. Part-time and evening/weekend programs available. Postbaccalaureate distance learning degree programs offered (minimal on-campus study). Electronic applications accepted.

University of Mississippi, Graduate School, School of Engineering, Oxford, University, MS 38677-9702. Offers computational engineering science (MS, PhD); engineering science (MS, PhD). *Faculty:* 42 full-time (4 women). *Students:* 116 full-time (21 women), 25 part-time (9 women); includes 9 minority (6 African Americans, 3 Asian Americans or Pacific Islanders), 100 international. In 1998, 46 master's, 12 doctorates awarded. *Degree requirements:* For master's, thesis required (for some programs), foreign language not required; for doctorate, dissertation required, foreign language not required. *Entrance requirements:* For master's, GRE General Test, TOEFL, minimum GPA of 3.0; for doctorate, GRE General Test, TOEFL. *Application deadline:* For fall admission, 8/1. Applications are processed on a rolling basis. Application fee: $0 ($25 for international students). Tuition, state resident: full-time $3,053; part-time $170 per credit hour. Tuition, nonresident: full-time $6,155; part-time $342 per credit hour. Tuition and fees vary according to program. *Financial aid:* Application deadline:3/1. *Unit head:* Dr. Allie M. Smith, Dean, 601-232-7407, Fax: 601-232-1287, E-mail: engineer@olemiss.edu.

Announcement: MS and PhD in engineering science programs feature emphases in the following general disciplines: chemical, civil, electrical, environmental, geological, materials, and mechanical engineering; computer science; computational hydroscience; computational engineering science; geology; materials science; and telecommunications. Specialized topical areas available for study in engineering science include soil mechanics, fluid mechanics, heat transfer, highway pavements, structural dynamics, electromagnetic theory, antennas, biochemical engineering, composites, minerals, artificial intelligence, computer networking, fault tolerance, formal methods, simulation, graphics, software engineering, energy conservation, hazardous-waste storage, coal research, combustion, hydraulics, marine minerals and mining technology, computational mechanics, remote sensing, geophysics, geohydrology, materials processing, and wireless communications.

University of Missouri–Columbia, Graduate School, College of Engineering, Columbia, MO 65211. Offers MS, PhD. Part-time programs available. *Faculty:* 124 full-time (9 women), 1 (woman) part-time. *Students:* 135 full-time (24 women), 141 part-time (20 women); includes 19 minority (8 African Americans, 6 Asian Americans or Pacific Islanders, 3 Hispanic Americans, 2 Native Americans), 173 international. 614 applicants, 53% accepted. In 1998, 98 master's, 31 doctorates awarded. *Degree requirements:* For doctorate, dissertation required. *Entrance requirements:* For master's and doctorate, GRE General Test, TOEFL. *Application deadline:* Applications are processed on a rolling basis. Application fee: $30 ($50 for international students). *Financial aid:* Fellowships, research assistantships, teaching assistantships, institutionally-sponsored loans available. *Unit head:* Dr. James Thompson, Dean, 573-882-4375.

See in-depth description on page 239.

University of Missouri–Rolla, Graduate School, School of Engineering, Rolla, MO 65409-0910. Offers MS, DE, PhD. Part-time and evening/weekend programs available. *Faculty:* 109 full-time (3 women), 1 (woman) part-time. *Students:* 356 full-time (62 women), 179 part-time (31 women); includes 35 minority (12 African Americans, 17 Asian Americans or Pacific Islanders, 6 Hispanic Americans), 224 international. Average age 28. 1130 applicants, 72% accepted. In 1998, 264 master's, 30 doctorates awarded. Terminal master's awarded for partial completion of doctoral program. *Degree requirements:* For doctorate, dissertation required, foreign language not required. *Application deadline:* Applications are processed on a rolling basis. Application fee: $25. Electronic applications accepted. *Financial aid:* In 1998–99, 71 fellowships with full and partial tuition reimbursements, 121 research assistantships with full and partial tuition reimbursements, 97 teaching assistantships with full and partial tuition reimbursements were awarded.; career-related internships or fieldwork, Federal Work-Study, institutionally-sponsored loans, traineeships, and tuition waivers (partial) also available. Aid available to part-time students. *Unit head:* Dr. O. Robert Mitchell, Dean, 573-341-4151, Fax: 573-341-4979, E-mail: mit@umr.edu. *Application contact:* Dr. Lokesh Dharani, Associate Dean, 573-341-4149, Fax: 573-341-4979, E-mail: dharani@umr.edu.

See in-depth description on page 241.

University of Nebraska–Lincoln, Graduate College, College of Engineering and Technology, Lincoln, NE 68588. Offers MS, PhD. *Faculty:* 75 full-time (5 women), 4 part-time (0 women). *Students:* 147 full-time (22 women), 102 part-time (14 women); includes 9 minority (3 African Americans, 6 Asian Americans or Pacific Islanders), 125 international. Average age 30. 530 applicants, 32% accepted. In 1998, 64 master's, 9 doctorates awarded. *Degree requirements:* For doctorate, dissertation, comprehensive exams required. *Entrance requirements:* For master's and doctorate, GRE General Test, TOEFL. *Average time to degree:* Doctorate–5.5 years full-time. Application fee: $35. Electronic applications accepted. *Financial aid:* In 1998–99, 16 fellowships with full tuition reimbursements, 123 research assistantships with full tuition reimbursements, 41 teaching assistantships were awarded.; Federal Work-Study also available. Aid available to part-time students. Financial aid application deadline: 2/15. *Unit head:* Dr. James L. Hendrix, Dean, 402-472-3181, Fax: 402-472-7792.

See in-depth description on page 243.

University of Nevada, Las Vegas, Graduate College, Howard R. Hughes College of Engineering, Las Vegas, NV 89154-9900. Offers MS, MSE, PhD. Part-time programs available. *Faculty:* 59 full-time (3 women). *Students:* 67 full-time (13 women), 79 part-time (20 women); includes 21 minority (1 African American, 14 Asian Americans or Pacific Islanders, 6 Hispanic Americans), 50 international. 153 applicants, 44% accepted. In 1998, 39 master's, 1 doctorate awarded. *Degree requirements:* For master's, comprehensive exam required, thesis optional, foreign language not required; for doctorate, dissertation required. *Entrance requirements:* For master's, minimum GPA of 3.0; for doctorate, minimum GPA of 3.5. *Application deadline:* For fall admission, 6/15; for spring admission, 11/15. Application fee: $40 ($95 for international students). *Financial aid:* In 1998–99, 25 research assistantships with full tuition reimbursements (averaging $7,968 per year), 50 teaching assistantships with partial tuition reimbursements (averaging $8,802 per year) were awarded.; fellowships, tuition waivers (full) also available. Financial aid application deadline: 3/1. *Unit head:* Dr. William Wells, Dean, 702-895-3699. *Application contact:* Graduate College Admissions Evaluator, 702-895-3320.

See in-depth description on page 245.

University of Nevada, Reno, Graduate School, College of Engineering, Reno, NV 89557. Offers MS, PhD. Terminal master's awarded for partial completion of doctoral program. *Degree requirements:* For master's, thesis optional, foreign language not required; for doctorate, dissertation required, foreign language not required. *Entrance requirements:* For master's, TOEFL (minimum score of 500 required), minimum GPA of 2.75; for doctorate, TOEFL (minimum score of 500 required), minimum GPA of 3.0.

University of New Brunswick, School of Graduate Studies, Faculty of Engineering, Fredericton, NB E3B 5A3, Canada. Offers M Eng, M Sc CS, M Sc E, PhD, Diploma. Part-time programs available. *Degree requirements:* For master's, thesis required, foreign language not required; for doctorate, dissertation, qualifying exam required, foreign language not required; for degree. *Entrance requirements:* For master's, TOEFL, TWE, minimum GPA of 3.0; for doctorate and Diploma, TOEFL, TWE.

University of New Hampshire, Graduate School, College of Engineering and Physical Sciences, Programs in Engineering, Durham, NH 03824. Offers chemical engineering (MS); civil engineering (MS); electrical and computer engineering (MS), including electrical engineering (MS, PhD); engineering (PhD), including chemical engineering, civil engineering, electrical engineering (MS, PhD), mechanical engineering, systems design engineering; mechanical engineering (MS); ocean engineering (MS). Part-time and evening/weekend programs available. *Faculty:* 63 full-time. *Students:* 52 full-time (11 women), 110 part-time (16 women); includes 8 minority (6 Asian Americans or Pacific Islanders, 2 Hispanic Americans), 28 international. Average age 30. 91 applicants, 89% accepted. In 1998, 31 master's, 6 doctorates awarded. *Degree requirements:* For doctorate, dissertation required, foreign language not required. *Application deadline:* For fall admission, 4/1 (priority date). Applications are processed on a rolling basis. Application fee: $50. Tuition, area resident: Full-time $5,750; part-time $319 per credit. Tuition, state resident: full-time $8,625. Tuition, nonresident: full-time $14,640; part-time $598 per credit. Required fees: $224 per semester. Tuition and fees vary according to course load, degree level and program. *Financial aid:* In 1998–99, 1 fellowship, 28 research assistantships, 36 teaching assistantships were awarded.; Federal Work-Study, scholarships, and tuition waivers (full and partial) also available. Aid available to part-time students. Financial aid application deadline: 2/15. *Unit head:* Dr. Roy B. Torbert, Dean, College of Engineering and Physical Sciences, 603-862-1781.

University of New Haven, Graduate School, School of Engineering and Applied Science, West Haven, CT 06516-1916. Offers MS, MSEE, MSIE, MSME, Certificate, MBA/MSIE. Part-time and evening/weekend programs available. *Students:* 61 full-time (18 women), 230 part-time (51 women); includes 25 minority (6 African Americans, 17 Asian Americans or Pacific Islanders, 1 Hispanic American, 1 Native American), 99 international. 138 applicants, 71% accepted. In 1998, 117 degrees awarded. *Degree requirements:* For master's, thesis or alternative required, foreign language not required. *Application deadline:* Applications are processed on a rolling basis. Application fee: $50. *Financial aid:* Federal Work-Study available. Aid available to part-time students. Financial aid application deadline: 5/1; financial aid applicants required to submit FAFSA. *Unit head:* Dr. M. Jerry Kenig, Dean, 203-932-7168.

See in-depth description on page 247.

University of New Mexico, Graduate School, School of Engineering, Albuquerque, NM 87131-2039. Offers ME, MS, MSME, PhD. Part-time and evening/weekend programs available. *Faculty:* 126 full-time (10 women), 48 part-time (3 women). *Students:* 210 full-time (45 women), 244 part-time (43 women); includes 79 minority (3 African Americans, 14 Asian Americans or Pacific Islanders, 54 Hispanic Americans, 8 Native Americans), 118 international. Average age 32. 367 applicants, 40% accepted. In 1998, 126 master's, 30 doctorates awarded. *Degree requirements:* For doctorate, dissertation required, foreign language not required, foreign language not required. *Entrance requirements:* For master's and doctorate, GRE General Test, minimum GPA of 3.0. *Application deadline:* For fall admission, 7/15; for spring admission, 11/14. Applications are processed on a rolling basis. Application fee: $25. *Financial aid:* In 1998–99, 40 fellowships (averaging $1,725 per year), 236 research assistantships with tuition reimbursements (averaging $3,142 per year), 36 teaching assistantships with tuition reimbursements (averaging $10,994 per year) were awarded.; career-related internships or fieldwork, Federal Work-Study, and grants also available. Total annual research expenditures: $50 million. *Unit head:* Dr. Paul A. Fleury, Dean, 505-277-5522, Fax: 505-277-1422, E-mail: pfleury@unm.edu.

See in-depth description on page 249.

University of New Orleans, Graduate School, College of Engineering, New Orleans, LA 70148. Offers MS, PhD, Certificate. Part-time and evening/weekend programs available. *Faculty:* 21 full-time (0 women), 3 part-time (0 women). *Students:* 85 full-time (12 women), 88 part-time (14 women); includes 13 minority (5 African Americans, 4 Asian Americans or Pacific Islanders, 3 Hispanic Americans, 1 Native American), 82 international. Average age 29. 361 applicants, 43% accepted. In 1998, 38 master's awarded. *Degree requirements:* For master's, thesis optional, foreign language not required; for doctorate, dissertation required. *Entrance requirements:* For master's, GRE General Test (minimum combined score of 1200 required), minimum GPA of 3.0; for doctorate, GRE General Test (minimum combined score of 1200 required). *Application deadline:* For fall admission, 7/1 (priority date). Applications are processed on a rolling basis. Application fee: $20. Tuition, state resident: full-time $2,362. Tuition, nonresident: full-time $7,888. Part-time tuition and fees vary according to course load. *Financial aid:* Fellowships, research assistantships, teaching assistantships, institutionally-sponsored loans available. *Faculty research:* Electrical, civil, environmental, mechanical, naval architecture, and marine engineering. Total annual research expenditures: $2.3 million. *Unit head:* Dr. John N. Crisp, Dean, 504-280-6825, Fax: 504-280-7413, E-mail: jncen@uno.edu. *Application contact:* Dr. Kazim Akyuzlu, Graduate Coordinator, 504-280-6186, Fax: 504-280-7413, E-mail: kmame@uno.edu.

See in-depth description on page 251.

University of North Carolina at Charlotte, Graduate School, The William States Lee College of Engineering, Charlotte, NC 28223-0001. Offers ME, MS, MSCE, MSE, MSEE, MSME, PhD. Part-time and evening/weekend programs available. *Faculty:* 81 full-time (6 women), 1 part-time (0 women). *Students:* 95 full-time (22 women), 156 part-time (26 women); includes 42 minority (7 African Americans, 33 Asian Americans or Pacific Islanders, 2 Hispanic Americans), 92 international. Average age 29. 230 applicants, 87% accepted. In 1998, 77 master's, 2 doctorates awarded. *Entrance requirements:* For master's, GRE General Test. *Application deadline:* For fall admission, 7/15; for spring admission, 11/15. Applications are processed on a rolling basis. Application fee: $35. Electronic applications accepted. *Financial aid:* In 1998–99, 76 research assistantships, 88 teaching assistantships were awarded.; fellowships, Federal Work-Study also available. Financial aid application deadline: 4/1. *Unit head:* Dr. Robert D. Snyder, Dean, 704-547-2301, Fax: 704-547-2352, E-mail: rds@email.uncc.edu. *Application contact:* Kathy Barringer, Assistant Director of Graduate Admissions, 704-547-3366, Fax: 704-547-3279, E-mail: gradadm@email.uncc.edu.

See in-depth description on page 253.

University of North Dakota, Graduate School, School of Engineering and Mines, Grand Forks, ND 58202. Offers M Engr, MA, MS, PhD. Part-time programs available. *Faculty:* 41 full-time (0 women). *Students:* 38 full-time (6 women), 4 part-time (1 woman). 30 applicants, 77% accepted. In 1998, 20 master's, 1 doctorate awarded. *Degree requirements:* For doctorate, dissertation required. *Entrance requirements:* For master's, GRE General Test, TOEFL (minimum score of 550 required), minimum GPA of 3.0 (MS), 2.5 (M Engr); for doctorate, GRE General Test, TOEFL (minimum score of 550 required), minimum GPA of 3.5. *Application deadline:* For fall admission, 3/1 (priority date). Applications are processed on a rolling basis. Application fee: $20. *Financial aid:* In 1998–99, 35 students received aid, including 12 research assistantships, 23 teaching assistantships; fellowships, career-related internships or fieldwork, Federal Work-Study, institutionally-sponsored loans, tuition waivers (full and partial), and unspecified assistantships also available. Financial aid application deadline: 3/15. *Unit head:* Dr. Don Richard, Dean, 701-777-3411, Fax: 701-777-4838, E-mail: dorichar@badlands.nodak.edu.

University of North Texas, Robert B. Toulouse School of Graduate Studies, College of Arts and Sciences, Department of Engineering Technology, Denton, TX 76203. Offers MS. Part-time programs available. *Faculty:* 10 full-time (0 women). *Students:* 10 full-time (3 women), 15 part-time; includes 2 minority (1 African American, 1 Asian American or Pacific Islander), 8 international. Average age 30. In 1998, 3 degrees awarded. *Degree requirements:* For master's, foreign language and thesis not required. *Entrance requirements:* For master's, GRE General Test (minimum score of 400 on each section, 1200 combined required), BS in related field. *Application deadline:* For fall admission, 7/17. Application fee: $25 ($50 for international students). *Financial aid:* Fellowships, research assistantships, teaching assistantships, career-related internships or fieldwork, Federal Work-Study, and institutionally-sponsored loans available. Financial aid application deadline: 4/1. *Faculty research:* Computer-aided design, robotics, computer-integrated manufacturing, automation, pattern recognition. *Unit head:* Dr. Albert B. Grubbs, Chair, 940-565-2022, Fax: 940-565-2666, E-mail: grubbs@unt.edu. *Application contact:* Dr. Michael Kozak, Graduate Adviser, 940-565-2022, Fax: 940-565-2666, E-mail: kozak@cas.unt.edu.

University of Notre Dame, Graduate School, College of Engineering, Notre Dame, IN 46556. Offers MS, PhD. Part-time programs available. *Faculty:* 106 full-time (3 women), 6 part-time (0 women). *Students:* 251 full-time (50 women), 15 part-time (2 women); includes 17 minority (7 African Americans, 5 Asian Americans or Pacific Islanders, 4 Hispanic Americans, 1 Native American), 159 international. 619 applicants, 27% accepted. In 1998, 55 master's, 30 doctorates awarded (47% entered university research/teaching, 47% found other work related to degree). Terminal master's awarded for partial completion of doctoral program. *Degree requirements:* For doctorate, dissertation required, foreign language not required, foreign language not required. *Entrance requirements:* For master's and doctorate, GRE General Test, TOEFL (minimum score of 600 required; 250 for computer-based). *Average time to degree:* Doctorate–5.8 years full-time. *Application deadline:* For fall admission, 2/1 (priority date). Applications are processed on a rolling basis. Application fee: $40. *Financial aid:* In 1998–99, 258 students received aid, including 48 fellowships with full tuition reimbursements available, 72 research assistantships with full tuition reimbursements available, 87 teaching assistantships with full tuition reimbursements available; scholarships, tuition waivers (full), and unspecified assistantships also available. Financial aid application deadline: 2/1. *Faculty research:* Aero/fluid dynamics, controls, hazardous waste, solid-state electronics, catalysis and surface dynamics. Total annual research expenditures: $9.8 million. *Unit head:* Dr. Frank P. Incropera, Dean, 219-631-5534, Fax: 219-631-8007, E-mail: fpi@nd.edu. *Application contact:* Dr. Terrence J. Akai, Director of Graduate Admissions, 219-631-7706, Fax: 219-631-4183, E-mail: gradad@nd.edu.

University of Oklahoma, Graduate College, College of Engineering, Norman, OK 73019-0390. Offers M Env Sc, M Nat Sci, MS, PhD. Part-time and evening/weekend programs available. *Faculty:* 91 full-time (8 women), 16 part-time (2 women). *Students:* 239 full-time (41 women), 271 part-time (52 women); includes 34 minority (7 African Americans, 14 Asian

Engineering and Applied Sciences—General

University of Oklahoma (continued)
Americans or Pacific Islanders, 6 Hispanic Americans, 7 Native Americans), 311 international. In 1998, 143 master's, 35 doctorates awarded. *Degree requirements:* For doctorate, dissertation, qualifying exam required. *Entrance requirements:* For master's and doctorate, TOEFL. Application fee: $25. Tuition, state resident: part-time $86 per credit hour. Tuition, nonresident: part-time $275 per credit hour. Tuition and fees vary according to course level, course load and program. *Financial aid:* In 1998–99, 171 research assistantships, 72 teaching assistantships were awarded.; fellowships, career-related internships or fieldwork, Federal Work-Study, institutionally-sponsored loans, and tuition waivers (full and partial) also available. Aid available to part-time students. *Unit head:* Dr. Arthur Porter, Dean, 405-325-2621.

University of Ottawa, School of Graduate Studies and Research, Faculty of Engineering, Ottawa, ON K1N 6N5, Canada. Offers M Eng, MA Sc, MCS, PhD. *Faculty:* 97 full-time, 23 part-time. *Students:* 244 full-time (46 women), 129 part-time (18 women), 44 international. Average age 32. In 1998, 76 master's, 19 doctorates awarded. *Degree requirements:* For master's, thesis or alternative required, foreign language not required; for doctorate, dissertation required, foreign language not required. *Entrance requirements:* For master's, honors degree or equivalent, minimum B average. *Application deadline:* For fall admission, 3/1. Applications are processed on a rolling basis. Application fee: $35. *Financial aid:* Fellowships, research assistantships, teaching assistantships, Federal Work-Study available. *Unit head:* Tyseer Aboulnasr, Dean, 613-562-5800 Ext. 6175, Fax: 613-562-5174. *Application contact:* Vivian Brazeau, Academic Administrator, 613-562-5800 Ext. 6171, Fax: 613-562-5174, E-mail: brazeau@genie.uottawa.ca.

University of Pennsylvania, School of Engineering and Applied Science, Philadelphia, PA 19104. Offers MS, MSE, PhD, M Arch/MSE, MD/PhD, MSE/MBA, MSE/MCP, VMD/PhD. M Arch/MSE offered jointly with the Master of Architecture Program. Part-time and evening/weekend programs available. Terminal master's awarded for partial completion of doctoral program. *Degree requirements:* For doctorate, dissertation required, foreign language not required. *Entrance requirements:* For master's and doctorate, TOEFL (minimum score of 600 required). Electronic applications accepted.

See in-depth description on page 255.

University of Pittsburgh, School of Engineering, Pittsburgh, PA 15260. Offers MS Ch E, MS Met E, MSBENG, MSCEE, MSEE, MSIE, MSME, MSMSE, MSMfSE, MSPE, PhD, Certificate, MS Ch E/MSPE. Part-time and evening/weekend programs available. Postbaccalaureate distance learning degree programs offered (no on-campus study). *Faculty:* 99 full-time (9 women), 11 part-time (2 women). *Students:* 297 full-time (73 women), 279 part-time (37 women); includes 45 minority (16 African Americans, 22 Asian Americans or Pacific Islanders, 7 Hispanic Americans), 200 international. 1090 applicants, 50% accepted. In 1998, 138 master's, 44 doctorates awarded. Terminal master's awarded for partial completion of doctoral program. *Degree requirements:* For doctorate, dissertation, comprehensive and final oral exams required, foreign language not required, foreign language not required. *Entrance requirements:* For master's and doctorate, TOEFL. *Average time to degree:* Master's–2 years full-time, 4 years part-time; doctorate–4 years full-time, 6 years part-time. *Application deadline:* For fall admission, 8/1 (priority date); for spring admission, 12/1 (priority date). Applications are processed on a rolling basis. Application fee: $30 ($40 for international students). *Financial aid:* In 1998–99, 246 students received aid, including 6 fellowships (averaging $16,972 per year), 162 research assistantships (averaging $15,435 per year), 64 teaching assistantships (averaging $16,065 per year); grants, scholarships, traineeships, and tuition waivers (full and partial) also available. Financial aid application deadline: 2/15. *Faculty research:* Artificial organs, biotechnology, signal processing, construction management, fluid dynamics. Total annual research expenditures: $15.5 million. *Unit head:* Dr. Gerald D. Holder, Dean, 412-624-9809, Fax: 412-624-0412, E-mail: holder@engrng.pitt.edu. *Application contact:* Office of Administration, 412-624-9800, Fax: 412-624-9808, E-mail: admin@engrng.pitt.edu.

See in-depth description on page 257.

University of Portland, Graduate School, Multnomah School of Engineering, Portland, OR 97203-5798. Offers MSCE, MSEE, MSME. Part-time and evening/weekend programs available. *Faculty:* 16 full-time (0 women). *Students:* 6 full-time (1 woman), 7 part-time. 18 applicants, 56% accepted. In 1998, 6 degrees awarded. *Degree requirements:* For master's, computer language required, foreign language and thesis not required. *Entrance requirements:* For master's, GRE General Test, TOEFL (minimum score of 550 required), minimum GPA of 3.0. *Application deadline:* For fall admission, 8/1 (priority date); for spring admission, 12/1. Applications are processed on a rolling basis. Application fee: $40. Tuition: Part-time $563 per semester hour. *Financial aid:* Teaching assistantships, career-related internships or fieldwork, Federal Work-Study, and institutionally-sponsored loans available. Aid available to part-time students. Financial aid application deadline: 3/15. *Unit head:* Dr. Zia Yamayee, Dean, 503-943-7314. *Application contact:* Dr. Khalid Khan, Graduate Program Director, 503-943-7276, E-mail: khan@up.edu.

University of Puerto Rico, Mayagüez Campus, Graduate Studies, College of Engineering, Mayagüez, PR 00681-5000. Offers M Ch E, M Co E, MCE, MEE, MME, MMSE, MS, PhD. Part-time programs available. *Degree requirements:* For master's, comprehensive exam required; for doctorate, one foreign language, dissertation required. *Entrance requirements:* For master's and doctorate, minimum GPA of 2.5, proficiency in English and Spanish.

University of Regina, Faculty of Graduate Studies and Research, Faculty of Engineering, Regina, SK S4S 0A2, Canada. Offers M Eng, MA Sc, PhD. *Faculty:* 22 full-time (3 women), 17 part-time (0 women). *Students:* 22 full-time (1 woman), 58 part-time (10 women). 76 applicants, 47% accepted. In 1998, 21 master's, 2 doctorates awarded. *Degree requirements:* For doctorate, dissertation required, foreign language not required, foreign language not required. *Entrance requirements:* For master's, TOEFL (minimum score of 550 required); for doctorate, TOEFL (minimum score of 550 required), master's degree. *Application deadline:* Applications are processed on a rolling basis. Application fee: $0. *Expenses:* Tuition and fees charges are reported in Canadian dollars. Tuition, state resident: full-time $1,688 Canadian dollars; part-time $94 Canadian dollars per credit hour. International tuition: $3,375 Canadian dollars full-time. Required fees: $65 Canadian dollars per course. Tuition and fees vary according to course load and program. *Financial aid:* In 1998–99, 1 fellowship, 16 research assistantships, 12 teaching assistantships were awarded.; career-related internships or fieldwork and scholarships also available. Financial aid application deadline: 6/15. *Unit head:* Dr. A. Chakma, Dean, 306-585-4159, Fax: 306-585-4855, E-mail: amit.chakma@uregina.ca.

University of Rhode Island, Graduate School, College of Engineering, Kingston, RI 02881. Offers MS, PhD. *Accreditation:* ABET (one or more programs are accredited). Part-time programs available.

University of Rochester, The College, School of Engineering and Applied Sciences, Rochester, NY 14627-0250. Offers MS, PhD. Part-time programs available. *Faculty:* 52. *Students:* 229 full-time (40 women), 54 part-time (8 women); includes 20 minority (5 African Americans, 14 Asian Americans or Pacific Islanders, 1 Hispanic American), 113 international. 774 applicants, 20% accepted. In 1998, 56 master's, 27 doctorates awarded. Terminal master's awarded for partial completion of doctoral program. *Degree requirements:* For doctorate, dissertation, preliminary and oral exams required, foreign language not required, foreign language not required. *Entrance requirements:* For master's and doctorate, GRE, TOEFL. *Application deadline:* For fall admission, 2/1 (priority date). Application fee: $25. *Financial aid:* Fellowships, research assistantships, teaching assistantships, scholarships and tuition waivers (full and partial) available. Financial aid application deadline: 2/1. *Unit head:* Kevin Parker, Dean, 716-275-4151.

See in-depth description on page 259.

University of St. Thomas, Graduate Studies, Graduate School of Applied Science and Engineering, St. Paul, MN 55105-1096. Offers MMSE, MS, MSDD, MSS, Certificate. *Accreditation:* ABET (one or more programs are accredited). Part-time and evening/weekend programs available. *Faculty:* 11 full-time (2 women), 39 part-time (2 women). *Students:* 45 full-time (22 women), 740 part-time (209 women); includes 120 minority (33 African Americans, 79 Asian Americans or Pacific Islanders, 7 Hispanic Americans, 1 Native American), 147 international. Average age 33. 333 applicants, 98% accepted. In 1998, 100 master's, 20 other advanced degrees awarded. *Application deadline:* For fall admission, 8/1 (priority date); for spring admission, 1/1 (priority date). Applications are processed on a rolling basis. Application fee: $30. Electronic applications accepted. Tuition: Part-time $437 per credit. Tuition and fees vary according to degree level, program and student level. *Financial aid:* In 1998–99, 78 students received aid; fellowships, research assistantships, grants and institutionally-sponsored loans available. Aid available to part-time students. Financial aid application deadline: 4/1; financial aid applicants required to submit FAFSA. *Application contact:* Dr. Miriam Q. Williams, Associate Vice President for Academic Affairs, 651-962-6032, Fax: 651-962-6930, E-mail: mqwilliams@stthomas.edu.

University of Saskatchewan, College of Graduate Studies and Research, College of Engineering, Saskatoon, SK S7N 5A2, Canada. Offers M Eng, M Sc, PhD, Diploma. *Degree requirements:* For master's and doctorate, thesis/dissertation required. *Entrance requirements:* For master's and doctorate, GRE, TOEFL.

University of South Alabama, Graduate School, College of Engineering, Mobile, AL 36688-0002. Offers MS Ch E, MSEE, MSME. Part-time programs available. *Faculty:* 22 full-time (0 women). *Students:* 53 full-time (5 women), 19 part-time; includes 5 minority (1 African American, 3 Asian Americans or Pacific Islanders, 1 Hispanic American), 43 international. 215 applicants, 66% accepted. In 1998, 16 degrees awarded. *Degree requirements:* For master's, project or thesis required. *Entrance requirements:* For master's, GRE General Test (minimum combined score of 1000 required), BS in engineering, minimum GPA of 3.0. *Application deadline:* For fall admission, 9/1 (priority date). Applications are processed on a rolling basis. Application fee: $25. Tuition, state resident: part-time $116 per semester hour. Tuition, nonresident: part-time $230 per semester hour. Required fees: $121 per semester. Part-time tuition and fees vary according to course load and program. *Financial aid:* In 1998–99, 4 research assistantships were awarded.; career-related internships or fieldwork and institutionally-sponsored loans also available. Aid available to part-time students. Financial aid application deadline: 4/1. *Unit head:* Dr. David T. Hayhurst, Dean, 334-460-6140. *Application contact:* Dr. Russell M. Hayes, Director of Graduate Studies, 334-460-6117.

See in-depth description on page 261.

University of South Carolina, Graduate School, College of Engineering and Information Technology, Columbia, SC 29208. Offers ME, MS, PhD. Part-time and evening/weekend programs available. Postbaccalaureate distance learning degree programs offered (minimal on-campus study). *Faculty:* 66 full-time (4 women), 1 (woman) part-time. *Students:* 211 full-time (37 women), 168 part-time (32 women); includes 28 minority (11 African Americans, 11 Asian Americans or Pacific Islanders, 4 Hispanic Americans, 2 Native Americans), 172 international. Average age 31. In 1998, 101 master's, 7 doctorates awarded. *Degree requirements:* For doctorate, dissertation required, foreign language not required, foreign language not required. *Entrance requirements:* For master's and doctorate, GRE General Test (minimum combined score of 1100 required), TOEFL. *Application deadline:* For fall admission, 3/1 (priority date); for spring admission, 11/1. Applications are processed on a rolling basis. Application fee: $35. Electronic applications accepted. Tuition, state resident: full-time $4,014; part-time $202 per credit hour. Tuition, nonresident: full-time $8,528; part-time $428 per credit hour. Required fees: $100; $4 per credit hour. Tuition and fees vary according to program. *Financial aid:* In 1998–99, 141 students received aid, including 55 research assistantships with partial tuition reimbursements available (averaging $12,000 per year), 86 teaching assistantships with partial tuition reimbursements available (averaging $12,000 per year); fellowships, career-related internships or fieldwork and institutionally-sponsored loans also available. *Faculty research:* Electrochemical engineering/fuel cell technology, fracture mechanics and nondestructive evaluation, virtual prototyping for electric power systems, wideband-gap electronics materials behavior/composites and smart materials. Total annual research expenditures: $11.9 million. *Unit head:* Dr. Craig A. Rogers, Dean, 803-777-4259, Fax: 803-777-9597. *Application contact:* Dr. J. H. Gibbons, Associate Dean, 803-777-4177, Fax: 803-777-0027, E-mail: gibbons@sc.edu.

See in-depth description on page 263.

University of Southern California, Graduate School, School of Engineering, Los Angeles, CA 90089. Offers MCM, MS, PhD, Engr, MBA/MS. Part-time programs available. *Students:* 1,120 full-time (179 women), 999 part-time (157 women); includes 365 minority (22 African Americans, 295 Asian Americans or Pacific Islanders, 48 Hispanic Americans), 1,364 international. Average age 24. 3709 applicants, 74% accepted. In 1998, 636 master's, 136 doctorates awarded. *Degree requirements:* For doctorate, dissertation required. *Entrance requirements:* For master's, doctorate, and Engr, GRE General Test. Application fee: $55. Tuition: Part-time $768 per unit. Required fees: $350 per semester. *Financial aid:* In 1998–99, 97 fellowships, 518 research assistantships, 180 teaching assistantships were awarded.; Federal Work-Study, institutionally-sponsored loans, and scholarships also available. Aid available to part-time students. Financial aid application deadline: 2/15; financial aid applicants required to submit FAFSA. *Unit head:* Dr. Leonard Silverman, Dean, 213-740-0017.

See in-depth description on page 265.

University of Southern Colorado, College of Applied Science and Engineering Technology, Pueblo, CO 81001-4901. Offers MS. Part-time and evening/weekend programs available. *Faculty:* 5 full-time (1 woman), 2 part-time (0 women). *Students:* 21 full-time (1 woman), 16 part-time (1 woman); includes 3 minority (all Hispanic Americans), 24 international. Average age 29. 23 applicants, 74% accepted. In 1998, 16 degrees awarded. *Degree requirements:* For master's, computer language required, thesis optional, foreign language not required. *Entrance requirements:* For master's, GRE General Test, TOEFL. *Average time to degree:* Master's–1.5 years full-time, 3.5 years part-time. *Application deadline:* For fall admission, 7/19 (priority date); for spring admission, 11/30 (priority date). Applications are processed on a rolling basis. Application fee: $15 ($30 for international students). *Financial aid:* In 1998–99, 5 teaching assistantships were awarded.; career-related internships or fieldwork, Federal Work-Study, institutionally-sponsored loans, and scholarships also available. Financial aid application deadline: 3/1; financial aid applicants required to submit FAFSA. *Faculty research:* Computer-integrated manufacturing, reliability, economic development, design of experiments, scheduling, simulation. Total annual research expenditures: $178,000. *Unit head:* Dr. Hector R. Carrasco, Dean, 719-549-2696, Fax: 719-549-2519, E-mail: carrasco@uscolo.edu. *Application contact:* Dr. Huseyin Sarper, Graduate Coordinator, 719-549-2889, Fax: 719-549-2519, E-mail: sarper@uscolo.edu.

University of Southern Indiana, Graduate Studies, School of Science and Engineering Technology, Evansville, IN 47712-3590. Offers MS. Part-time and evening/weekend programs available. *Degree requirements:* For master's, project required, foreign language and thesis not required. *Entrance requirements:* For master's, minimum GPA of 2.5, BS in engineering or engineering technology.

University of Southern Mississippi, Graduate School, College of Science and Technology, School of Engineering Technology, Hattiesburg, MS 39406-5167. Offers MS. Part-time programs available. *Faculty:* 19 full-time (1 woman), 1 part-time (0 women). *Students:* 25 full-time (4 women), 12 part-time (2 women); includes 14 minority (6 African Americans, 8 Asian Americans or Pacific Islanders) Average age 28. 43 applicants, 60% accepted. In 1998, 13 degrees awarded. *Degree requirements:* For master's, computer language, comprehensive exam required, thesis optional, foreign language not required. *Entrance requirements:* For master's, GMAT or GRE General Test, TOEFL, minimum GPA of 2.75. *Application deadline:*

Engineering and Applied Sciences—General

For fall admission, 8/6 (priority date). Applications are processed on a rolling basis. Application fee: $0 ($25 for international students). Tuition, state resident: full-time $2,250; part-time $137 per semester hour. Tuition, nonresident: full-time $3,102; part-time $172 per semester hour. Required fees: $602. *Financial aid:* Research assistantships, teaching assistantships, career-related internships or fieldwork and Federal Work-Study available. Financial aid application deadline: 3/15. *Faculty research:* Robotics; CAD/CAM; simulation; computer integrated manufacturing processes; construction scheduling, estimating, and computer systems. Total annual research expenditures: $1.9 million. *Unit head:* Dr. Ruth Ann Cade, Director, 601-266-4896, Fax: 601-266-5829. *Application contact:* Graduate Admissions, 601-266-5137.

University of South Florida, Graduate School, College of Engineering, Tampa, FL 33620-9951. Offers M Ch E, M Cp E, MCE, MCS, ME, MEE, MEVE, MME, MS, MS Ch E, MS Cp E, MSCE, MSCS, MSE, MSEE, MSEM, MSEV, MSIE, MSME, PhD, MS/MS. Part-time and evening/weekend programs available. *Faculty:* 89 full-time (7 women), 4 part-time (1 woman). *Students:* 269 full-time (58 women), 308 part-time (46 women); includes 102 minority (16 African Americans, 31 Asian Americans or Pacific Islanders, 53 Hispanic Americans, 2 Native Americans), 211 international. 623 applicants, 81% accepted. Terminal master's awarded for partial completion of doctoral program. *Degree requirements:* For doctorate, dissertation, 2 tools of research as specified by dissertation committee required, foreign language not required, foreign language not required. *Entrance requirements:* For master's, GRE General Test, minimum GPA of 3.0 during previous 2 years; for doctorate, GRE General Test. *Application deadline:* For fall admission, 6/1; for spring admission, 10/15. Applications are processed on a rolling basis. Application fee: $20. Electronic applications accepted. Tuition, state resident: part-time $148 per credit hour. Tuition, nonresident: part-time $509 per credit hour. *Financial aid:* In 1998–99, 268 students received aid, including 7 fellowships with full tuition reimbursements available, 195 research assistantships with full and partial tuition reimbursements available, 66 teaching assistantships with full tuition reimbursements available; career-related internships or fieldwork, Federal Work-Study, institutionally-sponsored loans, tuition waivers (partial), and unspecified assistantships also available. Aid available to part-time students. Financial aid applicants required to submit FAFSA. Total annual research expenditures: $4.3 million. *Unit head:* Michael G. Kovac, Dean, 813-974-3780, Fax: 813-974-5094, E-mail: kovac@eng.usf.edu.

Announcement: Master's and PhD degree programs are offered in chemical, civil, computer, electrical, environmental, industrial, and mechanical engineering; computer science; engineering management; and engineering science. Teaching and research assistantships are available to qualified students. Extensive student computing and laboratory facilities include vector and multiprocessor systems and advanced workstations.

See in-depth description on page 267.

University of Southwestern Louisiana, Graduate School, College of Engineering, Lafayette, LA 70504. Offers MS, MSE, MSET, MSTC, PhD. Part-time and evening/weekend programs available. *Faculty:* 60 full-time (6 women). *Students:* 115 full-time (25 women), 31 part-time (2 women); includes 6 minority (3 African Americans, 1 Asian American or Pacific Islander, 1 Hispanic American, 1 Native American), 112 international. Average age 25. 862 applicants, 64% accepted. In 1998, 97 master's, 15 doctorates awarded. Terminal master's awarded for partial completion of doctoral program. *Degree requirements:* For doctorate, computer language, dissertation, final oral exam required, foreign language not required, foreign language not required. *Entrance requirements:* For master's, GRE General Test; for doctorate, GRE General Test, minimum GPA of 3.0. *Application deadline:* For fall admission, 5/15. Application fee: $5 ($15 for international students). *Financial aid:* In 1998–99, 12 fellowships with full tuition reimbursements (averaging $15,417 per year), 102 research assistantships with full tuition reimbursements (averaging $6,319 per year) were awarded.; teaching assistantships, Federal Work-Study and tuition waivers (full and partial) also available. Aid available to part-time students. *Unit head:* Dr. Anthony B. Ponter, Dean, 318-482-6685.

University of Tennessee at Chattanooga, Graduate Division, School of Engineering, Chattanooga, TN 37403-2598. Offers MS. Part-time and evening/weekend programs available. *Faculty:* 17 full-time (2 women), 2 part-time (0 women). *Students:* 23 full-time (0 women), 86 part-time (20 women); includes 14 minority (8 African Americans, 4 Asian Americans or Pacific Islanders, 2 Hispanic Americans), 26 international. Average age 31. 96 applicants, 52% accepted. In 1998, 25 degrees awarded. *Degree requirements:* For master's, thesis required, foreign language not required. *Entrance requirements:* For master's, GRE General Test. *Application deadline:* Applications are processed on a rolling basis. Application fee: $25. *Financial aid:* Fellowships, research assistantships, Federal Work-Study and institutionally-sponsored loans available. Aid available to part-time students. Financial aid application deadline: 4/1. *Unit head:* Dr. Robert M. Desmond, Dean, 423-755-4121, Fax: 423-755-5229, E-mail: robert-desmond@utc.edu. *Application contact:* Dr. Deborah E. Arfken, Assistant Provost for Graduate Studies, 423-755-4667, Fax: 423-755-4478, E-mail: darfken@utcvm.utc.edu.

University of Tennessee, Knoxville, Graduate School, College of Engineering, Knoxville, TN 37996. Offers MS, PhD, MS/MBA. Part-time and evening/weekend programs available. Postbaccalaureate distance learning degree programs offered. *Faculty:* 134 full-time (5 women), 11 part-time (1 woman). *Students:* 301 full-time (60 women), 340 part-time (60 women); includes 48 minority (28 African Americans, 16 Asian Americans or Pacific Islanders, 1 Hispanic American, 3 Native Americans), 132 international. 575 applicants, 43% accepted. In 1998, 141 master's awarded (31% entered university research/teaching); 31 doctorates awarded. *Degree requirements:* For master's, thesis or alternative required, foreign language not required; for doctorate, dissertation required, foreign language not required. *Entrance requirements:* For master's and doctorate, TOEFL, minimum GPA of 2.7. *Application deadline:* For fall admission, 2/1 (priority date). Applications are processed on a rolling basis. Application fee: $35. Electronic applications accepted. *Financial aid:* In 1998–99, 8 fellowships, 143 research assistantships, 20 teaching assistantships were awarded.; career-related internships or fieldwork, Federal Work-Study, institutionally-sponsored loans, and unspecified assistantships also available. Financial aid application deadline: 2/1; financial aid applicants required to submit FAFSA. *Unit head:* Dr. Jerry Stoneking, Dean, 423-974-5321.

See in-depth description on page 269.

University of Tennessee Space Institute, Graduate Programs, Tullahoma, TN 37388-9700. Offers MS, PhD. Part-time programs available. Postbaccalaureate distance learning degree programs offered. *Faculty:* 41 full-time (1 woman), 8 part-time (0 women). *Students:* 87 full-time (24 women), 181 part-time (22 women); includes 21 minority (13 African Americans, 5 Asian Americans or Pacific Islanders, 2 Hispanic Americans, 1 Native American), 26 international. 180 applicants, 78% accepted. In 1998, 33 master's, 7 doctorates awarded. Terminal master's awarded for partial completion of doctoral program. *Degree requirements:* For doctorate, dissertation required, foreign language not required. *Application deadline:* Applications are processed on a rolling basis. Application fee: $35. *Financial aid:* In 1998–99, 8 fellowships, 50 research assistantships were awarded.; career-related internships or fieldwork, Federal Work-Study, and tuition waivers (full and partial) also available. Financial aid applicants required to submit FAFSA. *Faculty research:* Energy conversion, materials processing, computational fluid dynamics, aerodynamics, laser applications. Total annual research expenditures: $8 million. *Unit head:* Dr. T. Dwayne McCay, Vice President, 931-394-7213, Fax: 931-394-7211. *Application contact:* Dr. Edwin M. Gleason, Assistant Dean for Admissions and Student Affairs, 931-393-7432, Fax: 931-393-7346, E-mail: egleason@utsi.edu.

See in-depth description on page 271.

The University of Texas at Arlington, Graduate School, College of Engineering, Arlington, TX 76019. Offers M Engr, M Sw En, MCS, MS, PhD. Part-time programs available. *Faculty:* 90 full-time (6 women), 2 part-time (0 women). *Students:* 443 full-time (82 women), 447 part-time (68 women); includes 118 minority (11 African Americans, 84 Asian Americans or Pacific Islanders, 20 Hispanic Americans, 3 Native Americans), 487 international. 1222 applicants, 41% accepted. In 1998, 253 master's, 31 doctorates awarded. *Degree requirements:* For doctorate, dissertation required, foreign language not required. *Entrance requirements:* For master's and doctorate, GRE General Test, TOEFL. *Application deadline:* Applications are processed on a rolling basis. Application fee: $25 ($50 for international students). Tuition, state resident: full-time $1,368; part-time $76 per semester hour. Tuition, nonresident: full-time $5,454; part-time $303 per semester hour. Required fees: $66 per semester hour. $86 per term. Tuition and fees vary according to course load. *Financial aid:* Fellowships, research assistantships, teaching assistantships, career-related internships or fieldwork, Federal Work-Study, institutionally-sponsored loans, scholarships, and tuition waivers (partial) available. *Unit head:* Dr. J. Ronald Bailey, Dean, 817-272-2571, Fax: 817-272-2548, E-mail: bailey@uta.edu.

See in-depth description on page 273.

The University of Texas at Austin, Graduate School, College of Engineering, Austin, TX 78712-1111. Offers MSE, PhD, MBA/MSE, MP Aff/MSE. *Accreditation:* ABET (one or more programs are accredited). Part-time and evening/weekend programs available. *Faculty:* 236 full-time (21 women), 68 part-time (7 women). *Students:* 1,398 full-time (231 women), 358 part-time (59 women); includes 168 minority (20 African Americans, 93 Asian Americans or Pacific Islanders, 49 Hispanic Americans, 6 Native Americans), 901 international. 2859 applicants, 38% accepted. In 1998, 458 master's, 167 doctorates awarded. *Entrance requirements:* For master's and doctorate, GRE General Test. *Application fee:* $50 ($75 for international students). Electronic applications accepted. *Financial aid:* In 1998–99, 325 fellowships with partial tuition reimbursements (averaging $3,160 per year), 930 research assistantships with full tuition reimbursements (averaging $14,000 per year), 394 teaching assistantships with partial tuition reimbursements (averaging $13,000 per year) were awarded.; career-related internships or fieldwork, Federal Work-Study, institutionally-sponsored loans, scholarships, tuition waivers (partial), and academic assistantships, tutorships also available. Aid available to part-time students. Financial aid applicants required to submit FAFSA. Total annual research expenditures: $85 million. *Unit head:* Dr. Ben G. Streetman, Dean, 512-471-1166, Fax: 512-475-7072, E-mail: bstreet@mail.utexas.edu.

See in-depth description on page 275.

The University of Texas at Dallas, Erik Jonsson School of Engineering and Computer Science, Richardson, TX 75083-0688. Offers MS, MSEE, PhD. Part-time and evening/weekend programs available. *Faculty:* 37 full-time (3 women), 23 part-time (2 women). *Students:* 391 full-time (133 women), 332 part-time (74 women); includes 163 minority (7 African Americans, 142 Asian Americans or Pacific Islanders, 14 Hispanic Americans), 384 international. Average age 30. In 1998, 291 master's, 10 doctorates awarded. *Degree requirements:* For master's, minimum GPA of 3.0 required; for doctorate, dissertation required, foreign language not required. *Entrance requirements:* For master's, GRE General Test, TOEFL (minimum score of 550 required); for doctorate, GRE General Test, TOEFL (minimum score of 550 required), minimum GPA of 3.5. *Application deadline:* For fall admission, 7/15; for spring admission, 11/15. Applications are processed on a rolling basis. Application fee: $25 ($75 for international students). *Financial aid:* Fellowships, research assistantships, teaching assistantships, career-related internships or fieldwork, Federal Work-Study, grants, institutionally-sponsored loans, and scholarships available. Aid available to part-time students. Financial aid application deadline: 4/30; financial aid applicants required to submit FAFSA. *Unit head:* Dr. William P. Osborne, Dean, 972-883-2974, Fax: 972-883-2813, E-mail: wosborne@utdallas.edu. *Application contact:* Sheila R. Fleming, Student Development Specialist for Engineering and Computer Science, 972-883-6224, Fax: 972-883-2813, E-mail: fleming@utdallas.edu.

The University of Texas at El Paso, Graduate School, College of Engineering, El Paso, TX 79968-0001. Offers MEENE, MS, MSENE, PhD. Part-time and evening/weekend programs available. *Faculty:* 57 full-time (4 women), 17 part-time (1 woman). *Students:* 124 full-time (24 women), 124 part-time (21 women); includes 108 minority (1 African American, 4 Asian Americans or Pacific Islanders, 102 Hispanic Americans, 1 Native American), 99 international. Average age 28. 222 applicants, 50% accepted. In 1998, 66 master's awarded. *Degree requirements:* For doctorate, dissertation required, foreign language not required. *Entrance requirements:* For master's, GRE General Test, TOEFL (minimum score of 550 required); for doctorate, GRE General Test, TOEFL. *Application deadline:* Applications are processed on a rolling basis. Application fee: $15 ($65 for international students). Electronic applications accepted. Tuition, state resident: full-time $2,790. Tuition, nonresident: full-time $7,710. *Financial aid:* Fellowships, research assistantships, teaching assistantships, career-related internships or fieldwork, Federal Work-Study, institutionally-sponsored loans, and tuition waivers (partial) available. Financial aid applicants required to submit FAFSA. Total annual research expenditures: $3.7 million. *Unit head:* Dr. Andrew Swift, Interim Dean, 915-747-5460. *Application contact:* Susan Jordan, Director, Graduate Student Services, 915-747-5491, Fax: 915-747-5788, E-mail: sjordan@utep.edu.

The University of Texas at San Antonio, College of Sciences and Engineering, Division of Engineering, San Antonio, TX 78249-0617. Offers civil engineering (MS); electrical engineering (MS); mechanical engineering (MS). Part-time and evening/weekend programs available. *Faculty:* 22 full-time (2 women), 15 part-time (2 women). *Students:* 19 full-time (1 woman), 90 part-time (15 women); includes 38 minority (5 African Americans, 11 Asian Americans or Pacific Islanders, 21 Hispanic Americans, 1 Native American), 21 international. Average age 34. 69 applicants, 61% accepted. In 1998, 35 degrees awarded. *Degree requirements:* For master's, thesis optional, foreign language not required. *Entrance requirements:* For master's, GRE General Test. *Application deadline:* For fall admission, 7/1; for spring admission, 12/1. Applications are processed on a rolling basis. Application fee: $25. *Financial aid:* Research assistantships, teaching assistantships, career-related internships or fieldwork and institutionally-sponsored loans available. Aid available to part-time students. Financial aid application deadline: 3/31. *Faculty research:* Nonlinear digital signal processing, materials, environmental studies, digital systems, thermal systems. *Unit head:* Dr. Lex Akers, Director, 210-458-4490.

The University of Texas at Tyler, Graduate Studies, School of Engineering, Tyler, TX 75799-0001. Offers M Engr. Part-time programs available. *Faculty:* 9 full-time (0 women). 2 applicants, 100% accepted. *Degree requirements:* For master's, report required, foreign language and thesis not required. *Financial aid:* Application deadline: 7/1. *Unit head:* Dr. Leonard Hale, Dean, 903-566-7002, Fax: 903-566-7148, E-mail: lhale@mail.uttyl.edu.

University of Toledo, Graduate School, College of Engineering, Toledo, OH 43606-3398. Offers MS, MS Ch E, MSCE, MSEE, MSES, MSIE, MSME, PhD. Part-time and evening/weekend programs available. Postbaccalaureate distance learning degree programs offered (minimal on-campus study). *Faculty:* 84 full-time (9 women), 12 part-time (0 women). *Students:* 372 full-time (67 women), 131 part-time (18 women); includes 15 minority (8 African Americans, 5 Asian Americans or Pacific Islanders, 1 Hispanic American, 1 Native American), 357 international. Average age 26. 1471 applicants, 56% accepted. In 1998, 217 master's, 10 doctorates awarded. Terminal master's awarded for partial completion of doctoral program. *Degree requirements:* For doctorate, dissertation required, foreign language not required, foreign language not required. *Entrance requirements:* For master's and doctorate, GRE General Test, TOEFL (minimum score of 550 required). *Average time to degree:* Master's–2 years full-time; doctorate–4 years full-time. *Application deadline:* For fall admission, 5/31 (priority date). Applications are processed on a rolling basis. Application fee: $30. Electronic applications accepted. *Financial aid:* In 1998–99, 393 students received aid, including 2 fellowships with full tuition reimbursements available, 89 research assistantships with full tuition reimbursements available, 71 teaching assistantships with full tuition reimbursements available; Federal Work-Study, scholarships, tuition waivers (full), and unspecified assistantships also available. Aid available to part-time students. Financial aid application deadline: 4/1. *Faculty research:* Robotics, computer engineering, fluid dynamics, infrastructure, imaging. Total annual research expenditures: $5 million. *Unit head:* Dr. Phillip R. White, Interim Dean, 419-530-8020, Fax: 419-530-8026, E-mail: pwhite@eng.utoledo.edu. *Application contact:* Dr. Atam P. Dhawan, Assistant Dean for Graduate Studies, 419-530-7391, Fax: 419-530-7392, E-mail: adhawan@eng.utoledo.edu.

See in-depth description on page 277.

Engineering and Applied Sciences—General

University of Toronto, School of Graduate Studies, Physical Sciences Division, Faculty of Applied Science and Engineering, Toronto, ON M5S 1A1, Canada. Offers M Eng, M Sc, MA Sc, MH Sc, PhD. Part-time programs available. *Degree requirements:* For doctorate, dissertation required.

See in-depth description on page 279.

University of Tulsa, Graduate School, College of Engineering and Applied Sciences, Tulsa, OK 74104-3189. Offers ME, METM, MS, MSE, PhD, JD/MS. Part-time programs available. *Faculty:* 68 full-time (5 women), 1 part-time (0 women). *Students:* 143 full-time (35 women), 26 part-time (6 women); includes 6 minority (3 Asian Americans or Pacific Islanders, 1 Hispanic American, 2 Native Americans), 117 international. Average age 29. 212 applicants, 64% accepted. In 1998, 57 master's, 16 doctorates awarded. *Degree requirements:* For doctorate, dissertation required, foreign language not required. *Entrance requirements:* For master's and doctorate, GRE General Test, TOEFL. *Application deadline:* Applications are processed on a rolling basis. Application fee: $30. Electronic applications accepted. Tuition: Full-time $8,640; part-time $480 per hour. Required fees: $3 per hour. One-time fee: $200 full-time. Tuition and fees vary according to program. *Financial aid:* In 1998–99, 146 students received aid, including 5 fellowships with full and partial tuition reimbursements available (averaging $7,826 per year), 85 research assistantships with full and partial tuition reimbursements available (averaging $5,898 per year), 56 teaching assistantships with full and partial tuition reimbursements available (averaging $5,681 per year); career-related internships or fieldwork, Federal Work-Study, and tuition waivers (partial) also available. Aid available to part-time students. Financial aid application deadline: 2/1; financial aid applicants required to submit FAFSA. *Unit head:* Dr. Steve J. Bellovich, Dean, 918-631-2288.

University of Utah, Graduate School, College of Engineering, Salt Lake City, UT 84112-1107. Offers M Phil, ME, MEA, MS, PhD, EE. Part-time programs available. *Faculty:* 103 full-time (6 women), 155 part-time (11 women). *Students:* 317 full-time (66 women), 190 part-time (25 women); includes 15 minority (9 Asian Americans or Pacific Islanders, 3 Hispanic Americans, 3 Native Americans), 188 international. Average age 30. Terminal master's awarded for partial completion of doctoral program. *Degree requirements:* Foreign language not required. *Entrance requirements:* For master's and doctorate, TOEFL, minimum GPA of 3.0. Application fee: $30 ($50 for international students). *Financial aid:* In 1998–99, 85 teaching assistantships were awarded.; fellowships, research assistantships, career-related internships or fieldwork, Federal Work-Study, institutionally-sponsored loans, and traineeships also available. Aid available to part-time students. *Faculty research:* Biomaterials, wastewater treatment, computer-aided graphics design, semiconductors, polymers. *Unit head:* David Pershing, Dean, 801-581-6911, Fax: 801-581-8692, E-mail: david.pershing@dean.eng.utah.edu.

University of Vermont, Graduate College, College of Engineering and Mathematics, Burlington, VT 05405-0160. Offers MAT, MS, MST, PhD. Part-time programs available. *Degree requirements:* For doctorate, dissertation required, foreign language not required. *Entrance requirements:* For master's and doctorate, GRE General Test, TOEFL (minimum score of 550 required).

See in-depth description on page 281.

University of Victoria, Faculty of Graduate Studies, Faculty of Engineering, Victoria, BC V8W 2Y2, Canada. Offers M Eng, M Sc, MA, MA Sc, PhD. Part-time and evening/weekend programs available. *Faculty:* 63 full-time (7 women), 17 part-time (0 women). *Students:* 182 full-time (30 women), 8 part-time (1 woman), 79 international. Average age 27. 447 applicants, 12% accepted. In 1998, 31 master's, 13 doctorates awarded. *Degree requirements:* For doctorate, dissertation required. *Average time to degree:* Master's–2.5 years full-time; doctorate–4.5 years full-time. *Application deadline:* Applications are processed on a rolling basis. Application fee: $50. *Financial aid:* Fellowships, research assistantships, teaching assistantships, career-related internships or fieldwork, institutionally-sponsored loans, and awards available. Financial aid application deadline: 2/15. *Faculty research:* Computer-aided design, analog and digital filter design, VLSI system design, underwater acoustic systems, computer-aided manufacture. *Unit head:* Dr. M. Miller, Dean, 250-721-8612.

University of Virginia, School of Engineering and Applied Science, Charlottesville, VA 22903. Offers MAM, MCS, ME, MEP, MMSE, MS, PhD, ME/MBA. Part-time programs available. Postbaccalaureate distance learning degree programs offered (no on-campus study). *Faculty:* 162 full-time (21 women), 17 part-time (4 women). *Students:* 487 full-time (111 women), 23 part-time (4 women); includes 55 minority (19 African Americans, 24 Asian Americans or Pacific Islanders, 10 Hispanic Americans, 2 Native Americans), 137 international. Average age 27. 799 applicants, 39% accepted. In 1998, 159 master's, 53 doctorates awarded. Terminal master's awarded for partial completion of doctoral program. *Degree requirements:* For doctorate, dissertation, comprehensive exam required, foreign language not required. *Entrance requirements:* For master's and doctorate, GRE General Test. *Application deadline:* Applications are processed on a rolling basis. Application fee: $60. Electronic applications accepted. *Financial aid:* Fellowships with full tuition reimbursements, research assistantships with full tuition reimbursements, teaching assistantships with full tuition reimbursements, career-related internships or fieldwork available. Financial aid application deadline: 2/1. *Unit head:* Richard W. Miksad, Dean, 804-924-3593. *Application contact:* J. Milton Adams, Assistant Dean, 804-924-3897, E-mail: twr2c@virginia.edu.

See in-depth description on page 283.

University of Washington, Graduate School, College of Engineering, Seattle, WA 98195. Offers MAE, MS, MS Ch E, MS Civ E, MSAA, MSE, MSEE, MSIE, MSME, MSMSE, PhD, MBA/MSE. Part-time programs available. Postbaccalaureate distance learning degree programs offered (minimal on-campus study). *Faculty:* 185 full-time (22 women), 1 part-time (0 women). *Students:* 950 full-time (232 women), 383 part-time (68 women); includes 191 minority (22 African Americans, 133 Asian Americans or Pacific Islanders, 30 Hispanic Americans, 6 Native Americans), 384 international. Average age 29. 2254 applicants, 36% accepted. In 1998, 254 master's, 76 doctorates awarded. Terminal master's awarded for partial completion of doctoral program. *Degree requirements:* For doctorate, dissertation required, foreign language not required. *Entrance requirements:* For master's and doctorate, GRE, TOEFL. *Average time to degree:* Master's–2.1 years full-time, 4.5 years part-time; doctorate–5 years full-time, 7.75 years part-time. *Application deadline:* For winter admission, 11/1; for spring admission, 2/1. Applications are processed on a rolling basis. Application fee: $50. Electronic applications accepted. Tuition, state resident: full-time $5,196; part-time $475 per credit. Tuition, nonresident: full-time $13,485; part-time $1,285 per credit. Required fees: $387; $38 per credit. Tuition and fees vary according to course load. *Financial aid:* In 1998–99, 110 fellowships with full tuition reimbursements (averaging $10,807 per year), 460 research assistantships with full and partial tuition reimbursements, 204 teaching assistantships with full and partial tuition reimbursements were awarded.; career-related internships or fieldwork, Federal Work-Study, grants, institutionally-sponsored loans, scholarships, tuition waivers (full), unspecified assistantships, and stipend supplements also available. Aid available to part-time students. Financial aid application deadline: 2/28; financial aid applicants required to submit FAFSA. *Faculty research:* Advanced materials and manufacturing, biotechnology, computer systems and software, microelectronics, earthquake engineering. Total annual research expenditures: $39.5 million. *Unit head:* Dr. Denice D. Denton, Dean, 206-543-0340, Fax: 206-685-0666, E-mail: denton@engr.washington.edu. *Application contact:* Frank Ashby, Student and Community College Relations Manager, 206-543-1770, Fax: 206-616-8554, E-mail: engradv@engr.washington.edu.

University of Waterloo, Graduate Studies, Faculty of Engineering, Waterloo, ON N2L 3G1, Canada. Offers MA Sc, PhD. Part-time and evening/weekend programs available. Postbaccalaureate distance learning degree programs offered (no on-campus study). *Faculty:* 159 full-time (13 women), 82 part-time (9 women). *Students:* 456 full-time (109 women), 97 part-time (15 women). In 1998, 102 master's, 60 doctorates awarded. *Degree requirements:* For master's, research paper or thesis required; for doctorate, dissertation, comprehensive exam required. *Entrance requirements:* For master's, TOEFL (minimum score of 550 required), honors degree; for doctorate, TOEFL (minimum score of 550 required), master's degree. *Application deadline:* Applications are processed on a rolling basis. Application fee: $50. *Expenses:* Tuition and fees charges are reported in Canadian dollars. Tuition, state resident: full-time $3,168 Canadian dollars; part-time $792 Canadian dollars per term. Tuition, nonresident: full-time $8,000 Canadian dollars; part-time $2,000 Canadian dollars. Required fees: $45 Canadian dollars per term. Tuition and fees vary according to program. *Financial aid:* Fellowships, research assistantships, teaching assistantships, career-related internships or fieldwork, Federal Work-Study, and institutionally-sponsored loans available. *Unit head:* Dr. S. Chaudhuri, Dean, 519-888-4567 Ext. 3347, Fax: 519-746-1457. *Application contact:* Dr. A. Penlidis, Associate Dean of Graduate Studies, 519-888-4567 Ext. 3376, Fax: 519-746-1457, E-mail: penlidis@engmail.uwaterloo.ca.

The University of Western Ontario, Faculty of Graduate Studies, Physical Sciences Division, Faculty of Engineering Science, London, ON N6A 5B8, Canada. Offers M Eng, ME Sc, PhD. Part-time programs available. Terminal master's awarded for partial completion of doctoral program. *Degree requirements:* For master's and doctorate, thesis/dissertation required, foreign language not required. *Faculty research:* Wind, geotechnical, chemical reactor engineering, applied electrostatics, biochemical engineering.

University of Windsor, College of Graduate Studies and Research, Faculty of Engineering, Windsor, ON N9B 3P4, Canada. Offers MA Sc, PhD. Part-time programs available. *Degree requirements:* For doctorate, dissertation required. *Entrance requirements:* For master's, TOEFL, minimum B average; for doctorate, TOEFL, master's degree.

University of Wisconsin–Madison, Graduate School, College of Engineering, Madison, WI 53706-1380. Offers ME, MS, PhD, PDD. Part-time programs available. Postbaccalaureate distance learning degree programs offered (minimal on-campus study). *Faculty:* 187 full-time (17 women), 13 part-time (0 women). *Students:* 874 full-time (152 women), 148 part-time (17 women); includes 80 minority (14 African Americans, 36 Asian Americans or Pacific Islanders, 27 Hispanic Americans, 3 Native Americans), 510 international. 1639 applicants, 33% accepted. In 1998, 288 master's, 109 doctorates awarded. *Degree requirements:* For doctorate, dissertation required. Application fee: $45. Electronic applications accepted. *Financial aid:* Fellowships with full and partial tuition reimbursements, research assistantships with full tuition reimbursements, teaching assistantships with full tuition reimbursements, career-related internships or fieldwork, Federal Work-Study, institutionally-sponsored loans, scholarships, and unspecified assistantships available. Aid available to part-time students. Total annual research expenditures: $65 million. *Unit head:* John G. Bollinger, Dean, 608-262-3482, Fax: 608-262-6400, E-mail: boleng@engr.wisc.edu. *Application contact:* Graduate Admissions, 608-262-2433, Fax: 608-262-5134, E-mail: gradadmiss@mail.bascom.wisc.edu.

See in-depth description on page 285.

University of Wisconsin–Madison, Graduate School, Department of Engineering Professional Development, Madison, WI 53706-1380. Offers engineering (PDD); professional practice (ME); technical Japanese (ME).

University of Wisconsin–Milwaukee, Graduate School, College of Engineering and Applied Science, Milwaukee, WI 53201-0413. Offers MS, PhD, MUP/MS. Part-time programs available. *Faculty:* 57 full-time (3 women). *Students:* 89 full-time (23 women), 188 part-time (36 women); includes 25 minority (6 African Americans, 15 Asian Americans or Pacific Islanders, 4 Hispanic Americans), 108 international. 223 applicants, 61% accepted. In 1998, 42 master's, 12 doctorates awarded. *Degree requirements:* For master's, thesis or alternative required, foreign language not required; for doctorate, dissertation, internship required, foreign language not required. *Entrance requirements:* For master's, minimum GPA of 2.75; for doctorate, minimum GPA of 3.5. *Application deadline:* For fall admission, 1/1 (priority date); for spring admission, 9/1. Applications are processed on a rolling basis. Application fee: $45 ($75 for international students). *Financial aid:* In 1998–99, 5 fellowships, 30 research assistantships, 50 teaching assistantships were awarded.; career-related internships or fieldwork, Federal Work-Study, and unspecified assistantships also available. Aid available to part-time students. Financial aid application deadline: 4/15. *Unit head:* Dr. Chan Shih-Hung, Interim Dean, 414-229-5001.

See in-depth description on page 287.

University of Wyoming, Graduate School, College of Engineering, Laramie, WY 82071. Offers MS, PhD. Part-time programs available. *Faculty:* 74. *Students:* 87 full-time (11 women), 59 part-time (11 women); includes 4 minority (2 Asian Americans or Pacific Islanders, 1 Hispanic American, 1 Native American), 46 international. 233 applicants, 32% accepted. In 1998, 30 master's, 4 doctorates awarded. *Degree requirements:* For doctorate, dissertation required. *Entrance requirements:* For master's and doctorate, GRE General Test, TOEFL, minimum GPA of 3.0. *Application deadline:* Applications are processed on a rolling basis. Application fee: $40. Electronic applications accepted. Tuition, state resident: full-time $2,520; part-time $140 per credit hour. Tuition, nonresident: full-time $7,790; part-time $433 per credit hour. Required fees: $400; $7 per credit hour. Full-time tuition and fees vary according to course load and program. *Financial aid:* Fellowships, research assistantships, teaching assistantships, career-related internships or fieldwork, Federal Work-Study, and institutionally-sponsored loans available. Aid available to part-time students. Total annual research expenditures: $5.8 million. *Unit head:* Dr. Kynric Pell, Dean, 307-766-4253, Fax: 307-766-4444, E-mail: pell@uwyo.edu.

Utah State University, School of Graduate Studies, College of Engineering, Logan, UT 84322. Offers ME, MS, PhD, CE, EE. Part-time and evening/weekend programs available. *Students:* 179 full-time (20 women), 69 part-time (9 women); includes 4 minority (2 Asian Americans or Pacific Islanders, 1 Hispanic American, 1 Native American), 119 international. Average age 28. 445 applicants, 47% accepted. In 1998, 78 master's, 19 doctorates awarded. Terminal master's awarded for partial completion of doctoral program. *Degree requirements:* For doctorate, dissertation required, foreign language not required, foreign language not required. *Entrance requirements:* For master's and doctorate, GRE General Test (score in 40th percentile or higher required), TOEFL, minimum GPA of 3.0. *Application deadline:* For spring admission, 10/15. Applications are processed on a rolling basis. Application fee: $40. Tuition, state resident: full-time $1,492. Tuition, nonresident: full-time $5,232. Required fees: $434. Tuition and fees vary according to course load. *Financial aid:* Fellowships with partial tuition reimbursements, research assistantships with partial tuition reimbursements, teaching assistantships with partial tuition reimbursements, career-related internships or fieldwork, Federal Work-Study, institutionally-sponsored loans, and tuition waivers (partial) available. Aid available to part-time students. *Unit head:* A. Bruce Bishop, Dean, 435-797-2775.

Announcement: College of Engineering has graduate programs in biological and irrigation engineering, civil and environmental engineering, electrical and computer engineering, and mechanical, manufacturing, and aerospace engineering. Among the several research centers are the Utah Water Research Laboratory, Space Dynamics Laboratory, and Engineering Experiment Station. Research funding in the College averages $30 million per year.

See in-depth description on page 289.

Vanderbilt University, School of Engineering, Nashville, TN 37240-1001. Offers M Eng, MS, PhD, MBA/M Eng, MD/PhD. MS and PhD offered through the Graduate School. Part-time programs available. *Faculty:* 91 full-time (4 women), 10 part-time (3 women). *Students:* 291 full-time (59 women), 25 part-time (3 women); includes 16 minority (12 African Americans, 3 Asian Americans or Pacific Islanders, 1 Native American), 166 international. Average age 25. 536 applicants, 34% accepted. In 1998, 56 master's, 31 doctorates awarded. Terminal master's awarded for partial completion of doctoral program. *Degree requirements:* For doctorate, dissertation required, foreign language not required, foreign language not required. *Entrance requirements:* For master's and doctorate, GRE General Test. *Average time to degree:* Master's–2 years full-time; doctorate–3 years full-time. *Application deadline:* For fall admission, 1/

15. Application fee: $40. Electronic applications accepted. *Financial aid:* In 1998–99, 251 students received aid, including 16 fellowships with full tuition reimbursements available (averaging $15,000 per year), 130 research assistantships with full tuition reimbursements available (averaging $15,000 per year), 103 teaching assistantships with full tuition reimbursements available (averaging $11,250 per year); career-related internships or fieldwork, Federal Work-Study, institutionally-sponsored loans, scholarships, and tuition waivers (full and partial) also available. Aid available to part-time students. Financial aid application deadline: 1/15. *Faculty research:* Bio-optics, imaging, adsorption and surface chemistry, image processing, robotics, microelectronics, laser diagnostics. Total annual research expenditures: $11.2 million. *Unit head:* Kenneth F. Galloway, Dean, 615-322-0720, Fax: 615-343-8006, E-mail: kfg@vuse.vanderbilt.edu.

See in-depth description on page 291.

Villanova University, College of Engineering, Villanova, PA 19085-1699. Offers M Ch E, MCE, MME, MSCE, MSEE, MSTE, MSWREE, Certificate. Part-time and evening/weekend programs available. *Faculty:* 48 full-time (2 women), 23 part-time (0 women). *Students:* 58 full-time (19 women), 170 part-time (41 women); includes 14 minority (4 African Americans, 3 Asian Americans or Pacific Islanders, 7 Hispanic Americans), 36 international. Average age 26. 187 applicants, 73% accepted. In 1998, 73 degrees awarded. *Degree requirements:* For master's, thesis optional, foreign language not required. *Entrance requirements:* For master's, GRE General Test (for applicants with degrees from foreign universities), minimum GPA of 3.0. *Average time to degree:* Master's–2 years full-time, 4 years part-time. *Application deadline:* For fall admission, 8/1 (priority date); for spring admission, 12/1. Applications are processed on a rolling basis. Application fee: $40. *Financial aid:* In 1998–99, 43 students received aid, including 4 research assistantships with full tuition reimbursements available (averaging $9,215 per year), 28 teaching assistantships with full tuition reimbursements available (averaging $9,215 per year); Federal Work-Study, scholarships, and tuition waivers (full and partial) also available. Aid available to part-time students. *Faculty research:* Composite materials, economy and risk, heat transfer, signal detection. *Unit head:* Robert D. Lynch, Dean, 610-519-4940, Fax: 610-519-4941, E-mail: rlynch@email.vill.edu.

Virginia Commonwealth University, School of Graduate Studies, School of Engineering, Richmond, VA 23284-9005. Offers MS, PhD, MD/PhD. *Students:* 20 full-time (6 women), 3 part-time; includes 10 minority (all Asian Americans or Pacific Islanders) In 1998, 6 master's, 1 doctorate awarded. *Degree requirements:* For doctorate, dissertation, comprehensive oral and written exams required. *Entrance requirements:* For master's and doctorate, GRE General Test. *Application deadline:* For fall admission, 4/15. Application fee: $30. Tuition, state resident: full-time $4,031; part-time $224 per credit hour. Tuition, nonresident: full-time $11,946; part-time $664 per credit hour. Required fees: $1,081; $40 per credit hour. Tuition and fees vary according to campus/location and program. *Faculty research:* Artificial hearts, orthopedic implants, medical imaging, medical instrumentation and sensors, cardiac monitoring. *Unit head:* Dr. Henry A. McGee, Dean, 804-828-3636, Fax: 804-828-9866. *Application contact:* Dr. Gerald Miller, Associate Dean for Graduate Education, 804-828-7956, Fax: 804-828-4454, E-mail: gemiller@vcu.edu.

Virginia Polytechnic Institute and State University, Graduate School, College of Engineering, Blacksburg, VA 24061. Offers M Eng, MEA, MS, PhD. *Accreditation:* ABET (one or more programs are accredited). Part-time and evening/weekend programs available. *Faculty:* 263 full-time (19 women). *Students:* 997 full-time (168 women), 544 part-time (95 women); includes 161 minority (42 African Americans, 80 Asian Americans or Pacific Islanders, 34 Hispanic Americans, 5 Native Americans), 555 international. 2012 applicants, 43% accepted. In 1998, 401 master's, 91 doctorates awarded. Terminal master's awarded for partial completion of doctoral program. *Degree requirements:* Foreign language not required. *Entrance requirements:* For master's and doctorate, TOEFL. *Application deadline:* For fall admission, 12/1 (priority date). Applications are processed on a rolling basis. Application fee: $25. *Financial aid:* In 1998–99, 379 research assistantships, 172 teaching assistantships were awarded.; fellowships, career-related internships or fieldwork, Federal Work-Study, institutionally-sponsored loans, tuition waivers (full and partial), and unspecified assistantships also available. Aid available to part-time students. Financial aid application deadline: 1/15. *Unit head:* Dr. F. William Stephenson, Dean, 540-231-6641.

See in-depth description on page 293.

Washington State University, Graduate School, College of Engineering and Architecture, Pullman, WA 99164. Offers MS, PhD. *Faculty:* 107. *Students:* 219 full-time (49 women), 34 part-time (6 women); includes 12 minority (2 African Americans, 10 Asian Americans or Pacific Islanders), 146 international. In 1998, 174 master's, 23 doctorates awarded. Terminal master's awarded for partial completion of doctoral program. *Degree requirements:* For master's, oral exam required; for doctorate, dissertation, oral exam required. *Entrance requirements:* For master's and doctorate, minimum GPA of 3.0. *Average time to degree:* Master's–2 years full-time; doctorate–4 years full-time. *Application deadline:* For fall admission, 3/1 (priority date). Applications are processed on a rolling basis. Application fee: $35. *Financial aid:* In 1998–99, 83 research assistantships, 92 teaching assistantships were awarded.; fellowships, career-related internships or fieldwork, Federal Work-Study, institutionally-sponsored loans, tuition waivers (partial), and teaching associateships also available. Financial aid applicants required to submit FAFSA. Total annual research expenditures: $8.3 million. *Unit head:* Dr. Anjan Bose, Dean, 509-335-5593.

Washington University in St. Louis, School of Engineering and Applied Science, St. Louis, MO 63130-4899. Offers MA, MCE, MCE, MCM, MS, MSCE, MSE, MSEE, MSEE, D Sc, M Arch/MCM. Part-time and evening/weekend programs available. Terminal master's awarded for partial completion of doctoral program. *Degree requirements:* For master's, thesis optional, foreign language not required; for doctorate, dissertation required.

Wayne State University, Graduate School, College of Engineering, Detroit, MI 48202. Offers MS, PhD, Certificate. Part-time programs available. Terminal master's awarded for partial completion of doctoral program. *Degree requirements:* For master's, thesis optional, foreign language not required; for doctorate, dissertation required. *Faculty research:* Air/water quality, optoelectronic computing, manufacturing quality control, structural optimization, combustion.

Announcement: Wayne State University is located in the heart of one of the world's greatest manufacturing centers, and the College of Engineering maintains extensive relationships with the many large industries in the Detroit area. In fact, Wayne Engineering is a fully integrated part of the R&D infrastructure of the major manufacturers based in Detroit.

See in-depth description on page 295.

Western Michigan University, Graduate College, College of Engineering and Applied Sciences, Kalamazoo, MI 49008. Offers MS, MSE, PhD. Part-time programs available. *Faculty:* 111 full-time (9 women), 20 part-time (8 women). *Students:* 55 full-time (5 women), 310 part-time (34 women); includes 18 minority (6 African Americans, 10 Asian Americans or Pacific Islanders, 1 Hispanic American, 1 Native American), 160 international. 447 applicants, 67% accepted. In 1998, 66 degrees awarded. *Degree requirements:* For doctorate, dissertation, oral exam required. *Entrance requirements:* For master's, minimum GPA of 3.0; for doctorate, GRE General Test, minimum GPA of 3.0. *Application deadline:* For fall admission, 2/15 (priority date). Applications are processed on a rolling basis. Application fee: $25. *Financial aid:* Fellowships, research assistantships, teaching assistantships, career-related internships or fieldwork and Federal Work-Study available. Financial aid application deadline: 2/15; financial aid applicants required to submit FAFSA. *Unit head:* Dr. Leonard R. Lamberson, Dean, 616-387-4017. *Application contact:* Paula J. Boodt, Coordinator, Graduate Admissions and Recruitment, 616-387-2000, Fax: 616-387-2355, E-mail: paula.boodt@wmich.edu.

See in-depth description on page 297.

Western New England College, School of Engineering, Springfield, MA 01119-2654. Offers MSEE, MSEM, MSME. Part-time and evening/weekend programs available. *Faculty:* 17 full-time (0 women), 2 part-time (0 women). Average age 29. 20 applicants, 80% accepted. In 1998, 30 degrees awarded. *Degree requirements:* For master's, computer language, comprehensive exam required, thesis optional, foreign language not required. *Entrance requirements:* For master's, bachelor's degree in engineering or related field. *Application deadline:* Applications are processed on a rolling basis. Application fee: $30. *Financial aid:* Teaching assistantships available. Aid available to part-time students. Financial aid application deadline: 4/1; financial aid applicants required to submit FAFSA. *Faculty research:* Fluid mechanics, control systems. *Unit head:* Dr. Eric W. Haffner, Dean, 413-782-1273, E-mail: ehaffner@wnec.edu. *Application contact:* Harry F. Neunder, Coordinator, Continuing Education, 413-782-1750, Fax: 413-782-1779, E-mail: hneunder@wnec.edu.

West Texas A&M University, College of Agriculture, Nursing, and Natural Sciences, Department of Mathematics, Physical Sciences and Engineering Technology, Program in Engineering Technology, Canyon, TX 79016-0001. Offers MS. Part-time programs available. *Faculty:* 2 full-time (1 woman). *Students:* 4 full-time (1 woman), 17 part-time (3 women); includes 1 minority (Hispanic American), 5 international. Average age 34. 2 applicants, 0% accepted. In 1998, 3 degrees awarded. *Degree requirements:* For master's, comprehensive exam required, thesis optional, foreign language not required. *Entrance requirements:* For master's, GRE General Test (minimum combined score of 950 required; average 964). *Application deadline:* Applications are processed on a rolling basis. Application fee: $0 ($50 for international students). Electronic applications accepted. Tuition, state resident: full-time $1,152; part-time $48 per credit. Tuition, nonresident: full-time $6,336; part-time $264 per credit. Required fees: $1,063; $531 per semester. *Financial aid:* In 1998–99, research assistantships (averaging $6,500 per year), 1 teaching assistantship (averaging $6,500 per year) were awarded.; Federal Work-Study, institutionally-sponsored loans, and tuition waivers (partial) also available. Aid available to part-time students. Financial aid applicants required to submit FAFSA. *Faculty research:* Composites, firearms technology, small arms research and development. *Unit head:* Graduate Admissions, 608-262-2433, Fax: 608-262-5134, E-mail: gradadmiss@mail.bascom.wisc.edu. *Application contact:* Dr. Gerald Chen, Graduate Adviser, 806-651-2449, Fax: 806-651-2733, E-mail: gchen@mail.wtamu.edu.

West Virginia University, College of Engineering and Mineral Resources, Morgantown, WV 26506. Offers MS, MS Ch E, MSAE, MSCE, MSE, MSEE, MSEM, MSIE, MSME, MSPNGE, PhD. *Accreditation:* ABET (one or more programs are accredited). Part-time programs available. Terminal master's awarded for partial completion of doctoral program. *Degree requirements:* For master's, thesis optional, foreign language not required; for doctorate, dissertation, comprehensive exam required, foreign language not required. *Entrance requirements:* For master's and doctorate, TOEFL (minimum score of 550 required). *Faculty research:* Composite materials, software engineering, information systems, aerodynamics, vehicle propulsion and emission, manufacturing, longwall mining, transportation planning, biomedical engineering.

See in-depth description on page 299.

West Virginia University Institute of Technology, College of Engineering, Montgomery, WV 25136. Offers MS. Part-time programs available. *Faculty:* 13 full-time (1 woman). *Students:* 7 full-time (0 women), 5 part-time, 9 international. Average age 27. 52 applicants, 85% accepted. In 1998, 4 degrees awarded. *Degree requirements:* For master's, thesis or alternative, fieldwork required, foreign language not required. *Entrance requirements:* For master's, GRE General Test (minimum combined score of 1100 required), TOEFL (minimum score of 550 required), minimum GPA of 3.0. *Average time to degree:* Master's–2 years full-time. *Application deadline:* For fall admission, 3/15 (priority date). Applications are processed on a rolling basis. Application fee: $10. Tuition, state resident: full-time $2,816; part-time $312 per credit. Tuition, nonresident: full-time $6,964; part-time $774 per credit. *Financial aid:* In 1998–99, 7 teaching assistantships were awarded.; career-related internships or fieldwork, Federal Work-Study, and institutionally-sponsored loans also available. Financial aid application deadline: 3/15. *Unit head:* Dr. William Gregory, Dean, 304-442-3161, Fax: 304-442-1006. *Application contact:* Robert P. Scholl, Registrar, 304-442-3167, Fax: 304-442-3097, E-mail: rpscho@wvit.wvnet.edu.

Wichita State University, Graduate School, College of Engineering, Wichita, KS 67260. Offers MEM, MS, PhD. Part-time and evening/weekend programs available. *Faculty:* 46 full-time (4 women), 11 part-time (2 women). *Students:* 207 full-time (18 women), 202 part-time (25 women); includes 15 minority (1 African American, 10 Asian Americans or Pacific Islanders, 3 Hispanic Americans, 1 Native American), 272 international. Average age 32. 575 applicants, 74% accepted. In 1998, 99 master's, 14 doctorates awarded. Terminal master's awarded for partial completion of doctoral program. *Degree requirements:* For doctorate, dissertation, comprehensive exam required, foreign language not required. *Entrance requirements:* For master's and doctorate, GRE, TOEFL (minimum score of 550 required). *Application deadline:* For fall admission, 7/1 (priority date); for spring admission, 1/1. Applications are processed on a rolling basis. Application fee: $25 ($40 for international students). Electronic applications accepted. *Financial aid:* In 1998–99, 133 research assistantships (averaging $4,250 per year), 27 teaching assistantships with full tuition reimbursements (averaging $4,100 per year) were awarded.; fellowships, Federal Work-Study, institutionally-sponsored loans, and unspecified assistantships also available. Aid available to part-time students. Financial aid application deadline: 4/1; financial aid applicants required to submit FAFSA. *Faculty research:* Composite dynamics, controls, propulsion, optics, electronics manufacturing. Total annual research expenditures: $6.2 million. *Unit head:* Dr. William J. Wilhelm, Dean, 316-978-3400, Fax: 316-978-3853, E-mail: bwilhelm@engr.twsu.edu. *Application contact:* Dr. Mark M. Jong, Associate Dean, 316-978-3400, Fax: 316-978-3853, E-mail: mt.jong@engr.twsu.edu.

See in-depth description on page 301.

Widener University, School of Engineering, Chester, PA 19013-5792. Offers ME, ME/MBA. Part-time and evening/weekend programs available. *Faculty:* 27 part-time (4 women). *Students:* 18 full-time (2 women), 70 part-time (11 women); includes 15 minority (5 African Americans, 9 Asian Americans or Pacific Islanders, 1 Hispanic American), 24 international. 75 applicants, 88% accepted. In 1998, 43 degrees awarded. *Degree requirements:* For master's, thesis optional, foreign language not required. *Average time to degree:* Master's–2 years full-time, 4 years part-time. *Application deadline:* For fall admission, 8/1 (priority date); for spring admission, 12/1. Applications are processed on a rolling basis. Application fee: $25 ($300 for international students). *Financial aid:* In 1998–99, 5 teaching assistantships with full tuition reimbursements (averaging $7,500 per year) were awarded.; research assistantships, unspecified assistantships also available. Financial aid application deadline: 3/15. Total annual research expenditures: $82,500. *Unit head:* Dr. David H. T. Chen, Assistant Dean for Graduate Programs and Research, 610-499-4049, Fax: 610-499-4059, E-mail: david.h.chen@widener.edu.

See in-depth description on page 303.

Wilkes University, School of Science and Engineering, Wilkes-Barre, PA 18766-0002. Offers electrical engineering (MSEE). *Degree requirements:* Foreign language not required. *Entrance requirements:* For master's, GRE General Test.

Worcester Polytechnic Institute, Graduate Studies, Worcester, MA 01609-2280. Offers M Eng, MME, MS, PhD, Advanced Certificate, Certificate. Part-time and evening/weekend programs available. Postbaccalaureate distance learning degree programs offered (minimal on-campus study). *Faculty:* 178 full-time (23 women), 29 part-time (3 women). *Students:* 401 full-time (103 women), 376 part-time (77 women); includes 66 minority (8 African Americans, 40 Asian Americans or Pacific Islanders, 18 Hispanic Americans), 189 international. 1372 applicants, 57% accepted. In 1998, 208 master's, 27 doctorates awarded. Terminal master's awarded for partial completion of doctoral program. *Degree requirements:* For doctorate, dissertation required, foreign language not required, foreign language not required. *Entrance requirements:* For

Engineering and Applied Sciences—General–Cross-Discipline Announcements

Worcester Polytechnic Institute *(continued)*
master's, TOEFL (minimum score of 550 required). *Application deadline:* For fall admission, 2/15 (priority date); for spring admission, 10/15 (priority date). Applications are processed on a rolling basis. Application fee: $50. Electronic applications accepted. *Financial aid:* In 1998–99, 275 students received aid, including 33 fellowships with full tuition reimbursements available (averaging $14,200 per year), 119 research assistantships with full tuition reimbursements available (averaging $15,000 per year), 123 teaching assistantships with full tuition reimbursements available (averaging $11,970 per year); career-related internships or fieldwork, grants, institutionally-sponsored loans, scholarships, and tuition waivers (full) also available. Financial aid application deadline: 2/15; financial aid applicants required to submit FAFSA. *Faculty research:* Cryptography, space sciences, metals processing, computational modeling, bioengineering. Total annual research expenditures: $7.7 million. *Unit head:* Dianne E. Horgan, Director, 508-831-5561, Fax: 508-831-5717, E-mail: gao@wpi.edu. *Application contact:* Donna M. Johnson, Coordinator, 508-831-5248, Fax: 508-831-5717, E-mail: gao@wpi.edu.

Announcement: Known for academic excellence in technical education, WPI delivers more than 50 graduate science, engineering, and management programs to more than 1,000 full- and part-time students. WPI offers Master of Science, Master of Engineering, Master of Business Administration, and PhD degrees and graduate-level certificate programs. Graduate study is available in biology and biotechnology, biomedical and clinical engineering, biomedical sciences, chemical engineering, chemistry and biochemistry, civil and environmental engineering, computer and communications networks, computer science, electrical and computer engineering, fire protection engineering, management, manufacturing engineering, materials science and engineering, mathematical sciences, mechanical engineering, and physics. For more information, visit WPI's Web site at http://www.wpi.edu/

Wright State University, School of Graduate Studies, College of Engineering and Computer Science, Dayton, OH 45435. Offers MSCE, MSCS, MSE, PhD. Part-time and evening/weekend programs available. *Students:* 260 full-time (48 women), 140 part-time (26 women); includes 34 minority (7 African Americans, 24 Asian Americans or Pacific Islanders, 2 Hispanic Americans, 1 Native American), 189 international. Average age 29. 808 applicants, 58% accepted. In 1998, 165 master's, 1 doctorate awarded. *Degree requirements:* For master's, thesis optional, foreign language not required; for doctorate, dissertation, candidacy and general exams required. *Entrance requirements:* For master's, TOEFL (minimum score of 550 required); for doctorate, GRE General Test, TOEFL (minimum score of 550 required), minimum GPA of 3.3. Application fee: $25. *Financial aid:* In 1998–99, 65 fellowships, 60 research assistantships, 27 teaching assistantships were awarded.; Federal Work-Study, institutionally-sponsored loans, tuition waivers (full and partial), and unspecified assistantships also available. Aid available to part-time students. Financial aid applicants required to submit FAFSA. *Faculty research:* Robotics, heat transfer, fluid dynamics, microprocessors, mechanical vibrations. *Unit head:* Dr. James E. Brandeberry, Dean, 937-775-5001, Fax: 937-775-5009.

Announcement: College offers MS in engineering and PhD in engineering degree programs. The MS program has curricular tracks in electrical, mechanical, materials, biomedical, and human factors engineering. The PhD program supports research in the following 6 focus areas:

sensor signal and image processing, modern control and robotics, electronic and microwave circuits, processing and properties of high temperature and lightweight materials, computational design and optimization, and human interaction with complex systems.

See in-depth description on page 305.

Yale University, Graduate School of Arts and Sciences, Programs in Engineering and Applied Science, New Haven, CT 06520. Offers applied physics (MS, PhD); chemical engineering (MS, PhD); electrical engineering (MS, PhD); mechanical engineering (M Phil, MS, PhD), including applied mechanics and mechanical engineering. Part-time programs available. *Faculty:* 67. *Students:* 96 full-time (19 women), 3 part-time (1 woman); includes 6 minority (5 Asian Americans or Pacific Islanders, 1 Hispanic American), 62 international. 239 applicants, 21% accepted. In 1998, 3 master's, 18 doctorates awarded. Terminal master's awarded for partial completion of doctoral program. *Degree requirements:* For master's, foreign language and thesis not required; for doctorate, dissertation, exam required, foreign language not required. *Entrance requirements:* For master's and doctorate, GRE General Test, TOEFL. *Average time to degree:* Doctorate–5.3 years full-time. *Application deadline:* For fall admission, 1/4. Application fee: $65. *Financial aid:* Fellowships, research assistantships, teaching assistantships, Federal Work-Study and institutionally-sponsored loans available. Aid available to part-time students. *Unit head:* Director of Graduate Studies, 203-432-4250. *Application contact:* Admissions Information, 203-432-2770.

See in-depth description on page 307.

Youngstown State University, Graduate School, William Rayen College of Engineering, Youngstown, OH 44555-0001. Offers MSE. Part-time and evening/weekend programs available. *Faculty:* 21 full-time (1 woman), 1 (woman) part-time. *Students:* 39 full-time (8 women), 23 part-time (6 women); includes 2 minority (both African Americans), 19 international. 17 applicants, 82% accepted. In 1998, 24 degrees awarded. *Degree requirements:* For master's, computer language required, thesis optional, foreign language not required. *Entrance requirements:* For master's, TOEFL (minimum score of 550 required), minimum GPA of 2.75 in field. *Application deadline:* For fall admission, 8/15 (priority date); for winter admission, 11/15 (priority date); for spring admission, 2/15 (priority date). Applications are processed on a rolling basis. Application fee: $30 ($75 for international students). Tuition, state resident: part-time $97 per credit hour. Tuition, nonresident: part-time $219 per credit hour. Required fees: $21 per credit hour. $41 per quarter. *Financial aid:* In 1998–99, 14 students received aid, including 8 research assistantships with full tuition reimbursements available (averaging $7,500 per year), 1 teaching assistantship with full tuition reimbursement available (averaging $7,500 per year); Federal Work-Study, institutionally-sponsored loans, and scholarships also available. Aid available to part-time students. Financial aid application deadline: 3/1. *Faculty research:* Structural mechanics, water quality, wetlands engineering, control systems, power systems, heat transfer, kinematics and dynamics. *Unit head:* Dr. Charles A. Stevens, Dean, 330-742-3009, Fax: 330-742-1567. *Application contact:* Dr. Peter J. Kasvinsky, Dean of Graduate Studies, 330-742-3091, Fax: 330-742-1580, E-mail: amgrad03@ysub.ysu.edu.

Cross-Discipline Announcements

Carnegie Mellon University, Information Networking Institute, Pittsburgh, PA 15213-3891.

The MS in Information Networking is a cooperative endeavor of the Schools of Engineering, Computer Science, and Business, providing an alternative to the conventional one-year computer science or electrical engineering graduate program by integrating both and adding some features of an MBA program. Now in its eleventh year, the program carefully selects 35 people from the engineering and computer science disciplines to form each new class. The program provides technical electives aimed at several areas of specialization including (a) the telecommunications and computing industries, (b) wireless and mobile computing, (c) the financial services industry, and (d) systems integrating and consulting.

Carnegie Mellon University, School of Computer Science, Software Engineering Program, Pittsburgh, PA 15213-3891.

Master of Software Engineering is a unique 1-year program at Carnegie Mellon University. The program takes a hands-on approach to developing expertise under the guidance of experienced software engineers. Emphasizing practical results balanced by scientific underpinnings, the program concentrates on the engineering of superior software systems through the application of principles from computer science and related fields. The software development studio is a major component of the program. Working closely with a faculty member, student teams analyze a problem, plan and implement a software development project, and evaluate the outcome. Students have full access to the resources of the School of Computer Science, a world leader in computer science research, as well as the broad base of experience in the University's Software Engineering Institute, the only one of its kind.

Georgia Institute of Technology, Graduate Studies and Research, Ivan Allen College of Policy and International Affairs, Sam Nunn School of International Affairs, Atlanta, GA 30332-0001.

The Sam Nunn School of International Affairs at Georgia Tech offers an 18-month master's degree in international affairs that enables graduates to assume professional positions within business, government, and international organizations. The program is built around a core of 6 courses that provide strong theoretical and methodological skills and an understanding of the

major issues in international security and international political economy. Students also have the opportunity to tailor the program to their individual interests through elective offerings within the School and interdisciplinary work in economics, management, public policy, computer science, engineering, and other fields.

Union College, Graduate and Continuing Studies, Division of Engineering and Computer Science, Department of Electrical Engineering and Computer Science, Program in Computer Management Systems, Schenectady, NY 12308-2311.

Union College's Graduate Management Institute and the Department of Electrical Engineering and Computer Science offer a program leading to an MS in computer management systems. This program provides students with a technical education in applied computer science and an understanding of the managerial processes involved in the successful operation of modern computer-based information systems.

University of Tennessee, Knoxville, Graduate School, College of Human Ecology, Department of Consumer and Industry Services Management, Knoxville, TN 37996.

Students specializing in textile science enter a research-oriented area strongly based in mathematics, chemistry, physics, and engineering. Emphasis is placed on textile structural characterization, properties of textile materials, and textile processing. New textile materials, such as nonwoven webs and composites that expand the use of textiles, are emphasized.

University of Virginia, College and Graduate School of Arts and Sciences, Interdisciplinary Program in Biophysics, Charlottesville, VA 22903.

The Interdisciplinary Program in Biophysics at the University of Virginia offers training and research opportunities with more than 35 faculty members in the Schools of Graduate Arts and Sciences, Engineering, and Medicine. Macromolecular structure and physical biochemistry, membrane biophysics, and radiological physics are areas of specific research strength. All students are financially supported.

ARIZONA STATE UNIVERSITY

College of Engineering and Applied Sciences

Programs of Study

The College of Engineering and Applied Sciences (CEAS) offers opportunities for graduate study through the School of Engineering and the Del E. Webb School of Construction. The graduate programs in the School of Engineering are designed to bridge the gap between knowledge of engineering sciences and creative engineering practice, at the same time increasing students' depth and breadth of knowledge in their area of emphasis. The performance of scholarly research and the acceptance of professional responsibility for the documented results are considered to be essential requirements for graduate degrees and entrance into professional careers. Degrees offered include the Doctor of Philosophy (Ph.D.) in engineering and computer science; Master of Science in Engineering (M.S.E.); Master of Science (M.S.) with specialization in an area of engineering, computer science, or construction; and Master of Computer Science (M.C.S.). Arizona State University (ASU) is also collaborating with Northern Arizona University and the University of Arizona to offer the Tri-University Master of Engineering graduate degree program. This terminal master's program offers an opportunity to reflect the increasingly interdisciplinary nature of engineering practice.

Ph.D., M.S., and M.S.E. degrees are offered in aerospace, chemical, civil, electrical, industrial, and mechanical engineering and in engineering science. Students have the option of pursuing an interdisciplinary option in semiconductor processing and manufacturing that includes courses in electrical, chemical, and industrial engineering. Bioengineering offers M.S. and Ph.D. degrees, while computer science offers M.S., M.C.S., and Ph.D. degrees. An M.S. degree is offered in construction. Course work can be adapted to the needs and interests of each student, subject to certain minimum requirements in mathematics and science. Qualified students in engineering or selected fields, such as physics, biology, or chemistry, have an opportunity to specialize in particular subject areas within engineering. The M.S. degree requires a research thesis or an engineering report; those interested should check with the departments for specific requirements. The M.S.E. does not require a thesis or report, but students must take a written examination. Similar opportunities and requirements exist for qualified students who wish to pursue the Master of Computer Science degree. The Ph.D. program in engineering must be approved by a faculty supervisory committee and the faculty chairperson. Approved programs may include courses within one field of specialization or may include a rationally unified group in a combination of fields.

Research Facilities

Students have access to several outstanding experimental facilities, including a rich array of microscopy, synthesis, processing, and mechanical testing laboratories. Research facilities include capabilities for nanometer electronics, including state-of-the-art crystal growth, wafer processing, and materials/device characterization and analysis; biotechnology; biological and chemical analysis and treatment of water quality and organic wastes; conversion of organic wastes to high-quality fuels; development and evaluation of solar collectors; thermal sciences, including study of heating and cooling systems; antenna characterization in an anechoic chamber; signal processing; speech and audio processing; cryogenics; power systems analysis; power electronics, high-voltage, and insulation laboratories; investigation of heat transfer phenomena; fiber optics; radioisotope and neutron activation analysis; fluid mechanics; aerodynamics; structural and soils testing; microprocessor applications; research on photovoltaic cells and systems; computer engineering; engineering mechanics; bioengineering, biomechanics, bioinstrumentation, and motor dynamics; neurosciences, including neuromechanical control; materials science, including polymers and radiation damage; manufacturing processes; rapid fabrication in plastics and papers; computer-integrated manufacturing; statistical process control; quality control and reliability; and vehicle dynamics and control.

Engineering Technical Services has a staff of 24 full-time computer specialists who support the computing efforts of the College. The college computing environment is a large developing client/server enterprise that all faculty members, staff members, and students utilize. This enterprise consists of more than 3,000 desktop workstations that run either UNIX or the Windows environment, interconnected by an evolving 100MB Ethernet network to central enterprise-level application servers. Common baseline central services provided include e-mail, access to the World Wide Web on both UNIX and IIS, SQL servers, forums, and office productivity tools (MS Office). Other special services include support for high-performance computing, modeling/visualization, CAD/CAM, interactive mathematics, simulation, and general instrumentation.

Financial Aid

Approximately 480 graduate teaching or research assistantships are available, including sponsored research support. Students should write to the associate dean for academic affairs for further information. Convenient arrangements are made for the numerous graduate students who are employed full-time in local industries.

Cost of Study

For the 1999–2000 academic year, registration and tuition for 7 credit hours or more is $1094 per semester for Arizona residents. Nonresidents pay $389 per credit hour. The Graduate College considers 9 credit hours full-time enrollment.

Living and Housing Costs

Limited on-campus housing is available for unmarried students. For the 1999–2000 academic year, students living on campus can expect to pay about $3010, while students living off campus can expect to pay about $4950. Considering rent, food, personal expenses, fees, tuition, and books, Arizona residents living off-campus can expect to pay approximately $13,060 while nonresidents living off campus can expect to pay approximately $20,310.

Student Group

There are more than 43,000 students at Arizona State University, including almost 8,000 students pursuing a graduate degree program. More than 1,350 of the 5,547 students in the College of Engineering and Applied Sciences are pursuing graduate degrees. There are 374 students enrolled in doctoral programs within the College.

Location

Arizona is well known for its scenic attractions, which range from desert to mountain woodlands, lakes, and streams. The metropolitan community immediately surrounding Arizona State University thrives on high-technology industries that have large research and development staffs involved with computers, airborne electronics, semiconductors, turbines, energy production and conservation, food processing, and health services.

The College

The first bachelor's degree program in engineering was approved in 1956, largely through the efforts of valley civic and industry leaders who lobbied Arizona legislators. Today, there are approximately 180 tenure and tenure-track faculty members in six prominent engineering departments with the School of Engineering and the Del E. Webb School of Construction, which is one of the premier construction programs in the country. The College also supports six prestigious research centers, many of which are cooperative efforts between CEAS and other Arizona State University entities, other universities or research laboratories, and industry partners.

Applying

Students are required to complete an application form and to turn in certain documents to both the ASU Graduate College and the individual department of interest. Students may complete and submit the Graduate College application via the Web at http://www.asu.edu/graduate. Departments may require letters of reference, a statement of purpose, GRE scores, or other such documents not requested in the Graduate College application. Information regarding the necessary departmental application materials may be obtained by contacting the specific department of interest at the Web address listed below. Regular admission requires a grade point average of at least 3.0 (on a scale of 4.0) in the last two years of course work leading to the bachelor's degree. In special circumstances, provisional admission may be granted.

Correspondence and Information

Associate Dean for Academic Affairs
College of Engineering and Applied Sciences
Arizona State University
Tempe, Arizona 85287-5506

Telephone: 602-965-1726
E-mail: asuengr@asu.edu
World Wide Web: http://www.eas.asu.edu

Arizona State University

FACULTY HEADS AND RESEARCH AREAS

Aerospace Engineering: Don L. Boyer, Chair. Aerodynamics, design, dynamics and control, propulsion and structures. Research includes acoustic fatigue; aeroelasticity; aerospace vehicle dynamics, guidance, and control; aerospace structures; vehicle design and performance optimization; aircraft crashworthiness; boundary-layer transition; combustor modeling; composite materials; heat transfer in airbreathing and space propulsion systems; high-speed aerodynamics; laminar flow control; laser diagnostics in combustion and flows; orbital mechanics; rotorcraft aerodynamics and acoustics; structural optimization; and unsteady aerodynamics.

Bioengineering: Eric Guilbeau, Chair. Bioengineering topics include human movement and control of neuromuscular and neuroprosthetic systems; biosensors and neurostimulation; hard- and soft-tissue biomaterials, molecular/cellular/tissue, neural engineering, biomaterials, cardiac assist devices, and biocompatibility; orthopedic replacement devices and rehabilitation engineering; and artificial organs, medical device design, physiological transport, hybrid artificial organs, bioseparations, molecular and cellular bioengineering, and biosystems engineering. Interested students can refer to the in-depth description in the Biomedical Engineering section of *Peterson's Guide to Graduate Programs in Engineering and Applied Sciences.*

Chemical Engineering: Eric Guilbeau, Chair. Chemical engineering topics include bioseparations; thin-film, catalytic, and biochemical reactors; chemical vapor deposition and etching; pyrolysis of biomaterials and municipal wastes; atmospheric and groundwater modeling and hazardous-waste removal; environmentally conscious manufacturing, wet chemical surface treatment processes; thin-film characterization and colloid science; control theory, adaptive control, and batch and continuous control; and atmospheric transport and transport in porous media. Interested students can refer to the in-depth description in the Chemical Engineering section of *Peterson's Guide to Graduate Programs in Engineering and Applied Sciences.*

Civil and Environmental Engineering: Sandra L. Houston, Chair. Environmental/water resources engineering: water and wastewater treatment processes, hazardous wastes mitigation, water quality analysis, watershed management, risk assessment; hydraulic engineering, fluid mechanics, water resource systems, hydrology. Geotechnical/geoenvironmental engineering: soil mechanics, earthquake engineering, foundation analysis and design, hazardous waste transport and mitigation processes. Structures/materials engineering: analysis of reinforced concrete and steel structures, earthquake-resistant design, structural dynamics, optimization; composite materials, durability of materials, structural testing. Transportation/materials engineering: urban transportation planning, geometric design of facilities, traffic operations; mechanical properties, micromechanics, pavement materials, pavement management.

Computer Science and Engineering: Stephen Yau, Chair. Research groups include artificial intelligence (distributed planning systems, incremental planning, and applications), computer-aided geometric design/graphics (visual representation of data; surface, volume, and multidimensional image with biomedical applications and multiresolution flow visualization), databases (database and multimedia information systems for commercial and manufacturing applications), distributed processing and parallel high-performance systems and networks (high-speed computer networks, information highways, fault-tolerant computer networks and multiprocessors, and quality of service), microprocessors (hardware and software systems design and testing), and software engineering (software development and maintenance processes; component-based software development; quality assurance; and central, distributed, and parallel systems).

Del E. Webb School of Construction: William W. Badger, Director. The program includes concentrations in construction science, management, and facilities. The Alliance for Construction Excellence serves as a technical, scientific, and socioeconomic transfer center for construction industry problems and issues.

Electrical Engineering: Stephen Goodnick, Chair. Materials: epitaxial thin films, defects. Devices: discrete/integrated circuits, thin-film devices, VLSI analog and digital design, lasers, integrated optical circuits, optical interconnects, low-power electronics, nanoelectronics, optoelectronics. Characterization: electrical, optical, physical, and chemical measurements; SEM. Modeling: MOSFET and bipolar transistor modeling, charge carrier ballistics, Monte Carlo simulations, quantum transport. Power engineering: power systems, security assessment, software development for analysis and control, real-time computer control, large network analytical techniques. Transmission/distribution: design, load management, automation. Power electronics: rectifiers and inverter application, high-power switching devices, motor drivers, FACT devices, power quality studies, effects of harmonics, flicker control. High-voltage techniques: nonceramic insulators and insulation systems, insulation coordination. Control systems: analysis and design of multivariable, nonlinear, distributed, parameter, adaptive and intelligent, and digital control systems. Modeling, simulation, and real time control of dynamical systems. Neural networks: large systems, including manufacturing and biological systems. Signal processing: speech enhancement recognition and coding; image and video processing and coding; additive signal processing, detection, and estimation; filtering of stochastic processes; VLSI architectures for signal and image processing; mathematical signal analysis. Antennas: analysis, design and measurements for wireless communications; electromagnetic wave radiation, propagation, scattering, penetration, and reception. Computational electromagnetics: geometrical and physical theories of diffraction, moment method, finite element method, finite difference time domain method, and hybrid methods. Electronic packaging: modeling and measurements. Microwaves: circuits, devices, and systems; transient analysis of striplines and microstrips; solid-state circuits and devices, measurements. Communication: digital communications, wireless, networks, quality of service, systems, coding and modulation, multiple access. Lasers and coherent optics: fiber optics; communications, networks, components; holography. Interested students can refer to the in-depth description in the Electrical Engineering section of *Peterson's Guide to Graduate Programs in Engineering and Applied Sciences.*

Industrial Engineering: Gary Hogg, Chair. Manufacturing processes and controls, quality and reliability, enterprise information systems, management of technology, operations research, production systems, semiconductor manufacturing, human factors.

Materials Science and Engineering: Eric Guilbeau, Chair. Microelectronics: electrical ceramics, oxidation, metallization, thin-film growth, heteroepitaxy, high-temperature microelectronics, and computer modeling. Structural materials: composites, intermetallics, and high-temperature alloys. Interested students can refer to the in-depth description in the Materials Science and Engineering section of *Peterson's Guide to Graduate Programs in Engineering and Applied Sciences.*

Mechanical Engineering: Don L. Boyer, Chair. Design and manufacturing, dynamics and control, energy systems, engineering mechanics and thermosciences. Research includes laser diagnostics in combustion, solar energy systems, feature-based modeling, design automation, concurrent engineering, modeling and control of robots, failure analysis and life predictions, finite-element models, mechanics of thin films, fracture mechanics, metal cutting, hydrodynamic stability, turbulence modeling, two-phase flows, convective heat transfer in complex flows, rotating and stratified flows, pulverized-coal combustion, and pollutant formation and spray burning.

ARIZONA STATE UNIVERSITY EAST

College of Technology and Applied Sciences
Morrison School of Agribusiness and Resource Management

Programs of Study

The Arizona State University East Campus offers opportunities for graduate study through the Morrison School of Agribusiness and Resource Management and the College of Technology and Applied Sciences. Degrees offered include the Master of Science in agribusiness, the Master of Science in Technology, and through a cooperative agreement with the American School of International Management, a Master of International Management. The Master of Science in agribusiness programs offer graduates a broad choice of instruction tailored to the needs of students interested in careers in agribusiness or resource management. Focusing on the agribusiness sector allows the faculty to teach cutting-edge business ideas and methods within the unique economic environment of the human food chain. New biotechnologies, new products, new markets, and constantly changing global business conditions present a widening array of opportunities for well-trained managers.

The Master of Science in Technology degree is offered for students who seek technological preparation through the Departments of Aeronautical Management Technology, Electronics and Computer Engineering Technology, Information and Management Technology, and Manufacturing and Aeronautical Engineering Technology.

The graduate programs at ASU East are designed to bridge the development of theoretical knowledge, technological engineering sciences, agricultural principles, and creative industrial applications. The programs are formulated to increase students' depth and breadth of knowledge to facilitate their ability to execute integrative functions in the agribusiness and industrial environments. The performance of projects and applied research are used to develop the acceptance of professional responsibility for the documentation and management of essential functions within the industrial spectrum.

Research Facilities

The Morrison School of Agribusiness and Resource Management is currently involved in three major research projects: the Center for Agribusiness Policy Studies (CAPS), which is currently focusing on market reforms in Central Europe; the National Food and Agriculture Policy Project (NFAPP), which studies issues in fruit and vegetable markets; and the Food Industry Research Laboratory, which applies management, science, and technology to production, processing, packaging, distribution, and development of food products.

Research projects in the College of Technology and Applied Sciences reflect the varied interests of the faculties. Research is conducted in state-of-the-art facilities that support projects in aeronautical management technology, electronics and computer engineering technology, information and management technology, and manufacturing and aeronautical engineering technology. Specific areas of research are described on the reverse of this page.

Financial Aid

Financial aid from federal, state, local, institutional, and private sources is awarded in the form of scholarships, loans, grants, and work-study employment. A limited number of graduate teaching or research assistantships are available within the College of Technology and Applied Sciences, including sponsored research support. The Morrison School of Agribusiness and Resource Management offers about twenty such assistantships. Students should contact the Financial Aid Office of the Graduate College at 480-965-3521 for more information.

Cost of Study

For the 1999–2000 academic year, tuition for in-state, full-time (12 semester hours or more) students is approximately $2160; tuition for out-of-state, full-time students is $9112. Books and supplies are approximately $700.

Living and Housing Costs

Three academic villages provide a variety of living arrangements, including married and family housing and residence halls. Two-, three-, four-, and five-bedroom homes range from approximately $500 to $800 per month. Residence halls offer single or double occupancy, ranging in price from approximately $200 to $400 per month. Additional charges may include a security deposit, an application fee, a preparation fee, and, in some cases, utilities.

Student Group

Approximately one quarter of the ASU East student population is enrolled in graduate programs. Of the 291 graduate students at East, 86 are international students and 184 attend on a part-time basis. Two thirds of the graduate student population are men.

Location

ASU East is located in southeast Mesa, one of the fastest-growing communities in the Phoenix metropolitan area. Though located in a rural/suburban area, ASU East is just five minutes from a major shopping area with malls, restaurants, resorts, and entertainment venues. The beautiful Superstition Mountains and Sonoran Desert, with nearby lakes and skiing areas, create the contrast of Arizona's unique environment.

The University

ASU East opened in August 1996 in facilities conveyed by the Williams Air Force Base. Approximately 1,200 students are enrolled in undergraduate and graduate programs. ASU East is one of the educational partners of the Williams Campus. Maricopa Community Colleges, and Embry-Riddle University provide additional educational opportunities on the campus. The East Campus is located 23 miles from the ASU Main Campus. A shuttle service provides transportation between the Main and East Campuses for students taking courses at both locations.

Applying

Application forms may be obtained via the World Wide Web (http://asu.edu/graduate) or by writing to the Admissions Office, Graduate College, Arizona State University, Tempe, Arizona 85287. Students should submit an application for admission and two copies of all transcripts of undergraduate and graduate work to the Graduate College at least two months prior to the student's enrollment. It is recommended that scores on the General Test of the Graduate Record Examinations be submitted. Regular admission requires a grade point average of at least 3.0 (on a scale of 4.0) in the last two years of course work leading to the bachelor's degree. In special circumstances, provisional admission may be granted.

Correspondence and Information

Director of Academic Services
Arizona State University East
7001 East Williams Field Road, Building 20
Mesa, Arizona 85212

Telephone: 480-727-1028
E-mail: asueast@asu.edu
World Wide Web: http://www.asu.edu/east

Arizona State University East

FACULTY HEADS AND AREAS OF RESEARCH
AGRIBUSINESS AND RESOURCE MANAGEMENT

Agribusiness and Resource Management: Raymond Marquardt, Dean.

Agribusiness management, marketing, and finance; food industry; international agribusiness; resource management.

TECHNOLOGY AND APPLIED SCIENCES: Albert L. McHenry, Dean.

Department of Aeronautical Management Technology: Dr. William K. McCurry, Chair.

Applied research areas include human factors, air transportation, system requirements, airport management, fixed- and rotary-wing performance, National Airspace System infrastructure improvements, aviation safety, air traffic control, simulation and system planning, and aviation system management.

Department of Electronics and Computer Engineering Technology: Dr. Robert W. Nowlin, Chair.

Applied research areas include electronics communication systems (specialization in applications of DSP), digital/computer systems (specializations in microprocessor applications, computers integrated as a systems component, circuit design, and hardware design using state-of-the-art CAD/CAM), electrical systems (specializations in electronic system integration and hardware design using state-of-the-art CAD/CAM), microelectronics (specialization in state-of-the-art wafer processing, packaging, device characterization, and testing), and electronic technology education (specialization in electronic education, with special focus on the field of electronics).

Department of Information and Management Technology: Dr. Thomas Schildgen, Chair.

Applied research areas include graphics communication systems (specializations in digital imaging, gravure systems, color reproduction, and image acquisition), interactive computer graphics (specializations in interactive multimedia and computer animation), industrial management and supervision (specializations in management decision making, project management, total quality, technical communications, and industrial safety and health), hazardous materials and waste management (specializations in assessment of hazardous materials and management of hazardous waste.

Department of Manufacturing and Aeronautical Engineering Technology: Dr. Dale Palmgren, Chair.

Applied research areas include manufacturing engineering technology and aeronautical engineering technology. Specializations in the manufacturing engineering technology program are mechanical and material joining, statistical process control, robotics/automation, CAD/CAM/CAE, manufacturing processes, stress analysis, mechanical design, energy efficiency issues, and welding metallurgy. In the aeronautical engineering technology program, specializations are airplane design/performance, propulsion, wind tunnel testing, structures, aerodynamics, and controls.

BOSTON UNIVERSITY

College of Engineering

Programs of Study	The Doctor of Philosophy (Ph.D.) degree, Master of Science (M.S.) degree, and a post-bachelor's Ph.D. degree are offered. A combined Doctor of Philosophy/Doctor of Medicine (Ph.D./M.D.) program in biomedical studies is offered in cooperation with the School of Medicine. A combined Master of Science/Master of Business Administration (M.S./M.B.A.) program in manufacturing is offered in cooperation with the School of Management. Curricula and programs emphasize creative design and research in a major field of engineering specialization. The Ph.D. may be earned in aerospace, biomedical, computer, electrical, manufacturing, mechanical, and systems engineering. A minimum of 32 credits beyond the equivalent of an M.S., including a dissertation, is required for completion. An M.S. degree may be earned in aerospace, biomedical, computer systems, electrical, manufacturing, and mechanical engineering. Master's degree programs normally require one year of full-time study; part-time students should be able to complete a program within two calendar years. The Late Entry Accelerated Program (LEAP) enables students with at least a bachelor's degree in a liberal arts or other nonengineering field that does not qualify them for direct admission to any of the College's M.S. programs to apply and, ultimately, earn a master's degree in engineering in two years of full-time study or longer for part-time study.
Research Facilities	Modern research laboratories exist in each department and are supported by University centers, campus infrastructure, and specialized Boston area facilities. Engineers collaborate in the Biomolecular Engineering Research Center, the Center for Advanced Biotechnology, the Center for Space Physics, the Computational Sciences Center, the Hearing Research Center, the Center for Bio Dynamics, the Biomolecular Engineering Center, the Center for Photonics Research, and the Neuromuscular Research Center. The Scientific Instrumentation Facility provides precision fabrication, and the Power Challenge Array supercomputer supports modern computation. The Fraunhofer Research Institute has a manufacturing facility on campus, and the Center for Photonics Research has an industrial incubator facility. Students can pursue research at nearby hospitals and federal facilities such as Rome Labs and the Natick Army Research Center.
Financial Aid	A full range of financial aid is available, including Presidential University Graduate Fellowships, Dean's Fellows, research assistantships, and graduate teaching fellowships and scholarships. Teaching fellowships provide stipends of $12,500 per academic year in 1999–2000 and require approximately 20 hours a week of instructional duties. Recipients receive a tuition waiver for 8 to 10 credits per semester and up to 8 additional credits the following summer. Research assistantship stipend levels are comparable to those of teaching fellowships and are also supplemented by tuition waivers. University and Dean's Fellows scholarships range up to $37,830, including stipend and tuition, per academic year; the College encourages GEM scholars to apply. Federal Direct Student Loan and work-study applicants must submit a Free Application for Federal Student Aid (FAFSA) to the Federal Student Aid Programs Office. Work-study and FAFSA forms may be obtained from the Graduate Programs Office.
Cost of Study	In 1999–2000, tuition and fees for full-time study are $23,770. Part-time students pay $743 per credit hour.
Living and Housing Costs	Privately owned apartments or rooms are readily available. Living expenses for a single student are estimated at $10,730 for the nine-month academic year 1999–2000.
Student Group	The University has the distinction of having graduated the first woman Ph.D. in this country, as well as the first American Indian and one of the first blacks to receive the M.D. degree. Students of all races and cultural backgrounds continue to work and learn together in harmony and with mutual respect. In 1998–99, there were 10,223 graduate students, including 5,259 women, enrolled at the University. The College of Engineering has grown dramatically in recent years and has a current enrollment of more than 1,580.
Student Outcomes	Most of the Ph.D. program graduates obtained positions in their field of study within one month of graduation, with an average starting salary of $40,000 to $50,000. M.S. and Ph.D. program graduates have acquired consulting positions in high-technology industries, teaching positions, research positions in government laboratories, and various engineering positions in industry. M.S. program graduates have gone on to Ph.D. programs at Boston University, Stanford, MIT, and other major universities in the U.S. and abroad.
Location	Graduate students at Boston University enjoy the advantages of a sophisticated metropolitan area with world-renowned academic and scientific resources. Facilities of many other universities in the area, if needed, are easily accessible, and their excellent seminar and colloquium programs afford graduate students exceptional opportunities to participate at the cutting edge of current research. Access to nearby concentrations of high-technology industry provides another vital element in an exciting academically oriented region.
The University and The College	Boston University, incorporated in 1869, is a private, nonsectarian, coeducational, independent university fully open to men and women and members of all minority groups. Its approximately 23,500 full-time students and 3,130 faculty members make it one of the largest independent universities in the world. By granting its students access to almost every course offered within the institution, the University takes maximum advantage of its size and its range of resources. Boston University is responsive to the occupational needs of its students and the increasingly specialized demands they will face in the contemporary world. Boston University's College of Engineering has experienced remarkable growth over the last decade. The faculty of the College, which has tripled in size since 1980, seeks to build an ever-stronger research program while maintaining the traditional commitment to the educational experience of the students.
Applying	Early application is advisable. Applicants to the M.S., post-bachelor's Ph.D., and post-master's Ph.D. programs should have demonstrated a high degree of scholarship in an undergraduate program in engineering or science at an accredited college or university. Students desiring financial aid should apply by January 15 for fall admission and October 1 for spring. Applications without financial aid requests will be accepted until April 1 and October 15, respectively. Required admissions credentials include official transcripts and specific letters of recommendation. Forms are available on request from the address below. All applicants to the master's and Ph.D. programs must submit scores on the Graduate Record Examinations (GRE) General Test. Applicants to the master's program in manufacturing engineering must submit GRE General Test or Graduate Management Admission Test (GMAT) scores. All international students whose native language is not English must submit results of the Test of English as a Foreign Language (TOEFL) prior to acceptance; if they request financial aid in the form of a graduate teaching fellowship, a minimum TOEFL score of 600 is required.
Correspondence and Information	Graduate Programs College of Engineering Boston University 48 Cummington Street Boston, Massachusetts 02215 Telephone: 617-353-9760 Fax: 617-353-0259 E-mail: enggrad@bu.edu World Wide Web: http://www.bu.edu/eng/grad

Boston University

DEPARTMENT HEADS AND RESEARCH AREAS

The individual two-page descriptions for each department provide further information.

Aerospace and Mechanical Engineering. Professor Allan D. Pierce, Chairman. The department offers graduate programs leading to the Master of Science and Doctor of Philosophy degrees. Research is focused in four areas: waves and acoustics; dynamics, control, and robotics; fluid mechanics; and precision engineering. Some areas of specialization include aerodynamics, aeroelasticity, automatic control systems, biomechanics, noise control, photomechanical systems, structural mechanics, thermal processes, theoretical fluid dynamics, and turbulence. The departmental research facilities include the following laboratories for graduate research: Robotics Laboratory, Applied Acoustic Laboratory, Precision Engineering Research Laboratory, Mechanics of Materials Laboratory, Fluids and Instrumentation Laboratory, Boundary Layer Wind Tunnel, Computational Dynamics Facilities, Physical Acoustics Laboratory, and Nonlinear Acoustics Laboratory. In addition to these facilities, a number of interdisciplinary programs exist between faculty members in the aerospace and mechanical engineering department and faculty members in other departments within the University.

Biomedical Engineering. Professor Kenneth Lutchen, Chairman. The biomedical engineering graduate programs train students in the application of modern technology and quantitative engineering methods to biology and medicine. The department's approach to biomedical engineering is based on a cross-disciplinary approach, combining mechanical, electrical, and chemical engineering with physiology and biology. The programs combine empirical and theoretical approaches to the study of various systems, including sensorineural, neuromuscular, respiratory, cardiovascular, biomechanical, and molecular. Many faculty members have active collaborations with faculty from the Boston University School of Medicine and University Hospital or with faculty from the University's Sargent College of Allied Health Professions. In addition, departmental faculty collaborate or are affiliated with seven other hospitals in the Boston area—a region rich in hospitals, universities, and corporations involved with biomedical engineering. Graduate students may do their thesis projects as part of these collaborations.

Electrical, Computer, and Systems Engineering. Professor Bahaa Saleh, Chairman. Graduate programs within the department prepare students for the application of state-of-the-art analysis and design methods to problems in electrical, computer, and systems engineering. Master of Science degrees are offered in electrical engineering and computer systems engineering and the Ph.D. degree is offered in electrical, computer, or systems engineering. A thesis or project is required for the M.S. programs. Primary research areas in the department include speech, signal, and image processing; electronic materials and devices; photonics; reliable computing; electromagnetics; computer and communication networks; and high-performance computing and software engineering.

Manufacturing Engineering. Professor John Baillieul, Chairman. Graduate training at the M.S. and Ph.D. levels in the application of modern engineering methods to problems of manufacturing is the department's major objective. M.S. candidates can select among six concentration areas: manufacturing systems and operations research; computer-integrated design, analysis, and manufacture; manufacturing operations management; automation and control in manufacturing; engineered materials and processes; and process design. There are a variety of means of delivery to attain an M.S. in manufacturing engineering, including on-campus study, distance learning via interactive compressed video, accelerated low-resident programs for industry practitioners, and international exchange programs. A dual M.S./M.B.A. degree is an additional option for students selecting the operations management concentration area. The department's graduate program draws students from various disciplines, including mechanical, computer, electrical, systems, and industrial engineering. Laboratories in the areas of automated design and manufacturing systems, computer-integrated design, computer-aided manufacturing, computer-aided engineering (CAE), process controls, high-temperature oxidation, machining, microscopy, powder metallurgy, x-ray, production control, robotics, surface modification, control of discrete event systems (CODES), and microchip simulation currently support the graduate teaching and research activities. Doctoral work is ongoing in a number of externally funded research areas, including design and control of manufacturing systems and operations research, robotics and control of materials processing, engineered materials, computer-aided design and manufacture, and manufacturing of microelectronic components.

BRIGHAM YOUNG UNIVERSITY

College of Engineering and Technology

Programs of Study
The College of Engineering and Technology offers Master of Science (M.S.) degrees in chemical, civil, electrical, manufacturing, and mechanical engineering; computer-integrated manufacturing technology; and technology education. A typical course of study for a master's degree includes a minimum of 34 credit hours, including 6 to 9 hours of thesis credit. Nonthesis programs are also available in civil, electrical, and mechanical engineering and technology education.

An optional minor in engineering management at 9 credit hours may be elected for M.S. students in engineering or technology.

A cooperative program with the Marriott School of Management, a minimum two-year program in which students complete 78–81 credit hours of study, provides a dual M.B.A./M.S. degree.

Doctor of Philosophy (Ph.D.) degrees are offered in chemical, civil, electrical, and mechanical engineering. A typical course of study for a Ph.D. degree includes a minimum of 60 semester hours, at least 50 of which must be course work beyond the baccalaureate degree, plus 18 hours of dissertation.

Research Facilities
The College of Engineering and Technology occupies three major buildings at the Brigham Young University (BYU) campus. An extensive computing network provides high-speed computational capabilities to all students and faculty members. The NSF-sponsored Advanced Combustion Engineering Research Center (ACERC) develops and transfers to industry fundamental engineering knowledge and tools for solving critical national combustion problems, with emphasis on clean and efficient use of fossil fuels, including coal and natural gas, as well as combustion of waste materials. The state-sponsored Advanced Composites Manufacturing and Engineering Center (ACME) is established to promote the use and understanding of advanced materials, largely in support of the existing composite and plastic material companies operating in the state of Utah. The Catalysis Laboratory applies the principles of kinetics, chemistry, materials science, surface science, and chemical engineering to the understanding of catalyst properties and catalytic reactions. The Engineering Computer Graphics Laboratory (ECGL) is active in pursuing research for continued development of computer-aided engineering methodologies and tools. The Rapid Product Realization Center assists industry and individuals in developing new product prototypes and in launching new products. Other significant research activities include microwave remote sensing, signal processing, microstructure of materials, robotics, transportation, geotechnical engineering, water quality, computer architecture, and computer-aided engineering.

Financial Aid
Research and teaching assistantships are available in all departments within the College. Typical yearly assistantships pay full-time students up to $16,000 and require a 20-hour-per-week effort during the academic year and a 40-hour-per-week effort during the spring and summer terms. Fellowships, ranging from partial to full tuition, are also available from all departments.

Cost of Study
Semester tuition for the 1999–2000 academic year for full-time students is $1665 for Latter-day Saints (L.D.S.) students and $2500 for other students. Part-time tuition per credit hour is $185 for L.D.S. students and $278 for other students.

Living and Housing Costs
University housing for single students offers a two-semester room-only accommodation for $2000 and room and board accommodation for $4200. University family apartments for married students vary in cost between $370 and $550 per month depending upon apartment size. Early application for on-campus housing is recommended. Private, off-campus, academic-year housing is available beginning at $200 per month for singles and $430 per month for married couples. Rates may be significantly lower in the spring and summer terms.

Student Group
Total enrollment at BYU is approximately 30,000 students, of whom about 6 percent are graduate students. There are more than 225 graduate engineering students on campus.

Student Outcomes
There is nearly a 100-percent placement-rate for each department in the College. With respect to the various departments, examples of recent placements have been as follows: Chemical Engineering—Exxon, 3M Corporation, Eastmans Chemical, and Dow Chemical; Civil and Environmental Engineering—UDOT and many private businesses and governmental organizations; Electrical and Computer Engineering—Intel and Hewlett-Packard; School of Technology—Motorola, Allied Signal, and Boeing; Mechanical Engineering—Boeing, Exxon, Hewlett-Packard, and IBM. The average starting salary, depending on the program, ranges from $35,000 to $60,000.

Location
Brigham Young University is situated at the foot of the beautifully rugged Wasatch Range of the Rocky Mountains and bounded on the west by 23-mile-long Utah Lake. The campus is the focal point of a city of 105,000 and a valley of 318,000. Beyond it to the south and east are spectacular areas of vast sandstone canyons and monoliths, several of which are national parks. Forty-five miles north is the metropolitan area of Salt Lake City. Spectacular summer outdoor recreational locations and outstanding winter sports facilities are abundant.

The University
Established and sponsored by the Church of Jesus Christ of Latter-day Saints, Brigham Young University is the largest privately owned university in the United States. Founded in 1875 as Brigham Young Academy, the campus has grown from one building to 500 buildings on more than 600 acres. Nearly 1,400 full-time faculty members instruct more than 30,000 students. From its modest beginnings, Brigham Young University has grown to become one of the nation's distinguished institutions of private higher education. Faculty members and students work side by side in collegial scholarship enhanced by mutual commitment to the highest ideals of professional, ethical, and spiritual values. Engineering undergraduate programs were established in 1952, and master's degrees were begun in 1960. The engineering doctoral program was approved in 1968. Further information can be found by visiting Brigham Young's Web site at http://www.byu.edu.

Applying
Applications for admission to graduate study are available from the Brigham Young University Office of Graduate Studies, B-356 ASB, Provo, Utah 84602. A preliminary application, which includes all the international application deadlines for specific programs, can also be obtained from the Office of Graduate Studies. For specific U.S. and international student deadlines, students should check with the Graduate Studies Office (telephone: 801-378-4091; Web site: http://www.byu.edu/gradstudies).

Correspondence and Information
Graduate Advisor
Department of (specify Chemical, Civil and Environmental,
    Electrical and Computer, or Mechanical
    Engineering or School of Technology)
College of Engineering and Technology
Brigham Young University
Provo, Utah 84602

Brigham Young University

THE FACULTY AND THEIR RESEARCH

Chemical Engineering. Calvin H. Bartholomew, Professor; Ph.D., Stanford, 1972: catalysis, kinetics, air pollution. Merrill W. Beckstead, Professor; Ph.D., Utah, 1965: combustion, detonation transition, combustion instability. Thomas H. Fletcher, Professor; Ph.D., Brigham Young, 1983: coal devolatilization, combustion, laser diagnostics in combustion. John N. Harb, Associate Professor; Ph.D., Illinois, 1988: electrochemical engineering, learning styles, and teaching methodology. William C. Hecker, Associate Professor; Ph.D., Berkeley, 1982: chemical kinetics, coal chor reactivity, NO_x reduction, FTIR of absorbed surface species, heterogeneous catalysis. Paul O. Hedman, Professor; Ph.D., Brigham Young, 1973: jet engine combustion, laser diagnostics in combustion. John L. Oscarson, Professor; Ph.D., Michigan, 1985: thermodynamics of high-temperature aqueous solutions. William G. Pitt, Professor; Ph.D., Wisconsin, 1987: polymer science and engineering, biomaterials. Richard L. Rowley, Professor; Ph.D., Michigan State, 1978: thermophysical property prediction, correlation, and evaluation; maintenance of the DIPPR Thermophysical Properties Database. L. Douglas Smoot, Professor; Ph.D., Washington (Seattle), 1960: general combustion, pollutant formation, fossil fuels. Kenneth A. Solen, Professor; Ph.D., Wisconsin, 1974: biomedical engineering, artificial organs. Ronald E. Terry, Professor; Ph.D., Brigham Young, 1976: learning styles and teaching methodology, enhanced oil recovery, engineering ethics. W. Vincent Wilding, Associate Professor; Rice, 1985: thermodynamics and phase equilibria, environmental engineering.

Civil and Environmental Engineering. Richard J. Balling, Professor; Ph.D., Berkeley, 1982: optimization-based civil engineering design. Steven E. Benzley, Professor; Ph.D., California, Davis, 1971: finite element modeling and analysis. M. Brett Borup, Associate Professor; Ph.D., Clemson, 1985: industrial and hazardous waste control. W. Don Budge, Professor; Ph.D., Colorado, 1964: rigid and flexible pavement design. Henry N. Christiansen, Professor; Ph.D., Stanford, 1962: computer graphics, engineering management. Wayne Downs, Associate Professor; Ph.D., Florida, 1993: containment transport and biodegradation of hazardous waste. Fernando S. Fonseca, Assistant Professor; Ph.D., Illinois at Urbana-Champaign, 1997: structures. David W. Jensen, Associate Professor; Ph.D., MIT, 1986: advanced structural composite materials, smart structures. Norman L. Jones, Associate Professor; Ph.D., Texas, 1990: groundwater and terrain modeling, computer graphics. Warren K. Lucas, Instructor; M.S., Kansas, 1993: structural design, mechanics. LaVere B. Merritt, Professor; Ph.D., Washington (Seattle), 1970: water quality assessment and planning, modeling lakes and rivers. A. Woodruff Miller, Professor; Ph.D., Stanford, 1975: hydrologic modeling, satellite hydrology, hydraulic systems. E. James Nelson, Assistant Professor; Ph.D., Brigham Young, 1994: digital terrain modeling for watershed analysis. Kyle M. Rollins, Professor; Ph.D., Berkeley, 1987: collapsible soils, liquefaction. Mitsuru Saito, Associate Professor; Ph.D., Purdue, 1988: traffic engineering, control, and safety. Glen S. Thurgood, Professor; Ph.D., Texas A&M, 1975: street and highway design, traffic operations, highway safety. T. Leslie Youd, Professor; Ph.D., Iowa State, 1967: geotechnical engineering, earthquake engineering, liquefaction. Alan K. Zundel, Assistant Professor; Ph.D., Brigham Young, 1994: numerical hydrologic modeling, meshing techniques, free-form surfaces.

Electrical and Computer Engineering. James K. Archibald, Associate Professor; Ph.D., Washington (Seattle), 1987: computer architecture, parallel processing. David V. Arnold, Associate Professor; Ph.D., MIT, 1991: electromagnetic wave theory and microwave remote sensing. Randal Beard, Assistant Professor; Ph.D., Rensselaer, 1995: nonlinear control systems. LeRoy W. Bearnson, Associate Professor; Ph.D., Auburn, 1970: fault-tolerant computers/networks, data compression/security. Douglas M. Chabries, Professor; Ph.D., Brown, 1970: digital signal processing, image processing, speech processing. Richard W. Christiansen, Professor; Ph.D., Utah, 1976: digital signal processing, communication theory. David J. Comer, Professor; Ph.D., Washington State, 1966: communication circuits. Donald J. Comer, Professor; Ph.D., Santa Clara, 1968: VLSI. Richard L. Frost, Associate Professor; Ph.D., Utah, 1989: quantization, source and channel coding, signal processing. Brad L. Hutchings, Associate Professor; Ph.D., Utah, 1992: reconfigurable logic, VLSI design, tactile sensing. Brian D. Jeffs, Associate Professor; Ph.D., USC, 1989: digital image reconstruction, signal processing. Michael A. Jensen, Assistant Professor; Ph.D., UCLA, 1994: electromagnetics and wireless communications. David G. Long, Associate Professor; Ph.D., USC, 1989: microwave remote sensing, radar theory. Brent E. Nelson, Professor; Ph.D., Utah, 1984: computer systems performance analysis and modeling. Michael D. Rice, Associate Professor; Ph.D., Georgia Tech, 1991: digital communication theory. Richard H. Selfridge, Associate Professor; Ph.D., California, Davis, 1984: fiber-optic and electrooptic materials. Kevin B. Smith, Assistant Professor; Ph.D., Ohio State, 1996: noncontact meterology, design and control of motors. Wynn C. Stirling, Professor; Ph.D., Stanford, 1983: linear system theory, decision theory; signal processing. A. Lee Swindlehurst, Associate Professor; Ph.D., Stanford, 1991: estimation, control, and signal processing. Doran Wilde, Associate Professor; Ph.D., Oregon State, 1995: regular array design. Michael J. Wirthlin, Assistant Professor; Ph.D., Brigham Young, 1997: computer engineering and architecture.

Mechanical Engineering. W. Jerry Bowman, Associate Professor; Ph.D., Air Force Institute of Technology, 1987: heat transfer, heat pipes, regenerative heat exchangers, solar-powered aircraft. Kenneth W. Chase, Professor; Ph.D., Berkeley, 1972: computer-aided tolerancing and engineering, engineering design. Jordan J. Cox, Associate Professor; Ph.D., Purdue, 1991: advanced geometric modeling, reverse engineering. Russell Daines, Assistant Professor; Ph.D., Penn State, 1995: computational modeling of the flow fields of combined cycle engines. Paul F. Eastman, Associate Professor; Ph.D., Utah, 1965: materials applications, teaching methods. Mark Evans, Associate Professor; Ph.D., Rensselaer, 1987: dynamics and controls, robotics and automated systems. Larry L. Howell, Assistant Professor; Ph.D., Purdue, 1993: compliant mechanisms, engineering design. C. Gregory Jensen, Associate Professor; Ph.D., Purdue, 1993: computer-aided geometric design and manufacturing. Spencer P. Magleby, Associate Professor; Ph.D., Wisconsin–Madison, 1988: product development methods, engineering design. R. Daniel Maynes, Assistant Professor; Ph.D., Utah, 1997: fluid mechanics, rotating flows, turbulent flows and the behavior of fluid flow through microchannels. Tim McLain, Assistant Professor; Ph.D., Stanford, 1995: dynamic systems and controls. Mardson Q. McQuay, Professor; Ph.D., Carnegie Mellon, 1987: two-phase combustion, heat transfer. Kay S. Mortensen, Professor; Ph.D., Utah, 1967: engineering design, materials, teaching methods. Alan R. Parkinson, Professor; Ph.D., Illinois at Urbana-Champaign, 1982: optimization, engineering design. Predrag Radulovic, Associate Professor; Ph.D., Michigan, 1977: heat transfer, combustion, computer modeling. E. Max Raisor, Professor; M.S., Brigham Young, 1975: CAD, CAD/CAM integration, systems simulation. W. Edward Red, Professor; Ph.D., Arizona State, 1972: applied mechanics, robotics, manufacturing automation. Christopher A. Rotz, Associate Professor; Ph.D., MIT, 1978: damping in composites, design, testing and analysis of composites. Val E. Simmons, Associate Professor; Ph.D., Utah State, 1970: CAD/CAM and applications, heat transfer. Craig C. Smith, Professor; Ph.D., MIT, 1973: dynamic systems and controls, design of mechanical systems. Carl D. Sorensen, Associate Professor; Ph.D., MIT, 1985: analysis of manufacturing processes, materials. Robert H. Todd, Professor; Ph.D., Stanford, 1971: manufacturing process, MFG Proless Machine design and development. Dale R. Tree, Assistant Professor; Ph.D., Wisconsin, 1992: combustion engine design, emissions, engine fuels. Brent W. Webb, Professor; Ph.D., Purdue, 1986: heat transfer, thermodynamics, fluid mechanics.

School of Technology. Jeffery L. Campbell, Instructor; M.B.A., Phoenix, 1989: facilities and construction management. Perry W. Carter, Assistant Professor; Ph.D., Massachusetts, 1988: design of metal castings for quality and producibility. Kip W. Christensen, Associate Professor; Ph.D., Colorado State, 1991: education in manufacturing and construction technology. Jay P. Christofferson, Assistant Professor; Ph.D., Colorado State, 1996: computer integration and residential construction. Thomas L. Erekson, Professor; Ph.D., Illinois at Urbana-Champaign, 1979: hybrid and electrical vehicle development, administrative leadership in technology. Doran Gillie, Instructor; M.S., Brigham Young, 1997: technology and vocational education. Ronald F. Gonzales, Professor; Ph.D., Purdue, 1982: DOS-based applications and video graphics. Charles Harrell, Associate Professor; Ph.D., Denmark Technical, 1988: production and service system design and simulation. Val D. Hawks, Associate Professor; M.S., Lehigh, 1986: technical management, quality improvement, team building. Richard G. Helps, Associate Professor; M.S., Witwatersrand (Johannesburg), 1986: application of neural networks. D. Mark Hutchings, Instructor; M.S., Denver, 1977: construction company management—legal, accounting. A. Kent Johnson, Associate Professor; D.Sc., Stevens, 1965: active and passive filter analysis and design, passive networks. Kent E. Kohkonen, Assistant Professor; M.S., Brigham Young, 1976: cutting tools using polycrystalline diamonds. Barry M. Lunt, Assistant Professor; Ph.D., Utah State, 1993: electronics manufacturing and design, material properties and their low-frequency impedance. Loren Martin, Professor; Ed.D., Utah State, 1973: technology education/technical literacy. Tracy Nelson, Assistant Professor; Ph.D., Ohio State, 1998: welding and metallurgy. Jay Newitt, Professor; Ph.D., Colorado State, 1980: construction management. Earl F. Owen, Assistant Professor; M.S., Utah, 1972: microwave technology. Leon R. Rogers, Associate Professor; Ph.D., Texas A&M, 1988: construction management, estimating. Steven Shumway, Instructor; M.S., Utah State, 1993: computer integration, electronics control systems. Merrill J. Smart, Associate Professor; M.S., Utah, 1962: electronics, digital systems, single-board computers, embedded controllers, real-time computer systems. A. Brent Strong, Professor; Ph.D., Utah, 1971: composite materials manufacturing and properties. Gene A. Ware, Associate Professor; Ph.D., Utah State, 1980: infrared atmospheric measurements, signal processing.

BROWN UNIVERSITY

Division of Engineering

Programs of Study	The engineering program at Brown University is the oldest in the Ivy League and the third-oldest civilian program in the nation. In 1916, the separate Departments of Civil, Mechanical, and Electrical Engineering were combined into one Division of Engineering with a common core of basic engineering courses. This innovative change in the curriculum, which survives to this day, significantly increased the emphasis on engineering fundamentals shared by all disciplines. Today, the division is organized into the following four groups, which serve as focal points for faculty members and graduate students with common interests: electrical sciences and computer engineering; fluid, thermal, and chemical processes; materials science; and mechanics of solids and structures. The division offers Sc.M. and Ph.D. degrees in each of these four areas. The division also participates in cross-advising relationships on interdisciplinary graduate theses with the Division of Applied Mathematics, the Division of Biology and Medicine, and the Departments of Chemistry, Physics, and Computer Science. The lack of traditional departmental barriers within the Division of Engineering and a high faculty-student ratio are important factors in the high quality of the graduate experience at Brown. Graduate education at Brown takes place partly through formal courses and partly through lectures, seminars, and discussions with faculty members and other students, which form an intrinsic part of the graduate experience. For the Ph.D., no specified number of courses is required.

The focal point of a student's graduate education is the research done in close cooperation with a faculty adviser. Research areas in electrical sciences and computer engineering are solid-state materials and devices, quantum electronics, nonlinear optics and propagation, computer architecture and vision, VLSI design, digital signal and image processing, control systems and systems theory, and bioelectrical engineering. Areas of research in fluid, thermal, and chemical processes include energy and environmental technology, combustion and fire research, carbon materials, microfluidics and MEMS, turbulence and vortex dynamics, biomedical fluid mechanics, viscoelastic fluid dynamics, fluidization, chemical reaction engineering, and applied kinetics. Materials science is concerned with the processing, structure, and properties of metals, ceramics, and electronic materials, including fracture, with and without environmental effects; alloy design for strength; powder metallurgy; reaction sintering; crystalline defects; thin films; chemical vapor deposition; and semiconductor devices. The mechanics of solids and structures group emphasizes fracture of solids; stress-wave propagation; dynamic plasticity; mechanics of interfaces, thin films, and composites; structural dynamics; numerical methods in mechanics and mechanical testing of materials; and microscale and nanoscale approaches to the description of mechanical behavior. A number of faculty members and students participate in an NSF–sponsored materials research center (MRSEC) on micromechanics and nanomechanics of materials, centered in the Division of Engineering at Brown.

Research Facilities	The Barus and Holley Building houses the Division of Engineering and the Department of Physics. Faculty members and graduate students have proximate offices and laboratories, providing for close collaboration. Within the building are many smaller laboratories for experimental research. Large-scale experimental projects are contained in the adjoining Prince Engineering Laboratory where each group in the division has large research areas. This building houses the Fluids Mechanics Laboratory, including two low-turbulence wind tunnels, and the energy and environmental research laboratories, with various captive sample and flow reactors for combustion and chemical kinetics studies and extensive equipment for the chemical, physical, and thermal analysis of fuels, sorbents, and solid-waste materials. The laboratories for mechanics of solids and structures include facilities for studies on the fracture, fatigue, and creep of metals and plastics; soil mechanics; wave and impact phenomena; dislocation dynamics; and photoelasticity. Servo-controlled mechanical testing machines interfaced with PCs provide a modern system for data acquisition and reduction. There is an excellent central facility for electron microscopy that includes two scanning electron microscopes and two transmission electron microscopes, one equipped for in situ hot-stage experiments. The materials science laboratories offer equipment for quantitative microscopy, heat treatment, and sintering of metals and ceramics, as well as equipment for studies in physical metallurgy, micromechanisms of the strength of solids, and the thermomechanical behavior of solids. Extensive facilities exist for research in microelectronics, semiconductor properties, laser physics, infrared spectroscopy, FT-IR spectroscopy, optical preform fabrication and diagnostics, and the design and fabrication of microelectronic devices. In the field of digital systems, the Laboratory for Engineering Man/Machines Systems (LEMS) has facilities for speech recognition, image processing, and computer graphics. The Division maintains extensive, modern computing facilities, including a cluster of workstations and many PCs to support instructional activity. The LEMS laboratory supports its activities with a large network of Sun workstations and PC NT machines, as well as several special purpose machines. The Solid Mechanics group maintains a cluster of AlphaStations and AlphaServers for graduate student use. Subnets of the main, campuswide computer network interconnect all these machines. The Brown University Sciences Library houses an outstanding collection of books and periodicals in mathematics, physics, engineering, geology, medicine, biology, and chemistry.
Financial Aid	Most graduate students receive support in the form of fellowships, proctorships, or assistantships. Prize and industrial fellowships are available for qualified students. Additional fellowships are available for members of minority groups and women under the Graduate and Professional Opportunities Program. Stipends range from $14,500 to $17,000 for the 1999–2000 academic year, including two-month summer appointments. Most Ph.D. students do not pay any of their own tuition charges. Many graduate students devote 6 hours a week during the academic year as supplementary teaching assistants, which further increases the stipend level by about $1700 for each semester served.
Cost of Study	The tuition for students enrolled in the Division of Engineering for the 1999–2000 academic year is $24,624 for full-time study. Most Ph.D. students do not pay any of their own tuition charges.
Living and Housing Costs	Rooms are available at the Graduate Center, a modern complex of four 6-story residential towers connected to a centrally located commons building. Off-campus housing is also available.
Student Group	The total enrollment at Brown is 6,700, of whom 1,400 are in the Graduate School. The current graduate student population in engineering is 80, with 65 in the Ph.D. program and 15 in the Sc.M. program, including the integrated five-year program for Brown undergraduates. Graduate students represent all areas of the United States and many other countries.
Location	Metropolitan Providence has a population of about 800,000. Within a short driving distance are numerous ocean beaches; the historic resort community of Newport, Rhode Island; the city of Boston (one hour by car); and many locations for sailing and watersports on Narragansett Bay.
The Division	The total faculty of the division numbers 33, and the annual research budget is more than $7 million. Programs are integrated into a single organization unit, the Division of Engineering. This novel structure encourages collaboration by softening the boundaries between traditional engineering disciplines. Since many of the most exciting research problems today are interdisciplinary in nature, the structure at Brown offers distinct advantages.
Applying	The application deadline for fellowships is January 2; however, admission may be granted and research assistantships are often awarded after this date. The Division receives about 250 graduate applications per year, makes about 60 offers of admission, and matriculates 25 new graduate students.
Correspondence and Information	Graduate Program Coordinator Division of Engineering, Box DP Brown University Providence, Rhode Island 02912

Brown University

THE FACULTY AND THEIR RESEARCH

Electrical Sciences and Computer Engineering
R. Iris Bahar, Ph.D., Colorado. Computer engineering, CAD VLSI design, logic synthesis, low-power design.
J. Roderic Beresford, Ph.D., Columbia. Semiconductor heterostructure physics and devices, epitaxial crystal growth.
David B. Cooper, Ph.D., Columbia. Computer vision, pattern recognition, applied stochastic processes.
Gregory P. Crawford, Ph.D., Kent State. Liquid crystals, polymers, electrooptic materials, display devices.
Jerry D. Daniels, Ph.D., Berkeley. Biomedical engineering, neurophysiology, inhibition.
Benjamin B. Kimia, Ph.D., McGill. Computer vision, image processing, artificial intelligence, psychophysics.
Nabil M. Lawandy (research), Ph.D., Johns Hopkins. Dynamic instabilities in optically active media, laser theory, laser-induced transport.
Arto V. Nurmikko, Ph.D., Berkeley. Optical properties in semiconductors, quantum electronics.
Allan E. Pearson, Ph.D., Columbia. Systems theory, adaptive and optimal control systems, nonlinear systems.
Harvey F. Silverman, Ph.D., Brown. Computer architecture, digital signal processing, speech recognition and analysis.
William A. Wolovich, Ph.D., Brown. Multivariable and digital control systems.
Alexander Zaslavsky, Ph.D., Princeton. Physics and technology of semiconductor microstructures and devices.

Fluid, Thermal, and Chemical Processes
Joseph M. Calo, Ph.D., Princeton. Chemical kinetics and reaction engineering, atmospheric chemistry, carbon materials.
Kenneth S. Breuer, Ph.D., MIT. Microfluidics, MEMS, turbulence.
Bruce Caswell, Ph.D., Stanford. Viscoelastic fluids, heat transfer, computational fluid mechanics.
Robert H. Hurt, Ph.D., MIT. Energy and environmental technology, combustion, carbon materials.
Joseph T. C. Liu, Ph.D., Caltech. Coherent structure in turbulence, stratified and rotating flows, aerodynamic noise, fluidized beds.
Peter D. Richardson, Ph.D., D.Sc. (engineering), D.Sc. (physiology), London; FRS. Physiological fluid mechanics, heat and mass transfer, secondary flows, non-Newtonian flows.
Eric M. Suuberg, Sc.D., MIT. Fuel science, coal pyrolysis, combustion.

Materials Science
Clyde L. Briant, Sc.D., Columbia. Physical metallurgy, materials manufacturing, high-temperature superconductors, refractory metals.
Eric H. Chason, Ph.D., Harvard. Thin films, deposition processes, in situ optical diagnostics and spectroscopy.
Sharvan K. Kumar, Ph.D., Drexel. Composites, intermetallics, materials processing, high-temperature mechanical behavior.
David C. Paine, Ph.D., Stanford. Electronic materials, electron microscopy.
Janet Rankin, Ph.D., MIT. Sintering, in situ electron microscopy, energetics of interfaces in nanostructures and microstructures.
Brian W. Sheldon, Sc.D., MIT. Ceramic matrix composites, processing of advanced materials, chemical vapor deposition.

Mechanics of Solids and Structures
Janet A. Blume, Ph.D., Caltech. Continuum mechanics, elasticity, finite deformation theory.
Allan F. Bower, Ph.D., Cambridge. Tribology, plasticity, fracture.
Rodney J. Clifton, Ph.D., Carnegie Tech. Dynamic plasticity, rock mechanics, stress waves.
William A. Curtin, Ph.D., Cornell. Composite materials, microscopic modeling of materials, statistical mechanics.
L. Ben Freund, Ph.D., Northwestern. Fracture of solids, stress waves in solids.
Kyung-Suk Kim, Ph.D., Brown. Micromechanics of solids, adhesion, experimental mechanics.
Alan Needleman, Ph.D., Harvard. Structural stability, finite-element methods.
Rob Phillips, Ph.D., Washington (St. Louis). Microscopic modeling of materials, dynamics of dislocations, martensitic phase transformations.

CALIFORNIA INSTITUTE OF TECHNOLOGY

Programs in Engineering and Applied Science

Programs of Study

Graduate programs are offered leading to the Doctor of Philosophy degree, the Engineer degree, and the Master of Science degree in various fields of engineering. The program in chemical engineering is administered under the Division of Chemistry and Chemical Engineering, and an option in computation and neural systems is jointly administered by the Division of Engineering and Applied Science and two other divisions. All other graduate programs are under the Division of Engineering and Applied Science. Usually the requirements for the M.S. degree can be completed in one academic year. At least two years beyond the B.S. degree, including work on a thesis, are required for the Engineer degree. The Institute requirements for the Ph.D. degree include a minimum of at least three academic years of residence subsequent to the bachelor's degree, a thesis describing the results of independent research, and an oral examination based on the thesis and research.

Research Facilities

Engineering facilities are located in ten large laboratories on the campus. Aeronautics, applied mathematics, and jet propulsion are located in the Guggenheim, Firestone, and Kármán laboratories; applied mechanics, earthquake engineering, engineering design, mechanical engineering, soil mechanics, and structural engineering are in the Thomas Laboratory; electrical engineering and computation and neural systems are in the Moore Laboratory; control and dynamical systems are in the Steele Laboratory; chemical engineering in the Spalding Laboratory; environmental engineering science, materials science, and hydraulics and water resources are in the Keck Laboratories; computer science is in the Jorgensen Laboratory; computation and neural systems is in the Beckman Laboratories; and applied physics is in the Watson Laboratory. Computing facilities are available in the Engineering Computer Facility in the Thomas Lab and in the various laboratories.

Financial Aid

Fellowships, traineeships, and laboratory, research, and teaching assistantships are available. Those granted such aids may also receive a tuition award. In the academic year 1999–2000, stipends vary from $9038 to $14,748, depending upon duties, academic attainment, and experience. Teaching assistantships and research assistantships typically require 12 and 20 hours per week, respectively, permitting the holder to carry a full academic program of graduate work. Fellowships and traineeships do not require duties. Additional financial support may be provided during the summer months.

Cost of Study

Tuition, health-plan coverage, and other graduate student fees for the academic year 1999–2000 are $19,260. Books and supplies vary in cost; a typical figure might be $1026 a year.

Living and Housing Costs

Single-student dormitory-style accommodations are available in the on-campus Graduate Houses or Avery House at a cost of approximately $347 to $447 a month for a private room. Single-student apartment-style accommodations are available in the Catalina Complex I (four bedrooms, two bathrooms) for approximately $388 a month per student, plus utilities. Both married- and single-student apartment-style accommodations are available in the Catalina Complex II and III (two bedrooms, one bathroom) for approximately $836 a month for a family, or $418 a month per single student, plus utilities. Married student accommodations are available in the Catalina Complex III (one-bedroom apartments) at a cost of approximately $724 a month, plus utilities.

Student Group

There are 901 undergraduate students enrolled, including 256 women, and 957 graduate students, including 226 women. There are 412 graduate students enrolled in the engineering programs in the two divisions.

Student Outcomes

Typically, 95 percent of students graduating with a Ph.D. in one of the many engineering and applied science options have firm plans at graduation. Many students (36 percent) accept industrial employment. Employers of the 1998 class represented organizations in computers, electronics, chemicals, and pharmaceuticals; the national laboratories; nonprofit research organizations; and management consulting firms. Academic employment also figures prominently in the plans of graduates (51 percent). Students accept positions both as postdoctoral research fellows and tenure-track faculty members at institutions worldwide.

Location

Caltech is located in Pasadena, California, a city of approximately 125,000 inhabitants, about 10 miles northeast of Los Angeles. The Institute is situated in the center of a residential district but is within a few blocks of shopping facilities. Pasadena and the Metropolitan Los Angeles area provide abundant cultural and recreational opportunities.

The Institute

California Institute of Technology, more familiarly known as Caltech, is an independent, privately supported, and privately controlled institution, officially classed as a university, carrying on undergraduate and graduate instruction and research principally in the various fields of science and engineering. Throughout its history, Caltech has maintained a small, select student body and a faculty that is unusually active in research. The faculty in Engineering and Applied Science, including Chemical Engineering, consists of 86 professors. The Chairman of the Division of Engineering and Applied Science is Professor John H. Seinfeld.

Applying

Students are admitted only for September, and applications should be received by January 15. Application materials may be obtained from the office of the dean of graduate studies. Inquiries for further information will be forwarded to the appropriate group. Caltech welcomes applications from members of minority groups.

Correspondence and Information

Dean of Graduate Studies
California Institute of Technology
Pasadena, California 91125
Telephone: 626-395-6346
Fax: 626-577-9246
World Wide Web: http://www.cco.caltech.edu/~gradofc

California Institute of Technology

THE FACULTY AND RESEARCH AREAS

Aeronautics. The programs emphasize the fundamentals underlying aeronautics. Major groups include staff and graduate students interested in fluid mechanics, solid mechanics, and applied aerodynamics, with a close interrelationship with the programs of the Jet Propulsion Center, Applied Mathematics, and Applied Mechanics. Facilities consist of wind tunnels with velocities ranging from the very slow to hypersonic speeds, shock tubes, a free-surface and high-speed water tunnel, and instrumentation/equipment for experimentation in structures and solid mechanics related to fracture/failure of composites and metals.

Applied Mathematics. The program in applied mathematics emphasizes interdisciplinary studies and reflects the breadth of activity in applied mathematics at Caltech. In addition to basic and advanced study offered by the applied mathematics faculty, broad selections are available in mathematics, physics, engineering, and other such areas. Strong research programs cover nonlinear waves, perturbation theory, numerical analysis, fluid dynamics, computational fluid dynamics, scientific computation, computational materials science, and other areas.

Applied Mechanics. Advanced work is offered in elasticity, plasticity, dynamics, and vibrations. Research areas include linear and nonlinear dynamical systems; random vibrations; structural dynamics and control; fluid-structure interaction; wave propagation in solids; problems of large deformation and material instability in solids; phase transformations in solids; fracture mechanics, including elastic, viscoelastic, and plastic fracture; failure of composite materials; and finite-element analysis.

Applied Physics. Research opportunities are offered in a wide variety of fields, such as solid-state physics, lasers and integrated optics, photonics, metals and alloys, plasma physics, fluid dynamics, electronic materials, computer simulation of materials, and cryogenics. The program is intended for students with a strong interest in fundamental physics having technological applications.

Chemical Engineering. Research programs are conducted in air pollution and aerosol science, catalysis, zeolite synthesis, inorganic membranes, ceramics and electronic materials processing, statistical mechanics, dynamics of complex fluids, polymer rheology and polymer physics, biochemical engineering, multiphase flow, dynamics of gas-surface interactions, and nanoscale structures.

Civil Engineering. Advanced work is offered in structural engineering, earthquake engineering, hydrology, and coastal engineering. Research areas include the analysis of structures subjected to dynamic loadings, especially earthquake motions; the use of finite-element methods for structural analysis; seismic risk analysis and earthquake ground motion studies; system identification and control of structures; sediment transportation in streams; dispersion in turbulent shear flows; problems related to density-stratified fluids; generation, propagation, and coastal effects of tsunamis; pollutant transport in porous media; water reclamation; and the disposal of waste in the ocean.

Computation and Neural Systems. This integrated approach to graduate study is designed to promote a broad knowledge of relevant aspects of biophysics, neurophysiology, and psychophysics; computational devices; information theory; learning systems; neuronal modeling; and complex systems. The program encourages interdisciplinary approaches to problems in neurobiology, engineering, and information theory that share a common deep structure at the computational level. This approach is combined with an appropriate depth of knowledge in the particular field of thesis research. Students take classes in neurobiology, electrical engineering, and relevant mathematics. Ph.D. students only.

Computer Science. The department emphasizes a holistic approach to computer science research integrating: theory and systems development, programming and physical structures for computing, and computing fundamentals and applications. Research focuses on algorithms that scale; models of complex systems; graphics and human-computer interaction; concurrent computation; VLSI systems, with a specific emphasis on asynchronous and analog VLSI; ubiquitous computing; high-confidence systems, with a specific emphasis on fault tolerance, program verification and security; information theory; learning systems; and computer vision. Research projects frequently involve connections with other disciplines such as biology, mathematics, mechanics, control systems, and electrical engineering.

Control and Dynamical Systems. In this interdisciplinary graduate program, theoretical research is conducted in all aspects of control and dynamical systems, with emphasis on robustness, multivariable and nonlinear systems, optimal control, decentralized control, modeling, and system identification for robust control. Applications areas include large flexible structures; chemical process control; mixing and transport processes in fluids; turbomachines and complex combustion systems; global bifurcation analysis and control of nonlinear systems, including strongly nonlinear and chaotic dynamics; nonlinear flight dynamics for highly maneuverable aircraft; and robotic manipulation.

Electrical Engineering. Major research activities are solid-state electronics, including contacts to semiconductors, thin films, ion implantation, backscattering, and channeling analysis; microsensors and microactuators, including micromachining technology for accelerometers, infrared sensors, pressure sensors, microsurgical tools, and neuron probes; quantum electronics, including coherent radiation from submillimeter through ultraviolet wavelengths, nonlinear optics, and integrated devices in III–V semiconductors; electromagnetic waves, including integrated circuit antennas, dielectric waveguides, and optical signal and image processing; neural networks; communication systems, networks, switching, and traffic; coding and information theory and practice, wireless communications, including signal processing, data compression, and error-correcting coding techniques; power electronics, including the efficient conversion of electrical energy; digital signal processing, including adaptive and multirate filtering; wavelet transforms; and modern control theory, especially robust control.

Engineering Science. Programs are designed for students of subjects that might be called classical and semiclassical physics and mathematics, and engineering subjects that form the core of the new interdisciplinary sciences. Fields of research may include such topics as fluid mechanics (including applications to geophysical and biomechanical problems), physics of fluids, structure and properties of solids and liquids, dynamics of deformable bodies, and rheology of biological fluids.

Environmental Engineering Science. Research and instruction emphasize basic studies that underlie new solutions to challenging environmental problems, such as urban, regional, and global air quality; water supply and water quality control; effective management of hazardous substances in the environment; and maintenance of stable ecosystems. Among the academic disciplines central to the program are air pollution control engineering, atmospheric and aquatic chemistry, environmental fluid mechanics, applied microbiology, hydraulics, hydrology, and aerosol physics and chemistry. Extensive laboratories and modern instrumentation are available in all areas. Graduate students from a variety of backgrounds, including undergraduate majors in biology, chemical engineering, chemistry, civil engineering, mechanical engineering, and physics, among others, enter this program. A research report is available on request.

Jet Propulsion. The Guggenheim Jet Propulsion Center on the campus offers programs of instruction and research in the field of propulsion for advanced-degree candidates plus programs in other related engineering fields. Research topics include combustion in turbulent flow, combustion instability in propulsion systems and stationary power plants, and active feedback control of combustion systems.

Materials Science. Research is oriented toward fundamental phenomena that underlie the structure and properties of engineering materials. Emphasis is on the study of nonequilibrium metals, ceramics, and composites, but studies of other materials are encouraged through faculty members in other departments. Modern facilities for the preparation of materials include equipment for rapid quenching, physical vapor deposition, ion-beam bombardment, and shock-wave consolidation. Facilities for the characterization of materials include a 300 keV transmission electron microscope, a wide variety of X-ray diffraction equipment, Mössbauer spectrometers, and modern calorimetry equipment. Facilities for measuring the mechanical, magnetic, and electrical properties of materials are available. Information is also available on the World Wide Web (http://www.caltech.edu/~matsci/).

Mechanical Engineering. Graduate study emphasizes a broad coverage of such basic subjects as heat and energy transfer, thermodynamics, solid and fluid mechanics, kinematics, and dynamics, as well as interdisciplinary subjects, including engineering design, microelectromechanical systems (MEMS), and robotic systems. Typical research projects have been concerned with convective heat transfer in complex fluid-solid systems, turbulent flows, aeroacoustics, numerical methods for fluid dynamics, turbomachines, cavitation, hydrofoils in unsteady flow, two-phase flows, aerosols, combustion, synthesis of ceramics, diamond film, and other materials, automatic measurement of machine-part geometries, development of computer-aided engineering design (CAED) theory and methodologies, design of automobile chassis, microspacecraft propulsion systems, design of manipulator terminal devices, autonomous robotic systems, and control of mechanical systems.

CALIFORNIA POLYTECHNIC STATE UNIVERSITY, SAN LUIS OBISPO

College of Engineering

Programs of Study

The College of Engineering offers individualized graduate programs leading to the Master of Science degree in aeronautical engineering, civil and environmental engineering, computer science, electrical engineering, computer engineering, and mechanical engineering. In addition, M.S. degrees are offered with specializations in materials engineering, water engineering, and industrial engineering, and there are special interdisciplinary programs in biochemical engineering, bioengineering, and biomedical engineering. A joint program with the College of Business emphasizing the management of technology leads to an M.S. degree in industrial engineering and an M.B.A. The transportation planning specialization is a joint interdisciplinary program with the College of Architecture and Environmental Design, with successful completion resulting in the awarding of both the Master of City and Regional Planning (M.C.R.P.) and the M.S. in engineering.

The degrees require a minimum of 45 quarter units of course credit and culminate in a thesis, a graduate project, or a comprehensive examination. Each student designs his or her own program of study, subject to the approval of the Graduate Studies Committee. Students may specialize in any area supported by the faculty, either within a discipline or across disciplinary bounds. Full-time students will usually complete the master's degree in 1½ to 2 years. Cal Poly is particularly proud of its orientation toward industrial research and the strong emphasis placed on faculty-student interaction. Many students are eventually employed by companies that sponsored their graduate work. Twenty percent of master's graduates go on to study for the Ph.D. degree elsewhere. Cal Poly's College of Engineering has a holistic approach to graduate education. Faculty members, graduate students, and undergraduates are often teamed to attack real-world problems typical of what students will encounter in their professional careers.

Research Facilities

The College of Engineering is housed in eight buildings. Laboratory space occupies 160,000 square feet, of which approximately 20 percent is available for graduate research. Additional space will become available when the NSF-funded Advanced Technology Laboratories are completed. College of Engineering computing systems on all levels are available to all students for both classroom and research projects. The University Library has extensive collections in all engineering and applied science disciplines. The College has particular research emphasis (and facilities) in the areas of electric power, human factors, robotics, CAD, CIM, CAE, AI, transportation, nonmetallic materials, thermomechanical simulation, automotive electronics, CFD, FEA, nondestructive evaluation system design, mechatronics, and solar power.

Financial Aid

Cal Poly offers a number of financial aid programs funded by government, industry, and private sources, including research assistantships, teaching assistantships, grants, scholarships, and loans. Research assistantships are available to students assigned to research projects consistent with their preparation and interest. Cal Poly provides special funds (Graduate Equity Program) to support students who are members of underrepresented groups. Cal Poly has a unique Faculty Coordinated Graduate Internship Program that provides financial support and industrial experience for selected graduate students. Students should direct inquiries about financial aid of any sort to their department or program graduate coordinator.

Cost of Study

In 1999–2000, students who have been residents of California for at least one year pay $539 per quarter for up to 6 units or $761 per quarter for more than 6 units. Nonresidents, both American and international, pay $164 per unit in addition to registration fees. These costs do not include living expenses, books and supplies, parking, health fees, or miscellaneous expenses. Fees may increase in May 1999.

Living and Housing Costs

The campus residence halls accommodate about 2,800 students in comfortable double rooms. The cost of room and board for the 1998–99 academic year was $5000–$5200. Privately owned off-campus housing is available in private homes, condominiums, and apartment complexes; rents ranged from $236 to $600 per month for single rooms. The campus maintains a directory of available housing in the community (telephone: 805-756-5700).

Student Group

The University enrollment is more than 15,500 students, including about 4,000 undergraduates and 260 graduate students in the College of Engineering. Students come from every county in California, every state in the Union, and 105 other countries. The College of Engineering has a diversified population—in 1998–99, more than 20 percent of the students were women, 40 percent were members of ethnic minority groups, and roughly 5 percent were from other countries. Almost all students receive some form of financial aid. Virtually all 1998 graduates found employment in their areas of specialization or continued on to further graduate study.

Location

San Luis Obispo is a city of approximately 43,000 residents, situated 12 miles from the Pacific Ocean midway between San Francisco and Los Angeles on U.S. Highway 101. Several public beaches and lakes provide recreational opportunities for residents and tourists. The city was built around the site of Mission San Luis Obispo de Tolosa, the fifth of the twenty-one California missions. The scenic outdoor plaza of the mission provides a pleasant setting for many annual community events. Concerts and plays are presented throughout the county; the most notable of these is the series of concerts presented at the annual Mozart Festival during the first week of August. San Luis Obispo recently was selected as one of the top ten cities in the United States with a population of under 100,000.

The College

The College of Engineering at California Polytechnic State University, San Luis Obispo, has built a fine reputation with its learn-by-doing philosophy. Through its emphasis on laboratory classes, internships, and various research projects, the College has a history of successfully preparing its graduates for employment, and this, together with the desirable physical setting and geographical location of the University, has made Cal Poly, San Luis Obispo, one of the most popular campuses in the California State University System. The College is the only primarily undergraduate institution to be ranked in the top ten engineering schools in the country both for overall quality and for its ability to work with industry.

Applying

All applicants must file a complete application within the appropriate filing period. For admission under unclassified postbaccalaureate standing, a student must hold an acceptable baccalaureate degree from an institution accredited by a regional accrediting association or have completed equivalent academic preparation, as determined by an appropriate campus authority; must have attained a grade point average of at least 2.5 (on a 4.0 scale) in the last 60 semester (90 quarter) units attempted; and must have been in good standing at the last college attended. All students subject to degree requirements must demonstrate competence in writing skills as a requirement for graduation. Applications for financial aid must be received by March 1 for any quarter of the following academic year, summer through spring. For information regarding degree programs, students should address all correspondence to the specific faculty member indicated on the reverse of this page and send it to the address below.

Correspondence and Information

Graduate Program in (specify)
California Polytechnic State University
San Luis Obispo, California 93407

California Polytechnic State University, San Luis Obispo

FACULTY HEADS AND RESEARCH AREAS

Aeronautical Engineering
Dr. Jin Tso, Chair; Dr. Daniel Biezad, Coordinator.
The faculty consists of 5 professors, 1 assistant professor, and 2 lecturers.
The Master of Science program in aeronautical engineering prepares the student for entry into the well-established field of aerospace engineering. In addition, the areas of flight simulation and controls, structures, and aerothermal sciences have been integrated into the program. The program emphasizes engineering science and research activity. A student who has earned this degree has the capability for more complex research, development, and innovative design and is prepared for future graduate study in engineering leading to the Doctor of Engineering or Doctor of Philosophy degree. Active research involves the study of flows in compressible materials, CFD, FEA, flows at high Reynolds numbers, incidence effects, and stability of structures in space.

Biochemical Engineering
Dr. Nirupam Pal, Coordinator.
The interdisciplinary program emphasizes the fundamentals of bioprocess engineering, microbial growth kinetics and reactor design, enzyme kinetics, molecular genetics and control, specific process technology, and downstream processing of bioproducts.

Civil and Environmental Engineering
Dr. Robert Lang, Chair; Dr. Nirupam Pal, Coordinator.
The faculty consists of 14 professors, 1 associate professor, 4 assistant professors, and 8 lecturers.
The Master of Science program in civil and environmental engineering provides job-entry education for the more complex areas of engineering, such as research and development, innovative design, systems analysis and design, and managerial engineering; updating and upgrading opportunities for practicing engineers; graduate preparation for further study in engineering leading to the Doctor of Engineering or Doctor of Philosophy degree; and a base that allows graduates to maintain currency in their fields. Active research areas include transportation, hydrology, geotechnics, wood design, structures, soil micromechanics, hazardous waste treatment, bioremediation, and renewable energy production.

Computer Science
Dr. James Beug, Chair and Coordinator.
The faculty consists of 16 professors, 5 associate professors, and 3 assistant professors.
The department offers a program leading to a Master of Science in computer science with particular emphasis in the following areas: computer systems and software, computer graphics, numerical analysis, computer modeling and simulation, expert systems, information processing, and computer architecture. Special features of the program include its emphasis on the applications of computers to current industrial problems.

Electrical Engineering
Dr. Martin Kaliski, Chair; Dr. Donley Winger, Coordinator.
The faculty consists of 19 professors, 3 associate professors, and 7 lecturers.
The Master of Science program in electrical engineering provides job-entry education for the more complex areas of engineering, such as research and development, innovative design, systems analysis and design, and managerial engineering; updating and upgrading opportunities for practicing engineers; graduate preparation for further study in engineering leading to the Doctor of Engineering or Doctor of Philosophy degree; and a base that allows graduates to maintain currency in their fields. Active research areas include transportation electronics, computer networking, and polymer electronics.

Industrial Engineering
Dr. Sema Alptekin, Head and Coordinator.
The faculty consists of 11 professors, 3 associate professors, and 2 lecturers.
The Master of Science program in industrial engineering emphasizes the strong link between academia and industry. Active research areas include human factors, management of technology, facilities planning, rapid prototyping, CIM, stochastic processes, TQM, reliability theory, and forecasting.

Materials Engineering
Dr. Robert Heidersbach, Head and Coordinator.
The faculty consists of 3 professors, 1 associate professor, and 1 assistant professor.
The department emphasizes the fundamental relationships between atomic structure, microstructure, and the chemical and physical properties of materials. The department has research that emphasizes physical metallurgy, ceramics, polymeric materials, composites, and glasses. Active research programs include superconducting ceramics, electronic materials, semiconductor packaging, nondestructive evaluation of materials, microbiologically influenced corrosion, corrosion of rebar, the effect of minor elements on physical chemistry of molten metals, EVA welding, thermomechanical processing, and quantitative materialography.

Mechanical Engineering
Dr. Safwat Moustafa, Head; Dr. Saeed Niku, Coordinator.
The faculty consists of 18 professors, 4 associate professors, 2 assistant professors, and 1 lecturer.
Most of the faculty are registered professional engineers and are consultants to government, industry, and business. The program emphasizes fundamentals and engineering practice. Active research programs include composites, propulsion systems, thermal sciences, dynamic and solid mechanics, petroleum production, robotics, and engineering education.

CALIFORNIA STATE UNIVERSITY, LONG BEACH

College of Engineering

Programs of Study

Graduate programs leading to the degrees of Master of Science in Aerospace Engineering, Master of Science in Civil Engineering, Master of Science in Computer Science/Computer Engineering, Master of Science in Electrical Engineering, and Master of Science in Mechanical Engineering are offered by the College of Engineering at California State University, Long Beach (CSULB). An interdisciplinary Master of Science in Engineering is also offered to meet special needs. The professional degree of Civil Engineer is offered to persons who already have a master's degree in civil engineering.

A minimum of 30 semester credits in course work and research, including 24 credits earned at the University, is required for the master's degree. Proficiency in a foreign language is not required. Proficiency in English must be demonstrated by successful completion of the Writing Proficiency Examination.

A joint Ph.D. program in industrial and applied mathematics is offered with the Claremont Graduate University and currently has 40 students enrolled.

A minimum of 72 semester credits in course work, independent study, and research is required for the Ph.D. in industrial and applied mathematics. Proficiency in a foreign language and computer programming is required.

Research Facilities

Instructional and research equipment is housed in 205,000 square feet of laboratory space, of which approximately half is devoted to graduate research. Excellent computer systems are available to students for classroom and research projects. Much of the research program is supported by industry and federal agencies.

Financial Aid

For qualified graduate students, financial assistance is available in several forms: Federal Perkins Loans, federally insured student loans, emergency loans, Federal Work-Study awards, part-time employment, University scholarships, state graduate fellowships, educational and industrial scholarships, veterans' benefits, research-funded assistantships, graduate assistantships, and teaching assistantships. Research assistants are assigned to research projects consistent with their area of preparation and interest. Graduate and teaching assistantships are limited in number.

Cost of Study

Students who have been residents of California for more than one year pay $625 for 1 to 6 units or $991 for more than 6 units. Both domestic and international nonresidents pay $246 per unit, with a minimum of $811 and a maximum of $4588 for 15 units. These are tuition and mandatory fees for the fall 1999 semester.

Living and Housing Costs

The campus residence hall complex consists of eight halls with a maximum capacity of 2,000 students, equally divided between women and men. In 1999–2000, the cost of room, board, and utilities is $5000 per year. Students should apply early to secure a place in one of the campus halls of residence. Comparable off-campus housing is also available within reasonable commuting distance.

Student Group

The total fall 1998 University enrollment of 27,431 included 2,346 undergraduate and 670 graduate engineering students. Students come from every state in the Union and 109 other countries.

Location

The University is located a mile from the Pacific Ocean in the city of Long Beach (population 450,000). The climate is pleasantly semitropical, with an average winter low of 55°F. The city of Long Beach is the fifth largest in California and thirty-second largest in the United States.

Numerous recreational and cultural facilities and activities are available in Long Beach or within a reasonable driving distance, including several libraries, operas, theaters, museums, exhibits, convention centers, amusement parks, and professional football, baseball, basketball, soccer, and hockey teams. Every type of engineering employment is within easy driving distance of CSULB.

The University

CSULB was founded in 1949 as part of the twenty-five-campus California State University System. It is located on a beautifully landscaped 322-acre site, with eighty building complexes, on the southern edge of Long Beach.

Applying

Further information and application forms may be obtained from the associate dean for instruction of the College of Engineering (see below). Completed forms should be filed three months before the desired admission date. Applications for financial aid should be filed by March 2 for consideration for the fall semester. Current copies of the *University Bulletin* and *Schedule of Classes* may be obtained from the University bookstore.

Correspondence and Information

Dr. Mihir K. Das
Associate Dean for Instruction
College of Engineering
California State University
1250 Bellflower Boulevard
Long Beach, California 90840-8306
Telephone: 562-985-8032 or 5257
Fax: 562-985-7561
E-mail: mdas@engr.csulb.edu
     rprather@csulb.edu (International Education Center for students outside the U.S.)
World Wide Web: http://www.engr.csulb.edu

California State University, Long Beach

THE FACULTY AND AREAS OF RESEARCH

Aerospace Engineering. Dr. Tuncer Cebeci, Chairman; Dr. Orhan Kural, Graduate Adviser.

The faculty consists of 5 professors, 3 lecturers, and several adjunct and visiting faculty members of international reputation. The department is active in research and is equipped with modern laboratories and computer facilities. Recent sponsored research topics include unsteady flows, separated flows, high lift, aircraft icing, stability and transition, transonic flows, and manufacturing, testing, and analysis of composite materials. The department puts special emphasis on practical applications and interaction with the aerospace industry. Several scholarships and assistantships are available to the department's students.

Chemical Engineering. Dr. Lloyd Hile, Chairman; Dr. Shirley Tsai, Graduate Adviser.

The faculty consists of 5 professors and several part-time members from professional practice. The department is well equipped with modern laboratories for chemical engineering research, and several faculty members are actively involved in research. General areas of interest include rheology, two-fluid atomization, biochemical engineering, separation and purification, chemical reactor kinetics, safety and hazards engineering, solid-state device fabrication, polymer synthesis, computer simulation, and process control. The research projects supported by the National Science Foundation are: Rheology of Slurrys and Suspensions, Ultrasound-Modulated Two-fluid Atomization, and A Metal Ion Binding to Biopolymers. Students may pursue an interdisciplinary program in a Master of Science in Engineering with an emphasis in chemical engineering topics.

Civil Engineering. Dr. Chan-Feng (Steve) Tsai, Chairman; Dr. Peter Cowan, Graduate Adviser.

The faculty consists of 8 professors, 2 assistant professors, and engineers from the professional community as part-time lecturers. Most of the faculty members are registered professional engineers, some of whom are engaged in research and consulting activities for the federal government; others are consultants to federal, state, and/or local agencies. Areas of study include structural engineering, transportation and urban engineering, environmental engineering, construction engineering and management, water resources engineering, and geotechnical engineering. Recent funded research activities include a composite bridge deck study funded by the Federal Highway Administration, a structural health monitoring of a public safety building in southern California, an air condition metal ducting study for the thin gauge metal ducting that involves large displacements, a feasibility study for the reuse of slightly contaminated soils in parking lot and road constructions, and experimental studies for chemical flow diversions and modeling of sediment flow through a reservoir.

Computer Engineering and Computer Science. Dr. Sandra Cynar, Chairman; Dr. Dar-Biau Liu, Graduate Adviser.

The faculty is composed of 17 professors, 2 associate professors, and several temporary lecturers who are associated with high-technology companies in the area. The program is balanced in its coverage of hardware, software, and applications. Areas of faculty interest include computer networking, multimedia, object-oriented programming, embedded systems design, database management, artificial intelligence, distributed computer systems, software engineering, computer architecture, and simulation/modeling. The department has several laboratories to support its instructional and research programs. Computer systems include an SGI compute server, Sun workstations, and numerous networked microcomputers.

Electrical Engineering. Dr. Fumio Hamano, Chairman; Dr. Michael Chelian, Graduate Adviser.

The faculty is composed of 20 professors and several part-time faculty members from local high-tech industry. The programs are balanced, current, and industrially relevant. The focus is on sound fundamentals, applications, and comprehensive modern education. Several members are consultants to computer, electronics, and aerospace companies in the area. Areas of study include physical electronics and microelectronics, terrestrial and space power systems, networks and filters, electromagnetics and microwaves, robotics and control systems, communications, power electronics, digital design, microprocessor applications, image processing, biomedical engineering, and digital signal processing. The department has several laboratories to support both classroom and research activities. The department's computer facilities include Sun workstations, Windows NT, and PCs. There is also access to the college-wide Pentium labs and servers. Laboratory facilities include the following areas: analog and digital electronics, digital signal processing, digital image processing, robotics, neural networks, control and simulation, power systems, microwave, communication, biomedical engineering, network and filters, and solid-state and integrated circuits. Recent research includes computer applications in health-care systems and related applications in biomedical engineering; computer simulation of electrical power systems; computer communication network analysis and design, digital filters; computer control of robots; control theory; image processing; digital signal processing; microwave and communication systems; embedded controllers; microelectronics; and instrumentation systems.

Mechanical Engineering. Dr. Leonardo Pérez y Pérez, Chairman; Dr. C. Barclay Gilpin, Graduate Adviser.

The faculty consists of 16 professors, several adjunct professors and scholars, and engineers from the professional community as part-time lecturers. Most of the faculty members are registered professional engineers and are consultants to industry, government, and business. The program of study emphasizes both fundamentals and engineering practice. A student may choose to take either a broad program of study or a specialized program. Graduate courses are offered in the following subject areas: optimum design, CAD/CAM, materials and metallurgy, composite materials, integrated design, robotics and control systems, and advanced manufacturing engineering, including CIM, management engineering, mechanics of solids, fluid mechanics, heat transfer, thermodynamics, energy conversion, power systems, and energy conservation. The department offers a concentration of courses at the graduate level. These courses develop the concept of the integration of design and manufacturing and include the database of material selection, material forming, the interrelation of material properties and manufacturing processes, design for optimum properties and minimum cost, etc. The principles taught are common to all industries and will permit the graduate to apply these concepts to a variety of manufacturing processes. State-of-the-art engineering laboratories exist in manpowered vehicles, aerodynamics, automobile engine analysis, stress analysis, finite-element analysis, boundary-element analysis, solar energy, energy thermodynamics, acoustics, active noise control, integrated design, computer-integrated manufacturing, rapid prototyping, microanalysis (electromicroscope), fluid-thermoscience (holographic and laser-Doppler velocimetry system), and robotics. Some examples of recent thesis titles are Design of a Belt Mounted Air Purifying Respirator, Refueling Station Design Natural Gas Vehicles, Structural Instability of J-Stiffened Stitched Composite Shear Panel, Multiparameter Optimization Control of Distributed Parameter Systems, FEM Analysis of Flow Through a Bileaflet Heart Valve, Implementation of TQM, Superplastic Forming of Al-Li Alloys, Relaxation Spectrum of Polymers, Explicit Time Integration Method for FEA, Defense Industry Strategy in the 90's, and Dynamics of Rotor Bearing Systems. The department participates in NSF project SCCEME (Southern California Coalition for Education in Manufacturing Engineering).

CASE WESTERN RESERVE UNIVERSITY

The Case School of Engineering
Graduate Programs in Engineering

Programs of Study

The University offers M.S. and Ph.D. programs in biomedical engineering, chemical engineering, civil engineering, computer engineering and science, electrical engineering and applied physics, macromolecular science, materials science and engineering, mechanical and aerospace engineering, and systems and control engineering.

A Master of Engineering program that combines advanced study in business and the practice of engineering is also available for part-time students.

Special programs leading to the Ph.D./M.D. are available through the Department of Biomedical Engineering.

Research Facilities

Engineering facilities include seven buildings on campus. Each department has modern and fully equipped research and teaching laboratories. The basic computing platforms are powerful microcomputers, which are supplemented by campus workstations and off-campus supercomputers and specialized data resources. The University is at the leading edge of technology with respect to computer networking. Its network, CWRUnet, links more than eighty-five campus buildings and connects users to the most powerful supercomputers in the United States. It is the first network to use only fiber-optic cabling as wiring to each end-user computer. By employing both single-mode and multimode fiber-optic cabling in building premise wiring to the faceplate, CWRUnet transforms each advanced personal computer into a portal to vast repositories of electronic information. Specialized research is carried out in the following laboratories and centers: the Applied Neural Control Laboratory for development of orthotic and prosthetic systems; the Center for Applied Polymer Research, for fundamental research on polymers and interfaces; the Ernest B. Yeager Center for Electrochemical Sciences, for the study of energy conversion and storage, the conservation of natural resources, and biomedical applications; the Electronics Design Center, for research into applied neural control, microelectronics, biomedical electronics resource, advanced microsensors, and micromachined actuators; the Center for Automation and Intelligent Systems Research, where the "brains" of automated manufacturing systems-sensing devices, vision, speech, and intelligence are being developed; the Center for Advanced Liquid Crystalline Optical Materials, which is operated in conjunction with the University of Akron and Kent State University; the Cardiac Bioelectricity Research and Training Center, for the study of cardiac electrophysiology and electric fields generated by the heart using theoretical computer modeling and mapping of cardiac electrical activity; the Neural Engineering Center, for studies in computational neuroscience, engineering, and elctrophysiology to solve problems in the central and peripheral nervous systems; the Center for Cardiovascular Biomaterials, for development of new cardiovascular biomaterials, structure and property analysis, and biocompatibility; the FES Center, for the development and deployment of neural prostheses; and the National Center for Microgravity Research, for studies of transport phenomena and combustion in microgravity environments.

Financial Aid

Case Prime Fellowships provide full support plus discretionary funds for Ph.D.-bound students. Loans with deferred-repayment plans and low interest are available to U.S. citizens. Traineeships include NIH awards in biomedical engineering and macromolecular science and EPA awards in chemical and systems engineering.

Cost of Study

In 1999–2000, tuition is charged at the rate of $800 per semester hour. The insurance fee is $600 per academic year. To be considered full-time, a student must register for a minimum of 9 semester hours each semester.

Living and Housing Costs

The principal residence for unmarried graduate students is Clarke Tower, which accommodates 280 students. Graduate students are also housed in smaller University-sponsored residences. Room and board charges range from $6350 to $7030 for the 1999–2000 academic year. Students also live in off-campus apartments near the University.

Student Group

The total University enrollment is 9,753. Of this number, 2,171 are enrolled in the School of Graduate Studies, 3,962 in other graduate and professional programs, and 3,620 in the undergraduate colleges. There are 697 graduate students in the Case School of Engineering.

Location

Case Western Reserve University is located in University Circle, about 4 miles east of downtown Cleveland. There is considerable interaction between the University and other Circle institutions, such as the Cleveland Orchestra, the Cleveland Museum of Art, and the Cleveland Institute of Music. Metropolitan Cleveland offers a wide range of entertainment and national sports events.

The University and The School

Case School of Applied Science, founded in 1880, was renamed Case Institute of Technology in 1948. Master's work began in 1890, and doctoral work in the 1930s. In 1967, Case Institute of Technology joined with Western Reserve University, located on an adjoining campus, to form Case Western Reserve University. The Case School of Engineering has 108 faculty members, all of whom are active in the undergraduate and graduate programs.

Applying

Application for graduate study in engineering is effected through the particular department in which study is proposed. Departmental faculty members establish the prerequisites and make decisions on the basis of college records and recommendations. The GRE is required by the biomedical and computer engineering departments. Applicants for other departments are urged to submit scores if they have taken the exam. Completed applications must be filed at least thirty days prior to the beginning of the semester for which the student is applying. In addition to meeting the regular requirements, international students must demonstrate fluency in written and spoken English by taking the TOEFL (minimum score, 550). Applicants should indicate their proposed field of study to ensure an appropriate response.

Correspondence and Information

Dean, The Case School of Engineering
Glennan Building
Case Western Reserve University
University Circle
Cleveland, Ohio 44106-7220
Telephone: 216-368-4436

Case Western Reserve University

FACULTY HEADS AND DEPARTMENTAL AREAS OF RESEARCH

Biomedical Engineering (administered jointly with the School of Medicine). Professor Patrick E. Crago, Chairman.
Biomaterials/tissue engineering: Materials for implantation, including neural and cardiovascular tissue engineering, biomimetic materials, liposomal drug delivery, and biocompatible polymer surface modifications. Analysis of synthetic and biologic polymers by AFM, nanoscale structure-function relationships of orthopedic biomaterials. Biomedical image processing and analysis: MRI, PET, cardiac electrical potential mapping, human visual perception, image guided intervention. Neural prostheses and neural engineering: neural interfacing for stimulation and recording, neural prostheses for control of limb movement, bladder and bowel function, and respiration. Biomedical sensing: optical sensing and imaging, electrochemical and chemical fiber-optic sensors, chemical measurements in cells and tissues, endoscopy. Cardiac bioelectricity: cardiac electrophysiology (at ion-channel, cell, and tissue levels), models of cellular activity, mechanisms of cardiac arrhythmias, optical imaging of electrical propagation in the heart, noninvasive electrocardiographic imaging. Transport and metabolic systems engineering: modeling and analysis of tissue responses to heating (tumor ablation, implanted artificial heart) and of cellular metabolism related to organ and whole-body function in health (exercise) and disease (cardiac).

Chemical Engineering. Professor Nelson Gardner, Chairman.
Electrochemical engineering: microelectronic materials and sensors; modeling; electrodeposition and surface texturing; batteries and fuel cells; polymer-coated and diamond electrodes; electronic materials; alloys and compounds; corrosion protection; membranes for electrochemical applications; electrochemistry in surfactant systems. Light scattering: anemometric and quasi-elastic techniques, surface light scattering, Brewster-angle microscopy, statistical data analysis and parameter estimation. Materials engineering: low-pressure diamond growth, diamond-like carbon, wide band-gap nitrides, combustion and plasma synthesis of films, computation of phase diagrams, in situ spectroscopic techniques, aerosol synthesis, fine-particle processing strategies, dispersive mixing phenomena, microemulsion techniques for novel polymers, polymer nanocomposites; blends and nanoparticles; ultrathin films. Process control: control of multivariable, linear and nonlinear uncertain processes, virtual integrated prototyping. Reaction engineering: thermochemistry of redox systems, catalysis for environmental applications, reactive flow modeling. Separations: acoustic methods, process intensification, microemulsion techniques. Surface and colloidal phenomena: emulsion and coating rheology, microemulsions and micelles, polymeric surfactants and polymer-substrate interactions; Langmuir-Blodgett multilayers; monolayer structure and dynamics; spreading phenomena; image-force microscopy.

Civil Engineering. Professor Robert L. Mullen, Chairman.
Programs in environmental, structural, and geotechnical engineering and engineering mechanics. Theoretical and experimental research in environmental remediation; groundwater flow and contamination analysis; colloidal transport and aggregation; urban watershed and Great Lakes processes; joint design in prestressed concrete bridges; field load tests and slab bridges for rating purposes; reliability analysis and design of structural systems; random vibration, seismic analysis, and design of civil engineering systems; static and dynamic behavior of soils; the use of energy in predicting liquefaction; centrifuge modeling; bifurcation and shear banding phenomena in soils; highway pavement behavior; computational mechanics; and constitutive modeling, fracture, fatigue, and material failure analysis of microelectromechanical systems.

Electrical Engineering and Computer Science. Professor Robert V. Edwards, Acting Chairman.
Computer Engineering and Science. Computer-aided design and test of digital systems: VLSI, VHDL. Computer architecture and organization: autonomous agents and robotics, computational biology, and simulation of biological systems. Databases: design, data models, query processing, distributed databases, expert databases, statistical database systems. Design environments: software engineering, formal specifications, program verification and testing, design languages, design tools for hardware/software system development.
Electrical Engineering. Solid-state electronics, including semiconductor sensors and associated microcircuits; studies of semiconducting and insulating crystalline and polymer materials and their applications to devices; microelectromechanical systems, with emphasis upon sensors and actuators; materials for microelectromechanical systems, especially silicon carbide and TiNi shape memory alloys; high-frequency semiconductor device modeling and mixed-signal integrated circuit design and fabrication. Sensor systems, including MEMS sensor arrays, high-temperature sensors, and sensor arrays for aircraft applications. Automation, including expert systems, intelligent machine control and robotics, machine vision and industrial inspection, and agile manufacturing. Optical communications, especially optical fiber communications and wavelength-division multiplexing. Nonlinear and adaptive control of robotic systems. Control of micromechanical systems.
Systems and Control Engineering. Systems theory: optimization, stability of systems, algebraic systems theory. Estimation and control theory: identification, nonlinear filtering, stochastic control, adaptive control, process diagnostics, nonlinear control of robotic systems. Control of industrial systems: computers for real-time control and automation. Decision support systems. Production/manufacturing systems: process planning and machine setup, facility layout, scheduling and planning, integrated manufacturing systems. Global change modeling, simulation, and Internet-based education.

Macromolecular Science. Professor Alex M. Jamieson, Chairman.
Structure-property relationships in crystalline polymers, X-ray diffraction, light scattering, infrared spectroscopy, electron microscopy, dielectric relaxation, differential thermal analysis, high-resolution solid-state NMR, NMR imaging. Studies of polymer composites and blends; conducting polymers and polymer microdevices; the effects of crystalline and noncrystalline transitions on deformation processes; mechanisms of plastic and viscoelastic behavior; transport properties in macromolecular systems; polymer dynamics and polymer rheology; synthesis of unique membranes with unusual anisotropic mechanical properties; effect of processing history on polymer morphology and properties; synthesis, structure, and properties of liquid crystalline polymers; biological macromolecules; the assembly of fibrous proteins; connective-tissue structure and related aging processes; development of materials for artificial organs; theory of dispersive mixing and blending; reactive polymer processing; engineered microstructures through microlayer coextrusion; polymer composites.

Materials Science and Engineering. Professor Gary M. Michal, Chairman.
Relationships among processing, structure, properties, and performance of materials. Basic research in chemical thermodynamics and transport phenomena. Plastic deformation, fracture, materials characterization of metals, ceramics, and semiconductors. Core research areas include solidification applied to semiconductor single-crystal growth and metal casting; environmental effects encompassing corrosion, oxidation, coating performance and adhesion, aging infrastructure life predictions, reliability of electronic devices, and development of corrosion sensors; metal matrix and ceramic matrix composites focusing upon engineered interfaces; computational modeling evaluating near-net-shape manufacturing processes; development of rapid prototyping process technology; and experimental work evaluating materials under high hydrostatic pressures. Specific research topics include analytical and high-resolution electron microscopy; toughening mechanisms in ceramics, intermetallics, and alloys; lightweight materials for aerospace applications; biological models of ceramics; materials and coatings for elevated temperature use; defects in semiconductors; surface analysis of materials; and experimental investigation of metal-ceramic interface structures and fundamentals of epitaxy in bicrystals.

Mechanical and Aerospace Engineering. Professor Joseph M. Prahl, Chairman.
Fluid mechanics: aerospace mechanics, aerodynamics, airbreathing aircraft, stability and transition, drag reduction, turbomachinery, molecular dynamics, control of IC engines, tribology. Combustion: flame spread, microgravity combustion, catalytic ignition, ramjet combustion. Computational methods: unsteady incompressible flows, heat transfer, combustion, flows in deforming geometries. Dynamic systems: chaos, rotating machinery, roto/bearing/seal systems, control. Energy technology: heat rejection, porous media, heat pipes, engine combustion processes, natural convection. Engineering design: optimization and CAD, kinematic mechanisms, experimental stress analysis, failure analysis, product design. Microgravity research: crystal growth, containerless processing, g-jitter effects. Multiphase flow research: laser techniques, solid-liquid slurries, gas-solid flows, condensation. Orthopedic engineering: characteristics of the knee, hip, ankle, and spine and neuromuscular control; mechanics of injuries; gait analysis; prostheses; properties of bone and soft tissue; biomechanical measurements; tools and instrumentation. Robotics and space structures: flexible multijointed mechanical systems; maneuver, vibration, tracking problems.

THE CATHOLIC UNIVERSITY OF AMERICA

School of Engineering

Programs of Study

The School of Engineering offers M.S.E., M.S.C.S., D.Eng., and Ph.D. degree programs in the fields of biomedical, civil, computer science, electrical, and mechanical engineering. In addition, graduate programs are available in applied energy systems, biomechanics, cell/tissue engineering, computer engineering, construction management, control systems and mechatronics, engineering management, environmental engineering and management, fluid mechanics, intelligent control and robotics, mechanical design, medical imaging and signal processing, rehabilitation engineering, software engineering, solid mechanics, structures and structural mechanics, systems engineering, telemedicine, and thermal sciences.

Requirements for the master's degrees include a minimum of one academic year of residence and 30 semester hours of credit. Both thesis and nonthesis options are available.

The doctoral degrees normally require at least three years in residence, or the equivalent, beyond the bachelor's degree; a minimum of 53 semester hours of graduate course work must be completed. The doctoral dissertation is based on original research in the field of specialization. There is no foreign language requirement.

Research Facilities

Each department conducts research programs and has extensive research equipment and facilities.

The Computer Center maintains a DEC VAX 4100 and 4300, a large time-sharing system with approximately 400 to 500 network microcomputers located throughout the campus, including a cluster of facilities within the engineering school. Each of the departments in engineering also has a CAD laboratory.

The Mullen Library collection includes more than a million volumes housed in the main library and seven school and departmental libraries on campus. The libraries of the Consortium of Universities of the Washington Metropolitan Area extend borrowing privileges to graduate degree candidates. In addition, other library resources in the Washington area such as the Library of Congress and the National Library of Medicine are readily accessible.

Financial Aid

Financial aid is available in the form of research and teaching assistantships and tuition scholarships. In order to be eligible for the widest variety of aid programs, an applicant should take the Graduate Record Examinations (GRE) and submit a GAPSFAS application.

Cost of Study

Tuition for 1999–2000 is $18,350 per year for a full-time student. Additional fees total $776 per year.

Living and Housing Costs

On-campus dormitory facilities are available for unmarried graduate students. For 1999–2000, room costs vary from $3652 to $5850 per year. A meal plan costs an additional $1250 to $3480 per year depending on which plan is selected. There are numerous private rooms and apartments for rent in the nearby community.

Student Group

The Catholic University has an enrollment of about 6,100 full- and part-time students. The School of Engineering has about 200 undergraduate and 250 graduate students.

Location

The Catholic University is located in the nation's capital, a city that offers unsurpassed opportunities for study, research, and cultural enrichment. In addition to being the home of numerous government laboratories and research facilities, Washington is the center of a rapidly growing intellectual and research-minded community.

The University

The Catholic University of America was founded in 1889. It was originally intended for graduate instruction and research exclusively and has always retained a strong graduate orientation. The Institute of Technology was established in 1896, the forerunner of the School of Engineering, established in 1930, and of the School of Engineering and Architecture, established in 1936. Separate school status was achieved by the Schools of Engineering and Architecture in 1992. Although founded and partially supported by the Catholic Bishops of the United States, the University is essentially a free autonomous center of study and an agency serving the needs of human society.

Applying

Admission applications may be submitted at any time during the year; however, all supporting credentials should be received well in advance of the preferred semester of entry. Applicants requesting financial aid should ensure that completed applications, including GRE scores and the GAPSFAS, arrive by February 15.

Correspondence and Information

Chairman
(specify program)
School of Engineering
The Catholic University of America
Washington, D.C. 20064
Telephone: 202-319-5177
Fax: 202-319-4499
E-mail: cua-engineer@cua.edu
World Wide Web: http://www.ee.cua.edu

The Catholic University of America

FACULTY HEADS AND AREAS OF RESEARCH

Biomedical Engineering. Professor J. Winters, Chairman.

Major areas of graduate study include biotechnology, rehabilitation engineering, bioelectrical phenomena, and medical information technology. Teaching material is closely related to faculty research expertise and the resources available in neighboring biomedical establishments.

Civil Engineering. Professor T. W. Kao, Chairman.

Major areas of graduate study include structures and structural mechanics, environmental engineering, systems engineering, construction, management, and fluid and solid mechanics. The first four areas above are primarily professional in nature; the program in fluid and solid mechanics is heavily research oriented.

Electrical Engineering and Computer Science. Professor C. Nguyen, Chairman.

The faculty is engaged in several areas of research: signal and image processing, including digital signal, image, sonar, and neural network applications; communication systems, including computer, satellite, and nonsinusoidal systems; robotics and control, including adaptive and intelligent control and control of flexible space structures; electromagnetics, including antennas, properties of materials, and bioelectromagnetics; software engineering; artificial intelligence; and computer graphics.

Engineering Management. Professor S.A. Mohsberg III, Director.

This interdisciplinary ten-course (30 credit) degree program enhances professional performance and develops managerial skills through the completion of three specified courses in management theory (such as organizational theory and behavior), three specified courses in management tools (such as decision analysis), and four technical electives that may be chosen from any of the graduate courses offered by the School of Engineering (such as environmental engineering).

Mechanical Engineering. Professor S. Nieh, Chairman.

Major areas of graduate study are biomechanics, control mechatronics, energy systems, environmental engineering, thermofluid sciences, and mechanical design. Possible subdisciplines include vibration and structural mechanics, heat and mass transfer, HVAC, pollution control, active control, combustion and incineration, fluid mechanics, multiphase systems, intelligent systems and smart materials, computer-aided design, and advanced dynamics. Considerable flexibility is allowed students in making up their programs of study from the courses offered by the department and appropriate courses in science and engineering.

THE CITY COLLEGE AND
THE GRADUATE SCHOOL AND UNIVERSITY CENTER
OF THE CITY UNIVERSITY OF NEW YORK

School of Engineering

Programs of Study

The School of Engineering offers programs of study leading to Master of Science, Master of Engineering, and Doctor of Philosophy degrees in five areas: chemical engineering, civil engineering, computer science, electrical engineering, and mechanical engineering. The Ph.D. Program in Computer Science is located at the Graduate School and University Center. A new Ph.D. program in biomedical engineering and a twelve-credit advanced certificate program in all five departments are awaiting approval from the Board of Trustees.

The master's degree is awarded upon completion of 30 credits of approved graduate courses with a minimum average grade of B. Students are also required to complete any one of the following: a thesis (up to 6 credits), a project (3 credits), a report (no credit), or seminars (1 credit).

The doctoral degree requires satisfactory completion of 60 credits of approved graduate work beyond the bachelor's degree, of which at least 30 must be taken at the City University. It also requires the successful passing of the comprehensive first examination and the second examination in the field of dissertation research, proficiency in a tool for research, and the passing of an oral defense of the dissertation.

Research Facilities

Each department is equipped with substantial up-to-date research facilities to conduct its research programs. Additional interdisciplinary research facilities are also available. Some of the facilities and research groups are administered as research institutes, centers, and laboratories. These include the Benjamin Levich Institute for Physicochemical Hydrodynamics, the Institute for Biomedical Engineering, the Institute for Transportation Systems, the Center for Water Resources and Environmental Research, the Institute for Municipal Waste Research, the Earthquake Research Center, the Institute for Ultrafast Spectroscopy and Lasers, the Photonics Engineering Center, the Telecommunications Laboratory, the Optical Signal Processing Laboratory, the Biomechanics Laboratory, the Turbomachinery/Aerospace Laboratories, the Micro Electronic Heat Transfer and Materials Laboratory, the Theory of Machines Laboratory, the Solid Mechanics Research Laboratory, and the Roads and Airfields Materials Laboratory. The School of Engineering provides a wide range of networked computer facilities for both teaching and research. These total approximately 300 engineering workstations and various PC facilities.

Financial Aid

Many Ph.D. students are supported through University Fellowships, Gilleece Fellowships, college teaching or research assistantships, partial tuition waivers, the College Work-Study Program, Federal Stafford Loans, Magnet and Llewellyn Fellowships for minorities, and other scholarships, prizes, and awards.

Cost of Study

In 1999–2000 tuition for New York State residents is $185 to $245 per credit to a maximum of $2175 per semester. Non-state residents (including international students) pay $320 to $425 per credit to a maximum of $3800 per semester.

Living and Housing Costs

No dormitory is available for master's students, but limited accommodations are available for Ph.D. students. The rents vary from $5700 to $6900 per year.

Student Group

The average graduate enrollment in all engineering departments at the master's level is 500; at the doctoral level, not counting computer science, it is more than 120.

Student Outcomes

All of the Ph.D. graduates from 1994 to 1998 have found desirable positions. Approximately 20 percent are in academia, 50 percent in industry, 10 percent in government, and 20 percent in postdoctoral research positions.

Location

The City College occupies a 35-acre modern campus in historic St. Nicholas Heights, only a few blocks from the Hudson River. Some of its buildings are New York City landmarks.

The College and The School

City College was founded in 1847. In 1866, it was named College of the City of New York (CCNY). The College was committed to enlarge opportunities for all who sought a higher education, while maintaining and enhancing academic excellence. From 1944 to 1961, a wide range of master's programs were introduced. In 1961, doctoral-level programs were introduced. The City College is financed by the state of New York, student fees, and gifts.

The School of Engineering was formally established as a separate school in 1919 and was fully renovated in 1994. In fiscal year 1997–98, the School of Engineering had research expenditures of approximately $10.2 million.

Applying

The deadlines for application for admission to the master's program for the fall and spring semesters are May 15 and December 15 respectively.

The deadlines for application for admission to the Ph.D. program for the fall and spring semesters are April 15 and November 15 respectively.

Correspondence and Information

For master's program:
Office of Admissions
City College of New York
Convent Avenue at 138th Street
New York, New York 10031
Telephone: 212-650-6447
Fax: 212-650-6417

For doctoral program:
Office of Admissions
Graduate School and University
  Center
33 West 42nd Street
New York, New York 10036
Telephone: 212-642-2812
Fax: 212-642-2779

For additional information:
Dr. Mumtaz K. Kassir
  Associate Dean for Graduate
  Studies and Executive Officer
School of Engineering
The City College of New York
New York, New York 10031
Telephone: 212-650-8030
Fax: 212-650-8029
World Wide Web:
  http://www.engr.ccny.cuny.edu

The City College and the Graduate School and University Center of the City University of New York

THE FACULTY AND THEIR RESEARCH

Chemical Engineering

Professor Robert Graff, Chairman (telephone: 212-650-7232, fax: 212-650-6660). Areas of research include multiphase fluid mechanics (suspension rheology, sedimentation, hydrodynamic stability, interfacial phenomena, two-phase jet- and pore-flows, flow of near-critical binary mixtures, and high-velocity fluidization), effective properties of materials (flow in porous media, thermocapillarity, electrophoresis, and thermal and electronic properties of semiconductors), process control and simulation (combined cycle power generation), reaction engineering (multiphase reactor analysis), biomedical engineering (controlled drug release, study of the mechanism of artery disease, and liquid extraction of fermentation broths), polymer science (hydrogel technology, thin organic films, adsorption and self-assembly phenomena at solid-polymer interface, and viscoelasticity), powder technology (granulation, fluidization, and electrostatic effects of powders), and environmental engineering (air pollution control; soil remediation of organic, metallic, and radioactive contaminants).

Civil Engineering

Professor John Fillos, Chairman (telephone: 212-650-8010, fax: 212-650-6965). Areas of research in the three specialization options are structures (damage mechanics, wave propagation in solids, fracture mechanics, nonlinear behavior of structures, safety evaluation of bridges, structural reliability, earthquake engineering, soil-structure interaction and seismic risk assessment of nuclear power plants), transportation (traffic engineering; traffic safety; advanced traffic information systems; advanced traffic management systems; travel demand modeling; transportation infrastructure management; pavement design, construction, and maintenance; nondestructive testing of pavement; transportation finance; pricing; funding and institutions; transit planning, including rail, bus, ferry, and telecommuting; pedestrian and bicycle transportation; and sustainable urban transportation systems), and environmental engineering/water resources (municipal and industrial wastewater treatment process development, biological nutrient removal, drinking water quality and stability, disinfection by-products, modeling of groundwater flow and contaminant transport, non-point source pollution control, incinerator sludge ash utilization for shoreline protection, erosion controlled transport of radioactive and agrochemical pollutants, and modeling of contaminants in freezing/thawing soils).

Computer Science

Professor Stephen Lucci, Chairman (telephone: 212-650-6631, fax: 212-650-6184). Areas of instruction include computer systems (architecture, distributed systems, networks, and information management), theoretical computer science (formal systems, computability, complexity, and modeling), and scientific computing (numerical analysis, simulation, and optimization methods). Active areas of research include computer graphics, image processing, multimedia, virtual reality, computational geometry, mathematics of computation, cryptography, artificial intelligence, neural networks, mathematical fluid dynamics and simulation, networks, distributed computing, database systems, information management and virtual organization, economics of information, and computer–supported cooperative work.

Electrical Engineering

Professor Frederick E. Thau, Chairman (telephone: 212-650-7248, fax: 212-650-8249). The principal areas of research are photonics engineering (quantum optics and electronics, nonlinear optics, new laser sources, optical computing, ultrafast phenomena and devices, new optical materials, microstructures, laser remote sensing, and optical imaging), signal processing (signal detection and estimation theory, filter design, stability analysis, algorithms for extraction of parameters from radar and x-ray signals, development of fast algorithms, image processing, pattern recognition, and underwater acoustic signal processing), communication and networking engineering (data and digital communication, computer and local area networks, high-speed and multimedia communication networks, and optical communication), biomedical engineering (laser applications in medicine, biomedical image processing, and optical diagnostics instruments), and control and systems theory (adaptive, modal, nonlinear, and robust control in control of discrete event systems and in flight control applications).
A new management concentration in the master's degree became available in 1996.

Mechanical Engineering

Professor Yiannis Andreopoulos, Chairman (telephone: 212-650-5218, fax: 212-650-8013). Areas of research include fluid dynamics, turbulence and shock wave interaction, blast waves and composites interaction, advanced laser diagnostics, vortex dynamics, near-wall turbulence, spatio-temporal nonlinear dynamics, vibration and control, chaotic motions, hydrodynamics and interfacial phenomena, composite materials, fracture mechanics, adhesive bonding, elasticity of nonhomogeneous materials, biological materials, orthopedic biomechanics, bone remodeling, heat and mass transfer, bioheat transfer, image processing and electron microscopy, CAD/CAM/CIM, computational mechanics (including finite element and boundary element methods), machine dynamics, turbomachinery, wakes and corner flows, fish hydrodynamics, biofluid dynamics and transport phenomena, microvascular architecture.

CLARKSON UNIVERSITY

School of Engineering

Programs of Study

The School of Engineering, comprising departments of chemical, civil and environmental, electrical and computer, and mechanical and aeronautical engineering, offers programs of study leading to the Doctor of Philosophy, Master of Science, and Master of Engineering degrees. An interdisciplinary program allows the student to specialize in such areas as materials processing, robotics, or manufacturing. Interdisciplinary M.S. degrees are offered in computer science (joint program with the Departments of Mathematics and Computer Science and Electrical and Computer Engineering) and in engineering and manufacturing management (joint program with the School of Business). Descriptions of these programs can be found at http://www.clarkson.edu/~grad/gs/message.html.

The Master of Science degree is awarded upon completion of 30 credit hours of graduate work, including a thesis. The Master of Engineering degree can be obtained in one calendar year; it includes the completion of a design-oriented project. In addition, Clarkson has initiated a two-year, two-degree program whereby students may obtain an M.E. degree in one year and continue on for an additional year to obtain an M.B.A.

The Ph.D. is awarded upon completion of a minimum of 90 credit hours of graduate work, corresponding to a minimum of three academic years of full-time study beyond the bachelor's degree. A master's degree may be accepted in lieu of 30 credit hours. A comprehensive examination based on general preparation in the major field is required. Candidates for the Ph.D. are required to prepare an original dissertation in an advanced research area and defend it in an oral examination.

Research Facilities

The Department of Chemical Engineering houses modern research labs for plasma and laser processing, polymer fabrication and properties, crystal growth, rheology, bioengineering, fluid mechanics, heat and mass transfer, interfacial phenomena, nucleation, chemical metallurgy, chemical kinetics, separation process design, corrosion, and electrochemical engineering. The Department of Civil and Environmental Engineering has well-equipped environmental engineering laboratories with pilot plant facilities, walk-in constant-temperature rooms, and modern research instrumentation for organic and inorganic analyses; a hydraulics laboratory with a large automated tilting flume; temperature-controlled cold rooms and ice mechanics laboratories; concrete, asphalt, and structural testing laboratories, including a unique strong room; and soil mechanics and materials laboratories. The Department of Electrical and Computer Engineering has laboratories for distributed computing networks, intelligent information processing, microelectronics, motion control, robotics, power electronics, electric machines and drives, and liquid dielectric breakdown and a 1-million-volt high-voltage measurement laboratory. The Department of Mechanical and Aeronautical Engineering houses three wind tunnels, a clean room for microcontamination research, and labs for fluid mechanics, heat transfer, aerosol and multiphase flow, robotics, CAD, manufacturing, image processing, energy conversion, vibrations, combustion, and welding. The New York State Center for Advanced Materials Processing provides funding and facilities for interdisciplinary work on materials processing, including the only centrifuge facilities in the world dedicated to research on materials processing, welding, and related flow visualization at high gravity. Computing facilities within the School of Engineering include a variety of Sun and IBM RISC workstations, all interconnected to each office and laboratory by a high-speed wide-band network. Clarkson's Educational Resources Center houses modern information storage and retrieval facilities, the computing center, and the library.

Financial Aid

Several forms of financial assistance are available, which permit a full-time program of study and provide a stipend plus tuition. Instructional assistantships involve an obligation of 12 hours per week of assistance in courses or laboratories. Research assistantships require research activity that is also used to satisfy thesis requirements. Partial tuition scholarships are available for all degree programs. Industrial fellowships and traineeships are also available. In 1999–2000, the value of awards ranges up to $20,000 per year.

Cost of Study

Tuition for graduate work is $661 per credit hour in 1999–2000. Fees are about $215 per year.

Living and Housing Costs

Graduate students can find rooms or apartments near the campus. The University maintains single and married student housing units. Off-campus apartments for 2 students rent for $150 per month and up.

Student Group

There are approximately 150 students on campus pursuing graduate work in engineering. The total Graduate School enrollment is about 300, and the undergraduate enrollment is about 2,450.

Location

Potsdam, New York, is an attractive village located along the banks of the Raquette River on a rolling plain between the Adirondack Mountains and the St. Lawrence River. Three other colleges (one in Potsdam) provide a total college student body of 11,000 within a 12-mile radius. Potsdam is 100 miles from Montreal, 80 miles from Ottawa and Lake Placid, and 140 miles from Syracuse. The St. Lawrence Seaway, the Thousand Islands, and Adirondack resort areas are within a short drive. Opportunities for fishing, hiking, boating, golfing, camping, swimming, and skiing abound throughout the area.

The University

Clarkson University is a privately endowed school of science, engineering, and business. Master's degrees are offered in the engineering departments and in business administration, management systems, computer science, mathematics, physics, and chemistry; Ph.D. degrees are offered in physics, chemistry, mathematics, chemical engineering, civil and environmental engineering, electrical and computer engineering, mechanical engineering, and engineering science.

The academic year consists of two semesters of 15 weeks each. There is no formal summer session for graduate classes; graduate students and faculty devote the summer entirely to research.

Applying

It is recommended that applications be submitted by May 15 for the fall semester and October 15 for the spring. Study may begin in September, January, or June. Applicants requesting financial aid should submit their complete applications by March 1 for the fall semester and September 1 for the spring semester. Scores on the General Test of the GRE are required for all applications. TOEFL scores of at least 550 (paper-based test) or 213 (computer-based test) are required for all international applications.

Correspondence and Information

School of Engineering
Graduate Studies Office
Box 5700
Clarkson University
Potsdam, New York 13699-5700

Telephone: 315-268-7929
Fax: 315-268-3841
E-mail: schofeng@agent.clarkson.edu
World Wide Web: http://www.clarkson.edu/~eng/gradfram.html
                http://www.clarkson.edu/~grad/gs/gradapp.htm (downloadable application)

Clarkson University

THE FACULTY AND THEIR RESEARCH

Department of Chemical Engineering
S. V. Babu, Professor; Ph.D., SUNY at Stony Brook. Photolithography, plasma and laser processing, chemical-mechanical polishing.
Ruth E. Baltus, Associate Professor; Ph.D., Carnegie Mellon. Diffusion in porous media, chromatographic and membrane separations, immobilized enzyme processes.
Gregory A. Campbell, Professor and Dean of Engineering; Ph.D., Maine. Polymer properties and slurries.
Der-Tau Chin, Professor; Ph.D., Pennsylvania. Electrochemical engineering, corrosion, electroplating, batteries and fuel cells, waste treatment.
Sandra L. Harris, Associate Professor; Ph.D., California, Santa Barbara. Adaptive control, industrial applications of advanced control, process identification.
Richard J. McCluskey, Associate Professor; Ph.D., Minnesota. Chemical kinetics, microcontamination control.
John B. McLaughlin, Professor; Ph.D., Harvard. Computational fluid dynamics.
Richard J. Nunge, Professor; Ph.D., Syracuse. Transport phenomena.
Nsima T. Obot, Professor; Ph.D., McGill. Flue mechanics and turbulence, heat and mass flow, multiphase flow.
Don H. Rasmussen, Professor; Ph.D., Wisconsin–Madison. Nucleation and phase transformations, metal reduction, colloidal and interfacial phenomena.
R. Shankar Subramanian, Professor; Ph.D., Clarkson. Transport phenomena, colloidal and interfacial phenomena.
Ian Ivar Suni, Associate Professor; Ph.D., Harvard. Surface reaction, interfacial phenomena, solid-liquid interface.
Ross Taylor, Professor and Chair; Ph.D., Manchester. Multicomponent mass transfer, separation processes, symbolic computation.
Thomas J. Ward, Professor; Ph.D., Rensselaer. Process control.
William R. Wilcox, Professor and Co-Director, International Center for Gravity Materials Science and Applications; Ph.D., Berkeley. Materials processing, crystal growth.

Department of Civil and Environmental Engineering
Norbert L. Ackermann, Professor; Ph.D., Carnegie Tech. Mechanics of granular flow, river hydraulics.
John P. Dempsey, Professor; Ph.D., Auckland (New Zealand). Fracture mechanics, tribology, ice-structure interaction.
Stefan J. Grimberg, Assistant Professor; Ph.D., North Carolina at Chapel Hill. Bioremediation, bioavailability of organic environmental pollutants.
Christopher C. Higgins, Assistant Professor; Ph.D., Lehigh. Structural testing, passive structural control, earthquake engineering.
Thomas M. Holsen, Professor; Ph.D., Berkeley. Fate and transport of chemicals in the environment.
Boris Jeremic, Assistant Professor; Ph.D., Colorado. Elastoplastic behavior and constitutive modeling; computational methods.
Feng-Bor Lin, Professor; Ph.D., Carnegie Mellon. Modeling traffic operations, systems analysis.
Levon Minnetyan, Associate Professor; Ph.D., Duke. Structural analysis and design.
Dayakar Penumadu, Associate Professor; Ph.D., Georgia Tech. Calibration chamber testing of clays, artificial neurals.
Susan E. Powers, Associate Professor; Ph.D., Michigan. Multiphase fluid flow; hazardous-waste management.
Firas Sheikh-Ibrahim, Assistant Professor; Ph.D., Texas at Austin. Structural engineering, field testing of bridges.
Hayley H. Shen, Professor; Ph.D., Clarkson; Ph.D., Iowa. Sea ice dynamics.
Hung Tao Shen, Professor; Ph.D., Iowa. River and sea ice processes.
Thomas L. Theis, Professor and Chair; Ph.D., Notre Dame. Environmental chemistry, environmental systems analysis.
Jerry Yamamuro, Assistant Professor; Ph.D., UCLA. Mechanics of frictional materials, soil instability and liquefaction.
Poojitha Yapa, Associate Professor; Ph.D., Clarkson. Mathematical modeling of oil spills.
Thomas C. Young, Professor; Ph.D., Michigan State. Contaminant fate and transport modeling, applied environmental statistics.
Amy K. Zander, Associate Professor; Ph.D., Minnesota. Membrane systems in environmental processes.

Department of Electrical and Computer Engineering
James J. Carroll, Associate Professor; Ph.D., Clemson. High-performance motion control, nonlinear control, control strategies.
Susan E. Conry, Associate Professor and Chair; Ph.D., Rice. Multiagent systems, distributed problem solving, design of coordination strategies.
Abul N. Khondker, Associate Professor; Ph.D., Rice. Solid-state materials and device theory, modeling and characterization of semiconductor devices.
Jack Koplowitz, Associate Professor; Ph.D., Colorado. Image and signal processing, computer vision, pattern recognition.
Priyalal Kulasinghe, Assistant Professor; Ph.D., LSU. Parallel and distributed systems, design of effective bus-based architectures.
Milica Markovic, Research Assistant Professor; Ph.D., Colorado at Boulder. Quasi-optical systems for communications and other applications.
Paul B. McGrath, Professor; Ph.D., London. Dielectric materials and high-voltage engineering, insulation problems.
Robert A. Meyer, Professor; Ph.D., Rice. Artificial intelligence and distributed problem solving, verification of hardware designs, software engineering.
Rangaswamy Mukundan, Professor; Ph.D., Purdue. System theory, robotics, controller design, singular systems.
Thomas H. Ortmeyer, Professor; Ph.D., Iowa State. Power electronics, power quality, power system operation.
Pragasen Pillay, Professor; Ph.D., Virginia Tech. Modeling, analysis, design, and control of electric machines; electric motor drive systems.
Liya L. Regel, Research Professor and Director, International Center for Gravity Materials Science and Applications; Ph.D., Irkutsk State University (Russia); Doctorat, Ioffe Physical-Technical Institute (Russia). Materials science and its influence on properties and device performance.
Juergen Rilling, Assistant Professor; Ph.D., IIT. Software engineering, quality assurance and maintainability of software systems.
Robert J. Schilling, Professor; Ph.D., Berkeley. Control, nonlinear systems, robotics, active control of acoustic noise, motion planning.
Neeraj K. Sharma, Associate Professor; Ph.D., Akron. High-speed networking, multiprocessor interconnection, fault-tolerant systems.
James A. Svoboda, Associate Professor and Executive Officer; Ph.D., Wisconsin. Circuit theory, system theory, electronics, digital signal processing.

Department of Mechanical and Aeronautical Engineering
Goodarz Ahmadi, Professor; Ph.D., Purdue. Fluid mechanics, solid mechanics, turbulence modeling.
Daryush Aidun, Associate Professor; Ph.D., Rensselaer. Welding metallurgy, materials processing and solidification.
Ahmed A. Busnaina, Professor; Ph.D., Oklahoma State. Fluid mechanics, microcontamination control.
Frederick Carlson, Associate Professor; Ph.D., Connecticut. Heat transfer, crystal growth.
Cetin Cetinkaya, Assistant Professor; Ph.D., Illinois at Urbana-Champaign. Solid mechanics, stress wave propagation.
Mark Glauser, Professor; Ph.D., SUNY at Buffalo. Fluid mechanics, experimental turbulence.
James Kane, Associate Professor; Ph.D., Connecticut. Solid mechanics, boundary-element methods.
Ronald LaFleur, Associate Professor; Ph.D., Connecticut. Fluid mechanics, thermofluid design.
Sung P. Lin, Professor; Ph.D., Michigan. Fluid mechanics, fluid dynamic stability.
John Moosbrugger, Associate Professor; Ph.D., Georgia Tech. Solid mechanics, plasticity.
David Morrison, Associate Professor; Ph.D., Michigan. Materials science, fracture mechanics.
M. Sathyamoorthy, Professor and Chair; Ph.D., Indian Institute of Technology. Solid mechanics, nonlinear mechanics.
Eric Thacher, Associate Professor and Executive Officer; Ph.D., New Mexico State. Thermal sciences, solar energy.
Daniel Valentine, Associate Professor; Ph.D., Catholic University. Fluid mechanics, hydrodynamics.
Kenneth Visser, Assistant Professor; Ph.D., Notre Dame. Experimental aerodynamics.
Kenneth Willmert, Professor; Ph.D., Case Western Reserve. Solid mechanics, optimal design.
Steven W. Yurgartis, Associate Professor; Ph.D., Rensselaer. Solid mechanics, composite materials.

CLEMSON UNIVERSITY

College of Engineering and Science

Programs of Study

The College of Engineering and Science offers programs leading to both the M.S. and Ph.D. in twenty areas (see reverse side of this page). In addition, the College offers an M.Engr. program with work at an advanced level in courses that are professionally oriented; this degree is offered in seven areas. An Industrial Residency Program leading to the M.S. degree is available in certain departments. Further details can be found in the University's and departments' brochures.

Research Facilities

Research is an integral part of graduate study in engineering and science. Departmental research laboratories are located in each of the sixteen engineering and science buildings on the Clemson campus. Among facilities of special interest are a ceramic pilot lab, molecular simulation facilities, a 500-Mhz nuclear magnetic resonance facility, an X-ray diffraction facility, a high-resolution laser laboratory, a microelectronics clean-room facility, a wind engineering lab, a 15,000-gpm recirculating flume, computer-aided manufacturing facilities, machine shops, an optical flow measurements lab, a gas turbine flow facility, and a pressure casting facility. There is also a variety of labs for the study of experimental animal surgery and histopathology; artificial intelligence; microchemistry analysis; experimental thermodynamics; computer process control; polymer synthesis, characterization, and processing; polymeric resorbable biomaterials; biomaterials characterization; biomechanics; polymer composites; radioactive materials; hazardous and toxic materials analysis; bioorganic chemistry; fluorine chemistry; mass spectrometry; microelectronics/microcomputers; signal processing; wireless communications; power electronics; hydraulics; fluid mechanics; electrooptics and lasers; antennas and microwaves; semiconductor reliability; high-speed cinematography and stroboscopy; heat transfer; solar cell testing and reliability; materials testing; materials processing; robotics; aerodynamics; metallurgy; electrical fabrication, repair, and calibration; atmospheric phenomena through the use of radar, rocket, and satellite data; production of man-made and natural fibers; coloring agents and finishes; chemistry of composite systems; and textile technology.

The College of Engineering and Science is supported by multiple computer facilities at both the University and College levels. The University supports student course work and research through a series of network computer labs and on-campus computers. This network consists of an HDS AS/EX-100 mainframe computer with 320 megabytes of central storage and 192 megabytes of expanded storage running the MVS/ESA operating system and several Digital Equipment Corporation computers. More than 350 microcomputers are available through public-access computer labs. The College has recently finished an aggressive five-year, multimillion-dollar plan for enhancing computer facilities with 32-bit UNIX graphics workstations and advanced microcomputer facilities. More than 250 UNIX workstations and 125 microcomputers, as well as state-of-the-art design software, were added to the public-access College computer facilities. In addition, a substantial investment was made in the College networking backbone to ensure adequate bandwidth well into the future. A direct FDDI connection into the campus backbone allows fast access to University facilities and the Internet. Facilities for research computing range from a $750,000 Virtual-Reality Lab to an advanced networking lab with a gigabit backbone. Numerous other computers help complete a vast array of research computers. The College of Engineering and Science is also supported by the Robert Muldrow Cooper Library, which contains collections totaling more than 1.6 million books, periodicals, government publications, and microforms.

Financial Aid

Financial aid is available on a competitive basis in the form of full and partial fellowships and assistantships, with stipends that ranged from $2000 to $20,000 in fall 1998. Assistantships include teaching, research, and administrative duties. Graduate students can also apply for the Dean's Scholars Program, which provides a twelve-month departmental research teaching assistantship of at least $12,000; a fund of $1000 for conference attendance, books, etc.; and a supplement of $4000 from the Dean's Scholars Fund.

Cost of Study

For 1998–99, in-state academic fees for full-time graduate students (12 or more semester credit hours, with the addition of a technology fee) were $1577 per semester; part-time students (11 semester credit hours or fewer) paid $130 per semester credit hour. Out-of-state tuition was $3226 per semester for full-time students and $264 per semester credit hour for part-time students. Any graduate student holding an assistantship paid $493 per semester for an unlimited number of semester credit hours. All full-time students paid a $95 medical fee each semester. Part-time students taking more than 7 semester credit hours are required to pay the medical fee.

Living and Housing Costs

Rooms are available in University dormitories and apartments. In 1998–99, rents ranged from $810 to $1280. Married student apartments are available. Full cafeteria service is available on campus, with an unlimited meal plan for $937 per semester.

Student Group

Total enrollment at Clemson exceeds 16,500 students; approximately 25 percent are graduate students. The 1,052 graduate students in the College of Engineering and Science represent more than fifty universities and colleges.

Location

Clemson, South Carolina, is a residential community located approximately midway between Charlotte, North Carolina, and Atlanta, Georgia. Nearby Interstate 85 and the main line of Amtrak link Clemson with major cities in the region. The Greenville-Spartanburg airport is approximately 45 minutes away from Clemson's campus and is easily accessible from I-85. Many historic buildings, including Fort Hill, the home of John C. Calhoun, have been preserved in the community. The area has excellent facilities for all recreational activities, and many exciting cultural events are less than a 2-hour drive away.

The University

Clemson is both university and community. It is located in the scenic foothills of the Blue Ridge Mountains, on the 1,000-mile shoreline of Lake Hartwell. The campus consists of 1,400 acres and represents an investment of more than $200 million in academic buildings, student housing, and service facilities. It is surrounded by 24,000 acres of farms and forestry, agricultural, and engineering-research lands. Clemson University is the state land-grant institution of South Carolina.

Applying

For admission to the Graduate School, a degree-seeking student must submit an application (either electronically or via hard copy) and have a bachelor's degree from an institution with a scholastic rating satisfactory to the University, a satisfactory score on the General Test of the Graduate Record Examinations (not required for the M.Engr. degree program), and the approval of the head of the department in which the student plans to do his or her major work. International students are required to complete a self-managed application, available from the University's Office of International Programs and Services, and to submit TOEFL scores. For more information on graduate applications and policies or to complete an electronic application, students can visit Clemson's Graduate School Web Site (http://www.grad.clemson.edu).

Correspondence and Information

Associate Dean for Research and Graduate Studies
College of Engineering and Science
114 Riggs Hall
Box 340901
Clemson University
Clemson, South Carolina 29634-0901

Clemson University

FACULTY HEADS AND AREAS OF RESEARCH

Interested students may browse the College's World Wide Web server, which contains detailed information on each faculty member along with graduate course descriptions, by using the following address: http://www.eng.clemson.edu.

All inquiries should be addressed to the Graduate Program Coordinator and the appropriate box number, using the following: Clemson University, Clemson, South Carolina 29634.

Agricultural and Biological Engineering. Graduate Program Coordinator: Dr. R. E. Williamson, Box 340357. E-mail address: robert.williamson@ces.clemson.edu. Graduate degrees offered include the M.S., M.Engr., and Ph.D. Research areas: aquaculture, agricultural-waste management, biotechnology, crop production, mechanization, animal housing, residential housing, global climate change and air quality, soil and water management, crop processing, instrumentation and controls, and food and process engineering.

Bioengineering. Graduate Program Coordinator: Dr. R. A. Latour, Box 340905. E-mail address: robert.latour@ces.clemson.edu; World Wide Web: http://www.ces.clemson.edu/bio/. Graduate degrees offered include the M.S. and Ph.D. Research areas: biomaterials and biomechanics of living and nonliving materials used for the replacement of organs and tissues.

Ceramic and Materials Engineering. Graduate Program Coordinator: Dr. H. J. Rack, Box 340907. E-mail address: henry.rack@ces.clemson.edu. Graduate degrees offered include the M.S., M.Engr., and Ph.D. Research areas: traditional ceramics, ceramic fibers, ceramic matrix composites, glass, electronic ceramics, sol-gel processing, aerogels, and metallurgy.

Chemical Engineering. Graduate Program Coordinator: Dr. D. E. Hirt, Box 340909. E-mail address: che@ces.clemson.edu. Graduate degrees offered include the M.S. and Ph.D. Research areas: thermodynamics, polymers, kinetics, and bioseparations.

Chemistry. Graduate Program Coordinator: Dr. D. D. DesMarteau, Box 341905. E-mail address: arrickb@clemson.edu. Graduate degrees offered include the M.S. and Ph.D. Concentrations offered in analytical, inorganic, organic, and physical chemistry. Research areas: bioorganic chemistry, fluorine chemistry, photochemistry, polymer chemistry, organometallic chemistry, environmental sampling and analysis, mass spectrometry, and chemical physics.

Civil Engineering. Graduate Program Coordinator: Dr. D. V. Rosowsky, Box 340911. E-mail address: rdavid@ces.clemson.edu. Graduate degrees offered include the M.S., M. Engr., and Ph.D. Research areas: applied fluid mechanics, construction materials, construction project management, geotechnical engineering, structural engineering, and transportation engineering. Interdisciplinary research area: natural and man-made hazards.

Computer Engineering. Graduate Program Coordinator: Dr. David Lubkeman, Box 340915. E-mail address: david.lubkeman@ces.clemson.edu. Graduate degrees offered include the M.S. and Ph.D. Research areas: computer communications, computer systems architecture, communications/digital signal processing, and controls/robotics.

Computer Science. Graduate Program Coordinator: Dr. James M. Westall, Box 341906. E-mail address: westall@cs.clemson.edu. Graduate degrees offered include the M.S. and Ph.D. Research areas: computer architecture, database management systems, design and analysis of algorithms, operating systems, parallel computation, programming languages, graphical systems, networks, systems measurement and modeling, software engineering, and theory of computing.

Electrical Engineering. Graduate Program Coordinator: Dr. David Lubkeman, Box 340915. E-mail address: david.lubkeman@ces.clemson.edu. Graduate degrees offered include the M.S., M.Engr., and Ph.D. Research areas: computational electromagnetics, communications/digital signal processing, computer communications, control/robotics, electronics, and power systems.

Engineering Mechanics. Graduate Program Coordinator: Dr. M. W. Dixon, Box 340921. E-mail address: marvin.dixon@ces.clemson.edu. Graduate degrees offered include the M.S. and Ph.D. Research areas: mechanics of composite materials and structures.

Environmental Engineering and Science. Graduate Program Coordinator: Dr. C. P. L. Grady, Box 340919. E-mail address: les.grady@ces.clemson.edu. Graduate degrees offered include the M.S., M.Engr., and Ph.D. Research areas: environmental process engineering, hazardous and radioactive waste treatment, contaminant fate and transport, risk assessment and waste management, and environmental chemistry.

Hydrogeology. Graduate Program Coordinator: Dr. J. W. Castle, Box 341908. E-mail address: jcastle@ces.clemson.edu. Graduate degree offered is the M.S. Research areas: groundwater geology, subsurface geology and remediation, and numerical flow modeling.

Industrial Engineering. Graduate Program Coordinator: Dr. Anand Gramopadhye, Box 340920. E-mail address: agramop@ces.clemson.edu. Graduate degrees offered include the M.S. and Ph.D. Research areas: integrated manufacturing, quality engineering, human factors engineering, manufacturing systems, and operations research.

Materials Science and Engineering. Graduate Program Coordinator: Dr. R. Singh, Box 340915. E-mail address: raj.singh@ces.clemson.edu. Graduate degrees offered include the M.S. and Ph.D. Research areas: semiconductors, polymers, ceramics, metals, biomaterials, fast-cycle manufacturing, thermoelectrics, composite materials, atomistic simulations, mathematical modeling of materials processing, and ceramic manufacturing.

Mathematical Sciences. Graduate Program Coordinator: Dr. Doug Shier, Box 341907. E-mail address: shierd@math.clemson.edu. Graduate degrees offered include the M.S. and Ph.D. along with a joint Ph.D. in management science. Research areas: algebra/combinatorics, analysis, computational mathematics, operations research, and statistics/probability.

Mechanical Engineering. Graduate Program Coordinator: Dr. M. W. Dixon, Box 340921. E-mail address: marvin.dixon@ces.clemson.edu. Graduate degrees offered include the M.S., M.Engr., and Ph.D. Research areas: applied mechanics, mechanical and manufacturing systems, and thermal/fluid science.

Physics and Astronomy. Graduate Program Coordinator: Dr. M. F. Larsen, Box 341911. E-mail address: mlarsen@maxwell.phys.clemson.edu. Graduate degrees offered include the M.S. and Ph.D. Research areas: astrophysics, atmospheric physics, biophysics, computational physics, radiation physics, experimental and theoretical solid-state physics, and theoretical physics.

Textile Chemistry. Graduate Program Coordinator: Dr. Michael Drews, Interim Director, Box 341307. E-mail address: dmichae@clemson.edu. Graduate degree offered is the M.S. Research areas: fiber chemistry, polymer chemistry, chemistry of dyeing and/or finishing of fibers and textiles, and chemistry of composite systems.

Textiles, Fiber and Polymer Science. Graduate Program Coordinator: Dr. R. V. Gregory, Director, Box 341307. E-mail address: richar6@clemson.edu. Graduate degrees offered include the M.S. and Ph.D. Research areas: fiber chemistry, fiber physics, fiber science, polymer chemistry, polymer science, chemistry of dyeing and/or finishing of textile materials, chemistry of composite systems, fiber formation, textile structures, and textile technology.

Textile Science. Graduate Program Coordinator: Dr. Michael Drews, Interim Director, Box 341307. E-mail address: dmichae@clemson.edu. Graduate degree offered is the M.S. Research areas: fiber science, polymer science, and textile technology.

CLEVELAND STATE UNIVERSITY

Fenn College of Engineering

Programs of Study

Fenn College of Engineering at Cleveland State University (CSU) offers six Master of Science (M.S.) degree programs in chemical, civil, electrical, environmental, industrial, mechanics and materials, and mechanical engineering and an interdisciplinary Doctor of Engineering degree program. The Master of Science degree requires a minimum of 30 credit hours. Extensive course offerings are provided in the late afternoons and evenings for those who wish to study part-time while employed. The Doctor of Engineering degree requires a minimum of 60 credit hours beyond the master's degree. Of these credits, at least 31 must be earned in course work. A minimum of 30 credit hours is required for the doctoral dissertation. In most instances, the dissertation research is local industry or research center sponsored and involves an internship with the sponsoring company or institution.

Research Facilities

The University has a large library with more than 1 million volumes and computer networks for resource sharing that provide access to more than 20 million volumes. Each engineering department has state-of-the-art laboratories with specialized equipment for research and experimentation. Participation in ongoing externally funded research in industrial projects is available to qualified graduate students. Multidisciplinary research is also conducted through the auspices of the University's Center for Environmental Science, Technology and Policy and the Advanced Manufacturing Center. Total yearly research funding is more than $7 million.

Fenn College is equipped with distance learning facilities that allow the CSU faculty to simultaneously broadcast, via live interactive video and audio communication, advanced courses to multiple campuses throughout the state and to industry, as well as to receive specialized courses from other institutions.

Financial Aid

The engineering departments offer a substantial number of teaching assistantships to selected students who receive full tuition and a stipend. Doctoral students may also qualify for fellowships or instructorships. In addition, a number of tuition-only awards are offered to qualified master's and doctoral students. M.S. degree assistantships are also available through the Advanced Manufacturing Center for students interested in manufacturing-related studies. Faculty members award full research assistantships that are funded by their research grants to qualified students. Other financial aid programs, such as government-financed student loans as well as student employment opportunities, are administered by the University's Financial Aid Office. Most of these awards are made on a competitive basis; however, some programs take the student's financial need into consideration.

Cost of Study

Tuition for Ohio residents for the fall 1999 semester is $202 per credit hour up to 12 credit hours and a fixed $2626 for 13 to 16 credit hours. Tuition for out-of-state residents is $404 per credit hour up to 12 credit hours and a fixed $5252 for 13 to 16 credit hours. An additional technology fee of $7.50 per credit hour is also assessed.

Living and Housing Costs

Housing and food service for unmarried Cleveland State students are available in Viking Hall, which currently accommodates up to 600 students. They live in double, large double, or triple rooms, each with a private bath. CSU's Housing Bureau assists students in finding off-campus residences in the surrounding area.

Student Group

Enrollment at Cleveland State University is 15,735 students. Of these, about 30 percent are graduate students. Fenn College enrolled 775 undergraduates and 243 graduate students in 1997–98. In that same year, the College awarded 220 Bachelor of Engineering, 99 Master of Science, and 10 Doctor of Engineering degrees.

Location

CSU is located in the heart of downtown Cleveland, a highly industrialized metropolitan area that includes more than one fifth of the state's population. Local industries and organizations augment the scope of the University's academic programs by offering employment and internship possibilities.

The University

Cleveland State University, a state-assisted institution catering to an urban, primarily commuting student body, was established in 1964 through a merger with Fenn College, a private institution. The University's seven colleges offer sixty baccalaureate degree programs, three advanced degrees in law, and thirty-three graduate programs, including doctoral degrees in chemistry, biology, engineering, urban studies, and business administration. Long before the founding of CSU, Fenn College of Engineering had established a reputation for the excellence of its programs. The basic engineering degree programs are accredited by the ABET, as well as by the North Central Association of Colleges and Schools. For more information about CSU, students can visit the University's Web page (http://www.csuohio.edu).

Applying

Application forms must be filed at least one month prior to the academic semester for which admission is sought. Students must submit an official transcript from every college previously attended and two letters of recommendation from college professors, sent by the originating college directly to the Office of Graduate Admissions. Students with satisfactory academic records are generally not required to submit admission test scores. International students must follow special application and admission procedures.

Correspondence and Information

For information:
Chairperson, Department of (specify)
Fenn College of Engineering
Cleveland State University
Euclid Avenue at East 24th Street
Cleveland, Ohio 44115-2425

For application materials:
College of Graduate Studies
Cleveland State University
Euclid Avenue at East 24th Street
Cleveland, Ohio 44115-2440
Telephone: 216-687-3592
216-687-3593

For international applications:
Euclid Building 103
Cleveland State University
2344 Euclid Avenue
Cleveland, Ohio 44115-2407
Telephone: 216-687-3910

Cleveland State University

FACULTY HEADS AND AREAS OF RESEARCH

The faculty of the Fenn College of Engineering consists of 62 full-time faculty members and 36 part-time professors. All faculty members hold doctoral degrees and are active in research and scholarly activities. Areas of concentration in the various fields, together with the names of department chairpersons, are given below.

Chemical Engineering. Orhan Talu, Chairperson. The department offers advanced study in adsorption and adsorption processes, thermodynamics, fluid mechanics, biochemical engineering, materials science and engineering, and transport processes. Current faculty research topics include heterogeneous transport, adsorption equilibrium and dynamics, transfer process in non-Newtonian fluids, tribology, optimization of mammalian cell culture, and simulation and modeling. Advanced materials engineering studies are being offered in cooperation with the mechanics of materials activities in the Mechanical and Civil Engineering departments. In cooperation with the Civil Engineering department, environmental research studies are also conducted. A specialization in applied biomedical engineering within the Doctor of Engineering program is available to qualified students. This program is a partnership in graduate education and research with the renowned Cleveland Clinic Foundation.

Civil and Environmental Engineering. Paul A. Bosela, Chairperson. Areas of concentrated study and research are offered in structures and foundations, mechanics and materials, and environmental and water resources. Current faculty research interests include concrete and steel structures, concrete materials, structural dynamics, water resources, hydraulics, environmental engineering, solid-waste disposal, bioaugmentation, water filtration, groundwater decontamination, occupational health, industrial-waste treatment, soil mechanics, foundations, experimental stress analysis, stress wave propagation, composite materials, finite elements, nonlinear buckling, artificial intelligence in structures, fracture mechanics, and constitutive modeling.

Electrical and Computer Engineering. George L. Kramerich, Chairperson. Areas of concentrated study and research include communications, computer engineering, control systems, power systems, and power electronics. Among current faculty interests are asynchronous transfer mode networks, satellite/terrestrial integrated networks, field programmable gate arrays, spread spectrum communications, code division multiple access, knowledge-based controls systems, nonlinear systems control, system identification and adaptive control, industrial automation and mechatronics, power systems control in deregulated environments, flexible AC transmission systems, AC motor drives, and electromagnetic interference and compatibility.

Industrial and Manufacturing Engineering. Theodore Sheskin, Chairperson. Members of the department teach and pursue research in traditional industrial engineering topics, including ergonomics, statistical quality control, production and inventory control, and engineering management, as well as the contemporary topics of linear and nonlinear optimization, systems simulation, computer-aided facilities design, cellular manufacturing, industrial artificial intelligence applications, sensors, and manufacturing processes, sequencing, and scheduling.

Mechanical Engineering. Mounir Ibrahim, Chairperson. The department offers a broad program of studies covering the areas of fluid dynamics, thermal sciences, numerical and computational methods, machine systems and design optimization, instrumentation and controls, robotics, dynamics of machinery, and manufacturing processes. Current faculty research interests include self-diagnostic piezoelectric sensors, robot sensor design, automated laser-optical inspection techniques, mathematical simulation of forging and forming processes, multiphase flow and heat transfer, phase change heat transfer, design optimization, active control systems, non-Newtonian fluid mechanics and heat transfer, turbulent flows, and analysis of unsteady fluid flow and heat transfer processes.

Advanced Manufacturing Center. Frederick C. Schoenig Jr., Director. The Advanced Manufacturing Center (AMC) is a joint venture between Cleveland State's Fenn College of Engineering and AMP, Inc., a State of Ohio Edison Center. Under this agreement, AMC operates as a semiautonomous research center with a dedicated research staff and the ability to contract for the services of CSU faculty members and students. Areas of research include nondestructive testing techniques, wear and tribology phenomena, advanced electronic component design, laser welding, and process modeling and simulation. More than half of the multimillion-dollar research effort comes from industrial sponsors, including many of the Fortune 500 companies, such as DELPHI, Ford, FMC, GE, and Health-O-Meter. The AMC has received many prestigious awards for its light manufacturing automation designs. Research grants and assistantships are available for students interested in manufacturing. The AMC also operates the Manufacturing Learning Center (MLC). This federally supported effort helps regional manufacturers reposition products for commercial markets, make manufacturing practices more competitive, bring back offshore production, and train workers and engineering degree students through hands-on involvement in industry-sponsored manufacturing engineering projects. A number of scholarships are available to qualified undergraduate and graduate students interested in manufacturing engineering careers.

COLUMBIA UNIVERSITY

The Fu Foundation School of Engineering and Applied Science

Programs of Study

The Fu Foundation School of Engineering and Applied Science offers programs of study leading to the Master of Science degree, the Professional Engineering degree, and two doctoral degrees, the Ph.D. and the Eng.Sc.D., in the Departments of Applied Physics and Applied Mathematics, Biomedical Engineering, Chemical Engineering and Applied Chemistry, Civil Engineering and Engineering Mechanics, Computer Science, Earth and Environmental Engineering, Electrical Engineering, Industrial Engineering and Operations Research, and Mechanical Engineering. The faculty in the School consists of 127 full-time members.

These departments also provide the structure for graduate study leading to the master's and doctoral degrees in such interdisciplinary fields as applied mathematics, biomedical engineering, earth resources engineering, financial engineering, materials science, medical physics, plasma research, quantum electronics, solid-state science and engineering, and telecommunications engineering. A joint M.B.A./M.S. program is offered in cooperation with the Graduate School of Business in the fields of operations research, industrial engineering, and mineral economics.

The M.S. degree is awarded upon satisfactory completion of a minimum of 30 points of approved graduate study beyond the bachelor's degree. Programs leading to the Professional Engineering degree in engineering mechanics, computer science, and chemical, civil, electrical, industrial, mechanical, metallurgical, mineral, and mining engineering are available for engineers who wish advanced work beyond the level of the M.S. degree but do not wish to emphasize research. The Professional Engineering degree requires a minimum of 30 points of graduate work beyond the M.S. degree. Part-time programs leading to the master's and Professional Engineering degrees can be arranged in most departments. The minimum requirements for either the Ph.D. or the Eng.Sc.D. are the completion of 60 points of approved graduate work beyond the B.S., the passing of appropriate qualifying examinations as prescribed by a department or interdisciplinary committee, and the completion and oral defense of a dissertation based on original research.

Research Facilities

The Fu Foundation School of Engineering and Applied Science is housed in the Seeley W. Mudd Building, a fifteen-story classroom and laboratory building; the Engineering Terrace Building; and the Computer Science Building. In addition, the Schapiro Center for Engineering and Physical Science Research houses research programs in computers, microelectronics, telecommunications, and condensed matter physics as well as a 200-seat auditorium, seminar rooms, offices, and laboratories. Other research facilities with modern equipment are available in all departments. Among the facilities are laboratories for research in acoustics, artificial organs, advanced computer architecture, heat transfer, materials, microelectronics, lasers, telecommunications, nuclear measurements and technology, plasma physics, and fusion energy. The Monell Engineering Library contains approximately 200,000 volumes and more than a million technical reports. The University operates three IBM mainframes (a 3083, a 3090, and a 4381), a VAX 8700, three Sun-4 systems, a Macintosh laboratory, and extensive peripheral equipment. The campus network includes almost 1,000 terminals, many in dormitories and libraries. The University system is linked to nationwide computer networks such as ARPANET, New York State's Nysernet, and the National Supercomputer Centers. Network services are provided to School and department local area networks throughout the University. Students can communicate with their professors via the campus Ethernet. The School of Engineering maintains a large array of computers for the use of its students and faculty members. Included are more than a dozen VAX-11/750 and VAX-11/780 minicomputers, a DECSYSTEM-20, large numbers of Sun Workstations, MicroVAX's, IRIS graphic stations, HP workstations, more than 150 microcomputers for instruction and student use, CRT terminals, laser printers, and a Schoolwide LAN system—MUDDNET—connected to the University backbone. The School maintains a specially equipped interactive Graphics Laboratory with several dozen IBM PCs.

Financial Aid

Funds are available based on academic merit, financial need, and departmental requirements for research and teaching assistantships. Many students hold staff appointments, which form an integral part of students' training and allow rapid progress toward the degree. Loans and Federal Work-Study positions are available to needy students who are U.S. citizens, permanent resident aliens, or political refugees.

Cost of Study

Tuition for 1999–2000 is $24,150 (for 30 points), plus applicable fees.

Living and Housing Costs

University residence halls include traditional dormitory facilities, as well as suites and apartments for single and married students. On-campus living expenses for 1999–2000 are estimated at $10,800 (includes board). Limited graduate housing is available and is by application only. The cost averages between $512 and $965 per month, depending on the type of accommodations desired. Other off-campus rooms and apartments are also available.

Student Group

Attending the schools and colleges that constitute Columbia are 20,438 students, about 12,330 of whom are graduate students. The 1,000 graduate students attending the Fu Foundation School of Engineering and Applied Science represent almost 200 colleges and universities, about 50 of which are outside the United States.

Location

Columbia's Morningside Heights campus is located about 15 minutes from the heart of New York City. The student is offered a range of educational, cultural, and recreational opportunities.

The University and The School

Originally designated King's College, Columbia opened its doors in 1754 under a grant issued by King George II. Over the years, professional and graduate schools were added. In 1896, it was redesignated Columbia University. The Fu Foundation School of Engineering and Applied Science is the outgrowth of one of these professional schools, the School of Mines, established in 1864 as the first school of its kind in the United States.

The century-long tradition in the Fu Foundation School of Engineering and Applied Science has been the philosophy of combining a rich liberal education with the rigor of technical education. Many of the revolutionary advances of contemporary technology have been pioneered at Columbia.

Applying

The basic requirement for admission as a graduate student is a bachelor's degree in any field of engineering or a related field with a record that indicates the preparation and ability necessary for successful performance at Columbia. For the fall term, Ph.D., Eng.Sc.D., and financial aid applicants must apply by January 5. February 15 is the deadline for all other degrees. Applications that are received late are reviewed until August 1 if space is available. Those for the spring term should be submitted by October 1.

Correspondence and Information

Office of Graduate Student Services
The Fu Foundation School of Engineering and Applied Science
524 Seeley W. Mudd Building
Mail Code 4708
Columbia University
New York, New York 10027

Telephone: 212-854-6438
E-mail: seasgradmit@columbia.edu (admission)
        engradfinaid@columbia.edu (financial aid)
World Wide Web: http://www.seas.columbia.edu

Columbia University

AREAS OF RESEARCH

Applied Mathematics. The study of advanced mathematical and computational methods with applications to the physical and engineering sciences form the core of these activities. Areas of research, which are often of an interdisciplinary nature, include nonlinear dynamics and chaos, scientific computing, turbulence, and geophysical fluid dynamics.

Applied Physics. Faculty research interests fall into the following four general categories: plasma physics, solid-state physics, quantum electronics, and applied mathematics. Current research projects in plasma physics include equilibrium and stability studies of high-beta magnetized plasmas, theory of stellarators, innovative fusion confinement systems, coherent radiation from intense relativistic electron beams, free-electron lasers, space plasma physics, trapped-particle instability studies, computer simulation of plasmas, and fusion-energy engineering studies. In solid-state physics, projects include electronic transport in low-dimension semiconductor systems, the fractional quantum Hall effect, surface photophysics, properties of semiconductors at high pressure and temperature, semiconductor devices, and laser device processing. In quantum electronics, research includes nonlinear optics, laser physics, and optical scattering.

Center for Biomedical Engineering. The biomedical engineering program provides a curriculum in the fundamentals of engineering and biological sciences that serves as a foundation for three specialized tracks, one of which is chosen by each student. These three specialty areas, with their associated specific areas of faculty research interest, are biomechanics: biofluids, biomaterials, biotribology, biosolids, diffusion and mass transport, cardiac mechanics, computer-assisted surgery, implant technology, orthopaedics, pulmonary mechanics, rehabilitation engineering, and sports medicine; medical imaging: artificial vision, computed tomography (CT), computer graphics, computer-aided diagnosis, digital radiography, image compression, magnetic resonance (MR), pattern recognition, picture archival systems (PACS), quantitative analysis, signal processing, telemedicine, volume visualization, volume rendering, and 3-D ultrasound; and tissue and cellular engineering: artificial organs, bioelectricity, biosensors, body composition, cellular transducers, genomics, neurophysiology, molecular engineering, signal transduction, structure-function analysis, surgical metabolism, transport phenomenon, and tissue processes.

Chemical Engineering and Applied Chemistry. Emphasis is placed on the application of basic principles of chemistry, physics, mathematics, and engineering to the analysis, design, and operation of processes or systems that involve chemical reactions and the transport of momentum, energy, or mass. Areas of research include heat transfer, mass transfer, process dynamics and optimization, reaction kinetics, interfacial phenomena, electrochemical processes, polymer science, and thermodynamic properties. Interdisciplinary areas include bioengineering, environmental engineering, technology, nuclear heat transfer, and pharmaceutical engineering.

Civil Engineering, Engineering Mechanics, and Construction Engineering. The main faculty research focus is in structural mechanics, with applications to infrastructure deterioration, damage detection and monitoring, earthquake engineering, soil structure interaction, and active control of structural motion; mechanics of materials and continua, with applications to continuum damage mechanics, development of new construction materials such as concrete that uses waste glass, and to related areas in biomechanics; and fluid mechanics and geoenvironmental engineering, with applications in hydrogeology, contaminant transport, geosynthetics, and site remediation. Research in these areas is both experimental and analytical. There is extensive interaction with various New York City agencies on infrastructure-related problems.

Computer Science. Research areas include algorithmic analysis, collaborative work, computational complexity, computer-aided digital design, computer graphics, computer vision, databases and digital libraries, data mining and knowledge discovery, distributed systems, mobile computing, natural language processing, networking, real-time multimedia, robotics, and user interfaces.

Earth and Environmental Engineering/Earth Resources Engineering. Use of geostatistical methodology for environmental assessment of mines and brownfields, modeling of mine drainage, constructed pilot wetland for storage and treatment of harbor sediments, GIS-based models that superimpose multiple sets of data on flows and sources/sinks of contaminants, and treatment of land and sediments contaminated from mining and processing activities. Research on a variety of geoenvironmental issues, with the intent to quantify, assess, and ultimately manage adverse human effects on the environment, contaminant transport in the subsurface, flow phenomena in saturated and unsaturated soils, probabilistic assessment of the effects of human activities on the environment, geostatistical simulation, and numerical modeling of estuarine flow and transport processes. Many of the faculty members are associated with the Earth Engineering Center, the engineering arm of the Columbia Earth Institute. The Institute includes among its member units the Lamont-Doherty Earth Observatory, the Center for Environmental Research and Conservation, Biosphere 2, and the International Research Institute for climate prediction. Students have full opportunity to participate in the diverse activities of the Earth Institute.

Electrical Engineering. Students within the electrical engineering department can choose from a number of research specialities, including digital image processing, networking/communications, digital and analog circuit design, computer engineering, optoelectronics, semiconductor devices, electromagnetics, and plasmas. Research projects are grouped into four focus areas: communications and networking, image and signal processing, microelectronic circuits, and microelectronic devices, electromagnetics, optoelectronics, and plasma physics. Information on the specifics of many projects can be found on the Internet (http://www.ee.columbia.edu/Res/res.html). A survey of various student and professor Web sites can provide a sense of the extensive research facilities available at Columbia and the depth and breadth of the research that is in progress.

Financial Engineering. Faculty interests are in portfolio management; option pricing, including exotic and real options; computational finance, such as Monte Carlo simulation and numerical methods; and datamining.

Industrial Engineering. Faculty research interests are in the design, analysis, and control of production and service systems. Current studies involve flexible manufacturing systems, scheduling, supply chain management, production planning, inventory control, productivity, semiconductor manufacturing, and yield management.

Manufacturing Engineering. Engineers and managers are trained in new technologies, methods of production, and approaches to product design. Students receive hands-on training in an associated manufacturing engineering laboratory.

Materials Science and Minerals Engineering. Research in materials science includes microscopic study of interfaces, grain boundaries, and thin films; lattice defects and electrical properties of ceramics; laser processing and solidification of silicon; and optical and electrical properties of wide-band gap semiconductors. Research at the Langmuir Center for Colloids and Interfaces includes the enhancement of oil recovery by means of ficellar flooding of reservoirs, electroflotation of mineral particles, microbial interactions with minerals, selective flocculation of fine particles, and ultrafine grinding. There are many research projects in surface and colloid chemistry that involve both inorganic and organic materials, such as surfactants, polymers, and latexes.

Mechanical Engineering. Graduate study programs include fluid dynamics, mechanics of solids, kinematics, dynamics of machines, robotics, heat transfer, biomechanics, control theory, and manufacturing engineering. Major research areas are in fluid dynamics, convective heat transfer, analysis and synthesis of mechanisms, high-speed dynamics of machines, orthopedic biomechanics, manufacturing cryogenic machining, and control theory.

Medical Physics. Students are prepared for careers in medical physics and receive sufficient preparation for the ABMP certification exam. The program, offered in collaboration with faculty members from the College of Physicians and Surgeons, consists of a core curriculum of health physics and radiation physics courses and a practicum. Some opportunity for specialization exists.

Operations Research. In deterministic models, research includes linear, nonlinear, and integer programming algorithms, network flows, polyhedral combinatorics, and combinatorial optimization. In stochastic models, research is being conducted on Markovian decision processes, reliability theory, stochastic scheduling, statistical inference, simulation, queuing theory, queuing networks, risk analysis, and mathematical and computational finance.

Solid State Science and Engineering. The instructional program emphasizes basic physics and solid-state physics, with particular emphasis on the electrical, optical, and magnetic properties of solids. Research programs include studies of dielectric and optical properties of insulators, electronic materials and semiconductor devices, superconducting material, laser physics, high-pressure properties of solids, metallic thin films, inelastic light scattering, and the theory of exciton dynamics.

Telecommunications Engineering. Specialization is provided in a discipline in which the department has extensive offerings and a strong research effort.

Thermal Fluid Science. This M.S. degree program, which trains engineers in the design of thermal systems and computational methods, provides flexibility for interdisciplinary study for students with interests in thermofluid phenomena in chemical processes, environmental sciences, and oceanography.

CONCORDIA UNIVERSITY

Faculty of Engineering and Computer Science

Programmes of Study
The Faculty offers programmes leading to the Diploma in Computer Science; Graduate Certificates in Building Engineering, Environmental Engineering, and Mechanical Engineering; and the degrees of Master of/Magisteriate in Computer Science, Master of/Magisteriate in Applied Science, Master of/Magisteriate in Engineering, and Doctor of Philosophy. Programmes are available on a full- or part-time basis; all courses are scheduled in the evening.

The Diploma in Computer Science, a quota programme, provides expertise in computer science fundamentals for highly qualified university graduates with diverse backgrounds. A fully qualified candidate will be required to complete a minimum of 34 credits.

The Graduate Certificate in Building Engineering is a 15-credit programme that offers the following concentrations: building envelope, building science, construction management, energy efficiency, and indoor environment.

The new 15-credit (five-course) Graduate Certificate in Environmental Engineering focuses on industrial waste management and is offered through the Department of Building, Civil and Environmental Engineering.

Five new 15-credit (five-course) Graduate Certificates in Mechanical Engineering specialize in one of five dynamic areas: aerospace, composite materials, controls and automation, theoretical and computational fluid dynamics, and manufacturing systems.

The Master of/Magisteriate in Computer Science (M.Comp.Sc.) programme is offered in either of two options. The thesis option emphasizes research and may lead to further graduate studies. The course-oriented option provides a broad spectrum of computer science and application courses to allow students to gain depth in the areas of their choice. Applicants must hold an undergraduate degree with a concentration in computer science or a graduate diploma with high standing in computer science.

The Master of/Magisteriate in Applied Science (M.A.Sc.) programme is thesis oriented and is intended to provide a significant introduction to research. Consideration is given to candidates with a degree in a cognate area. Applicants holding a bachelor's degree in architecture may also be considered for admission to an extended programme. It will appeal primarily to the student interested in full-time study.

The Master of/Magisteriate in Engineering (M.Eng.) programme is course oriented, with the possibility of substituting a project for up to three courses, and it is designed to provide practicing engineers with an opportunity to strengthen and extend the knowledge they have obtained at the undergraduate level. Applicants must hold a Bachelor of Engineering degree or the equivalent, with high standing.

The minimum programme length for both the M.A.Sc. and M.Eng. is 45 credits.

Doctoral programmes are offered in all departmental research areas. Applicants must hold a master's degree or the equivalent, with high standing, in engineering, computer science, or a cognate discipline. The minimum programme length is 90 credits, including a 72-credit thesis. In addition, each student must complete the appropriate doctoral seminar and sit for a comprehensive examination (written and oral) on fundamentals in the research area and a written research proposal. The minimum residence requirement is two years of full-time study or the equivalent in part-time study.

Research Facilities
Well-developed laboratory facilities are available for research in many areas, including structures, experimental stress analysis, building enclosures, building materials, building acoustics, building aerodynamics, thermal environment, energy conservation, water resources, soil mechanics and geotechnical engineering, environmental engineering, microelectronics, micromachines, micromechanisms, circuits and systems, VLSI, communications, data processing, character recognition, computer graphics, distributed systems, databases and software engineering, computational mathematics, multimedia, power electronics, robotics, broadband switching, electrical machines, digital and analog instrumentation, CAD/CAM, vehicle dynamics, vehicle structure, computational fluid dynamics, rocket and air-breathing propulsion, fluid controls, multirobot workcells, automated guided vehicle with AI, fuel control systems, computer-integrated manufacturing, deep-hole machine tools, machining centres, sensor-based tool monitoring, high-speed workstations with parallel processors, interactive graphics, vibrations, hydraulic copying, shock tubes, heat transfer, heat pumps, combustion, composite materials processing, manufacturing, testing, and design of structures using composites.

Financial Aid
Sources of financial support available to full-time students (master's and doctoral) include scholarships, bursaries, research assistantships, teaching assistantships, and positions as laboratory instructors. The majority of these RA and TA positions are restricted to Canadian citizens and permanent residents.

Cost of Study
Tuition fees for 1999–2000 are Can$55.61 per credit for Quebec residents and Can$105.61 per credit for out-of-province Canadian citizens. For international students (visiting or independent), tuition fees are Can$275.61 per credit for the master's programme, graduate certificate programme, or graduate diploma programme and Can$248.61 per credit for the Ph.D. programme. Students must also pay term fees and student association, student services, and academic fees. The University reserves the right to change the published fees without notice. International students must also pay for medical insurance.

Living and Housing Costs
The University has no residence facilities for students, but off-campus rooms and apartments are available in the immediate vicinity. For a single student, the cost of living is about Can$11,000 per year in addition to tuition.

Student Group
Concordia University has an enrollment of about 26,000 full- and part-time students. There are about 239 students registered in the Diploma of Computer Science programme, 186 in the M.Comp.Sc. programme, 425 in the M.A.Sc. and M.Eng. programmes, and 150 in the Ph.D. programme.

Student Outcomes
Students who have completed their graduate studies in engineering and computer science have been hired as researchers and project engineers by the many industrial organizations located within the Montreal area and across North America. These organizations include Hydro-Quebec, MATROX, Pratt & Whitney Canada Inc., National Research Council of Canada, SPAR Aerospace Limited, Modern Engine Company, Bell Helicopter, Textron Canada, Canadair, CAE, Nortel, and Ford Motor Company.

Location
Montreal is an industrial, commercial, and transportation hub of Canada. The bilingual cultural life and the dynamic atmosphere of the city make it an exciting place in which to live.

The University
Concordia University was incorporated in 1974 through a merger of Sir George Williams University and Loyola College of Montreal. The Faculty's graduate programmes are offered on the downtown campus.

Applying
Applications, complete with a nonrefundable Can$30 fee, must be received in the Graduate Studies Office by June 1 for the fall term, October 1 for the winter term, and February 1 for the summer term when originating from within Canada and by February 15 for the fall term, June 15 for the winter term, and October 15 for the summer term when originating from outside Canada.

Correspondence and Information
Graduate Studies Programme
Concordia University
1455 de Maisonneuve Boulevard West, Room LB 1001
Montreal, Quebec H3G 1M8
Canada

Telephone: 514-848-3057
Fax: 514-848-8646
E-mail: infofac@encs.concordia.ca
World Wide Web: http://www.encs.concordia.ca/

Concordia University

THE FACULTY AND THEIR RESEARCH

Building, Civil and Environmental Engineering

S. Alkass, Ph.D.: construction management, building economics, construction plant, information systems in construction. B. Ashtakala, Ph.D.: transportation, planning, pavement design. A. Athienitis, Ph.D.: building thermal performance, solar engineering, HVAC, computer-aided building design. C. Bédard, Ph.D.: computer-aided building design, knowledge-based expert systems, information systems for construction. D. Derome, M.Eng.: building envelope, integrated building design, heat and moisture transfer performance. M. El-Badry, Ph.D.: bridge engineering, finite element analysis. M. Elektorowicz, Ph.D.: environmental engineering. P. Fazio, Ph.D.: building envelope, energy analysis, building design. D. Feldman, D.Sc.: polymers, polymers in energy conservation, building materials. R. W. Guy, Ph.D.: building acoustics, sound intensity management. K. H. Ha, Ph.D.: finite-element analysis, structural systems. F. Haghighat, Ph.D.: indoor air quality, HVAC. A. M. Hanna, Ph.D.: geotechnical engineering. C. Marsh, M. A.: seismic resistance, ultimate resistance, collapse analysis, elastic stability. O. Moselhi, Ph.D.: construction/project management, automated decision–support systems in construction, construction safety, and productivity. O. A. Pekau, Ph.D.: structural dynamics and earthquake engineering. H. B. Poorooshasb, Ph.D.: soil mechanics. A. S. Ramamurthy, Ph.D.: water resources and fluids engineering. T. Stathopoulos, Ph.D.: wind effects on building structures and environment. M. S. Troitsky, D.Tech.Sc.: bridge engineering, prestressed steel. M. Zaheeruddin, Ph.D.: building energy analysis, passive solar systems, control for energy systems, HVAC systems. Z. A. Zielinski, D.Tech.Sc.: precast prestressed concrete structures, structural connections. R. Zmeureanu, Ph.D.: building environment, energy analysis programs, computer-aided building design.

Electrical and Computer Engineering

M. O. Ahmad, Ph.D.: VLSI architecture for signal processing, parallel computer architecture for image processing. A. J. Al-Khalili, Ph.D.: VLSI. M. de Champlain, Ph.D.: software engineering. A. K. Elhakeem, Ph.D.: telecommunications networks. J. C. Giguère, Ph.D.: computer systems and networks. J. F. Hayes, Ph.D.: computer communications, local area networks. P. K. Jain, Ph.D.: power electronics and systems. G. Joós, Ph.D.: power electronics, motor drives. F. Khendek, Ph.D.: software engineering. K. Khorasani, Ph.D.: control, robotics, and neural networks. S. J. Kubina, Ph.D.: electromagnetic compatibility, antennas. L. M. Landsberger, Ph.D.: IC fabrication process, microelectronics, MEMS. T. Le-Ngoc, Ph.D.: communication systems. W. Lynch, Ph.D.: signal processing, image compression. M. K. Mehmet Ali, Ph.D.: performance of broadband networks. R. Paknys, Ph.D.: electromagnetics, electromagnetic compatibility, antennas, numerical electromagnetics. E. I. Plotkin, Ph.D.: signal processing. V. Ramachandran, Ph.D.: multivariable networks, digital filtering. R. Raut, Ph.D.: electronics and VLSI. M. R. Soleymani, Ph.D.: coding for broadband communication channels. M. N. S. Swamy, Ph.D.: digital signal processing. S. Tahar, Ph.D.: VLSI. C. W. Trueman, Ph.D.: electromagnetics and antennas, electromagnetic compatibility, numerical electromagnetics. C. Wang, Ph.D.: VLSI, image processing.

Mechanical Engineering

W. Ahmed, Ph.D.: dynamic systems, stability, vehicle systems, vibration isolation and control. S. Amiouny, Ph.D.: applied discrete optimization, production, distribution, and material handling systems. R. Bhat, Ph.D.: random vibrations, rotor dynamics, structural acoustics, micromechatronics. A. A. Bulgak, Ph.D.: modelling and performance analysis of flexible manufacturing/assembly systems, quality and productivity improvement. K. Demirli, Ph.D.: fuzzy logic–based reasoning and its application to manufacturing cells and autonomous mobile vehicles. R. Ganesan, Ph.D.: stochastic mechanics, stress analysis, finite element method, vibration, machinery monitoring and diagnostics. W. S. Ghaly, Ph.D.: computational fluid dynamics, analysis and design methods in turbomachinery, numerical aerodynamics. G. J. Gouw, Ph.D.: human factors engineering, safety engineering, biomedical engineering, tribology. W. G. Habashi, Ph.D.: computational fluid dynamics, finite-element methods, high-speed external and turbomachinery flows. E. Haseganu, Ph.D.: solid and structural mechanics, dynamics, vibration, computational solid mechanics, biomechanics. S. V. Hoa, Ph.D.: composite materials and structures, vibration, stress analysis, dynamics, finite-element method. H. Hong, Ph.D.: direct injection of alternative fuels for CI engines, electronic/hydromechanical fuel–control systems, microprocessor control of electromechanical systems. K. I. Krakow, M.S.: heat pumps, environmental control, refrigeration, thermodynamics, air conditioning. V. Latinovic, D.Eng.: computer-aided manufacturing, production technology, CADD. R. A. Neemeh, Ph.D.: shock wave physics and related phenomena, unsteady wave motion in compressible flow, high-speed aeroacoustics. J. Pegna, Ph.D.: computer-integrated design and manufacturing processes, free-form fabrication, rapid prototyping, precision engineering, machine design, metrology. M. Pugh, Ph.D.: ceramics, metals, and composites. R. Rajagopalan, Ph.D.: automation, real-time control, autonomous transit vehicles, parallel processing, robotics, robotic machining. S. Rakheja, Ph.D.: human response to vibration and shock, vehicle dynamics, optimization and vibration control in system design. I. G. Stiharu, Ph.D.: design and manufacturing, dynamics, micromachining and micromechanics, tribology. C. Y. Su, Ph.D.: control systems, mechatronics, precision engineering, robotics. J. V. Svoboda, D.Eng.: control systems, hydraulic systems, flight and vehicle simulators. G. H. Vatistas, Ph.D.: fluid mechanics, vortex dynamics and flow instabilities, experimental aerodynamics, finite difference in fluid dynamics, lubrication, microgravity fluid mechanics, thermodynamics. X. Xiao, Ph.D.: applied mechanics, composite materials, composite processing.

Computer Science

V. S. Alagar, Ph.D.: formal aspects of computing, robotics, database. J. W. Atwood, Ph.D.: specification, validation, implementation, and testing of communications protocols, distributed applications. S. Bergler, Ph.D.: AI, computational linguistics, cognitive science. T. D. Bui, Ph.D.: design and analysis of algorithms, numerical methods, mathematical modelling, CAD. G. Butler, Ph.D.: algebraic and combinatorial computing, object-oriented reusable software, deductive and object-oriented databases, formal methods, programming languages. B. C. Desai, Ph.D.: database systems, operating systems, applications of AI and intelligent systems, digital library. E. J. Doedel, Ph.D.: numerical analysis, differential equations, nonlinear analysis. T. Fancott, D.Sc.: parallel architectures, biomedical applications. D. Ford, Ph.D.: algorithms in computational algebra and computational number theory. G. Grahne, Ph.D.: databases and knowledge bases, knowledge representation, information systems, logic programming, computational complexity. P. Grogono, Ph.D.: programming languages, programming environments, compiler construction. R. Jayakumar, Ph.D.: VLSI algorithms and architectures, VLSI design automation, graph algorithms. S. L. Klasa, Ph.D.: algebraic K-theory, algebraic geometry, category theory, theoretical computer science, language and automata theory, applications of harmonic analysis to pattern recognition. A. Krzyzak, Ph.D.: pattern recognition, image processing, neural networks, computer vision, robotics and control. V. S. Lakshmanan, Ph.D.: relational databases, deductive databases, object-oriented databases, logic programming, knowledge representation, knowledge-base systems, parallel processing, graph algorithms, computational complexity. C. W. H. Lam, Ph.D.: combinatorial computing, algorithms, coding theory. H. F. Li, Ph.D.: parallel processing architectures and algorithms, VLSI systems and computation. T. Li, Ph.D.: parallel and distributed symbolic processing, VLSI architecture, testing and testable design for VLSI. G. Martin, M.Sc.: computer centre management, AI. J. McKay, Ph.D.: combinatorial and algebraic computation, numerical methods. L. Narayanan, Ph.D.: parallel and distributed computing, randomized algorithms. J. Opatrny, Ph.D.: interconnection networks, automata and formal languages, graph and network algorithms, parsing and compilers. D. K. Probst, Ph.D.: parallel and distributed systems and algorithms, multiprocessors architectures, hybrid systems. T. Radhakrishnan, Ph.D.: cooperating agents, man-machine communication, distributed processing, multimedia applications. M. Saksena, Ph.D.: real-time computing, distributed multimedia systems, computer networking. R. Shinghal, Ph.D.: artificial intelligence, pattern recognition. C. Y. Suen, Ph.D.: pattern recognition and machine intelligence, computational linguistics, large-scale computer input technologies. L. Tao, Ph.D.: parallel processing, distributed objects, combinatorial optimization.

DALHOUSIE UNIVERSITY

DalTech
Faculties of Engineering, Architecture, and Computer Science

Programs of Study

The University offers graduate work in architecture, engineering, naval architecture and marine engineering, engineering mathematics and computer science, food science, and urban and rural planning. Most programs in engineering and engineering mathematics lead to the Ph.D. degree.

Graduate research programs leading to the Master of Applied Science (M.A.Sc.) or Doctor of Philosophy (Ph.D.) degree are available in agricultural, chemical, civil, electrical, fisheries, industrial, mechanical, mining, and metallurgical engineering, and naval architecture and marine engineering. The graduate programs in engineering mathematics and food science lead to a Master of Science or Ph.D. degree. The graduate programs in computer science lead to a Master of Computer Science or Ph.D. degree. There are cooperative programs in urban and rural planning, metallurgical engineering, and computer science, involving both course work and on-the-job training.

For admission to the graduate programs in engineering, an undergraduate degree in engineering with high scholastic standing is the general requirement, but an honors degree in science or the equivalent will be accepted, in which case the candidate may be required to take essential undergraduate engineering courses. For admission to advanced work in engineering mathematics and computer science, a bachelor's degree in computer science, mathematics, physics, or engineering physics or its equivalent is accepted. The food science, chemical engineering, and metallurgical engineering programs have individual admission requirements.

For admission to the Master of Architecture program a professional degree in architecture with high scholastic standing is required. The Master of Urban and Rural Planning (M.U.R.P.) program accepts a range of undergraduate degrees in related professional fields.

Candidates for the Doctor of Philosophy degree are admitted at the recommendation of the department or school concerned. All undergraduate and graduate degrees submitted for consideration must be from a recognized university and be accompanied by a transcript of the candidate's academic record.

The course-work-only program for the Master of Engineering (M.Eng.) degree, which does not require the candidate to undertake research and write a thesis, represents an equivalent standard of achievement by requiring a larger amount of course work and a project.

Most graduate programs can be attempted on a part-time basis, provided that all requirements for a master's degree are completed within a five-year period.

The Ph.D. program is research oriented, with a seven-year limitation on the completion of all requirements for the degree.

Research Facilities

Each department or school has well-equipped laboratories for the type of research being done. There are adequate technical staff and facilities for the construction and maintenance of equipment. The University is surrounded by and interacts with a large research community. All laboratories have links to computing facilities both in house and in centralized locations. The DalTech library has the best collection of technical reference material in Maritime Canada and operates as part of the regional and national technical information network. Local research institutes and universities allow access to large-scale specialized equipment.

Financial Aid

A variety of scholarships, fellowships, and assistantships are available for students. Teaching assistantships can be part-time or full-time. Research assistantships are awarded from research funds. A graduate student can be awarded up to $16,000 per twelve-month period.

Cost of Study

Tuition and fees for full-time study in the engineering and computer science programs in 1999–2000 are $5302 for Canadian citizens and permanent residents and $8872 for visa students. Tuition and fees for full-time study in the architecture program are $5521 for Canadian citizens and permanent residents and $9091 for visa students.

Living and Housing Costs

Depending upon individual lifestyles, the yearly cost for an unmarried student living in residence is $7000 or more, excluding the cost of study.

Student Group

DalTech's enrollment of about 1,700 represents a senior-student atmosphere in an urban setting.

Location

The Halifax-Dartmouth area has a population of about 300,000 and has all the major cultural amenities of city life while being surrounded by the recreational facilities of the country and the seashore.

The University

The Technical University of Nova Scotia, formerly the Nova Scotia Technical College, was founded in 1907; it concentrated on the senior years of engineering education until 1961, when the School of Architecture was started, and 1981, when the School of Computer Science was formed. On April 1, 1997, the Technical University of Nova Scotia was amalgamated with Dalhousie University in what is commonly known as DalTech. Its downtown location makes the University very accessible to other universities, libraries, research centers, theaters, stores, and cultural activities.

Applying

All academic departments can supply specific information about programs and research assistantships, and they welcome inquiries. All requests for application forms and for residence information, including requests for University calendars, should be made to the Registrar. Formal application for graduate work must be made through the Registrar's Office. All students registering at Dalhousie University must provide proof of adequate health insurance coverage unless they are covered by a provincial health insurance plan.

Correspondence and Information

Graduate Studies and Research Office
DalTech
P.O. Box 1000
Halifax, Nova Scotia B3J 2X4
Canada
Telephone: 902-494-3951
Fax: 902-494-3149

Dalhousie University

FACULTY HEADS AND RESEARCH AREAS

F. Hamdullahpur, Associate Principal for Graduate Studies and Research; Ph.D., Technical University of Nova Scotia; P.Eng.
A. C. Bell, Dean of Engineering; Sc.D., MIT; P.Eng.
M. E. El-Hawary, Associate Dean of Engineering; Ph.D., Alberta; P.Eng.

Architecture and Planning
T. Emodi, Dean of Architecture; M.E.S., York.
Design studies take four major directions: adaptive reuse, architectural/constructional systems, housing, and urban form and place. Historical and theoretical research at the graduate level concentrates in urban design, planning, building-type studies, building performance studies, environmental controls.

Biological Engineering
N. Ben-Abdallah, Department Head; Ph.D., British Columbia; P.Eng., P. Ag.
Applications of computers to agri-food systems, non-point-source pollution, manure treatment and disposal, pollution control, fermentation technology, biomass and solar energy, drainage, irrigation, soil erosion, plant and animal environment, greenhouses, sprayer research, soil tillage, optimum selection of machinery, materials handling, on- and off-farm storage, processing and packaging of agricultural produce, food and bioprocess engineering, modeling thermal processing of solid foods, alcohol production from food processing waste, environmental engineering, agricultural engineering in developing countries.

Chemical Engineering
Y. P. Gupta, Department Head; Ph.D., Calgary; P.Eng.
Heat and mass transfer, gas-liquid contacting, processing of energy resources, development of alternative fuels, mathematical model simplification, process control, environmental engineering, nonintrusive measurement, fluidized-bed combustion, wood combustion, flame front propagation, dust explosions, materials engineering, runaway chemical reactions.

Civil Engineering
H. H. Vaziri, Department Head; Ph.D., British Columbia; P.Eng.
Bearing capacity of soils; buried flexible structures; soil dynamics; structural mechanics; structural impact assessment; fatigue and fracture mechanics; analysis of bridges; reinforced concrete; high-performance concrete; prestressed concrete, composite materials, and material properties; finite-element methods; computer-aided engineering; transportation systems analysis and design; pavement management and design; highway materials; wastewater treatment; on-site sewage disposal; lake-watersheds studies; stormwater management, groundwater modelling.

Electrical and Computer Engineering
S. T. Nugent, Department Head; Ph.D., New Brunswick; P.Eng.
Active and switched-capacitor networks, digital filters, signal processing, robotics, vehicle safety, control and optimization of power systems, cardiovascular dynamics, medical instrumentation, applications of computers in medicine, signal detection and estimation, underwater telemetry, spread spectrum systems, microcomputer-based systems, optoelectronic devices and systems, computer architecture, VLSI, communications, antennas, RF electronics, neural nets, electromagnetics and DSP.

Engineering Mathematics
W. J. Phillips, Department Head; Ph.D., British Columbia.
Computational fluid dynamics, hydrodynamics, wave loads on offshore structures, wave-ice-current interactions, applications of numerical linear algebra in digital signal processing, random field theory with applications to engineering problems, dynamical systems and the analysis of geophysical and medical time series data, piecewise regression and saddlepoint methods in statistics.

Food Science and Technology
Tom A. Gill, Director and Department Head; Ph.D., British Columbia.
Food science and fisheries engineering, postharvest handling, fish preservation and storage, food biochemistry, protein and fat deterioration, nontraditional species, fish oil utilization, fatty acid identification, fish-waste processing, improved protein utilization, quality control and assessment, rheology and texture of foods, structure of foods, food packaging, thermal processing of foods, beverage processing and handling.

Industrial Engineering
E. A. Gunn, Department Head; Ph.D., Toronto; P.Eng.
Development and application of job-shop scheduling rules, operations research studies in natural resource utilization, flexible manufacturing systems, computer-aided manufacture, mathematical programming, large-scale systems, economic modeling and optimization, workplace design, facilities design, manual materials handling, ergonomics, transportation, routing and scheduling, stochastic modeling, traffic control and resource management in telecommunications networks, manpower scheduling.

Mechanical Engineering
M. R. Kujath, Department Head; Ph.D., Warsaw (Poland); P.Eng.
Fluid power, energy conversion systems, fluidized-bed combustion, computational fluid dynamics, vibrations, CAD/CAM, CAE, process control theory and design, biomedical engineering, finite-element techniques, machine dynamics, solar energy, studies of ocean waves, wave energy, ship and marine hydrodynamics, composite materials, robotics, space mechanics.

Mining and Metallurgical Engineering
G. J. Kipouros, Department Head; Ph.D., Toronto; P.Eng.
Mine mechanization, simulation of mining systems, microseismic monitoring, rock mechanics, numerical modelling, pit wall and crown pillar stability, mine planning and design, mine feasibility studies, liquid/solid separation, mineral processing, industrial waste management, mine production optimization, horizontal drilling and completion, petroleum reservoir engineering, slag additions to cement, chemical desulfurization of coal, corrosion of superalloy materials, wear mechanisms and fracture in rail steels, ceramic coatings to resist refractory corrosion, welding metallurgy, fracture mechanics, fatigue crack growth, composite materials, cavitation erosion, Zn-based bearing materials, metal textures, electronic materials, molten-salt electrolysis, rare-earth metals, slag chemistry, structural ceramics.

Centre for Water Resources Studies
D. H. Waller, Director; Ph.D., Dalhousie; P.Eng.
The Centre for Water Resources Studies promotes research within the University and provides educational services to students and professionals. Current areas of interest include watershed management, lake and estuarine water quality, urban hydrology, stormwater quality and quantity, watershed modeling, rainwater cistern systems, on-site sewage disposal, landfill leachate, oxidation ponds, and air-pollutant transport.

Atlantic Industrial Research Institute
AIRI is an independent, self-supporting research institute on the campus of the University. It undertakes contract research primarily in the areas of industrial engineering and operations research and utilizes the capability, software packages, and other facilities available within the Department of Industrial Engineering. Interested parties should contact the institute directly.

Centre for Marine Vessel Design and Research
The Centre for Marine Vessel Design and Research is a nonprofit organization operating within the Mechanical Engineering Department. The centre is engaged in market-sensitive research and development for the marine industry in the areas of design, construction, operational performance, and safety of marine vessels and ocean structures.

Faculty of Computer Science
J. Slonim, Dean; Ph.D., Kansas State.
Artificial intelligence, CAD, computer architecture, computer vision, database systems, expert systems, graphical user interfaces, programming languages, robotics, software engineering, visual programming languages, VLSI.

DARTMOUTH COLLEGE

Thayer School of Engineering

Programs of Study

The Thayer School offers two programs of study: a professional project-based track leading to the degrees of Bachelor of Engineering and Master of Engineering in Engineering Management and a research thesis-based track leading to the Master of Science and Doctor of Philosophy degrees. Degrees are undifferentiated with respect to the engineering discipline, in order to emphasize interdisciplinary study that combines scholarship; creative, independent problem solving; and professional development.

The Bachelor of Engineering program is accredited by ABET. It serves as a first professional degree for students with A.B. degrees in engineering science, the physical sciences, mathematics, or computer science. Entrance requirements are the equivalent of an A.B. major in engineering sciences plus one science or math elective. Degree requirements are nine courses, including a design project. The Master of Engineering in Engineering Management (M.E.M.) program prepares the student for professional engineering practice and management and requires from four to six quarters of study (depending on previous education), with distributive requirements in engineering, design, and management and completion of an individual project defined by an industrial sponsor. An environmental option for the M.E.M. allows students to take courses at Vermont law school. A joint program of the School and the Tuck School of Business leads to the concurrent award of the Master of Science and the Master of Business Administration degrees. The Master of Science program prepares the student for a technical career in engineering and applied science and requires from three to four quarters of course work (depending on previous education) and completion of a thesis demonstrating ability to do research. The Doctor of Philosophy program is intended to provide preparation for a career in independent scientific research in engineering and applied science. Completion of a thesis that makes a significant contribution to knowledge or demonstrates advanced creative design is required. A joint M.D./Ph.D. program in biomedical engineering is offered in conjunction with the Dartmouth Medical School.

Research Facilities

Experimental facilities include advanced electron-optical, X-ray diffraction, and atomic force microscopy equipment (including the Dartmouth electron microscope facility) for microstructural examination; a clean room; a multiaxial testing system (MTS); and uniaxial testing systems (Instron and MTS), plus special laboratories for studies in materials, fabrication, and characterization of semiconducting devices; tribology; optics (including picosecond high-power laser system, gbit/s fiber-optic network); system dynamics and controls; numerical methods and visualization; ice physics and mechanics; fluid mechanics; biomedical engineering; and cold regions engineering. Computer facilities include on-campus superminicomputers and numerous laboratory microcomputers. A high-speed local area network hosts 150 general and special-purpose IBM, Hewlett-Packard, SGI, and Sun workstations. Biotechnology facilities support research in mammalian cell and virus culture, heterogeneous enzymology, biomass processing, and bioremediation.

Financial Aid

Students receive support from a number of sources, including research assistantships, fellowships, scholarships, and loans, and from such duties as course grading and assisting in laboratories. A number of fellowships that include a stipend and tuition are available to M.S. and Ph.D. students to cover the cost of the first two or three terms of residence. Students are expected to define their research topics by the end of their fellowship period, after which support is normally continued in the form of a research assistantship. Fellowship or research assistantship awards begin at $15,720, plus tuition, for the twelve-month year in 1999–2000.

Cost of Study

Tuition for the 1999–2000 academic year (nine months) is $24,624. Summer-term tuition for those in full-time residence during the previous nine months is $3283. All students must purchase the Dartmouth Student Group Health Insurance Plan ($1050 fee) unless granted a waiver.

Living and Housing Costs

Living expenses, including room and board, books, and personal expenses, are estimated at $9000 to $11,000 for the 1999–2000 academic year. These costs do not include tuition, travel, or summer living expenses.

Student Group

The graduate student body in engineering ranges from 120 to 150 students. They come from all areas of the country and abroad. Their average age is 24; 20 percent are women, and 5 percent are married. About 85 percent receive financial aid or are supported in some way.

Student Outcomes

More than 95 percent of Thayer School graduates are employed within three months of graduation. Recent M.E.M. management graduates are hired as design, project, and manufacturing engineers; consultants; strategic planners; and analysts. M.S. graduates not continuing for a doctorate enter technical positions as disciplinary engineers or in interdisciplinary areas involving product design and development. Ph.D. recipients pursue research in industry or at universities and large laboratories; some begin academic careers.

Location

The Dartmouth community is a center for the educational, scientific, medical, and cultural advantages most often associated with a metropolitan environment, yet it retains the small-town pleasantness of northern New England rural life. Recreational opportunities are excellent.

The College and The School

The Thayer School, founded in 1867, is affiliated with the College of Arts and Sciences and is joined by two other professional schools on campus, the Tuck School of Business and the Dartmouth Medical School. The medical school provides clinical facilities for Thayer School's Biomedical Engineering Program. Collaborative programs are maintained with other Dartmouth departments in computer science, geophysics, space and applied physics, biology, chemistry, and environmental science and with the U.S. Cold Regions Research and Engineering Laboratory, located in Hanover.

Applying

To receive priority consideration for financial aid, candidates for fall admission to the M.S. and Ph.D. programs should complete their application before January 15 of the year they wish to enter. Notification of admission is made as soon as possible thereafter. B.E. and M.E.M. applications are considered on a rolling basis. The GRE General Test is required of all applicants except those applying for admission to the B.E. program. Applicants to the B.E. and M.E.M. programs who request financial aid must complete the PROFILE form.

Correspondence and Information

Office of Graduate Admissions
Thayer School of Engineering
8000 Cummings Hall
Dartmouth College
Hanover, New Hampshire 03755-8000

Telephone: 603-646-2606
Fax: 603-646-3856
E-mail: thayer_admissions@dartmouth.edu
World Wide Web: http://thayer.dartmouth.edu

Dartmouth College

FACULTY RESEARCH AREAS

Graduate research in the following areas is sponsored at the Thayer School of Engineering by the School's 35 professors and 41 adjunct faculty members.

Biomedical Engineering. Biomaterials, orthopedic implant design, polymer bearing design and development, effects of sterilization on implants; microwave eye surgery; therapeutic and diagnostic bioelectromagnetics; bioinstrumentation and signal processing; image processing and reconstruction in tomography and microscopy; physiological system modeling; medical optics; biomedical informatics; neural engineering; information technologies in medicine.

Biotechnology and Biochemical Engineering. Bioreactor design, biochemical kinetics, evolutionary and viral biotechnology, biomimetics, bioremediation, metabolic engineering, biomass conversion and ethanol production, biosystems modeling.

Computational Methods. Finite difference, finite elements, spectral- and boundary-element methods; grid generation; coupling schemes for heterogeneous algorithms; parallel computation; visualization; applications in fluid mechanics, solid mechanics, heat transfer and diffusion, electromagnetic fields, oceanography, hydrology, geophysics, and space physics.

Computer Engineering. Computer architecture, parallel computing, simulation, mobile agents, wireless communications, distributed systems and networks, performance analysis, computer hardware design, VLSI design.

Environmental Science and Engineering. Environmental fluid mechanics, coastal and oceanic ecosystems, pollutant transport and fate, renewable resources, physics of ice and snow, hydrology, remote sensing.

Fluid Mechanics and Thermal Sciences. Two-phase flow, thermal spraying, energy conversion, biomass conversion, fuel cells, free-convection flow, basic fluid dynamics, magnetohydrodynamics.

Material Science. Plastic deformation and fracture of metals, intermetallic compounds, and ice; protein crystal growth; development of biomaterials; wear mechanisms in metals and polymers; microelectronic materials; magnetic materials; optical thin films and surface analysis; laser applications in materials; spectroscopy of glasses.

Microengineering and Electronics. Microelectronics; power electronics; electrooptics; microoptics; microfabrication; analog, digital, and mixed circuit design; VLSI design; microelectromechanical systems.

Optics and Signal Processing. Lasers and optical devices, optoelectronics, nonlinear optics, optical signal processing, image processing, 3-D microscopy, applications of optical waveguides, sensor development.

Solid Mechanics, Mechanical Design, Controls. Thermomechanical interactions in sliding contacts, stress analysis of knee-joint prostheses, tribology, control theory, smart structures, active noise control.

Space and Ionospheric Physics. Space environment and space-plasma physics; electromagnetic, magnetofluid, and plasma processes; physics of the magnetosphere and ionosphere; development of numerical simulation models; satellite data analysis techniques.

Experimental apparatus, which employs bacteria for continual production of fuel ethanol.

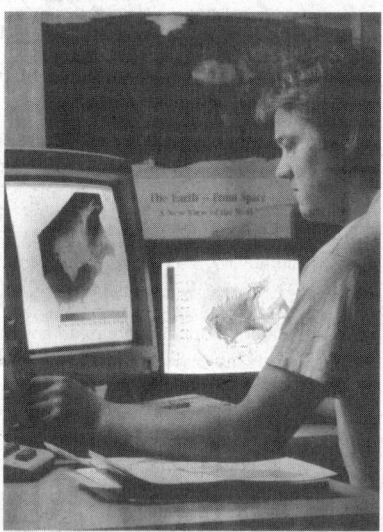

Environmental simulation is a major focus of the numerical methods laboratory. Shown here are animated computed trajectories of cod larvae in the Gulf of Maine—part of an international study of the Georges Bank ecosystem.

DREXEL UNIVERSITY

College of Engineering

Programs of Study

Proactive involvement in the global economy and NSF-recognized leadership in curriculum development are hallmarks of the College of Engineering, where chemical engineering, civil and architectural engineering, electrical and computer engineering, materials engineering, and mechanical engineering and mechanics can be studied at the Master of Science or the Doctor of Philosophy levels. The Master of Science degree is also offered in biochemical engineering, engineering geology, engineering management, software engineering, and telecommunications engineering. The Master of Engineering is available with a practice-oriented manufacturing option. Associated with the College of Engineering are two schools: the School of Biomedical Engineering, Science and Health Systems and the School of Environmental Science, Engineering, and Policy. These schools award the M.S and Ph.D. degrees.

Research Facilities

The College houses many state-of-the-art research facilities. For civil engineering, there are the Soil Mechanics, Structural Testing, Structural Models, Structural Dynamics, Construction Materials, Hydromechanics, and Environmental laboratories. The Geosynthetic Research Institute is a leading center for geosynthetic and geotechnical research. The Engineering Geology Laboratory houses more than 20,000 samples of earth materials; gravity, seismic, electrical, and magnetic geophysics equipment; and numerous geologic and topographic maps. For chemical engineering, there are the Animal Cell Bioreactor, Fermentation, Chemical Sensor, Semiconductor Processing, Multiphase Flow, Polymer Processing, and Control laboratories, as well as biochemical analytical instruments. For electrical and computer engineering, the laboratories are: Electrophysics; Power Systems; Applied Machine Intelligence and Robotics; Telecommunications; Signal Processing; Lightwave and Microwave Engineering; Computer-Based Interactive Systems (CBIS); Biomedical; Ultrasound; and Telemetry, Sensor, and Instrumentation. In addition, there is the Image Processing and Computer Vision Center, the Long Computer Center, and the Ben Franklin Superconductivity Center. In materials engineering, laboratory facilities are available for structural characterization, mechanical testing, deformation processing, spray-forming, plasma processing, powder metallurgy and particulate processing, ceramics and ceramic composites processing and testing, polymer processing, and the fabrication of advanced textile composites. The facilities for mechanical engineering and mechanics are the Biomechanics, CAD/CAM, Combustion and Fuel Chemistry, Combustion Emissions/Engine, Composite Mechanics, Heat and Mass Transfer, Fluid Mechanics, Stress Wave and Ballistics, Dynamic Systems and Controls, Robotics and Automation, Polymer Processing, and Experimental Mechanics laboratories. In 1991, the Center for Automation Technology (CAT) was opened, completing Drexel's Engineering Quadrangle. CAT houses the Rapid Product Development Center, which is a manufacturing extension service of the Delaware Valley Industrial Resource Center. LeBow Engineering Center houses the Calhoun Laboratory for Comparative Medical Science. Drexel recently established two new research centers—the Center for Telecommunications and Information Networking and the Center for Advanced Biomaterials and Tissue Engineering.

Financial Aid

In addition to the Federal Stafford Student and Federal Perkins Loan programs and work-study programs, aid is available in the form of fellowships and traineeships, research and/or teaching assistantships, cooperative education employment, and scholarships. Fellowships and traineeships enable the student to carry a full course load; work assignments are not required. Fellowships provide $15,000 in annual stipends and full graduate tuition of approximately $12,000 per year. The fellowships are awarded on a competitive basis to qualified Ph.D. applicants with M.S. degrees in engineering. The appointment carries up to 20 hours per week of teaching assignments determined by the department in which the student is pursuing his or her Ph.D. Assistantships enable a student to carry a full-time course load. Recipients work 20 hours per week assisting the teaching or research efforts of the College. Some appointments provide for a combination of teaching and research activities. In 1998–99, the College of Engineering provided funding for 80 full-time-equivalent teaching assistants. Appointments of research assistants vary by department. Stipends for these positions have an average range of $10,000 to $18,000 per year. Through Drexel's cooperative education program—the Career Integrated Education (C.I.E.) option—graduate students in certain engineering curricula can alternate periods of full-time study with periods of full-time employment in corporate, industrial, or governmental laboratories. Students earn competitive salaries during co-op work periods. The College awards scholarship funds based on financial need and/or academic achievement.

Cost of Study

Tuition is $585 per credit hour in 1999–2000. The general University fee is $125 per term for full-time students and $67 per term for part-time students.

Living and Housing Costs

Accommodations for single students are available in University residence halls. Ample housing is also available in the neighborhood bordering the campus. For the nine-month academic year, transportation and living expenses for a single student are estimated at $11,450.

Student Group

The College of Engineering enrolls 3,030 students; approximately 670 are involved in graduate study.

Location

Drexel is located in the University City section of Philadelphia, an area that consists of residential neighborhoods, two major universities, and several medical centers and research institutes. The campus is only minutes from downtown Philadelphia by public transportation or on foot, allowing students easy access to the city's museums, historic sites, and other attractions.

The University and The College

As a private university, Drexel builds on a century of experience in education. Drexel provides its graduate students the opportunity to study in a stimulating academic atmosphere promoted by both the University's long history and tradition of high-quality education and its firm commitment to excellence. The University's Career Services Center offers job placement services to all students. Assistance is provided in securing career information, writing resumes, interviewing, and planning job strategies.

Applying

Drexel operates on the quarter system, with terms beginning in September, January, March, and June. Candidates for full-time admission who are also applying for an assistantship must submit completed applications by February 1. Admission applications, with supporting transcripts, references, and requests for financial aid, should be submitted to the Office of Graduate Admissions.

Correspondence and Information

Office of Graduate Admissions, Box P
Drexel University
Philadelphia, Pennsylvania 19104
Telephone: 215-895-6700
E-mail: admissions-grad@post.drexel.edu
World Wide Web: http://www.coe.drexel.edu/coe.home/COE.home.html

Drexel University

THE FACULTY

The College of Engineering has 105 full-time faculty members in five major departments. College administrators, department heads, and program directors are listed below.

Raj Mutharasan, Ph.D., Frank A. Fletcher Professor of Chemical Engineering and Interim Dean.

Department of Chemical Engineering
Charles B. Weinberger, Ph.D., Professor and Department Head.

Department of Civil and Architectural Engineering
Joseph Martin, Ph.D., Associate Professor and Department Head.

Department of Electrical and Computer Engineering
Nihat M. Bilgutay, Ph.D., Professor and Department Head.

Department of Materials Engineering
Roger D. Doherty, Professor and Department Head.

Department of Mechanical Engineering and Mechanics
Nicholas P. Cernansky, Ph.D., Hess Chair Professor and Department Head.

Biochemical Engineering Program
Raj Mutharasan, Ph.D., Professor and Director.

Engineering Geology Program
Edward Doheny, Ph.D., Professor and Director.

Engineering Management Program
Steven Smith, Ph.D., Acting Director.

Geosynthetic Engineering Program
Robert Koerner, Ph.D., Professor and Director.

Telecommunications Engineering/Software Engineering
P. M. Shankar, Ph.D., Professor and Director.

School of Biomedical Engineering, Science and Health Systems
Banu Onaral, Ph.D., Professor and Director.

School of Environmental Science, Engineering, and Policy
Michael Gealt, Ph.D., Professor and Director.

RESEARCH CONCENTRATIONS

Chemical Engineering: Process dynamics and control, process design, semiconductor processing, biochemical engineering, tissue engineering, simulation and process modeling, biomaterials, polymer dynamics, computational fluid mechanics, drying.

Civil Engineering: Environmental engineering, geosynthetic engineering, water and wastewater treatment, highway engineering, building envelope studies, coastal engineering, hydrology, structural engineering, engineering geology, hazardous-waste engineering, construction materials.

Electrical and Computer Engineering: Electrophysics, microwave-lightwave engineering, ultrasonics and ultrasound, signal processing, communications, telecommunications engineering, networks, controls, circuits, electromagnetic fields, image processing, computer vision, power systems, artificial intelligence, optics, superconductivity, mobile communications.

Engineering Geology: Engineering geology, environmental geology, groundwater hydrology, hydrogeology, surface water hydrology, clay mineralogy, investigative geophysics.

Materials Engineering: Structure and properties of ceramics, metals, and polymers; composite materials; advanced textile composites; biomaterials; colloidal processing of ceramics; powder metallurgy; structural transformation in metallic alloys; plasma and spray-forming of materials; processing and properties of ultra-refractory composites and ceramics, smart materials, conversion of utility wastes into useful materials.

Mechanical Engineering and Mechanics: Structural dynamics, biomechanics, dynamics systems and controls, CAD/CAM, thermal sciences, nuclear engineering, fluid mechanics, combustion and fuels chemistry, manufacturing, robotics, thermodynamics, aerodynamics, robotics, smart materials and structures, composite materials, computational mechanics, solid mechanics.

School of Biomedical Engineering, Science and Health Systems: Biomaterials/biotechnology, biomechanics, biomedical signal processing, biosensors, bioelectrodes and biotelemetry, biophysics, biostatistics, cardiovascular dynamics and instrumentation, computer applications to health care, medical imaging and image processing, medical ultrasound, neural networks and systems, sensory systems, clinical engineering, rehabilitation engineering.

School of Environmental Science, Engineering, and Policy: Air pollution control, aquatic ecology, atmospheric chemistry, atmospheric science, biotechnology and bioremediation, biological unit operations, groundwater hydrology, hazardous waste treatment, microbial ecology, physical and chemical unit operations, physiological ecology, risk assessment, sludge treatment and disposal, unsaturated zone hydrology, water chemistry, water microbiology, water and wastewater treatment.

FLORIDA AGRICULTURAL AND MECHANICAL UNIVERSITY / FLORIDA STATE UNIVERSITY

College of Engineering

Programs of Study

The FAMU-FSU College of Engineering is a cooperative effort between two of Florida's more prominent state schools: Florida A&M University and Florida State University. Students choose one of the universities to enroll through but attend classes with students from both schools. The College offers Master of Science and Doctor of Philosophy programs in chemical, civil, electrical, industrial, and mechanical engineering. Current research consists of experimental and theoretical studies in the areas of fluid mechanics, aerodynamics, and heat transfer; computational mechanics, computer-aided design, and manufacturing and robotics; electromagnetics and communications, electronics (including microelectronic device design and fabrication), digital signal processing and control systems, and computer engineering; and polymer processing, biochemical engineering, materials research, semiconductor processing, macromolecule dynamics, molecular transport phenomena, biomedical devices, expert systems, thermodynamics, crystal growth, optimization and control of reactors, and air pollution control. Other areas of emphasis are bridge design, structural stability, and structural reliability; geotextiles, pavements, and soil dynamics; and transportation networks and multimodal systems, storm water, hazardous-waste and solid-waste management, water resource management, computer-aided design and planning, optimization, simulation, manufacturing process and system design, CAD/CAM, artificial intelligence in engineering, set-covering theory, predictive maintenance, simulation environment, and health-care delivery.

Completion of the M.S. degree requires a minimum of 30–33 semester credit hours of graduate work depending upon the degree type the student chooses: thesis or nonthesis. Up to 6 semester credit hours may be transferred from another accredited university. The Ph.D. degree program requires the completion of 45–54 semester credit hours beyond the M.S. degree. A minimum of 24 dissertation hours, passing a preliminary and/or qualifying examination, and successful dissertation defense are required for completion of the doctoral degree.

Research Facilities

Students in the FAMU-FSU College of Engineering have access to a large number and variety of computer systems. Course work and research are supported by a network of more than 400 computing devices in the College of Engineering building. More than 100 of these machines are managed by the College of Engineering Computing Services and are available for general student use. The remainder of the machines reside in departmental labs, research labs, faculty offices, and staff offices. Computing Services maintains several computer labs used to support undergraduate and graduate courses. These labs contain Sun color X-terminals as well as 486-based color PCs and are open more than 100 hours per week. All of the building's computers are connected to the College of Engineering LAN and to the Internet. They are supported by a cluster of Sun servers with about 60 Gbytes of general-use disk space. Network connectivity is available to machines and systems of the nearby Supercomputer Computations Research Institute (SCRI) and the FSU Academic Computing and Network Services (ACNS) department. SCRI has installed a Connection Machine, a CRAY YMP, and a cluster of IBM RS6000 and DEC Alpha workstations for supercomputer research. ACNS maintains a network of Sun Workstations and DEC equipment with access through more than 200 public terminals located throughout the Florida State University campus. In addition, ACNS provides 300 dial-up lines that are available to FAMU/FSU College of Engineering students at no charge.

Research laboratories include the Advanced Mechanics and Materials Laboratory, the Biomagnetic Fluid Dynamics Laboratory, Computational Fluid Dynamics at SCRI, the Electromagnetics Research Lab, the Finite Analytic Numerical Simulations Laboratory, the Fluid Mechanics Research Laboratory, the High-performance Computing and Simulation Research Laboratory, and polymer characterization, NMR imaging, and aerosol transport laboratories.

Financial Aid

Teaching and research assistantships are available on a competitive basis. Assistantship stipends range from $6000 to $14,000 for part-time duties for the nine-month academic year. In addition, tuition waivers are awarded on a competitive basis to qualified students. Students interested in departmental support should contact the chairperson of their major.

Cost of Study

Tuition in 1999–2000 is $139 per semester hour for Florida residents, and out-of-state students pay $483 per semester hour. Tuition costs are subject to change. The normal course load is 9 semester hours per semester for full-time students.

Living and Housing Costs

An estimate of the cost of attendance is $15,969 for the academic year (2 semesters) for in-state graduate students living in a University residence and participating in the University food plan. For nonresident graduate students, the cost of attendance is approximately $20,107. Rental rates for off-campus housing start at about $450 per month for a one-bedroom apartment.

Student Group

Current enrollment in the College exceeds 2,000 students. FAMU currently has more than 9,500 students, and enrollment at FSU exceeds 32,000. Total graduate enrollment at the FAMU-FSU College of Engineering is approximately 200. There are eleven professional and student groups at the College of Engineering.

Location

Florida A&M University and Florida State University are located in Tallahassee, the state capital. Although Tallahassee is among the state's fastest-growing cities, its natural beauty has been preserved. Five large lakes surrounding Tallahassee, as well as the nearby Gulf of Mexico, offer numerous recreational opportunities, including canoeing, fishing, waterskiing, boating, camping, hunting, and cycling. Among Florida cities, Tallahassee is distinguished by its gently rolling hills and roads canopied by majestic oaks.

The Universities

Florida A&M University and Florida State University are public coeducational institutions founded in 1887 and 1857, respectively. The Universities have great diversity in their cultural offerings and are rich in tradition. They have outstanding science departments and excellent schools and departments in such varied areas as architecture, business, education, law, music, religion, and theater. They are the homes of the renowned Rattlers and Seminoles.

Applying

Applications should be submitted as early as possible in the academic year prior to anticipated enrollment. The deadline for applications for the fall semester is July 1. International students should apply by March 1 because of visa procedures. Students seeking departmental assistantships are encouraged to apply at least one year in advance. Master's degree candidates must have a bachelor's degree in engineering or an allied field, have a minimum GPA of 3.0, and a minimum score of 1000 on the GRE verbal and quantitative sections. A score of 550 on the TOEFL exam is required of international applicants.

Correspondence and Information

For FSU admission application materials and catalog:

Graduate Admission Office
A2500 University Center
Florida State University
Tallahassee, Florida 32306
Telephone: 850-644-3420

For FAMU admission application materials and catalog:

Graduate Admission Office
G-9 Foote-Hilyer Administration Building
Florida A&M University
Tallahassee, Florida 32307
Telephone: 850-599-3796

For specific graduate program materials and College information:

Graduate Studies Coordinator
FAMU-FSU College of Engineering
2525 Pottsdamer Street
Tallahassee, Florida 32310-6046

Telephone: 850-487-6318
E-mail: gradstudy@eng.fsu.edu
World Wide Web: http://www.eng.
fsu.edu/research/gradstud.html

Florida Agricultural and Mechanical University/Florida State University

THE FACULTY AND THEIR RESEARCH

Department of Chemical Engineering
Rufina Alamo, Associate Professor; Ph.D., Complutense (Spain). Polymer science and engineering, crystallization, morphology, structure-properties relations. Pedro Arce, Associate Professor and Associate Faculty Member of MARTECH; Ph.D., Purdue. Molecular transport mechanics. Ravindran Chella, Associate Professor; Ph.D., Massachusetts. Polymer blends and composites. Stephen Gibbs, Assistant Professor and Faculty Member at NHMFL Center for Interdisciplinary Magnetic Resonance; Ph.D., Wisconsin. Applications of nuclear magnetic resonance. Eric Kalu, Assistant Professor; Ph.D., Texas A&M. Electrochemical engineering. Bruce R. Locke, Associate Professor; Ph.D., North Carolina State. Biochemical separations, biocatalysis, pollution control. Michael H. Peters, Professor and Chair; Ph.D., Ohio State. Macromolecular phenomena, brownian motion, aerosol science. Srinivas Palanki, Assistant Professor; Ph.D., Michigan. Nonlinear process control, process optimization, bioprocess engineering. John C. Telotte, Associate Professor; Ph.D., Florida. Biochemical processing, semiconductor processing. Jorge Viñals, Associate Professor and SCRI Research Scientist; Ph.D., Barcelona. Fluid mechanics. Dale Wesson, Assistant Professor; Ph.D., Michigan State. Liquid/liquid dispersions, hydrocyclones.

Department of Civil Engineering
Makola Abdullah, Assistant Professor; Ph.D., Northwestern. Structural dynamics, active control, earthquake and wind engineering. Andrew A. Dzurik, Professor; Ph.D., Cornell; PE. Water resources planning and management, environmental systems analysis. Gary Foose, Assistant Professor; Ph.D., Wisconsin. Geotechnical engineering, geotextiles. Millard W. Hall, Professor; Ph.D., Illinois. Water quality, water resources policy, environmental engineering. Wenrui Huang, Assistant Professor; Ph.D., Rhode Island. Modeling natural flow systems and pollution transport, coastal engineering analysis, time-series data analysis. Danuta Leszczynska, Associate Professor; Ph.D., Wroclaw Technical (Poland). Environmental engineering. Primus S. Mtenga, Assistant Professor; Ph.D., Wisconsin. Structural systems performance and reliability, wood structures. Renatus Mussa, Assistant Professor; Ph.D., Arizona State. Incident detection, traffic operations and control, highway safety and remedial measures, intelligent transportation systems. Sorronnadi Nnaji, Professor; Ph.D., Arizona. Water resources systems, hydrology, hydraulics, computer applications. Wei-Chou Virgil Ping, Associate Professor; Ph.D., Texas at Austin. Transportation engineering. John Sobanjo, Assistant Professor; Ph.D., Texas A&M. Transportation engineering. Lisa Spainhour, Assistant Professor; Ph.D., North Carolina State. Structural engineering. Kamal Tawfiq, Associate Professor; Ph.D., Maryland. Geotechnical engineering, soil dynamics, geotextiles. Jerzy Wekezer, Professor and Chair; Ph.D., Gdansk (Poland). Solid mechanics engineering. Nur Yazdani, Professor; Ph.D., Maryland; PE. Structural engineering, structural reliability, prestressed concrete.

Department of Electrical Engineering
Susan Allen, Professor and Vice President of Research; Ph.D., USC. Micromachining electronic and optical devices for chemical vapor deposition. Krishna Arora, Associate Professor; Ph.D., Indian Institute of Technology (Delhi). Optical interferometric systems, optical communications, underwater imaging, optical radiation effects in integrated circuits. Rajendra Arora, Professor; Ph.D., St. Andrews (Scotland). Electromagnetics (including microwaves and antennas), optical engineering and optoelectronics, high-temperature superconductivity. Juan Cockburn, Assistant Professor; Ph.D., Minnesota. Robust control theory and applications, computer-aided control systems design, signal and image processing. Steven M. Durbin, Assistant Professor; Ph.D., Purdue. Numerical simulation of semiconductor devices, solar cells, optoelectronics. Simon Foo, Associate Professor; Ph.D., South Carolina. Neural networks, CMOS digital and analog IC design, meta-stable and chaotic circuits. Frank B. Gross III, Associate Professor; Ph.D., Ohio State. SONAR, antennas, electromagnetic fields, RADAR, signal processing. Madhu S. Gupta, Professor and Chair; Ph.D., Michigan. Theory and design of microwave devices and circuits, RF multichip modules, fluctuation phenomenon, communication electronics. Thomas J. Harrison, Professor; Ph.D., Stanford. Personal computing in electrical engineering education, computer control systems and instrumentation, small computer architecture. Bruce A. Harvey, Associate Professor; Ph.D., Georgia Tech. Telecommunications, error-control coding, wireless communication, intelligent transportation systems. Bing W. Kwan, Associate Professor; Ph.D., Ohio State. Transient electromagnetic phenomenon, intelligent transportation systems and instrumentation, small computer architecture. Reginald Perry, Associate Professor and Associate Chair; Ph.D., Georgia Tech. CMOS optoelectric design, radiation effects of CMOS devices, ASIC system design methodologies. Rodney G. Roberts, Assistant Professor; Ph.D., Purdue. Robotics. Fred O. Simons Jr., Professor; Ph.D., Florida. Digital signal processing, optimized algorithms (DSP applications), signal and system theory, control systems. Norman Thagard, Professor and Director of College Relations; M.D., Texas Southwestern Medical Center at Dallas. Analog and digital electronic design, engineering science. Leonard Tung, Associate Professor; Ph.D., Texas Tech. Circuits and systems, automatic control, lightning protection, intelligent transportation systems. Jim Zheng, Professor; Ph.D., SUNY at Buffalo. Optoelectronic and energy storage devices, solid-state thin fluid.

Department of Industrial Engineering
Samuel A. Awoniyi, Associate Professor and Interim Dean; Ph.D., Cornell. Applied optimization in homotopy and related algorithms. Robert N. Braswell, Professor; Ph.D., Oklahoma State; PE. Mathematical methods in tolerancing, project analysis, supercomputing for engineering. Pi-Eih Lin, Courtesy Professor. Zoreh Moshir, Instructor; Ph.D., North Carolina. Engineering graphics, statistics. Okenwa Okoi, Visiting Assistant Professor; Ph.D., Warwick (England). Fiber-reinforced polymer composites and manufacturing. Yaw A. Owusu, Associate Professor; Ph.D., Penn State. Manufacturing processes and materials, waste materials processing. Joseph J. Pignatiello Jr., Associate Professor; Ph.D., Ohio. Quality control and improvement, robust design, simulation and engineering statistics. Brenda Rayco, Visiting Assistant Professor; Ph.D., Florida. Aggregation for location models, location theory, and optimization. H.-P. (Ben) Wang, Professor and Chair; Ph.D., Penn State. Integrated product/process design. C. (Chuck) Zhang, Assistant Professor; Ph.D., Iowa. Computer-integrated manufacturing.

Department of Mechanical Engineering
Farrukh Alvi, Visiting Assistant Professor; Ph.D., Penn State. Fluid dynamics, gasdynamics, optical fluid diagnostics. George Buzyna, Professor; Ph.D., Yale. Physical gasdynamics, geophysical fluid dynamics: convection in rotating stratified fluids and transition to turbulence. Namas Chandra, Professor; Ph.D., Texas A&M; PE. Finite-element methods, solid mechanics, superplastic metal forming, composite materials. C.-J. Chen, Professor and Dean of the College; Ph.D., Case Western Reserve. Heat transfer, bioengineering, and computational fluid mechanics. Emmanuel Collins, Associate Professor; Ph.D., Purdue. Controls and dynamics, numerical methods. Frederick Foreman, Assistant Professor; Ph.D., Florida A&M. Computational structure mechanics, design. Hamid Garmestani, Associate Professor; Ph.D., Cornell. Structural mechanics, materials science, composite materials. Peter Gielisse, Professor; Ph.D., Ohio State. Materials science, high-temperature composite materials, composite materials. Yousef Haif, Visiting Assistant Professor; Ph.D., Florida State. Computational fluid, heat transfer, magnetic drug delivery for cancer treatment. Patrick Hollis, Associate Professor; Ph.D., Cornell. Controls, nonlinear dynamical systems, robotics. Peter Kalu, Assistant Professor; Ph.D., London. Materials science, composites, deformation, recrystallization. Anjaneyulu Krothapalli, Professor and Chair; Ph.D., Stanford. Experimental fluid mechanics, aeroacoustics, aerodynamics. Luiz M. Lourenco, Professor; Ph.D., Brussels. Two-phase flows, laser diagnostics, instrumentation and heat transfer. Arthur Mutambara, Assistant Professor; Ph.D., Oxford. Robotics and controls. Simone Peterson, Assistant Professor; Ph.D., MIT. Ceramics and superconducting materials. Justin Schwartz, Associate Professor; Ph.D., MIT. Superconducting magnet engineering. John Seely, Adjunct Professor; M.S., Stevens. Heat transfer and microelectronics equipment. Chiang Shih, Associate Professor; Ph.D., USC. Unsteady aerodynamics, turbulent shear flows, laser diagnostics. Leon Van Dommelen, Professor; Ph.D., Cornell. Theoretical and computational fluid mechanics, computational mechanics. Steven Van Sciver, Professor; Ph.D., Washington (Seattle). Cryogenics and heat transfer.

GEORGE MASON UNIVERSITY

School of Information Technology and Engineering

Programs of Study

The School of Information Technology and Engineering awards the M.S. in computer science, electrical engineering, information systems, operations research and management science, software systems engineering, statistical sciences, systems engineering, and urban systems engineering. The Ph.D. is offered in information technology, as an integration and synthesis of the eight master's-level programs. A wide variety of specialties are available, including database management, human-computer interaction, cognitive science, knowledge engineering, decision support systems, computer architecture, controls, communications, robotics, software systems engineering, systems integration, command and control, computational analysis, parallel computing, distributed architectures, large-scale networks, systems management, total quality management and reengineering, transportation systems, operations research, applied statistics, and artificial intelligence. Programs in the School generally emphasize information and computing as major drivers of technological and organizational progress.

All M.S. degrees require 30 semester hours of graduate credit, often including a thesis or project. The Ph.D. requires at least 72 semester hours of credit beyond the bachelor's degree, including a dissertation of 24 semester hours. All programs require completion of a student-originated, approved plan of study.

Research Facilities

The computer and the Internet are essential tools for research in information technology and engineering. The School supports a network of four SPARC 1000 servers that provide computational support of student laboratories, student accounts, and Schoolwide network services. They serve as the backbone connecting the specialized facilities of a large number of research centers in the School, including high-performance computing in engineering education, computational probability and statistics, parallel computing, image analysis, robotics and automation, machine learning and inference, software systems engineering, secure computing, transportation systems, command control, and communication information systems.

Financial Aid

A variety of financial aid in the form of fellowships, teaching assistantships, research assistantships, and specially arranged cooperative work-study arrangements are available. These are awarded on a competitive basis. Stipends vary; in 1998–99, average teaching assistantship compensation was about $8500, plus an out-of-state tuition waiver for M.S. students and a full tuition waiver for doctoral students for half-time service during the nine-month academic year. Students on a half-time appointment typically take 9 semester hours of course work each term. Research assistants may register for additional thesis and dissertation credits. Summer appointments on a full- or part-time basis are available.

Cost of Study

Tuition for full-time study in 1998–99 was $4344 for Virginia residents and $12,504 for nonresidents. Part-time tuition and fees per credit hour were $181 for Virginia residents and $521 for nonresidents.

Living and Housing Costs

Living costs in northern Virginia are comparable to those in most large metropolitan areas; the benefits are generally much greater.

Student Group

In 1998–99, many of the School's 1,400 master's and 300 doctoral-level students were employed in a variety of information technology, computer science, and engineering professional activities in the greater Washington metropolitan area. The graduate student group is cosmopolitan and mature, and it contributes significantly to the intellectual environment at George Mason University.

Location

George Mason is located 15 miles west of the nation's capital on 677 acres in suburban Fairfax, Virginia. The area has a substantial number of private industries and federal research institutes, and participation in joint research and educational activities is much encouraged. This metropolitan region is the home of approximately 200,000 engineers and scientists, one of the largest such concentrations of technical talent in the nation. The area is culturally rich and offers many fine arts and performing arts opportunities. The four-season climate encourages outdoor activities that vary from swimming and mountain climbing in the summer to skiing in the winter. A wide range of recreational opportunities are within easy driving distance.

The University

George Mason University was founded more than thirty-five years ago. Student enrollment has more than quadrupled since 1972. The current enrollment is slightly more than 24,000 students; approximately one third of these are graduate and professional students. The University has six academic and professional schools, with approximately 700 full-time faculty members: Arts and Sciences, Management, Information Technology and Engineering, Law, Nursing, and Professional Studies. The development of the University has been shaped in response to the educational needs of its extraordinarily cosmopolitan constituency. One such development was the establishment in 1985 of the School of Information Technology and Engineering.

Applying

Students desiring admission to graduate study in information technology and engineering should have an undergraduate degree to enroll in the Master of Science degree program and a master's degree to enroll in the doctoral program, with background in an appropriate area of study: engineering, computer science, mathematics, or quantitatively oriented physical, social, and management sciences. A minimum undergraduate grade point average of 3.0 is desired, as is a composite score of 1200 on the verbal and quantitative sections of the General Test of the Graduate Record Examinations. A minimum score of 575 on the TOEFL is required from international students whose native language is not English. Applications should be submitted early, generally at least two months before the intended enrollment date. Further information, including application materials, may be obtained from the address below.

Correspondence and Information

Graduate Studies
School of Information Technology and Engineering
Mail Stop 3D5
George Mason University
Fairfax, Virginia 22030-4444
Telephone: 703-993-1512
Fax: 703-993-1633
E-mail: sitegrad@gmu.edu

George Mason University

THE FACULTY AND THEIR RESEARCH

Applied and Engineering Statistics
K. Bell, Ph.D., George Mason: statistical signal processing, statistical performance bounds. R. Bolstein, Ph.D., Purdue: sample surveys, mathematical analysis, deterministic operations research. D. Carr, Ph.D. Wisconsin–Madison: statistical graphics, data analysis management. D. Gantz, Ph.D., Rochester: computer performance evaluation, capacity assessment. J. Gentle, Ph.D., Texas A&M: Monte Carlo methods, random number generation, simulation, computational statistics. M. Habib, Ph.D., North Carolina: inference for stochastic processes, statistical information theory. J. Miller, Ph.D., Stanford: linear models, advanced statistical technology. C. Sutton, Ph.D., Stanford: geometric probability, gambling theory, probabilistic modeling. E. Wegman, Ph.D., Iowa: mathematical statistics, time-series analysis, computational statistics.

Civil, Environmental, and Infrastructure Engineering
T. Arciszewski, Ph.D., Warsaw University of Technology: engineering design, network computing. S. deMonsabert, Ph.D., Purdue: statistical modeling and analysis, environmental engineering, energy conservation, environmental protection. M. Houck, Ph.D., Johns Hopkins: water resource management, environmental engineering, systems analysis. D. Rathbone, Ph.D., Texas A&M: dynamic capacity, transportation planning, transportation infrastructure. T. Ryan, Ph.D., Illinois: construction management, decision support systems for construction and transportation.

Computer Science
R. Carver, Ph.D., North Carolina: software testing, concurrent systems. J. Chen, Ph.D., Central Florida: physically based modeling, real-time simulation, distributed interactive simulation, scientific visualization. K. DeJong, Ph.D., Michigan: adaptive systems design, artificial intelligence. P. Denning, Ph.D., MIT: concurrent systems, computer networks, system architecture. Z. Duric, Ph.D., Maryland: computer vision, video image processing, human-computer interaction. H. Hamburger, Ph.D., Michigan: natural-language processing. Y. Huang, Ph.D., Michigan State: Computer networking, distributed systems. T. Maney, D.A., George Mason: Intelligent tutoring, interactive multimedia. A. Marchant, Ph.D., Berkeley: history of computing, computer ethics. D. Menasce, Ph.D., UCLA: parallel computing, stochastic petri nets, heterogeneous supercomputing. R. Michalski, Ph.D., Silesia (Poland): machine learning, inference, data mining, intelligent systems cognition. D. Nordstrom, Ph.D., California: Multilevel secure systems. E. Norris, Ph.D., Florida: programming support environments, data flow architectures. M. Pullen, Ph.D., George Washington: distributed, parallel computing systems. D. Quammen, Ph.D., Pittsburgh: concurrent programming techniques. D. Richards, Ph.D., Illinois: comparisons of protein sequences, steimer tree algorithms, information dissemination in networks, parallel heuristics. D. Rine, Ph.D., Iowa: software engineering, software reuse, maintenance and object-oriented design, software metrics and formal methods. S. Setia, Ph.D., Maryland: parallel processing, performance evaluation, fault-tolerant computing. R. Simon, Ph.D., Pittsburgh: distributed multimedia systems, networks, computer-supported cooperative work. A. Sood, Ph.D., Carnegie Mellon: parallel, distributed processing, simulation, modeling, computer vision software. G. Tecuci, Ph.D., Paris: machine learning, knowledge acquisition, intelligent agents, artificial intelligence. P. Wang, Ph.D., Wisconsin–Milwaukee: applications of computer algorithms, expert systems. H. Wechsler, Ph.D., California, Irvine: vision and image analysis, knowledge-based information systems. E. White, Ph.D., Maryland: software architecture, control flow analysis, software verification, distributed computing.

Electrical and Computer Engineering
R. Athale, Ph.D., California, San Diego: optical computing and communications. A. Baraniecki, Ph.D., Windsor: digital signal processing, communications. G. Beale, Ph.D., Virginia: intelligent systems, control systems, digital simulation. A. Berry, Ph.D., Missouri–Columbia: semiconductors, electronic materials and devices. M. Black, Ph.D., Penn State: microwave processing of materials. P. Ceperly, Ph.D., Stanford: solar energy, heat transfer processes, electromagnetics. S. Chang, Ph.D., Hawaii: error control coding, networks, digital communications, information theory. G. Cook, Ph.D., MIT: robotics and automation, control systems engineering. Y. Ephraim, D.Sc., Technion (Israel): statistical signal processing, pattern recognition, source coding. K. Gaj, Ph.D., Warsaw University of Technology: Communication systems and networks, computer-network security, VLSI design and testing. J. Gertler, D.Sc., Hungarian Academy of Sciences: fault detection, systems control, estimation. L. Griffiths, Ph.D., Stanford: Space-time adaptive processing. M. Haney, Ph.D., Caltech: signal processing, communications. K. Hintz, Ph.D., Virginia: self-organizing machines, statistical pattern recognition, computer-based control. D. Ioannou, Ph.D., Manchester: solid-state devices, optical communications. B. Jabbari, Ph.D., Stanford: computer networks, digital communication, telecommunications. A. Levis, Sc.D., MIT: command and control, petri nets, organizational modeling. A. Manitius, Ph.D., Warsaw: math aspects of control theory, optimal control, numerical methods, computer simulation. R. Mulpuri, Ph.D., Michigan: electronic and microwave devices. P. Pachowicz, Ph.D., Stanislaw Staszic (Poland): intelligent autonomous systems and robotics, computer vision. P. Paris, Ph.D., Rice: communication systems, neural sciences, statistical signal processing. W. Sutton, Ph.D., Air Force Tech: semiconductor device physics, VLSI design. D. Tabak, Ph.D., Illinois at Urbana-Champaign: computer architecture, microcomputers, distributed systems. H. Van Trees, Sc.D., MIT: communication theory, command, control.

Information and Software Systems Engineering
P. Ammann, Ph.D., Virginia: software systems engineering, software testing, fault tolerance. D. Barbara, Ph.D., Princeton: data warehouses, data characterization, sequence searching. R. Baum, Ph.D., Michigan: Ada, programming languages, optimal control theory. P. Bose, Ph.D., USC: software architecture, formal specification and collaborative software design environments. A. Brodsky, Ph.D., Hebrew University: database and knowledge-based systems, constraints, geographic information systems. H. Gomaa, Ph.D., London: real-time systems, software design methods. S. Jajodia, Ph.D., Oregon: database management systems, software systems, parallel computing. L. Kerschberg, Ph.D., Case Western Reserve: expert database systems, information systems. A. Motro, Ph.D., Pennsylvania: relational databases, query systems. J. Offutt, Ph.D., Georgia Tech: automatic test data generation, software quality. A. Sage, Ph.D., Purdue: software process economics and quality assurance, software process management. R. Sandhu, Ph.D., Rutgers: information systems security, database management systems. E. Sibley, Sc.D., MIT: data dictionary systems, intelligent systems. X. Wang, Ph.D., USC: database query languages, theory and temporal databases.

Systems Engineering and Operations Research
L. Adelman, Ph.D., Colorado: decision support systems, human-computer interface, command and control. P. Brouse, Ph.D., George Mason: requirements engineering, multimedia technology, process improvement, decision support systems. D. Buede, Ph.D., Stanford: systems engineering methods, models, data fusion, decision analysis. K. C. Chang, Ph.D., Connecticut: distributed sensor networks, probabilistic data association. T. Friesz, Ph.D., Johns Hopkins: network analysis, optimization, spatial economics. I. Greenberg, Sc.D., NYU: statistical inference, stochastic systems. D. Gross, Ph.D., Cornell: applied probability, queuing theory, simulation. C. Harris, Ph.D., Polytechnic of Brooklyn: applied probability, queuing, simulation. K. Hoffman, D.Sc., George Washington: mathematical programming and modeling, computer analysis. K. Laskey, Ph.D., Carnegie Mellon: decision theory, information fusion, knowledge representation. D. Miller, Ph.D., Cornell: software reliability, queuing networks, telecommunications. S. Nash, Ph.D., Stanford: numerical analysis, mathematical programming. R. Polyak, Ph.D., USSR Academy of Sciences: linear and nonlinear programming, fixed points, mathematical economics. A. Sage, Ph.D., Purdue: decision support systems, systems design methodology, systems management. D. Schum, Ph.D., Ohio State: decision and inference analysis, human information processing. A. Sofer, D.Sc., George Washington: linear and nonlinear optimization, reliability. B. White, Ph.D., Virginia: systems engineering, decision analysis, risk assessment, transportation engineering.

THE GEORGE WASHINGTON UNIVERSITY

School of Engineering and Applied Science

Programs of Study

The School of Engineering and Applied Science (SEAS) offers the graduate degrees of Master of Science, Master of Engineering Management, Master of Science in computational science, Applied Scientist, Engineer, and Doctor of Science. A graduate-level certificate program is also offered. The seven fields of study are civil and environmental engineering, computer science, electrical engineering, engineering management, mechanical engineering, operations research, and telecommunications and computers. Interdisciplinary study is encouraged, especially at the doctoral level. Some interdisciplinary and cross-disciplinary programs are formalized at the master's level; for example, in electrical power and engineering management and in industrial engineering and engineering statistics. Within most fields, students may design their degree programs to pursue their own professional goals and academic interests, following curricular guidelines of the SEAS and in consultation with the academic adviser. The minimum master's program of 24 credit hours plus a 6-credit-hour thesis, or 30 or 33 credit hours without a thesis, culminates in some departments with the master's comprehensive examination. The professional degree programs (App.Sc. and Engr.) require a minimum of 30 hours of courses beyond the master's degree. A technical project may be required. Course work done for the professional degree may be transferred to the D.Sc., and vice versa, under some conditions. The D.Sc. program requires 30 hours of course work beyond the master's degree or 54 hours beyond the bachelor's degree, followed by a doctoral qualifying examination. The dissertation, requiring a minimum of 24 credit hours of work under the guidance of a thesis adviser selected by the student, is presented orally in the final examination. SEAS dissertations have had application in industry—such as an evaluation model of an oil spill contingency plan or an analysis of red blood cell flow; in government—for example, in projecting risks associated with nuclear reactors; and in higher education—through refined theoretical formulations of traditional problems.

Research Facilities

The University's three libraries hold 1.7 million volumes and more than 14,129 serials. Students have access to the Library of Congress and may consult GW's computerized catalog of holdings, those of six other local university libraries, and selected periodical indexes on line. SEAS students may use twenty-five laboratories that are well equipped as facilities for course work, experimentation, and research. Laboratories focus, for example, on artificial intelligence, biomedical systems, communications, computer-aided design, computer-aided manufacturing, decision support systems, digital electronics, energy technology, fiber optics and lasers, fluid mechanics, materials science, propulsion, robotics, soil mechanics, and VLSI design and testing. Research institutes organized by the faculty as sites for advancing their research are described on the reverse page. The Washington, D.C., area has the second-largest concentration of research and development activity in the United States, and many national laboratories are available to SEAS students for research.

Financial Aid

Teaching assistantships provide tuition for full-time graduate study and a salary of $1000 to $3500 for each section taught or supervised per semester in 1999–2000. Research assistants receive $8000 to $16,000 salaries for both academic and calendar year appointments. School Graduate Fellow, Dean's Fellow, and Department Fellow awards range from $7500 to $15,000 for eligible full-time students. Department Fellow awardees will also receive 18 semester hours tuition credit. Full-time students who are U.S. citizens or permanent residents may be eligible for half-tuition Graduate Engineering Honors Fellowships.

Cost of Study

In 1999–2000, tuition is charged at the rate of $701.50 per semester hour, calculated on a course-by-course basis. Mandatory University fees are $34.50 per semester hour to a maximum of $517.50, and books and supplies cost $600 to $1100, per year of full-time study.

Living and Housing Costs

Apartments for students attending George Washington University are available in the area at a range starting at about $600 a month. A fair estimate of a single student's living expenses for a full twelve-month year is $920 a month, including rent, food, local transportation, telephone, clothing, entertainment, and miscellaneous costs. GW's Off-Campus Housing Office assists registered students in finding rooms to share or to rent.

Student Group

Students in the School include graduates from most U.S. colleges and universities and from seventy countries around the world. Approximately 1,400 students are working on master's degrees, 50 on professional degrees, and 425 on doctorates. Approximately 40 percent of SEAS students are international, 17 percent are members of minority populations, and 17 percent are women. The Bureau of Labor Statistics projects that from 1990 to 2005, engineering figures in six of the top twenty fastest-growing careers.

Location

The School offers all SEAS programs at the main campus in the Foggy Bottom historic district of Washington, D.C., and selected programs at the GW Virginia campus in the nearby northern Virginia foothills. Within a couple of hours' drive east of the University lie the beaches of Virginia, Maryland, and Delaware, and west to the mountains are U.S. national forests and parks, where recreation includes skiing, fishing, hiking, and camping. Cultural activities in Washington, D.C., frequently free of charge, include the rich resources of the Smithsonian and other museums, performance art in numerous local and national theaters such as Wolf Trap Farm Park and the Kennedy Center for the Performing Arts, seasonal events and athletics along the Potomac River and the C&O Canal, and scores of restaurants featuring ethnic American and international foods.

The University and The School

George Washington University was chartered as a private university in 1821, in response to the hope of President George Washington that a national university be established in the federal city. GW has seven schools in addition to the School of Engineering and Applied Science. The others, which offer collaborative opportunities for engineering students, include arts and sciences, business and public management, education, international affairs, law, medicine, and public health and health services. Organized at GW in 1884, the engineering school is one of the oldest in America and was one of the first to admit women.

Applying

For all graduate programs except the D.Sc. in engineering management, applications are processed as they are received, and admission is granted while space is available; March 1 and October 1 are the priority deadlines for fall and spring. International applicants must submit scores on the TOEFL; 550 is the minimum for admission directly to graduate study. Submission of Graduate Record Examinations scores is recommended if the GRE is taken and is required when admission is sought to the D.Sc. program by an applicant whose highest degree is the baccalaureate or when financial aid is sought in the EECS Department.

Correspondence and Information

Office of Admissions and Student Records
School of Engineering and Applied Science
George Washington University
Washington, D.C. 20052

Telephone: 202-994-3096
　　　　　　800-537-7327 (toll-free)
Fax: 202-994-4522
E-mail: data:adms@seas.gwu.edu
World Wide Web: http://www.seas.gmu.edu

The George Washington University

AREAS OF RESEARCH

Civil and Environmental Engineering. Current research encompasses bedforms and their interaction with flow; adaptive parallel incompressible flow solvers and visualization tools; finite element modeling of convection-diffusion of sediments in flows; modeling of contaminants in groundwater; seismic retrofit of R/C beam to column connections using high-strength composite jackets; stochastic analysis of soil-structure interaction; and stochastic fatigue of welded steel joints and random vibration and transient response of payload.

Computer Science. Active research groups of the faculty and graduate students are working in adaptive learning image and signal analysis (ALISA)—scale and translation-invariant classification of subsymbolic and symbolic structures in images and signals; artificial intelligence—optical speech recognition, multivalued logics, genetic algorithms, genetic programming, speech and natural language processing, neural networks, and symbolic and logic programming; computer graphics and user interface—modeling, realistic-rendering, scientific visualization, and sound synthesis and synchronization; interactive multimedia and education technologies—cognitive paradigms, user interface design, expert systems, authoring languages, interactive multimedia, and the use of hypermedia in educational software; and VLSI—designing and testing VLSI digital/analog circuits, designing and testing novel architecture of VLSI circuits with applications in signal processing, and numerical device modeling.

Electrical Engineering. Active research groups of the faculty and graduate students are working in communications—theoretic problems of modulation and coding, mobile communication system design, analysis of all-optical networking; electrophysics—wave and boundary value, optics, remote sensing and parameter estimation, interaction of plasmas with electromagnetic fields, millimeter waves; energy conversion, power, and transmission—power systems operation, control and planning, insulation coordination, power electronics; medical imaging and image analysis, pattern recognition, imaging, and medical engineering—interdisciplinary research into medical imaging, from generation through interpretation, for clinical use in teaching and diagnosis as well as for developing software and hardware tools; and process automation, control, and robotics—systems theory, control systems, signal processing, robotics, manufacturing automation, image processing, and artificial intelligence.

Engineering Management and Systems Engineering. Active research areas of the faculty and graduate students encompass human factors studies; cost effectiveness: design-for-manufacturing methods for costing and process planning; decision support systems; engineering and project management; information technology applied to disaster and crisis management systems; modeling housing damage and resulting mass care needs in earthquakes; maritime safety systems and risk management; evaluating and monitoring risk; systems and software engineering; strategic planning; and systems engineering processes. Current research in stochastic modeling includes Bayesian statistics applied to decision analysis and quality assurance; practical concerns in reliability; improved assessment of software reliability; reliability and risk assessment; stress models for accelerated reliability testing; warranty analysis; reliability and quality control, with focus on software reliability and equipment warranties; and statistical and game theoretic aspects of the warranty problem. Other research embraces theory, algorithms, and applications related to optimization; sensitivity analysis for finitely constrained problems; integer, nonlinear, and nonconvex algorithmic development and applications, including health-care, defense, and economic modeling; nonparametric discriminant analysis via optimization; and game theoretic models and applications to societal problems.

Mechanical Engineering. Research encompasses design of mechanical engineering systems, including computer-aided design and manufacturing, computer-integrated manufacturing, and robotics; fluid mechanics, thermal sciences, and energy; aerospace engineering, including aeronautics and flight dynamics, astronautics, and propulsion; and solid mechanics and materials science. Major research investigations are under way in biomechanics and crash-related injuries, composite materials and testing, computational fluid dynamics, controls, finite-element analysis and ferrite element models for biomechanical analysis, fracture mechanics, neural networks, numerical optimization and artificial intelligence, robotics, structure and chemistry of thin films and surfaces, vehicle dynamics (including a space station modal identification experiment), computational aeroacoustics, high–Knudsen-number gasdynamics and wind tunnel design, laser diagnostics for supersonic and combustion flows (including experiments on fuel mixing and combustion in scramjets), microgravity two-phase flow, nonsteady flow induction, spacecraft control (including advanced flight deck technology research), strong vortex/boundary layer and launch vehicle/payload interactions, and vortex dynamics.

Telecommunications and Computers. Research groups in both the fields of computer science and electrical engineering incorporate aspects of the research that forms the foundation of this area of master's-level graduate study.

RESEARCH CENTERS, RESEARCH INSTITUTES, AND SPECIAL PROGRAMS

Civil, Mechanical, and Environmental Engineering Department
- Center for Intelligent Systems Research
- Center for the Study of Combustion and the Environment
- Environmental Systems Engineering Center
- GW Transportation Research Institute
- Institute for Materials Science
- Joint Institute for the Advancement of Flight Sciences
- National Crash Analysis Center
- PREST (Program for Research and Education in Space Technology)

Electrical Engineering and Computer Science Department
- Cyberspace Policy Institute
- Institute for Applied Space Research
- Institute for Magnetics Research
- Institute for Medical Imaging and Image Analysis

Engineering Management and Systems Engineering Department
- Center for Structural Dynamics
- Institute for Crisis and Disaster Management, Research, and Education
- Institute for Reliability and Risk Analysis
- Declassification Productivity Research Center

HARVARD UNIVERSITY

Division of Engineering and Applied Sciences

Programs of Study

The Division of Engineering and Applied Sciences (DEAS) offers versatile and broad training for students interested in developing and using mathematics, the engineering sciences, physics, chemistry, and biology to address technological and engineering problems. The Division's programs can be divided into four major overlapping categories: (1) investigations of the electronic and atomic properties of matter with concern for their applications (for example, lasers, semiconductors, superconductors, and amorphous solids); (2) studies of the macroscopic response of matter (for example, biomechanics, robotics, and mechanics of solids and fluids); (3) studies of the earth, its physical, chemical, and biological processes, and the effects of civilization upon them (for example, the physics and chemistry of atmospheres and oceans, climatology, geomechanics, and environmental science and engineering); and (4) studies of the control and organization of data and systems (for example, computer science and decision and control).

The Division's programs are typically interdisciplinary. Students follow programs and work on problems that, in a more subdivided faculty, might be pursued in a variety of departments. Thus, the first of the categories above covers topics often found in physics, chemistry, electrical and chemical engineering, and materials science departments. The second emphasizes topics often associated with departments of mechanical, chemical, civil, and materials engineering and applied mathematics, applied physics, and geophysics. The third covers topics taught elsewhere in departments of earth and space science, meteorology and oceanography, chemical engineering, environmental science and engineering, and civil engineering. The fourth involves topics often pursued in departments of computer science and engineering, operations research, electrical engineering, and statistics.

Close relations are maintained with many departments and schools of the University, and programs involving considerable work in these other departments are encouraged. Major emphasis is placed on the program leading to the Ph.D., the traditional degree of scholarship awarded for mastery of a significant field of knowledge. This mastery is demonstrated by work at honor grade level in the field of the candidate's research and at least one other related area and by the completion of an original research project, the results of which are submitted in a thesis. Programs leading to the degrees of Master of Science (S.M.) and Master of Engineering (M.E.) are also available. The S.M. is awarded for the successful completion of an integrated program of eight graduate half courses in a year or more. The M.E. is intended for students who wish to pursue more advanced courses without undertaking the long period of research and study required for the Ph.D. A minimum of two years in residence is required for this terminal degree.

Research Facilities

Extensive, modern research and computing facilities are available in the five laboratory buildings that house the experimental and theoretical research activities of the Division. Additional facilities in the Chemistry, Physics, Earth and Planetary Sciences, Biological, and Astronomical laboratories can be used by Division students.

Financial Aid

Students receive financial aid from a number of possible sources, such as employment in teaching and research, fellowships, traineeships, scholarships, and loans. Stipends are adjusted to meet the current living and educational expenses of the students.

Cost of Study

In 1999–2000, tuition for the two-term academic year is $22,054 plus $560 for health insurance and $711 for University health service. After the second year of study, tuition costs drop substantially.

Living and Housing Costs

Graduate dormitory accommodations are available to single students. Rates are approximately $3280–$5260 for the 1999–2000 academic year. Meals are available at the graduate commons. University-operated married students' apartments, from efficiency to three-bedroom suites, are available. The cost varies according to the accommodations. A wide range of private housing can be found within commuting distance.

Student Group

The graduate student body of the DEAS is composed of 167 men and women. Students are drawn from a great variety of intellectual and geographical backgrounds with approximately a quarter coming from abroad. About 90 percent of first-year students and most students beyond the first year receive some form of financial support through the University. Approximately 25 Ph.D. degrees are awarded each year.

Location

Cambridge, Massachusetts, the home of Harvard, is a city of approximately 96,000 and is just across the Charles River from Boston. The sea is 3 miles away, and 25 miles to the north lies the New Hampshire border. New England's historic countryside and ports, hills and beaches, and winter sports areas are close by, as are Boston's theaters and music and its cultural, economic, and sports activities. Cambridge itself has long been famous as a literary, scientific, and intellectual center. Within its borders are Radcliffe, MIT, and many other schools, colleges, and research institutions.

The University and The Division

Harvard is the oldest college in the United States. It was founded in 1636, sixteen years after the Pilgrims landed at Plymouth. The roots of technological education at Harvard date from 1847, when the Lawrence Scientific School was established. In 1951, with the generous support of the Gordon McKay bequest, the Division of Engineering and Applied Sciences was organized and charged with the responsibility of developing broad interdisciplinary educational and research programs in engineering and the applied sciences.

Applying

Students with a bachelor's degree in any field of engineering or the quantitative sciences are invited to submit applications. Admission and financial aid applications, along with supporting materials, must be received in the Graduate School Admissions Office before 5 p.m. on December 30.

Correspondence and Information

Information on the program:
Academic Office
Division of Engineering and Applied Sciences
Pierce Hall, 212b
Harvard University
Cambridge, Massachusetts 02138
Telephone: 617-495-2833
E-mail: admissions@deas.harvard.edu
World Wide Web: http://www.deas.harvard.edu

Application forms for admission and financial aid:
Admissions Office
Graduate School of Arts and Sciences
Byerly Hall
Harvard University
8 Garden Street
Cambridge, Massachusetts 02138
Telephone: 617-495-5315
E-mail: adm@hugsas.harvard.edu

Harvard University

THE FACULTY AND AREAS OF RESEARCH

ATOMS, MOLECULES, AND CONDENSED MATTER

Professors J. G. Anderson, M. Aziz, A. Dalgarno, H. Ehrenreich, D. S. Fisher, J. A. Golovchenko, B. Halperin, R. V. Jones, E. Kaxiras, C. Lieber, P. C. Martin, E. Mazur, V. Narayanamurti, D. R. Nelson, A. Pandiscio, W. Paul, P. S. Pershan, F. Spaepen, M. Tinkham, D. Weitz, R. Westervelt, T. T. Wu.

Condensed-Matter Theory. Equilibrium and transport theory of interacting many-body systems, critical phenomena and phase transitions, application of first-principles calculations to materials structure, crystal plasticity, and atomic transport kinetics.
Electronic Properties of Matter. Semiconductor physics, superconductivity, conduction in disordered systems.
Materials Science. Crystalline solids and their imperfections; amorphous materials, including semiconducting and metallic glasses; superconducting materials; kinetics of atomic diffusion, flow in solids, and crystal growth from solids, liquids, and gases; structure, properties, and atomic transport in thin films; surface science; synthesis and properties of new materials; materials under high pressure; pressure and stress effects on atomic transport; nanostructures.
Quantum Electronics and Atomic Physics. Electromagnetic phenomena, quantum electronics, optics, molecular physics, spectroscopy of condensed matter, nonlinear optics, laser chemistry, chemical kinetics.

COMPUTING, INFORMATION, AND CONTROL

Professors D. G. M. Anderson, W. H. Bossert, M. S. Brandstein, R. Brockett, U. O. Gagliardi, S. J. Gortler, B. J. Grosz, Y. C. Ho, R. V. Jones, A. Kaucic, H. T. Kung, H. R. Lewis, M. Mitzenmacher, A. G. Oettinger, M. O. Rabin, P. Rogers, M. I. Seltzer, S. Shieber, M. D. Smith, L. G. Valiant, W. Yang.

Applied Mathematics. Numerical methods, scientific computing, mathematical modeling.
Artificial Intelligence. Natural-language processing, distributed AI, vision, robotics, human-computer interfaces, optimization.
Computer Systems. Systems programming, operating systems, programming languages, information storage and retrieval, networking, computer architecture, compilers, graphics.
Control and Management of Energy and Environmental Systems.
Decision and Control. Formulation and analysis of problems of optimal decision and control under conditions of certainty and uncertainty with and without constraints; the methods are sometimes called operations research, systems analysis, management science, and game and control theory.
Information Policy.
Signal Processing. Implementation of VLSI systems, circuit design, computer vision and speech.
Theoretical Computer Science. Analysis of algorithms, computational complexity, computational learning, computer security, neural computation, parallelism.

CONTINUUM MODELS OF MATTER

Professors F. H. Abernathy, D. G. M. Anderson, D. S. Fisher, R. D. Howe, J. W. Hutchinson, R. V. Jones, R. E. Kronauer, R. O. O'Connell, J. R. Rice, F. Spaepen, H. A. Stone, T. T. Wu.

Bioengineering. Cardiac mechanics, mechanics of locomotion, pattern recognition in respiration and vision.
Electromagnetism. Radiation and scattering of electromagnetic waves.
Fluid Mechanics. Liquid dynamics and gasdynamics, especially in turbulent flow; geophysical fluid dynamics; ionized gas flow; wave propagation.
Heat Transfer and Thermodynamics. Convection, radiation, combustion.
Materials Science. Thermodynamic properties of condensed matter, phase transformations, mechanical properties.
Robotics.
Solid Mechanics. Elasticity of solids and structures, buckling and vibrations, shell theory, theory of plasticity, fatigue, fracture, micromechanics, geomechanics.

THE EARTH AND ITS RESOURCES

Professors J. G. Anderson, A. Barros, W. H. Bossert, J. N. Butler, B. F. Farrell, J. J. Harrington, H. Holland, D. J. Jacob, M. B. McElroy, R. Mitchell, R. O. O'Connell, J. R. Rice, A. R. Robinson, P. Rogers, P. Thaddeus, S. C. Wofsy.

Earth and Planetary Sciences. Theoretical, experimental, and observational studies of the physical properties of Earth and other planets; geophysics; physical oceanography; dynamic meteorology; atmospheric physics; climate dynamics; tectomophysics and earthquake dynamics.
Environmental Biology. Microbial ecology, water pollution biology, applied microbiology and population biology.
Environmental Chemistry. Atmospheric chemistry, electrochemistry, marine and water pollution chemistry and water treatment.
Environmental Engineering. Design and development of water resources and policies; solutions and suspensions in air and water; management, control, and treatment of chemical and radioactive pollutants; solid-waste management; environmental geomechanics.
Environmental Policy. The roles of technology, health, law, and economics in the formulation and assessment of policies for energy and the environment.

HOWARD UNIVERSITY

School of Engineering and Computer Science
College of Engineering, Architecture and Computer Sciences

Programs of Study

The School offers studies leading to the Master of Engineering (in civil, electrical, and mechanical engineering), Master of Science in Chemical Engineering, Master of Computer Science, and Doctor of Philosophy (in electrical and mechanical engineering). The Master of Engineering and the Master of Computer Science degree programs require from 1 to 2 years of study and may require a thesis. The Master of Science in Chemical Engineering degree program requires from 1 to 2 years and a thesis. The Master of Engineering in civil engineering requires 31 credit hours of study, which may include 6 credit hours for the thesis and a final oral examination on the thesis or a written comprehensive examination if no thesis is written. The Master of Engineering in electrical engineering involves 24 credit hours of study and 6 credit hours for the thesis option or 33 credit hours of study and comprehensive examinations for the nonthesis option. The Ph.D. in electrical engineering requires 72 credit hours beyond the bachelor's degree (60 credit hours of course work plus 12 credit hours for the doctoral dissertation); four semesters of residence and full-time study (two of the four semesters must be consecutive); a passing grade on the qualifying and preliminary examinations; and approval by the student's adviser. The thesis option for the Master of Engineering in mechanical engineering requires 24 credit hours of course work plus 6 credit hours for the thesis; the nonthesis option requires 33 credit hours, at least 3 credit hours of project work, and a comprehensive examination. The Ph.D. in mechanical engineering requires 72 credit hours beyond the bachelor's degree (60 credit hours of course work plus at least 12 credit hours for the doctoral thesis); four semesters of residence and full-time study (two of the four semesters must be consecutive); a passing grade on a comprehensive examination; and either demonstrated reading skill in French, German, Russian, or another language that the advisory committee considers useful or successful completion of approved courses in technical writing and speech. The Master of Science in Chemical Engineering requires 24 credit hours of course work, 6 credit hours for the thesis, and a final oral examination on the thesis. For the Master of Computer Science, 30 credit hours are required, but no thesis; however, a thesis may replace up to 6 of these credit hours. This program equips the student with a comprehensive knowledge of contemporary computer science through training that combines theory and practice.

In 1997, Howard University introduced two new interdisciplinary programs: the atmospheric sciences program (HUPAS) and the materials science and engineering program (MASEAD). These programs are offered in collaboration with departments in the College and with departments of other colleges. HUPAS offers studies leading to the M.S. and the Ph.D. in mechanical engineering/atmospheric sciences. MASEAD offers studies leading to the M.S. and the Ph.D. in physics/materials science, chemistry/materials science, electrical engineering/materials science, and mechanical engineering/materials science and the M.S. in chemical engineering/materials science. Individuals who are interested in these programs should contact the directors listed for admissions and application information.

Research Facilities

Within the School, facilities for conducting research include the following laboratories: energy conversion, mechanical measurement, system dynamics and control, air pollution, aerodynamics, nuclear engineering, water chemistry, structures and materials, soil mechanics, control systems, solid-state electronics (device fabrication, material and device characterization, and materials growth), electrical machinery, integrated circuits, microwave circuits, electronic simulation, fluid-thermal science, computer science, computer software design, computer engineering design, digital systems, microwave, communications and signal processing, and energy systems network. The Computer Learning and Design Center (CLDC), the College's centralized computing facility, provides the full spectrum of computer resources from high-end, engineering-oriented workstations to PCs and Macintoshes, which are housed in three labs. Peripherals include network system laser printers, color scanners, large format pen plotters, and communication devices that facilitate transmission between the CLDC, experimental labs, and offices in the building or beyond to the campuswide network and the Internet. Functions performed by clientele range from general presentation graphics, engineering drawing, computer-aided design, and animation to programming in all popular languages (Pascal, FORTRAN, UNIX, C++, HTML, Java, and others), engineering problem-solving (numerical and symbolic), large-scale simulation for research, mapping and broader geographic information system capability, literature searches, computer-based instruction and tutoring, reports and thesis publishing, e-mail, Internet and Web access, and authoring. Augmenting the CLDC are several special-purpose computer labs, including a multimedia center, a CAM lab with computer-controlled robot systems, and a virtual classroom for distance learning.

Financial Aid

Financial aid is available to qualified students through graduate teaching and research assistantships. Teaching assistantships provided a stipend of $8000 plus tuition in 1998–99. Some research assistantships are available at various salaries; they also provide summer support. The recipient is required to work up to 20 hours per week and is limited to 9 credit hours per semester.

Cost of Study

Comprehensive tuition for full-time graduate study was $10,200 for 1998–99. There is also a one-time enrollment fee of $150. These costs do not include books or housing.

Living and Housing Costs

Graduate students are not usually allocated accommodations on campus, as preference is given to undergraduate students. However, Howard Plaza Towers, a high-rise complex within convenient walking distance of the campus, has some accommodations for graduate and married students.

Student Group

There are approximately 500 undergraduate and 100 graduate students in the School of Engineering and Computer Science. Approximately 10,600 students are currently enrolled in all schools and colleges of Howard University.

Student Outcomes

Recent graduates have been employed by the U.S. Naval Academy, U.S.D.O.E., U.S.E.P.A., JPL, General Motors, General Electric, Xerox, Alcoa, and Bellcore, among others. Others have gone on to further study at this University or others, such as California, MIT, and Catholic.

Location

The University is located in Washington, D.C., a city that offers unsurpassed opportunities for study and research.

The University

Founded in 1867, Howard is a privately governed institution. It has sixteen schools and colleges, and its main campus occupies more than 89 acres. There is also the East Campus, home to the Divinity School, and the West Campus. Howard is a member of the National Consortium for Graduate Degrees for Minorities in Engineering.

Applying

Information and application forms are obtainable on request. Applicants should provide credentials prior to April 1 for August admission and before November 1 for January admission. Applicants requiring financial assistance should provide credentials as early as possible before the above deadlines. International students whose native language is not English must report scores from the TOEFL. International students applying for the computer science program must take the mathematics or engineering GRE Subject Test.

Correspondence and Information

Chair, Department of (specify by listing on reverse of page)
School of Engineering and Computer Science
College of Engineering, Architecture and Computer Sciences
Howard University
Washington, D.C. 20059

Howard University

THE FACULTY AND RESEARCH AREAS

Chemical Engineering

M. E. Aluko, Chair. J. N. Cannon, R. C. Chawla, W. E. Collins, H. M. Katz,[†] R. J. Lutz*, M. G. Rao, J. Tharakan.

The program offers advanced courses in chemical reaction engineering, thermodynamics, transport phenomena, process control, process design, and applied mathematics for chemical engineers. Research interests of the faculty include hazardous waste remediation, biopolymers, biochemical engineering, biomedical engineering, fluid and thermal sciences, environmental engineering, process control and design, numerical analysis, combustion and incineration, dynamics of reacting systems, and electronic materials processing.

Civil Engineering

L. Fleming, Chair. T. H. Broome Jr., S. Chellam*, R. Efimba, L. Fleming, D. Hampton,[†] J. H. Johnson Jr. (Dean), I. Jones,[†] J. Jones,[†] K. Jones, E. Martin, E. Noel, K. Ocran, G. Sabnis, B. Schimming, A. Shalaby, H, Wijeweera*, S. Zanganeh.

Major areas of study are environmental and water resources engineering, geotechnical engineering, structural engineering and mechanics, and transportation systems engineering. Current areas of research include beneficial aspects of wastewater sludges; mechanical behavior of large, repetitive lattice structures; behavior of concrete; turbulent open channel flow studies; geoenvironmental studies; soil-structure interactions; and treatment of contaminated soils and groundwater.

Electrical Engineering

J. Momoh, Chair. C. Bates, P. Bofah, H. T. Chieh,[†] A. K. Choudhury, M. Chouikha, T. L. Gill, J. Goshtashbi, G. Harris, S.N. Mohammad, S. Richardson, L. Rockett*, A. Rubaai, M. Spencer, Y. Wang, R. Yalamanchili.

On the master's level, major areas of study and research are solid-state electronics, control engineering, power systems, antennas, microwaves, and communications and signal processing. In the Ph.D. program, major areas of study are control engineering, power systems, communications, signal processing, solid-state electronics, and applied electromagnetics. Research interests of the faculty include artificial intelligence (expert systems), antennas, control systems, electrophysics, superconductivity, microwaves, microwave amplifiers, microwave solid-state devices, optimization, optical spectroscopy, fiber optics, semiconductor materials growth, characterization and device fabrication, mathematical physics, quantum field theory and magnetism, telecommunications, signal processing, fuzzy logic, optimization, and control theory for energy systems and power systems.

Mechanical Engineering

L. Thigpen, Chair. P. M. Bainum, D. N. Fan, E. Glakpe, K. Greenaugh*, M. Mosleh, R. Reiss, L. A. Scipio II,[†] S. Smith, N. R. Vira, M. L. Walker Jr., H. A. Whitworth.

Major areas of study at both the master's and Ph.D. levels are applied mechanics, aerospace engineering (dynamics and controls), CAD/CAM and robotics, and fluid-thermal sciences. With departmental approval, special interdisciplinary programs may be designed in such areas as atmospheric science and materials science. Ongoing research and development projects involve the study of attitude dynamics and control of flexible spacecraft; large space structures and tethered subsatellites; fatigue and failure of composite materials; thermal characteristics of phase-change materials in solar-heat receivers for space applications; theoretical and computational fluid dynamics applied to aeroacoustics and turbulence; structural optimization and sensitivity analysis; biomechanics; tribology of homocomposites, design and manufacturing, and robotics.

Systems and Computer Science

R. Leach (Acting Chair), M. Clark*, D. M. Coleman (on leave), W. Craven, H. Keeling*, P. Keiller, Y. Liang*, A. Paul, W. A. Semple, T. Shurn, J. Trimble.

The computer science program and the systems engineering program are offered through the Department of Systems and Computer Science. Computer science has an interdisciplinary structure that relies heavily on the pure sciences, the social sciences, engineering, and mathematics. The program seeks to provide students with knowledge basic to the computer scientist: the means by which information can be transferred, mathematical modeling of physical systems, development of efficient algorithms to represent processes, and the effective means by which information may be stored and retrieved. Major areas of study are computer design, programming and languages, mathematical analysis and theory of languages, and systems design and analysis.

Systems engineering is an interdisciplinary program utilizing the quantitative and social sciences. The problem-focused curriculum is based on systems analysis and operations research techniques applied to large-scale public systems problems. Course work and research may emphasize specific areas of application, such as transportation systems, health-care-delivery systems, solid-waste management systems, and other systems related to urban services. Current research interests include solid-waste collection, urban mass transit systems, urban jitney systems, and emergency ambulance services.

HUPAS Program

For information about this program, students should contact Professor Arthur N. Thorpe, Director, Center for Study of Terrestrial and Extraterrestrial Atmospheres, H.U. Program in Atmospheric Sciences, Howard University, Washington, D.C. 20059, telephone: 202-806-5172, fax: 202-806-4430.

MASEAD Program

For information about this program, students should contact Professor Clayton W. Bates Jr., Director, Interdisciplinary Graduate Program in Material Science and Engineering, School of Engineering and Computer Science, College of Engineering, Architecture and Computer Sciences, Howard University, Washington, D.C. 20059, telephone: 202-806-6147, e-mail: bates@negril.msrce.howard.edu.

---

*Part-time.
†Emeritus.

KANSAS STATE UNIVERSITY

College of Engineering

Programs of Study

The College of Engineering offers programs leading to the M.S. and Ph.D. degrees. The College has eight academic departments: Architectural Engineering and Construction Science, Biological and Agricultural Engineering, Chemical Engineering, Civil Engineering, Computing and Information Sciences, Electrical and Computer Engineering, Industrial and Manufacturing Systems Engineering, and Mechanical and Nuclear Engineering. The M.S. degree is offered in all departments. An M.S. degree in operations research and a Master of Software Engineering are also offered. The College offers an undesignated Ph.D. degree in any one of the academic departments except the Departments of Architectural Engineering and Construction Science and Computing and Information Sciences. A designated Ph.D. degree in computer science is offered by the Department of Computing and Information Sciences.

Candidates for the M.S. degree are normally required to spend one academic year in residence. Subject to the approval of the major department, the candidate may choose one of the following options: (1) a minimum of 30 semester hours of graduate credit including a master's thesis of 6–8 semester hours; (2) a minimum of 30 semester hours of graduate credit including a written report of 2 semester hours either of research or of problem work on a topic in the major field; or (3) a minimum of 30 semester hours of graduate credit in course work only, but including evidence of scholarly effort such as term papers, production of creative work, and so forth, as determined by the student's supervisory committee.

Candidates for the Ph.D. degree normally devote at least three years of two semesters each to graduate study, or about 90 semester hours beyond the bachelor's degree. A dissertation is required. Ph.D. candidates must complete a year of full-time study in residence at Kansas State University. Furthermore, a minimum registration of 30 hours in research is required, not including work done toward a master's degree. Each candidate also must have completed at least 24 hours of course work at the University. The foreign language requirement is determined as a matter of policy by the graduate faculty in each department.

Research Facilities

Each of the eight departments in the College of Engineering has modern and fully equipped teaching and research laboratories. In addition, the College has several centers and institutes, including the Testing Lab for Civil Infrastructure, the Advanced Manufacturing Institute, the Center for Hazardous Substance Research, the Pollution Prevention Institute, the Institute for Environmental Research, the Center for Transportation Research, and the National Gas Machinery Laboratory.

Financial Aid

The College of Engineering offers approximately 300 fellowships, traineeships, and assistantships each year. These awards are administered by individual departments.

Cost of Study

Fees for 1998–99 were $101 per credit hour for residents and $329.75 per credit hour for nonresidents. In addition, a $64 campus fee was charged for the first credit hour, with $17 charged for each additional credit hour, up to a maximum fee of $251 per semester. Students enrolled in the College of Engineering are assessed an engineering equipment fee of $85 (full-time student) or $42.50 (part-time student) per semester.

Living and Housing Costs

Residence hall rates for room and board were $1890 per semester per student in 1998–99; furnished apartments for married students were $273 per month for one bedroom and $312 per month for two bedrooms. These figures are subject to change. In addition, there are scholarship housing units, which function as cooperatives in which students provide their own services. There are numerous privately owned rooms and apartments with a wide range of rental rates in the community.

Student Group

Kansas State University enrolls about 20,000 students. The College of Engineering has approximately 2,600 undergraduate and 400 graduate students.

Location

The University's 315-acre campus is located in Manhattan, Kansas, a community of about 38,000 residents in addition to University students. Manhattan is in northeast Kansas and is easily accessible by I-70. Kansas City is 125 miles away, and Topeka, the state capital, is 54 miles to the east. Tuttle Creek Lake, 5 miles north of Manhattan, has picnic areas, boats for rent, and a large sand beach.

The University

Kansas State University was established in 1863 as the first land-grant institution under the Morrill Act. The University is composed of the Graduate School and the Colleges of Agriculture, Architecture and Design, Arts and Sciences, Business Administration, Education, Engineering, Human Ecology, Technology, and Veterinary Medicine. The College of Technology is located at Kansas State in Salina.

There are numerous cultural and entertainment activities associated with the University and the community. One of the most noteworthy is the Landon Lecture Series, which regularly brings outstanding speakers to the campus. The long list of notables has included Norman Borlaug, Tom Brokaw, George Bush, Jimmy Carter, Bob Dole, Gerald Ford, Billy Graham, Robert Kennedy, Walter Mondale, Richard Nixon, Sandra Day O'Connor, Colin Powell, Dan Rather, Ronald Reagan, Bernard Shaw, George Shultz, Sheikh Yamani, and many others.

The University is a member of the Big Twelve Conference and provides numerous facilities for athletic activities.

Applying

The Graduate School has an application fee of $25 for all international students. Requirements vary according to department. Students interested in graduate study in the College of Engineering should write to the Dean of Engineering, stating their area of interest, at the address given below.

Correspondence and Information

Terry S. King, Dean of Engineering
Kansas State University
Manhattan, Kansas 66506-5201

Telephone: 785-532-5590
Fax: 785-532-7810

Kansas State University

FACULTY HEADS AND AREAS OF RESEARCH

Architectural Engineering. David R. Fritchen, Head; M.S., Washington (Seattle). (13 faculty members) Structural, mechanical and electrical systems design for buildings: domestic water supply and sanitation systems, fire protection, heating and air-conditioning systems, lighting and electrical systems, environmental control systems in buildings, communication and energy management systems for buildings. Building design and construction: integration of structural, mechanical, and electrical systems in buildings.

Biological and Agricultural Engineering. James K. Koelliker, Head; Ph.D., Iowa State. (12 faculty members) Grain conditioning, handling, drying, and storage. Water and soil resources: irrigation systems, movement of pesticides and other chemicals in surface water and groundwater, improved water management techniques, erosion and sedimentation control, water quality and nonpoint pollution control, animal waste management. Tillage-vehicle mechanics: tillage and planting machines, soil compaction. Harvesting systems: performance modeling and analysis, online yield mapping, forage harvesting systems. Energy use in agriculture: efficient internal combustion engine operation. Control systems: instrumentation and controls, sensor development, image processing, chemical spray metering and control. Animal environment: air quality, environmental modification, ventilation-fan performance. Food engineering: process design, cereal-based product development, properties of biological products. Environmental engineering: constructed wetlands, vegetative filters, watershed modeling, bioremediation.

Chemical Engineering. Stevin H. Gehrke, Head; Ph.D., Minnesota. (10 faculty members) Transport phenomena: heat transfer, mass transfer, momentum transfer. Thermodynamics: phase equilibrium, adsorption equilibrium, thermodynamic efficiencies. Reactor engineering: catalytic reactions and reactors, polymerization reactions and reactors, enzyme-catalyzed reactors for gasification and liquefaction of coal and solid waste. Bioengineering: biological-waste treatment, hazardous-waste treatment, enzyme technology, biochemical engineering, food processing. Processes: fuel synthesis, energy resource conversion, air pollution control, water pollution control, thermodynamics, solid-waste conversion and recycling. Systems engineering: theory, simulation, optimization, process design and synthesis, process dynamics and control, artificial intelligence, knowledge engineering.

Civil Engineering. Stuart E. Swartz, Head; Ph.D., IIT. (13 faculty members) Hydrology and hydraulic engineering: hydraulic and hydrologic modeling, overland flow hydraulics. Environmental engineering: physical, chemical, and biological processes for water, wastewater, and hazardous waste treatment. Soil mechanics and foundation engineering: physical and mechanical properties of soil, soil stabilization, earth pressures and reactions, environmental geotechnology. Structural engineering: behavior and load-carrying capacity of steel and reinforced concrete members, fracture mechanics of concrete, finite-element methods, optimization applied to civil engineering structures, structural dynamics and earthquake engineering. Transportation engineering: urban transportation planning, transportation systems, analysis and simulation, geometric design of highways, highway safety, pavements and highway materials.

Computing and Information Sciences. Virgil E. Wallentine, Head; Ph.D. Iowa State. (16 faculty members) Parallel and distributed computing systems: distributed mutual exclusion; distributed shared memory; fault tolerant systems; synchronization and concurrency; construction, composition, and verification of distributed systems; algorithms and protocols; high-speed communication systems; operating systems; parallel programming languages and systems; real-time (embedded) computer systems. Programming languages: compiler construction, denotational semantics, operational semantics, partial evaluation, algebraic semantics, full abstraction, concurrency. Database systems: database design, object-oriented databases, database integrity and security. Software engineering: software life-cycle, software environments and tools, software metrics, software specification and verification, expert software systems, software testing, large software systems, computational science and engineering.

Electrical and Computer Engineering. David L. Soldan, Head; Ph.D., Kansas State. (23 faculty members) Bioengineering: physiological systems simulations. Communication systems: detection and estimation; digital modulation systems; analog/digital/RF circuits and systems; wireless telecommunications. Computer systems: computer vision; testing of digital systems; neural networks; computer architecture. Control systems: system theory and optimization; mobile autonomous robotics. Electromagnetics: device modeling and simulation; bioelectromagnetics. Instrumentation: computer-based instrumentation; intelligent instrumentation; microcontroller applications. Power systems: power system and stability; load management; distribution automation; power electronics; power devices, high-voltage circuits. Signal processing: adaptive signal processing, image processing, applications to spectroscopy. Solid-state electronics: sensors; device and process modeling; analog and digital integrated circuit design; infrared emitters and detectors.

Industrial and Manufacturing Systems Engineering. Bradley A. Kramer, Head; Ph.D., Kansas State. (11 faculty members) Operations research: mathematical programming, stochastic processes and queueing, fuzzy and uncertainty reasoning, discrete and intelligent systems, systems modeling and control, decision making. Manufacturing engineering: flexible manufacturing, numerically controlled machines, intelligent manufacturing and design, work design, facility layout, tool engineering, CAD/CAM, robotics and automation, quality control. Engineering management: artificial intelligence in management, project model, management decision making, organizational behavior, motivation of creative personnel. Ergonomics: man-machine interface, safety, work environment.

Mechanical and Nuclear Engineering. J. Garth Thompson, Head; Ph.D., Purdue. (22 faculty members) Heat and mass transfer: fluid mechanics, heat transfer, room air diffusion. Machine design and materials science: acoustics, dynamics, kinematics, rock mechanics, stress analysis, vibrations. Control systems: dynamic system modeling, fluid controls, fluidics, instrumentation and measurements, simulation and control. Heating, air conditioning, human comfort. Computer-assisted design and graphics. Nuclear reactor physics and engineering: radiation transport theory, neutron spectroscopy, neutron interactions with light nuclei, nuclear fuel management. Radiation detection and measurement: neutron activation analysis, X-ray and gamma-ray spectroscopy, nondestructive assay of fissile materials. Radiation effects on materials: photoluminescence of liquids, radiation-induced thermoluminescence. Radiation protection: radiation shielding, environmental monitoring. Controlled thermonuclear power: radiation damage and materials problems.

Advanced Manufacturing Institute. Farhad Azadivar, Director; Ph.D., Purdue. Expertise in computers, robotics, artificial intelligence, numerically controlled machines, flexible manufacturing systems, and instrumentation in the sensing, controlling, communicating, and decision-making processes in engineering design and manufacturing.

Center for Hazardous Substance Research. Larry E. Erickson, Director; Ph.D., Kansas State. Handling and processing hazardous waste/ materials, protection of water supplies: resource recovery, treatment, disposal, and storage of hazardous materials.

Pollution Prevention Institute. Jean S. Waters, Director; M.S., Kansas State. Provides technical assistance and training in source reduction and other environmentally sound practices to business, regulatory agencies, technical assistance groups, and private citizens throughout the Midwest. The institute also serves as a meeting ground for KSU faculty members involved in pollution prevention and other related activities.

Institute for Environmental Research. Mohammad H. Hosni, Director; Ph.D., Mississippi State. Study of the interaction of humans and their thermal environment: thermal comfort, humidification, air movement, clothing, physical activity, heat stress, cold stress, protective apparel, biothermal modeling.

Center for Transportation Research. Eugene R. Russell, Director; Ph.D., Purdue. Interdisciplinary mission-oriented research and training concerning national, regional, state, and local transportation problems. The K-TRAN program, started in 1991, is an ongoing, cooperative, and comprehensive research program addressing transportation needs of the state of Kansas utilizing academic and research resources from the Kansas Department of Transportation, KSU, and Kansas University.

National Gas Machinery Laboratory. Kirby S. Chapman, Director; Ph.D., Purdue. This laboratory provides the natural gas industry with independent testing and research capabilities, knowledge databases, and educational programs. A premier turbocharger test and research facility has been developed through acquisition of gas turbine engines, instrumentation, and a laboratory building.

Testing Lab for Civil Infrastructure. Hani G. Melhem, Director; Ph.D., Pittsburgh. The Testing Facility includes a pavement accelerated testing lab, a falling weight deflectometer state calibration station, and a shake-table for dynamic testing of model buildings. Future plans include structural testing of bridge components and prestressed concrete girders. The facility is a center for cooperation between academia, industry, and state departments of transportation. The pavement research and testing activity is sponsored by a consortium called the Midwest States Accelerated Testing Pooled Funds Program that fulfills the needs of the surrounding states for full-scale testing and addresses research topics of national and international importance.

LEHIGH UNIVERSITY

P. C. Rossin College of Engineering and Applied Science

Programs of Study

The College offers M.S. and Ph.D. degrees in the fields of chemical engineering, civil engineering, computational and engineering mechanics, computer science, electrical engineering, industrial engineering, materials science and engineering, and mechanical engineering. M.Eng. programs are offered in chemical engineering, civil engineering, electrical engineering, industrial engineering, materials science and engineering, and mechanical engineering. M.S. degrees may also be earned in computer engineering, quality engineering, and management science. Interdisciplinary graduate study is available in manufacturing systems engineering (M.S.) and polymer science and engineering (M.S., Ph.D.).

The master's degree is granted to properly qualified students who complete satisfactorily at least two full semesters of advanced work. The minimum required for award of a degree is 30 semester credit hours. In addition to a planned program of course study, students may be required to submit a thesis or project. Candidates for the master's degree must complete their programs within six years.

The Ph.D. is conferred on candidates who have demonstrated the capacity to carry on independent, original research as evidenced by an acceptable dissertation. A minimum of 72 semester credit hours beyond the bachelor's degree (48 credit hours beyond the master's) is required. Qualifying exams, a general exam in the major field (which may be partially oral), and a final oral exam are also required. Candidates must complete at least one full academic year of resident graduate study, and all post-baccalaureate work for the Ph.D. must be completed within ten years.

There is no University foreign language requirement. Part-time programs leading to the master's and Ph.D. degrees can be arranged in most departments.

Research Facilities

Lehigh's research labs include the Seeley G. Mudd Chemistry Building, a seven-level complex; Fritz Laboratory, which houses the civil engineering department; Whitaker Laboratory, for the materials science and engineering department, the interdisciplinary materials research center, and the world-class Laboratory for Electron Optics; Sinclair Laboratory, for surface chemistry and coatings research; Packard Laboratory, for electrical engineering and computer science and for mechanical engineering and mechanics; the Sherman Fairchild Laboratory, for solid-state studies; and Mohler Laboratory, for industrial engineering and manufacturing systems engineering. The chemical engineering department is housed at the Mountaintop Campus research complex. The civil engineering, electrical engineering and computer science, mechanical engineering and mechanics, and industrial engineering departments have extensive CAD/CAM/CAE facilities. The Herbert R. Imbt Laboratories house a state-of-the-art multidirectional structural testing laboratory for the Center of Advanced Technology for Large Structural Systems.

Lehigh's libraries include the Fairchild-Martindale and Mart Science and Engineering Library, which house computing facilities and more than 200,000 volumes in the natural and physical sciences, mathematics, and all branches of engineering. Resources of other libraries are available through an interlibrary loan system. The University has an InteCom digital PBX system that provides integrated voice and data communication services throughout the campus. The system provides access to the Computing Center mainframe computers, the Integrated Library System, and other computers located on the campus. The central computing facility contains a network of high-performance computers configured as centralized network servers, compute servers, and file servers. In addition to the centralized facilities, Lehigh maintains a total of forty public computing sites, computer classrooms, and computer-equipped lecture rooms across the campus. A wide variety of software, including word processing programs, spreadsheet programs, database management systems, mathematical and statistical applications, scientific and presentation graphics packages, terminal emulation programs, and high-level programming languages, are supported.

Financial Aid

For the 1999–2000 academic year, teaching assistants receive a tuition award and a stipend of $1250 to $1400 per month. Most appointments are half-time, require 20 hours of service per week, and allow registration for 10 credits per semester. Fellowships and research assistantships are awarded within departments. Stipends vary but are competitive.

Cost of Study

In 1999–2000, tuition is $860 per credit hour.

Living and Housing Costs

Housing is readily available in apartments in Lehigh's Saucon Valley Campus or within easy walking or driving distance of campus. Monthly rents for campus housing, exclusive of utilities, are approximately $375 for an efficiency apartment, $440 for a one-bedroom apartment, $475 to $500 for a two-bedroom apartment, and $510 for a three-bedroom apartment.

Student Group

In fall 1998, total enrollment at Lehigh was 6,233. Of these, there were 412 full-time and 374 part-time graduate students in the College of Engineering and Applied Science.

Student Outcomes

Lehigh maintains a strong tradition of placing its graduates in industry and academia. Recent graduates have joined companies such as AT&T, Ingersoll-Rand, IBM, and General Electric.

Location

Lehigh is on the north slope of South Mountain, overlooking Pennsylvania's Lehigh Valley (population 300,000) with its cities of Allentown, Bethlehem, and Easton. Bethlehem is a mostly residential community of 70,000. Its cultural heritage is seen in its buildings, providing a charming Colonial atmosphere. The community was founded in 1741, and many original buildings, remarkably preserved, remain in use. Bethlehem is 60 miles north of Philadelphia and 90 miles west of New York City. The Poconos are less than an hour away.

The University

Lehigh University, founded in 1865, was one of the first American institutions to offer a technical education. Its first five schools included a school of general literature and four scientific schools. The innovative concept of offering both technical and nontechnical courses of study has continued to be a successful formula at Lehigh. Today, Lehigh has a graduate school of education and three colleges: Engineering and Applied Science, Arts and Sciences, and Business and Economics. Thirty-five percent of the students are enrolled in the College of Engineering and Applied Science. Women have always been admitted to Lehigh at the graduate level, and in 1971 the University began admitting women at the undergraduate level. The campus recently expanded to 1,600 acres with the addition of the former Homer Labs of the Bethlehem Steel Corporation; the Mountaintop Campus houses several engineering departments and centers.

Applying

Although the deadline for regular admission is thirty days before graduate registration, those interested in financial aid must apply by January 15. Most assistantships are filled by May 1 for the fall semester.

Correspondence and Information

For application forms and additional information, students should contact the faculty graduate coordinator in the department of interest or:

Office of Graduate Studies
P. C. Rossin College of Engineering and Applied Science
Lehigh University
19 Memorial Drive West
Bethlehem, Pennsylvania 18015

Telephone: 610-758-6310
Fax: 610-758-5623
E-mail: ineas@lehigh.edu
World Wide Web: http://www.lehigh.edu

Lehigh University

DEPARTMENT CHAIRS AND AREAS OF FACULTY RESEARCH

Chemical Engineering. Professor Mohamed S. El-Aasser, Department Chairperson, and James Hsu, Faculty Graduate Coordinator, 111 Research Drive, Iacocca Hall, Mountaintop Campus. Major research thrusts are in biotechnology, polymer science, process modeling and control, and multiphase processing. Biotechnology research emphasizes reaction engineering, computer control, and separations technology of bioprocesses. Polymer science is concerned primarily with polymer colloids (their synthesis, characterization, and use), polymer interfaces, and multicomponent polymer materials (including interpenetrating network polymers). Process modeling and control is concerned with dynamic modeling of chemical processes and with advanced theory for process control, including multivariable and nonlinear control structures. Multiphase processing includes traditional and nontraditional areas of chemical engineering, such as research on fluidization, two-phase flow and heat transfer, mass transfer, catalysis and surface science, thin films and microelectronics processing, reactor design, and thermodynamic properties of fluid mixtures.

Civil and Environmental Engineering. Professor Arup K. Sengupta, Department Chairperson, and Professor John L. Wilson, Faculty Graduate Coordinator, 13 East Packer Avenue. The Fritz Engineering Laboratory offers complete facilities for research and instruction in environmental engineering, geotechnical engineering, hydraulic engineering, structural engineering, and related fields. Environmental laboratories are equipped with state-of-the-art analytical instruments and testing facilities for studying water and wastewater treatment, contaminant transport in groundwater, and bioremediation. Geotechnical testing equipment includes triaxial test apparatus with automatic data acquisition, shear and compressional wave velocity determination facilities, various consolidation equipment, multiple reactor cells for electrochemical soil treatment tests, flexible wall pemcameters, and standard equipment for soil property and index tests. Installations for testing models of spillways, open channels, and beach facilities are available in the H. R. Imbt Laboratory. Structural testing equipment includes dynamic testing machines, a 5-million-pound universal hydraulic testing machine, and other special loading apparatus. The Center for Advanced Technology and Large Structural Systems (ATLSS) on the Mountaintop Campus operates a world-class structural laboratory with multidirectional reaction walls and strong floors for performing static and dynamic tests.

Electrical Engineering and Computer Science. Professor Bruce D. Fritchman, Department Chairperson, and Professor Douglas R. Frey, Faculty Graduate Coordinator, 19 Memorial Drive West. The department's diversity in its research activities reflects the breadth of the field. One of its major research thrusts is in solid-state technology and devices, with particular emphasis on device modeling and characterization of submicron devices, MOS and MNOS structures, and defects; compound heterojunction semiconductor materials, device structure, and process modeling; packaging and interconnection modeling for high-speed digital and microwave circuits and inclusion in expert packaging design systems; thin-film silicon materials and transistors for matrix access to thin-film displays; and VLSI design and design automation. Other projects include light-wave technology, microprocessor technology, digital signal processing VLSI design, parallel and distributed computing, and networking; development of error-correcting codes and the verification and validation of large software systems; artificial intelligence within the context of knowledge-based expert systems for many applications, including large-scale building structures and device packaging; natural-language communication and man-machine interface; and manufacturing systems line simulation and optimization, robots, and CAD/CAM.

Industrial and Manufacturing Systems Engineering. Professor S. David Wu, Department Chairperson and Faculty Graduate Coordinator, 200 West Packer Avenue. The department has three primary areas of concentration: manufacturing systems and processes, operations research, and information systems and technology. Activities in manufacturing processes include traditional and nontraditional manufacturing processes, automation and robotics, electronics manufacturing, and computer-aided manufacturing. Manufacturing systems efforts include activities in production planning and scheduling, process planning, quality engineering, and traditional areas of industrial design. Operations research activities provide the opportunity for basic and applied research in areas such as neural networks, artificial intelligence and expert systems, simulation, and mathematical programming. The information systems and technology area includes designing, structuring, and using information systems based on structured files of a management information database for real-time applications. Extensive opportunities for industrial interaction are provided in each of the areas of concentration.

Materials Science and Engineering. Professor David Williams, Department Chairperson, and Professor Raymond A. Pearson, Faculty Graduate Coordinator, 5 East Packer Avenue. Special areas of research interest in the department include the formation, deformation, and fracture of materials; analytical electron microscopy and microstructural analysis of metals, ceramics, and glasses; electronic materials; processing and properties of ceramics; processing and characterization of thin films for corrosion protection and high-temperature environments; and theory and modeling of interfaces in metals, ceramics, and polymer composites. Electronic materials studies are also carried out jointly with other departments in the Sherman Fairchild Laboratory for Solid-State Studies. Welding of steels and other metals has been the focus of one of the longest research efforts, beginning in the 1930s and continuing today. Other areas of research are physical metallurgy of alloys, extractive metallurgy, magnetic properties of materials, and mechanical properties of polymers and ceramics.

Mechanical Engineering and Mechanics. Professor Charles R. Smith, Department Chairperson, and Professor Donald O. Rockwell, Faculty Graduate Coordinator. The department is particularly strong in four broad areas of study—fracture and solid mechanics, thermofluids, dynamic systems, and manufacturing/process modeling. Well-equipped laboratories, with supporting instrumentation, are maintained for a variety of disciplines, including heat transfer and combustion, turbulent fluid mechanics, solid mechanics, acoustics and system dynamics and control, computer-aided design, and integrated product development. Support facilities include a fully-staffed machine shop and electronics shop, superheated steam and high-pressure air laboratories, and data-acquisition systems connected to the CAD laboratory. The department has 29 regular faculty members with national and international reputations in their specialties. Most of the laboratories, classrooms, and offices are located in Packard Laboratory, although some faculty members and students who are engaged in interdisciplinary research use facilities in Whitaker Laboratory for materials and structural research and in Sinclair Laboratory for surface and coatings research.

GRADUATE PROGRAMS AND SPECIAL RESEARCH OPPORTUNITIES

In addition to studying for degrees in academic departments, graduate students may earn degrees in special interdisciplinary programs, such as polymer science and engineering and manufacturing systems engineering. The manufacturing systems engineering program, leading to an M.S. degree, is supported by the six engineering departments and by two departments in the College of Business and Economics. The program covers advanced manufacturing technologies from a strategic integrated systems perspective. Centers and institutes associated with the College of Engineering and Applied Science are devoted to molecular bioscience and biotechnology, environmental studies, surface and coatings research, design and manufacturing innovation, health services, social research, energy research, engineering, materials research, solid-state studies, polymer science and engineering, fracture and solid mechanics, metal forming, robotics, the study of the high-rise habitat, thermofluid engineering and science, biomedical engineering and mathematical biology, marine studies, innovation management studies, and chemical process modeling and control. Lehigh's Center for Advanced Technology for Large Structural Systems, a National Science Foundation engineering research center, was established in 1986 to study ways to improve the technological base for large-scale construction in cooperation with the Pennsylvania Ben Franklin Program and with industry.

MANHATTAN COLLEGE

Engineering Graduate Programs

Programs of Study

Programs offered by the School of Engineering that lead to a master's degree are intended to prepare professionals for advanced-level technical and administrative positions or admission to doctoral programs at other institutions. Since many students are pursuing a profession, engineering courses are generally offered in the evening. Master of Science degrees are offered in chemical, civil, computer, electrical, environmental, and mechanical engineering. A Master of Environmental Engineering degree is offered in addition to the Master of Science degree. Courses in computer, electrical, and mechanical engineering are also available through distance education via live two-way videoconferencing. Each department's curriculum includes both core and elective course work. All degree programs require the completion of 30 credit hours.

Research Facilities

There are more than forty scientific and engineering laboratories at Manhattan College. The environmental engineering laboratory complex in the Leo Engineering Building was completely renovated in 1992. The Research and Learning Center houses laser and microcomputer design labs as well as special rooms for robotics, graphics, CAD, and senior projects. All students are assigned e-mail accounts, and access to the Internet, the online library catalog, and research databases are supported in all computer labs. Manhattan belongs to major academic computer consortia and has license agreements with Lotus, SPSS, Microsoft, Corel, Digital, Borland, Novell, and Maple. The Grover H. Hermann Engineering Library, with more than 26,000 volumes, complements the approximately 200,000 volumes in the College's main library. Interlibrary loan agreements exist between Manhattan College and New York City and Westchester County public libraries.

Financial Aid

Financial aid is available through departmental fellowships, research and laboratory assistantships, and grants. The College offers limited loan opportunities to matriculated students, and some lab and research assistantships are available. To be eligible for any financial assistance, a student must be matriculated and attending classes at least half-time (6 credits a semester). Each student must complete a Manhattan College Financial Aid Application and the Free Application for Federal Student Aid (FAFSA).

Cost of Study

Tuition is $500 per credit. There is a laboratory fee of $100 (for a 2-hour lab) or $135 (for a 3-hour lab) per semester and a registration fee of $50 per semester.

Living and Housing Costs

Limited dormitory facilities are available for graduate students at the rate of $3675 per semester (double occupancy) or $4200 per semester (single occupancy), which includes a seven-day meal plan.

Student Group

Of the 487 students enrolled in Manhattan's Graduate Division, 205 are women, 282 are men, 403 are part-time students, and 84 study full-time, including 19 international students. A vast majority of students have prior professional experience in their field.

Student Outcomes

Graduate students are invited to participate in career services and campus recruitment. The College maintains a twelve-month resume referral service for participating students. Manhattan engineering graduates are employed by firms such as Con Edison, Fusco Corporation, IBM, Lockheed Martin, Merck, New York Power Authority, NYNEX, Pfizer, Prudential Securities Inc., Raytheon Electronic Systems, Silicon Valley Group, Sprint, Turner Construction Co., and the U.S. Department of Labor.

Location

Just 12 miles north of midtown Manhattan and 1 mile from Westchester County, the College is situated on the heights above Van Cortlandt Park in the Riverdale section of the Bronx. The Leo Engineering Building is located on Corlear Avenue. Riverdale, an upper-middle-class community, offers an ideal blend of the calm and quiet of a residential, suburban setting with easy access to the many advantages of New York City.

The College

Founded by the Christian Brothers in 1853 upon a Lasallian Catholic tradition of excellence in teaching, Manhattan College strives to provide a contemporary, person-centered education experience, characterized by high academic standards, reflection of values and principles, and lifelong career preparation. The School of Engineering was started in 1892 and is the oldest Catholic engineering program in the Northeast. Of the 250 faculty members, 163 are full-time, 87 are part-time, and 90 percent hold doctoral degrees. The student-faculty radio is 13:1.

Applying

Application forms for admission are provided by the Admission Center upon request. Filing for admission should be completed before May 2 for summer sessions, August 10 for the fall session, and January 7 for the spring session. Students seeking admission into the full-time engineering programs must have completed their applications by February 1 if they are applying for a fellowship or scholarship for the fall semester.

Correspondence and Information

William J. Bisset, Jr.
Dean of Admissions/Financial Aid
Manhattan College
Riverdale, New York 10471
Telephone: 800-MC-2-XCEL (toll-free)
Fax: 718-862-8019
E-mail: admit@manhattan.edu
World Wide Web: http://www.manhattan.edu

Manhattan College

THE FACULTY AND THEIR RESEARCH

Helen C. Hollein, Acting Dean; Eng.Sc.D., NJIT.

Chemical Engineering: Helen C. Hollein, Chairperson; Eng.Sc.D., NJIT.
Civil Engineering: Walter P. Saukin, Chairperson; Ph.D., CUNY, City College.
Electrical Engineering and Computer Engineering: Br. Henry Chaya, F.S.C., Chairperson; Ph.D., Princeton.
Environmental Engineering and Science: James A. Mueller, Chairperson; Ph.D., Wisconsin.
Mechanical Engineering: Daniel W. Haines, Chairperson; Eng.Sc.D., Columbia.

Faculty specializations: Structural analysis, hydrology, water and waste treatment, toxicology, analysis of natural water systems, air pollution control, stress analysis, computer-aided design, manufacturing, advanced separation processes, hazardous waste incineration, fields analysis, microwaves, heat transfer, telecommunications, multimedia technology, power systems, control systems, artificial intelligence, enzyme analysis, mathematical modeling, and heating, ventilating, and air conditioning.

MARQUETTE UNIVERSITY

College of Engineering

Programs of Study	Marquette University's College of Engineering, which dates back to 1908, is the largest Catholic college of engineering in the nation, blending technological training with an education in the liberal arts and Christian values. The College is a member of the American Society for Engineering Education, and its undergraduate programs are fully accredited by the Engineering Accreditation Commission of the Accreditation Board for Engineering and Technology.
	Through the College of Engineering, Marquette University's Graduate School offers the degrees of Master of Science and Doctor of Philosophy in five engineering disciplines: biomedical engineering, civil and environmental engineering, electrical and computer engineering, materials science and engineering, and mechanical engineering (including manufacturing systems engineering). A Master of Science in Engineering Management is also available through a joint program with the College of Business Administration. A Master of Science in health-care technologies management is available through a joint program with the Medical College of Wisconsin and the College of Business Administration.
	For Master of Science students, both thesis and nonthesis plans are available in most programs. The number of required semester hours varies from 30 to 36, depending on the program and the plan chosen. For the Doctor of Philosophy degree, a program of study must be prepared by the student in consultation with his or her doctoral adviser. Normally, the total course program, exclusive of the dissertation, is 60 semester hours beyond the baccalaureate degree, or 30 semester hours beyond the master's degree. Additional information may be obtained from Marquette University's Graduate School bulletin or by contacting either the College or the Graduate School.
Research Facilities	Numerous research facilities are available to graduate students within the College. Recognizing the multidisciplinary nature of most applied research activities, the College has created a number of research centers that capitalize on the strong technical skills of the faculty: Biomedical Engineering and Biomathematics, Energy Studies, Highway and Traffic Engineering, Industrial Processes and Productivity, Intelligent Systems, Sensor Technology, Materials Science and Technology, Rapid Prototyping, Signal Processing, and Water Quality. The College of Engineering is the recipient of research grants that total millions of dollars per year and are sponsored by government agencies, industry, and private foundations.
	In addition to its own PC and UNIX workstation laboratories and networks, the College makes use of Marquette's centralized computing facility and numerous public PCs and workstations. A fast-modem Internet link allows access to thousands of worldwide networks.
	The six-story Science Library, located next to the Engineering Building, provides comprehensive services in all scientific and engineering disciplines and complements the campus Memorial Library and the specialized Law Library.
Financial Aid	Many forms of financial aid are available to Marquette graduate students admitted to a degree program. Some types of aid are available from the Graduate School or through individual faculty research grants; other forms of aid are administered by the Office of Student Financial Aid. Qualified graduate students may be offered opportunities as teaching and research assistants, research fellows, and minority student fellows. Most financial aid decisions are based on academic performance, not financial need. Awards are determined during three annual competitions. Financial aid inquiries may be addressed to Mr. Thomas S. Marek, Financial Aid Coordinator, Graduate School (telephone: 414-288-5325 or 7137).
Cost of Study	The Graduate School tuition for the 1999–2000 academic year is $9180 (based on 9 credit hours per semester).
Living and Housing Costs	The University owns and operates two on-campus residences that are reserved for graduate and married students. Units vary from $1700 to $1900 per semester for efficiencies to $2300 to $3400 per semester for large one- or two-bedroom apartments. Numerous off-campus facilities are also available. Meal plans range from $320 (forty meals per semester) to $1196 (nineteen meals per week per semester).
Student Group	More than 7,100 undergraduates and 2,360 graduate students, in addition to approximately 1,000 students seeking professional degrees, are enrolled at Marquette University. The College of Engineering has more than 1,100 undergraduates and 325 graduate students.
Location	Marquette is an urban university located in midtown Milwaukee, Wisconsin. The city adds a special social and cultural dimension to the students' lives. Theaters, museums, and numerous restaurants are all within easy access of Marquette. Despite being a major metropolitan center, Milwaukee retains a small-town, friendly atmosphere. Famous for its festivals, cultural activities, and spectator sports, this city of 1.4 million people offers unlimited recreation possibilities.
The University	Marquette University is Wisconsin's largest independent institution of higher learning. The campus covers about 80 acres, and its fifty-one buildings house eleven colleges and schools. Founded in 1881 by members of the Society of Jesus, Marquette is committed to offering an education marked by intellectual excellence, the Judeo-Catholic tradition, and service to others.
Applying	Applicants must submit the following documents: a completed application form and application fee ($40), official transcripts from all current and previous colleges except Marquette, and three letters of recommendation. Doctoral applicants should write a statement of purpose and include GRE scores and copies of any published work, including theses and essays. International applicants must submit GRE and TOEFL scores. Application deadlines for financial aid awards are February 15 for the academic year or the fall semester, November 15 for the spring semester, and April 15 for the summer sessions.
Correspondence and Information	Admissions Coordinator Marquette University Graduate School P.O. Box 1881 Milwaukee, Wisconsin 53201-1881 Telephone: 414-288-7137 E-mail: mugs@marquette.edu World Wide Web: http://www.marquette.edu	Chairman, Department of (specify) Marquette University P.O. Box 1881 Milwaukee, Wisconsin 53201-1881 Telephone: 414-288-7079 (Engineering Office) E-mail: engrad@marquette.edu

Marquette University

FACULTY HEADS, RESEARCH AREAS, AND SPONSORED PROJECTS

Deans
G. E. Otto Widera, Professor, Chairman of the Department of Mechanical and Industrial Engineering, and Interim Dean of the College of Engineering; Ph.D., Wisconsin–Madison.
Jon K. Jensen, Associate Professor of Mechanical Engineering and Associate Dean for Undergraduate Affairs; Ph.D., Marquette.

BIOMEDICAL ENGINEERING
Dean C. Jeutter, Professor and Interim Chairman of the Department of Biomedical Engineering; Ph.D., Drexel.
Graduate Coordinator: Kristina M. Ropella, Associate Professor of Biomedical Engineering; Ph.D., Northwestern.

Research Areas
Artificial Limbs/Prostheses
Biomaterials/Biomechanics
Biomedical Optics
Biotelemetry
Cardiac Electrophysiology
Cell Transport and Metabolism
Functional Imaging (MR and X-Ray)
Hard and Soft Tissue Biomechanics
Head and Spinal Cord Trauma
Human Motion Analysis
Image Analysis/Processing
Mathematical Modeling
Physiological Signal Processing
Pulmonary Hemodynamics/Mass Transfer
Rehabilitation Engineering
Systems Physiology
Tissue Engineering
Transcutaneous Power Transfer

Recent Sponsored Projects
3-D Microvascular Imaging
Advanced Biotelemetry Systems for NASA
Bulk Soft Tissue in Trans-Tibial Amputees
Dental Biomechanics
Detection of Cardiac Arrythmias
Endothelial Cell Uptake and Metabolism
Functional MRI
Implantable Human Nerve Regeneration
Implantable Biotransceiver System
Microfocal Angiography
Neuromuscular Assessment in Children with Cerebral Palsy
Pediatric Motion Analysis
Pulmonary Hemodynamics/Hypertension
Radio Linked Tocodynamometer
Rocker Sole Shoe Kinematics in Diabetics
Soft Tissue Mechanics
Spinal Cord Injury Mechanisms/Kinematics
Vascular Biomechanics

CIVIL AND ENVIRONMENTAL ENGINEERING
Thomas H. Wenzel, Associate Professor and Chairman of the Department of Civil and Environmental Engineering; Ph.D., Northwestern.
Graduate Coordinator: Sriramulu Vinnakota, Professor of Civil Engineering; D.Sc., Swiss Federal Institute of Technology.

Research Areas
Accident Analysis
Asphalt/Concrete Pavement Performance
Biological Treatment of Groundwater
Concrete Materials
Diffuse Pollution
Drinking Water Treatment
Engineering Safety and Reliability
Environmental Control Systems
Pavement Textures
Public Works Management
Structural Stability
Structural Formwork
Theoretical and Applied Mechanics
Wastewater Treatment
Water Quality Modeling

Recent Sponsored Projects
Analysis and Design of Solder Joints in Microelectronics Applications
Accident Analysis of Abrasives and Salt for Snow and Ice Control
Cost Effective Concrete Pavement Cross Sections
Effects of Grinding on PCC Pavements
High Performance Rigid Pavements
Impacts of Pavement Surface Textures
Impacts Related to Pavement Texture Selection
Membrane Water Softening
Non-Linear Analysis of Steel Frames
Optimization of Milwaukee's Central Sewerage Control System
Public Perception of Midwest's Highway Pavements
Risk Based Urban Watershed Management
Stormwater Management and the Willingness to Pay for Environmental Quality
Treatability of Deicing Wastewater
Use of Micronutrients for the Control of Bio-Solids Bulking

ELECTRICAL AND COMPUTER ENGINEERING
Jeffrey L. Hock, Professor and Chairman of the Department of Electrical and Computer Engineering; Ph.D., Wisconsin–Milwaukee.
Graduate Coordinator: James E. Richie, Associate Professor of Electrical Engineering; Ph.D., Pennsylvania.

Research Areas
Acoustic Wave Devices
Antennas and Propagation
Artificial Intelligence and Expert Systems
Control Systems
Digital Signal Processing
Materials Science
Microwaves
Neural Networks
Power Systems and Devices
Sensors
Solid State Devices
Speech Processing

Recent Sponsored Projects
Characterization of Indoor Environmental and Cleanliness Sensors
Chemical Usage Monitoring for Dispensing Systems
Determination of Hexavalent Chromium in Liquid Environments Using Electrochemical Acoustic Wave Devices
Fluid Condition Monitoring
Graduate Assistance in Areas of National Need
Impact of Motor Loads on Power Electronic Building Blocks
Proximity Sensor Using Acoustic Waves
Reflection of Ultrasonic Lamb Waves: Theory and Applications
Shear Horizontal Surface Acoustic Wave Sensor Platform for Chemical and Biological Applications in Liquid Environments
Special Projects: GASDAY and Weathermaster
Study of Feature Extraction of Dental Disease Data Using Artificial Neural Networks
Surface Acoustic Wave (SAW) Sensor for Use in Ultrasonic Flow Measurements

MECHANICAL AND INDUSTRIAL ENGINEERING
G. E. Otto Widera, Professor and Chairman of the Department of Mechanical and Industrial Engineering; Ph.D., Wisconsin–Madison.
Graduate Coordinator: William E. Brower Jr., Professor of Mechanical Engineering; Ph.D., MIT.

Research Areas
Biomaterials and Biomechanics
Catalytic Materials
Composites
Deburring and Surface Finishing
Deformation Processing
Dynamic Systems and Control
Energy Conversion Processes
Ergonomics/Human Factors
Manufacturing Processes
Pressure Vessels and Piping
Quality and Reliability
Rapid Prototyping/Solidification
Solar Energy
Soldering in Electronics
Stress Analysis
Thermal Management of Electronic Assemblies
Thermodynamics/Heat Transfer
Thin Films
Vehicle Dynamics

Recent Sponsored Projects
Aircraft Subsystem Integration
Automated Deburring Process
Compliant Machine Tools for Surface Finishing Operations
Computational Fluid Dynamics Integration in Forging Processes
Electronics Thermal Management
Ergonomics Analysis of Line Mechanics in the Electric Power Industry
Ergonomics Evaluation and Design of Computer Keyboards
Ergonomics of Lawn and Garden Tools
Heat and Mass Transfer Characteristics of Indirect Evaporative Cooling
Performance of Deburring Systems
Pollution Control Equipment
Pressure Testing of Plastic Piping
Stiffness and Damping of Computer Keyboard Keys
Stress Analysis of Cylinder-Cylinder Intersections
Surface Mount Technology
Whole Body Vibration Isolation

MASSACHUSETTS INSTITUTE OF TECHNOLOGY

School of Engineering

Programs of Study

The MIT School of Engineering offers Master of Science (S.M.), Engineer, Ph.D., and Sc.D. programs in the Departments of Aeronautics and Astronautics, Chemical Engineering, Civil and Environmental Engineering, Electrical Engineering and Computer Science, Materials Science and Engineering, Mechanical Engineering, Nuclear Engineering, Ocean Engineering, and the Divisions of Bioengineering and Environmental Health and Engineering Systems. Master of Engineering (M.Eng.) degrees are offered in the Departments of Aeronautics and Astronautics, Civil and Environmental Engineering, Electrical Engineering and Computer Science, and Ocean Engineering. These eight departments and two divisions not only conduct advanced professional programs in their own disciplines but also provide the academic structure for graduate programs in interdisciplinary fields, which are administered by interdepartmental entities. The School also offers interdisciplinary programs leading to a master's degree or a Ph.D. in technology and policy; a program leading to a master's degree in transportation; joint master's programs with the Sloan School of Management in the management of technology, in transportation logistics, in systems design and management, and through the Leaders for Manufacturing program; and a joint doctoral program in polymer science and technology with the School of Science. Students studying in an area that encompasses more than one department may request the appointment of an interdepartmental committee to supervise their program. Descriptions of programs are included on the reverse side of this page; further information may be obtained by writing to the head of the appropriate department.

The Engineering Internship Program, available to students in the School, combines traditional on-campus academic programs with off-campus work experience in industry and government. It is a joint undergraduate and graduate program that leads to the simultaneous award of the bachelor's and master's degrees.

The master's degree is awarded upon the completion of an approved program of 66 subject units (approximately 22 semester credits) and presentation of an acceptable thesis. The Engineer degree requires an approved program of 162 subject units (approximately 55 semester credits) and an acceptable thesis. Students wishing to earn the doctorate must complete a program of advanced study, including a general examination, and perform an advanced research project, the results of which are reported in a thesis. In general, master's programs require a minimum of one year of full-time study, professional engineer's programs a minimum of two years, and doctorates a minimum of three years.

Research Facilities

Each department has an extensive complement of modern research facilities. The facilities of the Institute's interdepartmental laboratories are available to students. These include, but are not limited to, the Research Laboratory of Electronics; Laboratory for Information and Decision Systems; Artificial Intelligence Laboratory; Laboratory for Manufacturing and Productivity; Laboratory for Electromagnetic and Electronic Systems; Energy Laboratory; Laboratory for Computer Science; Microsystems Technology Laboratories; Center for Biotechnology Process Engineering; Center for Materials Science and Engineering; Materials Processing Center; Center for Technology, Policy, and Industrial Development; and Center for Transportation Studies. MIT's extensive computer facilities include departmental installations, large central time-sharing and batch services, a class 7 supercomputer, facilities for interactive graphics and computer-aided design, and the Athena facilities for educational computing.

Financial Aid

MIT makes financial support to graduate students available from a variety of sources in several different forms—fellowships, scholarships, traineeships, teaching and research assistantships, work-study (the Federal Work-Study Program), and loans (the Technology Loan Fund). More than half of the graduate students in the School are employed as teaching or research assistants as they progress toward their degree. These appointments give the student assistants opportunities for close interaction with faculty members.

Cost of Study

Tuition for graduate students in the School of Engineering for the two-term academic year is $25,000 in 1999–2000. Tuition for the summer is $8015 (maximum). Summer tuition may be waived for doctoral students engaged in research.

Living and Housing Costs

Dormitory accommodations for 466 single graduate students are available at a cost of $3508 to $4078 for the 1999–2000 academic year. Apartment accommodations for 594 single graduate students are available on a twelve-month basis (beginning September 1) at a cost of $376 to $576 per person per month (1999–2000). Meal service is offered seven days a week on either a contract or cash basis. For families, there are 406 apartments that range in price from $702 to $1060 per month, including utilities. Most residence assignments for new students are given for one year only for single students; there is a two-year option, however, for families. The majority of new graduate students applying for campus housing are assigned housing. Off-campus apartments are available within commuting distance of the campus; typical prices (excluding heat and utility costs) are $935 and up per month for a one-bedroom apartment and $1100 and up per month for a two-bedroom apartment. The quality of apartments is generally average, and the quantity is limited, so an early (one- to two-week) search for housing is advised.

Student Group

The engineering departments typically enroll 2,424 full-time graduate men and women, including 824 students from other countries. During the 1997–98 academic year, the School of Engineering awarded 791 master's degrees, 17 professional engineer's degrees, 222 Ph.D. degrees, and 17 Sc.D. degrees.

Location

MIT occupies a 154-acre campus on the north bank of the Charles River, with a fine view of Boston's skyline. The vacation areas of New Hampshire and Vermont, only a few hours from the campus, provide mountaineering, skiing, and hiking opportunities. The Boston area is world renowned for its cultural facilities. Performances of the Boston Symphony Orchestra, the Boston Pops, and the Boston Opera take place within a 5-minute drive or 30-minute walk of the campus. The area has many famous museums, restaurants, and historic sites, and more than a dozen universities and specialized schools are within a few miles of the campus.

The Institute

MIT, founded in 1861 as a private, coeducational, endowed institution committed to the extension of knowledge through teaching and research, is one of the foremost institutes of technology in the world. The Institute comprises five schools: Architecture and Planning, Engineering, Humanities and Social Science, Management, and Science. In 1998–99, 9,885 students were enrolled; more than half were in the Graduate School and the Whitaker College of Health Sciences and Technology. The campus provides excellent athletic facilities for indoor and outdoor sports, sailing, swimming, and crew. There are extensive cultural programs on campus.

Applying

Applications for entrance to the Graduate School in June or September should be made before January 15 for all engineering departments. Applicants seeking admission in February should file before November 1.

Correspondence and Information

Director of Admissions
Massachusetts Institute of Technology
77 Massachusetts Avenue
Cambridge, Massachusetts 02139
Telephone: 617-253-4791

Massachusetts Institute of Technology

DEPARTMENT HEADS AND RESEARCH AREAS

AERONAUTICS AND ASTRONAUTICS. Professor Edward F. Crawley, Head. Research is directed toward technologies for aircraft and space vehicles, their propulsion and information subsystems, and the communication, transportation, and exploration systems of which they are part. Areas of activity include composite materials and high-temperature structures; fracture and fatigue, structural dynamics, aeroelasticity, actively controlled structures, damage tolerance; aerodynamic design and optimization; active control applications in advanced propulsion systems; computational fluid dynamics; micromachines and devices; control-configured vehicles, fault tolerance in control systems, and control of space structures; concepts for space exploration; jet engine noise reduction; fluid mechanics of turbomachinery; turbine heat transfer; microjet and rocket engines; compression system stability; reduction of engine emissions; turbomachinery aeroelasticity; advanced concepts for high-performance and adaptation in space; international air transport, airline operation, air traffic control, airport planning; helicopter rotor dynamics and aerodynamics; navigation, guidance and control, autonomous vehicles, flight-critical software; space communication systems; complex product development and design for life-cycle cost and maintenance.

BIOENGINEERING AND ENVIRONMENTAL HEALTH. Professors Douglas A. Lauffenburger and Steven R. Tannenbaum, Co-Directors. Study and research focused on the interface between engineering and biology, combining quantitative, physical, integrative, and design-oriented principles with molecular and cellular biology and physiology. Major problem thrusts include biological imaging; biological microanalytics; biological transport processes; biomaterials; cell and tissue engineering; computational modeling and simulation of biological and physiological phenomena; delivery of molecular therapeutics; design and action of chemotherapeutic agents; electromechanical properties of tissues; environmental toxins, carcinogens, and pathogens; genetic toxicology and epidemiology; instrumentation for biological measurement; metabolism of exogenous compounds; molecular mechanisms of carcinogenesis and DNA damage/repair; and signal transduction, cell–cell communication, and extracellular matrix regulation of cell physiology.

CHEMICAL ENGINEERING. Professor R. C. Armstrong, Head. Study and research areas include biochemical engineering, emphasizing problems in protein engineering, cell culturing, and separations, as well as metabolic engineering for biological production of chemicals; biomedical engineering, with a focus on the analysis of pathophysiological phenomena, cell and tissue engineering, and the development of therapeutic devices; catalysis and chemical kinetics; chemical engineering systems and process control, including product design, process synthesis, process engineering, control, and optimization; combustion engineering; environmental engineering, employing chemical engineering fundamentals in air and water pollution control and solid-waste disposal; materials science and engineering of inorganic materials and organic polymers, including processing, rheology, structure-property relationships, and synthesis and characterization; thermodynamics, statistical mechanics, and molecular modeling; and transport processes. The School of Chemical Engineering Practice offers training in process engineering using an engineering internship in lieu of a master's thesis.

CIVIL AND ENVIRONMENTAL ENGINEERING. Professor R. L. Bras, Head. Research is carried out in two interdisciplinary laboratories. The focus is on the impact of human activities on the environment; on infrastructure development, renewal, management, and operation; and on the use of information technology to solve the engineering problems of the future. Research is carried out in the general areas of environmental chemistry and biology, environmental fluid dynamics and hydrodynamics, geoenvironment and geotechnology, surface and groundwater hydrology, materials and structures, transportation systems, information technology, and construction engineering and management. Interdisciplinary research is also promoted with the Technology Planning and Policy Program, the Center of Environmental Health Sciences, the Center for Global Change Science, the Center of Transportation Studies, the Woods Hole Oceanographic Institute, and the Center for Technology Policy and Industrial Development. Joint research is also ongoing with the Department of Architecture, the Media Laboratory, Lincoln Laboratory, Sloan School of Management, the Technology Development Program, and the Leaders for Manufacturing Program, among others.

ELECTRICAL ENGINEERING AND COMPUTER SCIENCE. Professor J. V. Guttag, Head. Most thesis research is conducted in laboratories sponsored by or affiliated with the department. The following partial list is indicative of the department's interests. **Laboratory for Information and Decision Systems:** modern control and system theory, computer communication networks. **Research Laboratory of Electronics:** radio astronomy, microwave electronics, digital signal processing, submicron technology, VLSI design, lasers and coherent optics, plasma physics and controlled fusion, bioengineering, communications biophysics, neurophysiology, speech communication, linguistics, cognitive information processing. **Laboratory for Computer Science:** knowledge-based application systems, computer languages, automata theory, complexity theory, computer networks, computer systems. **Artificial Intelligence Laboratory:** artificial intelligence, intelligent systems, vision, manipulation, robotics, natural language, research in learning. **Center for Materials Science and Engineering:** solid-state devices and MBE technology, microelectronic devices and integrated systems. **Laboratory for Electromagnetic and Electronic Systems:** electrohydrodynamics, production of high-energy particles and radiation, high-voltage transmission lines, power system modeling, application of semiconductor technology to power processing. **Microsystems Technology Laboratories:** silicon-integrated systems and process technology, compound semiconductor devices and processes, vacuum microelectronic devices.

MATERIALS SCIENCE AND ENGINEERING. Professor T. W. Eagar, Head. Research is conducted in the areas of synthesis and processing, structure, properties, and performance of materials. Programs are offered in ceramics, electronic materials, materials engineering, materials science, metallurgy, polymerics, and archaeological materials. A partial list of research areas includes the science and processing of electronic materials, physical ceramics, processing of high-technology ceramics, chemical and process metallurgy, polymer science and processing, composite materials, structure-processing-property relationships across classes of materials, structural materials, high-temperature materials, mathematical modeling, materials processing economics, manufacturing science, and industrial ecology. Examples of specific research topics include advanced spectroscopic and diffraction techniques, rapid solidification processing, high-strength polymer materials, computer modeling of structures and processing, high-temperature superconducting materials, high-temperature alloys, and photonic band gap materials.

MECHANICAL ENGINEERING. Professor N. P. Suh, Head. Major areas of research include disciplinary activities in mechanics and materials, fluid and thermal sciences, systems and control, and design and manufacturing. The department is also organized into research groups: manufacturing, product design, bioinstrumentation and bioengineering, information related to intelligent systems, and energy/transportation systems. Mechanics and materials research includes the study of acoustics, mechanics of sub-cellular biological systems, computational methodologies, continuum mechanics, nonlinear dynamics, waves, composites and the mechanical behavior of materials. Research in fluid and thermal sciences includes thermodynamics: combustion, IC engines, cryogenics, mesoscale heat engines, vortex methods; fluid mechanics: biofluids, non-Newtonian fluids, computational fluid dynamics, interfacial phenomena, molten microdrop deposition processes, flow imaging, and image processing; heat transfer: convection, extremely high heat fluxes, microscale transport, fluidized beds, electronics cooling, semiconductor processing, and nuclear thermohydraulics. Systems and control research includes control theory and implementation, bioengineering and living systems, robotics, classical and quantum information processing, precision mechatronics, and manufacturing systems. Design research includes axiomatic design, product design, environmentally conscious design, and web-based collaborative design, modeling, and optimization. Manufacturing systems research includes process innovation, DBM, 3-D printing, microcellular plastics, composites, precision engineering, and tribology.

NUCLEAR ENGINEERING. Professor Jeffrey Freidberg, Head. Research focus is on nuclear energy systems and radiation technology for medicine, industry, and the environment. Principal areas are reactor engineering, reactor safety analysis, nuclear materials engineering, applied plasma physics, applied radiation physics, biomedical nuclear technology, nuclear waste management, energy systems and policy analysis, radiation health physics, and radiological sciences. Research facilities include a 5,000-kw research reactor, an inelastic neutron spectrometer, a spatial NMR laboratory, two compact high-current accelerators, and various experimental facilities of the Plasma Fusion Center, including the ALCATOR C-Mod Tokamak and the Versatile Toroidal Facility (VTR).

OCEAN ENGINEERING. Professor C. Chryssostomidis, Head. Focus is on engineering for the ocean environment, including naval architecture, naval engineering design, ocean exploration and transportation, ocean sciences, marine robotics, and offshore engineering. A program in the utilization and management of ocean resources includes systems analysis, marine economics, policy, law, and management, with an emphasis on marine transportation and managing ocean-related environmental issues. The department conducts research in computer-aided engineering, marine hydrodynamics, marine resource and environmental management, ocean instrumentation and measurement for environmental monitoring, ocean-related acoustics, the response of ocean structures and ships to various loads, ship production, structural mechanics, systems analysis applied to marine operation, and underwater vehicles.

MICHIGAN STATE UNIVERSITY

College of Engineering

Programs of Study
Graduate programs leading to the M.S. and Ph.D. degrees are offered in biosystems engineering, chemical engineering, civil engineering, computer science, electrical engineering, environmental engineering, materials science, mechanical engineering, and mechanics. The College also offers programs leading to the M.S. degree in civil engineering–urban studies and environmental engineering–urban studies. The degree requirements vary by academic department.

Research Facilities
Each academic department has well-equipped laboratories for research and instruction. In addition, faculty and student researchers have access to a variety of College and University facilities.

The A. H. Case Center for Computer-Aided Engineering and Manufacturing conducts research activities that are responsive to industrial needs; operates the MSU Manufacturing Research Consortium, which focuses on environmentally responsible manufacturing; administers an International Technology Incubator to assist U.S. companies in tapping international expertise; conducts research on globally distributed teams; and co-administers the Genetic Algorithms Research and Applications Group (GARAGe).

The Composite Materials and Structures Center is an interdisciplinary research center supporting faculty and student research projects in composite materials and processing. Extensive state-of-the-art analytical, processing, fabrication, and characterization equipment is available. Equipment includes major groupings of instrumentation for surface and interfacial analysis, experimental mechanics and composite fabrication and characterization, as well as apparatus for characterization of chemical, thermal, and physical properties of materials. The center also houses the NSF State/Industry/University Cooperative Research Center for low-cost, high-speed polymer composite processing and operates the Advanced Materials Engineering Experiment Station in Midland, Michigan, for applied research and advanced development on polymeric composites.

Among the accessible University facilities are the Cyclotron Laboratory, the Center for Electron Optics, the Magnetic Resonance Imaging Facility, and the Laser Scanning Microscope Facility.

Financial Aid
Research and teaching assistantships and fellowships are available to qualified students. In 1998–99, annual half-time assistantship stipends were $15,192 for first-year M.S. students and $17,184 for senior M.S. students and Ph.D. students. Up to 6 credits of tuition per semester are waived for students with assistantships, and nonresident graduate assistants pay the resident tuition rate on all additional credits. Fellowships range from $100 per semester to $17,184 per calendar year.

Cost of Study
In 1998–99, graduate tuition was $222.50 per semester credit for Michigan residents and $450 per semester credit for nonresidents. Per-semester fees included a $288 registration fee for students enrolling for more than 4 credits ($238 for students enrolling for 4 or fewer credits), a $237 engineering program fee for students enrolling for more than 4 credits ($131 for students enrolling for 4 or fewer credits), a $3 FM radio tax, a $9 student information technology fee, a $4.25 student newspaper tax (for students enrolling for 10 or more credits), and a $4.50 Council of Graduate Students tax.

Living and Housing Costs
Owen Graduate Residence Hall offers comfortable living in an atmosphere conducive to advanced study and the exchange of ideas. The 1998–99 rates were $1928 per single room occupancy per semester and $1649 per double room occupancy per semester, including residence hall tax. Furnished University apartments are available for married students. The 1998–99 family monthly rates were $390 for a one-bedroom apartment and $432 for a two-bedroom.

Student Group
For the 1998 fall term, 43,189 students were enrolled on Michigan State University's East Lansing campus: 33,419 were undergraduates, 6,472 were graduate students, and 1,932 were professional students. Students come from every state in the Union and from 107 countries. Enrollment in the College of Engineering in the 1998 fall term was 4,480, of whom 3,888 were undergraduate students and 592 were graduate students.

Location
Michigan State University is located in the south-central part of the state in East Lansing, which is approximately 80 miles northwest of Detroit and about 210 miles northeast of Chicago. East Lansing is a city of about 30,000 people in a metropolitan area of more than 270,000. Lansing, the capital of Michigan, is located just 4 miles to the west. Many summer and winter recreational opportunities are located within driving distance of East Lansing.

The University and The College
Michigan State University, a land-grant A.A.U. institution, was founded in 1855. The curriculum includes more than 200 programs of undergraduate and graduate studies, all taught by more than 4,000 academic staff in fourteen degree-granting colleges. The property holdings at East Lansing number 5,239 acres. Some 2,100 acres are in existing or planned campus development; the remaining 3,139 acres are devoted to experimental farms, outlying research facilities, and natural areas. Major buildings number about 150 on the contiguous campus.

Applying
To ensure full consideration, applications for admission and all required documents must be received at least two months prior to the anticipated first semester of enrollment. If the applicant is also applying for financial aid, application materials must be received by January 15 for consideration for support commencing in the subsequent fall semester.

Correspondence and Information
Graduate Studies Coordinator
Department of (specify)
College of Engineering
Michigan State University
East Lansing, Michigan 48824

Michigan State University

FACULTY RESEARCH AREAS

Biosystems Engineering

Biosystems engineering combines the knowledge of physics, chemistry, and mathematics with engineering design, biology, and systems science. The discipline seeks sustainable solutions to support life on this planet. Biosystems engineers ensure an adequate and safe food supply, conserve the natural resources, and preserve the environment. The areas of study in biosystems engineering include systems engineering, food production, post-harvest technology, food engineering, natural resources, and the environment. Systems engineering research includes modeling of biologically based large-scale systems, resource optimization, and neural network applications. The food production area of emphasis includes soil physics and mechanics in tillage, traction, and soil compaction; crop planting, protection, and harvesting systems; computer vision guidance systems; global position systems (GPS); and precision agriculture. Projects in the post-harvest technology area encompass dehydration, automatic process control, product quality modeling, and product sorting, grading, and storage. The food engineering area includes a study of the rheological properties of food materials, supercritical fluid extraction, reverse osmosis, and extrusion processing and control. The projects in the natural resources and environment area include groundwater management, small watershed hydrology, supplemental irrigation, water quality, on-site wastewater treatment, composting, waste management, and odor control. Many other exciting and rewarding fields of study are available.

Chemical Engineering

Research interests within the department include programs or projects in separations, thermodynamics, reaction engineering, biochemical engineering, biomedical engineering, polymers, composite materials, ceramics, fluid mechanics/turbulence, transport properties, crystallization, surface and interfacial phenomena, adhesion, polymer processing, rheology, environmental remediation, and biomass conversion. Much of the research in the department is interdisciplinary, in cooperation with other investigators in the Colleges of Engineering and Natural Science, the Center for Composite Materials and Structures, the Crop and Food Bioprocessing Center, the Michigan Biotechnology Institute, and the Center for Microbial Ecology.

Civil and Environmental Engineering

The department offers graduate programs in the various focus areas of environmental, infrastructure, and transportation engineering with specific interests in transportation planning, traffic engineering, pavement engineering, geotechnical engineering, structural engineering, civil engineering materials, hydraulics, and environmental engineering. Current research includes computer-aided and reliability-based design in the following areas: geotechnical engineering; mechanistic analysis, performance prediction model development, nondestructive evaluation, and engineering management of pavement systems; truck-pavement interaction; field testing of instrumented pavement sections; groundwater contamination and dispersion processes; chemical and biological treatment of hazardous waste; remediation of chemical and petroleum spills in aquifers; cryogenic barrier technology to contain hazardous wastes; sorption phenomena; structural response to spatially varying ground motion; nonlinear random vibration; dynamic analysis on parallel computers; fiber-reinforced concrete; concrete durability and repair; recycling of waste materials and industrial by-products in concrete; highway safety-related geometric design; vehicle-driver characteristics and traffic operations; evaluation of highway safety programs; driver performance measurement; intelligent transportation systems; and highway safety in work zones. The department also conducts research in pavement engineering, traffic engineering, highway safety, and transit under the auspices of research center grants from the Michigan Department of Transportation.

Computer Science and Engineering

Advanced study is available in the areas of computer architecture, machine learning and robotics, high-performance computing, design automation, distributed systems, computer networks, artificial intelligence, knowledge-based systems, parallel systems and algorithms, pattern recognition, image processing, computer vision, software engineering, and theory of computing. The department operates a number of different laboratories with a variety of modern computing equipment. The department maintains state-of-the-art instructional and research facilities that are constantly upgraded. Instructional facilities include more than 200 color workstations, including Sun, Silicon Graphics, Pentium machines running NT, and Macintosh platforms, and servers. Facilities are networked campuswide and are available for student use 24 hours a day, seven days a week. Individual research groups in the department have their own laboratories consisting of specialized equipment. These include the Advanced Computing Systems Lab, the High-Speed Networking and Performance Lab, the Intelligent Systems Lab, the Pattern Recognition & Image Processing Lab, and the Software Engineering and Network Systems Lab. All computer science graduate students have a permanent account on the department network. In 1998–99, 121 students were supported as teaching and research assistants.

Electrical and Computer Engineering

The department has ongoing research projects in semiconductor physics and devices, integrated-circuits modeling, plasmas modeling and diagnostics; diamond and Si microsensors, high T_c superconductors, MOS technology; integrated-circuit fabrication, plasma-assisted etching and deposition; microwave processing of materials; transient electromagnetic scattering, antennas, radar target identification; analog electronics, macromodeling, computer-aided design, active filters; modeling and simulation of electronic circuits and systems; nonlinear control, singular perturbation methods, robust control, adaptive control using artificial neural networks; nonlinear systems and circuits, analog VLSI, adaptive and robust control; computer-aided design and manufacturing (CAD/CAM), interactive computer graphics; computer engineering, embedded processors, advanced architecture computers, performance measurement and visualization; VLSI architectures for enhanced control, VLSI design methodologies; digital circuit design, fault-tolerant design of microelectronic structures, logic synthesis for testability, and analog/digital circuits faults diagnosis; data acquisition, alternative energy systems; power systems stability, operation, and control; electrical machinery, finite-element methods for electromagnetic fields, electrical drives, power electronics; parameter estimation, system identification, adaptive signal processing; speech processing, system identification, biomedical signal processing.

Materials Science and Mechanics

Courses and research are available in the following areas: mechanics—continuum mechanics, micromechanics, linear and nonlinear elasticity, viscoelasticity, plasticity, experimental mechanics, optic methods of measurement, dynamics and vibrations, wave propagation, fatigue and fracture, biomechanics, mechanics of composites, computational mechanics and applied mathematics; materials science—mechanical metallurgy, composite materials, thermal and mechanical properties of metals, ceramics, polymers, and composites, phase transformations, electron microscopy, biomaterials, laser processing, high-temperature materials, ion-implantation, dynamic texture determination, superplastic forming, surface chemistry, and polymer physics. A majority of the faculty members in this department are associated with the Center for Composite Materials and Structures. Thus, multidisciplinary research opportunities in polymer-matrix, metal-matrix, and ceramic-matrix composites are also available.

Mechanical Engineering

Research programs in the Department of Mechanical Engineering include studies of energy conversion and utilization; optical measurements in heat transfer and fluid mechanics; enclosure heat transfer; fundamentals of fires and combustion; internal combustion engines; turbomachinery; fluid turbulence; systems modeling; heat transfer in biological materials; vibrations; dynamics; controls; photics; computational fluid mechanics, computational aeroacoustics, and heat transfer; active noise control; thermal properties of composite materials; nonlinear dynamics; shape optimization; material forming high-speed machinery; and smart composite materials. These studies involve analytical, computational, and experimental phases. Graduate courses help to prepare students for research and industrial practice. Students have the opportunity to conduct thesis research at several European universities, including Rheinisch-Westfälische Technische Hochschule in Aachen, Germany, and Technical University in Delft, Netherlands, under special exchange programs.

MICHIGAN TECHNOLOGICAL UNIVERSITY

College of Engineering

Programs of Study

The College of Engineering offers M.S. and Ph.D. degrees in all departments (see the reverse side of this page). Graduate programs generally have sufficient flexibility to encourage the tailoring of study plans to individual needs. Comprehensive examinations are given to Ph.D. candidates and, in some departments, to M.S. candidates before they begin their research activities. Thesis research is an important part of graduate study and is conducted under the guidance of the student's major adviser and graduate committee.

For students interested in a professional track, the College also offers a Master of Engineering degree.

In addition to the traditionally structured Ph.D., Michigan Technological University (MTU) offers a Ph.D. in engineering for study and research in the areas of computational science and engineering, environmental engineering, geotechnical engineering, sensing and signal processing, and structural engineering. The faculty members in these interdisciplinary areas come from two or more of the traditional engineering and science departments. Each interdepartmental grouping is charged with administration of the doctoral program in accordance with University rules and regulations for graduate studies and has the same degree of autonomy as an individual department in regulating its own programs. Interdepartmental activity of this type is conducted under the sponsorship of the College of Engineering, the Graduate School, and a University Interdepartmental Committee.

Research Facilities

The College of Engineering occupies seven modern buildings and has laboratories containing a wide variety of state-of-the-art research equipment. In addition to the academic department facilities, laboratory space and equipment for graduate students are provided by the Institute for Wood Research, the Institute for Materials Processing, the Keweenaw Research Center, and the Remote Sensing Institute.

Networked microcomputer labs and clusters of workstations are maintained by most academic departments. Departmental computer facilities provide faculty members and students with state-of-the-art hardware and software that not only meet the specific curricular and research objectives of the University, but also compare favorably with the computing environments students can expect to encounter in industry. Engineering departmental local area networks (LANs) are fully connected to each other and with almost all other nonengineering departmental LANs via a fiber campus backbone. The Michigan Tech fiber backbone is connected to the Internet via high-speed links through Michnet (Michigan's Regional Network) and NSFNET. Access to local and remote high-speed computing facilities is available along with an online library card catalog database.

The library maintains a modern open-stack system available to all students.

Financial Aid

Approximately 75 percent of all graduate students receive financial support through fellowships, traineeships, research assistantships, or graduate teaching assistantships. Although the amount of each stipend may vary, an M.S. student generally receives $2680 per quarter (plus tuition), and a Ph.D. student generally receives $3110 per quarter (plus tuition). Summer appointments are usually available.

Cost of Study

The quarterly tuition for 1998–99 was $2769 for out-of-state residents and $1292 for Michigan residents.

Living and Housing Costs

Apartments and dormitory rooms for single students are available in University-owned buildings. Room and board costs started at $4990 for the 1998–99 academic year. Housing for married students was available at $338 per month for a one-bedroom and $375 per month for a two-bedroom apartment. Utilities are included, and the apartments are partially furnished. These costs are approximate and subject to change.

Student Group

Michigan Technological University has 6,257 students enrolled University-wide, with 3,888 studying in the College of Engineering. The graduate program in engineering has 345 students in the M.S. and Ph.D. programs.

Location

The University is located in Houghton, Michigan, on the Keweenaw Peninsula, which protrudes into Lake Superior. This is an area of classic Precambrian geology, with many natural lakes, streams, and forests in the immediate vicinity. The area is known for its rugged natural beauty and offers numerous opportunities for outdoor recreation.

The local population of about 35,000 sponsors many cultural events. The major metropolitan centers of Detroit (550 miles away) and Minneapolis (325 miles) are readily accessible by regional airlines.

The University

Michigan Technological University has an excellent reputation in engineering education. By national standards, the University has superior laboratories and equipment and excels in several spheres of science and technology. Michigan Tech was founded in 1885 as a school of mining and metallurgical engineering. Although it maintains leadership in these areas, it has expanded its curriculum to include all areas of science and technology.

About eighty buildings are included in the main campus, housing administrative offices, classrooms, laboratories, computer facilities, lecture halls, living accommodations, a library, a student union, and a bookstore. The Walker Arts and Humanities Center is the campus focal point for cultural activities, including concerts, plays, lectures, and art exhibits. The A. E. Seaman Mineral Museum is known throughout the country for its outstanding mineral collections. In addition, the University operates an indoor ice arena, an eighteen-hole golf course, 8 kilometers of Nordic skiing and mountain biking trails, its own downhill ski hill with a chair lift, and an indoor tennis center. A major athletic-recreational complex houses a $\frac{1}{7}$-mile indoor track, two swimming pools, five handball courts, volleyball and basketball courts, dance rooms, a weight-lifting room, and a rifle range.

Applying

Applicants should have a baccalaureate degree from an accredited institution in an area of study that adequately prepares them for advanced study in their chosen field. They should have an undergraduate grade point average of 3.0 or better (on a 4.0 scale). Entrance exams may be required for some programs. Applications should be submitted as early as possible, preferably by February, in order to be considered for financial support. International applicants must indicate an ability to completely finance their own educational program for a period of twelve months and must submit TOEFL scores.

Correspondence and Information

Dean of the Graduate School
Michigan Technological University
1400 Townsend Drive
Houghton, Michigan 49931-1295
Telephone: 906-487-2327
World Wide Web: http://www.doe.mtu.edu

Michigan Technological University

FACULTY CHAIRS AND AREAS OF RESEARCH

Chemical Engineering. Associate Professor K. H. Schulz, Chair. Graduate research programs are available in process design and optimization, advanced process control, chemical process safety and risk assessment, polymer and composite materials processing, advanced materials/ceramics, surface science and catalysis, polymer rheology, thermodynamics and physical properties, biochemical engineering, and environmental engineering. The Process Simulation and Control Center integrates the Unit Operations Laboratory with advanced computer technologies for process simulation, evaluations, and control in an automated pilot plant setting.

Civil and Environmental Engineering. Professor C. R. Baillod, Chair. M.S. and Ph.D. degrees are offered in both civil engineering and environmental engineering. Areas of emphasis in the civil engineering graduate program include civil engineering materials, construction engineering/management, geotechnical engineering, transportation engineering, water resources engineering, and industrial by-product utilization. Areas of emphasis within environmental engineering include air-quality science and engineering, biological processes, environmental chemistry, groundwater and subsurface remediation, physical and chemical processes, pollution prevention, and surface water quality. The Master's International program allows students to combine graduate course work with the Peace Corp service. Twenty-six faculty members and more than 70 graduate students conduct more than $3 million of research per year. New and recently renovated laboratories located in Dillman Hall and the Dow Environmental Sciences and Engineering Building provide excellent facilities for analytical and experimental work. Computing facilities include more than forty Sun Workstations and more than seventy PCs. Additional information is available on the World Wide Web (http://www.cee.mtu.edu).

Electrical Engineering. Professor J. A. Soper, Interim Chair. Specialities are available in electrophysics, signal and image processing, power and energy conversion, and control systems. Research activities include fundamental properties of energetic materials; image recovery, with applications in astronomy and electron microscopy; imaging arrays; pattern recognition; adaptive optics; electric power systems analysis; intelligent systems; insulator failure measurements; power system transients; magnetic materials; collision avoidance systems; intelligent manufacturing; and image-based tracking. The department occupies five floors of the Electrical Energy Resources Center and has twenty laboratories for research and teaching, plus extensive computational facilities. A wide variety of sponsors are supported, including the U.S. Navy, the U.S. Air Force, the FAA, Ford, Northern States Power, Siemens, Consumers Power, and EPRI. The department's objective is to aggressively support both government and industrial sponsors and to push state-of-the-art engineering technology, while helping industry tackle some of its toughest current challenges. There are abundant opportunities for graduate students at both the M.S. and Ph.D. levels.

Geological Engineering and Sciences. Professor T. J. Bornhorst, Chair. The department offers graduate degrees in geological engineering, geophysics, and geology. Research projects cover a variety of areas in the general field of earth engineering and science. Research activities in geological engineering are concentrated in the fields of hydrogeology, multiphase groundwater systems, contaminant transport, soil and groundwater remediation, stability of clay liners, oil and gas production, and rock deformation. In geophysics, research activities focus on such topics as engineering geophysics and site investigations, electromagnetic and ground-penetrating radar, well logging and petrophysics, paleomagnetism and rock magnetism, and tectonics. Research in geology includes low-temperature aqueous geochemistry, basin evolution and sedimentology, volcanology, remote sensing and image analysis, clay mineralogy, petrology and water-rock interactions, mineral paragenesis and mineral deposits, structural geology, and petroleum geology. The department has modern, state-of-the-art research facilities, which are housed in the new Dow Environmental Sciences and Engineering Building. Major research laboratories include the Subsurface Visualization Laboratory, the Subsurface Remediation Laboratory, the Laboratory of Atmospheric Remote Sensing, and the Seaman Mineral Museum.

Mechanical Engineering and Sciences. Professor W. W. Predebon, Chair. Master of Science and doctoral degrees are offered. Department research efforts can be grouped into six broad areas: energy-thermal fluids, design, dynamic systems, manufacturing/industrial, solid mechanics, and bioengineering. An interdisciplinary option, manufacturing systems engineering, is offered. In addition to standard testing laboratories, special laboratories have been developed for work on automated manufacturing; metal removal; metal forming; stress analysis; fatigue, creep, vibration, and modal analysis; simulation of dynamic systems; acoustics and noise controls; controls; ferrography; internal combustion engine research (eight dynamometers); environmentally conscious manufacturing; high-strain rate behavior and impact response (split Hopkinson pressure bar); and plastics, composite materials, ceramics, and structures. Also available in the department are metallographic facilities, anechoic and reverberation chambers, an environmental chamber, and subsonic wind tunnels. Computing facilities include more than 110 Sun Workstations and more than seventy PCs.

Metallurgical and Materials Engineering. Professor C. L. White, Chair. The department is well equipped for research in all areas of metallurgy and materials. Complete optical microscopic facilities, multiple scanning and transmission electron microscopes, complete X-ray diffraction and spectrographic facilities, an electron microprobe, and a scanning Auger spectroscopy system are available. Materials preparation facilities are extensive and include single-crystal growing and metal hot working capabilities. Equipment for comminution, flotation, hydrometallurgy, and agglomeration is available for mining engineering research. The department, along with the University's Institute for Materials Processing (which is well equipped for pilot-scale processing of minerals and materials) is located in a 172,800-square-foot facility overlooking the Portage Waterway.

Mining Engineering. Professor O. F. Otuonye, Chair. Master of Science and doctoral degrees are offered in all areas of mining engineering. The department is a leader in resource engineering, the production of materials essential to human life from both natural and man-made sources. Mining, recycling, and waste utilization all provide useful resources, and the department research programs address how to mine and process these resources, regardless of their origin. In well-equipped rock mechanics, mineral and resource processing, underground, and computer laboratories, the department stresses resource engineering, geomechanics, tunneling, geotechnical site characterization, mechanical and explosive rock fragmentation, structural response to blast vibration, minerals and materials processing, drilling, construction of openings in rock, mine planning, and computer applications.

An aerial view of the campus. The College of Engineering buildings are in the background.

Michigan Technological University is an equal opportunity educational institution/equal opportunity employer.

MILWAUKEE SCHOOL OF ENGINEERING

Graduate Studies

Programs of Study

Milwaukee School of Engineering (MSOE) offers six master's degree programs. The Master of Science in Architectural Engineering (M.S.A.E.) offers a building design specialty that emphasizes application and design with functionality and economics. The Master of Science in Engineering (M.S.E.) enables the graduate engineering professional to solve problems by drawing from the fields of mechanical engineering, electrical engineering, computer engineering, and the physical sciences. The emphasis of this program is on the integration of technologies, rather than focusing on one narrow discipline. The Master of Science in Engineering Management (M.S.E.M.) is a technology-oriented management program designed to meet the needs of engineers, business managers, and other professional and technical personnel who are progressing into management. The Master of Science in Environmental Engineering (M.S.E.V.) provides practicing engineers with expertise in environmental systems design and environmental management in order to effectively address environmental regulations and issues. The Master of Science in Medical Informatics (M.S.M.I.) is a joint-degree program with the Medical College of Wisconsin that enables the medically based professional to combine medical science and computer information technologies to develop a body of knowledge and a set of techniques for the management of information in support of medical research, education, and patient care, which can contribute to improved medical care. The Master of Science in Perfusion (M.S.P.) is a hospital operating room discipline that is of great importance in invasive surgical techniques, particularly in open-heart surgery.

Courses in the M.S.A.E., M.S.E., M.S.E.M., and M.S.E.V. programs typically meet one evening per week during the regular academic year, and students attend on a part-time basis. The M.S.P. program is for full-time students only. It starts in September of each year, includes the summer, and ends in November of the following year. Upon completion of the program, students are eligible to sit for the Certified Clinical Perfusionist examination. The M.S.M.I. program is for full- and part-time students. Classes meet at MSOE and the Medical College of Wisconsin campus.

Research Facilities

The Applied Technology Center (ATC) is the research arm of MSOE. It conducts applied (strategic) research in conjunction with the university's various academic programs, utilizing faculty and student expertise as well as industrial-size laboratories to solve technological problems confronting business and industry. The close association between MSOE and the business and industrial community has long been one of the university's strengths. The ATC is heavily involved in the transferring of new technologies into real business practice through the Rapid Prototyping Center (MSOE is the only university in the world to possess the four leading rapid prototyping technologies), Fluid Power Institute (America's first fluid power research facility), Photonics and Applied Optics Center, High Impact Materials and Structures Center, Construction Science and Engineering Center, High Speed Video and Motion Analysis, and Center for BioMolecular Modeling. The ATC deals in research operations that include advanced manufacturing technologies, motion control and ultrafast videography, engineering and manufacturing consultation, and environmental areas. Both graduate and undergraduate students pursue research opportunities in the ATC.

Financial Aid

Most graduate students receive some type of tuition reimbursement from their employers. For those students who do not have this benefit, several financial loan options may be available. Nonimmigrant alien graduate students are not eligible for federal or state financial assistance or MSOE loan money. MSOE offers a limited number of graduate research assistantships.

Cost of Study

The 1999–2000 tuition for the M.S.A.E., M.S.E., M.S.E.M., M.S.E.V., and M.S.M.I. programs is $395 per credit hour. The cost of the M.S.P. program is $6000 per quarter (full-time enrollment).

Living and Housing Costs

MSOE operates three on-campus residence halls. Although undergraduate students comprise the largest segment of the resident population, the residence halls do offer an on-campus option to the graduate student. The Housing Office can provide more information on what is available and how personal needs might be accommodated. Renting off-campus housing from one of the many independently owned rental units near the university provides an alternative.

Student Group

In fall 1998, MSOE had 2,904 full- and part-time students representing thirty-six states and thirty-six countries. Of MSOE's 2,904 students, 413 were graduate students. Women constituted 18 percent of the overall student population. MSOE awarded 324 bachelor's degrees and 142 master's degrees in 1997–98. The majority of graduate students at MSOE have prior professional experience in their field.

Location

MSOE is located just a few blocks from Lake Michigan on the east side of downtown Milwaukee, which is approximately 90 miles north of Chicago. This central location is a short walk to the theater district, museums, sports arenas, shopping, and city festivals.

MSOE also offers classes at its Fox Valley Region Extension in Appleton, Wisconsin, for students who wish to pursue select graduate degrees in the evening on a part-time basis.

The University

Milwaukee School of Engineering was founded in 1903 to provide area industry with skilled mechanics and technicians. That one-room operation has grown into today's 13-acre university campus in downtown Milwaukee.

By focusing on applications-driven programs in engineering, business, technical communication, and health sciences, the university provides the best education in these specialty areas. Teaching is paramount, and faculty members have extensive experience in business and industry. No teaching assistants are used.

Applying

Applicants should submit an application, official transcripts of all undergraduate and graduate course work, letters of recommendation, a $30 fee, and appropriate test scores to the address given below.

Correspondence and Information

Graduate Admission
Milwaukee School of Engineering
1025 North Broadway
Milwaukee, Wisconsin 53202-3109
Telephone: 414-277-6763
         800-332-6763 (toll-free)
E-mail: explore@msoe.edu
World Wide Web: http://www.msoe.edu

Milwaukee School of Engineering

THE FACULTY AND THEIR RESEARCH

MSOE's faculty focus is on teaching, both in the classroom and in its world-class research facilities. Unlike many educational institutions, MSOE does not utilize teaching assistants. MSOE has 110 full-time faculty members and an additional 125 part-time nontenured faculty members. Small classes and a low 12:1 student-faculty ratio ensure that students receive personal attention.

The **Applied Technology Center (ATC)** serves as a technology transfer catalyst among academia, business and industry, and governmental agencies. The close association between MSOE faculty members and the business and industrial community has long been one of MSOE's strengths; applied research serves as a renewable resource in this linkage. The ATC undertakes more than 250 company-sponsored projects per year that involve faculty and staff members and students. Interdisciplinary capabilities provide a major advantage and can span fields such as engineering, science, business, computers, and technical communication. Modes of interaction include applied research and consulting by faculty members with industrial experience, often with graduate and undergraduate research assistants; projects in engineering and business disciplines, which are coordinated by company and faculty advisers; and referrals, which serve as an initial contact point for networking with others to optimize expertise and facilities for technology transfer.

The ATC, under the direction of Thomas Bray, Dean of Applied Research, is organized into several areas:

The **Rapid Prototyping Center (RPC)** is a joint effort between industry and MSOE that is dedicated to the application of proven technologies focused on reducing product development cycle times and biomedical applications. These technologies include computer-based rapid prototyping techniques and complementary processes leading up to full-scale manufacturing and modeling. The primary objectives of the RPC are integration of new technologies into real business practice and development of students who are ready to implement these practices upon entry into industry. The clients are industrial companies and educational institutions that desire to understand and advance rapid prototyping and related manufacturing methods and processes. Key to the RPC's vitality and success is a high level of industrial and biomedical parts design and fabrication activity using stereolithography (SLA), laminated object manufacturing (LOM), fused deposition modeling (FDM), and selective laser sintering (SLS) systems. Worldwide, MSOE is unique among universities in having this broad array of operating rapid prototyping systems. A twelve-module rapid prototyping curriculum is nearing completion under National Science Foundation (NSF) sponsorship. An NSF five-year program, Research Experience for Undergraduates, is active in this center. The RPC is also home to the Rapid Prototyping Consortium. This consortium, established in 1991, continues MSOE's tradition of building strong ties to business and industry. It includes industrial companies and educational institutions that cooperate in understanding and advancing rapid prototyping and manufacturing methods and processes.

RPC faculty members: John A. Choren, PE, Director of the Rapid Prototyping Consortium; Dr. Robert Crockett, Director, Academic Liaison; Professor Edward Howard, Program Manager, High School Outreach; Dr. Lisa Milkowski, Program Manager, Biomedical Applications.

The **Center for BioMolecular Modeling (CBMM)** was recently established within MSOE's RPC. This unique center focuses on the development of physical, hands-on models of proteins and other macromolecules for both research and educational purposes. In the coming years, MSOE hopes to become a recognized leader in the production of unique molecular models. Dr. Timothy Herman is Director of the CBMM.

The **Fluid Power Institute (FPI)** was established in 1962 as one of the first research facilities of its kind in the country and has remained a pioneer in motion control and fluid power education and technology transfer activities. It has expanded into electrohydraulic interface studies and currently has active programs in fluid power systems design, applications of fluid power to manufacturing, computational fluid dynamics, component evaluation, and filtration and contamination testing. Several laboratories with powerful variable fluid power sources and instrumentation support these programs. MSOE was recently selected by Caterpillar, Inc., together with Purdue University, as a partner in a long-term Master Sponsored Research Agreement in the field of electrohydraulics. This has allowed expansion of MSOE's research base and will improve Caterpillar's access to new technology. Thomas Wanke is Director of the FPI.

The **Photonics and Applied Optics Center** features state-of-the-art optical, laser, fiber-optic, sensor, and other photonics instrumentation. Holography, spectral analysis, communication and sensing, and many other optical and photonics applied research projects can be undertaken in the laboratory. Dr. A. James Mallman is Director of the Photonics and Applied Optics Center.

The **High Impact Materials and Structures Center** is developing concepts for blast-resistant cargo containers capable of mitigating explosions on aircraft flights for the Federal Aviation Administration (FAA). Participating faculty members have significant industrial experience and knowledge in the areas of shock-holing, dynamic structural analysis, materials testing, and advanced materials design. MSOE plans to consolidate design information and make it available to companies. Dr. Matthew Panhans is Director of the High Impact Materials and Structures Center.

The **Construction Science and Engineering Center** provides leadership and support to the construction industry through a variety of channels, including innovative applied research; testing, validating, and demonstrating new materials, processes, techniques, equipment, and methods; services to access information via the Internet—the latest in products, processes, and equipment; services for software evaluation and systems integration; and hosting continuing education programs needed by members of the industry. Dr. Douglas Stahl is Director of the Construction Science and Engineering Center.

The **High Speed Video and Motion Analysis** portable system has the ability to digitally capture and immediately play back events in the 1,000- to 12,000-frames-per-second range, enabling the user to analyze situations that would be impossible with conventional video or with the eye. Powerful motion analysis software can be used to track and graph up to nine points in the visual field. Contact for the system is Thomas Bray, Dean of Applied Research.

The Rapid Prototyping Center, through the Center for BioMolecular Modeling, is building world-class illustrations of life's building blocks, such as this protein.

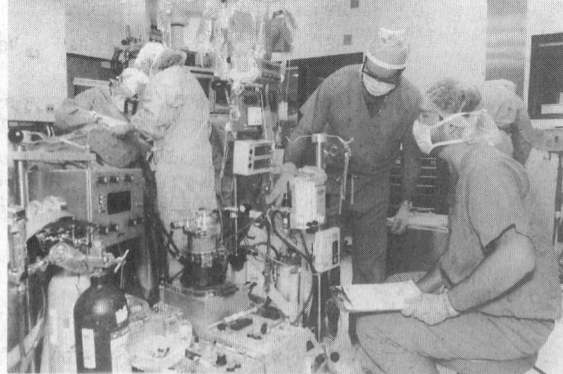

MSOE provides real-world learning experiences for all levels of students, including those in the Master of Science in Perfusion (M.S.P.) program.

MONTANA TECH OF THE UNIVERSITY OF MONTANA

Graduate Programs

Programs of Study	Montana Tech offers M.S. programs in geoscience with options in geochemistry, hydrogeology, hydrogeological engineering, geology, geological engineering, geophysical engineering, and mineral economics; engineering science; environmental engineering; industrial hygiene; metallurgical engineering; mineral processing engineering; mining engineering; petroleum engineering; and technical communication. The campus concentration of geotechnical and engineering expertise enables students to design individual programs such as hazardous wastes, mine waste, petroleum geology, geochemistry, and geophysics. Thesis and nonthesis options are available and normally take four semesters to complete. Students receive highly individualized instruction and thesis support.
Research Facilities	The engineering departments; the Montana Bureau of Mines and Geology, a research unit of the College; and the Center for Advanced Mineral and Metallurgical Processing each operate their own research facilities and programs. A wide variety of research is conducted in each program area (see reverse side of this page). Major efforts relate to geology; hydrogeology; mineral resources, including petroleum and coal; extractive metallurgy; geophysics; environmental geochemistry; engineering; bioremediation; and environmental controls. All programs have microcomputers and direct access to mainframe computers and the Internet on campus. Excellent analytical equipment is available to students, including several inductively coupled plasma mass spectrometers, a gas chromatograph/mass spectrometer, a scanning electron microscope, atomic absorption units, gas chromatographs, ion chromatographs, an electron microscope, seismic exploration apparatus, trace-gas analyzers for various gases, and other department-specific apparatus. There are two certified analytical laboratories on campus. The federal depository library on campus has 130,404 bound volumes, 155,210 microforms, 553 current periodicals, and an outstanding geosciences collection and is connected with the computerized Washington Library Network. There are approximately 600 PCs for student use, located in nine major computer labs on campus, as well as numerous laser and other printers.
Financial Aid	Teaching assistantships provided $6800 per academic year (taxable) in 1998–99 and required 20 hours per week. Some fee waivers for resident and nonresident fees are available, and there are various fellowships and research assistantships derived from contract research or research grants. These pay substantially more. Approximately 85 percent of graduate students receive some form of financial aid. Employment opportunities exist for students' spouses in local schools and businesses.
Cost of Study	In 1998–99, Montana residents paid $3107 for 12 credit hours. Nonresidents paid $9431 for 12 credit hours. Tuition and fees are estimated and are subject to change. Books may amount to $600 per academic year. The cost of thesis preparation ranges from $200 to $400. Sponsored research projects may pay most of the cost. Summer registration costs $702 for residents and $2190 for nonresidents for 6 credit hours.
Living and Housing Costs	In 1998–99, single students resided in the dormitory at a rate of $3629 for eight and a half months for a double room and board; single and duplex apartments were available for approximately $3900. The general cost of living is comparable to that of other towns in the Rocky Mountain area. Private automobile transportation is desirable if one lives off campus. Married student housing is available through the director of housing. Rents for unfurnished apartments range from $195 to $225 per month plus utilities; for furnished apartments, $225 to $257 per month plus utilities.
Student Group	Montana Tech has about 1,830 students, drawn largely from local areas and the state of Montana. The ratio of men to women in the student body is about 3:2. Out-of-state students account for 34 percent of the enrollment and international students for about 6 percent. Graduate students number 100, with about 6 percent international students.
Location	Located 5,750 feet above sea level, Butte has a dry climate that is cold in winter and cool in summer. About 35,000 people live in the area. Copper mining is conducted by Montana Resources, Inc. The Montana Power Company is headquartered in Butte, as are several subsidiary companies. Butte is also a regional medical center. Other major enterprises in the town include the DOE Resource Recovery Project and the EPA Mine Waste Technology Pilot Program, both operated by MSE; the National Center for Appropriate Technology; Special Resource Management; and research and development businesses such as Mycotech, Arco, and Pegasus Gold. The surrounding area supports ranching, tourism, and mining. Excellent hunting, camping, fishing, and skiing may be enjoyed in the mountainous countryside, much of which is national forestland. Served by Interstate Highways 15 and 90 and by Horizon Air and Skywest airlines, Butte is a distribution center for many commercial enterprises and is the site of the Port of Montana.
The College	Montana Tech is a member of the University of Montana System. The campus emphasizes engineering, science, and technology-based undergraduate and graduate programs with a strong orientation toward business and industry. Established in 1895 as the Montana School of Mines, the focus of the College historically has been on mineral- and energy-related professional engineering programs. While this focus has been maintained, degree programs have been added in the basic sciences that support the engineering programs and in fields related to the administration, application, and societal impact of the engineering programs. These programs derive special character and emphasis from the unusual setting of the institution and continue the tradition of high-quality education that has characterized Montana Tech since its founding. Montana Tech was ranked first among Western regional universities for the best educational value in the October 1994 issue of *U.S. News & World Report*. The College was also ranked in the top fifteen "Best Buys in College Education" (Southwest and Mountain States, 1995) in *Money* magazine and was listed in *America's 100 Best College Buys 1999*. The College houses the Montana Bureau of Mines and Geology, which conducts research, does public service work, and offers employment opportunities for students.
Applying	Semesters begin in early September and mid-January. The priority deadline for graduate admissions is April 1, but applications are received throughout the year. Financial aid, if available, is usually committed by late May; early application is advised. Admission is based on grade point index (B minimum) and references. GRE General Test scores are required for all prospective students. The TOEFL is required for non-English-speaking students, who must earn a minimum score of 525. All applications must be accompanied by a $30 (U.S. dollars) application fee. International applications must be accompanied by a financial statement.
Correspondence and Information	Graduate School Montana Tech 1300 West Park Street Butte, Montana 59701 Telephone: 406-496-4128 E-mail: cdunstan@mtech.edu World Wide Web: http://www.mtech.edu

Montana Tech of the University of Montana

THE FACULTY AND THEIR RESEARCH

Chemistry and Geochemistry Program
D. A. Coe, Professor and Department Head; Ph.D., Oregon State, 1974. Graph theory, environmental transport, chemical education.
D. A. Drew, Professor; Ph.D., Wyoming, 1971. Synthesis of organometallic compounds, molecular structure, spectroscopy.
D. B. Stierle, Professor; Ph.D., California, Riverside, 1977. Organic chemistry, natural products.
D. Cameron, Associate Professor; Ph.D., Purdue, 1979. Analytical environmental chemistry, chemical speciation, chelating and electrically conducting polymers.
A. Stierle, Research Professor; Ph.D., Montana State, 1988. Organic chemistry, natural products.
J. D. Hobbs, Assistant Professor; Ph.D., New Mexico, 1991. Computational chemistry and spectroscopy.

Engineering Science Program
D. Westine, Associate Professor and Department Head; Ph.D., North Dakota State, 1991. Systems control, bioengineering, electrical engineering.
L. L. Friel, Professor; Ph.D., Georgia Tech, 1973; PE. Civil engineering, optimum structural design, hydraulic backfill.
C. Hilpert, Professor; Ph.D., Oklahoma State, 1972; PE. Fluid power, machine design, theoretical stress analysis.
J. F. McGuire, Associate Professor; Ph.D., Utah, 1972. Engineering reports and graphic presentations.
R. Johnson, Associate Professor; M.S., Montana Tech, 1988. Concurrent engineering.
D. Trudnowski, Associate Professor; Ph.D., Montana State, 1991. Systems control, electrical engineering power, instrumentation.
N. K. Wahl, Associate Professor; M.S., Montana State, 1974; PE. Thermodynamics, heat transfer, fluid mechanics.
P. D. O'Leary, Director of Engineering Science Labs; B.S., Montana Tech, 1976. Welding engineering, nondestructive examination, welding metallurgy.

Environmental Engineering Program
K. Ganesan, Professor and Department Head; PE; Ph.D., Washington State, 1981. Air pollution measurement, modeling, and control.
R. Appleman, Professor; Ph.D., California, Irvine, 1978; PE. Environmental sampling and water quality modeling.
R. A. James, Professor; Ph.D., Montana State, 1973; PE. Pollution control, plant engineering, wetlands.
T. Waring, Professor; Ph.D., Pittsburgh, 1970. Environmental law.
W. J. Drury, Associate Professor; Ph.D., Montana State, 1992. Water and wastewater treatment, wetlands, solid-waste disposal.
H. Gerbrandt, Associate Professor; PE; Ph.D., New Mexico, 1993. Water quality, specifically lead and drinking water, wetlands reclamation.
H. Peterson, Associate Professor; Ph.D., Washington State, 1989. Atmospheric dispersions, instantaneous concentration predictions.
J. Larson, Director of Environmental Engineering Labs; B.S., Montana Tech, 1978. Environmental engineering.

Geological/Hydrogeological Engineering Program
D. Wolfgram, Associate Professor and Department Head; Ph.D., Berkeley, 1977. Mining geology, field geology, magmatic ore deposits.
M. Sholes, Professor; Ph.D., Texas at Austin, 1978. Coal geology and sedimentology, optical mineralogy.
W. Weight, Professor; Ph.D., Wyoming, 1989. Hydrogeology, groundwater, geostatistics, mathematical geology.
C. Gammons, Assistant Professor; Ph.D., Penn State, 1988. Hydrochemistry, ore deposits research, experimental geochemistry.
M. MacLaughlin, Assistant Professor; Ph.D., Berkeley, 1996. Rock mechanics, slope stability, earthquake engineering.
C. Elliott, Adjunct Assistant Professor; Ph.D., New Brunswick, 1988. Tectonics and structural geology.

Industrial Hygiene
J. Norman, Professor and Department Head; C.I.H., M.P.H., Houston, 1980. Industrial hygiene, toxicology, noise control.
B. Spath, Professor; Ph.D., Missouri–Columbia, 1979. Biomechanics, statistical analysis.
T. Spear, Professor; M.S., Minnesota, 1980. Industrial hygiene, health and safety management, particle aerosol sizing.
J. Amtmann, Associate Professor; M.S., Stroudsburg, 1990. Strength and conditioning, cardiac rehabilitation.

Metallurgical/Mineral Processing Engineering Program
C. Young, Associate Professor and Department Head; Ph.D., Utah, 1995. Mineral and coal processing, materials handling, electrochemistry, spectroscopy and control.
V. Griffiths, Professor; Sc.D., MIT, 1955. Materials problems, failure analysis, X-ray diffraction, scanning electron microscopy.
H. Haung, Professor; Ph.D., Stanford, 1975. Kinetics, hydrometallurgy, thermodynamics of aqueous solutions.
L. G. Twidwell, Professor; D.Sc., Colorado School of Mines, 1966. Thermodynamics of metallic solutions, extractive metallurgy, environmental problems.
S. A. Worcester, Professor; M.S., RPI, 1958. Materials science, plant design and operation, welding.
W. Huestis, Director of Metallurgy and Mineral Processing Engineering Laboratories; B.S., Montana Tech, 1969.
C. Anderson, Research Faculty, Center for Advanced Mineral and Metallurgical Processing; Ph.D., Idaho, 1988. Hydrometallurgy, electrometallurgy, pyrometallurgy, engineering design, X-ray diffraction.
C. Cross, Assistant Professor; PE; Ph.D., Colorado School of Mines, 1986. Welding metallurgy, weldability evaluation, modeling defect formation, filler alloy development, materials selection, foundry, solidification.

Mineral Economics Program
J. C. Brower, Professor and Program Director; Ph.D., Penn State, 1978. Mineral economics.
K. E. Burgher, Professor; Ph.D., Missouri–Rolla, 1985. Mineral economics, rock fragmentation.

Mining Engineering Program
H. P. Knudsen, Professor and Dean; Ph.D., Arizona, 1981. Geostatistics, mine planning.
T. E. Finch, Professor; Ph.D., Idaho, 1978; PE. Surface mining, ventilation, mining systems.
S. B. Patton, Assistant Professor; Ph.D., Alabama, 1993; PE. Ventilation, environmental aspects of mining.

Petroleum Engineering Program
T. Ahmed, Professor; Ph.D., Oklahoma, 1980. Phase behavior, reservoir simulation, EOR, reservoir engineering.
D. Bradley, Professor and Dean; Ph.D., Michigan State, 1977. Fluid properties, reservoir engineering.
G. V. Cady, Associate Professor; Ph.D., Stanford, 1969; PE. Well testing, management, economics.
J. G. Evans, Assistant Professor; M.S., Penn State, 1967. Oil production operations, reservoir engineering operations, well bore simulation.
M. Ziaja, Assistant Professor; Ph.D., University of Mining and Metallurgy (Poland), 1982. Rock mechanics, drill-bit design.

Physics/Geophysical Engineering Program
W. R. Sill, Professor and Department Head; Ph.D., MIT, 1967. Electrical methods, geothermal and exploration geophysics.
C. J. Wideman, Professor; Ph.D., Colorado School of Mines, 1974. Seismology, geothermal and exploration geophysics.
C. Link, Assistant Professor; Ph.D., Houston, 1993. Neural networks, exploration geophysics, crosshole seismology.
T. Moon, Assistant Professor; Ph.D., Washington (Seattle), 1985. Geophysics, paleomagnetism, remote sensing, image processing.
M. A. Speece, Assistant Professor; Ph.D., Wyoming, 1992. Engineering and environmental geophysics, high-resolution seismology, ground penetrating radar.

Project Engineering and Management Program (A joint degree between Montana Tech and Montana State University–Bozeman)
Montana Tech Program Director: K. Ganesan, Professor and Department Head of Environmental Engineering; Ph.D., Washington State, 1981. Air pollution measurement, modeling, and control.
R. A. James, Professor of Environmental Engineering; Ph.D., Montana State, 1973. Pollution control, plant engineering, wetlands.
MSU–Bozeman Program Director: J. T. Sears, Professor and Department Head of Chemical Engineering; Ph.D., Princeton, 1965. Kinetics, CVD, bacterial adsorption.
V. A. Cundy, Professor and Department Head of Mechanical and Industrial Engineering; Ph.D., Wyoming, 1979. Combustion, heat transfer.
W. P. Scarrah, Professor of Chemical Engineering; Ph.D., Montana State, 1973. Process development, process engineering, separations.

Technical Communication Program
J. Cortese, Professor and Department Head; Ed.D., Nova, 1990. Computer education, history, professional and technical communication.
D. Carter, Professor; Ph.D., California, Riverside, 1981. Psychology, quantitative research.
W. Macgregor, Professor; Ph.D., Colorado at Boulder, 1981. Literature, language and rhetoric, technical communication and production.
E. P. Munday, Associate Professor; Ph.D., Cornell, 1987. History and philosophy of scientific and technical communication.
P. Van der Veur, Associate Professor; Ph.D., Ohio, 1996. Media studies, multimedia productions, African studies.

NEW JERSEY INSTITUTE OF TECHNOLOGY

Newark College of Engineering

Programs of Study

New Jersey Institute of Technology (NJIT), America's most wired public university according to *Yahoo! Internet Life,* offers graduate programs of study in a wide range of engineering disciplines leading to the Doctor of Philosophy and Master of Science degrees through the Newark College of Engineering. Programs are available in applied chemistry, biomedical engineering, chemical engineering, civil engineering, computer engineering, electrical engineering, engineering management, engineering science, environmental engineering, environmental science, industrial engineering, manufacturing engineering, mechanical engineering, occupational safety and health engineering, occupational safety and industrial hygiene, power engineering, telecommunications, and transportation. Most programs may be completed on either a full- or part-time basis. Graduate certificate programs (12 credits of study) are also available in subjects that are in demand by business and industry.

Doctoral programs are designed to fill society's need for creative research scientists and engineers. The curriculum usually consists of a combination of course work and dissertation work for 60 credits beyond the master's degree. A unique Industry-Collaborative Doctoral Program also allows working professionals to pursue the Ph.D. while employed full-time.

Master's degree programs provide the advanced education needed by professionals in an era of rapidly expanding technology. These programs generally require 30 credits of course work with or without a 6-credit thesis or 3-credit project.

Research Facilities

NJIT is home to twenty-five major interdisciplinary research centers, many featuring partnerships with other leading research universities and industry. Graduate students may participate in cutting-edge research in areas such as environmental science and engineering, architecture and building science, manufacturing, electronics and communications, information technology, materials science and engineering, transportation, and infrastructure. More information may be obtained on the World Wide Web at http://www.njit.edu/Directory/Centers.html.

Financial Aid

Support is available to full-time students in the form of teaching, graduate, and research assistantships. Support for summer research is also available. Tuition remission is often included in assistantships. Students are urged to submit the application for admission no later than January 15, due to the highly competitive nature of securing financial support.

Cost of Study

Tuition for part-time students in 1998–99 was $368 per credit for residents of New Jersey and $509 per credit for nonresidents. Full-time tuition (12 to 19 credits) was $3476 per semester for residents and $4885 per semester for nonresidents. The tuition costs cited do not include student fees.

Living and Housing Costs

A limited amount of on-campus housing, the cost of which averages $4490 annually, is available for graduate students. Average room and board total $6382 per academic year. The Office of Residence Life assists in finding off-campus housing in Newark and the surrounding communities.

Student Group

Of the 8,200 students enrolled at New Jersey Institute of Technology, 3,000 are either full- or part-time graduate students.

Location

NJIT's 45-acre campus is located in the University Heights section of New Jersey's largest city, Newark. Its location offers many activities. It is home to the New Jersey Performing Arts Center, the New Jersey Symphony, and the Newark Museum. Branch Brook Park is just minutes from the campus. New York City is 10 miles away, and campuses of Rutgers University, the University of Medicine and Dentistry of New Jersey, Seton Hall Law School, and Essex County College are just a few blocks away. Many programs operate collaboratively with these institutions. Public transportation is available, and the New Jersey shore is close by.

The Institute

New Jersey Institute of Technology is the state's public technological research university. Founded in 1881, NJIT has maintained close ties with industry by preparing generations of students to assume leadership roles in an increasingly technological society. The university's first and largest school is Newark College of Engineering, which was established in 1919 to reflect the institution's evolution into a four-year college. When the School of Architecture was established in 1973, the name changed to New Jersey Institute of Technology in order to reflect a broadened mission. The College of Science and Liberal Arts, founded in 1982; the School of Management, founded in 1988; and the Albert Dorman Honors College, founded in 1993, round out the university's educational offerings.

Applying

Completed applications must be submitted to the Office of University Admissions no later than June 5 for fall admission and October 15 for spring admission. Students who wish to be considered for financial support must submit a completed application for admission by January 15. Students can also apply on line at the Web address listed below. GRE scores are required for admission to all doctoral programs and most master's degree programs, for all students seeking financial support, and for those whose last prior degree was from outside the United States. Official transcripts, letters of recommendation, and a $50 nonrefundable application fee are also required. International students must provide TOEFL scores as well as equivalent academic credentials from their countries of origin.

Correspondence and Information

Office of University Admissions
New Jersey Institute of Technology
University Heights
Newark, New Jersey 07102-1982
Telephone: 973-596-3300
Fax: 973-596-3461
World Wide Web: http://www.njit.edu

New Jersey Institute of Technology

ACADEMIC DEPARTMENTS AND RESEARCH CENTERS

Dr. Sheng-Taur Mau, Dean of Newark College of Engineering; telephone: 973-596-3222; Web site: http://www.njit.edu/Schools/NCE/nceadmin.html.

Chemical Engineering, Chemistry and Environmental Science: Dr. Gordon Lewandowski, Chairperson; telephone: 973-596-3568; Web site: http://www.njit.edu/Directory/Academic/Chem/Welcome.html.

Civil and Environmental Engineering: Dr. Edward Dauenheimer, Acting Chairperson; telephone: 973-596-2444; Web site: http://www.njit.edu/Directory/Academic/Civil/Welcome.html.

Electrical and Computer Engineering: Dr. Richard Haddad, Chairperson; telephone: 973-596-3512; Web site: http://www.njit.edu/Directory/Academic/ECE/Welcome.html.

Industrial and Manufacturing Engineering: Dr. Athanassios Bladikas, Acting Chairperson; telephone: 973-596-3185; Web site: http://www.njit.edu/Schools/NCE/imedept.html.

Mechanical Engineering: Dr. Dennis Siginer, Chairperson; telephone: 973-596-3331; Web site: http://www.njit.edu/Academic/ME/index.html.

RESEARCH CENTERS

Center for Communications and Signal Processing: Dr. Yeheskal Bar-Ness, Executive Director; telephone: 973-596-8474; Web site: http://www.njit.edu/Directory/Academic/ECE/Centers/ecelab.html#communications.

Otto H. York Center for Environmental Engineering and Science
    Center for Airborne Organics: Dr. Robert Pfeffer, Director; telephone: 973-642-7296.
    Hazardous Substance Management Research Center: Dr. Peter Lederman, Director; telephone: 973-596-3233; Web site: http://www.hsmrc.org.
    Northeast Hazardous Substance Research Center: Dr. Richard Magee, Director; telephone: 973-596-5883; Web site: http://www.njit.edu/NHSRC/.
    Sustainable Green Manufacturing Systems: Dr. Daniel Watts, Director; telephone: 973-596-3465.

Center for Manufacturing Systems: Dr. Donald Sebastian, Executive Director; telephone: 973-642-4869; Web site: http://www.cms.njit.edu.

Center for Membrane Technologies: Dr. Kamalesh Sirkar, Director and Sponsored Chair; telephone: 973-596-8447; Web site: http://www.njit.edu/Directory/Academic/Chem/faculty/sirkar.htm.

Center for Technology Studies: Dr. Nancy Jackson, Director; telephone: 973-596-8467; Web site: http://www.njit.edu/Directory/Centers/CPS/

Electronic Imaging Center: Dr. Haim Grebel, Director; telephone: 973-596-3533; http://www.njit.edu/Directory/Academic/ECE/Faculty/Grebel.html.

Institute for Transportation: Dr. Louis Pignataro, Executive Director; telephone: 973-596-3355; http://kimon.njit.edu/

Microelectronics Research Center: Dr. Kenneth Farmer, Executive Director; telephone: 973-596-5714; Web site: http://www.njit.edu/Directory/Centers/MRC/index.html.

Multi-Lifecycle Engineering Research Center: Dr. Reggie Caudill, Executive Director; telephone: 973-642-7198; Web site: http://www.njit.edu/MERC/.

National Center for Transportation and Industrial Productivity: Dr. Lazar Spasovic, Director; telephone: 973-596-3355; http://kimon.njit.edu/NCTIP/

New Jersey Center for Multimedia Research: Dr. Ali Akansu, Director; telephone: 973-596-5650; Web site: http://www.njcmr.org/.

New Jersey Center for Wireless Telecommunications: Dr. Richard Haddad, Executive Director; telephone: 973-596-8474.

New Jersey Technical Assistance Program: Michael Wallace, Director; telephone: 973-596-5864; Web site: http://www.njit.edu/njtap/.

Particle Technology Center: Dr. Robert Pfeffer, Director; telephone: 973-596-5829; Web site: http://ww-ec.njit.edu/ec_info/image2/ptc/.

Transportation Information and Decision Engineering Center: Dr. Louis Pignataro, Executive Director; telephone: 973-596-3355.

NORTH CAROLINA AGRICULTURAL AND TECHNICAL STATE UNIVERSITY

College of Engineering

Programs of Study

The College of Engineering offers the Doctor of Philosophy in electrical and mechanical engineering, the Master of Science in Engineering (M.S.E.), the Master of Science in Architectural Engineering (M.S.A.E.), the Master of Science in Chemical Engineering (M.S.Ch.E.), the Master of Science in Electrical Engineering (M.S.E.E.), the Master of Science in Industrial Engineering (M.S.I.E.), the Master of Science in Mechanical Engineering (M.S.M.E.), and the Master of Science in Computer Science (M.S.C.S.) degrees. The M.S.E. program is designed for students who wish to study topics of an interdisciplinary nature or to pursue graduate-level work in agricultural and biosystems or civil engineering. The other M.S. programs are for students who want to earn degrees in architectural, chemical, electrical, industrial, or mechanical engineering or computer science. Ph.D. programs require a Master of Science in Engineering and 24 credits of courses and 12 credits of dissertation research. M.S. programs have the thesis option, which includes 24 credits of courses and 6 credits of thesis work, or the project option, which includes 30 credits of courses and 3 credits of an independent project. Course work–only options are available and require 36 credits of courses. All thesis options are intended primarily for research-oriented students who are potential Ph.D. candidates and who wish to concentrate their efforts on a topic of publishable quality. Project options are intended for students wishing to investigate a design problem of current interest to industry or to pursue more applied study. Thesis and project option topics include, but are not limited to, HVAC and environmental systems design and structural systems (M.S.A.E.); facilities engineering; solid-state electronics and materials, digital systems and circuits, microelectronics, and communications (M.S.E.E.); energy management, human-machine systems, manufacturing engineering, and system analysis/simulation (M.S.I.E.); mechanics and materials, thermal sciences, and design and manufacturing (M.S.M.E.); and interdisciplinary topics in addition to those in chemical and civil engineering.

Ph.D.-level study is also available through an interinstitutional Ph.D. program with North Carolina State University (NCSU). The program is intended for Ph.D. degree candidates studying at North Carolina A&T and enrolled in a Ph.D. program at NCSU.

Research Facilities

The College's public facilities include five DEC Alpha computers, two SPARC computers, two MIPS computers, and fifty Intel PCs. The College's network consists of an FDDI ring that connects all buildings. General research equipment includes a Convex C-120 vector processor and a DEC 5400 minicomputer, both on a College-wide local area network. This network includes a number of workstations and personal computers and provides access to numerous computer facilities throughout the state and country. Other research facilities include a well-equipped shop; a scanning electron microscope with X-ray energy-dispersive analyzer; VLSI design and solid-state materials laboratories; a solar collector laboratory; a thermal-physical property laboratory with hot-wire/film, anemometry, and Schlieren systems; photoelasticity and interferometer facilities; a composite materials processing, fabrication, and testing center; solid-state device fabrication areas; integrated manufacturing and human-machine system laboratories; ergonomics and decision support systems laboratories; construction materials and structures laboratories; indoor environmental and HVAC laboratories; and the Center for Energy Research and Technologies. The College of Engineering's network provides access to the North Carolina Supercomputing Center and supports a teleclassroom/teleconferencing facility for transmitting and receiving courses, seminars, and conferences.

Financial Aid

Graduate students are generally considered for three types of support: research assistantships, teaching assistantships, and fellowships.

Cost of Study

The 1998–99 graduate tuition and required fees for North Carolina residents were $840 for 9 or more credit hours. Corresponding rates for nonresidents were $4475 for 9 or more credit hours.

Living and Housing Costs

Graduate students normally find reasonably priced housing near the campus or in the city of Greensboro. Typical apartment rents ranged from $350 to more than $450 per month.

Student Group

The engineering enrollment was 1,454 in the fall semester of 1998; the engineering graduate enrollment of full-time and part-time students was 293. Total enrollment for the University was approximately 7,354.

Location

Greensboro, Winston-Salem, High Point, and Burlington, all within a radius of 30 miles, are the major cities in the Piedmont Triad, the most heavily populated metropolitan area in North Carolina. Among the major industries and companies in the area are the corporate headquarters of Burlington Industries, Jefferson Standard Life Insurance Company, Analog Devices, Konica, AMP, Volvo-White Truck Corporation, and R F Micro Devices. Other universities in the area are the University of North Carolina at Greensboro, Bennett College, Wake Forest University, North Carolina School of the Arts, Winston-Salem State University, and Guilford College. To the east are Research Triangle Park (60 miles) and Raleigh (78 miles), the state capital; Charlotte is 85 miles west. Many professional and cultural activities flourish in the region. Greensboro has been noted for its outstanding quality of life in recent Rand-McNally surveys.

The University and The College

North Carolina A&T State University was originally organized as a land-grant college. The College of Engineering operates on a two-semester academic year, with two 5-week summer sessions at the graduate level.

Applying

Applications for graduate study are received the year round. For unconditional admission, the general requirements are an appropriate undergraduate engineering degree from an ABET-accredited program with a minimum overall average of 3.0 (out of 4.0). Students with other qualifications are considered on a case-by-case basis. International students are encouraged to apply early; a minimum of one semester in advance of the anticipated enrollment date is recommended.

Correspondence and Information

For the M.S.A.E.: Chairperson, Department of Architectural Engineering (336-334-7575)

For the M.S.E.: Chairperson, Department of Civil Engineering (336-334-7737)

For the M.S.Ch.E.: Chairperson, Department of Chemical Engineering (336-334-7564)

For the M.S.E.: Chairperson, Department of Agricultural and BioSystems Engineering (336-334-7787)

For the Ph.D. and M.S.E.E.: Chairperson, Department of Electrical Engineering (336-334-7760)

For the M.S.I.E.: Department of Industrial Engineering (336-334-7780)

For the Ph.D. and M.S.M.E.: Chairperson, Department of Mechanical Engineering (336-334-7620)

For the M.S.C.S.: Chairperson, Department of Computer Science (336-334-7245)

For graduate information:
Associate Dean for Graduate and Research Programs
NCA&TSU—College of Engineering
658 McNair Hall
Greensboro, North Carolina 27411

Telephone: 336-334-7594
Fax: 336-334-7540
College of Engineering e-mail: coegrad_info@ncat.edu
Graduate School e-mail: grad_info@ncat.edu
World Wide Web: http://www.eng.ncat.edu

North Carolina Agricultural and Technical State University

THE FACULTY

Agricultural and BioSystems Engineering. Peggy Fersner, Adjunct Assistant Professor; M.S., Virginia Tech. Godfrey A. Gayle, Professor and Chairperson; Ph.D., North Carolina State. Dick Phillips, Associate Professor; M.S., North Carolina State. Manuel R. Reyes, Assistant Professor; Ph.D., LSU. A. Shahbazi, Associate Professor and Graduate Program Coordinator; Ph.D., Penn State. Joan White, Adjunct Assistant Professor; M.S., Tufts.

Architectural Engineering. Ronnie S. Bailey, Associate Professor; M.U.P., Wisconsin–Madison. James M. Cox II, Adjunct Assistant Professor and Assistant to the Dean; M.S., Georgia Tech. Mike Ellis, Associate Professor; Ph.D., Virginia Tech. Sameer Hamoush, Assistant Professor; Ph.D., North Carolina State. Ronald N. Helms, Professor and Chairman; Ph.D., Ohio State. W. Mark McGinley, Associate Professor and Graduate Program Coordinator; Ph.D., Alberta. Peter Rojeski, Associate Professor; Ph.D., Cornell. Harmohindar Singh, Professor; Ph.D., Wayne State. Reginald C. Whitsett, Associate Professor; M.Arch., North Carolina State.

Chemical Engineering. Yusuf G. Adewuyi, Associate Professor; Ph.D., Iowa. Shamsuddin Ilias, Associate Professor; Ph.D., Queen's at Kingston. Vinayak N. Kabadi, Professor; Ph.D., Penn State. Franklin G. King, Professor and Chairperson; D.Sc., Stevens. Kenneth L. Roberts, Assistant Professor; Ph.D., South Carolina. Keith A. Schimmel, Associate Professor; Ph.D., Northwestern. Gary B. Tatterson, Professor; Ph.D., Ohio State.

Civil Engineering. Shoou-Yuh Chang, Professor and Graduate Coordinator; Ph.D., Illinois at Urbana-Champaign. Kenneth H. Murray, Professor; Ph.D., Virginia Tech. Emmanuel U. Nzewi, Associate Professor; Ph.D., Purdue. Miguel Picornell, Professor and Chairperson; Ph.D., Texas A&M. M. Reza Salami, Professor and Associate Dean for Graduate and Research Programs; Ph.D., Arizona. Andrew T. Watkin, Adjunct Associate Professor; Ph.D., Vanderbilt.

Computer Science. David Bellin, Associate Professor and Graduate Coordinator; Ph.D., CUNY. Sharon A. Brown, Adjunct Assistant Professor; M.S., Illinois at Urbana-Champaign. Albert Esterline, Assistant Professor; Ph.D., Minnesota. Raymond Hawkins, Adjunct Assistant Professor; M.S., SUNY at Albany. John C. Kelly, Associate Professor; Ph.D., Delaware State. Joseph Monroe, Professor and Chairperson; Ph.D., Texas A&M. Kenneth A. Williams, Assistant Professor; Ph.D., Minnesota. Anna Yu, Assistant Professor; Ph.D., Stevens.

Electrical Engineering. Ali Abul-Fadl, Associate Professor; Ph.D., Idaho. Marwan Bikdash, Assistant Professor; Ph.D., Virginia Tech. Eric Cheek, Adjunct Associate Professor and Associate Dean for Undergraduate Programs; Ph.D., Howard. Ward J. Collis, Associate Professor; Ph.D., Ohio State. Numan S. Dogan, Associate Professor; Ph.D., Michigan. Abdollah Homaifar, Associate Professor; Ph.D., Alabama. Esther A. Hughes, Assistant Professor; Ph.D., Cornell. Shanthi Iyer, Professor; Ph.D., Indian Institute of Technology (Delhi). Jung Kim, Professor; Ph.D., North Carolina State. Gary Lebby, Professor and Chairperson; Ph.D., Clemson. Clinton Lee, Associate Professor; Ph.D., North Carolina State. Jie Li, Adjunct Research Associate; Ph.D., Chinese Academy of Science. Robert Li, Associate Professor; Ph.D., Kansas. Harold L. Martin, Professor and Vice Chancellor for Academic Affairs; Ph.D., Virginia Tech. David E. Olson, Associate Professor; Ph.D., Utah. Earnest E. Sherrod, Associate Professor; M.S., Newark College of Engineering. David Song, Assistant Professor; Ph.D., Tennessee Tech. Feodorov Vainstein, Associate Professor; Ph.D., Boston University. Alvernon Walker, Associate Professor; Ph.D., North Carolina State. Chung Yu, Professor and Graduate Coordinator; Ph.D., Ohio State.

Industrial Engineering. Arup K. Mallik, Professor; Ph.D., North Carolina State. Ganelle Moultry-Gloster, Assistant Professor; Ph.D., Virginia Tech. Celestine A. Ntuen, Professor; Ph.D., West Virginia. Herbert Nwankwo, Assistant Professor; Ph.D., Texas at Austin. Eui Park, Professor and Chairman; Ph.D., Mississippi State. Bala Ram, Professor and Graduate Program Coordinator; Ph.D., SUNY at Buffalo. Sanjiv Sarin, Professor; Ph.D., SUNY at Buffalo. Silvanus Udoka, Associate Professor; Ph.D., Oklahoma State. Charles Vand de Zande, Adjunct Associate Professor; M.S., Rutgers. Samantha Wright, Research Associate; M.S., Texas A&M.

Mechanical Engineering. Vishnu S. Avva, Professor; Ph.D., Penn State. Suresh Chandra, Research Professor; Ph.D., Colorado State. Rajinder S. Chauhan, Assistant Professor; Ph.D., Auburn. William J. Craft, Professor and Chairperson; Ph.D., Clemson. DeRome Dunn, Assistant Professor; Ph.D., Virginia Tech. George J. Filatovs, Professor; Ph.D., Missouri–Rolla. Meldon Human, Associate Professor; Ph.D., Stanford. Ajit D. Kelkar, Professor; Ph.D., Old Dominion. David E. Klett, Professor; Ph.D., Florida. Carolyn W. Meyers, Professor; Ph.D., Georgia Tech. Tony C. Min, Professor Emeritus; Ph.D., Tennessee. Samuel P. Owusu-Ofori, Professor; Ph.D., Wisconsin–Madison. Devdas Pai, Associate Professor; Ph.D., Arizona State. Jagannathan Sankar, Professor; Ph.D., Lehigh. Mark Schulz, Assistant Professor; Ph.D., SUNY at Buffalo. Lonnie Sharpe Jr., Professor and Dean; Ph.D., Illinois at Urbana-Champaign. Kunigal Shivakumar, Research Professor; Ph.D., Indian Institute of Science (Bangalore). Shih-Liang Wang, Associate Professor; Ph.D., Ohio State.

SELECTED RESEARCH PROJECTS

Aerospace: Center for Aerospace Research—A NASA Center of Excellence. **Artificial Intelligence:** Development of Robust Algorithms for Threat/Explosive Recognition (FAA). Knowledge-Based Image Analysis for Object Identification (FAA). Model Analysis and Expert System Development for Planning and Scheduling (FNR/Wright-Patterson AFB). Testing an Expert System for Traffic Accident Analysis (USDOT). Feature Recognition for CAD/CAM Integration (AFOSR). **Communications:** Center of Excellence in Communications, Computer Vision, Statistical Pattern Recognition, and Neural Networks. **Computer Engineering:** Fault-Tolerant Computing, Self-Checking State Machine Realization in CMOS (ONR). Enhanced Computer System Performance and Reliability for Real-Time Object Recognition (ARPA). **Design and Manufacturing:** Product Realization Design and Prototype Implementation of Robotic Palletization in FMS (NSF). Vibration Reduction via Constrained Layer Technology (Army). Characterization of RTM Process (MLBC/Wright-Patterson AFB). Cylindrical Part Recognition and Modeling (MTIB/Wright-Patterson AFB). Bond Wrench Testing of Masonry Materials (Portland Cement Association). **Energy:** Energy-Efficient Dust Collection in North Carolina's Furniture Industry (North Alternative Energy Corporation). Center for Energy Research and Technologies (DOE). Diffusion of Carbon Dioxide and Iodine Through Topopah Spring Tuff (ORAU/DOE). Indoor Air Quality/Variable Ventilation (Honeywell). Alternative Cooling Technologies, Control Strategies to Enhance Energy Cost Reduction (North Carolina Energy Division). Coal Liquefaction/Processing, Heavy Oil Up-grading, High Temperature Membranes for Gas Separation, Novel Methods of Flux Enhancement in Membrane Filtration (DOE). **Energy/Thermo-Fluid Science:** Accessment of Cartesian/Cubic Grid Systems for CFD (Army). CFD High Speed Flows (NASA). **Environmental Engineering:** Waste Stream Life-Cycle Sequencing Model (Lockheed-Martin). The Communication Model and Environmental Justice (U.S. EPA). Pollution Prevention Opportunity Assessment for Radioactive Soil/Asphalt (Martin Marietta) (DOE-EM). Waste DOE, Samuel Massie, Chair of Excellence Project in Environmental Engineering. Air Force FAST Center for Environmental Remediation Fate and Transport of Hazardous Chemicals. Bioremediation of Hydrocarbons from Contaminated Soil (EG&G, Idaho). Recovery of Minerals (Bureau of Mines). Removal of Metal Ions from Wastewater Using CLM (ORNL/Martin Marietta). Removal of SOx/NOx for Flue Gas Using CLM (DOE). Solar Detoxification (ORNL/SNL). Tradeoff Analysis for Hydroelectric Development (Oak Ridge National Laboratories). Disposal Strategy of Low-Level Nuclear Waste (Oak Ridge National Laboratories). Effect of Ventilation on Indoor Environmental Quality (DOD). Process Solid Waste Assessment (Martin Marietta Energy Systems). Soot and NOx Emissions from Low Heat Rejection Diesels (U.S. ARO). **Geotechnical Engineering:** Multiaxial Testing and Constitutive Modeling (NSF). Mechanics of Gront Flow in Fractured Rock (U.S. Army–WES). **Human-Machine Systems:** Intelligent Human-Machine Interface for Mining Teleoperation (Bureau of Mines). **Mechanics and Materials:** High-Temperature Mechanical and Micro-Mechanical Characteristics of Ceramic Materials (Martin Marietta). Mars Mission Research Center (with North Carolina State University—NASA). Effects of Test Volume and Geometry on Ceramic Composites (Martin Marietta/DOE). Processing and Evaluation of Graphite/Polymides Tubes (NASA Lewis). Evaluation of Graphic/Copper Matrix Composite (NASA Lewis). Analysis of Low-Impact Damage to Composite Shells (WRDC/Wright-Patterson AFB). Development of Dry Stacked Masonry Systems (North Carolina Brick Association). Light Weight/Thermal Efficient/Structural Concrete (Carolina Power and Light–Kellogg). Combined Uplift and Shear on Masonry Wall Systems (NCBA). Rehabilitation of Concrete Masonry Walls with Fiber Composites (WES). **Object-Oriented (OO) Software Engineering:** Code Reuse, Distributed Communication, Smalltalk (NSA). Object Reuse, OO Class Testing, Source Code Metrics (IBM–Software Solutions Division). **Optical Electronics:** Brillouin Active Fiber Sensors (FAA). SBS-Induced Optical Switching in MIR Fibers (U.S. Army Research Office). **Solid-State Materials:** Material Growth and Characterization for Solid-State Devices (Langley Research Center, NASA). Light-Induced Effects in Amorphous Silicon (Solar Energy Research Institute). **Thermal Sciences/Energy Conversion:** Direct Contact Ice Making (ORNL).

NORTH CAROLINA STATE UNIVERSITY

College of Engineering

Programs of Study

The College of Engineering comprises ten degree-granting departments, all of which are authorized to award the Master of Science, the Master of Engineering in a designated field, and the Doctor of Philosophy. Programs of graduate study leading to the M.S. and Ph.D. offered are aerospace engineering, biological and agricultural engineering, chemical engineering, civil engineering, computer science, electrical and computer engineering, industrial engineering, materials science and engineering, mechanical engineering, and nuclear engineering; textile engineering offers the M.S. degree. Nonthesis master's degrees are also offered in most of the discipline areas and in the interdisciplinary program of integrated manufacturing systems engineering. Most nonthesis degrees require project or research work and a written technical report. Both M.S. and Ph.D. degrees are offered in the interdisciplinary program of operations research.

In most departments, the Master of Science degree is awarded for completing 30 credits of work, including a thesis, and passing a comprehensive examination. The Master of Engineering in a designated field is awarded for completing 30–36 course credits and passing a comprehensive examination. A Ph.D. degree is awarded for completing a program of work arranged by consultation between the student and his or her advisory committee, passing the written examination, completing a research dissertation, and passing a final examination on the dissertation.

Research Facilities

A new engineering graduate research center features more than 120,000 square feet of dedicated laboratory facilities. Special research facilities and equipment include transmission electron microscopes; computerized SEM with full X-ray and image analysis capabilities; electron beam–induced current and cathodoluminescence microscopy equipment; a scanning laser microscope; field emission electron beam lithography equipment; an imaging ion microscope for SIMS and 3-D ion imaging; a scanning Auger microprobe; an electron microprobe; complete X-ray analysis facilities including equipment for diffraction, topography, and radiography; a photoluminescence laboratory; MBE systems with in situ surface analysis; focused ion beam micromachinery; atomic resolution scanning tunneling microscopes; a precision engineering laboratory including diamond turning, ductile regime grinding, and surface metrology capabilities; a nuclear reactor with radiographic and neutron activation analysis; an applied energy laboratory; a plasma studies laboratory; a freon simulator of a PWR fission reactor; a synthesis laboratory for III-V semiconductor materials; an organometallic chemical vapor deposition system; a semiconductor device fabrication laboratory, a deep UV mask aligner, and oxidation diffusion furnaces; a plasma and chemical etching and vapor deposition facility; computer systems for research in communications and signal processing and in microelectronics; a commercial computer design system for large integrated circuits; an EPA automated pollution and combustion gas facility; anechoic and reverberation chambers; a computer-controlled gas chromatograph–mass spectrometer; a robotics and automation laboratory; state-of-the-art multimedia, voice I/O, and software engineering labs; DEC, HP, and Sun workstations linked through Ethernet; a large structures-testing system; pavement wheel-track testing; superpave asphalt testing; a shake table; and geotechnical test pits.

Financial Aid

Approximately half of the engineering graduate students are provided assistantships with full support for studies, including tuition and health insurance.

Cost of Study

Tuition and fees for full-time study in 1998–99 were $1185 per semester for North Carolina residents and $5768 per semester for nonresidents. Students taking fewer than 9 credits pay reduced amounts. Most students appointed as teaching or research assistants qualify for tuition and health insurance support.

Living and Housing Costs

On-campus dormitory facilities are provided for unmarried graduate students. In 1998–99, the rent for double rooms started at $1045 per semester. Accommodations in the newest residence hall for graduate students cost $1360 per semester. Apartments for married students in King Village rented for $360 per month for a studio, $385 for a one-bedroom apartment, and $420 for a two-bedroom apartment.

Student Group

The College of Engineering had an enrollment of 5,346 undergraduate students and 1,438 graduate students in 1998–99. Most graduate students find full- or part-time support through fellowships, assistantships, and special duties with research organizations in the area. During the 1997–98 academic year, the College conferred 1,351 degrees as follows: 94 doctoral degrees, 308 master's degrees, and 949 Bachelor of Science degrees.

Location

Raleigh, the state capital, has a population of about 280,000. Nearby is Research Triangle Park, one of the largest and fastest-growing research institutions of its type in the country. Sports and recreational facilities are also available.

The University and The College

North Carolina State University is the principal technological institution of the University of North Carolina System. Its largest schools are the Colleges of Engineering, Agriculture & Life Sciences, Physical & Mathematical Sciences, and Humanities & Social Sciences. Total enrollment is more than 27,000. A cooperative relationship with Duke University and the University of North Carolina at Chapel Hill contributes to a rich academic and research atmosphere, as does the University's association with the Research Triangle Park. The College of Engineering has more than 225 faculty members with professorial rank. Some of their current areas of interest are listed on the back of this page.

Applying

Applications may be submitted at any time. Although the GRE General Test is not always required, it is helpful in making decisions concerning financial aid. An applicant desiring to visit the campus may request information concerning travel allowances by writing to the graduate administrator of the preferred program of study. Students may apply for fellowships or assistantships in their application for admission. For application forms and further information, students should write to the address given below.

Correspondence and Information

Dean of the Graduate School
North Carolina State University
P.O. Box 7102
Raleigh, North Carolina 27695-7102
Telephone: 919-515-2872

North Carolina State University

THE FACULTY AND THEIR RESEARCH

Biological and Agricultural Engineering. D. B. Beasley, Department Head. Faculty: F. Abrams, D. Amatya, J. Barker, G. Baughman, S. Blanchard, R. Bottcher, C. Bowers, M. Boyette, J. Cheng, G. Chesehcir, J. Classen, C. Daubert, R. Evans, B. Farkas, S. Hale, A. Hassan, R. Huffman, F. Humenik, E. Humphries, G. Jennings, W. Johnson, K. Keener, G. Kriz, T. Losordo, W. McClure, S. Mohapatra, J. Parsons, G. Roberson, S. Roe, R. Rohrbach, A. Rubin, L. Safley, K. Sandeep, S. Seymour, R. Skaggs, R. Sowell, J. Spooner, L. Stikeleather, K. Swartzel, P. Westerman, T. Whitaker, D. Willits, J. Young.

Research areas: Bioinstrumentation, biomechanics, bioprocessing and materials handling, energy conservation and alternative fuels, environmental control, human engineering, mechanization, microprocessor applications, water and waste management, hydrology.

Chemical Engineering. R. Carbonell, Department Head. Faculty: J. DeSimone, P. Fedkiw, R. Felder, B. Freeman, C. Grant, J. Genzer, K. Gubbins, C. Hall, H. Hopfenberg, R. Kelly, S. Khan, P. Kilpatrick, H. Lamb, P. Lim, D. Ollis, M. Overcash, G. Parsons, S. Peretti, G. Roberts, C. Setzer, R. Spontak, H. Winston.

Research areas: Biotechnology, catalysis, chemical reaction engineering, electrochemical engineering, electronic materials, environmental studies, hazardous-waste management, interfacial phenomena, membrane transport processes, photochemical engineering, pollution prevention, polymer rheology, polymer sciences, polymer syntheses, separation processes, statistical and phase equilibrium thermodynamics, surface and colloid science, transport phenomena.

Civil Engineering. E. Downey Brill Jr., Department Head. Faculty: S. Ahmad, M. Barlaz, J. Baugh, L. Bernold, W. Bingham, R. C. Borden, R. H. Borden, A. Chao, J. Ducoste, J. Fisher, C. Frey, M. Gabr, G. Gilbert, A. Gupta, A. K. Gupta, J. Hanson, T. Hassan, K. Havner, Y. Horie, J. Hummer, D. Johnston, N. Khosla, Y. Kim, D. Knappe, M. Kowalsky, N. Krstulovic, P. Lambe, M. Leming, R. Malcom, V. Matzen, J. Nau, M. Overton, S. Rahman, S. Ranjithan, W. Rasdorf, N. Rouphail, J. Stone, A. Tayebali, C. Tung, P. Zia.

Research areas: Civil engineering systems; computer-aided engineering; construction engineering and management; construction materials; environmental engineering; geotechnical engineering; transportation systems and materials; solid mechanics; structural engineering; water resources.

Computer Science. A. L. Tharp, Department Head. Faculty: A. Anton, D. Bahler, D. Bitzer, F. Brglez, W. Chou, E. Davis, R. Dwyer, R. Fornaro, R. Funderlic, E. Gehringer, C. Healey, T. Honeycutt, P. Iyer, V. Jones, J. Lester, D. Martin, D. McAllister, H. Perros, D. Reeves, I. Rhee, W. Robbins, R. Rodman, J. Rossie, G. Rouskas, R. St. Amant, C. Savage, M. Singh, M. Stallmann, W. Stewart, K. Tai, M. Vouk, F. Wu, R. Young.

Research areas: Software systems (includes software engineering, AI, OOP, computer graphics, databases, concurrent systems, voice I/O, real-time systems, program methodology), computer networking, theory and algorithms, architecture.

Electrical and Computer Engineering. R. Kolbas, Department Head. Faculty: D. Agrawal, T. Alexander, W. Alexander, W. Allen, B. Baliga, M. Baran, S. Bedair, G. Bilbro, J. Brickley, B. Burger, G. Byrd, M. Chow, T. Conte, A. Duel-Hallen, A. Eichenberger, P. Franzon, E. Gehringer, T. Glisson, C. Gloster, J. Grainger, E. Grant, J. Hauser, C. Holton, B. Hughes, F. Kauffman, A. Kelley, K. Kim, R. Kolbas, H. Krim, R. Kuehn, M. Littlejohn, W. Liu, M. Masnari, T. Miller, J. Mink, V. Misra, A. Morazawi, T. Nagle, A. Nilsson, B. O'Neal, C. Osburn, H. Ozturk, M. Ozturk, S. Rajala, D. Reeves, W. Snyder, M. Steer, C. Townsend, J. Townsend, J. Trussell, I. Viniotis, M. White.

Research areas: Applied electromagnetics, microwaves/optics; communications; digital systems and computer architecture; microelectronics; power systems; signal and image processing; software engineering; solid-state devices and circuits; VLSI design.

Industrial Engineering. S. Roberts, Department Head. Faculty: M. Ayoub, R. Bernhard, D. Cormier, T. Culbreth, H. Damerdji, S. Elmaghraby, S. C. Fang, Y. Fathi, T. Hodgson, M. Kay, R. King, Y-S. Lee, W. Meier Jr., G. Mirka, R. Mittal, H. L. W. Nuttle, R. Pearson, E. Sanii, W. Smith, C. Sommerich, J. Taylor, J. Wilson, R. Young.

Research areas: Economic decision analysis, ergonomics, facilities design, inventory and scheduling theory, occupational safety, production planning and control, robotics, productivity improvement, material handling, simulation, concurrent engineering.

Integrated Manufacturing Systems Engineering. T. Hodgson, Executive Director, S. Jackson, Associate Director for Industrial Programs, L. Silverberg, Associate Director for Academic Programs. Associated Faculty: D. Bahler, P. Banks-Lee, R. Barker, M. Boyette, Y. Chen, T. Clapp, D. Cormier, T. Culbreth, Y. Fathi, T. Ghosh, P. Grady, G. Hodge, W. Jasper, T. Johnson, M. Kay, R. King, J. Leach, Y.-S. Lee, J. C. Lu, C. Maday, W. Meier, G. Mirka, H. Nuttle, M. Ramasubramanian, W. Rasdorf, P. Ro, R. Rodman, S. Roberts, J. Rust, E. Sanii, A. Seyam, W. Smith, K. Tai, J. Taylor, J. Wilson, R. Young.

Research areas: Automation, CAD, CAM, CIM and advanced information technology, logistics, material handling, mechatronics, part fabrication, quality assurance and testing, process and facilities planning, product assembly, product design, robotics, scheduling and operations management.

Materials Science and Engineering. J. Michael Rigsbee, Department Head. Faculty: K. Bachmann, C. M. Balik, R. Benson, D. Brenner, C. Chiklis, J. Cuomo, R. Davis, N. El-Masry, A. Fahmy, D. Griffis, T. Hare, J. Hren, J. Kasichainula, A. Kingon, C. Koch, D. Maher, K. Murty, J. Narayan, G. Rozgonyi, J. Russ, P. Russell, R. Scattergood, Z. Sitar, R. Spontak, H. Stadelmaier.

Research areas: Analytical techniques, ceramics and processing methods, composite materials, computer simulation techniques, corrosion, crystal structure and phase relations, electronic materials, irradiation effects, mechanical properties, metals, nonequilibrium processing, nuclear materials, polymers, stereology and image analysis, structure-property relations, superconducting materials, surface analysis, surface phenomena, thin-film processing and characterization, tribology.

Mechanical and Aerospace Engineering. F. R. DeJarnette, Department Head. Faculty: E. Afify, J. Bailey, A. Bayoumi, M. Boles, N. Chokani, J. David, T. Dow, H. Eckerlin, J. Edwards, J. R. Edwards, J. Eischen, M. Fikry, L. Franzoni, R. Gould, C. Hall, H. Hassan, T. Hodgson, R. Johnson, R. Keltie, E. Klang, C. Kleinstreuer, J. Leach, G. Lee, K. Lyons, C. Maday, D. McRae, J. Mulligan, R. Nagel, M. Ozisik, J. Perkins, M. Ramasubramanian, P. Ro, W. Roberts, L. Royster, L. Silverberg, Y. Sorrell, J. Strenkowski, R. Walberg, F. Yuan, M. Zikry.

Research areas: Heat transfer and fluid mechanics, energy conversion, combustion, environmental engineering, materials processing, manufacturing, precision engineering, solid mechanics, fracture mechanics, composite structures, fluid dynamics, aerothermodynamics and hypersonics, flight dynamics, propulsion, theoretical and structural acoustics, robotics, control systems, mechanical vibrations.

Nuclear Engineering. Donald J. Dudziak, Department Head. Faculty: M. Bourham, J. Doster, R. Gardner, J. Gilligan, O. Hankins, C. W. Mayo, R. Mayo, K. Murty, P. Turinsky, K. Verghese. M.-S. Yim.

Research areas: Computational reactor physics; fuel management; plasma engineering; radiation effects in nuclear materials; nuclear power systems modeling; plasma-surface interactions; radiation transport; reactor dynamics, control, and safety; computational thermal hydraulics; nuclear waste management; radiological engineering; industrial radiation applications; medical radiation physics.

Operations Research. W. Stewart, Director. Faculty: R. Bernhard, B. Bhattacharyya, J. Bishir, S. Campbell, W. Chou, H. Damerdji, J. Dunn, S. Elmaghraby, S. Fang, Y. Fathi, R. Funderlic, R. Hartwig, T. Hodgson, D. Holthausen, T. Honeycutt, C. Kelly, R. King, J. Lu, D. McAllister, C. Maday, C. Meyer, A. Nilsson, H. Nuttle, H. Perros, E. Peterson, T. Reiland, S. Roberts, J. Rodrigues, J. Roise, G. Rouskas, C. Savage, C. Smith, M. Stallmann, M. Suh, R. Taylor, H. Tran, I. Viniotis, M. Vasu, M. Vouk, J. Wilson.

Research areas: Mathematical programming, networks, production, queuing, project planning and control, routing and scheduling, simulation, stochastic processes, systems theory and optimal control.

Textile Engineering. M. Mohamed, Department Head. Faculty: D. Buchanan, T. Clapp, P. Grady, H. Hamouda, W. Jasper, G. Mock, J. Rust.

Research areas: Three-D composites fabrication, textile machinery design, process consumption modeling, biotextiles, robotics and automation.

NORTHEASTERN UNIVERSITY

Graduate School of Engineering

Programs of Study

The Graduate School of Engineering offers programs leading to the Master of Science degree, full-time and part-time, in chemical engineering, civil engineering, computer systems engineering, electrical engineering, engineering management, industrial engineering, information systems, mechanical engineering, and operations research. The Doctor of Philosophy degree is offered in chemical engineering, civil engineering, electrical engineering, industrial engineering, and mechanical engineering. An interdisciplinary program may be arranged at either the master's or doctoral level in most programs.

The programs leading to master's degrees in computer systems engineering (CSE) and in information systems (M.S.I.S.) are offered through the Graduate School of Engineering. The CSE program requires a background in engineering or a closely related field and has areas of specialization in CAD/CAM and engineering software design. The M.S.I.S. is for students with either technical or nontechnical backgrounds who seek career opportunities in the computer application and software fields. The Women in Information Systems program is designed for women who wish to make a career change. Finally, the Cooperative Education Plan, available to qualified full-time Master of Science degree students, presents the opportunity for integrating classroom theory with professional experience.

Research Facilities

The University Libraries system contains more than 870,000 volumes, 2 million microforms, 173,000 government documents, 8,400 serial titles, and 20,000 audio, video, and software titles. A large central library contains technologically sophisticated library services, including an online catalog and circulation system, an information gateway, and a network of CD-ROM optical disc databases. Graduate students also have access to other major research collections in the area through the Boston Library Consortium.

Departments maintain broad-based research facilities, with special emphases on, among other fields, sensing and imaging, signal processing, computer engineering, electron devices, thermofluids engineering, design and manufacturing, advanced materials engineering, and geotechnical/geoenvironmental engineering. The College of Engineering supports the Center for Electromagnetics Research, the Center for Communications and Digital Signal Processing, and the Center for Advanced Microgravity Materials Processing.

A high-speed intranet links users and facilities on the central campus and on three satellite campuses. The campus network is also connected via the Internet to computing resources around the world. At the university, students have access to a network of UNIX workstations, microcomputer labs, and an array of specialized departmental computing equipment.

Financial Aid

Northeastern awards need-based financial aid to graduate students through the Federal Perkins Loan, Federal Work-Study, and Federal Stafford Student Loan Programs. The University also offers a limited number of minority fellowships and Martin Luther King, Jr. Scholarships. The graduate school offers financial assistance through teaching, research, and administrative assistantship awards that include a stipend with tuition remission. In 1998–99, the stipend was $12,450 for nine months. These assistantships require a maximum of 20 hours of work per week. Also available are a limited number of tuition assistantships that provide partial or full tuition remission and require a maximum of 10 hours of work per week.

Cost of Study

In 1998–99, tuition in the Graduate School of Engineering was $465 per quarter hour of credit. Where applicable, there are special tuition charges for theses and dissertations. Other charges include the Student Center fee and health and accident insurance fee required of all full-time students.

Living and Housing Costs

On-campus and off-campus living expenses are estimated at $1200 to $1500 per month, with on-campus housing available on a limited basis to newly accepted students. A public transportation system services the Greater Boston area, and there are subway and bus services convenient to the University.

Student Group

In fall 1998, 24,027 students enrolled at Northeastern University, representing a wide variety of academic, professional, geographic, and cultural backgrounds. The Graduate School of Engineering had 1,011 students, 44 percent of whom attended on a full-time basis.

Student Outcomes

The Graduate School of Engineering offers degree programs designed to help students prepare themselves for technical positions in industrial organizations, government laboratories, research laboratories, and educational institutions. Graduates are employed around the world in a wide variety of fields, including aerospace, automotive, computer, electronics, environmental, and chemical, to name only a few. In addition, master's degree students at Northeastern University have the unique opportunity to participate in the Cooperative Education Plan, which in many cases has led to a full-time career upon graduation.

Location

Students at Northeastern University have access to the many cultural, educational, historical, and recreational offerings of Boston. Some of the city's cultural opportunities found near the main campus include the Museum of Fine Arts, Symphony Hall, and the Boston Public Library. Greater Boston is the home of sixty colleges and universities. The area's professional sports teams are the Red Sox, Celtics, Bruins, and the New England Patriots.

The University

Founded in 1898, Northeastern University is a privately endowed nonsectarian institution of higher learning. It is among the largest private universities in the country, with seven undergraduate schools and colleges, nine graduate and professional schools, and a number of continuing and special education programs and institutes.

Applying

Applicants must have obtained from a recognized institution a Bachelor of Science degree in engineering, or a field closely allied to engineering, with an acceptable quality of undergraduate work. Applicants to the M.S.I.S. program must have obtained a bachelor's degree from a recognized institution and must submit GRE scores. Applicants interested in full-time study should submit applications by February 15 to be considered for a graduate assistantship or fellowship award and by April 15 for fall admission. GRE and TOEFL scores are required of all applicants with undergraduate degrees from outside the United States.

Correspondence and Information

Director
Graduate School of Engineering
130 Snell Engineering Center
Northeastern University
Boston, Massachusetts 02115-5000
Telephone: 617-373-2711
E-mail: grad-eng@coe.neu.edu
World Wide Web: http://www.coe.neu.edu

Northeastern University

THE FACULTY AND THEIR RESEARCH

CHEMICAL ENGINEERING

Ralph A. Buonopane, Ph.D., Associate Professor and Chairman. **Faculty:** Gilda A. Barabino, Ph.D.; Albert Sacco Jr., Ph.D.; Ronald J. Willey, Ph.D.; Donald L. Wise, Ph.D. (Emeritus). The department offers a graduate program at both the M.S. and Ph.D. levels emphasizing process control, reactor and reaction modeling, transport processes, process design, biotechnology, and material science. The Ph.D. program involves advanced study in chemical engineering and an experimental thesis in a current research area. Research is in transport processes (innovative equipment for heat and mass transfer, multicomponent phase equilibrium, flow and structural properties of coal-water mixtures), computer simulation/control (continuous and discrete control system simulation), reaction kinetics (biochemical reactions, heterogeneous catalysis reactions, catalyst deactivation phenomena, aerogel catalysts), microgravity materials processing, and biochemical and biomedical (in vitro fluid mechanics) applications.

CIVIL AND ENVIRONMENTAL ENGINEERING

Mishac K. Yegian, Ph.D., PE, Professor and Chairman. **Faculty:** Akram N. Alshawabkeh, Ph.D.; Dionisio Bernal, Ph.D.; Frederic C. Blanc, Ph.D., PE; John J. Cochrane, Ph.D., PE; Peter Furth, Ph.D.; Constantine J. Gregory, Ph.D.; Fernando Miralles-Wilhelm, Ph.D., PE; Francis A. Oluokun, Ph.D.; Richard J. Scranton, M.S.; Thomas C. Sheahan, Sc.D., PE; Ali Touran, Ph.D., PE; Sara Wadia-Fascetti, Ph.D.; Irvine W. Wei, Ph.D. The department offers graduate programs at both the M.S. and Ph.D. levels in construction management, structures and materials, transportation, geotechnical/geoenvironmental engineering, and environmental engineering. Current research areas are earthquake-resistant design of structures, dynamic instability and nonlinear torsion in earthquake response, nonlinear system identification, condition assessment of existing structures, soil dynamics, liquefaction, seismic design of earthen dams and waste containment facilities, seismic hazard and risk analysis, dynamic interface properties of geosynthetics, geotechnical laboratory automation, cohesive soil rheological behavior, groundwater flow and contaminant migration in soils, numerical modeling of contaminant migration in groundwater systems, hazardous-waste-site remediation techniques, electrokinetic soil treatment, leakage through composite landfill liners, point-of-use water treatment, groundwater oxygenation by hydrogen peroxide injection, bioremediation of priority organic pollutants, simulation modeling of construction processes, bridge maintenance, heavy-equipment policy, transit data collection, traffic and transit modeling, travel demand estimation, and reduction of traffic congestion.

ELECTRICAL AND COMPUTER ENGINEERING

Fabrizio Lombardi, Ph.D., Professor and Chairman; Arvin Grabel, Sc.D., Professor and Associate Chairman. **Faculty:** David Brady, Ph.D.; Dana Brooks, Ph.D.; Soren Buus, Ph.D.; Chung Chan, Ph.D.; Jill Crisman, Ph.D.; Anthony Devaney, Ph.D.; Jack Hanania, Ph.D.; Jeffrey A. Hopwood, Ph.D.; Vinay Ingle, Ph.D.; David R. Kaeli, Ph.D.; Mieczyslaw Kokar, Ph.D.; Miriam Leeser, Ph.D.; Bradley Lehman, Ph.D.; Hanoch Lev-Ari, Ph.D.; Elias Manolakos, Ph.D.; Nicol McGruer, Ph.D.; Lisa McIlrath, Ph.D.; Stephen McKnight, Ph.D.; David McLaughlin, Ph.D.; Waleed Meleis, Ph.D.; Fred J. Meyer, Ph.D.; Eric Miller, Ph.D.; Sarma Mulukutla, Ph.D.; Sheila Prasad-Hinchey, Ph.D.; John G. Proakis, Ph.D.; Carey M. Rappaport, Ph.D.; Masoud Salehi, Ph.D.; Sheldon Sandler, Ph.D.; Martin Schetzen, Sc.D.; Philip Serafim, Sc.D.; Bahram Shafai, Sc.D.; Michael Silevitch, Ph.D.; Alex Stankovic, Ph.D.; Ioannis Stavrakakis, Ph.D.; Gilead Tadmor, Ph.D.; Man-Kuan Vai, Ph.D.; Carmine Vittoria, Ph.D.; Paul Zavracky, Ph.D. Six concentrations are offered: computer engineering; communications and signal processing; control systems and signal processing; electromagnetics, plasma, and optics; electronic circuits and semiconductor devices; and power systems. Research is in communications (advanced detection and estimation techniques for spread spectrum signals, local area networks with fiber-optic or atmospheric optical links, aerospace data transmission systems, adaptive filtering techniques for system identification and equalization of time-variant multipath channels), control systems (vibration control in space systems, discrete control of continuous processes), digital computers (microprocessor-based design and control, software engineering, theory of computation), digital signal image processing (recursive estimation of images, machine vision insights from biological systems, fast algorithms for linear filtering and prediction, VLSI architecture), electromagnetics (electromagnetic forces, strongly coupled plasma theory, high-voltage crossed-field discharge, dynamics of aurora ionospheric phenomena, RF intrusion sensing systems, antennas), instrumentation (telemetry systems), power systems (analysis, simulation, and optimum control of power systems; electromagnetic fields in electrical devices; special problems of electric machinery), and radar.

MECHANICAL, INDUSTRIAL, AND MANUFACTURING ENGINEERING

John W. Cipolla Jr., Ph.D., Professor and Chairman; Mohamad Metghalchi, Sc.D., Associate Professor and Associate Chairman. The department offers master's degrees in the fields of engineering management, industrial engineering, and mechanical engineering and Ph.D. degrees in both industrial engineering and mechanical engineering. In addition, the department offers an interdisciplinary master's degree in operations research in conjunction with the Department of Mathematics. The department also coordinates the M.S.I.S. program and the CAD/CAM and engineering software design options of the CSE program. **Industrial Engineering Faculty:** James Benneyan, Ph.D.; Thomas P. Cullinane, Ph.D.; Nasser Fard, Ph.D.; Surendra M. Gupta, Ph.D., PE; Sagar Kamarthi, Ph.D.; Shiwoo Lee, Ph.D.; Emanuel Melachrinoudis, Ph.D.; Ronald R. Mourant, Ph.D.; Ronald Perry, Ph.D.; Allen L. Soyster, Ph.D.; Gerard Voland, Ph.D. Current research includes production and manufacturing systems (design and analysis of flexible automated manufacturing systems, intelligent material handling, concurrent engineering, design of automated storage/retrieval systems using simulation), applied statistics/quality and reliability (steady-state estimation in simulation, validation of larger-scale simulation models, multivariate quality control development of a common-cause failure analysis method, analysis of dependent failures in redundancy optimization problems with multiple criteria), operation research (locating facilities using single or multiple criteria, simulations for tests of linear restrictions), mathematical programming and optimization, and computer systems engineering (decision-support systems development, artificial intelligence, design of a methodology for learning invariant functional descriptions, development and evaluation of models for designing). **Mechanical Engineering Faculty:** John W. Cipolla Jr., Ph.D.; George G. Adams, Ph.D.; Teiichi Ando, Ph.D.; Joseph T. Blucher, Ph.D.; Alexander M. Gorlov, Ph.D.; Hamid N-Hashemi, Ph.D.; Olusegun J. Ilegbusi, Ph.D.; Jacqueline A. Isaacs, Ph.D.; Gregory J. Kowalski, Ph.D.; Yiannis A. Levendis, Ph.D.; Achille Messac, Ph.D.; Mohamad Metghalchi, Sc.D.; Richard J. Murphy, Ph.D.; Uichiro Narusawa, Ph.D.; Welville B. Nowak, Ph.D. (Emeritus); John N. Rossettos, Ph.D.; Mohammad E. Taslim, Ph.D.; Bruce H. Wilson, Ph.D.; Yaman Yener, Ph.D.; Ibrahim Zeid, Ph.D. Current research includes thermofluids engineering (combustion of coal and coal-water slurries and pollution formation, gas turbine blade film cooling, heat transfer in rotating channels, incineration of waste plastics and hazardous materials, nonequilibrium thermodynamics, radiative transfer in high-temperature aerosols with thermophoresis, thermally stimulated nonlinear optics, thermofluids aspects of materials processing, processing and applications of metal-matrix composites, waste- and solar-power refrigeration), mechanics and design (modeling, analysis and simulation of the mechanics of lightweight flexible media, mechanics and tribology of magnetic recording systems, response of structures to moving loads, nonlinear finite-element analysis, CAD/CAM, design manufacturing, dynamic systems and control, automated modeling, design and control of drivetrain systems, modeling and analysis of composites, NDE of adhesive joints, fatigue and impact behavior of graphite/epoxy composite materials, Mode III fatigue crack propagation in turbine shafts), and materials science and engineering (thin films to resist corrosion, diffusion, and wear and for electronic applications; mechanical behavior of engineering materials; intelligent processing of materials; powder metallurgy; metal injection; molding; debinding and sintering; CAD of materials processes).

NORTHWESTERN UNIVERSITY

Robert R. McCormick School of Engineering and Applied Science

Programs of Study

The M.S. and Ph.D. degrees are offered in biomedical, chemical, civil, computer, electrical, industrial, and mechanical engineering; computer science; materials science and engineering; applied mathematics; and theoretical and applied mathematics. Professional master's degrees are also offered in engineering management, information technology, manufacturing engineering, and project management.

For the M.S. degree, in addition to a planned program of course study, a thesis or project may be required. A student can complete a professional degree in one academic year of full-time study for those programs, such as Master of Engineering Management, Master of Manufacturing Engineering, or Master of Project Management, in which a thesis is not required. When a thesis is required, the time of study is usually twelve to eighteen months. There also is a full-time, two-year M.S. program, Master of Management in Manufacturing, which is a joint venture between the J. L. Kellogg Graduate School of Management and the McCormick School of Engineering. The Ph.D. degree requires a minimum of three years beyond the B.S. degree; the average time is four to five years. A dissertation is always required for the Ph.D. degree. During the early phases of a Ph.D. program, while taking course work, the student is expected to determine an area of interest, select a dissertation adviser, and begin research.

Research Facilities

The University and departments provide extensive facilities to support research programs. Research and common equipment are housed in the Technological Institute, in the Adjacent Catalysis Center and Materials and Life Sciences Building, and in the Institute for Learning Sciences. Students conduct research in two kinds of laboratories—those assigned to faculty members and general user facilities that include an analytical services laboratory, electron probe instrumentation center, X-ray facility, surface science facility, and machine shop. The University also provides academic computing and network services.

Financial Aid

Most full-time graduate students in the Institute receive financial aid, which is available in a number of forms, such as fellowships, research assistantships, teaching assistantships, traineeships, and scholarships. Students are usually awarded a tuition scholarship as part of the appointment, or the stipend is sufficient to cover tuition and provide adequate living support. Many fellowships are supported by industry. Cabell Fellowships, which carry a higher stipend, are awarded to particularly outstanding students on a competitive basis. Research assistantships involve participation in research programs of the McCormick School, usually chosen in the area of the student's dissertation interest.

Cost of Study

Tuition totals $21,798 for the 1999–2000 academic year, with much lower rates applicable during the summer quarter and after the student has completed nine quarters of full-time registration.

Living and Housing Costs

In addition to on-campus housing, apartments and rooms nearby are usually available. There is an extensive complex of University-owned apartments for graduate housing. Food and living costs are close to the national average.

Student Group

Students at the McCormick School come from all parts of the United States, as well as from many other countries, with about 30 percent of the students coming directly from other countries. The graduate enrollment in the McCormick School is slightly more than 1,000. A placement office aids students in finding suitable jobs. A Ph.D. industrial internship is also offered.

Location

Northwestern University is located on the shores of Lake Michigan in the residential suburb of Evanston, immediately north of Chicago. A University beach, adjacent to the McCormick School, is used by students and their families. With its many museums, parks, zoos, musical and theatrical centers, night spots, and restaurants, Chicago offers remarkable cultural opportunities. For sports-minded students, there are all levels of activity, from participating in intramural sports to viewing collegiate and professional teams in nearly all sports. Many outstanding musical and theatrical performances take place on the University's campus.

The University and The School

The only privately supported university in the Big Ten, Northwestern University is a coeducational institution with two campuses, one on the lake shore in Evanston and the other on the Near North Side of Chicago. The McCormick School building, located on the Evanston campus, came into existence in 1939 through the generosity of Walter P. Murphy, who provided money to construct the original building and also left his residual estate as an endowment for the School. Expenditures for sponsored research exclusive of University funds in the School are approximately $45 million per year, and the University awards are in excess of $214 million.

Applying

Students with B.S. or M.S. degrees in any engineering discipline or with strong mathematics or science backgrounds and majors in other subjects are considered for graduate study provided that they have satisfactory overall records (at least a B average is required for admission). Northwestern University operates on a quarter system. Students may commence graduate study in any quarter of the year; however, most students enter at the beginning of the fall quarter. Application for admission and financial assistance for the forthcoming year is best completed by February 1 if a request for support for the fall (or summer) quarter is contemplated.

Correspondence and Information

Chairman, Department of (specify)
Robert R. McCormick School of Engineering and Applied Science
Northwestern University
Evanston, Illinois 60208
World Wide Web: http://www.tech.nwu.edu:80/gsroffice/

Northwestern University

FACULTY RESEARCH AREAS

Applied Mathematics. W.E. Olmstead, Admissions Officer. Coordinated study of mathematics, including mathematical methods and significant problems from various fields of application, is emphasized. Faculty research applies asymptotic analysis and perturbation theory, differential and integral equations, and numerical analysis to problems in extended dynamical systems, fiber optics, combustion, fluid and solid mechanics, geophysics, crystal growth, solidification, reaction-diffusion processes, pattern formation, and wave propagation. **Biomedical Engineering.** R.A. Linsenmeier, Chairman. Faculty members in the engineering and medical schools are involved in research in several areas: cardiopulmonary engineering (e.g., cell and tissue engineering and biomechanics of the vascular and pulmonary systems, magnetic resonance imaging of the heart, tissue microcirculation, and detection of cardiac arrhythmias), quantitative neuroscience (e.g., mechanisms of motor control, retinal information processing, auditory system biophysics, and neural microenvironment), rehabilitation engineering and movement biomechanics (e.g., limb prosthetics, joint biomechanics, and control of movement after stroke), medical devices (e.g., optical biosensors, laser-tissue interactions, intrathoracic artificial lung, and robotic microsurgery), biotechnology (e.g., macromolecular structure and kinetics of DNA and antibody reactions), and biomaterials (e.g., biomolecular engineering, self-assembly, phospholipid structures, and degradable polymers). **Chemical Engineering.** J. Ottino, Chairman. Emphasis is on biochemical and biomedical engineering, chemical reaction engineering and catalysis, multiphase flow, interfacial and chaotic phenomena, polymer science and engineering, computer-aided design and control of process systems, and pollution control engineering. Faculty members are actively involved in interdisciplinary research programs that include activities in the Materials Research Center, the Institute for Environmental Catalysis, the Center for Catalysis and Surface Science, and the Center for Biotechnology. Departmental facilities include computers, mass spectrometers, gas chromatographs, laser Raman spectroscopy, fluorescence and phosphorescence spectrometers, gel permeation chromatographs, and other analytical instruments. Many other facilities are available within the School, including the Polymer Characterization Laboratory, the Analytical Services Laboratory, and the various facilities of the Materials Research Center, the Catalysis Centers, and the Biotechnology Center. **Civil Engineering.** J. L. Schofer, Chairman. The fields of study and research include project management (planning, coordinating, controlling, and evaluating diverse activities on a civil engineering project), environmental health engineering (physical, chemical, and biological processes for water and waste treatment; environmental biotechnology; ecosystem-contaminant interactions; coal desulfurization; waste disposal; planning of environmental systems), environmental geotechnology (transport processes, groundwater flow, waste disposal, exploration), geotechnics (rock dynamics, TDR instrumentation, nondestructive evaluation of deep foundations, progressive failure in soils, ground modification, grouting, soil-structure interaction), mechanics of materials and solids (elasticity, finite elements and boundary elements, nondestructive evaluation, composites, geomechanics, micromechanics, dislocations, fracture), structural engineering and materials (analysis, stability, and dynamics of structures; fracture processes; damage mechanics and nonlinear constitutive laws; development of new cementitious materials; earthquake engineering; probabilistic methods and structural safety; chemistry and computer modeling of microstructure of concrete), transportation systems analysis and planning (urban transportation planning, transportation systems analysis, intercity travel modeling, logistics, systems design and evaluation, traffic engineering). **Computer Science.** L. Birnbaum, Chairman. Heavy emphasis on the development of theories and tools to enable the construction of large, distributed, usable, content-rich, interactive software systems. Core areas of research and study in support of this theme include artificial intelligence, cognitive science, computer vision, databases, graphics, human-computer interaction, natural language processing, software engineering, and systems. Applications include interactive intelligent learning environments for education and training, problem solvers for qualitative physics, mobile robots, distributed heterogeneous databases, case-based design tools, authoring tools, and networked environments for collaboration and learning. **Electrical and Computer Engineering.** P. Banerjee, Chairman. Fields of study include biomedical electronics, communication systems and networks, computational electromagnetics, computer architecture, control systems and robotics, detection and estimation, digital circuits, digital signal and image processing, electronic devices and materials, lasers, neural networks, nonlinear and quantum optics, numerical analysis, numerical optimization, optical communications, optical materials, quantum electronics, distributed and parallel computing, VLSI and CAD, and FPGA architectures. The department has well-equipped laboratories for electronic circuits, digital circuits, solid-state electronics, thin-film device development, biomedical electronics, microwave techniques, real-time control systems, guided-wave and nonlinear optics, fiber optics, biological control systems, digital systems design, digital signal processing, image and speech processing, MOCVD, MOMBE reactors for optoelectronic materials and devices fabrication, numerical analysis, computer architecture, distributed computing systems, robotics, VLSI/CAD, communication networks, and microprocessor systems design. The department's computing facilities include a 16-processor IBM SP-2 distributed-memory message-passing multicomputer, an 8-processor IBM J-10 shared-memory multiprocessor, an 8-processor SGI Origin 2000 distributed shared-memory multiprocessor, a network of fifty HP workstations, and a number of interconnected Sun SPARC teleservers and workstations. **Electronic Materials.** B. W. Wessels, Director. The program provides interdisciplinary education and research in the areas of semiconductor processing, semiconductor device physics, electronic materials characterization, preparation and properties of thin films, and optoelectronics. Collaborative research between Electrical and Computer Engineering and Materials Science and Engineering is offered. **Engineering Sciences and Applied Mathematics.** B. J. Matkowsky, Chairman. This department administers the program in applied mathematics. **Industrial Engineering and Management Sciences.** M. S. Daskin, Chair. Fields of study and research include applied probability, economics and production, optimization, organization theory, statistics and decision analysis, and systems analysis and design. Research in all areas ranges from abstract and theoretical studies to empirical investigations, experiments, and systems applications in industrial and government organizations. Specific faculty interests include large-scale linear, nonlinear, and combinatorial optimization; stochastic processes, quality control, and simulation; multiple objective decision making; equipment investment, facility location, and logistics; engineering economics; systems synthesis and design; management and economics of the R&D process; and modeling of manufacturing systems. **Materials Science and Engineering.** K. T. Faber, Chair. The department provides interdisciplinary education and research over a broad range of topics in metals, ceramics, polymers, electronic materials, biomaterials, and surfaces and interfaces. Research interests of the faculty extend to collaborative programs with chemistry, physics, geology, and chemical, civil, mechanical, biomedical, and electrical engineering and with the Medical School. Well-equipped research facilities are available for scanning and transmission electron microscopy, high-resolution electron microscopy with ultrahigh-vacuum capability, analytical electron microscopy, metallography, polymer characterization, atom-probe field-ion microscopy, mechanical behavior, X-ray diffraction, surface structure and composition, ceramics processing, optical characterization, crystal growth, and Mössbauer studies. Extensive use is made of minicomputers and microcomputers for control of experiments; data acquisition, including molecular beam epitaxy and organometallic vapor-phase epitaxy; and data processing. **Mechanical Engineering.** T. Belytschko, Chairman. Laboratories and facilities are available for work in lubrication, metal and polymer forming and cutting, mechanical design and CAD/CAM, automated assembly, robotics, combustion, fluid mechanics, and laser and optical interferometry. Current research activities involve combustion, fluid mechanics and fluid physics, turbulent and surface tension–dominated flows, tribology and surface engineering, dynamic systems, vibrations, finite elements, computational solid and fluid mechanics and radiative heat transfer, reliability analysis, control theory, robotics and man-machine systems, surgical robotics, design, manufacturing processes, metal cutting and forming, applied mechanics and stress analysis, composites and smart materials, fracture and porous media, and heat and mass transfer. **Theoretical and Applied Mechanics.** I. M. Daniel, Chairman. Departmental research areas include composite materials, computational mechanics, fracture and damage mechanics, geophysics, micromechanics, and nondestructive evaluation. Current research programs include nondestructive evaluation of aircraft airworthiness; size effects and scaling laws in fracture of materials; processing, characterization, and damage mechanics of composites; fiber-optic sensor development; computational structural dynamics; and mechanics of earthquakes.

OHIO UNIVERSITY

Fritz J. & Dolores H. Russ College of Engineering and Technology

Programs of Study

Programs of study leading to the M.S. degree are available in the Departments of Chemical, Civil, Industrial and Manufacturing Systems, and Mechanical Engineering and the School of Electrical Engineering and Computer Science. Programs leading to the Ph.D. degree are also offered in chemical and electrical engineering. An interdisciplinary program leading to the Ph.D. degree in integrated engineering with a focus in materials processing, geotechnical and environmental, or intelligent systems is also offered. Ohio's Individual Interdisciplinary Program is available in all majors at either the M.S. or Ph.D. degree levels. Normally, applicants to the M.S. program should have a B.S. degree in the respective departmental area. Applicants with a bachelor's degree in a related area can be admitted and take undergraduate courses to make up deficiencies. Students in this category should contact the chair of the department's graduate committee. Thesis and nonthesis options are available in some M.S. programs; prospective students should contact the specific departmental chair concerning the thesis/nonthesis policy. Applicants wishing to pursue the Ph.D. degree in chemical or electrical engineering should have an M.S. degree in the respective area. In some cases, a student with an M.S. in a related area can be admitted to the Ph.D. program. An M.S. degree in an engineering or related field is required for entry into the Ph.D. program in integrated engineering. The graduate programs in engineering are enhanced by an endowment now worth more than $16 million provided by distinguished alumni, the late Dr. C. Paul Stocker and his wife, Beth. The Stocker Endowment provides two endowed chairs, scholarships, and faculty enrichment programs. Exciting research has been initiated as a result of this remarkable legacy, and existing research has been significantly augmented.

Research Facilities

All departments are housed in the C. Paul and Beth K. Stocker Engineering and Technology Center. This is a state-of-the-art facility with modern research equipment. More than $8 million of endowment funds have been expended in support of research and graduate education. Approximately $10 million of equipment has been purchased over the last twelve years to support the College's programs. The University and College computer facilities provide students the microcomputers, minicomputers, workstations, and mainframes for educational and research purposes. An Intergraph CAD/CAM system provides exceptional facilities for students and faculty members. Computers are interconnected by a local area network (LAN) in the Stocker Center, providing access to Internet and other national networks. In addition to the general College computer labs, there are a large number of computers and workstations in departmental labs. The Large Scale Software Systems Development Laboratory provides opportunities for "real-world" software projects utilizing state-of-the-art Silicon Graphics workstations. The Avionics Engineering Center is internationally known for its research in avionics and provides exceptional research opportunities. The College also has the Ohio Research Institute for Transportation and the Environment, the Ohio Coal Research Center, the Center for Automatic Identification Education and Research, the Center for Corrosion of Multiphase Systems, the Center for Advanced Materials, and the Center for Advanced Software Systems Integration.

Financial Aid

Several types of competitive, merit-based financial support are available, ranging from tuition scholarships to part-time faculty appointments. Graduate Teaching Associateships are teaching-related appointments that normally require 20 hours per week of assigned duties. Graduate Research Associateships entail work on supported research contracts and grants; a commitment of 20 hours per week is required. Associateships require 20 hours per week; duties are normally associated with teaching activities. Tuition scholarships are also available. Stocker Fellowships carry stipends of $11,000–$16,000 for the academic year with no required duties. Stocker Research Associateships provide stipends of $8000–$10,000 for the academic year and $4000–$5000 for the summer; these assignments require one-half-time service for the academic year and full-time service for the summer. Recipients of Stocker Fellowships and Research Associateships must be U.S. citizens and meet departmental academic standards. Other financial support programs are available; see the *Graduate Bulletin*. Stipends range upward from a minimum of $5600 for three quarters. Appointments are made on a competitive basis.

Cost of Study

In 1998–99, tuition (9–18 credit hours) was $1542 per quarter for in-state students and $2955 per quarter for out-of-state students. Tuition costs are subject to change without notice.

Living and Housing Costs

The 1998–99 quarterly dormitory rates were $902 for a standard single and $733 for a standard double room. Board costs ranged from $515 per quarter for the one-meal plan to $758 for the three-meal-a-day plan. University-owned apartments are also available. A one-bedroom unfurnished unit was $463 per month (furnished, $526); a two-bedroom unfurnished unit was $546 per month (furnished, $610). All apartments have a stove and refrigerator; utilities are included in the costs. Students interested in these apartments should apply by January. Many private apartments, varying in cost, are available close by. These costs are subject to change without notice.

Student Group

The University enrollment is approximately 19,000, including students from more than ninety other countries. The College enrollment is about 1,700 students, including approximately 300 graduate students.

Location

Ohio University is located in Athens, Ohio, a city of more than 40,000 people, located 75 miles southeast of Columbus, Ohio, the state capital. Situated in the heart of scenic southeastern Ohio, Athens is a short traveling distance from many state and national parks and forests. Restaurants and stores are within walking distance of the campus.

The University

Ohio University was the first university in the Northwest Territory. Chartered in 1804 by the First General Assembly of Ohio, it is a symbol of America's early realization of the importance of education. The University's commitment is perhaps best indicated by the motto above the main gate to the campus: "So enter that daily thou mayest grow in wisdom, knowledge, and love." The multimillion-dollar C. Paul and Beth K. Stocker Endowment provides many benefits for faculty enrichment, student scholarships, endowed chairs, and funds for advanced research and equipment for the College. The University and its colleges bring many distinguished speakers and performers to campus. There are both social and professional student organizations on campus. Many recreational facilities are available for both indoor and outdoor sports and activities.

Applying

Normally, the admission deadline is March 1. Application materials may be obtained from the Office of Graduate Student Services, Wilson Hall, Ohio University, Athens, Ohio 45701 (telephone: 614-593-2800).

Correspondence and Information

Associate Dean for Research and Graduate Studies
Russ College of Engineering and Technology
Ohio University
Athens, Ohio 45701

Telephone: 614-593-1482
World Wide Web: http://www.ent.ohiou.edu

Ohio University

DEPARTMENT CHAIRS AND AREAS OF RESEARCH

Warren K. Wray, Dean. (e-mail: wray@bobcat.ent.ohiou.edu)
Jerrel R. Mitchell, Associate Dean for Research and Graduate Studies. (e-mail: mitchell@bobcat.ent.ohiou.edu)
Roger Radcliff, Associate Dean for Academic Affairs. (e-mail: radcliff@ouvaxa.cats.ohiou.edu)
Pamela Parker, Associate Dean for Development. (e-mail: parker@bobcat.ent.ohiou.edu)

Chemical Engineering. Dr. Michael Prudich, Chair (e-mail: prudich@bobcat.ent.ohiou.edu); Dr. Kendree Sampson, Graduate Chair (e-mail: sampson@bobcat.ent.ohiou.edu). Programs leading to M.S. and Ph.D. degrees are offered with research particularly in the areas of coal conversion and utilization, flue gas desulfurization, thin-film materials, polymerization reaction engineering, process control and dynamics, biochemical engineering, corrosion, environmental assessment, separation processes, transport phenomena, applied mathematics, aerosols, and thermodynamics. The Ohio Coal Research Center, Center for Advanced Materials, and the NSF Corrosion in Multiphase Systems Center emphasize research projects in the above.

Civil Engineering. Dr. Gayle Mitchell, Chair (e-mail: gmitchell@bobcat.ent.ohiou.edu); Dr. Glenn Hazen, Graduate Chair (e-mail: ghazen@bobcat.ent.ohiou.edu). Programs of study leading to the M.S. degree may be formulated in the areas of environmental engineering, geotechnical engineering, geo-environmental, hydrology, mechanics, structures, or transportation. Research activities include studies of structural and environmental response of pavements, testing and structural performance of pipes, flow through porous media, soil-structure interaction, constitutive relations for soils and rocks, nondestructive testing, computational methods in soils and structures, dynamic response of structures, transportation noise impacts and assessment, reclamation of mined lands, subsurface investigations, landfills, stochastic flood and drought analyses, and zebra mussels. Most of the research is conducted through the Ohio Research Institute for Transportation and the Environment. Specialized research equipment includes a 25-ton Cone Penetrometer Test truck, a 200-gravities geophysical test centrifuge, a 1-million pound load frame, and an indoor accelerated pavement load facility.

Electrical Engineering and Computer Science. Dr. Dennis Irwin, Chair (e-mail: irwin@bobcat.ent.ohiou.edu); Dr. Douglas Lawrence, Graduate Chair (e-mail: dal@bobcat.ent.ohiou.edu) Programs leading to the M.S. and Ph.D. degrees are offered. Areas of interest include computers and computer science, control systems, VLSI design, communications, information theory, solid-state electronics, energy conversion and power systems, electromagnetics, avionics, and signal processing. One of the most distinctive features of the School of Electrical Engineering and Computer Science at Ohio University is its Avionics Engineering Center. Initiated in 1963, this center provides educational opportunities for graduate students. The center participates in NASA's Tri-University Program with Princeton University and the Massachusetts Institute of Technology. Research projects at the center include instrument landing technology, airborne data collection, communications, and navigational systems, including GPS.

Industrial and Manufacturing Systems Engineering. Dr. Charles M. Parks, Chair (e-mail: cparks@bobcat.ent.ohiou.edu); Dr. Thomas A. Lacksonen, Graduate Program Chair (e-mail: lacksonen@bobcat.ent.ohiou.edu). The department offers the M.S. in industrial and systems engineering and offers the Ph.D. in intelligent systems through the College's Integrated Ph.D. Program. The primary research and teaching thrust of the Ohio University IMSE department is manufacturing systems engineering. Within this context, students can focus their graduate studies in the areas of specialization, including human factors/ergonomics, manufacturing systems design and control, manufacturing information systems, and quality systems. A unique feature of the OU IMSE department is its Center for Advanced Software Systems Integration, which manages funded research projects in manufacturing software systems integration. With the School of EECS, the IMSE department jointly operates a state-of-the-art Large Scale Software Systems Development Laboratory populated with commercial-grade manufacturing software tools, CASE tools, and Silicon Graphics engineering workstations. For more details, students can access the IMSE department entry on the Russ College of Engineering's World Wide Web home page at http://www.ent.ohiou.edu.

Mechanical Engineering. Dr. Jay Gunasekera, Chair (e-mail: gsekera@bobcat.ent.ohiou.edu); Dr. M. Khairul Alam, Graduate Chair (e-mail: alam@bobcat.ent.ohiou.edu). Graduate work leading to an M.S. degree in mechanical engineering can be formulated with specialization in mechanical systems, manufacturing, CAD/CAM, thermo-fluid sciences, or product management. Areas of interest include computer-aided design and manufacturing, microcomputer control and data acquisition systems, automated manufacturing systems, finite-element analysis, polymer processing, robotics, combustion, thermal process engineering, thermo-fluid systems, composite processing, heat transfer, fluid mechanics, and mechanical design. A number of students in mechanical engineering are completing their Ph.D. degrees in the integrated Ph.D. program.

Integrated Engineering. An interdisciplinary Ph.D. is offered with three specialty tracks: (1) materials processing, (2) geotechnical and environmental, and (3) intelligent systems. Many research topics are available in these areas. Students can tailor a program of study to address problems of national and international importance. An M.S. degree in engineering or a related field is required for admission. A plan of study is developed on an individual basis by the student's adviser and the special committee. That plan must include a set of designated core courses and courses appropriate to the chosen specialty track. At least 15 credit hours from each of two departments in the Russ College of Engineering and Technology or 12 credit hours from each of three departments must be included in the plan of study. Cross-disciplinary research is facilitated by the College's six centers and institute: Avionics Engineering Center, Ohio Coal Research Center, Center for Automatic Identification, Corrosion in Multiphase Systems Research Center, Center for Advanced Materials Processing Research, the Center for Advanced Software Systems Integration, and the Ohio Research Institute for Transportation and the Environment. Many interesting research opportunities are available in the three designated specialty tracks. For more information about this program, students should contact the Associate Dean for Research and Graduate Studies.

Test of antenna array system.

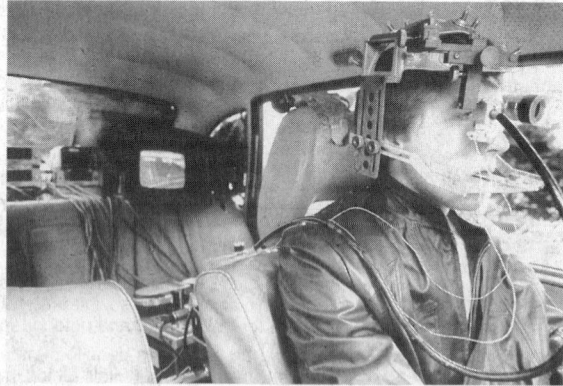

Human factors and driver eye-scanning research in the Department of Industrial and Manufacturing Systems Engineering.

OKLAHOMA STATE UNIVERSITY

College of Engineering, Architecture and Technology

Programs of Study

Through its various schools, the College of Engineering, Architecture and Technology offers Master of Science, Master of Engineering, and Doctor of Philosophy degree programs in agricultural, chemical, civil, electrical, industrial, and mechanical engineering. The School of Architecture offers Master of Architectural Engineering and Master of Architecture degrees. Master of Science and Master of Engineering degrees are offered in environmental engineering by the School of Civil Engineering. Also offered are multidisciplinary programs leading to the Master of Manufacturing Engineering and the Master of Telecommunication Management degrees.

A Master of Engineering degree is offered for students who want an emphasis on design rather than research. It requires 32 semester credit hours plus a creative design project of up to 8 credit hours, usually done as part of an internship with industry that is of three to seven months' duration. It builds upon an accredited B.S. degree in the same discipline.

The Master of Science degree can consist of 24 credit hours of course work plus 6 for a thesis or 32 credit hours (35 credit hours for the Master of Architectural Engineering) including a creative component (independent study) of at least 2 hours.

The master's degrees in the School of Architecture require two semesters of course work beyond the usual five-year undergraduate professional degree in architecture or architectural engineering.

The Doctor of Philosophy degree requires 90 semester credit hours beyond the B.S. or 60 beyond the master's, including 18 to 30 thesis credits. Preliminary, qualifying, and final examinations are required.

The master's degrees can normally be completed in twelve months. The Ph.D. requires from two to three years beyond the master's. The University offers two 16-week semesters plus an 8-week summer session each year.

Research Facilities

The College of Engineering's annual research budget is approximately $12 million. A wide variety of computer equipment and numerous minicomputers, microcomputers, and microprocessors, with interactive graphics capability, are available within the College. Extensive laboratory space and research equipment are available for use by students. The College's Office of Engineering Research provides administrative support services for more than 200 active research projects. These services include budget preparation, proposal editing, and production and fiscal reporting on contracts and grants. The University has a large central research library covering more than 6 acres of floor space, with a substantial amount devoted to engineering and physical science volumes. The Advanced Technology Research Center was dedicated in October 1997.

Financial Aid

Financial aid for graduate students includes fellowships, scholarships, and teaching and research assistantships. Stipends for half-time assistantships for the 1999–2000 year range from $5500 to $9400 for master's students and from $6300 to $11,000 for Ph.D. students. Nonresident tuition is waived for graduate assistants. The fellowship and scholarship application deadline is March 1 for the following academic year. The University has a large number of clerical job openings for students' spouses.

Cost of Study

For 1999–2000, tuition for graduate-level credit is $95.91 per semester credit hour for Oklahoma residents and $270.91 for nonresidents. In addition, a health center fee of $46 per semester ($7 for those who take 6 credit hours) and a records maintenance fee of $5 per semester are charged to all full-time students. All engineering students also pay a technology fee of $19 per semester credit hour to a maximum of $200 each semester.

Living and Housing Costs

The cost of dormitory room and board was about $1960 per semester for 1999–2000, including local telephone service. OSU owns seventy-seven low-cost, furnished, two-bedroom apartments for married students. Numerous private houses and apartment buildings are located near the campus. A nonresident single student can expect to spend $3200 per semester for dormitory housing, food, books, and miscellaneous expenses.

Student Group

The OSU enrollment is about 19,400 full-time on-campus students; approximately 4,300 of these are graduate students. Men constitute 54 percent and women, 46 percent of the student body. The College of Engineering, Architecture and Technology has 3,085 students, including 540 graduate students in engineering and architecture.

Student Outcomes

Students receiving graduate degrees in engineering at Oklahoma State University are recruited in the national marketplace. Increasing numbers of employment opportunities within the geographical area for these graduates include General Motors/Delphi in Wichita Falls, Texas; Koch Industries in Wichita, Kansas; Texas Instruments in the Dallas area; American Airlines, The Williams Companies, and the Corps of Engineers in Tulsa; and General Motors, AT&T, Seagate, and Xerox in Oklahoma City.

Location

Stillwater, located 65 miles from both Tulsa and Oklahoma City, has a population of 40,000 and is essentially a university town. There are a number of large lakes and recreational areas nearby that are usually uncrowded. The climate is mild and pleasant, typical of the Sun Belt area.

The University

Founded in 1890, Oklahoma State University is a land-grant institution with eight colleges: Agriculture, Arts and Sciences, Business Administration, Education, Engineering, Graduate Studies, Home Economics, and Veterinary Medicine. The Stillwater campus has 100 buildings situated on 415 acres, plus the nearby Lake Carl Blackwell area of 19,364 acres. Cultural and recreational facilities are provided by the Seretean Center for the Performing Arts, the Valerie Colvin Physical Education Center, and the award-winning Student Union. The University participates in all major intercollegiate sports and ranks third nationally in the number of NCAA championships in varsity sports. OSU is a member of the Big Twelve Conference.

Applying

Application forms are available from the Graduate College and must be submitted in duplicate with official transcripts of all academic work completed. Requests for financial aid should be made directly to the school of interest by March 1 for summer or fall semester admission and by October 1 for spring admission.

Correspondence and Information

Graduate Adviser
School of (specify)
Oklahoma State University
Stillwater, Oklahoma 74078

Dean of the Graduate College
202 Whitehurst Hall
Oklahoma State University
Stillwater, Oklahoma 74078

Oklahoma State University

FACULTY HEADS AND RESEARCH AREAS

Dean, College of Engineering, Architecture and Technology: Karl N. Reid.

Architecture and Architectural Engineering. Professor Randy Seitsinger, Head. Research concentration is in design, construction, high-rise buildings, architectural history (especially medieval), and computer applications in architecture.

Biosystems and Agricultural Engineering. Professor Billy J. Barfield, Head. Areas of emphasis include soil erosion, sediment control, nonpoint-source pollution, irrigation, hydrology, groundwater, water quality, hydraulics, crop processing, grain storage, food processing, physical properties of biological materials, sensors and control technology, machine vision, image processing, energy conservation, expert systems, ground and aerial pesticide application, equipment design, low-input agricultural systems, and precision farming.

Chemical Engineering. Professor Russell Rhinehart, Head. The School of Chemical Engineering is involved in a variety of research. The major focus of the School is toward industrially relevant research, but fundamental research is also undertaken. Students have the opportunity to pursue graduate research in traditional areas of chemical engineering, including vapor-liquid equilibrium thermodynamics, adsorption thermodynamics, and rheology. In addition, graduate research is available in more recent areas such as computer-assisted process design, polymer processing and engineering, membrane separations, ultrapure water processing, industrial ion exchange, artificial intelligence applied to process control and monitoring, and ocean thermal energy conversion. Finally, the School has several focus areas in the environmental area, including groundwater and contaminant transport modeling, development of design strategies for environmentally benign processes, and industrial waste treatment.

Civil and Environmental Engineering. Professor Robert K. Hughes, Head. Research interests include expansive soils, dynamic compaction, biological and chemical treatment of industrial domestic and hazardous wastes, air and water pollution, groundwater pollution, aquifer restoration, geosynthetics, flexible and rigid pavements, construction material, offshore structures, computational mechanics, composites, lightweight concrete, construction scheduling and estimating, and alternate disputes resolution.

Electrical and Computer Engineering. Professor Michael A. Soderstrand, Head. Graduate and research fields of specialization include energy/power/renewable energy systems, power economics, power electronics, computer systems, signal/image/speech processing, VLSI, parallel processing, analog CMOS electronics, control systems, robotics, neural networks, communication theory, estimation theory, data transportation and protection, engineering reliability, telecommunication systems, electromagnetics/antennas/radar, optical engineering, fiber optics, photonics, optoelectronics, quantum electronics, and laser systems, including ring lasers. Exceptional research facilities are available in the optics/laser area. Modern computer facilities that include a large number of networked Sun Workstations are available for use by graduate students and faculty members. Facilities are also available for a variety of signal-processing research activities.

Industrial Engineering. Professor William J. Kolarik, Head, and Professor Allen C. Schuermann, Graduate Program Director. The Computer Integrated Manufacturing Center, the Oklahoma Industrial Analysis and Diagnostic Center, the Center for Local Government Technology, and the Hazardous Material and Waste Management Program are among the School's research facilities. Teaching and research interests include computer-integrated manufacturing, automation, scheduling, object-oriented modeling, performance measurement, robotics, total quality management, operations research, stochastic systems analysis and systems simulation, economic analysis, ergonomics, group technology, information systems, management and organizational behavior, productivity management and energy, water and hazardous waste management.

Mechanical and Aerospace Engineering. Professor Larry L. Hoberock, Head. The School of Mechanical and Aerospace Engineering is a major participant in the interdisciplinary Manufacturing Laboratory and the Web Handling Research Center, with close ties to the Laser Center. The graduate program is suitable for those who have undergraduate degrees in engineering, physics, or applied mathematics, but mathematics students may need prerequisite work at the undergraduate level. The research program covers a wide variety of topics, and brief descriptions of these follow. (Much more information can be found on the World Wide Web at http://www.mae.okstate.edu). The research for manufacturing processes and materials includes ultraprecision machining and grinding, nontraditional machining, forming of metals and composites, synthesis of diamond films and coatings, advanced ceramic finishing and processing, machine tool monitoring and sensing. For web handling, the research involves nip mechanics, roll defects, roll structure analysis and measurement, constituent properties of web stacks, winding mechanics, air entrainment, roll buckling and wrinkling, viscoelastic and hygroscopic material effects, online tension measurement and control, traction, web flutter, lateral dynamics and control. The research for fluid mechanics and aerodynamics focuses on computational fluid dynamics, boundary layer transition, flow-induced vibrations, flow characterization through filters, and high-speed air films. The thrust areas for heat transfer, thermal, and environmental systems involve thermal system simulation and optimization, building simulation and load calculation, energy conservation, radiative energy propagation, static and dynamic laser light scattering measurement of dense slurries, electronic cooling, combustion, and high-temperature flame impingement. Design research includes creative and conceptual design, geometric modeling, interactive computer-aided design, manufacturing process design, and robotic systems design. The research topics for solid mechanics, structures, and material behavior are structural dynamics, finite element methods, plasticity modeling, material characterization, corrosion, fatigue, and environmental assisted cracking. For dynamics and control, the thrust areas involve real-time distributed control, control of nonlinear uncertain systems, sliding-mode control, system identification and parameter estimation, robotic systems control, fuzzy systems control, neural networks, and dynamic synchronization. Vibrations and acoustics research focuses on vibration of heat-exchanger tubes, signal analysis, acoustic emission sensors, sonic measurement techniques, and vibratory conveying.

OLD DOMINION UNIVERSITY

College of Engineering and Technology

Programs of Study

The College of Engineering and Technology offers the Master of Science, Master of Engineering, and Doctor of Philosophy degrees in aerospace, civil, electrical, environmental, and mechanical engineering and in engineering mechanics. The degrees of Master of Engineering Management, Master of Computer Engineering, Master of Engineering in operations research and systems analyses and in experimental methods, and M.S. and Ph.D. in engineering management are also offered.

Research Facilities

The College has extensive research facilities. The Department of Aerospace Engineering provides graphics workstations for computational fluids and structural analysis. Students can directly access supercomputers at off-campus locations, notably NASA facilities at Langley and Ames. The department operates a state-of-the-art computational laboratory, a large low-speed wind tunnel, a supersonic wind tunnel, two other smaller tunnels, an acoustics and vibrations laboratory, a controls and robotics laboratory, and the 30-foot by 60-foot full-scale wind tunnel at Langley. The Department of Civil and Environmental Engineering has well-equipped and state-of-the-art laboratories in the areas of civil and environmental engineering computations (including geographical information systems), environmental, hydraulics/water resources, geotechnical/soil dynamics, structures/earthquake, and coastal engineering. The Department of Electrical and Computer Engineering provides computer laboratories for speech processing, image processing, neural networks, and architectures for specialized computations. A clean room allows for research in thin films using diamond, and the Physical Electronics Research Institute facilities house laboratories for research in laser applications, laser switch technology, E-Beam switching, semiconductor fabrication and characterization, and ultrafast science. The Department of Mechanical Engineering has an extensive CAD/CAM computer laboratory for structural analysis as well as materials testing laboratories. The Department of Engineering Management operates a computer integration lab for both automated manufacturing and systems engineering applications. The University's Computer Center houses an IBM 3090 processor complex, which is locally networked with high-speed data links nationwide. Departments are equipped with state-of-the-art Sun-based workstations that are networked to the University's Computer Center.

Financial Aid

Department financial assistance is available in the form of research or teaching assistantships, fellowships, and unfunded tuition scholarships. Information is available through the department graduate program director. Student support varies up to $18,000 for the academic year (including federal loans); the average federal loan is $11,000. To apply for federal or state financial assistance, students should complete the Free Application for Federal Student Aid (FAFSA). Information regarding Federal Direct Student Loans is available through the Office of Enrollment Services/Student Financial Aid, Rollins Hall, Old Dominion University, Norfolk, Virginia, 23529-0052; telephone: 757-683-3683; fax: 757-683-5920.

Cost of Study

For 1999–2000, tuition per credit hour is $180 for Virginia residents and $477 for nonresidents, plus a $38 health service fee and a $22 transportation fee. (These fees are subject to change.)

Living and Housing Costs

The University is located in an urban residential area and is closely surrounded by a great variety of living accommodations. Monthly rents range from $200 to $300 for comfortable quarters. Single rooms can be found for less. Students are eligible to apply for on-campus housing, which costs $3024 per academic year for an apartment without meals or $5000 per academic year for a residence hall room with meals. Questions concerning on-campus housing should be referred to the Office of Housing Services, 4701 Powhatan Avenue, Suite G-1, Norfolk, Virginia 23508-1850; telephone: 757-683-4283 or 800-766-0833 (toll-free).

Student Group

The University's enrollment for fall 1998 was 17,800, including 1,749 engineering majors, both graduate and undergraduate. There were 1,375 undergraduates, 261 master's-level students, and 113 Ph.D. students in the College of Engineering and Technology. The College's graduate student body is composed of approximately 14 percent women, 12 percent minority students, and 32 percent international students.

Location

Norfolk is a major international commercial and tourist destination. It is the hub of Hampton Roads, Virginia, whose industrial base provides students with the chance to interact with local industries, meet practicing engineers, and study current projects. Nearby are Virginia Beach, museums, theaters, concert halls, the Virginia Air and Space Center in Hampton, and historic Colonial Williamsburg. Centrally located on the Eastern Seaboard, Norfolk is just 4 hours south of Washington, D.C., and just a few hours from the hiking trails of the Blue Ridge Mountains to the west.

The University

Founded in 1930 as the Norfolk Division of the College of William and Mary, Old Dominion University gained independent status in 1962 and became a university in 1969. One of five academic colleges in the University, the College of Engineering and Technology has the largest number of majors. In addition to its facilities on the main campus in Norfolk, the College operates the Peninsula Graduate Engineering Center in Hampton, Virginia. This center, located a short distance from NASA Langley Research Center, provides classrooms, research facilities, and staff to support the burgeoning engineering community on the Virginia Peninsula. A separate nonprofit corporation, the Old Dominion University Research Foundation, facilitates research projects by administering the grants and contracts obtained by the faculty.

Applying

Applicants must have earned a bachelor's degree from an accredited university or an equivalent degree from an institution overseas and have a minimum 3.0 cumulative grade point average (on a 4.0 scale). A completed application must be sent to the Admissions Office, along with an application fee of $30. Application and credential deadlines are July 1 for the fall semester, November 1 for the spring semester, and April 1 for the summer semester. Most programs require GRE test scores, letters of recommendation, and written essays.

Correspondence and Information

For more information:
Dr. Griffith R. McRee, Associate Dean
College of Engineering and Technology
Old Dominion University
Norfolk, Virginia 23529-0236
Telephone: 757-683-3789
World Wide Web: http://www.eng.odu.edu

For admissions information:
Admissions Office
Old Dominion University
Norfolk, Virginia 23529-0050
Telephone: 757-683-3637
    800-348-7926
E-mail: aos100s@shawnee.oa.odu.edu
World Wide Web: http://www.odu.edu/gnusers/sbh/admit.htm

Old Dominion University

FACULTY HEADS AND AREAS OF RESEARCH

Aerospace Engineering Department. Osama A. Kandil, Ph.D., Chair of the Department, and Thomas E. Alberts, Ph.D., Graduate Program Director. The department emphasizes graduate instruction and research in all aspects of aerospace engineering and engineering mechanics. Well over $1.5 million is generated annually in externally funded research. The department has exceptionally strong ties with the nearby NASA Langley Research Center, and many of the department's graduate students have the opportunity to work on collaborative projects with NASA, frequently on-site at Langley. Special areas of research interest in the department include high-speed flows, computational aerodynamics and fluid dynamics, vortex flows, unsteady flows, flow stability and control, turbulence modeling, aerodynamic design optimization, multidisciplinary design optimization, aeroacoustics, sonic fatigue, space structures, finite-element analysis, structural mechanics, composite structures, mechanics of smart structures, controls, robotics, magnetic suspensions, and experimental aerodynamics and fluid mechanics.

Civil and Environmental Engineering Department. A. Osman Akan, Ph.D., Chair of the Department, and Isao Ishibashi, Ph.D., Graduate Program Director. Graduate studies and research in civil and environmental engineering cover five distinct but related principal disciplines, varying from solid to fluid phenomena. Structural engineering includes structural dynamics, nonlinear structural response, structural optimization, passive damping, lifeline earthquake engineering, and computational structural mechanics, using concurrent processing. Geotechnical engineering includes soil dynamics, soil-structure interactions, earthquake engineering, and the study of fabric in granular materials. Coastal engineering studies include monitoring of directional wave shoreline data, groin field improvement, and dredging management. Environmental engineering includes water and wastewater treatment, environmental chemistry, modeling and management of natural aquatic systems, and hazardous waste. Water resources includes hydraulics and hydrology, groundwater, watershed modeling, GIS applications, and urban stormwater management.

Electrical and Computer Engineering Department. Stephen A. Zahorian, Ph.D., Chair of the Department, and Amin Dharamsi, Ph.D., Graduate Program Director. Graduate studies and research are directed toward computer engineering, controls, signal processing, pulsed power, semiconductor devices, and physical electronics. Recent funded research projects have included laser diodes and materials; pumping and injection control of solid-state lasers; diamond film properties; guidance and control for aerospace vehicles; devices for flow measurements; electron beam– and laser-controlled semiconductor switches; concurrent processing in data-driven architectures; nonequilibrium radiation additions to equilibrium codes, including AOTV applications; visual speech display for the deaf; speaker-independent isolated word recognition; ambulatory fetal heart-rate monitoring; automated task learning, decomposition, and scheduling; applications of neural networks; and ultrafast science.

Engineering Management Department. Ralph V. Rogers, Ph.D., Chair of the Department, and Resit Unal, Ph.D., Graduate Program Director. Research and professional activities include computer-integrated management, cost and scheduling control systems, human performance engineering, knowledge systems, cybernetics and systems theory, logistics systems, quality assurance management, group decision making and team building, statistical process control for job-shop manufacturing, and facilitation of access to space by commercial entities.

Mechanical Engineering Department. Sushil Chaturvedi, Ph.D., Chair of the Department, and J. K. Huang, Ph.D., Graduate Program Director. The department is substantially involved in research, generating more than $1.5 million in externally funded research projects each year. It has a variety of ongoing collaborative research projects with researchers at NASA Langley Research Center, located less than 22 miles from the main campus. Special areas of research interest in the department are fluids—high-speed flows, computational fluid dynamics, mixing and reacting flows, turbulence modeling, experimental fluid mechanics, and aerodynamic optimization; thermal—heat transfer instrumentation, heat and mass transfer in energy systems, radiation, combustion, and free and forced convection; solids—structural dynamics, composite materials, and experimental mechanics; and mechanical systems—controls, robotics, optimization, mechanisms, computer-aided design, and vibrations. An emphasis area in design/manufacturing is offered at the master's level.

ENTERPRISE CENTERS

Applied Research Center (ARC). The center is an advanced materials science engineering and laser technology research center. Staffed with industry/University teams utilizing the Jefferson Lab technologies, ARC provides commercial product-related research. (E-mail: mgupta@odu.edu)

The Center for Advanced Ship Repair and Maintenance (CASRM). Tom Fox, Ph.D., Executive Director. The center is a partnership between the private ship repair yards of Hampton Roads, Virginia; Old Dominion University; CIT; and the city of Norfolk. The goal is to make ship repair operations more effective while meeting or exceeding environmental requirements. (World Wide Web: http://www.odu.edu/gnusers/miatc_v/casrm.htm)

Center for Continuing Engineering Education (C²E²). John Gawne, Executive Director. The center provides customized individual and group training in engineering and computers. Activities include engineering management, information technology, network administration, network engineering, and PC specialist studies. (World Wide Web: http://www.eng.odu.edu/~ccee)

Industrial Assessment Center (IAC). Sid Roberts, Ph.D., Director. The center utilizes student and faculty expertise to reduce companies' energy expenditure. At no cost to the company, an IAC team will provide practical recommendations to increase efficiency and reduce operating costs. (World Wide Web: http://www.eng.odu.edu/~iac)

The Langley Full-Scale Wind Tunnel (LFST). The tunnel is a full-scale facility for aerodynamic testing of ground, air, and sea vessel structures. LFST is the second-largest wind tunnel in the U.S. in terms of test section size and is one of the largest in the world. (World Wide Web: http://www.lfst.com)

The Technology Applications Center (TAC). Helen Madden, Ph.D., Director. The center identifies and focuses University resources on engineering and management problems. Activities include prototyping, customized testing, manufacturing process improvements, product development, sales and marketing, strategic planning, and performance benchmarking. (World Wide Web: http://www.tac.odu.edu)

Virginia Modeling, Analysis and Simulation Center (VMASC). Thomas Mastaglio, Ph.D., Executive Director. The center applies Department of Defense simulation and training technology to improving corporate and human performance. Recent programs include Port Simulation, Diagnostic Models, Interactive Decision Making Models, and End-to-End Process Simulations. (World Wide Web: http://www.vmasc.odu.edu)

Virginia Space Flight Center (VSFC). Billie Reed, Ph.D., Executive Director. The center is a Virginia, NASA, and Old Dominion University initiative to develop a commercial spaceport and to bring about a center for excellence in space operations. Suborbital and orbital missions with payloads of up to 9,000 pounds are projected for commercial uses. (World Wide Web: http://www.va-spaceflightcenter.org)

OREGON STATE UNIVERSITY

College of Engineering

Programs of Study

The College of Engineering, one of eleven colleges in Oregon State University, offers the Master of Science (M.S.) and Doctor of Philosophy (Ph.D.) degrees in the Departments of Bioresource Engineering; Chemical Engineering; Civil, Construction, and Environmental Engineering; Computer Science; Electrical and Computer Engineering; Industrial Engineering; Mechanical Engineering; and Nuclear Engineering. Master of Engineering degrees are also offered in ocean engineering, manufacturing engineering, and environmental engineering. M.S. degrees can be completed in a minimum of one year, but normally take one and a half to two years, and can include a project or a thesis. The Ph.D. degree entails the successful completion of at least three academic years of residence subsequent to the bachelor's degree, a thesis describing the results of independent research, and an oral examination based on the thesis and research. Most graduate students become involved with funded research projects.

Research Facilities

Extensive research facilities and resources are located on campus. These include bioconversion, bioseparation, biomedical materials, remote sensing, and surface and groundwater laboratories in bioresource engineering; biochemical, high-temperature ceramics, membrane separation process, combustion and gasification, and process simulation and control facilities in chemical engineering; environmental, construction materials, wave research, geotechnical, structures, and hydrology and hydraulics facilities in civil engineering; many high-end workstations, several parallel computers, and a Meiko supercomputer in computer science; the Network for Engineering and Research in Oregon (NERO), power electronics, integrated circuits and devices, epitaxial growth including an MBE system, optical and electrical characterization, and electrooptical and microwave measurements facilities in electrical and computer engineering; computer-integrated manufacturing, robotics, and human factors laboratories in industrial engineering; computer-aided design, materials testing including nondestructive evaluation, wind tunnel, robotics, engine test cells, combustion, heat transfer, and fluidized bed laboratories in mechanical engineering; and nuclear thermal hydraulics and safety, TRIGA reactor, and fabric composite facilities in nuclear engineering. Extensive computer facilities are available in the Milne Computing Center as well as in the various departments.

Financial Aid

Fellowships and research and teaching assistantships are available from various departments. Those students who are granted such support also receive a tuition waiver. In the academic year 1998–99, stipends varied from $6500 to $15,000, depending upon duties and experience. Teaching and research assistantships typically require 13 to 20 hours a week. Fellowships do not require duties. Additional financial support may be provided during the summer months.

Cost of Study

Estimated tuition and other fees for the 1999–2000 academic year are $6012 for residents and $10,230 for nonresidents. The cost of books and supplies range from about $700 to $1000. (Figures are subject to change without notice.)

Living and Housing Costs

For 1998–99, the rate for double occupancy in University dormitories and a full meal plan was about $4600 for the academic year. A few University apartments are available for married students with rents starting at about $300 per month. Off-campus one-bedroom apartments rent for about $350–$450 per month.

Student Group

Of the 14,618 students enrolled at Oregon State University in fall 1998, 2,824 were in the College of Engineering. Of these, 259 were enrolled for the M.S. degree and 144 for the Ph.D. degree.

Student Outcomes

Graduates of the College's programs find employment in industry (e.g., Intel, Tektronix, Hewlett Packard, and Boeing), government (e.g., EPA, Federal Laboratories, and the Department of Defense), and at universities.

Location

Oregon State University is located in Corvallis, Oregon, approximately 80 miles south of Portland. Corvallis is a university community of about 50,000 and is located on the Willamette River, about 50 miles from the Pacific Ocean and 60 miles from the Cascade Mountain Range. Outdoor activities, such as skiing, hiking, and fishing, are available. The community and the University offer concerts, plays, movies, and art exhibits.

The University and The College

Oregon State University is a land-grant, sea-grant, and space-grant university. Since its origin in 1893, the College has graduated more than 22,000 engineers. Each year it awards about 400 B.S., 140 M.S., and 40 Ph.D. degrees. The faculty currently numbers about 130. Interdisciplinary programs with the Departments of Science, Forestry, and Agriculture are also available.

Applying

Application forms may be obtained by writing to the admissions office. The application deadline for the fall is February 15. For admission at other times, applications should be received two months before the beginning of the term except for electrical and computer engineering, which accept applications for fall term only. Regular admission requires a grade point average of at least 3.0 (on a 4.0 scale) for the last 90 hours leading to the bachelor's degree. Some departments require that scores on the Graduate Record Examinations also be submitted.

Correspondence and Information

Office of Admissions and Orientation
Kerr Administration Building 104
Oregon State University
Corvallis, Oregon 97331-2106
Telephone: 541-737-4411
          800-291-4192 (toll-free)
Fax: 541-737-2482
E-mail: osuadmit@orst.edu
World Wide Web: http://osu.orst.edu

College of Engineering
101 Covell Hall
Oregon State University
Corvallis, Oregon 97331-2409
Telephone: 541-737-3101
Fax: 541-737-1805
E-mail: info@engr.orst.edu
World Wide Web: http://www.engr.orst.edu

Oregon State University

DEPARTMENT HEADS AND RESEARCH AREAS

Bioresource Engineering
Department Head: James A. Moore, Ph.D., Minnesota. Thirteen faculty members. The program focuses on the interface between biological sciences and engineering. Emphasis areas include bioprocess engineering (bioseparations, bioconversions, food engineering); biomedical engineering; water resources and quality (watershed analysis, hydrologic systems modeling, groundwater monitoring and modeling irrigation management); biological systems modeling; and remote sensing and geographical information systems. Inquiries: 541-737-2041; fax: 541-737-2082; e-mail: info-bre@bre.orst.edu; WWW: http://www.bre.orst.edu

Chemical Engineering
Interim Department Head: Shoichi Kimura, Ph.D., Osaka. Eight faculty members. Major areas for study and research include chemical reactor engineering, chemical recovery, fine particle processing, transport phenomena, heat transfer, electronic materials processing, fluidization, process optimization and control, protein adsorption, biofilm development, rheology, characterization of polymers, and biochemical reaction engineering. Inquiries: 541-737-4791; fax: 541-737-4600; e-mail: mail@che.orst.edu; WWW: http://www.che.orst.edu

Civil, Construction, and Environmental Engineering
Department Head: Wayne C. Huber, Ph.D., MIT. Twenty-four faculty members. Programs of study and research include construction engineering management in the areas of equipment and methods, productivity, and project control and management; environmental engineering in the areas of management of hazardous substances, bioremediation, water quality, wastewater treatment, and mitigation of environmental impacts; geotechnical engineering in the areas of slope stability, cold regions engineering, earthquake engineering, and innovative earth structures; ocean engineering in the areas of coastal structures, ocean outfall behavior, and ocean structures; structural engineering in the areas of dynamics, wood and steel structures, seismic response of structures, and advanced computational analysis; transportation engineering in the areas of materials, pavement evaluation and performance, traffic engineering, truck size and weight impacts, mobility of the handicapped and elderly, and safety; and water resources in the areas of hydrology and hydraulic issues related to environmental problems and stormwater management. Inquiries: 541-737-4934; fax: 541-737-3052; e-mail: civil@engr.orst.edu; WWW: http://www.ccee.orst.edu

Computer Science
Department Head: Michael J. Quinn, Ph.D., Washington State. Fifteen faculty members. Major areas of study include artificial intelligence (especially machine learning, real-time reasoning, and reasoning with uncertainty), parallel and distributed computing, programming languages, software engineering, and user interface design. Most basic research includes an application component in areas such as ecological modeling, climate change, oceanography, molecular biology, population biology, engineering design, speech recognition, air-traffic control, and quantum physics. Inquiries: 541-737-3273; fax: 541-737-3014; e-mail: bernie@cs.orst.edu; WWW: http://www.cs.orst.edu/

Electrical and Computer Engineering
Interim Department Head: Alan K. Wallace, Ph.D., Sheffield (England). Twenty-five faculty members. Areas of study and research include energy systems in the areas of power electronics, power quality–converters and drives; computer engineering in the areas of digital system design, microprocessor systems, computer arithmetic, ASIC and VHDL design, parallel system architecture, and processing; data security and cryptology; systems and control with emphases on system identification and detection, adaptive control and biomodeling, and communication and signal processing; and electronics and electronic materials with emphases in advanced integrated circuit design, including RF, IF, analog, mixed signal IC, and VLSI design, solid-state materials and devices, electroluminescent display technologies, electronic packaging, applied electromagnetics, microwaves, and optoelectronics. Inquiries: 541-737-3617; fax: 541-737-1300; e-mail: graduatesecretary@ece.orst.edu; WWW: http://www.ece.orst.edu

Industrial and Manufacturing Engineering
Department Head: Sabah U. Randhawa, Ph.D., Arizona State. Twelve faculty members. Areas of focus include computer-integrated manufacturing, robotics, engineering economics and multicriteria decision analysis, human factors engineering, management systems engineering, production control, quality control and management, simulation languages and modeling, operations research, and artificial intelligence. Inquiries: 541-737-2365; fax: 541-737-5241; e-mail: phyllis.helvie@orst.edu; WWW: http://www.ie.orst.edu

Mechanical Engineering
Department Head: Gordon M. Reistad, Ph.D., Wisconsin. Twenty-two faculty members. Areas of emphasis are design, materials, mechanics, and thermal/fluids engineering. Current research includes design—concurrent engineering, design verification and methods, CAD/CAM, expert systems, diagnostics, self-tuning sensors, and experimental modal analysis; materials—composite micromechanical and thermophysical properties, creep, fracture, fatigue, alloy design, thin films, phase diagrams, dimensional stability of solids, composite design for superconductors, microstructure in superconductors, and cryogenic properties; mechanics—composites, finite-element modeling, continuum damage, laser/material interactions, dynamics and stochastic loading, impact dynamics, sports mechanics, predictive monitoring and diagnosis, optimal and stochastic control, artificial intelligence, and mechatronics; thermal/fluids engineering—convection, including fluidized beds, circuit board cooling, and non-Newtonian fluids; radiant energy transport; and microscale energy conversion; thermodynamics—energy, second law analysis, and thermophysical properties; fuel burning, flame stabilization, and quenching; synthesis of solid materials; ionic and atomic species; enhanced mixing of diffusion flames; modeling of reactive flows; and biomass and municipal solid waste combustion; and fluid mechanics—aerodynamics of wind turbines, buoyant jets, fluidization of particle beds, and microscale fluid mechanics. Inquiries: 541-737-3441; fax: 541-737-2600; e-mail: info-me@engr.orst.edu; WWW: http://www.me.orst.edu

Nuclear Engineering
Department Head: Andrew C. Klein, Ph.D., Wisconsin. Ten faculty members. Graduate degrees are offered in nuclear engineering and radiation health physics. Areas of emphasis in nuclear engineering include reactor engineering, nuclear power generation, thermal hydraulics, space nuclear power, nuclear waste management, artificial intelligence, in-core fuel management, nuclear instrumentation, radioisotope production, radiation shielding, fuel cladding corrosion, and neutron radiography. Areas of emphasis in radiation health physics include radioactive material transport, research reactor health physics, radiation instrumentation, radiation shielding, environmental modeling and monitoring, emergency response planning, waste management, risk assessment, boron neutron capture therapy, radiation dosimetry, nuclear rules and regulations, and radiation detection methods. Inquiries: 541-737-2343; fax: 541-737-0480; e-mail: nuc_engr@ne.orst.edu; WWW: http://www.ne.orst.edu

PENNSTATE

PENNSYLVANIA STATE UNIVERSITY

College of Engineering

Programs of Study

Programs of study offer M.S., M.Eng., and Ph.D. programs in acoustics, aerospace engineering, architectural engineering, computer science and engineering, engineering science and mechanics, and civil, environmental, industrial, mechanical, and nuclear engineering. M.S. and Ph.D. programs are offered in bioengineering and in agricultural, chemical, and electrical engineering. M.S. programs are offered in engineering mechanics and engineering science. A one-year M.M.M. degree program is offered in manufacturing management. Many programs are interdisciplinary and are available in air pollution control, bioengineering, acoustics, materials engineering, transportation, computational fluid dynamics, microelectronics, manufacturing, signal processing, propulsion engineering, construction, and science, technology, and society.

All master's programs require a minimum of 30 graduate semester credits. The M.S. degree requires the presentation of an acceptable thesis or paper, and the M.Eng. requires a scholarly paper. The Ph.D. is awarded upon completion of a program of advanced study that includes at least a minimum period of residence, a satisfactory thesis, and the passing of comprehensive and final oral examinations as determined by the student's doctoral committee.

Research Facilities

General University facilities used by engineering faculty members and students include IBM RS-6000 workstation labs. College facilities include College and departmental NT Domain Servers, DEC Alphas, Sun Workstations, and Power PCs; they are all networked via the fiber-optic–linked College Ethernet in the Center for Electronic Design, Communications and Computing (CEDCC). CEDCC also maintains the College's communications and computing infrastructure, Web Server, Mail Server, and other support systems. Other College centers work on research programs in materials, robotics, manufacturing, chemistry, construction, electronics, aerospace, transportation, and other specialized areas. Facilities include a Triga 1-MW pulsing nuclear reactor, a robotics lab, a Co-60 gamma-ray pool, hot cells, and a vehicle test track; transmission and scanning electron microscopes; ESR and NMR systems; a variety of low- and high-speed aerodynamic and propulsion facilities; a clean room for electronic materials processing; and a separate machine shop facility.

Financial Aid

Graduate students are supported by a variety of government and industrial fellowships and traineeships and by research assistantships and teaching assistantships. In 1999–2000, stipends for fellows and trainees range from $5000 to $14,000. Stipends for half-time research or teaching assistantships are $13,000 to $20,000 (including summer). Grants-in-aid covering tuition accompany these stipends. Federal Stafford Student Loans and Federal Perkins Loans are available upon evidence of need. There is also a graduate work-study program available, paying up to $2400 per semester.

Cost of Study

Tuition in 1999–2000 is estimated at $3600 per semester for Pennsylvania residents and $7000 per semester for nonresidents, with two semesters to the academic year. Thesis fees are $17 for the master's and $70 for the doctorate.

Living and Housing Costs

University and private housing is available to graduate students. The University's facilities range from dormitory rooms for single students to two-bedroom apartments for families. Dormitory rooms cost $1300 and up per semester, room and full board approximately $2300 to $2555, and apartments $1467 and up per semester in 1999–2000. Privately owned apartments are also available in the community.

Student Group

There are currently 74,977 students enrolled for credit at Pennsylvania State University, with 41,050 on the University Park campus. There are 6,163 graduate students enrolled at University Park, with 1,273 in the College of Engineering. Additional engineering students are in the College of Earth and Mineral Sciences. Students come from all parts of the United States and from many countries. Most students are financially assisted by fellowships or assistantships.

Location

Penn State's main campus, University Park, is located in the center of the state in the borough of State College. The town and its surrounding area, with a population of about 80,000, are located in low, rolling mountain country and offer a variety of recreational activities. The community and the University present a wide array of cultural and athletic events.

The University and The College

Penn State is a land-grant university founded in 1855. Graduate work began in 1862 with 2 students. Today, the Graduate School has 2,340 faculty members and grants degrees in 155 majors. It awards about 2,034 master's degrees and 593 doctorates each year. The University has extensive research programs in many areas, and research expenditures in the College of Engineering exceed $49 million annually.

Applying

Qualified students may be admitted at the beginning of any semester—in August, January, or June. Admission is granted by the dean of the Graduate School after approval of the application by the department or program in which the student plans to major. Prospective students may write directly to any of the departments listed on the reverse side of this page.

Correspondence and Information

David N. Wormley, Dean
College of Engineering
101 Hammond Building
Pennsylvania State University
University Park, Pennsylvania 16802
World Wide Web: http://www.engr.psu.edu

Head of (specify department)
College of Engineering
Pennsylvania State University
University Park, Pennsylvania 16802

Pennsylvania State University

FACULTY HEADS AND RESEARCH AREAS

Acoustics. Professor A. A. Atchley, Head. Research areas include acoustical diagnostics, acoustic imaging and holography, active control of sound and vibration, adaptive signal processing, aeroacoustics, architectural acoustics, atmospheric acoustics, boundary- and finite-element techniques, chaos, computational acoustics, hydroacoustics, intensity technique, noise control, nondestructive evaluation, nonlinear acoustics, ocean acoustics, optoacoustics, physical acoustics, signal processing sonar engineering, speech intelligibility, structural acoustics, transducers, ultrasonics, underwater acoustics, wave propagation and scattering, thermoacoustics, sound quality, and virtual reality.

Aerospace Engineering. Professor D. K. McLaughlin, Head. Research areas include astrodynamics, analytical and computational fluid dynamics, experimental fluid mechanics, flight science and vehicle dynamics, structures, space propulsion, turbomachinery, and air-breathing propulsion. Research facilities include subsonic and supersonic wind tunnels, a large water channel, turbomachinery compressor and turbine facilities, an advanced composites (fabrication and testing) laboratory, an electric propulsion facility, a vacuum chamber space simulator, a supersonic shear layer facility, an anechoic jet noise facility, massively parallel computer systems, and other modern computer facilities.

Agricultural and Biological Engineering. Professor R. E. Young, Head. Programs are available in food engineering, soil and water resource management and conservation, properties of biological materials, environmental control, expert systems, particulate materials, agricultural structures, systems engineering, machinery systems, safety engineering, horticultural engineering, microclimate modifications, and wood engineering. Facilities include modern laboratories for food engineering, geographic information systems, controlled environmental studies, water quality, electronics instrumentation, waste management, hydraulic power and engines, physical properties of biological materials, structural component testing, computer vision, and machine research and design.

Architectural Engineering. Professor Richard A. Behr, Head. Graduate study and research are performed in four main subject areas: construction–modeling of organizations and information modeling, productivity, contracting and delivery methods; illumination engineering–modeling and visualization, daylighting, optical design, photometry, human factor issues; mechanical and energy systems–district energy, thermal storage, indoor air quality, solar, controls, system modeling; building structural systems–advancement of analysis and design methods, optimization, seismic evaluation, structural control, envelope systems, historic preservation, cable and membrane structures.

Bioengineering. Professor H. H. Lipowsky, Head. This is an interdisciplinary program. Current areas of research include development of artificial organs (e.g., heart and lung), electrocardiography, hemodynamics, ultrasonic imaging and transducers, pulmonary measurements, blood rheology, cellular biomechanics, medical imaging, electrophysiology, and function of the microcirculation.

Chemical Engineering. Professor J. L. Duda, Head. Research areas include applied thermodynamics; biotechnology—protein separations, plant cell cultures, bioreactor design; catalysis/surface science—catalyst preparation/characterization, metal-support effects, molecular dynamics/Monte Carlo simulations; physiological transport—flow/diffusion in lung, cardiovascular fluid mechanics/mass transfer, cellular biomechanics; polymers/colloids—diffusion, organized molecular assemblies; transport phenomena—turbulent reacting flows, surface tension driven flows; and tribology—lubricant rheology, oxidation stability, high-temperature lubrication.

Civil Engineering. Professor P. P. Jovanis, Head. Programs include civil and building construction, project management, civil engineering materials (geotechnical engineering, portland cement and asphalt concrete, pavement design), hydrosystems (watershed models, groundwater modeling, systems analysis, hydraulics of open channels), structures (earthquake, blast, abnormal loadings; bridges, buildings, and the building enclosure; off-shore structures, structural control, reliability, and rehabilitation), and transportation (traffic engineering, transportation planning, facilities design, network optimization algorithms, and intelligent transportation systems).

Computer Science and Engineering. Professor D. Miller, Head. Faculty research interests include computer architecture, digital system design, embedded processors, fault-tolerant computing, interconnection networks, parallel processing, performance evaluation, VLSI design, computer security, database systems, distributed computing, programming language design, semantics and implementation, software engineering, computer networks, data communications, computer vision, document image analysis, machine learning algorithms, pattern recognition, algorithmic design, computational complexity, numerical computing, scientific visualization, and computational molecular biology.

Electrical Engineering. Professor J. D. Mitchell, Interim Head. Research areas include antennas, propagation, microwaves, FDTD methods, radar and lidar, radiometry, in situ and remote sensing of the ionosphere; digital communications, optical switching, image and signal processing, multidimensional signals, signal recognition, reconstruction, neural networks; nonlinear optics, fibers, optical storage, computing; silicon, III-V, organic, wide bandgap semiconductors; ceramic, ferroelectric, and quantum devices, processing techniques; linear systems, active vision, control systems; power system planning and control, drive systems, power electronics.

Engineering Science and Mechanics. R. P. McNitt, Head. Research focuses on optical, electronic, or mechanical material property control; advanced material fabrication and processing; and material, device, and structure response simulation. Specific research areas include composite materials; rheological and biological materials; continuum mechanics; powdered materials engineering; fatigue and fracture; failure analysis; micromechanics; surface engineering; CVD and ion implantation; microelectronic materials and devices; ESR; thin films; solid-state devices; display materials and devices; nanofabrication; diamond films; NDE; sensors and actuators; adaptive control; smart materials; wave-material interactions; ultrasonics; structural dynamics; chaos; acoustics; boundary and finite elements; condition monitoring; and artificial intelligence. Facilities are available for SEM, SAM, SFM, X-ray clean rooms, metallography, shock and vibration, and fatigue.

Environmental Engineering. Professor P. P. Jovanis, Head. Research areas include water and wastewater treatment processes, aquatic chemistry and microbiology, solid and hazardous waste treatment, industrial pollution prevention and waste minimization, air pollution treatment and control, water resource systems, and soil bioremediation.

Industrial and Manufacturing Engineering. Professor A. "Ravi" Ravindran, Head. Programs are available in human factors—ergonomics engineering, human/machine interface design, safety; manufacturing systems—metal-cutting theory, plastic deformation and welding processes, group technology, design of production systems, CAPP, engineering for production, automation, robotics, control, micromachining, CAD/CAM, flexible manufacturing systems, machine tool sensing and diagnostics, tolerancing; operations research—applied stochastic processes, decision analysis, mathematical programming, graph theory and networks, engineering economy, artificial intelligence, expert systems; and systems design—quality assurance, reliability, experimental design, and systems simulation techniques.

Mechanical Engineering. Professor R. C. Benson, Head. Major fields of specialization include acoustics, automatic control, biomechanics, compressible and incompressible flow, computer vision, computational fluid dynamics, transport phenomena, combustion and flame kinetics, space propulsion, dynamics of machines, turbulence theory, simulation and modeling, transportation and vehicles, mechatronics, heat and mass transfer, design analysis, optimization and synthesis, simulation of mechanical systems, computer-aided design, robotics, smart materials, structural dynamics, vibrations and noise control, tribology, laser machining, and heat exchanger design. Departmental facilities include wind tunnels, laser facilities, advanced diagnostics for combustion and fluid flow studies, solid rocket propellants, engines, control and mechatronics, vibrations, robotics activities, and computer-based diagnostic facilities.

Nuclear Engineering. Professor A. J. Baratta, Program Chair. Graduate studies and research are offered in reactor safety—advanced reactor design, thermal-hydraulic modeling, transient analysis, accident analysis; reactor theory—computational methods, transport theory; reactor control—advanced control methods, use of artificial intelligence; reactor operations—fuel management, radiation instrumentation, radiation monitoring and dosimetry; materials research—radiation effects, plant-life extension issues, hyperfine probes for defects in solids; and radiation applications—neutron radiography. Facilities include a 1-MW TRIGA reactor, hot cells, a thermal-hydraulic test facility, and gamma irradiation, neutron radiography, neutron activation analysis, reactor simulation, nuclear materials engineering, and low-level radiation monitoring laboratories.

Quality and Manufacturing Management Program. Professor M. P. Hottenstein and Professor C. O. Ruud, Co-Directors. This is a one-year, interdisciplinary program leading to a Master in Manufacturing Management (M.M.M.). Topics covered are business concepts in manufacturing, engineering design, quality management, statistical process control and experimental design, manufacturing systems planning and control, design practice for manufacturing, manufacturing processes and materials, manufacturing strategy, and communication and leadership skills.

Science, Technology, and Society. Professor Hector E. Flores, Director. This is an interdisciplinary, intercollege program with research interests in technological literacy, science-technology education and policy, and engineering and environmental ethics.

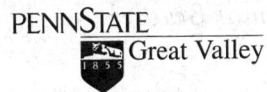

PENNSYLVANIA STATE UNIVERSITY GREAT VALLEY SCHOOL OF GRADUATE PROFESSIONAL STUDIES

Graduate Programs in Engineering

Programs of Study

Penn State Great Valley offers the Master of Engineering (M.Eng.) degree in environmental engineering and in systems engineering, the Master of Environmental Pollution Control (M.E.P.C.), and the Master of Software Engineering (M.S.E.). The environmental curriculum offers a multidisciplinary approach using a blend of theory and practice. Systems engineering is designed to integrate the traditional engineering practices of electrical, industrial, and mechanical methodologies to design and manage complex systems. The software engineering degree prepares students to develop the next generation of software products and services for industry and government.

Graduate study is designed to meet the needs of professionals who wish to obtain a master's degree in a part-time evening or weekend format. To earn a master's degree, engineering students complete a 30- to 36-credit program, which may include a professional paper.

Research Facilities

The computer center provides laboratories and classroom networks of more than 150 Intel-based microcomputer workstations for student use. These workstations are connected to the campus's local area network (LAN) and to the University's wide area network (WAN). The latter allows student access to the Internet and the World Wide Web. Students are provided with e-mail accounts and dial-up access to facilitate remote use of University-wide computer resources, libraries, and the World Wide Web.

The research library houses more than 24,000 books; 360 current professional, trade, and popular periodicals; and a collection of government publications, microfiche, CD-ROMs, and books on audiotape. Drawing on the resources of the entire University, the library at Great Valley is part of Penn State's University Libraries system, one of the leading academic research library organizations in the nation. Students have access to more than 4 million cataloged volumes, 1.4 million government publications, and 32,000 current journals and serials, plus a number of informational materials in various formats, from maps to microforms. Other accessible resources include materials at all Big Ten university libraries, other national research centers, and the Tri-State College Library Cooperative, an organization that provides members with access to the library resources of more than thirty colleges in the Philadelphia area.

Financial Aid

Financial assistance at Penn State Great Valley exists in the form of scholarships, grants, Federal Stafford Student Loans, graduate research assistantships, a minority fellowship, and Federal Work-Study.

Cost of Study

Part-time tuition for 1998–99 was $343 per graduate semester credit for Pennsylvania residents. Tuition for non-Pennsylvania residents was $611 per graduate semester credit.

Living and Housing Costs

Most graduate students are enrolled at Penn State Great Valley on a part-time basis and take evening or Saturday courses. They live and are employed in the greater Philadelphia region. Penn State Great Valley does not offer on-campus housing.

Student Group

More than 500 students are enrolled in graduate engineering programs. They are a diverse group of students employed full-time in business, computer science, engineering, health and human services, information, and scientific fields. These students pursue graduate study to keep up with the latest technological challenges, build academic credentials to advance their careers, and redirect their career paths into emerging, growing fields.

Location

The campus is situated in the Great Valley Corporate Center, along Route 202, the region's high-technology corridor. It is the nation's first university facility permanently housed in a corporate park, alongside world-class companies.

The University and The College

Penn State Great Valley is designed specifically for adult learners. To meet the needs of working adults, most courses meet two nights a week in seven-week sessions, allowing students to take one course at a time and complete two courses each semester. With six sessions offered each year (in the fall, spring, and summer semesters), students may complete as many as six graduate courses (18 credits) in one year and earn their degree in about two years on a part-time basis. Evening or Saturday classes enable students to participate in the program while maintaining full-time professional positions.

Applying

To receive admission consideration, applicants must hold a bachelor's degree from a regionally accredited U.S. institution or a comparable degree from a recognized college or university outside the United States. Four further information and applications, interested students should contact the Admissions Office.

Correspondence and Information

Admissions Office
Pennsylvania State University Great Valley School of Graduate Professional Studies
30 East Swedesford Road
Malvern, Pennsylvania 19355
Telephone: 610-648-3243
Fax: 610-889-1334
E-mail: gvinfo@psu.edu
World Wide Web: http://www.gv.psu.edu

Pennsylvania State University Great Valley School of Graduate Professional Studies

THE FACULTY AND THEIR RESEARCH

Penn State Great Valley engineering faculty members have established national and international reputations as scholars, consultants, practitioners, and researchers, and they bring a broad range of backgrounds and experiences to the classroom. They publish in premier journals and refereed conference proceedings. Faculty members present their research at national and international conferences and often thread their consulting experience into classroom instruction. The faculty is made up of full-time members as well as affiliate and part-time members and adjunct professors, many of whom hold full-time corporate positions or manage their own businesses. The following is a list of the full-time engineering faculty members at Penn State Great Valley.

James Alpigini, Assistant Professor of Electrical Engineering; Ph.D., Wales. System modeling, simulation, and visualization; control of ill-defined and nonlinear systems; artificial intelligence in real-time control; chaos theory and dynamics.

Mohamad Ansari, Associate Professor of Mechanical Engineering; Ph.D., Michigan. Operator theory, reductive and transitive operator algebras.

Robert Hartman, Associate Professor of Mechanical Engineering; Ph.D., Delaware. Computations and measurements of separated, reattaching, turbulent flows.

Kathryn Lilly Jablokow, Associate Professor of Mechanical Engineering; Ph.D., Ohio State. Dynamic simulation and real-time control of complex robotic systems, such as walking machines, flexible mechanisms, and dexterous hands; creativity and innovation, especially creative problem solving.

Eugene Kozik, Associate Professor of Industrial and Manufacturing Engineering; Ph.D., Pittsburgh. Systems engineering, database technology, information systems design, computer-integrated manufacturing, data communications technology, computer network security.

John McCool, Associate Professor of Industrial and Manufacturing Engineering; Ph.D., Temple. Problems related to bearing design and application, including life testing and reliability; the analysis of surface roughness effects, fluid traction, and metrology. Dr. McCool established the Surface Microtopography Laboratory at Penn State Great Valley.

John S. Mullin, Lecturer in Software Engineering; M.Eng., Penn State. Software engineering, project management, information systems, computer performance analysis, database and expert systems, human-computer interaction.

Colin Neill, Assistant Professor of Software Engineering; Ph.D., Wales. Real-time software engineering, software process assessment, communications systems, bioelectronics, manufacturing systems.

Michael J. Piovoso, Associate Professor of Electrical Engineering; Ph.D., Delaware. AI-based systems, digital signal processing, control engineering, neural nets, fuzzy logic, expert systems, data mining.

David Russell, Professor of Electrical Engineering and Academic Division Head for Graduate Programs in Engineering; Ph.D., Council for National Academic Awards (Liverpool Polytechnic, Manchester University, and the United Kingdom Atomic Energy Authority, U.K.). Application of artificial intelligence in the control of real-time, ill-defined systems; factory information systems; the philosophy of machine intelligence.

Lily Sehayek, Assistant Professor of Environmental Engineering; Ph.D., Rutgers. Groundwater modeling, informatics, the application of artificial neural networks to environmental problems.

James Weisbecker, Associate Professor of Computer Science; Ph.D., Temple. Parallel computing, programming languages, system development methods.

PORTLAND STATE UNIVERSITY

School of Engineering and Applied Science

Programs of Study

The School of Engineering and Applied Science at Portland State University (PSU) offers programs leading to master's and doctoral degrees and several specialized graduate certificates. Master of Science (M.S.) programs are offered in civil engineering, computer science, electrical and computer engineering, engineering management, and mechanical engineering.Master of Engineering (M.Eng.) programs are offered in civil engineering, civil engineering management, electrical and computer engineering, engineering management: project management, engineering management: technology management, manufacturing engineering, and systems engineering. Ph.D. programs are available in electrical and computer engineering, systems science (civil engineering, mechanical engineering, or engineering management), and environmental sciences and resources–civil engineering. The M.S. programs in civil engineering, computer science, electrical and computer engineering, and mechanical engineering require a minimum of 45 quarter hours after the completion of the bachelor's degree in the discipline. The M.S. in engineering management requires a minimum of 54 quarter hours beyond the bachelor's degree. Programs with or without thesis are available in all departments. The M.Eng. requires a minimum of 45 quarter hours beyond the bachelor's degree, possibly including credit for an industry internship. No thesis is required. A maximum of 15 quarter hours may be transferred in as part of a master's degree. The Ph.D. in electrical and computer engineering requires a minimum of 54 quarter hours beyond the bachelor's degree. The single-discipline option (civil engineering, mechanical engineering, or engineering management) of the systems science doctoral program requires a minimum of 72 quarter hours beyond the bachelor's degree. The environmental sciences and resources—civil engineering program requires a minimum of 78 quarter hours beyond the bachelor's degree. For all doctoral programs, candidates must complete a dissertation containing a real contribution to knowledge, based on their own investigation.

Research Facilities

Research and computing facilities of the School are housed in six buildings. These buildings contain departmental laboratories, computing facilities, and networking infrastructure that support research in the following areas: Civil Engineering—structures and materials, concrete, surveying and mapping, geotechnical, computational water quality/resources, hydraulics, environmental, and transportation; Computer Science—databases, parallel processing, and software engineering; Electrical and Computer Engineering—computer architecture, VLSI design, IC test, laser systems, analog and digital circuit design, signal and image processing, controls, microcomputer systems, parallel processing, power electronics, and neural networks; Engineering Management—decision modeling, cognitive sciences, and simulation for technology management; and Mechanical Engineering—computational fluid dynamics, structural engineering, robotics and control, CNC machining, thermal systems, and building science. Several labs are strongly supported by industry and provide state-of-the-art equipment for research: the Seismic Testing and Applied Research (STAR) Lab, the Materials Testing Systems (MTS) Lab, the Mechanical Computer Aided Engineering (MCAE) Lab, the IC Design and Test Lab, and the Intel Microcomputer Engineering Lab.

Financial Aid

Graduate assistantships for teaching or research are available on a competitive basis. Graduate assistants also receive a substantial tuition reduction. Students wishing to apply for a graduate assistantship must correspond directly with the appropriate academic department or program. Graduate students are eligible to apply for educational loans through the Federal Perkins Loan program, the Federal Stafford Student Loan program, the Federal Supplemental Loans for Students program, and the Federal Work-Study program.

Cost of Study

For 1998–99, tuition and fees for full-time graduate students were $2033.50 per quarter for Oregon residents and $3481.50 per quarter for nonresidents.

Living and Housing Costs

More than 1,100 units in eleven buildings on campus and four buildings off campus are available to PSU students. The buildings offer a variety of units and house approximately 1,600 students and their spouses and dependents. For 1998–99, monthly rents for accommodations in the campus buildings ranged from $200 to $273 for a sleeping room (shared bath facilities), from $315 to $399 for a studio apartment, from $405 to $515 for a one-bedroom unit, and from $594 to $697 for a two-bedroom unit.

Student Group

The enrollment at Portland State University is in excess of 14,000. The School of Engineering and Applied Science has more than 1,300 majors, of whom about 23 percent are full- and part-time graduate students.

Student Outcomes

Graduates of these programs are in high demand with regional agencies and industry, and many obtain research and academic positions nationally. Electrical and computer engineers find employment with electronics and software companies.Engineering management graduates find jobs in technology-driven companies, and some have started companies of their own. Civil engineers find employment in government agencies and structural and environmental consulting firms. Mechanical engineers' employment includes energy analysis in buildings, product design and manufacture in material handling, and electronics industries. Computer science graduates obtain positions in software engineering, database management, and software quality assurance.

Location

The Portland metropolitan area, with a population of 1.2 million, is Oregon's major cultural, commercial, and industrial center. Portland has been a major West Coast transportation center and international seaport for the past century. Cultural and recreational opportunities are plentiful. Theater, concerts, opera, symphony, and professional sports are popular in the city. Ninety minutes away by car are the Cascades. Equally close is the rugged and spectacular Oregon coast. Because of the recent phenomenal growth of high-technology industries, the Portland metropolitan area has been referred to as the "Silicon Forest."

The University

Portland State University is the urban university in the Oregon State System of Higher Education and is among the largest of the doctoral-degree-granting institutions in the Northwest. Located in Oregon's major metropolitan area, the University has the mission of serving the educational and professional needs of the region through excellence in teaching, scholarly research, and community service. Portland State has been designated as one of the leading institutions for the development of high-technology education and research in Oregon, and the School of Engineering and Applied Science is playing a major role in this development.

Applying

Applicants should have a baccalaureate degree from an accredited institution in a field related to the intended area of graduate study. GRE scores are required of applicants for all doctoral degree programs and of applicants to the master's degree program in computer science. A minimum TOEFL score of 550 is normally required of all candidates whose native language is not English and who have not completed an undergraduate degree at an accredited U.S. institution. Two copies of the Application for Admission to Graduate Study, accompanied by a nonrefundable $50 application fee, should be submitted to the Office of Admissions. Two official transcripts of all previous college or university work must be mailed directly to the Office of Admissions by each institution the applicant has attended. In addition, three letters of recommendation should be submitted to the appropriate department.

Correspondence and Information

For program information:
Office of the Dean
School of Engineering and Applied Science
Portland State University
P.O. Box 751
Portland, Oregon 97207
Telephone: 503-725-4631
Fax: 503-725-4298
World Wide Web: http://www.eas.pdx.edu/

For admission information:
Office of Admissions
Portland State University
P.O. Box 751
Portland, Oregon 97207
Telephone: 800-547-8887 (toll-free)
World Wide Web: http://www.pdx.edu

Portland State University

DEPARTMENTS AND THEIR RESEARCH

Civil Engineering Department

Franz N. Rad, Chair (telephone: 503-725-4282, fax: 503-725-5950, World Wide Web: http://www.ce.pdx.edu/). The department has 8 resident and numerous part-time faculty members who are involved in instruction and research at the M.S. and Ph.D. levels. Social, economic, and environmental challenges in the U.S. and around the globe are strongly interrelated to the quality of the physical infrastructure. The quality of air, water, land, transportation, housing, and power and communication systems are paramount in the socioeconomic/environmental well being of any region in the world. The graduate programs aim at educating leaders to meet challenges related to enhancing infrastructure. Students learn about conducting research and solve real problems that exist in Oregon and the Northwest. The following agencies/programs located in the metropolitan area are recent past or current sources of research funding: Bonneville Power Administration, Corps of Engineers, City of Portland, Metro, U.S. Geological Survey, U.S. Department of Transportation, Oregon Department of Transportation, and the Department of Geology and Mineral Industries. Faculty and their research areas follow. Bill Fish: heavy metals and other toxic materials in aquatic systems, process characterization, remediation, and treatment. Manouchehr Gorji: structural mechanics, plates and shells, composite materials, plasticity. Roy W. Koch: hydrologic modeling, forecasting, water resources systems analysis. B. Kent Lall: transportation planning and design, video imaging technologies, traffic engineering and management, pavement design. Shu-Guang Li: groundwater modeling, contaminant transport in heterogeneous porous media, numerical and probabilistic methods, environmental and hydrologic systems. Wendelin H. Mueller: computer and microprocessor applications to engineering problems, post-elastic member behavior, full-scale structural testing, seismic qualification and testing using a shake table. Franz N. Rad: reinforced/prestressed concrete, limit-states design; earthquake engineering. Trevor D. Smith: laterally loaded foundation piles, field in situ testing, pressuremeters, collapse arid soils. Scott A. Wells: physical/chemical/water/waste water treatment processes, water quality and hydrodynamic monitoring and modeling of surface-water systems.

Computer Science Department

Cynthia A. Brown, Chair (telephone: 503-725-4036, fax: 503-725-3211, e-mail: cmps@cs.pdx.edu, World Wide Web: http://www.cs.pdx.edu/). The department has 16 full-time faculty members. Emphasis is on interaction with local software industry and individualized graduate programs. Educational programs and research interests include formal methods, programming languages, software engineering, database systems, software metrics, software tools, parallel programming environments, languages and compilers, software testing, computer security, logic programming, and functional programming. Details on faculty members and their interests are available on the departmental Web page.

Electrical and Computer Engineering Department

Rolf Schaumann, Chair (telephone: 503-725-3806, fax: 503-725-3807, e-mail: grad_studies@ee.pdx.edu, World Wide Web: http://www.ee.pdx.edu/). The department's 15 full-time and numerous part-time faculty members support research and study programs at the M.S., M.Eng., and Ph.D. levels. Graduate certificates are offered in several specialties in electrical and computer engineering. Faculty research is sponsored by industry, including Intel, Tektronix, Credence, Triquint, Hewlett-Packard, Cadence, Mentor Graphics, and NEC America, and national agencies such as the National Science Foundation, the Department of Defense, and the National Institutes of Health. The booming high-tech industry in the Portland area provides ample opportunities for student internships and full- and part-time employment. Extensive laboratory and computer facilities are available within the department. Faculty members and their specific research areas follow. Lee Casperson: optical phenomena, devices and systems; short-pulse laser dynamics, light propagation, wave guides, and resonators. W. Robert Daasch: digital and analog VLSI circuit design and design automation, design for test, computational science and engineering, distributed computing and computer networks. Michael A. Driscoll: high-performance CPU design; computer architecture; hardware and software aspects of parallel and distributed computing. Andy Fraser: image processing; pattern recognition; information theory; nonlinear dynamics; chaos; neural networks and fuzzy systems. Douglas Hall: digital circuit design; computer systems; distributed computing. Y.-C. Jenq: digital and adaptive signal processing; theory and applications of non-uniformly sampled signals; A/D and D/A converters; high-speed digital networks and wireless communications. Malgorzata Chrzanowska-Jeske: VLSI physical layout design automation; field-programmable gate arrays; low-power design; layout and logic synthesis; testing of digital systems; low-temperature devices and phenomena. George Lendaris: neural networks; control systems. Fu Li: signal and image processing, video and wireless communications. Branimir Pejcinovic: microwave circuits; photo detectors and noise. Marek Perkowski: design automation: logic synthesis; high-level synthesis; fast prototyping with field programmable gate arrays; formal methods of design and verification; machine learning: constructive induction. Rolf Schaumann: analog and digital integrated circuits and filters. Xiaoyu Song: design automation, physical design, synthesis, test and verification, formal methods. Richard Tymerski: power electronics, high frequency switching power converters; analog CAD, simulation; control theory applications. Paul Van Halen: device design, device modeling and parameter extraction.

Engineering Management Program

Dundar F. Kocaoglu, Director (telephone: 503-725-4660, fax: 503-725-4667, e-mail: info@emp.pdx.edu, World Wide Web: http://www.emp.pdx.edu/). The Engineering Management Program has 3 full-time faculty members supplemented with adjuncts and interdisciplinary faculty members from other departments. Educational programs and research interests include management of innovation, creativity, technical organizations, technical people, team building, organizational culture, concurrent engineering, technology planning, technology evaluation, selection and acquisitions, strategic management of technology, management of R&D and engineering projects, decision theory, multicriteria decisions, group decision making, Analytic Hierarchy Process, operations research, single and multiobjective optimization, mathematical programming, data envelopment analysis (DEA), productivity management, scheduling models, multivariate statistical analysis, traditional and advanced manufacturing management, manufacturing simulations, quality management, decision support systems, management of information technology, artificial intelligence, and expert systems for technology management.

Mechanical Engineering Department

Graig Spolek, Chair (telephone: 503-725-4290, fax: 503-725-8255, World Wide Web: http://www.me.pdx.edu/). The department has 11 faculty members. Research and applications are in energy systems, building science, advanced design methods, manufacturing engineering, and engineering science, with emphasis in thermo-fluid sciences and design and manufacturing, including energy systems, building science, control systems, design methods, and manufacturing systems. Specialized areas of study include fundamental heat and mass transfer, control of energy systems, HVAC systems design, moisture transport in buildings, energy-efficient building design, indoor air quality, appliance combustion testing, computer-aided design, mechanical tolerancing, computational geometry, design automation, expert systems, optimization, finite-element methods, analysis and design of high-speed mechanical systems, design for manufacturing, statistical process improvement, robotics application, computer-controlled machining, inspection and quality assurance, motion synthesis for manufacturing, computer-aided manufacturing, automatic controls, advanced manufacturing processes, computational and experimental fluid dynamics, heat transfer in electronic equipment, depth-averaged models of the convection, iterative procedures on massively parallel computers, and algorithm applications and implementation on parallel-processing computers. The department also administers the M.Eng. in manufacturing engineering program.

Systems Engineering Directorate

Herman Migliore, Director (telephone: 503-725-4288, fax: 503-725-4298, World Wide Web: http://www.eas.pdx.edu/systems/). The systems engineering approach is presented in short courses and graduate courses as applied to development of interdisciplinary products and processes. A Master of Engineering degree is planned for fall 1999, with core courses planned in a distance learning format. Discipline-specific courses serve as electives. The directorate participates in the interinstitutional Systems Engineering Center of Excellence as part of the International Council on Systems Engineering.

PURDUE UNIVERSITY

Schools of Engineering

Programs of Study

The Schools of Engineering at Purdue University offer M.S. (with specialty designation), M.S.E., M.S., and Ph.D. degrees in the following ten disciplines: aeronautics and astronautics, agricultural and biological engineering, biomedical engineering, chemical engineering, civil engineering, electrical engineering, industrial engineering, materials engineering, mechanical engineering, and nuclear engineering. All programs are open to qualified students with a bachelor's degree in engineering, but students with degrees in such areas as science and mathematics may be admitted also. For the master's degree to have "engineering" in its title, the student must have an undergraduate degree in engineering; otherwise, the M.S. without designation is awarded. Entering students negotiate their programs with the school or department of engineering in which study is proposed. Brief descriptions of each unit's research activities are included on the reverse side of this page. Further information may be obtained by contacting the Graduate Program Administrator in the discipline of interest (see address at bottom of page).

The Master of Science degree, with or without specialty designation, typically requires a year of full-time study. The Ph.D. program requires approximately two years of full-time study beyond the master's degree. Each program is set up on an individual basis with the approval of a graduate adviser and an advisory committee. The doctoral dissertation is based on original research in the field of specialization. Students with part-time teaching or research appointments should expect to take longer than full-time students to complete their degrees.

Research Facilities

Annual research expenditures in the Schools of Engineering exceed $80 million. The principal research facilities are housed in ten buildings on the central campus, with five major laboratories on or near the campus. Each school maintains its own laboratories, as well as computer facilities and specialized equipment for research. Students also have ready access to the Engineering Computer Network, the University Computer Center, and the University and Engineering Libraries.

The Institute for Interdisciplinary Engineering Studies, located in the A. A. Potter Engineering Center, was established for interdisciplinary research and education in such areas as biomedical engineering, energy, environment, manufacturing, and transportation. Research activities include biotissue engineering; magnetic resonance imaging; energy policy and conservation; effects of electric utility deregulation; automotive safety; computer-integrated design, manufacturing, and automation; and biomass utilization as a fuel, food, or chemical source. The Engineering Research Center for Collaborative Manufacturing, a center of excellence funded by the National Science Foundation, is a cross-disciplinary center involving industry, faculty members, and graduate and undergraduate students from several disciplines.

Further information on research laboratories and centers may be found at the Schools of Engineering Web site listed at the bottom of this page.

Financial Aid

Numerous fellowships, scholarships, and traineeships awarded by Purdue University, various state and federal agencies, and industrial sponsors are available each year. Teaching and research assistantships are also available, with stipends that ranged from $1000 to $2700 per month in 1998–99. The stipends also carry exemptions from University fees and tuition, except for approximately $320 per semester. Further information may be obtained from the school or department in which the student wishes to enroll.

Cost of Study

In 1998–99, Indiana residents paid $1882 per semester for tuition and fees, and nonresidents paid $5992 per semester. Summer session charges are one half of the semester rates.

Living and Housing Costs

University-supervised graduate residences cost from $275 to $575 monthly in 1998–99. The University operates more than 1,300 married student apartments that rent at a reasonable rate. A variety of off-campus housing is available in the Lafayette and West Lafayette communities.

Student Group

There are 36,000 students on the main campus at West Lafayette, of whom 8,000 are in engineering. Graduate students total 6,400, with 1,600 of these in engineering. The student body represents all fifty states in the U.S. and many other countries.

Location

Purdue's West Lafayette campus is situated across the Wabash River from Lafayette. Population of the combined metropolitan area (excluding students) is approximately 130,000; the area offers an exceptional variety of cultural activities, historic landmarks, and recreational attractions. The campus is located 60 miles north of Indianapolis and 130 miles south of Chicago.

The University

Purdue is a comprehensive research university with an international reputation. Established as a land-grant institution in 1869, Purdue has grown from one campus with 39 students and 6 instructors to four campuses with a total enrollment of more than 65,000 students and more than 3,500 faculty members. The undergraduate and graduate programs in the Schools of Engineering have long been recognized as leaders in their disciplines, based on the scope and quality of the curriculum, faculty, facilities, and student body.

Applying

Applications and supporting materials should be submitted at least four months prior to the beginning of the semester for which admission is sought. International students whose native language is not English are required to take the Test of English as a Foreign Language. Students wishing information should contact the Graduate Program Administrator in their selected engineering discipline as indicated below. Any message to the e-mail address below generates an automatic response with current contact information for all of the disciplines.

Correspondence and Information

Graduate Program Administrator
(specify engineering discipline)
Purdue University
West Lafayette, Indiana 47907
E-mail: graduate@ecn.purdue.edu
World Wide Web: http://www.ecn.purdue.edu

Purdue University

FACULTY HEADS AND RESEARCH

School of Aeronautics and Astronautics. Professor T. N. Farris, Head. Special emphasis is placed on the advanced technology associated with the aerospace field. Research and study programs are offered in aerodynamics of propellers, wind energy machines, and buildings; boundary-layer theory; computational and unsteady aerodynamics; hypersonic aerodynamics; laser-Doppler velocimetry; control of flexible vehicles; flight and orbit mechanics; guidance; handling quality; linear and bilinear large-scale systems theory; large space structures; man-vehicle integration; model reduction; parameter identification; pilot modeling; simulation; combustion diagnostics; rocket propulsion; low gravity fluid mechanics; aeroelasticity; composite materials; experimental stress analysis; failure and damage assessment; fracture mechanics; finite elements; tribology; and structural dynamics and design. Excellent laboratory and computer facilities are available within the School.

Department of Agricultural and Biological Engineering. Professor V. F. Bralts, Head. Discovery of new technologies to produce and process food, fiber, and biological materials is the broad mission. Exceptional faculty and facilities enable engineering research programs concerning environmental and natural resources, geographic information systems, erosion mechanics, hydrologic modeling, water resource management, irrigation systems, bioprocessing, separations, biochemical reactor design and kinetics, biomass conversion, enzyme genetics, machine systems engineering, precision agriculture, robotics and sensors, electrohydraulic power control systems, mechatronics, off-highway vehicle design, grain handling and storage, post-harvest engineering, air quality, animal housing, and waste processing.

Department of Biomedical Engineering. Professor G. R. Wodicka, Head. This new department, established July 1, 1998, provides a formal administrative focus for graduate study in this interdisciplinary field. Major areas of emphasis include biological signal processing, biomaterials, biomechanics, biomedical acoustics, biomedical optics, cardiovascular devices, diagnostic and therapeutic devices, drug delivery systems, medical imaging, and tissue engineering.

School of Chemical Engineering. Professor G. V. Reklaitis, Head. Students and faculty are engaged in research in applied mathematics, artificial intelligence, biochemical engineering, biomedical engineering, chemical process research and development, colloid and interfacial phenomena, computer-aided design, catalysis and surface science, fluid mechanics, particulate systems, polymer science and engineering, process control, separation processes, surface science, process systems engineering and optimization, thermodynamics and statistical mechanics, transport phenomena, and a number of other areas. Excellent state-of-the-art instrumentation and computer facilities are available within the School.

School of Civil Engineering. Professor V. P. Drnevich, Head. A wide variety of research and study opportunities exists in all areas of civil engineering, including environmental and hydraulic engineering: wastewater and water process systems, environmental chemistry and microbiology, pollution prevention, air pollution control, and fluid mechanics and hydrology; geotechnical engineering: soil and rock mechanics, foundations, dams, levees, highways and airfields, and geoenvironmental projects; materials civil engineering: construction and transportation materials, including Portland cement, conventional and high-performance concretes, aggregates, bituminous materials, other nonmetallic materials, superpave technology, instrumentation, and highway and airport pavement design; structural engineering: design in concrete and/or steel, earthquake engineering; classical and computer analysis of structures, solid mechanics; geomatics engineering: surveying, geodesy, GPS, photogrammetry, digital mapping, remote sensing, automated/semiautomated spatial image analysis, cartography, and geographic information systems; transportation and infrastructure systems engineering: transportation systems planning and management, traffic engineering, public infrastructure, intelligent transportation systems, airport planning and design; construction and engineering management: technical and managerial responsibilities of the constructor, scheduling, utilization of construction equipment, and management of fiscal and human resources. Excellent research facilities are available in all areas, including experimental laboratories and a broad range of computational facilities.

School of Electrical and Computer Engineering. Professor W. K. Fuchs, Head. Research may be undertaken in robotics, optimal control theory, variable structure control, analog fault diagnosis, chaos, nonlinear system theory, digital process control, discrete event systems, neural networks, fuzzy systems, data mining, expert systems, artificial intelligence, advanced automation, microelectromechanical systems, MEMS for biomedical applications, biomedical acoustics, biological signal processing, bioelectromagnetics, computational imaging of biological structures, medical imaging, telemedicine, haptic human-machine interfaces, digital signal processing, speech processing, natural language processing, pattern recognition, image and video processing, inverse problems, image and video databases, communication theory, mobile communications, analog and digital communications systems, information theory, radar, microprocessor design and applications, computer architecture and systems, multiprocessor architectures, memory system design, computer graphics, distributed computing, computer networks, wireless networks, network interface architectures, multimedia computing, high-performance computing, parallel processing, fault-tolerant computing, parallel algorithms, automata, formal languages, software compilers, operating systems, computer-aided circuit design, VLSI, low-power electronics, integrated circuits, integrated circuit testing, semiconductor materials and devices, microwave solid-state devices, nanoelectronics and mesoscopic systems, molecular electronics, quantum electronics, holography, optics, integrated nonlinear optics, optical bistability, molecular-beam epitaxy, magnetics, numerical electromagnetics, electroceramics, bulk power system analysis, electromechanical motion devices and control, power electronics, electric machinery and drive systems, large power-electronics–based power distribution systems, electric propulsion systems, solar cells, amorphous materials, optimization and control of power systems, remote sensing, and numerous interdisciplinary programs. Excellent laboratory and computer facilities are available within the School.

School of Industrial Engineering. Professor W. D. Compton, Interim Head. Major areas of study are human factors engineering, manufacturing systems engineering, operations research, and production systems engineering, with research opportunities in combinatorial optimization, mathematical programming, stochastic systems, computer simulation, ergonomics, man-computer intelligence, decision support systems, management of technology, production control, material handling, computer-integrated manufacturing, industrial automation, robotics, manufacturing processes, and precision engineering. Excellent laboratory and computer facilities are available.

School of Materials Engineering. Professor G. L. Liedl, Head. Research programs encompass the structure, properties, processing, and performance of materials. Programs range from basic studies of properties and phenomena to the consideration of complex materials systems. Opportunities for research and studies include the processing and characterization of compound semiconductors, superconductors, high-temperature alloys, metallic alloys, and ceramics; the mechanical, electrical, or magnetic properties and mass transport of a variety of materials; the specialized extractive processes; and numerous interdisciplinary programs. Excellent central facilities exist in electron microcopy and X-ray diffraction, as well as computers to support the programs.

School of Mechanical Engineering. Professor E. D. Hirleman, Head. Major research and study programs within the School provide opportunity for graduate student and faculty research in automatic control and microprocessors; design and computer-aided design; robotics; manufacturing and material processing; heat and mass transfer; fluid mechanics, gasdynamics, and turbomachinery and propulsion; combustion and thermodynamics; energy conversion and utilization; acoustics and noise control; environmental control; refrigeration and air conditioning; kinematics; dynamics, vibration, and stress analysis; fatigue and fracture of materials; theoretical and applied mechanics; tribology; applied optics and laser diagnostics; and bioengineering. Extensive computational, electronics, design, and machine shop facilities are available, and seven major buildings accommodate the broad spectrum of graduate and research programs in the School.

School of Nuclear Engineering. Professor A. L. Bement Jr., Head. Research and study programs are available in reactor analysis, reactor safety, thermal hydraulics, complex adaptive systems, nuclear-fuel management, fusion technology, nuclear materials, nuclear waste, two-phase flow, magnetofluidmechanics, and heat transfer in liquid metals. These are supported by extensive laboratory facilities. Particularly noteworthy are the thermal-hydraulic and reactor safety laboratory, the reactor materials research laboratory, and the magnetofluidmechanics laboratory. Advanced computing technologies such as fuzzy logic, neural networks, and genetic algorithms are applied in several research areas. Although nuclear power reactors and fusion reactors receive emphasis in these areas, significant attention is also given to space and medical applications of nuclear technology.

RENSSELAER POLYTECHNIC INSTITUTE

School of Engineering

Programs of Study

The School of Engineering at Rensselaer Polytechnic Institute is ranked seventeenth among engineering schools nationally according to *U.S.News & World Report* and is ranked in the top ten by practicing engineers. The School offers the Master of Engineering (M.Eng.), the Master of Science (M.S.), the Doctor of Engineering (D.Eng.), and the Doctor of Philosophy (Ph.D.). Programs include aeronautical engineering, biomedical engineering, chemical engineering, civil engineering, computer and systems engineering, decision sciences and engineering systems, electrical engineering, electric power engineering, environmental engineering, materials engineering, mechanical engineering, nuclear engineering, transportation engineering, engineering physics, industrial and management engineering, manufacturing systems engineering, and operations research and statistics. There is no foreign language requirement for any of these programs.

The M.Eng. program is a nonthesis degree intended primarily as preparation for professional practice. A student with an accredited engineering B.S. degree or with equivalent experience can typically complete this degree in one year. Students with no engineering background may require more time.

The Master of Science program in engineering encompasses more diverse educational needs. Admission to the program requires a baccalaureate degree in an area appropriate to the individual's proposed plan of study but not necessarily an engineering degree. Some additional course work not qualifying for graduate credit may be necessary, however, depending on the individual's background and intended study plan. A thesis is generally required.

The Doctor of Engineering program qualifies the graduate to accept professional responsibility at the frontiers of engineering practice, while the Doctor of Philosophy program is intended primarily for those planning research-oriented careers. Both programs require completion of an approved program, typically 90 credit hours beyond the bachelor's degree, including a dissertation.

Research Facilities

Research is supported by state-of-the-art facilities, such as the George M. Low Center for Industrial Innovation; the Rensselaer Libraries, whose electronic information system provides access to collections, databases, and the Internet from campus and remote terminals; the Rensselaer Computing System, which permeates the campus with a coherent array of advanced workstations, a shared toolkit of applications for interactive learning and research, and high-speed Internet connectivity; one of the country's largest academically-based clean room facilities; a visualization laboratory for scientific computation; and a high-performance computing facility. In addition, the academic departments have extensive research capabilities and equipment. There are also several central research support units and numerous centers and institutes, including Composite Materials and Structures, Integrated Electronics and Electronic Manufacturing; Multiphase Research; Glass Research Center; Scientific Computation Research; Automation Technologies; Image Processing; Digital Video and Media Research; Geotechnical Centrifuge; and Infrastructure and Transportation Studies; Services Research and Education; Statistical Consulting; and Bioseparation.

Financial Aid

Financial aid is available in the form of fellowships, research or teaching assistantships, and scholarships. The stipend generally range from $10,000 to $12,000 for the nine-month 1999–2000 academic year. In addition, full tuition is usually granted. Additional compensation for study during the summer months may also be available. Outstanding students may qualify for University-supported Rensselaer Scholar Fellowships, which carry a stipend of $15,000 plus 30 hours of tuition credit, including fees. Low-interest, deferred-repayment graduate loans are also available to U.S. citizens with demonstrated need.

Cost of Study

Tuition for 1999–2000 is $665 per credit hour. Other fees amount to approximately $535 per semester. Books and supplies cost about $1700 per year.

Living and Housing Costs

The cost of rooms for single students in residence halls or apartments ranges from $3356 to $5298 for the 1999–2000 academic year. Family student housing, with a monthly rent of $592 to $720, is available.

Student Group

There are about 4,300 undergraduates and 1,750 graduate students representing all fifty states and more than eighty countries at Rensselaer.

Student Outcomes

Eighty-eight percent of Rensselaer's 1998 graduate students were hired after graduation, with starting salaries that averaged $56,269 for master's degree recipients and $57,000–$75,000 for doctoral degree recipients.

Location

Rensselaer is situated on a scenic 260-acre hillside campus in Troy, New York, across the Hudson River from the state capital of Albany. Troy's central northeast location provides students with a supportive, active, medium-sized community in which to live and an easy commute to Boston, New York, Montreal, and some of the country's finest outdoor recreation, including Lake George, Lake Placid, and the Adirondack, Catskill, Berkshire, and Green Mountains. The Capital Region has one of the largest concentrations of academic institutions in the United States. Sixty thousand students attend fourteen area colleges and benefit from shared activities and courses.

The University

Founded in 1824 and the first American college to award degrees in engineering and science, Rensselaer Polytechnic Institute is accredited by the Middle States Association of Colleges and Schools and is a private, nonsectarian, coeducational university. Rensselaer has five schools—Architecture, Engineering, Management, Science, and Humanities and Social Sciences—that offer a total of ninety-eight graduate degrees in forty-seven fields.

Applying

Admissions applications and all supporting credentials should be submitted well in advance of the preferred semester of entry to allow sufficient time for departmental review and processing. The application fee is $35. Since the first departmental awards are made in February and March for the next full academic year, applicants requesting financial aid are encouraged to submit all required credentials by February 1 to ensure consideration.

Correspondence and Information

For written information about graduate study:
Department of (specify)
Rensselaer Polytechnic Institute
Troy, New York 12180-3590
Telephone: 518-276-6000
World Wide Web: http://www.eng.rpi.edu/
    www/welcome/html

For applications and admissions information:
Director of Graduate Academic and Enrollment
    Services, Graduate Center
Rensselaer Polytechnic Institute
Troy, New York 12180-3590
Telephone: 518-276-6789
E-mail: grad-services@rpi.edu

Rensselaer Polytechnic Institute

FACULTY HEADS AND RESEARCH AREAS

Biomedical Engineering. Dr. Robert L. Spilker, Chair. Rensselaer's programs provide a thorough integration of advanced knowledge in a traditional engineering discipline with appropriate experience and knowledge in the life sciences so that meaningful advances in the life sciences can be made using engineering principles and techniques. Areas of advanced study and research include biomaterials, biomechanics, computational bioengineering, biofluidmechanics, computer applications and systems design, cellular tissue bioengineering, biomedical imaging, systems physiology, clinical medicine and anesthesiology, and bioinstrumentation and medical devices.

Center for Composite Materials and Structures. Dr. Sanford S. Sternstein, Director. The center covers all aspects of metal, ceramic, and polymer matrix composites, including materials-related areas such as matrix synthesis and characterization; CVD and ceramic precursor chemistry, fiber and matrix characterization; mechanics-related areas such as micromechanics and modeling of failure, fatigue, and damage mechanisms; and structural mechanics such as laminate analysis, joints, and vibrations. Composites-related undergraduate activities include the sailplane and a hybrid electric automobile.

Center for Image Processing Research (CIPR). Dr. James W. Modestino, Director. Interdisciplinary research and academic programs in image and video processing, including, but not limited to, image transmission, compression, coding, enhancement, interpretation, and applications.

Center for Infrastructure and Transportation Studies (CITS). Dr. George F. List, Director. Research emphasizes application of advanced technologies to the design, maintenance, and operation of civil infrastructure systems such as roadways, bridges, buildings, pipelines, and power distribution networks. Current projects include infrastructure management systems, remote sensing condition assessment, deterioration modeling, real-time risk analysis, capital investment planning, on- and off-road vehicle mobility modeling, advanced traffic management, emergency management, and electric roadways.

Center for Integrated Electronics and Electronics Manufacturing (CIEEM). Dr. Don L. Millard, Director. The center emphasizes leading-edge research and the education and training of students capable of performing high-quality interdisciplinary research and development in integrated electronics and electronics manufacturing. The center's research focuses on on-chip and off-chip interconnect technology, including high-conductivity metals and low-dielectric constant interlayer dielectrics (reliability, process integration, characterization, modeling, simulation, metrology, etc.), low- and high-power devices, compound semiconductor materials and devices, optoelectronic materials and interconnections, advanced packaging research, full integration of design and manufacturing incorporating the art-to-part concept, basic and applied research in commercialization and technology transfer, and electronic agile manufacturing. CIEEM facilities include complete characterization and testing facilities, a 9,000-square-foot Class 100 microfabrication clean room, design and modeling facilities, and several faculty and staff laboratories in one centralized location to bring synergism in faculty and staff from several departments in the Schools of Engineering and Science.

Center for Multiphase Research (CMR). Dr. Michael Z. Podowski, Director. Interdisciplinary research is conducted in such major areas as nucleation phenomena, phase change and interfacial phenomena, particulate and multiphase flow and heat transfer, analytical and computational methods and instrumentation and measurement.

Chemical Engineering. Dr. Jonathan Dordick, Chair. The chemical conversion of resources into new, more useful forms, together with new developments in biochemical engineering and semiconductor processing, pose challenges that are being addressed by the department. Research areas include nonlinear diffusion; interfacial phenomena; thermodynamics; combustion and high-temperature kinetics; generation of air pollutants; polymer engineering; biomass conversion and biochemical engineering; membrane and chromatographic separations; processing of semiconductors and other advanced materials; process control and design; drug formulation and delivery; and meso/nanoscale engineering.

Civil Engineering. Dr. George F. List, Chair. Research programs include geotechnical engineering, mechanics of composite materials and structures, computational mechanics, earthquake engineering, structural engineering, transportation engineering, infrastructure engineering, and computational techniques with direct ties to simulation and state-of-the-art experimentation.

Decision Sciences and Engineering Systems. Dr. Charles Malmborg, Acting Chair. This interdisciplinary department offers programs in manufacturing systems engineering, industrial and management engineering, operations research and statistics, and information systems. It prepares its students to model complex systems and to use analytical techniques in design problem solving and support of engineering and business and computational decision making.

Electrical, Computer and Systems Engineering. Dr. Joe H. Chow, Acting Chair. The department offers M.Eng., M.S., and Ph.D. programs in electrical engineering and computer and systems engineering. The research strengths of the department include microelectronics technology and design, automation and robotics, advanced image processing, digital signal processing, computer communication networks, agile manufacturing, plasma diagnostics and electromagnetics, and the development of multimedia educational materials.

Electric Power Engineering. Dr. J. Keith Nelson, Chair. Electric energy technology is a mature technology in the sense that some problems and solutions have existed for many years. However, today's environmental constraints, combined with demands for flexibility, increased reliability, and competitive costs, force new solutions to old problems while introducing entirely new considerations. Current research is in the areas of transmission and distribution, compaction of equipment, large electrical apparatus design, machine analysis, circuit interruption technology, economic studies of systems, modeling of power systems and component devices, and insulation systems. A new thrust into power electronics, started in 1991–92, has generated both new laboratories and research into drive systems, harmonic effects, and power quality issues.

Environmental and Energy Engineering. Dr. Don Steiner, Chair. The department offers degrees in environmental engineering, nuclear engineering, and engineering physics. Research areas include bioremediation and filtration of water systems, fission and fusion reactor technology, nuclear data measurements, health physics, multiphase phenomena and applied radiation. Facilities include an electron linear accelerator and a reactor critical facility.

Materials Science and Engineering. Dr. Richard W. Siegel, Chair. The department offers major research programs in electronic materials, nanostructural and composite materials, materials processing, and glass and ceramic science. Processing research includes metal, ceramic, and polymer systems. A new thrust in nanoscale technology augments and unites these areas of strength. Individual faculty research areas include solidification and crystal growth, welding and joining, corrosion, mechanical metallurgy and surface and interface studies.

Mechanical Engineering, Aeronautical Engineering and Mechanics. Dr. John Tichy, Chair. Major areas of graduate study and research are aeronautics, applied mechanics/mechanics of materials, design, manufacturing, thermofluid sciences, and tribology. Subdisciplines include automatic control, flight mechanics, propulsion, computational and experimental fluid mechanics, gasdynamics, solid mechanics, plasticity, viscoplasticity, continuum physics, creep, fatigue and fracture, advanced composite materials, noise control, structural dynamics, multidisciplinary design optimization, biomechanics, finite-element methods, kinematic synthesis, optics, thermodynamics, heat and mass transfer, energy systems and conversion, thermal design, materials processing, vehicle dynamics, kinematics and robots, CAD/CAM, and helicopter technology.

New York State Center for Advanced Technology (CAT) in Automation Technology. Dr. Harry E. Stephanou, Director. This center conducts interdisciplinary research in close collaboration with industrial sponsors. Research is focused in four major areas: manufacturing automation (precision assembly, handling of soft goods, welding, vision and inspection, rapid tooling, rapid product prototyping), robotics (bilateral manipulation, intelligent vehicles, teleoperation and shared control, reconfigurable systems, remote inspection in nuclear power plants, surgical robotics), materials processing (smart materials, aluminum extrusion, freeform power molding), and information processing (product data modeling and transfer standards, collaborative design, metadatabases, optical character recognition, text and document processing).

Scientific Computation Research Center (SCOREC). Dr. Mark S. Shephard, Director. Research focuses on high-performance computing strategies to improve understanding of physical phenomena, to provide new modeling and simulation techniques, and to support computational experimentation. Current projects include automated adaptive techniques for solving PDEs, parallel computation techniques, and procedures for critical applications. Computing facilities include a state-of-the-art IBM SP2 parallel computer and advanced workstations from Apple, IBM, Silicon Graphics, and Sun.

R·I·T

ROCHESTER INSTITUTE OF TECHNOLOGY

Kate Gleason College of Engineering

Programs of Study	The Kate Gleason College of Engineering offers programs that include traditional Master of Science (M.S.) degrees in the fields of applied statistics, computer engineering, electrical engineering, industrial engineering, mechanical engineering, and microelectronics manufacturing engineering, as well as Master of Engineering (M.Eng.) degrees in engineering management, industrial engineering, manufacturing engineering, mechanical engineering, microelectronics manufacturing engineering, and systems engineering. Two Master of Science degrees, manufacturing management and leadership and materials science and engineering, are offered as interdisciplinary degrees with the College of Business and the College of Science, respectively. The Master of Science degree may lead to employment in engineering in an industrial environment or to further graduate study at the doctoral level. The Master of Engineering degree is designed as a terminal master's program leading to industrial employment. In the Master of Engineering program, an industrial internship or an engineering case study replace the traditional thesis.
Research Facilities	The Center for Integrated Manufacturing Studies (CIMS) is a university-industry-government collaboration designed to increase the competitiveness of U.S. manufacturing companies in the global marketplace. In 1993, CIMS broke ground for a $22-million facility that greatly expanded the center's ability to provide hands-on training. The center is a showcase for manufacturing technology in electronics, imaging, printing and publishing, mechantronics, and advanced materials. Each of the five one-of-a-kind industry-focused teaching factories within the facility is dedicated to demonstrating the transfer of new and proven technologies into the manufacturing process. In addition, the Kate Gleason College of Engineering has numerous state-of-the-art facilities, including the Center for Microelectronic and Computer Engineering.
Financial Aid	A limited number of teaching assistantships, research assistantships, and tuition scholarships are available to graduate students in most degree programs on a competitive basis. Federal, state, and institutional aid is also available to those who qualify. More detailed information may be obtained from the individual academic departments. Applicants seeking financial aid must have all documents submitted to the Office of Financial Aid by February 15 to be considered for entry with support the following September.
Cost of Study	In 1999–2000, tuition and fees are $546 per credit hour (1–11 credits) and $6487 per quarter (12–18 credits). Fees for internships in the master of engineering degree programs are $290 per credit hour. An estimated cost for books and supplies for full-time students ranges from approximately $500 to $2500. Depending on the number of courses, part-time students' books and supplies may cost approximately $300–$450. All full-time graduate students are required to pay a student activities fee of $48 per quarter. All fees are subject to change.
Living and Housing Costs	Room and board for full-time students for 1999–2000 is $2284 per quarter for a standard meal plan and double room occupancy. A variety of residence hall and apartment housing options and meal plans are available, and costs vary according to options selected. Housing within the surrounding community is plentiful and moderately priced.
Student Group	Current enrollment for the Institute is approximately 13,000, including 8,000 full-time and 2,700 part-time undergraduates and 2,200 graduate students. The Kate Gleason College of Engineering has approximately 1,100 undergraduates and 250 graduate students, 255 of whom are women.
Location	The campus occupies 1,300 acres in suburban Rochester, the third-largest city in New York state. Rochester boasts a thriving arts community that includes the International Museum of Photography, the Memorial Art Gallery, the Rochester Philharmonic Orchestra, the Eastman School of Music, and the Rochester Museum and Science Center and Planetarium. Also close by are the vineyards and wine-tasting region of the Finger Lakes and Lake Ontario. The population of the metropolitan area is just under 1 million, with industries such as Eastman Kodak, Xerox, and Bausch & Lomb providing the area's economic base.
The University	Founded in 1829 and emphasizing career education, the Institute is a privately endowed, coeducational university consisting of seven colleges. RIT is the fourth oldest and one of the largest cooperative education institutions in the world, annually placing 2,600 students in co-op positions with approximately 1,300 employers. Enrolled students represent all fifty states and more than eighty other countries. The National Technical Institute for the Deaf has a current enrollment of approximately 1,200 students.
Applying	Any student who wishes to become a candidate for a master's degree must first be formally admitted to the appropriate graduate program. Applicants who have successfully completed a bachelor's degree from a regionally accredited school or who are currently completing a bachelor's degree are eligible for admission providing they meet grade, degree, and other credential requirements, including a grade point average of 3.0 out of 4.0 To be considered for admission it is necessary to submit an Application for Admission to Graduate Study accompanied by the appropriate undergraduate and graduate transcripts and two letters of recommendation. Applications are accepted on a rolling basis and must be accompanied by a $40 application fee. In addition, TOEFL scores may be required of students whose native language is not English.
Correspondence and Information	Rochester Institute of Technology Office of Part-time and Graduate Enrollment Services 58 Lomb Memorial Drive Rochester, New York 14623-5604 Telephone: 716-475-2229 E-mail: opes@rit.edu

Rochester Institute of Technology

THE FACULTY

Paul E. Petersen, Professor and Dean; Ph.D., Michigan State.
Richard Reeve, Professor and Associate Dean; Ph.D., SUNY at Buffalo.
Margaret M. Urckfitz, Assistant Dean for Student Services; B.S., RIT.
Donald D. Baker, Professor and Director, Center for Quality and Applied Statistics; M.B.A., Ed.D., Rochester.

Roy S. Czernikowski, Professor and Department Head, Computer Engineering; Ph.D., Rensselaer.
Lynn F. Fuller, Motorola Professor and Department Head, Microelectronic Engineering; Ph.D., SUNY at Buffalo.
Charles W. Haines, Professor and Department Head, Mechanical Engineering; Ph.D., Rensselaer.
Michael J. Lutz, Professor and Coordinator, Software Engineering; M.S., SUNY at Buffalo.
Jasper E. Shealy, Professor and Department Head, Industrial and Manufacturing Engineering; Ph.D., SUNY at Buffalo.
Raman M. Unnikrishnan, Professor, Computer Engineering Department, and Department Head, Electrical Engineering; Ph.D., Missouri.

Anne M. Barker, Assistant Professor; M.S., RIT.
Thomas B. Barker, Associate Professor; M.S., RIT.
Richard G. Budynas, Professor; Ph.D., Massachusetts; PE.
John T. Burr, Assistant Professor; Ph.D., Purdue.
Tony Chang, Professor; Ph.D., Chinese Academy of Science (Peking).
Edward Chung, Assistant Professor; Ph.D., Ohio University.
Soheil A. Dianat, Professor; Ph.D., George Washington.
Robert A. Ellson, Professor; Ph.D., Rochester, PE.
Jon Freckleton, Associate Professor; M.S., Nazareth College; PE.
Hany A. Ghoneim, Associate Professor; Ph.D., Rutgers.
Amitabha Ghosh, Associate Professor; Ph.D., Mississippi State.
Surendra K. Gupta, Professor; Ph.D., Rochester.
Roger E. Heintz, Professor; Ph.D., Syracuse.
Michael P. Hennessey, Assistant Professor; Ph.D., Minnesota.
Richard B. Hetnarski, James E. Gleason Professor; Dr. Tech.Sci., Polish Academy of Sciences; PE.
Karl D. Hirschmann, Visiting Assistant Professor; M.S., RIT.
Mark A. Hopkins, Associate Professor; Ph.D., Virginia Tech.
John D. Hromi, Professor Emeritus; D.Engr., Detroit.
Kenneth W. Hsu, Professor; Ph.D., Marquette; PE.
Michael A. Jackson, Associate Professor; Ph.D., SUNY at Buffalo.
Satish Kandlikar, Professor; Ph.D., Indian Institute of Technology.
Bhalchandra V. Karlekar, Professor; Ph.D., Illinois; PE.
Mark Kempski, Professor; Ph.D., SUNY at Buffalo.
Kevin Kochersberger, Visiting Assistant Professor; Ph.D., Virginia Tech.
Santosh K. Kurinec, Professor; Ph.D., Delhi (India).
Richard L. Lane, Professor; Ph.D., Alfred.
Daniel R. Lawrence, Assistant Professor; Ph.D., Toronto.
Guifang Li, Gleason Professor of Photonics and Assistant Professor; Ph.D., Wisconsin.
Michael J. Lutz, Professor; M.S., SUNY at Buffalo.
Swaminathan Madhu, Professor; Ph.D., Washington (Seattle).
Athimoottil V. Mathew, Professor; Ph.D., Queens (Canada).
Norman A. Miller, Lecturer; B.S., London.
Jacqueline Reynolds Mozrall, Assistant Professor; Ph.D., SUNY at Buffalo.
P. R. Mukund, Associate Professor; Ph.D., Tennessee.
Madhu R. Nair, Visiting Instructor; M.S., Lehigh.
Nabil Nasr, Associate Professor; Ph.D., Rutgers.
Jose Fernando Naveda, Assistant Professor; Ph.D., Minnesota.
Chris Nilsen, Professor; Ph.D., Michigan State; PE.
Alan H. Nye, Professor; Ph.D., Rochester.
Ali Ogut, Associate Professor; Ph.D., Maryland.
Sudhakar R. Paidy, Professor; Ph.D., Kansas State.
James E. Palmer, Professor; Ph.D., Case Tech.
Robert E. Pearson, Associate Professor; Ph.D., SUNY at Buffalo.
David Perlman, Associate Professor; M.S., Cornell.
Amedeo Qualich, Students' Adviser; M.S., RIT.
Mysore R. Raghuveer, Associate Professor; Ph.D., Connecticut.
Sannasi Ramanan, Associate Professor; Ph.D., Indian Institute of Technology.
V.C.V. Pratapa Reddy, Professor; Ph.D., Indian Institute of Technology (Madras).
Marietta R. Scanlon, Assistant Professor; Ph.D., Johns Hopkins.
Edward G. Schilling, Professor; Ph.D., Rutgers.
Frank Sciremammano Jr., Professor; Ph.D., Rochester.
Muhammad E. Shaaban, Assistant Professor; Ph.D., USC.
Bruce W. Smith, Associate Professor; Ph.D., RIT.
Robert L. Snyder, Professor; Ph.D., Iowa State; PE.
Paul H. Stiebitz, Assistant Professor; M.E., RIT.
David A. Sumberg, Associate Professor; Ph.D., Michigan State.
Brian K. Thorn, Associate Professor; Ph.D., Georgia Tech.
Albert H. Titus, Assistant Professor; Ph.D., Georgia Tech.
Fung-I Tseng, Professor; Ph.D., Syracuse.
David G. Tomer, Senior Lecturer; M.E., Penn State.
Josef S. Torok, Associate Professor; Ph.D., Ohio State.
I. Renan Turkman, Associate Professor, The John D. Hromi Center for Quality and Applied Statistics; Docteur-Ingenieur, Institut Nationale des Sciences Appliques (France).
Jayanti Venkataraman, Professor; Ph.D., Indian Institute of Science (Bangalore).
Panchapakesan Venkataraman, Assistant Professor; Ph.D., Rice.
Joseph G. Voelkel, Associate Professor; Ph.D., Wisconsin–Madison.
Wayne W. Walter, Professor Microelectronic Engineering Department; Ph.D., Rensselaer; PE.
Mason E. Wescott, Professor Emeritus, Statistics; Ph.D., Northwestern.
Hubert D. Wood, Assistant Professor; M.S., Rochester.
Kathryn Woodcock, Visiting Assistant Professor; M.A.Sc., Waterloo (Canada).

ROSE-HULMAN INSTITUTE OF TECHNOLOGY

Faculty of Engineering and Applied Sciences

Programs of Study Graduate programs leading to the degree of Master of Science are authorized in applied optics and in biomedical, chemical, civil, electrical, environmental, and mechanical engineering, as well as in chemistry, mathematics, physics, and engineering management. Applicants are currently being accepted only for work leading to the M.S. degree in applied optics, biomedical engineering, chemical engineering, civil engineering, electrical engineering, engineering management, environmental engineering, and mechanical engineering.

A thesis is considered to be a normal part of the master's degree program. The degree program is tailored to the needs of each of the students, who develop a specific plan of study jointly with their individual advisory committee. A minimum of 51 quarter credits and an average grade of B are required. A maximum of 12 quarter credits may be earned by thesis research.

Research Facilities Moench Hall, Olin Hall, and Crapo Hall house laboratories, classrooms, and offices for all the academic departments. The Waters Computing Center supports three major logical networks: an AFS-based Unix network, a Novell NetWare–based PC/Mac network, and an Open VMS Cluster.

Financial Aid In general, Rose-Hulman has three types of financial aid available to graduate students. Assistantships—both graduate and research appointments—are awarded by the president of the Institute upon the recommendation of the Graduate Studies Committee. Tuition grants are also awarded by the president upon the recommendation of the Graduate Studies Committee. Loans are applied for through the Office of Student Life. Assistantships carry a stipend for the academic year and normally include a full tuition grant; payment is made in installments beginning in September.

Cost of Study The 1998–99 tuition for a full-time graduate student was $6014 per quarter. Part-time students enrolled at $505 per quarter hour. Book costs and expenses for reproducing the thesis are normally the responsibility of the student.

Living and Housing Costs Although there are no special facilities for housing graduate students on campus, space is frequently available in the undergraduate residence halls. Students living on campus must purchase the meal plan. The cost of double-occupancy accommodations is $910 per quarter. The board cost is $825 per quarter.

Student Group There are approximately 1,450 students at the Institute, all majoring in engineering or science. There are about 90 students in the on-campus graduate program and 60 in Rose-Hulman's off-campus graduate program.

Location The city of Terre Haute is located approximately 70 miles west of Indianapolis, near the Indiana-Illinois state line. The campus consists of 200 rolling acres in a suburban/residential setting east of Terre Haute. It is within easy driving distance of many major midwestern cities and universities. Numerous parks and recreational areas are also close by.

The University Rose-Hulman Institute of Technology is one of the select few independent colleges of engineering and science in the United States. It was founded in 1874 by Chauncey Rose, a pioneer industrialist and entrepreneur. The college's mission is to educate Renaissance scientists and engineers—people who can change the times, who understand the importance of their technical knowledge in relation to society, and who maintain their appreciation for the arts and humanities. Some of the most academically talented students in the nation are enrolled at Rose-Hulman. The campus recently has undergone extensive renovation, and the oldest laboratory now in use dates to 1984.

Applying Applications for admission to graduate work and information may be obtained from the address given below. Regular admission requires that the applicant have both a bachelor's degree in an appropriate undergraduate field from an accredited educational institution and at least a B average. International applicants who are not graduates of an accredited bachelor's program offered by a U.S. college or university must provide official copies of the results of the Graduate Record Examinations and the Test of English as a Foreign Language (TOEFL). The minimum acceptable score on the TOEFL is 580 divided equally among test sections.

Correspondence and Information Dean for Research and Graduate Studies
Rose-Hulman Institute of Technology
5500 Wabash Avenue
Terre Haute, Indiana 47803
Telephone: 812-877-8403

Rose-Hulman Institute of Technology

THE FACULTY AND THEIR RESEARCH

Applied Optics. Robert M. Bunch, Professor; Ph.D., Kansas, 1981: laser illuminated elastic light scattering, fiber-optic sensor applications, optical testing and instruments. Richard Ditteon, Professor; Ph.D., UCLA, 1981: geometrical optics, computer-aided optical system design. Daniel L. Hatten, Assistant Professor; Ph.D., Maryland, 1996: laser physics, laser-matter interaction. Charles Joenathan, Associate Professor; Ph.D., Indian Institute of Technology, 1986: speckle techniques, holography, fiber-optic sensors, phase-measuring interferometry. Brij M. Khorana, Professor, Director of the Center for Applied Optic Studies; Ph.D., Case Western Reserve, 1968: fiber-optic sensors; interferometry; image processing, including machine vision applications and fringe analysis. Sudipa Mitra-Kirtley, Assistant Professor; Ph.D., Kentucky, 1991: UV-visible absorption and fluorescence studies of organic molecules, synchrotron radiation. Michael F. McInerney, Professor; Ph.D., Kent State, 1978: image processing, neural nets, videodiscs, and computer-assisted instruction. Michael J. Moloney, Professor; Ph.D., Maryland, 1966: physics of solid-state devices, data acquisition via PCs, interaction with area high school teachers. Azad Siahmakoun, Associate Professor; Ph.D., Arkansas, 1987: nonlinear optics, laser physics, applications of optical phase conjugation, photorefractive materials, chaos. Jerome F. Wagner, Professor; Ph.D., Ohio, 1971: infrared detectors coupled with integrated optics. Arthur B. Western Jr., Professor and Head of Physics and Applied Optics; Ph.D., Montana State, 1976: optical holographic interferometry, solid-state physics.

Chemical Engineering. Carl F. Abegg, Professor; Ph.D., Iowa State, 1966: chemical reaction engineering, kinetics of crystallization processes, controlled release nitrogen fertilizer technology. Ronald S. Artigue, Professor; D.E., Tulane, 1980: fermentation technology, distillation control, optimization. Jerry A. Caskey, Professor; Ph.D., Clemson, 1965; PE: applications of water-soluble polymers in flocculation processes and wastewater treatment, polymerization process development for the manufacture of polyacrylamide. M. Hossein Hariri, Professor and Head of the Department of Chemical Engineering; Ph.D., Manchester, 1979: vapor-liquid equilibrium, air-pollution control, petroleum engineering, sorption of NOx and SOx from gas emissions, molecular modeling, agglomeration in fluidized beds, mass transfer in bubble columns. Stuart Leipziger, Professor; Ph.D., IIT, 1964: measurement and prediction of properties of fluid mixtures, thermodynamic analysis of processes, interfacial instabilities during mass transfer operations.

Civil Engineering. Thomas J. Descoteaux, Assistant Professor; Ph.D., Connecticut, 1992; PE: structural analysis and design in timber, building science engineering. Robert J. Houghtalen, Associate Professor; Ph.D., Colorado State, 1988; PE: urban stormwater management, groundwater and surface water models. Sharon A. Jones, Assistant Professor; Ph.D., Carnegie Mellon, 1996: solid and hazardous waste, decision making for the environment. James L. McKinney, Professor and Head of the Department of Civil Engineering; Ph.D., Purdue, 1980; PE: pavement design, pavement rehabilitation, construction management. Martin J. Thomas, Professor; Ph.D., Notre Dame, 1975: aquatic chemistry, hazardous waste, environmental engineering.

Electrical and Computer Engineering. Frank E. Acker, Professor; Ph.D., Carnegie Mellon, 1967; PE: robotics, simulation and control, manufacturing automation, reliability of high-voltage equipment. Bruce A. Black, Professor; Ph.D., California, 1971; PE: spread spectrum communications, computer-assisted instruction. Frederick C. Brockhurst, Associate Professor; Ph.D., Missouri, 1973; PE: variable speed drive applications, integration of computers into the teaching process, dynamic modeling of electric machines. Edward R. Doering, Assistant Professor; Ph.D., Iowa State, 1992: image and signal processing, nondestructive evaluation, technology-enabled instruction. William J. Eccles, Professor; Ph.D., Purdue, 1965: microcomputer systems, engineering education. Barry J. Farbrother, Professor and Head of the Department of Electrical and Computer Engineering; Ph.D., Hertfordshire (England), 1977; C.Eng. (England): embedded computer systems, imaging systems. Jeffrey E. Froyd, Professor; Ph.D., Minnesota, 1979: computer-aided control system design, VLSI computer-aided design tools, computer-based instrumentation, electrical characterization of interconnection technologies, neural networks and fuzzy systems. Clifford H. Grigg, Professor; Ph.D., Manchester Institute of Science and Technology (England), 1977; PE: power system operation and planning related to reliability, control, and economics. Keith E. Hoover, Associate Professor; Ph.D., Illinois, 1976: development of digital radio direction-finding and radio-locating techniques, ionospheric modeling, digital signal processing, neural networks, microcomputer applications. Wayne T. Padgett, Assistant Professor; Ph.D., Georgia Tech, 1994: digital signal processing, detection and estimation, pattern recognition, aerial robotics. Niusha Rostamkolai, Associate Professor; Ph.D., Virginia Tech, 1986: power system dynamics and control, power system protection, flexible AC transmission, high-voltage direct-current transmission and static var compensators. David R. Voltmer, Professor; Ph.D., Ohio State, 1970; PE: microwave CAD and metrology, computational electromagnetics, computer-aided instruction, GPS applications. Ruth I. Waite, Associate Professor; Ph.D., Iowa State, 1987: VLSI design and testing, neural networks. Mark A. Yoder, Associate Professor; Ph.D., Purdue, 1984: symbolic algebra systems and multimedia in engineering education, image and speech processing.

Mechanical Engineering. M. Patricia Brackin, Associate Professor; Ph.D., Georgia Tech, 1996; PE: design processes and optimization, quality management. Christine A. Buckley, Assistant Professor; Ph.D. Northwestern, 1994: corrosion, evaluation and development of biomaterials, medical device design. Phillip J. Cornwell, Associate Professor; Ph.D., Princeton, 1989: structural dynamics of periodic structures with parameter uncertainties, dynamics of large flexible space structures. Don L. Dekker, Professor; Ph.D., Stanford, 1973; PE: engineering design education, engineering design processes and creativity. Robert E. Dillon Jr., Associate Professor; Ph.D., Rensselaer, 1983; PE: computational fluid dynamics, gasdynamics, ballistics, internal combustion engines, mechanical design. James R. Eifert, Professor, Vice President for Academic Affairs, and Dean of the Faculty; Ph.D., Ohio State, 1973: heat treatment of steel, phase transformation in iron alloys. Jerry M. Fine, Associate Professor; Ph.D., Texas at Austin, 1984: development of Runge-Kutta-Nystrom methods with interpolants, finite-element analysis applicable to thermomechanical analysis of the aluminum die-casting process. J. Darrell Gibson, Professor; Ph.D., New Mexico, 1968; PE: noise and vibration analysis. Samuel F. Hulbert, Professor and President of the University; Ph.D., Alfred, 1964; PE: ceramic engineering, biomaterial development and evolution, design and evaluation of artificial organs and prosthetic devices, evaluation of biocompatibility of biomaterials and proposed biomaterials, use of ceramics and carbons in surgical implants. Andrew R. Mech, Associate Professor; Ph.D., Illinois, 1986; PE: system simulation and optimization, gas turbines, fluid science learning center. Donald G. Morin, Professor; Ph.D., Maryland, 1972: machine dynamics and controls, design and application of robotics, computer-automated machinery. C. Mallory North, Professor; Ph.D., Alabama, 1969; PE: mathematical and computer simulation of dynamic/vibratory systems, structural dynamics. William G. Ovens Jr., Professor; Ph.D., Connecticut, 1971: materials processing and design, manufacturing process development, concurrent engineering. David J. Purdy, Professor and Head of the Department of Mechanical Engineering; Ph.D., Purdue, 1981: dynamics, simulation, computer-aided instruction, control system design and analysis. Donald E. Richards, Associate Professor; Ph.D., Ohio State, 1981; PE: heat exchanger design with enhanced heat transfer surfaces, augmentation techniques, absorption heat pump technology, multifluid heat exchanges. A. T. Roper, Professor, Vice President for Planning and Data Systems, and Director of the Center for Technology and Policy Studies; Ph.D., Colorado State, 1967: social impact assessment, low Reynolds number aerodynamics, computational fluid mechanics, wind tunnel testing. L. Wayne Sanders, Professor; Ph.D., SMU, 1974; PE: internal combustion engines, finite-element computational heat transfer, waste heat recovery in gas turbines. Leland K. Shirely, Professor; Ph.D., Brown, 1964: technology for analyzing fluid flow in diesel engine systems. David Stienstra, Associate Professor; Ph.D., Texas A&M, 1990: fracture mechanics and fatigue, mechanical behavior of materials. Lee R. Waite, Associate Professor; Ph.D., Iowa State, 1987; PE: analysis and modeling of arterial blood flow, design of biomedical instrumentation.

SAN JOSE STATE UNIVERSITY

College of Engineering

Programs of Study

The College of Engineering offers opportunities to qualified students for postgraduate education and research leading to the M.S. degree in nine programs, including aerospace, chemical, civil and environmental, computer, electrical, industrial and systems, interdisciplinary, mechanical, and materials engineering. These programs are designed with flexibility to accommodate the special needs of working students as well as full-time students.

Night and early morning classes are available. Each student is expected to plan a program of study with the help of an advisory committee. Students have the option to work on a thesis or project based on their career goals. Industry projects are encouraged. The M.S. degree is usually awarded upon completion of 30 semester units of course work and a comprehensive examination.

Research Facilities

The College has extensive research laboratories in orbital mechanics and demonstration, flow visualization, TQM, signal processing, networking, DSP, semiconductor devices, integrated circuits, digital control, neural networks, parallel computing, information engineering in manufacturing, design for manufacturability, robotics, vision systems, environmental engineering, process design materials characterization, earthquake engineering, structures, geotechnical, transportation, construction, mechatronics, micromachining, electronical cooling, and control systems. The research expenditure was $3.5 million in 1998–99 and involved 36 faculty members. The College also has the following resources: Applied Technology Institute; Institute for Environmental Research; Client Server Group; Center for Electronic Materials and Devices; Center for Productivity and Quality; Biometrics Group; and Signal Processing Group.

Financial Aid

One quarter of San Jose State University students receive financial aid, totaling more than $26 million per year. Aid is available through scholarships; grants, including Pell, Supplemental Educational Opportunity, State University, and Educational Opportunity; loans, including Perkins and Guaranteed Student Loans; and College Work-Study. A number of graduate assistantships and scholarships are available to qualified students. Students should contact individual departments for more information.

Cost of Study

Tuition is charged at the rate $652 per semester for 6 units or fewer. Books and supplies cost approximately $320 per semester.

Living and Housing Costs

The University maintains residence halls for students; rates are approximately $2670 per semester for food and housing. There are one- and two- bedroom apartments on campus at rates ranging from $318 to $575 per month. Many rooms and apartments are available in the surrounding community. Meals are served at several on-campus dining halls.

Student Group

The campus enrollment is about 27,500 students, approximately 14 percent of whom are in the College of Engineering. There are 830 graduate students and 3,500 undergraduate among the various engineering departments.

Student Outcomes

It is estimated that 10 percent of California State System engineering graduates come from San Jose State University. A sampling of employers of recent graduates includes Intel, Applied Materials, Lockheed, NASA–Ames Research Center, Bechtel Corporation, EPA, National Semiconductor, Signetics, Apple Computer, TRW, General Electric, Sun Microsystems, Hewlett Packard, IBM, Cisco, Quantum, Seagate, Stryker, city of San Jose, Cal Trans, Santa Clara Valley Water District, and Pacific Gas and Electric.

Location

San Jose State University is located at the hub of one of the most vigorous and vibrant areas in the world—Silicon Valley, which is known internationally as a center for innovation and creativity in the realm of high-technology research and development. Just a short drive away is San Francisco, beaches in Santa Cruz, Carmel, Monterey Bay, and Coastal Redwoods.

The University

Founded in 1857, San Jose State University is California's oldest institution of public higher education. In 1972 San Jose State achieved university status and become a part of the twenty-three-campus California State University System, the largest university system in the U.S. Today, the multicultural, multi-ethnic student body prepares for careers in the professions, business, social work, engineering, science, technology, education, social science, the arts, and the humanities.

Applying

Admissions applications and all supporting credentials should be submitted well in advance of the application dates. The fall date is July 15, and the spring date is December 15. International deadlines can be found on the University's Web site. The application fee is $59.

Applicants should have a baccalaureate degree in engineering or science, an appropriate GPA, and, in some cases, a GRE score. Additional course work may be assigned to help prepare students for graduate studies.

Correspondence and Information

Department of (specify)
San Jose State University
One Washington Square
San Jose, California 95192
Telephone: 408-924-3800 (Dean Donald Kirk)
E-mail: geneng@email.sjsu.edu
World Wide Web: http://www.engr.sjsu.edu/graduate

San Jose State University

THE FACULTY AND THEIR RESEARCH

Aerospace Engineering
Dr. Fred Barez, Chair.
Areas of study: gas and fluid dynamics, hypersonic flow and aerodynamics, aircraft design, propulsion, spacecraft design, microsatellite technology, guidance and control, and remote sensing.

Chemical Engineering
Dr. Michael Jennings, Chair.
Areas of study: kinetics, catalysis, reactor design, biochemical engineering, environmental engineering, electrochemistry, semiconductor processing, surface engineering.

Civil and Environmental Engineering
Dr. Thalia Anagnos, Chair.
Areas of study: concrete technology, structural dynamics, earthquake engineering, design methods, traffic engineering, traffic safety, congestion pricing, transportation economics, intelligent transportation systems, water resources engineering, groundwater, hydrology, hydraulic modeling, open-channel hydraulics, hydraulic structures, hydraulics of sediment flow, geotechnical engineering, soil mechanics, rock mechanics, geological engineering, educational software development, environmental engineering, treatment plant design, bioreactor design, hazardous-waste treatment and management, applied limnology, watershed management, feasibility studies, environmental regulations, international construction, international standards, delay analysis, forensic engineering, computer applications for construction, privatization.

Computer Engineering
Dr. Lee Chang, Director.
Areas of study: computer architecture, software engineering, memory design, microcomputers, embedded systems, multimedia computing, computer networks, distributed objects, parallel computing, performance analysis, computer vision and robotics, computer applications, operating systems, secure operating systems.

Electrical Engineering
Dr. Belle Wei, Chair.
Areas of study: signal processing, video/audio processing, multimedia systems, sound localization and separation, statistical pattern recognition, neural networks, digital signal processing and architectures, digital communication systems, wireless communication systems, digital systems, hardware description languages and synthesis, telecommunication networks, expert systems, semiconductor devices and circuits, micromachining, microfabrication, motion control, modeling of semiconductor processing, microelectromechanical (MEMS) devices, analog and digital circuit design, VLSI, solid-state circuits, electromagnetics, antennas, microwaves, optoelectronic devices and systems, power electronics and applications, automatic control systems, microcontroller systems and applications, lossless data compression/decompression circuits, video compression algorithms and architectures.

General Engineering (Interdisciplinary)
Dr. Ahmed Hambaba, Director.
Areas of study: engineering education, software development processes, data structures, software, software productivity, software tools and processes, software quality assurance, software testing, object-oriented technology, systems engineering, engineering management (ISO 9001), client server, biometrics identification, design for manufacturability, CAD, CAM, CIM, composites engineering, product realization, information systems engineering, artificial intelligence, expert system development, multi-agent architecture, self-organization design, distributed production systems, adaptive problem solving, real-time problem solving, fuzzy logic systems, neural networks, learning paradigms, reinforcement learning, genetic algorithm, open system scalability and interoperability, image/signal analysis and understanding, integrated distributed intelligent systems in manufacturing.

Industrial and Systems Engineering
Dr. Ernie Unwin, Director.
Areas of study: human factors/ergonomics, simulation of business and management systems, work measurement, quality control and total quality management.

Materials Engineering
Dr. Michael Jennings, Chair.
Areas of study: semiconductor materials and processing, thin-film deposition methods, ion-beam processing, ion implantation, solid-state diffusion, silicon integrated circuits and devices, corrosion, mechanical testing methods, ceramics engineering, materials characterization, microelectronic encapsulation and interconnection, magnetic materials and biomaterials.

Mechanical Engineering
Dr. Fred Barez, Chair.
Areas of study: computer-aided design, solids modeling, finite-element analysis, mechanical design and optimization, machine and mechanism design, mechatronics and mechatronic systems design, robotics and control, electronics packaging and thermal management of electronics, high-vacuum systems design, disk drive mechanics and technology, fluid mechanics and heat transfer, HVAC, noise and vibration.

Student in research laboratory.

The College of Engineering at San Jose State University.

SANTA CLARA UNIVERSITY

School of Engineering

Programs of Study	Santa Clara University (SCU) offers Master of Science, Engineer, and Doctor of Philosophy degree programs in addition to specialized engineering certificate programs. M.S. degrees are offered in applied mathematics, civil engineering, computer engineering, electrical engineering, engineering management and leadership, mechanical engineering, and software engineering. The program leading to the degree of Engineer is designed particularly for the ongoing education of the practicing engineer. It is offered in computer, electrical, and mechanical engineering. The degree is granted on completion of an approved academic program and on demonstration of a record of acceptable technical achievement in the candidate's field of engineering. Courses are selected to advance engineers' competence in specific areas related to their professional work. Evidence of technical achievement must include a paper written principally by the candidate and accepted for publication by a recognized engineering journal prior to the granting of the degree. (In certain cases, departments may accept as publication the proceedings of an appropriate conference.) The degree of Doctor of Philosophy is conferred by the School of Engineering primarily in recognition of competence in a subject field and in the ability to investigate engineering problems independently. The work for the degree consists of engineering research and preparation of a thesis based on that research, acceptance for publication of one or more refereed articles based on the thesis, and a program of advanced studies in engineering, mathematics, and related physical sciences. The student's work is directed by a specific department and is subject to the general supervision of the School of Engineering. The School grants the Ph.D. in electrical engineering, computer engineering, and mechanical engineering. Graduate engineering courses are primarily offered one day a week from 7 to 9 a.m., with additional courses offered in the evening.
Research Facilities	The School's research facilities include laboratories in the areas of computer graphics, object-oriented database systems, distributed computer systems, biomechanics, robotics and controls, circuits, image processing, digital systems, geotechnical testing, materials, water quality engineering, microwaves and communications, semiconductor fabrication, thermofluids, dynamics and control, and mechanical instrumentation. Computer facilities include a DEC VAX 6610; more than seventy-five state-of-the-art engineering workstations, most of which are manufactured by Hewlett-Packard and Sun Microsystems; and many Pentium-based personal computers. Most computers are connected to the campuswide Ethernet, which is also linked to the Internet for international access. The Institute for Information Storage Technology, organized under the auspices of the University, serves as a key technical resource to the information storage industry and provides outstanding educational opportunities for students in the relevant areas of data storage.
Financial Aid	Teaching and research assistantships are available and are administered by respective departments. Students must apply separately to a specific department for these assistantships at the time they apply to the graduate program. Because the University does not maintain its own scholarship program for students enrolled in the graduate programs of the School of Engineering, students applying for aid may find that the most advantageous method of financing their education is through loan programs. Among those available to students in the School are the Federal Perkins Loan, Federal Stafford Student Loan, and Supplemental Loans for Students programs. Application forms and further information may be obtained from the Financial Aid Office. International students requesting Immigration Form I-20 must submit a financial statement showing their ability to completely finance the entire program.
Cost of Study	For 1999–2000, tuition charges are $458 a unit. In addition, there is a registration fee of $10 per quarter and an incremental tuition fee of $100 per quarter. Books and supplies for the academic year cost about $750.
Living and Housing Costs	Living expenses for a single student, including housing, food, and moderate entertainment costs, range from $1000 to $1500 per month. Typical two-bedroom apartments in the area rent for $850 to $1000.
Student Group	The SCU student body consists of 3,800 undergraduates and 4,000 graduate and law students. There are about 700 undergraduates in the School of Engineering. Approximately 1,100 graduate students are in the graduate engineering programs. About 60 percent of these students are employed in Silicon Valley and attend SCU on a part-time basis.
Location	Santa Clara University is 46 miles from San Francisco, near the southern tip of San Francisco Bay, in an area rich in opportunities for learning. The campus is situated in the midst of the nation's great concentrations of high-tech industry and professional and scientific activity. Many nearby firms and social agencies are world leaders in the search for solutions to man's technological problems. The cultural and entertainment centers of San Francisco, Berkeley, Oakland, and Marin County are within an hour's travel by bus, train, or car. In the opposite direction, the Pacific beaches of Santa Cruz are about 30 minutes away; the world-famous Monterey Peninsula and Carmel are 2 hours away. Santa Clara has a moderate Mediterranean climate. Over a period of sixty-seven years, the average maximum temperature was 71 degrees and the average minimum, 42 degrees. The sun shines approximately 293 days per year, and the average annual rainfall is 15 inches.
The University	Santa Clara, the first institution to offer classes in higher learning in California, is a private Jesuit university. Graduate programs are conducted in the Leavey School of Business and Administration, the Division of Counseling Psychology and Education, the Division of Pastoral Ministries, and the School of Law, as well as in the School of Engineering.
Applying	Applications are accepted for the fall, winter, and spring quarters; deadlines for completed applications are June 1, October 1, and January 1, respectively. Scores on the GRE General Test are required of all applicants except for applicants majoring in engineering management or applying to a certificate program. Scores on the TOEFL are required of all students who are not United States citizens or have not earned a degree in the United States. A statement of research interests and purpose and three letters of recommendation are required of applicants to the Engineer degree and Ph.D. programs.
Correspondence and Information	Graduate Admissions Office School of Engineering Santa Clara University 500 El Camino Real Santa Clara, California 95053-0583 Telephone: 408-554-4313 Fax: 408-554-5474 E-mail: engr-grad@sunrise.scu.edu World Wide Web: http://www.engr.scu.edu

Santa Clara University

THE FACULTY AND THEIR RESEARCH

Applied Mathematics Department
G. Fegan, Associate Professor and Chair of Applied Mathematics; Ph.D., Oregon, 1973. Modern algebra, linear algebra, probability and statistics.

J. Leader, Assistant Professor; Ph.D., Brown, 1989. Numerical analysis, chaos.

Civil Engineering Department
S. Chiesa, Associate Professor and Assistant Dean, Undergraduate Services; Ph.D., Notre Dame, 1982; PE. Environmental engineering, biological wastewater treatment, biological nutrient control techniques, solid-waste management.

E. J. Finnemore, Professor; Ph.D., Stanford, 1970; PE. Hydraulic engineering, hydrology, hydrogeology, water quality.

S. Nsour, Assistant Professor; Ph.D., Clemson, 1991. Transportation engineering, construction engineering.

R. Serrette, Associate Professor; Ph.D., Cornell, 1992. Structural steel design, timber design.

S. Singh, Professor and Chair of Civil Engineering; Ph.D., Berkeley, 1979; PE. Geotechnical engineering, experimental soil mechanics, earthquake engineering.

W. Taniwangsa, Assistant Professor; Ph.D., Berkeley, 1996. Design and analysis of reinforced concrete structures, earthquake-resistant design and analysis, earthquake protective systems.

Computer Engineering Department
R. Danielson, Associate Professor; Ph.D., Illinois, 1975. Image processing, impact of information technology on business.

R. Davis, Professor; Ph.D., California, Santa Cruz, 1979. Logic programming.

M. Ketabchi, Professor; Ph.D., Minnesota, 1985. Database theory.

D. Lewis, Associate Professor and Chair of Computer Engineering; Ph.D., Syracuse, 1975. Systems programming.

Q. Li, Associate Professor; Ph.D., Florida International, 1989. Parallel processing and architecture, operating system, simulation, network.

N. Ling, Associate Professor; Ph.D., Southwestern Louisiana, 1989. Computer architecture, VLSI, parallel architecture.

L. Seiter, Assistant Professor; Ph.D., Northeastern, 1996. Software engineering.

W. Shang, Associate Professor; Ph.D., Purdue, 1990. Parallel computing.

Electrical Engineering Department
T. Healy, Professor; Ph.D., Colorado, 1966. Communications, microwaves.

A. Hoagland, Adjunct Professor and Director, Institute for Information Storage Technology; Ph.D., Berkeley, 1954. Magnetic recording, data storage.

S. Mourad, Professor; Ph.D., North Carolina State, 1970. Digital testing and reliability of large-scale computer systems.

T. Ogunfunmi, Associate Professor; Ph.D., Stanford, 1990. Signal processing, neural networks, VLSI.

G. Okamoto, Assistant Professor; Ph.D., Texas at Austin, 1998. Wireless communications.

M. Rahman, Associate Professor; Ph.D., Tokyo Institute of Technology, 1984. Microelectronics.

D. Siljak, Professor; Ph.D., Belgrade, 1963. Controls and systems.

S. Wood, Professor and Chair of Electrical Engineering; Ph.D., Stanford, 1978. Graphics and digital image processing.

C. Yang, Professor; Ph.D., Pennsylvania, 1975. Semiconductors.

A. Zecevic, Assistant Professor; Ph.D., Santa Clara, 1993. Circuits and systems.

Engineering Management and Leadership Department
R. Parden, Professor and Chair of Engineering Management and Leadership; Ph.D., Iowa, 1953. Engineering management for employed technical professionals.

Mechanical Engineering Department
M. Ardema, Professor; Ph.D., Berkeley, 1974. Nonlinear dynamic systems, optimal control, differential games, singular perturbations, aerospace applications.

T. Hight, Associate Professor and Chair of Mechanical Engineering; Ph.D., Stanford, 1977. Design, computer-aided design, finite-element analysis, biomechanics.

L. Hornberger, Associate Professor and Associate Dean, Graduate Services; Ph.D., Utah, 1986. Materials, manufacturing processes, plastics, product design.

J. Ma, Adjunct Professor; Ph.D., Iowa, 1959. Thermoscience, tribology, dynamics, servomechanics and control electromechanical systems.

M. Saad, Professor; Ph.D., Michigan, 1956. Thermodynamics, heat transfer, nonconventional energy, compressible flow, combustion.

L. Sanchez, Assistant Professor; Ph.D., Houston, 1988. Mechanical design, kinematics, biomechanics, holographic interferometry.

T. Shoup, Professor and Dean, School of Engineering; Ph.D., Ohio State, 1969. Mechanisms, biomechanics, numerical methods.

SOUTHERN ILLINOIS UNIVERSITY CARBONDALE

College of Engineering

Programs of Study

Graduate programs are offered leading to the Ph.D. in engineering science, with concentrations in mechanics, fossil energy, and electrical systems, and to the Master of Science in civil engineering, electrical engineering, mechanical engineering, mining engineering, and manufacturing systems.

Course offerings and research activities in the Department of Civil Engineering include numerical fluid and solid mechanics, mechanics of composite materials, computational mechanics, stability, water quality control, hazardous-waste treatment and disposal, hydraulic design, viscous and inviscid flow, wave motion, turbulence, structural dynamics, nonlinear structural analysis, structural design, and geotechnical and geoenvironmental engineering. In the Department of Electrical Engineering, course work and research include circuits and systems theory, electronics, solid-state devices and materials, digital systems, computer architecture, instrumentation, biomedical engineering, electromagnetics, optics, communication theory, image processing, robust systems, control theory, robotics, neural networks, energy conversion, power systems, power electronics, expert systems, and distributed processing. In the Department of Mechanical Engineering and Energy Processes, the areas of study and research include air pollution control, mass and heat transfer, coal conversion, electrochemical processes, catalysis, thermal science, thermal systems design, combustion, internal combustion engines, chemical and biochemical processes, mechanical systems, computer-aided design, and materials science. Course offerings and research activities in the Department of Mining Engineering include rock mechanics, coal mining, coal utilization and mine environment control, mine planning, fly ash utilization, and ground control problems. Areas of study and research activities in manufacturing systems include computer-aided manufacturing, reliability theory, robotics, and automated factory systems.

To be admitted to the Ph.D. program, students must have a master's degree, or the equivalent, in engineering. To earn the Ph.D., a minimum of 32 semester hours of course work and 24 semester hours of dissertation research is required. The course work must be completed in two areas: the area of concentration and the program core. A student must complete a minimum of 15 semester hours of course work relevant to an area of concentration and 17 hours of courses in the program core, which is required of all students. A dissertation must be completed in the student's area of research interest, with the approval of the dissertation committee. Written candidacy exams, covering all course work taken, are required, and an oral defense of the dissertation must be accomplished.

Students who choose the M.S. thesis option must complete a minimum of 30 semester hours of acceptable graduate credit, including 18 semester hours within the major department. Each candidate must also pass a comprehensive examination covering all graduate work including the thesis. Students who choose the M.S. nonthesis option must complete a minimum of 36 semester hours of acceptable graduate credit, including at least 21 semester hours within the major department. Of these 21 semester hours, 3 should be in a course that can be devoted to the preparation of a research paper. In addition, each candidate is required to successfully complete a research paper and a written comprehensive examination.

Research Facilities

The air-conditioned laboratory facilities are furnished with modern equipment and are located in the College of Engineering building complex near the 26-acre campus lake. Laboratories are available for research in the areas of biomedical engineering, circuits and systems, electrochemical engineering, electronics, electric machines, information processing, lasers, power systems and power electronics, microelectronics, microprocessors and digital systems, microwaves, optics, control and robotics, signal processing and pattern recognition, intelligent and expert systems, fluid mechanics and hydraulics, heat transfer, holography, materials testing and stress analysis, materials science and metallurgy, rate processes, combustion, internal combustion engines, soil mechanics, systems, ultrasonics and catalytic research, air and water quality, coal conversion and utilization, rock and soil mechanics, mine ventilation, blasting and fragmentation, and environmental laboratories.

Financial Aid

Graduate teaching and/or research assistantships that carry a stipend of approximately $9200 for the 1999–2000 academic year, in addition to a tuition waiver, are available in the departments. A few fellowships with stipends ranging from $9900 to $16,500 per calendar year and a tuition waiver are also available on a competitive basis.

Cost of Study

The tuition and fee charge for students enrolled in 1998–99 was $3998 for Illinois residents and $9925 for nonresidents.

Living and Housing Costs

During the 1998–99 academic year, the cost for on-campus room and board was $3777. There are 571 furnished and unfurnished apartments available for married students; rents ranged from $325 to $385 a month, including utilities.

Student Group

In 1998–99, there were 22,252 students at the University. The engineering departments had 164 graduate students working toward the master's degree and 27 working toward the Ph.D., as well as 984 undergraduate students.

Location

The city of Carbondale is approximately 100 miles southeast of St. Louis, Missouri, in Jackson County, the western border of which is the Mississippi River. Immediately south of Carbondale is some of the most rugged and picturesque terrain in Illinois. The region immediately surrounding Carbondale is noted for its large peach and apple orchards. Within 10 miles of the campus are two state parks and four lakes, and much of the area is a part of the Shawnee National Forest.

The University

Southern Illinois University is in its second 100 years of providing high-quality education. Graduate studies were first offered in 1943, and the first doctoral degree was granted in 1959.

Applying

Students interested in graduate studies in engineering should seek admission to the Graduate School and acceptance in a degree program offered by one of the four engineering departments. The applicant must have a bachelor's degree with a major in engineering, mathematics, physical science, or life science and demonstrate competence in mathematics. A student whose undergraduate training is deficient may be required to take course work without graduate credit.

Correspondence and Information

For information about M.S. programs:
Chairman, Department of (specify)
College of Engineering
Southern Illinois University
Carbondale, Illinois 62901-6603

For information about Ph.D. programs:
Associate Dean
College of Engineering
Southern Illinois University
Carbondale, Illinois 62901-6603

Southern Illinois University Carbondale

THE FACULTY AND THEIR RESEARCH

Civil Engineering

Rolando Bravo, Associate Professor; Ph.D., Houston, 1990; PE, PH: hydraulics and hydraulic design, dewatering and ground subsidence. Lizette Chevalier, Associate Professor; Ph.D., Michigan State, 1993; EIT: transport and remediation of nonaqueous phase liquids (NAPL). James N. Craddock, Associate Professor; Ph.D., Illinois at Urbana-Champaign, 1979; PE: finite-element stress analysis, mechanics of composite materials, solid mechanics, numerical analysis. Bruce DeVantier, Associate Professor; Ph.D., California, Davis, 1983; PE: water resources, environmental engineering. William F. Eichfeld, Assistant Professor; M.S., Wisconsin–Madison, 1973; PE: structural analysis, strength of materials, highways and highway construction materials. Roy R. Frank Jr., Assistant Professor; M.S., Southern Illinois at Carbondale, 1983; IPLSA: surveying, photogrammetry, GIS, GPS, heavy construction. Nader Ghafoori, Professor; Ph.D., Miami (Florida), 1986; PE: structural analysis, reinforced and prestressed concrete, properties of concrete, pavement design. Aslam Kassimali, Professor; Ph.D., Missouri–Columbia, 1976: nonlinear structural analysis, structural dynamics and stability, analysis of fiber-composite structures. Sanjeev Kumar, Assistant Professor; Ph.D., Missouri–Rolla, 1996; PE: geotechnical earthquake engineering, dynamic soil structure analysis, piles under lateral loads, hydraulic conductivity of clays, landfill designs. John W. Nicklow, Assistant Professor; Ph.D., Arizona State, 1998; PE: hydraulics, hydrology, optimal control of water resources systems. Vijay K. Puri, Associate Professor; Ph.D., Missouri–Rolla, 1984: geotechnical engineering, soil dynamics, machine foundations, liquefaction of soils. Billy T. Ray, Associate Professor; Ph.D., Missouri–Rolla, 1985; PE: nitrification in activated sludge process; anaerobic digestion, using fluidized-bed reactors; hazardous-materials handling and disposal. Sedat Sami, Professor, Chairman, and Acting Dean; Ph.D., Iowa, 1966; PE: fluid mechanics, turbulence, hydraulics and hydrology. Shing-Chung Yen, Professor; Ph.D., Virginia Tech, 1984: analysis of composite materials and structures, solid mechanics, structural dynamics and vibrations.

Electrical Engineering

Nazeih M. Botros, Associate Professor; Ph.D., Oklahoma, 1985: digital hardware design, artificial intelligence, bioengineering. David P. Brown, Professor; Ph.D., Michigan State, 1961: circuit and system theory, discrete system stability. Morteza Daneshdoost, Professor; Ph.D., Drexel, 1984: power systems analysis, expert systems, man-machine interface, neural network applications to power. Shirshak Dhali, Professor; Ph.D., Texas Tech, 1984: gas discharges, plasma processing electronics. Ralph R. Etienne-Cummings, Assistant Professor; Ph.D., Pennsylvania, 1994: computer engineering, VLSI, analog VLSI neural systems. V. K. Feiste, Associate Professor; Ph.D., Missouri Columbia, 1966; PE: power systems analysis, distribution automation. Glafkos D. Galanos, Professor and Chairman; Ph.D., Manchester (England), 1970: power systems, control, power electronics, AC/DC systems. Charles A. Goben, Professor; Ph.D., Iowa State, 1965: solid-state electronics, materials, surface electromagnetic wave effects, radiation effects in semiconductors, optoelectronics and fiber optics. L. Gupta, Associate Professor; Ph.D., SMU, 1986: computer vision, pattern recognition, digital signal processing. Frances J. Harackiewicz, Associate Professor; Ph.D., Massachusetts at Amherst, 1990: electromagnetics, antennas, microwaves and millimeterwaves, ferrites and microstrip-phased arrays. Constantine I. Hatziadoniu, Associate Professor; Ph.D., West Virginia, 1987: power systems; modeling, simulation, and control; high-voltage DC transmission; power electronics. C. J. Hu, Professor; Ph.D., Colorado at Boulder, 1966: microwave systems, nonlinear systems and neural controllers, electrooptics, bioengineering. Dimitros N. Kagaris, Assistant Professor; Ph.D., Dartmouth, 1994: VLSI design automation, digital circuit testing, communication networks. Mahmoud A. Manzoul, Associate Professor; Ph.D., West Virginia, 1985: fuzzy logic hardware, computer architecture, microprocessor-based design. Farzad Pourboghrat, Associate Professor; Ph.D., Iowa, 1984: control, robotics, mechatronics, adaptive control systems, neural networks. Mohammad R. Sayeh, Associate Professor; Ph.D., Oklahoma State, 1985: neural networks, optical information processing. R. Viswanathan, Professor; Ph.D., SMU, 1983: spread spectrum systems, wireless communications, detection and estimation theory.

Manufacturing Systems

Serge Abrate, Professor; Ph.D., Purdue, 1983: structures, structural dynamics, vibrations, design, composite materials. Gary Butson, Associate Professor and Chairman; Ph.D., Illinois, 1981: mechanical response of wire rope. Feng-Chang Roger Chang, Associate Professor; Ph.D., Ohio State, 1985: computer-integrated manufacturing systems, production planning and control, decision support systems, knowledge-based system. Jefferson F. Lindsey, Professor; D.Engr., Lamar, 1976; PE: electrical aspects of robots, research methods, electronic materials. James P. Orr, Associate Professor; Ph.D., Southern Illinois at Carbondale, 1983: human aspects of automation. Yiming Rong, Associate Professor; Ph.D., Kentucky, 1989: CAD, automated systems, manufacturing processes. Julie K. Spoerre, Assistant Professor; Ph.D., Florida State, 1995: neural networks, composites manufacturing, design optimization, concurrent engineering. Marek L. Szary, Assistant Professor; D.Engr., Wroclaw Technical (Poland), 1977: mechanical aspects of robots, acoustics. Tomas Velasco, Assistant Professor; Ph.D., Arkansas, 1991: quality control, statistics, reliability, artificial intelligence. Alan J. Weston, Assistant Professor; Ph.D., Southern Illinois at Carbondale, 1991: chemical processes, process control.

Mechanical Engineering and Energy Processes

Om P. Agrawal, Associate Professor; Ph.D., Illinois at Chicago, 1984: computer-aided analysis and design of rigid/flexible multibody systems, numerical analysis, finite element methods, continuum mechanics. James Blackburn, Associate Professor; Ph.D., Tennessee, 1988: biokinetics, bioremediation, biotechnology and pollution prevention. Christopher Byrne, Assistant Professor; Ph.D., Johns Hopkins, 1996: composite materials, nondestructive evaluation, friction materials. Philip Chu, Associate Professor; Ph.D., South Carolina, 1982: CAD/CAM composite materials, NDE, FEA. Jarlen Don, Associate Professor; Ph.D., Ohio State, 1982: materials creep and creep fatigue, surface phenomena, carbon-carbon composites. Kambiz Farhang, Professor; Ph.D., Purdue, 1989: CAD, controls, vibrations. Edwin J. Hippo, Professor; Ph.D., Penn State, 1977: coal conversion and cleaning, solid-carbon materials, scanning tunneling microscopy. Michael M. Khonsari, Professor and Chairman; Ph.D., Texas at Austin, 1983: tribology, heat transfer, numerical analysis, machinery performance analysis. Rasit Koc, Professor; Ph.D., Missouri–Rolla, 1989: ceramic materials, powder processing. Manohar Kulkarni, Assistant Professor; Ph.D., Missouri–Columbia, 1986; PE: thermal analysis of materials, thermal modeling, energy management. Shashi B. Lalvani, Professor; Ph.D., Connecticut, 1982; PE: electrochemical processes, materials and environmental sciences. Alay Mahajan, Associate Professor; Ph.D., Tulane, 1994: robotics, controls, intelligent and autonomous systems. Charles B. Muchmore, Professor; Ph.D., Southern Illinois at Carbondale, 1969; PE: biological, physical, and chemical aspects of water quality control; coal research; mass transfer operations. S. Rajan, Professor; Ph.D., Illinois at Urbana-Champaign, 1970: internal combustion engines, combustion and energy utilization. Dale E. Wittmer, Professor; Ph.D., Illinois at Urbana-Champaign, 1980: high-temperature-resistant materials, carbon fiber production and composites, ceramics. Maurice Wright, Professor; Ph.D., Wales, 1962: fiber-reinforced composites, carbon-carbon composites, fracture mechanics, friction materials, brake systems.

Mining Engineering

Yoginder Paul Chugh, Professor and Chairman; Ph.D., Penn State, 1971: rock mechanics and ground control, production engineering in surface and underground coal mines, mining subsidence, management of coal combustion residues. R. Honaker, Associate Professor; Ph.D., Virginia Tech, 1992: coal processing. Bradley C. Paul, Associate Professor; Ph.D., Utah, 1989: solution mining, minerals processing, underground mining, management of coal combustion residues. H. Sevim, Professor and Acting Associate Dean; D.Eng.Sc., Columbia, 1983: mineral economics and operations research, materials handling, experimental design. Atmesh K. Sinha, Professor; Ph.D., Sheffield (England), 1963: coal processing, mine electrical engineering, mine health and safety.

STATE UNIVERSITY OF NEW YORK AT BINGHAMTON

Thomas J. Watson School of Engineering and Applied Science

Programs of Study
The Watson School offers graduate programs leading to the M.Eng., with specializations in computer, electrical, industrial, and mechanical engineering; to the M.S. in electrical engineering, industrial engineering, mechanical engineering, computer science, and systems science; and to the Ph.D. in electrical engineering, mechanical engineering, systems science, computer science, and systems science with a specialization in manufacturing systems.

The M.Eng. is a practice-oriented graduate degree that requires eight courses plus a two-course project.

All M.S. degree programs in the Watson School require that the student complete at least eight courses plus a thesis or, in approved cases, additional course work and a termination requirement. The normal period for completion of a master's degree is 1½ years of full-time study.

Graduation requirements for the Ph.D. degree include satisfactory completion of a comprehensive examination, based on an individual learning contract, and satisfactory defense of a dissertation. There is a 24-credit-hour residence requirement. The normal period for completion of the degree is two years beyond the master's degree. The learning contract permits the development of a highly individualized course of study in close cooperation with senior members of the academic faculty.

Doctoral programs in the Watson School focus on studies at the forefront of science and technology. Creative approaches to state-of-the-art problems in areas of faculty research interest are emphasized.

Research Facilities
The Watson School has one of the most advanced college-campus computer systems in the nation, consisting of an IBM 9121 linked to computers at each faculty or staff desk in addition to a Watson School facility with microcomputers and workstations for student use. Numerous Sun Workstations, a Symbolics computer, and a Hypercube computer are available for research. Watson School members have access to the supercomputer at Cornell for research. In addition, students have access to the University computer facilities for specialized applications.

Although the Watson School was established as recently as 1983, it has developed an international reputation in the multidisciplinary research specialty of electronic packaging. This research is housed in the Watson School's Integrated Electronics Engineering Center (IEEC). The IEEC is also a designated National Science Foundation state/industry university cooperative research center, and in 1993 it became a New York State Center for Advanced Technology (CAT). State-of-the-art research laboratories support the efforts of faculty members to develop research programs. Two other research centers under development are the Center for Computing Technology and the Center for Intelligent Systems.

The University library system consists of the Glenn G. Bartle Library, housing materials in the social sciences and the humanities, a Science Library, and a Fine Arts/Music Library with a total collection of more than 2 million items. Online access to other SUNY collections exists through the campus computer network. Resources are supplemented by membership in academic library consortia, notably the Research Libraries Group, Inc.

Financial Aid
Many students hold fellowships, traineeships, or graduate, research, or teaching assistantships. Most awards include a full waiver of tuition. Other sources of financial aid include the New York State Tuition Assistance Program, the Federal Stafford Student Loan Program, the graduate and professional school College Work-Study Program, and campus jobs.

Cost of Study
For full-time matriculated graduate students, tuition in 1998–99 cost $2550 per semester for state residents and $4208 per semester for nonresidents.

Living and Housing Costs
A recently completed apartment complex, the Graduate Community, has 3- and 4-person apartments, with living room, dining area, kitchen, and bath. Based on a 1998–99 academic-year lease, the semester rate for a single bedroom was $2050, and for a double bedroom, $1775 per person and $3085 per couple. The cost of meal plans per semester is as follows: basic, $787; standard, $1022; and ultra, $1097. Assistance in locating off-campus housing is provided by the listing services of Off-Campus College.

Student Group
Of the 12,259 students enrolled at Binghamton University, 2,700 are graduate students. In the Watson School, there are 876 undergraduates and 368 graduate students. Many obtain jobs in local high-technology enterprises during their enrollment at the School and after graduation.

Location
The University's 606-acre campus is in a suburban setting just west of Binghamton. More than 300,000 people live within commuting distance of the campus. Cultural offerings in the community include the museum and programs of the Roberson Center for the Arts and Sciences, as well as performances by the Binghamton Symphony, Tri-Cities Opera, Civic Theater, and other groups. The University's Art Gallery has a permanent collection representing all periods and also displays works from special loan exhibitions. The annual concert series of the Anderson Center brings a wide variety of performing artists to campus. The Department of Theater stages more than twenty-five productions each year.

The University and The School
The State University of New York at Binghamton is one of the four university centers in the State University of New York System. The faculty numbers about 700. Graduate programs were initiated in 1961 with the establishment of Master of Arts programs in English and mathematics.

The Watson School was created in 1983 by combining the established graduate programs in computer science and systems science from the School of Advanced Technology with new programs in electrical, industrial, and mechanical engineering.

Applying
Holders of an appropriate bachelor's degree from any recognized college or university are eligible to apply. Application forms should be requested from the Office of Graduate Admissions at gradad@binghamton.edu. Applicants should submit GRE General Test scores. International applicants must submit TOEFL scores and provide proof of their ability to meet academic expenses. All credentials should be on file at least one month prior to anticipated enrollment. To ensure consideration for assistantship and fellowship awards, admission credentials should be received by February 15. An online application is available at http://www.gradschool.binghamton.edu.

Correspondence and Information
Associate Dean for Academic Affairs and Administration
Thomas J. Watson School of Engineering and Applied Science
State University of New York at Binghamton
P.O. Box 6000
Binghamton, New York 13902-6000
World Wide Web: http://www.watson.binghamton.edu

State University of New York at Binghamton

THE FACULTY AND THEIR RESEARCH

Department of Computer Science
Nael B. Abu-Ghazaleh, Assistant Professor; Ph.D., Cincinnati. Parallel and distributed processing, computer architecture, parallel discrete event simulation.
Sudhir Aggarwal, Professor and Department Chair; Ph.D., Michigan. Computer networks, distributed systems, protocols, information retrieval from the World Wide Web, simulation, networks, real-time systems.
Michal Cutler, Associate Professor; Ph.D., Weizmann (Israel). Design automation, information retrieval, expert systems.
Richard Eckert, Associate Professor; Ph.D., Kansas. Computer graphics, human-computer interaction, computer architecture, microprocessor-based systems, computer science education.
Dennis Foreman, Lecturer; M.S., SUNY at Binghamton. Design and development of operating systems and computers.
Kanad Ghose, Associate Professor; Ph.D., Iowa State. Parallel processing, computer architecture, VLSI architectures, distributed systems, operating systems.
Margaret E. Iwobi, Lecturer; M.S., SUNY at Binghamton. Software engineering principles, software development environments.
Walker Land, Lecturer; M.S., George Washington. Neural networks, evolutionary computing, object-oriented design, systematic design and applications.
Leslie Lander, Associate Professor; Ph.D., Liverpool. Formal aspects of software engineering, programming languages and paradigms.
Michael J. Lewis, Assistant Professor; Ph.D., Virginia. Distributed computing, metasystems, parallel computing, object-orientation, component-based software development, operating systems.
Patrick H. Madden, Assistant Professor; Ph.D., UCLA. VLSI computer-aided design, computational geometry, optimization for NP-hard problems.
Weiyi Meng, Associate Professor; Ph.D., Illinois at Chicago. Internet-based information retrieval, heterogeneous database systems, query optimization and translation.
Walter G. Piotrowski, Associate Professor; Ph.D., SUNY at Binghamton. Operating systems, distributed systems and networks.
Stephen Y. H. Su, Professor; Ph.D., Wisconsin–Madison. Fault-tolerant computing, design automation, computer architecture.
William L. Ziegler, Associate Professor; M.S., Syracuse. Programming languages and paradigms, computer architecture, university-industry collaboration

Department of Electrical Engineering
Craig Bergman, Assistant Professor; M.S., Illinois at Urbana-Champaign. Digital design, microprocessors, human factors.
Nikolaos Bourbakis, Professor; Ph.D., Patras (Greece). Applied AI, robotics, knowledge-based VLSI design, computer vision, text and image processing, multiprocessor system architectures, neural nets, automated software environment.
Monish Chatterjee, Associate Professor; Ph.D., Iowa. Nonlinear wave phenomena, nonlinear modeling, quantum electronics, fiber, acousto-optics, optics and optical communications.
James Constable, Professor; Ph.D., Ohio State. Instrumentation, cryogenics, electrical noise, contact resistance, electronics packaging.
Jose Delgado-Frias, Associate Professor; Ph.D., Texas A&M. Computer engineering, VLSI/WSI design, parallel computer architectures, reconfigurable computing, interconnection networks, novel computing paradigms.
Lyle D. Feisel, Professor and Dean; Ph.D., Iowa State. Physical electronics, thin films, semiconductors, continuing education.
Mark Fowler, Assistant Professor; Ph.D., Penn State. Digital signal processing, video compression.
Harry Kroger, Professor; Ph.D., Cornell. Electronics packaging, physics and fabrication of superconductor and semiconductor devices, superconductor-semiconductor hybrid circuits.
James Morris, Professor; Ph.D., Saskatchewan. Thin films, semiconductor devices, automotive electronics, engine sensors, electronics packaging.
Dhananjay Phatak, Assistant Professor; Ph.D., Massachusetts at Amherst. Computer architectures, computer arithmetic, neural networks and applications.
Richard Plumb, Professor and Department Chair; Ph.D., Syracuse. Electromagnetics, ground-penetrating radar, scattering theory.
George Sackman, Professor; Ph.D., Stanford. Signal processing, acoustic space-time array processing, microwave electronics.
Richard Schwartz, Professor Emeritus; Ph.D., Pennsylvania. Microwave theory, antennas and propagation, acoustics, signal processing.
Victor A. Skormin, Professor; Ph.D., Moscow. Control engineering, operations research, computer simulation.
Douglas Summerville, Assistant Professor; Ph.D., SUNY at Binghamton. Computer engineering, parallel computer architectures, interconnection networks.
Charles Taylor, Associate Professor; M.S., SUNY at Binghamton. Automatic controls, microprocessor applications, robotics.
Peter E. Wagner, Professor; Ph.D., Berkeley. Semiconductor circuit elements, microwave resonance, surface electricity, applied optics.
Eva Wu, Associate Professor; Ph.D., Minnesota. Approximation, optimization, and stabilization of distributed parameter systems; robust control synthesis theory; control of robotic manipulators; signal processing.

Department of Mechanical Engineering
Frank Cardullo, Associate Professor; M.S., SUNY at Binghamton. Vehicle simulation, vehicle dynamics, man-machine systems.
Richard Culver, Professor; Ph.D., Cambridge. Dynamic instabilities in metal deformation, engineering education.
John Fillo, Professor and Associate Dean; Ph.D., Syracuse. Thermal fluid analysis, mathematical modeling, heat transfer in electronics, advanced technology.
Robert Frey, Lecturer; M.S., Syracuse. Experimental methods, instrumentation, vibration testing.
James Geer, Bartle Professor; Ph.D., NYU. Perturbation methods, nonlinear problems, slender body theory, symbolic computation.
Gary Lehmann, Associate Professor; Ph.D., Clarkson. Fluid dynamics, numerical and experimental heat transfer, cooling of electronics.
Ronald Miles, Professor and Department Chair; Ph.D., Washington (Seattle). Vibrations, acoustics, fatigue, noise, biomechanics.
Bruce Murray, Assistant Professor; Ph.D., Arizona. Thermal and fluid sciences, computational fluid dynamics, materials processing.
James Pitarresi, Associate Professor; Ph.D., SUNY at Buffalo. Computational mechanics, vibration modeling and testing, electronic packaging.
Chittaranjan Sahay, Associate Professor; Ph.D., Indian Institute of Technology (Delhi). Solid mechanics, manufacturing and design.
Bahgat Sammakia, Professor and Director of the Integrated Electronics Engineering Center; Ph.D., SUNY at Buffalo. Thermal and fluid sciences, electronic packaging.
Timothy Singler, Associate Professor; Ph.D., Rochester. Experimental and analytical fluid mechanics, geophysical fluid mechanics, interfacial fluid mechanics, interfacial stability, applied mathematics.
D. C. Sun, Professor; Ph.D., Princeton. Mechanics, fluid and mechanical systems, tribology.

Department of Systems Science and Industrial Engineering
Robert Emerson, Professor and Department Chair; Ph.D., Purdue. Integrated manufacturing, quality assurance, decision support systems.
David Enke, Assistant Professor; Ph.D., Missouri–Rolla. Neural networks, artificial vision, optimization, applied statistics, decision support systems, cognitive modeling.
Donald Gause, Bartle Professor; M.S., Michigan State. General design processes, user-oriented systems design, problem resolution processes, adaptive programming.
George Klir, Distinguished Professor and Director, Center for Intelligent Systems; Ph.D., Czechoslovak Academy of Sciences. General systems methodology, logic design and computer architecture, information theory, fuzzy systems.
Sarah Lam, Assistant Professor; Ph.D., Pittsburgh. Intelligent systems, statistical analysis and design of experiments, neural networks modeling.
Harold W. Lewis III, Associate Professor; Ph.D., SUNY at Binghamton. Fuzzy expert systems, approximate reasoning.
Howard Pattee, Professor Emeritus; Ph.D., Stanford. Theoretical biology, evolutionary models, linguistic control of dynamic systems.
Daryl Santos, Assistant Professor; Ph.D., Houston. Production scheduling and control, engineering economics, engineering management, simulation.
Krishnaswami Srihari, Professor; Ph.D., Virginia Tech. Manufacturing systems, computer-aided process planning, expert systems, computer-integrated manufacture.

STATE UNIVERSITY OF NEW YORK AT BUFFALO

School of Engineering and Applied Sciences

Programs of Study

Graduate programs leading to the M.S., M.Eng., and Ph.D. degrees are offered in aerospace, chemical, civil, electrical, industrial, and mechanical engineering. An M.S. degree in engineering science is available for studies in environmental science. Programs for specialization in such areas as engineering mechanics, engineering materials, systems engineering, operations research, and environmental engineering are available. Requirements for the M.S. degree include a minimum of one academic year of full-time study and the completion of 30 semester hours of credit. No more than 6 hours of transfer credit from another institution are accepted. Other specific requirements vary with the department. The general requirement for the M.Eng. degree is completion of 30 semester hours of credit in a program, designed by the student and the department, that includes 15 credit hours in regular graduate-level engineering courses (excluding project) in a major and 3 to 6 credit hours of an engineering project or thesis. The program is designed for completion in one year of residence. The M.Eng. degree with an engineering management concentration prepares students for management positions within an engineering organization. Students enrolled on a full-time basis can complete the program in one year; part-time students generally take three to four years. All students are required to complete a project linked to industry and related to practical problems. Students may pursue the M.S. or Ph.D. in three areas of concentration: human factors and ergonomics, operations research, or production systems and manufacturing engineering. Distance learning programs are available via EngiNet™.

The general requirements for the Ph.D. degree include at least two full years of study and research beyond the bachelor's or master's degree and the completion of a dissertation. Each candidate is assigned an adviser for his or her study program. The candidate usually is required to pass a preliminary qualifying examination and a comprehensive examination following completion of all course work and to defend the dissertation orally at the end of the program. Applicants should consult the department of interest regarding other regulations. Graduate programs are flexible. Independent creative work is emphasized, and students may integrate educational and research activities from more than one department. Graduates are prepared for careers in teaching and research in a university, in government, or in industry.

Research Facilities

Each of the engineering departments has its own modern, fully equipped laboratories. The University libraries serve the teaching, educational, and advanced research needs of the faculty and graduate students and provide direct access to 2.6 million volumes and more than 23,300 periodical and serial subscriptions. The library resources of primary interest to engineering faculty members and students are housed in the Science and Engineering Library, which holds more than 530,000 books and bound periodicals, 2,350 periodical and serial subscriptions, 1.5 million microforms, 200,000 maps, and extensive technical reports, government documents, and electronic resources.

Faculty members, students, and staff have direct access to University-maintained computers, software libraries, and peripheral facilities. Central computing services offers access to clusters of Sun Workstations and time-sharing computers. Engineering Computing Services offers access to engineering-specific software running on a Sun-based system that includes servers, time-share resources and individual workstations located in both public and faculty laboratories. Also, department-specific PC laboratories offer a wide variety of engineering and writing software.

Financial Aid

A variety of research appointments are available, as are University-supported assistantships and Graduate School fellowships. Tuition scholarships are available. Fellowships and assistantships carry stipends that ranged from $10,700 to $18,000 for the 1998–99 academic year. Summer support is available for most research appointments. Work done as a research assistant is generally applicable to the student's thesis or dissertation.

Cost of Study

The tuition in 1998–99 for full-time graduate studies was $2550 per semester for state residents and $4208 per semester for nonresidents. Part-time students who were state residents paid $213 per semester hour and nonresidents $351 per semester hour. Tuition scholarships in the form of fellowships or assistantships are normally provided to students who receive financial aid. Mandatory fees are approximately $435 per semester.

Living and Housing Costs

The School is located on the North Campus, a 1,200-acre site in Amherst, New York. This campus has modern residence halls that accommodate more than 4,000 students. The University also maintains five residence halls for more than 1,000 students on the South Campus. New apartment-style University housing is also available for graduate students. Students seeking housing should write to the University Housing Office, Richmond Building 4, Ellicott Complex, Amherst, New York 14261-0009 or telephone 716-645-2171. Adequate off-campus housing, informally inspected and approved by the University, is available near the South Campus. The Housing Office maintains a file to assist single and married students in finding suitable accommodations. The cost of living is typical of medium-sized cities in the Northeast. Meals may be bought in campus cafeterias at reasonable cost.

Student Group

State University of New York (SUNY) at Buffalo has approximately 24,000 students; 2,400 are in the School of Engineering and Applied Sciences. Of these, 700 are graduate students enrolled either full- or part-time.

Location

Buffalo, located along Lake Erie and the Niagara River, is New York State's second-largest city and the dynamic capital of its expanding Niagara Frontier. Metropolitan Buffalo has nearly 1 million inhabitants. It ranks eighth nationally in expenditures for research and supports more than 100 laboratories. The city's cultural facilities include the world-famous Albright-Knox Art Gallery, the Buffalo Philharmonic Orchestra, the State University of New York at Buffalo's programs for the performing arts, the Studio Arena Theatre, and numerous historical and science museums. Recreational facilities are available in and around the metropolitan area for both summer (swimming, boating, and fishing) and winter (skating, sleighing, and skiing) sports. A major sports stadium and an auditorium house spectator sports. There are several city parks for picnics and outings, including a zoological park and a children's zoo. Niagara Falls, the major scenic attraction in the area, is within a half-hour drive of downtown Buffalo. The network of highways in and around the city makes all areas accessible to the University, which is situated in a residential area at the Buffalo city limits.

The University

The State University of New York at Buffalo was founded in 1846 as the University of Buffalo. The merger in 1962 with the State University system signaled a period of dramatic development; today it is the largest single unit and most comprehensive undergraduate and graduate center of the State University, enrolling about 24,000 students (about 19,000 full-time). The University offers 92 undergraduate, 111 master's, 99 doctoral, and 4 professional programs. It is a member of the American Association of Universities (AAU).

Applying

Applications for graduate work and other information may be obtained from the address given below. Students should indicate their department of interest.

Correspondence and Information

Office of Graduate Education
School of Engineering and Applied Sciences
412 Bonner Hall
State University of New York at Buffalo
Buffalo, New York 14260-1900
World Wide Web: http://www.eng.buffalo.edu/

State University of New York at Buffalo

AREAS OF RESEARCH CONCENTRATION

Chemical Engineering
Triantafillos J. Mountziaris, Director of Graduate Studies; Ph.D., Princeton.

The department offers research projects in many of the fundamental areas of chemical engineering, including transport phenomena, biochemistry, fluid dynamics, thermodynamics, process design and control, surface chemistry, catalysis, and reaction kinetics. These fundamentals also underlie specialized projects in other areas such as biochemical engineering, thermodynamics and electric properties of polymers, polymer processing, fracture mechanics and the physical chemistry of adhesion, reactor stability and reaction engineering, and ceramics engineering. The department is able to draw on the attendant strengths of a large and diverse university center. This environment encourages interdisciplinary research programs not only with other engineering departments but also with other scientific disciplines.

Civil Engineering
Joseph F. Atkinson, Director of Graduate Studies; Ph.D., MIT.

The department offers graduate studies in five defined areas of concentration and in new, developing areas. Defined areas include structural and earthquake engineering, environmental engineering and science, computational engineering mechanics, construction engineering and management, and geomechanics, geotechnical, and geoenvironmental engineering. The programs are designed to develop professional leaders in areas of specialization or to create an environment of interdisciplinary learning in new, developing fields of civil engineering.

Unique courses in vibration reduction using structural control, base isolations, energy dissipation systems, and seismic experimentation, along with theoretical subjects in boundary elements, finite elements, and structural plasticity, are taught on annual or biannual cycles. Laboratory facilities include one of the most advanced experimentation facilities in the U.S. for seismic simulation (a large shaking table) and vibration testing, environmental and geoenvironmental labs, and an environmental hydraulics laboratory that includes a rotating lab for geophysical fluid flow.

The department is host of and participates in three major research centers: the National Center for Earthquake Engineering Research, the Great Lakes Program (in environmental engineering), and the New York State Hazardous Waste Management Center. All centers sponsor and maintain research programs and experimental facilities used by students and faculty members.

Computer Science and Engineering
Raj Acharya, Director of Graduate Studies; Ph.D., Minnesota.

The department offers two master's-level tracks, one in computer science and the other in computer engineering. At the master's and Ph.D. levels, areas of specialization include networking, complexity theory, parallel computing, algorithms, computer vision, cognitive science, knowledge representation, and image analysis. Many students and faculty members are involved in interdisciplinary research through the Center for Cognitive Science, the Center of Excellence for Document Analysis and Recognition, the National Center for Geographic Information Analysis, and others.

Electrical Engineering
David M. Benenson, Director of Graduate Studies; Ph.D., Caltech.

The department offers graduate studies in the research focus areas of communications and signal processing, materials and electronics, lasers and photonics, and plasma and high-power electronics. These areas include mobile computing and networking; reliability and diagnostic reasoning; multimedia computing, visualization, and fractals; telecommunications, mobile and personal communication systems, and high-speed and optical networks; digital signal and image processing; machine vision and neural networks; adaptive signal processing, detection, and estimation; adaptive antennas and radar arrays; inverse scattering and wave-propagation and diffraction theory; electronic instrumentation and sensors; medical electronics and biomedical imaging; microelectronics and quantum electronics; semiconductors, heterostructures of III-V compounds, and photovoltaics; semiconductor photonic devices; computational photonics; laser spectroscopy; microscopy, microtomography, and lithography; holographic and laser-ablation techniques; superconducting and nanophase materials; plasma physics, processing, and diagnostics; metal vapor plasmas, arc plasmas, and switching; electromagnetic compatibility, RF, and microwaves; energy systems and high-power electronics, pulsed power, dielectrics, and insulation; and industrial automation and energy conservation.

Industrial Engineering
Colin G. Drury, Director of Graduate Studies; Ph.D., Birmingham (England).

Students may pursue three areas of concentration: human factors and ergonomics, operations research, and production systems and manufacturing engineering. Human factors and ergonomics focuses on applications of engineering, psychology, biomechanics, and physiology to the modeling, analysis, and design of equipment and environments for industry, the service sector, transportation, and personal living. Operations research applies mathematics, statistics, computer science, and engineering principles to the formulation and solution of mathematical models for problems in long-range planning, energy and urban systems, health systems, and manufacturing. Production systems and manufacturing engineering focuses on production planning and scheduling, computer-integrated manufacturing, enterprise management, quality assurance, facilities location and design, robotic systems, concurrent engineering, control of flexible manufacturing systems, materials handling, and storage systems.

Mechanical and Aerospace Engineering
Dale B. Taulbee, Director of Graduate Studies; Ph.D., Illinois.

The department offers courses of study and research opportunities in the basic areas of fluid and thermal sciences, mechanics and materials, and systems and design. The current research interests of the faculty span a wide range of topics in mechanical and aerospace engineering, including bioengineering and biomechanics; combustion and propulsion; composite materials; compressible flows; computational fluid dynamics; computer-aided design and optimization; dynamics and control systems; electronic materials and packaging; fluid mechanics; fracture and fatigue and creep; heat transfer; manufacturing systems; materials science and engineering; metals, ceramics, and polymers; solid mechanics; thermodynamics and energy systems; turbulence; two-phase flows; and vibration analysis. Modern, well-equipped laboratories that support the research and graduate study activities of the department are located on the Amherst campus. These laboratories include the Assistive Device Design Laboratory; the CFD Laboratory; a Combustion Research Laboratory; the Heat Transfer Laboratory; the Hemodynamics Laboratory; the Materials Laboratory, with equipment for materials processing and testing; the Multidisciplinary Optimization and Design Laboratory; and the Turbulence Research Laboratory, with a low-speed wind tunnel and a plume facility.

STEVENS INSTITUTE OF TECHNOLOGY

Charles V. Schaefer, Jr. School of Engineering

Programs of Study	Stevens Institute of Technology offers graduate programs of study in the Charles V. Schaefer, Jr. School of Engineering leading to the degrees of Master of Engineering and Doctor of Philosophy in a broad range of engineering disciplines. These include chemical engineering; civil, environmental, and ocean engineering; computer engineering; electrical engineering; engineering physics; materials engineering; mechanical engineering; and polymer engineering. Interdisciplinary programs in concurrent engineering, construction management, and telecommunications management and the degree of Engineer in chemical, civil, computer, electrical, and mechanical engineering are also offered. In addition, a broad range of graduate certificate programs are available for industrialists and practicing engineers.

The mission of the School of Engineering is to provide high-quality education beyond the undergraduate degree by promoting a rigorous and scholarly environment with strong cross-disciplinary links supporting world-class education, research, and technical applications. It strives to ensure the continual relevance of its programs and keeps the community keenly aware of national and international needs, developments, and trends in education, research, and technology.

The Master of Engineering program is intended to extend and broaden undergraduate education. Strong emphasis is placed on providing the flexibility required for responding to a rapidly changing technological environment. It requires 30 credits of approved course work with or without a thesis. The program may be completed in one year of full-time study or in longer periods of part-time study. The Engineer degree requires an additional 30 credits of course work beyond that required for the master's degree, including an in-depth design project.

The Ph.D. program in engineering is for students who are primarily interested in a research or teaching career. The program aims to prepare the students to make important contributions at the frontiers of their disciplines and is granted in recognition of superior academic preparation and creative scholarly research. A Ph.D. student must pass a qualifying examination at the beginning of the program and is formally admitted to candidacy after passing a comprehensive examination in a major field. The Ph.D. degree requires 60 credits beyond the master's degree (including 30 thesis credits), and a candidate for the degree must complete and defend an acceptable dissertation.

Research Facilities
All departments have state-of-the-art research facilities. In addition, Stevens has developed "Industrial Alliances" (Steeples of Excellence), which strive to identify problems impacting U.S. competitiveness and develop solutions that industry can implement. Major areas of interdisciplinary research addressed by the Industrial Alliances include automated concurrent engineering, environmental and coastal engineering, highly filled materials technologies, the development of polymer processes, telecommunications, the effective management of manufacturing, and laser physics and quantum electronics technologies.

The support services include the Stevens Computer Center, which has a six-processor minisupercomputer with extensive preprocessing and postprocessing capabilities; a fully networked academic infrastructure; and the Samuel C. Williams Library, which offers a wide range of information-gathering tools, such as an information-retrieval system of computer databases containing references to millions of documents and rapid delivery of engineering publications.

Financial Aid
Assistantships, fellowships, scholarships, loan and deferred-payment plans, work-study, and employer tuition benefits are available to qualified students. Assistantships include nine-month academic-year stipends ranging form $11,000 to $13,000 for 1999–2000 plus remission of tuition and fees; recipients devote 20 hours per week to teaching or research.

Cost of Study
Tuition is $695 per credit for the 1999–2000 academic year.

Living and Housing Costs
Residence costs for 1999–2000 range from $3300 to $5000 per academic year for an off-campus room to $5670 per academic year for on-campus married and graduate student apartments. Additional living expenses are approximately $3500 for thirty-eight weeks. Books and supplies cost about $700 per year.

Student Group
There is an exceptionally diverse group of about 2,000 graduate students at Stevens, nearly 70 percent of whom are enrolled part-time.

Location
Stevens is located on the west bank of the Hudson River in Hoboken, New Jersey, a community that has undergone a remarkable renaissance and become a popular residential, recreational, and cultural center. The campus is 15 minutes from New York City by bus or subway. World-famous year-round resort areas and beaches are less than 2 hours away.

The Institute
Founded in 1870, Stevens is a pioneer in technical education and a highly regarded independent center of study and research accredited by MSACS, ABET, and CSAB. Total enrollment is approximately 3,500, and there are 120 full-time faculty members, more than 95 percent of whom hold a doctorate. A leader in integrating computers into engineering education, Stevens has excellent computing facilities and a campuswide computer network that greatly expands the capabilities of the entire college community.

Applying
An application, a transcript, two recommendations, and a $45 application fee should be filed with the office of the Dean of the Graduate School two to four weeks before the beginning of the semester for domestic applicants and two to four months before the beginning of the semester for international applicants. The fall semester begins in late August and the spring semester in mid-January. Registration should be completed about a week before the term opening.

Correspondence and Information
For written information about graduate study:
Department of (specify)
Stevens Institute of Technology
Castle Point on Hudson
Hoboken, New Jersey 07030
Telephone: 201-216-5105

For applications and admission:
Dr. Charles Suffel
Dean of the Graduate School
Stevens Institute of Technology
Castle Point on Hudson
Hoboken, New Jersey 07030
Telephone: 201-216-5234
Fax: 201-216-8044

Stevens Institute of Technology

ACADEMIC DEPARTMENTS AND FACULTY RESEARCH

Bernard M. Gallois, Dean of Engineering

Chemical Engineering. Dr. Traugott E. Fischer, Interim Director. Current research activities within the department include polymer reaction engineering, reactive extrusion, and devolatization; polymer rheology, processing, simulation, and dispersive and distributive mixing; characterization of highly filled suspensions; biochemical reaction engineering; mass transfer; bioprocess control, modeling, and identification; process control, identification, and modeling; waste treatment; crystallization; combustion; and process synthesis and analysis. Contact: Professor S. Koven (telephone: 201-216-5519; e-mail: skoven@stevens-tech.edu)

Civil, Environmental, and Ocean Engineering. Dr. Richard I. Hires, Director. Major research activities of the department are in the following three programs: **Civil Engineering:** analysis and optimization of structural systems and concrete technology, nonlinear dynamics and stochastic processes, flow-induced vibration, soil mechanics, and geoenvironmental engineering. Contact: Professor K. Y. Billah (telephone: 201-216-5344; e-mail: billah@stevens-tech.edu) **Ocean Engineering:** analysis of beach erosion and evaluation of remedial alternatives, observation and analysis of ocean surface waves and currents and their interaction with coastal structures, marine craft hydrodynamics, numerical modeling and analysis, and applied oceanography. Contact: Professor Michael Bruno (telephone: 201-216-5338; e-mail: mbruno@stevens-tech.edu) **Environmental Engineering:** hydrolysis and biodegradation of explosives and rocket propellants, denitrification of high nitrite mixtures, nutrient recovery from inedible plant biomass for application in space, surface enhancement of industrial wastewater filtration media, transport and fate of nonaqueous liquids in saturated and unsaturated soils, in situ and ex situ contaminant removal, beneficial use of contaminated soils and industrial waste by-products, chemical fixation and immobilization in soils and sludges, management of contaminated dredged sediment, and development of leaching protocols; numerical and statistical modeling of contaminant fate and transport, activated sludge process, bioregenerative life support systems. Contact: Professor David A. Vaccari (telephone: 201-216-5570; e-mail: dvaccari@stevens-tech.edu)

Electrical and Computer Engineering. Dr. Stuart K. Tewksbury, Director. Faculty research areas include wavelength division multiplexed all-optical networks; network reliability; optical fiber communication systems; integrated voice, data, and video networks; neural networks; digital image/signal processing; VLSI circuits and systems; design methodologies for synthesis of ASICs; statistical signal processing; wireless communications; high-speed multimedia networks; and asynchronous transfer mode networks. Contact: Graduate Program Coordinator (telephone: 201-216-5623; e-mail: fflaniga@stevens-tech.edu; World Wide Web: http://www.ece.stevens-tech.edu)

Materials Science and Engineering. Dr. Traugott E. Fischer, Director. The faculty research interests include chemical vapor deposition of structural coatings and electronic ceramic thin films; high-temperature corrosion of ceramics, applications of self-assembled molecules; thin-film deposition and characterization, plasma cleaning; advanced electron optical methods, especially applications to polymer characterization and holography; surface electron spectroscopy; ultra-thin magnetic films; electrochemical aspects of materials, including corrosion and electrodeposition; and tribology and tribochemical polishing, particularly of ceramics. Contact: Professor H. Du (telephone: 201-216-5262; e-mail: hdu@stevens-tech.edu; World Wide Web: http://www.stevens-tech.edu)

Mechanical Engineering. Dr. Siva Thangam, Director. Current areas of research include acoustical source characterization, noise control, duct acoustics; kinematic and dynamic characteristics of mechanisms, finite-element methods, modeling of biomechanical systems, composite materials, fracture mechanics; integrated product design, design for manufacturability, development of expert systems, computer-aided manufacturing; parameter sensitivity reduction in control systems, control of robot end effectors, studies on dynamic balancing; computational fluid mechanics and heat transfer, hydrodynamic stability, thermomechanical analysis of electronic systems, heat transfer in manufacturing processes, dynamic behavior of turbomachinery, stall characteristics of turbomachinery, aeroelastic tailoring of turbine/compressor blades; diagnostics for monitoring submicron particulates, partitioning of heavy metals, alternative fuels, generation of particulates. Contact: Professor S. Thangam (telephone: 201-216-5558; e-mail: sthangam@stevens-tech.edu; World Wide Web: http://www.me.stevens-tech.edu)

Physics and Engineering Physics. Dr. Edward A. Whittaker, Director. Faculty research areas include laser spectroscopy, atomic and molecular physics, electron-atom collisions, physical kinetics, nonlinear phenomena, and solid-state device modeling. Interdisciplinary activities include active collaborations with environmental and materials engineering faculty members at Stevens and solid-state device researchers at several industrial organizations and national laboratories. Contact: Professor H. L. Cui (telephone: 201-216-5637; e-mail: hcui@stevens-tech.edu; World Wide Web: http://attila.stevens-tech.edu/physics)

RESEARCH CENTERS

Center for Environmental Engineering (CEE). Dr. George P. Korfiatis, Director. The Center for Environmental Engineering is dedicated to applied, interdisciplinary research for the solution of pressing, real-world environmental problems. Through advanced knowledge and in-depth professional expertise, CEE is recognized as a leader in the development, evaluation, and implementation of new environmental technologies. CEE research specialties include physical, chemical, and biological waste treatment processes; soil and groundwater remediation; computer modeling of contaminant transport and fate in surface, coastal, and ground waters; ocean and estuary environmental hydrodynamic measurements; and commercial, industrial, and residential water conservation.

Davidson Laboratory. Dr. Michael S. Bruno, Director. Since 1935, the Davidson Laboratory has been an international leader in the fields of ocean engineering and naval architecture. It conducts basic and applied research in hydrodynamics, ocean engineering, and the environment, making use of facilities that include a towing tank that is 313 feet long, 12 feet wide, and 6 feet deep and an oblique-sea basin (one of only two in the nation) that is 75 feet long, 75 feet wide, and 5 feet deep. Three research vessels include one 30-foot and two 25-foot vessels that are fully equipped to conduct coastal and estuary field measurements.

Design and Manufacturing Institute (DMI). Dr. Souran Manoochehri, Director. DMI is an interdisciplinary center integrating product design, materials processing, and manufacturing expertise with modern computer software technology. DMI has been integrating design and manufacturing technology for polymeric and composite parts manufactured using injection molding and resin molding processes. The DMI's research has resulted in a knowledge-based engineering design system called the Automated Concurrent Engineering Software (ACES) that guides a product developer concurrently through the conception and production of a part. The system allows for concurrent optimization and trade-off studies with respect to cost and performance, considering manufacturability, reliability, affordability, and other influences affecting the product's life cycle early on in the design process. Currently, DMI is extending this technology to large-scale system assemblies and to other manufacturing processes such as metal casting, forming, and machining. The "Learning Factory" at DMI is a computer-integrated facility containing a fully equipped modern numerically controlled environment configured to serve as a molding and tool production facility; a state-of-the-art rapid prototyping environment; computer-controlled coordinate measuring equipment for product quality control and verification; and a well-equipped laboratory for material characterization and mechanical testing.

Highly Filled Materials Institute (HFMI). Dr. Dilhan M. Kalyon, Director. The center specializes in materials filled with solids at concentrations that approach the maximum packing fraction used in energetics, personal care, battery, ceramic, magnetic construction, and oil drilling industries. Capabilities include specialized source codes and experimental facilities, especially for extrusion processing, rheological analysis, microstructural distributions, and ultimate mechanical, electrical, and magnetic properties of highly filled materials.

Center for Product Lifecycle Management (CPLM). Dr. Stephen L. Wythe, Director. The Center for Product Lifecycle Management is a focal point for both information and technology on plastics products over their lifecycle: design, manufacture, use, and disposal. Working with industry and government, the center emphasizes the development of products and fabrication processes that reduce the potential for significant environmental problems and risks. The center is an alliance of two existing organizations at Stevens: the Center for Environmental Engineering and the Polymer Processing Institute, and its activities include contract product and process research, engineering studies, educational and training programs, and technology transfer.

AFFILIATED RESEARCH CENTER

Polymer Processing Institute (PPI). Dr. Costas G. Gogos, Director. The Polymer Processing Institute is an independent research corporation hosted by Stevens Institute of Technology, with which it maintains close ties. Its mission is to serve industry by advancing the scientific underpinnings of polymer technology through industry-sponsored research, development, and education and to disseminate information pertaining thereto through technology transfer. PPI operates its own extension center for the plastics industry (NJPEC), offers advanced-level short courses annually for industrial engineers and scientists, and supports the education and research needs of graduate and undergraduate students at Stevens.

STEVENS INSTITUTE OF TECHNOLOGY

School of Applied Sciences and Liberal Arts

Programs of Study

Stevens offers graduate programs in applied mathematics (M.S.), applied statistics (M.S.), chemical biology (M.S. and Ph.D.), chemistry (M.S. and Ph.D.), computer science (M.S. and Ph.D.), engineering physics optics and solid state (M.Eng.), materials science (M.S. and Ph.D.), mathematics (M.S. and Ph.D.), and physics (M.S. and Ph.D.).

Stevens also offers interdisciplinary science and engineering programs (M.S., M.E., and Ph.D.) that allow for program flexibility. With the assistance of a graduate adviser, students may design a program suited to their individual needs.

Graduate Certificate programs are available in specialized areas, including applied optics, computer mathematics, biomedical chemistry, and programming for critical applications.

A variety of off-campus and on-site corporate programs are also available, some of which are accessible using distance learning technologies.

Research Facilities

All departments have state-of-the-art research facilities and equipment, and Stevens supports a powerful computing environment that includes a wide range of information-gathering tools for bibliographic, database, design, and modeling uses. Research facilities and equipment include various advanced lasers and spectrometers, the Center for Quantum Electronics (a high-performance computing facility for semiconductor device modeling), a material characterization facility with advanced electron microscopes and spectrometers, the Barasch mass spectrometer facility, an IBM-Brucker superconducting 200-MHz NMR spectrometer, an Enraf-Nonius X-ray spectrometer with automated diffractometer, a computational facility with Silicon Graphics R10000 Octane computer, a kinetics and bioseparation facility, a multimedia research laboratory, and other resources of the Advanced Telecommunications Institute.

Financial Aid

Assistantships, fellowships, scholarships, loan and deferred-payment plans, and work-study are available to qualified students. In 1998–99, assistantships included a minimum stipend of $9800 plus remission of tuition and fees; recipients devote 20 hours per week to teaching or research. Additional information on financial aid may be obtained from the Dean of the Graduate School.

Cost of Study

Tuition is $695 per credit for the 1999–2000 academic year. The fee for a typical 2½-credit graduate course is $1737.50. An enrollment fee of $80 is charged per semester. Books and supplies cost about $500 per year.

Living and Housing Costs

Residence costs for 1998–99 ranged from $3300 to $5300 per academic year for an off-campus room; on-campus married and graduate student apartments cost approximately $4150 per academic year. Additional living expenses were approximately $3900 for thirty-eight weeks.

Student Group

An exceptionally diverse group of about 2,000 graduate students (400 full-time and 1,600 part-time) pursue Stevens graduate programs. Four hundred forty students study the natural sciences, computer science, and mathematics.

Location

Stevens is located in Hoboken, New Jersey, on the west bank of the Hudson River across from mid-town Manhattan. Hoboken has undergone a remarkable renaissance and has become a popular residential, recreational, and cultural center. The campus is 15 minutes from the center of New York City. World-famous year-round resort areas and beaches are less than 2 hours away.

The Institute

Founded in 1870, Stevens is a pioneer in scientific, engineering, and management research and education. Total graduate and undergraduate enrollment is approximately 3,400, and there are 120 full-time faculty members, more than 90 percent of whom hold a doctorate. A campuswide computer network greatly expands the capabilities of the entire community.

Applying

An application, transcript, two recommendations, and a $45 application fee should be filed with the Dean of the Graduate School at least two weeks before the beginning of a semester for domestic applicants and two to four months before for international applicants. The fall semester begins in late August and the spring semester in mid-January. Registration should be completed at least one week before the term opening.

Correspondence and Information

For information about programs:
Department of (specify)
Stevens Institute of Technology
Castle Point on Hudson
Hoboken, New Jersey 07030
Telephone: 201-216-5105 or 5107

For applications and admission information:
Dr. Charles Suffel
Dean of The Graduate School/G-2
Stevens Institute of Technology
Castle Point on Hudson
Hoboken, New Jersey 07030
Telephone: 201-216-5234
Fax: 201-216-8044
E-mail: thegradschool@stevens-tech.edu
World Wide Web: http://www.stevens-tech.edu

Stevens Institute of Technology

THE FACULTY AND THEIR RESEARCH

School of Applied Sciences and Liberal Arts
Patrick Flanagan, Dean; Ph.D., McGill, 1968. (telephone: 201-216-8220).

Department of Chemistry and Chemical Biology
Director: W. C. Ermler, Director; Ph.D., Ohio State, 1972. Electronic structure of atoms and molecules, relativistic effects in heavy atoms. (e-mail: wermler@stevens-tech.edu; telephone: 201-216-5544).

A. Aguanno, Visiting Assistant Professor; Ph.D., NYU, 1991. Regulation of genes encoding the catecholamine biosynthetic enzymes.

A. K. Bose, Professor; Sc.D., MIT, 1950. Synthesis of penicillins, steroids.

J. Carroll, Visiting Assistant Professor; Ph.D., Wisconsin–Madison, 1995. Physical chemistry.

P. W. Flanagan, Ph.D., McGill, 1968. Global terrestrial ecosystems, microbial genetics and ecology.

F. T. Jones, Professor; Ph.D., Polytechnic of Brooklyn, 1960. Mass spectrometry, reaction kinetics, radiation and photochemistry, decomposition of halogenated compounds.

N. Kumbaraci, Associate Professor; Ph.D., Columbia, 1977. Neuromuscular physiology.

M. S. Manhas, Professor Emeritus; Ph.D., Allahabad (India), 1951. Synthesis of heterocyclic compounds, penicillins and cephalosporins.

M. M. Marino, Ph.D., Stevens, 1990. Theoretical chemistry, psuedopotentials.

S. S. Stivala, Research Professor; Ph.D., Pennsylvania, 1960. Physical and polymer chemistry.

C. Stone, Associate Professor; Ph.D., Indiana, 1989. Enzymology and protein chemistry, kinetics of alcohol oxidoreduction.

Department of Computer Science
Director: Stephen L. Bloom, Professor; Ph.D., MIT, 1968. Semantics. (e-mail: bloom@menger.eecs.stevens-tech.edu; fax: 201-216-8246; telephone: 201-216-5439).

E. Angelopoulou, Assistant Professor; Ph.D., Johns Hopkins, 1997. Computer vision.

A. Banerjee, Assistant Professor; Ph.D., Kansas State, 1995. Programming languages and compilers.

D. Chatziantoniou, Assistant Professor; Ph.D., Columbia, 1990. Databases.

A. Compagnoni, Assistant Professor; Ph.D., Katholieke (Nijmegen), 1995. Programming languages and type theory, with applications for theorem provers.

D. Duggan, Associate Professor; Ph.D., Massachusetts, 1990. Programming languages and environments.

D. Klappholtz, Associate Professor; Ph.D., Pennsylvania, 1974. Parallel compilers.

P. Morreale, Associate Professor; Ph.D., IIT, 1991. Network design and performance.

D. Naumann, Assistant Professor; Ph.D., Texas at Austin, 1992. Data refinement in imperative languages with higher-order, object-oriented, and polytypic features.

A. Satyanaryana, Professor; Ph.D., Jawaharlal Nehru (New Delhi), 1981. Network reliability.

Department of Mathematical Sciences
Director: Robert H. Gilman, Professor; Ph.D., Columbia, 1969. Geometric group theory, symbolic computation. (e-mail: rgilman@stevens-tech.edu; fax: 201-216-8231; telephone: 201-216-5449).

D. I. Bauer, Professor; Ph.D., Stevens, 1978. Hamiltonian cycle theory and network reliability.

M. Dostal, Professor; Ph.D., Mathematical Institute of the Czechoslovak Academy of Sciences, 1966. Asymptotic expansions, Fourier analysis, probability.

B. Hayes, Assistant Professor; Ph.D., NYU (Courant), 1994. Partial differential equations.

P. Miller, Assistant Professor; Ph.D., Massachusetts Amherst, 1994. Differential equations, nonlinear dynamics, geophysics, reaction-diffusion equations.

K. Khashana, Visiting Assistant Professor; Ph.D., Delaware, 1994. Operator theory, singular perturbations, inverse problems, nonlinear theory.

L. Levine, Professor; Ph.D., Maryland, 1968. Applied mathematics, differential equations, educational technology.

J. Manogue, Associate Professor; M.S., Michigan, 1955. Ordinary differential equations.

V. Mazmanian, Senior Lecturer; M.S., Stevens, 1971. Pedagogy.

R. S. Pinkham, Professor; Ph.D., Harvard, 1955. Statistics, probability, numerical methods, vision research, mathematics education.

C. L. Suffel, Professor; Ph.D., Polytechnic of Brooklyn, 1969. Network and graph theory.

Department of Physics and Engineering Physics
Director: Edward A. Whittaker, Professor; Ph.D., Columbia, 1982. High-sensitivity laser absorption spectroscopy, frequency modulation spectroscopy, quantum optics. (e-mail: ewhittak@stevens-tech.edu; fax: 201-216-5638; telephone: 201-216-5665).

K. Becker, Professor; Ph.D., Saarlandes (Germany), 1981. Collision cross section measurements, optical and mass spectroscopy, elementary process in discharges, plasmas, and planetary atmospheres.

E. B. Brucker, Professor; Ph.D., Johns Hopkins, 1959. Experimental high-energy physics, optics, optical holography.

W. E. Carr, Professor; Ph.D., Illinois, 1967. Plasma physics, electron and positive ion beams, computational physics.

H.-L. Cui, Associate Professor; Ph.D., Stevens, 1987. Solid-state theory, semiconductor electronic and optoelectronic device modeling.

N. J. Horing, Professor; Ph.D., Harvard, 1964. Quantum many-body theory, solid-state and surface physics, high magnetic field effects, collective modes in nanostructures, quantum transport theory.

E. E. Kunhardt, Professor; Ph.D. Polytechnic of New York, 1976. Nonequilibrium and nonlinear behavior of electrons in gaseous and condensed matter.

H. Salwen, Professor; Ph.D., Columbia, 1956. Fluid dynamics, kinetic theory, quantum mechanics.

Advanced Telecommunications Institute (ATI)
Director: Patricia A. Morreale; Ph.D., IIT, 1991. (e-mail: pat@ati.stevens-tech.edu; telephone: 201-216-8072).

SYRACUSE UNIVERSITY

L. C. Smith College of Engineering and Computer Science

Programs of Study	The L. C. Smith College of Engineering and Computer Science offers programs leading to the following graduate degrees: aerospace engineering, M.S.; bioengineering, M.S.; chemical engineering, M.S. and Ph.D.; civil engineering, M.S. and Ph.D.; computational science, M.S.; computer engineering, M.S., Ph.D., and Computer Engineer; computer and information science, Ph.D.; computer science, M.S.; electrical engineering, M.S., Ph.D., and Electrical Engineer; engineering management, M.S.; environmental engineering, M.S.; environmental engineering science, M.S.; hydrogeology, M.S.; manufacturing engineering, M.S.; mechanical engineering, M.S.; mechanical and aerospace engineering, Ph.D.; neuroscience, M.S. and Ph.D.; solid-state science and technology, M.S.; and systems and information science, M.S. In general, the Master of Science degree requires a minimum of 30 credit hours except for the Master of Science degrees in engineering management and bioengineering, which require 36 credit hours. The thesis option requires 24 credits of course work plus 6 hours accounted for by a research-related thesis. Most departments offer a nonthesis option that substitutes 6 hours of course work for the thesis and requires a comprehensive examination. Requirements for the Ph.D. vary among the academic units. In general, a minimum of 78 credit hours beyond the baccalaureate is required, including graduate course work, independent study, and a dissertation. The College also offers the following certificates: computational science, M.S. and Ph.D. levels; and computational neuroscience, graduate level.
Research Facilities	The L. C. Smith College of Engineering and Computer Science occupies two modern buildings on the main campus quadrangle and shares space in the Center for Science and Technology. The Institute for Sensory Research is located just 2 miles away on the South Campus in a facility uniquely suited to sensory research. Each of the buildings has modern, fully equipped laboratories for research, as well as study laboratories, classrooms, and seminar rooms. Major research laboratory facilities include a high-performance distributed computing laboratory; distributed information systems laboratory; scalable concurrent processing laboratory; the solid-state laboratory; the plasma laboratory; a composite materials manufacturing and testing laboratory; the microwave laboratory; a printed-circuits facility; a VLSI design laboratory; a signal processing laboratory; a robotics laboratory; two anechoic chambers; low-speed, supersonic, and hypersonic wind tunnels; a biomechanics laboratory; and eleven fully equipped, computerized neuroscience laboratories for physiological, psychophysical, biophysical, and neuroanatomical studies of the auditory, tactile, and visual systems. In addition, the College maintains special laboratories for membrane engineering and science, supercritical extraction and chemical reaction, absorption and filtration, liquid extraction, polymer processing, catalyst preparation, and testing magnetic properties of thin films. The College of Engineering and Computer Science (ECS) maintains excellent computer facilities available for use by graduate students. The College's Computer and Information Technologies Group provides each student with both an NT LAN account and a UNIX account. Within ECS, there are four NT clusters (108 Pentium PCs) and five UNIX clusters (sixty Sun Ultra 5s). These facilities complement dedicated computer labs maintained by individual faculty members for their research teams. ECS computer resources include four Sun Microsystems E450s that support file and application services, Web page development and hosting, and a general purpose timesharing compute server. A rich suite of software that is unique to ECS studies is provided on the respective systems. In addition, students in the College can readily access facilities operated by the University's central Computing and Media Services (CMS) organization, the Center for Advanced Technology in Computer Applications and Software Engineering (CASE), and the Northeast Parallel Architecture Center (NPAC). Detailed information is available via the World Wide Web concerning computer facilities available at the University (http://cms.syr.edu or http://netsys.syr.edu/sunix/), the CASE Center (http://www.cat.syr.edu), and NPAC (http://www.npac.syr.edu). Campuswide computer resources include two dual-processor Ultra-2 timesharing systems and three 2-processor SPARCserver 10s. NPAC resources include an 8-processor SGI Challenge L; a dedicated cluster of workstations connected by high-performance networks, including eight Sun UltraSPARC2s connected by an OC3 ATM switch; a cluster of SGI Indy and O2 workstations; and a cluster of dual Pentium II PCs.
Financial Aid	Aid is available in the form of scholarships, fellowships, and teaching and research assistantships. The number of scholarships and assistant positions varies by department. Inquiries should be directed to the specific department chair. Scholarships provide full tuition for 30 credits an academic year. Assistantships carry a stipend and require an average of 20 hours a week of instruction or research responsibilities. Assistants may be awarded a scholarship and may also apply for one of a limited number of fellowship awards to support summer study or research. Fellowships are available for superior students. These are awarded on a competitive basis and carry a stipend of $11,384 for the 1999–2000 academic year and remitted tuition.
Cost of Study	Tuition in 1999–2000 for graduate students is $583 per credit hour.
Living and Housing Costs	University-owned furnished apartments for single students are available for $3500 per student per semester. Married students may rent one- or two-bedroom apartments at a monthly cost of $560 to $675 and two-bedroom apartments at $645 to $715 per month.
Student Group	Undergraduate enrollment averages 10,000, and graduate enrollment is more than 4,000. Students are drawn from every state in the Union and more than 100 countries. About 1,600 international students are registered each year in the undergraduate and graduate programs. Current full-time graduate enrollment on campus in engineering is more than 400 students.
Student Outcomes	Ultimately, most graduates enter academic or commercial careers. Ph.D. graduates commonly accept postdoctoral research positions.
Location	Syracuse, New York, is a city of approximately 150,000 people. The population of the greater metropolitan Syracuse area is more than 450,000. Located in the center of the state, the city serves as a focal point for many business, cultural, and entertainment activities for the central New York area. To the west is the famed Finger Lakes region; to the north and northeast are the Thousand Islands and the Adirondack Mountains.
The University	Syracuse University is a medium-sized, private, coeducational university. Founded in 1870, it is a comprehensive research university that has thirteen degree-granting schools and colleges and several interdisciplinary and continuing education programs. The University's 640-acre campus is beautifully situated among the hills of central New York State. The University has a growing stature in the sciences and engineering and maintains outstanding traditions in music, art, drama, communications, and public affairs.
Applying	Students are admitted for the semesters beginning in both September and January. Applications for aid must be received by January 1. Assistantship applications should be submitted before January 1. GRE scores are required for admission to some of the programs and are highly recommended for others, especially for international students. It is suggested that applicants take the General and Subject (engineering or mathematics) tests. International applicants without degrees from English-speaking universities are required to demonstrate English proficiency by examination. Application forms may be obtained from the Graduate School, 303 Bowne Hall.
Correspondence and Information	Director of Graduate Programs L. C. Smith College of Engineering and Computer Science 227E Link Hall Syracuse University Syracuse, New York 13244-1240 E-mail: gradinfo@ecs.syr.edu World Wide Web: http://www.ecs.syr.edu

Syracuse University

THE FACULTY AND AREAS OF RESEARCH

Bioengineering and Neuroscience

S. C. Chamberlain (Chair), S. J. Bolanowski Jr., G. A. Engbretson, J. M. Gilbert, K. M. Hiiemae, E. M. Relkin, N. B. Slepecky, R. L. Smith, R. T. Verrillo (Emeritus Professor), and J. Zwislocki (Emeritus Professor).

The Bioengineering and Neuroscience faculty pursues advanced multidisciplinary research, much of it at the Institute for Sensory Research. The program combines engineering and life sciences in the study of the sensory systems of hearing, touch, and vision as well as the mechanics and materials of biological systems. Current research interests include intensity coding in the auditory and tactile systems, visual information processing in the retina, cochlear biophysics, regeneration and repair in the inner ear, mechanisms of transduction in peripheral sense organs, parallel processing of information from sensory receptors to the brain, sensory-motor aspects of chewing, biomechanics of extremities and spine, modeling of bone, joint force, and motion, development and behavior of biomaterials and self-reinforced composites, degradation and corrosion in biological environments, micromechanisms of viable biological tissues, tissue engineering, and development of novel medical devices.

Chemical Engineering and Materials Science

G. C. Martin (Chair), A. J. Barduhn (Emeritus Professor), J. C. Heydweiller, C. J. Kelly, P. A. Rice, A. S. Sangani, K. Schröder (Emeritus Professor), J. A. Schwarz, S. A. Stern (Emeritus Professor), L. L. Tavlarides, and C. Tien (Emeritus Professor).

The department offers a broad spectrum of research opportunities in chemical engineering. Current research interests are in the areas of chemical equilibria and kinetics, supercritical extraction and chemical reaction of hazardous wastes, process optimization, chemical reaction and transport in biological systems, bioremediation, fluid mechanics in multiphase systems, magnetic phenomena in thin films, catalysis and surface chemistry, and chemistry-property relations in polymers and polymer-based composites.

Civil and Environmental Engineering

S. K. Bhatia (Chair), R. Aboutaha, S. P. Clemence, A. Costello, C. T. Driscoll Jr., A. A. Friedman (Emeritus Professor), C. E. Johnson, R. D. Letterman, E. M. Lui, J. A. Mandel (Emeritus Professor), D. Negussey, and E. M. Owens (Research Professor).

Major areas of study are environmental engineering, geotechnical engineering, and structural engineering. Current research activities are in the areas of aquatic chemistry, acidic precipitation, acid waters and their chemistry, problems related to water-treatment and waste-treatment systems, soil dynamics, geotextiles, geofoams, uplift of helical anchors, fracture mechanics of composite materials, structural dynamics, probability applications to structural engineering, and computer-aided analysis and design of structural systems.

Electrical Engineering and Computer Science (see Peterson's Guide, Section 9)

C. R. P. Hartmann (Chair), E. Arvas, H. Blair, P. Brinch Hansen, R. Chen, S. K. Chin, E. Ercanli (Temporary-term Assistant Professor), J. W. Fawcett (Part-time Associate Professor), G. Foster, G. Fox, P. Ghosh, A. Goel, S. Hariri, C. Isik, K. Jabbour, Y. Jia (Part-time Instructor), D. V. Keller Jr. (Research Professor), P. G. Kornreich, J. K. Lee, H. F. Mattson Jr. (Research Professor), K. Mehrotra, C. K. Mohan, S. Older, D. J. Pease, F. Phelps, J. Royer, R. Sargent (Research Professor), T. Sarkar, E. Sibert, Q. W. Song, S. Taylor, P. K. Varshney, H. Wang, and D. D. Weiner (Research Professor). In addition, the department has five affiliated faculty members from other units of the University, namely mathematics, NPAC, political science, and the School of Information Studies. Several research scientists from NPAC also collaborate in the research activities of the department.

Current faculty member research interests include RF/wireless engineering, electromagnetic theory, computational electromagnetic analysis, microwave transmission, satellite communication, digital signal processing, communication theory, robotics, control systems, microelectronic devices, optics, optoelectronic devices, software engineering, computer architecture, software migration, artificial intelligence, expert systems, neural networks, fuzzy logic, evolutionary computing, computer networks, VLSI design, computer-aided design, design automation and verification, formal methods for hardware/software design, parallel and distributed computing, concurrent programming languages, computational science, distributed and multimedia information systems, semantics of concurrency, logic programming, computational logic, structural computational complexity, plasma simulation, and error-correcting codes.

Mechanical, Aerospace, and Manufacturing Engineering

J. E. LaGraff (Chair), E. A. Bogucz Jr., T. Q. Dang, B. Davidson, H. Higuchi, A. Levy, J. Lewalle, F. A. Lyman (Emeritus Professor), Y. Moon, V. Murthy, R. W. Perkins, U. Roy, E. F. Spina, T. Vedder, and V. Weiss (Emeritus Professor).

Major areas of study are fluid dynamics, solid mechanics, and manufacturing systems. Current research projects are focused in the areas of experimental aerodynamics, turbulence modeling, computational fluid dynamics, gas turbine flows, hypersonic aerothermodynamics, composite materials, applied mechanics, fracture mechanics, biomechanics, manufacturing processes, geometric tolerancing, intelligent manufacturing systems, and helicopter rotor dynamics.

Solid State Science and Technology Program

K. Schröder (Director), R. R. Birge, J. Chaiken, B. Davidson, J. H. Fendler, P. K. Ghosh, J. Goodisman, A. Honig, P. Kornreich, A. Levy, M. C. Marchetti, A. Miller, E. A. Schiff, J. A. Schwarz, J. T. Spencer, G. Vidali, R. W. Vook, and V. Weiss.

Current faculty research interests are in the areas of thin-film science and technology, surface science and catalysis, mechanical behavior of solids, membranes, theoretical and experimental solid-state physics and chemistry, microelectronics and optoelectronics, electronic and magnetic materials, and design and characterization of new materials, such as composites.

TENNESSEE TECHNOLOGICAL UNIVERSITY

College of Engineering

Programs of Study

The College of Engineering offers programs leading to the degrees of Master of Science and Doctor of Philosophy.

The Master of Science, a research-oriented degree program, is offered with majors in chemical engineering, civil engineering, electrical engineering, industrial engineering, and mechanical engineering. Areas of research are indicated on the reverse side of this page. The Master of Science degree requires a minimum of 30 semester credits, of which 6 may be counted toward the thesis requirement. No foreign language is required for the master's programs, which can typically be completed in two calendar years.

The Doctor of Philosophy in engineering is an interdisciplinary degree program under the direction of advisory committees that are interdepartmental in nature. The degree requires a minimum of 24 semester hours of course work beyond the master's degree (48 beyond the baccalaureate), 8 semester hours of communication skills, and the equivalent of 24 semester hours of doctoral research and dissertation. A highly qualified student possessing an M.S. degree in engineering normally needs at least three years of full-time study to complete the degree. Current areas of doctoral research include acoustics, computer engineering, control systems, environmental engineering, fluid mechanics, lasers, lightning, machine design, material sciences, neural networks, physical electronics, plasmas, polymers, power systems, process design, robotics, signal and image processing, solid mechanics, structural mechanics, telecommunications, and thermal sciences.

Research Facilities

The College of Engineering operates state-supported research Centers of Excellence in Manufacturing Research, Water Resources, and Electric Power. The Electric Power Center has computer and laboratory facilities to perform engineering and economic modeling for the design of power plants and electrical distribution and transmission systems. The Manufacturing Center includes extensive computer-aided design (CAD) and computer-aided manufacturing (CAM) capabilities. In addition to computer modeling capabilities, the Water Resources Center has an EPA-certified water analysis laboratory.

The Chemical Engineering Department maintains research facilities in energy conservation, mass transfer, computer-aided process design, distillation, polymers, and physical properties. The Civil and Environmental Engineering Department and the Water Resources Center have facilities for water and industrial-waste treatment research, chemical analyses, soils and structural engineering, stress analysis, and transportation materials. The Electrical and Computer Engineering Department and the Electric Power Center have laboratories for antennas, digital systems, plasmas, lasers, power-system simulation and training, high-voltage engineering, robotics, telecommunication and signal processing, gaseous electronics, and nuclear engineering. The Industrial and Manufacturing Engineering Department has facilities for research in ergonomics, manufacturing systems, and the simulation of industrial activities. The Mechanical Engineering Department, the Electric Power Center, and the Manufacturing Center have laboratories for the study of noise control, combustion engines, computer-aided design, fluid dynamics, heat transfer, machine design, and material sciences.

Financial Aid

Aid is available through individual departments and centers in the form of teaching or research assistantships. Full master's-level assistantships pay tuition and fees plus a stipend of between $1800 and $5400 per semester. Full doctoral-level assistantships also pay tuition and fees plus a stipend of between $2700 and $6750 per semester. Partial assistantships, which pay a prorated share of tuition, fees, and a stipend, are sometimes awarded. A limited amount of support is available during the summer months. Approximately 85 percent of all graduate students received aid during 1997–98.

Cost of Study

The 1998–99 tuition and fees for full-time graduate students were $2960 per academic year (two semesters) for Tennessee residents and $7786 per academic year for nonresidents. A typical annual figure for books and other supplies is $1000.

Living and Housing Costs

In 1998–99, dormitory accommodations were available at $850 per semester for double occupancy. The rooms are fully furnished with all utilities paid. Off-campus housing is also available. The Tech Village consists of more than 300 apartments, which have been available at approximately $225 per month. Meals on campus averaged $1080 per semester in 1998–99.

Student Group

The University enrolls about 7,100 undergraduate and 1,100 graduate students, of whom 51 percent are men and 49 percent are women. The College of Engineering has about 1,500 undergraduate, 130 master's, and 30 doctoral students. In 1998–99, forty-seven master's and three doctoral degrees were awarded.

Location

Cookeville is a city of about 25,000 people, located halfway between Nashville and Knoxville. The surrounding area abounds in natural beauty and includes several state parks and large lakes. The University is within 80 miles of the Tennessee Valley Authority Headquarters, the Oak Ridge National Laboratory, and the Arnold Engineering Development Center of the U.S. Air Force.

The University and The College

Tennessee Technological University, founded in 1915, is a coeducational, state-supported university occupying a 235-acre main campus. The College of Engineering is the largest of the six undergraduate colleges: Agriculture and Human Ecology, Arts and Sciences, Business Administration, Education, Engineering, and Nursing. All members of the graduate faculty in the College hold an earned doctorate.

Applying

Students can be admitted for any term, but most are admitted for the fall term, which begins in August. Completed applications, an official transcript of the student's undergraduate and previous graduate work, and three letters of recommendation should be submitted at least four weeks (six months for international students) prior to the anticipated date of enrollment. GRE General Test scores are required (minimum total score: 1500). TOEFL scores (minimum score: 550) are required for international students. Financial aid applications received by March 1 are given priority.

Correspondence and Information

For application requests:

Dean
Graduate School
Box 5036
Tennessee Technological University
Cookeville, Tennessee 38505
Telephone: 931-372-3233
Fax: 931-372-3497
E-mail: g_admissions@tntech.edu
World Wide Web: http://www.tntech.edu/

For other information:

Associate Dean for Graduate Studies
College of Engineering
Box 5005
Tennessee Technological University
Cookeville, Tennessee 38505
Telephone: 931-372-3834
Fax: 931-372-6172
E-mail: engr_grad@tntech.edu

Tennessee Technological University

THE GRADUATE FACULTY AND THEIR RESEARCH

Chemical Engineering

Joseph J. Biernacki, Associate Professor; D.Eng., Cleveland State, 1988; PE: materials processing, properties and performance, modeling process dynamics. J. R. Booth, Adjunct Associate Professor; Ph.D., Clemson, 1965; PE: polymers, controls. Patricia J. M. Dycus, Assistant Professor; Ph.D., Tennessee Tech, 1997: thermodynamics, transport properties. Clayton P. Kerr, Professor; Ph.D., LSU, 1968; PE: air pollution control, process design, optimization. Ted Shutov, Professor; D.Sc., Mendeleev (Moscow), 1985: polymer processing, plastic composites, plastics recycling, biodegradable plastics. David W. Yarbrough, Professor and Chairperson; Ph.D., Georgia Tech, 1966; PE: thermodynamics, transport properties, energy conservation.

Civil and Environmental Engineering

Daniel A. Badoe, Assistant Professor; Ph.D., Toronto, 1994: modeling of urban travel demand and travel behavior. William P. Bonner, Professor and Chairperson; Ph.D., Florida, 1967: water quality, industrial-waste treatment, oil shale effects. George R. Buchanan, Professor; Ph.D., Virginia Tech, 1966; PE: solid mechanics, wave propagation, seismic analysis. L. K. Crouch, Associate Professor; Ph.D., Missouri–Rolla, 1990; PE: pavement design, construction materials. Dennis B. George, Professor and Director, Water Resources Center; Ph.D., Clemson, 1976; PE: water quality, water and wastewater treatment, hazardous-waste treatment. John A. Gordon, Professor; Ph.D., Purdue, 1970; PE: water-quality modeling, waste heat, reservoirs, environmental assessments. R. Craig Henderson, Assistant Professor; Ph.D., Tennessee, 1994; PE: structural design, wind and seismic analysis. Sharon Huo, Assistant Professor; Ph.D., Nebraska–Lincoln, 1997; PE: high-performance concrete, earthquake-resistant structures. Roy C. Loutzenheiser, Professor and Assistant Dean for Undergraduate Affairs; Ph.D., Texas A&M, 1972; PE: traffic, transportation engineering. Richard W. Lowhorn, Assistant Professor; Ph.D., Tennessee Tech, 1994; PE: environmental engineering, biological treatment of hazardous waste. Vincent S. Neary, Assistant Professor; Ph.D., Iowa, 1995; PE: river mechanics and fluvial hydraulics, movable-bed model studies. K. Larry Roberts, Professor; Ph.D., Tennessee, 1980; PE: physical-chemical water treatment, systems analysis. Edmond P. Ryan, Associate Professor; Ph.D., New Mexico, 1974; PE: foundations, structural design. Dallas G. Smith, Professor; Ph.D., Virginia Tech, 1969; PE: composite materials, fracture mechanics. R. Noel Tolbert, Professor; Ph.D., Vanderbilt, 1974; PE: viscoelastic materials, polymor concrete. Lenly J. Weathers, Assistant Professor; Ph.D., Iowa, 1995; EIT: bioremediation, wastewater treatment.

Electrical and Computer Engineering

Mohamed Abdelrahman, Assistant Professor; Ph.D., Idaho State, 1996: intelligent robust control, instrumentation and measurement systems, nuclear engineering. Ali T. Alouani, Professor; Ph.D., Tennessee, 1986; EIT: estimation theory, signal processing, nonlinear control of power systems, fuzzy logic control, systems. Joseph N. Anderson, Professor; Ph.D., Tennessee Tech, 1976: systems, robotics. Jeffrey R. Austen, Associate Professor; Ph.D., Illinois at Urbana-Champaign, 1991; EIT: communications, digital systems, signal processing, remote sensing. Charles L. Carnal, Professor; Ph.D., Tennessee, 1984; PE: system identification and state estimation. A. Chandrasekaran, Professor; Ph.D., Indian Institute of Technology–Madras, 1973: power systems, electrical machines. Pritindra Chowdhuri, Professor; D.Eng., RPI, 1966; PE: electric power-systems analysis, electromagnetic transients, transmission. Jeffrey Frolick, Assistant Professor; Ph.D., Michigan, 1995: telecommunication systems, signal processing. Roger L. Haggard, Associate Professor; Ph.D., Georgia Tech, 1991: digital systems, computer engineering, parallel computer architectures, numerical simulations. Charles E. Hickman, Professor and Interim Dean; Ph.D., Tennessee, 1966; PE: power plant performance. Satish M. Mahajan, Professor; Ph.D., South Carolina, 1987: high-voltage devices, optoelectronics. Wagdy H. Mohmoud, Assistant Professor; Ph.D., Alabama, 1997: computer engineering, integrated circuit design, VLSI system design, image processing. Sundaram Natarajan, Professor; Ph.D., Concordia (Montreal), 1979: electronic circuits and signal processing. Joseph O. Ojo, Associate Professor; Ph.D., Wisconsin–Madison, 1987: power systems, electric machines and drives. Esther T. Ososanya, Associate Professor; Ph.D., Bradford (England), 1984: VLSI, supercomputers, microprocessors, computer networks. Ghadir Radman, Associate Professor; Ph.D., Tennessee Tech, 1983; PE: power system control, control system design. Periasamy K. Rajan, Professor and Chairperson; Ph.D., Indian Institute of Technology–Madras, 1975: digital signal processing, image processing. Carl A. Ventrice, Professor; Ph.D., Penn State, 1962: lasers, plasmas, antennas, thin films.

Industrial and Manufacturing Engineering

Kenneth R. Currie, Associate Professor; Ph.D., West Virginia, 1988; PE: simulation, operations research, neural networks. S. Deivanayagam, Professor; Ph.D.,Texas Tech, 1973; PE: ergonomics, work design, safety. David W. Elizandro, Professor; Ph.D., Arkansas, 1974; PE: systems engineering. Jessica O. Matson, Professor and Chairperson; Ph.D., Georgia Tech, 1982; PE: operational research, material handling, storage systems. James R. Smith, Professor; Ph.D., Virginia Tech, 1971; PE: quality control, applied statistics. R. Meenakshi Sundaram, Professor; Ph.D., Texas Tech, 1976; PE: manufacturing systems, computer-aided manufacturing, technology transfer and assessment.

Industrial Technology

Delbert Stone, Associate Professor; Ed.D., Northern Colorado, 1978: welding, CNC machining, robotics and programmable controllers.

Mechanical Engineering

Stephen L. Canfield, Assistant Professor; Ph.D., Virginia Tech, 1997: kinematics and dynamics in design, robotics and parallel manipulators. John C. Chai, Assistant Professor; Ph.D., Minnesota, 1994: advanced computational methods for analyzing transport phenomena, especially radiative heat transfer. Glenn T. Cunningham, Associate Professor; Ph.D., Tennessee Tech, 1990; PE: heat transfer, solid mechanics. Corinne Darvennes, Associate Professor; Ph.D., Texas at Austin, 1989: acoustics, noise control and nondestructive evaluation. Edwin I. Griggs, Professor and Chairperson; Ph.D., Purdue, 1970; PE: thermodynamics, heat transfer, solar energy. Samuel Sang Moo Han, Professor; Ph.D., Alabama in Huntsville, 1977: fluid dynamics, numerical heat transfer. Darrell E. P. Hoy, Professor; Ph.D., North Carolina State, 1985: digital imaging, stress analysis, solid mechanics. Stephen A. Idem, Professor; Ph.D., Purdue, 1986: heat transfer, fluid mechanics. Jeffrey A. Marquis, Associate Professor; Ph.D., Kentucky, 1978: heat transfer, turbomachinery, small-scale hydro. Sastry S. Munukutla, Professor; Ph.D., Iowa, 1981: power-plant modeling, heat transfer, fluid mechanics. John Peddieson Jr., Professor; Ph.D., Virginia Tech, 1969: fluid mechanics, vibration, combustion instability. Kenneth R. Purdy, Professor and Interim Director, Electric Power Center; Ph.D., Georgia Tech, 1963; PE: heat transfer, thermochemical conversion of biomass, combustion instability. Ahmad A. Smaili, Associate Professor; Ph.D., Tennessee Tech, 1986: mechanical design, finite element analysis, robotics. Robert A. Smoak, Professor; D.Sc., Virginia, 1966: power plant control, simulation, optimal control. George M. Swisher, Professor; Ph.D., Ohio State, 1969; PE: systems, instrumentation, computer simulation. Kwun-Lon Ting, Professor; Ph.D., Oklahoma State, 1982: mechanisms, robotics. Marie B. Ventrice, Professor and Associate Dean for Graduate Studies and Research; Ph.D., Tennessee Tech, 1974; PE: heat and mass transfer, cogeneration, combustion. Christopher D. Wilson, Assistant Professor; Ph.D., Tennessee, 1997: engineering materials, stress, fatigue, fracture. Dale A. Wilson, Professor; Ph.D., Missouri–Columbia, 1978; PE: fracture mechanics, fatigue design.

TEXAS A&M UNIVERSITY

College of Engineering

Programs of Study

The Texas A&M University College of Engineering offers programs leading to the Master of Engineering in twelve major subject areas, the Master of Science in sixteen major areas, the Master of Computer Science, the Doctor of Philosophy in thirteen disciplines, and the Doctor of Engineering in twelve disciplines. Joint degree approvals under consideration include the M.D./Ph.D. in engineering and the M.D./Master of Engineering. Research concentrations and areas of emphasis in the various departments are briefly described on the reverse of this page.

The Ph.D. requires a minimum of 64 credit hours beyond the M.S. The Doctor of Engineering program requires a minimum of 64 credit hours beyond the M.S., including a one-year internship. The M.S. degree program requires completion of a minimum of 32 credit hours. The M.E. and M.C.S. require completion of a minimum of 36 hours.

Research Facilities

In addition to the laboratories provided by the various departments, a number of outstanding University research facilities are used by Texas A&M graduate students in engineering. Included are laboratories for research in biomedical engineering, food science and engineering, solid-state electronics, energy and propulsion, remote sensing, gasdynamics, materials and structural mechanics, nuclear science, petroleum engineering, space technology, ocean engineering, terramechanics, thermodynamics, highway safety, environmental engineering, aerodynamics and flight mechanics, and many other areas. The University library contains more than 1 million books and journals.

Financial Aid

A variety of fellowships and traineeships provide support for graduate students. Teaching and research appointments are also available, as are a number of assistantships provided by the University through grants from industry and other sources.

Assistantships in a number of significant research programs within the Texas Engineering Experiment Station and the Texas Transportation Institute are other means by which engineering graduate students receive support. These assignments bring students into direct contact with active researchers. Inquiries to department heads should be made as early as possible.

Cost of Study

Current tuition for residents of Texas is $72 per semester credit hour with a $120 minimum; for others it is $285 per semester credit hour. Additional fees ($450–$550 per semester) are required for student services, engineering equipment and computer access, and laboratory use. Out-of-state tuition and fees are waived for eligible students receiving fellowships and/or assistantships.

Living and Housing Costs

A limited number of University-owned apartments, both furnished and unfurnished, are available for married students. Rents range from $230 to $360 per month, plus electricity. A large number of private apartment complexes with one- to three-bedroom apartments are available, with rents ranging from $442 to $774 per month.

Student Group

In 1998–99, graduate enrollment in the College of Engineering was 1,646. The majority of students are young engineers who have begun graduate study immediately after receiving bachelor's degrees from universities throughout this country and internationally.

Location

The Bryan–College Station area is a progressive community with a population of 140,000. It is centrally located in relation to Houston, Dallas, Fort Worth, and Austin. The area provides many recreational and cultural activities.

The University

Texas A&M University was founded as a land-grant college in 1876 and is the state's oldest public institution of higher education. Because it was conceived as a school of agriculture and engineering, the institution maintains the expected emphases, but offerings in many other areas have been added to broaden the scope of instruction. The campus covers 5,500 acres. In 1998–99, the total enrollment was 43,389, including 6,746 graduate students. Sponsored research for 1998 totaled $366.8 million, leading universities in the Southwest.

Applying

A formal application, including a transcript and GRE General Test and Subject Test scores, is required. Forms, available from the director of admissions, should be filed as early as possible to be considered for financial aid.

Correspondence and Information

For application:
Director of Admissions
College of Engineering
Texas A&M University
College Station, Texas 77843
Telephone: 409-845-1031

For program information:
Head, (specify department)
College of Engineering
Texas A&M University
College Station, Texas 77843

Texas A&M University

FACULTY HEADS AND RESEARCH AREAS

Aerospace Engineering. Dr. K. T. Alfriend, Head. Research programs are being conducted in low-speed, transonic, and hypersonic aerodynamics; materials research; composites; fracture mechanics; active materials and smart structures; guidance; dynamics and control of satellites; astrodynamic microsatellite technology; stability and control of aircraft; aeroelasticity; and computational fluid dynamics. Low-speed and supersonic wind tunnels provide facilities for aerodynamic research in fundamental fluid-flow problems and in three-dimensional testing of complete airplane and other models. A flight mechanics laboratory with five aircraft is used for full-scale testing. Structures and materials research in composites, high-temperature metals, and active materials is supported by laboratories for mechanical testing, wave propagation, and nondestructive testing. An intelligent systems laboratory supports research on smart structures and uninhabited aerial vehicles (UAV).

Agricultural Engineering. Dr. James R. Gilley, Head. Research programs are available in water resource development, air and water quality, solute transport in groundwater, alternate energy systems, agricultural machinery systems, bioprocessing of food and fiber products, biochemical and food engineering, forest and wood products engineering, microelectronic and systems applications in food and fiber production, and engineering applications in biotechnology.

Biomedical Engineering. Dr. W. A. Hyman, Chair. The program offers a wide range of research and academic opportunities involving the application of engineering principles to living systems: biomechanics of tissues and bone, physiological fluid mechanics, artificial organs, bioinstrumentation, lasers in medicine, biosensors, medical imaging systems, and rehabilitation engineering. Considerable laboratory, fabrication, and computing facilities support the program.

Chemical Engineering. Dr. R. G. Anthony, Head. The program offers a wide variety of research topics in thermodynamics, catalysis, chemical reaction engineering, biochemical engineering, separation processes, polymers, colloids and interfaces, asphaltic materials, process simulation and control, mathematical modeling, enhanced oil recovery, rheology, and environmental engineering. Specialized equipment includes fermentation and cell culture facilities, a laser-Doppler electrophoresis instrument, a laser light scattering instrument, an X-ray diffractometer, and an NMR spectroscopy system. (Department Web Site: http://www-chen.tamu.edu)

Civil Engineering. Dr. J. M. Niedzwecki, Head. The department offers a wide range of research and academic opportunities in civil systems engineering; coastal, hydraulic, and ocean engineering; construction engineering and project management; environmental engineering; geotechnical engineering; hydrology; materials engineering; mechanics and materials; public works engineering and management; structural engineering and structural mechanics; transportation engineering; and water resources engineering.

Computer Engineering. (Administered by the Departments of Computer Science and Electrical Engineering.) This new program began operation in the spring of 1997, offering master's and Ph.D. degrees.

Computer Science. Dr. W. Zhao, Head. The department offers excellent opportunities for research in software engineering, computer networks, knowledge systems, real-time systems, computer architecture, fault-tolerant computing, digital libraries, hypermedia, information retrieval, computer vision, computer graphics, robotics, computational science, and parallel computation and, in conjunction with the Department of Electrical Engineering, computer engineering and microelectronics design and testing. Extensive computer and graphics workstations and a large set of parallel and supercomputer facilities support the instructional and research activities of the department.

Electrical Engineering. Dr. C. Singh, Head. The department offers work in semiconductor materials and devices, optics, automatic control systems, electromagnetic fields, microwaves, digital systems, microprocessors, computer engineering, communication systems, digital-signal processing, power systems, power electronics, digital and analog VLSI circuits, and biomedical. In addition to well-equipped general instructional and research laboratories, special laboratory facilities are available for research in solid-state electronics, microwaves, antennas, digital systems, microprocessors, power electronics and power system protection and automation, telecommunications, and control systems simulation. Interdisciplinary research programs include computer engineering, manufacturing, transportation, biomedical imaging, electronic materials, and wavelets.

Health Physics. Dr. A. E. Waltar, Head. Active research involves both internal and external dosimetry, with emerging emphasis in medical physics, plutonium disposition, nuclear waste site clean-up, and nuclear waste disposal. Both applied and research degrees are offered.

Industrial Engineering. Dr. Way Kuo, Head. The department offers excellent opportunities for research in manufacturing and production systems, operations research, simulation modeling, quality and reliability, transportation systems, robotics and automation, and knowledge engineering.

Industrial Hygiene. Dr. A. E. Waltar, Head. Research is dedicated to providing safe and healthful means to use necessary hazardous materials and hazardous processes in Texas industry. Emphases are in sampling strategies and methods, ventilation, and acoustics and noise control. The program has links to toxicology and epidemiology programs in the College of Medicine.

Interdisciplinary Engineering. Dr. Karan L. Watson, Head. Study and research are provided in this field for those with educational aims that do not identify with the classical engineering departments at the University or that encompass several disciplines. Examples of such programs are systems engineering, materials, structural mechanics, medicine, and activation analysis.

Mechanical Engineering. Dr. S. Jayasuriya, Head. The department is housed in an engineering building with modern office and laboratory space and has additional space in other A&M buildings. Active research facilities exist for fracture testing, composite materials, plastics, metallurgical studies, corrosion, experimental stress analysis, ultrasonics, nondestructive testing, vibration, shock, rotating machinery, turbomachinery, seals, fluid dynamics, gasdynamics, turbulence, heat transfer, power generation, tribology, combustion, in situ lignite gasification, solar energy, wind tunnel studies, aerosols, manufacturing processes, robotics and intelligent machinery, and computer-aided design and manufacturing. Computer facilities available to graduate students include the University's Cray and Silicon Graphics supercomputers, the College of Engineering's VAX cluster, and the department's PC lab, with more than seventy-five high-performance PCs and twenty high-performance workstations. There are also several graphics devices, including an Evans & Sutherland PS 300 system.

Nuclear Engineering. Dr. A. E. Waltar, Head. The department conducts experimental and computational research on advanced nuclear reactors for terrestrial and space systems, applied and experimental health physics, industrial hygiene, safety engineering, charged-particle interactions with matter, and fusion engineering. Department facilities include a training reactor, a large research reactor, counting and spectroscopy laboratories, accelerators and thermal hydraulics labs, and access to a 300-MeV cyclotron, tomography systems, and rectilinear scanners. (Department Web site: http://trinity.tamu.edu)

Ocean Engineering (Administered by the Civil Engineering Department). The program provides an opportunity to solve engineering problems in the coastal and deep ocean environments. Areas of research interest include marine hydrodynamics, ocean wave mechanics, wave-structure interaction, dredging, offshore pipelines, diving technology, and coastal processes. Laboratory facilities include a dredge pump and pipe test loop system, a large shallow wave basin, a state-of-the-art deep-water wave basin, and several long open-channel flumes with wave-making capabilities.

Petroleum Engineering. Dr. C. H. Bowman, Head. Advanced-level research and course study are offered in drilling, production, well stimulation, reservoir engineering, reservoir characterization, and reservoir management. Specific teaching and research areas include core analysis and flow imaging, drilling and drilling fluids, economic evaluation, field-scale reservoir studies, formation damage, formation evaluation, horizontal and extended-reach wells, improved oil and gas recovery, production and well test data analysis, reservoir characterization, reservoir management, reservoir simulation, rock mechanics, well logging, and well stimulation. New areas of research include multilateral wells, nonparametric optimization (including neural networks), streamtube flow simulation, and well performance modeling and interpretation.

Safety Engineering. Dr. A. E. Waltar, Head. The program conducts experimental, computational, and policy research in systems safety, industrial safety, fire protection, and product safety. The emphasis is on explosion suppression, electrostatic fire safety, and improved safety by process substitution.

TEXAS A&M UNIVERSITY–KINGSVILLE

College of Engineering
Graduate Engineering Programs

Programs of Study

The College of Engineering offers several interdisciplinary courses of study leading to Master of Science and Master of Engineering degrees in chemical engineering, civil engineering, computer science, electrical engineering, environmental engineering, industrial engineering, industrial technology, mechanical engineering, and natural gas engineering. For Master of Science degrees, both Plan I (24 semester hours of course work and 6 semester hours of thesis) and Plan II (36 semester hours of course work) are available. The Master of Engineering degree is a special program intended to provide practicing engineers with the opportunity for advanced studies, for which 36 hours of course work are required (registration as a Professional Engineer in Texas may qualify a person to complete the program in 30 hours). Students intending to pursue a Ph.D. are urged to complete the Plan I requirements. Comprehensive oral exams are required for all degrees.

Chemical and natural gas research areas include gross error detection and data reconciliation, polar-nonpolar vapor liquid phase equilibrium, and microbial desulfurization of coal and gas. Civil engineering research includes structural dynamics, concrete design, and groundwater hydrology. Electrical engineering and computer science research involves compiler design, real-time systems, control system synthesis, digital signal processing, communications, and integrated electronics. Environmental engineering research explores air pollution control, wastewater treatment, bioremediation, solid/hazardous waste design, industrial health/safety, and product safety/toxicology. Mechanical and industrial engineering research includes network flows, computer-integrated manufacturing, advanced dynamics, and intelligent control.

Research Facilities

The College of Engineering's research facilities are divided into a number of laboratories in three engineering buildings: the Kleberg Engineering Building, the Dotterweich Engineering Laboratory, and the McNeil Engineering Laboratory. Advanced specialty equipment is housed in dedicated labs for each discipline for materials testing, electronic design, chemical separation and distillation, and environmental analysis. Extensive computer facilities utilizing 486 PC and Sun SPARCstation machines are available to all students. In addition, research facilities in other areas of the University (e.g., biology, chemistry, earth sciences), such as an advanced scanning electron microscope, are available for engineering usage.

Financial Aid

University teaching and research assistantships provide annual stipends, while fellowships and internships supported by federal, state, and University funds provide stipends of $1000 to $14,000. International students are required to guarantee their support but are eligible for in-state tuition rates if awarded competitive fellowships. Inquiries about the availability of support should be addressed to the graduate coordinator of the various disciplines (see address listed below).

Cost of Study

Tuition and fees (15 semester hours) for Texas residents are $1121 per semester. The cost for non-Texas residents and international students is $4271 per semester.

Living and Housing Costs

The cost of living in south Texas is relatively low when compared to that of other areas of the country. Apartments and rental homes in the $300 to $400 per month range can easily be found. Several apartment complexes and rental properties are within walking distance or short driving distance of the University. Dormitory rooms for unmarried students and apartments for married students and their families are available through the University. Dorm rooms with optional meal plans start at about $1700 per semester. Student family apartments rent for about $250 per month.

Student Group

The total University enrollment is about 6,000. Of the College of Engineering's 950 students, 200 are enrolled as graduate students. International students make up 64 percent of the graduate student enrollment, but the student population is very diverse.

Location

Texas A&M University–Kingsville is located in semitropical south Texas. Kingsville is situated about 40 miles southwest of Corpus Christi, 153 miles southeast of San Antonio, and 120 miles north of Mexico. Kingsville, a town of about 25,000 people, is in a semirural area but is easily accessible to urban areas. The coastline is nearby, making water sports and activities popular. The King Ranch, one of the largest commercial ranches in the world, provides tours and other community services. Nearby Corpus Christi features many social and cultural activities, including museums, concerts, plays, and minor league baseball.

The University

Texas A&M University–Kingsville (formerly Texas A&I University) was established in 1925 as Texas State Teacher's College and has since evolved into a comprehensive institution of higher learning. Its name and scope have undergone major changes, most recently in 1989, when the University became a part of the Texas A&M University System, and in 1993, when the name was changed to Texas A&M University–Kingsville.

Applying

Candidates must submit admissions applications to the Admissions Office and must be admitted to the College of Graduate Studies and the engineering program of their choice. Students must have a GRE score of at least 1000 (verbal plus quantitative); environmental engineering applicants must have a GRE score of at least 1100 and an undergraduate GPA of at least 3.0. Students whose native language is not English must supply a minimum score of 550 on the Test of English as a Foreign Language. Application deadlines are for the fall, July 1; spring, November 15; and summer, April 15.

Correspondence and Information

Graduate Coordinator
(Department of interest)
College of Engineering
MSC 188
Texas A&M University–Kingsville
Kingsville, Texas 78363
Telephone: 361-593-2001
Fax: 361-593-2106

Texas A&M University–Kingsville

THE FACULTY AND THEIR RESEARCH

Hayder Abdul-Razzak, Professor of Mechanical and Industrial Engineering; Ph.D., IIT, 1988. Design, analysis, and modeling of fluid/thermal systems; numerical computations.

Faleh T. Al-Saadoon, Professor of Chemical and Natural Gas Engineering; Ph.D., Pittsburgh, 1970; PE. Unconventional gas resources, profile control/modification using cross-linked polymer.

Fred C. Benson, Associate Professor of Civil Engineering; Ph.D., Texas A&M, 1988. Healing characteristics of pavements using asphalt cement binders, inherent molecular properties of different asphaltic cements.

T. Joe Boehm, Associate Professor of Electrical Engineering and Computer Science; Ph.D., Oklahoma State, 1975. Ultra-low-level and ultra-stable empirical studies, fluctuation phenomena.

Rajab Challoo, Professor of Electrical Engineering and Computer Science; Ph.D., Wichita State, 1988; PE. Modeling, analysis, and design of linear and nonlinear dynamic systems/control systems; intelligent control of robotic systems.

John L. Chisholm, Assistant Professor of Chemical and Natural Gas Engineering; Ph.D., Oklahoma, 1992. Microbially enhanced oil recovery, reservoir simulation, time-series analysis.

Yousri Elkassabgi, Professor of Mechanical and Industrial Engineering; Ph.D., Houston, 1986; PE. Heat transfer with or without phase change, fluid mechanics, energy conservation.

Andrew N. S. Ernest, Associate Professor and Chairman of Environmental Engineering; Ph.D., Texas A&M, 1991. Surface and subsurface water quality modeling; particle and particle-mediated contaminant transport, sediment-water column interactions, parameter estimation and process optimization, in situ bioremediation.

Kambiz Farahmand, Associate Professor of Mechanical and Industrial Engineering; Ph.D., Texas at Arlington, 1992; PE. Ergonomics and occupational health, quality control and ISO 9000 Standard Series, production and inventory control and simulation.

Mohammed A. Faruqi, Assistant Professor of Civil Engineering; Ph.D., Arkansas, 1996. Structural materials, concrete structures, structural dynamics, model analysis, and mathematical modeling of composites.

Jerry W. Hedrick, Associate Professor of Industrial Technology; M.S., Texas A&M, 1965. Industrial safety.

William A. Heenan, Professor and Chairman of Chemical and Natural Gas Engineering; D.Engr., Detroit, 1969; PE. Chemical process control and optimization, error detection/data reconciliation.

Farzin Heidari, Assistant Professor of Industrial Technology; Ph.D., Idaho, 1990. Automation, cellular manufacturing, flexible manufacturing systems (FMS).

Kuruvilla John, Assistant Professor of Environmental Engineering; Ph.D., Iowa, 1996. Regional/urban tropospheric ozone, photochemical air quality modeling, air pollution monitoring, emissions inventory assessment, environmental impact assessment.

Pat T. Leelani, Professor and Chairman of Civil Engineering; Ph.D., Akron, 1980. Geotechnical engineering, static and dynamic properties of soil.

Chung S. Leung, Associate Professor of Electrical Engineering and Computer Science; Ph.D., Florida Atlantic, 1989; PE. Speech recognition, data compression, pattern recognition, neural networks, fuzzy logic.

John S. Linder, Professor of Electrical Engineering and Computer Science; Ph.D., Arizona, 1967. Modeling and experimental verification of single-event effects using synthesis-based approaches to IC design in multiple technologies.

Bruce Marsh, Assistant Professor of Industrial Technology; D.I.T., Northern Iowa, 1996. Dimensional metrology, quality control and fluid power.

Ronald D. Matthys, Associate Professor of Civil Engineering; Ph.D., Texas at Arlington, 1984. Structural analysis and design of structural steel and reinforced concrete structures.

Robert A. McLauchlan, Professor and Chairman of Mechanical and Industrial Engineering; Ph.D., Texas, 1978; PE. Dynamics; sensor fusion/integration and intelligent control of systems, processes, devices, and machines; neurocomputing; higher-order spectra; fuzzy logic and fractals/multiresolution techniques; hybrid neuro-fuzzy-genetic/evolutionary control and learning systems; process improvement/process innovation applied to engineering curriculum change.

Mark Miller, Associate Professor and Acting Chairman of Industrial Technology; Ph.D., Texas A&M, 1993. Computer-aided design.

Frank M. Mullen, Associate Professor of Industrial Technology; Ed.D., East Texas State, 1965. Industrial materials, 1983 industrial controls.

Reza Nekovei, Assistant Professor of Electrical Engineering and Computer Science; Ph.D., Rhode Island, 1994. Massive parallel processing, rapid prototyping, medical imaging, distributed systems.

Syed Iqbal Omar, Professor of Electrical Engineering and Computer Science; Ph.D., Carleton, 1971. Intelligent systems, image processing, telerobotic protocols.

Selahattin Ozcelik, Assistant Professor of Mechanical and Industrial Engineering; Ph.D., RPI, 1996. Adaptive control, robust and nonlinear control, robotics, intelligent control, flexible structures and active vibration control.

Sung-won Park, Associate Professor and Acting Chairman of Electrical Engineering and Computer Science; Ph.D., New Mexico, 1985; PE. Spectral analysis, data compression, speech processing, image processing.

Ali A. Pilehvari, Associate Professor of Chemical and Natural Gas Engineering; Ph.D., Tulsa, 1984; PE. Rheology of non-Newtonian fluids, including drilling fluids, cement slurries, solid-liquid, and gas-liquid.

Charanjit Rai, Professor of Chemical and Natural Gas Engineering; Ph.D., IIT, 1960; PE. Kinetic modeling, gas processing–enhanced oil recovery, microbial desulfurization.

Joseph O. Sai, Associate Professor of Civil Engineering; Ph.D., Texas A&M, 1982. Water resources, liner systems design and evaluation, infiltrometers.

Barbara Schreur, Professor of Electrical Engineering and Computer Science; Ph.D., Florida State, 1979. Real-time, systems software reliability, Oort cloud, presentation software–based instruction.

Dale L. Schruben, Professor of Chemical and Natural Gas Engineering; Ph.D., Carnegie-Mellon, 1973; PE. Fluid mechanics of wind turbines, particle-polymer composites, cold petroleum rheology.

Robert Serth, Professor of Chemical and Natural Gas Engineering; Ph.D., SUNY at Buffalo, 1968; PE. Gross-error detection/data reconciliation.

Robert F. Tucker, Professor of Mechanical and Industrial Engineering and Assistant Dean; Ph.D., Louisiana Tech, 1976; PE. Product design and development, numerical applications.

Gary R. Weckman, Assistant Professor of Mechanical and Industrial Engineering; Ph.D., Cincinnati, 1996. Reliability optimization, modeling, work measurements, production and inventory management.

TUFTS UNIVERSITY

College of Engineering

Programs of Study	The College of Engineering offers programs of study leading to the Master of Science (M.S.) and Doctor of Philosophy (Ph.D.) degrees in the Departments of Chemical Engineering, Civil and Environmental Engineering, Electrical Engineering and Computer Science, and Mechanical Engineering. In addition to the usual areas of concentration within these disciplines, other departments offer programs in distinctive specialties, including biomedical technology, biotechnology, environmental health, hazardous materials management, product engineering, human factors, and electrooptics. The Graduate School also offers an interdisciplinary doctoral program that allows students to devise their course of study.

Applicants for graduate degrees in engineering are required to have a suitable background in mathematics and/or engineering sciences and the prerequisite understanding for the advanced engineering courses to be taken. Requirements for the M.S. degree are ten courses including a thesis—except in civil and environmental engineering and electrical engineering and computer science, where a master's report or a design project may be elected. There is an oral examination covering the thesis research. For computer science, an undergraduate major in computer science is expected, but other backgrounds will be considered.

Candidates for the Ph.D. degree normally have completed requirements for the M.S. degree in their discipline and must pass a qualifying examination. The candidate must satisfactorily complete a program of course work established by a faculty committee, write a dissertation on the research effort, and defend the dissertation orally.

Full-time students ordinarily take five courses per term or four courses plus thesis research. Students with research or teaching assistantships take the equivalent of two or three courses per term. Part-time students take one or two courses per term, and some programs can be completed in the evening. One year of residence is required for the master's degree; another two years beyond the master's degree for the Ph.D. |
Research Facilities	The College maintains an environment that makes research attractive by providing modern facilities and instrumentation. In electrical engineering, the Electro-Optics Technology Center has attracted support from more than twenty corporations, which supply equipment and visiting faculty. Special facilities are available for microwave engineering and for VLSI design in the electrical engineering and computer science department. Chemical engineering has complete molecular biology cell culture fermentation/bioprocessing and biophysical characterization laboratories associated with the Biotechnology Center, a fully equipped heterogeneous catalysis laboratory, a teaching/research laboratory for pollution prevention projects, and the Laboratory for Materials and Interfaces with state-of-the-art equipment, including an environmental scanning electron microscope and an atomic force microscope. In mechanical engineering, several laboratory facilities are used for both teaching and research, including specialized facilities dedicated to the investigation of turbulent flow, comparative biomechanics, and thermal manufacturing processes. The Department of Civil and Environmental Engineering maintains modern and well-equipped teaching and research laboratories in the areas of environmental, geotechnical, materials, structures, and water resources engineering.
Financial Aid	Full and partial tuition scholarships are available to many M.S. and most Ph.D. students. Teaching and research assistantships are available on a competitive basis for students with good academic standing.
Cost of Study	Tuition for students beginning in the academic year 1999–2000 is $24,804 for full-time students, plus a mandatory health service fee of $445. The tuition for part-time students is $2480 per course. Scholarships are available.
Living and Housing Costs	Graduate students tend to find housing near the campus, often sharing apartments. Monthly rentals for a 2-bedroom apartment are between $450 and $650 per person per month. A limited amount of graduate housing is available on campus and ranges from $4510 to $4780 for a single for the academic year. Meal plans are available at University dining rooms. The cost of the full meal plan is $3460 for the 1999–2000 academic year.
Student Group	Approximately 300 graduate students are enrolled in the College of Engineering, including 60 Ph.D. candidates and about 100 part-time students. About 20 percent are international students, and 32 percent are women.
Location	Tufts University is located in the Boston suburbs, just 7 miles from the city center and 2 miles from Cambridge. Because of the high density of educational institutions, many distinguished scholars who reside in or visit the Boston area present seminars and confer with colleagues at Tufts. Students with specialized research needs may obtain Boston Library Consortium privileges, enabling them to use the library facilities of other local universities. The area is a center for historic points of interest and some of the finest cultural offerings in the world, including the Boston Symphony, the Museum of Fine Arts, the Museum of Science, the New England Aquarium, and many performing groups and theaters that stage professional productions.
The University	Since its designation as Tufts College in 1852 and as Tufts University in 1955, Tufts has grown to comprise seven primary faculties: the Faculty of Arts and Science; the Fletcher School of Law and Diplomacy; the Schools of Medicine, Dental Medicine, and Veterinary Medicine; the Sackler Graduate School of Biomedical Science; and the School of Nutrition. At present, the total enrollment in all schools is about 7,690 students, of whom approximately 2,950 are graduate and professional students.
Applying	For September enrollment, applications should be submitted by February 15 for chemical engineering, civil and environmental engineering, and mechanical engineering. The electrical engineering and computer science application deadlines are March 15. For January enrollment, applications should be submitted by October 15. However, applications received after these dates are given consideration. TOEFL scores are required for international students. Scores on the GRE General Test are required for applicants to chemical engineering and electrical engineering and computer science and recommended for applicants to other departments.
Correspondence and Information	Graduate School of Arts and Sciences Tufts University 120 Packard Avenue Medford, Massachusetts 02155 Telephone: 617-627-3395 E-mail: gsas@infonet.tufts.edu

Tufts University

THE FACULTY AND THEIR RESEARCH

Ioannis Miaoulis, Professor of Mechanical Engineering and Dean; Ph.D., Tufts. Thermal-fluid processes, microscale heat transfer, science education.
G. Kim Knox, Adjunct Assistant Professor and Assistant Dean; M.S., Tufts; PE. Mechanics, structural engineering.

Chemical Engineering

Jerry H. Meldon, Associate Professor and Chairman; Ph.D., MIT. Mass transfer, membrane processes, reaction-separation coupling.
Gregory D. Botsaris, Professor; Ph.D., MIT. Crystallization, applied surface science, nucleation.
Eliana DeBernardez-Clark, Associate Professor; Ph.D., Littoral (Argentina). Biochemical engineering, protein folding and aggregation, fermentation processes.
Aurelie Ewards, Visiting Assistant Professor; Ph.D., MIT. Physiological modeling.
Maria Flytzani-Stephanopoulos, Associate Professor and Raytheon Professor of Pollution Prevention; Ph.D., Minnesota. Catalysis, clean energy transportation technologies, pollution prevention.
Walter Juda, Adjunct Professor; Ph.D., Lyons (France). Electrochemistry and chemical reaction engineering.
David Kaplan, Associate Professor and Director of Biotechnology Center; Ph.D., Syracuse. Biotechnology, biomaterials.
Brian Kelley, Adjunct Assistant Professor; Ph.D., MIT. Protein purification, large-scale purifications, high-density bacterial fermentation.
Kim Lewis, Research Associate Professor; Ph.D., Moscow. Multidrug resistance, drug discovery, metabolic engineering.
Daniel F. Ryder, Associate Professor; Ph.D., Worcester Polytechnic. Glass and ceramic materials processing, sol-gel processing technology, electronic ceramics, neural networks.
Howard Saltsburg, Research Professor; Ph.D., Boston University. Solid-state electrochemistry, nanoscale structures.
Nak-Ho Sung, Professor; Ph.D., MIT. Polymers and composites, interface science, polymer diffusion, surface modification.
Kenneth A. Van Wormer Jr., Professor; Sc.D., MIT. Optimization, nucleation, reaction kinetics, VLSI fabrication.
Gordana Vunjak-Novakovic, Adjunct Professor; Ph.D., Belgrade. Transport phenomena in multiphase systems, tissue engineering.

Civil and Environmental Engineering

Stephen H. Levine, Associate Professor and Chairman; Ph.D., Massachusetts at Amherst. Environmental, ecological, and economic systems modeling.
Linfield C. Brown, Professor; Ph.D., Wisconsin–Madison. Water quality modeling, environmental engineering and statistics.
Steven C. Chapra, Professor; Ph.D., Michigan. Surface water-quality modeling.
Wayne Chudyk, Associate Professor; Ph.D., Illinois at Urbana-Champaign. Drinking water quality and toxic materials, groundwater monitoring.
Larry Cohen, Lecturer; M.S., MIT. Project management, risk assessment, hazardous waste treatment technologies.
Anne Marie Desmarais, Lecturer; M.S., Michigan. Environmental health.
John Durant, Assistant Professor; Ph.D., MIT. Environmental engineering, hazardous materials management.
Lewis Edgers, Professor; Ph.D., MIT. Geotechnical engineering, environmental geotechnology.
David M. Gute, Associate Professor; Ph.D., Yale. Environmental and occupational epidemiology.
Rachid Hankour, Lecturer; Ph.D., Tufts. Geotechnical construction materials and hydraulics.
Ronald Hirschfeld, Lecturer; Ph.D., Harvard. Soil mechanics.
Daniel Jansen, Assistant Professor; Ph.D., Northwestern. Materials, mechanical behavior of concrete.
Paul Kirshen, Research Associate Professor; Ph.D., MIT. Water resources research.
Kim Knox, Adjunct Assistant Professor; M.S., Tufts. Structural engineering, bridge design.
Lee Minardi, Lecturer; M.S., Tufts. Computer-aided design, geometric modeling.
Masoud Olia, Lecturer; Ph.D., Northeastern. Damage in composites, systems dynamics, stress analysis.
Masoud Sanayei, Associate Professor; Ph.D., UCLA. Structural engineering, finite element analysis.
Christopher Swan, Assistant Professor; Ph.D., MIT. Environmental geotechnology, geotechnical engineering.
Richard M. Vogel, Professor; Ph.D., Cornell. Water resources, statistics, engineering economics.
Richard Weber, Lecturer; M.S., Florida. Geotechnical engineering.
Mark Woodin, Lecturer; D.Sc., Harvard. Epidemiology and public health.

Electrical Engineering and Computer Science

Robert A. Gonsalves, Professor and Chairman; Ph.D., Northeastern. Digital image processing, communications.
Mohammed N. Afsar, Professor; Ph.D., London. Microwaves and submillimeter waves, design and applications.
Anselm C. Blumer, Associate Professor; Ph.D., Illinois. Data compression, machine learning.
David Boas, Research Assistant Professor; Ph.D., Pennsylvania. Biomedical optics.
Chorng Hwa Chang, Associate Professor; Ph.D., Drexel. Computer engineering, digital systems.
Alva Couch, Associate Professor; Ph.D., Tufts. Parallel computing, computer graphics.
Mark Cronin-Golomb, Associate Professor; Ph.D., Caltech. Photorefractive devices, nonlinear optics.
Denis W. Fermental, Associate Professor; Ph.D., Northeastern. Control theory, computer engineering.
Ronald B. Goldner, Professor; Ph.D., Purdue. Electrooptics including materials, devices and systems, solar energy conversion.
Soha Hassoun, Assistant Professor; Ph.D., Washington (Seattle). CAD, VLSI.
Robert J. K. Jacob, Assistant Professor, Ph.D., Johns Hopkins. Human-computer interaction.
Paul Kelley, Research Professor and Director, Electro-Optics Technology Center; Ph.D., MIT. Photonics, optical communications, optical networking.
David W. Krumme, Associate Professor; Ph.D., Berkeley. Parallel computing.
Karen P. Lentz, Assistant Professor; Ph.D., Northeastern. Digital simulation, multimedia computer architecture.
Carolyn L. McCreary, Associate Professor; Ph.D., Colorado. Graph visualization, parallel processing.
Joseph F. Noonan, Associate Professor; Ph.D., Tufts. Statistical communication, digital signal processing, information theory.
Douglas Preis, Professor; Ph.D., Utah State. Electromagnetics, signal processing.
James Schmolze, Associate Professor; Ph.D., Massachusetts. Artificial intelligence.
Diane L. Souvaine, Associate Professor; Ph.D., Princeton. Computational geometry.
Van Toi Vo, Associate Professor; Ph.D., Swiss Federal Institute of Technology. Biomedical electromechanical engineering, microtechnology design, mathematical modeling.

Mechanical Engineering

Vincent P. Manno, Associate Professor and Chairman; Sc.D., MIT. Computational thermal-fluid dynamics.
Behrouz Abedian, Associate Professor; Ph.D., MIT. Electrokinetics, fluid dynamics.
I. Melvin Bernstein; Professor and Vice-President of Arts, Sciences, and Technology; Ph.D., Columbia. Metallurgy.
Allan H. Clemow, Adjunct Assistant Professor; M.S., Penn State. Consumer product analysis, products liability.
William J. Crochetiere, Professor; Ph.D., Case Western Reserve. Machine design, mechatronics.
Martha N. Cyr, Research Assistant Professor; Ph.D., Worcester Polytechnic. Thermal-fluid sciences, education outreach.
Charalabos C. Doumanidis, Associate Professor; Ph.D., MIT. Controls, thermal manufacturing.
Robert Greif, Professor; Ph.D., Harvard. Applied mechanics, vibration control, composite materials.
Mark Kachanov, Professor; Ph.D., Brown. Fracture mechanics, micromechanics of materials.
John G. Kreifeldt, Professor; Ph.D., Case Western Reserve. Engineering psychology, operations research, product design.
Frederick C. Nelson, Professor; Ph.D., Harvard. Vibrations and noise control.
James P. O'Leary, Associate Professor; M.S., West Virginia. Machine design, manufacturing, bioengineering.
Oscar Orringer, Research Associate; Sc.D., MIT. Aircraft and ground vehicle analysis.
A. Benjamin Perlman, Professor; Ph.D., Lehigh. Vehicle dynamics, applied mechanics, finite-element analysis.
Livia M. Racz, Assistant Professor; Ph.D., MIT. Transport phenomena, materials processing.
Chris Rogers, Associate Professor; Ph.D., Stanford. Experimental fluid dynamics, science education.
Anil Saigal, Professor; Ph.D., Georgia Tech. Advanced materials, manufacturing processes, quality control.
Leslie S. Schneider, Research Assistant Professor; Ph.D., Stanford. Networking of manufacturing facilities.
Michael E. Wiklund, Adjunct Associate Professor; M.S., Tufts. Software interfaces.
Peter Y. Wong, Research Assistant Professor; Ph.D., Tufts. Thermal materials processing, radiative heat transfer.
Michael Zimmerman, Adjunct Associate Professor; Ph.D., Pennsylvania. Thermal manufacturing.

UNIVERSITY OF BRIDGEPORT

School of Engineering and Design
Programs in Engineering

Programs of Study

Intended for individuals who wish to enhance their expertise with an emphasis on professional applications in the engineering disciplines, the Master of Science (M.S.) is offered in computer engineering, electrical engineering, management engineering, and mechanical engineering. The program in computer engineering requires completion of 33 credit hours of study; the programs in electrical engineering, management engineering, and mechanical engineering require 30 credit hours. Completion of all of these programs requires at least one of the following: a comprehensive examination, a written thesis based on independent research, or completion of an appropriate special project.

The computer engineering curriculum includes computer architecture, advanced programming in C, computer networking, digital signal processing, VLSI design, logic synthesis with VHDL, and electives featuring more advanced topics.

The electrical engineering program features communications, controls, electronics, signal processing, and electromagnetics.

The mechanical engineering program focuses on solid mechanics, structural dynamics, fluid mechanics, heat transfer, composite materials, and aerospace-related subjects.

The management engineering program features a flexible curriculum tailored to individual needs, with half of the courses involving elective study in the sciences, engineering, technology management, business concepts, and organizational systems.

In addition, the School also offers the M.S. in the allied field of computer science, which includes the study of advanced C programming, advanced algorithms and data structures, operating systems, database design, and theory of computation. The M.S. can also be pursued on weekends in fewer than 18 months at the University of Bridgeport–Stamford (UB-Stamford) campus.

Research Facilities

The computing facilities at the University of Bridgeport are among the best available. SPARC, digital design, digital signal processing, mixed-signal, image sequence, and microprocessor laboratories are among those available to students in the programs. Laboratory facilities for research provide a complement and include the following areas of focus: communication systems, microwaves, power electronics, digital signal processing, computer-aided design, materials testing, and thermofluids. The University's Wahlstrom Library contains approximately 275,000 bound volumes (including bound journals and indexes) and more than 1 million microforms and subscribes to more than 1,700 periodicals and other serials. Online databases available from off campus via the library's Internet Web site include more than sixty databases in OCLC's FirstSearch that cover all subject areas, EBSCOhost's Academic Search FullTEXT Elite, UMI's ProQuest Direct ABI Global, and MANTIS. Online databases available throughout campus from the library's Web site include LEXIS-NEXIS Academic Universe, Moody's Company Data Direct, Britannica Online, and STAT-USA. CD-ROM databases in the library include ERIC, MEDLINE, ALT HealthWatch, Allied & Alternative Medicine, and Index to Chiropractic Literature. All students have access to e-mail, Netscape, and word processing. Residence halls are wired for individual computer hook-ups. An extension library is maintained at the UB-Stamford campus, with more than 1,000 volumes, more than forty-five periodicals, and extensive electronic access.

Financial Aid

Financial aid is available in the form of the Federal Work-Study Program, Federal Perkins Loans, Federal Stafford Student Loans, graduate assistantships, and internships. The University also hires graduate students as residence hall directors and assistant hall directors. Additional information can be obtained from the Financial Aid Office (203-576-4568). The University also has a long-standing partnership with local corporations, who provide employees with excellent educational opportunities that lead to degrees and career advancement.

Cost of Study

In 1999–2000, tuition is $380 per credit hour in the School of Engineering and Design for students taking up to 12 credits per semester.

Living and Housing Costs

Graduate students may reside either in the University's on-campus residence halls or in private (off-campus) apartments or rooms. The cost of off-campus living varies widely. Additional information related to on-campus residence may be obtained from the Office of Residential Life (203-576-4395).

Student Group

As of fall 1998, there were 65 students enrolled in the engineering programs in the School of Engineering and Design among approximately 1,400 graduate students enrolled at the University. Of the total University graduate population, approximately 51 percent are women, 33 percent are international, and 15 percent are members of minority groups.

Location

The University of Bridgeport's 86-acre campus is situated on Long Island Sound. Both the Bridgeport and Stamford campuses are within easy reach of Westchester County, New York City, and northern New Jersey. Sixty-five percent of Connecticut's largest corporations are located in Fairfield County; these companies provide students with excellent opportunities for jobs, both before and after graduation. University faculty members maintain close relationships with area corporations, school systems, and agencies.

The University

Founded in 1927, the University of Bridgeport is a private, nonsectarian, urban, and comprehensive university. Professional accreditations include those from the ADA, ABA, ACBSP, NASAD, CCE, and ABET. The University's campus is composed of ninety-one buildings of diverse architectural styles. The Bernhard Arts and Humanities Center is a cultural hub, and the Wheeler Recreation Center is a complete recreation and physical fitness facility.

The University's Stamford Center provides convenient access from southern Fairfield County; Westchester County, New York; and northern New Jersey.

Applying

Students are encouraged to apply well in advance of the term they expect to enter but no later than thirty days before the beginning of the semester. Applications are accepted for fall, spring, and summer semesters. Electronic applications are welcome through the University's Web site (address below).

Correspondence and Information

For general inquiries:
Office of Admissions
University of Bridgeport
126 Park Avenue
Bridgeport, Connecticut 06601
Telephone: 203-576-4552
      800-EXCEL-UB (toll-free)
Fax: 203-576-4941
E-mail: admit@bridgeport.edu
World Wide Web: http://www.bridgeport.edu

For additional information:
School of Engineering and Design
Charles A. Dana Hall of Science
University of Bridgeport
169 University Avenue
Bridgeport, Connecticut 06601
Telephone: 203-576-4111
Fax: 203-576-4766
E-mail: kristie@bridgeport.edu
      deptcse@bridgeport.edu

University of Bridgeport

THE FACULTY AND THEIR RESEARCH

Paul Bauer, Associate Professor of Mechanical and Management Engineering (Chair); Ph.D., Oklahoma State. Technology management, strategic planning, marketing, energy studies, and the thermal sciences.

Julius Dichter, Assistant Professor of Computer Science and Engineering; Ph.D., Connecticut. Neural networks, parallel processing.

Stephen Grodzinsky, Professor of Computer Science and Engineering (Chair); Ph.D., Illinois. Digital design, logic synthesis, VLSI design, and microelectronics.

Lawrence Hmurcik, Associate Professor of Electrical Engineering; Ph.D., Clarkson. Digital design, electronic properties of materials, and fiber optics.

Wenelin Janeff, Professor of Electrical Engineering (Chair); Ph.D., Dresden (Germany). Power electronics, communications, and signal processing.

Gonhsin Liu, Associate Professor of Computer Science and Engineering; Ph.D., SUNY at Buffalo. Signal processing, image processing, computer vision, and UNIX programming.

Douglas Lyon, Assistant Professor of Computer Science and Engineering; Ph.D., RPI. Computer-generated music, diffraction rangefinding, image sequencing, and signal processing.

Austif Mahmood, Associate Professor of Computer Science and Engineering; Ph.D., Washington State. Algorithms, computer architecture, parallel VLSI simulation, compiler design, and numerical methods.

Steven Malary, Associate Professor of Mechanical Engineering and Director of the School; Ph.D., Brown. Combustion chemistry and diffusion flames.

Valluru Rao, Professor of Computer Science; D.Sc., Washington (St. Louis). Fuzzy logic, neural networks, and programming languages.

Tarek Sobh, Associate Professor of Computer Science and Engineering; Ph.D., Pennsylvania. Control and simulation of electromechanical systems, parallel architecture, reverse engineering, and robotics.

Tienko Ting, Visiting Professor of Mechanical Engineering (Acting Chair); Ph.D., Michigan. Structural dynamics and damage detection.

UNIVERSITY OF CALIFORNIA, BERKELEY

Graduate Group in Applied Science and Technology

Programs of Study

The program in Applied Science and Technology operates under the auspices of the College of Engineering's Interdisciplinary Studies Center. The program has three major areas of emphasis: applied physics, engineering science, and mathematical sciences. Faculty members associated with the program are drawn from several departments within the College of Engineering as well as from the Departments of Physics, Chemistry, Chemical Engineering, and Mathematics. Topics of interest include the novel properties and applications of nanostructures, thin films and interface science, microelectromechanical systems (MEMS), short wavelength coherent radiation, X-ray microimaging for the life and physical sciences, plasma physics and plasma-assisted materials processing, laser-induced chemical processes, laser probing of complex reacting systems, ultrafast phenomena, particle accelerators, nonlinear dynamics, chaotic systems, numerical methods, and topics in computational fluid mechanics and reacting flows. This program awards Ph.D. degrees.

Research Facilities

Graduate research in the AS&T program profits greatly from the multitude of state-of-the-art experimental facilities on the UC Berkeley campus and at the adjacent Lawrence Berkeley National Laboratory. Among these facilities are the National Center for Electron Microscopy, with the world's highest resolution high-voltage microscope and a microfabrication lab for student work involving lithography, ion-implantation, and thin-film deposition; an integrated sensors laboratory; femtosecond laser laboratories; optical, electrical, and magnetic resonance spectroscopies; short wavelength laser and X-ray research laboratories; an unparalleled variety of material, chemical, and surface science analytic equipment; and a soft X-ray synchrotron radiation facility dedicated to materials, chemical, and biological research using high-brightness and partially coherent X-rays. The interdisciplinary, collaborative nature of the AS&T Program provides ample opportunity to develop new research directions by making the best use possible of these facilities and of the other research instrumentation available to AS&T faculty members.

Financial Aid

Students are encouraged to apply for extramural (e.g., NSF) fellowships as well as for various others administered by the University. Research assistantships are available in projects supported by extramural grants or contracts. An effort is made to align the research assistant's interests with those of a faculty member working with extramural support.

Cost of Study

Fees, insurance, and tuition for 1999–2000 total $4408.50 for California residents and $14,730.50 for nonresidents.

Living and Housing Costs

Room and board in the San Francisco Bay area for the 1999–2000 nine-month academic year average $8802. Books and supplies total about $578, and entertainment and miscellaneous expenses are about $2200. Costs are proportionately higher for the twelve-month academic year and are subject to change.

Student Group

Approximately 30,000 students, including close to 9,000 graduate students, are enrolled at Berkeley. The cosmopolitan student body includes about 1,900 international students from nearly ninety countries and about 3,500 students from states other than California.

Location

The University is located at the base of the Berkeley hills, looking across the San Francisco Bay to San Francisco. The San Francisco Bay Area has a tremendous variety of cultural and entertainment activities to suit all tastes and interests. Students have ready access to the Pacific Coast and beaches, and excellent skiing areas can be found 3½ hours east in the Sierra Nevada. The climate is cool in the summer, with no rainfall. Winters are mild, with intermittent rainfall and sunny days. It is an excellent working climate.

The University

The Berkeley campus of the University of California is the system's parent campus. It has an enrollment of 30,000, with 8,700 graduate students in 100 fields of study, and is noted for the academic distinction of its faculty, the high quality and wide scope of its research activities, and the variety and vitality of student activities. It is generally ranked by its academic peers as one of the best graduate institutions in the United States.

Applying

Complete applications, including official transcripts, GRE scores, three letters of reference, and a statement of academic and professional goals, are due January 3 for the following fall. To obtain application forms, students should contact the address below.

Correspondence and Information

Applied Science and Technology
230 Bechtel Engineering Center, 1708
University of California
Berkeley, California 94720-1708
Telephone: 510-642-8790
Fax: 510-643-6103
E-mail: ast-program@coe.berkeley.edu
World Wide Web: http://www.coe.berkeley.edu/ast/

University of California, Berkeley

FACULTY AND THEIR RESEARCH

Paul Alivisatos, Chemistry (UCB). Semiconductor nanocrystals, synthesis of crystallites consisting of a few hundred to tens of thousands of atoms, spectroscopy of clusters. (alivis@uclink4.berkeley.edu)

David T. Attwood, Applied Science and Technology (UCB), Center for X-ray Optics (LBNL). Partially coherent radiation at short wavelengths; synchrotrons; undulators; X-ray optics, microscopes, and holography; applications for the life and physical sciences. (attwood@eecs.berkeley.edu)

Stanley A. Berger, Mechanical Engineering (UCB). Physiological fluid mechanics, particular flow in curved blood vessels and in the microcirculation. (saberger@me.berkeley.edu)

Charles K. Birdsall, Electrical Engineering (UCB). Plasma theory and simulation, including plasma processing of materials; relativistic electron beam interactions with slow and fast waveguides. (birdsall@eecs.berkeley.edu)

Jeffrey Bokor, Electrical Engineering and Computer Science (UCB), Center for X-ray Optics (LBNL). Nanostructure device physics and fabrication, X-ray optics and lithography. (jbokor@eecs.berkeley.edu)

Van P. Carey, Mechanical Engineering (UCB), Applied Science Division (LBNL). Thermophysics of multiphase systems, molecular simulation modeling of microscale transport processes, statistical mechanics of microscale systems. (vcarey@me.berkeley.edu)

C. J. Chang-Hasnain, Electrical Engineering and Computer Science (UCB). Semiconductor lasers and optoelectronic devices, III-V compound epitaxy and processing, applications in novel optical systems. (cch@eecs.berkeley.edu)

Daniel Chemla, Physics (UCB), Materials Sciences Division (LBNL). Quantum size effects and optical properties of semiconductor nanostructures, time-resolved scanning tunneling microscopy, near-field optical microscopy. (dschemla@lbl.gov)

Nathan W. Cheung, Electrical Engineering and Computer Science (UCB). Integrated-circuit processing and electronic materials. (cheung@eecs.berkeley.edu)

Daryl Chrzan, Materials Science and Mineral Engineering (UCB). Computational materials science, with emphases on dislocation dynamics and the growth of thin films. (dcchrzan@socrates.berkeley.edu)

Leon O. Chua, Electrical Engineering and Computer Science (UCB). Nonlinear circuits, nonlinear systems, bifurcation theory, chaos, neural networks, CAD and nonlinear electronics. (chua@eecs.berkeley.edu)

Paul Concus, Mathematics (UCB), Physics Division (LBNL). Mathematical, computational, and experimental study of fluid interfaces under microgravity conditions. (concus@lbl.gov)

Didier de Fontaine, Materials Science and Mineral Engineering (UCB), Materials Sciences Division (LBNL). Thermodynamics of solids, quantum and statistical mechanics of alloy phase stability. (ddf@isis.berkeley.edu)

Lutgard C. De Jonghe, Materials Science and Mineral Engineering (UCB), Center for Advanced Materials (LBNL). Ceramic processing; particulate composites of ceramics with polymer, metal, or ceramic matrix; microstructure characterization. (dejonghe@lbl.gov)

Robert W. Dibble, Mechanical Engineering (UCB). Combustion diagnostics, laser diagnostics in reactive flows, gas-phase chemical kinetics, pollution studies. (rdibble@me.berkeley.edu)

Roger W. Falcone, Physics (UCB). Quantum electronics and short wavelength coherent light sources applied to atomic physics, solid-state physics, and plasma physics. (rwf@physics.berkeley.edu)

Michael Frenklach, Mechanical Engineering (UCB). Chemical vapor deposition of thin films, diamond synthesis, nucleation of interstellar dust. (myf@me.berkeley.edu)

Ronald Gronsky, Materials Science and Mineral Engineering (UCB), National Center for Electron Microscopy (LBNL). Identification of atomic structure of interfaces and defects in materials using techniques of electron microscopy. (rgronsky@socrates.berkeley.edu)

F. Alberto Grunbaum, Mathematics (UCB). Image reconstruction; tomographic methods in medicine, geophysics, and nondestructive evaluation; solitons; signal processing. (grunbaum@math.berkeley.edu)

T. Kenneth Gustafson, Electrical Engineering and Computer Science (UCB), Center for X-ray Optics (LBNL). Modern optics and quantum electronic techniques; nonlinear phenomena, thresholding, and logic devices; X-rays and nanostructure fabrication. (tkg@eecs.berkeley.edu)

Eugene E. Haller, Materials Science and Mineral Engineering (UCB), Materials Sciences Division (LBNL). Semiconductors: thin-film growth, electronic and structural properties of defects and impurities. (eehaller@lbl.gov)

Charles Harris, Chemistry (UCB), Chemical Sciences Division (LBNL). Transport properties and femtosecond dynamics of electrons in disordered media. (harris@socrates.berkeley.edu)

Roger T. Howe, Electrical Engineering and Computer Science (UCB). Silicon micromachining processes, modeling of microdynamic devices and systems. (howe@eecs.berkeley.edu)

Tsu-Jae King, Electrical Engineering and Computer Science (UCB). Accelerator Fusion Research (LBNL). Integrated circuit devices and technology and thin-film technology for larger area electronics; microelectromechanical devices (MEMS). (tking@eecs.berkeley.edu)

Harold Lecar, Molecular and Cell Biology (UCB). Neurobiology, ionic channels and cell membranes. (hlecar@uclink4.berkeley.edu)

Allan J. Lichtenberg, Electrical Engineering and Computer Science (UCB). Nonlinear dynamics; plasma confinement, heating, and fusion; plasma discharges for materials processing. (ajl@eecs.berkeley.edu)

Michael A. Lieberman, Electrical Engineering and Computer Science (UCB). Plasma-assisted materials processing and processing discharges. (lieber@eecs.berkeley.edu)

Dorian Liepmann, Mechanical Engineering (UCB). Experimental biofluid mechanics, vortex dynamics, free-surface flows, hydroacoustics. (liepmann@me.berkeley.edu)

Roya Maboudian, Chemical Engineering (UCB). Surface science of semiconductor materials, micromachining (MEMS). (maboudia@socrates.berkeley.edu)

Philip S. Marcus, Mechanical Engineering (UCB). Bifurcations and development of chaotic flows, numerical simulation of three-dimensional fluid flow, vortex dynamics, numerical algorithms for applications to astrophysics and geophysics. (phil@marpc.berkeley.edu)

C. Bradley Moore, Chemistry (UCB). Unimolecular reaction dynamics of highly vibrational excited molecules—transition states and energy flow using laser, spectroscopic and dynamical methods; dynamics in chemical and biological systems. (dbmoore@socrates.berkeley.edu)

Richard S. Muller, Electrical Engineering and Computer Science and Director, Berkeley Sensor & Actuator Center (UCB). Solid-state microsensors and microactuators, materials and processes for micromechanics, micromechanical structures. (muller@eecs.berkeley.edu)

Andrew R. Neureuther, Electrical Engineering and Computer Science (UCB). Photolithography and integrated circuit process technology simulation. (neureuth@eecs.berkeley.edu)

William G. Oldham, Electrical Engineering and Computer Science (UCB) and Director, Electronics Research Laboratory. Integrated circuit process technology, microstructure fabrication, and modeling. (oldham@eecs.berkeley.edu)

Kristofer S. J. Pister, Electrical Engineering and Computer Science (UCB). Microelectromechanical systems (MEMS), microrobotics, CAD for MEMS. (pister@eecs.berkeley.edu)

Jeffrey Reimer, Chemical Engineering (UCB). Magnetic resonance studies of surfaces, polymers and defects in semiconductors. (reimer@socrates.berkeley.edu)

Timothy D. Sands, Materials Science and Mineral Engineering (UCB). Thin-film materials and devices, nanostructures, artificially structured materials, information storage media and devices. (sands@uclink.berkeley.edu)

Peter G. Schultz, Chemistry (UCB). Synthesis of novel materials.

Charles V. Shank, Chemistry, Electrical Engineering and Computer Science, and Physics (UCB) and Director, Lawrence Berkeley National Laboratory. Investigation of ultrafast phenomena in physics, chemistry, and biology using femtosecond optical pulse techniques. (cvshank@lbl.gov)

Andrew J. Szeri, Mechanical Engineering (UCB). Convective/diffusive transport, nonlinear dynamics, perturbation methods. (aszeri@me.berkeley.edu)

Eicke R. Weber, Materials Science and Mineral Engineering (UCB), Materials Sciences Division (LBNL). Defects in bulk and thin-film semiconductors, electronic and structural properties and their influence on device performance. (weber@socrates.berkeley.edu)

Alan Weinstein, Mathematics (UCB). Symplectic geometry, Hamiltonian structures, stability, connections between classical and quantum mechanics. (alanw@math.berkeley.edu)

K. Birgitta Whaley, Chemistry (UCB). Theory of atomic and molecular clusters, nanostructures, and high-energy density materials. (whaley@holmium.cchem.berkeley.edu)

Richard M. White, Electrical Engineering and Computer Science (UCB). Ultrasonics, micromachined sensing and actuation devices for physical, chemical, and biological applications. (rwhite@eecs.berkeley.edu)

UNIVERSITY OF CALIFORNIA, IRVINE

School of Engineering

Programs of Study

The School of Engineering offers programs of study leading to the degrees of M.S. and Ph.D. in the following fields: chemical and biochemical engineering; civil engineering; electrical and computer engineering with concentrations in computer networks and distributed computing, computer systems and software, and electrical engineering; engineering with concentrations in environmental engineering and materials science and engineering; and mechanical and aerospace engineering. An interdepartmental concentration in biomedical engineering began in fall 1999.

Chemical and biochemical engineering focuses on bioremediation, cellular growth kinetics and regulation, materials science and engineering, optimization and control of reactors, protein engineering, recombinant DNA technology, and separations. Civil engineering emphasizes structural and earthquake engineering, transportation systems engineering, and water resources and environmental engineering. Electrical and computer engineering includes study in the areas of computer engineering, optical and solid-state devices, and systems engineering and signal processing. Environmental engineering addresses the development of strategies to control anthropogenic emissions of pollutants to the atmosphere, waterways, and terrestrial environment; remediation of polluted natural systems; design of technologies to treat waste; and the evaluation of contaminant fate in urban environments. Materials science and engineering focuses on electronic and photonic materials, structure of materials, mechanics of solids, and chemical processing of materials. Areas emphasized in mechanical and aerospace engineering include fluid and thermal sciences, combustion and propulsion, systems and design (including control and robotics), and aerospace engineering. Biomedical engineering involves biophotonics, biomedical MEMS, biomedical nanoscale systems (including DNA microchip technology), and biomedical computational technology.

Faculty research activities are noted on the reverse side of this page. Further information may be obtained by writing the chairperson of the appropriate department or, for materials science and engineering, environmental engineering, and biomedical engineering, contacting the Graduate Affairs Office at the address below.

Interdisciplinary research units affiliated with the School of Engineering are the Institute for Combustion and Propulsion Science and Technology, the Institute of Transportation Studies, and the National Fuel Cell Research Center.

The faculty of the School is a well-blended group of distinguished senior faculty—18 members have achieved Fellow grade in their respective professional societies—and promising junior faculty—7 members have been named National Science Foundation (NSF) Young Investigators, or have received the NSF Career Development Award. As evidence of the School's quality and innovation, *U.S. News & World Report* ranked the School "Up and Coming" for three years in a row and more recently listed the School as one of the top fifty graduate schools of engineering in the nation.

Research Facilities

The School, which occupies eight buildings, has approximately 82,000 square feet of well-equipped research space, including a bioreactor laboratory, a bioseparation laboratory, a recombinant cell laboratory, a molecular beam epitaxial growth facility, a wind tunnel facility, and laboratories for structural dynamics, hydraulics, soil/water physics, water quality, VLSI design automation, parallel distributed computing, image processing, quantum electronics and optics, integrated optics, microfabrication, distributed real-time microcomputing, combustion and propulsion, turbulence, robotics, controls, and materials. All the departments have facilities for analytic and computational research using networked Apollo Workstations, Sun workstations, microcomputer labs, microprocessor development systems, multiprocessor systems, a central VAX/VMS system, a Convex high-performance computer, and the University of California supercomputer.

Financial Aid

Fellowships and teaching and research assistantships are available on a competitive basis. Except for students on visas, there are opportunities for part-time work in the engineering community of Orange County. With the same exception, financial aid may be obtained from UCI's Financial Aid Office.

Cost of Study

In 1999–2000, student fees are $1726 per quarter for California residents and an additional $3441 per quarter for nonresidents. These fees are subject to change.

Living and Housing Costs

On-campus housing is available. In 1999–2000, monthly apartment rents are from $277 to $725 for single students and from $554 to $1202 for married students and families. Early application is advised for on-campus housing. Privately owned apartments are available close to the campus, and many types of housing can be found in the surrounding communities of Santa Ana, Newport Beach, Costa Mesa, Irvine, Tustin, and Laguna Beach.

Student Group

Current campus enrollment is 18,209, including 1,263 undergraduate and 309 graduate students in the School of Engineering.

Student Outcomes

Graduates of the degree programs offered within the School of Engineering hold positions in academia or in the industrial or governmental sectors that involve the development of new technologies for the benefit of society. Many graduates take various professional positions in local, national, and international high-tech companies. Others obtain faculty appointments or research and development positions in their areas of specialization.

Location

The 1,510-acre UCI campus is in Orange County, 40 miles south of Los Angeles. Irvine is one of the nation's fastest-growing residential, industrial, and business areas, yet within view of the campus is a wildlife sanctuary; Pacific Ocean beaches are nearby. Residential areas range from the beach communities of Newport Beach and Laguna Beach to the socially and economically diverse urban centers of Santa Ana, Tustin, and Costa Mesa.

The University

One of the nine campuses in the University of California system, UCI now enrolls 3,638 graduate and professional students. The University offers graduate degrees through the Schools of Biological Sciences, Engineering, Fine Arts, Humanities, Physical Sciences, Social Ecology, and Social Sciences; the Graduate School of Management; the College of Medicine; and the Department of Information and Computer Science. The Department of Education offers courses and training leading to California teaching credentials.

Applying

Application forms and general information may be obtained by writing to the School of Engineering. The deadlines for applications are May 1 for the fall quarter, October 15 for the winter quarter, and January 15 for the spring quarter. Applicants who wish to be considered for fellowships or for teaching or research assistantships should apply by February 1. Applicants must submit official records covering all postsecondary academic work, three letters of recommendation, and official scores on the General Test of the Graduate Record Examinations. International students whose native language is not English must submit the results of the Test of English as a Foreign Language (TOEFL).

Correspondence and Information

Department of (specify)
School of Engineering
University of California
Irvine, California 92697

Graduate Student Affairs
School of Engineering
104 Engineering and Computing Trailers
University of California
Irvine, California 92697-2750

Telephone: 949-824-3562
E-mail: jdsommer@uci.edu
World Wide Web: http://www.eng.uci.edu

University of California, Irvine

THE FACULTY AND THEIR RESEARCH

Department of Chemical and Biochemical Engineering and Materials Science

Ying C. Chang, Ph.D.: molecular engineering, biophysical chemistry, polymer interfacial chemistry and physics, organic vapor deposition and thin film fabrication. Nancy A. DaSilva, Ph.D.: recombinant cell technology. James C. Earthman, Ph.D.: fatigue behavior and cyclic damage, defect monitoring techniques. Steven C. George, M.D., Ph.D.: physiological modeling, tissue engineering, pulmonary gas exchange, pulmonary metabolism and transfer properties of nitric oxide. G. Wesley Hatfield, Ph.D.: molecular mechanisms of biological control systems. Juan Hong, Ph.D.: biochemical reaction processes and separation processes. Enrique J. Lavernia, Chair, Ph.D.: manufacturing, composite materials, nanostructured materials, modeling, synthesis by nonequilibrium methods. Henry C. Lim, Ph.D.: bioreactor and bioreaction engineering. Martha L. Mecartney, Ph.D.: microstructure of materials. Farghalli A. Mohamed, Ph.D.: mechanical behavior of materials, creep, superplasticity, strengthening mechanisms. Frank G. Shi, Ph.D.: semiconductor processing and modeling, interconnect and packaging polymers, amorphous and nanocrystalline materials, thermodynamics and kinetics of nucleation/crystallization/glass transition processes. Vasan Venugopalan, Ph.D.: biomedical laser applications; fundamentals of laser-induced thermal, mechanical, and radiative transport phenomena.

Department of Civil and Environmental Engineering

Alfredo H-S. Ang, Ph.D.; PE: structural and earthquake engineering, risk and reliability engineering. Constantinos V. Chrysikopoulos, Ph.D.: subsurface solute transport, nonaqueous phase liquid dissolution in porous media, mathematical modeling. Maria Q. Feng, Ph.D.: structural engineering, intelligent control of structural systems. Stanley B. Grant, Ph.D.: environmental engineering, coagulation and filtration of colloidal contaminants, environmental microbiology. Gary L. Guymon, Ph.D.; PE: water resources, groundwater, modeling uncertainty. Medhat A. Haroun, Ph.D.; PE: numerical and experimental modeling of the seismic behavior of structural systems for design, retrofit and repair; systems evaluated are liquid storage tanks, bridge-supporting columns and piers, and concrete masonry/composite buildings. R. Jayakrishnan, Ph.D.: transportation systems analysis. Michael G. McNally, Ph.D.: travel behavior, activity-based approaches, transportation planning and modeling, transportation and land use, computer applications in transportation. Terese M. Olson, Ph.D.: environmental aquatic chemistry, colloidal processes, pollutant transformation processes in natural and water treatment systems. Gerard C. Pardoen, Ph.D.; PE: structural dynamics, including ambient and forced vibration testing of large- and small-scale structures; static and dynamic testing of timber, steel, and concrete components with applications to earthquake engineering; nonlinear finite element analysis of lightweight, composite material structures. Wilfred W. Recker, Ph.D.: transportation engineering, demand analysis, intermodal transfer analysis, urban planning and transportation interaction. Amelia C. Regan, Ph.D.: logistics, freight and fleet management, intermodal transportation systems. Stephen G. Ritchie, Chair, Ph.D.: transportation systems engineering, advanced traffic management and control systems, development and application of emerging technologies in transportation. Brett F. Sanders, Ph.D.: environmental and computational fluid dynamics, flood mitigation, urban runoff, adjoint methods. Jan Scherfig, Ph.D.; PE: water resources and water quality, effect of low temperature on waste treatment processes; modeling, water reclamation and reuse. Robin Shepherd, Ph.D., D.Sc.; PE: structural engineering, structural dynamics analysis of buildings, bridges, and aerospace structures. Roberto Villaverde, Ph.D.; PE: earthquake engineering, nonlinear structural mechanics; passive structural control; seismic response of nonstructural components; earthquake ground motion characterization. Jann N. Yang, D.Sc.; PE: fatigue, reliability, maintainability, and control of structures.

Department of Electrical and Computer Engineering

Nicolaos G. Alexopoulos, Dean; Ph.D.: integrated microwave and millimeter-wave circuits and antennas, substrate materials and thin films, electromagnetic theory. Nader Bagherzadeh, Chair, Ph.D.: parallel processing, computer architecture, VLSI design. Casper W. Barnes Jr., Ph.D.: digital signal processing. Harut Barsamian, M.S.: computer architectures, software engineering. Neil J. Bershad, Ph.D.: communication and information theory, signal processing. Lubomir Bic, Ph.D.: parallel processing, distributed systems, database machines. Douglas M. Blough, Ph.D.: parallel processing, fault-tolerant computing, computer architecture. Lynn Choi, Ph.D.: computer architecture, microprocessor design, optimizing compilers. Rui J. P. de Figueiredo, Ph.D.: intelligent sensing and control, applied mathematics. Franco De Flaviis, Ph.D.: wireless communications devices and systems. Nikil Dutt, Ph.D.: design modeling, languages and synthesis, CAD tools, computer architecture. Daniel D. Gajski, Ph.D.: parallel algorithms and architectures, design methodology, design science, CAD algorithms and tools, software/hardware codesign. Hideya Gamo, D.Sc.: quantum electronics, electromagnetics, optics. Michael Green, Ph.D: analog integrated circuit design, circuit simulation, nonlinear circuits. Glenn E. Healey, Ph.D.: machine vision, computer engineering, image processing, computer graphics, intelligent machines. Daniel Hirschberg, Ph.D.: analysis of algorithms, data structures, models of computation. K. H. (Kane) Kim, Ph.D.: ultrareliable distributed and parallel computing, real-time object-based system engineering. Fadi J. Kurdahi, Ph.D.: VLSI system design, design automation of digital systems. Tomas Lang, Ph.D.: numerical processors and multiprocessors, parallel computer systems. Chin C. Lee, Ph.D.: electronic packaging, thermal management, integrated optics, photonics. Henry P. Lee, Ph.D.: optoelectronics, semiconductor materials and devices. Guann-Pyng Li, Ph.D.: high-speed semiconductor technology, optoelectronic devices, integrated circuit fabrication and testing. Kwei-Jay Lin, Ph.D.: real-time systems, distributed systems. Orhan Nalcioglu, Ph.D.: nuclear magnetic resonance imaging and spectroscopy, digital radiography. Richard D. Nelson, Ph.D.: sensors, microelectronics, photonics, medical imaging. Alexandru Nicolau, Ph.D.: architecture, parallel computation, programming languages and compilers. Robert M. Saunders, Dr.Eng.; PE: electromechanics, power systems. Issac D. Scherson, Ph.D.: parallel computing architectures, massively parallel systems, parallel algorithms, interconnection networks, performance evaluation. Roland Schinzinger, Ph.D.; PE: electromagnetics, power systems, operations research. Phillip C.-Y. Sheu, Ph.D.: robotics, database systems. Jack Sklansky, D.Sc.; PE: pattern recognition, machine vision, medical imaging, neural learning, computer engineering. Keyue M. Smedley, Ph.D.: power electronics. Allen R. Stubberud, Ph.D.; PE: control systems, digital signal processing, estimation and optimization. Tatsuya Suda, Ph.D.: computer networks, distributed systems, performance evaluations. Harry H. Tan, Ph.D.: communication systems, information theory, coding theory, stochastic processes. Chen S. Tsai, Ph.D.: integrated and fiber-optic devices and materials, acoustooptics, magnetooptics, acoustic microscopy. Wei Kang (Kevin) Tsai, Ph.D.: data communication networks, neural networks, parallel algorithms and architectures, CAD for VLSI systems engineering.

Department of Mechanical and Aerospace Engineering

James E. Bobrow, Ph.D.: dynamic systems and control, robotics, fluid power. Haris J. Catrakis, Ph.D.: turbulence and mixing at high Reynolds numbers, flow control for aerospace and marine vehicles. Donald Dabdub, Ph.D.: mathematical modeling of air pollution dynamics by parallel computation. Derek Dunn-Rankin, Ph.D.: combustion, aerosol sizing and transport, laser diagnostics and spectroscopy. Donald K. Edwards, Ph.D.; PE: heat and mass transfer. Said E. Elghobashi, Chair, Ph.D.: direct simulation of turbulent, chemically reacting, and dispersed two-phase flows. Carl A. Friehe, Ph.D.: fluid mechanics, turbulence, micrometeorology, instrumentation. Faryar Jabbari, Ph.D.: robust and nonlinear control theory, adaptive parameter identification. John C. LaRue, Associate Dean for Student Affairs, Ph.D.: fluid mechanics, heat transfer, turbulence, microelectromechanical systems, instrumentation. Feng Liu, Ph.D.: computational fluid dynamics, aerodynamics, and turbo machines. J. Michael McCarthy, Ph.D.: kinematic theory of spatial motion, design of mechanical systems, cooperating robots. Kenneth D. Mease, Ph.D.: air and spacecraft guidance and control, geometric nonlinear control. Melissa E. Orme, Ph.D.: droplet dynamics, fluid mechanics of materials synthesis, net-form manufacturing. Dimitri Papamoschou, Ph.D.: compressible mixing and turbulence, supersonic jet noise reduction, diagnostics for compressible flow, acoustics in moving media, respiratory fluid mechanics. Roger H. Rangel, Ph.D.: fluid mechanics and heat transfer of multiphase systems including spray combustion, atomization, and metal spray solidification; applied mathematics. David J. Reinkensmeyer, Ph.D.: biomedical engineering robotics. G. Scott Samuelsen, Ph.D.; PE: energy, propulsion, combustion, and environmental conflict; turbulent transport in complex flows, spray physics, NO_x and soot formation, laser diagnostics, and experimental methods; application of engineering science to practical propulsion and stationary systems; environmental ethics. William E. Schmitendorf, Associate Dean for Academic Affairs, Ph.D.: control theory and applications. Athanasios Sideris, Ph.D.: control systems, neural networks. William A. Sirignano, Ph.D.: combustion theory and computational methods, multiphase flows, turbulent reacting flows, flame spread.

UNIVERSITY OF CALIFORNIA, LOS ANGELES

School of Engineering and Applied Science

Programs of Study

The School of Engineering and Applied Science offers courses leading to the degrees of Master of Science, Engineer, and Doctor of Philosophy. The School is divided into six departments, which encompass the major engineering disciplines. These are chemical engineering; civil and environmental engineering; computer science; electrical engineering; materials science and engineering; and mechanical and aerospace engineering.

The Interdepartmental Degree Program in Biomedical Engineering also offers the Master of Science and Doctor of Philosophy degrees. The Master of Science degree program consists of nine courses of graduate and upper-division work in addition to a thesis or a comprehensive examination. The basic program of study for the Ph.D. is usually built around one major field and two minor fields. Ph.D. candidates are permitted to propose ad hoc fields when the established fields do not meet their educational objectives. A five-course graduate program leading to a Certificate of Specialization is offered in forty-five fields of engineering and applied science.

Research Facilities

Laboratory facilities are available for research in such areas as computer design and applications, circuits, electronics, solid-state devices, holography and modern optics, electromagnetics, plasma engineering, air pollution, metal corrosion, chemical processes, catalysis, combustion, cryogenics, electrochemistry, heat transfer, nuclear engineering, city and regional planning, biotechnology, sanitary engineering, control systems, communications, ceramics, electron microscopy, materials, metallography, welding and X-ray studies, earthquake engineering, fluid mechanics, high-speed aerodynamics and propulsion, soil mechanics, structures, subsonic wind tunnels, medical engineering, and electrical and mechanical standards. The School of Engineering and Applied Science operates an advanced computer network of UNIX machines built around a variety of IBM, Sun, and HP servers and workstations. Combined with dedicated NFS servers, this UNIX-based network provides cycle, file services, and general network services for more than 1,000 workstations and servers of all types. The School also maintains a computer classroom and five open computer labs with approximately 175 workstations (Pentium PCs and X-terminals) where computerized instruction occurs. The Office of Academic Computing supports a cluster of IBM RISC System/6000 servers and a multiprocessor SP/2 system. Network access to campus, regional, and national networks is also provided. The UCLA Library, which ranks in the top three nationally, has more than 6.6 million volumes. One of its specialized branches, the Science and Engineering Library (SEL), contains more than 460,000 volumes and receives more than 7,000 serials, as well as more than 1.9 million technical reports. SEL provides major access to library materials through ORION, the UCLA online information system, and Melvyl, the UC (nine campuses) online system.

Financial Aid

Fellowships are available for students devoting full time to study and research. Stipends range from $7500 to $27,000 in 1999-2000 Teaching assistantships, starting at $13,329 for approximately half-time teaching assistance for the academic year, are also offered. Research assistantships are available; although the stipends vary, the net earnings are approximately the same as for teaching assistantships. In addition, financial support is available through government and privately sponsored fellowships.

Cost of Study

For 1999–2000, California residents pay $1518 per quarter in registration and incidental fees. These fees include mandatory health insurance. Nonresidents pay an additional fee of $3128 per quarter. The academic year consists of three quarters.

Living and Housing Costs

A typical budget for a California resident living in an off-campus apartment is $14,082 per year and includes required books and supplies, board and room for the period classes are in session during the three quarters, and a minimum allowance for miscellaneous items.

Student Group

UCLA's total enrollment is about 35,000. The School of Engineering and Applied Science has approximately 1,000 graduate students and about 2,300 undergraduates.

Location

The greater Los Angeles area is a major site of industrial research, engineering, and business activity—for example, in aerospace, electronics, and petrochemicals. It is a noted center for the performing arts, and it provides cultural and recreational facilities to meet varied tastes and budgets. Skiing, water sports, boating, and hiking are all available nearby.

The University and The School

Ranking academically among the leading universities in the United States, UCLA has attracted distinguished scholars and researchers from all over the world. It shares with eight other campuses the history and prestige of the statewide University of California System. The UCLA campus is located in the Westwood area, set against the Santa Monica Mountains, a few miles from the ocean. Numerous cultural programs, including musical and dramatic productions, are presented on the campus. Among the recreational facilities is the Sunset Canyon Recreation Center, located in the hills of the west campus adjacent to the residence halls. It features two outdoor swimming pools, a clubhouse, multipurpose playing fields, an outdoor amphitheater, and picnic areas.

The School of Engineering and Applied Science, established during the academic year 1968–69, is an outgrowth of the College of Engineering, founded in 1945. The curriculum of the School, covering the basic concepts of science and technology, provides a resource for all UCLA students and makes possible close interaction with the behavioral sciences, humanities, and other professional schools. Dr. A. R. Wazzan is Dean of the School. The faculty has 126 full-time professors and 70 visiting and part-time professorial appointees. The School of Engineering and Applied Science also offers frequent seminars by noted scientists and engineers from the United States and abroad.

Applying

Application for admission to graduate standing, with complete credentials and the application fee, must be filed with Graduate Admissions. Applicants should contact the department for the deadlines for the fall, winter, and spring quarters. Information concerning departmental offerings and applications may be obtained from department chairs. Applicants may obtain graduate information and may submit graduate applications on line at the Web site (http://www.seas.ucla.edu/).

Correspondence and Information

Associate Dean, Student Affairs
School of Engineering and Applied Science
6426 Boelter Hall
University of California
P.O. Box 951601
Los Angeles, California 90095-1601
Telephone: 310-825-1704
E-mail: diane@ea.ucla.edu

University of California, Los Angeles

DEPARTMENT HEADS AND RESEARCH AREAS

Chemical Engineering. S. Senkan, Chair. Areas include (1) reaction engineering–molecular modeling of petroleum processing, gas-to-particle conversion, catalytic oxidation, detailed chemical kinetic mechanisms, chemical vapor deposition, and computational chemistry; (2) statistical mechanics—structure and function of small biological molecules, structure determination of microcrystalline materials, diffusion in disordered microporous materials, and field theories for transport-limited reactions; (3) materials processing in polymers, superconductors, ultrafine particle synthesis, and semiconductor manufacturing; (4) combustion kinetics and hydrodynamics, nonintrusive optical diagnostics in combustion monitoring, chemical kinetics of combustion and incineration of hazardous materials, formation of trace combustion by-products, experimental studies of flames using molecular beam sampling coupled to mass spectrometry and gas chromatography/mass spectrometry, and photoionization time-of-flight mass spectrometry; (5) transport phenomena, movement of pollutants in the environment (i.e., air, water, and soil), and transport at interfaces and at cryogenic temperatures; (6) electrochemistry, reduction of corrosion in marine environments, and electrochemical manufacturing processes; (7) biochemical engineering, biomedical engineering, metabolic engineering, DNA microtechnology, electroenzymology, biosensors, biochemical analysis of useful chemicals, biochemical degradation of contaminants in the environment, and studies of *Archaea* for the treatment of toxic chemicals and for the synthesis of new biomaterials; (8) process design and control, applied mathematics, mass exchange networks, process design for waste minimization, integrated design and control, robust and optimal control of nonlinear systems, model reduction and control of nonlinear distributed parameter systems, simulation and control of fluid flows and particulate process modeling and control; and (9) separations, selective membranes, and adsorption of organic and inorganic chemicals for the treatment of wastewaters.

Civil and Environmental Engineering. M. Stenstrom, Chair. Research areas include (1) geotechnical engineering—analysis of soil behavior under static and dynamic loads, including stress-strain relationships, shear strength, and volume change characteristics; design of retaining structures and embankments; foundation engineering; static and cyclic soil testing; response of soil deposits and foundations to earthquake loads; and soil-structure interaction; (2) structures and mechanics—static and dynamic analysis of structural systems; structural design and optimization; finite-element methods and computational techniques for linear and nonlinear structural analysis, including vibrations, dynamic response, and stability effects; earthquake response and seismic design of structural systems; retrofitting and strengthening of existing structures: criteria assessment and design; experimental structural analysis, including full-scale component testing; failure analysis and structural system identification and damage assessment of structures; analysis of disjoint particle assemblies; damage mechanics, mechanics of composites, micromechanics, fatigue and fracture, plasticity, nondestructive testing and experimental mechanics, service life predictions, constitutive modeling and nonlinear computational mechanics; structural reliability, random vibrations, probabilistic structural dynamics, and stochastic finite elements; (3) water resources—water resources engineering, surface and groundwater hydrology, optimization of water resources systems, contaminant transport in the subsurface environment, remediation of contaminated soil and groundwater, hydroclimatology; and (4) environmental engineering—water and wastewater treatment, physiocochemical and biological processes, environmental organic and inorganic chemistry, water quality assessment, stormwater management, solid and groundwater characterization and remediation, environmental fluid mechanics, environmental microbiology and biotechnology, and mathematical modeling of environmental processes.

Computer Science. R. Muntz, Chair. Theory, design, implementation, and use of computer systems, including (1) theoretical foundations—design and analysis of algorithms, computational complexity, models and algorithms for parallel and distributed computation, online and randomized algorithms, combinatorial optimization algorithms and VLSI/CAD, and automata and formal languages; (2) computer networking—data and multimedia communications; local, campus, and wide area networks; Internet, ATM, optical, and wireless architectures and protocols; performance modeling, analysis, and optimization; network simulation and measurements; computer scheduling and resource allocation; and distributed systems; (3) computer system design—computer system architecture; digital systems; logic design; memory, arithmetic; control; data transmission, reliable fault-tolerant systems, VLSI system design, and input-output systems design; (4) programming languages and systems—general- and special-purpose languages; compilers; systems programming; syntax, semantics, and pragmatics of programming languages; implementation; database-management systems; and parallel and distributed systems; (5) scientific computing—simulation; numerical analysis; optimization; physical systems; biological systems; and interactive systems; and (6) artificial intelligence—natural-language processing, problem solving, knowledge modeling, pattern recognition, cognitive systems, machine perception, and neural models.

Electrical Engineering. William J. Kaiser, Chair. Study covers (1) communications and telecommunications—satellite, spread spectrum, and digital communication systems; estimation, detection, and optimization algorithms; stochastic modeling of telecommunication systems; and information theory; (2) integrated circuits (ICs) and systems—analog and digital ICs, radio-frequency circuits, digital signal processors, microsensors, and application-specific computer ICs; (3) signal processing, speech and image processing systems, multirate signal processing, adaptive filtering, biomedical speech and image processing, and communications signal processing; (4) solid-state electronics—new devices with picosecond switching times and bandwidths up to submillimeter wave ranges; hot-electron and heterojunction transistors, quantum effect devices, and other more conventional devices such as scaled-down MOSFETs and SOI devices; basic materials for solid-state devices; and VLSI; (5) control systems; linear and nonlinear systems analysis; stochastic estimation and filtering; adaptive distributed, nonlinear, optimal, and robust control; guidance and tracking; (6) photonics and optoelectronics—solid-state, gas, and semiconductor lasers; nonlinear optics; optoelectronics; ultrafast optics; photonic devices and systems; fiber optics; and optical communication; (7) electromagnetics—integrated microwave and millimeter-wave circuits, fiber optics, optical signal processing, antenna theory and design, satellite and personal communication antennas, biological interactions, modem antenna measurement and diagnostic techniques, scattering, and electromechanics and nonlinear electrodynamics; (8) plasma electronics—plasma production, confinement, and heating; applications of plasma physics to laser plasma accelerators, plasma materials processing, free-electron lasers, RF accelerators, and large-scale computer simulation of plasmas; (9) operations research—optimization theory, linear and nonlinear programming, numerical methods, nonconvex programming; and stochastic processes and applications to communications and telecommunications.Students should see the two-page description in Section 8 for more details.

Materials Science and Engineering. K.-N. Tu, Chair. Areas are (1) ceramics and ceramic processing—mechanical properties of oxides, glass science, electrical and optical properties of inorganic materials, sol-gel processing and thin films, hybrid organic-inorganic solids, and structure-property relationships of ceramic-matrix composite materials; (2) electronic materials—materials science of thin films, phase transformation in solids, electron microscopy, high-resolution X-ray diffraction, electromigration, irradiation effects on structural materials and strengthening mechanisms in solids, structure of glass and semiconductors, and structure and properties of polymers; and (3) structural materials—physical metallurgy, composite materials, fatigue and high-temperature fracture of metal and carbon-matrix composite materials, mechanical metallurgy, thin films and metallization on microelectronic devices, vapor deposition and materials synthesis, structure-property relationships, nondestructive testing, and materials characterization.

Mechanical and Aerospace Engineering. V. Dhir, Chair. Areas include (1) dynamics—systems and control of physical systems, such as spacecraft, helicopters, aircraft, industrial robots, machines and manipulators, large space structures, and helicopter rotor dynamics; (2) fluids—theoretical and experimental studies on steady and unsteady aerodynamics, microgravity fluid physics, hypersonic flow, computational fluid mechanics, combustion and propulsion, thermal convection, flow instabilities, and turbulence and aeroacoustics; (3) manufacturing and design—CAD/CAM applications to mechanical engineering, aerospace and electronics, numerical control, robotics and artificial intelligence, flexible manufacturing, and design methodology; (4) solids—theoretical and numerical studies in the mechanics of solids, including fracture mechanics, micromechanics, mechanics of composite materials, nondestructive evaluation, and wave propagation in solids; (5) structures—structural dynamics, fluid-structure interaction, fixed-wing and rotary-wing aeroelasticity, finite-element methods, structural optimization with static and dynamic constraints, and integrated design and optimization of actively controlled aerospace structures, such as composite wings and large space structures; (6) thermal science and engineering—laminar and turbulent, free and forced convection, including flow instability, boiling, condensation, evaporation, and two-phase flow; and hazardous-waste incineration; and (7) Microelectromechanical systems (MEMS)—focuses on science, engineering, and fabrication issues ranging in size from nanometers to millimeters, including both experimental and theoretical studies covering fundamentals to industrial applications.

Interdepartmental Graduate Degree Program in Biomedical Engineering. J. D. Mackenzie, Interim Chair. Seven subfields leading to M.S./Ph.D. degrees in: (1) biomedical signal and image processing—digital imaging modalities such as MR, CT, and PET and design of intelligent monitoring and recording devices; (2) bioacoustics, speech, and hearing—quantitative methods of engineering, physical and biological sciences to solve problems in speech and hearing; (3) biomedical instrumentation—use of lasers in surgery, sensors for detection and monitoring of disease and controlled drug delivery system; (4) biomechanics, biomaterials, and tissue engineering—properties of bone, muscles and tissues, the replacement of natural materials with artificial compatible and functional materials; (5) biochemical engineering—development and analysis of processes involving biological materials and biocatalysts; (6) biocybernetics—properties or behavior of living systems, including regulation, control, communication, and visualization; and (7) neuroengineering—locomotion and pattern generation, central control of movement, processing of sensory information, MEMS, signal processing, and photonics.

UNIVERSITY OF CENTRAL FLORIDA

College of Engineering

Programs of Study

The College of Engineering offers master's and doctoral degrees in civil, computer, electrical, environmental, industrial, and mechanical engineering. The master's programs are designed to provide competent students who have a baccalaureate in engineering or other selected fields (e.g., mathematics, computer science, physics, biology) with an opportunity to specialize in a particular subject area within engineering. A master's degree with a thesis requires a minimum of 30 semester credits, including 6 hours of thesis. Some departments offer a course work–only master's degree program requiring 36 semester credits with a final comprehensive examination. Graduate programs provide direct support for the emergence of the central Florida area as one of the national centers of high-technology industry. The Ph.D. program is primarily intended for those with a master's degree in engineering; but, with appropriate articulation courses, those who hold a master's degree in a related discipline are able to use the program to study engineering disciplines in depth. A student's Ph.D. program consists of a minimum of 81 semester hours of graduate work; a maximum of 36 hours may be transferred from a master's program.

Research Facilities

The college has modern shop facilities and well-equipped laboratories devoted to research in mechanics and materials, heat transfer, internal combustion, engine optimization, laser machining, manufacturing, minicomputer applications, microprocessor development systems, digital signal processing, software engineering, microelectronics, materials engineering, rapid prototyping, optical communication systems, environmental engineering, and bridge expansion joint test track. The College also maintains a close liaison with the Florida Solar Energy Center (FSEC), located at Brevard; the Institute for Simulation and Training (IST); the Center for Research in Electro-Optics & Lasers (CREOL); and the Space Education and Research Center (SERC), located in the Central Florida Research Park that adjoins the main campus. Computer support is provided by numerous PCs and workstations throughout the College. A local area network connects all College computers and provides gateways to external systems and the Internet. The total number of volumes in the University library exceeds 750,000, with 600 periodical titles in engineering.

Financial Aid

Teaching and research assistantships require one-quarter-time to one-half-time work loads, with compensation up to $14,200 for twelve months in 1998–99. Assistantships provide substantial tuition waivers. The College offers eight fellowships yearly honoring Dr. Robert D. Kersten, the College's founding dean, which include a $3250 stipend, full tuition waiver, and a teaching or research assistantship. The application deadline is March 1. There are also four similar doctoral fellowships offered yearly. The University's Student Financial Assistance Office administers long-term loans and institutional emergency short-term loans.

Cost of Study

Tuition in 1999–2000 is $136.89 per semester hour for in-state students; out-of-state students pay $480.45 per semester hour. General fees paid by all students are $47.30 per term plus lab fees where applicable. Fees are subject to change.

Living and Housing Costs

The cost of living is at or below the national average. There is no state income tax in Florida. University housing is not available for graduate students. Many apartments are available near campus with rent ranging from $400 for one-bedroom units to $600 for two-bedroom units.

Student Group

The University's enrollment exceeds 28,000, with the student body almost equally divided between men and women. The enrollment of the College of Engineering is approximately 3,500, which includes about 900 graduate students. The majority of the graduate students are industrial employees studying part-time. Most of the College's graduate courses are recorded on videotape and made available to local branch campuses and industrial sites.

Student Outcomes

UCF College of Engineering graduates are on the faculty of universities such as Old Dominion University, Arizona State University, U.S. Military Academy, Tuskegee University, Southwestern Louisiana University, and American University in Beirut. Engineering graduates are working in their own firms and in industry and are conducting research and development at Lockheed-Martin, Loral, U.S. Army STRICOM, Harris Corporation, NASA-Kennedy Space Center, Westinghouse, Anderson Consulting, Deloitte-Touche, Walt Disney World, AT&T, U.S. Geological Survey, Intel Corporation, and the city of Orlando.

Location

UCF is located 15 miles east of downtown Orlando. Central Florida has a high percentage of simulation and training, high-technology laser, aerospace, communications, electronics, and electrooptics industries. The Atlantic Ocean, the Gulf of Mexico, and numerous rivers and spring-fed lakes provide many opportunities for outdoor recreation.

The University and The College

Established as a state university in 1963, UCF admitted its first students in 1968. Today, the campus covers 1,300 wooded acres. The University's central location makes it accessible from all parts of the state. In addition, there are branch campuses located in Cocoa, Daytona Beach, and south Orlando.

Applying

Prospective students should apply to the University Graduate Studies Office at least six weeks (five months for international students) before the start of classes for the term in which they plan to enroll. A $20 application fee, official transcripts from an accredited college, and GRE General Test scores are required. The minimum admission requirement is based on an average of B or better in the last 60 attempted semester hours of the baccalaureate program and a minimum combined score of 1000 on the verbal and quantitative portions of the GRE General Test. International students must score a minimum of 550 on the TOEFL in order to be accepted by the University.

Correspondence and Information

Director of Graduate Affairs
College of Engineering
University of Central Florida
Orlando, Florida 32816-2993
Telephone: 407-823-2455
World Wide Web: http://pegasus.cc.ucf.edu/~acad

University of Central Florida

THE FACULTY AND THEIR RESEARCH

M. Wanielista, Dean, Ph.D., Cornell; PE: water resources systems, stormwater management systems.

Civil and Environmental Engineering (CEE): A. E. Radwan, Chair; Ph.D., Purdue; PE: intelligent transportation systems, traffic flow modeling. Mohamed A. Abdel-Aty, Ph.D., California, Davis: transportation demand analysis, traffic safety. H. M. Al-Deek, Ph.D., Berkeley: intelligent transportation systems, traffic management. D. L. Block, Director of Florida Solar Energy Center (FSEC); Ph.D., Virginia Tech; PE: solar energy. M. B. Chopra, Assistant Chair; Ph.D., SUNY at Buffalo: geoenvironmental problems, nonlinear soil consolidation. C. David Cooper, Ph.D., Clemson; PE: air pollution control and modeling, incineration. J. D. Dietz, Ph.D., Clemson; PE: water and wastewater, industrial waste. S. El-Tawil, Ph.D., Cornell: seismic inelatic analysis, hybrid systems. S. C. Hagen, Ph.D., Notre Dame: water resources. C. M. Head, Ph.D., Georgia; PE: beach erosion, geomorphology. S. K. Hong, Ph.D., UCLA: electrical engineering, drinking water. S. K. Kunnath, Ph.D., SUNY at Buffalo; PE: reinforced concrete, nonlinear structural dynamics and structural analysis software. S. Kuo, Ph.D., Michigan State; PE: asphalt/concrete mix, pavement design, geotechnical studies. F. N. Nnadi, Ph.D., Queen's at Kingston: storm water management, control of toxic heavy metal resuspension, wetland conservation. A. Oloufa, Ph.D., Berkeley. O. Onyemelukwe, Ph.D., Pittsburgh: statically and dynamically loaded structural systems, wind engineering. A. A. Randall, Ph.D., Auburn; PE: biological nutrient removal, industrial wastewater treatment, bioremediation. D. Reinhart, Ph.D., Georgia Tech; PE: solid and hazardous waste. R. L. Wayson, Graduate Program Coordinator; Ph.D., Vanderbilt; PE: air quality, noise pollution due to mobile sources.

Electrical and Computer Engineering (ECE): W. B. Mikhael, Chair, Ph.D., Concordia: digital signal processing. I. E. Batarseh, Assistant Dean; Ph.D., Illinois at Chicago: electronic circuits, power electronics. C. S. Bauer, Ph.D., Florida; PE: software engineering, computer graphics, real-time simulation. M. Belkerdid, Ph.D., Central Florida: communications, electrooptics. R. F. DeMara, Ph.D., USC: digital systems, computer architecture. M. Georgiopoulos, Ph.D., Connecticut: communications networks, spread spectrum systems. A. Gonzalez, Ph.D., Pittsburgh; PE: computer-aided instruction, knowledge-based systems verification. F. Gonzalez, Ph.D., Illinois at Urbana-Champaign: intelligence control of discreet-event systems. M. G. Haralambous, D.Sc., George Washington: optimal control, communications. L. Jones, Ph.D., Virginia Polytechnic: remote sensing. T. Kasparis, Ph.D., CUNY, City College: digital signal and image processing, electronics. H. I. Klee, Ph.D., Polytechnic of Brooklyn; PE: systems engineering, flight simulation. D. G. Linton, Ph.D., Florida; PE: software engineering and application of CASE tools, simulation. J. J. Liou, Graduate Coordinator; Ph.D., Florida: microelectronics, solid-state devices. D. C. Malocha, Ph.D., Illinois at Urbana-Champaign: surface waves, acoustooptics, RF communications, microelectronics. R. N. Miller, Associate Dean of Undergraduate Affairs; Ph.D., SUNY at Buffalo; PE: high-power switching, instrumentation, alternative energy sources. H. R. Myler, Ph.D., New Mexico State: machine intelligence, knowledge engineering. B. E. Petrasko, D.Eng., Detroit: computer organization and architecture, digital communications. R. L. Phillips, Ph.D., Arizona State: optical communications, optical propagation. Z. Qu, Ph.D., Georgia Tech: controls, robotics. S. Richie, Ph.D., Central Florida: surface waves, digital systems. W. Shu, Ph.D., Illinois at Urbana-Champaign: software engineering. K. B. Sundaram, Ph.D., Indian Institute of Technology: microelectronics, thin films. N. S. Tzannes, Ph.D., Johns Hopkins: communication systems, information theory, signal processing. P. Wahid, Ph.D., Indian Institute of Science: electromagnetics. G. Walton, Ph.D., Tennessee: software engineering, software specification, software testing, software quality measurement and models. A. D. Weeks, Ph.D., Central Florida: embedded microprocessor, digital image processing. W. Wu, Ph.D., Santa Clara: parallel processing multimedia systems. J. S. Yuan, Ph.D., Florida: microelectronics. J. Zalewski, Ph.D., Warsaw: software engineering.

Industrial Engineering and Management Systems (IEMS): C. Reilly, Chair; Ph.D., Purdue: mathematical programming, evaluating optimization solution methods, applications of mathematical programming, simulation, applied operations research. R. L. Armacost, D.Sc., George Washington: operations research, decision analysis, resource constrained project scheduling. A. Elshennawy, Ph.D., Penn State: manufacturing, robotics, quality and reliability. R. Hoekstra, Ph.D., Cincinnati: alternative fuels, engine optimization, design theory, design for assembly. Y. A. Hosni, Ph.D., Arkansas; PE: productivity, operations research, logistics. D. Kolunda, Ph.D., North Carolina State: management systems engineering, organizational learning, organizational change and transitions, strategic planning and project management. T. Kotnour, Ph.D., Virginia Tech: management systems engineering, organizational learning, project management. G. Lee, Ph.D., Texas Tech; PE: human factors, productivity, ergonomics. L. Malone, Graduate Program Coordinator; Ph.D., Virginia Tech: response surface modeling, regression modeling, statistical aspects of simulation. P. McCauley-Bell, Ph.D., Oklahoma: expert systems, fuzzy logic, knowledge acquisition, simulation, failure analysis, ergonomics. M. Mollaghasemi, Ph.D., Louisville: operations research, simulation, multicriteria optimization. M. Mullens, Ph.D., Georgia Tech: manufacturing systems, warehousing, materials handling, automation, simulation. J. Pet-Edwards, Ph.D., Case Western Reserve: systems engineering, risk assessment and management, decision support systems, decision making, project cost estimation. M. Proctor, Ph.D., North Carolina State: simulation, distributed interactive simulation, military applications. J. Ragusa, D.B.A., Florida State: multimedia applications, artificial intelligence. G. Schrader, Ph.D., Illinois at Urbana-Champaign; PE: manufacturing, engineering management. J. A. Sepúlveda, Ph.D., Pittsburgh; PE: health operations research, simulation, economic analysis. K. Stanney, Ph.D., Purdue: human-computer interaction, human factors, ergonomics. W. J. Thompson, Ph.D., Arizona State: facilities design, cost engineering, production control. G. Whitehouse, Provost; Ph.D., Arizona State; PE: computer modeling, operations research, engineering management. K. E. Williams, Ph.D., Connecticut: cognitive modeling, intelligent simulation, intelligent tutoring training systems engineering design, human-computer interface design, intelligent agents, decision support systems.

Mechanical and Aerospace Engineering (MAE): L. C. Chow, Chair; Ph.D., Berkeley: two-phase heat transfer, electronic packaging. P. J. Bishop, Interim Associate Vice President for Graduate Studies; Ph.D., Purdue; PE: heat transfer, laser-materials interactions. R. H. Chen, Ph.D., Michigan: combustion, propulsion, experimental methods. L. Chew, Ph.D., Washington (Seattle): fluid mechanics, turbulence, vortex dynamics. V. H. Desai, Ph.D., Johns Hopkins: materials corrosion, fractography. B. E. Eno, Ph.D., Cornell; PE: fluid mechanics, energy conversion, HVAC. R. M. Evan-Iwanowski, Research Professor; Ph.D., Cornell: dynamics, nonstationary vibrations, chaos. L. A. Giannuzzi, Ph.D., Penn State: electron microscopy, structure/properties of materials. A. H. Hagedoorn, Ph.D., Cornell; PE: finite elements, structures, computer-aided design. E. R. Hosler, Ph.D., Illinois at Urbana-Champaign; PE: two-phase flow, power cycles, flow visualization. R. W. Johnson, Ph.D., UCLA; PE: control, astrodynamics system design. J. S. Kapat, Sc.D., MIT: fluid mechanics and heat transfer, turbulence and transition, gas turbines, transport of flow in material processing, MEMS for flow control. A. Kar, Ph.D., Illinois: materials processing, laser-aided manufacturing. A. J. Kassab, Graduate Program Coordinator; Ph.D., Florida: computational fluid mechanics and heat transfer, boundary element methods. K. Lin, Ph.D., Michigan: controls, flight simulation. J. D. McBrayer, D.Sc., Washington (St. Louis); PE: heat transfer, applied aerodynamics. A. Minardi, Ph.D., Central Florida: heat transfer, laser machining, energy conversion. F. A. Moslehy, Ph.D., South Carolina; PE: applied and experimental mechanics, vibrations, nondestructive evaluation. J. F. Nayfeh, Associate Chair; Ph.D., Virginia Tech: dynamics, composite structures, computer-aided design. D. Nicholson, Ph.D., Yale: finite elements, fracture mechanics, contact mechanics. C. E. Nuckolls, Ph.D., Oklahoma; PE: mechanisms, design, vehicle dynamics. S. Seal, Ph.D., Wisconsin: surface science, methods and instrumentation, high-temperature corrosion, coatings, nitrides, thin films, biomaterials, nanoparticles. W. F. Smith, Sc.D., MIT: PE: mechanical properties of alloys. G. G. Ventre, Director of Space Education and Research Center, Ph.D., Cincinnati; PE: space systems, orbital mechanics, photovoltaics. D. Zhou, Ph.D., Arizona: diamond films, carbon nanotubes, nanowhiskers, electron field emission, plasma processing, electrochemical processing, microstructural characterizations.

UNIVERSITY OF CINCINNATI
College of Engineering

Programs of Study	The College of Engineering has 149 tenure-track faculty members in six departments: Aerospace Engineering and Engineering Mechanics; Civil and Environmental Engineering; Chemical Engineering; Electrical and Computer Engineering and Computer Science; Materials Science and Engineering; and Mechanical, Industrial and Nuclear Engineering.
	The Doctor of Philosophy degree is offered in aerospace engineering, chemical engineering, civil engineering, computer science and engineering, engineering mechanics, environmental engineering, environmental science, electrical engineering, industrial engineering, materials science, mechanical engineering, metallurgical engineering, and nuclear and radiological engineering. The Master of Science degree is offered in all of the above areas, plus computer engineering, computer science, health physics, and solid-state electronics.
Research Facilities	The new Engineering Research Center consists of 175,000 square feet of space with new or expanded research facilities in the areas of computing, microelectronics, materials characterization, polymer research, energy, structures, biomedical engineering, membrane technology, infrastructure research, digital systems, intelligent vision, and high-temperature thin-film superconductivity. In addition, the facility houses an advanced materials processing center. Graduate students also conduct research in special research centers within the College of Engineering: Advanced Manufacturing Sciences, Aerosol Processes, Noyes-Giannestras Biomechanics Labs, Fossil Fuel Research, Composite Materials and Structures, Intelligent Vision and Information Processing, Hazardous Waste Research and Education, Geoenvironmental Science and Technology, Computer-Aided Molecular Design, Cincinnati Infrastructure Institute, Accelerated Life Testing and Environmental Research (ALTER), Ergonomics Research Lab, Groundwater Research, Microelectronics Sensors and Microstructures, Computational Fluid Dynamics, Computational Simulation of Aerospace Propulsion Structures, Nanoscience and Technology, Basic and Applied Nuclear Research Lab, Robotics Research, Structural Dynamics Research Lab, the Health and Environmental Risk Institute, and the Computer-Aided Manufacturing Lab.
Financial Aid	The College provides 150 teaching and 350 to 400 research assistantships, traineeships, and fellowships. Stipends for the academic year range from $10,200 to $18,000 plus a tuition scholarship. In 1998–99, the College awarded 525 teaching and research assistantships and provided University graduate scholarships to 750 students.
Cost of Study	Graduate tuition is on a quarterly basis. For 1999–2000, full-time tuition is $1862 per quarter for Ohio residents and $3508 per quarter for out-of-state students; there is also a general fee of $185 per quarter.
Living and Housing Costs	Campus housing for graduate students enrolled in the College of Engineering at the University of Cincinnati is available in apartments; 1998–99 rates ranged from $393 to $544 per month.
Student Group	Enrollment at the University is 35,000, including 22,000 full-time students. College enrollment includes 1,850 full-time undergraduate students and 750 full-time graduate students.
Location	Cincinnati is situated on the banks of the Ohio River in the tristate area of Ohio, Indiana, and Kentucky. It is a major midwest crossroads of air, highway, and rail transportation. The population of greater Cincinnati is 1.7 million; more than 38,000 companies drive the city's economy. Cincinnati has been named as one of the most liveable cities in the United States.
The University and The College	The University of Cincinnati (UC) is a Research I institution with particular strengths in the quality of its faculty members and their teaching and research capabilities, especially in the developmental and applied nature of their work. The College of Engineering provides a high-quality learning environment, an internationally recognized faculty, and challenging research projects.
	The University traces its origin to 1819, when Cincinnati College was founded. UC became a state institution in 1977; its eighteen colleges and schools offer 445 undergraduate, graduate, doctoral, and professional degrees. Conducting more than $100 million annually in externally funded research, UC is designated a Research 1 University by the Carnegie Commission and is ranked in the top 3 percent of leading research universities by the National Science Foundation. The College maintains excellent cooperative relations with more than 600 engineering firms, manufacturing industries, research laboratories, public utilities, and government agencies.
Applying	Inquiries should be sent to the major department. The department supplies application materials. The application fee is $40. The GRE General Test is required; the Subject Test is required in some departments and recommended by all. International applicants must also submit minimum TOEFL scores ranging from 550 to 580; the TSE is required by some departments.
Correspondence and Information	Associate Dean for Academic and Administrative Affairs College of Engineering University of Cincinnati Cincinnati, Ohio 45221-0018 Telephone: 513-556-2739 E-mail: gloria.cole@uc.edu World Wide Web: http://www.eng.uc.edu

University of Cincinnati

DEPARTMENT HEADS AND RESEARCH AREAS

Aerospace Engineering and Engineering Mechanics (Mail Location 0070, telephone: 513-556-3219, e-mail: ase_em@uc.edu, World Wide Web: http://www.ase.uc.edu/). Gary L. Slater, Ph.D., Head; Stanley G. Rubin, Ph.D., Graduate Studies Director. Research areas—fluid mechanics: computational fluid dynamics; grid generation, separation, and bifurcation; turbulent and transitional flows; jet mixing; plume interaction; subsonic, supersonic, and hypersonic internal and external flows; fluid–plasma interactions; aeroacoustics; fluid-aerosol interaction and environmental impact; propulsion: two-phase reacting flows, steady and unsteady flow in turbomachines, heat transfer, compressor performance, propulsion system transients, unsteady inlet-compressor interactions, liquid fuel combustion, fuel sprays, turbine blade film cooling, cascade flow, pressure- and temperature-sensitive paint diagnostics, particulate flow, erosion, aging effects, coatings; controls: fault tolerance, optimal trajectories, modern control design, parameter estimation, neural and intelligent control; structures and controls: large space structures, flexible aircraft; astrodynamics: satellite orbital mechanics, space-vehicle attitude dynamics and control, spacecraft proximity operations, nonlinear dynamical systems, hybrid numerical simulation procedures; solid mechanics: inelastic, anisotropic constitutive modeling, high-temperature alloys, thermomechanical coupling, fibrous composites, thick composite shells, structural similitude and size effects, structural optimization, dynamic stability, aeroelasticity, wave propagation in various media, nondestructive evaluation; biomechanics: tissue engineering, properties of ligaments and tendons, cell mechanics, knee and ankle mechanics.

Chemical Engineering (Mail Location 0171, telephone: 513-556-2761, e-mail: gradp@alpha.che.uc.edu, World Wide Web: http://www.chemical.uc.edu/). Joel R. Fried, Ph.D., Interim Head; Y. S. (Jerry) Lin, Ph.D., Graduate Studies Director. Research areas—process dynamics, process synthesis, process simulation, process control, optimization, reactive distillation, reactor modeling; transport in polymeric systems, polymer blends, composites, polymer thermal analysis, molecular modeling, computer-aided molecular design; membrane separations, membrane reactors, catalytic membranes, inorganic membranes, pervaporation; transport phenomena, flow through porous media and dense solids; gas separations, characterization of nanoporous solids; energy and fuels research; digital computation, energy and environmental engineering, chemical vapor deposition; advanced materials for separation, chemical vapor deposition/sol-gel processing, adsorption separation; adsorption and ion exchange, chromatography, chemical sensors on a chip; water and wastewater treatment; hazardous waste treatment, waste minimization, waste materials as sources of chemical feedstocks and carbons; aerosol science, flame synthesis of materials, materials processing; heterogeneous catalysis, environmental catalysis, development and optimization of refining processes, oxygenated fuel additives.

Civil and Environmental Engineering (Mail Location 0071, telephone: 513-556-3648). Mark T. Bowers, Ph.D., P.E., Interim Head; Frank E. Weisgerber, Ph.D., P.E., Graduate Studies Director. Research areas in environmental engineering and science—water resources, gaseous pollutant control, flue gas desulfurization, aerosol science and engineering, combustion, landfill leachate treatment, landfill design and waste incineration, water chemistry, microbiology, water and wastewater treatment systems, natural treatment systems, surface and groundwater flows and contamination, hydroclimatology, geoenvironmental systems, and water distribution systems. Research areas in civil infrastructure systems—condition assessment, deterioration mechanisms, renewal engineering, and optimum management of the highway transportation systems, including traffic, pavements, and bridges; innovative and intelligent structural systems, members, and materials; and seismic vulnerability evaluation and retrofit of buildings and bridges. Advanced tools developed and used in research include destructive and nondestructive testing of large-scale laboratory models and actual buildings, bridges, and pavements in the field; instrumented monitoring and local nondestructive probing (including modal testing) in the field; laboratory-accelerated life testing; and linear and nonlinear system identification through finite-element simulation.

Electrical and Computer Engineering and Computer Science (Mail Location 0030, telephone: 513-556-4769). Thomas Mantei, Ph.D., Head; Dieter Schmidt, Ph.D., Graduate Program Director. Research areas in systems engineering—computer vision, image processing, artificial neural network, fuzzy expert systems, automatic factory control, process control systems, inspection and quality control, control theory, modeling and identification of dynamical systems, intelligent control systems, adaptive and nonlinear systems, digital and adaptive signal processing, pattern recognition, I/O and real-time system design, signal processing for communication, multiuser communication, information theory and coding, biomedical systems, and complex systems. Research areas in computer science—artificial intelligence, logic systems, computer science theory, design and analysis of algorithms, fuzzy logic systems, distributed systems, communication networks, distributed artificial intelligence, database theory, computer graphics, human-computer interface, multimedia systems, VLSI algorithms, parallel architectures, graph algorithms, programming languages theory, and machine learning. Research areas in computer engineering—computer architecture, operating systems, parallel and distributed processing, database systems, compilers, embedded systems, Web-based computing, network-enabled computing, wireless networking, mobile computing, active networks, software engineering, formal methods, formal verification, VLSI design, design and test automation, and reconfigurable and adaptive computing. Research areas in electronic materials and devices—smart sensors and systems for biology; implantation in medicine, aerospace, and industry; micromotors and microfluidics; micromechanical structures, integrated circuits, and sensors; materials and devices in elemental semiconductors, III-V compound semiconductors, wide bandgap semiconductors, semiconducting glasses, and high TC superconductors; electronic devices (MESFETs, HEMTs, HBTs) and optical devices (modulators and photodetectors); optical information processing; engineering of materials with novel properties for electronic and optoelectonic devices, nanoelectronics, optoelectronics, integrated optics, optical sensors, focused ion beam technology, quantum well structures, atomic probe microscopy, plasma etching, and plasma deposition.

Materials Science and Engineering (Mail Location 0012, telephone: 513-556-3096). Raj Singh, Ph.D., Interim Head; R. J. Roe, Ph.D., Graduate Studies Director. Research areas—surface and interface properties; nonlinear optical properties of polymers; metal-polymer adhesive bonding; inorganic polymers; cyclization in polymeric systems; conducting composites; effect of aging and environment on polymer composites; liquid crystalline polymers; mobility in ordered polymers; thermodynamics of polymer blends; spectroscopy; small angle X-ray scattering; analytical TEM; high-temperature materials; intermetallic compounds; metal, ceramic, and polymer matrix composites; joining, welding, and solid-state bonding; creep-fatigue-environment interactions; deformation, damage evolution, and fracture; experimental micromechanics; hydrogen embrittlement; principles of microstructural evolution; solidification processing; combustion synthesis; advanced ceramic fiber development; molten salt physical chemistry; processing of ceramics and composites; advanced materials characterization techniques; solution processing of ceramic materials; and glass and glass ceramics.

Mechanical, Industrial, and Nuclear Engineering (Mail Location 0072, telephone: 513-556-5157). Urmila Ghia, Ph.D., Head; Frank M. Gerner, Ph.D., Director of Graduate Studies in Mechanical, Industrial, and Nuclear Engineering; Jay Kim, Ph.D., Graduate Studies Director in Mechanical Engineering; Sam Anand, Ph.D., Graduate Studies Director in Industrial Engineering; Henry Spitz, Ph.D., Graduate Studies Director in Nuclear and Radiological Engineering. Mechanical engineering research areas—applied mechanics, machine design and analysis, vibrations, robotics, active vibration and acoustic control, phase-change and convection heat transfer, heat-transfer augmentation, computational fluid dynamics, experimental fluid mechanics, control systems, engineering design, acoustics, finite-element methods, modal analysis, signature analysis, and machinery monitoring, computer-aided design and manufacturing, and manufacturing processes. Industrial engineering research areas—quality control, reliability, computer-aided manufacturing, manufacturing systems and processes, computer-integrated manufacturing, feature-based manufacturing, tolerancing and metrology, machine vision inspection, simulation, mathematical modeling, operations research, production planning and control, material handling, facilities design, ergonomics, injury prevention and control, cumulative trauma and engineering controls, training, occupational biomechanics, hazard evaluation, and occupational safety and robotics. Nuclear and Radiological engineering research areas—environmental radiological measurements and risk assessment, radiation detection and measurement, occupational and environmental radiation dosimetry, human health risk assessment, probabilistic risk assessment, radiological engineering, health physics, biomedical radiation measurements, nuclear reactor analysis and control, and reactor analysis computer programs.

UNIVERSITY OF COLORADO AT BOULDER

College of Engineering and Applied Science

Programs of Study

Graduate study in engineering and applied science at the University of Colorado at Boulder is offered in three degree programs. Programs leading to the Master of Science and Doctor of Philosophy degrees are provided for students who seek academic or research careers. A professional course of study leading to the Master of Engineering degree is intended primarily for students who wish to pursue further study in the professional disciplines. All three degrees are offered through the Graduate School, whose faculty is organized into fields of instruction.

Each student plans and carries out a program with the assistance of an advisory committee chaired by the faculty member with whom the student wishes to work. Minimal requirements concerning course level, credit hours, grades, residence, examinations, and the thesis are specified by the Graduate School for all three degrees offered. In general, the Master of Engineering degree requires satisfactory completion of 30 semester hours, 15 of which must be advanced-level engineering courses, and a report. The Master of Science involves successful completion of not fewer than 24 semester hours of thesis and/or course work. The Doctor of Philosophy is awarded for proficiency and originality in a field where the recipient has made significant research contributions. A minimum of 30 semester hours of advanced course work and 30 semester hours of dissertation credit are required for this degree.

Research Facilities

The Engineering Center on the Boulder campus is a complex of classrooms, faculty offices, computing facilities, a library, and more than fifty research laboratories that cover over 10 acres of floor space. In addition to the facilities necessary to provide for a wide variety of research programs, the center houses a significant number of mass and electronic spectrometers, electron and other microscopes, the nation's most powerful geotechnical centrifuge, a structural testing facility, a unique commercial building HVAC testing laboratory, a class-10,000 clean room facility, and ion-implantation and microwave-propagation equipment. Fourteen research centers (listed on the reverse of this page) offer additional opportunities for interdisciplinary study and support.

Near the campus are government laboratories, including the National Institute of Standards and Technology, the National Oceanic and Atmospheric Administration, the National Center for Atmospheric Research, and the National Renewable Energy Lab. National laboratories and numerous industrial firms involve the College's students and faculty in many joint research projects.

Each engineering department is extensively supported by microcomputers, minicomputers, and mainframe computers. Terminals for student use are available throughout the Engineering Center.

Financial Aid

Available financial aid includes research fellowships, research assistantships, teaching assistantships, and traineeships as well as other forms of support. Many appointments cover tuition and fees; they are comparable to those offered at other first-rate graduate institutions. Loan funding is also available.

Cost of Study

In 1999–2000, graduate tuition for Colorado residents is $1254 for 6 credit hours and $1871 for 9 to 18 credit hours per semester; nonresident graduate tuition is $5094 for 6 credit hours and $7641 for 9 to 18 credit hours per semester. Student fees are approximately $350.

Living and Housing Costs

On-campus housing is $3013 per semester for a single room with nineteen meals a week in 1999–2000. Information about single-student housing may be obtained from the Supervisor of Reservations, Campus Box 154, University of Colorado at Boulder, Boulder, Colorado 80310-0154. Family housing information may be obtained from the Family Housing Office, University of Colorado at Boulder, 1350 20th Street, Boulder, Colorado 80302. A separate housing application is required.

Student Group

Total University enrollment is 25,125, including 4,530 graduate students. Enrollment in engineering is 3,526, including 1,091 graduate students.

Student Outcomes

Primary employment for Ph.D. graduates is in research and development positions in the engineering and computing industries, government laboratories, and academia (both in teaching and research). Primary employment for M.S. graduates is in technical, development, and research positions in the engineering, computing, and telecommunications industries, as well as in government laboratories and academia.

Location

The Boulder campus is located along the Front Range of the Rocky Mountains. Outdoor recreation is near at hand, offering celebrated skiing, backpacking, fishing, mountain climbing, and cycling in a health- and fitness-oriented area. Boulder is a community of 96,000; metropolitan Denver, 30 miles away, offers all the cultural amenities of a large city and is easily accessible from Boulder by public transportation.

The University

Graduate study in engineering and applied science at the University of Colorado at Boulder is conducted within the framework of a large and diverse university with an international reputation. The mission of the University is to lead in discovery, communication, and use of knowledge through instruction, research, and service to the public. As a comprehensive university, it provides for each graduate student an educational experience that is distinguished by the scope of its programs and course offerings, the outstanding quality of its research facilities, the diversity of its student body, and the professionalism and dedication of its faculty.

Applying

Requests from U.S. citizens and permanent U.S. residents for admission to the Graduate School should be sent to Graduate Admissions in care of the engineering department in which the applicant wishes to study. An application packet will be sent in return. Individual departments should be contacted directly for information concerning application deadlines. Qualified students are recommended for admission to regular degree status by the major department. International students have a December 1 deadline for the fall semester and an October 1 deadline for the spring semester.

Correspondence and Information

Graduate Admissions
(Name of Department and
  C.B.# from list at right)
College of Engineering and Applied Science
University of Colorado
Boulder, Colorado 80309
World Wide Web: http://www.colorado.edu/engineering

Aerospace Engineering Sciences, C.B. 429
Chemical Engineering, C.B. 424
Civil, Environmental, and Architectural
  Engineering, C.B. 428
Computer Science, C.B. 430
Electrical and Computer Engineering, C.B. 425
Engineering Management, C.B. 435
Mechanical Engineering, C.B. 427
Interdisciplinary Telecommunications, C.B. 530

University of Colorado at Boulder

THE FACULTY AND THEIR RESEARCH

Aerospace Engineering Sciences. A. Richard Seebass, Chairman. B. Argrow: fluid dynamics. P. Axelrad: global positioning. M. Balas: controls. S. Biringen: fluids. G. H. Born: astrodynamics. C. Y. Chow: fluids. R. D. Culp: astrodynamics. J. Curry: atmospheric sciences. W. Emery: satellite oceanography, remote sensing. C. Farhat: structural dynamics. C. Felippa: structural dynamics. J. Forbes: atmospheric sciences and remote sensing. P. Freymuth: experimental fluids. A. Hoehn: space bioregenerative life support. L. Kantha: ocean modeling. J. Koster: experimental fluids. K. Larson: geophysics. D. Lawrence: systems and controls. M. Leben: remote sensing. M. Lesoinne: computational fluid structural dynamics. J. Maslanik: remote sensing. M. Mikulas: structural design. G. Morgenthaler: systems. K. C. Park: structural dynamics. L. Peterson: structures and controls. G. Rosborough: astrodynamics. A. R. Seebass: aerodynamics/gasdynamics. H. Snyder: cryogenics and low-g fluid mechanics. L. Stodieck: space life sciences experimentation.

Chemical Engineering. Robert H. Davis, Chairman. K. S. Anseth: biomedical engineering, biomaterials. V. A. Barocas: biomedical engineering, biomechanics. C. Bowman: polymers. D. E. Clough: multivariable adaptive process control. R. H. Davis: biotechnology, membrane separations, suspension fluid mechanics. J. L. Falconer: heterogeneous catalysis and surface analysis. R. I. Gamow: biomedical engineering and biotechnology, cell-wall physiology. C. M. Hreyna: fluidization, granular flows, turbulence. D. S. Kompala: biochemical engineering. W. B. Krantz: membranes, thin films, combustion geophysics. R. D. Noble: chemical complexation, membranes. W. F. Ramirez: control, optimization. T. W. Randolph: biotechnology, supercritical fluids. R. L. Sani: fluid dynamics. P. Todd: biochemical separations. A. W. Weimer: ceramics, particle processes.

Civil, Environmental, and Architectural Engineering. Hon-Yim Ko, Chairman. B. Amadei: rock mechanics. G. Amy: environmental engineering. A. Bielefeldt: bioremediation, environmental engineering. M. J. Brandemuehl: building energy systems. H. Brown: construction management. S. C. Chapra: environmental systems. R. Corotis: structural engineering, design and optimization. R. Davis: illumination engineering. J. E. Diekmann: construction engineering. D. DiLaura: illumination engineering. J. O. Dow: structural dynamics. D. M. Frangopol: structural reliability and optimization. K. Gerstle: steel and concrete structures. V. Gupta: hydrological science. M. Halek: surveying and photogrammetry. J. Heaney: environmental engineering and water resources. G. Hearn: nondestructive evaluation of structures. M. Hernandez: environmental engineering. H.-Y. Ko: geotechnical modeling and testing. M. Krarti: building energy systems. J. F. Kreider: building systems, renewable energy. D. McKnight: environmental engineering. R. Muehleisen: acoustics, architectural engineering. R. Pak: soil-structure interaction. R. Qualls: water resources, hydrology. H. Rajaram: environmental engineering. J. Ryan: environmental engineering. V. Saouma: computer-aided design, fracture mechanics. P. S. B. Shing: earthquake engineering and structural dynamics. J. Silverstein: biological wastewater treatment processes. A. Songer: construction management. E. Spacone: nonlinear analysis of structures and computational methods. K. M. Strzepek: water resources planning. S. Sture: geotechnical engineering and mechanics. L. H. Summers: building energy management. R. S. Summers: environmental engineering. K. J. Willam: computational mechanics. Y. Xi: civil engineering materials. D. Znidarcic: soil mechanics and foundations. J. Zornberg: environmental geotechnics.

Computer Science. Karl Winklmann, Chairman. K. Anderson: software engineering, hypermedia, human-computer interaction. E. Bradley: scientific computation/artificial intelligence. R. Byrd: numerical computation. X. C. Cai: numerical analysis. A. Ehrenfeucht: theory of computation, artificial intelligence. M. Eisenberg: artificial intelligence, human-computer interaction. C. Ellis: systems, groupware. G. Fischer: artificial intelligence, human-computer interaction. L. Fosdick: parallel computation, numerical computation. H. Gabow: algorithms. D. Grunwald: parallelizing compilers, parallel systems. E. Jessup: numerical computation. H. F. Jordan: parallel architectures and algorithms. R. King: databases. C. Lewis: artificial intelligence, user interface design. M. Main: theory of computation. J. Martin: natural-language processing, knowledge representation, machine learning. O. McBryan: numerical and parallel computation. M. Mozer: cognitive science, neural networks. E. Nemeth: networking, combinatorics. G. Nutt: computer systems, performance measurement and modeling. R. Osborne: heterogenous distributed database systems. L. Palen: human-computer interaction, computer-supported cooperative work, social analysis of information technologies. A. Repenning: visual programming, interactive simulation, computers in education, agents. R. Schnabel: numerical and parallel computation. S. Singh: machine learning, adaptive systems, artificial intelligence. W. M. Waite: compiler construction. K. Winklmann: theory of computation. A. Wolf: software engineering, object databases. B. Zorn: programming languages, multiprocessing.

Electrical and Computer Engineering. Renjeng Su, Chairman. J. Avery: microprocessors. S. Avery: radar remote sensing. F. Barnes: microwaves, quantum electronics, biology. D. Beeman: computational neuroscience. T. X. Brown: telecommunications systems, networking, neural networks, novel computing. W. Cathey: nonconventional imaging systems. J. Dunn: electromagnetics, microwaves. R. Erickson: power electronics. D. Etter: digital signal processing. E. Fuchs: varaible-speed devices, renewable and alternative energy, power quality. K. Gupta: microwave circuits, electromagnetic fields. G. Hachtel: computer-aided design. J. Hauser: systems, control theory. R. Hayes: semiconductor devices, optoelectronics. V. Heuring: programming language design. H. Hinton: photonic telecommunications systems. R. Hooker: optoelectronic systems. K. Johnson: optical systems. H. Jordan: parallel optical computing systems. E. Kuester: fields, radio propagation. M. Lightner: computer-aided design. A. Majerfeld: devices, materials, quantum electronics. D. Maksimovic: power electronics. P. Mathys: communication theory. W. May: integrated circuits. D. Meyer: control theory, manufacturing. A. Mickelson: optics, electromagnetics. R. Mihran: biomedical engineering. G. Moddel: optoelectronic thin-film materials and devices. C. Mullis: communications, digital signal processing. J. Pankove: solid-state devices, materials. L. Pao: control systems. M. Piket-May: computational electromagnetics. A. Pleszkun: computer architecture. Z. Popovic: experimental microwave active devices. L. Scharf: communications, digital signal processing. F. Somenzi: computer-aided design. R. Su: nonlinear control, robotics. B. Van Zeghbroeck: optoelectronic devices, integrated circuits. H. Varanasi: communication theory. H. Wachtel: biomedical engineering, bioelectromagnetics, neurophysiology and cancer therapeutics. K. Wagner: adaptive, nonlinear optical computing systems. W. Waite: programming language design. M.-Y. Wu: systems, control theory.

Interdisciplinary Telecommunications. Frank S. Barnes, Director. Gary Bardsley, Associate Director. J. H. Alleman: telecommunications economics, policy and management. G. L. Bardsley: cable TV, economics and policy aspects of telecommunications, strategic planning for telecommunications. F. S. Barnes: fiber optics and advanced optical application. S. Black: telecommunications law and regulation. R. S. Bloomfield: telecommunications performance and standardization. T. X. Brown: wireless communications, ATM, switching. S. E. Bush: digital wide area networks and protocols, ISDN/BISDN, signaling system 7, frame relay. H. M. Gates: data communications. K. Klingenstein: computer network management. S. B. McCray: legal and regulatory issues in telecommunications. G. Mitchell: telecommunications systems, telecommunications theory and applications, traffic and queuing theory, satellite communications. J. R. Sauer: fiber optics, optical switching systems and software.

Mechanical Engineering. S. K. Datta, Chairman. M. C. Branch: experimental combustion. V. Bright: microelectromechanical systems. L. E. Carlson: design of prosthetic devices. J. W. Daily: combustion processes. S. K. Datta: wave propagation in elastic media. M. Dunn: solid mechanics. K. Gall: fatigue and fracture of metals. T. L. Geers: structure-medium interaction. A. R. Greenberg: polymeric and biological materials. J. Hertzberg: combustion. D. R. Kassoy: analytical fluid mechanics, combustion. Y. C. Lee: intelligent electronics manufacturing. R. L. Mahajan: electronics manufacturing and packaging. S. Mahalingam: computational fluid mechanics, combustion. J. Milford: air quality modeling. S. L. Miller: indoor air quality, exposure assessment. R. Raj: ceramics, metal-organic chemical vapor deposition. G. Subbarayan: optimal design, electronic packaging. C. H. Suh: computer-aided design. P. D. Weidman: fluid dynamics. P. Zoller: polymer science/engineering.

Engineering Research Centers. For further information, students should contact the director in care of the appropriate department.
Center for Advanced Decision Support in Water and Environmental Systems (CADSWES): Jacqueline F. Sullivan, Managing Director, c/o CEAE.
Center for Aerospace Structures (CAS): Charbel Farhat, Director, c/o AES.
Center for Applied Parallel Processing (CAPP): Oliver McBryan, Director, c/o CS.
Colorado Center for Astrodynamics Research (CCAR): George H. Born, Director, c/o AES.
Bioserve Space Technologies Center: George Morgenthaler, Director, c/o AES.
Center for Combustion Research (CCR): John W. Daily, Director, c/o ME.
Joint Center for Energy Management (JCEM): Michael J. Brandemuehl, Director, c/o CEAE.
Center for Advanced Manufacturing and Packaging of Microwave, Optical and Digital Electronics: R. Mahajan, Director, c/o ME.
Optoelectronic Computing Systems Center (OCSC): John A. Neff, Director, c/o ECE.
Process Biotechnology Program (PBP): Robert H. Davis, Co-Director, c/o CHE.
Center for Separations Using Thin Films (CSTF): Richard Noble, Co-Director, William Krantz, Co-Director, c/o CHE.
Center for Software Systems Science (CS^3): Gary Nutt, Director, c/o CS.
Center for Pharmaceutical Biotechnology (CPB): Theodore Randolph, Director, c/o CHE.
Center for Acoustics, Mechanics, and Materials (CAMM): Thomas L. Geers, Director, c/o ME.
Colorado Center for Information Storage: Renjeng Su, Director, c/o ECE.
Center for Drinking Water Optimization (CDWO): H. Scott Summers, Director, c/o CEAE.
Center for Fundamentals and Applications of Photopolymerizations: Chrsitopher N. Bowman, Director, c/o CHE.

UNIVERSITY OF COLORADO AT DENVER

College of Engineering and Applied Science

Programs of Study

Graduate engineering programs currently offered by the University of Colorado at Denver (CU-Denver) include the Master of Science (M.S.) in civil engineering, electrical engineering, mechanical engineering, and computer science. The Master of Engineering (M.Eng.) degree is also offered in civil engineering, electrical engineering, and mechanical engineering and permits the combination of graduate courses in engineering with those in other fields such as business administration, computer science, and public administration. Up to 12 of the required 30 semester hours for the degree may be selected from courses outside of engineering. A Ph.D. program is also offered in civil engineering.

Two degree plans are available for Master of Science students: Plan I requires a minimum of 24 semester hours of graduate work and a research thesis. Plan II requires a minimum of 30 semester hours of graduate work, including an independent study and report. Master of Engineering candidates follow the Plan II model but may select 12 semester hours from approved nonmajor courses. Ph.D. students require a minimum of 30 semester hours of graduate course work and a research dissertation. The actual number of hours of formal course work normally exceeds the minimum of 30.

Most engineering graduate courses are offered in the evening and are well suited to the needs of the practicing professional as well as the full-time student.

Research Facilities

The College of Engineering and Applied Science is located in downtown Denver. Engineering facilities include laboratories for bioengineering, computer engineering, electric power, robotics, controls, communications and microwave studies, VLSI and computer development, computer-aided design, materials science, geotechnical engineering (soil dynamics), and transportation engineering. Students and faculty members have access to the World Wide Web, the Internet, and on-campus computing in more than twenty-one departmental and general-purpose computing laboratories located throughout campus. CU-Denver is a member of Internet2 and is a fully participating member of WestNet. CU-Denver Computing, Information and Network Services supports general-purpose computing on DEC Alpha and Sun minicomputers. These machines provide access to many specialized software packages, including ADINA, SAS, SPSS, MatLab, and Lindo, as well as support for programming languages (FORTAN, Pascal, LISP, etc.). The three general-purpose computing facilities on campus support word processing, World Wide Web access, and many classroom-specific applications on both Windows- and Macintosh-based machines. The remaining departmental labs provide access to general-purpose software and specialized (discipline-specific) applications and data. PPP dial-up access to the Internet and the World Wide Web is provided at low cost to all CU-Denver students and faculty and staff members.

Financial Aid

Colorado Doctoral Fellowship and Colorado Graduate Grant Fellowship funds are available to residents of the state on a limited basis. In addition, teaching and research assistantships are available. Other fellowships, scholarships, and traineeships are available from various state and federal agencies and industrial sponsors.

Cost of Study

For 1998–99, Colorado residents paid $212 per semester hour for tuition, and nonresidents paid $764. Fees average about $135 per semester, and books and supplies average about $300 per semester. Tuition rates are subject to change at any time.

Living and Housing Costs

There is no University housing, and all students make their own arrangements for housing in the Denver metro area. Costs vary widely, depending upon individual requirements and choices.

Student Group

The College of Engineering and Applied Science has approximately 600 undergraduate and 280 graduate students from all parts of the United States and many other countries. Approximately 16 percent are women, and 8 percent are members of ethnic minority groups. The average age of graduate students is 32. Many are working professional engineers who have chosen to combine a working and educational schedule designed to meet their career goals.

Location

Denver is the central city of a rapidly growing metropolitan area that is spreading along the Front Range of the Rocky Mountains. The national emphasis on high-tech and energy industries has spurred economic, population, and government growth in Colorado. The campus is located in downtown Denver and is at the heart of a comprehensive public transportation system, making it relatively easy to reach regardless of where one lives in the metro area. Areas for skiing, backpacking, hiking, and fishing are relatively close and are easily reached by public and private transportation.

The University

Since its founding in 1876, the University of Colorado has served as a major research and teaching institution in the western United States. The University awarded its first advanced degree in 1893. It has grown from one campus in Boulder to four campuses (Boulder, Colorado Springs, and two in Denver). As of spring 1998, the Denver campus had 10,526 students, 47 percent of whom were graduate students. The College of Engineering has provided engineering instruction in Denver since 1914. Graduate engineering degrees were offered in Denver through the system-wide Graduate School beginning in the 1930s.

Applying

Applications for admission should be requested from the Graduate School or the appropriate department, as shown below. Programs are open to students who hold a bachelor's degree in a curriculum that includes the necessary prerequisites with above-average grades for the engineering degree program in which they wish to specialize. Applications and supporting materials should be received in the office of the major department at least sixty days prior to the term for which admission is sought, or earlier if required by the major department. A fee of $50 is required and must be paid before the application is processed.

Correspondence and Information

Dean
Graduate School
University of Colorado at Denver
P.O. Box 173364
Denver, Colorado 80217-3364
Telephone: 303-556-2663

Chairperson
Department of (specify Civil Engineering,
    Computer Science and Engineering, Electrical
    Engineering, or Mechanical Engineering)
College of Engineering and Applied Science
University of Colorado at Denver
P.O. Box 173364
Denver, Colorado 80217-3364
Telephone: 303-556-2870

SECTION 1: ENGINEERING AND APPLIED SCIENCES

University of Colorado at Denver

THE FACULTY AND THEIR RESEARCH

Civil Engineering
Nien-Yin Chang, Professor; Ph.D., Ohio State, 1976; PE. Geotechnical engineering, earthquake engineering, ground motion amplification, seismic dam safety, soil dynamics, soil liquefaction, dynamic soil-pile-bridge interaction, constitutive modeling of soils, expansive soil foundation designs, application of statistics and probability to geotechnical problems.
James Guo, Associate Professor; Ph.D., Illinois, 1982; PE. Hydraulic engineering, groundwater hydrology, mathematical modeling of fluid systems, urban drainage.
David W. Hubly, Associate Professor and Chair; Ph.D., Iowa State, 1976; PE. Design of water distribution and sewage collection systems, water and wastewater treatment, water quality management, treatment of groundwater.
William C. Hughes, Professor; Ph.D., New Mexico, 1969; PE. Hydrology and hydraulic engineering.
Bruce N. Janson, Associate Professor; Ph.D., Illinois, 1981. Transportation engineering, planning, and management; dynamic network equilibrium modeling and design; travel demand forecasting; real-time traffic simulation; environmental impacts and safety; traveler information systems; geographic information systems.
Lynn E. Johnson, Professor; Ph.D., Cornell, 1980; PE. Geographic information systems, water resource systems modeling and management (hydrology, reservoir operations, and water quality), computer-aided planning and design, river basin and urban water system management, flood forecasting, decision support systems.
Sarosh Khan, Assistant Professor; Ph.D., California, Irvine, 1995. Advanced traffic management for intelligent transportation systems, traffic operations and control, traffic simulation modeling, application of advanced software techniques, including artificial intelligence techniques to transportation and infrastructure management.
John R. Mays, Professor; Ph.D., Colorado, 1967; PE. Structural engineering, dynamic response of nonlinear structural systems, finite element analysis of structural systems.
Dunja Peric, Assistant Professor; Ph.D., Colorado, 1990. Constitutive modeling of soils, groundwater, finite element analysis, geotechnical engineering.
Anuhurada Ramaswami, Assistant Professor; Ph.D., Carnegie Mellon, 1994. Environmental engineering, multimedia transport and fate of pollutants, hazardous waste site remediation, integration of physicochemical and biological phenomena for clean up of contaminated soils and wastewaters.
Kevin L. Rens, Assistant Professor; Ph.D., Iowa State, 1994; PE. Structural engineering, nondestructive evaluation techniques, inspection and rating of infrastructure, forensic engineering, maintenance and repair of infrastructure, computer-aided structural engineering, structural dynamics and mechanics.
Judith Stalnaker, Associate Professor; Ph.D., Colorado, 1985; PE. Structural engineering, design of timber structures, concrete and steel structures, application of finite elements to concrete structures.
Jonathan Wu, Professor; Ph.D., Purdue, 1980. Soil engineering, finite element applications, ground improvements, seepage and groundwater flow, simulation of soil behavior, geosynthetics.

Computer Science and Engineering
Gita Alaghband, Professor and Chair; Ph.D., Colorado, 1986. Parallel processing involving algorithms, languages, and architecture; performance measurement; operating systems.
Tom Altman, Professor; Ph.D., Pittsburgh, 1984. Complexity theory and optimization.
John R. Clark, Professor; Ph.D., MIT, 1971. Information theory, random processes, artificial life, chaos.
Elain Eschen, Assistant Professor; Ph.D., Vanderbilt, 1997. Design and analysis of algorithms (graph algorithms, combinatorial optimization, approximate algorithms, data structures), discrete structures (graph theory, perfect graphs, partial orders).
Ross McConnell, Assistant Professor; Ph.D., Colorado, 1994. Algorithms, graph and network theory, combinatorial optimization, pattern matching in text and images.
John Noll, Assistant Professor; Ph.D., USC, 1997. Software engineering.
Christopher Smith, Assistant Professor; Ph.D., Minnesota, 1996. Robotics, computer vision, intelligent transportation systems, artificial intelligence.
Boris Stilman, Professor; Ph.D., National Research Institute for Electrical Engineering (Russia), 1984. Artificial intelligence, complex intelligent systems, linguistic geometry, software engineering.
William J. Wolfe, Associate Professor; Ph.D., CUNY, 1976. Computer vision, robotics, artificial intelligence, neural networks, expert systems, automated planning and scheduling.

Electrical Engineering
Brian Atkinson, Senior Instructor; M.S., Colorado, 1965. Audio electronics and system design, loudspeaker design, analog low-frequency electronics, control system implementation using analog techniques and microprocessor designs.
Jan Bialasiewicz, Associate Professor; Ph.D., 1966, D.Sc., 1972, Silesian Technical (Poland); PE. Optimal control, adaptive systems, stochastic systems, neural networks, identification and control of flexible structures, wavelets.
Tamal Bose, Associate Professor; Ph.D., Southern Illinois, 1988. Digital signal processing, communications, image and speech processing.
Hamid Fardi, Associate Professor; Ph.D., Colorado, 1986. Microelectronics, VLSI simulation and modeling, analog/digital electronics, optoelectronics, photovoltaics, measurement and characterization.
Shelly Goggin, Assistant Professor; Ph.D., Colorado, 1992. Pattern recognition, image processing, fuzzy systems, biomedical processing, processing for power systems, neural networks.
Joseph L. Hibey, Professor; D.Sc., Washington (St. Louis), 1976. Application of stochastic processes to estimation, decision, and control; nonlinear systems; stability theory.
Gary Leininger, Professor; Ph.D., SUNY at Buffalo, 1970. Adaptive systems in control and signal processing, intelligent control, statistical process control, manufacturing systems, management of technology.
Miloje (Mike) Radenkovic, Associate Professor; Sc.D., Belgrade (Yugoslavia), 1986. Robust control systems, stochastic control and system identification, adaptive systems in control and signal processing, control of large-scale systems, intelligent control.
Pankaj K. Sen, Professor and Chair; Ph.D., Technical of Nova Scotia, 1974; PE. Electric power systems, machines and energy.

Mechanical Engineering
William Clohessy, Professor; Ph.D., Cornell, 1948. Thermodynamics, primarily liquid drop formation and evaporation; statistical mechanics.
Joseph F. Cullen Jr., Senior Instructor; M.S., Colorado, 1985. Renewable energy, thermomechanical fluid theory, water resources engineering, aeronautical sciences.
James C. Gerdeen, Professor and Chair; Ph.D., Stanford, 1965; PE. Computer-aided design and manufacturing, sheet metal forming, design with polymers and composites.
Peter E. Jenkins, Professor and Dean; Ph.D., Purdue, 1974; PE. Turbomachinery (pumps, compressors, and turbines), engine technology, propulsion systems, energy conversion systems, basic and applied thermal sciences.
J. Kenneth Ortega, Professor; Ph.D., Colorado, 1976. Bioengineering: fluid mechanics in biological systems and stress-strain in biomaterials (plant and fungal walls), fluid mechanics: pulsatile flow, viscous flow, and convective heat and mass transfer.
Richard S. Passamaneck, Associate Professor; Ph.D., USC, 1978; PE. Fluid mechanics, thermodynamics, IC engines, high-energy well fracturing.
Ronald A. L. Rorrer, Assistant Professor; Ph.D., Virginia Tech, 1991; PE. Tribology, thermomechanical properties of polymers, friction-induced vibration, nonlinear dynamics, fatigue of polymers and composites, vibration damping and isolation, adhesion.
L. Rafael Sanchez, Associate Professor; Ph.D., Michigan Tech, 1987. Sheet metal forming, computer-aided design and manufacturing, experimental mechanics, bulk metal forming.
John A. Trapp, Professor; Ph.D., Berkeley, 1970. Two-phase flow modeling, computational fluid dynamics, thermal hydraulics, computational mechanics.
Samuel Welch, Assistant Professor; Ph.D., Colorado, 1993. Computational fluid dynamics.

UNIVERSITY OF CONNECTICUT

School of Engineering

Programs of Study

The University of Connecticut offers M.S. and Ph.D. programs in each of the six departments within the School of Engineering—Chemical Engineering, Civil and Environmental Engineering, Computer Science and Engineering, Electrical and Systems Engineering, Mechanical Engineering, and Metallurgy and Materials Engineering—and through four interdisciplinary programs—Biomedical Engineering, Environmental Engineering, Materials Science, and Polymer Science—affiliated with the School. The programs offer considerable flexibility for students who wish to tailor their studies to satisfy personal interests and professional requirements.

The M.S. degree program offers two options; one emphasizes research and requires the completion of a thesis, and the other stresses comprehensive understanding of professional practice in a field by successful fulfillment of course work and engineering projects.

The Ph.D. program prepares individuals with outstanding ability to become creative contributors in engineering or a related field. Award of the highest degree testifies to broad mastery of an established subject area, acquisition of research skills, and a concentration of knowledge in a specific field. A dissertation contributing to the body of knowledge in a chosen area of research is required for the degree. An individual plan of study is prepared with an advisory committee. The residence requirement is at least two consecutive semesters.

Research Facilities

The School and affiliated centers are housed in eight buildings on the Storrs and adjoining Depot campuses. Laboratories within academic departments and research centers are well-equipped and maintained by a professional staff. These laboratories provide graduate students with up-to-date facilities for their research in fundamental and applied investigations.

Graduate students are provided with ample research opportunities, particularly through multidisciplinary research centers with University-wide participation: the Biotechnology Center, the Taylor L. Booth Center for Computer Applications and Research, the Environmental Research Institute (which includes the Pollution Prevention Research and Development Center), the Institute of Materials Science (which includes the Electric Insulation Research Center and the Polymer Compatibilization Consortium), the Photonics Research Center, the Precision Manufacturing Center (which includes the Center for Grinding Research and Development), and the Transportation Institute. These research centers provide the most modern equipment and facilities in their respective area of specialization. The research centers enjoy strong external support from federal and state agencies and industry. Working in synergy with industry in cooperative research and development tasks, research is usually interdisciplinary in nature with specific applications in mind.

The School maintains its own network of computing systems in addition to the University-wide computing and communications facilities. Offices and laboratories are equipped with PCs and workstations that are connected by local area networks. The School's network is connected to the Internet for accessing computing resources throughout the world. Engineering students have access to sophisticated engineering graphics and imaging workstations, CAD/CAM, and parallel computers.

Financial Aid

Financial aid is available through fellowships, assistantships, and other types of support by the University, industry, and federal and state agencies. Research and teaching assistantships require part-time service for sponsored research projects and/or for the University. Assistantships may be continued through the summer on a part-time or full-time basis. Currently, the levels of support per nine-month academic year are $14,155 for graduate assistants with a baccalaureate, $14,895 for experienced graduate assistants with a master's degree, and $16,555 for experienced graduate assistants who have a master's degree or its equivalent and who have passed the general examination for the Ph.D. In addition, health benefits are provided, and tuition is waived for students who hold at least half-time appointments of 10 hours per week.

Cost of Study

For the 1999–2000 academic year, students who are classified as Connecticut residents pay tuition and fees of $3065 per semester. Out-of-state students pay tuition and fees of $7155 per semester. Tuition is prorated for students registering for fewer than 9 credits per semester. Student fees are prorated if registration is for fewer than 12 credits.

Living and Housing Costs

University housing for married and unmarried graduate students is available at a reasonable cost, but is limited. Apartments are also available in the surrounding area. The University Housing Office offers assistance in locating suitable housing on and off campus. Limited residential hall space is available for $1605 per semester, as is an optional meal plan for $1400 per semester.

Student Group

The University has about 22,000 students, and there are about 6,200 graduate students in eighty different fields of study. The School of Engineering has about 400 graduate students; more than 65 percent are full-time, and about 51 percent are in a Ph.D. program.

Location

The University of Connecticut is situated in picturesque northeastern Connecticut, which offers rural living with ready access to major cities. Boston is 1½ hours away by car; New York City, 2½ hours. Fishing and boating are popular in neighboring lakes and beaches. Excellent skiing is within easy driving distance.

The University

Founded in 1881, the University of Connecticut is the flagship and major research university in Connecticut. In addition to the main campus at Storrs and the adjoining Depot campus, there are five regional campuses, the UConn Health Center, the Law School, and the School of Social Work located throughout the state. The main campus in Storrs has 120 major buildings and covers 3,520 acres.

Applying

Application materials may be obtained from the graduate admission office. Applicants must hold a bachelor's degree or its equivalent and must submit all previous transcripts, three letters of recommendation, and a personal letter of application. The General Test of the GRE is strongly recommended, and international students from non-English-speaking countries are required to submit acceptable TOEFL scores.

Correspondence and Information

Graduate Admissions Office
University of Connecticut, U-6
438 Whitney Road Extension
Storrs, Connecticut 06269-1006

Telephone: 860-486-3617
E-mail: gradinfo@engr.uconn.edu
World Wide Web: http://www.engr.uconn.edu

University of Connecticut

THE GRADUATE FACULTY AND THEIR RESEARCH

Chemical Engineering
Luke E. Achenie, Thomas F. Anderson, James P. Bell, Douglas J. Cooper, Robert W. Coughlin, Michael B. Cutlip, Can Erkey, James M. Fenton, Joseph Helble, Jeffrey T. Koberstein, Patrick Mather, Montgomery Shaw, Robert A. Weiss, Thomas K. Wood.

This faculty offers research and advanced studies in polymer science, an interdisciplinary area cooperating with the Institute of Materials Science, with many active research projects, particularly in polymer compatibilization; in biochemical engineering and biotechnology (in close cooperation with the Department of Molecular and Cell Biology, the Department of Pharmacy, and the Biotech Center), with a wide range of research projects from water and wastewater treatment to fermentation and separation; in environmental research and hazardous wastes, which cooperates with the Environmental Research Institute studying hazardous materials, remediation, and pollution prevention; in catalysis/surface science and chemical reaction engineering, which conducts joint research with the chemistry faculty; in process control and optimization, which includes research in control, optimization, and large-scale simulations; in fuel cells and electrochemical engineering, which has considerable cooperative research with industry; and in thermodynamics and phase equilibrium, which has interests in the application of thermodynamics and phase equilibrium, pollution prevention, process design, materials development, energy technology, and biochemical engineering (http://www.engr.uconn.edu/cheg).

Civil and Environmental Engineering
Nelly M. Abboud, Michael L. Accorsi, Emmanouil N. Anagnostou, Christian F. Davis, John T. DeWolf, Kenneth R. Demars, Howard I. Epstein, Gregory C. Frantz, Norman W. Garrick, Domenico Grasso, George E. Hoag, John N. Ivan, John W. Leonard, Allison Mackay, Ramesh B. Malla, Rusk Y. Masih, Erling Murtha-Smith, Nikolaos P. Nikolaidis, Fred L. Ogden, Charles S. Sawyer, Barth F. Smets.

This faculty offers advanced studies in environmental and water resources engineering, which has research interests in environmental biotechnology, fate and transport of contaminants, environmental interfacial and colloidal phenomena, abiotic transformations, hazardous waste remediation technologies, filtration processes, transport and fate of sediments, hydrologic modeling and scaling, and remote sensing applications in hydrometeorology and hydoclimatology and uncertainty analysis of hydrologic processes; in structures and applied mechanics, which has research interests in finite element analysis and structural optimization, theory of poroelasticity, biomechanics, bridge vibrations and monitoring systems, fluid-structure interactions, dynamics and vibration of structures, structural integrity and progressive collapse, nonlinear analysis, structural connections, structural design code issues in steel and concrete, and geotechnical mechanics and processes; and in transportation systems engineering, which has research interests in livable urban street and highway design, social and environmental impacts of transportation decisions, pavement design and evaluation, highway crash reduction measure evaluation, highway crash prediction, and temporal/spatial link traffic volume forecasting (http://www.engr.uconn.edu/cee).

Computer Science and Engineering
Reda Ammar, Keith Barker, Steven A. Demurjian, Ian R. Greenshields, Lester Lipsky, Fred. J. Maryanski, Marios Marronicolas, Robert McCartney, Thomas J. Peters, John Roulier, Eugene Santos Jr., Mallory Selfridge, Dong-Guk Shin, Howard A. Sholl (Emeritus), Alexander Shvartsman, T. C. Ting.

This faculty offers advanced studies and research in artificial intelligence, which has interests in cognition and learning, case-based reasoning, intelligent human system interface and communications, expert systems, robotics, intelligent manufacturing systems and control, and distributed decision making; in graphic and imaging systems, which has interests in graphics generation, processing and editing, imaging processing and pattern recognition, medical imaging systems, computational geometry, feature-based CAD/CAM, and georeferenced information systems; in software design and engineering, which has research interests in software system design, development and evaluation, software system performance and evaluation, data models and database systems, object-oriented systems design, data security, human-system interactions and interface systems, information engineering, and distributed systems; in architecture and distributed systems, which has interests in distributed and network systems, distributed systems design, development and performance evaluation, querying models and data flow, control, parallel systems and architecture, and internetworking; and in theoretical foundations, which has interests in automatic querying models, data structure, discrete mathematics and structures, information theory, and graph theory (http://www.engr.uconn.edu/cse).

Electrical and Systems Engineering
Douglas Abraham, A. F. M. Anwar, John E. Ayers, Rajeev Bansal, Yaakov Bar-Shalom, Steven Boggs (Research Faculty), Joseph Bronzino (Professor-in-Residence), H. Chen, Peter K. Cheo, Eric Donkor, John Enderle, Martin D. Fox, Faquir Jain, Bahram Javidi, David Jordan (Emeritus), T. Kirubarajan (Research Faculty), David Kleinman (Emeritus), Peter B. Luh, Robert B. Northrop (Emeritus), Krishna Pattipati, Geoff Taylor, Peter Willett, Quing Zhu.

This faculty offers advanced studies and research in electromagnetics and physical electronics, which includes research in acoustical and optical holography and imaging, quantum electronics, optoelectronics, lasers, photonics, solar cells, semiconductor heterojunctions with application to integrated circuits, electron transport theory, antenna design, microwave technology, and dielectrics; in control and communication systems, which includes research in distributed decision making, adaptive control, digital control, optimal control, combination optimization, manufacturing systems, power systems, communication theory, stochastic communication and control, information theory, communication networks, fault-tolerant systems, signal analysis and processing, optical pattern recognition, and neural networks; and in biomedical engineering, which serves as a multidisciplinary program with faculty members in physiology, neurobiology, psychology, materials science, medicine, and dental medicine (http://www.ee.uconn.edu).

Mechanical Engineering
Matthew Begley, John C. Bennett, Theodore L. Bergman, Zbigniew M. Bzymek, Baki M. Cetegen, Eli K. Dabora (Emeritus), Amir Faghri, Robert G. Jeffers, Eric Jordan, Kazem Kazerounian, Herbert A. Koenig, Lee S. Langston, Kevin Murphy, Nejat Olgac, Ranga Pitchumani, Roman Solecki (Emeritus), Marios Soteriou, Bi Zhang.

This faculty offers advanced studies and research in design, manufacturing, and systems, which includes design of mechanisms and robotics systems, design methodology, computer-aided design and manufacturing, design of machine elements, design of precision manufacturing processes, computer-integrated manufacturing, precision grinding, ceramic and metal cutting, control systems theory, design and automatic control, robust control algorithms in robotics, and automated manufacturing systems; in applied mechanics, which has active research in stress analysis and vibrations of solids with cutouts/cavities/inclusions, mechanical behavior of materials including the modeling of the micromechanics of viscoplastic composites and polycrystalline material, numerical modeling and finite element analysis, active control of structural response, and the identification of complex systems; and in fluids and thermal engineering, which has research in compressible and heated flows, laminar and turbulent flow, two-phase flow and heat transfer, combustion, computational fluid dynamics, classical and statistical thermodynamics, conduction, convection and radiation heat transfer, combustion, fluid and thermal engineering devices and systems, and thermal manufacturing (http://www.engr.uconn.edu/me).

Metallurgy and Materials Engineering
Mark Aindow, Harold D. Brody, Philip C. Clapp, Owen F. Devereux (Emeritus), James M. Galligan, Norbert D. Greene, Trevor D. Howes, Theo Z. Kattamis, Harris Marcus, Arthur J. McEvily (Emeritus), John E. Morral, Nitin Padture, Donald I. Potter (Emeritus), Leon Shaw, Peter R. Strutt (Emeritus).

This faculty offers advanced studies and research in alloy science, corrosion, mechanical properties, physical and process metallurgy, and ceramics. Faculty research interests include thermal barrier coatings, ceramic composites, mechanically activated synthesis of powder, electronic structure of alloys, diffusion in high-temperature alloys, directional solidification mechanisms, novel casting technologies, quantitative prediction of solidification behavior, fracture toughness, martensitic transformations, electrochemical/corrosion measurement, dislocation drag mechanisms, photoplasticity and defects in electronic materials, fatigue crack growth, metallurgical surfaces by ion implantation, analytical transmission electron microscopy, thin-film high-temperature superconductors, and synthesis of nanostructured materials (http://www.ims.uconn.edu/metal).

UNIVERSITY OF DAYTON

School of Engineering

Programs of Study

The School offers programs leading to the degrees of Doctor of Philosophy in aerospace engineering, electrical engineering, electrooptics, materials engineering, and mechanical engineering and Doctor of Engineering in aerospace engineering, electrical engineering, materials engineering, and mechanical engineering, as well as a Master of Science degree in the following areas: aerospace engineering, chemical engineering, civil engineering, electrical engineering, electrooptics, engineering, engineering management, engineering mechanics, management science, materials engineering, and mechanical engineering.

Research Facilities

The modern Eugene W. Kettering Engineering and Research Laboratories building, a six-story, air-conditioned facility, provides 211,000 square feet and contains eighty-eight laboratories, fourteen classrooms, 115 faculty offices, and eight seminar rooms. Practically all of the engineering, technology, and research activities are housed in this building. Laboratories are available in such areas as man-machine simulation, energy conversion, electronics, electrooptics, digital systems, systems and human performance, microwaves, holographics, manufacturing, metallurgy, arc plasma, high-temperature materials, materials analysis, structural mechanics, civil engineering, spectroscopy, environmental engineering, soil mechanics, and aerospace, instrumentation, and fluid mechanics. Campus computer facilities are networked with all the engineering and research activities. Many high-performance and personal computers are available for faculty and student use.

Financial Aid

Graduate students are eligible for financial aid in several forms, including teaching and research assistantships, fellowships, traineeships, and scholarships. Fellowships provide for tuition plus a stipend. These fellowships involve full-time summer employment and half-time employment during the academic year. Teaching and research assistantships provide tuition plus a stipend. Appointments in these categories permit half-time study.

Cost of Study

Tuition in 1999–2000 for graduate students in engineering working toward a master's degree is $424 per registered credit hour; for those working toward a doctoral degree, the cost is $478 per registered credit hour.

Living and Housing Costs

There are no dormitory facilities for graduate students. However, there are many apartments and rooming facilities adjacent to the campus.

Student Group

In 1998–99, there were approximately 400 students from many states and various European and far eastern countries enrolled in the School's graduate programs.

Student Outcomes

The most recent graduates have been successful in gaining employment in a variety of positions with companies such as Procter and Gamble, Texas Instruments, Ford, Kodak, Bell Aerospace, General Dynamics, Lockheed-Martin, Hewlett-Packard, General Motors, Wright Laboratories, Motorola, SRL, SAIC, BDM, Spectra-Physics, EDS, Cummings Engine, Allied Signal, and G. E. Aircraft and with various government agencies like the U.S. Army Corps of Engineers and NASA. Some master's-level graduates begin their doctoral studies at major universities, and some doctoral-level graduates gain faculty appointments at other universities.

Location

Greater Dayton is a metropolitan area of about 933,500 people, situated on the Miami River in southwestern Ohio. It is approximately an hour's drive from Metropolitan Cincinnati and Columbus. As a result, a variety of cultural and recreational activities is available. The Dayton area is also in a favorable academic climate, with fifteen colleges and universities—seven of which offer graduate work—located within 75 miles. The Dayton area has one of the highest ratios of engineers and scientists to total population in the nation, and there is a heavy concentration of industrial and government engineering activity.

The University and The School

The University of Dayton, a privately endowed institution, was established in 1850. The School of Engineering was started in 1910, and graduate engineering activities were initiated in 1961. The University annually engages in approximately $45 million of engineering and scientific research through the sponsorship of private industry and government agencies.

Applying

Entrance requirements vary with each degree program. Applicants for admission must present evidence of their undergraduate and/or graduate preparation, as well as letters of recommendation. Applications for admission must be filed by August 1 for the first term, December 1 for the second term, and April 1 for the third term.

Correspondence and Information

For applications and graduate catalogs:
Office for Graduate Studies
200 St. Mary's Hall
University of Dayton
Dayton, Ohio 45469-0001
Telephone: 513-229-2343
World Wide Web: http://www.engr.udayton.edu

For additional information:
Office for Graduate Engineering Programs
   and Research
261 Kettering Laboratories Building
University of Dayton
Dayton, Ohio 45469-0227
Telephone: 513-229-2241
E-mail: gradinfo@engr.udayton.edu

University of Dayton

THE FACULTY AND THEIR RESEARCH

M. Atiquzzaman, Assistant Professor; Ph.D., Manchester (England), 1987. Computer networks, communication.
D. R. Ballal, Professor; Ph.D., Cranfield (England), 1972. Gas turbine combustion, jet propulsion.
L. I. Boehman, Professor; Ph.D., IIT, 1967. Advanced fluid dynamics, combustion, internal combustion engines.
F. K. Bogner, Professor; Ph.D., Case Tech, 1967. Numerical methods, structural response, composite materials, finite elements.
R. J. Brecha, Assistant Professor; Ph.D., Texas, 1990. Quantum optics.
R. A. Brockman, Professor; Ph.D., Dayton, 1979. Structural dynamics, solid mechanics, finite-element analysis, energy management.
C. E. Browning, Adjunct Professor; Ph.D., Dayton, 1976. Composites.
R. P. Chartoff, Professor; Ph.D., Princeton, 1968. Engineering polymers.
D. V. Chase, Assistant Professor; Ph.D., Kentucky, 1993. Water resources, numerical modeling, computer simulation, optimization.
H. N. Chuang, Professor; Ph.D., Carnegie Tech, 1966. Thermodynamics, energy conversion systems, solar heating systems analysis.
B. A. Craver, Associate Professor; Ph.D., Purdue, 1976. Nonlinear optics, optical hardening, optical chaos.
M. Daniels, Assistant Professor; Ph.D., Strathclyde (Scotland), 1982. Power, control systems, electrical machinery.
R. Deep, Associate Professor; Ph.D., Florida State, 1976. Artificial intelligence.
J. A. Detrio, Associate Professor; M.S., Alabama, 1966. Optical and electronic materials.
V. G. Dominic, Assistant Professor; Ph.D., USC, 1993. Nonlinear optics, beam agility.
G. R. Doyle Jr., Professor; Ph.D., Akron, 1973. Dynamics, vibrations, acoustics, controls.
B. D. Duncan, Associate Professor; Ph.D., Virginia Tech, 1991. Fiber optics, waveguides, electrooptic sensors, Fourier optics.
F. E. Eastep, Professor; Ph.D., Stanford, 1968. Aerodynamics and aeroelasticity.
C. E. Ebeling, Associate Professor; Ph.D., Ohio State, 1973. Reliability engineering.
J. P. Eimermacher, Professor; Ph.D., Cincinnati, 1973. Manufacturing, mechanical design, mechanical analysis, concurrent engineering.
T. E. Endres, Associate Professor; M.S.M.E., Dayton, 1969. Automatic assembly.
E. Ervin, Assistant Professor; Ph.D., Michigan, 1993. Computational fluid dynamics.
J. Ervin, Associate Professor; Ph.D., Michigan, 1991. Experimental heat transfer, multiphase flow.
D. Eylon, Professor; D.Sc., Technion (Israel), 1972. Microstructure and fracture.
L. Flach, Associate Professor; Ph.D., Colorado at Boulder, 1989. Process modeling, dynamics and control.
D. L. Flannery, Associate Professor; Ph.D., MIT, 1968. Coherent optics, electrooptic devices, optical computing.
J. P. Gallagher, Professor; Ph.D., Illinois, 1968. Fracture mechanics.
S. C. Gustafson, Associate Professor; Ph.D., Duke, 1974. Optical processing/computing, statistical optics, laser radar.
K. P. Hallinan, Associate Professor; Ph.D., Johns Hopkins, 1988. Capillary flow, multiphase flow, phase change heat transfer.
R. C. Hardie, Assistant Professor; Ph.D., Delaware, 1992. Signal and image processing, pattern recognition, remote sensing.
R. S. Harmer, Associate Professor; Ph.D., Illinois, 1971. Materials engineering, ceramics.
M. Hayat, Assistant Professor; Ph.D., Wisconsin–Madison, 1994. Optical communications, image/signal processing.
N. L. Hecht, Associate Professor; Ph.D., Alfred, College of Ceramics, 1972. Structural ceramics.
V. K. Jain, Professor; Ph.D., Iowa State, 1980. Machine design, materials, manufacturing, tribology.
K. V. Jata, Adjunct Professor; Ph.D., Minnesota, 1981. Physical metallurgy.
G. E. Johnson, Professor; Ph.D., Vanderbilt, 1978. Design, manufacturing.
A. R. Kashani, Associate Professor; Ph.D., Wisconsin–Madison, 1989. Control of manufacturing processes, structural dynamics and control.
R. J. Kee, Assistant Professor; D.E., Dayton, 1989. Power systems, power electronics, energy conversion.
J. K. Kissock, Assistant Professor; Ph.D., Texas A&M, 1993. Commercial and residential energy use.
S. Koh, Assistant Professor; Ph.D., Cincinnati, 1994. Computer engineering, microelectronics, optical interconnects.
B. Kumar, Associate Professor; Ph.D., Penn State, 1976. Optical materials, superconductivity.
C. W. Lee, Professor; Ph.D., Ohio State, 1982. Transport phenomena and polymer processing.
A. J. Lightman, Associate Professor; Ph.D., Weizmann (Israel), 1971. Optical measurement systems and machine vision.
G. R. Little, Assistant Professor; Ph.D., Ohio State, 1979. Optical computing, infrared fiber optics, electrooptic devices and systems.
J. S. Loomis, Professor; Ph.D., Arizona, 1980. Optical design, interferometry, computational optics, image processing.
C. C. Lu, Associate Professor; Ph.D., Texas, 1972. Transport phenomena, polymer science, thermodynamics.
D. L. Moon, Professor; Ph.D., Ohio State, 1974. Digital systems, computer architecture, fiber optics, electrooptic displays.
A. P. Murray, Assistant Professor; Ph.D., California, Irvine, 1996. Robotics and kinematic synthesis.
P. T. Murray, Associate Professor; Ph.D., North Carolina, 1979. Laser solid interaction.
K. J. Myers, Professor; D.Sc., Washington (St. Louis), 1986. Agitation, chemical reaction engineering.
J. M. O'Hare, Professor; Ph.D., SUNY at Buffalo, 1966. Theoretical physics, solid-state physics, optical properties.
K. M. Pasala, Professor; Ph.D., Bangalore (India), 1974. Antenna theory, electromagnetic scattering, microwave theory, signal processing.
L. Pedrotti, Associate Professor; Ph.D., New Mexico, 1986. Laser gyro, quantum optics, optical metrology.
R. Penno, Assistant Professor; Ph.D., Dayton, 1987. Antenna theory, electromagnetic scattering, microwave theory.
J. C. Petrykowski, Associate Professor; Ph.D., Illinois, 1981. Fluid mechanics, mathematical methods, mechanics and heat transfer.
A. E. Ray, Professor Emeritus; Ph.D., Iowa State, 1959. Metallurgy.
D. B. Rogers, Professor Emeritus; Ph.D., Dayton, 1978. Computer architecture, logical design, systems modeling, solid-state optical devices.
S. J. Ryckman, Distinguished Service Professor; M.S., Missouri, 1942. Environmental engineering: water treatment and pollution control.
S. I. Safferman, Assistant Professor; Ph.D., Cincinnati, 1994. Environmental engineering, hydraulics.
J. E. Saliba, Professor; Ph.D., Dayton, 1983. Structural analysis and design, finite-element analysis.
T. E. Saliba, Professor; Ph.D., Dayton, 1986. Process control, composite materials.
S. S. Sandhu, Professor; Ph.D., Imperial College (London), 1973. Combustion, thermodynamics, transport phenomena, process modeling.
F. Scarpino, Associate Professor; Ph.D., Dayton, 1987. Control systems, signal processing, microelectronics, avionics systems.
J. J. Schauer, Professor; Ph.D., Stanford, 1964. Heat and mass transfer, advanced engineering analysis, acoustics.
B. M. Schmidt, Distinguished Service Professor; Ph.D., Ohio State, 1963. Solid-state devices and quantum electronics.
R. A. Servais, Professor; D.Sc., Washington (St. Louis), 1969. Transport phenomena, process modeling.
J. A. Snide, Professor; Ph.D., Ohio State, 1975. Materials engineering, failure analysis.
G. Subramanyam, Assistant Professor; Ph.D., Cincinnati, 1993. Tunable components for communication systems.
P. J. Sweeney, Professor; Ph.D., Dayton, 1977. Engineering management.
F. Takahashi, Associate Professor; Dr.Eng., Keio (Japan), 1982. Combustion.
P. Taylor, Assistant Professor; Ph.D., Penn State, 1984. Laser-induced fluorescence/laser-induced photolysis.
D. Thebert-Peeler, Assistant Professor; Ph.D., Dayton, 1992. Surface modification and analysis.
G. A. Thiele, Tait Professor; Ph.D., Ohio State, 1968. Electromagnetics, antennas, radar scattering.
J. Ullett, Adjunct Professor; Ph.D., Dayton, 1992. Polymer chemistry.
T. M. Weeks, Adjunct Professor; Ph.D., Syracuse, 1965. Turbulent boundary layers in hypersonic flow, numerical analysis.
J. J. Westerkamp, Associate Professor; Ph.D., Purdue, 1985. Biomedical applications, statistical signal processing.
J. M. Whitney, Professor; Ph.D., Ohio State, 1968. Mechanics of composites.
T. L. Williamson, Associate Professor; Ph.D., Ohio State, 1975. Communications, electrooptics, holography.
J. C. Wurst, Professor Emeritus; Ph.D., Illinois, 1971. Ceramics, metallurgy.
P. Yaney, Professor; Ph.D., Cincinnati, 1963. Spectroscopic optical probe techniques, including Raman scattering, and absorption processes.
M. Zoghi, Associate Professor; Ph.D., Cincinnati, 1988. Soil mechanics, geotechnical engineering, earthquake engineering.

UNIVERSITY OF DELAWARE

College of Engineering

Programs of Study

The College offers master's and Ph.D. degrees through the Departments of Chemical Engineering, Civil and Environmental Engineering, Electrical and Computer Engineering, Materials Science and Engineering, and Mechanical Engineering. Interdisciplinary degree programs can be structured through concentration in bioelectronics; biomedical, environmental, and ocean engineering; energy conversion; engineering mechanics; materials science; and telecommunications. Part-time graduate students can attend classes as graduate nondegree students through the Engineering Outreach Program. These courses may apply to a graduate degree in engineering upon regular admission to a specific degree program in the College of Engineering.

Research Facilities

A recent renovation and addition provides state-of-the-art facilities for chemical engineering research. The chemical engineering laboratories include laboratories for process control, thermodynamics, mass transfer, polymer physics, polymer processing, biochemical research, surface science, colloids, protein adsorption, and photovoltaic unit operations. Computer facilities include several Sun Enterprise RISC server systems and a computer lab containing twenty networked Dell Pentium computers. Catalysis Center labs include high- and low-pressure reaction facilities, surface spectroscopy labs, synthetic labs, and equipment for other in situ spectroscopic techniques, such as FT-IR and solid-state NMR. Computational equipment for molecular modeling is also available. The civil and environmental engineering department has large research facilities with separate labs for each research area, which are equipped with computers and sophisticated research equipment to maximize each faculty's area of research. The ocean engineering lab contains a large directional wave basin, a precision wavetank, and several minor wavetanks, flumes, and other research facilities. It is one of the largest and best equipped in the country. The departmental computer system consists of more than thirty Sun Workstations networked to a central file server and to University mainframes. The electrical and computer engineering labs consist of several Sun Ultra 10/20 and Sun Ultra450 server platforms that support more than 200 desktop Ultra and SPARCstations and high-performance HP stations across nine TCP/IP Ethernet/Fast Ethernet subnetworks that maintain more than 80 GB of central disk storage. Fast Ethernet/FDDI switches provide seamless access to the more than 300 GB of distributed disk storage. Specialized computing equipment includes an ATM-based laboratory with twenty-five-seat Silicon Graphics workstations for image and HDTV research. The department is also home to a distributed computing NSF-funded high-performance cluster of twenty 4-processor Sun Ultra450s networked with 1GB/s Myrinet interfaces to form an 80-process/10-GB main-memory super computer cluster. Academic use of the SGI lab and Sun Ultra2 is facilitated by a twenty-seat lab of Windows-NT workstations and server housed in a lecture/presentation-capable room. There are labs for optoelectronic fabrication and a state-of-the-art clean room for device fabrication and semiconductor processing. Materials science and engineering facilities include thermal analysis; electron, optical, and atomic-force microscopes; X-ray diffraction and fluorescence, synthesis, and processing equipment; and FTIR, Raman, and FT-Raman spectroscopic instrumentation. The mechanical engineering labs have facilities for research in polymer, metal, and ceramic matrix composites manufacturing and nondestructive evaluation; precision forming and design of trim panels for aircraft interior noise suppression; robotics, real-time planning, and optimization of dynamic systems and rehabilitation robotics; musculoskeletal modeling and the study of rheumatoid arthritis; impact response of solids using high-speed photography and infrared thermography; manufacturing applications; atmospheric aerosols related to global warming and urban smog; and turbulence and particle-laden flows on SGI/Sun UNIX computers and remote supercomputers. Six interdisciplinary centers provide additional research facilities and opportunities. They are the Center for Advanced Coastal Research, the Center for Catalytic Science and Technology, the Center for Composite Materials, the Center for Molecular and Engineering Thermodynamics, the Center for Nanomachined Surfaces, and the Orthopedic and Biomechanical Engineering Center.

Financial Aid

Fellowships, assistantships, and tuition scholarships are available. In 1999–2000, twelve-month stipends range from $13,200 to $18,200. Teaching assistantships also cover tuition. International teaching assistants are required to attend a free four-week training program. A stipend is paid for attending. Minority fellowships and support for summer research are available.

Cost of Study

The tuition fees for 1999–2000 for full-time graduate students are $2190 per semester for Delaware residents and $6375 per semester for nonresidents. Part-time study is $243 and $708 per credit hour for resident and nonresident students, respectively.

Living and Housing Costs

University-owned graduate housing is offered on a first-come, first-served basis. In 1999–2000, monthly apartment rents range from $340 to $680 per month. Privately owned apartments, both furnished and unfurnished, are available in the surrounding area.

Student Group

More than 21,000 students are enrolled at the University. The College has approximately 1,400 students from all over the world. Women constitute 22 percent of the engineering students at Delaware. In 1998–99, 209 students who were members of minority groups were enrolled in engineering programs.

Student Outcomes

Approximately 85 percent of graduates choose employment in private industry, government laboratories and agencies, and nonprofit research centers. Another 15 percent pursue careers in academia.

Location

The College of Engineering is on the main campus of the University in the suburban community of Newark, Delaware, within an hour's drive of Philadelphia, Baltimore, Wilmington, Dover, and the Chesapeake Bay. New York City and Washington, D.C., are only 2 hours away. The College's central location provides ready access to major corporate centers and facilities throughout the mid-Atlantic region.

The University

The University was founded in 1743 as a small liberal arts school. Today, as one of the oldest land-grant institutions in the nation, it is recognized as a major state-assisted, private university. It is situated in an area rich in technical talent and interests. The cultural and technical interaction between students and faculty and the surrounding community provides a stimulating environment.

Applying

Applications for admission must be submitted by July 1 for the fall semester and by December 1 for the spring semester. Students seeking financial aid should complete their applications by January 1. International students interested in teaching assistantships must pass the TOEFL and an oral proficiency test and, if selected, attend the University's International TA Training Program.

Correspondence and Information

Department of (specify)
College of Engineering
University of Delaware
Newark, Delaware 19716
World Wide Web: http://www.udel.edu/engg/

University of Delaware

DEPARTMENT CHAIRS AND AREAS OF FACULTY RESEARCH

Chemical Engineering. Professor Eric W. Kaler, Chair. Major research thrusts are in biochemical and biomedical engineering; catalysis, surface science, chemical reaction engineering; colloid science; electronic materials processing; expert systems and process control; polymer science and engineering and fluid mechanics; thermodynamics and phase equilibria. Biochemical and biomedical engineering research includes the development of artificial tissues and organs, properties and purification of biological fluids, protein folding, molecular and cellular engineering, bioreactor control, and the study of the effects of environmental factors on protein activity. The department's Center for Catalytic Science and Technology provides an intense focus on catalysis and surface chemistry. Development of molecular-level understanding of catalytic reaction mechanisms through both experiments and computational methods; synthesis and characterization of novel catalytic materials, including nanoporous solids; and the development and application of chemical reaction engineering approaches are key components of the effort to develop new, environmentally benign catalytic processes. Electronic materials processing includes the development of semiconductors for photovoltaic and other electronic device applications, chemical vapor deposition for the production of advanced electronic materials with semiconductor applications, and fundamental research in the chemical synthesis of high-temperature superconductors. Expert systems and process control research involves the application of expert systems and artificial intelligence methodology to fault diagnosis and control in chemical production and process simulation. Polymer science and engineering and fluid mechanics research includes the study of fabrication operations, such as polymer extrusion and injection molding, detailed numerical analysis of the complex flow of composite materials during manufacture, flow properties of polymeric liquid crystals, flow of highly concentrated ceramic suspensions, and new techniques for the control of the flow of ceramic pastes and the orientation of any suspended fibers. The department's Center for Molecular and Engineering Thermodynamics focuses on the experimental and theoretical studies of the thermodynamics of phase behavior of complex fluids, polymeric materials and proteins, high-temperature aqueous solutions and biological compounds in water, molecular simulations, and the chemistry of environmental problems.

Civil and Environmental Engineering. Professor Chin-Pao Huang, Chair. Four major areas of research interest in the department are environmental engineering, ocean engineering, structural and geotechnical engineering, and transportation engineering. Specific research in environmental engineering includes sediment and soil quality standards, remediation of heavy metal contamination, fate of chemical warfare agents, biological sludge treatment and toxic waste biodegradation, analysis of the chemical and biological processes in landfills, photocatalytic oxidation of organic wastes, and treatment of contaminated aquifers. The Center for Applied Coastal Research provides a focus for research of nearshore circulation and offshore breakwaters; computer modeling of the water-wave spectrum; wave breaking; and the generation of infragravity waves, such as surf beat and edge waves. The transport and fate of spilled oil are also among the research topics of the faculty. The department conducts active research in nonlinear dynamic analyses of building frameworks and steel structures, application of composite materials to civil engineering structures, development of efficient models for geomaterials, behavior of composite or reinforced soil structures to seismic loadings, bearing capacity of floating ice plates, analysis and design of railway tracks, analytical modeling of concrete pavements for airports and highways, and thermal buckling of concrete pavements, three-dimensional stability analysis of inhomogeneous slopes, design of inflated fabric-reinforced concrete structures, and random vibration of arch dam reservoirs during earthquakes. Transportation research efforts include evaluation of multilane design alternatives for improving suburban highways, application of fuzzy set theory and artificial intelligence to transportation problems, an integrated traffic monitoring system for the state of Delaware, evaluation of rail-highway crossing safety programs, and pavement design optimization with reinforced roller-compacted concrete.

Electrical and Computer Engineering. Professor Gonzalo R. Arce, Interim Chair. The special research interests of the faculty fall into four broad groups: signal processing and communications, electronic devices and materials, computer systems engineering, and optics and electromagnetics. The signal processing and communications thrust emphasizes robust signal processing, multimedia communications, wireless communication networks, source coding, image and video signal processing, wavelets, and multirate signal processing. Research includes the development of authentication digital image watermarks and multiuser communication systems. The electronic devices and materials area emphasizes photonic and microwave frequency devices, optoelectronic systems, and electronic materials. Research includes the fabrication and characterization of a new generation of photonic devices made from alloy films of silicon, germanium, and carbon (SiGeC). Computer systems engineering emphasizes the operation of high-performance networks, computation with analog neural networks, and special purpose computers for image and signal processing applications. Research includes human-computer interaction systems and methodology and the development of computers that mimic the function of biological systems. The activities of the electromagnetics group include ultrafast electronic circuits for terahertz spectroscopic sensors and the design of subwavelength diffractive optical elements for optoelectric sensing and communications. The ECE department is home to the Center for Nanomachined Surfaces, a State of Delaware Advanced Technology Center that is a combined industry/academic effort to develop tools and processes for producing extremely smooth and flat surfaces for the semiconductor industry. The multifunctional scanned probe microscopes and other surface analytical tools are developed as part of the department's research activities in electromagnetics. Research is under way to greatly improve the routing and scheduling of high-bandwidth networks with minimal internal storage as well as time-keeping and synchronization in very wide area computer networks.

Materials Science and Engineering. Professor John F. Rabolt, Chair. The major focus of this department's research efforts is in polymers, materials chemistry, composites, electronic and magnetic materials, physical metallurgy, catalysts, gas sensors, and inorganic and organic thin films. Polymers represents a growth area for the department and includes studies of crystallization, morphology, and the synthesis and characterization of advanced polymeric materials. Metal matrix and fiber reinforced composites are being investigated (in conjunction with the Center for Composite Materials) for applications that range from lightweight armor to civilian infrastructure (bridge) programs. Research in electronic and magnetic materials focuses on advanced organic films for flat panel displays, electroluminescent polymers for LED applications, the development of materials useful in the conversion of solar energy into electricity, and the magnetic properties of high-temperature superconductors. The area of physical metallurgy brings together research in thermal conductivity and corrosion, while that in organic thin films focuses on novel methods of deposition, including Langmuir-Blodgett (LB), self-assembly (SA), and hybrid LB/SA techniques.

Mechanical Engineering. Professor Suresh G. Advani, Acting Chair. The research of the mechanical engineering department is focused in six areas: composites, air pollution, biomedical engineering, manufacturing science, robotics and control, and advanced materials. The department faculty members are also involved in cross-disciplinary research programs. The Center for Composite Materials' work includes the mechanics and manufacture of advanced composite materials for the study of smart structures. Air pollution research involves high-performance computing techniques and advanced instrumentation to study particulate air pollutants and transport phenomena in combustion. The Center for Biomedical Engineering Research provides a framework for interdisciplinary research in the general area of bioengineering. Topics include the generation of force and motion in the human body, orthopedic and rehabilitation engineering, and the study of pulmonary and renal fluid mechanics. Manufacturing research is concerned with all aspects of flow in thin liquid films, including paints and other coatings, the behavior of fibers in concentrated suspensions, resin transfer mold–filling processes in composites manufacturing, rapid tooling, and lubrication and cooling during machining. Current research areas in robotics and control are design of novel robotic systems, coordination and control of multi-degree-of-freedom robot systems, smart materials and intelligent structures, and optimization of dynamic manufacturing processes. Advanced materials engineering is concerned with characterization and modeling of engineering materials, including polymer, metal, and ceramic matrix composites and high-strain-rate deformation and Split Hopkinson Pressure Bars for high-strain-rate testing.

UNIVERSITY OF DENVER

Department of Engineering

Programs of Study

The University of Denver's (DU) Department of Engineering offers four traditional and two nontraditional graduate programs. The department offers programs leading to the Master of Science in electrical engineering and the Master of Science in mechanical engineering. The purpose of these programs is to serve the profession of engineering and the Colorado community through advanced study in electrical or mechanical engineering and related fields. The specializations offered within the electrical engineering program are electromagnetics, quantum optics, semiconductors, signal processing, and communications. The specializations offered within the mechanical engineering program are structure and behavior of materials, fluid mechanics, heat transfer, mechanical design, and analysis/robotics. Both programs also prepare the students for academic and industrial advancements. Each program offers a thesis and a nonthesis option.

The Master of Science in management and engineering, a unique joint degree offered by the Daniels College of Business and the Department of Engineering, is designed primarily for the professional engineer who is in midcareer (five to twelve years post-B.S.). Many engineers advance to leadership positions within their businesses. They are designated team or group leaders or promoted to section, branch, or division heads. In these roles, they are called upon to exercise managerial skills that often have not been fully developed. Furthermore, with technology constantly undergoing rapid change, skills acquired in undergraduate education may be in need of upgrading and expanding to maintain mastery of the profession. The purpose of this program is to serve the engineering professional and the Colorado industrial community through advanced study in both management and engineering, thus meeting both needs in one comprehensive program. The program prepares its graduates for industrial advancement.

The Master of Science in computer science and engineering, a joint degree, offers the opportunity to concentrate on the key engineering areas of interest and to obtain a solid set of core knowledge in computer science. The typical person seeking this degree is an engineer who is highly interested in computer science. Such individuals are involved in the intimate use and integration of computers in their everyday work.

The Master of Science and the Doctor of Philosophy in materials science degree programs are designed to prepare the student for research and development work in the materials field. The programs are multidisciplinary and involve the Departments of Physics, Chemistry, and Engineering, with Engineering as the administering department. The programs reflect this multidisciplinary nature by providing a thorough grounding in each of the basic disciplines of the field. Depth in specialized areas is achieved through the research interests of faculty members in each of the participating departments. These faculty members constitute the Materials Science Faculty Group, which is responsible for implementing and administering the programs.

Research Facilities

The University of Denver possesses excellent media and computer facilities. Penrose Library has online search capabilities that include the Colorado Alliance of Research Libraries (CARL), LEXIS-NEXIS, Dialog, and other business and humanities indexes. Research areas in which the department currently has strength are finite element modeling of fiber composite failure and mechanical behavior; detection and measurement of tropospheric and stratospheric aerosols; acoustic emission from materials, including composites; residual stress determination in composites; finite element modeling of electromagnetic fields in wave guides; and robotic vision/image processing/networking.

Financial Aid

Two kinds of financial aid are available for graduate students: need-based and merit-based aid. Students applying for either kind of aid must be accepted into an eligible graduate program at the University. Need-based financial aid consists of Federal Perkins Loans, the Federal Work-Study Program, and Colorado Graduate Grants. Merit-based financial aid consists of graduate tuition scholarships, graduate teaching and research assistantships, Colorado graduate fellowships, and minority student fellowships. A student who seeks any kind of financial assistance to pursue graduate studies at the University must file the Free Application for Federal Student Aid (FAFSA). For more information, students should contact the graduate school or department in which they wish to enroll.

Cost of Study

For the 1998–99 academic year, the tuition rates were as follows: $506 per quarter hour and $6072 per quarter (maximum charge for 12 to 18 quarter hours). Students registering for more than 18 quarter hours, including noncredit courses, were charged the hourly rate of $506 per quarter hour in excess of 18. Tuition charges are the same for all students, whether residents of Colorado or nonresidents. The University reserves the right to make changes in tuition charges or refund policies without advance notice.

Living and Housing Costs

The estimated housing cost per year for a one-bedroom apartment was $4305, effective autumn quarter 1998. The cost per quarter was $1096. Off-campus apartments are available nearby and range from $300 to $500 per month. A liberal estimate of total monthly living expenses (rent, board, books, and personal spending) is $1000.

Student Group

Approximately 3,000 graduate students attend the University and comprise a little more than half of the 5,818 traditional students enrolled. Another 2,892 nontraditional (mostly graduate) students are enrolled, bringing the University's total student enrollment to 8,710 students. Approximately 14 percent of the total student population consists of students who are members of ethnic minority groups.

Student Outcomes

Most graduate students in the University of Denver engineering department find employment within a year of graduation, most of them within ninety days of graduation.

Location

The 125-acre University Park campus is located in a residential neighborhood 8 miles southeast of downtown Denver. The Rocky Mountains are 1 hour west by car. The Park Hill campus, 9 miles northeast of the main campus, is home to the Lamont School of Music, the College of Law, and the Women's College.

The University

The University of Denver is an independent institution offering undergraduate and graduate degree programs. Founded in 1864 by John Evans, who was appointed by Abraham Lincoln as the second governor of the Colorado Territory, it is the largest independent university in the state and one of four research universities in Colorado. As the oldest and largest private university in the Rocky Mountain region, the University of Denver draws students from all fifty states and eighty-eight countries.

Applying

Applications are accepted and reviewed on a rolling admissions basis. This means that prospective students can expect an admission decision within three to seven weeks after the application file is complete. Most students apply between November and February for the subsequent fall quarter. The application fee is $40. All graduate applicants must take the GRE and have an acceptable score. If applying for the joint degree with the Daniels College of Business, the student must also take the GMAT.

Correspondence and Information

Graduate Program Coordinator
University of Denver
2199 South University Boulevard
Denver, Colorado 80210
Telephone: 303-871-2305
Fax: 303-871-4566
E-mail: gfac@denver.du.edu
World Wide Web: http://www.du.edu/grad/gradadm.html

Dr. Paul Predecki
Engineering Program Coordinator
University of Denver
2390 South York Street, Room 200
Denver, Colorado 80210
Telephone: 303-817-2107
Fax: 303-871-4450
E-mail: ppredeck@du.edu

University of Denver

THE FACULTY

Professors
M. A. Hamstad, Ph.D., Berkeley; PE.
P. K. Predecki, Ph.D., MIT.
A. J. Rosa, Ph.D., Illinois at Urbana-Champaign.
E. R. Tuttle, Ph.D., Colorado at Boulder; PE.
J. C. Wilson, Ph.D., Minnesota.

Associate Professors
R. R. Delyser, Ph.D., Colorado at Boulder.
M. M. Kumosa, Ph.D., Wroclaw Technical (Poland).
R. E. Salters, Ph.D., New Mexico.

Assistant Professors
G. Edwards, Ph.D., South Florida.
K. Meehan, Ph.D., Illinois at Urbana-Champaign.
D. Megherbi, Ph.D., Brown.
P. Rullkoetter, Ph.D., Purdue.
R. Whitman, Ph.D., Colorado at Boulder.

Lecturers
A. Boulenouar, D.E., Northeastern.
F. Long, Ph.D., Iowa State.
J. D. Mote, Ph.D., Berkeley; PE.

Adjunct Faculty
A. Czanderna, Ph.D., Purdue; National Renewable Energy Laboratory.
I. Hindash, Ph.D., Colorado at Boulder; Aerovehicles.
A. Hinrichs, E.E.E., USC; Hughes Electronics.
D. Kishoni, Ph.D., Cornell.
K. E. Mahrer, Ph.D., Stanford; Denver Research Institute.
W. V. McCullough, Ph.D., Carnegie Mellon; Hughes Aircraft Corporation.
D. C. Rolley, M.S., Denver; Hughes Aircraft Corporation.
J. Sebesta, M.S., Colorado State; United Engineers and Technologists.

UNIVERSITY OF DETROIT MERCY

College of Engineering and Science

Program of Study	Graduate programs leading to the degrees of Master of Engineering (M.Engr.) and Doctor of Engineering (D.Engr.) are offered through the departments of chemical engineering, civil and environmental engineering, electrical engineering, and mechanical engineering. The chemistry department offers programs leading to the M.S. and the Ph.D. degrees in chemistry as well as a Master of Science in Economic Aspects of Chemistry (M.S.E.C.). The biology department offers programs leading to the M.S. in biology. The programs of the department of mathematics and computer science lead to either the M.A. in mathematics or Master of Arts in Teaching of Mathematics (M.A.T.M.) and the Master of Science in Computer Science (M.S.C.S.). The College, with support from the College of Business Administration, offers programs leading to the Master of Engineering Management (M.E.M.) and Master of Science in Product Development (M.S.P.D.). The M.P.D. cohort program has been developed in collaboration with Massachusetts Institute of Technology and five industrial partners as well as with the U.S. Navy and NSF.
Research Facilities	The Center of Excellence in Environmental Engineering and Science is an interdisciplinary research and educational center that supports studies in a wide variety of environmental areas, with a special emphasis on polymers. A research and administration center houses a variety of new instrumentation and equipment for environmental research, including a Haake rheometer; extruders; shredding, grinding, and pulverizing equipment; high-pressure reactors; and several chromatographs and spectrometers. The Polymer Institute is a center for research and development in both fundamental and applied polymer areas. The institute has extensive equipment for the characterization of polymers (chemical, mechanical, thermal, spectroscopic, and environmental). Polymer process equipment includes RIM machines, an injection machine, a foam machine, extruders and presses, and specialized equipment for the testing of coatings, sealants, and adhesives. Complete rubber compounding and molding facilities are also available. The recently expanded Manufacturing Laboratory conducts research to advance the state of the art in four dominant areas: metal cutting, surface finish, computer-aided manufacturing, and quality control. Major capital resources are available for research projects, including CNC Milling and Turning Centers, a coordinate measuring machine, machine tool dynamometers, an X-ray fluorescence machine, and surface finish characterization equipment. A $500,000 renovation project (with NSF support) was recently completed, creating the Polymer Synthesis Laboratory, the Polymer Characterization and Testing Laboratory, and the Physical Measurement and Computation Laboratory. The Advanced Computing Laboratory supports both the teaching and research of the College, providing advanced computer workstations and a wide variety of software for such functions as geometric and solid modeling; design; thermal, kinetic, and finite element modeling; modeling of injection molding, cooling, and warpage; NC program generation; chemical process modeling and control; and modeling of electrical systems.
Financial Aid	Teaching and research assistantships are limited and are normally obtained only after a student has spent at least one term in the program. The possibility of assistantships is explored through consultations with individual faculty members. In addition, the Scholarship and Financial Aid Office accepts applications for grants, loans, and work-study assistance. Aid available includes the Michigan Tuition Grant (for Michigan residents only), various work-study programs, and a wide variety of loans. The University also accepts third-party payments from employers and government agencies and offers its own payment plans. For information regarding financial aid programs, students should call 313-993-3350.
Cost of Study	Tuition in 1999–2000 is $545 per credit hour. Registration fees are $100 per term for full-time students.
Living and Housing Costs	Housing is available on both the McNichols and Outer Drive campuses. Double-occupancy rates range from $1410 to $3280, single-occupancy rates from $2420 to $2720. The University offers meal plans at costs ranging from $595 to $1085. All rates are for a sixteen-week term. For more information, students should contact Residence Life at 313-993-1230.
Student Group	In fall 1998, the total enrollment of the University was approximately 6,700. International students constituted 15 percent of the graduate student body; 55 percent of the graduate students were women. That same year, there were 180 graduate students in the College of Engineering and Science.
Location	The campus and the city offer a variety of cultural programs. The University has on-campus performances, and the city has a symphony orchestra, an art institute, excellent library facilities, resident theater companies, and programs of visiting opera, ballet, and other events. The metropolitan area has many lakes. Ski resorts are within a half-hour drive. Detroit supports several professional sports teams.
The University	The University of Detroit Mercy was formed in 1990 through consolidation of the University of Detroit (founded in 1877) and Mercy College of Detroit (founded in 1941). The University is an independent institution operated under the auspices of Jesuit educators and the Religious Sisters of Mercy. The University comprises three campuses. McNichols Campus houses the Colleges of Business Administration, Engineering and Science, and Liberal Arts and the School of Architecture. Outer Drive Campus houses the College of Education and Human Services, the College of Health Professions, and the Dental School. There is also a Law School campus in downtown Detroit.
Applying	Applications for admission should be completed at least six weeks before the beginning of a term. Official transcripts are required from all colleges attended. Applications for financial aid should be submitted by March 1. International students are urged to complete their applications at least three months before classes begin. Admission requirements are a bachelor's degree from an accredited college, a B average in the total undergraduate program and in the proposed field of study, and, normally, an undergraduate major or the equivalent in the proposed field. Applicants with less than a B average who present other evidence of ability to perform graduate-level work may be admitted as probationary students upon the recommendation of the director of the program concerned. The application fee is $30.
Correspondence and Information	Records Office College of Engineering and Science University of Detroit Mercy P.O. Box 19900 Detroit, Michigan 48219-0900 Telephone: 313-993-3336 Fax: 313-993-1187 E-mail: engineering@udmercy.edu

University of Detroit Mercy

FACULTY RESEARCH AREAS

Biology
The Department of Biology offers graduate programs specifically designed to meet the needs of a variety of students in the life sciences. This includes, for example, students preparing themselves for more advanced studies in health-related careers, students wishing to delve more deeply into one of the more specialized areas of biology, teachers and others wishing to upgrade their biology background, and adults contemplating or working toward career changes that require a deeper knowledge of the life sciences. Areas of research emphasis include cellular-molecular biology and microbiology. Recent research has focused on such areas as bioremediation of dangerous waste materials and chromosome fingerprinting.

Chemical Engineering
Research interests within the department include programs or projects focused mainly on three broad areas. Chemical and polymer processes: fluid flow, polymers, composite manufacturing and processing, separations, polymer charging (electrets), polymerization and curing kinetics, high-conversation kinetics and RIM, injection molding, blow-film extrusion, and application of artificial neural networks. Electrochemical engineering: plating, batteries, and electrolytical recovery of materials from waste streams. Environmental solutions: filtration, aerosol and electrostatics charging, and plastics recycling and reuse. Interdisciplinary research is routinely conducted in cooperation with investigators from other engineering departments, the Polymer Institute, the Center of Excellence in Environmental Engineering and Science, and the chemistry department.

Chemistry and Biochemistry
Research programs in the Department of Chemistry and Biochemistry include the following areas: regulation of cerebral endothelial $(Na^+ +K^+)$-ATPase, hypertension, heavy-metal toxicity, high-affinity choline uptake, photochemistry in organized media, asymmetric synthesis, analytical techniques for complex polymers, size-exclusion chromatography, study of dynamical processes in disordered systems using electron-spin probes and ^2H NMR, effects of high-energy radiation on clathrate compounds and stabilization of reactive intermediates and electron pairs, synthesis, transition-metal catalysis on polymeric supports, water-soluble organo-metallic catalysts, atmospheric chemistry and quantum theory, carborane monomers and polymers, free-radical ligand chemistry, magnetic organo-based compounds, and regulation of cytochrome P-450.

Civil and Environmental Engineering
The department engages in research in the following areas. Geotechnical: constitutive properties of cohesive soils, effect of chemicals on the hydraulic conductivity of clay liners, and groundwater modeling. Traffic: traffic safety, smart cars on highways, and transit maintenance. Structural engineering: soil-structure interaction analysis, structural optimization, and finite-element modeling. Environmental engineering: performance of waste tire–asphalt mix, analytical and experimental study on containment of hazardous wastes, and solidification of waste particulates using isocyanate-based binders.

Electrical and Computer Engineering
Faculty members are involved in research in the areas of: automotive electronics, image and speech processing (including vision systems), digital signal processing, computer architecture and networking, sensor analysis, microstrip antennas, electromagnetic interference (EMI) and electromagnetic compatibility (EMC), and digital control systems.

Mathematics and Computer Science
Faculty research interests cover a broad range of mathematical and computer sciences, including holography, functional analysis, actuarial mathematics, abelian groups, modules and rings, and mathematical modeling applied to industrial problems, health sciences, statistics, and computing.

Mechanical Engineering
Faculty members work closely with industry and government agencies to conduct research focused mainly on the department's three areas of concentration: manufacturing systems, automotive engineering, and thermal fluid systems. Topics of research include modeling of manufacturing processes and systems, laser welding and machining, nondestructive identification and electronic cataloging of fasteners, design of composite parts, multimedia tools for engineering education, automobile emissions, sensors and actuators for automotive electronics, computational fluid dynamics and heat transfer, and two-phase flow.

Product Development
Faculty research interests focus on various aspects of "end-to-end-to-end" product development process. The faculty works closely with industry leaders, government agencies, and academics/research institutions, such as MIT and the Center for Innovation in Product Design. The thrust of the research is to drastically enhance competitiveness of the U.S. automotive and related industries.

UNIVERSITY OF FLORIDA

College of Engineering
Graduate Engineering and Research Center

Programs of Study	The College of Engineering's Graduate Engineering and Research Center (GERC) at the University of Florida offers Master of Science (M.S.), Master of Engineering (M.E.), Engineer (Engr.), and Doctor of Philosophy (Ph.D.) degrees in aerospace engineering, electrical and computer engineering, engineering mechanics, and industrial and systems engineering. In addition to traditional degree programs, certificate programs are offered in test and evaluation and in environmental policy and management.
	A minimum of 30 semester hours is required for the M.S. and M.E. with thesis, 6 of which represent work on the thesis. The nonthesis M.S. and M.E. require 32 to 34 semester hours. (An accredited bachelor's degree in engineering or its equivalent is a prerequisite for the M.E.) The Engineer degree requires 30 semester hours beyond the master's. The Ph.D. degree requires 90 semester hours (including dissertation) beyond the bachelor's; 30 semester hours in one calendar year or 36 semester hours in no more than four semesters within a period of two calendar years must be earned on a University of Florida campus. The language requirement for the Ph.D. degree varies by department.
	Programs of study may be tailored to emphasize the individual interests of graduate students. Master's and Ph.D. candidates who are preparing theses or dissertations work closely with individual faculty members on theoretical or applied research problems. An articulation program of foundation courses is required if the student's background is deficient.
	The University operates on a schedule of two semesters plus one summer term.
Research Facilities	The Center has 15,000 square feet of research laboratory space. Facilities include those for microwave, photonics, laser, holography, shock tubes, and material testing and characterization research. A high-speed network connection to the computer resources of the Eglin Air Force Base, including the Eglin AFB Super Computer, is available for research purposes. Specialized research laboratories on the Eglin AFB are also available for joint University–Air Force projects.
Financial Aid	Research assistantships are available from the Center. The University of Florida Graduate School may provide additional funding for fellowships. Assistantship stipends begin at $10,440 for the 1999–2000 academic year. Research fellowships are available for applicants seeking a Ph.D. degree.
	Financial assistance is available through the University. Employment benefits are the primary source of tuition payments for many students. Full-time students supported by assistantships participate primarily in advanced defense-related research projects and have access to laboratory facilities at the nearby Eglin AFB.
Cost of Study	The 1998–99 tuition was $137.75 per graduate semester credit hour for Florida residents and $481.31 for nonresidents. Students on assistantships are normally eligible for full tuition fee waivers.
Living and Housing Costs	On-campus housing is not available at this location. The area has a moderate cost of living. The cost of condominiums and single-family homes in the area begins at $45,000, while rentals begin at $325 per month.
Student Group	Part-time students constitute 94 percent of the graduate student population; 83 percent are male, 21 percent are military officers, and 45 percent are government scientists and engineers.
Student Outcomes	Most students attend the Center to enhance their knowledge and credentials for career advancement with their current employers. Students have typically been promoted within their organizations or quickly hired by local and national industrial firms or government agencies after receiving a master's or Ph.D. degree.
Location	Shalimar, Florida, is located in the Fort Walton Beach area on the Emerald Coast of northwest Florida. The area has several military bases, a variety of high-technology industries, and an abundance of educational and recreational advantages. In 1997, this area was listed by *Money* magazine as the tenth-best place to live among the 100 largest U.S. metropolitan areas.
The Center	The Graduate Engineering and Research Center was established in 1969 as a satellite campus of the University of Florida in Gainesville, Florida, to support the graduate engineering education needs of the northwest Florida technical community. The academic degree programs offered at the GERC are identical to those offered on the main campus. Courses are offered at the Center as well as in distance learning formats. Students enjoy a family-like atmosphere; flexibility in classes, course work, and research; and close contact with faculty and support staff members.
Applying	Application requirements and forms are the same as those of the main campus. However, applications should be submitted via the Graduate Engineering and Research Center. For degree-seeking students applying for admission, a B.S. degree in an engineering discipline is recommended. Other requirements include a minimum upper-division GPA of 3.0 and GRE verbal and quantitative minimum scores of 1000. International students should have a TOEFL score of at least 550.
Correspondence and Information	Graduate Coordinator Graduate Engineering and Research Center University of Florida 1350 North Poquito Road Shalimar, Florida 32579-1163 Telephone: 850-833-9350 Fax: 850-833-9366 E-mail: reginfo@gerc.eng.ufl.edu World Wide Web: http://www.gerc.eng.ufl.edu/

University of Florida

THE FACULTY AND THEIR RESEARCH

The Center is in a growth phase. Additional faculty members are currently being considered. Courses are taught by resident faculty members; faculty members located on the University of Florida, Gainesville, campus via distance learning media; and highly qualified adjunct faculty members from the Eglin AFB technical community. A list of resident faculty members for the 1999–2000 academic year follows.

Aerospace Engineering
James E. Milton, Ph.D., Florida, 1966. Aerodynamics, terradynamics, flight dynamics.
Pasquale M. Sforza, Ph.D., Polytechnic of Brooklyn, 1965. Fluid mechanics, heat transfer, energy conversion, propulsion.

Electromagnetics
Henry Zmuda, Ph.D., Cornell, 1984. Optical processing of microwave signals, microwave system and component design, photonics.

Engineering Mechanics
C. Allen Ross, Ph.D., Florida, 1971. Aerospace structures and materials, composite materials, experimental stress analysis, high-strain material properties.

Photonics
Christopher S. Anderson, Ph.D., North Carolina State, 1991. Optical instrumentation, high-speed photography, optical signal processing, holography/interferometry, laser communications.

System Theory
Kenneth E. Dominiak, Ph.D., Polytechnic of Brooklyn, 1971. System and signal theory, digital signal processing.

UNIVERSITY OF FLORIDA

College of Engineering

Programs of Study

The College of Engineering offers Master of Science, Master of Engineering, Engineer, and Doctor of Philosophy degrees in aerospace engineering, agricultural and biological engineering, chemical engineering, civil engineering, coastal and oceanographic engineering, computer and information science, electrical and computer engineering, engineering mechanics, engineering science, environmental engineering, environmental sciences, industrial and systems engineering, materials science and engineering, mechanical engineering, and nuclear engineering sciences. A Master of Civil Engineering professional degree is also offered. The Departments of Computer and Information Science and Engineering, Electrical and Computer Engineering, Industrial and Systems Engineering, Materials Science and Engineering, and Mechanical Engineering offer a multidisciplinary certificate program in manufacturing systems engineering in conjunction with the Master of Engineering degree program. Several departments offer the opportunity for a five-year combined Bachelor of Science/Master of Science program. The Department of Materials Science and Engineering also offers a joint M.D./Ph.D. program in biomaterials. Master of Science, Master of Engineering, and Ph.D. degrees in engineering management are offered by the Department of Industrial and Systems Engineering in conjunction with the College of Business Administration. A joint program in health physics is offered between the Departments of Environmental Engineering Sciences and Nuclear and Radiological Engineering. Interdisciplinary Master of Science, Master of Engineering, and Ph.D. programs in biomedical engineering are also offered.

A minimum of 30 semester hours is required for the M.S. or M.E. with thesis, 6 of which represent work on the thesis. The nonthesis M.S. or M.E. requires 32 to 34 semester hours. (An accredited bachelor's degree in engineering or its equivalent is a prerequisite for the M.E.) The Engineer degree requires 30 hours beyond the master's, with an optional thesis. The Ph.D. degree requires 90 credit hours (including dissertation) beyond the bachelor's; 30 semester hours in one calendar year or 36 semester hours in no more than four semesters within a period of two calendar years must be earned on the Gainesville campus. The language requirement for the Ph.D. degree varies by department.

With twelve engineering departments and many programs in cooperation with the other twenty University colleges and schools, including the large on-campus health science center, the opportunities for interdisciplinary research are numerous. The College ranks among the top thirteen U.S. engineering colleges in research funding.

The University operates on a schedule of two semesters plus two summer terms. Graduate courses are offered all year.

Research Facilities

The College currently performs about $60 million per year in research and has extensive research laboratories in a broad range of disciplines in modern buildings. The Engineering and Industrial Experiment Station, the research arm of the College, is well recognized nationally and internationally for the quality and breadth of its programs, and faculty members are at the forefront of their fields. The College of Engineering and the University have extensive computational facilities, including parallel processing, computer graphics, and minicomputer and personal computer laboratories. The library system consists of several major units, including a modern science and engineering library. The library includes 2.4 million volumes and more than 2 million microforms.

Financial Aid

Nearly 750 graduate assistantships with competitive stipends are available in research and teaching for one-fourth to three-fourths-time work loads. There are also traineeships and fellowships that provide from $4000 to $20,000 and are supported by NSF, NIH, NDEA Title IV, NANT, USDOE, NASA, the University of Florida Graduate School, and the College of Engineering.

Cost of Study

The registration fee for most graduate course work is $144.20 per credit hour for Florida residents and $504.93 per credit hour for out-of-state students in 1999–2000. The nonresident fee can usually be waived for students who hold graduate assistantships, fellowships, or traineeships.

Living and Housing Costs

The rent for one semester in dormitories or in the modern, air-conditioned Twin Towers residence halls (one for men and one for women), including utilities, is $1087 per student in 1999–2000. The University operates six apartment villages for married students. Privately owned rooms and apartments are readily available.

Student Group

The total enrollment at the University is approximately 43,300, including 7,571 graduate students on campus. The College of Engineering graduate enrollment is more than 1,500. The percentages of men and women in the College for graduate programs are 81.5 percent and 18.5 percent, respectively. Thirty-seven percent of the College enrollment is international.

Student Outcomes

Employment opportunities for graduates are available in a variety of government agencies, consulting firms, industries, and other businesses. A wide variety of employers recruit each year at the University of Florida. Recent employers include Andersen Consulting; CH2M Hill; Hewlett-Packard Company; Microsoft; Motorola, Inc.; Pratt & Whitney; Texas Instruments; Entergy; Lockheed Martin; and Westinghouse Corporation.

Location

The University of Florida is located in Gainesville, a city of approximately 100,000, situated in north-central Florida. Gainesville lies midway between the Atlantic Ocean and the Gulf of Mexico, each of which is within a 2-hour drive. A University golf course is adjacent to the campus, and there are opportunities for swimming and boating at nearby lakes, springs, and rivers. Gainesville is served by several airlines and bus lines and is located along I-75, 1 hour south of I-10 and 2 hours north of Orlando.

The University

A combined state university and land-grant college, the University of Florida has sixteen upper-division colleges and schools and four professional colleges (Dentistry, Law, Medicine, and Veterinary Medicine). The College of Engineering consists of twelve degree-granting departments occupying seventeen buildings and has 279 full-time faculty members. The University of Florida is a member of the Association of American Universities.

Applying

Application forms may be obtained from the director of admissions. A baccalaureate degree from an accredited college and an average grade of B or better for the junior and senior years are generally required. All students must submit satisfactory scores on the verbal and quantitative portions of the General Test of the Graduate Record Examinations for admission to the Graduate School. Applications and transcripts should be submitted up to one year but not later than sixty days before registration; students may be admitted for any semester.

Correspondence and Information

Requests for information and application forms:

Chairman, Department of (specify)
College of Engineering
University of Florida
Gainesville, Florida 32611

Telephone: 352-392-0943
Fax: 352-392-9673
E-mail: academics@eng.ufl.edu
World Wide Web: http://www.eng.ufl.edu

University of Florida

FACULTY HEADS AND RESEARCH

Aerospace Engineering. Professor Wei Shyy, Chairman. Graduate research involves scientific and engineering problems related to flight. Major areas include gasdynamics, combustion, composite structures, micro air vehicles, flight dynamics, guidance and control, microelectromechanical systems, structural and system optimization, aerodynamics, turbulence, aeropropulsion, computational fluid and solid mechanics, and automated design. Equipment includes subsonic and supersonic tunnels, subsonic and supersonic combustion rigs, lasers, and modern instrumentation.

Agricultural and Biological Engineering. Professor C. Direlle Baird, Chairman. A broad and active research program emphasizes engineering solutions to problems associated with biological and agricultural systems, often related to renewable natural resources. Areas of study and research include irrigation and drainage, water quality, energy management, systems analysis, machinery design, food engineering and bioprocessing, remote sensing, postharvest technology, structures, environmental control, aquaculture, nonpoint pollution control, mathematical modeling of plant and animal systems, information technology/systems, and management of agricultural and municipal wastes.

Chemical Engineering. Professor Timothy J. Anderson, Chairman. Graduate study and research emphasize chemical engineering science: transport phenomena, thermodynamics, kinetics and catalysis, interfacial phenomena, and materials science; chemical engineering systems: chemical and biochemical reaction engineering, computer-aided design, optimization, and control; and interdisciplinary chemical engineering: electrochemical engineering, microelectronics, biomedical interfaces, cell and tissue engineering, and particle science and technology.

Civil Engineering. Professor Paul Y. Thompson, Chairman. Graduate study and research areas include transportation and traffic operations, hydraulics, hydrology, and construction engineering; geotechnical engineering, foundations, structures, and highway engineering; groundwater hydraulics; open-channel flow; sediment transport; hydraulic models and analogues; hydraulic transients; hydraulics of stratified flow; coastal hydraulics; harbor hydraulics; advanced metal structures; advanced soil mechanics; foundations; centrifugal modeling; in situ techniques for soils; structure and engineering properties of soils; soil dynamics; structural dynamics; advanced reinforced concrete; statically indeterminate structures; structural analysis; framed structures, folded plates, and shells; nonlinear structural analysis; prestressed concrete; surveying and mapping, remote sensing, geographic information systems, laser swath mapping, geodesy, and global systems; materials and recycling of waste products; construction engineering; and public works.

Coastal and Oceanographic Engineering. Professor Paul Y. Thompson, Chairman. Research areas include coastal and offshore structures, modeling, coastal sediment and pollutant transport, and estuary and lake hydrodynamics. Facilities are available for hydraulic model studies. A wave tank is available for studies of combined effects of wind, wave, and current. Models of large bays and estuaries can be constructed in a 165-foot by 125-foot basin that has a snake-type wave generator and a variety of pump arrangements to simulate conditions in nature. Other facilities include a tilting flume, a wave channel, and two wave tanks. Field equipment includes several boats, an amphibian (LARC), and a wide range of monitoring instruments.

Computer and Information Science and Engineering. Professor Gerhard X. Ritter, Chairman. Graduate research is in theoretical and applied areas of computer science, computer engineering, information science and information engineering, specifically, computer vision and visualization, database systems, software engineering, parallel and distributed computing systems, high-performance computing, computer networking, computer communications, artificial intelligence, computer algorithms, simulation, management information systems, and other areas. Students have access to a variety of computers: a 14-processor IBM SP-2 parallel supercomputer; a high-availability cluster of Sun E3500s; Silicon Graphics workstations; numerous Sun, IBM, and HP workstations; and numerous Windows NT PCs.

Electrical and Computer Engineering. Professor Martin A. Uman, Chairman. The ten major areas of teaching and research are communications, computer systems and networks, device and physical electronics, digital signal processing, electric energy systems, electromagnetics, electronic circuits, intelligent and information systems, photonics, and systems and controls. The research facilities include laboratories devoted to IC processing, VLSI circuits, photonics, robotics, digital design processing, lightning, power systems, communication systems, and speech analysis and synthesis. Research computing facilities are excellent and include eighty workstations, several hundred PCs, and extensive industrial software that run a wide variety of educational software packages. Teaching facilities include UNIX workstations and Pentium PCs. Computers are networked to other University computing resources and to national and international networks.

Engineering Science and Engineering Mechanics. Professor Wei Shyy, Chairman. Areas of study and research are composite materials, materials processing, biomechanics, dynamic materials, plasticity, viscoelasticity, fatigue, fracture, creep, vibrations, structural dynamics, multiphase systems, photoelasticity and optics, space mechanics, elastic and hydrodynamic stability, modeling, optimal control, estimation theory, plates and shells, engineering optimization, wave propagation in solids, numerical techniques, and applied mathematics. There are laboratories for study of materials, mechanics and physics of fluids, dynamics and vibrations, stress waves, stress analysis, computational mechanics, and biomechanics.

Environmental Engineering Sciences. Professor John J. Warwick, Chairman. The areas of concentration are air pollution, environmental biology, radiological health, solid and hazardous wastes, water chemistry, water resources management, water supply, water pollution control, and wetlands and systems ecology, directed by a multidisciplinary 18-member faculty. A modern four-story building contains 34,000 square feet of space for research and teaching. The Center for Wetlands and the Center for Environmental Policy are associated with the department.

Industrial and Systems Engineering. Professor Donald W. Hearn, Chairman. Graduate work emphasizes engineering management, manufacturing systems engineering, health systems, operations research, and quality and reliability assurance. The engineering management program is offered in conjunction with the College of Business Administration. Ph.D. concentrations include decision support systems, intelligent manufacturing, engineering management, energy efficiency, facilities layout and location, operations research, and production planning and control. Extensive computing resources are available in computing laboratories and the Center for Applied Optimization.

Materials Science and Engineering. Professor Reza Abbaschian, Chairman. Research programs emphasize processing, properties, structure, and application of biomaterials, ceramics, electronic materials, magnetic materials, metals, and polymers and include biopolymers, colloid and interfacial chemistry, composites, corrosion, crystal growth, diamond thin films, electronic materials, electrotransport, fracture, glasses, grain boundaries and interfaces, ion implantation, laser processing, mechanical behavior, metals and minerals processing, molecular beam epitaxy, optical properties, optoelectronics, physical ceramics, polymers, prosthetic materials, sol-gel processing, solidification, stereology, surface characterization, and thin films. Analytical facilities include Auger, EMP, ESCA, scanning Auger, SEM, STEM, TEM, XPS, and X-ray equipment. The faculty is internationally recognized.

Mechanical Engineering. Professor William G. Tiederman, Chairman. Programs of study and research are available in energy conversion, solar energy, combustion, transportation systems, heat transfer, robotics, machine design, manufacturing, experimental and computational fluid dynamics, gasdynamics, HVAC, two-phase flow and heat transfer, microscale heat transfer, and biomedical engineering. Facilities include a solar research park, high-speed machine tools, robots, and a MEMS laboratory. A full range of modern diagnostic and test equipment supports the activities of the 31 faculty members.

Nuclear and Radiological Engineering. Professor James S. Tulenko, Chairman. Graduate study and research are conducted in nuclear power technology, nuclear space power reactors and energy conversion, radiological sciences (health physics and medical physics), bionuclear engineering, radiation imaging, engineering physics, robotics and expert systems as applied to nuclear operations and maintenance, environmental aspects of nuclear power generation, neutron and reactor physics, nuclear waste management, nuclear and radiation chemistry, and high-temperature materials. Facilities include a 100-kW training and research reactor with associated analytical laboratory and neutron radiography facilities, an ultrahigh-temperature fluids and materials lab, a plasma irradiation facility, a high-temperature diagnostic lab, and a laboratory for mobile robots for hazardous environments, including four mobile robots for experimental purposes.

UNIVERSITY OF HOUSTON

Cullen College of Engineering

Programs of Study	The Cullen College of Engineering offers programs leading to the degrees of Master of Engineering (M.E.), Master of Science (M.S.), and Doctor of Philosophy (Ph.D.) and has five departments: chemical engineering, civil and environmental engineering, electrical and computer engineering, industrial engineering, and mechanical engineering. The M.E. and M.S. are offered by all departments and the M.S. is offered in graduate-level interdisciplinary programs, which include aerospace, biomedical, computer and systems, environmental, materials, and petroleum engineering. Ph.D. degrees are offered in each department and in the following engineering interdisciplinary programs: aerospace, computer and systems, environmental, and materials.
Research Facilities	Laboratories in the chemical engineering department include a catalyst development laboratory; a two-phase flow laboratory where research is conducted on gas-liquid mixtures flowing in channels; biochemical and biomedical laboratories where research is done on protein separations, antibody technology, and mammalian cell signal transduction mechanisms; electronic materials processing laboratories where research is performed on etching and deposition of thin films for microelectronic device fabrication; an Improved Oil Recovery Institute with research interests in reservoir simulation, transport in porous media, and improved oil recovery techniques; and a reaction engineering laboratory where work focuses on chemical reactor design and control, catalysis, and multiphase reactors. In civil and environmental engineering, specialized facilities include a wave and fluid-structure interaction laboratory; a structural research laboratory capable of full-scale testing of panels, beams, and columns; a national geotechnical experimentation site for studying in situ behavior of soils; a Geographical Information Systems Laboratory; an environmental engineering research laboratory where studies on the analysis and control of organic and inorganic contaminants in drinking water and waste water are conducted; and geotechnical laboratories containing a fully instrumented pile-driving facility and soil modification equipment. The electrical and computer engineering department has an expert-systems laboratory for signal and image understanding, focusing on biomedical engineering applications; a microwave anechoic chamber with ancillary positioning equipment for the measurement of antenna radiation patterns, using a computer-controlled wideband receiver; a thin-film processing laboratory for research in microelectronic materials and semiconductor processing, containing advanced deposition, patterning annealing equipment; an optoelectronics laboratory equipped with a wide range of gas, solid-state, and semiconductor lasers for generating continuous wave to femto-second pulses; a seismic acoustics laboratory focusing on reflection seismology; an API nuclear calibration facility with a gamma pit, a neutron pit, and an electronically operated jib crane for standardizing nuclear logging radiation units; and a well-logging laboratory where research is done in the electromagnetic characteristics of well-logging tools. The industrial engineering department has extensive research facilities for work in CAD/CAM, AI/expert systems, human factors/ergonomics, and other areas. The facilities of the mechanical engineering department include laboratories for the study of aerodynamics; turbulence; shear flow; aeroacoustics; holographic velocimetry; computational fluid dynamics and combustion; the behavior of metals, polymers, ceramics, composites, and superconducting materials; the effects of high temperature on mechanical properties of dynamic fracture and the nondestructive characterization of elastic and mechanical properties of metallic and metal-matrix composites; mechanics; boiling and turbulent convective heat transfer; computer-based tools for designing systems and their components; and vibration and control of dynamic systems, including fault-tolerant intelligent structures.
Financial Aid	All graduate students who are U.S. citizens and who have at least a B average from an accredited university in the United States and high GRE scores may be considered for a research or teaching assistantship. Assistantships ranged from $9600 to $16,000 a year in 1998–99, depending on qualifications. Financial assistance is also available to exceptional international students.
Cost of Study	In 1999–2000, Texas residents pay $1560 per semester in required tuition and fees for 12 semester credit hours; nonresidents pay $3983. Students on fellowship or holding a half-time teaching or research assistantship automatically qualify for resident tuition. A graduate assistant tuition fellowship is available covering 9 hours of resident tuition.
Living and Housing Costs	On-campus housing in 1999–2000 is approximately $2763 per academic year. Meal plans range from $960 to $2260 per academic year. Off-campus apartments begin at about $560 per month. Most graduate students enrolled at the University live off campus.
Student Group	About 2,280 students are enrolled in the Cullen College of Engineering, including 646 graduate students.
Location	Houston is the fourth-largest city in the nation and maintains a high rate of economic growth. The University's 390-acre campus is only a short drive from downtown Houston. Within the 200-acre area of the internationally known Texas Medical Center complex are the Baylor and University of Texas colleges of medicine, the M. D. Anderson Cancer and Tumor Institute, and the School of Public Health, to name but a few. These institutions, together with the Johnson Space Center and other organizations, offer numerous interdisciplinary research opportunities. The Alley Theatre, the Museum of Fine Arts, the Contemporary Arts Museum, the Houston Ballet, the Houston Symphony, the Houston Grand Opera Association, and other cultural resources offer the finest in classical and contemporary entertainment.
The College	The Cullen College of Engineering is one of fourteen colleges of the University. It is housed in two buildings with about 280,000 square feet. Several large College laboratories contain major research facilities.
Applying	Applications for admission and two copies of all transcripts must be submitted to the major department before July 3 for the fall semester and by October 1 for the spring semester. International students and those seeking financial assistance are urged to apply early in the calendar year for fall semester admission. TOEFL and GRE scores are required of international students applying for admission. U.S. students are required to take the GRE General Test.
Correspondence and Information	Graduate Studies Coordinator (indicate department) Cullen College of Engineering University of Houston 4800 Calhoun Houston, Texas 77204-4814 Telephone: 713-743-4200 Fax: 713-743-4214 World Wide Web: http://www.egr.uh.edu/

University of Houston

THE FACULTY AND THEIR RESEARCH

Chemical Engineering

N. R. Amundson, Cullen Distinguished Professor; Ph.D., Minnesota: chemical reaction analysis. C. Arnold, Adjunct Associate Professor; Ph.D., Texas at Austin; PE: enhanced oil recovery, reservoir engineering. V. Balakotaiah, Professor; Ph.D., Houston: reaction engineering. D. Economou, Moores' University Scholar, Associate Chair, and Professor; Ph.D., Illinois: electronic materials. E. J. Henley, Professor; D.Sc., Columbia; PE: reliability, computer-aided design. R. Krishnamoorti, Assistant Professor; Ph.D., Princeton: thermodynamic interactions in polymer blends. D. Luss, Cullen Distinguished Professor; Ph.D., Minnesota; PE: reaction engineering. K. Mohanty, Associate Professor; Ph.D., Minnesota: multiphase transport through porous media. M. Nikolaov, Associate Professor; Ph.D., UCLA: computer-aided process engineering, control of microelectronics processes. R. Pollard, Professor; Ph.D., Berkeley: electrochemical engineering. H. W. Prengle Jr., Professor Emeritus; D.Sc., Carnegie Mellon; PE: thermodynamics. J. T. Richardson, Professor and Chairman; Ph.D., Rice: catalysis. F. M. Tiller, M. D. Anderson Distinguished Professor Emeritus; Ph.D., Cincinnati; PE: filtration, coal liquefactions. R. Willson, Associate Professor; Ph.D., MIT: biochemical engineering. F. L. Worley Jr., Professor Emeritus; Ph.D., Houston; PE: air pollution.

Civil and Environmental Engineering

T. G. Cleveland, Associate Professor; Ph.D., UCLA: environmental systems modeling. D. A. Clifford, Professor and Chairman; Ph.D., Michigan; PE: ion exchange and membrane processes. O. I. Ghazzaly, Professor; Ph.D., Texas at Austin; PE: soil-structure interaction behavior of expansive clays. K. J. Han, Associate Professor; Ph.D., Washington (St. Louis): structural mechanics. T. A. Helwig, Assistant Professor; Ph.D., Texas at Austin: steel structures. T. C. Hsu, Professor; Ph.D., Cornell; PE: reinforced- and prestressed-concrete structures, structural mechanics. S. T. Mau, Professor; Ph.D., Cornell; PE: concrete structures. M. W. O'Neill, Cullen Distinguished Professor; Ph.D., Texas at Austin; PE: foundations. Hanadi S. Rifari, Assistant Professor; Ph.D., Rice; PE: surface water, groundwater hydrology, contaminant transport. W. G. Rixey, Assistant Professor; Ph.D., Berkeley: hazardous waste treatment. D. J. Roberts, Assistant Professor; Ph.D., Alberta: applied microbiology. J. R. Rogers, Associate Professor; Ph.D., Northwestern; PE: water resources and quality, stormwater runoff. S. W. Tabsh, Assistant Professor; Ph.D., Michigan; PE: structural safety, bridge structures. C. Vipulanandan, Professor; Ph.D., Northwestern; PE: geomechanics experimental methods, waste disposal. K. H. Wang, Associate Professor; Ph.D., Iowa: coastal/estuary hydrodynamics. A. N. Williams, Professor; Ph.D., Reading: hydrodynamics, water waves.

Electrical and Computer Engineering

W. L. Anderson, Professor and Chairman; Sc.D., New Mexico: optics, biomedical engineering, pattern recognition and nondestructive testing. B. J. Barr, Assistant Professor; Ph.D., Houston: applied mathematics. G. Chen, Associate Professor; Ph.D., Texas A&M: computations, dynamics and control of nonlinear systems, spline approximations in engineering. O. Crisan; Ph.D., Timisoara (Romania): power systems modeling. J. R. Glover Jr., Professor; Ph.D., Stanford; PE: adaptive systems, digital and biomedical signal processing. T. J. Hebert, Associate Professor, Ph.D., USC: image and signal processing. M. Herbordt, Assistant Professor; Ph.D., Massachusetts at Amherst: computer architecture, parallel architectures, algorithms and languages for massively parallel arrays. D. R. Jackson, Associate Professor; Ph.D., UCLA: applied electromagnetics, printed antennas, leaky-wave antennas. B. H. Jansen, Professor; Ph.D., Amsterdam: biomedical signal processing, pattern recognition, knowledge-based systems, chaos theory and nonlinear modeling. N. Karayiannis, Associate Professor; Ph.D., Toronto: neural networks. T. King, Assistant Professor; Ph.D., Illinois at Urbana-Champaign: control systems, electronic circuit design, instrumentation. P. Y. Ktonas, Professor; Ph.D., Florida: bioelectrical signal analysis and applications in clinical medicine, random data analysis. R. Liu, Associate Professor; Ph.D., Jiatong: well-logging. S. A. Long, Professor and Associate Dean; Ph.D., Harvard; PE: applied electromagnetics, printed-circuit and millimeter-wave antennas, applied superconductivity. P. Markenscoff, Associate Professor; Ph.D., Minnesota: computer architecture, distributed processing. H. Ogmen, Associate Professor; Ph.D., Laval: neural networks, biological and machine vision. P. S. Ong, Associate Professor; D.Sc., Delft; PE: X-ray and electron optics, scanning electron microscopy. D. M. Pai, Associate Professor; Ph.D., British Columbia: exploration seismics, electrical well-logging methods. G. F. Paskusz, Professor; Ph.D., UCLA; PE: computer-aided circuit analysis and design. S. Pei, Professor; Ph.D., SUNY at Stony Brook: optoelectronic devices, compound semiconductors. W. P. Schneider, Professor; S.M., MIT; PE: controls and electronic instrumentation. D. P. Shattuck, Associate Professor; Ph.D., Duke: acoustic imaging, well-logging. L. C. Shen, Professor; Ph.D., Harvard; PE: subsurface sensing, well-logging. L. S. Shieh, Professor; Ph.D., Houston; PE: control and power systems, identification and design. L. P. Trombetta, Associate Professor; Ph.D., Lehigh: electrical properties of semiconductor materials, electron devices. J. T. Williams, Associate Professor; Ph.D., Arizona: applied electromagnetics, numerical analysis, microwave and millimeter-wave antennas. D. R. Wilton, Professor; Ph.D., Illinois: computational electromagnetics, electromagnetic scattering radiation and penetration. J. C. Wolfe, Professor and Dean; Ph.D., Rochester: materials research, electron devices, microfabrication. W. Zagozdzon-Wosik, Associate Professor; Ph.D., Warsaw: semiconductor-integrated circuit-processing technology, electron devices.

Industrial Engineering

J. Chen, Associate Professor and Chairman; Ph.D., Oklahoma; PE: ergonomics/safety, artificial intelligence, manufacturing systems, automation. C. A. Chung, Assistant Professor; Ph.D., Pittsburgh: engineering management, manufacturing systems, management training and education simulators, simulation. E. E. Deal, Associate Professor; Ph.D., Houston: operations research, scheduling, heuristic optimization, probability. C. E. Donaghey, Professor; Ph.D., Pittsburgh; PE: facility layout, applied statistics, reliability, simulation. B. Goldberg, Adjunct Professor; Ph.D., New School: engineering economics, engineering management. M. Heller, Assistant Professor; Ph.D., Johns Hopkins: artificial intelligence, applied operations research, environmental application. J. L. Hunsucker, Associate Professor; Ph.D., LSU; PE: engineering management, productivity, system design, risk management, scheduling. B. Ostrofsky, Professor; Ph.D., UCLA; PE: system management, system engineering, integrated logistics support. L. J. H. Schulze, Associate Professor; Ph.D., Texas A&M; PE: ergonomics, safety engineering, system design and evaluation. T. Yang, Associate Professor; Ph.D., Penn State: scheduling and sequencing, expert systems, optimization, manufacturing systems.

Mechanical Engineering

R. B. Bannerot, Professor and Chairman; Ph.D., Rice; PE: thermal sciences. H. F. Brinson, Professor; Ph.D., Stanford: mechanical behavior of adhesives, composites, and polymers. Y.-C. Chen, Associate Professor; Ph.D., Cornell: solid mechanics, composite materials. C. Dalton, Professor and Associate Dean; Ph.D., Texas at Austin; PE: fluid mechanics. R. Eichhorn, Professor; Ph.D., Minnesota; PE: heat transfer, fluid mechanics. R. D. Finch, Professor; Ph.D., Imperial College (London): acoustics. R. Glowinski, Professor; Ph.D., Paris: scientific computing, applied mathematics. K. M. Grigoriadis, Assistant Professor; Ph.D., Purdue: control systems analysis and design. D. Keith Hollingsworth, Associate Professor; Ph.D., Stanford; PE: convective heat transfer. A. K. M. F. Hussain, Cullen Professor; Ph.D., Stanford: fluid mechanics. S. J. Kleis, Associate Professor; Ph.D., Michigan State: fluid mechanics. I. Kunin, Professor; Ph.D., Leningrad Polytechnic: microcontinuum mechanics. J. H. Lienhard, M. D. Anderson Distinguished Professor; Ph.D., Berkeley: heat transfer, fluid mechanics. R. W. Metcalfe, Professor; Ph.D., MIT: computational fluid dynamics, spectral methods. A. Powell, Professor; Ph.D., Southampton: aeroacoustics, flow-induced oscillations. J. R. Rao, Associate Professor; Ph.D., Michigan: optimization theory, design methodology, systems modeling. K. Ravi-Chandar, Professor; Ph.D., Caltech: mechanics of fracture. K. Salama, Professor; Ph.D., Cairo: materials science, metallurgy, superconductivity. N. Shamsundar, Associate Professor; Ph.D., Minnesota: thermal sciences. W. E. Van Arsdale, Associate Professor; Ph.D., Cornell: applied mechanics, polymer rheology. S. Wang, Distinguished University Professor; Sc.D., MIT.; solid mechanics, materials science and engineering, superconductivity. L. T. Wheeler, Professor; Ph.D., Caltech; PE: solid mechanics, elasticity. K. White, Associate Professor; Ph.D., Washington (Seattle): materials science, ceramics, metallurgy. L. C. Witte, Professor and Associate Dean; Ph.D., Oklahoma State; PE: thermal sciences. D. C. Zimmerman, Associate Professor; Ph.D., SUNY at Buffalo: vibrations and control.

UIC

UNIVERSITY OF ILLINOIS AT CHICAGO

College of Engineering

Programs of Study

The College of Engineering offers programs leading to M.S. and Ph.D. degrees in the Departments of Chemical Engineering; Civil and Materials Engineering; Electrical Engineering and Computer Science; and Mechanical Engineering (includes industrial engineering) and in the bioengineering program.

The chemical engineering department offers programs leading to M.S. and Ph.D. degrees. The department has its own building containing all instructional and research facilities. Research facilities include state-of-the-art equipment for combustion kinetics, catalyst preparation, microelectronic materials processing, fluid-particle processes, and synthesis in supercritical fluids. Faculty members also have interests in thermodynamics, statistical methods, transport phenomena, and process optimization.

The civil and materials engineering department offers programs leading to M.S. and Ph.D. degrees in two major areas: civil engineering and materials engineering. Students may select courses in the areas of environmental engineering, structural engineering, transportation, mechanics, and materials science. The courses include hydraulics and hydrology, groundwater modeling and pollutant transport, soil mechanics and foundations, environmental remediation and geotechnology, structures, plates and shells, earthquake engineering, design in reinforced concrete and steel, prestressed concrete, traffic, transportation facilities, elasticity, plasticity, viscoelasticity, wave motion, fracture mechanics, composite materials, chemical and process metallurgy, stress corrosion cracking, and solidification and welding.

The electrical engineering and computer science department offers programs leading to M.S. and Ph.D. degrees in several areas, including computer science (algorithms, artificial intelligence, computational complexity, computer architecture, computer graphics, computer vision, database systems, human-computer interaction, natural-language processing, net theory of concurrent systems, programming languages, software engineering, and theoretical computer science) and electrical engineering (active networks, applied graph theory, broadband matching networks, controls theory, digital communication systems, digital filters, electromagnetic and ultrasonic imaging, electromagnetic theory and antennas, microwave electronics, noise in electronic devices, optical communications and waveguides, process control, remote sensing, robotics, solid-state–device theory, VLSI, digital systems, and microprocessors).

The mechanical engineering department offers degree programs leading to the M.S. in mechanical engineering and in industrial engineering and the Ph.D. in mechanical engineering and in industrial engineering and operations research. The programs cover mechanical, industrial, and aerospace engineering and related fields. The primary areas include mechanical analysis and design, thermal sciences, fluids engineering, robotics, manufacturing, and industrial engineering. Students may select courses dealing with topics in fluid mechanics, stress analysis, mechanisms, dynamics and vibration, mechanical design, computer-aided design and manufacturing, heat transfer, combustion, multiphase flow and heat transfer, industrial automation, and energy conversion. Interdisciplinary programs are encouraged, especially in the biological, environmental, electrical engineering, and computer science fields. The department has laboratories for work in biomechanics, mechanical design, manufacturing, mobile robots, controls, modal analysis, tribology, quality control, CAD/CAM, CIM, internal-combustion engines, fluid mechanics, two-phase flow, heat transfer, mass transfer, rheology, combustion, materials processing, interferometry, and virtual reality applications in manufacturing.

The bioengineering department offers M.S. and Ph.D. degrees. Faculty interests focus on the areas of cell and tissue engineering, neural engineering, biomechanics and rehabilitation engineering, biomedical imaging and visualization, bioMEMS, and biomaterials. The department closely interacts with other departments within the College of Engineering as well as with the Colleges of Medicine, Dentistry, and Pharmacy.

Research Facilities

Departmental laboratories have modern equipment and facilities.

Financial Aid

Some full-time graduate students receive teaching assistantships or research assistantships. Aid is available throughout the graduate program, including summer sessions.

Cost of Study

In 1998–99, Illinois residents registering for 12 or more semester hours paid $5476 for two semesters; nonresidents, $12,112. For programs of fewer than 12 hours, the fee is less. (All fees are subject to change.)

Living and Housing Costs

In addition to UIC residence halls, off-campus apartments and rooms are available. Meals may be taken in campus dining facilities or at a variety of nearby commercial establishments. Food, housing, transportation, medical care, personal items, clothing, and incidentals are estimated to cost $11,000 per calendar year.

Student Group

UIC enrolls more than 25,000 students from throughout Illinois, the United States, and overseas. Thirty-five percent of the total enrollment is at the graduate and professional level.

Location

UIC is located on the Near West Side, 5 minutes by public transit from the center of Chicago. The campus is accessible via all major transportation routes serving the Greater Chicago area.

The University

The University of Illinois at Chicago is the largest institution of higher learning in the Chicago area, one of the top seventy research universities in the United States, and an increasingly significant center for international education and research. It comprises fifteen colleges, which offer bachelor's degrees in more than ninety fields, master's degrees in eighty-nine areas, and doctorates in fifty-two specializations.

Applying

Application forms are sent upon request. The completed forms should be submitted as early as possible, preferably before March 1. Scores on the Graduate Record Examinations are required by some departments. Students from abroad must submit TOEFL scores.

Correspondence and Information

Lawrence A. Kennedy, Dean
College of Engineering
University of Illinois at Chicago
851 South Morgan Street
Chicago, Illinois 60607-7043

The University of Illinois at Chicago

THE FACULTY AND THEIR RESEARCH

Bioengineering

F. Amirouche: biomechanics, human locomotion, microdynamics, robotics. D. Carley: cardiorespiratory control and dynamics, computational neurobiology. T. Desai: tissue engineering, cell encapsulation, bioMEMS. D. Graupe: adaptive filtering, signal processing, muscle stimulation. B. He: modeling of physiological function, electrocardiology. P. Hesketh: biosensors and microfabrication. J. Hetling: electrophysiology of vision and retinal prosthetics. J. Karlsson: system modeling and control. J. Lin: noninvasive radio frequency and microwave sensing of physiological signatures, cardiovascular diseases, hyperthermia treatment for cancer. F. Loth: biofluid dynamics in using experimental simulations to model the hemodynamics flow field. R. Magin: magnetic resonance imaging. A. Masud: computational mechanics. R. Natarajan: biomechanics. W. O'Neill: information theory. B. Zuber: Physiological control and neurobiology.

Chemical Engineering

K. Brezinsky: chemical kinetics in automotive engine emissions, combustion synthesis of ceramics, reactions in supercritical fluids. J. H. Kiefer: kinetics of gas reactions, combustion. A. A. Linninger: product and process design, computer-aided modeling and simulation, pollution prevention in pharmaceutical and specialty chemical batch manufacturing, logic-based control and process operations, process optimization. G. A. Mansoori: applied statistical mechanics and thermodynamics, supercritical extraction retrograde condensation, asphaltene deposition, bioseparation. S. Murad: computer modeling of reverse osmosis separations. L. C. Nitsche: transport phenomena in microstructured media, fluid mechanics, Brownian motion, colloidal and macromolecular separations (especially membrane separations), multiphase flow, applied mathematics. J. R. Regalbuto: catalyst preparation, characterization of solid catalysts, heterogeneous reaction kinetics. H. Reyes: enzymatic hydrolysis and synthesis of polymers, enzymatic catalysis in nonaqueous media, deep culture fermentation of mycobacteria. S. Szepe: chemical and catalytic reaction engineering, optimization. C. G. Takoudis: microelectronic materials and processing, chemical vapor deposition, microfabrication techniques, heterogeneous catalysis, reaction kinetics, in situ surface spectroscopies at interfaces, heterogeneous catalysis-reaction engineering. R. M. Turian: complex fluid and coal-water mixture flows, rheology, slurry transport, hydrodynamics and heat transfer, microbial treatment of coal. L. E. Wedgewood: transport phenomena, rheology, computational non-Newtonian fluid mechanics, macromolecular kinetic theory, Brownian dynamics, stochastic simulations.

Civil and Materials Engineering

F. Ansari: fiber-optic sensors for construction materials and structures. J. Botsis (Adjunct): polymer science and engineering, fracture. D. Boyce: urban transportation planning. R. H. Bryant: structural mechanics and optimization. A. Chudnovsky: fracture and damage, thermodynamics. R. D. Crago: hydraulics and hydrology, environmental engineering and remote sensing. S. Ghosh (Adjunct): reinforced and prestressed concrete. S. Harren: micromechanics, crystal plasticity. J. E. Indacochea: solidification, welding. M. Issa: reinforced and prestressed concrete. D. Lemke: stress analysis and dynamics. A. Masud: computational mechanics, smart materials, fluid structure interaction. M. J. McNallan: chemical and process metallurgy, thermodynamics and kinetics of metallurgical processes, high-temperature corrosion. K. Reddy: geotechnical and geoenvironmental engineering, earthquake engineering, groundwater and contaminant hydrology, engineering geology. T. C. T. Ting: wave propagation, viscoelasticity, continuum mechanics. M. L. Wang: structural dynamics, smart and composite structures, monitoring systems for long-span bridges. C.-H. Wu: linear and nonlinear elasticity, fracture.

Electrical Engineering and Computer Science

G. C. Agarwal: automatic control. R. Ansari: image/video processing. D. Babic: physical principles of nanoscale Si particles. R. Beigel: HCI, data mining, algorithms, complexity. J. Ben-Arie: object and target recognition. W. M. Boerner: electromagnetics, inverse scattering. U. A. Buy: software engineering, concurrency and real-time analysis. C. K. Chang: software engineering. W.-K. Chen: broadband matching, applied graph theory. T. A. DeFanti: computer graphics and video animation, virtual reality. B. Di Eugenio: artificial intelligence, natural language processing. S. Dutt: parallel and distributed computing, VLSI CAD, fault-tolerant computing and computer architecture. A. D. Feinerman: miniaturization of the scanning electron microscope. G. D. Friedman: electromagnetics. D. Graupe: control systems. B. He: imaging systems, signal and image processing. P. J. Hesketh: solid-state sensors and actuators. J. Hummel: compilers, programming languages, parallelism. A. Johnson: collaborative virtual reality, human-computer interaction, educational applications. S. Kasif: high-performance intelligent systems machine learning, data mining, computational biology, computational neuroscience, parallel computation. R. V. Kenyon: human visual and motor systems. A. Kshemkalyani: computer networks, distributed computing. S. R. Laxpati: antennas, electromagnetic theory. J. Lillis: computer-aided design for VLSI circuits and combinatorial optimization. J. C. Lin: electromagnetics in biology and medicine. J. Lobo: knowledge representation. T. G. Moher: human-computer interaction. T. Murata: petri net modeling and analysis of concurrent computer systems. D. L. Naylor: development and application of microfabricated photonic materials and devices. A. Nehorai: signal and image processing. P. C. Nelson: AI research in heuristic search and applied AI research in the areas of transportation. W. D. O'Neill: signal and image processing, bioengineering, information theory. R. Priemer: optimal and adaptive digital signal processing. F. K. H. Quek: human-computer interaction. D. Schonfeld: signal and image processing. S. M. Shatz: distributed computing systems. K. Shenai: solid-state electronics and microelectronics. A. P. Sistla: distributed systems, semantics and verification of concurrent systems. R. H. Sloan: design and analysis of algorithms, computational learning theory. J. A. Solworth: computer architecture. B. Super: computer and biological vision. J. J. P. Tsai: knowledge-based software systems. P. L. E. Uslenghi: electromagnetics, scattering theory. O. Wolfson: database systems. H.-Y. Yang: applied electromagnetics in ferromagnetic integrated circuits and antennas. C. T. Yu: database management. O. Yu: telecommunications and mobile commmunications networking.

Mechanical Engineering

S. K. Aggarwal: combustion, turbulent sprays, computational methods. F. M. L. Amirouche: multibody dynamics, CAD/CAE, robotics, human body vibration, biomechanics. P. Banerjee: virtual manufacturing, software environment design, interactive system design. J. G. Boyd: microelectromechanical and microfluidic devices. S. Cetinkunt: automatic control, robotics, microprocessor. S. Cha: optical measurements, radiative/convective transfer, spray and combustion. W. Chen: CAD, robust design, optimization. M. Y. Choi: combustion: radioactive emission measurements and modeling of turbulent diffusion flames, optical and intrusive techniques for flame characterization and particulate measurements, fundamental droplet combustion characteristics. P. M. Chung: turbulence, combustion, heat transfer. B. D. Coller: nonlinear dynamics and control, bifurcation theory, chaos, fluids, aeroelasticity. D. M. France: heat transfer in multiphase flows, including boiling and particulate flows and hydrodynamic instabilities. A. A. Fridman: plasma processes, chemical kinetics, non-equilibrium processes, gas discharges, emission controls, aerosols and catalysis. C. George: numerical heat transfer, plasma processes. Y. G. Gogotsi: ceramic composites, mechanical dynamics of materials. S. I. Guceri: ceramic composites, rapid prototyping. K. C. Gupta: mechanisms, robotic manipulators, optimization. J. P. Hartnett: energy resources, heat transfer, non-Newtonian fluids. J. O. Karlsson: cryobiology, tissue engineering. L. A. Kennedy: combustion, non-equilibrium processes, emission control, fluid mechanics, optical methods, heat transfer. K. Kim: computer-aided manufacturing, automation and robotics, machining control, mechanical vibrations, time series analysis. F. L. Litvin: gear theory and applications, robot manipulators, spatial mechanisms. F. Loth: biofluid engineering. C. M. Megaridis: spray combustion, one- and two-phase heat and mass transfer. F. G. Miller: industrial and manufacturing engineering, plant and equipment facility analysis. W. J. Minkowycz: heat transfer, two-phase and porous media flows, numerical heat transfer. I. K. Puri: combustion, fluid dynamics of multiphase media, chemical kinetics of flames. T. J. Royston: sound and vibration control, electromechanical systems. A. A. Shabana: computer-aided design, dynamic systems, vibrations, finite-element methods. S. M. Song: kinematics, robotics, walking machines, computer-aided design. J. J. Stukel: electrohydrodynamics of multiphase systems, aerosol deposition. W. M. Worek: heat and mass transfer, optical techniques, advanced energy systems.

UNIVERSITY OF ILLINOIS AT URBANA–CHAMPAIGN

College of Engineering

Programs of Study

The departments of the College of Engineering offer master's and doctoral degree programs in aeronautical and astronautical, agricultural, ceramic, chemical, civil and environmental, electrical and computer, general, mechanical and industrial, and nuclear engineering; computer science; materials science and engineering; physics; and theoretical and applied mechanics. Interdisciplinary specialties include bioengineering, computational science and engineering, engineering design, environmental control, manufacturing systems, and polymer science.

Graduate degree requirements vary by department; typical programs are listed below. For the master's degree, 8 units (1 unit equals 4 semester hours) of approved graduate credit, or 9 units if no thesis is rendered, are required; 3 of the 8 units must be in the 400 series, and 2 of these 3 must be in the major field. There is no foreign language requirement, but a student must spend a minimum of two semesters in residence. The doctoral degree program can be divided into three stages, two of which must be completed in residence: (1) a master's degree or an equivalent number of credits (8 units or 32 semester hours); (2) 8 units of work and a preliminary examination; and (3) 8 units of thesis research, a dissertation, and a final examination. Most departments do not have a language requirement. Graduate students often do interdisciplinary research work in the Materials Research Laboratory, Microelectronics Laboratory, Coordinated Science Laboratory, National Center for Supercomputing Applications, and Beckman Institute for Advanced Science and Technology. Faculty members in these laboratories hold joint appointments in the degree-granting departments.

Research Facilities

The College of Engineering had separately budgeted research expenditures in the 1998 fiscal year that totaled more than $90 million. Special laboratories and unusual facilities, such as the facility for submicron fabrication, multiport molecular-beam epitaxy machine, metal organic chemical vapor deposition system, nuclear reactor, radio and optical telescopes, electron microscopy center, and radioisotope lab, are available. The College has an extraordinary computing environment, equipped with everything from supercomputers to a wide variety of workstations. NCSA Mosaic™, a distributed hypermedia browser that provides a unified interface to the Internet, was developed at the National Center for Supercomputing Applications. The College is noted for its dedication to interdisciplinary studies, and centers have been established for the Advanced Transportation Research and Engineering Laboratory, the Air Conditioning and Refrigeration Center, bioengineering, the Center for Compound Semiconductor Microelectronics, the Center for Computational Electromagnetics, the Center for Fluorescence Dynamics, the Center for Machine Tool Analysis, the Center for Microanalysis of Materials, the Center for Reliable and High-Performance Computing, the Center for Simulation of Advanced Rockets, computational science and engineering, the Coordinated Science Laboratory, the Machine Tool Agile Manufacturing Research Institute, the Materials Research Laboratory, the Microelectronics Laboratory, the Mid-America Earthquake Center, the Science and Technology Center for Cement-Based Materials, and the Science and Technology Center for Superconductivity. The University Library, with a collection of more than 9 million volumes, is the third-largest academic library in the world. The Grainger Engineering Library Information Center holds more than 254,500 bound volumes and approximately 2,825 serial titles.

Financial Aid

Financial aid includes federally sponsored traineeships and fellowships and University and industry fellowships. Also available are part- and full-time research and teaching assistantships. The stipend for a half-time assistantship varies by department and includes exemption from tuition and service fees.

Cost of Study

Tuition and fees vary according to the number of semester hours taken. For a full program (3 or more units), tuition and fees per semester in 1998–99 were $2819 for Illinois residents and $6257 for nonresidents. Summer session charges for a full program (2.5 or more units) were $1778 for Illinois residents and $3927 for nonresidents.

Living and Housing Costs

University graduate residence halls have single rooms for $2696 and $3008 and double rooms for $2416 and $2872 per academic year. Board contracts are available for between $1435 and $3368. University family housing rents for between $350 and $510 per month. Privately owned rooms and apartments are available at similar and higher rents.

Student Group

Enrollment at the Urbana-Champaign campus is 36,303 students. The College of Engineering has 5,285 undergraduates and 2,007 graduate students. Most graduate students receive some form of financial aid. For example, during 1998–99, there were 225 fellows and trainees and 1,187 part-time research assistants.

Student Outcomes

Top corporations target graduates from the University of Illinois at Urbana-Champaign's (UIUC) College of Engineering. A recent trend includes an increase in the number of smaller, high-technology companies seeking graduates with a computer background. Many of the graduates accept offers from large corporations, such as Intel, General Electric, Motorola, Ford Motor, Caterpillar, Andersen Consulting, Texas Instruments, Procter & Gamble, IBM, Ernst & Young, AT&T, Lucent, and Hewlett-Packard.

Location

The campus is 130 miles south of Chicago in the twin cities of Urbana and Champaign (population 110,000). The area is primarily a university community, with excellent schools and parks and modern shopping facilities.

The University and The College

The University of Illinois at Urbana-Champaign is in its second century of operation and is recognized as a major national center of excellence in graduate education. The College of Engineering, founded in 1868, has grown and prospered with the University, establishing itself as a productive center of engineering research and education. It offers both graduate and undergraduate programs. In recent years, one out of every twenty-nine doctoral degrees awarded by colleges of engineering has been awarded by UIUC. The graduate program of the College has been ranked among the very best in the United States by the Conference Board of Associated Research Councils. The strength and accomplishments of the 420 faculty members are exemplified by their achievements. The faculty includes 24 members of the National Academy of Engineering, 10 members of the National Academy of Sciences, 13 members of the American Academy of Arts and Sciences, and 3 National Medal of Science winners.

Cultural activities on the campus are fostered by the Krannert Center for the Performing Arts, which contains five separate theaters for orchestra, opera, choral groups, theater, and dance. The 16,000-seat Assembly Hall serves as the stage for some of the nation's foremost entertainers. In addition to Big Ten football and basketball, the University offers a broad program of athletics using the Intramural Physical Education Building. The University's 1,500-acre Allerton Park is only one of a number of nearby parks and recreation areas.

Applying

Graduate application forms and information are available upon request from the heads of the individual departments (see reverse side of this page). Applications should be returned as early as possible. On-campus registration is in mid-August for the fall semester and in mid-January for the spring semester.

Correspondence and Information

Dean of Engineering
2-114 Engineering Sciences Building
University of Illinois at Urbana-Champaign
1101 West Springfield Avenue
Urbana, Illinois 61801

World Wide Web: http://www.engr.uiuc.edu/

University of Illinois at Urbana-Champaign

DEPARTMENT HEADS AND RESEARCH AREAS

Aeronautical and Astronautical Engineering. M. B. Bragg, Head. Aerodynamics, aerospace vehicle flight simulation, astrodynamics, combustion, computational fluid dynamics, dynamical systems, flight vehicle synthesis, lasers, materials and structures, propulsion and combustion, space systems design, structural dynamics, structural mechanics, systems and control, two-phase flow.

Agricultural Engineering. L. E. Bode, Head. Computer vision applications, mechatronics, biosensing, precision agriculture, cropping/tillage systems, chemical application systems, irrigation/drainage, watershed hydrology, soil erosion, water quality, bioenvironmental engineering, structural design/materials, animal waste bioprocessing, grain quality, grain drying, corn milling, biophysical properties, food engineering, bioprocess control, accident reconstruction, electronic applications, engines, alternative fuels and energy.

Chemical Engineering. C. F. Zukoski, Head. Applied mathematics, catalysis, cellular and molecular bioengineering, colloid and interfacial phenomena, computational biology, computer simulations, electrochemical engineering, fluid mechanics, heat and mass transfer, high-pressure studies, kinetics, modeling entire chemical industries, nucleation phenomena, plasma materials processing in microelectronics, polymer science and engineering, process dynamics, reactor design, semiconductor growth, solid-state physics, thermodynamics, tribology.

Civil and Environmental Engineering. D. E. Daniel, Head. Computer-aided engineering systems; construction management; construction materials; environmental engineering and science in civil engineering; environmental hydrology and hydraulic engineering; geotechnical engineering; structural engineering and structural dynamics; system safety, reliability, and design; transportation facilities; transportation systems.

Computer Science. D. A. Reed, Head. Artificial intelligence, communications, computer-aided design of digital systems, computer architecture and systems, databases and information systems, distributed systems, human-computer interactions and interfaces, numerical and scientific computing, operating systems, parallel computing, programming languages, real-time systems, software engineering, theoretical computing.

Electrical and Computer Engineering. S. M. Kang, Head. Advanced automation, advanced processing and circuits, aeronomy, analog and digital circuits, bioacoustics, Center for Reliable and High-Performance Computing, communications, decision and control, digital signal and image processing, electromagnetic communication and electronics packaging, electromagnetics, electrophysics, engineering education, gaseous electronics, high-frequency devices, magnetic resonance, photonic systems, power and energy systems, quantum electronics, remote sensing and wave propagation, semiconductor lasers, semiconductor physics, semiconductors, solid-state devices, supercomputing research and development, thin films and charged particles, tunneling microscopy.

General Engineering. H. E. Cook, Head. Biomechanics and rehabilitation engineering, communications networks, computer-aided design, control systems, design and manufacturing systems, design theory and methodology, engineering graphics and geometry, evolutionary computation, integrated mechanical and structural design, nondestructive evaluation and testing, operations research, robotics, tribology, vehicle dynamics.

Materials Science and Engineering. J. Economy, Head. Bioceramics; cementitious materials; ceramic and glassy solids; composites; computer simulations of materials; electrical ceramics; interfaces; materials chemistry; mechanical behavior of solids; microelectronics packaging materials and processes; phase transformation and microcharacterization; polymers; processing; properties of rapidly solidified metals; radiation damage in metals; structural ceramics; superconductors; surface studies, coatings, and laser processing; thin-film physics.

Mechanical and Industrial Engineering. R. O. Buckius, Head. Automotive systems, bioengineering, combustion and propulsion, computational science and engineering, control systems, design methodology and tribology, dynamic systems, energy systems and thermodynamics, engineering mechanics, engineering statistics, environmental engineering, fluid dynamics and gasdynamics, heat transfer, human factors, manufacturing systems, materials behavior, materials processing, MEMS, operations research, production management.

Nuclear Engineering. J. F. Stubbins, Head. Applied plasma physics; computational mechanics; controlled nuclear fusion; health physics, radiological, and medical applications; neutron activation and environmental engineering; nuclear materials, radiation effects, and waste management; nuclear power, operations, and control; reactor physics and reactor kinetics; space propulsion and power systems; thermal hydraulics and reactor safety.

Physics. D. K. Campbell, Head. Theoretical, computational, and experimental research: condensed-matter physics, including superconductivity, magnetism, and other cooperative phenomena, magnetic resonance, thin-film growth and properties, surface physics, and mesoscopic systems; high-energy particle physics, including strong and electroweak interactions; theoretical astrophysics, including cosmology and general relativity; low-temperature physics; intermediate-energy physics, including the structure of nucleons; biological physics, including fluorescence spectroscopies, visualization, and simulations; nonlinear and complex systems.

Theoretical and Applied Mechanics. H. Aref, Head. Applied mathematics; behavior of engineering materials (creep, fracture, localization, damage, phase transformations, failure analysis); computational mechanics; dynamics, vibrations, and waves; mechanics of fluids (turbulence, chaos, particle-image velocimetry, boundary layers, combustion, computational fluid dynamics); mechanics of solids (structures, composites, experimental stress analysis, elasticity, plasticity, optimization).

Bioengineering. L. A. Frizzell, Chairman, Bioengineering Faculty. Bioelectric phenomena; bioelectromagnetics; bioenvironmental engineering; bioinstrumentation; biological modeling; biomaterials; biomechanics; biomedical imaging; bioultrasonics; cellular bioengineering; food and bioprocessing; heat and mass transfer; ionizing radiation; microelectomechanical systems; physiological modeling; rehabilitation engineering; retinal neurophysiology; signal processing.

Coordinated Science Laboratory. W. K. Jenkins, Director. Advanced automation; analog and digital circuits; applied computation theory; communications; decision and control; digital signal and image processing; electromagnetic communication, radiation, and scattering; heterostructure devices and physics; information retrieval; quantum electronics; reliable and high-performance computing; semiconductor materials and devices; semiconductor physics; supercomputing research and development; surface studies; thin-film electronics.

Manufacturing Systems. S. G. Kapoor, Director. Simultaneous engineering (design methods, CAD/CAM, and factory models/simulation); management/systems integration (human factors/man-machine systems, organization/management, and database design/information management); decision and control (process/adaptive control, production control, and strategic planning); data acquisition, processing, and distribution (sensing/signal processing, interfaces, and networking/communications); materials processing, finishing, and materials handling (unit processes, robotics, and storage and retrieval systems); agile manufacturing; environmentally conscious manufacturing.

Materials Research Laboratory. H. K. Birnbaum, Director. Aspects of solid-state and materials science that support materials technology, including crystal structure; microstructure and microchemistry in metals and ceramics; diffusion of atoms and ions; superconductivity; properties of liquid helium; semiconductors and semiconducting heterojunctions; high-pressure phenomena; crystal defects and radiation damage; magnetic properties; optical properties, including infrared, far-ultraviolet absorption, and Raman scattering; phase transformations; thin films; crystal growth; molecular beam epitaxy; properties of heterophase and other modulated structures; phase transitions; organic and polymeric solids. Emphasis is on the science of synthesis and processing.

Microelectronics Laboratory. S. G. Bishop, Director. Growth, processing, and characterization of optoelectronic materials; design, simulation, fabrication, and testing of monolithically integrated devices, circuits, and systems based on compound semiconductors; Center for Compound Semiconductor Microelectronics; optoelectronic integrated circuits; optical interconnects; metal-organic chemical vapor deposition, molecular and chemical beam epitaxy, plasma-enhanced CVD; high-resolution electron-beam lithography and chemically assisted ion-beam etching; scanning, tunneling microscopy; infrared spectroscopy and photoluminescence.

THE UNIVERSITY OF IOWA

College of Engineering

Programs of Study

The College of Engineering at the University of Iowa offers M.S. and Ph.D. programs in biomedical engineering, chemical and biochemical engineering, civil and environmental engineering, electrical and computer engineering, industrial engineering, and mechanical engineering. Master's candidates must maintain a 3.0 grade point average and may choose either a thesis or nonthesis program. Students must also successfully complete a minimum of 30 semester hours, 24 of which must be taken at the University of Iowa. Doctoral candidates must complete three years beyond the bachelor's degree, with a minimum of 72 semester hours. One academic year must be in residence. Research tools may be required as specified by the individual program. Those interested should contact the specific department for additional requirements.

Research Facilities

In addition to the research activities conducted within the six collegiate departments, there are three major research units within the College. The Iowa Institute of Hydraulic Research (IIHR) is one of the nation's premier and oldest fluids research and engineering laboratories. Its activities encompass the broad spectrum of fluid mechanics and hydraulics. The IIHR conducts programs of teaching, together with basic and applied research, in fluid mechanics (turbulent shear flows, vortex dynamics, ship hydrodynamics, and computational fluid dynamics), hydraulics (river hydraulics, computational hydraulics, hydraulic structures, and environmental hydraulics), cold-regions engineering (ice-related river hydraulics, ice mechanics, winter highway maintenance, and ice modeling), water resources (hydrometeorology, integrated watershed modeling, and environmental hydrology), and history of hydraulics and fluid mechanics. Specialized facilities include the IIHR supercomputer facility, a large low-turbulence wind tunnel, vortex-dynamics laboratory, ship-model towing tank with a wavemaker and particle image velocimeter (PIV), ice-engineering laboratories (including an ice towing tank/model ice basin), computational fluids/hydraulics laboratory, a mobile hydrometeorological laboratory (with surface energy balance, radiometric, and meteorological stations), a three-dimensional scanning elastic lidar, a differential absorption lidar (DIAL), and a number of PIV and fiber-optics-based LDV systems. High-level involvement of graduate students is a hallmark of most IIHR projects. This involvement provides unique opportunities for valuable research and engineering experience to students and postdoctoral trainees as part of their educational programs. The Iowa Spine Research Center (ISRC) assesses clinical effectiveness and outcome in diagnosing and treating various spinal diseases. It also provides guidance in spinal research and patient care. The center is a unique effort involving the Colleges of Engineering and Medicine. Teams of investigators at the center include engineers, economists, surgeons, research scientists, nurses, therapists, and students. The Center for Computer-Aided Design (CCAD) was founded for the purpose of enhancing research and development of simulation-based design methods utilizing modern computer technology and simulation tools. The center's research programs are focused on the dynamic and kinematic analyses and design of mechanical systems, robotics, structural optimization, meshless methods, fatigue analysis, and probabilistic analysis as well as operator-in-the-loop design and analysis, human factor considerations, and simulation techniques. Within CCAD are the Iowa Driving Simulator (IDS) as well as an HP S-class sixteen-processor supercomputer complemented with an extensive network of computer servers, X terminals, workstations, and personal computers to meet the computing needs for simulation-based design technology evolution. At present, the IDS is the most advanced facility of its kind in the United States. It is composed of a Harris Nighthawk 6808 real-time computer, an Evans & Sutherland ESIG 2000 Image Generator, a wrap-around sound capability, and a six-degree-of-freedom motion platform. As a result of the center's groundbreaking research and its commitment to state-of-the-art simulation technology, the U.S. Department of Transportation will build the National Advanced Driving Simulator (NADS) on the University of Iowa Oakdale research campus. The Iowa Computer-Aided Engineering Network (ICAEN) supports the curricular and research computing needs of the College through state-of-the-art hardware, the same commercial software used by engineers in the industry, and a dedicated professional support staff. All engineering students receive computer accounts and maintain those accounts throughout their college careers. Full Internet and Web access complement local educational resources, which include enhanced classroom instruction, online classes, engineering design and simulation packages, programming languages, and productivity software.

The Engineering Library is a center of College activity. Major interdisciplinary research centers have been established by the College faculty members in collaboration with faculty members from other colleges on campus. Prominent centers include the Center for Biocatalysis and Bioprocessing, the Center for Global and Regional Environmental Research, the Center for Health Effects of Environmental Contamination, and the Center for International and Rural Environmental Health.

Financial Aid

Financial aid is available to graduate students in the form of research and teaching assistantships as well as fellowships from federal agencies and industry. Support includes a competitive stipend and reduction in tuition. Specific information may be obtained from the individual departments.

Cost of Study

For 1999–2000, tuition is $2786 for state residents and $10,228 for nonresidents. There is a computer fee of $175 per semester, which allows students use of the Iowa Computer-Aided Engineering Network. In addition, there is a mandatory student health fee of $53 per semester.

Living and Housing Costs

Housing is available in apartments or private homes within walking distance of campus.

Student Group

Total enrollment at the University for fall 1998 was approximately 28,700 students. Students come from all fifty states, two U.S. territories, and 105 other countries. Engineering enrollment for fall 1998 was 1,135 undergraduate students and 305 graduate students.

Student Outcomes

Nearly half of the graduates accept positions in Iowa and Illinois, though companies from across the country present offers. Recent graduates have taken positions with companies such as 3M, Anderson Consulting, Caterpillar, Deere & Co., G. D. Searle, Hewlett-Packard, Intel, Motorola, PriceWaterhouseCoopers, Rockwell International, and Trane Company.

Location

The University is located in Iowa City, which lies in the Iowa River valley among the rolling hills of east-central Iowa. Iowa City is known as the "Athens of the Midwest" because of the many cultural, intellectual, and diverse opportunities available. The Iowa City metropolitan area is a community of 86,000 people approximately 25 miles from Cedar Rapids, Iowa's second-largest city, with more than 125,000 people.

The University

The University of Iowa, established in 1847, comprises ten colleges. The University was the first state university to admit women on an equal basis with men. The University founded the first law school west of the Mississippi River, established one of the first university-based medical centers in the Midwest, and was the first state university in the nation to establish an interfaith school of religion. It was an innovator in accepting creative work—fine art, musical compositions, poetry, drama, and fiction—for academic credit. The University established Iowa City as a national college-prospect testing center. It was a leader in the development of actuarial science as an essential tool of business administration. As a pioneering participant in space exploration, it has become a center for education and research in astrophysical science.

Applying

The application fee is $30 ($50 for international students). Admission requirements differ in each department; students should contact the department in which they are interested for additional requirements.

Correspondence and Information

Admissions
107 Calvin Hall
The University of Iowa
Iowa City, Iowa 52242
World Wide Web: http://www.uiowa.edu/~gradcoll

The University of Iowa

DEPARTMENTS, CHAIRS, AND AREAS OF FACULTY RESEARCH

College Administration
P. Barry Butler, Interim Dean.
A. Jacob Odgaard, Associate Dean for Research and Graduate Studies.
Forrest M. Holly, Interim Associate Dean for Academic Programs.

Biomedical Engineering
Chair: Krishnan B. Chandran, Professor.
Research areas: biomechanics of the spine, low back pain and scoliosis, biomechanics of tractor roll-over injuries, the physiology and kinematics of swallowing, articular joint contact mechanics, total joint replacement, in vitro and in vivo studies of heart valve prosthesis and total artificial heart, regional vascular reactivity in arterial segments with atherosclerosis or hypertension, hemodynamics of arterial disease, mechanical properties of diseased arteries, characterization of natural and synthetic composites, solution-perfused tubes for preventing blood-materials interaction, control and coordination of the cardiovascular and respiratory systems, theory and applications of equilibrium and nonequilibrium hydrosel phase transitions to biomaterials, controlled drug delivery, tissue engineering, medical image acquisition, processing and quantitative analysis, development of thermal seeds for hyperthermia in prostate, wire coil–reinforced bone cement.

Chemical and Biochemical Engineering
Interim Chair: Gregory Carmichael, Professor.
Research areas: atmospheric chemistry; air pollution engineering; artificial organs; biocatalysis, biochemical engineering, biomaterials, bioremediation, and biotechnological applications of extremophiles; catalysis and kinetics; cell culture; dilute separation technology; high-speed computing; membrane separation; polymer science; crystallization.

Civil and Environmental Engineering
Interim Chair: Robert Ettema, Professor.
Research areas: air entrainment during steel casting, air pollution, applied environmental chemistry and process design, bioremediation, computational solid mechanics, constitutive modeling, design of hydraulics structures, design simulation, hydropower, ice mechanics, optimal control of nonlinear systems, optimal design of nonlinear structures, diverse aspects of water resources engineering, rainfall and flood forecasting, thermal pollution/power plant operation, transport modeling, water quality modeling, winter highway maintenance.

Electrical and Computer Engineering
Chair: Sudhakar M. Reddy, Professor.
Research areas: acoustooptics, medical image processing, communication systems and computer networks, controls and robotics, signal processing, reliable computing systems, image processing, parallel and distributed computing systems, photonics, plasma waves, software engineering, very large scale integrated circuits.

Industrial Engineering
Chair: Peter J. O'Grady, Professor.
Research areas: concurrent engineering, engineering design, engineering economics, ergonomics/human factors, manufacturing, manufacturing process control, mathematics programming, operations research, telerobotics, quality control and reliability.

Mechanical Engineering
Interim Chair: James Andrews, Professor.
Research areas: biomechanics, combustion and chemically reactive flows, computer-aided analysis and design, dynamics, fatigue and fracture mechanics, fluid mechanics, heat transfer, materials processing and behavior, product liability, reliability-based design, robotics, structural mechanics, system simulation, thermal systems, vehicle dynamics and simulation, virtual prototyping.

The Center for Computer-Aided Design focuses on mechanical system analysis and design, structural analysis and optimization, reliability-based design, robotics, simulation-based design, human factors analysis, and driving simulation. It houses the Iowa Driving Simulator, used for a wide range of experimental studies related to traffic safety, highway and vehicle design, and basic human factors.

The Iowa Institute of Hydraulic Research, one of the nation's premier and oldest fluids research and engineering laboratories, encompasses research activities that extend into the international domain. It is renown for its expertise in fluid mechanics, hydraulics, river engineering, cold-regions engineering, and water resources.

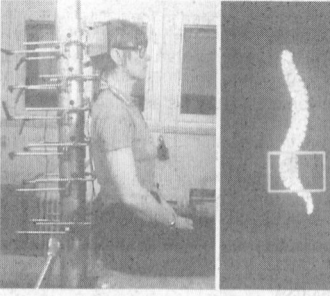

The Iowa Spine Research Center assesses clinical effectiveness and outcome in diagnosing and treating various spinal diseases. It provides guidance in spinal research and patient care and fosters interdisciplinary research on prevention, diagnosis, and treatment of low back pain as well as mathematically modeling and testing prosthetic devices used in helping speed recovery.

UNIVERSITY OF KANSAS

School of Engineering

Programs of Study

The School of Engineering offers the Master of Science degree in thirteen areas: aerospace engineering, architectural engineering, chemical engineering, civil engineering, computer engineering, computer science, electrical engineering, engineering management, environmental engineering, environmental science, mechanical engineering, petroleum engineering, water resources engineering, and water resources science.

The School also offers Ph.D. degrees in aerospace engineering, chemical and petroleum engineering, civil engineering, computer science, electrical engineering, environmental engineering, environmental science, and mechanical engineering.

For students interested in a career in advanced engineering design, the School offers a program leading to the Doctor of Engineering (D.E.) degree in aerospace engineering, civil engineering, electrical engineering, and mechanical engineering. The School also offers a program leading to a Master of Engineering (M.E.) degree in aerospace engineering.

Candidates for the M.S. or M.E. degrees are required to complete between 30 and 39 hours of course work, including a thesis or a special project.

The M.S. degree programs on the Kansas City and Topeka campuses are oriented toward the part-time graduate student.

Research Facilities

Each department in the School has access to modern, fully equipped facilities and laboratories. The University's West Campus contains the Space Technology Center, a focal point of NASA-sponsored research; the Center for Excellence in Computer Aided Systems Engineering; the Radar Systems and Remote Sensing Laboratory; the Telecommunications and Information Sciences Laboratory; the Kurata Thermodynamics Laboratory; and the Flight Research Laboratory.

Research facilities on the main campus include the Transportation Center, Structural and Materials Laboratory, Environmental Engineering Laboratory, Water Resources Laboratory, Scanning Electron Microscope Laboratory, a variety of computer-integrated manufacturing and computer-aided engineering facilities, Biomedical Engineering Laboratory, Design Technologies Lab, Tertiary Oil Recovery Project, and the Intelligent Systems and Automation Laboratory.

Financial Aid

A variety of fellowships and assistantships are available for graduate students in the School of Engineering. These awards are administered through individual departments and the Center for Research, Inc.

Cost of Study

Tuition is paid by the credit hour at a cost of $101 per credit hour for a Kansas resident and $329.75 per credit hour for a nonresident. Staff rates (TAs and most RAs) are $101 per credit hour, but up to 100 percent of tuition is waived for TAs. There is also a $15 per engineering credit hour equipment fee. International students must show minimum financial resources of $15,500 to qualify for a visa.

Living and Housing Costs

The University offers 300 apartment units that are available for married students and their families. Rents range from approximately $200 a month for a one-bedroom apartment to $225 a month for a two-bedroom apartment. Residence hall fees range from $3080 to $3850 per academic year.

Student Group

Approximately 26,400 students attend the Lawrence campus each year; approximately 6,000 of those students enroll in graduate school. The School of Engineering enrolls about 1,400 undergraduates and 650 graduate students each semester.

Student Outcomes

The primary goal of the various M.S. and M.E. degree programs is to prepare graduates for professional practice in engineering. The M.S. degree programs on the Lawrence campus also prepare the best graduate students to continue on to Ph.D. degree programs and careers in academia or industrial research. The D.E. degree program is for the practicing engineer who wants to emphasize project management skills along with course work beyond the M.S. or M.E. degree.

Location

The University of Kansas is situated in Lawrence, a community of 75,000 people. The University's main campus occupies about 1,000 acres on and around Mount Oread, a limestone bluff that overlooks the city's verdant surroundings. Lawrence is situated 21 miles east of Topeka, the state capital, and 40 miles west of Kansas City, Missouri. Around Lawrence there are numerous lakes and parks for recreational use and many theaters and museums for cultural enjoyment.

The University

The University of Kansas was founded in 1866 and is a charter member of the Association of American Universities, which was founded in 1900. The KU School of Engineering was founded in 1891, although the University has been offering engineering courses since 1869.

Applying

Application forms should be requested from, and returned to, the department in which the applicant wishes to study. A $30 nonrefundable application fee, two copies of official transcripts of all undergraduate and graduate work, and three letters of recommendation are required. GRE scores are required by some departments, and international students must submit TOEFL scores.

Correspondence and Information

Graduate Program
Department of (specify)
School of Engineering
Learned Hall
University of Kansas
Lawrence, Kansas 66045

University of Kansas

FACULTY HEADS AND RESEARCH AREAS

Aerospace Engineering

Mark Ewing, Chair; Ph.D., Ohio State. Eight faculty members. Computer-aided aircraft engineering and design, airplane stability and control, automatic flight control design and development, composite structures, computational aerodynamics, flight testing, powered-lift aerodynamics, propulsion integration, structural design and testing. Research facilities: two subsonic wind tunnels, a supersonic wind tunnel, a propulsion-test cell, two aircraft, and aircraft structural testing and computer facilities.

Architectural Engineering

Thomas E. Glavinich, Chair; D.E., Kansas. Six faculty members. Building environmental systems: analysis and design, interior and exterior lighting design, building power, control and communication systems, HVAC building-systems design, active and passive solar systems, building acoustics; structural analysis and design; construction engineering and management: building construction, cost estimating, planning, scheduling, project management, site engineering.

Chemical and Petroleum Engineering

Don Green, Chair; Ph.D., Oklahoma. Eleven faculty members. Enhanced oil recovery: micellar/polymer/thermal process; polymers: characterization, adaptation for drug-delivery systems; thermodynamics: phase equilibria, equations of state; natural gas reservoir engineering: mathematical and physical modeling; transport: boiling heat transfer; reservoir engineering: characterization of porous media, mathematical modeling, transient-test analysis; kinetics and catalysis: supercritical reactions, pollution-control catalysis for internal-combustion engines and wood-burning devices; process engineering: computer-aided design; process dynamics and control: optimal control, computer applications to control.

Civil Engineering

Steve McCabe, Chair; Ph.D., Illinois at Urbana-Champaign. Eighteen faculty members. Fluid mechanics; water resources management; water resources policy; hydrograph design methodologies; flood and drought disaster mitigation; hazardous-waste materials; biological-waste treatment; industrial waste biomonitoring and toxicity studies; biological wastewater treatment, aerobic and anaerobic; fracture mechanics, fatigue, fracture control plans for structures, structural mechanics; theoretical and experimental stress analysis; reinforced concrete design and analysis; plain concrete, composite construction; earthquake engineering; soil dynamics and soil-structure interaction; traffic flow theory, traffic control, transportation planning; geotechnical engineering, tunnels, earth dams and foundations; highway engineering.

Electrical Engineering and Computer Science

Search pending for Chair. Twenty-eight faculty members. Telecommunication and information systems: statistical communications theory, digital signal processing, image processing, pattern recognition, microwave communications, general systems theory, stochastic modeling, estimation theory; remote sensing and radar systems: radio wave propagation, antennas, radar imaging and backscatter systems, microwave remote sensing; computer engineering: embedded systems, microprocessor design, artificial/machine intelligence, computer-aided design, computer-based information systems, computer networking and protocols, computational methods and algorithms, computer graphics and modeling; electronics: circuit design for communications and control, pulse circuits, biomedical applications using digital signal processing; electromagnetic theory: antenna arrays, plasma propagation, acoustic waves, applied mathematics.

Mechanical Engineering

Terry Faddis, Chair; D.E., Kansas. Eleven faculty members. Computer-integrated manufacturing and behavior of materials, computational mechanics and finite-element analysis, robotics and control systems, mechanical system design and analysis, thermal-fluid sciences (thermodynamics, fluid mechanics, and heat and mass transfer) and their applications (momentum machines, heat power, air conditioning, solar energy, and aspects of environmental changes). Biomechanics of human balance, gait and motor control.

Engineering Management

Robert Zerwekh, Director; Ph.D., Iowa State. Five faculty members. Simulation, project management, and venture analysis.

RESEARCH CENTERS

Kansas University Energy Research Center
Transportation Center

UNIVERSITY OF LOUISVILLE

Speed Scientific School

Programs of Study

The Speed Scientific School is organized as a professional school of engineering. The professional program consists of five years of integrated academic course work, cooperative internships, and individual thesis research leading to the Master of Engineering degree. The Master of Engineering professional programs are offered in the following areas of specialization: chemical engineering, civil engineering, electrical engineering, engineering mathematics and computer science, industrial engineering, and mechanical engineering. A postbaccalaureate Master of Engineering professional program in engineering management is offered during evening hours for practicing engineers.

The Speed Scientific School also offers advanced programs through the Graduate School leading to the Master of Science and Doctor of Philosophy degrees. These programs are offered in the following areas of specialization: chemical engineering (M.S., Ph.D.), civil engineering (M.S., Ph.D.), computer science (M.S.), electrical engineering (M.S. in electrical engineering, Ph.D. in computer science and engineering), engineering mathematics and computer science (M.S. in computer science, Ph.D. in computer science and engineering), industrial engineering (M.S., Ph.D.), and mechanical engineering (M.S.).

Research Facilities

Research facilities and equipment supporting strong graduate and professional programs are present throughout the Speed Scientific School. The Henry Vogt Building is the Center for Computer-Aided Engineering, Rapid Prototyping, and Factory Automation. Paul C. Lutz Hall houses research laboratories devoted to air quality and emissions, pollution prevention, large-scale testing, industrial ergonomics, biomechanics, bioreactors, bioseparations, reactive mixing, food processing, artificial neural systems, communications, multimedia, computer vision/image processing, electrooptics, materials research, and a class-100 microfabrication cleanroom.

Financial Aid

Financial aid is available through industrial foundation fellowships, research and teaching assistantships, and appointments to sponsored projects in the School's research program. There are also graduate fellowships. These sources of financial support usually provide sufficient funds for living expenses and tuition fees. If necessary, however, long-term loans can be made to qualified students.

Cost of Study

In 1998–99, tuition and fees for full-time graduate professional students were $1460 per semester for state residents and $4140 per semester for nonresidents. Part-time resident students paid $175 per semester hour, and part-time nonresidents paid $502 per semester hour.

Living and Housing Costs

Attractive dormitory facilities for men and women are available on campus. Application for rooms and meals is made through the Housing Office.

Student Group

Speed School's undergraduate student body totals 1,299. The postbaccalaureate population totals 586 students: 260 enrolled in the Graduate School and 326 in the graduate professional programs of the Speed School.

Student Outcomes

Alumni of all graduate and graduate professional programs of the Speed School have been extremely successful in finding employment at, or shortly after, graduation. In this time frame, 95–98 percent of students who earn Master of Engineering or Master of Science degrees have found employment as practicing engineers. A similar proportion of Ph.D. graduates have also found positions either in academe or in industry, both international and domestic.

Location

Louisville is a beautiful city, encompassing the mixed cultures of the North and South. The Louisville Symphony Orchestra, Kentucky Opera Association, Louisville Ballet, and Actor's Theatre are well-known. There are four other colleges and two seminaries in the area. Historic sites abound, and recreational activities are many and varied.

The University

The University of Louisville, organized in 1798, is one of the oldest municipal institutions in the country. Since 1970, it has been a state university in the Kentucky System of Higher Education. There is a total student body of 20,894 in the University's twelve schools and colleges. The Graduate School has an enrollment of 1,590 full-time and 2,440 part-time students within these schools and colleges. The Speed Scientific School, the professional school of engineering of the University, is located in an engineering complex on the University of Louisville's Belknap campus.

Applying

Students are admitted to the individual departments after careful consideration of their undergraduate records, letters of recommendation, and seriousness of purpose. Excellent cooperation is received from local industry in permitting employees to enroll in early morning and late afternoon classes.

Correspondence and Information

For admission to the Speed School:
Professor Donald L. Cole, Assistant Dean
Speed Scientific School
University of Louisville
Louisville, Kentucky 40292
Telephone: 502-852-6194
Fax: 502-852-8890

For admission to the Graduate School:
Dr. Paul D. Jones, Associate Dean
Graduate School
University of Louisville
Louisville, Kentucky 40292
Telephone: 502-852-8372
Fax: 502-852-8375

University of Louisville

DEPARTMENT CHAIRS AND RESEARCH AREAS

Chemical Engineering. D. Collins, Acting Chair. Advanced study and research in chemical engineering includes basic applied and multidisciplinary areas, with particular emphasis on materials research, polymer processing, catalysis, process control, pollution prevention, reactive mixing, and the environment. Facilities include a Materials Research and Characterization Laboratory for biotechnology separation processes, interdisciplinary focus on thin-film science, CVD of diamond and diamond-like films, and microelectronic materials.

Civil and Environmental Engineering. L. F. Cohn, Chair. The department offers advanced study in five subdisciplines, which are structural, geotechnical, transportation, and water resources. Scholarship through research is a top priority of the department. Individual faculty members are currently active in the following research projects: earthquake resistant design, bridge safety analysis, reinforced concrete design, structural dynamics, bridge scour, streambank erosion, landslide stabilization, urban hydrology, regionalization of flood data, water quality modeling, groundwater modeling, rural water distribution systems, transportation noise, and computer graphics. Emphasis is placed on computer applications in analysis and design. Graduate students play major roles in each of these projects.

Electrical Engineering. D. L. Chenoweth, Chair. The graduate and professional programs include study and research in the following areas: circuit theory, integrated electronics, microelectronic devices and sensors, computer design and architectures, adaptive control and system identification, microprocessor systems (including applications for the blind), computer vision, communications systems, optical computing, computational electromagnetics, microwave and antenna components, neural networks and algorithms, analog and digital VLSI circuits and architectures, software tools for design automation, and automated logic design at the register transfer level. Departmental laboratories for graduate-level research include laboratories for CAE, signal and image processing, communications systems, VLSI circuit design, microdevices, microprocessor systems development and interfacing, electrooptics, automated testing, and microwave components.

Engineering Mathematics And Computer Science. K. A. Kamel, Chair. The graduate and professional programs of this department include study and research in operating systems design, compiler design, database-management systems, artificial intelligence, computer graphics, distributed processing, performance evaluation, software engineering, microcomputer systems, computer architecture, image processing, computer networking, computer vision, program testing, large-scale models, optimization, numerical analysis, computer control and automation, and simulation of engineering applications. Excellent computing facilities are available, including SG and HP workstations and IBM mainframes in a campus ISN/Ethernet network. The department has vision systems, a multimedia laboratory, robotics and computer control systems, graphic equipment, AI Symbolics and UNIX workstations, and microsystems in an LAN connected to the campus network. Robotics and automation research facilities are available through the Vogt Center for Computer-Aided Engineering and Factory Automation.

Industrial Engineering. S. M. Alexander, Chair. Advanced study is offered in human factors/ergonomics engineering, manufacturing engineering, operations research, and production systems engineering. Faculty research areas include optimization, simulation, manufacturing decision support systems, human performance and reliability, human aspects of advanced manufacturing technology, quality assurance and reliability, fuzzy set theory, robotics and flexible manufacturing system design, facilities planning and material handling, manufacturing planning and control, and artificial intelligence applications in industrial engineering. Laboratories are available in factory automation and manufacturing processes, industrial ergonomics/human factors, and microcomputer systems.

Mechanical Engineering. G. Prater, Chair. Advanced study is offered in the areas of heat transfer, thermodynamics, fluid mechanics, system dynamics, mechanical design, computer-aided engineering, and instrumentation. Departmental laboratories are available to support faculty research interests, which include environmental engineering, fluid power simulation and design, biomedical modeling and testing, design software development, experimental stress analysis, analytical and experimental modal analysis, acoustics, factory automation, and reverse engineering. Opportunities for interdisciplinary research are available through the School's biotechnology, manufacturing, materials, and environmental research focus groups.

RESEARCH

Speed Scientific School conducts engineering and related applied science research on a contract basis for industry and government. The research program furnishes technical assistance to industry and government, provides research opportunities for graduate students and faculty, and gives the faculty a background knowledge so that their teaching will relate more directly to engineering as it is practiced professionally. Opportunities for interdisciplinary research exist in the School's five research focus groups. Following are examples of active research projects involving faculty members and students.

Classification schedules for soil elastic properties.
Seismic response of self-strained structures.
Modeling and computer control of distillation columns.
Microprocessor implementation of direct digital control systems.
Digital signal processing.
Renewable energy resources.
Active network research.
Municipal and industrial water and wastewater treatment.
Software engineering.
Earthquake engineering.
Image processing and pattern recognition.
Computer-aided facility-location analysis.
Error rates in manual and electronic time studies.
Adaptive control of unattended machine tools.
Sensory fusion.
Psychophysical scaling of performance rating.
Materials research and analysis.
Neural networks.
Database design and text processing.
Computer-aided modeling and simulation.
Computer graphics.
Knowledge-based engineering.
Speech processing.
Computer-aided-design software.
Antenna array design.
Computer process control.
Modeling of polymer processing.
Monitoring of acid precipitation.

Heterogenous catalysis and surface chemistry.
Expert systems in civil engineering.
Highway and airport noise.
Remote short-range communications equipment.
Computer-aided circuit design and analysis.
VLSI systems.
Expert systems for manual lifting.
Human factors in design of automated manufacturing systems.
Expert systems for process quality control, production control, robot selection.
Adaptive control of machine tools.
Computer vision for microcirculatory applications.
Mixing in bioreactors.
Gas-liquid mass transfer in agitated vessels.
Electrooptics.
Parallel computing systems.
Waste minimization and pollution prevention.
Horizontal boring technology.
Acoustic tuning of refrigeration compressors.
Form optimization of composite components.
Design software development.
Design and testing of medical appliances.
Industrial assessments.
Modeling and control of MSF and RO desalination processes.
CFD simulation of mixing processes.
Rheological characterization of biochemical systems.
Thin-film science.
CVD of diamond and diamond-like films.
Microelectronic materials.

UNIVERSITY OF MARYLAND, BALTIMORE COUNTY

College of Engineering

Programs of Study

Graduate programs leading to the Master of Science and Doctor of Philosophy degrees are offered in chemical and biochemical engineering, computer science, electrical engineering, and mechanical engineering. In each of these areas, a program can be designed to meet the interests and requirements of the individual student. A Master of Science degree in engineering management is also offered through a joint program with the University of Maryland University College and the University of Baltimore. A civil engineering program is being developed with structural and environmental tracks in conjunction with the mechanical and chemical engineering departments, respectively. M.S. and Ph.D. degrees in civil engineering are expected to be offered in fall 2000.

Candidates for the M.S. degree may select either the thesis or nonthesis option. For students choosing the thesis option, a minimum of 24 semester hours of graduate-level courses and six semester hours of thesis research credit are required in addition to the thesis. For students selecting the nonthesis option, the requirements are 30 semester hours of graduate-level courses, one or more scholarly papers, and a final comprehensive examination.

Candidates for the Ph.D. degree are required to complete the equivalent of three years of full-time graduate study and research beyond the B.S. degree. The number of credit hours required varies, depending on the background of the candidate and the program in question. The ability to do independent scholarly research must be demonstrated in a dissertation on a topic connected with the major subject. The student is also required to provide a public oral defense of the dissertation.

Research Facilities

Research facilities are located in the Engineering and Computer Science Building and in the Technology Research Center, which is located approximately ½ mile from the center of campus. Class 100 and Class 1000 clean rooms and other state-of-the-art laboratories are available for research in such areas as machine tool dynamics, mechatronics, robotics, biorheology, soft tissue, orthopedics, connective tissue, inelastic impact dynamics, nonlinear materials characterization, heat transfer, diode and picosecond lasers, upstream and downstream biochemical processing, tissue engineering, drug delivery, biosensors, biooptics, protein engineering, remote sensing signal and image processing laboratory, metal organic chemical vapor deposition (MOCVD), chemically assisted ion beam etching (CAIBE), communications and signal processing, optical communication, intelligent information systems, digital libraries, parallel and distributed processing, computer networks, artificial intelligence, and computer graphics and animation.

Financial Aid

Research and teaching assistantships are available for well-qualified applicants. Graduate assistantships carry a basic stipend ranging from $7000 to $18,000 per academic year. Health insurance coverage and tuition for the required 7 graduate credit hours of enrollment per semester are included. Graduate School fellowships are offered, as are fellowships and grants in aid for eligible students from minority groups.

Cost of Study

In 1999–2000, tuition is $268 per credit hour for Maryland residents and $470 per credit hour for nonresidents. Additional fees are approximately $47 per credit hour. The cost of books and supplies varies according to courses taken.

Living and Housing Costs

A very limited number of spaces are set aside for graduate students in the University apartment complexes on campus. Room and board are approximately $3000 per semester, plus utilities. Off-campus housing is available, with costs ranging from $200 to $300 per month for rooms to share in private homes and from $600 to $900 per month for a three-bedroom house. Many of the housing units are on the University shuttle bus route.

Student Group

There are approximately 250 graduate students and approximately 1,300 undergraduates in the College of Engineering. Total campus enrollment is approximately 10,000.

Location

The University of Maryland, Baltimore County (UMBC) campus is located on a 477-acre suburban site 10 minutes away by car from the Inner Harbor and Oriole Park at Camden Yards in downtown Baltimore. The campus is strategically located where Interstate 95 intersects the southern part of the Baltimore Beltway in the midst of one of the greatest concentrations of research facilities and talent in the nation. The campus is only 30 miles from the nation's capital and is near theaters, museums, and Baltimore/Washington International Airport.

The College

The College of Engineering was started in 1984 as part of the University of Maryland College Park School of Engineering. In 1992, the UMBC College of Engineering became an independent engineering program within the University System of Maryland. The College's innovative research program focuses on the areas of bioprocess engineering, photonics, signal processing and communications, biomechanical engineering, advanced manufacturing, robotics and automation, materials, intelligent information systems, high-performance computing and communications, computer graphics, and algorithms and theory.

Applying

Applications for admission to the University of Maryland Graduate School, Baltimore must be submitted directly to the Graduate School by July 1 for admission in the fall semester and December 1 for admission in the spring semester. Deadlines for international students are January 1 for the fall semester and May 1 of the prior academic year for the spring semester. The application fee is $40.

Correspondence and Information

Graduate Adviser
Department of (specify)
College of Engineering
University of Maryland, Baltimore County
1000 Hilltop Circle
Baltimore, Maryland 21250
Telephone: 410-455-3400 (Chemical and Biochemical)
            410-455-3500 (Computer Science and Electrical)
            410-455-3330 (Mechanical)
Fax: 410-455-1049 (Chemical and Biochemical)
     410-455-1048 (Computer Science and Electrical)
     410-455-1052 (Mechanical)

University of Maryland, Baltimore County

THE FACULTY AND THEIR RESEARCH

Chemical and Biochemical Engineering

Duane F. Bruley, Professor; Ph.D., Tennessee, 1962. Bioprocess dynamics and control, biodownstream processing.

Douglas Frey, Associate Professor and Interim Chair; Ph.D., Berkeley, 1984. Chromatographic separations, electrophoresis.

Kyung Kang, Assistant Professor; Ph.D., California, Davis, 1991. Biomedical and bioprocess engineering, bioseparations.

Mark Marten, Assistant Professor; Ph.D., Purdue, 1991. Cellular response to dynamic environments, rheology in viscous cell broth.

Antonio R. Moreira, Professor and Associate Provost; Ph.D., Pennsylvania, 1977. Biotransformations, expression of proteins.

Gregory F. Payne, Professor; Ph.D., Michigan, 1984. Enzyme-based strategies for waste management, metabolic engineering of Streptomyces.

Govind Rao, Professor; Ph.D., Drexel, 1987. Oxygen toxicity, noninvasive sensors.

Julia Ross, Assistant Professor; Ph.D., Rice, 1994. Molecular mechanisms of cell adhesion, biomaterials, and biocompatibility.

Michael R. Sierks, Associate Professor; Ph.D., Iowa State, 1988. Structural and functional interactions of proteins and substrates.

Computer Science Electrical Engineering

Tulay Adali, Associate Professor; Ph.D., North Carolina State, 1992. Analysis of adaptive signal-processing algorithms and their application.

Alan J. Baumgarten, Lecturer; M.S., Johns Hopkins, 1986. C++, software development.

Susan Bogar, Lecturer; M.S., Maryland, Baltimore County, 1997. Information retrieval, Internet, intelligent agents.

Gary Burt, Lecturer; M.S., Bowie State, 1997. Design, development, and support of communications.

Gary M. Carter, Professor; Ph.D., MIT, 1975. Optoelectronics, diode lasers, nonlinear optics, coherent optical communications.

Chein-I Chang, Associate Professor; Ph.D., Maryland, 1986. Information theory, image processing, digital signal processing.

Richard Chang, Associate Professor; Ph.D., Cornell. Computational complexity theory, structural complexity, analysis of algorithms.

Jyh-Chia Chen, Associate Professor; Ph.D., SUNY at Buffalo, 1989. Optoelectronic materials and devices, thin film technology.

Yung-Jui Chen, Professor; Ph.D., Pennsylvania, 1976. Integrated optics and optoelectronics, optical and electronic properties of materials, ultrashort optical pulse spectroscopy.

Fow-sen Choa, Associate Professor; Ph.D., SUNY at Buffalo, 1988. Semiconductor lasers, optoelectronic integrated circuits.

David Ebert, Associate Professor; Ph.D., Ohio State, 1994. Visualization, procedural modeling, advanced rendering and animation techniques, volumetric rendering.

Tim Finin, Professor; Ph.D., Illinois at Urbana-Champaign, 1980. Artificial intelligence, knowledge representation, intelligent agents.

Dennis Frey, Lecturer; M.E.S., Loyola (Baltimore), 1998. Real-time transaction processing, proprietary database design, relation databases.

Anupam Joshi, Assistant Professor; Ph.D., Purdue, 1993. Distributed, networked, and mobile computing.

Konstantinos Kalpakis, Assistant Professor; Ph.D., Maryland, 1994. Parallel and distributed systems, digital libraries and electronic commerce, database systems, data structures, combinatorial optimization.

Hank Katz, Lecturer; M.S., Florida, 1962. Networking.

Samuel Lomonaco, Professor; Ph.D., Princeton, 1964. Algebraic coding theory, cryptography.

Curtis R. Menyuk, Professor; Ph.D., UCLA, 1981. Light propagation, optical fibers, nonlinear phenomena.

Ethan Miller, Assistant Professor; Ph.D., Berkeley, 1995. Massive storage systems, parallel file systems, multimedia, computing.

Joel M. Morris, Professor; Ph.D., Johns Hopkins, 1975. Communications and signal processing, signal detection and estimation, information theory.

Charles Nicholas, Associate Professor; Ph.D., Ohio State, 1988. Electronic document processing, software engineering, intelligent information systems.

Yun Peng, Associate Professor; Ph.D., Maryland College Park, 1985. Artificial intelligence, neural network computing, medical applications, automated diagnostic systems, reasoning under uncertainty, expert systems.

John Pinkston, Professor and Chair; Ph.D., MIT, 1967. Coding theory, data compression, information security.

James Plusquellic, Assistant Professor; Ph.D., Pittsburgh, 1997. Microelectronic device testing, modeling and simulation of digital devices and fabrication processes, neural network architectures and algorithms.

Penny Rheingans, Assistant Professor; Ph.D., North Carolina at Chapel Hill, 1993. Interactive computer graphics for the representation of information, manipulation of virtual objects.

Alan T. Sherman, Associate Professor; Ph.D., MIT, 1987. Discrete algorithms, cryptology, algebraic and security properties of cryptographic functions.

Deepinder Sidhu, Professor; Ph.D., SUNY at Stony Brook, 1979. Computer networks, distributed systems, distributed and heterogeneous databases.

Brooke Stephens, Associate Professor and Associate Dean; Ph.D., Maryland College Park, 1962. Numerical analysis, computational fluid dynamics and resource allocation problems relating to distributed systems.

Andrew Veronis, Visiting Professor; Ph.D., Manchester, 1968. Computer architectures, speech processing.

Li Yan, Associate Professor; Ph.D., Maryland, 1989. Quantum electronics, ultrashort pulse formation, ultrafast nonlinear optics.

Yaacov Yesha, Professor; Ph.D., Weizmann (Israel), 1979. Algorithms, computational complexity, source coding, speech and image compression.

Yelena Yesha, Professor; Ph.D., Ohio State, 1989. Distributed systems, database systems, digital libraries, electronic commerce, performance modeling.

Mechanical Engineering

Muniswamappa Anjanappa, Associate Professor; Ph.D., Maryland College Park, 1986. Mechatronics, controls, sensors, actuators.

Dwayne Arola, Assistant Professor; Ph.D., Washington (Seattle), 1996. Net shape machining of composite materials, tool design.

Stephen M. Belkoff, Assistant Professor (Joint with Orthopedic Biomechanics Laboratory, University of Maryland School of Medicine); Ph.D., Michigan State, 1990. Hard and soft tissue biomechanics, orthopedic implant mechanics, microstructural modeling.

Shlomo Carmi, Professor and Dean; Ph.D., Minnesota, 1968. Fluid mechanics, rheology computational mechanics, stability and bifurcation.

Panos Charalambides, Associate Professor; Ph.D., Illinois at Urbana-Champaign, 1986. Fracture mechanics, finite elements, mechanics of composites and biomechanics.

Charles Eggleton, Assistant Professor; Ph.D., Stanford, 1994. Computational biofluid mechanics, blood flow, non-Newtonian flows.

Tony Farquhar, Assistant Professor; Ph.D., Cornell, 1991. Mechanical analysis of living structures: mechanical factors in osteoarthritis and cartilage.

Akhtar S. Khan, Professor; Ph.D., Johns Hopkins, 1972. Dynamic plasticity, constitutive modeling of finite plastic behavior, fracture mechanics.

Severino L. Koh, Professor; Ph.D., Purdue, 1962. Theoretical and applied mechanics, continuum mechanics, composite materials.

Uri Tasch, Associate Professor; Ph.D., MIT, 1983. Automatic controls, robotics, manufacturing, biomechanics.

L. D. T. Topoleski, Associate Professor; Ph.D., Pennsylvania, 1990. Biomaterials and mechanics, fracture and fatigue.

Christian H. von Kerczek, Associate Professor and Interim Chair; Ph.D., Johns Hopkins, 1973. Theoretical and computational fluid mechanics.

William Wood III, Assistant Professor; Ph.D., Berkeley, 1996. Decision support and collaborative design, formal methods for conceptual engineering design, education design, information systems.

Neil T. Wright, Associate Professor; Ph.D., Pennsylvania, 1992. Boiling heat transfer, thermal contact conductance, microelectronic cooling.

Liang Zhu, Assistant Professor; Ph.D., CUNY, City College, 1995. Bioheat transfer, thermal regulation in tissue, bioheat transfer modeling.

UNIVERSITY OF MASSACHUSETTS AMHERST

College of Engineering

Programs of Study

The College of Engineering at the University of Massachusetts offers seven M.S. and five Ph.D. programs in four departments. These are the Departments of Chemical Engineering, Civil and Environmental Engineering, Electrical and Computer Engineering, and Mechanical and Industrial Engineering. The programs offer considerable flexibility to the student who wishes to develop a curriculum tailored to his or her own professional requirements. A student majoring in any of the academic curricula must have a strong background in the physical sciences and engineering.

A wide variety of programs in mathematics, computer sciences, and the physical and biological sciences are available on the campus, allowing students a wide range of offerings that strongly support engineering work. A program in polymer science and engineering is offered in the College of Natural Sciences and Mathematics.

Research Facilities

The College is housed in six buildings designed for both instruction and research. Research laboratories and field and demonstration facilities are well equipped for their respective functions. Supporting facilities include well-staffed and well-equipped electronics, instrument, machine, welding, and carpentry shops. Engineering Computer Services provides network and central computer services to the faculty and students of the College of Engineering. These services include central file and compute servers as well as e-mail and Internet access. Several public access computer rooms, including a 48-seat PC lab/classroom, two additional PC labs, and a UNIX workstation area, provide access to these services to the students. Multimedia PCs and Macintosh computers, scanners, and writeable CD-ROMs are also available for use by faculty members and students. Students may develop their own World Wide Web page, converse with family and friends via e-mail, explore the Internet, or do course work from any of the College's public rooms or via modem from their dorm or home.

Financial Aid

Financial aid is available in the form of fellowships and traineeships supported by the University and NSF. Also available are teaching assistantships supported by the University and research assistantships supported by sponsored research projects. The amount of stipends for fellowships and traineeships depends on the supporting agency and the recipient's number of dependents. Research and teaching assistantships require part-time work for the University during the academic year. An assistant is permitted to carry up to 12 semester hours of courses. Frequently, assistantships may be continued through the summer on either a part-time or a full-time basis. Remuneration is commensurate with the experience of the holder and averages $12,000 to $17,000 per year. Tuition charges are usually waived for students whose stipends are greater than $1735 per semester.

Cost of Study

In the spring of 1999, tuition for Massachusetts residents was $119 per credit hour (up to $1428 per semester); for nonresidents, it was $375.75 per credit hour (up to $4509 per semester). Full-time graduate students were charged general fees of $1428 per semester and an engineering fee of $150.

Living and Housing Costs

University housing for graduate students—married and unmarried—is available but not adequate to meet the demand. Rooms and apartments in the surrounding area are adequate and comparatively reasonable in cost. The University maintains a housing office that offers assistance in finding suitable housing on or off campus.

Student Group

About 6,000 students were pursuing graduate degrees in fifty academic departments during 1998–99. There were more than 400 graduate engineering students in the four major departments. The engineering student body is divided evenly between U.S. and non-U.S. residents.

Location

The University of Massachusetts is situated on 1,200 acres in one of the most picturesque sections of New England. The University at its Amherst campus joins with its academic neighbors—Amherst, Hampshire, Smith, and Mount Holyoke colleges—in maintaining the rich tradition of educational and cultural activity associated with the beautiful Connecticut Valley region. The area is small town and rural in character but is within 4 hours' drive of any East Coast location from New York City to Portland, Maine. Skiing is available locally, and the large ski areas of Vermont and New Hampshire are 2 to 4 hours away by car.

The University

One of the leading centers of public higher education in the Northeast, the University of Massachusetts at Amherst was established in 1863 under the original Land Grant Act. In recent decades, it has achieved a growing reputation for excellence in an increasing number of disciplines, for the breadth of its academic offerings, and for the expansion of its historical roles in education, research, and public service.

Within its ten schools, colleges, and faculties, the University offers bachelor's degrees in more than ninety areas, master's degrees in seventy, and the doctoral degree in forty-eight. There are approximately 25,000 students enrolled at the University—19,000 undergraduates and 6,000 graduates.

Applying

Application for graduate study in the College of Engineering for fall enrollment should be made by February 1, February 15, or March 1, depending on the department. An undergraduate cumulative grade point average of 2.75 or better is normally required for admission to degree status. Applicants who fail to meet this requirement must present other substantial evidence of the capacity to do satisfactory graduate work. They may then be admitted with provisional status. The application fee is $25 for residents and $40 for nonresidents.

Correspondence and Information

Graduate School
University of Massachusetts
Amherst, Massachusetts 01003
Telephone: 413-545-0721

University of Massachusetts Amherst

THE FACULTY AND THEIR RESEARCH

Chemical Engineering
Professors S. M. Auerbach (Adjunct), W. C. Conner Jr., A. K. Dillow, M. F. Doherty, J. M. Douglas (Emeritus), J. W. Eldridge (Emeritus), R. J. Farris (Adjunct), V. Haensel (Emeritus), D. A. Hoagland (Adjunct), R. S. Kirk (Emeritus), R. L. Laurence, R. W. Lenz (Adjunct), M. F. Malone (Head), P. A. Monson, K. M. Ng, S. C. Roberts, M. Tsapatsis, M. Vanpee (Emeritus), D. Vlachos, J. J. Watkins, P. R. Westmoreland, H. H. Winter, Z. Q. Zheng. Graduate research may be pursued in key areas of chemical engineering, including process design and control, applied molecular and materials modeling, transport phenomena, biochemical engineering, and chemical reaction engineering. Opportunities for research in interdisciplinary areas such as polymer science and engineering are available. A few examples of current research problems are synthesis and surface studies of new catalysts; process design methodology; diffusion in polymers; polymer rheology; azeotropic distillation; separation processes; statistical thermodynamics; plasma-enhanced chemical vapor deposition; combustion kinetics; nanophase materials synthesis and modeling; and supercritical fluid processing.

Civil and Environmental Engineering
Professors D. P. Ahlfeld, B. B. Berger (Emeritus), A. Chajes, C.-S. Chang, S. A. Civjan, D. DeGroot, J. K. Edzwald, S. J. Ergas, K. L. Hancock, W. H. Highter, C. L. Ho, J. M. LaFave, T. J. Lardner, S. C. Long, A. J. Lutenegger, D. A. Noyce, D. W. Ostendorf, E. Parkany, D. A. Reckhow, E. T. Selig (Emeritus), P. W. Shuldiner, F. D. Stockton, M. S. Switzenbaum, J. E. Tobiason. Graduate students may specialize in the following areas: structural engineering and mechanics, geotechnical engineering (subconcentration in environmental geotechnology is available), transportation, and environmental engineering. Research, sponsored and unsponsored, is being pursued in water supply and treatment, water quality and pollution, wastewater treatment, hazardous wastes, transportation of sediment in fluids, structural dynamics, stability of structures, finite-element analysis, fracture mechanics, transportation technology transfer, applications of geographic information systems, intelligent transportation systems, pavement design, pavement management, highway traffic operations, soil dynamics, soil-structure interaction, railroad track performance, buried pipelines, earth retaining structures, soil behavior, in situ testing techniques, soil constitutive relationships, and computer analysis in geotechnical engineering.

Electrical and Computer Engineering
Professors N. G. Anderson, L. S. Bobrow, W. P. Burleson, K. R. Carver, M. Ciesielski, A. P. DeFonzo, H. Dorin, S. B. Desu (Head), T. F. Djaferis, L. E. Franks (Emeritus), S. J. Frasier, A. Ganz, D. L. Goeckel, W. Gong, I. G. Harris, F. S. Hill, C. V. Hollot, R. W. Jackson, P. A. Kelly, W. L. Kilmer, I. Koren, C. M. Krishna, K. M. Lau, D. P. Looze, D. H. Navon (Emeritus), C. M. Pozar, S. C. Reising, D. H. Schaubert, K. D. Stephan (Associate Head), C. T. Swift, T. W. Tang, R. Tessier, K. S. Yngvesson. The department offers advanced study and research in communications, signal processing, and systems and control engineering (including wireless communication theory, image processing and analysis, coding for multimedia systems, network modeling and control, intelligent vehicle systems, control theory, and large-scale optimization); computer systems engineering (including VLSI computer-aided design, architecture, and testing; parallel, real-time, and fault-tolerant computing systems; computer communication networks; and computer graphics); and microwave, solid-state, and optical electronics (including microwave systems and remote sensing; antennas and electromagnetic analysis; microwave, millimeter wave, and ultra–high speed circuits; compound semiconductor materials and devices; superconducting materials and devices; optoelectronic devices and materials; and numerical modeling of semiconductor microdevices).

Environmental Engineering
Professors D. P. Ahlfeld, B. B. Berger (Emeritus), J. K. Edzwald, S. J. Ergas, S. C. Long, D. W. Ostendorf, D. A. Reckhow, M. S. Switzenbaum (Coordinator), J. E. Tobiason. The environmental engineering curriculum within the civil and environmental engineering department prepares students for engineering careers related to the protection of the environment as it pertains to public well-being and health. Specialization is offered in water and wastewater treatment, water resources, groundwater, hazardous wastes, and environmental chemistry and microbiology. Sponsored research is being carried out in biological, physical, and chemical treatment of waters and wastewater; drinking water supply systems; wastewater collection systems; water resources and systems analysis; groundwater pollution and modeling; unconfined aquifer containment; and modeling and treatment of hazardous contaminants in drinking waters, groundwater, soils, and wastewaters.

Industrial Engineering
Professors A. Deshmukh, D. L. Fisher, R. J. Giglio, D. Kazmer, D. S. Kim, L. M. Seiford, J. M. Smith, J. Terpenny. This program focuses on operating person-machine systems of society: factories and other productive systems, service systems, and policy and control of systems. In the programs of teaching, research, and service, the problems of such systems are studied in order to understand their behavior and to be able to solve problems that arise in them. In particular, the skills of operations research, productivity improvement, total quality management, and human-factors science are applied to such fields as automated factories and robotics, business systems, health-care systems, and government operating systems. Much of the work of industrial engineering and operations research is interdisciplinary and proceeds with the cooperation of other departments inside and outside the College of Engineering.

Manufacturing Engineering
Professors K. Danai, A. Deshmukh, R. Gao, D. Kazmer, B. H. Kim, S. Malkin, L. E. Murch, C. R. Poli, L. M. Seiford. This interdisciplinary program is concerned with the planning and selection of methods of manufacture, the design of equipment for manufacture, the improvement of established manufacturing techniques, and the development of new ones. Automatic assembly, automation in manufacturing, robots, grinding, intelligent control of machining, injection molding, and design for ease of manufacture are emphasized.

Mechanical Engineering
Professors L. L. Ambs, T. R. Blake, Y. Chait, K. Danai, A. Deshmukh, J. Donovan, R. Gao, J. I. Goldstein (Dean), W. P. Goss, I. R. Grosse, K. Jakus, D. Kazmer, B. H. Kim, R. H. Kirchhoff, S. Krishnamurty, S. Malkin, J. G. McGowan, L. E. Murch, S. Nair, B. Perot, C. R. Poli (Head), J. R. Rinderle, J. E. Ritter, G. A. Russell, J. E. Sunderland, J. Terpenny, G. E. Zinsmeister. Faculty members conduct research in areas of control theory and applications, mechanical design, design theory and methodology, manufacturing engineering, materials engineering, renewable energy, energy conservation, and thermal-fluids engineering. This research, which is sponsored by both industry and government, involves experimental, analytical, and computational activities. Examples of current research are robust controls, fault diagnosis, automation of mechanical design, design for producibility, CAD/CAM, injection molding, CNC machining, grinding, mechanical behavior and fracture of brittle materials, mechanical behavior of polymers, multiphase fluid mechanics, combustion, industrial and residential energy conservation, and solar and wind energy systems.

Polymer Science and Engineering
Professors R. J. Farris (Head), S. P. Gido, R. B. Hallock (Adjunct), D. A. Hoagland (Graduate Program Director), S. L. Hsu, F. E. Karasz, R. W. Lenz (Emeritus), A. J. Lesser, C. P. Lillya (Adjunct), W. J. MacKnight, T. J. McCarthy, M. Muthukumar, J. Penelle, T. P. Russell, K. Schmidt-Rohr, R. S. Stein (Emeritus Adjunct), H. H. Strey, H. H. Winter (Adjunct). This interdisciplinary Ph.D.-granting department offers research opportunities across the polymer field, spanning the chemistry, physics, and engineering of all types of polymeric materials. Multi-investigator programs exist in biosynthesis, interfaces, mechanical properties, morphology, transport, blends, networks, and block copolymers.

UNIVERSITY OF MASSACHUSETTS LOWELL

James B. Francis College of Engineering

Programs of Study

The education of engineers in state-of-the-art areas of advanced technology and the University's commitment to national and regional economic development are the major premises upon which the graduate programs in the James B. Francis College of Engineering are based. These programs are intended to produce engineers whose education not only develops expertise in the design, development, and production of products but also an understanding of the management involved in the creation of new products, companies, and service organizations. The graduate programs in engineering are thus intended to educate engineers capable of keeping abreast with the rapidly changing technology that characterizes the high-technology economy of the Northeast. The programs lead to degrees of M.S.Eng. in the areas of chemical, civil, computer, electrical, energy, materials, mechanical, and plastics engineering; the M.S. in environmental studies, and work environment; the Sc.D. in work environment; the D.Eng. in electrical, mechanical, and plastics engineering; and, through the College of Arts and Sciences, the Ph.D. in applied chemistry (with options in environmental science and polymer science/plastics engineering) and the Ph.D. in applied physics (with options in energy engineering and engineering mechanics).

Research Facilities

Each department has modern laboratory equipment supporting its specific research areas. The College has a collection of laboratories available to students for study in the areas of acoustics and vibrations, atmospheric measurements, modal analysis and controls, plastics and composite material processing and evaluation, solar energy and photovoltaics, structural ceramics, and water resources–environmental, and transportation. In addition, there are a number of interdepartmental laboratories in support of interdisciplinary faculty team initiatives and University-wide centers to support multidisciplinary programs in electronics technologies, telecommunications, materials and materials processing, atmospheric research, imaging and optical computing, sustainable production, high-speed computing, computer-aided engineering and design, data visualization, ergonomics, manufacturing, environmental engineering and toxics use reduction, and radiation and submillimeter technology. The College has extensive computer facilities that are completely networked for system integration in computer-aided engineering, testing, design, and manufacturing using state-of-the-art engineering application programs.

Financial Aid

Approximately 65 teaching assistantships and 65 research fellowships are available through the Graduate School, individual faculty members, and the Research Foundation. Grants for research projects during the summer are available from the Graduate Office. Federal Perkins Loans and other forms of aid may be obtained through the Financial Aid Office.

Cost of Study

In 1998–99, tuition and fees for full-time graduate students were $4869 per academic year for residents of Massachusetts and $10,278 per academic year for all others. Health insurance and thesis preparation fees are extra. In some programs, reduced tuition is available to residents of the New England region.

Living and Housing Costs

In 1998–99, on-campus graduate housing cost about $390 per month. The cost of married student housing ranged from $440 to $490 a month. Furnished rooms near the campus cost $250 to $400 a month, and unfurnished apartments are available.

Student Group

In fall 1998, the enrollment of the College of Engineering totaled 1,466 day and early evening students. Of the total enrollment, 576 were graduate students; 361 of these were studying part-time.

Student Outcomes

UMass Lowell graduates are employed in consulting firms, varying industries, and state and federal research and regulatory agencies. These include the EPA, Stone & Webster, FAA, state transportation agencies, U.S. Air Force, Raytheon, DEC, M/A COM, Polaroid, AT&T, NYNEX, and Foster Miller. Recent graduates have gone on to graduate programs at many prestigious institutions, including MIT, Berkeley, Rensselaer, Cornell, Illinois, Stanford, Case Western Reserve, Georgia Tech, and Northwestern.

Location

The University is in Lowell, a city of about 95,000, 30 miles northwest of Boston. The campus lies on both sides of the Merrimack River. Within an hour's drive are the cultural, educational, and recreational activities of Boston and many national historic sites. Lowell is also ideally situated for outdoor activities. The lake and mountain regions of New Hampshire and the Atlantic beaches are easily accessible.

The University

The former University of Lowell became part of the University of Massachusetts five-campus system in 1991.

Applying

Applications must be submitted no later than June 1 preceding the fall term or November 1 preceding the spring term in which the applicant wishes to enroll. Scores from the GRE General Test, GMAT, or MAT and from the TOEFL for international students whose native language is not English; transcripts in duplicate; and three letters of reference are required.

Correspondence and Information

Dr. Jerome L. Hojnacki
Dean of the Graduate School
University of Massachusetts Lowell
Lowell, Massachusetts 01854
Telephone: 508-934-2380
Fax: 508-934-3010

University of Massachusetts Lowell

DEPARTMENT HEADS AND RESEARCH AREAS

Krishna Vedula, Dean.

Chemical Engineering and Materials Engineering. Professor Alfred Donatelli, Head. Research areas within the department include biotechnology (biomaterials and bioprocesses), computer-assisted process control and engineering; surface and interfacial engineering (colloids, tribology, gels, and membranes), advanced engineering materials (high-temperature applications, ceramics, composites, paper-like products, and "smart" optical materials); energy and thermofluids (solar fuel cells, solar distillation, correlation of physical properties, heat transfer, mass and momentum transport in non-Newtonian fluids, energy storage, and innovative combustion cycles); paper and pulp (delignification in Kraft pulping, enzymatic hydrolysis of cellulose, recycling, and testing), and environmental protection and site remediation. Many projects are carried out in collaboration with faculty members from other departments and programs.

Civil Engineering and Environmental Studies. Professor William Moeller, Head. Research within the Department of Civil Engineering is conducted in four speciality areas—environmental, geotechnical, structural, and transportation engineering. A list of research interests in the environmental area includes transport and remediation of groundwater contaminants, use of surfactants to enhance organic contaminant mobilization in soils, bioremediation of soil contaminants, soil vapor extraction and air sparging of groundwater contaminants, electrokinetic technology for the removal of organics and metals from fine-grained soils, measurement of atmospheric deposition of toxic materials and excess nutrients onto Massachusetts and Cape Cod bays, incineration of CFC-containing foams, statistical modeling of ozone concentrations, design of power plants to facilitate CO_2 capture, and studies of biological treatment of industrial wastewater. In the geotechnical area, research includes static and dynamic behavior of deep foundations, interfacial friction, photoelastic modeling and discrete element analysis of granular materials, anchorages for deployable structures, engineering properties, and behavior of nonwoven geotextiles. In the structural area, research includes studies of quick deployment structures, methods to evaluate the structural capacity of existing bridges and buildings, use of polymer films to strengthen masonry walls, and seismic engineering. In the transportation area, research includes simulation and optimization models for transportation networks, real-time traffic-adaptive signal control systems, and dynamic route guidance for Intelligent Vehicle Highway Systems (IVHS).

Electrical Engineering. Professor Michael A. Fiddy, Head; Professor R. Holmstrom, Coordinator. Research areas in the department include remote sensing of Earth's ionosphere, radar studies of the mesosphere, electromagnetic wave propagation in plasmas, meteor scatter communications, and scattering and inverse scattering from objects. There is materials- and device-related research in semiconductors and organics, photovoltaics, VLSI design and fabrication, optical computing, solid-state power electronics, and microwave circuits. Information and telecommunications research is conducted in the areas of probabilistic models and characterization of video and voice traffic, networks, wireless communications, image and signal processing, pattern and speech recognition, acoustics, optical image processing, and spectral estimation. Systems research is in linear and nonlinear systems, circuit simulations, air traffic control problems, stability and chaotic dynamics. Compute-intensive modeling and studies are in computer architecture with application to signal processing.

Energy Engineering. Professor Alfred Donatelli, Head; Professor Gilbert J. Brown, Coordinator. Energy engineering serves as an umbrella for graduate study in nuclear engineering and in solar energy utilization, and it promotes research in optics applied to energy generation and the protection of the environment. Research interests in nuclear engineering include reactor physics methods development, radiation damage, thermal-hydraulic systems studies, cross-section measurements, numerical analysis, fuel and waste management, operations, and plant life extension. In solar thermal engineering, interests include high-concentration systems, power plant evaluation, passive solar systems, and solar fuels and chemicals. The photovoltaics effort includes work on design method development, PV-assisted lighting, battery evaluation, and solar spectrum splitting. Other interests include energy economics, safety, and site remediation. Major research facilities include a 1-Mw research reactor, a 5-MeV Van de Graaf accelerator, and a 1-megacurie Cobalt-60 gamma irradiation facility. Within the University of Massachusetts Lowell Photovoltaic Program, there are a testing laboratory, a solar simulator, and photovoltaic arrays with a peak generating capacity of more than 10 kw.

Mechanical Engineering. Professor Struan Robertson, Head; Professor J. Sherwood, Coordinator. Major research is conducted in four general areas of concentration: mechanics and materials, thermo/fluids/energy, vibrations and dynamics, and design and manufacturing. Specific areas of specialization include processing, analysis and testing of composite materials; computational mechanics; finite element analysis; computational heat transfer; thermal plasma systems; solar systems; aerodynamics; modal analysis and structural dynamics; dynamic systems and controls; automation, robotics, and workcell design; theory and practice of engineering design; and manufacturing systems and controls. Departmental laboratories and computer facilities supporting this research include the Center for Computer-aided Engineering and Design (CAEDC), the Advanced Composite Materials and Textile Research Laboratory, the Modal Analysis and Controls Laboratory, the Heat Transfer Laboratory, the Materials Testing and Instrumentation Laboratory, and the Automated Manufacturing and Assembly Laboratory.

Plastics Engineering. Professor Robert Nunn, Head. Major areas of research are polymer process development, polymer materials engineering, and composite structures and evaluation. The department has complete facilities to process polymer materials and to fabricate composite structures. Equipment is available to characterize the molecular structure and to determine the mechanical, electrical, and thermal properties of polymeric materials. The Institute for Plastics Innovation was established at the University of Massachusetts Lowell to work with industry and government institutions and to provide practical services to industry. Flexible research programs can be individually tailored to meet the specific needs and objectives of the participant.

Work Environment. Professor David H. Wegman, Head. Master's and doctoral programs and research activities are centered in industrial hygiene, occupational ergonomics, occupational epidemiology, and work environment policy. Industrial hygiene research interests include modeling and design of exhaust ventilation systems and development of methods for real-time exposure monitoring and innovative exposure assessment in industrial settings (currently in the automotive and plastics industries). Occupational ergonomics research addresses development and application of biomechanical models, work analysis through laboratory simulations and field studies, and work reorganization for skill-based productivity, innovation, and health. Occupational epidemiology research focuses on selected health risks (acute respiratory hazards, musculoskeletal problems, acute occupational injuries, psychosocial strain, and workplace cancer) and epidemiological methods. Policy research includes investigations of social factors in occupational health, especially the recognition of occupational disease; economic analysis of injury causation, prevention, and control; politics of regulation; and cross-national studies of environment and occupational health policy. The Toxics Use Reduction Institute is a multidisciplinary research, education, and policy center that conducts research and provides technical assistance and training to promote reduction in the use of toxic and hazardous chemicals in industry and in commerce.

UNIVERSITY OF MIAMI

College of Engineering

Programs of Study

The College of Engineering offers courses of graduate study leading to the M.S., Ph.D., and Doctor of Arts (D.A.) degrees. M.S. programs are offered in architectural, biomedical, civil, electrical and computer, industrial, mechanical, and ocean engineering. A dual M.S. degree in engineering and business administration is offered through a joint program with the School of Business Administration. Ph.D. programs are currently available in biomedical, civil, electrical and computer, industrial, and mechanical engineering and industrial ergonomics. The D.A. degree, designed primarily to prepare students for college teaching, is offered with the cooperation of the School of Education. The D.A. is available in civil engineering and mechanical engineering. Graduate programs offered in conjunction with other schools or units of the University include biomedical engineering, with the School of Medicine; engineering management, with the School of Business Administration; environmental safety and health, with the School of Medicine; a joint M.S.I.E./M.B.A. weekend executive program and management of technology program, with the School of Business Administration; and ocean engineering, with the Rosenstiel School of Marine and Atmospheric Science. The Departments of Electrical and Computer, Industrial, and Mechanical Engineering offer interdisciplinary studies in computer-integrated manufacturing leading to the award of the M.S. degree. The M.S. and Ph.D. programs in interdepartmental graduate studies permit highly qualified students to pursue a privileged, individualized program of studies that cuts across disciplinary lines. Approval of the Graduate Council is required to enroll in the interdepartmental studies program. Master's degree programs are available with and without a thesis in most areas of specialization. Requirements for the M.S. with a thesis include an approved integrated program of a minimum of 30 semester credits with an average grade of B or better and no grade below C; at least 9 course credits on the 600 level; an approved thesis of 6 credits; and an oral examination in defense of the thesis. Requirements for the M.S. without a thesis include an approved integrated program of a minimum of 36 semester credits with an average grade of B or better and no grade below C; at least 12 course credits on the 600 level; and an oral examination taken after completion of a minimum of 18 credits of graduate work. For the Ph.D. degree, at least 24 credits in courses and seminars must be taken beyond the requirements for the M.S. degree. All Ph.D. candidates are expected to complete an appropriate integrated program of studies in preparation for the comprehensive qualifying examination. Such preparation normally requires two years after the bachelor's degree. One or two years beyond the qualifying examination will usually be needed for the completion of an acceptable dissertation and the remaining course work; the student will then be admitted to the final oral examination. For the D.A. degree, 78 credits beyond the baccalaureate are required. Of these, at least 48 credits must be in the major or cognate fields, 9 credits in higher education, and 12 credits in an internship and a project.

Research Facilities

The University's Ungar Computing Center is equipped with an IBM 9672-R42 with PR/SM feature, an IBM AS/400 system, two IBM RISC/6000-570s, and an open VMS cluster with AlphaServer 4100. In addition, the College has a variety of computing facilities. Numerous Sun and DEC workstations are distributed throughout the College and are interconnected with local area networks. These networks are connected to the University-wide network and the global Internet. There are several PC facilities within the College and in the University.

Financial Aid

Teaching and research assistantships are available. Graduate assistants are expected to spend 15–20 hours per week assisting in instructional or research activities. An internship cooperative program allows the student to combine graduate studies with professional employment in local engineering firms.

Cost of Study

Tuition for the master's degree as well as for post-master's and doctoral work is $852 per credit hour in 1999–2000.

Living and Housing Costs

Single men and women admitted to the Graduate School may be accommodated in single or double rooms in one of the residential colleges or may share one of a limited number of apartments. There is limited housing available to married or single-parent families. Application requests should be directed to the Assignments Office (for single students) or to the Married Family Housing Office, Department of Residence Halls, P.O. Box 248044, Coral Gables, Florida 33124 (telephone: 305-284-4505 and 305-284-4634, respectively). The cost of accommodations in the community varies with the needs and resources of the individual.

Student Group

There are 3,111 graduate students on the University campus; 150 are in the College of Engineering.

Location

Coral Gables, Miami, and south Florida offer many advantages. The Miami area is subtropical, coastal, urban, and bilingual, and it is located at an international crossroads.

The University

Students in the University number 13,422 and come from forty-nine states and more than 109 countries. Students of all religions, races, and nationalities participate in the objective of the University, which is to produce graduates who can make significant contributions to the society in which we live. Founded in 1925, the University is nonprofit, nondenominational, and coeducational. It is free from religious and political control and derives its funds from tuition and from direct contributions from alumni and others interested in its work in teaching and research.

Applying

Applicants for admission must supply transcripts, scores on the GRE General Test, and three letters of recommendation. International students are required to submit TOEFL scores, and those who are graduates of U.S. institutions must also submit GRE scores. Applications for financial assistance should be submitted before March 1.

Correspondence and Information

Office of the Associate Dean
College of Engineering
University of Miami
P.O. Box 248294
Coral Gables, Florida 33124-0620
Telephone: 305-284-2942

University of Miami

THE FACULTY AND THEIR RESEARCH

Biomedical Engineering. O. Ozdamar, Professor and Chairman; Ph.D., Northwestern: auditory instrumentation, biomedical signal processing. S. S. Asfour, Secondary Professor; Ph.D., Texas Tech: ergonomics and rehabilitation. J. Augenstein, Professor; M.D., Ph.D., Miami (Florida): medical informatics, trauma. H. Baier, Professor; M.D., Frankfurt: respiratory physiology. N. Block, Professor; M.D., NYU: artificial bladder, urologic oncology. Redmond P. Burke, Adjunct Instructor; M.D., Harvard: cardiac surgery. H. S. Cheung, Secondary Professor; Ph.D., USC: cellular engineering. D. de Azevedo, Adjunct Professor; Ph.D., Miami (Florida): image processing, instrumentation for vision. R. E. Delgado, Adjunct Assistant Professor; Ph.D., Miami (Florida): hearing instrumentation. W. Gu, Assistant Professor; Ph.D., Columbia: soft tissue biomechanics.B. Hurwitz, Associate Professor; Ph.D., Florida: physiological psychology. N. H-C. Hwang, James L. Knight Professor; Ph.D., Colorado State: fluid mechanics. P. Jayakar, Adjunct Professor; M.D., Bombay (India); Ph.D., Manitoba: electrophysiological techniques in neurology, epilepsy monitoring. J. W. Keller, Adjunct Associate Professor; M.S., Arkansas: automatic controls in biomedical engineering. K. Khalaf, Post Doctoral Associate; Ph.D., Ohio State: orthopedic engineering. J. Kline, Professor Emeritus; Ph.D., Iowa State: instrumentation, artificial organs, dental electrosurgery. R. Knighton, Research Associate Professor; Ph.D., Michigan: biophysics. D. Larnard, Adjunct Assistant Professor; Ph.D., Purdue: simulation, automatic control. L. Latta, Professor; Ph.D., Miami (Florida): orthopedic instrumentation and devices. C. Lavernia, Assistant Professor; M.D., Puerto Rico: orthopedics. Jean-Marc Legeais, Adjunct Instructor; M.D., Paris: ophthalmology. M. Lenart, Adjunct Professor; Ph.D., Stuttgart (Germany): computer-aided design. C.-C. Lu, Research Assistant Professor; Ph.D., Miami (Florida): medical instrumentation. Fabrice Manns, Research Assistant Professor; Ph.D., Miami: ophthalmic physics and devices. H. Mayrovitz, Adjunct Professor; Ph.D., Pennsylvania: cardiovascular physiology, microcirculation. P. M. McCabe, Professor; Ph.D., Illinois: neuroscience. J. Nagel, Adjunct Professor; Ph.D., Erlangen-Nuremberg (Germany): imaging, instrumentation and signal analysis. J. M. Parel, Research Associate Professor; Eng. ETS-G, Geneva: ophthalmic physics and devices. L. Pinchuk, Adjunct Assistant Professor; Ph.D., Miami (Florida): biomaterials and devices. D. Popovic, Adjunct Professor; Ph.D., Belgrade: electrical engineering, motor functions and paralysis. J. Radovich, Adjunct Professor; Ph.D., Washington (St. Louis): polymers, artificial dialysis. J. Raines, Associate Professor; Ph.D., MIT: cardiovascular fluid mechanics. R. Rol, Adjunct Assistant Professor; Ph.D., Swiss Federal Institute of Technology: ophthalmic physics and devices. N. Salansky, Adjunct Assistant Professor; Ph.D., Russia: electrostimulation and biophysics. R. Schmitt, Adjunct Professor; Ph.D., Bremen (Germany): medical ultrasound, manufacturing. N. Schneiderman, Professor; Ph.D., Indiana: psychophysiological systems. T. Shipley, Adjunct Professor; Ph.D., Johns Hopkins: psychophysiology of vision. P. Soderberg, Secondary Faculty; Ph.D., Karolinska Institutet (Sweden): ophthalmic research. P. P. Tarjan, Professor; Ph.D., Syracuse: instrumentation, implantable and assistive devices. F. Teleschi, Secondary Associate Professor; M.D., Miami (Florida): cochlear implants, neurotological monitoring. D. Tepevac, Research Assistant Professor; Ph.D., Belgrade; rehabilitation engineering, microcomputer-based medical instrumentation. Frank L. Villain, Adjunct Instructor; Ph.D., Rouen: polymer chemistry. A. Wanner, Professor; M.D., Basel: pulmonary function mechanics. B. Watson, Professor; Ph.D., Florida State: physics, photochemistry, photobiology.

Civil, Architectural, and Environmental Engineering. D. A. Chin, Professor and Chairman; Ph.D., Georgia Tech; PE: water resources and environmental engineering. J. D. Englehardt, Associate Professor; Ph.D., California, Davis: environmental planning, risk analysis, physical and chemical treatment. A. H. Namini, Associate Professor; Ph.D., Maryland: computational mechanics, bridge structures, computer-aided design. M. K. Phang, Professor; Ph.D., Rensselaer; PE: structural engineering, building architectural engineering, construction materials. H. M. Solo-Gabriele, Assistant Professor; Ph.D., MIT; PE: water resources, environmental engineering. M. Soltani, Associate Professor; Ph.D., Johns Hopkins; PE: architectural engineering, structural reliability and optimization, and computer-aided design. W. Suaris, Associate Professor; Ph.D., Northwestern; PE: structural engineering, fracture mechanics. F. H. Tinoco, Lecturer; Ph.D., Iowa State: geotechnical engineering, rock mechanics, experimental testing, foundations, shear strength and compressibility of soils. T. D. Waite, Professor; Ph.D., Harvard; PE: environmental engineering, water quality control. R. F. Zollo, Professor, Ph.D., Carnegie Mellon; PE: fiber-reinforced concrete, concrete materials testing and experimental analysis.

Electrical and Computer Engineering. T. Y. Young, Professor and Chairman; D.Eng., Johns Hopkins: computer vision and image processing. C. Douligeris, Associate Professor; Ph.D., Columbia: communication and computer networks. N. G. Einspruch, Professor and Senior Fellow in Science and Technology; Ph.D., Brown: solid-state electronics, management of technology. G. Gonzalez, Professor; Ph.D., Arizona: electronics, electromagnetics. E. I. Jury, Research Professor Emeritus; Ph.D., Columbia: signal processing, control theory and multidimensional systems. M. R. Kabuka, Professor; Ph.D., Virginia: computer vision, medical informatics, robotics. C. S. Lindquist, Professor; Ph.D., Oregon State: signal processing; passive, active, and digital filters. P. S. Liu, Professor; Ph.D., Purdue: VLSI systems, computer architecture. S. Negahdaripour, Professor; Ph.D., MIT: computer vision, undersea images. K. Premaratne, Associate Professor; Ph.D., Miami (Florida): control and signal processing. A. A. Recio, Associate Professor; D.C.A., Havana: power and illumination. M. Scordilis, Associate Professor; Ph.D., Clemson: speech processing, audio engineering. M. A. Tapia, Professor; Ph.D., Notre Dame: digital systems, compiler optimization and fault-tolerant concurrent systems. M. R. Wang, Assistant Professor; Ph.D., California, Irvine: electrooptics, integrated optics. K. Yacoub, Professor; Ph.D., Pennsylvania: statistical communications.

Industrial Engineering. N. G. Einspruch, Professor, Chairman, and Senior Fellow in Science and Technology; Ph.D., Brown: solid-state electronics, management of technology. E. M. Abdel-Moty, Research Associate Professor; Ph.D., Miami (Florida): rehabilitation, ergonomics. R. Aboudi, Associate Professor; Ph.D., Cornell: operations research, applied statistics. S. Asfour, Professor; Ph.D., Texas Tech: work physiology, ergonomics, biomechanics. N. Boubekri, Associate Professor; Ph.D., Nebraska: automated assembly, industrial robotics, computer-aided manufacturing. S. Czaja, Professor; Ph.D., SUNY at Buffalo: aging, human-computer interaction, human factors. Mohamed Aly Fahmy, Research Associate; Ph.D., Miami (Florida): biomechanics, ergonomics, finite element modeling. E. Iakovou, Associate Professor; Ph.D., Cornell: operations research, manufacturing systems, production planning and control. T. M. Khalil, Professor; Ph.D., Texas Tech: ergonomics, technology management. C. Kurucz, Professor; Ph.D., SUNY at Buffalo: operations research, applied statistics. V. K. Omachonu, Associate Professor; Ph.D., Polytechnic of New York: health systems, production control. J. Sharit, Research Associate Professor; Ph.D., Purdue: human-computer interaction, human factors. D. J. Sumanth, Professor; Ph.D., IIT: productivity engineering and management.

Mechanical Engineering. S. S. Rao, Professor and Chairman; Ph.D., Case Western Reserve: engineering optimization, reliability-based design, vibrations, design for manufacturability, fuzzy systems. J. Catz, Professor; Sc.D., MIT: measurement uncertainty, design. A. Hsu, Associate Professor; Ph.D., Georgia Tech: computational fluid dynamics, turbulence, biomedical fluid dynamics. S. Kakac, Professor; Ph.D., Manchester (England): heat transfer, nuclear engineering. S. S. Lee, Professor and Associate Dean; Ph.D., Berkeley: fluid mechanics and heat transfer, mathematical modeling of thermal discharge. H. T. Liu, Assistant Professor; Ph.D., Miami (Florida): fuel cells, solar energy, heat transfer, and two-phase flow. R. Narasimhan, Research Assistant Professor and Assistant Dean; Ph.D., Miami (Florida): fluid mechanics, heat transfer. A. Shahin, Assistant Professor; Ph.D., Purdue: controls and dynamics. N. Simha, Assistant Professor; Ph.D., Minnesota: continuum and fracture mechanics, shape memory, alloy, soft tissues. M. R. Swain, Associate Professor; Ph.D., Miami (Florida): conventional and alternative fueled internal combustion engines, hydrogen safety analysis. T. N. Veziroglu, Professor and Director, Clean Energy Research Institute; Ph.D., London: two-phase fluid flows, thermal contact conductance, solar energy, hydrogen energy. K.-F. V. Wong, Professor; Ph.D., Case Western Reserve: fluid/mass transfer in porous media, environment-energy systems, expert systems, air pollution, ocean pollution, solid wastes.

UNIVERSITY OF MICHIGAN

College of Engineering

Programs of Study

The College of Engineering offers graduate programs leading to the Master of Science, Master of Science in Engineering, Master of Engineering, and Doctor of Philosophy degrees in the most current and vital areas of engineering, including aerospace, automotive, chemical, civil and environmental, nuclear, and industrial and operations engineering; applied physics; atmospheric, oceanic, and space sciences; bioengineering; electrical engineering and computer science; macromolecular science and engineering; manufacturing; materials science; mechanical engineering; naval architecture and marine engineering; radiological health; technical communication; and transportation. The College also offers a Doctor of Engineering degree in manufacturing. The specific degree requirements for the master's are 30 credit hours of approved graduate course work, with at least a B average. There are no foreign language requirements. The residency period is at least two terms of 8 credit hours each or the equivalent of full-time or summer half-term work.

Departmental doctoral course requirements vary, but students must pass a comprehensive examination in a major field or specialization before being recommended for candidacy. A dissertation describing an independent research project, under the direction of a dissertation committee, is required. Requirements for foreign languages and nontechnical course work are determined by individual departments or programs and the Graduate School. At least two terms of 8 credit hours each must be completed on campus. Equivalent summer or part-time work may be substituted upon recommendation of the department.

Research Facilities

The College's laboratories are housed in modern facilities with equipment that has been brought up to the state of the art within the last few years. Some of the unique research facilities of the College are for space physics research for survey of the atmosphere with rocket, orbiting satellite, and planetary probes instrumentation; weather forecasting and meteorological laboratories; solid-state electronics laboratories for fabrication of ultrafast and ultrasmall electronic components; ion beam laboratories for surface modification of metals; remote sensing and robotics laboratories for automated manufacturing and research; a ship hydrodynamics laboratory; nonlinear optics research laboratories; cellular biotechnology laboratories; and the Tauber Manufacturing Institute. Over the past nine years, the College of Engineering has created the Computer Aided Engineering Network (CAEN), which is now recognized as a model distributed computing environment for engineering and computer science instruction and research. CAEN consists of more than 5,000 computer workstations (Apollo, Sun, Apple, IBM, HP, DEC) richly connected to a variety of local and remote supercomputers (CRAY Y-MP, NCUBE, IBM ES 9000/720), file and print servers, implementation servers, and national networks. Students have full access to this entire environment, including a huge array of software for analysis, design (CAD), communication, and personal productivity.

Financial Aid

The College offers an array of fellowships, teaching assistantships, and research assistantships associated with its large sponsored research program. The College tries to offer support for all well-qualified students with the exception that teaching assistantships are not normally available for international students during their first year.

Cost of Study

For 1999–2000, tuition is approximately $6042 per term for state residents and approximately $11,173 per term for nonresidents. There is also a modest fee that covers the registration fee, health service, and student assembly and government. (Costs are subject to change each year.)

Living and Housing Costs

The University of Michigan's Housing Information Office serves as an information source and provides assistance to graduate students seeking accommodations to suit their individual needs. In addition to on-campus single-student residence halls, adequate married student housing is available. The Housing Information Office administers University-operated facilities, including residence halls, co-ops, apartments, and suites, and coordinates information concerning privately owned housing. Food and living costs are about equivalent with the national level.

Student Group

Of the total enrollment of 37,197 at the University of Michigan, there are 6,684 undergraduate and graduate students in the College of Engineering. Of that number, there are 1,002 master's students and 1,043 Ph.D. students.

Location

The University of Michigan is located in the heart of Ann Arbor, just 40 miles west of Detroit. There are exceptional cultural and educational advantages in an environment rich with private art galleries, theatrical productions, concerts, ballet, opera, lectures, recitals, discussions, and colloquia and seminars. Every season features numerous cultural and sports events, both collegiate and professional, in Ann Arbor and the Detroit area.

The University

The University of Michigan was founded in 1817 and was the nation's first public university.

Applying

For applications forms and information on financial aid, students should write to the graduate program chair of the department in which they are interested.

Correspondence and Information

Graduate Program Chair
Department of (specify)
College of Engineering
University of Michigan
Ann Arbor, Michigan 48109

University of Michigan

FACULTY HEADS AND RESEARCH

Aerospace Engineering: David C. Hyland, Chair. Turbulent flows, computational fluid dynamics, space propulsion, turbulent combustion, laser diagnostics, dust and spray detonations, optimal structural design, composite analysis and measurements, fracture mechanics, microstructure measurements and analysis, trajectory optimization, spacecraft dynamics and control, control of flexible structures, control theory and applications, including optimal, digital, and robust control.

Atmospheric, Oceanic and Space Sciences: Lennard A. Fisk, Chair. Global change, air pollution meteorology, atmospheric evolution, climatology, comet dynamics, computational fluid dynamics, interstellar chemical physics, physics and chemistry of planetary atmospheres, radiative transfer, remote sensing, synoptic meteorology, upper-atmospheric physics and chemistry, wave-turbulence interaction, weather forecasting, atmospheric chemistry, biogeochemical cycles, mesoscale meteorology, space plasmas, spacecraft, ground-based instrumentation, and applications of computer technology to K–12 earth science curriculum.

Biomedical Engineering: Charles Cain, Chair. Signal processing, biomechanics, image processing, biological chemistry, biological sensors, bioacoustics, biotechnology, tissue engineering, prosthetics, rehabilitation engineering, physiology, medical imaging, biomaterials and bioelectrical sciences.

Chemical Engineering: Ralph T. Yang, Chair. Catalysis, cellular bioengineering, bioseparations, reaction engineering, simulation and mathematical modeling, environmental engineering, colloids and interfaces, flow in porous media, fluid mechanics, materials processing, process control, sensing, microelectronic materials, electrochemical engineering.

Civil and Environmental Engineering: Richard D. Woods, Chair. Construction engineering and management: cost engineering, human resources in construction, construction methods and equipment, construction decision support systems, knowledge-based systems, scheduling and layout; environmental and water resources engineering: environmental chemistry and microbiology, hazardous waste management, water resource policy and risk-benefit analysis, groundwater contaminant hydrology, water quality modeling and simulation, water supply and waste treatment; geotechnical engineering: foundation design, physicochemical properties, soil dynamics, soil properties, soil and site improvement, soil stabilization and slope stability; hydraulics and hydrologic engineering: coastal engineering, steady and unsteady open-channel flow, surface and subsurface hydrology, environmental fluid mechanics; materials and highway engineering: advanced cement-based materials, fibrous and particulate composites, materials/structure interactions, material durability, micromechanics and fracture mechanics of materials, microstructural analysis, pavement materials and geotextiles, materials for infrastructure rehabilitation; structural engineering: earthquake engineering, elastic and inelastic analysis/design, material and member behavior, repair and strengthening of structures, reliability and risk analysis, structural dynamics.

Electrical Engineering and Computer Science: Pramod Khargonekar, Chair. Computer Science and Engineering (CSE): artificial intelligence, hardware systems, software design, database design, machine learning, machine vision, computer languages and semantics, distributed systems, VLSI systems, robotics. Electrical Science and Engineering (ESE): solid-state electronics, electromagnetics, optics, sensors and integrated circuits, optoelectronics, high frequency microelectronics, remote sensing, terahertz devices and systems, antennas, VLSI design, vehicular engineering. System Science and Engineering (SSE): communications and signal processing, control systems, biosystems, manufacturing systems, integrated vehicle highway systems, signal processing and detection, medical imaging.

Industrial and Operations Engineering: John R. Birge, Chair. Operations research, linear and nonlinear optimization, stochastic processes, control of stochastic systems, production and manufacturing systems analysis, facility design, materials handling, sequencing and scheduling, equipment replacement, ergonomics, human factors, human-computer interaction, occupational safety, information systems development, process optimization, computational methods, quality control, engineering management, financial engineering.

Materials Science and Engineering: Albert F. Yee, Chair. Composite materials, polymer alloys, structural ceramics, electronic materials, magnetic materials, optical materials, materials synthesis, processing and manufacturing, mechanical behavior of materials, surface modification, theoretical modeling and computer simulation, materials characterization, materials chemistry and physics.

Mechanical Engineering and Applied Mechanics: A. Galip Ulsoy, Chair. Crystal growth, frost formation, combustion processes, convection, characteristics of porous media, heat recovery, thin-film measurements, structural and design optimization, AI in design, formal design methodologies, computational geometry and geometric modeling, kinematics, CAD/CAM, process control and sensing, quality control, machining control, mechanics of materials analysis, composite materials, tribology, robotics and mobile robots, biomechanics, automotive engineering, vibrations and acoustics, contact phenomena, laser materials processing, laser measurements, CMM technology, precision machining, machine monitoring and diagnostics, welding, engineering ethics, technology and society, microelectromechanical systems, computational mechanics, dynamics, turbulent flow, multiphase flows, cavitation, high- and low-gravity heat transfer, fracture mechanics and fatigue, solids and fluids of polymeric materials, electrorheological materials, metallurgical failure analysis, dynamic systems modeling.

Naval Architecture and Marine Engineering: Michael M. Bernitsas, Chair. System and structural reliability, nonlinear seakeeping analysis, analysis of advanced propulsors, remote sensing of ship wakes and ocean surface processes, nearshore coastal hydrodynamics, design and analysis of offshore structures, offshore mooring system dynamics, dynamic positioning, computer-aided marine design, free-form surface design and scientific visualization, virtual reality, wave mechanics, turbulent flow, ice mechanics, advanced ship production planning, small-craft resistance and dynamics, marine transportation systems optimization, probabilistic modeling and management, port and inland waterway planning.

Nuclear Engineering and Radiological Sciences: Gary S. Was, Chair. Nuclear reactor analysis and design, inherently safe nuclear reactors, plasma physics, plasma processes, measurement of neutron cross sections, intelligent reactor controls, computational physics, transport theory, neutron activation analysis, radiation measurements and imaging, internal and external dosimetry, medical radiation applications, plasma kinetic theory, intense particle beams, neutron spectroscopy, reactor kinetics, computational transport on advanced computer architecture, Monte Carlo methods, advanced reactor design, photon transport, electron transport, fusion reactor design, space nuclear power, material modification, neutron scattering, radiation effects on materials, ion implantation, metastable materials, environmental effects on materials, corrosion, stress corrosion cracking, hydrogen embrittlement.

Programs

Applied Physics, Program in: Roy Clarke, Director. Interdisciplinary studies in materials physics, optical sciences, plasma science and engineering, solid-state electronics, synchrotron radiation, environmental physics, quantum structures and devices

Macromolecular Science and Engineering, Program in: Richard E. Robertson, Director. Science and technology of synthetic and natural macromolecules, modeling, characterization, retardation, mechanical behavior, molecular physics, morphology, melts, solutions, spectroscopic analysis, deformation and failure, structure and properties of polymers, conjugated polymers, organometallic polymers, glassy polymers, plasma protein-polymer interactions, polymerization kinetics, hybrid inorganic/organic polymers, dental composite processing, liquid crystalline polymers, block copolymers, biomaterials and tissue engineering, polymer glasses, polermic alloys, liquid crystal and electrooptic biopolymers.

Manufacturing, Program in: Debasish Dutta, Director. Manufacturing design/process integration, manufacturing process/production control, quality control, organizational behavior, accounting and finance, marketing and strategy, optimal structural design, process control in the chemical industries, microelectronics process technology, optimal control, robotics, simulation, plant flow systems, theory of scheduling, safety management, human factors in engineering systems, computer control of manufacturing systems, ship production.

Technical Communication, Program in: Leslie Olsen, Director. This program is structured to produce graduates with advanced technical skills and strong communication expertise. Graduates are in a position to design, write, produce, and manage today's sophisticated documentation and other forms of technical communication required by modern high-technology companies.

UNIVERSITY OF MICHIGAN–DEARBORN

College of Engineering and Computer Science

Programs of Study

The postbaccalaureate programs in engineering at the University of Michigan–Dearborn (U of M–Dearborn) are geared to the demands of the student and the needs of industry and are designed to further the theoretical and technical background of the engineer. The College offers programs leading to the Master of Science in Automotive Engineering, Master of Science in Computer Engineering, Master of Science in Computer and Information Science, Master of Science in Electrical Engineering, Master of Science in Engineering Management, Master of Science in Industrial and Systems Engineering, Master of Science in Manufacturing Systems Engineering, Master of Science in Mechanical Engineering, and joint M.S./M.B.A. degrees through the School of Management. The College also provides education and research opportunities for students in the Doctor of Engineering Program offered through the Ann Arbor campus. Working students are accommodated by course offerings late in the afternoon and evening in automotive engineering, computer engineering, computer and information science, engineering management, electrical and computer engineering, industrial and systems engineering, manufacturing systems engineering, and mechanical engineering.

All programs in graduate studies in engineering at the University of Michigan–Dearborn, except the M.B.A. part of the I.S.E./M.B.A. degrees, are offered through the Horace H. Rackham School of Graduate Studies in Ann Arbor.

A master's degree is awarded after completion of a minimum of 30 credit hours, although each program has individual requirements that must be met prior to graduation.

Research Facilities

The College of Engineering and Computer Science built the Engineering Complex in 1997, adding 53,000 square feet of laboratory, classroom, office, and study space. The complex houses a rapid prototype laboratory, a human factors laboratory, a design studio, a hypermedia laboratory, and CAD, PC, Macintosh, networking, and Sun computer laboratories. The Manufacturing Systems Engineering Laboratory building is equipped with laboratories that include metrology, machine dynamics and diagnostics, precision machining, and computer-integrated manufacturing. This component of manufacturing research is supplemented by an extensive array of computers dedicated to the engineering disciplines. In addition, the College has several other experimental laboratories available for research: machine vision and intelligence, design and fatigue, acoustics and vibrations, combustion engines and fuels, vehicle electronics, plastics and composites, circuits, electronic control systems, energy conversion, manufacturing simulation, 3-D imaging, applied thermodynamics, fluid mechanics, heat transfer, computer automation, robotics, data communications, and digital systems. Combined, the College's facilities provide effective and comprehensive areas for teaching, student projects, research, and faculty projects that impact curriculum and build strong partnerships with industry, government, and the community.

Financial Aid

Scholarships, fellowships, and other grants-in-aid, as well as financial assistance through departmental employment, are often available to qualified students in engineering. In keeping with University practice and policy, such assistance is available without regard to race, color, creed, sex, or national origin.

The number of awards available each year is variable, as is the amount of the stipend. Recipients are appointed by, or upon the nomination of, the departments in which the applicants are enrolled. Application forms for students who will be registered at the University of Michigan–Dearborn can be obtained from the College of Engineering. When submitted to the College of Engineering, this application form, entitled "Application for Graduate School Fellowship, Teaching, or Research Assistantship," serves as the vehicle for obtaining consideration for all awards administered by the College.

Cost of Study

The tuition in 1998–99 for full-time graduate students (8 credit hours) working toward a master's degree was $2092 for Michigan residents and $5898 for out-of-state residents. In addition, students were assessed a $30 fee per credit hour, not to exceed $100 per term, and a per semester computer-user fee of $80.

Living and Housing Costs

The local living costs are somewhat dependent upon the availability of housing. An estimation of living costs beyond the cost of study is $1200 per month.

Student Group

Enrollment at the University of Michigan–Dearborn in fall 1998 was 8,219. Of this figure, 2,195 students were enrolled in the College of Engineering and Computer Science: 1,487 undergraduate and 708 graduate students.

Location

The University of Michigan–Dearborn is located in the heart of Michigan's largest urban area, just 10 miles from downtown Detroit and a wide variety of cultural, athletic, and recreational opportunities. Many outdoor recreation facilities, including rivers, lakes, beaches, and ski areas, are within a short driving distance.

The University

The University of Michigan–Dearborn is one of three campuses governed by the University of Michigan Board of Regents. As a regional campus of the University of Michigan system, it shares in the tradition of excellence in teaching, research, and service. The campus, which is located on 202 acres of the former estate of the late Henry Ford, is primarily a commuter campus. It was founded in 1959 as a senior-level institution offering only junior, senior, and graduate courses. Since 1971, the Dearborn campus has offered full four-year degree programs and expanded its graduate offerings. As part of the University of Michigan system, U of M–Dearborn enjoys the resources of a large multiuniversity and the advantages of moderate size.

Applying

Applications for graduate admission, accompanied by a $55 nonrefundable fee, transcripts, and letters of recommendation, should reach the department by August 1 for the fall term, December 1 for the winter term, or April 1 for the spring term.

Correspondence and Information

For information about the various engineering programs at the University, students should contact the following faculty members: Dr. P. K. Mallick, Automotive Engineering (313-593-5119); Dr. K. Modesitt, Computer and Information Science (313-593-5680); Dr. M. Shridhar, Electrical and Computer Engineering (313-593-5420); Dr. S. Kachhal, Industrial and Manufacturing Systems Engineering (313-593-5361); Dr. P. K. Mallick, Manufacturing Systems Engineering (313-593-5119); and Professor C. Chow, Mechanical Engineering (313-593-5241).

College of Engineering and Computer Science
University of Michigan–Dearborn
4901 Evergreen Road
Dearborn, Michigan 48128-1491
Telephone: 313-593-0897
Fax: 313-593-9967

University of Michigan–Dearborn

THE FACULTY AND THEIR RESEARCH

Alan Argento, Associate Professor of Mechanical Engineering; Ph.D., Michigan, 1989. Structural dynamics, elastic wave propagation.

A. Adnan Aswad, Professor of Industrial and Manufacturing Systems Engineering; Ph.D., Michigan, 1972. Decision theory, production processes.

Selim Awad, Professor of Electrical and Computer Engineering; Ph.D., Polytechnic Institute of Grenoble (France), 1983. Communications, speech analysis, microprocessor applications.

G. Fredric Bolling, Professor of Mechanical Engineering; Ph.D., Toronto, 1957. Materials; manufacturing, quality; management decision making and strategies.

Charu Chandra, Assistant Professor of Industrial and Manufacturing Systems Engineering; Ph.D., Arizona State, 1994. Information design and modeling for enterprise/supply chain management and integration.

Chia-hao Chang, Professor of Industrial and Manufacturing Systems Engineering; Ph.D., Oregon State, 1978. Application of artificial intelligence to knowledge engineering and expert systems, database systems.

Yubao Chen, Associate Professor of Industrial and Manufacturing Systems Engineering; Ph.D., Wisconsin–Madison, 1986. Time-series analysis, CAD/CAM, CIM, process monitoring, diagnosis and control.

John Cherng, Associate Professor of Mechanical Engineering; Ph.D., Tennessee, 1978. Computer-aided design, computer graphics, thermal/fluid sciences, vibrations and acoustics.

Chi L. Chow, Professor and Chairperson, Mechanical Engineering; Ph.D., London, 1965. Fatigue, fracture and damage mechanics.

Bruce Elenbogen, Associate Professor of Computer and Information Science; Ph.D., Northwestern, 1981. Theory of computation, parallel algorithms, CAD, fractals.

Hugh E. Huntley, Assistant Professor of Mechanical Engineering; Ph.D., Michigan, 1992. Solid mechanics, computational methods, design theory and finite element.

Swatantra K. Kachhal, Professor and Chairperson, Industrial and Manufacturing Systems Engineering; Ph.D., Minnesota, 1974. Automated material handling and warehousing, statistical process control, statistical design and analysis of experiments, health-care systems.

Roberto R. Kampfner, Associate Professor of Computer and Information Science; Ph.D., Michigan, 1981. Organizational information systems, decision support systems, artificial intelligence and expert systems, adaptive systems and biological information processing.

Ali Kamrani, Associate Professor of Industrial and Manufacturing Systems Engineering; Ph.D., Louisville, 1991. Robotics and automation, computer integrated manufacturing, mathematical modeling and computer simulation.

Rajgopal Kannan, Assistant Professor of Computer and Information Science; Ph.D., Denver, 1996. High-speed networks, optical computing and communications, ATM switching networks.

James W. Knight, Associate Professor of Industrial and Manufacturing Systems Engineering; Ph.D., Ohio State, 1977. Ergonomics, statistical design of experiments.

Ghassan Kridli, Assistant Professor of Industrial and Manufacturing Systems Engineering; Ph.D., Missouri–Columbia, 1997. Manufacturing processes, computer-aided design and manufacturing (CAD/CAM), CNV machining.

Shridhar Lakshmanan, Associate Professor of Electrical and Computer Engineering; Ph.D., Massachusetts at Amherst, 1991. Signal and image processing.

Robert E. Little, Professor of Mechanical Engineering; Ph.D., Michigan, 1963. Reliability, modes of failure, fatigue, mechanical design.

P. K. Mallick, Professor of Mechanical Engineering and Director of Interdisciplinary Programs; Ph.D., IIT, 1973. Materials and manufacturing processes, solid mechanics, failure analysis/design, composites.

Bruce Maxim, Associate Professor of Computer and Information Science; Ph.D., Michigan, 1982. Artificial intelligence, comparative programming languages, data structures, file processing.

John Miller, Associate Professor of Electrical and Computer Engineering; Ph.D., Toledo, 1983. Machine vision systems, digital signal processing, logic and computer design.

William J. Mitchell, Assistant Professor of Mechanical Engineering; M.S., Michigan, 1960. Materials and manufacturing processes.

Kenneth L. Modesitt, Professor and Chairperson, Computer and Information Science; Ph.D., Washington State, 1972. Expert knowledge-based systems, software engineering, computer-based learning.

Yi Lu Murphey, Associate Professor of Electrical and Computer Engineering; Ph.D., Michigan, 1989. Character segmentation and recognition.

Syed Murtuza, Professor of Electrical and Computer Engineering; Ph.D., Purdue, 1967. Digital control, optimal control systems.

Tsung Y. Na, Professor of Mechanical Engineering; Ph.D., Michigan, 1964. Thermal/fluid sciences, fluid mechanics, numerical/computer methods.

Natarajian Narasimhamurthi, Associate Professor of Electrical and Computer Engineering; Ph.D., Berkeley, 1979. Systems theory, power systems, computer science and mathematics.

Elsayed A. Orady, Associate Professor of Industrial and Manufacturing Systems Engineering; Ph.D., McMaster, 1982. Manufacturing processes, computer-aided manufacturing, robotics.

Eric Ratts, Assistant Professor of Mechanical Engineering; Ph.D., MIT, 1993. Thermodynamics, heat transfer and cryogenics.

Indrajit Ray, Assistant Professor of Computer and Information Science; Ph.D., George Mason, 1997. Advanced database technology and formal methods.

Indrakshi Ray, Assistant Professor of Computer and Information Science; Ph.D., George Mason, 1997. Distributed database systems, transaction processing, workflows, information security.

Sibabrata Ray, Assistant Professor of Computer and Information Science; Ph.D., Nebraska–Lincoln, 1995. Networking, parallel and distributed computing, computer architecture.

Subrata Sengupta, Professor of Mechanical Engineering and Dean of the College of Engineering and Computer Science; Ph.D., Case Western Reserve, 1974. Waste heat management and utilization.

Tariq Shamim, Assistant Professor of Mechanical Engineering; Ph.D., Michigan, 1997. Combustion, flame modeling, emission control, computational fluid dynamics.

Adnan K. Shaout, Associate Professor of Electrical and Computer Engineering; Ph.D., Syracuse, 1987. Computer design, image processing and pattern recognition.

Malayappan Shridhar, Professor and Chairperson, Electrical and Computer Engineering; Ph.D., Aston (England), 1969. Speech and image processing, pattern recognition.

Louis Y. Tsui, Associate Professor of Computer and Information Science; Ph.D., Michigan, 1984. Production scheduling, simulation, software engineering.

Onur Ulgen, Professor of Industrial and Manufacturing Systems Engineering; Ph.D., Texas Tech, 1979. Discrete and continuous simulation, modeling of production systems, material handling systems, ARIMA and other time-series techniques.

Keshav S. Varde, Professor of Mechanical Engineering and Associate Dean of the College of Engineering and Computer Science; Ph.D., Rochester, 1971. Thermal/fluid sciences, thermodynamics, combustion, alternative energy sources/fuels.

Paul Watta, Assistant Professor of Electrical and Computer Engineering, Ph.D., Wayne State, 1994. Artificial neural networks, dynamical systems, automata networks, optimization, control systems design, pattern recognition.

David Yoon, Associate Professor of Computer and Information Science; Ph.D., Wayne State, 1989. Computational geometry, computer-aided geometric modeling, theoretical computer science.

Armen Zakarian, Assistant Professor of Industrial and Manufacturing Systems Engineering; Ph.D., Iowa, 1997. Modeling and analysis of manufacturing systems, intelligent simulation environments.

Yi Zhang, Assistant Professor of Mechanical Engineering; Ph.D., Illinois, 1989. Design and manufacturing of gearing systems.

Dongming Zhao, Associate Professor of Electrical and Computer Engineering; Ph.D., Rutgers, 1990. Image processing and analysis, machine vision, parallel real-time image processing.

Qiang Zhu, Assistant Professor of Computer and Information Science; Ph.D., Waterloo, 1995. Databases, information systems, software engineering, scientific computation.

UNIVERSITY OF MISSOURI–COLUMBIA

College of Engineering

Programs of Study

The College of Engineering at the University of Missouri–Columbia offers Master of Science (M.S.), and Master of Engineering (M.Eng.), and Doctor of Philosophy (Ph.D.) programs in the fields of biological and agricultural engineering, chemical engineering, civil engineering, computer engineering and computer science, electrical engineering, industrial engineering, mechanical engineering, and nuclear engineering.

The Master of Science program is designed to prepare students for research and advanced engineering careers. Core courses may be taken to supplement graduate work. The program requires completion of a minimum of 30 hours beyond the bachelor's degree, including a thesis in some departments.

The Master of Engineering program is designed for professional working students. It is a nonthesis option that requires completion of a minimum of 36 hours of course work beyond the bachelor's degree and is available from all departments in the College.

The Doctor of Philosophy program qualifies the graduate to accept professional responsibility of the engineering practice and is intended primarily for those planning research-oriented careers. The program requires a qualifying examination, a minimum of 72 hours past a bachelor's degree, individual research through a dissertation, and a comprehensive exam. All departments have different requirements for admission and attainment of degrees.

Research Facilities

Research is supported by state-of-the-art facilities. Ellis Library, the main library on campus, occupies an entire city block, houses approximately 1.5 million volumes, and allows access to collections, databases, an online catalog, LUMIN, and ERIC. The Engineering Library, which encompasses 10,000 square feet holds almost 80,000 volumes, and is equipped with an online database and catalog and a seminar room. Campus Computing provides access to the computing resources of the University and offers free training classes; MU's mainframe computing environments are provided through machines that offer UNIX, VM/CMS, and MVS operating systems. The Engineering Computer Network provides advanced engineering computation, CAD/CAM and graphics, artificial intelligence, multiple high-level programming languages, and computational and simulation libraries. The Science Instrument Shop and Glassblowing Service was facilitated to help design and build sophisticated research equipment. The Electronic Instrument Laboratory is a versatile facility with the capability of designing, building, and maintaining complex electronic research systems, instruments, and computers. In addition, the academic departments have extensive research equipment, laboratories, and facilities.

Financial Aid

Students are eligible for a number of financial aid packages, including teaching and research assistantships administered by the College and by individual departments, graduate school fellowship programs, and national and industrial fellowships. Programs include a monthly stipend, and most include a waiver of tuition. Typical financial packages (including stipends and tuition waivers) range from $10,000 to 20,000 per year.

Cost of Study

In-state tuition for the 1999–2000 term is $167.80 per credit hour, plus a $36.80-per-hour supplemental fee for engineering courses. Out-of-state tuition is $504.80 per credit hour. Most students in the program receive tuition waivers as part of their financial aid package. Students also pay an activities fee of $9.53 per credit hour, a computer fee of $8.30 per credit hour, and a health fee of $57.50 per semester.

Living and Housing Costs

On-campus, single-student room and board cost $4655 per year, including twenty-one meals per week. Married student on-campus housing ranges from $286 to $380 per month. Private accommodations are available throughout Columbia at modest rates due to the low cost of living.

Student Group

There are about 25,000 students at the University of Missouri–Columbia. MU's Graduate School enrolls more than 4,600 graduate students in more than ninety graduate degree programs. In its history, the University has granted a total of more than 50,000 master's degrees and approximately 10,000 doctoral degrees. The graduate programs in the College average more than 300 graduate students, representing a diverse and balanced group of ethnic and cultural backgrounds. Approximately 150 are Ph.D. students.

Location

The University of Missouri–Columbia is located in Columbia, Missouri, on Interstate 70, halfway between Kansas City and St. Louis. The campus is near downtown Columbia and offers the advantages of a rural setting in a community of more than 75,000 people. Columbia is an educational and medical center and offers a wide range of cultural activities. It is routinely included among the top twenty best places to live by *Money* magazine.

The University

MU, established in 1839, is the oldest state university west of the Mississippi River. The Columbia campus is the largest of the four campuses of the University of Missouri System. Other campuses are in St. Louis, Kansas City, and Rolla. A member of the American Association of Universities (AAU) and a university classified Research I by the Carnegie Foundation for the Advancement of Teaching, MU is a premier provider of graduate and professional education.

Applying

Admission to a program requires an undergraduate degree, preferably in engineering at an institution accredited by the Accreditation Board for Engineering and Technology, Inc. (ABET), Graduate Record Examinations (GRE) scores, and a 3.0 grade point average in the last 60 hours of undergraduate study. TOEFL scores of 500 or better are required of international students whose native language is not English. Applicants should contact the Director of Graduate Studies in the particular department of interest for more specific information.

Correspondence and Information

Director of Graduate Studies
(Program of Interest)
College of Engineering
University of Missouri–Columbia
Columbia, Missouri 65211

E-mail: whmiller@risc1.ecn.missouri.edu
World Wide Web: http://www.ecn.missouri.edu/index.html

University of Missouri–Columbia

DEPARTMENTS, CENTERS, FACULTY HEADS, AND RESEARCH AREAS

DEPARTMENTS AND PROGRAMS

Biological and Agricultural Engineering: Dr. Neil Meador, Chair. The department offers graduate programs leading to the degrees of Master of Engineering, Master of Science, and Doctor of Philosophy in biological engineering. Thesis research may emphasize bioprocessing, food engineering, environmental engineering, biochemical engineering, biomedical engineering, or precision agriculture. Laboratories are well equipped for research in bioprocessing, bioreactor design, value-added processes and products, properties of biological and food materials, food extrusion, water quality, wetlands, process control, computer vision, GIS, precision agriculture, chemical application technology, soil physics, hydrology, and renewable energy.

Chemical Engineering: Dr. Sunggyu Lee, Chair. The department offers graduate work leading to the degrees of Master of Science and Doctor of Philosophy. Areas of study are non-ideal fluid mechanics, rheology, process control and optimization, reaction kinetics, catalysis (vibrating and nonvibrating systems), thermodynamics, transport, properties of gases, heat- and mass-transfer, supercritical fluid technology, air pollution monitoring and control, energy resource and reproduction, and biochemical engineering research. There are excellent facilities for student research, including an equation of state and transport properties lab, a heterogeneous catalysis and reaction lab, a heat and mass transport lab, an air pollution monitoring and control lab, a biochemical engineering lab, a non-Newtonian fluid mechanics lab, a computation lab, a process engineering lab, and a transport properties phenomena lab.

Civil and Environmental Engineering: Dr. Sam Kiger, Chair. Graduate programs offered by the department prepare students for research and advanced engineering careers. Major program areas include structural mechanics, structural engineering, explosion resistant design, and material and geotechnical engineering. The department has well-equipped laboratories for extensive research and several computer-controlled electrohydraulic testing machines. The laboratories are served by a 5-ton overhead crane. The geotechnical lab houses porosimetry, permeable, consolidation, and trivial testing. The environment labs are supplied with analytical equipment for the complete physical, chemical, and microbiological analysis of water and wastewater. The hydraulics lab has two medium-sized flumes and a 3-foot by 20-foot wind tunnel.

[L+2,1]**Computer Engineering and Computer Science:** Dr. Su-Shing Chen, Chair. The department prepares students for careers as computer professionals in industry and academia. Faculty research interests include distributed systems, networked information systems, databases, digital libraries, parallel algorithms, computational intelligence, neural networks, fuzzy logics, computer vision, image processing, speech understanding, sensor-based robots, computer graphics, visualization, networking, and performance analysis. The department has excellent computing facilities including SGI, Sun, DEC, and Intel workstations and servers.

Electrical Engineering: Dr. Kai-Fong Lee, Chair. The department prepares students to apply the knowledge of electrical phenomena to real-world problems. Research areas include semiconductor devices, digital systems, laser applications, robotics, environmental and biomedical research, signal and image processing, microwave and antenna systems, telecommunications, and power electronics. The department has excellent computer equipment and other laboratory facilities, which are used for applied research sponsored by various sources of research funding.

Industrial and Manufacturing Systems Engineering: Dr. Larry G. David, Chair. The graduate program provides a scholarly environment in which highly qualified, creative students develop the necessary skills to solve complex industrial, governmental, and societal system design problems. These systems are required to operate within increasingly complex constraints, thus requiring the use of sophisticated and creative designs. The industrial engineer responsible for such designs must be capable of applying a broad spectrum of scientific tools if the most effective systems are to be obtained. Areas of research include manufacturing systems, production planning and control, mathematical programming, discrete optimization, fuzzy set theory, statistical data analysis and response surface technologies, integrated production systems, material flow systems, stochastic processes, scheduling, quality assurance techniques, facilities design, and health-care delivery services.

Mechanical and Aerospace Engineering: Dr. Robert D. Y. Tzou, Chair. The department prepares students for advanced professional engineering careers. Areas of concentration include AI/expert systems, automation, bioengineering, combustion, computational fluid dynamics, control, creep and plasticity, design optimization, finite and boundary element methods, fluid and aerosol mechanics, fracture mechanics, heat transfer, intelligent systems, interactive computer graphics, laser diagnostics, manufacturing processes, materials science, mechanical synthesis, mechatronics, microprocessor applications, nonlinear structural mechanics, orbital mechanics, parallel computation, residual stress, robotics, thermal systems design, ultrasonic nondestructive evaluation, and vehicle dynamics. The department has a number of specialized laboratories in the areas of aerosol mechanics, combustion, computer control, creep, fluid mechanics, fracture mechanics, heat transfer, manufacturing, materials, and structural dynamics.

Nuclear Engineering: Dr. William Miller, Chair. The graduate-level nuclear engineering program is one of only thirty-seven such university programs nationwide. The program works closely with the MU Research Reactor, which is the most powerful university research reactor in the country. Nuclear engineering offers several options, including basic nuclear engineering, nuclear power engineering, health physics, and medical physics. Area research topics include nuclear material management, aerosol mechanics, reactor safety analysis, nuclear energy conversion, reactor physics, reactor design, nondestructive testing and measurement, radiative heat transfer, neutron spectrometry, neutron and gamma ray transport, neutron activation analysis, nuclear waste management, nuclear plasma research, health physics, magnetic resonance imaging, radiation therapy, and alternative and renewable energy concepts.

RESEARCH CENTERS AND OTHER RESEARCH OPPORTUNITIES

Capsule Pipeline Research Center. The center began operating in fall 1991 and is the only National Science Foundation–sponsored center in Missouri. The center's mission during its first four years focused on the development of coal log pipeline technology for transporting coal. The center has successfully won NSF and state support for a second four-year term and plans to broaden its mission to cover other types of capsule pipelines for freight transport, including solid waste and hazardous waste and grain.

The Center for Surface Science and Plasma Technology. The center deals with interfaces, or boundaries, between plasma, the fourth state of matter, and conventional matter such as gases, liquids, or solids. Research projects include applying the technology to make a car frame more corrosion resistant, improving composite materials for orthopedic applications, making membranes for biochemical reactors, and modifying the surface of powders.

Industrial and Technological Development Center. The research facility is designed to enhance industrial productivity and America's competitive position in the world market through fundamental engineering research. The ITDC has recently succeeded in stimulation research relating to the product delivery process for increased productivity, quality of design and manufacturing, and environment protection.

The John M. Dalton Cardiovascular Research Center. The center has excellent facilities for research in all biomedical disciplines, including bioengineering. Four research areas are emphasized: exercise and cardiovascular function, cardiovascular consequences of diabetes, neurohumoral control of the circulation, and membrane biology. Research spans molecular biology to integrated systems function and basic science to clinical application.

The Particulate Systems Research Center. The center provides a focus for researchers from chemical, civil, mechanical, and nuclear engineering. Ongoing funded research includes single-particle and integral experiments, indoor air pollution, clean-room technology, radon measurements and mitigation, nuclearation and condensation, particle location in viscous fluids, materials manufacturing using aerosol reactors, and large-scale computer code development.

The University of Missouri Research Reactor Center. The center provides MU with opportunities for research and graduate education in the neutron-related sciences and engineering that are unmatched at any other U.S. university. The scientific research at the reactor spans a broad spectrum of disciplines and techniques and includes archaeometry, elastic and inelastic neutron scattering, gamma-ray scattering, neutron interferometry, neutron activation analysis, human and animal nutrition, epidemiology and immunology, radiation effects, radioisotope studies, health physics, and nuclear engineering.

Coordinated Engineering Program (CEP). Administered in conjunction with the University of Missouri–Kansas City, the CEP offers Master of Science degrees in mechanical, civil, and electrical engineering. An interdisciplinary Ph.D. in engineering is also available.

Oak Ridge Associated Universities. ORAU is a consortium of colleges and universities and an operating contractor for the U.S. Department of Energy in Oak Ridge, Tennessee. ORAU works with its member institutions to help students and faculty members gain access to federal research facilities throughout the country; to keep its members informed about opportunities for fellowship, scholarship, and research appointments; and to organize research alliances among members.

UNIVERSITY OF MISSOURI–ROLLA

School of Engineering and School of Mines and Metallurgy

Programs of Study

The University of Missouri–Rolla (UMR) offers Master of Science degrees in aerospace engineering, ceramic engineering, chemical engineering, civil engineering, computer engineering, electrical engineering, engineering management, engineering mechanics, environmental engineering, geological engineering, geology and geophysics, mechanical engineering, metallurgical engineering, mining engineering, nuclear engineering, and petroleum engineering. The M.S. with a thesis requires 30 credits; without a thesis, the M.S. requires 33 credits.

The Ph.D. is offered in aerospace engineering, ceramic engineering, chemical engineering, civil engineering, computer engineering, electrical engineering, engineering management, engineering mechanics, geological engineering, geology and geophysics, mechanical engineering, metallurgical engineering, mining engineering, nuclear engineering, and petroleum engineering. The Ph.D. requires a minimum of 72 credits, and a one-year residency is required. Students must pass a qualifying exam to become Ph.D. candidates, a comprehensive exam over course work, and a final oral exam after the research dissertation is completed.

The Doctor of Engineering degree is available in ceramic engineering, chemical engineering, civil engineering, computer engineering, electrical engineering, engineering management, geological engineering, mechanical engineering, mining engineering, nuclear engineering, and petroleum engineering. The D.E. degree has requirements similar to the Ph.D., except that 65 total credits of course work are required, with three areas of emphasis and a one-year industrial internship that forms the basis for the dissertation, which may be in a design-related area.

Research Facilities

Each department has an extensive complement of modern research facilities. Within the academic departments the following laboratory facilities exist: biochemical processing, thermodynamic properties, computer vision, electronic materials, environmental research, cold-formed steel testing, computer-integrated, packaging, robotics, structural vibration, engine testing, composite materials, manufacturing, large wind tunnel, smart structures, and flexible manufacturing. Specialized research and service facilities include the Biochemical Processing Institute, the Center for Cold-Formed Steel Structures, the Center for Infrastructural Engineering Studies, the Design Engineering Center, the Electronic Materials Processing and Characterization Institute, the Energy Analysis and Diagnostic Center, the Center for Environmental Science and Technology, the Environmental Research Center, the Experimental Mine, the Generic Mineral Technology Center for Pyrometallurgy, the Aerosol and Cloud Physics Laboratory, the Electronic Materials Processing and Characterization Institute, the Graduate Center for Materials Research, the High Pressure Waterjet Laboratory, the Institute for Chemical and Extractive Metallurgy, the Institute of River Studies, the Institute of Thin Film Processing Science, the Intelligent Systems Center, the Rock Mechanics and Explosive Research Center, the Transportation Institute, and the Electronic Materials Applied Research Center.

Financial Aid

Financial assistance is available for graduate students at UMR in the forms of graduate teaching and research assistantships, instructorships, and fellowships.

Half-time graduate assistants devote approximately 20 hours per week to laboratory supervision or other assigned duties, including research, and receive a stipend of $12,985 on a nine-month basis in 1999–2000. Chancellor's and Dean's Fellowships, which cover fees, are available for outstanding students.

Also available are federal and industrial fellowships and traineeships, which cover tuition and fees. In all cases, the department where the student would study should be contacted for information regarding financial aid.

Cost of Study

In 1999–2000, fees total $2097 for 9 credits per semester for residents and $5130 for 9 credits per semester for nonresidents. Holders of graduate assistantships and fellowships are considered residents for the purpose of fee payment. Costs are subject to change.

Living and Housing Costs

University residence hall rates, which include room and board, range from $3590 to $4935 per academic year in 1999–2000. Students can obtain off-campus apartments for approximately $300 to $450 per month. Information is available from the Residential Life Office, 104 Norwood Building (telephone: 573-341-4218).

Student Group

The enrollment at the University of Missouri–Rolla for fall 1998 was 4,918, including 640 graduate students. Of this number, 356 were enrolled in graduate studies in the School of Engineering; 115 were enrolled in graduate studies in the School of Mines and Metallurgy.

Student Outcomes

Graduates of UMR graduate programs in engineering have found jobs in all areas of technology and business. M.S. graduates typically take jobs at the same companies but at higher entry levels than the B.S. graduates. These are typically in management of manufacturing facilities, technical support of manufacturing, applied research, and technical marketing. At the Ph.D. level, graduates have entered academic teaching and research, government laboratory research, and corporate research.

Location

Rolla, a community of about 15,000 people, is located in south-central Missouri, 100 miles southwest of St. Louis and 130 miles northeast of the Springfield-Branson area. Entertainment is available through local musical and theatrical groups, the University's cultural programs, movie theaters, athletic events, and recreational facilities. Nearby state parks, rivers, and lakes offer excellent fishing, hunting, and canoeing.

The University

The University of Missouri–Rolla, founded as the University of Missouri School of Mines and Metallurgy in 1871, is one of the four campuses of the University of Missouri. Today, UMR maintains its specializations in engineering; the physical, earth, and computer sciences; and mathematics as well as offering undergraduate degrees in the humanities and social sciences. UMR is the only university to ever win the Missouri Quality Award, which is patterned after the Malcolm Baldrige National Quality Award.

Applying

The completed application form, official transcripts, and GRE scores should be sent directly to the Office of Admissions, which forwards the application to the department of interest. No application is considered complete until all official documents are received by admissions; however, informal consideration begins immediately. Applications for financial aid should be sent directly to the academic department at least six months, and preferably nine months, in advance. Applications may also be made electronically via the World Wide Web at the address listed below.

Correspondence and Information

Office of Admissions
106 Parker Hall
1870 Miner Circle
University of Missouri–Rolla
Rolla, Missouri 65409-1060
Telephone: 800-522-0936 (toll-free)
Fax: 573-341-4082
E-mail: umrolla@umr.edu
World Wide Web: http://www.umr.edu/enrol

University of Missouri–Rolla

THE FACULTY AND THEIR RESEARCH

CERAMIC ENGINEERING

Professors: Wayne Huebner (Chair), Harlan Anderson, Richard Brow, Delbert Day, Douglas Mattox, Robert Moore, P. Darrell Ownby, Mohamed Rahaman. **Research:** Mass transport, mechanical damping, high-temperature creep, piezoelectrics, dielectrics, mixed conductors, solid electrolytes, glass, ceramic-to-metal seals, refractories, slag resistance, wetting, processing and sintering.

CHEMICAL ENGINEERING

Professors: Douglas Ludlow (Chair), James Johnson, A. I. Liapis, David Manley, Nicholas Morosoff, Gary Patterson, X. B. Reed Jr., Stephen Rosen. **Associate Professors:** Neil Book, Daniel Forciniti, Parthasahka Neogi, Oliver Sitton. **Assistant Professor:** Dennis Sourlas. **Research:** Mixing, fluid mechanics, bioengineering, adsorption, freeze-drying, extraction, distillation, polymers, process control, surface phenomena, energy storage, kinetics, vapor-liquid equilibria, electrochemical kinetics, bioprocesses, membrane technology.

CIVIL AND ENVIRONMENTAL ENGINEERING

Professors: Paul Munger (Interim Chair), Franklin Cheng, Roger LaBoube, Antonio Nanni, Thomas Petry, Shamsher Prakash, Richard Stephenson, Jerome Westphal. **Associate Professors:** Craig Adams, Abdeldjelil Belarbi, Rodney Lentz, Charles Morris, David Richardson, Gary Spring, Purush TerKonda. **Assistant Professors:** Jerry Bayless, Genda Chen, Joel Burken, Mark Fitch. **Research:** Environmental and sanitary engineering, pollution abatement, water resources, potamology, cold-formed steel structures, composite materials, highway drainage, curtain-wall structures, urban watershed modeling, wind-structure interaction phenomena, structural optimization and control systems.

ELECTRICAL ENGINEERING

Professors: E. Keith Stanek (Chair), Max Anderson, Jack Boone, Jack Bourquin, Gordon Carlson, David Cunningham, Darrow Dawson, O. Robert Mitchell, Randy Moss, Vittal Rao, Paul Stigall, John Stuller, Thomas VanDoren, C. H. Wu. **Associate Professors:** Levant Acar, Badrul Chowdhury, Mariesa Crow, James Drewniak, Richard DuBroff, Kelvin Erickson, James Hahn, Thomas Herrick, Nancy Hubing, Todd Hubing, Kurt Kosbar, Hardy Pottinger, Steve Watkins. **Assistant Professors:** Daryl Beetner, Norman Cox, Stephen Pekarek, Ashok Tikku. **Research:** Circuits-electronics; communications and digital signal processing; computer engineering; computer vision; electronic materials, grounding, and shielding; control systems; digital system design; electromagnetics; intelligent industrial systems; physical electronics.

ENGINEERING MANAGEMENT

Professors: Henry Wiebe (Interim Chair), Cihan Dagli, Madison Daily, Donald Myers, Yildirim Omurtag, Kenneth Ragsdell. **Associate Professors:** Bahador Ghahramani, Raymond Kluczny, Henry Metzner, Stephen Raper. **Assistant Professors:** Venkat Allada, Kevin Hubbard, Halvard Nystrom, Susan Murray, Peter Schmidt, David Shaller. **Research:** Manufacturing engineering, packaging engineering, industrial engineering, quality engineering, management of technology, computer-integrated manufacturing, value analysis, total quality management, smart engineering system design, optimization, scheduling, plant layout, solid-waste management, packaging materials and machinery, project management, neural networks, work design, safety, management information systems, databases.

GEOLOGICAL ENGINEERING

Professors: John Rockaway (Chair), David Barr, Dale Elifrits, Allen Hatheway. **Associate Professor:** Jeffrey Cawlfield. **Assistant Professors:** Norbert Maerz, Paul Santi, Merrill Stevens. **Research:** Digital imaging processing and geographic information systems, probabilistic modeling and geostatistics, groundwater hydrology and contaminant transport, environmental geology and hazardous-waste management and cleanup, engineering geology and geotechnics.

GEOLOGY AND GEOPHYSICS

Professors: Richard Hagni (Chair), Neil Anderson, Jay Gregg, Robert Laudon. **Associate Professor:** Francisca Oboh-Ikuenobe. **Assistant Professors:** Steven Cardimona, David Wronkiewicz. **Research:** Mineralogy, igneous and sedimentary petrology-geochemistry, sedimentology, stratigraphy, palynology, biostratigraphy, exploration, engineering and environmental geophysics, exploration and environmental geochemistry, economic geology, petroleum geology, basin studies.

MECHANICAL AND AEROSPACE ENGINEERING AND ENGINEERING MECHANICS

Professors: Ashok Midha (Chair), Darryl Alofs, Bassem Armaly, Xavier Avula, S. N. Balakrishnan, Clark Barker, Victor Birman, K. Chandrashekhara, Ta-Shen Chen, Donald Cronin, Alfred Crosbie, Lokesh Dharani, Walter Eversman, Virgil Flanigan, Leslie Koval, K. Krishnamurthy, Shen Lee, Terry Lehnhoff, D. C. Look Jr., H. Fred Nelson, Harry Sauer Jr., John Sheffield, Hai-Ling Tsai. **Associate Professors:** James Drallmeier, Fathi Finaish, K. M. Issac, Frank Liou, Wen Lu, Gearoid MacSithigh, J. Keith Nisbett, Anthony Okafor, David Riggins. **Assistant Professors:** Nancy Ma, Samit Roy. **Research:** Aerodynamics; turbulent flows; computational fluid dynamics; space mechanics; hypersonics; acoustics; combustion; aerospace propulsion; aerospace systems/structures; flight simulation; stability of aerospace vehicles; modern control theory; aeroelasticity; composites; optimization; fracture mechanics; composite materials; metal forming; continuum mechanics; experimental mechanics; vibrations; sonic fatigue; biomechanics; finite-element methods; modal analysis; shear banding; penetration mechanics; convective, radiative, and aerosol mechanics; two-phase phenomena; acoustics; CAD/CAM/CAE; dynamics and controls; flexible manufacturing and robotics.

METALLURGICAL ENGINEERING

Professors: John L. Watson (Chair), Donald Askeland, Ronald Kohser, Thomas O'Keefe, David Robertson. **Associate Professors:** Joseph Newkirk, Christopher Ramsay, Mark Schlesinger, David Van Aken. **Assistant Professor:** Kent Peaslee. **Teaching Associate:** F. Scott Miller. **Research:** Physical, manufacturing, chemical, and extractive metallurgy; mineral processing.

MINING ENGINEERING

Professors: John Wilson (Chair), Richard Bullock, Tad Golosinski, Marian Mazurkiewicz, Lee Saperstein, David Summers. **Associate Professors:** Jerry Tien, Paul Worsey. **Assistant Professor:** Karl Zipf. **Research:** Mine health and safety, modeling of mining operations, blasting, mechanical rock breaking, strata control, waterjet cutting, mine ventilation, mineral economics, computer-aided mine design, rock mechanics, materials transportation, high-level explosives research.

NUCLEAR ENGINEERING

Professors: Arvind Kumar (Chair), Ray Edwards, Nicholas Tsoulfanidis. **Associate Professors:** Albert Bolon, Shahla Keyvan, Gary Mueller. **Research:** Nuclear radiation transport, power plant diagnostics and control using artificial intelligence, radiation effects, heat transfer, fluid mechanics, radiation protection, fuel cycle, radioactive waste management.

PETROLEUM ENGINEERING

Professors: Daopu Numbere (Program Head), Leonard Koederitz. **Associate Professors:** Shari Dunn-Norman, Philip Schenewerk. **Research:** Well-completion design, production engineering, reservoir engineering and simulation, well testing, numerical modeling, natural-gas engineering, reservoir characterization, environmental applications of oil recovery technology, abandoned-well research.

UNIVERSITY OF NEBRASKA–LINCOLN

College of Engineering and Technology

Programs of Study	The M.S. and Ph.D. degrees are granted by the Graduate College. Master's programs are available in computer science; in engineering mechanics; and in biological systems, chemical, civil, electrical, industrial and management systems, manufacturing systems, and mechanical engineering. Eight doctoral fields are available: agricultural and biological systems engineering; chemical and materials engineering; civil engineering; computer engineering; electrical engineering; engineering mechanics; industrial, management systems, and manufacturing engineering; and mechanical engineering. These fields are described on the reverse side of this page.
Research Facilities	The College of Engineering and Technology and the Engineering Research Centers maintain modern laboratories for research and teaching in all of the fields and academic disciplines listed above. Technician-staffed machine shops—including foundry and carpentry facilities—and a technician-staffed electronics shop repair, maintain, and develop instrumentation for the research and teaching activities of the College. In addition, the biological systems engineering field operates the Nebraska Power Laboratory.
	Extensive computational facilities include various networked microcomputers, minicomputers, and superminicomputers and a CRAY J90 supercomputer with access to central mainframes and the Internet.
	The chemistry, physics, and engineering libraries contain the major archival journals, references, and texts. The University libraries also provide computer literature searching and participate in an interlibrary loan system for rapid access to references not available locally.
Financial Aid	Scholarships, fellowships, and teaching and research assistantships are available on a competitive basis. Teaching and research assistants receive a full tuition grant plus stipends ranging from $7200 to $14,800 per academic year for a half-time appointment.
Cost of Study	Current graduate tuition for Nebraska residents is $103.75 per credit hour. For nonresident students, it is $256.25 per credit hour. Lab fees range from $15 to $55 per semester. Program and facilities fees are $207 for full-time students.
Living and Housing Costs	Graduate students can arrange housing on campus through the University Housing Office. Room and board costs for the academic year 1999–2000 are $4070. There are a few on-campus apartments for married students; privately owned rental units are readily available in Lincoln.
Student Group	The University of Nebraska–Lincoln enrolls approximately 24,000 students, including 4,546 graduate students. The College of Engineering and Technology has a total enrollment of 2,209, including 423 graduate students.
Student Outcomes	During 1996–97, the College of Engineering and Technology graduated 459 students. Of those, 43 percent participated in a survey on outcomes, in which 30 reported going on to graduate school and 195 reported taking jobs related to engineering. The average wage of those who graduated with a bachelor's degree was $35,291; those with a master's, $43,282; and those with doctorates, $50,833.
Location	Lincoln—Nebraska's capital and second-largest city—is easily accessible by all kinds of transportation. The "All-America City," population 200,000, offers abundant cultural and recreational opportunities. The Lincoln and University communities offer many concerts, recitals, plays, art exhibits, and other cultural activities through excellent museums, galleries, and performing arts facilities. Numerous recreational opportunities are provided by the city's many parks and Salt Valley lakes.
The University and The College	The University of Nebraska–Lincoln is the largest component of the University of Nebraska System, which also includes the University of Nebraska at Kearney, the University of Nebraska at Omaha, and the University of Nebraska Medical Center, also located in Omaha. Beginning as a land-grant college chartered in 1869, the University of Nebraska–Lincoln now has thirteen colleges. The College of Engineering and Technology administers programs on the East and City campuses in Lincoln as well as on the Omaha campus.
Applying	Students who wish to pursue a graduate degree should write to the addresses below. Application materials must be received by the Graduate College two months before the first semester of planned registration. A $35 application fee is required.

Correspondence and Information

Regarding doctoral programs (except for computer science):

Coordinator of Engineering Doctoral Studies
W181 Nebraska Hall
College of Engineering and Technology
University of Nebraska–Lincoln
Lincoln, Nebraska 68588-0501

Regarding master's programs and the Ph.D. in computer engineering:

Chair, Computer Engineering Curriculum
Department of Computer Science
  and Engineering
University of Nebraska–Lincoln
Lincoln, Nebraska 68588-0115

University of Nebraska–Lincoln

THE FACULTY AND AREAS OF RESEARCH

Agricultural and Biological Systems Engineering. Darrel L. Martin, Field Chair. Graduate Faculty Fellows: L. Bashford, R. Brand, L. D. Clements, M. Dahab, E. Dickey, D. Edwards, D. Eisenhauer, J. Gilley, R. Grisso, M. Hanna, G. Hoffman, T. Howell, D. Jones, M. Kocher, M. Meagher, G. Meyer, J. Nienaber, D. Schulte, D. Shelton, L. Stetson, D. Vanderholm, D. Watts, C. Weller, W. Woldt. Agricultural power and machinery systems, plant sensors, irrigation system design, ground and surface water management, water quality, environmental engineering, solid and hazardous waste management, food and industrial materials handling and processing systems, extrusion processing, bioprocess reactor engineering, monitoring and controlling biological systems, decision support systems, geographical information systems, biomedical engineering.

Chemical and Materials Engineering. Robert DeAngelis, Field Chair. Graduate Faculty Fellows: D. Alexander, L. D. Clements, L. Lauderback, B. Robertson, D. Timm, H. Viljoen, J. Woollam. Participating faculty members are from chemical, electrical, and mechanical engineering, with interests in chemical, metallurgical, and materials engineering operations and the production, development, and utilization of materials. Renewable fuels and fuel oxygenates; biomass hydrolysis; fermentation, pyrolysis, and gasification; computer-aided process design; hazardous waste minimization in the process industries; surface science; electrochemical processes; corrosion of thin films; analytical electron microscopy; powder metallurgy; biomaterial separation and biochemical processes; thermodynamics; polymeric materials and composites; electrical and optical materials; graphite science; phase transitions in solids; ellipsometry; light-scattering spectroscopy.

Civil Engineering. Mohamed Dahab, Field Chair. Graduate Faculty Fellows: A. Azizinamini, I. Bogardi, P. McCoy, M. Moussavi, B. Rosson, J. Sherrad, D. Sicking, M. Tadros, C. Tuan, T. Zhang. Structures, transportation, environmental studies, water resources, and geotechnical research; development of energy-efficient building materials, such as high-strength concrete, concrete sandwich panels, and ultra-light concrete masonry units; research on culverts, bridges, and high-rise buildings; full-scale crash testing of roadside safety appurtenances; removal of nitrate from drinking water by biological and chemical processes; development of a geophysical monitoring network for groundwater contamination; risk analysis and risk management hydrological impacts of global change; hazardous-waste management and remediation; strategies for improving safety of elderly drivers; geometric design of highways.

Computer Engineering. Sharad C. Seth, Field Chair. Graduate Faculty Fellows: P. Bhattacharya, J.-C. Birget, J. Deogun, D. Klarner, J. Leung, S. Magliveras, S. Margolis, S. Reichenbach, A. Samal, S. Shende, D. Stinson, A. Surkan, S. Wiedenbeck. Coding and information theory, combinatorial optimization and combinatorics, computer architecture design and fault tolerance, computer networks, computer vision and image processing, databases, data security and encryption, human-computer interaction, information retrieval, multimedia, neural networks, parallel and distributed processing, scheduling theory and real-time systems software engineering, theory of computation, VLSI.

Electrical Engineering. Ram Narayanan, Field Chair. Graduate Faculty Fellows: D. Alexander, M. Algrain, E. Bahar, D. P. Billesbach, J. Boye, R. Dilion, N. Ianno, M. Hoffman, S. R. Liberty, R. Maher, R. Narayanan, D. Nelson, R. Palmer, K. St. Germain, K. Sayood, P. Snyder, R. Soukup, R. Throne, F. Ullman, H. Vakilzadian, J. Varner, R. Voelker, F. Williams, J. Woollam. Aerosols, biomedical engineering, communication, control system, digital system, electromagnetics, electronic circuits, environmental remediation, gaseous electronics and plasmas, nanostructures, power systems, remote sensing, solid-state materials and devices. World Wide Web: http://www.engr.unl.edu/ee/

Engineering Mechanics. Mao S. Wu, Field Chair. Graduate Faculty Fellows: M. F. Beatty, S. Chou, Y. Dzenis, M. Negahban, Y. C. Pao. Analytical mechanics—dynamics, vibrations, and nonlinear mechanics; computational mechanics—finite-element and boundary-element methods; mechanics of materials—metals, polymers, ceramics, biomaterials, and composites; mechanics of solids—linear and nonlinear elasticity, plasticity, viscoelasticity, damage and fracture mechanics, micromechanics, piezoelectricity, stress waves; experimental mechanics—acoustic emission and ultrasonic techniques, impact and dynamic materials characterization.

Industrial, Management Systems, and Manufacturing Engineering. Ram R. Bishu, Field Chair. Graduate Faculty Fellows: J. L. Ballard, F. Choobineh, D. Cochran, M. S. Hallbeck, R. O. Hoffman, K. P. Rajurkar, M. Riley. Ergonomics, manufacturing, manufacturing systems, computer-integrated manufacturing, traditional and nontraditional processes, fuzzy logic controllers, operations research, simulation, optimization, decision theory, statistical analyses.

Mechanical Engineering. Lorraine G. Olson, Field Chair. Graduate Faculty Fellows: J. P. Barton, K. D. Cole, R. J. DeAngelis, G. Gogos, D. Y. S. Lou, A. R. Peters, J. D. Reid, B. R. Robertson, S. L. Rohde, C. W. S. To. Acoustics and laser beam/particle interactions; diffusion theory for hot film sensors; numerical heat transfer and fluid flow in energy systems, combustion, manufacturing, and materials processing; aerodynamics; vehicle crashworthiness; probabilistic design; dynamics of machinery and robotics; mechatronics; sound and vibration studies; mechanical alloying; X-ray diffraction methods; electron microscopy; thin film deposition; failure analysis. World Wide Web: http://www.engr.unl.edu/me

ENGINEERING RESEARCH CENTERS

Center for Communication and Information Science. Sharad Seth, Director. This center combines faculty members from engineering, computer science, and mathematics. Research encompasses accessing, transmitting, sharing, and disseminating information and protecting such information from unauthorized use. Specific areas of research include the theory of networks and their efficient operation and management, coding theory, computer security, data compression, recognition, parallel and distributed systems, and multimedia systems.

Center for Electro-Optics. Dennis Alexander, Director. Research is in the general area of the application of electromagnetic radiation (lasers) for remote in situ measurements; specific programs are in applied optical measurements, modeling of electromagnetic radiation interactions with matter, remote sensing, radar cross sections, computer graphics and vision, particle or surface metrology, optical communications, particle sizing, ultrafast optical phenomena, measurement techniques, femtosecond laser nano-machining and nanostructures, environmental remediation, and atmospheric boundary layer research.

Center for Ergonomics and Safety Research. Michael W. Riley, Director. Research is focused on multidisciplinary activities such as musculoskeletal injuries, cumulative trauma disorders, work place design, hand tools, human-computer interaction, farm safety and injury risk factors. Ten different departments participate in the Center.

Center for Infrastructure Research. Paul Seaburg, Interim Director. Basic and applied research is aimed at improving the safety, efficiency, and productivity of the physical infrastructure, including all types of public facilities. Specific research programs include those on highways, roads, bridges, railroads, and mass transit systems; water supply and treatment systems; sewer and wastewater treatment systems; solid and hazardous-waste treatment and resource recovery systems; and public buildings.

Center for Laser-Analytical Studies of Trace Gas Dynamics. Shashi Verma, Director. This center has the broad-range objective of developing a trace gas measurement capability in Nebraska. Research focuses on the dynamics of trace gases, such as methane, carbon dioxide, and nitrous oxide, in the atmosphere; air quality issues in the livestock industry; problems that arise in materials processing; and the detection of fuel consumption by-products.

Center for Microelectronic and Optical Materials. John Woollam, Director. Research is conducted in the general areas of electrical and optical materials, with specific programs in magnetic and protective coating materials and vapor-deposited diamond and diamondlike carbon films, advanced compound semiconductors for microelectronics and optoelectronics, thin-film high-temperature superconductors, evaluation of environmental and corrosion protection of surfaces by coatings, materials for magneto-optic recording, materials for fast high-current switching, and optical instrumentation development.

Center for Nontraditional Manufacturing Research. K. P. Rajurkar, Director. Research is focused on nontraditional manufacturing/processing methods (such as electrodischarge, electrochemical, ultrasonic, and hybrid), with a mission of targeting the existing and future needs for software and hardware related to machinability, surface integrity, adaptive control, and agile manufacturing systems in the processing of new materials such as ceramics, superalloys, and composites for the automobile, aerospace, and electronics industries.

Mid-America Transportation Center. Patrick McCoy, Director. The U.S. Department of Transportation selected the College as a regional university transportation center in 1995. The Center pools research from Midwestern universities and state transportation departments. Research focuses on designing and operating transportation facilities, with the objective of building better cooperation between agencies in the region and working with industry to develop marketable products.

Midwest Roadside Safety Facility (MwRSF). Dean Sicking, Director. The MwRSF is internationally known for its work in the testing of roadside safety hardware. Full-scale crash testing is completed at the facility test site. Research focuses on developing roadside hardware with the objective of increasing the safety of the nation's highway travelers. The MwRSF is one of three test facilities approved by the federal government to certify that roadside hardware meets national safety guidelines.

National Bridge Research Organization (NaBRO). Atorod Azizinamini, Director; Dimitri Rizos, Assistant Director. The establishment of NaBRO was made possible through initial funding provided by the Nebraska Research Initiative (NRI) program, which aims in fostering the state's science infrastructure. NaBRO acts as a liaison between academia, government agencies, and design professionals for bridge-related issues. Research focuses on general bridge engineering with specific programs in the areas of innovative designs, nondestructive evaluation techniques, high-performance steel materials, advanced rating methodologies, seismic design and analysis, soil structure interaction, health monitoring of bridges, and full-scale testing, among others.

UNIVERSITY OF NEVADA, LAS VEGAS

Howard R. Hughes College of Engineering

Programs of Study

Programs of study lead to the degrees of Master of Science in Engineering (M.S.E.) in civil, electrical, and mechanical engineering; the Master of Science (M.S.) in computer science; and the Doctor of Philosophy (Ph.D.) in civil engineering, electrical engineering, mechanical engineering, and computer science. Areas of study and research include civil and environmental engineering (fluid mechanics and hydraulics, structural analysis and design, geotechnical and geological, transportation, and environmental), electrical and computer engineering (power systems, communications, computer engineering, electronics, solid state devices, and controls), mechanical engineering (robotics, controls, vibrations, and acoustics; heat transfer, fluid flow, and energy systems; mechanics and materials), and computer science (algorithms, artificial intelligence, software engineering, operating systems, programming languages, and graphics). There are two M.S. programs: 30 credits of graduate course work plus a 3-credit design project and 24 credits of graduate course work plus a 6-credit thesis. The Ph.D. degree requires the M.S. degree or equivalent for admission to the engineering or computer science programs. A minimum of 27 credits of course work is required beyond the M.S. plus a minimum of 18 credits for the dissertation. A qualifying oral and written examination is required after the completion of at least 12 but not more than 18 credits of required course work. A preliminary examination is required upon presentation of a dissertation research proposal.

Research Facilities

The College of Engineering is housed in the Thomas E. Beam Engineering Complex. This facility contains fifteen laboratories, seven classrooms, an auditorium, several College computer and terminal rooms, machine shops, a supercomputer center, and office space.

Each department has laboratories with virtually new equipment. The Tiberti Research Laboratory is a large, high-profile lab containing three large acoustical chambers, two 100-foot wind tunnel ducts, piping and controls for a 3,600-GPM flume, and a state-of-the-art MTS machine for a variety of materials testing applications. Other laboratories are devoted to soil mechanics and construction materials, water and wastewater treatment, fluid mechanics and hydraulics, heat transfer and thermal fluids, energy conversion and robotics, machine design, electric power systems, electronics and communications, digital systems and controls, computer-aided design, geographical information systems, computer architecture, and computer graphics.

Every department has access to Sun and Digital Equipment Corporation computers, graphics workstations, CAD/CAM stations, more than a dozen GIS stations, and many IBM-compatible personal computers. The University Computer Center is available for student and faculty use. The National Supercomputing Center for Energy and the Environment supports an Origin 2000 system, a Cray EL92, a Convex C-220, a Storage Tek Silo, and a variety of software, all of which are housed in the College of Engineering. Additional research opportunities are available through the Transportation Research Center, the Information Science Research Institute (ISRI), and the Center for Mechanical and Environmental Systems Technology.

Financial Aid

Each department has teaching and research assistantships that pay student tuition plus half-time monthly stipends. Fellowships, scholarships, and grants are available through the Graduate College. Students may apply for employment and a number of loan programs (work-study, Federal Perkins Loans, Federal Stafford Student Loans, and Incentive Grants) through Student Financial Services.

Cost of Study

Tuition in 1999–2000 is $96.50 per credit hour for Nevada residents, with an additional fee of $3173.50 per semester for out-of-state students. Health and other fees are $81 per semester.

Living and Housing Costs

Eight dorms are available, with room and board prices ranging from $2600 to $2925 per semester in 1999–2000. One hall is reserved for graduate students. Off-campus rental housing costs range from $500 to $700 per month, depending on utility and other costs. Students often live together in order to cut costs. Living costs are quite reasonable in the Las Vegas community.

Student Group

In 1998, about 1,000 undergraduate and 160 graduate students were enrolled in engineering and computer science. Most graduate students are part-time students who work full-time.

Location

The University's campus of about 335 acres is located on Maryland Parkway in Las Vegas about a mile north of McCarran International Airport and 3 miles east of I-15. The Las Vegas community has a population in excess of 1 million people. The city is considered one of the entertainment capitals of the world. Outdoor activities include boating on Lake Mead and skiing on Mount Charleston.

The University

In 1957, the University of Nevada, Las Vegas, constructed its first building on a 60-acre plot of desert and opened its doors to an inaugural class of 300 students. Today, enrollment exceeds 21,000, and some 700 professors bring degrees and teaching experience from all over the world.

Applying

Applicants to the M.S. programs should have earned a bachelor's degree in the appropriate discipline (ABET-accredited for engineering programs) or a closely related field with a minimum GPA of 3.0 (on a 4.0 scale). Provisional admission may be granted to promising applicants, but any deficiencies must be made up prior to the start of graduate work. A doctoral candidate must obtain satisfactory scores on the General Test of the GRE. The complete application consists of the application form, official transcripts of all previous university work, and three letters of recommendation. International students must file a financial statement and complete the TOEFL with a score of at least 550. They must also take the UNLV English as a Second Language Placement Test and may be required to take English language courses.

Correspondence and Information

For program information:
Dr. Walter C. Vodrazka, Associate Dean
Howard R. Hughes College of Engineering
University of Nevada, Las Vegas
4505 Maryland Parkway
Box 454005
Las Vegas, Nevada 89154-4005
Telephone: 702-895-3699
E-mail: dodgers@ce.unlv.edu

For application forms and admission questions:
Graduate College
University of Nevada, Las Vegas
4505 Maryland Parkway
Box 451017
Las Vegas, Nevada 89154-1017
Telephone: 702-895-3320
E-mail: hoefle@ccmail.nevada.edu

University of Nevada, Las Vegas

THE FACULTY AND RESEARCH AREAS

William R. Wells, Dean; Ph.D., Virginia Tech.
Walter C. Vodrazka, Associate Dean; Ph.D., Purdue.

Civil and Environmental Engineering

Jacimaria Batista, Assistant Professor; Ph.D., Penn State. James A. Cardle, Associate Professor; Ph.D., Minnesota. Gerald Frederick, Professor; Ph.D., Purdue. John Gambatese, Assistant Professor; Ph.D., Washington (Seattle). David E. James, Associate Professor and Chairman; Ph.D., Caltech. Moses Karakouzian, Professor; Ph.D., Ohio State. Mohammed Kaseko, Associate Professor; Ph.D., California, Irvine. Samaan G. Ladkany, Professor; Ph.D., Wisconsin. Barbara A. Luke, Assistant Professor; Ph.D., Texas at Austin. Edward S. Neumann, Professor; Ph.D., Northwestern. Neil Opfer, Associate Professor; M.S., Purdue. Thomas Piechota, Assistant Professor; Ph.D., UCLA. Shashi K. Sathisan, Professor; Ph.D., Berkeley. Jaeho Son, Assistant Professor; Ph.D., Purdue. Walter C. Vodrazka, Professor and Associate Dean; Ph.D., Purdue. Herbert C. Wells, Professor Emeritus; M.S., Berkeley. Richard V. Wyman, Professor Emeritus; Ph.D., Arizona.

Areas of current research interests include concrete structures, finite element analysis, robotics, materials, computational methods of fluid mechanics, nuclear waste engineering, solid and hazardous waste disposal, air pollution control, chemical and biochemical processes, pollutant distribution and movement, remediation of contaminated sites, containment technology, remote sensing, nuclear and hazardous wastes transportation, urban transportation planning, traffic engineering, air quality modeling, air transportation, geographical information systems (GIS) applications, public transportation, transportation safety, transportation technology, construction engineering, and construction management.

Computer Science

Wolfgang Bein, Assistant Professor; Ph.D., Berkeley. Hal Berghel, Professor and Chairman; Ph.D., Nebraska. Yonina S. Cooper, Associate Professor; Ph.D., New Mexico State. Ajoy K. Datta, Professor; Ph.D., Jadavpur (Calcutta). Laxmi P. Gewali, Professor; Ph.D., Texas at Dallas. Clinton L. Jeffery, Assistant Professor; Ph.D., Arizona. Lawrence Larmore, Professor; Ph.D., California, Irvine. John T. Minor, Associate Professor; Ph.D., Texas at Austin. Maria L. Misch, Lecturer; M.S., Nevada, Las Vegas. Thomas Nartker, Professor and Director, ISRI; Ph.D., Texas A&M. Roy Ogawa, Associate Professor; Ph.D., Berkeley. Kazem Taghva, Professor; Ph.D., Iowa. Evangelos A. Yfantis, Professor; Ph.D., Wyoming.

Areas of current research interests include document analysis, text and document retrieval, artificial intelligence (including automated deduction, rule-based knowledge systems, and natural language processing), compilers and supercompilers, computer graphics, image processing, online algorithms, parallel algorithms and systems, distributed systems, database systems, algorithm analysis, and computational geometry.

Electrical and Computer Engineering

Yahia Baghzouz, Professor; Ph.D., LSU. William L. Brogan, Professor Emeritus; Ph.D., UCLA. Lori Bruce, Assistant Professor; Ph.D., Alabama in Huntsville. Ashok Iyer, Professor and Chairman; Ph.D., Texas Tech. Abdol R. Khoie, Associate Professor; Ph.D., Pittsburgh. Shahram Latifi, Professor; Ph.D., LSU. Ramon J. Martinez, Associate Professor; M.S.E.E., Worcester Polytechnic. Eugene E. McGaugh, Associate Professor; Ph.D., Kansas. Robert Schill, Associate Professor; Ph.D., Wisconsin. Henry Selvaraj, Assistant Professor; Ph.D., Warsaw Technical. Sahjendra Singh, Professor; Ph.D., Johns Hopkins. Peter A. Stubberud, Associate Professor; Ph.D., UCLA. Rama Venkatasubramanian, Professor; Ph.D., Purdue.

Areas of current research activity include computer-aided analysis of power systems, power system harmonics, electric drives, control theory, filtering, estimation theory, application of analog and digital control laws, interactive software for control systems design, compound semiconductors, CAD/solid state devices, optoelectronics, solar cells, crystal growth, computer networks, fault tolerance computing, parallel processing, digital systems design, neural networks, digital signal processing, speech analysis, control systems, digital filters, systems analysis, modeling device physics, and ultrafast devices.

Mechanical Engineering

Robert F. Boehm, Professor; Ph.D., Berkeley. William G. Culbreth, Associate Professor; Ph.D., California, Santa Barbara. Bingmei May Fu, Assistant Professor; Ph.D., CUNY. Georg F. Mauer, Professor; Ph.D., Berlin Technical. Samir F. Moujaes, Associate Professor; Ph.D., Pittsburgh. Brendan J. O'Toole, Associate Professor; Ph.D., Delaware. Darrell W. Pepper, Professor and Chairman; Ph.D., Missouri–Rolla. Douglas D. Reynolds, Professor; Ph.D., Purdue. Robert L. Skaggs, Professor Emeritus; Ph.D., Iowa State. Mohammed B. Trabia, Associate Professor; Ph.D., Arizona State. Zhiyong Wang, Assistant Professor; Ph.D., Harbin Institute of Technology (China). William R. Wells, Professor and Dean; Ph.D., Virginia Tech. Woosoon Yim, Associate Professor; Ph.D., Wisconsin.

Areas of current research interest include computational fluid mechanics, heat transfer, two-phase flows, energy conservation, thermo comfort systems, solar energy utilization, performance and sound characteristics of ventilation systems, acoustics, mechanical vibration, vibration of fan systems, elastic robot arms, robot sensor control, engine diagnostics, nuclear waste engineering, nuclear waste container design, flow-in porous media, and composite materials.

UNIVERSITY OF NEW HAVEN

School of Engineering and Applied Science
Graduate Studies

Programs of Study	The University of New Haven (UNH) offers Master of Science degree programs in computer and information science, electrical and computer engineering, environmental engineering, industrial engineering (M.S.I.E.), mechanical engineering (M.S.M.E.), and operations research. A dual-degree program allows students to earn both the Master of Business Administration (M.B.A.) and the M.S.I.E. All master's degree candidates must complete a minimum of 30 credit hours in residence at the University of New Haven. Degree requirements differ in each program. Prospective students should consult with the chair of the engineering department to ascertain specific program requirements.
Research Facilities	The School of Engineering and Applied Science provides extensive, modern laboratory facilities designed for graduate instruction and research, including a state-of-the-art multimedia laboratory, the Computer-Aided Engineering Center (CAEC), the Digital & Image Processing Laboratory, the Control Systems Laboratory, and the Manufacturing and Robotics Laboratory. The UNH Center for Computing Services provides both administrative and academic computing support. Administrators, faculty members, and students have access to the latest in computer technology. Personal computers for student use are spread throughout the campus, with the largest concentration located at the Center for Computing Services. The Computer-Aided Engineering Center laboratory in the School of Engineering and Applied Science houses Windows NT Pentium workstations and microcomputers networked to the campus backbone as well as the Internet, printing and plotting devices, laser printing, and a wide variety of software and simulation packages.
Financial Aid	Financial aid is available for graduate students through a wide variety of sources, including assistantships, fellowships, need-based grants-in-aid, loans, and work-study programs.
Cost of Study	Tuition for master's degree students for the 1999–2000 academic year is $390 per graduate credit, or $1170 per course, for most graduate courses. Engineering courses are subject to a $50-per-credit-hour lab differential fee. Registration/Graduate Student Council fees total $15 per term. All charges and fees are subject to change.
Living and Housing Costs	Limited campus housing is available for graduate students. The Resident Services Office maintains a partial listing of apartments in the local area at a variety of costs.
Student Group	Most students are from Connecticut, but a significant number come from other states and many other countries. The graduate student body of about 2,400 ranges from recent college graduates to professionals with several years of experience in their fields. About 46 percent of the graduate students are women, about 10 percent receive some sort of financial aid, approximately 14 percent are international students, and about 18 percent are members of minority groups.
Location	The University of New Haven maintains a close relationship with the surrounding community. Although the campus is located in West Haven, it is less than 3 miles from downtown New Haven, and students can easily take advantage of the cultural offerings of the city. New Haven has rail, bus, and air service, and its location at the junction of two major interstate highways places the school within easy driving distance of New York, Boston, and Providence.
The University	The University of New Haven was founded in 1920 and is accredited as a general purpose institution by the New England Association of Schools and Colleges. A number of graduate classes are held at several off-campus locations across the state. Most graduate classes are held in early evening to accommodate both part-time and full-time students. An ELS Center is located on the main campus, which provides international students with English language preparation.
Applying	Applicants must hold a baccalaureate degree from an accredited college or university. An applicant must submit the following before the initial registration: a formal application, a nonrefundable $50 application fee, two letters of recommendation, final official transcripts (in English) of all previous college work, a TOEFL score (for students whose native language is not English), and certified financial support forms (for all international students). Late applicants may register as nonmatriculated students. All correspondence and requests for material should be directed to the Graduate School. Descriptions of programs and procedures are available in the *Graduate Catalog*. Information about the University of New Haven is available on the Internet via the World Wide Web at the address listed below.
Correspondence and Information	Joseph F. Spellman Director of Graduate Admissions University of New Haven 300 Orange Avenue West Haven, Connecticut 06516 Telephone: 203-932-7133 800-DIAL-UNH (toll-free) Fax: 203-932-7137 E-mail: gradinfo@charger.newhaven.edu World Wide Web: http://www.newhaven.edu/

University of New Haven

THE FACULTY

The faculty consists of 560 full-time and part-time professors. The coordinators for the various graduate programs in the School of Engineering and Applied Science are listed below.

Computer and Information Science: Tahany A. Fergany, Ph.D., Connecticut.
Electrical Engineering: Bijan Karimi, Ph.D., Oklahoma State.
Environmental Engineering: Agamemnon D. Koutsospyros, Ph.D., Polytechnic.
Industrial Engineering: Ronald N. Wentworth, Ph.D., Purdue.
Mechanical Engineering: Konstantine C. Lambrakis, Ph.D., Rensselaer.
Operations Research: Ronald N. Wentworth, Ph.D., Purdue.

UNIVERSITY OF NEW MEXICO

School of Engineering

Programs of Study

The School of Engineering offers M.S. and Ph.D. programs in chemical, civil, electrical, mechanical, and nuclear engineering and in computer science and M.E. degrees in manufacturing and hazardous-waste engineering. A graduate certificate in scientific and engineering computation may be earned in conjunction with any of the graduate degrees. A dual M.E. in manufacturing/M.B.A. program is available.

Students wishing to pursue graduate programs in engineering must meet both the requirements for admission to graduate study at the University of New Mexico (UNM) and the prerequisites of the department through which the desired program is offered. Admission decisions are made by departments. Deficiencies may need to be corrected in individual cases.

Requirements for the M.S. degree include a minimum of 30 semester hours with a minimum academic average of B. Six of these hours may be thesis work. A total of 30 to 36 hours is required for the M.E. degrees.

Requirements for the Ph.D. degree include a minimum of 48 semester hours of courses beyond the B.S. degree or 24 semester hours beyond the M.S. degree and evidence of superior scholarship and ability as an independent investigator. The M.S. and M.E. degrees require eighteen to twenty-four months of full-time study; the Ph.D., thirty-six to forty-eight months.

Departmental research activities are summarized on the reverse of this page; further information may be obtained by writing to the graduate adviser of the appropriate department.

Research Facilities

The principal facilities for the School of Engineering are located in the Farris Engineering Center and various department buildings. Additional research laboratories are located in the UNM Research Park, about 1 mile from campus. Research organizations housed in and/or closely affiliated with the School of Engineering include New Mexico Engineering Research Institute, Institute for Space and Nuclear Power Studies, Center for High Technology Materials, Center for Micro-Engineered Materials, Microelectronics Research Center, Center for Autonomous Control Engineering, High Performance Computing Education and Research Center, Center for Radioactive Waste Management, Alliance for Transportation Research Institute, Waste-management Education and Research Consortium, Advanced Materials Laboratory, Training and Research Institute for Plastics, and Manufacturing Training and Technical Center. Total research expenditures for the School of Engineering exceeded $40 million in 1997–98. Research facilities at nearby Sandia National Laboratories, Los Alamos National Laboratory, and the Air Force Research Laboratory are often used by graduate students.

The Centennial Science and Engineering Library has approximately 350,000 volumes and about 2,000 current journal subscriptions. Computer-searchable database services are available.

Financial Aid

Numerous research assistantships paying from $12,000 to $19,000 per year are available through the departments, along with teaching assistantships paying from $9000 to $12,000. Assistantships also pay the cost of tuition. Almost all full-time graduate students receive financial aid.

Cost of Study

In 1999–2000, tuition is $1221 per semester for full-time New Mexico residents and $4345 per semester for nonresident students. There are also a Graduate Student Association fee of $16 per semester and an equipment fee of $10 per credit hour.

Living and Housing Costs

In 1998–99, minimum full-time expenses, excluding tuition, were approximately $7000 on campus and $9000 off campus per semester. This included health and accident insurance, books, supplies, board, room, clothing, laundry, and miscellaneous expenses. University housing is available for both individuals and families.

Student Group

The University of New Mexico has a total undergraduate and graduate enrollment of approximately 30,000. The School of Engineering has about 1,400 undergraduate students and 470 graduate students, of whom 66 percent are master's candidates and 34 percent are doctoral candidates. Women constitute 19 percent of the graduate engineering enrollment. About half of the graduate students are studying on a part-time basis.

Student Outcomes

Recent recipients of master's and doctoral degrees from the University of New Mexico have found employment in a wide variety of organizations, including Sandia and Los Alamos National Laboratories; manufacturing firms such as Intel and General Motors; consulting firms such as CH2M–Hill and International Technology Corporation; government agencies such as the U.S. Department of Energy and the New Mexico Environment Department; and at universities both in the U.S. and abroad.

Location

The University is situated in Albuquerque, which has a metropolitan population of more than 500,000. The city is a mile above sea level, overlooking the Rio Grande, and it abuts the Sandia Mountains, which reach to 10,678 feet and offer skiing and hiking opportunities. Although Albuquerque undergoes seasonal changes, the dry, sunny climate rarely exhibits temperature extremes. Santa Fe, the first North American capital city, is nearby. The setting is rich with the traditions of Indian, Spanish, and Anglo cultures.

The University

The University of New Mexico is one of six universities in the U.S. that is both a Carnegie Research I school and is dedicated to serving members of minority groups. It is the flagship university of the state. UNM's mission includes offering comprehensive educational programs, conducting research and participating in other scholarly activities, and contributing to the quality of life in the state.

Applying

Admission decisions are made by each department and apply only for the semester for which a student applies. Applicants must hold an accredited bachelor's degree and have above a B average in their last two undergraduate years and in their major field. Departments may have more rigorous admission requirements. All applicants must submit the results of the Graduate Record Examinations General Test to the appropriate department prior to admission. Applications for financial aid and forms for letters of reference are available from the Office of Graduate Studies. Application deadlines for financial aid and assistantships vary, depending on departmental application deadlines. The Office of Graduate Studies fellowship deadline is April 16.

Correspondence and Information

Graduate Adviser
Department of (specify; see other side)
University of New Mexico
Albuquerque, New Mexico 87131
Telephone: 505-277-5521
Fax: 505-277-1422
World Wide Web: http://www.unm.edu

University of New Mexico

DEPARTMENT HEADS AND RESEARCH AREAS

Chemical and Nuclear Engineering. Professor Joseph L. Cecchi, Chairman. Advanced study leads to the M.S. in chemical or nuclear engineering and the Ph.D. in engineering. The department provides education and research in diverse fields of advanced technologies. Faculty and students participate in a number of special programs at the Institute for Space Nuclear Power Studies, the Center for Micro-Engineered Materials, and the Center for Radioactive Waste Management. Research facilities are located in the Farris Engineering Center and the Nuclear Engineering, the Space Technology, the Thermal-Hydraulics, and the Advanced Materials laboratories. State-of-the-art equipment is available in the research laboratories. Current research topics include biotechnology, biomedical sensors, radioactive waste management, health physics, space nuclear power and propulsion systems engineering and design, heat pipes technology, nuclear reactor safety and two-phase flow, reactor thermal-hydraulics, accelerator physics and engineering, interaction of radiation with matter, radiation measurement diagnostics, in situ fractional solidification, application of new techniques in powder and porous materials characterization, microstructure of ceramics, catalysis by metals on model supports, chemical and plasma processing of semiconductors, colloid transport, vapor-phase synthesis of materials, catalytic-waste degradation, inorganic membranes, biomedical sensors and instrumentation, and the use of self-assembled monolayers to study interfacial phenomena. Close collaboration with researchers at Sandia National, the Los Alamos National, and the Air Force Research laboratories provides access to an extremely large local scientific and engineering community. Graduates are intensively recruited by employers, with local and national opportunities that range from operation of high-technology facilities to research and development on the cutting edge. Continued professional opportunities for graduates are ensured due to close ties between the department's research programs and technologies of the future.

Civil Engineering. Professor Timothy J. Ward, Chairman. Advanced study leads to Ph.D. and M.S. degrees with specialization in construction engineering (M.S. only), environmental engineering, geotechnical engineering, structural engineering and structural mechanics, transportation engineering (M.S. only), and water resources engineering. Research facilities include a 2-kip electromagnetic shaker table in the Structural Dynamics Laboratory; a 320-KV X-ray system and a 110-kip INSTRON biaxial testing machine in the Meso-Mechanics Laboratory; a fast scanning electron microscope, a pulsed ruby laser, and a high-speed streak camera for dynamic interferometry in the Micro-Mechanics Laboratory; a high-vacuum vapor deposition unit and E-3 environmental scanning electron microscope in the Environmental SEM Laboratory; a 100-ton servohydraulic testing machine and 220-kip INSTRON axial and torsional loading system in the Materials Response Laboratory; a dynamic triaxial testing system in the Geotechnical Laboratory; complete water quality analytic instrumentation in the Environmental Laboratory; a 50-foot tilting table in the Hydraulics Laboratory; and Sun SPARCstations and PC systems in the computer laboratory. There is research collaboration with the Sandia and Los Alamos national laboratories, the Phillips Laboratory, the Army Corps of Engineers, the Alliance for Transportation Research Institute, New Mexico Engineering Research Institute, Waste-management Education and Research Consortium, NASA, and state agencies and private enterprises. Research topics include finite-element analysis, micromechanics, mesomechanics, geomechanics, reinforced earth, industrial and hazardous waste, highway safety and design, shock analysis and testing, pavement testing, expert systems, construction project organization, groundwater quality, open-channel hydraulics, erosion and sediment transport, random vibrations, structural reliability, system identification, and arid regions hydrology.

Computer Science. Professor Deepak Kapur, Chairman. The department offers M.S. and Ph.D. degrees in computer science. There are active research programs in adaptive computation, artificial intelligence, automated reasoning, computer security, data mining, distributed operating systems, experimental algorithms, and scientific visualization. Research programs have national recognition with funding from the Advanced Research Project Agency, National Science Foundation, Office of Naval Research, national laboratories, and other agencies. The department was recently awarded a five-year NSF Research Infrastructure grant to support computer science research fundamental to creation of a national information infrastructure and to establish a National Information Infrastructure (NII) Experimental Laboratory. Computing facilities are excellent. The department maintains a network of workstations (primarily SGI, Sun, Macintosh, and Linux PCs) with access to the Internet. Students also have access to additional visualization facilities in the Engineering and Science Computer Pod, as well as the facilities of the NII Experimental Laboratory. The department has close research relationships with the Santa Fe Institute and the national laboratories at Los Alamos and Sandia. Supercomputing facilities at the national labs as well as at the Maui High Performance Computing and Visualization Center (UNM is the prime contractor for the Maui Center) are available. A collaboration between faculty in computer science at UNM and researchers at Sandia National Laboratory has led to the development of the SUNMOS (Sandia and UNM OS) operating system. It served as the compute node operating system in the 1994 Gordon Bell award-winning submission and in establishing the Intel Paragon as the world's fastest computer (281 gflops on MP LINPACK, 329 gflops on LU factorization code).

Electrical and Computer Engineering. Professor Christos Christodoulou, Chairman. The department offers advanced study leading to the M.S. in computer or electrical engineering and the Ph.D. in engineering. Research facilities are located in the EECE building, the Engineering Annex Building, the Center for High Technology Materials (CHTM) Building, and the Microelectronics Research Center (MRC) facilities. Both CHTM and MRC are located in the Research Park. Faculty members and students participate in research in computer engineering, manufacturing engineering (interdisciplinary M.S. program), microelectronics, optoelectronics, plasma science, RF and microwaves, antennas, systems and controls (part of which research is done through the Center for Autonomous Control Engineering), signal processing and communications, high-performance computing, real-time systems, parallel algorithms, and image processing and visualization. State-of-the-art equipment is available in the research laboratories, including equipment for lasers and electrooptics, microprocessors, robotics, solid-state fabrication, a class-100 clean room, RF generation and characterization, pulsed power and plasma science, and high-performance and parallel computing. In addition, there is a clean room for instruction in microelectronics fabrication. The department maintains a large collection of high-end PC and UNIX workstations, multimedia output devices, and industry-standard engineering software for student use. There is close collaboration with researchers at Los Alamos National Laboratory, Sandia National Laboratories, the Air Force Research Laboratory, and the UNM-administered High Performance Computing Education and Research Centers in Albuquerque and Maui. Opportunities for graduates continue to grow as the demand for trained professionals in high-technology fields increases.

Mechanical Engineering. Professor David Thompson, Chairman. Programs of study and research are offered in the areas of thermal science, computational mechanics, fluid mechanics, solid mechanics, materials science, biomechanics, dynamic systems and control, robotics, and manufacturing engineering. Recent and ongoing research areas are thermal sciences—stratified thermal storage, energy conservation, radiative heat transfer, second law analysis of thermal systems, heat transfer engineering in periodically varying phenomena, electronic cooling, thermal shock wave, friction heating, transient burning of granular energetic materials, light-sensitive explosive initiation and thermophysical properties, dynamical systems and stochastic processes, bifurcation theory, and thermal convection; fluid mechanics—boundary-element method, finite difference/finite volume, turbulence modeling, turbulence measurement, and aeroptics; computational mechanics—finite-element features and constraint algorithms; solid mechanics—anisotropic constitutive equations for plasticity and damage, fracture mechanics and composite failure, mesocracking damage, irreversibility and macroscopic damage, structure damage diagnosis, dynamic properties of materials, transducer response from extreme inputs, finite-element method, boundary finite-element method, seismic isolation studies, optimization, and nonlinear system and chaotic motion; robotics—robot dynamics, flexible robots, dexterous end effectors, smart/intelligent materials, and structures and systems (in particular, ionic polymeric gels or artificial muscles); manufacturing—computer-integrated manufacturing, enterprise simulation, clean-room automation, sensors, microfabrication, and computer-based training for semiconductor manufacturing; and materials science—thermomechanical integrity of microelectronic devices and packages, mechanical behavior of thin films and composite materials, and modeling of materials across length scales. Substantial computing and laboratory facilities are available in the ME building to support research in these areas.

UNIVERSITY OF NEW ORLEANS

College of Engineering

Programs of Study

The College of Engineering offers programs of graduate study leading to the degrees of Doctor of Philosophy in engineering and applied sciences, Master of Science in Engineering (with concentrations in civil, electrical, mechanical, and naval architecture and marine engineering), and Master of Science in Engineering Management.

After admission to a master's program, the student selects one of two options: the thesis (or research) option, which requires 30 hours of graduate work, including 6 hours of thesis research, or the nonthesis (or courses only) option, which requires 33 hours of graduate credit.

Students entering the Doctor of Philosophy program in engineering and applied science choose an Advisory Department. The Ph.D. committee chairman resides in that department, and the department administers the qualifying examination. (Most students with a specialized master's degree in the Advisory Department's area should be able to take the qualifying examination shortly after enrolling.) After passing the qualifying examination, a Dissertation Advisory Committee is established for the student. Although a majority of the Dissertation Advisory Committee members may be from the Advisory Department, the committee reflects the interdisciplinary nature of the program.

Research Facilities

The $20-million nine-story Engineering Building was dedicated at UNO in 1987. This facility includes classrooms, instructional and research laboratories, and departmental offices. It features a number of state-of-the-art specialized facilities, including a specially designed towing tank for deepwater and shallow-water testing. The 125-foot by 15-foot by 7-foot-deep towing tank enables students and faculty members to perform experiments with ships and offshore structures in waves and in other extreme conditions.

A number of special laboratories for instruction and research are also housed in the Engineering Building. These specialized facilities include an anechoic chamber designed to absorb microwave reflections, a structures laboratory capable of structures testing of 65-foot-long structural members such as beams and trusses used for offshore platforms and bridges, a hydraulic and water resource laboratory, an optics laboratory, and a state-of-the-art computer-aided design/computer-aided manufacturing (CAD/CAM) laboratory.

A modern environmental engineering laboratory, consisting of adjoining areas for the wet laboratory and instrumentation rooms, is utilized for environmental studies, including waste management and research. The existing equipment includes laboratory-scale pilot units for waste treatability studies, e.g., activated sludge, granular activated carbon, and sand filtration. Analytic instrumentation and equipment capabilities include a GC and GC/MS with data systems, total organic carbon analyzer, Alpkem Auto Analyzer, atomic absorption spectrophotometer, infrared analyzers, and others. Automatic samplers, flow monitors, and other field survey equipment and instrumentation are also available for engineering characterization studies for the structural integrity of waste used in a recycling mode or for the geotechnical laboratory for the investigation of landfills.

The Engineering Building has a comprehensive internal system of fiber-optic computer cabling and is connected to the University's central computer facility, which houses several systems, including a Cray supercomputer. The system's networking capability permits extensive sharing of data within the building as well as high-speed access to the Internet. Facilities for transmitting and receiving closed-circuit television have been provided in all classrooms and selected laboratories, and these have also been connected to the campuswide television system.

Financial Aid

Teaching assistantships are available to qualified graduate students. Research assistantships supported by grant funds of individual faculty members are also available. The stipend is identical to that of teaching assistants. Summer support is available in each category. The amount paid is proportionally scaled to the academic year stipend.

Cost of Study

The full-time graduate tuition in 1998–99 was $2362 per semester for residents and $7888 per semester for nonresidents. The corresponding estimated fees for the summer session were $690 for residents and $1045 for nonresidents.

Living and Housing Costs

In 1998–99, the cost of double-occupancy rooms for single students ranged from $775 to $1160 per semester. Unfurnished married student apartments were available for $350 to $400 per month. In the newly constructed Privateer Place, housing ranged from $1025 to $2338 per semester. Off-campus apartments close to the University are plentiful.

Student Group

The total enrollment in 1998–99 at the New Orleans campus was 15,629; graduate enrollment totaled 3,991. Graduate enrollment in the College of Engineering for the fall 1998 term was 173 students. Students came to the University from forty-eight states and more than sixty-nine other countries. In 1998, the College of Engineering awarded forty-nine master's degrees.

Location

As a major urban institution located in one of Louisiana's most exciting cities, the University of New Orleans has a history of providing an exemplary education for its students. Greater New Orleans is a metropolitan area of more than 1 million people. The University is located on the 300-acre tract of land on the southern shore of Lake Pontchartrain. New Orleans is an ideal place to participate as well as spectate. The city's semitropical climate and thousands of acres of public parks allow year-round jogging, bicycling, golfing, and tennis. Surrounded by lakes, rivers, and bayous, the city is outstanding for fishing, sailing, and other water sports, and the surrounding marshlands are a hunter's paradise.

The University and The College

The University of New Orleans is a state institution that was formally opened in 1958. It was initially established as the Louisiana State University in New Orleans, and the name was changed to the University of New Orleans in 1974. The School of Engineering was established in 1973 and became the College of Engineering in 1980. The University is the second-largest university in Louisiana.

Applying

Qualified students may be admitted for any semester and begin in August, January, or June. Applications are reviewed continually; applicants seeking admission to a graduate program in engineering must have received a bachelor's degree in a field of engineering, mathematics, or the sciences; or, in the case of international students, they must present evidence of an equivalent academic preparation. Students who do not have an engineering undergraduate degree must complete a core program of 30 semester hours of undergraduate engineering courses or pass equivalent credit examinations.

Correspondence and Information

For catalogs and application forms:
Office of Graduate Admission
University of New Orleans
Lakefront
New Orleans, Louisiana 70148
Telephone: 504-280-6595

For master's program information:
Dr. Alim P. Hannoura
Graduate Coordinator
College of Engineering
University of New Orleans
Lakefront
New Orleans, Louisiana 70148
Telephone: 504-280-6283

For Ph.D. program information:
Dr. Alim P. Hannoura
Program Director
College of Engineering
University of New Orleans
Lakefront
New Orleans, Louisiana 70148
Telephone: 504-280-6283

University of New Orleans

FACULTY HEADS AND RESEARCH AREAS

Civil and Environmental Engineering. Kenneth McManis, Chairman. Faculty interest and current research programs include reliability-based analysis and design, nonlinear analysis of frames, biaxial bending of reinforced-concrete columns, design of prestressed bridge girders with high-strength concrete, waste management, groundwater modeling and aquifer remediation, water quality, radioactive waste containment systems, expert system applications, wave interaction with rubble-mound breakwater foundation engineering, pavement systems, soil stabilization, pile foundations, geosynthetics, and gravimetric study of areal subsidence. The Civil and Environmental Engineering Department also encompasses the Urban Waste Management and Research Center, an EPA-funded center with research in the impact of urban runoff surface water quality, regional waste management systems, biofilter systems for treating landfill leachate, waste-to-energy, expert systems computer models, and other related areas. Research assistantships are available. The department also houses the Freeport-McMoRan Endowed Chair in Environmental Modeling. As chairman, Dr. J. Alexander McCorquodale heads the Freeport-McMoRan Center for Environmental Modeling, made up of multidisciplinary teams pursuing environmental research. Research assistantships are available.

Electrical Engineering. Russell Trahan, Chairman. Current research topics include ellipsometry optics, thin-film optics, development of new entirely solid-state light-polarization sensors, optical coatings design, design of polarizing optical elements, determination of optical properties of materials in the visible and near-visible spectrum—the materials may be transparent or absorbing, isotropic or crystalline, in bulk or thin-film form, theory of light reflection, biomedical applications of microwaves, biomedical signal processing, digital circuits, instrumentation, microcomputers—design and application, interface, optical computing, communication and signal processing, wave propagation and antennas, radiation in plasma, microwave radiation, analog circuit design, electromagnetic theory, RF, microwave, and ionizing radiation bioeffects and medical applications, bioelectric current and ion transport in isolated tissues, radioactive tracer studies, ocular effects of RF radiation on the eyes of Rhesus monkeys, disposal of hospital and research center radioactive waste, transport of ions across isolated membranes, RF heating of hypothermic patients, artificial intelligence, neutral networks, expert systems, image processing, power systems (neutral net and intelligent system applications, transmission and load modeling, fault location, system harmonies), acoustics, communications, signal processing (image, audio, biomedical), application of digital signal processing techniques to audio engineering and biomedical signal analysis, Kalman filtering estimation, adaptive filtering, digital filters, transient response, control theory—optimization, data acquisition—real time, signal processing, optimal control of gas turbines by computer simulation, optimal calibration of transponder arrays using the Kalman filter, optimal control of a distillation process, fault detection in control schemes, polarization optics, ellipsometry, photopolarimetry, lasers, holography, fiber optics, computer-aided control system design, analog-digital/digital-analog package, Wavelet transform, optical signal processing, automatic target recognition, and nondestructive testing. The Electrical Engineering Department also encompasses the Center for the Industrial Applications of Electric Power and Instrumentation, with research in the improvement of the quality of power, improvement of motor performance, continuity of service, shut-down schemes, computerized energy monitoring, fault location, power factor improvement, new motor and drive technology, effects of magnetic fields, and other related areas. Research assistantships are also available in this area of study.

Mechanical Engineering. Edwin P. Russo, Chairman. Research interests fall into four major areas: solid mechanics, fluid dynamics, heat transfer, and system dynamics and design. Topics in solid mechanics include buckling and vibrations of laminated plates and shells, dynamic behavior of laminated imperfect flat or cylindrical panels under axial impact, constitutive equations for engineering materials, and lifting eye analysis. In fluid dynamics and heat transfer, topics include transient two-phase flow with constant wall temperature, cryogenics, thermal-hydrodynamic modeling of shallow lakes, submerged towing cable configuration analysis, supersonic diffusers and ejectors, and incompressible viscous flow in ducts and pipes. Topics in system dynamics and design include surface finish imperfections from workpiece vibration, CNC machine code generation, and chaotic motion of an inertially asymmetric circular wheel. The Mechanical Engineering Department encompasses the Composites Engineering Research Laboratory, with research in impact, buckling, and vibration behavior of laminated panels. This rapidly developing research area is expected to continue growing, and research assistantships are available. The department also has an Energy Conversion and Conservation Research Center, with current research in energy from waste, solar energy, and energy conservation in buildings. A new building for energy management is planned that will provide a laboratory facility, which will double the College's current research space. Research assistantships are available for this research area and in the cryogenics laboratory, which has been developed over the last five years.

National Biodynamics Laboratory. The National Biodynamics Laboratory, a division of the College of Engineering, conducts research aimed at preventing injury and enhancing performance in humans subjected to external motion forces by ships, airplanes, automobiles, offshore oil structures, etc. The research uses devices such as acceleration sleds, a ship-motion simulator, vibration equipment, and desensitization devices in conjunction with comprehensive data acquisition systems.

Naval Architecture and Marine Engineering. William Vorus, Chairman. Research interests fall into four major areas: design of ship and offshore platforms, marine hydrodynamics, ship and offshore platform structure design, and marine propulsion plant design. Topics in design of ship and offshore platform design include high-speed marine craft design and database development, design of offshore platforms, design of river ships, computer-integrated manufacturing, and ship production procedures. In marine hydrodynamics, topics include computational fluid dynamics, calculation of ship roll motions and of the flow around ship hulls, prediction of large-amplitude (nonlinear) motions and ship capsize using dynamical systems analysis techniques, investigation of ship towing and steering in shallow water, investigation of the flow around the ship bow in shallow water, experimental evaluation of ships and offshore structures performance using computer simulations and model tests conducted in the 38.3-meter long by 4.6-meter wide by 2.1-meter deep towing tank. Topics on ship and offshore platform structure design include study of the use of fiberglass in ship structures, analysis of high-speed craft structure, structural optimization, analysis of complex structures using FEM, and analyses of welded plate behavior under variable loading. Topics in marine engineering include computer simulation of marine diesel engine performance, shafting and propeller vibration analysis, design of marine auxiliary machinery, and reliability and maintenance of marine propulsion and auxiliary equipment using the UNO-RAM database. The School of Naval Architecture and Marine Engineering also supports the activities of the Gulf Coast Regional Maritime Technology Center, which focuses on advanced structure design, environmental ship design, hydrodynamics of marine vehicles, vehicle maintenance and operation, and advanced materials and production processes.

UNO–Avondale Maritime Technology Center (UAMTCE). The UAMTCE's mission is to foster technological progress in the maritime industry, including advanced research and enhancements to UNO's NAME curriculum. Specialties include concurrent engineering, IPPD/IPDE and simulation-based design, including 3-D visualization and virtual reality.

Urban Waste Management and Research Center (UWMRC). The UWMRC addresses the environmental interactions between air, water, and land as they relate to preventive and corrective measures employed to eliminate sources of pollution in urban areas.

UNIVERSITY OF NORTH CAROLINA AT CHARLOTTE

The William States Lee College of Engineering

Programs of Study

The William States Lee College of Engineering offers Master of Science degrees in civil engineering (M.S.C.E.), electrical engineering (M.S.E.E.), mechanical engineering (M.S.M.E.), and computer science (M.S.). The College offers a Master of Science in Engineering (M.S.E.) for students with nonengineering backgrounds and a Master of Engineering (M.E.) on an off-campus basis. The College offers Ph.D. degree programs in electrical engineering and mechanical engineering and also offers study and research in civil engineering and computer science leading to the Ph.D. degree through an interinstitutional program with North Carolina State University. Doctoral study in certain areas of computer science (logic, optimization, computer integrated manufacturing, and computer engineering) can be pursued through one or another of the doctoral programs in electrical engineering, mechanical engineering, applied mathematics, or the Ph.D. in information technology.

The primary purpose of all these programs is to provide an opportunity for people in the engineering and computer science professions to obtain graduate-level education that will improve on-the-job skills, provide for total career development, and prepare them for additional advanced education and research. The programs are structured to be responsive to the ever-changing needs of the engineering and computer science professions and to foster a high degree of university-industry interaction. The programs are available in the evening.

The master's degrees in engineering and computer science require satisfactory completion of at least 30 semester hours of approved graduate course work, no more than 6 of which can be earned for a research thesis. For an engineering degree, a minimum of 18 semester hours of graduate course work must be in a major area of engineering, and at least 6 hours must be in mathematics, science, or a different engineering area. For the computer science degree, 12 hours of core courses and a minimum of 9 semester hours in an area of concentration are required. Upon completion of all degree requirements, the candidate stands for a final comprehensive examination, which may be written, oral, or both. All work for the master's and Ph.D. degrees must be completed within a six-year period. Ph.D. degree requirements include at least 45 hours of course work beyond the baccalaureate and at least 18 hours of research credit for work on a publishable dissertation.

Research Facilities

The Sheldon P. Smith Engineering Building and the Woodford A. Kennedy Building housing the Department of Computer Science contain more than 45,000 square feet of engineering and computer laboratories that have approximately $10 million worth of equipment to support the educational research and development activities of the College. Major research laboratory facilities are available in each of the following areas: precision engineering and design; thermal fluids and power; materials science and metallurgy; robotics/manufacturing; biomedical and biotechnical engineering; electron-microscopy; CAD; biotechnology; geotechnical engineering; structural engineering; transportation engineering; environmental engineering; electronics; electric power; optoelectronics; microelectronics design, fabrication, packaging, measurements, and testing; intelligent systems; software and computer engineering; and computer vision. The College has a fully integrated system of networked workstations, file servers, printers, and plotters to support the full range of academic and research computing. Access to outside specialized computing capabilities including superminis and supercomputers is also available. The adjacent 75,000-square-foot C. C. Cameron Applied Research Center contains a state-of-the-art microelectronics clean room, optoelectronics, a precision machine shop, semiconductor device characterization, and a metrology laboratory.

Financial Aid

Approximately 200 half-time teaching and research assistantships starting at $9000 and several Giles Graduate Fellowships at $12,000 plus tuition are available, as well as a limited number of out-of-state tuition awards.

Cost of Study

Tuition and fees for full-time study in 1999 are approximately $975 per semester for North Carolina residents and approximately $5000 per semester for nonresidents. Students taking fewer than 9 credit hours pay reduced rates.

Living and Housing Costs

On-campus dormitory facilities are provided for unmarried graduate students at costs ranging from about $900 to $1300 per semester. Various meal plans ranging from about $500 to $1000 per semester are available. Ample housing is available near the campus; typical apartment rents range from $500 to $900 per month.

Student Group

The William States Lee College of Engineering enrolls 1,121 undergraduate students and 181 graduate students in engineering and 563 undergraduate students and 87 graduate students in computer science. Most of the full-time graduate students receive some form of financial support.

Student Outcomes

All of the College's graduate programs, at both the M.S. and Ph.D. levels, are oriented toward professional practice. Employers include small and large consulting engineering firms; electronics firms such as Motorola, Intel, Texas Instruments, and Northern Telecom; manufacturing firms such as Ford, Cummings, and Caterpillar; computer-related firms such as Nationsbank, IBM, and Pacific Scientific; and government agencies and laboratories.

Location

The University of North Carolina at Charlotte has a modern campus, located just outside Charlotte, the largest city and urban center in the Carolinas. The location of a number of industries and a major research park near the campus provide excellent opportunities for university-industry interaction, particularly with regard to research and graduate education.

The University and The College

The University of North Carolina at Charlotte was established in 1965 as one of the sixteen campuses of the University of North Carolina System. The William States Lee College of Engineering is one of six colleges that constitute the University.

Applying

Applications for graduate study, which must be submitted on a form supplied by the dean of the Graduate School, are received the year round. Admission to the Ph.D. programs requires a master's degree in engineering or a closely allied field. General admission requirements are satisfactory undergraduate preparation in engineering, computer science, or a related area, usually manifested by the possession of a baccalaureate degree with a grade point average of 3.0 on a 4.0 scale; submission of two official transcripts of all academic work attempted beyond high school; submission of satisfactory scores on the Graduate Record Examinations or, for engineering programs, on the Miller Analogies Test and the Doppelt Test of Mathematics Reasoning; at least three evaluations from persons familiar with the applicant's personal and professional qualifications; an essay describing the applicant's experience and objective in undertaking graduate study; and, for international students, financial certification and acceptable scores on the Test of English as a Foreign Language or the Michigan test.

Correspondence and Information

Dr. Robert J. Mundt
Associate Vice Chancellor for Graduate Programs
    and Dean of Graduate School
University of North Carolina at Charlotte
9201 University City Boulevard
Charlotte, North Carolina 28223-0001
Telephone: 704-547-3371
Fax: 704-547-3279

Chairman, Department of (specify)
University of North Carolina at Charlotte
9201 University City Boulevard
Charlotte, North Carolina 28223-0001
Telephone: 704-547-2301
Fax: 704-547-2352

University of North Carolina at Charlotte

THE FACULTY AND THEIR RESEARCH

Civil Engineering
David M. Bayer, Professor; Ph.D., Vanderbilt, 1968. Structural analysis and design, steel structures, computer applications in structures.
James D. Bowen, Assistant Professor; Ph.D., MIT, 1990. Surface water quality modeling, microscale fluid mechanics, constructed wetlands, discharge diffuser design.
Rachael Davidson, Assistant Professor; Ph.D., Stanford, 1998. Structural engineering, earthquake engineering, disaster risk analysis.
Jack B. Evett, Professor; Ph.D., Texas A&M, 1968. Hydraulics, hydrology, surveying, water resources and water conservation.
Janos Gergely, Assistant Professor; Ph.D., Utah, 1998. High-performance materials, composites, structural retrofit.
Johnny R. Graham, Associate Professor; Ph.D., North Carolina State, 1990. Transportation engineering and traffic studies.
Helene A. Hilger, Assistant Professor; Ph.D., North Carolina at Charlotte/North Carolina State, 1998. Environmental engineering, biotechnology, microbiology, pollution prevention.
Rajaram Janardhanam, Professor; Ph.D., Virginia Tech, 1981. Geotechnical and geoenvironmental engineering, substructure analysis and design, pavement materials.
Martin R. Kane, Assistant Professor; Ph.D., Michigan State, 1995. Transportation engineering/planning, transit, highway safety, human factors.
L. Ellis King, Professor; D.Eng., Berkeley, 1967. Transportation and traffic engineering, human factors.
Karl Linden, Assistant Professor; Ph.D., California, Davis, 1995. Environmental engineering, biotechnology.
Alan T. Stadler, Assistant Professor; Ph.D., Kansas, 1995. Geotechnical engineering, geosynthetics.
Jy S. Wu, Professor; Ph.D., Rutgers, 1980. Environmental engineering, water resources, biotechnology.
David T. Young, Associate Professor and Chairman; Ph.D., Virginia Tech, 1985. Structural engineering, concrete/masonry materials, optimization.

Computer Science
C. Michael Allen, Professor; Ph.D., SUNY at Buffalo, 1968. Computer engineering, medical imaging, embedded controllers.
Keh-Hsun Chen, Associate Professor; Ph.D., Duke, 1976. AI systems, heuristic search, computer Go, theory of computing.
Bei-Tseng Bill Chu, Associate Professor; Ph.D., Maryland, 1988. Computer-integrated manufacturing, intelligent diagnostic and control systems, system integration, human-computer interfaces.
Mirsad Hadzikadic, Associate Professor and Chairman; Ph.D., SMU, 1987. AI, cognitive science, machine learning, expert systems.
Junshong Long, Associate Professor; Ph.D., Illinois at Urbana-Champaign, 1992. Object-oriented system programming, distributed and parallel processing, computer architecture.
Zbigniew Michalewicz, Professor; Ph.D., Polish Academy of Sciences, 1981. Genetic algorithms, statistical databases.
M. Taghi Mostafavi, Associate Professor; Ph.D., Oklahoma State, 1986. Computer engineering, VLSI architecture and design.
Zbigniew W. Ras, Professor; Ph.D., Warsaw, 1973. Intelligent information systems, logic for artificial intelligence.
Hassan Modaress Razavi, Associate Professor; Ph.D., West Virginia, 1978. Computer organization and architecture, fault-tolerant computing.
Gyorgy E. Revesz, Professor; Ph.D., Eötvös Loránd (Budapest), 1968. Formal languages and automata theory, compiler design, semantics of programming languages, functional programming, symbolic computing, parallel programming.
Kalpathi R. Subramanian, Associate Professor; Ph.D., Texas at Austin, 1990. Computer graphics, scientific and engineering visualizations.
William J. Tolone, Assistant Professor; Ph.D., Illinois at Urbana-Champaign, 1996. Computer-supported cooperative work, groupware, meta-level architectures, object-oriented design, specification environments, agent technologies, software engineering.
Anthony Barry Wilkinson, Professor; Ph.D., Manchester (England), 1974. Multiprocessor system design and networks.
Jing Xiao, Associate Professor; Ph.D., Michigan, 1990. Robotics, computer vision, artificial intelligence.
Jan Zytkow, Professor; Ph.D., Warsaw, 1972. Artificial intelligence, machine learning, knowledge discovery.

Electrical and Computer Engineering
Falih H. Ahmad, Assistant Professor; Ph.D., Mississippi State, 1990. Electromagnetics and communications.
Stephen Bobbio, Professor; Ph.D., William and Mary, 1972. Microelectronics, plasma processing and microstructure.
Robert Coleman, Professor; Ph.D., Auburn, 1970. Electromagnetics.
Teresa Dahlberg, Assistant Professor; Ph.D., North Carolina State, 1993. Communications and computer networking.
Kasra Daneshvar, Professor; Ph.D., Illinois at Urbana-Champaign, 1979. Laser optics, sensors, vacuum microelectronics, optoelectronics.
Arthur H. Edwards, Associate Professor; Ph.D., Lehigh, 1981. Band structure theory and device modeling.
Michael R. Feldman, Associate Professor; Ph.D., California, San Diego, 1989. Optical interconnects, computer-generated holography, optoelectronic.
Richard F. Greene, Professor; Ph.D., Pennsylvania, 1951. Semiconductor science, vacuum microelectronics, nanometrics.
M. A. Hasan, Assistant Professor; Ph.D., Sweden, 1990. Thin film materials and processing.
Irvin Jones, Assistant Professor; Ph.D., Colorado at Boulder, 1998. Computer architecture and engineering.
Yogendra P. Kakad, Professor; Ph.D., Florida, 1975. Control system theory, spacecraft controls, CIM, operations research.
Fouad Kiamilev, Associate Professor; Ph.D., California, San Diego, 1992. Optoelectronics.
Vasilije P. Lukic, Professor; Ph.D., Belgrade (Yugoslavia), 1968. Power system controls.
Rafic Z. Makki, Professor; Ph.D., Tennessee Tech, 1983. VLSI design and test.
Mehdi Miri, Associate Professor; Ph.D., Ohio State, 1987. Electrical power and breakdown, robotics.
Howard Phillips, Professor; Ph.D., New Mexico, 1972. Artificial vision and digital systems.
Barry Sherlock, Associate Professor; Ph.D., Cape Town (London), 1989. Signal processing, image compression, fluid flow modeling, biomedical.
Farid M. Tranjan, Professor and Chairman; Ph.D., Kentucky, 1984. Device processing, display technology.
Raphael Tsu, Professor; Ph.D., Ohio State, 1960. Quantum devices, device materials, materials growth and characterization.
Sheng-Guo Wang, Associate Professor; Ph.D., Houston, 1994. Systems and control, robust control, communications, computer networks.
Thomas Weldon, Assistant Professor; Ph.D., Penn State, 1995. Signal processing.
Dian Zhou, Associate Professor; Ph.D., Illinois at Urbana-Champaign, 1990. VLSI routing and layout.

Mechanical Engineering and Engineering Science
Harish Cherukuri, Assistant Professor; Ph.D., Illinois at Urbana-Champaign, 1994. Computational mechanics, plasticity, creep, deformation processing.
Robin Coger, Assistant Professor; Ph.D., Berkeley, 1993. Thermosciences, bioengineering, cryopreservation, tissue engineering.
Paul H. DeHoff, Professor; Ph.D., Purdue, 1965. Viscoelasticity, finite-element analysis, stress analysis, compatibility of metal/ceramic systems.
Horacio V. Estrada, Associate Professor; Ph.D., RPI, 1983. Semiconductor characterization, sensors and transducers.
Yogeshwar Hari, Associate Professor; Ph.D., Purdue, 1969; PE. Electromechanical systems, CAD, vibration analysis.
Robert J. Hocken, Professor; Ph.D., SUNY at Stony Brook, 1973. Metrology, precision machinery, nanotechnology, automated manufacturing, optical physics.
Robert E. Johnson, Professor and Chairman; Ph.D., Caltech, 1977. Materials processing, fluid mechanics, applied mathematics.
Russell G. Keanini, Assistant Professor; Ph.D., Berkeley, 1992. Computational fluid dynamics, materials processing, bioheat and biomass transfer.
Rhyn H. Kim, Professor; Ph.D., Michigan State, 1965; PE. CAD of energy conversion devices, visualization of fluid phenomena, fluid bed combustion, vibrations and noise.
Harry J. Leamy, Professor; Ph.D., Iowa State, 1967. Metallurgy, materials science.
Ganesh P. Mohanty, Professor; Ph.D., IIT, 1961. Metallurgy, materials processing and characterization.
Edgar G. Munday, Associate Professor; Ph.D., Virginia Tech, 1984; PE. Electromechanical design, automatic control.
John A. Patten, Ph.D., North Carolina State, 1996. Manufacturing, CAD/CAM, CIM, robotics, precision engineering, energy conservation, electric vehicles, solar charging.
Steven R. Patterson, Professor; Ph.D., California, Davis, 1987. Precision machine design.
Richard D. Peindl, Associate Professor; Ph.D., Ohio State, 1984. Experimental mechanics, digital control, orthopedic biomechanics.
Jay Raja, Professor; Ph.D., Indian Institute of Technology (Madras), 1980. CAM, metrology, precision engineering.
J. William Shelnutt, Associate Professor; M.S., Air Force Tech, 1966. Total quality systems, teamwork, statistically designed experiments, statistical process control.
Scott Smith, Professor; Ph.D., Florida, 1987. High speed and advanced machining, dynamics.
Stuart Smith, Associate Professor; Ph.D., Warwick (England), 1987. Precision engineering and instrumentation design.
Robert G. Wilhelm, Assistant Professor; Ph.D., Illinois at Urbana-Champaign, 1992. Geometric modeling, tolerance and design theory, manufacturing systems, CAD/CAM, precision engineering.

UNIVERSITY OF PENNSYLVANIA

School of Engineering and Applied Science

Programs of Study	The School of Engineering and Applied Science (SEAS) offers M.S.E. and Ph.D. programs through seven graduate groups: Bioengineering, Chemical Engineering, Computer and Information Science, Electrical Engineering, Materials Science and Engineering, Mechanical Engineering and Applied Mechanics, and Systems Engineering. The School also offers three interdisciplinary master's programs in biotechnology, telecommunications and networking engineering, and virtual environments. The M.S.E. degree is awarded upon successful completion of 10 course units of work, including a 2-course-unit thesis in some graduate groups. One year of full-time study, including summer course work, is generally required, although part-time study is also possible. The Ph.D. degree is awarded upon successful completion of at least 20 course units of work, a dissertation showing a high capacity for independent research that represents a concrete contribution to human knowledge, and the passing of preliminary and final examinations. Several joint-degree programs are offered between the School of Engineering and Applied Science and other schools within the University, including the School of Arts and Sciences (Master of Biotechnology), Wharton School (M.B.A./M.S.E.), the School of Medicine (M.D./Ph.D.), and the Graduate School of Fine Arts (M.S.E./M.Arch., M.S.E./M.C.P.).
Research Facilities	Research programs in the School of Engineering and Applied Science are necessary components of graduate education. Individual faculty members implement their research in conjunction with the graduate students. Such research may involve experimentation, for which appropriate and well-equipped laboratories are essential, or theoretical analysis and synthesis, in which libraries and computational facilities are essential, or both. Appropriate facilities are described on the reverse side of this page.
Financial Aid	A number of fellowships, assistantships, and scholarships are available on a yearly competitive basis. Provisions of these awards vary; the maximum benefits include payment of tuition and the general and technology fees plus a stipend. No financial aid is offered to part-time students.
Cost of Study	Tuition for the academic year 1999–2000 is $23,670, and there are a general fee of $1546 and a technology fee of $445 for full-time study. For part-time study, the tuition is $2996 per course unit (one course), the general fee is $188, and the technology fee is $60.
Living and Housing Costs	On-campus housing is available for both single and married students. Residences for single students cost $675 and up per month; for a shared-living situation, $510 and up per month; and for a private apartment, $815 and up per month. Housing costs for married couples range from $675 to $1065 per month. There are numerous privately owned apartments in the immediate area.
Student Group	There are approximately 20,000 students at the University, more than 10,000 of whom are enrolled in graduate and professional schools. Of these, 746 are in graduate engineering programs.
Location	The University is located in West Philadelphia, just a few blocks from the heart of the city. Philadelphia is a twentieth-century city with seventeenth-century origins. Renowned museums, concert halls, theaters, and sports arenas provide cultural and recreational outlets for students. Fairmount Park extends through large sections of Philadelphia, occupying both banks of the Schuylkill River. Not far away are the Jersey shore to the east, Pennsylvania Dutch country to the west, and the Poconos to the north. Equidistant from New York City and Washington, D.C., the city of Philadelphia is a patchwork of distinctive neighborhoods ranging from Colonial Society Hill to Chinatown.
The School	The School of Engineering and Applied Science has a distinguished reputation for the quality of its programs. Its alumni have achieved international distinction in research, higher education, management, entrepreneurship and industrial development, and government service. Its faculty leads a research program that is at the forefront of modern technology and has made major contributions in a wide variety of fields.
Applying	Candidates may apply for admission by submitting an application to the address below or may apply directly through an online express application system (http://sentry.isc.upenn.edu/ws/expressap). Ph.D. applications for fall matriculation must be received by January 2 to ensure consideration for financial aid. Admission is based on the student's past record as well as on letters of recommendation. Scores on the Graduate Record Examinations are required. All students whose native language is not English must arrange to take the Test of English as a Foreign Language (TOEFL) prior to making application; the minimum score accepted is 600.
Correspondence and Information	For additional information about programs, applications, and admissions: Office of Graduate Admissions School of Engineering and Applied Science Towne Building University of Pennsylvania Philadelphia, Pennsylvania 19104-6391 Telephone: 215-898-4542 E-mail: engadmis@seas.upenn.edu

University of Pennsylvania

AREAS OF RESEARCH

Bioengineering. The nation's first Ph.D. in bioengineering was granted at the University of Pennsylvania, and today the department consists of 12 primary faculty members and more than 60 secondary and associated faculty members. The Bioengineering Ph.D. Program is designed to train individuals for academic, government, or industrial research careers. Research interests include cellular biomechanics, bioactive biomaterials, cell and tissue engineering, neuroengineering, orthopedic bioengineering, neurorehabilitation, respiratory mechanics and transport, molecular and cellular aspects of bioengineering, and biomedical imaging. Penn's interdisciplinary research training laboratories are in the Department of Bioengineering, the School of Engineering and Applied Science, and the new Institute for Medicine and Engineering; the University's medical, dental, and veterinary schools; and four research-oriented hospitals, all of which are located on campus. The professional master's in bioengineering prepares students for leadership in the rapidly developing fields of industrial, entrepreneurial, and governmental biomedical engineering. The program is one of the first to recognize the growing need for biomedical engineering professionals whose expertise is in creating and managing new technology. Specially designed professional business courses set this program apart from most other biomedical engineering master's programs. Courses provide students with a full understanding of the process involved in taking a new technology from the laboratory to the biomedical marketplace and a complete awareness of the procedures and strategies needed to develop a company around that new technology. (World Wide Web: http://www.seas.upenn.edu/be)

Biotechnology. The Master of Biotechnology degree is offered jointly by the School of Arts and Sciences and the School of Engineering and Applied Science. This interdisciplinary program prepares both full- and part-time students for productive and creative careers in the biotechnology and pharmaceutical industries. Students can specialize in basic biotechnology, engineering biotechnology, or computational biology/bioinformatics. These tracks, in combination with core courses, ensure that the students get a uniquely broad exposure to the entire field of biotechnology. (World Wide Web: http://www.seas.upenn.edu/biotech)

Chemical Engineering. The department was one of the first in the United States to offer a degree in chemical engineering. Courses and research programs are offered in applied mathematics, adsorption, biochemical and biomedical engineering, computer-aided design, transport and interfacial phenomena, thermodynamics, polymer engineering, semiconductor and ceramic materials processing, reaction kinetics, catalysis, artificial intelligence, and process control. Many research projects are collaborative and take advantage of other strong programs in the University. Ongoing research includes joint projects with faculty from the medical school and Wistar Institute, from the Department of Biology, from the Department of Chemistry, from Computer Science and Engineering, and from Materials Science and Engineering. (World Wide Web: http://www.seas.upenn.edu/ohome)

Computer and Information Science. The program is intended for students with undergraduate training in any one of a broad range of disciplines related to modern information processing, including engineering, computer science, mathematics, physical sciences, business management science, philosophy (logic), linguistics, and psychology. Major current research areas are computer architecture, computational linguistics, database and information systems, operating systems, programming languages, artificial intelligence, graphics and image processing, theory of computation, algorithms, robotics, computer vision, massively parallel systems, computational biology, medical informatics, and networking. (World Wide Web: http://www.cis.upenn.edu)

Electrical Engineering. The graduate program in electrical engineering encompasses the physical, device, telecommunications, and signal processing aspects of electrical engineering. The course work and research are coordinated into four areas of specialization: electromagnetic field phenomena, including diffraction scattering, propagation, guided waves, electromagnetic wave interaction with complex media (such as chiral, bianisotropic, fractal, and knotted media), imaging sciences, mathematical aspects of electromagnetic theory (such as fractional calculus and symmetry in electromagnetism), remote sensing of the environment, microwave and long-wavelength holographic imaging, and electrooptics; nonlinear dynamics, neurodynamics, and communication theory, spectrum estimation and adaptive techniques, image processing, statistical techniques, digital signal processing, and neural networks; telecommunications, including packet and circuit switching, IP and ATM networks and protocols, network design, wireless communication, performance modeling in particular quality-of-services issues, communication architectures, and protocols; and solid-state electronics, including integrated sensors, interface phenomena, mixed analog and digital integrated-circuit and other devices, and electronic properties of materials and their applications. The department also offers a Professional Master of Science Program that involves an internship in industry. (World Wide Web: http://www.seas.upenn.edu/ee)

Materials Science and Engineering. The department conducts an extensive program of graduate education and research aimed at understanding the physical origins of the behavior of ceramics, polymers, metals, and alloys in electronic, structural, magnetic, and interfacial applications. Students have access to a broad range of state-of-the-art instrumentation in the department and the Laboratory for Research on the Structure of Matter (LRSM), which is housed in the same building. The LRSM is one of the largest NSF-supported Materials Research Science and Engineering Centers in the country and includes central facilities for surface studies, ion scattering, electron microscopy, X-ray diffraction, computer simulation, mechanical testing, and materials synthesis and processing. Access to synchrotron radiation (X-ray and UV) and neutron-scattering facilities is also available at nearby National Labs. Research within the department can be grouped under four general headings: surfaces and interfaces (polymer-polymer, metal-ceramic, and grain boundaries in metallic materials), complex materials (carbon-based nanotubes, copolymers, intermetallic alloys, and nanomaterials), failure mechanisms (plastic deformation, fatigue, embrittlement, corrosion, and predictive modeling), and novel electronic ceramics (ferroelectrics, microwave materials, batteries and fuel cells, superconductors, and catalysts). (World Wide Web: http://www.seas.upenn.edu/mse)

Mechanical Engineering and Applied Mechanics. The research in the department combines applications and theory. It is often interdisciplinary in nature and is done in collaboration with material sciences, computer sciences, electrical engineering, and the medical school. The areas of focus are thermal and fluid sciences, mechanics of materials, mechanical systems, and biomechanics. Research in thermal fluids focuses on energy conversion, combustion, water desalination, microelectronic device fabrication and cooling, inorganic and organic (macromolecular) crystal growth, active control of flow patterns, transport processes associated with mesodevices and microdevices and with sensors, material processing, multiphase flows, and computational fluid dynamics. The research in mechanics of materials focuses on crystal plasticity, effective properties of nonlinear composites, intermetallic compounds, localization studies, metal-forming processes, interfacial fracture, fatigue and high-temperature fracture, soft material, phase transitions in thermoelastic solids, and cell mechanics. Research in mechanical systems focuses on robotics, dynamics, controls, design, mechanisms, optimization, virtual and rapid prototyping, and microelectromechanical systems (MEMS). Biomechanics research spans scales from the tissue level through the molecular, with major efforts in cell mechanics, tendon and ligament properties, biomolecular simulation, and gravity effects on cells and tissues. (World Wide Web: http://www.seas.upenn.edu/meam)

Systems Engineering. The department's programs focus on design and operation of large-scale systems. The core of the systems engineering discipline consists of the tools and methods necessary to deal quantitatively and qualitatively with large-scale systems. Research activities are directed toward development of the core systems tools and methodology and toward integrated application of these tools and methodology in several interdisciplinary fields. The core systems concepts, tools, and methodology areas include system design, system integration, modeling and simulation, network analysis, optimization, human factors, system evaluation, reliability, and control and automation. Application areas include computer systems, environmental and resources systems, logistics systems, manufacturing systems, structural and construction systems, telecommunications systems, and transportation systems. Most of the research activities are interdisciplinary in nature, and systems engineering faculty and students typically interact or collaborate with professors and students from other departments of SEAS, the School of Arts and Sciences, and the Wharton School. (World Wide Web: http://www.seas.upenn.edu/sys)

Telecommunications. This unique interdisciplinary program draws its faculty and courses from several SEAS departments—computer and information science, electrical engineering, and systems engineering—and from the Wharton School of Business. The program's interdisciplinary approach gives both full- and part-time students the flexibility to tailor the curriculum to their specific interests, backgrounds, and career goals. Students typically have an undergraduate degree in engineering, but students with other degrees will be considered. The program's three required telecommunication courses cover the theory and practice of modern data and voice networking and future broadband integrated networking. Electives address the increasingly complex demands placed on current and future telecommunications managers. Such electives range from computer operating systems and digital signal processing to engineering economics and queuing theory. (World Wide Web: http://www.seas.upenn.edu/profprog/tcom)

Virtual Environments. The Master of Science in Engineering in virtual environments is a ten-course-unit program in computer science and related technology that focuses on the digital media of computer graphics, visualization, modeling, and virtual reality. Such synthetic imagery has become an integral part of human experience through 3-D images, games, movie special effects, architectural walk-throughs, educational software, Internet-based virtual communities, computer-aided design and engineering, event reconstructions, entertainment, and real-time simulations. (World Wide Web: http://www.cis.upenn.edu)

UNIVERSITY OF PITTSBURGH

School of Engineering

Programs of Study

The School offers the M.S. and Ph.D. in chemical engineering, civil engineering, electrical engineering, industrial engineering, materials science and engineering, mechanical engineering, and metallurgical engineering and an interdisciplinary Ph.D. in bioengineering in collaboration with the School of Medicine. M.S. degree programs are offered in bioengineering, manufacturing systems engineering, and petroleum engineering. Dual-degree programs are offered in chemical and petroleum engineering and electrical engineering in collaboration with the Department of Mathematics and Statistics. Interdepartmental programs in biomedical engineering and bioengineering, ceramics, engineering management, environmental engineering, manufacturing systems engineering, operations research, polymer engineering, systems management, and other specialized studies can be structured. The School offers a professional master's degree program requiring 30 credits with no comprehensive examination or thesis requirement. For students who are pursuing research and will be seeking a Ph.D. degree, a comprehensive examination and a thesis (or acceptable nonthesis option) are required. Course requirements for the Ph.D. degree are determined by the Graduate Faculty and vary with programs; the total number of credits covering subjects in the major and cognate areas is typically 55 to 65. The minimum requirement is 72 credits beyond the bachelor's degree. A dissertation, graduate preliminary examinations, comprehensive examinations, and thesis/dissertation oral examinations are required. Additional requirements vary with departments.

Research Facilities

The Michael L. Benedum Hall of Engineering is one of the country's most modern and best-equipped engineering buildings, offering a variety of state-of-the-art laboratories. The off-campus facilities of the University's Applied Research Center are also available. Engineering students have access to the mainframe and personal computers operated by Computing and Information Systems (CIS). The CIS academic mainframe computing facilities include the VAX family of computers. The VAXcluster consists of two VAX (6420 and 6540) and two Alpha 3000 processors. The UNIX environment is built from four DECsystem 5000 processors. CIS also offers a Sun UNIX service consisting of two SPARCserver 10 processors. Each of these systems supports a wide range of software applications, including database management systems, graphic analysis programs, mathematical and statistical program libraries, simulation packages, tape utilities, file transfer facilities, and text processors. Academic computing timesharing services are provided to students and faculty members. In addition to supporting and providing access to the University's computing systems, CIS supports and provides access to the Pittsburgh Supercomputing Center, a joint venture of the University of Pittsburgh, Carnegie Mellon University, and Westinghouse Electric Corporation. The center's CRAY Y-MP supercomputer is one of the fastest processors in the world. CIS maintains and supports eleven public computing labs on the campus, three of which are in Benedum Hall. Of particular interest is the Advanced Graphics Laboratory (one of the three in Benedum Hall), which offers advanced computer graphics systems and services. Equipment includes a networked cluster of Sun Workstations, IBM workstations for video digitizing, image analysis, and computer art/paint applications. The George M. Bevier Engineering Library holds 65,262 volumes, 83,504 microforms, and 1,014 journals covering most engineering topics. Computerized database searching by subject or author of such files as the Engineering Index, Chemical Abstracts, Science Abstracts, and Metals Abstracts is available. The library is part of the University Library System, which maintains collections totaling more than 7.3 million volumes, including microtext. The University is a member of the Center for Research Libraries, through which uncommon materials on many subjects are available via interlibrary loan. Through the University's membership in the Pittsburgh Regional Library Center, cooperative arrangements have been developed between the University and neighboring institutions, including Carnegie Mellon and Duquesne universities, for borrowing and research privileges.

Financial Aid

Teaching and research assistantships are available; stipends ranged from $1286 to $1338 per month in 1998–99. Graduate scholarships and fellowships, supported by the School, alumni, and industrial and foundation gifts, are awarded on a competitive basis.

Cost of Study

In 1998–99, resident tuition was $4602 per term (spring and fall) for full-time (9–15 credits) students and $442 per credit for part-time (1–8 credits). Nonresident tuition was $9476 per term for full-time and $902 per credit for part-time. Students registered during the summer are billed on a per-credit basis regardless of the number of credits taken. Books, supplies, and incidental fees are additional.

Living and Housing Costs

For 1998–99, living expenses for students attending the University of Pittsburgh were approximately $850 per month for housing, meals, and miscellaneous expenses. The cost of books is approximately $650 per term.

Student Group

More than 26,000 students are enrolled at the Pittsburgh campus; of these, 7,364 are pursuing graduate degrees. The School of Engineering's graduate enrollment is 576.

Location

Pittsburgh is a growing cultural center and the heart of one of the world's large industrial regions, with more than fifty industrial research laboratories and the headquarters of some of the nation's major corporations located in the area. Culturally and socially, Pittsburgh offers the world-renowned Pittsburgh Symphony, the opera, and Carnegie Institute with its library, museum, art galleries, and music hall. The 500-acre Schenley Park, adjacent to the University's main campus, provides opportunities for outdoor recreation.

The University

Founded in 1787, the University of Pittsburgh is the oldest institution of higher education west of the Allegheny Mountains. It is an independent, state-related, nonsectarian, coeducational institution offering a variety of undergraduate, graduate, and adult education programs. The main campus occupies fifty-five buildings, among them the forty-two-story Cathedral of Learning. Graduate degrees have been conferred since 1836, and the first Ph.D. program was developed in 1884. The first M.S. degree in engineering was awarded in 1919; the first Ph.D. in engineering, in 1929. The University was elected in 1974 to the Association of American Universities, an organization of the fifty-six most respected graduate and research institutions.

Applying

To apply for the fall term, an applicant should file the application and complete credentials by August 1; for the spring term, by December 1; and for the summer term, by April 1. Applicants for financial aid must file five months prior to these dates. The TOEFL is required of all applicants whose native language is not English.

Correspondence and Information

Office of Administration
School of Engineering
253 Benedum Hall
University of Pittsburgh
Pittsburgh, Pennsylvania 15261

University of Pittsburgh

DEPARTMENT CHAIRMEN/PROGRAM DIRECTORS AND AREAS OF FACULTY RESEARCH

Bioengineering. Professor Jerome S. Schultz, Chairman. The aim of the program is to prepare individuals for careers in academic and industrial research. Students are directed toward graduate engineering education and research, with particular emphasis on the Ph.D. This program enlists faculty members from the School of Engineering, the School of Medicine, and the School of Health and Rehabilitation Sciences. A large number of the engineering faculty members associated with the program also have appointments and research laboratories within the School of Medicine, which provides an unusual opportunity for students to develop research projects in the basic and directly associated departments. The scope of the program includes the application of engineering principles, methods, and technology to three broad areas: fundamentals of molecular and cellular biology; development of instrumentation, products, processes, and systems for potential industrial application; and improvement of health-care delivery and assist systems. The bioengineering faculty is applying various forms of engineering principles, technology, and methodology to a broad variety of medical and life science problems. The major research areas include image analysis, heart assist pumps, blood oxygenators, rehabilitation engineering, tissue biomaterial interactions, biomechanics, prosthetic devices, ergonomics, cardiac dynamics, Doppler ultrasound, NMR imaging and spectroscopy, gene therapy, bioseparations, enzyme engineering, biosensors, and metabolic engineering. (WWW: http://www.engrng.pitt.edu/bioeng/)

Chemical and Petroleum Engineering. Professor Alan J. Russell, Chairman. The department offers the Master of Science degree in petroleum engineering and the Master of Science and Doctor of Philosophy degrees in chemical engineering. The department maintains strong research programs in the areas of catalysis; bioengineering; energy and environmental engineering; supercritical fluids; polymerization and polymer processing; molecular modeling and simulation; thermodynamics; gas-solid, gas-liquid, and gas-liquid-solid flow processes; and advanced separation processes. Research thrusts have been established in the areas of biochemical and biomedical engineering, in collaboration with the School of Medicine; and environmental engineering, in collaboration with the Civil and Environmental Engineering Department and the Graduate School of Public Health. (WWW: http://www.engrng.pitt.edu/~chewww/)

Civil and Environmental Engineering. Professor Rafael G. Quimpo, Chairman. More than any other of the engineering professions, civil engineering is intimately involved with public, private, and governmental activities that have an impact on the quality of life. Whether dealing with the condition of the infrastructure or assessing the needs for clean air and water resources, civil engineering exerts a primary influence through all levels of society. The department has particular strengths in the areas of structural engineering, environmental engineering, and hydrology and water resources, with strong support in geotechnical engineering, geomechanics, and transportation planning and analysis. One of the two thrust areas of the department is environmental engineering, which is enhanced by cooperative efforts with the Center for Hazardous Materials and Research and the Biotechnology Center of the University of Pittsburgh. These activities are complemented by the Center for Environment and Energy, which is housed in the department. Close involvement with the Graduate School of Public Health and the School of Medicine further expands the scope of the environmental and biomechanics/fluid mechanics efforts. The other thrust area is structural engineering, emphasizing bridge and building structures, materials performance, fatigue and fracture, risk analysis, and methods of advanced analysis and design. The department also emphasizes a broad interdisciplinary program in construction management. (WWW: http://www.engrng.pitt.edu/~civwww/index.htm)

Electrical Engineering. Professor Joel Falk, Chairman. Areas of study and research include computer engineering (algorithm development, parallel computer architectures, optical computing, VLSI architectures, computer-aided design for VLSI, microprocessor systems, computer and communication networks); control (control of artificial organs, model reduction and realization of multidimensional and uncertain systems, knowledge-based control, statistical process control); electronics (microelectronics, semiconductor device modeling, analog circuit design, linear and nonlinear optical devices; optoelectronic integrated devices, semiconductor lasers, semiconductor materials and devices); image processing/computer vision (parallel algorithms and architectures for image processing, digital topology, pattern recognition, biomedical image processing); power (electrical transients in power systems, pulse power components and systems, real-time computer control of power systems); and signal processing/communications (knowledge-based signal processing, statistical signal processing, time-frequency, multidimensional system theory, processing of speech signals, spectral estimation, neural networks, communications, optical processing, and biomedical signal processing. Major facilities located in the department include computer vision and pattern recognition laboratory, laser laboratory, optoelectronics laboratory, optical computing systems laboratory, Pittsburgh Integrated Circuits and Analysis (PICA) laboratory, Pitt parallel computer laboratory, network communications laboratory, and applied signal and system analysis laboratory. (WWW: http://ee.pitt.edu/)

Industrial Engineering. Professor Harvey Wolfe, Chairman. The department has three thrust areas: manufacturing systems, operations research, and information systems engineering. Manufacturing systems research emphasizes systems integration with respect to CAD/CAM, vision systems, and robotics. There is also a strong research component in manufacturing design; in particular with respect to biodevices. The Manufacturing Assistance Center is a 34,000-square-foot off-campus laboratory. In the operations research area, research emphasis is in the applied areas of mathematical programming; statistical modeling, including reliability; inventory systems; stochastic systems; and health systems. Information systems engineering research is being carried out in wireless communications, databases, and bar codes, as they apply to warehouse management, ERP, supply chain management, and manufacturing execution systems. The Automatic Data Collection Research Center features the latest bar code, radio frequency, voice, and CCD technology. There is also a strong emphasis on intelligent computing, including neural networks, genetic algorithms, and fuzzy logic. The department offers a strong research program in human factors both in the cognitive sciences with respect to applications in engineering education and in ergonomics in conjunction with the Department of Bioengineering and the School of Medicine. (WWW: http://www.pitt.edu/~pittie/)

Manufacturing Systems Engineering. Professor John H. Manley, Director. This interdisciplinary M.S. program offers courses developed and taught by School of Engineering faculty members. Using video teleconferencing, courses presented in a live classroom setting on the Pittsburgh campus can also be delivered simultaneously to other locations. Areas of study and research include aspects of modern manufacturing from a systems perspective with emphasis on applied research. All manufacturing systems engineering students are required to perform on-site, thesis-level research in manufacturing companies. Research topical areas supported by the faculty include computer-aided manufacturing, agile manufacturing, high-precision robotics, quality and productivity improvement, manufacturing management, materials processing, safety engineering, hazardous materials reduction, design for manufacture and maintenance, manufacturing information systems, machine intelligence, real-time process control, modeling and simulation, and applications of operations research to manufacturing systems.

Materials Science and Engineering. Professor William A. Soffa, Chairman. The department has several outstanding research programs in the areas of physical and process metallurgy, materials degradation, high temperature materials, advanced electronic and magnetic ceramic materials, and composite materials. Synthesis, processing, and microstructure-property relationships form the nucleus of these research programs, many of which are highly interdisciplinary. A number of research programs have significant research interactions with industry. Highly sophisticated instrumentation for microstructure characterization is available. This includes state-of-the-art XRD, TEM, STEM and a field emission SEM, an image analysis system, and an atom probe facility. In addition, facilities also exist for the characterization of electronic and magnetic properties of materials. Excellent research equipment for the determination of mechanical properties of materials is also available.
(WWW: http://www.engrng.pitt.edu/~msewww/mse.html)

Mechanical Engineering. Professor Frederick S. Pettit, Interim Chairman. The department offers a strong integrated program of studies in fluid mechanics, heat transfer, dynamics, vibrations, biomechanics, composite materials, mechanical design, fracture mechanics, thermal systems, acoustics, and modern computational techniques. Specific areas of research include combustion, friction/vibration interaction, composite materials, thermal systems analysis, fracture mechanics, vibration control, smart materials and structures, vibration modeling, integrated product and process development, multiphase flows, plasticity, viscoelasticity, and rotating machinery. Numerous opportunities for Industry-University Partnerships have been and are currently under development in these areas. The department also offers a strong interdisciplinary research program in vascular and musculoskeletal mechanics in conjunction with the Department of Bioengineering and the School of Medicine.
(WWW: http://www.engrng.pitt.edu/~mewww/me.html)

UNIVERSITY OF ROCHESTER

School of Engineering and Applied Sciences

Programs of Study

The School of Engineering and Applied Sciences offers programs leading to M.S. and Ph.D. degrees in biomedical engineering, chemical engineering (M.S. only), electrical and computer engineering, materials science, mechanical engineering, and optics. The M.S. degree requires a minimum of 30 semester hours of graduate credit and may be earned with or without a thesis. Thesis research can involve up to 12 hours of graduate credit. The M.S. option without a thesis may include up to 6 credits of independent study or project work and requires a comprehensive final examination. Programs are open to both full-time and part-time students. A special part-time master's program designed to accelerate the time period required to earn the degree is available for selected employees in local industries. The Ph.D. degree is offered to prepare individuals for careers in research and teaching. The requirements include 90 semester hours of credit beyond the bachelor's degree and at least one academic year of full-time study in residence. A typical academic program is divided between course work and research credits to provide Ph.D. candidates with a broad exposure to their fields of interest, the requisite training for mastery of their area of specialization, and experience in conducting scholarly research. Ph.D. students must pass a preliminary examination and an oral qualifying examination and must present and defend an original thesis that contributes to knowledge in the field.

Research Facilities

The four academic departments of the School are located on the River Campus, the University's main campus. Each department has extensive laboratories with modern equipment for research and instruction. Research centers within the academic units include the Center for Optoelectronics Research, the Center for Superconducting Electronics, the Rochester Center for Biomedical Ultrasound, and the Diagnostic Radiology Research Laboratory. Off-campus facilities include the Laboratory for Laser Energetics, which houses one of the world's most powerful laser systems and conducts controlled thermonuclear fusion and laser physics research, and the Center for Optoelectronics and Imaging, where the research efforts of the Center for Optics Manufacturing and the Center for Electronic Imaging Systems are performed.

The University's library system contains more than 2.75 million volumes, and the Carlson Science and Engineering Library maintains complete collections in the research areas of the School. A campuswide network connects the many computing facilities including numerous Sun file servers and computation servers. The School has more than seventy monochrome and color UNIX workstations. In addition, a variety of microcomputers and related peripherals are available throughout the School.

Financial Aid

Research and teaching assistantships, department fellowships, and other fellowships and scholarships are available. Graduate assistantships provide a basic stipend of $13,000 to $15,000 per year plus a full tuition scholarship. This support includes the summer months. University fellowships for outstanding candidates are also available with stipends in excess of $15,000 for twelve months. Special honors fellowships are also offered for up to $20,000 per year. Nearly all full-time graduate students receive financial aid.

Cost of Study

Tuition for 1998–99 was $672 per credit. The mandatory health fee for 1998–99 was $336. Costs are subject to change for the 1999–2000 academic year.

Living and Housing Costs

University-owned apartments near the River Campus accommodate about 40 percent of the graduate students. This housing ranges from studio apartments to three-bedroom town houses, both furnished and unfurnished. Rents range from about $320 per month for a room adjoining campus to about $630 per month for a furnished two-bedroom, two-bathroom apartment in an on-campus housing complex. Off-campus (private) housing is plentiful. Food and other living costs in Rochester are moderate.

Student Group

The University's total enrollment is about 8,500, including 4,500 full-time undergraduates, 2,600 full-time graduate students, and 1,400 part-time students. There are approximately 245 full-time graduate students and 45 part-time graduate students in the School of Engineering and Applied Sciences.

Student Outcomes

Students completing engineering graduate studies at University of Rochester (UR) have many options available to them. Recent graduates have found positions in a wide range of the industrial and business sectors. Examples of companies where graduates work include Cisco, Xerox, Hewlett-Packard, Allied Chemical, IBM, 3M, Corning, Fujitsu Corporation, Merck Pharmaceutical, and General Electric. Some graduates have gone on to faculty positions in prestigious engineering departments and medical centers around the country, including Rensselaer, Duke, Cornell, and Lehigh. Opportunities for UR graduate students remain abundant.

Location

The Rochester metropolitan area has a population of just under 1 million. Its economy is based primarily on high-technology industries. Eastman Kodak and Xerox are major employers. The area is unusually strong in the quality of its public institutions and cultural life and is the home of the Rochester Philharmonic Orchestra, the Eastman School of Music, the Memorial Art Gallery, the Museum and Science Center, and the International Museum of Photography. Recreational opportunities include boating and fishing on Lake Ontario and the nearby Finger Lakes, skiing in the Bristol Hills, touring the Finger Lakes wineries, and camping and hiking in the Adirondacks (only a 4-hour drive from Rochester).

The University

Founded in 1850, the University of Rochester is an independent, nonsectarian, coeducational institution of higher learning and research. It is one of the nation's smallest distinguished universities. Academic and research programs are conducted by seven schools and colleges on three campuses. Programs ranging from the undergraduate to the postdoctoral level are offered in the humanities, social sciences, natural sciences, and professional fields of business, education, engineering, medicine, music, and nursing.

The River Campus, which includes the College (the School of Engineering and Applied Sciences as well as Arts and Sciences), is situated on the tree-lined bank of the Genesee River about 3 miles south of downtown Rochester. The Medical Center is adjacent to the River Campus; the Eastman School of Music is in the heart of the cultural district of downtown Rochester. The University offers excellent facilities for sports and recreation, including the multimillion-dollar Zornow Sports Center.

Applying

Admission to graduate study normally begins in the fall semester. Applicants seeking financial aid beginning in the fall semester should submit complete applications by the preceding February 1. Students not requesting financial aid should submit applications by August 1 for fall admission and December 1 for spring admission. The application fee is $25. TOEFL scores are required for international students whose native language is not English. GRE scores are strongly recommended for all applicants. Direct contact with the department of interest is encouraged.

Correspondence and Information

School of Engineering and Applied Sciences
Box 270076
University of Rochester
Rochester, New York 14627-0076

Telephone: 716-275-4151
Fax: 716-461-4735
E-mail: graddean@seas.rochester.edu (general information)
bme_gradinfo@seas.rochester.edu (biomedical engineering)
gradinfo@che.rochester.edu (chemical engineering)
gradinfo@ece.rochester.edu (electrical and computer engineering)
matsci_gradinfo@seas.rochester.edu (materials science)
gradinfo@me.rochester.edu (mechanical engineering)
gradinfo@optics.rochester.edu (the Institute of Optics)
World Wide Web: http://www.seas.rochester.edu:8080/

University of Rochester

THE FACULTY AND THEIR RESEARCH

BIOMEDICAL ENGINEERING. Abramowicz, Carstensen, X. Chen, Clark, Dalecki, Fauchet, Fenton, Flessner, Foster, Frame, Gingrich, Gracewski, Haake, Hocking, Jorne, Kutulakos, Lerner, Levinson, Maciunas, Maurer, Moore, Morris, Mottley, Ning, Palmer, Parker, Pentland, Peruccchio, Puzas, Rivers, Rubens, Sarelius, Schwartz, Seidman, Sheu, Shrager, Totterman, Waag, Ward, Waugh (Program Director), Williams, Wu, Yang, Yu, and Zhong.
Molecular cell and tissue engineering: cellular mechanics, mechanics of growth and development of heart muscle and bone, oxygen transport and blood rheology, advanced cell culture systems for growing cells from bone marrow. **Biomedical imaging and optics:** applications and technological developments of ultrasound in clinical settings, medical image processing and functional imaging, optical devices incorporating state-of-the-art optical technology, photodynamic methods for cancer treatment, vision research.

CHEMICAL ENGINEERING. S. H. Chen, Chimowitz, Heist, Jenekhe, Jorne, Notter, Palmer (Chair), Sotirchos, and Wu.
Biochemical and biomedical engineering: molecular biology of biocatalysts, genetic and tissue engineering, fermentation, lung surfactants, interfacial phenomena in biological systems. **Chemical reaction and process dynamics:** complex reaction systems, optimal reactor design, reaction engineering, catalysis, gas-solid reactions, electrochemical reaction engineering, process dynamics, mathematical and computational modeling. **Polymer science:** liquid crystalline polymers; block copolymers; conducting polymers; biomedical polymers; structure-property relationships; electronic, photonic, and optoelectronic polymers; photophysics of polymers; polymer nanocomposites. **Materials synthesis and processing:** chemical vapor deposition, ceramic materials, nanostructured catalysts, nanostructured fuel cells, electronic and optoelectronic materials processing, chemistry of advanced materials, microelectronics processing, porous silicon, liquid crystals, nucleation, aerosols, ultrafine particles. **Interfacial and transport processes:** interfacial dynamics, mixing with chemical reaction, transport in porous media, interfacial transport, supercritical extraction, critical phenomena, statistical mechanics of fluids, air pollution control.

ELECTRICAL AND COMPUTER ENGINEERING. Albicki, Albonesi, Bocko, Carstensen, C. W. Chen, Dalecki, Derefinko, Dwarkadas, Erdem, Everbach, Fauchet (Chair), Feldman, Friedman, Hsiang, Jones, Kadin, Kinnen, Kriss, Levinson, Merriam, Mottley, Ning, Parker, Pentland, Saber, Sobolewski, Tekalp, Titlebaum, Tsybeskov, Waag, and Ward.
Computer systems and microelectronics: logic design, electric power quality, data communications, microprocessor design, high-speed adaptive architectures, multi-processor systems, computer systems performance analysis, digital systems design, VLSI circuits and systems, CMOS circuits, synchronization, clock distribution, pipelining, signal integrity, speed/power/area tradeoffs, WSI, VLSI systems, routing and placement, CAD tools, computer architecture, computer architecture, computer organization, programming languages. **Optoelectronics:** optoelectronics and photonic materials and devices, semiconductor physics, light-emitting porous and nanoscale silicon, femtosecond lasers, optical diagnostics, ultrafast phenomena, superconductivity, electronic noise, solid-state and quantum electronics, nonequilibrium and ultrafast phenomena in condensed matter, nanodevices and nanotechnologies, microwaves and millimeter waves. **Signal processing and biomedical imaging:** biomedical ultrasound, bioelectric phenomena, studies of the interaction of acoustic and electric fields with biological materials, quantitative ultrasonic tissue and materials characterization, anisotropy of ultrasonic parameters, contractile-state-dependence of ultrasonic parameters, ultrasonic contrast agents, medical imaging, Doppler imaging techniques, digital halftoning, 3D/4D medical imaging, image processing, digital image and video processing, object mmodeling and tracking, video compression, video filtering, video editing, content-based video indexing, 3D motion estimation and modeling, multiple-access communications, radar, sonar, signal design and coding psychoacoustics, echolocation, computer languages, ultrasonic scattering, imaging. **Superconductivity and quantum electronics:** quantum electronics, superconducting electronics, quantum noise, quantum computing, electromechanical transducers, quantum electronic circuits and devices, digital analog and hybrid superconducting integrated circuits, radio astronomy instrumentation, quantum computation, superconducting thin films and devices, nonequilibrium effects, high-temperature superconductors, thin-film fabrication and processing, optoelectronic switching, magnetic thin films, infrared detection. **Electromechanics and electrostatics:** electromechanics of particles, microelectromechanics, biological dielectrophoresis, industrial electrostatic hazards and nuisances.

MATERIALS SCIENCE. Basu, Bazan, Bigelow, Bocko, Boyd, Burns (Program Director), S. H. Chen, Eisenberg, Farrar, Fauchet, Funkenbusch, Gage, Gao, Hall, Heist, Houde-Walter, Hsiang, Jacobs, Jenekhe, Jorne, Kadin, Lambropoulos, Li, Moore, Mukamel, Myers, Quesnel, Shapir, Sobolewski, Sotirchos, Teegarden, Teitel, Watson, and Wicks.
Materials synthesis: new nonlinear optical crystals, glasses and polymers, liquid crystal polymers, organic and inorganic polymers made with transition metal catalysts, block and conjugated copolymers, magnetooptical materials, polymer LEDUs, fluorescent materials, nanocomposites, epitaxial semiconductors and optoelectronic devices, III–V and group IV compounds and semiconductors. **Materials processing:** chemical vapor deposition of ceramics, deterministic optical fabrication and manufacturing, microgrinding, mechanical and electrochemical polishing, sputtering and thin film deposition and processing, lithography, powder ceramics and metals, bulk and thin film ceramic superconductors and gradient index materials. **Materials characterization:** X-ray diffraction, scanning and transmission electron microscopy, atomic force microscopy, nanoindentation, near-field and Nomarski optical microscopy, Raman spectroscopy and extensive facilities for optical properties. **Materials testing:** nanoindentation and scratching, AFMUs for indentation, impression creep, fatigue and recovery, acoustic and optical damping, laser damage testing, thermomechanical and fracture toughness facilities. **Analytic and computational studies:** deformation, dislocation mechanics, nucleation, phase transitions, cyclic and statistical thermodynamics of solids, adhesion, fluctuations (especially in superconductors), molecular dynamics, laser physics, fracture mechanics and failure analysis.

MECHANICAL ENGINEERING. Betti, Burns, Clark, Conners, Funkenbusch, Gans, Genberg, Gracewski, Lambropoulos (Chair), Lerner, Li, McCrory, McKinstrie, Meyerhofer, Peruccchio, Quesnel, Ramesh, Ronald, Simon, Thomas, and Waugh.
Biomechanics: elasticity, poroelasticity, material properties, and nonlinear finite-element modeling of the heart; mechanics of cardiac growth and development; biomedical ultrasound; mechanical properties of biomembranes; bone growth and orthopedics. **Computational mechanics:** chaos in mechanical systems, mechanical properties of materials, geometric modeling and adaptive algorithms for finite-element analysis, modeling of crystal growth. **Fluid mechanics:** astrophysical fluid dynamics and magnetohydrodynamics, hydrodynamic lubrication, low-dimension system modeling. **Solid mechanics:** elastic waves in layered media, laser generation of GHz-frequency acoustics, mechanics of bonded interfaces and solder mechanics, thermal and mechanical properties of thin films, expansion and collapse of bubbles, fracture mechanics and failure of biological stones. **Mechanical testing and failure analysis:** fracture and fatigue, mechanical damping, scratch resistance and impression creep, thermomechanical deformation of materials, indentation. **Microstructure characterization:** atomic force microscopy; nanoindentation; optical, scanning, and transmission electron microscopy; surface profilometry; friction and wear; X-ray and electron diffraction. **Materials processing:** laser damage of materials, mechanics and material problems in deterministic microgrinding and optical manufacturing, processing of powder materials, thermal and deformation processing of metals and alloys, crystal growth of semiconductors. **Special materials:** biomedical materials, nonlinear optical materials and superconducting ceramics. **Plasma and laser physics:** high-field atomic physics, inertial confinement fusion, laser-matter interactions, magnetic confinement fusion, plasma-based particle acceleration, short pulse X-ray sources.

THE INSTITUTE OF OPTICS. Agrawal, Boyd, Brown, Eberly, Erdogan, Fauchet, George, Hall (Director), Houde-Walter, Jacobs, Moore, Morris, Seka, Smith, Stroud, Teegarden, Walmsley, Wicks, Williams, and Wolf.
Component technologies: diffractive optics, gradient-index optics, holographic optical elements. **Guided-wave optics:** fiber optics, integrated optics, optical waveguide phenomena, fiber gratings, fiber amplifiers. **Imaging:** diffractive theory, electronic imaging, pattern recognition, Fourier optics. **Lasers:** solid-state lasers, semiconductor lasers, fiber lasers, laser instabilities, high-power lasers, laser fusion, ultrafast laser physics and engineering. **Medical optics:** human visual system, mechanisms of vision, laser-tissue interactions. **Nonlinear optics:** nonlinear interactions, phase conjugation, nonlinear optical materials. **Optical materials:** glasses, III–V and group IV semiconductors, epitaxial growth, nonlinear materials, liquid crystals, materials processing, polishing science, optical thin-film coatings. **Optical system design:** design algorithms, novel optical systems, optical aberration theory, optical systems without symmetry, nonimaging optics. **Photonics:** optoelectronics, optical communications, quantum electronics. **Quantum optics:** resonant interaction of light with matter, Rydberg atoms, electron and atomic wavepackets, multiphoton processes, ultrafast phenomena. **Theoretical foundations:** coherence theory, quantum and classical electrodynamics, propagation of light, statistical optics, radiation theory.

UNIVERSITY OF SOUTH ALABAMA

College of Engineering

Programs of Study

The College of Engineering offers programs leading to the Master of Science degree in chemical, electrical, and mechanical engineering. The Department of Civil Engineering offers graduate-level course work but does not offer a graduate degree program at present. A minimum of 33 semester hours of credit are required to complete the degree requirements, with thesis, project, and course options available. The thesis option requires a minimum of 6 semester hours of research, and the project option requires a minimum of 3 semester hours of project work.

Research Facilities

The College of Engineering is housed on the main University in three buildings on campus that cover more than 88,000 square feet of space, of which approximately 20 percent is devoted to graduate research. Each academic department maintains an extensive portfolio of state-of-the-art research equipment. The facilities are reflective of the strength of individual faculty members. The College maintains a wide range of networked computer facilities, both for teaching and for research. Access to supercomputing is available through the Alabama Supercomputing Authority. Facilities for materials research are particularly extensive, including laser ablation chambers, scanning electron microscopy, X-ray diffractometry, and a variety of other techniques for surface characterization. There is close collaboration with the College of Medicine in a variety of research areas, with a strong focus on health-care engineering. This collaboration also provides access to the extensive research facilities found in the University's three teaching and research hospitals. Other research facilities focus on molecular sieve adsorption, reclamation of scrap tires, environmental remediation and wastewater treatment, transportation engineering in the severe weather coastal environment water resources, computation fluid mechanics, computer engineering, VLSI, electromagnetic radiation, photo optics, advanced process control, and magnetic storage of data.

Financial Aid

The University offers assistantships in all fields of the master's program. Assistantships provide stipends of $4000 to $12,000 for the academic year plus remission of course fees. Assistants are expected to pay other specific fees. Applications for assistantships should be made to the respective Department Chair.

Cost of Study

The basic fees for fall 1999 amounted to $146.50 per semester plus course fees of $125.50 per semester hour; therefore, a student carrying a 9-semester-hour load paid a total of $1276 per semester or a total of $2552 for the academic year. Out-of-state rates were $263.50 per semester hour; therefore an out-of-state student who carried a 9-semester-hour load paid a total of $2518 per semester or a total of $5036 for the academic year.

Living and Housing Costs

The University has extensive housing near the campus for single and married students; rent is about $215 to $265 per month. Single students may live in the dormitories; the cost is about $745 per semester for a suite to about $850 per semester for a two-person efficiency apartment. A board plan is available, with options from $699 to $796 per semester. The cost of living in Mobile is slightly below the national average.

Student Group

The University of South Alabama enrolled approximately 12,600 students in its most recent academic year; 1,800 of them were graduate students. Seventy-one percent of these students are from Alabama, 23 percent come from other states, and 6 percent come from other countries.

Location

Mobile is a thriving port city located on the Gulf of Mexico with a rich international heritage and a population of slightly more than 500,000. The Gulf coast beaches, as well as New Orleans and Atlanta, are easily accessible by automobile. Mobile offers numerous museums and historic sites as well as fine and performing arts, such as ballet, opera, theater, and symphony. The climate in Mobile is semitropical, making outdoor activities enjoyable year-round. The summers are warm and the winters mild, with average temperatures ranging from 51.4°F in January to 81.8°F in August. Mobile has 219 days of sunshine every year (60% of the year).

The University

Founded in 1964, the University includes the Graduate School; the Colleges of Allied Health Professions, Arts and Sciences, Business, Education, Engineering, Medicine, and Nursing; the School of Continuing Education and Special Programs; and the School of Computer and Information Science. There are three specialized departments: Cooperative Education, Military Science, and Aerospace Studies. The University has three major teaching hospitals in Mobile. All facilities are entirely modern.

Applying

The deadline for applications and all supporting documents is August 1 for fall, December 15 for spring, and May 20 for summer. The admission decision is based on the applicant's previous academic record and evidence of the applicant's ability to pursue work on the graduate level.

Correspondence and Information

For admission information:
Director of Admissions
Administration Building 182
University of South Alabama
Mobile, Alabama 36688-0002
Telephone: 800-872-5247 (toll-free)
World Wide Web: http://www.usouthal.edu

For engineering program information:
Dr. Keith Harrison
College of Engineering
Graduate Program Director
EGLB Room 242
Mobile, Alabama 36688-0002
Telephone: 334-460-6160

University of South Alabama

THE FACULTY AND THEIR RESEARCH

Chemical Engineering
W. Crews Askew, Professor; Ph.D., Florida. Pollution control strategies and chemical reactor analysis.
Jagdish C. Dhawan, Professor; Ph.D., Mississippi. Supercritical fluid extraction technology, coal liquefaction.
B. Keith Harrison, Professor and Chair, Ph.D., Missouri–Rolla. Process simulation and thermodynamics.
David T. Hayhurst, Professor and Dean of the College; Ph.D., Worcester Polytechnic. Zeolites, molecular sieves, separation processes.
H. Ted Huddleston, Professor; Ph.D., Case Western Reserve. Expert system control, process modeling and simulation.
Stephen J. Morisani, Assistant Professor; Ph.D., Rhode Island. Mass transfer and unit operations.
Nicholas D. Sylvester, Professor; Ph.D., Carnegie Mellon. Petroleum engineering.
Nelson L. Smith, Visiting Associate Professor; Ph.D., Columbia. Engineering management.

Civil Engineering
Scott M. Douglass, Professor; Ph.D., Drexel. Coastal sediment processes, littoral zone processes and hydraulics.
James H. Lane, Associate Professor, Ph.D., North Carolina State. Severe weather construction techniques.
Joseph M. Olsen, Professor and Chair; Ph.D., MIT. Geotechnical, soil mechanics.
Husam Omar, Associate Professor; Ph.D., Manitoba.
Kevin D. White, Associate Professor; Ph.D., Virginia Tech. Constructed wetlands, wastewater management, environmental remediation, water quality.

Electrical and Computer Engineering
Peter C. Byrne, Associate Professor; Ph.D., University College (Dublin). Digital control systems, nonlinear control systems.
Michael Hamid, Professor; Ph.D., Toronto. Electromagnetics, antennas and propagation, microwave techniques and devices.
Ashok Jumar, Assistant Professor; Ph.D., North Carolina State. Deposition, microstructural, and electrical characterization of superconducting; semiconducting; ferroelectric and diamond thin films; VLSI technology and fabrication; laser and plasma-assisted processing of advanced electronic materials; device applications of superconducting and semiconducting thin films.
Martin Parker, Professor and Chair; Ph.D., Salford (England). Magneto resistance and electromigration.
Arifur Rahman, Associate Professor; Ph.D., Kentucky. Design of electric machines, power electronics, solid-state drives, instrumentation in power systems, magnetic materials.
Vahid Riasati, Assistant Professor; Ph.D., Alabama in Huntsville. Optical pattern recognition.
Adel Sakla, Professor; Ph.D., Illinois. High-performance computer architecture and hardware design language; design of digital, active, and switched capacitor filters; digital image processing.
William Stapleton, Assistant Professor; Ph.D., Alabama. Digital image compression, high-performance computer architecture.
Thomas G. Thomas, Assistant Professor; Ph.D., Alabama in Huntsville. Intelligent instrumentation, neural networks, environmental remediation with molecular sieves, gas chromotography.

Mechanical Engineering
Lanier S. Cauley, Associate Professor; Ph.D., Clemson. Heat transfer and thermal fluid mechanics.
F. Carroll Dougherty, Assistant Professor; Ph.D., Stanford. Computational biofluid mechanics.
Francis M. Donovan, Professor; Ph.D., Purdue. Fluid dynamics and instrumentation.
Ali E. Engin, Professor and Chair; Ph.D., Michigan. Biomechanics, health-care engineering.
J. Marcus Hollis, Adjunct Professor; Ph.D., California, San Diego. Biomechanical orthopedic reconstruction.
Eugene I. Odell Professor; Ph.D., Ohio State. Finite-element analysis, machine design.
Cecil H. Rarnage, Associate Professor; Ph.D., Texas at Austin. Machine design.
Edmund Tsang, Associate Professor; Ph.D., Iowa State. Service learning, pre-engineering education.
Charlie Zheng, Assistant Professor; Ph.D., Old Dominion. Computational mechanics, aircraft wake vortices.

UNIVERSITY OF SOUTH CAROLINA

College of Engineering and Information Technology

Programs of Study	The College offers programs leading to both Master of Science and Doctor of Philosophy degrees in the following fields: chemical engineering, civil and environmental engineering, computer engineering, electrical engineering, and mechanical engineering. In addition, the College offers a Master of Engineering program in the aforementioned disciplines that consists of work at an advanced level in professionally oriented courses. Further details and requirements are outlined in University and department brochures. The College has created a new School of Information Technology, with the intent of establishing initial degree programs in software engineering, network and communications engineering, and information systems engineering. The software engineering program offers Master of Science, Master of Engineering, and Doctor of Philosophy degree programs. The network and communications engineering program offers Master of Science, Master of Engineering, and Doctor of Philosophy degree programs. The information systems engineering program offers Master of Science and Master of Engineering degrees. Students should contact the College for further information. The College also offers many courses, fully creditable to graduate degree programs, off campus in various locations in South Carolina through A Program of Graduate Engineering Education (APOGEE), which utilizes a combination of videotapes and live closed-circuit television lectures featuring student talk-back capability. Faculty members have highly diverse educational and professional backgrounds, and their research projects span a wide range of interests. This research enhances the overall academic program and provides excellent thesis and dissertation topics for students. Three faculty members are chaired professors and 2 others hold University Research Professorship appointments.
Research Facilities	Research is an integral part of the graduate programs of the College. Students and faculty members work together closely on research projects, usually on a one-to-one basis. The College emphasizes original research to help solve major problems in such areas as energy and energy storage, environmental engineering, pollution prevention and waste minimization, high-voltage engineering, geotechnical engineering, engineering materials, electronic materials, photonics, hazardous-waste management, nuclear- and hazardous-waste processing, advanced measurements, fracture mechanics, nondestructive evaluation, computer systems, machine intelligence, robotics, polymers, corrosion, catalysis, separations, materials, process modeling and control, applied mathematics and numerical methods, and electrochemical engineering. The College generated more than $15 million in research in 1997–98—an average of more than $250,000 per faculty member. Two modern buildings offer more than 500,000 square feet of space that is fully wired, with comprehensive computing capabilities. More than 100 modern research laboratories offer support for an aggressive research enterprise. The University library was recently recognized as one of the top fifty libraries in the U.S., with more than 3 million volumes. The College conducts extensive research in the following centers and laboratories: the Center for Electrochemical Engineering, the Center for Information Technology, the Center for Mechanics of Materials and Non-destructive Evaluation, the Filtration Research Engineering Demonstration, the Microelectronics and Photonics Laboratory, and the Virtual Test Bed Project.
Financial Aid	Financial aid for graduate students in the College is available in the form of full and partial fellowships and teaching and research assistantships. Assistantships are awarded through individual departments.
Cost of Study	In 1999–2000, in-state full-time graduate students not holding assistantships pay $1947 per semester; out-of-state tuition is $4057. Graduate students who hold assistantships of twenty hours per week pay a flat rate of $710 per semester. Fees may change without notice.
Living and Housing Costs	Most rooms in University residence halls are designed for double occupancy; a few rooms accommodate three. Single-occupancy rooms are available at higher rates. Charges for summer sessions are lower. Monthly charges for family housing range from $385 to $616. An initial deposit of $100 and an application fee of $20 are required.
Student Group	The Columbia campus of the University of South Carolina system enrolled 25,250 students in fall 1998. Undergraduates numbered 15,907, while graduate and professional students totaled 9,343. Seventy-seven percent of the students are from South Carolina, with all other states and 116 other countries represented. Members of minority groups make up 23 percent of the student body.
Location	Columbia, South Carolina, is one of the fastest-growing metropolitan areas in the Southeast and currently has a population of 475,000. Columbia is served by several major airlines, major train and bus lines, and three interstate highways. There are outstanding cultural and recreational opportunities in the area, including intercollegiate sports, semiprofessional baseball, such outdoor activities as hunting and boating, and programs offered by various cultural organizations. The famous South Carolina beaches and the beautiful Blue Ridge Mountains are only a few hours away. The average temperature in the winter is in the mid-40s(°F), with the summertime average in the upper 80s.
The University and The College	Founded in 1801, the University of South Carolina has grown to include eight regional campuses across the state and offers bachelor's degrees in 123 fields through fifteen colleges on the Columbia campus. Master's degrees are offered in 182 fields, and doctoral degrees are offered in sixty-one fields. The University houses a medical school, law school, and pharmacy college. The College of Engineering and Information Technology celebrated its 100th anniversary in 1994. It currently offers both master's and doctoral degrees in five disciplines (chemical, civil/environmental, computer, electrical, and mechanical engineering) and is planning to add programs of study in software engineering, network and communications engineering, and information systems engineering. The current faculty of 62 is scheduled to grow to 125 as the new programs of study are established. More than 100 master's degrees and seven doctoral degrees were awarded in 1997–98.
Applying	To be admitted to the Graduate School, a degree-seeking student must have a bachelor's degree from an accredited engineering program, with satisfactory scholastic standing. Students from nonaccredited undergraduate engineering programs must have a satisfactory score on the General Test of the Graduate Record Examinations. Other requirements are imposed by individual departments in the College. These are specified in the admission packet mailed to prospective students at the time of application. In addition, all applicants whose native language is not English must submit Test of English as a Foreign Language (TOEFL) scores. Applications should be received by early March for fall admission, mid-September for spring admission, and early February for summer admission.
Correspondence and Information	Office of Graduate Studies College of Engineering and Information Technology University of South Carolina Columbia, South Carolina 29208 Telephone: 803-777-4177 E-mail: info@engr.sc.edu World Wide Web: http://www.engr.sc.edu

University of South Carolina

THE FACULTY AND RESEARCH AREAS

Craig A. Rogers, Dean; Ph.D., Virginia Tech.
Joseph H. Gibbons, Associate Dean for Academics; Ph.D., Pittsburgh
William F. Ranson, Associate Dean for Extended Learning; Ph.D., Illinois.
Michael M. Reischman, Associate Dean for Research and Faculty Development; Ph.D., Oklahoma State.

Chemical Engineering
Michael D. Amiridis, Assistant Professor; Ph.D., Wisconsin. Perla Balbuena, Assistant Professor; Ph.D., Texas. Francis A. Gadala-Maria, Associate Professor; Ph.D., Stanford. Joseph H. Gibbons, Professor; Ph.D., Pittsburgh. Karlene Kosanovich, Assistant Professor; Ph.D., Notre Dame. Michael A. Matthews, Associate Professor; Ph.D., Texas A&M. Thanasis Papathanasiou, Associate Professor; Ph.D., McGill. Harry J. Ploehn, Associate Professor and Graduate Director; Ph.D., Princeton. Branko N. Popov, Research Professor; Ph.D., Zagreb. James A. Ritter, Assistant Professor; Ph.D., SUNY at Buffalo. Thomas G. Stanford, Assistant Professor; Ph.D., Michigan. Vincent Van Brunt, Professor; Ph.D., Tennessee. John W. Van Zee, Associate Professor; Ph.D., Texas A&M. John W. Weidner, Associate Professor; Ph.D., North Carolina State. Ralph E. White, Professor and Chair; Ph.D., Berkeley.

Research areas include adsorption cycles; pressure swing adsorption; adsorption equilibria; adsorber dynamics; diffusion and mass transfer in porous media; heat transfer; fluid dynamics; fluid separations modeling; cross-flow filtration; extraction chemistry; isotopic separations; electrochemical engineering; electrodeposition of new materials; corrosion science and engineering; battery and fuel cell design; electrochemical sensors, waste treatment, and reactor design; rheology; composite materials; polymer processing; sol-gel processing and materials; supercritical fluid technology; phase equilibria and complex mixture thermodynamics; kinetics and reaction engineering; catalysis; photocatalysis; photochemical reactor design; environmental engineering; environmental remediation; environmentally benign manufacturing; nuclear- and hazardous-waste treatment; process safety; process modeling and control; neural networks; and applied mathematics and numerical methods.

Civil and Environmental Engineering
Ronald L. Baus, Professor and Graduate Director; Ph.D., Penn State. J. Hugh Bradburn, Associate Professor; Ph.D., North Carolina State. Hanif Chaudhry, Professor and Chair; Ph.D., British Columbia. Adrienne T. Cooper, Assistant Professor; Ph.D., Florida. John Dickerson, Associate Professor; Ph.D., Caltech. Joseph Raymond Flora, Assistant Professor; Ph.D., Cincinnati. Sarah L. Gassman, Assistant Professor; Ph.D., Northwestern. Molly M. Gribb, Assistant Professor; Ph.D., Wisconsin–Milwaukee. Kent Harries, Assistant Professor; Ph.D., McGill. Jasim Imran, Assistant Professor; Ph.D., Minnesota. Anthony S. McAnally, Associate Professor; Ph.D., Auburn. Michael E. Meadows, Associate Professor; Ph.D., Tennessee. Michael F. Petrou, Assistant Professor; Ph.D., Case Western Reserve. Charles Pierce, Assistant Professor; Ph.D., Northwestern. Richard P. Ray, Associate Professor; Ph.D., Michigan. J. David Waugh, Professor; M.S., Yale.

Areas of research activity include pavement management systems, civil engineering materials, computer-controlled laboratory testing, soil dynamics, characterization of soil physical and hydraulic properties in the unsaturated zone, application of neural networks for parameter estimation in multivariate functions, contaminated soil and groundwater remediation, in situ measurement of soil and hydraulic parameters, lead and copper corrosion in water systems, electrolytically enhanced biological degradation, stormwater runoff and quality management, water treatment and wastewater treatment, hazardous wastes, finite-element analysis, boundary-element analysis, nondestructive testing of bridges, structural modeling, nonlinear structural analysis, statics and dynamics of cable systems, and soil-structure interaction.

Electrical and Computer Engineering
Benjamin Beker, Associate Professor; Ph.D., Illinois. Ronald D. Bonnell, Professor; M.S.M.E., Kentucky. John B. Bowles, Associate Professor; Ph.D., Rutgers. Charles W. Brice III, Associate Professor; Ph.D., Georgia Tech. George J. Cokkinides, Associate Professor; Ph.D., Georgia Tech. Roger Dougal, Professor; Ph.D., Texas Tech. Edward W. Ernst, Allied Signal Professor; Ph.D., Illinois. Jerry L. Hudgins, Associate Professor and Interim Chair; Ph.D., Texas Tech. Michael N. Huhns, Professor; Ph.D., USC. Asif Khan, Professor; Ph.D., MIT. Hideaki Kobayashi, Associate Professor; Ph.D., Waseda (Tokyo). Robert O. Pettus, Professor; Ph.D., Auburn. Enrico Santi, Assistant Professor; Ph.D., Caltech. Ted L. Simpson, Professor; Ph.D., Harvard. Larry M. Stephens, Professor; Ph.D., Johns Hopkins. Tangali S. Sudarshan, Professor and Graduate Director; Ph.D., Waterloo. Juan E. Vargas, Associate Professor; Ph.D., Vanderbilt. Jose M. Vidal, Assistant Professor; Ph.D., Michigan.

Research efforts in the department are broadly classified into six categories: systems analysis (digital signal processing, reliability, electric power, electric machinery and drives), VLSI design (ASIC design, CAD tools, VLSI architecture, design methodology, silicon compilation), artificial intelligence (expert systems, distributed AI, machine learning, neural networks, object-oriented programming, intelligent control, knowledge-based management systems, intelligent tutoring systems), robotics and intelligent systems (machine vision, image processing, signal processing, robot control, robot sensing, industrial inspection), electromagnetics (modern antenna and microwave integrated circuits, anisotropic and inhomogeneous media, scattering, radiation), and physical electronics (pulsed power, surface flashover, high-voltage insulation, high field effects in gap wideband semiconductors and dielectric materials, gaseous electronics, electro-optics, power electronic devices, device physics, programming environments).

Mechanical Engineering
Sarah Baxter, Assistant Professor; Ph.D., Virginia. Y. J. Chao, Professor; Ph.D., Illinois. Xiaomin Deng, Associate Professor; Ph.D., Caltech. Sandip Dutta, Assistant Professor; Ph.D., Texas A&M. Victor Giurgiutiu, Associate Professor; Ph.D., London. D. A. Keating, Associate Professor; M.S., Dayton. Jamil Khan, Associate Professor; Ph.D., Clemson. Jed S. Lyons, Associate Professor; Ph.D., Georgia Tech. Stephen R. McNeill, Associate Professor and Graduate Director; Ph.D., South Carolina. Jeff Morehouse, Associate Professor; Ph.D., Auburn. Walter Hamilton Peters, Professor; Ph.D., Virginia Tech. William Ranson, Tamper Chair Professor; Ph.D., Illinois. Anthony P. Reynolds, Assistant Professor; Ph.D., Virginia. Curtis A. Rhodes, Professor; Ph.D., Carnegie Mellon. David N. Rocheleau, Assistant Professor; Ph.D., Florida. Craig A. Rogers, Professor; Ph.D., Virginia Tech. Michael A. Sutton, Professor and Interim Chair; Ph.D., Illinois.

Areas of current research activity include computer vision in motion analysis; nondestructive analysis; combustion and solar heating; solidification; bioreactors; thermodynamics; heat transfer; compressible fluid flow; analysis of pressure vessels and piping; fracture mechanics; theoretical and experimental mechanics; laser-speckle interferometry; coherent optics in experimental mechanics; 3-D vision systems; mechanical testing of electronic components; robotics; composite processing; microstructural characterization of metallic and polymeric components through SEM, TEM, and optical microscopic evaluation; and computer vision methods for noncontacting strain and deformation measurements in general surfaces.

UNIVERSITY OF SOUTHERN CALIFORNIA

School of Engineering

Programs of Study

The School of Engineering—ranked twelfth in the nation in graduate programs in the current *U.S. News & World Report* and sixth nationally in active faculty who are members of the National Academy of Engineering—offers graduate programs leading to the M.S., Engineer, and Ph.D. degrees in aerospace engineering, applied mechanics, biomedical engineering, chemical engineering, civil engineering, computer engineering, computer science, electrical engineering, engineering management, environmental engineering, industrial and systems engineering, manufacturing engineering, materials engineering, materials science, mechanical engineering, operations research, petroleum engineering, and systems architecture and engineering. The Master of Construction Management and the Master of Engineering in Computer-Aided Engineering are also offered. Areas of emphasis such as biomedical imaging and telemedicine, computer networks, multimedia and creative technologies, robotics and automation, software engineering, and VLSI design are available within certain majors.

Requirements for the master's degree are satisfied by completion of a minimum of 27 units of graduate course work; a thesis may be written as part of this requirement. The Engineer degree requires 30 units of course work beyond the M.S. degree, with concentration in two areas of engineering, and a qualifying examination. The Ph.D. requires 60 units beyond the bachelor's degree and a qualifying examination. Students must demonstrate an ability to make an original contribution to a chosen field through independent study and research. A minimum of one year of research is required after passing the qualifying examination. The student must defend the dissertation before a suitably constituted committee. A minimum grade point average of 3.0 must be earned in all graduate programs.

The USC Interactive Instructional Television System broadcasts regular University courses via a microwave network to companies in the Los Angeles area and in Orange and Ventura counties and via a satellite system to companies outside the Los Angeles area. Students at these companies may take their degree courses through this system without having to commute frequently to campus.

Research Facilities

The School of Engineering occupies thirteen buildings on the west side of the campus at University Park in Los Angeles and has extensive research facilities in areas that include biomedical and neural engineering, composite materials, computer communications, information sciences, multimedia technologies, software engineering, signal and image processing, robotics and intelligent systems, and advanced transportation technologies. The Engineering and Sciences Library in the Seaver Science Center has extensive collections in all areas of engineering. Computer facilities for instruction and research are provided by University Computing Services. The Student Computing Facility, dedicated to instruction, houses three large-scale systems to support student computing needs: a Sun Ultra Enterprise 4000 time-sharing server with twelve processors and 4 GB of memory; a Sun SparcServer 1000 time-sharing server with eight processors; and a Sun Ultra Enterprise 4000 file server with six processors and 300 GB of disk storage. The Research Computing Facility has four powerful systems: a Sun Ultra Enterprise 4000 time-sharing and file server; a Silicon Graphics Origin 2000 time-sharing and file server with thirty-two processors; and IBM SP2 with thirty-three processors; and a Convex Exemplar SPP-2000 with sixteen processors. USC ranks among the top dozen research universities in the nation in available computing power.

Financial Aid

Financial assistance is available in the form of teaching assistantships, research assistantships, fellowships, and work-study programs. The stipend for assistantships ranges from approximately $6200 to $14,000 per academic year; awards also carry tuition remission. Graduate fellowships cover 24 units of tuition and all mandatory fees and carry stipends ranging from $14,000 to $20,000 per academic year for three years. Several local industries participate in work-study programs that provide for part-time study every semester or full-time study during alternate semesters. For information and applications for all assistantships and fellowships, students should contact the relevant departmental office and the Graduate School.

Cost of Study

Tuition is $768 per semester unit in 1999–2000. Full-time students typically take 9–12 units per semester. Mandatory fees are $350 per semester.

Living and Housing Costs

University housing averages $1900 per semester in 1999–2000. Various University meal plans can be purchased for approximately $1550 per semester. Privately owned off-campus apartments can be rented in the immediate vicinity of campus for $500 to $800 per month. Housing is available in the Greater Los Angeles area at all levels of cost.

Student Group

The University of Southern California has a total enrollment of approximately 28,000 students, 46 percent of whom are women. The School of Engineering has 3,900 students, of whom 2,200 are graduate students. More than 30,000 alumni hold engineering degrees.

Location

Only 5 minutes by freeway from the center of Los Angeles, USC is secluded on an extensive, landscaped campus with a quiet, academic atmosphere. Students may take advantage of the broad cultural offerings of a major metropolis or participate in solving the problems of the inner city. Situated less than an hour from both the mountains and the sea, USC offers students a choice of many outdoor diversions.

The University and The School

USC is private, nonsectarian, and coeducational. It ranks in the top 10 among private research universities in the United States in federal research and voluntary support. It is a member of the Association of American Universities, which is a group that represents the top 1 percent of the nation's accredited universities. Attracted by the variety and quality of the University's programs and facilities, faculty and students come from every corner of the globe to engage in teaching, learning, and research. At USC, freedom and responsibility characterize the atmosphere for everyone from the freshman to the postdoctoral student. Convinced that student activities are an important part of an engineer's training in human relationships, the School of Engineering has encouraged organizations that bring students together in professional or social groups. Thirteen professional societies that emphasize the different branches of engineering have student chapters at the University, including the Society of Women Engineers; there are eight student chapters of engineering honor societies.

Applying

Early application is advisable, and applications should be mailed to the Office of Admissions only. To be admitted to full graduate standing, the student must have a bachelor's degree and must have taken at least the General Test of the Graduate Record Examinations. Some degree programs require a GRE Subject Test as well. Students can be admitted conditionally pending receipt of their GRE results; when all records are complete, they may qualify for regular graduate status. The application fee is $55.

Correspondence and Information

Graduate Study Office
School of Engineering
University of Southern California
Los Angeles, California 90089-1454

Telephone: 213-740-6241
E-mail: berti@mizar.usc.edu
World Wide Web: http://www.usc.edu/engineering

University of Southern California

FACULTY CHAIRMEN AND RESEARCH

Aerospace Engineering. Professor R. E. Kaplan, Chairman. Three major areas of graduate study and research: astronautics and planetary science; theoretical, experimental, and computational fluid mechanics; and transportation systems and technologies. Current research involves studies of planetary atmospheres, magnetospheres and the heliosphere, electric propulsion, high-energy gas flows and plasmas, studies of highly nonequilibrium flows using Monte Carlo Direct Simulation, interaction of energetic atmospheric species with gases and surfaces, high-altitude exhaust plume phenomena, hypersonic and rarefied gas dynamics, optical diagnostics of gas flows, flow instabilities, transition to turbulence, turbulent boundary layers and flow control, computational fluid mechanics, mixing layers and jets, stratified wakes, unsteady separation, hydrodynamic wave interactions, chaos in dynamic systems, nonstationary problems related to oscillating wings and to short-takeoff aircraft, low drag vehicles, ground transportation aerodynamics, numerical simulations of engineering and geophysical turbulent flows, rotating and stratified fluids, geophysical and planetary flow phenomena, investigations of micromechanical systems, flow in porous media, the effect of flow on solidification processes, material processing, and satellite systems, power, and operations.

Biomedical Engineering. Professor David D'Argenio, Chairman. Current research is primarily focused on advanced methods of physiological system modeling and simulation with applications to neurophysiological processes, neurophysiological mechanisms of memory and learning, cardiopulmonary control, pharmacokinetics and pharmacodynamics, neuromuscular control, visuomotor coordination, sensory systems, orthopedic research, and biofluid mechanics. Additional research involves emerging medical technologies utilizing lasers and various imaging modalities (MRI, PET, NMI, etc.), spectroscopic techniques, microelectronics, neural prostheses, and artificial neural networks.

Chemical Engineering and Petroleum Engineering. Professor Katherine S. Shing, Chairman. Major areas of research are chemical reaction engineering, involving membrane reactors and heterogeneous catalysis; polymer science and engineering, including the study of polymer composites, irradiation, and morphology of polymers; acoustic emission, micromechanical, and finite-element studies of materials; adhesion and adsorption of polymers on solid surfaces; applied statistical physics and thermodynamics, including computer simulation of molecular fluids, the study of phase behavior, criticality, aggregation, and percolation phenomena. Research in petroleum engineering involves the development of improved methods for predicting gas and oil reservoir performance; fluid flow through porous materials, particularly as it relates to methods for increasing oil recovery; flow regimes for displacement processes, thermal methods, and flow of foams and polymer solutions. Other areas of research include reservoir characterization studies of naturally fractured reservoirs, gas reservoirs and their optimization, and geothermal engineering.

Civil Engineering and Environmental Engineering. Professor L. Carter Wellford, Chairman. Major areas of research are earthquake engineering (wave propagation, measurement and analysis of strong ground motion, and soil-structure interaction), geotechnical engineering (finite-element applications, computational micromechanics, constitutive models for cyclic behavior of soils, liquefaction, seismic design of bridge foundations, and ground remediation), structural engineering and design (performance of structures during earthquakes, suspension bridges, shake table testing, and seismic analysis and design of reinforced concrete structures), performance of structural systems (experimental methods, advanced materials, and advanced simulation methods); structural control and health monitoring (passive, hybrid, and active control); finite-element modeling of structures; probabilistic methods; optimal design of nonlinear structural systems; and computer tomography), transportation engineering (mathematical programming for urban transportation policy alternatives, artificial intelligence applications, transportation system management, and urban transportation planning), water resources and coastal engineering (wave-structure interaction, surface runoff, slurry transport, groundwater basins, backflow preventers, laser-Doppler velocimetry, elliptical jets, and analysis and field investigation of tsunamis). Research in environmental engineering includes combustion research for air pollution reduction; modeling, control, and management of landfills; modeling and simulation of pollutant transport in ground water; and biofilters, absorber columns, and oxidation processes for waste water treatment.

Computer Science. Professor Ellis Horowitz, Chairman. Research is carried out in active areas of software, applications, and theory. Typical but not exclusive fields are programming languages, operating systems, software engineering, program specification, concurrency, artificial intelligence (natural language, knowledge representation, distributed problem solving, and intelligent robots), brain theory and neural computation, database management systems (distributed databases, deductive databases, database semantics, and user interfaces), networks, design and analysis of algorithms and data structures, theory of computation, semantics, complexity, automata, formal languages, robotics, manufacturing, cryptography, computer-aided design, graphics, geometric modeling, and computer animation.

Electrical Engineering. Professor Robert A. Scholtz, Co-Chairman, Systems; Professor Hans H. Kuehl, Co-Chairman, Electrophysics. Research in Systems includes coding theory, communication systems, computer architecture, computer communication networks, computer design, control theory, data compression, design automation, digital computer systems, distributed databases, estimation theory, fuzzy logic, image understanding, integrated multimedia systems, microprocessors, neural nets, nonlinear stability theory, numerical aspects of optimal control, optical computing, optical signal processing, parallel processing, robotics, signal design, signal synchronization, spread spectrum systems, stochastic and adaptive computer systems, system identification, system theory, and VLSI. Areas in Electrophysics include electromagnetics, microwaves, advanced accelerator and radiation sources, pulsed power, antennas, plasma electronics, electric power systems, pollution abatement with pulsed plasmas, applied plasma science, environmental applications of plasma science, pulsed power switching, pulse generators for pulsed power, high-performance active and passive filters, coupling and interconnect parasitics, VLSI, microchips, neural probe arrays, hybrid analog/digital circuits, smart camera circuitry, integrated optoelectronic devices and systems, lasers, photonic devices, optics, photonic device processing, nonlinear optics, phase-conjugate optics, photonic crystals, vertical cavity surface emitting lasers (VCSELs), multidimensional laser arrays, high-bandwidth optical modulators, fiber optics, photorefractive devices, sensors, optical networks, and optoelectronic VLSI systems.

Industrial and Systems Engineering. Professor F. Stan Settles, Chairman. Major areas of research include production systems, intelligent simulation environments, operations research, operations management, transportation planning and intelligent vehicle highway systems, performance measurement and evaluation, rapid prototyping, skill acquisition and transfer, concurrent product and process design, management of technology, and total quality systems.

Materials Science. Professor Florian Mansfeld, Chairman. The basic core courses in this area are quantum mechanics, solid state, chemical thermodynamics, crystals and anisotropy, diffusion and phase equilibria, and dislocation theory and applications; interdisciplinary topics include materials preparation and purification, electronic and photonic materials, metals, ceramics, semiconductor physics, solid-state devices, composite materials, electrochemistry of solids, electrochemical engineering, corrosion, science and technology, mechanical behavior, phase transformations, and electron-optical and cathodoluminescent examination of material.

Mechanical Engineering. Professor R.E. Kaplan, Chairman. Four major areas of graduate courses and research: thermal and fluid systems, mechanical design and manufacturing systems, material behavior, and mechanical systems. Current research includes intelligent manufacturing systems, concurrent engineering, design automation, propellant combustion, droplet burning, turbulent combustion, microgravity phenomena, two-phase fluid systems, bubbly flow dynamics, dynamics of granular media, evaporation of spray drops, thermal-contact resistance, advanced kinematics, dynamics and control of rigid and/or flexible systems, kinematics and control of robots, wave propagation in layered media, long-period microtremors, identification and analysis of mechanical and structural systems, alloy design, and geophysical fluid mechanics.

School of Engineering Research Centers and Institutes.
Asia Pacific Institute for Industrial Leadership; Biomedical Simulations Resource; Center for Advanced Transportation Technologies; Center for Composites Research; Center for Electron Microscopy and Micro-Analysis; Center for Neural Engineering; Center for Photonic Technology; Center for Research on Applied Signal Processing; Center for Research on Environmental Sciences, Policy and Engineering; Center for Software Engineering; Communication Sciences Institute; Foundation for Cross-Connection Control and Hydraulic Research; Information Sciences Institute; Institute for Robotics and Intelligent Systems; Integrated Media Systems Center; International Institute for Innovative Risk Reduction Research on Civil Infrastructure Systems; Signal and Image Processing Institute; USC Engineering Technology Transfer Center; Western Research Application Center.

UNIVERSITY OF SOUTH FLORIDA

College of Engineering

Programs of Study

Graduate programs leading to master's and doctoral degrees are offered in chemical engineering, civil and environmental engineering, computer science and engineering, electrical engineering, engineering management, engineering science, environmental engineering, industrial engineering, and mechanical engineering. These programs cover engineering principles as well as applications at an advanced level and prepare students for careers in such fields as engineering design, operations, research, development, and teaching. Special interdisciplinary study is available through the engineering science graduate program. Advanced work in a major field of specialization may be combined with one or more supporting fields to develop individuals capable of creative, interdisciplinary work in engineering science and applications. Students are assigned to a major professor at the start of studies, and a supervisory committee guides them in their research activities.

Master's degrees require a minimum of two year's full-time study (30–36 semester hours) beyond the bachelor's degree. Programs with or without a thesis can be pursued in all departments. A final written or oral examination is required of each student. A maximum of 8 semester-hour transfer credits or three courses may be accepted as part of a master's degree program.

The Ph.D. degree normally requires three to four years of full-time study and research beyond the bachelor's degree, or two to three beyond the master's. Ordinarily, at least two full years are spent on course work. The remainder of the time is spent on a searching and authoritative investigation of a special area of the candidate's choice, culminating in the dissertation, which demonstrates the student's capacity for considerable original thought, talent for significant research and/or design, and ability to organize and present findings.

Students must achieve and maintain a minimum grade point average of 3.0 (on a 4.0 scale) in all courses taken for graduate credit. Courses with grades below 2.0 will not be accepted toward a graduate degree.

Research Facilities

Graduate students at the University of South Florida have access to extensive computing capabilities. The University Computing Center operates an IBM 3090 vector facility. The College of Engineering facilities include Sun Enterprise servers, Sun and SGI Workstations, and Intel Pentium II- and III-based PC laboratories for student use. The Department of Computer Science and Engineering also operates an Intel Parallel Hypercube and Sun machines. Ethernet and ISN networks provide connections to laboratories and offices. An extensive CAD facility is available for VLSI and mechanical CAD, supported by analytical software with data visualization capability and automated test and data acquisition instrumentation. Laboratory inventories include compound semiconductor materials processing, thin-film and hybrid circuits facilities, extensive analytical and testing equipment, a software engineering laboratory, industrial robots, a Cyberwave Tyangulation Range Scanner, and a well-equipped machine shop. A scanning electron microscope with surface analytical capability is available.

Financial Aid

Teaching and research assistantships, which require 10 to 20 hours per week, are available for well-qualified full-time students. Stipends for graduate assistantships range from $6825 to $25,000 for the nine-month academic year. Graduate assistants also receive a tuition credit worth up to $7000. Supplemental support for the summer term is often available. In addition, several prestigious fellowships are awarded each year to outstanding students.

Cost of Study

The in-state matriculation fee was $142 (1998–99) per semester credit hour; out-of-state students paid an additional tuition fee of $343 per semester credit hour.

Living and Housing Costs

The cost of living in the Tampa Bay area compares favorably with that of most other parts of the United States. Limited facilities for unmarried students are available on campus. A wide range of off-campus housing is available immediately adjacent to the University. Meal plans are available in the University's dining halls. Excluding tuition, books, and transportation, living costs are estimated at $2500 per semester for a single student.

Student Group

There are more than 36,000 students at the University of South Florida, including about 4,500 enrolled in graduate programs. The College of Engineering has an enrollment of about 2,690, including more than 650 graduate students.

Location

The Tampa Bay area, with a population of more than 1.5 million, is rich in recreational, cultural, and athletic activities. The University is located approximately 10 miles from downtown Tampa and 35 miles from Clearwater and St. Petersburg. Symphonies and concerts in diverse musical genres, professional theater, and professional sports events are regularly available. Fine beaches and parks make Tampa Bay an elite resort and vacation spot. The area is undergoing rapid industrial growth and houses divisions of many internationally known high-technology industries.

The University and The College

The University of South Florida opened its doors in 1960 and now has more than 36,000 students on four campuses. The Tampa campus is the largest, with approximately 25,000 students. It comprises the Colleges of Engineering, Business, Education, Fine Arts, Arts and Sciences, Medicine, and Nursing and the School of Public Health. The campus is modern, airy, and attractive and provides a wide range of recreational facilities, including a golf course and a riverfront park.

The College of Engineering maintains close contact with local industry, where many of its students find part-time or full-time employment. The College is housed in two modern buildings.

Applying

All applicants for a master's program in engineering must have a bachelor's degree in an accredited engineering or related program, with a grade point average of 3.0 or higher during the last two years of their undergraduate work and/or a GRE General Test score of 1000 or better (verbal and quantitative scores combined). All applicants must submit valid GRE scores. Individual programs may set higher minimum qualifications. Applicants for the doctoral program are expected to have an academic record that exceeds these minimum requirements. Applications for admission should be received by the University's Office of Graduate Admissions at least three months prior to first expected enrollment. Applications from abroad should be submitted at least five months prior to first expected enrollment.

Correspondence and Information

Associate Dean for Academic Affairs
College of Engineering
University of South Florida
Tampa, Florida 33620

Telephone: 813-974-3782
Fax: 813-974-5092
World Wide Web: http://www.eng.usf.edu

University of South Florida

FACULTY HEADS AND RESEARCH AREAS

Chemical Engineering
L. Garcia-Rubio, Ph.D., Chairman. E-mail: voulgari@eng.usf.edu. The Chemical Engineering Department offers graduate programs leading to M.S., M.E. (nonthesis), and Ph.D. degrees in chemical engineering and engineering science. Faculty research interests cover a broad range of topics in reacting systems, thermodynamics, transport phenomena, system design and control, and fundamentals as well as applications in such domains as biomedical, materials, and environmental. The department has well-equipped laboratories for polymer synthesis and characterization, supercritical fluid technology, biomedical engineering, sensor development and instrumentation, process design and control, unit operations, and phase equilibrium.

Civil and Environmental Engineering
William C. Carpenter, Ph.D., PE, Chairman. E-mail: carpente@eng.usf.edu. The department offers master's and Ph.D. programs in the areas of environmental engineering, including water supply, wastewater treatment, air quality management, solid and hazardous waste management, and environmental risk assessment; geotechnical engineering; materials engineering, including properties of materials, corrosion, and microscopic and macroscopic properties of concrete; structural engineering and structural mechanics; transportation engineering; and water resources engineering, including urban hydrology, computer modeling of aquatic systems, and geographic information systems. Some of the current research projects are development and application of computer models of surface and subsurface water systems, physical modeling of complex hydraulic systems, computer analysis of structures subjected to dynamic loading, experimental stress analysis of bridge models, monitoring systems for corrosion protection and corrosion control, cooling tower operations, reverse osmosis water purification, landfill liners, solid-waste landfill separation and degradation processes, organic soil characterization, computational studies of pavement surfaces, and impact of materials characterization on durability of concrete structures.

Computer Science and Engineering
A. Kandel, Ph.D., Chairman. E-mail: msphd@csee.usf.edu. Master's and Ph.D. programs in computer science and engineering emphasize the design, analysis, and application of computer systems as well as theoretical aspects of computing. Research areas include fault analysis and diagnosis of digital systems, fault-tolerant distributed computing, artificial intelligence and expert systems, neural networks, fuzzy set theory, computer vision and image processing, parallel processing, database and computer networks, VLSI design, robotics, microprogramming and emulation, programming methodology, user interface design, and computer graphics, including color graphics, CAI, and CAD/CAM applications. Research facilities include several SUN-based, research-oriented local area networks; more than 150 SPARC workstations; PCs; Macintoshes; and IBM and DEC workstations. The department's facilities also include a microprocessor laboratory, a hardware/architecture laboratory, and a PC-compatible laboratory for instructional purposes. College of Engineering facilities available to the department include a second network of Sun Workstations. In addition, the University operates a large IBM mainframe, which is available for the department's instructional and research purposes.

Electrical Engineering
E. Stefanakos, Ph.D., PE, Chairman. E-mail: gayla@eng.usf.edu. Faculty research as an integral part of the graduate programs includes externally and internally funded studies related to microelectronic design, fabrication, and testing (VLSI, VHSIC, MIMIC, and ASIC design, microwave and high-frequency analog and digital circuit modeling and testability, interconnection systems, electrical noise, reliability and failure mode studies on Si, III-V and II-VI processing); communications and signal processing (networks, packet switching, digital video and HDTV, ISDN, satellite communications, comm-software, comm-terminals); systems and controls; solid-state material and device processing and characterization; electromagnetics, microwave and millimeter-wave engineering (antennas, devices, systems); CAD and microprocessors; power systems; electrical vehicles; and photovoltaics.

Industrial and Management Systems Engineering
M. Weng, Ph.D., Graduate Director. E-mail: brett@eng.usf.edu. The department offers graduate work in the areas of operations research, applied statistics, engineering management, reliability and maintainability engineering, simulation, engineering economy, production and inventory control, manufacturing systems, and human factors engineering. Research interests cover a broad spectrum of industrial engineering techniques. The department also houses its own Sun Workstation computer laboratory that is used for simulation, artificial intelligence, research work, industrial automation, manufacturing, and human factors laboratories.

Mechanical Engineering
R. V. Dubey, Ph.D., Chairman. E-mail: rahman@eng.usf.edu. Graduate research and course work are available in the areas of aerodynamics, computational fluid dynamics, heat transfer, energy systems for space applications, HVAC, building energy conservation, thermal power generation, robot sensors, robotics teleoperation, composite materials, structural analysis, modeling and simulation, controls, tribology, and vibrations. Students have access to a large array of computers and software for data acquisition, analysis, graphical display, and artificial intelligence. The department has a subsonic wind tunnel and laboratories for heat transfer, fluid mechanics, vibrations, stress analysis, computer-aided design, computational fluid dynamics, and HVAC.

RESEARCH CENTERS

Center for Microelectronics Research (CMR)
Robert P. Carnahan, Ph.D., PE, Interim Director. CMR is a state-funded research center with 19 full-time tenured and tenure-track faculty and research staff members. The Center executes state-of-the-art research in microelectronics materials, devices and processes, design, prototyping, and testing and promotes the transfer of this technology to industry and government.

Research focuses on the architecture, design, and test methodologies for VLSI/ULSI/WSI/MCM systems, high-speed test and interconnect technologies for wafer scale and hybrid ULSL defect engineering in semiconductor materials, and development of in-line monitoring techniques for next-generation IC fabrication. Strong industrial interaction and partnerships are part of the CMR research strategy, with several cooperative research ventures having already been established. CMR currently has seven research laboratories: Microelectronics Design Laboratory, Microelectronics Test Laboratory, Device Engineering Laboratory, Microelectronics Processing Laboratory, Defect Engineering Laboratory, Rapid Prototyping Laboratory, and Reliability and Accelerated Life Test Laboratory.

Center for Urban Transportation Research (CUTR)
Gary L. Brosch, M.S., Director. The Center has a full-time professional staff of 37 research associates with backgrounds in economics, engineering, urban planning, public administration, and geography. CUTR's research program focuses on a variety of transportation issues, including suburban mobility, public transportation, high-speed rail, intelligent transportation systems, growth management, transportation finance, transportation safety, geographic information systems, specialized transportation, access management, alternative fuels, and public policy. CUTR was named a National Urban Transit Institute by Congress in 1991. CUTR serves as the focus organization in an interdisciplinary transportation program for students in economics, public administration, and civil engineering. Through technical support, policy analysis, and research support and by identifying innovative solutions to transportation problems in the state and the nation and through publications, presentations, and seminars, CUTR also serves as an information exchange for transportation agencies on state-of-the-art solutions.

Southern Technology Applications Center (STAC)
W. L. Cahoon, Ph.D., Director. E-mail: cahoon@eng.usf.edu. In 1977, NASA and the State University System of Florida combined resources to form the Southern Technology Applications Center to disseminate and transfer NASA technology, products, and processes to the private sector. In January 1992, STAC was appointed the Southeast Regional Technology Transfer Center (RTTC) with responsibility for nine Southeastern states. Its effective network of experts and resources are located at the Colleges of Engineering at six of the SUS universities.
The cornerstone of STAC's technology transfer success is a professional staff trained and experienced in engineering, physical and biological sciences, medicine, social and behavioral sciences, business planning, marketing, training, library science, and government. STAC is the connection to access the information, technology, inventions, equipment, facilities, and expertise that reside within NASA, more than 700 federal laboratories, and the SUS universities.

THE UNIVERSITY OF TENNESSEE, KNOXVILLE

College of Engineering

Programs of Study

The College of Engineering at the University of Tennessee endeavors to provide an educational experience that imparts the technical competence, proficiency of inquiry, and social understanding that will enable graduates to fulfill their responsibilities as professional engineers. Graduate programs in the College of Engineering provide opportunities for advanced study leading to Master of Science and Doctor of Philosophy degrees in the fields of aerospace, biosystems, chemical, civil, electrical, mechanical, metallurgical, nuclear, and polymer engineering and engineering science. In addition, the M.S. degree is offered in environmental engineering and industrial engineering. At the University of Tennessee Space Institute near Tullahoma, graduate-level courses are offered in engineering fields such as aerospace, chemical, electrical, engineering science and mechanics, industrial, mechanical, and engineering management as well as in the engineering-related fields of mathematics and physics. Programs in all areas lead to the M.S. degree, and Ph.D. programs are available in several areas.

Research Facilities

The on-campus research facilities, which are housed in an eight-building complex, are complemented by those at the nearby Oak Ridge National Laboratory (ORNL). In addition, the Centers of Excellence Program, which is funded through Tennessee's Better Schools Program, enables the University to strengthen teaching and research programs and to achieve national recognition in specific areas. Faculty members participate in three major Centers of Excellence—Science Alliance, the Center for Materials Processing, and the Waste Management Research and Education Institute. Science Alliance is a collaborative research effort between the Colleges of Engineering and Arts and Sciences and ORNL. The Center for Materials Processing is an interdisciplinary program aimed at solving important materials-processing problems. The Waste Management Research and Education Institute is an interdisciplinary organization involving the faculties and facilities of the Colleges of Engineering, Business, and Arts and Sciences, along with the Institute of Agriculture, to provide an engineering focus on waste-treatment technologies.

The College also participates heavily in the Transportation Center; the Center for Environmental Biotechnology; the Energy, Environment, and Resources Center; the Measurement and Controls Engineering Center; and the Maintenance and Reliability Center, the last two being administered within the College of Engineering with major support from industrial sponsors. These centers frequently involve facilities and personnel from other colleges, universities, and research laboratories. Finally, applied and theoretical work in computational mechanics is conducted in the state-of-the-art College Center for Computer Integrated Engineering (C^3IE), a multivendor complex connected by the Internet to remote supercomputing facilities.

Financial Aid

Teaching and research assistantships are available through each department, as are some scholarships and fellowships. The University's financial aid office administers work-study and loan programs and may assist with on-campus part-time employment.

Cost of Study

For 1998–99, academic fees for in-state students were $160 per semester hour, with a maximum of $1437 per semester. Out-of-state fees were $441 per semester hour, with a maximum of $3965 per semester. All full-time students paid a programs/services fee of $140 and a technology fee of $100 per semester.

Living and Housing Costs

Single student housing is available in both University dormitories and apartments. Rates for 1998–99 ranged from $930 to $1630 per semester. The University also provides excellent apartment facilities in several locations for married students with or without families. Apartments not required to house married students are made available to single graduate and professional students. Rates ranged from $280 to $395 per month. University Food Services provides meal plans that ranged from $946 for ten meals per week to $1165 for a seven-day plan per semester.

Student Group

In 1998–99, a total of 26,064 students were enrolled at the University of Tennessee, Knoxville. Of this number, 5,638 were graduate students. During the 1997–98 academic year, the College of Engineering awarded 339 B.S., 136 M.S., and 32 Ph.D. degrees.

Location

Knoxville is located on the banks of the Tennessee River in eastern Tennessee in the foothills of the beautiful Great Smoky Mountains and is within easy reach of several TVA lakes. Close by are many technology-based companies, with the immediate area claiming more than 2,000 residents who hold Ph.D. degrees. Knoxville is an area hub for employment, commerce, and health care. It is within extended reach of the Atlantic Ocean, the Gulf of Mexico, Atlanta, Memphis, and Washington, D.C.

The University

Spread over 528 acres, the University of Tennessee, Knoxville, one of the nation's oldest institutions of higher education, traces its origins back to 1794 when George Washington was President of the United States. In 1869, under terms of the Morrill Act passed by Congress in 1862, the state legislature selected the University as the state's federal land-grant institution, thereby enabling the University to broaden its offerings by establishing an agricultural and mechanical college.

Applying

Initial inquiry and application should be made to the Graduate School. To be considered for admission, the Graduate School requires a minimum overall GPA of 2.7 on a 4.0 scale (minimum 3.0 for electrical engineering) or a minimum GPA of 3.0 for the senior year of undergraduate study. A minimum B average is required of international students. Departmental requirements may be higher than those of the Graduate School. A substantial number of requisite courses may be required for applicants with nonengineering and nonmajor backgrounds. For students whose native language is not English, a minimum TOEFL score of 550 is required. GMAT scores are not required; GRE scores are required for international applicants in most programs. Application may be made at any time; however, international students have deadlines well in advance of the desired date of entry.

Correspondence and Information

Graduate Admissions and Records
218 Student Services Building
The University of Tennessee
Knoxville, Tennessee 37996-0220
Telephone: 423-974-3251
Fax: 423-974-6541
E-mail: gsinfo@utk.edu
World Wide Web: http://web.utk.edu/~gsinfo

Associate Dean for Graduate Studies
College of Engineering
101 Perkins Hall
The University of Tennessee
Knoxville, Tennessee 37996-2011
Telephone: 423-974-2454
E-mail: engracad@utk.edu
World Wide Web: http://www.engr.utk.edu

The University of Tennessee, Knoxville

ACADEMIC UNITS AND AREAS OF RESEARCH

J. E. Stoneking, Dean

All inquiries should be directed to the head of the specific department with the following address: The University of Tennessee, College of Engineering, Knoxville, Tennessee 37996.

Biosystems Engineering. C. R. Mote, Head. Soil and water engineering: soil erosion prediction and control, irrigation, water management, water quality, waste management and utilization, nonpoint source pollution control, watershed hydrology, domestic wastewater renovation, disturbed land reclamation. Power and machinery: mechanization, machine design, traction mechanics, soil dynamics, soil-machine interaction, power unit and implement performance, chemical application technology, conservation tillage and planting, forage harvesting and processing. Food and process engineering: properties of biological materials, drying and curing, osmotic dehydration, materials handling, storage and packaging, heat and mass transfer. Instrumentation: sensor development and application, online analytical sensors, foreign matter detection, biological activity detection, automatic process control, environmental sampling and monitoring.

Chemical Engineering. J. R. Collier, Head. Chemical engineering applications in biology and biochemistry, thermodynamics, chemical process design and optimization, industrial pollution prevention, environmental control technology, industrial automation and control, deterministic chaos applications, statistical process analysis and control, chemometrics, kinetics and combustion, materials chemistry, crystal structure, molecular rheology, polymer morphology and processing, analytical measurement and sensor development, statistical mechanics, molecular simulation, mixed-solvent electrolyte systems, supercritical fluids, computational fluid dynamics, neutron scattering, applied mathematics, numerical methods, industrial monitoring, industrial safety, product life-cycle analysis.

Civil and Environmental Engineering. G. D. Reed, Head. Transportation system planning and design; transportation safety; traffic operations; transportation air quality; investment/financial analysis; bridge testing and evaluation; testing of masonry infills; lateral load tests of driven piles; testing, behavior, and modeling of highway and construction materials; stability of slopes, embankments, and tunnels; nondestructive evaluation of pavement systems; railroad engineering; risk assessment; automated highway systems; intelligent vehicle systems; GPS/GIS applications; air pollution control technologies; air pollution dispersion modeling; water and wastewater treatment; hazardous waste management; environmental restoration; mixed and radioactive waste management; bioremediation; aquatic chemistry; fate and transport of contaminants; surface and groundwater hydrology; erosion and sediment transport; soil and geosynthetic hydraulic barriers; remediation of mines; characteristics of fractures in soil and rock.

Electrical Engineering. M. A. Karim, Head. Communications: telecommunications, spread spectrum, coding. Computer engineering: networks, architecture, systems. Computer vision and robotics: pattern recognition, navigation, virtual reality. Controls: state-space estimators and observers, multivariable and adaptive approaches. Electromagnetics: waves, media, antennas. Electrooptics: processing, communication, systems, displays. Image processing: spatial and transform processing, enhancement, restoration, segmentation, filtering. Industrial plasma engineering: applications. Information processing: security, computing, data visualization. Intelligent control: sensor fusion, process control, automation, real-time systems. Microelectronics: devices, systems, mixed-signal VLSI, VHDL, monolithic sensors. Power electronics and systems: inverters, drives, control. Signal processing: filtering, instrumentation.

Industrial Engineering. T. E. Shannon, Interim Head. Manufacturing systems engineering: design/redesign of manufacturing systems, improvement of production processes, lean production methodologies, environmental impact of manufacturing, computer simulation of manufacturing processes and systems, and other technologies focused on manufacturing as part of an integrated value stream that must be managed as a total system. Human factors engineering: workplace and human-machine systems design, cumulative and sudden trauma workplace disorders, biomechanics, and accident reconstruction. Maintenance and reliability engineering: management and improvement of the maintenance/reliability function and maintenance/reliability strategies. Information systems engineering: design of information systems, managing information technology as a change enabler, and expert systems. Operations research: optimization techniques in industrial engineering and computer simulation modeling. General industrial engineering: engineering economical analysis of capital investments, workflow analysis and measurement, quality assurance, planning and scheduling methodologies, inventory management, materials handling, and facilities layout and design. Engineering management: management of technology, improvement/change processes, and redesign of organizational systems. Dual-degree program (M.S./M.B.A.) in manufacturing: management, improvement/change, redesign of manufacturing and production systems.

Materials Science and Engineering. J. E. Spruiell, Head. Specialized graduate programs exist in both metallurgical engineering and polymer engineering. Research areas include high-temperature materials, high-temperature creep-fatigue behavior, high-temperature coatings, solidification and crystal growth, welding and joining, corrosion, physical metallurgy, thin-film deposition and processing, laser processing, ion implantation and plasma processing, microwave processing of ceramics, composite materials (metal, ceramic, and polymer matrix), fatigue and fracture behavior of ceramic and intermetallic composites, interface behavior of ceramic composites, time-dependent fracture mechanics, life prediction technology, forced chemical vapor infiltration fabrication of ceramic composites, polymer crystallization and morphology, rheology, polymer processing and structure development, mathematical modeling, high-performance polymers, biomedical applications of polymers, and thermal and electrical aging of polymers.

Mechanical and Aerospace Engineering and Engineering Science. D. W. Dareing, Head. Aerodynamic drag on flapping high-aspect cylinders, unsteady motion of flexible body driven by aerodynamic forces, thermal analysis of fine fiber attenuation in plane jet flow, reduction of structural vibrations using piezoelectric actuators, numerical modeling of intelligent structures, advanced space mission planning, hypersonics, robotics, flexible automation, product development, product development of commercial devices, electro/mechanical devices, vibrations/dynamics, nonlinear dynamics of mechanical systems, phase-change heat transfer in thermal energy storage systems, noncontact temperature measurement of opaque and semitransparent materials, computational heat transfer and fluid mechanics, alternative fuels for light-duty vehicles, emission controls for light-duty vehicles, on-road emission testing of light-duty vehicles, thermal response of decomposing polymers, heat transfer in spent nuclear fuel canisters, cooling of electronics packages, experimental and theoretical studies of performance of three-way catalyst, deterministic chaos in fluidized bed, controlling chaos in engine emission, compressible fluid flow and heat transfer, phase change in heat transfer, numerical methods in heat transfer, thermoacoustic convection. Biomedical engineering: biomechanics, impact trauma, ergonomics, cellular biomechanics, interaction of mechanical stress and pathology. Fundamental and applied research and applications in computational fluid dynamics: weak statement discrete approximation theory, finite elements, compressible and incompressible flows, buoyancy, radiation, turbulence and reacting flows. Computational solid mechanics: linear and nonlinear finite-element analysis, computational constitutive modeling, 3-D static and dynamic stress analysis, application to real-world problems. Dynamics and vibrations: dynamic system simulation, vibration analysis, vehicle dynamics. Fracture mechanics: ductile fracture, transition fracture, fracture testing of polymers. Composite materials: polymeric and ceramic composites, environmental effects, time-dependent response, interfacial and interphase properties, fabrication and experimental mechanics of composites.

Nuclear Engineering. H. L. Dodds, Head. Reactor physics, criticality safety, reactor instrumentation and control, applied artificial intelligence, plant surveillance and diagnostic technology, radiological engineering (health physics and medical physics), shielding, reactor safety, risk assessment, thermal hydraulics, radioactive waste management, maintenance and reliability engineering, radiation risk analysis, charged particle transport, neutron source and optics, nuclear nonproliferation systems, radiochemical engineering, space radiation protection, energy and waste policy, spallation neutron source development.

UNIVERSITY OF TENNESSEE SPACE INSTITUTE

Programs in Engineering and Applied Science

Programs of Study

The University of Tennessee Space Institute (UTSI) offers graduate programs leading to the Doctor of Philosophy and the Master of Science degrees in various fields of engineering and applied science. The accredited academic programs and educational policies of the Institute have their origins in appropriate departments of the University of Tennessee, Knoxville. M.S. and Ph.D. programs are available in aerospace engineering, electrical engineering, engineering science, mechanical engineering, metallurgical engineering, and physics. M.S. programs are offered in aviation systems, chemical engineering, computer science, engineering management, industrial engineering, and mathematics. M.S. programs in aerospace engineering, aviation systems, engineering management, and mechanical engineering are available to off-campus students via interactive video and/or videotape.

Research Facilities

Aerodynamics facilities include water tunnels, a high–Reynolds number transonic wind tunnel, a supersonic tunnel, and a low-speed wind tunnel. Propulsion research is performed in a building that includes five bunkered test cells. Computational research is conducted using numerous Institute computing facilities or on supercomputers through high-speed communication links with the DOD High Performance Computer Centers. Students in aviation systems conduct flight research in specially equipped aircraft, including a T-39 Sabreliner, a Piper Saratoga, a Piper Navajo, a Piper Cub, two Bell OH-58C (Bell Jet Ranger) helicopters, and a sailplane, as well as two Navion variable stability aircraft and a STOL jetwing aircraft.

The Center for Laser Applications (CLA) is a multidisciplinary group of faculty members, staff, and students that conducts research in nonlinear optics, nonequilibrium plasma, combustion and fluid physics, and laser materials processing. The CLA facilities include more than 2,400 square meters of modern laboratory space. Major laser systems include very high power excimer, Nd-YAG, and Ti-sapphire nanosecond, picosecond, and femtosecond lasers; a ring-jet dye laser; and copper vapor and argon-ion lasers. Laser materials processing research utilizes 3kW, 2kW, and 650W Nd-YAG and 3kW CO_2 laser systems. Materials-processing and metallurgical analysis facilities are also available. Diagnostic instrumentation includes picosecond streak cameras, a 20-MHz framing camera, a 5-kHz video imaging system, image analysis computers, two-dimensional image intensifiers, optical multichannel analyzers, spectrometers, interferometers, photon-counting systems, a holographic camera, scanning electron microscopes, and an electron probe microanalyzer.

Energy conversion research is carried out on a large scale in a complete combustion heat recovery and gas clean-up laboratory. A parallel program includes environmental research. Current energy conversion activities concentrate on coal combustion and coal combined cycles. Materials research and advanced instrumentation are carried out on combustion-driven experiments in modern, well-equipped laboratories. Chemical process engineering is related to energy conversion processes and environmental issues. Research in these areas requires the application of advanced concepts involving acoustics, optics, gasdynamics, heat transfer, materials science, mechanical design, electronics, and computer science. A high-temperature superconductor development program is also being carried out.

Financial Aid

Graduate research assistantships (GRAs), which include a stipend and waiver of tuition and maintenance fees, are available for outstanding students. For 1999–2000, stipends range from $14,750 to $18,800 for twelve months and may be renewed based on satisfactory academic and research progress, subject to the availability of funds. Industrial internships and fellowships are also available.

Cost of Study

Maintenance and other fees for in-state full-time students are $1728 per semester in 1999–2000. For out-of-state full-time students, maintenance fees, tuition, and other fees are $4762 per semester.

Living and Housing Costs

A limited number of single- and double-occupancy rooms are available on campus and are equipped with efficiency kitchens. Housing costs for 1999–2000 are $260 per month for a single and $140 per month for a double; a $75 deposit is required to reserve a room. Apartments in the local area may be rented for approximately $300 per month plus utilities.

Student Group

There are approximately 300 graduate students enrolled in UTSI courses; about 90 are full-time students.

Student Outcomes

UTSI has a telecommunication link with the UT Knoxville Placement Office that permits UTSI students to interview with corporation recruiters visiting the main campus. Recent graduates have been employed by Jet Propulsion Lab, Johnson Controls, Lockheed-Martin, Boeing, Texas Instruments, General Electric, Toyota, TRW, Lucent Technology, McDonnell Douglas, Pratt & Whitney, Teledyne Brown Engineering, Procter & Gamble, Aero Astro, Sverdrup Technology, Coleman Company, United Defense, Hewlett-Packard, Raytheon, NASA, DOE, and others.

Location

UTSI is located on a picturesque, 365-acre lakeshore campus near the U.S. Air Force's Arnold Engineering Development Center, one of the world's largest concentrations of aerospace test facilities. The Space Institute's location in middle Tennessee, near Tullahoma, is the center of a triangle formed by Nashville, Chattanooga, and the NASA Marshall Space Flight Center at Huntsville, Alabama.

The Institute

The Institute is devoted solely to graduate education and research. UTSI has awarded more than 1,500 M.S. and Ph.D. degrees since it was established in 1964.

Applying

Students may enter in the fall (late August), spring (January), or summer (June). GRE scores are required for admission by some departments and for assistantships by all departments.

Correspondence and Information

Admissions
University of Tennessee Space Institute
Tullahoma, Tennessee 37388-9700
Telephone: 931-393-7432
        888-822-8874 Ext. 432 (toll-free)
Fax: 931-393-7346
E-mail: admit@utsi.edu
World Wide Web: http://www.utsi.edu

University of Tennessee Space Institute

PROGRAM HEADS AND RESEARCH AREAS

Aerospace Engineering. Degree Program Chairman: Dr. Ahmad Vakili. Graduate programs leading to the Master of Science and doctoral degrees are available to qualified graduates of recognized undergraduate curricula in either aerospace or mechanical engineering and to qualified graduates of other curricula who satisfy necessary prerequisites. The major areas of study and research include air and space vehicle design, flight mechanics, propulsion, fluid mechanics, aerodynamics, and space engineering. Study of these areas usually involves a combination of analytical, computer modeling, and experimental approaches. The primary goals of the program are to provide the student with a strong background in fundamentals and to offer a research environment representative of activity in the aerospace engineering profession.

Aviation Systems. Degree Program Co-Chairmen: Dr. Frank Collins and Dr. Ralph Kimberlin. The aviation systems program is designed for students who possess a bachelor's degree in engineering or science and who wish to study under a "systems philosophy" toward careers in research and development or administration in areas pertinent to aviation. Current areas in aviation that receive emphasis in the program include flight testing, aircraft design, aviation safety, air traffic control, and aviation technical management. Opportunities for research exist at UTSI's flight research facility at the Tullahoma Municipal Airport. Students who are unable to study in residence at the Space Institute may be able to complete the program by videotape at one of the off-campus locations.

Chemical Engineering. Degree Program Chairman: Dr. Atul Sheth. The M.S. degree with a major in chemical engineering is offered. Research opportunities are in the areas of fossil fuel utilization, high-temperature superconductor-related process development, hazardous- and nonhazardous-waste management, and related topics, with a major emphasis on coal-fired power generation. Particular subjects include aspects of air pollution control, recovery of chemicals, combustion, gasification, and heat and mass transfer.

Computer Science. Degree Program Chairman: Dr. Bruce Whitehead. UTSI offers an M.S. in computer science with emphasis in scientific/engineering software design—the theory, algorithms, data structures, and software engineering principles of scientific (as opposed to business) applications of computer science. The application areas currently emphasized by the program are VLSI computer-aided design, computer graphics, neural networks, and genetic algorithms. In each of these areas, course work focuses on the underlying theory of the application area and the translation of this theory into well-engineered software. The goal is that the student will graduate with an understanding of principles and design techniques that will remain valid long after current commercial products in these areas have become obsolete. Classes are small, the student-to-faculty ratio is quite favorable, faculty members are readily available to advise students, and there is ample opportunity for students to become involved in ongoing research of the faculty and to develop related thesis topics. Computing facilities include a network of PCs and SPARC-5, SPARC-10, and ultrasparc workstations.

Electrical Engineering. Degree Program Chairman: Dr. Roy D. Joseph. The M.S. with a major in electrical engineering is offered to qualified graduates with a bachelor's degree in electrical engineering. The M.S. thesis option includes 6 semester hours of required course work in electrical engineering (EE); 6 semester hours in approved mathematics, EE, or non-EE courses; 12 semester hours in approved EE electives; and 6 hours of thesis work. A nonthesis option replaces the 6 hours of thesis work with 6 hours of EE courses plus a 3-hour project course. The faculty is involved with research and graduate instruction in analog and digital signal processing, digital hardware design, advanced control theory, image processing, and electrooptics. Under special circumstances the Ph.D. may be earned at UTSI; interested students should consult with the department to determine specific requirements.

Engineering Management. Degree Program Chairman: Dr. Max Hailey. The University of Tennessee's Engineering Management (EM) program is headquartered at UTSI. One of the largest in the country, the program features a concentration of behavioral and management courses balanced with quantitative methods courses. The EM program, while popular on campus, is available by a variety of media to students across the state and nation. EM is a freestanding concentration of UT's MSIE. Experience as a practicing engineer or engineering manager is required. International students must have at least two years of U.S. industrial work experience before being admitted to the program.

Engineering Science. Degree Program Chairman: Dr. Ahmad Vakili. Graduate programs leading to M.S. and Ph.D. degrees with a major in engineering science are available to graduates of recognized curricula in engineering, mathematics, or one of the physical or biological sciences. Program concentrations include structural mechanics, fluid mechanics, computational mechanics, laser materials processing, plasma physics, microgravity science, acoustics, and optical engineering. In each of these concentrations, interdisciplinary programs are arranged to meet individual needs or interests. The flexibility and interdisciplinary aspect of the program is intended to be of particular interest to prospective students currently employed in research, development, or design activities and whose interests in continuing education (either full-time or part-time) lie at one of the interfaces between science and engineering or can best be met by interdisciplinary study in engineering.

Industrial Engineering. Degree Program Chairman: Dr. Max Hailey. Systems thinking and its influence on management behaviors, analysis and redesign of organizational systems, development of redesign templates for change (manufacturing, financial, service), design of manufacturing systems, improvement in quality and productivity through focused attention on reduced variation, process management using experimental design, designing organizations to support high-performance work teams, management of technology, biomechanics, cumulative and sudden trauma workplace disorders, activity-based management, application of demand-flow manufacturing techniques in manufacturing and nonmanufacturing areas, information technology as a change enabler, expert systems and artificial intelligence, product development, lean production systems design and implementation, evaluation of environmental impact on manufacturing, process modeling for the product development process, development of tools relating to determining of strategic enablers in product development and assessment of risks of employing or not employing them, development of error profiles and compensation strategies in precision manufacturing, cognitive ergonomics, accident reconstruction.

Mathematics. Degree Program Chairman: Dr. K. C. Reddy. A graduate study program with emphasis on interdisciplinary studies in mathematical sciences, leading to an M.S., is offered. Research topics include numerical analysis, computational fluid dynamics, Hamiltonian mechanics, integrable systems, signal processing, and calculus of variations. Opportunities exist to carry out research on mathematical problems of practical interest and application in the aerospace industry.

Mechanical Engineering. Degree Program Chairman: Dr. Ahmad Vakili. Graduate programs leading to the Master of Science and Doctoral degrees are available to qualified graduates of recognized undergraduate curricula in either mechanical or aerospace engineering and to qualified graduates of other curricula who satisfy necessary prerequisites. The major areas of study and research include thermal sciences (e.g., thermodynamics, heat transfer, energy conversion, and combustion) and propulsion (air breathing engines, liquid and solid rocket motors, and futuristic flight propulsion systems). Programs in mechanics, structures, and vibrations are offered on a limited basis.

Metallurgical Engineering. Degree Program Chairman: Dr. Mary Helen McCay. Graduate programs leading to M.S. and Ph.D. degrees with a major in metallurgical engineering are available to qualified graduates who hold a bachelor's degree in either metallurgical engineering or materials science and to qualified graduates of other curricula who satisfy necessary prerequisites. The major areas of study and research include physical metallurgy, phase transformations and structure-properties correlations with concentration on laser materials interactions, solidification physics, and high-temperature superconductor processing. Opportunities for research also exist in the area of processing and characterization of advanced materials. The interdisciplinary aspect of the program is intended to provide students with a strong background in the fundamentals of science and engineering related to materials.

Physics. Degree Program Chairman: Dr. Horace Crater. Graduate programs in physics leading to Master of Science and Doctor of Philosophy degrees are available to qualified students who have completed an undergraduate major in physics or its equivalent. The minimum courses prerequisite to graduate study are one semester of thermodynamics and two semesters each of classical mechanics, electricity and magnetism, and introduction to quantum mechanics or modern physics. First-year graduate students are required, for advising purposes only, to take a qualifying examination in undergraduate physics during the fall semester registration period. The Graduate Record Examinations should be submitted for admission evaluation. Participation in research is a basic requirement for advanced degrees in physics. Active research areas at UTSI include atomic and molecular spectroscopy, theoretical particle physics, chemical and laser physics, optics, and medical physics. The Center for Laser Applications has ongoing physics research in laser-related studies of plasma and combustion phenomena of propulsion systems, ultrasensitive spectroscopy, nonlinear optics, ultrafast phenomena, nonequilibrium fluid physics, and biomedical applications.

THE UNIVERSITY OF TEXAS AT ARLINGTON

College of Engineering

Programs of Study

The College of Engineering at the University of Texas at Arlington (UTA) offers master's and doctoral degrees in the areas of aerospace, biomedical, civil, computer science, electrical, industrial, and mechanical engineering; computer science; manufacturing; materials science; and management of technology. UTA's biomedical engineering is a joint program with the University of Texas Southwestern Medical Center at Dallas, materials science and engineering is an interdisciplinary program with UTA's College of Science, environmental science and engineering is an interdisciplinary program with the College of Science and the School of Urban and Public Affairs, and management of technology is an interdisciplinary program with the College of Business. A manufacturing engineering option is also available through the Departments of Computer Science and Engineering, Electrical Engineering, Industrial and Manufacturing Systems Engineering, and Mechanical and Aerospace Engineering. A Master of Software Engineering degree is also offered.

Research Facilities

The College of Engineering has state-of-the-art equipment for research housed in seven buildings. Central computing facilities as well as distributed nodes are available for both student and faculty member use. Specialized laboratories and equipment are found in the following institutes and research centers: the Automation and Robotics Research Center (ARRI); Agile Aerospace Manufacturing Research Institute; Human Performance Institute; Center for Advanced Engineering Systems and Automation Research (CAESAR); Energy Systems Research Center; Wave Scattering Research Center; Center for Electron Materials, Devices, and Systems; Software Engineering Center for Telecommunications; Center for Composite Materials; Fluid Control Development Center; Aerodynamics Research Center; Computational Fluid Dynamics Center; Center for Hypersonic Research; Construction Research Center; ElectroOptics Research Center; and Center for Transportation Studies.

Financial Aid

Research and teaching assistantships, fellowships, scholarships, and loans are available to students. Faculty have acquired substantial state, federal, and private funding for research with stipends for graduate students. In addition, there are part-time and full-time engineering employment opportunities in the Dallas–Fort Worth metroplex.

Cost of Study

The estimated tuition rate for a 12-hour course load is expected to be $1918 per semester for Texas residents and $4632 for nonresidents in 2000–01. Half-time teaching and research assistants and holders of competitive fellowships are entitled to Texas resident rates.

Living and Housing Costs

The University has residence halls and apartments available. Residence hall rates range from $1514 to $2574 for a single student for the two long semesters (nine months) in 1999–2000. One-bedroom apartments start at $285 per month. There are also many privately owned apartments in Arlington available at reasonable rates.

Student Group

There are more than 18,000 students at UT Arlington. Roughly 2,800 are in the College of Engineering, of which about half are pursuing graduate degrees in engineering. The College has a rich cultural diversity and active and competitive student organizations in every department. Many graduate engineering classes are also Web-enabled or televised and transmitted to students employed at high-technology companies throughout the metroplex and north Texas.

Location

The University of Texas at Arlington is in the center of the Dallas–Fort Worth metroplex, one of the largest high-technology regions in the country. The area provides, in addition to campus activities, a full range of cultural and recreational opportunities, including museums, ballet, theater, amusement parks, and professional sports. The region is served by Dallas/Fort Worth International Airport, one of the largest and busiest in the country.

The University and The College

Founded in 1895, the University has experienced most of its growth in the past thirty-five years. It is located on a beautiful, modern 347-acre campus. The first baccalaureate class in engineering graduated in 1961 and the first Ph.D.'s were awarded in 1971. It attracts outstanding faculty members, who guide student research for many local industries, state and federal agencies, and national and international institutions. The College has seventeen organized research centers in automation and robotics, environment, construction, transportation, software, physical electronics, energy systems, electron devices and systems, human performance, wave scattering, composite materials, fluid controls, aerodynamics, computational fluid dynamics, and hypersonic research. According to the Engineering Manpower Commission, UTA ranks in the nation's top 10 percent in the number of master's degrees and in the top 15 percent in the number of Ph.D.'s awarded in engineering. UTA has earned Carnegie Foundation Doctoral University I status.

Applying

Students holding a B.S. degree with a minimum upper-level grade point average of 3.0 (on a scale of 4.0) and a minimum combined GRE General Test score of 1100 on the verbal and quantitative sections are invited to apply for admission. Applications should be submitted by April 1 for the fall semester and October 15 for the spring semester. Admission materials can be obtained from the Graduate School office. Applications for financial aid should be submitted directly to the applicable department by March 1 to be considered for fall awards, and they should include copies of transcripts and GRE scores. TOEFL scores for international students must be at least 550. (TSE is 560).

Correspondence and Information

For catalogs and application forms:
Graduate School
Box 19167
The University of Texas at Arlington
Arlington, Texas 76019-0167

Telephone: 817-272-2681
E-mail: graduate.school@uta.edu

For other information:
Dean of Engineering
Box 19019
The University of Texas at Arlington
Arlington, Texas 76019-0019

Telephone: 817-272-2571
World Wide Web: http://www-eng.uta.edu

University of Texas at Arlington

THE FACULTY AND RESEARCH AREAS

Biomedical Engineering. R. Eberhart, Chairman; Ph.D., Berkeley. K. Behbehani, Ph.D., Toledo. C.-J. Chuong, Ph.D., California, San Diego. H. Liu, Ph.D., Wake Forest. K. D. Nelson, Ph.D., Texas Southwestern Medical Center at Dallas. Biological signal processing; biosensors; neuroscience engineering; soft- and hard-tissue mechanics; wave mechanics and lithotripsy; artificial- and hybrid-organ design; biomaterials; recombinant DNA technology; medical imaging with ultrasound, gamma ray, magnetic resonance imaging, electron and confocal microscopy, and fluorometer sensing modalities; clinical and rehabilitation engineering. Program offers training in five different tracks: bioinstrumentation, medical imaging, biomechanics/orthopedics, biomaterials/tissue engineering, and molecular engineering. (World Wide Web: http://www.uta.edu/biomed_eng/bme.htm).

Civil Engineering. C. E. Parker, Chair; Ph.D., Arizona. S. A. Ardekani, Ph.D., Texas at Austin. D. Clark, M.S., SMU. E. C. Crosby, Ph.D., Tennessee. S. Govind, Ph.D., Texas at Austin. T. Huang, Ph.D., Illinois. A. P. Kruzic, Ph.D., California, Davis. J. H. Matthys, Ph.D., Texas at Austin. W. H. Nedderman, Ph.D., Iowa State. A. Puppala, Ph.D., LSU. S. R. Qasim, Ph.D., West Virginia. M. Spindler, Ph.D., Northwestern. J. C. Williams, Ph.D., Texas at Austin. R. L. Yuan, Ph.D., Illinois. Environmental engineering; construction; infrastructure; transportation planning; hazardous- and toxic-waste abatement; asbestos abatement; hydrology; structural analysis; analytical methods in structural dynamics; shell structures and marine riser mechanics; biological and chemical processes in water quality control; water reclamation and reuse; natural systems for wastewater treatment; structural analysis and design of reinforced concrete, steel, timber, and masonry systems; soil mechanics; soil stabilization; foundation engineering; traffic flow theory; traffic engineering; highway capacity analysis; transportation systems analysis; operations research; properties and behavior of structural concrete; experimental stress analysis; composite structural materials; in situ soil testing. (World Wide Web: http://www-ce.uta.edu).

Computer Science and Engineering. B. D. Carroll, Chairman; Ph.D., Texas at Austin. C. T. Bruggeman, Ph.D., Indiana. D. J. Cook, Ph.D., Illinois. R. A. Elmasri, Ph.D., Stanford. L. Fegaras, Ph.D., Massachusetts. P. Gmytrasiewicz, Ph.D., Michigan. L. Holder, Ph.D., Illinois. P. Hsia, Ph.D., Texas at Austin. F. A. Kamangar, Ph.D., Texas at Arlington. D. Kung, Ph.D., Norwegian Institute of Technology. L. L. Peterson, Ph.D., Texas Health Science Center at Dallas. B. Shirazi, Ph.D., Oklahoma. L. D. Umbaugh, Ph.D., Ohio State. R. L. Walker, Ph.D., Texas at Austin. B. P. Weems, Ph.D., Northwestern. L. Welch, Ph.D., Ohio State. R. Yerraballi, Ph.D., Old Dominion. H.-Y. Youn, Ph.D., Massachusetts. Computer system architecture and modeling, fault-tolerant computing, parallel systems, interconnection networks, simulation and performance evaluation, mobile computing, telecommunications, computer security, parallel processing, distributed systems, temporal databases, system integration, schema versioning, object databases, query optimization, intelligent databases, knowledge-based systems, artificial intelligence, multi-agent systems, knowledge engineering, image processing, robotics, neural networks, machine learning, object-oriented systems, software engineering, software testing, object-oriented testing, software development methodologies, reverse engineering, software and hardware systems specification, distributed real-time systems. (World Wide Web: http://www-cse.uta.edu).

Electrical Engineering. R. Magnusson, Chairman; Ph.D., Georgia Tech. K. Alavi, Ph.D., MIT. J. Bredow, Ph.D., Kansas. R. L. Carter, Ph.D., Michigan State. M. S. Chen, Ph.D., Texas at Austin. M. P. Chwialkowski, Ph.D., Warsaw Tech. W. A. Davis, Ph.D., Michigan. V. Devarajan, Ph.D., Texas at Arlington. W. E. Dillon, Ph.D., Texas at Arlington. J. Fitzer, D.Sc., Washington (St. Louis). A. K. Fung, Ph.D., Kansas. G. V. Kondraske, Ph.D., Texas at Arlington/Texas Southwestern Medical Center. W. J. Lee, Ph.D., Texas at Arlington. F. L. Lewis, Ph.D., Georgia Tech. T. A. Maldonado, Ph.D., Georgia Tech. M. T. Manry, Ph.D., Texas at Austin. J. H. McElroy, Ph.D., Catholic University. V. Prabhu, Ph.D., MIT. K. R. Rao, Ph.D., New Mexico. R. R. Shoults, Ph.D., Texas at Arlington. C. V. Smith, Ph.D., MIT. S. Tijuata, Ph.D., Texas at Arlington. K. S. Yeung, Dr.Ing., Karlsruhe (Germany). Photoconductive devices, pulsed power high voltage, electrical insulation, power electronics, space power systems, holography, integrated optics, quantum well devices, microwave and millimeter-wave integrated circuits, molecular beam epitaxy, electrooptics, power systems, remote sensing and wave scattering, pollution monitoring, robotics, robust control, signal processing, flight simulation, utility deregulation issues, neural networks, computer vision, telecommunications, fiber optics, low-power digital VLSI, microwave communications, instrumentation. The faculty includes 5 IEEE Fellows. (World Wide Web: http://www-ee.uta.edu).

Industrial and Manufacturing Systems Engineering. D. H. Liles, Chairman; Ph.D., Texas at Arlington. Oklahoma State. H. W. Corley, Ph.D., Florida. B. Huff, Ph.D., Texas at Arlington. S. N. Imrhan, Ph.D., Texas Tech. J. H. McElroy, Ph.D., Catholic University. F. A. Meier, Ph.D., Washington (St. Louis). J. W. Priest, Ph.D., Texas at Arlington. J. Rogers, Ph.D., Texas at Arlington. G. T. Stevens Jr., Ph.D., Oklahoma State. Design for producibility and reliability, manufacturing systems, automation, CAD/CAM, robotics, engineering design and development process, ergonomics, computer-integrated enterprise, enterprise design and analysis, statistical process control, manufacturing error analysis, linear models, work sampling, discrete event computer simulation, economic decision making, engineering economy, production and inventory control, material and project control, production and quality control. Major areas of interaction with industry: manufacturing, logistics, enterprise engineering. (World Wide Web: http://www.uta.edu/ie/).

Materials Science and Engineering. R. L. Elsenbaumer, Chairman; Ph.D., Stanford. P. Aswath, Ph.D., Brown. W. S. Chan, Ph.D., Purdue. R. D. Goolsby, Ph.D., Berkeley. R. M. Johnson, Ph.D., Oklahoma. C. Kim, Ph.D., Berkeley. Thirty-six other professors from the College of Science and the College of Engineering also participate. Interdisciplinary research is carried out in eight departments in these Colleges, including the Departments of Mechanical and Aerospace Engineering, Electrical Engineering, Biomedical Engineering, Civil Engineering, Physics, Chemistry, Biology, and Mathematics. Semiconductor processing, metal matrix composites, positron annihilation, electrically conductive polymers, structural materials, biomaterials, electronic devices, biocompatible materials, polymeric materials, ceramics, ceramic matrix composites, fatigue and fracture mechanics, surface physics, intermetallic systems, materials modeling, optoelectronics, advanced composites and applied mathematics. (World Wide Web: http://mse.uta.edu).

Mechanical and Aerospace Engineering. D. R. Wilson, Chairman; Ph.D., Texas at Arlington. R. Bailey, Dean of Engineering; Ph.D., Southampton (England). R. D. Goolsby, Associate Dean of College of Engineering; Ph.D., Berkeley. D. A. Anderson, Vice President for Research and Dean of Graduate School; Ph.D., Iowa State. P. B. Aswath, Ph.D., Brown. W. S. Chan, Ph.D., Purdue. J. H. Gaines, Ph.D., Texas at Austin. A. Haji-Sheikh, Ph.D., Minnesota. K. T. Harris, Ph.D., Mississippi. D. A. Hullender, Ph.D., MIT. R. M. Johnson, Ph.D., Oklahoma. S. P. Joshi, Ph.D., Purdue. T. J. Lawley, Ph.D., SMU. K. L. Lawrence, Ph.D., Arizona State. F. K. Lu, Ph.D., Penn State. T. S. Lund, Ph.D., Stanford. J. J. Mills, Director, Automation and Robotics Research Institute; Ph.D., Durham (England). D. Musielak, Ph.D., Alabama. S. Nomura, Dr.Engr., Tokyo; Ph.D., Delaware. F. R. Payne, Ph.D., Penn State. D. D. Seath, Ph.D., Iowa State. P. Shiakolas, Ph.D., Texas at Arlington. A. Y. Tong, Ph.D., Carnegie Mellon. D. G. Tuckness Jr., Ph.D., Texas at Austin. B. P. Wang, Ph.D., Virginia. R. L. Woods, Ph.D., Oklahoma State. S. M. You, Ph.D., Minnesota. Automatic control, dynamic system analysis, flight controls, simulators, fluidics, optimal control, parameter estimation; design and manufacturing; structural analysis, multidisciplinary design optimization, CAE, CAD/CAM adaptive FEM analysis, application of microprocessors in mechanical systems, composites, smart structures damage tolerance, robotics, and integrated machining systems; thermal science: solar energy, energy storage systems, heat transfer, fluid mechanics, thermal curing of composites, electronic cooling, combustion of fuel droplets; computational fluid dynamics, aerodynamics, guidance, navigation and control system, finite-element methods, supersonic and hypersonic flow, turbulence modeling and measurement, low-speed aerodynamic analysis and testing, experimental fluid dynamics, aerospace propulsion systems, pulse detonation wave engines, composite structures, damage mechanics, smart structures. The department has a full complement of wind tunnel facilities, from subsonic to hypersonic. (World Wide Web: http://mae.uta.edu).

Automation & Robotics Research Institute (ARRI). J. J. Mills, Director; Ph.D., Durham (England). D. J. Cook, Ph.D., Illinois. V. Devarajan, Ph.D., Texas at Arlington. R. Elmasri, Ph.D., Stanford. J. Fitzer, D.Sc., Washington (St. Louis). B. Huff, Ph.D., Texas at Arlington. T. J. Lawley, Ph.D., SMU. F. L. Lewis, Ph.D., Georgia Tech. D. H. Liles, Ph.D., Texas at Arlington. J. W. Priest, Ph.D., Texas at Arlington. P. Shiakolas, Ph.D., Texas at Arlington. C. V. Smith, Ph.D., MIT. ARRI research to advance manufacturing processes is interdisciplinary and includes five College of Engineering departments and the College of Business. ARRI's mission is to help U.S. industry become more competitive in the global marketplace through world-class manufacturing, technology transfer, manufacturing, concurrent engineering, liquid metal jetting, reconfigurable controls, assembly and information systems, robotics, adaptive systems, controls, singular systems, enterprise engineering, vision systems. (World Wide Web: http://arri.uta.edu).

UNIVERSITY OF TOLEDO

College of Engineering

Programs of Study

The College of Engineering (COE) at the University of Toledo (UT) offers Master of Science (M.S.) and Doctor of Philosophy (Ph.D.) degrees through the Departments of Bioengineering; Chemical and Environmental Engineering; Civil Engineering; Electrical Engineering and Computer Science; and Mechanical, Industrial, and Manufacturing Engineering.

The Master of Science program is designed to prepare students for research and advanced engineering careers. The program requires the completion of 30 semester hours beyond the bachelor's degree for the thesis/project option.

The Doctor of Philosophy program is designed for those planning research-oriented industrial or academic careers to pursue professional engineering practice for the advancement of science, engineering, and technology. The program requires a qualifying examination, a minimum of 90 semester credit hours (60 semester credit hours beyond the master's degree) of course work and dissertation, and a successful oral defense of the dissertation work.

Research Facilities

Research is supported by state-of-the-art research and computing laboratories. Multidisciplinary collaborative research facilities include the Polymer Institute, International Blast Mitigation Research Center, Advanced MicroMachining Laboratory, Center for Integrated Manufacturing, Center for Abrasive Micromachining, Macromolecular Crystallization Laboratory, Medical Imaging and Informatics Laboratory, Center for Intelligent Transportation, and the Pavement Management Center. The College of Engineering has an overall research expenditure of $6 to $8 million from research grants and contracts.

Financial Aid

Nearly all full-time graduate students receive some financial support. COE fellowships, teaching assistantships, and research assistantships, which include a stipend and a tuition waiver, are available for qualified students on a competitive basis.

Cost of Study

The tuition rate for graduate tuition for the 1998–99 school year was $212.60 per semester credit hour for in-state and $459.60 per semester credit hour for out-of-state students. Full-time tuition (between 12 and 16 hours) was $2551.20 for in-state students and $5551.20 for out-of-state students.

Living and Housing Costs

The University of Toledo offers off-campus shuttle service to many off-campus housing and apartment complexes. Many students can find affordable, high-quality housing within easy walking distance of campus. The shuttle routes extend as far as 5 miles away from campus.

Student Group

There are approximately 21,000 students at the University of Toledo. Approximately 4,400 are graduate students. Of those, approximately 500 are graduate students in the College of Engineering. The University has a rich mixture of diverse student organizations. Students join groups that are organized around common cultural, religious, athletic, and educational interests.

Location

The University of Toledo has several campus sites in the city of Toledo. All engineering graduate students take classes on the Bancroft campus, which is located in suburban western Toledo near Ottawa Hills. Toledo is the fiftieth largest city in the U.S. It is located on the western shores of Lake Erie and within a 2-hour drive of Cleveland and Detroit.

The University and The College

The University of Toledo was founded by Jessup W. Scott in 1872 as a municipal institution and became part of the state of Ohio's system of higher education in 1967. The College of Engineering was founded in 1934 and began offering graduate degrees in 1947. The College of Engineering is housed in a three-building, $21-million complex, which is composed of Nitschke Hall, Palmer Hall, and North Engineering. A new multimedia state-of-the-art auditorium is scheduled to open in fall 1999 for computer-aided instructions and distance learning initiatives.

Applying

Students with a Bachelor of Science in engineering or one of the physical, mathematical, or biological sciences are encouraged to apply. Applicants should have a 3.0 grade point average (on a 4.0 scale) or better, but exceptions are made for those who demonstrate ability for graduate study. Applications should be completed by March 1 for full consideration for the fall semester. Admission materials can be obtained from the graduate school office or from the Web site listed below.

Correspondence and Information

Graduate School
University of Toledo
2801 West Bancroft Street
Toledo, Ohio 43606
Telephone: 419-530-4723
E-mail: grdsch@utnet.utoledo.edu

Office of Graduate Studies
College of Engineering
University of Toledo
Toledo, Ohio 43606
Telephone: 419-530-7391
E-mail: mharlow@eng.utoledo.edu
World Wide Web: http://eng.utoledo.edu/grad

University of Toledo

THE FACULTY AND THEIR RESEARCH

Bioengineering: Ronald L. Fournier, Chairperson; Ph.D., Toledo: transport processes in biological systems, bioartificial organs, tissue engineering, genetic engineering. Krzysztof J. Cios, Ph.D., Mining and Metallurgy (Poland): neurocomputing, machine and human learning, medical informatics. Alam P. Dhawan, Ph.D., Manitoba: medical imaging, intelligent image analysis, genetic algorithms, neural networks, adaptive learning. Jeffrey D. Johnson, Ph.D., Cincinnati: intelligent systems for imaging and control, computational neuroscience. Vik J. Kapoor, President; Ph.D., Lehigh: biosensors, bioelectronics, biocompatible integrated circuits, implantable signal processing. Frank J. Kollarits, Ph.D., Ohio State: biosensors, biomaterials, bioelectric instrumentation. Demetrios D. Raftopoulos, Ph.D., Penn State: biomechanics, biomaterials, gait analysis. Patricia A. Relue, Ph.D., Michigan: fluorescence microscopy, mathematical modeling, biomedical/tissue engineering. Susan Sharfstein, Ph.D., Berkeley: biochemical engineering, mammalian cell culture, bioreactors.

Chemical and Environmental Engineering: Steven LeBlanc, Chairperson; Ph.D., Michigan: environmental applications of chemical engineering, flue gas cleanup using membrane absorbers, photocatalytic membrane reactors for air and water treatment. Martin Abraham, Ph.D., Delaware: environmental reaction engineering and catalysis, supercritical water oxidation, heterogeneous catalysis for solventless processing. Maria Coleman, Ph.D., Texas: development of gas separation membranes, transport mechanism of small molecules within a glassy polymer matrix, immobilized metal affinity membrane protein separation. Kenneth DeWitt, Ph.D., Northwestern: fluid-thermal physics and transport phenomena. John Dismukes, Ph.D., Illinois: accelerated innovation in technology, novel materials synthesis and process development, fluid convection on growth striations in semiconductors. Saleh Jabarin, Ph.D., Massachusetts: polymer materials and orientation, crystallization and morphology, permeation and transport behavior of polymers. Glenn Lipscomb, Ph.D., Berkeley: separating gas or liquid mixtures, development of improved polymeric materials, membrane separation systems. Arunan Nadarajah, Ph.D., Florida: transport and growth processes in fluid systems, semiconductor materials, protein crystallization, protein adsorption. Bruce Poling, Ph.D., Illinois: physical property estimation of environmentally important compounds, phase equilibrium in systems with environmental applications. Constance Schall, Ph.D., Rutgers: crystallization and precipitation processes, bioseparations, protein crystallization. Sashidar Varanasi, Ph.D., SUNY at Buffalo: polymers at interfaces and immobilized enzyme technology.

Civil Engineering: Brian Randolph, Chairperson; Ph.D., Ohio State: subsurface instrumentation, geosynthetics, soil testing, flow modeling. Donald Angelbeck, Ph.D., Purdue: reliability of conventional biological and chemical wastewater treatment processes, utilization of waste products of sludge stabilization. Eddie Chou, Ph.D., Texas A&M: transportation facilities design, systems analysis, engineering material properties, pavement performance evaluation, infrastructure management. Kuan-Chen Fu, Ph.D., Notre Dame: structural analysis, finite element and boundary element methods, optimal systems design. Jiwan Gupta, Ph.D., Waterloo: transportation engineering and intelligent transportation system (ITS). Andrew Heydinger, Ph.D., Houston: foundation engineering, laboratory testing, field instrumentation and mathematical modeling, analysis of deep foundations, geoenvironmental engineering, testing of pavement base and subbase materials. Ashok Kumar, Ph.D., Waterloo: air pollution, risk analysis and environmental information technology. Blair McDonald, Ph.D., Utah: soil mechanics and geotechnical engineering. Naser Mostaghel, Ph.D., Berkeley: structural mechanics, earthquake engineering, blast-proof structures. George Murnen, Ph.D., Notre Dame: matrix analysis of structures and finite elements. Douglas Nims, Ph.D., Berkeley: passive seismic control of buildings, shape memory energy dissipators, seismic restraint of piping. Azadeh Parvin, D.Sc., George Washington: structural engineering. Mark Pickett, Ph.D., Connecticut: extreme load safety analysis of structures and mechanical components.

Electrical Engineering and Computer Science: Adel A. Ghandakly, Chairperson; Ph.D., Calgary: computer control of dynamic systems, adaptive control systems, electric glass melting modeling and control. Mansoor Alam, Ph.D., Indian Institute of Science: fault tolerance and reliability, high-performance computer networking. Dr. Adel H. Eltimsahy, Ph.D., Michigan: robotics, robotic control, neural networks and fuzzy logic, flexible manipulators. Sammie Giles Jr., Ph.D., Michigan: electromagnetic field theory, EMP signal processing. Gerald R. Heuring, Ph.D., Illinois at Urbana-Champaign: physically based modeling and animation, modeling natural phenomena, file system design, object-oriented design/programming. Mohsin M. Jamali, Ph.D., Windsor: real-time digital systems, automotive electronics. Anthony D. Johnson, Director, VLSI Design Tool Laboratory; Ph.D., Belgrade: VLSI and ASIC system design. Vik J. Kapoor, President; Ph.D., Lehigh: biosensors, bioelectronics, signal processing. Devinder Kaur, Ph.D., Wayne State: computer architecture, design of microcontrolled systems, fuzzy logic. Junghwan Kim, Ph.D., Virginia Tech: communications and networking, mobile communications, digital signal processing. Roger J. King, Ph.D., Toledo: power electronics, switching power converters, modeling and control. Subhash C. Kwatra, Ph.D., South Florida: satellite communications, data compression, intelligent transportation systems. Henry F. Ledgard, Ph.D., MIT: human factors, software engineering, programming languages. Richard G. Molyet, Ph.D., Toledo; Automatic control, robotics. Ezzatollah Salari, Ph.D., Wayne State: data compression and coding, image/video/signal processing, multimedia. Gursel Serpen, Ph.D., Old Dominion: artificial intelligence, artificial neural networks, fuzzy systems and evolutionary computing, real-time systems. Edwyn D. Smith, Ph.D., Arizona: electronic circuits, solid-state electronics, VLSI. Hilda M. Standley, Ph.D., Toledo: compiler design, high-performance architecture. Thomas A. Stuart, Ph.D., Iowa State: electrical power systems, power electronics.

Mechanical, Industrial and Manufacturing Engineering: Robert J. Abella, Ph.D., Toledo: manufacturing systems, numerical control, production management, computer-aided manufacturing. Abdollah A. Afjeh, Toledo: fluid dynamics, propulsion systems, computational methods, energy conversion systems. Robert A. Bennett, Ph.D., Wayne State: plastics recycling, materials, product design, engineering economy, applied statistics. Frank Chen, Ph.D., Missouri–Columbia: manufacturing cell design and cell control systems, machining and tooling technologies, planning and operation modules for flexible manufacturing systems. Ali Fatemi, Ph.D., Iowa: solid mechanics, mechanical design, materials mechanical behavior, composite materials, fatigue, fracture mechanics. M. Samir Hefzy, Ph.D., Cincinnati: orthopedic biomechanics and rehabilitation engineering, with a focus on the knee joint. Theo G. Keith, Ph.D., Maryland: tribology, computational fluid dynamics, fluid mechanics, heat transfer, aeroelasticity. Steven N. Kramer, Ph.D., Rensselaer: mechanisms, robotics, dynamics of mechanical systems. K. Cyril Masiulaniec, Ph.D., Toledo: phase change heat transfer, convective heat transfer from roughened surfaces, high-temperature heat exchangers and industrial furnaces. Ioan D. Marinescu, Ph.D., Galatzi: manufacturing processes, tribology. Roger J. McNichols, Ph.D., Ohio State: reliability, applied statistics and experimental design, environmental monitoring, manufacturing processes. Nagi G. Naganathan, Ph.D., Oklahoma State: smart materials, dynamics, vibrations, robotics. Tsung-Ming Terry Ng, Ph.D., Berkeley: aerodynamics, stability, turbulence, experiment. Keytack H. Oh, Ph.D., Ohio State: engineering economy and decision strategy, human factors, production management systems. Douglas L. Oliver, Ph.D., Washington State: heat transfer, mathematical modeling, microgravity flow dynamics. Walter W. Olson, Ph.D., Rensselaer. Mehdi Pourazady, Ph.D., Cincinnati: finite element methods, computer-aided design and manufacturing. Phillip R. White, Ph.D., Purdue: computer graphics, solid modeling, computer-aided design and manufacturing.

UNIVERSITY OF TORONTO

Faculty of Applied Science and Engineering

Programs of Study

Graduate programs are offered leading to the degrees of Master of Applied Science, Master of Engineering, and Doctor of Philosophy in aerospace science and engineering, biomedical engineering, chemical engineering and applied chemistry, civil engineering, electrical and computer engineering, environmental engineering, mechanical and industrial engineering, metallurgy and materials science, and welding engineering.

Candidates for the M.A.Sc. degree should hold a B.A.Sc. degree from the University of Toronto or an equivalent degree in engineering. An applicant with a bachelor's degree in science or applied mathematics may be admitted on the recommendation of the department concerned and subject to the approval of the School of Graduate Studies. The program includes course work and the preparation of a research thesis, the latter being the major requirement. The minimum residence requirement for the degree is one academic session, and the requirements for the degree must be completed within three calendar years.

Similarly, candidates for the M.Eng. degree should hold a B.A.Sc. degree from Toronto or an equivalent degree in engineering. An applicant with a bachelor's degree in science or applied mathematics may be admitted on the recommendation of the department concerned and subject to the approval of the School of Graduate Studies. The M.Eng. program may include a project in addition to lecture and laboratory courses. The degree program must be completed within six calendar years of registration. The program can usually be completed in one year of full-time study.

The Ph.D. degree program, which normally requires a minimum of two years of full-time study beyond the master's degree, includes advanced course work in addition to the major requirement of a research thesis.

Research Facilities

Each department has laboratories with modern equipment for research and instruction, as well as shop facilities for the construction of special equipment. In addition, researchers have access to such instrumentation as electron microscopes and X-ray equipment, wind tunnels, a towing channel, data acquisition systems, and extensive computing facilities.

The University's library resources consist of approximately 5 million volumes, 800,000 microtexts, 100,000 maps, 7,000 manuscript titles, and 310,000 other items.

Financial Aid

Most research assistantships are available only to Canadian citizens and landed immigrants. In 1998–99, research and teaching assistantships provided up to Can$15,000 for a calendar year. Residents of Ontario should apply for Ontario graduate scholarships, and applicants eligible for awards of the Natural Sciences and Engineering Research Council of Canada (NSERC) should apply through the University or directly to NSERC early in the fall term.

Cost of Study

For 1998–99, tuition and incidental fees for full-time study were Can$4070 per year for Canadian citizens and landed immigrants and Can$9077 per year, including health insurance, for those on student visas. A limited number of differential fee waivers are available to visa students.

Living and Housing Costs

The minimum annual cost of living in Toronto for a single student at a medium standard (in a University residence) is estimated to be $15,000, excluding academic and incidental fees.

Student Group

The University of Toronto has an enrollment of about 52,200 full- and part-time students. The Faculty of Applied Science and Engineering has about 3,200 undergraduate and 1,100 full- and part-time graduate students. Students are drawn from every province of Canada and all parts of the world.

Location

Toronto is a city of nearly 3 million people. It is the capital of the province of Ontario and is situated on the north shore of Lake Ontario. Culturally rich, it is the headquarters of the Canadian Opera Company, the National Ballet Guild of Canada, the National Youth Orchestra, and the Ontario Drama Festival and the home of the internationally known Toronto Symphony Orchestra. It has an extensive number of live theaters. In addition to the world-famous Royal Ontario Museum, there is the Ontario Art Gallery and the Ontario Science Centre.

Fifty to 100 miles north of Toronto are lakes, forests, conservation areas, and ski runs.

The University

Founded by Royal Charter in 1827, the University of Toronto is one of the oldest universities in Canada, with a staff of approximately 6,000 and an enrollment of 35,000 full-time students. The resources of the University permit it to offer an environment conducive to challenging research. The main campus of the University, known as the St. George campus, occupies about 52 hectares of land (about 129 acres) in downtown Toronto. In 1998, Maclean's magazine rated the University of Toronto highest among research-intensive universities in Canada for the fifth consecutive year.

Applying

Before making a formal application, prospective students are advised to visit or write to the appropriate department, supplying all pertinent information. Application forms may be obtained on request from the Secretary, School of Graduate Studies. It is suggested that admission applications be forwarded before February 1 for the session commencing in September. All students who are accepted for admission and whose applications have been received before February 1 are automatically considered for University of Toronto fellowship support.

Correspondence and Information

Coordinator of Graduate Studies
Department of (specify)
University of Toronto
Toronto, Ontario M5S 1A4
Canada
World Wide Web: http://www.ecf.utoronto.ca/apsc

University of Toronto

FACULTY HEADS AND RESEARCH AREAS

Chemical Engineering
D. G. B. Boocock, Professor and Chair of the Department; Ph.D., London; CIC, DIC, FCIC.
M. T. Kortschot, Professor and Coordinator of Graduate Studies; Ph.D., Cambridge; PE.

Civil Engineering
B. J. Adams, Professor and Chair of the Department; Ph.D., Northwestern; PE.
S. A. Sheikh, Professor and Coordinator of Graduate Studies; Ph.D., Toronto; PE.

Electrical and Computer Engineering
S. G. Zaky, Professor and Chair of the Department; Ph.D., Toronto; PE.
A. N. Venetsanopoulos, Associate Chair of Graduate Studies; Ph.D., Yale; D.Eng., Athens; FEIC, FIEEE, PE.

Mechanical and Industrial Engineering
J. S. Wallace, Professor and Chair of the Department; Ph.D., Michigan; PE.
A. Mandelis, Professor, Associate Chair, and Coordinator of Graduate Studies; Ph.D., Princeton; FAPS.

Metallurgy and Materials Science
D. D. Perovic, Professor and Chair of the Department; Ph.D., Toronto.
Z. Wang, Professor and Associate Chair of the Department; Ph.D., Polytechnic.

Environmental Engineering (Collaborative Program)
P. H. Byer, Professor of Civil Engineering and Chair of the Division of Environmental Engineering; Ph.D., MIT; PE.

Institute for Aerospace Studies
A. A. Haasz, Professor and Director; Ph.D., Toronto; AFCASI, PE.
L. D. Reid, Professor and Associate Director; Ph.D., Toronto; FCASI, PE.

Institute of Biomaterials and Biomedical Engineering
H. Kunov, Professor and Acting Director of the Institute; Ph.D., Copenhagen.
A. M. Dolan, Associate Professor and Coordinator of Graduate Studies; M.Sc., Missouri; PE.
K. H. Norwich, Professor and Associate Director; Ph.D., Toronto.

CURRENT AREAS OF RESEARCH INTEREST

Aerospace Studies: dynamics and control of aircraft and spacecraft; space robotics; compressible flows and shock waves; combustion and propulsion; low-speed aerodynamics; mechanics of gases, plasmas, and molecular beams; ultrahigh vacuum technology; mechanics of solids and structures; air-cushion and air-bearing technology; fusion reactor materials and technology; smart structure and optical-fiber sensor technology; helicopter flight testing; man-machine systems; hypersonic and high-temperature gasdynamics; flight and ground vehicle simulators; computational fluid dynamics. For further information, students can refer to the Web site (http://www.utias.utoronto.ca).

Biomedical Engineering: metabolic control systems, biomaterials, vascular disease detection, microprocessors in medicine, nuclear medicine engineering, neurophysiology, hearing and acoustics, ultrasound, microcirculation, vision research, drug delivery device, orthopaedic implant design, bone-material interface, biodegradable implants, tissue mechanics, load-bearing biomaterials, microencapsulation of mammalian cells, drug delivery systems, biomechanics, tissue engineering, implants and tissue repair/regeneration, dental implant design, dental restorative materials, degradation of biomaterials, mechanical characteristics of biomaterials, blood-interfacing biomaterials, surface modification, composite biomaterials, metallic biomaterials, bioceramics, biopolymers, hydrogels, trace element analysis, processing and properties of biomaterials, neural grafts, cell-biomaterial interactions, implant-related bone remodeling, device retrieval and analysis, surface science. For further information, students should refer to the Institute's Web site (http://www.ibbme.utoronto.ca).

Chemical Engineering and Applied Chemistry: fluid mechanics and transport, rheology, polymers and materials science, advanced ceramics, biomaterials, pulp and paper, nuclear engineering, radiochemical technology, food engineering, biotechnology, environmental engineering, optimization, process control and modelling, chemical reactor design and analysis, electrochemistry, catalysis and surface science, hydrometallurgy. For further information and research links, students should refer to the department's Web site (http://www.chem-eng.utoronto.ca).

Civil Engineering: structural engineering (concrete, steel, and composite structures; materials engineering; building science; rehabilitation and restoration of structures; construction engineering and management), geomechanics (rock mechanics and engineering, soil mechanics, mining applications, transportation planning and engineering, intelligent infrastructure systems, environmental and water resource engineering, environmental remediation, environmental planning). For further information, students should refer to the department's Web site (http://www.civ.toronto.edu).

Electrical and Computer Engineering: communications—digital signal processing, digital communications, communication networks, data compression, information theory and coding, satellite communications, image processing and multimedia, spread spectrum systems, wireless communications, and radio networks; computers—computer architecture, systems programming, computer networks, distributed systems, trusted systems, array processors, nonbinary logic, digital circuits, computer-aided design, and computer applications; FPGA applications and systems; electronics—semiconductor device physics, integrated circuit design, network theory, linear and digital circuits, transport and optical properties of semiconductors, VLSI design and technology, filters, computer-aided circuit design and testing, and solid-state transducers; power devices and systems—electric power systems, high-voltage phenomena, energy conversion, power modulators, power semiconductor systems, induction heating, electromagnetic field–fluid interaction, magnetic materials, linear motors, electric propulsion systems, machine systems stability, and electromechanical devices; electromagnetics—electromagnetic theory and measurement, antennas, electromagnetic compatibility, nonlinear wave interactions, quasi-optics, microwave circuits, and bioelectromagnetics; systems control—foundations of control theory, control of multivariable systems, control of discrete-event systems, process modelling and identification, stochastic control, adaptive control, control of queuing systems, microprocessor control systems, large-scale system theory, optimization and simulation, metallurgical control, urban traffic control; photoelectronics—crystal optics, fiber sensors, solar cells, integrated optics, lightwave technology, optoelectronic devices, nonlinear optics, optical properties of semiconductors, semiconductor laser and optoelectronic devices, and short-wavelength lasers. For further information about the department, including degree programs, students should refer to the department's Web site (http://www.ece.utoronto.ca/grad/handbook/).

Environmental Engineering (Collaborative Program): air pollution control, clean manufacturing processes, environmental fluid mechanics and dynamics, environmental planning, fate and transport of contaminants, hydrogeology, incineration and combustion, materials recycling and recovery, municipal and industrial wastewater treatment, nuclear engineering and waste disposal, preventive engineering, pulp and paper processes, soil and groundwater remediation, solid-waste management, stormwater management, sustainable transportation systems, urban infrastructure rehabilitation, water resources analysis, water and air quality modelling, water treatment. For further information, students can refer to the department's Web site (http://www.ecf.utoronto.ca/apsc/misc/envir.html)

Mechanical and Industrial Engineering: applied mechanics, biomedical engineering, computer-aided engineering, energy studies, fluid mechanics and hydraulics, human factors engineering, management information systems, management science, materials, manufacturing, operations research, robotics, automation and control, social impact of technology, systems design and optimization, surface sciences, thermodynamics and heat transfer, plasma processing, vibration, computational fluid dynamics, environmental engineering, coatings, finite element methods, internal combustion engines, spray-forming processes, laser photothermal and optoelectronic diagnostic science and instrumentation, ultrasonic nondestructive evaluation. For further information, students should refer to the Web site (http://www.mie.utoronto.ca).

Metallurgy and Materials Science: nonferrous pyrometallurgy, iron and steel making, process metallurgy, welding, phase transformations, mechanical properties, corrosion, ceramics, biomaterials, amorphous materials, composite materials, nuclear materials, electronic materials, microstructural science, nanotechnology. For further information, students should refer to the department's Web site (http://www.ecf.utoronto.ca/apsc/mms).

UNIVERSITY OF VERMONT

College of Engineering and Mathematics

Programs of Study
Small and vibrant graduate programs are offered in six curricular areas and in three interdisciplinary areas.

Master of Science degree programs are offered in biomedical engineering, civil and environmental engineering, electrical engineering, mechanical engineering, computer science, materials science, mathematics, statistics, and biostatistics. In most programs, thesis and nonthesis options are available. All programs require a minimum of 30 semester hours of approved graduate credit; the civil and environmental engineering program requires 36 hours for the nonthesis option. In thesis options, the number of credit hours earned in thesis research varies between 6 and 12 credits; these credits are included in the minimum required for the degree. Ph.D. programs are offered in the areas of civil and environmental engineering, electrical engineering, mechanical engineering, mathematical sciences (pure and applied), and materials science. At least 75 credit hours must be earned in course work and dissertation research. Satisfactory performance on a comprehensive examination, completion of an acceptable dissertation, and an oral defense of the dissertation are required.

Research Facilities
The College is especially strong in the area of computational research. The EMBA Computer Facility provides a sophisticated UNIX computing environment for the networking of workstations, microcomputers, and superminicomputers throughout the College for use by faculty and staff members and students. This network includes two parallel processing machines from Silicon Graphics, an Encore, five Sun Microsystems servers, and an IBM RS6000/950.

Each department has research laboratories for its specialized areas. The computer science department has research expertise in communications, algorithms, and networks. The electrical and computer engineering department has research capabilities in areas of digital signal processing, optical fiber sensors, semiconductor processing and device characterization, electromagnetic field theory, speech recognition, and VLSI design and testing. In civil and environmental engineering, laboratories support soil and groundwater remediation, surface-water hydrology, inhalation toxicology, hydraulics, structures, and geotechnical research. Mechanical engineering laboratories support research in manufacturing, metallurgy, heat transfer, fluid mechanics, biomechanics, materials processing and testing, robotics, and vibrations. The biomedical engineering, mathematics, and biostatistics programs closely collaborate and share facilities with the College of Medicine. Sixty workstations, fifty X-terminals, three microcomputer laboratories, and two graphics laboratories support instructional and research needs.

Additional academic support includes the Bailey/Howe Library, which houses the University's main collection of more than 1 million volumes; a physics and chemistry library located in the Cook Physical Sciences Building; and collections in medicine and health sciences located in the Dana Medical Library.

Financial Aid
Most full-time graduate students receive a graduate teaching fellowship or a graduate research fellowship. Graduate teaching fellows are generally appointed for nine months and receive stipends averaging $11,275 and tuition scholarships covering a maximum of 10 credit hours per semester. Graduate research assistants, generally appointed for nine or twelve months, receive stipends ranging from $17,250 (nine months) to more than $21,250 (twelve months), including tuition at in-state rates. Students requesting financial support must submit scores from the General Test of the Graduate Record Examinations. Possibilities exist for summer support.

Cost of Study
Tuition rates for 1999–2000 are $311 per credit hour for Vermont residents. For nonresidents, the rates are $778 per credit hour. The New England Regional Student Program provides reduced tuition in selected programs for students who are residents of Connecticut, Maine, Massachusetts, New Hampshire, or Rhode Island.

Living and Housing Costs
Up-to-date listings maintained by the Department of Residential Life are available for many apartments, houses, and rooms for rent at or near the University. Rents in the area vary from approximately $75 per week for a single room to $600 or more for a two-bedroom apartment. A single student should expect minimum overall living expenses to be about $750 per month.

Student Group
There are 159 graduate students in the College of Engineering and Mathematics. The Graduate College has 1,112 students, and the campus population numbers about 10,000.

Student Outcomes
The College's graduate students get field-related jobs without exception; some are in Vermont. In civil and environmental engineering, recent employers include IBM, Camp Dresser and McKee, Vollmer Associates, Research Engineers Inc., and GEO Engineering. Electrical engineers and computer scientists have recently joined IBM, Hewlett-Packard, IDX, and Raytheon. Likewise, recent mechanical engineering graduates have joined IBM, NASA, Lockheed-Martin, Corning, and Air Quality Associates. Some mathematics and statistics employers include IBM, VT-Life, and Sun Microsystems.

Location
Burlington is Vermont's largest city, with a population of approximately 39,000. Greater Burlington, with some 132,000 inhabitants, is divided between pleasant suburbs, farms, and woodland. Burlington enjoys magnificent views of Lake Champlain and the Adirondack Mountains to the west, plus Vermont's Green Mountains to the east. Outdoor activities include swimming, boating, hiking, climbing, and skiing. Burlington is easily accessible by air, bus, train, and car.

The University
The University of Vermont was founded by charter in 1791, the same year that Vermont became the fourteenth of the United States. It consists of the Colleges of Agriculture and Life Sciences, Arts and Sciences, Engineering and Mathematics, Education and Social Services, and Medicine; the Graduate College; the Schools of Allied Health Sciences, Business Administration, Natural Resources, and Nursing; and the Office of Continuing Education.

Applying
Students seek formal admission to the Graduate College and must apply on an official form obtained from the Graduate College Admissions Office. Scores from the Graduate Record Examinations are required. Completed applications and supporting materials must be received in the Graduate College Admissions Office by April 1 for admission to the fall semester (by March 1 for financial aid consideration) for most programs and by February 1 for the civil and environmental engineering program. Applicants for admission must hold a baccalaureate degree prior to the date of first enrollment or have completed work equivalent to that required for a baccalaureate. International students must submit results from the Test of English as a Foreign Language (TOEFL).

Correspondence and Information
Graduate Coordinator for (area)
University of Vermont
Burlington, Vermont 05405

Telephone: 802-656-3390
E-mail: emdean@emba.uvm.edu
World Wide Web: http://www.emba.uvm.edu

University of Vermont

THE FACULTY AND THEIR RESEARCH

Civil and Environmental Engineering
Jean-Guy Beliveau, Ph.D., Princeton, 1974. Structural dynamics, system identification.
Robert F. Dawson, Ph.D., Purdue, 1964. Aggregates, material reliability, pavement design, stochastic modeling.
David E. Dougherty, Ph.D., Princeton, 1985. Transport in porous media, computational methods, optimization.
Richard N. Downer, Ph.D., Colorado State, 1967. Hydraulics, hydrology, open-channel flow.
Nancy J. Hayden, Ph.D., Michigan State, 1992. Remediation of hazardous sites, chemical and physical processes.
David R. Hemenway, Ph.D., North Carolina, 1974. Aerosol generation and monitoring, inhalation toxicology, air pollution.
† Jeffrey P. Laible, Ph.D., Cornell, 1973. Dynamic and finite-element modeling of structures, hydrodynamics.
James P. Olson, Ph.D., North Carolina State, 1969. Geotechnics, soil behavior, groundwater.
George F. Pinder, Ph.D., Illinois, 1968. Groundwater behavior, numerical modeling, experimental studies of multiphase media.
Donna M. Rizzo, Ph.D., Vermont, 1994. Groundwater hydrology, numerical methods, optimization.
Adel W. Sadek, Ph.D., Virginia, 1998. Transportation, ITS, modeling, simulation.

Computer Science
Charles J. Colbourn, Ph.D., Toronto, 1980. Combinational designs, network design and analysis.
Byung Lee, Ph.D., Stanford, 1991. Database systems.
Robert R. Snapp, Ph.D., Texas at Austin, 1987. Pattern recognition, neural networks, learning algorithms, image analysis.
Guo-Liang Xue, Ph.D., Minnesota, 1991. Numerical optimization, parallel and distributed computing.

Electrical and Computer Engineering
† Richard G. Absher, Ph.D., Duke, 1967. Speech processing, digital control, digital logic design and test.
Peter L. Fuhr, Ph.D., Johns Hopkins, 1985. Electrooptical systems, laser communications, fiber-optic sensors.
Gagan Mirchandani, Ph.D., Cornell, 1968. Digital signal processing, parallel processing, hardware implementation.
Kurt Oughstun, Ph.D., Rochester, 1978. Electromagnetic wave/optical field theory, dispersive pulse propagation.
* Stephen Titcomb, Ph.D., Lehigh, 1983. Semiconductor device physical electronics, cryoelectronics, engineering education.
* Walter J. Varhue, Ph.D., Virginia, 1984. Plasma processing of electronic materials.
Ronald W. Williams, Ph.D., Iowa State, 1966. Microprocessors-based design, VLSI circuit design.

Mathematics
Daniel S. Archdeacon, Ph.D., Ohio State, 1980. Combinatorics, graph theory.
Daniel E. Bentil, Ph.D., Oxford, 1990. Biomathematics.
James W. Burgmeier, Ph.D., New Mexico, 1969. Numerical analysis, modeling, software.
Roger L. Cooke, Ph.D., Princeton, 1966. Fourier analysis.
Jeffrey H. Dinitz, Ph.D., Ohio State, 1980. Combinatorics.
David Dummit, Ph.D., Princeton, 1980. Algebraic number theory.
Richard M. Foote, Ph.D., Cambridge, 1976. Group theory, algebraic number theory.
Kenneth I. Golden, Ph.D., Paris (Sorbonne), 1964. Applied mathematics, plasma, electrodynamics and statistical mechanics.
Kenneth I. Gross, Ph.D., Washington (St. Louis), 1966. Applied mathematics, asymptotics, mechanics, biomathematics.
† William D. Lakin, Ph.D., Chicago, 1968. Applied mathematics, asymptotics, mechanics, biomathematics.
Jonathan W. Sands, Ph.D., California, San Diego, 1982. Algebraic number theory, computational number theory.
J. Michael Wilson, Ph.D., UCLA, 1981. Fourier analysis.
Robert K. Wright, Ph.D., Columbia, 1966. Approximation theory, computational linear algebra.
Jianke Yang, Ph.D., MIT, 1994. Applied mathematics, asymptotics, nonlinear waves.
† Jun Yu, Ph.D., Washington (Seattle), 1988. Applied computational mathematics, biomathematics.

Mechanical Engineering
† Bruce D. Beynnon, Ph.D., Vermont, 1991. Biomechanics.
Naomi C. Chesler, Ph.D., MIT, 1996. Cardiovascular fluid dynamics, assistive technology design.
* Delcie R. Durham, Ph.D., Vermont, 1981. Mechanical metallurgy, material characterization, product and process design.
Ted B. Flanagan, Ph.D., Washington (Seattle), 1955. Physical chemistry, thermodynamics.
Gerald P. Francis, Ph.D., Cornell, 1964. Fluid mechanics.
Clarke E. Hermance, Ph.D., Princeton, 1963. Transient- and steady-state combustion and heat transfer.
Darren L. Hitt, Ph.D., John Hopkins, 1997. Experimental, computational, and theoretical aspects of fluid dynamics, acoustics, and biophysics.
Mahendra S. Hundal, Ph.D., Wisconsin, 1964. Computer-aided design, product development.
† Dryver R. Huston, Ph.D., Princeton, 1986. Electromechanical design, smart structures, biomechanics.
*† Tony S. Keller, Ph.D., Vanderbilt, 1988. Biomechanics, stereology, fracture mechanics.
Ian A. Stokes, Ph.D., Polytechnic of Central London, 1975. Biomechanics of the spine.
* Branimir F. von Turkovich, Ph.D., Illinois, 1961. Manufacturing, high-speed material removal, materials.
* JunRu Wu, Ph.D., UCLA, 1985. Ultrasound, lasers.

Statistics and Biostatistics
John Aleong, Ph.D., Iowa State, 1975. Experimental design, reliability, quality control, statistical methods.
Takamaru Ashikaga, Ph.D., UCLA, 1973. Discriminant analysis, pattern recognition, program evaluation.
Jeffrey Buzas, Ph.D., North Carolina State, 1993. Measurement error models, instrumental variables, statistical computing.
Peter Callas, Ph.D., Massachusetts, 1994. Epidemiology, medical biostatistics.
Michael C. Costanza, Ph.D., UCLA, 1977. Biostatistics, multivariate analysis, survival analysis.
† Larry D. Haugh, Ph.D., Wisconsin, 1972. Quality control, reliability, time series, applied statistics, medical biostatistics.
Ruth M. Mickey, Ph.D., UCLA, 1983. Categorical data analysis, biostatistical methods in epidemiology and physiology.
Mun S. Son, Ph.D., Oklahoma State, 1984. Time series, regression, econometrics, sequential analysis.

Biomedical Engineering
Biomedical Engineering is an interdisciplinary program that includes faculty members from Mechanical Engineering, Electrical Engineering, and the College of Medicine. The symbol (†) indicates participating faculty members from the College of Engineering and Mathematics.

Materials Science
Materials Science is an interdisciplinary program that includes faculty members from Mechanical Engineering, Electrical Engineering, Chemistry, and Physics. The symbol (*) indicates participating faculty members from the College of Engineering and Mathematics.

UNIVERSITY OF VIRGINIA

School of Engineering and Applied Science

Programs of Study

The University of Virginia School of Engineering and Applied Science offers Master of Science, Master of Engineering, and Doctor of Philosophy programs in ten curricula: applied mechanics (no Ph.D.), biomedical engineering, chemical engineering, civil engineering, computer science, electrical engineering, engineering physics, materials science and engineering, mechanical and aerospace engineering, and systems engineering.

The Master of Science degree requires the completion of a minimum of 24 semester hours of graduate courses and 6 semester hours of research and the submission of a thesis. The Master of Engineering degree requires a minimum of 30 semester hours of graduate course work; no thesis is required. Well-prepared students require a minimum of one calendar year for degree completion, including a nine-week summer session. For the Ph.D. degree, students typically complete 24 semester hours of courses and 24 semester hours of research beyond the master's degree, pass a comprehensive written and oral examination, and submit and defend a dissertation.

Research Facilities

Each department maintains extensive modern research facilities pertinent to its own research programs and has organized interdisciplinary research centers as well as departmental laboratories. Interdisciplinary research is carried out through research centers in which graduate students in two or more disciplines work together on a research project. There are thirty-one such centers currently in operation: the Aerogel Research Laboratory; the Aerospace Research Laboratory; the A. H. Small Center for Computer Aided Engineering; the Applied Electrophysics Laboratory; the Center for Advanced Computational Technology; the Center for Bioprocess Development; the Center for Electrochemical Science and Engineering; the Center for Engineering of Wound Prevention and Repair; the Center for Genetic Engineering Targeting Vascular Disease; the Center for High Temperature Composites; the Center for Magnetic Bearings; the Center for Risk Management of Engineering Systems; the Center for Semicustom Integrated Systems; the Center for Survivable Information Systems; the Center for Transportation Studies; the Communications, Control, and Signal Processing Laboratory; the Composite Mechanics Laboratory; the Injury Prevention Program; the Institute for Technology in Medicine; the Intelligent Processing of Materials Laboratory; the Internet Technology Innovation Center; the Legion–Meta-Computing facility; the Light Metals Center; the Mathematical-Computational Modeling Laboratory; the Microscale Heat Transfer Laboratory; the Networking Multimedia facility; the Next-Generation Real-Time Systems Laboratory; the Rotating Machinery and Controls Industrial Program; the Institute for Parallel Computation; the Light Aerospace Alloy and Structure Technology Program; The University of Virginia's Institute for Microelectronics; and the Virginia Laboratory for Engineering and Automated Design. The University's Information Technology and Communication Organization supports a wide variety of computing environments ranging from single-user workstations to supercomputing. The majority of these are housed throughout the University and within the ITC computing center. The major supported environments are: UNIX computing servers and desktop workstations, including IBM RiscSystem/6000; Silicon Graphics and Sun computers; Windows-based microcomputers and Apple Macintoshes; and supercomputing and parallel computing resources that are available through affiliation with the National Partnership for Advanced Computational Infrastructure. As a participant in the National Science Foundation's Internet 2 Project, the University provides very high-speed (ATM) network connections to other top-tier research universities and national laboratories.

Financial Aid

Financial aid is available in the form of fellowships and research and teaching assistantships, which usually carry stipends of $16,000 and up, plus tuition and fees for the calendar year.

Cost of Study

Full-time tuition and fees for two semesters in 1998–99 were approximately $4900 for Virginia residents and $15,900 for out-of-state students. Part-time tuition and fees were proportionately less. University fees are subject to change.

Living and Housing Costs

Dormitory facilities are available for single students for $1960–$3110 for the academic year. University-operated accommodations for student families consist of furnished one-bedroom apartments for $467 per month, two-bedroom apartments for $505 per month, and three-bedroom apartments for $577 per month (including utilities). An off-campus housing bureau maintains a listing of privately owned rooms, apartments, and houses for students and student families. The cost of living is comparable to that in most medium-to-large cities in the United States. Housing costs are subject to change.

Student Group

Approximately 18,000 students, including about 6,200 graduate and professional students, are currently enrolled in the University. There are about 2,400 students in the School of Engineering and Applied Science—approximately 1,800 undergraduates and 600 graduate students. In a typical year, approximately 400 bachelor's degrees, 200 master's degrees, and 80 Ph.D. degrees are awarded to students in the School of Engineering and Applied Science.

Location

Charlottesville, including its environs, is a community of approximately 100,000 situated in the foothills of the Blue Ridge Mountains. The climate is relatively mild, and the opportunity for outdoor recreation extends throughout most of the year. The Shenandoah National Park is 20 miles away, Wintergreen Ski Resort is 38 miles, and Washington, D.C., is 110 miles. The Tuesday Evening Concerts present a number of outstanding concerts each year, and there are frequent popular concerts and art exhibitions in both the University and the community. An active players group and visiting troupes offer further entertainment. The area contains many places of historic interest, including Monticello and Ash Lawn, the homes of Jefferson and Monroe, respectively.

The University and The School

The University of Virginia, founded in 1819 by Thomas Jefferson, is a state-aided institution. Its graduate student body is nationally and internationally diverse. It is widely known for its effective student-controlled honor system and its outstanding programs in a variety of areas, including engineering and applied science. Approximately 180 full-time faculty members are active in teaching, research, and public service in the School of Engineering and Applied Science.

Applying

Applicants must have a baccalaureate degree from a recognized college or university, should have at least a B average, and should furnish three letters of reference, transcripts of all academic course work, and the results from the Graduate Record Examinations. Applications from students whose native language is other than English must include the results of the Test of English as a Foreign Language. Students may apply for January, June, or September admission. Those applying for the fall semester who desire financial aid must submit a complete application by February 1. Applications, however, are accepted up to one month prior to each possible entrance date. International students who are holders of a visa other than a Permanent Resident or Immigrant Visa must submit complete applications by August 15 for January admission or by April 15 for September admission. For additional information, students should write to the chairperson of the appropriate department or curriculum or to the address below.

Correspondence and Information

Graduate Studies
School of Engineering and Applied Science
A 115 Thornton Hall
University of Virginia
Charlottesville, Virginia 22903-2442
Telephone: 804-924-3897
   804-982-HEAR (TDD)
E-mail: seas-grad-admission@Virginia.edu

University of Virginia

RESEARCH AREAS

Applied Mechanics. Graduate students in applied mechanics conduct research on fundamental problems in solids, fluids, and motion. In most cases, financial support is available in the form of research assistantships, teaching assistantships, and/or fellowships. Research areas include mechanics of composite materials, shell theory, nonlinear elasticity, fracture mechanics, structural mechanics, random vibrations, continuum mechanics, high-temperature test methods, thermal structures, fluid mechanics, nonlinear dynamical systems and chaos, biomechanics, optimization, finite elements, anistropic elasticity, vibrations, galactic dynamics, planetary systems, fatigue, and micromechanics. (804-924-3605)

Biomedical Engineering. Research specialties include biomechanics, biotransport, cellular and tissue engineering, genetic engineering, medical imaging, neural engineering, and orthopedic engineering. Active research areas include microvascular mechanics, vascular remodeling, transport across blood-brain barrier, blood volume control, adhesion biomechanics, targeted gene delivery, molecular mechanics, ultrasound imaging, digital radiographic detection, rapid MRI acquisition, MRI and ultrasonic contrast agents, ion-channel dysfunction, orthopedic mechanics, and motion analysis. (804-924-5101)

Chemical Engineering. Fundamental research areas include catalysis, diffusion and mass transfer, electrochemical systems, interfacila characterization, thermodynamics and physical properties of solutions and surfactants, molecular modeling, magnetic resonance imaging, and nonlinear dynamics of chemical reactions. Applications-oriented programs include separations technology such as adsorption, chromatography and ion exchange, and crystallization; reaction engineering, including electrochemical and solid-state processing; bioprocessing technology and bioseparations; bioremediation; and materials synthesis. (804-924-7778)

Civil Engineering. The three principal areas of research and graduate study are water resources and environmental engineering, transportation engineering and management, and structural and applied mechanics. The environmental engineering program emphasizes research in watershed management, stormwater and nonpoint pollution control, surface and groundwater hydrology, contaminant transport, water quality modeling, groundwater remediation, sorption of organic pollutants to natural soil, and planning of environmental systems. Research in transportation engineering focuses on traffic systems management, traffic flow theory, intelligent transportation systems, transportation planning, safety, public transportation, and intermodal freight. Research in structural and applied mechanics focuses on dynamic response of structures, numerical techniques in structural mechanics, random vibrations and stochastic processes, dynamic stability of structural systems, field testing of highway bridges, composite material systems, functionally graded materials, and optimization in design. (804-924-7464)

Computer Science. Research in computer science includes operating systems, analysis of algorithms and computation theory, programming languages, compilers, database systems, software engineering, graph theory, real-time systems, computer vision, computer networks, distributed multimedia, electronic commerce, computer architecture, computer graphics, and human-computer interfaces. An underlying theme in much of this research is distributed and parallel processing systems. (804-924-2200)

Electrical Engineering. Graduate work and research are offered in three principal laboratories: the Communication, Control, and Signal Processing Laboratory; the Applied Electrophysics Laboratories; and the Center for Semicustom Integrated Systems (Computer Engineering). Areas of research in these laboratories include digital communications and communication networks, computer systems, VLSI system design, fault tolerance, test technology, system design methodologies, control systems, wireless communication, pattern recognition, signal and image processing, microwave devices and circuits, semiconductor properties, device development and fabrication processes, superconducting electronics and fabrication technology, solid-state devices, integrated optics, and optical signal processing. Interdisciplinary activities include magnetic bearings and rehabilitation engineering. (804-924-6077)

Engineering Physics. Current research includes electronic materials, advanced material design, atomistic-level computations in material science, surface modifications and interactions, plasma physics, atomic collisions, medical physics, space science instrumentation, computational methods in fluid physics, nonlinear dynamics and chaos, accelerator design, and low temperature physics. A flexible program allows students wide latitude in the selection of course work and research areas. (804-924-7237)

Materials Science and Engineering. Graduate education and research in materials science and engineering are concerned with the structure, properties, performance, and processing of materials. Graduate programs include fundamental scientific investigations as well as applied research in the development of new engineering materials. Research programs exist in the areas of applied electrochemistry, electronic materials, computational modeling, crystal defects, electrical brushes, friction and wear, mechanical properties, fracture, materials processing, phase transformations, polymer physics, dielectric properties, metallic glasses, catalysis, surface and interfacial studies, and high-resolution analytical electron microscopy. (804-982-5640)

Mechanical and Aerospace Engineering. Training and research are offered in the areas of solid and structural mechanics, thermal sciences, system dynamics and vibrations, fluid mechanics and gasdynamics, control systems, and manufacturing and design. Current research includes studies of turbomachinery flows, aerodynamic flows, supersonic combustor flows, boundary layer stability, solar energy thermal systems performance analysis, rotating machinery, spacecraft, very large structures in space, automobile crashworthiness, shock isolation, integrative techniques for industrial automation, modeling and simulation of manufacturing and material processes, computer-aided design, magnetic bearings, design optimization with regard to robotics and heat-treating processes, and application to structural design, design of improved-performance wheelchairs and other rehabilitation devices, biomechanics and biomaterials problems, automotive safety systems for the handicapped, thermal structural analysis, biological- and chemical-agent sensing, microgravity fluid mechanics, microgravity combustion, fire suppression, and insect and fish locomotion. (804-924-7425)

Systems Engineering. Studies and research have two orientations, basic and applied, and are invariably interdisciplinary. Basic studies focus on the development of theoretical and methodological foundations for problem-solving and decision-making systems. These draw on disciplines such as information technology, modeling and simulation, mathematical systems theory, decision theory, control theory, economics, operations research, management science, risk analysis, computer science, artificial intelligence, cognitive science, and human factors. Applied studies aim at the objective of the systems approach: comprehensive solutions to complex problems that require the integration of technological, informational, organizational, behavioral, human, and economic factors. The department offers ample opportunities for engaging in challenging basic or applied research on significant contemporary topics. Current research topics include (1) total quality control and reliability in manufacturing; (2) risk management of engineering and environmental systems; (3) analysis and design of large-scale communications systems; (4) statistical decision analysis and its application to civilian, military, and large-scale systems; (5) control of discrete-event systems (e.g., manufacturing, communication, computer, and transportation systems); (6) intelligent systems, data fusion, and data analysis; (7) systems design methodologies that incorporate ethics and sustainability; (8) financial engineering and financial services operations management; (9) information and decision support systems; and (10) hierarchical-multiobjective analysis of large-scale systems. (804-924-5393)

UNIVERSITY OF WISCONSIN–MADISON

College of Engineering

Programs of Study	Both M.S. and Ph.D. degrees are offered in the Departments of Chemical, Civil and Environmental, Electrical and Computer, Industrial, and Mechanical Engineering; Materials Science and Engineering; Nuclear Engineering and Engineering Physics; and Engineering Mechanics. Students can also pursue an M.S. or Ph.D. in several multidisciplinary and interdisciplinary areas: materials science, oceanography and limnology, geological engineering, water chemistry, and biomedical engineering. An interdisciplinary M.S. program in manufacturing systems engineering is offered for engineers currently working in industry as well as those just completing the bachelor's degree. Special committee degrees can be arranged for students who wish to combine other areas of study with engineering while pursuing an M.S. or Ph.D., such as computational science and engineering.
	The completion of 24 to 30 credits, 5 to 8 of which may be in work on a thesis, is required for the M.S. degree; an M.S. thesis is not required in all departments. A student whose undergraduate degree is not in the same disciplinary area as the department may need additional credits to obtain the M.S.
	The Ph.D. degree requires six full semesters of credit, half of which may be transferred from an approved institution upon recommendation of the student's adviser. Foreign language requirements are determined by each department. A student is formally admitted to candidacy for the Ph.D. degree after passing a comprehensive preliminary examination in the major field. A thesis is required as well as a final oral examination that covers the thesis and the general fields of major and minor studies. For further information, students should see the *Graduate School Catalog*.
Research Facilities	Each department maintains extensive research facilities. The College has access to many University facilities and services, including the Internet; it also operates a computer network within the College for research, instruction, and administration. The Engineering Research Building houses laboratories for research in interdisciplinary areas, such as applied superconductivity, biomedical engineering, engine research, fusion and plasma physics, materials science, surface science and thin films, and solar energy. Excellent library and information-search facilities are housed in the Kurt F. Wendt Engineering and Physical Sciences Library.
Financial Aid	Financial aid available to graduate students includes fellowships, traineeships, scholarships, research assistantships, teaching assistantships, and loans. Application forms for all except the loans can be obtained from the student's major department. Nearly all include payment of nonresident tuition; full tuition is paid by some awards.
	Many on-campus jobs are available for students and their spouses. The Student Financial Services Office can provide further information.
Cost of Study	Tuition for 1998–99 was $2463 per semester for state residents and $7594 per semester for nonresidents.
Living and Housing Costs	University-owned apartments are available for married graduate students; rents ranged from $380 to $563 per month as of July 1998. Single graduate students may apply for single student housing; the 1998–99 rates for single rooms ranged from $2194 to $2860 per academic year. Numerous off-campus rooms and apartments are also available.
Student Group	More than 40,000 students attend the University of Wisconsin–Madison; they represent every state in the Union and more than 100 other countries. The College of Engineering has an enrollment of about 4,500 students, including more than 1,000 graduate students. The size and diversity of the student body provide opportunities for interaction with people of varied backgrounds and in many areas of study.
Student Outcomes	Many recent graduates of M.S. and Ph.D. programs remain in Madison, primarily as research associates with the University. Ph.D. graduates are likely to accept assistant professorships or research positions at American universities or federal laboratories, and many work in private industry at companies such as Intel, Motorola, and Bell Labs. Finally, many international students return to their country of citizenship, thus enhancing the strong global network of College of Engineering alumni.
Location	Madison is the capital of and second-largest city in Wisconsin, with a population of 192,262. The University and the city and state governments are Madison's largest employers. Skiing and skating are popular winter sports, and, in the summer, Madison's four lakes provide facilities for swimming, fishing, and sailing. Restaurants in the city offer varied cuisines. A bus transit system connects all areas of the city, and Madison has excellent air and bus transportation to Chicago, Milwaukee, Minneapolis–St. Paul, and other major metropolitan areas.
The University and The College	The University of Wisconsin–Madison is located along the south shore of Lake Mendota, and its campus stretches for 1½ miles west from the center of Madison. There are more than 125 departments offering more than 4,500 courses in ten major schools and colleges. The University is noted for the high quality of its research in many areas.
Applying	Application materials should be submitted to the Graduate School at least six weeks before the anticipated start of graduate study. International students whose native language is not English must include results of the Test of English as a Foreign Language (TOEFL). Applications for scholarships, fellowships, and assistantships must be received by the appropriate department's office no later than March 15 for the following fall semester. GRE scores and letters of recommendation are required to complete the financial aid application. Scores on the GRE General Test are required for fellowship support and are recommended in all cases. Notification of admission and all financial aid awards is given by March. Individual departments and programs may have additional requirements.

Correspondence and Information	For information about departmental degree programs:	For information about nondepartmental degree programs:
	Chairman (name of department) College of Engineering University of Wisconsin–Madison Madison, Wisconsin 53706	(name of program) College of Engineering University of Wisconsin–Madison Madison, Wisconsin 53706 World Wide Web: http://www.engr.wisc.edu

University of Wisconsin–Madison

FACULTY HEADS AND RESEARCH AREAS

Chemical Engineering. J. Dumesic, Chair. Transport phenomena in chemical reactors and separation processes; analysis of biological and environmental systems; bioinformatics; biomass and food processing; kinetic theory of polymeric fluids, structure-property relations of polymers; high-performance parallel computing; electrolytic and electrogenerative processes; electrochemical reaction engineering; hydrometallurgy; chromatography; colloidal suspension rheology; protein-protein interactions, applied immunology, fermentation technology; in situ determination of heterogeneous catalyst structure and surface properties, Raman and infrared spectroscopic studies of catalytic systems; mathematical modeling of fixed-bed reactors; design and control of single processes and entire industrial systems, dynamic behavior of reactors, and real-time computer control of industrial processes; computer-aided process design; thermodynamics and statistical physics of fluids.

Civil and Environmental Engineering. R.L. Smith, Chair. Composite material structures; structural engineering; reinforced/precast/prestressed concrete structures; earthquake engineering; wood structures; bridge design; aseismic concrete structures; structural dynamics; behavior of concrete, asphalt, and wood-based materials; rheology of asphalt binders; asphalt concrete mixture design; environmental engineering; water chemistry; limnology; ceramic membranes and catalysis; photocatalysis; biofilms; small-scale wastewater treatment systems; oxygen transfer; solid-wastes engineering and management; water; wastewater and industrial waste treatment; water supply; water quality; computer-aided process modeling and design; hazardous wastes remediation and disposal; pollutant transport modelings; hydraulic engineering; sedimentation; ground water; turbulence; fluid mechanics; water resources systems; hydrologic modeling; stormwater modeling; environmental transport and mixing; in situ groundwater bioremediation; air quality management; geo-spatial information engineering; spatial data acquisition; satellite geopositioning; remote sensing; land and geographic information systems; digital photogrammetry; geotechnical engineering; soil mechanics; environmental geotechnics; landfill design; engineering geology; subsurface exploration and in situ testing; soil-structure interaction; landslides; computer-aided geotechnical construction; intelligent transportation systems; traffic engineering; highway safety; highway engineering; transportation planning; city planning; transit planning and operations; pavement analysis and design; construction engineering and management; robotics and automation; decision analysis; contractor analysis; decision support systems; computer integrated construction; constructibility analysis; expert systems; information engineering; infrastructure and facilities management.

Electrical and Computer Engineering. W. J. Tompkins, Chair. Computer control, optimization, system analysis and synthesis, medical instrumentation, biomedical digital signal processing, microelectromechanical systems (MEMS), wireless communications, X-ray lithography, computer-aided design, testing and built-in testing, fault-tolerant computing, real time systems, computer architecture, microwave systems and devices, antennas and radiation, wave propagation and scattering, information theory, digital signal processing, pattern recognition, speech and image analysis, magnetic confinement, plasma processing of materials, radio frequency heating, modeling of AC and DC power systems (power electronics), rotating machine applications, superconductor materials and devices, VLSI circuit design, fabrication, and processing, lasers, optoelectronic devices and systems, holography.

Engineering Mechanics. G. A. Emmert, Chair. Analytical dynamics; astronautics, including space dynamics, satellite dynamics, celestial mechanics, and space applications of robotics; continuum mechanics and rheology; engineering materials and composite materials; experimental mechanics, including experimental stress analysis, fatigue, and biomechanics; fracture mechanics, including both analytical and experimental work; geotechnical engineering, including rock mechanics and geophysics; solid mechanics and structural mechanics, including elasticity, plasticity, elastic stability, plates, and shells; computational mechanics and numerical methods, including finite-element analysis for dynamics, solid mechanics, and fluid mechanics applications; vibrations and wave propagation.

Industrial Engineering. H. J. Steudel, Chair. CAD/CAM, large-scale simulation, facilities planning, computer-aided process planning, statistical process control, online system modeling and control, production planning and control, work design, application of computer-based decision-analytic methodologies to delivery systems, medical technology assessment, multiple-criteria evaluation, design/evaluation of medical systems, computer applications, workplace design, human-computer interaction, human disabilities and aging, ergonomics, sociotechnical systems, occupational health and safety, biomechanics, decision analysis, statistical decision theory, multiple objective decision making, stochastic processes, quantitative methods for economic planning, mathematical programming, applications of data analysis and designs of experiments to problems of industrial quality, problems of robust design, process control, variance reduction, multifactors, and experimentation.

Materials Science and Engineering. E. E. Hellstrom, Chair. Materials processing (melting and casting, welding, powder metallurgy), physical metallurgy (phase transformation, rapid solidification, oxidation, radiation damage, point defects, superconducting materials, surface science), ceramics and electronic materials (metallization, surface science, thermodynamics, kinetics, oxidation, electrical and mechanical characterization), mechanical behavior (metals and ceramics), rock mechanics, rock drilling, in situ stress and rock property determination, rock chamber design, underwater mining, tight gas sands, nuclear-waste disposal.

Mechanical Engineering. K. W. Ragland, Chair. Fundamental studies and applications of machine and mechanical system design, CAD, controls (applications of electromechanical, computer control, and estimation theory to automation devices), solid and virtual reality, modeling robotics, kinematics, system dynamics, structural dynamics and vibrations, biomechanics, tribology, vehicle design, fluid power systems, product and process modeling and design, internal combustion engines (fuel properties and ignition, emissions, optical diagnostics applied to engines, computational fluids modeling of in-cylinder events), energy conversion (biomass combustion, energy utilization, solar energy, refrigeration, cryogenics, building HVAC system analysis and control for buildings), manufacturing processes and systems, polymer processing, statistical methods for improved product quality and process productivity, computational mechanics, experimental mechanics (moiré techniques, holography, modal analysis, straingages, transducer design), structural mechanics (composite materials, plates, shells, and pressure vessels).

Nuclear Engineering and Engineering Physics. G. A. Emmert, Chair. Engineering physics, with particular attention to nuclear fission and fusion. Fission reactor safety, including thermal hydraulics and operational aspects; development of fission reactors for terrestrial and space applications. High-temperature plasmas for controlled thermonuclear fusion and low-temperature plasmas for industrial plasma-processing and plasma/ion-implantation applications. Design of terrestrial and space reactors for converting fusion energy to useful power. Nuclear materials, including radiation damage to materials. Applied superconductivity and cryogenics, with applications to energy storage, superconducting magnets, and the cooling of instrumentation in space. Use of supercomputers in theoretical plasma physics, radiation hydrodynamics, and general engineering design.

NONDEPARTMENTAL GRADUATE DEGREE PROGRAMS

Biomedical Engineering. R. G. Radwin, Chair. Biomaterials, biomechanics, biomedical computing and signal processing, bioinstrumentation, biosensors and electrodes, ergonomics, health infomatics, imaging, medical decision making and health services, micro-electromechanical systems, minimally invasive medicine and surgery, molecular computing, molecular engineering, neurophysiology, radiological engineering, rehabilitation engineering, robotics and computer vision, and targeted drug delivery.

Geological Engineering. B. C. Haimson, Chair. Geoenvironmental engineering, groundwater hydrology, hydrogeology, surficial geology, rock physics, rock mechanics, rock fracture mechanics, in situ stress measurements, soil mechanics, geotechnics, numerical methods, polluted site remediation, polluted aquifer restoration, waste disposal in rocks and soils.

Manufacturing Systems Engineering. R. Suri, Director. Design, development, implementation, operation, and management of modern manufacturing systems; computer-aided design; manufacturing and engineering; issues affecting manufacturing competitiveness.

Materials Science. E. E. Hellstrom, Chair. High-technology materials: applied superconductivity, ceramics, metals, semiconductors, microelectronics, polymers, surface science, interfaces, thin films, X-ray lithography, photoresists, submicron circuits, micron-size machines.

Oceanography and Limnology. W. C. Sonzogni, Chair. Interdisciplinary studies including fate of organic compounds in aqueous systems, primary productivity in northern Wisconsin lakes, cool-water fish culture, microorganisms and aquatic plants, organic compounds in lakes, nutrient dynamics in lake systems, lake-groundwater interaction, acoustic studies of fish, wetland formation processes, predator-prey ecology of aquatic organisms, freshwater symbioses, animals and algae, behavior of sediment-laden river plumes, measurement and design of fluid flows, aquatic ecosystems, acid rain and lake ecosystems, microseismicity and teleseismic studies, coastal engineering and wave dynamics, heat balance of small lakes, long-term ecological research.

Water Chemistry. D. Armstrong, Chair. Applications of chemical principles to aquatic systems: chemistry of colloids and solid-solution interfaces, atmospheric chemistry, chemistry of surface and groundwaters, sediment-water interactions, biogeochemical cycles in lakes, aquatic photochemistry, contaminated water and sediment remediation.

UNIVERSITY OF WISCONSIN–MILWAUKEE

College of Engineering and Applied Science

Programs of Study

The College of Engineering and Applied Science (CEAS) offers the Master of Science (M.S.) in engineering, the M.S. in computer science, and a Ph.D. in engineering. The M.S. program provides breadth by requiring a program of 21 credits of course work and a minimum of 3 credits of thesis work. For those with prior engineering or scientific work experience, the program offers a nonthesis option that requires 9 additional credits of course work in lieu of a thesis. Each student, in consultation with faculty members, has the flexibility to put together a program of study that is compatible with the student's objectives. Up to 9 credits of business school courses are allowed for those students contemplating a career in management.

The M.S. program in engineering currently offers seven areas of concentration: civil engineering, electrical and computer engineering, engineering mechanics, industrial and management engineering, manufacturing engineering, materials engineering, and mechanical engineering.

The M.S. in computer science is a departmental program offered by the Department of Electrical Engineering and Computer Science. The program is designed to provide the student with a broad background in various aspects of computer science. Areas of concentration include software engineering, computer systems, database organization, data security, and artificial intelligence.

There are six major areas in the Ph.D. program: civil engineering, computer science, electrical engineering, industrial engineering, materials engineering, and mechanical engineering. The Ph.D. degree requires a minimum of 66 credits beyond the baccalaureate, including a dissertation. Many of the courses leading toward the various master's degrees and the doctoral degree are offered in the late afternoon or evening; thus students can complete much of their course work on a part-time basis.

Research Facilities

The College of Engineering and Applied Science has 61 full-time faculty members, all of whom hold doctorates, and about 50 staff personnel. The College has some of the finest research centers and laboratories in the nation. Faculty research interests include combustion diagnostics, computer-aided manufacturing, composite materials, computer architecture, data security, digital communications, electromagnetics, energy analysis and diagnostics, environmental studies, electronics cooling, foundry and solidification processing, high-speed computing, human factors, image and signal processing, nonlinear optical device modeling, radon abatement, rheology, structural research, thermal engineering, transportation, tribology, and two-phase flow. The College is also the home of several excellent research centers, including the Center for By-Product Utilization; the Center for Urban Transportation Studies; the Center for Cryptography, Computer and Network Security; the Energy Analysis and Diagnostics Center; and the Wisconsin Alternative Fuels Test Center. Members of the College faculty also participate in the Center for Great Lakes Studies and the Laboratory for Surface Studies, two University of Wisconsin–Milwaukee (UWM) centers not housed within the College.

Financial Aid

Most full-time graduate students in the College are supported by research and teaching assistantships. Students should contact the respective department for specific application information. Fellowships and scholarships are also available. All qualified students can defray their educational expenses by applying for financial aid, including fellowships; teaching, program, and research assistantships; loans; work-study; nonresident tuition remission; and student employment programs. For more information, students may contact the Graduate School's Fellowship Coordinator, (telephone: 414-229-6267; e-mail: gradschool@uwm.edu; World Wide Web: http://www.uwm.edu/Dept/Grad_Sch).

Cost of Study

Per credit fees vary for enrollment below 8 credits. For Wisconsin residents, the 1998–99 fee for 8 to 12 credits was $2498, and for nonresidents it was $7608. Doctoral students who have reached dissertator status are assessed at a special lower rate.

Living and Housing Costs

Although many graduate students live in off-campus housing (apartments), some on-campus dormitory housing (with meal plan) is available for single students at approximately $4000 per year. The campus also provides a variety of food services for off-campus students. Estimated living costs for a single graduate student are $7900 per academic year, exclusive of tuition and fees.

Student Group

Total University enrollment in 1998–99 was approximately 23,000 students from more than forty states and ninety countries. Engineering and computer science graduate enrollment was 276. Sixty-four percent were part-time students.

Student Outcomes

Because the University of Wisconsin–Milwaukee is a comprehensive premier urban research university with applied research at the federal and industrial levels, employment opportunities for its graduates are outstanding. Local companies such as Kohler, Eaton, Briggs & Stratton, and Harley-Davidson recruit from the College. CEAS graduate students are also recruited nationally and internationally by companies such as Motorola, Hewlett Packard, Sundstrand, John Deere, General Electric, Rockwell, Andersen Consulting, Siemens, Underwriters Laboratory, and Caterpillar.

Location

Located 90 miles north of Chicago, Milwaukee (the seventeenth-largest city in the U.S.) is an ethnically diverse city with an abundance of parks, museums, art galleries, major-league sports, and fine restaurants. Educational and cultural institutions include the Milwaukee Art Museum, the Performing Arts Center, the Milwaukee Symphony Orchestra, the Florentine Opera Company, the Milwaukee Ballet Company, and the Milwaukee Repertory Theater.

The University and the College

Near Lake Michigan on Milwaukee's northeast side, UWM's beautiful 92-acre campus is surrounded by natural areas that offer scenic beauty and outdoor activity all year long. Since 1964, the College has developed a full complement of undergraduate and graduate programs in engineering and computer science. To accommodate the needs of students, courses are offered on a day or evening schedule. UWM combines the excellence of the University of Wisconsin System with resources only a major metropolitan area can offer.

Applying

Applicants for admission to the Graduate School must meet the admission requirements of the Graduate School and the requirements of the graduate program to which they are applying. Applicants should have a 2.75 minimum undergraduate GPA on a 4.0 scale, and international applicants must have a minimum TOEFL score of 550. Applicants may be admitted with no more than two course deficiencies. A $45 fee must accompany applications. International student applications require an additional $30 fee for credential evaluation.

Correspondence and Information

Graduate Program Coordinator
College of Engineering and Applied Science
University of Wisconsin–Milwaukee
3200 North Cramer Street
Milwaukee, Wisconsin 53211

Telephone: 414-229-6169
Fax: 414-229-6958
E-mail: ceas_adv@engr.uwm.edu
World Wide Web: http://www.ceas.uwm.edu

For information and applications:
UWM Graduate School Admissions
261 Mitchell Hall
P.O. Box 340
Milwaukee, Wisconsin 53201

E-mail: gradschool@uwm.edu
World Wide Web: http://www.uwm.edu/
Dept/Grad_Sch

University of Wisconsin–Milwaukee

THE FACULTY AND THEIR RESEARCH

Civil Engineering

Edward A. Beimborn, Professor; Ph.D., Northwestern. Transportation engineering.
Theodore Bratanow, Professor; Ph.D., Karlsruhe (West Germany). Aerodynamics.
Hector Bravo, Associate Professor; Ph.D., Iowa. Hydraulics.
Erik R. Christensen, Associate Professor; Ph.D., California, Irvine. Environmental studies.
Verne C. Cutler, Professor; Ph.D., Wisconsin–Madison. Fatigue.
Al Ghorbanpoor, Professor; Ph.D., Maryland. Nondestructive testing.
Sam Helwany, Assistant Professor; Ph.D., Colorado at Boulder. Geotechnical engineering.
Alan J. Horowitz, Professor; Ph.D., California, Los Angeles. Transportation development.
Kwang K. Lee, Professor and Department Chair; Ph.D., Cornell. Hydrodynamics.
Tarun R. Naik, Associate Professor and Director of the Center For By-Products Utilization; Ph.D., Wisconsin–Madison. Waste recycling.
A. Fattah Shaikh, Professor; Ph.D., Iowa. Reinforced and prestressed concrete.
Michael P. Wnuk, Professor; Ph.D., Technical University of Krakow (Poland). Nonlinear mechanics.

Computer Science

John Boyland, Assistant Professor; Ph.D., Berkeley. Programming languages.
George I. Davida, Professor; Ph.D., Iowa. Cryptography and data security.
Yvo G. Desmedt, Professor; Ph.D., Catholic University of Leuven (Belgium). Cryptography.
Sumanta Guha, Associate Professor; Ph.D., Michigan. Computational geometry.
Peter Haddawy, Associate Professor; Ph.D., Illinois at Urbana-Champaign. Artificial intelligence.
S. Hossein Hosseini, Associate Professor; Ph.D., Iowa. Computer architecture.
Leonard P. Levine, Professor; Ph.D., Syracuse. Computer graphics.
Susan McRoy, Assistant Professor; Ph.D., Toronto. Artificial intelligence and computational linguistics.
Ethan V. Munson, Assistant Professor; Ph.D., Berkeley. Software engineering.
Rene Peralta, Associate Professor; Ph.D., Berkeley. Theory of computation.
Ichiro Suzuki, Professor; Ph.D., Osaka (Japan). Computational geometry.
K. Vairavan, Professor and Department Chair; Ph.D., Notre Dame. Computer structures.
Adam Webber, Assistant Professor; Ph.D., Cornell. Compilers.

Electrical Engineering

Carolyn R. Aita, Wisconsin Distinguished Professor of Materials Engineering; Ph.D., Northwestern. Sputter deposition of thin films.
Brian Armstrong, Assistant Professor; Ph.D., Stanford. Robotics.
Keith Corzine, Assistant Professor; Ph.D., Missouri–Rolla, Power electronics.
George W. Hanson, Assistant Professor; Ph.D., Michigan State. Electromagmetics.
S. Hossein Hosseini, Associate Professor; Ph.D., Iowa. Computer architecture.
Ivan Howitt, Assistant Professor; Ph.D., California, Davis. Wireless communication.
Robert J. Krueger, Associate Professor; Ph.D., Wisconsin–Madison. Network theory.
Chiu-Tai Law, Assistant Professor; Ph.D., Purdue. Nonlinear optics.
Devendra K. Misra, Associate Professor and Department Chair; Ph.D., Michigan State. Imaging technology.
Ali M. Reza, Associate Professor; Ph.D., Wyoming. Signal processing.
David C. Yu, Associate Professor; Ph.D., Oklahoma. Power systems analysis.
Jun Zhang, Associate Professor; Ph.D., Rensselaer. Image processing.

Industrial and Manufacturing Engineering

Nidal Abu-Zahra, Assistant Professor; Ph.D., Cleveland State. Metal cutting, polymer processing.
Robert D. Borchelt, Associate Professor; Ph.D., Missouri–Rolla. Manufacturing.
Tsong-How (Phil) Chang, Associate Professor; Ph.D., Wisconsin–Madison. Statistical methods.
Arun Garg, Professor; Ph.D., Michigan. Ergonomics.
Minnie H. Patel, Associate Professor; Ph.D., Georgia Tech. Operations research.
Umesh K. Saxena, Professor; Ph.D., Wisconsin–Madison. Economic analysis.
Hamid Seifoddini, Associate Professor and Department Chair; Ph.D., Oklahoma State. Cellular manufacturing.

Materials

Carolyn R. Aita, Wisconsin Distinguished Professor of Materials Engineering; Ph.D., Northwestern. Sputter deposition of thin films.
Tery L. Barr, Professor; Ph.D., Oregon. ESCA and AUGER surface analysis.
Hugo F. Lopez, Associate Professor; Ph.D., Ohio State. Mechanical properties of materials.
Joachim P. Neumann, Professor; Ph.D., Berkeley. Alloy development.
Pradeep K. Rohatgi, Ford/Briggs & Stratton Professor; D.Sc., MIT. Composite materials, foundry technology, failure analysis.
Dev Venugopalan, Associate Dean; Ph.D., McMaster. Heat treatment, corrosion, casting, welding.

Mechanical Engineering

Ryoichi S. Amano, Professor; Ph.D., California, Davis. Computational fluid dynamics.
S. H. Chan, Wisconsin Distinguished Professor and Interim Dean; Ph.D., Berkeley. Combustion, heat transfer, genetic engineering.
Anoop K. Dhingra, Associate Professor; Ph.D., Purdue. Design optimization.
Suresh V. Garimella, Cray-Research Associate Professor; Ph.D., Berkeley. Electronic cooling, solidification and jet impingement heat transfer.
Tien-Chien Jen, Assistant Professor; Ph.D., California, Los Angeles. Grinding and machining.
Yong Se Kim, Associate Professor; Ph.D., Stanford. Computer-aided engineering.
Dilip Kohli, Professor and Department Chair; Ph.D., Oklahoma State. Robotics, locomotion kinematics.
Gunol Kojasoy, Professor; Ph.D., Georgia Tech. Two-phase flows.
Ronald A. Perez, Associate Professor; Ph.D., Purdue; PE. Control theory and applications.
John Reisel, Assistant Professor; Ph.D., Purdue. Combustion diagnostics, chemical kinetics.
Kevin J. Renken, Associate Professor; Ph.D., Illinois at Chicago. Radon abatement, heat-transfer augmentation, convection in porous media.

UTAH STATE UNIVERSITY

College of Engineering

Programs of Study

Graduate programs leading to the M.E., M.S., and Ph.D. degrees are offered in biological and irrigation engineering, civil and environmental engineering, electrical and computer engineering, and mechanical and aerospace engineering.

A minimum of 30 semester credits is required for the M.E. and M.S. degrees. The M.E. program requires a design problem or laboratory work and the M.S. a research thesis or an engineering report.

At least 60 semester credits beyond the master's degree, or 90 credits beyond the B.S., are required for the Ph.D.; about 35 credits are normally granted for the dissertation. No foreign language is required for any degree.

Research Facilities

The Utah Water Research Laboratory provides 80,000 square feet of highly flexible space for a wide range of experimental work in fluid mechanics and hydraulics, water quality analysis, and instrumentation. Digital, analog, and hybrid computer equipment is available for experimental applications.

The Space Dynamics Laboratory and the Center for Space Engineering conduct research projects involving space exploration, upper-atmospheric studies using rockets, and infrared engineering and is currently involved in space shuttle experiments. It offers numerous opportunities for creative research. Assistantships are available for qualified electrical engineering, mechanical and aerospace engineering, computer science, and physics students. The major facilities for the Center are located in Logan, Utah, and Bedford, Massachusetts.

The Center for Atmospheric and Space Sciences, although formally within the College of Science, shares many research projects with the College of Engineering faculty. Basic research is conducted on atmospheric and space plasma physics with a strong blend of theory and field experiments (including the space shuttle). Student opportunities range from designing and building Getaway Special shuttle payloads to fieldwork in both polar regions and participation in numerous research projects as a research assistant.

Utah State University (USU) computer facilities available for faculty and student use include a VMS cluster of four DEC 3000/400S (RISC Alpha) computers, an IBM RS6000/550 numeric intensive processor, twenty DECstation/ 5000 RISC graphics workstations, and microcomputer laboratories that include more than 550 PC and Macintosh computers. The campus network consists of a Fiber Distributed Digital Interface (FDDI) backbone transmitting 100 megabits/second. Each building on campus is connected to the backbone with 10 megabits/second Ethernet service to offices and labs. The campus network is connected to UTAHNET, Internet, NSI/DECnet, and other regional networks. College facilities include two graphics and computational intensive computer laboratories with twenty IBM RS6000 workstations and sixteen Sun SPARCstations. In addition, more than 100 high-performance microcomputers and minicomputers are located in various laboratories throughout the College.

During the past year, the total extramural research funding in the College of Engineering was more than $31 million. Students participate and play an important role in these funded projects.

Financial Aid

Aid for qualified students is available in the form of teaching and research assistantships. Assistantships carried a nine-month stipend of up to $1600 per month in 1998–99. Research assistants are assigned to a project consistent with their area of major interest.

A number of research organizations provide funding for graduate students, including the Engineering Experiment Station, the Utah Center for Water Resources Research, the Utah Water Research Laboratory, the Space Dynamics Laboratory, the Center for Atmospheric and Space Sciences, and the Center for Space Engineering.

Cost of Study

In 1999–2000, tuition and fees are $1250.32 per semester for Utah residents and $3874.69 per semester for nonresidents and international students. Out-of-state tuition is waived for students holding an assistantship.

Living and Housing Costs

In 1999–2000, the cost of on-campus housing for single students ranges from $779 to $1257 per semester for an apartment with cooking facilities. A dormitory room and board cost $1105 per semester. Married student apartments on campus range from $414 to $554 per month. Off-campus housing ranges from $130 to $370 per month.

Student Group

Total University enrollment is 19,322, including 2,815 graduate students. Undergraduate engineering enrollment is 1,378, and graduate engineering enrollment is 238.

Location

Utah State University is located in Logan, a university community with a population of approximately 37,000. The Wasatch Mountains, just east of the campus, rise to an altitude of 9,800 feet. Fishing, hunting, skiing, hiking, and boating are some of the recreational activities available within several miles of the University. Salt Lake City is 80 miles south of Logan.

Summers in northern Utah are characterized by warm, dry days and cool nights. In winter, heavy snows fall in the mountains surrounding Cache Valley, where the University is located.

The University

Established in 1888 as a land-grant agricultural college, the University today comprises nine colleges.

Applying

Students should apply directly to the School of Graduate Studies. A catalog may be obtained from the Distribution Center.

Correspondence and Information

Head, Department of (specify)
College of Engineering
Utah State University
Logan, Utah 84322-4100

Utah State University

THE FACULTY AND RESEARCH AREAS

Biological and Irrigation Engineering
Wynn R. Walker, Head. The faculty consists of 5 professors, 2 associate professors, 1 assistant professor, 3 adjunct professors, and 6 emeritus professors. Graduate research is available in bioprocess systems engineering (biowaste and biotechnology areas), food engineering, and soil and water resource engineering (irrigated agricultural systems).
Bioprocess Systems Engineering: a new area of specialization that focuses on the scale-up commercialization of biotechnology products; processing traditional agricultural materials into nonfood products, such as biodegradable plastics, paints, pharmaceuticals, and biofuels; and biowaste recycling.
Food engineering: land application of food processing wastes, extrusion of dairy-based foods, multistage anaerobic digestion of biological materials, functional properties of foods, and biological detoxification of metals.
Soil and Water Resource Engineering: improving irrigation system and project management through irrigation conveyance network analyses, remote sensing and geographical information system applications, electronic data transfer and manipulation, remote computer automation of delivery systems, and on-farm water management; crop production parameters; low-energy sprinkle irrigation systems; synthetic envelope materials for agricultural drains; salt balance research as it relates to both drainage and water management; saline water for irrigation; optimizing regional and local scale groundwater utilization; groundwater quality management; modeling for water management; and coordinating the use of groundwater and surface-water resources.
The department and the International Irrigation Center are also directly involved in international technical assistance, adaptive research, and training activities concerned with on-farm water management, design and evaluation of irrigation and drainage systems, irrigation scheduling, water resource development, management of water delivery systems, irrigation project management, drainage of agricultural lands, computer applications in irrigation and drainage, design and maintenance of wells and pumps for irrigation, farmer participation in irrigation project management, and technology transfer.

Civil and Environmental Engineering
Loren R. Anderson, Head. The faculty includes 13 professors, 16 associate professors, 5 assistant professors, and 12 emeritus professors. Graduate research is available in environmental engineering, geotechnical engineering, structural engineering and mechanics, transportation engineering, and water engineering (fluid mechanics and hydraulics, groundwater, hydrology, and water resources). In addition, the department has an international division which coordinates international research and training activities.
Environmental Engineering: bioremediation, phytoremediation, on-site waste disposal, nonpoint pollution control, environmental management, natural systems modeling, hazardous waste management, eutrophication in lakes and reservoirs, numerical determination of hydraulic and quality characteristics, fish habitat studies, pollution prevention, and water and wastewater treatment.
Geotechnical Engineering: computer applications in geotechnical engineering, landslide control, liquefaction analysis and mapping, soil reinforcement methods, risk assessment of dams, earthquake engineering, pile foundations, pavement design, and geoenvironmental engineering.
Structural Engineering and Mechanics: computer-aided design of structures, structural optimization, soil-structure interaction, fiber-reinforced composites, constitutive modeling of reinforced concrete, earthquake engineering, load and resistance factor design, structural dynamics, bridge design, damping of composite systems and base isolation systems for earthquake loading.
Transportation Engineering: traffic systems modeling, geographical information systems (GIS), evaluation of intelligent transportation systems, transportation investment appraisal, public transit systems planning, and intermodal terminal design applications.
Water Engineering: fluid mechanics and hydraulics, groundwater, hydrology, and water resources engineering; model studies of hydraulic structures; sedimentation; erosion control; cavitation; subsurface contaminant transport; climate modeling; flood hydrology; snow hydrology; drought management; terminal lake analyses; use of remote sensing and GIS for water management and forecasting; dam safety and risk assessment; optimal river basin management; and international water resources planning and management.
International Division: Some of the most challenging opportunities for applied research in water resources development and systems operation around the world are overseas. The Utah Water Research Laboratory and the USU Civil and Environmental Engineering Department have considerable expertise to contribute, and this expertise can be enhanced through experiences gained in meeting these international challenges. Through the international division, research is funded which provides opportunities for participation of USU faculty and graduate students in the planning and operation of major water resources projects overseas. As these opportunities are realized, the division supplements and strengthens the research and academic programs in water resources engineering at USU. Current or recently completed contracts for training, technical assistance, and research involve Egypt, India, Thailand, and West Africa.

Electrical and Computer Engineering
G. S. Stiles, Acting Head. The faculty is composed of 8 professors, 6 associate professors, and 2 assistant professors. Graduate research areas include computer-based systems, controls, signal processing, communications, space science and engineering, RF systems, VLSI and optical computing, and the design of molecular function.
Computer-Based Systems: the Center for Self-Organizing and Intelligent Systems (CSOIS) develops real-time and embedded applications of single and multiple processor systems in robotics and controls. Other groups develop systems for image and audio processing, spacecraft payloads, and high-performance data acquisition and analysis.
Controls: CSOIS staff applies controls to problems in robotics, including autonomous vehicles, navigation, obstacle avoidance, and telepresence.
Signal Processing: the compression group works on hardware and software projects in video and audio compression, including satellite data compression and PC and network-based video phone and video conferencing systems.
Communications: video and audio communications, network systems.
Space Science and Engineering: the department, the Space Dynamics Laboratory, and the Center for Atmospheric and Space Sciences conduct ionospheric and magnetospheric experiments from ground instruments and space probes utilizing ionospheric HF radar, optical measurements, photonic systems, theoretical modeling, and extensive facilities for infrared Fourier spectroscopy; faculty members and students design, construct, and operate payloads for rocket and space shuttle missions.
RF systems: design at HF and microwave frequencies, antennas for wireless communication, numerical simulation of biological, geophysical, and other RF applications.
VLSI and Optical Computing: the VLSI research group focuses on device development and design techniques for implementing optical computing elements on semiconductor substrates. Other areas of interest are architecture synthesis, microsystems (combining electronics, mechanics, fluidics, and optics), and special-purpose microchip architecture and design.
Molecular Function Design: the National Center for the Design of Molecular Function (NCDMF), associated with the ECE department, develops methods to tailor the function of molecules for applications in biotechnology, biomedical photonics, and environmental monitoring. NCDMF electrical, computer, and optical engineers work on new, high-performance real-time analysis techniques, including optical absorption, fluorescence, dual-wavelength pulsed-laser photothermal spectroscopy, and photothermal imaging spectroscopy.

Mechanical and Aerospace Engineering
J. C. Batty, Head. The faculty includes 6 professors, 6 associate professors, 1 research associate professor, 3 assistant professors, 2 adjunct professors, and 4 emeritus professors. Graduate research is ongoing in several broad areas.
Fluid Mechanics/Aerodynamics: experimental and computational fluid dynamics (CFD) and heat transfer. Current research involves parafoil, airfoil, bluff body, and wing-tip vortex studies; aircraft design; vorticity dynamics; algorithm development; and flow visualization techniques. Major facilities include the Aerodynamics Research Laboratory, which has a 4-foot by 4-foot subsonic wind tunnel with a data acquisition and signal processing center and a CFD laboratory with a multiprocessor Silicon Graphics Origin 2000 deskside supercomputer and twenty-one Silicon Graphics 02 workstations.
Manufacturing/Automation: life cycle design, sustainable manufacturing, manufacturing process development, mechatronics, neural networks, control logic, intelligent systems, CAD/CAM, flexible manufacturing, design theory, methodology, concurrent engineering, process control, manufacturing cell development, and economics of product/process interaction.
Mechanics/Dynamics/Materials: advanced strength of materials, finite element applications, mechanics of composite materials, material science, continuum mechanics, elasticity, plasticity, dynamics, advanced dynamics, orbital mechanics, advanced vibrations, material image analysis system, MTS tensile/compression test system, fatigue testing, buried structures lab, vibration test facility, and I-DEAS solid modeling/FEM/visualization software on twenty-one SGI workstations, space structures, composite materials damping, plastic pipe design and testing, and accident reconstruction.
Space Engineering: cryogenics, hydrogen storage and handling in space, thermal management of orbiting and interplanetary space systems, miniaturization of space-deployed instruments, fiber support technology, thermal link development, orbital mechanics, space systems design, and attitude control systems. Major facilities at the Space Dynamics Laboratory include cryogenic laboratories, clean rooms, thermal/vacuum systems, high bay integration areas, and machine shops.

VANDERBILT UNIVERSITY

School of Engineering

Programs of Study	The Graduate School offers advanced study leading to the degrees of Master of Science and Doctor of Philosophy in the following engineering fields: biomedical engineering, chemical engineering, civil engineering, computer science, electrical engineering, environmental and water resources engineering, materials science and engineering, and mechanical engineering. Individualized, interdisciplinary programs of study (such as Management of Technology) leading to the master's and Ph.D. degrees are also possible for students with special goals that require combining courses in several disciplines.
	Paralleling the research-oriented M.S. degree is the Master of Engineering, an advanced professional degree awarded by the School of Engineering. Practice-oriented, the degree stresses engineering technology and is offered in chemical engineering, civil engineering, computer science, electrical engineering, environmental and water resources engineering, management of technology, materials science and engineering, and mechanical engineering. Thesis and residence requirements are flexible, and either full-time or part-time enrollment can be arranged. Tuition is the same for the M.Eng. degree as for the M.S., but application for admission is made to the School of Engineering.
Research Facilities	Research facilities are housed in three engineering buildings. In addition, engineering students have access to exceptional support facilities in the Stevenson Science Center and the Vanderbilt Medical Center.
	The Jean and Alexander Heard Library is one of the important research libraries in the South, with more than 2 million volumes in nine divisions and three special units. Access to materials is provided through its integrated, automated system—ACORN. The Sara Shannon Stevenson Science Library in the Stevenson Center contains more than 260,000 volumes in the fields of chemistry, engineering, general biology, geology, mathematics, molecular biology, and physics. The central computing facility provides enterprise-wide services, such as access to the World Wide Web, e-mail, dialup access, Web publishing, and kerberos-based user authentication.
	The computing facilities within the School of Engineering include 150 Sun workstations, 700 Pentium-based computers that primarily run either Windows NT or Linux, and smaller numbers of computers that each run one of the following: HP-UX, AIX, IRIX, CLIX, Windows 95/98, or MacOS. An ATM backbone with switched 10/100 Ethernet at the edges connects all of the computers to the campus LAN.
	Engineering facilities include laser instrumentation with eximer and free-electron lasers for fluid flow studies and combustion dynamics employing Raman scattering, fluorescence, and phased Doppler anemometry. Acoustic levitation apparatus for fluids studies are located in the Microgravity Research and Applications Center. Facilities are available for conventional and containerless melting, splat cooling, and directional solidification of metals, glasses, and ceramics, as well as state-of-the-art characterization employing SEM and TEM microscopy, X-ray, electron paramagnetic resonance, dielectric measurements, potentiostat, ellipsometry, and tribological and mechanical testing. Materials modifications are undertaken employing CO_2, eximer and free-electron lasers, and ion implantation with a 300-kv accelerator. Other resources of special note include an intelligent robotics laboratory; a physical acoustics laboratory; a structural laboratory with static and dynamic testing equipment; a computer-oriented transportation laboratory; a soil mechanics laboratory; environmental laboratories with equipment for analysis of water, air, and waste; a hydraulics laboratory; extensive materials processing equipment; subsonic and supersonic wind tunnels; a visually integrated soft-arm robot; and facilities for cell cultures, cellular bioengineering, and quantitative physiology.
Financial Aid	Financial support is available to qualified graduate students pursuing an engineering major. Assistantships, fellowships, and traineeships offer tuition grants and/or a salary varying from $9000 to $20,000 per year. Most carry teaching or research duties. Vanderbilt University is a member of the National Consortium for Graduate Degrees for Minorities in Engineering and Science, Inc. Fellowship programs are available to members of minority groups who are pursuing master's and Ph.D. degrees.
Cost of Study	Tuition for full-time study (12 semester hours) is $11,518 per semester. Minimum tuition for students not taking courses is $200 per semester (in residence) or $50 per semester (not in residence).
Living and Housing Costs	Although ample, privately owned rental accommodations are available in the area, graduate students may apply for housing in University apartments. There are accommodations for both single and married students, with studio and bedroom units ranging from $2800 to $3900 per semester. Food service is available for approximately $300 to $1500 per semester.
Student Group	In 1998–99, there were 4,292 graduate students and professional school students in the student body of 10,110 at Vanderbilt. Engineering enrollment was 1,298 undergraduate and 314 graduate students. Twenty-five percent of the engineering students were women. A total of 232 bachelor's degrees, 56 master's degrees, and 31 doctorates were awarded.
Location	Vanderbilt is near the center of Nashville, a metropolitan area of half a million people in the heart of the Tennessee Valley Authority lake country. The rolling hills and waterways of beautiful Tennessee have made Nashville a mecca for lovers of outdoor life. Nashville is the cultural, commercial, and financial center of the mid-South and is known as the "Athens of the South" and "Music City U.S.A."
The University	Vanderbilt is an independent, privately supported university, founded in 1873 by Commodore Cornelius Vanderbilt "to strengthen the ties which should exist between all geographical sections of our common country." It cooperates with other local private universities and colleges (Fisk and Meharry) through the Nashville University Center, a consortium that permits cross-registration for courses and full credit exchange. The University is a member of the Association of American Universities and is accredited by the Southern Association of Colleges and Schools. Various undergraduate engineering curricula are accredited by the Accreditation Board for Engineering and Technology (ABET).
Applying	Applicants should submit complete credentials to the Graduate School prior to January 15; late applications are reviewed as space and time permit. A fee of $25 should accompany the application. Qualified graduates with a bachelor's degree from an ABET-accredited or equivalent curriculum who have maintained at least a B average in undergraduate work are considered. Applicants are required to take the GRE; most programs now require submission of scores before enrollment.
Correspondence and Information	The Graduate School Vanderbilt University Nashville, Tennessee 37240

Vanderbilt University

THE FACULTY AND THEIR RESEARCH

The faculty of the School of Engineering consists of 132 full-time and 39 part-time members with professorial rank. Of the full-time faculty, all hold doctorates and 45 percent are registered engineers.

Major fields of study and areas of concentration are listed below. Directors of graduate studies named in each area may be contacted for further information.

Kenneth F. Galloway, Dean
Edward J. White, Associate Dean for Academic Affairs.
Thomas A. Cruse, Associate Dean for Research and Graduate Affairs.
Robert E. Stammer Jr., Assistant Dean for Student Affairs.
John R. Veillette, Associate Dean for Preparatory Academics.

Biomedical Engineering
Professor T. R. Harris, Chair; Robert J. Roselli, Director of Graduate Studies.
Areas of study: Quantitative physiology of the cardiopulmonary, musculoskeletal, and neurological systems and biomedical optics; cellular bioengineering, physiological transport phenomena, and medical imaging; medical computing and expert systems; and vision research.

Chemical Engineering
Professor Douglas LeVan, Chair; G. Kane Jennings, Director of Graduate Studies.
Areas of study: Adsorption and surface chemistry, biochemical engineering and biotechnology, chemical reaction engineering, environment, materials, and process modeling and control. These are united through study of the fundamental topics of transport phenomena, chemical reaction kinetics, and thermodynamics.

Civil Engineering
Professor Edward Thackston, Chair; S. Mahadevan, Director of Graduate Studies.
Areas of study: Structural engineering, structural mechanics, finite-element analysis, transportation safety facility design, structural dynamics, stochastic method design, transportation analysis design, traffic and transit engineering, and hazardous material transport.

Computer Science
Professor E. J. White, Acting Chair; Douglas Fisher, Director of Graduate Studies.
Areas of study: Artificial intelligence, machine learning, model-based reasoning, qualitative reasoning, intelligent tutoring systems, database theory, database systems, knowledge-based systems, design and analysis of algorithms, graph theory, computational complexity, operating systems, performance evaluation, parallel and distributed computing, computer networks, fault diagnosis, supercomputers, numerical software, atomic structure software, image processing, medical imaging, concurrency/parallelism, software engineering, parallel architectures, and program testing.

Electrical Engineering
Professor S. E. Kerns, Chair; Francis Wells, Director of Graduate Studies.
Areas of study: Computer engineering, intelligent systems, robotics, control systems, microelectronics, solid-state theory and devices, VLSI design, solid-state sensors, signal processing, systems and circuits, artificial intelligence, expert systems, automatic control, and innovative computer test methodologies.

Environmental and Water Resources Engineering
Professor Edward Thackston, Chair and Director of Graduate Studies.
Areas of study: Water-quality engineering, physical/chemical/biological treatment of priority pollutants, transport of pollutants in surface and groundwater, handling and cleanup of hazardous/toxic/radioactive wastes, nuclear facility decontamination, environmental chemistry and microbiology, and modeling the impact of combined sewer overflows on rivers.

Materials Science and Engineering
Professor Robert Bayuzick, Program Director and Director of Graduate Studies.
Areas of study: Analytical electron microscopy, containerless processing, solidification, crystal growth, fluid mechanics, dynamics of drops and bubbles, physical acoustics, living cell encapsulation, wear of metals and ceramics, erosion, transfer-layer formation, mechanical properties of coatings and welds, rolling contact deformation and fatigue, surface modification, fretting, fracture mechanics, cyclic stress/strain behavior and crack growth, structure of glasses, radiation effects on optical properties, ion implantation and laser modification of glasses, stress corrosion cracking, secondary emission, structure of liquid alloy surfaces, sputtering and desorption, surface analytical techniques, diamond films, direct fabrication, and development of superconducting materials.

Mechanical Engineering
Professor Robert W. Pitz, Chair; A. M. Mellor, Director of Graduate Studies.
Areas of study: Acoustics; active vibration control; adaptive nonlinear control; aeroelasticity; biomechanics; combustion in gas turbines, diesel engines, and natural gas appliances; computational fluid dynamics; computer animation; controls; cooling of electronics; crack growth and arrest; design synthesis; dual-fueled engines; dynamics of flexible robots; environment-induced fracture; finite element and boundary element stress analysis as applied to fracture; gas dynamics of rarefied hypervelocity flows; haptic interfaces; inverse computational methods; laminar and turbulent combustion; laser diagnostics of combustion; mesoscale energy conversion; microgravity engineering; micromanufacturing; microrobotics; microscale heat transfer; optimal design; rigid body kinematics and dynamics; robot design and control; rolling contact studies; smart structures; solar energy and energy storage studies; spacecraft tethers; surface-enhanced convection; telerobotics; thermomolecular flow phenomena; turbulence scale effects; unsteady and turbulent flows; vibrations.

Virginia Tech

VIRGINIA POLYTECHNIC INSTITUTE AND STATE UNIVERSITY

College of Engineering

Programs of Study

The Virginia Tech College of Engineering offers both thesis and nonthesis M.S. degrees, as well as the more practice- and professional-oriented M.Eng. degree. Programs are available in aerospace, agricultural, chemical, civil, computer, and electrical engineering; engineering mechanics; environmental engineering; environmental sciences and engineering; industrial and systems engineering; materials science and engineering; mechanical and mining and minerals engineering; and systems engineering. Residence and other requirements vary with programs. Ph.D. programs are available in aerospace, agricultural, chemical, civil, computer, electrical, mechanical, and mining engineering; engineering mechanics; environmental sciences and engineering; industrial and systems engineering; and materials science and engineering. The M.S. with thesis requires 30 semester credits, including 6 to 10 semester credits of thesis work; the nonthesis M.S. requires a minimum of 30 course credits. The M.Eng. requires 30 credits, including 3 to 6 credits for a project and report. The Ph.D. requires a minimum of 90 credits beyond the B.S., including 30 to 60 credits of research and dissertation. Proficiency in a foreign language may be required. A full year of residence on campus is required. Up to 42 credits of the Ph.D. program may be transferred from accredited institutions.

Research Facilities

Aerospace and Ocean Engineering facilities include a full-scale motion-based aircraft simulator, a subsonic stability wind tunnel with a 6-foot by 6-foot test section with a dynamic motion model mount; a 9-inch by 9-inch supersonic wind tunnel for Mach numbers 2 through 4; a boundary-layer research tunnel; a towing tank; a cavitation test rig; a structures lab, including a large-space structures simulation facility and a fatigue-rated axial-torsional load frame (MTS). Facilities for Biological Systems Engineering include small watersheds for hydrologic experiments, an artificial intelligence and geographic information systems lab, a water quality lab, animal environment systems watershed assessment lab, a physical properties lab, a fluid power lab, a bioprocess engineering lab, and solar and wind energy systems. Chemical Engineering has labs in polymer characterization and polymer processing, water soluble polymers and colloids, catalysts, fluid particle systems, novel separation techniques, environmental engineering, high-vacuum surface science, and biotechnology. Civil Engineering has labs for research in hydraulics, structural and materials testing, soil mechanics, and environmental engineering. Environmental Sciences and Engineering facilities include controlled-temperature rooms; instruments for the analysis of organic and halogenated compounds, heavy metals, and particle size; and light incubators. Electrical Engineering facilities include a satellite-tracking station and labs for image processing, acoustooptics, fiber optics, electrooptics, computer-aided design, hybrid microelectronics, power electronics, time-domain measurements, digital signal processing, optical image processing, computer engineering, alternative energy, electronic materials, motion controls, power systems, mobile/portable radio, robotics, and VLSI design. The facilities in Engineering Science and Mechanics consist of axial-torsional loading systems, materials testing equipment, a scanning acoustic microscope, a photomechanics lab, a nondestructive evaluation lab, several composite materials labs, an autoclave and clean room, a fluid mechanics lab with a towing basin and carriage, a nonlinear vibrations lab, a biofluids lab, a fracture mechanics lab, minicomputers, and modern high-speed data acquisition systems. Industrial and Systems Engineering facilities include metal-casting, machining, welding, CNC, automation, and robotics labs; telephotometric, microdensitometric, and microphotometric units; an auto simulator; an aircraft simulator; an ergonomics lab; a rehabilitation assessment lab; a displays lab; high-temperature impact and shock testing facilities; and environmental and acoustical chambers. Research equipment and precision instrumentation for Mechanical Engineering include hot-film and laser-Doppler anemometer systems, a calorimeter room, two anechoic chambers, a reverberation chamber, fatigue-testing machines, an experimental structural imaging and modal analysis lab, a tribology lab, a network of workstations for engineering design, a CAD/CAM lab containing an IBM 980 Power Server and twenty-one IBM 41-W CAD/CAM workstations, a rotor dynamics lab, a gas turbine lab with two research gas turbine engines, a fuel-cell systems lab, a flexible-structure controls lab, and a smart materials and structures lab. Materials Science and Engineering facilities include a scanning transmission electron microscope; an environmental scanning electron microscope; optical microscopes and metallographic facilities, including hot stages; X-ray diffraction equipment; ESCA surface analysis facilities; metallurgical and ceramic processing equipment, including a melt spinner and hot isostatic press; mechanical testing equipment; instruments for measuring the thermal response of materials, including thermal expansion, thermal diffusivity by a laser pulse technique, and differential thermal analysis. Mining and Minerals Engineering facilities include labs for rock mechanics, surface chemistry, electrochemistry, comminution, size analysis, image analysis, electrostatic separation, environmental control, mining ventilation, systems analysis, health and safety, and chemical analysis, and a pilot plant for mineral processing and coal preparation. There are minicomputers and microcomputers.

Financial Aid

Various fellowships, traineeships, and numerous corporation and foundation grants are available, and students may obtain University loans. Teaching and research assistantships are also available, with a monthly salary of $1335 to $1750 for half-time appointments in 1999–2000.

Cost of Study

In 1999–2000, students who are state residents pay $229 per semester credit hour ($385 for out-of-state students). For any one semester, these charges are not less than $793 ($1261 for nonresidents) and not more than $2475 ($3879 for nonresidents).

Living and Housing Costs

There are many apartments in the area, with monthly rents ranging from $350 to $800. For additional information, graduate students planning to attend Virginia Tech should contact the Off-Campus Housing Office.

Student Group

Of the 25,608 students enrolled in the fall of 1998, 4,193 were graduate students. The total engineering enrollment was 6,646, including 1,755 graduate students. In May 1998, 570 master's degrees and 99 doctoral degrees were awarded in engineering.

Location

The Virginia Tech campus is in Blacksburg, located in the Appalachian Mountains in southwest Virginia.

The University

Virginia Tech, Virginia's land-grant university, was founded in 1872 and awarded its first M.S. degree in 1892. The first Ph.D. was awarded in 1942. Today, graduate degrees are offered in approximately seventy fields. The campus provides facilities for outdoor and indoor sports. Many cultural programs are held on campus.

Applying

Applicants must submit a completed application form, three letters of recommendation, two official transcripts of undergraduate and graduate records, and, for certain departments, GRE scores.

Correspondence and Information

Dean of the Graduate School
Virginia Polytechnic Institute and State University
Blacksburg, Virginia 24061-0325

Telephone: 540-231-6691
Fax: 540-231-2727
World Wide Web: http://www.rgs.vt.edu

Virginia Polytechnic Institute and State University

DEPARTMENTS AND RESEARCH PROGRAMS

Aerospace and Ocean Engineering Department. Professor B. Grossman, Head. Research and graduate study specializations are available in aerodynamics, hydrodynamics, computational fluid dynamics, reentry physics, propulsion, flight and space mechanics, trajectory optimization, guidance and control, multidisciplinary design optimization, aeroelasticity, structural dynamics, composite structures, large-space structures, structural optimization, ocean wave mechanics, and ocean engineering. World Wide Web: http://www.aoe.vt.edu/

Biological Systems Engineering Department. Professor J. V. Perumpral, Head. The department offers graduate studies and research opportunities in waste management, alternative energy sources and energy management, bioprocessing, food engineering, wood engineering, soil and water conservation engineering, watershed engineering, nonpoint source pollution control, surface water and groundwater quality modeling, knowledge-based systems, geographic information systems, and sustainable agriculture. World Wide Web: http://www.eng.vt.edu/eng/bse/

Bradley Department of Electrical and Computer Engineering. Professor L. A. Ferrari, Head. Areas of graduate study and research include applied electromagnetics, wave propagation and scattering, microwave remote sensing, antennas, communication integrated circuits, RF/microwave/mixed-signal systems, satellite communications, computer engineering, embedded systems, computer vision, fault-tolerant systems, digital signal processing, digital/optical image processing, machine vision, holography, electric power systems, alternative energy, automatic control, motor drives, electronic materials, multichip modules and microelectronics packaging, semiconductor devices, computer networks and computer architecture, time-domain and RF measurements, fiber optics and sensors, power electronics, wireless telecommunication systems and applications, robotics, VLSI design/test, intelligent systems, configurable computing, and neural and fuzzy systems. World Wide Web: http://www.ee.vt.edu/home.html

Charles Edward Via Department of Civil and Environmental Engineering. Professor W. R. Knocke, Head. Research and graduate study in environmental engineering and hazardous-waste management; construction engineering and management; hydrology and hydraulic engineering; geotechnical and geoenvironmental engineering; structural engineering; transportation systems engineering, pavement design, and civil engineering materials; infrastructure engineering and assessment; and geographical information systems. World Wide Web: http://www.ce.vt.edu/

Chemical Engineering Department. Professor W. L. Conger, Head. Research and graduate study are conducted in polymer science and engineering, solid-state surface science, catalysts, kinetics, microelectronic circuit processing, reaction engineering, biochemical engineering, hazardous-waste treatment and environmental engineering, applied surface chemistry, and computer-aided design. The department emphasizes applied science and the application of new science to important problems. World Wide Web: http://www.eng.vt.edu/eng/che/

Engineering Science and Mechanics Department. Professor E. G. Henneke, Head. Graduate study and research are offered in the areas of mechanics of solids and fluids, dynamics, vibrations, biomedical engineering, controls, properties of engineering materials, and computational mechanics. Current research interests include a wide range of activities related to the areas of focus. Examples are laminar-flow control, finite-element analyses, structural optimization, reliability and probability, nonlinear vibrations, nondestructive testing, composite materials and structures, adhesive and sealant science, nonlinear dynamics, fracture mechanics, failure and damage of composites, hydrodynamic stability, and adhesion science. World Wide Web: http://www.esm.vt.edu/esm.html

Industrial and Systems Engineering Department. Professor J. G. Casali, Head. Offers graduate study and research opportunities in human factors engineering, industrial engineering, manufacturing systems engineering, management systems engineering, operations research, and engineering administration, as well as an interdisciplinary program in systems engineering. Current research interests include auditory displays, automation, computer simulation, display system evaluation, hearing protection, ergonomics, manufacturing, human factors, human-computer interaction, vehicle simulation and displays, flexible manufacturing systems, vehicle guidance systems, manual-control theory, optimization, process planning, organization theory, queuing theory, reliability, scheduling, robotics, electronics manufacturing, and information technology. World Wide Web: http://www.ise.vt.edu/

Materials Science and Engineering Department. Professor R. S. Gordon, Head. Interdisciplinary and cross-disciplinary graduate study and research are offered in theoretical, computational, and applied subjects related to the processing, properties, structure, and performance of engineering materials, including ceramics, metals, polymers, electronic and optical materials, composites, and biomaterials. Qualified students with appropriate backgrounds in engineering and physical science participate in the following research programs: electrical, dielectric, optical, thermal, and mechanical property characterization of ceramics, metals, polymers, and electronic/optical materials; processing of all material forms, including composites and glasses; computer simulation and modeling in materials systems; mechanical alloying in metal systems; phase transformations; ceramic packaging in high-power integrated circuits; sintering of ceramics and powder metallurgy; electron and optical microscopy, X-ray diffraction, crystallography, and crystal chemistry; corrosion, biomaterials synthesis, and characterization; fracture, deformation, fatigue, thermal shock, and micromechanics; adhesion and surface studies and modeling and visualization of materials structures and properties. World Wide Web: http://www.eng.vt.edu/eng/materials/mse.html

Mechanical Engineering Department. Professor W. F. O'Brien, Head. Graduate studies and research are offered in acoustics; automatic control; combustion; emission control; internal combustion engines; vehicle dynamics and control; dynamics and vibration of machinery; design; computer-aided design and computer graphics, and rapid prototyping; experimental model analysis; laser imaging; robotics; spatial mechanisms; fluid mechanics; tribology; heat and mass transfer; microwave processing; kinematics; boundary-layer flow; multiphase flow; propulsion; atmospheric radiation; energy management; turbomachinery; finite-element analyses; smart materials systems and structures; and structural dynamics. World Wide Web: http://www.me.vt.edu/

Mining and Minerals Engineering Department. Professor S. C. Suboleski, Head. Graduate study and applied and theoretical research include rock mechanics, ground control, equipment evaluation, systems analysis, health and safety, mineral processing, coal preparation, conservation, mining environment, mining ventilation, and mineral economics. Research projects include control of unstable ground conditions in coal mines, design and mine layout, computer modeling of mining systems, and mine ventilation. Mineral processing projects include production of superclean coal, fines dewatering, and electrochemistry of sulfide minerals. World Wide Web: http://www.eng.vt.edu/eng/mining/minehome.html

OFF-CAMPUS PROGRAMS

Commonwealth Graduate Engineering Program. Virginia Tech is one of five universities that participate in the Commonwealth Graduate Engineering Program, utilizing Net.Work.Virginia, the state's broadband, high-speed asynchronous mode transfer (ATM) network. The students take classes at numerous locations throughout the state, from any of the five participating universities, but must have at least five classes from their home institution to graduate.

The Virginia Consortium of Engineering Schools (VCES). VCES is a consortium of the colleges of engineering at William and Mary, Old Dominion, the University of Virginia, and Virginia Tech and maintains facilities in Hampton, Virginia. Ph.D. and M.S. courses are transmitted between the member universities and the Hampton classrooms. Students may take up to 50 percent of their courses from VCES universities other than the one in which they are registered for the degree. **The Alexandria Research Institute** provides research facilities to help ensure the state's global preeminence in research intensive technology fields.

OTHER MAJOR RESEARCH ORGANIZATIONS OF THE COLLEGE

Advanced Vehicle Dynamics Center
Alexandria Research Center
Antennas and Satellite Communications Center
Center for Advanced Ceramic Materials
Center for Automotive Fuel Cell Systems
Center for Biomedical Engineering
Center for Coal and Minerals Processing
Center for Composite Materials and Structures
Center for Construction Performance Improvement
Center for Energy and Global Environment
Center for Geotechnical Practice and Research
Center for Infrastructure Assessment and Management
Center for Intelligent Material Systems and Structures
Center for Wireless Telecommunications Systems
Energy Management Institute
Fiber and Electro-Optics Research Center
Human Factors Engineering Center
Manufacturing Systems Engineering Center
Materials Response Group

Mining and Minerals Resources Research Institute
Mobile and Portable Radio Research Group
Multidisciplinary Analysis and Design (MAD) Center for Advanced Vehicles
NASA–Virginia Tech Composites Program
NSF Center for Power Electronics Systems
Pharmaceutical Engineering Institute
Polymer Materials and Interfaces Laboratory
Software Technologies Laboratory
Systems Performance Laboratory
Transportation Research Center
Turbomachinery and Propulsion Research Center
University Visualization and Animation Group
Vibration and Acoustics Laboratories
Virginia Active Combustion Control Group
Virginia Center for Coal and Energy Processing
Virginia Tech Information Systems Center
Visualization and Animation Laboratory of Engineering

WAYNE STATE UNIVERSITY

College of Engineering

Programs of Study

Education of engineers for leadership in urban society involves an expanded treatment of the traditional engineering disciplines. Interaction among the pure sciences, the engineering sciences, and the social sciences, together with unusual opportunities for research, prepares the graduate for his or her profession.

The College of Engineering, under the leadership of Dean Chin Kuo, includes the Departments of Chemical Engineering and Materials Science, Civil and Environmental Engineering, Electrical and Computer Engineering, Industrial and Manufacturing Engineering, and Mechanical Engineering and the Division of Engineering Technology. Special programs include automotive engineering, bioengineering, environmental and infrastructure engineering, noise and acoustics, quality control, reliability engineering, energy research, transportation, vehicle systems, engineering management, manufacturing, control systems, combustion, photonics, microelectronics, smart sensors, and interdisciplinary programs.

A master's degree is awarded after completion of the minimum 32 semester hours of credit. Ph.D. candidates complete a minimum of 60 semester hours beyond the master's degree, of which 30 semester hours apply to the dissertation requirement.

Graduates of the College have found career opportunities in industry, research laboratories, universities, and government agencies.

Research Facilities

The College of Engineering has a wide range of general undergraduate, departmental, and research laboratories as well as support facilities. The Center for Automotive Research provides a focus for a broad range of interdisciplinary research in diesel and gasoline engines, automotive electronics, engine and vehicle noise, alternative fuels, and automobile safety. The Bioengineering Center operates in close collaboration with the Wayne Medical School on a range of programs, with emphasis on biomechanics studies. Computer-aided manufacturing and robotics are studied through the industrial engineering and mechanical engineering departments and the Division of Engineering Technology. Additional research laboratories include an acoustics and vibrations laboratory; a laboratory for the study of gasoline and diesel combustion characteristics; a structures and earthquake systems laboratory; a composite materials testing laboratory; an environmental kinetics laboratory; materials, fluids, and metallurgical laboratories; a solid-state electronics laboratory; a microprocessor laboratory; a smart-sensor laboratory; and molecular beams and laser light scattering laboratories.

Excellent computer facilities are available to students and faculty through both the College's Computer Aided Design Laboratory and the University's Computing Services Center. In the College, networked Sun computer systems accommodate extensive CAD/CAM/graphics software. For larger studies, the University's center features a Cray J916 Vector Parallel supercomputer, which is accessible through remote terminals over the MERIT network.

Financial Aid

Many full-time students are supported through sponsored research programs or by individual departments. Fellowships, traineeships, and graduate assistantships for teaching or research typically provide stipends of $9550 to $16,000 per academic year for part-time work, tuition waivers, and housing allowances.

Cost of Study

In 1998–99, tuition for a full-time resident graduate student was $163 per credit hour; the nonresident rate was $355 per credit hour. Eight credits is considered a minimum full-time load. (Tuition is subject to change by action of the Board of Governors.)

Living and Housing Costs

The University offers housing for single and married graduate students. The Housing Office provides a list of approved off-campus accommodations.

Student Group

Approximately 29,200 students are enrolled at Wayne State University. More than 50 percent of these students are women, 15,700 attend the University full-time, and 13,500 attend part-time. There are 12,500 graduate students enrolled in a wide variety of programs. The College of Engineering currently enrolls 1,405 undergraduates and 1,190 graduate students. Approximately 50 percent of engineering graduate students come from the state of Michigan.

Location

Located in the heart of Detroit, the University is surrounded by an area that offers many industrial, business, and cultural opportunities. Nearby are major research and production centers devoted to chemicals, pharmaceuticals, computers, telecommunications, fuels, metallurgy, and automobiles and automotive equipment. The area is also a center for highway and construction industries and has one of the largest hospital complexes in the nation. Among the cultural attractions are professional and university theaters; opera and symphony orchestras; art galleries and institutes; museums; classical, jazz, and soul music; and top international and Broadway entertainment. Detroit is the home of the exciting Renaissance Center. The Henry Ford Museum and Greenfield Village, where American history is reenacted, are only a short drive away. The Detroit River, Lake St. Clair, and numerous smaller lakes attract water sports enthusiasts. Michigan, a tourist state, has many lakes, beaches, forests, and skiing areas.

The University and The College

Wayne State University evolved from a one-year premedical program incorporated into Central High School to serve the forty-five-year-old Detroit Medical College. Successive educational needs led to junior college, city college, and city university status. In 1956, Wayne became a state university. It is now composed of eleven colleges and schools plus the Graduate School and the Center for Urban Studies. Its buildings have been designed by some of the finest architects in the world and are surrounded with beautiful landscaping to serve human needs. The University spirit is one of intellectual curiosity.

The University is accredited by the North Central Association of Colleges and Schools. In fiscal year 1998, the College's research programs attracted more than $16 million in gifts, grants, and contracts from government agencies and public and private institutions.

Applying

Applications for graduate admission, accompanied by a $20 fee ($30 for international students) and transcripts must reach the Office of Admissions by July 1 for the fall term, November 1 for the winter term, or March 15 for the spring term. The deadline for fellowships, assistantships, and scholarships is May 1. The GRE examination is not required for admission. A TOEFL score of at least 550 is necessary for international students.

Correspondence and Information

Research and Graduate Programs
College of Engineering
Wayne State University
Detroit, Michigan 48202
Telephone: 313-577-3861

Wayne State University

THE FACULTY AND THEIR RESEARCH

CHEMICAL ENGINEERING AND MATERIALS SCIENCE

Professors. Esin Gulari, Chair; Ph.D., Caltech: thermodynamics and transport properties of polymer solutions and melts, processing of polymers with supercritical fluids, light scattering–based particle and drop sizing techniques. Ralph H. Kummler, Ph.D., Johns Hopkins: environmental fate and transport, hazardous-waste management. Charles Manke, Ph.D., Berkeley: polymer rheology and processing, polymer kinetic theory. K. Y. Simon Ng, Ph.D., Michigan: heterogeneous catalysis, polymerization kinetics, spectroscopic characterization techniques. Erhard W. Rothe, Ph.D., Michigan: chemical kinetics, laser imaging and spectroscopy of combusting systems.

Associate Professors. Yinlun Huang, Ph.D., Kansas State: pollution prevention and waste minimization, process design and synthesis. Susil Putatunda, Ph.D., Indian Institute of Technology (Bombay): physical metallurgy, fatigue and fracture toughness of metals. Steven O. Salley, Ph.D., Detroit: artificial internal organs and biomedical engineering.

Assistant Professors. Rangaramanujam Kannan, Ph.D., Caltech: dynamics of polymeric systems and interfaces, rheo-optical spectroscopy and scattering techniques. Guang-Zhao Mao, Ph.D., Minnesota: interfacial science of complex fluids, self-assembly and thin films of polymers and surfactants. Howard Matthew, Ph.D., Wayne State: tissue engineering and artificial organs. Gina Shreve, Ph.D., Michigan: biocatalysis, multiphase transport in biological systems. Paul VanTassel, Ph.D., Minnesota: shape selective catalysis, protein adsorption and bioseparations.

CIVIL AND ENVIRONMENTAL ENGINEERING

Professors. Haluk Aktan, Ph.D., Michigan: civil engineering structures. Tapan K. Datta, Ph.D., Michigan State: transportation. Snehamay Khasnabis, Ph.D., North Carolina State: transportation. Chin Kuo, Dean; Ph.D., Princeton: water resources engineering. Carol J. Miller, Ph.D., Michigan: engineering hydraulics and environmental engineering. Mumtaz Usmen, Chairman; Ph.D., West Virginia: geotechnical and construction engineering.

Associate Professors. Gongkang Fu, Ph.D., Case Western Reserve: civil engineering structures. Thomas M. Heidtke, Ph.D., Michigan: water resources and environmental engineering. Takaaki Kagawa, Ph.D., Berkeley: geotechnical engineering.

Assistant Professors. Nazli Yesiller, Ph.D., Wisconsin: geotechnical and geoenvironmental engineering. H. C. Wu, Ph.D., MIT: infrastructure materials and structural mechanics.

ELECTRICAL AND COMPUTER ENGINEERING

Professors. Raymond Arrathoon (Emeritus), Ph.D., Stanford: optical computing. Robert D. Barnard (Emeritus), Ph.D., Case Tech: estimation theory, stability theory, optimal control. Zinovi Gribnikov, Ph.D., USSR Academy of Sciences: pn-junctions, heterojunctions, transistors, thyristors, injection lasers, hot electrons, quantum wells, terahertz generators, couplers, real-space transfer. Mohamad H. Hassoun, Ph.D., Wayne State: artificial neural systems, soft computing, pattern recognition. Jerome Meisel, Ph.D., Case Western Reserve: power systems, electromechanics, power electronics. Vladimir Mitin, Ph.D., Ukranian Academy of Sciences: theory of solid-state physics and semiconductors. Melvin P. Shaw (Emeritus); Ph.D., Case Tech: solid-state physics, semiconductor devices. Donald J. Silversmith, Ph.D., MIT: solid-state devices, microstructures fabrication. Harpreet Singh, Ph.D., Roorkee (India): computers, controls, and systems. Franklin Westervelt, Chair; Ph.D., Michigan: computers.

Associate Professors. Gregory W. Auner, Ph.D., Wayne State: solid-state devices and materials. Jatinder Bedi, Ph.D., Roorkee (India): computer communications and networks. Vipin Chaudhary, Ph.D., Texas at Austin: parallel and distributed systems, computer vision. Robert Erlandson, Ph.D., Case Western Reserve: bioengineering systems. Feng Lin, Ph.D., Toronto: discrete event systems, manufacturing systems, control systems. Syed M. Mahmud, Ph.D., Washington (Seattle): computer architecture, parallel processing, digital system design, microprocessor-based instrumentation. Pepe Siy, Ph.D., Akron: pattern recognition. Le Y. Wang, Ph.D., McGill: robust control system design. James Woodyard, Ph.D., Delaware: solid-state electronics. Yang Zhao, Ph.D., Penn State: nonlinear optics.

Assistant Professors. Guy Edjlali, Ph.D., Paris VI (Curie): run-time and operating system support for parallel and distributed computational intensive programs, Internet security. Ryan Jin, Ph.D., Utah: molecular dynamics simulations of polymeric materials. Dmitri Romanov, Ph.D., USSR Academy of Sciences: solid-state physics and electronics, low-dimensional systems, theoretical and computational physics. Loren J. Schwiebert, Ph.D., Ohio State: computers. Shishir K. Shah, Ph.D., Texas at Austin: computer vision, scene understanding, statistical modeling, pattern recognition. Cheng-Zhong Xu, Ph.D., Hong Kong: parallel and distributed systems, high-performance computing.

INDUSTRIAL AND MANUFACTURING ENGINEERING

Professors. Kenneth R. Chelst, Ph.D., MIT: decision and risk analysis in engineering management, emergency services, optimal retail networks. Donald R. Falkenburg, Ph.D., Case Western Reserve: concurrent engineering, management of technology. Frank Plonka, Ph.D., Michigan: manufacturing engineering, computer-integrated engineering. Nanua Singh, Ph.D., Rajasthan (India): manufacturing systems engineering, concurrent engineering.

Associate Professors. Olugbenga Mejabi, Ph.D., Lehigh: process management, flexible manufacturing, simulation, factory control. Gary S. Wasserman, Ph.D., Georgia Tech: reliability, robust engineering, quality assurance systems. Kai Yang, Ph.D., Michigan: operations research, robust engineering, quality engineering, reliability.

Assistant Professors. Darin Ellis, Ph.D., Penn State: human factors, gerontology. Leslie F. Monplaisir, Ph.D., Missouri–Rolla: collaborative engineering, group decision support systems.

MECHANICAL ENGINEERING

Professors. Victor Berdichevsky, Ph.D., Moscow: mechanics. Ronald F. Gibson, Ph.D., Minnesota: mechanical behavior of composites. Naeim A. Henein, Ph.D., Michigan: combustion, energy conversion. Raouf A. Ibrahim, Ph.D., Edinburgh: nonlinear vibrations, fluid dynamics. Albert I. King, Ph.D., Wayne State: bioengineering. Kenneth A. Kline, Chair; Ph.D., Minnesota: structural dynamics. Ming-Chia Lai, Ph.D., Penn State: laser diagnostics. Golam M. Newaz, Ph.D., Illinois: composites. Eugene Rivin, D.Sc., Moscow: vibration, noise, and structural dynamics. Trilochan Singh, Associate Chair; Ph.D., Berkeley: combustion, energy conversion. Dinu Taraza, Ph.D., Bucharest: engine dynamics and simulation. Alan B. Whitman, Ph.D., Minnesota: classical mechanics.

Associate Professors. Emmanuel O. Ayorinde, Ph.D., Nottingham (England): vibrations, composites. John Cavanaugh, M.D., Michigan State: impact biomechanics. Nabil G. Chaloub, Ph.D., Michigan: dynamic modeling, controls, robotics. Jerry C. Ku, Ph.D., SUNY at Buffalo: thermal-fluid sciences. Sheng Liu, Ph.D., Stanford: electronic packaging. Chin-An Tan, Ph.D., Berkeley: dynamics, vibrations. Sean F. Wu, Ph.D., Georgia Tech: acoustics, vibrations. King Hay Yang, Ph.D., Wayne State: biomechanics. Sheng-Tao Yu, Ph.D., Penn State: computational fluid dynamics.

Assistant Professors. Theresa Atkinson, Ph.D., Michigan State: biomechanics. Michele J. Grimm, Ph.D., Pennsylvania: biomechanics. Xin Wu, Ph.D., Michigan: metal forming.

DIVISION OF ENGINEERING TECHNOLOGY

Professor. Mulchand S. Rathod, Director; Ph.D., Mississippi State: artificial heart, HVAC, heat transfer, energy, electronic packaging.

Associate Professors. Vladimir Sheyman, Ph.D., Academy of Sciences Minsk (Russia): heat transfer, electronic packaging, IC engines, energy conversion. Mukasa E. Ssemakula, Ph.D., Manchester (England): manufacturing processes, CIM, process planning. Chih-Ping Yeh, Ph.D., Texas A&M: image processing, computer vision, signal processing, communication.

Assistant Professors. Shamala A. Chickamenahalli, Ph.D., Kentucky: power, controls, electronic systems. Akihiko Kumagai, Ph.D., Wisconsin: manufacturing, CAD/CAM, robotics. Ece Yaprak, Ph.D., Wayne State: computer network and communications.

WESTERN MICHIGAN UNIVERSITY

College of Engineering and Applied Sciences

Programs of Study Men and women with advanced degrees in engineering and technical fields are ideal candidates for leadership positions in industry and education. At the doctoral level, Western Michigan University offers the Ph.D. in industrial engineering, mechanical engineering, and paper and imaging science and engineering. At the master's level, WMU offers the Master of Science in Engineering degree in computer, electrical, industrial, and mechanical engineering and the Master of Science degree in construction management, engineering management, manufacturing engineering, materials science and engineering, operations research, and paper science and engineering.

To earn a Master of Science degree, a student must complete 30 to 33 graduate semester credit hours. Fifteen of these hours must be in courses at the 600 or 700 level. A typical program consists of ten courses beyond the prerequisite requirements. Thesis, project, or course work–only options are available, depending on the master's program selected.

Research Facilities The College of Engineering and Applied Sciences has well-equipped research facilities. Tribology, the study of friction and wear, is addressed in three laboratories: a fluid mechanics laboratory for experimental and computational analysis of fluid flow; a materials science laboratory with atomic force and scanning-tunneling microscopes; and a tribology laboratory focused primarily on mechanical seals using a variety of seal test stands and analytical equipment, including a vertical interference microscope for surface topography measurement. Applied mechanics laboratories support fatigue, impact, and high-capacity torsional and axial load testing. The thermal systems laboratory supports research in heat exchanger design and in efficient refrigeration systems. The coherent optics laboratory uses optical methods for noncontact full-field strain measurements. Materials research is supported by facilities for the testing of mechanical, thermal, and physical properties of materials, including Mössbauer spectroscopy, electron microscopy, and X-ray diffractometry. There are fabrication and processing facilities for metals, polymers, and woods, including modern computer-controlled manufacturing capabilities. The Human Performance Institute conducts research in ergonomics, motion analysis, and workplace injury prevention. The Engineering Management Research Laboratory is a blend of faculty members, students, and sponsoring organizations focused on the investigation of organizational problems and issues. WMU offers unparalleled capabilities in its paper and printing pilot plants, with facilities in pulping, papermaking, and printing. Research is conducted on pulping, papermaking, coating, and printing, with particular emphasis on environmentally benign processing. A new Center for Non-Wood Fibers has been established to investigate alternatives to wood as a fiber source for paper products. The electrical and computer engineering laboratories house an anechoic chamber, two RF-shielded rooms, a digital signal processing laboratory, and Mentor Graphics tools for automated design of electronic circuits and systems. Modern computing facilities are available through both the College and the University.

Financial Aid Grants and fellowships available at Western Michigan University include assistantships and fellowships for master's students ranging from approximately $8200 to $12,000 and doctoral associateships starting at $9780. Students with appointments are required to pay regular tuition fees, but a tuition grant is often provided to nonresident students to supplement an appointment. Small grants are available through the Graduate Student Research Fund to encourage research by currently enrolled graduate students and to assist them in presenting their findings to professional groups.

The Financial Aid Office administers financial aid programs, such as loans and work-study programs, that are supported directly by federal and state sources.

Cost of Study In 1998–99, the cost for graduate study was $154.24 per credit hour for Michigan residents and $372.30 for nonresidents. Other fees were $602 for full-time students.

Living and Housing Costs Western offers a number of housing options. The University provides 585 apartments for students. Priority assignments are often given to graduate assistants, associates, and fellows. Rent for an unfurnished one-bedroom apartment ranges from $395 to $643. The University also offers a variety of options in its residence halls on campus.

Student Group Western Michigan University offers sixty-five master's degree programs, two specialist programs, and twenty-five doctoral programs. Approximately 54 percent of the graduate students in the University receive some form of financial aid. Of the students enrolled in graduate programs in the College of Engineering and Applied Sciences, approximately 85 percent are men and 15 percent are women. Thirty-three percent are international students.

Location Western Michigan University is located in Kalamazoo, Michigan, a metropolitan area of 200,000 midway between Detroit and Chicago on Interstate 94. Kalamazoo offers a combination of big-city attractions and a warm, small-town atmosphere. There are eighty lakes within 20 miles of downtown and more than sixty parks in the surrounding area.

The University and The College Founded in 1903, Western Michigan University is a large, comprehensive university with a Carnegie Doctoral I classification. Its enrollment of 26,575 includes students from all over the world. Graduate students make up approximately 22 percent of the enrollment.

The College of Engineering and Applied Sciences has six departments. The College offers undergraduate engineering programs in aeronautical, chemical, computer, construction, electrical, industrial, manufacturing, materials, mechanical, and paper engineering. The programs in aeronautical engineering, computer engineering, electrical engineering, industrial engineering, and mechanical engineering are EAC/ABET accredited. Bachelor of Science degrees are offered in manufacturing-related fields, such as engineering graphics and design technology, industrial design, printing, and engineering management. The programs in engineering graphics and design technology and manufacturing engineering technology are TAC/ABET accredited.

Applying Applicants for graduate programs in the College of Engineering and Applied Sciences should have a B.S. in engineering or a related field and a grade point average of 3.0 (A = 4.0) in their undergraduate work or in the last two undergraduate years for manufacturing engineering applicants. In addition to satisfying the general admission requirements of the Graduate College, applicants for the master's degree in electrical engineering or manufacturing engineering should submit results of the GRE. Applicants with degrees in other related disciplines may be considered for admission after they have satisfactorily completed the necessary undergraduate prerequisite courses. Application deadlines are July 1 for the fall semester, November 1 for the winter semester, March 1 for the spring session, and May 1 for the summer session. However, recommended deadlines for full consideration and financial aid are February 15 for the fall semester and October 15 for the winter semester.

Correspondence and Information Graduate Programs
College of Engineering and Applied Sciences
Western Michigan University
Kalamazoo, Michigan 49008-5062
Telephone: 616-387-4017
Fax: 616-387-4024
World Wide Web: http://www.wmich.edu/engineer

Western Michigan University

THE FACULTY AND THEIR RESEARCH

Department of Electrical and Computer Engineering
Ikhlas Abdel-Qader, Assistant Professor; Ph.D., North Carolina State. Digital image processing, control systems.
Johnson Asumadu, Assistant Professor; Ph.D., Missouri–Columbia. Power electronics, fuzzy logic control.
Sanjeev Baskiyar, Assistant Professor; Ph.D., Minnesota. Parallel processing, computer architecture.
Raghvendra Gejji, Associate Professor; Ph.D., Notre Dame. Control systems, computers.
John W. Gesink, Associate Professor; Ph.D., Michigan. Biomedical engineering, instrumentation.
Janos L. Grantner, Associate Professor; Ph.D., Budapest Technical. Fuzzy logic, finite-state machines/controllers.
Garrison W. Greenwood, Assistant Professor; Ph.D., Washington (Seattle). Computer architecture, logic design.
Dean R. Johnson, Associate Professor; Ph.D., Michigan State. Computer systems, fiber optics.
Joseph A. Kelemen, Associate Professor; M.S.E.E., St Louis; PE. Power systems.
John L. Mason, Associate Professor; Ph.D., Michigan; PE. Electromagnetics, microwaves.
Damon A. Miller, Assistant Professor; Ph.D., Louisville. Neural networks, electronics.
S. Hossein Mousavinezhad, Professor and Chair; Ph.D., Michigan State. Electromagnetics, communications.
Frank L. Severance, Associate Professor; Ph.D., Michigan State; CDP. Control systems, computer systems.
Lambert R. Vander Kooi, Professor, Ph.D., Michigan. Computer systems, communications.

Department of Construction Engineering, Materials Engineering, and Industrial Design
Osama Abudayyeh, Assistant Professor; Ph.D., North Carolina State; PE. Integrated project control systems, database management systems.
Pnina Ari-Gur, Associate Professor; D.Sc., Technion (Israel). Ceramic matrix composites, texture analysis.
Mohammed E. Haque, Assistant Professor; Ph.D., NJIT; PE. Fracture mechanics of engineering materials, composite materials.
Anil Sawhney, Assistant Professor; Ph.D., Alberta-Edmonton (Canada). Design and analysis of construction operations, construction bidding.
Vladimir Tsukruk, Professor and Chair; Ph.D., D.Sc., Ukrainian Academy of Sciences. Polymer materials science, surface engineering.

Department of Industrial and Manufacturing Engineering
Michael B. Atkins, Professor and Chair; Ed.D., Texas A&M. Computer-aided design, computer graphics.
Kailash M. Bafna, Professor; Ph.D., Purdue; PE. Facilities design, quality control.
Steven Butt, Assistant Professor; Ph.D., Pennsylvania State. Operations research, facilities location and vehicle routing.
Paul V. Engelmann, Associate Professor; Ed.D., Western Michigan. Plastics injection molding, extrusion.
Tycho K. Fredericks, Assistant Professor; Ph.D., Wichita State. Industrial ergonomics, work measurement and design.
Tarun Gupta, Associate Professor; Ph.D., Wisconsin–Milwaukee. Computer simulation, mechatronics.
Abdolazim Houshyar, Professor; Ph.D., Florida. Facilities design, simulation.
Mitchel Keil, Assistant Professor; Ph.D., Virginia Tech. Computer aided design and manufacturing, design for manufacturability.
Leonard R. Lamberson, Professor; Ph.D., Texas A&M; PE. Reliability.
David M. Lyth, Professor; Ph.D., Michigan State; CQE. Production/operations management, quality management.
Larry A. Mallak, Associate Professor; Ph.D., Virginia Tech. Engineering management, strategic planning and analysis.
William R. Peterson, Assistant Professor; Ph.D., Ohio State. Engineering management, production/operations management.
Sam N. Ramrattan, Associate Professor; Ph.D., Iowa State. Quality control/process control, industrial materials and processes.
Jorge Rodriguez, Assistant Professor; Ph.D., Wisconsin–Madison. Computer-aided design, finite element analysis.
Frederick Z. Sitkins, Professor; M.A., Eastern Michigan; CmfgE. Metrology, ISO 9000.
Ralph Tanner, Associate Professor; Ph.D., Oakland; PE; CmfgE. Automated manufacturing systems, robotics.
James VanDePolder, Associate Professor; M.S., Colorado State. Automotive engine testing, pollution control.
Bob E. White, Professor; Ph.D., Iowa State; PE. Engineering economy, ergonomics.
Robert M. Wygant, Professor; Ph.D., Houston; CMfgE. Human performance/ergonomics, work design.

Department of Mechanical and Aeronautical Engineering
Judah Ari-Gur, Professor; D.Sc., Technion (Israel). Static and dynamic response of composite materials, static and dynamic response of structures.
Kasim Biber, Assistant Professor; Ph.D., Wichita State. Applied aerodynamics, stability and control.
Christopher Cho, Professor; Ph.D., SUNY at Stony Brook; PE. Two-phase flow, heat transfer.
Jay Easwaran, Associate Professor; Ph.D., Toronto. Design, fabrication, and testing of resin-based composites; lost-foam casting.
Meshulam Groper, Professor; D.Sc., Technion (Israel). Experimental stress analysis.
Philip J. Guichelaar, Professor; Ph.D., Michigan; PE. Tribology, manufacturing processes.
Jerry H. Hamelink, Professor; Ed.D., Western Virginia; PE. Experimental design, thermodynamics.
Richard B. Hathaway, Professor; Ph.D., Oakland; PE. Vehicle structures and dynamics, application of optical methods.
Arthur W. Hoadley, Professor; M.S., Ohio State; PE. Applied aerodynamics, flight test instrumentation.
James Kamman, Assistant Professor; Ph.D., Cincinnati. Applied mechanics, dynamics.
Daniel Kujawski, Associate Professor; D.Sc., Warsaw Technical; Ph.D., Polish Academy of Sciences (Warsaw). Testing, fatigue, and fracture of advanced materials.
William W. Liou, Assistant Professor; Ph.D., Penn State. Analytical and computational fluid dynamics in propulsion, turbulence modeling and simulations.
Parviz Merati, Professor and Chair; Ph.D., Illinois at Urbana-Champaign; PE. Experimental and numerical fluid mechanics and heat transfer, tribology.
Koorosh Naghshineh, Associate Professor; Ph.D., Penn State. Noise and vibration control.
Iskender Sahin, Professor; Ph.D., Virginia Tech; PE. Incompressible fluid mechanics, water waves.
Rameshwar Sharma, Associate Professor; Ph.D., Wayne State. Control systems.
Dennis VandenBrink, Associate Professor; Ph.D., Virginia Tech; PE. Finite-element analysis.
Molly W. Williams, Professor and Associate Dean; Ph.D., Berkeley; PE. Tribology, failure analysis.

Department of Paper and Printing Science and Engineering
Raja G. Aravamuthan, Professor; Ph.D., Washington (Seattle). Pulping methods, process simulation.
John H. Cameron, Associate Professor; Ph.D., Michigan State. Heat and mass transfer, specialty papers.
Paul D. Fleming, Associate Professor; Ph.D., Harvard. Digital printing, simulation.
Margaret K. Joyce, Assistant Professor; Ph.D., North Carolina State. Paper coatings, rheology.
Thomas W. Joyce, Professor and Chair; Ph.D., Purdue; PE. Environmental engineering, biotechnology.
Peter E. Parker, Associate Professor; Ph.D., Michigan. Process modeling, towel and tissue creping.
David K. Peterson, Associate Professor; Ph.D., Denver; PE. Environmental control, conservation process control.
Dewei Qi, Assistant Professor; Ph.D., Waterloo. Paper surface properties, fiber physics.

WEST VIRGINIA UNIVERSITY

College of Engineering and Mineral Resources

Programs of Study

The College of Engineering and Mineral Resources offers nine Ph.D. programs in engineering: aerospace, chemical, civil, computer, electrical, industrial, mechanical, mining, and petroleum and natural gas and a Ph.D. in computer science. Designated master's degrees are offered in aerospace, chemical, civil, computer science, electrical, industrial, mechanical, mining, petroleum and natural gas, and software engineering. M.S. degrees are offered in occupational hygiene and occupational safety and safety management; both programs are accredited by the Related Accredited Commission of the Accreditation Board for Engineering and Technology (ABET). In addition, the College offers an M.S. degree in engineering to qualified students whose baccalaureate work was done in a field other than engineering. The College offers a certificate in bioengineering, manufacturing systems engineering, materials engineering, and software engineering. Programs of study are structured to fit the goals of individual students within a discipline. The graduate faculty is composed of scholars who are highly competent in their fields of expertise. Many have national and international reputations. The College of Engineering and Mineral Resources is dedicated to providing its undergraduate and graduate students with an excellent engineering education and to preparing students to become tomorrow's leaders.

For all M.S. degrees, each candidate, with the approval of his or her advisory and examining committee, follows a planned program that must contain a minimum of 30 semester credit hours. If a thesis or a problem report is part of the candidate's program, not more than 6 semester credit hours of research leading to an acceptable thesis nor more than 3 semester credit hours of work for an acceptable problem report may be applied toward the semester credit hour requirement.

The Ph.D. degree is awarded upon completion of a program of advanced study that includes at least two semesters of continuous on-campus residence, submission of an acceptable dissertation, and the successful completion of comprehensive and final oral examinations.

Research Facilities

The College of Engineering and Mineral Resources is housed principally in three buildings—Engineering Sciences (131,000 square feet), Mineral Resources (89,000 square feet), and Engineering Research (37,000 square feet)—with an additional 43,000 square feet of research space in other locations. The College houses several externally supported centers: Alternative Transportation Fuels, Constructed Facilities, Energy Utilization and Assessment, Entrepreneurial Studies and Development, Microelectronics Systems Research, NASA Space Consortium, Longwall Mining and Ground Control Research, and Transportation. It also operates a fluid-particle science facility, a microelectronics fabrication facility, three experimental aircraft, a bridge testing facility, an engine emissions testing laboratory, several alternate-fuel vehicles, a mine-navigating robot, wind tunnels, and a number of other facilities that provide the opportunity to investigate materials, evaluate designs, and perform engineering research. West Virginia University's (WVU) computer-controlled intercampus transportation system serves as an engineering laboratory as well.

The College is served by a multitude of modern computer systems, most of which are accessible by a college-wide local area network. It has a computer laboratory consisting of personal computers, workstations, terminals, plotters, and printers reserved for graduate research. Each department has additional facilities as well. Fiber optic communications pathways connect engineering to the computer center, other campus locations, and the Internet. The West Virginia Network for Telecomputing is housed within the University, providing access to machines throughout West Virginia and the world.

Financial Aid

Aid for qualified students is available in the form of teaching and research assistantships. Assistantships carry a stipend that ranges from $5040 to $18,000 in 1999–2000, and tuition is waived for these students. Research assistants are assigned to a project consistent with their major area of interest.

Cost of Study

Tuition for graduate students is $2930 per year for West Virginia residents and $8354 per year for out-of-state residents for the 1999–2000 academic year. These costs are subject to change. Tuition is waived for students holding research or teaching assistantships.

Living and Housing Costs

The cost of an efficiency apartment in University-owned housing is $353 per month during 1999–2000. There are a limited number of apartments available for married students. Privately owned apartments and dormitories in the Morgantown area rent from a low of approximately $350 per month.

Student Group

Total University enrollment for fall 1998 was 22,238, including 7,063 graduate and professional students. There are currently 629 graduate students enrolled in the College of Engineering and Mineral Resources, both full- and part-time.

Location

West Virginia University is located in Morgantown, a university community with a population of approximately 26,000. Monongalia County is one of the largest deep-mine, coal-producing counties in the nation, and WVU is the county's largest single employer. Located on the east bank of the Monongahela River, Morgantown is situated on the rugged terrain of the Appalachian highlands; its altitude varies from 800 to 1,150 feet above sea level. The northern West Virginia area has a growing high-technology industry, anchored by the Federal Energy Technology Center, the NASA-WVU Software Independent Validation and Verification Center, the National Institute for Occupational Safety and Health (NIOSH), and the FBI Fingerprint Center, which offers a unique opportunity to develop partnerships in research and economic development. Morgantown is within easy traveling distance to Washington, D.C., and Pittsburgh, Pennsylvania.

The Personal Rapid Transit (PRT) System, which was built by the U.S. Department of Transportation as a national research and demonstration project, is perhaps the largest project of its kind ever built on a university campus. It consists of computer-directed electric-powered cars that operate on a concrete-and-steel guideway without drivers on board and is used by students for transportation between the three campuses.

The University

Since its founding in 1867, WVU has developed into the center of graduate and professional education, research, and extension programs in West Virginia. It is one of only forty-three state universities in the nation that serve their states as research and land-grant institutions.

Applying

Students should apply to the WVU Office of Admissions and Records, P.O. Box 6009, Morgantown, West Virginia 26506-6009, and are encouraged to correspond directly with the Department they wish to enroll in. Requirements include an application fee of $45, transcripts of all previous academic work, and three letters of reference. GRE General Test scores are required by some departments. International students must submit a TOEFL score of at least 550 and satisfactory GRE scores.

Correspondence and Information

Chairman (specify department)
College of Engineering and Mineral Resources
West Virginia University
P.O. Box 6101
Morgantown, West Virginia 26506-6101

Telephone: 304-293-4821
Fax: 304-293-5024
E-mail: info@cemr.wvu.edu
World Wide Web: http://www.cemr.wvu.edu

West Virginia University

FACULTY HEADS AND AREAS OF RESEARCH

Chemical Engineering. Dr. Dady B. Dadyburjor, Chair.

The graduate program in chemical engineering at both the M.S. and Ph.D. levels is a research degree. The department offers a graduate program of medium size that is focused on student involvement in applied research in a modern, interactive learning environment. The total student experience reflects a balance of course work and thesis/dissertation research. Faculty members have active research programs in catalysis, reaction engineering, electronic materials, fluidization, fluid particle and coating science, polymer rheology, composite materials, surface science, mineral processing, and environmental, biochemical, and biomedical engineering. Typical projects currently include coal liquefaction and gasification processes; particle coating and adhesion technology, cell bioreactor dynamics; carbon products from coal; rheological characterizations of foods, emulsions, and composites; and biotransport phenomena. Faculty members receive funding for research to support students at competitive assistantship levels from various government and corporate agencies and sponsors. Faculty members are active in publication and participate regularly in professional societies and technical meetings.

Civil and Environmental Engineering. Dr. David Martinelli, Chair.

There are currently 19 full-time graduate faculty members and more than 100 graduate students specializing in one or more of the following subdisciplines: environmental and hydrotechnical engineering; structural engineering; geotechnical, geoenvironmental, and materials engineering; and transportation systems engineering. Current research in the department includes in situ bioremediation, formation of in situ barriers using jet/permeation grouts for waste containment, acid mine drainage remediation, removal of heavy metals using activated carbon, shoreline erosion protection methods, fly ash seals to control acid mine drainage, subsidence prediction and control over abandoned mine land, landfill design, asphalt paving technology, evaluation of wastewater treatment filters, dynamics of granular flows, application of image-analysis techniques to define watershed drainage paths, innovative design and diagnostics applied to constructed facilities, design and monitoring of timber bridges, fiber-reinforced plastic building components, environmental geotechnology, a pilot plant for manufacturing new composite materials, advanced technologies for traffic monitoring, traffic engineering, highway design, maintenance and operations, and infrastructure management.

Computer Science and Electrical Engineering. Dr. George Trapp, Chair.

The faculty in Computer Science and Electrical Engineering has an excellent research reputation, particularly in the areas of high performance computing, software engineering, software verification and validation, microelectronics systems, control systems, digital communications and signal and image processing, electromechanical systems, and power systems. While these are currently the areas of greatest emphasis, the department also has strengths in other areas of computer science, computer engineering, and electrical engineering. The number of faculty members and students in the graduate program is well balanced, and a graduate student can realistically expect to work in close cooperation with the faculty member who serves as the research adviser. The atmosphere is congenial and friendly, while achieving the professional purpose of maintaining an excellent research environment. Current major sponsored projects are in the areas of software engineering, software IV and V, computer engineering, electromechanical systems, image processing, and microelectronic systems, photonic systems, and power systems.

Industrial and Management Systems Engineering (IMSE). Dr. Ralph W. Plummer, Chair.

Graduate studies and research programs are offered in industrial engineering, occupational hygiene and occupational safety (OHOS), and safety and environmental management (SEM). The IMSE Department consists of 17 full-time graduate faculty members and more than 150 graduate students. Academic programs and research opportunities are offered in the areas of ergonomics, decision sciences, manufacturing systems, operations research/simulation, industrial hygiene, and safety management. The OHOS and SEM programs are accredited by ABET/RAC in industrial hygiene and safety management, respectively. Current research topics include biomechanics, computer-aided design, expert systems, computer simulation, decision theory, energy conservation, human factors, industrial hygiene, inventory control, mathematical programming and optimization, pollution prevention, quality control, safety in systems design, scheduling, sequencing, training, transportation systems, information systems, and management of systems. Graduates of these programs have a broad variety of career opportunities.

Mechanical and Aerospace Engineering. Dr. Donald W. Lyons, Chair.

Graduate studies and research are offered in fluid mechanics, aerodynamics, thermal sciences, theoretical and applied mechanics, machine design, materials, structures, manufacturing systems engineering, robotics, CAD/CAM, automation and controls, and bioengineering. Current research includes projects in multiphase flow, aircraft stability and control, fluidized-bed combustion, internal combustion engine development, alternative fueled vehicles, laser velocimetry, emissions of fossil fuel combustion, bearing loads, fracture mechanics of coal, composite materials, space structures, robot work-cell design, robot safety, computer-aided design methodologies, interactive graphics, rehabilitation and medical devices, and microfracture of biomedical materials. Research is experimental, theoretical, and computational, and extensive laboratory and computer facilities are available.

Mining Engineering. Dr. Syd S. Peng, Chair.

The mining engineering graduate program offers both M.S. and Ph.D. degrees. The program is housed in a new building with state-of-the-art laboratories such as rock mechanics, mine ventilation, health and safety, mine design, mine equipment, longwall, computer, coal crushing, and froth flotation laboratories, and a mineral dressing pilot plant. The faculty members are well-established and well-known nationally and internationally in their fields of expertise. They are very active in coal mining research, including mountaintop removal mining, coal bed methane, application of mathematical optimization to mine ventilation, respirable coal dust (generation and distribution), longwall mining and management systems, ground control, rock mechanics, surface subsidence, coal and mineral preparation, and environments. Being in the center of one of the richest coalfields in the world, all research projects utilize local coal mines for verification of techniques developed in the laboratory. The majority of research is funded by the mining industry and some is co-funded by state agencies and thus involves technology transfer.

Petroleum and Natural Gas Engineering. Professor Sam Ameri, Chair.

The department's research programs are focused on the areas of natural gas production and storage, formation evaluation, reservoir characterization and modeling, flow through porous media, low permeability gas reservoirs, drilling engineering, and environmental applications. Most of the department's research is sponsored by the U.S. Department of Energy. Several departmental faculty members are working on application of petroleum engineering technology to environmental protection and increased well productivity and well life. Major research programs include use of artificial intelligence for well productivity enhancement and application of petroleum technology for environmental protection/remediation. Much of the work done in the department is specific to petroleum and natural gas industry problems in the state. Research is experimental, theoretical, and computational. There are also extensive laboratory and computer facilities available.

WICHITA STATE UNIVERSITY

College of Engineering
Graduate Engineering Programs

Programs of Study

The College of Engineering offers graduate programs leading to the Master of Science (M.S.) and the Doctor of Philosophy (Ph.D.) in aerospace engineering (aerodynamics and fluid mechanics; structures, solid mechanics, and composites; flight dynamics and control; and multidisciplinary design, analysis, and optimization), electrical engineering (control systems, communications, signal processing, computers and digital systems, and energy and power systems), industrial and manufacturing engineering (industrial ergonomics/human factors, manufacturing systems engineering, and engineering systems), and mechanical engineering (engineering materials properties and failure modes; controls, robotics, and automation; multibody and impact dynamics; mechanical engineering design and manufacturing; thermodynamics and transport processes; combustion; and heating, ventilating, and air-conditioning (HVAC) and energy conservation). The College also offers the Master of Engineering Management (M.E.M.) degree in the industrial and manufacturing engineering department.

For the M.S. degree, three degree options are available in each department. The thesis option requires a minimum of 30 credit hours, including 6 hours of thesis; the directed project option requires a minimum of 33 credit hours, including 3 hours of project; and the all-courses option requires a minimum of 33 credit hours of course work. All options involve a written or oral concluding examination. The M.E.M. degree requires a minimum of 33 credit hours of course work and a written exit examination.

For the Ph.D. degree, approximately 60 credit hours beyond the M.S. degree are required. This includes a minimum of 24 hours of dissertation and a distribution of remaining courses between major, minor, and mathematics. The student's advisory committee administers required examinations, including the dissertation defense.

Research Facilities

The College has significant research facilities supporting programs in all four departments. The graduate programs are enhanced by the presence of the industrial complex in Wichita and the National Institute for Aviation Research (NIAR) on the Wichita State campus. Major facilities available in engineering and the NIAR include subsonic and supersonic wind tunnels, structures, crash dynamics, CIM, manufacturing processes, graphics, metrology, propulsion, computational mechanics, materials research, composites, ergonomics, shock and vibration, and vehicle dynamometer laboratories. Appropriate library and computer facilities are also available to engineering graduate students.

Financial Aid

Fellowships, teaching assistantships, and research assistantships are available on a competitive basis. The awards range from $4000 to more than $10,000 per academic year, with some possibility of summer support in teaching and research. Students awarded assistantships pay in-state tuition and fees. Teaching assistants receive an additional partial waiver of in-state tuition based on the number of credit hours taught.

Cost of Study

Estimated tuition and fees for the 1999–2000 academic year are $118 per credit hour for Kansas residents and $348 per credit hour for nonresidents. Engineering courses are subject to a $15 per credit hour engineering equipment fee, the proceeds of which are used to purchase laboratory equipment for instructional and student research purposes.

Living and Housing Costs

Many graduate students live in apartments and prepare their own meals. University apartments range from $195 per month for double occupancy with shared bath and kitchen to $550 per month for a single-occupancy studio-type apartment with kitchen and bath. Average costs for single-bedroom off-campus housing are estimated to be $350 per month.

Student Group

There is great diversity in the students pursuing graduate degrees in the College of Engineering. Many are part-time evening students who work for local industries. More than 50 percent of the full-time students receive some type of financial support. International students, who comprise 50 percent of the College of Engineering graduate student enrollment, are considered an asset since engineering is now clearly a worldwide profession.

Student Outcomes

Graduates from these programs have moved into the international marketplace. Many graduates find employment in the Wichita area; others find positions in major engineering centers throughout the nation. Many of the international students return home for job opportunities with international companies.

Location

Wichita, the largest city in Kansas and part of a metropolitan area of 400,000, offers the cultural and economic advantages of a big city but maintains the friendly atmosphere of a smaller town. It is a regional medical center and the focus of the majority of industry in Kansas. These industries include the Boeing Company, Cessna Aircraft, Raytheon Aircraft, Bombardier-Learjet Aircraft, Koch Industries, Case Industries, Evcon Industries, Brite-Voice Communications, LSI Logic, and many others. Native American, Hispanic, Asian, and Middle Eastern groups are typical of Wichita's multicultural and ethnic diversity.

The University and The College

Wichita State University is an urban university and as such offers the "Metropolitan Advantage" to its students. This advantage includes not only the arts, cultural, and social activities associated with the city and university but also, for engineering students, the advantages of a location in the major manufacturing area of the state. College of Engineering students are involved in industrial research projects and co-op positions with local companies; many are employed locally. The College is large enough to offer full programs at the B.S., M.S., and Ph.D. levels but small enough for students to get to know the faculty members on a personal basis.

Applying

Full admission into an M.S. program requires the equivalent of an undergraduate degree in that engineering discipline or a related area, with a minimum grade point average of 3.0 for the last two years of undergraduate work, all engineering courses, and mathematics and physical science courses. Each applicant's academic record is evaluated prior to admission to the program to determine their potential for success in graduate study. The M.E.M. program requires two years of professional experience.

Full admission into a Ph.D. program requires a completed M.S. degree in engineering or physical science. GRE General Test scores must be submitted. Some students may find it necessary to take prerequisite courses to be able to meet the course breadth requirements. Each applicant's record is reviewed and evaluated by the department receiving the application.

Correspondence and Information

Graduate Coordinator
College of Engineering
Wichita State University
Wichita, Kansas 67260-0044

Telephone: 316-978-3400
Fax: 316-978-3853
World Wide Web: http://www.engr.twsu.edu/

Wichita State University

THE FACULTY AND THEIR RESEARCH

Aerospace Engineering

Ramesh K. Agarwal, Bloomfield Distinguished Professor and Executive Director of the National Institute for Aviation Research; Ph.D., Stanford, 1975. Computational fluid dynamics, electromagnetics and acoustics.

Klaus A. Hoffmann, Professor and Ph.D. Graduate Coordinator; Ph.D., Texas at Austin, 1983. Computational fluid dynamics, aerothermodynamics, hypersonics.

Steven J. Hooper, Associate Professor; Ph.D., Iowa State, 1983. Solid mechanics, structures, structural dynamics, composites.

Walter J. Horn, Professor and Chair; Ph.D., Texas at Austin, 1972. Solid mechanics, structures, composites, aeroelasticity.

Thomas E. Lacy, Assistant Professor; Ph.D., Georgia Tech, 1998. Solid mechanics, aircraft structures, fracture mechanics, damage tolerance.

L. Scott Miller, Associate Professor; Ph.D., Texas A&M, 1988. Theoretical and experimental aerodynamics.

Roy Y. Myose, Associate Professor; Ph.D., USC, 1991. Experimental aerodynamics and structures.

M. Gawad Nagati, Associate Professor; Ph.D., Iowa State, 1984. Flight dynamics and control.

Michael Papadakis, Associate Professor and Bombardier-Learjet Fellow; Ph.D., Wichita State, 1986. Computational, experimental, and theoretical aerodynamics.

Kamran Rokhsaz, Associate Professor and Master's Graduate Coordinator; Ph.D., Missouri–Rolla, 1988. Aerodynamics and design, flight mechanics.

Bert L. Smith, Professor; Ph.D., Kansas State, 1966. Solid mechanics, composites.

John S. Tomblin, Assistant Professor; Ph.D., West Virginia, 1994. Solid mechanics, structures, composites.

Electrical Engineering and Computer Engineering

Robert I. Egbert, Professor and Director of the Center for Energy Studies; Ph.D., Missouri–Rolla, 1976. Power systems, control systems.

Elmer A. Hoyer, Professor and Director of the Federal Rehabilitation Engineering Research Center; Ph.D., Missouri–Columbia, 1967. Digital signal processing, speech signal processing, rehabilitation engineering.

Ward T. Jewell, Professor and Director of Power Quality Center; Ph.D., Oklahoma State, 1986. Power systems.

Everett L. Johnson, Professor and Chair; Ph.D., Kansas, 1969. Digital design, computer engineering.

Mark T. Jong, Professor and Associate Dean; Ph.D., Missouri–Columbia, 1967. Circuit/system theory, signal processing, energy.

Raj Katti, Associate Professor; Ph.D., Washington State, 1991. Digital systems, VLSI, VHDL, computer architecture.

Hyuck M. Kwon, Associate Professor; Ph.D., Michigan, 1984. Digital, wireless, and satellite communications systems.

Larry D. Paarmann, Associate Professor; Ph.D., IIT, 1983. Signal processing.

Ravindra Pendse, Assistant Professor and Director of CISCO CCIE Lab; Ph.D., Wichita State, 1994. Computer engineering, VHDL, network systems.

M. Edwin Sawan, Professor and Graduate Coordinator; Ph.D., Illinois at Urbana-Champaign, 1979. Control theory.

Steven R. Skinner, Associate Professor; Ph.D., Iowa, 1991. Nonlinear optics, optical neural networks and computing, computer engineering.

Asrat Teshome, Associate Professor; Ph.D., Cornell, 1980. Power systems, power electronics, control theory, fiber optics.

Paul K. York, Professor; Ph.D., Texas A&M, 1967. Signal processing, navigational systems and avionics systems.

Industrial and Manufacturing Engineering

S. Hossein Cheraghi, Associate Professor; Ph.D., Penn State, 1992. Precision measurement, automated inspection, computer vision, tolerancing, GD&T, CAD/CAM, manufacturing systems.

Jeffrey E. Fernandez, Professor; Ph.D., Texas Tech, 1986. Ergonomics, human factors engineering, safety, biomechanics.

Mark Kaiser, Assistant Professor; Ph.D., Purdue, 1991. Operations research, geometric analysis and metrology, energy systems, engineering optimization.

Krishna Krishnan, Assistant Professor; Ph.D., Virginia Tech, 1994. Manufacturing systems, CAD/CAM, free-form surfaces, process planning, design for manufacturability, process automation and control, flexible manufacturing, material handling.

Viswanathan Madhavan, Assistant Professor; Ph.D., Purdue, 1996. Traditional/nontraditional machining, cutter technology, forming processes, tool and jig design.

Don E. Malzahn, Professor; Ph.D., Oklahoma State, 1975. Operations management, decision analysis, project management, rehabilitation engineering, engineering management.

Abu S. M. Masud, Professor, Graduate Coordinator, and Chair; Ph.D., Kansas State, 1978. Operations research, multiple-objective decision making, forecasting, expert systems, decision analysis, QFD, modeling and optimization of productions and service systems.

Jamal Y. Sheikh-Ahmad, Assistant Professor; Ph.D., North Carolina State, 1993. Metal and composite manufacturing; computer-aided manufacturing; tools, jig, and fixture design.

Janet M. Twomey, Assistant Professor; Ph.D., Pittsburgh, 1994. Simulation, artificial neural networks, statistical analysis intelligent data analysis/processing.

Lawrence E. Whitman, Assistant Professor; Ph.D., Texas at Arlington, 1998. Enterprise engineering, supply-chain management, CAD/CAM, manufacturing systems engineering.

Heecheon You, Assistant Professor; Ph.D., Penn State, 1999. Biomechanics, industrial and cognitive ergonomics, CTD, human factors engineering.

Mechanical Engineering

Behnam Bahr, Associate Professor; Ph.D., Wisconsin–Madison, 1988. Manufacturing, robotics, automation, artificial intelligence in manufacturing.

Jharna Chaudhuri, Professor; Ph.D., Rutgers, 1982. Materials science and engineering, X-ray diffraction, electronic materials, wide-band gap semiconductors, composite materials, fatigue, fracture mechanics, impact.

Mahesh S. Greywall, Professor and Graduate Coordinator; Ph.D., Berkeley, 1962. Fluid dynamics, thermal science, computational fluid dynamics, environmental engineering.

Richard T. Johnson, Professor and Chair; Ph.D., Iowa, 1968. Internal-combustion engines, alternative fuels, design, manufacturing automation.

David N. Koert, Assistant Professor; Ph.D., Drexel, 1990. Combustion and combustion chemistry, alternative fuels, air pollution control, high-temperature processes and material properties.

Hamid M. Lankarani, Associate Professor and Bombardier-Learjet Fellow; Ph.D., Arizona, 1988. Multibody and impact dynamics, biomechanics, mechanical engineering design.

Julie A. Mathis, Associate Professor; Ph.D., LSU, 1989. Fluid dynamics, thermal science, structural engineering.

T. S. Ravigururajan, Associate Professor; Ph.D., Iowa State, 1986. Heat transfer enhancement, HVAC, microscale heat transfer.

James E. Steck, Associate Professor; Ph.D., Missouri–Rolla, 1989. Artificial neural networks, dynamics and control, FEM, acoustics.

George E. Talia, Professor; Ph.D., Case Western Reserve, 1980. Mechanical properties of materials, microscopy, defects in materials, erosion, composites.

C. Charles Yang, Assistant Professor; Ph.D., LSU, 1993. Application and manufacture of composite materials, mechanical engineering design.

WIDENER UNIVERSITY
Graduate Programs in Engineering

Programs of Study
Widener University awards the degree of Master of Engineering (M.E.) in chemical engineering, civil engineering, computer and software engineering, electrical/telecommunication engineering, mechanical engineering, and engineering management. The environmental engineering option is available in chemical and civil engineering and engineering management.

The dual-degree program, Master of Engineering/Master of Business Administration (M.E./M.B.A.), is offered in conjunction with the School of Business Administration.

The Master of Engineering programs require a minimum of 33 credits without a thesis or 30 credits with a thesis. The M.E./M.B.A. program credit requirements vary according to the undergraduate business courses completed. A saving of two or three courses results from the combining of the two degree programs.

Research Facilities
Advanced laboratory facilities are available to engineering students for problem solving and project activities, including a Tektronix-computer–aided design laboratory and laboratories for microprocessors, digital signal processing computers, energy conversion, electronics and microwaves, automatic controls, structures and mechanics, transport phenomena, unit operations, soil mechanics, and fluids. A machine shop is also available.

The Wolfgram Memorial Library has a collection of 300,000 printed volumes and 2,205 periodical titles. Services include online access to bibliographic information, CD-ROM databases, audiovisual-media collections and facilities, and access to other libraries' resources through interlibrary loans.

Financial Aid
Students in graduate programs can apply for financial aid programs through the Financial Aid Office on the Main Campus. A limited number of graduate assistantships are available to full-time students in graduate programs, and a number of loan programs are available to all eligible students.

Cost of Study
For the 1999–2000 year, tuition in the Master of Engineering program is $530 per credit. Students in the M.E./M.B.A. program pay $490 per credit for M.B.A. courses.

Living and Housing Costs
Affordable rental apartments are available within a 3-mile radius of the campus.

Student Group
Classes are small. The average class size is fewer than 20. Approximately 4,200 students are pursuing graduate or professional degrees at the University. About 55 percent are women. The student population is largely drawn from the mid-Atlantic region; about 5 percent of the graduate and professional students are from other countries.

Location
Widener's Main Campus, occupying more than 100 acres in Chester, Pennsylvania, is easily accessible from Interstate 95, which runs north-south along the Delaware River. Located in Delaware County, one of the oldest counties in Pennsylvania, the campus is near historic and commercial areas; Philadelphia is just 15 miles north.

The 40-acre Delaware Campus (15 miles southwest of the Main Campus) is located on Route 202 (Concord Pike) north of Wilmington and is only a short distance from Interstate 95. The Widener School of Law is located on the Delaware Campus. Another branch of the School of Law is located on a 14-acre campus in Harrisburg, Pennsylvania. The Harrisburg Campus is located on Interstate 81 at the Progress Avenue exit.

The University
A private, accredited university founded in 1821, Widener offers doctoral, master's, baccalaureate, and associate degrees through its eight schools and colleges. Incorporated in both Pennsylvania and Delaware, the University has three campuses, which are located in Chester and Harrisburg, Pennsylvania; and Wilmington, Delaware. All three campuses serve both resident and commuter students. The total University enrollment is 7,082.

Applying
Candidates should hold a bachelor's degree in engineering in an EAC/ABET accredited program. Scores on the GRE, although not required, are recommended for admission to the graduate engineering program. Scores on the GMAT are required for the M.E./M.B.A. program. All transcripts of records covering all academic work beyond high school must be submitted to the University. An application fee is required.

Correspondence and Information
Graduate Programs In Engineering
Widener University
One University Place
Chester, Pennsylvania 19013
Telephone: 610-499-4198

Widener University

THE FACULTY AND THEIR RESEARCH

Fred A. Akl, Professor and Dean of Civil Engineering; Ph.D., Calgary; PE (Ohio). Vibration analysis, finite element methods, computational mechanics, experimental mechanics.

Charles L. Bartholomew, Professor of Civil Engineering; Ph.D., Illinois; PE (Colorado, Illinois, Pennsylvania). Geotechnical engineering, transportation engineering, dams and water supply.

Vicki L. Brown, Associate Professor of Civil Engineering; Ph.D., Delaware; PE (Pennsylvania). Structural analysis and design, reinforced concrete structures.

David H. T. Chen, Associate Professor of Chemical Engineering; Ph.D., Rochester; PE (Delaware). Process design, process development, heat and mass transfer, environmental engineering.

John F. Davis, Assistant Professor of Civil Engineering; Ph.D., Penn State. Environmental engineering, water and wastewater treatment, water quality modeling and assessment.

Balaur S. Dhillon, Associate Professor of Engineering; Ph.D., Colorado; PE (Colorado, New Jersey, Pennsylvania). Finite elements and numerical analysis, structural analysis and design, reinforced concrete design.

Rocco A. Di'Taranto, Professor of Engineering; Ph.D., Pennsylvania; PE (Pennsylvania). Analysis of damped vibrations of beams and plates, theory of vibratory bending, structural vibrations and dynamics of laminated beams and cylinders, composite materials.

Raymond P. Jefferis III, Professor of Engineering; Ph.D., Pennsylvania. Systems engineering, software engineering, modeling and control, direct digital control of industrial processes, microprocessors, and distributed hierarchal control.

Alfred T. Johnson, Professor of Engineering; Ph.D., Pennsylvania; PE (Pennsylvania). Digital signal processing, circuit and system theory.

Nathaniel R. Kornfield, Professor Emeritus of Engineering; Ph.D., Pennsylvania. Systems engineering, computer engineering, computers in engineering, bioengineering.

JoAnn B. Koskol, Associate Professor of Electrical Engineering; Ph.D., Delaware. Signal and image processing.

Anastas Lazaridis, Professor of Mechanical Engineering; Sc.D., Columbia; PE (Delaware). Heat transfer, thermodynamics, thermal engineering, energy conservation, solar engineering, fluid mechanics, solidification-melting, metal castings, centrifugal casting, textile engineering.

Bryen E. Lorenz, Associate Professor of Engineering; Ph.D., Drexel. Electromagnetics.

Gennaro J. Maffia, Associate Professor of Chemical Engineering; M.B.A., NYU; D.E., Dartmouth. Process design, process development, biotechnology, environmental engineering.

James R. May, Associate Professor of Law; J.D., Kansas; LL.M., Pace. Environmental law, law of hazardous waste and substances.

Kenneth M. McNeil, Associate Professor of Engineering; Ph.D., Cambridge; PE (Pennsylvania). Fluidization, kinetics and catalysis, mass transfer, process development, reactor design.

Thomas G. McWilliams Jr., Professor Emeritus of Engineering; Ph.D., Maryland. Thermodynamics, materials engineering.

Ronald L. Mersky, Associate Professor of Civil Engineering; Ph.D., Pennsylvania; PE (Pennsylvania). Environmental engineering, water resources, resources analysis, waste management.

John E. Molyneux, Professor of Mechanical Engineering; Ph.D., Pennsylvania; PE (New York). Statistical theory of continua, numerical solution of partial differential equations, geophysical applications in inverse scattering theory.

Gopalakrishna Nadig, Associate Professor of Engineering; Ph.D., Cornell. Dynamics and vibrations, stress analysis (theoretical and experimental), machine design.

Charles R. Nippert, Associate Professor of Engineering; Ph.D., Lehigh; PE (Pennsylvania). Mechanical agitation, process development, computer simulation.

John A. Okolowski, Associate Professor of Engineering; Ph.D., Drexel. Electricity and magnetism, electromagnetics.

Sohail Sheikh, Associate Professor of Electrical Engineering; Ph.D., Syracuse. Multiprocessor interconnection networks, fault-tolerant computing and optical computing.

Maria Slomiana, Associate Professor of Mechanical Engineering; Ph.D., Drexel. Solid mechanics, composite materials, theoretical and experimental fracture mechanics.

Michael P. Smyth, Professor Emeritus of Engineering; Ph.D., Pennsylvania; PE (Pennsylvania). Linear engineering systems, network theory, author of *Linear Engineering Systems: Tools and Techniques*, coauthor of *You and Technology*.

Suk-chung Yoon, Associate Professor of Computer Science; Ph.D., Northwestern. Computer science.

Adjunct Faculty

Dale A. Brandreth, Ph.D., Toronto. Thermodynamics, surface chemistry, building insulation.

Constantine G. Fountzoulas, Ph.D., Northeastern. Thermodynamics, fluid mechanics, material engineering.

Amalie J. Frank, Ph.D., Columbia. Telecommunications and computer networks, image coding, multimedia, computer science, combinatorial optimization.

John W. Hoopes Jr., Ph.D., Columbia; PE (Delaware). Process design, process development, heat and mass transfer, multiphase flow.

Amy M. Jakubowski Murphy, Ph.D., Temple. Linear systems and control.

Forouza Pourki, Ph.D., Purdue. Controls, optimization.

Murali Rao, M.S.E.E., Drexel.

Vincent R. Rice, J.D., Widener. Business law, engineering management.

Mahendra J. Shah, Ph.D., Kentucky. Structures.

Kirkbride Hall.

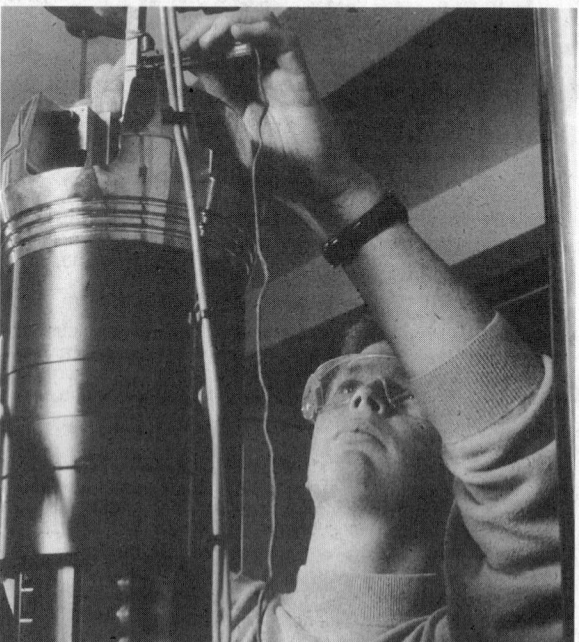

A wide variety of laboratory facilities are available to Widener University graduate engineering students.

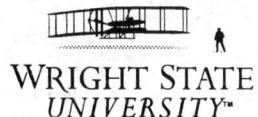

**WRIGHT STATE
UNIVERSITY™**

WRIGHT STATE UNIVERSITY

College of Engineering and Computer Science

Programs of Study

The College offers graduate programs leading to the Master of Science (M.S.) in engineering and Doctor of Philosophy (Ph.D.) in engineering degrees. The M.S. program offers curricular tracks in electrical, mechanical, materials, biomedical, and human factors engineering.

The M.S. program is broad in scope and emphasizes portable concepts in the design and analysis of complex physical systems, such as modeling, synthesis, and optimization, and provides a bridge for interdisciplinary areas such as control systems, composite materials, and robotics. The Ph.D. program supports research in the following six focus areas: sensor signal and image processing, modern control and robotics, electronic and microwave circuits, processing and properties of high-temperature and lightweight materials, computational design and optimization, and human interaction with complex systems.

M.S. students must complete a minimum of 45 graduate credit hours in an approved program of study. Students accepting graduate assistantships are required to complete a thesis, whereas others may elect to pursue a course option. Most graduate courses are offered in the late afternoon or early evening to allow employed students to complete the program on a part-time basis.

Ph.D. students must complete at least 135 graduate credit hours beyond the bachelor's degree in engineering in an approved program of study or 90 hours beyond the master's degree in engineering. At least 30 of these hours must be for graduate course work beyond a master's degree. Students must pass the Ph.D. qualifying examinations and candidacy examination and successfully complete and defend a dissertation.

Research Facilities

Modern laboratory facilities provide equipment for research in a number of areas. Graduate students have access to a wide range of computer systems interconnected by local and wide-area networks. Equipment includes DEC Alpha servers and workstations, a Silicon Graphics (SGI) Onyx 2, and SGI, DEC, and Sun workstations as well as numerous networked PCs and X-Windowing terminals. Access is also available to the Ohio Supercomputer via the Ohio Academic and Research Network (OARNET). The Paul Laurence Dunbar Library and the Fordham Health Sciences Library provide excellent services for research and have online electronic access to more than 4,000 other libraries in the United States and abroad. A modern machine shop is also available for research support. Laboratories specifically dedicated to student and faculty research exist in the areas of robotics, heat transfer, fluid dynamics, microprocessors, mechanical vibrations, signal processing, analog and digital electronics, microwave devices, VLSI circuit design, materials testing, materials processing, electron microscopy, augmentative communications, diagnostic ultrasonics, medical imaging, human-machine systems, and visual displays. Wright-Patterson Air Force Base and several Dayton industries participate in sponsored research, and their facilities are frequently available for graduate students and faculty members. Wright State's Information Technology Research Institute, a collaboration of academia, industry, and government, provides opportunities for research in all areas of modern information technology. In addition, the Edison Materials Technology Center provides an opportunity for collaborative research with a timely applied focus.

Financial Aid

Financial aid available to graduate students includes research assistantships, teaching assistantships, Federal Perkins Loans, Federal Stafford Student Loans, short-term loans, and academic fellowships. Scholarships are also available through the Dayton Area Graduate Studies Institute. Competitive stipends for research and teaching assistantships are available.

Cost of Study

Tuition in 1998–99 for residents of Ohio was $161 per quarter hour for part-time study (1–10½ hours) and $1703 for full-time study (11–18 hours). Tuition for nonresident students was $282 per quarter hour for part-time study and $3013 for full-time study.

Living and Housing Costs

On-campus housing is available, and there is a variety of housing available in the local area. Room and board cost about $1600 per quarter. The cost of living in Dayton is low compared with that in most other metropolitan areas.

Student Group

More than 3,200 graduate students are enrolled at Wright State University in forty graduate degree programs that lead to master's, Ed.S., Ph.D., M.D., and Psy.D. degrees.

Student Outcomes

Graduates are employed in a variety of professional positions both in-state and out-of-state. Typical positions include research scientist, avionics systems engineer, acquisition project officer, product design engineer, senior engineer, and project engineer.

Location

Although Wright State has a suburban setting, it is closely tied to Dayton, the fourth-largest metropolitan area in Ohio. Dayton is a manufacturing center and is also the home of Wright-Patterson Air Force Base, the center of Air Force research and procurement. This combination of industrial and military development has produced an unusual concentration of scientific, technical, and research activity. Dayton has a considerable and varied cultural life. In the surrounding area there are at least fifteen other colleges and universities.

The University

Wright State University is an exciting and expanding university that continues in the innovative spirit of aviation pioneers Orville and Wilbur Wright. The University has modern buildings that facilitate access for all, and the 557-acre main campus has accommodations for academics, support programs, sports, housing, and arts activities. Wright State's phenomenal growth as an institution—from one building, 92 employees, and 3,200 students in 1964 to forty-two modern buildings, more than 2,100 employees, and more than 15,000 students in 1998—has stimulated a comparable growth in the areas that surround the University.

Applying

The basic requirement for admission to the master's degree program is a bachelor's degree in engineering or a related area with an overall undergraduate grade point average of at least 2.7 (on a 4.0 scale) or an overall undergraduate grade point average of at least 2.5 with an average of 3.0 or better in the major field. The minimum admission requirements for the Ph.D. program are a B.S. from an ABET–accredited program, with a minimum 3.0 grade point average, or an M.S. from an engineering program, with a minimum 3.5 grade point average. Scores on the GRE General Test are required of all Ph.D. applicants.

Correspondence and Information

James E. Brandeberry, Ph.D., PE, Dean
College of Engineering and Computer Science
Wright State University
Dayton, Ohio 45435

Telephone: 937-775-5001
Fax: 937-775-5009
E-mail: dean_ecs@cs.wright.edu
World Wide Web: http://www.cs.wright.edu/

Wright State University

THE FACULTY AND THEIR RESEARCH

Maher S. Amer, Assistant Professor; Ph.D., Drexel, 1995. Raman spectroscopy, polymers, composites, micromechanics of multiphase materials.

Richard J. Bethke, Associate Professor and Chair, Mechanical and Materials Engineering; Ph.D., Wisconsin–Madison, 1971. Signal and systems modeling and analysis, stochastic processes.

James E. Brandeberry, Professor and Dean, College of Engineering and Computer Science; Ph.D., Marquette, 1969; PE. Circuit and interface design, microprocessors, digital control, robotics, computer-aided design.

Chien-In Chen, Associate Professor; Ph.D., Minnesota, 1989. VLSI design, design testability, computer-aided design automation.

Kenneth C. Cornelius, Associate Professor; Ph.D., Michigan State, 1978. Aerodynamics; turbomachinery; measurement, computational analysis, and modeling of turbulence in transonic flows; vortex flows on low-aspect-ratio wings; boundary-layer control.

Parviz Dadras, Professor; Ph.D., Delaware, 1972. Solid mechanics, carbon-carbon composites, manufacturing processes.

Billy W. Friar, Assistant Professor Emeritus; Ph.D., Ohio State, 1970; PE. Thermodynamics, heat transfer, fluid mechanics.

Jennie J. Gallimore, Associate Professor; Ph.D., Virginia Tech, 1989. Visualization, virtual environments, spatial orientation, adaptive displays.

Fred D. Garber, Associate Professor; Ph.D., Illinois at Urbana-Champaign, 1983. Decision theory and pattern recognition, communication theory.

Ramana V. Grandhi, Distinguished Professor; Ph.D., Virginia Tech, 1984. Structural optimization, mechanical vibrations, finite-element methods.

Thomas N. Hangartner, Professor; Ph.D., Swiss Federal Institute of Technology, 1978. Biomedical engineering, medical imaging, CT scanning, instrumentation, computers, quantitative bone measurements.

Wilbur L. Hankey Jr., Professor Emeritus; Ph.D., Ohio State, 1962. Computational fluid dynamics, aerodynamics, aerothermodynamics.

Russell A. Hannen, Associate Professor Emeritus; Ph.D., Ohio State, 1960. Electronic systems, control theory, stochastic processes.

Ping He, Professor; Ph.D., Drexel, 1984. Biomedical engineering, medical imaging, ultrasonics, instrumentation, computers.

Lang Hong, Professor; Ph.D., Tennessee, 1989. Stochastic control systems, computer vision, image processing and pattern recognition, robotics, multiple sensor integration and target tracking.

Marian K. Kazimierczuk, Professor; Ph.D., Warsaw Technical, 1978. Electronic circuit analysis, high-frequency tuned power amplifiers, power electronics.

Richard J. Koubek, Professor and Chair, Biomedical and Human Factors Engineering; Ph.D., Purdue, 1987. Usability, human-computer interaction, training, human aspects of manufacturing.

Junghsen Lieh, Associate Professor; Ph.D., Clemson, 1990. Dynamics and control of mechanical systems, manufacturing, impact mechanics, biomechanics, vehicle systems, instrumentation.

William S. McCormick, Professor; Ph.D., Wisconsin–Madison, 1967. Communication theory, bioengineering, electromagnetics, electrooptics.

James A. Menart, Assistant Professor; Ph.D., Minnesota, 1996. Thermal sciences, heat transfer, plasma science.

Pradeep Misra, Associate Professor; Ph.D., Concordia, 1987. Multivariable control theory, decentralized system theory, robotics, applied numerical analysis, two-dimensional discrete-time systems, robust control theory.

Sharmila M. Mukhopadhyay, Assistant Professor; Ph.D., Cornell, 1989. Surface and interface phenomena, composites, photoelectron spectroscopy, bonding in solids.

Krishna Naishadham, Associate Professor; Ph.D., Mississippi, 1986. Electromagnetics, antennas and microwaves, high-frequency asymptotic techniques, numerical methods for wave propagation, scattering and radiation problems, and microwave characterization of materials.

S. Narayanan, Associate Professor; Ph.D., Georgia Tech, 1994. Modeling and simulation of complex technological systems, cognition, computational representation, interactive systems.

Chandler A. Phillips, Professor; M.D., USC, 1969; PE. Mathematical modeling of biomechanics, fuzzy decision making in rehabilitation, functional electrical stimulation.

L. Rai Pujara, Professor; Ph.D., Ohio State, 1971. Multivariable control systems, systems analysis, robust control theory.

Kuldip S. Rattan, Professor; Ph.D., Kentucky, 1975. Computer-aided design, digital signal processing and control, bioengineering, robotics.

David B. Reynolds, Associate Professor; Ph.D., Virginia, 1978. Biomedical engineering, biofluid mechanics, engineering approaches to respiratory/pulmonary physiology and to orthopedics and prosthetics.

Blair A. Rowley, Professor; Ph.D., Missouri–Columbia, 1970; PE. Biomedical engineering, rehabilitation engineering, computer applications to augmentative communication, instrumentation, bioelectric effects of low-level electrical currents on tissue growth and healing.

Arnab K. Shaw, Associate Professor; Ph.D., Rhode Island, 1987. Communication theory and stochastic processes, estimation and detection, signal modeling and signal processing, simulation of communication systems.

Belle A. Shenoi, Professor; Ph.D., Illinois, 1962. Network theory, active and digital filters, communication circuits, digital signal processing.

Raymond E. Siferd, Professor and Chair, Electrical Engineering; Ph.D., Air Force Tech, 1977. Integrated circuits, signal processing, microelectromechanical systems.

Joseph C. Slater, Associate Professor; Ph.D., SUNY at Buffalo, 1993. Structure dynamics and control.

George R. Spalding, Associate Professor; Ph.D., Lehigh, 1974. Systems identification, robotics, dynamics and control.

Raghavan Srinivasan, Associate Professor; Ph.D., SUNY at Stony Brook, 1983. High-temperature deformation, materials behavior modeling, materials engineering.

Joseph F. Thomas Jr., Professor and Dean, School of Graduate Studies and Associate Vice President for Research; Ph.D., Illinois at Urbana-Champaign, 1968. Materials engineering, mechanical behavior.

Scott K. Thomas, Associate Professor; Ph.D., Dayton, 1993. Heat and mass transfer, fluid mechanics, two-phase flow.

Isaac Weiss, Professor; Ph.D., McGill, 1978. Materials engineering, thermomechanical processing, powder metallurgy.

J. Mitch Wolff, Assistant Professor; Ph.D., Purdue, 1995. Fluid mechanics, turbomachinery, unsteady aerodynamics, CFD, aeroelasticity, instrumentation.

Kefu Xue, Associate Professor; Ph.D., Penn State, 1987. Image processing and computer vision, stochastic processes and filtering, computer and communication systems, control and estimation theory.

Xudong Zhang, Assistant Professor; Ph.D., Michigan, 1997. Occupational biomechanics, computer-aided ergonomics, ergonomics.

YALE UNIVERSITY

Programs in Engineering and Applied Science

Programs of Study

All research and instructional programs in engineering and applied science are coordinated by the Faculty of Engineering, consisting of the Departments of Chemical, Electrical, and Mechanical Engineering and Applied Physics. These four units have autonomous faculty appointments and instructional programs, and students may obtain degrees designated according to different disciplines. A single Director of Graduate Studies oversees all graduate student matters. Students have considerable freedom in selecting programs to suit their interests and may choose programs of study that draw upon the resources of science departments that are not within the Faculty of Engineering, including the Departments of Physics, Chemistry, Mathematics, Statistics, Astronomy, Geology and Geophysics, and Computer Science and departments of the School of Medicine and of the School of Organization and Management.

There are no general qualifying examinations. Students who perform well in their first year of course work and have been accepted by one of the members of the faculty as research assistants in their second year are admitted to candidacy; they take an area examination designed to correct possible weaknesses in their preparation. Students are required to submit an acceptable dissertation containing a substantial research contribution. The minimum residence requirement is three years of full-time study, and students are encouraged to finish their degree within four years. Most first-year work consists of courses designed to give the student basic graduate-level instruction. Each student is urged to engage in research as soon as possible. Since graduate study in engineering continues throughout the year, students usually stay on the campus during the summer to continue their research. Graduate courses are not offered during the summer.

M.S. degrees are offered in applied mechanics, applied physics, chemical engineering, and electrical engineering and require the successful completion of at least eight term courses, two of which may be special projects. Although this program can normally be completed in one year of full-time study, a part-time M.S. program is available for practicing engineers and others. Its requirements are the successful completion of eight term courses in a time period not to exceed four calendar years.

Research Facilities

Located in the middle of the campus, the Becton Center of Engineering and Applied Science offers outstanding research facilities, which include modern laboratories for work in solid-state, surface, low-temperature, atomic, molecular, and plasma physics; semiconductor devices; optical and mechanical properties; and biomechanics and chemical engineering, together with extensive facilities for construction of apparatus. This center is adjacent to the Departments of Mathematics and Computer Science and is near the complex of facilities for physics, chemistry, and the biological sciences. The Faculty of Engineering has a rich computing environment, including servers, UNIX workstations, and Macintosh and Microsoft Windows personal computers. A high-speed data network interconnects engineering and extends to the campus network. Yale has been long connected to the Internet and is now participating in vBNS and the emerging Internet II. In addition, advanced instrumentation, computing, and networking are combined in a number of laboratories.

Financial Aid

Almost all first-year Ph.D. students receive a University fellowship paying full tuition and an adjusted stipend. Support thereafter is generally provided by research assistantships, which pay $16,200 plus full tuition in 1998–99. A number of prize fellowships paying larger-than-normal stipends are available to exceptional students. Students may supplement their income with various teaching fellowships and project assistantships. Fellowship support is normally not available for master's degree or part-time students.

Cost of Study

Tuition is $23,670 for the 1999–2000 academic year.

Living and Housing Costs

In 1998–99, single students in University housing paid from $3700 to $4130 for the academic year. Housing units are available for married students.

Student Group

Yale has 10,800 students—5,300 are undergraduates and the remainder are graduate and professional students. About 160 graduate students are in engineering, most of them working toward the Ph.D.

Location

Situated on Long Island Sound, among the scenic attractions of southern New England, New Haven provides outstanding cultural and recreational opportunities. The Greater New Haven area has a population of more than 350,000 and is only 1½ hours from New York by train or car.

The University

Yale is the third-oldest university in the United States, and its engineering program is also one of the oldest. The administrative organization for engineering has changed through the years, but the intention has always been to give students a high degree of flexibility in arranging their programs, with close interaction between individual students and faculty members.

Applying

Students with a bachelor's degree in any field of engineering or in mathematics, physics, or chemistry may apply for admission to graduate study, as may other students prepared to do graduate-level work in any of the study areas of the chosen department, regardless of their specific undergraduate field. Normally, students are admitted only at the beginning of the fall term. Application should be initiated about a year in advance of desired admission, and the application should be filed preferably before December 25; the file, including letters of reference, should be completed before January 2. Notifications of admission and award of financial aid are sent by April 1. Applicants must take the General Test of the Graduate Record Examinations; the exam should be taken in October. International applicants must submit scores on the TOEFL. The brochure *Graduate Bulletin of the Faculty of Engineering* provides information on the faculty and degree programs and can be obtained from the address below.

Correspondence and Information

Director of Graduate Studies
Programs in Engineering and Applied Science
Yale University, Dunham Lab
P.O. Box 208267
New Haven, Connecticut 06520-8267
Telephone: 203-432-4250
Fax: 203-432-2797

Yale University

THE FACULTY AND AREAS OF RESEARCH

APPLIED MECHANICS/MECHANICAL ENGINEERING. R. E. Apfel, I. B. Bernstein, J. Cholewicki, B. T. Chu, N. Delson, J. Fernández de la Mora, A. Gomez, R. B. Gordon, R. H. Kraichnan, A. Liñan-Martinez, M. B. Long, E. T. Onat, L. Pfefferle, G. L. Povirk, D. E. Rosner, B. Saltzman, R. B. Smith, M. D. Smooke, K. R. Sreenivasan, W. Tong, G. Veronis, F. A. Williams, D. T. Wu.

Mechanics of Fluids. Acoustics and bioeffects of ultrasound, bulk and surface properties of liquids (including metastable liquids, radiation-induced bubble formation, and surfactant-induced effects); dynamics and stability of drops and bubbles; experimental, theoretical, and computational studies of turbulence; chaos; fractals; aerodynamics; kinetic theory of gases and mixtures; electrospray theory and characterization; combustion and flames; computational methods for fluid dynamics and reacting flows; laser diagnostics of reacting and nonreacting flows; atmospheric turbulence, climate, theoretical and laboratory modeling of large-scale ocean circulation.

Mechanics of Solids/Material Science. Mechanics of deformation, mass transport, and nucleation within material systems through experimental, analytic, and computational studies; mechanical testing of small-scale structures; characterization of microscale inhomogeneities in plastic flow; impact loading of materials; diffusion of dopants within semiconductor films; evolution of surface roughness during plastic deformation; ion implantation–induced disorder in crystalline films; incorporation of microstructural information into constitutive laws; biomechanics of the heart; electromigration in metallic interconnects; transient nucleation in multicomponent systems.

APPLIED PHYSICS. R. K. Chang, R. D. Grober, V. E. Henrich, D. E. Prober, K. M. Rabe, N. Read, R. D. Schoelkopf, A. D. Stone, W. P. Wolf. Joint appointments (designation of primary departments is as follows: DR—Diagnostic Radiology; ME—Mechanical Engineering; EE—Electrical Engineering; CH—Chemistry; PH—Physics): A. W. Anderson (DR), R. C. Barker (EE), S. E. Barrett (PH), J. C. Gore (DR), L. J. Guido (EE), M. B. Long (ME), T. P. Ma (EE), S. Sachdev (PH), R. Shankar (PH), M. A. Reed (EE), M. D. Smooke (ME), K. R. Sreenivasan (ME), J. C. Tully (CH). Adjunct Faculty: J. F. Dillon Jr., P. C. Hohenberg. Emeritus Faculty: W. R. Bennett Jr., A. Herzenberg, R. G. Wheeler.

Condensed-Matter Physics and Materials Science. Experimental and theoretical investigations of crystal surfaces and material interfaces by electron spectroscopy; optical, X-ray, Auger, and electron tunneling spectroscopy; superconductivity and quantum transport phenomena (in ultrasmall metal and semiconductor structures); superconductor tunneling microwave detectors and X-ray detectors; optical properties of microcavities and micro-objects, such as optical fibers and droplets; theoretical studies of magnetic insulators, interactions and phase transitions, development and application of ab initio methods in studying structural and electronic properties of solids.

Lasers, Nonlinear Optical Spectroscopy, Atoms, and Molecules. Use of laser techniques in atomic and molecular spectroscopy, tunable minilasers, laser spectroscopy, optical scattering; optical filters for telecommunication; Raman scattering of gaseous molecules, liquids, and semiconductors; nonlinear optics of microparticles; interaction of laser radiation with atoms/molecules on metallic surfaces; cavity quantum electrodynamic effects of emission inside microcavities; application of laser diagnostics to combustion and turbulence.

Physics in Medicine. Development and applications of NMR imaging and spectroscopic techniques and applications in biology and medicine; design and study of new NMR techniques and instrumentation for medical diagnostic applications; development of functional imaging of brain and heart; proton NMR relaxation mechanisms in heterogeneous media; cross-relaxation and magnetization transfer in biological tissues and tissue models; susceptibility contrast mechanisms; diffusion and restricted diffusion of water in tissues and other media. Quantification of flow and perfusion by NMR imaging; studies of turbulent and complex flow by NMR. Improvements in NMR signal processing; image restoration, segmentation, and analysis. Web site: http://www.eng.yale.edu/aphy.

CHEMICAL ENGINEERING. E. I. Altman, M. Elimelech, R. Ely, M. Grant, G. L. Haller, C. G. Horvath, M. Loewenberg, L. D. Pfefferle, D. E. Rosner, C. Walker (Emeritus), J. Walz. Adjunct Faculty: F. P. Bocr, D. M. Crothers, L. S. Ettre, T. E. Graedel, J. Levitzky, R. Weber, L. L. Wikstrom, K. W. Zilm.

Biochemical Engineering. Enzyme reactor technology; medical applications; biopolymer separation and purification; transport phenomena in living systems; hydrophobic interactions; properties of biological substances via high-performance liquid chromatography; chromatographic enzyme reactors; biochemical reactors and processes for product recovery, enzymatic catalysis in nonaqueous solvents.

Heterogeneous Catalysis. Diffusion and reaction in pulsed and steady-flow reactors, XPS and EXAFS studies of bimetallic catalysts, chemical energy accommodation, catalysis with state-selected reactants and/or products using molecular beams.

High-Temperature Chemical Reaction Engineering. Energy and mass transfer from combustion gases; dynamics of nonisothermal multiphase flow systems; gasification kinetics of solids and liquids; resistance-relaxation studies of gas absorption-desorption; reactivity of refractory nonmetals; chemical energy accommodation; diffusion limitations in flow reactors; heterogeneous combustion; chemical vapor deposition/etching, aerosol nucleation, growth, transport.

Separation Science and Technology. Liquid and gas chromatography; free jet separation methods for macromolecules and aerosols; partial condensation; subsieve particle column theory; displacement chromatography; adsorption from liquid mixtures; adsorption and ion exchange processes, thermal (Soret) diffusion, gas separation; statistical mechanics of competitive adsorption.

Molecular-Beam Chemical Engineering. Scattering cross sections of organic molecules, IR emission kinetics; nozzle beam: IR techniques for pollution monitoring, vibration and translational excitation in surface-reaction macromolecule detection, scattering from liquid surfaces; molecular beam: gas chromatography interfacing, energy relaxation in nozzle expansions and translational-to-vibrational energy transfer in cross beam interactions.

Environmental Engineering. Physical and chemical processes for water quality control; aquatic and environmental chemistry; transport and fate of chemical substances in the environment; colloidal and interfacial phenomena in aquatic systems; environmental engineering biotechnology; membrane separation processes; aerosol science and technology; incineration of toxic wastes; industrial ecology; geochemical cycles and the global environment; chemical reactions at the mineral-water interface.

ELECTRICAL ENGINEERING. R. C. Barker, P. N. Belhumeur, R. Chang, J. Duncan, G. Hager, D. Henry, P. J. Kindlmann, D. Kriegman, R. Kuc, B. Kuszmaul, T. P. Ma, A. S. Morse, K. S. Narendra, J. Pan, M. Pinto, A. Rangarajan, M. A. Reed, L. Staib, H. D. Tagare, D. Updegrove, J. R. Vaisnys, J. Woodall, S. Zucker.

Computer Engineering. VLSI system design and architecture, timing analysis and optimization, parallel algorithms, fault-tolerant design, high-level testing and simulation.

Microelectronics/Photonics. Physics and technology of semiconductor devices, electronic and photonic materials, heterojunction structures and devices, electron tunneling, quantum device physics, nanoelectronics, molecular electronics.

Systems and Information Sciences. Signal and image processing, computer vision, robot navigation, intelligent sensors, adaptive and learning systems, neural networks, control of flexible structures, biomedical engineering.

INTERDISCIPLINARY STUDIES

Biomedical Engineering. Formation of anatomical and functional medical images, especially magnetic resonance; analysis and processing of medical image data for recovering quantitative information; development of various types of enzyme reactors for use in therapy, tubular enzymic wall reactors for clinical analyzers; modeling of physiological systems; pharmacokinetics; quantitative structure-activity relationships for antitumor agents; analysis of physiological fluids; development of techniques for clinical analysis of biological signals, especially neurological signals; decompression sickness investigation; biocontrols; development of ultrasonic probing systems for the diagnosis of liver disease; characterization and separation of blood and other biomaterials.

Atmospheric Sciences and Geophysical/Geochemical Studies. Atmospheric aerosols, interpretation of seismic data, continental drift, marine sedimentation, effect of intense pressures and temperatures on geological materials.

Section 2
Aerospace/Aeronautical Engineering

This section contains a directory of institutions offering graduate work in aerospace/aeronautical engineering, followed by in-depth entries submitted by institutions that chose to prepare detailed program descriptions. Additional information about programs listed in the directory but not augmented by an in-depth entry may be obtained by writing directly to the dean of a graduate school or chair of a department at the address given in the directory.

For programs offering related work, see also in this book Engineering and Applied Sciences and Mechanical Engineering and Mechanics. In Book 4, see Geosciences and Physics.

CONTENTS

Aerospace/Aeronautical Engineering

Air Force Institute of Technology, School of Engineering, Department of Aeronautics and Astronautics, Program in Aeronautical Engineering, Wright-Patterson AFB, OH 45433-7765. Offers MS, PhD. *Accreditation:* ABET (one or more programs are accredited). Part-time programs available. *Faculty:* 13 full-time (0 women), 1 part-time (0 women). *Students:* 20 full-time, 6 part-time. In 1998, 8 master's awarded (100% found work related to degree); 1 doctorate awarded (100% found work related to degree). *Degree requirements:* For master's and doctorate, thesis/dissertation required, foreign language not required. *Entrance requirements:* For master's, GRE General Test (minimum score of 500 on verbal section, 600 on quantitative required), minimum GPA of 3.0, must be military officer or U.S. citizen; for doctorate, GRE General Test (minimum score of 550 on verbal section, 650 on quantitative required), minimum GPA of 3.0, must be military officer or U.S. citizen. *Average time to degree:* Master's–1.5 years full-time; doctorate–3 years full-time. Application fee: $0. *Faculty research:* Computational fluid dynamics, experimental aerodynamics, computational structural mechanics, experimental structural mechanics, aircraft stability and control. *Unit head:* Dr. Paul King, Head, 937-255-3636 Ext. 4628.

Air Force Institute of Technology, School of Engineering, Department of Aeronautics and Astronautics, Program in Astronautical Engineering, Wright-Patterson AFB, OH 45433-7765. Offers MS, PhD. *Accreditation:* ABET (one or more programs are accredited). Part-time programs available. *Faculty:* 13 full-time (0 women). *Students:* 7 full-time. In 1998, 2 master's awarded (100% found work related to degree); 1 doctorate awarded (100% found work related to degree). *Degree requirements:* For master's and doctorate, thesis/dissertation required, foreign language not required. *Entrance requirements:* For master's, GRE General Test (minimum score of 500 on verbal section, 600 on quantitative required), minimum GPA of 3.0, must be military officer or U.S. citizen; for doctorate, GRE General Test (minimum score of 550 on verbal section, 650 on quantitative required), minimum GPA of 3.0, must be military officer or U.S. citizen. *Average time to degree:* Master's–1.5 years full-time. Application fee: $0. *Faculty research:* Orbital mechanics, spacecraft design, spacecraft control. *Unit head:* Dr. Bill Weisel, Head, 937-255-3636 Ext. 4312.

Air Force Institute of Technology, School of Engineering, Department of Aeronautics and Astronautics, Program in Space Operations, Wright-Patterson AFB, OH 45433-7765. Offers MS, PhD. Part-time programs available. *Faculty:* 13 full-time (0 women). *Students:* 15 full-time. *Degree requirements:* For master's and doctorate, thesis/dissertation required, foreign language not required. *Entrance requirements:* For master's, GRE General Test (minimum score of 500 on verbal section, 600 on quantitative required), minimum GPA of 3.0, must be military officer or U.S. citizen; for doctorate, GRE General Test (minimum score of 550 on verbal section, 650 on quantitative required), minimum GPA of 3.0, must be military officer or U.S. citizen. Application fee: $0. *Faculty research:* Orbital mechanics, spacecraft design, spacecraft control. *Unit head:* Dr. Bradley S. Liebst, Head, Department of Aeronautics and Astronautics, 937-255-3069, Fax: 937-656-7621, E-mail: bliebst@afit.af.mil.

Arizona State University, Graduate College, College of Engineering and Applied Sciences, Department of Mechanical and Aerospace Engineering, Tempe, AZ 85287. Offers aerospace engineering (MS, MSE, PhD); engineering science (MS, MSE, PhD); mechanical engineering (MS, MSE, PhD). *Faculty:* 35 full-time (3 women), 2 part-time (0 women). *Students:* 83 full-time (10 women), 39 part-time (6 women); includes 16 minority (1 African American, 11 Asian Americans or Pacific Islanders, 3 Hispanic Americans, 1 Native American), 64 international. Average age 28. 521 applicants, 49% accepted. In 1998, 37 master's, 8 doctorates awarded. *Degree requirements:* For master's, computer language, thesis or alternative required; for doctorate, computer language, dissertation required. *Entrance requirements:* For master's and doctorate, GRE General Test. Application fee: $45. *Financial aid:* Fellowships available. *Faculty research:* Aerodynamics, fluid mechanics, propulsion and space power, advanced structures and materials, robotics and automation. *Unit head:* Dr. Don L. Boyer, Chair, 480-965-3291, E-mail: mae@asu.edu. *Application contact:* Graduate Secretary, 480-965-4979.

See in-depth description on page 370.

Arizona State University East, College of Technology and Applied Sciences, Department of Aeronautical Management Technology, Mesa, AZ 85212. Offers MS. Part-time and evening/weekend programs available. *Faculty:* 3 full-time (0 women). *Students:* 4 full-time (0 women), 7 part-time (2 women); includes 2 minority (1 Asian American or Pacific Islander, 1 Hispanic American), 3 international. Average age 34. 5 applicants, 80% accepted. *Degree requirements:* For master's, thesis or applied project and oral defense required. *Entrance requirements:* For master's, minimum GPA of 3.0. *Application deadline:* Applications are processed on a rolling basis. Application fee: $45. *Financial aid:* In 1998–99, 7 research assistantships with partial tuition reimbursements (averaging $11,895 per year) were awarded.; career-related internships or fieldwork, Federal Work-Study, grants, scholarships, tuition waivers (full and partial), and unspecified assistantships also available. Aid available to part-time students. Financial aid application deadline: 3/1; financial aid applicants required to submit FAFSA. *Faculty research:* Approach lighting, cooperative and collaborative learning, distance education, human factors, situation awareness. Total annual research expenditures: $583,147. *Unit head:* Dr. William K. McCurry, Chair, 602-727-1998, Fax: 602-727-1730, E-mail: mccurry@asu.edu. *Application contact:* Dr. William K. McCurry, Chair, 602-727-1998, Fax: 602-727-1730, E-mail: mccurry@asu.edu.

Arizona State University East, College of Technology and Applied Sciences, Department of Manufacturing and Aeronautical Engineering Technology, Mesa, AZ 85212. Offers MS. Part-time and evening/weekend programs available. *Faculty:* 2 full-time (0 women), 1 part-time (0 women). *Students:* 21 full-time (14 women), 64 part-time (25 women); includes 5 minority (3 Asian Americans or Pacific Islanders, 2 Hispanic Americans), 15 international. Average age 33. 42 applicants, 79% accepted. In 1998, 30 degrees awarded. *Degree requirements:* For master's, thesis or applied project and oral defense required. *Entrance requirements:* For master's, minimum GPA of 3.0. *Average time to degree:* Master's–2.8 years full-time, 5.4 years part-time. *Application deadline:* Applications are processed on a rolling basis. Application fee: $45. *Financial aid:* In 1998–99, 16 students received aid, including 5 research assistantships with partial tuition reimbursements available (averaging $8,556 per year); teaching assistantships, career-related internships or fieldwork, Federal Work-Study, grants, scholarships, and tuition waivers (full and partial) also available. Aid available to part-time students. Financial aid application deadline: 3/1; financial aid applicants required to submit FAFSA. *Faculty research:* Robotics, wind tunnel testing, propulsion systems, statistical process control. *Unit head:* Dr. Dale Palmgren, Chair, 602-727-1584, Fax: 602-727-1549, E-mail: palmgren@asu.edu. *Application contact:* Dr. Dale Palmgren, Chair, 602-727-1584, Fax: 602-727-1549, E-mail: palmgren@asu.edu.

Auburn University, Graduate School, College of Engineering, Department of Aerospace Engineering, Auburn, Auburn University, AL 36849-0002. Offers MAE, MS, PhD. Part-time programs available. *Faculty:* 14 full-time (1 woman). *Students:* 9 full-time (0 women), 19 part-time (2 women), 3 international. 15 applicants, 40% accepted. In 1998, 9 master's, 1 doctorate awarded. *Degree requirements:* For master's, thesis (MS), exam required; for doctorate, dissertation, exams required, foreign language not required. *Entrance requirements:* For master's, GRE General Test; for doctorate, GRE General Test (minimum score of 400 on each section required). *Application deadline:* For fall admission, 9/1; for spring admission, 3/1. Applications are processed on a rolling basis. Application fee: $25 ($50 for international students). Tuition, state resident: full-time $2,760; part-time $76 per credit hour. Tuition, nonresident: full-time $8,280; part-time $228 per credit hour. *Financial aid:* Fellowships, research assistantships, teaching assistantships, Federal Work-Study available. Aid available to part-time students. Financial aid application deadline: 3/15. *Faculty research:* Aerodynamics, flight dynamics and simulation, propulsion, structures and aeroelasticity, aerospace smart structures. *Unit head:*

Dr. John E. Cochran, Head, 334-844-4874. *Application contact:* Dr. John F. Pritchett, Dean of the Graduate School, 334-844-4700.

See in-depth description on page 321.

Boston University, College of Engineering, Department of Aerospace and Mechanical Engineering, Boston, MA 02215. Offers aerospace engineering (MS, PhD); mechanical engineering (MS, PhD). Part-time programs available. *Faculty:* 26 full-time (3 women), 4 part-time (1 woman). *Students:* 41 full-time (8 women), 5 part-time (1 woman); includes 3 minority (1 African American, 2 Asian Americans or Pacific Islanders), 19 international. Average age 26. 178 applicants, 17% accepted. In 1998, 9 master's, 6 doctorates awarded. Terminal master's awarded for partial completion of doctoral program. *Degree requirements:* For master's, thesis or alternative required, foreign language not required; for doctorate, dissertation required, foreign language not required. *Entrance requirements:* For master's, GRE General Test, TOEFL (minimum score of 500 required; 213 for computer-based); for doctorate, GRE General Test, TOEFL. *Application deadline:* For fall admission, 4/1; for spring admission, 10/1. Applications are processed on a rolling basis. Application fee: $50. Tuition: Full-time $23,770; part-time $743 per credit. Required fees: $220. Tuition and fees vary according to class time, course level, campus/location and program. *Financial aid:* In 1998–99, 7 fellowships with full tuition reimbursements (averaging $13,000 per year), 19 research assistantships with full tuition reimbursements (averaging $11,500 per year), 20 teaching assistantships with full tuition reimbursements (averaging $11,500 per year) were awarded.; career-related internships or fieldwork, Federal Work-Study, institutionally-sponsored loans, and scholarships also available. Financial aid application deadline: 12/15; financial aid applicants required to submit FAFSA. *Faculty research:* Waves and acoustics, dynamics and controls, fluid mechanics, materials processing, precision engineering. Total annual research expenditures: $2.1 million. *Unit head:* Dr. Allan Pierce, Chairman, 617-353-2877, Fax: 617-353-5866. *Application contact:* Cheryl Kelley, Graduate Programs Director, 617-353-9760, Fax: 617-353-0259, E-mail: enggrad@bu.edu.

See in-depth description on page 323.

Brown University, Graduate School, Division of Engineering, Program in Aerospace Engineering, Providence, RI 02912. Offers Sc M, PhD. *Degree requirements:* For doctorate, dissertation, preliminary exam required, foreign language not required, foreign language not required.

California Institute of Technology, Division of Engineering and Applied Science, Option in Aeronautics, Pasadena, CA 91125-0001. Offers MS, PhD, Engr. *Faculty:* 11 full-time (0 women). *Students:* 60 full-time (6 women), 37 international. 61 applicants, 30% accepted. In 1998, 10 master's, 7 doctorates, 4 other advanced degrees awarded. *Degree requirements:* For master's, foreign language and thesis not required; for doctorate, dissertation required, foreign language not required. *Application deadline:* For fall admission, 1/15. Application fee: $0. *Faculty research:* Computational fluid dynamics, technical fluid dynamics, structural mechanics, mechanics of fracture, aeronautical engineering and propulsion. *Unit head:* Dr. Hans G. Hornung, Director, GALCIT, 626-395-4551. *Application contact:* Dr. Wolfgang G. Knauss, Representative, 626-395-4751.

California Polytechnic State University, San Luis Obispo, College of Engineering, Department of Aeronautical Engineering, San Luis Obispo, CA 93407. Offers MSAE. Part-time programs available. *Faculty:* 6 full-time (0 women), 4 part-time (1 woman). *Students:* 3 full-time (1 woman), 1 part-time. 6 applicants, 67% accepted. In 1998, 5 degrees awarded (33% entered university research/teaching, 67% found other work related to degree). *Degree requirements:* For master's, thesis required, foreign language not required. *Entrance requirements:* For master's, GRE General Test, minimum GPA of 2.5 during last 2 years of course work. *Average time to degree:* Master's–2 years full-time, 4 years part-time. *Application deadline:* For fall admission, 5/31 (priority date); for spring admission, 12/31. Applications are processed on a rolling basis. Application fee: $55. Tuition, nonresident: part-time $164 per unit. Required fees: $531 per quarter. *Financial aid:* Research assistantships, teaching assistantships, career-related internships or fieldwork available. Financial aid application deadline: 3/2; financial aid applicants required to submit FAFSA. *Faculty research:* Aerodynamics, fluid dynamics, computational fluid dynamics, aircraft structures, propulsion, flight simulation and control. *Unit head:* Dr. Jin Tso, Chair, 805-756-2562, Fax: 805-756-2376, E-mail: jtso@calpoly.edu. *Application contact:* Dr. Daniel J. Biezad, Professor, 805-756-5126, Fax: 805-756-2376, E-mail: dbiezad@calpoly.edu.

Announcement: The MS program in aeronautical engineering emphasizes engineering science and research activity, with students required to complete a research thesis. The department's faculty members offer applied research specializations in traditional aeronautics disciplines, including aerodynamics, fluid dynamics, aerospace structures, stability and control, propulsion systems, and aircraft design. Many students conduct research at government or industry laboratories. Further information may be obtained from the above address.

California State University, Long Beach, Graduate Studies, College of Engineering, Department of Aerospace Engineering, Long Beach, CA 90840. Offers MSAE. Part-time programs available. *Faculty:* 7 full-time (1 woman), 6 part-time (0 women). *Students:* 4 full-time (0 women), 6 part-time; includes 1 minority (Native American), 5 international. Average age 33. 10 applicants, 20% accepted. In 1998, 3 degrees awarded. *Degree requirements:* For master's, thesis or alternative required, foreign language not required. *Entrance requirements:* For master's, TOEFL (minimum score of 550 required). *Application deadline:* For fall admission, 8/1; for spring admission, 12/1. Application fee: $55. Electronic applications accepted. Tuition, nonresident: part-time $246 per unit. Required fees: $569 per semester. Tuition and fees vary according to course load. *Financial aid:* Career-related internships or fieldwork, Federal Work-Study, grants, institutionally-sponsored loans, and unspecified assistantships available. Financial aid application deadline: 3/2. *Faculty research:* Aerodynamic flows, ice accretion, stability and transition. *Unit head:* Dr. Tuncer Cebeci, Chair, 562-985-1503, Fax: 562-985-1669, E-mail: cebeci@engr.csulb.edu. *Application contact:* Dr. Hamid Hefazi, Graduate Adviser, 562-985-1502, Fax: 562-985-1669, E-mail: hefazi@engr.csulb.edu.

California State University, Northridge, Graduate Studies, College of Engineering and Computer Science, Department of Civil, Industrial and Applied Mechanics, Department of Mechanical Engineering, Northridge, CA 91330. Offers aerospace engineering (MS); applied engineering (MS); machine design (MS); mechanical engineering (MS); structural engineering (MS); thermofluids (MS). Part-time and evening/weekend programs available. *Faculty:* 8 full-time, 4 part-time. *Students:* 3 full-time (0 women), 44 part-time (4 women); includes 17 minority (14 Asian Americans or Pacific Islanders, 2 Hispanic Americans, 1 Native American), 3 international. *Degree requirements:* For master's, thesis or alternative required, foreign language not required. *Entrance requirements:* For master's, GRE General Test, TOEFL, minimum GPA of 2.5. *Application deadline:* For fall admission, 11/30. Application fee: $55. Tuition, nonresident: part-time $246 per unit. International tuition: $7,874 full-time. Required fees: $1,970. Tuition and fees vary according to course load. *Unit head:* Dr. William J. Rivers, Chair, 818-677-2187. *Application contact:* Dr. Tom Mincer, Graduate Coordinator, 818-677-2007.

Carleton University, Faculty of Graduate Studies, Faculty of Engineering and Design, Department of Mechanical and Aerospace Engineering, Ottawa, ON K1S 5B6, Canada. Offers aerospace engineering (M Eng, PhD); materials engineering (M Eng); mechanical engineering (M Eng, PhD). *Faculty:* 23 full-time (1 woman). *Students:* 51 full-time (4 women), 11 part-time (2 women). Average age 29. In 1998, 18 master's, 1 doctorate awarded. *Degree requirements:* For master's, thesis optional; for doctorate, dissertation required. *Entrance requirements:* For master's, TOEFL (minimum score of 550 required), honors degree; for doctorate, TOEFL (minimum score of 550 required), MA Sc or M Eng. *Average time to degree:* Master's–2.2 years full-time; doctorate–6.3 years full-time. *Application deadline:* For fall admission, 3/1. Applications are processed on a rolling basis. Application fee: $35. *Financial aid:* In 1998–99, 1 student received aid. Application deadline: 3/1. *Faculty research:* Thermal fluids engineering, heat transfer, vehicle engineering. Total annual research expenditures: $1.9 million. *Unit head:* Paul Straznicky, Director, 613-520-2600 Ext. 5684, Fax: 613-520-5715, E-mail: pstrazni@mae.carleton.ca. *Application contact:* Ata M. Khan, Associate Dean of Engineering, 613-520-5659, Fax: 613-520-5682, E-mail: ata_khan@carleton.ca.

Concordia University, School of Graduate Studies, Faculty of Engineering and Computer Science, Program in Aerospace Engineering, Montréal, PQ H3G 1M8, Canada. Offers M Eng. *Students:* 18 full-time (2 women), 12 part-time. *Degree requirements:* For master's, computer language, thesis or alternative required. *Application deadline:* For fall admission, 5/1; for spring admission, 10/1. Application fee: $50. *Faculty research:* Aeronautics and propulsion avionics and control, structures and materials, space engineering. *Unit head:* Dr. J. V. Svoboda, Director, 514-848-4171, Fax: 514-848-3175.

Cornell University, Graduate School, Graduate Fields of Engineering, Field of Aerospace Engineering, Ithaca, NY 14853-0001. Offers M Eng, MS, PhD. *Faculty:* 24 full-time. *Students:* 19 full-time (2 women); includes 1 minority (Hispanic American), 13 international. 71 applicants, 31% accepted. In 1998, 4 master's, 4 doctorates awarded. *Degree requirements:* For master's, thesis (MS) required; for doctorate, dissertation required. *Entrance requirements:* For master's and doctorate, GRE General Test, TOEFL (minimum score of 550 required). *Application deadline:* For fall admission, 1/15. Application fee: $65. Electronic applications accepted. *Financial aid:* In 1998–99, 17 students received aid, including 6 fellowships with full tuition reimbursements available, 8 research assistantships with full tuition reimbursements available, 3 teaching assistantships with full tuition reimbursements available; institutionally-sponsored loans, scholarships, tuition waivers (full and partial), and unspecified assistantships also available. Financial aid applicants required to submit FAFSA. *Faculty research:* Aerodynamics, fluid mechanics, turbulence, combustion/propulsion, aeroacoustics. *Unit head:* Director of Graduate Studies, 607-255-5250, Fax: 607-255-1222. *Application contact:* Graduate Field Assistant, 607-255-5250, Fax: 607-255-1222, E-mail: maegrad@cornell.edu.

École Polytechnique de Montréal, Graduate Programs, Department of Mechanical Engineering, Montréal, PQ H3C 3A7, Canada. Offers aerothermics (M Eng, M Sc A, PhD); applied mechanics (M Eng, M Sc A, PhD); tool design (M Eng, M Sc A, PhD). Part-time and evening/weekend programs available. *Degree requirements:* For master's and doctorate, one foreign language, computer language, thesis/dissertation required. *Entrance requirements:* For master's, minimum GPA of 2.75; for doctorate, minimum GPA of 3.0. *Faculty research:* Noise control and vibration, fatigue and creep, aerodynamics, composite materials, biomechanics, robotics.

Embry-Riddle Aeronautical University, Daytona Beach Campus Graduate Program, Department of Aeronautical Science, Daytona Beach, FL 32114-3900. Offers MAS. Part-time and evening/weekend programs available. *Faculty:* 8 full-time (0 women), 2 part-time (0 women). *Students:* 27 full-time (4 women), 26 part-time (6 women); includes 5 minority (2 African Americans, 1 Asian American or Pacific Islander, 2 Hispanic Americans), 27 international. Average age 30. 30 applicants, 83% accepted. In 1998, 23 degrees awarded. *Degree requirements:* For master's, thesis optional, foreign language not required. *Entrance requirements:* For master's, TOEFL (minimum score of 550 required), minimum GPA of 2.5. *Application deadline:* Applications are processed on a rolling basis. Application fee: $30 ($50 for international students). Electronic applications accepted. Tuition: Full-time $8,190; part-time $455 per credit. Required fees: $105 per semester. Tuition and fees vary according to program. *Financial aid:* In 1998–99, 2 teaching assistantships with tuition reimbursements (averaging $8,640 per year) were awarded.; fellowships, research assistantships, career-related internships or fieldwork, Federal Work-Study, and unspecified assistantships also available. Aid available to part-time students. Financial aid application deadline: 4/15; financial aid applicants required to submit FAFSA. *Faculty research:* Child and infant safety seats for airliners, global positioning satellite navigation, aircraft accidents caused by unapproved parts, cockpit checklists, improving fire suppression techniques at airports. *Unit head:* Dr. Charles E. Richardson, Program Chair, 904-226-6442, Fax: 904-226-6012, E-mail: richard@cts.db.erau.edu. *Application contact:* Ginny Tait, Graduate Admissions Specialist, 904-226-6115, Fax: 904-226-6299, E-mail: taitg@cts.db.erau.edu.

See in-depth description on page 327.

Embry-Riddle Aeronautical University, Daytona Beach Campus Graduate Program, Department of Aerospace Engineering, Daytona Beach, FL 32114-3900. Offers MSAE. Part-time and evening/weekend programs available. *Faculty:* 6 full-time (0 women). *Students:* 17 full-time (6 women), 8 part-time (1 woman); includes 1 minority (African American), 10 international. Average age 26. 41 applicants, 73% accepted. In 1998, 11 degrees awarded. *Degree requirements:* For master's, thesis optional, foreign language not required. *Entrance requirements:* For master's, TOEFL (minimum score of 550 required), BS in aeronautical engineering or equivalent; minimum GPA of 3.0 in junior and senior years, 2.5 overall. *Application deadline:* Applications are processed on a rolling basis. Application fee: $30 ($50 for international students). Electronic applications accepted. Tuition: Full-time $8,820; part-time $490 per credit. Required fees: $105 per semester. Tuition and fees vary according to program. *Financial aid:* In 1998–99, 4 research assistantships with tuition reimbursements (averaging $8,640 per year), 11 teaching assistantships with tuition reimbursements (averaging $8,640 per year) were awarded.; fellowships, career-related internships or fieldwork, Federal Work-Study, and unspecified assistantships also available. Aid available to part-time students. Financial aid application deadline: 4/15; financial aid applicants required to submit FAFSA. *Faculty research:* Flight testing and simulation; nondestructive testing, acoustic emissions, and neural networks; acoustics, dynamics and vibrations; nonlinear dynamics, chaos control and acoustics; structural dynamics and vibrations. Total annual research expenditures: $1.3 million. *Unit head:* Dr. Yechiel Crispin, Program Coordinator, 904-226-6257, Fax: 904-226-6747, E-mail: crispiny@cts.db.erau.edu. *Application contact:* Ginny Tait, Graduate Admissions Specialist, 904-226-6115, Fax: 904-226-6299, E-mail: taitg@cts.db.erau.edu.

See in-depth description on page 325.

Embry-Riddle Aeronautical University, Extended Campus, Graduate Resident Centers, Department of Aeronautical Science, Daytona Beach, FL 32114-3900. Offers MAS. Part-time and evening/weekend programs available. Postbaccalaureate distance learning degree programs offered (minimal on-campus study). *Students:* 26 full-time (2 women), 1,251 part-time (142 women); includes 166 minority (64 African Americans, 28 Asian Americans or Pacific Islanders, 61 Hispanic Americans, 13 Native Americans), 16 international. Average age 35. 314 applicants, 94% accepted. In 1998, 713 degrees awarded. *Degree requirements:* For master's, thesis optional, foreign language not required. *Application deadline:* Applications are processed on a rolling basis. Application fee: $30 ($50 for international students). Electronic applications accepted. Tuition: Full-time $6,897; part-time $314 per credit. Tuition and fees vary according to course load and campus/location. *Financial aid:* Available to part-time students. Applicants required to submit FAFSA. *Unit head:* Dr. Stephen O'Brien, Chair, 850-689-8050, E-mail: obriens@cts.db.erau.edu. *Application contact:* Pam Thomas, Director of Admissions and Records, 904-226-6910, Fax: 904-226-6984, E-mail: ecinfo@ec.db.erau.edu.

Florida Institute of Technology, Graduate School, College of Engineering, Division of Engineering Science, Department of Aerospace Engineering, Melbourne, FL 32901-6975. Offers MS, PhD. Part-time programs available. *Faculty:* 6 full-time (0 women), 1 part-time (0 women). *Students:* 1 (woman) full-time, 5 part-time, 2 international. Average age 28. 18 applicants, 50% accepted. In 1998, 1 master's awarded. *Degree requirements:* For master's, thesis optional, foreign language not required; for doctorate, dissertation, comprehensive exam required, foreign language not required. *Entrance requirements:* For master's, GRE General Test, GRE Subject Test, minimum GPA of 3.0; for doctorate, GRE General Test, GRE Subject Test, minimum GPA of 3.2. *Application deadline:* Applications are processed on a rolling basis. Application fee: $50. Electronic applications accepted. Tuition: Part-time $575 per credit hour. Required fees: $100. Tuition and fees vary according to campus/location and program. *Financial aid:* Research assistantships with full and partial tuition reimbursements, teaching assistantships with full and partial tuition reimbursements, tuition remissions available. Financial aid application deadline: 3/1; financial aid applicants required to submit FAFSA. *Faculty research:* Aerodynamics and fluid dynamics, aerospace structures and materials, combustion and propulsion. Total annual research expenditures: $98,590. *Unit head:* Dr. C. S. Subramanian, Chair, 407-674-8092, Fax: 407-674-8813, E-mail: subraman@zach.fit.edu. *Application contact:* Carolyn P. Farrior, Associate Dean of Graduate Admissions, 407-674-7118, Fax: 407-723-9468, E-mail: cfarrior@fit.edu.

Florida Institute of Technology, Graduate School, School of Aeronautics, Melbourne, FL 32901-6975. Offers aviation (MSA); aviation human factors (MS). Part-time and evening/weekend programs available. *Faculty:* 11 full-time (1 woman), 4 part-time (0 women). *Students:* 8 full-time (2 women), 13 part-time (5 women); includes 2 minority (both Hispanic Americans), 4 international. Average age 30. 20 applicants, 50% accepted. In 1998, 5 degrees awarded. *Degree requirements:* For master's, thesis optional, foreign language not required. *Entrance requirements:* For master's, GRE General Test, minimum undergraduate GPA of 3.0. *Application deadline:* For fall admission, 8/1; for spring admission, 12/1. Applications are processed on a rolling basis. Application fee: $50. Electronic applications accepted. Tuition: Part-time $575 per credit hour. Required fees: $100. Tuition and fees vary according to campus/location and program. *Financial aid:* In 1998–99, 2 students received aid; research assistantships, teaching assistantships, career-related internships or fieldwork and tuition remission available. Financial aid application deadline: 3/1. *Faculty research:* Three dimensional airspace analysis, cooperative learning with educational simulations, education technology in the classroom. Total annual research expenditures: $318,650. *Unit head:* Dr. Nathaniel Villaire, Program Chairman of Graduate Studies, 407-674-8120, Fax: 407-725-6974, E-mail: villaire@fit.edu. *Application contact:* Carolyn P. Farrior, Associate Dean of Graduate Admissions, 407-674-7118, Fax: 407-723-9468, E-mail: cfarrior@fit.edu.

See in-depth description on page 329.

Florida Institute of Technology, Graduate School, School of Extended Graduate Studies, Program in Aerospace Engineering, Melbourne, FL 32901-6975. Offers MS. Part-time programs available. Average age 29. 1 applicants, 0% accepted. In 1998, 1 degree awarded (100% found work related to degree). *Degree requirements:* For master's, thesis optional, foreign language not required. *Entrance requirements:* For master's, GRE General Test, GRE Subject Test, minimum GPA of 3.0. *Average time to degree:* Master's–1 year full-time, 3 years part-time. *Application deadline:* Applications are processed on a rolling basis. Application fee: $50. Electronic applications accepted. Tuition: Part-time $270 per credit hour. Part-time tuition and fees vary according to campus/location. *Financial aid:* Applicants required to submit FAFSA. *Application contact:* Carolyn P. Farrior, Associate Dean of Graduate Admissions, 407-674-7118, Fax: 407-723-9468, E-mail: cfarrior@fit.edu.

The George Washington University, School of Engineering and Applied Science, Department of Mechanical and Aeronautical Engineering, Washington, DC 20052. Offers MS, D Sc, App Sc, Engr. Part-time and evening/weekend programs available. *Degree requirements:* For master's, thesis optional, foreign language not required; for doctorate, computer language, dissertation, final and qualifying exams required, foreign language not required; for other advanced degree, foreign language and thesis not required. *Entrance requirements:* For master's, TOEFL (minimum score of 550 required; average 580) or George Washington University English as a Foreign Language Test, appropriate bachelor's degree, minimum GPA of 3.0; for doctorate, TOEFL (minimum score of 550 required; average 580) or George Washington University English as a Foreign Language Test, appropriate bachelor's or master's degree, minimum GPA of 3.4, GRE required if highest earned degree is BS; for other advanced degree, TOEFL (minimum score of 550 required; average 580) or George Washington University English as a Foreign Language Test, appropriate master's degree, minimum GPA of 3.0. *Application deadline:* For fall admission, 3/1 (priority date); for spring admission, 10/1. Applications are processed on a rolling basis. Application fee: $55. Tuition: Full-time $17,328; part-time $722 per credit hour. Required fees: $828; $35 per credit hour. Tuition and fees vary according to campus/location and program. *Financial aid:* Fellowships, research assistantships, teaching assistantships, career-related internships or fieldwork and institutionally-sponsored loans available. Financial aid application deadline: 3/1; financial aid applicants required to submit FAFSA. *Unit head:* Dr. Michael K. Myers, Chair, 202-994-6749, Fax: 202-994-0238. *Application contact:* Howard M. Davis, Manager, Office of Admissions and Student Records, 202-994-6158, Fax: 202-994-0909, E-mail: data:adms@seas.gwu.edu.

Georgia Institute of Technology, Graduate Studies and Research, College of Engineering, School of Aerospace Engineering, Atlanta, GA 30332-0001. Offers aerospace engineering (MS, MSAE, PhD); biomedical engineering (MS Bio E). Part-time programs available. Terminal master's awarded for partial completion of doctoral program. *Degree requirements:* For master's, computer language required, thesis optional, foreign language not required; for doctorate, computer language, dissertation required, foreign language not required. *Entrance requirements:* For master's, GRE, TOEFL (minimum score of 550 required), minimum GPA of 3.0; for doctorate, GRE, TOEFL (minimum score of 550 required), minimum GPA of 3.25. *Faculty research:* Structural mechanics and dynamics, fluid mechanics, flight mechanics and controls, combustion and propulsion, system design and optimization.

See in-depth description on page 331.

Howard University, College of Engineering, Architecture, and Computer Sciences, School of Engineering and Computer Science, Department of Mechanical Engineering, Washington, DC 20059-0002. Offers aerospace engineering/dynamics and controls (M Eng, PhD); applied mechanics (M Eng, PhD); CAD/CAM and robotics (M Eng, PhD); fluid and thermal sciences (M Eng, PhD). Part-time programs available. *Faculty:* 9 full-time (1 woman). *Students:* 17 full-time (7 women), 2 part-time; includes 7 African Americans, 1 Asian American or Pacific Islander, 7 international. Terminal master's awarded for partial completion of doctoral program. *Degree requirements:* For master's, computer language, comprehensive exam required; for doctorate, one foreign language, computer language, dissertation, 2 terms of residency required. *Entrance requirements:* For master's and doctorate, GRE General Test, TOEFL, minimum GPA of 3.0. *Application deadline:* For fall admission, 4/1 (priority date); for spring admission, 11/1. Applications are processed on a rolling basis. Application fee: $45. Electronic applications accepted. *Unit head:* Dr. Lewis Thigpen, Chair, 202-806-6600, Fax: 202-806-5258, E-mail: lthigpen@scs.howard.edu. *Application contact:* Dr. Sonya Smith, Graduate Director, 202-806-4837.

Illinois Institute of Technology, Graduate College, Armour College of Engineering and Sciences, Department of Mechanical, Materials and Aerospace Engineering, Mechanical and Aerospace Engineering Division, Chicago, IL 60616-3793. Offers MMAE, MS, PhD. Part-time programs available. *Faculty:* 18 full-time (2 women), 8 part-time (0 women). *Students:* 39 full-time (0 women), 66 part-time (7 women); includes 17 minority (5 African Americans, 9 Asian Americans or Pacific Islanders, 3 Hispanic Americans), 41 international. 206 applicants, 44% accepted. In 1998, 18 master's, 3 doctorates awarded. Terminal master's awarded for partial completion of doctoral program. *Degree requirements:* For master's, thesis (for some programs), comprehensive exam required, foreign language not required; for doctorate, dissertation, comprehensive exam required, foreign language not required. *Entrance requirements:* For master's, GRE General Test (minimum combined score of 1200 required), TOEFL (minimum score of 550 required), undergraduate GPA of 3.0; for doctorate, GRE (minimum score of 1200 required), TOEFL (minimum score of 550 required), undergraduate GPA of 3.0 required. *Application deadline:* For fall admission, 7/1; for spring admission, 11/1. Applications are processed on a rolling basis. Application fee: $30. Electronic applications accepted. *Financial aid:* In 1998–99, 2 fellowships, 11 research assistantships, 14 teaching assistantships were awarded.; Federal Work-Study, institutionally-sponsored loans, scholarships, and graduate assistantships also available. Financial aid application deadline: 3/1. *Faculty research:* Solid and structural mechanics, fluid dynamics, thermal sciences, transportation engineering, design and manufacturing. *Unit head:* Dr. Kevin Meade, Associate Chair, Mechanical Engineering,

Aerospace/Aeronautical Engineering

Illinois Institute of Technology (continued)
312-567-3175, Fax: 312-567-7230, E-mail: meade@mae.iit.edu. *Application contact:* Dr. S. Mohammad Shahidehpour, Dean of Graduate College, 312-567-3024, Fax: 312-567-7517, E-mail: grad@minna.cns.iit.edu.

Iowa State University of Science and Technology, Graduate College, College of Engineering, Department of Aerospace Engineering and Engineering Mechanics, Ames, IA 50011. Offers aerospace engineering (M Eng, MS, PhD); engineering mechanics (M Eng, MS, PhD). *Faculty:* 41 full-time. *Students:* 39 full-time (5 women), 14 part-time (3 women), 41 international. 79 applicants, 43% accepted. In 1998, 12 master's, 9 doctorates awarded. *Degree requirements:* For master's, thesis required (for some programs); for doctorate, dissertation required. *Entrance requirements:* For master's and doctorate, GRE General Test, TOEFL. *Application deadline:* For fall admission, 3/1 (priority date); for spring admission, 10/1 (priority date). Application fee: $20 ($50 for international students). Electronic applications accepted. Tuition, state resident: full-time $3,308. Tuition, nonresident: full-time $9,744. Part-time tuition and fees vary according to course load, campus/location and program. *Financial aid:* In 1998–99, 15 research assistantships with partial tuition reimbursements (averaging $11,076 per year), 25 teaching assistantships with partial tuition reimbursements (averaging $9,892 per year) were awarded.; fellowships, scholarships also available. *Unit head:* Dr. Thomas J. Rudolphi, Chair, 515-294-0095, E-mail: aeem_info@iastate.edu. *Application contact:* Dr. Ambar Mitra, Director of Graduate Education, 515-294-2694, E-mail: aeem_info@iastate.edu.

Massachusetts Institute of Technology, School of Engineering, Department of Aeronautics and Astronautics, Cambridge, MA 02139-4307. Offers M Eng, SM, PhD, Sc D, EAA. *Faculty:* 30 full-time (2 women). *Students:* 196 full-time (31 women), 3 part-time (2 women); includes 16 minority (2 African Americans, 11 Asian Americans or Pacific Islanders, 3 Hispanic Americans), 88 international. Average age 26. 207 applicants, 62% accepted. In 1998, 58 master's, 15 doctorates awarded. *Degree requirements:* For master's and EAA, thesis required, foreign language not required; for doctorate and EAA, dissertation, comprehensive exams required, foreign language not required. *Entrance requirements:* For master's, TOEFL (minimum score of 600 required); for doctorate, MS. *Application deadline:* For fall admission, 1/15; for spring admission, 11/1. Application fee: $55. *Financial aid:* In 1998–99, 190 students received aid, including 27 fellowships, 155 research assistantships, 5 teaching assistantships; institutionally-sponsored loans and scholarships also available. Financial aid application deadline: 3/1; financial aid applicants required to submit FAFSA. *Faculty research:* Composite materials, structural dynamics, aerodynamic design and optimization, computational fluid dynamics, micromachines and devices. Total annual research expenditures: $18.4 million. *Unit head:* Dr. Edward F. Crawley, Head, 617-253-7570, E-mail: crawley@mit.edu. *Application contact:* Jaime Peraire, Chairman of the Graduate Committee, 617-253-1981, Fax: 617-253-0823.

See in-depth description on page 333.

Middle Tennessee State University, College of Graduate Studies, College of Basic and Applied Sciences, Department of Aerospace, Murfreesboro, TN 37132. Offers aerospace education (M Ed); airport/airline management (MS); asset management (MS). *Accreditation:* NCATE (one or more programs are accredited). Part-time and evening/weekend programs available. *Faculty:* 4 full-time (0 women), 3 part-time (0 women). *Students:* 12 full-time (6 women), 9 part-time (2 women); includes 5 minority (3 African Americans, 1 Asian American or Pacific Islander, 1 Hispanic American) Average age 30. 6 applicants, 17% accepted. In 1998, 9 degrees awarded. *Degree requirements:* For master's, one foreign language, comprehensive exams required, thesis not required. *Entrance requirements:* For master's, GRE, MAT (minimum score of 38 required). *Application deadline:* For fall admission, 8/1 (priority date). Application fee: $25. *Financial aid:* Application deadline: 5/1. *Unit head:* Ronald J. Ferrara, Chair, 615-898-3515, E-mail: rferrara@frank.mtsu.edu.

Mississippi State University, College of Engineering, Department of Aerospace Engineering, Mississippi State, MS 39762. Offers aerospace engineering (MS); engineering mechanics (MS). Part-time programs available. *Students:* 8 full-time (0 women), 2 part-time; includes 6 minority (all Asian Americans or Pacific Islanders), 2 international. Average age 29. 7 applicants, 29% accepted. In 1998, 12 degrees awarded. *Degree requirements:* For master's, computer language required, foreign language not required. *Entrance requirements:* For master's, GRE General Test, TOEFL (minimum score of 550 required), minimum GPA of 2.75. *Application deadline:* For fall admission, 7/1; for spring admission, 11/1. Applications are processed on a rolling basis. Application fee: $25 for international students. *Financial aid:* Federal Work-Study, institutionally-sponsored loans, and unspecified assistantships available. Financial aid applicants required to submit FAFSA. *Faculty research:* Computational fluid dynamics, flight mechanics, aerodynamics, composite structures, prototype development. Total annual research expenditures: $3.4 million. *Unit head:* Dr. John C. McWhorter, Head, 662-325-3623, Fax: 662-325-7730, E-mail: mcwho@ae.msstate.edu. *Application contact:* Jerry B. Inmon, Director of Admissions, 662-325-2224, Fax: 662-325-7360, E-mail: admit@admissions.msstate.edu.

Naval Postgraduate School, Graduate Programs, Department of Aeronautics and Astronautics, Monterey, CA 93943. Offers MS, D Eng, PhD, Eng. Program only open to commissioned officers of the United States and friendly nations and selected United States federal civilian employees. *Accreditation:* ABET (one or more programs are accredited). Part-time programs available. Postbaccalaureate distance learning degree programs offered (minimal on-campus study). *Students:* 48 full-time, 7 international. In 1998, 46 master's, 7 other advanced degrees awarded. *Degree requirements:* For master's and Eng, computer language, thesis required, foreign language not required; for doctorate, one foreign language, computer language, dissertation required. *Unit head:* Dr. Gerald H. Lindsay, Chairman, 831-656-2311. *Application contact:* Theodore H. Calhoon, Director of Admissions, 831-656-3093, Fax: 831-656-2891, E-mail: tcalhoon@nps.navy.mil.

Naval Postgraduate School, Graduate Programs, Program in Space Systems, Monterey, CA 93943. Offers MS. Program only open to commissioned officers of the United States and friendly nations and selected United States federal civilian employees. Part-time programs available. *Students:* 60 full-time. In 1998, 15 degrees awarded. *Degree requirements:* For master's, computer language, thesis required, foreign language not required. *Unit head:* Dr. Rudy Panholzer, Academic Group Chairman, 831-656-2278. *Application contact:* Theodore H. Calhoon, Director of Admissions, 831-656-3093, Fax: 831-656-2891, E-mail: tcalhoon@nps.navy.mil.

North Carolina State University, Graduate School, College of Engineering, Department of Mechanical and Aerospace Engineering, Program in Aerospace Engineering, Raleigh, NC 27695. Offers MS, PhD. *Faculty:* 1 full-time (0 women), 1 part-time (0 women). *Students:* 21 full-time (2 women), 7 part-time (1 woman); includes 4 minority (2 African Americans, 2 Asian Americans or Pacific Islanders), 4 international. Average age 30. 30 applicants, 33% accepted. In 1998, 4 master's, 3 doctorates awarded. *Degree requirements:* For master's, thesis, oral exam required, foreign language not required; for doctorate, dissertation, oral and preliminary exams required, foreign language not required. *Entrance requirements:* For master's and doctorate, GRE General Test. *Average time to degree:* Master's–2 years full-time, 3 years part-time; doctorate–3 years full-time, 4 years part-time. *Application deadline:* For fall admission, 7/15; for spring admission, 12/15. Applications are processed on a rolling basis. Application fee: $45. *Financial aid:* Fellowships, research assistantships, teaching assistantships, career-related internships or fieldwork and institutionally-sponsored loans available. *Faculty research:* Vibration and control, fluid dynamics, thermal sciences, structure and materials, aerodynamics acoustics. *Unit head:* Dr. James C. Mulligan, Director of Graduate Programs, 919-515-2856, Fax: 919-515-7968, E-mail: mulligan@eos.ncsu.edu.

The Ohio State University, Graduate School, College of Engineering, Department of Aerospace Engineering, Applied Mechanics, and Aviation, Program in Aeronautical and Astronautical Engineering, Columbus, OH 43210. Offers MS, PhD. *Faculty:* 12 full-time. *Students:* 31 full-time (4 women), 7 part-time; includes 2 minority (both Asian Americans or Pacific Islanders), 12 international. 98 applicants, 18% accepted. In 1998, 2 master's, 1 doctorate awarded. *Degree requirements:* For master's, computer language required, thesis optional, foreign language not required; for doctorate, computer language, dissertation required, foreign language not required. *Entrance requirements:* For master's and doctorate, GRE General Test or minimum GPA of 3.0. *Application deadline:* For fall admission, 8/15. Applications are processed on a rolling basis. Application fee: $30 ($40 for international students). *Financial aid:* Fellowships, research assistantships, teaching assistantships, career-related internships or fieldwork, Federal Work-Study, institutionally-sponsored loans, and unspecified assistantships available. *Unit head:* Richard Bodonyi, Graduate Studies Committee Chair, 614-292-7354, Fax: 614-292-8290, E-mail: bodonyi.1@osu.edu.

Old Dominion University, College of Engineering and Technology, Department of Aerospace Engineering, Norfolk, VA 23529. Offers aerospace engineering (ME, MS, PhD); engineering mechanics (ME, MS, PhD). Part-time and evening/weekend programs available. *Faculty:* 11 full-time. *Students:* 34 full-time (1 woman), 28 part-time (2 women); includes 4 minority (2 African Americans, 1 Asian American or Pacific Islander, 1 Hispanic American), 23 international. Average age 29. In 1998, 11 master's, 8 doctorates awarded. *Degree requirements:* For master's, computer language, thesis, comprehensive exam required, foreign language not required; for doctorate, computer language, dissertation, candidacy exam required, foreign language not required. *Entrance requirements:* For master's, TOEFL (minimum score of 550 required), minimum GPA of 3.0; for doctorate, TOEFL (minimum score of 550 required), minimum GPA of 3.25. *Average time to degree:* Master's–6 years part-time; doctorate–8 years part-time. *Application deadline:* For fall admission, 7/1; for spring admission, 10/1. Applications are processed on a rolling basis. Application fee: $30. Electronic applications accepted. *Financial aid:* In 1998–99, 59 students received aid, including 28 research assistantships (averaging $14,379 per year); fellowships, teaching assistantships, career-related internships or fieldwork, grants, and tuition waivers (partial) also available. Aid available to part-time students. Financial aid application deadline: 2/15; financial aid applicants required to submit FAFSA. *Faculty research:* Computational fluid dynamics, experimental fluid dynamics, structural mechanics, dynamics and control, multidisciplinary problems. Total annual research expenditures: $850,000. *Unit head:* Dr. Thomas E. Alberts, Chair, 757-683-3736, Fax: 757-683-3200, E-mail: talberts@aero.odu.edu. *Application contact:* Dr. Thomas E. Alberts, Chair, 757-683-3736, Fax: 757-683-3200, E-mail: talberts@aero.odu.edu.

Pennsylvania State University University Park Campus, Graduate School, College of Engineering, Department of Aerospace Engineering, State College, University Park, PA 16802-1503. Offers M Eng, MS, PhD. *Students:* 50 full-time (4 women), 13 part-time. In 1998, 17 master's, 7 doctorates awarded. *Degree requirements:* For master's and doctorate, thesis/dissertation required, foreign language not required. *Entrance requirements:* For master's and doctorate, GRE General Test. Application fee: $50. *Unit head:* Dr. Dennis K. McLaughlin, Head, 814-865-2569.

See in-depth description on page 335.

Polytechnic University, Brooklyn Campus, Department of Mechanical, Aerospace and Manufacturing Engineering, Major in Aeronautics and Astronautics, Brooklyn, NY 11201-2990. Offers MS. Part-time programs available. *Students:* 1 full-time (0 women), 2 part-time (1 woman); includes 1 minority (Asian American or Pacific Islander) Average age 33. 5 applicants, 20% accepted. *Entrance requirements:* For master's, BS in aerospace or mechanical engineering. *Application deadline:* Applications are processed on a rolling basis. Application fee: $45. Electronic applications accepted. *Financial aid:* Fellowships, research assistantships, teaching assistantships, institutionally-sponsored loans available. Aid available to part-time students. Financial aid applicants required to submit FAFSA. *Faculty research:* UV filter, fuel efficient hydrodynamic containment for gas core fission, turbulent boundary layer research. Total annual research expenditures: $583,050. *Unit head:* Wanda Frederick, Graduate Coordinator, 206-616-1113, Fax: 206-543-0217, E-mail: wanda@aa.washington.edu. *Application contact:* John S. Kerge, Dean of Admissions, 718-260-3200, Fax: 718-260-3446, E-mail: admitme@poly.edu.

Polytechnic University, Farmingdale Campus, Graduate Programs, Department of Mechanical, Aerospace and Manufacturing Engineering, Major in Aeronautics and Astronautics, Farmingdale, NY 11735-3995. Offers MS. 5 applicants, 40% accepted. In 1998, 1 degree awarded. *Degree requirements:* For master's, computer language required. *Application deadline:* Applications are processed on a rolling basis. Application fee: $45. Electronic applications accepted. *Financial aid:* Institutionally-sponsored loans available. Aid available to part-time students. Financial aid applicants required to submit FAFSA. *Unit head:* Joan P. Smith, Graduate Student Adviser, 435-797-0330, Fax: 435-797-2417, E-mail: jpsmith@mae.usu.edu. *Application contact:* John S. Kerge, Dean of Admissions, 718-260-3200, Fax: 718-260-3446, E-mail: admitme@poly.edu.

Princeton University, Graduate School, School of Engineering and Applied Science, Department of Mechanical and Aerospace Engineering, Princeton, NJ 08544-1019. Offers applied physics (M Eng, MSE, PhD); computational methods (M Eng, MSE); dynamics and control systems (M Eng, MSE, PhD); energy and environmental policy (M Eng, MSE, PhD); energy conversion, propulsion, and combustion (M Eng, MSE, PhD); flight science and technology (M Eng, MSE, PhD); fluid mechanics (M Eng, MSE, PhD). *Faculty:* 26 full-time (2 women). *Students:* 46 full-time (6 women); includes 5 minority (4 Asian Americans or Pacific Islanders, 1 Hispanic American), 17 international. Average age 23. 173 applicants, 22% accepted. In 1998, 6 master's awarded (83% found work related to degree, 17% continued full-time study); 14 doctorates awarded (29% entered university research/teaching, 71% found other work related to degree). *Degree requirements:* For master's and doctorate, thesis/dissertation required, foreign language not required. *Entrance requirements:* For master's and doctorate, GRE General Test. *Average time to degree:* Master's–2 years full-time; doctorate–5.53 years full-time. *Application deadline:* For fall admission, 1/3. Electronic applications accepted. *Financial aid:* Fellowships, research assistantships, teaching assistantships, Federal Work-Study and institutionally-sponsored loans available. Financial aid application deadline: 1/3. Total annual research expenditures: $5.3 million. *Unit head:* Prof. Richard B. Miles, Director of Graduate Studies, 609-258-4683, Fax: 609-258-6109, E-mail: maegrad@princeton.edu. *Application contact:* Etta Recke, Graduate Administrator, 609-258-4683, Fax: 609-258-6109, E-mail: etta@princeton.edu.

See in-depth description on page 337.

Purdue University, Graduate School, Schools of Engineering, School of Aeronautics and Astronautics, West Lafayette, IN 47907. Offers MS, MSAAE, MSE, PhD. *Faculty:* 22 full-time (1 woman). *Students:* 110 full-time (11 women), 16 part-time; includes 8 minority (3 Asian Americans or Pacific Islanders, 4 Hispanic Americans, 1 Native American), 76 international. 167 applicants, 78% accepted. In 1998, 25 master's, 9 doctorates awarded. *Degree requirements:* For master's, foreign language and thesis not required; for doctorate, variable foreign language requirement, dissertation required. *Entrance requirements:* For master's and doctorate, GRE General Test, TOEFL (minimum score of 550 required). Application fee: $30. Electronic applications accepted. *Financial aid:* In 1998–99, 13 fellowships, 57 research assistantships, 21 teaching assistantships were awarded.; career-related internships or fieldwork also available. Aid available to part-time students. Financial aid applicants required to submit FAFSA. *Faculty research:* Structures and materials, propulsion, aerodynamics, dynamics and control. *Unit head:* Dr. Thomas N. Farris, Head, 765-494-5117, Fax: 765-494-0307, E-mail: farrist@ecn.purdue.edu. *Application contact:* Linda Flack, Administrative Assistant, 765-494-5152, Fax: 765-494-0307, E-mail: flack@ecn.purdue.edu.

See in-depth description on page 339.

Rensselaer Polytechnic Institute, Graduate School, School of Engineering, Department of Mechanical Engineering, Aeronautical Engineering and Mechanics, Program in Aeronautical Engineering, Troy, NY 12180-3590. Offers M Eng, MS, D Eng, PhD, MBA/M Eng. Part-time and evening/weekend programs available. *Faculty:* 31 full-time (2 women), 5 part-time (0 women). *Students:* 13 full-time (2 women), 2 part-time; includes 1 minority (Asian American or

Pacific Islander), 10 international. 27 applicants, 37% accepted. In 1998, 2 master's, 2 doctorates awarded. *Degree requirements:* For master's, computer language, thesis required (for some programs), foreign language not required; for doctorate, computer language, dissertation required, foreign language not required. *Entrance requirements:* For master's and doctorate, GRE, TOEFL (minimum score of 550 required). *Application deadline:* For fall admission, 2/1 (priority date). Applications are processed on a rolling basis. Application fee: $35. *Financial aid:* Fellowships, research assistantships, teaching assistantships, career-related internships or fieldwork, institutionally-sponsored loans, and tuition waivers (partial) available. Financial aid application deadline: 2/1. *Faculty research:* Vehicular performance and flight mechanics, gas dynamics, aerodynamics, structural dynamics, advanced propulsion. Total annual research expenditures: $2.5 million. *Application contact:* Dr. Michael Jensen, Associate Head, 518-276-6432, Fax: 518-276-6025, E-mail: burged@rpi.edu.

See in-depth description on page 341.

Rutgers, The State University of New Jersey, New Brunswick, Graduate School, Program in Mechanical and Aerospace Engineering, New Brunswick, NJ 08903. Offers computational fluid dynamics (MS, PhD); design and dynamics (MS, PhD); fluid mechanics (MS, PhD); heat transfer (MS, PhD); solid mechanics (MS, PhD). Part-time and evening/weekend programs available. *Faculty:* 29 full-time (0 women), 5 part-time (0 women). *Students:* 59 full-time (12 women), 21 part-time (4 women); includes 9 minority (1 African American, 6 Asian Americans or Pacific Islanders, 2 Hispanic Americans), 38 international. Average age 25. 214 applicants, 29% accepted. In 1998, 8 master's, 9 doctorates awarded. *Degree requirements:* For master's, thesis required (for some programs), foreign language not required; for doctorate, dissertation required, foreign language not required. *Entrance requirements:* For master's, GRE General Test, BS in mechanical/aerospace engineering or related field; for doctorate, GRE General Test, MS in mechanical/aerospace engineering or related field. *Application deadline:* For fall admission, 6/1. Application fee: $50. *Financial aid:* In 1998–99, 57 students received aid, including 8 fellowships, 25 research assistantships, 24 teaching assistantships; tuition waivers (full) also available. Financial aid application deadline: 3/1; financial aid applicants required to submit FAFSA. Total annual research expenditures: $8 million.

Saint Louis University, Graduate School, Department of Aerospace and Mechanical Engineering, St. Louis, MO 63103-2097. Offers MS, MS(R). *Faculty:* 13 full-time (2 women), 4 part-time (0 women). *Students:* 4 full-time (2 women), 11 part-time (2 women); includes 2 minority (1 Asian American or Pacific Islander, 1 Hispanic American), 9 international. Average age 27. 10 applicants, 70% accepted. In 1998, 10 degrees awarded. *Degree requirements:* For master's, comprehensive oral exam required, thesis optional, foreign language not required. *Entrance requirements:* For master's, GRE General Test. *Application deadline:* For fall admission, 7/1; for spring admission, 11/1. Applications are processed on a rolling basis. Application fee: $40. Tuition: Full-time $20,520; part-time $507 per credit hour. Required fees: $38 per term. Tuition and fees vary according to program. *Financial aid:* In 1998–99, 10 students received aid, including 1 research assistantship, 4 teaching assistantships Financial aid application deadline: 4/1; financial aid applicants required to submit FAFSA. *Faculty research:* Flight dynamics/control, structural dynamics, experimental aerodynamics, aircraft design/optimization. *Unit head:* Dr. Krishnaswamy Ravindra, Chairman, 314-977-8438, Fax: 314-977-8403, E-mail: ravindrak@slu.edu. *Application contact:* Dr. Marcia Buresch, Assistant Dean of the Graduate School, 314-977-2240, Fax: 314-977-3943, E-mail: bureschm@slu.edu.

See in-depth description on page 343.

San Diego State University, Graduate and Research Affairs, College of Engineering, Department of Aerospace Engineering and Engineering Mechanics, San Diego, CA 92182. Offers aerospace engineering (MS); engineering mechanics (MS); engineering sciences and applied mechanics (PhD); flight dynamics (MS); fluid dynamics (MS). *Students:* 4 full-time (0 women), 2 part-time; includes 1 minority (Asian American or Pacific Islander), 3 international. 8 applicants, 50% accepted. In 1998, 3 degrees awarded. Terminal master's awarded for partial completion of doctoral program. *Degree requirements:* For doctorate, dissertation required, foreign language not required. *Entrance requirements:* For master's, GRE General Test (minimum combined score of 950 required), TOEFL (minimum score of 550 required). *Application deadline:* For fall admission, 7/1 (priority date); for spring admission, 12/1. Applications are processed on a rolling basis. Application fee: $55. *Faculty research:* Organized structures in post-stall flow over wings/three dimensional separated flow, airfoil growth effect, probabilities, structural mechanics. Total annual research expenditures: $35,000. *Unit head:* Joseph Katz, Chair, 619-594-6074, Fax: 619-594-6005, E-mail: jkatz@mail.sdsu.edu. *Application contact:* Allen Plotkin, Graduate Adviser, 619-594-7019, Fax: 619-594-6005, E-mail: allen.plotkin@sdsu.edu.

San Jose State University, Graduate Studies, College of Engineering, Department of Mechanical and Aerospace Engineering, Program in Aerospace Engineering, San Jose, CA 95192-0001. Offers MS. *Students:* 1 full-time (0 women), 11 part-time (5 women); includes 4 minority (2 Asian Americans or Pacific Islanders, 2 Hispanic Americans), 2 international. Average age 30. 8 applicants, 88% accepted. In 1998, 4 degrees awarded. *Application deadline:* For fall admission, 6/1. Applications are processed on a rolling basis. Application fee: $59. Tuition, nonresident: part-time $246 per unit. Required fees: $1,939; $1,309 per year. *Unit head:* Dr. Marcia Buresch, Assistant Dean of the Graduate School, 314-977-2240, Fax: 314-977-3943, E-mail: bureschm@slu.edu. *Application contact:* Dr. Richard Desautel, Coordinator, 408-924-3840.

Stanford University, School of Engineering, Department of Aeronautics and Astronautics, Stanford, CA 94305-9991. Offers MS, PhD, Eng. *Faculty:* 16 full-time (0 women). *Students:* 144 full-time (19 women), 63 part-time (8 women); includes 29 minority (3 African Americans, 13 Asian Americans or Pacific Islanders, 13 Hispanic Americans), 79 international. Average age 27. 165 applicants, 75% accepted. In 1998, 38 master's, 20 doctorates awarded. Terminal master's awarded for partial completion of doctoral program. *Degree requirements:* For master's, foreign language and thesis not required; for doctorate, dissertation required, foreign language not required; for Eng, thesis required. *Entrance requirements:* For master's, doctorate, and Eng, GRE General Test, GRE Subject Test, TOEFL. *Application deadline:* For fall admission, 2/1. Applications are processed on a rolling basis. Application fee: $65 ($80 for international students). Electronic applications accepted. Tuition: Full-time $24,588. Required fees: $152. Part-time tuition and fees vary according to course load. *Financial aid:* Fellowships, research assistantships, teaching assistantships, Federal Work-Study and institutionally-sponsored loans available. Financial aid application deadline: 2/1. *Unit head:* George Springer, Chairman, 650-723-4135, Fax: 650-723-0062, E-mail: springer@sierra.stanford.edu. *Application contact:* Graduate Admissions Coordinator, 650-723-2757.

See in-depth description on page 345.

State University of New York at Buffalo, Graduate School, School of Engineering and Applied Sciences, Department of Mechanical and Aerospace Engineering, Buffalo, NY 14260. Offers aerospace engineering (M Eng, MS, PhD); mechanical engineering (M Eng, MS, PhD). Part-time programs available. *Faculty:* 23 full-time (2 women), 8 part-time (0 women). *Students:* 74 full-time (2 women), 126 part-time (6 women); includes 12 minority (3 African Americans, 9 Asian Americans or Pacific Islanders), 91 international. Average age 24. 216 applicants, 76% accepted. In 1998, 40 master's, 8 doctorates awarded. Terminal master's awarded for partial completion of doctoral program. *Degree requirements:* For master's, comprehensive exam, project, or thesis required; for doctorate, dissertation required, foreign language not required. *Entrance requirements:* For master's and doctorate, GRE General Test, GRE Subject Test, TOEFL (minimum score of 550 required). *Average time to degree:* Master's–2 years full-time, 4 years part-time; doctorate–4 years full-time, 6 years part-time. *Application deadline:* For fall admission, 2/1; for spring admission, 10/1. Applications are processed on a rolling basis. Application fee: $35. Tuition, state resident: full-time $5,100; part-time $213 per credit hour. Tuition, nonresident: full-time $8,416; part-time $351 per credit hour. Required fees: $870; $75 per semester. Tuition and fees vary according to course load and program. *Financial aid:* In 1998–99, 63 students received aid, including 2 fellowships with tuition reimbursements available, 37 research assistantships with tuition reimbursements available (averaging $10,500 per year), 24 teaching assistantships with tuition reimbursements available (averaging $10,350 per year); Federal Work-Study, institutionally-sponsored loans, tuition waivers (full), and unspecified assistantships also available. Financial aid application deadline: 2/1; financial aid applicants required to submit FAFSA. *Faculty research:* Fluid and thermal sciences, systems anddesign, mechanics and materials. Total annual research expenditures: $924,273. *Unit head:* Dr. Christina L. Bloebaum, Chairman, 716-645-2593 Ext. 2231, Fax: 716-645-3875, E-mail: clb@eng.buffalo.edu. *Application contact:* Dr. Dale B. Taulbee, Director of Graduate Studies, 716-645-2593 Ext. 2307, Fax: 716-645-3875, E-mail: trldale@eng.buffalo.edu.

Syracuse University, Graduate School, L. C. Smith College of Engineering and Computer Science, Department of Mechanical and Aerospace Engineering, Program in Aerospace Engineering, Syracuse, NY 13244-0003. Offers MS, PhD. *Students:* 6 full-time (2 women), 4 part-time, 8 international. Average age 29. 8 applicants, 100% accepted. In 1998, 1 master's, 1 doctorate awarded. *Degree requirements:* For master's, project or thesis required; for doctorate, computer language, dissertation required, foreign language not required. *Entrance requirements:* For master's and doctorate, GRE General Test, GRE Subject Test. *Application deadline:* Applications are processed on a rolling basis. Application fee: $40. Tuition: Full-time $13,992; part-time $583 per credit hour. *Financial aid:* Fellowships, research assistantships, teaching assistantships, Federal Work-Study and tuition waivers (partial) available. Financial aid application deadline: 3/1. *Unit head:* Alan Levy, Graduate Director.

Texas A&M University, College of Engineering, Department of Aerospace Engineering, College Station, TX 77843. Offers M Eng, MS, PhD. *Faculty:* 19 full-time (1 woman). *Students:* 65 full-time (4 women); includes 5 minority (2 Asian Americans or Pacific Islanders, 2 Hispanic Americans, 1 Native American), 40 international. Average age 29. 116 applicants, 96% accepted. In 1998, 16 master's, 5 doctorates awarded. *Degree requirements:* For master's, thesis (MS) required; for doctorate, dissertation required, foreign language not required. *Entrance requirements:* For master's and doctorate, GRE General Test, TOEFL. *Application deadline:* For fall admission, 1/15 (priority date); for spring admission, 9/15. Applications are processed on a rolling basis. Application fee: $50 ($75 for international students). Electronic applications accepted. *Financial aid:* In 1998–99, 8 fellowships, 39 research assistantships, 14 teaching assistantships were awarded. Financial aid application deadline: 3/1; financial aid applicants required to submit FAFSA. *Faculty research:* Materials and structures, aerodynamics and CFD, flight dynamics and control. *Unit head:* Dr. Terry Alfriend, Head, 409-845-7541. *Application contact:* Dr. Dimitris Lagoudas, Graduate Adviser, 409-845-5520, Fax: 409-845-6051, E-mail: karer@aero.tamu.edu.

Université Laval, Faculty of Graduate Studies, Faculty of Sciences and Engineering, Department of Mechanical Engineering, Program in Aerospace Engineering, Sainte-Foy, PQ G1K 7P4, Canada. Offers M Sc. *Students:* 5 full-time (0 women), 2 part-time. 8 applicants, 75% accepted. In 1998, 5 degrees awarded. *Application deadline:* For fall admission, 3/1. Application fee: $30. *Unit head:* Yuan Maciel, Director, 418-656-2131 Ext. 7967, Fax: 418-656-7415, E-mail: yuan.maciel@gmc.ulaval.ca.

The University of Alabama, Graduate School, College of Engineering, Department of Aerospace Engineering and Mechanics, Tuscaloosa, AL 35487. Offers MSAE, MSESM, PhD. Part-time programs available. Postbaccalaureate distance learning degree programs offered (no on-campus study). *Faculty:* 14 full-time (0 women), 3 part-time (0 women). *Students:* 26 full-time (2 women), 42 part-time (2 women); includes 14 minority (1 African American, 13 Asian Americans or Pacific Islanders), 2 international. Average age 26. 40 applicants, 40% accepted. In 1998, 14 master's, 1 doctorate awarded. *Degree requirements:* For master's, thesis or alternative required, foreign language not required; for doctorate, dissertation required, foreign language not required. *Entrance requirements:* For master's, GRE General Test (minimum combined score of 1500 on three sections required). *Application deadline:* For fall admission, 7/6 (priority date). Applications are processed on a rolling basis. Application fee: $25. *Financial aid:* In 1998–99, 25 students received aid, including 5 fellowships, 12 research assistantships, 8 teaching assistantships; Federal Work-Study and institutionally-sponsored loans also available. Financial aid application deadline: 7/6. *Faculty research:* Flight simulation, advanced mechanical behavior in materials, fluid and solid computational mechanics, hypersonic aerodynamics, intelligent systems. Total annual research expenditures: $5 million. *Unit head:* Dr. Tom E. Novak, Head, 205-348-7300, Fax: 205-348-2094, E-mail: tnovak@coe.eng.ua.edu. *Application contact:* Dr. Amnon Katz, 205-348-7300.

The University of Arizona, Graduate College, College of Engineering and Mines, Department of Aerospace and Mechanical Engineering, Program in Aerospace Engineering, Tucson, AZ 85721. Offers MS, PhD. Part-time programs available. *Students:* 19 full-time (2 women), 2 part-time, 16 international. Average age 26. 39 applicants, 67% accepted. In 1998, 2 master's, 1 doctorate awarded. *Degree requirements:* For master's, thesis or alternative required, foreign language not required; for doctorate, dissertation required. *Entrance requirements:* For master's and doctorate, GRE General Test, GRE Subject Test, TOEFL (minimum score of 550 required), minimum GPA of 3.0. *Application deadline:* For fall admission, 7/23. Applications are processed on a rolling basis. Application fee: $35. *Financial aid:* Fellowships, research assistantships, teaching assistantships available. *Faculty research:* Fluid mechanics, structures, computer-aided design, stability and control, combustion. *Application contact:* Barbara Heefner, Graduate Secretary, 520-621-4692, Fax: 520-621-8191.

See in-depth description on page 347.

University of California, Davis, Graduate Studies, College of Engineering, Program in Mechanical and Aeronautical Engineering, Davis, CA 95616. Offers aeronautical engineering (M Engr, MS, D Engr, PhD, Certificate); mechanical engineering (M Engr, MS, D Engr, PhD, Certificate). *Faculty:* 27 full-time (1 woman), 5 part-time (0 women). *Students:* 81 full-time (10 women), 4 part-time; includes 14 minority (11 Asian Americans or Pacific Islanders, 3 Hispanic Americans), 19 international. 125 applicants, 82% accepted. In 1998, 17 master's, 5 doctorates awarded. *Degree requirements:* For master's, thesis optional, foreign language not required; for doctorate, dissertation required, foreign language not required. *Entrance requirements:* For master's and doctorate, GRE General Test, minimum GPA of 3.0. *Application deadline:* For fall admission, 3/15. Application fee: $40. Electronic applications accepted. *Financial aid:* In 1998–99, 17 fellowships with full and partial tuition reimbursements, 36 research assistantships with full and partial tuition reimbursements, 18 teaching assistantships with full and partial tuition reimbursements were awarded. Financial aid application deadline: 1/15; financial aid applicants required to submit FAFSA. *Unit head:* Bahram Ravani, Chairperson, 530-752-0581, Fax: 530-752-4158. *Application contact:* Susan Fann, Academic Assistant, 530-752-0581, Fax: 530-752-4158, E-mail: sfann@ucdavis.edu.

University of California, Irvine, Office of Research and Graduate Studies, School of Engineering, Department of Mechanical and Aerospace Engineering, Irvine, CA 92697. Offers MS, PhD. Part-time programs available. *Faculty:* 18 full-time (1 woman). *Students:* 53 full-time (7 women), 8 part-time; includes 20 minority (1 African American, 15 Asian Americans or Pacific Islanders, 4 Hispanic Americans), 20 international. 134 applicants, 49% accepted. In 1998, 18 master's, 9 doctorates awarded. Terminal master's awarded for partial completion of doctoral program. *Degree requirements:* For doctorate, dissertation required, foreign language not required. *Entrance requirements:* For master's, GRE General Test, minimum GPA of 3.0; for doctorate, GRE General Test. *Application deadline:* For fall admission, 1/15 (priority date). Applications are processed on a rolling basis. Application fee: $40. Electronic applications accepted. *Financial aid:* Fellowships, research assistantships, teaching assistantships, institutionally-sponsored loans and tuition waivers (full and partial) available. Financial aid application deadline: 3/2; financial aid applicants required to submit FAFSA. *Faculty research:* Thermal and fluid sciences, combustion and propulsion, control systems, robotics. *Unit head:* Dr. Said Elghobashi, Chair, 949-824-8451, Fax: 949-824-8585, E-mail: selghoba@uci.edu. *Application contact:* Dorothy Miles, Graduate Coordinator, 949-824-5469, Fax: 949-824-8585, E-mail: djmiles@uci.edu.

Aerospace/Aeronautical Engineering

University of California, Los Angeles, Graduate Division, School of Engineering and Applied Science, Department of Mechanical and Aerospace Engineering, Program in Aerospace Engineering, Los Angeles, CA 90095. Offers MS, PhD. *Students:* 28 full-time (5 women); includes 4 minority (all Asian Americans or Pacific Islanders), 12 international. 32 applicants, 66% accepted. In 1998, 7 master's, 2 doctorates awarded. *Degree requirements:* For master's, comprehensive exam or thesis required; for doctorate, dissertation, qualifying exams required, foreign language not required. *Entrance requirements:* For master's, GRE General Test, GRE Subject Test (required for foreign students), minimum GPA of 3.0; for doctorate, GRE General Test, GRE Subject Test (required for foreign students), minimum GPA of 3.25. *Application deadline:* For fall admission, 1/5; for spring admission, 12/31. Application fee: $40. Electronic applications accepted. *Financial aid:* Fellowships, research assistantships, teaching assistantships, Federal Work-Study, institutionally-sponsored loans, and tuition waivers (full and partial) available. Financial aid application deadline: 1/5; financial aid applicants required to submit FAFSA. *Application contact:* Student Affairs Officer, E-mail: maeapp@ea.ucla.edu.

University of California, San Diego, Graduate Studies and Research, Department of Applied Mechanics and Engineering Sciences, Program in Aerospace Engineering, La Jolla, CA 92093-5003. Offers MS, PhD. Part-time programs available. *Students:* 5 full-time (0 women), 2 international. 33 applicants, 24% accepted. In 1998, 1 doctorate awarded. *Degree requirements:* For master's, comprehensive exam or thesis required; for doctorate, dissertation, qualifying exam required. *Entrance requirements:* For master's and doctorate, GRE General Test, TOEFL (minimum score of 550 required), minimum GPA of 3.0. *Application deadline:* For fall admission, 5/31. Application fee: $40. *Financial aid:* In 1998–99, fellowships with full tuition reimbursements (averaging $15,000 per year), research assistantships with full tuition reimbursements (averaging $15,000 per year), teaching assistantships with partial tuition reimbursements (averaging $13,000 per year) were awarded.; career-related internships or fieldwork and scholarships also available. Financial aid application deadline: 1/31; financial aid applicants required to submit FAFSA. *Faculty research:* Aerospace structures, turbulence, gas dynamics and combustion. *Unit head:* Dr. Amnon Katz, 205-348-7300. *Application contact:* AMES Graduate Student Affairs, 619-534-4387, Fax: 619-534-1730, E-mail: bwalton@ames.ucsd.edu.

University of Central Florida, College of Engineering, Department of Mechanical, Materials, and Aerospace Engineering, Orlando, FL 32816. Offers aerospace systems (MSME, PhD); materials science and engineering (MSME, PhD); mechanical systems (MSME, PhD); mechanical, materials, and aerospace engineering (Certificate); thermofluids (MSME, PhD). Part-time and evening/weekend programs available. *Faculty:* 18 full-time, 6 part-time. *Students:* 61 full-time (14 women), 30 part-time (4 women); includes 12 minority (3 African Americans, 6 Asian Americans or Pacific Islanders, 3 Hispanic Americans), 42 international. Average age 31. 48 applicants, 58% accepted. In 1998, 23 master's, 2 doctorates awarded. *Degree requirements:* For master's, thesis or alternative required, foreign language not required; for doctorate, dissertation, departmental qualifying exam required, foreign language not required. *Entrance requirements:* For master's, GRE General Test (minimum combined score of 1000 required), TOEFL (minimum score of 550 required; 213 computer-based), minimum GPA of 3.0 in last 60 hours; for doctorate, TOEFL (minimum score of 550 required; 213 computer-based), minimum GPA of 3.5 in last 60 hours. *Application deadline:* For fall admission, 7/15; for spring admission, 12/15. Application fee: $20. Tuition, state resident: full-time $2,054; part-time $137 per credit. Tuition, nonresident: full-time $7,207; part-time $480 per credit. Required fees: $47 per term. *Financial aid:* In 1998–99, 71 students received aid, including 17 fellowships with partial tuition reimbursements available (averaging $2,824 per year), 36 teaching assistantships with partial tuition reimbursements available (averaging $3,509 per year); research assistantships with partial tuition reimbursements available, career-related internships or fieldwork, Federal Work-Study, institutionally-sponsored loans, tuition waivers (partial), and unspecified assistantships also available. Financial aid application deadline: 3/1; financial aid applicants required to submit FAFSA. *Faculty research:* Aerospace systems, computation of methods, dynamics and control, laser applications, materials science. *Unit head:* Dr. Louis Chow, Chair, 407-823-2333. *Application contact:* Dr. A. J. Kassab, Coordinator, 407-823-2416, Fax: 407-823-0208.

University of Cincinnati, Division of Research and Advanced Studies, College of Engineering, Department of Aerospace Engineering and Engineering Mechanics, Program in Aerospace Engineering, Cincinnati, OH 45221-0091. Offers MS, PhD. *Students:* 53 full-time (8 women), 31 part-time (1 woman); includes 4 minority (1 African American, 3 Asian Americans or Pacific Islanders), 52 international. In 1998, 19 master's, 8 doctorates awarded. *Degree requirements:* For master's, project or thesis required; for doctorate, one foreign language, dissertation required. *Entrance requirements:* For master's and doctorate, GRE General Test, TOEFL (minimum score of 525 required). *Average time to degree:* Master's–2.7 years full-time; doctorate–5.9 years full-time. *Application deadline:* For fall admission, 2/1 (priority date). Application fee: $40. *Financial aid:* Fellowships, tuition waivers (full) available. Aid available to part-time students. Financial aid application deadline: 2/1. *Faculty research:* Computational fluid mechanics/propulsion, large space structures, dynamics and guidance of VTOL vehicles. *Unit head:* AMES Graduate Student Affairs, 619-534-4387, Fax: 619-534-1730, E-mail: bwalton@ames.ucsd.edu. *Application contact:* Dr. Stan Rubin, Director of Graduate Studies, 513-556-3711, Fax: 513-556-5038, E-mail: srubin@uceng.uc.edu.

University of Colorado at Boulder, Graduate School, College of Engineering and Applied Science, Department of Aerospace Engineering Sciences, Boulder, CO 80309. Offers ME, MS, PhD. *Degree requirements:* For master's, thesis or alternative, comprehensive exam required, foreign language not required; for doctorate, dissertation required, foreign language not required. *Entrance requirements:* For master's, GRE General Test (minimum combined score of 1700 on three sections required), minimum undergraduate GPA of 3.25. *Faculty research:* Systems and control, fluids, bioengineering, structures, astrodynamics.

University of Colorado at Colorado Springs, Graduate School, College of Engineering and Applied Science, Program in Aerospace Engineering, Colorado Springs, CO 80933-7150. Offers ME. Part-time and evening/weekend programs available. *Faculty:* 2 full-time (0 women), 15 part-time (1 woman). *Students:* 43 full-time (6 women), 26 part-time (4 women); includes 8 minority (2 African Americans, 4 Asian Americans or Pacific Islanders, 2 Hispanic Americans), 3 international. Average age 29. 43 applicants, 100% accepted. In 1998, 44 degrees awarded. *Degree requirements:* For master's, computer language required, thesis optional, foreign language not required. *Entrance requirements:* For master's, GRE General Test (minimum combined score of 1200 required), TOEFL (minimum score of 550 required), bachelor's degree in engineering or related degree, minimum GPA of 3.0. *Average time to degree:* Master's–2 years full-time, 4 years part-time. *Application deadline:* For fall admission, 7/15; for spring admission, 12/10. Applications are processed on a rolling basis. Application fee: $40 ($50 for international students). Tuition, state resident: full-time $2,768; part-time $118 per credit. Tuition, nonresident: full-time $10,392; part-time $425 per credit. Required fees: $265; $7.5 per credit. One-time fee: $28. Tuition and fees vary according to program and student level. *Financial aid:* In 1998–99, 2 students received aid. Application deadline: 5/1. *Faculty research:* Neural networks, artificial intelligence, robust control, space operations, space propulsion. Total annual research expenditures: $40,000. *Unit head:* Dr. Charles E. Fosha, Director, 719-262-3573, Fax: 719-262-3589, E-mail: cfosha@mail.uccs.edu. *Application contact:* Karen Clevenger, Academic Adviser, 719-262-3243, Fax: 719-262-3589, E-mail: kcleveng@mail.uccs.edu.

University of Connecticut, Graduate School, School of Engineering, Department of Mechanical Engineering, Storrs, CT 06269. Offers aerospace engineering (MS, PhD); biomedical engineering (MS, PhD); mechanical engineering (MS, PhD); ocean engineering (MS, PhD). Terminal master's awarded for partial completion of doctoral program. *Degree requirements:* For master's, thesis or alternative required; for doctorate, dissertation required. *Entrance requirements:* For master's and doctorate, GRE General Test, GRE Subject Test. *Faculty research:* Design, applied mechanics, dynamics and control, energy and thermal sciences, manufacturing.

University of Dayton, Graduate School, School of Engineering, Department of Mechanical and Aerospace Engineering, Dayton, OH 45469-1300. Offers aerospace engineering (MSAE, DE, PhD); mechanical engineering (MSME, DE, PhD). Part-time programs available. *Faculty:* 20 full-time (0 women), 7 part-time (0 women). *Students:* 34 full-time (2 women), 27 part-time (3 women); includes 9 minority (3 African Americans, 3 Asian Americans or Pacific Islanders, 3 Hispanic Americans), 12 international. Average age 26. In 1998, 15 master's, 5 doctorates awarded. *Degree requirements:* For master's, foreign language and thesis not required; for doctorate, dissertation, departmental qualifying exam required. *Entrance requirements:* For master's, TOEFL. *Application deadline:* For fall admission, 8/1 (priority date). Applications are processed on a rolling basis. Application fee: $30. *Financial aid:* In 1998–99, 18 students received aid, including 1 fellowship with full tuition reimbursement available (averaging $18,000 per year), 15 research assistantships with full tuition reimbursements available (averaging $13,500 per year), 2 teaching assistantships with full tuition reimbursements available (averaging $9,000 per year); institutionally-sponsored loans and tuition waivers (full and partial) also available. *Faculty research:* Turbine blade convection, jet engine combustion, energy storage, heat pipes surface transfer, surface coating friction and wear. Total annual research expenditures: $400,000. *Unit head:* Dr. Glen E. Johnson, Chairperson, 937-229-2835, Fax: 937-229-2756, E-mail: gjohnson@engr.udayton.edu. *Application contact:* Dr. Donald L. Moon, Associate Dean, 937-229-2241, Fax: 937-229-2471, E-mail: dmoon@engr.udayton.edu.

University of Florida, Graduate School, College of Engineering, Department of Aerospace Engineering, Mechanics, and Engineering Science, Program in Aerospace Engineering, Gainesville, FL 32611. Offers ME, MS, PhD, Certificate, Engr. *Students:* 38 full-time (4 women), 11 part-time (4 women); includes 3 minority (1 African American, 1 Asian American or Pacific Islander, 1 Hispanic American), 24 international. In 1998, 9 master's, 1 doctorate awarded. *Degree requirements:* For master's and other advanced degree, thesis optional; for doctorate, dissertation required. *Entrance requirements:* For master's and doctorate, GRE General Test, TOEFL, minimum GPA of 3.0; for other advanced degree, GRE General Test. *Application deadline:* For fall admission, 6/1 (priority date). Applications are processed on a rolling basis. Application fee: $20. Electronic applications accepted. *Financial aid:* Fellowships available. *Application contact:* Dr. Chen-Chi Hsu, Graduate Coordinator, 352-392-9823, Fax: 352-392-7303, E-mail: cch@aero.ufl.edu.

University of Florida, Graduate School, Graduate Engineering and Research Center (GERC), Gainesville, FL 32611. Offers aerospace engineering (ME, MS, PhD, Engr); electrical and computer engineering (ME, MS, PhD, Engr); engineering mechanics (ME, MS, PhD, Engr); industrial and systems engineering (ME, MS, PhD, Engr). Part-time programs available. Postbaccalaureate distance learning degree programs offered. *Faculty:* 6 full-time (0 women), 16 part-time (1 woman). *Students:* 13 full-time (6 women), 159 part-time (29 women); includes 21 minority (8 African Americans, 8 Asian Americans or Pacific Islanders, 4 Hispanic Americans, 1 Native American) Terminal master's awarded for partial completion of doctoral program. *Degree requirements:* For master's, computer language required (for some programs), thesis optional, foreign language not required; for doctorate, computer language (for some programs), dissertation required; for Engr, computer language (for some programs), thesis required, foreign language not required. *Entrance requirements:* For master's, GRE General Test (minimum combined score of 1000 required), TOEFL, minimum GPA of 3.0; for doctorate, GRE General Test (minimum combined score of 1200 required), written and oral qualifying exams, TOEFL, minimum GPA of 3.0, master's degree in engineering; for Engr, GRE General Test (minimum combined score of 1000 required), TOEFL, minimum GPA of 3.0, master's degree in engineering. *Application deadline:* For fall admission, 6/1; for spring admission, 10/1. Applications are processed on a rolling basis. Application fee: $20. Electronic applications accepted. *Unit head:* Dr. Pasquale M. Sforza, Director, 850-833-9355, Fax: 850-833-9366, E-mail: sforza@gerc.eng.ufl.edu. *Application contact:* Judi Shivers, Program Assistant, 850-833-9350, Fax: 850-833-9366, E-mail: reginfo@gerc.eng.ufl.edu.

University of Houston, Cullen College of Engineering, Program in Aerospace Engineering, Houston, TX 77004. Offers MSAER, PhD. Part-time and evening/weekend programs available. *Students:* 2 full-time (0 women), 8 part-time (2 women); includes 1 minority (Hispanic American), 2 international. Average age 30. 5 applicants, 80% accepted. In 1998, 2 master's awarded. Terminal master's awarded for partial completion of doctoral program. *Degree requirements:* For master's, thesis required, foreign language not required; for doctorate, dissertation, departmental qualifying exam required, foreign language not required. *Entrance requirements:* For master's and doctorate, GRE General Test, TOEFL. *Application deadline:* For fall admission, 7/3 (priority date); for spring admission, 12/4. Applications are processed on a rolling basis. Application fee: $25 ($75 for international students). *Financial aid:* Research assistantships, teaching assistantships, career-related internships or fieldwork and Federal Work-Study available. Financial aid application deadline: 2/15. *Faculty research:* Smart structures, composites, polymers, fracture, fluid mechanics. *Unit head:* Dr. K. Ravi-Chandar, Director, 713-743-4500, Fax: 713-743-4503, E-mail: ravi@uh.edu. *Application contact:* Sheena Paul, Graduate Admissions Analyst, 713-743-4505, Fax: 713-743-4503, E-mail: megrad@mail.me.uh.edu.

University of Illinois at Urbana–Champaign, Graduate College, College of Engineering, Department of Aeronautical and Astronautical Engineering, Urbana, IL 61801. Offers MS, PhD. *Faculty:* 19 full-time (1 woman). *Students:* 71 full-time (9 women); includes 8 minority (5 Asian Americans or Pacific Islanders, 3 Hispanic Americans), 29 international. 113 applicants, 19% accepted. In 1998, 12 master's, 9 doctorates awarded. *Degree requirements:* For master's and doctorate, thesis/dissertation required, foreign language not required. *Entrance requirements:* For master's, GRE General Test, TOEFL (minimum score of 607 required; 253 for computer-based); for doctorate, GRE General Test, TOEFL (minimum score of 607 required, 253 for computer-based). *Application deadline:* For fall admission, 2/20; for spring admission, 11/5. Applications are processed on a rolling basis. Application fee: $40 ($50 for international students). Tuition, state resident: full-time $4,616. Tuition, nonresident: full-time $11,768. Full-time tuition and fees vary according to course load. *Financial aid:* In 1998–99, 5 fellowships, 34 research assistantships, 13 teaching assistantships were awarded. Financial aid application deadline: 2/20. *Unit head:* Dr. Wayne C. Solomon, Head, 217-333-2651. *Application contact:* Dr. Michael B. Bragg, Director of Graduate Studies, 217-333-2651, Fax: 217-244-0720, E-mail: mbragg@uiuc.edu.

See in-depth description on page 349.

University of Kansas, Graduate School, School of Engineering, Department of Aerospace Engineering, Lawrence, KS 66045. Offers ME, MS, DE, PhD. *Faculty:* 8 full-time. *Students:* 7 full-time (1 woman), 15 part-time (1 woman), 13 international. In 1998, 6 master's, 2 doctorates awarded. *Degree requirements:* For master's, thesis or alternative, exam required, foreign language not required; for doctorate, dissertation, comprehensive exam required. *Entrance requirements:* For master's and doctorate, Michigan English Language Assessment Battery, TOEFL, minimum GPA of 3.0. *Application deadline:* For fall admission, 7/1. Application fee: $30 ($45 for international students). *Financial aid:* Fellowships, research assistantships, teaching assistantships, career-related internships or fieldwork available. *Faculty research:* Control systems, aerodynamics, propulsion. *Unit head:* David Downing, Chair, 785-864-4267. *Application contact:* Jan Roskam, Graduate Director.

University of Maryland, College Park, Graduate School, A. James Clark School of Engineering, Department of Aerospace Engineering, College Park, MD 20742-5045. Offers M Eng, MS, PhD. Part-time and evening/weekend programs available. Postbaccalaureate distance learning degree programs offered. *Faculty:* 26 full-time (0 women), 12 part-time (0 women). *Students:* 70 full-time (12 women), 40 part-time (6 women); includes 16 minority (6 African Americans, 8 Asian Americans or Pacific Islanders, 2 Hispanic Americans), 37 international. 131 applicants, 35% accepted. In 1998, 15 master's, 4 doctorates awarded. *Degree requirements:* For master's, thesis optional, foreign language not required; for doctorate, dissertation required. *Entrance requirements:* For master's, minimum GPA of 3.2. *Application deadline:* Applications are processed on a rolling basis. Application fee: $50 ($70 for international students). Tuition, state resident: part-time $272 per credit hour. Tuition, nonresident: part-time $475 per credit hour. Required fees: $632; $379 per year. *Financial aid:* In 1998–99, 4 fellowships with full tuition reimbursements (averaging $18,000 per year), 64 research assistantships with tuition reimbursements (averaging $15,377 per year), 10 teaching assistantships with tuition reimbursements (averaging $10,526 per year) were awarded.; Federal Work-Study, grants, and scholarships also available. Aid available to part-time students. Financial aid applicants required to submit FAFSA. *Faculty research:* Aerodynamics and propulsion, structural mechanics, flight dynamics, rotor craft, space robotics. Total annual research expenditures: $5.4 million. *Unit head:* Dr. William Fourney, Chairman, 301-405-1129, Fax: 301-314-9001. *Application contact:* Trudy Lindsey, Director, Graduate Admission and Records, 301-405-4198, Fax: 301-314-9305, E-mail: grschool@deans.umd.edu.

University of Maryland, College Park, Graduate School, A. James Clark School of Engineering, Professional Program in Engineering, College Park, MD 20742-5045. Offers aerospace engineering (M Eng); chemical engineering (M Eng); civil engineering (M Eng); electrical engineering (M Eng); fire protection engineering (M Eng); materials science and engineering (M Eng); mechanical engineering (M Eng); reliability engineering (M Eng); systems engineering (M Eng). Part-time and evening/weekend programs available. Postbaccalaureate distance learning degree programs offered. *Faculty:* 11 part-time (0 women). *Students:* 20 full-time (3 women), 205 part-time (42 women); includes 58 minority (27 African Americans, 25 Asian Americans or Pacific Islanders, 5 Hispanic Americans, 1 Native American), 20 international. *Degree requirements:* For master's, foreign language and thesis not required. *Application deadline:* Applications are processed on a rolling basis. Application fee: $50 ($70 for international students). Tuition, state resident: part-time $272 per credit hour. Tuition, nonresident: part-time $475 per credit hour. Required fees: $632; $379 per year. *Unit head:* Dr. Patrick Cunniff, Associate Dean, 301-405-5256, Fax: 301-314-9477. *Application contact:* Trudy Lindsey, Director, Graduate Admission and Records, 301-405-4198, Fax: 301-314-9305, E-mail: grschool@deans.umd.edu.

University of Michigan, Horace H. Rackham School of Graduate Studies, College of Engineering, Department of Aerospace Engineering, Ann Arbor, MI 48109. Offers M Eng, MS, MSE, PhD, Aerospace E. Part-time programs available. Terminal master's awarded for partial completion of doctoral program. *Degree requirements:* For master's, thesis not required; for doctorate, dissertation, oral defense of dissertation, preliminary exams required. *Entrance requirements:* For master's, GRE General Test; for doctorate, GRE General Test, master's degree.

See in-depth description on page 351.

University of Michigan, Horace H. Rackham School of Graduate Studies, College of Engineering, Department of Atmospheric, Oceanic, and Space Sciences, Ann Arbor, MI 48109. Offers atmospheric and space sciences (M Eng, MS, PhD); oceanography: physical (MS, PhD); remote sensing and geoinformation (M Eng); space and planetary physics (PhD); space systems (M Eng). Part-time programs available. *Faculty:* 21 full-time (2 women), 9 part-time (0 women). *Students:* 61 full-time (26 women), 1 part-time; includes 2 minority (both African Americans), 19 international. Terminal master's awarded for partial completion of doctoral program. *Degree requirements:* For master's, thesis required (for some programs); for doctorate, dissertation, oral defense of dissertation, preliminary exams required. *Entrance requirements:* For master's and doctorate, GRE General Test (combined average 2000 on three sections), TOEFL (minimum score of 600 required). *Application deadline:* For fall admission, 1/15 (priority date). Applications are processed on a rolling basis. Application fee: $60. *Unit head:* Lennard Fisk, Chair, 734-764-3335, Fax: 734-764-4585, E-mail: lafisk@umich.edu. *Application contact:* Susan Schreiber, Academic Services Assistant, 734-764-3336, Fax: 734-764-4585, E-mail: aoss.um@umich.edu.

University of Minnesota, Twin Cities Campus, Graduate School, Institute of Technology, Department of Aerospace Engineering and Mechanics, Minneapolis, MN 55455-0213. Offers aerospace engineering (M Aero E, MS, PhD); mechanics (MS, PhD). Part-time programs available. *Faculty:* 18 full-time (2 women), 2 part-time (0 women). *Students:* 56 full-time (9 women); includes 15 minority (all Asian Americans or Pacific Islanders), 19 international. Average age 24. 109 applicants, 41% accepted. In 1998, 9 master's awarded (44% found work related to degree, 56% continued full-time study); 8 doctorates awarded (13% entered university research/teaching, 87% found other work related to degree). Terminal master's awarded for partial completion of doctoral program. *Degree requirements:* For master's, foreign language and thesis not required; for doctorate, dissertation required, foreign language not required. *Average time to degree:* Master's–2 years full-time, 5 years part-time; doctorate–5 years full-time. *Application deadline:* For fall admission, 7/15. Applications are processed on a rolling basis. Application fee: $40 ($50 for international students). *Financial aid:* In 1998–99, 27 research assistantships with full tuition reimbursements, 24 teaching assistantships with full tuition reimbursements were awarded. Financial aid application deadline: 2/1. *Faculty research:* Fluid mechanics, solid and continuum mechanics, dynamical systems and control therapy. Total annual research expenditures: $4.4 million. *Unit head:* William L. Garrard, Head, 612-625-8000, Fax: 612-626-1558, E-mail: dept@aem.umn.edu. *Application contact:* Ellen K. Longmire, Director of Graduate Studies, 612-625-8000, Fax: 612-626-1558, E-mail: dgs@aem.umn.edu.

See in-depth description on page 353.

University of Missouri–Columbia, Graduate School, College of Engineering, Department of Mechanical and Aerospace Engineering, Columbia, MO 65211. Offers MS, PhD. *Faculty:* 21 full-time (1 woman). *Students:* 21 full-time (1 woman), 17 part-time (1 woman); includes 4 minority (2 African Americans, 1 Asian American or Pacific Islander, 1 Native American), 19 international. 23 applicants, 52% accepted. In 1998, 18 master's, 3 doctorates awarded. *Degree requirements:* For master's, thesis required, foreign language not required; for doctorate, dissertation required. *Entrance requirements:* For master's and doctorate, GRE General Test, TOEFL (minimum score of 550 required), minimum GPA of 3.0. *Application deadline:* Applications are processed on a rolling basis. Application fee: $30 ($50 for international students). *Financial aid:* Research assistantships, teaching assistantships, institutionally-sponsored loans available. *Unit head:* Dr. Uee Wan Cho, Director of Graduate Studies, 573-882-3778.

University of Missouri–Rolla, Graduate School, School of Engineering, Department of Mechanical and Aerospace Engineering and Engineering Mechanics, Program in Aerospace Engineering, Rolla, MO 65409-0910. Offers MS, PhD. Part-time programs available. *Faculty:* 6 full-time (0 women). *Students:* 11 full-time (1 woman), 2 part-time (1 woman), 5 international. Average age 27. 21 applicants, 71% accepted. In 1998, 2 master's awarded (100% found work related to degree); 1 doctorate awarded. *Degree requirements:* For master's, thesis required (for some programs), foreign language not required; for doctorate, dissertation required, foreign language not required. *Entrance requirements:* For master's, GRE General Test (minimum combined score of 1150 required), TOEFL (minimum score of 570 required), minimum GPA of 3.0; for doctorate, GRE General Test (minimum combined score of 1150 required), TOEFL (minimum score of 570 required), minimum GPA of 3.5. *Average time to degree:* Master's–1.9 years full-time, 6 years part-time; doctorate–3.4 years full-time. *Application deadline:* For fall admission, 7/1; for spring admission, 12/1. Applications are processed on a rolling basis. Application fee: $25. Electronic applications accepted. *Financial aid:* In 1998–99, 3 fellowships with full and partial tuition reimbursements (averaging $9,739 per year), 6 research assistantships with full and partial tuition reimbursements (averaging $12,985 per year), 1 teaching assistantship with full and partial tuition reimbursement (averaging $9,774 per year) were awarded. Financial aid application deadline: 3/1. *Faculty research:* Aerodynamics, stability and control, fluid dynamics and propulsion, acoustics, radiative transfer. *Unit head:* Dr. Michael B. Bragg, Director of Graduate Studies, 217-333-2651, Fax: 217-244-0720, E-mail: mbragg@uiuc.edu. *Application contact:* Dr. James A. Drallmeier, Associate Professor, Chair for Graduate Studies, 573-341-4710, Fax: 573-341-4607, E-mail: grad-students@gearbox.maem.umr.edu.

University of New Haven, Graduate School, School of Public Safety and Professional Studies, Program in Aviation Science, West Haven, CT 06516-1916. Offers MS. *Degree requirements:* For master's, thesis or alternative required, foreign language not required. Application fee: $50. *Financial aid:* Application deadline: 5/1. *Unit head:* John Kelly, Director, 203-932-7472.

University of Notre Dame, Graduate School, College of Engineering, Department of Aerospace and Mechanical Engineering, Program in Aerospace Engineering, Notre Dame, IN 46556. Offers MS, PhD. Part-time programs available. *Students:* 25 full-time (2 women), 2 part-time, 18 international. 43 applicants, 56% accepted. In 1998, 1 master's, 3 doctorates awarded. Terminal master's awarded for partial completion of doctoral program. *Degree requirements:* For master's, thesis or alternative required, foreign language not required; for doctorate, dissertation required, foreign language not required. *Entrance requirements:* For master's and doctorate, GRE General Test, TOEFL (minimum score of 600 required; 250 for computer-based). *Average time to degree:* Master's–2 years full-time; doctorate–5 years full-time. *Application deadline:* For fall admission, 2/1 (priority date); for spring admission, 10/15. Applications are processed on a rolling basis. Application fee: $40. *Financial aid:* In 1998–99, fellowships with full tuition reimbursements (averaging $16,000 per year), research assistantships with full tuition reimbursements (averaging $11,500 per year), teaching assistantships with full tuition reimbursements (averaging $11,500 per year) were awarded.; tuition waivers (full) and unspecified assistantships also available. Financial aid application deadline: 2/1. *Faculty research:* Acoustics, structural design, particle dynamics, aerodynamics, compressible flows. *Unit head:* Margaret A. Fillion, Student Services Assistant, 734-764-3311, Fax: 734-763-0578, E-mail: mafn@engin.umich.edu. *Application contact:* Dr. Terrence J. Akai, Director of Graduate Admissions, 219-631-7706, Fax: 219-631-4183, E-mail: gradad@nd.edu.

See in-depth description on page 355.

University of Oklahoma, Graduate College, College of Engineering, School of Aerospace and Mechanical Engineering, Program in Aerospace Engineering, Norman, OK 73019-0390. Offers MS, PhD. *Students:* 8 full-time (0 women), 4 part-time; includes 3 minority (2 Hispanic Americans, 1 Native American), 7 international. Average age 28. 12 applicants, 92% accepted. In 1998, 6 master's awarded. *Degree requirements:* For master's, thesis or alternative, comprehensive exam required, foreign language not required; for doctorate, dissertation, comprehensive exam, qualifying exam required, foreign language not required. *Entrance requirements:* For master's, GRE General Test, TOEFL (minimum score of 550 required), BS in engineering or physical sciences; for doctorate, GRE General Test, TOEFL (minimum score of 550 required), MS in aerospace engineering or equivalent. *Application deadline:* For fall admission, 6/1 (priority date). Applications are processed on a rolling basis. Application fee: $25. Tuition, state resident: part-time $86 per credit hour. Tuition, nonresident: part-time $275 per credit hour. Tuition and fees vary according to course level, course load and program. *Financial aid:* Fellowships, research assistantships, teaching assistantships, Federal Work-Study, institutionally-sponsored loans, and tuition waivers (partial) available. Financial aid application deadline: 3/1. *Faculty research:* Aerodynamics, aerospace structures and propulsion, composite materials, aircraft design. *Unit head:* Jan Roskam, Graduate Director. *Application contact:* Jan Roskam, Graduate Director.

See in-depth description on page 357.

University of Ottawa, School of Graduate Studies and Research, Faculty of Engineering, Ottawa-Carleton Institute for Mechanical and Aerospace Engineering, Ottawa, ON K1N 6N5, Canada. Offers M Eng, MA Sc, PhD. *Faculty:* 57 full-time, 3 part-time. *Students:* 113 full-time (9 women), 20 part-time (2 women), 24 international. Average age 31. In 1998, 30 master's, 6 doctorates awarded. *Degree requirements:* For master's, thesis or alternative required, foreign language not required; for doctorate, dissertation required, foreign language not required. *Entrance requirements:* For master's, honors degree or equivalent, minimum B average; for doctorate, master's degree, minimum B+ average. *Application deadline:* For fall admission, 3/1. Application fee: $35. *Financial aid:* Fellowships, research assistantships, teaching assistantships, Federal Work-Study available. *Faculty research:* Fluid mechanics, heat transfer, solid mechanics, design, manufacturing. *Unit head:* David Redekor, Director, 613-562-5800 Ext. 6290, Fax: 613-562-5177. *Application contact:* Solange Lamontagne, Academic Assistant, 613-562-5834, Fax: 613-562-5177, E-mail: gradinfo@eng.uottawa.ca.

University of Southern California, Graduate School, School of Engineering, Department of Aerospace Engineering, Los Angeles, CA 90089. Offers MS, PhD, Engr. Part-time programs available. *Faculty:* 13 full-time (0 women), 5 part-time (0 women). *Students:* 32 full-time (3 women), 45 part-time (4 women); includes 17 minority (14 Asian Americans or Pacific Islanders, 3 Hispanic Americans), 24 international. Average age 30. 88 applicants, 65% accepted. In 1998, 10 master's, 12 doctorates awarded. *Degree requirements:* For master's, thesis optional; for doctorate, dissertation required. *Entrance requirements:* For master's, 0RE General Test, GRE Subject Test; for doctorate and Engr, GRE General Test, GRE Subject Test. *Application deadline:* For fall admission, 2/15 (priority date); for spring admission, 11/1. Application fee: $55. Tuition: Part-time $768 per unit. Required fees: $350 per semester. *Financial aid:* In 1998–99, 2 fellowships, 8 research assistantships, 9 teaching assistantships were awarded.; Federal Work-Study, institutionally-sponsored loans, and scholarships also available. Aid available to part-time students. Financial aid application deadline: 2/15; financial aid applicants required to submit FAFSA. *Unit head:* Dr. Richard Kaplan, Chairman, 213-740-5353.

See in-depth description on page 359.

University of Tennessee, Knoxville, Graduate School, College of Engineering, Department of Mechanical and Aerospace Engineering and Engineering Science, Program in Aerospace Engineering, Knoxville, TN 37996. Offers MS, PhD. Part-time programs available. *Students:* 16 full-time (3 women), 14 part-time; includes 1 minority (African American), 7 international. 36 applicants, 69% accepted. In 1998, 5 master's awarded. *Degree requirements:* For master's, thesis or alternative required, foreign language not required; for doctorate, dissertation required, foreign language not required. *Entrance requirements:* For master's and doctorate, TOEFL (minimum score of 550 required), minimum GPA of 2.7. *Application deadline:* For fall admission, 2/1 (priority date). Applications are processed on a rolling basis. Application fee: $35. Electronic applications accepted. *Financial aid:* Application deadline: 2/1; *Application contact:* Dr. Allen Yu, Graduate Representative, 923-974-4159, E-mail: nyu@utk.edu.

University of Tennessee, Knoxville, Graduate School, Intercollegiate Programs, Program in Aviation Systems, Knoxville, TN 37996. Offers MS. Part-time programs available. Postbaccalaureate distance learning degree programs offered (no on-campus study). *Students:* 7 full-time (0 women), 42 part-time (1 woman), 3 international. 27 applicants, 19% accepted. In 1998, 17 degrees awarded. *Degree requirements:* For master's, thesis optional, foreign language not required. *Entrance requirements:* For master's, TOEFL (minimum score of 550 required), minimum GPA of 2.7. *Application deadline:* For fall admission, 2/1 (priority date). Applications are processed on a rolling basis. Application fee: $35. Electronic applications accepted. *Financial aid:* Application deadline: 2/1; *Unit head:* Dr. Ralph Kimberlin, Head, 931-393-7411, Fax: 931-393-7409.

University of Tennessee Space Institute, Graduate Programs, Program in Aerospace Engineering, Tullahoma, TN 37388-9700. Offers MS, PhD. Part-time programs available. *Faculty:* 8 full-time (0 women). *Students:* 10 full-time (3 women), 13 part-time; includes 1 minority (African American), 4 international. 28 applicants, 79% accepted. In 1998, 3 master's, 2 doctorates awarded. *Degree requirements:* For master's, thesis required (for some programs), foreign language not required; for doctorate, dissertation required. *Entrance requirements:* For master's and doctorate, GRE General Test. *Application deadline:* Applications are processed on a rolling basis. Application fee: $35. *Financial aid:* Fellowships, research assistantships, career-related internships or fieldwork, Federal Work-Study, and tuition waivers (full and partial) available. Financial aid applicants required to submit FAFSA. *Unit head:* Dr. Ahmad Vakili, Degree Program Chairman, 931-393-7483, Fax: 931-393-7530, E-mail: avakili@utsi.edu. *Application contact:* Dr. Edwin M. Gleason, Assistant Dean for Admissions and Student Affairs, 931-393-7432, Fax: 931-393-7346, E-mail: egleason@utsi.edu.

Aerospace/Aeronautical Engineering–Cross-Discipline Announcement

The University of Texas at Arlington, Graduate School, College of Engineering, Department of Mechanical and Aerospace Engineering, Program in Aerospace Engineering, Arlington, TX 76019. Offers M Engr, MS, PhD. *Students:* 12 full-time (0 women), 13 part-time (1 woman); includes 1 minority (Asian American or Pacific Islander), 8 international. 17 applicants, 29% accepted. In 1998, 2 master's, 3 doctorates awarded. *Degree requirements:* For master's and doctorate, thesis/dissertation required, foreign language not required. *Entrance requirements:* For master's, GRE General Test (minimum combined score of 1000 required), TOEFL (minimum score of 550 required); for doctorate, GRE General Test (minimum combined score of 1250 required), TOEFL (minimum score of 550 required). *Application deadline:* Applications are processed on a rolling basis. Application fee: $25 ($50 for international students). Tuition, state resident: full-time $1,368; part-time $76 per semester hour. Tuition, nonresident: full-time $5,454; part-time $303 per semester hour. Required fees: $66 per semester hour. $86 per term. Tuition and fees vary according to course load. *Financial aid:* Research assistantships, teaching assistantships available. *Unit head:* Jan Roskam, Graduate Director. *Application contact:* Dr. Frank K. Lu, Graduate Adviser, 817-272-2603, Fax: 817-272-2538, E-mail: lu@mae.uta.edu.

The University of Texas at Austin, Graduate School, College of Engineering, Department of Aerospace Engineering and Engineering Mechanics, Program in Aerospace Engineering, Austin, TX 78712-1111. Offers MSE, PhD. *Students:* 107 (13 women); includes 7 minority (3 Asian Americans or Pacific Islanders, 3 Hispanic Americans, 1 Native American) 40 international. 102 applicants, 20% accepted. In 1998, 14 master's, 15 doctorates awarded. *Entrance requirements:* For master's and doctorate, GRE General Test (minimum combined score of 1000 required). *Application deadline:* For fall admission, 2/1 (priority date); for spring admission, 10/1. Applications are processed on a rolling basis. Application fee: $50 ($75 for international students). Electronic applications accepted. *Financial aid:* Fellowships, research assistantships, teaching assistantships available. Financial aid application deadline: 2/1. *Unit head:* Dr. Frank K. Lu, Graduate Adviser, 817-272-2603, Fax: 817-272-2538, E-mail: lu@mae.uta.edu. *Application contact:* Graduate Coordinator, 512-471-7595, E-mail: gradprog@orion.ae.utexas.edu.

See in-depth description on page 361.

University of Toronto, School of Graduate Studies, Physical Sciences Division, Faculty of Applied Science and Engineering, Institute for Aerospace Science and Engineering, Toronto, ON M5S 1A1, Canada. Offers M Eng, MA Sc, PhD. Part-time programs available. *Degree requirements:* For master's, thesis required (for some programs); for doctorate, dissertation required.

University of Virginia, School of Engineering and Applied Science, Department of Mechanical and Aerospace Engineering, Charlottesville, VA 22903. Offers ME, MS, PhD. *Faculty:* 30 full-time (2 women). *Students:* 65 full-time (4 women), 81 part-time (8 women); includes 8 minority (1 African American, 5 Asian Americans or Pacific Islanders, 2 Hispanic Americans), 17 international. Average age 27. 76 applicants, 18% accepted. In 1998, 25 master's, 4 doctorates awarded. *Degree requirements:* For master's, thesis (MS) required; for doctorate, dissertation, comprehensive exam required, foreign language not required. *Entrance requirements:* For master's, GRE General Test; for doctorate, GRE General Test (minimum score of 700 on quantitative section required), TOEFL (minimum score of 600 required). *Application deadline:* For fall admission, 8/1 (priority date). Applications are processed on a rolling basis. Application fee: $60. *Financial aid:* Fellowships, research assistantships, teaching assistantships available. Financial aid application deadline: 2/1. *Unit head:* Dr. Jeffrey Morton, Graduate Director, 804-924-6224, E-mail: jbm@virginia.edu. *Application contact:* J. Milton Adams, Assistant Dean, 804-924-3897, E-mail: twr2c@virginia.edu.

University of Washington, Graduate School, College of Engineering, Department of Aeronautics and Astronautics, Seattle, WA 98195. Offers MAE, MSAA, PhD. Part-time programs available. Postbaccalaureate distance learning degree programs offered (minimal on-campus study). *Faculty:* 18 full-time. *Students:* 57 full-time (5 women), 18 part-time (3 women). Average age 25. 95 applicants, 55% accepted. In 1998, 17 master's awarded (76% found work related to degree, 24% continued full-time study); 5 doctorates awarded (100% found work related to degree). *Degree requirements:* For master's, thesis optional, foreign language not required; for doctorate, dissertation required, foreign language not required. *Entrance requirements:* For master's, GRE General Test (minimum score of 450 on verbal section, 700 on quantitative, 650 on analytical required), TOEFL (minimum score of 580 required), minimum GPA of 3.0; for doctorate, GRE General Test (minimum score of 450 on verbal section, 700 on quantitative, 650 on analytical required), TOEFL (minimum score of 580 required), minimum GPA of 3.35. *Average time to degree:* Master's–1.5 years full-time, 3 years part-time; doctorate–5.25 years full-time, 8 years part-time. *Application deadline:* For fall admission, 7/1; for winter admission, 11/1; for spring admission, 2/1. Applications are processed on a rolling basis. Application fee: $50. Electronic applications accepted. Tuition, state resident: full-time $5,196; part-time $475 per credit. Tuition, nonresident: full-time $13,485; part-time $1,285 per credit. Required fees: $387; $38 per credit. Tuition and fees vary according to course load. *Financial aid:* In 1998–99, 12 fellowships (averaging $2,292 per year), 25 research assistantships with full tuition reimbursements (averaging $1,135 per year), 21 teaching assistantships with full tuition reimbursements (averaging $1,123 per year) were awarded.; Federal Work-Study and tuition waivers (full) also available. Financial aid application deadline: 2/1. *Faculty research:* Space system design, aircraft systems, energy systems, composites/structures, fluid dynamics. Total annual research expenditures: $4 million. *Unit head:* Dr. Adam P. Bruckner, Chair, 206-543-1950, Fax: 206-543-0217, E-mail: bruckner@aa.washington.edu. *Application contact:* Wanda Frederick, Graduate Coordinator, 206-616-1113, Fax: 206-543-0217, E-mail: wanda@aa.washington.edu.

Utah State University, School of Graduate Studies, College of Engineering, Department of Mechanical and Aerospace Engineering, Logan, UT 84322. Offers aerospace engineering (MS, PhD); mechanical engineering (ME, MS, PhD). *Faculty:* 14 full-time (1 woman). *Students:* 31 full-time (2 women), 7 part-time; includes 1 minority (Asian American or Pacific Islander), 12 international. Average age 27. 74 applicants, 58% accepted. In 1998, 13 master's, 2 doctorates awarded. Terminal master's awarded for partial completion of doctoral program. *Degree requirements:* For master's, computer language, thesis required (for some programs), foreign language not required; for doctorate, computer language, dissertation required, foreign language not required. *Entrance requirements:* For master's, GRE General Test (score in 40th percentile or higher required), TOEFL (minimum score of 550 required), minimum GPA of 3.0; for doctorate, GRE General Test (score in 40th percentile or higher required), GRE Subject Test, TOEFL (minimum score of 550 required), minimum GPA of 3.0. *Application deadline:* For fall admission, 3/15 (priority date); for spring admission, 10/15. Applications are processed on a rolling basis. Application fee: $40. Tuition, state resident: full-time $1,492. Tuition, nonresident: full-time $5,232. Required fees: $434. Tuition and fees vary according to course load. *Financial aid:* In 1998–99, 20 students received aid, including 14 research assistantships with partial tuition reimbursements available (averaging $12,000 per year), 5 teaching assistantships with partial tuition reimbursements available (averaging $9,000 per year); fellowships with partial tuition reimbursements available, Federal Work-Study and institutionally-sponsored loans also available. Financial aid application deadline: 3/15. *Faculty research:* In-space instruments, cryogenic cooling, thermal science, space structures, composite materials. *Unit head:* J. Clair Batty, Head, 435-797-2868, Fax: 435-797-2417. *Application contact:* Joan P. Smith, Graduate Student Adviser, 435-797-0330, Fax: 435-797-2417, E-mail: jpsmith@mae.usu.edu.

Virginia Polytechnic Institute and State University, Graduate School, College of Engineering, Department of Aerospace and Ocean Engineering, Program in Aerospace Engineering, Blacksburg, VA 24061. Offers M Eng, MS, PhD. *Students:* 68 full-time (6 women), 21 part-time (2 women); includes 4 minority (1 African American, 1 Asian American or Pacific Islander, 2 Hispanic Americans), 41 international. 100 applicants, 44% accepted. In 1998, 11 master's, 10 doctorates awarded. *Degree requirements:* For master's, thesis required (for some programs), foreign language not required; for doctorate, dissertation required, foreign language not required. *Entrance requirements:* For master's and doctorate, GRE (non-native speakers only), TOEFL. *Application deadline:* For fall admission, 12/1 (priority date). Applications are processed on a rolling basis. Application fee: $25. *Financial aid:* Application deadline: 4/1. *Unit head:* Dr. Bernard Grossman, Head, Department of Aerospace and Ocean Engineering, 540-231-6611, E-mail: grossman@aoe.vt.edu.

Webster University, School of Business and Technology, Department of Business, St. Louis, MO 63119-3194. Offers business (MA, MBA); computer resources and information management (MA, MBA); computer science/distributed systems (MS); finance (MA, MBA); health care management (MA); health services management (MA, MBA); human resources development (MA, MBA); human resources management (MA); international business (MA, MBA); management (MA, MBA); marketing (MA, MBA); procurement and acquisitions management (MA, MBA); public administration (MA); real estate management (MA, MBA); security management (MA, MBA); space systems management (MA, MBA, MS); telecommunications management (MA, MBA). *Faculty:* 5 full-time (1 woman). *Students:* 3,474 full-time (1,390 women), 1,592 part-time (603 women); includes 1,465 minority (1,021 African Americans, 169 Asian Americans or Pacific Islanders, 255 Hispanic Americans, 20 Native Americans), 397 international. *Degree requirements:* For master's, foreign language and thesis not required. *Application deadline:* Applications are processed on a rolling basis. Application fee: $25 ($50 for international students). *Unit head:* Lucille Berry, Chair, 314-968-7022, Fax: 314-968-7077, E-mail: berrylm@webster.edu. *Application contact:* Dr. Beth Russell, Director of Graduate Admissions, 314-968-7089, Fax: 314-968-7166, E-mail: russelmb@webster.edu.

West Virginia University, College of Engineering and Mineral Resources, Department of Mechanical and Aerospace Engineering, Program in Aerospace Engineering, Morgantown, WV 26506. Offers aerospace engineering (MSAE); engineering (MSE, PhD). Part-time programs available. Terminal master's awarded for partial completion of doctoral program. *Degree requirements:* For master's, thesis required, foreign language not required; for doctorate, dissertation, comprehensive exam required, foreign language not required. *Entrance requirements:* For master's and doctorate, GRE General Test, TOEFL (minimum score of 550 required), minimum GPA of 3.0. *Faculty research:* Transonic aerodynamics, viscous/inviscid interactions, combustion, aerospace structures, space mechanics.

Wichita State University, Graduate School, College of Engineering, Department of Aerospace Engineering, Wichita, KS 67260. Offers MS, PhD. Part-time programs available. *Faculty:* 12 full-time (0 women), 5 part-time (1 woman). *Students:* 39 full-time (3 women), 49 part-time (8 women); includes 4 minority (3 Asian Americans or Pacific Islanders, 1 Native American), 47 international. Average age 33. 45 applicants, 67% accepted. In 1998, 20 master's, 2 doctorates awarded. Terminal master's awarded for partial completion of doctoral program. *Degree requirements:* For master's, oral or written exam required, thesis optional, foreign language not required; for doctorate, one foreign language (computer language can substitute), dissertation, comprehensive exam required. *Entrance requirements:* For master's and doctorate, GRE, TOEFL (minimum score of 550 required). *Application deadline:* For fall admission, 7/1 (priority date); for spring admission, 1/1. Applications are processed on a rolling basis. Application fee: $25 ($40 for international students). Electronic applications accepted. *Financial aid:* In 1998–99, 77 research assistantships (averaging $5,000 per year), 2 teaching assistantships with full tuition reimbursements (averaging $5,000 per year) were awarded.; fellowships, Federal Work-Study, institutionally-sponsored loans, and unspecified assistantships also available. Financial aid application deadline: 4/1. *Faculty research:* Composite materials and structures, electro-impulse de-icing, computational fluid dynamics, stall-spin aerodynamics and simulation, water droplet trajectories on and around aircraft surfaces. Total annual research expenditures: $5.4 million. *Unit head:* Dr. Walter Horn, Chairperson, 316-978-3410, Fax: 316-978-3307. *Application contact:* Dr. Klaus A. Hoffmann, Graduate Coordinator, 316-978-3410, Fax: 316-978-3307, E-mail: hoffmann@twsuvm.uc.twsu.edu.

Cross-Discipline Announcement

Carnegie Mellon University, School of Computer Science, Robotics Institute, Pittsburgh, PA 15213-3891.

Carnegie Mellon's MS and PhD programs in robotics are highly research oriented and interdisciplinary in nature, drawing from such fields as computer science, electrical and computer engineering, and mechanical engineering. Students may specialize in machine perception, artificial intelligence, manipulation, autonomous vehicles, manufacturing automation, or other areas. See in-depth description in Section 8: Computer Science and Information Technology.

ARIZONA STATE UNIVERSITY

Department of Mechanical and Aerospace Engineering

Programs of Study	The faculty of the department offers a broad spectrum of opportunities in graduate education. Emphasis is placed on providing a thorough background in fundamentals leading to specialization in one area. Typical programs of study include courses and seminars given by the Department of Mechanical and Aerospace Engineering, other engineering departments, and the Departments of Mathematics, Physics, and Chemistry.
	Graduate programs leading to the Master of Science, Master of Science in Engineering, and Doctor of Philosophy degrees are offered with majors in aerospace engineering and mechanical engineering. Within any of these majors, the following areas of emphasis may be pursued: design and manufacturing, dynamics and controls, energy systems, engineering mechanics, fluid and aerodynamics, heat transfer, and thermodynamics.
	The Master of Science degree program requires a minimum of 30 semester hours, including 6 hours of thesis work. The Master of Science in Engineering degree program is intended primarily for part-time students working in local industry. It requires a minimum of 30 semester hours, all of which may be in course work.
	The Doctor of Philosophy degree program requires a minimum of 84 semester hours beyond the bachelor's degree. At least 30 hours of classwork must be completed at ASU, and the student must spend a minimum of two regular semesters of full-time study on campus. The department requires Ph.D. students to complete a set of qualifying courses before beginning their research. Students are admitted to candidacy after passing a comprehensive examination, which is given near the completion of the course work in the student's program of study.
Research Facilities	Specialized laboratory facilities are available for research in the areas of acoustics, aerodynamics, environmental fluid dynamics, fluids mechanics, heat transfer, combustion, laser diagnostics, metal grinding and forming, solar energy, vibrations, composite materials, robotics, CAD/CAM, and control systems. ASU operates an IBM ES9000-732, an IBM 3090-300E, a VAX 6000-634, a MASPAR-MP-2, and a cluster of four IBM RISC-6000 substations, which are available to support graduate research. The College of Engineering and Applied Sciences provides Engineering Technical Services, with a staff of 24 full-time computer specialists who support the computing efforts of the College. The College computing environment is a large, developing client-server enterprise that all faculty members, staff members, and students utilize. This enterprise consists of more than 3,000 desktop workstations that run either UNIX or the Windows environment and are interconnected by an evolving 100-MB Ethernet network to central enterprise-level application servers. Common baseline central services provided include e-mail, World Wide Web access on both UNIX and IIS, SQL servers, forums, and office productivity tools (e.g., MS Office). Other special services include support for high-performance computing, modeling/visualization, CAD/CAM, interactive mathematics, simulation, and general instrumentation.
	The University library system, which includes the Daniel E. Noble Science and Engineering Library adjacent to the Engineering Research Center, houses more than 2 million volumes, 1.7 million microforms, 24,000 subscriptions to journals and serials, and an extensive collection of federal reports and has a full-time staff of 250. Microfiche cataloging and computerized bibliographic search capabilities facilitate the use of the library.
Financial Aid	A wide range of financial support is available for graduate students, including teaching and research assistantships, University and industrial fellowships, and a variety of scholarships. Awards are made on a competitive basis, depending on the student's past experience and scholastic record. One-half-time assistantships carry stipends ranging from $9500 to $12,700 per academic year, in addition to a waiver of nonresident tuition. Fellowships ranging up to $5000 are also available for outstanding students and supplement the regular stipends. Summer salaries for research assistants are available also.
Cost of Study	The tuition rate during the 1999–2000 academic year for 7 or more credit hours is $1094 for Arizona residents. Nonresidents pay $389 per credit hour; however, those nonresident students holding appointments as graduate assistants (quarter time or more) pay the resident tuition rate. Books, supplies, and incidental fees run about $300 per semester.
Living and Housing Costs	Students living on campus pay about $3010 for the 1999–2000 academic year. Information can be obtained from Residential Life, Student Services Building, Room A131. Students living off campus pay approximately $4950 for the academic year. Considering rent, food, personal expenses, fees, tuition, and books, Arizona residents living off campus pay approximately $13,060 per academic year, while nonresidents living off campus pay approximately $20,310 per academic year.
Student Group	The graduate student enrollment at Arizona State University is approximately 8,000, with 1,400 in the graduate programs within the College of Engineering and Applied Sciences. There are approximately 120 students enrolled in graduate programs in the Department of Mechanical and Aerospace Engineering; 50 of these are Ph.D. candidates. There are currently 80 research and teaching assistants among the department's graduate students.
Location	The Phoenix metropolitan area is the population, economic, and industrial center of the state of Arizona. Entertainment centers, art and anthropological museums, and sports arenas are among the wide variety of facilities available in the Valley of the Sun. The area's mild climate also provides opportunities for numerous outdoor activities.
The University and The Department	Arizona State University was founded in 1885 and is the largest and oldest institution of higher learning in the state of Arizona. Its present enrollment of more than 43,200 places it as the third-largest central-campus university in the United States.
	The Department of Mechanical and Aerospace Engineering provides a stimulating environment for graduate study. In addition to a firm foundation of regularly scheduled classes, seminars frequently offer advanced study in engineering topics of current interest.
Applying	Applications for admission to graduate programs are reviewed continually; however, it is recommended that they be submitted six months in advance of registration. Submission of GRE General Test scores is required of all applicants. Students seeking financial assistance should submit their applications for admission and for financial aid by February 15 for the following academic year.
Correspondence and Information	For admission: Graduate College Arizona State University Box 871003 Tempe, Arizona 85287-1003	For admission and financial aid: Dr. Don L. Boyer, Chair Department of Mechanical and Aerospace Engineering Arizona State University Box 876106 Tempe, Arizona 85287-6106 E-mail: mae@asu.edu

Arizona State University

THE FACULTY AND THEIR RESEARCH

D. L. Boyer, Professor; Ph.D., Johns Hopkins, 1965. Geophysical fluid dynamics.

L. Chapsky, Assistant Professor; Ph.D., Berkeley, 1998. Heat transfer, thermodynamics, antifreeze proteins.

A. Chattopadhyay, Associate Professor; Ph.D., Georgia Tech, 1984. Design and manufacturing, solid mechanics and dynamics.

K. P. Chen, Associate Professor; Ph.D., Minnesota, 1990. Multiphase flows, melt fracture, coextrusion and codrawing of polymeric liquids, bifurcation and transition of hydrodynamic instabilities.

J. K. Davidson, Professor; Ph.D., Ohio State, 1965. Kinematic geometry, dynamics, robotics, mechanical design.

D. L. Evans, Professor; Ph.D., Northwestern, 1967. Energy studies, system simulation, thermodynamics, heat transfer.

H. J. S. Fernando, Professor; Ph.D., Johns Hopkins, 1983. Fluid mechanics, turbulence, diffusive phenomenon.

D. F. Jankowski, Professor; Ph.D., Michigan, 1964. Fluid mechanics, hydrodynamic stability.

D. A. Kouris, Associate Professor; Ph.D., Northwestern, 1987. Micromechanics, composite materials, elasticity.

D. Krajcinovic, Professor; Ph.D., Northwestern, 1968. Micromechanics, fracture and damage mechanics, composites.

C. Y. Kuo, Associate Professor; Ph.D., Berkeley, 1984. Robotics, vibration control, system simulation.

D. H. Laananen, Associate Professor; Ph.D., Northeastern, 1968. Composite materials, aircraft crashworthiness.

T.-W. Lee, Assistant Professor; Ph.D., Michigan, 1990. Combustion science, fluid mechanics.

D. D. Liu, Professor; Ph.D., Southampton, 1974. High-speed aerodynamics, unsteady aerodynamics, aeroelasticity.

A. Mahalov, Assistant Professor; Ph.D., Cornell, 1991. Fluid mechanics.

B. W. McNeill, Assistant Professor; Ph.D., Stanford, 1976. Engineering/design process.

M. P. Mignolet, Associate Professor; Ph.D., Rice, 1987. Random and deterministic vibrations, stochastic process theory.

R. E. Peck, Professor; Ph.D., California, Irvine, 1976. Combustion and thermal systems.

P. Peralta, Assistant Professor; Ph.D., Pennsylvania, 1996. Nanomechanics.

P. E. Phelan, Assistant Professor; Ph.D., Berkeley, 1990. Thermal phenomena in superconducting devices, cryogenics, microscale applications, environmental control engineering.

R. L. Rankin, Associate Professor; Ph.D., Rice, 1971. Structural dynamics, continuum mechanics, finite elements.

H. L. Reed, Professor; Ph.D., Virginia Tech, 1981. Fluid mechanics, hydrodynamic stability, transition.

R. P. Roy, Professor; Ph.D., Berkeley, 1975. Two-phase flow, boiling heat transfer, conduction and convection heat transfer.

W. S. Saric, Professor; Ph.D., IIT, 1968. Hydrodynamic stability, boundary-layer transition, nonlinear waves.

J. J. Shah, Professor; Ph.D., Ohio State, 1984. CAD/CAM, expert systems, mechanical design, automation.

M. C. Shaw, Professor Emeritus; Sc.D., Cincinnati, 1942. Engineering design, materials behavior, tribology, machine-tool technology.

K. Sieradzki, Associate Professor; Ph.D., Syracuse, 1978. Fracture of solids, environmental effects on the mechanical behavior of materials, mechanics and physics of thin films and nanostructured materials, thin-film deposition and growth processes, phase transformations.

K. Squires, Associate Professor; Ph.D., Stanford, 1990. Large eddy simulation, sub-grid-scale turbulence modeling, particle-laden turbulent flows, numerical techniques for computational fluid mechanics.

A. A. Tseng, Professor; Ph.D., Georgia Tech, 1978. Metal and plastic forming, sensing, microfabrication, numerical analysis.

V. L. Wells, Associate Professor; Ph.D., Stanford, 1985. Rotary-wing aeroacoustics, aerodynamics.

B. Wie, Professor; Ph.D., Stanford, 1981. Spacecraft dynamics, attitude control, control theory.

L. S. Yao, Professor; Ph.D., Berkeley, 1974. Heat and mass transfer, fluid mechanics.

AREAS OF CURRENT RESEARCH

Aerodynamics and Fluid Mechanics. Mixing of swirling flows, hydrodynamic stability of laminar flows, numerical methods, rough surface flow, aerodynamics of gas turbine combustor flows, turbulence modeling, fluid mechanics of crystal growth, unsteady aerodynamics, transition to turbulence, mixing in stratified flows, rotating flows, double diffusive phenomena, rotary-wing aerodynamics.

Design and Manufacturing. Computer-aided design, expert systems, design methods, robotics, design optimization, laser diagnostics, optical inspection, feature technology, geometric modeling, computer graphics, manufacturing automation, structural shape synthesis, aerospace structural design, vehicle dynamics, crashworthiness.

Heat Transfer Thermodynamics and Energy Systems. Solar energy systems, boiling and condensation, heat transfer in complex flows, combustion emissions and diagnostics, numerical heat transfer, gas turbine engine heat transfer and combustion, thermal control of electronic equipment, heat transfer and combustion in porous media, radiation heat transfer, combined heat and mass transfer, two-phase flows.

Solid Mechanics, Structures, and Vibrations. Structural optimization, structural dynamics, biomechanics, composite materials, rotor dynamic systems, computer-aided testing, finite elements, acoustic fatigue, noise control, fluid structure interaction, earthquake excitation, plate and shell structures, buckling, damage mechanics, random vibrations.

System Dynamics and Control. Dynamics and control of aircraft and spacecraft, robotics, nonlinear mechanics, vehicle dynamics, aircraft performance optimization, orbit mechanics, optical sensors, control theory, digital implementation.

AUBURN UNIVERSITY

Department of Aerospace Engineering

Programs of Study

The Department of Aerospace Engineering offers three graduate degree programs: the Master of Science (M.S.), Master of Aerospace Engineering (M.A.E.), and Doctor of Philosophy (Ph.D.). The graduate program prepares students for careers in the aerospace industry, in government laboratories, and in academia. Studies for the Ph.D. also are designed to produce research scholars.

For the M.S., the student must complete an approved program of at least 45 credit hours in aerospace engineering or closely related supporting subjects, with a minimum of 30 credit hours at the 0600 level or above. The M.S. degree requirements include the completion of a thesis under the supervision of a major professor and an advisory committee.

The M.A.E. is a nonthesis degree for which the student must complete an approved program of at least 48 credit hours of course work with a minimum of 33 credit hours at the 0600 level or above. A suitable project in aerospace engineering, culminating in a final written report approved by the students advisory committee, may be substituted for 3 credit hours of course work. An oral presentation is also required for the M.A.E. degree.

For the Ph.D., the student must complete a minimum of 92 credit hours beyond the bachelor's degree. A plan of study is arranged on an individual basis, and students may elect to specialize in the general areas of aerodynamics, astrodynamics, control theory, flight dynamics, propulsion, structures, or structural dynamics. A written qualifying examination and a general doctoral examination, with both written and oral parts, are required of all doctoral candidates. An oral defense of the doctoral dissertation is also required of each student. There is no language requirement for the master's or Ph.D. degrees.

Research Facilities

Specialized equipment and facilities available for student use include a low-speed (130 mph) wind tunnel with a 3-foot by 4.5-foot test section, including a dedicated computer with data acquisition software; a Mach 2 wind tunnel with a 1-inch by 3-inch test section; an instructional Mach 1.5-3.5 supersonic tunnel with a four-inch by four-inch test section; a seven-inch by seven-inch supersonic wind tunnel capable of testing discrete Mach numbers from 1.01 to 3.28; a structures and structural dynamics laboratory for experiments in structures, dynamics, and vibration classes; a propulsion laboratory containing gas turbine engines and a supersonic nozzle flow test bed; a composites and materials laboratory that houses freezers for composite materials storage and a four-foot by five-foot computer-controlled oven for composites processing; a design laboratory housed in two rooms for computer aided design and presentations in senior design courses; two departmental computing laboratories, with PCs and Macintosh computers; and a flight simulation laboratory for students interested in learning to fly. In addition, the local airport is operated by the University, with airport facilities available for aircraft familiarization, orientation, and test flights. Students may include flight training in their program of study, and a professional flight option is available for engineering students who wish to combine engineering and flight careers.

Financial Aid

Students with superior records may qualify as graduate research assistants (GRAs) or graduate teaching assistants (GTAs). GRAs work with faculty members on sponsored research projects. GTAs assist faculty members with teaching duties, teach in laboratory sections, and occasionally teach elementary courses. GRA and GTA stipends vary with the student's stage in graduate school and other qualifications. Out-of-state tuition differentials are waived for students receiving GTA and GRA appointments of one-quarter time or more. An outstanding doctoral student may qualify for one of the Presidential Tuition Fellows. Many job opportunities for students' spouses exist at the University and with other firms in the surrounding community.

Cost of Study

For the 1998–99 academic year, tuition for full-time study was $920 per quarter for residents. For others, the tuition was approximately $2760 per quarter for full-time study. Books and supplies cost about $300 per quarter.

Living and Housing Costs

A limited number of University-owned apartments are available for married as well as single students. There are also a large number of private apartment complexes in the area, ranging from one- to three-bedroom apartments. A listing of off-campus housing facilities may be obtained by writing the Offices of Housing and Residence Life, Admissions, or Student Affairs.

Student Group

The University, a comprehensive research institution, has an enrollment of more than 21,000, of which more than 3,500 are in the College of Engineering, making it the second-largest college in the University. The graduate enrollment is 2,633 for the University, 456 for the College of Engineering, and 20 for the Department of Aerospace Engineering.

Student Outcomes

Graduates of the master's and doctorate programs have been employed by virtually every major aerospace firm in the United States and by most of the technically oriented government agencies, including NASA, the U.S. Army, and the U.S. Air Force. Many graduates of the master's program have risen to high-level management positions in these organizations. Other graduates of the master's program have gone on to doctoral programs both at Auburn and at other schools. Graduates of the doctoral program have been employed at high levels of responsibility by aerospace companies, universities, and government laboratories as engineers, managers, faculty members, and administrators.

Location

Referred to as "the loveliest city on the plains", the city of Auburn is a growing community that retains the relaxing small-town atmosphere it is known for. Auburn and Auburn University have always had a warm relationship in which both the city and the University benefit from each other. The community of Auburn has a population of 34,000 and, at 723 feet above sea level, enjoys a moderate climate. Recreation of all types, including a variety of amateur and nationally prominent intercollegiate teams, is a part of campus life. Major golf courses blanket the area. There is also a full calendar of concerts, plays, and nationally recognized guest speakers, which provides a strong cultural stimulus. For those who enjoy a metropolitan atmosphere, the cities of Montgomery and Birmingham in Alabama and historical Atlanta and Columbus in Georgia are all within a 120-mile radius. The Gulf of Mexico, with its miles of beaches and deep-sea fishing, is about four hours away by automobile. Atlanta's Hartsfield International Airport is less than two hours from Auburn.

The University

The University was originally founded in 1856 and was later rechartered in 1872 to become the Agricultural and Mechanical College of Alabama, the state's land-grant university. Over the years, Auburn University gained recognition for its high academic standards and competitive spirit. Student enrollment increased, and award-winning faculty members from around the world were added. In 1960, the University adopted the name of its home, and Auburn University came to be. The University is on the quarter system, with quarters starting in January, March, June, and September. It will change to the semester system in fall 2000.

Applying

Further information and applications forms are provided upon request.

Correspondence and Information

Dr. Roy J. Hartfield, Associate Professor
Aerospace Engineering Department
211 Aerospace Engineering Building
Auburn University, Alabama 36849-5338
Telephone: 334-844-6819
Fax: 334-844-4803
E-mail: aegrad@eng.auburn.edu
World Wide Web: http://www.eng.auburn.edu/department/ae

Auburn University

THE FACULTY AND THEIR RESEARCH

A. Ahmed, Ph.D. (aerospace engineering), Wichita State. Experimental aerodynamics as fluid, experimental aerodynamics, model vortex dynamics, boundary layers.

R. Barrett, Ph.D. (aerospace engineering), Kansas. Aerodynamics, design adaptive aerostructures.

J. E. Burkhalter, Ph.D. (civil engineering), Texas at Austin. Aerodynamics, fluid mechanics.

D. A. Cicci, Ph.D. (aerospace engineering), Texas at Austin. Astrodynamics, orbit determination, guidance and control, numerical analysis.

J. E. Cochran Jr., Ph.D. (aerospace engineering), Texas at Austin. Dynamics, flight dynamics and control, spacecraft attitude dynamics and control, legal aspects of engineering.

W. A. Foster Jr., Ph.D., (aerospace engineering), Auburn. Numerical structure analysis, solid rocket propulsion, engineering computer applications.

R. S. Gross, Ph.D. (engineering mechanics), Clemson. Elasticity, experimental mechanics, composite materials.

R. J. Hartfield Jr., Ph.D. (mechanical and aerospace engineering), Virginia. Nonintrusive flow diagnostics and propulsion.

R. M. Jenkins, Ph.D. (aeronautics, astronautics, and engineering science), Purdue. Propulsion and aerodynamics.

J. B. Lundberg, Ph.D. (aerospace engineering), Texas at Austin. Astronautics orbit determination, navigation, numerical analysis, GPS.

Aerodynamics: Auburn University actively conducts both basic and applied research in areas of interest to government agencies and private companies. Aerodynamics encompasses the kinematics and kinetics of fluid motion and its effects on objects in contact with the fluid. Recent research efforts at Auburn include theoretical studies of innovative lifting devices and aerodynamic optimization using neural networks and genetic algorithms. Experimental aerodynamic research includes characterization of lifting surfaces at low Reynolds numbers, including unsteady phenomena; the generation of vorticity; and high-speed mixing using both computational fluid dynamics and nonintrusive optical diagnostic techniques.

Adaptive Aerostructures and Composite Materials: Auburn is currently conducting several major research efforts related to the use of adaptive structural components and composite materials in aerospace vehicles. Specific projects include the development of a guided bullet, the development of a new type of helicopter rotor blade, and the development of a new class of adaptive hypersonic interceptor flight control devices.

Astrodynamics: Astrodynamics (astronautics) research includes attitude of dynamics and control of spacecraft orbital mechanics, mission analysis, and the design of devices and equipment to be used in space. Current projects focus on the development of mathematical models and algorithms for the detection, slate estimation, and motion prediction; multibody modeling of spacecraft; development of algorithms for utilizing the Global Positioning Satellite System (GPSS) in navigation and in orbit and attitude determination; tethered satellite detection; remote sensing; satellite-geodesy; and numerical methods for solving ill-conditioned slate estimation problems.

Flight Dynamics and Control: The mathematical modeling, simulation, stability analysis, and control of the motion of flight vehicles are areas that provide many opportunities for research. Projects include the development of multiple rigid-body models of the main rotor head of a helicopter and simulation of its operation; the modeling, control, and simulation of towed targets; and the development and testing of guidance algorithms for missiles using six-degree-of-freedom simulations.

Intermodal Transportation: A new research thrust has been initiated in the area of intermodal (airway, railway, highway, and waterway) transportation and, in particular, in the simulation of the terminals that provide the interfaces between the four modes of transportation. Systems theory and the application of resources such as the GPSS are used in the development of real-time simulations of terminal operations and the development of control systems for future intermodal terminals.

Propulsion: The generation of propulsive forces can be accomplished using a wide array of devices, including liquid, solid, and hybrid chemical rockets; various aerodynamic devices, including propellers, turbojets, ramjets, and scramjets; and various electric and thermal devices proposed for extended space travel and other applications. Recent propulsion research at Auburn has included studies of ignition phenomena for large solid-rocket motors, combustion stability in hybrid motors, experimental and analytical investigations of various electric thrust generators mixing "o" combustion in scramjets, and innovative vibration suppression techniques for gas turbine-based propulsion devices.

Solid Mechanics, Structures, and Structural Dynamics: Solid mechanics research at Auburn includes finite-element modeling of aerospace structures, the development of composite materials and composite material applications, adaptive aerostructures, aeroelasticity, and vibrations.

BOSTON UNIVERSITY

College of Engineering
Department of Aerospace and Mechanical Engineering

Programs of Study	The department offers graduate programs leading to the Ph.D. and M.S. degrees. Research is focused in four principal areas: waves and acoustics; dynamics, control, and robotics; fluid mechanics; and precision engineering. Some areas of specialization are aerodynamics, aeroelasticity, automatic control systems, biomechanics, noise control, photomechanical systems, structural mechanics, thermal processes, theoretical fluid dynamics, and turbulence. Both M.S. and Ph.D. degrees are available in aerospace engineering and mechanical engineering. To receive the M.S. degree, a student must complete 32 credits. A thesis is recommended but not required. A cumulative average of at least 3.0 (B) is required for graduation. The postbachelor's and post-master's Ph.D. programs require a minimum of 64 and 32 credits, respectively, and at least two consecutive semesters of residence, written and oral examinations, and a dissertation.
Research Facilities	The Robotics Laboratory contains a variety of fixed and mobile robots, a network of computer workstations, and two computer systems for simultaneous simulation and real-time control of nonlinear mechanical systems. The Precision Engineering Research Laboratory houses advanced equipment for the measurement and fabrication of engineering materials to submicrometer tolerances. The Mechanics of Materials Laboratory is equipped for computer-aided mechanical testing of composite and biomedical materials. The Aerodynamic and Fluid Mechanics Laboratory contains three wind tunnels instrumented with electronic and computer-assisted data acquisition systems, including a boundary-layer wind tunnel for fundamental turbulent transport studies in three-dimensional boundary layers. The Applied Acoustics Laboratory contains instrumentation and supplies for studies involving underwater, industrial, and medical acoustics and acoustic cavitation. The Nonlinear Acoustics Laboratory is instrumented for experiments in high-intensity sound and includes a research lithotripter (a medical device that uses shock waves to break kidney stones). The Physical Acoustics Laboratory possesses instrumentation devoted to experiments probing science and engineering aspects of acoustic levitation, bubble dynamics, droplet dynamics, and the acoustics and rheology of foam. There are extensive computer resources available to the department, including clusters of graphics workstations, multiprocessor VAX machines, and a massively parallel TMC CM-5 supercomputer.
Financial Aid	A full range of financial aid opportunities is available, including Presidential University Graduate Fellowships, Dean's Fellowships, graduate research assistantships, teaching fellowships, and scholarships. In 1999–2000, teaching fellowships provide stipends of $12,500 per academic year and require 20 hours a week of instructional and other duties. Recipients receive a tuition waiver for 8 to 10 credits per semester and up to 8 additional credits the following summer. Research assistantship stipend levels are comparable to those of teaching fellowships and are also supplemented by tuition waivers. University and Dean's Fellows scholarships range up to $37,830 (including stipend and tuition). The department encourages GEM scholars to apply. Federally funded work-study applications may be obtained from the Graduate Programs Office. Federal Direct Student Loan and Work-Study applicants must send a Free Application for Federal Student Aid (FAFSA) to the Federal Student Aid Programs Office. Work-Study and FAFSA forms may be obtained from the Graduate Programs Office.
Cost of Study	For the 1999–2000 academic year, tuition and fees for full-time study are $23,770. Part-time students pay $743 per credit hour.
Living and Housing Costs	Privately owned apartments or rooms are readily available. Living expenses for a single student are estimated at $10,730 for the nine-month 1999–2000 academic year.
Student Group	The department has 26 students in the M.S. programs and 16 students pursuing the Ph.D. In addition, 4 students are enrolled in a B.S./M.S. program that lasts four to five years. Students with nonscience bachelor's degrees can participate in the LEAP program, which qualifies them to apply to the M.S. program after only about one year of full-time remedial science study.
Student Outcomes	Department graduates are highly competitive in today's marketplace. Recent graduates have started their own businesses, found jobs consulting for high-technology industries, obtained teaching positions at the university level, and been hired to work in government laboratories. M.S. students have gone on to Ph.D. programs at Boston University, Stanford, MIT, Michigan, and other schools.
Location	Boston University offers a sophisticated metropolitan environment with world-renowned academic and scientific resources. The facilities of other area universities are easily accessible, and excellent seminar and colloquium programs afford graduate students unique opportunities to participate at the cutting edge of research. Access to high-technology industry provides another vital element in an academically oriented region. A vast range of cultural, social, and recreational activity is available in the Greater Boston area. The Boston Symphony Orchestra and the Boston Pops perform only a 15-minute drive away from campus, and the Museum of Fine Arts and the Museum of Science are close by. Virtually every kind of professional athletic team is represented in Boston.
The University and The Department	Boston University is an independent nonsectarian university, open to women and all minorities. Its approximately 23,500 full-time students and 3,130 faculty members make it one of the largest universities in the world. The Department of Aerospace and Mechanical Engineering is one of four departments in the College of Engineering.
Applying	Applicants to the M.S. or post-bachelor Ph.D. programs should have demonstrated a high degree of scholarship in an undergraduate program in science at an accredited college or university. Post-master's Ph.D. applicants must hold an M.S., M.A., or the equivalent from an accredited institution. Students desiring financial aid should send completed applications, including official transcripts and GRE General Test scores, by January 15 for fall admission and by October 1 for spring. Applications without financial aid requests are accepted until April 1 and October 15 for fall and spring admission, respectively.
Correspondence and Information	For program information: Graduate Programs Department of Aerospace and Mechanical Engineering Boston University 110 Cummington Street Boston, Massachusetts 02215 Telephone: 617-353-2814 World Wide Web: http://www.bu.edu/eng/ame	For admission application forms: Graduate Programs College of Engineering Boston University 44 Cummington Street Boston, Massachusetts 02215 Telephone: 617-353-9760 Fax: 617-353-0259 E-mail: enggrad@bu.edu World Wide Web: http://www.bu.edu/eng/grad

Boston University

THE FACULTY AND THEIR RESEARCH

John Baillieul, Ph.D., Harvard. Directs and conducts research in robotics, the control of mechanical systems, and mathematical system theory. Research projects in the Robotics Laboratory are aimed at the design and implementation of sensor-driven control systems that involve closing feedback loops, using a variety of sensing technologies. Also interested in the nonlinear dynamics of complex mechanisms featuring multiply articulated and elastic components.

Paul Barbone, Ph.D., Stanford. Research focuses on aspects of dynamic fluid-structure interaction and ultrasonic medical imaging. Analytical and numerical methods based on asymptotics and perturbation theory are used to describe basic interaction phenomena such as surface wave excitation and propagation and modal coupling. In medical ultrasound, these techniques are used to suggest practical means to overcome diffraction limitations in imaging. Fundamental research includes development problems in hybrid analytical/numerical methods and exponential asymptotics.

Thomas Bifano, Ph.D., North Carolina State. Research includes pioneering development of ultraprecise machines for ductile-regime microgrinding of brittle materials, resulting in the first published analytical models and empirical results describing this important process. Applications to production of strategically and commercially important components include ceramic space-based mirrors, ultrathin quartz wafers, and computer hard disk substrates. Other research interests include deformable mirror fabrication, development of micromechanical devices for precision actuation, and optical storage disk manufacturing.

Robin Cleveland, Ph.D., Texas at Austin. Research focuses on nonlinear acoustics, cavitation, shock-wave lithotripsy, shock-wave propagation.

Debora Compton, Ph.D., Stanford. Conducts fundamental experimental studies of the physical structure and statistical behavior of boundary-layer turbulence. Focus is on boundary layers subject to streamline curvature. Recently designed an innovative laser-Doppler anemometer for the measurement of Reynolds stresses in the near-wall region. Applications include active control of stall.

Pierre Dupont, Ph.D., RPI. Research is in kinematics, dynamics, and control, especially as applied to robotics. Recent contributions include modeling low-velocity dynamic friction phenomena in lubricated metals and providing associated lower bounds on closed-loop controller gains for machine stability. Current topics include the identification of friction models based on surface physics, determination of standards for friction parameter identification in assembled machines, and the creation of novel adaptive and robust compensators for precise motion control.

Leopold Felsen, D.E.E., Polytechnic; Member, National Academy of Engineering. Research areas include wave propagation and diffraction in various disciplines, high-frequency asymptotics, wave-oriented data processing and imaging, complexity and deterministic-stochastic wave interactions, ultrawideband (short pulse) phenomena, spectral and phase space methods.

Donald Fraser, Sc.D., MIT; Director of the Boston University Center for Photonics Research. Directs and oversees research conducted in this center, which was established with a $29-million federal award. Research addresses a broad range of photonic technologies, including III-V semiconductors, near-field spectroscopy, biomolecular photonic materials, optical wideband parallel data communications, optical data storage, and environmental and medical sensing applications. Emphasizes both basic research and prototype product development. Technology transfer to the marketplace occurs via licensing agreements with existing companies or spin-off ventures from the University.

Sheryl Grace, Ph.D., Notre Dame. Research focuses on unsteady aerodynamics/hydrodynamics, aeroacoustics, inverse problems, computational fluid dynamics, and applied mathematics.

R. Glynn Holt, Ph.D., Mississippi. Research focuses on physical acoustics. Projects include rheology of foam, acoustic levitation, bubble and droplet dynamics, and sonoluminescence in space.

Michael S. Howe, Ph.D., Imperial College (London). Research involves noise generation in turbomachines and turbulent, two-phase boundary-layer flows, and the theory of wave propagation in fluid-loaded, composite elastic media.

Harley Johnson, Ph.D., Brown. Research projects include mixed continuum/atomistic modeling and strain effects on confinement in Quantum wires.

J. Gregory McDaniel, Ph.D., Georgia Tech. Primary focus on the development of innovative models of structural acoustics and vibration phenomena, including wave propagation and reflection, damping mechanisms, acoustic radiation and scattering, and causality constraints.

Ali Nadim, Ph.D., MIT. Research includes fundamental studies of multiphase fluid dynamics, focusing on the dynamics of gas bubbles, liquid drops, solid particles, biological cells, and polymer molecules suspended in viscous fluids. Recent work has examined the nonlinear oscillations of gas bubbles under ultrasonic excitation, the effects of surfactants and thermocapillary phenomena on the migration of drops and bubbles, and the non-Newtonian rheology of polymeric solutions.

Raymond Nagem, Ph.D., MIT. Conducts fundamental research in solid mechanics and structural dynamics, numerical and analytical methods in acoustic and elastic wave propagation, image processing applied to acoustic and optical signals, and design of structures for extraterrestrial habitats.

Allan Pierce, Ph.D., MIT; Department Chair. Author of graduate text, "Acoustics: An Introduction to its Physical Principles and Applications." Research interests include supersonic flight, sonic booms, wave-based applications involving elastic structures and fluids, identification and measurement of statistical descriptors of dynamical systems with high complexity, optoacoustic and electroacoustic devices, acoustic detection systems, bioacoustics of aneurysms, active materials in control of vibration and sound radiation, and wave propagation through turbulence.

Ronald Roy, Ph.D., Yale. Research focuses on physical, underwater, industrial, and medical acoustics, with an emphasis on acoustic cavitation and bubble-related acoustical processes, such as noise production, sea-surface scattering, sonoluminescence, sonochemistry, and high-intensity focused ultrasound for therapeutic medicine.

Charles Speziale, Ph.D., Princeton. Research is in the theoretical analysis and modeling of turbulent flows. The ultimate goal is to develop new models for the reliable prediction of turbulent flows. Basic research on turbulence theory and modeling relevant to aerodynamic and hydrodynamic applications such as flow around naval vessels.

Ann Stokes, Ph.D., Harvard. Research focuses on nonlinear dynamics and theoretical issues in the control of mechanical systems. Applications include the design of robotic manipulators and active noise control.

Ananias Tomboulides, Ph.D., Princeton. Development and implementation of numerical methods for both direct and large-eddy simulations of incompressible flows using spectral-type methods. Applications include bluff-body, pipe, and crystal-melt flows.

Daniel Udelson, Ph.D., Harvard. Current research is in biomechanics, focusing on the urodynamics and hemodynamics of the human urinary system. The former is for the purpose of providing new methods to diagnose disorders of the urinary tract; the latter involves modeling to enable better understanding of male potency problems.

Donald Wroblewski, Ph.D., Berkeley. Forging a fundamental understanding of scalar transport processes in turbulent flows, his efforts center on experimental investigations into a variety of complex and unsteady turbulent flow situations, including streamwise-vortex-dominated and unsteady boundary layers, oscillatory flows in bifurcating geometries like the upper respiratory tract, and pulsatile pipe flows. Improved experimental techniques for turbulent scalar flux measurements and for time-resolved wall heat flux measurements is also an active area of research.

Victor Yakhot, Ph.D., Moscow State. Renormalization group for turbulence and turbulence modeling, long-wave stability of three-dimensional flows, theory of probability distributions in random flow, generalized statistical hydrodynamics, theory of flame-front propagation in turbulence.

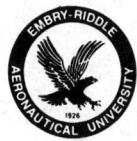

EMBRY–RIDDLE AERONAUTICAL UNIVERSITY

Office of Graduate Programs and Research
Department of Aerospace Engineering

Programs of Study

Embry-Riddle Aeronautical University's Master of Aerospace Engineering (M.A.E.) and Master of Science in Aerospace Engineering (M.S.A.E.) degree programs provide formal postbaccalaureate study with emphasis on hands-on and applied knowledge required by engineers engaged in aircraft- and aerospace-oriented research, development, and design. Each degree program is planned to augment the individual student's previous engineering and science background in areas of aeroacoustics, acoustic emission nondestructive testing, aerodynamics, propulsion, optimal systems, aerospace structures, or other areas of aerospace engineering. Candidates for the degrees can select courses with the goal of building a graduate program that supports their interests in the aerospace engineering profession or that prepares them to continue on to their doctoral studies. The M.A.E. program (all course work option) requires a minimum of 33 credit hours of graduate course work. The M.S.A.E. program (thesis option) requires 24 credit hours of course work and 9 credit hours of thesis.

Research Facilities

Aerospace engineering students have access to an acoustics laboratory; several subsonic wind and smoke tunnels, including an anechoic wind tunnel; a composite materials laboratory; a flight dynamics and control laboratory; a material testing laboratory; a computer-aided manufacturing laboratory; a stereolithographic machine; a thermal sciences laboratory; two engine test cells; and a flight test airplane, which provide opportunities for study and research. In addition, the department computer resources include IBM RS6000s, IBM PowerSeries AIX's, Sun Workstations with Ethernet connections to IBM PowerPCs, PowerMacs, and Pentium PCs. Commercial software is available for static, dynamic, linear, and nonlinear structural analysis, limited computational fluid mechanics, and computer-aided drafting and design works. All students have access to the Airway Science Simulation Laboratory, which simulates the various elements of the National Airspace System: weather, airports, airways, air traffic control, flow control, and pilot and aircraft performance. Currently, the lab houses several flight simulators, a meteorological center, an air traffic control simulation system (nonradar trainers), an air traffic control intelligent simulation training system, and a computer-based instructional system center. The Gill Robb Wilson Memorial Aeronautical Science Center houses a flight simulator laboratory, an aviation weather center, and a flight dispatch center for single and multiengine aircraft. The Aviation Maintenance Technology Center is equipped with both reciprocating and jet engine laboratories, an avionics laboratory, and jet engine test cells.

Financial Aid

Embry-Riddle makes every effort, within the limitations of the financial resources available, to ensure that no qualified student is denied the opportunity to obtain an education because of inadequate funds. However, the primary responsibility for financing an education must be assumed by the student. A number of graduate assistantships that provide a stipend and a tuition waiver are available to qualified students on a competitive basis each year. Other financial aid programs are Federal Stafford Student Loans, short-term loans, scholarship and fellowship programs, and the Embry-Riddle Student Employment Program. All graduate programs are approved for Veterans' Affairs education benefits.

Cost of Study

In 1999–2000, tuition costs are $490 per semester hour. Books and supplies cost approximately $300 per semester.

Living and Housing Costs

Some on-campus housing is available to graduate students. The cost of a standard double-occupancy room is $1400 per semester. Off-campus housing is reasonably priced. Single students who are sharing rental and utility expenses with someone can expect off-campus room and board yearly expenses to be $4000. Expenses for married students are higher.

Student Group

The graduate programs currently enroll 250 students on the Daytona Beach campus. The College of Career Education enrolls more than 3,000 students in graduate degree programs off campus. On the Daytona Beach campus, 40 percent are from other countries, 41 percent are women, and 42 percent are members of minority groups. More than 10 percent of the campus-based graduate students are employed full-time.

Student Outcomes

Employment opportunities are steadily improving for graduates of the M.S.A.E./M.A.E. degree programs. Recent graduates were in high demand, with multiple offers from companies such as Boeing, Lockheed-Martin, Sikorsky, and Dassault Falcon.

Location

The Daytona Beach campus is adjacent to the Daytona Beach International Airport and is 10 minutes from the Daytona beaches. Within an hour's drive are Disney World and EPCOT, Universal Studios, the Kennedy Space Center, Sea World, and St. Augustine.

The University

The University comprises the main campus at Daytona Beach; a western campus in Prescott, Arizona; and the College of Career Education. Within the field of aviation, Embry-Riddle Aeronautical University has built a reputation for the high quality of instruction in its programs since its founding in 1926.

Applying

The desired minimum undergraduate cumulative grade point average is 2.5 on a 4.0 scale and a minimum of 3.0 in the discipline during the junior and senior years. Applications from U.S. citizens and permanent residents should be received at least 30 days prior to the first day of the term in which the applicant plans to enroll. International students should submit all of their documents at least ninety days prior to the first day of the term in which they plan to enroll.

Correspondence and Information

Graduate Admissions
Embry-Riddle Aeronautical University
600 S. Clyde Morris Boulevard
Daytona Beach, Florida 32114-3900
Telephone: 904-226-6115
       800-388-3728 (toll-free)
Fax: 904-226-7050
E-mail: admit@db.erau.edu
World Wide Web: http://www.db.erau.edu

Embry-Riddle Aeronautical University

THE FACULTY AND THEIR AREAS OF SPECIALIZATION

The following are faculty members at the Daytona Beach campus.

Yechiel Crispin, Professor; D.Sc., Israel. Multidisciplinary design optimization.
Howard D. Curtis, Professor; Ph.D., Purdue; PE. Numerical structural analysis, engineering mechanics.
Habib Eslami, Professor; Ph.D., Old Dominion. Anisotropic theory, solid mechanics, composite materials.
Tej R. Gupta, Professor; Ph.D., Roorkee (India); Ph.D., Virginia Tech. Aerodynamics, numerical techniques, CFD.
Eric v.K. Hill, Professor; Ph.D., Oklahoma. Aerostructural mechanics and acoustic emissions.
Ira D. Jacobson, Professor; Ph.D., Virginia. Design, flight dynamics, human factors.
T. David Kim, Professor; Ph.D., Georgia Tech. Flight mechanics, testing, and simulation.
James G. Ladesic, Professor; Ph.D., Florida; PE. Aerostructural design, analysis, experimental methods.
Lakshmanan L. Narayanaswami, Professor; Ph.D., Georgia Tech. Thermal systems, propulsion.
Allen I. Ormsbee, Professor; Ph.D., Caltech. Aerodynamics.
Howard Patrick, Associate Professor; Ph.D., North Carolina State. Acoustics and aeroacoustics.
Frank J. Radosta, Professor; Ph.D., Florida. Structural mechanics, elastic/plastic stress analysis.
R. Luther Reisbig, Professor; Ph.D., Michigan State. Thermal sciences.

EMBRY–RIDDLE AERONAUTICAL UNIVERSITY

Office of Graduate Programs and Research
Department of Applied Aviation Sciences and
Department of Aeronautical Science

Program of Study

Embry-Riddle Aeronautical University's Master of Aeronautical Science (M.A.S.) degree program is designed to guarantee the aviation professional a generalist education operations perspective. It provides an opportunity for flight crew members, air traffic control personnel, flight operations specialists, industry management and technical representatives, and aviation educators to enhance their knowledge and pursue additional career opportunities. Although aviation experience is desired for entry into the M.A.S. program, pilot qualifications are not necessary.

All students must complete a 12-credit advanced aviation/aerospace science core. Students then complete 12 credits in one of seven specializations: aeronautics, aviation/aerospace operations, aviation/aerospace education, aviation/aerospace management, aviation/aerospace safety, human factors in aviation systems, and space studies. The remaining credits consist of electives and either a thesis or a graduate research project. The program can generally be completed in three or four semesters of full-time study.

The program is offered in three modes: a residential program through the Department of Applied Aviation Sciences on the Daytona Beach campus, a classroom program at more than 100 resident centers throughout the United States and in Europe through the Extended Campus' College of Career Education, and a World Wide Web–based distance learning program through its Center for Distance Learning.

Research Facilities

Aeronautical science students have access to the Airway Science Simulation Laboratory, which simulates the various elements of the National Airspace System: weather, airports, airways, air traffic control, flow control, and pilot and aircraft performance. The lab houses a meteorological center, an air traffic control simulation system, an air traffic control intelligent-simulation training system, and a computer-based instructional system center.

The Air Traffic Management Research Laboratory's core capabilities are built around the Total Airspace and Airport Modeler (TAAM). This sophisticated modeling and simulation platform is a graphical fast-time simulation tool that offers full functionality for airspace and airport applications. Its uses include configuration and design, delay analysis, operations planning, and predeployment safety and economic decision making. Embry-Riddle has developed proprietary tools to enhance the functionality of TAAM and to address added complexities of airspace and airport modeling.

The controller-in-the-loop laboratory, which was developed under contract to the National Aviation Research Institute and NASA's Ames Research Center, is used to conduct experiments in the areas of air traffic control human factors and the implications for controllers of new technologies and procedures being considered for the modernized National Airspace system. The lab provides realistic functionality that is similar to what is available in en route and terminal ATC facilities.

Financial Aid

Embry-Riddle makes every effort, within the limitations of the financial resources available, to ensure that no qualified student is denied the opportunity to obtain an education because of inadequate funds. However, the primary responsibility for financing an education must be assumed by the student. A number of graduate assistantships providing a stipend and a tuition waiver are available on a competitive basis each year. Other financial aid programs are Federal Stafford Student Loans, short-term loans, scholarship and fellowship programs, and the Embry-Riddle Student Employment Program, available on the Daytona Beach campus. All graduate programs are approved for Veterans' Affairs education benefits.

Cost of Study

In 1999–2000, tuition costs on the Daytona Beach campus are $455 per semester hour. The tuition costs for 1999–2000 are $238 per semester hour at Extended Campus resident centers and $280 per semester hour through the Center for Distance Learning.

Living and Housing Costs

Some on-campus housing is available to graduate students. The cost of a standard double-occupancy room is $1450 per semester. Off-campus housing is reasonably priced. Single students who are sharing rental and utility expenses with someone can expect off-campus room and board yearly expenses of $4000.

Student Group

The graduate programs currently enroll 250 graduate students on the Daytona Beach campus. The College of Career Education enrolls more than 3,000 students in graduate degree programs off campus, and the Center for Distance Learning enrolls an additional 900 students. On the Daytona Beach campus, 40 percent are from other countries, 41 percent are women, and 42 percent are members of minority groups. More than 10 percent of the campus-based graduate students are employed full-time.

Student Outcomes

Employment opportunities are excellent for graduates of the M.A.S. program. In an era of industry consolidation, aviation/aerospace firms continue to seek the best aviation educated people. Higher education is an important criteria in the selection of company personnel. Typical employers of recent graduates of the M.A.S. in safety specialization program are the FAA, NTSB, TSI, Sikorsky and Cessna Aircraft, and major airlines. Some M.A.S. graduates continue to fly for major and commuter airlines in the U.S. and South America. International students have returned to more advanced positions of authority in their country's companies or governments. Some M.A.S. graduates choose to further their education and are pursuing postgraduate degrees at schools that include MIT, Penn State, Indiana, and Florida State. Military officers are furthering their active duty careers with advanced positions after completing their M.A.S. degrees.

Location

The Daytona Beach campus is adjacent to the Daytona Beach International Airport and 10 minutes from the Daytona beaches. Within an hour's drive are Disney World and EPCOT, the Kennedy Space Center, SeaWorld, and St. Augustine. Extended Campus resident centers are located in thirty-seven states and seven other nations. A map of these sites can be found on the Web site listed below.

The University

The University comprises the eastern campus at Daytona Beach; a western campus in Prescott, Arizona; and the Extended Campus with off-campus programs. Within the field of aviation, Embry-Riddle Aeronautical University has built a reputation for high quality of instruction in its programs since its founding in 1926.

Applying

Entry into the M.A.S. program requires a demonstrated knowledge in the areas of college-level mathematics, introduction to computers, economics, behavioral science, and aviation rules and regulations. The minimum undergraduate cumulative GPA is 2.5 out of a possible 4.0. Applications from U.S. citizens and permanent residents should be received at least thirty days prior to the first day of the term in which the student plans to enroll. International students should submit all of their documents at least ninety days prior to the first day of the term in which they plan to enroll. Applications for graduate admission are handled separately for the Daytona Beach campus, for the College of Career Education, and for the Center for Distance Learning. Applicants should contact the admissions office that corresponds to their preferred mode of study.

Correspondence and Information

Daytona Beach Campus:

Graduate Admissions
Embry-Riddle Aeronautical University
600 South Clyde Morris Boulevard
Daytona Beach, Florida 32114-3900

Telephone: 904-226-6115
        800-388-3728 (toll-free)
Fax: 904-226-7050
E-mail: admit@db.erau.edu
World Wide Web: http://www.db.erau.edu

Extended Campus:

Admissions, Records and Registration
College of Career Education and Center for Distance
  Learning
Extended Campus
Embry-Riddle Aeronautical University
600 South Clyde Morris Boulevard
Daytona Beach, Florida 32114-3900

Telephone: 940-226-6910
        800-522-6787 (toll-free)
Fax: 904-226-6984
World Wide Web: http://www.ec.erau.edu

Embry-Riddle Aeronautical University

THE FACULTY AND THEIR AREAS OF SPECIALIZATION

The following are faculty members at the Daytona Beach campus:

Bishop Blackwell, Professor; Ed.D., Florida. Aviation education, aircraft systems.

Tim Brady, Professor; Ph.D., St. Louis. Safety.

Thomas J. Connolly, Professor; Ed.D., Nova. Aviation education, simulation systems.

Lance Erickson, Professor; Ph.D., Florida. Control and communication systems, meteorology, space studies.

John Ernst, Associate Professor; Ph.D., St. Louis. Meteorology.

Donald B. Hunt, Associate Professor, M.A.S., Embry-Riddle. Aviation aircraft investigation, safety systems.

Leslie L. Kumpula, Professor; M.S.A.E., Minnesota. Aerodynamics, performance.

James E. Lewis, Associate Professor; M.A.S., Embry-Riddle. Rotorcraft operations, aerodynamics.

William A. Martin, Professor; M.A.S., Embry-Riddle. Safety.

Bradley M. Muller, Assistant Professor; Ph.D., Florida State. Meteorology.

Franklin D. Richey, Professor; D.B.A., Nova. Air carrier operations, airport simulation.

Marvin Smith, Associate Professor; Ed.D., Nova. Air traffic control, educational technology.

The following are faculty members at the Extended Campus:

Armando Alvear, Assistant Professor; M.A., Texas A&M. Aviation/aerospace communications control systems.

Jon Anderson, Assistant Professor; M.A.S., Embry-Riddle. Accident investigation, airport operations safety.

Paul Bankit, Professor; Ph.D., Michigan State. Air carrier operations, air transportation systems, corporate aviation operations.

Thomas S. Barker, Assistant Professor; Ph.D., North Texas. Economics, logistics.

Alan R. Bender, Associate Professor; Ph.D., Berkeley. Research methods, aviation planning.

John F. Bollinger, Associate Professor; M.A. Central Michigan. Air transportation, aircraft and spacecraft development, aircraft maintenance management.

Thomas Brown, Associate Professor, M.A.M., Embry-Riddle. Aviation communication systems, airport operations.

Richard P. Buchtmann, Associate Professor; M.A.M., Embry-Riddle. Aviation communication systems, airport operations.

Gene E. Burton, Professor; Ph.D., North Texas State. Research methods.

Larry S. Carlton, Associate Professor; M.S., Golden Gate. Air transportation systems, aircraft and spacecraft development, research methods.

Glenn A. Chaffee, Assistant Professor; M.S., Golden Gate. Corporate aviation operations, human factors.

Ronald E. Clark, Associate Professor; Ed.D., Nova. Research methods, aviation psychology.

Charles M. Court, Professor; Ph.D., St. Louis. Air transportation systems, aircraft and spacecraft development, human factors, logistics, research methods.

Ernest Dammier, Associate Professor; Ed.D., Nova Southeastern. Aviation safety, research methods, simulation systems.

Orin Godsey, Associate Professor; M.A., Rensselaer. Air carrier operations, aviation safety, human factors.

Wayne R. Harsha, Assistant Professor; Ed.D., Montana State. Technical management, research methods.

Daniel E. Johnson, Assistant Professor; Ed.D., USC. Human factors, aviation safety.

Kenneth J. Kovach, Associate Professor; Ed.D., Nova Southeastern. Air transportation, corporate aviation operations, research methods.

Michael J. Kraus, Assistant Professor; M.S., USC. Aviation maintenance management.

Mary Landers, Assistant Professor; Ed.D., East Texas State. Research methods.

George Langhorne, Assistant Professor; Ed.D., United States International. Research methods.

Harold J. Maloney, Associate Professor; M.B.A., Pepperdine. Air transportation, logistics.

William L. March, Professor; Ed.D., Indiana. Aviation operations, human factors, research methods.

James Marlow, Associate Professor; Ph.D., Tennessee. Aviation legislation.

Frederick E. McNally, Assistant Professor; Ed.D., San Francisco. Air carrier operations, research methods.

David A. Miramonti, Assistant Professor; Ed.D., Western Michigan. Management, organizational behavior.

Vance F. Mitchell, Professor; Ph.D., Berkeley. Aviation psychology, human factors.

Thomas Moe, Assistant Professor; J.D., North Dakota. Aviation insurance, aviation labor relations, aviation legislation.

John L. Neff, Associate Professor; Ed.D., Indiana. Economics, research methods.

Gerald P. Nicoletta, Assistant Professor; M.A., Alabama. Corporate aviation, economics.

Donald M. Nixon, Professor; Ed.D., Auburn. Research methods, airline operations.

Hubert C. Pate, Associate Professor; M.S., Troy State. Simulation systems.

Robert W. Reed, Assistant Professor; Ph.D., Michigan. Operations research, research methods.

David Rollins, Assistant Professor; D.B.A., United States International. Logistics, management.

John. A. Rooke, Assistant Professor; M.A., Pepperdine. Aviation safety, insurance.

Franz G. Rosenhammer, Assistant Professor; D.B.A., Tennessee. Air transpiration, logistics, research methods.

Bruce A. Rothwell, Assistant Professor; D.P.A., Alabama. Logistics, research methods.

Kent W. Rowe, Assistant Professor; M.A., Wichita State. Aircraft development, aviation safety.

James T. Schultz, Associate Professor; Ed.D., USC. Human factors, marketing, research methods.

Clarence Schumaker, Assistant Professor; Ph.D., Catholic University. Research methods, psychology.

Larry W. Shadow, Assistant Professor; Ph.D., United States International. Simulation systems, research methods.

Guy Smith, Assistant Professor; Ed.D., Montana State. Human factors, research methods, transportation systems.

David F. Soutamire, Associate Professor; M.B.A., Troy State. Aircraft development, aviation safety, human factors.

Michael D. Warner, Assistant Professor; M.S., Naval Postgraduate College. Aviation safety, accident investigation.

Sidney Earl Wheeler, Professor; Ph.D., Florida. Aviation economics, human factors, research methods.

Mary K. Whitmire, Associate Professor; M.B.A., Corpus Christi State. Aviation marketing, economics.

Joseph Wildinger, Assistant Professor; M.M.S., M.S., Troy State. Air carrier operations, corporate aviation.

Jack Wrinkle, Assistant Professor; M.A.S., Embry-Riddle. Aviation safety, human factors.

FLORIDA INSTITUTE OF TECHNOLOGY
School of Aeronautics
Division of Aviation Studies
Master of Science in Aviation Program
Master of Science in Aviation Human Factors Program

Program of Study

The Master of Science in Aviation (M.S.A.) is designed to help meet the professional growth needs of persons interested in a wide range of aviation careers. The degree is especially relevant for those who have earned a baccalaureate degree in aviation and those who have worked in the aviation field and now require more specialized knowledge.

Two areas of emphasis are currently being offered: the airport development and management option and the aviation science option. Persons interested in careers in airport or airline management, airport consulting, and governmental organizations involved in the management or regulation of airports will be interested in the airport development and management option, which is designed to offer specialization in the management and operation of airports. Modern airport management requires a unique set of skills that cross traditional corporate lines into government regulation and oversight, public finance, public administration, urban planning, environmental protection, security, flight safety, contract management, airspace management, and a host of other disciplines. This degree option helps place all of these disciplines in context and prepares professionals for demanding but uniquely rewarding careers.

The aviation science option (safety) places emphasis on aviation safety, accident investigation, technical aviation consulting, and educational, regulatory, or investigative positions in governmental or trade organizations. This option is designed to broaden knowledge in a variety of technical areas of aviation. Modern accident investigators, consultants, government employees, teachers, and researchers often need to be able to assimilate and interpret aviation information from a variety of specialists and to apply this information to the resolution of complex problems and issues. The degree option prepares professionals to understand, integrate, and use information derived from such diverse fields as aviation psychology, human factors, avionics, electronics, meteorology, and aviation physiology.

Research Facilities

The M.S.A. program offers modern computer laboratories for weather, air traffic control, advanced planning applications, statistical analyses, and word processing. The School of Aeronautics is housed in the Skurla Building located on the main campus of Florida Tech. Modern conference rooms and classrooms are available for meetings, program courses, and content-specific seminars. All rooms have access to audiovisual equipment, including projectors, VCR equipment, projection screens, movie projectors, and engine mock-up and avionics equipment. Library resources contain major, general-purpose magazines and newspapers as well as professional reference books, dictionaries, and indexes. There is also electronic library search capability on the Internet and access to the World Wide Web.

A complete flight training facility consisting of fifty single and multiengine aircraft, flight simulators, classrooms with appropriate flight instructional aids, and a comprehensive maintenance facility are located minutes away at the Melbourne International Airport for students' use. A new multiengine simulator is also available on the main campus.

Financial Aid

Awards are based on academic promise, need, college costs, and the availability of funding. Inquiries should be sent to the Director of Financial Aid. Students eligible for Veterans Administration (VA) benefits may contact the VA representative on the Melbourne campus. Some flight, airport, and human factors internship may be available.

Cost of Study

Tuition for the academic year 1998–99 was $550 per credit hour. Graduate student teaching and research awards may include tuition remission and a stipend.

Living and Housing Costs

Room and board on campus cost approximately $2200 per semester in 1998–99. On-campus housing (dormitories and apartments) is available for full-time single and married graduate students, but priority for dormitory rooms is given to undergraduate students. Many apartment complexes and rental houses are available near the campus.

Student Group

Florida Tech has an active student government that acts as a vital link between the administration and the student body, as the liaison between the university and the community, and as the catalyst for social events. The organization promotes new ideas and encourages student participation at all levels of university activity. The School of Aeronautics specifically has a student-organized Aeronautics Committee, the National Association of Women in Aviation (NAWA), Florida Institute of Technology School of Aeronautics Alumni organization (FITSA), Alpha Eta Rho, the Falcons Flight Team, and the American Association of Airport Executives (AAEE).

Student Outcomes

Graduates of the program obtain positions in various areas and companies such as airport management; FAA; Airborne Express; K-C Aviation; Molex Inc.; UXB International; Collier County Airport Authority; U.S. Air Force Civilian Personnel; Pittsburgh International Airport; City of Houston Department of Aviation; San Francisco International Airport; NASA; Hoyle, Tanner, & Associates; Flight Data Inc.; Kenton County Airport Board; Hanover County Municipal Airport; Greiner Inc.; and The New Piper Aircraft.

Location

Melbourne, Florida, is a medium-sized community with a subtropical climate. Shopping centers and a major hospital are nearby. Melbourne International Airport serves the community and campus with flights from all major cities. There are beaches approximately 3 miles from the campus for surfing, sailing, skin diving, and water-skiing. The Kennedy Space Center and Disney World are also nearby.

The Institute and The School

Florida Tech is an accredited, coeducational, independent university. Since its founding in 1958, along with the U.S. space program, the university has grown rapidly. Today nearly 4,000 students are enrolled in undergraduate and graduate programs. The university offers doctoral degrees in eighteen disciplines, while master's degrees are offered in more than forty areas of study. The School of Aeronautics is also accredited by the Council for Aviation Accreditation.

Applying

A strong background in aviation or its related fields is recommended. Scores on the GRE General Test are required. Applications should be received at the Graduate Admissions Office by early January. Students are selected on the basis of undergraduate records and interviews with several members of the faculty.

Correspondence and Information

Dr. Nathaniel Villaire
Program Chairman of Graduate Studies
School of Aeronautics
Florida Institute of Technology
150 West University Boulevard
Melbourne, Florida 32901-6988
Telephone: 407-674-8120
Fax: 407-984-8461
E-mail: villaire@fit.edu
World Wide Web: http://www.fit.edu/AcadRes/aero/
index2.html

For Catalog and Application:
Graduate Admissions Office
150 West University Boulevard
Melbourne, Florida 32901
Telephone 407-674-8027
    800-944-4348 (toll-free)
Fax: 407-723-9468
World Wide Web: http://www.fit.edu

Florida Institute of Technology

THE FACULTY AND THEIR RESEARCH

Rich Adams, Assistant Professor; Ph.D. candidate, Central Florida. Systems analysis of human error accidents in aviation and the role of cognitive information processing in error reduction.

Ballard Barker, Associate Professor and Associate Dean; Ph.D., Oklahoma. Planning, design, operation, and management of airports and other aviation facilities.

John Cain, Instructor; Ed.S., Ph.D. candidate, Florida Tech. Developing curriculum and instructing aviation science and aviation management academics. Instructs aeronautical science courses in aerodynamics, computer systems, aviation math, aircraft systems, aircraft performance, and accident investigation.

Kenneth Crooks, Assistant Professor and Director, Division of Aviation Studies; J.D., Florida; M.P.A., Golden Gate. Corporate finance, decision theory, investments, business law, management, transportation, and labor relations with an emphasis on the legal environment of aviation management.

Paul Davis, Assistant Professor and Industry Intern Coordinator; M.B.A., Florida Tech. Aviation studies, including cockpit/crew resource management, aviation management, flight safety, human factors, and ergonomics.

William Graves, Assistant Professor and Program Chair, Aviation Management; M.B.A., Alabama. Airport planning, development, and design. Extensive experience in operational test and evaluation of aircraft and program management of aircraft and associated weapons development.

Richard Lanier, Associate Professor; M.A., Florida; Ph.D., Central Florida. Sensation and perception, decision making, NASA space systems human factors.

James W. McIntyre, Associate Professor; M.S., Rensselaer. Flight training, fixed-base operations, operations research.

N. Thomas Stephens, Professor and Dean, School of Aeronautics; Ph.D., Florida. Numerous types of single engine and multiengine aircraft. Commanded operational radar surveillance missions; conducted research and development involving monitoring air pollutant dispersion from power plant plumes from an instrument aircraft. Consultant to the World Health Organization, University Pertanian Malaysia, the Emirate of Abu Dhabi, the Pan American Health Organization, and the Federal Highway Administration's National Highways Institute.

Nathaniel Villaire, Associate Professor and Program Chair, Graduate Studies, School of Aeronautics; Ed.D., William and Mary, M.P.A., Golden Gate. High-altitude pulmonary physiology, human factors in ATC, airspace management and safety.

Michael Wilson, Associate Professor; M.S., Oregon State. Captain in various single engine and multiengine jet and helicopter aircraft. Management experience and special training includes flight safety international type, six-month recurrent for Learjet, Falcon 10, and Falcon Fanjet aircraft; cockpit resource management; international operations; medical emergency training; Loran instructor; and VLF navigation systems training.

Donna Forsyth Wilt, Assistant Professor; M.S.E.E., Ph.D., Florida Tech. Situational awareness, avionics, navigation and communications systems, physical science, flight training (CFII).

Michael Witiw, Associate Professor; M.Pr.M., Saint Louis; Ph.D., Florida Tech. Aviation, synoptic, and satellite meteorology; climatology; use of multiple-regression techniques to compact and store climatological data.

GEORGIA INSTITUTE OF TECHNOLOGY
A Unit of the University System of Georgia

School of Aerospace Engineering

Programs of Study

The School of Aerospace Engineering encompasses a wide variety of disciplines related to flight. As new opportunities and problems are explored, the boundaries of the field of aerospace engineering continue to expand. Graduate studies in aerospace engineering expose the student to advanced course work, research, and design. While the courses provide specialization and depth in individual areas, research and design pose unique challenges that require the student to venture far and wide in search of new solutions.

The School offers the Master of Science and the Ph.D. degrees. The Master of Science degree requires either 33 semester hours of course work beyond the bachelor's degree, of which up to 3 semester hours may be a research project, or 24 semester hours of course work plus a thesis. Depending on the student's background, this program can be completed in three to four semesters.

The Doctor of Philosophy degree requires 50 semester hours of course work beyond the bachelor's degree, admission to candidacy based on a qualifying examination and a dissertation proposal, and successful completion and defense of a doctoral thesis on original research. Most students complete the Ph.D., degree in three to five years.

Courses are offered and research is conducted in the areas of aeroacoustics, aerodynamics, aeroelasticity, astrodynamics, combustion, computational fluid dynamics, composite materials, computer-aided design, design optimization, experimental diagnostics, flight controls, flight mechanics, gasdynamics, propulsion, rotorcraft technology, structural mechanics, and turbulence simulations.

Research Facilities

The research program at Georgia Tech ranks among the top five aerospace programs nationwide and enjoys a worldwide reputation. Research projects in a variety of areas are performed under contract to federal agencies that include the U.S. Army, Navy, and Air Force; NASA; the Department of Energy; the National Science Foundation; the Environmental Protection Agency; and the Federal Aviation Administration, as well as several industries. The major laboratory facilities in the School are the acoustics laboratory, aeroelastic rotor test chamber, composites laboratory, crashworthiness facility, flight dynamics laboratory, low-turbulence wind tunnel, chemical propulsion research laboratory, laboratory for pulse combustion processing, rotor static thrust facility, rotorcraft simulator, solid propellant combustion laboratory, system identification laboratory, turbulent combustion diagnostics laboratory, and John J. Harper 7 foot by 9 foot wind tunnel. Many of these facilities are equipped with state-of-the-art laser diagnostics and computerized data acquisition.

In addition, the facilities of the Computer-aided Design Laboratory and the Symbolic Computation Laboratory of the College of Engineering are available for use by the aerospace faculty members and students. A variety of mainframe computers, dedicated minicomputers, workstations, and microcomputers are available, and several research projects are conducted on off-campus supercomputers using nationwide networks. This strong research program enables students to work with state-of-the-art research facilities and ensures that aeronautical engineering courses are taught by faculty members who are actively engaged in research at the cutting edge of their fields.

Financial Aid

Financial assistance for highly qualified applicants is available through research or teaching assistantships, and Presidential Fellowships. Financial awards are based on academic qualifications and potential for research productivity. Stipends can range from $14,500 to $16,000 per year, plus all tuition and fees. International students are required by faculty members to provide certified documentation of financial support but are eligible to compete for financial aid.

Cost of Study

Tuition and fees for the 1999–2000 academic year are $6193 per semester for nonresidents of Georgia and $1849 per semester for residents. Part-time students are charged prorated amounts. The rates for tuition and fees are subject to change at the end of any semester.

Living and Housing Costs

Dormitory rooms for unmarried students and apartments for married students and their families are available at reasonable cost through the Institute. Most of these rooms were upgraded for the 1996 Olympics. Rooms and apartments in privately-owned dwellings within walking distance or a short driving distance are available for a range of prices. Assistance in locating housing is offered by the Institute's Housing Office.

Student Group

Approximately 13,000 students are enrolled at Georgia Tech. About 9,300 of these are undergraduate students, and 3,700 are graduate students. The School of Aerospace Engineering has an approximate enrollment of 250 undergraduate and 210 graduate students.

Student Outcomes

Graduates have found employment in industry, government laboratories, and universities in this country and abroad.

Location

Atlanta is one of the most beautiful and exciting cities in the United States. Situated at an altitude of 1,050 feet above sea level, Atlanta is the second-highest major city in the country. Its topography is responsible for a favorable climate of moderate summers and mild winters. Atlanta's location has been an important factor in its development into the transportation, financial, and communications hub of the Southeast. Within a 2-hour drive of Atlanta are the beautiful Great Smoky Mountains; much closer are Lake Lanier and Lake Allatoona, noted for their swimming, boating, fishing, and picnicking opportunities. Atlanta is also the home of the Atlanta Braves, the Falcons, and the Hawks. Atlanta, site of the 1996 Summer Olympics, is the headquarters of many large corporations and is the financial hub of the rapidly growing southeastern United States.

The School

The Daniel Guggenheim School of Aeronautics was established in 1930. It has continuously been at the forefront of the rapidly changing field of aerospace engineering. The first master's degree was awarded in 1934, and the Ph.D. program was started in 1961. The School currently has 29 faculty members and nearly 40 professional staff members. Over the five-year period from 1992 to 1997, the School awarded a total of 296 M.S. degrees and 88 Ph.D.'s. The high graduate-degree productivity is the result of the very active research programs throughout all disciplines of the School. Most of the research conducted is performed by students as part of their M.S. or Ph.D. programs.

Applying

To assure full consideration for financial aid, submit a complete application by the end of January of the year of planned enrollment.

Correspondence and Information

Dr. J. I. Jagoda
Associate Chair for Research and Graduate Studies
School of Aerospace Engineering
Georgia Institute of Technology
Atlanta, Georgia 30332-0150

Telephone: 800-738-3359 (toll-free)

Georgia Institute of Technology

THE FACULTY AND THEIR RESEARCH

David S. Lewis, Jr. Chair
Ben T. Zinn, Regents' Professor; Ph.D., Princeton, 1965. Combustion instability, combustion control, propulsion, pulse combustion.

Regents' Professor
N. L. Sankar, Ph.D., Georgia Tech, 1977. Computational fluid dynamics, helicopter aerodynamics.

Regents' Professors Emeriti
Robin B. Gray, Ph.D., Princeton, 1957. Aerodynamics.
Edward W. Price, B.A. (math), B.A. (physics), UCLA, 1948. Propulsion, combustion.

Professors
Krishan K. Ahuja, Ph.D., Syracuse, 1976. Aeroacoustics, fluid mechanics. (Joint appointment with the Georgia Tech Research Institute)
E. A. Armanios, Ph.D., Georgia Tech, 1985. Composite and structures, fracture mechanics, design.
Olivier Bauchau, Ph.D., MIT, 1981. Structural dynamics, multi-body dynamics, experimental dynamics.
Anthony J. Calise, Ph.D., Pennsylvania, 1968. Flight mechanics and controls.
James I. Craig, Ph.D., Stanford, 1968. Structural mechanics, experimental mechanics, design.
Wassim M. Haddad, Ph.D., Florida Tech, 1987. Stochastic modeling, robust multivariable control, structural dynamic control.
S. V. Hanagud, Ph.D., Stanford, 1963. Structural mechanics and materials, flexible body control, nonlinear dynamics.
D. H. Hodges, Ph.D., Stanford, 1973. Nonlinear structural mechanics, computational mechanics and dynamics, rotorcraft dynamics and aeroelasticity.
J. I. Jagoda, Associate Director for Research and Graduate Studies; Ph.D., Imperial College (London), 1976. Experimental combustion, optical diagnostics, chemical propulsion systems.
M. P. Kamat, Ph.D., Georgia Tech, 1972. Nonlinear structural analysis and optimization, computational methods.
George A. Kardomateas, Ph.D., MIT, 1985. Mechanics of materials and structures, composite structures, fracture mechanics.
N. M. Komerath, Ph.D., Georgia Tech, 1982. Experimental fluid mechanics, aerodynamics.
Robert G. Loewy, Chair; Ph.D., Pennsylvania, 1962. Helicopter structure dynamics, aeroelasticity, composite structures for aircraft and spacecraft, unsteady aerodynamics.
David J. McGill, Ph.D., Kansas, 1960. Dynamics. (Joint appointment with the School of Civil and Environmental Engineering)
Suresh Menon, Ph.D., Maryland, 1984. Combustion/propulsion, computational fluid dynamics, turbulence and turbulent mixing.
J. V. R. Prasad, Ph.D., Georgia Tech, 1985. Applied mechanics, flight mechanics and controls.
Daniel P. Schrage, D.Sc., Washington (St. Louis), 1978. Rotorcraft and aircraft design, aeroelasticity, flight mechanics and controls, concurrent engineering.
Ramesh R. Talreja, Ph.D., Denmark Technical, 1974. Composite materials and structures.

Professors Emeriti
Robert L. Carlson, Ph.D., Ohio State, 1962. Structural mechanics, fatigue in structures.
Donnell W. Dutton, M.S., Georgia Tech, 1940; PE. Systems engineering design, stress analysis, aerodynamic vehicle structures.
Wilfred F. Horton, B.Sc., University College (England), 1940. Structures, design.
James E. Hubbartt, M.S., Case Tech, 1950. Fluid mechanics, boundary layer control and propulsion.
Howard M. McMahon, Ph.D., Caltech, 1958. Fluid mechanics, turbulent boundary layers, helicopters, V/STOL aerodynamics.
G. Alvin Pierce, Ph.D., Ohio State, 1966. Aeroelasticity, unsteady aerodynamics.
James C. Wu, Ph.D., Illinois at Urbana-Champaign, 1957. Unsteady aerodynamics, viscous flow, computational aerodynamics, turbulence.

Associate Professors
Stanley C. Bailey, Ph.D., Stanford, 1967. Structural mechanics, solar energy.
Oliver McGee, Ph.D., Arizona, 1988. Composite structures, structural dynamics, turbomachinery.
Stephen Ruffin, Ph.D., Stanford, 1993. Computational fluid dynamics, high-speed propulsion, hypersonics and nonequilibrium flows, aerodynamics.
C. V. Smith, Sc.D., MIT, 1962. Structural mechanics, dynamics.
P. K. Yeung, Ph.D., Cornell, 1989. Turbulence and turbulent mixing, computational fluid dynamics.

Assistant Professors
Dimitri N. Mavris, Ph.D., Georgia Tech, 1988. Aircraft and notercraft design, air breathing propulsions system design, aerodynamics.
John Olds, Ph.D., North Carolina State, 1993. Multidisciplinary design optimization, orbital mechanics, space launch vehicle design.
Amy R. Pritchett, Sc.D., MIT, 1997. Flight simulation, avionics and cockpit design.
Jerry M. Seitzman, Ph.D., Stanford, 1991. Experimental combustion, propulsion, laser diagnostics.
Marilyn Smith, Ph.D., Georgia Tech, 1994. Computational aeroelasticity, computational fluid mechanics, aeroacoustics.

Lecturer
Michael W. M. Jenkins, C.Eng., Gloucester College (England), 1958. Aerospace vehicle design, stability/control and handling qualities, flight and tunnel testing, advanced concepts.

Adjunct Professors
David A. Peters, Ph.D., Stanford, 1974. Aeroelasticity, vibrations and helicopter dynamics.
Brian L. Stevens, Ph.D., Manchester (England), 1966. Controls, nonlinear simulations.

MASSACHUSETTS INSTITUTE OF TECHNOLOGY

School of Engineering
Department of Aeronautics and Astronautics

Programs of Study

The MIT Department of Aeronautics and Astronautics offers Master of Engineering in aeronautics and astronautics, Master of Science in aeronautics and astronautics, Master of Science in engineering and management, Doctor of Philosophy, and Doctor of Science degrees. Graduates with an aerospace engineering degree find career opportunities in commercial and military aircraft and spacecraft engineering, space exploration, airlines and the air transportation industry, teaching, research, military service, and in many related technology-intensive fields, such as transportation, information, and the environment. The demanding technical education has a strong emphasis on understanding complex systems and is excellent preparation for careers in business, law, medicine, and public service. Critical disciplines and technologies with which the aerospace engineer must be familiar include mechanics and physics of fluids; structures and materials; propulsion and energy conversion; information, control, and estimation; humans and automation; avionics; and aerospace systems. The department believes that a strong research program is essential to a challenging and relevant educational environment; research currently ranges from jet aircraft and rotorcraft to rockets and spacecraft and includes, as a key aspect, the worldwide information and navigation systems in which these sophisticated vehicles must operate. In addition, the department is currently in the process of developing a broad educational initiative, which has as its engineering context the conception, design, implementation, and operation (CDIO) of aerospace and related complex high-performance systems.

Research Facilities

The Department of Aeronautics and Astronautics has an extensive complement of modern research facilities. These include the Active Materials and Structures Laboratory, the Fluid Dynamics Research Laboratory, the Gas Turbine Laboratory, the Information and Control Engineering Laboratory, the International Center for Air Transportation, the Lean Aerospace Initiative, the Man Vehicle Laboratory, the Space Systems Laboratory, the Technology Laboratory for Advanced Composites, and the Wright Brothers Wind Tunnel. The department is also a participant in the Laboratory for Information and Design Systems. In addition, as part of its CDIO initiative, the department is constructing a Learning Laboratory for Complex Systems, which will bring together undergraduate and graduate students, faculty members, and industrial partners in an integrated design-build environment.

Financial Aid

MIT makes financial support to graduate students available from a variety of sources in several different forms, including fellowships, scholarships, teaching and research assistantships, the Federal Work-Study Program, and loans (the Technology Loan Fund). More than three fourths of the graduate students in the department are employed as research or teaching assistants as they progress toward their degree. These jobs give the students financial assistance and opportunities for close interaction with faculty members.

Cost of Study

Tuition for graduate students in the School of Engineering for the two-term academic year is $25,000 in 1999–2000. Tuition for the summer is $8335.

Living and Housing Costs

Dormitory and apartment accommodations for 1060 single graduate students are available at a cost of $3258 to $8415 for the 1999–2000 academic year. Meal service is offered seven days a week on either a contract or cash basis. For families, there are 407 apartments that ranged in price from $675 to $1019 per month, including utilities. Most residence assignments for new, single students are given for one year only; there is a two-year option, however, for families. The majority of new graduate students who apply for campus housing are assigned housing. Off-campus apartments are available within commuting distance of the campus; typical prices (excluding heat and utility costs) were $900 and up per month for a one-bedroom apartment and $1200 and up per month for a two-bedroom apartment. The quality of apartments is generally average, and the quantity is limited, so an early (one- to two-week) search for housing is advised.

Student Group

A total of 238 applications were received for the fall 1998 term; of this number, 134 were admitted, and 67 accepted the offer of admission. Enrollment for spring 1998 included 117 Master of Science, 63 Ph.D., and 10 M.Eng. students. There were 8 members of minority groups and 24 women.

Location

MIT occupies a 125-acre campus on the north bank of the Charles River with a fine view of Boston's skyline. The vacation areas of New Hampshire and Vermont, only a few hours from the campus, provide mountaineering, skiing, and hiking opportunities. The Boston area is world-renowned for its cultural facilities. Performances of the Boston Symphony Orchestra, the Boston Pops, and the Boston Opera take place within a 5-minute drive or 30-minute walk of the campus. The area has many famous museums, restaurants, historic sites, and more than a dozen universities and specialized schools within a few miles of the campus.

The Institute

MIT, founded in 1861 as a private, coeducational, endowed institution committed to the extension of knowledge through teaching and research, is one of the foremost institutes of technology in the world. The Institute is composed of five schools: Architecture and Planning, Engineering, Humanities and Social Science, Management, and Science. In 1997, 9,880 students were enrolled; more than half were in the Graduate School and the Whitaker College of Health Science and Technology. The campus provides excellent athletic facilities for indoor and outdoor sports, sailing, swimming, and crew. There are extensive cultural programs on campus.

Applying

Applications for entrance to the Graduate School in June or September should be made before January 15. Applicants seeking admission in February should file before November.

Correspondence and Information

Mark D'Avila
Graduate Student Services Coordinator
Massachusetts Institute of Technology
125 Massachusetts Avenue, 33-208
Cambridge, Massachusetts 02139
Telephone: 617-253-2260
Fax: 617-253-0823
World Wide Web: http://web.mit.edu/aeroastro/www/core

Massachusetts Institute of Technology

THE FACULTY AND THEIR RESEARCH

Edward F. Crawley, Sc.D., Professor and Department Head. Space structures, structural dynamics, turbomachine structures, aeroelasticity, control structure interaction, aeroservoelasticity.

Edward M. Greitzer, Ph.D., Professor and Associate Department Head. Gas turbine engines, turbomachinery, fluid dynamics, propulsion, active control of turbomachinery flows.

Professors

Vincent Chan, Ph.D. Optical communications, wireless communications, space communications, networks.

John J. Deyst Jr., Sc.D. Avionics and controls, flight guidance, cockpit instrumentation, critical systems, fault tolerance.

Alan H. Epstein, Ph.D. Propulsion, turbomachinery, engine controls, microelectrical and mechanical systems.

R. John Hansman Jr., Ph.D. Aircraft/atmospheric interaction, aviation safety, cockpit human factors and information management, air traffic control, airline operations instrumentation.

Wesley L. Harris, Ph.D. Unsteady aerodynamics, aeroacoustics, computational fluid dynamics, biomedical fluid dynamics (hemodynamics), sustainment, logistics, government acquisition policy.

Daniel E. Hastings, Sc.D. Spacecraft-environmental interactions, plasma physics, power generation in space, space systems, space policy.

Walter M. Hollister, Sc.D. Air traffic control, aircraft operations, cockpit displays, flight guidance, navigation systems.

Jack L. Kerrebrock, Ph.D. Aircraft engines, compressors, turbine cooling.

Paul A. Lagace, Ph.D. Composite materials, fracture and fatigue (longevity), damage tolerance, manufacturing technology, system engineering, management issues.

Nancy G. Leveson, Ph.D. Safety of software-controlled systems, software engineering, system engineering, system safety engineering, human-computer interaction.

Winston R. Markey, Sc.D. Automatic geometric measurements, automatic control systems.

Manuel Martinez-Sanchez, Ph.D. Rocket propulsion, space power, space tethers, dynamics of turbomachinery.

Earll M. Murman, Ph.D. Systems engineering, product development, engineering education, aerodynamics.

Amedeo R. Odoni, Ph.D. Airport planning and design, air traffic control, applied probability theory, operations research, transportation systems analysis.

Jaime Peraire, Ph.D. Computational fluid mechanics, aerodynamics, unstructured mesh technology, parallel computing.

Wallace E. VanderVelde, Sc.D. Automatic control, estimation, interial systems, navigation, fault-tolerant systems.

Sheila E. Widnall, Sc.D. Space and aeronautical systems, acquisition reform, defense aerospace management.

Laurence R. Young, Sc.D. Human factors, space life science, flight simulation, expert systems, biomechanics.

Associate Professors

Mark Drela, Ph.D. Computational fluid dynamics, transonic aerodynamics, low-speed aerodynamics, boundary layers, design methodology.

Eric M. Feron, Ph.D. Automatic control, numerical algorithms for optimization.

Nesbitt Hagood, Sc.D. Adaptive structures, structural dynamics and control, space structures, actuator/sensor development, active materials, micromotors and generators.

Steven R. Hall, Sc.D. Automatic control, control of helicopters, control of structures.

Dava J. Newman, Ph.D. Aerospace medicine, bioengineering, physiology, dynamics and control, teleoperation and virtual reality, robotics.

S. Mark Spearing, Ph.D. Layered and composite materials, high-temperature materials, fracture and fatigue, micromechanics, electronic packaging, ceramics, microelectromechanical systems.

Ian A. Waitz, Ph.D. Propulsion, supersonic combustion, fluid dynamic mixing, aeroacoustics, hydroacoustics, environmental aerospace engineering.

Brian Williams, Ph.D. Space and aerial robotics, cognitive robotics, automated reasoning and artificial intelligence, automation for operations and design, hybrid control systems.

Assistant Professors

Carlos E. S. Cesnik, Ph.D. Structural mechanics, structural dynamics, aeroelasticity, computational mechanics.

John-Paul Clarke, Sc.D. Air traffic control; airline and airport operations; modeling, design, and operation of complex systems; environmental impact of aviation (noise and emissions); application of advanced technology to aircraft operations and air traffic control.

David L. Darmofal, Ph.D. Computational fluid dynamics, probabilistic aerothermal design, fluid dynamics, scientific visualization.

Daniel Frey, Ph.D. Probabilistic design, manufacturing.

James K. Kuchar, Ph.D. Information management, probabilistic systems modeling, hazard alerting and avoidance, cockpit displays, flight simulation.

David Miller, Sc.D. Distributed satellite systems, precision telescope structures, formation flying satellites.

Eytan Modiano, Ph.D. Communication networks and protocols, satellite and hybrid networks, high-speed networks.

Professors Emeriti

Richard H. Battin, Ph.D. Celestial navigation, optimal mechanics.

Eugene E. Covert, Ph.D. Steady and unsteady aerodynamics, aircraft power plant integration (performance), boundary layers, three-dimensional flow, wind tunnel.

John Dugundji, Ph.D. Aeroelasticity, structural dynamics, composite materials.

Norman D. Ham, Sc.D. Vibration control in helicopters through control of rotors.

Thomas B. Sheridan, Ph.D. Humans and automation, cognitive engineering, teleoperation and virtual reality.

Leon Trilling, Ph.D. Gas-surface interactions, rarefied gasdynamics, science, technology and society.

Senior Lecturers

Charles W. Boppe, Ph.D. Student projects, product development projects.

Fredric F. Ehrich, Sc.D. Microengine research, aircraft gas turbine technology.

Deborah Nightingale, Ph.D. Manufacturing systems design, implementation, and operation; industrial and systems engineering.

Charles Oman, Ph.D. Flight simulation, spatial orientation, instrumentation, space physiology, manual control, human factors.

Rudrapatna V. Ramnath, Ph.D. Modern dynamics and its application to athletics.

Joyce M. Warmkessel, Ph.D. Systems engineering; spacecraft, aircraft, and ground systems, with particular expertise in the establishment of systems engineering processes and tools.

Peter W. Young, M.S. Space systems management, United States Air Force.

PENNSYLVANIA STATE UNIVERSITY

Department of Aerospace Engineering

Programs of Study

Graduate students in Penn State's Department of Aerospace Engineering can pursue studies leading to the M.S., M.Eng., and Ph.D. degrees. A high-performance computing graduate minor has recently been created to educate students in scientific and high-performance computing with emphasis on the uses of parallel computers. Course work and research projects are offered in the following general areas of specialization: analytical and computational fluid dynamics, aeroacoustics, experimental fluid dynamics, dynamics and control, flight science and vehicle dynamics, rotorcraft engineering, structures and materials, structural dynamics, space propulsion, and turbomachinery. The course requirements for students pursuing the M.S. degree total 30 credits, including 6 thesis credits and 24 course credits. In addition to completing the specified core courses, a wide variety of courses may be selected in the areas of specialization. A thesis is required for the M.S. degree. The M.Eng. degree requires 30 course credits, including a 2-credit scholarly paper and a 1-credit seminar; no thesis is required for the M.Eng. degree. Candidates for the Ph.D. degree take a candidacy exam within two semesters after identified entry into the doctoral program. There is no foreign language requirement for the Ph.D.; however, the University requires each Ph.D. candidate to demonstrate satisfactory proficiency in written and spoken English. When a Ph.D. student has substantially completed the course work, a comprehensive written and oral examination covering both the major program and the minor field of study is required. After the doctoral candidate has satisfied all other requirements for the degree, the final oral examination is held.

Research Facilities

A major thrust of the department involves advances in computer usage, both for analysis and for experimental data acquisition and processing. For such activity, excellent computer facilities and equipment are available, including an IBM 9000 with three vector processors; a 75-node IBM SP2; a 16-node SGI Power Challenge; numerous DEC, IBM, Sun, and Silicon Graphics workstations; X-terminals; and a massively parallel Connection Machine (CM-200). There is also excellent access to NSF and NASA supercomputer facilities.

Other major facilities include a large, low-turbulence subsonic wind tunnel with a 3.25-foot by 5-foot test section, speed range to 150 mph, and a floor-mounted six-component strain gauge balance; two additional low-turbulence wind tunnels (one with a 2-foot by 3-foot test section and the second with a 3-foot diameter axisymmetric test section); a supersonic wind tunnel (6-inch by 6-inch test section, speed range from Mach 1.4 to Mach 4.0); a supersonic free shear layer facility for mixing studies; a laminar flow water channel (2.5-foot by 1.5-foot test section); an axial flow turbine facility with heavily instrumented blading to measure unsteady pressures, heat transfer, and shear stresses; a multistage compressor facility driven by a 500-hp motor; a single-stage axial flow compressor (3-foot diameter) facility; an automotive torque converter facility; a heat-transfer facility to simulate turbine flow; a linear turbine cascade for heat-transfer research; a real-time color image processing system for the post-processing of liquid crystal images in convective heat transfer research; various probe calibration jets; several laser-Doppler anemometers, including a subminiature semiconductor model; an ATC/510G flight simulator; aeroacoustic research facilities—a small jet noise facility, an anechoic chamber, and a reverberant room; a vacuum tank facility for low-density flow (pressure range to 10^{-4} Torr, pumping approximately 5,000 cfm at 5×10^{-3} Torr) and associated instrumentation; an unsteady propellant combustion facility and a variable power microwave generator; and a compressed air flow facility (300-psi reservoir). Also available are an autoclave, a pultrusion machine, a filament winder, a braiding machine for composite materials manufacturing, a thermal analysis system, an ultrasonic inspection system, an acoustic emission system, a servohydraulic materials testing machine, a high-temperature biaxial tension/torsion testing facility, a fiber-optic interferometer, a reflection polariscope used in material fabrication and characterization, a structural dynamics laboratory, a space environmental simulator, a spectrometer, and a CW Nd laser for space-propulsion research.

Financial Aid

Graduate students are supported by a variety of government and industrial fellowships and by research and teaching assistantships. In 1998–99, stipends for half-time research assistants ranged from $14,946 to $16,030 for the year, including the summer. Teaching assistants earned from $11,205 to $12,015, excluding the summer. Stipends for fellows averaged $16,000. Federal Perkins and Pennsylvania Higher Education Assistance Agency loans are available upon evidence of need.

Cost of Study

Tuition for 1998–99 was $3267 per semester for Pennsylvania residents and $6730 per semester for nonresidents, with two semesters to the academic year.

Living and Housing Costs

University and private housing is available to graduate students. In 1998–99, University dormitory rooms cost $1255 and up per semester, room and full board $2080 per semester, and apartments $1270 and up per semester. Privately owned apartments are also available in the community.

Student Group

The Aerospace Engineering Department enrolls about 70 graduate students, approximately one half of whom are at the Ph.D. level.

Student Outcomes

Graduates with advanced degrees from Penn State's Department of Aerospace Engineering have a 100 percent employment record during the past ten years in various industry, governmental, and university positions. Examples of employers include Lockheed Martin Missiles and Space Center, GE Aircraft Engines, Naval Air Warfare Center, Knolls Atomic Power Laboratory, and General Motors Institute.

Location

Penn State's main campus, University Park, is located in the center of the state on U.S. Highway 322 near Interstate 80. The town and its surrounding area has a community population of about 74,000.

The University

Pennsylvania State University is a land-grant university founded in 1855. The Graduate School awards about 1,965 master's degrees and 545 doctorates each year. Research expenditures in aerospace engineering are approximately $2.5 million annually.

Applying

Qualified students may be admitted at the beginning of any semester—in August, January, or June. Admission is granted by the dean of the Graduate School after approval of the application by the aerospace engineering department.

Correspondence and Information

Dr. Dennis K. McLaughlin, Professor and Head
Department of Aerospace Engineering
233 Hammond Building
Pennsylvania State University
University Park, Pennsylvania 16802
E-mail: dkmaer@engr.psu.edu
World Wide Web: http://www.psu.edu/dept/aerospace

Pennsylvania State University

THE FACULTY AND THEIR RESEARCH

Anthony K. Amos, Professor of Aerospace Engineering; Ph.D. (civil engineering), Princeton. Structural mechanics, dynamics, and control: analysis of flexible space structures, aeroelasticity problems in V/STOL aircraft. (aka@psu.edu)

Cengiz Camci, Associate Professor of Aerospace Engineering; Ph.D. (convective heat transfer and fluid dynamics), Von Karman Institute and K. U. Leuven (Belgium). Aerothermodynamics of turbomachinery, convective heat transfer, short-duration wind tunnel techniques, finite-element techniques for flow and heat-transfer calculations, laser-Doppler anemometry, liquid crystal imaging for heat transfer studies. (c-camci@psu.edu)

George S. Dulikravich, Associate Professor of Aerospace Engineering; Ph.D. (mechanical and aerospace engineering), Cornell. Computational fluid dynamics, including analysis and multidisciplinary inverse design and optimization; solutions of Euler and Navier-Stokes equations; hypersonic flows; design of coolant flow passage shapes; adaptive computational grids; artificial dissipation concepts; magnetohydrodynamics and electrohydrodynamics; flows with solidification and melting. (ft7@email.psu.edu)

Farhan S. Gandhi, Assistant Professor of Aerospace Engineering; Ph.D. (aerospace engineering), Maryland. Computational structural mechanics, structural dynamics, viscoelasticity and damping, smart structures, helicopter dynamics and aeroelasticity, advanced bearingless rotor systems. (fgandhi@euler.aero.psu.edu)

J. William Holl, Professor Emeritus of Aerospace Engineering; Ph.D. (mechanical engineering), Penn State. Hydrodynamics and cavitation: effect of microbubbles on cavitation inception, influence of surface deviations on cavitation performance.

Budugur Lakshminarayana, Evan Pugh Professor of Aerospace Engineering and Director, Center for Gas Turbines and Power; Ph.D. (mechanical engineering), D.Eng., Liverpool (England). Experimental, computational, and analytical fluid dynamics; measurement of three-dimensional unsteady flow in turbomachinery; three-dimensional unsteady Navier-Stokes solver development for the prediction of turbomachinery flows; turbulence modeling for curvature and rotation effects. (b1laer@engr.psu.edu)

George A. Lesieutre, Professor of Aerospace Engineering, Director of Graduate Studies, and Associate Director, Center for Acoustics and Vibration; Ph.D. (aerospace engineering), UCLA. Structural dynamics and vibration damping, composite structures, material damping modeling and characterization, piezoceramic actuation, structural control. (g-lesieutre@psu.edu)

Lyle N. Long, Professor of Aerospace Engineering, Director of the Institute for High Performance Computing Applications, and Co-Director of the Penn State Rotorcraft Center of Excellence; D.Sc. (aerospace engineering), George Washington. Computational fluid dynamics and hypersonics: massively parallel processing, aeroacoustics, and molecular dynamics. (lnl@psu.edu)

Mark D. Maughmer, Associate Professor of Aerospace Engineering; Ph.D. (aeronautical and astronautical engineering), Illinois at Urbana-Champaign. Analytical, computational, and experimental aerodynamics: aircraft design, performance, stability, and control; airfoil design and analysis; low Reynolds number aerodynamics. (mdm@euler.aero.psu.edu)

Barnes W. McCormick, Professor Emeritus of Aerospace Engineering; Ph.D. (aeronautical engineering), Penn State. Subsonic aerodynamics: advanced concepts in V/STOL, flight mechanics, and wake turbulence. (bwmaer@engr.psu.edu)

Dennis K. McLaughlin, Professor and Head of Aerospace Engineering; Ph.D. (aeronautics and astronautics), MIT. Experimental fluid dynamics and aeroacoustics: turbulent structure in supersonic shear layers, the aerodynamics and aeroacoustics of supersonic free jets and of centrifugal turbomachinery. (dkmaer@engr.psu.edu)

Robert G. Melton, Associate Professor of Aerospace Engineering and Director of Undergraduate Studies; Ph.D. (engineering physics), Virginia. Astrodynamics, spacecraft dynamics and control: trajectory optimization, perturbation analysis of low-thrust orbital motion, orbit determination, dynamics and control of multibody spacecraft. (rgmaer@engr.psu.edu)

Michael M. Micci, Professor of Aerospace Engineering; Ph.D. (aerospace engineering), Princeton. Rocket propulsion: experimental and analytical work on oscillatory burning of solid and liquid propellants; rocket motor instabilities; advanced propulsion concepts, particularly the heating of propellant gases to high temperature by the absorption of microwave radiation; optical diagnostics of nozzle flows expanding into a vacuum. (micci@henry2.aero.psu.edu)

Philip J. Morris, Boeing/A. D. Welliver Professor of Aerospace Engineering; Ph.D. (aeronautical engineering), Southampton (England). Computational and analytical fluid dynamics, hydrodynamic stability, computational aeroacoustics: turbulence modeling in high-speed flows, aerodynamics and acoustics of jets, boundary layer stability. (pjm@psu.edu)

Blaine R. Parkin, Professor Emeritus of Aerospace Engineering; Ph.D. (aeronautics), Caltech. Hydrodynamics, subsonic aerodynamics and aeroelasticity/hydroelasticity: boundary layer analysis and unsteady cavitation.

Edward C. Smith, Associate Professor of Aerospace Engineering and Co-Director of the Penn State Rotorcraft Center of Excellence; Ph.D. (aerospace engineering), Maryland. Composite structures, rotorcraft dynamics: aeroelastic and aeromechanical tailoring of composite rotor blades, composite beam modeling, elastomeric materials. (ecs@rcoe.psu.edu)

Hubert C. Smith, Associate Professor of Aerospace Engineering; Ph.D. (aerospace engineering), Virginia. Aircraft design, performance, and operations: low-speed aerodynamics and air transportation. (hcsaer@engr.psu.edu)

PRINCETON UNIVERSITY

School of Engineering and Applied Science
Department of Mechanical and Aerospace Engineering

Programs of Study

The Department of Mechanical and Aerospace Engineering provides three programs of graduate study and research leading to the following degrees: Master of Engineering, Master of Science in Engineering, and Doctor of Philosophy. The M.E. is a nonthesis degree program that requires eight courses. An M.S.E. candidate must successfully complete a coordinated series of departmental courses and independent study and submit an acceptable research thesis. A Ph.D. candidate is required to pass the departmental general examination (usually taken during the second year), submit an acceptable dissertation, and pass a final oral examination.

Coordinated programs of graduate course work and research training are available in a broad spectrum of topics that fall within the following major areas of departmental activity: combustion and energy conversion, computational methods, dynamics and control systems, energy and environmental technology assessment, flight science and technology, fluid mechanics, lasers and applied physics, and materials and mechanical systems. A typical graduate program in any of these areas includes a group of fundamental courses taken both in the Department of Mechanical and Aerospace Engineering and in fields closely related to the area of major interest, as well as an extensive involvement with experimental or theoretical research problems leading to an M.S.E. or Ph.D. thesis.

Typically, two years of full-time resident graduate study are necessary to fulfill the requirements for the M.S.E. degree; four years are normally required for the Ph.D. degree; and the M.E. degree requires nine to twelve months.

Research Facilities

The department offers outstanding facilities to conduct research in the Engineering Quadrangle and at the Forrestal campus. Laboratories are available for research in propulsion, combustion, engines, gasdynamics, flight mechanics, instrumentation, aeroelasticity, lasers, and materials.

Financial Aid

A substantial number of fellowships, such as the Daniel and Florence Guggenheim, the H. C. Phillips, and the Gordon Wu, are awarded on a competitive basis. Additional fellowships are available to qualified students through the support of the National Science Foundation, the Hertz Foundation, and Zonta International. In 1999–2000, fellowships provide a stipend of up to $17,400 plus tuition. In addition to these awards, assistantships in research and in instruction are available; these provide a stipend plus full tuition. Summer support is normally available to augment both fellowship and assistantship awards.

Cost of Study

Tuition for 1999–2000 is $25,050, which includes use of University facilities and medical insurance. There is a graduation fee of $15 for the M.E. and M.S.E. degrees and $65 for the Ph.D. degree.

Living and Housing Costs

Rooms for single men and women at the Graduate College range from $2589 to $4027 in 1999–2000; board costs range from $2529 to $3638. Students living off campus may purchase meal tickets at these rates. University housing is available for married students at $485 to $755 per month; utilities are sometimes included. Jobs are generally available at the University or nearby for graduate students' spouses.

Student Group

Enrollment in the department stands at 60 registered students, of whom 53 are Ph.D. candidates. Thirty-seven American and nineteen other universities are represented, as are the Army, Air Force, and Naval Academies. The average age of students is 23. Most M.E. and M.S.E. graduates go into industry or government; Ph.D. graduates usually follow careers in industrial research or university teaching.

Location

Princeton is a residential community about an hour's drive from New York City, Philadelphia, and the Atlantic coast. The town has preserved its historic character by maintaining many original Colonial buildings and homes in an area in which there are a large and growing number of industrial research facilities. Athletic facilities are extensive and include those for golf, tennis, sailing, swimming, and squash. The Pocono Mountains, about 75 miles away, offer excellent skiing and camping areas. There is a rich variety of performing arts, including concerts, dance, classic and international films, and professional theater. The art museum possesses and displays distinguished paintings, drawings, and sculptures.

The Department

The Department of Mechanical and Aerospace Engineering features intensive programs in both mechanical and aeronautical engineering. It was founded in the 1940s as the Department of Aeronautical Engineering, but, because of a basic shift in student interests, it merged with the Department of Mechanical Engineering in 1963. Although some major facilities are still based at the Forrestal campus, many elements of research are located at the Engineering Quadrangle, ensuring a good balance between theoretical and experimental programs. By design, the department is kept small; emphasis is placed on close student-faculty relationships and on each student's developing to his or her full capacity.

Applying

Transcripts of grades and three letters of recommendation are required; TOEFL scores are required for international students whose native language is not English. The GRE General Test is required, and it is strongly suggested that students take the Subject Test in engineering, mathematics, or physics. Applications that are postmarked by December 1, 1999, have an application fee of $40. Applications that are postmarked after this date have an application fee of $70. The deadline for filing 2000–01 applications is 5 p.m. on January 3, 2000. Official notification of application decisions is mailed during the month of March.

Correspondence and Information

Director of Graduate Studies—GCA
Department of Mechanical and Aerospace Engineering
D228 Engineering Quadrangle
Princeton University
Princeton, New Jersey 08544
Telephone: 609-258-4683
World Wide Web: http://www.princeton.edu/~mae/MAE.html

Princeton University

THE FACULTY AND THEIR RESEARCH

F. V. Bracco, Professor; Ph.D., Princeton, 1970. Theoretical-experimental energy-pollution combustion research.

G. L. Brown, Professor; D.Phil., Oxford, 1967. Experimental fluid mechanics, turbulence, rotational flows, compressible flows.

E. Y. Choueiri, Assistant Professor; Ph.D., Princeton, 1991. Spacecraft propulsion, space plasma physics, plasma dynamics and astronautics.

F. L. Dryer, Professor; Ph.D., Princeton, 1972. Fuels, combustion, energy conservation, environment, waste reduction/recovery.

A. G. Evans, Professor; Ph.D., Imperial College (London), 1967. Thermomechanical performance of structural materials used in aerospace, automotive, and electronic applications; design and durability models, multilayers, ultralight structures, and thin films.

P. J. Holmes, Professor; Ph.D., Southampton, 1974. Nonlinear mechanics, dynamical systems, bifurcation theory.

R. G. Jahn, Professor; Ph.D., Princeton, 1955. Electric propulsion, plasmadynamics, ionized gases, human-machine anomalies.

N. J. Kasdin, Assistant Professor; Ph.D., Stanford, 1991. Optimal control and estimation, nonlinear systems, space systems engineering, attitude control and orbital mechanics, stochastic processes.

C. K. Law, Professor; Ph.D., California, San Diego, 1973. Combustion and propulsion, heat and mass transfer.

N. E. Leonard, Associate Professor; Ph.D., Maryland, 1994. Nonlinear control, control of mechanical systems, autonomous vehicles.

M. G. Littman, Associate Professor; Ph.D., MIT, 1977. Quantum computing, atomic spectroscopy, laser technology, robotics.

L. Martinelli, Assistant Professor; Ph.D., Princeton, 1987. Computational fluid dynamics for engineering analysis and design.

R. B. Miles, Professor; Ph.D., Stanford, 1972. Laser imaging of complex flows, laser control of plasmas, radiatively driven phenomena.

D. M. Nosenchuck, Associate Professor; Ph.D., Caltech, 1982. Design experimental fluid dynamics.

M. Q. Phan, Assistant Professor; Ph.D., Columbia, 1989. System identification, adaptive control, learning control, control of flexible structures.

B. S. H. Royce, Professor; Ph.D., King's College (London), 1957. Materials, mechanical properties.

A. J. Smits, Professor and Chairman; Ph.D., Melbourne, 1975. Structure of turbulence in subsonic and supersonic flow, scaling of turbulent flow.

W. O. Soboyejo, Professor; Ph.D., Cambridge, 1988. Mechanical behavior of materials, solid mechanics, and alternative energy systems.

R. H. Socolow, Professor; Ph.D., Harvard, 1964. Global environmental issues, energy conservation technology and policy.

D. J. Srolovitz, Professor; Ph.D., Pennsylvania, 1981. Computer simulation and theoretical modeling of microstructural evolution, thin-film deposition and mechanical behavior.

R. F. Stengel, Professor; Ph.D., Princeton, 1968. Optimal control and estimation, robotics and intelligent systems, aircraft dynamics, space flight.

S. Suckewer, Professor; Ph.D., 1966, D.Sc., 1971, Warsaw. Plasma physics and engineering, X-ray lasers and microscopes, spectroscopy, atomic processes, nonlinear optics, laser applications in biomaterials, new ignition systems for internal combustion.

Z. Suo, Professor; Ph.D., Harvard, 1989. Solid mechanics, mechanical properties, microstructural evolution, stress-related phenomena in microelectronics.

M. J. Wornat, Assistant Professor; Sc.D., MIT, 1988. Combustion, pyrolysis, environment, aerosols, heterogeneous reactions.

PURDUE UNIVERSITY

School of Aeronautics and Astronautics

Programs of Study
The School offers the degrees of M.S., Master of Science in Aeronautics and Astronautics (M.S.A.A.), Master of Science in Engineering (M.S.E.), and Ph.D. Instruction is available in many areas, such as subsonic, supersonic, and hypersonic aerodynamics; computational and experimental aerodynamics; V/STOL, noise, and turbulence; flight and orbit mechanics; systems analysis, guidance and control, optimization, and control of flexible aerospace vehicles; rocket propulsion, propellant combustion, and combustion instability in ramjets and rockets; and elasticity, fracture mechanics, waves, finite elements, material properties, fatigue, aeroelasticity, composite materials, structural dynamics, experimental mechanics, and tribology. Thesis and nonthesis options are available for the master's degree programs. Candidates must complete a minimum of 30 semester hours of work, of which 9 hours may be allocated for the thesis. A master's degree program can be completed in two semesters of full-time study or in approximately three semesters with a half-time teaching or research assistantship. A minimum of 48 semester hours of course work beyond the B.S. degree (exclusive of language requirements) is required for the Ph.D. program. The Ph.D. plan of study consists of a primary area and two related areas, one of which must be applied mathematics, with at least 6 hours of course work completed outside the School. A written qualifying examination and oral preliminary and final exams based primarily on the thesis are required.

Research Facilities
The Aerospace Sciences Laboratory has a 4-foot by 6-foot dual-purpose subsonic wind tunnel, a 4-inch Mach 4 quiet-flow Ludwieg tube, a 2-inch Mach 2.5 supersonic wind tunnel, a 4-inch shock tube, two 18-inch by 24-inch low-speed wind tunnels, an impinging jet facility, extensive machine shop facilities, computers, a laser-Doppler velocimeter, and other instrumentation. The Photomechanics Laboratory is equipped for modern methods of stress analysis, including two- and three-dimensional photoelasticity, birefringent coatings, brittle lacquer and moiré techniques, and static strain measurements. The Structural Dynamics Laboratory is exceptionally well equipped for dynamic signal processing. Of particular note is the 1-million-frames-per-second Cordin 116 camera. The Composite Materials Laboratory contains facilities for general materials testing and for curing composite laminates. An autoclave specially designed for curing epoxy-matrix composites is available for laminate fabrication. Three complete MTS material and fatigue testing machines (55-kip, 22-kip, and 11-kip capacities) and associated equipment are used to perform ultimate strength, stiffness, and fatigue tests. The Fatigue and Fracture Laboratory has two computerized servohydraulic fatigue machines (22-kip and 11-kip) and all associated equipment required to study fatigue crack formation, propagation, and fracture in aerospace materials. The Tribology and Materials Processing Laboratory contains a 22-kip computer-controlled electrohydraulic test machine and associated equipment for fretting and fatigue testing, infrared sensors for temperature measurements, mechanical and optical profilometers, an atomic force microscope, a nanoindenter, and a variety of instrumented machine tools. The Propulsion and Power Laboratory contains two instrumented rocket and turbine test cells, propellant and explosive storage facilities, and hydrogen peroxide concentration apparatus. The Energy Conversion Laboratory is a unique area to study fuel cells, heat engines, and direct energy conversion techniques. The University Computing Center provides computer services through remote terminals and has an IBM 3090 vector computer, a compute cluster consisting of IBM RISC 6000 workstations, and a 142-node Intel Paragon and an eighteen-node IBM SP2 parallel processing supercomputer. The School also has Sun Workstation access to a network of superminicomputers. The Aerospace Post-processing and Visualization Lab has three SGI Indigo 2 workstations, two Indy workstations, and video processing equipment.

Financial Aid
Graduate teaching assistants receive $1200 per month for a half-time appointment. Graduate research assistants can receive from $990 to $1500 depending on the research funding available. All tuition is waived, but students pay a fee of $380 per semester and $155 for the summer session. A few University fellowships, available for highly qualified students, pay $15,000 for the calendar year, with tuition waived; these may be supplemented by a one-quarter-time assistantship.

Cost of Study
In 1999–2000, full-time graduate students who are Indiana residents pay fees of $1750 per semester; nonresidents pay $5860 per semester. Summer session rates are half the semester rates.

Living and Housing Costs
Graduate dormitory rooms are available for between $2520 and $3360 per ten-month academic year, without meals. Privately owned rooms and apartments are also available near the campus.

Student Group
The School currently has 130 graduate students. Most are continuing their studies immediately after completing their undergraduate work, but about 10 to 15 percent have returned after several years of employment. Employment opportunities in the aerospace field are excellent in industry or government laboratories and in academic institutions.

Student Outcomes
M.S. graduates are placed in a wide variety of industrial and government positions as well as Ph.D. programs. Ph.D. graduates are placed in a wide variety of university, industrial, and government positions. The duties of these positions span production, development, teaching, and research.

Location
Purdue University is located in West Lafayette, Indiana, across the Wabash River from Lafayette. The area population is about 100,000, including 36,000 students. Chicago is about 125 miles northwest, and Indianapolis is about 60 miles southeast. The area has excellent schools and other facilities.

The University and The School
Purdue University has nine schools of engineering and three divisions representing nearly all fields of engineering. These schools have about 5,800 undergraduate and 1,500 graduate students. Sponsored research amounts to about $45 million annually. The School of Aeronautics and Astronautics has 21 full-time faculty members. The research and teaching program is supplemented by a colloquium series that features noted speakers in the aerospace field.

Applying
Graduate study in aeronautics and astronautics is based on a background equivalent to the undergraduate courses at Purdue. Students from other engineering fields or the sciences may be admitted but may be required to take certain undergraduate courses. The GRE General Test is required for applicants seeking a fellowship and is desirable for others. Graduate study may begin in January, June, or August, but fellowships begin only in the fall, and there are usually more assistantships available then. Assistantship or fellowship applications for the fall semester should be submitted by February 1.

Correspondence and Information
Head
School of Aeronautics and Astronautics
1282 Grissom Hall
Purdue University
West Lafayette, Indiana 47907-1282
Telephone: 765-494-5152

Purdue University

THE FACULTY AND THEIR RESEARCH

D. Andrisani II, Associate Professor; Ph.D., SUNY at Buffalo, 1979. Estimation, control, dynamics.

G. A. Blaisdell, Associate Professor; Ph.D., Stanford, 1990. Computational fluid mechanics, transition and turbulence.

S. H. Collicott, Associate Professor; Ph.D., Stanford, 1990. Experimental and low-gravity fluid dynamics, optical diagnostics, applied optics.

M. Corless, Professor; Ph.D., Berkeley, 1984. Dynamics, systems, and control.

W. A. Crossley, Assistant Professor; Ph.D., Arizona State, 1995. Optimization, rotorcraft and aircraft design, structure design.

J. F. Doyle, Professor; Ph.D., Illinois, 1977. Structural dynamics, experimental mechanics, photomechanics, wave propagation.

H. D. Espinosa, Associate Professor; Ph.D., Brown, 1992. Micromechanics of ceramics and composites, experimental and computational mechanics.

T. N. Farris, Professor and Head; Ph.D., Northwestern, 1966. Tribology, manufacturing processes, fatigue and fracture.

A. E. Frazho, Professor; Ph.D., Michigan, 1977. Control systems.

A. F. Grandt Jr., Professor; Ph.D., Illinois, 1971. Damage-tolerant structures and materials, fatigue and fracture, aging aircraft.

S. D. Heister, Associate Professor; Ph.D., UCLA, 1988. Rocket propulsion, liquid propellant injection systems.

K. C. Howell, Professor; Ph.D., Stanford, 1983. Orbit mechanics, spacecraft dynamics, control; trajectory optimization.

J. M. Longuski, Professor; Ph.D., Michigan, 1979. Spacecraft dynamics, orbit mechanics, control, orbit decay and reentry.

A. S. Lyrintzis, Associate Professor; Ph.D., Cornell, 1988. Computational aeroacoustics, aerodynamics for rotorcraft and jet flows.

M. A. Rotea, Associate Professor; Ph.D., Minnesota, 1990. Robust and nonlinear multivariable control, modeling and identification.

J J. Rusek, Assistant Professor; Ph.D., Case Western Reserve, 1983. Experimental energy conversion and rocket propulsion.

S. P. Schneider, Associate Professor; Ph.D., Caltech, 1989. Experimental fluid mechanics, high-speed laminar-turbulent transition.

J. P. Sullivan, Professor; Sc.D., MIT, 1973. Experimental aerodynamics, propellers, laser-Doppler velocimetry.

C. T. Sun, Professor; Ph.D., Northwestern, 1967. Composites, fracture and fatigue, structural dynamics.

T. A. Weisshaar, Professor; Ph.D., Stanford, 1971. Aircraft structural mechanics, aeroelasticity, integrated design.

M. H. Williams, Professor and Associate Head; Ph.D., Princeton, 1975. Aerodynamics, computational fluid mechanics.

RENSSELAER POLYTECHNIC INSTITUTE

Department of Mechanical Engineering, Aeronautical Engineering and Mechanics

Programs of Study

The department offers four degrees (M.S., M.Eng., and Ph.D.) in each of three curricula (aeronautical engineering, mechanical engineering, and mechanics). The curricula are flexible and can accommodate individual interests.

Programs of study are organized under aeronautics, applied mechanics, design and manufacturing, and energy and thermal/fluid systems. Additional areas of study include structures and composites, tribology, rotorcraft technology, space technology, mechanics of materials, computational mechanics, and applied physics.

A Master of Science degree requires 30 credit hours, usually eight courses plus 6 credit hours of thesis or project work. The M.Eng. program is a course-based degree program without a thesis requirement intended for professional practice. A student with an accredited B.S. or its equivalent can typically complete the M.Eng. in one year. Admission to the doctoral program is based on a review of the academic and research record by the Graduate Program Committee of the department. Completion of the doctoral program requires 60 credit hours beyond the master's degree, about half in the form of courses, with the remaining being thesis research. An oral candidacy exam and a public thesis defense, conducted by the student's Doctoral Committee, are also required.

Research Facilities

Department facilities include an extensive array of PCs and workstations, which are located in the research laboratories of individual faculty members. There are more than thirty laboratories that support research. In aeronautics/space technology, these include Sonic, Transonic, and Hypersonic Wind Tunnels; Transatmospheric Propulsion; Multiphase Aerodynamics; and Composite Materials and Structures laboratories. In applied mechanics/computational mechanics/biomechanics/mechanics of materials, there are Computational Biomechanics, Computational Mechanics, Mechanics of Materials, Noise and Vibration Control Research, Experimental Mechanics, Symbolic Computation, and Nonlinear Optics laboratories. In thermal fluid and energy sciences, laboratories are available for the study of gas turbines, electronics cooling, heat transfer, laser processing, fouling research, and phase change heat transfer. In manufacturing/design, there are Design Optimization, Robotics and Mechanism, High Speed Machining, Mechatronics, Rapid Prototype, and Smart Materials and Structures laboratories. In tribology, laboratories for the study of fluid systems, tribology, and boundary analysis are available.

Research is supported by such state-of-the-art facilities as the computing facilities in the Center for Industrial Innovation, which provides graduate students with walk-in access to advanced workstations and to programs ranging from personal productivity aids to advanced computer-aided design and analysis packages; the Rensselaer Libraries, whose electronic information systems provide access to collections, databases, and Internet resources from campus and remote terminals; the Rensselaer Computing System, which saturates the campus with a coherent array of advanced workstations, a shared toolkit of applications for interactive learning and research, and high-speed Internet connectivity; a visualization laboratory for scientific computation; and a high-performance computing facility that includes a 36-node SP2 parallel computer.

Financial Aid

Most support is in the form of research or teaching assistantships. Stipends ranged up to $11,000 for the 1998–99 academic year. In addition, full tuition is granted. Additional compensation for summer months is often available from funded research grants and contracts. Outstanding students may qualify for Rensselaer Scholar Fellowships ($15,000 plus full waiver of tuition and fees), specially designated fellowship awards (stipends up to $11,000 plus tuition), or supplemental fellowships. Low-interest, deferred-repayment graduate loans are also available to U.S. citizens with demonstrated need.

Cost of Study

Tuition for 1999–2000 is $665 per credit hour. Other fees amount to approximately $535 per semester. Books and supplies cost about $1700 per year.

Living and Housing Costs

The cost of rooms for single students in residence halls or apartments ranges from $3356 to $5298 for the 1999–2000 academic year. Family student housing, with a monthly rent from $592 to $720, is available.

Student Group

There are about 4,300 undergraduates and 1,750 graduate students representing all fifty states and more than eighty countries at Rensselaer.

Student Outcomes

Eighty-eight percent of Rensselaer's 1998 graduate students were hired after graduation with starting salaries that averaged $56,259 for master's degree recipients and $57,000 to $75,000 for doctoral degree recipients.

Location

Rensselaer is situated on a scenic 260-acre hillside campus in Troy, New York, across the Hudson River from the state capital of Albany. Troy's central Northeast location provides students with a supportive, active, medium-sized community in which to live; an easy commute to Boston, New York, and Montreal; and some of the country's finest outdoor recreation, including Lake George, Lake Placid, and the Adirondack, Catskill, Berkshire, and Green Mountains. The Capital Region has one of the largest concentrations of academic institutions in the United States. Sixty thousand students attend fourteen area colleges and benefit from shared activities and courses.

The University

Founded in 1824 and the first American college to award degrees in engineering and science, Rensselaer Polytechnic Institute today is accredited by the Middle States Association of Colleges and Schools and is a private, nonsectarian, coeducational university. Rensselaer has five schools—Architecture, Engineering, Management, Science, and Humanities and Social Sciences. The School of Engineering is ranked among the top twenty engineering schools in the nation by the *U.S. News & World Report* survey and is ranked in the top ten by practicing engineers.

Applying

Admissions applications and all supporting credentials should be submitted well in advance of the preferred semester of entry to allow sufficient time for departmental review and processing. The application fee is $35. Applicants requesting financial aid are encouraged to submit all required credentials by February 1 to ensure consideration, since the first departmental awards are made in February and March for the next full academic year.

Correspondence and Information

For written information about graduate study:
Professor Michael K. Jensen
Associate Chair for Graduate Studies
Department of Mechanical Engineering,
　Aeronautical Engineering and Mechanics
Rensselaer Polytechnic Institute
110 8th Street
Troy, New York 12180-3590
Telephone: 518-276-6432
World Wide Web: http://www.meche.rpi.edu

For applications and admissions information:
Director of Graduate Academic and
　Enrollment Services, Graduate Center
Rensselaer Polytechnic Institute
110 8th Street
Troy, New York 12180-3590
Telephone: 518-276-6789
E-mail: grad-services@rpi.edu
World Wide Web: http://www.rpi.edu

Rensselaer Polytechnic Institute

THE FACULTY AND THEIR RESEARCH

Kurt S. Anderson, Assistant Professor; Ph.D., Stanford. Computational methods in multibody dynamics and dynamics.

Thierry A. Blanchet, Assistant Professor; Ph.D., Dartmouth. Tribology, solid lubrication, surface science, contact mechanics.

Luciano Castillo, Assistant Professor; Ph.D., Buffalo. Fluid mechanics, turbulence.

Kevin C. Craig, Associate Professor; Ph.D., Columbia. Mechanics of granular material flow, dynamic system modeling and control, viscoelastic material behavior.

Marcelo R. M. Crespo da Silva, Professor; Ph.D., Stanford. Nonlinear vibrations, helicopter dynamics, nonlinear structural dynamics, perturbation analysis, computerized symbolic manipulation.

Stephen J. Derby, Associate Professor; Ph.D., Rensselaer. Simulation, kinematics, applications of robotics.

George J. Dvorak, William Howard Hart Professor of Rotational and Technical Mechanics; Ph.D., Brown. Mechanics of solids, composite materials and structures, fracture and fatigue.

Gary A. Gabriele, Professor and Associate Dean of the School of Engineering; Ph.D., Purdue. Design automation, design optimization.

Henrik J. Hagerup, Associate Professor and Associate Chair of the Department (Undergraduate); Ph.D., Princeton. Fluid mechanics, airplane dynamics.

Prabhat Hajela, Professor; Ph.D., Stanford. Computer-aided optimum design of structural and mechanical systems, aeroelasticity and structural dynamics.

Amir Hirsa, Associate Professor; Ph.D., Michigan. Fluid mechanics, experimental gas dynamics.

Kenneth Jansen, Assistant Professor; Ph.D., Stanford. Computational mechanics, parallel computing, computational fluid dynamics.

Michael K. Jensen, Professor and Associate Chair for Graduate Studies; Ph.D., Iowa State; PE. Heat transfer, fluid mechanics, two-phase flow, numerical analysis, enhanced heat transfer.

Deborah A. Kaminski, Associate Professor and Director of Core Engineering; Ph.D., Rensselaer. Radiation, heat transfer, fluid flow, lasers, solar power, numerical methods.

Erhard Krempl, Rosalind and John J. Redfern Jr. Professor of Engineering; Dr.Ing., Munich Technical (Germany). Mechanics of materials, theoretical and experimental studies, plasticity, creep, fatigue, fracture mechanics.

Daeyong Lee, Professor; Sc.D., MIT. Mechanics of materials, materials processing, CAM, manufacturing processes, computer software systems for manufacturing, materials testing.

Andrew Z. Lemnios, Research Professor; Ph.D., Connecticut. Rotorcraft technology, aeroelasticity, unsteady aerodynamics, structural dynamics.

C. James Li, Associate Professor; Ph.D., Wisconsin. Control of manufacturing processes and equipment, machine condition monitoring, nonlinear system identification.

Antoinette M. Maniatty, Clare Boothe Luce Associate Professor; Ph.D., Cornell. Continuum mechanics, mechanics of materials.

Leik N. Myrabo, Associate Professor; Ph.D., California, San Diego. Advanced propulsion and power systems, energy conversion, space technology, beamed energy.

Henry T. Nagamatsu, Professor Emeritus; Ph.D., Caltech. Hypersonic gas dynamics, transonic plasma dynamics, fluid dynamics, heat transfer.

Aleksandar G. Ostrogorsky, Associate Professor; Sc.D., MIT. Heat and mass transfer, fluid mechanics, solidification phenomena, opto-electronic materials processing.

Catalin Picu, Assistant Professor; Ph.D., Dartmouth. Mechanics of solids, micromechanics and nanomechanics of crystalline defects, atomistic simulations.

Zvi Rusak, Associate Professor; D.Sc., Technion (Israel). Theoretical aerodynamics, fluid mechanics.

Henry A. Scarton, Associate Professor; Ph.D., Carnegie Mellon. Vibrations, acoustics, noise control, machine dynamics, sensors, biomechanics.

Ting-Leung Sham, Associate Professor; Ph.D., Brown. Mechanics of solids, fracture mechanics, finite-element methods.

Richard N. Smith, Professor; Ph.D., Berkeley. Thermal/fluid sciences, melting and solidification, energy systems, materials processing.

Burt L. Swersey, Senior Lecturer; B.S., Cornell. Creativity in design, design methodology.

Matt P. Szolwinski, Assistant Professor; Ph.D., Purdue. Fatigue and fracture, tribology, aerospace structures and materials.

Brian E. Thompson, Associate Professor; Ph.D., Imperial College (London). Aerodynamics, fluid mechanics, multiphase flow.

John A. Tichy, Professor and Chair of the Department; Ph.D., Michigan. Tribology, friction, wear, rheology, fluids.

Harry F. Tiersten, Professor; Ph.D., Columbia; PE. Continuum mechanics, continuum physics, electromechanical devices, structures.

Daniel F. Walczyk, Assistant Professor; Ph.D., MIT. Mechanical design, machine design, integration of CAD/CAM, manufacturing, rapid tooling methods.

Areas of research include but are not limited to flight mechanics and aerodynamics, propulsion, fluid mechanics, gas dynamics, helicopter technology, advanced composite structures, structural dynamics, vibrations and dynamics, acoustics and noise control, solid mechanics, mechanics of materials, optics, finite-element analysis, tribology, computer-aided design, automation and manufacturing, kinematics and robotics, automatic controls, energy conversion, heat and mass transfer, advanced energetics, thermal design and control, multiphase phase phenomena, and computational methods.

Opportunities for interdisciplinary activity are available through involvement with a number of independent centers and multidepartment programs, such as the Design and Manufacturing Institute (DMI) and Centers for Integrated Electronics, Multiphase Research, Rotorcraft Technology, Tribology and Wear Control, Composite Structures and Materials, and Advanced Technology in Robotics and Automation.

SAINT LOUIS UNIVERSITY

Parks College of Engineering and Aviation
Department of Aerospace and Mechanical Engineering

Program of Study

The Department of Aerospace and Mechanical Engineering offers an evening program that leads to the Master of Science (M.S.) degree. Research and nonresearch options are offered. Typical areas of specialization include theoretical and experimental aerodynamics, aeroelasticity, flight vehicles design, aircraft flight mechanics, stability and control, nonlinear system modeling and dynamics, mechanics and structure of composites, and structural mechanics.

Candidates for the M.S. degree are required to complete 30 credit hours of graduate work, 15 of which are core credits. In the research option, 6 credits should be devoted to thesis work. In the nonresearch option, 3 credits should be devoted to independent study. Each student plans his or her study program in consultation with a departmental adviser.

The Department of Aerospace Engineering provides a stimulating environment for graduate study. The program is applications oriented. The adjunct faculty members are drawn from the Boeing Company to teach specialized elective courses.

Research Facilities

The modern aerospace engineering laboratories contain a variety of facilities. The aerodynamic laboratory includes one low-speed wind tunnel with a six-component balance and a computerized data acquisition system, one supersonic wind tunnel with a four-component balance and computerized data acquisition system, and one Eidetic 1520 flow visualization water tunnel with a one-dimensional LDV system and a PIV system. The structures laboratory is equipped with several universal testing machines and strain-gauge units. A composite materials laboratory is under development. A state-of-the-art, real-time, man-in-the-loop Engineering Flight Simulator is available for research in flight simulation.

The University Computing Center provides computer services through personal computers and remote terminals networked to Micro VAXs, SUN stations, Silicon Graphics Machines, and a DECstation. Access to remote computers is available through the Internet.

Financial Aid

Qualified students may obtain financial assistance in the form of Saint Louis University fellowships, tuition waiver scholarships, and teaching and research assistantships.

Cost of Study

Full-time students pay tuition and fees of $552 per credit hour in 1999–2000. (Figures are subject to change by the Board of Trustees).

Living and Housing Costs

For the 1999–2000 academic year, a single room in the University's graduate residence hall is $2076 per semester.

Student Group

There are approximately 2,000 students in the Graduate School at Saint Louis University. Forty percent of those students are women.

Location

Parks College of Engineering and Aviation is located on the Frost campus of Saint Louis University. The greater St. Louis area is a cosmopolitan region with a wealth of cultural and recreational opportunities that include an art museum, the world-famous Missouri Botanical Gardens, the St. Louis Zoo, the symphony, theaters, and major-league baseball, football, and hockey.

The University

During its 170-year history, Saint Louis University has developed into a major, urban, higher-education complex. The University now embraces the following administrative units: College of Arts and Sciences (1818); Graduate School (1832); School of Medicine (1836); School of Law (1842); College of Philosophy and Letters (1889); School of Business and Administration (1910); School of Nursing (1928); School of Social Service (1930); Parks College (1946); and School of Allied Health Professionals (1979).

The Medical campus is on South Grand Avenue in St. Louis. Saint Louis University has a campus in Madrid, Spain. Saint Louis University is a private university under Roman Catholic auspices. The University is sponsored and assisted by the Society of Jesus (Jesuits), a Catholic religious order founded by Saint Ignatius Loyola in 1540. The University is private by charter and Catholic in its philosophy and commitments; it is not church-related in the sense of receiving financial support from a church body.

Applying

Applicants should have a background equivalent to an undergraduate degree in aerospace engineering from Parks College. Students from other disciplines may be admitted, provided that they take certain undergraduate prerequisites. Prospective graduate students may register for a limited number of credits while awaiting notice of admission. Applicants for regular admission must take the GRE General Test before final admission is granted. Successful completion of the Engineering In Training (EIT) examination may be acceptable in lieu of the GRE. Assistantships and fellowships start in the fall semester, and application for these forms of aid should be submitted sufficiently in advance. For exceptions of the regular admission procedures, please refer to the *Graduate Bulletin*.

Correspondence and Information

Dean of Graduate School
110 O'Donnell Hall
3663 Lindell Boulevard
Saint Louis University
St. Louis, Missouri 63108
Telephone: 314-977-2244

Graduate Program Coordinator
Department of Aerospace Engineering
Parks College of Engineering and Aviation
Saint Louis University
3450 Lindell Boulevard
St. Louis, Missouri 63103-0907
Telephone: 314-977-8213

Saint Louis University

THE FACULTY AND THEIR RESEARCH

Salahuddin Ahmed, Assistant Professor; Ph.D., Iowa State. Computational fluid dynamics, viscous flow and turbulence modeling, ultrasonic testing.

Richard M. Andres, Professor; Ph.D., Saint Louis. Flight vehicle design, stability of control, wind tunnel testing.

Patricia A. Benoy, Associate Professor; Ph.D., Rensselaer. Tribology, heat transfer, thermodynamics.

William W. Bower, Associate Professor; Ph.D., Purdue. Viscous flow.

Alan Cain, Adjunct Associate Professor; Ph.D., Stanford. Computational fluid dynamics, design optimization.

Sridhar S. Condoor, Assistant Professor; Ph.D., Texas A&M. Machine design, solid mechanics, mechatronics.

Paul A. Czysz, Professor; M.S., Ohio State. Hypersonic propulsion, system integration.

Marty A. Ferman, Professor; Ph.D., California Coast. Aeroelasticity, structural dynamics.

John A. George, Professor; Ph.D., Saint Louis. Space dynamics, propulsion, fluid and aerodynamics.

Swami Karunamoorthy, Professor; D.Sc., Washington (St. Louis). Composite materials, aircraft structures.

Amy E. Lang, Assistant Professor; Ph.D., Caltech. Experiments in fluid mechanics, engineering science.

Mortaza Mani, Adjunct Associate Professor; Ph.D., Old Dominion. Computational fluid dynamics, heat transfer.

David Manor, Professor; Ph.D., Wichita State. High angle of attack aerodynamics, flight vehicle design, flight testing.

Bellur L. Nagabhushan, Professor; Ph.D., Virginia Tech. Stability and control of flight vehicles, design and synthesis of automatic controls.

Ray N. Nitzsche, Associate Professor; Ph.D., Illinois. Finite-element methods, aircraft structures.

David A. Peters, Adjunct Professor; Ph.D., Stanford. Dynamics, vibration, aeroelasticity, applied thermodynamics.

Krishnaswamy Ravindra, Professor and Chairman; Ph.D., Penn State. Cavitation, flight vehicle dynamics, wind tunnel testing.

STANFORD UNIVERSITY

Department of Aeronautics and Astronautics

Programs of Study	The Department of Aeronautics and Astronautics covers the areas of acoustics, aerodynamics, aerospace robotics, aerospace structures, aircraft and space design, composite structures, computational fluid dynamics, flight mechanics, fluid mechanics, guidance and control, hypersonic and physical gasdynamics, navigation systems (including GPS), optical diagnostics, optimal control design, propulsion, and transportation systems. The Department of Aeronautics and Astronautics has a close affiliation with the Department of Mechanical Engineering. An important outcome of this affiliation is that the students in the two departments have many opportunities to study with faculty members in both departments, as well as to take additional courses from the distinguished faculties of mathematics, chemical engineering, materials sciences and engineering, electrical engineering, computer sciences, physics, and chemistry and from elsewhere in the University.
	The Master of Science degree requires no thesis and can be completed in three academic quarters (one year). In addition, the degrees of Engineer and Ph.D. are offered to post-master's students.
Research Facilities	There are excellent facilities for research in fluid mechanics; heat transfer; combustion; laser diagnostics; robotics; navigation systems; flexible space structures; guidance, control, and precision instrumentation; manufacturing automation and biomechanics; structures and composites; and aircraft and spacecraft system design. A number of these laboratories are unique in this country, and several are closely coupled with the Stanford Institute for Manufacturing and Automation, the NASA–Ames Research Center, the Center for Integrated Systems, and the Center for Turbulence Research.
Financial Aid	Each year, the department awards several graduate fellowships, primarily to entering master's degree candidates. Fellowships normally provide full tuition and a substantial living-expense stipend for a three-quarter period of study. The department also nominates outstanding applicants for Stanford Graduate Fellowships, which provide three years of full support. Research and teaching assistantships are available to students at the post-master's degree level and occasionally for master's degree candidates. These assistantships provide a substantial living-expense stipend and a tuition grant for up to 9 units per quarter. The required work serves as the basis for the Engineer or Ph.D. thesis. For teaching assistantships, students who have taken Stanford courses are usually preferred.
Cost of Study	In 1999–2000, tuition is $24,588 for the academic year ($8196 per quarter). Books and supplies cost about $1250 for the academic year. Students in the School of Engineering are expected to be full-time students.
Living and Housing Costs	Room and board for the nine-month 1999–2000 academic year average $10,300; miscellaneous costs may add several hundred dollars. Additional funds are needed if summer enrollment is planned. Couples need at least $2250 per month for living expenses, plus $300 per month for each child. University apartments rent for $350 to $900 per month, including utilities. Off-campus housing costs are considerably higher close to campus.
Student Group	The University enrollment is approximately 13,800, including 7,260 graduate students. About one fourth of the graduate students and one half of the undergraduates are women. The graduate students of the department number approximately 250, primarily recent graduates of other universities from all parts of the nation and the world.
Student Outcomes	Recent graduates are employed in the aerospace industry, at research labs, and as university faculty members. Several graduates have also joined or founded start-up companies in various fields.
Location	The campus extends from the wooded area surrounding Palo Alto to the foothills of the Coast Range, offering a great variety of recreational activities. The University is surrounded by the Stanford Industrial Park and the Silicon Valley's busy suburbs. The cities of San Francisco, San Jose, and Oakland are each within 30 miles of the campus. There is boating on nearby San Francisco Bay, and Pacific beaches are a 45-minute drive to the west. The Sierra Nevada snow country is a 4-hour drive away. The wine-producing areas of the state, the Gold Rush country, and Monterey, Carmel, and the Big Sur are within easy reach of the campus.
The University	Stanford University was founded in 1885 by Senator and Mrs. Leland Stanford and has an international reputation as an outstanding educational institution. Stanford has a long-standing tradition of academic excellence in the engineering and physical science fields and has produced many prominent engineers and scientists. The University's atmosphere is an unusual blend of a pleasant and uncrowded environment, a dynamic and diverse student body and faculty, and unswerving standards of academic excellence.
Applying	Students can be admitted for any quarter; most are admitted for autumn. Most fellowship support is allocated before April 15; assistantships are often decided somewhat later. To be considered for a fellowship, applicants should see that completed applications, including GRE scores, are received by the department before January 15 and that the GRE General Test is taken no later than December. Applicants for a fellowship are encouraged to take the GRE Subject Test in engineering by December.

Correspondence and Information

For applications and admission information:
Office of Graduate Admissions
Stanford University
Stanford, California 94305-3052

Telephone: 650-723-4291
E-mail: ck.gaa@forsythe.stanford.edu
World Wide Web: http://www.stanford.edu/dept/
    registrar/admissions

For further program information:
Department of Aeronautics and Astronautics
Stanford University
Stanford, California 94305-4035

Telephone: 650-723-2757
E-mail: aero.grad.office@forsythe.stanford.edu
World Wide Web: http://aa.stanford.edu

Stanford University

THE FACULTY AND THEIR RESEARCH

Aerospace Systems Analysis, Guidance, Control, and Aerospace Robotics
Holt Ashley, Professor Emeritus of Aeronautics and Astronautics and of Mechanical Engineering; Sc.D., MIT, 1951.
Arthur E. Bryson Jr., Professor Emeritus of Aeronautics and Astronautics and of Mechanical Engineering; Ph.D., Caltech, 1951.
Robert H. Cannon, Professor Emeritus of Aeronautics and Astronautics; Sc.D., MIT, 1950.
Daniel B. DeBra, Professor Emeritus of Aeronautics and Astronautics and of Mechanical Engineering; Ph.D., Stanford, 1962.
Per Enge, Professor of Aeronautics and Astronautics; Ph.D., Illinois, 1983.
Jonathan How, Assistant Professor of Aeronautics and Astronautics; Ph.D., MIT, 1993.
Ilan Kroo, Professor of Aeronautics and Astronautics; Ph.D., Stanford, 1983.
Bradford W. Parkinson, Professor of Aeronautics and Astronautics; Ph.D., Stanford, 1966.
J. David Powell, Professor Emeritus of Aeronautics and Astronautics and of Mechanical Engineering; Ph.D., Stanford, 1970.
Stephen M. Rock, Associate Professor of Aeronautics and Astronautics; Ph.D., Stanford, 1978.
Claire J. Tomlin, Assistant Professor of Aeronautics and Astronautics; Ph.D., Berkeley, 1998.
Robert J. Twiggs, Consulting Professor of Aeronautics and Astronautics; M.S., Stanford, 1963.

Research problems: aeroelasticity; aerospace and deep underwater robotics; aircraft and spacecraft vehicle systems design; guidance, control, and precision instrumentation; multidisciplinary design optimization; navigation systems (especially GPS).

Fluid Mechanics
Juan Alonso, Assistant Professor of Aeronautics and Astronautics; Ph.D., Princeton, 1997.
Donald Baganoff, Professor of Aeronautics and Astronautics; Ph.D., Caltech, 1964.
Peter Bradshaw, Professor Emeritus of Mechanical Engineering and of Aeronautics and Astronautics; B.A., Cambridge, 1957.
Brian J. Cantwell, Professor of Aeronautics and Astronautics and of Mechanical Engineering; Ph.D., Caltech, 1976.
Joel H. Ferziger, Professor of Mechanical Engineering and of Aeronautics and Astronautics; Ph.D., Michigan, 1962.
Ronald K. Hanson, Professor of Mechanical Engineering and of Aeronautics and Astronautics; Ph.D., Stanford, 1968.
Lambertus Hesselink, Professor of Aeronautics and Astronautics and of Electrical Engineering; Ph.D., Caltech, 1977.
Antony Jameson, Professor of Aeronautics and Astronautics; Ph.D., Cambridge, 1963.
Sanjiva Lele, Assistant Professor of Mechanical Engineering and of Aeronautics and Astronautics; Ph.D., Cornell, 1985.
Robert W. MacCormack, Professor of Aeronautics and Astronautics; M.S., Stanford, 1967.
Parvis Moin, Professor of Mechanical Engineering and of Aeronautics and Astronautics; Ph.D., Stanford, 1978.
William C. Reynolds, Professor of Mechanical Engineering and of Aeronautics and Astronautics; Ph.D., Stanford, 1958.
Milton Van Dyke, Professor Emeritus of Mechanical Engineering and of Aeronautics and Astronautics; Ph.D., Caltech, 1949.

Research problems: aerodynamics, computational aeroacoustics, computational fluid dynamics, hypersonic and supersonic flow, numerical simulation, physical gasdynamics, turbulence.

Solid Mechanics
Fu-Kuo Chang, Associate Professor of Aeronautics and Astronautics; Ph.D., Michigan, 1983.
Richard Christensen, Professor (Research) of Aeronautics and Astronautics; Ph.D., Yale, 1961.
George S. Springer, Professor and Chairman of Aeronautics and Astronautics; Ph.D., Yale, 1962.
Charles R. Steele, Professor of Mechanical Engineering and of Aeronautics and Astronautics; Ph.D., Stanford, 1960.
Stephen Tsai, Professor (Research) of Aeronautics and Astronautics; D.Eng., Yale, 1961.

Research problems: aerospace structures, biomechanics, composite structures, computational mechanics, fracture and defect mechanics, materials system optimization, mechanics of deformable solids, smart structures.

THE UNIVERSITY OF ARIZONA

Department of Aerospace and Mechanical Engineering

Programs of Study

The Department of Aerospace and Mechanical Engineering (AME) offers graduate programs that lead to M.S. and Ph.D. degrees in aerospace engineering, mechanical engineering, and nuclear engineering. Research in aerospace engineering covers the fields of aeronautics and space technology. Research in mechanical engineering includes the four fundamental areas of fluid mechanics, solid mechanics, computational mechanics, and thermosciences. The department also offers vigorous research programs in two applications areas of mechanical engineering: reliability engineering and biomedical engineering. It maintains close relations with the Applied Mathematics Program, with which several faculty members are formally associated, and the Biomedical Interdisciplinary Program.

The Master of Science degree requires a minimum of 32 units, with thesis, report, and course options.

The Ph.D. program must contain a minimum of 57 units of course work and 18 units of dissertation study. To satisfy the residence requirement, the student must spend a minimum of two regular semesters of full-time study on campus. Students are admitted to candidacy after passing a qualifying examination.

Research Facilities

The department has well-equipped laboratories for research in aerodynamics, combustion, composite materials, computational fluid mechanics, computer-aided engineering, control of manufacturing processes, control systems, convective heat transfer, double-diffusive convection, hydrodynamics, interactive graphics, rocketry, solar energy, space technologies, and turbulent flow. Of particular note are an 80-foot-long low-speed wind tunnel with very low background turbulence levels and a 2-foot by 3-foot cross section; wind tunnels for convective heat transfer; a natural convection test cell; an air-jet facility; a turbulent gas combustor for flame studies; a composite materials fabrication laboratory; high-quality graphics terminals, including Silicon Graphics IRIS and Sun workstations; a 50-foot-long, closed-return water tunnel; a boundary-layer transition water tunnel; several laser-Doppler velocimetry systems; and a variety of digital data-acquisition systems.

The department owns and operates open-access computer laboratories that feature more than fifty state-of-the-art Windows- and UNIX-based workstations and dozens of department-specific software packages. In addition, the University operates, within the AME building, an open-access laboratory that features approximately fifty state-of-the-art Windows-based workstations and a large number of general-purpose software titles. The department separately operates eleven graduate computer laboratories that feature approximately fifty workstations, most of them UNIX-based, and special-purpose research software for exclusive research purposes. Work is also under way to build special-purpose, computer-aided instruction rooms for a variety of courses. All the computers in the department have direct access to the Internet over a state-of-the-art, high-speed, fiber-optic data network installed in the AME building. Research computers in certain laboratories are configured to operate as local-area multicomputers. The department also has access to University-operated, special-purpose computational facilities, which include an IBM SP multiprocessor, minisupercomputers, and clusters of IBM RS-6000 and HP 900 series workstations. The University participates in the Cornell National Supercomputing Facility's Smart Node Program and has privileged access to the facilities operated by the San Diego Supercomputing Center.

Financial Aid

Fellowships, traineeships, research and teaching assistantships, and tuition scholarships are available for qualified students. Assistantships provide a stipend of $15,713 for the 1999–2000 academic year (nine months) plus a waiver of out-of-state tuition. Recipients devote 20 hours per week to research or teaching duties during the academic year and may significantly supplement their stipends through full-time work in the department during the summer.

Cost of Study

The registration fee for full-time Arizona residents is $2162 for the 1999–2000 academic year. Students who have not yet established Arizona residence pay $9114 in out-of-state tuition; this is normally waived for students supported by an assistantship, fellowship, or traineeship.

Living and Housing Costs

The average cost of a room in the residence halls is $2396 for the 1999–2000 academic year. Comparably priced off-campus housing is available within easy walking distance.

Student Group

There are 109 graduate students in the department. Of these, 58 are Ph.D. candidates, and 76 are being supported by research or teaching assistantships, fellowships, and traineeships.

Location

The University is located in Tucson, where the excellent climate, clean air, and mountain vistas act as a magnet for visitors and new residents. The metropolitan area has a population of 750,000. Tucson offers the best of two worlds; it is a modern, progressive city, yet it retains the culture and flavor of its Spanish and Old West past.

The University

The University of Arizona was founded in 1885 as the land-grant institution of Arizona. The University has thirteen colleges and an enrollment of 34,327 students; of these, 819 are enrolled in the eight graduate departments of the College of Engineering and Mines.

Applying

Applications for admission and all supporting material should be submitted to the Graduate College at least three months prior to registration. Applicants are required to submit GRE General Test scores. All students whose native language is other than English must submit TOEFL scores. All Ph.D. applicants and M.S. applicants must submit three letters of reference.

Correspondence and Information

Graduate Program
Department of Aerospace and Mechanical Engineering
The University of Arizona
Tucson, Arizona 85721
Telephone: 520-621-4692
Fax: 520-621-8191
World Wide Web: http://www.ame.arizona.edu

The University of Arizona

THE FACULTY AND THEIR RESEARCH

Ara Arabyan, Associate Professor; Ph.D., USC, 1986. Dynamics of rigid and deformable bodies, advanced computational methods in mechanics, simulation of multisegmented biosystems.

Thomas F. Balsa, Professor and Head; Ph.D., Princeton, 1970. Analytical fluid mechanics, nonlinear waves in shear flows, flow stability, jet noise.

Francis F. Champagne, Professor; Ph.D., Washington (Seattle), 1966. Experimental fluid mechanics, turbulent shear flows, laminar-turbulent transition, drag reduction.

Cho Lik Chan, Associate Professor; Ph.D., Illinois, 1986. Heat transfer, laser materials-processing, boundary element method.

Chuan F. Chen, Professor; Ph.D., Brown, 1960. Double-diffusive convection, stability of convection.

Weinong W. Chen, Assistant Professor; Ph.D., Caltech, 1995. Experimental solid mechanics, mechanical behavior of materials at high strain rates, fatigue of materials.

Hermann F. Fasel, Professor; Ph.D., Stuttgart, 1974. Computational fluid dynamics, hydrodynamic stability, laminar-turbulent transition, flow control, turbulent flows.

Ernest D. Fasse, Assistant Professor; Ph.D., MIT, 1992. Dynamic systems and control, control of spatial mechanisms and structures, energetic human-machine systems, biomechanics, robotics and automation.

Rocco Fazzolari, Associate Professor Emeritus; Ph.D., UCLA, 1967. Energy management, HVAC, renewable energy systems, direct energy for conversion.

Barry Ganapol, Professor; Ph.D., Berkeley, 1971. Particle transport theory, radiative transfer, satellite remote sensing.

Juan C. Heinrich, Professor; Ph.D., Pittsburgh, 1975. Solidification of alloys, lubrication, finite element analysis.

David L. Hetrick, Professor Emeritus; Ph.D., UCLA, 1954. Dynamics and safety of nuclear reactors, critical assemblies, processing plants.

Jeffrey W. Jacobs, Associate Professor; Ph.D., UCLA, 1986. Fluid mechanics, hydrodynamic stability.

Matthew R. Jones, Assistant Professor; Ph.D., Illinois, 1993. Radiative heat transfer, application of lasers in biomedicine, inverse problems.

H. A. Kamel, Professor Emeritus; Ph.D., Imperial (London), 1964. Finite-element methods, computer-aided design, interactive graphics.

Dimitri B. Kececioglu, Professor; Ph.D., Purdue, 1953; PE. Reliability, maintainability, and availability engineering; operational readiness; system effectiveness; accelerated testing; reliability growth; burn-in testing; environmental stress screening; software reliability.

Edward J. Kerschen, Professor; Ph.D., Stanford, 1978. Fluid mechanics and applied mathematics, hydrodynamic stability and receptivity, aeroacoustics, unsteady aerodynamics.

Erdogan Madenci, Professor and Associate Head; Ph.D., UCLA, 1987. Fabrication, analysis, and testing of composite structures, thermal and mechanical analysis of electronic packages.

Parviz E. Nikravesh, Professor; Ph.D., Tulane, 1976. Computational methods in kinematics and dynamics, computer-aided design, optimization, vehicle dynamics, biomechanics.

Alfonso Ortega, Associate Professor; Ph.D., Stanford, 1986. Experimental and numerical convective heat transfer in complex laminar and turbulent flows, conjugate heat transfer, natural convection, electronics cooling, gas turbine heat transfer, microsensors.

Edwin K. Parks, Professor Emeritus and Adjunct Professor; Ph.D., Toronto, 1952. Flight testing and modeling the dynamics of airplanes, wind-shear-related airplane accidents.

Henry C. Perkins Jr., Professor; Ph.D., Stanford, 1963. Thermodynamics, energy, air pollution.

Kumar Ramohalli, Professor; Ph.D., MIT, 1971. Space technologies, propulsion and power, combustion, acoustics, solid propellant rockets.

Lawrence B. Scott Jr., Professor; Ph.D., Stanford, 1967. Rarefied gasdynamics, wind energy conversion systems, computer-aided aircraft design and optimization, wind tunnel testing.

Robert L. Seale, Professor Emeritus; Ph.D., Texas, 1953. Nuclear reactor and criticality safety, reactor thermal-hydraulic performance, energy systems.

William R. Sears, Professor Emeritus; Ph.D., Caltech, 1938. Aerodynamics, adaptive-wall wind tunnels.

Bruce R. Simon, Professor; Ph.D., Washington (Seattle), 1971. Finite element analysis, solid and structural mechanics, transport in soft tissues, biomechanics, thermal stress analysis.

K. R. Sridhar, Associate Professor; Ph.D., Illinois, 1990. Heat transfer, multiphase flow, space technology, microsensors, fuel cells.

Thomas L. Vincent, Professor; Ph.D., Arizona, 1963. Nonlinear dynamical systems, optimal control and game theory, evolution and adaptation of biological systems.

Morton E. Wacks, Professor Emeritus; Ph.D., Utah, 1958. Nuclear waste management, activation analysis and isotope applications, health physics, radiation detection and measurement.

John G. Williams, Professor; Ph.D., Imperial College (London), 1971. Nuclear reactor instrumentation and control, dosimetry, adaptive control and learning systems, artificial neural networks and cognitive systems modeling.

Paul H. Wirsching, Professor Emeritus; Ph.D., New Mexico, 1970. Structural reliability and probabilistic mechanics, random vibration and design, metal fatigue and fracture reliability.

Israel J. Wygnanski, Professor; Ph.D., McGill, 1964. Aerodynamics, turbulent shear flows, drag reduction, control of turbulent mixing, boundary layers, laminar-turbulent transition, hydrodynamic stability.

Ming de Zhou, Research Professor; Ph.D., North-Western University Tech (China), 1967. Aerodynamics, dynamics of flight, turbulent shear flows, control of transition and separation, hydrodynamic stability.

AREAS OF RESEARCH

Biomedical Engineering and Biomechanics: Computational activities in biomotor control, constitutive modeling of soft tissues, finite-element analysis of the spine and cardiovascular system, models of the head and neck, orthopedic mechanics, evolution and adaptation of biosystems, microsensing, and design and evaluation of medical implants.

Computational Mechanics: Activities focus on multibody systems, parallel computing, highly flexible bodies, articulated joints, vibration analysis, damage detection, and graphics and animation.

Controls and Flight Mechanics: Theoretical and experimental activities on nonlinear and optimal controls, differential game theory, and control of manufacturing processes and flexible structures.

Fluid Mechanics and Aerodynamics: Activities span the experimental, diagnostic, computational, and analytical aspects of laminar and turbulent shear flows, receptivity and transition, hydrodynamic stability, control of flows, aeroacoustics, lubrication, thermal and double-diffusive convection, and particle-laden flow.

Manufacturing Technology: Activities range from analysis, design-synthesis, and sensitivity studies to real-time identification and control of processes applicable to forming, joining, and machining; laser welding; directional solidification of alloys; residual stresses; and damage nucleation.

Nuclear Engineering: Research includes fission reactor systems, with emphasis in kinetics, control, dosimetry, neutral particle transport theory, power plant operations, and waste management.

Reliability Engineering: Research includes preventive maintenance policies and their optimization, accelerated tests, reliability optimization and growth, environmental stress screening and optimization, probabilistic mechanics and structural reliability, and fatigue and fracture reliability.

Solid Mechanics: Analytical and numerical research covers microdamage in metals, fabrication and failure of composites, constitutive laws for finite strains, poroelastic-transport field problems, thermal stresses in electronic packages, and finite- and boundary-element techniques.

Space Technologies: The UA–NASA Space Engineering Center provides a focus for most of the space technology activities, including debris removal from orbits, indigenous materials utilization and in situ processing, rocketry, hybrid combustion, smart composites, microgravity heat transfer, and thermal control.

Thermosciences: Experimental and computational research spans a range of activities in laminar and turbulent heat transfer, convective stability and conduction phenomena applicable to propulsion, electronic cooling, laser welding, multiphase flows, and gas turbines.

UNIVERSITY OF ILLINOIS AT URBANA–CHAMPAIGN

College of Engineering
Department of Aeronautical and Astronautical Engineering

Programs of Study

Graduate programs in aeronautical and astronautical engineering lead to the degrees of Master of Science and Doctor of Philosophy. Typical areas of specialization include computational fluid dynamics, multiphase flow, rarefield gas dynamics, transonic and hypersonic aerodynamics, theoretical and experimental steady and unsteady aerodynamics, biofluids, flight vehicle design, aircraft flight mechanics, orbital and celestial mechanics, spacecraft optimization and control, nonlinear system modeling and dynamics, combustion, detonation, chemical and electric propulsion, chemical lasers, mechanics and structures of composites, structural mechanics, aeroelasticity, and wind and solar energy.

Candidates for the M.S. degree are required to have a minimum of 8 units of graduate work, including a thesis. In special circumstances, the M.S. thesis requirement may be waived, in which case 9 units of graduate work are required. All M.S. students must complete a mathematics requirement and a core requirement by taking one course from each of the following three core areas: (1) aerodynamics, fluid mechanics, combustion, and propulsion; (2) astrodynamics, control, and dynamics; and (3) structural mechanics and materials. Candidates for the Ph.D. must pass a qualifying examination, complete a minimum of 16 units of graduate courses including mathematics, submit a thesis, pass an oral comprehensive examination prior to starting thesis research, and pass a final examination after completing the thesis. Proficiency in a foreign language is not required. Admission to the doctoral program is based upon course work, the qualifying exam, and an assessment of the candidate's ability to do independent research.

The Faculty of the Department of Aeronautical and Astronautical Engineering has international stature. All faculty members teach and conduct research, working closely with students. Class sizes are typically 4 to 10 students and research groups are typically 3 to 6 students. Each student plans his or her study and research program in consultation with a departmental adviser.

Research Facilities

The aeronautical and astronautical engineering department has excellent facilities for research in computational fluid dynamics, computational aerodynamics, low-speed and high-speed experimental aerodynamics, chemical lasers, electric propulsion, high-pressure combustion, structures, structural dynamics, composite materials and structures, and dynamics and control. Cooperative research programs are also carried out jointly with the Department of Computer Science and the Department of Mechanical and Industrial Engineering. The department is adjacent to the new $30-million Granger Engineering Library. All students have access to fiber-optic-linked computers in their offices.

The Computing Services Office provides computer access through remote terminals, workstations, and microcomputer laboratories. Faculty members and students also have access to the most powerful computers at the National Center for Supercomputing Applications at the University. The College of Engineering and the Department of Aeronautical and Astronautical Engineering operate several workstation laboratories, and a computer-aided-engineering facility is available for interactive graphics and computations.

Financial Aid

Qualified students may obtain financial assistance in the form of University of Illinois fellowships, tuition waiver scholarships, and teaching and research assistantships. The usual stipend for teaching and research assistants is a minimum of $11,907 per academic year for 50 percent appointments. Summer research appointments are also available.

Cost of Study

Full-time students who are Illinois residents paid tuition and fees of $2819 per semester in 1998–99; nonresidents paid $6257 per semester. (Tuition and fees are subject to change by the Board of Trustees.)

Living and Housing Costs

For the 1998–99 academic year (nine months), single rooms in the University's graduate residence halls rented for $2696 and $3008, and double rooms cost $2416 and $2872 per student. Optional board contracts cost from $1435 to $3368. University family housing rented for $350 to $510 per month, usually with additional charges for utilities. (Housing and board changes are subject to change by the Board of Trustees.) Privately owned rooms and apartments in Champaign-Urbana are available at comparable and higher rents.

Student Group

Most students begin graduate work directly after their undergraduate studies, although some return from industry. The majority of the students are U.S. citizens, and approximately 17 percent are women. Currently, all students are full-time.

Student Outcomes

Master's graduates either enter the Ph.D. program or seek industrial full-time positions; doctoral graduates enter academic careers or obtain jobs with industry in research and development.

Location

The University of Illinois is located 130 miles south of Chicago in the twin cities of Urbana and Champaign, whose total population is about 110,000. The cities have an excellent public school system composed of thirty-four elementary and secondary schools, as well as excellent private schools.

The University

The Department of Aeronautical and Astronautical Engineering has been part of the College of Engineering since 1944. Aeronautical and astronautical engineering students are a socially cohesive group and partake in numerous department and self-organized activities throughout the academic year. The $20-million Krannert Center for the Performing Arts attracts outstanding professional artists to the campus. Other entertainers perform in the 16,000-seat Assembly Hall, which is also the arena for all varsity basketball games. A $9-million intramural sports building is also available. The University maintains a large indoor ice rink, and an active Sierra Club promotes outings for outdoor enthusiasts. Many University conferences and symposia are held at Allerton Park, a 1,500-acre estate owned by the University.

Applying

The prerequisite for graduate study normally is the equivalent of the undergraduate courses required at the University of Illinois for a bachelor's degree in aeronautical and astronautical engineering; however, applicants holding bachelor's degrees in other fields of engineering, the physical sciences, or mathematics may also be admitted to advanced study. The GRE General Test is required for all graduate students applying to either the M.S. or Ph.D. program. For international students, the number of assistantships and fellowships available is limited compared to the large number of highly qualified applicants. International applicants must meet the departmental English proficiency requirement on the TOEFL (minimum 607, paper-based; minimum 253, computer-based). The TOEFL must be taken within two years of the proposed term of entry.

Applications for admission to the Graduate College are accepted throughout the year, and students may begin study at the start of the fall or spring semester. Applications must be received by February 1 for the fall semester and by November 1 for the spring semester.

Correspondence and Information

Sandee G. Moore
Graduate Program Coordinator
Department of Aeronautical and Astronautical Engineering
University of Illinois at Urbana–Champaign
104 South Wright Street
Urbana, Illinois 61801-2935

Telephone: 217-333-2651
E-mail: "sandeemoore" <aae@uiuc.edu>
World Wide Web: http://www.aae.uiuc.edu/

University of Illinois at Urbana–Champaign

THE FACULTY AND THEIR RESEARCH

Professors

Michael B. Bragg, Head of the Department; Ph.D., Ohio State, 1981. Aerodynamics, flight mechanics, aircraft icing, unsteady aerodynamics.

Rodney L. Burton, Associate Head of the Department; Ph.D., Princeton, 1966. Electric and advanced chemical rocket propulsion, hypersonic flows, hypervelocity accelerators.

Lawrence A. Bergman, Ph.D., Case Western Reserve, 1980. Structural dynamics and control, stochastic dynamics, system identification, smart structures.

Charles E. Bond, Ph.D., Michigan, 1964. Magnetohydrodynamics, aerodynamics, solar and wind energy.

John D. Buckmaster, Ph.D., Cornell, 1969. Combustion, fluid mechanics, applied mathematics.

Bruce A. Conway, Ph.D., Stanford, 1981. Celestial mechanics, optimal control.

J. Craig Dutton, Ph.D., Illinois, 1979. Gasdynamics, fluid mechanics, propulsion.

Ki D. Lee, Ph.D., Illinois, 1976. Computational fluid dynamics, aerodynamics, transonic flows, vortex flows, design optimization.

N. Sri Namachchivaya, Ph.D., Waterloo, 1984. Nonlinear dynamical systems, bifurcation theory, stability analysis, stochastic processes, structural dynamics (deterministic and probabilistic).

John E. Prussing, Sc.D., MIT, 1967. Orbital mechanics, spacecraft trajectories, optimal control.

Lee H. Sentman, Ph.D., Stanford, 1965. Chemical lasers, nonequilibrium flow modeling, molecular energy transfer, kinetic theory and statistical mechanics, fluid dynamics, space environmental effects on satellite motion.

Wayne C. Solomon, Ph.D., Oregon, 1963. Gas-phase kinetics, air-breathing and rocket propulsion.

Associate Professors

Robert A. Beddini, Ph.D., Rutgers, 1981. Fluid dynamics, computational aerothermochemistry, propulsion.

Victoria L. Coverstone-Carroll, Ph.D., Illinois, 1992. Aerospace system design, robotics, spacecraft control.

Eric Loth, Ph.D., Michigan, 1988. Experimental and computational studies of supersonic flow and two-phase flow, propulsion fluid dynamics.

Michael S. Selig, Ph.D., Penn State, 1992. Computational and experimental aerodynamics; airfoil design and analysis; wind energy systems; aircraft design, performance, stability, and control.

Petros G. Voulgaris, Ph.D., MIT, 1991. Robust control of time-varying and nonlinear systems, general systems theory, estimation and identification of complex systems, emphasis on aerospace applications.

Scott R. White, Ph.D., Penn State, 1990. Manufacturing of composites, solid mechanics, composite materials, smart structures and materials.

Assistant Professor

Philippe H. Geubelle, Ph.D., Caltech, 1993. Theoretical and computational solid mechanics, fracture mechanics, constitutive behaviors of solids.

Professors Emeriti

Harry H. Hilton, Ph.D., Illinois, 1951. Solid mechanics, viscoelasticity, composites, structures, dynamics, numerical analysis, computer-aided engineering.

Allen I. Ormsbee, Ph.D., Caltech, 1955. Aerodynamics.

Shee-Mang Yen, Ph.D., Illinois, 1951. Rarefied gas dynamics, computational fluid dynamics, hypersonic aerodynamics.

Adam R. Zak, Ph.D., Purdue, 1961. Structures, dynamics, solid mechanics.

Associate Professors Emeriti

Harold O. Barthel, Ph.D., Illinois, 1957. Unsteady waves in gasdynamics.

Kenneth R. Sivier, Ph.D., Michigan, 1967. Experimental and applied aerodynamics, aircraft flight mechanics, aircraft design education, system design and engineering.

UNIVERSITY OF MICHIGAN

Department of Aerospace Engineering

Programs of Study

The Department of Aerospace Engineering offers graduate programs leading to the degrees of Master of Science in Engineering, Master of Engineering in Aerospace Engineering, Master of Engineering in Space Systems, Master of Science, Aerospace Engineer, and Doctor of Philosophy.

The Master of Science in Engineering and the Master of Science degrees require a minimum of 30 credit hours of work beyond the bachelor's degree, including at least five advanced courses in aerospace engineering and at least two courses in mathematics beyond advanced calculus. A master's thesis is not required; however, a student may elect to complete a thesis, usually equivalent to 6 credit hours. For the Master of Science in Engineering program, the student must present the equivalent of a bachelor's degree in engineering at the University of Michigan, while a bachelor's degree in physics or mathematics is required for the Master of Science program.

The Master of Engineering in Aerospace Engineering provides an emphasis on engineering practice for those students who desire a systems- and applications-focused engineering education. The Master of Engineering in Space Systems provides a broad interdisciplinary education in the scientific, engineering, and management aspects of complex space systems. Both degrees require 30 credit hours of course work beyond the bachelor's degree.

The degree of Aerospace Engineer provides advanced training beyond the master's degree without the research emphasis of the Doctor of Philosophy degree. The equivalent of a master's degree plus a program of 30 additional credit hours is required. The total program includes at least 24 credit hours in aerospace engineering; 6 hours devoted to a research, design, or development problem, including a written report; 9 hours in mathematics beyond advanced calculus; and three courses in cognate fields other than mathematics. Students must also pass a written examination in their field of specialization.

There are no specific course requirements for the Doctor of Philosophy degree. However, each student must pass an oral and written examination in his or her area of specialization and in the broader field of aerospace engineering. The student must then carry out an original research investigation and must present the results in a written doctoral dissertation.

Combined master's and doctoral programs with other departments can be easily arranged.

Research Facilities

The Gas Dynamics Laboratory includes wind tunnels and shock tubes for studying phenomena over a wide range of fluid velocity and temperature. Two low-speed, low-turbulence tunnels are used for studying turbulence, boundary-layer control, unsteady flows, and automotive and architectural aerodynamics. Several shock tubes are available and have been used for the study of gaseous and two-phase detonation, droplet breakup and ignition, and metal and coal dust combustion. Facilities for the study of supersonic flames and turbulent spray flames are available. Extensive laser diagnostics include laser-induced fluorescence imaging and particle imaging velocimetry.

The W. M. Keck Foundation Laboratory for Computational Fluid Dynamics is home to a number of high-powered RISC-based workstations that serve as a front end to a 16-processor SGI Origin 2000 computer server. There is also access to extensive collegewide computing facilities, including a 64-processor IBM SP2. Students and faculty members working in the laboratory also make use of off-site computing facilities at NASA and NSF supercomputing centers.

There are several laboratories that specialize in structural and solid mechanics. These include laboratories with extensive computer facilities and equipment for structural and material testing. Laboratory activities include development of computational procedures (e.g., finite element and optimization methods), the study of advanced materials (e.g., composites and active materials), the creation of novel structural systems (e.g., inflatables), and methods in structural control. In the area of flight dynamics and control systems, extensive digital computing facilities are available for simulation, optimization, and flight mechanics studies.

The University of Michigan Library System is the sixth largest in North America. The Media Union Library has a collection of more than 500,000 volumes that cover all fields of engineering. The University has extensive computing facilities, including a network of more than 2,000 HP, Sun, and NT workstations, as well as Apple Power Macintosh PCs. These machines are available for use by all students.

Financial Aid

Sources of financial aid include the Milo B. Oliphant Fellowship in Aerospace Engineering, the Boeing Fellowship, the Regents Fellowship, and the François-Xavier Bagnoud Fellowship, as well as other University and industrial fellowships paying annual stipends of up to $18,000 plus the cost of tuition. A number of research and teaching assistantships also provide financial support.

Cost of Study

The 1998–99 tuition for full-time study was $5841 per term for Michigan residents and $10,940 per term for nonresidents.

Living and Housing Costs

The 1998–99 cost of living for a single student was estimated to be $10,800, which was intended to cover food, housing, and personal needs. Married students and those with children should plan on additional living, insurance, and child-care costs. University and private housing for single and married students is readily available. All costs are subject to change for the 1999–2000 academic year.

Student Group

Approximately 35,000 students are enrolled at the University of Michigan. The College of Engineering has about 4,500 undergraduate and 2,000 graduate students. In the Department of Aerospace Engineering, about 220 undergraduate and 115 graduate students are enrolled.

Location

Ann Arbor, a university town of about 145,000 (including students), is an hour's drive from the heart of Detroit. Ann Arbor is on the Huron River and is set in rolling rural countryside. The river and a series of lakes and parks to the north provide opportunities for outdoor recreation. The city is a center of cultural and artistic activities, and a number of public and private research laboratories are located there. Ann Arbor has excellent public schools, and nursery schools and day-care facilities are also available.

The University

The University of Michigan was founded in 1817 and moved from Detroit to Ann Arbor in 1837. The University consists of nineteen schools and colleges, including the Colleges of Engineering (founded in 1895) and Literature, Science, and the Arts; the Medical and Law Schools; and the Schools of Business Administration, Dentistry, Natural Resources, and Music. There have been many research and degree programs involving cooperation between different units of the University. The breadth of the programs and the diversity of the student body create a cosmopolitan atmosphere.

Applying

There is no specific deadline for application for admission to graduate study. However, students requesting financial assistance should have their completed application submitted by January 15. The same application is used for all forms of financial aid. The General Test of the Graduate Record Examinations is required for admission. Applications for admission and financial assistance, catalogs, and a brochure describing the graduate program in detail may be obtained by writing to the Chair of the department.

Correspondence and Information

Chair, Department of Aerospace Engineering
François-Xavier Bagnoud Building
University of Michigan
1320 Beal Avenue
Ann Arbor, Michigan 48109-2140
Telephone: 734-764-3311
Fax: 734-763-0578
World Wide Web: http://www.engin.umich.edu/dept/aero/

University of Michigan

THE FACULTY AND THEIR RESEARCH

Professors

William J. Anderson, Ph.D., Caltech. Finite-element analysis, structural design, structural dynamics.

Dennis S. Bernstein, Ph.D., Michigan. Systems, control theory, space structures.

Werner J. A. Dahm, Ph.D., Caltech. Laser diagnostics, fluid mechanics, turbulent combustion.

James F. Driscoll, Ph.D., Princeton. Propulsion, fluid mechanics, turbulent combustion.

Gerard M. Faeth, Head, Gas Dynamics Laboratory; Ph.D., Penn State. Turbulent flames, multiphase combustion processes.

Peretz P. Friedmann, François-Xavier Bagnoud Professor; Sc.D., MIT. Aeroelasticity of helicopters and fixed-wing aircraft, structural dynamics and its control, adaptive structures.

Paul B. Hays, Ph.D., Michigan. Remote sensing of the geophysical environment and aeronomy.

David C. Hyland, Chairman; Ph.D., MIT. Robust fixed gain, system identification and adaptive control, noise and vibration control.

Pierre Kabamba, Ph.D., Columbia. Systems theory, control theory, dynamics.

C. William Kauffman, Ph.D., Michigan. Airplane designs, experimental methods, dust explosions, combustion.

N. Harris McClamroch, Ph.D., Texas at Austin. Linear and nonlinear system theory, stability, optimal control.

Philip L. Roe, B.A., Cambridge. Computational fluid dynamics, applied aerodynamics.

Nicolas Triantafyllidis, Ph.D., Brown. Continuum mechanics, plasticity theory, structural stability, finite elements, fracture mechanics.

Bram Van Leer, Ph.D., Leiden (Netherlands). Computational fluid dynamics, compressible flows.

Associate Professors

Luis Bernal, Ph.D., Caltech. Fluid mechanics, turbulent shear flows, instrumentation.

Iain D. Boyd, Ph.D., Southampton. Spacecraft propulsion, hypersonics, MEMS, computational modeling.

Alec D. Gallimore, Ph.D., Princeton. Spacecraft propulsion, plasma dynamics, plasma diagnostics.

Kenneth G. Powell, Sc.D., MIT. Computational fluid dynamics.

Anthony Waas, Ph.D., Caltech. Composites, experimentation, biomechanics.

Peter D. Washabaugh, Ph.D., Caltech. Fracture mechanics, composites, experimentation.

Assistant Professors

Daniel J. Scheeres, Ph.D., Michigan. Astrodynamics, spacecraft navigation, orbit determination.

John A. Shaw, Ph.D., Texas at Austin. Active materials and intelligent structures, material and structural instabilities, thermal-mechanical constitutive modeling.

Research Scientist

Donald E. Geister, M.S., Michigan. Computer programming, computer graphics, experimental fluid dynamics, experimental diagnostics for dentistry.

UNIVERSITY
OF MINNESOTA

UNIVERSITY OF MINNESOTA

Department of Aerospace Engineering and Mechanics

Programs of Study

The University of Minnesota offers graduate study programs leading to master's and doctoral degrees in the major fields of mechanics and aerospace engineering. These programs emphasize the basic engineering sciences of fluid mechanics, solid and continuum mechanics, and dynamical systems and control theory. Faculty and student research includes experimental, computational, and theoretical projects.

The Master of Science degree requires 20 quarter credits, or about six courses. A minimum of 14 credits must be earned in the major field, and at least 6 credits must be in one or more related fields outside the major. If a student chooses to write a thesis (Plan A), the thesis research counts for 10 credits. A 3-credit individual project is required as part of the program for students who earn the degree primarily through course work (Plan B).

The Master of Aerospace Engineering program offers professional training beyond the bachelor's degree. Twenty credits are required for the degree. The program may include a project or consist entirely of course work.

The Ph.D. program in the two major fields, mechanics and aerospace engineering, requires approximately two years of course work (or one year beyond the master's). Each student must pass written and oral preliminary examinations and complete a doctoral thesis based on an original research topic. Most students finish the Ph.D. after about five years of study beyond the bachelor's degree.

Research Facilities

The Department of Aerospace Engineering and Mechanics (AEM) possesses extensive experimental facilities and diagnostic equipment. Mechanics of materials facilities include a Bridgeman-StockBarger crystal growth apparatus and an arc-furnace capable of producing single crystals of custom metallic alloys that exhibit shape-memory, magnetostrictive, or other active behaviors. Further facilities include a one-of-a-kind biaxial tension test machine, a 1-Tesla electromagnet for magnetostriction studies, and a vibration-isolated Instron mechanical testing machine with an in situ moire interferometer for measuring surface displacements. The department also has a 50-ton composites press for manufacturing composite plates.

In addition to sensors and standard laboratory equipment, the dynamics and controls program has recently acquired four 7-pound LabWorks shakers, an UmHoltz-Dickie shaker, and a Spectral Dynamics controller. Data acquisition equipment includes B&K structural analyzers and Macintosh computers with DSP processors.

Fluid mechanics facilities include wind tunnels, water channels and tanks, jets, fluidized beds, and a shock tube. Hot-wire and laser-Doppler anemometers, a PIV system, high-power lasers, extensive equipment for high- and low-speed video and still photography, and numerous computers are available for diagnostics.

The department maintains an active machine shop that contains, in addition to standard machines, a wire EDM machine and a computer-controlled mill.

In addition, a high-speed network connects the department to the University of Minnesota Supercomputer Institute and the Army High Performance Computing Center. The current resources of the institute include a 256-processor IBM SP supercomputer with 192 gigabytes of memory, a seventy-two-processor cluster of IBM workstations, and a thirty-two processor SGI Origin 2000. The resources of the center include a 256-processor Cray T3E-1200 and a state-of-the-art Graphics and Visualization Laboratory.

Financial Aid

The primary form of financial aid for first-year graduate students is the teaching assistantship. A half-time appointment as a teaching assistant carries a stipend of approximately $12,500 for the nine-month academic year. Tuition for students with half-time appointments is waived. Students with outstanding records are offered Graduate School Fellowships, and a few incoming students are offered research assistantships. Ph.D. students are normally supported by research assistantships once they begin their research.

Cost of Study

Quarterly tuition during the 1998–99 academic year for full-time enrollment was $1710 (for residents) and $3358 (for nonresidents) for 7 to 14 credits. Each credit over 14 was $216 (for residents) and $432 (for nonresidents). The quarterly student services fee was about $160. All graduate students who take one or more courses are required to pay a $100 computing fee.

Living and Housing Costs

Living expenses range from $12,000 to $19,000 for a twelve-month period. These expenses include housing (the largest individual expense), utilities, food, "settling in" costs, and miscellaneous costs. University residence halls and apartments, as well as a variety of privately owned apartments and houses, are available near the campus or University express bus lines.

Student Group

In 1998–99, 56 students (84 percent men and 16 percent women) were enrolled in AEM graduate programs. Ninety-eight percent were full-time students, and about 91 percent received financial assistance. About 54 percent were master's degree students, and 46 percent were Ph.D. students.

Location

With a population of more than 2.2 million, the Twin Cities metropolitan area is a lively urban center with a full range of entertainment, including what is widely regarded as the most active theater scene between the East and West Coasts. There is no shortage of fine museums, galleries, night clubs, and classical music and dance. Minneapolis is an attractive city of lakes and parks, tree-lined streets, and a thriving downtown—the kind of city where a contemporary skyline can be viewed from a shady lakeshore. As the capital city, St. Paul has a strong historical feel, with an abundance of both old and new architectural gems. The countryside bordering the Twin Cities offers many parks and preserves and smaller towns to explore on day trips. Clear winter days bring skiers, skaters, hikers, and joggers to the lakes and parkways. The same areas accommodate swimmers, sailors, cyclists, golfers, and tennis players in the spring and summer. The Twin Cities area supports a strong and diversified economy, including many Fortune 500 companies as well as numerous high-tech start-ups.

The University

The University of Minnesota, Twin Cities, is a classic Big Ten campus in the heart of the Minneapolis–St. Paul metropolitan area. The largest of the four University of Minnesota campuses, it comprises nineteen colleges and offers 161 bachelor's, 218 master's, 114 doctoral, and five professional degree programs. With a host of nationally recognized, highly ranked programs, the University's Twin Cities campus provides a world-class setting for lifelong learning.

Applying

A four-year B.S. degree in an engineering, basic science, or mathematics program is required. Admission depends primarily on the undergraduate record and letters of recommendation. The GRE is not required, although submitted scores are considered in admissions and financial aid decisions. A TOEFL score of 550 is required of international students whose native language is not English. All prospective students for the fall quarter are urged to apply by January 15 in order to receive maximum consideration for financial aid. Applications are accepted through the spring and summer, but financial aid is not necessarily guaranteed.

Correspondence and Information

Director of Graduate Studies
Department of Aerospace Engineering and Mechanics
107 Akerman Hall
University of Minnesota
110 Union Street, SE
Minneapolis, Minnesota 55455

Telephone: 612-625-8000
E-mail: dgs@aem.umn.edu
World Wide Web: http://www.aem.umn.edu

University of Minnesota

THE FACULTY AND THEIR RESEARCH

Amy E. Alving, Associate Professor; Ph.D. (mechanical and aerospace engineering), Princeton, 1988. Experimental research in turbulence and fluid mechanics.

Gary J. Balas, Associate Professor; Ph.D. (aeronautics), Caltech, 1989. Aerospace control systems, experimental and theoretical.

Gordon S. Beavers, Professor; Ph.D. (mechanical science), Cambridge, 1963. Experimental fluid mechanics, rheological fluid mechanics.

Graham V. Candler, Associate Professor; Ph.D. (aeronautics and astronautics), Stanford, 1988. Hypersonic aerodynamics, computational fluid dynamics, high-temperature gas physics, thermochemical nonequilibrium flows.

Dale F. Enns, Adjunct Associate Professor; Ph.D. (aeronautics and astronautics), Stanford, 1984. Controls, dynamics, aeroelasticity, flight mechanics, dynamical systems.

Roger L. Fosdick, Professor; Ph.D. (applied mathematics), Brown, 1963. Broad spectrum of problems in thermodynamics and continuum mechanics at both the applied and foundation levels, nonlinear material behavior using the methods of applied mathematics.

William L. Garrard, Professor; Ph.D. (engineering mechanics), Texas at Austin, 1968. Dynamics and control of aerospace vehicles, stability and control of nonlinear systems, control of gas turbines, parachute dynamics.

Richard D. James, Professor; Ph.D. (mechanical engineering), Johns Hopkins, 1979. Thermodynamics of solids, phase transformations, thermomechanical behavior of shape-memory materials, micromagnetics, materials with large magnetostriction.

Daniel D. Joseph, Professor; Ph.D. (mechanical engineering), IIT, 1963. Two-phase flow, rheology, fluid mechanics, stability bifurcation.

Perry H. Leo, Associate Professor; Ph.D. (metallurgical engineering and materials science), Carnegie Mellon, 1987. Phase transformations, micromechanics of defects in solids, biological materials, composites.

Ellen K. Longmire, Associate Professor; Ph.D. (mechanical engineering), Stanford, 1991. Experimental fluid mechanics, particle-laden and multiphase flow, turbulence, vortex dynamics.

Thomas S. Lundgren, Professor; Ph.D. (fluid mechanics), Minnesota, 1960. Vortex dynamics, turbulence, two-phase flows, tube transportation systems.

Ivan Marusic, Assistant Professor; Ph.D. (mechanical engineering), Melbourne, 1992. Experimental and theoretical study of turbulent boundary layers.

Thomas W. Shield, Associate Professor; Ph.D. (mechanical engineering), Berkeley, 1988. Experimental solid mechanics, mechanics of materials, single crystal plasticity, shape-memory and magnetostrictive materials, fracture mechanics, elasticity.

Lev Truskinovsky, Associate Professor; Ph.D. (mechanics of solids), USSR Academy of Sciences, 1984. Nonlinear continuum mechanics, thermodynamics, fracture, phase transformations, geophysics.

Andrew Vano, Akerman Professor of Design; B.A.E. (aeronautical engineering), Minnesota, 1963; FAA DER (flight analyst, structures, systems and equipment, powerplant installation, and test pilot). Aircraft and spacecraft design, flight testing, project management.

Theodore A. Wilson, Professor; Ph.D. (aerospace engineering), Cornell, 1962. Respiratory mechanics: modeling lung structure and deformation, respiratory flow, chest wall mechanics.

Yiyuan Zhao, Associate Professor; Ph.D. (aeronautics and astronautics), Stanford, 1989. Guidance/control, optimization, dynamics, air-traffic management, rotorcraft flight trajectories.

UNIVERSITY OF NOTRE DAME

College of Engineering
Department of Aerospace and Mechanical Engineering

Programs of Study

The Department of Aerospace and Mechanical Engineering offers graduate programs leading to the Doctor of Philosophy degree in aerospace and in mechanical engineering. The research activities of the department are divided into three main areas: aerospace sciences, thermal and fluid sciences, and design or manufacturing and solid mechanics. A student is admitted to the Ph.D. degree program upon passing the qualifying examination. He or she must also pass a candidacy examination and then write and defend a dissertation. A total of 72 credit hours are required for the degree; of these, a minimum of 39 must be course credits, including 12 in departmental core courses, 9 in an area different from the area of research, and 6 outside the department. There is no foreign language requirement. Students are expected to complete their degree requirements in about four years. Those entering with a master's degree can have up to 24 graduate credits transferred and need about a year less to complete requirements.

Doctoral students who have successfully completed the candidacy examination may be recommended by the department for a master's degree. A separate master's degree program, requiring 30 credit hours of work (including a thesis, required of most students), is also offered. This degree can be completed in a year and a half.

The department encourages a close working relationship between faculty and students, both inside and outside the classroom. The Annual Graduate Student Research Conference, in which students present their work to the faculty and to other students, is part of this interaction.

Research Facilities

There are excellently equipped laboratories in the department. The Hessert Center for Aerospace Research contains some of the finest experimental research equipment in the country, including subsonic, transonic, supersonic, atmospheric, and anechoic wind tunnels. The Particle Dynamics Laboratory includes a vacuum/pressure tank facility and two HEPA-filtered circular wind tunnels. Additional laboratories include the Environmental Fracture Laboratory, the Tribology/Manufacturing Laboratory, the Metrology Laboratory, the Solid Mechanics Laboratory, the Thermal Sciences Laboratory, the Design Automation Laboratory, and the Automation and Robotics Laboratory.

The department also maintains a state-of-the-art computer graphics laboratory and a number of Sun, Silicon Graphics, DEC, and IBM workstations. The University operates three mainframe computers: an 8-processor IBM SP2, a 16-processor IBM SP1, and an IBM ES/9000. About 160 powerful Sun SPARCstations are available on campus, with 100 workstations located within the College of Engineering.

Financial Aid

Nearly all students admitted are given some sort of financial aid. In addition to several prestigious University fellowships for outstanding students, the department offers financial aid in the form of assistantships. Graduate assistants are paid an average of $16,000 for the twelve-month calendar year, with tuition and fees provided separately by the University. The stipend increases after successful completion of the candidacy examination. Additional support is available for the summer months.

Cost of Study

The tuition in the Graduate School is $10,400 per semester. (Notre Dame's semester system does not include the summer, which the student is expected to devote to full-time research.)

Living and Housing Costs

Some living accommodations are available on campus for both single and married students. Most students, however, prefer to live off campus in rented rooms and apartments close to the University. On-campus housing averages $1213 to $1464 per semester for single students and $250 to $355 per month plus utilities for married couples. The cost of meals in the University dining halls ranges from $530 to $1420 per semester, depending on the meal option chosen.

Student Group

The University enrollment is about 7,800 undergraduates and 2,400 graduate students. Enrollment in the College of Engineering is about 800 undergraduates and 250 graduate students. There are currently about 60 graduate students and several postdoctoral fellows in the Department of Aerospace and Mechanical Engineering.

Student Outcomes

Most students graduate with a doctoral degree and find employment in academia, research laboratories, industry, or postdoctoral positions. For example, of those students who graduated with a Ph.D. in the last two years, 4 individuals have academic positions, 2 are involved in postdoctoral work, 3 are conducting research at a national laboratory, and 6 found employment in industry.

Location

The University is situated immediately north of the city of South Bend, Indiana, an industrial center of about 110,000 people, approximately 90 miles east of Chicago. The Century Center hosts the road-show companies of Broadway plays and is the home of a first-rate symphony orchestra. Notre Dame is only 2 hours by toll road from downtown Chicago, with its many attractions.

The University

The University of Notre Dame was founded in 1842 by the Reverend Edward Frederick Sorin and 6 brothers of the French religious community known as the Congregation of Holy Cross. It was chartered as a university by a special act of the Indiana legislature in 1844, and engineering studies were begun in 1873. The University is the cultural center of the northern Indiana–southwestern Michigan area known as "Michiana." It houses the renowned Snite Museum of Art and offers a wide variety of cultural, social, sports, and political events throughout the year. The twin lakes and many wooded areas of the 1,250-acre campus provide a setting of natural beauty for more than ninety-five University buildings.

Applying

Applicants should arrange for the results of the Graduate Record Examinations (GRE), transcripts of previous academic credits and degrees earned, and three letters of recommendation from college teachers to be sent to the Graduate School at least six months prior to the beginning of the academic session in which enrollment is sought. Prospective candidates are encouraged to correspond with the director of graduate studies at the address below to obtain further information about the current research interests of the department.

Correspondence and Information

Director of Graduate Studies
Department of Aerospace and Mechanical Engineering
University of Notre Dame
Notre Dame, Indiana 46556

Telephone: 219-631-5430
Fax: 219-631-8341
E-mail: ame.amedept.1@nd.edu
World Wide Web: http://www.nd.edu/~ame

University of Notre Dame

THE FACULTY AND THEIR RESEARCH

Hafiz Atassi, Professor; Ph.D., Paris, 1963. Hydrodynamics, aerodynamics, wing theory, aeroelasticity, aeroacoustics and noise control, turbomachinery, kinetic theory and gasdynamics, unsteady flows, applied mathematics, numerical analysis.

Stephen M. Batill, Professor; Ph.D., Notre Dame, 1972. Experimental aerodynamics, aeroelasticity, aerospace structures, aerospace systems design, structural dynamics and vibrations.

Raymond Brach, Professor; Ph.D., Wisconsin–Madison, 1965. Engineering mechanics, dynamics, vibrations, acoustics, engineering design, engineering applications of statistics.

Edmundo Corona, Associate Professor; Ph.D., Texas at Austin, 1990. Structural/solid mechanics, experimental mechanics, cyclic plasticity, structural stability, composite materials.

Patrick F. Dunn, Professor; Ph.D., Purdue, 1974. Aerosol science, propulsion, aerodynamics, magnetohydrodynamic flows and heat transfer, fluid mechanics.

Mohamed Gad-el-Hak, Professor; Ph.D., Johns Hopkins, 1973. Fluid mechanics, turbulence, flow control, unsteady aerodynamics, experimental methods, geophysical flows.

J. William Goodwine Jr., Assistant Professor; Ph.D., Caltech, 1998. Nonlinear control theory, robotics and mechanics.

Robert A. Howland, Associate Professor; Ph.D., North Carolina State, 1974. Mechanics, analytical dynamics, nonlinear mechanics, applied mathematics and perturbation methods.

Nai-Chien Huang, Professor, Ph.D., Harvard, 1963. Stability of structures, micromechanics, fracture mechanics, applied mathematics.

Eric J. Jumper, Professor; Ph.D., Air Force Tech, 1975. Steady and unsteady aerodynamics, hypersonics, real gas phenomena, plasma dynamics and laser physics, fluid mechanic transport phenomena and mixing, homogeneous and heterogeneous reaction kinetics and surface catalysis, aero-optical phenomena.

John W. Lucey, Associate Professor; Ph.D., MIT, 1965. Nuclear engineering, numerical methods, radiation shielding, educational technology.

James J. Mason, Assistant Professor; Ph.D., Caltech, 1993. Experimental solid mechanics, materials science, dynamic deformation, shear instabilities, fracture, fatigue, micromechanics.

Stuart McComas, Professor; Ph.D., Minnesota, 1964. Thermodynamics, convective heat transfer, fluid mechanics, energy systems.

Thomas J. Mueller, Professor; Ph.D., Illinois, 1961. Gasdynamics of separated and wake flows, low Reynolds number aerodynamics, boundary-layer transition, propulsion, aerodynamics and flow visualization.

Victor W. Nee, Professor; Ph.D., Johns Hopkins, 1967. Fluid mechanics, transport phenomena, heat transfer, applied mathematics.

Robert C. Nelson, Professor and Chairman; Ph.D., Penn State, 1974. Aerodynamics, fluid mechanics, aircraft stability and control.

Samuel Paolucci, Associate Professor; Ph.D., Cornell, 1979. Fluid mechanics, natural convection in enclosures, nonlinear dynamical systems, two-phase flow, analytical and computational methods.

Joseph M. Powers, Associate Professor; Ph.D., Illinois at Urbana-Champaign, 1988. Gasdynamics, two-phase flow, detonation and combustion theory, numerical methods, applied mathematics.

Francis H. Raven, Professor; Ph.D., Cornell, 1958. Robotics, kinematics, control systems.

John E. Renaud, Associate Professor; Ph.D., Rensselaer, 1992. Mechanical systems design, design optimization methods, design for manufacture, computer-aided design and manufacturing automation.

Steven R. Schmid, Assistant Professor; Ph.D., Northwestern, 1993. Tribology, manufacturing process simulation and optimization, surface generation, measurement and modeling, tribo-characteristics and wear of tool materials.

Mihir Sen, Professor; Ph.D., MIT, 1975. Natural convection in enclosures, forced convection, two-phase systems, chemically reacting flows, stability of flows, numerical and analytical methods, energy applications.

Steven B. Skaar, Professor; Ph.D., Virginia Tech, 1982. Dynamics and controls, control of holonomic and nonholonomic systems and of elastic and viscoelastic systems, camera space manipulation.

Michael M. Stanisic, Associate Professor; Ph.D., Purdue, 1986. Kinematic control, design of dextrous manipulators, canonical kinematics, design optimization.

Albin A. Szewczyk, Professor; Ph.D., Maryland, 1961. Measurements in turbulent flows with buoyancy, unsteady effects on flows past bluff bodies, effects of freestream shear on heated bluff bodies.

Flint O. Thomas, Associate Professor; Ph.D., Purdue, 1983. Experimental fluid dynamics/aerodynamics, turbulent wall layers, free-shear layer stability and transition, high-speed compressible shear layers, shock wave/turbulent boundary layer interactions, aeroacoustics of high-speed jet screech.

James P. Thomas, Assistant Professor, Ph.D., Lehigh, 1989. Continuum mechanics and thermodynamics, mechanics of materials, fatigue and fracture of solids, mechanics of damage and defects in solids, applied mathematics.

UNIVERSITY OF OKLAHOMA

School of Aerospace and Mechanical Engineering

Programs of Study

Programs of advanced study and research are offered leading to the degrees of Master of Science and Doctor of Philosophy in the fields of aerospace and mechanical engineering. A variety of specializations are available. Mechanical engineering students may select programs in theory and applications of solid mechanics, fluid mechanics, and thermal sciences or the design and analysis of advanced systems and devices. The aerospace engineering program offers opportunities for specialization in aerodynamics, aeroelasticity, aerospace structures, aeroacoustics, aero-vehicle design, and other aero-related applications of engineering mechanics. These coordinated graduate programs also offer a major concentration on techniques of advanced engineering analysis, with current emphases on finite-element methods, nonlinear analysis, variational calculus, perturbation methods, differential quadrature, and multidisciplinary design optimization.

The M.S. degree with thesis requires 30 credit hours, including a maximum 6 hours of thesis research credit. A nonthesis M.S. degree requiring 36 credits is available under restricted circumstances. Nonthesis M.S. candidates must pass a comprehensive final examination covering their fields of academic study. Typical M.S. programs require from fifteen to thirty months, depending upon background and level of involvement.

The doctoral programs consist of at least 90 hours beyond the baccalaureate, normally including about 30 hours from the M.S. program and at least 42 hours of dissertation research. Doctoral candidates must pass a comprehensive qualifying exam in the first two semesters of their work and complete a general exam on their individual program when they are ready to begin their dissertation. Doctoral programs typically require two to four years beyond the M.S. (or equivalent).

Research Facilities

The School's research laboratories are located in Felgar Hall and the Engineering Research Center on North Campus. Specialized laboratories have been developed for combustion and propulsion, composite materials and structures, dynamics, stress analysis and mechanical behavior of materials, fatigue, fluid flow and heat transfer, aerodynamics, laser velocimetry and fluid measurements, nondestructive evaluation techniques, thermal imaging, radiative heat transfer, systems control, computer-aided vision, and CAD/CAE/CAM. A departmental network, consisting of Sun SPARCstations, links faculty offices and research laboratories. Network users have access to the College of Engineering and campus computing facilities, consisting of Alliant, Encore, and IBM mainframes. A variety of PC and workstation laboratories are also available on campus. A branch of the University library located in Felgar Hall serves the primary needs of engineering students and faculty members.

Financial Aid

Fellowships, research and teaching assistantships, grading positions, and tuition waiver scholarships are available to qualified students. Academic year stipends vary from $6300 to $12,000 for half-time appointments. Students appointed half time also receive a waiver of the nonresident portion of tuition plus a 4-hour waiver each semester (fall and spring) of the resident portion of tuition. Half-time students also receive a health insurance payment totaling $485 for fall and spring. A majority of the full-time graduate students receive some departmental support as well.

Cost of Study

In 1998–99, per-credit-hour tuition for the graduate-level courses was $86.65 for Oklahoma residents and $261.15 for nonresidents. Charges for health, recreation, and other student services are not included in these amounts.

Living and Housing Costs

The cost of living in Norman is below the national average. University dormitories provide room and board for single students for $3800 per calendar year. University apartments rent for $359 to $529 a month. Off-campus rents are slightly higher. The University bus system provides transportation from University residences and from many other apartment complexes to academic buildings.

Student Group

There are almost 25,000 students at the University of Oklahoma. Approximately 3,700 students are enrolled in the Graduate College on the Norman campus. The School of AME has about 90 graduate students; most are full-time students, and about 30 percent are in doctoral studies.

Location

Norman is a suburban university town with a population of about 80,000. It is located 18 miles south of Oklahoma City, which provides diverse cultural, recreational, and sports entertainment. Lake Thunderbird, 10 miles east of Norman, offers boating, waterskiing, fishing, and camping. The Oklahoma Art Center; Oklahoma Historical Society Museum; Western Heritage Center; Oklahoma Theatre Center; professional baseball, hockey, and rodeo; and a broad menu of campus events and activities provide an abundance of recreational opportunities.

The University

The University of Oklahoma was chartered in 1890, eighteen years before Oklahoma achieved statehood. The main campus occupies 1,000 acres in Norman. The University also maintains a Health Sciences Center in Oklahoma City, with a branch in Tulsa. The University's nineteen colleges offer 107 bachelor's, seventy-six master's, and fifty-one doctoral programs.

Applying

To be admitted to a graduate program, the applicant must have earned a bachelor's degree with a superior record from an accredited college or university or an institution approved by the graduate studies coordinator. The application deadlines for admission to fall, spring, and summer terms are July 15, December 1, and May 1, respectively, for U.S. citizens and April 1, September 1, and February 1 for international students. Applicants who wish to be considered for teaching support positions must provide GRE scores and letters of recommendation. All applicants are asked to specify their field of interest. All students from non-English-speaking countries must present a TOEFL score of at least 550.

Correspondence and Information

Graduate Studies Coordinator
School of Aerospace and Mechanical Engineering
University of Oklahoma
865 Asp Avenue, Room 212
Norman, Oklahoma 73019-0601
Telephone: 405-325-5011

University of Oklahoma

THE FACULTY AND THEIR RESEARCH

Ajay K. Agrawal, Assistant Professor; Ph.D., Miami (Florida). Combustion, gas turbines, flow imaging and diagnostics, computational fluid mechanics.

M. Cengiz Altan, Associate Professor; Ph.D., Delaware. Advanced material processing, composite materials manufacturing, computational fluid dynamics and flow modeling, non-Newtonian fluids, rheology.

Robert P. Anex, Assistant Professor; Ph.D., California, Davis. Industrial ecology, green design, decision making under uncertainty, technology policy; control of aerospace, environmental, and economic systems.

James David Baldwin, Assistant Professor; Ph.D., Virginia. Mechanical design, metal fatigue analysis, probabilistic design methods, structural analysis of injection-molded composites, corrosion fatigue analysis for aging aircraft.

Charles W. Bert, Benjamin H. Perkinson Chair, George Lynn Cross Research Professor; Ph.D., Ohio State. Solid mechanics: composite-material structures, nonlinear structural dynamics, impact energy absorption, design synthesis, pressure vessels, material damping, rotor dynamics.

Kuang-Hua Chang, Assistant Professor; Ph.D., Iowa. Concurrent design and manufacturing, design for mechanical reliability, design for structural fatigue, CAD/CAE/CAM, design optimization.

Davis M. Egle, Professor; Ph.D., Tulane. Nondestructive evaluation, acoustic emission, ultrasonics, computer-aided instruction, wave propagation in solids, biomechanics.

George Emanuel, Professor; Ph.D. Stanford. Fluid dynamics: diffuser, ejector, and hydraulic valve dynamics; chemical kinetics; mathematical modeling of systems; condensation theory; scramjet and space-plane modeling; retrograde fluids; critical point theory; aerodynamics.

S. R. Gollahalli, Lesch Centennial Professor; Ph.D., Waterloo. Internal combustion engines, combustion, propulsion.

Kurt Gramoll, Professor, Hughes Professorship; Ph.D., Virginia Tech. Multimedia technology and software, composites, pressure vessels design, thermoviscoelastic materials.

F. C. Lai, Associate Professor; Ph.D., Delaware. Convective heat transfer, heat transfer in porous media, electrohydrodynamics, HVAC.

Ramkumar N. Parthasarathy, Assistant Professor; Ph.D., Michigan. Laminar and turbulent flows, two-phase flows, laser and optical diagnostics for particle-laden flows, atomization of liquids and slurries.

Harold L. Stalford, Professor and Director; Ph.D., Berkeley. Nonlinear, robust, autonomous, optimal, and real-time control of systems and vehicles (theoretical and experimental).

Alfred G. Striz, Professor; Ph.D., Purdue. Aeroelasticity, numerical structural analysis and optimization, composites, biomechanics, multidisciplinary design optimization.

William H. Sutton, Associate Professor; Ph.D., North Carolina State. Theoretical and experimental methods in radiative heat transfer, combined modes of heat transfer, energy systems, thermal imaging, automotive alternative fuels.

CURRENT RESEARCH AREAS

The AME faculty members and graduate students are involved in an active, diversified program of fundamental research and engineering development activities. Many of these programs are carried on with external sponsorship from interested state or federal agencies or corporate sponsors. Other programs are funded through a variety of internally administered research grants and funds. Many research programs, particularly those involving the investigation of new fields and analytical concepts, are initiated as unsponsored departmental programs. Taken all together, these activities provide a vital and supportive research environment.

Solid Mechanics, Material Behavior, and Aeroelasticity. Current activities cluster in broad areas of experimental mechanics, plates and shells, buckling, mechanical behavior of materials with emphasis on plasticity and fracture mechanics, fatigue, composite materials and structures, structural dynamics, elastic and inelastic wave propagation, structural and multidisciplinary design optimization, and nondestructive testing and evaluation. Examples of ongoing research programs include the characterization of cure state in polymer composites, research in hybrid numerical-experimental elastic-plastic stress, studies of the dynamics of flexible pipelines, weight optimization of aerospace structures with strength and aeroelastic constraints, corrosion-fatigue interaction in aging aircraft structures (AFOSR), structural modeling of sheet fiber reinforced composite disk drive components, numerical modeling of small crack mechanics in airframe fastener holes, materials processing under microgravity conditions, and the development of efficient methodologies in computational mechanics and structural optimization.

Fluid Mechanics and Aerodynamics. Current research includes studies in compressible flow, viscous flow, non-Newtonian fluids, reentry aerophysics and hypersonic flow, investigations of inviscid supersonic flow over a compressive ramp, aerodynamics of supersonic lifting bodies, work on an integrated aerodynamic/propulsion study for generic aerospace planes based on waverider concepts, development and application of CFD algorithms to incompressible and compressible flow problems, rheology of polymer processing, flow of fiber suspensions, anisotropic fluids, studies of laminar and turbulent two-phase jets and jet break-up, aerodynamic optimization of flow passages.

Thermosciences, Heat Transfer, and Combustion. Research is carried on in theoretical and applied studies of radiative, conductive, and convective heat transfer; thermal properties of materials; combustion and flame dynamics; propulsion; solar energy; and thermography and rainbow schlieren imaging. Typical ongoing projects include combustion characteristics of synthetic fuels, combustion in microgravity, studies of radiative properties for participating media, droplet combustion in cross flow, stability and flame structure of turbulent gas jet flames, analysis and prevention of abnormal combustion in jet engine cartridge starters, the thermo-chemical processes of pollutant formation in combustion, applications of thermal imaging, and automotive application of alternative fuels.

Engineering Design and Development. The design and development activities tend to cluster into several major areas: aircraft design and evaluation, computer-aided design and engineering, control of engineering systems, concurrent design and manufacturing, and industrial product improvement. Specific ongoing projects include the development and evaluation of computer-automated ultrasonic NDT systems; development of airfoil edge devices for improved performance of conventional aircraft; development of sensors and a microcomputer-based monitoring system for evaluation of a large facility HVAC system; the development and application of computer-aided vision systems; the evaluation and development of assistive technology for rehabilitation engineering; design of commercial autogyros; SII pulsejet, pulsefan, pulseprop concept; tool and information integration for concurrent design and manufacturing; biomechanics; and revolutionary general aviation aircraft configurations.

Interdisciplinary Research. AME faculty members and students are involved in interdisciplinary programs with other groups, such as the Center for Engineering Optimization, both inside and outside of the University

Controls Engineering. The School has more than 30 Ph.D. and M.S. degree candidates in the field of controls. Funded research includes structural control (NSF, FHWA), automotive dynamics and control (GM), and the flight control of helicopters. Recent projects include semiactive vibration control for bridges, the development of an automotive seat motion simulator, and the testing of novel automotive suspension designs (DOT). Recent dissertations completed include work on model reduction, nonlinear control design for hydraulic systems, robust control analysis of motion simulators, control techniques for distributed parameter systems, and the control of seismic structures. Hardware development includes embedded controllers, wireless instrumentation, and specialized mechanical systems. Recently initiated projects include research on bridge/truck vibration control systems and automatic weather information technology using GPS.

UNIVERSITY OF SOUTHERN CALIFORNIA

Department of Aerospace Engineering

Programs of Study

The Department of Aerospace Engineering offers Master of Science, Engineer, and Doctor of Philosophy degrees in aerospace engineering. Instruction is available in the full spectrum of areas related to both atmospheric and space flight. Joint studies with other engineering departments permit specialties in aerodynamics, flight mechanics, propulsion, controls, structures, and materials on the M.S. and Engineer levels. The department places particular emphasis on flow phenomena and their applications from hydrodynamics to plasmas, with intensive theoretical, computational, and experimental programs.

Thesis and nonthesis options are available for the M.S. program, which comprises 27 units (nine classes). Both the Engineer and Ph.D. degrees require 60 units of graduate study; the Ph.D. is the more research-oriented degree. Admission to the Ph.D. program follows a screening procedure that includes examinations and an extensive review of the candidate's academic background. Candidates for the Ph.D. are expected to complete substantial amounts of directed research, supervised by a faculty guidance committee, and pass a qualifying written and oral examination prior to starting a dissertation. The completed dissertation must be publicly defended.

Research Facilities

The principal aerospace engineering research laboratories are quite modern and contain a variety of facilities. The labs include two low-turbulence wind tunnels, a low-density hypervelocity facility, a large mixing layer tunnel, a water channel, and an unsteady water tunnel. A large bay houses many smaller, temporary facilities, including jets, a micromachine laboratory, and a curved channel flow. A geophysical fluid-dynamics laboratory contains several experimental setups for exploring the behavior of rotating and stratified flows. An aerodynamic sound laboratory includes a large anechoic chamber with both subsonic and supersonic jet flows. There are an electric propulsion laboratory, a hypersonic reacting flow wind tunnel, and two space sciences laboratories. Another laboratory is dedicated to the study of net-shape manufacturing techniques.

The computer facilities in the Department of Aerospace Engineering provide the means for doing computational work in-house as well as access to the University computers and a variety of supercomputers in the United States. The department operates a computer laboratory equipped with a number of Sun and Silicon Graphics workstations that are used for data processing, program development, and graphics. Assorted workstations and PCs serve as laboratory data-acquisition computers. All computers are linked together and are connected via Ethernet to the University network, which consists of two parallel minisupercomputers and a large number of Sun workstations and Sun file servers. The University provides links to the NSF supercomputer centers and other supercomputer installations in the U.S. Since USC is a member of the San Diego Supercomputer Center Consortium, most large-scale numerical calculations are performed at SDSC's supercomputers (CRAY C-90, Intel, and CRAY T3D).

Financial Aid

Graduate teaching and research assistantships provide tuition remission for up to 12 units per semester and an annual stipend that ranges from $12,509 to $13,863 in 1999–2000 for a full-time appointment, depending upon experience (or a quarter-time appointment with half the stipend and tuition remission for 8 units). Additional summer support of $6253 to $6930 is available for students engaged in continuing research programs.

Cost of Study

Tuition for 1999–2000 is $11,099 per semester for 15–18 units or $748 per unit for students who take fewer than 15 or more than 18 units.

Living and Housing Costs

The University maintains a complex of apartments near campus that vary in rent from $500 to $800 per person per month in 1999–2000. Accommodations range from rooms and studios to one- and two-bedroom units. Additional housing is available in areas a brief bus ride from campus. The cost of living in Los Angeles is comparable to that in most other major urban areas.

Student Group

The department currently has more than 120 graduate students. Ten percent are international students. More than 60 percent of the graduate students are part-time, but most students pursuing the Ph.D. are full-time. Many M.S. students and a number of Engineer and Ph.D. degree candidates are currently employed in the southern California aerospace industry. Opportunities for graduate assistantships and/or part-time employment in the aerospace industry are perhaps the best in the nation.

Location

The University of Southern California (USC) is located in Los Angeles, California, adjacent to the Los Angeles Coliseum and three major museums and 7 miles southeast of Hollywood, in the center of the southern California industrial, entertainment, and recreational capital. As one of the largest urban centers in America, Los Angeles has excellent cultural and recreational facilities.

The University and The School

The University of Southern California, the oldest major university in the West, is private, nonsectarian, and coeducational. The variety and quality of USC's programs and facilities attract faculty, students, and visitors from around the world. The School of Engineering is one of the University's twenty-three schools and colleges. Aerospace Engineering is one of eleven departments within the School of Engineering. The School of Engineering operates the Instructional Television Center, which enables part-time students in the surrounding industry to enroll in graduate courses at their place of employment or in regional classrooms.

Applying

Applicants should have an undergraduate degree in engineering, mathematics, or physics, or an equivalent background. Prospective graduate students may register for a limited number of units while awaiting notice of admission. All applicants must take the GRE General Test before final admission is granted. The Subject Test in engineering, mathematics, or physics is recommended. Assistantships and fellowships start in the fall semester, and applications for these forms of aid should be submitted prior to February 1.

Correspondence and Information

Graduate Admissions Officer
Department of Aerospace Engineering
University of Southern California
Los Angeles, California 90089-1191

Telephone: 213-740-5353
Fax: 213-740-7774
E-mail: ae@usc.edu
World Wide Web: http://ae-www.usc.edu

University of Southern California

THE FACULTY AND RESEARCH AREAS

Ron F. Blackwelder, Professor; Ph.D., 1970; Fellow, APS.
Frederick K. Browand, Professor; Ph.D., 1968; Fellow, APS.
H. K. Cheng, Professor Emeritus; Ph.D., 1952; Fellow, AIAA and APS, and Member, National Academy of Engineering.
J. A. Domaradzki, Professor; Ph.D., 1978.
Daniel A. Erwin, Associate Professor; Ph.D., 1986.
Henryk Flashner, Professor (Mechanical Engineering); Ph.D., 1979.
Michael Gruntman, Professor; Ph.D., 1984.
Gerald Hintz, Adjunct Professor; Ph.D., 1969.
Richard E. Kaplan, Professor; Sc.D., 1964.
Joseph A. Kunc, Professor; Ph.D., 1974; Fellow, APS.
Robert Liebeck, Adjunct Professor; Ph.D., 1964; Fellow, AIAA, and Member, National Academy of Engineering.
Peter B. S. Lissaman, Adjunct Professor; Ph.D., 1965.
Tony Maxworthy, Professor (Mechanical Engineering); Ph.D., 1960; Fellow, APS, and Member, National Academy of Engineering.
John E. McIntyre, Adjunct Professor; Ph.D., 1964.
Eckart H. Meiburg, Professor; Ph.D., 1986.
E. Phillip Muntz, Professor (Radiology); Ph.D., 1961; Fellow, AIAA and APS, and Member, National Academy of Engineering.
Paul K. Newton, Professor; Ph.D., 1985.
Larry G. Redekopp, Professor; Ph.D., 1969; Fellow, APS.
Paul Ronney, Associate Professor; Ph.D., 1983.
Donald E. Shemansky, Professor; Ph.D., 1965.
Geoffrey Spedding, Associate Professor; Ph.D., 1981.
Costas Synolakis, Associate Professor (Civil Engineering); Ph.D., 1986.

RESEARCH AREAS

Aerodynamics. Theoretical work is under way in which aerodynamic effects on bodies operating at high angles of incidence are studied. Unsteady and separated flows in two and three dimensions are studied theoretically, numerically, and experimentally. Analytical and numerical studies on hypersonic aerodynamics related to aerospace transports are ongoing.

Aerodynamics of Ground Vehicles. Recent wind tunnel measurements of drag forces on a platoon of vehicles has demonstrated substantial drag savings with commensurate fuel savings and pollution abatement. The work is continuing by means of wind tunnel tests and full-scale road tests utilizing several vehicles connected by a unique instrumented tow bar designed by the department. A new wind tunnel containing a moving ground plane–simulated road surface is currently under design for the laboratory.

Computational Fluid Dynamics. Supercomputers are being used to numerically simulate three-dimensional fluid flows. Current research involves investigation of energy transfer in homogeneous turbulence, numerical simulation of transition to turbulence in boundary layers and free shear flows, simulations of convective and stably stratified flows with applications to atmospheric and oceanographic phenomena, and numerical modeling of fluid flows interacting with structures made of composite materials.

Computational Techniques for Energetic Flows. Work in progress is aimed at improving the ability of computations to accurately model nonequilibrium, high-speed flow. Under study is the way in which real-gas effects (excitation, dissociation, and ionization) are taken into account and the transition of expanding (nozzle) flows from the continuum to the rarefied regime.

Electric Propulsion. Investigations are being made of a number of types of electric thrusters that will be of great importance to space-based work. This involves experimental investigation of the thruster plasma dynamics and an effort to accurately model the propellant acceleration mechanisms.

Filtering and Control Theory. Work is being done on nonlinear optimum filtering; estimation and control of random fields; stochastic solutions of partial differential equations; multiple target tracking and associated problems of direction finding; and filtering for processes on Riemannian manifolds.

Geophysical Fluid Mechanics. This area of research involves the study of fluid motion in which rotation and/or stratification are important dynamical influences. Research includes the use of rotating-flow experiments to simulate the flows observed in intense storms, in planetary atmospheres, and in ocean currents. The motions of fluids having vertical density stratification are also examined in the laboratory and in field studies.

Kinetic Processes in Gas Flows. Theoretical and experimental research dealing with microscopic properties of flowing gases is conducted on the level of atomic and molecular interactions. Statistical Monte Carlo methods are used for determination of the nonequilibrium properties of the gases. Energy and momentum transfer during interaction of fast streams of neutral particles is also studied. Transition probabilities for exchange of translational, rotational, and vibrational energy in molecular collisions are being analyzed.

Material Processing. Experiments in composites and net-shape manufacturing are conducted. Research on the efficient formation of designer composite materials and coatings is done based on the kinetic processes taking place in gas flows. A significant research program in the one-step formation of structural machine parts is in progress.

Optical Diagnostic Techniques. Methods are being developed to study nonequilibrium and mixing processes in hypersonic, rarefied flows. These involve electron-beam and laser-induced fluorescence and allow for the study of fast transient processes in the flows as well as instantaneous visualization of flowfields containing mixing species.

Planetary and Space Science. A research program in theory and practice is carried out in solar systems and heliosphere phenomena. Areas of interest include upper atmospheric, ionospheric, and magnetospheric plasma phenomena. Models are constrained by measurements from major spacecraft and laboratory experiments. The work includes laboratory astrophysics and gas- and solid-phase physical chemistry.

Plasma Dynamics. Nonequilibrium and non–steady state phenomena in partially ionized gases (plasmas) are studied by using detailed microscopic models. Research includes simulation of phenomena in high-temperature gases, flames, interplanetary and interstellar mediums, and ionospheres. Interactions of electrons, ions, and radiation with atoms and molecules—and with solids—are examined, as is behavior of plasma streams in electromagnetic fields. Work is also under way on chemical processes in partially ionized gases.

Transonic Flow and Unsteady Aerodynamics. Nonlinear and linear problems arising from transonic flight are studied by combined asymptotic and numerical approaches. Attention is given to three-dimensional flow structure and to separation phenomena related to trailing- and leading-edge stall.

Turbulence Research. Two low-speed wind tunnels are being used for experimental determination of mechanisms by which energy is transferred from the free stream to the boundary layer. Additional facilities are in use for the study of two- and three-dimensional jets, mixing layers, and curved boundary layers. Hot-wire anemometry and flow visualization are used extensively in these facilities.

Waves and Instabilities in Fluids. Experimental and theoretical studies are conducted on linear and nonlinear aspects of wave motion and fluid stability. Bifurcations and transitions to chaos and their relevance to the onset of turbulence are studied theoretically from the viewpoint of nonlinear dynamical systems.

UNIVERSITY OF TEXAS AT AUSTIN

Department of Aerospace Engineering and Engineering Mechanics

Programs of Study

The Department of Aerospace Engineering and Engineering Mechanics offers advanced study and research leading to Master of Science and Doctor of Philosophy degrees. (Major areas of study are shown on the reverse side of this page.) The normal prerequisite for graduate study is a bachelor's degree in aerospace engineering or a related field of engineering. However, for those with degrees in science or mathematics, graduate study in orbital mechanics, computational mechanics, computational and applied mathematics, and theoretical mechanics is possible.

The M.S. degree program is available with three options: 30 semester hours, including 6 credit hours for a thesis; 33 semester hours, including 3 credit hours for a report; or 36 semester hours of course work. In all options, students are required to take 6 hours of supporting course work outside the major. Students receiving financial aid are expected to choose the thesis option. The M.S. degree can be completed within three semesters and a summer session of full-time study.

Students seeking a Ph.D. degree are expected to take at least 24 semester hours of course work beyond the M.S. degree, although no specific courses are required. To be admitted to candidacy for the Ph.D., students must pass both written and oral examinations covering technical material relevant to their major area of study. Separate regulations apply to the Computational and Applied Mathematics (CAM) program.

Research Facilities

Computational facilities available to graduate students include the Academic Center Workstation Laboratory, the TICAM Distributed Computing Laboratory, the Data Visualization Laboratory, and the High-Performance Computing Facility. These facilities include high-performance VMS- and UNIX-based computer graphics and engineering workstations. Parallel processing efforts are supported through clusters of workstations connected via high speed fiber-optic networks and scalable parallel multicomputers. Internet access and World Wide Web publishing are also provided.

There are extensive experimental laboratories for aerospace engineering students. These include supersonic and subsonic wind tunnels, an extensive modern laboratory for mechanics of solids and materials, structural dynamics test facilities, and a composite materials fabrication and test laboratory. The major research laboratories are the Center for Space Research; Texas Institute of Computational and Applied Mathematics; Center for Aeromechanics Research; Computational Fluid Dynamics Laboratory; Structural Dynamics Laboratory; Experimental Aerodynamics Laboratory; Composite Materials Laboratory; Center for Research in the Mechanics of Solids, Structures and Materials; and Flow Imaging Laboratory.

Financial Aid

Financial aid is available to students in the form of fellowships, research assistantships, and teaching assistantships. In 1998–99, fellowships provided stipends between $2000 and $15,000 per calendar year. Assistantships require the recipient to work approximately 20 hours per week. Research assistantships usually involve programs that satisfy thesis or dissertation needs and often do not delay the degree. In 1999–2000, stipends range upward from $11,000 plus tuition and fees for half-time appointments. With the exception of part-time students, all students are required to take a minimum of 9 course hours per semester. A number of CAM fellowships at $25,000 per calendar year are available for U.S. citizens.

Cost of Study

For 9 semester hours, Texas residents pay an estimated $1535 per semester in tuition and fees in 1999–2000, while nonresidents pay $3900 per semester. Nonresident students holding a half-time assistantship automatically qualify for resident tuition. Tuition is subject to change without notice.

Living and Housing Costs

University dormitories and apartments are available at reasonable rates. For reservations, students should contact the Division of Housing and Food Service. Off-campus housing is available in Austin, and a shuttle bus system serves areas densely populated by students.

Student Group

The University of Texas at Austin has approximately 48,800 students on campus. Total graduate student enrollment is about 11,650; graduate enrollment in engineering is approximately 1,900, with 160 in aerospace engineering and engineering mechanics programs. Every region of the United States and many other countries are represented in the student body.

Location

The population of Austin, the capital of Texas, and the surrounding metropolitan area is approximately 1 million. A wide spectrum of cultural and entertainment activities is available in both the University and city communities. The Colorado River has been dammed to form the chain of Highland Lakes, two of which are located partially within the city limits. Because of the lakes and the pleasant climate, water-oriented activities such as waterskiing, sailing, fishing, and swimming are easily accessible, and many outdoor sports may be enjoyed the year round. The cities of Dallas and Houston are approximately 4 hours by car from Austin, and excellent air service to the major cities of the United States is available.

The University

The University of Texas at Austin was formally opened in 1883. The College of Engineering was established in 1894 and has earned a position of importance among the major universities of the country. The University of Texas System, with campuses located around the state, has an enrollment of approximately 85,000. Academic programs cover many fields, including engineering, law, business, and liberal arts. The University is organized on an academic year consisting of two 16-week semesters and a 12-week summer session.

Applying

Application forms for admission and financial aid can be obtained from the graduate coordinator in aerospace engineering and engineering mechanics or the graduate adviser in computational and applied mathematics.

Students seeking financial aid must submit their material by February 1 for the summer and fall semesters and by October 1 for the spring semester. For all other students, the application file must be completed by March 1 for the summer and fall semesters and by October 1 for the spring semester.

Correspondence and Information

Graduate Coordinator
Department of Aerospace Engineering
    and Engineering Mechanics
University of Texas at Austin
Austin, Texas 78712-1085
Telephone: 512-471-7595
E-mail: gradprog@mail.ae.utexas.edu
World Wide Web: http://www.ae.utexas.edu/

Graduate Adviser
Computational and Applied Mathematics Program
Taylor Hall 2.312
University of Texas at Austin
Austin, Texas 78712
Telephone: 512-471-7386

University of Texas at Austin

THE FACULTY AND THEIR RESEARCH

I. M. Babuška. Computational mechanics and applied mathematics.
E. B. Becker. Finite-element methods in solid and fluid mechanics.
A. M. Bedford. Wave propagation, impact mechanics.
J. K. Bennighof. Computation in structural dynamics, control of flexible structures.
R. H. Bishop. Guidance, navigation, and control.
R. A. Broucke. Dynamics, celestial mechanics, satellite theory.
G. F. Carey. Finite elements, computational fluid mechanics, transport phenomena.
N. T. Clemens. Compressible/reacting turbulent flows, laser diagnostics.
R. R. Craig. Analytical and experimental structural dynamics, control of flexible structures, flexible multibody dynamics.
C. N. Dawson. Computational mathematics, modeling of surface and subsurface flows.
L. F. Demkowicz. Computational mechanics and applied mathematics.
D. S. Dolling. Unsteady flows, shock/boundary-layer interactions, hyperoclocity projectile aerodynamics.
W. T. Fowler. Flight testing, orbital mechanics, spacecraft/mission design.
D. B. Goldstein. Computational fluid dynamics, aerodynamics, turbulence.
L. J. Hayes. Finite-element methods in heat transfer and biomedical applications.
D. G. Hull. Flight mechanics, optimization, trajectory optimization, guidance.
J. Kallinderis. Fluid mechanics and computational fluid dynamics.
S. Kyriakides. Stability of solids, structures, and materials; plasticity, composite materials.
K. M. Liechti. Composite materials, fracture mechanics, adhesive bonding.
H. M. Mark. Engineering science, spacecraft design, transport physics, astrophysics.
M. E. Mear. Solid mechanics, plasticity.
R. S. Nerem. Orbit determination, satellite geodesy, satellite remote sensing.
J. T. Oden. Finite element methods in solid and fluid mechanics.
G. J. Rodin. Solid mechanics, fracture of materials.
R. A. Schapery. Composite materials, damage and fracture mechanics, viscoelasticity.
B. E. Schutz. Orbital mechanics, dynamics, satellite geodesy, numerical methods.
R. O. Stearman. Aeroelasticity, structural dynamics, experimental mechanics, safety and reliability.
M. Stern. Continuum mechanics, solid mechanics, boundary-element methods.
B. D. Tapley. Satellite applications, geodesy, optimal estimation and control.
P. L. Varghese. High-temperature gasdynamics, nonequilibrium flows, laser diagnostics of combustion and plasmas, laser sensors.
M. F. Wheeler. Fluid mechanics, numerical analysis.

RESEARCH AREAS

Aerothermodynamics and Fluid Mechanics. Research projects are both computational and experimental in nature and cover a broad spectrum of problems associated with subsonic and supersonic flight. Current computational projects include 3-D Navier-Stokes calculations of flow around aircraft using hybrid grids, study of algorithms and data structures for efficient computation on massively parallel computers, unsteady hypersonic cavity flows, direct numerical simulation of turbulent boundary layers, rarefied gas flow on the Jovian moon Io, and flows with thermal and chemical nonequilibrium. Experimental projects include the study of shock–boundary layer interactions, laser flow visualization of supersonic wakes, development and application of laser-based sensors to high-speed flows, flames, and plasmas. Experimental facilities include a Mach 5 supersonic wind tunnel, a supersonic wake and turbulent flame facility with laser flow visualization capabilities, and a laser-based sensors laboratory.

Computational and Applied Mathematics. The department is directly involved with the interdisciplinary CAM program. Graduate students in the program are expected to develop proficiency in applicable mathematics, numerical analysis, and mathematical modeling. Current research topics include adaptive finite and boundary element methods, error estimation, large-scale parallel computing, modeling of semiconductors, computational fluid mechanics, structural acoustics, electromagnetics, parallel data structures, simulation of microstructure of composite materials, surface water flow and transport, environmental modeling and remediation, flows in permeable media.

Engineering Mechanics. Engineering mechanics is concerned with the development of a fundamental framework for analyzing and predicting the behavior of a wide variety of engineered systems. Graduate study and research is directed primarily toward the areas of mechanics of solids, structures and materials, fluid mechanics, and computational mechanics. Research areas include fracture mechanics; mechanics of offshore structures; mechanics of electronic materials; stability at the structural and material levels; micromechanics of composites, ceramics, ice, and polymers; flow through porous media; computational fluid dynamics; wave propagation; and plasticity.

Guidance, Navigation, Control, and Flight Mechanics. This area involves study and research in system theory, control theory, estimation theory, and optimal control theory and their applications to guidance, navigation, control, and flight mechanics associated with aircraft and spacecraft. Current research topics include adaptive Kalman filtering and data fusion, spacecraft navigation and rendezvous, guidance of maneuverable reentry vehicles, second variation conditions for optimal control problems, and numerical methods for solving optimization problems.

Mechanics of Solids, Structures, and Materials. This program transcends the aerospace and mechanics branches of the department. Emphasis is on investigation of problems that require fundamental understanding of mechanical behavior of solids and materials. Research is approached through detailed mathematical analysis, numerical techniques, and careful examination. Current investigations include fracture mechanics, behavior of composite structures, and structural stability. Applications come from the aerospace, automotive, and offshore industries. Research on the mechanical behavior of materials deals with the development of models for their constitutive behavior and failure mechanisms. The materials covered include polymer matrix/fiber composites, ceramic and metal matrix particulate composites, solid propellants, cellular materials, shape memory alloys, elastomers, ice, and others. The majority of these investigations involve a combination of experimental, analytical, and numerical work. Laboratories equipped with modern testing facilities are one of the strengths of the group.

Satellite Applications. This broad research program addresses a variety of satellites, satellite sensors, and ground-tracking networks. Research is conducted in the earth, ocean, and space sciences: oceanography; meteorology; geodesy; gravity field modeling; atmospheric modeling; and data processing techniques. Heavy emphasis is placed on satellite remote sensing, gravity field determination, and multisensor integration. Satellite altimetry, scatterometer, radiometer, and laser ranging data are being processed and analyzed. Analysis of Global Positioning System (GPS) data is being conducted for centimeter accuracy positioning as well as other applications such as kinematic positioning and attitude determination. Pattern recognition, image enhancement, and other advanced graphics techniques are being used in the earth and ocean research efforts. Data analysis techniques are being developed to study tectonic plate motion, regional subsidence, the earth's rotation, and polar motion. Research is also being conducted on uses of the Hubble Space Telescope.

Spacecraft/Mission Design and Planning. This program includes research in all aspects of spacecraft and space mission planning and design. Research efforts address spacecraft systems, characteristics, attitude dynamics, guidance, control, communications, sensors, propulsion, consumables, life support, radiation shielding, mission requirements, and trajectory design and analysis. Space station, lunar, and martian missions are being investigated. Improved mission processes are also being studied. Current projects are supported by the NASA Johnson Space Center and Jet Propulsion Laboratory. Current research focuses on Mars mission planning, precision landing, and the uses of low thrust propulsion to support planetary exploration.

Structural Dynamics, Aeroelasticity, and Wave Propagation. This area includes analytical, computational, and experimental research on a range of topics, including modeling, identification, and control of flexible structures; multilevel adaptive computational structural dynamics algorithms; aeroelasticity; reliability; wave propagation; and flexible multibody dynamics. Well-equipped laboratories support experimental programs in aeroelasticity and structural dynamics, and a computer lab supports research in parallel processing for structural dynamics applications.

Section 3
Agricultural Engineering

This section contains a directory of institutions offering graduate work in agricultural engineering, followed by in-depth entries submitted by institutions that chose to prepare detailed program descriptions. Additional information about programs listed in the directory but not augmented by an in-depth entry may be obtained by writing directly to the dean of a graduate school or chair of a department at the address given in the directory.

For programs offering related work, see also in this book Bioengineering, Biomedical Engineering, and Biotechnology; Civil and Environmental Engineering; Engineering and Applied Sciences; and Management of Engineering and Technology. In Book 3, see Biological and Biomedical Sciences; Ecology, Environmental Biology, and Evolutionary Biology; Marine Biology; Nutrition; and Zoology. In Book 4, see Agricultural and Food Sciences and Natural Resources.

CONTENTS

Program Directory

Announcements

In-Depth Descriptions

Agricultural Engineering

Auburn University, Graduate School, College of Agriculture, Department of Agricultural Engineering, Auburn, Auburn University, AL 36849-0002. Offers MS, PhD. Part-time programs available. *Faculty:* 11 full-time (0 women). *Students:* 3 full-time (0 women), 3 part-time. In 1998, 2 master's awarded. *Degree requirements:* For master's and doctorate, thesis/dissertation required, foreign language not required. *Entrance requirements:* For master's, GRE General Test; for doctorate, GRE General Test (minimum score of 400 on each section required), GRE Subject Test, master's degree. *Application deadline:* For fall admission, 9/1; for spring admission, 3/1. Applications are processed on a rolling basis. Application fee: $25 ($50 for international students). Tuition, state resident: full-time $2,760; part-time $76 per credit hour. Tuition, nonresident: full-time $8,280; part-time $228 per credit hour. *Financial aid:* Fellowships, research assistantships, teaching assistantships available. Financial aid application deadline: 3/15. *Faculty research:* Power and machinery, environmental engineering, forest engineering, waste management, food engineering. *Unit head:* Dr. Clifford A. Flood, Acting Head, 334-844-4180. *Application contact:* Dr. John F. Pritchett, Dean of the Graduate School, 334-844-4700.

Clemson University, Graduate School, College of Agriculture, Forestry and Life Sciences, School of Applied Science and Agribusiness, Department of Agricultural and Biological Engineering, Clemson, SC 29634. Offers biosystems engineering (M Engr, MS, PhD). Part-time programs available. *Students:* 7 full-time (0 women), 4 part-time, 4 international. 6 applicants, 50% accepted. In 1998, 3 master's, 1 doctorate awarded. *Degree requirements:* For master's, thesis required (for some programs), foreign language not required; for doctorate, dissertation required, foreign language not required. *Entrance requirements:* For master's and doctorate, GRE General Test, TOEFL (minimum score of 530 required), minimum GPA of 3.0. *Application deadline:* For fall admission, 6/1. Application fee: $35. *Financial aid:* Fellowships, research assistantships available. Financial aid applicants required to submit FAFSA. *Faculty research:* Natural resources management, biotechnical engineering; agricultural structure with animal and plant environment control; aquaculture, fruit, vegetable, and tobacco crop mechanics. Total annual research expenditures: $1.4 million. *Unit head:* Dr. John C. Haynes, Chair, 864-656-4061, Fax: 864-656-0338. *Application contact:* Dr. R. E. Williamson, Graduate Coordinator, 864-656-4074, Fax: 864-656-0338, E-mail: bwllmsn@clemson.edu.

Colorado State University, Graduate School, College of Engineering, Department of Chemical and Bioresource Engineering, Program in Bioresource and Agricultural Engineering, Fort Collins, CO 80523-0015. Offers MS, PhD. Part-time programs available. *Faculty:* 9 full-time (0 women). *Students:* 12 full-time (3 women), 9 part-time (1 woman), 10 international. Average age 30. 26 applicants, 54% accepted. In 1998, 7 master's, 2 doctorates awarded. Terminal master's awarded for partial completion of doctoral program. *Degree requirements:* For master's, thesis or alternative required, foreign language not required; for doctorate, dissertation required, foreign language not required. *Entrance requirements:* For master's and doctorate, GRE General Test, TOEFL (minimum score of 550 required), minimum GPA of 3.0. *Application deadline:* For fall admission, 2/1 (priority date). Applications are processed on a rolling basis. Application fee: $30. Electronic applications accepted. *Financial aid:* In 1998–99, 1 fellowship, 5 research assistantships, 3 teaching assistantships were awarded.; career-related internships or fieldwork, Federal Work-Study, and institutionally-sponsored loans also available. *Faculty research:* Irrigation, water quality, environmental engineering, groundwater, farm machinery. *Unit head:* Dr. Jim C. Loftis, Chair, Department of Chemical and Bioresource Engineering, 970-491-5252, Fax: 970-491-7369, E-mail: loftis@engr.colostate.edu.

Cornell University, Graduate School, Graduate Fields of Agriculture and Life Sciences, Field of Agricultural and Biological Engineering, Ithaca, NY 14853-0001. Offers biological engineering (M Eng, MPS, MS, PhD); energy (M Eng, MPS, MS, PhD); environmental engineering (M Eng, MPS, MS, PhD); environmental management (MPS); food processing engineering (M Eng, MPS, MS, PhD); international agriculture (M Eng, MPS, MS, PhD); local roads (M Eng, MPS, MS, PhD); machine systems (M Eng, MPS, MS, PhD); soil and water engineering (M Eng, MPS, MS, PhD); structures and environment (M Eng, MPS, MS, PhD). *Faculty:* 28 full-time. *Students:* 45 full-time (13 women); includes 5 minority (1 African American, 3 Asian Americans or Pacific Islanders, 1 Hispanic American), 19 international. 58 applicants, 62% accepted. In 1998, 17 master's, 9 doctorates awarded. *Degree requirements:* For master's, thesis (MS) required; for doctorate, dissertation required, foreign language not required. *Entrance requirements:* For master's and doctorate, GRE General Test, TOEFL (minimum score of 550 required). *Application deadline:* For fall admission, 1/15. Applications are processed on a rolling basis. Application fee: $65. Electronic applications accepted. *Financial aid:* In 1998–99, 25 students received aid, including 5 fellowships with full tuition reimbursements available, 14 research assistantships with full tuition reimbursements available, 6 teaching assistantships with full tuition reimbursements available; institutionally-sponsored loans, scholarships, tuition waivers (full and partial), and unspecified assistantships also available. Financial aid applicants required to submit FAFSA. *Unit head:* Director of Graduate Studies, 607-255-2173, Fax: 607-255-4080. *Application contact:* Graduate Field Assistant, 607-255-2173, Fax: 607-255-4080, E-mail: abengradfield@cornell.edu.

Announcement: Diversity of interests and integration of engineering and physical and biological sciences characterize graduate study and research in agricultural and biological engineering at Cornell. The 28 graduate faculty members work in subject areas of energy, environmental engineering and management, food and biological engineering, international development, local roads, machine systems, soil and water engineering, and structures and environments. Students are drawn from all areas of engineering and from other sciences with equivalent engineering backgrounds. The concepts of academic freedom, interdisciplinary study, and responsibility for one's own program provide a foundation for the graduate experience at Cornell. Ithaca is a small city in a rural area and has a rich cultural, recreational, and intellectual diversity. Applications accepted any time, but University fellowship deadline is 1/15.

Dalhousie University, Faculty of Graduate Studies, DalTech, Faculty of Engineering, Department of Biological Engineering, Halifax, NS B3H 3J5, Canada. Offers M Eng, MA Sc, PhD. *Faculty:* 5 full-time (0 women), 1 part-time (0 women). *Students:* 11 full-time (2 women), 1 part-time. Average age 33. 11 applicants, 64% accepted. In 1998, 3 doctorates awarded (100% entered university research/teaching). *Degree requirements:* For master's and doctorate, thesis/dissertation required, foreign language not required. *Entrance requirements:* For master's and doctorate, TOEFL (minimum score of 580 required). *Application deadline:* For fall admission, 6/1; for winter admission, 10/1; for spring admission, 2/1. Applications are processed on a rolling basis. Application fee: $55. *Financial aid:* In 1998–99, 1 research assistantship (averaging $1,600 per year), 4 teaching assistantships (averaging $4,000 per year) were awarded.; fellowships, scholarships and unspecified assistantships also available. *Faculty research:* Waste management, energy and environment, bio-machinery and robotics, soil and water, aquacultural and food engineering. *Unit head:* Dr. N. Ben-Abdallah, Head, 902-494-6003, Fax: 902-423-2423, E-mail: bio.engineering@dal.ca. *Application contact:* Shelley Parker, Admissions Coordinator, Graduate Studies and Research, 902-494-1288, Fax: 902-494-3149, E-mail: shelley.parker@dal.ca.

Instituto Tecnológico y de Estudios Superiores de Monterrey, Campus Monterrey, Graduate and Research Division, Program in Agriculture, Monterrey, 64849, Mexico. Offers agricultural parasitology (PhD); agricultural sciences (MS); farming productivity (MS); food processing engineering (MS); phytopathology (MS). Part-time programs available. *Degree requirements:* For master's and doctorate, thesis/dissertation required. *Entrance requirements:* For master's, PAEG, TOEFL; for doctorate, GMAT or GRE, TOEFL, master's in related field. *Faculty research:* Animal embryos and reproduction, crop entomology, tropical agriculture, agricultural productivity, induced mutation in oleaginous plants.

Iowa State University of Science and Technology, Graduate College, College of Engineering, Department of Agricultural and Biosystems Engineering, Ames, IA 50011. Offers M Eng, MS, PhD. *Faculty:* 24 full-time, 4 part-time. *Students:* 37 full-time (9 women), 13 part-time (3 women); includes 3 minority (2 African Americans, 1 Native American), 29 international. 24 applicants, 25% accepted. In 1998, 7 master's, 4 doctorates awarded. *Degree requirements:* For master's, computer language, thesis required (for some programs), foreign language not required; for doctorate, one foreign language, computer language, dissertation required. *Entrance requirements:* For master's and doctorate, TOEFL (minimum score of 550 required). *Application deadline:* For fall admission, 5/1 (priority date); for spring admission, 10/1. Applications are processed on a rolling basis. Application fee: $20 ($50 for international students). Electronic applications accepted. Tuition, state resident: full-time $3,308. Tuition, nonresident: full-time $9,744. Part-time tuition and fees vary according to course load, campus/location and program. *Financial aid:* In 1998–99, 29 research assistantships with partial tuition reimbursements (averaging $9,999 per year) were awarded.; fellowships, teaching assistantships, scholarships also available. *Faculty research:* Grain processing and quality, tillage systems, simulation and controls, water management, environmental quality. *Unit head:* Dr. Stewart W. Melvin, Head, 515-294-0462, Fax: 515-294-6633, E-mail: ageng@iastate.edu. *Application contact:* Ramesh Kanwar, Director of Graduate Education, 515-294-4913, E-mail: ageng@iastate.edu.

Kansas State University, Graduate School, College of Engineering, Department of Biological and Agricultural Engineering, Manhattan, KS 66506. Offers MS, PhD. *Faculty:* 12 full-time (0 women). *Students:* 21 full-time (3 women), 5 part-time, 18 international. 15 applicants, 20% accepted. In 1998, 7 master's awarded (100% found work related to degree); 1 doctorate awarded. Terminal master's awarded for partial completion of doctoral program. *Degree requirements:* For master's and doctorate, thesis/dissertation required, foreign language not required. *Entrance requirements:* For master's and doctorate, TOEFL (minimum score of 500 required). *Average time to degree:* Master's–2.5 years full-time; doctorate–4 years full-time. *Application deadline:* For fall admission, 3/1; for spring admission, 11/1. Applications are processed on a rolling basis. Application fee: $0 ($25 for international students). Electronic applications accepted. *Financial aid:* In 1998–99, 5 fellowships (averaging $8,649 per year), 17 research assistantships (averaging $8,649 per year) were awarded. Total annual research expenditures: $843,000. *Unit head:* Dr. James K. Koelliker, Head, 785-532-5580, Fax: 785-532-5825. *Application contact:* Naiqian Zhang, Graduate Coordinator, 785-532-5580, Fax: 785-532-5825.

Louisiana State University and Agricultural and Mechanical College, Graduate School, College of Agriculture, Department of Biological and Agricultural Engineering, Baton Rouge, LA 70803. Offers biological and agricultural engineering (MSBAE); engineering science (MS, PhD). Part-time programs available. *Faculty:* 12 full-time (2 women), 1 part-time (0 women). *Students:* 6 full-time (1 woman), 6 part-time (2 women); includes 2 minority (1 African American, 1 Native American), 5 international. Average age 28. 5 applicants, 60% accepted. In 1998, 2 master's awarded. Terminal master's awarded for partial completion of doctoral program. *Degree requirements:* For master's and doctorate, thesis/dissertation required, foreign language not required. *Entrance requirements:* For master's and doctorate, GRE General Test (minimum combined score of 1000 required), minimum GPA of 3.0. *Application deadline:* For fall admission, 1/25 (priority date). Applications are processed on a rolling basis. Application fee: $25. *Financial aid:* In 1998–99, 4 research assistantships with partial tuition reimbursements (averaging $9,937 per year) were awarded.; fellowships, teaching assistantships with partial tuition reimbursements, career-related internships or fieldwork also available. Financial aid application deadline: 7/1. *Faculty research:* Machine development, aquaculture, environmental engineering, microprocessor applications, ergonomics engineering, bioprocessing, hydrology, biosensors, food engineering. Total annual research expenditures: $1.9 million. *Unit head:* Dr. Lalit Verma, Head, 225-388-3153, Fax: 225-388-3492, E-mail: lverma@gumbo.bae.lsu.edu. *Application contact:* Dr. Thomas Lawson, Graduate Coordinator, 225-388-3153, Fax: 225-388-3492, E-mail: hawson@gumbo.bae.lsu.edu.

McGill University, Faculty of Graduate Studies and Research, Faculty of Agricultural and Environmental Sciences, Department of Agricultural and Biosystems Engineering, Montréal, PQ H3A 2T5, Canada. Offers computer applications (M Sc, M Sc A, PhD); food engineering (M Sc, M Sc A, PhD); grain drying (M Sc, M Sc A, PhD); irrigation and drainage (M Sc, M Sc A, PhD); machinery (M Sc, M Sc A, PhD); pollution control (M Sc, M Sc A, PhD); postharvest (M Sc, M Sc A, PhD); soil dynamics (M Sc, M Sc A, PhD); structure and environment (M Sc, M Sc A, PhD); vegetable and fruit storage (M Sc, M Sc A, PhD). Part-time programs available. *Faculty:* 12 full-time (2 women). *Students:* 52 full-time (9 women), 15 international. Average age 26. 24 applicants, 79% accepted. In 1998, 10 master's, 3 doctorates awarded. *Degree requirements:* For master's and doctorate, thesis/dissertation required. *Entrance requirements:* For master's, TOEFL (minimum score of 550 required), minimum GPA of 3.0; for doctorate, TOEFL (minimum score of 550 required), M Sc. *Application deadline:* For fall admission, 1/1 (priority date); for winter admission, 5/1 (priority date); for spring admission, 9/1 (priority date). Applications are processed on a rolling basis. Application fee: $60. *Financial aid:* Fellowships, research assistantships, teaching assistantships, institutionally-sponsored loans available. *Faculty research:* Postharvest technology, geotextiles for soil-lined manure reservoirs, groundwater transport of contaminants, insect protein for human food. Total annual research expenditures: $1.3 million. *Unit head:* Dr. G. S. V. Raghavan, Chair, 514-398-7773, Fax: 514-398-8387, E-mail: raghavan@agreng.lan.mcgill.ca. *Application contact:* 514-398-7708, Fax: 514-398-7968, E-mail: grad@macdonald.mcgill.ca.

Michigan State University, Graduate School, College of Agriculture and Natural Resources, Department of Agricultural Engineering, Program in Agricultural Technology and Systems Management, East Lansing, MI 48824-1020. Offers MS, PhD. *Students:* 2 (1 woman). Average age 32. *Degree requirements:* For master's, foreign language and thesis not required; for doctorate, dissertation required. *Entrance requirements:* For master's, GRE; for doctorate, GRE, MS. *Application deadline:* Applications are processed on a rolling basis. Application fee: $30 ($40 for international students). *Financial aid:* Applicants required to submit FAFSA. Total annual research expenditures: $37,000. *Unit head:* Dr. Ajit K. Srivastava, Chairperson, Department of Agricultural Engineering, 517-353-7268, Fax: 517-432-2892, E-mail: srivasta@pilot.msu.edu.

Michigan State University, Graduate School, College of Agriculture and Natural Resources, Department of Agricultural Engineering, Program in Biosystems Engineering, East Lansing, MI 48824-1020. Offers MS, PhD. *Students:* 9 full-time (0 women), 5 international. Average age 32. *Degree requirements:* For master's, foreign language and thesis not required; for doctorate, dissertation required. *Entrance requirements:* For master's, GRE; for doctorate, GRE, MS. *Application deadline:* Applications are processed on a rolling basis. Application fee: $30 ($40 for international students). *Financial aid:* Applicants required to submit FAFSA. *Unit head:* Shelley Parker, Admissions Coordinator, Graduate Studies and Research, 902-494-1288, Fax: 902-494-3149, E-mail: shelley.parker@dal.ca.

North Carolina Agricultural and Technical State University, Graduate School, College of Engineering, Department of Agricultural and Biosystems Engineering, Greensboro, NC 27411. Offers MSE. *Degree requirements:* Foreign language not required. *Application deadline:* For fall admission, 7/1 (priority date); for spring admission, 1/9. Applications are processed on a rolling basis. *Unit head:* Dr. Godfrey Gayle, Chairperson, 336-334-7543, Fax: 336-334-7844, E-mail: gayle@garfield.ncat.edu.

North Carolina State University, Graduate School, College of Agriculture and Life Sciences, Department of Biological and Agricultural Engineering, Raleigh, NC 27695. Offers MBAE, MS, PhD. Part-time programs available. *Faculty:* 31 full-time (1 woman), 11 part-time (2 women). *Students:* 40 full-time (12 women), 10 part-time (5 women); includes 6 minority (5 African Americans, 1 Asian American or Pacific Islander), 13 international. Average age 30. 18 applicants, 61% accepted. In 1998, 7 master's, 3 doctorates awarded. *Degree requirements:* For master's, thesis or alternative required, foreign language not required; for doctorate, dissertation required, foreign language not required. *Entrance requirements:* For master's and doctorate, TOEFL (minimum score of 550 required), GRE (international students only). *Application deadline:* For fall admission, 6/25. Applications are processed on a rolling basis. Application fee: $45. *Financial aid:* In 1998–99, 4 fellowships (averaging $6,381 per year), 26 research assistantships (averaging $4,899 per year) were awarded.; teaching assistantships, career-related internships or fieldwork also available. Financial aid application deadline: 2/28. *Faculty research:* Bioinstrumentation, biomechanics, processing of biological materials, water table management, animal waste management. Total annual research expenditures: $8.3 million. *Unit head:* Dr. David B. Beasley, Head, 919-515-2694, Fax: 919-515-6772, E-mail: beasley@eos.ncsu.edu. *Application contact:* Dr. James H. Young, Director of Graduate Programs, 919-515-6710, Fax: 919-515-6772, E-mail: jim_young@ncsu.edu.

North Dakota State University, Graduate Studies and Research, College of Engineering and Architecture, Department of Agricultural and Biosystems Engineering, Fargo, ND 58105. Offers agricultural and biosystems engineering (MS); natural resource management (MS). Part-time programs available. *Faculty:* 7 full-time (0 women). *Students:* 6 full-time (0 women), 6 part-time. Average age 25. 10 applicants, 20% accepted. In 1998, 7 degrees awarded (100% found work related to degree). *Degree requirements:* For master's, computer language, thesis required, foreign language not required. *Entrance requirements:* For master's, TOEFL (minimum score of 525 required). *Average time to degree:* Master's–2 years full-time, 4 years part-time. *Application deadline:* For fall admission, 7/1. Applications are processed on a rolling basis. Application fee: $25. *Financial aid:* In 1998–99, 8 students received aid, including 8 research assistantships with full tuition reimbursements available (averaging $8,600 per year); career-related internships or fieldwork, Federal Work-Study, institutionally-sponsored loans, and tuition waivers (full) also available. Financial aid application deadline: 4/15. *Faculty research:* Agricultural power and machine systems, irrigation, crop processing, food engineering, environmental resources. *Unit head:* E. C. Stegman, Chair, 701-231-7261, Fax: 701-231-1008, E-mail: stegman@plains.nodak.edu.

The Ohio State University, Graduate School, College of Food, Agricultural, and Environmental Sciences, Department of Food, Agricultural, and Biological Engineering, Columbus, OH 43210. Offers MS, PhD. *Faculty:* 17 full-time, 7 part-time. *Students:* 17 full-time (4 women), 2 part-time, 12 international. 16 applicants, 38% accepted. In 1998, 3 master's, 3 doctorates awarded. *Degree requirements:* For master's, computer language required, thesis optional, foreign language not required; for doctorate, computer language, dissertation required, foreign language not required. *Entrance requirements:* For master's, GRE General Test, GRE Subject Test, or minimum GPA of 3.0 (international students); for doctorate, GRE General Test, GRE Subject Test, or minimum GPA of 3.5 (international students). *Application deadline:* For fall admission, 8/15. Applications are processed on a rolling basis. Application fee: $30 ($40 for international students). *Financial aid:* Fellowships, research assistantships, teaching assistantships, career-related internships or fieldwork, Federal Work-Study, and institutionally-sponsored loans available. Aid available to part-time students. *Unit head:* Thomas L. Bean, Chairman, 614-292-6131, Fax: 614-292-9448, E-mail: bean.3@osu.edu.

Oklahoma State University, Graduate College, College of Agricultural Sciences and Natural Resources, School of Biosystems and Agricultural Engineering, Stillwater, OK 74078. Offers M Bio E, MS, PhD. *Faculty:* 19 full-time (1 woman). *Students:* 19 full-time (10 women), 17 part-time (4 women); includes 5 minority (1 African American, 1 Asian American or Pacific Islander, 2 Hispanic Americans, 1 Native American), 11 international. Average age 28. In 1998, 7 master's, 2 doctorates awarded. *Degree requirements:* For master's and doctorate, thesis/dissertation required, foreign language not required. *Entrance requirements:* For master's and doctorate, TOEFL (minimum score of 550 required). *Application deadline:* For fall admission, 6/1 (priority date). Application fee: $25. *Financial aid:* In 1998–99, 28 students received aid, including 24 research assistantships (averaging $11,417 per year), 2 teaching assistantships (averaging $11,400 per year); career-related internships or fieldwork, Federal Work-Study, and tuition waivers (partial) also available. Aid available to part-time students. Financial aid application deadline: 3/1. *Unit head:* Bill Barfield, Head, 405-744-5431.

Pennsylvania State University University Park Campus, Graduate School, College of Engineering, Department of Agricultural and Biological Engineering, State College, University Park, PA 16802-1503. Offers agricultural engineering (MS, PhD). *Students:* 27 full-time (6 women), 16 part-time (2 women). In 1998, 9 master's, 5 doctorates awarded. *Degree requirements:* For master's and doctorate, thesis/dissertation required, foreign language not required. *Entrance requirements:* For master's and doctorate, GRE General Test. Application fee: $50. *Unit head:* Dr. Roy E. Young, Head, 814-865-7792.

Purdue University, Graduate School, Schools of Engineering, School of Agricultural and Biological Engineering, West Lafayette, IN 47907. Offers MS, MSABE, MSE, PhD. Part-time programs available. *Faculty:* 23 full-time (1 woman), 8 part-time (1 woman). *Students:* 43 full-time (10 women), 4 part-time; includes 3 minority (1 African American, 1 Asian American or Pacific Islander, 1 Hispanic American), 19 international. 44 applicants, 20% accepted. In 1998, 11 master's, 1 doctorate awarded. Terminal master's awarded for partial completion of doctoral program. *Degree requirements:* For doctorate, dissertation required, foreign language not required, foreign language not required. *Entrance requirements:* For master's and doctorate, GRE General Test, TOEFL (minimum score of 550 required). Application fee: $30. Electronic applications accepted. *Financial aid:* In 1998–99, 4 fellowships, 34 research assistantships, 3 teaching assistantships were awarded.; career-related internships or fieldwork and instructorships also available. Aid available to part-time students. Financial aid applicants required to submit FAFSA. *Faculty research:* Food engineering, environmental engineering, robotics, biotechnology, machine intelligence. Total annual research expenditures: $2.8 million. *Unit head:* Dr. V. F. Bralts, Head, 765-494-1162, Fax: 765-496-1115, E-mail: bralts@ecn.purdue.edu. *Application contact:* Kathy Lucas, Graduate Secretary, 765-494-1166, Fax: 765-496-1115, E-mail: lucask@ecn.purdue.edu.

Rutgers, The State University of New Jersey, New Brunswick, Graduate School, Program in Bioresource Engineering, New Brunswick, NJ 08903. Offers MS. Part-time programs available. *Faculty:* 13 full-time (2 women), 2 part-time (0 women). *Students:* 3 full-time (0 women), 7 part-time (2 women); includes 1 minority (Asian American or Pacific Islander), 5 international. Average age 27. 11 applicants, 55% accepted. In 1998, 1 degree awarded. *Degree requirements:* For master's, thesis, seminar required, foreign language not required. *Entrance requirements:* For master's, GRE General Test. *Average time to degree:* Master's–4 years full-time. *Application deadline:* For fall admission, 5/1 (priority date); for spring admission, 12/1. Applications are processed on a rolling basis. Application fee: $50. *Financial aid:* In 1998–99, 2 students received aid, including 2 research assistantships; teaching assistantships Financial aid application deadline: 3/1; financial aid applicants required to submit FAFSA. *Faculty research:* Greenhouse engineering, energy and environment, machine vision, flexible automation and robotics, systems analysis. Total annual research expenditures: $300,000. *Unit head:* Gene Giacomelli, Director, 732-932-9753, Fax: 732-932-7931.

South Dakota State University, Graduate School, College of Engineering, Department of Agricultural Engineering, Brookings, SD 57007. Offers MS, PhD. PhD offered jointly with Iowa State University of Science and Technology. *Degree requirements:* For master's, thesis, oral exam required, foreign language not required; for doctorate, dissertation, preliminary oral and written exams required. *Entrance requirements:* For master's, TOEFL (minimum score of 550 required); for doctorate, TOEFL. *Faculty research:* Water resources, machine visions, food engineering, environmental engineering, machine design.

Texas A&M University, College of Agriculture and Life Sciences, Department of Agricultural Engineering, College Station, TX 77843. Offers M Agr, M Eng, MS, DE, PhD. Part-time programs available. *Faculty:* 28 full-time (3 women), 8 part-time (0 women). *Students:* 37 full-time (13 women), 13 part-time (3 women); includes 6 minority (1 African American, 1 Asian American or Pacific Islander, 4 Hispanic Americans), 14 international. Average age 30. 21 applicants, 67% accepted. In 1998, 12 master's, 3 doctorates awarded. *Degree requirements:* For master's, thesis (MS), preliminary and final exams required; for doctorate, dissertation, preliminary and final exams required, foreign language not required. *Entrance requirements:* For master's and doctorate, GRE General Test (minimum score of 450 on verbal section, 600 on quantitative required), TOEFL. *Application deadline:* For fall admission, 2/1 (priority date); for spring admission, 10/1. Applications are processed on a rolling basis. Application fee: $50 ($75 for international students). Electronic applications accepted. *Financial aid:* In 1998–99, 4 fellowships (averaging $12,000 per year), 14 research assistantships (averaging $14,250 per year), 14 teaching assistantships (averaging $14,271 per year) were awarded.; career-related internships or fieldwork and tuition waivers (partial) also available. Financial aid application deadline: 3/1; financial aid applicants required to submit FAFSA. Total annual research expenditures: $1.2 million. *Unit head:* James R. Gilley, Head, 409-845-3931, Fax: 409-862-3442, E-mail: gilley@zeus.tamu.edu. *Application contact:* Jeana Goodson, Academic Programs Assistant, 409-845-0609, Fax: 409-862-3442, E-mail: j-goodson@tamu.edu.

Université Laval, Faculty of Graduate Studies, Faculty of Agricultural and Food Sciences, Department of Soils and Agricultural Engineering, Programs in Agricultural Engineering, Sainte-Foy, PQ G1K 7P4, Canada. Offers M Sc. *Students:* 9 full-time (1 woman), 2 part-time. 12 applicants, 50% accepted. In 1998, 2 degrees awarded. *Application deadline:* For fall admission, 3/1. Application fee: $30. *Unit head:* Alfred Marquis, Director, 418-656-2131 Ext. 3798, Fax: 418-656-3723, E-mail: alfred.marquis@sga.ulaval.ca.

The University of Arizona, Graduate College, College of Agriculture, Department of Agricultural and Biosystems Engineering, Tucson, AZ 85721. Offers MS, PhD. *Faculty:* 15. *Students:* 24 full-time (6 women), 5 part-time; includes 1 minority (African American), 17 international. Average age 33. 7 applicants, 57% accepted. In 1998, 3 master's, 3 doctorates awarded. Terminal master's awarded for partial completion of doctoral program. *Degree requirements:* For master's and doctorate, computer language, thesis/dissertation required, foreign language not required. *Entrance requirements:* For master's and doctorate, TOEFL (minimum score of 550 required), minimum GPA of 3.0 in last 2 years of undergraduate study. *Application deadline:* For fall admission, 3/1. Applications are processed on a rolling basis. Application fee: $45. *Financial aid:* Fellowships, research assistantships, teaching assistantships, Federal Work-Study, institutionally-sponsored loans, scholarships, and tuition waivers (full and partial) available. Financial aid application deadline: 5/1. *Faculty research:* Irrigation system design, energy-use management, equipment for alternative crops, food properties enhancement. *Unit head:* Donald Slack, Head, 520-621-3691. *Application contact:* Kathleen Crist, Graduate Secretary, 520-621-1607.

University of Arkansas, Graduate School, College of Engineering, Department of Biological and Agricultural Engineering, Fayetteville, AR 72701-1201. Offers MSBAE, MSE, PhD. *Faculty:* 10 full-time (0 women). *Students:* 9 full-time (1 woman), 1 (woman) part-time; includes 1 minority (Hispanic American), 6 international. 5 applicants, 40% accepted. In 1998, 3 master's, 1 doctorate awarded. *Degree requirements:* For master's, thesis required, foreign language not required; for doctorate, one foreign language, dissertation required. Application fee: $40 ($50 for international students). Tuition, state resident: full-time $3,186. Tuition, nonresident: full-time $7,560. Required fees: $378. *Financial aid:* In 1998–99, 9 research assistantships were awarded.; career-related internships or fieldwork and Federal Work-Study also available. Aid available to part-time students. Financial aid application deadline: 4/1; financial aid applicants required to submit FAFSA. *Unit head:* Ivan L. Berry, Head, 501-575-2351.

University of British Columbia, Faculty of Graduate Studies, Faculty of Applied Science, Department of Bioresource Engineering, Vancouver, BC V6T 1Z2, Canada. Offers M Sc, MA Sc, PhD. *Degree requirements:* For master's, thesis required, foreign language not required. *Entrance requirements:* For master's, TOEFL (minimum score of 550 required). *Faculty research:* Environmental management, water resource development, bio-process engineering, waste utilization (biomass conversions).

University of California, Davis, Graduate Studies, College of Engineering, Program in Biological and Agricultural Engineering, Davis, CA 95616. Offers M Engr, MS, D Engr, PhD, M Engr/MBA. Part-time programs available. *Faculty:* 27 full-time (3 women). *Students:* 43 full-time (15 women); includes 7 minority (5 Asian Americans or Pacific Islanders, 2 Hispanic Americans), 22 international. Average age 25. 37 applicants, 68% accepted. In 1998, 6 master's awarded (100% found work related to degree); 7 doctorates awarded. Terminal master's awarded for partial completion of doctoral program. *Degree requirements:* For master's and doctorate, thesis/dissertation required, foreign language not required. *Entrance requirements:* For master's, minimum GPA of 3.0; for doctorate, GRE, minimum graduate GPA of 3.25. *Application deadline:* For fall admission, 4/1 (priority date). Application fee: $40. Electronic applications accepted. *Financial aid:* In 1998–99, 28 students received aid, including 9 fellowships with full and partial tuition reimbursements available, 13 research assistantships with full and partial tuition reimbursements available, 4 teaching assistantships with full and partial tuition reimbursements available Financial aid application deadline: 1/15; financial aid applicants required to submit FAFSA. *Faculty research:* Forestry, irrigation and drainage, power and machinery, structures and environment, information and energy technologies. *Unit head:* David J. Hills, Graduate Adviser, 530-752-0102, Fax: 530-752-2640, E-mail: bioageng@ucdavis.edu.

University of Dayton, Graduate School, School of Engineering, Department of Civil Engineering, Dayton, OH 45469-1300. Offers engineering mechanics (MSEM); environmental engineering (MSCE); soil mechanics (MSCE); structural engineering (MSCE); transport engineering (MSCE). Part-time programs available. *Faculty:* 9 full-time (0 women), 2 part-time (0 women). *Students:* 7 full-time (1 woman), 6 part-time (1 woman); includes 1 minority (African American), 4 international. *Degree requirements:* For master's, thesis or alternative required, foreign language not required. *Entrance requirements:* For master's, TOEFL. *Application deadline:* For fall admission, 8/1. Applications are processed on a rolling basis. Application fee: $30. *Unit head:* Dr. Joseph Saliba, Chairperson, 937-229-3847. *Application contact:* Dr. Donald L. Moon, Associate Dean, 937-229-2241, Fax: 937-229-2471, E-mail: dmoon@engr.udayton.edu.

University of Florida, Graduate School, College of Engineering, Department of Agricultural and Biological Engineering, Gainesville, FL 32611. Offers agricultural engineering (ME, MS, PhD, Engr); agricultural operations management (MS, PhD). Part-time programs available. *Faculty:* 39. *Students:* 27 full-time (4 women), 7 part-time (2 women); includes 6 minority (4 African Americans, 1 Asian American or Pacific Islander, 1 Native American), 16 international. 27 applicants, 59% accepted. In 1998, 9 master's, 3 doctorates awarded. Terminal master's awarded for partial completion of doctoral program. *Degree requirements:* For master's and Engr, thesis optional; for doctorate, dissertation required. *Entrance requirements:* For master's and doctorate, GRE General Test, minimum GPA of 3.0; for Engr, GRE General Test. *Application deadline:* For fall admission, 6/1 (priority date). Applications are processed on a rolling basis. Application fee: $20. Electronic applications accepted. *Financial aid:* In 1998–99, 23 students received aid, including 6 fellowships, 8 research assistantships, 1 teaching assistantship; unspecified assistantships also available. *Faculty research:* Soil and water engineering, structures and environments, power and machinery, biological processing, food engineering, remote sensing hydrology, geographic information systems. *Unit head:* Dr. C. Direlle Baird, Chair, 352-392-7655, Fax: 352-392-4092, E-mail: baird@agen.ufl.edu. *Application contact:* Dr. Kenneth Campbell, Graduate Coordinator, 352-392-8534, Fax: 352-392-4092, E-mail: klc@agen.ufl.edu.

University of Georgia, Graduate School, College of Agricultural and Environmental Sciences, Department of Biological and Agricultural Engineering, Athens, GA 30602. Offers agricultural engineering (MS); biological and agricultural engineering (PhD); biological engineering (MS). *Faculty:* 30 full-time (1 woman). *Students:* 17 full-time (4 women), 1 part-time. 37 applicants, 32% accepted. In 1998, 8 master's, 2 doctorates awarded. *Degree requirements:* For master's, thesis required, foreign language not required; for doctorate, one foreign language (computer language can substitute), dissertation required. *Entrance requirements:* For master's and doctorate, GRE General Test. *Application deadline:* For fall admission, 7/1 (priority date); for spring admission, 11/15. Application fee: $30. Electronic applications accepted. *Financial aid:* Fellowships, research assistantships, teaching assistantships, unspecified assistantships avail-

Agricultural Engineering

University of Georgia (continued)

able. *Unit head:* Dr. Brahm Verma, Graduate Coordinator, 706-542-0862, Fax: 706-542-8806, E-mail: bverma@bae.uga.edu.

University of Hawaii at Manoa, Graduate Division, College of Tropical Agriculture and Human Resources, Department of Biosystems Engineering, Honolulu, HI 96822. Offers MS. Part-time programs available. *Faculty:* 16 full-time (0 women). *Students:* 10 full-time (4 women), 6 part-time (2 women); includes 6 minority (1 African American, 5 Asian Americans or Pacific Islanders), 4 international. 6 applicants, 83% accepted. In 1998, 4 degrees awarded. *Degree requirements:* For master's, computer language, thesis required, foreign language not required. *Application deadline:* For fall admission, 3/1; for spring admission, 9/1. Application fee: $25 ($50 for international students). *Financial aid:* In 1998–99, 10 research assistantships (averaging $15,017 per year), 1 teaching assistantship (averaging $12,786 per year) were awarded.; fellowships, Federal Work-Study, institutionally-sponsored loans, and tuition waivers (full) also available. *Faculty research:* Mechanization, agricultural systems, waste management, water management, cell culture. *Unit head:* Dr. Charles M. Kinoshita, Chairperson, 808-956-8867, Fax: 808-956-9269, E-mail: kinoshi@wiliki.eng.hawaii.edu. *Application contact:* Dr. John Grove, Chairman, 808-956-5779, Fax: 808-956-4585, E-mail: jgrove@hawaii.edu.

University of Idaho, College of Graduate Studies, College of Agriculture, Department of Biological and Agricultural Engineering, Moscow, ID 83844-4140. Offers M Engr, MS, PhD. *Faculty:* 10 full-time (0 women). *Students:* 7 full-time (1 woman), 8 part-time; includes 1 minority (African American), 6 international. In 1998, 1 master's, 2 doctorates awarded. *Degree requirements:* For master's, thesis or alternative required, foreign language not required; for doctorate, dissertation required. *Entrance requirements:* For master's, minimum GPA of 2.8; for doctorate, minimum undergraduate GPA of 2.8, 3.0 graduate. *Application deadline:* For fall admission, 8/1; for spring admission, 12/15. Application fee: $35 ($45 for international students). *Financial aid:* In 1998–99, 9 research assistantships (averaging $11,651 per year), 2 teaching assistantships (averaging $10,097 per year) were awarded.; career-related internships or fieldwork also available. Financial aid application deadline: 2/15. *Faculty research:* Irrigation, soil and water conservation, agricultural mechanization. *Unit head:* Dr. James De Shazer, Head, 208-885-6182.

University of Illinois at Urbana–Champaign, Graduate College, College of Engineering, Department of Agricultural Engineering, Urbana, IL 61801. Offers MS, PhD. *Faculty:* 20 full-time (0 women). *Students:* 42 full-time (12 women); includes 3 minority (1 African American, 1 Asian American or Pacific Islander, 1 Hispanic American), 23 international. 18 applicants, 33% accepted. In 1998, 7 master's, 6 doctorates awarded. *Degree requirements:* For master's and doctorate, thesis/dissertation required, foreign language not required. *Application deadline:* Applications are processed on a rolling basis. Application fee: $40 ($50 for international students). Tuition, state resident: full-time $4,616. Tuition, nonresident: full-time $11,768. Full-time tuition and fees vary according to course load. *Financial aid:* In 1998–99, 3 fellowships, 35 research assistantships, 1 teaching assistantship were awarded.; tuition waivers (full and partial) also available. Financial aid application deadline: 2/15. *Unit head:* Loren E. Bode, Head, 217-333-3570. *Application contact:* Marvin Paulsen, Director of Graduate Studies, 217-333-7926, Fax: 244-0323, E-mail: mrp@sugar.age.uiuc.edu.

University of Kentucky, Graduate School, Graduate School Programs from the College of Engineering, Program in Agricultural Engineering, Lexington, KY 40506-0032. Offers MSAE, PhD. Part-time programs available. *Degree requirements:* For master's, comprehensive exam required, thesis optional, foreign language not required; for doctorate, dissertation, comprehensive exam required, foreign language not required. *Entrance requirements:* For master's, GRE General Test, minimum undergraduate GPA of 2.5; for doctorate, GRE General Test, minimum graduate GPA of 3.0. *Faculty research:* Machine systems, food engineering, fermentation, hydrology, water quality.

University of Maine, Graduate School, College of Natural Sciences, Forestry, and Agriculture, Department of Biosystems Science and Engineering, Program in Bio-Resource Engineering, Orono, ME 04469. Offers MS. *Faculty:* 7 full-time (0 women). *Students:* 2. 3 applicants, 33% accepted. *Degree requirements:* For master's, thesis required (for some programs), foreign language not required. *Entrance requirements:* For master's, GRE General Test, GRE Subject Test, TOEFL (minimum score of 550 required), undergraduate major in agricultural or forest engineering. *Application deadline:* For fall admission, 2/1 (priority date); for spring admission, 10/15. Applications are processed on a rolling basis. Application fee: $50. *Financial aid:* Federal Work-Study available. Financial aid application deadline: 3/1. *Faculty research:* Aquaculture, irrigation and drainage, forest machinery, biomass energy. *Unit head:* Dr. Hayden Soule, Chair, 207-581-2718, Fax: 207-581-2725. *Application contact:* Scott G. Delcourt, Director of the Graduate School, 207-581-3218, Fax: 207-581-3232, E-mail: graduate@maine.edu.

University of Manitoba, Faculty of Graduate Studies, Faculty of Engineering, Department of Agricultural Engineering, Winnipeg, MB R3T 2N2, Canada. Offers M Eng, M Sc, PhD. *Unit head:* N. R. Bulley, Head.

University of Maryland, College Park, Graduate School, College of Agriculture and Natural Resources, Department of Biological Resources Engineering, College Park, MD 20742-5045. Offers MS, PhD. *Faculty:* 13 full-time (0 women), 6 part-time (1 woman). *Students:* 17 full-time (10 women), 12 part-time (3 women); includes 7 minority (4 African Americans, 3 Asian Americans or Pacific Islanders), 6 international. 37 applicants, 43% accepted. In 1998, 6 master's, 1 doctorate awarded. *Degree requirements:* For master's, thesis optional, foreign language not required; for doctorate, dissertation required, foreign language not required. *Entrance requirements:* For master's, minimum GPA of 3.0. Application fee: $50 ($70 for international students). Tuition, state resident: part-time $272 per credit hour. Tuition, nonresident: part-time $475 per credit hour. Required fees: $632; $379 per year. *Financial aid:* In 1998–99, 17 research assistantships with tuition reimbursements (averaging $13,288 per year), 7 teaching assistantships with tuition reimbursements (averaging $11,818 per year) were awarded.; fellowships with full tuition reimbursements, career-related internships or fieldwork also available. Financial aid applicants required to submit FAFSA. *Faculty research:* Engineering aspects of production; harvesting, processing, and marketing of terrestrial and aquatic food and fiber. Total annual research expenditures: $385,057. *Unit head:* Dr. Frederick Wheaton, Chairman, 301-405-2223, Fax: 301-314-9023. *Application contact:* Trudy Lindsey, Director, Graduate Admission and Records, 301-405-4198, Fax: 301-314-9305, E-mail: grschool@deans.umd.edu.

Announcement: Housed in one of the University's newest buildings, the Department of Biological Resources Engineering provides graduate students with state-of-the-art research facilities, including specialized laboratories for work on instrumentation, cellular engineering, physiology, ecological engineering, biomechanics, geographic information systems, water quality, aquacultural engineering, and water resources engineering. The graduate program provides qualified students with the interdisciplinary study and research experience they need to contribute to this exciting field. Under the personal guidance of outstanding faculty members, graduate students design educational programs leading to both Master of Science and Doctor of Philosophy degrees. Programs of study are tailored to meet the individual research interests and career ambitions of each graduate student. For additional information, see the in-depth description in Section 5.

University of Minnesota, Twin Cities Campus, Graduate School, College of Agricultural, Food, and Environmental Sciences, Department of Biosystems and Agricultural Engineering, Minneapolis, MN 55455-0213. Offers MBAE, MSBAE, PhD. Part-time programs available. *Faculty:* 16 full-time (1 woman). *Students:* 18 full-time (8 women), 10 part-time (1 woman). 22 applicants, 73% accepted. In 1998, 5 master's awarded (100% found work related to degree); 1 doctorate awarded (100% found work related to degree). *Degree requirements:* For master's, thesis (for some programs), seminar required, foreign language not required; for doctorate, dissertation, seminar required, foreign language not required. *Entrance requirements:* For master's and doctorate, TOEFL (minimum score of 550 required; average 590), BS in engineering or related field. *Average time to degree:* Master's–2.5 years full-time; doctorate–5 years full-time. *Application deadline:* For fall admission, 7/15; for spring admission, 12/15. Applications are processed on a rolling basis. Application fee: $40 ($50 for international students). *Financial aid:* In 1998–99, 18 students received aid, including 2 fellowships (averaging $17,000 per year), 16 research assistantships (averaging $14,500 per year) Aid available to part-time students. Financial aid application deadline: 2/15. *Faculty research:* Water quality, bioprocessing, food engineering, terramechanics, process and machine control. Total annual research expenditures: $2.3 million. *Unit head:* Dr. R. Vance Morey, Head, 612-625-7733, Fax: 612-624-3005. *Application contact:* Kevin A. Janni, Director of Graduate Studies, 612-625-3108, Fax: 612-624-3005, E-mail: keven.a.janni-1@umn.edu.

University of Missouri–Columbia, Graduate School, College of Engineering, Department of Biological Engineering, Columbia, MO 65211. Offers agricultural engineering (MS); biological engineering (MS, PhD). *Faculty:* 13 full-time (0 women), 1 (woman) part-time. *Students:* 7 full-time (2 women), 15 part-time (4 women), 19 international. 1 applicants, 0% accepted. In 1998, 4 master's, 1 doctorate awarded. *Degree requirements:* For master's and doctorate, thesis/dissertation required. *Entrance requirements:* For master's and doctorate, GRE General Test, TOEFL, minimum GPA of 3.0. *Application deadline:* For fall admission, 5/1 (priority date). Applications are processed on a rolling basis. Application fee: $30 ($50 for international students). *Financial aid:* Research assistantships, teaching assistantships, institutionally-sponsored loans available. *Unit head:* Dr. Jinglu Tan, Director of Graduate Studies, 573-882-7778.

University of Nebraska–Lincoln, Graduate College, College of Engineering and Technology, Department of Biological Systems Engineering, Lincoln, NE 68588. Offers agricultural and biological systems engineering (MS); agricultural science (MS), including mechanized systems management; engineering (PhD). *Faculty:* 21 full-time (1 woman), 6 part-time (0 women). *Students:* 17 full-time (3 women), 24 part-time (5 women); includes 15 minority (1 African American, 11 Asian Americans or Pacific Islanders, 3 Hispanic Americans) Average age 25. 41 applicants, 12% accepted. In 1998, 12 master's, 3 doctorates awarded. *Degree requirements:* For master's, thesis (for some programs), teaching experience required; for doctorate, dissertation, comprehensive exams required. *Entrance requirements:* For master's and doctorate, GRE General Test, TOEFL (minimum score of 550 required). *Average time to degree:* Master's–2.5 years full-time, 4 years part-time; doctorate–3.5 years full-time, 6 years part-time. *Application deadline:* Applications are processed on a rolling basis. Application fee: $35. Electronic applications accepted. *Financial aid:* In 1998–99, 13 students received aid, including 4 fellowships with full tuition reimbursements available (averaging $17,000 per year), 8 research assistantships with full tuition reimbursements available (averaging $12,600 per year), 1 teaching assistantship (averaging $12,000 per year) *Faculty research:* Food and biochemical engineering, environmental engineering, precision agriculture, sensors and controls, soil and water. Total annual research expenditures: $2.5 million. *Unit head:* Dr. Glenn J. Hoffman, Head, 402-472-1413, Fax: 402-472-6338, E-mail: bsen001@unlvm.onl.edu.

University of Saskatchewan, College of Graduate Studies and Research, College of Engineering, Department of Agricultural Engineering, Saskatoon, SK S7N 5A2, Canada. Offers M Eng, M Sc, PhD. *Degree requirements:* For master's, computer language, thesis required, foreign language not required; for doctorate, dissertation required, foreign language not required. *Entrance requirements:* For master's and doctorate, GRE, TOEFL.

University of Tennessee, Knoxville, Graduate School, College of Agricultural Sciences and Natural Resources, Department of Agricultural and Biosystems Engineering, Program in Biosystems Engineering, Knoxville, TN 37996. Offers MS, PhD. *Students:* 5 full-time (2 women), 4 part-time, 3 international. 3 applicants, 0% accepted. In 1998, 3 master's, 1 doctorate awarded. *Degree requirements:* For master's and doctorate, thesis/dissertation required, foreign language not required. *Entrance requirements:* For master's and doctorate, GRE General Test, TOEFL (minimum score of 550 required), GRE Subject Test, minimum GPA of 2.7. *Application deadline:* For fall admission, 2/1 (priority date). Applications are processed on a rolling basis. Application fee: $35. Electronic applications accepted. *Financial aid:* Application deadline: 2/1; *Unit head:* Kevin A. Janni, Director of Graduate Studies, 612-625-3108, Fax: 612-624-3005, E-mail: keven.a.janni-1@umn.edu. *Application contact:* Dr. Ronald Yoder, Graduate Representative, E-mail: ryoder@utk.edu.

University of Tennessee, Knoxville, Graduate School, College of Agricultural Sciences and Natural Resources, Department of Agricultural and Biosystems Engineering, Program in Biosystems Engineering Technology, Knoxville, TN 37996. Offers MS. *Students:* 4 full-time (1 woman), 7 part-time (2 women). 2 applicants, 50% accepted. In 1998, 4 degrees awarded. *Degree requirements:* For master's, thesis or alternative required, foreign language not required. *Entrance requirements:* For master's, TOEFL (minimum score of 550 required), GRE General Test, minimum GPA of 2.7. *Application deadline:* For fall admission, 2/1 (priority date). Applications are processed on a rolling basis. Application fee: $35. Electronic applications accepted. *Financial aid:* Application deadline: 2/1; *Unit head:* Dr. Constance L. Wood, Associate Dean, 606-257-4613, Fax: 606-323-1928. *Application contact:* Dr. Ronald Yoder, Graduate Representative, E-mail: ryoder@utk.edu.

University of Wisconsin–Madison, Graduate School, College of Agricultural and Life Sciences, Department of Biological Systems Engineering, Madison, WI 53706-1380. Offers MS, PhD. Part-time programs available. Terminal master's awarded for partial completion of doctoral program. *Degree requirements:* For master's, computer language required, thesis optional, foreign language not required; for doctorate, computer language, dissertation required, foreign language not required. *Entrance requirements:* For master's, GRE, TOEFL. Electronic applications accepted. *Faculty research:* Waste systems, food engineering, power and machinery, structures and environment, construction management.

Utah State University, School of Graduate Studies, College of Engineering, Department of Biological and Irrigation Engineering, Logan, UT 84322. Offers biological and agricultural engineering (MS, PhD); irrigation engineering (MS, PhD). Part-time programs available. *Faculty:* 10 full-time (0 women). *Students:* 26 full-time (4 women), 11 part-time (4 women); includes 1 minority (Hispanic American), 27 international. Average age 28. 21 applicants, 76% accepted. In 1998, 7 master's, 8 doctorates awarded. Terminal master's awarded for partial completion of doctoral program. *Degree requirements:* For master's, computer language, thesis required (for some programs), foreign language not required; for doctorate, computer language, dissertation required, foreign language not required. *Entrance requirements:* For master's and doctorate, GRE General Test (score in 40th percentile or higher required), TOEFL (minimum score of 550 required), minimum GPA of 3.0. *Application deadline:* For fall admission, 6/15 (priority date); for spring admission, 10/15. Applications are processed on a rolling basis. Application fee: $40. Tuition, state resident: full-time $1,492. Tuition, nonresident: full-time $5,232. Required fees: $434. Tuition and fees vary according to course load. *Financial aid:* In 1998–99, 15 fellowships with partial tuition reimbursements, 15 research assistantships with partial tuition reimbursements, 2 teaching assistantships with partial tuition reimbursements were awarded.; tuition waivers (partial) also available. Aid available to part-time students. *Faculty research:* Surge flow, on-farm water management, crop-water yield modeling, drainage, groundwater modeling. *Unit head:* Wynn Walker, Head, 435-797-2788, Fax: 435-797-1248, E-mail: wynnwalk@cc.usu.edu. *Application contact:* Lyman Willardson, Graduate Adviser, 435-797-2789, Fax: 435-797-1248, E-mail: fath8@cc.usu.edu.

Virginia Polytechnic Institute and State University, Graduate School, College of Engineering, Department of Biological Systems Engineering, Blacksburg, VA 24061. Offers M Eng, MS, PhD. *Faculty:* 12 full-time (0 women), 8 part-time (0 women). *Students:* 24 full-time (7 women), 1 part-time; includes 1 minority (Asian American or Pacific Islander), 12 international. Average age 25. 13 applicants, 62% accepted. In 1998, 7 master's, 3 doctorates awarded. *Degree requirements:* For master's and doctorate, thesis/dissertation required, foreign language not required. *Entrance requirements:* For master's and doctorate, GRE General Test, TOEFL (minimum score of 600 required). *Application deadline:* For fall admission, 12/1 (priority date). Applications are processed on a rolling basis. Application fee: $25. *Financial aid:* In 1998–99, 4 fellowships, 17 research assistantships were awarded.: career-related internships or fieldwork, institutionally-sponsored loans, tuition waivers (full and partial), and unspecified assistantships also available. Aid available to part-time students. Financial aid application deadline: 4/1. *Faculty research:* Soil and water engineering, alternative energy sources for agriculture and agricultural mechanization. *Unit head:* Dr. John V. Perumpral, Head, 540-231-6615, E-mail: perump@vt.edu. *Application contact:* Dr. S. Mostaghimi, Chairman, 540-231-7605, E-mail: smostagh@vt.edu.

Section 4
Architectural Engineering

This section contains a directory of institutions offering graduate work in architectural engineering, followed by in-depth entries submitted by institutions that chose to prepare detailed program descriptions. Additional information about programs listed in the directory but not augmented by an in-depth entry may be obtained by writing directly to the dean of a graduate school or chair of a department at the address given in the directory.

For programs offering related work, see also in this book Engineering and Applied Sciences and Management of Engineering and Technology. In Book 2, see Applied Arts and Design (Industrial Design and Interior Design), Architecture (Environmental Design), Political Science and International Affairs, and Public, Regional, and Industrial Affairs (City and Regional Planning and Urban Studies). In Book 4, see Environmental Sciences and Management.

CONTENTS

Architectural Engineering

Illinois Institute of Technology, Graduate College, Armour College of Engineering and Sciences, Department of Civil and Architectural Engineering, Chicago, IL 60616-3793. Offers M Geoenv E, M Trans E, MCEM, MGE, MPW, MS, MSE, PhD. Part-time and evening/weekend programs available. *Faculty:* 9 full-time (0 women), 11 part-time (0 women). *Students:* 36 full-time (8 women), 60 part-time (6 women); includes 13 minority (2 African Americans, 7 Asian Americans or Pacific Islanders, 4 Hispanic Americans), 55 international. 108 applicants, 63% accepted. In 1998, 23 master's, 4 doctorates awarded. Terminal master's awarded for partial completion of doctoral program. *Degree requirements:* For master's, thesis (for some programs), comprehensive exam required, foreign language not required; for doctorate, dissertation, comprehensive exam required, foreign language not required. *Entrance requirements:* For master's and doctorate, GRE (minimum score of 1200 required), TOEFL (minimum score of 550 required), undergraduate GPA of 3.0 required. *Application deadline:* For fall admission, 7/1; for spring admission, 11/1. Applications are processed on a rolling basis. Application fee: $30. Electronic applications accepted. *Financial aid:* In 1998–99, 1 fellowship, 11 teaching assistantships were awarded.; research assistantships, Federal Work-Study, institutionally-sponsored loans, scholarships, and graduate assistantships also available. Financial aid application deadline: 3/1. *Faculty research:* Structural engineering, construction management, geotechnical engineering, transportation engineering, optimum design of civil structures. Total annual research expenditures: $51,000. *Unit head:* Dr. J. Mohammadi, Chairman, 312-567-3540, Fax: 312-567-3519, E-mail: cemohammadi@minna.cns.iit.edu. *Application contact:* Dr. S. Mohammad Shahidehpour, Dean of Graduate College, 312-567-3024, Fax: 312-567-7517, E-mail: grad@minna.cns.iit.edu.

Kansas State University, Graduate School, College of Engineering, Department of Architectural Engineering and Construction Science, Manhattan, KS 66506. Offers architectural engineering (MS). *Faculty:* 7 full-time (0 women). *Students:* 16 full-time (3 women), 1 part-time. 9 applicants, 100% accepted. In 1998, 6 degrees awarded (100% found work related to degree). *Degree requirements:* For master's, thesis or alternative required, foreign language not required. *Average time to degree:* Master's–1.5 years full-time. *Application deadline:* Applications are processed on a rolling basis. Application fee: $0 ($25 for international students). Electronic applications accepted. *Financial aid:* In 1998–99, 6 research assistantships (averaging $5,067 per year) were awarded.; career-related internships or fieldwork and Federal Work-Study also available. *Faculty research:* Building systems, structures. Total annual research expenditures: $155,000. *Unit head:* David Fritchen, Head, 785-532-5964, Fax: 785-532-6944.

Milwaukee School of Engineering, Department of Architectural Engineering and Building Construction, Program in Architectural Engineering, Milwaukee, WI 53202-3109. Offers MS. Part-time programs available. *Faculty:* 4. *Application deadline:* For fall admission, 8/15. Applications are processed on a rolling basis. Application fee: $30. Electronic applications accepted. *Unit head:* Dr. John Zachar, Director, 414-277-7307, Fax: 414-277-7479, E-mail: zachar@msoe.edu. *Application contact:* Graduate Admissions, 800-332-6763, Fax: 414-277-7475.

Announcement: MSOE offers a Master of Science in Architectural Engineering (MSAE) degree. The MSAE program has a building structural design specialty and is ideal for individuals who have an undergraduate degree in architectural, civil, or structural engineering. The program is offered on a part-time basis.

North Carolina Agricultural and Technical State University, Graduate School, College of Engineering, Department of Architectural Engineering, Greensboro, NC 27411. Offers MSAE. Part-time programs available. *Faculty:* 6 full-time (0 women). *Students:* 6 full-time (3 women), 1 part-time; includes 4 minority (all African Americans) Average age 24. 10 applicants, 70% accepted. In 1998, 2 degrees awarded. *Degree requirements:* For master's, thesis defense required. *Entrance requirements:* For master's, GRE General Test, GRE Subject Test (recommended). *Application deadline:* For fall admission, 7/1; for spring admission, 1/9. Applications are processed on a rolling basis. Application fee: $35. *Financial aid:* In 1998–99, 1 fellowship, 3 research assistantships, 2 teaching assistantships were awarded.; career-related internships or fieldwork and unspecified assistantships also available. *Faculty research:* Lightning, indoor air quality, material behavior HVAC controls, structural masonry systems. Total annual research expenditures: $250,000. *Unit head:* Dr. Ronald N. Helms, Chairperson, 336-344-7575, Fax: 336-344-7126. *Application contact:* Dr. W. Mark McGinley, Graduate Coordinator, 336-334-7575, Fax: 336-334-7126, E-mail: mcginley@garfield.ncat.edu.

Oklahoma State University, Graduate College, College of Engineering, Architecture and Technology, School of Architecture, Program in Architectural Engineering, Stillwater, OK 74078. Offers M Arch E. *Degree requirements:* For master's, thesis or alternative required, foreign language not required. *Entrance requirements:* For master's, TOEFL (minimum score of 550 required). *Application deadline:* For fall admission, 7/1 (priority date). Application fee: $25. *Financial aid:* Career-related internships or fieldwork, Federal Work-Study, and tuition waivers (partial) available. Aid available to part-time students. Financial aid application deadline: 3/1.

Pennsylvania State University University Park Campus, Graduate School, College of Engineering, Department of Architectural Engineering, State College, University Park, PA 16802-1503. Offers M Eng, MAE, MS, PhD. *Students:* 25 full-time (5 women), 2 part-time. In 1998, 10 master's, 1 doctorate awarded. *Degree requirements:* For master's and doctorate, thesis/dissertation required, foreign language not required. *Entrance requirements:* For master's and doctorate, GRE General Test. Application fee: $50. *Unit head:* Dr. Richard Behr, Head, 814-865-6394. *Application contact:* Graduate Program Officer, 814-863-2078.

See in-depth description on page 371.

Rensselaer Polytechnic Institute, Graduate School, School of Architecture, Program in Lighting, Troy, NY 12180-3590. Offers MS. *Faculty:* 11 full-time (3 women), 15 part-time (5 women). *Students:* 20 full-time (10 women), 2 part-time (1 woman); includes 4 minority (2 Asian Americans or Pacific Islanders, 1 Hispanic American, 1 Native American), 10 international. 18 applicants, 67% accepted. In 1998, 4 degrees awarded. *Degree requirements:* For master's, thesis required, foreign language not required. *Entrance requirements:* For master's, GRE General Test, TOEFL (minimum score of 550 required), portfolio. *Application deadline:* For fall admission, 4/1 (priority date). Applications are processed on a rolling basis. Application fee: $35. *Financial aid:* In 1998–99, 4 students received aid; research assistantships, career-related internships or fieldwork and institutionally-sponsored loans available. Financial aid application deadline: 4/1. *Faculty research:* Lighting and human performance, energy-efficient lighting, lighting product development, lighting design demonstration. Total annual research expenditures: $3.5 million. *Unit head:* Dr. Mark Rea, Director, Lighting Research Center, 518-276-8701. *Application contact:* Daniel Frering, Manager, Outreach Education, 518-276-8716, Fax: 518-276-4835, E-mail: frerid@rpi.edu.

University of Colorado at Boulder, Graduate School, College of Engineering and Applied Science, Department of Civil, Environmental, and Architectural Engineering, Boulder, CO 80309. Offers building systems (MS, PhD); construction engineering and management (MS, PhD); environmental engineering (MS, PhD); geoenvironmental engineering (MS, PhD); geotechnical engineering (MS, PhD); structural engineering (MS, PhD); water resource engineering (MS, PhD). *Degree requirements:* For master's, thesis or alternative, comprehensive exam required; for doctorate, dissertation required. *Entrance requirements:* For master's, GRE General Test, minimum undergraduate GPA of 3.0.

University of Kansas, Graduate School, School of Engineering, Department of Architectural Engineering, Lawrence, KS 66045. Offers MS. *Faculty:* 4 full-time, 1 part-time. *Students:* 4 full-time (1 woman), 9 part-time (4 women); includes 1 minority (African American), 2 international. In 1998, 2 degrees awarded. *Degree requirements:* For master's, thesis required, foreign language not required. *Entrance requirements:* For master's, Michigan English Language Assessment Battery, TOEFL, minimum GPA of 3.0. *Application deadline:* For fall admission, 7/1. Application fee: $30 ($45 for international students). *Financial aid:* Fellowships, research assistantships, teaching assistantships, career-related internships or fieldwork available. *Faculty research:* Construction management, building mechanical systems, illumination engineering, building power systems. *Unit head:* Thomas Glavinich, Chair, 785-864-3434.

The University of Memphis, Graduate School, Herff College of Engineering, Department of Engineering Technology, Memphis, TN 38152. Offers architectural technology (MS); electronics engineering technology (MS); manufacturing engineering technology (MS). Part-time programs available. *Faculty:* 6 full-time (2 women). *Students:* 19 full-time (4 women), 8 part-time (2 women); includes 3 minority (2 African Americans, 1 Asian American or Pacific Islander), 21 international. *Degree requirements:* For master's, comprehensive exam required. *Entrance requirements:* For master's, GRE General Test (minimum combined score of 1000 required) or MAT, interview, minimum undergraduate GPA of 2.5. *Application deadline:* For fall admission, 8/1; for spring admission, 12/1. Applications are processed on a rolling basis. Application fee: $25 ($50 for international students). Electronic applications accepted. Tuition, state resident: full-time $3,410; part-time $178 per credit hour. Tuition, nonresident: full-time $8,670; part-time $408 per credit hour. Tuition and fees vary according to program. *Unit head:* Ronald L. Day, Chairman, 901-678-2238, Fax: 901-678-5145, E-mail: rday@memphis.edu. *Application contact:* Dr. Dean L. Smith, Coordinator of Graduate Studies, 901-678-3300, Fax: 901-678-5145, E-mail: dlsmith@cc.memphis.edu.

University of Miami, Graduate School, College of Engineering, Department of Civil, Architectural, and Environmental Engineering, Coral Gables, FL 33124. Offers architectural engineering (MSAE); civil engineering (MSCE, DA, PhD). Part-time and evening/weekend programs available. *Faculty:* 9 full-time (1 woman), 8 part-time (1 woman). *Students:* 17 full-time (7 women), 2 part-time (1 woman); includes 8 minority (5 Asian Americans or Pacific Islanders, 3 Hispanic Americans) Average age 25. 43 applicants, 74% accepted. In 1998, 4 master's awarded. *Degree requirements:* For master's, thesis required, foreign language not required; for doctorate, dissertation, oral and qualifying exams required, foreign language not required. *Entrance requirements:* For master's and doctorate, GRE General Test (minimum score of 500 on verbal section, 500 on quantitative required), TOEFL (minimum score of 550 required), minimum GPA of 3.0. *Average time to degree:* Master's–1.5 years full-time, 3 years part-time; doctorate–3 years full-time, 4 years part-time. *Application deadline:* For fall admission, 4/1 (priority date); for spring admission, 11/1 (priority date). Applications are processed on a rolling basis. Application fee: $35. Tuition: Full-time $15,336; part-time $852 per credit. Required fees: $174. Tuition and fees vary according to program. *Financial aid:* In 1998–99, 11 students received aid, including 3 research assistantships, 9 teaching assistantships; fellowships, institutionally-sponsored loans and tuition waivers (partial) also available. *Faculty research:* Electron beam, wastewater treatment, water management, structural reliability, structural dynamics, wind engineering. Total annual research expenditures: $2.3 million. *Unit head:* Dr. David A. Chin, Chairman, 305-284-3391, Fax: 305-284-3492. *Application contact:* Dr. Ahmad H. Namini, Adviser, 305-284-3391, Fax: 305-284-3492.

The University of Texas at Austin, Graduate School, College of Engineering, Program in Architectural Engineering, Austin, TX 78712-1111. Offers MSE. Part-time programs available. *Faculty:* 8 full-time (0 women), 2 part-time (1 woman). *Students:* 11 (2 women); includes 2 minority (1 Asian American or Pacific Islander, 1 Hispanic American) 6 international. 10 applicants, 50% accepted. In 1998, 5 degrees awarded. *Degree requirements:* For master's, thesis required, foreign language not required. *Entrance requirements:* For master's, GRE General Test (minimum combined score of 1000 required). *Application deadline:* For fall admission, 4/1; for spring admission, 11/1. Applications are processed on a rolling basis. Application fee: $50 ($75 for international students). *Financial aid:* In 1998–99, 2 fellowships, 6 research assistantships, 1 teaching assistantship were awarded.; career-related internships or fieldwork also available. Aid available to part-time students. Financial aid application deadline: 2/1. *Faculty research:* Materials engineering, structural engineering, construction engineering, project management. Total annual research expenditures: $2 million. *Unit head:* Dr. James O. Jirsa, Chairman, Department of Civil Engineering, 512-471-4921, Fax: 512-471-0592, E-mail: jirsa@uts.cc.utexas.edu. *Application contact:* James T. O'Connor, Graduate Adviser, 512-471-4645, Fax: 512-471-3191, E-mail: jtoconnor@mail.utexas.edu.

PENNSYLVANIA STATE UNIVERSITY

College of Engineering
Department of Architectural Engineering

Programs of Study

The Department of Architectural Engineering in the College of Engineering offers graduate study and research leading to the M.Eng., M.S., and Ph.D. degrees. Teaching and research are carried out in four major areas: building mechanical and energy systems engineering, building illumination engineering, building structures engineering, and building construction engineering. The building mechanical and energy systems area focuses on HVAC simulation and optimization, building energy analysis, building automation and control, solar and alternate energy sources, indoor air quality control, and vibration and noise control. Building illumination engineering focuses on lighting design, luminaire optics, photometry, lighting system modeling and visualization, psychological and behavioral issues, and daylighting. Building structures engineering focuses on masonry structural design, steel building structural design, reinforced concrete structures for buildings, and glass and aluminum curtain wall (i.e., building envelope) systems. Building construction focuses on project delivery systems; integration of the design, build, and operating processes; process and information modeling; domestic and international construction project organization and control; and advanced technology projects.

The M.Eng. course requirements are 30 course credits and a paper. The course requirement for students pursuing the M.S. degree is 30 credits, which include 24 course credits and 6 thesis credits. A thesis is required for the M.S. degree. Students entering the doctoral program take the candidacy exam within two semesters of enrollment. There is no foreign language requirement for the doctoral degree. When a Ph.D. student has substantially completed the course work, a comprehensive written and oral examination is administered by the student's doctoral committee. After the doctoral candidate has satisfied all other requirements for the degree, the final oral examination is held. This exam is related primarily to the thesis, but it may cover the candidate's entire program of study.

Research Facilities

A wide variety of facilities are available to the architectural engineering student in support of graduate study and research. Within the department, these facilities include the Computer Aided Design Laboratory (CAD Lab), the Computer Integrated Construction Laboratory, the Illumination Laboratory, the Building Enclosure Testing Laboratory, and the Building Thermal and Mechanical Systems Laboratory. Other facilities provided on a College- or University-wide basis include the Engineering Acoustics Laboratory, the Engineering Computer Laboratory, the Civil Engineering Structures Laboratory, and the University libraries. A special facility for the study of buildings and building systems is the readily available laboratory provided by the University Physical Plant. The ongoing construction and operation of this extensive campus facility provides opportunities for many firsthand experiences in support of graduate study and research. In addition, through an advisory group, the Facilities Engineering Institute, the department assists with design, operation, and maintenance problems in state-owned buildings throughout Pennsylvania.

Financial Aid

Opportunities exist for financial support for M.S. and Ph.D. students through teaching and research assistantships, which are normally held for a nine-month period. M.Eng. students do not qualify for department support. Graduate assistantships provide for tuition as well as a stipend, which depends upon the specific level of time commitment. Students offered assistantships are selected on the basis of academic potential and the ability to contribute to the ongoing activities of the program.

Offers of financial aid are made in March for assignments beginning in the following fall semester.

Cost of Study

Tuition in 1998–99 was $3267 per semester for Pennsylvania residents and $6730 per semester for nonresidents, with two semesters per academic year. There is a mandatory $90 computer fee, a $36 activity fee, and a $233 engineering surcharge each semester.

Living and Housing Costs

University and privately owned housing is available to graduate students. The University's facilities range from dormitory rooms for single students to two-bedroom apartments for families. Dormitory rooms were $1375 per semester and up, and one-bedroom apartments were $325 per month and up plus utilities in 1998–99. Privately owned apartments are also available in the community.

Student Group

Graduate enrollment in the department stands at 28 registered students, 6 of whom are Ph.D. candidates. The size of the graduate student body promotes considerable interaction with the faculty and fellow students. The student body is culturally diverse, with varied academic backgrounds in architectural engineering, civil engineering, mechanical engineering, architecture, or science.

Location

Penn State's main campus, University Park, is located in the center of the state in the borough of State College. The University may be reached via public bus or air transportation through the University Park Airport. The town and its surrounding area, with a population of about 75,000, are located in low, rolling mountain country that offers a variety of recreational activities. The community and the University present a wide array of cultural and athletic events.

The University and The Department

Pennsylvania State University is a land-grant university founded in 1855. Graduate work began in 1862 and has grown to the point where approximately 1,400 master's degrees and 470 doctorates are awarded University-wide each year. Architectural engineering at Penn State was initiated in 1913 and is the only architectural engineering department offering the Ph.D. degree.

Applying

Qualified students may be admitted for either the fall or the spring semesters, which begin in August and January, respectively. Applications are reviewed and accepted on a continuous basis. Results from the GRE are required prior to review of any M.S. or Ph.D. application. The GRE exam is not required for M.Eng. applicants. A separate application for assistantships is required. Decisions on fellowships and scholarships are made about February 28, and on teaching and research assistantships about March 15.

Correspondence and Information

Graduate Program Officer
Department of Architectural Engineering
Pennsylvania State University
225 Engineering Unit A
University Park, Pennsylvania 16802

Telephone: 814-863-2078
Fax: 814-863-4789
E-mail: aegrad@psu.edu

Pennsylvania State University

THE FACULTY AND THEIR RESEARCH

William P. Bahnfleth, Assistant Professor; Ph.D., Illinois, 1989; PE. Industrial experience: U.S. Army Construction Engineering Research Laboratory, ZBA Inc. Consulting Engineers. Professional affiliations: ASHRAE, ASME, Sigma Xi. Research areas: stratified chilled water storage, district heating and cooling, numerical heat transfer, building energy modeling and analysis.

Richard A. Behr, Professor and Head; Ph.D., Texas Tech, 1982; PE. Industrial experience: Atlantic Richfield Corporation, Dravo Corporation. Research areas: structural performance and durability of building envelope systems under severe windstorm, earthquake, and accelerated weathering effects; development of innovative laboratory facilities and teaching methods for college-level structural engineering and mechanics courses.

Craig A. Bernecker, Associate Professor; Ph.D., Penn State, 1989. Industry positions: The Ballinger Company, Peerless Electric Company. Professional affiliations: IESNA Fellow, Research Committee, Nomenclature Committee; CIE, U.S. National Committee Member, Technical Expert, Division 3 and 7; IALD. Research areas: illumination engineering, engineering psychology, human factors, distance learning.

Thomas E. Boothby, Associate Professor; Ph.D., Washington, 1991; PE, RA. Industry positions: Design Professionals, Inc.; Cibola Energy Corp., Wilson and Co. Professional affiliations: ASCE, ASEE, TMS, British Masonry Society. Research areas: laboratory and field monitoring of full-scale structures, evaluation of old and historic structures, plasticity applied to unreinforced masonry structures, fiber-reinforced plastics for repair of concrete and masonry structures, sound barriers.

Eric F. P. Burnett, Bernard and Henrietta Hankin Professor; Ph.D., Imperial College (London), 1969; PE, Fellow ACI and CSCE. Industrial experience: Technical Director, Trow Consulting Engineers; Manager of Building Research, Canada Mortgage and Housing Corporation; Ove Arup Associates. Professional affiliations: ASCE, Concrete Society, Masonry Society, CIB. Research areas: housing, enclosure design and performance, structural concrete, the building process.

Louis F. Geschwindner, Professor; Ph.D., Penn State, 1977; PE. Professional affiliations: Chair, ASCE Committee on Design of Steel Building Structures and Vice Chair, Committee on Load and Resistance Factor Design, Standards Committee for Tensioned Fabric Structures. Research areas: design of steel buildings, drift control, semirigid connections, leaning columns, fabric and tension structure analysis and design, masonry, computer applications, computer-aided design.

Linda M. Hanagan, Assistant Professor; Ph.D., Virginia Tech, 1994; PE. Industrial experience: Duffield Associates, Clifton Theobald Associates, URS Dalton. Professional affiliations: ASCE, AE Division Education Committee, Structural Control Committee; NSAE; ASEE. Research areas: structural control, experimental testing, serviceability of building structures, structural design and analysis.

Ali M. Memari, Assistant Professor; Ph.D., Penn State, 1989; PE. Industrial experience: PMB Systems Engineering, Gannett Fleming, Inc. Professional affiliation: Earthquake Engineering Research Institute, ASCE, ACI. Research areas: seismic assessment and strengthening of buildings; full-scale dynamic testing of buildings; seismic performance of partition, infill and curtain walls, and cladding elements; structural optimization.

Richard G. Mistrick, Associate Professor; Ph.D., Penn State, 1991; PE. Previous employment: Lighting Technologies. Professional affiliations: IESNA Fellow; Chair of Papers Committee, Calculation Procedures Committee, Economics Committee; CIE, TC 3-31, Lighting Analysis for Real Interiors, U.S. National Committee Member. Research areas: illumination engineering, advanced lighting system modeling and visualization, daylighting, and photometry.

Stanley A. Mumma, Professor; Ph.D., Illinois, 1974; PE, ASHRAE Fellow. Industrial experience: General Motors Corporation, U.S. Army Construction Engineering Research Laboratory. Professional affiliations: ASHRAE Planning Council, Scholarship Trustee, Nominating Committee, Publishing Council, Technology Council, Society Program Committee, Innovative Research Ideas Committee, Government Affairs Committee. Chairman, Research and Technical Committee; Environmental Health Committee; Standards Committee; Chairman, TC 6.7 Solar Energy Utilization; TCs. Research areas: mathematical modeling-simulation and optimization of building mechanical components and systems, solar and alternate energy, demand side management, control of ventilation systems to provide acceptable indoor air quality, direct digital controls.

M. Kevin Parfitt, Associate Professor; M.E.C.E., Cornell, 1979; PE. Previous employment: Architects Hansen Ling Meyer, Raymond A. DePasquale and Associates, and BASCO Associates. Professional affiliations: ASCE, Technical Council on Computer Practices, Architectural Engineering Institute Education and Publication Committee; National Society of Professional Engineers; Pennsylvania Society of Professional Engineers; and National Computer Graphics Association. Research areas: structural engineering, computer-aided design, automated quality control, professional practice.

Victor E. Sanvido, Professor; Ph.D., Stanford, 1984. Previous employment: Sanvido & Sons (Pty), Ltd.; Clifford Harris (Pty), Ltd.; Sohio Construction Company. Professional affiliations: DBIA, ASCE. Research areas: project delivery systems, process and information modeling, productivity improvement, international project delivery, advanced technology project delivery.

Architectural Engineering at Penn State

Architectural engineering is a specialized academic field of study and research aimed at developing leading engineers for the building industry. Architectural engineering graduates provide the following engineering services to the building industry: conceptualization, analysis, synthesis, budgeting, research, development, design, management, construction, and facility utilization. Architectural engineers are responsible for designing, integrating, and constructing the major systems within a building, including the electrical, communications, illumination, envelope, mechanical, acoustics, plumbing, fire protection, and structural systems. Graduate study at the master's and Ph.D. levels provides an outstanding opportunity for in-depth engineering specialization within the context of buildings. Penn State has the only architectural engineering department in the United States presently offering the Ph.D. The department offers four areas of specialization at the graduate level: building construction, building illumination systems, building mechanical and energy systems, and building structural systems. Graduate students generally enter this program of study with an undergraduate degree in mechanical engineering, electrical engineering, civil engineering, architectural engineering, architecture, or a related science field.

Section 5
Bioengineering, Biomedical Engineering, and Biotechnology

This section contains a directory of institutions offering graduate work in bioengineering, biomedical engineering, and biotechnology, followed by in-depth entries submitted by institutions that chose to prepare detailed program descriptions. Additional information about programs listed in the directory but not augmented by an in-depth entry may be obtained by writing directly to the dean of a graduate school or chair of a department at the address given in the directory.

For programs offering related work, see also in this book Aerospace/Aeronautical Engineering, Engineering and Applied Sciences, Engineering Design, Engineering Physics, Management of Engineering and Technology, and Mechanical Engineering and Mechanics. In Book 3, see Biological and Biomedical Sciences and Physiology; in Book 4, Mathematical Sciences (Biometrics and Biostatistics); and in Book 6, Allied Health.

CONTENTS

Bioengineering

Arizona State University, Graduate College, College of Engineering and Applied Sciences, Department of Chemical, Bio and Materials Engineering, Program in Bioengineering, Tempe, AZ 85287. Offers MS, PhD. *Degree requirements:* For doctorate, dissertation required. *Entrance requirements:* For master's and doctorate, GRE General Test. Application fee: $45. *Financial aid:* Fellowships, research assistantships, teaching assistantships available. *Faculty research:* Biotechnology, biocontrol, biomechanics, bioinstrumentation and materials, biosystems engineering/biotransport. *Application contact:* Graduate Secretary, 480-965-9707.

See in-depth description on page 387.

Carnegie Mellon University, Carnegie Institute of Technology, Department of Civil and Environmental Engineering, Pittsburgh, PA 15213-3891. Offers civil engineering (MS, PhD); civil engineering and industrial management (MS); civil engineering and robotics (PhD); civil engineering/bioengineering (PhD); civil engineering/engineering and public policy (MS, PhD). Part-time programs available. *Faculty:* 21 full-time (5 women). *Students:* 56 full-time (20 women), 11 part-time (2 women); includes 3 minority (1 African American, 1 Hispanic American, 1 Native American), 44 international. Terminal master's awarded for partial completion of doctoral program. *Degree requirements:* For master's, thesis required (for some programs), foreign language not required; for doctorate, dissertation, qualifying exam required, foreign language not required. *Entrance requirements:* For master's and doctorate, GRE General Test, TOEFL. *Application deadline:* For fall admission, 2/1 (priority date); for spring admission, 10/15. Application fee: $45. *Unit head:* Chris Hendrickson, Head, 412-268-2941, Fax: 412-268-7813. *Application contact:* Maxine A. Leffard, Graduate Program Administrator, 412-268-8712, Fax: 412-268-7813, E-mail: ce-addmissions+@andrew.cmu.edu.

Case Western Reserve University, School of Medicine, Graduate Programs in Medicine, Department of Physiology and Biophysics, Cleveland, OH 44106. Offers biophysics and bioengineering (PhD); cell physiology (PhD); physiology and biophysics (PhD); systems physiology (PhD). *Faculty:* 64. *Students:* 57 full-time (19 women); includes 2 minority (1 African American, 1 Asian American or Pacific Islander), 28 international. *Degree requirements:* For doctorate, dissertation required, foreign language not required. *Entrance requirements:* For doctorate, GRE General Test, TOEFL (minimum score of 550 required). *Application deadline:* For fall admission, 3/15 (priority date). Applications are processed on a rolling basis. Application fee: $25. Electronic applications accepted. *Unit head:* Dr. Antonio Scarpa, Chairman, 216-368-5298, Fax: 216-368-5586. *Application contact:* M. Wendy Schneider, Administrator, 216-368-5517, Fax: 216-368-5586.

Clemson University, Graduate School, College of Engineering and Science, School of Chemical and Materials Engineering, Department of Bioengineering, Clemson, SC 29634. Offers MS, PhD. Part-time programs available. *Students:* 43 full-time (15 women), 8 part-time (3 women); includes 2 minority (both Hispanic Americans), 8 international. Average age 23. 61 applicants, 69% accepted. In 1998, 13 master's, 2 doctorates awarded. *Degree requirements:* For master's, thesis optional, foreign language not required; for doctorate, dissertation required, foreign language not required. *Entrance requirements:* For master's and doctorate, GRE General Test, TOEFL. *Application deadline:* For fall admission, 6/1; for spring admission, 11/1. Application fee: $35. *Financial aid:* Fellowships, research assistantships, teaching assistantships, career-related internships or fieldwork available. Financial aid application deadline: 2/15; financial aid applicants required to submit FAFSA. *Faculty research:* Biomaterials, biomechanics. *Unit head:* Dr. R. Larry Dooley, Chair, 864-656-3051, Fax: 864-656-4466, E-mail: dooley@eng.clemson.edu. *Application contact:* Dr. Robert Latour, Graduate Student Coordinator, 864-656-5552, Fax: 864-656-4466, E-mail: latourr@eng.clemson.edu.

See in-depth description on page 393.

Colorado State University, Graduate School, College of Engineering, Department of Mechanical Engineering, Program in Bioengineering, Fort Collins, CO 80523-0015. Offers MS, PhD. *Faculty:* 18 full-time (0 women). *Degree requirements:* For doctorate, dissertation required, foreign language not required, foreign language not required. *Entrance requirements:* For master's and doctorate, GRE General Test (minimum combined score of 1850 on three sections required; average 1872), TOEFL (minimum score of 550 required; average 596), minimum GPA of 3.0. *Application deadline:* For fall admission, 2/1 (priority date). Applications are processed on a rolling basis. Application fee: $30. Electronic applications accepted. *Faculty research:* Orthopedic biomechanics, instrumentation for assessment of cardiovascular function, modeling biological systems. *Unit head:* Dr. Susan James, Assistant Professor, 970-491-3573, Fax: 970-491-3827, E-mail: sjames@lamar.colostate.edu. *Application contact:* Dr. Susan James, Assistant Professor, 970-491-3573, Fax: 970-491-3827, E-mail: sjames@lamar.colostate.edu.

Cornell University, Graduate School, Graduate Fields of Agriculture and Life Sciences, Field of Agricultural and Biological Engineering, Ithaca, NY 14853-0001. Offers biological engineering (M Eng, MPS, MS, PhD); energy (M Eng, MPS, MS, PhD); environmental engineering (M Eng, MPS, MS, PhD); environmental management (MPS); food processing engineering (M Eng, MPS, MS, PhD); international agriculture (M Eng, MPS, MS, PhD); local roads (M Eng, MPS, MS, PhD); machine systems (M Eng, MPS, MS, PhD); soil and water engineering (M Eng, MPS, MS, PhD); structures and environment (M Eng, MPS, MS, PhD). *Faculty:* 28 full-time. *Students:* 45 full-time (13 women); includes 5 minority (1 African American, 3 Asian Americans or Pacific Islanders, 1 Hispanic American), 19 international. 58 applicants, 62% accepted. In 1998, 17 master's, 9 doctorates awarded. *Degree requirements:* For master's, thesis (MS) required; for doctorate, dissertation required, foreign language not required. *Entrance requirements:* For master's and doctorate, GRE General Test, TOEFL (minimum score of 550 required). *Application deadline:* For fall admission, 1/15. Applications are processed on a rolling basis. Application fee: $65. Electronic applications accepted. *Financial aid:* In 1998–99, 25 students received aid, including 5 fellowships with full tuition reimbursements available, 14 research assistantships with full tuition reimbursements available, 6 teaching assistantships with full tuition reimbursements available; institutionally-sponsored loans, scholarships, tuition waivers (full and partial), and unspecified assistantships also available. Financial aid applicants required to submit FAFSA. *Unit head:* Director of Graduate Studies, 607-255-2173, Fax: 607-255-4080. *Application contact:* Graduate Field Assistant, 607-255-2173, Fax: 607-255-4080, E-mail: abengradfield@cornell.edu.

Dalhousie University, Faculty of Graduate Studies, DalTech, Faculty of Engineering, Department of Biological Engineering, Halifax, NS B3H 3J5, Canada. Offers M Eng, MA Sc, PhD. *Faculty:* 5 full-time (0 women), 1 part-time (0 women). *Students:* 11 full-time (2 women), 1 part-time. Average age 33. 11 applicants, 64% accepted. In 1998, 3 doctorates awarded (100% entered university research/teaching). *Degree requirements:* For master's and doctorate, thesis/dissertation required, foreign language not required. *Entrance requirements:* For master's and doctorate, TOEFL (minimum score of 580 required). *Application deadline:* For fall admission, 6/1; for winter admission, 10/1; for spring admission, 2/1. Applications are processed on a rolling basis. Application fee: $55. *Financial aid:* In 1998–99, 1 research assistantship (averaging $1,600 per year), 4 teaching assistantships (averaging $4,000 per year) were awarded.; fellowships, scholarships and unspecified assistantships also available. *Faculty research:* Waste management, energy and environment, bio-machinery and robotics, soil and water, aquacultural and food engineering. *Unit head:* Dr. N. Ben-Abdallah, Head, 902-494-6003, Fax: 902-423-2423, E-mail: bio.engineering@dal.ca. *Application contact:* Shelley Parker, Admissions Coordinator, Graduate Studies and Research, 902-494-1288, Fax: 902-494-3149, E-mail: shelley.parker@dal.ca.

Georgia Institute of Technology, Graduate Studies and Research, College of Engineering, School of Biomedical Engineering, Atlanta, GA 30332-0001. Offers MS Bio E, PhD, Certificate, MD/PhD. MD/PhD offered jointly with Emory University. MS Bio E offered jointly with the Schools of Aerospace Engineering, Chemical Engineering, Civil and Enviro nmental Engineering, Materials Science and Engineering, Mechanical Engineeri ng, and Textile Engineering. Terminal master's awarded for partial completion of doctoral program. *Degree requirements:* For master's and doctorate, thesis/dissertation required, foreign language not required. *Entrance requirements:* For master's and doctorate, TOEFL (minimum score of 550 required). *Faculty research:* Biomechanics and tissue engineering, bioinstrumentation and medical imaging.

See in-depth description on page 401.

Kansas State University, Graduate School, College of Engineering, Department of Biological and Agricultural Engineering, Manhattan, KS 66506. Offers MS, PhD. *Faculty:* 12 full-time (0 women). *Students:* 21 full-time (3 women), 5 part-time, 18 international. 15 applicants, 20% accepted. In 1998, 7 master's awarded (100% found work related to degree); 1 doctorate awarded. Terminal master's awarded for partial completion of doctoral program. *Degree requirements:* For master's and doctorate, thesis/dissertation required, foreign language not required. *Entrance requirements:* For master's and doctorate, TOEFL (minimum score of 500 required). *Average time to degree:* Master's–2.5 years full-time; doctorate–4 years full-time. *Application deadline:* For fall admission, 3/1; for spring admission, 11/1. Applications are processed on a rolling basis. Application fee: $0 ($25 for international students). Electronic applications accepted. *Financial aid:* In 1998–99, 5 fellowships (averaging $8,649 per year), 17 research assistantships (averaging $8,649 per year) were awarded. Total annual research expenditures: $843,000. *Unit head:* Dr. James K. Koelliker, Head, 785-532-5580, Fax: 785-532-5825. *Application contact:* Naiqian Zhang, Graduate Coordinator, 785-532-5580, Fax: 785-532-5825.

Kansas State University, Graduate School, College of Engineering, Department of Electrical and Computer Engineering, Manhattan, KS 66506. Offers bioengineering (MS, PhD); communications (MS, PhD); computer engineering (MS, PhD); control systems (MS, PhD); electric energy systems (MS, PhD); instrumentation (MS, PhD); signal processing (MS, PhD). Postbaccalaureate distance learning degree programs offered (no on-campus study). *Faculty:* 21 full-time (3 women). *Students:* 26 full-time (2 women), 24 part-time (2 women), 18 international. *Degree requirements:* For master's, thesis optional; for doctorate, dissertation required. *Entrance requirements:* For master's, GRE General Test (minimum score of 400 on verbal section, 600 on quantitative, 600 on analytical required); for doctorate, GRE General Test (minimum score of 400 on verbal section, 600 on quantitative required). *Application deadline:* For fall admission, 3/1; for spring admission, 9/1. Applications are processed on a rolling basis. Application fee: $0 ($25 for international students). Electronic applications accepted. *Unit head:* Dr. David Soldan, Head, 785-532-5600, E-mail: grad@eece.ksu.edu.

Louisiana State University and Agricultural and Mechanical College, Graduate School, College of Agriculture, Department of Biological and Agricultural Engineering, Baton Rouge, LA 70803. Offers biological and agricultural engineering (MSBAE); engineering science (MS, PhD). Part-time programs available. *Faculty:* 12 full-time (2 women), 1 part-time (0 women). *Students:* 6 full-time (1 woman), 6 part-time (2 women); includes 2 minority (1 African American, 1 Native American), 5 international. Average age 28. 5 applicants, 60% accepted. In 1998, 2 master's awarded. Terminal master's awarded for partial completion of doctoral program. *Degree requirements:* For master's and doctorate, thesis/dissertation required, foreign language not required. *Entrance requirements:* For master's and doctorate, GRE General Test (minimum combined score of 1000 required), minimum GPA of 3.0. *Application deadline:* For fall admission, 1/25 (priority date). Applications are processed on a rolling basis. Application fee: $25. *Financial aid:* In 1998–99, 4 research assistantships with partial tuition reimbursements (averaging $9,937 per year) were awarded.; fellowships, teaching assistantships with partial tuition reimbursements, career-related internships or fieldwork also available. Financial aid application deadline: 7/1. *Faculty research:* Machine development, aquaculture, environmental engineering, microprocessor applications, ergonomics engineering, bioprocessing, hydrology, biosensors, food engineering. Total annual research expenditures: $1.9 million. *Unit head:* Dr. Lalit Verma, Head, 225-388-3153, Fax: 225-388-3492, E-mail: lverma@gumbo.bae.lsu.edu. *Application contact:* Dr. Thomas Lawson, Graduate Coordinator, 225-388-3153, Fax: 225-388-3492, E-mail: hawson@gumbo.bae.lsu.edu.

Massachusetts Institute of Technology, School of Engineering, Division of Bioengineering and Environmental Health, Cambridge, MA 02139-4307. Offers bioengineering (PhD); toxicology (SM, PhD, Sc D). *Faculty:* 7 full-time (1 woman). *Students:* 33 full-time (20 women), 1 (woman) part-time; includes 10 minority (1 African American, 7 Asian Americans or Pacific Islanders, 2 Hispanic Americans), 6 international. Average age 26. 42 applicants, 24% accepted. Terminal master's awarded for partial completion of doctoral program. *Degree requirements:* For master's, thesis required, foreign language not required; for doctorate, dissertation, oral and written qualifying exams required, foreign language not required. *Entrance requirements:* For master's and doctorate, GRE General Test, TOEFL (minimum score of 600 required). *Application deadline:* For fall admission, 1/15. Application fee: $55. *Financial aid:* In 1998–99, 34 students received aid, including 16 fellowships, 12 research assistantships, 9 teaching assistantships; Federal Work-Study, grants, institutionally-sponsored loans, and scholarships also available. Financial aid application deadline: 1/15; financial aid applicants required to submit FAFSA. *Faculty research:* Biological imaging, biological microanalytics, biological transport process, biomaterials, cell and tissue engineering. Total annual research expenditures: $7.1 million. *Unit head:* Dr. Steven R. Tannenbaum, Co-Director, 617-253-3729, E-mail: srt@mit.edu. *Application contact:* Debra A. Luchanin, Academic Administrator, 617-253-5804.

See in-depth description on page 407.

McMaster University, School of Graduate Studies, Faculty of Engineering, Bioengineering Committee, Hamilton, ON L8S 4M2, Canada. Offers M Eng, PhD. *Degree requirements:* For doctorate, dissertation, comprehensive exam required, foreign language not required. *Application deadline:* For fall admission, 3/1 (priority date). Applications are processed on a rolling basis. Application fee: $50. *Unit head:* Dr. J. L. Brash, Chair, 905-525-9140 Ext. 24946.

MCP Hahnemann University, School of Medicine, Biomedical Graduate Programs, Program in Bioengineering, Philadelphia, PA 19102-1192. Offers PhD, MD/PhD. Offered jointly with Lehigh University. *Degree requirements:* For doctorate, dissertation, qualifying exam required, foreign language not required. *Entrance requirements:* For doctorate, GRE General Test, TOEFL.

Mississippi State University, College of Engineering, Department of Agricultural and Biological Engineering, Mississippi State, MS 39762. Offers biological engineering (MS). *Students:* 7 full-time (2 women), 4 part-time (1 woman). Average age 26. 7 applicants, 57% accepted. *Degree requirements:* Foreign language not required. *Entrance requirements:* For master's, GRE General Test (minimum combined score of 1050 required), TOEFL (minimum score of 550 required), minimum GPA of 2.75. *Application deadline:* For fall admission, 7/1; for spring admission, 11/1. Applications are processed on a rolling basis. Application fee: $25 for international students. *Financial aid:* Federal Work-Study, institutionally-sponsored loans, and unspecified assistantships available. Financial aid applicants required to submit FAFSA. *Faculty research:* Bioenvironmental engineering, bioinstrumentation, biomechanics/biomaterials, chemical application in agriculture, biological modeling. Total annual research expenditures:$950,000. *Unit head:* Dr. Jerome A. Gilbert, Head, 662-325-3280, Fax: 662-325-3853, E-mail: jgilbert@abe.msstate.edu. *Application contact:* Jerry B. Inmon, Director of Admissions, 662-325-2224, Fax: 662-325-7360, E-mail: admit@admissions.msstate.edu.

North Carolina State University, Graduate School, College of Agriculture and Life Sciences, Department of Biological and Agricultural Engineering, Raleigh, NC 27695. Offers MBAE, MS, PhD. Part-time programs available. *Faculty:* 31 full-time (1 woman), 11 part-time (2 women). *Students:* 40 full-time (12 women), 10 part-time (5 women); includes 6 minority (5 African Americans, 1 Asian American or Pacific Islander), 13 international. Average age 30. 18 applicants, 61% accepted. In 1998, 7 master's, 3 doctorates awarded. *Degree requirements:*

For master's, thesis or alternative required, foreign language not required; for doctorate, dissertation required, foreign language not required. *Entrance requirements:* For master's and doctorate, TOEFL (minimum score of 550 required), GRE (international students only). *Application deadline:* For fall admission, 6/25. Applications are processed on a rolling basis. Application fee: $45. *Financial aid:* In 1998–99, 4 fellowships (averaging $6,381 per year), 26 research assistantships (averaging $4,899 per year) were awarded.; teaching assistantships, career-related internships or fieldwork also available. Financial aid application deadline: 2/28. *Faculty research:* Bioinstrumentation, biomechanics, processing of biological materials, water table management, animal waste management. Total annual research expenditures: $8.3 million. *Unit head:* Dr. David B. Beasley, Head, 919-515-2694, Fax: 919-515-6772, E-mail: beasley@eos.ncsu.edu. *Application contact:* Dr. James H. Young, Director of Graduate Programs, 919-515-6710, Fax: 919-515-6772, E-mail: jim_young@ncsu.edu.

The Ohio State University, Graduate School, College of Food, Agricultural, and Environmental Sciences, Department of Food, Agricultural, and Biological Engineering, Columbus, OH 43210. Offers MS, PhD. *Faculty:* 17 full-time, 7 part-time. *Students:* 17 full-time (4 women), 2 part-time, 12 international. 16 applicants, 38% accepted. In 1998, 3 master's, 3 doctorates awarded. *Degree requirements:* For master's, computer language required, thesis optional, foreign language not required; for doctorate, computer language, dissertation required, foreign language not required. *Entrance requirements:* For master's, GRE General Test, GRE Subject Test, or minimum GPA of 3.0 (international students); for doctorate, GRE General Test, GRE Subject Test, or minimum GPA of 3.5 (international students). *Application deadline:* For fall admission, 8/15. Applications are processed on a rolling basis. Application fee: $30 ($40 for international students). *Financial aid:* Fellowships, research assistantships, teaching assistantships, career-related internships or fieldwork, Federal Work-Study, and institutionally-sponsored loans available. Aid available to part-time students. *Unit head:* Thomas L. Bean, Chairman, 614-292-6131, Fax: 614-292-9448, E-mail: bean.3@osu.edu.

Oklahoma State University, Graduate College, College of Agricultural Sciences and Natural Resources, School of Biosystems and Agricultural Engineering, Stillwater, OK 74078. Offers M Bio E, MS, PhD. *Faculty:* 19 full-time (1 woman). *Students:* 19 full-time (10 women), 17 part-time (4 women); includes 5 minority (1 African American, 1 Asian American or Pacific Islander, 2 Hispanic Americans, 1 Native American), 11 international. Average age 28. In 1998, 7 master's, 2 doctorates awarded. *Degree requirements:* For master's and doctorate, thesis/dissertation required, foreign language not required. *Entrance requirements:* For master's and doctorate, TOEFL (minimum score of 550 required). *Application deadline:* For fall admission, 6/1 (priority date). Application fee: $25. *Financial aid:* In 1998–99, 28 students received aid, including 24 research assistantships (averaging $11,417 per year), 2 teaching assistantships (averaging $11,400 per year); career-related internships or fieldwork, Federal Work-Study, and tuition waivers (partial) also available. Aid available to part-time students. Financial aid application deadline: 3/1. *Unit head:* Bill Barfield, Head, 405-744-5431.

Oregon State University, Graduate School, College of Engineering, Department of Bioresource Engineering, Corvallis, OR 97331. Offers M Agr, MAIS, MS, PhD. *Faculty:* 13 full-time (2 women), 7 part-time (2 women). Average age 26. In 1998, 2 degrees awarded (100% found work related to degree). Terminal master's awarded for partial completion of doctoral program. *Degree requirements:* For master's, thesis or alternative required, foreign language not required; for doctorate, dissertation required, foreign language not required. *Entrance requirements:* For master's and doctorate, TOEFL (minimum score of 550 required), minimum GPA of 3.0 in last 90 hours. *Application deadline:* For fall admission, 3/1. Applications are processed on a rolling basis. Application fee: $50. *Financial aid:* Fellowships, research assistantships, teaching assistantships, Federal Work-Study and institutionally-sponsored loans available. Aid available to part-time students. Financial aid application deadline: 2/1. *Faculty research:* Bioengineering, water resources engineering, food engineering, cell culture and fermentation, vadose zone transport, regional hydrology modeling, bioseparations, post-harvest processing, biomedical engineering, nonpoint pollution abatement, drug formulation and delivery, waste management, stochastic hydrology, biological modeling. *Unit head:* Dr. James A. Moore, Head, 541-737-2041, Fax: 541-737-2082, E-mail: info@pandora.bre.orst.edu.

See in-depth description on page 415.

Pennsylvania State University University Park Campus, Graduate School, Intercollege Graduate Programs, Intercollege Graduate Program in Bioengineering, State College, University Park, PA 16802-1503. Offers MS, PhD. *Students:* 20 full-time (3 women), 10 part-time (6 women). Terminal master's awarded for partial completion of doctoral program. *Degree requirements:* For master's and doctorate, thesis/dissertation required. *Entrance requirements:* For master's and doctorate, GRE General Test, TOEFL. Application fee: $50. *Unit head:* Dr. Herbert H. Lipowsky, Head, 814-865-1407.

See in-depth description on page 417.

Purdue University, Graduate School, Schools of Engineering, School of Chemical Engineering, West Lafayette, IN 47907. Offers biomedical engineering (MS Bm E, PhD); chemical engineering (MS Ch E, PhD). *Faculty:* 18 full-time (3 women), 6 part-time (0 women). *Students:* 89 full-time (21 women), 6 part-time (4 women); includes 9 minority (3 African Americans, 4 Asian Americans or Pacific Islanders, 2 Hispanic Americans), 50 international. *Degree requirements:* For master's and doctorate, thesis/dissertation required, foreign language not required. *Entrance requirements:* For master's and doctorate, TOEFL (minimum score of 550 required). *Application deadline:* Applications are processed on a rolling basis. Application fee: $30. Electronic applications accepted. *Unit head:* Dr. G. V. Reklaitis, Head, 765-494-4075. *Application contact:* Linda Hawkins, Graduate Administrator, 765-494-4057.

Purdue University, Graduate School, Schools of Engineering, School of Electrical and Computer Engineering, West Lafayette, IN 47907. Offers biomedical engineering (MS Bm E, PhD); computer engineering (MS, PhD); electrical engineering (MS, PhD). Part-time programs available. Postbaccalaureate distance learning degree programs offered (no on-campus study). *Faculty:* 60 full-time (4 women), 9 part-time (0 women). *Students:* 330 full-time (68 women), 4 part-time (3 women); includes 6 African Americans, 11 Hispanic Americans *Degree requirements:* For master's, thesis optional, foreign language not required; for doctorate, dissertation required, foreign language not required. *Entrance requirements:* For master's and doctorate, GRE General Test (combined average 2070 on three sections), TOEFL (minimum score of 575 required; average 635). *Application deadline:* For fall admission, 1/15 (priority date); for spring admission, 9/1. Applications are processed on a rolling basis. Application fee: $30. Electronic applications accepted. *Unit head:* Dr. W. K. Fuchs, Head, 765-494-3539, Fax: 765-494-3544, E-mail: fuchs@purdue.edu. *Application contact:* Dr. A. M. Weiner, Director of Admissions, 765-494-3392, Fax: 765-494-3393, E-mail: ecegrad@ecn.purdue.edu.

Purdue University, Graduate School, Schools of Engineering, School of Mechanical Engineering, West Lafayette, IN 47907. Offers biomedical engineering (MS Bm E, PhD); mechanical engineering (MS, MSE, MSME, PhD). *Faculty:* 45 full-time (2 women), 6 part-time (0 women). *Students:* 230 full-time (21 women), 22 part-time (1 woman). *Degree requirements:* For doctorate, dissertation required. *Entrance requirements:* For master's and doctorate, TOEFL (minimum score of 575 required). Application fee: $30. Electronic applications accepted. *Unit head:* Dr. F. Dan Hirleman, Head, 765-494-5688.

Rice University, Graduate Programs, George R. Brown School of Engineering, Department of Bioengineering, Houston, TX 77251-1892. Offers MS, PhD. *Entrance requirements:* For master's and doctorate, GRE General Test, TOEFL.

Rice University, Graduate Programs, George R. Brown School of Engineering, Department of Electrical and Computer Engineering, Houston, TX 77251-1892. Offers bioengineering (MS, PhD); circuits, controls, and communication systems (MS, PhD); computer science and engineering (MS, PhD); electrical engineering (MEE); lasers, microwaves, and solid-state electronics (MS, PhD). Part-time programs available. *Degree requirements:* For master's, thesis required (for some programs), foreign language not required; for doctorate, dissertation required, foreign language not required. *Entrance requirements:* For master's and doctorate, GRE General Test, GRE Subject Test, TOEFL (minimum score of 550 required), minimum GPA of 3.0. *Faculty research:* Physical electronics.

Rutgers, The State University of New Jersey, New Brunswick, Graduate School, Program in Bioresource Engineering, New Brunswick, NJ 08903. Offers MS. Part-time programs available. *Faculty:* 13 full-time (2 women), 2 part-time (0 women). *Students:* 3 full-time (0 women), 7 part-time (2 women); includes 1 minority (Asian American or Pacific Islander), 5 international. Average age 27. 11 applicants, 55% accepted. In 1998, 1 degree awarded. *Degree requirements:* For master's, thesis, seminar required, foreign language not required. *Entrance requirements:* For master's, GRE General Test. *Average time to degree:* Master's–4 years full-time. *Application deadline:* For fall admission, 5/1 (priority date); for spring admission, 12/1. Applications are processed on a rolling basis. Application fee: $50. *Financial aid:* In 1998–99, 2 students received aid, including 2 research assistantships; teaching assistantships Financial aid application deadline: 3/1; financial aid applicants required to submit FAFSA. *Faculty research:* Greenhouse engineering, energy and environment, machine vision, flexible automation and robotics, systems analysis. Total annual research expenditures: $300,000. *Unit head:* Gene Giacomelli, Director, 732-932-9753, Fax: 732-932-7931.

Syracuse University, Graduate School, L. C. Smith College of Engineering and Computer Science, Department of Bioengineering and Neuroscience, Syracuse, NY 13244-0003. Offers bioengineering (MS). *Degree requirements:* Foreign language not required. *Entrance requirements:* For master's, GRE General Test, GRE Subject Test. *Application deadline:* Applications are processed on a rolling basis. Application fee: $40. Tuition: Full-time $13,992; part-time $583 per credit hour. *Financial aid:* Application deadline: 3/1. *Unit head:* Steve Chamberlain, Chair, 315-443-9711, Fax: 315-443-1184. *Application contact:* Norma Slepecky, Contact, 315-443-9749, Fax: 315-443-1184.

Texas A&M University, College of Engineering, Department of Industrial Engineering, Division of Biomedical Engineering, College Station, TX 77843. Offers M Eng, MS, D Eng, PhD. Part-time programs available. *Students:* 40 full-time (8 women), 7 part-time (1 woman); includes 7 minority (1 African American, 3 Asian Americans or Pacific Islanders, 3 Hispanic Americans), 28 international. Average age 27. 60 applicants, 57% accepted. In 1998, 9 master's, 4 doctorates awarded. *Degree requirements:* For master's, computer language, thesis (MS) required; for doctorate, computer language, dissertation (PhD) required. *Entrance requirements:* For master's, GRE General Test (combined average 1100), TOEFL; for doctorate, GRE General Test (combined average 1250), TOEFL. *Application deadline:* Applications are processed on a rolling basis. Application fee: $50 ($75 for international students). *Financial aid:* In 1998–99, 20 students received aid; fellowships, research assistantships, teaching assistantships, career-related internships or fieldwork available. Financial aid application deadline: 4/1; financial aid applicants required to submit FAFSA. *Faculty research:* Biological models, medical lasers, optical biosensors, medical instrumentation, artificial heart. Total annual research expenditures: $400,000. *Unit head:* Dr. William Hyman, Program Head, 409-845-5532. *Application contact:* S. Rastegar, Graduate Adviser, 409-845-5532.

University of Arkansas, Graduate School, College of Engineering, Department of Biological and Agricultural Engineering, Fayetteville, AR 72701-1201. Offers MSBAE, MSE, PhD. *Faculty:* 10 full-time (0 women). *Students:* 9 full-time (1 woman), 1 (woman) part-time; includes 1 minority (Hispanic American), 6 international. 5 applicants, 40% accepted. In 1998, 3 master's, 1 doctorate awarded. *Degree requirements:* For master's, thesis required, foreign language not required; for doctorate, one foreign language, dissertation required. Application fee: $40 ($50 for international students). Tuition, state resident: full-time $3,186. Tuition, nonresident: full-time $7,560. Required fees: $378. *Financial aid:* In 1998–99, 9 research assistantships were awarded.; career-related internships or fieldwork and Federal Work-Study also available. Aid available to part-time students. Financial aid application deadline: 4/1; financial aid applicants required to submit FAFSA. *Unit head:* Ivan L. Berry, Head, 501-575-2351.

University of British Columbia, Faculty of Graduate Studies, Faculty of Applied Science, Department of Bioresource Engineering, Vancouver, BC V6T 1Z2, Canada. Offers M Sc, MA Sc, PhD. *Degree requirements:* For master's, thesis required, foreign language not required. *Entrance requirements:* For master's, TOEFL (minimum score of 550 required). *Faculty research:* Environmental management, water resource development, bio-process engineering, waste utilization (biomass conversions).

University of California, Berkeley, Graduate Division, Group in Bioengineering, Berkeley, CA 94720-1708. Offers PhD. *Faculty:* 117 full-time (13 women). *Students:* 50 full-time (16 women); includes 15 minority (5 African Americans, 7 Asian Americans or Pacific Islanders, 3 Hispanic Americans), 2 international. Average age 26. 146 applicants, 18% accepted. In 1998, 4 degrees awarded. *Degree requirements:* For doctorate, dissertation, qualifying exam required, foreign language not required. *Entrance requirements:* For doctorate, GRE General Test, minimum GPA of 3.0. *Application deadline:* For fall admission, 1/3. Application fee: $40. *Financial aid:* Fellowships, research assistantships, Federal Work-Study, institutionally-sponsored loans, and traineeships available. Financial aid application deadline: 1/3. *Faculty research:* Imaging, biomechanics, biomems modeling, neuroscience, biomedical computing, vision. *Unit head:* Dr. Rajendra Bhatnagar, Chair, E-mail: bhatnag@itsa.ucsf.edu. *Application contact:* Dr. Rajendra Bhatnagar, Chair, E-mail: bhatnag@itsa.ucsf.edu.

See in-depth description on page 429.

University of California, Davis, Graduate Studies, College of Engineering, Program in Biological and Agricultural Engineering, Davis, CA 95616. Offers M Engr, MS, D Engr, PhD, M Engr/MBA. Part-time programs available. *Faculty:* 27 full-time (3 women). *Students:* 43 full-time (15 women); includes 7 minority (5 Asian Americans or Pacific Islanders, 2 Hispanic Americans), 22 international. Average age 25. 37 applicants, 68% accepted. In 1998, 6 master's awarded (100% found work related to degree); 7 doctorates awarded. Terminal master's awarded for partial completion of doctoral program. *Degree requirements:* For master's and doctorate, thesis/dissertation required, foreign language not required. *Entrance requirements:* For master's, minimum GPA of 3.0; for doctorate, GRE, minimum graduate GPA of 3.25. *Application deadline:* For fall admission, 4/1 (priority date). Application fee: $40. Electronic applications accepted. *Financial aid:* In 1998–99, 28 students received aid, including 9 fellowships with full and partial tuition reimbursements available, 13 research assistantships with full and partial tuition reimbursements available, 4 teaching assistantships with full and partial tuition reimbursements available Financial aid application deadline: 1/15; financial aid applicants required to submit FAFSA. *Faculty research:* Forestry, irrigation and drainage, power and machinery, structures and environment, information and energy technologies. *Unit head:* David J. Hills, Graduate Adviser, 530-752-0102, Fax: 530-752-2640, E-mail: bioageng@ucdavis.edu.

University of California, San Diego, Graduate Studies and Research, Department of Bioengineering, La Jolla, CA 92093-5003. Offers MS, PhD. *Students:* 65 (18 women). 183 applicants, 36% accepted. In 1998, 18 master's, 10 doctorates awarded. *Entrance requirements:* For master's and doctorate, GRE General Test, TOEFL (minimum score of 550 required), minimum GPA of 3.0. Application fee: $40. *Unit head:* Shu Chien, Chair. *Application contact:* Graduate Coordinator, 619-534-6884.

See in-depth description on page 431.

University of California, San Francisco, Graduate Division, Program in Bioengineering, San Francisco, CA 94143. Offers PhD. *Faculty:* 117 full-time (13 women). *Students:* 50 full-time (16 women); includes 15 minority (5 African Americans, 7 Asian Americans or Pacific Islanders, 3 Hispanic Americans), 2 international. Average age 26. 146 applicants, 18% accepted. In 1998, 4 degrees awarded. *Degree requirements:* For doctorate, dissertation, qualifying exam required, foreign language not required. *Entrance requirements:* For doctorate, GRE General Test, minimum GPA of 3.0. *Application deadline:* For fall admission, 1/3. Application fee: $40. *Financial aid:* Fellowships, research assistantships, Federal Work-Study, institutionally-sponsored loans, and traineeships available. Financial aid application deadline: 1/3; financial

Bioengineering

University of California, San Francisco (continued)
aid applicants required to submit FAFSA. *Faculty research:* Imaging, biomechanics, modeling, neuroscience, biomedical computing, vision. *Unit head:* Rajendra Bhatnagar, Chair. *Application contact:* Rajendra Bhatnagar, Chair.

See in-depth description on page 429.

University of Connecticut, Graduate School, School of Engineering, Field of Electrical and Systems Engineering, Storrs, CT 06269. Offers biological engineering (MS); control and communication systems (MS, PhD); electromagnetics and physical electronics (MS, PhD). Terminal master's awarded for partial completion of doctoral program. *Degree requirements:* For master's, thesis or alternative required; for doctorate, dissertation required. *Entrance requirements:* For master's and doctorate, GRE General Test, TOEFL.

University of Georgia, Graduate School, College of Agricultural and Environmental Sciences, Department of Biological and Agricultural Engineering, Athens, GA 30602. Offers agricultural engineering (MS); biological and agricultural engineering (PhD); biological engineering (MS). *Faculty:* 30 full-time (1 woman). *Students:* 17 full-time (4 women), 1 part-time. 37 applicants, 32% accepted. In 1998, 8 master's, 2 doctorates awarded. *Degree requirements:* For master's, thesis required, foreign language not required; for doctorate, one foreign language (computer language can substitute), dissertation required. *Entrance requirements:* For master's and doctorate, GRE General Test. *Application deadline:* For fall admission, 7/1 (priority date); for spring admission, 11/15. Application fee: $30. Electronic applications accepted. *Financial aid:* Fellowships, research assistantships, teaching assistantships, unspecified assistantships available. *Unit head:* Dr. Brahm Verma, Graduate Coordinator, 706-542-0862, Fax: 706-542-8806, E-mail: bverma@bae.uga.edu.

See in-depth description on page 435.

University of Guelph, Faculty of Graduate Studies, College of Physical and Engineering Science, School of Engineering, Guelph, ON N1G 2W1, Canada. Offers biological engineering (M Sc, PhD); environmental engineering (M Eng, M Sc, PhD); water resources engineering (M Eng, M Sc, PhD). Part-time programs available. *Faculty:* 18 full-time (1 woman), 29 part-time (4 women). *Students:* 54 full-time (15 women), 13 part-time (2 women); includes 22 minority (2 African Americans, 18 Asian Americans or Pacific Islanders, 2 Hispanic Americans), 7 international. *Degree requirements:* For master's, thesis required (for some programs); for doctorate, dissertation required. *Entrance requirements:* For master's, minimum B- average during previous 2 years; for doctorate, minimum B average. *Application deadline:* For fall admission, 8/1 (priority date); for winter admission, 11/1 (priority date); for spring admission, 4/1 (priority date). Applications are processed on a rolling basis. Application fee: $60. *Expenses:* Tuition and fees charges are reported in Canadian dollars. Tuition, area resident: Full-time $4,725 Canadian dollars; part-time $1,055 Canadian dollars per term. International tuition: $6,999 Canadian dollars full-time. Required fees: $295 Canadian dollars per term. *Unit head:* Dr. Lambert Otten, Director, 519-824-4120 Ext. 2043, Fax: 519-836-0227, E-mail: lotten@uoguelph.ca. *Application contact:* Dr. Ramesh P. Rudra, Graduate Coordinator, 519-824-4120 Ext. 2110, Fax: 519-836-0227, E-mail: rrudra@uoguelph.ca.

University of Hawaii at Manoa, Graduate Division, College of Tropical Agriculture and Human Resources, Department of Biosystems Engineering, Honolulu, HI 96822. Offers MS. Part-time programs available. *Faculty:* 16 full-time (0 women). *Students:* 10 full-time (4 women), 6 part-time (2 women); includes 6 minority (1 African American, 5 Asian Americans or Pacific Islanders), 4 international. 6 applicants, 83% accepted. In 1998, 4 degrees awarded. *Degree requirements:* For master's, computer language, thesis required, foreign language not required. *Application deadline:* For fall admission, 3/1; for spring admission, 9/1. Application fee: $25 ($50 for international students). *Financial aid:* In 1998–99, 10 research assistantships (averaging $15,017 per year), 1 teaching assistantship (averaging $12,786 per year) were awarded.; fellowships, Federal Work-Study, institutionally-sponsored loans, and tuition waivers (full) also available. *Faculty research:* Mechanization, agricultural systems, waste management, water management, cell culture. *Unit head:* Dr. Charles M. Kinoshita, Chairperson, 808-956-8867, Fax: 808-956-9269, E-mail: kinoshi@wiliki.eng.hawaii.edu. *Application contact:* Dr. John Grove, Chairman, 808-956-5779, Fax: 808-956-4585, E-mail: jgrove@hawaii.edu.

University of Illinois at Chicago, Graduate College, College of Engineering, Bioengineering Program, Chicago, IL 60607-7128. Offers MS, PhD, MD/PhD. *Students:* 32 full-time (10 women), 11 part-time (3 women); includes 14 minority (1 African American, 11 Asian Americans or Pacific Islanders, 1 Hispanic American, 1 Native American), 13 international. Average age 24. 84 applicants, 40% accepted. In 1998, 8 master's, 4 doctorates awarded. Terminal master's awarded for partial completion of doctoral program. *Degree requirements:* For master's and doctorate, computer language, thesis/dissertation required, foreign language not required. *Entrance requirements:* For master's and doctorate, GRE Subject Test, TOEFL (minimum score of 550 required), minimum GPA of 4.0 on a 5.0 scale. *Application deadline:* For fall admission, 7/3; for spring admission, 11/8. Application fee: $40 ($50 for international students). *Financial aid:* In 1998–99, 23 students received aid; fellowships, research assistantships, teaching assistantships, career-related internships or fieldwork available. *Faculty research:* Imaging systems, bioinstrumentation, electrophysiology, biological control, laser scattering. *Unit head:* Dr. Richard Magin, Head, 312-996-2331.

See in-depth description on page 437.

University of Illinois at Urbana–Champaign, Graduate College, College of Engineering, Program in Bioengineering, Urbana, IL 61801.

See in-depth description on page 439.

University of Maryland, College Park, Graduate School, College of Agriculture and Natural Resources, Department of Biological Resources Engineering, College Park, MD 20742-5045. Offers MS, PhD. *Faculty:* 13 full-time (0 women), 6 part-time (1 woman). *Students:* 17 full-time (10 women), 12 part-time (3 women); includes 7 minority (4 African Americans, 3 Asian Americans or Pacific Islanders), 6 international. 37 applicants, 43% accepted. In 1998, 6 master's, 1 doctorate awarded. *Degree requirements:* For master's, thesis optional, foreign language not required; for doctorate, dissertation required, foreign language not required. *Entrance requirements:* For master's, minimum GPA of 3.0. Application fee: $50 ($70 for international students). Tuition, state resident: part-time $272 per credit hour. Tuition, nonresident: part-time $475 per credit hour. Required fees: $632; $379 per year. *Financial aid:* In 1998–99, 17 research assistantships with tuition reimbursements (averaging $13,288 per year), 7 teaching assistantships with tuition reimbursements (averaging $11,818 per year) were awarded.; fellowships with full tuition reimbursements, career-related internships or fieldwork also available. Financial aid applicants required to submit FAFSA. *Faculty research:* Engineering aspects of production; harvesting, processing, and marketing of terrestrial and aquatic food and fiber. Total annual research expenditures: $385,057. *Unit head:* , Dr. Frederick Wheaton, Chairman, 301-405-2223, Fax: 301-314-9023. *Application contact:* Trudy Lindsey, Director, Graduate Admission and Records, 301-405-4198, Fax: 301-314-9305, E-mail: grschool@deans.umd.edu.

See in-depth description on page 441.

University of Missouri–Columbia, Graduate School, College of Engineering, Department of Biological Engineering, Columbia, MO 65211. Offers agricultural engineering (MS); biological engineering (MS, PhD). *Faculty:* 13 full-time (0 women), 1 (woman) part-time. *Students:* 7 full-time (2 women), 15 part-time (4 women), 19 international. 1 applicants, 0% accepted. In 1998, 4 master's, 1 doctorate awarded. *Degree requirements:* For master's and doctorate, thesis/dissertation required. *Entrance requirements:* For master's and doctorate, GRE General Test, TOEFL, minimum GPA of 3.0. *Application deadline:* For fall admission, 5/1 (priority date). Applications are processed on a rolling basis. Application fee: $30 ($50 for international students). *Financial aid:* Research assistantships, teaching assistantships, institutionally-sponsored loans available. *Unit head:* Dr. Jinglu Tan, Director of Graduate Studies, 573-882-7778.

University of Nebraska–Lincoln, Graduate College, College of Engineering and Technology, Department of Biological Systems Engineering, Lincoln, NE 68588. Offers agricultural and biological systems engineering (MS); agricultural science (MS), including mechanized systems management; engineering (PhD). *Faculty:* 21 full-time (1 woman), 6 part-time (0 women). *Students:* 17 full-time (3 women), 24 part-time (5 women); includes 15 minority (1 African American, 11 Asian Americans or Pacific Islanders, 3 Hispanic Americans) Average age 25. 41 applicants, 12% accepted. In 1998, 12 master's, 3 doctorates awarded. *Degree requirements:* For master's, thesis (for some programs), teaching experience required; for doctorate, dissertation, comprehensive exams required. *Entrance requirements:* For master's and doctorate, GRE General Test, TOEFL (minimum score of 550 required). *Average time to degree:* Master's–2.5 years full-time, 4 years part-time; doctorate–3.5 years full-time, 6 years part-time. *Application deadline:* Applications are processed on a rolling basis. Application fee: $35. Electronic applications accepted. *Financial aid:* In 1998–99, 13 students received aid, including 4 fellowships with full tuition reimbursements available (averaging $17,000 per year), 8 research assistantships with full tuition reimbursements available (averaging $12,600 per year), 1 teaching assistantship (averaging $12,000 per year) *Faculty research:* Food and biochemical engineering, environmental engineering, precision agriculture, sensors and controls, soil and water. Total annual research expenditures: $2.5 million. *Unit head:* Dr. Glenn J. Hoffman, Head, 402-472-1413, Fax: 402-472-6338, E-mail: bsen001@unlvm.onl.edu.

University of Notre Dame, Graduate School, College of Engineering, Department of Civil Engineering and Geological Sciences, Notre Dame, IN 46556. Offers bioengineering (MS); civil engineering (MS); civil engineering and geological sciences (PhD); environmental engineering (MS); geological sciences (MS). *Faculty:* 13 full-time (1 woman). *Students:* 35 full-time (10 women), 4 part-time; includes 4 minority (2 African Americans, 1 Asian American or Pacific Islander, 1 Hispanic American), 11 international. Terminal master's awarded for partial completion of doctoral program. *Degree requirements:* For master's and doctorate, thesis/dissertation required, foreign language not required. *Entrance requirements:* For master's and doctorate, GRE General Test, TOEFL (minimum score of 600 required; 250 for computer-based). *Application deadline:* For fall admission, 2/1 (priority date); for spring admission, 10/15. Applications are processed on a rolling basis. Application fee: $40. *Unit head:* Dr. Billie F. Spencer, Director of Graduate Studies, 219-631-5381, Fax: 219-631-9236, E-mail: cegeos@nd.edu. *Application contact:* Dr. Terrence J. Akai, Director of Graduate Admissions, 219-631-7706, Fax: 219-631-4183, E-mail: gradad@nd.edu.

University of Pennsylvania, School of Engineering and Applied Science, Department of Bioengineering, Philadelphia, PA 19104. Offers MSE, PhD, MD/PhD, VMD/PhD. Terminal master's awarded for partial completion of doctoral program. *Degree requirements:* For master's, computer language required, thesis optional, foreign language not required; for doctorate, computer language, dissertation required, foreign language not required. *Entrance requirements:* For master's, GRE General Test (minimum combined score of 1900 on three sections required; average 2100), TOEFL (minimum score of 600 required; average 630); for doctorate, GRE General Test, TOEFL (minimum score of 600 required; average 630). Electronic applications accepted. *Faculty research:* Biomaterials and biomechanics, biofluid mechanics and transport, bioelectric phenomena, computational neuroscience.

See in-depth description on page 455.

University of Pittsburgh, School of Engineering, Department of Bioengineering, Pittsburgh, PA 15260. Offers MSBENG, PhD. Part-time and evening/weekend programs available. *Faculty:* 2 full-time (0 women). *Students:* 45 full-time (20 women), 10 part-time (2 women); includes 8 minority (1 African American, 4 Asian Americans or Pacific Islanders, 3 Hispanic Americans), 7 international. 58 applicants, 47% accepted. In 1998, 7 master's, 3 doctorates awarded. Terminal master's awarded for partial completion of doctoral program. *Degree requirements:* For master's, computer language, thesis required, foreign language not required; for doctorate, computer language, dissertation, comprehensive and final oral exams required, foreign language not required. *Entrance requirements:* For master's and doctorate, GRE General Test, TOEFL (minimum score of 550 required), minimum QPA of 3.0. *Average time to degree:* Master's–2 years full-time, 3 years part-time; doctorate–5 years full-time, 7 years part-time. *Application deadline:* For fall admission, 8/1 (priority date); for spring admission, 12/1. Applications are processed on a rolling basis. Application fee: $30 ($40 for international students). *Financial aid:* In 1998–99, 42 students received aid, including 5 fellowships (averaging $12,368 per year), 13 research assistantships (averaging $10,160 per year), 8 teaching assistantships (averaging $9,088 per year); grants, scholarships, and traineeships also available. Financial aid application deadline: 2/15. *Faculty research:* Artificial organs, biomechanics, biomaterials, signal processing, biotechnology. Total annual research expenditures: $4.5 million. *Unit head:* Dr. Jerome S. Schultz, Director, 412-383-9713, Fax: 412-383-9710, E-mail: jssbio@vms.cis.pitt.edu.

See in-depth description on page 459.

University of Toledo, Graduate School, College of Engineering, Department of Bioengineering, Toledo, OH 43606-3398. Offers MS, PhD. *Faculty:* 10 full-time (2 women), 1 part-time (0 women). *Students:* 42 full-time (13 women), 7 part-time (1 woman); includes 1 minority (African American), 28 international. Average age 26. 69 applicants, 49% accepted. In 1998, 23 master's awarded (100% found work related to degree). Terminal master's awarded for partial completion of doctoral program. *Degree requirements:* For master's, computer language required, thesis optional, foreign language not required; for doctorate, computer language, dissertation required, foreign language not required. *Entrance requirements:* For master's, GRE General Test (minimum combined score of 1800 on three sections required; average 1900), TOEFL (minimum score of 550 required), minimum GPA of 3.3; for doctorate, GRE General Test, TOEFL (minimum score of 550 required). *Average time to degree:* Master's–2 years full-time; doctorate–4 years full-time. *Application deadline:* For fall admission, 5/31 (priority date). Applications are processed on a rolling basis. Application fee: $30. Electronic applications accepted. *Financial aid:* In 1998–99, 44 students received aid, including 17 research assistantships with full tuition reimbursements available, 12 teaching assistantships with full tuition reimbursements available; scholarships, tuition waivers (full), and unspecified assistantships also available. Financial aid application deadline: 4/1. *Faculty research:* Artificial organs, biochemical engineering, bioelectrical systems, biomechanics, cellular engineering. Total annual research expenditures: $374,096. *Unit head:* Dr. Ronald L. Fournier, Chairman, 419-530-8030, Fax: 419-530-8076, E-mail: rfourni@uoft02.utoledo.edu. *Application contact:* Heather M. Kohler, Academic Program Coordinator, 419-530-8078, Fax: 419-530-8076, E-mail: hkohler@eng.utoledo.edu.

University of Utah, Graduate School, College of Engineering, Department of Bioengineering, Salt Lake City, UT 84112-1107. Offers ME, MS, PhD. *Faculty:* 9 full-time (0 women), 26 part-time (5 women). *Students:* 65 full-time (18 women), 17 part-time (7 women); includes 2 minority (both Asian Americans or Pacific Islanders), 16 international. Average age 28. In 1998, 9 master's, 6 doctorates awarded. Terminal master's awarded for partial completion of doctoral program. *Degree requirements:* For master's, comprehensive exam, thesis (MS), written project and oral presentation (ME) required; for doctorate, dissertation required, foreign language not required. *Entrance requirements:* For master's and doctorate, GRE, TOEFL (minimum score of 500 required), minimum GPA of 3.0. *Application deadline:* For fall admission, 5/1. Application fee: $30 ($50 for international students). *Financial aid:* In 1998–99, 8 fellowships, 47 research assistantships were awarded.; traineeships also available. Financial aid application deadline: 5/1. *Faculty research:* Bioinstrumentation, biomaterials, ultrasonic bioinstrumentation, medical imaging, neuroprosthesis. Total annual research expenditures: $2.1 million. *Unit head:* Richard A. Normann, Chair, 801-581-7645, Fax: 801-585-5361. *Application contact:* Richard D. Rabbit, Admissions Coordinator, 801-581-8559.

See in-depth description on page 467.

University of Washington, School of Medicine, Graduate Programs in Medicine, Department of Bioengineering, Seattle, WA 98195. Offers MS, MSE, PhD. *Faculty:* 29 full-time (2 women),

34 part-time (4 women). *Students:* 111 full-time (37 women), 2 part-time; includes 20 minority (3 African Americans, 11 Asian Americans or Pacific Islanders, 4 Hispanic Americans, 2 Native Americans), 21 international. 246 applicants, 27% accepted. In 1998, 3 master's, 10 doctorates awarded. Terminal master's awarded for partial completion of doctoral program. *Degree requirements:* For master's, thesis required; for doctorate, dissertation, qualifying exam, general exam required. *Entrance requirements:* For master's and doctorate, GRE General Test, TOEFL, minimum GPA of 3.0. *Average time to degree:* Master's–2.5 years full-time; doctorate–5 years full-time. *Application deadline:* For fall admission, 1/15. Application fee: $45. Tuition, state resident: full-time $5,196; part-time $475 per credit. Tuition, nonresident: full-time $13,485; part-time $1,285 per credit. Required fees: $387; $38 per credit. Tuition and fees vary according to course load. *Financial aid:* In 1998–99, 7 fellowships (averaging $16,800 per year), 96 research assistantships (averaging $16,800 per year) were awarded.; Federal Work-Study, institutionally-sponsored loans, traineeships, and tuition waivers (full) also available. Aid available to part-time students. Financial aid application deadline: 2/28. *Unit head:* Dr. Yongmin Kim, Chair, 206-685-2000, Fax: 206-685-3300. *Application contact:* Dr. Gerald Pollack, Chair, Student Admissions Committee, 206-685-2021, Fax: 206-685-3300.

See in-depth description on page 469.

Virginia Polytechnic Institute and State University, Graduate School, College of Engineering, Department of Biological Systems Engineering, Blacksburg, VA 24061. Offers M Eng, MS, PhD. *Faculty:* 12 full-time (0 women), 8 part-time (0 women). *Students:* 24 full-time (7 women), 1 part-time; includes 1 minority (Asian American or Pacific Islander), 12 international. Average age 25. 13 applicants, 62% accepted. In 1998, 7 master's, 3 doctorates awarded. *Degree requirements:* For master's and doctorate, thesis/dissertation required, foreign language not required. *Entrance requirements:* For master's and doctorate, GRE General Test, TOEFL (minimum score of 600 required). *Application deadline:* For fall admission, 12/1 (priority date). Applications are processed on a rolling basis. Application fee: $25. *Financial aid:* In 1998–99, 4 fellowships, 17 research assistantships were awarded.; career-related internships or fieldwork, institutionally-sponsored loans, tuition waivers (full and partial), and unspecified assistantships also available. Aid available to part-time students. Financial aid application deadline: 4/1. *Faculty research:* Soil and water engineering, alternative energy sources for agriculture and agricultural mechanization. *Unit head:* Dr. John V. Perumpral, Head, 540-231-6615, E-mail: perump@vt.edu. *Application contact:* Dr. S. Mostaghimi, Chairman, 540-231-7605, E-mail: smostagh@vt.edu.

Biomedical Engineering

Arizona State University, Graduate College, College of Engineering and Applied Sciences, Department of Chemical, Bio and Materials Engineering, Program in Bioengineering, Tempe, AZ 85287. Offers MS, PhD. *Degree requirements:* For doctorate, dissertation required. *Entrance requirements:* For master's and doctorate, GRE General Test. Application fee: $45. *Financial aid:* Fellowships, research assistantships, teaching assistantships available. *Faculty research:* Biotechnology, biocontrol, biomechanics, bioinstrumentation and materials, biosystems engineering/biotransport. *Application contact:* Graduate Secretary, 480-965-9707.

See in-depth description on page 387.

Baylor College of Medicine, Medical School, Biomedical Engineering Program, Houston, TX 77030-3498. Offers MD/PhD. *Students:* 4 full-time (1 woman); includes 1 minority (Asian American or Pacific Islander) Average age 27. 262 applicants, 16% accepted. *Application deadline:* For fall admission, 11/1 (priority date). Applications are processed on a rolling basis. Application fee: $35. Tuition, state resident: full-time $6,550. Tuition, nonresident: full-time $19,650. *Financial aid:* Federal Work-Study, institutionally-sponsored loans, and tuition waivers (full and partial) available. Financial aid application deadline: 3/29. *Unit head:* Dr. James R. Lupski, Director, 713-798-5264, Fax: 713-798-6325, E-mail: mstp@bcm.tmc.edu. *Application contact:* 713-798-4842, Fax: 713-798-5563, E-mail: melodym@bcm.tcm.edu.

Boston University, College of Engineering, Department of Biomedical Engineering, Boston, MA 02215. Offers MS, PhD, MD/PhD. Part-time programs available. *Faculty:* 25 full-time (5 women), 7 part-time (1 woman). *Students:* 75 full-time (26 women), 1 part-time; includes 6 minority (all Asian Americans or Pacific Islanders), 22 international. Average age 25. 139 applicants, 39% accepted. In 1998, 9 master's, 5 doctorates awarded. Terminal master's awarded for partial completion of doctoral program. *Degree requirements:* For master's and doctorate, thesis/dissertation required, foreign language not required. *Entrance requirements:* For master's, GRE General Test, TOEFL (minimum score of 500 required; 213 for computer-based); for doctorate, GRE General Test, TOEFL. *Application deadline:* For fall admission, 4/1; for spring admission, 10/1. Applications are processed on a rolling basis. Application fee: $50. Tuition: Full-time $23,770; part-time $743 per credit. Required fees: $220. Tuition and fees vary according to class time, course level, campus/location and program. *Financial aid:* In 1998–99, 6 fellowships with full tuition reimbursements (averaging $13,000 per year), 56 research assistantships with full tuition reimbursements (averaging $11,500 per year), 11 teaching assistantships with full tuition reimbursements (averaging $11,500 per year) were awarded.; career-related internships or fieldwork, Federal Work-Study, institutionally-sponsored loans, and scholarships also available. Financial aid application deadline: 12/15; financial aid applicants required to submit FAFSA. *Faculty research:* Biotechnological and human genome engineering, sensory biophysics, auditory neurophysiology, respiratory physiology and mechanics, microscale biomechanics. Total annual research expenditures: $7.5 million. *Unit head:* Dr. Kenneth R. Lutchen, Chairman, 617-353-2805, Fax: 617-353-6766. *Application contact:* Cheryl Kelley, Graduate Programs Director, 617-353-9760, Fax: 617-353-0259, E-mail: enggrad@bu.edu.

See in-depth description on page 389.

Brown University, Graduate School, Division of Biology and Medicine, Program in Artificial Organs/Biomaterials/Cellular Technology, Providence, RI 02912. Offers MA, Sc M, PhD. *Students:* 12 full-time (5 women); includes 1 minority (Asian American or Pacific Islander), 2 international. 27 applicants, 15% accepted. In 1998, 2 degrees awarded. Terminal master's awarded for partial completion of doctoral program. *Degree requirements:* For doctorate, dissertation, preliminary exam required, foreign language not required. *Entrance requirements:* For master's and doctorate, GRE General Test, GRE Subject Test. *Application deadline:* For fall admission, 1/2 (priority date). Applications are processed on a rolling basis. Application fee: $60. *Financial aid:* In 1998–99, 4 fellowships, 1 research assistantship, 4 teaching assistantships were awarded. Financial aid application deadline: 1/2. *Unit head:* Dr. Edith Mathiowitz, Director, 401-863-3262.

Brown University, Graduate School, Division of Engineering, Program in Biomedical Engineering, Providence, RI 02912. Offers Sc M. *Degree requirements:* For master's, thesis required, foreign language not required.

California State University, Northridge, Graduate Studies, College of Engineering and Computer Science, Department of Electrical and Computer Engineering, Northridge, CA 91330. Offers biomedical engineering (MS); communications/radar engineering (MS); control engineering (MS); digital/computer engineering (MS); electronics engineering (MS); microwave/antenna engineering (MS). Part-time and evening/weekend programs available. *Faculty:* 17 full-time, 3 part-time. *Students:* 20 full-time (2 women), 77 part-time (8 women); includes 33 minority (3 African Americans, 24 Asian Americans or Pacific Islanders, 6 Hispanic Americans), 9 international. *Degree requirements:* For master's, thesis or alternative required, foreign language not required. *Entrance requirements:* For master's, GRE General Test, TOEFL, minimum GPA of 2.5. *Application deadline:* For fall admission, 11/30. Application fee: $55. Tuition, nonresident: part-time $246 per unit. International tuition: $7,874 full-time. Required fees: $1,970. Tuition and fees vary according to course load. *Unit head:* Dr. Nagwa Bekir, Chair, 818-677-2190. *Application contact:* Nagi El Naga, Graduate Coordinator, 818-677-2180.

Carnegie Mellon University, Carnegie Institute of Technology, Department of Electrical and Computer Engineering, Concentration in Biomedical Engineering, Pittsburgh, PA 15213-3891. Offers MS, PhD. Part-time programs available. *Degree requirements:* For master's, thesis required, foreign language not required; for doctorate, computer language, dissertation, qualifying exam, teaching experience required, foreign language not required. *Entrance requirements:* For master's and doctorate, GRE General Test, TOEFL. *Application deadline:* For fall admission, 2/1; for spring admission, 10/15. Application fee: $45. *Financial aid:* Fellowships, research assistantships, teaching assistantships available. Financial aid application deadline: 1/15. *Unit head:* Michael M. Domach, Head, 412-268-2246. *Application contact:* Lynn E. Philibin, Assistant Head for Graduate Studies, 412-268-3291, Fax: 412-268-2860, E-mail: lynn@ece.cmu.edu.

Carnegie Mellon University, Carnegie Institute of Technology, Interdisciplinary Biomedical Engineering Program, Pittsburgh, PA 15213-3891. Offers MS, PhD. *Faculty:* 3 part-time (0 women). *Students:* 8 full-time (2 women), 1 part-time; includes 2 minority (1 African American, 1 Asian American or Pacific Islander), 1 international. Average age 25. In 1998, 1 master's, 4 doctorates awarded (100% found work related to degree). Terminal master's awarded for partial completion of doctoral program. *Degree requirements:* For master's, computer language, thesis required, foreign language not required; for doctorate, computer language, dissertation, qualifying exam required, foreign language not required. *Entrance requirements:* For master's and doctorate, GRE General Test (minimum score of 549 on verbal section, 740 on quantitative section, 680 on analytical section required), TOEFL (minimum score of 550 required). *Application deadline:* For fall admission, 2/1 (priority date). Applications are processed on a rolling basis. Application fee: $45. *Financial aid:* In 1998–99, 3 research assistantships were awarded.; fellowships, Federal Work-Study also available. *Faculty research:* Cellular and molecular systematics, signal and image processing, materials and mechanics. *Unit head:* Michael M. Domach, Director, 412-268-2246. *Application contact:* Hilda Diamond, Associate Director, 412-268-2521.

Announcement: PhD and MS degrees in bioengineering prepare students for careers in teaching, basic/clinical research, and industrial research/development. Research areas include tissue engineering, cellular and molecular systematics, signal and image processing, and materials and mechanics. Fellowships, traineeships, and research assistantships are available. Laboratories include state-of-the-art computing facilities.

Case Western Reserve University, School of Graduate Studies, The Case School of Engineering, Department of Biomedical Engineering, Cleveland, OH 44106. Offers biomedical engineering (MS, PhD); clinical engineering (MS). *Faculty:* 16 full-time (1 woman), 37 part-time (2 women). *Students:* 60 full-time (15 women), 63 part-time (5 women). Average age 25. 302 applicants, 9% accepted. In 1998, 26 master's, 7 doctorates awarded. Terminal master's awarded for partial completion of doctoral program. *Degree requirements:* For master's, thesis required (for some programs), foreign language not required; for doctorate, dissertation required, foreign language not required. *Entrance requirements:* For master's and doctorate, GRE, TOEFL (minimum score of 600 required). *Application deadline:* For fall admission, 2/1 (priority date). Applications are processed on a rolling basis. Application fee: $25. *Financial aid:* In 1998–99, 35 fellowships with full tuition reimbursements (averaging $15,000 per year), 56 research assistantships with full and partial tuition reimbursements (averaging $13,200 per year), 16 teaching assistantships (averaging $5,400 per year) were awarded.; traineeships also available. Financial aid application deadline: 2/15. *Faculty research:* Image processing and analysis, cardiac bioelectricity, biomaterials and tissue engineering, neuroprosthesis and neural engineering, chemical and optical diagnostic devices. Total annual research expenditures: $5.3 million. *Unit head:* Dr. Patrick E. Crago, Chairman, 216-368-3977, Fax: 216-368-4969, E-mail: pec3@po.cwru.edu. *Application contact:* Clarressa Phillips, Admissions Coordinator, 216-368-4094, Fax: 216-368-4969, E-mail: cmp13@po.cwru.edu.

See in-depth description on page 391.

The Catholic University of America, School of Engineering, Department of Mechanical Engineering, Program in Biomedical Engineering, Washington, DC 20064. Offers MBE, MS Engr, PhD. Part-time and evening/weekend programs available. *Students:* 2 full-time (both women), 7 part-time (3 women); includes 1 minority (African American), 2 international. Average age 28. 10 applicants, 100% accepted. In 1998, 2 degrees awarded. *Degree requirements:* For master's, comprehensive exam required, thesis optional, foreign language not required; for doctorate, dissertation, comprehensive and oral exams required, foreign language not required. *Entrance requirements:* For master's, minimum GPA of 3.0; for doctorate, minimum GPA of 3.5. *Application deadline:* For fall admission, 8/1 (priority date); for spring admission, 12/1. Applications are processed on a rolling basis. Application fee: $50. *Financial aid:* Research assistantships, teaching assistantships, career-related internships or fieldwork, Federal Work-Study, institutionally-sponsored loans, and tuition waivers (full and partial) available. Aid available to part-time students. Financial aid application deadline: 2/1. *Faculty research:* Cell and tissue engineering, biomechanics, rehabilitation engineering, neural engineering, medical instrumentation. Total annual research expenditures: $100,000. *Unit head:* Dr. Aydin Tozeren, Director, 202-319-5181.

Clemson University, Graduate School, College of Engineering and Science, School of Chemical and Materials Engineering, Department of Bioengineering, Clemson, SC 29634. Offers MS, PhD. Part-time programs available. *Students:* 43 full-time (15 women), 8 part-time (3 women); includes 2 minority (both Hispanic Americans), 8 international. Average age 23. 61 applicants, 69% accepted. In 1998, 13 master's, 2 doctorates awarded. *Degree requirements:* For master's, thesis optional, foreign language not required; for doctorate, dissertation required, foreign language not required. *Entrance requirements:* For master's and doctorate, GRE General Test, TOEFL. *Application deadline:* For fall admission, 6/1; for spring admission, 11/1. Application fee: $35. *Financial aid:* Fellowships, research assistantships, teaching assistantships, career-related internships or fieldwork available. Financial aid application deadline: 2/15; financial aid applicants required to submit FAFSA. *Faculty research:* Biomaterials, biomechanics. *Unit head:* Dr. R. Larry Dooley, Chair, 864-656-3051, Fax: 864-656-4466, E-mail: dooley@eng.clemson.edu. *Application contact:* Dr. Robert Latour, Graduate Student Coordinator, 864-656-5552, Fax: 864-656-4466, E-mail: latourr@eng.clemson.edu.

See in-depth description on page 393.

Cleveland State University, College of Graduate Studies, Fenn College of Engineering, Doctoral Program in Applied Biomedical Engineering, Cleveland, OH 44115-2440. Offers D Eng.

Biomedical Engineering

Cleveland State University (continued)
Unit head: Dr. Orhan Talu, Chairperson, Department of Chemical Engineering, 216-687-2571, Fax: 216-687-9220, E-mail: talu@csvax.csuohio.edu.

See in-depth description on page 395.

Colorado State University, Graduate School, College of Engineering, Department of Mechanical Engineering, Program in Bioengineering, Fort Collins, CO 80523-0015. Offers MS, PhD. *Faculty:* 18 full-time (0 women). *Degree requirements:* For doctorate, dissertation required, foreign language not required, foreign language not required. *Entrance requirements:* For master's and doctorate, GRE General Test (minimum combined score of 1850 on three sections required; average 1872), TOEFL (minimum score of 550 required; average 596), minimum GPA of 3.0. *Application deadline:* For fall admission, 2/1 (priority date). Applications are processed on a rolling basis. Application fee: $30. Electronic applications accepted. *Faculty research:* Orthopedic biomechanics, instrumentation for assessment of cardiovascular function, modeling biological systems. *Unit head:* Dr. Susan James, Assistant Professor, 970-491-3573, Fax: 970-491-3827, E-mail: sjames@lamar.colostate.edu. *Application contact:* Dr. Susan James, Assistant Professor, 970-491-3573, Fax: 970-491-3827, E-mail: sjames@lamar.colostate.edu.

Columbia University, Fu Foundation School of Engineering and Applied Science, Department of Biomedical Engineering, New York, NY 10027. Offers MS, Eng Sc D. Part-time programs available. *Faculty:* 5 full-time (0 women), 4 part-time (0 women). *Students:* 14 full-time (3 women), 3 part-time. Average age 24. 71 applicants, 31% accepted. In 1998, 3 master's awarded (33% found work related to degree, 67% continued full-time study); 2 doctorates awarded (50% entered university research/teaching, 50% continued full-time study). *Degree requirements:* For master's, thesis required, foreign language not required; for doctorate, dissertation, qualifying exam required. *Entrance requirements:* For master's and doctorate, GRE General Test, TOEFL. *Average time to degree:* Master's–1 year full-time; doctorate–5.5 years full-time. *Application deadline:* For fall admission, 1/5; for spring admission, 10/1. Application fee: $55. *Financial aid:* In 1998–99, 3 fellowships with full tuition reimbursements (averaging $16,500 per year), 11 research assistantships with full tuition reimbursements (averaging $16,000 per year), 3 teaching assistantships with full tuition reimbursements (averaging $12,000 per year) were awarded.; Federal Work-Study also available. Financial aid application deadline: 1/5; financial aid applicants required to submit FAFSA. *Faculty research:* Artificial organs, orthopedic and musculoskeletal biomechanics, cellular and tissue engineering, artificial organs, cardiovascular biomechanics, auditory biophysics, algorithms for quantitative analysis, image enhancement, 3-D diagnostic imaging, multiresolution representations, image compression. Total annual research expenditures: $442,000. *Unit head:* Dr. Van C. Mow, Head, 212-854-8458, Fax: 212-854-8725, E-mail: vcm1@columbia.edu. *Application contact:* X. Edward Guo, Assistant Professor, 212-854-6196, Fax: 212-854-3304, E-mail: exgl@columbia.edu.

Cornell University, Graduate School, Graduate Fields of Engineering, Field of Biomedical Engineering, Ithaca, NY 14853-0001. Offers MS, PhD. *Faculty:* 19 full-time. *Students:* 2 full-time (1 woman). 40 applicants, 18% accepted. *Degree requirements:* For doctorate, dissertation required, foreign language not required, foreign language not required. *Entrance requirements:* For master's, TOEFL; for doctorate, GRE General Test, TOEFL (minimum score of 550 required). *Application deadline:* For fall admission, 1/15. Application fee: $65. Electronic applications accepted. *Financial aid:* In 1998–99, 2 students received aid, including 1 fellowship with full tuition reimbursement available, 1 research assistantship with full tuition reimbursement available; institutionally-sponsored loans, scholarships, tuition waivers (full and partial), and unspecified assistantships also available. *Faculty research:* Biomaterials, biomedical instrumentation, and diagnostics; biomedical mechanics; drug delivery, design, and metabolism. *Unit head:* Director of Graduate Studies, 607-255-1003. *Application contact:* Graduate Field Assistant, 607-255-1003, E-mail: biomedeng@cornell.edu.

Dartmouth College, Thayer School of Engineering, Program in Biomedical Engineering, Hanover, NH 03755. Offers MS, PhD, MD/PhD. *Degree requirements:* For master's, thesis required; for doctorate, dissertation, candidacy oral exam required. *Entrance requirements:* For master's and doctorate, GRE General Test. *Application deadline:* For fall admission, 1/15 (priority date). Application fee: $20 ($40 for international students). *Financial aid:* Fellowships, research assistantships, teaching assistantships, career-related internships or fieldwork, Federal Work-Study, institutionally-sponsored loans, and tuition waivers (full and partial) available. Financial aid application deadline: 1/15. *Faculty research:* Imaging and image processing, biomaterials and orthopedics, physiological modeling, cancer hyperthermia and radiation therapy, bioelectromagnetics, biomedical optics. Total annual research expenditures: $1.2 million. *Unit head:* Lynn E. Philibin, Assistant Head for Graduate Studies, 412-268-3291, Fax: 412-268-2860, E-mail: lynn@ece.cmu.edu. *Application contact:* Candace S. Potter, Admissions Coordinator, 603-646-3844, Fax: 603-646-3856, E-mail: candace.potter@dartmouth.edu.

Drexel University, Graduate School, School of Biomedical Engineering, Science and Health Systems, Philadelphia, PA 19104-2875. Offers biomedical engineering (MS, PhD); biomedical science (MS, PhD); biostatistics (MS); clinical/rehabilitation engineering (MS). *Faculty:* 6 full-time, 4 part-time. *Students:* 32 full-time (12 women), 50 part-time (17 women); includes 13 minority (7 African Americans, 5 Asian Americans or Pacific Islanders, 1 Native American), 34 international. Average age 30. 146 applicants, 62% accepted. In 1998, 16 master's, 3 doctorates awarded. *Degree requirements:* For master's, thesis required (for some programs); for doctorate, dissertation, 1 year of residency, qualifying exam required. *Entrance requirements:* For master's, TOEFL (minimum score of 570 required), minimum GPA of 3.0; for doctorate, TOEFL (minimum score of 570 required), minimum GPA of 3.0, MS. *Application deadline:* For fall admission, 8/21. Applications are processed on a rolling basis. Application fee: $35. Tuition: Full-time $15,795; part-time $585 per credit. Required fees: $375; $67 per term. Tuition and fees vary according to program. *Financial aid:* In 1998–99, 9 research assistantships, 2 teaching assistantships were awarded.; career-related internships or fieldwork, Federal Work-Study, institutionally-sponsored loans, tuition waivers (full and partial), and unspecified assistantships also available. Financial aid application deadline: 2/1. *Faculty research:* Cardiovascular dynamics, diagnostic andtherapeutic ultrasound. *Unit head:* Dr. Banu Onaral, Director, 215-895-2215. *Application contact:* Dr. William Freedman, Graduate Adviser, 215-895-2225.

See in-depth description on page 397.

Duke University, Graduate School, School of Engineering, Department of Biomedical Engineering, Durham, NC 27708-0586. Offers MS, PhD. *Faculty:* 36 full-time, 2 part-time. *Students:* 75 full-time (25 women); includes 10 minority (1 African American, 8 Asian Americans or Pacific Islanders, 1 Hispanic American), 10 international. 176 applicants, 20% accepted. In 1998, 9 master's, 11 doctorates awarded. *Degree requirements:* For doctorate, dissertation required, foreign language not required. *Entrance requirements:* For master's and doctorate, GRE General Test. *Application deadline:* For fall admission, 12/31; for spring admission, 11/1. Application fee: $75. *Financial aid:* Fellowships, research assistantships, teaching assistantships, Federal Work-Study available. Financial aid application deadline: 12/31. *Unit head:* Dr. Gregg Trahey, Director of Graduate Studies, 919-660-5132, Fax: 919-684-4488, E-mail: kwb@acpub.,duke.edu.

See in-depth description on page 399.

École Polytechnique de Montréal, Graduate Programs, Institute of Biomedical Engineering, Montréal, PQ H3C 3A7, Canada. Offers M Eng, M Sc A, PhD, DESS. M Sc A and PhD offered jointly with Université de Montréal. Part-time programs available. *Degree requirements:* For master's and doctorate, one foreign language, computer language, thesis/dissertation required. *Entrance requirements:* For master's, minimum GPA of 2.75; for doctorate, minimum GPA of 3.0. *Faculty research:* Cardiac electrophysiology, biomedical instrumentation, biomechanics, biomaterials, medical imagery.

Georgia Institute of Technology, Graduate Studies and Research, College of Engineering, School of Biomedical Engineering, Atlanta, GA 30332-0001. Offers MS Bio E, PhD, Certificate, MD/PhD. MD/PhD offered jointly with Emory University. MS Bio E offered jointly with the Schools of Aerospace Engineering, Chemical Engineering, Civil and Enviro nmental Engineering, Materials Science and Engineering, Mechanical Engineeri ng, and Textile Engineering. Terminal master's awarded for partial completion of doctoral program. *Degree requirements:* For master's and doctorate, thesis/dissertation required, foreign language not required. *Entrance requirements:* For master's and doctorate, TOEFL (minimum score of 550 required). *Faculty research:* Biomechanics and tissue engineering, bioinstrumentation and medical imaging.

See in-depth description on page 401.

Harvard University, Graduate School of Arts and Sciences, Department of Physics, Cambridge, MA 02138. Offers experimental physics (AM, PhD); medical engineering/medical physics (PhD, Sc D), including applied physics (PhD), engineering sciences (PhD), medical engineering/medical physics (Sc D), physics (PhD); theoretical physics (AM, PhD). *Students:* 80 full-time (26 women). *Degree requirements:* For doctorate, dissertation, final exams, laboratory experience required, foreign language not required. *Entrance requirements:* For master's, GRE General Test, TOEFL (minimum score of 550 required); for doctorate, GRE General Test, GRE Subject Test, TOEFL (minimum score of 550 required). *Application deadline:* For fall admission, 12/14. Application fee: $60. *Unit head:* Dr. David Nelson, Chairperson, 617-495-2866. *Application contact:* Office of Admissions and Financial Aid, 617-495-5315.

Harvard University, Graduate School of Arts and Sciences, Division of Engineering and Applied Sciences, Cambridge, MA 02138. Offers applied mathematics (ME, SM, PhD); applied physics (ME, SM, PhD); computer science (ME, SM, PhD); computing technology (PhD); engineering science (ME); engineering sciences (SM, PhD); medical engineering/medical physics (PhD, Sc D), including applied physics (PhD), engineering sciences (PhD), medical engineering/medical physics (Sc D), physics (PhD). *Students:* 143 full-time (31 women); includes 10 minority (1 African American, 9 Asian Americans or Pacific Islanders), 57 international. Terminal master's awarded for partial completion of doctoral program. *Degree requirements:* For master's, foreign language and thesis not required; for doctorate, dissertation required, foreign language not required. *Entrance requirements:* For master's and doctorate, GRE General Test, GRE Subject Test, TOEFL (minimum score of 550 required). Application fee: $60. *Unit head:* Dr. Paul C. Martin, Dean, 617-495-2833. *Application contact:* Office of Admissions and Financial Aid, 617-495-5315.

Harvard University, Medical School, Division of Health Sciences and Technology, Program in Medical Engineering/Medical Physics, Cambridge, MA 02138. Offers applied physics (PhD); engineering sciences (PhD); medical engineering/medical physics (Sc D); physics (PhD). Offered jointly with Massachusetts Institute of Technology. *Degree requirements:* For doctorate, dissertation, oral and written qualifying exams required, foreign language not required.

See in-depth description on page 403.

Indiana University–Purdue University Indianapolis, School of Engineering and Technology, Biomedical Engineering Program, Indianapolis, IN 46202-2896. Offers MS Bm E, PhD. *Degree requirements:* For master's, thesis optional, foreign language not required. *Entrance requirements:* For master's, GRE, TOEFL (minimum score of 550 required), minimum B average; for doctorate, GRE General Test, TOEFL. *Application deadline:* For fall admission, 5/1. Application fee: $30 ($50 for international students). *Financial aid:* Fellowships, research assistantships, teaching assistantships available. Financial aid application deadline: 3/1. *Unit head:* Dr. Edward Berbari, Chair, 317-274-9721. *Application contact:* Mary DeBruicker, Graduate Office, 765-494-3649, Fax: 765-494-6440, E-mail: bmeprogram@ecn.purdue.edu.

Johns Hopkins University, G. W. C. Whiting School of Engineering, Department of Biomedical Engineering, Baltimore, MD 21205. Offers MSE, PhD, MD/PhD. *Faculty:* 17 full-time (1 woman), 4 part-time (0 women). *Students:* 100 full-time (31 women), 2 part-time (1 woman); includes 18 minority (4 African Americans, 14 Asian Americans or Pacific Islanders), 29 international. Average age 23. 280 applicants, 22% accepted. In 1998, 17 master's, 7 doctorates awarded. *Degree requirements:* For master's and doctorate, thesis/dissertation required, foreign language not required. *Entrance requirements:* For master's and doctorate, GRE General Test, TOEFL (minimum score of 560 required). *Application deadline:* For fall admission, 1/10. Application fee: $50. Tuition: Full-time $23,660. Tuition and fees vary according to program. *Financial aid:* In 1998–99, fellowships (averaging $12,150 per year), 8 teaching assistantships (averaging $12,150 per year) were awarded.; research assistantships, training grants also available. Financial aid application deadline: 1/10. *Faculty research:* Biomedical instrumentation, cardiovascular system, neutral encoding, biomedical sensors, biomaterials and imaging. Total annual research expenditures: $6.4 million. *Unit head:* Dr. Murray B. Sachs, Director, 410-955-3131, Fax: 410-955-0549, E-mail: msachs@eureka.wbme.jhu.edu. *Application contact:* Dr. Lawrence P. Schramm, Director, PhD Degree Program, 410-955-3277, Fax: 410-955-9826, E-mail: lschramm@bme.jhu.edu.

See in-depth description on page 405.

Louisiana Tech University, Graduate School, College of Engineering and Science, Department of Biomedical Engineering, Ruston, LA 71272. Offers MS, PhD. Part-time programs available. Terminal master's awarded for partial completion of doctoral program. *Degree requirements:* For master's and doctorate, thesis/dissertation required, foreign language not required. *Entrance requirements:* For master's, GRE General Test (minimum combined score of 1070 required), TOEFL (minimum score of 550 required), minimum GPA of 3.0 in last 60 hours; for doctorate, TOEFL (minimum score of 550 required), minimum graduate GPA of 3.25 (with MS) or GRE General Test (minimum combined score of 1270 required without MS). *Faculty research:* Microbiosensors and microcirculatory transport, speech recognition, artificial intelligence, rehabilitation engineering, bioelectromagnetics.

Marquette University, Graduate School, College of Engineering, Department of Biomedical Engineering, Milwaukee, WI 53201-1881. Offers bioinstrumentation/computers (MS, PhD); biomechanics/biomaterials (MS, PhD); functional imaging (PhD); healthcare technologies management (MS); systems physiology (MS, PhD). Part-time and evening/weekend programs available. *Faculty:* 12 full-time (2 women), 31 part-time (4 women). *Students:* 41 full-time (15 women), 31 part-time (5 women); includes 9 minority (6 Asian Americans or Pacific Islanders, 3 Hispanic Americans), 11 international. 41 applicants, 73% accepted. In 1998, 5 master's awarded (100% found work related to degree). Terminal master's awarded for partial completion of doctoral program. *Degree requirements:* For master's, computer language, thesis, comprehensive exam required, foreign language not required; for doctorate, computer language, dissertation defense, qualifying exam required. *Entrance requirements:* For master's and doctorate, GRE General Test, TOEFL (minimum score of 575 required). *Application deadline:* For fall admission, 2/15 (priority date); for spring admission, 11/15 (priority date). Applications are processed on a rolling basis. Application fee: $40. Tuition: Part-time $510 per credit hour. Tuition and fees vary according to program. *Financial aid:* In 1998–99, 12 fellowships with full tuition reimbursements, 9 research assistantships with full tuition reimbursements, 6 teaching assistantships with full tuition reimbursements were awarded.; scholarships also available. Financial aid application deadline: 2/15. *Faculty research:* Cell and organ physiology, signal processing, gait analysis, radio frequency instrument design. Total annual research expenditures: $1.5 million. *Unit head:* Dr. Dean C. Jeutter, Acting Chairman, 414-288-3375, Fax: 414-288-7938, E-mail: jeutterd@ums.csd.mu.edu.

Massachusetts Institute of Technology, Whitaker College of Health Sciences and Technology, Division of Health Sciences and Technology, Medical Engineering/Medical Physics Program, Cambridge, MA 02139-4307. Offers medical engineering (PhD); medical engineering and medical physics (Sc D); medical physics (PhD). Offered jointly with Harvard University. *Degree* *requirements:* For doctorate, dissertation, oral and written departmental qualifying exams required, foreign language not required.

See in-depth description on page 403.

Mayo Graduate School, Graduate Programs in Biomedical Sciences, Program in Biomedical Engineering, Rochester, MN 55905. Offers PhD. *Faculty:* 33 full-time (1 woman). *Students:* 25 full-time (7 women); includes 4 minority (1 African American, 2 Asian Americans or Pacific Islanders, 1 Hispanic American), 4 international. In 1998, 4 degrees awarded. *Degree requirements:* For doctorate, oral defense of dissertation, qualifying oral and written exam required. *Entrance requirements:* For doctorate, GRE, TOEFL, 2 years of chemistry; 1 year of biology, calculus, and physics. *Application deadline:* For fall admission, 12/31 (priority date). Applications are processed on a rolling basis. Application fee: $0. *Financial aid:* In 1998–99, 25 fellowships were awarded. *Unit head:* Dr. Richard A. Robb, Education Coordinator, 507-284-4937, E-mail: rar@mayo.edu. *Application contact:* Sherry Kallies, 507-266-0122, Fax: 507-284-0999, E-mail: phd.training@mayo.edu.

See in-depth description on page 409.

McGill University, Faculty of Graduate Studies and Research, Faculty of Medicine, Department of Biomedical Engineering, Montréal, PQ H3A 2T5, Canada. Offers M Eng, PhD. *Faculty:* 9 full-time (1 woman), 10 part-time (1 woman). *Students:* 33 full-time (9 women). Average age 24. 26 applicants, 69% accepted. In 1998, 8 master's, 3 doctorates awarded. *Degree requirements:* For master's and doctorate, thesis/dissertation required, foreign language not required. *Entrance requirements:* For master's, GRE (score in 10th percentile or higher required), TOEFL (minimum score of 600 required), minimum GPA of 3.3; for doctorate, GRE (score in 10th percentile or higher required), TOEFL (minimum score of 600 required). *Average time to degree:* Master's–3.5 years full-time; doctorate–4.5 years full-time. *Application deadline:* For fall admission, 3/1 (priority date); for winter admission, 11/1 (priority date). Application fee: $60. *Financial aid:* In 1998–99, 10 fellowships, 2 research assistantships were awarded.; institutionally-sponsored loans and tuition waivers (full) also available. *Faculty research:* Medical image processing, pulmonary and auditory mechanics, neuromuscular control, eye/head control (real and artificial), software and instrumentation. *Unit head:* Dr. R. E. Kearney, Chair, 514-398-6737, Fax: 514-398-7461, E-mail: rob@cortex.biomed.mcgill.ca. *Application contact:* Pina Sorrini, Coordinator, Graduate Program, 514-398-6736, Fax: 514-398-7461, E-mail: pina@cortex.biomed.mcgill.ca.

MCP Hahnemann University, School of Medicine, Biomedical Graduate Programs, Program in Bioengineering, Philadelphia, PA 19102-1192. Offers PhD, MD/PhD. Offered jointly with Lehigh University. *Degree requirements:* For doctorate, dissertation, qualifying exam required, foreign language not required. *Entrance requirements:* For doctorate, GRE General Test, TOEFL.

Mercer University, School of Engineering, Macon, GA 31207-0003. Offers biomedical engineering (MSE); electrical engineering (MSE); engineering management (MSE); mechanical engineering (MSE); software engineering (MSE); software systems (MS); technical management (MS). Part-time and evening/weekend programs available. *Faculty:* 23 full-time (1 woman), 6 part-time (0 women). *Degree requirements:* For master's, computer language, thesis or alternative required, foreign language not required. *Entrance requirements:* For master's, GRE, minimum undergraduate GPA of 3.0. *Application deadline:* For fall admission, 7/1; for spring admission, 11/15. Applications are processed on a rolling basis. Application fee: $35 ($50 for international students). *Unit head:* Dr. Benjamin S. Kelley, Dean, 912-752-2459, Fax: 912-752-5593, E-mail: kelley_bs@mercer.edu. *Application contact:* Kathy Olivier, Coordinator, Special Programs, 912-752-2196, E-mail: oliver_kh@mercer.edu.

Milwaukee School of Engineering, Department of Electrical Engineering and Computer Science, Program in Perfusion, Milwaukee, WI 53202-3109. Offers cardiovascular perfusion (MS). *Students:* 5 full-time (1 woman), 1 international. Average age 25. *Degree requirements:* For master's, thesis defense required. *Entrance requirements:* For master's, GRE General Test, BS in biology, chemistry, engineering, or related sciences. *Average time to degree:* Master's–1 year full-time. *Application deadline:* For fall admission, 8/15. Applications are processed on a rolling basis. Application fee: $30. Electronic applications accepted. *Financial aid:* Career-related internships or fieldwork available. Aid available to part-time students. *Unit head:* Dr. Vincent Canino, Director, 414-277-7331, Fax: 414-277-7465, E-mail: canino@msoe.edu. *Application contact:* Helen Boomsma, Director, Lifelong Learning Institute, 800-321-6763, Fax: 414-277-7475, E-mail: boomsma@msoe.edu.

Announcement: Milwaukee School of Engineering offers an MS in perfusion. Perfusion is the science of supporting or replacing a patient's circulatory or respiratory function during invasive surgery. Extensive clinical experience and an emphasis on the technology involved in perfusion make the full-time program especially suited to biomedical engineering graduates. Other MS degrees include architectural engineering, engineering, environmental engineering, engineering management, and medical informatics.

New Jersey Institute of Technology, Office of Graduate Studies, Department of Chemical Engineering, Chemistry and Environmental Science, Interdisciplinary Program in Biomedical Engineering, Newark, NJ 07102-1982. Offers MS. Part-time and evening/weekend programs available. *Degree requirements:* For master's, thesis required, foreign language not required. *Entrance requirements:* For master's, GRE General Test (minimum score of 450 on verbal section, 600 on quantitative, 550 on analytical required). Electronic applications accepted. *Faculty research:* Medical instrumentation, prosthesis design, biodegradation of hazardous waste, orthopedic biomechanics, image processing.

Northwestern University, The Graduate School, Robert R. McCormick School of Engineering and Applied Science, Department of Biomedical Engineering, Evanston, IL 60208. Offers MS, PhD. Admissions and degrees offered through The Graduate School. Part-time programs available. *Faculty:* 28 full-time (0 women). *Students:* 69 full-time (28 women); includes 16 minority (4 African Americans, 11 Asian Americans or Pacific Islanders, 1 Hispanic American), 8 international. 117 applicants, 41% accepted. In 1998, 14 master's, 10 doctorates awarded. Terminal master's awarded for partial completion of doctoral program. *Degree requirements:* For master's, thesis or alternative required, foreign language not required; for doctorate, dissertation required, foreign language not required. *Entrance requirements:* For master's, GRE General Test (combined average 1980 on three sections), TOEFL (minimum score of 560 required); for doctorate, GRE General Test (combined average 1995 on three sections), TOEFL (minimum score of 560 required). *Application deadline:* For fall admission, 8/30. Application fee: $50 ($55 for international students). *Financial aid:* In 1998–99, 8 fellowships with full tuition reimbursements (averaging $11,673 per year), 16 research assistantships with partial tuition reimbursements (averaging $16,285 per year), 8 teaching assistantships with full tuition reimbursements (averaging $12,042 per year) were awarded.; career-related internships or fieldwork, Federal Work-Study, institutionally-sponsored loans, and tuition scholarships, traineeships also available. Financial aid application deadline: 1/15; financial aid applicants required to submit FAFSA. *Faculty research:* Biomechanics and transport, rehabilitation engineering, neuroscience, biomaterials, cellular engineering. Total annual research expenditures: $3.3 million. *Unit head:* Robert Linsenmeier, Chair, 847-491-3043, Fax: 847-491-4928. *Application contact:* Mary Anne Peruchini, Coordinator, 847-491-5635, Fax: 847-491-4928, E-mail: mperuchini@nwu.edu.

See in-depth description on page 411.

The Ohio State University, Graduate School, College of Engineering, Program in Biomedical Engineering, Columbus, OH 43210. Offers MS, PhD. Evening/weekend programs available. *Faculty:* 47 full-time, 29 part-time. *Students:* 45 full-time (17 women), 6 part-time (4 women); includes 8 minority (3 African Americans, 3 Asian Americans or Pacific Islanders, 2 Hispanic Americans), 15 international. 157 applicants, 20% accepted. In 1998, 10 master's, 8 doctorates awarded. *Degree requirements:* For master's, computer language required, thesis optional, foreign language not required; for doctorate, computer language, dissertation required, foreign

language not required. *Entrance requirements:* For master's and doctorate, GRE General Test, GRE Subject Test. *Application deadline:* For fall admission, 8/15. Applications are processed on a rolling basis. Application fee: $30 ($40 for international students). *Financial aid:* Fellowships, research assistantships, career-related internships or fieldwork, Federal Work-Study, and institutionally-sponsored loans available. Aid available to part-time students. *Unit head:* Morton H. Friedman, Acting Director, 614-292-7165, Fax: 614-292-7301, E-mail: friedman.1@osu.edu.

See in-depth description on page 413.

Pennsylvania State University University Park Campus, Graduate School, Intercollege Graduate Programs, Intercollege Graduate Program in Bioengineering, State College, University Park, PA 16802-1503. Offers MS, PhD. *Students:* 20 full-time (3 women), 10 part-time (6 women). Terminal master's awarded for partial completion of doctoral program. *Degree requirements:* For master's and doctorate, thesis/dissertation required. *Entrance requirements:* For master's and doctorate, GRE General Test, TOEFL. Application fee: $50. *Unit head:* Dr. Herbert H. Lipowsky, Head, 814-865-1407.

See in-depth description on page 417.

Purdue University, Graduate School, Interdisciplinary Biomedical Engineering Program, West Lafayette, IN 47907. Offers MS Bm E, PhD. *Faculty:* 40 full-time (10 women). *Students:* 3. 48 applicants, 6% accepted. *Application deadline:* For fall admission, 1/15 (priority date). Application fee: $30. Electronic applications accepted. *Financial aid:* Fellowships, research assistantships, teaching assistantships available. Aid available to part-time students. Financial aid applicants required to submit FAFSA. *Faculty research:* Biomaterials, biomechanics, medical image and signal processing, medical instrumentation, tissue engineering. *Unit head:* Dr. G. R. Wodicka, Chairman, Graduate Committee, 765-494-0637, Fax: 765-494-6440, E-mail: wodicka@ecn.purdue.edu. *Application contact:* Biomedical Engineering Graduate Office, 765-494-3649, Fax: 765-494-6440, E-mail: bmeprogram@ecn.purdue.edu.

Rensselaer Polytechnic Institute, Graduate School, School of Engineering, Department of Biomedical Engineering, Troy, NY 12180-3590. Offers M Eng, MS, D Eng, PhD, MBA/M Eng. Part-time programs available. *Faculty:* 8 full-time (2 women), 11 part-time (0 women). *Students:* 36 full-time (9 women), 4 part-time; includes 7 minority (6 Asian Americans or Pacific Islanders, 1 Hispanic American), 10 international. 102 applicants, 52% accepted. In 1998, 10 master's, 3 doctorates awarded. *Degree requirements:* For master's, thesis required (for some programs), foreign language not required; for doctorate, dissertation required, foreign language not required. *Entrance requirements:* For master's and doctorate, GRE, TOEFL (minimum score of 550 required). *Application deadline:* For fall admission, 2/1 (priority date). Applications are processed on a rolling basis. Application fee: $35. *Financial aid:* In 1998–99, 2 fellowships (averaging $4,000 per year), 7 research assistantships with full and partial tuition reimbursements (averaging $5,500 per year), 12 teaching assistantships with full and partial tuition reimbursements (averaging $11,000 per year) were awarded.; career-related internships or fieldwork and institutionally-sponsored loans also available. Financial aid application deadline: 2/1. *Faculty research:* Internal and electrical impedence, computational mechanics, biomechanics and biomaterials, cellular and tissue bioengineering. Total annual research expenditures: $400,000. *Unit head:* Dr. Robert Spilker, Chair, 518-276-6548, Fax: 518-276-3035. *Application contact:* Lorrie Citarella, Coordinator of Student Affairs, 518-276-6547, Fax: 518-276-3035, E-mail: bme_coord@rpi.edu.

See in-depth description on page 419.

Rice University, Graduate Programs, George R. Brown School of Engineering, Department of Bioengineering, Houston, TX 77251-1892. Offers MS, PhD. *Entrance requirements:* For master's and doctorate, GRE General Test, TOEFL.

Rose-Hulman Institute of Technology, Faculty of Engineering and Applied Sciences, Interdisciplinary Program in Biomedical Engineering, Terre Haute, IN 47803-3920. Offers MS, MD/MS. Part-time programs available. *Faculty:* 13 full-time (3 women). *Students:* 11 full-time (3 women), 1 international. Average age 22. 18 applicants, 61% accepted. In 1998, 3 degrees awarded. *Degree requirements:* For master's, thesis required, foreign language not required. *Entrance requirements:* For master's, GRE, TOEFL (minimum score of 580 required), minimum GPA of 3.0. *Average time to degree:* Master's–2 years full-time. *Application deadline:* For fall admission, 2/1 (priority date). Applications are processed on a rolling basis. Application fee: $0. *Financial aid:* In 1998–99, 6 students received aid, including 3 fellowships (averaging $6,000 per year); research assistantships, teaching assistantships, grants, institutionally-sponsored loans, and tuition waivers (full and partial) also available. Financial aid application deadline: 2/1. *Unit head:* Dr. Lee Waite, Chair, 812-877-8404, Fax: 812-877-3198, E-mail: lee.waite@rose-hulman.edu. *Application contact:* Dr. Buck F. Brown, Dean for Research and Graduate Studies, 812-877-8403, Fax: 812-877-8102, E-mail: buck.brown@rose-hulman.edu.

Rutgers, The State University of New Jersey, New Brunswick, Graduate School, Program in Biomedical Engineering, New Brunswick, NJ 08903. Offers MS, PhD. Part-time programs available. *Faculty:* 63 full-time (7 women), 2 part-time (0 women). *Students:* 52 full-time (22 women), 40 part-time (16 women); includes 17 minority (7 African Americans, 8 Asian Americans or Pacific Islanders, 2 Hispanic Americans), 24 international. 133 applicants, 35% accepted. In 1998, 9 master's, 7 doctorates awarded. Terminal master's awarded for partial completion of doctoral program. *Degree requirements:* For master's and doctorate, thesis/dissertation required, foreign language not required. *Entrance requirements:* For master's and doctorate, GRE General Test, minimum GPA of 3.0. *Average time to degree:* Master's–3.7 years full-time; doctorate–5.5 years full-time. *Application deadline:* For fall admission, 4/1. Applications are processed on a rolling basis. Application fee: $50. *Financial aid:* In 1998–99, 22 fellowships with full tuition reimbursements, 17 research assistantships with full tuition reimbursements, 11 teaching assistantships with full tuition reimbursements were awarded.; career-related internships or fieldwork, Federal Work-Study, and tuition waivers (partial) also available. Financial aid application deadline: 3/1; financial aid applicants required to submit FAFSA. *Faculty research:* Biomedical instrumentation, biomechanics and biomaterials, vision and imaging. Total annual research expenditures: $1.5 million. *Unit head:* Stanley M. Dunn, Director, 732-445-3706, Fax: 732-445-3753, E-mail: smd@occlusal.rutgers.edu. *Application contact:* Graduate Secretary, 732-445-3706, Fax: 732-445-3753, E-mail: graduate@biomed.rutgers.edu.

Announcement: The Graduate Program in Biomedical Engineering offers MS and PhD programs in 2 options: physiological systems and biomedical instrumentation, and biomechanics and biomaterials. This joint program with the University of Medicine and Dentistry of New Jersey allows for excellent collaboration between engineering, life science, and clinical faculty members, with outstanding research opportunities. There are strong research interests in cardiovascular system dynamics; automated diagnostic devices; bioinstrumentation, including biotelemetry and implants; cardiac-assist devices; signal processing; neural networks and modeling; eye movements, pattern recognition, and visual information processing; respiratory control; neurophysiological responses in cell culture systems; computer image processing, ultrasonic imaging, and nuclear magnetic resonance; biosensors; tissue engineering; and biomaterials, biomechanics, and orthopedic implants.

Stanford University, School of Engineering, Department of Mechanical Engineering, Program in Biomechanical Engineering, Stanford, CA 94305-9991. Offers MS. *Entrance requirements:* For master's, GRE General Test, TOEFL. *Application deadline:* For fall admission, 1/15. Application fee: $65 ($80 for international students). Electronic applications accepted. Tuition: Full-time $24,588. Required fees: $152. Part-time tuition and fees vary according to course load. *Financial aid:* Application deadline: 1/15. *Unit head:* Kathy Kelly, Director of Admissions, 973-596-3300, Fax: 973-596-3461, E-mail: admissions@njit.edu. *Application contact:* Admissions Office, 650-723-3148.

Biomedical Engineering

State University of New York at Stony Brook, Graduate School, Program in Biomedical Engineering, Stony Brook, NY 11794. Offers MS, PhD, Certificate. *Faculty:* 53. 15 applicants, 13% accepted. In 1998, 4 degrees awarded. *Degree requirements:* For doctorate, dissertation, qualifying exams required, foreign language not required, foreign language not required. *Entrance requirements:* For master's and doctorate, GRe General Test, TOEFL. *Application deadline:* For fall admission, 1/15. Application fee: $50. Total annual research expenditures: $225,059. *Unit head:* Dr. Clint T. Rubin, Director, 516-444-2302, Fax: 516-444-7671, E-mail: clint@bone.ortho.sunysb.edu. *Application contact:* Anne Marie Dusatko, Administrative Secretary, 516-444-2302, Fax: 516-444-7671, E-mail: anne@bone.ortho.sunysb.edu.

Syracuse University, Graduate School, L. C. Smith College of Engineering and Computer Science, Department of Bioengineering and Neuroscience, Syracuse, NY 13244-0003. Offers bioengineering (MS). *Degree requirements:* Foreign language not required. *Entrance requirements:* For master's, GRE General Test, GRE Subject Test. *Application deadline:* Applications are processed on a rolling basis. Application fee: $40. Tuition: Full-time $13,992; part-time $583 per credit hour. *Financial aid:* Application deadline: 3/1. *Unit head:* Steve Chamberlain, Chair, 315-443-9711, Fax: 315-443-1184. *Application contact:* Norma Slepecky, Contact, 315-443-9749, Fax: 315-443-1184.

Announcement: The Department of Bioengineering and Neuroscience offers a master's program in bioengineering. The program is designed to prepare students with undergraduate backgrounds in engineering or physics for careers in the biomedical industry. The curriculum provides the opportunity for students to take a 4-course specialization in technology transfer, engineering management, or manufacturing engineering and the flexibility for the curriculum to be customized to meet the career objectives of each student. Research areas include systems neuroscience, bioinstrumentation, imaging, and skeletal muscular biomechanics; research involves laboratories at the Institute for Sensory Research at Syracuse University and the Orthopedic Research Laboratory at the adjacent SUNY Health Science Center, Syracuse.

Texas A&M University, College of Engineering, Department of Industrial Engineering, Division of Biomedical Engineering, College Station, TX 77843. Offers M Eng, MS, D Eng, PhD. Part-time programs available. *Students:* 40 full-time (8 women), 7 part-time (1 woman); includes 7 minority (1 African American, 3 Asian Americans or Pacific Islanders, 3 Hispanic Americans), 28 international. Average age 27. 60 applicants, 57% accepted. In 1998, 9 master's, 4 doctorates awarded. *Degree requirements:* For master's, computer language, thesis (MS) required; for doctorate, computer language, dissertation (PhD) required. *Entrance requirements:* For master's, GRE General Test (combined average 1100), TOEFL; for doctorate, GRE General Test (combined average 1250), TOEFL. *Application deadline:* Applications are processed on a rolling basis. Application fee: $50 ($75 for international students). *Financial aid:* In 1998–99, 20 students received aid; fellowships, research assistantships, teaching assistantships, career-related internships or fieldwork available. Financial aid application deadline: 4/1; financial aid applicants required to submit FAFSA. *Faculty research:* Biological models, medical lasers, optical biosensors, medical instrumentation, artificial heart. Total annual research expenditures: $400,000. *Unit head:* Dr. William Hyman, Program Head, 409-845-5532. *Application contact:* S. Rastegar, Graduate Adviser, 409-845-5532.

Thomas Jefferson University, College of Graduate Studies, Program in Cell and Tissue Engineering, Philadelphia, PA 19107. Offers PhD. *Faculty:* 4 full-time (1 woman). *Students:* 1 full-time (0 women). *Degree requirements:* For doctorate, dissertation required, foreign language not required. *Entrance requirements:* For doctorate, TOEFL (minimum score of 550 required); GRE General Test (minimum combined score of 1100 required), minimum GPA of 3.2. *Application deadline:* Applications are processed on a rolling basis. Application fee: $40. Tuition: Full-time $12,670. Tuition and fees vary according to degree level and program. *Financial aid:* In 1998–99, 1 fellowship with full tuition reimbursement was awarded. Financial aid application deadline: 5/1. *Unit head:* Dr. Rocky S. Tuan, Director, 215-955-5479, Fax: 215-955-9159. *Application contact:* Jessie F. Pervall, Director of Admissions, 215-503-4400, Fax: 215-503-3433, E-mail: cgs-info@mail.tju.edu.

See in-depth description on page 421.

Tulane University, School of Engineering, Department of Biomedical Engineering, New Orleans, LA 70118-5669. Offers MS, MSE, PhD, Sc D. MS and PhD offered through the Graduate School. *Faculty:* 10 full-time (1 woman), 27 part-time (7 women). *Students:* 44 full-time (12 women), 2 part-time (1 woman). Average age 23. 88 applicants, 42% accepted. In 1998, 11 master's awarded. Terminal master's awarded for partial completion of doctoral program. *Degree requirements:* For master's and doctorate, thesis/dissertation required, foreign language not required. *Entrance requirements:* For master's and doctorate, GRE General Test (minimum combined score of 1000 required; average 1250), TOEFL, minimum B average in undergraduate course work. *Average time to degree:* Master's–1.9 years full-time; doctorate–5.6 years full-time. *Application deadline:* For fall admission, 7/15 (priority date); for spring admission, 12/1 (priority date). Applications are processed on a rolling basis. Application fee: $35. *Financial aid:* In 1998–99, 35 students received aid, including 8 fellowships, 20 research assistantships, 10 teaching assistantships; career-related internships or fieldwork, Federal Work-Study, institutionally-sponsored loans, and tuition waivers (full and partial) also available. Financial aid application deadline: 2/1. *Faculty research:* Pulmonary and biofluid mechanics and biomechanics of bone, biomaterials science, finite element analysis, electric fields of the brain. *Unit head:* Dr. Richard T. Hart, Chairman, 504-865-5897, Fax: 504-865-8779. *Application contact:* Dr. E. Michaelides, Associate Dean, 504-865-5764.

See in-depth description on page 423.

Université de Montréal, Faculty of Medicine, Graduate Programs in Medicine, Institute of Biomedical Engineering, Montréal, PQ H3C 3J7, Canada. Offers M Sc A, PhD. *Faculty:* 7 full-time (0 women). *Students:* 10 full-time (4 women). 3 applicants, 33% accepted. In 1998, 1 master's awarded (100% entered university research/teaching); 1 doctorate awarded. *Degree requirements:* For master's, computer language, thesis required; for doctorate, computer language, dissertation, general exam required. *Entrance requirements:* For master's and doctorate, proficiency in French, knowledge of English. *Application deadline:* For fall admission, 4/1 (priority date). Applications are processed on a rolling basis. Application fee: $30. *Financial aid:* Career-related internships or fieldwork available. *Faculty research:* Electrophysiology, biomechanics, instrumentation, imaging, simulation. *Unit head:* A. Robert LeBlanc, Director, 514-343-6357. *Application contact:* A. Robert LeBlanc, Director, 514-343-6357.

The University of Akron, Graduate School, College of Engineering, Department of Biomedical Engineering, Akron, OH 44325-0001. Offers MS, MSE, PhD, MD/PhD. Part-time and evening/weekend programs available. *Faculty:* 8 full-time, 2 part-time. *Students:* 12 full-time (7 women), 2 part-time (1 woman); includes 4 minority (3 African Americans, 1 Hispanic American), 5 international. Average age 25. In 1998, 11 degrees awarded. Terminal master's awarded for partial completion of doctoral program. *Degree requirements:* For master's, thesis required, foreign language not required; for doctorate, variable foreign language requirement (computer language can substitute for one), dissertation, candidacy exam, qualifying exam required. *Entrance requirements:* For master's, GRE General Test, TOEFL (minimum score of 590 required), minimum GPA of 3.0; for doctorate, GRE General Test, TOEFL (minimum score of 590 required), minimum graduate GPA of 3.3. *Average time to degree:* Master's–2 years full-time, 4 years part-time. *Application deadline:* For fall admission, 3/1. Applications are processed on a rolling basis. Application fee: $25 ($50 for international students). Tuition, state resident: part-time $189 per credit. Tuition, nonresident: part-time $353 per credit. Required fees: $7.3 per credit. *Financial aid:* Fellowships with full tuition reimbursements, research assistantships with full tuition reimbursements, teaching assistantships with full tuition reimbursements, career-related internships or fieldwork, Federal Work-Study, scholarships, and tuition waivers (full) available. Financial aid application deadline: 3/1. *Faculty research:* Signal and image processing, gait analysis, biomechanics, biocontrols instrumentation, vascular dynamics. *Unit head:* Dr. Mary Verstraete, Chair, 330-972-7691, E-mail: mverstraete@uakron.edu. *Application contact:* Dr. Mary Verstraete, Chair, 330-972-7691, E-mail: mverstraete@uakron.edu.

See in-depth description on page 425.

The University of Alabama at Birmingham, Graduate School, School of Engineering, Department of Biomedical Engineering, Birmingham, AL 35294. Offers MSBE, PhD, DMD/PhD, MD/PhD. *Students:* 51 full-time (18 women), 8 part-time (2 women); includes 8 minority (7 African Americans, 1 Hispanic American), 8 international. 56 applicants, 59% accepted. In 1998, 10 master's, 4 doctorates awarded. *Degree requirements:* For master's, computer language, thesis or alternative, oral exam required, foreign language not required; for doctorate, computer language, dissertation, comprehensive exam required, foreign language not required. *Entrance requirements:* For master's, GRE General Test (minimum score of 600 on each section required), TOEFL (minimum score of 550 required); for doctorate, GRE General Test (minimum score of 650 on each section required), TOEFL (minimum score of 550 required). *Application deadline:* For fall admission, 4/15. Applications are processed on a rolling basis. Application fee: $30 ($60 for international students). Electronic applications accepted. *Financial aid:* In 1998–99, 48 students received aid; fellowships with full tuition reimbursements available, research assistantships, career-related internships or fieldwork, Federal Work-Study, and institutionally-sponsored loans available. *Unit head:* Dr. Linda C. Lucas, Chair, 205-934-8420, Fax: 205-975-4919.

Announcement: The Department of Biomedical Engineering has an undergraduate honors program in BME and a graduate program offering MS and PhD degrees. Three broad areas of research focus have been designated: (1) biomedical implants, (2) imaging of the brain, and (3) cardiac electrophysiology. Primary faculty members currently have annual grant support of more than $2.5 million.

See in-depth description on page 427.

University of Alberta, Faculty of Medicine and Dentistry, Graduate Programs in Medicine, Department of Biomedical Engineering, Edmonton, AB T6G 2E1, Canada. Offers biomedical engineering (M Sc); medical sciences (PhD). *Faculty:* 7 full-time (0 women). *Students:* 14 full-time (3 women), 3 international. Average age 23. 22 applicants, 23% accepted. In 1998, 2 master's awarded (50% found work related to degree, 50% continued full-time study); 1 doctorate awarded (100% entered university research/teaching). *Degree requirements:* For master's and doctorate, thesis/dissertation required, foreign language not required. *Average time to degree:* Master's–2.5 years full-time; doctorate–4.5 years full-time. *Application deadline:* For fall admission, 7/1. Applications are processed on a rolling basis. *Financial aid:* In 1998–99, 7 students received aid, including 2 research assistantships (averaging $14,000 per year), 4 teaching assistantships (averaging $14,000 per year); institutionally-sponsored loans and scholarships also available. Financial aid application deadline: 3/1. *Faculty research:* NMR spectroscopy, NMR imaging, rehabilitation engineering, brain electromagnetic activity, biomaterials, functional neuromuscular stimulation, MRI. Total annual research expenditures: $1.6 million. *Unit head:* Dr. P. S. Allen, Chair, 780-492-6397, Fax: 780-492-8259, E-mail: allen@fourier.bme.med.ualberta.ca. *Application contact:* Dr. R. E. Snyder, Graduate Coordinator, 780-492-6343, Fax: 780-492-8259, E-mail: r.snyder@ualberta.ca.

University of California, Berkeley, Graduate Division, Group in Bioengineering, Berkeley, CA 94720-1708. Offers PhD. *Faculty:* 117 full-time (13 women). *Students:* 50 full-time (16 women); includes 15 minority (5 African Americans, 7 Asian Americans or Pacific Islanders, 3 Hispanic Americans), 2 international. Average age 26. 146 applicants, 18% accepted. In 1998, 4 degrees awarded. *Degree requirements:* For doctorate, dissertation, qualifying exam required, foreign language not required. *Entrance requirements:* For doctorate, GRE General Test, minimum GPA of 3.0. *Application deadline:* For fall admission, 1/3. Application fee: $40. *Financial aid:* Fellowships, research assistantships, Federal Work-Study, institutionally-sponsored loans, and traineeships available. Financial aid application deadline: 1/3. *Faculty research:* Imaging, biomechanics, biomems modeling, neuroscience, biomedical computing, vision. *Unit head:* Dr. Rajendra Bhatnagar, Chair, E-mail: bhatnag@itsa.ucsf.edu. *Application contact:* Dr. Rajendra Bhatnagar, Chair, E-mail: bhatnag@itsa.ucsf.edu.

See in-depth description on page 429.

University of California, Davis, Graduate Studies, College of Engineering, Program in Biomedical Engineering, Davis, CA 95616. Offers MS, PhD. *Faculty:* 39 full-time (6 women), 1 part-time (0 women). *Students:* 54 full-time (17 women); includes 12 minority (10 Asian Americans or Pacific Islanders, 2 Hispanic Americans), 2 international. Average age 31. 64 applicants, 70% accepted. In 1998, 6 master's, 2 doctorates awarded. *Degree requirements:* For master's and doctorate, thesis/dissertation required, foreign language not required. *Entrance requirements:* For master's and doctorate, GRE General Test, minimum GPA of 3.25. *Application deadline:* For fall admission, 1/15 (priority date). Applications are processed on a rolling basis. Application fee: $40. Electronic applications accepted. *Financial aid:* In 1998–99, 13 fellowships, 24 research assistantships, 11 teaching assistantships were awarded.; career-related internships or fieldwork, Federal Work-Study, institutionally-sponsored loans, and tuition waivers (full and partial) also available. Financial aid application deadline: 1/15; financial aid applicants required to submit FAFSA. *Faculty research:* Orthopedic biomechanics, cell/molecular biomechanics and transport, biosensors and instrumentation, human movement, biomedical image analysis, spectroscopy. *Unit head:* Maury L. Hull, Graduate Adviser, 530-752-2611, Fax: 530-752-2123. *Application contact:* Cassandra Fong, Administrative Assistant, 530-752-2611, Fax: 530-752-2123, E-mail: csfong@ucdavis.edu.

University of California, Irvine, Office of Research and Graduate Studies, School of Engineering, Department of Chemical and Biochemical Engineering and Materials Science, Program in Biomedical Engineering, Irvine, CA 92697. Offers engineering (MS, PhD). Part-time programs available. *Faculty:* 10 full-time (2 women). Terminal master's awarded for partial completion of doctoral program. *Degree requirements:* For doctorate, dissertation required, foreign language not required, foreign language not required. *Entrance requirements:* For master's, GRE General Test, minimum GPA of 3.0; for doctorate, GRE General Test. *Application deadline:* For fall admission, 1/15 (priority date). Applications are processed on a rolling basis. Application fee: $40. Electronic applications accepted. *Financial aid:* Fellowships, research assistantships, teaching assistantships, institutionally-sponsored loans available. Financial aid application deadline: 3/2; financial aid applicants required to submit FAFSA. *Unit head:* Steve George, Director, 949-824-3941, Fax: 949-824-3440, E-mail: scgeorge@uci.edu. *Application contact:* Admissions Assistant, 949-824-3562, Fax: 949-824-3440.

University of California, Los Angeles, Graduate Division, School of Engineering and Applied Science, Interdepartmental Graduate Program in Biomedical Engineering, Los Angeles, CA 90095. Offers MS, PhD. *Students:* 7 full-time (1 woman); includes 3 minority (2 Asian Americans or Pacific Islanders, 1 Hispanic American), 1 international. 13 applicants, 54% accepted. *Degree requirements:* For master's, comprehensive exam or thesis required; for doctorate, dissertation, qualifying exams required, foreign language not required. *Entrance requirements:* For master's, GRE General Test, minimum GPA of 3.0; for doctorate, GRE General Test, minimum GPA of 3.25. *Application deadline:* For fall admission, 1/15. Application fee: $40. Electronic applications accepted. *Financial aid:* In 1998–99, 5 fellowships, 1 research assistantship were awarded. *Unit head:* Dr. John Mackenzie, Interim Chair, 310-825-3539. *Application contact:* Stacey Tran, Student Affairs Officer, 310-794-5945, Fax: 310-794-5956, E-mail: stacey@ea.ucla.edu.

University of California, San Diego, Graduate Studies and Research, Department of Bioengineering, La Jolla, CA 92093-5003. Offers MS, PhD. *Students:* 65 (18 women). 183 applicants, 36% accepted. In 1998, 18 master's, 10 doctorates awarded. *Entrance requirements:* For master's and doctorate, GRE General Test, TOEFL (minimum score of 550 required), minimum GPA of 3.0. Application fee: $40. *Unit head:* Shu Chien, Chair. *Application contact:* Graduate Coordinator, 619-534-6884.

See in-depth description on page 431.

University of California, San Francisco, Graduate Division, Program in Bioengineering, San Francisco, CA 94143. Offers PhD. *Faculty:* 117 full-time (13 women). *Students:* 50 full-time (16 women); includes 15 minority (5 African Americans, 7 Asian Americans or Pacific Islanders, 3 Hispanic Americans), 2 international. Average age 26. 146 applicants, 18% accepted. In 1998, 4 degrees awarded. *Degree requirements:* For doctorate, dissertation, qualifying exam required, foreign language not required. *Entrance requirements:* For doctorate, GRE General Test, minimum GPA of 3.0. *Application deadline:* For fall admission, 1/3. Application fee: $40. *Financial aid:* Fellowships, research assistantships, Federal Work-Study, institutionally-sponsored loans, and traineeships available. Financial aid application deadline: 1/3; financial aid applicants required to submit FAFSA. *Faculty research:* Imaging, biomechanics, modeling, neuroscience, biomedical computing, vision. *Unit head:* Rajendra Bhatnagar, Chair. *Application contact:* Rajendra Bhatnagar, Chair.

See in-depth description on page 429.

University of Connecticut, Graduate School, School of Engineering, Department of Mechanical Engineering, Field of Biomedical Engineering, Storrs, CT 06269. Offers MS, PhD. Terminal master's awarded for partial completion of doctoral program. *Degree requirements:* For master's, thesis or alternative required; for doctorate, dissertation required. *Entrance requirements:* For master's and doctorate, GRE General Test, TOEFL.

University of Florida, Graduate School, College of Engineering, Department of Biomedical Engineering, Gainesville, FL 32611. Offers MS, PhD. *Faculty:* 21. *Students:* 11 full-time (3 women); includes 5 minority (2 African Americans, 3 Asian Americans or Pacific Islanders), 2 international. *Degree requirements:* For master's, thesis optional; for doctorate, dissertation required. *Entrance requirements:* For master's, GRE General Test, minimum GPA of 3.0; for doctorate, GRE General Test. *Application deadline:* For fall admission, 6/1 (priority date). Applications are processed on a rolling basis. Application fee: $20. Electronic applications accepted. *Financial aid:* In 1998–99, 9 research assistantships were awarded. *Unit head:* Dr. Christopher Batich, Director, 352-392-6630, Fax: 352-392-7303, E-mail: cbati@ufl.edu. *Application contact:* Dr. Roger Tran-Son-Tay, Graduate Coordinator, 352-392-6229, Fax: 352-392-7303, E-mail: rtst@aero.ufl.edu.

See in-depth description on page 433.

University of Houston, Cullen College of Engineering, Program in Biomedical Engineering, Houston, TX 77004. Offers MS. Part-time and evening/weekend programs available. *Students:* 5 full-time (2 women), 3 part-time (1 woman); includes 1 minority (Asian American or Pacific Islander), 2 international. Average age 31. 62 applicants, 5% accepted. In 1998, 1 degree awarded. *Degree requirements:* For master's, thesis required, foreign language not required. *Entrance requirements:* For master's, GRE General Test, TOEFL. *Application deadline:* For fall admission, 7/3 (priority date); for spring admission, 12/4. Applications are processed on a rolling basis. Application fee: $25 ($75 for international students). *Financial aid:* Research assistantships, teaching assistantships, career-related internships or fieldwork, Federal Work-Study, institutionally-sponsored loans, and tuition waivers (partial) available. Financial aid application deadline: 7/1. *Faculty research:* Bioelectromagnetic signal analysis and modeling, antibody technology, biological neural networks, drug transport in tissues, medical imaging. *Unit head:* Dr. Pericles Y. Ktonas, Director, 713-743-4403, Fax: 713-743-4444, E-mail: ktonas@jetson.uh.edu. *Application contact:* Mylyssa McDonald, Graduate Analyst, 713-743-4403, Fax: 713-743-4444, E-mail: mmm05866@jetson.uh.edu.

University of Illinois at Chicago, Graduate College, College of Engineering, Bioengineering Program, Chicago, IL 60607-7128. Offers MS, PhD, MD/PhD. *Students:* 32 full-time (10 women), 11 part-time (3 women); includes 14 minority (1 African American, 11 Asian Americans or Pacific Islanders, 1 Hispanic American, 1 Native American), 13 international. Average age 24. 84 applicants, 40% accepted. In 1998, 8 master's, 4 doctorates awarded. Terminal master's awarded for partial completion of doctoral program. *Degree requirements:* For master's and doctorate, computer language, thesis/dissertation required, foreign language not required. *Entrance requirements:* For master's and doctorate, GRE Subject Test, TOEFL (minimum score of 550 required), minimum GPA of 4.0 on a 5.0 scale. *Application deadline:* For fall admission, 7/3; for spring admission, 11/8. Application fee: $40 ($50 for international students). *Financial aid:* In 1998–99, 23 students received aid; fellowships, research assistantships, teaching assistantships, career-related internships or fieldwork available. *Faculty research:* Imaging systems, bioinstrumentation, electrophysiology, biological control, laser scattering. *Unit head:* Dr. Richard Magin, Head, 312-996-2331.

See in-depth description on page 437.

University of Illinois at Urbana–Champaign, Graduate College, College of Engineering, Program in Bioengineering, Urbana, IL 61801.

See in-depth description on page 439.

The University of Iowa, Graduate College, College of Engineering, Department of Biomedical Engineering, Iowa City, IA 52242-1316. Offers MS, PhD. *Faculty:* 7 full-time, 4 part-time. *Students:* 23 full-time (5 women), 12 part-time (1 woman); includes 4 minority (1 African American, 2 Asian Americans or Pacific Islanders, 1 Hispanic American), 14 international. 53 applicants, 70% accepted. In 1998, 7 master's, 8 doctorates awarded. *Degree requirements:* For master's, thesis optional; for doctorate, dissertation, comprehensive exam required. *Entrance requirements:* For master's and doctorate, GRE, TOEFL. *Application deadline:* Applications are processed on a rolling basis. Application fee: $30 ($50 for international students). *Financial aid:* In 1998–99, 4 fellowships, 24 research assistantships, 6 teaching assistantships were awarded. Financial aid applicants required to submit FAFSA. *Faculty research:* Automated image analysis of heart, mathematical modelling of eye control, cardiorespiratory control, computer algorithms in health, systems physiology. *Unit head:* , Krishnar B. Chandran, Chair, 319-384-0504, Fax: 319-335-5631.

University of Kentucky, Graduate School, Program in Biomedical Engineering, Lexington, KY 40506-0032. Offers MSBE, PhD. *Degree requirements:* For master's, comprehensive exam required, thesis optional, foreign language not required; for doctorate, dissertation, comprehensive exam required, foreign language not required. *Entrance requirements:* For master's, GRE General Test, minimum undergraduate GPA of 2.5; for doctorate, GRE General Test, minimum graduate GPA of 3.0. *Faculty research:* Signal processing and dynamical systems, cardiopulmonary mechanics and systems, bioelectromagnetics, neuromotor control and electrical stimulation, biomaterials and musculoskeletal biomechanics.

University of Massachusetts Worcester, Graduate School of Biomedical Sciences, Program in Biomedical Engineering and Medical Physics, Worcester, MA 01655-0115. Offers PhD. *Faculty:* 20 full-time (1 woman). *Students:* 1 (woman) full-time; minority (Asian American or Pacific Islander) Average age 26. *Degree requirements:* For doctorate, dissertation required, foreign language not required. *Entrance requirements:* For doctorate, GRE General Test, 1 year of calculus, physics, organic chemistry and biology. *Application deadline:* For fall admission, 2/1 (priority date). Applications are processed on a rolling basis. Application fee: $25 ($50 for international students). *Financial aid:* Unspecified assistantships available. *Unit head:* Dr. Peter Grigg, Director, 508-856-2457.

See in-depth description on page 443.

University of Medicine and Dentistry of New Jersey, Graduate School of Biomedical Sciences, Graduate Programs in Biomedical Sciences, Program in Biomedical Engineering, Newark, NJ 07107-3001. Offers MS, PhD. *Degree requirements:* For master's, thesis, qualifying exam required; for doctorate, dissertation, qualifying exam required, foreign language not required. *Entrance requirements:* For master's and doctorate, GRE General Test, TOEFL. *Application deadline:* For fall admission, 4/1 (priority date); for spring admission, 10/1. Application fee: $40. *Financial aid:* Fellowships, research assistantships, teaching assistantships available. Financial aid application deadline: 5/1. *Unit head:* Stanley M. Dunn, Director, 732-445-3706, Fax: 732-445-3753.

The University of Memphis, Graduate School, Herff College of Engineering, Program in Biomedical Engineering, Memphis, TN 38152. Offers MS, PhD. *Faculty:* 17 full-time (1 woman), 2 part-time (0 women). *Students:* 52 full-time (16 women), 11 part-time, 31 international. Average age 26. 155 applicants, 48% accepted. In 1998, 11 degrees awarded. *Degree requirements:* For master's, thesis or alternative, level A/oral exam required; for doctorate, dissertation, level A and level B exams required. *Entrance requirements:* For master's, GRE General Test (minimum score of 500 on each of three sections required) or MAT, minimum undergraduate GPA of 3.0; for doctorate, GRE General Test (minimum combined score of 1800 on three sections required), minimum undergraduate GPA of 3.25 required or master's degree in biomedical engineering. *Application deadline:* For fall admission, 8/1 (priority date); for spring admission, 12/1. Applications are processed on a rolling basis. Application fee: $25 ($50 for international students). Tuition, state resident: full-time $3,410; part-time $178 per credit hour. Tuition, nonresident: full-time $8,670; part-time $408 per credit hour. Tuition and fees vary according to program. *Financial aid:* In 1998–99, 38 fellowships with full tuition reimbursements, 8 research assistantships with full tuition reimbursements, 4 teaching assistantships with full tuition reimbursements were awarded.; career-related internships or fieldwork and tuition waivers (full and partial) also available. Financial aid application deadline: 4/1. *Faculty research:* Biomechanics, including orthopedic implants, prosthetic devices and design; cell and tissue engineering with a focus on the cardiovascular system, artificial organs, biomaterials, and hemodynamics; electropysiology, including measurement methods, modeling and computation, signal analysis, and biosensors and microfabrication; imaging including novel medical image-acquisition devices, computational image processing and quantitative analysis techniques. *Unit head:* Dr. Vincent T. Turitto, Chairman, 901-678-4299. *Application contact:* Dr. Michael R. T. Yen, Coordinator of Graduate Studies, 901-678-3263.

Announcement: The joint program offers master's and doctoral training in biomedical engineering. Participation by the University of Memphis and the University of Tennessee strengthens and augments available courses. Faculty disciplines include biomechanics, cell and tissue, electrophysiology, medical imaging, and sensors. The program is adding a new emphasis in pediatric biomedical engineering.

See in-depth description on page 445.

University of Miami, Graduate School, College of Engineering, Department of Biomedical Engineering, Coral Gables, FL 33124. Offers MSBE, PhD. Part-time programs available. *Faculty:* 8 full-time (1 woman), 24 part-time (1 woman). *Students:* 43 full-time (14 women), 12 part-time (4 women); includes 15 minority (4 African Americans, 1 Asian American or Pacific Islander, 9 Hispanic Americans, 1 Native American), 10 international. Average age 29. 129 applicants, 83% accepted. In 1998, 5 master's, 3 doctorates awarded. *Degree requirements:* For master's, thesis or alternative, oral exam or thesis required, foreign language not required; for doctorate, dissertation, oral and qualifying exams required, foreign language not required. *Entrance requirements:* For master's and doctorate, GRE General Test (minimum combined score of 1000 required), TOEFL (minimum score of 550 required), minimum GPA of 3.0. *Average time to degree:* Master's–2.5 years full-time, 4 years part-time; doctorate–4 years full-time, 5 years part-time. *Application deadline:* For fall admission, 5/1 (priority date); for spring admission, 10/1. Applications are processed on a rolling basis. Application fee: $35. Tuition: Full-time $15,336; part-time $852 per credit. Required fees: $174. Tuition and fees vary according to program. *Financial aid:* In 1998–99, 43 students received aid, including 3 fellowships with tuition reimbursements available, 22 research assistantships with tuition reimbursements available, 1 teaching assistantship with tuition reimbursement available; career-related internships or fieldwork also available. Aid available to part-time students. *Faculty research:* Instrumentation for auditory, cardiovascular, neural, orthopedic, and pulmonary systems; flow cytometry; artificial bladders; fluid dynamics. Total annual research expenditures: $440,062. *Unit head:* Dr. Ozcan Ozdamar, Chairman, 305-284-2442, Fax: 305-284-6494, E-mail: ptarjan@engine01.msmail.miami.edu. *Application contact:* Sandra Perdomo, Staff Associate, 305-284-2445, Fax: 305-284-6494, E-mail: sperdomo@miami.edu.

University of Michigan, Horace H. Rackham School of Graduate Studies, College of Engineering, Department of Biomedical Engineering and Center for Biomedical Engineering Research, Ann Arbor, MI 48109. Offers MS, PhD. *Faculty:* 38 full-time (5 women), 65 part-time (9 women). *Students:* 97 full-time (27 women); includes 20 minority (4 African Americans, 14 Asian Americans or Pacific Islanders, 2 Hispanic Americans), 25 international. Average age 25. 265 applicants, 48% accepted. In 1998, 19 master's, 7 doctorates awarded. *Degree requirements:* For master's, thesis optional, foreign language not required; for doctorate, dissertation, oral defense of dissertation, preliminary exams required, foreign language not required. *Entrance requirements:* For master's, GRE General Test (combined average of 2000 on three sections), TOEFL (minimum score of 560 required); for doctorate, GRE General Test (combined average of 2000 on three sections), TOEFL (minimum score of 660 required), master's degree. *Average time to degree:* Master's–3 years full-time; doctorate–4 years full-time. *Application deadline:* For fall admission, 2/1 (priority date); for winter admission, 11/1 (priority date); for spring admission, 3/1 (priority date). Applications are processed on a rolling basis. Application fee: $55. Electronic applications accepted. *Financial aid:* In 1998–99, 10 fellowships with full tuition reimbursements (averaging $17,600 per year), 60 research assistantships with full tuition reimbursements (averaging $17,600 per year), 5 teaching assistantships with full tuition reimbursements (averaging $17,000 per year) were awarded.; Federal Work-Study, scholarships, and traineeships also available. Financial aid application deadline: 2/1; financial aid applicants required to submit FAFSA. Total annual research expenditures: $20 million. *Unit head:* Dr. Matthew O'Donnell, Chair, Biomedical Engineering, 734-764-9588, Fax: 734-936-1905, E-mail: biomede@umich.edu. *Application contact:* Maria E. Steele, Student Services Assistant, 734-764-9588, Fax: 734-936-1905, E-mail: msteele@umich.edu.

See in-depth description on page 447.

University of Minnesota, Twin Cities Campus, Graduate School, Institute of Technology, Department of Biomedical Engineering, Minneapolis, MN 55455-0213. Offers MS, PhD, MD/PhD. Part-time programs available. *Faculty:* 41 full-time (2 women), 14 part-time (4 women). *Students:* 50 full-time (14 women), 13 part-time (1 woman); includes 6 minority (5 Asian Americans or Pacific Islanders, 1 Native American), 22 international. Average age 25. 102 applicants, 43% accepted. In 1998, 6 master's, 2 doctorates awarded. Terminal master's awarded for partial completion of doctoral program. *Degree requirements:* For master's, thesis optional; for doctorate, dissertation required. *Entrance requirements:* For master's, GRE General Test (minimum score of 500 on verbal sectinos, 1350 combined required); for doctorate, GRE General Test (minimum score of 500 on verbal section, 1350 combined required). *Average time to degree:* Master's–2 years full-time; doctorate–4.5 years full-time. *Application deadline:* For fall admission, 1/15 (priority date); for spring admission, 7/15. Applications are processed on a rolling basis. Application fee: $40 ($50 for international students). *Financial aid:* In 1998–99, 5 fellowships with full tuition reimbursements (averaging $14,500 per year), 20 research assistantships with full tuition reimbursements (averaging $14,500 per year) were awarded.; teaching assistantships *Faculty research:* Biomedical microelectromechanical systems, tissue engineering, biomechanics and blood/fluid dynamics, biomaterials, soft tissue mechanics, biomedical imaging. Total annual research expenditures: $8 million. *Unit head:* Dennis L. Polla, Head, 612-624-9603, Fax: 612-626-6583, E-mail: polla@lenti.med.umn.edu. *Application contact:* Anne Longo, Graduate Coordinator, 612-624-9603, Fax: 612-626-6583, E-mail: bmengp@maroon.tc.umn.edu.

See in-depth description on page 449.

University of Nevada, Reno, School of Medicine, Graduate Programs in Medicine, Program in Biomedical Engineering, Reno, NV 89557. Offers MS, PhD. Offered jointly with the College

Biomedical Engineering

University of Nevada, Reno (continued)
of Engineering. *Degree requirements:* For doctorate, dissertation required, foreign language not required. *Entrance requirements:* For master's, GRE General Test, minimum GPA of 2.75; for doctorate, GRE General Test, minimum GPA of 3.0.

The University of North Carolina at Chapel Hill, School of Medicine, Graduate Programs in Medicine, Department of Biomedical Engineering, Chapel Hill, NC 27599. Offers MS, PhD. *Faculty:* 13 full-time (1 woman), 14 part-time (2 women). *Students:* 71 full-time (11 women); includes 12 minority (4 African Americans, 5 Asian Americans or Pacific Islanders, 3 Hispanic Americans), 11 international. In 1998, 10 master's, 4 doctorates awarded. Terminal master's awarded for partial completion of doctoral program. *Degree requirements:* For master's, thesis, comprehensive exam required, foreign language not required; for doctorate, dissertation, qualifying exam required. *Entrance requirements:* For master's and doctorate, GRE General Test (minimum combined score of 1000 required; average 1359), minimum GPA of 3.0, average 3.2. *Average time to degree:* Master's–2.5 years full-time; doctorate–5 years full-time. *Application deadline:* For fall admission, 2/1. Application fee: $55. *Financial aid:* In 1998–99, 19 fellowships, 33 research assistantships were awarded.; teaching assistantships, career-related internships or fieldwork and unspecified assistantships also available. *Faculty research:* Medical imaging, medical informatics, biomaterials, microelectronics instrumentation, neuroscience engineering. *Unit head:* Dr. Carol L. Lucas, Chair, 919-966-1175, Fax: 919-966-2963, E-mail: chairman@bme.unc.edu. *Application contact:* Dr. Henry S. Hsiao, Director of Graduate Studies, 919-966-1175, Fax: 919-966-2963, E-mail: hsiao@med.unc.edu.

Announcement: Flexible MS/PhD program in biomedical engineering welcomes applications from talented students with backgrounds in physical or biological sciences. Training areas include biomaterials, instrumentation, imaging, informatics, microelectronics, signal processing, and systems physiology. Major research opportunities exist in medical imaging, medical informatics, dental materials, electrocardiology, and systems neuroscience.

See in-depth description on page 453.

University of Pennsylvania, School of Engineering and Applied Science, Department of Bioengineering, Philadelphia, PA 19104. Offers MSE, PhD, MD/PhD, VMD/PhD. Terminal master's awarded for partial completion of doctoral program. *Degree requirements:* For master's, computer language required, thesis optional, foreign language not required; for doctorate, computer language, dissertation required, foreign language not required. *Entrance requirements:* For master's, GRE General Test (minimum combined score of 1900 on three sections required; average 2100), TOEFL (minimum score of 600 required; average 630); for doctorate, GRE General Test, TOEFL (minimum score of 600 required; average 630). Electronic applications accepted. *Faculty research:* Biomaterials and biomechanics, biofluid mechanics and transport, bioelectric phenomena, computational neuroscience.

See in-depth description on page 455.

University of Pittsburgh, School of Engineering, Department of Bioengineering, Pittsburgh, PA 15260. Offers MSBENG, PhD. Part-time and evening/weekend programs available. *Faculty:* 2 full-time (0 women). *Students:* 45 full-time (20 women), 10 part-time (2 women); includes 8 minority (1 African American, 4 Asian Americans or Pacific Islanders, 3 Hispanic Americans), 7 international. 58 applicants, 47% accepted. In 1998, 7 master's, 3 doctorates awarded. Terminal master's awarded for partial completion of doctoral program. *Degree requirements:* For master's, computer language, thesis required, foreign language not required; for doctorate, computer language, dissertation, comprehensive and final oral exams required, foreign language not required. *Entrance requirements:* For master's and doctorate, GRE General Test, TOEFL (minimum score of 550 required), minimum QPA of 3.0. *Average time to degree:* Master's–2 years full-time, 3 years part-time; doctorate–5 years full-time, 7 years part-time. *Application deadline:* For fall admission, 8/1 (priority date); for spring admission, 12/1. Applications are processed on a rolling basis. Application fee: $30 ($40 for international students). *Financial aid:* In 1998–99, 42 students received aid, including 5 fellowships (averaging $12,368 per year), 13 research assistantships (averaging $10,160 per year), 8 teaching assistantships (averaging $9,088 per year); grants, scholarships, and traineeships also available. Financial aid application deadline: 2/15. *Faculty research:* Artificial organs, biomechanics, biomaterials, signal processing, biotechnology. Total annual research expenditures: $4.5 million. *Unit head:* Dr. Jerome S. Schultz, Director, 412-383-9713, Fax: 412-383-9710, E-mail: jssbio@vms.cis.pitt.edu.

See in-depth description on page 459.

University of Rochester, The College, School of Engineering and Applied Sciences, Program in Biomedical Engineering, Rochester, NY 14627-0250. Offers MS, PhD. Part-time programs available. *Students:* 11 full-time (3 women), 2 part-time; includes 1 minority (Asian American or Pacific Islander), 4 international. 66 applicants, 18% accepted. Terminal master's awarded for partial completion of doctoral program. *Degree requirements:* For master's, foreign language and thesis not required; for doctorate, dissertation, qualifying exam required, foreign language not required. *Entrance requirements:* For doctorate, GRE General Test, TOEFL. *Application deadline:* For fall admission, 2/1. Application fee: $25. *Financial aid:* Fellowships, research assistantships, teaching assistantships, tuition waivers (full and partial) available. Financial aid application deadline: 2/1. *Unit head:* Dr. Richard Waugh, Director, 716-275-4151. *Application contact:* Mary Wallman, Graduate Program Secretary, 716-275-3891.

See in-depth description on page 461.

University of Saskatchewan, College of Graduate Studies and Research, College of Engineering, Division of Biomedical Engineering, Saskatoon, SK S7N 5A2, Canada. Offers M Eng, M Sc, PhD. *Degree requirements:* For master's and doctorate, thesis/dissertation required. *Entrance requirements:* For master's and doctorate, GRE, TOEFL.

University of Southern California, Graduate School, School of Engineering, Department of Biomedical Engineering, Los Angeles, CA 90089. Offers biomedical engineering (MS, PhD); biomedical imaging and telemedicine (MS). *Faculty:* 54 full-time (23 women), 1 (woman) part-time. *Students:* 43 full-time (16 women), 12 part-time (3 women); includes 13 minority (11 Asian Americans or Pacific Islanders, 2 Hispanic Americans), 25 international. Average age 26. 113 applicants, 64% accepted. In 1998, 13 master's, 8 doctorates awarded. *Degree requirements:* For master's and doctorate, thesis/dissertation required. *Entrance requirements:* For master's and doctorate, GRE General Test, GRE Subject Test. *Application deadline:* For fall admission, 7/1 (priority date); for spring admission, 12/1. Application fee: $55. Tuition: Part-time $768 per unit. Required fees: $350 per semester. *Financial aid:* In 1998–99, 7 fellowships, 20 research assistantships, 10 teaching assistantships were awarded.; Federal Work-Study, institutionally-sponsored loans, and scholarships also available. Aid available to part-time students. Financial aid application deadline: 2/15; financial aid applicants required to submit FAFSA. *Faculty research:* Respiratory mechanics, signal processing, pharmacokinetics, mathematical modeling, basic medical imaging. *Unit head:* Dr. David D'Argenio, Chairman, 213-740-7237.

University of Tennessee, Knoxville, Graduate School, College of Engineering, Department of Mechanical and Aerospace Engineering and Engineering Science, Program in Engineering Science, Knoxville, TN 37996. Offers applied artificial intelligence (MS); biomedical engineering (MS, PhD); composite materials (MS, PhD); computational mechanics (MS, PhD); engineering science (MS, PhD); fluid mechanics (MS, PhD); industrial engineering (MS, PhD); optical engineering (MS, PhD); solid mechanics (MS, PhD). Part-time programs available. *Students:* 33 full-time (9 women), 16 part-time (1 woman); includes 4 minority (1 African American, 2 Asian Americans or Pacific Islanders, 1 Native American), 10 international. *Degree requirements:* For master's, thesis or alternative required, foreign language not required; for doctorate, dissertation required, foreign language not required. *Entrance requirements:* For master's and doctorate, TOEFL (minimum score of 550 required), minimum GPA of 2.7. *Application deadline:* For fall admission, 2/1 (priority date). Applications are processed on a rolling basis. Application fee: $35. Electronic applications accepted. *Application contact:* Dr. Allen Yu, Graduate Representative, 923-974-4159, E-mail: nyu@utk.edu.

University of Tennessee, Memphis, College of Graduate Health Sciences, School of Biomedical Engineering, Memphis, TN 38163-0002. Offers MS, PhD. Part-time programs available. Terminal master's awarded for partial completion of doctoral program. *Degree requirements:* For master's, thesis, oral and written comprehensive exams required, foreign language not required; for doctorate, dissertation, oral and written preliminary and comprehensive exams required, foreign language not required. *Entrance requirements:* For master's and doctorate, GRE General Test (minimum combined score of 1200 required), TOEFL (minimum score of 525 required), minimum B average; bachelor's degree in engineering, physics, chemistry, computer or mathematical science, biology, or a closely related field.

Announcement: The joint program offers master's and doctoral training in biomedical engineering. Participation by the University of Memphis and the University of Tennessee strengthens and augments available courses. Faculty disciplines include biomechanics, cell and tissue, electrophysiology, medical imaging, and sensors. The program is adding a new emphasis in pediatric biomedical engineering.

See in-depth description on page 445.

The University of Texas at Arlington, Graduate School, College of Engineering, Biomedical Engineering Program, Arlington, TX 76019. Offers MS, PhD. Part-time programs available. *Faculty:* 4 full-time (1 woman), 1 part-time (0 women). *Students:* 11 full-time (2 women), 14 part-time (3 women); includes 2 minority (1 African American, 1 Asian American or Pacific Islander), 9 international. 43 applicants, 30% accepted. In 1998, 8 master's, 1 doctorate awarded. *Degree requirements:* For master's, computer language, comprehensive exam required, thesis optional, foreign language not required; for doctorate, computer language, dissertation, qualifying exam required, foreign language not required. *Entrance requirements:* For master's, GRE General Test (minimum combined score of 1100 required), TOEFL (minimum score of 575 required), minimum GPA of 3.0; for doctorate, GRE General Test (minimum combined score of 1150 required), TOEFL (minimum score of 575 required), minimum GPA of 3.4. *Application deadline:* Applications are processed on a rolling basis. Application fee: $25 ($50 for international students). Tuition, state resident: full-time $1,368; part-time $76 per semester hour. Tuition, nonresident: full-time $5,454; part-time $303 per semester hour. Required fees: $66 per semester hour. $86 per term. Tuition and fees vary according to course load. *Financial aid:* Research assistantships, teaching assistantships, career-related internships or fieldwork, institutionally-sponsored loans, and tuition waivers (partial) available. Financial aid application deadline: 6/30. *Faculty research:* Instrumentation, mechanics, materials. *Unit head:* Dr. Robert C. Eberhart, Chair, 817-272-2249, Fax: 817-272-2251. *Application contact:* Dr. Charles Chuong, Graduate Adviser, 817-272-2249, Fax: 817-272-2251, E-mail: chuong@uta.edu.

The University of Texas at Austin, Graduate School, College of Engineering, Program in Biomedical Engineering, Austin, TX 78712-1111. Offers MSE, PhD. Part-time programs available. *Students:* 10 full-time (3 women), 7 international. Average age 25. 62 applicants, 58% accepted. In 1998, 13 master's, 9 doctorates awarded. *Degree requirements:* For master's, foreign language and thesis not required; for doctorate, dissertation required, foreign language not required. *Entrance requirements:* For master's and doctorate, GRE General Test (minimum combined score of 1000 required). *Application deadline:* For fall admission, 2/1; for spring admission, 10/1. Applications are processed on a rolling basis. Application fee: $50 ($75 for international students). Electronic applications accepted. *Financial aid:* In 1998–99, 3 fellowships were awarded.; research assistantships, teaching assistantships, Federal Work-Study and tuition waivers (partial) also available. Financial aid application deadline: 2/1. *Faculty research:* Biomechanics, bioengineering, tissue engineering, tissue optics, biothermal studies. *Unit head:* Dr. Kenneth R. Diller, Director, 512-471-7167, Fax: 512-471-0616, E-mail: kdiller@mail.utexas.edu. *Application contact:* Dr. John A. Pearce, Graduate Adviser, 512-471-4984, Fax: 512-471-0616, E-mail: jpearce@mail.utexas.edu.

See in-depth description on page 463.

The University of Texas Southwestern Medical Center at Dallas, Southwestern Graduate School of Biomedical Sciences, Biomedical Engineering Program, Dallas, TX 75235. Offers MS, PhD. *Faculty:* 29 full-time (2 women), 26 part-time (3 women). *Students:* 27 full-time (11 women); includes 17 minority (1 African American, 15 Asian Americans or Pacific Islanders, 1 Hispanic American) Average age 25. 120 applicants, 58% accepted. In 1998, 7 master's awarded. *Degree requirements:* For master's, computer language, comprehensive exam or thesis required; for doctorate, computer language, dissertation, comprehensive exam required, foreign language not required. *Entrance requirements:* For master's, GRE General Test (minimum combined score of 1000 required; average 1260), minimum GPA of 3.0; for doctorate, GRE General Test (minimum combined score of 1150 required; average 1290), TOEFL (minimum of 575 required), minimum GPA of 3.4. *Application deadline:* For fall admission, 5/15 (priority date); for spring admission, 10/15. Application fee: $0. *Financial aid:* In 1998–99, 1 fellowship with partial tuition reimbursement (averaging $900 per year), 24 research assistantships (averaging $774 per year) were awarded.; career-related internships or fieldwork, grants, institutionally-sponsored loans, and tuition waivers (partial) also available. Financial aid application deadline: 3/15; financial aid applicants required to submit FAFSA. *Faculty research:* Noninvasive image analysis, biomaterials development, rehabilitation engineering, biomechanics, bioinstrumentation. *Unit head:* Dr. Robert Eberhart, Chair, 214-648-2052, Fax: 214-648-2979, E-mail: eberhart@email.swmed.edu. *Application contact:* Helen Leonhardt, 214-648-3512, Fax: 214-648-2979, E-mail: helen.leonhardt@email.swmed.edu.

See in-depth description on page 465.

University of Toronto, School of Graduate Studies, Physical Sciences Division, Faculty of Applied Science and Engineering, Institute of Biomedical Engineering, Toronto, ON M5S 1A1, Canada. Offers biomedical engineering (M Eng, M Sc, MA Sc, PhD); clinical biomedical engineering (MH Sc). Part-time programs available. *Degree requirements:* For master's, thesis required (for some programs); for doctorate, dissertation required.

University of Utah, Graduate School, College of Engineering, Department of Bioengineering, Salt Lake City, UT 84112-1107. Offers ME, MS, PhD. *Faculty:* 9 full-time (0 women), 26 part-time (5 women). *Students:* 65 full-time (18 women), 17 part-time (7 women); includes 2 minority (both Asian Americans or Pacific Islanders), 16 international. Average age 28. In 1998, 9 master's, 6 doctorates awarded. Terminal master's awarded for partial completion of doctoral program. *Degree requirements:* For master's, comprehensive exam, thesis (MS), written project and oral presentation (ME) required; for doctorate, dissertation required, foreign language not required. *Entrance requirements:* For master's and doctorate, GRE, TOEFL (minimum score of 500 required), minimum GPA of 3.0. *Application deadline:* For fall admission, 5/1. Application fee: $30 ($50 for international students). *Financial aid:* In 1998–99, 8 fellowships, 47 research assistantships were awarded.; traineeships also available. Financial aid application deadline: 5/1. *Faculty research:* Bioinstrumentation, biomaterials, ultrasonic bioinstrumentation, medical imaging, neuroprosthesis. Total annual research expenditures: $2.1 million. *Unit head:* Richard A. Normann, Chair, 801-581-7645, Fax: 801-585-5361. *Application contact:* Richard D. Rabbitt, Admissions Coordinator, 801-581-8559.

See in-depth description on page 467.

University of Vermont, Graduate College, College of Engineering and Mathematics, Department of Computer Science and Electrical Engineering, Program in Biomedical Engineering, Burlington, VT 05405-0160. Offers MS. *Degree requirements:* For master's, thesis required, foreign language not required. *Entrance requirements:* For master's, GRE General Test, TOEFL (minimum score of 550 required).

University of Virginia, School of Engineering and Applied Science, Department of Biomedical Engineering, Charlottesville, VA 22903. Offers ME, MS, PhD. *Faculty:* 8 full-time (1 woman), 1 part-time (0 women). *Students:* 49 full-time (14 women); includes 12 minority (1 African American, 7 Asian Americans or Pacific Islanders, 3 Hispanic Americans, 1 Native American), 1 international. Average age 24. 55 applicants, 56% accepted. In 1998, 15 master's, 4 doctorates awarded. *Degree requirements:* For master's, project or thesis required; for doctorate, dissertation, comprehensive exam required. *Entrance requirements:* For master's and doctorate, GRE General Test. *Application deadline:* For fall admission, 8/1; for spring admission, 12/1. Application fee: $60. *Financial aid:* Fellowships, research assistantships, teaching assistantships available. Financial aid application deadline: 2/1. *Faculty research:* Cardiopulmonary and neural engineering, cellular engineering, image processing, orthopedics and rehabilitation engineering. *Unit head:* Dr. J. S. Lee, Chairman, 804-924-5101.

University of Washington, School of Medicine, Graduate Programs in Medicine, Department of Bioengineering, Seattle, WA 98195. Offers MS, MSE, PhD. *Faculty:* 29 full-time (2 women), 34 part-time (4 women). *Students:* 111 full-time (37 women), 2 part-time; includes 20 minority (3 African Americans, 11 Asian Americans or Pacific Islanders, 4 Hispanic Americans, 2 Native Americans), 21 international. 246 applicants, 27% accepted. In 1998, 3 master's, 10 doctorates awarded. Terminal master's awarded for partial completion of doctoral program. *Degree requirements:* For master's, thesis required; for doctorate, dissertation, qualifying exam, general exam required. *Entrance requirements:* For master's and doctorate, GRE General Test, TOEFL, minimum GPA of 3.0. *Average time to degree:* Master's–2.5 years full-time; doctorate–5 years full-time. *Application deadline:* For fall admission, 1/15. Application fee: $45. Tuition, state resident: full-time $5,196; part-time $475 per credit. Tuition, nonresident: full-time $13,485; part-time $1,285 per credit. Required fees: $387; $38 per credit. Tuition and fees vary according to course load. *Financial aid:* In 1998–99, 7 fellowships (averaging $16,800 per year), 96 research assistantships (averaging $16,800 per year) were awarded.; Federal Work-Study, institutionally-sponsored loans, traineeships, and tuition waivers (full) also available. Aid available to part-time students. Financial aid application deadline: 2/28. *Unit head:* Dr. Yongmin Kim, Chair, 206-685-2000, Fax: 206-685-3300. *Application contact:* Dr. Gerald Pollack, Chair, Student Admissions Committee, 206-685-2021, Fax: 206-685-3300.

See in-depth description on page 469.

Vanderbilt University, School of Engineering, Department of Biomedical Engineering, Nashville, TN 37240-1001. Offers MS, PhD, MD/PhD. *Faculty:* 11 full-time (2 women). *Students:* 37 full-time (13 women); includes 5 minority (4 African Americans, 1 Asian American or Pacific Islander), 1 international. Average age 26. 63 applicants, 37% accepted. In 1998, 12 master's awarded (50% found work related to degree, 50% continued full-time study); 1 doctorate awarded. *Degree requirements:* For master's and doctorate, thesis/dissertation required, foreign language not required. *Entrance requirements:* For master's and doctorate, GRE General Test (minimum combined score of 1000 required; average 1250). *Average time to degree:* Master's–2 years full-time; doctorate–5 years full-time. *Application deadline:* For fall admission, 1/15; for spring admission, 11/1. Application fee: $40. *Financial aid:* In 1998–99, 19 fellowships with full tuition reimbursements, 14 research assistantships with full tuition reimbursements, 10 teaching assistantships with full tuition reimbursements were awarded.; institutionally-sponsored loans, scholarships, and tuition waivers (partial) also available. Aid available to part-time students. Financial aid application deadline: 1/15. *Faculty research:* Quantitative physiology, bio-optics, imaging, medical informatics and computing, medical instrumentation, physiological transport phenomena. Total annual research expenditures: $1.8 million. *Unit head:* Dr. Thomas R. Harris, Chair, 615-322-0842, Fax: 615-343-7919. *Application contact:* Dr. Robert J. Roselli, Director of Graduate Studies, 615-322-2602, Fax: 615-343-7919, E-mail: roselli@vuse.vanderbilt.edu.

See in-depth description on page 471.

Virginia Commonwealth University, School of Graduate Studies, School of Engineering, Department of Biomedical Engineering, Richmond, VA 23284-9005. Offers MS, PhD, MD/PhD. *Students:* 20 full-time (6 women), 3 part-time; includes 10 minority (all Asian Americans or Pacific Islanders) In 1998, 6 master's, 1 doctorate awarded. *Degree requirements:* For doctorate, dissertation, comprehensive oral and written exams required. *Entrance requirements:* For master's and doctorate, GRE General Test. *Application deadline:* For fall admission, 4/15. Application fee: $30. Tuition, state resident: full-time $4,031; part-time $224 per credit hour. Tuition, nonresident: full-time $11,946; part-time $664 per credit hour. Required fees: $1,081; $40 per credit hour. Tuition and fees vary according to campus/location and program. *Faculty research:* Clinical instrumentation, mathematical modeling, neurosciences, radiation physics and rehabilitation. *Unit head:* Dr. Gerald Miller, Chair, 804-828-7956, Fax: 804-828-4454, E-mail: gemiller@vcu.edu. *Application contact:* Dr. Jennifer S. Wayne, Graduate Program Director, 804-828-2595, Fax: 804-828-4454, E-mail: jswayne@vcu.edu.

See in-depth description on page 473.

Wake Forest University, School of Medicine, Graduate Programs in Medicine, Program in Medical Engineering, Winston-Salem, NC 27109. Offers PhD. *Faculty:* 8 full-time (0 women). *Students:* 8 full-time (1 woman), 1 part-time; includes 2 minority (1 African American, 1 Asian American or Pacific Islander), 2 international. Average age 26. 11 applicants, 45% accepted. *Degree requirements:* For doctorate, dissertation required, dissertation required. *Entrance requirements:* For doctorate, GRE General Test, GRE Subject Test. *Application deadline:* For

fall admission, 2/15 (priority date). Applications are processed on a rolling basis. Application fee: $25. Electronic applications accepted. *Financial aid:* In 1998–99, 4 fellowships, 4 research assistantships were awarded.; grants and tuition waivers (full) also available. Financial aid application deadline: 2/15. Total annual research expenditures: $57,478. *Unit head:* Dr. Pete Santago, Director, 336-716-2703.

Washington University in St. Louis, School of Engineering and Applied Science, Department of Biomedical Engineering, St. Louis, MO 63130-4899. Offers MS, D Sc. *Faculty:* 6 full-time (0 women), 78 part-time (2 women). *Students:* 27 full-time (8 women); includes 10 minority (all Asian Americans or Pacific Islanders), 4 international. Average age 23. 80 applicants, 26% accepted. In 1998, 1 degree awarded (100% found work related to degree). Terminal master's awarded for partial completion of doctoral program. *Degree requirements:* For master's, thesis optional; for doctorate, dissertation required. *Entrance requirements:* For master's, GRE, TOEFL (minimum score of 600 required), minimum GPA of 3.0; for doctorate, GRE General Test (minimum score of 2000 on 3 sections required), TOEFL (minimum score of 600 required), minimum GPA of 3.5. *Average time to degree:* Master's–2.5 years full-time. *Application deadline:* For fall admission, 2/1 (priority date). Application fee: $20. *Financial aid:* In 1998–99, 22 students received aid, including 5 fellowships with full tuition reimbursements available (averaging $15,000 per year), 13 research assistantships with full tuition reimbursements available (averaging $15,000 per year), 1 teaching assistantship with full tuition reimbursement available (averaging $15,000 per year); Federal Work-Study, institutionally-sponsored loans, traineeships, and tuition waivers (partial) also available. Financial aid application deadline: 2/1; financial aid applicants required to submit FAFSA. *Faculty research:* Biomedical and biological imaging, cardiovascular engineering, cell and tissue engineering, computational molecular biology, computational neuroscience. *Unit head:* Dr. Frank C. P. Yin, Chairman, 314-935-6164, Fax: 314-935-7448, E-mail: yin@biomed.wustl.edu. *Application contact:* Beverly Jane Spudlich, Department Secretary, 314-935-6164, Fax: 314-935-7448, E-mail: beej@biomed.wustl.edu.

See in-depth description on page 475.

Wayne State University, Graduate School, College of Engineering, Department of Biomedical Engineering, Detroit, MI 48202. Offers MS, PhD. *Degree requirements:* For master's, thesis optional, foreign language not required; for doctorate, dissertation required.

Worcester Polytechnic Institute, Graduate Studies, Department of Biomedical Engineering, Worcester, MA 01609-2280. Offers biomedical engineering (M Eng, MS, PhD, Certificate); clinical engineering (MS). Part-time and evening/weekend programs available. *Faculty:* 5 full-time (0 women), 2 part-time (1 woman). *Students:* 35 full-time (13 women), 8 part-time (3 women); includes 8 minority (1 African American, 5 Asian Americans or Pacific Islanders, 2 Hispanic Americans), 12 international. 87 applicants, 64% accepted. In 1998, 10 master's, 6 doctorates awarded. Terminal master's awarded for partial completion of doctoral program. *Degree requirements:* For master's, thesis optional, foreign language not required; for doctorate, dissertation required, foreign language not required. *Entrance requirements:* For master's and doctorate, GRE General Test (combined average 1837 on three sections), TOEFL (minimum score of 550 required; average 607). *Application deadline:* For fall admission, 2/15 (priority date); for spring admission, 10/15 (priority date). Applications are processed on a rolling basis. Application fee: $50. Electronic applications accepted. *Financial aid:* In 1998–99, 16 students received aid, including 11 fellowships with full tuition reimbursements available (averaging $14,940 per year), 5 research assistantships with full tuition reimbursements available (averaging $15,000 per year), 1 teaching assistantship with full tuition reimbursement available (averaging $11,970 per year); career-related internships or fieldwork, grants, institutionally-sponsored loans, and scholarships also available. Financial aid application deadline: 2/15; financial aid applicants required to submit FAFSA. *Faculty research:* Biomedical sensors, medical imaging, biomechanics/biomaterials, somatosensory system analysis, cardiac electrophysiology. Total annual research expenditures: $524,161. *Unit head:* Dr. Robert A. Peura, Head, 508-831-5447, Fax: 508-831-5541, E-mail: rapeura@wpi.edu.

See in-depth description on page 477.

Wright State University, School of Graduate Studies, College of Engineering and Computer Science, Programs in Engineering, Program in Biomedical and Human Factors Engineering, Dayton, OH 45435. Offers biomedical engineering (MSE); human factors engineering (MSE). Part-time programs available. *Students:* 22 full-time (7 women), 27 part-time (10 women); includes 8 minority (2 African Americans, 6 Asian Americans or Pacific Islanders), 7 international. Average age 27. 75 applicants, 69% accepted. In 1998, 18 degrees awarded. *Degree requirements:* For master's, thesis or course option alternative required. *Entrance requirements:* For master's, TOEFL (minimum score of 550 required). *Application deadline:* For fall admission, 5/30. Applications are processed on a rolling basis. Application fee: $25. *Financial aid:* Fellowships, research assistantships, teaching assistantships, Federal Work-Study, institutionally-sponsored loans, and unspecified assistantships available. Aid available to part-time students. Financial aid application deadline: 3/15; financial aid applicants required to submit FAFSA. *Faculty research:* Medical imaging, functional electrical stimulation, implantable aids, man-machine interfaces, expert systems. *Unit head:* Dr. Richard J. Koubek, Chair, 937-775-5044, Fax: 937-775-7364.

Biotechnology

Brown University, Graduate School, Division of Biology and Medicine, Program in Artificial Organs/Biomaterials/Cellular Technology, Providence, RI 02912. Offers MA, Sc M, PhD. *Students:* 12 full-time (5 women); includes 1 minority (Asian American or Pacific Islander), 2 international. 27 applicants, 15% accepted. In 1998, 2 degrees awarded. Terminal master's awarded for partial completion of doctoral program. *Degree requirements:* For doctorate, dissertation, preliminary exam required, foreign language not required. *Entrance requirements:* For master's and doctorate, GRE General Test, GRE Subject Test. *Application deadline:* For fall admission, 1/2 (priority date). Applications are processed on a rolling basis. Application fee: $60. *Financial aid:* In 1998–99, 4 fellowships, 1 research assistantship, 4 teaching assistantships were awarded. Financial aid application deadline: 1/2. *Unit head:* Dr. Edith Mathiowitz, Director, 401-863-3262.

Dartmouth College, Thayer School of Engineering, Program in Biotechnology and Biochemical Engineering, Hanover, NH 03755. Offers MS, PhD. *Degree requirements:* For master's, thesis required; for doctorate, dissertation, candidacy oral exam required. *Entrance requirements:* For master's and doctorate, GRE General Test. *Application deadline:* For fall admission, 1/15 (priority date). Application fee: $20 ($40 for international students). *Financial aid:* Fellowships, research assistantships, teaching assistantships, career-related internships or fieldwork, Federal Work-Study, institutionally-sponsored loans, and tuition waivers (full and partial) available. Financial aid application deadline: 1/15. *Faculty research:* Biomass processing, metabolic engineering, bioplastic synthesis, kinetics and reactor design, applied microbiology. Total annual research expenditures: $400,000. *Unit head:* Graduate Field Assistant, 607-255-1003, E-mail: biomedeng@cornell.edu. *Application contact:* Candace S. Potter, Admissions Coordinator, 603-646-3844, Fax: 603-646-3856, E-mail: candace.potter@dartmouth.edu.

East Carolina University, Graduate School, College of Arts and Sciences, Department of Biology, Greenville, NC 27858-4353. Offers biology (MS); molecular biology/biotechnology (MS). Part-time programs available. *Faculty:* 19 full-time (5 women). *Students:* 28 full-time (15

women), 56 part-time (26 women); includes 11 minority (6 African Americans, 3 Asian Americans or Pacific Islanders, 2 Native Americans), 1 international. *Degree requirements:* For master's, one foreign language (computer language can substitute), thesis, comprehensive exams required. *Entrance requirements:* For master's, GRE General Test, GRE Subject Test, TOEFL. *Application deadline:* For fall admission, 6/1 (priority date); for spring admission, 10/15. Applications are processed on a rolling basis. Application fee: $40. Tuition, state resident: full-time $1,012. Tuition, nonresident: full-time $8,578. Required fees: $1,006. Part-time tuition and fees vary according to course load. *Unit head:* Dr. Gerhard W. Kalmus, Director of Graduate Studies, 252-328-6722, Fax: 252-328-4178, E-mail: kalmusg@mail.ecu.edu. *Application contact:* Dr. Paul D. Tschetter, Senior Associate Dean, 252-328-6012, Fax: 252-328-6071, E-mail: grad@mail.ecu.edu.

Florida Institute of Technology, Graduate School, College of Science and Liberal Arts, Department of Biological Sciences, Program in Biotechnology, Melbourne, FL 32901-6975. Offers MS. Part-time programs available. *Students:* 2 full-time (1 woman), 1 (woman) part-time. Average age 28. 6 applicants, 83% accepted. In 1998, 2 degrees awarded. *Degree requirements:* For master's, internship required, foreign language and thesis not required. *Entrance requirements:* For master's, GRE General Test, minimum GPA of 3.0. *Application deadline:* Applications are processed on a rolling basis. Application fee: $50. Electronic applications accepted. Tuition: Part-time $575 per credit hour. Required fees: $100. Tuition and fees vary according to campus/location and program. *Financial aid:* Career-related internships or fieldwork available. Financial aid application deadline: 3/1; financial aid applicants required to submit FAFSA. *Faculty research:* Marine microbiology, drugs from the sea, natural products, drug discovery. *Unit head:* Candace S. Potter, Admissions Coordinator, 603-646-3844, Fax: 603-646-3856, E-mail: candace.potter@dartmouth.edu. *Application contact:* Carolyn P. Farrior, Associate Dean of Graduate Admissions, 407-674-7118, Fax: 407-723-9468, E-mail: cfarrior@fit.edu.

Biotechnology

Howard University, College of Medicine, Department of Biochemistry and Molecular Biology, Washington, DC 20059-0002. Offers biochemistry and molecular biology (PhD); biotechnology (MS). *Faculty:* 12 full-time (3 women), 5 part-time (3 women). *Students:* 9 full-time (5 women); all minorities (all African Americans) *Degree requirements:* For master's, externship required, foreign language and thesis not required; for doctorate, dissertation, oral and written comprehensive exams required, foreign language not required. *Entrance requirements:* For master's and doctorate, GRE General Test, minimum GPA of 3.0. *Application deadline:* For fall admission, 4/1 (priority date). Application fee: $45. *Unit head:* Dr. Matthew George, Interim Chair, 202-806-6289, Fax: 202-806-5784, E-mail: mgeorge@fac.howard.edu. *Application contact:* Dr. Cynthia K. Abrams, Director of Graduate Studies, 202-806-6289, Fax: 202-806-5784, E-mail: cabrams@fac.howard.edu.

Illinois Institute of Technology, Graduate College, Armour College of Engineering and Sciences, Department of Biological, Chemical and Physical Sciences, Biology Division, Chicago, IL 60616-3793. Offers biochemistry (MS); biology (PhD); biotechnology (MS); cell biology (MS); microbiology (MS). Part-time and evening/weekend programs available. *Faculty:* 8 full-time (0 women), 3 part-time (1 woman). *Students:* 26 full-time (7 women), 77 part-time (45 women); includes 27 minority (16 African Americans, 7 Asian Americans or Pacific Islanders, 4 Hispanic Americans), 33 international. Terminal master's awarded for partial completion of doctoral program. *Degree requirements:* For master's, thesis (for some programs), comprehensive exam required, foreign language not required; for doctorate, dissertation, comprehensive exam required, foreign language not required. *Entrance requirements:* For master's and doctorate, GRE (minimum score of 1200 required), TOEFL (minimum score of 550 required), undergraduate GPA of 3.0 required. *Application deadline:* For fall admission, 7/1; for spring admission, 11/1. Applications are processed on a rolling basis. Application fee: $30. Electronic applications accepted. *Unit head:* Dr. Benjamin Stark, Associate Chair, 312-567-3980, Fax: 312-567-3494, E-mail: bstark@charlie.iit.edu. *Application contact:* Dr. S. Mohammad Shahidehpour, Dean of Graduate College, 312-567-3024, Fax: 312-567-7517, E-mail: grad@minna.cns.iit. edu.

Instituto Tecnológico y de Estudios Superiores de Monterrey, Campus Monterrey, Graduate and Research Division, Program in Natural and Social Sciences, Monterrey, 64849, Mexico. Offers biotechnology (MS); chemistry (MS, PhD); communications (MS); education (MA). Part-time programs available. *Degree requirements:* For master's and doctorate, thesis/dissertation required. *Entrance requirements:* For master's, PAEG, TOEFL; for doctorate, PAEG, TOEFL, master's in related field. *Faculty research:* Cultural industries, mineral substances, bioremediation, food processing, CQ in industrial chemical processing.

Manhattan College, Graduate Division, School of Engineering, Program in Biotechnology, Riverdale, NY 10471. Offers MS. Offered jointly with the College of Mount Saint Vincent. Admissions temporarily suspended. *Degree requirements:* Foreign language not required. *Faculty research:* Tissue culture, protein structure, molecular biochemistry.

McGill University, Faculty of Graduate Studies and Research, Faculty of Agricultural and Environmental Sciences, Institute of Parasitology, Ste. Anne de Bellevue, PQ H9X 3V9, Canada. Offers biotechnology (Certificate); parasitology (M Sc, PhD). *Faculty:* 9 full-time (2 women), 4 part-time (1 woman). *Students:* 29 full-time (11 women), 12 international. Terminal master's awarded for partial completion of doctoral program. *Degree requirements:* For master's and doctorate, thesis/dissertation required; for Certificate, thesis not required. *Entrance requirements:* For master's, TOEFL (minimum score of 570 required), minimum GPA of 3.2; for doctorate, TOEFL (minimum score of 570 required), M Sc; for Certificate, TOEFL (minimum score of 550 required), minimum GPA of 3.0, B Sc in biological sciences. *Application deadline:* For fall admission, 1/1 (priority date); for winter admission, 5/1 (priority date); for spring admission, 9/1 (priority date). Applications are processed on a rolling basis. Application fee: $60. *Unit head:* Dr. Marilyn E. Scott, Director, 514-398-7722, Fax: 514-398-7857. *Application contact:* 514-398-7708, Fax: 514-398-7968, E-mail: grad@macdonald.mcgill.ca.

North Carolina State University, Graduate School, College of Management, Program in Management, Raleigh, NC 27695. Offers biotechnology (MS); computer science (MS); engineering (MS); forest resources management (MS); general business (MS); management information systems (MS); operations research (MS); statistics (MS); telecommunications systems engineering (MS); textile management (MS); total quality management (MS). Part-time programs available. *Faculty:* 40 full-time (9 women), 4 part-time (0 women). *Students:* 48 full-time (15 women), 156 part-time (43 women); includes 33 minority (16 African Americans, 15 Asian Americans or Pacific Islanders, 1 Hispanic American, 1 Native American), 4 international. *Degree requirements:* For master's, computer language required, foreign language and thesis not required. *Entrance requirements:* For master's, GRE or GMAT, TOEFL (minimum score of 550 required), minimum undergraduate GPA of 3.0. *Application deadline:* For fall admission, 6/25; for spring admission, 11/25. Applications are processed on a rolling basis. Application fee: $45. *Unit head:* Dr. Jack W. Wilson, Director of Graduate Programs, 919-515-4327, Fax: 919-515-6943, E-mail: jack_wilson@ncsu.edu. *Application contact:* Dr. Steven G. Allen, Director of Graduate Programs, 919-515-6941, Fax: 919-515-5073, E-mail: steve_allen@ncsu.edu.

Northwestern University, The Graduate School, Division of Interdepartmental Programs, Interdepartmental Biological Sciences Program (IBiS), Concentration in Biotechnology, Evanston, IL 60208. Offers PhD. *Degree requirements:* For doctorate, dissertation, 2 quarters of teaching experience required, foreign language not required. *Entrance requirements:* For doctorate, GRE General Test, TOEFL (minimum score of 600 required), TSE. *Application deadline:* For fall admission, 1/15. Applications are processed on a rolling basis. Application fee: $50 ($55 for international students). *Financial aid:* Fellowships, research assistantships, teaching assistantships, Federal Work-Study, institutionally-sponsored loans, and traineeships available. Financial aid application deadline: 1/15; financial aid applicants required to submit FAFSA. *Unit head:* Dr. Steven G. Allen, Director of Graduate Programs, 919-515-6941, Fax: 919-515-5073, E-mail: steve_allen@ncsu.edu. *Application contact:* Dr. Steven G. Allen, Director of Graduate Programs, 919-515-6941, Fax: 919-515-5073, E-mail: steve_allen@ncsu.edu.

Northwestern University, The Graduate School, Division of Interdepartmental Programs, Program in Biotechnology, Evanston, IL 60208. Offers MS. Admissions and degrees offered through The Graduate School. Part-time programs available. *Students:* 32 full-time (17 women); includes 6 minority (1 African American, 5 Asian Americans or Pacific Islanders), 4 international. In 1998, 31 degrees awarded. *Entrance requirements:* For master's, GRE or MCAT (strongly recommended), TOEFL (minimum score of 560 required), bachelor's degree in biology, chemistry, engineering, or related area. *Average time to degree:* Master's–1 year full-time, 2 years part-time. *Application deadline:* For fall admission, 8/7 (priority date); for spring admission, 2/28. Applications are processed on a rolling basis. Application fee: $50 ($55 for international students). *Financial aid:* Career-related internships or fieldwork and institutionally-sponsored loans available. Financial aid applicants required to submit FAFSA. *Faculty research:* Genetic engineering, cell biology/immunology, medicinal chemistry/bioinformatics, bioengineering. *Unit head:* Alicia Loffler, Director, 847-467-1453, E-mail: a-loffler@nwu.edu. *Application contact:* Jeanne Sheppard, Admission Contact, 847-467-1453, Fax: 847-467-2180, E-mail: j-sheppard@nwu.edu.

Salem-Teikyo University, Graduate School, Department of Bioscience, Salem, WV 26426-0500. Offers biotechnology/molecular biology (MS). *Faculty:* 5 full-time (2 women). *Students:* 4 full-time (2 women); includes 2 minority (1 African American, 1 Asian American or Pacific Islander), 1 international. *Degree requirements:* For master's, thesis required, foreign language not required. *Entrance requirements:* For master's, GRE, minimum undergraduate GPA of 3.0. *Application deadline:* Applications are processed on a rolling basis. Application fee: $25. Electronic applications accepted. Tuition: Full-time $10,000; part-time $165 per credit hour. Required fees: $55; $55 per year. *Unit head:* Dr. Patrick Lai, Chair, 304-782-5575, Fax: 304-782-5579, E-mail: lai@salem.wvnet.edu. *Application contact:* Carolyn Sue Ritter, Director of Admissions, 304-782-5336, Fax: 304-782-5592, E-mail: admissions@stunix.salem-teikyo. wvnet.edu.

Stephen F. Austin State University, Graduate School, College of Sciences and Mathematics, Department of Chemistry, Program in Biotechnology, Nacogdoches, TX 75962. Offers MS. *Students:* 13 full-time (6 women), 9 part-time (6 women); includes 7 minority (1 African American, 3 Asian Americans or Pacific Islanders, 3 Native Americans), 1 international. 9 applicants, 78% accepted. *Degree requirements:* For master's, comprehensive exam required, foreign language and thesis not required. *Entrance requirements:* For master's, GRE General Test (minimum combined score of 1000 required), TOEFL, minimum GPA of 2.8 in last 60 hours, 2.5 overall. *Application deadline:* For fall admission, 8/1 (priority date); for spring admission, 12/15. Applications are processed on a rolling basis. Application fee: $0 ($50 for international students). Tuition, state resident: full-time $1,792. Tuition, nonresident: full-time $6,880. *Financial aid:* In 1998–99, research assistantships (averaging $6,750 per year) Financial aid application deadline: 3/1. *Unit head:* Dr. Beatrice Clack, Director, 409-468-3606.

Thomas Jefferson University, College of Graduate Studies, Program in Cell and Tissue Engineering, Philadelphia, PA 19107. Offers PhD. *Faculty:* 4 full-time (1 woman). *Students:* 1 full-time (0 women). *Degree requirements:* For doctorate, dissertation required, foreign language not required. *Entrance requirements:* For doctorate, TOEFL (minimum score of 550 required); GRE General Test (minimum combined score of 1100 required), minimum GPA of 3.2. *Application deadline:* Applications are processed on a rolling basis. Application fee: $40. Tuition: Full-time $12,670. Tuition and fees vary according to degree level and program. *Financial aid:* In 1998–99, 1 fellowship with full tuition reimbursement was awarded. Financial aid application deadline: 5/1. *Unit head:* Dr. Rocky S. Tuan, Director, 215-955-5479, Fax: 215-955-9159. *Application contact:* Jessie F. Pervall, Director of Admissions, 215-503-4400, Fax: 215-503-3433, E-mail: cgs-info@mail.tju.edu.

See in-depth description on page 421.

Tufts University, Division of Graduate and Continuing Studies and Research, Professional and Continuing Studies, Biotechnology Engineering Program, Medford, MA 02155. Offers Certificate. Part-time and evening/weekend programs available. Average age 27. 1 applicants, 100% accepted. In 1998, 1 degree awarded. *Average time to degree:* 1 year part-time. *Application deadline:* For fall admission, 8/15 (priority date); for spring admission, 12/12. Applications are processed on a rolling basis. Application fee: $40. *Financial aid:* Available to part-time students. Application deadline: 5/1; *Unit head:* Jessie F. Pervall, Director of Admissions, 215-503-4400, Fax: 215-503-3433, E-mail: cgs-info@mail.tju.edu. *Application contact:* Jessie F. Pervall, Director of Admissions, 215-503-4400, Fax: 215-503-3433, E-mail: cgs-info@mail.tju.edu.

Tufts University, Division of Graduate and Continuing Studies and Research, Professional and Continuing Studies, Biotechnology Program, Medford, MA 02155. Offers Certificate. Part-time and evening/weekend programs available. Average age 25. 0 applicants, 100% accepted. In 1998, 1 degree awarded. *Average time to degree:* 1 year part-time. *Application deadline:* For fall admission, 8/15 (priority date); for spring admission, 12/12. Applications are processed on a rolling basis. Application fee: $40. *Financial aid:* Available to part-time students. Application deadline: 5/1; *Unit head:* Jeanne Sheppard, Admission Contact, 847-467-1453, Fax: 847-467-2180, E-mail: j-sheppard@nwu.edu. *Application contact:* Jeanne Sheppard, Admission Contact, 847-467-1453, Fax: 847-467-2180, E-mail: j-sheppard@nwu.edu.

University of Alberta, Faculty of Graduate Studies and Research, Department of Biological Sciences, Edmonton, AB T6G 2E1, Canada. Offers environmental biology and ecology (M Sc, PhD); microbiology and biotechnology (M Sc, PhD); molecular biology and genetics (M Sc, PhD); physiology and cell biology (M Sc, PhD); systematics and evolution (M Sc, PhD). Terminal master's awarded for partial completion of doctoral program. *Degree requirements:* For master's and doctorate, thesis/dissertation required. *Entrance requirements:* For master's and doctorate, TOEFL (minimum score of 600 required).

University of Connecticut, Graduate School, College of Liberal Arts and Sciences, Biological Sciences Group, Storrs, CT 06269. Offers ecology and evolutionary biology (MS, PhD), including botany, ecology, entomology, systematics, zoology; molecular and cell biology (MS, PhD), including biochemistry, biophysics, biotechnology (MS), cell and developmental biology, genetics, microbiology, plant molecular and cell biology; physiology and neurobiology (MS, PhD), including neurobiology, physiology. *Degree requirements:* For doctorate, dissertation required. *Entrance requirements:* For master's and doctorate, GRE General Test, GRE Subject Test, TOEFL.

University of Connecticut, Graduate School, College of Liberal Arts and Sciences, Biological Sciences Group, Department of Molecular and Cell Biology, Field of Biotechnology, Storrs, CT 06269. Offers MS. *Entrance requirements:* For master's, GRE General Test, GRE Subject Test, TOEFL.

Announcement: The MS in biotechnology is administered by the Department of Molecular and Cell Biology and uses the facilities of the Biotechnology Center, with participation by the College of Agriculture and Natural Resources (animal sciences, nutritional sciences, pathobiology, plant sciences, environmental health), the School of Pharmacy, and the Departments of Chemical Engineering, Chemistry, Ecology and Evolutionary Biology, and Physiology and Neurobiology. Biotechnology students may enter PhD programs in any of these fields. Prospective students should contact Dr. Robert Vinopal, 860-486-4886.

University of Delaware, Delaware Biotechnology Institute, Newark, DE 19716. Offers PhD.

University of Massachusetts Boston, Graduate Studies, College of Arts and Sciences, Faculty of Sciences, Program in Biotechnology and Biomedical Science, Boston, MA 02125-3393. Offers MS. *Degree requirements:* For master's, thesis, comprehensive exams required, foreign language not required. *Entrance requirements:* For master's, GRE General Test, GRE Subject Test, minimum GPA of 2.75.

University of Massachusetts Lowell, Graduate School, College of Arts and Sciences, Department of Biological Sciences, Lowell, MA 01854-2881. Offers biochemistry (PhD); biological sciences (MS); biotechnology (MS). Part-time programs available. *Faculty:* 12 full-time (3 women). *Students:* 17 full-time (8 women), 45 part-time (30 women); includes 9 minority (2 African Americans, 5 Asian Americans or Pacific Islanders, 2 Hispanic Americans), 9 international. *Degree requirements:* For master's, thesis required, foreign language not required; for doctorate, computer language, dissertation required. *Entrance requirements:* For master's and doctorate, GRE General Test. *Application deadline:* For fall admission, 4/1 (priority date); for spring admission, 10/1. Applications are processed on a rolling basis. Application fee: $20 ($35 for international students). *Unit head:* Dr. Robert Lynch, Chair, 978-934-2891, E-mail: robert_lynch@woods.uml.edu. *Application contact:* Dr. Ilze Skare, Coordinator, 978-934-2885, E-mail: ilze_skare@woods.uml.edu.

University of Minnesota, Twin Cities Campus, Medical School, Graduate Programs in Medicine, Program in Microbial Engineering, Minneapolis, MN 55455-0213. Offers MS. Part-time programs available. *Degree requirements:* For master's, computer language, thesis or alternative required, foreign language not required. *Entrance requirements:* For master's, GRE General Test (combined average 1800 on three sections), TOEFL (minimum score of 550 required; average 600). *Faculty research:* Microbial genetics, oncogenesis, gene transfer, fermentation, bioreactors, genetics of antibiotic biosynthesis.

See in-depth description on page 451.

University of Missouri–St. Louis, Graduate School, College of Arts and Sciences, Department of Biology, St. Louis, MO 63121-4499. Offers biology (MS, PhD), including animal behavior (MS), biochemistry, biotechnology (MS), conservation biology (MS), development (MS), ecology (MS), environmental studies (PhD), evolution (MS), genetics (MS), molecular biology and biotechnology (PhD), molecular/cellular biology (MS), physiology (MS), plant systematics, population biology (MS), tropical biology (MS); biotechnology (Certificate); tropical biology and conservation (Certificate). Part-time programs available. *Faculty:* 47. *Students:* 20 full-time (12 women), 73 part-time (47 women); includes 18 minority (3 African Americans, 4 Asian Americans or Pacific Islanders, 11 Hispanic Americans), 22 international. *Degree requirements:* For master's, thesis or alternative required, foreign language not required; for doctorate, one foreign language, dissertation, 1 semester of teaching experience required. *Entrance requirements:* For doctorate, GRE General Test. *Application deadline:* For fall admission, 7/1 (priority date); for spring admission, 11/1 (priority date). Applications are processed on a rolling basis. Application fee: $25 ($40 for international students). Electronic applications accepted. *Unit head:* Director of Graduate Studies, 314-516-6203, Fax: 314-516-6233, E-mail: icte@umsl.edu. *Application contact:* Graduate Admissions, 314-516-5458, Fax: 314-516-6759, E-mail: gradadm@umsl.edu.

University of Pennsylvania, School of Engineering and Applied Science, Program in Biotechnology, Philadelphia, PA 19104. Offers MS. Part-time programs available. *Entrance requirements:* For master's, GRE General Test, TOEFL (minimum score of 600 required), bachelor's degree in science or undergraduate course work in molecular biology. Electronic applications accepted.

See in-depth description on page 457.

University of Tennessee, Knoxville, Graduate School, College of Arts and Sciences, Program in Life Sciences, Knoxville, TN 37996. Offers biotechnology (MS); plant physiology and genetics (MS, PhD). *Students:* 7 full-time (3 women), 6 part-time (3 women); includes 1 minority (Hispanic American), 4 international. *Degree requirements:* For master's, thesis required (for some programs), foreign language not required; for doctorate, dissertation required. *Entrance requirements:* For master's and doctorate, GRE General Test, TOEFL (minimum score of 550 required), minimum GPA of 2.7. *Application deadline:* For fall admission, 2/1 (priority date). Applications are processed on a rolling basis. Application fee: $35. Electronic applications accepted. *Unit head:* Dr. Frank Harris, Chairperson, 423-974-6841, Fax: 423-974-4057, E-mail: rmattingly@utk.edu.

The University of Texas at San Antonio, College of Sciences and Engineering, Division of Life Sciences, Programs in Biology and Biotechnology, San Antonio, TX 78249-0617. Offers biology (MS); biotechnology (MS). Part-time programs available. *Degree requirements:* For master's, comprehensive exam required, thesis optional. *Entrance requirements:* For master's, GRE General Test, minimum GPA of 3.0. *Faculty research:* Plant ecology, bioremediation, neurophysiology, neurotoxicology, neuroendocrinology, neural circuit analysis.

University of the Sciences in Philadelphia, College of Graduate Studies, Program in Cell Biology and Biotechnology, Philadelphia, PA 19104-4495. Offers MS. *Degree requirements:* For master's, thesis required (for some programs), foreign language not required. *Entrance requirements:* For master's, GRE General Test (minimum combined score of 1500 on three sections required), TOEFL (minimum score of 550 required). *Application deadline:* For fall admission, 5/1; for spring admission, 10/1. Applications are processed on a rolling basis. Application fee: $30. *Financial aid:* Application deadline: 5/1. *Unit head:* Dr. John Porter, Director, 215-596-8917, E-mail: j.porter@usip.edu.

University of Washington, School of Medicine, Graduate Programs in Medicine, Department of Molecular Biotechnology, Seattle, WA 98195. Offers PhD. *Faculty:* 9 full-time (3 women), 5 part-time (2 women). *Students:* 13 full-time (5 women); includes 1 minority (Asian American or Pacific Islander), 1 international. 37 applicants, 19% accepted. In 1998, 1 degree awarded (0% continued full-time study). *Degree requirements:* For doctorate, computer language, disserta-

tion, qualifying, general and final exams required. *Entrance requirements:* For doctorate, GRE General Test, GRE Subject Test, TOEFL, minimum GPA of 3.0. *Average time to degree:* Doctorate–3.5 years full-time. *Application deadline:* For fall admission, 1/15. Application fee: $45. Electronic applications accepted. Tuition, state resident: full-time $5,196; part-time $475 per credit. Tuition, nonresident: full-time $13,485; part-time $1,285 per credit. Required fees: $387; $38 per credit. Tuition and fees vary according to course load. *Financial aid:* In 1998–99, 13 students received aid, including 12 fellowships with full tuition reimbursements available (averaging $16,800 per year), 1 research assistantship with full tuition reimbursement available (averaging $12,600 per year); grants, traineeships, and tuition waivers (full) also available. Financial aid application deadline: 2/28; financial aid applicants required to submit FAFSA. *Faculty research:* Technologies for genome analysis, molecular cytogenetics, protein sequencing, mass spectrometry, molecular immunology, cellular activation. Total annual research expenditures: $5 million. *Unit head:* Dr. Leroy E. Hood, Chairman, 206-685-7367. *Application contact:* Colbey Harris, Program Manager, 206-616-7297, Fax: 206-221-5661, E-mail: gradprog@u.washington.edu.

William Paterson University of New Jersey, College of Science and Health, Department of Biology, Program in Biotechnology, Wayne, NJ 07470-8420. Offers MS. Part-time and evening/weekend programs available. *Degree requirements:* For master's, exit exam required. *Entrance requirements:* For master's, GRE General Test, GRE Subject Test (biology), minimum GPA of 2.75. *Faculty research:* DNA cloning, genetic engineering, planttissue culture, molecular genetics, *Drosophila* gene expression.

Worcester Polytechnic Institute, Graduate Studies, Department of Biology and Biotechnology, Worcester, MA 01609-2280. Offers biology (MS); biomedical sciences (PhD); biotechnology (MS, PhD). *Faculty:* 11 full-time (4 women). *Students:* 25 full-time (13 women), 2 part-time (1 woman); includes 1 minority (Hispanic American), 3 international. 40 applicants, 35% accepted. In 1998, 18 master's, 1 doctorate awarded. *Degree requirements:* For master's, thesis required, foreign language not required; for doctorate, dissertation, qualifying exam required, foreign language not required. *Entrance requirements:* For master's and doctorate, GRE General Test (combined average 1963 on three sections), TOEFL (minimum score of 550 required; average 640). *Application deadline:* For fall admission, 2/15 (priority date); for spring admission, 10/15 (priority date). Applications are processed on a rolling basis. Application fee: $50. Electronic applications accepted. *Financial aid:* In 1998–99, 27 students received aid, including 7 fellowships with full tuition reimbursements available (averaging $14,400 per year), 9 research assistantships with full tuition reimbursements available (averaging $15,000 per year), 11 teaching assistantships with full tuition reimbursements available (averaging $11,970 per year); career-related internships or fieldwork, grants, institutionally-sponsored loans, scholarships, and tuition waivers (full and partial) also available. Financial aid application deadline: 2/15; financial aid applicants required to submit FAFSA. *Faculty research:* Genetic engineering, microbial genetics, immunology, DNA technology, fermentation genetics, pharmaceutical production. Total annual research expenditures: $465,825. *Unit head:* Dr. Ronald D. Cheetham, Head, 508-831-5582, Fax: 508-831-5936, E-mail: cheetham@wpi.edu.

Worcester State College, Graduate Studies, Program in Biotechnology, Worcester, MA 01602-2597. Offers MS. Part-time and evening/weekend programs available. *Degree requirements:* For master's, one foreign language (computer language can substitute), thesis or alternative, oral comprehensive exam required. *Entrance requirements:* For master's, GRE General Test or MAT, minimum undergraduate QPA of 3.0 in biology. *Faculty research:* Effects of insulin in invertebrates, ecology of freshwater turtles, symbiotic relations of plants and animals.

Cross-Discipline Announcements

Case Western Reserve University, School of Medicine, Graduate Programs in Medicine, Department of Physiology and Biophysics, Cleveland, OH 44106.

The Departments of Biomedical Engineering and Physiology and Biophysics offer a joint program in Biophysics/Bioengineering with areas of concentration including molecular biophysics, biological imaging, cellular electrophysiology, modeling of biological processes, membrane biophysics, and biosensors. A complete description of this multidisciplinary graduate program is located in Book 3, Biological Sciences, Section 17: Physiology.

Dartmouth College, Thayer School of Engineering, Hanover, NH 03755.

Thayer School offers MS and PhD programs emphasizing biotechnology and biochemical engineering. Active interdisciplinary collaborations with Departments of Chemistry and Biology and the medical school provide opportunities for study and research in environmental, biomolecular, and bioprocess engineering. See in-depth description in this volume.

Drexel University, Graduate School, School of Biomedical Engineering, Science and Health Systems, Philadelphia, PA 19104-2875.

Graduate programs in biomedical engineering and biomedical science at Drexel University offer several cross-disciplinary specialization areas, including biomedical ultrasound and biomedical optics, biosensing, biomedical imaging, biomedical signal processing, biomechanics, biomaterials, tissue and cellular engineering, human performance and neuroengineering, medical informatics, computational biomedicine, genome sciences and bioinformatics, systems physiology, and biostatistics.

Georgetown University, Graduate School, Programs in Biomedical Sciences, Department of Biochemistry and Molecular Biology, Washington, DC 20057.

The department has 2 unique programs in biotechnology that provide opportunities for full- and part-time students. The MS degree in biochemistry and molecular biology, with specialization in biotechnology, provides a comprehensive background suitable for the career development of students interested in biotechnology, medicine, and the biomedical sciences. The Certificate Program is divided between theory and methods. The Certificate Program is for individuals who have interest in pursuing careers in the biotechnology industry. Participants usually have experience in a variety of allied areas, such as the sciences, business, or law. See in-depth description in the Biochemistry section in Book 3 of this series.

Harvard University, Graduate School of Arts and Sciences, Program in Biological and Biomedical Sciences, Boston, MA 02115.

Established program leading to doctorate in biological and biomedical sciences offers students a wide range of research opportunities through laboratories at Harvard Medical School and affiliated research institutes and hospitals. Interdisciplinary structure leads to breadth of research

activities. Program allows development of individual interests while encouraging students to establish a firm grasp of modern biology. For other related programs at Harvard University, students should see the index.

Texas A&M University, College of Medicine, Graduate Program in Medical Sciences, Department of Medical Physiology, College Station, TX 77843.

The department is part of an interdisciplinary doctoral program in basic medical science, preparing individuals for academic careers in research and teaching. Departmental research focuses on cardiovascular science, ranging in scope from molecular biology to systems integration. Specific areas of expertise include microcirculation, vascular cell biology/electrophysiology, angiogenesis, coronary circulation, lymphatics, physiological imaging, and integrative cardiovascular physiology. The college program previously had about 50 students, 95% of whom had fellowships or assistantships. Students with training in engineering or biological or physical sciences are encouraged to apply. See in-depth description in the Physiology section of Book 3.

University of California, Los Angeles, School of Medicine, Graduate Programs in Medicine, Department of Biomathematics, Los Angeles, CA 90095.

University of California, Los Angeles Department of Biomathematics offers a graduate program leading to the PhD degree that trains creative, fully independent investigators who can initiate research in mathematical biology/applied mathematics and their chosen biomedical specialty, including genetics, molecular biology, neurosciences, physiology, pharmacology, and immunology. See the in-depth description in Book 4, Section 7 of these guides.

University of California, San Francisco, School of Pharmacy, Department of Pharmaceutical Chemistry, San Francisco, CA 94143.

The PhD degree program in pharmaceutical chemistry is directed toward research at the interface between chemistry and biology. The program offers research in areas of bioorganic chemistry, macromolecular structure and function, computer-aided drug design, medicinal chemistry, drug metabolism and biochemical toxicology, drug delivery, molecular parasitology, molecular pharmacology, and pharmacokinetics/pharmacodynamics.

University of Delaware, Delaware Biotechnology Institute, Newark, DE 19716.

A cross-disciplinary Institute specializing in plant biology, cell and molecular biology, protein biochemistry, biologically-oriented engineering, and bioinformatics. As part of the Institute's graduate program it offers a unique PhD training in plant molecular biology and related skills, including biotechnology, in collaboration with DuPont. One rotation at the University, one in a DuPont lab. Student can choose to complete PhD thesis at either institution. Information and application available online (http://www.udel.edu/plants/index.html). Biotechnology combined admissions program under development. Application also possible through chemical engineer-

Cross-Discipline Announcements

University of Delaware (continued)
ing, electrical engineering, biology, chemistry, plant and soil sciences, animal science, and marine science.

University of Pittsburgh, School of Health and Rehabilitation Sciences, Department of Rehabilitation Science and Technology, Pittsburgh, PA 15260.

Students may pursue an MS or PhD in bioengineering or rehabilitation science. Certificates are available in rehabilitation engineering and assistive technology service delivery. The department offers strong research, clinical, and academic programs. Students come from a variety of academic and professional disciplines. Specialized tracks are available. See in-depth description in Book 6 of this series.

Virginia Polytechnic Institute and State University, Center for Biomedical Engineering, Blacksburg, VA 24061.

Virginia Tech offers opportunities for interdisciplinary graduate study through the Center for Biomedical Engineering. Led by a core group of faculty members, students may pursue programs leading to the MS and PhD in chemical engineering, electrical and computer engineering, engineering science and mechanics, industrial and systems engineering, materials science and engineering, and mechanical engineering. A graduate certificate in biomedical engineering is offered in conjunction with these degrees. For more information, contact Cathy Hill, Graduate Program Advisor (telephone: 540-231-7460; e-mail: bmegrad@vt.edu).

ARIZONA STATE UNIVERSITY

Department of Chemical, Bio and Materials Engineering
Bioengineering Program

Programs of Study

The Bioengineering Program at Arizona State University (ASU) is in the Department of Chemical, Bio and Materials Engineering (BME) and offers graduate programs that lead to the Master of Science and Doctor of Philosophy degrees in bioengineering. Established in 1966, the BME program at ASU was one of the early pioneers in biomedical engineering education and research. The BME program awards M.S. and Ph.D. degrees in bioengineering that are highly flexible and offers students a broad selection of research topics. In addition, a graduate transition program in bioengineering is available to non-bioengineering majors, including those with backgrounds in the life and physical sciences.

Seven full-time tenure-track faculty members, 2 research faculty members, and 19 associated and adjunct faculty members administer the graduate bioengineering program. Some research projects are in collaboration with medical researchers in the Phoenix area. The graduate BME program is broadly based, with core research activities in biocontrols (optimal strategies for human movement, gait analysis, anthrorobotic/neuroprosthetic systems), bioinstrumentation (bioelectricity, bioelectronic devices, medical diagnostic and therapeutic instrumentation, implantable bioelectric devices), neuromuscular stimulation (cardiac assistance, electrophysiology, applied neural control, neural protheses), neuroscience (auditory neurophysiology, neural modeling), biomaterials (tissue replacement, surface modification, biocompatibility), biomechanics (rehabilitation engineering, musculoskeletal modeling), biosystems/biotransport engineering (artificial organs, cardiovascular engineering, bioseparations), and molecular and cellular engineering (hybrid molecular and cellular-based devices, tissue engineering).

The Master of Science degree program requires a minimum of 30 credit hours, including 6 hours of thesis credit. All students are required to enroll in a 1-credit seminar during each semester in residence. The Doctor of Philosophy degree program requires a minimum of 84 hours beyond the bachelor's degree. To satisfy the residency requirements, the students must spend a minimum of two semesters of full-time study on campus. All Ph.D. candidates must take oral and written qualifying exams during their first year in the program. These exams cover mathematical, life sciences, and engineering fundamentals. Students are admitted to candidacy after passing a comprehensive exam, which is given near the completion of course work. The exam consists of an oral and written dissertation prospectus. Students with a B.S. degree in a field other than bioengineering are encouraged to apply. The Graduate Committee determines, on an individual basis, which undergraduate courses must be taken to ensure success in the graduate program.

Research Facilities

With approximately 15,000 square feet of modern and specialized in vitro and in vivo laboratory space, the bioengineering facilities at Arizona State University are among the best among engineering colleges in the United States at providing the capability to carry a research project from concept to preclinical trials. Core faculty research laboratories include artificial organs, biocontrols, bioinstrumentation and measurements, biomaterials, experimental and computational biomechanics, biotechnology, cardiovascular engineering, cellular and tissue engineering, neuroengineering, neurosensory, and neurostimulation. In addition, the Whitaker Center for Neuromechanical Control is a multidisciplinary joint research facility between bioengineering faculty members and neuroscientists at the Barrow Neurological Institute. Associated bioengineering research facilities include the Science and Engineering Materials facility for multipurpose synthesis and characterization and the multidisciplinary Scanning Tunneling and Atomic Force Microscopy facility.

Engineering Computer Services has a staff of 35 full-time computer specialists who support the wide array of computers and electronic equipment within the College. The College also maintains complete machine, carpentry, electrical, and paint shops to support graduate research. The University libraries house more than 2 million volumes, 1.7 million microfilms, and 24,000 journal titles. The Noble Science and Engineering Library, a designated patent depository, houses the entire U.S. patent collection.

Financial Aid

A wide variety of financial support is available for graduate students, including teaching and research assistantships and University and industrial fellowships. Out-of-state tuition is waived for all graduate assistants and fellows. In addition, there are a number of Graduate Tuition Scholarships (GTS) that cover out-of-state tuition and Graduate Academic Scholarships (GAS) that cover in-state registration fees available on a highly competitive basis.

Cost of Study

In 1998–99, the registration and tuition fee for 12 hours or more was $1044 per semester for Arizona residents; nonresidents paid an additional $3476 (prorated for fewer than 12 hours).

Living and Housing Costs

Limited on-campus housing is available for unmarried students. In 1998–99, room rates for the academic year ranged from $2180 to $3805. Thirteen different meal plans are available (per week or semester). Moderately priced apartments are available within walking distance of the campus.

Student Group

There are more than 47,000 students at Arizona State, including more than 11,000 graduate students. More than 1,800 of the 6,300 students in the College of Engineering and Applied Sciences are enrolled in graduate programs. In fall 1998, there were 140 graduate students in the department, with 47 of them enrolled in the bioengineering program.

Location

Tempe is a suburb that borders the city of Phoenix. The metropolitan area is the population, economic, and industrial center of the state of Arizona. Entertainment centers, art and anthropology museums, and sports arenas are among the wide variety of facilities available in the Valley of the Sun. The area's mild climate also provides for numerous year-round outdoor activities. The urban community surrounding ASU thrives on high-technology industries and is one of the fastest-growing communities in the country.

The University

Arizona State University traces its origin to 1885 and is the largest and oldest institution of higher learning in the state of Arizona. Its present enrollment of 47,000 places it as the fifth-largest central-campus university in the United States. ASU's campus comprises 700 acres and offers outstanding physical facilities to support the University's research and educational mission. Included within the more than 125 buildings are twelve colleges and schools, a University-wide computer system, seven libraries (including the Noble Science and Engineering Library), and more than two dozen specialized centers of research. ASU was recently awarded Research I status as a result of its past and continuing commitment to academic research.

Applying

All students must apply for admission through the Graduate College. For application forms, students should contact Graduate Admissions, Arizona State University, Tempe, Arizona 85287-1003 or call 602-965-6113. Submission of Graduate Record Examinations (GRE) scores and a statement of purpose is required. International students whose native language is not English must submit a Test of English as a Foreign Language (TOEFL) score of 580 or better. Applications are reviewed continuously, but for consideration for financial support, should be received by February 1 for the following academic year.

Correspondence and Information

Chair of Graduate Committee
Chemical, Bio and Materials Engineering
Box 876006
Arizona State University
Tempe, Arizona 85287-6006

Telephone: 602-965-3313
E-mail: cbmerec@asuvax.eas.asu.edu

Arizona State University

THE FACULTY AND THEIR RESEARCH

Bioengineering

Eric J. Guilbeau, Ph.D., Louisiana Tech. Biomedical engineering, biomaterials, skeletal muscle cardiac assist, biological transport phenomena, development of biosensors, physiological systems analysis and simulation, artificial internal organs.

Jiping He, Ph.D., Maryland. Biomechanics, robotics, computational neuroscience, optimal control, system dynamics and control.

Daryl R. Kipke, Ph.D., Michigan. Computational neuroscience, auditory neurophysiology, electrical stimulation, speech recognition, signal processing, biomedical instrumentation, neural modeling using parallel computers and analog VLSI technology.

Steve Massia, Ph.D, Texas at Austin. Biomaterials, biomimetic surface modifications, cell-biomaterial interactions, vascular implants, tissue engineering, local drug delivery, cell-extracellular matrix interactions.

Alyssa Panitch, Ph.D., Massachusetts Amherst. Bioorganic polymer chemistry, biomaterials, cell-material interactions, vascular grafts, tissue engineering.

Vincent B. Pizziconi, Ph.D., Arizona State. Biomedical engineering, molecular and cellular engineering/biological engineering/tissue engineering, artificial organs, biomaterials, biosensors, biotechnology, bioseparations, scanning tunneling microscopy (STM) and atomic force microscopy (AFM) of natural and synthetic biomaterials, fractals/nonlinear dynamical systems.

James D. Sweeney, Ph.D., Case Western Reserve. Biomedical engineering, rehabilitation engineering, applied neural control, neurophysiology, motor physiology, mathematical modeling, cardio assistance, cardiac defibrillation, medical devices.

Bruce C. Towe, Ph.D., Penn State. Bioinstrumentation, implantable microelectronic devices, medical ultrasound, bioelectric phenomena, cardiac defibrillation.

Gary T. Yamaguchi, Ph.D., Stanford. Biomechanics and rehabilitation engineering design, including joint mechanics; computer modeling of muscle, tendon, and joints; optimal control and dynamic analysis/simulation of movement, coordination, and functional neuromuscular stimulation.

Materials Science Engineering

Terry L. Alford, Ph.D., Cornell. Microelectronic metallization and reliability, silicide formation, ion-beam modification of materials.

Ray W. Carpenter, Ph.D., Berkeley. Atomic structure and chemistry of interfaces and boundaries in solids, phase transformation mechanisms in metals and ceramics, electron microscopy methods and instrumentation.

Sandwip K. Dey, Ph.D., Alfred, College of Ceramics. Thin-film processing science of electroceramics; characterization of electrical, microstructural, and microchemical properties; high-permittivity dielectrics for ULSI DRAMs and microelectronic packages.

Lester E. Hendrickson (Emeritus), Ph.D., Illinois. Corrosion, fracture and failure analysis, physical and chemical metallurgy.

Dean L. Jacobson, Ph.D., UCLA. Thermionic energy conversion, high-temperature materials, erosion, heat pipes, thermodynamics.

Stephen L. Krause, Ph.D., Michigan. Ordered polymers, composite materials, electronic materials, X-ray diffraction, electron X-ray diffraction, electron microscopy.

James W. Mayer, Ph.D., Purdue. Electronic materials and metallization of integrated circuits; development of new semiconductor materials, such as the ternary alloy SiGeC grown on silicon; development of new metal systems for interconnectors; interdiffusion and reactions in thin films; analysis of paint pigments, art media, and metallic artifacts; ion-beam analysis and Rutherford Backscattering analysis.

James T. Stanley (Emeritus), Ph.D., Illinois. Phase transformations, radiation effects on materials, erosion-corrosion.

Chemical Engineering

Stephen P. Beaudoin, Ph.D., North Carolina State. Transport phenomena; surface science concerning pollution prevention, waste minimization, and air pollution remediation.

James R. Beckman, Ph.D., Arizona. Unit operations, applied mathematics, crystallization control and nucleation, process simulation, data analysis, energy storage, solar cooling.

Lynn Bellamy, Ph.D., Tulane. Cognitive science, learning theory, learning style preferences, constructivist learning, classroom assessment, quality management principles, distancing theory (and representational competence), team dynamics, active learning and other classroom management strategies, curriculum (or course) development, design, specification, and assessment.

Neil S. Berman, Ph.D., Texas at Austin. Regional air pollution and global warming physical and numerical models, turbulent mixing experiments and calculations in atmospheric boundary layers and in liquid flows with polymeric additives, experimental measurements using laser-induced fluorescence, transformations and deposition of air toxic materials.

Veronica A. Burrows, Ph.D., Princeton. Surface science; environmental sensors; semiconductor processing; interfacial chemical and physical processes in sensor processing, lubrication, and composite materials.

Antonio A. Garcia, Ph.D., Berkeley. Protein purification, acid-base molecular interactions in separations, solid-liquid interfacial phenomena, scanning probe microscopy, biocolloid chemistry, chromatography, biosensor immobilization.

James L. Kuester, Ph.D., Texas A&M. Chemical reactor analysis, thermochemical conversion processes, complex reaction systems, catalytic processes, process instrumentation and control, optimization, applied statistics, applied mathematics.

Gregory B. Raupp, Ph.D., Wisconsin. Gas-solid surface reaction mechanisms and kinetics, interaction between surface reactions and simultaneous transport processes, semiconductor materials processing, thermal and plasma-enhanced chemical vapor deposition (CVD), environmental pollution remediation and control, photocatalytic oxidation.

Daniel E. Rivera, Ph.D., Caltech. Control systems engineering, computer-aided design.

Vernon E. Sater, Ph.D., IIT. Heavy metal removal from wastewater, online process instrumentation.

Robert S. Torrest, Ph.D., Minnesota. Multiphase flow, filtration, polymer solution flow in porous media, in situ processes for energy and mineral recovery, pollution control.

Imre Zwiebel (Emeritus), Ph.D., Yale. Mass transfer, adsorption, surface phenomena, adsorption of proteins and macromolecules, protein engineering, separations and bioseparations, molecular sieves, applied mathematics.

BOSTON UNIVERSITY

College of Engineering
Department of Biomedical Engineering

Programs of Study

The Boston University Biomedical Engineering graduate programs train students in the application of modern technology and quantitative engineering methods to biology and medicine. The Ph.D., M.D./Ph.D., and M.S. degrees in biomedical engineering are offered. Areas of concentration are auditory research, biomechanics of movement, biotechnological and human genome engineering, cellular biomechanics, computational biology, computational vision, electromechanical and electrokinetic interactions in tissue, mathematical and computer modeling, molecular engineering, motor control, neuroengineering, neuromuscular signal processing, neurophysiology, nonlinear biodynamics, pulmonary biomechanics, and pulmonary instrumentation. (For a more complete listing of faculty research activity and interests, see the reverse of this page.) The M.S. in biomedical engineering requires 36 credits (nine semester courses, including a thesis), of which at least 28 credits must be earned at Boston University. The Ph.D. program requires at least two consecutive semesters of full-time resident commitment to the discipline, written and oral qualifying examinations, research, the dissertation, and a final oral examination. A cumulative grade point average of at least 3.0 (B) is required.

Research Facilities

Laboratories for research are located primarily in the Engineering Research Building, part of the Metcalf Center for Science and Engineering. Additional laboratories are located in the University's Medical Center, approximately 3 miles from the Charles River campus. Students and faculty members have access over a campus network to a variety of powerful computers. Specialized equipment is available in the biomedical engineering teaching and research laboratories. These include the Biomolecular Engineering Research Center, the Center for Advanced Biotechnology, the Fields and Tissues Laboratory, the Hearing Research Center, the Brain and Vision Laboratory, the NeuroMuscular Research Center, the Respiratory Research Laboratories, the Neural Dynamics Laboratory, and the Molecular Engineering Laboratory.

Financial Aid

A full range of financial aid opportunities is available, including the Presidential University Graduate Fellowships, Dean's Fellowships, departmental fellowships, graduate teaching fellowships, research assistantships, and scholarships. In 1999–2000, teaching fellowships provide stipends of $12,500 per academic year and require approximately 20 hours a week of instructional and other duties. Recipients receive a tuition waiver for 8 to 10 credits per semester and up to 8 additional credits for the following summer. Research assistantship stipend levels are comparable to those of teaching fellowships and are also supplemented by tuition waivers. University and Dean's Fellowships range up to $37,830 (including stipend and tuition) per year; the department encourages GEM scholars to apply. Applicants for the Federal Direct Student Loan Program and work-study must send a Free Application for Federal Student Aid (FAFSA) to the Federal Student Aid Programs Office. Work-study and FAFSA forms may be obtained from the Graduate Programs Office.

Cost of Study

In 1999–2000, tuition and fees for full-time study are $23,770. Part-time students pay $743 per credit hour.

Living and Housing Costs

Privately owned apartments or rooms are readily available. Living expenses for a single student are estimated at $10,730 for the nine-month 1999–2000 academic year.

Student Group

The department has 76 graduate students (M.S. and Ph.D.). About two thirds are working toward the Ph.D. degree; 34 percent of these are women. Students come from many areas of the United States and from several countries. In 1998–99, 92 percent of the M.S. and Ph.D. students were partially or fully supported by teaching fellowships, research assistantships, scholarships, or awards from governments, foundations, or industry.

Student Outcomes

Some graduate students continue in academia at the Ph.D. or postdoctoral level at Boston University or at other institutions. Others enter the biomedical industry as research and development engineers. Some academic institutions that recent graduates have joined in addition to Boston University include Cornell Medical College, Johns Hopkins, MIT, Penn State, and the University of Chicago. Companies where program graduates now work include Brigham and Women's Hospital, Bose Corp., Guidant Corp., and SRI International.

Location

Boston University's location beside the Charles River in Boston's Back Bay provides easy access to the many resources of the Greater Boston area. Approximately sixty colleges and universities are located in this area, adding an atmosphere of vitality, scholarship, and fun to the resources of one of the world's great cities, which include hearing the Boston Symphony; sampling Italian cuisine in the North End; sailing on the Charles; walking the historic Freedom Trail; browsing in Faneuil Hall's Quincy Market; visiting numerous museums, nightclubs, and theaters; and cheering the Red Sox, the Bruins, and the Celtics.

The University and The Department

Boston University is an independent nonsectarian university open to women and all minorities. Its approximately 23,500 full-time students and 3,130 faculty members help to make it one of the largest independent universities in the world. The Department of Biomedical Engineering is a rapidly developing department within the University's College of Engineering, which has experienced remarkable growth in the past several years. The department maintains internationally prominent research programs and is deeply committed to the education of its students. Through the department's interactions with local industries and hospitals, the faculty and students maintain an appreciation for the relevance of various engineering issues.

Applying

Applicants to the Ph.D. and M.S. programs should have attained a high degree of scholarship in an undergraduate program in engineering or science at an accredited college or university. Applications for admission with financial aid consideration must be submitted by January 15 for the fall semester and by October 1 for the spring semester. Required credentials for both the M.S. and Ph.D. programs include a B+ average and GRE quantitative and analytical scores in the 80th-percentile level. A minimum TOEFL score of 625 is expected of international applicants.

Correspondence and Information

For program information:

Graduate Admissions
Biomedical Engineering Department
College of Engineering
Boston University
44 Cummington Street
Boston, Massachusetts 02215
Telephone: 617-353-7609
World Wide Web: http://www.bu.edu/eng/bme

For admission application forms:

Graduate Programs
College of Engineering
Boston University
48 Cummington Street
Boston, Massachusetts 02215
Telephone: 617-353-9760
Fax: 617-353-0259
E-mail: enggrad@bu.edu
World Wide Web: http://www.bu.edu/eng/grad

Boston University

THE FACULTY AND THEIR RESEARCH

Mark W. Bitensky, Research Professor; M.D., Yale. G-protein signal transduction, erythrocyte biology, macromolecular ensembles.

Stephen Burns, Research Associate; Ph.D., MIT. Telecommunications, quantitative exercise instrumentation, medical instrumentation.

Charles R. Cantor, Professor of Biomedical Engineering and Biochemistry and Director, Center for Advanced Biotechnology; Ph.D., Berkeley. Human genome analysis, molecular genetics, new biophysical tools and methodologies, genetic engineering.

Laurel Carney, Associate Professor and Associate Chair of Graduate Studies; Ph.D., Wisconsin–Madison. Auditory research, neural processing of complex sounds.

H. Steven Colburn, Professor and Director, Hearing Research Center; Ph.D., MIT. Hearing research, particularly binaural interaction; virtual acoustic environments.

James Collins, Professor; D.Phil., Oxford (England). Developing and implementing nonlinear dynamics and statistical physics to study the biomechanics and neural control of posture and locomotion.

Douglas A. Cotter, Adjunct Associate Professor; Ph.D., North Carolina State. Hospital information systems, medical instrumentation and diagnostics, pattern recognition, research and development management.

Charles DeLisi, Professor, Biomedical Engineering, and Dean, College of Engineering; Ph.D., NYU. Analysis of DNA function, protein structure, optimization algorithms, neural net applications to molecular biology, drug and vaccine design, membrane biophysics.

Carlo J. DeLuca, Professor, Biomedical Engineering; Research Professor, Neurology; and Director, NeuroMuscular Research Center; Ph.D., Queen's at Kingston. Neuromuscular signals and controls, rehabilitation engineering.

Micah Dembo, Professor; Ph.D., Cornell. Statistical mechanics in biological systems, cellular information processing and signal transduction, thermodynamics and mechanics of cell adhesion, biophysics of cell deformation, active motility.

William F. Dolphin, Research Assistant Professor; Ph.D., Boston University. Temporal coding in the perception of complex acoustic signals in the mammalian auditory system.

Nathaniel I. Durlach, Visiting Scientist; M.A., Columbia. Sensory communication: application areas include communication aids for the deaf and deaf-blind, teleoperator systems, and virtual-environment systems.

Solomon Eisenberg, Associate Professor of Biomedical Engineering and Electrical Engineering and Associate Dean of Undergraduate Studies, College of Engineering; Sc.D., MIT. Electrically mediated phenomena in tissues and biopolymers.

Evan A. Evans, Professor; Ph.D., California, San Diego. Nanomicroscale biomechanics, ultrasensitive force probes and extreme resolution optical techniques, material properties of cellular structure, role of structural forces in cell biochemistry.

Maxim Frank-Kamenetskii, Professor; Ph.D., Moscow Physico-Technical Institute (Russia); Sc.D., Academy of Sciences (Russia). DNA structures, DNA topology, DNA functioning, new drugs interacting with DNA.

Federico Girosi, Research Assistant Professor; Ph.D., Genoa (Italy). Study and use of artificial neuronal networks, models for visual motion processing.

Franco Giulianini, Research Associate; Ph.D., Northeastern. Development and implementation of computer programs to statistically analyze and model experimental data.

Ary L. Goldberger, Adjunct Associate Professor; M.D., Yale. Erratic time series formed by the human heartbeat, sequencing of nucleotides appearing in DNA.

Stephen Grossberg, Wang Professor of Cognitive and Neural Systems and Professor of Biomedical Engineering, Mathematics, and Psychology; Ph.D., Rockefeller. Neural models of adaptive behavior, vision, audition, speech, learning, memory, and recognition.

Ji. C. He, Research Associate; Ph.D., Chicago. Psychological optics: optical quality of the eye, myopia, accommodation, Stiles Crawford effect, color vision and color deficiency, and spatial vision.

Allyn E. Hubbard, Associate Professor, Electrical Engineering and Biomedical Engineering; Ph.D., Wisconsin. Auditory physiology, experiments and modeling, neurocomputing, VLSI on biomedical applications, biosensors.

Andrew C. Jackson, Professor; Ph.D., Mississippi. Respiratory physiology; respiratory mechanics.

Hernan J. Jara, Adjunct Professor; Ph.D., Illinois at Chicago. Atomic, molecular, and laser physics.

W. Clement Karl, Assistant Professor, Biomedical Engineering and Electrical Engineering; Ph.D., MIT. Multiresolution statistical signal and image processing, geometric estimation.

David Kennedy, Adjunct Assistant Professor; Ph.D., MIT. Computerized methods of in vivo morphometric analysis and magnetic resonance pulse sequence analysis, neurologic imaging and chemical shift imaging.

Ron Kikinis, Research Assistant Professor; M.D., Zurich Medical School (Switzerland). 3-D MRI reconstruction of the human brain from 2-D MRI images; neuroradiology, MRI, and medical image processing.

Kenneth R. Lutchen, Professor and Chairman; Ph.D., Case Western Reserve. Airway and lung tissue mechanics and ventilation, heart rate variability, linear and nonlinear systems identification, optimal design, minimally invasive diagnostics.

Roberto Merletti, Associate Professor, Biomedical Engineering and NeuroMuscular Research Center; Ph.D., Ohio State. Neuromuscular system performance, electrical stimulation.

Marvin Minsky, Adjunct Professor; Ph.D., Princeton. Artificial intelligence and the "Society of the Mind" theory of intelligence and psychology.

David C. Mountain Jr., Professor, Biomedical Engineering, and Research Professor, Otolaryngology; Ph.D., Wisconsin. Auditory information processing, sensory biophysics, computer simulation, biomedical electronics, biomedical signal processing.

Koichi Nakano, Research Associate; Ph.D., Osaka Medical School. Increased expression of CD44 variants in differentiated thyroid cancers, identification of NGF-response element in the rat neuropeptide Y gene, induction of the binding proteins.

S. Hamid Nawab, Associate Professor, Electrical Engineering and Biomedical Engineering; Ph.D., MIT. Digital signal processing, knowledge-based signal processing, auditory scene analysis, signal processing for low-power and communications applications.

J. Philip Saul, Adjunct Assistant Professor; M.D., Duke. Signal processing, information theory, and cardiovascular physiology.

Robert Sekuler, Research Professor; Ph.D., Brown. Human visual psychophysics; effects of global motion in random dot patterns.

Barbara Shinn-Cunningham, Assistant Professor of Cognitive and Neural Systems and Biomedical Engineering; Ph.D., MIT. Binaural hearing, localization, modeling of psychoacoustic phenomena, resolution and bias in psychoacoustic tasks.

Cassandra Smith, Professor, Biomedical Engineering, Biology, and Biochemistry, and Deputy Director, Center for Advanced Biotechnology; Ph.D., Texas A&M. Development of novel methods for mapping and sequencing of large and small genomes.

Temple F. Smith, Professor and Director, Biomolecular Engineering Research Center; Ph.D., Colorado. The syntactic and semantic structure of the genetic information in biomolecular sequences, structures and their evolution.

Dimitrije Stamenovic, Associate Professor; Ph.D., Minnesota. Respiratory mechanics, rheology of soft tissues, mechanics of foamlike structures and cell mechanics.

Bela Suki, Assistant Professor; Ph.D., Jozef Attila (Hungary). Mechanical properties of living tissues, the ensemble behavior of complex biological systems, nonlinearities in biological systems.

George E. Tarr, Research Associate; Ph.D., Massachusetts Amherst. Development of general and specific structural analysis of proteins and peptides, current emphasis on the union of chemistry and enzymology with MALDI-TOF mass spectrometry.

Malvin C. Teich, Professor of Biomedical Engineering and Electrical Engineering; Ph.D., Cornell. Wavelet analysis of fractal biological signals, neural coding, auditory and visual psychophysics, quantum optics.

Lucia M. Vaina, Professor of Biomedical Engineering and Associate Research Professor, Neurology; and Director, Brain and Vision Research Laboratory; Ph.D., Sorbonne; Dr. ès Science, Toulouse Polytechnic. Computational visual neuroscience, biological and computational learning.

Sandor Vajda, Associate Professor; Ph.D., Hungarian Academy of Science. Scientific computing, computational chemistry, combinatorial optimization, molecular biology, protein and peptide structure determination.

Herbert F. Voigt, Associate Professor and Associate Chair of Undergraduate Studies, Biomedical Engineering, and Associate Research Professor, Otolaryngology; Ph.D., Johns Hopkins. Auditory neurophysiology, neural circuitry, neural modeling.

Zhiping Weng, Instructor of Biomedical Engineering; Ph.D., Boston University. Bioinformatics, DNA and protein sequence analysis.

John White, Assistant Professor; Ph.D., Johns Hopkins. Nonlinear membrane conductances in neurons, modulation of ion channels, dynamics of neuronal networks.

Joyce Y. Wong, Clare Booth Luce Assistant Professor; Ph.D., MIT. Biomaterials, tailoring cell-material interfaces for drug delivery and tissue engineering applications, direct measurement of biological interactions.

Tatsuro Yoshida, Research Associate; Ph.D., Michigan. Regulation of G-protein-coupled signal transduction in retinal rod cells, refrigerated storage of human red blood cells, quantitative studies and computer simulation of metabolic pathways in cells, bioenergetics.

CASE WESTERN RESERVE UNIVERSITY

Department of Biomedical Engineering

Programs of Study

Many exceptional and innovative educational programs are offered to provide career opportunities for biomedical engineering (BME) research, development, and design that relate to diagnostic and therapeutic methods in industry, medical centers, and academic institutions. Graduate programs lead to the M.S., Ph.D., and M.D./Ph.D. in BME. Programs of study, based on individual needs and interests, allow the student to develop strength in an engineering specialty and apply this expertise to an important biomedical problem under the supervision of a faculty Guidance Committee. Graduate students can choose from more than twenty-five courses regularly taught in BME, as well as many courses taught in other departments of Case Western Reserve University (CWRU). Typically, an M.S. program consists of seven to nine courses and a Ph.D. program consists of about fourteen courses beyond the B.S.

Students can select research projects from among the many strengths of the department, including applied neural control, neuromuscular prostheses, neural engineering, biomaterials, tissue engineering, biomedical imaging, chemical microsensors, optical diagnostics, cardiac bioelectricity, mass and heat transport, and metabolic systems. Opportunities for collaborative research and training in engineering with application to basic biomedical sciences as well as to clinical and commercial developments are available through primary faculty members, associated faculty members, and researchers in the nearby major medical centers.

Research Facilities

The primary faculty members have laboratories focusing on applied neural control, cardiovascular and skeletal biomaterials, cardiovascular and neural tissue engineering, drug and gene delivery, biomedical image processing, biomedical optical imaging, cellular and tissue cardiac bioelectricity, ion channel function, electrochemical and fiber optic sensors, neural engineering and brain electrophysiology, and neural prostheses. BME faculty members and students also make extensive use of campus research centers for special purposes such as microelectronic fabrication and material analyses. Associated faculty members have labs devoted to eye movement control, gait analysis, implantable sensors/actuators, magnetic resonance imaging (MRI), PET imaging, metabolism, and tissue pathology. These are located at five major medical centers and teaching hospitals that (with one exception) are next to the campus or within easy walking distance.

Financial Aid

Graduate students may receive financial support from faculty members as research assistants (M.S. and Ph.D.), from the department as NIH and Whitaker Foundation research trainees (Ph.D. only), from the Case School of Engineering as Case Prime Fellows, and from the School of Medicine as NIH medical scientist trainees (combined M.D./Ph.D. only). These positions are awarded on a competitive basis. There are also opportunities for part-time work in various laboratories of the University and nearby medical institutions.

Cost of Study

Tuition in 1999–2000 at CWRU for graduate students is $800 per credit hour. A full load for graduate students is a minimum of 9 credits per semester. Fees for health insurance and activities are approximately $305 per semester.

Living and Housing Costs

Graduate housing is available on campus for unmarried students. In 1999–2000, the costs for single housing range from $3810 to $4640. Within a 2-mile radius of the campus, numerous apartments are available for married and single students at rents ranging from $400 to $900 per month.

Student Group

The Department of Biomedical Engineering has approximately 125 graduate students, of whom about 60 percent are advancing toward the Ph.D. At Case Western Reserve University, approximately 3,600 students are enrolled as undergraduates, 2,200 in graduate studies, and 4,000 in the professional schools.

Location

CWRU is located on the eastern boundary of Cleveland in University Circle, which is the city's cultural center. The area includes Severance Hall (home of the Cleveland Orchestra), the Museum of Art, the Museum of Natural History, the Garden Center, the Institute of Art, the Institute of Music, the Western Reserve Historical Society, and the Crawford Auto-Aviation Museum. Metropolitan Cleveland has a population of almost 2 million. The Cleveland Hopkins International Airport is 30 minutes away by rail transit. A network of parks encircles the greater Cleveland area. Opportunities are available for sailing on Lake Erie and for hiking and skiing nearby in Ohio, Pennsylvania, and New York. Major-league sports, theater, and all types of music provide a full range of entertainment.

The University and the Department

The Department of Biomedical Engineering at Case Western Reserve University is part of both the Case School of Engineering and the School of Medicine, which are located on the same campus. Established in 1967, the department is one of the pioneers in biomedical engineering education and is currently among the nation's largest and highest rated (according to *U.S. News & World Report*). Case Western Reserve University was formed in 1967 by a federation of Western Reserve College and Case Institute of Technology. Numerous interdisciplinary programs exist with the professional Schools of Medicine, Dentistry, Nursing, Law, Social Work, and Management.

Applying

Applications that request financial aid should be submitted before February 1. The completed application requires official transcripts, scores on the GRE General Test, and three letters of reference. Application forms are available from the BME admissions coordinator or can be downloaded from the CMRU Web site (http://www.cwru.edu). M.D./Ph.D. applicants should apply directly to the School of Medicine.

Correspondence and Information

Admissions Coordinator
Department of Biomedical Engineering
Wickenden Building 501
Case Western Reserve University
10900 Euclid Avenue
Cleveland, Ohio 44106-7207

Telephone: 216-368-4094
Fax: 216-368-4969
E-mail: xx220@po.cwru.edu
World Wide Web: http://convolve.ebme.cwru.edu/bmehome.html

Case Western Reserve University

THE FACULTY AND THEIR RESEARCH

Primary Faculty

Ravi V. Bellamkonda, Ph.D., Assistant Professor. Biomaterials, neural and vascular cell and tissue engineering, nerve regeneration and vascular grafts, and controlled drug and gene delivery.

Patrick E. Crago, Ph.D., Professor and Chairman. Neuroprostheses for restoration of motor function, modeling of neuromuscular control.

Jianmin Cui, Ph.D., Assistant Professor. Molecular and biophysical mechanisms of ion channel function and modulation, the role of ion channels in cardiac excitation and arrhythmias.

Dominique M. Durand, Ph.D., Professor. Neural engineering, neural prostheses, magnetic and electric stimulation of the nervous system, electrophysiology of epilepsy, computational neuroscience.

Steven J. Eppell, Ph.D., Assistant Professor. Biomaterials instrumentation, nanoscale structure-function analysis of orthopaedic biomaterials, and scanning probe microscopy and spectroscopy of skeletal tissues.

Janie Fouke, Ph.D., Professor. Transducers, sensors, and medical devices; instrumentation systems; mechanical and optical properties of tissues and organs.

Jinming Gao, Ph.D., Assistant Professor. Biomaterials, organ-targeted drug delivery, imaging biopolymers, vascular tissue engineering.

Miklos Gratzl, Ph.D., Associate Professor. Electrochemical and optical biosensors and cost-effective diagnostic devices, measurements of cellular neurotransmitter release and cancer cell drug resistance.

Joseph A. Izatt, Ph.D., Assistant Professor. Biomedical optical imaging and spectroscopy, fiber-optic biosensors, minimally invasive diagnostic technology, optical coherence tomography.

J. Lawrence Katz, Ph.D., Professor. Structure-property relationships in bone, osteophilic biomaterials, ultrasonic studies of tissue anisotropy, scanning acoustic microscopy.

Dmitri E. Kourennyi, Ph.D., Assistant Professor. Signal processing in the retina, ion channel biophysics.

Roger E. Marchant, Ph.D., Associate Professor and Director of the Center for Cardiovascular Biomaterials. Biopolymers, biosynthetic surfactants, polymer surface modification for implants and sensors, protein-surface interactions by AFM.

J. Thomas Mortimer, Ph.D., Professor and Director of the Applied Neural Control Laboratory. Neural control and prostheses, electrical activation of neural tissue, membrane properties and electrodes.

Niels F. Otani, Ph.D., Associate Professor. Computer models of cardiac action potential wave propagation, nonlinear dynamical properties of excitable tissues.

P. Hunter Peckham, Ph.D., Professor and Director of the Functional Electrical Stimulation Center. Motor function restoration with neural prostheses, control of orthotic and prosthetic systems.

Yoram Rudy, Ph.D., Professor and Director of the Cardiac Bioelectricity Research and Training Center. Models of cardiac cellular activity and cardiac excitation, cardiac electric mapping, mechanisms of cardiac arrhythmias, electrocardiographic imaging.

Gerald M. Saidel, Ph.D., Professor. Mass transport, heat transport, and metabolism in cells, tissues, and organs; modeling and nonlinear parameter estimation of dynamic systems.

David L. Wilson, Ph.D., Associate Professor. Biomedical image processing: digital processing and quantitative image quality of X-ray fluoroscopy images; interventional MRI.

Associated Faculty

James M. Anderson, M.D./Ph.D., Professor (Pathology, University Hospitals). Biocompatibility of implants, human vascular grafts.

Marco E. Cabrera, Ph.D., Assistant Professor (Pediatric Cardiology, RB&C, University Hospitals). Modeling and control of metabolic processes, metabolic regulation in hypoxia, ischemia, and exercise.

Brian Davis, Ph.D., Adjunct Assistant Professor (Cleveland Clinic Foundation). Human locomotion, diabetic foot pathology, spaceflight-induced osteoporosis and biomedical instrumentation.

David Dean, Ph.D., Assistant Professor (Neurological Surgery, University Hospitals). 3-D medical imaging and morphometrics; skull, brain, soft tissue face.

Louis F. Dell'Osso, Ph.D., Professor (Neurology, VA Medical Center). Neurophysiological control, ocular motor control and oscillations.

Jeffrey L. Duerk, Ph.D., Associate Professor (Radiology, University Hospitals). MRI, flow visualization, interventional MRI.

Igor R. Efimov, Ph.D., Assistant Professor (Cardiology, Cleveland Clinic Foundation). Imaging and modeling of cardiac arrhythmias.

Mark D. Grabiner, Ph.D., Adjunct Assistant Professor (Cleveland Clinic Foundation). Human neuromotor control and strength and endurance.

Hiroaki Harasaki, M.D./Ph.D., Adjunct Associate Professor (Cleveland Clinic Foundation). Artificial hearts and circulatory assists, heart valve prostheses, vascular grafts, cardiovascular biomaterials.

Karl J. Jepsen, Ph.D., Assistant Professor (Orthopaedics, University Hospitals). Skeletal fragility, bone damage mechanics, influence of mechanical loading on osteoblast differentiation and matrix organization.

Jill S. Kawalec, Ph.D., Adjunct Assistant Professor (Ohio College of Podiatric Medicine). Biomaterials, small joint replacement, orthopaedic fixation devices.

Kevin Kilgore, Ph.D., Adjunct Assistant Professor (MetroHealth Medical Center). Function electrical stimulation, restoration of hand function in quadriplegics, hand biomechanics.

Robert Kirsch, Ph.D., Researcher (MetroHealth Medical Center). Human movement control, restoration by functional stimulation.

Kandice Kottke-Marchant, M.D./Ph.D., Adjunct Associate Professor (Cleveland Clinic Foundation). Interaction of blood and materials, endothelial cell function on biomaterials.

Kenneth R. Laurita, Ph.D., Assistant Professor (Medicine/Cardiology, University Hospitals). Cardiac electrophysiology, arrhythmia mechanisms, intracellular calcium homeostasis, fluorescence imaging, instrumentation and software for potential mapping.

R. John Leigh, M.D., Professor (Neurology, VA Medical Center). Normal and abnormal motor control, eye movements.

Raymond Muzic, Ph.D., Assistant Professor (Radiology, University Hospitals). Physiologic modeling and experiment design for PET.

David S. Rosenbaum, M.D., Associate Professor (Medicine/Cardiology, University Hospitals). High-resolution cardiac optical mapping, arrhythmia mechanisms, ECG signal processing.

Mark S. Rzeszotarski, Ph.D., Assistant Professor (Radiology, PHS Mt. Sinai Medical Center). Computer applications in radiology: magnetic resonance imaging, computed tomography, nuclear medicine, and ultrasound.

Ronald Triolo, Ph.D., Assistant Professor (Orthopaedics, VA Medical Center). Rehabilitation engineering, neural control of motion, lower-extremity neuroprostheses, orthopaedic biomechanics and prosthetic/orthotic design.

Clayton L. Van Doren, Ph.D., Assistant Professor (Orthopaedics, MetroHealth Medical Center). Sensory motor control and mechanical properties of the hand, tactile sensation, electrocutaneous stimulation, and artificial sensory feedback.

Ivan Vesely, Ph.D., Adjunct Associate Professor (Cleveland Clinic Foundation). Artificial heart valves, soft tissue biomechanics, tissue engineering.

D. Geoffrey Vince, Ph.D., Adjunct Assistant Professor (Cleveland Clinic Foundation). Imaging, quantitative microscopy, vascular pathology.

Albert L. Waldo, M.D., Professor (Medicine/Cardiology, University Hospitals). Cardiac electrophysiology, cardiac excitation mapping, mechanisms of cardiac arrhythmias and conduction.

Michael Wendt, Ph.D., Assistant Professor (Radiology, University Hospitals). Interventional MRI, fast MRI sequences.

Guang H. Yue, Ph.D., Adjunct Assistant Professor (Cleveland Clinic Foundation). Neural control of movement, electrophysiology, fMRI.

Maciej Zborowski, Ph.D., Adjunct Assistant Professor (Cleveland Clinic Foundation). High-speed magnetic cell sorting.

Nicholas P. Ziats, Ph.D., Assistant Professor (Pathology, University Hospitals). Vascular grafts, cell-material interactions, extracellular matrix, tissue engineering, blood compatibility.

CLEMSON
U N I V E R S I T Y

CLEMSON UNIVERSITY

College of Engineering and Science
Department of Bioengineering

Program of Study	The bioengineering program at Clemson is devoted to the application of engineering science and technology to the problems of medicine, spanning the range from the mechanics of health-care delivery systems to investigations of fundamental physiological processes. The principal thrust of this program is in the areas of biomaterials, biomechanics, and cellular biology, including the development and evaluation of living and nonliving prosthetic materials, the physical and mechanical behavior of tissues treated as engineering materials, and the physiological response of the host to foreign matter. Specific areas of research include synthesis and evaluation of biocompatible polymers; friction, wear, and lubrication of artificial and natural joints; cell biology and tissue engineering; application of artificial intelligence techniques to the computer-aided design of prosthetic devices; interaction of material surfaces with cells and biomolecules; biomolecular modeling; and mathematical modeling of biomechanical structures.
	M.S. and Ph.D. degrees in bioengineering can be earned under the direction of faculty members in the Department of Bioengineering. Close working relationships exist between the Department of Bioengineering and other engineering and life sciences departments.
	The Bioengineering Alliance of South Carolina, a consortium consisting of Clemson University, the University of South Carolina, and the Medical University of South Carolina, promotes unified biomedical engineering education and research in South Carolina. Graduate students in the bioengineering program have an opportunity to participate in consortium research programs and receive valuable clinical experience. A clinical internship course is available that permits a student to study under a selected clinical preceptor and gain experience in clinical applications of bioengineering. Agreements between Clemson University, the Greenville Hospital System, and other health-care facilities also provide expanded opportunities for clinical internships and research.
Research Facilities	The Department of Bioengineering is housed in the Rhodes Engineering Research Center. Fully equipped laboratories for polymer synthesis and characterization, mechanical testing, histopathology, tribology, surface analysis, image analysis, cell culture, and computer-aided analysis and design are located in this building. Other on-campus facilities include an electron microscopy lab, a state-of-the-art animal research facility, clean rooms, and machine shops. The College of Engineering and Science maintains computational facilities that include more than 250 UNIX workstations and 125 microcomputers, as well as design and analysis software. Departmental resources include a recently acquired virtual reality computing lab with a dedicated supercomputer and a biotribology lab that includes knee-joint simulators.
Financial Aid	A limited number of University assistantships requiring up to 20 hours of work per week are available. These assistantships carry a stipend of up to $15,000 per year and a reduction of tuition fees. There are opportunities for exceptional students to obtain additional University fellowships of up to $5000 for twelve months.
Cost of Study	The tuition and fees for full-time students in 1998–99 were $1577 per semester for South Carolina residents and $3072 for nonresidents. All graduate assistants received a reduction in tuition and fees and paid $493 per semester and $165 per summer session.
Living and Housing Costs	Rooms are available in dormitories for $810 to $1185 per semester or in University apartments for $1010 to $1280 per semester. Housing for married students is available for $290 to $495 a month; graduate assistants and fellows are given priority. Applications must be received at the housing office before May 1 for August housing and before November 1 for January housing.
Student Group	There are about 16,700 students at Clemson University, of whom approximately 25 percent are graduate students. There are approximately 1,000 graduate students in the College of Engineering and Science. The Department of Bioengineering has approximately 50 graduate students. No undergraduate degree is offered in bioengineering, although an undergraduate minor is available.
Student Outcomes	About one third of the students completing an M.S. degree go on to work toward a Ph.D. degree. After completion of an M.S. degree, recent graduates of the program have obtained employment as research/project engineers in the medical device industry, federal research laboratories, or academic laboratories. After completion of the Ph.D. degree, students go on to upper-level research positions in the medical device industry or to postdoctoral or faculty positions in academia.
Location	Clemson, South Carolina, is a residential community located approximately midway between Charlotte, North Carolina, and Atlanta, Georgia. Nearby Interstate Highway 85, Amtrak, and the Greenville-Spartanburg Airport link Clemson with major cities in the region. Clemson is 240 meters above sea level and has an average temperature of 17°C; the annual rainfall is about 130 centimeters. It is located in the scenic foothills of the Blue Ridge Mountains on the sprawling 1,600-kilometer shoreline of Lake Hartwell.
The University	The 2.4-square-kilometer campus represents an investment of approximately $270 million in academic buildings, student housing, and service facilities and is surrounded by 81 square kilometers of farms and research lands for forestry, agriculture, and engineering. Clemson University is the state land-grant institution of South Carolina.
Applying	Applicants should have a B.S. degree from an accredited college or university in any of the major engineering disciplines. Some background in general biology and physiology is recommended but is not a prerequisite. Students with degrees in physics, chemistry, or life sciences are also considered for admission if they can demonstrate proficiency in certain prescribed engineering courses. Ph.D. candidates should have an M.S. in engineering and should preferably have completed an M.S. thesis. All applicants must take the General Test of the Graduate Record Examinations.
Correspondence and Information	Graduate Student Coordinator Department of Bioengineering Rhodes Engineering Research Center Clemson University Clemson, South Carolina 29634-0905 Telephone: 864-656-5559 Fax: 864-656-4466 E-mail: ruth.watkins@ces.clemson.edu

Clemson University

THE FACULTY AND THEIR RESEARCH

J. Black, Professor Emeritus; Ph.D. Biomaterials, elastomeric composites, corrosion.
T. Boland, Ph.D. Protein-biomaterial interactions, molecular films, surface and interface engineering, AFM.
R. L. Dooley, Department Head; Ph.D. Bioinstrumentation, medical computer applications, advanced manufacturing and design.
V. M. Gharpuray, Ph.D. Biomechanics, solid mechanics, spine mechanics.
M. LaBerge, Ph.D. Tribology, orthopedic implants.
R. A. Latour, Ph.D. Biomaterials, biomolecule-surface interactions, biomolecular modeling.
P. L. Mente, Ph.D. Cell mechanics, musculoskeletal diseases.
S. Saha, Ph.D. Biomechanics, bone-cement, bioinstrumentation, bone.

Adjunct Faculty
G. Acres, M.D. Hemodialysis materials.
B. L. Allen Jr., M.D. Orthopedic surgery.
Y. An, M.D. Orthopedics.
L. S. Bowman, M.D. Orthopedic surgery.
J. Burns, Ph.D. Cardiovascular replacements.
R. Christensen, M.D. Temporomandibular joint replacements.
G. Cooper IV, M.D. Cardiology.
B. Cuddy, M.D. Neurosurgery, spine.
R. A. Draughn, Ph.D. Dental materials, mechanics.
H. I. Friedman, M.D., Ph.D. Plastic surgery.
R. J. Friedman, Ph.D. Orthopedic surgery.
G. Heimke, Ph.D. Ceramics, dental materials.
W. C. Hutton, Ph.D. Biomechanics.
E. M. Langan, M.D. Vascular surgery.
B. J. Love, Ph.D. Polymeric materials.
S. Martin, M.D. Orthopedics.
B. B. Michniak, Ph.D. Drug delivery.
P. M. Murray, M.D. Orthopedic surgery.
T. B. Pace, M.D. Orthopedic surgery.
F. J. Pearce, Ph.D. Combat trauma and casualty research.
G. L. Picciolo, Ph.D. Cell physiology.
D. L. Powers, D.V.M. Veterinary medicine.
W. Ramp, Ph.D. Cell and tissue physiology.
C. D. Riddle, M.D. Orthopedic surgery.
R. M. Sade, M.D. Cardiac surgery.
S. W. Shalaby, Ph.D. Biopolymers.
G. Sherouse, Ph.D. Radiation oncology.
F. G. Spinale, Ph.D. Cardiovascular physiology.
P. Stasikelis, M.D. Orthopedic surgery.
R. Straup, D.M.D. Dental materials.
T. M. Sullivan, M.D. Vascular surgery.
L. Terracio, Ph.D. Cellular biomechanics.
J. D. Thompson, M.D. Pediatric orthopedics.
F. A. Young, Ph.D. Dental materials, metallurgy.
A. F. von Recum, D.V.M., Ph.D. Tissue/material interface.
M. R. Zile, M.D. Cardiology.

Representative Thesis Titles
The Formation and Surface Modification of Absorbable Glasses.
A Mathematical Model to Enhance Surgical Treatment of Scoliosis.
Gas Phase Phosphorylation of Thermoplastic Polymer.
Evaluation of Closed Chondromalacia Patellae Using Ultrasound: A Feasibility Study.
In Vitro Tribological Assessment of Compliant Orthopaedic Bearing Surfaces in Pathological Conditions.
Structural, Mechanical, and Tribological Response of Lapine Articular Cartilage to Intra-Articular Ketorolac Tromethamine.
The Influence of Implant Porosity on Epithelial and Soft Tissue Response in the Percutaneous Location.
Cellular Response to Heterogeneously Charged Polymers.
Soft Tissue Response to Porous Implants of Varying Bending Stiffness.
Effects of Reinforcement and Irradiation on Thermal and Mechanical Properties of Ultrahigh Molecular Weight Polyethylene.
Intramedullary Tantalum Foam Fixation in an Intact Tibia.
Collagen Types Found at the Material-Tissue Interface.
Microsurface Texturing: An Industrial Approach for Soft Tissue Implants.
An Analytical Model of Human Intervertebral Disc Prolapse.
The Synthesis, Characterization, and Study of the Hydrolytic Degradation Characteristics of Oxalate-Based Polyesters.
Immobilization of Bioreagents and Conductive or Fluorescent Adjuvants to Phosphonylated Polyethylene.
Effect of Orientation on the Physicochemical and Morphological Changes in Absorbing Polylactide Films.
Biocompatability of Prosthesis Intraocular Lens Implants: A Review and Histological Method.
Absorption Profile and Tissue Distribution of a Methoxypropyl-Cyanoacrylate/Oxalate System as a Tissue Adhesive.
In Vivo Strains in Bone Near Transcortical Implants.
The Effects of Cervical Fusion on the Behavior and Composition of Adjacent Discs: An In Vivo Model.
The Use of Modulus-Graded Polyurethane Elastomers as Orthopaedic Implant Bearing Surfaces.
Hip Joint Simulator Design.
3D Reconstruction of 2D Medical CT Scans: An Application to Biomechanical Analysis.
Reorienting Parts in Stereolithography Using a Rotational Degree of Freedom.
Mechanical Behavior of the Human Lumbar Spine: A Finite Element Study.
Bone Remodeling Around Transcortical Implants.
Application of an Iterative Finite Element Analytical Technique in the Determination and Comparison of Mechanical Properties of Isolated Mammalian Cardiocytes.
The Effect of Orthopaedic Metals on Osteoblastic Expression of Osteocalcin.
Development and Characterization of a PEG-based Polymer as a Biliary Stent Material.
Effect of Bovine Serum Concentration on the Lubrication of UHMWPE/Cobalt-Chromium Alloy Tribosystem.
Controlled Surface Modification of Thermoplastic Polymers for Use in Biomedical Applications.
In Vivo Spinal Stiffness Measurement System for the Optimal Surgical Correction of Scoliosis.
A Finite Element Study of Stresses in Bone Near Transcortical Implants.
Mechanical and Physical Characterization of Sonicated Acrylic Bone Cement.
Dextran as a Biomaterial Coating for Reduced Cell Adhesion (MS).
Synthesis and Characterization of a Novel PLA/Dextran Copolymer for Biomaterials Applications (MS).

CLEVELAND STATE UNIVERSITY

Fenn College of Engineering
Chemical Engineering Department
Doctoral Program in Applied Biomedical Engineering

Program of Study

The Doctoral Program in Applied Biomedical Engineering is an interdisciplinary program and a specialization within the Doctor of Engineering degree program. It is a partnership in graduate education and research between the Fenn College of Engineering at Cleveland State University (CSU) and the world-renowned Cleveland Clinic Foundation. The program is housed in the Chemical Engineering department.

Research Facilities

Researchers and doctoral students in the program have access to state-of-the-art facilities at Cleveland State University and at the Cleveland Clinic Foundation. These facilities include equipment for surface analysis, tissue culture, inspection and image analysis, electronics, histopathology and electron microscopy, mechanical prototyping, heart valve studies, biomechanics, and image processing and a host of PCs, workstations, and mainframe computers. In addition, Cleveland State's Advanced Manufacturing Center has numerous facilities that supplement research needs, including a modern machine shop complete with CNC equipment and solid-modeling software, laser welding for fabrication of very small pieces, wire-cutting equipment, and various testing equipment.

Financial Aid

The Cleveland Clinic Foundation currently supports 10 doctoral students involved in research, providing them with stipends of $12,000, which Cleveland State University matches with full tuition. Faculty members award full research assistantships, which are funded by their research grants, to qualified students. Other financial aid programs are available through the University's Financial Aid Office.

Cost of Study

Tuition for Ohio residents for the fall 1999 semester is $202 per credit hour for up to 12 credit hours and a fixed $2626 for 13 to 16 credit hours. Tuition for out-of-state residents is twice these amounts. An additional technology fee of $7.50 per credit hour is also assessed.

Living and Housing Costs

Housing and food service for unmarried Cleveland State students are available in Viking Hall, which accommodates up to 600 students. These students live in double, large double, or triple rooms, each with a private bath. The University's Housing Bureau assists students in finding off-campus residences in the surrounding area.

Student Group

The applied biomedical engineering program began in the fall 1998 semester and had 12 doctoral students enrolled. It is expected that another 12 will be admitted to the program for the 1999–2000 academic year.

Location

CSU is located in the heart of downtown Cleveland on the shore of Lake Erie. This highly cosmopolitan area is the home for more than one fifth of the state's population. The Northeastern Ohio region has the presence of a strong local health-care industry, which includes some well-known biomedical companies, such as Steris, Picker International, Invacare, and AcroMed. The urban setting gives students easy access to the many cultural, scientific, sports-oriented, entertainment, and dining advantages of the city.

The University

Cleveland State University, a state-assisted institution, was established in 1964 through a merger with Fenn College, a private institution offering engineering education since 1923. The University's seven colleges offer sixty baccalaureate degree programs, three advanced degrees in law, and thirty-six graduate programs, including doctoral degrees in chemistry, biology, engineering, urban studies, and business administration. For more information about CSU, prospective students are invited to visit the University's Web page (http://www.csuohio.edu).

Applying

Applicants must have a master's degree in an engineering discipline from an accredited institution and present evidence of the ability to pursue graduate work, as exemplified by high scholastic achievement and strong recommendations. Applicants holding an M.S. in related sciences or an M.D. may be admitted, provided that their undergraduate degree is in an engineering discipline. Admission forms may be obtained from the College of Graduate Studies and must be filed no later than July 1; however, applicants are advised to apply early to be considered for financial aid. International students must follow special application and admission procedures.

Correspondence and Information

Information:
Chairman, Chemical Engineering Department
Cleveland State University
Euclid Avenue at East 24th Street
Cleveland, Ohio 44115-2425
E-mail: o.talu@csuohio.edu

Application materials:
College of Graduate Studies
Cleveland State University
Euclid Avenue at East 24th Street
Cleveland, Ohio 44115-2440
Telephone: 215-687-3592
E-mail: p.bellini@csuohio.edu

International applications:
International Admissions
Euclid Building 103
2344 Euclid Avenue
Cleveland, Ohio 44115-2407
Telephone: 216-687-3910
E-mail: b.turner@csuohio.edu

Cleveland State University

THE FACULTY AND THEIR RESEARCH

The focus of the program is to educate engineers who have the ability to do applied research; bring the results of basic research to the development of products, materials, and new technologies; and develop and perform prototype testing and evaluation for commercial viability.

Joanne M. Belovich, Ph.D., Michigan. Bioartifical pancreas, cell separation, bioseparations.

J. Fredrick Cornhill, D.Phil., Oxford. Image processing, atherosclerosis, cardiovascular devices.

Brian L. Davis, Ph.D., Penn State. Orthopedic biomechanics, space flight–induced osteoporosis, instrumentation.

Jorge E. Gatica, Ph.D., SUNY at Buffalo. Kinetics, mathematical modeling.

Rama S. R. Goria, Ph.D., Toledo. Blood flow, fluid mechanics.

Mark D. Grabiner, Ph.D., Illinois at Urbana-Champaign. Neuromotor control, skeletal muscle strength and endurance.

Geofrey R. Lockwood, Ph.D., Toronto. Ultrasound imaging systems, ultrasound transducer design.

Cahir McDevitt, Ph.D., London. Articular cartilage biochemistry, type VI collagen, adhesion proteins.

Majid Rashidi, Ph.D., Case Western Reserve. Living tissue mechanics, artificial joints and cardiac assist pump design.

William Smith, D.Eng., Cleveland State. Ventricular assist devices, medical device design.

Orhan Talu, Ph.D., Arizona State. Adsorption, bioseparations, thermodynamics.

Ivan Vesely, Ph.D., Western Ontario. Artificial heart valves, soft-tissue biomechanics, tissue engineering.

D. Geoffrey Vince, Ph.D., Liverpool. Vascular pathology, intravascular ultrasound, cell quantification.

Fuqin Xiong, Ph.D., Manitoba. Imaging, bioelectronics, communication networks.

DREXEL UNIVERSITY

School of Biomedical Engineering, Science and Health Systems

Programs of Study

The School of Biomedical Engineering, Science and Health Systems offers M.S. and Ph.D. programs in biomedical engineering and biomedical science. The biomedical engineering programs are intended for individuals with prior training in engineering, physical science, or mathematics. The biomedical science programs are designed for students with a life science background or with professional or advanced degrees in the health sciences or in biology. A special curriculum is available for students who wish to transfer from biomedical science to biomedical engineering. The completion of 45 quarter credits is required for the M.S. degree in biomedical engineering and the M.S. degree in biomedical science. A thesis is highly recommended but a nonthesis option is offered. The program of study for the M.S. degree usually takes two years to complete. Admission to the Ph.D. program is open to qualified students with training in engineering or natural or physical sciences and mathematics and to individuals with academic or professional degrees in a medical science discipline. Requirements for the Ph.D. include one full year of residency, 45 credits of course work beyond the M.S. degree, and dissertation research; a qualifying examination that covers both the program's core subjects and the student's areas of specialization is required. The Ph.D. candidacy exam is intended to examine the student's preparation and his or her ability to undertake dissertation research and the defense of the dissertation.

Research Facilities

The School operates core research and computing laboratories for both educational and research purposes. The School also operates the Calhoun Comparative Medicine Laboratories, which are Drexel's central facilities for the care and use of laboratory animals. The laboratories occupy about 9,000 square feet in the modern Lebow Engineering Center. Facilities are available for sterile surgery, radiography, and other research procedures. Networked computing equipment includes a number of IBM-compatible computers as well as Sun Systems, Silicon Graphics systems, a MicroVAX system, and Apple Macintosh computers. In addition to the core facilities, laboratories designed for specific research projects are operated by individuals or by teams of faculty members. These laboratories provide facilities for research in artificial organs, biocontrols, bioelectrochemistry, bioelectrodes, biomaterials, biomechanics, biomedical imaging and signal processing, biosensors, biostatistics, cardiac assist devices, cardiovascular system dynamics, cell culture, chronobiology, dental implants, diagnostic ultrasound, electrophysiology, modeling of physiological systems, neural networks and systems, and rehabilitative engineering.

Financial Aid

A limited number of teaching and research assistantships are available, with stipends that ranged from $800 to $1300 a month in 1998–99. Graduate fellowships supported by the Calhoun Endowment as well as various grants, scholarships, loans, employment opportunities, and internships are also available.

Cost of Study

Tuition for 1998–99 for students in the School of Biomedical Engineering, Science and Health Systems was $565 per credit hour. The general fee is $121 per term.

Living and Housing Costs

A limited number of accommodations for single graduate students are available in University residence halls. Apartment listings and housing assistance for graduate students are also available. Additional information can be obtained from the Residential Living Office, Kelly Hall, 203 North 34th Street.

Student Group

Drexel enrolls approximately 12,000 students, including more than 3,000 graduate students. Enrollment in the School of Biomedical Engineering, Science and Health Systems is composed of approximately 75 percent biomedical engineering students and 25 percent biomedical science students.

Location

Philadelphia, a city of 2 million people, is a center of science, industry, and culture. The Drexel campus is only a few minutes' walk from the heart of the city, with its many theaters and museums and its opera, orchestra, and ballet. Drexel is located in one of the major industrial and medical research centers in the nation. Five medical schools and more than a dozen major research and teaching hospitals are in the area. The School of Biomedical Engineering, Science and Health Systems has a cooperative agreement with Thomas Jefferson University. In addition, there are close working relationships with most of the area medical schools and hospitals, especially Temple University and Allegheny University.

The University and The School

Drexel, founded in 1891, consists of the Colleges of Engineering, Arts and Science, Business and Administration, Information Studies and the Nesbitt College of Design Arts, and the Schools of Biomedical Engineering, Science and Health Systems; Environmental Science, Engineering and Policy; and Education.

The School of Biomedical Engineering, Science and Health Systems (formerly the Biomedical Engineering and Science Institute), originally formed as a "program" in 1959, is one of the earliest efforts of its kind. It has always addressed the dual role of teaching and research in the application of scientific and technical means to the solution of medical and biological problems. The biomedical engineering academic program and participation in research activity provide students with strong engineering training as well as the basic medical and biological knowledge needed to function in a multidisciplinary environment. The biomedical science program provides biological, medical, mathematical, and physical tools needed to function in the clinical health-care field and in basic biomedical research. Through training in engineering, life science, and administration as well as an internship, the clinical/rehabilitation engineering program prepares students to transfer technology to the clinical environment.

The school recently developed an undergraduate program in biomedical engineering and enrolled its first freshman class in fall 1998.

Applying

Students may be admitted at any time, but most matriculate in September. Admission in any term other than fall requires special permission from the Graduate Advisor. Applicants are selected on the basis of previous academic work, letters of recommendation, and, whenever possible, a personal interview. Verbal and quantitative scores on the GRE General Test are recommended for admission and required for assistantship consideration.

Correspondence and Information

For departmental information:
School of Biomedical Engineering, Science
  and Health Systems
Drexel University
Philadelphia, Pennsylvania 19104
Telephone: 215-895-2215
Fax: 215-895-4983
World Wide Web: http://www.biomed.drexel.edu

For applications:
Dean of Admissions
Graduate School
Drexel University
Philadelphia, Pennsylvania 19104

Drexel University

THE FACULTY

Banu Onaral, Director and Professor, School of Biomedical Engineering, Science and Health Systems; Ph.D., Pennsylvania.

Anthony Addison, Professor, Chemistry; Ph.D., Kent at Canterbury (England).

Sorin Adrian, School of Biomedical Engineering, Science and Health Systems; Ph.D.

Tayfun Akgul, Visiting Assistant Professor, Electrical and Computer Engineering; Ph.D., Pittsburgh.

Leon Bahar, Professor, Electrical and Computer Engineering; Ph.D., Lehigh.

Richard Beard, Professor Emeritus, Electrical and Computer Engineering; Ph.D., Pennsylvania.

Nihat Bilgutay, Department Head, Electrical and Computer Engineering; Ph.D., Purdue.

Philip Bloomfield, Research Professor, School of Biomedical Engineering, Science and Health Systems; Ph.D., Chicago.

Jean-Claude Bradley, Assistant Professor, Chemistry; Ph.D., Ottawa.

Joseph Cammarota, Research Associate Professor, School of Biomedical Engineering, Science and Health Systems; Ph.D., Drexel.

Robin Carr, Director of Engineering Labs, School of Biomedical Engineering, Science and Health Systems; Ph.D., Toronto.

Douglas Chute, Professor, Psychology/Sociology/Anthropology; Ph.D., Missouri.

Fernand Cohen, Professor, Electrical and Computer Engineering; Ph.D., Brown.

Stephen Dubin, Clinical Professor, School of Biomedical Engineering, Science and Health Systems; Ph.D., Columbia Pacific.

Bruce Eisenstein, Professor, Electrical and Computer Engineering; Ph.D., Pennsylvania.

Mahmoud El-Sherif, Research Professor, Materials Engineering; Ph.D., Drexel.

Frank Ferrone, Professor, Physics and Atmospheric Sciences; Ph.D., Princeton.

Leonard Finegold, Professor, Physics and Atmospheric Sciences; Ph.D., London.

William Freedman, Associate Director and Associate Professor, School of Biomedical Engineering, Science and Health Systems; Ph.D., Drexel.

Richard Foulds, Research Professor, School of Biomedical Engineering, Science and Health Systems; Ph.D., Tufts.

Eli Fromm, Professor and Vice Provost; Ph.D., Jefferson Medical.

Vladimir Genis, Research Associate Professor, School of Biomedical Engineering, Science and Health Systems; Ph.D., Kiev.

Allon Guez, Associate Professor, Electrical and Computer Engineering; Ph.D., Florida.

Stephen Hartley, Assistant Professor, Math and Computer Sciences; Ph.D., Virginia.

Nira Herrmann, Department Head, Math and Computer Sciences; Ph.D., Stanford.

Thomas Hewett, Associate Professor, Psychology/Sociology/Anthropology; Ph.D., Illinois at Urbana–Champaign.

Leonid Hrebien, Professor, Electrical and Computer Engineering; Ph.D., Drexel.

Dov Jaron, Professor, School of Biomedical Engineering, Science and Health Systems; Ph.D., Pennsylvania.

Surya Kalidindi, Associate Professor, Materials Engineering; Ph.D., MIT.

Kylie Keshav, Assistant Professor, Bioscience and Biotechnology; Ph.D., Wollongong (Australia).

Frank Ko, Professor, Materials Engineering; Ph.D., Georgia Tech.

Alan Lau, Associate Professor, Mechanical Engineering; Ph.D., MIT.

Young Lee, Professor, Chemical Engineering; Ph.D., Purdue.

Peter Lewin, Professor, School of Biomedical Engineering, Science and Health Systems; Ph.D., Denmark.

Anthony Lowman, Assistant Professor, Chemical Engineering; Ph.D., Purdue.

Michele Marcolongo, Assistant Professor, Materials Engineering; Ph.D., Pennsylvania.

Katherine McCain, Professor, IST; Ph.D., Drexel.

Donald McEachron, Associate Director and Research Professor, School of Biomedical Engineering, Science and Health Systems; Ph.D., California, San Diego.

Alexander Meystel, Professor, Electrical and Computer Engineering; Ph.D., ENIMS (Russia).

Bahram Nabet, Associate Professor, Electrical and Computer Engineering; Ph.D., Washington (Seattle).

Vernon Newhouse, Professor Emeritus, School of Biomedical Engineering, Science and Health Systems; Ph.D., Leeds (England).

Jonathan Nissanov, Research Associate Professor, School of Biomedical Engineering, Science and Health Systems; Ph.D., Colorado.

Michael O'Connor, Assistant Professor, Bioscience and Biotechnology; M.D., Johns Hopkins.

Scott Overmyer, Assistant Professor, IST; Ph.D., George Mason.

Athina Petropulu, Associate Professor, Electrical and Computer Engineering; Ph.D., Northeastern.

Kambiz Pourezzaei, Associate Professor, Electrical and Computer Engineering; Ph.D., Rensselaer.

John Reid, Calhoun Professor Emeritus, School of Biomedical Engineering, Science and Health Systems; Ph.D., Pennsylvania.

Arye Rosen, Research Professor, School of Biomedical Engineering, Science and Health Systems; Ph.D., Drexel.

Kevin Scoles, Associate Professor, Electrical and Computer Engineering; Ph.D., Dartmouth.

Rahamim Seliktar, Professor, School of Biomedical Engineering, Science and Health Systems; Ph.D., Strathclyde (Scotland).

P. M. Shankar, Professor, Electrical and Computer Engineering; Ph.D., Indian Institute of Technology.

Wei-Heng Shih, Associate Professor, Materials Engineering; Ph.D., Ohio State.

Sorin Siegler, Associate Professor, Mechanical Engineering; Ph.D., Drexel.

James Spotila, Professor of Bioscience and Biotechnology; Ph.D., Arkansas.

Hun H. Sun, Professor Emeritus, Biomedical and Electrical and Computer Engineering; Ph.D., Cornell.

Oleh Tretiak, Professor, Electrical and Computer Engineering; Sc.D., MIT.

Thomas Twardowski, Assistant Professor, Materials Engineering; Ph.D., Illinois at Urbana-Champaign.

Yen Wei, Professor, Chemistry; Ph.D., CUNY.

Margaret Wheatley, Associate Professor, School of Biomedical Engineering, Science and Health Systems; Ph.D., Toronto.

Jack Zhou, Assistant Professor, Mechanical Engineering; Ph.D., NJIT.

Stanley Zietz, Associate Professor, Math and Computer Sciences; Ph.D., Berkeley.

Adjunct Faculty

David Bell, Ph.D., Thomas Jefferson University Hospital, Department of Biomedical Instrumentation.

David Brooks, Ph.D., Research Professor, Math and Computer Science Department.

Britton Chance, Ph.D., Department of Biochemistry and Biophysics, University of Pennsylvania.

Robert Clancy, M.D., Division of Neurology, Children's Hospital of Philadelphia.

Rebecca Craik, Ph.D., Professor of Physiology, Beaver College.

Michael Dellavecchia, Ph.D., M.D., Wills Eye Hospital.

Alberto Esquenazi, M.D., Director, Gait and Motion Analysis Laboratory, Moss Rehabilitation Hospital.

Scott Faro, M.D., Director of MRI and MRI Physics Group, Allegheny University of the Health Sciences.

Flemming Forsberg, Ph.D., Thomas Jefferson University Hospital.

Lawrence Gessman, M.D., Electrophysiology Department, Deborah Heart and Lung Center.

Barry Goldberg, M.D., Director, Division of Diagnostic Ultrasound Research and Education Institute.

Howard Hillstrom, Ph.D., Director of Gait Study Center, Pennsylvania College of Podiatric Medicine.

John Holmes, Ph.D., Center for Clinical Epidemiology and Biostatistics, University of Pennsylvania School of Medicine.

Ziwei Huang, Ph.D., Kimmel Cancer Center, Jefferson Medical College.

Jeffrey Joseph, D.O., Director, The Artificial Pancreas Center at Jefferson.

Philip Katz, Ph.D., Vice President and CIO, Health Systems International.

Cato Laurencin, M.D., Associate Professor of Orthopaedic Surgery, Allegheny University of the Health Sciences.

Andrew Maidment, Ph.D., Assistant Professor of Radiology, Thomas Jefferson University Hospital.

Suzane Maxian, Ph.D., Director, Biomaterials Study Center, Pennsylvania College of Podiatric Medicine.

Feroze Mohamed, Ph.D., Department of Radiology, Allegheny University of the Health Sciences.

Rodney Murray, Ph.D., Director of Academic Computing and Instructional Technology, Thomas Jefferson Medical Center.

David Nash, M.B.A./M.D., Director of Health Policy and Clinical Outcomes, Thomas Jefferson University Hospital.

Igal Nevo, M.D., Director, Laboratory of Medical Informatics, Albert Einstein Medical Center.

William Santamore, Ph.D., M.D., Professor of Surgery, University of Louisville School of Medicine.

Mark Schafer, Ph.D., Sonic Technologies.

Lisa Selby-Silverstein, Ph.D., PT, Director, Human Performance Laboratory, Thomas Jefferson University.

Barry Shender, Ph.D., Crew Systems Engineering Department/Human Performance Branch, Navy.

David Soll, M.D., Soll Eye Associates.

Samuel Steinberg, M.B.A., President, Graduate Hospital and Mt. Sinai Hospital.

Gerald Sterling, Ph.D., Department of Pharmacology, Temple University School of Medicine.

Ira Tackel, M.S., Department of Biomedical Instrumentation, Thomas Jefferson University.

Ronald Tallarida, Ph.D., Department of Pharmacology, Temple University School of Medicine.

Rocky Tuan, Ph.D., Director of Orthopaedic Research, Thomas Jefferson University.

Simon Vinitski, Ph.D., Professor of Radiology, Thomas Jefferson University.

Arjun Yodh, Ph.D., Department of Physics and Astronomy, University of Pennsylvania.

DUKE UNIVERSITY

School of Engineering
Department of Biomedical Engineering

Programs of Study

The Department of Biomedical Engineering offers programs leading to the M.S. and Ph.D. degrees. Graduate programs in biomedical engineering are organized primarily for students who are working toward the Ph.D. degree. There is no single list of required graduate courses. The best course choices must necessarily take into account the diverse backgrounds of entering students and the different objectives of various research projects. Each student's program is supervised by a committee of graduate faculty members. The committee meets with the student at specified intervals to review all aspects of his or her program. The committee's chairman advises the graduate student and the director of graduate studies in biomedical engineering, in writing, of the decisions of the committee.

A minimum of 30 credit hours is required for the M.S. degree; 6 of these may be for the thesis. Both thesis and nonthesis M.S. programs are offered. Students who are working toward a Ph.D. often earn an M.S. degree through the thesis option as a first step. A minimum of 60 credit hours is required for the Ph.D. degree, including 12 hours for the doctoral dissertation. The doctoral dissertation, accomplished as an independent investigation, should demonstrate significant and original contributions to an interdisciplinary topic. Every biomedical engineering student is required to serve as a teaching assistant as part of the graduate training. One semester is required of master's candidates, and three semesters are required of Ph.D. candidates.

Research Facilities

The Department of Biomedical Engineering has about 17,500 square feet of office and laboratory space in the Engineering Building. Numerous facilities for biomedical engineering research are available on campus.

The Tissue Properties and Orthopaedic Research Laboratory includes equipment for biomechanical studies of tissue properties, trauma mechanics, and orthopaedic appliance development. The Interventional Cardiac Catheterization Facility is available for the study of mechanical and laser methods for treating atherosclerosis. The Optics and Biosensors Laboratory is used to study polymer optical waveguides and their interaction with adsorbed proteins and the development of fiber-optic biosensors.

The Ultrasonic Laboratories are housed within the Department of Biomedical Engineering as well as in the area of the Cardiac Diagnostic Unit of Duke Medical Center. Facilities include a variety of modern electronic test equipment and ultrasound transducer measurement equipment. There is also an ultrasound transducer fabrication facility and several clinical phased array scanners. The Medical Imaging Laboratory resources range from well-equipped instrumentation laboratories to state-of-the-art digital angiography. PET imagers, animal surgical suites, CO_2 and Yag lasers, and a wide variety of computational facilities are available. The research SPECT Laboratory and workstation rooms are equipped with computers, a cyclotron, a two-ring PET tomograph, and chemistry laboratories.

The Cardiac Stimulation and Simulation Laboratory, part of the Engineering Research Center, includes animal and electronics laboratories and a computer room. Animal experiments are performed in a well-equipped 800-square-foot electrophysiology animal laboratory.

The Medical Informatics Laboratory is concerned with the acquisition, storage, processing, analysis, and presentation of medical data; linkage of knowledge to data; and conversion of data into new knowledge in order to satisfy educational, patient care, research, administrative, and financial requirements.

The Experimental Electrophysiology Laboratory is well equipped with the various manipulators, electrode systems, and stimulators used in electrophysiological studies and is particularly well equipped for extracellular, as well as intracellular, waveform measurements. Duke researchers and graduate students also have access to a CRAY Y-MP/432 supercomputer at the North Carolina Supercomputing Center at Research Triangle Park.

There is unrestricted access to the Duke University Library System, which includes the William R. Perkins Library and its seven branches on campus and the independently administered libraries of Law, Medicine, and Business. The Engineering Library is adjacent to the School of Engineering.

Financial Aid

The Graduate School and School of Engineering offer fellowships to outstanding students. Support is sometimes available from research grants made to the faculty. In 1999–2000, the average stipend level for biomedical engineering graduate students is $1400 per month.

Cost of Study

For the 1999–2000 academic year, tuition for all full-time students is a flat rate of $8760 per semester plus a $1250 registration fee per semester. After six semesters, only the registration fee is charged. For part-time students, tuition is $730 per credit plus an additional $1250 registration fee per semester. All students are required to pay the student health fee of $220 per semester and the student government fee of $9.50 per semester. There is a one-time transcript fee of $30. All full-time students are required to register for 12 units per semester until their unit requirement is met. Tuition and fees are subject to change for the 2000–01 academic year.

Living and Housing Costs

Graduate students share off-campus apartments or houses, paying rents of about $300–$375 per month depending on size and location.

Student Group

In 1998–99, the department had 80 full-time graduate students. Many countries were represented. Women made up about 29 percent of the group. The School of Engineering enrolls about 250 of the approximately 2,260 graduate students who attend the University. The average grade point of matriculants over the past five years has exceeded 3.5. Average GRE scores for the verbal and quantitative portions of the test have exceeded 1350.

Location

Duke University and the Duke Medical Center are located at one of the vertices of the triangle formed by Raleigh, Durham, and Chapel Hill. Within the boundaries of that triangle are three major universities and the Research Triangle Park.

The University

Duke University, a privately supported institution, has a competitively selected student body.

Applying

Applications for admission should be completed by December 31 for the fall semester and by October 1 for the spring semester. Funding is limited for spring semester awards. Applications must include transcripts, GRE General Test scores, and recommendations before they can be considered by the department.

Correspondence and Information

Gregg E. Trahey
Director of Graduate Studies
Department of Biomedical Engineering
Duke University
Box 90281
Durham, North Carolina 27708
World Wide Web: http://bme-www.mc.duke.edu/bme.html

Duke University

THE FACULTY AND THEIR RESEARCH

Roger C. Barr, Professor of Biomedical Engineering and Associate Professor of Pediatrics; Ph.D., Duke. Cardiac electrophysiology and computing.

Alan H. Baydush, Assistant Research Professor of Radiology and Biomedical Engineering; Ph.D., Duke. Bayesian image processing.

Ashutosh Chilkoti, Assistant Professor of Biomedical Engineering; Ph.D., Washington. Protein engineering and molecular surface engineering.

James T. Dobbins, Associate Professor of Radiology and Biomedical Engineering; Ph.D., Wisconsin. Advanced digital radiographic imaging.

Carey E. Floyd Jr., Research Professor of Radiology and Biomedical Engineering; Ph.D., Duke. Digital radiography.

David R. Gilland, Assistant Research Professor of Radiology and Biomedical Engineering; Ph.D., North Carolina at Chapel Hill. Nuclear medicine.

Farshid Guilak, Assistant Professor of Orthopaedic Surgery and Biomedical Engineering; Ph.D., Columbia. Orthopaedic biomechanics, cell and tissue engineering.

Joseph W. Hales, Assistant Research Professor of Community and Family Medicine and Biomedical Engineering; Ph.D., Utah. Medical informatics: medical concept representation and computer-based patient records.

William E. Hammond, Professor of Community and Family Medicine and of Biomedical Engineering; Ph.D., Duke. Biomedical computing.

Craig S. Henriquez, Associate Professor of Biomedical Engineering; Ph.D., Duke. Modeling cardiac propagation and arrhythmogenesis.

Edward W. Hsu, Assistant Professor of Biomedical Engineering; Ph.D., Johns Hopkins. Magnetic resonance imaging.

Ronald J. Jaszczak, Associate Professor of Radiology and Biomedical Engineering; Ph.D., Florida. Single-photon emission computed tomography.

G. Allan Johnson, Professor of Radiology, Physics and Biomedical Engineering; Ph.D., Duke. Nuclear magnetic resonance microscopy.

David F. Katz, Professor of Biomedical Engineering and Obstetrics and Gynecology; Ph.D., Berkeley. Reproduction and fertility.

Wanda Krassowska, Associate Professor of Biomedical Engineering; Ph.D., Duke. Electrical stimulation of the heart.

Joseph Y. Lo, Assistant Research Professor of Radiology and Biomedical Engineering; Ph.D., Duke. Computer-aided image analysis.

David F. Lobach, Assistant Research Professor of Community and Family Medicine and Biomedical Engineering; M.D., Ph.D., Duke. Development and evaluation of decision support systems for medical care.

James R. MacFall, Associate Professor of Radiology and Biomedical Engineering; Ph.D., Maryland. Functional MRI and NMR microscopy high-speed imaging.

James H. McElhaney, Professor of Biomedical Engineering and Experimental Orthopaedics, Department of Surgery; Ph.D., West Virginia. Biomechanics and prostheses.

Barry S. Myers, Associate Professor of Biomedical Engineering and Orthopaedic Surgery; M.D., Ph.D., Duke. Orthopaedic biomechanics.

David Needham, Associate Professor of Mechanical Engineering and Materials Science and of Biomedical Engineering; Ph.D., Nottingham (England). Material properties of biological cells and lipid bilayer membranes, structure at liquid/"soft" solid interfaces and the balance and range of intersurface forces with applications in microcarrier drug delivery.

Laura E. Niklason, Assistant Professor of Anesthesia and Biomedical Engineering; Ph.D., Chicago, M.D., Michigan. Tissue-engineered vascular grafts.

William M. Reichert, Associate Professor of Biomedical Engineering and Biochemical Engineering; Ph.D., Michigan. Biosensors.

Lori A. Setton, Assistant Professor of Biomedical Engineering; Ph.D., Columbia. Soft-tissue mechanics.

Stephen W. Smith, Professor of Biomedical Engineering and Adjunct Professor of Radiology; Ph.D., Duke. Acoustical imaging and instrumentation.

Gregg E. Trahey, Professor of Biomedical Engineering and Assistant Professor of Radiology; Ph.D., Duke. Acoustical imaging.

George A. Truskey, Associate Professor of Biomedical Engineering and Biochemical Engineering; Ph.D., MIT. Lipoprotein transport in arterial wall, mammalian cell culture.

Olaf T. von Ramm, Professor of Biomedical Engineering and Assistant Professor of Medicine; Ph.D., Duke. Acoustical imaging and instrumentation.

Patrick Wolf, Assistant Professor of Biomedical Engineering; Ph.D., Duke. Instrumentation for the diagnosis and treatment of cardiac arrhythmias.

Fan Yuan, Assistant Professor of Biomedical Engineering; Ph.D., CUNY. Drug delivery to and within tumors.

GEORGIA INSTITUTE OF TECHNOLOGY
A Unit of the University System of Georgia

Graduate Studies in Bioengineering

Program of Study	Graduate M.S. and Ph.D. degrees in bioengineering are coordinated through the College of Engineering's Multidisciplinary Committee for Bioengineering. Academic units involved in bioengineering education and research include the Schools of Aerospace Engineering, Biomedical Engineering, Chemical Engineering, Electrical and Computer Engineering, Materials Engineering, Mechanical Engineering, and Textile and Fiber Engineering and the College of Computing. Biomedical engineering is a joint academic department with the Emory University School of Medicine. In addition, qualified graduate students may apply for the M.D./Ph.D. program offered jointly with Emory University and the Medical College of Georgia. Bioengineering faculty members at Georgia Tech participate in collaborative research programs with a variety of medical schools across the country, including Emory University, Harvard University, the University of Alabama at Birmingham, the University of Texas Health Science Center at San Antonio, the University of Chicago, and the Medical College of Georgia.
Research Facilities	The bioengineering program has an extensive array of specialized equipment to facilitate research in biomechanics, tissue engineering, cardiovascular dynamics, bioprocessing, noninvasive diagnostic methods, and rehabilitation engineering. A variety of computing facilities are available, including state-of-the-art parallel supercomputers and single-user workstations. The Price Gilbert Memorial Library's collection includes 1.7 million volumes, 2.1 million microtexts, and the largest collection of patents in the Southeast. Georgia Tech currently receives more than 14,000 serials. The recently completed, $30-million Petit Institute for Bioengineering and Bioscience now houses the Bioengineering Program, the School of Biomedical Engineering, and extensive research facilities.
Financial Aid	Four sources of financial aid are available to graduate students: teaching assistantships, research assistantships, fellowships, and out-of-state tuition waivers. Graduate assistantships carry a twelve-month stipend and a reduction of fees. A wide variety of federal, industrial, and private financial aid packages are available. Two highlights of these packages are the Medtronic Scholarships and the Industrial Internship Program. For outstanding Ph.D. applicants, a special traineeship program funded by a recent educational grant from the Whitaker Foundation is also available. International students are required to guarantee their support but are eligible to compete for financial assistance.
Cost of Study	In 1999–2000, estimated total fees for graduate students carrying a full academic load are $1847 per semester for residents of Georgia and $6173 for nonresidents. Part-time students are charged prorated amounts. Fees are subject to change without notice.
Living and Housing Costs	Dormitory rooms for unmarried students and apartments for married students and their families are available at reasonable costs through the Institute. Rooms and apartments in privately owned dwellings within walking distance or a short driving distance are available in several price categories. Students should write to the Housing Office for details. Unmarried students will find that the minimum necessary yearly expenses, exclusive of tuition and fees, are estimated at $12,722.
Student Group	In the fall of 1998, there were 13,959 students at Georgia Tech, of whom 3,655 were graduate students. Of the 8,459 engineering students, 2,282 were graduate students. Many graduate students received financial aid. Forty-five percent of Georgia Tech's student body come from outside the state of Georgia.
Location	Atlanta is a cosmopolitan city, the center of a metropolitan area of more than 2 million inhabitants. Located high on the Piedmont Plateau, the area is one of beautiful homes, rolling terrain, and dense stands of woods. There are numerous cultural facilities in town and at the nineteen colleges and universities in the area. There are numerous cultural facilities in town and at the nineteen colleges and universities in the area. Professional football, baseball, basketball, and hockey are played in Atlanta's municipal stadium and at other sports facilities. A mecca for young professionals, Atlanta has a vibrant night life and many eclectic cultural pockets. These include Buckhead, Little Five Points, Virginia Highlands, the Sweet Auburn Area, and the Lenox Area.
The Institute	The Georgia Institute of Technology is located on a 330-acre campus near downtown Atlanta, the business and communications center of the Southeast. The quality of the student body is high; the Institute enrolls the highest percentage of freshmen National Merit Scholars and National Achievement Scholars among publicly supported U.S. colleges. Georgia Tech also ranks first among public universities in engineering research and development expenditures, with a total annual research budget exceeding $130 million. In addition to the academic departments, three research centers participate in bioengineering programs: the Bioengineering Center and the Center for Rehabilitation Technology at Georgia Tech and the Emory University–Georgia Tech Biomedical Technology Research Center. The Bioengineering Center provides leadership and coordination for biomedical research both on campus and with off-campus organizations. The Emory–Georgia Tech Center was established to create research and educational opportunities in which the Emory Medical School in Atlanta and Georgia Tech both participate. This center provides seed funds for cooperative research and coordinates the M.D./Ph.D. programs between the two schools.
Applying	Application forms for admission and financial aid may be obtained by writing to the address given below and should be returned, together with letters of recommendation and official transcripts of previous academic work, before January 31 by those applying to enter the next fall. Applications to the M.D./Ph.D. program should be submitted simultaneously to the director of the Bioengineering Center at Georgia Tech and to the Admissions Office at Emory University Medical School or the Admissions Office of the Medical College of Georgia.
Correspondence and Information	Professor Ajit P. Yoganathan Director, Bioengineering Center–Petit Institute for Bioengineering and Bioscience School of Biomedical Engineering Georgia Institute of Technology Atlanta, Georgia 30332-0535 Telephone: 404-894-7063 Fax: 404-894-4243

Georgia Institute of Technology

THE FACULTY AND THEIR RESEARCH

Paul J. Benkeser, Associate Professor; Ph.D., Illinois. Biomedical ultrasonics, medical image processing.

Gregory Berns, Assistant Professor; Ph.D., California, Davis. Cognitive neuroscience and imaging.

Robert S. Cargill II, Assistant Professor; Ph.D., Pennsylvania. Cell biomechanics, neural cell trauma, smooth muscle tissue development, tissue engineering.

Elliot L. Chaikof, Assistant Professor; M.D., Johns Hopkins; Ph.D., MIT. Biomolecular materials, cell motility, bioartificial organs.

Stephen P. DeWeerth, Associate Professor; Ph.D., Caltech. Neuromorphic systems, neural prostheses, neuro-microelectric interfacing.

William Ditto, Professor; Ph.D., Clemson. Research into the biodynamics and chaotic behavior of fibrillation in hearts and seizures in brains.

Robert L. Eisner, Associate Professor; Ph.D., Purdue. Noninvasive cardiac imaging.

Norberto Ezquerra, Associate Professor; Ph.D., Florida State. Medical informatics, medical imaging, computer vision, visualization, artificial intelligence.

Zorina Galis, Assistant Professor of Medicine; Ph.D., McGill. Pathology of blood vessels.

Andres J. Garcia, Assistant Professor; Ph.D., Pennsylvania. Musculoskeletal tissue engineering, biomaterials, cellular engineering, integrin adhesion receptors and signaling.

Ernest V. Garcia, Professor; Ph.D., Miami (Florida). Noninvasive cardiac imaging, positron emission tomography.

Don P. Giddens, Professor and Chair; Ph.D., Georgia Tech. Lawrence L. Gellerstedt Jr., Chair, Georgia Tech. Professor of Bomedical Engineering, Emory. Biofluid dynamics, atherosclerosis, biomedical modeling and computing, vascular biomechanics.

Robert E. Guldberg, Assistant Professor; Ph.D., Michigan. Orthopedic biomechanics, image-based modeling, bone-tissue engineering.

William D. Hunt, Associate Professor; Ph.D., Illinois. Transducer development for intravascular ultrasound, biobased chemical sensors, thrombometer.

William C. Hutton, Professor; D.Sc., London. Orthopedic mechanics, lumbar spine, osteoarthritis, foot function.

Arthur Koblasz, Associate Professor; Ph.D., Caltech. Rehabilitation engineering and medical diagnostics.

David N. Ku, Regent's Professor; Ph.D., Georgia Tech; M.D., Emory. Hemodynamics, MRI, thrombosis, tissue vascular grafts, biomaterials.

Michelle LaPlaca, Assistant Professor; Ph.D., Pennsylvania. Biomechanics of traumatic neural injury, molecular mechanisms of cell death, cellular engineering.

Joseph LeDoux, Assistant Professor; Ph.D., Rutgers. Genetic therapies and tissue engineering.

Marc Levenston, Assistant Professor; Ph.D., Stanford. Problems in mechanical engineering, effects of mechanical environment on orthopedic soft tissue biology.

Peter J. Ludovice, Assistant Professor; Ph.D., MIT. Computer simulation and molecular modeling of biomolecular systems.

Shamkant B. Navathe, Professor; Ph.D., Michigan. Database modeling and design, user interfaces and query languages for information retrieval, knowledge-based systems, genome data management.

Robert M. Nerem, Parker H. Petit Distinguished Professor; Ph.D., Ohio State. Atherosclerosis, hemodynamics, vascular endothelial cells, cellular and tissue engineering.

Roderic I. Pettigrew, Associate Professor; Ph.D., MIT; M.D., Miami (Florida). Magnetic resonance and nuclear imaging of the cardiovascular system.

Athanassios Sambanis, Associate Professor; Ph.D., Minnesota. Tissue engineering, cellular engineering, intracellular protein trafficking and secretion, modeling of the cell.

Raymond P. Vito, Professor; Ph.D., Cornell. Soft tissue mechanics, biomechanical design.

Timothy M. Wick, Associate Professor; Ph.D., Rice. Cell-cell interactions, blood-cell rheology, blood-cell adhesion, large-scale cell culture, cellular engineering, tissue engineering.

Ajit P. Yoganathan, Regents Professor; Ph.D., Caltech. Cardiovascular fluid mechanics, artificial heart valves, Doppler ultrasound, MRI, blood rheology.

Guotong Zhou, Assistant Professor; Ph.D., Virginia. Biosignal processing, medical image processing.

Cheng Zhu, Associate Professor; Ph.D., Columbia. Cell/molecular mechanics, application to immunology and tumor biology.

HARVARD UNIVERSITY/MASSACHUSETTS INSTITUTE OF TECHNOLOGY

Division of Health Sciences and Technology
Medical Engineering/Medical Physics Program

Program of Study

This curriculum is a five- to six-year program leading to the Ph.D. or Sc.D. degree awarded by MIT or the Harvard Faculty of Arts and Sciences. The objective of this program is to educate individuals who will be well qualified as engineers or physicists, and who will have extensive knowledge of the medical sciences such that they may engage in productive independent investigation of important problems at the interface of technology and clinical medicine. During the initial phase of the program, students are registered in both a graduate department and HST and concentrate on developing strength in their basic engineering or physical science disciplines. In some departments this involves completing departmental requirements for an S.M. degree, including thesis. Students also take approximately six subjects in human anatomy, pathology, and pathophysiology together with HST medical students. In order to continue to the second phase of the program, Medical Engineering/Medical Physics (MEMP) students must satisfactorily pass a qualifying examination administered by their department and by HST. This examination tests competence in engineering or physics and also reviews the student's progress in independent research.

Students next participate in two intensive six-week clinical courses. One is held primarily during January, and the second occurs immediately following the end of the spring semester. Students acquire the skills of physical examination and history taking and participate in the longitudinal care of patients. Later in the program, usually in association with their doctoral research, students participate in an individually arranged interdisciplinary clinical preceptorship in which they study, in depth, a particular biomedical technology and its application to patient care or clinical research. These clinical experiences provide the student with an intimate understanding of the world of medicine and of the practical constraints and the opportunities for applying science and technology to health needs. A new MEMP track emphasizes the application of cellular and molecular biology to problems in clinical medicine. This track gives students the option of substituting some of the clinical and preclinical courses with advanced training in cellular and molecular biology.

Doctoral thesis research is conducted under the direction of faculty members from MIT or Harvard and should focus on a fundamental problem of clinical relevance. Thesis research may be done on the MIT or Harvard campus or at one of the teaching hospitals.

Research Facilities

The research facilities are extensive and include laboratories at MIT, Harvard Medical School, the Harvard Division of Applied Sciences, the Harvard Faculty of Arts and Sciences, and the affiliated hospitals. Students have access to the libraries of Harvard, MIT, and the Harvard Medical School.

Ph.D. candidates in the Medical Engineering/Medical Physics Program are eligible to conduct their Ph.D. research in the laboratories of faculty members at MIT, Harvard University, Harvard Medical School, and the affiliated hospitals. MEMP students are thus able to access the extraordinary diversity and richness of the combined Harvard-MIT resources in biomedical engineering.

Financial Aid

Financial aid to Ph.D. candidates is made available from a variety of sources and in several different forms—fellowships, traineeships, scholarships, teaching and research assistantships, and loans. While the Division of Health Sciences and Technology makes every effort to secure resources that will provide financial aid to deserving and needy students, it is important that every prospective student investigate other sources of aid that may be available from the National Science Foundation and other agencies. The Division provides MEMP students with three semesters of fellowship and/or research assistantship support. Support may also come from the research laboratory where students conduct their Ph.D. research.

Cost of Study

For students registered at MIT, tuition is $25,000 (plus $636 in fees) for the 1999–2000 academic year and $8335 for the summer. For students registered at Harvard, tuition is $23,325 for the academic year.

Living and Housing Costs

Living expenses, including room, meals, and miscellaneous expenses, average $1450 per month for single students and slightly more for married students. Single graduate students at MIT may live in Ashdown House or Tang Hall. Married graduate students at MIT may live at Westgate or Eastgate. Five dormitories at Harvard University are reserved for single men and women graduate students. Married student graduate housing at Harvard University is available in several apartment complexes owned by Harvard. Many graduate students at both institutions live in apartments and houses in the surrounding communities.

Student Group

The Medical Engineering/Medical Physics Program enrolled 73 students in 1998–99, including 19 women. All students receive financial aid.

Location

Harvard and MIT are located in Cambridge, just across the Charles River from Boston.

The Institutions and The Division

Harvard is the oldest college in the United States, founded in 1636. MIT, founded in 1861 as a private, endowed institution committed to the extension of knowledge through teaching and research, has grown to be one of the foremost institutes of technology in the world.

The HST Division, established in 1977, formalized a major collaborative effort in the health sciences begun in 1970 as a program. This effort, designed to focus science and technology on human health needs, draws on the complementary strengths of the two institutions.

Applying

Applicants generally have a baccalaureate degree in engineering or physical science. At least one undergraduate course in each of the following is strongly recommended: biology, organic chemistry, biochemistry, and advanced calculus.

Students who apply through MIT must also apply as regular graduate students in one of the engineering departments, in physics, or in chemistry (normally the department most closely related to the undergraduate major). Interested individuals should write to the Office of Admissions, Room 3-103, Massachusetts Institute of Technology, Cambridge, Massachusetts 02139. The application deadline for September admission at MIT is January 15.

Students who apply through Harvard must apply as candidates for the Ph.D. degree in engineering or applied physics under the Harvard Division of Applied Sciences or for the Ph.D. in physics under the Harvard Department of Physics. Applications may be obtained from the Admissions Office, Graduate School of Arts and Sciences, Byerly Hall 203, Harvard University, 8 Garden Street, Cambridge, Massachusetts 02138. Applications for September admission at Harvard are due by December 15.

Correspondence and Information

Dr. Martha Gray
Room E25-519
Massachusetts Institute of Technology
Cambridge, Massachusetts 02139
Telephone: 617-253-1445

Russell E. Berg
Byerly Hall, 2nd Floor
Harvard University
Cambridge, Massachusetts 02138
Telephone: 617-495-5396

Harvard University / Massachusetts Institute of Technology

RESEARCH PROGRAMS

Research and development programs under the aegis of the Division are focused on major medical and health problems. The Division brings together multidisciplinary teams—physicians, engineers, and scientists—from Harvard Medical School, its teaching hospitals, and MIT to work on a scale appropriate to the goals of the research. The programs span the range from fundamental scientific research to applied research and development. Some active programs within the Division are listed here.

Harvard-MIT Arteriosclerosis Center. Robert S. Lees, M.D., Director. The Center's programs include development of telemedicine technology; new approaches to drug and extracorporeal therapy of hypercholesterolemia; ultrasound, angiographic, and magnetic resonance studies of cardiovascular function; and new diagnostic agents for radionuclide imaging of atherosclerosis. The faculty includes Anibal A. Arjona, Ph.D.; Robert E. Dinsmore, M.D.; Linda Hemphill, M.D.; Fred Holmvang, M.D.; Simon W. Law, Ph.D.; Ann M. Lees, M.D.; and Robert S. Lees, M.D.

Harvard-MIT Center for Biomedical Engineering. Elazer R. Edelman, M.D., Ph.D., Director. The focus of the center is to apply principles of physics and engineering to basic biomedical research and to solve practical problems in clinical medicine. The center has particular strengths in signal processing, physiological modeling, transducer technology, and development of microcomputer-based electronic instrumentation. Among the more than forty current projects within the center are the following: Electroporation: Theory of Basic Mechanisms: James C. Weaver, Ph.D.; Cardiac Arrhythmia Analysis: Roger G. Mark, M.D., Ph.D.; Medical Instrumentation and Physiologic Monitoring: H. F. Bowman, Ph.D.; Optimal Control Modeling in Respiratory Physiology: Chi-Sang Poon, Ph.D.; Imaging Cardiac Electrical Activity: Richard J. Cohen, M.D.; Mechanism of Atherosclerosis: Elazer R. Edelman, M.D., Ph.D.

Hyperthermia: Clinic, Biology, Physics. H. Frederick Bowman, Ph.D., Program Director. Superficial and deep tumor hyperthermia utilizing steered, focused ultrasound. Thermal models, thermal dosimetry, thermal physics, thermometry, perfusion measurements, 3-D visualization of thermal parameters, correlation with clinical results.

Regulation of Protein Synthesis and Erythropoiesis. Irving M. London, M.D., Principal Investigator. The objectives of this research are the elucidation of the mechanisms controlling the synthesis of hemoglobin and other proteins in eukaryotic cells, and the definition of the determinants of erythroid cell differentiation and maturation. The program is applying this new knowledge to the development of gene therapy for sickle cell anemia and thalassemia.

Cartilage Degradation in Arthritis and Cartilage Tissue Engineering. Alan J. Grodzinsky, Sc.D., Principal Investigator. Study of effects of mechanical loading forces on connective tissue cell metabolism. Study of the pathological effects of matrix proteases that degrade human cartilage; relevance to arthritis and normal cartilage repair. Design and development of a sensor probe for detection of early osteoarthritic disease via electromechanical spectroscopic imaging; synthesis of a mechanically functional cartilage substitute tissue from chondrocytes in gel matrices.

RNA Structure and RNA-Protein Interactions in mRNA Translation, Virus Replication, and Virus Assembly. Lee Gehrke, Ph.D., Principal Investigator. Biochemical and biophysical methods are used to define sequences and structures required for formation of specific ribonucleoprotein complexes. Translational control mechanisms involving 3-foot untranslated sequences are studied.

Magnetic Resonance Imaging of Cartilage. Martha Gray, Ph.D., and Deborah Burstein, Ph.D., Principal Investigators. NMR spectroscopy techniques are used to investigate composition and transport in cartilage in order to assess nondestructively the functional status of cartilage in vitro and in vivo.

Lasers and Medicine. Michael S. Feld, Ph.D. The Laser Biomedical Research Center and NIH Biotechnology Resource Center for Research in Lasers and Medicine is a user-oriented facility set up in the MIT Spectroscopy Laboratory to study fundamental light-tissue interactions that have potential biomedical applications. Research areas of interest include the physics of laser tissue removal for advanced microsurgical applications and the use of laser spectroscopy to type tissue and diagnose disease.

Biomedical Engineering. Ernest G. Cravalho, Ph.D. The Department of Surgery, West Roxbury Veterans Administration Hospital, has ties to both the Harvard Medical School and MIT. In addition to providing patient care, the department is responsible for the safe use of technology in patient care and is engaged in the design and development of special devices and instruments for both clinical and research applications.

Medical Computer Science. Peter Szolovits, Ph.D. Research within the Clinical Decision Making Group of the MIT Laboratory for Computer Science focuses on the application of artificial intelligence and database techniques to the study of medical reasoning, the development of computer programs to aid medical personnel in diagnosis and therapy, the creation of integrated medical record systems, and development of lifelong personal health information systems (http://www.medg.lcs.mit.edu)

Drug Delivery Systems. Robert S. Langer Jr., Sc.D. Research efforts focus on studying and developing new approaches for controlled drug administration, particularly for the delivery of large molecules such as proteins (MW>1000kD) produced by genetic engineering.

Humans in Space. Laurence C. Young, Sc.D. Research efforts focus on the survival and productivity of humans in space, including the balance system, vestibuloocular reflexes, and motor control. Human factors research includes the use of artificial intelligence to assist astronaut decision making. Dr. Young also directs the National Space Biomedical Research Institute, of which MIT and Harvard are members.

Auditory Physiology. William T. Peake, Ph.D.; Dennis M. Freeman, Ph.D.; Thomas F. Weiss, Ph.D.; and Bertrand Delgutte, Ph.D. The fundamental objective is to further understanding of the mechanisms that underlie sensory communication. Current research focuses on the study of transmission and coding of stimulus information in normal and pathological auditory nervous systems. A strong collaboration exists with the Eaton-Peabody Laboratory at the Massachusetts Eye and Ear Infirmary.

Uses of Radiation in Medicine. Jacquelyn C. Yanch, Ph.D. The Laboratory for Accelerator Beam Applications houses a high current 4 MeV proton/deuteron accelerator used in various research applications of radiation therapy. In the Biomedical Imaging and Computation Laboratory, applications of radiation in medicine (both diagnostic and therapeutic) are explored using computational tools, such as Monte Carlo simulation of photons, electrons, and neutrons or medical image analysis and processing. Computational research is coupled with experimental data acquisition in such areas as radiation dosimetry, radiation beam design, and image acquisition in nuclear medicine.

HST Center for Experimental Pharmacology and Therapeutics. Robert H. Rubin, M.D., Director. Research focuses on the development of innovative quantitative techniques for the assessment of physiological function, pathophysiological derangement, and pharmacological effects. Applications include positron emission tomography, magnetic resonance imaging and advanced ultrasound to assess patients before and after therapy.

Auditory Perception and Aids for the Deaf. Louis Braida, Ph.D. Program focuses on hearing aid research; functional models of hearing impairment; perception of speech by listeners with normal and impaired hearing; and acoustic and visual aids to speech reading.

Noninvasive Techniques and Electroporation. James C. Weaver, Ph.D. Basic and applied research for new noninvasive technologies, particularly transdermal drug delivery and chemical sensing. Theory of electric and magnetic field effects in biological systems, emphasizing electroporation of cells and tissues.

Mechanism of Atherosclerosis. Elazer R. Edelman, M.D., Ph.D., Director. Research in this lab melds clinical interests in unstable coronary syndromes with scientific studies in pharmacology, biomaterials science, high resolution microscopy and image analysis, polymeric drug delivery, tissue engineering, cell and molecular biology, and biochemistry.

Laboratory of Surgical Science and Engineering (LSSE). Martin L. Yarmush, M.D., Ph.D.; Ronald Tompkins, M.D., Sc.D.; Mehmet Toner, Ph.D.; Jeffery Morgan, Ph.D. Engineers, scientists, and physicians collaborate within an interdisciplinary environment to solve clinically relevant problems at the interface of engineering and medicine. Research topics include: tissue engineering and artificial organs; applications of microfabrication in biology and medicine; gene therapy and nucleic acid biotechnology; bioseparations; protein engineering; antibody targeted photolysis; preservation of tissues and organs; genetically modified skin; cell injury and repair; metabolism and metabolic engineering.

Biomedical Instrumentation. Stephen Burns, Ph.D., Technical Director, Harvard-MIT Biomedical Engineering Center. Electronics and microprocessor-based instruments, intelligent instruments, model-based measurement-control procedures, medical equipment in developing countries, telecommunications.

JOHNS HOPKINS UNIVERSITY

Biomedical Engineering Program

Programs of Study	Since 1966 the Biomedical Engineering Program has provided advanced training in engineering, physics, chemistry, and mathematics coupled with a thorough understanding of the basic biological sciences. Students may seek either the Ph.D. degree or a combined M.D./Ph.D. degree. Both programs include courses in the School of Medicine, the School of Engineering, and the School of Arts and Sciences. Strong emphasis is placed on the original research that leads to the doctoral dissertation. The interdisciplinary nature of the program is overseen and maintained by its interdivisional governing body, the Committee on Biomedical Engineering. The students' breadth of outlook is maintained through periodic meetings with advisory committees representing both the biomedical and engineering sciences.
	There are no formal course requirements and no fixed number of semester credit hours for Ph.D. candidates. However, under the supervision of their advisory committees, students usually select two years of formal course work in the engineering sciences, mathematics, and biomedicine. A high degree of competence in physiology, biochemistry, and anatomy is expected. Most Ph.D. students and all M.D./Ph.D. students obtain this competence in the course work required for first-year medical students. Ph.D. students spend much of their second year taking advanced courses in engineering and the physical sciences; M.D./Ph.D. students take these courses after they have completed the second year of medical school course work. Students spend their summers in biomedical research laboratories conducting research preliminary to their thesis investigation.
	When students have completed most of their course work, they must pass an oral preliminary exam, after which they concentrate on their thesis research. Although this research is usually experimental in nature and requires biological techniques, the student can emphasize experiment and theory as desired. Many research opportunities are available within the Department of Biomedical Engineering, located in the School of Medicine. However, in order that they may have the widest range of research opportunities, students are free to work in laboratories throughout the University. All students write and defend a dissertation describing their original research. M.D./Ph.D. students usually need to finish required clinical clerkships upon completion of the requirements for the Ph.D. degree. Suitable scheduling of elective and clerkship periods may result in the simultaneous completion of requirements for both degrees.
Research Facilities	Although students in the Biomedical Engineering Program use laboratories and other facilities throughout the University, the majority work within the Department of Biomedical Engineering, which occupies four floors of the Traylor Research Building. In addition to 11,000 square feet of research laboratory space, the department has lecture and seminar rooms, a complete library, and a computer laboratory that is devoted largely to student use. Other laboratories participating in the program and located within the University occupy another 15,000 square feet. There are specialized facilities for studies on the encoding of information in the nervous system; neuroengineering; motor control; biological and molecular analysis of ion channels; imaging; image analysis and reconstruction; cellular, molecular, and systems analysis of cardiovascular function; neural control of autonomic function; large-scale computational models of biological systems; polymeric materials for biological applications; optical analysis of biological motors, biosensors, and bioinstrumentation.
Financial Aid	For 1999–2000, each student in the program is provided with medical insurance, a stipend of $16,800 per year, and a tuition grant of $23,660 per year. Current sources for funds include NIH systems and integrative physiology traineeships, NIH Medical Scientist Training Program awards, and University fellowships.
Cost of Study	In 1998–99, all first-year graduate students were charged a matriculation fee of $540, covered by the program. Books and supplies cost approximately $575 a year.
Living and Housing Costs	University apartment costs range from $370 per month for unfurnished single rooms to $685 per month for unfurnished two-bedroom apartments in 1999–2000. Many privately owned apartments are available at comparable costs.
Student Group	There are 70 graduate students in the program, of whom about one fifth are seeking the combined M.D./Ph.D. degree. Between 10 and 15 students are accepted each year. The majority of graduates are engaged in basic research and teaching in academic departments of biomedical engineering, physiology and biophysics, electrical engineering, computer science, and otolaryngology.
Location	Baltimore is a vital city long known for its ethnic neighborhoods and, more recently, for its innovations in the preservation and restoration of urban homes. The city abounds with cultural opportunities; it has four major museums, superb orchestras, numerous choral and chamber music organizations, an opera company, ballet, and theaters. Baltimore's excellent park system and its proximity to the Appalachian region and the Chesapeake Bay provide ample opportunity for outdoor activities.
The University	Johns Hopkins University includes the School of Arts and Sciences, the School of Medicine, the Whiting School of Engineering, the School of Hygiene and Public Health, and the Peabody Conservatory. Additional resources of importance to the Biomedical Engineering Program are provided by the Johns Hopkins Applied Physics Laboratory, which is currently engaged in more than fifty collaborative research and development projects with the Johns Hopkins medical institutions.
Applying	Applicants should hold a bachelor's degree from an accredited college and have taken at least one semester of organic chemistry, one year of college biology, and mathematics, including differential equations and linear algebra. Undergraduate fields of study most appropriate to the program are engineering, physics, mathematics, and biology.
	The application deadline for fall admission to the Ph.D. program is January 10, and the deadline for applying for admission to the combined M.D./Ph.D. program is November 1. Applicants to the combined-degree program should write directly to the Office of Admissions of Johns Hopkins University School of Medicine.
Correspondence and Information	Dr. Eric D. Young, Director Doctoral Training Program Department of Biomedical Engineering Johns Hopkins University School of Medicine 720 Rutland Avenue, Room 606 Baltimore, Maryland 21205-2195 Telephone: 410-955-3277 Fax: 410-955-9826 E-mail: grad_phd@bme.jhu.edu

Johns Hopkins University

THE FACULTY AND THEIR RESEARCH

Rita Alevriadou, Assistant Professor of Biomedical Engineering; Ph.D., Rice, 1992. Fluid mechanics, cell and molecular biology of the cardiovascular system, thrombosis/atherosclerosis, metabolism of endothelial cells.

Christopher Chen, Assistant Professor of Biomedical Engineering; Ph.D., MIT, 1997; M.D., Harvard, 1999. Regulatory effects of cell–extracellular matrix interactions, cellular mechanosensation, tissue engineering, cancer biology, microfabrication.

Andrew S. Douglas, Associate Professor of Mechanical Engineering (joint appointment with Biomedical Engineering); Ph.D., Brown, 1982. Nonlinear solid mechanics, dynamic fracture, shear localization; cardiac mechanics.

Jerry D. Glickson, Professor of Radiology and Biological Chemistry; Ph.D., Columbia, 1969. NMR spectroscopy of cancer and heart disease.

William C. Hunter, Associate Professor of Biomedical Engineering; Ph.D., Pennsylvania, 1977. Cardiac dynamics: muscle, chambers, coupling to vasculature.

Kenneth O. Johnson, Professor of Neuroscience and Biomedical Engineering; Ph.D., Johns Hopkins, 1970. Study of sensory information processing in the cortex.

David A. Kass, Associate Professor of Medicine and Biomedical Engineering; M.D., Yale, 1980. Mechanical and hormonal regulation of systolic and diastolic dysfunction in experimental heart failure; organ and cell physiology of vascular response to increased pulsatile perfusion.

Scot C. Kuo, Assistant Professor of Biomedical Engineering; Ph.D., Berkeley, 1988. Mechanism of biological force generation, laser-induced optical forces.

Kam W. Leong, Assistant Professor of Biomedical Engineering (joint appointment with Materials Science and Engineering); Ph.D., Pennsylvania, 1982. Biological applications of synthetic polymers and pharmacokinetics.

W. Lowell Maughan, Associate Professor of Medicine and Biomedical Engineering; M.D., Washington (Seattle), 1970. Cardiovascular hemodynamics and coupling.

Elliot R. McVeigh, Assistant Professor of Biomedical Engineering (joint appointment with Radiology); Ph.D., Toronto, 1988. Magnetic resonance imaging, image analysis with computers, cardiac mechanics.

Richard A. Meyer, Professor of Neurosurgery and Biomedical Engineering; M.S., Johns Hopkins, 1971. Peripheral neural mechanisms of pain sensation.

Michael I. Miller, Professor and Director, Center for Imaging Science, Professor of Biomedical Engineering and Electrical and Computer Engineering; Ph.D., Johns Hopkins, 1983. Image understanding, computational anatomy, medical imaging, computational neuroscience.

Wayne A. Mitzner, Professor of Environmental Health Sciences; Ph.D., Johns Hopkins, 1972. Cardiopulmonary mechanics and fluid exchange, lung structure, development.

John Murphy, Professor of Biomedical Engineering; Ph.D., Catholic University, 1971. Electroporation of membranes, mechanisms of microwave interaction with tissue, minimally invasive photoacoustic spectroscopy of tissue.

Aleksander S. Popel, Professor of Biomedical Engineering (joint appointment with Mechanical Engineering); Ph.D., Moscow, 1972. Fluid mechanics, mathematical modeling of microcirculatory exchange.

Murray B. Sachs, Massey Professor and Director, Department of Biomedical Engineering, and Professor of Neuroscience and Otolaryngology; Ph.D., MIT, 1966. Neurophysiology, auditory neurophysiology, psychophysics.

Lawrence P. Schramm, Professor of Biomedical Engineering and Neuroscience and Director, Doctoral Training Program; Ph.D., Rochester, 1970. Neural control of metabolism, autonomic control systems.

Reza Shadmehr, Assistant Professor of Biomedical Engineering; Ph.D., USC, 1991. Application of robotics to human motor control and learning, computational neuroscience.

Mark Shelhamer, Assistant Professor of Otolaryngology–Head and Neck Surgery and Biomedical Engineering; Sc.D., MIT, 1990. Vestibular and oculomotor analysis and modeling, nonlinear dynamics in physiology.

Artin A. Shoukas, Professor of Biomedical Engineering; Ph.D., Case Western Reserve, 1972. Overall systems analysis of circulatory systems, systems physiology.

Nitish V. Thakor, Associate Professor of Biomedical Engineering (joint appointments with Materials Science and Engineering and with Mechanical Engineering); Ph.D., Wisconsin–Madison, 1981. Medical instrumentation, cardiovascular and neurologic signal processing and patient monitoring, microcomputers and robotics applications.

Leslie Tung, Assistant Professor of Biomedical Engineering; Ph.D., MIT, 1978. Electrophysiology and contractile force of single heart cells, analysis of multicellular structure.

Xiaoqin Wang, Assistant Professor of Biomedical Engineering; Ph.D., Johns Hopkins, 1992. Neurophysiology of the auditory cortex, neural mechanisms of speech perception and learning, computational neuroscience.

Raimond L. Winslow, Assistant Professor of Biomedical Engineering and Computer Science; Ph.D., Johns Hopkins, 1985. Auditory and visual physiology, computational neuroscience, nonlinear dynamical systems, high-performance computing.

Eric D. Young, Professor of Biomedical Engineering, Neuroscience, and Otolaryngology; Ph.D., Johns Hopkins, 1972. Auditory neurophysiology, sensory processes, computers.

David T. Yue, Assistant Professor of Biomedical Engineering; M.D., Ph.D., Johns Hopkins, 1987. Ion channels in the heart, biophysical aspects of electrophysiology.

MASSACHUSETTS INSTITUTE OF TECHNOLOGY

Division of Bioengineering and Environmental Health
Graduate Program in Bioengineering

Programs of Study

A program leading to the Ph.D. in bioengineering is offered within the recently formed Division of Bioengineering and Environmental Health. The purpose of this program is to educate engineers in applying their measurement and modeling perspectives to understanding how biological systems operate, especially when perturbed by genetic, chemical, or materials interventions or subjected to pathogens or toxins, and in applying their design perspective to create innovative technologies in biology-based diagnostics, therapeutics, and devices or in manufacturing industries for non-health-related markets. The program is designed to provide training in problem solving using modern biotechnology, emphasizing an ability to measure, model, and rationally manipulate biological systems.

Students admitted to the program typically have a B.S. or an M.S. degree in engineering. During their first year they pursue a unified core curriculum, in which approaches from the various engineering disciplines are used to examine mechanical phenomena, rates of transport, and the kinetics of various processes in biological materials and organisms over a wide range of length and time scales. Topics covered in bioengineering electives include the structure and properties of natural and synthetic biomaterials, biological instrumentation and measurement, and bioinformatics and computational biology. The core and elective courses in bioengineering are supplemented by training in biochemistry and cell biology and in a traditional engineering discipline, as appropriate for the background of a given student. The written part of the doctoral qualifying examinations, centered on the core curriculum, is taken after the second semester. The students select a research adviser and begin research before the end of the first year. The oral part of the doctoral qualifying examinations, which focuses on the student's area of research, is taken during the second year. Approximately five years of total residence are needed to complete the doctoral thesis and other degree requirements.

The faculty members associated with the program have a wide range of research interests within bioengineering. Areas in which students may specialize include biological imaging; biological transport phenomena; biomaterials; cell engineering; cell, tissue, and fluid biomechanics; drug delivery; electromechanical properties of tissues; instrumentation for biological measurements; metabolism of toxic substances; microfabrication biotechnology; physiological modeling and simulation; and tissue engineering. Most of the faculty members are associated with one or more interdisciplinary research centers at Massachusetts Institute of Technology (MIT), including the Biotechnology Process Engineering Center, the Center for Biomedical Engineering, and the Center for Environmental Health Sciences.

Research Facilities

Laboratories of faculty members are equipped in accordance with their particular research interests. Shared resources include cell and tissue culture facilities associated with quantitative microscopy and imaging equipment (including fluorescence, two-photon, and laser-confocal) and equipment for studying chemical and mechanical interactions among cells and biomolecules (BIACore, Cytosensor, optical trap, radiolabeling facility).

Financial Aid

Financial support (full tuition and stipend) is provided for all admitted students. The stipend for 1999–2000 is $1500 per month. Employment opportunities are usually available for students' spouses both on campus and in the community. International students and their spouses may find work permits difficult to obtain and should plan accordingly.

Cost of Study

In 1999–2000, tuition is $12,500 per regular term ($25,000 for fall and spring); summer tuition may be waived in the future for "research only" registration. Hospital and accident insurance is approximately $640 per year, with additional coverage available for dependents. The cost of textbooks is estimated at $650 per year. All students in the program receive a stipend and full tuition support.

Living and Housing Costs

On-campus rooms are available for single graduate students. Rates range from approximately $375 to $900 per month. Rents for Institute housing for married students vary from $675 to $1025 per month, including utilities, for efficiency and one- and two-bedroom apartments. Off-campus apartments are available at considerably higher rates.

Location

MIT is located on the banks of the Charles River, which separates Cambridge and Boston, cities with a combined metropolitan population of 2.5 million. Numerous concerts, museums, exhibits, and other cultural resources are readily available. The large concentration of universities in the metropolitan area—Harvard, MIT, Boston University, Tufts, and others—provides a stimulating intellectual environment. MIT offers the combined advantages of New England's largest metropolitan center with easy access to the Atlantic Ocean beaches and the New England countryside. As one of the earliest communities established in the United States, Boston also has a wide variety of historical sites and traditions.

The Institute

Massachusetts Institute of Technology was founded in 1861 to teach "exactly and thoroughly the fundamental principles of positive science." Its goals include the education of men and women with sound capabilities in science, but whose abilities and goals are shaped by nonscientific influences as well. These objectives are maintained for the current enrollment of about 4,500 undergraduate and 5,300 graduate students. At present the Institute's involvement in the community is growing through expanding tutoring and community service projects. Its primary objectives, however, remain the education of its students and the advancement of scientific research.

Applying

Applicants must have a B.S. degree or the equivalent in engineering, biology, chemistry, physics, or mathematics. Transcripts, three letters of recommendation, and scores from the GRE General Test must be submitted. In addition, a minimum score of 600 on the TOEFL is required of applicants whose first language is not English. Applications should be completed by January 15 for entrance in September. Applicants are generally notified of the outcome by April 1.

Correspondence and Information

Debra A. Luchanin, Academic Administrator
Bioengineering and Environmental Health
Academic Office, Room 56-651
Massachusetts Institute of Technology
Cambridge, Massachusetts 02139-4307

Telephone: 617-253-1712
Fax: 617-258-8676
World Wide Web: http://www.mit.edu/beh/

Massachusetts Institute of Technology

THE BIOENGINEERING FACULTY AND THEIR RESEARCH

William M. Deen, Professor of Chemical Engineering and Bioengineering; Ph.D., Stanford. Physiological transport and kinetics.

C. Forbes Dewey, Professor of Mechanical Engineering and Bioengineering; Ph.D., Caltech. Cell mechanics, biological imaging.

Linda G. Griffith, Associate Professor of Chemical Engineering and Bioengineering; Ph.D., Berkeley. Tissue engineering, biomaterials.

Alan J. Grodzinsky, Professor of Electrical Engineering and Computer Science and Bioengineering; Sc.D., MIT. Electromechanical properties of tissues.

Neville J. Hogan, Professor of Mechanical Engineering and Bioengineering; Ph.D., MIT. Neuromuscular system dynamics.

Ian W. Hunter, Professor of Mechanical Engineering and Bioengineering; Ph.D., Auckland, Micromechanical devices.

Roger D. Kamm, Professor of Mechanical Engineering and Bioengineering; Ph.D., MIT. Cell and tissue mechanics.

Paul E. Laibinis, Assistant Professor of Chemical Engineering; Ph.D., Harvard. Biosensors, microfabrication.

Robert S. Langer, Professor of Chemical Engineering and Biomedical Engineering; Sc.D., MIT. Biomaterials, drug delivery, tissue engineering.

Douglas A. Lauffenburger, Professor of Chemical Engineering and Bioengineering; Ph.D., Minnesota. Cell engineering.

Harvey F. Lodish, Professor of Biology and Bioengineering; Ph.D., Rockefeller. Molecular and cell therapeutics.

L. Mahadevan, Assistant Professor of Mechanical Engineering; Ph.D., Stanford. Molecular biomechanics.

Paul T. Matsudaira, Professor of Biology and Bioengineering; Ph.D., Dartmouth. Biological imaging, cytoskeletal structure, genomics biotechnology.

Ram Sasisekharan, Assistant Professor of Toxicology and Bioengineering; Ph.D., Harvard. Extracellular matrix regulation of cell function.

Peter T. So, Assistant Professor of Mechanical Engineering; Ph.D., Princeton. Biological imaging.

Ioannis Yannas, Professor of Mechanical Engineering, Materials Science and Engineering, and Bioengineering; Ph.D., Princeton. Tissue engineering.

MAYO GRADUATE SCHOOL

Graduate Program in Biomedical Engineering

Programs of Study

The Mayo Graduate School offers a graduate program leading to the Ph.D. and M.D./Ph.D. in biomedical sciences with specialization in the basic science disciplines of biochemistry, biomedical engineering, immunology, molecular biology, molecular neuroscience, pharmacology, and physiology. These programs are designed to provide highly motivated students with an educational background and laboratory experience that will prepare them for careers as independent research investigators.

Each program of study emphasizes acquisition of proficiency in a specialized area such as biomedical engineering, pharmacology, immunology, biochemistry, molecular biology, or molecular neuroscience. In addition, each program is rounded out by course work in two or more areas of study other than the area of specialization. The formal academic portion of the program can be completed in two years. An additional two to three years are usually needed for carrying out a research-based thesis project. Mayo provides an outstanding environment for research in the basic sciences.

Further details concerning the individual programs of study listed above are provided in separate descriptions.

Research Facilities

Each area of specialization is well equipped with state-of-the-art facilities for research and education in the life sciences. The campus in Rochester contains more than 965,000 square feet of space devoted to research and related activities. Research opportunities in molecular neuroscience and pharmacology are also available at the campus in Jacksonville, Florida, and opportunities in biochemistry and molecular biology are available at the Scottsdale, Arizona, campus. A full-time faculty of 100 scientists and an annual budget of $100 million support current research. Mayo Medical Library contains 413,000 volumes and subscribes to 4,300 medical and scientific journals.

Financial Aid

Full-time graduate students are provided with Mayo Graduate School predoctoral research fellowships. They provide a yearly stipend (approximately $17,000 in 1999–2000) in addition to medical insurance and other benefits. Receipt of fellowship stipends does not obligate students to work as laboratory assistants or as teaching assistants. Since stipends are provided by the Graduate School, students are not constrained in their choice of research advisers by funding limitations.

Cost of Study

Tuition costs for course work taken at Mayo's Rochester, Minnesota; Jacksonville, Florida; and Scottsdale, Arizona, campuses and off campus are provided in addition to the yearly stipend. Students pay no tuition or ancillary fees.

Living and Housing Costs

Living costs in Rochester are comparable to those in cities of similar size within the upper Midwest and generally lower than those in urban areas of the East and West Coasts. A single student can live comfortably in Rochester on the stipend provided.

Student Group

Approximately 25 students are admitted each year into the Ph.D. in biomedical sciences program, and 6 are admitted into the M.D./Ph.D. program. Enrollment in other Mayo educational programs includes 1,000 clinical residents and fellows and 275 postdoctoral research fellows.

Student Outcomes

Approximately two thirds of students enrolled in the Mayo Ph.D. programs graduate with a Ph.D. degree. Students take just over five years to complete a Ph.D. at Mayo. Most students have authored several publications by the time they graduate.

Location

The city of Rochester combines the best of two worlds—the warmth and friendliness of a small town with the bustling commerce, entertainment, and conveniences of a metropolis. Lectures, symphony concerts, art exhibits, and a civic theater contribute to a cosmopolitan atmosphere that is unusual for a city this size. The city and its environs provide an extensive four-season calendar of recreation. Attractions in Minneapolis and St. Paul, a major metropolitan area just 80 miles north of Rochester, include professional sports and a variety of cultural opportunities.

The Graduate School

The Mayo Graduate School is a division of the Mayo Foundation, which includes Mayo Clinic, Mayo Graduate School of Medicine, Mayo Medical School, Mayo School of Health-Related Sciences, and affiliated hospitals. Mayo Foundation is accredited by the Commission on Institutions of Higher Education of the North Central Association of Colleges and Schools.

Applying

The general requirements of the Mayo Graduate School for admission to the Ph.D. program include a bachelor's degree from an accredited college or university and scores on the GRE General Test, or the equivalent, indicating strong academic ability. Specific course prerequisites include two years of college chemistry (including organic), one year of calculus, one year of biology, and one year of physics, with evidence of superior performance in all courses. A foundation course in biochemistry or molecular biology is strongly recommended. Certain program tracks have additional requirements such as course work in physical chemistry and advanced mathematics. Official transcripts from schools attended, three letters of recommendation, and a summary of the student's scientific interests and career goals are also required. Interviews are suggested. Applications to the M.D./Ph.D. program are initiated via the AMCAS system and must be submitted by November 15. The Ph.D. portion of the M.D./Ph.D. application should be submitted by December 21. Applications for the Ph.D. program should be submitted by December 31. Mayo Foundation is an affirmative action and equal opportunity educator and employer.

Correspondence and Information

Graduate Program in Biomedical Engineering
Mayo Graduate School
200 First Street SW
Rochester, Minnesota 55905
Telephone: 507-284-4356
E-mail: phd.training@mayo.edu

Mayo Graduate School

THE FACULTY AND THEIR RESEARCH

Kai-Nan An, Ph.D. Musculoskeletal biomechanics, orthopedic and rehabilitation engineering.

Marek Belohlavek (Internal Medicine and Biomedical Engineering), M.D., Ph.D. Parametric analysis of cardiovascular perfusion and function with ultrasound.

Michael Camilleri, M.D. Image-based contraction/relaxation studies of human digestive tract; accommodation, transit profile, and residue of meals.

Stephen Carmichael, Ph.D. Anatomy innovative methods in anatomic education, asynchronous learning.

Thoms Dousa (Physiology and Medicine/Nephrology), M.D., Ph.D. Intracellular signal transduction mechanisms in health and disease, pathogenesis of glomerulonephritis.

Richard Ehman, M.D. Magnetic resonance imaging physics, motion correction algorithms, vascular imaging techniques for MRI, tissue characterization, adaptive technique for high-definition MR imaging of moving structures, magnetic resonance elastography.

Bradley Erickson (Radiology), M.D., Ph.D. Application of computers to solve neuroradiology problems, radiology informatics.

Gianrico Farrugia (Gastroenterology and Hepatology, Physiology and Biophysics), M.D. Gastrointestinal motility, ion channels, interstitial cells of Cajal.

Joel Felmlee, Ph.D. Magnetic resonance imaging (MRI) improvements in safety and image quality.

Julio Fernandez, Ph.D. Mechanisms of exocytotic secretion studied with biophysical techniques; tension in secretory granule membranes caused by extensive membrane transfer through the exocytotic fusion pore; structural basis of protein elasticity, using molecular biology and atomic-force microscopy.

Barry Gilbert, Ph.D. Design of computer hardware and computer-aided design (CAD) software, development of gallium arsenide digital integrated circuit technology for high-performance digital signal processors, studies of mathematical algorithms for biomedical signal and image processing and analysis, computation-bound problems in biomedical research, packaging of GaAs signal processors on multichip modules.

James Greenleaf (Physiology and Biophysics), Ph.D. Ultrasound imaging and therapy devices.

Nicholas Hangiandreou, Ph.D. Digital imaging and management systems, image quality assessment.

Michael Herman, Ph.D. Use of digital imaging for treatment planning, treatment verification, and visualization in oncology.

Rolf Hubmayr, M.D. Biomechanics of the lung, lung deformation injury, cell mechanics.

John Huston III (Radiology), M.D. Neuro MRI and MR angiography.

Clifford Jack, M.D. Functional MRI, speech and language mapping in the brain.

Michael Joyner, M.D. Autonomic nervous system interactions, factors that regulate arterial pressure and blood flow.

Kenton Kaufman (Biomechanics Laboratory), Ph.D. Gait analysis, lower extremity biomechanics, muscle modeling.

Bernard King, M.D. MRI, CT, and ultrasound of the genitourinary system.

Robert Kline, Ph.D. Three-dimensional radiation therapy planning, radiation physics and imaging.

Armando Manduca, Ph.D. Image processing, image analysis, neural networks, texture analysis.

Piotr Marszalek, Ph.D. Single molecule mechanics by atomic-force microscopy techniques.

David Masters, Ph.D. Fabrication and characterization of local drug delivery systems for pharmacotherapy and tissue engineering, using biodegradable natural and synthetic polymers.

Cynthia McCollough, Ph.D. X-ray–computed tomography, radiation dosimetry, spiral and electron-beam CT.

Edwin McCullough (Radiation Oncology), Ph.D. Photons, electrons, and radionuclides in the treatment/diagnosis of cancer.

Virginia Miller (Surgery and Physiology), Ph.D. Effects of hormonal and mechanical stimuli on vascular remodeling.

Andres Oberhauser (Molecular Biophysics), Ph.D. Molecular mechanisms of exocytosis, atomic-force microscopy analysis of protein folding reactions.

Michael O'Connor (Radiology/Nuclear Medicine), Ph.D. SPECT reconstruction techniques (attenuation and scatter correction and OSEM), multi-isotope imaging techniques.

Hal Ottesen (Biomedical Engineering), Ph.D. Efficient methods in digital signal and image processing, real-time digital control with fuzzy logic.

Luo Zong Ping, Ph.D. Musculoskeletal biomechanics, cellular and molecular biomechanics in orthopedics.

James Rae, Ph.D. Structure and function of ion channels in ocular epithelia.

Stephen Riederer, Ph.D. Medical imaging systems, magnetic resonance imaging (MRI), high-speed MRI.

Erik Ritman, M.D., Ph.D. Whole-body and microcomputed tomography for structure-to-function relationships.

Richard Robb, Ph.D. Computer-aided visualization and analysis of biological systems; 3-D image reconstruction and display; image-guided surgery and clinical interventions; image segmentation, registration, volume rendering, and image modeling; virtual reality and high-performance graphics.

Juan Romero (Physiology—Division of Hypertension, Division of Nephrology and Internal Medicine), M.D. Characterization of renal function in Goldblatt hypertension with electron beam–computed tomography.

Frank Rusnak, Ph.D. Metalloproteins, mechanisms of protein phosphatases and immunosuppressant drugs.

Gary Sieck, Ph.D. Plasticity in neuromotor control, including alterations in motor-unit mechanical and metabolic properties, alterations in motoneuron morphometry, and alterations at neuromuscular junction (models include postnatal development, neural inactivation, compensatory loading, and steroidal treatment); neuromuscular interactions, muscle fiber mechanics, and energetics.

Steven Sine, Ph.D. Structure-function relationships in nicotinic acetylcholine receptors (AchR); mechanistic bases of human AchR channelopathies; mutagenesis, protein structure, and single-channel kinetics.

Joseph Szurszewski, Ph.D. Gastrointestinal motility and electrophysiology, 3-D volume reconstruction of single-cell function.

Stuart Taylor, Ph.D. Digital imaging microscopy; imaging Ca2+ in EC coupling; X-ray diffraction and imaging microscopy of muscle contraction.

Russell Turner, Ph.D. Bone fracture site remodeling, tissue repair.

Richard Vetter, Ph.D. Radiation biophysics and dosimetry.

David Warner (Anesthesiology), M.D. Airway smooth muscle physiology; anesthetic mechanisms of action; respiratory muscle function.

Michael Yaszemski, M.D., Ph.D. Tissue biomechanics and engineering.

NORTHWESTERN UNIVERSITY

McCormick School of Engineering and Applied Science
Department of Biomedical Engineering

Program of Study

The objective of the Ph.D. program in the Department of Biomedical Engineering is to train independent researchers and provide them with a solid foundation in engineering and a strong background in the life sciences. During the first year, students develop engineering skills and initiate their research. First-year class work typically includes quantitative systems physiology, advanced mathematics, statistics for experimenters, and advanced engineering classes within the student's area of interest. The second year is devoted to completion of research for the master's thesis, which requires an oral defense. The master's thesis defense serves, in part, as the Ph.D. qualifying exam. For those who enter the program with an M.S. degree, a Ph.D. proposal exam serves as the research component of the qualifying exam. Subsequent years are devoted to advanced study of engineering and the life sciences as well as selection, completion, and oral defense of the Ph.D. dissertation research. A total of 15 one-quarter courses is required for the Ph.D. Completion of the Ph.D. requirements typically takes from four to six years. The Ph.D. program prepares students for careers of independent research in solving important medical and biological problems in areas requiring engineering skills and life-science knowledge; such research is currently undertaken in industry, hospitals, government laboratories, and academia.

The objective of the M.S.-only program is to train individuals to use engineering principles to solve biological and medical problems by providing them with a foundation in engineering and a background in the life sciences. Students are required to complete twelve one-quarter courses. The program is designed to be completed within nine to twelve months and must include a significant project, typically from a list of options provided upon matriculation. Class work includes quantitative systems physiology, advanced mathematics, statistics for experimenters, and advanced engineering classes within the student's area of interest. The M.S.-only program prepares students for careers in industry as well as in hospital, government, and academic laboratories. Other students obtain further training in graduate and medical school.

The Department of Biomedical Engineering has close ties with other departments in the engineering school as well as with most other schools within the University, including the Medical and Dental schools, the School of Speech, and the College of Arts and Sciences. Areas of research include sensory physiology (especially visual and auditory physiology); biomechanics and transport (including fluid and solid mechanics); orthopedic biomechanics; rehabilitation engineering; lasers and optics; image processing; electronic instrumentation; and biotechnology.

Research Facilities

State-of-the-art research facilities support substantial efforts in each of the above-mentioned areas and include laboratories within the McCormick School of Engineering and Applied Science, the Medical School, and affiliated hospitals. Computing facilities ranging from microcomputers to mainframes are readily available. The Science and Engineering Library and the Medical Library have extensive collections and online catalog and literature-search capabilities.

Financial Aid

Most full-time graduate students receive financial aid, which is available in a number of forms, such as fellowships and research and teaching assistantships. All aid is awarded on a competitive basis. Generally, first-year fellowship support is available only to U.S. and Canadian citizens and U.S. permanent residents. Teaching assistantships are not given to first-year students. Research assistantships, whose holders are selected by individual faculty members, involve participation in a research program and are rarely given to first-year students. While the Biomedical Engineering Department makes every effort to provide financial aid to deserving students, all prospective students are encouraged to investigate aid from NSF and other agencies.

Cost of Study

Tuition for the 1999–2000 academic year is $21,798; it is $7266 for the summer.

Living and Housing Costs

Most students live in off-campus apartments or houses where rent is approximately $725 per month for a single and $425 per month per person for those who share a house or an apartment. Most students live in Evanston, in the neighboring suburbs, or in residential northern Chicago. Students who have a fellowship can live reasonably without taking out loans; other students should plan on spending approximately $31,300 per year for the costs of living and education.

Student Group

There were 69 graduate students in the Biomedical Engineering Department in 1999–2000, including 28 women. There were 657 graduate students in engineering and 2,387 students in the Graduate School.

Location

Northwestern University is located on the scenic shores of Lake Michigan, with campuses in suburban Evanston and downtown Chicago. The University beach, sailing club, recreational facilities, and the beautiful lakefront areas are used by students and their families for athletics as well as leisurely walks and picnics. Chicago, with its many museums, parks, zoos, musical and theater centers, night spots, and restaurants, offers remarkable cultural opportunities. For athletically minded students, there are all levels of activity, from swimming, track, and aerobic/workout facilities in the aquatics/sport center to viewing collegiate and professional teams in nearly all sports.

The University

The only privately supported university in the Big Ten, Northwestern University was founded in 1851 and currently has approximately 2,100 full-time faculty members. Northwestern University is a full-service institution with renowned schools of music, speech, and journalism and has all the attendant cultural facilities, including a concert hall and several theaters, which are an integral part of the Evanston campus.

Applying

A B.S. in either engineering or a quantitative science is required. Applicants are judged on previous academic performance, GRE scores (the General Test is required, the Subject Test in engineering is not required), letters of recommendation, and the student's statement of purpose. An application for admission can be considered for any quarter of the year, but matriculation in the fall is recommended. Applications for fall quarter should be completed and received by January 15 in order to receive prime consideration. Fellowship appointments are awarded only to students entering in the fall.

Correspondence and Information

Admissions Officer
Biomedical Engineering Department
Northwestern University
2145 Sheridan Road
Evanston, Illinois 60208-3107

Telephone: 847-491-5635
Fax: 847-491-4928
E-mail: bmeadmit@nwu.edu
World Wide Web: http://www.nwu.edu/bme

Northwestern University

THE FACULTY AND THEIR RESEARCH

Ernest Byrom, Ph.D., Adjunct Assistant Professor; Director, Computing Cardiology, Evanston Hospital. Digital image processing in cardiac angiography, computer-aided analysis of invasive cardiac electrophysiology studies, signal averaging for surface ECG, modeling conduction abnormalities.

Dudley S. Childress, Ph.D., Professor; Director, Prosthetics Research Laboratory and Rehabilitation Engineering Research Program. Limb prosthetics and orthotics, biomechanics, walking and aided ambulation, control of artificial arms, engineering in rehabilitation.

Peter Dallos, Ph.D., Professor. Biophysics and physiology of the auditory system, bioacoustics, sensory neurobiology.

Christina Enroth-Cugell, M.D., Ph.D., Professor Emerita. Neurophysiology of vision, especially recording from single retinal, ganglion cells in the cat.

Max Epstein, Ph.D., Professor Emeritus. Developments in microendoscopic instrumentation, medical imaging.

Matthew Glucksberg, Ph.D., Associate Professor. Pulmonary mechanics, pathophysiology of pulmonary edema, lung liquid transport and microvessel-wall permeability, robotic microsurgery, blood-pressure and flow measurements of the retinal microcirculation.

Thomas K. Goldstick, Ph.D., Professor Emeritus. Mass transport, especially of oxygen in the retina and in blood; regulation and control of transport; effect on the tissue transport systems of diseases such as diabetes and AIDS; measurement of oxygen tension and tissue blood flow; characterization of blood coagulation.

Kevin E. Healy, Ph.D., Associate Professor. Surface engineering and characterization of implant materials, analysis of protein adsorption and cell response to artificial materials with defined surface physical-chemical properties, development of biodegradable polymer implants used for tissue and organ regeneration.

James C. Houk, Ph.D., Professor and Chairman of Physiology. Brain mechanisms of movement control, neural network models of information processing in sensorimotor systems, microelectrode studies of nerve signals in awake animals, in vitro studies of neural circuits that produce motor programs, biomechanics of movement.

Mark Johnson, Ph.D., Associate Professor. Biotransport phenomena, pathogenesis of glaucoma, morphometric studies of connective tissues, biophysical characterization of normal and dysfunctional behavior of lung surfactant, Respiratory Distress Syndrome, chaos.

Robert M. Judd, Ph.D., Assistant Professor. Use of MRI to study the heart, using MR contrast agents to examine myocardial perfusion and viability, use of NMR spectroscopy techniques to examine myocardial metabolism.

David M. Kelso, Ph.D., Associate Professor. Medical instrumentation, biosensors and microminiature analytical devices.

Andrew E. Kertesz, Ph.D., Professor. Binocular vision, eye movements and perception, normal function as well as disorders, physiological control systems, clinical instrumentation.

Francis J. Klocke, Ph.D., Professor and Director, Feinberg Cardiovascular Research Institute. Endothelial mechanisms contributing to control of the coronary circulation and myocardial contractile function, downregulation of myocardial metabolic demand in the presence of coronary flow limitation, angiogenesis in the coronary and peripheral vascular beds, flow-limiting effects of sulfonylureas in the coronary circulation.

Debiao Li, Ph.D., Associate Professor. Magnetic resonance imaging, heart and blood vessels, imaging of coronary arteries, blood oxygenation evaluations.

Robert A. Linsenmeier, Ph.D., Professor and Chairman of Biomedical Engineering. Microenvironment of the mammalian retina, including ionic balance and transport of nutrients, especially oxygen; electrophysiology of the mammalian retina.

Shu Liu, Ph.D., Assistant Professor. Mechanisms of remodeling and growth of blood vessels and lung tissues under the influence of mechanical forces at the molecular, cellular, and organ levels; roles of mechanical forces in the development of engineering approaches to prolong vein graft patency.

Phillip B. Messersmith, Ph.D., Assistant Professor. Biomaterials, particularly the development of temperature-sensitive liposomes for drug delivery.

Lyle F. Mockros, Ph.D., Professor. Fluid mechanics, analytical and experimental investigations of mass transfer processes in flowing blood, development of artificial lungs, blood processing, thrombogenesis in prosthetic devices, cardiac mechanics, hemodynamics.

David J. Mogul, Ph.D., Assistant Professor. Cellular electrophysiology of the brain, ion channels, neuromodulation, synaptic plasticity, cellular mechanisms of memory, computational neuroscience, nonlinear networks, epilepsy.

Todd B. Parrish, Ph.D., Assistant Professor. Neuroimaging utilizing MRI methods to investigate brain activation including imaging processing and statistical analysis, investigating cerebral vascular blood flow dynamics with MRI.

Barry W. Peterson, Ph.D., Professor. Central nervous system physiology, specifically motor learning and sensorimotor transformation in the vestibular system.

William Z. Rymer, M.D., Ph.D., Professor. Neural control and biomechanics of movement studied in human and animal models; movement disturbances induced by neurological disorders, including stroke and spinal cord injury.

Alan V. Sahakian, Ph.D., Associate Professor. Electrophysiology of the heart, especially atrial arrhythmias; pacing and automatic cardioversion and defibrillation; signal and image processing and instrumentation for medical and aerospace applications.

Kenneth G. Spears, Ph.D., Professor. Therapeutic applications of lasers in ophthalmology, imaging with coherent optical radiation, assessment of the properties of biomolecules using ultrafast spectroscopy probing techniques.

Melody A. Swartz, Ph.D., Assistant Professor. Interstitial mechanics and transport; mechanotransduction; lymphatic transport; tissue fluid balance.

John B. Troy, Ph.D., Associate Professor. Processing and analysis of images by the mammalian visual system, functional circuitry of the retina, application of signal theory to visual physiology, optic nerve regeneration.

Jeffrey S. Vender, M.D., Professor and Associate Chair of Anesthesiology. Blood gas monitoring, pulmonary system monitoring, hemodynamic monitoring.

Joseph T. Walsh Jr., Ph.D., Associate Professor. Analytic and experimental analysis of the interaction of laser radiation with tissue, development of diagnostic and therapeutic applications of lasers.

Harvey A. Wigdor, D.D.S., M.S., Adjunct Associate Professor and Chief of Dental Service, Ravenswood Hospital. Laser dental applications.

Tai T. Wu, M.D., Ph.D., Professor. Structure and function of proteins and DNA, especially those pertaining to the immune system.

Li-Qun Zhang, Ph.D., Assistant Professor. Systems identification of reflex and intrinsic properties of spastic and normal limb muscles, compensatory mechanisms and rehabilitation of knee injuries, mechanical actions and load sharing of individual muscles.

THE OHIO STATE UNIVERSITY

Biomedical Engineering Center

Programs of Study	The Biomedical Engineering Center offers graduate programs leading to a Master of Science and a Doctor of Philosophy degree for students with academic backgrounds in engineering or the physical or life sciences. Students without engineering degrees are required to complete additional undergraduate courses to develop competence in engineering. Graduates are prepared to pursue hospital, industrial, or academic careers in biomedical engineering. Areas of research include artificial intelligence, atherosclerosis, biocompatibility, biofluid mechanics, biomaterials, biomedical product design, bioMEMS, biomolecular spectroscopy, boron neutron capture therapy, cardiovascular device evaluation, cardiovascular modeling and simulation, cognitive modeling, diagnostic imaging (including SPECT), distance learning, Doppler echocardiography, educational robotics, electrocardiography, electrogastrography, electromyography, expert/ intelligent systems, functional muscle stimulation, gait analysis, glaucoma, human-computer interactions, human movement, image processing with cardiovascular and computer vision applications, instrumentation, laser/optical applications, lumbar epidural virtual reality simulator, machine perception, magnetic cell separation, magnetic resonance applications, mass transport, medical physics, modeling, motion biomechanics, motor control, neural networks, neurosciences, occupational biomechanics, ophthalmology, orthopaedics and orthopaedic biomechanics, pathophysiology of heart failure, pulmonary mechanics, sensory systems and rehabilitation, signal processing, tongue biomechanics and prosthesis, ultrasound, vascular physiology, visual information systems for the blind, visual psychophysics, and visualization and volume rendering. A strong focus in biomedical image engineering, with cardiovascular, orthopedic, and neuroscience applications, is being developed under a Special Opportunity Award from the Whitaker Foundation. The M.S. requires 36 quarter credit hours of courses and 9 quarter credit hours of thesis research, with course work from both engineering and the life sciences, and a minimum cumulative GPA of 3.0. A thesis is required, with a final oral examination. The Ph.D. requires 75 quarter credit hours of courses (master's work may be applied) and 50 quarter credit hours of research, with a cumulative GPA of 3.0. The program of study for the doctorate includes one major area in biomedical engineering and one or two minor areas in related engineering fields or the life sciences. Written and oral candidacy examinations must be taken successfully. Research for the dissertation is supervised by the student's academic adviser, who also chairs the committee that administers the final oral examination. Exceptional applicants may be admitted directly to the Ph.D. program. A combined M.D./Ph.D. program is available for selected highly qualified students through the College of Medicine and Graduate School.
Research Facilities	The combined technical and clinical research facilities of the College of Medicine, the College of Engineering, and the College of Veterinary Medicine, in proximity to the medical center of the Ohio State University Hospitals, provide extensive resources for biomedical engineering research. Research facilities include the Balance Disorders Laboratory, Biochemical Engineering Laboratory, Biodynamics Laboratory, Biomedical Microdevices Laboratory (including microfabrication lithography, biohybrid microMD, biochemistry/surface modification, and polymer processing), Biosonar Laboratory, Biospectroscopy Laboratory, Cardiac Catherization Laboratory, Cardiovascular Simulations Laboratory, Computer Graphics Laboratory, Electrogastrography Laboratory, Gait Analysis Laboratory, Laboratory for AI Research in Perception and Neurodynamics, Laboratory for Electrophysiology, Laboratory for Research in Vision, Laboratory of Experimental Atherosclerosis, Laboratory of In-Vivo Electron Spin Resonance Spectroscopy, Laboratory of Vascular Diseases, Laser-Tissue Interactions Laboratory, Magnetic Resonance Imaging and Spectroscopy Research Facility, Nuclear Medicine Laboratory, Nuclear Reactor Laboratory, Ohio Lions Eye Research Laboratory, Orthopaedic BioMaterials Laboratory, Physical Therapy Human Movement Laboratory, Restorative Materials Laboratory, Sensory and Cognitive Engineering Research and Design Laboratory, Ultrasonics and Fiber Optics Laboratory, Ultrasound Imaging Laboratory, Vestibular Research Laboratory, Vision Rehabilitation Laboratory, Visual Neuroscience Laboratory, and X-ray Microtomography Laboratory. Extensive computing facilities are available, including the workstation-based Biomedical Engineering Computing Center, state-of-the-art biomedical image processing facilities, and access to the CRAY Y-MP supercomputer at the Ohio Supercomputer Center. An affiliation with the Cleveland Clinic Foundation affords additional research opportunities in the cardiovascular, orthopaedic, and imaging areas, including the Doris Alburn Laboratory for Musculoskeletal Research, the Whitaker Biomedical Imaging Laboratory, the Cardiovascular Imaging Center, and the Cardiovascular Device Research and Development Center.
Financial Aid	Most students are supported by research associateships and fellowships. Associateship support includes tuition and fees as well as a monthly stipend and requires a 20-hour-per-week commitment.
Cost of Study	Tuition and fees in spring 1999 were $2029 per quarter for Ohio residents; for nonresidents, tuition and fees were $4844 per quarter.
Living and Housing Costs	For 1998–99, single rooms in graduate and professional student housing cost $300 per month, with optional food service costing up to $786 per quarter. Family student housing in 1998–99 cost $375 per month for a one-bedroom unit and $480 per month for a two-bedroom unit. A wide variety of near-campus and farther-off-campus housing is also available, with a broad range of monthly rents.
Student Group	The Biomedical Engineering Center had an enrollment of 70 students in 1998–99, with approximately 60 percent in the Ph.D. program and 25 percent women. A strong student chapter of the Biomedical Engineering Society is involved in professional, social, and campus activities.
Student Outcomes	Graduates of the program are successfully employed in research institutions (e.g., Battelle Memorial Institute, NASA), hospitals (e.g., The Cleveland Clinic Foundation), industry (e.g., Boeing, Ford Motor Co.), and academia. Since 1989, nearly 50 percent of the Ph.D. graduates have assumed faculty positions in colleges and universities worldwide. Other graduates have become technical consultants; clinical, design, and software engineers; programmer analysts; and research scientists.
Location	The Ohio State University is in Columbus, the state capital, which offers the advantages of a major metropolitan area, including excellent restaurants, a symphony, ballet and opera, theaters, museums, night spots, and a nationally known zoo. The park system features many facilities and shelters for outdoor activities.
The University and The Center	The Ohio State University, Ohio's land-grant institution, is attended by more than 50,000 students, with 1,500 graduate students in the College of Engineering alone. The Biomedical Engineering Center, officially established in 1971, has expanded with the renovation of 15,000 square feet of research and office space and the assignment of other resources to targeted areas of research. More than 60 participating faculty members throughout the University contribute to the activities of the Center.
Applying	Completed applications, official transcripts, and test score reports should be sent to Graduate Admissions at the address below. An autobiographical statement and three letters of reference should be sent directly to the Biomedical Engineering Center. Ph.D. applicants should include a copy of their master's thesis in English. GRE General Test scores are required of all applicants. The deadline for an application to be considered for a University fellowship is January 15. Other applications are reviewed throughout the year.
Correspondence and Information	Director, Biomedical Engineering Center The Ohio State University 270 Bevis Hall 1080 Carmack Road Columbus, Ohio 43210-1002 Telephone: 614-292-7152

Graduate Admissions
3rd Floor
Lincoln Tower
The Ohio State University
1800 Cannon Drive
Columbus, Ohio 43210

The Ohio State University

THE FACULTY AND THEIR RESEARCH

James O. Alben, Professor, Medical Biochemistry; Ph.D., Oregon. Biomolecular spectroscopy, mathematical modeling, respiratory proteins.

Jessie Au, Professor, Pharmacy and Medicine; Ph.D., California. Pharmacokinetics, pharmacodynamics and toxicity of anticancer drugs.

Kamran Barin, Assistant Professor, Otolaryngology; Ph.D., Ohio State. Human postural control, vestibular neurophysiology.

Lawrence J. Berliner, Professor, Chemistry; Ph.D., Stanford. Diagnostic imaging by electron spin resonance and in vivo spectroscopy.

Necip Berme, Professor, Mechanical Engineering; Ph.D., Case Western Reserve. Orthopedic biomechanics, gait analysis.

Philip F. Binkley, Associate Professor, Medicine; M.D., Ohio State. Clinical hemodynamics, aortic impedance, heart rate variability, vascular biology.

Thomas E. Blue, Associate Professor, Nuclear and Mechanical Engineering; Ph.D., Michigan. Boron neutron capture therapy, radiation therapy.

Kim L. Boyer, Professor, Electrical Engineering; Ph.D., Purdue. Computer vision.

William A. Brantley, Professor, Restorative Dentistry, Prosthodontics, and Endodontics; Ph.D., Carnegie Mellon. Dental biomaterials.

Robert S. Brodkey, Professor Emeritus, Chemical Engineering; Ph.D., Wisconsin. Arterial blood flow, particle tracking velocimetry.

Martin Caffrey, Professor, Chemistry; Ph.D., Cornell. Membrane structure, drug delivery systems, liquid crystal diffraction.

Jeff Chalmers, Associate Professor, Chemical Engineering; Ph.D., Cornell. Effects of hydrodynamic forces on cells, magnetic cell separation.

Shive Chaturvedi, Associate Professor, Civil Engineering; Ph.D., Indian Institute of Technology.

Peter Clarkson, Research Scientist, Biomedical Engineering; Ph.D., Southampton. Signal processing, ECG analysis.

Bradley D. Clymer, Associate Professor, Electrical Engineering; Ph.D., Stanford. Image capture and image and signal processing.

J. Fredrick Cornhill, Professor, Surgery, and Staff, Cleveland Clinic; D.Phil., Oxford. Atherosclerosis, image processing, cardiovascular devices.

Roger A. Crawfis, Assistant Professor, CIS; Ph.D., California, Davis. Computer graphics, scientific visualization, virtual reality.

Brian L. Davis, Assistant Professor, Mechanical Engineering, and Staff, Cleveland Clinic; Ph.D., Penn State. Gait analysis, orthopedic biomechanics, diabetes, space flight-induced osteoporosis.

Roger Dzwonczyk, Assistant Professor, Anesthesiology; M.S., Ohio State. Cardiopulmonary resuscitation, ECG signal processing.

Mauro Ferrari, Professor, Internal Medicine and Mechanical Engineering, and Director, Biomedical Engineering Center; Ph.D., Berkeley. BioMEMS, biomedical nanotechnology for drug delivery, cell transplantation, bioseparation, biological micromechanics.

Elizabeth Fisher, Assistant Professor, Radiology and Staff, Cleveland Clinic; Ph.D., Rutgers. 3-D image analysis, neuroimaging, MRI.

Morton H. Friedman, Professor, Biomedical and Chemical Engineering and Pathology, and Associate Director, Biomedical Engineering Center; Ph.D., Michigan. Vascular hemodynamics and physiology, intravascular device evaluation, angiographic image processing.

Donald L. Fry, Professor, Internal Medicine and Pathology; M.D., Harvard. Experimental atherosclerosis, arterial mass transport.

Osamu Fujimura, Professor, Speech and Hearing Science; D.Sc., Tokyo. Speech production, tongue prosthesis, phonetics, X-ray microbeam.

Reinhard Gahbauer, Professor and Director, Radiation Oncology; M.D., Ludwig Maximilian (Germany). Radiation therapy.

Somnath Ghosh, Associate Professor, Applied Mechanics; Ph.D., Michigan. Computational mechanics, composite/heterogeneous materials.

Leonard Golding, Professor, Surgery and Staff, Cleveland Clinic; M.D., Sydney (Australia). Cardiovascular device evaluation.

Ernesto Goldman, Assistant Professor, Anesthesiology; M.D., Ohio State. Electromyography, cardiopulmonary interactions, paralysis.

Mark Grabiner, Staff Scientist, Cleveland Clinic. Human aging and motor control.

Robert L. Hamlin, Professor, Veterinary Biosciences; D.V.M., Ph.D., Ohio State. Pathophysiology of heart failure, pulmonary mechanics.

Mardi C. Hastings, Associate Professor, Mechanical Engineering and Biomedical Engineering; Ph.D., Georgia Tech. Bioacoustics, ultrasonics, fiber-optic sensors, instrumentation.

Deborah G. Heiss, Assistant Professor, Physical Therapy Division; Ph.D., Iowa; PT, OCS. Motor control, rehabilitation, orthopaedic biomechanics.

Hooshang Hemami, Professor, Electrical Engineering; Ph.D., Ohio State. Dynamics, control, and simulation of human movement.

C. Russell Hille, Professor, Medical Biochemistry; Ph.D., Rice. Enzymology of metalloproteins, biological electron transfer.

Richard D. Howell, Associate Professor, Special Education; Ph.D., New Mexico. Educational robotics, assistive technologies.

Hsiung Hsu, Professor Emeritus, Electrical Engineering; Ph.D., Harvard. Nonlinear optics, noncontact tonometry.

Carl R. Ingling, Associate Professor, Biophysics and Zoology; Ph.D., Rochester. Psychophysical study of human visual system.

Jogikal M. Jagadeesh, Associate Professor, Electrical Engineering and Pharmacy; Ph.D., Ohio State. Instrumentation, image processing, visual neuroscience, wavelet analysis.

William M. Johnston, Professor, Restorative Dentistry, Prosthodontics, and Endodontics; Ph.D., Michigan. Restorative materials.

P. Ewen King-Smith, Professor, Optometry; Ph.D., Cambridge. Visual sensitivity in optic nerve disorders, tear film spectrophotometry.

Lawrence E. Leguire, Clinical Associate Professor, Ophthalmology; Ph.D., Vanderbilt. Visual function improvement in infants and preverbal children.

Richard Lembach, Professor, Ophthalmology; M.D., Ohio State. Corneal topography, corneal transplantation, refractive surgery.

Alan S. Litsky, Associate Professor, Surgery and Biomedical Engineering; Sc.D., MIT; M.D., Columbia. Natural and synthetic biomaterials, orthopedic implant fixation, novel implant materials.

Geoffrey R. Lockwood, Assistant Professor, Mechanical Engineering, and Staff, Cleveland Clinic; Ph.D., Toronto. Ultrasound imaging systems, ultrasound, transducer design.

Adolf V. Lombardi Jr., Clinical Assistant, Professor, Orthopaedic Surgery; M.D. Temple. Orthopaedic biomechanics.

Thomas H. Mallory, Clinical Assistant Professor, Orthopaedic Surgery; M.D., Ohio State. Stress analysis of total joint prosthesis, biomechanics.

William S. Marras, Professor, Industrial and Systems Engineering; Ph.D., Wayne State. Occupational biomechanics.

W. Mitchell Masters, Associate Professor, Zoology; Ph.D., Cornell. Auditory information processing, sensory ecology and physiology.

John S. McDonald, Professor and Chairman, Anesthesiology. Lumbar epidural virtual reality simulator.

Don W. Miller, Professor, Nuclear and Mechanical Engineering; Ph.D., Ohio State. Digital radiography and radiation imaging.

Philip T. Nowicki, Associate Professor, Pediatrics and Physiology; M.D., Tufts. Developmental vascular physiology, endothelial cell biology.

Mohamad Parnianpour, Associate Professor, Industrial and Systems Engineering; Ph.D., NYU. Spine and occupational biomechanics, rehabilitation.

William S. Pease, Associate Professor, Physical Medicine; M.D., Cincinnati. Gait analysis, electromyography, rehabilitation.

Kimerly A. Powell, Assistant Professor, Radiology and Staff, Cleveland Clinic; Ph.D., Ohio State. Quantitative image processing, detection and diagnosis of breast cancer, myocardial mechanics, magnetic resonance imaging.

Cynthia J. Roberts, Associate Professor, Biomedical Engineering, Ophthalmology, and Surgery; Ph.D., Ohio State. Laser and optical applications in ophthalmology, ophthalmic imaging, corneal topography.

Pierre-Marie Robitaille, Associate Professor, Radiology and Medical Biochemistry; Ph.D., Iowa State. In vivo NMR spectroscopy, MRI.

Douglas W. Scharre, Assistant Professor; M.D., Georgetown. Diagnostic imaging in neuropsychiatric and dementing disorders.

Petra Schmalbrock, Assistant Professor, Radiology; Ph.D., Münster (Germany). Vascular and high-resolution magnetic resonance imaging.

Robert H. Small, Assistant Professor, Anesthesiology; M.D., Ohio State. Functional MRI, microvascular physiology and pharmacology, virtual reality.

William Smead, Associate Professor, Surgery; M.D., Vanderbilt. Noninvasive evaluation of arterial and venous disease.

Philip J. Smith, Professor, Industrial and Systems Engineering; Ph.D., Michigan. Human-computer interaction, cognitive systems engineering.

Bradford T. Stokes, Professor, Physiology and Surgery; Ph.D., Rochester. Fetal transplantation, spinal cord injury, biomechanics.

Roger P. Stradley, Assistant Professor, Veterinary Physiology; Ph.D., Ohio State. Gastrointestinal motility.

Richard D. Tallman, Associate Professor, Allied Health and Physiology; Ph.D., Ohio State. Cardiopulmonary interaction and artificial organs.

James D. Thomas, Professor, Medicine, and Staff, Cleveland Clinic; M.D., Harvard. Doppler echocardiography, cardiac mechanics, image and signal processing, numerical modeling, echocardiography in manned space flight.

Jean A. Tkach, Associate Professor, Radiology and Staff, Cleveland Clinic; Ph.D., Case Western Reserve. MR angiography, functional MRI, flow quantification, magnetization transfer, MRI and steady-state MR techniques.

Douglas B. VanFossen, Associate Professor, Clinical Medicine; M.D., Ohio State. Intravascular devices, vascular hemodynamics, digital imaging.

D. Geoffrey Vince, Assistant Professor, Pathology, and Staff, Cleveland Clinic; Ph.D., Liverpool (United Kingdom). Vascular pathology, intravascular ultrasound, cell quantification.

Andreas von Recum, Professor, Surgery and Oral Surgery; Ph.D., Colorado State. Biocompatibility of medical implants.

Manjula B. Waldron, Professor, Biomedical Engineering and Engineering Graphics; Ph.D., Stanford. Neural networks, communication and education aids for deaf persons, cognitive modeling, life cycle design.

DeLiang L. Wang, Associate Professor, CIS; Ph.D., USC. Neural networks, machine perception, cognitive modeling.

Paul A. Weber, Chair and Professor, Ophthalmology; M.D. Glaucoma.

Herman R. Weed, Professor, Electrical Engineering and Preventive Medicine; M.S.E.E., Ohio State. Functional muscle stimulation, vision information for the blind, rehabilitation, physiological control systems.

Richard White, Associate Professor, Radiology and Staff, Cleveland Clinic; M.D., Duke. Cardiovascular magnetic resonance imaging.

OREGON STATE UNIVERSITY

Bioresource Engineering Department

Programs of Study

The department offers programs leading to the M.S. and Ph.D. degrees. The objective of the bioresource engineering program is to serve at the interface between life sciences and engineering. Bioresource engineering is the application of engineering and life-science principles and problem-solving techniques to the optimum use and sustainability of biological resources. The curriculum is engineering-based, with equal emphasis on the life sciences and engineering. Courses focus on biological systems modeling, theoretical and applied aspects of fermentation and bioseparation processes, cell culture engineering, metabolic engineering, biomedical engineering, microscale systems engineering, regional hydrologic analysis, groundwater systems, irrigation, water resource optimization, remote sensing, image analysis, and instrumentation. The department has concentrated its research effort toward two major thrusts: bioprocess engineering and water resource engineering. Specific research topics in bioprocess engineering include bacterial biofilm development, protein and surfactant interactions at interfaces, downstream processing in biotechnology, purification of proteins and peptides at high loadings, biological modeling, quality preservation and energy conservation in fresh fruit storages, quality retention in frozen seafood, thermal process design for foods, and thermophysical property modeling. Research topics in water resource engineering include constructed wetland treatment systems, crop growth modeling, optimum irrigation, management, crop water requirements, groundwater and subsurface contaminant transport, hydrologic modeling, watershed and regional scale hydrologic analysis, image processing and artificial intelligence technologies to manage water resources, livestock production odor control, nonpoint source water pollution control, livestock waste management, and use of remote sensing.

Research Facilities

Between 1991 and 1993, the department's facilities (Gilmore and Gilmore Annex) were completely remodeled, with nine research laboratories established. In addition, the facilities were expanded to include four new graduate student offices and a 1,000-square-foot graduate teaching laboratory. The department completed construction of a new student computer lab/classroom, with fifteen computers available for student use. These computers are networked to a departmental server, the campus network, and the Web. A new conference room/classroom is now available in Gilmore Hall. The department has excellent fabrication facilities for constructing experimental apparatus and electronic devices and more than fifty personal computers and workstations networked to a departmental server and the campus network. These networks allow local, national, and international communications as well as access to the University's library holdings. Because OSU has had a Land Grant tradition for 125 years, its library has considerable holdings in the life sciences and engineering.

Financial Aid

Eighty-six percent of students are supported by assistantships or fellowships. The department offers graduate assistantships to qualified applicants on a competitive basis. M.S. and Ph.D. graduate assistants on half-time appointment are paid $14,570 and $16,224 per annum, respectively, in addition to tuition remission.

Cost of Study

Total tuition for full-time graduate students is $1910 per quarter for residents and $3570 for nonresident graduate students. Graduate students on assistantships qualify for tuition remission but must pay fees of $400 per term.

Living and Housing Costs

The estimated cost of living for a single student, including room and board, books and supplies, and personal expenses, is about $8400 per year. The University provides graduate dormitory housing and student family housing. Off-campus housing is available at a wide range of prices. The University Student Activities Office is available to assist students with housing issues.

Student Group

Oregon State University has an enrollment of more than 14,500, including more than 3,000 graduate students. The student body is diverse in its representation of gender, race, and religion. Of OSU's student population, approximately 58 percent are men, 42 percent are women, 77 percent are Oregonians, and 12 percent are students of color. There are approximately 40 graduate students enrolled in the department, with about 40 percent in the Ph.D. program.

Student Outcomes

Training in the graduate programs continues to prepare students for challenging positions in industry, government, and academia at highly competitive salaries. Recent graduates have joined consulting engineering firms, food and pharmaceutical companies, and prestigious research groups at the Battelle Pacific Northwest and Lawrence Berkely Laboratories. Others have gone on to further study in science and engineering at U.S. universities and in medical schools.

Location

Oregon State University's 500-acre main campus is in Corvallis, a city of approximately 45,000 that retains the friendliness and convenience of a small town. Corvallis lies in the heart of the beautiful Willamette Valley, about an hour's drive from the spectacular Oregon coast to the west and the rugged Cascade Mountains to the east. There are abundant opportunities for year-round outdoor recreation nearby. The University also has many facilities outside the Corvallis area, including the OSU Mark O. Hatfield Marine Science Center at Newport, the OSU Seafoods Laboratory at Astoria, ten agricultural research stations, and offices of the OSU Extension Service in every county.

The University and The Department

Oregon State University was designated as Oregon's Land Grant, state-assisted college in 1868. In 1968, Oregon State University became one of the nation's first three Sea Grant Colleges. Space Grant designation followed in 1991, making OSU one of only a handful of universities in the United States to hold all three designations.

Applying

Admission is open to graduates of fields related to bioresource engineering (engineering and science). Admission is possible in every quarter, although starting in the fall term is preferred. Assistantships can be awarded at any time. Applicants must submit to the Office of Admissions two copies of the application form, a $50 nonrefundable application fee, official sealed transcripts from every college or university attended, and a letter indicating the student's objectives and particular fields of interest. A third copy of the application form must be sent directly to the major department, along with copies of transcripts, GRE scores, a copy of the letter of interest, and three letters of reference addressed to the department. Applicants from other countries are required to provide copies of scores on the TOEFL.

Correspondence and Information

For program information:

Graduate Committee Chair
Department of Bioresource Engineering
116 Gilmore
Oregon State University
Corvallis, Oregon 97331-3906

Telephone: 541-737-2041
Fax: 541-737-2082
E-mail: info-bre@orst.edu
World Wide Web: http://www.bre.orst.edu

For application forms:

Director of Admissions
The Graduate School
Administrative Services A300
Oregon State University
Corvallis, Oregon 97331

Telephone: 541-737-4881

Oregon State University

THE FACULTY AND THEIR RESEARCH

Bioprocess Engineering

The bioprocess engineering group addresses a wide range of topics, from fundamental studies at the molecular level to value-added processing. Research areas include using genetically altered enzymes as protein analogues to model competitive molecular events at interfaces, theoretical and experimental studies on large-scale chromatography, the study of biologically mediated changes in food quality, the processing of fruits and vegetables to increase their quality and value, and the use of biological processes to convert biological materials into fuel, chemicals, and other useful products. Equipment to support research and teaching efforts in bioprocess engineering includes walk-in environmental chambers, laboratory fermenters, a high-speed centrifuge, dialysis and ion exchange equipment for protein isolation, gas chromatographs, HPLCs, two fully automated in situ ellipsometers, a contact angle goniometer, a liquid surface tensiometer, a differential scanning calorimeter, high-temperature/pressure reactors, a tissue culture hood, and CO_2 incubators.

Michelle Bothwell, Assistant Professor, Bioresource Engineering Department; Ph.D., Cornell, 1994. Biointerfacial phenomena with applications in biotechnology and biomedicine.

Frank Chaplen, Assistant Professor, Bioresource Engineering Department; Ph.D., Wisconsin–Madison, 1996. Cellular and metabolic pathway engineering, improvement of animal cell function in industrial cell culture, development of tools and techniques for bioprocess prediction and control, microscale systems for bioengineering and biological research.

Andrew G. Hashimoto, Professor and Associate Provost for Academic Affairs, Bioresource Engineering Department; Ph.D., Cornell, 1972. Bioprocessing, conversion of biomass into methane and ethanol, biomass pretreatment, agricultural waste management, methane emissions related to global climate change.

Martin L. Hellickson, Associate Professor, Bioresource Engineering Department; Ph.D., Minnesota, 1975. Postharvest preservation, handling, storage, and transportation of fresh fruits and vegetables; energy conservation in refrigerated storages.

Edward R. Kolbe, Professor, Bioresource Engineering Department/Portland; Ph.D., New Hampshire, 1975. Measuring and modeling thermal properties of food and biological materials; food freezing; modeling quality change in frozen storage; cooking/gelation processes, including ohmic heating; seafood technology.

Joseph McGuire, Professor, Bioresource Engineering Department; Ph.D., North Carolina State, 1987. Colloidal and interfacial phenomena, protein and surfactant adsorption, biofilm development, biomedical engineering.

Michael T. Morrissey, Associate Professor, Food Science and Technology Department and Director of OSU Seafoods Lab (Astoria); Ph.D., Oregon State, 1982. Seafood processing, quality standards in seafood, biochemical/microbial changes in seafood.

Jae Park, Associate Professor, Food Science and Technology Department/Seafoods Lab (Astoria); Ph.D., North Carolina State, 1985. Rheological and chemical characterization of seafood muscle proteins and their interaction with other ingredients and/or physicochemical components, surimi production and utilization, ohmic heat processing.

Michael H. Penner, Associate Professor, Food Science and Technology Department; Ph.D., California, Davis, 1984. Enzyme technology of glycan hydrolases, cellulases, side-chain cleaving hemicellulases.

Ajoy Velayudhan, Assistant Professor, Bioresource Engineering Department; Ph.D., Yale, 1990. Downstream processing in biotechnology, design and scale-up of nonlinear adsorptive and chromatographic separations, optimization of separation sequences, biological modeling.

Water Resource Engineering

The water resource engineering group addresses a wide range of contemporary issues of hydrologic analysis and water quality and quantity. The topic area focuses on simulation modeling and decision support, application of remote sensing and GIS to water resource management, regional hydrologic modeling, optimum irrigation management, animal waste management, nonpoint source pollution management, constructed wetlands water treatment, and groundwater quality. Research includes field campaigns in Africa and Canada and across the United States, with a strong emphasis on applied research. The Hydrologic Engineering group in the Bioresource Engineering Department has put a priority on maintaining state-of-the-art research facilities. Equipment dedicated to the area of watershed hydrology includes a spectral radiometer, three fully automated TDR sensor systems, three tension infiltrometers, numerous electronic rain gauges, three generators with six pumping systems, several ISCO automated samplers, a 3/4-ton four-wheel-drive truck, and a digital SIR-10 GPR system. For laboratory investigations, the University maintains two dedicated flow visualization laboratories, a 20Ghz Tektronix TDR system, and analysis tools, including a gas chromatograph, a Waters HPLC, spectrophotometers, ion-specific bromide and chloride meters, a fluorometer, fraction collectors, and numerous digitally networked balances and pH meters. Numerical modeling tools include state-of-the-art software for multiphase flow (e.g., STOMP, HYDRUS2D) and GGID (e.g., ARCINFO), and Sun and linked IBM RS/6000 workstations for numerically intensive operations. The Hydrologic Engineering group aggressively maintains the computing facilities at a state-of-the-art level.

Dominique Bachelet, Assistant Professor, Bioresource Engineering Department; Ph.D., Colorado State, 1983. Ecosystem modeling, biogeochemical cycling, methane emission, global climate change effects on vegetation distribution, GIS application of models.

John P. Bolte, Associate Professor, Bioresource Engineering Department; Ph.D., Auburn, 1987. Ecological systems modeling, decision support, artificial intelligence.

Richard H. Cuenca, Professor, Bioresource Engineering Department; Ph.D., California, Davis, 1978. Hydrologic system analysis, irrigation system design, evapotranspiration analysis, soil water measurement and physical property analysis.

Marshall J. English, Professor, Bioresource Engineering Department; Ph.D., California, Davis, 1978. Water resources systems analysis, irrigation management, optimum irrigation system design and operation.

Derek C. Godwin, Extension Agent, Bioresource Engineering Department/Curry County; M.S., Oregon State, 1993. Watershed management as Extension Sea Grant agent; educational programs related to improving salmon populations and habitat, water quality and conservation, and ecosystem health; assists in the formation and operation of community-based watershed councils sanctioned by state legislation.

Jonathan D. Istok, Professor, Civil Engineering; Ph.D., Oregon State, 1986. Soil and water engineering, aquifer testing, numerical modeling of groundwater flow, bioremediation.

Danny Marks, Assistant Professor, EPA; Ph.D., California, Santa Barbara, 1988. Global climate change, large-scale spatial modeling, regional hydrology, snow and alpine hydrology, watershed hydrology, precipitation chemistry, remote sensing, digital terrain analysis, surface climate simulation, geobase information systems, remote data collection systems.

J. Ronald Miner, Professor, Bioresource Engineering Department; Ph.D., Kansas State, 1967; Extension Water Quality Specialist. Animal waste management, odor control, water quality.

James A. Moore, Professor and Department Head, Bioresource Engineering Department; Ph.D., Minnesota, 1975. Agricultural waste management, water quality, constructed wetlands, nonpoint source pollution.

John S. Selker, Associate Professor, Bioresource Engineering Department; Ph.D., Cornell, 1990. Groundwater quality hydrology and modeling research, study of water and contaminant transport through unsaturated natural media (vadose zone transport), including the development of numerical models, laboratory scale flow cells, and field experiments.

R. Thomas Wykes, Energy Agent, Bioresource Engineering Department/Deschutes County; B.S., Oregon State, 1984. Home energy conservation, renewable energy (solar), telecommuting, boiler maintenance and operation.

PENNSTATE

PENNSYLVANIA STATE UNIVERSITY

Bioengineering Program

Programs of Study

Bioengineering, an intercollege graduate degree–granting program offering M.S. and Ph.D. degrees, is staffed by 5 full-time faculty members and by 24 other faculty members whose primary appointments are in the Colleges of Engineering, Medicine, Health and Human Development, and Science. About twenty courses in bioengineering provide a firm foundation in the application of engineering methods to the solution of problems in medicine and biology. The core curriculum consists of introductory courses on structure and function of the cardiovascular system, physiology, and biomedical instrumentation; in the internship program, students are exposed to an overview of clinical procedures and laboratory research in the College of Medicine of Penn State University at the Milton S. Hershey Medical Center.

Entering students begin graduate study with the core curriculum and are encouraged to become actively involved in a research project. In subsequent years, research activities become intensified while students take technical electives to strengthen their academic skills in traditional engineering disciplines and the life sciences. A minimum of 30 credits of graduate study is required for M.S. and Ph.D. programs. All students attend a weekly seminar series with lectures by faculty members and distinguished researchers from outside the University and are required periodically to give seminars on their research and to develop effective communication skills in lecture and classroom settings. Ph.D. students are expected to complete a balanced program of courses in bioengineering, the life sciences, and traditional engineering in addition to fulfilling degree requirements by taking a written and oral qualifying examination, taking a comprehensive oral examination after a minimum of one year of research, and publicly defending the dissertation before the faculty. Students accepted into both the Bioengineering Program and the Penn State University Medical School may pursue a combined M.D./Ph.D. program.

The Bioengineering Program maintains strong research efforts on the biophysical basis of cardiovascular function and the development of artificial organs. The widely recognized collaboration between engineering and clinical faculty, aimed at development of an artificial heart, continues to set new milestones in the development of artificial organs. The Penn State Heart Assist Pump was recognized in 1990 as an International Historic Mechanical Engineering Landmark by the American Society of Mechanical Engineers. Major research programs facilitate funding of graduate research assistantships in various areas of faculty research interests.

Research Facilities

The Bioengineering Program maintains more than 8,545 square feet of research space devoted to the Artificial Heart, Microcirculation, Neuroelectrophysiology, Cell Biomechanics, and Ultrasound Imaging laboratories in the Hallowell Building at the University Park campus. Numerous other laboratories are located on this campus and at the Hershey Medical Center. State-of-the-art research equipment, microcomputers and mainframe computers, and graphics workstations are available to support all research projects.

Financial Aid

Graduate research assistantships, College of Engineering and departmental fellowships, and University-sponsored minority fellowships are available to qualified applicants for support of all tuition and fees and a stipend starting at $15,911 per year. Stipends may be supplemented, through the Office of the Dean of the College of Engineering, by a special award to exceptional and outstanding students. All financial aid is awarded on the basis of academic merit and availability of funds.

Cost of Study

Full-time tuition for 1998–99 was $3267 per semester for Pennsylvania residents and $6730 per semester for nonresidents. Each student was required to pay $637 per year for medical insurance.

Living and Housing Costs

On-campus housing is available for single and married graduate students. Details may be obtained by writing to the Assignment Office–Campus Residences, 101 Shields Building. Off-campus apartments are also available.

Student Group

Thirty-six graduate students are currently enrolled, two thirds of whom are studying for the Ph.D. degree and the remainder for the M.S. Approximately one fourth of the students are in residence full-time at the Hershey Medical Center, where they perform their research following completion of the core curriculum of bioengineering courses at the University Park campus. About 25 percent of the students are women. Students share a camaraderie and mutually supportive social experience during their residence. The Bioengineering Program's highly active student chapter of the national Biomedical Engineering Society won national recognition by receiving the Meritorious Achievement Award from the society in 1991, 1992, and 1993.

Location

The University Park campus of Pennsylvania State University is situated in central Pennsylvania, within a few hours' drive of New York City, Philadelphia, Pittsburgh, and Washington, D.C. The Milton S. Hershey Medical Center is located in Hershey, Pennsylvania, 1¾ hours by car from University Park. Communications between the Medical Center and the main campus are maintained electronically by closed-circuit TV link, which is used for course offerings common to the two campuses.

The University and The Department

Pennsylvania State University consists of the main campus at University Park (38,219 students) and seventeen branch campuses (33,651 students) throughout the state. The College of Engineering offers one of the largest undergraduate programs in the country, leading to B.S. degrees in ten engineering disciplines. Due to the split of its parent administrative body between the Graduate School and the College of Engineering, the Bioengineering Program is beyond departmental status and is considered an intercollege program. As such, it enjoys an academic freedom that allows it to respond rapidly to the ever-changing needs of biomedical research and of the medical community and its supporting industry.

Applying

The Bioengineering Program welcomes applications from students with diverse backgrounds in engineering and the life sciences. Application forms for admission and financial aid may be obtained from the program office and must be submitted along with supporting documents (official transcripts, three letters of recommendation, and GRE scores) by February 15 of the academic year preceding admission to ensure consideration for financial aid. International applicants are required to take the TOEFL.

Correspondence and Information

Dr. Herbert H. Lipowsky
Bioengineering Program
233 Hallowell Building
Pennsylvania State University
University Park, Pennsylvania 16802
Telephone: 814-865-1407
Fax: 814-863-0490
E-mail: bsb1@psu.edu

Pennsylvania State University

THE FACULTY AND THEIR RESEARCH

Herbert H. Lipowsky, Ph.D., Professor and Chairman of Bioengineering. Pressure and flow relationships in microcirculation, in vivo rheology blood flow in sickle-cell disease and other hematological disorders. Image enhancement of the in vivo leukocyte-endothelium contact zone using optical sectioning microscopy. *Ann. Biomed. Eng.* 25:521–35, 1997.

Professors

Harry R. Allcock, Ph.D., Evan Pugh Professor of Chemistry. Synthesis of biomedical and bioactive materials. Use of polyphosphazenes for skeletal tissue regeneration and cardiovascular devices. Novel polyphosphazene/poly(lactideco-glycolide) blends: Miscibility and degradation studies. *Biomaterials* 18:1565–9, 1997.

James G. Brasseur, Ph.D., Professor of Mechanical Engineering. Biofluid mechanics, neuromuscular mechanics, turbulent flows, graphical imaging. The impact of fundoplication on bolus transit across the gastoesophageal junction. *Am. J. Physiol.* 275:G1386–93, 1998.

Paul W. Brown, Ph.D., Professor of Materials Science and Engineering. Biomaterials and synthetic bone. The effects of magnesium on hydroxyapatite formation in vitro at 37.4°C. *Calcif. Tissue Int.* 60:538–46, 1997.

Peter R. Cavanagh, Ph.D., Distinguished Professor of Kinesiology, Medicine, Orthopaedics and Rehabilitation, and Biobehavioral Health. Mechanics of the lower extremities in normal and diabetic subjects. The role of cutaneous information in a contact control task of the leg in humans. *Hum. Movement Sci.* 17:95–120, 1998.

Andris Freivalds, Ph.D., Professor of Industrial Engineering. Biomechanics, ergonomics. A graphic model of the human hand using CATIA. *Int. J. Ind. Ergonomics* 11:255–64, 1994.

David B. Geselowitz, Ph.D., Distinguished Alumni Professor of Bioengineering and Professor of Medicine. Theoretical cardiac electrophysiology and electrocardiography, artificial hearts and cardiac-assist devices. Use of surface Laplacian to extract local activation from epicardial electrograms. *Computers in Cardiology*, pp. 129–32. IEEE Computer Society Press, 1996.

William E. Higgins, Ph.D., Professor of Electrical and Computer Engineering. 3-D/4-D medical image analysis and visualization, virtual endoscopy. System for analyzing true three-dimensional angiograms. *IEEE Trans. Med. Imaging* 15(3):377–85, 1996.

Edward S. Kenney, Ph.D., Professor Emeritus of Nuclear Engineering. Industrial and medical radiation imaging, nuclear instrumentation, nuclear reactor control. A high speed Compton scatter pipe wall imaging system. Nuclear instruments and methods. *Phys. Res. A* 353:334–7, 1994.

William Pierce, M.D., Professor Emeritus of Surgery. Artificial heart and mechanical circulatory assistance. An electrically powered total artificial heart: Over 1 year survival in the calf. *ASAIO J.* 42(5):M342–6, 1996.

Joseph L. Rose, Ph.D., Paul Morrow Professor of Engineering Design and Manufacturing. Development of ultrasound imaging and guided wave devices. A computer model for simulating ultrasonic scattering in biological tissues with high scatterer concentration. *J. Ultrasound Med. Biol.* 20(9):903–13, 1994.

Gerson Rosenberg, Ph.D., Jane A. Fetter Professor of Surgery and Bioengineering. Mechanical circulatory assistance, the electric artificial heart, artificial organs. Dynamic in vitro and in vivo performance of a permanent total artificial heart. *Artif. Organs* 22(1):87–94, 1998.

James P. Runt, Ph.D., Professor of Polymer Science. Poly(urethaneureas) as blood contacting materials, biostability, microphase-separated structure, improved barrier properties. An investigation of the *in vivo* stability of poly(ether urethaneurea) blood sacs. *J. Biomed. Mater. Res.* 44:371–80, 1999.

K. Kirk Shung, Ph.D., Professor of Bioengineering and Director, Center for Ultrasound Transducer Engineering. Ultrasonic imaging, transducers, and tissue characterization. Ultrasonic transducers and arrays. *IEEE Eng. Med. Biol. Mag.* 15:20–30, 1996.

John M. Tarbell, Ph.D., Distinguished Professor of Chemical Engineering. Biomolecular transport dynamics, cardiovascular fluid dynamics, artificial heart fluid mechanics, blood damage. Mean velocity and Reynolds stress measurements in the regurgitant jets of tilting disk heart valves in an artificial heart environment. *Ann. Biomed. Eng.* 26:146–56, 1998.

James S. Ultman, Ph.D., Professor of Chemical Engineering. Biomass and heat transfer, pulmonary physiology, health effects of air pollutants. Longitudinal distribution of ozone absorption in the lung: Simulation with a single-path model. *Toxicol. Appl. Pharmacol.* 140:219–26, 1996.

Robert Zelis, M.D., Professor of Medicine and Cellular and Molecular Physiology. Analysis of cardiac function. Compartmental analysis of norepinephrine kinetics in congestive heart failure. *Am. J. Cardiol.* 75:299–301, 1995.

Associate Professors

Abdellaziz Ben-Jebria, Ph.D., Associate Professor of Chemical Engineering. Kinetics of air pollutants in lung airways. Pulmonary drug delivery with controlled release. *J. Appl. Physiol.* 84:379–85, 1998; 81:1651–7, 1996. *Science* 276:1868–71, 1997.

Wenwu Cao, Ph.D., Associate Professor of Mathematics and Materials Science. Ultrasound transducer design. Finite element analysis and experimental studies on piezoelectric composite transducers and arrays. *Ultrason. Imaging* 18:1–9, 1996.

Cheng Dong, Ph.D., Associate Professor of Bioengineering and Engineering Science and Mechanics. Biomechanics, cellular mechanics, cellular response to physical and chemical signals in biological systems, tissue mechanics, computer modeling. Two phases of pseudopod protrusion in tumor cells revealed by a micropipette. *Microvasc. Res.* 47:55–67, 1994.

John F. Gardner, Ph.D., Associate Professor of Mechanical Engineering. Control systems and modeling. An empirically based aortic pressure observer for use with electro-mechanical circulatory assist devices. *ASME J. Biomed. Eng.* 115:187–94, 1993.

Roger P. Gaumond, Sc.D., Associate Professor of Bioengineering. Neural signals and neuromuscular stimulation, control of respiration and of cardiovascular function. *Artificial Hearts and Other Organs. Encyclopedia of Electrical and Electronics Engineering,* New York: John Wiley and Sons, 1999.

Kane High, M.D., Associate Professor of Anesthesia. Anesthesia management and respiratory assist devices. Modeling gas exchange through a blind-ended microporous fiber. *ASAIO Abstracts*, p. 62, 1994.

Joseph J. McInerney, Ph.D., Associate Professor of Medicine and Bioengineering. Diagnostic imaging of the cardiovascular system. Real-time ventricular volumes and cardiac function in the working rat heart. *Am. J. Med. Sci.* 307:92–6, 1994.

Neil A. Sharkey, Ph.D., Associate Professor of Kinesiology and Orthopaedics and Rehabilition and Associate Director of the Center for Locomotion Studies. Musculoskeletal research, including normal, pathologic, and reconstructed function of bones and joints; mechanisms of injury; internal biomechanical behavior of the foot and ankle, knee, hip, and shoulder. *Clin. Biomechan.* 13:420–33, 1998.

Michael B. Smith, Ph.D., Associate Professor of Radiology, Center for NMR Research. Magnetic resonance imaging. Three-dimensional mapping of the static magnetic field inside the human head. *Magn. Reson. Med.* 36:705–14, 1996.

Alan J. Snyder, Ph.D., Associate Professor of Bioengineering and Research Associate in Surgery. Artificial heart, circulatory assist, electronic design for implantable devices. Non-invasive control of cardiac output for alternately ejecting dual pusherplate pumps. *Artif. Organs* 16:189–94, 1992.

Assistant Professors

Norman R. Harris, Ph.D., Assistant Professor of Bioengineering. Effects of leukocyte endothelial cell adhesion and cardiac risk factors on microvascular permeability. Age-dependent responses of the mesenteric vasculature to ischemia-reperfusion. *Am. J. Physiol.* 274:H1509–15, 1998.

Christopher R. Jacobs, Ph.D., Assistant Professor of Orthopaedics and Rehabilitation, Bioengineering, and Engineering Science. Orthopaedic biomechanics, implants, tissue adaptation, cellular response to biophysical signals, large-scale finite-element modeling. Adaptive bone remodeling incorporating simultaneous density and anisotropy considerations. *J. Biomech.* 30(6):603–13, 1997.

Nadine B. Smith, Ph.D., Assistant Professor of Bioengineering. Noninvasive focused ultrasound surgery and hyperthermia. MRI temperature monitoring. *Int. J. Radiat. Oncol. Biol. Phys.* 43:217–25, 1999.

William J. Weiss, Ph.D., Assistant Professor of Bioengineering and Research Associate in Surgery. Implantable circulatory support devices, transcutaneous energy transmission. Recent improvements in the completely implanted total artificial heart. *ASAIO J.* 42(5):M342–6, 1996.

Research Associate

Arnold A. Fontaine, Ph.D., Research Associate, Applied Research Laboratory. Biofluid dynamics, turbulence, drag reduction. Chordal force distribution determines systolic mitral leaflet configuration and severity of functional mitral regurgitation. *JACC*, in press.

RENSSELAER POLYTECHNIC INSTITUTE

Department of Biomedical Engineering

Programs of Study

Biomedical engineering is both an education and career choice. Qualified students are admitted to the Department of Biomedical Engineering for graduate work. Degrees offered are Master of Science (M.S.), Master of Engineering (M.Eng.), Doctor of Philosophy (Ph.D.), and Doctor of Engineering (D.Eng.). The M.S./Ph.D. track is oriented toward research, while the M.Eng./D.Eng. track is oriented toward design and is recommended for students who are interested in a career in industry. The M.Eng. Program is a nonthesis degree intended for professional practice. Each graduate student develops an individual plan of study in consultation with a faculty adviser. The master's program is typically completed in two years; the doctoral program requires five to six years.

Three concentrations for graduate study are available. The electrical concentration focuses on computing and signal processing applications to such problems as online processing of patient data and instrumentation for measurement of physiological parameters. In the materials area, students address problems associated with the relationships between biomaterials properties and cellular responses at the tissue-biomaterial interface. In the biomechanics area, key research problems include biomechanics of cells and tissues in the vascular and musculoskeletal systems, stress transfer at the bone-implant interface, and computer modeling of the biomechanics of synovial joints.

Research Facilities

Departmental research facilities include a bioinstrumentation teaching laboratory and seven other research labs in the Jonsson Engineering Center. The laboratories are oriented toward studies of cardiopulmonary transport and dynamics, physiological control, image processing, biomaterials, cellular/tissue engineering, computational biomechanics, biomechanical testing, and stress analysis of implants and interfaces. Rensselaer is involved in collaborative research activities with area medical institutions, including Albany Medical College, Albany Veterans Administration Hospital, Wadsworth Center for Laboratories and Research, New York State Department of Health, and Columbia University.

Research is supported by state-of-the-art facilities such as the computing facilities in the Center for Industrial Innovation, which provides graduate students with walk-in access to advanced workstations and to programs ranging from personal productivity aids to advanced computer-aided design and analysis packages; the Rensselaer Libraries, with electronic information systems that provide access to collections, databases, and Internet resources from campus and remote terminals; the Rensselaer Computing System, which permeates the campus with a coherent array of advanced workstations, a shared toolkit of applications for interactive learning and research, and high-speed Internet connectivity; a visualization laboratory for scientific computation; and a high-performance computing facility, which includes a 36-node SP2 parallel computer.

Financial Aid

Students are eligible for financial aid of various types, including fellowships, teaching assistantships, tuition scholarships, and research assistantships. The amount of aid varies with the source and type of support and the progress of the student. Outstanding students may qualify for University-supported Rensselaer Scholar Fellowships, which carry a stipend of $15,000 and a full waiver of tuition and fees. Low-interest, deferred-repayment graduate loans are also available to U.S. citizens with demonstrated need.

Cost of Study

Tuition for 1999–2000 is $665 per credit hour. Other fees amount to approximately $535 per semester. Books and supplies cost about $1700 per year.

Living and Housing Costs

The cost of rooms for single students in residence halls or apartments ranges from $3356 to $5298 for the 1999–2000 academic year. Family student housing, with a monthly rent of $592 to $720, is available.

Student Group

There are about 4,300 undergraduates and 1,750 graduate students representing all fifty states and more than eighty countries at Rensselaer.

Student Outcomes

Eighty-eight percent of Rensselaer's graduate students were hired after graduation with starting salaries that averaged $56,259 for master degree recipients and $57,000 to $75,000 for doctoral degree recipients in 1998.

Location

Rensselaer is situated on a scenic 260-acre hillside campus in Troy, New York, across the Hudson River from the state capital of Albany. Troy's central Northeast location provides students with a supportive, active, medium-sized community in which to live and an easy commute to Boston, New York, Montreal, and some of the country's finest outdoor recreation, including Lake George, Lake Placid, and the Adirondack, Catskill, Berkshire, and Green Mountains. The Capital Region has one of the largest concentrations of academic institutions in the United States. Sixty thousand students attend fourteen area colleges and benefit from shared activities and courses.

The University

Founded in 1824 and the first American college to award degrees in engineering and science, Rensselaer Polytechnic Institute today is accredited by the Middle States Association of Colleges and Schools and is a private, nonsectarian, coeducational university. Rensselaer has five schools: Architecture, Engineering, Management, Science, and Humanities and Social Sciences. The School of Engineering ranks nationally among the top twenty engineering schools by the *U.S. News & World Report* survey and is ranked in the top ten by practicing engineers.

Applying

Admissions applications and all supporting credentials should be submitted well in advance of the preferred semester of entry to allow sufficient time for departmental review and processing. GRE General Test scores are required. The application fee is $35. Since the first departmental awards are made in February and March for the next full academic year, applicants requesting financial aid are encouraged to submit all required credentials by February 1 to ensure consideration.

Correspondence and Information

For written information about graduate study:

Coordinator of Student Affairs
Department of Biomedical Engineering
Rensselaer Polytechnic Institute
110 8th Street
Troy, New York 12180-3590
Telephone: 518-276-6547
E-mail: citarl@rpi.edu
World Wide Web: http://www.rpi.edu/dept/
   biomed/WWW/

For applications and admissions information:

Director of Graduate Academic and Enrollment
   Services, Graduate Center
Rensselaer Polytechnic Institute
110 8th Street
Troy, New York 12180-3590
Telephone: 518-276-6789
E-mail: grad-services@rpi.edu
World Wide Web: http://www.rpi.edu

Rensselaer Polytechnic Institute

THE FACULTY AND THEIR RESEARCH

R. Bizios, Professor; Ph.D., MIT, 1979. Cellular and tissue bioengineering, cell-biomaterial interactions, biomaterials.

J. B. Brunski, Professor; Ph.D., Pennsylvania, 1977. Dental biomechanics and implants, biomaterials.

N. DePaola, Assistant Professor; Ph.D., MIT–Harvard, 1991. Biofluidmechanics, cellular bioengineering.

J. C. Newell, Professor; Ph.D., Albany Medical College, 1974. Cardiopulmonary physiology, systems modeling, electrical impedance imaging.

L. E. Ostrander, Associate Professor; Ph.D., Rochester, 1966. Information processing, biomedical signal analysis, human factors in medical equipment design.

R. J. Roy, Professor Emeritus; D.Eng.Sci., Rensselaer, 1962; M.D., Albany Medical College, 1976. Biocontrol systems, digital signal processing, pattern recognition, adaptive processes.

B. Roysam, Associate Professor of Electrical, Computer, and Systems Engineering; D.Sc., Washington (St. Louis), 1989. Intelligent imaging at low SNR, parallel computation, biomedical applications.

R. L. Spilker, Professor and Chairman of the Department; Sc.D., MIT, 1974. Computational mechanics and biomechanics.

A. Zelman, Professor; Ph.D., Berkeley, 1971. Membrane transport phenomena, design and construction of toys and assistive devices to aid persons who are physically and mentally challenged.

Affiliated Faculty

M. Cheney, Associate Professor of Mathematical Sciences; Ph.D., Indiana, 1982. Applied mathematics, differential equations, mathematical physics, analysis.

R. H. Doremus, Professor of Glass and Ceramics Science; Ph.D., Illinois, 1953; Ph.D., Cambridge, 1956. Physical chemistry, solutions of polyelectrolytes and proteins.

D. G. Gisser, Professor Emeritus of Electrical, Computer, and Systems Engineering; D.Eng, Rensselaer, 1944. Electronic devices and circuits, measurement instrumentation, biomedical applications.

D. Isaacson, Professor of Mathematics and Computer Science; Ph.D., NYU, 1977. Electric current–computed tomography.

J. W. Modestino, Professor of Electrical, Computer, and Systems Engineering; Ph.D., Princeton, 1969. Stochastic processes in communication and control, information theory and coding, detection and estimation theory, digital signal and image processing.

M. Savic, Professor of Electrical, Computer, and Systems Engineering; Eng.D.Sc., Belgrade, 1965. Controlled cryodestruction, signal processing.

H. A. Scarton, Associate Professor of Mechanical Engineering and Mechanics; Ph.D., Carnegie Mellon, 1970. Biomechanics, wave phenomena, acoustics, noise control.

Adjunct Faculty

S. S. Bowser Jr., Adjunct Assistant Professor; Ph.D., SUNY at Albany, 1984. Cell structure and function, particularly cell motility and cytoskeleton-membrane interactions; effects of mechanical forces on cell physiology; biology of benthic foraminifera.

J. P. Cousins, Adjunct Assistant Professor; Ph.D., Johns Hopkins, 1988. Magnetic resonance imaging and spectroscopy.

P. Del Vecchio, Adjunct Associate Professor; Ph.D., Fordham, 1975. Biology, vascular endothelium.

P. M. Edic, Adjunct Assistant Professor; Ph.D., Rensselaer, 1994. Electrical impedance imaging and magnetic resonance imaging computation.

P. Feustel, Adjunct Assistant Professor; Ph.D., Albany Medical College, 1978. Cerebral circulation and regulation of respiration.

R. L. Jacobs, Adjunct Professor; M.D., Iowa State, 1956. Orthopedics, physiology and biochemistry of bone.

B. Lee, Adjunct Professor; M.D., Seoul, 1951. Surgical research, peripheral vascular surgery.

B. Rangert, Adjunct Associate Professor; Ph.D. Chalmers, 1974. Dental implants, biomaterials, and biomechanics.

B. Rosenblatt, Adjunct Assistant Professor; D.M.D., Tufts, 1975. Dental materials research.

T. M. Saba, Adjunct Professor; Ph.D., Tennessee, 1967. Physiology of the reticuloendothelial system, cardiovascular and pulmonary function during shock, host defense mechanisms.

J. N. Turner, Adjunct Professor; Ph.D., SUNY at Buffalo, 1973. Biophysics, anatomic pathology, quantitative light microscopy.

RESEARCH AREAS

Bioinstrumentation and Medical Devices. Research and advance study include work on respiratory transducers and monitors, computer-controlled devices for physiological and medical studies, cardiac output monitors, viscoelastic testing of biological materials, development of noninvasive/nontraumatic diagnostic instrumentation for bone and vascular abnormalities, design and manufacturing of unique, non-commercially available toys and assistive devices to aid persons who are physically and/or mentally challenged.

Biomaterials. Efforts center on the design, construction, implantation, and evaluation of implants for use in bone. Cell cultures are used to elucidate bone/implant interactions; the results are incorporated into the design of biomaterials, which elicit specific and desirable responses from cells and tissues. Light microscopy plus X-ray and electron optical techniques are available for studies of variations in structure and composition of bones and teeth.

Biomechanics. Research in biomechanics includes studies of the physical properties of hard and soft connective tissues at bone-implant interfaces. Research is also under way to predict and measure biting forces on endosseous dental implants and to relate implant-tissue biomechanics to interfacial tissue responses. In cell culture studies, the strength of the attachment of cells to biomaterial cells is being investigated.

Impedance Imaging. Electrical impedance imaging is the technology for creating images of the internal structures of a body from measurements made at electrodes on the body's surface. Electrical currents are applied to the body from an array of electrodes, and from the resulting voltages an image can be reconstructed of the electrical conductivity of the body.

Computing and Signal Processing Applications. Research in imaging includes studies that apply engineering methods of computer-assisted tomography to image reconstruction for microscopy. Major efforts are in three-dimensional visualization of fluorescence micrographs from a conventional light microscope. Such methods allow biologists to observe the activity of chromosomes throughout various phases of the cell cycle. Other research projects are in new methods of image reconstruction for X-ray-computed tomography.

Systems Physiology and Clinical Medicine and Surgery. The techniques of systems modeling are applied to a variety of physiological systems to elucidate their function and behavior, to determine the etiology of disease, to suggest improved methods of diagnosis and treatment, and to predict courses of response and recovery. Cardiopulmonary mechanics and computer control of the administration of vasoactive drugs are actively studied.

Cellular Bioengineering. Cultured mammalian cells are used to study, in vitro and at the cellular/molecular level, systems of biomedical interest. Experimental projects in progress include investigations of the mechanisms of osteoblast interactions with orthopedic/dental implant materials; the structure and biochemistry of the cell/biomaterial interface; and the effects of mechanical stresses on cellular function and morphology.

Computational Bioengineering. The level of complexity inherent in the study of human systems such as musculoskeletal or cardiovascular systems frequently dictates the need for numerical methods of solution. High performance computational methods to the study of diathrodial joint mechanics, cardiovascular mechanics, dental mechanics, and imaging are being developed and applied. Projects involving the development of computational methods for bioengineering applications are done in collaboration with Rensselaer's Scientific Computation Research Center.

Other Research. Theoretical studies of membrane transport phenomena involving chemical, electrical, and mechanical properties of ionic membranes. Biomedical engineering research at Rensselaer involves three schools within the Institute and interactions with Albany Medical College.

THOMAS JEFFERSON UNIVERSITY

College of Graduate Studies
Graduate Program in Cell and Tissue Engineering

Program of Study

The Ph.D. Graduate Program in Cell and Tissue Engineering (CTE) provides an innovative and rigorous research training and education program in the application of contemporary biological and engineering principles to study normal and abnormal tissue and cells for the ultimate goal of designing functional tissue substitutes. The program emphasizes strengths in molecular biology, cell biology, physiology, biomaterials sciences, and clinical sciences to provide innovative approaches to understanding, imaging, and ameliorating disease states through the use of state-of-the-art tools in biotechnology and informatics. The program in cell and tissue engineering provides a multidisciplinary approach to train and engage students in such innovative biomedical research areas as genetic engineering of animal models of human diseases, tissue engineering, biomaterials, cellular and functional bioimaging, cellular bioreactor technology, cellular and molecular phenotyping technologies, and biological/medical information technology. The program prepares students for careers in academia and in the private sector in such areas as biopharmaceuticals, biotechnology, biomedical materials and devices, and bioinstrumentation.

The interdepartmental training faculty consists of researchers actively engaged in the research areas above. The student population is composed of students from many backgrounds, including biology, biochemistry, chemistry, biophysics, and bioengineering. In addition, through an academic alliance with the Drexel University School of Biomedical Engineering Sciences and Health Systems, research interests and training in biomaterials, biomedical computing, bioinformatics, and biomechanics are expanded.

Research Facilities

The laboratories of the program faculty are housed in modern research buildings that are fully equipped for investigations in all aspects of cellular, molecular, and clinical analyses. Students in the CTE program have access to state-of-the-art facilities for nucleic acid sequencing, recombinant DNA technology, PCR analysis, protein purification and analysis, cell and tissue culture, and transgenic facilities, as well as confocal and electron microscopes and ultrasound and MRI facilities. In addition, the University houses a computer-automated biomedical research library.

Financial Aid

Financial support is available to the majority of Ph.D. students in the form of University fellowships, teaching and research assistantships, and training grants. In 1999–2000, students granted full fellowship support receive funds for tuition remission and a stipend of $16,600. University loan programs are also available to qualified students.

Cost of Study

Tuition and fees for full-time Ph.D. students are $12,670 per year in 1999–2000.

Living and Housing Costs

Living expenses in Philadelphia are quite reasonable in comparison to that of other East Coast metropolitan areas. Affordable student housing is available on campus and in the immediate area.

Student Group

The University community totals approximately 2,500 students, of whom 150 are currently full-time Ph.D. candidates. The student body is diverse, with members from across the country and abroad. Many activities are coordinated by the Graduate Student Association, and a variety of cultural and recreational activities are available within and around the University.

Location

The University is located in Center City Philadelphia, within walking distance of shopping, restaurants, concert and theater halls, museums, and numerous historic sites.

The University

Thomas Jefferson University, founded in 1824, is an academic health center devoted to research and teaching in the biomedical sciences and is recognized as a leader in biomedical research and education.

Applying

Completed application forms may be obtained from and should be returned to the Office of Admissions, College of Graduate Studies, Thomas Jefferson University, 1020 Locust Street, Philadelphia, Pennsylvania 19107 (telephone: 215-955-4400). Prospective students are encouraged to visit the University for interviews and to meet with members of the faculty. The application fee is $40.

Correspondence and Information

For applications:
Jessie Pervall
Director of Admissions and Recruitment
College of Graduate Studies
Thomas Jefferson University
1020 Locust Street, M46
Philadelphia, Pennsylvania 19107-6799
Telephone: 215-503-4400
Fax: 215-503-3433
E-mail: cgs-info@mail.tju.edu
World Wide Web: http://www.tju.edu

For program information:
Dr. Rocky S. Tuan
Director, Cell and Tissue Engineering Graduate Program
Thomas Jefferson University
501 Curtis Building
1015 Walnut Street
Philadelphia, Pennsylvania 19107
Telephone: 215-955-5479
Fax: 215-955-9159
E-mail: rocky.s.tuan@mail.tju.edu
World Wide Web: http://jeffline.tju.edu/CWIS/DEPT/CTE/

Thomas Jefferson University

THE FACULTY AND THEIR RESEARCH

Nicholas A. Abidi, M.D., Pennsylvania, 1991. Molecular and cellular biology of bone and cartilage regeneration; growth factor enhancement of tissue healing; biomechanical testing of tissue repair techniques.

Robert Akins Jr., Ph.D., Pennsylvania, 1992. Cell and molecular biology of mammalian striated muscle; tissue engineering, gene therapy, and surgical implantation for the treatment of muscular diseases.

George C. Brainard, Ph.D., Texas at San Antonio, 1982. Photobiology and neuroendocrine regulation; control of melatonin in humans and animals; effects of light on behavior and mood.

Mon-Li Chu, Ph.D., Florida, 1975. Molecular biology of extracellular matrix proteins; function and molecular interactions of collagens and basement membrane proteins; regulation of extracellular matrix gene expression; connective tissue diseases.

Matt During, M.D., Auckland (New Zealand), 1982. Gene therapy of neurological disorders; gene therapy of metabolic disorders; molecular basis of learning and memory; development of gene transfer methods; immunological interactions with the nervous system; novel therapeutic approaches to neurodegenerative disorders.

Vicky L. Funanage, Ph.D., Delaware, 1981. Cell and molecular biology of musculoskeletal development, growth, and disease; DNA diagnostics for neuromuscular and musculoskeletal diseases; tissue engineering of bone and muscle; molecular pathophysiology of spinal muscular atrophy and myotonic dystrophy; placental leptin and its role in growth and development.

David J. Hall, Ph.D., Minnesota, 1984. Cell and molecular biology of cell-cycle transit; regulation of cell proliferation in three-dimensional cell culture; utilization of yeast for development of therapeutic drugs; alteration of cell/tissue/organ shape and effects on gene expression; control of cell-cycle transit by extracellular matrix/protein; subcellular protein targeting detected by confocal imaging.

Robert W. Harrison, Ph.D., Yale, 1985. Molecular modeling and rational drug design, antiviral agents; HIV protease; bioinformatics; protein structure and its relationship to function.

Noreen Hickok, Ph.D., Brandeis, 1981. Bone-biomaterial interaction and implant design; tissue engineering of bone and cartilage; mechanisms of chondrocyte hypertrophy; regulation of cell proliferation via cytoskeletal/oncogene interactions; confocal laser imaging of cells and extracellular matrix.

William J. Hozack, M.D., McGill, 1981. Factors controlling successful hip and knee replacements.

Ziwei Huang, Ph.D., California, San Diego, 1993. Chemical and structural approaches to immunology and cell biology; structure and function of proteins, peptides, and small molecular mimetics; structure-based drug design for cancer, AIDS, and immune diseases.

Jeffrey I. Joseph, D.O., Philadelphia College of Osteopathic Medicine, 1983. Implantable optical blood chemistry sensors; tissue engineering for stable vascular interface; biomaterials; neovascularization; implantable drug delivery systems; artificial endocrine pancreas; implantable cardiovascular sensor.

Jaspal Khillan, Ph.D., Punjab (India), 1980. Gene regulation during development and differentiation; chondrocyte differentiation; molecular biology of skeletal formation during normal and diseased state; tissue engineering of cartilage; transgenic animal models of skeletal disorders; embryonic stem cells for tissue regeneration.

Karen A. Knudsen, Ph.D., Pennsylvania, 1978. Cell-cell adhesion; cadherins and catenins; cadherin-mediated signaling; skeletal myogenesis; cardiomyocyte biology.

Thomas B. Knudsen, Ph.D., Thomas Jefferson, 1981. Experimental and molecular toxicology of the embryo; regulation and function of the mitochondrial DNA genome; cell-signaling pathways in diseases with a developmental basis; environmental responsive genes associated with birth defects.

Devendra M. Kochhar, Ph.D., Florida, 1964. Mammalian embryology; cartilage and bone cell differentiation; genetic and experimental analysis of skeletal dysphasias; in vitro bioassays to screen environmental agents; computer-based design of retinoid molecules; functional analyses of agonists and antagonists of retinoid receptors; natural and synthetic retinoids in skin and bone diseases.

Yefu Li, M.D., Beijing Union Medical College, 1986; Ph.D., Brigham Young, 1991. Molecular study of human genetic skeletal disorders; molecular genetic basis of spinal bifida; signal pathways controlling chondrogenesis.

Sue A. Menko, Ph.D., Pennsylvania, 1978. Cell-adhesion receptors in embryonic development; integrins and the regulation of lens cell differentiation; matrix compartmentalization and the regulation of lens development; cadherins and lens morphogenesis; integrin matrix interactions in skeletal muscle differentiation; adhesion molecules and salivary gland biogenesis.

Pamela A. Norton, Ph.D., Tufts, 1986. Transcriptional regulation of the fibronectin gene in cells and transgenic mice; mechanisms of fibronectin alternative mRNA splicing; extracellular matrix in liver development and disease; mesenchymal cell-cell and cell-matrix interactions.

Joan Overhauser, Ph.D., Purdue, 1983. Molecular basis of chromosome disorders, elucidation of haplo-insufficiency syndromes; molecular cytogenetics of chromosome abnormalities, molecular elucidation of psychiatric disorders; physical mapping of chromosome 5 and 18.

Richard H. Rothman, M.D., Pennsylvania, 1962; Ph.D., Thomas Jefferson, 1965. Design and optimization of total joint arthroplasty; biology of implant wear; osteoarthritis and degenerate joint diseases.

James D. San Antonio, Ph.D., Pennsylvania, 1987. Mechanisms of angiogenesis and chondrogenesis; vascular smooth muscle cell growth and differentiation; atherosclerosis and restenosis; biochemistry and functions of proteoglycan-collagen interactions; tissue engineering; development of autologous organ implants.

Peter F. Sharkey, M.D., SUNY Health Science Center at Syracuse, 1984. Bone ingrowth into biomaterials; osteoarthritis; particulate disease; pulmonary embolism; clinical results of total hip arthroplasty/total knee arthroplasty.

J. Bruce Smith, M.D., Wake Forest, 1965. Immunopathogenesis of rheumatoid arthritis; development of cell-based vaccines for treatment of autoimmune disease; maternal-fetal cell trafficking; roles of persistent fetal lymphocytes in the later development of autoimmune disease, especially systemic sclerosis (scleroderma); pregnancy-related clinical changes in patients with autoimmune diseases; cytokine production and regulation in first-trimester placenta.

Marla Steinbeck, Ph.D., Iowa State, 1987. Production of reactive oxygen species (ROS) by inflammatory cells and osteoclasts; mechanism of ROS-mediated bone resorption; elucidation of signaling pathways and transcriptional regulators involved in ROS-mediated differentiation.

Rocky S. Tuan, Ph.D., Rockefeller, 1977. Cell and molecular biology of skeletal development, growth, and disease; bone-biomaterial interaction and implant design; tissue engineering of bone and cartilage; confocal laser imaging of cells and extracellular matrix; animal models of skeletal diseases; gene-based infection diagnostics.

Jouni Uitto, M.D./Ph.D., Helsinki, 1970. Molecular biology of the cutaneous basement membrane zone; genetic disorders of the skin; heritable connective tissue disorders.

David A. Wenger, Ph.D., Temple, 1968. Biochemical, molecular, pathological, and genetic studies on certain lysosomal storage diseases; development of improved diagnostic methods; use of animal models for treatment studies; development of vectors for treating Krabbe disease by ex vivo and in vivo gene therapy methods.

Eric Wickstrom, Ph.D., Berkeley, 1972. Antisense oligonucleotides for cancer therapy; antigenomic oligonucleotides for viral therapy; gene insertion at defined sequences for congenital diseases.

Edward Winter, Ph.D, SUNY at Stony Brook, 1985. Role of signal transduction pathways in morphogenesis and development; mechanisms regulating gene expression.

Charlene J. Williams, Ph.D., Rutgers, 1983. Genetic linkage analysis of osteo- and inflammatory arthropathies; positional cloning of disease susceptibility genes for osteo- and inflammatory arthropathies; isolation and characterization of matrix genes in normal and diseased cartilage; molecular ecology.

TULANE UNIVERSITY

School of Engineering
Department of Biomedical Engineering

Programs of Study	Tulane's biomedical engineering (BME) department provides a comprehensive engineering science approach to research and education in the health sciences and medicine. The goal is to train highly qualified researchers and educators with general competence in all branches of engineering science as well as specific in-depth expertise in particular fields of biomedical research. Most of the students in the department choose to study in one of the areas represented by the full-time faculty. These include biosolid mechanics and biofluid mechanics, the mechanics and physiology of the pulmonary system, cardiovascular fluid mechanics, biomaterials science, brain physics and neurophysiology, cardiac electrophysiology, bioinstrumentation, and tissue engineering. In addition, many students find other projects that suit their interests under the supervision of the adjunct faculty.
	Since students enter the program with varied academic backgrounds and course preparation, degree requirements are specified in terms of a level of competence expected of the student rather than by the completion of particular courses. Nevertheless, it is normally expected that M.S.E. candidates will complete a minimum of 24 semester hours of approved graduate-level course work and will write a scholarly thesis describing original research. These courses must meet prescribed distribution criteria. Students pursuing the D.Sc. degree must also carry out a further 24 semester hours of graduate course work, pass a written/oral qualifying examination, and successfully defend their doctoral dissertation. All students pursuing a graduate degree must demonstrate teaching competence as part of their training.
Research Facilities	The Department of Biomedical Engineering is located in the Lindy Claiborne Boggs Center, a facility built in 1988 to foster biotechnology research. The BME department has more than 25,000 square feet of office and laboratory space. Among the laboratories are a biomaterials laboratory, a biomechanical testing laboratory, a computational cardiac electrophysiology and imaging laboratory, cardiovascular and pulmonary mechanics testing laboratories, a neurosignal analysis laboratory, and a cell and tissue engineering laboratory. The department has recently been awarded a $1 million Special Opportunity Award from the Whitaker Foundation; this award money will help develop a laboratory for computational modeling of cells and tissues.
	Notable pieces of equipment include an MTS axial-torsion testing machine, EG&G PAR electrochemical and polarographic measurement systems, an Electronetics pulsating bubble surfactometer, a Cahn surface tension balance, video analysis hardware and software, a water-jacketed CO_2-controlled incubator, an Olympus IX-50 inverted microscope, an IEC refrigerated centrifuge, and a laminar flow hood and an autoclave. One of the department's strengths involves the use of computational methods toward the understanding of biomedical problems. To complete these studies, the department has substantial computational facilities, including a Cray Y-EL94 supercomputer, and an SGI Origin 2000 multiprocessor server.
Financial Aid	Nearly all full-time students in the department receive competitive, merit-based financial aid. The department typically offers teaching assistantships to new students. These provide a stipend of approximately $12,000 and a tuition waiver for nine months of service. Research assistantships are also occasionally awarded to new students. The department frequently has several four-year fellowships ($17,000-per-year stipend and tuition waiver). Graduate students are encouraged to apply for individual fellowships, such as those awarded by the National Science Foundation, the Whitaker Foundation, the American Heart Association, and Tau Beta Pi. In recent years, students in the department have been very successful in obtaining these awards.
Cost of Study	In 1998–99, tuition was $22,160, subject to a maximum increase of 4 percent in the next year. The estimated cost of books and supplies was $750. Student health insurance was $750 per year, and health center fees were $304 per year. Full-time students were also assessed $280 for activities and the use of the Reilly athletic center. All of these costs are revised annually.
Living and Housing Costs	A small number of on-campus dormitory housing openings are available each year, but most students elect to live off campus. Monthly rents range between $250 and $500. The total cost of rent, utilities, and food for single students is approximately $6500–$7500 annually.
Student Group	In 1998–99, of the approximately 40 graduate students in the department, approximately 35 percent were women. Roughly 40 percent of the graduate students were pursuing doctoral degrees, while the remainder were pursuing master's degrees. The graduate students come from many parts of the United States and a variety of other countries. Applications are welcomed from interested students with academic backgrounds in any physical or biologic science.
Student Outcomes	Graduates have found employment in the biomedical device industry, hospitals, government agencies, and many research and academic areas.
Location	Tulane University is located in a beautiful residential section of New Orleans, a city that is famous for its restaurants. Plays and concerts are numerous in New Orleans, while the Mardi Gras celebration and its preliminary pageants offer festivities unparalleled in the United States. New Orleans is also the site of the Louisiana Superdome, which is home to Tulane football and the New Orleans Saints. Sailing and boating are popular on Lake Pontchartrain, located within 15 minutes of the campus, and the beaches of the Mississippi Gulf Coast are less than an hour away.
The University and The Department	Tulane is a private, nonsectarian university that offers a wide range of undergraduate, graduate, and professional courses of study. There are approximately 6,600 undergraduate and 3,600 graduate students enrolled in the University, with roughly 10 percent of the student body attending the School of Engineering. The Department of Biomedical Engineering is one of the premier departments at Tulane and has consistently been ranked in the top eight departments nationally in recent Gourman surveys.
Applying	Applications for admission with financial aid for the fall semester are due by February 1. All applications must be submitted by July 15 for the fall semester and December 1 for the spring semester. Applicants must hold, or be in the process of obtaining, a bachelor's degree, and all applications must include transcripts of prior work, three letters of recommendation, and GRE test scores. International applicants must also include TOEFL scores.
Correspondence and Information	For program information and application forms: Administrative Assistant Department of Biomedical Engineering Tulane University New Orleans, Louisiana 70118 Telephone: 504-865-5897 E-mail: amills@mailhost.tcs.tulane.edu World Wide Web: http://www.bmen.tulane.edu

Tulane University

THE FACULTY AND THEIR RESEARCH

Ronald C. Anderson, Associate Professor; Ph.D., Tulane, 1987. Biomechanics, orthopedic materials.

Kirk J. Bundy, Professor; Ph.D., Stanford, 1975. Biomaterials, corrosion, bioadhesion, environmental science.

Kay C. Dee, Assistant Professor; Ph.D., Rensselaer, 1996. Tissue engineering, cellular bioengineering.

Donald P. Gaver III, Associate Professor and Assistant Chairman; Ph.D., Northwestern, 1988. Biofluid mechanics, pulmonary mechanics, bioremediation, interfacial flows, surfactant physicochemical hydrodynamics, cell and tissue modeling, computational fluid mechanics.

Richard T. Hart, Professor and Chairman of the Department; Ph.D., Case Western Reserve, 1983. Mechanics of bone, finite-element analysis, functional adaptation.

Glen A. Livesay, Assistant Professor; Ph.D., Pittsburgh, 1996. Experimental and theoretical mechanics, soft tissue mechanics, optimization.

Paul L. Nunez, Professor; Ph.D., California, San Diego, 1969. Electroencephalography, signal processing, neocortical dynamics.

David A. Rice, Associate Professor; Ph.D., Purdue, 1974; PE. Physiological modeling, cardiopulmonary mechanics, bioacoustics, instrumentation and signal processing.

Natalia A. Trayanova, Associate Professor; Ph.D., Sofia (Bulgaria), 1986. Theoretical and computational electrophysiology, cardiac pacing and defibrillation, scientific visualization.

Cedric F. Walker, Professor; Ph.D., Duke, 1978; PE. Telemedicine, neural prostheses and electrical stimulation, clinical engineering.

Affiliated Faculty

Richard B. Ashman, Director of Biomechanics Laboratory, Texas Scottish Rite Hospital for Crippled Children; Ph.D., Tulane, 1982. Biomechanics.

Richard Baratta, Assistant Professor of Orthopedic Surgery, Louisiana State University School of Medicine; Ph.D., Tulane, 1989. Rehabilitation engineering.

Andrew J. Buda, Chief, Cardiology Section, Department of Medicine, Tulane University School of Medicine; M.D., Toronto, 1973. Cardiology.

Claude F. Burgoyne, Assistant Professor of Ophthalmology, Louisiana State University Medical Center; M.D., Minnesota, 1987. Mechanical properties of soft tissue.

Dzung Hong Dinh, Assistant Professor of Neurosurgery, Tulane University School of Medicine; M.D., Iowa, 1985. Spinal mechanics.

Lawrence M. Gettleman, Professor of Reconstructive Dentistry, University of Louisville; D.M.D., Harvard, 1966. Dental biomaterials.

Henry Glindmeyer, Research Professor of Medicine, Tulane University School of Medicine; D.Eng., Tulane, 1976. Environmental influences on pulmonary mechanics.

Salvadore Guccione, Head of Mathematics Division, University of New Orleans National Biodynamics Laboratory; Ph.D., Missouri, 1977. Statistical modeling of head-neck impact response, mechanics of injury.

Robert G. Heath, Professor of Psychiatry and Neurology, Tulane University School of Medicine; M.D., Pittsburgh, 1938; D.M.Sci., Columbia, 1949. Neural stimulation, biological psychiatry.

Jean Jacob, Assistant Professor of Ophthalmology, Louisiana State University Eye Center; Ph.D., Tulane, 1988. Biomedical research.

Jack Kent, Professor and Head of Oral and Maxillofacial Surgery, Louisiana State University School of Dentistry; D.D.S., Nebraska, 1963. Dental implants and alloplastic materials for maxillofacial reconstruction.

Bahram Khoobehi, Assistant Professor of Ophthalmology, Louisiana State University Eye Center; Ph.D., North Texas State, 1982. Laser-liposome techniques for studying retinal blood flow.

Stephen D. Klyce, Professor of Ophthalmology, Louisiana State University Eye Center; Ph.D., Yale, 1971. Corneal physiology and topology.

J. Monroe Laborde, Chairman, Department of Orthopedic Surgery, Touro Infirmary; M.D., Tulane, 1973; M.B.E., Case Western Reserve, 1976. Orthopedics.

Peter V. Moulder, Professor of Surgery; M.D., Chicago, 1945. Cardiovascular surgery and systems modeling.

Donald R. Owen, President, BioSouth Research Labs; Ph.D., Houston, 1973. Polymeric biomaterials.

Ken Pilgreen, Alabama Neurophysics Laboratory; M.D., LSU, 1982. Electroencephalography.

Janet C. Rice, Associate Professor of Biostatistics, Tulane University School of Public Health; Ph.D., Purdue, 1974. Statistical analysis and experimental design.

Donald E. Richardson, Professor and Chairman of Neurosurgery, Tulane University School of Medicine; M.D., Tulane, 1957. Neurosurgery and electrical stimulation for control of pain.

Kevin A. Thomas, Assistant Professor of Orthopedic Surgery, Louisiana State University Medical Center; Ph.D., Tulane, 1985. Biomaterials in orthopedics.

Karen H. Watanabe, Adjunct Assistant Professor of Biomedical Engineering; Ph.D., California, 1993. Heat transfer and transport in porous media.

Marc Weiss, Ph.D., Rochester, 1969. Human impact tolerance.

Robert Zone, Vice President, Vital Assist Inc.; Ph.D., Tulane, 1993. Polymeric biomaterials.

UNIVERSITY OF AKRON

Department of Biomedical Engineering

Programs of Study

The graduate programs in the Department of Biomedical Engineering are designed to develop the student's understanding of the application of engineering principles to medical problems and medical research. Current graduate programs in the department lead to the M.S. and Ph.D. degrees. An integrated M.D./Ph.D. program is available in cooperation with the Northeastern Ohio Universities College of Medicine (NEOUCOM).

The department is involved in many collaborative research projects with industry, regional hospitals, and NEOUCOM. This strong interaction between academia and the medical community creates unusual opportunities for training that involves basic sciences and their applications.

Biomedical engineering courses are offered in the areas of biomedical instrumentation; soft connective tissue biomechanics; orthopedic implants; math modeling in biomedicine; physiological control systems; cardiovascular diagnostic and therapeutic techniques; cardiovascular dynamics; experimental methods in biomechanics; imaging devices, biosensors, and detectors; biomedical signal processing; image processing; biometry; mechanics in physiology and medicine; rehabilitation engineering; biomedical computing; artificial organs; hard tissue biomechanics; sensory systems analysis; physiological systems; neural networks; muscle mechanics and optimization; kinematics of the human body; and finite elements in biomechanics.

Research Facilities

The Department of Biomedical Engineering occupies 7,400 square feet of laboratory and office space in the Sidney L. Olson Research Center. Research programs in the department are carried out not only in the facilities of the department but also in laboratories in Akron City Hospital, Akron General Medical Center, Children's Hospital Medical Center of Akron, Saint Thomas Hospital Medical Center, Edwin Shaw Rehabilitation Hospital, the Cleveland Clinics, NEOUCOM, and local industrial research centers. There are laboratories involved in image analysis; image detectors, devices, and biosensors; biostereometrics; motion analysis; biomaterials; vascular dynamics; musculoskeletal biomechanics; human interface and rehabilitation engineering; and medical instrumentation in the department.

Financial Aid

The University awards a number of graduate assistantships to qualified students. Assistantships are normally awarded for up to two years of master's degree study and up to four years of doctoral degree study. These assistantships provided a stipend of $7500 for 1997–98, plus remission of tuition and fees, and are available in the Department of Biomedical Engineering. In addition, externally funded fellowships are available from the grants and contracts received for work conducted in the department.

Cost of Study

In 1998–99, tuition was $178 per semester hour for Ohio residents and $333.10 per semester hour for out-of-state students. There was also a general service fee of $6.85 per semester hour.

Living and Housing Costs

Apartment suites are available for graduate students and range from $250 to $300 per month per student. For details, students should contact the Off-Campus Housing Office, Spanton Hall, Room 109 (telephone: 330-972-6936).

Student Group

At the present time, there are more than 30 full-time graduate students in the Department of Biomedical Engineering, with applications from potential additional graduate students currently being reviewed. There is an active chapter of the Biomedical Engineering Society on campus. Some of the interests of graduate students are sports medicine, biomechanics, biotransport, biostereometrics, rehabilitation engineering, biomaterials, biomedical image science, flow modeling, noninvasive diagnostics, and artificial organs, to name a few.

Location

Akron, with a rich heritage of industrial leadership in rubber and related products, has evolved into a center for high technology and service-oriented businesses. The city provides northeastern Ohio with a stimulating blend of educational, cultural, recreational, commercial, research, and industrial resources.

The University

Founded in 1870 as Buchtel College by the Unitarians, the University of Akron became a municipal college in 1913 and a state university in 1967. The student body and campus have grown rapidly in recent years. With more than 26,000 students, the University of Akron is the third-largest university in Ohio. In 1979, the Institute for Biomedical Engineering Research was founded in order to tie together the research at Akron hospitals and the engineering expertise at the University. Within the College of Engineering, the Department of Biomedical Engineering was formed in 1984.

Applying

The application deadline for admission to the Graduate School is two months before fall or spring registration. The assistantship application deadline is March 1. TOEFL scores of 590 or higher are required for international students whose native language is not English, and GRE General Test and TSE scores are required for all applicants. Students should apply directly to the Graduate Office, University of Akron.

Correspondence and Information

Admissions Committee
Department of Biomedical Engineering
University of Akron
Akron, Ohio 44325
E-mail: info@biomed.uakron.edu
World Wide Web: http://www.biomed.uakron.edu

University of Akron

THE FACULTY AND THEIR RESEARCH

Ted Conway, Ph.D. (theoretical and applied mechanics), Illinois at Urbana-Champaign. Biomechanics, elastic and viscoelastic mechanics, mechanical design, analytical and computational nonlinear mechanics of polymers, wire rope and polymer cord design and analysis.

George C. Giakos, Ph.D. (electrical engineering), Marquette. Medical imaging, imaging devices and detectors, spectroscopical imaging, biosensors, solid-state devices, fiber-optical systems and biooptics.

Irving Miller, Ph.D. (chemical engineering), Michigan. Biomaterials, biotransport phenomenon, drug delivery.

Glen O. Njus, Ph.D. (mechanical engineering), Iowa. Orthopedic biomechanics, prosthetic design, soft tissue implants, wound healing, reconstructive surgery.

Narender Reddy, Ph.D. (biomedical engineering), Texas A&M. Medical devices, rehabilitation engineering, human interface technology, virtual reality, noninvasive measurement, neural networks, fuzzy logic and computer modeling.

Stanley Rittgers, Ph.D. (biomedical engineering), Ohio State. Cardiovascular flow modeling, ultrasound Doppler color flow imaging, laser Doppler anemometry, quantitative flow visualization, computational fluid mechanics, MRI and noninvasive vascular diagnostic techniques.

Daniel Sheffer, Ph.D. (exercise physiology), Texas A&M; Director, Biostereometrics Laboratory, University of Akron Laboratory at Akron City Hospital. Biostereometrics, breast cancer detection, human ergonomics, human physiology, whole-body stereophotogrammetry.

Bruce Taylor, Ph.D. (physiology and biomedical engineering), Kent State. Biomedical instrumentation—blood flow and blood pressure measurements, computer simulations of cardiovascular dynamics, clinical engineering.

Mary C. Verstraete, Ph.D. (applied mechanics), Michigan State. Gait analysis, human motion analysis, sports biomechanics, optimization, finite-element modeling.

Faculty with Joint Appointments in Biomedical Engineering

Frank Harris, Ph.D., Polymer Science; Eberhard Meinecke, Ph.D., Polymer Science; Richard Mostardi, Ph.D., Biology; Dale Mugler, Ph.D., Mathematical Sciences; Thomas Price, Ph.D., Mathematical Sciences; Jonathon Rakich, Ph.D., Management; Daniel Smith, Ph.D., Chemistry; Max Willis, Ph.D., Mechanical Engineering.

Research Fellow

Robert Pierson, B.S. (chemical engineering), Princeton. Polymer research—synthetic rubbers; artificial organs, with emphasis on the artificial heart.

Adjunct Faculty (Institute for Biomedical Engineering Research)

Michael Askew, Ph.D., Director of Musculoskeletal Research, Summa Health System.

Gordon Bennett, M.D.; Orthopedics, Akron General Medical Center.

Enrique Canilang, M.D.; Physical Medicine Department, Edwin Shaw Rehabilitation Hospital.

Surendra Chawla, Ph.D.; Engineering, Goodyear Tire & Rubber Company.

William Cruce, Ph.D.; Neurobiology, NEOUCOM.

William Davros, Ph.D.; Radiology, Cleveland Clinic Foundation.

Michael Domski, B.S.; Heart Center, Akron General Medical Center.

Duane Donovan, M.D.; Vascular Surgery, Akron City Hospital.

Elizabeth Dumont, Ph.D.; Anatomy, NEOUCOM.

Robert Flora, M.D.; Obstetrics and Gynecology, NEOUCOM.

Ernie Freeman, Ph.D.; Cardiovascular, Akron General Medical Center.

William Gardner, M.D.; Internal Medicine, Akron General Medical Center.

Lowell Gerson, Ph.D.; Epidemiology, NEOUCOM.

Ivan Gradisar, M.D.; Orthopedics, Orthopaedic Surgeons, Inc.

Walter Hoyt, M.D., Emeritus; Orthopedics, Akron City Hospital.

Richard Josephson, M.D.; Cardiology, Akron City Hospital.

Anand Kantak, M.D.; Associate Director of Neonatology, Children's Medical Center.

David Kay, M.D.; Surgery, Akron General Medical Center.

Richard Klich, Ph.D.; School of Speech Pathology and Audiology, Kent State University.

William Landis, Ph.D.; Biochemistry, NEOUCOM.

Mark Leeson, M.D.; Orthopedics, Akron General Medical Center.

Robert Liebelt, M.D., Ph.D.; Director of Medical Education, Saint Thomas Hospital.

Steven Lippitt, M.D.; Orthopedics, Akron General Medical Center.

C. William Loughry, M.D.; Surgery, Akron City Hospital.

John McCulloch, M.D.; Orthopedics, Orthopaedic Surgeons, Inc.

Arne Melby III, M.D.; Orthopedics, Orthopaedic Surgeons, Inc.

Amy Milsted, Ph.D.; Biology, University of Akron.

Nina Njus, M.D.; Surgery, Summit Hand Center, Inc.

Anthony Passalaqua, M.D.; Nuclear Medicine, Akron City Hospital.

Richard Pepe, D.O.; Radiology, Akron City Hospital.

Gregory F. Powell, Ph.D.; Imaging, Picker International.

Steve Reger, Ph.D.; Musculoskeletal Research, Cleveland Clinic Foundation.

Donna Richardson, Ph.D.; Signal and Image Processing.

Kenneth Rosenthal, Ph.D.; Microbiology/Immunology, NEOUCOM.

Bruce Rothschild, M.D.; Rheumatology, St. Elizabeth Hospital.

Steven Schmidt, Ph.D.; Director of Falor Center for Vascular Study, Akron City Hospital.

Buel Smith, M.D., Emeritus; Orthopedics, Akron General Medical Center.

Paul Steurer, M.D.; Orthopedics, Akron General Medical Center.

Sherry Steusse, Ph.D.; Neurobiology, NEOUCOM.

Jack Summers, M.D.; Urology, Akron City Hospital.

Timothy Teyler, Ph.D.; Neurobiology, NEOUCOM.

William D. Timmons, Ph.D.; Consultant.

Richard Varga, Ph.D.; Director, Institute for Computational Mathematics, Kent State University.

Douglas Wagner, M.D.; Surgery, Plastic and Reconstructive Surgeons, Inc.

Steven Ward, Ph.D.; Anatomy, NEOUCOM.

Jeffrey Wenstrup, Ph.D.; Neurobiology, NEOUCOM.

THE UNIVERSITY OF ALABAMA AT BIRMINGHAM

School of Engineering
Department of Biomedical Engineering

Programs of Study

The Department of Biomedical Engineering offers the Master of Science and Doctor of Philosophy degrees. The Master of Science may be taken as a terminal degree, or the program may be pursued as an ideal background for the qualified student seeking entry into the doctoral program. With the terminal M.S. degree, most graduates obtain employment either in medical centers and hospitals, in industry with manufacturers of medical products, or with government agencies, health-care groups, and computer application groups. Doctoral candidates are usually preparing for careers in research, teaching, or advanced design and development.

Biomedical engineering students should have a foundation in the basic sciences and engineering. The three research areas of emphasis are brain imaging, biomedical implants, and cardiac electrophysiology. Other areas of research include biofluids, biocontrols, and tissue engineering.

Research Facilities

The biomedical engineering program at The University of Alabama at Birmingham (UAB) is one of few such graduate programs with an internationally recognized medical center located on its campus. Some 5,600 square feet of laboratory space is dedicated to the programs, and additional space is available at the University Medical Center. The Biomaterials Laboratory has complete facilities for materials characterization, including FTIR, DCA, SEM, STEM, Auger spectroscopy, image analysis, X-ray diffraction, NMR, and atomic and absorption spectroscopy. Corrosion testing, mechanical testing, and cell-culture analyses are also available. The Biomechanics Laboratories contain experimental and analytical capabilities, including finite-element analysis–based research utilizing Marc, Cosmos, NASTRAN, ANSYS, and PDA/PATRAN-G (for finite-element analysis) programs as well as traditional stress analysis capability. The Alabama CRAY X-MP supercomputer is also available. A biofluids lab with computerized LDV and Doppler ultrasound equipment is also available. A DASH CAD system, several microprocessor development systems, HP PC instruments, numerous Sun and SGI workstations, and PCs are available. A 4-Tesla whole-body MR facility offers opportunities for collaborative work in medical imaging and spectroscopy. Laboratories for the cardiac electrophysiology program are located in a newly renovated 15,000-square-foot space in Volker Hall. Equipment for experiments includes optical and electrical recording systems for multichannel mapping of heart tissue.

Financial Aid

For certain qualified graduate students, stipends of $15,000 per year (plus tuition and fees) are available in the form of research assistantships in a clinical or research environment. Graduate students may also qualify for College Work-Study awards or Stafford Loans or for veterans' benefits through the Veterans Administration. Financial aid applications are available through the Office of Student Financial Aid, University Center, Room 250.

Cost of Study

In 1999–2000, tuition for state residents is $99 per semester hour; for out-of-state graduate students, tuition is $198 per semester hour.

Living and Housing Costs

Some University-owned and -operated housing is available on campus at reasonable rates. Apartments or modest rooming accommodations and a variety of inexpensive dining establishments are available within a few blocks of the campus. The University's cafeterias also provide excellent food at reasonable prices.

Student Group

The majority of the full-time M.S. and Ph.D. students in the department received B.S. degrees in biomedical, electrical, materials, or mechanical engineering or physics. One third of the students are women, and most students receive some type of financial aid. One half of the students are pursuing the Ph.D. degree.

Student Outcomes

Since the inception of the UAB biomedical engineering program, 31 percent of graduates have found employment in medical device companies, 10 percent in government agencies, 45 percent in hospitals and universities, and 7 percent in professional schools or private practice. Only about 7 percent go into non-biomedical engineering industry jobs. Examples of employers include Collagen Corporation, Eli Lily, W. L. Gore, C. R. Bard, Wright Technologies, Stryker, and Intermedics.

Location

The 4,034-square-mile Birmingham metropolitan area (population 903,000) has an average high temperature of 73.6° F and an average low of 51.2° F. The University's 70-square-block campus is located six blocks from the downtown area of Birmingham, within walking distance of restaurants, entertainment facilities, and shopping areas.

The University

UAB is an autonomous campus within The University of Alabama System, with an enrollment exceeding 16,000 students (3,154 in the Graduate School). The University is composed of thirteen schools, including a large medical center. UAB faculty members are currently involved in more than $225 million worth of externally funded research grants and contracts. For graduate students, this level of research means availability of financial support, access to research laboratories, and the opportunity to interact with faculty members who are recognized nationwide by their colleagues.

Applying

Students should apply to the Graduate School, Room 511, University Center. A completed application form, transcripts, three letters of reference, GRE scores, and a letter of intent are required. Although the record of each applicant is considered as a whole, successful applicants should have maintained at least a B average and usually present GRE General Test scores (verbal and quantitative portions) of at least 600 on each section to enter the M.S. program and 650 on each to enter the Ph.D. program. Applicants are encouraged (but not required) to take the GRE Subject Test. Applicants whose native language is not English are required to present a TOEFL score of at least 550 and a minimum score of 3.5 on the TWE. A personal interview is normally required as part of the application.

Correspondence and Information

Graduate Program Director
Department of Biomedical Engineering, Hoehn 370
The University of Alabama at Birmingham
Birmingham, Alabama 35294-4440
World Wide Web: http://bmewww.eng.uab.edu

The University of Alabama at Birmingham

THE FACULTY AND THEIR RESEARCH

Jorge E. Alonzo, Assistant Professor; M.D., Salamanca (Spain), 1978. Biomechanics of orthopedic trauma.

Franklin Amthor, Associate Professor; Ph.D., Duke, 1979. Neurophysiology of vision, computer graphics.

Andreas Anayiotos, Assistant Professor; Ph.D., Georgia Tech, 1991. Biofluids, dynamics of the vascular system.

Gary Barnes, Professor; Ph.D., Wayne State, 1970. Medical imaging, X-ray physics.

James M. Cuckler, Professor; M.D., NYU, 1975. Orthopedic surgery, skeletal implants, effects on cells of mechanical stress.

Jan Anthonie den Hollander, Professor; Ph.D., Leiden (Netherlands), 1976. Magnetic resonance imaging.

Allan C. Dobbins, Assistant Professor; Ph.D., McGill, 1992. Human and machine vision, neural computation, brain imaging, scientific visualization.

Kenneth Doblar, Assistant Professor; M.D., Rutgers, New Brunswick, 1978. Control of ventilation, cerebral flow monitoring, anesthesiology.

Mark Doyle, Associate Professor; Ph.D., Nottingham (England), 1983. Processing and acquisition of magnetic resonance images.

Alan Eberhardt, Assistant Professor; Ph.D., Northwestern, 1991. Solid mechanics, analytical and numerical methods in biomechanics.

Evangelos Eleftheriou, Assistant Professor; Ph.D., Tennessee Tech, 1991. Mechanical systems, automated manufacturing, and mechanical design.

William T. Evanochko, Associate Professor; Ph.D., Auburn, 1979. Magnetic resonance spectroscopy, medical imaging.

Vladimir G. Fast, Research Assistant Professor; Ph.D., Moscow Institute of Physics and Technology, 1992. Cardiac electrophysiology.

Dale S. Feldman, Associate Professor; Ph.D., Clemson, 1982. Biomaterials, soft-tissue biomechanics, polymeric implants.

Timothy Jerner Gawne, Assistant Professor; Ph.D., Uniformed Services Health Sciences, 1984. Physiological optics, neural dynamics of form perception.

Richard A. Gray, Assistant Professor; Ph.D., Virginia, 1993. Optical mapping of re-entry fibrillation and defibrillation.

Gary J. Grimes, Professor; Ph.D., Colorado at Boulder, 1973. Telecommunications and fiber-optic data transmission, virtual reality, telemedicine.

Deliah Huelsing, Research Assistant Professor; Ph.D., Tulane, 1998. Cardiac electrophysiology.

James W. Hugg, Assistant Professor; Ph.D., Stanford, 1978. Magnetic resonance imaging, imaging of the brain, spectroscopic MRI.

Raymond E. Ideker, Professor; M.D./Ph.D., Tennessee, 1974. Electrophysiology, study of cardiac arrythmia, cardioversion and electrical ablation for treatment of arrythmia.

Marjorie K. Jeffcoat, Professor; D.M.D., Harvard, 1976. Periodontics.

Robert Lee Jeffcoat, Research Assistant Professor; Ph.D., MIT, 1975. Dentistry biomaterials, quantitative diagnostic techniques and instrumentation.

John S. Kirkpatrick, Assistant Professor; M.D., Wake Forest, 1985. Orthopedic surgery.

Martin Klinger, Research Assistant Professor; Ph.D., Alabama at Birmingham, 1998. Implants, biochemistry.

Stephen B. Knisley, Associate Professor; Ph.D., North Carolina, 1988. Myocardial electrophysiology, study of membrane potentials using laser-excited fluorescent dyes.

William Lacefield, Associate Professor; Ph.D., Florida, 1980. Ceramic biomaterials and coatings for dental and orthopedic applications.

Jack E. Lemons, Professor; Ph.D., Florida, 1968. Biomaterials, biological tissue reaction to synthetic materials, biomechanics.

Linda C. Lucas, Professor; Ph.D., Alabama at Birmingham, 1982. Biomaterials, biocompatibility, surgical alloys, corrosion resistance of implant materials.

Martin J. McCutcheon, Professor; Ph.D., Arkansas, 1967. Medical instrumentation, speech physiology, signal processing.

Andrew E. Pollard, Associate Professor; Ph.D., Duke, 1988. Simulation and modeling of electrical signals of the heart.

Charles W. Prince, Professor; Ph.D., Alabama at Birmingham, 1984. Dental nutrition, bone biochemistry, vitamin D, calcium and phosphorus metabolism.

Firoz Rhaemtulla, Professor; Ph.D., Umea (Sweden), 1974. Connective tissue biochemistry.

E. Douglas Rigney, Assistant Professor; Ph.D., Alabama at Birmingham, 1990. Coatings for biomaterials, ion-beam sputtering.

Jack M. Rogers, Assistant Research Professor; Ph.D., California, San Diego, 1993. Computer simulations of re-entry, signal analysis of cardiac arrhythmias.

Rosalia N. Scripa, Professor; Ph.D., Florida, 1976. Biomaterials, ceramics/glass, extractive metallurgy, semiconducting materials.

William M. Smith, Professor; Ph.D., Duke, 1970. Bioinstrumentation, multichannel cardiac mapping, ECG mapping and signal analysis.

Ernest M. Stokely, Professor; Ph.D., SMU, 1972. 3-D medical imaging, 3-D computer graphics, digital imaging.

J. Anthony Thompson, Associate Professor; Ph.D., North Carolina, 1977. Molecular mechanisms of angiogenesis and fibroblast growth factor.

Donald B. Twieg, Associate Professor; Ph.D., SMU, 1977. Medical imaging, magnetic resonance imaging (MRI) techniques, functional MRI of brain and heart.

Gregg L. Vaughn, Associate Professor; Ph.D., Alabama, 1974. Imaging, digital signal processing, applications of microprocessors.

Ramakrishna Venugopalan, Research Assistant Professor; Ph.D., Alabama at Birmingham, 1998. Biocorrosion, orthopedic and cardiovascular devices.

Adjunct Faculty

Martha W. Bidez, Associate Professor; Ph.D., Alabama at Birmingham, 1987. Biomechanics of hard tissues, mechanical systems, computer-aided design, stress analysis.

Glenn S. Fleisig, Director of Research, American Sports Medicine Institute; Ph.D., Alabama at Birmingham, 1994. Injury prevention and proper mechanics for baseball pitching and other throwing activities, exercise, and golf.

Steven D. Girouard, Research Scientist; Ph.D., Case Western Reserve, 1996. Cellular mechanisms of initiation, maintenance, and termination of arrhythmias.

Richard J. Holl, Director of Product Development, Southern BioSystems, Inc.; Ph.D., Auburn, 1989. Biodegradable polymers, drug delivery systems, microencapsulation, drug and device regulations, transport phenomena, fluid dynamics.

Louis C. Sheppard, Ph.D., London, 1966. Control systems, expert systems.

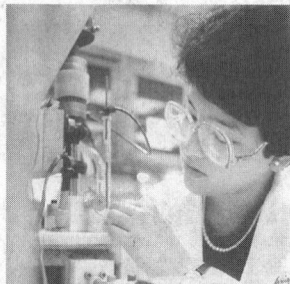

Prosthetic device corrosion research.

Auger Electron Microscope for surface analyses.

Biomedical instrumentation.

UNIVERSITY OF CALIFORNIA, BERKELEY / UNIVERSITY OF CALIFORNIA, SAN FRANCISCO

Graduate Program in Bioengineering

Program of Study

The University of California campuses at Berkeley and San Francisco offer a joint graduate program in bioengineering. This program permits students to benefit from both the strong clinical and health sciences resources available on the San Francisco campus and the strong engineering and basic life sciences resources available on the Berkeley campus.

The program is interdepartmental as well as intercampus. It formally combines related interests and research activities of faculty from five of the seven engineering departments and from several nonengineering departments (e.g., Biology, Vision Science, Integrative Biology) at Berkeley with those of the faculty from all four professional schools (Dentistry, Medicine, Nursing, and Pharmacy) at San Francisco.

All students in the program are simultaneously enrolled in the Graduate Divisions of both campuses and are free to take advantage of the courses and research opportunities of both.

The program awards Ph.D. degrees that carry the names of both campuses. M.S. degrees are awarded while students work toward Ph.D. completion.

Research Facilities

Berkeley and San Francisco combine to offer state-of-the-art research facilities for engineering, life sciences, and clinical research with a wide variety of laboratories, computing and fabrication facilities, and clinical settings. (See the reverse of this page for the list of faculty and research areas.)

Financial Aid

Students are encouraged to apply for extramural (e.g., NSF) fellowships as well as for various other fellowships administered by the University. Research assistantships are available in projects supported by extramural grants or contracts. An effort is made to align the research assistant's interests with those of a faculty member working with extramural support. Many research assistants work half-time during the academic year and full-time during part of the summer. Some traineeships are available to bioengineering students who are U.S. citizens through a training grant from the National Institutes of Health. Two 2-year Whitaker traineeships (including tuition and fees) may become available per year to students interested in the application of bioengineering tools to study the problems of vision. Information about other types of financial aid is available from the Financial Aid Office on each campus.

Cost of Study

Fees, insurance, and tuition for 1999–2000 total $4408.50 for California residents and $14,730.50 for nonresidents. Books and supplies cost about $500.

Living and Housing Costs

Room and board in the San Francisco Bay Area for the nine-month academic year 1999–2000 average $4725 to $7700. Books and supplies average $578. Entertainment and miscellaneous expenses total $1600. Costs are proportionately higher for the twelve-month academic year.

Student Group

Approximately 31,000 students, including close to 9,000 graduate students, are enrolled at Berkeley. The cosmopolitan student body includes about 1,900 international students from nearly ninety countries and about 3,500 students from states other than California.

Approximately 3,500 students are enrolled at San Francisco, including almost 940 graduate students. Graduate students come from all over the United States, and approximately 10 percent are from other countries.

Location

The Berkeley campus is surrounded by wooded hills and by the business and residential districts of Berkeley (population 105,000). Despite its rapid growth, the campus retains much natural beauty, with wooded glens, spacious plazas, and picturesque Strawberry Creek running the length of the campus.

The San Francisco campus is located in the center of San Francisco near Golden Gate Park and commands an impressive view of the city, surrounded by the Pacific Ocean, the Golden Gate, and San Francisco Bay. The campus community enjoys all the social and cultural advantages of a cosmopolitan, metropolitan area as well as easy access to the beaches and redwood groves of Marin County and San Mateo County and the ski slopes of the Sierra.

The University

In 1868, the legislature of California approved an act creating the University. Following a move from temporary facilities at Oakland, Berkeley became the site of the first permanent campus in 1873. As enrollment increased, architecturally varied structures were added, and the present campus gradually took shape. Eight other campuses, each maintaining a separate administrative organization and style of academic life, were later established throughout the state. Today the University of California is one of the largest in the world. The University also maintains research and field stations, extension centers, and other facilities for research and instruction in more than eighty locations throughout California. With such extensive resources, the University enjoys a leading position among universities, offering advancement of knowledge in virtually every field of modern human endeavor and serving as an indispensable force in the development and growth of American society.

The University of California at San Francisco is another of the nine campuses of the statewide university system. The Medical Sciences Building and the Health Sciences Research and Instruction Buildings at UCSF have about 800,000 square feet of floor space on sixteen levels devoted to graduate and professional education in pharmacy, dentistry, nursing, medicine, and the basic health sciences. The campus also includes two hospitals and the Millberry Student Union, which is the cultural, social, and recreational center of the campus. The library contains more than 628,000 volumes, and more than 4,000 leading international periodicals in the health sciences are received.

Applying

All applications are processed at Berkeley. The application deadline is January 3 for the following fall. To obtain application forms, students should contact the address below.

Correspondence and Information

Bioengineering Graduate Group
230 Bechtel Engineering Center, #1708
College of Engineering
University of California
Berkeley, California 94720
Telephone: 510-642-8790
E-mail: bioeng@coe.berkeley.edu
World Wide Web: http://www.coe.berkeley.edu/bioengineering/

University of California, Berkeley/University of California, San Francisco

THE FACULTY AND THEIR RESEARCH

Katherine Andriole, Ph.D.; Radiology (UCSF). Medical image processing and analysis.
Ronald Arenson, M.D.; FACR; Radiology (UCSF). Medical picturing archiving and communication, medical imaging.
David M. Auslander, Ph.D.; Mechanical Engineering (UCB). Automatic control, motion control and robotics.
Marty Banks, Ph.D.; Optometry (UCB). Visual space perception, stereopsis, motion, vestibular system, virtual reality.
Christopher Benz, M.D.; Medicine (UCSF). Breast cancer, molecular therapeutics and diagnostic receptors, oncogenes.
Stanley A. Berger, Ph.D.; Engineering Science (UCB). Biofluid dynamics, blood flow, sperm-mucus transport.
Rajendra S. Bhatnagar, Ph.D.; Biochemistry (UCSF). Connective tissue biophysics and biophysical chemistry.
Harvey Blanch, Ph.D.; Chemical Engineering (UCB). Biochemical engineering, applied enzymology, bioproduct recovery.
Bernhard Boser, Ph.D.; EECS (UCB). Integrated circuit design, signal processing, medical imaging, communication.
David S. Bradford, M.D.; Orthopaedic Surgery (UCSF). Spinal reconstructive surgery.
Thomas F. Budinger, M.D., Ph.D.; Electrical Engineering and Computer Sciences (UCB). PET, NMR, signal processing.
Christopher E. Cann, Ph.D.; Radiology (UCSF). Quantitative CT, bone calcium metabolism.
Gary Caputo, M.D.; Radiology (UCSF). Cardiovascular imaging and radiological informatics.
Theodore B. Cohn, Ph.D.; Vision Science (UCB). Psychoneurophysiology of vision.
Roger Cooke, Ph.D.; Biochemistry and Biophysics (UCSF). Mechanical aspects and changes occurring during fatigue in muscle contraction.
David Copenhagen, Ph.D.; Ophthalmology and Physiology (UCSF). Retina, synaptic transmission, vision.
Chris Cullander, Ph.D.; Biopharmaceutical Sciences (UCSF). Drug transport and permeation enhancement in epithelia.
Yang Dan, Ph.D.; Molecular and Cell Biology (UCB). Information processing in the visual system.
Stephen E. Derenzo, Ph.D.; EECS (UCB). Radioisotope detection and imaging, PET.
Edward Diao, M.D.; Orthopaedic Surgery (UCSF). Wrist anatomy, tendon healing, nerve syndromes, hand arthritis.
Chris J. Diederich, Ph.D.; Radiation Oncology (UCSF). Ultrasound devices for hyperthermia cancer.
Claire Farley, Ph.D.; Integrative Biology (UCB). Biomechanics of locomotion and skeletal muscle.
John Featherstone, D.D.S.; Restorative Dentistry (UCSF). Demineralization of dental enamel and laser effects on dental hard tissues.
Thomas E. Ferrin, Ph.D.; Pharmaceutical Chemistry (UCSF). Computer-aided drug design.
Dan Fried, Ph.D.; Restorative Dentistry (UCSF). Laser therapeutics and optical diagnostic methods in dentistry and medicine.
Stanton A. Glantz, Ph.D.; Medicine (UCSF). Cardiac mechanics, computers in medicine, health policy.
Robert Gould, Ph.D.; Radiology (UCSF). CT, digital radiography.
Randall Hawkins, M.D., Ph.D.; Radiology (UCSF). Metabolic imaging (PET), MH processing and modeling, applications to cancer.
Bruce Hasegawa, Ph.D.; Radiology (UCSF). Medical imaging, computerized tomography.
Harriet Hopf, M.D., Ph.D.; Anesthesia (UCSF). Increasing tissue oxygen in acute and chronic wounds and in critically ill patients.
Serena A. Hu, M.D.; Orthopaedic Surgery (UCSF). Spine biomechanics, including instrumentation and motion analysis; osteoporosis.
H. K. (Bernie) Huang, D.Sc.; Radiology (UCSF). Image processing, picture archiving, computer applications in medicine.
C. Anthony Hunt, Ph.D.; Biopharmaceutical Sciences (UCSF). Macromolecular engineering, targeted delivery of drugs, biomimetics.
Nola Hylton, Ph.D.; Radiology (UCSF). Magnetic resonance imaging of breast cancer.
Thomas James, Ph.D.; Pharmaceutical Chemistry and Radiology (UCSF). NMR spectroscopy.
Hami Kazerooni, Sc.D.; Mechanical Engineering (UCB). Robotic systems worn by humans, orthotic devices, virtual exercise machines.
Jay D. Keasling, Ph.D.; Chemical (UCB). Metabolic engineering of microorganisms and bioremediation.
Tony Keaveny, Ph.D.; Mechanical Engineering (UCB). Orthopaedic biomechanics, total joint prostheses, bone fracture.
Stanley Klein, Ph.D.; Vision Science (UCB). Nonlinear analysis of neural responses.
Rodger Kram, Ph.D.; Integrative Biology (UCB). Biomechanics and energetics of locomotion.
Steven L. Lehman, Ph.D.; Integrative Biology (UCB). Neuromuscular control, mathematical models of muscle.
Michael Lesh, M.D.; Medicine (UCSF). Computer modeling of cardiac arrhythmias, catheter ablation.
Dorian Liepmann, Ph.D.; Mechanical Engineering (UCB). Experimental biofluid mechanics.
Stephen Lisberger, Ph.D.; Physiology (UCSF). Computational, behavioral, and neurophysiological analysis of the vestibulo-ocular reflex.
Jeffery Lotz, Ph.D.; Orthopaedic Surgery (UCB). Biomechanics of the spine.
Andrew Lou, Ph.D.; Radiology (UCSF). Intelligent radiological image display workstation, CT/MR mapping of prostate gland.
Sharmila Majumdar, Ph.D.; Radiology (UCSF). Quantitative magnetic resonance imaging and spectroscopy, image analysis.
Grayson W. Marshall Jr., D.D.S., Ph.D.; Restorative Dentistry (UCSF). Microscopy and X-ray microanalysis of biomaterials.
Sally J. Marshall, Ph.D.; Restorative Dentistry (UCSF). Structure-property relationships in dental materials, metals, ceramics.
Richard A. Mathies, Ph.D.; Chemistry (UCB). Development of microfabricated integrated DNA analysis systems on glass chips.
Michael M. Merzenich, Ph.D.; Physiology (UCSF). Neural prosthesis, cortical plasticity and the origins of higher brain functions.
Kenneth Miller, Ph.D.; Physiology (UCSF). Theoretical and computational neuroscience, analysis of cerebral cortical function.
Sheldon Miller, Ph.D.; Vision Science (UCB). Cell and molecular analysis of structure, function, and regulation of epithelia.
Susan Muller, Ph.D.; Chemical (UCB). Polymer processing, non-Newtonian fluid mechanics, polymer dynamics, and rheology.
Mohandas Narla, D.Sc.; Laboratory Medicine (UCSF). Red cells and hematology.
Sarah J. Nelson, Ph.D.; Radiology (UCSF). NMR, quantification of changes in metabolites by in vivo spectroscopy.
Lisa Pruitt, Ph.D.; Mechanical Engineering (UCB). Ultrahigh molecular weight polyethylene, total joint replacements.
Stanley Prussin, Ph.D.; Nuclear Engineering (UCB). Radionuclear chemistry and applications, radiation detection and measurements.
David Rempel, M.D., M.P.H.; Medicine (UCSF). Biomechanical evaluation of work tasks, preventing musculoskeletal disorders.
Robert Ritchie, Ph.D.; Materials Science (UCB). Prosthetic devices.
Stephen Robinovitch, Ph.D.; Orthopaedic Surgery (UCSF). Orthopaedic biomechanics, age-related changes in balance and mobility.
Stephen Rothman, Ph.D.; Physiology (UCSF). X-ray optics to analyze cell processes.
Boris Rubinsky, Ph.D.; Mechanical Engineering (UCB). Cryosurgery, cryobiology, biological heat transfer.
David Saloner, M.D.; Radiology (UCSF-VAMC). Magnetic resonance imaging, visualization and quantitation of blood flow.
Clifton Schor, Ph.D.; Vision Science (UCB). Computation and investigating behavioral measurements of human eye movements.
Christoph Schreiner, Ph.D.; Otolaryngology (UCSF). Representation of complex auditory signals in the central auditory system.
Vernon Smith, Ph.D.; Radiation Oncology Physics (UCSF). Magnetic resonance spectroscopy, image processing.
Robert C. Spear, Ph.D.; Environmental Health (UCB). Occupational and environmental health, mathematical modeling.
Paul R. Stauffer, M.S.E.E., M.S.C.E.; Radiation Oncology (UCSF). Medical applications of RF and microwave.
Robert M. Stroud, Ph.D.; Biochemistry and Biophysics (UCSF). Biological structure and function at the molecular level.
Francis C. Szoka Jr., Ph.D.; Biopharmaceutical Sciences (UCSF). Drug delivery systems.
Frank Tendick, Ph.D.; Electrical Engineering and Computer Science (UCB). Biorobotics, human interfaces, virtual environments for surgery.
Lynn Verhey, Ph.D.; Radiation Oncology (UCSF). Radiotherapy delivery, clinical implementation of new technologies.
Daniel Vigneron, Ph.D.; Radiology (UCSF). MRI and spectroscopy for assessment of tumors and their response to therapy.
Jasmina Vujic, Ph.D.; Nuclear Engineering (UCB). Advanced numerical methods in radiation transport.
Michael W. Weiner, M.D., Ph.D.; Medicine and Radiology (UCSF). Development of magnetic resonance for neurological disease.
Frank Werblin, Ph.D.; Neurobiology (UCB). Processing of visual information in the vertebrate retina.
Richard M. White, Ph.D.; Electrical Engineering (UCB). Electronic, ultrasonic, and optical microsensors.
Charles D. Yingling, Ph.D.; Medical Psychology (UCSF). EEG and ERP, surgical monitoring, CNS stimulation.

UNIVERSITY OF CALIFORNIA, SAN DIEGO

Department of Bioengineering

Programs of Study

The Department of Bioengineering offers graduate instruction leading to the M.S. and Ph.D. degrees. The bioengineering graduate program started in 1966, and the department was established in the Jacobs School of Engineering in 1994. The interdisciplinary Program in Bioengineering, which is a joint program with the School of Medicine, provides an excellent graduate education, integrating the fields of engineering and the biomedical sciences. Students with an undergraduate education in engineering or biological sciences will learn how to use engineering concepts and methodology to study biomedical problems associated with genes, molecules, cells, tissues, organs, and systems, with applications to clinical medicine. Education and research in bioengineering has been facilitated by the establishment of an Institute for Biomedical Engineering and the receipt of Whitaker Foundation Development and Leadership Awards.

The M.S. program is intended to equip the student with fundamental knowledge in bioengineering. The degree may be terminal or obtained on the way to earning the Ph.D. Two plans of study are offered, both requiring successful completion of 48 quarter units of credit: Plan I is a combination of course work and research, culminating in a thesis; Plan II involves course work only and culminates in a comprehensive examination. In addition to the existing M.S. degree, the department is proposing to offer a Master of Engineering (M.Eng.) degree, pending approval. The purpose of this new degree is to prepare design and project engineers for careers in the biomedical and biotechnology industries within the framework of the graduate program of the bioengineering department. It will be a terminal professional degree in engineering. With the introduction of the M.Eng degree, the M.S. Plan II option will no longer be available to new M.S. students.

The Ph.D. program is designed to prepare students for a career in research and/or teaching in bioengineering. Each student, in conjunction with a faculty adviser, develops a course program that prepares him or her for the departmental Ph.D. qualifying examination, which tests students' capabilities in three areas of specialization and ascertains their potential for independent study and research. The degree requires the completion of a dissertation and defense of that research.

There are also M.D./M.S. and M.D./Ph.D. degrees offered in conjunction with the UCSD Medical School, pending independent admission to the Medical School.

Research Facilities

The research laboratories in the department are fully equipped for modern bioengineering research. Among the equipment available are electronic, video, ultrasonic, and imaging instruments for investigating cell rheology, tissue viscoelasticity, orthopedic biomechanics, cardiac mechanics, and microcirculatory dynamics; implantable sensors for monitoring blood glucose concentration and oxygen tension; and facilities for DNA sequencing, genetic engineering, and molecular, cellular, and tissue bioengineering. The department maintains excellent computer facilities, including many advanced graphics workstations and a confocal laser scanning microscope. There are CRAY T3E and IBM SP2 supercomputers at the San Diego Supercomputer Center on campus. The School of Engineering has excellent scanning and transmission electron microscopic facilities.

Financial Aid

The department supports most full-time graduate students at the Ph.D. level. Financial support is available in the form of fellowships, traineeships, teaching assistantships, and research assistantships. Awarding of financial support is competitive, and stipends average $16,000 for the academic year, plus tuition and fees. Sources of funding include University fellowships and traineeships from an NIH training grant and the Whitaker Foundation Development Award. Funds for support of international students are extremely limited, and the selection process is highly competitive.

Cost of Study

In 1999–2000, full-time students who are California residents pay $1629.50 per quarter in registration and incidental fees. Non-California residents pay a total of $5071.50 per quarter in registration and incidental fees. There is a reduced-fee structure for students enrolled on a half-time basis. In addition, effective fall 1997, the annual nonresident tuition fee is reduced by 75 percent for graduate doctoral students who have advanced to candidacy. This reduced rate is available for a maximum of three years, depending upon the date of advancement. Fees are subject to change.

Living and Housing Costs

UCSD provides 1,400 apartments for graduate students. Current monthly rates range from $405 for a single student to $900 for a family. There is also a variety of off-campus housing in the surrounding communities. Prevailing rents range from $398 per month for a room in a private home to $1250 or more per month for a two-bedroom apartment. Information may be obtained from the UCSD Residential Housing Office at 619-824-0850.

Student Group

Current campus enrollment is 17,227 students, of whom 14,021 are undergraduates and 2,145 are graduate students. The Bioengineering Department has an undergraduate enrollment of about 591 and a graduate enrollment of 66.

Location

The 2,040-acre campus spreads from the coastline, where the Scripps Institution of Oceanography is located, across a large wooded portion of the Torrey Pines Mesa overlooking the Pacific Ocean. To the east and north lie mountains, and to the south are Mexico and the almost uninhabited seacoast of Baja California.

The University

One of nine campuses in the University of California System, UCSD comprises the General Campus, the School of Medicine, and the Scripps Institution of Oceanography. Established in La Jolla in 1960, it is one of the newer campuses but in this short time has become one of the major universities in the country.

Applying

A minimum GPA of 3.4 (on a 4.0 scale) is required for admission. The average GPA for students offered support in 1998–99 was 3.75. All applicants are required to take the GRE General Test. International applicants whose native language is not English are required to take the TOEFL and obtain a minimum score of 550. In addition to test scores, applicants must submit a completed Graduate Admission and Award Application, all official transcripts (English translation must accompany official transcripts written in other languages), a statement of purpose, and three letters of recommendation. The deadline for international students to request application materials is November 1. The deadline for filing applications for both international students and U.S. residents is January 15. Applicants are considered for admission for the fall quarter only.

Correspondence and Information

Department of Bioengineering 0419
University of California, San Diego
La Jolla, California 92093-0419

Telephone: 619-822-0006
E-mail: be-gradinfo@bioeng.ucsd.edu
World Wide Web: http://www-bioeng.ucsd.edu/

University of California, San Diego

THE FACULTY AND THEIR RESEARCH

Sangeeta Bhatia, M.D., Ph.D., Assistant Professor. Microscale tissue engineering; microfabrication techniques for control of cellular microenvironment; role of homotypic and heterotypic cell-cell interactions; cell-ECM interactions in tissue function; hepatic physiology, pathophysiology, and replacement therapies; cell-based biosensors and tailored biomaterials.

Pao Chau, Ph.D., Associate Professor of Chemical Engineering and of Bioengineering. Integrative protein systems, cell-cycle dependent events, bioreactor and mammalian cell cultures.

Shu Chien, M.D., Ph.D., Professor of Bioengineering and of Medicine. Effects of mechanical forces on endothelial gene expression and signal transduction; circulatory regulation in health and disease; transendothelial transport of macromolecules; energy balance and molecular basis of leukocyte-endothelial interactions; vascular tissue engineering.

John A. Frangos, Ph.D., Professor. Mechanisms of shear-induced activation of endothelial membrane function, bone tissue engineering, muscle tissue engineering, signaling by mechanical stimulation in plants.

Yuan-Cheng Fung, Ph.D., Professor Emeritus. Biomechanics of circulation, respiration, and muscle; stress-growth law of blood vessels; determination of the zero-stress state and constitutive equations of components of blood vessel and the whole vessel; morphometry of systemic and pulmonary blood vessels; continuum mechanics in pulmonary physiology; integration of morphology, mechanical properties, rheology, and boundary conditions into pressure-flow relations.

David A. Gough, Ph.D., Professor. Implantable glucose sensor for diabetes; glucose and oxygen transport through tissues, sensor biocompatibility; dynamic models of the natural pancreas on based glucose input and insulin output.

Anne Hoger, Ph.D., Associate Professor of Engineering Science and of Bioengineering. Continuum mechanics, constitutive theory, residual stress, mechanics of biological tissues.

Marcos Intaglietta, Ph.D., Professor. Development of plasma expanders and artificial blood, theory of tissue oxygenation at the microvascular level, optical methods for the study of microcirculation.

Paul C. Johnson, Ph.D., Adjunct Professor. Local and neural mechanisms of blood flow regulation in the microcirculation, including cellular mechanisms; flow properties of blood in vivo; oxidative metabolism-blood flow-relations; development of instrumentation for microcirculatory research; image analysis of in vivo flow patterns.

Deidre A. MacKenna, Ph.D., Adjunct Assistant Professor. Integrin cell and molecular biology, extracellular matrix biology, mechanotransduction, integrin function in inflammatory diseases, cardiovascular pathobiology, quantitative morphometry.

Andrew D. McCulloch, Ph.D., Professor. Biomechanics of normal and diseased ventricular myocardium, mechanics of cardiac connective tissue matrix, mechanics of ventricular growth and remodeling, effects of mechanical strain on cardiac fibroblasts, cardiac electromechanics, mechanics of ischemic myocardium.

Bernhard Palsson, Ph.D., Professor. Hematopoietic tissue engineering, stem cell technology, bioreactor design, retroviral gene transfer for gene therapy, metabolic dynamics and regulation, whole cell simulators, metabolic engineering, genetic circuits.

Thomas D. Pollard, M.D., Adjunct Professor. Biophysical analysis of the mechanical properties of the actin cytoskeleton.

Robert L. Y. Sah, M.D., Sc.D., Charles Lee Powell Associate Professor. Cartilage tissue engineering at the molecular, cellular, and tissue levels; cartilage biophysics in health and arthritis: mechanical-chemical-electrical transduction, poroelasticity, and electrokinetics; macromolecular partitioning and transport; cartilage growth, repair, and degeneration; biophysical and biochemical regulation.

Geert W. Schmid-Schönbein, Ph.D., Professor. Analysis of blood flow, exchange, and lymphatic transport in the microcirculation in health and disease; roles of erythrocytes, leukocytes, platelets, and endothelium in microvascular blood flow, capillary entrapment, and transendothelial migration; mechanisms for cell activation in the microcirculation; development of a model of blood flow in skeletal muscle, with applications to hypertension, ischemia, and shock.

Sidney S. Sobin, M.D., Ph.D., Adjunct Professor. Biochemical changes in the microvasculature in aging; composition, source, mechanisms of formation, rate of accumulation, and distribution of microvascular constituents; cellular mechanisms of hypoxic pulmonary hypertension and their role in the prediction of established pulmonary hypertension and disease.

Lanping Amy Sung, Ph.D., Associate Professor. Molecular structure and control of gene expression of membrane skeletal proteins in relation to the mechanical properties of cells and tissues in differentiation, aging, and disease; molecular defects of membrane skeletal proteins in hereditary diseases; protein 4.2 as a pseudozyme in maintaining the stability and flexibility of erythrocyte membranes; mechanical function of tropomodulin (a tropomyosin-binding protein) in the heart, muscles, and erythrocytes.

Lewis K. Waldman, Ph.D., Associate Adjunct Professor. Measurement of multidimensional finite strains in the beating heart, stress analysis, estimation of ventricular material properties and wall stresses, catheter design, medical imaging and image analysis.

John T. Watson, Ph.D., Adjunct Professor. Heart failure and mechanical circulatory support; biomaterials; medical implant design; bioimaging; creativity, innovation, and techology transfer.

School of Medicine Affiliates

Kenneth R. Chien, Ph.D., Professor of Medicine and of Bioengineering. Utilizing state-of-the-art technology for genetic engineering of animal models of human cardiovascular disease, with an emphasis on congenital heart disease, cardiac hypertrophy, and heart failure; cardiovascular cell type restricted gene knockouts utilizing CRE-LOX technology; inducible cardiac restricted gene targeting with a tet-off system; conditional ventricular restricted transgenesis with CRE as a genetic switch.

James W. Covell, M.D., Professor of Medicine and of Bioengineering. Cardiovascular physiology and pharmacology; biomedical computing; mechanisms of diseased cardiac muscle contraction in the intact animal, the function of ischemic and hypertrophied cardiac muscle, and the role of the extracellular matrix in hypertrophy and heart failure; structure-function relationships of ventricular myocardium, using direct determinations of myocardial structure with confocal and electron microscopy, high-resolution measurements of finite deformation, and finite-element modeling to explore these relationships; role of the extracellular matrix linking adjacent myocardial laminae in normal hearts; changes in the extracellular matrix in ischemia and heart failure that contribute to loss of normal cardiac function.

Mark H. Ellisman, Ph.D., Professor of Neurosciences and of Bioengineering. Development and application of advanced imaging technologies to obtain new information about cell structure and function, structural correlates of nerve impulse conduction and axonal transport, cellular interactions during nervous system regeneration, cellular mechanisms regulating transient changes in cytoplasmic calcium, aging in the central nervous system.

Arnost Fronek, M.D., Ph.D., Professor Emeritus of Surgery and of Bioengineering. Physiology and pathophysiology of peripheral circulation, use of noninvasive techniques to improve diagnosis and therapy of vascular diseases, development of calibrated photoplethysmography and its utilization to evaluate the arterial venous system.

Richard L. Lieber, Ph.D., Professor of Orthopaedics and of Bioengineering. Musculoskeletal system design and plasticity; skeletal muscle architecture and its relation to tendon transfer surgery, development of intraoperative and rehabilitative measuring devices, skeletal muscle mechanics, sarcomere length measurement in isolated fibers and whole muscles, myosin expression in skeletal muscle after exercise-induced injury, immobilization, spinal cord injury and electrical stimulation.

Jeffrey H. Omens, Ph.D., Assistant Adjunct Professor of Medicine and of Bioengineering. Regional mechanics of the normal and diseased heart; miniaturization of functional measurement techniques for rat and mouse hearts; role of mechanical factors in cardiac hypertrophy, remodeling and growth; residual stress in the heart; computer-assisted analysis of cardiac mechanics.

K-L. Paul Sung, Ph.D., Professor-In-Residence of Orthopaedics and of Bioengineering. Regulation of cell adhesion by microfilament capping and associated proteins; energy balance and molecular mechanisms of cell interactions in immune response; intracellular ion (Ca^{2+}) and signal pathway dependence during cell adhesion, migration, and activation; molecular organization of cell membranes; biophysical properties of blood cells and endothelial cells in inflammatory response; gene activities, migration, mechanical-stressed, and healing processes of ligament and vertebrate dick cells with and without inflammation.

Peter D. Wagner, M.D., Professor of Medicine and of Bioengineering. Theoretical and experimental basis of oxygen transport in the lungs and skeletal muscles; muscle capillary growth regulation using molecular biological approaches in integrated systems—the role of 02, microvascular hemodynamics, physical factors, and inflammatory mediators; mechanisms of exercise limitation in health and disease, especially the role of muscle dysfunction in heart failure, emphysema, and renal failure.

John B. West, M.D., Ph.D., D.Sc., Professor of Medicine and of Bioengineering. Bioengineering aspects of the lung; stress failure and physiology of pulmonary capillaries when exposed to high transmural pressures; distribution of ventilation and blood flow in the lung; effect of gravity on the lung; measurements of pulmonary function during sustained weightlessness on Spacelab; distortion of the lung resulting from its weight; regulation of the extracellular matrix of capillary walls, including changes of gene expression as the result of stress; high-altitude physiology, especially extreme altitude and intermittent exposure to altitude.

UNIVERSITY OF FLORIDA

College of Engineering
Biomedical Engineering Program

Programs of Study

The mission of the Biomedical Engineering (BME) Program is to educate students who have strong engineering and science backgrounds for M.S. and/or Ph.D. degrees in biomedical engineering so that they can productively apply their education and training to the solution of engineering problems in the fields of medicine and biology and related fields. BME graduate students are educated with in-depth, specialized knowledge and research experience in at least one of four core areas (biomaterials; biomechanics; molecular, cellular, and tissue engineering; and biomedical imaging and signal processing); receive firsthand experience and understanding about the variety of opportunities and constraints that occur in a clinical and biological science environment, including technology transfer; become familiar with state-of-the-art techniques in imaging and interactions between cells and biomaterials; and are educated in a very broad-based and synergistic medical and engineering environment. Existing BME activities in the College of Engineering (COE) are extensive and involve collaborations with faculty members in the College of Medicine (COM). These collaborations have produced numerous publications and patents and have created enabling technologies in health-related engineering that have been commercialized.

Core courses include Introduction to Biomedical Engineering and Physiology, Clinical Shadowing for Engineers, and a regular seminar. These courses should be taken during the first year. Students should select a core area by the end of the first semester. Each core area requires one introductory course, 6 credits of option courses from the core area, and a proposed 2-credit Advanced Topics Course.

Research Facilities

The University of Florida Brain Institute is designed to enhance collaborative research and training opportunities in all aspects of head and spinal cord injury, paralysis, neurodegenerative diseases, and general neuroscience. It is housed in a newly constructed 200,000-square-foot building and involves extensive BME activities. The National High Magnetic Field Laboratory develops the highest magnetic fields and provides a national users facility near Tallahassee, with its main medical activities (MRI) centered at the University of Florida (UF). The Major Analytical Instrumentation Center, which houses large state-of-the-art instruments such as electron microscopes and X-ray equipment, is available within the COE. The NSF Engineering Research Center for Particle Science and Technology focuses on key aspects of particle technology: the production, characterization, modification, handling, and utilization of organic and inorganic powders in both dry and wet conditions. Currently funded activities include bacterial adhesion to surfaces and viral filters. A new building has recently been constructed to house these facilities.

Financial Aid

Financial aid is available in the form of graduate research assistantships. For the 1999–2000 terms, full-time assistantships range from $12,500 to $15,000 for twelve months. Students with assistantships have been granted tuition waivers in past years.

Cost of Study

For 1999–2000, the cost of tuition is approximately $138 per credit hour for Florida residents and $481 per credit hour for out-of-state students. Books and supplies cost approximately $700 per year. Many graduate assistantships provide funds for tuition and fees.

Living and Housing Costs

Living expenses, such as food and local transportation, are approximately $2630 per student per semester. For single students, a room in the coeducational residence halls is available for $853 to $1744 per student per semester. Married students and their families can also live on campus in apartments arranged in village communities at a cost of $261 to $464 per month. Off-campus housing can be found within walking distance of the campus, and ample accommodations are available locally at a wide range of prices.

Student Group

The enrollment for the program is currently around 25 full-time graduate students, as the program has just begun (1998). The total University enrollment is more than 40,000 students. There are about 150 graduate students engaged in biomedical engineering activities in the various graduate departments.

Location

The University campus is located in Gainesville, a city of more than 90,000 in north-central Florida, midway between the Atlantic Ocean and the Gulf of Mexico. There are opportunities for swimming and boating at nearby lakes, springs, and rivers. Airlines, Amtrak, and bus lines serve Gainesville. The city is located along I-75, 1 hour south of I-10 and 2 hours north of Orlando.

The University and The Program

The University of Florida, a combined state university and land-grant college, was founded in 1906. It is the state's oldest, largest, and most comprehensive university and is among the nation's most academically diverse public universities. The COE administers the BME Program using a nontraditional departmental structure. The current program at UF is broadly based and highly interdisciplinary, with faculty participation from most engineering disciplines as well as the COM. Thus, a traditional departmental configuration, with all BME faculty members in a single unit, is modified to accommodate the faculty members' discipline-affiliated departments and, simultaneously, their interdisciplinary biomedical engineering focus.

Applying

Admission requires a baccalaureate degree in engineering or science from an accredited college, an excellent upper-division grade point average, and satisfactory scores on the GRE and TOEFL (international students only). Application forms, transcripts, and test scores should be submitted up to one year but no later than six months prior to the term of admission. For more information, students may visit the Web site at the address listed below.

Correspondence and Information

For questions and correspondence regarding admission:
Biomedical Engineering Program
University of Florida
P.O. Box 116131
Gainesville, Florida 32608
E-mail: admin@bme.ufl.edu

For program information and application forms:
Graduate Faculty Advisor
Biomedical Engineering Program
University of Florida
P.O. Box 116131
Gainesville, Florida 32611-6131
Telephone: 352-392-9790
Fax: 352-392-9791
E-mail: admin@bme.ufl.edu
World Wide Web: http://www.bme.ufl.edu

University of Florida

THE FACULTY AND THEIR RESEARCH

Aerospace Engineering, Mechanics, and Engineering Sciences

R. Hirko. Engineering design course to aid the handicapped; brain-actuated control, utilizing tactile and auditory stimulus; automatic model-based segmentation of knee prosthesis silhouettes in fluoroscopic image sequences; real-time simultaneous data acquisition of 3-D range of motion and grip strength of the human wrist.

U. Kurzweg. High-frequency pulmonary ventilation.

R. Tran-Son-Tay. Rheological studies of sickle hemoglobin polymerization, shear sensitivity of stem cells, biorheology of leukocytes, rheological measurements of biological fluids, left ventricular assist device design.

E. Walsh. Intracranial pressure effects in the anesthesiology human patient simulator.

Chemical Engineering

R. Dickinson. Adhesion-mediated biased cell migration in three-dimensional matrices, receptor-mediated bacterial adhesion.

M. Orazem. Transdermal drug delivery.

D. Shah. Contact lenses and biolubrication.

Computer and Information Science Engineering

A. Laine. Wavelet processing for digital mammography, wavelet representation for digital mammography, quantitative methods of multiscale analysis for digital mammography.

S. Sahni. Algorithms for compression and registration of brain MRI.

B. Vemuri. Multiresolution stochastic 3-D shape models for segmentation of shapes from brain MRI.

Electrical and Computer Engineering

J. Anderson. Reconstruction algorithms for PET.

D. Childers. Modeling vocal disorders, interactive model of the vocal folds and turbulent noise for speech synthesis.

R. Duensing. Optimized reconstruction of images from MRI arrays, electrical receiver noise characterization in high-field MRI coil arrays.

J. Principé. Localization and recognition of sound signatures, using biologically plausible sensors and dynamic neural networks; recurrent neural networks for the processing of nonlinear nonstationary signals; neurocomputational models for describing feature extraction phenomena in the auditory cortex; automated volumetric analysis of the auditory cortex, using self-organizing principles.

F. Taylor. Overcomplete enhancement of digital mammograms.

Materials Science and Engineering

C. Batich. Physical modeling of arteriovenous malformations, drug delivery systems, implant degradation, detection of caries beneath dental restoration.

A. Brennan. Composites for dental restorations, surface modifications, resorbable biomaterials.

L. Hench (Adjunct). Bioactive glass research and development, modeling of bioactive glass interfaces.

E. Goldberg. Hydrophilic polymers for prevention of surgical adhesions, intraocular lens (IOL) implant project, IOL surface characterization and biocompatibility studies, protein microspheres for vaccines and cancer therapy, viscosurgical polymers for ophthalmic surgery, properties of silicones for implants, hydrograft surface modification of implants and devices, laser-biopolymer systems.

Mechanical Engineering

J. Klausner. New nasal continuous positive airway pressure device.

K. Reisinger. Influence of age on standing/balance stability, myoelectric control of above-knee prosthesis, synthesis and stability of bipedal locomotion.

A. Seireg. Optimal design of orthopedic implants; bone remodeling and prescription of exercise; prosthesis design, with consideration of muscle forces.

Nuclear and Radiological Engineering

S. Anghaie. Spectral sensitivity of radiotherapy procedures and measurement of beam spectral characteristics.

W. Bolch. Estimates of organ dose from pediatric radiological examinations, MRI-based bone dosimetry model and its application to probabilistic dose assessment, electron transport in internal dosimetry, dosimetric model of the head and brain for nuclear medicine neuroimaging studies, small-scale dosimetry techniques for radioimmunotherapy, use of radioactive stents for the prevention of restenosis following balloon angioplasty.

F. Bova. Frameless stereotactic applications, advances in stereotactic radiosurgery of high-precision radiotherapy.

J. Fitzsimmons, Center for the Study of the Psychology of Emotion and Attention. Optic neuritis: oxidative injury.

D. Hintenlang. Dosimetry and biological effects of high-field magnetic resonance imaging.

K. Scott. NMR imaging and spectroscopy for in vivo applications, H-1 and P-31 NMR of arm and leg lesions, cardiac NMR spectroscopy.

UNIVERSITY OF GEORGIA

Department of Biological and Agricultural Engineering

Programs of Study	The department offers three graduate degrees: the M.S., with a major in biological engineering; the M.S., with a major in agricultural engineering; and the Ph.D. The objectives of the graduate program are to foster and ensure unification of the diverse knowledge bases of the biological, agricultural, physical, social, and engineering sciences to develop a synergistic knowledge and to use the new knowledge base to educate students in engineering analysis, problem solving, and research methods within the complex domain of biological and agricultural systems. Graduate students are extended a wide latitude in the selection of a field of study and research from the broad domain of biological and agricultural engineering.
	The M.S. degrees are granted in recognition of competency in the selected area of study, in the development of synergistic knowledge, in rational problem-solving skills, and in conducting independent work. A minimum of 24 semester hours of course work and an acceptable thesis are required for the completion of the degree.
	The Ph.D. degree is granted in recognition of research, breadth and soundness of scholarship, and thorough knowledge of the field of study rather than upon completion of any specific work prescribed in advance. In general, a student takes 48–55 semester hours of course work beyond the B.S. The selection of courses is guided to have students acquire proficiency in the selected area of science, the ability to integrate diverse knowledge, creative thinking ability for defining problems, and the ability to conduct original research. The comprehensive examination tests students' proficiency in the area of science and scholarship, and the final oral examination evaluates students' dissertation research for original contribution to scientific literature.
	Faculty interests are diverse but are generally related to bioprocess, electrical phenomena in biological materials, biomechanics, bioremediation, biosensors/instrumentation, and modeling biological and ecological processes for biological, agricultural, food, and forest systems. The program offers students opportunities to integrate several disciplines and gain experience in biological engineering approaches to developing new systems and solutions.
Research Facilities	The department has extensive field and laboratory facilities. These include modern instrumentation and laboratories for biochemical, biomechanical, biosensor, biophysical, bioseparation, enzyme, and fermentation engineering; supercritical extraction; electrostatics; environmental physiology; water quality; and materials testing research. In addition, extensive facilities are available for nondestructive sensing with an X-ray CT scanner and NMR and for image analysis, materials handling, food engineering, and other technologies. A pilot plant for evaluating newly developed alternatives with enzymes, ozone, and biotechnology for bleaching pulp for the paper industry is a unique facility in the country. The department collaborates with the College of Veterinary Medicine, the Division of Biological Sciences, and the Artificial Intelligence Center, all of which have outstanding laboratory equipment and facilities.
	The affiliated research facilities of the USDA laboratories in Georgia, the Savannah River Research Laboratory, and the Agricultural Experiment Stations at Griffin and Tifton are valuable assets. The University of Georgia's library is one of the top twenty among the nation's 106 best university, government, and private research libraries.
Financial Aid	The department offers graduate assistantships to qualified applicants on a competitive basis. M.S. and Ph.D. graduate assistants on regular appointment are paid $14,533 and $15,511 per annum, respectively. Graduate assistants pay only a $40 matriculation fee per semester in lieu of tuition, plus student activity fees of $298.
Cost of Study	Total tuition and fees for full-time students are $1399 per semester for Georgia residents and $4119 per semester for nonresidents.
Living and Housing Costs	The estimated cost of living for a single student, including room and board, books and supplies, and personal expenses, is about $5500 per year (two semesters). The University provides graduate dormitory housing, ranging from $2350 to $3200 per year, and married student housing at $250 to $350 per month for one- and two-bedroom apartments; all include the costs of water and cable television.
Student Group	In 1998–99, of the 26 graduate students in the program, 42 percent were enrolled in the Ph.D. program and 31 percent were women. The Graduate Club is very active in organizing academic and social events. The Graduate School has nearly 5,600 students.
Student Outcomes	Graduates of the program are employed as university faculty members, in private companies as environmental engineers for designing and managing remediation systems, in the computer software industry, and work to develop information systems, design and manufacture machinery, and develop food processing systems. Graduates of the program also go to other graduate schools for additional degrees.
Location	Athens is a college town of about 75,000, located about 65 miles east of Atlanta. The Appalachian Mountains are a 2-hour drive to the north, and the seacoast is approximately 5 hours to the southeast. Athens is the recreational and cultural center of northeast Georgia, with a lively community of local artists and musicians. Popular annual local events include the Athens Criterium (a bicycle race), the annual HCA Tennis Tournament, the Special Olympics, the Athens Human Rights Festival, the Earth Day Festival, the North Georgia Folk Festival, and the Athens New Jazz Festival.
The University and The Department	The 1785 charter to establish the University of Georgia was the first charter for a state-supported university in the United States. As the state's land-grant and sea-grant university, the University provides the state's most comprehensive education, research, and service programs from its thirteen schools and colleges. The University is particularly noted for its programs in agricultural and environmental sciences, biological sciences, business, chemistry, ecology, education, law, and veterinary science. In 1992, the University made environmental literacy a mandatory requirement for all undergraduate students. In 1990, the initiation of biological engineering added a new dimension to the department's long-established engineering programs for agricultural systems. New faculty members and facilities in bioseparation, biosensors, biomechanics, nondestructive sensing, computer modeling, and enzyme engineering are contributing to the development of biological engineering in collaboration with the excellent programs in the biological and veterinary sciences.
Applying	Admission is open to graduates of any engineering discipline. Graduates with nonengineering degrees (the biological, physical, and chemical sciences; mathematics; computer science; and the agricultural and food sciences) are considered for provisional admission and are required to take 15 to 21 semester hours of undergraduate engineering science courses before they receive full admission. Admission is possible in the fall or spring semester, but fall semester admission is preferred. Applicants seeking financial assistance should apply by January 31, although assistantships are sometimes awarded at other times. Applicants must submit an admission form, official transcripts, GRE scores, and a $30 application fee to the Graduate School. The application for financial assistantship and three letters of reference should be sent directly to the Graduate Coordinator of the department. TOEFL scores are required for applicants from non-English-speaking countries. The application for admission can be submitted electronically by visiting the Graduate School Web site.
Correspondence and Information	For program information: Dr. Brahm P. Verma, Graduate Coordinator Biological and Agricultural Engineering 117 Driftmier Engineering Center University of Georgia Athens, Georgia 30602 Telephone: 706-542-0860 Fax: 706-542-6063 E-mail: gradprog@bae.uga.edu World Wide Web: http://www.bae.uga.edu	For application forms: Director of Admissions The Graduate School 534 Boyd Graduate Studies Research Center University of Georgia Athens, Georgia 30602 Telephone: 706-542-1789 E-mail: gradadm@uga.cc.uga.edu World Wide Web: http://www.gradsch.uga.edu

University of Georgia

THE FACULTY AND THEIR RESEARCH

Gerald F. Arkin, Professor; Ph.D., Illinois, 1971. Modeling crop growth and interactions of meteorological conditions in production of crops.

Michael J. Bader, Assistant Professor; Ph.D., Kentucky, 1991. Extension and continuing education programs in chemical applications, harvesting, processing, drying, storing, and grading in peanut and cotton production.

M. Bruce Beck, Adjunct Professor and Eminent Scholar of Environmental Technology; Ph.D., Cambridge, 1973. Environmental systems analysis, engineering process control, and policy analysis using concepts of filtering theory, adaptive control, fuzzy logic and expert systems.

David D. Bosch, Adjunct Research Scientist; Ph.D., Arizona, 1990. Fate and transport of agricultural chemicals by numerical modeling of the transport process and evaluation of models under laboratory and field conditions.

Manjeet S. Chinnan, Professor; Ph.D., North Carolina State, 1976. Modified atmosphere packaging of food materials, development and evaluation of edible films for foods, processing of peanuts and cereal legumes, mathematical modeling and computer simulation of food processes.

Rex L. Clark, Professor; Ph.D., Mississippi State, 1968. Controlled-traffic agricultural systems using wide-span vehicles, automatic weight distribution for improved traction, automatic guidance and spatial location systems, and applications of GPS and GIS to enable prescription farming operations.

K. C. Das, Assistant Research Scientist; Ph.D., Ohio State, 1995. Development and analysis of biological techniques for the treatment of municipal, industrial, and agricultural wastes; image analysis; spectrometric applications; artificial intelligence applications.

Mark A. Eiteman, Assistant Professor; Ph.D., Virginia, 1991. Recovery and purification of compounds from biological processes and from agricultural materials and heterogeneous ozone treatment of hazardous waste.

Mark D. Evans, Assistant Professor; Ph.D., Texas A&M, 1987. Application of image processing to engineering measurements and quality detection of biological and agricultural materials.

Timothy L. Foutz, Assistant Professor; Ph.D., North Carolina State, 1988. Analysis of the cellular and material responses of bone to mechanical usage in order to identify control systems that initiate bone remodeling.

John W. Goodrum, Associate Professor; Ph.D., Georgia Tech, 1974. Chemical or physical processing of biological materials, with focus on design and testing of biomass-based liquid fuels; computer vision for process analysis and control; supercritical solvent extraction.

Takoi Hamrita, Assistant Professor; Ph.D., Georgia Tech, 1993. Applications of automatic control in agricultural systems, modeling and optimal control of biological systems.

Gerrit Hoogenboom, Assistant Professor; Ph.D., Auburn, 1985. Development and application of crop simulation models to study the soil-plant-atmosphere continuum and the effect of the environment and meteorological conditions on crop growth, development, and yield.

James R. Kastner, Assistant Professor; Ph.D., Georgia Tech, 1993. Design and scale-up of bioreactors, use of microorganisms to produce/develop environmentally benign products/processes, production of specialty chemicals using enzymes and/or microorganisms.

William S. Kisaalita, Assistant Professor; Ph.D., British Columbia, 1986. Biomolecular engineering (neural cell tissue-culture processes) and cross-disciplinary development of biomimetic systems for biomedical and biotechnological sensing applications.

Andrzej W. Kraszewski, Adjunct Research Scientist; D.Sc., Polish Academy of Sciences, 1973. Microwave measurements, particularly as related to permittivity of materials, microwave aquametry, and interaction of electromagnetic waves with water and moist substances.

S. Edward Law, D. W. Brooks Distinguished Professor; Ph.D., North Carolina State, 1968. Basic and applied studies of electrostatics and its development for beneficial uses in agricultural and biological systems, including aerodynamic/electrostatic crop spraying, polyphase traveling field conveyance and separation of biomaterials, pollen management, and electric-discharge-generated ozone for pulp bleaching and pathogen and ethylene control in food storage.

Bryan W. Maw, Associate Professor; Ph.D., North Carolina State, 1981. Crop modeling, expert systems for irrigation, machine design for plant propagation and harvesting, controlled-traffic systems in crop production.

Ronald W. McClendon, Professor; Ph.D., Mississippi State, 1974. Application of artificial intelligence and operations research in the development of decision support systems for the management of biological systems.

Bailey W. Mitchell, Adjunct Professor; Ph.D., Purdue, 1969. Development of computer-based instrumentation and controls for managing environment as it relates to the development and transmission of poultry diseases.

Shree Nath, Assistant Research Scientist; Ph.D., Oregon State, 1996. Development and application of artificial intelligence and computer tools for sustainable use of natural resources.

Stuart O. Nelson, Adjunct Professor; Ph.D., Iowa State, 1972. Dielectric properties of materials, methods of measurement, and their use in agricultural applications of radio-frequency and microwave energy for developing sensors.

Stanley E. Prussia, Professor; Ph.D., California, Davis, 1980. Postharvest engineering for horticultural crops based on principles from systems theory, quality management, ergonomics, and materials science.

Glen Rains, Assistant Professor; Ph.D., Virginia Tech, 1992. Sensor development for the measurement of soil properties for use in precision farming management, pest detection using sensors that respond to plant stress indicators, measurement of local spatial and temporal rainfall variation.

L. Mark Risse, Public Service Assistant; Ph.D., Purdue, 1994. Extension and applied research in agricultural pollution prevention, animal waste management, control and prevention of nonpoint source pollution, and hydrologic and soil erosion modeling.

Matthew C. Smith, Assistant Professor; Ph.D., Florida, 1988. Mathematical modeling of the fate and transport of agricultural chemicals in the environment, field and laboratory-scale measurements for the validation of transport models.

David E. Stooksbury, Assistant Professor and State Climatologist; Ph.D., Virginia, 1992. Applied climatology, climate impacts on agricultural and environmental systems, wind and solar energy, wind resource assessment.

Chi N. Thai, Associate Professor; Ph.D., California, Davis, 1983. Development of instrumentation and decision support systems for the management of postharvest systems of fresh agricultural produce based on physical and sensory quality attributes.

Daniel L. Thomas, Associate Professor; Ph.D., Purdue, 1984. Hydrologic, water quality, and crop-production models for improved water and chemical management in small and large watersheds and agricultural production fields.

Sidney A. Thompson, Associate Professor; Ph.D., Kentucky, 1981. Structural design of buildings, materials handling, physical properties of granular materials, on-farm waste management systems.

E. Dale Threadgill, Professor and Department Head; Ph.D., Auburn, 1968. Systems and procedures for effective and safe distribution of chemicals through irrigation systems.

Ernest W. Tollner, Professor; Ph.D., Auburn, 1981. Modeling root growth and other ecological processes, X-ray tomography for visualizing and quantifying soils and biological materials, nuclear magnetic resonance spectrometry for quantifying soil-water-plant systems and properties of food products.

Bobby L. Tyson, Professor; Ph.D., Clemson, 1979. Extension and continuing education programs in farm safety; hay harvesting, handling, and storage; and energy program coordination.

Garrett L. VanWicklen, Associate Professor; Ph.D., Cornell, 1982. Physiologic response of animals to different stressors of the ambient environment, including air temperature, humidity, and airborne contaminants.

George Vellidis, Assistant Professor; Ph.D., Florida, 1989. Impacts of agricultural production systems on water quality, including nutrient and pesticide transport through wetland systems; environmental risk assessment; development of instrumentation.

Brahm P. Verma, Professor; Ph.D., Auburn, 1968. Application of theory of models and fuzzy set theory to biological and agricultural systems for developing decision support systems for economic development and environmental sustainability.

John W. Worley, Assistant Professor; Ph.D., Virginia Tech, 1990. Structures, waste handling systems, electrical systems and controls.

UIC

UNIVERSITY OF ILLINOIS AT CHICAGO

Bioengineering Department

Programs of Study
The department's research and curriculum focus in the areas of cell and tissue engineering, neural engineering, biomechanics and rehabilitation engineering, biomedical imaging and visualization, biomicroelectromechanical systems, and biomaterials.

The Ph.D. program in bioengineering is interdisciplinary in nature and designed to train students to solve problems at the interface of biology and engineering. The Bioengineering Department closely interacts with other departments within the College of Engineering and the Colleges of Medicine, Dentistry, and Pharmacy.

The M.S. program in bioengineering offers specializations in all of the areas mentioned above.

Both the M.S. and Ph.D. programs require the completion of a thesis. The M.S. program takes approximately two calendar years to complete, while the Ph.D. program takes approximately four years beyond the bachelor's degree. Both of these programs are designed to be fairly flexible, so that a student can pursue his or her own interests.

Research Facilities
Approximately 11,000 square feet of laboratory space is devoted to bioengineering. The University operates a mainframe computer with appropriate peripherals. In addition, the Bioengineering Department has several minicomputers that are available for student use with interfacing to provide analog-to-digital and digital-to-analog capability. The Research Resources Center occupies approximately 115,000 square feet of laboratory space and includes an animal laboratory with 7,500 animals. The library has 1.3 million volumes in its collection.

Financial Aid
Most full-time graduate students receive financial aid in the form of fellowships, teaching assistantships, tuition and fee waivers, or research assistantships. This aid is available throughout the year, including the summer session. Stipends for University fellows are $1300 per month for up to twelve months, while those for teaching and research assistants are approximately $12,000 for the academic year. All fellowships and assistantships carry remission of tuition and of most fees.

Cost of Study
For students who register for 12 or more credit hours of study per semester, tuition and fees for two semesters in 1999–2000 total $5492 for Illinois residents and $12,128 for nonresidents. For programs of fewer than 12 hours, the cost is less. (All fees are subject to change.)

Living and Housing Costs
In addition to University of Illinois at Chicago (UIC) dormitory accommodations, off-campus apartments and rooms are available for $400 per month and up. Meals may be taken on or off campus. The cost of food, transportation, medical care, clothing, and incidentals is estimated at $9500 per calendar year.

Student Group
The total enrollment on the campus is about 25,000, including approximately 6,100 graduate students. The current enrollment in the Bioengineering Department is 24 M.S. students and 19 Ph.D. students. In addition, there are about 90 undergraduate students in bioengineering.

Location
The University of Illinois at Chicago is located in the heart of the city. Chicago offers a multitude of recreational and cultural facilities within a short distance of the campus. Beaches situated on the southwest bank of Lake Michigan are frequented by thousands of Chicagoans every year. Virtually every kind of athletic team is represented in the city, and it has many museums with international reputations, such as the Art Institute, the Field Museum of Natural History, and the Museum of Science and Industry. Chicago's symphony orchestra is considered one of the world's finest.

The University
The University of Illinois at Chicago is the largest institution of higher learning in the Chicago area, one of seventy Research I universities in the United States, and an increasingly significant center for education, research, and health-care and public service. It offers doctorates in fifty-two specializations, master's degrees in eighty-nine fields, and bachelor's degrees in ninety-eight areas.

Applying
Application forms to apply for admission and financial assistance are sent on request. The completed forms should be returned as early as possible, preferably before March 1. Students from abroad must submit TOEFL scores. All applicants must submit GRE General Test scores and three letters of recommendation.

Correspondence and Information
Director
UIC Bioengineering Department (M/C 063)
851 South Morgan Street, Room 218
University of Illinois at Chicago
Chicago, Illinois 60607-7052
Telephone: 312-996-2335

University of Illinois at Chicago

THE FACULTY AND THEIR RESEARCH

College of Engineering

G. Agarwal, Professor of Electrical Engineering and Computer Science; Ph.D., Purdue, 1965. Control of human movements, analysis of EMG signal.

H. Alkan-Onyuksel, Associate Professor of Pharmaceutics; Ph.D., London, 1978. Drug delivery and transport, solubilization through biological membranes.

N. Alperin, Assistant Professor of Medical Physics; Ph.D., Chicago, 1992. Magnetic resonance imaging.

F. Amirouche, Associate Professor of Mechanical Engineering and of Bioengineering; Ph.D., Cincinnati, 1984. Biomechanics, human locomotion, microdynamics, and robotics.

R. Anderson, Associate Professor of Obstetrics and Gynecology; Ph.D., Illinois at Chicago, 1973. Biomedical instrumentation.

A. Ansari, Assistant Professor of Physics; Ph.D., Illinois at Urbana-Champaign, 1988. Molecular biophysics, protein-DNA interactions, kinetics of short DNA.

D. Braddock, Professor of Human Development; Ph.D., Texas at Austin, 1973. Developmental disabilities and assistive technology.

D. Carley, Research Associate Professor of Medicine and Pharmacology; Ph.D., Cambridge, 1985. Cardiorespiratory control and dynamics, computational neurobiology.

W. Chen, Assistant Professor of Dermatology; Ph.D., Temple, 1988. Bioelectromagnetic field, drug delivery, electrical injuries.

M. Cooper, Associate Professor of Emergency Medicine; M.D., Michigan State, 1974. Electrical injuries.

M. Damaser, Adjunct Assistant Professor of Urology; M.D., Berkeley, 1994. Biomechanics and rehabilitation of the lower urinary tract.

J. Daugirdas, Professor of Medicine; M.D., Northwestern, 1973. Vascular physiology.

T. Desai, Assistant Professor of Bioengineering; Ph.D., Berkeley, 1998. Artificial organs, pharmacokinetics.

J. Drummond, Professor of Restorative Dentistry and of Bioengineering; Ph.D., D.D.S., Illinois, 1979. Biomaterials, dental materials, tissue structure bonding, fracture mechanics.

C. Evans, Associate Professor of Orthodontics; D.M.Sc., Cambridge, 1975. Facial growth and development connective tissue.

D. Fiat, Professor of Physiology and Biophysics; D.Sc., Israel Institute of Technology (Haifa), 1960. Biophysics, nuclear magnetic resonance imaging.

D. Graupe, Professor of Electrical Engineering and Computer Science; Ph.D., Liverpool (England), 1963. Adaptive filtering, signal processing, muscle stimulation.

L. Hanley, Associate Professor of Chemistry; Ph.D., SUNY at Stony Brook, 1988. Molecular ion-surface interactions and biomaterials.

B. He, Assistant Professor of Electrical Engineering and of Bioengineering; Ph.D., Tokyo Institute of Technology, 1988. Modeling of physiological function, electrocardiology.

G. Hedman, Coordinator of Institute of Disability and Human Development; M.S., Virginia, 1983. Rehabilitation engineering.

P. Hesketh, Associate Professor of Electrical Engineering and Computer Science; Ph.D., Pennsylvania, 1987. Biosensors and microfabrication.

J. Hetling, Assistant Professor of Bioengineering; Ph.D., Illinois at Chicago, 1997. Electrophysiology of vision and retinal prosthetics.

D. Hier, Professor of Neurology; M.D., Cambridge, 1973. Neural networks and multi-imaging modalities.

A. Hopfinger, Professor of Medicinal Chemistry; Ph.D., Case Western Reserve, 1969. Computer-assisted molecular design.

D. Hurwitz, Assistant Professor of Orthopedic Surgery; Ph.D., Illinois at Chicago, 1994. Biomechanics and gait analysis.

J. Karlsson, Assistant Professor of Mechanical Engineering; Ph.D., MIT, 1994. System modeling and control.

S. Lavender, Assistant Professor of Orthopedic Surgery; Ph.D., Ohio State, 1990. Spine mechanics.

J. Lin, Professor of Electrical Engineering and Computer Science; Ph.D., Washington (Seattle), 1971. Noninvasive radio frequency and microwave sensing of physiological signatures, cardiovascular diseases, hyperthermia treatment for cancer.

F. Loth, Assistant Professor of Mechanical Engineering; Ph.D., Georgia Tech, 1993. Biofluid dynamics in using experimental simulations to model the hemodynamics flow field.

R. Magin, Professor and Head of Bioengineering; Ph.D., Rochester, 1976. Magnetic resonance imaging.

J. Marko, Assistant Professor of Physics; Ph.D., MIT, 1989. Protein-DNA interactions and molecular dynamics.

A. Masud, Assistant Professor of Computational Mechanical Engineering; Ph.D., Stanford, 1992. Computational mechanics.

R. Natarajan, Adjunct Professor of Bioengineering; Ph.D., London, 1971. Biomechanics.

A. Nehorai, Professor of Electrical Engineering and Computer Science; Ph.D., Stanford, 1983. Biomedical signal processing, biomedical imaging.

S. Olson, Associate Professor of Biology of Oral Diseases; Ph.D., Michigan, 1979. Engineering of novel anticoagulant, antiapoptotic and antifibrinolytic proteins.

W. O'Neill, Professor of Electrical Engineering and Computer Science; Ph.D., Notre Dame, 1965. Neuroscience and pupillography.

A. Patwardhan, Visiting Professor of Bioengineering; Ph.D., Oklahoma State, 1980. Biomechanics, spinal disorders, bone and soft tissue mechanics.

D. Pavel, Professor of Radiology; M.D., Northwestern, 1957. Medical imaging.

D. Pepperberg, Research Assistant Professor of Ophthalmology; Ph.D., Cambridge, 1973. Retinal physiology.

J. Rothstein, Professor of Physical Therapy and of Bioengineering; Ph.D., NYU, 1983. Clinical measurement, kinesiology.

L. Sadler, Associate Professor and Head of Biomedical Visualization and Associate Professor of Bioengineering; M.Sc., Michigan, 1972. Morphometrics, human body databases.

D. Schonfeld, Associate Professor of Electrical Engineering and Computer Science; Ph.D., Johns Hopkins, 1990. Neural networks, image processing.

B. Sposato, Project Coordinator of Disability and Human Development; M.S., Virginia, 1990. Rehabilitation engineering.

J. Sychra, Associate Professor of Radiology; Ph.D., Charles (Prague), 1968. Medical imaging and propagation of radar waves.

K. Thorne, Assistant Professor of Restorative Dentistry; Ph.D., UCLA, 1994. Bioceramics and advanced processing solutions.

D. Yeates, Research Professor of Medicine and Professor of Bioengineering; Ph.D., Toronto, 1975. Pulmonary physiology and clearance.

B. Zuber, Professor Emeritus of Chemical Engineering and of Bioengineering; Ph.D., MIT, 1965. Physiological control, neurobiology.

UNIVERSITY OF ILLINOIS
AT URBANA–CHAMPAIGN

Bioengineering Program

Program of Study

The University of Illinois at Urbana-Champaign (UIUC) Bioengineering Program was founded in 1973. The program is interdisciplinary with participating faculty members in the Colleges of Engineering, Medicine, Liberal Arts and Sciences, Agricultural Engineering, Applied Life Studies, and Veterinary Medicine. Graduate students must enroll in the primary department that most closely corresponds to their specific area of interest. The faculty adviser in the primary department is usually selected from the list of Bioengineering faculty members. The Bioengineering Program can assist prospective students in selecting an appropriate primary department and subsequently offer guidance in the selection of courses and in the availability and use of campuswide bioengineering research facilities. Each student completes graduate training course work in a traditional graduate subject (e.g., electrical engineering, physiology, or biomaterials) through the departmental qualifying exam. Subsequent interdisciplinary research training, under the direction of the Ph.D. committee, closely monitors the student's progress through the preliminary exam to the final Ph.D. defense. Bioengineering faculty members make up a significant portion of each student's Ph.D. committee. In this manner, the student receives the flexibility of graduate training in bioengineering within the formal degree requirements of a specific graduate department. The Bioengineering Program at UIUC coordinates graduate study in bioelectric phenomena, bioelectromagnetics, bioinstrumentation, biological modeling, biomaterials, biomechanics, biomedical imaging, bioultrasonics, cellular bioengineering, food and bioprocessing, heat and mass transfer, ionizing radiation, physiological modeling, and rehabilitation engineering.

The University of Illinois at Urbana-Champaign also has an established Medical Scholars Program (MSP) that provides for simultaneous pursuit of the M.D. and Ph.D. degrees administered through the Medical School. The Bioengineering Program is an active participant in the recruitment of MSP students and tries to integrate this program with existing bioengineering areas.

Research Facilities

A wide variety of state-of-the-art research facilities are available on the UIUC campus in each of the colleges that participate in the Bioengineering Program. In addition, the Beckman Institute, the Biotechnology Center, the National Center for Supercomputing Applications, and a number of NSF- and NIH-funded centers on campus provide extensive support for bioengineering research. Research laboratories participating in bioengineering studies exist in the areas of bioacoustics, magnetic resonance imaging, electron spin resonance, biomaterials, bioprocessing technology, fluorescence dynamics, veterinary nuclear medicine, radiology, and rehabilitation engineering. The science, engineering, and medical libraries provide technical support for bioengineering research through extensive collections of bioengineering-related publications.

Financial Aid

Financial aid is available in the form of federally sponsored traineeships and fellowships as well as University and industry fellowships. Recipients are exempt from tuition and fees. Also available are part- and full-time research and teaching assistantships. The stipend for a half-time assistantship varies by department and includes exemption from tuition and service fees.

Cost of Study

Tuition and fees vary according to the number of semester hours taken. For a full program (3 or more units), tuition and fees per semester in 1998–99 were $2856 for Illinois residents and $6294 for nonresidents. Summer session charges for a full program (2.5 or more units) were $1778 for Illinois residents and $3927 for nonresidents. For students with fellowships or assistantships, only the insurance fee must be paid.

Living and Housing Costs

University graduate residence halls have single rooms for $2696 and $3008 and double rooms for $2416 and $2872 per academic year. Optional board contracts are available for $1435 to $3368. University family housing rents start at $350 per month. Privately owned rooms and apartments are available at similar and higher rents.

Student Group

Enrollment at the Urbana-Champaign campus in 1998 was 36,303. The Bioengineering Program has an undergraduate enrollment of approximately 200 students, and 100 graduate students are affiliated with the program through participating departments. A student chapter of the IEEE Engineering in Medicine and Biology Society is sponsored by the program.

Location

The campus is 130 miles south of Chicago in the twin cities of Urbana and Champaign (population 110,000). The area is primarily a university community, with excellent public school, mass transit, and park systems and modern shopping facilities.

The University

The University of Illinois at Urbana-Champaign is in its second century of operation and is recognized as a major national center of excellence in graduate education. Cultural activities on the campus are fostered by the Krannert Center for the Performing Arts, which contains five separate theaters for orchestra, opera, choral groups, theater, and dance. The 16,000-seat Assembly Hall serves as the stage for some of the nation's foremost entertainers. In addition to Big Ten football and basketball, the University offers a broad program of athletics, utilizing the Intramural Physical Education Building. The University's 1,500-acre Allerton Park is only one of a number of nearby parks and recreation areas.

Applying

Applications are accepted on a rolling basis. Students must submit GRE scores and letters of recommendation. International students must take the TOEFL. All students must satisfy the application requirements of the specific department that contains the area in which they wish to specialize. Inquiries concerning the Bioengineering Program should be sent to the address below.

Correspondence and Information

Leon A. Frizzell
Chair of Bioengineering
53 Everitt Laboratory
University of Illinois at Urbana-Champaign
1406 West Green Street
Urbana, Illinois 61801
Telephone: 217-333-1867
E-mail: bioen@uiuc.edu
World Wide Web: http://www.ece.uiuc.edu/~bioen/

University of Illinois at Urbana-Champaign

THE FACULTY AND THEIR RESEARCH INTERESTS

All faculty members have appointments in the Bioengineering Program and in the departments listed below.

Thomas Anastasio, Associate Professor, Department of Molecular and Integrative Physiology. Computational neuroscience.

Lloyd Barr, Professor, Department of Molecular and Integrative Physiology. Excitable cells.

David J. Beebe, Assistant Professor, Department of Electrical and Computer Engineering and Department of Mechanical and Industrial Engineering. Microelectromechanical systems, microfabrication.

Philip M. Best, Professor, Department of Molecular and Integrative Physiology. Ion channels.

Yoram Bresler, Professor, Department of Electrical and Computer Engineering. Imaging.

John C. Chato, Professor Emeritus, Department of Mechanical and Industrial Engineering and Department of Electrical and Computer Engineering. Heat transfer.

John W. Chow, Assistant Professor, Department of Kinesiology. Biomechanics.

Leslie L. Christianson, Professor, Department of Agricultural Engineering. Microenvironment control.

Robert B. Clarkson, Associate Professor, Department of Veterinary Clinical Medicine and Department of Veterinary Biosciences. Imaging, magnetic resonance.

Edward R. Damiano, Assistant Professor, Department of Mechanical and Industrial Engineering. Continuum mechanics, mixture theory, perturbation methods, microvascular and vestibular mechanics.

M. Joan Dawson, Associate Professor, Department of Molecular and Integrative Physiology, Department of Radiology (UIC), and Department of Obstetrics and Gynecology. NMR spectroscopy and spectroscopy imaging.

Mrinal K. Dewanjee, Professor, Department of Veterinary Biosciences. Physiology, nuclear medicine, mechanism of thrombosis and cancer: diagnosis and prevention.

Howard S. Ducoff, Professor Emeritus, Department of Molecular and Integrative Physiology. Radiation biophysics.

Floyd Dunn, Professor Emeritus, Department of Electrical and Computer Engineering and Department of Molecular and Integrative Physiology. Ultrasonic bioengineering/biophysics.

Albert S. Feng, Professor, Department of Molecular and Integrative Physiology. Auditory physiology.

Raymond M. Fish, Adjunct Assistant Professor, Department of Electrical and Computer Engineering. Analog circuits, medicine.

Leon A. Frizzell, Professor, Department of Electrical and Computer Engineering. Ultrasonic biophysics/bioengineering.

John G. Georgiadis, Professor, Department of Mechanical and Industrial Engineering. Heat/mass transfer and fluid mechanics related to process and energy industries, advanced diagnostics for multiphase flows, magnetic resonance imaging.

David R. Gross, Professor and Head, Department of Veterinary Biosciences. Cardiovascular physiology, physiological fluid dynamics.

Bruce M. Hannon, Professor, Department of Geography. Ecological modeling.

Sandy I. Helman, Professor, Department of Molecular and Integrative Physiology. Electrophysiology of epithelial.

Kenneth R. Holmes, Associate Professor, Department of Veterinary Biosciences. Tissue blood flow and heat transfer.

Eric G. Jakobsson, Professor, Department of Molecular and Integrative Physiology. Biomolecular and biophysical theory and computation.

Ann L. Johnson, Professor, Department of Veterinary Clinical Medicine. Orthopedic surgery.

Paul C. Lauterbur, Professor, Department of Medical Information Science, Department of Chemistry, Department of Molecular and Integrative Physiology, and the Center for Advanced Study. Magnetic resonance, imaging, image processing, and neuroscience.

Zhi-Pei Liang, Associate Professor, Department of Electrical and Computer Engineering. Biomedical imaging.

J. Bruce Litchfield, Professor, Department of Agricultural Engineering. Food and bioprocessing.

Norman R. Miller, Associate Professor, Department of Mechanical and Industrial Engineering. Biomechanics.

Manssour H. Moeinzadeh, Associate Professor, Department of General Engineering, Department of Kinesiology, and Division of Rehabilitation-Education Services. Biomechanics.

Mark E. Nelson, Assistant Professor, Department of Molecular and Integrative Physiology. Computational neuroscience.

Burks Oakley II, Associate Professor, Department of Electrical and Computer Engineering and Biophysics Program. Bioinstrumentation.

William D. O'Brien Jr., Professor, Department of Electrical and Computer Engineering, Department of Medical Information Science, and Division of Nutritional Sciences. Ultrasonic biophysics, acoustic microscopy, and imaging.

William A. Olson, Assistant Professor, Department of Veterinary Clinical Medicine. Clinical pharmacology.

Marvin R. Paulsen, Professor, Department of Agricultural Engineering. Machine vision, NIR spectroscopy.

Adrienne L. Perlman, Associate Professor, Department of Speech and Hearing Science, and Adjunct Associate Professor, College of Medicine. Swallowing.

Gerald J. Pijanowski, Associate Professor, Department of Veterinary Biosciences. Orthopedic biomechanics.

John F. Reid, Professor, Department of Agricultural Engineering. Machine vision.

Nikolaos V. Sahinidis, Assistant Professor, Department of Mechanical and Industrial Engineering. Biomolecular design and optimization.

Mark R. Simon, Associate Professor, Department of Veterinary Biosciences. Cartilage growth.

Shankar Subramaniam, Professor, Department of Molecular and Integrative Physiology. Protein modeling and design.

Jonathan V. Sweedler, Associate Professor, Department of Chemistry. Neurotransmitter distribution and release.

John C. Thurmon, Professor, Department of Veterinary Clinical Medicine. Cardiopulmonary physiology-anesthesiology, anesthesiology.

A. Robert Twardock, Professor, Department of Veterinary Biosciences and Department of Molecular and Integrative Physiology, and department affiliate, Department of Nuclear Engineering. Nuclear medicine.

Alexander F. Vakakis, Assistant Professor, Department of Mechanical and Industrial Engineering. Nonlinear dynamics and modal analysis.

Tony G. Waldrop, Professor, Department of Molecular and Integrative Physiology, Neuroscience Program, Department of Kinesiology, and Biophysics Program. Effects of hypoxia on brain stem neurons, developmental neurobiology, and hypertension.

Andrew Webb, Assistant Professor, Department of Electrical and Computer Engineering. Magnetic resonance.

Bruce C. Wheeler, Associate Professor, Department of Electrical and Computer Engineering. Neural engineering.

Erik C. Wiener, Assistant Professor, Department of Nuclear Engineering and Department of Medical Information Sciences. Functional and medical imaging, targeted drug delivery, and the implications of tumor heterogeneity for therapy and diagnosis.

K. Dane Wittrup, Assistant Professor, Department of Chemical Engineering. Molecular bioengineering.

James F. Zachary, Associate Professor, Department of Veterinary Pathobiology. Ultrasound-induced bioeffects and in vivo ultrasonic microprobe for tumor diagnosis.

Charles F. Zukoski, Professor, Department of Chemical Engineering. Protein and colloidal.

UNIVERSITY OF MARYLAND, COLLEGE PARK

Department of Biological Resources Engineering

Programs of Study

Biological resources engineers improve societies, ecosystems, and the lives and health of individuals. Specializing in systems made from, used with, or applied to living organisms, they engineer solutions involving human and animal health and safety, environmental quality, and sustainable food production. The graduate program of the Department of Biological Resources Engineering at the University of Maryland, College Park (UMCP), provides qualified students with the multidisciplinary study and research experience they need to contribute to this exciting field. Under the personal guidance of an outstanding faculty, graduate students design educational programs leading to Master of Science (M.S.) and Doctor of Philosophy (Ph.D.) degrees. They develop these programs within the framework of three areas of graduate study: bioengineering, bioenvironmental systems engineering, and ecological engineering. All of the programs are tailored to meet the individual research interests and career ambitions of each graduate student.

Research Facilities

Housed in one of UMCP's newest buildings, the Department of Biological Resources Engineering provides graduate students with state-of-the-art research facilities, including specialized laboratories for work on instrumentation, cellular engineering, exercise physiology and biomechanics, wetland and plant ecology, geographic information systems, water quality, aquacultural engineering, and water resources engineering. Equipment includes walk-in environmental and growth chambers, treadmills, a digital plant canopy analyzer, light measurement equipment, liquid-gas transfer and reverse osmosis units, an ultrahigh-temperature processing unit, a particle analyzer, a gas chromatograph, video and fluorescent microscopes, an ion auto-analyzer, a UV-VIS spectrophotometer, a mass spectrometer, a blood gas analyzer, a portable rainfall simulator, and a tilting hydraulic flume. Graduate students also have access to computer facilities that feature engineering workstations, computer-aided design and modeling equipment, and customized data acquisition systems.

Financial Aid

The Department of Biological Resources Engineering provides financial support for the majority of its graduate students through assistantships and fellowships. Both teaching and research assistantships are available. Assistantships are provided as part of ongoing research grants by the University and through cooperative agreements with surrounding federal agencies. The research activities associated with these assistantships are usually part of ongoing faculty research and may contribute to thesis or dissertation research.

Cost of Study

In the 1998–99 academic year, tuition was $272 per credit hour for Maryland residents and $400 per credit hour for nonresidents. Other costs include mandatory fees that range from $171 to $282, depending on the number of credits taken.

Living and Housing Costs

University-owned apartments are available for graduate students. Monthly rates for this housing start at $525 for an efficiency apartment, $629 for a one-bedroom apartment, and $752 for a two-bedroom apartment. Off-campus housing ranges from approximately $300 per month for rooms in private homes to $600 to $1100 per month for a two-bedroom apartment. Additional information is available from the University's housing bureau.

Student Group

The University of Maryland has an enrollment of approximately 24,500 undergraduates and 8,300 graduate students. There are approximately 30 graduate students (40 percent women) in the Department of Biological Resources Engineering. More than 40 percent of these students are enrolled in the Ph.D. program, with the other 60 percent working toward an M.S.

Student Outcomes

Opportunities for the graduates in biological resources engineering are as diverse as the program. They range from positions with biotechnology and environmental companies to medical research. In addition, students have the opportunity to continue their studies in the sciences or engineering and in veterinary or medical school.

Location

Located near Washington, D.C., Annapolis, and Baltimore, UMCP offers easy access to some of the world's most important scientific organizations and a variety of cultural and recreational opportunities. These include the Smithsonian, the National Institutes of Health, the USDA's Beltsville Laboratories, the National Agricultural Library, the Library of Congress, the National Medical Library, the Kennedy Center, and the National Archives.

The University and The Department

As the flagship institution of the University of Maryland System, UMCP is one of the original land-grant institutions and is a Carnegie Foundation Research I institution. The Department of Biological Resources Engineering is a leader in its dynamic and innovative approaches to combining biology and engineering and is part of both the A. James Clark School of Engineering and the College of Agriculture and Natural Resources.

Applying

Outstanding graduates from engineering backgrounds and biological and physical science backgrounds are encouraged to apply. Admission to the master's program requires a bachelor's degree from an accredited institution. Although admission to the Ph.D. program normally requires an M.S. degree, outstanding students with a B.S. may directly enter the Ph.D. program. Program requirements vary with the background of the students. The application fee is $50 for U.S. citizens and permanent residents and $70 for international applicants.

Correspondence and Information

For program information:
Graduate Coordinator
Department of Biological Resources Engineering
University of Maryland, College Park
College Park, Maryland 20742
Telephone: 301-405-1198
Fax: 301-314-9023
E-mail: bioresengr@runoff.umd.edu
World Wide Web: http://www.bre.umd.edu

For application forms:
Graduate School
Admissions Office
Lee Building
University of Maryland, College Park
College Park, Maryland 20742
World Wide Web: http://www.inform.umd.edu/grad

University of Maryland, College Park

THE FACULTY AND THEIR RESEARCH

DEPARTMENTAL RESEARCH

Research within the department offers an integrated systems approach to solving diverse problems that involve biological and engineered components. Fundamental to this approach is development of techniques to manipulate and control biological systems. Engineering studies at UMCP emphasize a systems approach to problem solving. Students learn to adapt and apply engineering techniques to living systems of all scales.

On the smallest of these scales, bioengineers at UMCP are involved in research that includes the cryopreservation of living cells, the development of methods to remove biofilms from food-processing equipment, biological transport modeling, evaluation of biofilms for water treatment, the reduction of salmonella contamination in poultry and other products, and the evaluation of microscale damage within the lungs. On a somewhat larger scale, engineers at UMCP have the opportunity to work with entire organisms, instrumentation design, and food-processing equipment. Some of the research opportunities are the examination of physiological control during exercise; the evaluation of respiratory stress caused by protective mask-wear; the discovery of ways to harvest, transport, process, store, and deliver superior-quality aquatic food products; robotics; and computer vision.

On the largest scale, the program trains graduate students to use the same multidisciplinary systems approach to solving societal and environmental problems through the use of designed living systems. Current ecological engineering and bioenvironmental systems engineering research projects focus on such issues as restoration of damaged ecosystems and creation of new habitats; closed, semiclosed, and open aquacultural production systems; mathematical modeling of aquatic systems; water treatment using vegetation-based systems; the use of waste to fertilize agricultural land and natural habitats; nonpoint source pollution assessment using an integrated watershed approach, and the pollution potential of wastewater treatment systems.

THE FACULTY

Lowell Adams, Lecturer; Ph.D. (zoology), Ohio State, 1976. Wildlife ecology and management of urban open spaces, human-wildlife interactions in the metropolitan environment.

Andrew H. Baldwin, Assistant Professor; Ph.D. (botany), LSU, 1996. Ecological engineering and natural resources management; ecology of natural, restored, and created wetlands; wetlands for wastewater treatment; disturbance and regeneration processes in wetland vegetation.

Herbert L. Brodie, Professor Emeritus; M.S. (agricultural engineering), Maryland, 1972; PE. Bioenvironmental and water resources engineering, water and waste systems (industrial, municipal, and agricultural).

Lewis E. Carr, Instructor; Ph.D. (agricultural engineering), Maryland, 1987. Bioenvironmental and water resources engineering, poultry production and processing, environmental control of structures, composting and waste management, food safety.

Gary K. Felton, Water Quality Extension Specialist; Ph.D. (agricultural engineering), Texas A&M, 1987. Bioenvironmental and water resources engineering, nonpoint source pollution, urban nutrient management, groundwater hydrology and quality, field sampling of water flow and quality.

Theodore H. Ifft, Faculty Extension Associate; agricultural engineering, Penn State, 1958; PE. Bioenvironmental and water resources engineering, geographic information systems, global positioning systems, satellite image processing and computer modeling.

Arthur T. Johnson, Professor; Ph.D. (agricultural engineering), Cornell, 1969. Bioengineering, instrumentation, transport processes, modeling, exercise physiology and respiratory stress.

Patrick Kangas, Associate Professor; Ph.D. (environmental engineering sciences), Florida, 1983. Ecological engineering and natural resources management, tropical ecosystems.

Hubert Montas, Assistant Professor; Ph.D. (agricultural and biological engineering), Purdue, 1996. Bioenvironmental and water resources engineering, numerical methods in environmental biodynamics, transport modeling, information and decision support systems.

David S. Ross, Associate Professor; Ph.D. (agricultural engineering), Penn State, 1973. Bioengineering, horticultural engineering, controlled-environment agriculture, trickle irrigation, postharvest cooling.

Paul D. Schreuders, Assistant Professor and Graduate Coordinator; Ph.D. (biomedical engineering), Texas at Austin, 1989. Bioengineering, biological transport phenomena, cellular engineering, biofilms, Web-based teaching, cryopreservation and cryobiology.

Adel Shirmohammadi, Professor; Ph.D. (biological and agricultural engineering), North Carolina State, 1982. Bioenvironmental and water resources engineering, water quality/transport modeling, nonpoint source pollution assessment using an integrated watershed approach.

Fredrick W. Wheaton, Professor and Chair; Ph.D. (agricultural engineering), Iowa State, 1968. Bioengineering, aquacultural engineering, closed-cycle aquaculture, automation, seafood processing, modeling of aquacultural systems, seafood plant waste disposal and use.

UNIVERSITY OF MASSACHUSETTS WORCESTER AND WORCESTER POLYTECHNIC INSTITUTE

Medical School
Joint Program in Biomedical Engineering and Medical Physics

Program of Study

The joint program in biomedical engineering and medical physics of the Graduate School of Biomedical Sciences at the University of Massachusetts Medical School and of Worcester Polytechnic Institute offers a program of study and research that leads to a Ph.D. degree in biomedical engineering. The faculties of the program have varied research interests, including biomedical imaging, biosensors and biomedical instrumentation, tissue and cell engineering, and biomechanics. The program emphasizes basic and applied research in an environment that promotes strong interaction between biomedical engineers, researchers, and medical professionals.

Incoming students should have good preparation in the fundamentals of mathematics, physics, and engineering. Formal course requirements for the Ph.D. include a core curriculum of graduate-level courses in biomedical engineering and biomedical science. Students should undertake three to four laboratory rotations prior to selecting a laboratory for thesis research. In addition to taking formal courses and conducting research, students are expected to attend programmatic seminars on both campuses and to participate in a weekly seminar program. Admission to candidacy occurs following successful completion of a qualifying examination administered prior to the beginning of the third year. The Ph.D. degree generally requires four to five years to obtain and is awarded following completion and successful defense of the Ph.D. dissertation. The program employs the advanced technical knowledge and expertise of engineering and medical faculty members to provide students with the knowledge and skills to apply engineering principles to medically related problems.

Research Facilities

The Medical School was constructed in 1974 and is generously endowed with office, classroom, and laboratory space. Research facilities and equipment are available in all modern areas of biomedical research. Major facilities include a large experimental animal care facility, a virus laboratory, a cell and tissue culture facility, a protein chemistry laboratory, a Diabetes Research Center, a Cancer Center, and electronics and machine shops. The Medical School and Worcester Polytechnic Institute are at the forefront of the development of new technologies in biomedical engineering, with recognized leadership in the area of biomedical imaging, having major facilities in three-dimensional reconstruction microscopy, X-ray crystallography, and NMR spectroscopy. Facilities at the Medical School, including the Center for Advanced Clinical Technologies–Smith & Nephew Center for Research and Endoscopic Minimally Invasive Surgery, integrate into the community of pharmaceutical and biotechnology companies in the Biotechnology Research Park.

Financial Aid

Beginning full-time students are offered a graduate assistantship (in 1999–2000, the annual stipend is $17,500). Research assistantships paying $17,500 per year are available to students starting thesis research. Application for assistantships is made in the application for admission.

Cost of Study

Tuition waivers are available for both graduate and research assistants.

Living and Housing Costs

Apartments and rooms are available in the Worcester area and nearby communities. No on-campus housing is provided at the Medical School. The large number of students in the area makes the sharing of apartments feasible; local transportation is available and quite economical. The housing office assists students in finding appropriate places to live.

Student Group

This new program currently has 7 graduate students enrolled. Anticipated total enrollment will be 30 to 40 students. Current enrollment in the Graduate School of Biomedical Sciences' Ph.D. programs is approximately 200 students.

Location

The second-largest city in New England, Worcester is 40 miles west of Boston. Worcester and the surrounding towns offer varied living conditions and a wide range of educational and cultural activities. Located in the city are numerous other educational institutions, including Clark University and Holy Cross, Assumption, and Worcester State Colleges. The University of Massachusetts Medical School and Worcester Polytechnic Institute are less than 3 miles apart.

The University and The Medical School

The University of Massachusetts has five campuses: the main one at Amherst, about 70 miles west of Worcester; the Medical School in Worcester; and campuses in Boston, Dartmouth, and Lowell. The Medical School opened in 1970; there are now 100 medical students in each class. The Medical School joins regional institutions and the University of Massachusetts Amherst in the training of health science professionals. A Graduate School of Nursing opened in 1985. A program in molecular medicine recently opened at Massachusetts Biotechnology Research Park, with 52,250 square feet of laboratory and office space and 200 scientists, graduate students, and staff members. The Goff Learning Center, completed in 1997, offers a state-of-art teaching facility. A Cancer Center offers both basic and clinical approaches to improve the diagnosis and therapy of patients with cancer.

Applying

The program accepts graduate students through a joint admissions committee of the Graduate School of Biomedical Sciences and Worcester Polytechnic Institute. Applicants should have a bachelor's degree in a physical science or engineering. Students who do not have a degree in engineering can apply but are expected to take undergraduate engineering course(s). Applicants must take both the Graduate Record Examinations General Test and an appropriate advanced Subject Test. Three letters of recommendations and an application fee ($25 in-state, $50 out-of-state) are required. Completed applications (including transcripts and letters of recommendation) should be received no later than January 15 for both summer and fall admission.

The University of Massachusetts is an Equal Opportunity/Affirmative Action employer. The University of Massachusetts Medical School recruits minority and women candidates, considers all applicants, and, once students are accepted, ensures that they will not be discriminated against in any area.

The graduate catalog and application forms may be obtained from the address below. Potential applicants may wish to contact Dr. Peter Grigg (Graduate Director) at 508-856-2101 for any additional information about the program.

Correspondence and Information

Graduate School of Biomedical Sciences
University of Massachusetts Medical School
55 Lake Avenue North
Worcester, Massachusetts 01655-0116
Telephone: 508-956-4135
            888-860-2334 (toll-free)
E-mail: gsbs@umassmed.edu
World Wide Web: http://www.umassmed.edu/GSBS

University of Massachusetts Worcester and Worcester Polytechnic Institute

THE GRADUATE FACULTY AND THEIR RESEARCH

Fred Anderson, Ph.D., Professor of Surgery and Director of the School for Outcomes Research. Electrical impedance plethysmography, physician practices in the prevention and management of venous thromboembolism, quality assurance program to improve outcomes for patients at risk for venous thromboembolism.

Holly K. Ault, Ph.D., Assistant Professor of Mechanical Engineering. Computer-aided design, engineering design, graphics, biomechanics of joints and soft tissues.

Lawrence Bonassar, Ph.D., Assistant Professor of Anesthesiology. Tissue engineering, soft tissue biomechanics, mechanical regulation of connective tissue metabolism, structure-function relationships in cartilage extracellular matrix.

Walter Carrington, Ph.D., Associate Professor of Physiology. Application of mathematical and computational methods to fluorescence microscopy, especially deconvolution methods, that provide three-dimensional images of cells with substantially greater resolution than other approaches to light microscopy can provide.

David Cyganski, Ph.D., Professor of Electrical and Computer Engineering. Model-based 3-D object identification and pose estimation from linear signal decomposition and direction of arrival analysis.

Michael Davis, M.D., Sc.D., Professor of Radiology. Synthesis and biologic evaluation of diagnostic imaging agents and therapeutic radiopharmaceuticals; design, synthesis, and biologic evaluation of radiographic contrast agents for diagnostic radiology.

David DiBiasio, Ph.D., Associate Professor of Chemical Engineering. Biological reactor engineering, biochemical kinetics.

Carl J. D'Orsi, M.D., Professor of Radiology. Breast imaging: new technologies for light scanning of the breast, including digital breast imaging and interventional breast ultrasound.

William W. Durgin, Ph.D., George I. Alden Professor of Mechanical Engineering. Biomedical fluid mechanics, aerodynamics, hydrodynamics, flow-induced vibration, microgravity fluid dynamics.

Michael A. Gennert, Sc.D., Associate Professor of Computer Sciences. Computer vision and image processing, including computer theories of vision and applications; tomographic reconstruction and image processing of reconstructed imagery; artificial intelligence and neural networks, especially applied to vision problems.

Stephen Glick, Ph.D., Associate Professor of Nuclear Medicine. Single photon emission computed tomographic imaging of the brain and cardiovascular system, methods for reconstructing three-dimensional data from its two-dimensional projections.

Peter Grigg, Ph.D., Professor of Physiology. Stress-sensitive responses of stretch-sensitive neurons in mouse skin; constitutive properties of skin measured with both static and dynamic stimuli; water movement in soft tissues under tensile loading, measured with NMR spectroscopy.

Karl G. Helmer, Ph.D., Assistant Professor of Biomedical Engineering. Nuclear magnetic resonance imaging and theory of biological systems, evaluation and treatment of stroke and cancer, impedance spectroscopy, biomedical instrumentation, noninvasive tissue ischemia.

Allen H. Hoffman, Ph.D., Professor of Mechanical Engineering. Biomechanics, biomaterials, biomedical engineering, dynamics, fluid mechanics, continuum mechanics.

Andrew Karellas, Ph.D., Associate Professor of Radiology. Developing novel techniques for improving lesion detectability in diagnostic X-ray imaging, with particular emphasis on mammography; research and development in small-format digital mammography using charge-coupled devices for stereotactic localization and high-resolution spot views; full-breast digital mammography and its adaptation to tomographic imaging of the breast.

Michael King, Ph.D., Professor of Nuclear Medicine. Clinical applications of radionuclides: improving both the detection of disease and the quantitation of function through the application of the principles of medical physics and biomedical engineering; tomographic reconstruction; compensation for the physical degradations present in imaging; estimation of attenuation maps; image segmentation; numerical and human observer detection studies; physiological monitoring; functional imaging.

Sean Kohles, Ph.D., Assistant Professor of Biomedical Engineering. Biomechanics, biomaterials, tissue engineering, statistics, experimental mechanics techniques applied to soft and hard connective tissues and orthopedic devices.

Stevan Kun, Ph.D., Assistant Professor of Biomedical Engineering. Impedance spectroscopy, biomedical instrumentation, noninvasive tissue ischemia.

Larry Lifshitz, Ph.D., Assistant Professor of Physiology. Computer vision and computer graphics; algorithms to identify features of interest in 2-D and 3-D images generated with a 3-D wide-field digital imaging microscope; developer of the Data Analysis and Visualization Environment, which is used to analyze images; scale-space differential geometry for feature extraction, deformable models for image segmentation, spatial statistics for image analysis, and interactive graphics for visualization; design and implementation of cell simulation models.

Fred J. Looft III, Ph.D., Professor of Electrical and Computer Engineering. Instrumentation, biomedical signal processing, digital and analog systems, biomedical engineering.

George Mardirossian, Ph.D., Assistant Professor of Nuclear Medicine. Dosimetry analysis, radiolabeling agents for radioimmunoscintigraphy, pharmacokinetic evaluation and modeling techniques; application of infusional brachytherapy techniques to deliver a maximum radiation dose of antibody labeled with therapeutic radionuclides by direct infusion into prostate cancers.

Yitzhak Mendelson, Ph.D., Associate Professor of Biomedical Engineering. Biosensors, microcomputer-based biomedical instrumentation, invasive and noninvasive blood gas monitoring, applications of optics to biomedicine.

Peder Pedersen, Ph.D., Professor of Electrical and Computer Engineering. Inverse methods in ultrasonics, biomedical applications of ultrasound, transducer characterization techniques, ultrasound Doppler measurements, ultrasound tissue characterization.

Robert A. Peura, Ph.D., Professor and Chair, Biomedical Engineering, Worcester Polytechnic Institute, and Program Co-Director. Biomedical instrumentation; noninvasive biosensors; optical bioinstrumentation; blood pressure, blood glucose, and blood gas monitoring.

Richard S. Qimby, Ph.D., Associate Professor of Physics. Optics, laser spectroscopy, photoacoustic spectroscopy, optics of thin metal films, nondestructive testing, biomedical applications of lasers, fiber optics.

Brian J. Savilonis, Ph.D., Associate Professor of Mechanical Engineering. Fluid mechanics, biofluid mechanics, heat transfer.

Babs R. Soller, Ph.D., Assistant Professor of Surgery. Development of noninvasive and minimally invasive sensors based on optical spectroscopy with determined blood and tissue chemistry of surgical and critical-care patients.

Christopher H. Sotak, Ph.D., Associate Professor of Biomedical Engineering. Application of nuclear magnetic resonance spectroscopy (NMRS) and nuclear magnetic resonance imaging (NMRI) in medicine and biology, evaluation and treatment of stroke and cancer, detection of biological metabolites.

John Sullivan, Ph.D., Associate Professor of Mechanical Engineering. Computational mechanics, biomechanics, finite-element analysis, heat transfer.

Richard A. Tuft, Ph.D., Associate Professor of Physiology. Development of new optical/electronic microscopes and their application to the problems of imaging rapid changes in the concentration of intracellular ions and molecules; a microscope incorporating fast piezoelectric focusing (at 1 micrometer per millisecond), a low-noise high-quantum-efficiency solid-state CCD camera, and ultraviolet and visible laser illumination used to study intracellular [Ca2+] near the plasma membrane of smooth muscle cells, the role of Ca2+ and other second messengers in the chemotaxis of white blood cells, and [Ca2+] dynamics in secretory cells.

Yu-Li Wang, Ph.D., Professor of Physiology. Responses of cultured cells to mechanical input, fluorescence imaging of cells and molecules, cell motility and cell division.

David E. Wolf, Ph.D., Professor of Physiology, University of Massachusetts Medical School, and Program Co-Director. Role of membrane organization in cellular differentiation systems, such as epididymal maturation and capacitation of mammalian spermatozoa and early mammalian embryogenesis.

THE UNIVERSITY OF MEMPHIS
UNIVERSITY OF TENNESSEE, MEMPHIS

Joint Graduate Program in Biomedical Engineering

Programs of Study

The UM/UT Joint Graduate Program offers M.S. and Ph.D. degrees in biomedical engineering that stress the application of engineering and physical science to biomedical problems, including research and development of new technologies. The curriculum seeks to provide each student with integrated skills in life science, applied mathematics, and engineering and offers research specialization in four major subdisciplines: biomechanics and rehabilitation engineering, including orthopedic implants, prosthetic devices and design engineering; cell and tissue engineering, focusing on the cardiovascular system and including artificial organs, biomaterials, and hemodynamics; electrophysiology, including measurement methods, modeling and computation, signal analysis, and biomedical sensors and microfabrication; and imaging, including novel medical image–acquisition devices, computational image processing, and quantitative analysis techniques.

The M.S. degree is offered with a thesis option (requiring completion of 30 semester hours) and a nonthesis option (33 semester hours). The Ph.D. degree requires completion of 57 semester hours (credits) in graded courses and research beyond an M.S. degree or 90 credits total when proceeding from a bachelor's degree. The elective curriculum and research of each student is guided by a faculty committee. On completion of requirements, each M.S. thesis or Ph.D. dissertation candidate is examined by the faculty committee during a final oral presentation of the research. Candidates for the M.S. degree without thesis are examined orally on results from work on a 3- or 6-credit project and must also pass a written exam that assesses the degree to which academic course work has been integrated into practical knowledge.

Research Facilities

Extensive facilities for biomedical engineering research exist on both university campuses. The laboratories of the UT School of Biomedical Engineering include research facilities in biomechanics, cell and tissue engineering, cardiac electrophysiology, computer modeling of biosystems, medical imaging, and rehabilitation engineering. Research facilities of the UM Department of Biomedical Engineering include laboratories in biofluid dynamics, biomaterials testing, biomechanics, biorheology, electrocardiology, and microfabrication of biomedical sensors. The new Memphis Pediatric Biomedical Engineering Center has laboratory space in the LeBonheur Children's Hospital and offers clinical rotations and projects in technology-intensive clinical areas. External facilities that are available to students for collaborative research through the Joint Graduate Program include the combined resources of the UM Herff College of Engineering, the UT Health Sciences Center, and affiliations with Campbell Clinic, St. Jude Children's Research Hospital, Veteran Affairs Medical Center, other area hospitals, and leading health industry corporations.

Financial Aid

Financial aid in varying amounts in the form of full- and part-time fellowships, assistantships, and tuition remission is regularly awarded on a competitive basis. The program attempts to support all full-time M.S. and Ph.D. students with some form of assistance, although students may be admitted without any financial aid. Assistance in the form of loans is also available for qualified applicants.

Cost of Study

Full-year tuition and fees are approximately $4229 for Tennessee resident students and $10,598 for out-of-state students. Students who receive tuition remission are responsible for mandatory health insurance unless proof of other coverage is provided.

Living and Housing Costs

Total annual living expenses for a single student are estimated at $8710, which comprises room and board, $5200; books and supplies, $1100; personal expenses, $1650; and mandatory health insurance, $760. Graduate dormitories and neighborhood housing at comparable rates are available convenient to both campuses. City bus service is available between the two campuses (about 5 miles apart); however for reasonable convenience, personal transportation may be desirable. Carpooling is commonly used.

Student Group

Approximately 65 students are enrolled in the Joint Graduate Program in Biomedical Engineering. Combined enrollment at both universities averages more than 22,000, including 4,650 graduate students. A large fraction of the graduate student body is composed of women and international students.

Student Outcomes

Since 1990, 83 students have received master's degrees, and 13 have received doctoral degrees in biomedical engineering from the two universities. The majority of graduates are currently employed in industry, with such corporations as Medtronic, Sofamor Danek, Smith & Nephew Richards Orthopaedics, Celcore, Picker X-ray, Stryker, Massachusetts General Hospital, Alabama Vocational Rehab Agency, University of Mississippi Medical Center, Andrews University, University of Tennessee, and the Mayo Clinic, among others. Seven others are pursing their Ph.D.'s (at Johns Hopkins, Tulane, and the University of Memphis), and 2 others are in medical school. All doctoral graduates are now employed as research associates or postdoctorals (Brigham and Women's Hospital; University of California, San Diego; and University of Tennessee).

Location

The Memphis metropolitan area has more than 1 million residents and is the educational, commercial, and cultural center of the mid-South. Situated in the southwestern corner of Tennessee, Memphis is convenient to a variety of prime vacation spots, such as Hot Springs, Arkansas; the Gulf Coast; and the Great Smoky Mountains. Within minutes of either campus are a variety of cultural attractions, including theater, symphony, opera, ballet, and museums.

The Universities and The Program

The University of Memphis, the largest of six 4-year universities governed by the Tennessee Board of Regents, is a comprehensive urban university. It is composed of the Colleges of Arts and Sciences, Business and Economics, Communication and Fine Arts, Education, and Engineering as well as Schools of Nursing and Law.

The University of Tennessee, Memphis, is one of four instructional campuses of the University of Tennessee. As the University's academic health sciences center, the Memphis facilities include the Colleges of Allied Health Sciences, Dentistry, Graduate Health Sciences, Medicine, Nursing, and Pharmacy; the School of Biomedical Engineering; and the 160-bed Bowld Hospital. The two universities have a combined faculty of more than 1,700 and annually receive combined research funding of more than $50 million.

Applying

Applicants are sought who have B.S. or M.S. degrees in any branch of engineering, physics, or mathematics (including computer science). Students with degrees in other appropriate physical (and certain biological) sciences may be accepted on condition that they complete specified prerequisite work in mathematics and engineering needed for graduate study in biomedical engineering. Requirements for admission include official submission of all undergraduate and graduate transcripts and GRE General Test scores, with scores of at least 500 on each of the verbal, quantitative, and analytical sections. In addition, international students must submit minimum TOEFL scores of 550.

Correspondence and Information

Director of Graduate Studies
School of Biomedical Engineering
University of Tennessee, Memphis
899 Madison Avenue, Suite 801
Memphis, Tennessee 38163
Telephone: 901-448-7343
Fax: 901-448-7387
E-mail: ljordan@utmem.edu
World Wide Web: http://memphis.mecca.org/BME/

Coordinator of Graduate Studies
Department of Biomedical Engineering
The University of Memphis
Memphis, Tennessee 38152
Telephone: 901-678-3263
Fax: 901-678-5281
E-mail: mryen@cc.memphis.edu

The University of Memphis/University of Tennessee, Memphis

THE FACULTY AND THEIR RESEARCH

Endowed Chairs/Professors

*A. U. (Dan) Daniels, Wilhelm Professor of Orthopedic Surgery; Ph.D., Utah, 1966. Orthopedic implant materials, device properties and performance.

*Frank A. DiBianca, Crippled Children's Foundation Chair of Excellence; Ph.D., Carnegie Mellon, 1970. Medical imaging devices, digital image analysis, computer simulations.

*Eugene C. Eckstein, Hyde Chair of Excellence; Ph.D., MIT, 1975. Transport in flowing blood, blood-material interactions, design and testing of implantable and prosthetic devices, biomechanics.

*Michael R. Neuman, Herff Chair of Excellence in Biomedical Engineering; Ph.D., Case Tech, 1966; M.D., Case Western Reserve, 1974. Biomedical sensors, clinical instrumentation, pediatrics, and perinatology.

*Vincent T. Turitto, Herff Chair of Excellence in Biomedical Engineering; Sc.D., Columbia, 1972. Fluid flow and cellular response, biomaterials, thrombosis.

Professors

Lloyd D. Partridge (Emeritus), Ph.D., Michigan, 1953. Physiology, nervous systems, control theory.

*Michael R. T. Yen, Ph.D., California, San Diego, 1973. Tissue perfusion, biomechanics, orthopedic engineering.

Associate Professors

*Jack W. Buchanan, M.S.E.E., M.D., Kentucky, 1975. Cardiac electrophysiology and modeling, physiological instrumentation, high-performance distributed computing.

*Frank J. Claydon, Ph.D., Duke, 1987. Cardiac electrophysiology, mapping of myocardial infarction, ischemia.

Lawrence M. Jordan, Ph.D., Princeton, 1964. Digital X-ray imaging, computer modeling, applications of control systems.

Erno Lindner, Ph.D., Budapest Technical, 1985. Chemical sensors, microfabrication, analytical chemistry.

Mohammad F. Kiani, Ph.D., Louisiana Tech, 1990. Effects of ionizing radiation on microvasculature, hemodynamics, tumor biology.

*Herbert D. Zeman, Ph.D., Stanford, 1972. Medical imaging, intravenous angiography, image-processing algorithms.

Assistant Professors

Semahat S. Demir, Ph.D., Rice, 1995. Docent, Republic of Turkey, 1998. Mathematical modeling and computer simulation in cardiac electrophysiology, computational neuroscience.

Denis J. DiAngelo, Ph.D., McMaster, 1993. Spine and orthopedic biomechanics, design of multiaxis testing equipment, mechanics of the musculoskeletal system, design of lower extremity sports prosthetics.

Douglas J. Goetz, Ph.D., Cornell, 1995. Cell adhesion, tumor biology, drug delivery.

Robert A. Malkin, Ph.D., Duke, 1993. Electrophysiology, instrumentation, cardiac activation mapping, signal processing.

Jae-Young Rho, Ph.D., Texas, 1991. Bone and cartilage biomechanics, biomaterials, orthopedics, tissue engineering related to orthopedics.

Steven M. Slack, Ph.D., Washington (Seattle), 1989. Biomaterials, cellular responses to mechanical forces.

Tenured faculty

UNIVERSITY OF MICHIGAN

Department of Biomedical Engineering and Center for Biomedical Engineering Research

Program of Study

Biomedical engineering at the University of Michigan is an interdisciplinary program jointly sponsored by the College of Engineering and the Medical School. Faculty members for the program come not only from Engineering and Medicine but also from the School of Dentistry; the School of Public Health; the College of Literature, Science, and the Arts; the Veterans Administration Hospital in Ann Arbor; and the Henry Ford Hospital in Detroit. The variety and quality of research pursued by this diverse faculty provide a rich environment for graduate study in biomedical engineering. Specific areas of study include biomechanics, biotechnology, tissue engineering, bioelectrical systems, medical imaging, biomaterials, and rehabilitation engineering. Such depth and diversity are rare among research universities.

The graduate program in biomedical engineering grants both the M.S. and Ph.D. degrees. The specific degree requirements for a master's are 30 credit hours of approved graduate course work, with at least a B in each course. Directed research work is required to familiarize the student with the problems associated with biomedical engineering research. Students usually select a program of study in one of five formal options: biomechanics, biomaterials, biotechnology, bioelectrical engineering (including medical imaging), and rehabilitation engineering.

A second master's degree may be obtained for a minimum of 24 additional credit hours, provided that the admission and degree requirements are satisfied for the department awarding the second degree. A master's thesis option exists for students who wish to undertake individual research.

The Ph.D. is conferred in recognition of marked ability and scholarship in the field of biomedical engineering. Students must pass qualifying and preliminary examinations before being recommended for candidacy. A dissertation describing an independent research project, carried out under the direction of a dissertation committee, is required. Students complete course work as needed to prepare them for doctoral research.

The University of Michigan has an established M.D./Ph.D. degree program, which is administered through the Medical School with funds provided by the National Institutes of Health. The Biomedical Engineering Program has been an active participant in recruiting and training M.D./Ph.D. candidates. In addition to the M.D./Ph.D. degree program, students may obtain an M.D./M.S. degree.

Research Facilities

The Engineering College and the Medical School have very impressive state-of-the-art laboratories and teaching facilities. Because they are in an interdisciplinary program, biomedical engineering students have access to all of the facilities of both schools. The biomedical engineering faculty members have research laboratories located in their respective schools.

The University has a solid-state electronics laboratory in the Electrical Engineering Building. This laboratory is one of the best university facilities in the country for the design and manufacture of microelectronic circuits and is already being utilized extensively for the fabrication of implantable biosensor transducers.

Three Medical Science Research Buildings provide more than 300,000 square feet of prime basic laboratory and support space for the Medical Center.

Financial Aid

The program offers an array of fellowships and research assistantships associated with its large sponsored research program. All admitted students are considered for financial aid.

Cost of Study

In 1998–99, tuition was $6042 per term for state residents and $10,943 per term for nonresidents. There is also a modest fee that covers registration, health service, and student assembly and government. (Costs are subject to change each year.)

Living and Housing Costs

The University's Housing Information Office serves as an information source and provides assistance to graduate students seeking accommodations to suit their individual needs. In addition to on-campus single-student residence halls, adequate married student housing is available. The Housing Information Office administers University-operated facilities, including residence halls, co-ops, apartments, and suites, and coordinates information concerning privately owned housing. Food and living costs are close to the national average.

Student Group

There are approximately 115 graduate students in the Biomedical Engineering Program. Of that number, there are approximately 40 Master of Science students and 75 Ph.D. students.

Location

The University of Michigan is located in the heart of Ann Arbor, just 40 miles west of Detroit. There are exceptional cultural and educational advantages in an environment rich with private art galleries, theatrical productions, concerts, ballet, opera, lectures, recitals, discussions, and colloquia and seminars. Every season features numerous cultural and sports events, both collegiate and professional, in Ann Arbor and the Detroit area.

The University

The University of Michigan was founded in 1817 and was the nation's first public university.

Applying

For application forms, students should write to the address below. The application deadline for the fall term is February 1.

Correspondence and Information

Maria Steele
Department of Biomedical Engineering
3304A GG Brown
University of Michigan
Ann Arbor, Michigan 48109-2125
Telephone: 734-764-9588
E-mail: biomede@umich.edu
World Wide Web: http://www.bme.umich.edu

University of Michigan

THE FACULTY AND THEIR RESEARCH

Neal Alexander, M.D., Assistant Professor. Mobility in older adults, focused on rising from a bed and a chair, maintenance of standing balance.
David Anderson, Ph.D., Professor. Auditory and vestibular neurophysiology, vestibular physiology and space medicine.
Larry Antonuk, Ph.D., Associate Professor. Design, development, and clinical implementation of diverse imagers.
Thomas Armstrong, Ph.D., Professor. Physiology and biomechanics of human performance, physical stresses.
James Ashton-Miller, Ph.D., Research Scientist. Human musculoskeletal biomechanics, effects of aging.
Robert Bartlett, M.D., Professor. Blood surface interaction, organ replacement, prolonged extracorporeal circulation.
Spencer BeMent, Ph.D., Professor. Integrated circuit microprobes, neuromuscular signal processing, robotic mobility, orthotic device control.
Jeff Bonadio, M.D., Associate Research Scientist. Contribution of extracellular matrix to tissue function.
Fred Bookstein, Ph.D., Research Scientist. Morphometrics, shape and growth forecasting, image detection and reconstruction.
Susan Brooks, Ph.D., Assistant Professor. Skeletal muscle mechanics.
David Burke, Ph.D., Assistant Professor. Genetics and genome analysis, technology development for genetic testing and DNA sequencing.
Mark Burns, Ph.D., Associate Professor. Biochemical separations, magnetic enhancement separations, membrane separations.
Charles A. Cain, Ph.D., Professor. Medical applications of ultrasound, hyperthermia cancer therapy, biological effects of ultrasound.
Bruce Carlson, M.D., Ph.D., Professor. Regeneration and transplantation of mammalian skeletal muscle.
James Carpenter, M.D., Associate Professor. Biomechanics, sports medicine, shoulder joints.
Paul Carson, Ph.D., Professor. Diagnostic ultrasound, digital analysis of pulse echo ultrasound waveforms, correlation of tissue properties.
Kenneth Casey, M.D., Professor. Central nervous system pain pathways and their controlled alterations.
Steven Ceccio, Ph.D., Associate Professor. Fluid mechanics, with emphasis on multiphase flows.
Don Chaffin, Ph.D., Professor. Musculoskeletal biomechanics, occupational task and stress analysis, muscle physiology and control.
Thomas Chenevert, Ph.D., Associate Professor. MRI, diagnostic NMR imaging and spectroscopy.
David Dawson, Ph.D., Professor. Ion transport mechanisms, fluctuation analysis, and impedance analysis.
Robert Dennis, Ph.D., Research Investigator. Skeletal muscle denervation, contraction-induced injury and aging in muscle groups.
John Faulkner, Ph.D., Professor. Muscle physiology, muscle transplant, aging effects on muscle, muscle injury.
Jeffrey Fessler, Ph.D., Assistant Professor. Statistical methods for medical image processing and tomographic image reconstruction.
Michael Flynn, Ph.D., Adjunct Assistant Professor. Diagnostic imaging and image analysis, tissue morphology and function.
Brian Fowlkes, Ph.D., Research Investigator. Diagnostic and therapeutic ultrasound, acoustic cavitation, image processing.
Benedick A. Fraass, Ph.D., Professor. Computer-controlled radiotherapy, 3-D conformal treatment planning, clinical radiation therapy.
Ari Gafni, Ph.D., Professor. Protein folding.
Steven Goldstein, Ph.D., Professor. Bone and joint biomechanics, total joint arthroplasty, hard and soft tissue mechanics, tissue adaptation.
Daniel Green, Ph.D., Professor. Physiology of vision, neural networks, neural basis for visual adaptations and control, visual prosthetics.
Karl Grosh, Ph.D., Assistant Professor. Mechanics and response of biological systems.
Melissa Gross, Ph.D., Assistant Professor. Biomechanics and motor control, movement in the elderly.
James Grotberg, M.D., Ph.D., Professor. Biofluid mechanics, respiratory flows and transport, pulmonary drug delivery, and surfactant replacement therapy.
Carl Hanks, D.D.S., Ph.D., Professor. Biocompatibility, dental materials and biomaterials, bone and connected tissue healing.
Alfred Hero, Ph.D., Professor. Mathematical and statistical modeling of imaging systems, image reconstruction and signal processing.
Scott Hollister, Ph.D., Assistant Professor. Bone mechanics and bone adaption during aging and in response to total joint replacement.
Bret Hughes, Ph.D., Assistant Professor. Retinal physiology, ion transport mechanisms.
David Humes, M.D., Professor. Developing tissue engineering constructs, bioartificial kidney, hemocompatible vascular grafts.
Alan Hunt, Ph.D., Assistant Professor. Biomechanics of mitotic chromosome movements.
Janice Jenkins, Ph.D., Professor. Cardiology and signal processing, computerized diagnostic electrocardiography.
Tibor Juhasz, Ph.D., Associate Research Scientist. Biomedical applications of lasers.
Peter Kaufman, Ph.D., Professor. Plant cell biology, gravitropic response mechanisms.
Marc Kessler, Ph.D., Assistant Professor. Medical image processing, multiselector data fusion, computer-controlled radiation therapy.
Denise Kirshner, Ph.D., Assistant Professor. Computational pathogenesis.
Glenn Knoll, Ph.D., Professor. Nuclear medicine imaging, radioisotope imaging, unconventional aperture design.
David Kohn, Ph.D., Associate Professor. Orthopedic and dental biomechanics and biomaterials.
Arthur Kuo, Ph.D., Assistant Professor. Neuromuscular biomechanics, motor control of movement, human balance, dynamics.
Ron Kurtz, M.D., Assistant Professor. Ultrafast laser tissue removal in treatment of eye diseases.
William M. Kuzon Jr., M.D., Ph.D., Assistant Professor. Skeletal muscle mechanics.
Simon Levine, Ph.D., Associate Professor. Assistive technology for handicapped people, rehabilitation robotics.
Jennifer Linderman, Ph.D., Associate Professor. Receptor mediated cell phenomenon, mathematical modeling, digital fluorescence imaging.
Robert MacDonald, M.D., Professor. Neurophysiology, neuropharmacology, synaptic transmission.
Beth Malow, M.D., Assistant Professor. Epilepsy, sleep, EEG signal analysis.
Bernard Martin, Ph.D., D.S., Associate Professor. Psychophysiology sensorimotor control, neurophysiology.
Carlos Mastrangelo, Ph.D., Assistant Professor. VLSI and micromechanical fabrication, pressure sensors, microfluidic DNA sequencer.
Larry Matthews, Ph.D., Professor. Musculoskeletal biomechanics, joint prosthesis, fracture healing and fixation, tissue adaptation.
Daniel McShan, Ph.D., Associate Professor. Computer-controlled radiotherapy, 3-D conformal treatment planning.
Joseph Metzger, Ph.D., Assistant Professor. Contractile mechanics.
Charles Meyer, Ph.D., Professor. Diagnostic ultrasound, image reconstruction, ultrasonic microprobe transducer development.
Rees Midgley, M.D., Professor. Communication and control of cellular function, biosensors, intercellular signaling.
Josef Miller, Ph.D., Professor. Tissue engineering of the inner ear.
Maria Moalli, D.V.M., Research Investigator. Signal transduction mechanisms that regulate bone adaptation to physical forces.
David Mooney, Ph.D., Assistant Professor. Cellular and tissue engineering.
Michael Morris, Ph.D., Professor. Raman microprobe spectroscopy and spectral imaging.
Doug Noll, Ph.D., Associate Professor. Magnetic Resonance Imaging (MRI), Functional MRI, medical imaging, image processing.
Matthew O'Donnell, Ph.D., Professor. Medical imaging, biomedical ultrasonics, NMR imaging.
Mathilde Peters, D.M.D., Ph.D., Professor. Biomechanical research applied to restorative dentistry.
Malini Raghavan, Ph.D., Assistant Professor. Molecular interactions involved in the invasion of host cells by viruses.
Jonathan Raz, Ph.D., Associate Professor. Time series and repeated measures analysis, applications to biomedical signals and images.
W. Leslie Rogers, Ph.D., Professor. Nuclear imaging, coded apertures, emission computed tomography, positron imaging.
Ann Marie Sastry, Ph.D., Assistant Professor. Composite and fibrous mechanics.
Michael Savageau, Ph.D., Professor. Function, design, and evolution of cellular and molecular networks; nonlinear dynamics.
Christoph Schmidt, Ph.D., Assistant Professor. Microscopic mechanics of cytoskeletal and motor proteins.
Lawrence Schneider, Ph.D., Research Scientist. Impact biomechanics, anthropometry, ergonomics of seating.
Paul A. Sieving, M.D., Ph.D., Professor. Neurophysiology of vision, retinal electrophysiology, clinical vision testing.
Vijendra Singh, Ph.D., Senior Research Scientist. Role of immune targeting.
Duncan Steele, Ph.D., Senior Research Scientist. Biophysical methodology for protein structure and dynamics.
Gregory Wakefield, Ph.D., Associate Professor. Signal and image processing, psychoacoustics, sensory system analysis.
Henry Wang, Ph.D., Professor. Industrial biotechnology, biosensing and bioscreening, cell cultivation technology.
Roger Wiggins, M.D., Professor. Renal elastic and transplant.
William Williams, Ph.D., Professor. Electrophysiology and signal processing, communication and control theory, human perception.
Alan Wineman, Ph.D., Professor. Solid mechanics.
Kensall Wise, Ph.D., Professor. Development and characterization of solid-state sensors, microelectrode development.
J. S. Wolf, M.D., Assistant Professor. Endourology and laparoscopic urology.
Wen-Jei Yang, Ph.D., Professor. Hyperthermia, hemodynamics in flows through prosthetic heart valves, biofluid mechanics.

UNIVERSITY OF MINNESOTA

Institute of Technology and Medical School
Department of Biomedical Engineering

Programs of Study
The Biomedical Engineering Graduate Program is administered by the Department of Biomedical Engineering (BMEn). Study in biomedical engineering leads to the M.S. or Ph.D. degree and provides for students a broad familiarity with the interactions among the engineering, biological, and medical sciences plus in-depth training in at least one of these disciplines. Thesis research might typically be in one of the following areas: biocompatibility, biomaterials and biointerfacial science, biomedical imaging, biomicroelectromechanical devices, blood fluid mechanics, cardiovascular engineering, human factors engineering, instrumentation, medical informatics, membranes and mass transfer, microbial population dynamics, neurological devices, prosthetic devices, simulation of physiological systems, and tissue engineering. Further information on current research areas is available from the Director of Graduate Studies.

Minors are offered, and students may design a program tailored to their special interests with the aid of their faculty committees.

Research Facilities
Investigative facilities for the biological, medical, physical, and engineering sciences are located for the most part in a cluster on the Minneapolis campus. Some participating laboratories are located on the St. Paul campus.

Financial Aid
The Department of Biomedical Engineering provides $16,000 fellowships plus tuition for qualified first-year graduate students in biomedical engineering. Outstanding students may receive Graduate School fellowships. There are also research and teaching assistantships available that provide a stipend and a tuition scholarship. Biomedical engineering faculty members conduct research programs supported by federal and state agencies and industry, and these programs employ students for research that may also be used for the thesis.

Cost of Study
In 1998–99, resident tuition was $1710 per quarter (7–14 credits), and nonresident tuition was $3358 per quarter. Student service fees were $266.55 per quarter. Tuition and fees for 1999–2000 may change pending action by the Board of Regents. Graduate teaching or research assistants holding quarter-time appointments pay half the tuition rate, and those with half-time appointments pay no tuition. Detailed information on tuition and service fees can be obtained from the Office of the Registrar. The University of Minnesota is converting to the semester system beginning in fall 1999. Tuition and fees will be prorated for the semester conversion.

Living and Housing Costs
Housing is available near campus at reasonable cost. There is a broad mix of both University campus housing and rental apartments. University and Metropolitan Transit systems link the campus with all parts of the Twin Cities. The cost of living is, on the average, less than that in major metropolitan areas on the East and West Coasts.

Student Group
There are approximately 9,000 graduate students on the Twin Cities campus. The student body consists of students from all sections of the country and from nearly every continent. There is a very active Biomedical Engineering Student Society chapter on campus.

Student Outcomes
Recent biomedical engineering Ph.D. and M.S. graduates have gone on to take positions as postdoctoral fellows, research engineers, staff scientists, chemists, doctors, knowledge engineers, and faculty members at such institutions as the University of Minnesota; United HealthCare Corporation, Minneapolis; Children's Hospital Medical Center, Cincinnati; Medtronic, Minneapolis; Johns Hopkins, Baltimore; St. Jude Medical Center, Minneapolis; and the Brain Research Institute, University of Zurich.

Location
The Twin Cities area offers outstanding opportunities for cultural and recreational activities. Both the Minnesota Orchestra and the St. Paul Chamber Orchestra are nationally known, and there are several fine chamber music groups as well as many theater groups (including the Guthrie Theatre), four major art museums, and an excellent public radio network. There are major-league teams in baseball, football, and basketball and opportunities for swimming, fishing, canoeing, and skiing. The Twin Cities area has a vigorous and rapidly growing biomedical engineering industry, which is one of the largest concentrations of biomedical engineering activity in the country.

The University
The University of Minnesota, founded in 1851 by the Minnesota Territorial Legislature, now functions both as a state land-grant institution and as the state's major urban university. The proximity of the Medical School and the Institute of Technology and the collaboration of many departments provide an unusually stimulating environment for graduate study. A strong seminar program brings many well-known scientists and engineers to the campus and serves to unite the academic and industrial communities. The University's Department of Biomedical Engineering is an interdisciplinary focal point for the support of education, research, and innovation in biomedical engineering.

Applying
Students with baccalaureate degrees in engineering, the sciences, or mathematics are encouraged to apply. An applicant with a baccalaureate degree in physical or biological science may be accepted but will be required to complete course work through the junior year of an undergraduate engineering curriculum before being admitted as a candidate for the M.S. or Ph.D. degree. In most cases, this course work is not considered part of the graduate degree program.

Correspondence and Information
Professor Stanley M. Finkelstein
Director of Graduate Studies
Biomedical Engineering Graduate Program
University of Minnesota
420 Delaware Street, SE, Box 368 Mayo
Minneapolis, Minnesota 55455

Telephone: 612-624-9603
E-mail: bmengp@tc.umn.edu
World Wide Web: http://www.bmei.umn.edu

University of Minnesota

THE FACULTY AND THEIR RESEARCH

The departmental affiliations of the faculty members associated with the Program in Biomedical Engineering are shown in parentheses at the end of each listing.

Dennis L. Polla, Earl E. Bakken Professor of Biomedical Engineering and Head. BioMEMS and medical instrumentation. (Biomedical Engineering)

Stanley M. Finkelstein, Director of Graduate Studies. Instrumentation, medical computing, patient home monitoring. (Laboratory Medicine and Pathology)

Jerome H. Abrams. Modeling nonlinear physiologic systems. (Surgery)

Robert J. Bache. Cardiology, biorheology, NMR, vascular biology. (Medicine)

Joan E. Bechtold. Biomechanics, mechanical testing, fracture fixation, joint prostheses. (Orthopaedic Biomechanics Laboratory, Hennepin County Medical Center)

David G. Benditt. Cardiac electrophysiology. (Medicine)

John C. Bischof. Measurement and modeling of heat and mass transfer in biological tissues. (Mechanical Engineering)

Perry L. Blackshear Jr. Biorheology, soft-tissue mechanics, artificial organs. (Mechanical Engineering)

Henry Buchwald. Artificial organs, biomaterials. (Surgery)

Dennis D. Caywood. Experimental cardiopulmonary and vascular surgery. (Small Animal Clinical Sciences)

Frank B. Cerra. Biomedical applications in human intensive care practice. (Surgery)

Jay N. Cohn. Cardiology, biorheology. (Medicine)

Max Donath. Orthopedic prostheses, robotics, ergonomics. (Mechanical Engineering)

William K. Durfee. Muscle biomechanics, electrical stimulation, medical product design, virtual reality. (Mechanical Engineering)

Arthur G. Erdman. Medical device and product design, kinematics and dynamic applications, MEMS devices in medicine. (Mechanical Engineering)

Martha Flanders. Musculoskeletal modeling, neuromuscular control of arm movement. (Neuroscience)

John E. Foker. Prostheses, bioprosthetic cardiac valves. (Surgery)

Leo T. Furcht. Cell adhesion and migration, synthetic peptides, biomaterials, biocompatibility. (Laboratory Medicine and Pathology)

James R. Gage. Clinical applications of gait analysis, biomechanics. (Orthopaedic Surgery)

Michael G. Garwood. In vivo magnetic resonance spectroscopy and imaging. (Radiology)

William B. Gleason. Biomaterials, computer modeling, drug delivery. (Laboratory Medicine and Pathology)

Bruce E. Hammer. NMR, imaging and spectroscopy, bioreactors. (Radiology)

Linda K. Hansen. Regulation of cell growth and function by adhesion to extracellular matrix and biomaterials. (Laboratory Medicine and Pathology)

Robert P. Hebbel. Endothelial biology and tissue engineering, iron and oxidant damage to membrane molecules. (Medicine)

James E. Holte. Integration of instrumentation and modeling techniques. (Electrical and Computer Engineering)

Wei-Shou Hu. Tissue engineering, biotechnology. (Chemical Engineering and Materials Science)

Xiaoping Hu. Medical imaging, image processing and reconstruction. (Radiology)

Allison Hubel. Culture and cryopreservation of engineered tissues and cellular therapeutics. (Laboratory Medicine and Pathology)

Paul A. Iaizzo. Skeletal and cardiac muscle physiology, pathophysiology of inherited muscle disorders, human thermoregulation and biomedical correlates of wound prevention. (Anesthesiology and Physiology)

Mostafa Kaveh. Statistical signal processing, image processing, computerized tomography. (Electrical and Computer Engineering)

Seong-Gi Kim. Neural MRI and MRS, blood flow measurement. (Radiology)

Tarald O. Kvalseth. Human factors engineering. (Mechanical Engineering)

Paul C. Letourneau. Mechanisms of cell motility, neuronal guidance and nervous system development and regeneration. (Neuroscience)

David G. Levitt. Biophysics, determination of protein structure using X-ray crystallography. (Physiology)

Jack L. Lewis. Musculoskeletal biomechanics, orthopedic engineering. (Orthopaedic Surgery and Mechanical Engineering)

Rex Lovrien. Thermodynamics, cells, macromolecules, proteins. (Biochemistry)

David Masters. Biodegradable drug delivery systems for local and systemic applications, biomaterials and in vitro and in vivo analyses. (Pharmaceutics and Anesthesiology)

James B. McCarthy. Characterization of cell adhesion molecules, cell motility and adhesion receptor function, control of inflammation. (Laboratory Medicine and Pathology)

Ronald C. McGlennen. Molecular diagnostics procedures for genetic testing, microfabricated bioanalytic microchips. (Laboratory Medicine and Pathology)

Wilmer G. Miller. Biorheology, polymers. (Chemistry)

Daniel L. Mooradian. Cardiovascular drug delivery, tissue engineering. (Biomedical Engineering)

David A. Nelson. Audition, clinical psychoacoustics, electrically stimulated hearing. (Otolaryngology and Communication Disorders)

Robert P. Patterson. Medical instrumentation, bioelectric impedance measurements, bioelectric potentials, therapeutic electrical stimulation. (Physical Medicine and Rehabilitation)

Richard E. Poppele. Neurobiology, sensorimotor integration and control of posture and movement. (Neuroscience)

Gundu H. R. Rao. Platelet biochemistry, physiology, and pharmacology. (Laboratory Medicine and Pathology)

Michael Schwartz. Biomechanics, gait analysis. (Orthopaedic Surgery)

Ronald Siegel. Drug delivery, modeling of transport, pharmacodynamics. (Pharmaceutics)

Carl S. Smith. Neural network modeling, chaos/fractals in urologic physiology. (Surgery)

Clark M. Smith II. Blood rheology; white-cell, red-cell, and platelet hematology. (Pediatrics)

John F. Soechting. Biomechanics, movement control. (Neuroscience)

Ephraim M. Sparrow. Heat transfer and fluid mechanics. (Mechanical Engineering)

Ahmed H. Tewfik. Signal and image processing, high-speed MRI, ultrasound imaging, image databases. (Electrical and Computer Engineering)

Matthew V. Tirrell. Biomolecular materials, bioadhesion, biosurface science. (Chemical Engineering and Materials Science)

Robert T. Tranquillo. Cell motility, wound healing, nerve regeneration, bioartificial tissues. (Chemical Engineering and Materials Science)

Neal F. Viemeister. Auditory psychophysics. (Psychology)

Jay Zhang. Cardiology, NMR spectroscopic study of myocardial bioenergetics, cardiac applications of MRS and MRI, heart energetics. (Medicine)

Cheryl L. Zimmerman. Pharmacokinetics, in vivo drug transport. (Pharmaceutics)

UNIVERSITY OF MINNESOTA

Graduate Program in Microbial Engineering

Program of Study

The Master of Science degree or Ph.D. minor in microbial engineering at the University of Minnesota is offered through an interdisciplinary program integrating basic microbiology, molecular biology, chemical engineering, and related sciences. Graduates of the program are trained as professionals in the industrial application of microorganisms, cultured cells, and immunologic agents. They know both modern basic microbiology and biological engineering and can either proceed to a Ph.D. degree program in a related discipline or work directly with research and development staff in biotechnology industries. The microbial engineering program is coordinated by the Biological Process Technology Institute (BPTI). BPTI also administers an NIGMS Training Grant in Biotechnology used to support predoctoral students in related disciplines. Faculty members from eleven departments and seven centers or institutes of the University provide creative and stimulating research direction.

The two-year program comprises course work in a specialized program of microbiology, molecular biology, immunobiology, and chemical engineering. In addition, students present one seminar and instruct one laboratory course in advanced microbiology, molecular biology, immunobiology, or chemical engineering. Students may choose supporting course work from specified fields including biochemistry, food science, pharmacognosy, genetics, and cell biology, and they must demonstrate proficiency in computer programming and one computer language before completing the program. The presentation of an original laboratory research thesis to the graduate faculty is required at the end of the second and final year.

Research Facilities

The University of Minnesota offers up-to-date research facilities and libraries. Departments participating in the program include the Departments of Agronomy and Plant Genetics; Biochemistry and Biophysics; Chemical Engineering and Materials Science; Chemistry; Food Science and Nutrition; Laboratory Medicine and Pathology; Medicinal Chemistry; Microbiology, Immunology and Molecular Pathobiology; Molecular, Cellular, Developmental Biology and Genetics; Oral Sciences; and Soil Sciences. Research institutes include the Biological Process Technology Institute (BPTI), the Center for Biodegradation Research and Informatics (CBRI), the Center for Interfacial Engineering (CIE), the Center for Microbial Physiology and Metabolic Engineering (CMPME), the Industry/University Cooperative Research Center for Biocatalytic Processing, the Institute of Human Genetics, and the Plant Molecular Genetics Institute.

Financial Aid

Students compete for a limited number of research assistantships. The basic stipend is $1104 per month, with tuition and health insurance paid. Applicants are notified of acceptance and financial commitments by June 1 of each year. For detailed information, students should contact the director of graduate studies.

Cost of Study

The projected 1999–2000 semesterly (7–12 credits) full-time tuition for state residents is $2490. Nonresidents not holding fellowships or assistantships pay $3260. The semesterly student service fee amounts to $290 for residents and nonresidents alike.

Living and Housing Costs

Health insurance is provided to single students holding assistantships. Family coverage is provided with a copayment. University residence halls are within easy walking distance. The 1999–2000 costs average $3237 per semester for room (double occupancy) and board for graduate students. A variety of other housing facilities off campus are also available. For details, students should contact the housing office (Comstock Hall East, 210 Delaware Street SE, Minneapolis, Minnesota 55455; 612-624-2994).

Student Group

Students come from all over the United States and other countries. Graduates have pursued diverse and attractive careers in industry or further study for a Ph.D. in allied areas. Approximately 8 students participate in the program.

Location

The Minneapolis and St. Paul area is the cultural and industrial center of the upper Midwest, with a population of about 2 million. The Twin Cities are rich in what makes cities exciting but have few of the drawbacks that can be associated with urban life. Beautiful lakes, rivers, and city and state parks combined with the cold winter, cool spring, warm summer, and colorful autumn provide excellent opportunities for a variety of outdoor activities, including biking, camping, swimming, fishing, and skiing.

The University and The Program

The University of Minnesota was established in 1851 and has grown to include eighteen major colleges, divisions, and institutes. A major research and high-technology center, it regularly ranks among the top U.S. universities in receipt of federal grants for research and development. The microbial engineering program began in 1984 to meet the demand for students with expertise in biotechnology.

Applying

Applicants with a bachelor's degree in biological sciences, biochemistry, chemistry, or chemical engineering are encouraged to apply. Recommended academic preparation includes one year each of calculus, organic chemistry, physics, microbiology, and basic chemical engineering and a background in basic biology, physical chemistry, biochemistry, and genetics. If necessary, deficiencies can be made up during the first year of graduate work.

Application forms may be obtained from the Biological Process Technology Institute. Students should submit their applications as early as possible since most selections are made in February and March. To be evaluated, applications must be complete, including scores on the GRE General Test, TOEFL scores for international applicants, an autobiographical and goals statement, three letters of recommendation from professors or employers able to judge the applicant's capability for advanced study, and transcripts from institutions previously attended. Personal interviews may be scheduled once the application has been reviewed.

Correspondence and Information

Michael J. Sadowsky, Ph.D., Director of Graduate Studies
Program in Microbial Engineering
Biological Process Technology Institute
University of Minnesota
1479 Gortner Avenue, Suite 240
St. Paul, Minnesota 55108-6106
Telephone: 612-625-0212
Fax: 612-625-1700
E-mail: mic_eng@cbs.umn.edu
World Wide Web: http://cbs.umn.edu/bpti/bpti.html

University of Minnesota

THE FACULTY AND THEIR RESEARCH

Robert J. Brooker, Professor; Ph.D., Yale, 1983. Biology of cell surfaces, molecular genetics, transport.

Peter W. Carr, Professor; Ph.D., Penn State, 1969. Analytical chemistry, bioanalytical chemistry, chromatography.

Pat Cleary, Professor; Ph.D., Rochester, 1972. Molecular genetics of pathogenic streptococci, vaccine development.

Gary M. Dunny, Professor; Ph.D., Michigan State, 1978. Microbial genetics, microbial development, intercellular chemical communication.

Lynda B. Ellis, Associate Professor; Ph.D., Brandeis, 1971. Health informatics, computational biology.

Anthony J. Faras, Professor; Ph.D., Colorado at Boulder, 1970. Tumor viruses, oncogenesis, gene transfer.

Michael C. Flickinger, Professor; Ph.D., Wisconsin–Madison, 1977. Fermentation, animal cell culture technology, kinetics of protein synthesis and regulation.

James A. Fuchs, Professor; Ph.D., Texas A&M, 1970. Prokaryotic molecular biology, genetic engineering, protein engineering, gene regulation.

Richard S. Hanson, Professor; Ph.D., Illinois at Urbana-Champaign, 1962. Biochemistry and genetics of bacteria that degrade single carbon compounds, biodegradation of xenobiotic chemicals.

Alan B. Hooper, Professor; Ph.D., Johns Hopkins, 1964. Mechanisms of nitrogen metabolism in bacteria.

Wei-Shou Hu, Professor; Ph.D., MIT, 1983. Biochemical engineering, bioreactors, animal cell culture, plant cell culture.

Theodore P. Labuza, Professor; Ph.D., MIT, 1965. Kinetics and utilization of plant cell culture systems for food, food flavors, and drugs.

R. Scott McIvor, Professor; Ph.D., Minnesota, 1982. Mammalian expression, molecular virology and gene therapy.

Larry L. McKay, Professor; Ph.D., Oregon State, 1969. Plasmid biology and gene transfer systems in lactococci.

Daniel O'Sullivan, Assistant Professor; Ph.D., University College, Cork (Ireland), 1990. Gene regulation, genetic fingerprinting, bacteriocins, bacteriophage.

Bernard E. Reilly, Associate Professor; Ph.D., Case Western Reserve, 1965. Virus morphology and coat protein assembly mechanisms.

Palmer Rogers, Professor; Ph.D., Johns Hopkins, 1957. Enzymes and regulation of solvent production of *Clostridium acetobutylicum*.

Michael J. Sadowsky, Professor and Director of Graduate Studies; Ph.D., Hawaii, 1983. Identification and regulation of bacterial genes, proteins, and metabolic pathways involved in the biodegradation of chlorinated herbicides.

Janet L. Schottel, Professor; Ph.D., Washington (St. Louis), 1977. Regulation of gene expression, mechanism of mRNA turnover, plant-pathogen interactions, genetics of phytotoxin production.

David H. Sherman, Associate Professor; Ph.D., Columbia, 1981. *Streptomyces* genetics, genetics of antibiotic biosynthesis.

W. Thomas Shier, Professor; Ph.D., Illinois, 1970. Mammalian cell culture, antibiotic resistance production, novel transfection reagents.

David A. Somers, Professor; Ph.D., Washington State, 1983. Plant molecular genetics, plant genetic engineering, development in crop improvement.

Peter J. Southern, Associate Professor; Ph.D., Edinburgh, 1978. Regulation of gene expression and molecular basis of virus-induced diseases.

Friedrich Srienc, Professor; Ph.D., Graz Technical (Austria), 1980. Biochemical engineering, cell cycle kinetics, biopolymers, flow cytometry.

Robert T. Tranquillo, Associate Professor; Ph.D., Pennsylvania, 1986. Cell and tissue engineering, wound healing.

Dan Urry, Professor; Ph.D., Utah, 1964. Basic and applied studies on elastic model proteins, with emphasis on energy conversion.

Lawrence P. Wackett, Professor; Ph.D., Texas at Austin, 1984. Biodegradation of hazardous wastes, biotransformation for specialty chemical production.

Carston R. Wagner, Associate Professor; Ph.D., Duke, 1987. Biocatalysis and protein engineering.

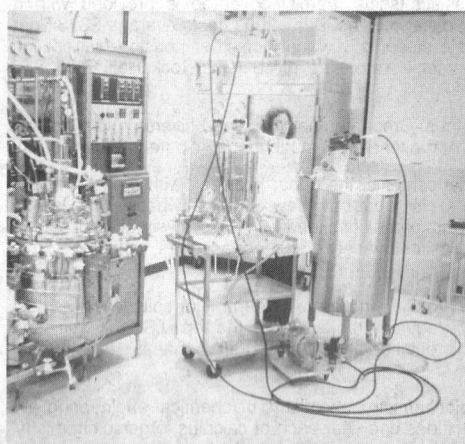

Animal-cell culture pilot plant in the Central Fermentation Research Facility of the Biological Process Technology Institute.

Staff members discuss run data at the Rosemount System 3 DCS control console at the BPTI Bioprocessing Pilot Facility.

THE UNIVERSITY OF NORTH CAROLINA AT CHAPEL HILL

Department of Biomedical Engineering

Programs of Study

The Department of Biomedical Engineering is an academic graduate department within the School of Medicine that stresses the application of science and engineering, mathematical analysis, and computer techniques to biomedical problems. Graduate education leading to the M.S. and Ph.D. degrees in biomedical engineering is offered, and a combined M.D./Ph.D. program is available to students who also gain admission to the School of Medicine. Students may enter the program with an undergraduate degree in engineering or in the physical or life sciences. A core curriculum consisting of courses in advanced mathematics and statistics, biomedical computer applications, biomaterials, biomedical instrumentation, and physiology provides a strong foundation for biomedical engineering students. In addition, students are encouraged to focus on one of the following six areas: medical imaging; medical informatics; biomaterials and biomechanics; instrumentation, microelectronics, and telemedicine; digital systems; and biosystems analysis (modeling and control theory). Students may also enrich their programs by taking courses at either Duke University or North Carolina State University, the other two universities included in the Research Triangle. The Microelectronics Center of North Carolina, situated within the Research Triangle Park, has established a complete two-way television and data communications network within the state of North Carolina, and a number of graduate-level teleclasses in topics related to microelectronics, computer science, and engineering are available to students at each institution. Furthermore, the network interconnects statewide campuses with the North Carolina Supercomputing Center, which houses a Cray processor. The principal areas of research within the biomedical engineering department are in medical imaging, neuroscience engineering, medical informatics, biomedical computer communications, bioelectronics and sensors, physiological system modeling, biomaterials, and real-time computer systems. The Department of Biomedical Engineering is a participant in the Duke–North Carolina Engineering Research Center for Emerging Cardiovascular Technologies. In addition, largely due to the talented pool of affiliated faculty members, students may conduct their research at North Carolina State University, Duke University, the Research Triangle Institute, or industrial sites within the Research Triangle area.

Research Facilities

The biomedical engineering department has extensive research facilities located within the University of North Carolina (UNC) School of Medicine. The department directly supports the six program areas listed above through a fully networked group of computers. The department maintains a network of Sun Ultra 2 and DEC Alpha UNIX workstations. These are connected to a high-speed campuswide backbone and then over high-speed links to the North Carolina Supercomputer Center, North Carolina State University, Duke University, and the Internet. A lab of networked PCs as well as PCs in various labs are available for use in courses and research. Further computing facilities are available at the UNC Computation Center, which maintains two large computational servers, and at the North Carolina Supercomputer Center, which maintains several Cray supercomputers and several Silicon Graphics workstations. Major software packages are available in biomedical engineering, including ACSL and SPICE for system modeling, matlab, Mathematica, Labview, AVS, ANSYS, and ABEL. Computing facilities within the Department of Biomedical Engineering are available to all students and faculty members.

Financial Aid

Fellowships are awarded on a competitive basis, and assistantships are usually available in a variety of research areas beginning in the second semester of study. Financial aid in the form of fellowships, research assistantships, or affiliated industrial research positions is generally available to all students enrolled in the program. Stipends are typically $14,000 to $15,000, with a half-time effort required for assistantships. Support for half-time industrial research positions is ordinarily greater than $15,000.

Cost of Study

Tuition and fees for the 1998–99 year were $2250 for North Carolina residents and $11,420 for nonresidents (two semesters). Books and supplies are estimated to be $500 (two semesters). Tuition remission to in-state levels may be granted to research assistants for a maximum of four semesters. Students may also qualify for full tuition waivers.

Living and Housing Costs

The cost of dormitory rooms (double occupancy) was $2110, and meals were about $2390 for the 1997–98 year (two semesters); married student housing is available. Apartment rents ranged from $200 to $800 per month in the town of Chapel Hill.

Student Group

The University of North Carolina has more than 22,000 students enrolled at the Chapel Hill campus, including approximately 5,500 graduate students. The biomedical engineering department had 70 graduate students, including 13 women and 17 international students, enrolled for the 1998–99 year.

Location

The University is located in the town of Chapel Hill. With the support of several nearby universities, the area provides excellent theater, music, sports, and special attractions. Research and development laboratories of many leading corporations are located in the Research Triangle Park, including the microelectronics facilities of General Electric, IBM, and Data General as well as the research laboratories of Becton-Dickinson, Glaxo Burroughs Wellcome, and the U.S. Environmental Protection Agency. Both the mountains and the coast are easily accessible, with the Great Smoky Mountains National Park, the Blue Ridge Mountains, and the Carolina beaches all less than half a day away.

The University

The first state university in the nation to admit students, the University of North Carolina was chartered in 1789 and formally opened in 1795. The University offered graduate course work as early as 1853, and several graduate degrees were awarded before the turn of the century. The Graduate School encompasses seven divisions: Fine Arts, Humanities, Biological Sciences, Social Sciences, Physical Sciences and Mathematics, and two professional divisions. The University has approximately 2,000 faculty members and is a member of the Association of American Universities.

Applying

Requirements for admission to the biomedical engineering and mathematics graduate program include GRE General Test scores (a minimum combined score of 1000 on the verbal and quantitative portions is required), a personal statement defining career goals, three letters of recommendation, and an undergraduate transcript for at least seven semesters of work toward a B.S. or B.A. degree in engineering, physics, chemistry, computer science, mathematics, or biology. The average combined verbal and quantitative GRE scores of enrolled students is about 1360 with a GPA of 3.2. The deadline for applying is February 1. Late applications are considered on a space-available basis. The application fee is $55.

Correspondence and Information

Director of Admissions
Department of Biomedical Engineering
School of Medicine
152 MacNider Hall, CB #7575
The University of North Carolina at Chapel Hill
Chapel Hill, North Carolina 27599-7575
Telephone: 919-966-1175
Fax: 919-966-2963
E-mail: chairman@bme.unc.edu

The University of North Carolina at Chapel Hill

THE FACULTY AND THEIR RESEARCH

Teaching Faculty

Steven M. Downs, Assistant Professor of Pediatrics and Biomedical Engineering; M.D., Stanford. Decision analysis and its applications in clinical guidelines development and computer-based medical decision support systems.

Oleg Favorov, Assistant Professor of Biomedical Engineering; Ph.D., North Carolina at Chapel Hill. Somatosensory cortical physiology and neural network modeling of cortical information processing.

Eric Frey, Assistant Professor of Biomedical Engineering; Ph.D., North Carolina at Chapel Hill. Nuclear medicine imaging, corrective reconstruction techniques in emission computer tomography, applications of high-speed computers to image reconstruction.

Daniel S. Fritsch, Research Assistant Professor of Radiation Oncology and Biomedical Engineering; Ph.D., North Carolina at Chapel Hill. Image processing, display for radiation therapy treatment planning.

Richard L. Goldberg, Instructor of Biomedical Engineering; Ph.D., Duke. Medical instrumentation, assistive technology for the disabled.

John E. Hammond, Professor of Pathology, Medicine, Biochemistry, and Biomedical Engineering; Ph.D., Florida. Medical systems and informative interfacing microprocessors to laboratory instruments, laboratory medicine.

Henry S. Hsiao, Associate Professor of Biomedical Engineering; Ph.D., Berkeley. Medical instrumentation, computer applications, cardiovascular dynamics.

Timothy A. Johnson, Associate Professor of Biomedical Engineering and Medicine; Ph.D., North Carolina at Chapel Hill. Cardiac electrophysiology, real-time computer applications, digital signal processing, control theory.

Robert P. Kusy, Professor of Dentistry and Biomedical Engineering; Ph.D., Drexel. Biomedical materials, applied mechanics, structure-property relationships.

David S. Lalush, Research Assistant Professor of Biomedical Engineering; Ph.D., North Carolina at Chapel Hill. Tomographic reconstruction algorithms, processing and display of nuclear medicine images.

John W. Loonsk, Assistant Professor of Biomedical Engineering and Assistant Dean and Director, School of Medicine; M.D., SUNY at Buffalo. Medical informatics, medical information delivery, medical information management, medical education, information technology, clinical decision support.

Carol L. Lucas, Professor of Biomedical Engineering, Surgery, and Applied and Material Science; Ph.D., North Carolina at Chapel Hill. Hemodynamics, pulmonary circulation, digital signal processing, mathematical modeling.

H. Troy Nagle Jr., Professor of Electrical and Computer Engineering, North Carolina State, and Professor of Biomedical Engineering, North Carolina at Chapel Hill; M.D., Miami (Florida); Ph.D., Auburn. Fault-tolerant microelectronic circuits for implantable devices, biosensors, and neural modeling.

Stephen M. Pizer, Professor of Computer Science, Biomedical Engineering, Radiology, and Radiation Oncology; Ph.D., Harvard. Medical image processing.

Stephen R. Quint, Research Associate Professor of Biomedical Engineering, Applied and Material Science, and Neurology; Ph.D., Virginia. Digital signal processing, systems analysis, real-time computer applications, cardiopulmonary physiology, cerebral blood flow.

Jeffrey Y. Thompson, Assistant Professor of Dentistry and Biomedical Engineering; Ph.D., Florida. Dental biomaterials, ceramics.

Mark A. Tommerdahl, Assistant Professor of Biomedical Engineering; Ph.D., North Carolina at Chapel Hill. Neurophysiology, mechanisms of neurocomputation in the somatosensory nervous system.

Benjamin M. W. Tsui, Professor of Biomedical Engineering and Radiology; Ph.D., Chicago. Digital medical imaging, theory and instrumentation in nuclear medicine and in nuclear magnetic resonance.

Paul Weinhold, Research Assistant Professor of Biomedical Engineering and Orthopaedics; Ph.D., North Carolina State. Biomechanics.

Barry L. Whitsel, Professor of Physiology and Biomedical Engineering; Ph.D., Illinois. Elucidation of the normal mammalian somatosensory nervous system and the neural mechanisms that underlie its computational capacities, determination of the effects on the computational capacities of centrally acting drugs and of nervous system damage due to disease or trauma.

Affiliated Faculty

Craig A. Branch, Associate Professor of Physics and Biomedical Engineering; Director of Laboratory Research, Medical Physics, Advanced Brain Imaging, Nathan Kline Institute for Psychiatric Research; Ph.D., Oakland. Dynamic magnetic resonance imaging (MRI) techniques for measurement of cerebral function and their application for the study of pathophysiology of cerebral trauma.

James John Brickley Jr., Assistant Dean, Engineering, North Carolina State; Ph.D., Virginia. Clinical neurophysiology, biomechanics, computer applications, digital process control.

Norman A. Coulter Jr., Professor Emeritus of Biomedical Engineering and Surgery; M.D., Harvard. Prevention of nuclear war.

John D. Charlton, Adjunct Research Associate Professor of Biomedical Engineering; Ph.D., Virginia. Intracranial pressure, intensive care monitoring, control theory.

Edward Chaney, Professor of Radiation Oncology and Biomedical Engineering, School of Medicine, North Carolina at Chapel Hill; Ph.D., Tennessee. Medical informatics, medical imaging.

Anthony J. Hickey, Professor of Pharmacy and Biomedical Engineering; Ph.D., Aston (Birmingham). Pulmonary drug delivery, aerosol formulations.

David W. Hislop, Research Associate Professor of Biomedical Engineering, Program Manager, U.S. Army Research Center; Ph.D., Illinois. Neural networks and pattern recognition.

William T. Krakow, Associate Professor of Biomedical Engineering, Director of IC Design and Test, MCNC, Research Triangle Park; Ph.D., North Carolina at Chapel Hill. IC and system design, development of VLSI system and CAD.

Lester Kwock, Professor of Radiology and Biomedical Engineering; Ph.D., California, Santa Barbara. NMR imaging and spectroscopy of the lung; NMR markers for tumor hypoxia.

Michael L. McCartney, Adjunct Associate Professor of Biomedical Engineering, Senior Scientist, Research Triangle Institute; Sc.D., Virginia. Noninvasive instrumentation, mathematical modeling, sensory prostheses.

Thomas K. Miller, Professor of Electrical and Computer Engineering, North Carolina State; Ph.D., North Carolina at Chapel Hill. Biomedical signal processing, computer architectures, VLSI systems.

Paul R. Moran, Professor of Radiology and Biomedical Engineering, Bowman Gray School of Medicine; Ph.D., Cornell. Medical imaging, MR flow imaging, MR phase contrast angiography.

Julian Roseman, Professor of Radiation Oncology and Biomedical Engineering, School of Medicine, North Carolina at Chapel Hill; M.D., Southwestern Medical School (Texas); Ph.D., Texas. Medical informatics, medical imaging.

Robert Rutledge, Professor of Surgery and Biomedical Engineering, UNC Hospitals; M.D., Florida. Medical informatics.

Barbara Wildemuth, Associate Professor of Information and Library Science and Biomedical Engineering, North Carolina at Chapel Hill; Ph.D., Drexel. Medical informatics.

Lloyd R. Yonce, Professor Emeritus of Physiology and Biomedical Engineering; Ph.D., Michigan. Cardiovascular physiology.

UNIVERSITY OF PENNSYLVANIA

School of Engineering and Applied Science
Department of Bioengineering

Programs of Study	The Department of Bioengineering offers two independent and streamlined graduate degree programs—the Doctor of Philosophy in bioengineering and the professional Master of Science in Engineering (M.S.E.). The nation's first Ph.D. in bioengineering was granted at the University of Pennsylvania, and the department consists of 12 primary faculty members and more than 60 secondary and associated faculty members. The bioengineering Ph.D. program is designed to train individuals for academic, government, or industrial research careers. Research interests include cellular biomechanics, bioactive biomaterials, cell and tissue engineering, neuroengineering, orthopedic bioengineering, neurorehabilitation, respiratory mechanics and transport, molecular and cellular aspects of bioengineering, and biomedical imaging. With the establishment of an interschool Institute for Medicine and Engineering in 1996, Penn provides a unique, integrated, campuswide, interdisciplinary learning and research environment at the cutting edge of biomedical engineering. The professional master's degree in bioengineering prepares students for leadership in the rapidly developing fields of industrial, entrepreneurial, and governmental biomedical engineering. The program is one of the first to recognize the growing need for biomedical engineering professionals whose expertise is in creating and managing new technology. Specially designed professional business courses set this program apart from other biomedical engineering master programs. Courses provide students with a full understanding of the process involved in taking a new technology from the laboratory to the biomedical marketplace and a complete awareness of the procedures and strategies needed to develop a company around that new technology. Joint-degree programs with the Schools of Medicine and Veterinary Medicine and with the Pennsylvania College of Podiatric Medicine are also available.
Research Facilities	State-of-the-art equipment for microscopy, imaging, computational simulations, materials development and testing is available, as are electronics labs and machine shops and cell and tissue culture facilities. For more information about specialized laboratories for biotransport, biointerfaces, tissue engineering and biomaterials, cellular and tissue biomechanics, computational neuroscience and neuroengineering, vision studies, electron optics, materials testing, and surface analysis students may consult the Web site (http://www.seas.upenn.edu/be/orgs.html).
Financial Aid	A limited number of fellowships and scholarships are available on a competitive basis to full-time Ph.D. students.
Cost of Study	Tuition and general fees for the academic year 1999–2000 are $25,216 plus a technology fee of $458 for full-time study or $2997 per course plus a $189 general fee and a $58 technology fee for part-time study.
Living and Housing Costs	On-campus housing is available for both single and married students. In 1998–99, residences for single students cost $815 per month for a one-bedroom apartment with living room, kitchen, and bath or $510 per month per person for a two-bedroom apartment with kitchen and bath. The cost for married student housing ranged from $830 per month for a one-bedroom apartment with living room, kitchen, and bath to $1050 per month for a two-bedroom apartment with living room, kitchen, and bath. There are also numerous apartments in the west Philadelphia and Center City areas available within walking distance of campus.
Student Group	More than 10,000 students are enrolled in Penn's twelve graduate and professional schools, and many are leaders in their fields. The Department of Bioengineering has more than 60 students in the graduate program.
Student Outcomes	More than 90 percent of recent graduates have continued in research or postdoctoral positions at universities and research institutes. A number of students have combined the Ph.D. in bioengineering with an M.D. degree. Other recent positions have been found in biology-related companies both nationally and internationally.
Location	The University is located in west Philadelphia, just a few blocks from the heart of the city. Philadelphia is a twentieth-century city with seventeenth-century origins. Renowned museums, concert halls, theaters, and sports arenas provide cultural and recreational outlets for students. Fairmount Park extends through large sections of Philadelphia, occupying both banks of the Schuylkill River. Not far away are the New Jersey shore to the east, the Pennsylvania Dutch country to the west, and the Pocono Mountains to the north. The city of Philadelphia is a patchwork of distinctive neighborhoods ranging from Colonial Society Hill to Chinatown.
The University and The School	The University of Pennsylvania was founded by Benjamin Franklin in 1740. It is an Ivy League institution of twelve schools and colleges occupying a 262-acre campus. The School of Engineering and Applied Science has a distinguished reputation for the high quality of its programs. Its alumni have achieved international distinction in research, management, industrial development, government service, and engineering education. Its faculty leads a research program that is at the forefront of modern technology and has made major contributions in a wide variety of fields. The School is the birthplace of the modern computer: it was at Penn that ENIAC, the world's first electronic, large-scale, general-purpose digital computer, was created.
Applying	Admissions are among the most selective in the country. An undergraduate degree in engineering or the physical sciences is preferred. Minimum course requirements are two years of calculus through differential equations and one year of physics with calculus and laboratory. Applicants holding an M.Sc. in an engineering discipline can apply for academic credit in the program. The application deadline is January 2 for Ph.D. applicants and July 1 for professional master's applicants. consideration with financial aid. Scores on the General Test of the Graduate Record Examinations are required. All international students whose native language is not English must arrange to take the Test of English as a Foreign Language (TOEFL) prior to making application; the minimum acceptable score is 600. To obtain application forms, students should write to the Office of Graduate Education and Research, School of Engineering and Applied Science, Room 112 Towne Building, University of Pennsylvania, Philadelphia, Pennsylvania 19104-6391 or visit the Web site at http://www.seas.upenn.edu/be/admin/process.html.
Correspondence and Information	Dr. David F. Meaney, Graduate Group Chair Department of Bioengineering University of Pennsylvania 120 Hayden Hall 3320 Smith Walk Philadelphia, Pennsylvania 19104-6392 Telephone: 215-898-8501 E-mail: begrad@eniac.seas.upenn.edu

University of Pennsylvania

THE FACULTY AND THEIR RESEARCH

This list is subject to change.

Martin D. Altschuler, Ph.D. 3-D treatment planning, optimization and decision analysis for radiation therapy, surface mensuration by laser beams.

Leon Axel, Ph.D., M.D. MRI: basic approaches to image formation and applications to blood flow and heart wall motion analysis.

Norman Badler, Ph.D. Computer graphics, human movement simulation and animation, AI, integrated graphics and language system.

Navin Bansal, Ph.D. Development and application of 23Na magnetic resonance imaging and spectroscopy.

Peter Bloch, Ph.D. Nuclear MR characterization of tumors, tumor localization using MRI X-ray fluorescence to determine heavy metals in situ.

Kwabena A. Boahen, Ph.D. Neuromorphic engineering, neurobiology, synthesizing integrated electronic circuits, asynchronous communications.

Daniel K. Bogen, Ph.D., M.D. Pediatric cognitive rehabilitation, biomedical instrumentation.

Dawn Bonnell, Ph.D. Mechanical/electrical interface properties, electronic structure variation occurring during brittle fracture.

Raymond C. Boston, Ph.D. Dynamic modeling in epidemiology, modeling biological systems.

Charles R. Bridges, M.D., Sc.D. Experimental models of chronic coronary ischemia, development of theoretical and mathematical models of cardiac function; magnetic resonance imaging of the heart; application of gene therapy to treatment of cardiac disease.

Carl T. Brighton, M.D., Ph.D. Bone and cartilage growth, maintenance, and repair; electrically-induced osteogenesis.

Gershon Buchsbaum, Ph.D. Visual signal processing and image coding, modeling of retinal and visual system architecture and function.

Donald G. Buerk, Ph.D. Microcirculatory studies of ocular blood flow and oxygen delivery, carotid body chemosensory transduction mechanisms.

Artur Cideciyan, Ph.D. Understanding functional deficits in hereditary retinal degenerations caused by known gene mutations; imaging of the retina.

Mortimer M. Civan, M.D. Electrophysiology of ion transport, molecular basis, and regulation of aqueous humor formation in the eye.

Peter F. Davies, Ph.D. Molecular mechanisms of atherosclerosis; mechanisms of interaction of hemodynamic forces with vascular endothelium and vascular cell-cell interactions; cell and molecular biology and membrane biophysics, biomechanics, and computational fluid dynamics.

Scott Diamond, Ph.D. Endothelial cell mechanobiology, drug and gene delivery, thrombosis and thrombolysis, biotransport phenomena.

Dennis E. Discher, Ph.D. Biomembrane mechanics—blood cells and muscle cells, structural molecule mechanochemistry—biochemical and biophysical methods; statistical mechanics of networks and polymers—computational emphasis, biomechanics.

Paul Ducheyne, Ph.D. Tissue engineering, biomaterials, bioactive ceramics, biocompatibility of metallic materials, prosthesis design.

David M. Eckmann, Ph.D., M.D. Pulmonary and cardiovascular biofluid mechanics and biotransport phenomena.

Nader Engheta, Ph.D. Biologically inspired polarization-difference imaging, fractional calculus in electrodynamics, wave interaction with complex media.

S. Walter Englander, Ph.D. Protein and nucleic acid structure and function; soluble and membrane proteins and polypeptides, ribosomes, polynucleotides, RNA and DNA, glycoproteins and oligosaccharides, sickle cell hemoglobin problem.

Aron B. Fisher, Ph.D. Lung surfactant metabolism, mechanisms of surfactant secretion, oxidative injury to the lung.

Kenneth Foster, Ph.D. Nonionizing radiation from audio through microwave frequency ranges, dielectric properties of tissue and biological molecules.

George L. Gerstein, Ph.D. Electrical activity of neuronal assemblies, computer simulations of neuronal networks.

Robert Goldman, M.D. Applying known electronic fields similar to postulated endogenous fields in wounds; obtaining metabolic responses.

Yale E. Goldman, M.D., Ph.D. Molecular mechanism of muscle contraction; structural, mechanical, and biochemical events in contractile proteins.

Keith Gooch, Ph.D. Tissue-engineered microvascular networks for gene therapy.

David J. Graves, Sc.D. DNA arrays for analysis by hybridization, cellular adhesion processes, production of recombinant proteins, bioseparations.

Joel H. Greenberg, Ph.D. Cerebral blood flow and metabolism, stroke and ischemia, PET.

Daniel Hammer, Ph.D. Cellular bioengineering, cell adhesion, virus-cell interactions, membrane dynamics and structure, biomaterials, biomimetics.

Gabor T. Herman, Ph.D. Computer technology for transaxial tomography, 3-D display of medical objects.

Paul A. Janmey, Ph.D. Defining phosphoinositide-binding sites on proteins; cytoskeletal functions; interactions of different cytoskeletal filaments.

Peter M. Joseph, Ph.D. X-ray, MRI, physiologic imaging.

Frederick Kaplan, M.D. Molecular genetics of disorders of ossification, topographic organization of repetitive elements in the human genome.

Jonathan Kaufman, Ph.D. Human thermoregulation and effect of external conditions on homeostasis.

John S. Leigh, Ph.D. Development and application of MR techniques, optical imaging in highly scattering media, protein folding and structure.

Robert M. Lewitt, Ph.D. Formulation and computer implementation of algorithms for image reconstruction from data derived from transmitted X-rays.

Brian Litt, M.D. Electromagnetism and the brain, including neuronal networks, source localization in the brain, and signal processing in epilepsy; nonlinear dynamics; frequency and domain methods applied to seizure prediction.

Mitchell Litt, D.Eng.Sc. Biorheology of body fluids, biotransport.

David Longnecker, M.D. Microcirculation, peripheral circulatory control, tissue oxygenation, endothelial-vascular smooth muscle cell-cell interactions.

Edward J. Macarak, Ph.D. Mechanisms of gene activation and phenotypic modulation in cells responding to mechanical forces.

Susan Margulies, Ph.D. Cell and tissue biomechanics; ventilator-induced injury; spinal cord injury; pediatric and adult brain injury.

Bryan E. Marshall, M.D., F.R.C.P. Mechanisms of pulmonary circulation regulation studied from organism to vascular smooth-muscle cell level.

Joseph McGowan, Ph.D. MR: applications to neuroradiology—MS, head trauma, and neonatal brain development; quantitative imaging.

Tracy McIntosh, Ph.D. Molecular and cellular sequelae of central nervous system injury, development of novel strategies to treat brain injury.

David Meaney, Ph.D. Biomechanics of CNS injury, experimental and computational modeling of brain and spinal cord injury mechanics.

Dimitri Metaxas, Ph.D. Physics-based modeling and simulation, computer graphics and animation, computer vision, scientific visualization.

David M. Nunamaker, M.B., Ch.B., Ph.D. Biomechanics, fracture healing, internal fixation, bone remodeling, bone fatigue.

Allan Pack, M.D. Neural control of respiration during sleep, spectral analysis, pattern generators, computational neuroscience.

Warren Pear, Ph.D. Tumor biology; signal transduction; leukemogenesis, including chronic myelogenous leukemia and myeloproliferative diseases.

Solomon R. Pollack, Ph.D. Bioelectrical properties of bone and connective tissue; electrical stimulation of bone growth and fracture healing.

John A. Quinn, Ph.D. Biochemical engineering, novel applications of membranes in bioprocessing, cell motility and chemotaxis.

Ramesh Raghupathi, Ph.D. Development of cardiovascular MR imaging and analysis methods—applications to the investigation of cardiac mechanics.

Ravinder Reddy, Ph.D. Development of sodium and proton MRI-based diagnostic tools for detecting early degenerative changes in cartilage.

Virginia M. Richards, Ph.D. Auditory perception, psychophysics, mathematical psychology.

Kathryn Saatman, Ph.D. Using in vivo and in vitro models of central nervous system injury to study changes in neuronal cytoskeleton, calcium-mediated neuropathology, axonal injury and roles for trophic factors.

James C. Saunders, Ph.D. Structure and function of peripheral auditory system, micromechanics and macromechanics of middle and inner ear.

Peter Scherer, Ph.D., M.D. Biofluid mechanics; interest in liquid, gas, particle, and heat flow in the human nose and respiratory system; capnography.

Ralph Schumacher, M.D. Cell biology, particle-induced inflammation, particle-protein-cell interactions, drug effects on particle-associated inflammation.

Chandra Sehgal, Ph.D. 3-D sonographic imaging, mathematical modeling and design of echo contrast agents.

Irving M. Shapiro, Ph.D. Mechanism of bone and cartilage mineralization.

Henry Shuman, Ph.D. Lipid transport, molecular mechanism of anesthesia.

Douglas Smith, M.D. Modeling focal and diffuse brain injury, posttraumatic cognitive dysfunction, traumatic axonal injury, magnetic resonance techniques for diagnosis of brain trauma, development of brain injury therapies.

Louis J. Soslowsky, Ph.D. Orthopedic biomechanics, soft-tissue mechanics, tissue engineering; biomechanics of the shoulder.

Bernard Steinberg, Ph.D. High-resolution imaging using microwaves and ultrasonics, adaptive signal processing, image-quality enhancement.

Jayaram K. Udupa, Ph.D. Visualization and analysis of multidimensional biomedical images, computer graphics for medical applications, kinematics of joints from image sequences, volume rendering.

Edward Vresilovic, M.D., Ph.D. Spinal cord injury; artificial discs and disc prostheses of the lumbar spine; biomaterials, including artificial grafts.

Felix W. Wehrli, Ph.D. Quantitative NMR imaging of connective tissue microstructure and function in vitro and in vivo.

Flaura K. Winston, Ph.D., M.D. Study of pediatric trauma using principles of biomechanics, pediatrics, and epidemiology; injury prevention strategies.

UNIVERSITY OF PENNSYLVANIA

School of Engineering and Applied Science
Master of Biotechnology Program

Program of Study

The Master of Biotechnology Program prepares students, both full- and part-time, for leadership in the biotechnology and pharmaceutical industries. Strongly interdisciplinary, this innovative professional master's program is a joint program of the School of Arts and Sciences (biology and chemistry departments) and the School of Engineering and Applied Science (chemical engineering, bioengineering, and computer and information science departments). Penn's world-class biomedical research centers, renowned science departments, and position at the hub of the largest pharmaceutical/biotechnology corridor in the United States place this program at the forefront of biotechnology education.

Three parallel curriculum tracks, basic biotechnology, engineering biotechnology, and computational biology/bioinformatics, give students flexibility to tailor their degree to their background, interests, and current career or career goals. Basic biotechnology emphasizes molecular biology, engineering biotechnology stresses bioprocess engineering central to pharmaceutical manufacturing, and computational biology/bioinformatics prepares students for analyzing ever-expanding genomic databases. These tracks, in combination with core courses, ensure that students get a uniquely broad exposure to the entire field of biotechnology. All students learn about topics that range from the latest developments in recombinant DNA technology to nontechnical issues that they must understand in order to bring biotechnology products to market, such as bioethics, government regulatory and drug-approval policies, and patent law.

The curriculum for all tracks consists of eleven courses, six core courses and five courses from the chosen track. The program starts in the fall semester. Full-time students can complete the program's eleven course units by the end of the following fall. Late afternoon classes accommodate students who want to earn their degree while working and are willing to take longer to complete their degree. Core courses include Biochemistry, Biotechnology I, Biotechnology II, Laboratory in Biotechnology and Genetic Engineering, Statistics, and Biotechnology Seminar I and II. Biotechnology I covers molecular biology, recombinant DNA technology, transgenic organisms, protein engineering, combinatorial chemistry, and molecular modeling; Biotechnology II examines production of biological molecules using cell culture technologies and introduces bioreactor design and control, molecular and cellular bioseparations, and tissue and cellular engineering. The Biotechnology Seminars review current scientific, regulatory, and ethical issues in biotechnology.

Research Facilities

Unlike many other major universities, the University of Pennsylvania has its medical, dental, nursing, and veterinary schools on the same campus as the School of Engineering and Applied Science and the School of Arts and Sciences. Another rich resource is Penn's own biomedical research center, the Institute for Medicine and Engineering (IME), which includes the Center for Bioinformatics. The IME, a collaboration between the School of Medicine and the School of Engineering and Applied Science, focuses on fundamental research in areas that are likely to develop practical application. The Center for Bioinformatics applies computational science to the analysis of the informational content of biological databases, such as the Human Genome Project.

Financial Aid

Although financial aid is not available, students can apply for a variety of both subsidized and nonsubsidized loans with the assistance of University Student Financial Services. Part-time students are also encouraged to check their company's plan for tuition reimbursement. Many companies cover all or part of the cost of the program.

Cost of Study

Tuition for the academic year 1999–2000 is $23,670, and there are a general fee of $1546 and a technology fee of $445 for full-time study. Per semester, the cost is $11,358 for tuition, $742 for the general fee, and $220 for the technology fee. Part-time, the tuition is $2996 per course unit, the general fee is $188, and the technology fee is $60.

Living and Housing Costs

On-campus housing is available for both single and married students. Residences for single students cost $625 and up per month ($510 and up per month for shared living); private apartments cost $815 and up per month. Housing costs for married couples range from $675 to $1065 per month. Many privately owned apartments exist in the immediate area.

Student Group

There are approximately 20,000 students at the University. More than 10,000 are enrolled in graduate and professional schools. Of these, 746 are in the graduate engineering program. The biotechnology program has a mix of full- and part-time students. Many part-time students already work for pharmaceutical and biotechnology companies. Students have undergraduate degrees in biology, biochemistry, chemistry, and many of the engineering disciplines, including chemical engineering, computer science, and bioengineering.

Location

The University is located in west Philadelphia, close to the heart of the city. Philadelphia's renowned museums, concert halls, theaters, sports arenas, and parks provide students with rich cultural and recreational outlets. Equidistant from New York City and Washington, D.C., Philadelphia is also close to the Jersey shore, Pennsylvania Dutch country, and the Poconos.

The School

In 1997, using Penn's well-established tradition of interdisciplinary teaching, the School of Engineering and Applied Science and the School of Arts and Sciences created this innovative professional master's program. The program combines the strengths of the School of Engineering and Applied Science's chemical engineering, bioengineering, and computer and information science departments and the School of Arts and Sciences' biology and chemistry departments.

Applying

Admission is competitive and on a rolling basis. For the spring semester, applications are due by November 1; for fall, applications are due by June 1. Candidates may apply by submitting an application to the address below or through Penn's online express application at http://sentry.isc.upenn.edu/ws/expressapp. Options for transfer credit, part-time study, and an industrial internship are available. Admission is based on the student's past record and on letters of recommendation. Students should have an undergraduate degree in science (such as biology, biochemistry, or chemistry) or engineering (such as chemical engineering, computer science, or bioengineering) and should have taken an undergraduate course in molecular biology and a year of college-level calculus. Scores on the Graduate Record Examinations (GRE) are required. All students whose native language is not English must take the Test of English as a Foreign Language (TOEFL); the minimum accepted score is 600.

Correspondence and Information

For information:
SEAS Professional Education Programs
Towne Building, Room 119
University of Pennsylvania
Philadelphia, Pennsylvania 19104-6391
Telephone: 215-898-0696
Fax: 215-573-9673
E-mail: biotech@pobox.upenn.edu
World Wide Web: http://www.upenn.edu/biotech

For admission:
Office of Graduate Admissions
School of Engineering and Applied Science
Towne Building, Room 111
University of Pennsylvania
Philadelphia, Pennsylvania 19104-6391
Telephone: 215-898-9246

University of Pennsylvania

THE FACULTY AND THEIR RESEARCH

SCHOOL OF ENGINEERING AND APPLIED SCIENCE

Bioengineering

Keith J. Gooch, Assistant Professor; Ph.D., Penn State, 1995. Tissue engineering and gene therapy.

David F. Meaney, Assistant Professor; Ph.D., Pennsylvania, 1991. Biomechanics of central nervous system injury, evaluation of automotive crash dynamics and occupant restraints, experimental and computational modeling of brain and spinal cord injury mechanics.

Chemical Engineering

Eric T. Boder, Assistant Professor; Ph.D., Illinois, 1998. Protein engineering, biophysics of protein recognition, biopharmaceuticals, engineering the immune response.

Scott L. Diamond, Associate Professor; Ph.D., Rice, 1990. Endothelial cell mechanobiology, thrombosis and thrombolytics, endothelial gene therapy.

David Graves, Associate Professor; Sc.D., MIT, 1967. Bioproduct recovery from living cells, chemical conversions in living cells, DNA identification with immobilized hybridization.

Daniel A. Hammer, Professor, Director of the Master of Biotechnology Program, Director of the Program's Engineering Track, and Member of the Institute for Medicine and Engineering; Ph.D., Pennsylvania, 1987. Biophysics of cell adhesion, virus-cell interactions, soft-bio materials.

Computer and Information Science

Susan B. Davidson, Professor, Director of the Computational Biology/Bioinformatics Track, and Co-director of the Center for Bioinformatics; Ph.D., Princeton, 1982. Distributed systems, database systems, real-time systems.

G. Christian Overton, Adjunct Professor, Computer and Information Sciences, and Associate Professor, Genetics; Ph.D., Johns Hopkins. Databases, knowledge bases, knowledge discovery in databases, the human genome project and control of gene expression.

Tandy Warnow, Associate Professor; Ph.D., Berkeley, 1991. Algorithms, graph theory, combinatorics, computational biology.

SCHOOL OF ARTS AND SCIENCES

Biology

Andrew N. Binns, Professor; Ph.D., Princeton, 1979. Cell biology of *Agrobacterium*-mediated plant transformation.

Anthony R. Cashmore, Professor and Director of the Basic Biotechnology Track; Ph.D., Auckland (New Zealand). Molecular biology, control of genes critical to photosynthesis.

Warren J. Ewens, Ph.D., Australian National, 1964. Mathematical and statistical methods in genetics.

David Roos, Associate Professor; Ph.D., Rockefeller, 1984. Molecular genetics and cell biology of protozoan parasites: *Toxoplasma* and *plasmodium* (malaria), eukaryotic evolution.

Chemistry

David Christianson, Professor; Ph.D., Harvard, 1987. Protein engineering of transition metal–binding sites, structural and mechanistic studies of hydrolytic enzymes, structural basis of terpenoid biosynthesis.

UNIVERSITY OF PITTSBURGH

School of Engineering
Bioengineering Department

Programs of Study	This program, which enlists faculty members from both the School of Engineering and the School of Medicine, is directed toward graduate engineering education and research, with particular emphasis on the Ph.D. The bioengineering faculty members thus apply engineering principles, technology, and methodology to a broad variety of medical and life science problems. Active, externally funded areas of research include computer processing of biologically derived signals; computer analysis of radiographic, ultrasonic, and nuclear magnetic resonance images; development of prostheses, artificial organs, and implantable sensors; development of medically related instrumentation; health operations research; biomaterials; biomechanics of soft tissues, tendons, ligaments, and articular cartilage; tissue engineering; metabolic engineering; pharmacokinetics; enzyme engineering; bioseparation; and protein engineering of biocatalysts. Degrees awarded are the Ph.D. in bioengineering, the M.S. in bioengineering, the Certificate in Rehabilitation Engineering, and the Certificate in Clinical Cardiovascular Engineering.
	The curriculum in bioengineering is designed to provide a flexible framework to accommodate students while emphasizing in-depth preparation in a student area of special interest. Doctoral-level examinations and entry examinations are individually designed for each student. Because the program expects students to meet graduate-level performance in both engineering and biological focus areas, students should have a high level of motivation for this graduate program.
	A large number of the engineering faculty members in the program also have appointments and research laboratories within the School of Medicine and the School of Rehabilitation Sciences, thus providing an unusual opportunity for students to develop research projects in the basic science and clinical departments.
Research Facilities	Facilities for research include the Center for Biotechnology and Bioengineering, the Center for Sports Medicine and Rehabilitation, the Joint Replacement Center, the Transplantation Institute, the Musculoskeletal Research Laboratories, the Section of Biomedical Informatics, the Pittsburgh Cancer Institute, the Pittsburgh NMR Center for Biomedical Research, the Pittsburgh Supercomputing Center, the Center for Balance Disorders, the Center for Clinical Neurophysiology, Rehabilitation Technology and Engineering Laboratories, and the Functional Imaging Laboratory.
Financial Aid	Graduate students may receive financial support in the form of Dean's Fellowships or research, training, or teaching assistantships; all are awarded on a competitive basis. There are also related internships available, as well as many opportunities for part-time employment on or around the campus.
Cost of Study	In 1998–99, Pennsylvania resident tuition was $4602 per term (spring and fall) for full-time students (9–15 credits) and $442 per credit for part-time students. Nonresident tuition was $9476 per term for full-time students and $902 per credit for part-time students. Students registered for the summer are billed on a per-credit basis regardless of the number of credits taken. Books, supplies, and incidental fees are additional.
Living and Housing Costs	In 1998–99, living expenses were approximately $850 per month for housing, meals, and miscellaneous expenses. The cost of books was approximately $650 per term. Pitt students find that public transportation in Pittsburgh is convenient and affordable. City buses run regularly on campus and provide ready access to area attractions, shopping, parks, and nightlife.
Student Group	Students in this program come from all over the world. Most of those admitted have an undergraduate QPA of at least 3.3; preferred applicants have a bachelor's degree in one of the engineering disciplines. Entrance into the program is very competitive; so far this year, less than 68 percent of program applicants have been accepted.
	Graduates of the bioengineering programs usually go on to have successful careers in academia; in medical equipment, pharmaceutical, or insurance companies; or in clinical engineering units within hospitals.
Location	Rich in history, culture, and ethnic diversity, Pittsburgh was recently hailed as "one of the world's most livable cities." Pittsburghers enjoy an exciting artistic culture that ranges from the Pittsburgh Ballet Theatre and the Pittsburgh Symphony Orchestra to local bands and artists at the leading edge of their genres. The city is home to several national sports teams, including the 1991 and 1992 Stanley Cup champion Pittsburgh Penguins. A myriad of recreational activities are available, from skiing in the Allegheny Mountains to visiting Kennywood amusement park, the "Roller Coaster Capital of the World."
The University and The School	In addition to the Bioengineering Department, a number of other departments from both the School of Engineering (Chemical and Petroleum Engineering, Civil and Environmental Engineering, Electrical Engineering, Industrial Engineering, and Mechanical Engineering) and the School of Medicine (Molecular Genetics and Biochemistry, Neurobiology, Surgery, Orthopedic Surgery, Radiology, and Physiology) are involved in the growing interest in research in bioengineering and biotechnology. Research faculty members from both schools collaborate on related research in bioengineering and biotechnology, including biomechanics, cardiac dynamics, gene therapy, ergonomics, artificial organs, enzyme engineering, biosensors, and rehabilitation engineering.
Applying	For students applying for admission to the fall term with financial aid, applications must be received by February 15; for admission to the spring term, applications must be received by November 15. All applicants must supply two letters of recommendation. The TOEFL is required of all applicants whose native language is not English. GRE scores are not required but are preferred; the GRE taken in a subject area is necessary to qualify for certain grants and/or fellowships. Applications and fees should be submitted to Office of Administration, 253 Benedum Engineering Hall, School of Engineering, University of Pittsburgh, Pittsburgh, Pennsylvania 15261.
Correspondence and Information	Bioengineering Department 749 Benedum Hall University of Pittsburgh Pittsburgh, Pennsylvania 15261 Telephone: 412-383-9713 E-mail: bioeng@engrng.pitt.edu

University of Pittsburgh

THE FACULTY AND THEIR RESEARCH

J. F. Antaki, Ph.D., Assistant Professor of Surgery. Cardiac biomechanics and dynamics, artificial heart research, blood rheology, computational fluid dynamics, image analysis.

M. M. Ataai, Ph.D., Associate Professor of Chemical and Petroleum Engineering. Mathematical modeling of biological systems, models for recombinant cells, bioreactor design, continuous culture, affinity techniques for protein purification.

M. J. Bloom, M.D./Ph.D., Assistant Professor of Anesthesia. Interneuronal interactions, particularly within the auditory cortex.

H. S. Borovetz, Ph.D., Professor of Surgery and of Civil Engineering. Cardiovascular bioengineering, hemodynamics and lipoproteins uptake, rheological abnormalities associated with heart assist devices, biomechanical properties of cardiac tissue.

J. R. Boston, Ph.D., Visiting Assistant Professor of Electrical Engineering. Clinical applications of sensory-evoked potentials and computer techniques, use of semi-automated continuous evoked potentials in the operating room.

D. M. Brienza, Ph.D., Assistant Professor, Schools of Health and Rehabilitation Sciences and of Electrical Engineering. Biomedical instrumentation for the measurement and assessment of biomedical properties of soft tissues.

C. E. Brubaker, Ph.D., Professor and Dean, School of Health and Rehabilitation Sciences and Professor of Industrial Engineering. Rehabilitative engineering, disability and assistive devices, technologies for disabilities of mobility and seating.

K. C. Chung, Ph.D., Associate Professor, Schools of Health and Rehabilitative Sciences and of Mechanical Engineering. Biomechanical modeling and MRI measurements of weight-bearing buttock's soft tissues in response to contoured supports related to pressure sore prevention.

R. A. Cooper, Ph.D., Associate Professor of Rehabilitation Science and Technology. Rehabilitation engineering, wheelchair design.

W. Federspiel, Ph.D., Associate Professor of Surgery and of Chemical Engineering. Cardiopulmonary bioengineering, biomedical transport phenomena, blood rheology, gas exchange, artificial lung research.

K. J. Fischer, Ph.D., Assistant Professor of Orthopaedic Surgery and of Mechanical Engineering. Computational analysis of bone adaptation to mechanical stimulus, bone tissue mechanics, hand and forearm biomechanics, total joint replacement design analysis.

J. M. R. Furman, M.D./Ph.D., Associate Professor and Director of the Vestibular Section of the Department of Otolaryngology and Associate Professor of Neurology and of Electrical Engineering. Digital signal processing systems analysis for human vestibular physiology and pathophysiology; vestibulo-ocular, vestibulospinal, and oculomotor functions.

L. G. Gilbertson, Ph.D., Assistant Professor of Orthopaedic Surgery and of Mechanical Engineering. Spine biomechanics, including role of mechanical factors in disc degeneration, mechanisms of injury, and effects of surgery.

D. Gur, Sc.D., Professor of Radiology and of Radiation Health. Digital radiography and mammography, xenon-enhanced imaging of cerebral blood flow, picture archiving and communications systems (PACS) in radiology.

G. D. Holder, Ph.D., Professor of Chemical Engineering and Dean of the School of Engineering. Molecular thermodynamics and phase equilibria, supercritical fluids.

T.-K. Hung, Ph.D., Professor of Civil Engineering and of Neurological Surgery. Theoretical and computational analysis of blood flow in normal and abnormal arteries, heart valves, intra-aortic balloon pumping, microcirculation, biomechanics of spinal cord injury in an animal model using NMR spectroscopy.

P. Loughlin, Ph.D., Assistant Professor of Electrical Engineering. Signal processing, time-frequency analysis, biomedical and industrial applications, control systems for biomedical applications.

I. J. Lowe, Ph.D., Professor of Physics and Astronomy. NMR imaging and spectroscopy; NMR instrumentation; pulse sequences for measuring T_1, T_2 diffusion, perfusion, and flow.

J. F. Patzer II, Ph.D., Research Assistant Professor of Chemical Engineering. Portable dialysis regeneration system, development of improved tissue culture reactors and an artificial liver.

M. S. Redfern, Ph.D., Assistant Professor of Otolaryngology and of Industrial Engineering. Human movement, with regard to the balance system and the biomechanics of posture.

A. J. Russell, Ph.D., Associate Professor of Chemical and Petroleum Engineering. Protein engineering in enzymology, surface charge on modification of enzyme catalysis, properties and applications of enzymes in organic solvents.

G. Salama, Ph.D., Associate Professor of Physiology. Imaging apparatus for high temporal and spatial resolution, fundamental problems in normal and abnormal cardiac rhythm.

J. S. Schultz, Ph.D., Professor of Chemical Engineering and Director of the Center for Biotechnology and Bioengineering. Biosensors using biomolecules and fiber optics, culturing and storage of hematopoietic cells, new cell-based therapies.

R. J. Sclabassi, M.D./Ph.D., Professor of Neurosurgery, of Electrical Engineering, of Mechanical Engineering, of Psychiatry, and of Behavioral Neuroscience. Neurophysiological modeling of the neurosurgical patient.

L. J. Shuman, Ph.D., Professor of Industrial Engineering and Associate Dean of School of Engineering. Operations research as applied to the problems of health-care delivery systems and emergency medical services.

P. Smolinski, Ph.D., Associate Professor of Mechanical Engineering. Computational modeling and finite element methods, new materials for bone implants, implant-induced bone remodeling, swelling of casted limbs.

J.-K. Suh, Ph.D., Assistant Professor of Orthopaedic Surgery and of Musculoskeletal Research Laboratories. Theoretical and experimental biomechanics of soft tissues, electromechanical properties of articular cartilage, repair of articular cartilage, biomechanical aspects of osteoarthritis, contact problems in musculoskeletal joints.

D. A. Vorp, Ph.D., Assistant Professor of Surgery and of Mechanical Engineering. Mechanical factors in the genesis and progression of vascular disease, effect of disease on the mechanical properties of vascular tissue, patency rate of vascular prostheses.

W. R. Wagner, Ph.D., Assistant Professor of Surgery and Adjunct Assistant Professor of Chemical Engineering. Biomaterial-associated thrombosis and thromboembolism, in vivo thrombotic mechanisms associated with ventricular assist devices.

H. Wolfe, Ph.D., Professor and Chairman of Industrial Engineering. Operations research applications to health delivery systems, health-care reimbursement, medical decision making, nurse staffing, hospital microcosting, and emergency medical services.

S. L.-Y. Woo, Ph.D., Professor of Orthopaedic Surgery and of Mechanical Engineering; Vice Chairman for Research in Orthopaedic Surgery; and Director of the Musculoskeletal Research Laboratories. Nonlinear material properties of biological tissues; measurements of soft tissue behavior by video methods; homeostatic responses and remodeling of bones, ligaments, and tendons; healing and repair of articular cartilage, meniscus tendons, and ligaments.

UNIVERSITY OF ROCHESTER

Program in Biomedical Engineering

Program of Study	The program in biomedical engineering is a free-standing interdisciplinary graduate program leading to a doctoral degree in biomedical engineering. Students are admitted directly to the program and are subject only to the degree requirements established by the program's faculty members. The program focuses primarily on two areas of biomedical engineering research: molecular, cell and tissue engineering and medical imaging and optics. Recommended programs of study are tailored to these areas as well as the specific needs and interests of the student. Students are encouraged to begin research as soon as possible, typically by the summer following the first year of study. Course work is generally completed by the end of the second year of study, after which the student focuses all of their time on original research leading to the dissertation. Formal requirements include written and oral examinations, a thesis, and an oral defense. The expected time to complete the degree is five to six years. The program includes more than 40 faculty members in the departments of chemical engineering, electrical engineering, mechanical engineering, and the Institute of Optics, as well as in the clinical and basic science departments in the Medical School. The wide range of research opportunities in which students may engage includes investigations into the physical and biochemical mechanisms regulating microvascular flow and interstitial transport, development of culture systems for production of cells and tissues, single cell micromechanical testing of cellular rheological properties and adhesive interactions, modeling of tissue mechanics in the developing heart, imaging, tissue characterization and treatment using ultrasound, medical image processing, and application of optical methods in clinical diagnosis and treatment.
Research Facilities	The School of Engineering and Applied Sciences is housed in more than 110,000 square feet of laboratory, classroom, and office space, approximately 41,500 square feet of which are research laboratories. In addition, there are extensive research facilities in the Medical Center, which is a 5-minute walk from the engineering departments. In addition to research labs in the Engineering School, the participating faculty members occupy a total of approximately 40,000 square feet of laboratory and office space in the Medical Center, which is available for student research activities.
Financial Aid	All students admitted to the Ph.D. program receive financial aid, including a full tuition scholarship and a stipend. Students should contact the program for the current stipend. This financial support does not require teaching or laboratory assistant duties, although some teaching experience is required of all degree recipients. No stipend support is available for M.S. degree candidates, but some partial tuition scholarships are available. Federal work-study program funds, government or personal loans, or part-time employment may sometimes be used to meet expenses. Employment opportunities for student spouses are available in the community and at the University.
Cost of Study	Full-time doctoral students generally receive full tuition assistance and health insurance in addition to their stipend. Students are responsible for purchasing books, the cost of which varies depending on the specific program of study.
Living and Housing Costs	The University of Rochester offers a variety of housing accommodations for married and single graduate students ranging in cost from $300 per month for a studio apartment to $634 per month for a three-bedroom town house. Other housing options are available in the nearby community, many within walking distance of the campus. Cost of food and sundries generally ranges from $150 to $250 per month for a single person.
Student Group	In its second year, the program includes 11 graduate students. Of these students, 3 are women and 7 are U.S. citizens. The program seeks to admit 6 students each year, and a steady enrollment of about 30 students is anticipated by the year 2001. These will join more than 30 students enrolled in other programs who are presently working in the laboratories of the participating faculty members. Current graduate enrollment in the entire School of Engineering and Applied Sciences includes about 45 percent international students and approximately 20 percent women.
Student Outcomes	Because this is a new program, no students have yet graduated. Doctoral students from similar programs at the University frequently go on to further studies as post-doctoral fellows in industry, government, or at other academic institutions, or find positions in industry in research and product development.
Location	Rochester is located in the Genesee River Valley on the shores of Lake Ontario. The city is in the heart of the Finger Lakes region and close to the Adirondack wilderness region. With no heavy industry or refineries, the city's commercial base is primarily technological. A small city in size, with a population for the metropolitan area of about 1 million, the community has big-city offerings in museums and other cultural resources and in professional sports. The area has numerous parks, fields, woods, streams, and lakes that provide opportunities for water sports, hiking, fishing, and skiing.
The University and The Program	The University of Rochester is a privately endowed, nonsectarian, coeducational institution founded in 1850. It consists of seven colleges and the Eastman Theater, the Memorial Art Gallery, and the Eastman House, a museum of photography. The total full-time enrollment of the University is approximately 7,500, of whom more than 2,300 are graduate students. The faculty consists of about 1,200 full-time members. The Program in Biomedical Engineering was established at the University of Rochester in the 1960s when the University received one of the first Biomedical Engineering Training Grants awarded by the National Institutes of Health. Formerly, students were admitted to and given degrees in the traditional engineering departments. In 1996, an independent, interdepartmental degree program in biomedical engineering was established.
Applying	The program is designed primarily for students who have received a four-year baccalaureate degree in engineering or applied physics, but students from other baccalaureate degree programs may apply. Undergraduate training should include courses in calculus through differential equations, inorganic chemistry, and physics, as well as in-depth training in engineering or the physical sciences appropriate for one of the main focus areas of our program. Applicants are required to take the Graduate Record Examinations (GRE), and foreign applicants whose native language is not English must achieve a minimum score of 600 on the Test of English as a Foreign Language (TOEFL), unless they are graduates of a U.S. undergraduate program. Completed applications should be received by February 1.
Correspondence and Information	Program in Biomedical Engineering School of Engineering and Applied Sciences Box 270168, Gavett Hall University of Rochester Rochester, New York 14627-0168 Telephone: 716-275-3891 Fax: 716-756-7771 E-mail: bme_gradinfo@seas.rochester.edu World Wide Web: http://www.seas.rochester.edu:8080/bme

University of Rochester

THE FACULTY AND THEIR RESEARCH

J. S. Abramowicz, Associate Professor; M.D., Tufts, 1975. Prenatal diagnosis of fetal anomalies; use of contrast media in enhancing sonographic imaging of the placenta.

E. L. Carstensen, Professor; Ph.D., Pennsylvania, 1955. Biophysics and biological effects of acoustic, electric, and magnetic fields.

X. Chen, Assistant Professor; Ph.D., Yale, 1991. Ultrasound imaging; acoustic radiation and scattering.

A. Clark Jr., Professor; Ph.D., MIT, 1963. Oxygen transport in the microcirculation.

D. Dalecki, Assistant Professor; Ph.D., Rochester, 1993. Biomedical ultrasound; acoustics; lithotripsy; biological effects of ultrasound.

P. M. Fauchet, Professor; Ph.D., Stanford, 1984. Optoelectronic and photonic materials and devices, with particular emphasis on developing applications of novel technology for improving health care and reducing its cost.

B. M. Fenton, Associate Professor; Ph.D., California, San Diego, 1980. Tumor vascular structure, oxygenation and radiosensitivity.

M. F. Flessner, Associate Professor; Ph.D., Michigan, 1981; M.D., Maryland, 1985. Transport of water, small solutes, and macromolecules across the peritoneum and through the underlying tissue.

T. H. Foster, Associate Professor; Ph.D., Rochester, 1990. Photodynamic therapy of cancer, optical spectroscopy and imaging of tissue.

M. D. S. Frame, Assistant Professor; Ph.D., Missouri–Columbia, 1990. Vascular communications and the control of peripheral blood flow.

K. J. Gingrich, Assistant Professor; M.D., Pittsburgh, 1984. Biophysics, structure-function, and pharmacology of ion channels.

S. M. Gracewski, Associate Professor; Ph.D., Berkeley, 1984. Methods of stone fragmentation during clinical lithotripsy; cavitation in response to ultrasound pulses.

A. R. Haake, Assistant Professor; Ph.D., South Carolina, 1985. Cell-signaling mechanisms controlling proliferation and apoptosis during skin development and aging.

D. C. Hocking, Assistant Professor; Ph.D., Albany, 1992. Regulation of cell behavior by the extracellular matrix.

J. Jorne, Professor; Ph.D., Berkeley, 1972. Electrochemistry; bioelectrochemistry; microelectronics processing and sensors; modeling.

K. Kutulakos, Assistant Professor; Ph.D., Wisconsin–Madison, 1994. Dermatological application of image processing, including computer vision and robotics, silhouette-based shape recovery, augmented reality, and image-based rendering.

A. L. Lerner, Assistant Professor; Ph.D., Michigan, 1996. Biomechanics of bone growth.

S. F. Levinson, Assistant Professor; Ph.D. Purdue, 1981; M.D. Indiana, 1983. Ultrasonic elasticity imaging (sonoelastography) of human skeletal muscle.

R. J. Maciunas, Professor; M.D., Illinois, 1980. Image-guided neurosurgery, stereotaxy, stereotactic radiosurgery, gene therapy, neurooncology, movement disorders.

C. R. Maurer Jr., Assistant Professor; Ph.D., Vanderbilt, 1996. Medical imaging and image processing, image registration; image-guided therapy; medical applications of augmented reality.

D. T. Moore, Professor; Ph.D., Rochester, 1974. Design of endoscopic instruments for the visible and the infrared; optical metrology.

T. W. Morris, Associate Professor; Ph.D., Michigan, 1972. Characterization of vascular lesions, blood flow, and tissue perfusion using X-ray and MRI pharmaceuticals.

J. G. Mottley, Associate Professor; Ph.D., Washington (St. Louis), 1985. Biomedical applications of ultrasound, including ultrasonic tissue characterization and contrast agents.

R. Ning, Assistant Professor; Ph.D., Utah, 1989. Three-dimensional medical imaging, including image processing and feature-detection techniques applied to clinical image acquisition modalities.

Harvey J. Palmer, Professor; Ph.D., Washington (Seattle), 1971. Interfacial phenomena, heat, and mass transfer applied to biological systems.

K. J. Parker, Professor; Ph.D., MIT, 1981. Medical imaging; medical ultrasound; elasticity imaging; Doppler imaging; image processing.

A. P. Pentland, Professor; M.D., Michigan, 1978. Phospholipases and cyclooxygenases in epidermal function and their role in carcinogenesis and cell differentiation; digital imaging and virtual reality for dermatology.

R. Perucchio, Associate Professor; Ph.D., Cornell, 1981; D.Engr., Pisa (Italy), 1977. Finite element biomechanical modeling of the embryonic heart; computational geometric modeling from biological data; large scale computation.

J. E. Puzas, Professor; Ph.D., Rochester, 1976. Molecular and cellular biology of bone.

R. J. Rivers, Associate Professor; M.D., Ph.D., Virginia, 1990. Integration of the regulators of blood flow in the microcirculation.

D. J. Rubens, Associate Professor; M.D., Rochester, 1979. Ultrasound sonelasticity; three-dimensional imaging; contrast agents.

I. H. Sarelius, Professor; Ph.D., Auckland (New Zealand), 1978. Vascular cell communication and microvascular function.

K. Q. Schwarz, Associate Professor; M.D., Rochester, 1983. Contrast echocardiography; basic science and clinical applications.

S. H. Seidman, Senior Instructor; Ph.D., Case Western, 1993. Neural control and mathematical modeling of reflex eye movements; physiological control systems.

S.-S. Sheu, Professor; Ph.D., Chicago, 1979. Molecular mechanisms of intracellular Ca^{2+} signaling.

P. G. Shrager, Professor; Ph.D., Berkeley, 1969. Ionic channels in neurons and glial cells.

S. M. S. Totterman, Associate Professor; M.D., Oulu (Finland), 1967; Ph.D., Bergman (Norway), 1983. Magnetic resonance imaging for clinical diagnosis, with particular application to evaluation and treatment of orthopedic injury and disease.

R. C. Waag, Professor; Ph.D., Cornell, 1965. Ultrasonic scattering and propagation problems in medical imaging and other applications.

D. S. Ward, Professor; M.D., Miami, 1976; Ph.D., UCLA, 1975. Systems modeling of respiratory processes.

R. E. Waugh, Professor; Ph.D., Duke, 1977. Mechanical and thermodynamic properties of biological membranes; cellular mechanics and function of cytoskeletal proteins.

D. R. Williams, Professor; Ph.D., California (San Diego), 1979. Physiological optics; visual instrumentation and retinal imaging; color and spatial vision.

J. H. D. Wu, Associate Professor; Ph.D., MIT, 1987. Bone marrow tissue engineering and microenvironment studies; lymphocyte culture; molecular and biochemical engineering.

J. Yang, Associate Professor; M.D., Brown, 1982; M.D., Washington (St. Louis), 1986. Molecular pharmacology of ion channel and viral vector–mediated neuronal receptor engineering.

Y. Yu, Associate Professor; Ph.D., London (England), 1986. Automatic real-time segmentation of transrectal ultrasound for intraoperative dosimetry of prostate brachytherapy.

J. Zhong, Associate Professor; Ph.D., Brown, 1988. Development and medical application of magnetic resonance imaging.

THE UNIVERSITY OF TEXAS AT AUSTIN

College of Engineering
Biomedical Engineering Graduate Program

Program of Study	Biomedical engineering is the development and application of engineering principles to medical problems. The Biomedical Engineering Graduate Program offers the M.S. and Ph.D. degrees. Many graduates have found the program to be excellent preparation for entry into medical school. The program is interdisciplinary, with a faculty that includes members of several departments as well as practicing physicians. Typical course requirements for the M.S. degree are 15 hours of biomedical engineering core courses, 9 hours of upper-level or graduate supporting work in a related area, and 6 hours of thesis work. No specific number of hours is required for the Ph.D., although 48 to 60 hours of multidisciplinary course work beyond the B.S. is suggested.
Research Facilities	The Biomedical Engineering Graduate Program occupies more than 15,000 square feet in the Engineering Science Building, with offices and many specially equipped laboratories in various locations on campus. Collaboration with the University of Texas (UT) Health Science Centers at Dallas and San Antonio, the UT MD Anderson Cancer Center in Houston, and the UT Medical Branch in Galveston also offers excellent opportunities for student research. The University computer facilities are among the finest in the nation, with Cray, IBM SP2, and RS600 servers and workstations available for graduate research. Smaller machines, such as the Alpha and high-end PCs, are also available within the program.
Financial Aid	Financial aid in the form of teaching assistantships, research assistantships, and fellowships is available for a limited number of students. Awards are made on a competitive basis. To be considered, applications must be submitted by February 1. Some students find part-time employment in local industry, but they must negotiate such arrangements individually.
Cost of Study	For 1998–99, graduate student fees, including tuition and other mandatory fees, were approximately $1300 (for 9 credit hours) per semester for residents of Texas and approximately $3200 for nonresidents. Complete information may be obtained from the *General Information* bulletin of the University of Texas. In some instances, fellowship recipients and students who are employed at least 20 hours per week as research or teaching assistants are considered state residents for tuition purposes.
Living and Housing Costs	The cost of room and board in University dormitories ranges from $4537 to $5883 for nine months. University apartments (for families) rent for $367 to $540 per month. A shuttle-bus system provides quick access to apartments in many areas of the city.
Student Group	During the academic year 1998–99, there were approximately 1,800 graduate students enrolled in the College of Engineering, with 85 in the Biomedical Engineering Graduate Program. The ratio of M.S. to Ph.D. candidates is about 1:2. Upon completion of the M.S. degree, students take positions in industry, medical research laboratories, and hospitals; enter medical, dental, or veterinary school; or continue study toward the Ph.D.
Location	Austin is the capital of Texas and has a population of approximately 500,000. Its main industries have traditionally been education and government. However, the city also has a number of major "clean" industries concentrating on high-level technology (e.g., Texas Instruments, Motorola, Tracor, IBM, MCC, Sematech, and Samsung). The climate is mild, encouraging year-round outdoor activities. There are many cultural and recreational opportunities available. Water sports are especially popular, and Austin is within an hour of seven highland lakes.
The University	The University of Texas at Austin is a state-supported institution consisting of fourteen schools and colleges. Although there is no medical school at Austin, the University of Texas system has medical schools at Dallas, San Antonio, Houston, and Galveston. The Austin campus has a student population of approximately 48,000; but, students need not fear being swallowed up in anonymity. The great size and diversity of the University provide a rich variety of programs and activities.
Applying	A degree in engineering or physics is required to apply to the Biomedical Engineering Graduate Program. Students without a degree in engineering or physics need to alleviate any deficiencies by enrolling in supplementary course work prior to applying for admission to the program. These courses may be taken at UT Austin (nondegree) or at any accredited institution of higher learning. Interested students should contact the program for assistance in selecting the necessary course work. Students accepted into the program typically have a combined verbal and quantitative score on the General Test of the GRE of 1300 for those applying to the M.S. program and 1400 for those applying to the Ph.D. program. A GPA of at least 3.3 on a 4.0 scale is expected for students applying to the M.S. program; a GPA of at least 3.5 is expected for students applying to the Ph.D. program. Students whose native language is not English must also submit a score of at least 550 on the Test of English as a Foreign Language (TOEFL). Also required are a statement of career objectives and three letters of recommendation from previous professors and/or employers who can evaluate the applicant's potential for graduate work and independent research. An online application is available (http://dpweb1.dp.utexas.edu/adappw/apply.wb).
Correspondence and Information	Graduate Information Biomedical Engineering Graduate Program Engineering Science Building 610 The University of Texas at Austin Austin, Texas 78712-1084 Telephone: 512-471-4679 Fax: 512-471-0616 E-mail: burks@mail.utexas.edu World Wide Web: http://www.ece.utexas.edu/bme/ http://www.utexas.edu/student/giac (general graduate studies information)

The University of Texas at Austin

THE FACULTY AND THEIR RESEARCH

Acoustics
Mark F. Hamilton, Ph.D., Penn State, 1983. Physical acoustics, ultrasonics. E-mail: hamilton@mail.utexas.edu
Elmer L. Hixson, Ph.D., Texas at Austin, 1960. Bioacoustics, sensors. E-mail: ehixson@mail.utexas.edu

Bioethics
Margaret N. Maxey, Ph.D., Union Theological Seminary (New York), 1971. Bioethical dilemmas generated by biotechnological innovations. E-mail: dr.maxey@mail.utexas.edu

Biological Control Systems
Robert H. Flake, Ph.D., Washington (St. Louis), 1962. Biological control systems, environment systems, pattern classification. E-mail: flake@ece.utexas.edu
Baxter F. Womack, Ph.D., Purdue, 1963. Control and modeling of biological systems. E-mail: womack@ece.utexas.edu
Hao Ying, Ph.D., Alabama, 1990. Fuzzy control, modeling and systems in biomedicine. E-mail: hying@utmb.edu

Biomechanics
C. Mauli Agrawal, Ph.D., Duke, 1989; PE. Biomaterials, tissue engineering, bone mechanics. E-mail: agrawal@uthscsa.edu
K. A. Athanasiou, Ph.D., Columbia, 1989. Tissue engineering and biomechanics. E-mail: athanasiou@uthscsa.edu
Lawrence D. Abraham, Ed.D., Columbia, 1975. Neural control of movement, human biomechanics. E-mail: l.abraham@mail.utexas.edu
Ronald E. Barr, Ph.D., Marquette, 1975. Biosignal analysis, computer graphics, biomechanics. E-mail: rbarr@mail.utexas.edu
William L. Buford, Ph.D., LSU, 1984. Computer graphics simulation of musculoskeletal kinematics. E-mail: william.buford@utmb.edu
Richard J. Lagow, Ph.D., Rice, 1967. Organometallic chemistry, polylithium organic compounds, organic and inorganic fluorine chemistry, synthetic bone materials, fluorocarbon and inorganic polymer chemistry, synthesis of new forms of carbon, synthesis of new sigma-bonded xenon-carbon compounds. E-mail: rjlagow@mail.utexas.edu
Clarence L. Nicodemus, Ph.D., California, Davis, 1993. Investigation of the kinematic of the in vivo human lumbar spine. E-mail: clnicode@utmb.edu
Marcus G. Pandy, Ph.D., Ohio State, 1987. Musculoskeletal biomechanics. E-mail: pandy@mail.utexas.edu
S. V. Sreenivasan, Ph.D., Ohio State, 1994. Biomechanics/rehabilitation machines, robotics. E-mail: sv.sreeni@mail.utexas.edu

Cardiovascular System
Lee E. Baker, Ph.D., Baylor College of Medicine, 1965. Cardiopulmonary physiology, noninvasive acquisition of physiological data. E-mail: leb@mail.utexas.edu
Thomas M. Runge, M.D., Texas Medical Branch, 1947. Cardiology, extracorporeal blood pumps, pulsatile hemodialysis blood substitutes. E-mail: runge@mail.utexas.edu

Molecular Bioengineering/Tissue Engineering
C. Mauli Agrawal, Ph.D., Duke, 1989; PE. Biomaterials, tissue engineering, bone mechanics. E-mail: agrawal@uthscsa.edu
K. A. Athanasiou, Ph.D., Columbia, 1989. Tissue engineering and biomechanics. E-mail: athanasiou@uthscsa.edu
George Georgiou, Ph.D., Cornell, 1987. Biotechnology. E-mail: gg@che.utexas.edu
David G. Gorenstein, Ph.D., Harvard, 1969. NMR spectroscopy of proteins and nucleic acids; biochemistry, biophysical chemistry, computational biochemistry, enzymology, organophosphorus chemistry. E-mail: david@nmr.utmb.edu
Christine Schmidt, Ph.D., Illinois, 1995. Cellular and tissue engineering. E-mail: schmidt@che.utexas.edu

Bioheat Transfer
Kenneth R. Diller, Sc.D., MIT, 1972. Low-temperature biology, tissue banking, burn injury, computer vision. E-mail: kdiller@mail.utexas.edu
Linda Hayes, Ph.D., Texas at Austin, 1979. Finite-element processing of heat transfer in human tissues. E-mail: lhayes@mail.utexas.edu
John A. Pearce, Ph.D., Purdue, 1980. Thermal damage in tissue, electrosurgery, bioelectric phenomena. E-mail: jpearce@mail.utexas.edu
Jonathan W. Valvano, Ph.D., MIT, 1981. Bioinstrumentation, measurement of perfusion, tissue thermal properties, real-time temperature measurements, thermal modeling in tissue, embedded applications of microcomputers. E-mail: valvano@uts.cc.utexas.edu

Image Processing
Alan C. Bovik, Ph.D., Illinois, 1984. Biomedical image processing, computer vision. E-mail: bovik@ece.utexas.edu
John A. Pearce, Ph.D., Purdue, 1980. Thermal damage in tissue, electrosurgery, bioelectric phenomena. E-mail: jpearce@mail.utexas.edu
Bugao Xu, Ph.D., Maryland, 1992. Applications of computervision technology to measure size and shape of the human body, image processing. E-mail: bxu@mail.utexas.edu

Instrumentation
Jonathan W. Valvano, Ph.D., MIT, 1981. Bioinstrumentation, measurement of perfusion, tissue thermal properties, real-time temperature measurements, thermal modeling in tissue, embedded applications of microcomputers. E-mail: valvano@uts.cc.utexas.edu

Laser-Tissue Interaction/Tissue Optics
Thomas E. Milner, Ph.D., Arizona, 1991. Optical-based therapeutics and diagnostic imaging, biomedical fiber sensors. E-mail: milner@ece.utexas.edu
Massoud Motamedi, Ph.D., Texas at Austin, 1988. Lasers in medicine. E-mail: mmotamedi@utmb.edu
Rebecca Richards-Kortum, Ph.D., MIT, 1990. Laser spectroscopy for the study of pathophysiology and diagnosis of disease. E-mail: kortum@mail.utexas.edu
A. J. Welch, Ph.D., Rice, 1964. Optical and thermal interactions of laser light with tissue, medical applications of lasers. E-mail: welch@mail.utexas.edu

Physiology and Electrophysiology
Lee E. Baker, Ph.D., Baylor College of Medicine, 1965. Cardiopulmonary physiology, noninvasive acquisition of physiological data. E-mail: leb@mail.utexas.edu
Akhil Bidani, M.D., Texas Medical Branch, 1981. Physiology and biophysics. E-mail: abidani@utmb.edu
Harvey M. Fishman, Ph.D., Berkeley, 1968. Biophysics. E-mail: hfishman@utmb.edu
Glen E. Journeay, M.D., Texas Medical Branch, 1960. Toxicology. E-mail: journeay@mail.utexas.edu
George C. Kramer, Ph.D., Texas Medical Branch, 1979. Physiology, burn and trauma research. E-mail: gkramer@utmb.edu
James F. Leary, Ph.D., Penn State, 1977. Molecular cytometry to isolate and characterize rare cells. E-mail: james.leary@utmb.edu
H. Grady Rylander III, M.D., Texas Health Science Center at San Antonio, 1974. Vision research, biomedical sensors, laser applications, neuroprosthesis design. E-mail: rylander@mail.utexas.edu
Louis C. Sheppard, Ph.D., London, 1976. Closed-loop medical delivery systems.

Robotics
S. V. Sreenivasan, Ph.D., Ohio State, 1994. Biomechanics/rehabilitation machines, robotics. E-mail: sv.sreeni@mail.utexas.edu
Delbert Tesar, Ph.D., Georgia Tech, 1964. Robotics of microsurgery, othotics. E-mail: tesar@mail.utexas.edu

Signal Processing
Ronald E. Barr, Ph.D., Marquette, 1975. Biosignal analysis, computer graphics, biomechanics. E-mail: rbarr@mail.utexas.edu
Benito Fernández-Rodriguez, Ph.D., MIT, 1988. Systems engineering, neural networks. E-mail: benito@mail.utexas.edu
R. Joe Thornhill, Ph.D., Texas at Austin, 1980. Digital signal processing. E-mail: thornhill@mail.utexas.edu

Spectroscopy
Thomas E. Milner, Ph.D., Arizona, 1991. Optical-based therapeutics and diagnostic imaging, biomedical fiber sensors. E-mail: milner@ece.utexas.edu
Massoud Motamedi, Ph.D., Texas at Austin, 1988. Lasers in medicine. E-mail: mmotamedi@utmb.edu
Rebecca Richards-Kortum, Ph.D., MIT, 1990. Laser spectroscopy for the study of pathophysiology and diagnosis of disease. E-mail: kortum@mail.utexas.edu
A. J. Welch, Ph.D., Rice, 1964. Optical and thermal interactions of laser light with tissue, medical applications of lasers. E-mail: welch@mail.utexas.edu

Vision
Wilson S. Geisler, Ph.D., Indiana, 1975. Psychophysics, physiology and modeling of vision systems, computer vision. E-mail: geisler@psy.utexas.edu
Alan C. Bovik, Ph.D., Illinois, 1984. Biomedical image processing, computer vision. E-mail: bovik@ece.utexas.edu
H. Grady Rylander III, M.D., Texas Health Science Center at San Antonio, 1974. Vision research, biomedical sensors, laser applications, neuroprosthesis design. E-mail: rylander@mail.utexas.edu

SOUTHWESTERN **UTA**

THE UNIVERSITY OF TEXAS SOUTHWESTERN MEDICAL CENTER AT DALLAS / THE UNIVERSITY OF TEXAS AT ARLINGTON

Joint Program in Biomedical Engineering

Program of Study

The graduate program in biomedical engineering is administered jointly by the University of Texas Southwestern Medical Center at Dallas (UT Southwestern) and the University of Texas at Arlington (UTA). Degrees are awarded jointly by both institutions. The program prepares students for careers in industry, medicine, and academe. The 37-credit-hour M.S. program normally takes two years to complete, including a project or a master's thesis. The Ph.D. program normally takes four years to complete beyond the M.S. degree, including dissertation research, writing, and defense. Course work and research are organized in six tracks: bioinstrumentation, biomaterials and tissue engineering, biomechanics, medical imaging, molecular engineering, and orthopedic engineering. All tracks emphasize interaction with the life science and clinical faculties at on-site facilities. Industrial and clinical internships are available to M.S. candidates.

Research Facilities

Well-equipped laboratories exist at both campuses for all six research tracks. State-of-the-art facilities exist for positron, NMR, ultrasound, and X-ray imaging and near-infrared spectroscopy. Instrumentation facilities include a modern controls lab, a bioinstrumentation development lab, and a lab that develops high-throughput instrumentation for the human genome project and beyond. Facilities for wireless microsphere integrated circuit biosensor development are available through industry collaboration. A haptic workstation facility for simulation of minimally invasive surgery is available. Robotics-based surgical instrumentation is planned in conjunction with the UTA Automation and Robotics Research Institute. Well-equipped polymer fabrication labs are available. Major analytical equipment is available to support the polymer labs. Extensive computer facilities are available at both campuses to support all aspects of research. Facilities for in vitro and in vivo experimental studies are available. Specialized labs are available for research in patient monitoring; cardiac and muscle metabolism; structure and function of many cell systems; pulmonary, cardiac, orthopedic, and muscle mechanics; brain imaging and MRI spectroscopy; genomic engineering; and cell therapy, stent, and other tissue engineering topics.

Financial Aid

Financial aid is available for eligible students in the form of research assistantships, a work-study program, part-time jobs, and loans that provide $400–$1000 per month.

Cost of Study

In 1998–99, tuition is $72 per credit hour for Texas residents and $285 per credit hour for out-of-state students, not including fees. Research assistants pay Texas resident rates.

Living and Housing Costs

Dormitories are available at UTA for $210 per month. Private apartments are available near each institution and in adjacent communities. Housing costs for multiple occupancy begin at $450 per person per month for an unfurnished apartment.

Student Group

There are 60 full-time students in the graduate biomedical engineering program. More than 60 percent are supported on graduate assistantships. There is an active BME Society chapter.

Location

The Dallas–Fort Worth metroplex is an international center for technology-intensive industries. A wide variety of cultural and recreational opportunities are available.

The Universities

UT Southwestern is one of the nation's premier biomedical institutions. Its facilities include the Southwestern Graduate School of Biomedical Sciences, Southwestern Medical School, and Zale-Lipshy University Hospital. It is affiliated with Parkland Memorial Hospital, Children's Medical Center, St. Paul Medical Center, and the Veterans Administration Medical Center. Three research buildings on a 30-acre campus are newly completed.

The University of Texas at Arlington's current enrollment exceeds 19,000, including 2,800 students in the College of Engineering, which ranks in the top fifty engineering colleges in the nation. The engineering classrooms and laboratories are modern and well equipped.

Applying

A formal application, transcripts of all undergraduate and graduate work, three letters of recommendation, scores on the GRE General Test, and a letter stating personal objectives must be submitted. The sum of the GRE verbal and quantitative scores must be at least 1000 for master's degree applicants and at least 1150 for Ph.D. applicants. International students must achieve a score of 575 or better on the TOEFL. Although not required, a personal or telephone interview is recommended. Applications should be filed at either school before April 15 for fall admission or before September 1 for spring admission.

For catalogs and application forms from UT Southwestern, students should write to the program chairman; for UTA, students may write to the Graduate School (Box 19167). For department information, students should contact the program chairman at either of the addresses listed below.

Correspondence and Information

Biomedical Engineering Program
The University of Texas Southwestern Medical
  Center at Dallas
5323 Harry Hines Boulevard, G8.248
Dallas, Texas 75235-9130
Telephone: 214-648-2052
Fax: 214-648-2979
World Wide Web: http://www.swmed.edu/home_pages/
  bme/index.bme.html

Biomedical Engineering Program
The University of Texas at Arlington
Box 19138
Arlington, Texas 76019
Telephone: 817-272-2249
Fax: 817-272-2252
World Wide Web: http://www.uta.edu/biomed_eng/
  bme.htm

The University of Texas Southwestern Medical Center at Dallas/The University of Texas at Arlington

THE FACULTY AND THEIR RESEARCH AREAS

The University of Texas Southwestern Medical Center at Dallas

Nadir Alikacem, Assistant Professor; Ph.D., Sussex, 1991. Intravitreal drug release, MRI in diabetic retinopathy.

Peter Antich, Professor; Ph.D., Johns Hopkins, 1971. Tumor and bone imaging.

Loren Bertocci, Assistant Professor; Ph.D., Washington State, 1986. Energy metabolism studied by magnetic resonance and optical spectroscopy.

C. Gunnar Blomqvist, Professor; M.D., Karolinska (Stockholm), 1967. Space physiology and instrumentation.

Michael J. Bolesta, Assistant Professor; M.D., Missouri, 1981. Minimally invasive surgery of the cervical spine.

Michael D. Devous Sr., Professor; Ph.D., Texas A&M, 1976. PET, SPECT imaging in psychiatric and neurologic disorders.

Robert C. Eberhart, Professor; Ph.D., Berkeley, 1965. Biomaterials, circulatory assist devices, minimally invasive surgery.

Maureen A. Finnegan, Associate Professor; M.D., British Columbia, 1974. Orthopedics.

Kenneth P. Gall, Associate Professor; Ph.D., Boston University, 1989. Radiotherapy dosimetry, radiographic instrumentation, conformal 3-D radiotherapy.

Harold (Skip) Garner Jr., Professor; Ph.D., Wisconsin, 1982. Human genome project, computational biology instrumentation.

Cole A. Giller, Associate Professor; M.D./Ph.D., UCLA, 1984. Signal analysis of cerebral blood flow, neurologic function.

Herbert Hagler, Associate Professor; Ph.D., SMU, 1975. Analytical electron microscopy, distance learning.

Jureta W. Horton, Professor; Ph.D., Texas Health Science Center at Dallas, 1981. Burn and trauma surgery (shock).

Michael Jessen, Associate Professor; M.D., Manitoba, 1981. Circulatory assist devices, cardiac metabolism.

Robert L. Johnson Jr., Professor; M.D., Northwestern, 1951. Pulmonary physiology, critical-care medicine.

Padmakar Kulkarni, Professor; Ph.D., Rensselaer, 1973. Nuclear chemistry.

Vladislav S. Markin, Professor; D.Sc./Ph.D., Moscow, 1964. Biophysics (membranes).

Ralph P. Mason, Associate Professor; Ph.D./C.Chem., Cambridge, 1986. 19F NMR, PET-MR, MRI, bold MRI, υMRI.

Roderick McColl, Assistant Professor; Ph.D., Warwick, 1992. υMRI, DICOM.

Jere Mitchell, Professor; Ph.D., Texas Southwestern Medical Center at Dallas, 1954. Cardiovascular physiology.

George Ordway, Associate Professor; Ph.D., Kentucky, 1979. Regulation of responses and adaptations of skeletal muscle to contractile activity.

Ronald Peshock, Professor and Assistant Dean for Informatics; M.D., Texas Southwestern Medical Center at Dallas, 1976. MRI.

W. Matthew Petroll, Associate Professor; Ph.D., Virginia, 1989. Wound healing, confocal microscopy, cell mechanics.

Jeffrey T. Potts, Assistant Professor; Ph.D., North Texas Health Science at Fort Worth, 1993. Central neurotransmission in cardiovascular function.

Morton Prager, Professor; Ph.D., Purdue, 1951. Biomaterials, tissue engineering.

A. Dean Sherry, Professor; Ph.D., Kansas State, 1971. Lanthanide, 23 Na and 13C NMR to study intermediary metabolism.

Richard Srebro, Professor; M.D., Washington State, 1959. Ophthalmology, signal processing.

Gaylord Throckmorton, Professor; Ph.D., Chicago, 1974. Craniofacial biomechanics.

Jihong Wang, Assistant Professor; Ph.D., Colorado, 1994. Human perception, picture achieving, υMRI imaging.

Robert Sanders Williams, Professor; M.D., Duke, 1974. Molecular genetics of muscle, ischemic heart disease, isolation of cardiogenic cells.

The University of Texas at Arlington

Khosrow Behbehani, Professor; Ph.D., Toledo, 1979. Signal processing, control theory, instrumentation.

C. J. Charles Chuong, Professor; Ph.D., California, San Diego, 1981. Cardiopulmonary biomechanics, shock-wave lithotripsy, finite-element applications.

Diane J. Cook, Associate Professor; Ph.D., Illinois, 1990. Artificial intelligence, machine planning, parallel AI algorithms, machine learning.

Jerome Eisenfeld, Professor; Ph.D., Chicago, 1966. Biomathematics.

Ronald L. Elsenbaumer, Professor; Ph.D., Stanford, 1978. Biomaterials/polymers.

George V. Kondraske, Professor; Ph.D., Texas at Arlington and Texas Health Science Center at Dallas, 1982. Human performance theory, sensors.

Hanli Liu, Assistant Professor; Ph.D., Wake Forest, 1994. Biomedical optics for tissue characterization using near-infrared light.

Michael T. Manry, Professor; Ph.D., Texas at Austin, 1976. Signal/image processing, estimation theory, neural networks.

Kevin D. Nelson, Assistant Professor; Ph.D., Texas Southwestern Medical Center at Dallas, 1995. Biomaterials, tissue engineering.

Lynn Peterson, Associate Dean of Engineering for Academic Affairs; Ph.D., Texas Southwestern Medical Center at Dallas, 1978. Knowledge representation, artificial intelligence applications.

Richard B. Timmons, Professor; Ph.D., Catholic University, 1962. Biomaterials, surface coatings.

Industrial Adjunct Faculty

Jeanette E. Ahrens, Instructor; Ph.D., Tulane, 1996. Orthopedics.

Navin Bansal, Assistant Professor; Ph.D., Delaware, 1988. MRI spectroscopy.

Tracy L. Cameron, Senior Research Scientist; Ph.D., Queen's at Kingston, 1996. Neural prosthesis and neural modulation.

Kenneth Diller, Adjunct Professor and Director; Sc.D., MIT, 1972. Bioheat transfer.

Thomas D. Franklin Jr., Executive Director; Ph.D., Illinois, 1972. Spine biomechanics, ultrasonic imaging, minimally invasive surgery.

Maruta Ram Gudavalli, Research Scientist; Ph.D., Cincinnati, 1989. Biomechanics, robotics, computer-aided design.

Morley Herbert, Adjunct Assistant Professor; Ph.D., Toronto, 1972. Orthopedic biomechanics.

Millard M. Judy, Adjunct Assistant Professor; Ph.D., Colorado School of Mines, 1969. Lasers in surgery, photochemistry in medicine and surgery.

Edgar A. Lucas, Director; Ph.D., UCLA, 1972. Instrumentation.

Fabian E. Pollo, Director, Motion and Performance Lab; Ph.D., Texas A&M, 1992. Orthopedics, motion and performance, human motor control.

Michael J. Torma, Director, Center for Biomedical Technology and Innovation; M.D., Alabama, 1968. Minimally invasive surgery instrumentation.

John J. Triano, Director, Chiropractic Division; D.C./Ph.D., Michigan, 1998. Joint biomechanics, bioinstrumentation, clinical trials.

David B. Wallace, Vice President, Technology Development; Ph.D., Texas at Arlington, 1987. Microfabrication for medical treatment.

UNIVERSITY OF UTAH

Department of Bioengineering

Program of Study	The Department of Bioengineering offers an interdisciplinary program leading to the Master of Engineering (M.E.), Master of Science (M.S.), and Doctor of Philosophy (Ph.D.) degrees. Instruction is provided in numerous areas, including applied neuroscience, biomedical polymers, biobased engineering, bioinstrumentation, biosensors, biomaterials, biomechanics, tissue engineering, micromachined medical instrumentation, and medical imaging. Graduate education is enhanced by close interaction with colleagues in the School of Medicine, College of Science, College of Pharmacy, and College of Engineering.
	Students in the M.S. program must complete master's-level core curriculum and bioengineering track elective courses (minimum 5 credit hours) in one of the following areas: biomaterials, bioinstrumentation/imaging, biomechanics, or neural interfaces. Additionally, all M.S. students are required to pass a comprehensive exam and defend their thesis in a public forum.
	Based on qualification at the time of admission, students may enter the Ph.D. program directly. Ph.D. students must successfully complete the bioengineering graduate core curriculum or its equivalent and take additional advanced graduate courses. Students must also pass a written qualifying exam, write a research proposal on their dissertation topic, and publicly defend their dissertation.
	All graduate students are encouraged to select a research director and to begin thesis research as soon as they begin their studies. The program, which is individually tailored to meet the specific objectives of each candidate, may involve collaboration with faculty members in other departments. The Ph.D. degree program normally takes about four to five years.
Research Facilities	The Department of Bioengineering has a strong research orientation. In addition to having access to the research laboratories of the individual faculty members, students have use of the research facilities of the College of Engineering, including campuswide access to the University Computer Center, and of the collaborating departments. Bioengineering faculty members and students also make use of the orthopedic bioengineering, orthopedic histomorphometry, optical bioinstrumentation, advanced imaging methods, surface analysis, vision research, medical imaging, microelectronics research, Langmuir-Blodgett trough, anesthesia bioinstrumentation, neuroprosthetics, and interfacial spectroscopy laboratories, as well as the Center for Biopolymers at Interfaces and the Center for Engineering Design.
Financial Aid	The majority of the students receive support from research assistantships that are funded by grants and contracts to faculty members from outside agencies. A limited number of scholarships and distinguished fellowships are also available.
Cost of Study	Graduate students receiving financial support through the University of Utah are given full tuition waivers. Tuition and fees for 1999–2000 for 12 credit hours are approximately $1280 for state residents and $3897 for nonresidents per semester.
Living and Housing Costs	On-campus housing for unmarried graduate students is available for $1657 to $2405 (depending on the type of room chosen) per academic year. University housing for married students ranges from $340 to $582 per month, including utilities, depending on the size of the apartment. Medical Plaza housing costs for both married and single students range from approximately $433 to $680 per month. Off-campus housing near the University is also available.
Student Group	The University of Utah has a student population of 25,000, representing fifty states and fifty other countries. The Department of Bioengineering welcomes approximately 20 new students each year and maintains an average total enrollment of 80.
Location	Salt Lake City is the center of a metropolitan area of nearly a million people. It lies in a valley with an elevation varying between 4,200 and 5,500 feet and is surrounded by mountain peaks reaching nearly 12,000 feet in elevation. The city is the cultural center of the intermountain area, having resident ballet and modern dance companies, theater and opera companies, and a symphony orchestra. It also supports professional basketball, hockey, and baseball teams and is within 30 minutes of half a dozen of the best ski areas in the country. A major wilderness area is less than 2 hours away, and ten national parks are within a day's drive.
The University	Although the University of Utah is one of the oldest state universities west of the Missouri River, having been founded in 1850, the 1,500-acre campus, nestled at the foothills of the Wasatch Mountains, is characterized by modern buildings, open malls, fountains, and attractive landscaping. An international faculty of 3,600 provides comprehensive instruction and research in disciplines ranging from medicine and law to fine arts and business. The University has an excellent library system and an outstanding computer center.
Applying	Instructions for applying to the program and a packet of application materials may be obtained by writing to the address given below. In addition to the application form and fee, students must submit three letters of recommendation, scores on the General Test of the Graduate Record Examinations, and a written statement of interests and goals.
	Detailed information on the various aspects of the Department of Bioengineering at the University of Utah can be obtained by accessing the department's home page (http://www.bioen.utah.edu).
Correspondence and Information	Dr. Vladimir Hlady Director of Graduate Admissions Department of Bioengineering 50 South Central Campus Drive, Room 2480 University of Utah Salt Lake City, Utah 84112-9202

University of Utah

THE FACULTY AND THEIR RESEARCH

*T. L. Allinger, Ph.D., Calgary. Orthopedics biomechanics, specifically sports biomechanics.

J. D. Andrade, Ph.D., Denver. Interfacial biochemistry, biochemical sensors, proteins engineering, integrated science education, bioluminescence.

*K. N. Bachus, Ph.D., Utah. Bone biomechanics, fracture analysis, implant failure mechanisms.

M. J. Berggren, Ph.D., Stanford. Advanced imaging techniques, algorithm development.

R. D. Bloebaum, Ph.D., Western Australia. Orthopedic implants.

*D. Bloswick, Ph.D., Michigan. Biomechanics, ergonomics.

S. C. Bock, Ph.D., California, Irvine. Antithrombin III heparin cofactor activity-function, development of medically useful serpins, regulation of glycoprotein N-glycosylation.

*J. N. B. Bridge, Ph.D., UCLA. Cardiac muscle biophysics.

G. Burns, D.V.M., Colorado State; Ph.D., Washington State. Biomaterial implant pathology, biomaterial-related immune response, total artificial heart.

K. D. Caldwell, Ph.D., Uppsala (Sweden). Separation and characterization of biopolymers, subcellular particles, and cells.

D. A. Christensen, Ph.D., Utah. Optical/ultrasonic bioinstrumentation.

*R. Clackdoyle, Ph.D., Dalhousie. Medical imaging, 3-D image reconstruction.

*E. B. Clark, M.D., Albany Medical College. Cardiovascular development in humans.

G. A. Clark, Ph.D., California, Irvine. Neurobiology, biological basis of behavior, cellular neurophysiology, computational neuroscience, cellular mechanisms of learning.

*K. Dusek, Ph.D., Czechoslovak Academy of Sciences. Formation-structure-properties relations of polymers.

*R. S. Eidens, Ph.D., Utah. Medical imaging, ultrasonic bioinstrumentation, application of VSI technology to acoustic transducers/imaging systems.

*E. P. France, Ph.D., Wright State. Biomechanics.

A. Bruno Frazier, Ph.D., Georgia Tech. Biobased micromachining technologies and applications.

*R. D. Gesteland, Ph.D., Harvard. Genetics, DNA sequencing.

*J. M. Harris, Ph.D., Purdue. Laser-based bioinstrumentation, interfacial spectroscopy.

*T. G. Henderson, Ph.D., Texas at Austin. Artificial intelligence, computer vision, robotics.

J. N. Herron, Ph.D., Illinois. Protein engineering, molecular graphics, biosensors.

V. Hlady, Ph.D., Zagreb. Biochemistry/biophysics at interfaces, solid/liquid interface of biomaterials, proteins as engineering.

K. W. Horch, Ph.D., Yale. Neuroprostheses, biomedical instrumentation, information processing in the somatosensory system, tactile aids.

*D. T. Hutchinson, M.D., Jefferson Medical. Orthopedics implants for the hand.

S. C. Jacobsen, Ph.D., MIT. Prosthesis design, microelectromechanical systems, control theory, robotics.

J. Janatova, Ph.D., Czechoslovak Academy of Sciences. Protein chemistry, biocompatibility.

C. R. Johnson, Ph.D., Utah. Theoretical/computational electrophysiology, inverse electrocardiography, applications of dynamical systems theory in medicine/biology.

S. A. Johnson, Ph.D., Stanford. Ultrasonic, X-ray, NMR tomographic, and 3-D imaging.

*S. C. Johnson, Ph.D., Utah. Exercise physiology, energy metabolism, orthopedics biomechanics.

*S. W. Kim, Ph.D., Utah. Blood compatibility, drug-delivery systems.

S. E. Kern, Ph.D., Utah. Pharmacokinetics and pharmacodynamics modeling and control.

*K. Knutson, Ph.D., Utah. Biopolymers, biomembranes, controlled delivery.

*C. Konak, Ph.D., Karlova (Prague). Physical properties of polymer solutions and gels as studied by light scattering methods.

J. Kopecek, Ph.D., Czechoslovak Academy of Sciences. Biomaterials, chemistry/biochemistry of macromolecules, drug-delivery systems.

*P. Kopeckova, Ph.D., Czechoslovak Academy of Sciences. Bioorganic polymer chemistry, biodegradability of polymers, drug-delivery systems.

*J. K. Leypoldt, Ph.D., California, San Diego. Artificial kidney, biological transport phenomena, membrane applications in biotechnology.

R. S. MacLeod, Ph.D., Dalhousie. Cardiac bioelectric modeling, body surface potential mapping, cardiac electrophysiology, scientific visualization.

*J. C. McRea, Ph.D., Utah. Medical device development.

*S. G. Meek, Ph.D., Utah. Prosthetic design and control, EMG signal processing, biomechanics.

*C. Moncur, Ph.D., Utah. Physical therapy for rheumatic diseases, biomechanics of disability.

*J. R. Nelson, Ph.D., Utah. Microbiology, immunology.

R. A. Normann, Ph.D., Berkeley. Cell physiology, bioinstrumentation, neuroprosthetics.

D. B. Olsen, D.V.M., Colorado State. Artificial heart/assist devices, design, control, surgical implantation/physiologic interfaces, other organ replacement devices.

G. W. Pantalos, Ph.D., Ohio State. Circulatory mechanical support, aerospace physiology.

*D. L. Parker, Ph.D., Utah. Medical imaging, applications of physics in medicine.

*W. G. Pitt, Ph.D., Wisconsin–Madison. Polymers and composite materials for biomedical applications, surface chemistry.

*G. D. Prestwich, Ph.D., Stanford. Bioorganic chemistry.

A. Pungor, Ph.D., Technical University (Hungary). Scanning force microscopy, near-field optical microscopy, bioinstrumentation, mechanical property measurement on surfaces with SFM.

R. D. Rabbitt, Ph.D., RPI. Biomechanics, hearing/vestibular mechanisms, computational mechanics, computational neuroscience.

*N. Rapoport, Ph.D., Moscow State; D.Sc., Academy of Science (USSR). Polymeric materials, biological magnetic resonance.

*R. B. Roemer, Ph.D., Stanford. Heat transfer, thermodynamics, design, optimization to biomedical problems.

*W. A. Sands, Ph.D., Utah. Biomechanics of sport, exercise, sports science.

M. K. Sharp, Sc.D., MIT. Biofluid mechanics, biotransport processes.

*G. D. Smith, Ph.D., Utah. Property-structure relationships in soft condensed matter, especially polymer structure-dynamics.

*R. S. Smith, Ph.D., Arizona State. Immunology, immunosensing, wave-guided immunodiagnostics.

*K. W. Spitzer, Ph.D., SUNY at Buffalo. Cardiac cellular electrophysiology, intracellular pH regulation.

R. J. Stewart, Ph.D., California, Santa Barbara. Protein engineering, biological energy transduction, structure/activities of cytoskeletal proteins, molecular motors.

*C. L. Thomas, Ph.D., Drexel. Imaging, modeling, applications of ultrasound.

P. A. Tresco, Ph.D., Brown. Cellular-based molecular delivery systems, synthetic membrane fabrication, neurodegenerative/neuroendocrine/endocrine deficiency disorders.

*E. G. Vajda, Ph.D., Utah. Bone biomechanics, orthopedic implants.

*R. A. VanWagenen, Ph.D., Utah. Biosensors, bioluminescence, biomaterials.

J. A. Weiss, Ph.D., Utah. Biomechanics, mechanics of normal/healing soft tissues, evaluation of injury mechanics/treatment regimens.

D. L. Wells, Ph.D., Utah. Microfabrication, minimally invasive medical systems, microsensors, microactuators.

D. R. Westenskow, Ph.D., Utah. Bioinstrumentation, microprocessor applications in medicine.

J. W. Wiskin, Ph.D., Utah. Mathematical modeling/numerical techniques, inverse scattering.

---

*Adjunct faculty.

UNIVERSITY OF WASHINGTON

Department of Bioengineering

Programs of Study

The Department of Bioengineering provides comprehensive, multidisciplinary programs of research and education. The concepts and techniques of the physical sciences and engineering are applied to challenges in the health sciences and medicine. Likewise, the concepts of biology are applied to problems in engineering. There are options for study and research leading to the M.S. in bioengineering, the M.S.E., and the Ph.D. degrees.

The objective of the Ph.D. program is to train qualified persons for careers in bioengineering research and teaching. The training has three major components. First, students acquire a breadth of knowledge in engineering and medicine and in the interdisciplinary interface between these quite disparate fields. Second, they develop a depth of knowledge and expertise in a particular scientific specialty. Finally, each student develops and demonstrates the potential for independent research. These objectives are met through a combination of didactic, research, and teaching experiences. The program places rigorous expectations on students' performance while maintaining sufficient flexibility regarding specific requirements to accommodate qualified students with diverse backgrounds. An M.D./Ph.D. program is also available. The master's degree programs provide course work and research experience that prepare students for careers in academic, industrial, and hospital settings.

Areas of research include bioinstrumentation (biosensors, optical instrumentation, cardiovascular instrumentation, cochlear prosthesis, biological laboratory instrumentation), biomechanics (muscle mechanics, robotic surgery, prosthetics, and soft tissue implants), biosystems (microcirculatory transport and exchange, simulation methodologies, cellular biomechanics), biomaterials (blood compatibility, cell growth, drug delivery systems, surface modification, biomolecule separations), cellular bioengineering (muscle contraction, cell motility, filament motion, phase transition in cellular process, enzyme moiety), biomedical imaging (PET, MRI, spectroscopy, image processing, CT, ultrasound, tomography), and molecular bioengineering (molecular modeling, nanotechnology, protein engineering, self organization, chemical sensors, biosensors). The research program is very active and is supported by more than $10 million per year from external grants.

Research Facilities

Offices and laboratories are located in the College of Engineering and the School of Medicine, which are adjacent to each other on the University campus. Laboratories have state-of-the-art equipment; computers are used extensively. Students have access to the University Hospital, vivarium, Primate Center, computer center, and libraries, as well as to all engineering and health science departments and facilities.

Financial Aid

Financial aid in the form of traineeships, fellowships, and assistantships is available for qualified graduate students. Nearly all full-time students are supported. The funding is derived from federal research and training programs, the Graduate School Research Fund, and sponsored research projects. Predoctoral students are encouraged to apply individually for fellowships sponsored by such agencies as the National Institutes of Health; information concerning these fellowships is available from the Department of Bioengineering.

Cost of Study

Tuition and fees for the academic year 1999–2000 are $1861 per quarter for Washington residents and $4624 per quarter for out-of-state residents. All holders of assistantships qualify for resident status.

Living and Housing Costs

The cost of room and board in University residences varies depending upon the type of accommodations and meal plan; rates for room and board are approximately $1830 per quarter. Married student housing is available for qualified applicants. Numerous off-campus rooms, apartments, and houses are also available nearby. Their cost varies according to proximity to campus and demand.

Student Group

There are 114 students in the program. Of these, 86 percent are pursuing the Ph.D., 33 percent are women, and 23 percent are from countries other than the United States. Undergraduate backgrounds include all branches of engineering, physics, chemistry, and biological sciences.

Location

The University of Washington is located in Seattle, a major city of the Pacific Northwest. It has a moderate climate conducive to academic and recreational pursuits. Shopping and entertainment centers are near the University. Downtown Seattle is a few minutes away by public transportation. The area is known for its boating, camping, fishing, hiking, mountain climbing, and skiing opportunities as well as its rock, folk, and classical music and visual and performing arts.

The University

The University of Washington was founded in 1861 and is the oldest state-assisted institution of higher education on the Pacific Coast. There are approximately 35,000 students studying in a variety of fields, including arts and sciences, business administration, dentistry, education, engineering, fisheries, forest resources, law, medicine, nursing, pharmacy, and public health and community medicine. Comprehensive intercollegiate and intramural athletic programs, as well as a range of musical and cultural programs, are offered. The beautiful green campus encompasses 680 acres and is bordered on the east by Lake Washington and on the south by Lake Union. The University of Washington is ranked among the top universities in the United States and is an outstanding center of academic excellence in the Northwest. A particularly strong research institution, the University is first among all universities in receipt of federal research funding.

Applying

Applicants holding a B.S. degree in engineering, physical sciences, or biological sciences are encouraged to apply. A 3.0 minimum grade point average (on a 4.0 scale) and scores from the General Test of the Graduate Record Examinations are required for admission. International applicants are required to take the TOEFL and to obtain a minimum score of 580; if scores fall between 500 and 580, ESL classes are required. Graduate admissions and assistantship applications, a statement of purpose, and three letters of recommendation are also needed in order to be considered for admission. Admission is very competitive. In recent years, 10 percent of the applicants have been offered admission. The application deadline for fall quarter is January 15. Further information may be obtained from the address below.

Correspondence and Information

Academic Counselor
Department of Bioengineering, 357962
University of Washington
Seattle, Washington 98195-7962
Telephone: 206-685-2021

University of Washington

THE FACULTY AND THEIR RESEARCH

Core Faculty

J. B. Bassingthwaighte, M.D., Ph.D. Cardiovascular mass transport and ion exchanges, simulation analysis of integrated systems, PET imaging, cardiac metabolism, fractal physiology.

D. G. Castner, Ph.D. Surface analysis, surface modification, biomaterials, organic thin films.

L. Crum, Ph.D. Biomedical acoustics.

D. M. Foster, Ph.D. Biomathematics and modeling methodology, simulation analysis, lipid and lipoprotein metabolism, gluconeogenesis, population kinetics.

C. Giachelli, Ph.D. Characterization of cellular responses to cytokines and other proteins, biomaterials.

A. S. Hoffman, Sc.D. Polymeric biomaterials, immobilized biomolecules, drug delivery, diagnostics, bioseparations, bioprocesses.

L. E. Hood, M.D., Ph.D. T-cell receptor and MHC recognition: tolerance, autoimmunity, and genetic predisposition; T-cell development; genome sequencing of T-cell receptor loci.

T. A. Horbett, Ph.D. Interactions of cells and proteins with foreign materials, insulin delivery devices, glucose sensors.

L. L. Huntsman, Provost, Ph.D. Mechanics of heart and heart muscle, cardiovascular system assessment, new measurement techniques.

M. J. Kushmerick, M.D., Ph.D. Imaging.

Y. Kim, Ph.D. Medical imaging, diagnostic ultrasound, telemedicine, instrumentation, system modeling, media processing and processors.

H. C. Lai, Ph.D. Cellular effects of electromagnetic fields.

Z. Li, Ph.D. Cardiac metabolism, PET imaging, kinetic modeling.

T. P. Lybrand, Ph.D. Molecular modeling.

R. W. Martin, Ph.D. Bioinstrumentation, ultrasonic Doppler, echo, tissue characterization, signal processing, 3-D ultrasonic imaging and reconstruction.

D. Martyn, Ph.D. Regulation and mechanical properties of contraction in skeletal and cardiac muscle.

J. J. Medina, Ph.D. Science education.

G. H. Pollack, Ph.D. Muscular contractions, cardiac dynamics, optical image processing.

B. D. Ratner, Ph.D. Synthesis and characterization of biomaterials for cardiovascular, ophthalmologic, and drug delivery applications; surface analysis by ESCA, SIMS, STM, and FR-1R-ATR; drug release.

M. Regnier, Ph.D. Mechanical and kinetic measurements of skeletal and heart muscle contractions and their regulation by calcium.

J. E. Sanders, Ph.D. Tissue mechanics, external prosthetics, novel measurement techniques, tissue adaptation to mechanical stress.

N. Singh, Ph.D. Cellular effects of electromagnetic fields.

F. A. Spelman, Ph.D. Local control of peripheral circulation, biophysics of the implanted cochlea, bioinstrumentation for primate research.

P. S. Stayton, Ph.D. Protein engineering, design proteins for biosensors.

S. Vaezy, Ph.D. Biomedical applications of high-intensity focused ultrasound.

P. Verdugo, M.D. Microrheology and control of ciliary and flagellar motion, polymer physics of mucin secretion, instrumentation in dynamic laser scattering spectroscopy.

P. Vicini, Ph.D. Biomathematics and modeling methodology, simulation analysis, population kinetics.

V. Vogel, Ph.D. Molecular assemblies, surface modification, Langmuir-Blodgett technique, microscopy, nonlinear optics.

P. Yager, Ph.D. Physical chemistry and applications of biomembranes.

Adjunct Faculty

The departmental affiliation for adjunct faculty members is indicated in parentheses following each listing.

M. A. Afromowitz, Ph.D. Instrumentation (Electrical Engineering).

D. Baker, Ph.D. Protein folding.

F. Baneyx, Ph.D. Protein engineering (Chemical Engineering).

G. Bashein, M.D., Ph.D. Anesthetic delivery (Anesthesiology).

K. W. Beach, M.D., Ph.D. Cardiovascular instrumentation (Surgery).

J. H. Caldwell, M.D. PET imaging (Medicine/Cardiology).

J. B. Callis, Ph.D. Instrumentation (Chemistry).

R. P. Ching, Ph.D. Orthopedic biomechanics (Orthopedics).

K. E. Conley, Ph.D. In vivo muscle metabolism and energetics, magnetic resonance spectroscopy, muscle structure and function, muscle physiology, exercise physiology.

S. R. Dager, M.D. Biosystems (Radiology).

C. H. Daly, Ph.D. Biomechanics (Mechanical Engineering).

M. M. Graham, M.D., Ph.D. Imaging (Radiation Oncology).

B. Hannaford, Ph.D. Robotics (Electrical Engineering).

R. M. Haralick, Ph.D. Imaging (Electrical Engineering).

M. P. Hlastala, Ph.D. Systems (Respiratory Diseases and Physiology and Biophysics).

W. Hol, Ph.D. Protein crystallography (Biomolecular Structure).

T. Hunkapiller, Ph.D. Biological computation (Molecular Biotechnology).

I. J. Kalet, Ph.D. Imaging (Oncology).

T. K. Lewellen, Ph.D. Imaging (Radiology).

D. T. Linker, M.D. Cardiac ultrasound (Medicine/Cardiology).

F. A. Matsen III, M.D. Biomechanics (Orthopaedics).

D. Meldrum, Ph.D. Robotics (Electrical Engineering).

D. Nickerson, Ph.D. Molecular genetics (Molecular Biotechnology).

T. L. Richards, Ph.D. Imaging (Radiology).

E. Riskin, Ph.D. Imaging (Electrical Engineering).

S. M. Schwartz, M.D., Ph.D. Systems (Pathology).

M. Soma, Ph.D. Instrumentation (Electrical Engineering).

B. Stewart, Ph.D. Imaging (Radiology).

B. Trask, Ph.D. Cytogenetics (Molecular Biotechnology).

G. J. van den Engh, Ph.D. Quantitative cytogenetics (Molecular Biotechnology).

J. Yates, Ph.D. Sequencing (Molecular Biotechnology).

C. Yuan, Ph.D. Imaging (Radiology).

VANDERBILT UNIVERSITY

Department of Biomedical Engineering

Programs of Study

Vanderbilt University offers Master of Science and Doctor of Philosophy degrees with a major in biomedical engineering. The program is interdisciplinary in scope. Advanced courses in engineering specialties, life sciences, and biomedical engineering are combined with training in the conception and performance of biomedical research. The goal of the program is to provide advanced education and research training in quantitative biology, in biomedical information and instrumentation systems, and in the scientific principles underlying the creation of therapeutic devices and processes. The program is specifically concerned with the interface between the engineering, physical, computing, and mathematical sciences and biology. Current graduate research is concentrated on circulatory-tissue mass transfer and mechanics in the heart and lungs, cellular bioengineering, magnetic resonance imaging, interactive image-guided surgery, vision research, biosensors, biomedical optics, and applications of advanced computing methods to biomedical systems.

Candidates for the degree of Master of Science must complete 24 semester hours of approved graduate-level courses and present a research thesis. Requirements for the Doctor of Philosophy degree are 48 semester hours of course work beyond the bachelor's degree, distributed between life sciences and engineering; passage of a written examination covering basic knowledge in biomedical engineering; completion of a qualifying examination consisting of presentation of a proposal for doctoral research; and presentation of a dissertation presenting the results of original research in biomedical engineering. Students wishing to combine study for the M.D. degree with that for a Ph.D. in biomedical engineering may apply to the School of Medicine for admission to the Medical Scientist Training Program.

Research Facilities

Extensive facilities in the School of Engineering and the School of Medicine are available for graduate research in biomedical engineering. These include laboratories for the study of cardiopulmonary function (the Vanderbilt Lung Center Laboratories and the Biomedical Engineering Cardiopulmonary Laboratory), laboratories for the study of cell function and biotechnology (the Cellular Engineering Laboratory and the Cell Transport Laboratory), the laboratory for stereotactic neurosurgery in the Department of Neurosurgery, the Free Electron Laser Laboratory, the laboratory for the development of computerized intraoperative monitoring, the Biomedical Computing Laboratory, and the Vision Research Laboratory. In addition, the Department of Radiology and Radiological Sciences, which cooperates fully with the graduate program in biomedical engineering, has extensive computer and data-analysis systems including a magnetic resonance imaging system, digital radiography facilities, CAT scanners, positron emission tomography facilities, and gamma scanning systems.

Financial Aid

The department intends, within its resources, to provide adequate financial assistance to graduate students with high academic potential who need help in meeting expenses. Both stipends and full-tuition scholarships are available for students on a competitive basis. Stipends are given as research assistantships, teaching assistantships, and service-free federal traineeships. Special supplementary fellowship awards additive to these stipends are available for exceptionally qualified students.

Cost of Study

Tuition for graduate studies in 1998–99 was $914 per semester hour during the academic year and $914 per semester hour during the summer session. A minimum tuition charge of $200 was required of all enrolled graduate students regardless of course load. Books, supplies, and health insurance are approximately $1800 per year.

Living and Housing Costs

On-campus housing is available for both single and married students. Married student apartments cost $400–$600 per month on campus; similar prices are available in the residential areas surrounding the University.

Student Group

The total undergraduate, graduate, and professional school enrollment at Vanderbilt is approximately 9,000 students. The School of Engineering enrolls about 1,200 undergraduates and 310 graduate students in all programs. The Department of Biomedical Engineering has an undergraduate enrollment of 270 students and a graduate enrollment of 43, most of whom are pursuing the Ph.D. degree. Approximately 80 percent of the graduate students are U.S. citizens; the remainder come from a variety of other countries.

Location

Nashville offers professional, cultural, and recreational opportunities. Modern buildings and restored historic structures give graphic evidence of the city's vitality. More than a dozen colleges and universities attract more than 30,000 students. These institutions together form a broad, influential community of higher education and earn Nashville the nickname "Athens of the South." The surrounding area offers a great variety of outdoor activities, including the Great Smoky Mountains National Park, which is about 5 hours distant.

The University and The Department

Vanderbilt University, founded in 1873, is an independent institution offering a full range of instruction. Its units include the Graduate School, the School of Engineering, and the School of Medicine. The medical school and hospital are located on the main campus, approximately one block from the School of Engineering. This unusual juxtaposition has made possible a wide variety of collaborative efforts between faculty and students of the two schools. Biomedical Engineering is a separately budgeted full department of the School of Engineering, with 10 primary faculty members and 10 others appointed jointly with other departments. All cooperate in graduate student research and teaching.

Applying

Applications should be made to the Graduate School by January 15 for fall admission and November 1 for spring admission. Applications are accepted after these deadlines, but competition for financial aid is positively influenced by early application. Students should have an undergraduate degree in engineering or natural science and at least a B average in undergraduate work and must take the General Test of the Graduate Record Examinations. International students must take the TOEFL. Life science students may need to make up deficits in mathematics and engineering subjects.

Correspondence and Information

Director of Graduate Studies
Department of Biomedical Engineering
Vanderbilt University
P.O. Box 1631, Station B
Nashville, Tennessee 37235
Telephone: 615-322-3521

Vanderbilt University

THE FACULTY AND THEIR RESEARCH

Alfred B. Bonds III, Professor of Electrical and Biomedical Engineering; Ph.D. (electrical/biomedical engineering), Northwestern; postdoctoral training, Berkeley. Physiology and bioengineering of the visual system, advanced biomedical instrumentation associated with the neurological and optical systems, neural computing.

Kenneth L. Brigham, Joe and Morris Werthan Professor of Experimental Medicine, Professor of Biomedical Engineering, Associate Professor of Molecular Physiology and Biophysics, Director of the Division of Pulmonary Medicine (Department of Medicine), and Director of the Vanderbilt Lung Center; M.D., Vanderbilt; postdoctoral training, Johns Hopkins and Cardiovascular Research Institute, California, San Francisco. Pulmonary physiology and medicine, cellular and molecular physiology of endothelial cells, lung vascular transport.

Jerry C. Collins, Research Associate Professor of Biomedical Engineering and Director of the Biomedical Computing Laboratory, Clinical Research Center; Ph.D. (biomedical/electrical engineering), Duke; postdoctoral training, Vanderbilt. Lung fluid balance, coronary physiology, medical informatics, nutrition transport physiology.

Michael Fitzpatrick, Associate Professor of Computer Science and of Radiology and Radiological Sciences; Ph.D. (physics), Florida State; postdoctoral training, North Carolina at Chapel Hill. Imaging science, computer-enhanced determinants of coronary flow from digital subtraction angiography.

Robert L. Galloway Jr., Associate Professor of Biomedical Engineering and Neurological Surgery; Ph.D. (biomedical engineering), Duke; postdoctoral training, Vanderbilt. Medical applications of magnetic resonance imaging, ultrasound imaging, stereotactic neurosurgery and infrared biosensors.

Todd D. Giorgio, Associate Professor of Biomedical Engineering and Chemical Engineering; Ph.D. (chemical engineering), Rice. Effects of shear on cellular function, platelet physiology, biorheology, biosensors of cellular function.

Thomas R. Harris, Professor of Biomedical and Chemical Engineering and of Medicine, Chair of the Department of Biomedical Engineering (School of Engineering), and Director of the Division of Biomedical Engineering and Computing (Department of Medicine); Ph.D. (chemical engineering), Tulane; M.D., Vanderbilt. Physiological transport phenomena, computer simulation of cardiopulmonary function, infrared biosensors of cardiopulmonary function.

Frederick R. Haselton, Associate Professor of Biomedical Engineering; Ph.D. (bioengineering), Pennsylvania; postdoctoral training, Vanderbilt. Cellular bioengineering, endothelial cell function, physiological transport phenomena.

Stanley B. Higgins, Research Assistant Professor of Medicine (Biomedical Engineering); Ph.D. (mathematics), Texas Christian. Medical computing and informatics, expert systems, medical research database management.

E. Duco Jansen, Assistant Professor of Biomedical Engineering; Ph.D. (biomedical engineering), Texas at Austin; postdoctoral training, Texas at Austin. Biomedical optics, laser-tissue interaction.

Paul H. King, Associate Professor of Biomedical and Mechanical Engineering and of Anesthesiology; Ph.D. (mechanical/biomedical engineering), Vanderbilt. Medical instrumentation, computerized operating-room monitoring, expert systems in anesthesiology.

Anita Mahadevan-Jansen, Assistant Professor of Biomedical Engineering; Ph.D. (biomedical engineering), Texas at Austin. Biomedical optics, spectroscopy.

K. A. Overholser, Professor of Biomedical and Chemical Engineering; Ph.D. (chemical engineering), Wisconsin; postdoctoral training, London and Cardiovascular Research Institute, California, San Francisco. Cardiovascular fluid mechanics, biorheology, physiological transport phenomena, cellular endocytosis.

C. Leon Partain, Professor of Radiology and Biomedical Engineering and Director of the Division of Imaging (Department of Radiology and Radiological Sciences); M.D., Washington (St. Louis); Ph.D. (nuclear engineering), Purdue; postdoctoral training, North Carolina at Chapel Hill. Magnetic resonance imaging, nuclear medicine, neurological radiology.

Cynthia B. Paschal, Assistant Professor of Biomedical Engineering and of Radiology and Radiological Sciences; Ph.D. (biomedical engineering), Case Western Reserve. Magnetic resonance imaging of the cardiovascular system.

David R. Pickens, Associate Professor of Radiology and Radiological Sciences and of Biomedical Engineering; Ph.D. (mechanical/biomedical engineering), Vanderbilt. Magnetic resonance imaging, measurement of blood flow by magnetic resonance.

Robert J. Roselli, Professor of Biomedical Engineering; Ph.D. (biomedical/mechanical engineering), Berkeley; postdoctoral training, Vanderbilt. Lung fluid balance, physiological transport phenomena, functional imaging of the lung circulation, biosensors of cardiopulmonary function, biomechanics.

Richard G. Shiavi, Professor of Biomedical and Electrical Engineering and Assistant Professor of Orthopaedics and Rehabilitation Medicine; Ph.D. (biomedical engineering), Drexel. Quantitative kinesiology, analysis of human gait and joint motion, physiological signal processing, electromyography.

VIRGINIA COMMONWEALTH UNIVERSITY

Biomedical Engineering Program

Programs of Study

Biomedical engineering at Virginia Commonwealth University offers graduate study leading to the M.S., Ph.D., and M.D./Ph.D. degrees. The program offers state-of-the-art research and development opportunities in orthopedic biomechanics, biomaterials, cellular and tissue engineering, cardiac electrophysiology, medical imaging systems, cardiovascular dynamics, rehabilitation engineering, and man-machine interfacing. The M.S. requires 32 credit hours of course work in addition to research leading to a thesis. The Ph.D. requires 41 hours of course work above the B.S. (or 9 above the M.S. in biomedical engineering) in addition to research leading to a dissertation. A Ph.D. qualifying examination is required at the conclusion of course work and a dissertation defense at the conclusion of the research phase. The Biomedical Engineering (BME) Program is housed on the University's Medical College of Virginia (MCV) campus, the nation's fourth-largest medical academic campus. The primary and affiliate faculty members in biomedical engineering maintain appointments and interconnections in medicine, pharmacy, dentistry, nursing, allied health, science, and engineering.

Research Facilities

Teaching and research laboratories are located throughout the Virginia Commonwealth University campus, with the majority located on the Medical College of Virginia campus. The Biomedical Engineering Program maintains research laboratories in orthopedic biomechanics with a servo-hydraulic Instron system; cardiac electrophysiology with pacemaker and defibrillator testing systems; heart assist with various artificial heart technologies; medical imaging with MRI, CAT, and ultrasonic scanners; cellular mechanics and tissue engineering with the latest biotechnology equipment, evoked potentials; and rehabilitation engineering with voice- and eye-tracking hardware. Numerous PC, minicomputer, and mainframe computers are linked with both engineering and University servers and to the Internet. Custom microelectronic fabrication is available through the School of Engineering Microelectronics Center with a class 1000 clean room. Custom hardware and device fabrication is available through the BME machine and electronics shops with 3 full-time staff members.

Financial Aid

Graduate students may receive support through teaching assistantships, research assistantships, fellowships, or traineeships. Support includes a stipend, tuition, and fees for a twelve-month period, with renewals based upon satisfactory academic performance. These awards are determined on a competitive basis. There are also several opportunities for employment in the School of Medicine, the School of Engineering, and numerous Medical College of Virginia Hospital clinics and laboratories.

Cost of Study

In 1998, tuition and fees for full-time study were $4800 for in-state students and $12,500 for out-of-state students.

Living and Housing Costs

Room and board fees for medical dormitories are approximately $4500 per year, with additional housing in apartments and rental houses throughout greater Richmond.

Student Group

Biomedical engineering graduate enrollment has grown considerably over the last several years, and the 1998–99 enrollment was expected to reach 30 full-time students, with approximately 50 percent M.S., 40 percent Ph.D., and 10 percent M.D./Ph.D. students. All full-time students receive financial support through teaching and research assistantships or fellowships.

Student Outcomes

Students graduating with the M.S. degree have been employed at leading biomedical device firms and medical facilities throughout the nation. Students graduating with the Ph.D. degree have been employed in academia, medical research laboratories, and medical device industries. Students completing the M.D./Ph.D. degree program continue into medical residencies or internships.

Location

Virginia Commonwealth University is located in downtown Richmond, Virginia. The University maintains two campuses located only 2 miles apart: the academic campus, where the School of Engineering has opened a new $40-million teaching and research building, and the Medical College of Virginia campus, with the University's Schools of Medicine, Pharmacy, Dentistry, Nursing, and Allied Health, along with three hospitals and two clinics. The Virginia Biotechnology Research Park is located at the edge of the MCV campus, and the city is ringed by high-technology industry in microelectronics, computing, biotechnology, and chemical-processing facilities. Richmond is 2 hours south of Washington, D.C. and 2 hours west of Virginia Beach.

The University and The Program

Virginia Commonwealth University is a large, state-supported university with an enrollment of 23,000 for 1998–99. There are more than 1,600 faculty members with an annual research budget of more than $100 million and an overall University budget near $1 billion.

Biomedical engineering is a rapidly growing program with Virginia's only BME program offering all three degrees (B.S., M.S., Ph.D.) in BME. The program is located in the heart of the MCV campus, with strong interaction in virtually all clinical and research programs throughout the hospitals, clinics, and medical academic departments.

Applying

Applications should be made prior to March 1 for the upcoming fall semester. Those applicants seeking financial support should submit applications prior to February 1. The application requires official transcripts, scores on the GRE General Test, and scores on the TOEFL for international students.

Correspondence and Information

Graduate Admissions Coordinator
Biomedical Engineering Program
220 McGuire Hall Annex
Virginia Commonwealth University
1112 East Clay Street
Richmond, Virginia 23298-0694
Telephone: 804-828-7263
Fax: 804-828-4454
E-mail: jwayne@gems.vcu.edu
World Wide Web: http://www.vcu.edu/egrweb/bme/bmehome.html

Virginia Commonwealth University

THE FACULTY AND THEIR RESEARCH

Robert Adelaar, Ph.D., Professor of Orthopedic Surgery. Orthopedic and joint biomechanics.

Clive Baumgarten, Ph.D., Professor of Physiology. Properties of excitable membranes, cardiac electrophysiology.

Thomas Biber, Ph.D., Professor of Physiology. Cellular transport mechanisms.

Gary Bowlin, Ph.D., Assistant Professor of Biomedical Engineering. Cellular engineering, tissue engineering, biosensors, biomaterials compatibility, endothelial cell seeding.

Peter Byron, Ph.D., Professor of Pharmacy and Pharmaceutics. Instrumentation and formulations for aerosols.

John Cardea, M.D., Professor and Chairman of Orthopedic Surgery. Biomechanics and surgical procedures of total joint replacement.

Marcus Carr, Ph.D., Professor of Internal Medicine. Platelet function, coagulation, thrombosis.

Walter Carter, Ph.D., Professor of Biostatistics. Response surfaces, optimization, design of experiments.

Yin Chang, Ph.D., Assistant Professor of Orthopedic Surgery. Molecular biology of soft tissues, including cartilage and tendon.

Stephen Cleary, Ph.D., Professor of Physiology. EMF effects on biological systems, radiation biophysics.

Ralph Damiano, M.D., Assistant Professor of Surgery. Cardiac waveform analysis.

John DeSimone, Ph.D., Professor of Physiology. Biomembranes, chemosensory physiology and biophysics.

Panos Fatouros, Ph.D., Professor of Radiology. Cephalometric radiography, magnetic resonance imaging.

Ding-Yu Fei, Ph.D., Associate Professor of Biomedical Engineering. Biomedical instrumentation, medical imaging systems, Doppler ultrasound, hemodynamic measurements.

Thomas Haas, Ph.D., Professor of Mechanical Engineering. Biomedical materials, polymers.

Stephen Harkins, Ph.D., Professor of Gerontology. Pain and age, psychophysiology of sensation, perception and cognitive processing.

Rosalyn Hobson, Ph.D., Assistant Professor of Electrical Engineering. Neural networks, artificial intelligence.

Peng-Wie Hsia, Ph.D., Associate Professor of Biomedical Engineering. Physiological signal processing, cardiac electrophysiology, cardiac defibrillation techniques, microcomputer applications in medicine.

William Jiranek, Ph.D., Assistant Professor of Orthopedic Surgery. Molecular mechanisms of bone resorption.

H. Thomas Karnes, Ph.D., Professor of Pharmacy. Optical spectroscopy, laser-induced fluorescence.

Glen Kellogg, Ph.D., Assistant Professor of Medicinal Chemistry. Molecular graphics and software development.

Monty Kier, Ph.D., Professor of Medicinal Chemistry. Cellular automata, dynamic simulations.

Kenneth Kraft, Ph.D., Associate Professor of Radiology. Medical imaging, magnetic resonance imaging.

Robert Lamb, Ph.D., Professor and Chairman of Physical Therapy. Muscle biomechanics.

Martin Lenhardt, Ph.D., Professor of Biomedical Engineering. Auditory systems, noise analysis, sensory systems, ultrasonic systems.

Anthony Marmarou, Ph.D., Professor of Surgery. Modeling of neurological function following cerebral trauma.

Gerald Miller, Ph.D., Professor of Biomedical Engineering and Department Chairman. Biofluid dynamics, artificial organs, man-machine interfacing, rehabilitation engineering, human factors.

Gregory Miller, Ph.D., Professor of Pathology. Biosensors.

Peter Moon, Ph.D., Associate Professor of Dentistry. Dental bonding and mechanical testing of dental materials and biological reactions of materials.

Alfred Ochs, Ph.D., Associate Professor of Biomedical Engineering. Ocular motion in cerebella disease, visual-evoked potentials in multiple sclerosis.

Joseph Ornato, M.D., Professor of Internal Medicine. Computer applications in cardiology and emergency medicine.

Raphael Ottenbrite, Ph.D., Professor of Chemistry. Design and synthesis of polymers and copolymers.

Roland Pittman, Ph.D., Professor of Physiology. Microcirculation, oxygen delivery.

David Sarrett, Ph.D., Professor of Dentistry. Composites, adhesion, dental materials.

Jennifer Wayne, Ph.D., Associate Professor of Biomedical Engineering. Biomechanics, orthopedic mechanics, soft tissue mechanics, biomaterials testing, limb motion analysis.

Gary Wnek, Ph.D., Professor and Chairman of Chemical Engineering. Polymeric materials, hydrogels.

Wilhelm Zuelzer, M.D., Associate Professor of Orthopedic Surgery. Biomechanics of knee and ankle joints.

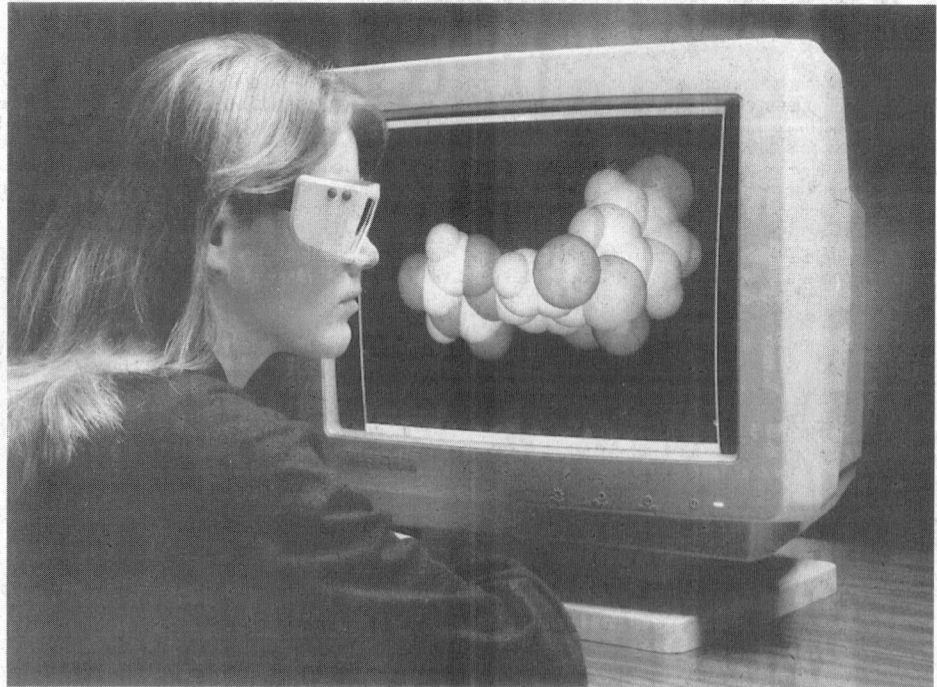

Molecular modeling is used to design new materials and biopolymers.

WASHINGTON UNIVERSITY IN ST. LOUIS

School of Engineering and Applied Science
Department of Biomedical Engineering

Programs of Study

The School of Engineering and Applied Science, through its professional division, the Sever Institute of Technology, offers instruction and research opportunities that lead to the M.S. and D.Sc. degrees in biomedical engineering. The primary focus of the Department of Biomedical Engineering is doctoral training. Graduate students form the core of an interdepartmental network, the Institute of Biological and Medical Engineering, that brings together faculty members from the School of Engineering, the School of Medicine, and the School of Arts and Sciences through the training process. The Department of Biomedical Engineering offers rigorous programs that are designed to develop leaders in molecular engineering, neural engineering, and the engineering of growth and remodeling. This is a new vision of biological engineering as a field, and this new formulation defines a role for which Washington University is ideally suited.

All students in the M.S. and D.Sc. degree programs must fulfill the requirements of a core curriculum in biomedical engineering. Thirty units of graduate credit are required for the M.S. degree. These may consist entirely of course work or a combination of course work and a thesis. Students electing the thesis option undertake an individual research project in lieu of 6 units of course work.

The D.Sc. degree programs emphasize creative research work that leads to a dissertation. Total credit requirements beyond the baccalaureate degree are 36–48 units of course work (including acceptable transfer credit from an M.S. program) and 24–36 units of research. In addition to course work in the core curriculum phase, students participate in a weekly seminar series and in research rotations. A D.Sc. candidate must pass a comprehensive qualifying examination with both written and oral sections, submit a satisfactory dissertation that involves independent creative work in the area of specialization, and defend the dissertation during a final oral examination.

Research Facilities

Many of the other departments in the School of Engineering and Applied Science are also involved in biomedical engineering–related research and have extensive, modern research facilities. Many additional research facilities are available at the School of Medicine.

Financial Aid

Graduate students at the Sever Institute are supported by a variety of traineeships, fellowships, and assistantships. Loan funds, made available on the basis of an applicant's academic promise and need, are offered by regular University sources and by the Federal Stafford Loan Program.

Cost of Study

In the Sever Institute of Technology, tuition is charged on a per-unit basis. The rate for the 1998–99 academic year was $925 per unit. The tuition for a student taking the normal full-time load of 12 units each semester was $22,200 for the year.

Living and Housing Costs

On-campus housing is not provided for graduate students, although numerous apartments are available within both walking and commuting distance of the campus. Costs are commensurate with those of similar urban areas.

Student Group

The Sever Institute enrolls about 650 full- and part-time graduate students. About 29 percent of these students are from other countries.

Location

Among the major cultural and educational institutions in the area are the St. Louis Symphony Orchestra, the St. Louis Art Museum, the Missouri Botanical Garden, the Jefferson National Expansion Memorial, the St. Louis Science Center, the Municipal Opera, and the St. Louis Zoo. Numerous concerts, lectures, and film series are presented on the campus and in the city.

The University

Washington University in St. Louis is an independent, privately endowed and supported institution that was founded in 1853. The campus, which adjoins the largest park in St. Louis, is in a pleasant residential area. The University has approximately 9,500 full-time students and more than 2,060 full-time faculty members. The University ranks among the top ten in the nation in the number of Nobel laureates associated with it.

Applying

The deadline for filing applications for financial aid is February 1 preceding the academic year for which aid is requested. There is no deadline for filing for admission. The typical cumulative GRE score (verbal + quantitative + analytical) of recently admitted students is above 2000. For international students whose native language is not English, a minimum TOEFL score of 600 is required. A minimum score of 5.0 is required on the TWE. All international students are required to take an English proficiency examination upon arrival and may be required to enroll at their expense in English as a foreign language courses. The application deadline for international students is February 1 for fall admission and November 1 for spring admission.

Correspondence and Information

Frank C.-P. Yin, M.D., Ph.D.
Chairman, Department of Biomedical Engineering
School of Engineering and Applied Science
Campus Box 1097
Washington University in St. Louis
One Brookings Drive
St. Louis, Missouri 63130-4899
Telephone: 314-935-6164
Fax: 314-935-7448
E-mail: admissioncoordinator@biomed.wustl.edu

Washington University in St. Louis

THE FACULTY AND THEIR RESEARCH

Full-Time and Joint Faculty

Amir A. Amini, Ph.D., Assistant Professor of Medicine. Medical imaging, cardiac MRI, medical computer vision.
R. Martin Arthur, Ph.D., Professor of Electrical Engineering. Ultrasonic imaging, electrocardiography.
Philip V. Bayly, Ph.D., Assistant Professor of Mechanical Engineering. Cardiac dynamics.
Elliot L. Elson, Ph.D., Professor of Biochemistry and Molecular Biophysics. Cellular mechanics and molecular biophysics.
Bijoy K. Ghosh, Ph.D., Professor of Systems Science and Mathematics. Biomedical imaging, microcircuits in turtle visual systems.
Sandor J. Kovacs, M.D., Ph.D., Associate Professor of Medicine. Cardiovascular biophysics, systems physiology.
Christine H. Lorenz, Ph.D., Assistant Professor of Medicine and Biomechanics. Cardiovascular, MRI.
James G. Miller, Ph.D., Professor of Physics. Ultrasonic imaging.
William F. Pickard, Ph.D., Professor of Electrical Engineering. Biological transport, electrobiology, plant electrophysiology.
Marcus E. Raichle, M.D., Professor of Radiology and Neurology. Functional brain imaging.
William D. Richard, Ph.D., Associate Professor of Electrical Engineering. Ultrasonic imaging, medical instrumentation.
Jin-Yu Shao, Ph.D., Assistant Professor of Biomedical Engineering. Cell mechanics, receptor and ligand interactions.
Joseph Smith, M.D., Assistant Professor of Medicine. Electrocardiography, modeling and quantitative analysis of arrhythmias.
Donald L. Snyder, Ph.D., Professor of Electrical Engineering. Communication theory, random process theory, signal processing.
David J. States, M.D., Ph.D., Associate Professor of Biomedical Computing. Computational molecular biology and genome analysis.
Salvatore P. Sutera, Ph.D., Professor of Biomedical Engineering. Hemorheology, mechanically assisted circulation.
Larry A. Taber, Ph.D., Professor of Biomedical Engineering. Mechanics of cardiovascular development.
David C. Van Essen, Ph.D., Professor of Neurobiology. Brain mapping.
Robert H. Waterston, M.D., Ph.D., Professor of Genetics. Mapping and sequencing the *C. elegans* genome.
Samuel A. Wickline, M.D., Professor of Medicine. Cardiovascular biophysics, acoustics, magnetic resonance imaging.
Frank C.-P. Yin, M.D., Ph.D., Professor and Chair of Biomedical Engineering. Biomechanics, cell mechanics.
George I. Zahalak, Eng.Sc.D., Professor of Mechanical Engineering. Mechanics of muscle and cells.

Affiliated Faculty

Charles H. Anderson, Ph.D., Research Professor of Neurobiology. Neuroscience of the visual system.
Kyongtae Ty Bae, M.D., Ph.D., Assistant Professor of Radiology. Magnetic resonance and CT imaging.
G. James Blaine III, D.Sc., Professor of Radiology. Digital electronic radiology.
John P. Boineau, M.D., Professor of Surgery (Cardiothoracic). Mapping and surgical treatment of arrhythmias.
Gary H. Brandenburger, D.Sc., Adjunct Professor of Biomedical Engineering. Ultrasonic contrast agents.
Paul C. Bridgman, Ph.D., Associate Professor of Neurology. Cellular properties of developing nerve and muscle.
Andreas Burkhalter, Ph.D., Associate Professor of Neurobiology. Synaptic mechanisms and organization of visual cortex.
Harold Burton, Ph.D., Professor of Neurobiology. Anatomy and physiology of the somatosensory system.
Michael E. Cain, M.D., Professor of Internal Medicine. Electrocardiography, cardiac electrophysiology.
Jose-Angel Conchello, Ph.D., Assistant Professor of Biomedical Computing in the IBC. Microscopy systems.
Thomas E. Conturo, M.D., Ph.D., Assistant Professor of Radiology. Magnetic resonance imaging.
P. Duffy Cutler, Ph.D., Assistant Professor of Radiology. Positron emission tomography imaging.
Ron K. Cytron, Ph.D., Associate Professor of Computer Science. DNA mapping and biological simulation.
Robert H. Deusinger, Assistant Professor of Physical Therapy. Biomechanical bases of musculoskeletal coordination and balance.
Michael L. Dustin, Ph.D., Assistant Professor of Pathology. Dynamic regulation of integrin interactions in the immune system.
Sean Eddy, Ph.D., Assistant Professor of Genetics. Computational molecular biology and genome analysis.
Jack R. Engsberg, Research Associate Professor of Neurological Surgery. Biomechanics, gate analysis, and dynometry.
Stefan Fischer, Ph.D., Clinical Scientist. Magnetic resonance imaging of myocardial deformation and perfusion.
William A. Frazier III, Ph.D., Professor of Biochemistry and Molecular Biophysics. Regulation of cellular phenotype by extracellular matrix.
Daniel R. Fuhrmann, Ph.D., Associate Professor of Electrical Engineering. Automated DNA mapping and sequencing.
Will D. Gillett, Ph.D., Associate Professor of Computer Science. Compiler theory and implementation, algorithm analysis.
Warren Gish, Ph.D., Assistant Professor of Genetics. Computational molecular biology and genome analysis.
Robert J. Gropler, M.D., Associate Professor of Radiology. Positron emission tomography, magnetic resonance imaging.
E. Mark Haacke, Ph.D., Professor of Radiology. Magnetic resonance imaging, MRI of coronary vasculature.
Stephen Highstein, M.D., Ph.D., Professor of Otolaryngology. Motor learning in the vestibulo-ocular reflex of squirrel monkeys.
James Huettner, Ph.D., Assistant Professor of Cell Biology and Physiology. Physiology of glutamate receptor-mediated signaling.
Joseph Klaesner, Ph.D., Research Assistant Professor of Physical Therapy. Factors contributing to injury in patients with chronic disabilities.
Jeff W. Lichtman, M.D., Ph.D., Professor of Neurobiology. Optical methods to visualize synaptic structure and function.
Weili Lin, Ph.D., Assistant Professor of Radiology. Magnetic resonance imaging.
Garland R. Marshall, Ph.D., Professor of Molecular Biology and Pharmacology. Molecular design.
Robert P. Mecham, Ph.D., Professor of Cell Biology and Physiology. Extracellular matrix and its influence on the phenotype of cells.
Thomas R. Miller, M.D., Ph.D., Professor of Radiology. Nuclear medicine imaging.
Scott D. Minor, Ph.D., Assistant Professor of Physical Therapy. Isokinetic muscle performance testing and injury prevention.
Stanley Misler, M.D., Ph.D., Associate Professor of Medicine. Modulation of quantal release by neuroendocrine cells.
Michael J. Mueller, Ph.D., Associate Professor of Physical Therapy. Mechanical factors to prevent amputations in diabetics.
Joseph A. O'Sullivan, Ph.D., Associate Professor of Electrical Engineering. Tomographic imaging.
Ruth Okamoto, D.Sc., Assistant Professor of Mechanical Engineering. Regional mechanical properties of myocardium.
John M. Ollinger, Ph.D., Research Assistant Professor of Radiology. Positron emission and magnetic resonance imaging.
Michael K. Pasque, M.D., Professor of Surgery (Cardiothoracic). Regional mechanical properties of myocardium.
Steven Petersen, Ph.D., Professor of Neurology (Neuropsychology). Functional imaging of vision, attention, memory, and language.
Jay Ponder, Ph.D., Assistant Professor of Biochemistry and Molecular Biophysics. Molecular design and molecular biophysics.
Fred U. Rosenberger, D.Sc., Associate Professor of Biomedical Computing in IBC. Image processing.
Carl M. Rovainen, Ph.D., Professor of Cell Biology and Physiology. Brain blood vessels.
John Schotland, M.D., Ph.D., Assistant Professor of Electrical Engineering. Optical imaging.
Richard B. Schuessler, Ph.D., Research Associate Professor of Surgery (Cardiothoracic). Cardiac arrhythmias.
Matthew J. Silva, Ph.D., Assistant Professor of Orthopedic Surgery. Bone mechanics, tendon mechanics and repair.
Stefano Soatto, D.Ing., Ph.D., Assistant Professor of Electrical Engineering. Dynamic vision, modeling of human and computer vision.
Joe Henry Steinbach, Ph.D., Professor of Anesthesiology. Channels gated by neurotransmitters, synaptic transmission.
Subhash Suri, Ph.D., Associate Professor of Computer Science. Computational geometry, algorithms.
Barna A. Szabo, Ph.D., Professor of Mechanical Engineering. Finite element analysis of regional heart mechanical properties.
Tzyh-Jong Tarn, D.Sc., Professor of Systems Science and Mathematics. Modeling, identification, and control of robotic systems.
Alan R. Templeton, Ph.D., Professor of Biology. Computational genomics, evolutionary genetics, genotype/phenotype complexity.
W. Thomas Thach, M.D., Professor of Pediatrics. Neurophysiology and modeling of learning and control of movements.
Jerold W. Wallis, M.D., Ph.D., Associate Professor of Radiology. Nuclear medicine imaging.
Michael J. Welch, Ph.D., Professor of Radiology. Radiopharmaceutical chemistry for imaging agents.
M. Victor Wickerhauser, Ph.D., Professor of Mathematics. Wavelets and image compression.
Thomas A. Woolsey, M.D., Professor of Experimental Neurological Surgery. Structure, function, and development of nervous system.
Dequan Zou, D.Sc., Research Assistant Professor of Physical Therapy. Biomechanics, FEA modeling, application of computer technology.
Michael Zuker, Ph.D., Associate Professor of Biomedical Computing. Algorithms for RNA structure prediction, sequence analysis.

WORCESTER POLYTECHNIC INSTITUTE

Biomedical Engineering Department

Programs of Study

The goal of the biomedical engineering programs is to apply engineering principles and technology to significant biomedical problems. A B.S. degree in engineering, physics, or computer science or the equivalent is required for admission to all programs. Special programs are available for outstanding graduates in the life sciences to accelerate their competence in engineering. All students are expected to achieve some basic knowledge in engineering, life sciences, and biomedical engineering. Full- and part-time programs are available.

The department offers a Master of Science (M.S.) degree in biomedical engineering and Master of Engineering (M.E.) degrees in biomedical engineering and clinical engineering. A minimum of 30 credit hours is required for the M.S. degree; at least 6 credit hours must be a thesis. Course requirements include 6 credits each of life science, biomedical engineering, advanced engineering–math, and electives. The clinical engineering M.S. program is for individuals interested in employment in hospitals and other clinical environments. The M.E. program differs from the M.S. program in that it is especially suited for those students with general-practice career goals and for those not planning subsequent pursuit of the Ph.D. degree. A minimum of 33 credit hours is required for the M.E. degree. Course requirements include 6 credits of life science, 12 credits of biomedical engineering, 6 credits of advanced engineering–math, and 9 credits of electives. Students can substitute 3–6 credits of directed research for 3 credits of biomedical engineering and/or 3 credits of electives. Students who are enrolled in the M.E. degree program in clinical engineering take a clinical engineering internship at an affiliated hospital. Students without an engineering or related undergraduate degree must take thematically related undergraduate and graduate engineering, math, and science courses in addition to the required biomedical engineering courses and/or thesis. Full-time graduate students with an engineering or equivalent background should expect to spend about two years in the master's program. The Ph.D. requires a minimum of 90 credits beyond the bachelor's degree, with at least one year of full-time residence at WPI. A Ph.D. qualifying exam is required and is normally taken following the first year of study. A candidacy exam to test the student's knowledge within a chosen specialization is given within two years following the qualifying exam. The Ph.D. program does not require a foreign language exam.

The joint WPI–University of Massachusetts Medical Center (UMMC) Ph.D. program combines the resources of WPI with those of UMMC, a medical school and biomedical research center located 2 miles from WPI. The faculty members in this program have varied research interests, including biomedical imaging, biosensors and biomedical instrumentation, tissue and cell engineering, and biomechanics. The program emphasizes basic and applied research in an environment that promotes strong interaction between biomedical engineers, researchers, and medical professionals.

Research Facilities

Research projects are conducted in the Salisbury Laboratories at WPI. The facilities include a Biosensors Laboratory comprised of four labs for basic and applied medical optics research. Other research is carried out in the physiology, medical imaging, and electrophysiology labs. The facilities also contain animal surgery and holding quarters. An additional medical imaging facility has been established at the Central Massachusetts Magnetic Imaging Center as a part of a joint research project between the Biomedical Engineering Department and the Department of Radiology at the University of Massachusetts Medical Center (UMMC). Close cooperation with UMMC, Tufts University Veterinary School (TUVS), and St. Vincent Hospital (SVH) makes their staff and facilities available for thesis projects and internships. Research projects are also available at other affiliated institutions, such as Clark University and UMass–Memorial Health Care. The Gordon Library provides complete library service, including an extensive microfiche collection, resource videotapes, and a computerized literature search service. Access is also available to other libraries in the area, including the library at the University of Massachusetts Medical School (UMMS).

Financial Aid

Assistantships are available for teaching or research. In 1999–2000, these generally carry a stipend of $1370 per month and remission of tuition for 20 credits per academic year; some assistantships have a higher basic stipend. WPI's Goddard Fellowship and various research grants and employment opportunities also provide financial support. Additional assistance may be available during the summer months.

Cost of Study

Graduate tuition for the 1999–2000 academic year is $661 per credit hour. There are nominal extra charges for theses and health insurance.

Living and Housing Costs

Although graduate students do not generally live in dormitory rooms, they may eat at the Institute's dining facilities. Apartments and rooms in private homes near the campus are available at varying costs.

Student Group

WPI enrolls about 3,500; approximately 1,000 are graduate students. The department has about 50 full- and part-time graduate students; 25 percent of the students are women, and 30 percent are international students. About 60 percent of the full-time students in the department receive some financial aid.

Location

WPI is located on a 62-acre campus in Worcester. Flanked by city parks, it has a feeling of openness though located in the second-largest city in New England. Boston, Cape Cod, and the Berkshires are easily reached for recreation. The city is well known for its many colleges, museums, the Science Center, outstanding shopping malls, and the Centrum, a 15,000-seat concert hall and multifunction civic center.

The Institute and The Department

The Institute, founded in 1865, is the third-oldest U.S. college of engineering and science. Graduate study has been a part of the Institute's activity for more than eighty years. Classes are small and provide for close student-faculty relationships. Graduate students frequently interact in research with undergraduates.

The department began as the Biomedical Engineering Program in 1962. It has strong cooperative programs with UMMC, TUVS, and SVH. Students may take courses through the Worcester Consortium Colleges.

Applying

Application for admission and financial assistance should be made by February 15. Applications for students not requesting financial aid are considered until June 1. The deadline for Goddard Fellowship applications is February 15. Results of the GRE General Test are also required for admission.

Correspondence and Information

For additional information:
Christopher H. Sotak, Ph.D.
Professor and Department Chairman
Biomedical Engineering Department
Worcester Polytechnic Institute
Worcester, Massachusetts 01609

For application materials:
Office of Graduate Admissions
Worcester Polytechnic Institute
100 Institute Road
Worcester, Massachusetts 01609

Worcester Polytechnic Institute

THE FACULTY AND THEIR RESEARCH

Department Faculty

Christopher H. Sotak, Professor and Department Chairman; Ph.D., Syracuse, 1983. Nuclear magnetic resonance (NMR) spectroscopy and imaging in biology and medicine, nonmedical applications of high-resolution ("Microscopic") NMR imaging, hardware and software aspects of NMR instrumentation.

Karl G. Helmer, Research Assistant Professor; Ph.D., Rochester, 1992. Application of porous media methods to biological systems using nuclear magnetic resonance spectroscopy (NMRS) and imaging (NMRI), evaluation and treatment of stroke and tumor.

Sean S. Kohles, Associate Professor; Ph.D., Wisconsin–Madison, 1994. Tissue engineering, biomechanics, solid mechanics, experimental mechanics, injury mechanics, biomaterials, composite materials, fatigue.

Stevan Kun, Research Assistant Professor; Ph.D., Worcester Polytechnic, 1990. Biomedical instrumentation, impedance imaging and spectroscopy, impedance-based noninvasive biosensors, cardiac electrophysiology, analytical and numerical modeling of electric fields.

Yitzhak Mendelson, Associate Professor; Ph.D., Case Western Reserve, 1983. Biosensors, microcomputer-based biomedical instrumentation, invasive and noninvasive blood gas monitoring, applications of optics to biomedicine.

Robert A. Peura, Professor; Ph.D., Iowa State, 1969. Biomedical instrumentation and biosensors; noninvasive monitoring of blood glucose and urea; impedance imaging and spectroscopy.

Ross D. Shonat, Assistant Professor; Ph.D., Pennsylvania, 1991. Development and application of biomedical imaging techniques to in vivo studies of physiology in inbred, transgenic, and knockout mouse models; studies of oxygen and blood-flow dynamics in neural tissues and the eye.

WPI Faculty Associated with the Program

Holly K. Ault, Associate Professor (mechanical engineering); Ph.D., Worcester Polytechnic, 1988. Computer-aided design, engineering design, graphics; biomechanics of joints and soft tissues.

David Cyganski, Professor (electrical and computer engineering); Ph.D., Worcester Polytechnic, 1981. Bioimaging, multivariate classification, signal decomposition.

David DiBiasio, Associate Professor (chemical engineering); Ph.D., Purdue, 1980. Biological reactor engineering, biochemical kinetics.

William W. Durgin, George I. Alden Professor (mechanical engineering); Ph.D., Brown, 1970. Biomedical fluid mechanics, aerodynamics, hydrodynamics, flow-induced vibration, microgravity fluid dynamics.

Michael A. Gennert, Associate Professor (computer science); Sc.D., MIT, 1987. Computer vision and image processing, including computational theories of vision and applications; tomographic reconstruction and image processing of reconstructed imagery; artificial intelligence and neural networks, especially applied to vision problems.

Daniel G. Gibson III, Assistant Professor (biology and biotechnology); Ph.D., Boston University, 1983. Neurobiology, cell culture.

Allen H. Hoffman, Professor (mechanical engineering); Ph.D., Colorado, 1970. Biomechanics, biomaterials, biomedical engineering, dynamics, fluid mechanics, continuum mechanics.

Fred J. Looft III, Professor (electrical engineering); Ph.D., Michigan, 1979. Instrumentation, biomedical signal processing, digital and analog systems, biomedical engineering.

Peder C. Pedersen, Professor (electrical engineering); Ph.D., Utah, 1976. Applications of inverse methods for early detection of arteriosclerotic plaque; imaging of the skin with ultrasound for assessment of cancer, burns, skin grafts, and effects of therapy in dermatology; ultrasound transducer characterization techniques; ultrasound attenuation mapping technique for soft tissue; detection of deep-vein thrombosis with Doppler ultrasound.

Richard S. Quimby, Associate Professor (physics); Ph.D., Wisconsin–Madison, 1979. Optics, laser spectroscopy, photoacoustic spectroscopy, optics of thin metal films, nondestructive testing, biomedical applications of lasers, fiber optics.

Brian J. Savilonis, Professor (mechanical engineering); Ph.D., SUNY at Buffalo, 1976. Fluid mechanics, biofluid mechanics, heat transfer.

Satya Shivkumar, Associate Professor (mechanical engineering); Ph.D., Stevens, 1987. Biomaterials, polymers, materials science and engineering.

John Sullivan, Associate Professor (mechanical engineering); Ph.D., Dartmouth, 1986. Computational mechanics, biomechanics, finite-element analysis, heat transfer.

Matthew O. Ward, Associate Professor (computer science); Ph.D., Connecticut, 1979. Computer graphics, visualization, biomolecular pattern recognition.

Adjunct Faculty

Shyue-Ling Chen, M.S., Assistant Professor, Engineering in the Clinical Environment. Director of Medical Engineering, New England Medical Center, Boston.

M. A. King, Ph.D., Professor of Nuclear Medicine, University of Massachusetts Medical School.

Michael Leal, Instructor, Medical Device Regulation and Quality Control. Engineer, U.S. Food and Drug Administration.

John McCracken, Ph.D., Professor, Laboratory Animal Surgery. Senior Scientist, Worcester Foundation for Experimental Biology.

Affiliate Faculty

F. A. Anderson Jr., Ph.D., Associate Professor, Division of Surgical Research, University of Massachusetts Medical School.

S. Aronow, Ph.D., Certified Radiological Physicist, Technology in Medicine, Inc.

G. Cho, M.S., Senior Vice President, Cynosure, Inc.

B. S. Cutler, M.D., Professor and Chairman, Division of Vascular Surgery, University of Massachusetts Medical School.

M. A. Davis, M.D., Sc.D., Professor of Radiology and Director, Radiological Research Laboratories, University of Massachusetts Medical School.

R. M. Dunn, M.D., Associate Professor, Division of Plastic and Reconstructive Surgery, University of Massachusetts Medical Center.

C. L. Feldman, Sc.D., Professor, Brigham & Women's Hospital, Boston.

M. Fisher, M.D., Chief of Neurology, Medical Center of Central Massachusetts–Memorial.

R. M. Giasi, M.D., Associate Professor of Anesthesiology, University of Massachusetts Medical School.

P. Grigg, Ph.D., Professor, Department of Physiology, University of Massachusetts Medical School.

J. B. Hermann, M.D., Professor, Division of Vascular Surgery, University of Massachusetts Medical School.

A. Karellas, Ph.D., Associate Professor, Department of Radiology, University of Massachusetts Medical School.

M. T. Kirber, Ph.D., Assistant Professor, Department of Physiology, University of Massachusetts Medical School.

P. W. Kotilainen, Ph.D., Product Manager, Hewlett-Packard Co.

K. Kraus, D.V.M., Professor of Surgery, Tufts University School of Veterinary Medicine.

J. P. Lock, M.D., Assistant Professor of Medicine, University of Massachusetts Medical School.

G. Majno, M.D., Professor and Chairman, Department of Pathology, University of Massachusetts Medical School.

I. S. Ockene, M.D., Professor of Medicine; Director of Preventative Cardiology Program; and Associate Director of Division of Cardiovascular Medicine, University of Massachusetts Medical School.

J. A. Paraskos, M.D., Professor and Director of Noninvasive Cardiology, Cardiovascular Medicine, and Associate Chairman, Department of Medicine, University of Massachusetts Medical School.

N. A. Patwardhan, M.D., Associate Professor of General Surgery, University of Massachusetts Medical School.

M. J. Rohrer, M.D., Assistant Professor of Surgery, University of Massachusetts Medical School.

A. Shahnarian, Ph.D., Consultant, Biomedical Anesthesia Associates.

J. Singer, Ph.D., Professor of Physiology, University of Massachusetts Medical School.

T. J. Vander Salm, M.D., Professor, Division of Cardiothoracic Surgery, University of Massachusetts Medical School.

J. V. Walsh, M.D., Professor of Physiology, University of Massachusetts Medical School.

H. B. Wheeler, M.D., Harry M. Haidak Distinguished Professor and Chairman, Department of Surgery, University of Massachusetts Medical School.

Section 6
Chemical Engineering

This section contains a directory of institutions offering graduate work in chemical engineering, followed by in-depth entries submitted by institutions that chose to prepare detailed program descriptions. Additional information about programs listed in the directory but not augmented by an in-depth entry may be obtained by writing directly to the dean of a graduate school or chair of a department at the address given in the directory.

For programs offering related work, see also in this book Engineering and Applied Sciences; Geological, Mineral/Mining, and Petroleum Engineering; Management of Engineering and Technology; and Materials Sciences and Engineering. In Book 2, see Home Economics and Family Studies (Clothing and Textiles); in Book 3, Biochemistry; and in Book 4, Chemistry and Geosciences (Geochemistry and Geology).

CONTENTS

Biochemical Engineering

California Polytechnic State University, San Luis Obispo, College of Engineering, Program in Engineering, San Luis Obispo, CA 93407. Offers biochemical engineering (MS); industrial engineering (MS); integrated technology management (MS); materials engineering (MS); mechanical engineering (MS); water engineering (MS). *Faculty:* 98 full-time (8 women), 82 part-time (14 women). *Students:* 25 full-time (3 women), 6 part-time. *Degree requirements:* Foreign language not required. *Entrance requirements:* For master's, GRE General Test, minimum GPA of 2.5 in last 90 quarter units. Application fee: $55. Tuition, nonresident: part-time $164 per unit. Required fees: $531 per quarter. *Unit head:* Dr. Paul E. Rainey, Associate Dean, 805-756-2131, Fax: 805-756-6503, E-mail: prainey@calpoly.edu. *Application contact:* Dr. Paul E. Rainey, Associate Dean, 805-756-2131, Fax: 805-756-6503, E-mail: prainey@calpoly.edu.

Cornell University, Graduate School, Graduate Fields of Engineering, Field of Chemical Engineering, Ithaca, NY 14853-0001. Offers advanced materials processing (M Eng, MS, PhD); applied mathematics and computational methods (M Eng, MS, PhD); biochemical engineering (M Eng, MS, PhD); chemical reaction engineering (M Eng, MS, PhD); classical and statistical thermodynamics (M Eng, MS, PhD); fluid dynamics, rheology and biorheology (M Eng, MS, PhD); heat and mass transfer (M Eng, MS, PhD); kinetics and catalysis (M Eng, MS, PhD); polymers (M Eng, MS, PhD); surface science (M Eng, MS, PhD). *Faculty:* 18 full-time. *Students:* 71 full-time (17 women); includes 9 minority (1 African American, 7 Asian Americans or Pacific Islanders, 1 Hispanic American), 29 international. *Degree requirements:* For master's, thesis (MS) required; for doctorate, dissertation required, foreign language not required. *Entrance requirements:* For master's and doctorate, GRE General Test, TOEFL (minimum score of 580 required). *Application deadline:* For fall admission, 1/15. Application fee: $65. Electronic applications accepted. *Unit head:* Director of Graduate Studies, 607-255-4550, Fax: 607-255-9166. *Application contact:* Graduate Field Assistant, 607-255-4550, Fax: 607-255-9166, E-mail: dgs@cheme.cornell.edu.

Dartmouth College, Thayer School of Engineering, Program in Biotechnology and Biochemical Engineering, Hanover, NH 03755. Offers MS, PhD. *Degree requirements:* For master's, thesis required; for doctorate, dissertation, candidacy oral exam required. *Entrance requirements:* For master's and doctorate, GRE General Test. *Application deadline:* For fall admission, 1/15 (priority date). Application fee: $20 ($40 for international students). *Financial aid:* Fellowships, research assistantships, teaching assistantships, career-related internships or fieldwork, Federal Work-Study, institutionally-sponsored loans, and tuition waivers (full and partial) available. Financial aid application deadline: 1/15. *Faculty research:* Biomass processing, metabolic engineering, bioplastic synthesis, kinetics and reactor design, applied microbiology. Total annual research expenditures: $400,000. *Unit head:* Graduate Field Assistant, 607-255-1003, E-mail: biomedeng@cornell.edu. *Application contact:* Candace S. Potter, Admissions Coordinator, 603-646-3844, Fax: 603-646-3856, E-mail: candace.potter@dartmouth.edu.

Drexel University, Graduate School, College of Engineering, Department of Chemical Engineering, Program in Biochemical Engineering, Philadelphia, PA 19104-2875. Offers MS. Part-time and evening/weekend programs available. *Students:* 2 full-time (1 woman), 1 part-time, 2 international. Average age 26. 21 applicants, 52% accepted. In 1998, 1 degree awarded. *Degree requirements:* For master's, thesis required, foreign language not required. *Entrance requirements:* For master's, TOEFL (minimum score of 570 required), minimum GPA of 3.0 in chemical engineering or biological sciences. *Application deadline:* For fall admission, 8/21. Applications are processed on a rolling basis. Application fee: $35. Tuition: Full-time $15,795; part-time $585 per credit. Required fees: $375; $67 per term. Tuition and fees vary according to program. *Financial aid:* In 1998–99, 2 research assistantships, 1 teaching assistantship were awarded.; career-related internships or fieldwork, Federal Work-Study, tuition waivers (full and partial), and unspecified assistantships also available. Financial aid application deadline: 2/1. *Faculty research:* Monitoring and control of bioreactors, sensors for bioreactors, large-scale production of monoclonal antibodies. *Unit head:* Dr. Rajakkannu Mutharasan, Director, 215-895-2236. *Application contact:* Kelli Kennedy, Director of Admissions, 215-895-6706, Fax: 215-895-5939, E-mail: crowlka@duvm.ocs.drexel.edu.

Rutgers, The State University of New Jersey, New Brunswick, Graduate School, Program in Chemical and Biochemical Engineering, New Brunswick, NJ 08903. Offers MS, PhD. Part-time and evening/weekend programs available. *Faculty:* 18 full-time (2 women), 3 part-time (1 woman). *Students:* 48 full-time (26 women), 20 part-time (5 women); includes 8 minority (2 African Americans, 6 Asian Americans or Pacific Islanders), 21 international. Average age 25. 163 applicants, 10% accepted. In 1998, 12 master's awarded (33% found work related to degree, 67% continued full-time study); 5 doctorates awarded (40% entered university research/teaching, 60% found other work related to degree). Terminal master's awarded for partial completion of doctoral program. *Degree requirements:* For master's, thesis required (for some programs), foreign language not required; for doctorate, dissertation required, foreign language not required. *Entrance requirements:* For master's, GRE General Test (minimum combined score of 1100 required; average 1320), TOEFL (minimum score of 550 required); for doctorate, GRE General Test (minimum combined score of 1100 required; average 1320). *Application deadline:* For fall admission, 1/15 (priority date); for spring admission, 10/15. Applications are processed on a rolling basis. Application fee: $50. *Financial aid:* In 1998–99, 48 students received aid, including 21 fellowships with full tuition reimbursements available (averaging $16,500 per year), 12 research assistantships with full tuition reimbursements available (averaging $16,500 per year), 10 teaching assistantships with full tuition reimbursements available (averaging $16,500 per year); unspecified assistantships also available. Financial aid application deadline: 1/15; financial aid applicants required to submit FAFSA. *Faculty research:* Biotechnology, environmental engineering, statistical thermodynamics, polymers, pharmaceutical engineering. Total annual research expenditures: $4 million. *Unit head:* Dr. Henrik Pedersen, Graduate Director, 732-445-4950, Fax: 732-445-2421, E-mail: cbemail@sol.rutgers.edu.

University of California, Irvine, Office of Research and Graduate Studies, School of Engineering, Department of Chemical and Biochemical Engineering and Materials Science, Irvine, CA 92697. Offers biomedical engineering (MS, PhD), including engineering; chemical and biochemical engineering (MS, PhD). Part-time programs available. *Faculty:* 10 full-time (2 women). *Students:* 30 full-time (9 women), 4 part-time (1 woman); includes 11 minority (10 Asian Americans or Pacific Islanders, 1 Hispanic American), 17 international. 72 applicants, 46% accepted. In 1998, 11 master's, 6 doctorates awarded. Terminal master's awarded for partial completion of doctoral program. *Degree requirements:* For doctorate, dissertation required, foreign language not required, foreign language not required. *Entrance requirements:* For master's, GRE General Test, minimum GPA of 3.0; for doctorate, GRE General Test. *Application deadline:* For fall admission, 1/15 (priority date). Applications are processed on a rolling basis. Application fee: $40. Electronic applications accepted. *Financial aid:* Fellowships, research assistantships, teaching assistantships, institutionally-sponsored loans available. Financial aid application deadline: 3/2; financial aid applicants required to submit FAFSA. *Faculty research:* Bioreactors, recombinant cells, separation operations. *Unit head:* Dr. Enrique J. Lavernia, Chair, 949-824-8277, Fax: 949-824-2541, E-mail: lavernia@uci.edu. *Application contact:* Nancy Carter-Fields, Graduate Coordinator, 949-824-3786, Fax: 949-824-2541, E-mail: nvcarter@uci.edu.

See in-depth description on page 509.

The University of Iowa, Graduate College, College of Engineering, Department of Chemical and Biochemical Engineering, Iowa City, IA 52242-1316. Offers MS, PhD. *Faculty:* 5 full-time, 1 part-time. *Students:* 16 full-time (6 women), 13 part-time (2 women); includes 2 minority (both African Americans), 16 international. 124 applicants, 12% accepted. In 1998, 4 master's, 4 doctorates awarded. *Degree requirements:* For master's, thesis optional; for doctorate, dissertation, comprehensive exam required. *Entrance requirements:* For master's and doctorate, GRE, TOEFL. *Application deadline:* Applications are processed on a rolling basis. Application fee: $30 ($50 for international students). *Financial aid:* In 1998–99, 3 fellowships, 22 research assistantships, 8 teaching assistantships were awarded. Financial aid applicants required to submit FAFSA. *Faculty research:* Catalysis and reactor design, fine-particle morphology and behavior, air pollution modelling. *Unit head:* Dr. Gregory Carmichael, Interim Chair, 319-335-1414.

University of Maryland, Baltimore County, Graduate School, College of Engineering, Department of Chemical and Biochemical Engineering, Baltimore, MD 21250-5398. Offers MS, PhD. *Faculty:* 8 full-time (4 women), 8 part-time (0 women). *Students:* 37 full-time (15 women), 8 part-time (1 woman); includes 4 minority (all Asian Americans or Pacific Islanders), 18 international. 78 applicants, 9% accepted. In 1998, 1 master's, 2 doctorates awarded. *Entrance requirements:* For master's, GRE General Test, minimum GPA of 3.0; for doctorate, GRE General Test, GRE Subject Test, TOEFL, minimum GPA of 3.0. *Application deadline:* For fall admission, 7/1. Applications are processed on a rolling basis. Application fee: $45. *Financial aid:* Fellowships, research assistantships, teaching assistantships available. *Faculty research:* Bioengineering, mammalian cell culture, protein purification, adsorptive separation. *Unit head:* Dr. Douglas D. Frey, Interim Chair, 410-455-3400. *Application contact:* Director, Graduate Program, 410-455-3693.

Chemical Engineering

Arizona State University, Graduate College, College of Engineering and Applied Sciences, Department of Chemical, Bio and Materials Engineering, Program in Chemical Engineering, Tempe, AZ 85287. Offers MS, MSE, PhD. *Degree requirements:* For doctorate, dissertation required. *Entrance requirements:* For master's and doctorate, GRE General Test. Application fee: $45. *Financial aid:* Fellowships, research assistantships, teaching assistantships available. *Faculty research:* Biomedical and clinical engineering, chemical process engineering, energy and materials conversion, environmental control. *Application contact:* Graduate Secretary, 480-965-9707.

See in-depth description on page 495.

Auburn University, Graduate School, College of Engineering, Department of Chemical Engineering, Auburn, Auburn University, AL 36849-0002. Offers M Ch E, MS, PhD. Part-time programs available. *Faculty:* 14 full-time (0 women). *Students:* 56 full-time (6 women), 11 part-time (2 women); includes 9 minority (6 African Americans, 2 Asian Americans or Pacific Islanders, 1 Hispanic American), 28 international. 70 applicants, 23% accepted. In 1998, 7 master's, 16 doctorates awarded. *Degree requirements:* For master's, thesis (MS) required; for doctorate, dissertation, comprehensive exams required, foreign language not required. *Entrance requirements:* For master's, GRE General Test; for doctorate, GRE General Test (minimum score of 400 on each section required). *Application deadline:* For fall admission, 9/1; for spring admission, 3/1. Applications are processed on a rolling basis. Application fee: $25 ($50 for international students). Tuition, state resident: full-time $2,760; part-time $76 per credit hour. Tuition, nonresident: full-time $8,280; part-time $228 per credit hour. *Financial aid:* Fellowships, research assistantships, teaching assistantships, Federal Work-Study available. Aid available to part-time students. Financial aid application deadline: 3/15. *Faculty research:* Coal liquefaction, asphalt research, pulp and paper engineering, surface science, biochemical engineering, microfibrous materials manufacturing. *Unit head:* Dr. Robert P. Chambers, Head, 334-844-4827. *Application contact:* Dr. John F. Pritchett, Dean of the Graduate School, 334-844-4700.

Brigham Young University, Graduate Studies, College of Engineering and Technology, Department of Chemical Engineering, Provo, UT 84602-1001. Offers chemical engineering (MS); engineering (PhD). *Faculty:* 13 full-time (0 women), 2 part-time (0 women). *Students:* 36 full-time (3 women); includes 1 minority (Hispanic American), 11 international. Average age 25. 56 applicants, 16% accepted. In 1998, 8 master's awarded (100% found work related to degree); 3 doctorates awarded (100% found work related to degree). *Degree requirements:* For master's and doctorate, thesis/dissertation required, foreign language not required. *Entrance requirements:* For master's, minimum GPA of 3.0 in upper-division major classes; for doctorate, minimum GPA of 3.3. *Average time to degree:* Master's–2.5 years full-time; doctorate–5.25 years full-time. *Application deadline:* For fall admission, 5/1. Application fee: $30. Tuition: Full-time $3,330; part-time $185 per credit hour. Tuition and fees vary according to program and student's religious affiliation. *Financial aid:* In 1998–99, 12 fellowships (averaging $5,000 per year), 14 research assistantships (averaging $4,400 per year), 8 teaching assistantships (averaging $9,450 per year) were awarded.; career-related internships or fieldwork, institutionally-sponsored loans, and scholarships also available. Financial aid application deadline: 3/15. *Faculty research:* Kinetics-catalysis, combustion, thermodynamics, environmental engineering, biomedical engineering. Total annual research expenditures: $1.5 million. *Unit head:* Dr. Kenneth A. Solen, Chair, 801-378-2586, Fax: 801-378-7799. *Application contact:* Thomas H. Fletcher, Graduate Coordinator, 801-378-6236, Fax: 801-378-7799, E-mail: tom@harvey.et.byu.edu.

Brown University, Graduate School, Division of Engineering, Program in Fluid Mechanics, Thermodynamics, and Chemical Processes, Providence, RI 02912. Offers Sc M, PhD. *Degree requirements:* For doctorate, dissertation, preliminary exam required, foreign language not required, foreign language not required.

Bucknell University, Graduate Studies, College of Engineering, Department of Chemical Engineering, Lewisburg, PA 17837. Offers MS, MS Ch E. *Faculty:* 6 full-time. *Students:* 5 (1 woman). *Degree requirements:* For master's, thesis required, foreign language not required. *Entrance requirements:* For master's, GRE General Test (minimum combined score of 1000 required), GRE Subject Test, TOEFL (minimum score of 550 required), minimum GPA of 2.8. *Application deadline:* For fall admission, 6/1 (priority date); for spring admission, 12/1 (priority date). Applications are processed on a rolling basis. Application fee: $25. Tuition: Part-time $2,600 per course. Tuition and fees vary according to course load. *Financial aid:* Research assistantships, teaching assistantships, unspecified assistantships available. Financial aid application deadline: 3/1. *Faculty research:* Computer-aided design, software engineering, applied mathematics and modeling, polymer science, digital process control. *Unit head:* Dr. Jeffrey Cscernica, Chairman, 570-577-1114.

Chemical Engineering

California Institute of Technology, Division of Chemistry and Chemical Engineering, Program in Chemical Engineering, Pasadena, CA 91125-0001. Offers MS, PhD. Terminal master's awarded for partial completion of doctoral program. *Degree requirements:* For master's and doctorate, thesis/dissertation required, foreign language not required. *Faculty research:* Chemical reaction engineering, transport phenomena, biochemical engineering, catalysis, ceramics and electronic materials.

Carnegie Mellon University, Carnegie Institute of Technology, Department of Chemical Engineering, Pittsburgh, PA 15213-3891. Offers chemical engineering (M Ch E, MS, PhD); colloids, polymers and surfaces (MS). Part-time and evening/weekend programs available. *Faculty:* 21 full-time (2 women), 1 (woman) part-time. *Students:* 71 full-time (17 women), 29 part-time (9 women). Average age 26. In 1998, 15 master's, 12 doctorates awarded. Terminal master's awarded for partial completion of doctoral program. *Degree requirements:* For doctorate, dissertation, qualifying exam required, foreign language not required, foreign language not required. *Entrance requirements:* For master's and doctorate, GRE General Test, GRE Subject Test, TOEFL. *Average time to degree:* Doctorate–4.7 years full-time. *Application deadline:* For fall admission, 2/1; for spring admission, 10/15. Application fee: $50. *Financial aid:* Fellowships, research assistantships, teaching assistantships, Federal Work-Study available. *Faculty research:* Computer-aided design in process engineering, biomedical engineering, biotechnology, complex fluids. Total annual research expenditures: $2.1 million. *Unit head:* Ignacio Grossman, Head, 412-268-2230. *Application contact:* Amanda Mottorn, Graduate Admissions Coordinator, 412-268-2243, Fax: 412-268-7139.

Case Western Reserve University, School of Graduate Studies, The Case School of Engineering, Department of Chemical Engineering, Cleveland, OH 44106. Offers ME, MS, PhD. Part-time and evening/weekend programs available. Postbaccalaureate distance learning degree programs offered. *Faculty:* 11 full-time (0 women), 1 part-time (0 women). *Students:* 7 full-time (1 woman), 34 part-time (8 women). Average age 25. 100 applicants, 5% accepted. In 1998, 12 master's, 4 doctorates awarded. Terminal master's awarded for partial completion of doctoral program. *Degree requirements:* For master's, thesis required (for some programs), foreign language not required; for doctorate, dissertation, qualifying exam, research proposal required, foreign language not required. *Entrance requirements:* For master's and doctorate, TOEFL (minimum score of 600 required). *Application deadline:* For fall admission, 3/1 (priority date); for spring admission, 11/1. Applications are processed on a rolling basis. Application fee: $25. *Financial aid:* In 1998–99, 10 fellowships with full and partial tuition reimbursements (averaging $15,000 per year), 13 research assistantships with full and partial tuition reimbursements (averaging $15,000 per year), 41 teaching assistantships with full and partial tuition reimbursements (averaging $11,250 per year) were awarded.; Federal Work-Study, institutionally-sponsored loans, and tuition waivers (full and partial) also available. Financial aid application deadline: 3/1. *Faculty research:* Advanced materials (diamond thin-film ceramics), surface chemistry and engineering, electrochemistry and engineering, advanced separation processes. Total annual research expenditures: $1.5 million. *Unit head:* Nelson Gardner, Chairman, 216-368-4182, Fax: 216-368-3016, E-mail: nxg3@po.cwru.edu. *Application contact:* Seymour Birnbaum, Business Manager, 216-368-3840, Fax: 216-368-3016, E-mail: sxb33@po.cwru.edu.

City College of the City University of New York, Graduate School, School of Engineering, Department of Chemical Engineering, New York, NY 10031-9198. Offers ME, MS, PhD. Part-time programs available. *Students:* 1 (woman) full-time, 16 part-time (3 women). In 1998, 8 degrees awarded. *Degree requirements:* For master's, computer language required, thesis optional, foreign language not required; for doctorate, one foreign language (computer language can substitute), dissertation, comprehensive exams required. *Entrance requirements:* For master's, TOEFL (minimum score of 500 required); for doctorate, GRE General Test, TOEFL. *Application deadline:* Applications are processed on a rolling basis. Application fee: $40. *Financial aid:* Fellowships, research assistantships available. *Faculty research:* Theoretical turbulences, bio-fluid dynamics, polymers, fluidization, transport phenomena. *Unit head:* Robert Graff, Chairman, 212-690-8136. *Application contact:* Graduate Admissions Office, 212-650-6977.

Clarkson University, Graduate School, School of Engineering, Department of Chemical Engineering, Potsdam, NY 13699. Offers ME, MS, PhD. *Faculty:* 13 full-time (2 women). *Students:* 25 full-time (7 women), 1 (woman) part-time, 24 international. Average age 25. 121 applicants, 12% accepted. In 1998, 7 master's, 4 doctorates awarded. *Degree requirements:* For master's, thesis required, foreign language not required; for doctorate, dissertation, departmental qualifying exam required, foreign language not required. *Entrance requirements:* For master's, GRE, TOEFL. *Application deadline:* For fall admission, 5/15 (priority date); for spring admission, 10/15 (priority date). Applications are processed on a rolling basis. Application fee: $25 ($35 for international students). Tuition: Part-time $661 per credit hour. Required fees: $215 per semester. *Financial aid:* In 1998–99, 17 research assistantships, 6 teaching assistantships were awarded.; fellowships *Faculty research:* Separation techniques, fluid mechanics, computational thermodynamics, corrosion, surface studies, electronic manufacturing. Total annual research expenditures: $885,134. *Unit head:* Dr. Ross Taylor, Chairman, 315-268-6652, Fax: 315-268-6654, E-mail: taylor@clarkson.edu. *Application contact:* Dr. Philip K. Hopke, Dean of the Graduate School, 315-268-6447, Fax: 315-268-7994, E-mail: hopkepk@clarkson.edu.

Clemson University, Graduate School, College of Engineering and Science, School of Chemical and Materials Engineering, Department of Chemical Engineering, Clemson, SC 29634. Offers M Engr, MS, PhD. *Students:* 29 full-time (5 women), 6 part-time (2 women); includes 1 minority (Asian American or Pacific Islander), 9 international. Average age 25. 58 applicants, 19% accepted. In 1998, 5 master's, 5 doctorates awarded. *Degree requirements:* For master's and doctorate, thesis/dissertation required, foreign language not required. *Entrance requirements:* For master's and doctorate, GRE General Test, TOEFL. *Application deadline:* For fall admission, 6/1. Application fee: $35. *Financial aid:* Fellowships, research assistantships, teaching assistantships, career-related internships or fieldwork available. Financial aid applicants required to submit FAFSA. *Faculty research:* Polymer processing, catalysis, process automation, thermodynamics, separation processes. Total annual research expenditures: $800,000. *Unit head:* Dr. Charles H. Gooding, Chair, 864-656-3055, Fax: 864-656-0784, E-mail: chgdng@clemson.edu. *Application contact:* Dr. Doug Hirt, Graduate Coordinator, 864-656-0822, Fax: 864-656-0784, E-mail: hirtde@clemson.edu.

Cleveland State University, College of Graduate Studies, Fenn College of Engineering, Department of Chemical Engineering, Cleveland, OH 44115-2440. Offers MS, D Eng. Part-time programs available. *Faculty:* 7 full-time (1 woman). *Students:* 6 full-time (1 woman), 25 part-time (2 women); includes 2 minority (1 African American, 1 Asian American or Pacific Islander), 19 international. Average age 25. 125 applicants, 79% accepted. In 1998, 14 master's awarded. *Degree requirements:* For master's, project or thesis required; for doctorate, dissertation, candidacy and qualifying exams required, foreign language not required. *Entrance requirements:* For master's, GRE General Test, GRE Subject Test, TOEFL, minimum GPA of 2.75; for doctorate, GRE General Test, GRE Subject Test, TOEFL, minimum GPA of 3.25. *Application deadline:* For fall admission, 7/15 (priority date). Applications are processed on a rolling basis. Application fee: $25. *Financial aid:* In 1998–99, 5 research assistantships, 3 teaching assistantships were awarded.; fellowships, career-related internships or fieldwork, Federal Work-Study, institutionally-sponsored loans, and unspecified assistantships also available. Financial aid application deadline: 4/30. *Faculty research:* Heterogeneous transport, absorption equilibrium and dynamics, transfer process in non-Newtonian fluids, tribology, optimization of mammalian cell culture, simulation and modeling. *Unit head:* Dr. Orhan Talu, Chairperson, 216-687-2571, Fax: 216-687-9220, E-mail: talu@csvax.csuohio.edu. *Application contact:* Dr. Niluter Dural, Coordinator, 216-687-2569.

Colorado School of Mines, Graduate School, Chemical Engineering and Petroleum Refining Department, Golden, CO 80401-1887. Offers ME, MS, PhD. Part-time programs available. *Faculty:* 24 full-time (3 women), 4 part-time (0 women). *Students:* 42 full-time (14 women), 14 part-time (4 women); includes 2 minority (1 Asian American or Pacific Islander, 1 Hispanic American), 29 international. 87 applicants, 32% accepted. In 1998, 12 master's awarded (92% found work related to degree); 1 doctorate awarded (100% found work related to degree). *Degree requirements:* For master's, thesis required (for some programs), foreign language not required; for doctorate, dissertation, comprehensive exam required, foreign language not required. *Entrance requirements:* For master's and doctorate, GRE General Test, minimum GPA of 3.0. *Application deadline:* Applications are processed on a rolling basis. Application fee: $40. Electronic applications accepted. *Financial aid:* In 1998–99, 37 students received aid, including 26 research assistantships, 8 teaching assistantships; fellowships, unspecified assistantships also available. Aid available to part-time students. Financial aid applicants required to submit FAFSA. *Faculty research:* Liquid fuels for the future, responsible management of hazardous substances, surface and interfacial engineering, advanced computational methods and process control, gas hydrates. Total annual research expenditures: $1 million. *Unit head:* Dr. Robert M. Baldwin, Head, 303-273-3720, E-mail: rbaldwin@mines.edu. *Application contact:* John Dorgan, Assistant Professor, 303-273-3539, Fax: 303-273-3730, E-mail: jdorgan@mines.edu.

Colorado State University, Graduate School, College of Engineering, Department of Chemical and Bioresource Engineering, Program in Chemical Engineering, Fort Collins, CO 80523-0015. Offers MS, PhD. Part-time programs available. *Faculty:* 10 full-time (1 woman). *Students:* 14 full-time (7 women), 13 part-time (4 women); includes 2 minority (1 African American, 1 Native American), 11 international. Average age 28. 92 applicants, 38% accepted. In 1998, 8 master's, 2 doctorates awarded. Terminal master's awarded for partial completion of doctoral program. *Degree requirements:* For master's, thesis or alternative required, foreign language not required; for doctorate, dissertation required, foreign language not required. *Entrance requirements:* For master's and doctorate, GRE General Test, TOEFL (minimum score of 550 required), minimum GPA of 3.0. *Application deadline:* For fall admission, 2/1 (priority date). Applications are processed on a rolling basis. Application fee: $30. Electronic applications accepted. *Financial aid:* In 1998–99, 9 research assistantships, 2 teaching assistantships were awarded.; fellowships, career-related internships or fieldwork, Federal Work-Study, and institutionally-sponsored loans also available. *Faculty research:* Heat and mass transfer, semiconductor materials, biochemical engineering, advanced materials, environmental engineering. *Unit head:* Dr. Jim C. Loftis, Chair, Department of Chemical and Bioresource Engineering, 970-491-5252, Fax: 970-491-7369, E-mail: loftis@engr.colostate.edu.

Columbia University, Fu Foundation School of Engineering and Applied Science, Department of Chemical Engineering and Applied Chemistry, New York, NY 10027. Offers chemical engineering (MS, Eng Sc D, PhD, Engr). PhD offered through the Graduate School of Arts and Sciences. *Faculty:* 7 full-time (0 women), 6 part-time (0 women). *Students:* 21 full-time (6 women), 10 part-time (2 women). 153 applicants, 43% accepted. In 1998, 7 master's, 4 doctorates awarded. *Degree requirements:* For master's, thesis required, foreign language not required; for doctorate, dissertation, qualifying exam required, foreign language not required. *Entrance requirements:* For master's, doctorate, and Engr, GRE General Test, TOEFL. *Application deadline:* For fall admission, 1/5; for spring admission, 10/1. Application fee: $55. *Financial aid:* In 1998–99, 31 students received aid, including fellowships (averaging $18,000 per year), 10 research assistantships (averaging $18,000 per year), 10 teaching assistantships (averaging $18,000 per year); Federal Work-Study also available. Financial aid application deadline: 1/5; financial aid applicants required to submit FAFSA. *Faculty research:* Electrochemistry, synthetic membrane applications, chemical process analysis, applied polymer physics. Total annual research expenditures: $2.6 million. *Unit head:* Dr. Carl Gryte, Head, 212-854-4453, Fax: 212-854-3054, E-mail: ccg2@columbia.edu. *Application contact:* Mary Ferraro, Departmental Administrator, 212-854-4453, Fax: 212-854-3054, E-mail: mf224@columbia.edu.

Cornell University, Graduate School, Graduate Fields of Engineering, Field of Chemical Engineering, Ithaca, NY 14853-0001. Offers advanced materials processing (M Eng, MS, PhD); applied mathematics and computational methods (M Eng, MS, PhD); biochemical engineering (M Eng, MS, PhD); chemical reaction engineering (M Eng, MS, PhD); classical and statistical thermodynamics (M Eng, MS, PhD); fluid dynamics, rheology and biorheology (M Eng, MS, PhD); heat and mass transfer (M Eng, MS, PhD); kinetics and catalysis (M Eng, MS, PhD); polymers (M Eng, MS, PhD); surface science (M Eng, MS, PhD). *Faculty:* 18 full-time. *Students:* 71 full-time (17 women); includes 9 minority (1 African American, 7 Asian Americans or Pacific Islanders, 1 Hispanic American), 29 international. 340 applicants, 49% accepted. In 1998, 20 master's, 9 doctorates awarded. *Degree requirements:* For master's, thesis (MS) required; for doctorate, dissertation required, foreign language not required. *Entrance requirements:* For master's and doctorate, GRE General Test, TOEFL (minimum score of 580 required). *Application deadline:* For fall admission, 1/15. Application fee: $65. Electronic applications accepted. *Financial aid:* In 1998–99, 55 students received aid, including 17 fellowships with full tuition reimbursements available, 25 research assistantships with full tuition reimbursements available, 13 teaching assistantships with full tuition reimbursements available; institutionally-sponsored loans, scholarships, tuition waivers (full and partial), and unspecified assistantships also available. Financial aid applicants required to submit FAFSA. *Faculty research:* Biomedical engineering, catalysis, materials design and processing, polymer science and engineering. *Unit head:* Director of Graduate Studies, 607-255-4550, Fax: 607-255-9166. *Application contact:* Graduate Field Assistant, 607-255-4550, Fax: 607-255-9166, E-mail: dgs@cheme.cornell.edu.

Dalhousie University, Faculty of Graduate Studies, DalTech, Faculty of Engineering, Department of Chemical Engineering, Halifax, NS B3H 3J5, Canada. Offers M Eng, MA Sc, PhD. *Faculty:* 6 full-time (0 women), 2 part-time (0 women). *Students:* 20 full-time (3 women). 22 applicants, 36% accepted. In 1998, 3 master's awarded (100% found work related to degree). *Degree requirements:* For master's and doctorate, thesis/dissertation required, foreign language not required. *Entrance requirements:* For master's and doctorate, TOEFL (minimum score of 580 required). *Application deadline:* For fall admission, 6/1; for winter admission, 10/1; for spring admission, 2/1. Applications are processed on a rolling basis. Application fee: $55. *Financial aid:* Fellowships, research assistantships, teaching assistantships available. *Faculty research:* Coal science and dispersion, dust explosions, separation processes, air and water pollution control, process control. *Unit head:* Dr. Yash Gupta, Head, 902-494-3953, Fax: 902-420-7639. *Application contact:* Shelley Parker, Admissions Coordinator, Graduate Studies and Research, 902-494-1288, Fax: 902-494-3149, E-mail: shelley.parker@dal.ca.

Drexel University, Graduate School, College of Engineering, Department of Chemical Engineering, Program in Chemical Engineering, Philadelphia, PA 19104-2875. Offers MS, PhD. *Students:* 13 full-time (2 women), 30 part-time (9 women); includes 8 minority (1 African American, 7 Asian Americans or Pacific Islanders), 22 international. 101 applicants, 53% accepted. In 1998, 5 master's, 2 doctorates awarded. *Degree requirements:* For doctorate, dissertation required. *Entrance requirements:* For master's, TOEFL (minimum score of 570 required), minimum GPA of 3.0; for doctorate, TOEFL (minimum score of 570 required), minimum GPA of 3.5, MS in chemical engineering. *Application deadline:* For fall admission, 8/21. Applications are processed on a rolling basis. Application fee: $35. Tuition: Full-time $15,795; part-time $585 per credit. Required fees: $375; $67 per term. Tuition and fees vary according to program. *Financial aid:* In 1998–99, 1 research assistantship, 6 teaching assistantships were awarded.; unspecified assistantships also available. Financial aid application deadline: 2/1. *Unit head:* John Dorgan, Assistant Professor, 303-273-3539, Fax: 303-273-3730, E-mail: jdorgan@mines.edu. *Application contact:* Kelli Kennedy, Director of Admissions, 215-895-6706, Fax: 215-895-5939, E-mail: crowlka@duvm.ocs.drexel.edu.

Announcement: The department offers MS and PhD programs in chemical engineering, with concentrations in areas such as computer science, engineering management, environmental engineering, biochemical engineering, and materials engineering. In addition, it offers an MS program in biochemical engineering, customized to each student's background. Visit the Web site at http://www.chemeng.drexel.edu/.

École Polytechnique de Montréal, Graduate Programs, Department of Chemical Engineering, Montréal, PQ H3C 3A7, Canada. Offers M Eng, M Sc A, PhD, DESS. Part-time and

Chemical Engineering

École Polytechnique de Montréal (continued)
evening/weekend programs available. *Degree requirements:* For master's and doctorate, one foreign language, computer language, thesis/dissertation required. *Entrance requirements:* For master's, minimum GPA of 2.75; for doctorate, minimum GPA of 3.0. *Faculty research:* Polymer engineering, biochemical and food engineering, reactor engineering and industrial processes pollution control engineering, gas technology.

Florida Agricultural and Mechanical University, Division of Graduate Studies, Research, and Continuing Education, FAMU-FSU College of Engineering, Department of Chemical Engineering, Tallahassee, FL 32310-6046. Offers MS, PhD. *Students:* 4 (2 women); includes 3 minority (all African Americans) *Entrance requirements:* For master's, GRE General Test (minimum combined score of 1000 required), minimum GPA of 3.0. *Application deadline:* For fall admission, 7/1. Application fee: $20. *Financial aid:* Research assistantships, teaching assistantships, tuition waivers (full) available. *Faculty research:* Macromolecular transport, polymer processing, biochemical engineering, process control, environmental engineering. *Unit head:* Graduate Field Assistant, 607-255-4550, Fax: 607-255-9166, E-mail: dgs@cheme.cornell.edu. *Application contact:* Dr. Bruce R. Locke, Graduate Coordinator.

Florida Institute of Technology, Graduate School, College of Engineering, Division of Engineering Science, Program in Chemical Engineering, Melbourne, FL 32901-6975. Offers MS, PhD. Part-time programs available. *Faculty:* 5 full-time (1 woman), 1 part-time (0 women). *Students:* 5 full-time (0 women), 2 part-time, 5 international. Average age 26. 68 applicants, 66% accepted. In 1998, 3 master's awarded. Terminal master's awarded for partial completion of doctoral program. *Degree requirements:* For master's and doctorate, thesis/dissertation required, foreign language not required. *Entrance requirements:* For master's, TOEFL (minimum score of 550 required), minimum GPA of 3.0; for doctorate, GRE General Test, minimum GPA of 3.2. *Application deadline:* Applications are processed on a rolling basis. Application fee: $50. Electronic applications accepted. Tuition: Part-time $575 per credit hour. Required fees: $100. Tuition and fees vary according to campus/location and program. *Financial aid:* In 1998–99, 3 students received aid, including 1 research assistantship (averaging $5,229 per year), 2 teaching assistantships (averaging $2,945 per year); tuition remissions also available. Financial aid application deadline: 3/1; financial aid applicants required to submit FAFSA. *Faculty research:* Space technology, biotechnology, materials synthesis and processing, supercritical fluids, water treatment, process control. Total annual research expenditures: $44,500. *Unit head:* Dr. Paul A. Jennings, Chair, 407-674-7561, Fax: 407-984-8461, E-mail: jennings@fit.edu. *Application contact:* Carolyn P. Farrior, Associate Dean of Graduate Admissions, 407-674-7118, Fax: 407-723-9468, E-mail: cfarrior@fit.edu.

Florida State University, Graduate Studies, FAMU/FSU College of Engineering, Department of Chemical Engineering, Tallahassee, FL 32310-6046. Offers MS, PhD. Part-time programs available. *Faculty:* 11 full-time (1 woman), 1 part-time. *Students:* 23 full-time (7 women); includes 3 minority (2 African Americans, 1 Hispanic American), 15 international. Average age 25. 50 applicants, 20% accepted. In 1998, 5 master's awarded (100% found work related to degree); 2 doctorates awarded (100% found work related to degree). *Degree requirements:* For master's, computer language, thesis required, foreign language not required; for doctorate, computer language, dissertation, preliminary exam, qualifying exam required, foreign language not required. *Entrance requirements:* For master's, GRE General Test (minimum combined score of 1800 on three sections required), BS in chemical engineering, minimum GPA of 3.2; for doctorate, GRE General Test. *Average time to degree:* Master's–2 years full-time; doctorate–5 years full-time. *Application deadline:* For fall admission, 3/1 (priority date). Applications are processed on a rolling basis. Application fee: $20. *Financial aid:* In 1998–99, 23 students received aid, including 1 fellowship, 10 research assistantships with tuition reimbursements available (averaging $16,000 per year), 10 teaching assistantships with tuition reimbursements available (averaging $16,000 per year); scholarships also available. Financial aid application deadline: 3/1. *Faculty research:* Macromolecular transport, polymer processing, biochemical engineering, process control, environmental engineering, thermodynamics. Total annual research expenditures: $500,000. *Unit head:* Dr. Srinivas Palanki, Graduate Coordinator, 850-410-6170, Fax: 850-410-6150, E-mail: palanki@eng.fsu.edu. *Application contact:* 850-410-6149, Fax: 850-410-6150.

Georgia Institute of Technology, Graduate Studies and Research, College of Engineering, School of Chemical Engineering, Atlanta, GA 30332-0001. Offers biomedical engineering (MS Bio E); chemical engineering (MS Ch E, PhD); polymers (MS Poly); pulp and paper engineering (Certificate). *Degree requirements:* For master's and doctorate, computer language, thesis/dissertation required, foreign language not required; for degree, foreign language not required. *Entrance requirements:* For master's and doctorate, GRE, TOEFL (minimum score of 550 required), minimum GPA of 2.7. Electronic applications accepted. *Faculty research:* Biochemical engineering; process modeling, synthesis, and control; polymer science and engineering; thermodynamics and separations; surface and particle science.

Graduate School and University Center of the City University of New York, Graduate Studies, Program in Engineering, New York, NY 10036-8099. Offers chemical engineering (PhD); civil engineering (PhD); electrical engineering (PhD); mechanical engineering (PhD). *Faculty:* 68 full-time (1 woman). *Students:* 105 full-time (16 women), 11 part-time (2 women); includes 12 African Americans, 5 Asian Americans or Pacific Islanders, 4 Hispanic Americans *Degree requirements:* For doctorate, dissertation required, dissertation required. *Entrance requirements:* For doctorate, GRE General Test. *Application deadline:* For fall admission, 4/15. Application fee: $40. *Unit head:* Dr. Mumtaz Kassir, Acting Executive Officer, 212-650-8030.

Howard University, College of Engineering, Architecture, and Computer Sciences, School of Engineering and Computer Science, Department of Chemical Engineering, Washington, DC 20059-0002. Offers MS. Offered through the Graduate School of Arts and Sciences. Part-time programs available. *Faculty:* 6 full-time (0 women), 1 part-time (0 women). *Students:* 3 full-time (0 women); all minorities (all African Americans) Average age 25. 15 applicants, 7% accepted. *Degree requirements:* For master's, computer language, thesis required, foreign language not required. *Entrance requirements:* For master's, GRE General Test, TOEFL, minimum GPA of 3.0. *Average time to degree:* Master's–2.5 years full-time. *Application deadline:* For fall admission, 4/1; for spring admission, 11/1. Applications are processed on a rolling basis. Application fee: $45. *Financial aid:* In 1998–99, 1 teaching assistantship with full tuition reimbursement (averaging $8,000 per year) was awarded.; research assistantships with full tuition reimbursements, tuition waivers (partial) also available. Financial aid application deadline: 4/1. *Faculty research:* Bioengineering, computational fluid mechanics, process control, reaction kinetics, reactor modeling. Total annual research expenditures: $500,000. *Unit head:* Dr. Mobolaji E. Aluko, Chair, 202-806-6624.

Illinois Institute of Technology, Graduate College, Armour College of Engineering and Sciences, Department of Chemical and Environmental Engineering, Chemical Engineering Division, Chicago, IL 60616-3793. Offers M Ch E, MS, PhD. Part-time and evening/weekend programs available. *Faculty:* 13 full-time (0 women), 5 part-time (0 women). *Students:* 67 full-time (11 women), 68 part-time (12 women); includes 23 minority (5 African Americans, 13 Asian Americans or Pacific Islanders, 5 Hispanic Americans), 63 international. 175 applicants, 75% accepted. In 1998, 19 master's, 7 doctorates awarded. Terminal master's awarded for partial completion of doctoral program. *Degree requirements:* For master's, thesis (for some programs), comprehensive exam required, foreign language not required; for doctorate, dissertation, comprehensive exam required, foreign language not required. *Entrance requirements:* For master's and doctorate, GRE (minimum score of 1200 required), TOEFL (minimum score of 550 required), undergraduate GPA of 3.0 required. *Application deadline:* For fall admission, 7/1; for spring admission, 11/1. Applications are processed on a rolling basis. Application fee: $30. Electronic applications accepted. *Financial aid:* In 1998–99, 25 research assistantships, 15 teaching assistantships were awarded.; fellowships, Federal Work-Study, institutionally-sponsored loans, and scholarships also available. Financial aid application deadline: 3/1. *Faculty research:* Multiphase flow, polymers, bioengineering, process control, energy conversion. *Unit head:* Dr. Ali Cinar, Associate Chair, 312-567-3040, Fax: 312-567-8874, E-mail: cinar@charlie.iit.edu. *Application contact:* Dr. S. Mohammad Shahidehpour, Dean of Graduate College, 312-567-3024, Fax: 312-567-7517, E-mail: grad@minna.cns.iit.edu.

See in-depth description on page 497.

Institute of Paper Science and Technology, Graduate Programs, Program in Chemical Engineering, Atlanta, GA 30318-5794. Offers MS, PhD. Part-time programs available. Terminal master's awarded for partial completion of doctoral program. *Degree requirements:* For master's, industrial experience, research project required, foreign language and thesis not required; for doctorate, dissertation required, foreign language not required. *Entrance requirements:* For master's and doctorate, GRE (score in 50th percentile or higher required), minimum GPA of 3.0. *Application deadline:* For fall admission, 3/1 (priority date). Application fee: $0. *Financial aid:* Fellowships, career-related internships or fieldwork and institutionally-sponsored loans available. Financial aid applicants required to submit FAFSA. *Unit head:* Dr. William James Frederick, Director, Chemical Recovery and Corrosion Division, 404-894-5303, Fax: 404-894-5752. *Application contact:* Dana Carter, Student Development Counselor, 404-894-5745, Fax: 404-894-4778, E-mail: dana.carter@ipst.edu.

Instituto Tecnológico y de Estudios Superiores de Monterrey, Campus Monterrey, Graduate and Research Division, Programs in Engineering, Monterrey, 64849, Mexico. Offers applied statistics (M Eng); artificial intelligence (PhD); automation engineering (M Eng); chemical engineering (M Eng); civil engineering (M Eng); electrical engineering (M Eng); electronic engineering (M Eng); environmental engineering (M Eng); industrial engineering (M Eng, PhD); manufacturing engineering (M Eng); mechanical engineering (M Eng); systems and quality engineering (M Eng). M Eng offered jointly with the University of Waterloo; PhD (industrial engineering) offered jointly with Texas A&M University. Part-time and evening/weekend programs available. Terminal master's awarded for partial completion of doctoral program. *Degree requirements:* For master's and doctorate, one foreign language, computer language, thesis/dissertation required. *Entrance requirements:* For master's, PAEG, TOEFL; for doctorate, GRE, TOEFL, master's in related field. *Faculty research:* Flexible manufacturing cells, materials, statistical methods, environmental prevention, control and evaluation.

Iowa State University of Science and Technology, Graduate College, College of Engineering, Department of Chemical Engineering, Ames, IA 50011. Offers M Eng, MS, PhD. *Faculty:* 15 full-time. *Students:* 41 full-time (14 women), 6 part-time (3 women), 26 international. 148 applicants, 6% accepted. In 1998, 4 master's, 11 doctorates awarded. *Degree requirements:* For master's, thesis required (for some programs), foreign language not required; for doctorate, dissertation required, foreign language not required. *Entrance requirements:* For master's and doctorate, GRE General Test (foreign students), TOEFL. *Application deadline:* For fall admission, 2/1 (priority date); for spring admission, 10/1. Application fee: $20 ($50 for international students). Electronic applications accepted. Tuition, state resident: full-time $3,308. Tuition, nonresident: full-time $9,744. Part-time tuition and fees vary according to course load, campus/location and program. *Financial aid:* In 1998–99, 42 research assistantships with partial tuition reimbursements (averaging $13,506 per year) were awarded.; teaching assistantships, scholarships also available. *Unit head:* Dr. Charles E. Glatz, Chair, 515-294-7643, Fax: 515-294-2689, E-mail: chemengr@iastate.edu.

Johns Hopkins University, G. W. C. Whiting School of Engineering, Department of Chemical Engineering, Baltimore, MD 21218-2699. Offers MS, MSE, PhD. Part-time programs available. *Faculty:* 10 full-time (1 woman), 4 part-time (0 women). *Students:* 48 full-time (18 women); includes 10 minority (2 African Americans, 6 Asian Americans or Pacific Islanders, 2 Hispanic Americans), 13 international. Average age 25. 111 applicants, 16% accepted. In 1998, 2 master's, 6 doctorates awarded. *Degree requirements:* For master's and doctorate, thesis/dissertation required, foreign language not required. *Entrance requirements:* For master's, GRE General Test, TOEFL (minimum score of 560 required); for doctorate, GRE General Test, TOEFL (minimum score of 600 required). *Average time to degree:* Master's–2 years full-time; doctorate–5 years full-time. *Application deadline:* For fall admission, 2/1. Application fee: $50. Tuition: Full-time $23,660. Tuition and fees vary according to program. *Financial aid:* In 1998–99, 4 fellowships (averaging $11,250 per year), 28 research assistantships (averaging $13,122 per year), 7 teaching assistantships (averaging $14,067 per year) were awarded.; Federal Work-Study, institutionally-sponsored loans, and training grants also available. Aid available to part-time students. Financial aid application deadline: 3/14. *Faculty research:* Biotechnology, fluid mechanics, phase transitions and equilibria, separations. Total annual research expenditures: $1.7 million. *Unit head:* Dr. Michael Paulaitis, Chair, 410-516-7170, Fax: 410-516-5510, E-mail: michaelp@jhunix.hcf.jhu.edu. *Application contact:* Lynn Johnson, 410-516-5455, Fax: 410-516-5510, E-mail: che@jhu.edu.

Kansas State University, Graduate School, College of Engineering, Department of Chemical Engineering, Manhattan, KS 66506. Offers MS, PhD. *Faculty:* 11 full-time (0 women). *Students:* 22 full-time (7 women), 4 part-time; includes 1 minority (African American), 18 international. 48 applicants, 13% accepted. In 1998, 3 master's, 2 doctorates awarded. Application fee: $0 ($25 for international students). *Financial aid:* In 1998–99, 1 teaching assistantship was awarded.; research assistantships Total annual research expenditures: $759,000. *Unit head:* Dr. L. T. Fan, Head, 785-532-5584. *Application contact:* James H. Edgar, Graduate Coordinator, 785-532-5584.

Lamar University, College of Graduate Studies, College of Engineering, Department of Chemical Engineering, Beaumont, TX 77710. Offers ME, MES, DE. *Faculty:* 7 full-time (0 women), 1 part-time (0 women). *Students:* 18 full-time (3 women), 10 part-time (1 woman); includes 2 minority (both Asian Americans or Pacific Islanders), 24 international. Average age 25. 52 applicants, 58% accepted. In 1998, 7 master's awarded (86% found work related to degree, 14% continued full-time study); 1 doctorate awarded (100% found work related to degree). *Degree requirements:* For master's, thesis required (for some programs), foreign language not required; for doctorate, computer language, dissertation required, foreign language not required. *Entrance requirements:* For master's, GRE General Test (minimum combined score of 950 required), TOEFL; for doctorate, GRE General Test (minimum combined score of 950 required), TOEFL (minimum score of 530 required). *Average time to degree:* Master's–2 years full-time, 5 years part-time; doctorate–3 years full-time, 5 years part-time. *Application deadline:* For fall admission, 5/15 (priority date); for spring admission, 10/1 (priority date). Applications are processed on a rolling basis. Application fee: $0. *Financial aid:* In 1998–99, 24 students received aid, including 1 fellowship with partial tuition reimbursement available, 24 research assistantships with partial tuition reimbursements available, 4 teaching assistantships with partial tuition reimbursements available; tuition waivers (full and partial) also available. Financial aid application deadline: 4/1. *Faculty research:* Industrial waste water treatment, air pollution, soil bio-remediation, fluidization and fluidized bed, combustion, metal and sulfur emission control. Total annual research expenditures: $4.1 million. *Unit head:* Dr. Jack Hopper, Chair, 409-880-8784, Fax: 409-880-2197, E-mail: che_dept@hal.lamar.edu. *Application contact:* Sandy Drane, Coordinator, International Students and Graduate Studies, 409-880-8349, Fax: 409-880-8414, E-mail: dranesl@lub002.lamar.edu.

Lehigh University, College of Engineering and Applied Science, Department of Chemical Engineering, Bethlehem, PA 18015-3094. Offers M Eng, MS, PhD. Part-time programs available. Postbaccalaureate distance learning degree programs offered (no on-campus study). *Faculty:* 17 full-time (2 women). *Students:* 60 full-time (14 women), 55 part-time (13 women); includes 2 minority (1 Asian American or Pacific Islander, 1 Hispanic American), 42 international. Average age 26. 199 applicants, 17% accepted. In 1998, 6 master's, 7 doctorates awarded. Terminal master's awarded for partial completion of doctoral program. *Degree requirements:* For master's, thesis not required; for doctorate, dissertation, qualifying and general exams required. *Entrance requirements:* For master's, GRE General Test, TOEFL (minimum score of 550 required; average 587); for doctorate, GRE General Test, TOEFL (minimum score of 550 required; average 600). *Application deadline:* For fall admission, 7/15; for spring admission, 12/1. Applications are processed on a rolling basis. Application fee: $40. *Financial aid:* In 1998–99, 42 students received aid, including 7 fellowships, 24 research assistantships, 8 teaching assistantships; career-related internships or fieldwork and institutionally-sponsored loans also available. Financial aid application deadline: 1/15. *Faculty research:* Emulsion polymers, process control, biochemical engineering, biotechnology, electronic materials processing, multiphase processing. Total annual research expenditures: $4 million. *Unit head:* Dr. Mohamed S. El-Aasser, Chairman, 610-758-4260, Fax: 610-758-5057. *Application contact:* Dr. Mohamed S. El-Aasser, Chairman, 610-758-4260, Fax: 610-758-5057.

Louisiana State University and Agricultural and Mechanical College, Graduate School, College of Engineering, Department of Chemical Engineering, Baton Rouge, LA 70803. Offers MS Ch E, PhD. Part-time and evening/weekend programs available. *Faculty:* 19 full-time (1 woman), 1 part-time (0 women). *Students:* 43 full-time (4 women), 12 part-time (1 woman); includes 1 minority (African American), 39 international. Average age 27. 126 applicants, 28% accepted. In 1998, 14 master's, 5 doctorates awarded. Terminal master's awarded for partial completion of doctoral program. *Degree requirements:* For master's, comprehensive exam or thesis required; for doctorate, dissertation, general exam, qualifying exam required, foreign language not required. *Entrance requirements:* For master's and doctorate, GRE General Test, minimum GPA of 3.0. *Application deadline:* For fall admission, 1/25 (priority date). Applications are processed on a rolling basis. Application fee: $25. *Financial aid:* In 1998–99, 3 fellowships, 35 research assistantships with partial tuition reimbursements (averaging $10,868 per year) were awarded. Financial aid application deadline: 4/15. *Faculty research:* Reaction engineering, control, thermodynamic and transport phenomena, polymer processing and properties, biochemical engineering, environmental engineering. Total annual research expenditures: $860,000. *Unit head:* Dr. F. Carl Knopf, Chair, 225-388-1426, Fax: 225-388-1476, E-mail: knopf@che.lsu.edu. *Application contact:* Dr. Martin A. Hjortso, Graduate Adviser, 225-388-1426.

See in-depth description on page 499.

Louisiana Tech University, Graduate School, College of Engineering and Science, Department of Chemical Engineering, Ruston, LA 71272. Offers MS, D Eng. Part-time programs available. Terminal master's awarded for partial completion of doctoral program. *Degree requirements:* For master's and doctorate, thesis/dissertation required, foreign language not required. *Entrance requirements:* For master's, GRE General Test (minimum combined score of 1070 required), TOEFL (minimum score of 550 required), minimum GPA of 3.0 in last 60 hours; for doctorate, TOEFL (minimum score of 550 required), minimum graduate GPA of 3.25 (with MS) or GRE General Test (minimum combined score of 1270 required without MS). *Faculty research:* Artificial intelligence, biotechnology,hazardous waste process safety.

Manhattan College, Graduate Division, School of Engineering, Program in Chemical Engineering, Riverdale, NY 10471. Offers MS. Part-time and evening/weekend programs available. *Degree requirements:* For master's, computer language, thesis or alternative required, foreign language not required. *Entrance requirements:* For master's, GRE, TOEFL, minimum GPA of 3.0. *Faculty research:* Advanced separation processes, environmental management, combustion.

Massachusetts Institute of Technology, School of Engineering, Department of Chemical Engineering, Cambridge, MA 02139-4307. Offers SM, PhD, Sc D. *Faculty:* 29 full-time (4 women). *Students:* 198 full-time (47 women), 1 (woman) part-time. Average age 26. 435 applicants, 17% accepted. In 1998, 43 master's, 42 doctorates awarded. Terminal master's awarded for partial completion of doctoral program. *Degree requirements:* For master's, thesis required (for some programs); for doctorate, dissertation, comprehensive exams required. *Entrance requirements:* For master's and doctorate, GRE General Test. *Average time to degree:* Master's–1.5 years full-time; doctorate–5 years full-time. *Application deadline:* For fall admission, 1/15; for spring admission, 11/1. Application fee: $55. *Financial aid:* In 1998–99, 192 students received aid, including 50 fellowships, 145 research assistantships, 33 teaching assistantships Financial aid applicants required to submit FAFSA. *Faculty research:* Biochemical engineering, metabolic engineering, biomedical engineering, catalysis and chemical kinetics, combustion engineering. Total annual research expenditures: $19.2 million. *Unit head:* Dr. Robert C. Armstrong, Head, 617-253-4581, E-mail: rca@mit.edu. *Application contact:* Janet E. Fischer, Administrator, 617-253-4579, E-mail: info@chemegrad.mit.edu.

McGill University, Faculty of Graduate Studies and Research, Faculty of Engineering, Department of Chemical Engineering, Montréal, PQ H3A 2T5, Canada. Offers M Eng, PhD. Part-time programs available. *Faculty:* 16 full-time (1 woman), 14 part-time (0 women). *Students:* 73 full-time (18 women). 60 applicants, 37% accepted. In 1998, 23 master's, 12 doctorates awarded. Terminal master's awarded for partial completion of doctoral program. *Degree requirements:* For master's, computer language required, thesis optional; for doctorate, computer language, dissertation required. *Entrance requirements:* For master's, TOEFL (minimum score of 550 required), minimum GPA of 3.0; for doctorate, TOEFL (minimum score of 550 required). *Application deadline:* For fall admission, 3/1 (priority date); for winter admission, 9/15; for spring admission, 1/31. Applications are processed on a rolling basis. Application fee: $60. *Financial aid:* Fellowships, research assistantships, teaching assistantships available. Financial aid application deadline: 3/1. *Faculty research:* Polymers, plasmas, pulp and paper, biotechnology, separation processes. *Unit head:* R. J. Munz, Head, 514-398-4277, Fax: 514-398-6678. *Application contact:* J. H. Vera, Graduate Director, 514-398-4274, Fax: 514-398-6678.

McMaster University, School of Graduate Studies, Faculty of Engineering, Department of Chemical Engineering, Hamilton, ON L8S 4M2, Canada. Offers M Eng, PhD. *Faculty:* 14 full-time, 1 part-time. *Students:* 44 full-time, 11 part-time. *Degree requirements:* For master's, thesis required, foreign language not required; for doctorate, dissertation, comprehensive exam required, foreign language not required. *Application deadline:* For fall admission, 3/1 (priority date). Applications are processed on a rolling basis. Application fee: $50. *Financial aid:* In 1998–99, teaching assistantships (averaging $7,722 per year); fellowships, research assistantships, career-related internships or fieldwork also available. *Unit head:* Dr. John L. Brash, Chair, 905-525-9140 Ext. 24949.

McNeese State University, Graduate School, College of Engineering and Technology, Lake Charles, LA 70609-2495. Offers chemical engineering (M Eng); civil engineering (M Eng); electrical engineering (M Eng); mechanical engineering (M Eng). Part-time and evening/weekend programs available. *Faculty:* 13 full-time (1 woman). *Students:* 5 full-time (0 women), 3 part-time. *Degree requirements:* For master's, computer language, thesis or alternative required, foreign language not required. *Entrance requirements:* For master's, GRE General Test, TOEFL, minimum undergraduate GPA of 3.0. *Application deadline:* For fall admission, 7/15 (priority date). Applications are processed on a rolling basis. Application fee: $10 ($25 for international students). *Unit head:* Dr. O. C. Karkalits, Dean, 318-475-5875.

Michigan State University, Graduate School, College of Engineering, Department of Chemical Engineering, East Lansing, MI 48824-1020. Offers chemical engineering (PhD), including environmental toxicology. Part-time programs available. *Faculty:* 14. *Students:* 33 full-time (7 women), 15 part-time (3 women); includes 10 minority (1 African American, 9 Asian Americans or Pacific Islanders), 16 international. Average age 26. 165 applicants, 23% accepted. In 1998, 9 master's, 11 doctorates awarded. Terminal master's awarded for partial completion of doctoral program. *Degree requirements:* For master's, foreign language and thesis not required; for doctorate, dissertation required. *Entrance requirements:* For master's, GRE, TOEFL, minimum GPA of 3.0; for doctorate, GRE, TOEFL. *Application deadline:* For fall admission, 1/15 (priority date); for spring admission, 11/1. Applications are processed on a rolling basis. Application fee: $30 ($40 for international students). *Financial aid:* In 1998–99, 27 research assistantships (averaging $12,722 per year), 13 teaching assistantships (averaging $11,624 per year) were awarded.; fellowships Financial aid applicants required to submit FAFSA. Total annual research expenditures: $1.6 million. *Unit head:* Dr. Bruce E. Dale, Chairperson, 517-355-5135, Fax: 517-432-1105, E-mail: bdale@egr.msu.edu.

Michigan Technological University, Graduate School, College of Engineering, Department of Chemical Engineering, Houghton, MI 49931-1295. Offers MS, PhD. Part-time programs available. *Faculty:* 14 full-time (2 women). *Students:* 29 full-time (7 women), 15 international. Average age 26. 133 applicants, 8% accepted. In 1998, 12 master's, 1 doctorate awarded. *Degree requirements:* For master's and doctorate, thesis/dissertation required, foreign language not required. *Entrance requirements:* For master's, GRE General Test (combined average 1769 on three sections), TOEFL (minimum score of 600 required); for doctorate, GRE General Test (combined average 1837 on three sections), TOEFL (minimum score of 600 required; average 619). *Average time to degree:* Master's–3.2 years full-time; doctorate–5.7 years full-time. *Application deadline:* For fall admission, 3/15 (priority date). Applications are processed on a rolling basis. Application fee: $30 ($35 for international students). Tuition, state resident: full-time $4,377. Tuition, nonresident: full-time $9,108. Required fees: $126. Tuition and fees vary according to course load. *Financial aid:* In 1998–99, 4 fellowships (averaging $2,472 per year), 11 research assistantships (averaging $8,965 per year), 8 teaching assistantships (averaging $7,260 per year) were awarded.; career-related internships or fieldwork, Federal Work-Study, institutionally-sponsored loans, and unspecified assistantships also available. Aid available to part-time students. Financial aid application deadline: 3/1; financial aid applicants required to submit FAFSA. *Faculty research:* Polymer engineering, thermodynamics, chemical process safety, surface science/catalysis, environmental chemical engineering. Total annual research expenditures: $1.5 million. *Unit head:* Dr. Kirk Schulz, Chair, 906-487-3132, Fax: 906-487-3213, E-mail: khschulz@mtu.edu. *Application contact:* Dr. Kirk Schulz, Chair, 906-487-3132, Fax: 906-487-3213, E-mail: khschulz@mtu.edu.

Mississippi State University, College of Engineering, Department of Chemical Engineering, Mississippi State, MS 39762. Offers MS. Part-time programs available. Postbaccalaureate distance learning degree programs offered (minimal on-campus study). *Students:* 13 full-time (5 women), 5 part-time (2 women); includes 6 minority (all Asian Americans or Pacific Islanders), 1 international. Average age 26. 18 applicants, 28% accepted. In 1998, 3 degrees awarded. *Degree requirements:* For master's, thesis, comprehensive oral or written exam required, foreign language not required. *Entrance requirements:* For master's, GRE General Test (minimum combined score of 1200 required), TOEFL (minimum score of 550 required), minimum GPA of 2.75. *Application deadline:* For fall admission, 7/1; for spring admission, 11/1. Applications are processed on a rolling basis. Application fee: $25 for international students. *Financial aid:* Fellowships, Federal Work-Study, institutionally-sponsored loans, and unspecified assistantships available. Financial aid applicants required to submit FAFSA. *Faculty research:* Kinetics of ozone oxidation, waste reduction gas recovery, thermodynamics, composite materials, treatment of hazardous waste. Total annual research expenditures: $951,434. *Unit head:* Dr. Donald Hill, Head, 662-325-2480, Fax: 662-325-2482. *Application contact:* Jerry B. Inmon, Director of Admissions, 662-325-2224, Fax: 662-325-7360, E-mail: admit@admissions.msstate.edu.

Montana State University–Bozeman, College of Graduate Studies, College of Engineering, Department of Chemical Engineering, Bozeman, MT 59717. Offers chemical engineering (MS); engineering (PhD); environmental engineering (MS, PhD); project engineering and management (MPEM). Part-time programs available. *Students:* 17 full-time (6 women), 12 part-time (8 women). Average age 28. 16 applicants, 38% accepted. In 1998, 6 master's, 1 doctorate awarded. *Degree requirements:* For master's, thesis or alternative required, foreign language not required; for doctorate, dissertation required, foreign language not required. *Entrance requirements:* For master's and doctorate, GRE General Test, TOEFL (minimum score of 550 required). *Application deadline:* For fall admission, 6/1 (priority date); for spring admission, 11/1. Applications are processed on a rolling basis. Application fee: $50. *Financial aid:* In 1998–99, 24 students received aid, including 2 teaching assistantships; fellowships, research assistantships, Federal Work-Study and scholarships also available. Financial aid application deadline: 3/1. *Faculty research:* Biofilms, materials, separations, composites. Total annual research expenditures: $624,080. *Unit head:* Dr. John T. Sears, Head, 406-994-2221, Fax: 406-994-5308, E-mail: cheme@coe.montana.edu.

National Technological University, Programs in Engineering, Fort Collins, CO 80526-1842. Offers chemical engineering (MS); computer engineering (MS); computer science (MS); electrical engineering (MS); engineering management (MS); hazardous waste management (MS); health physics (MS); management of technology (MS); manufacturing systems engineering (MS); materials science and engineering (MS); software engineering (MS); special majors (MS); transportation engineering (MS); transportation systems engineering (MS). Part-time programs available. *Faculty:* 600 part-time (20 women). *Entrance requirements:* For master's, BS in engineering or related field. *Application deadline:* Applications are processed on a rolling basis. Application fee: $50. *Unit head:* Lionel V. Baldwin, President, 970-495-6400, Fax: 970-484-0668, E-mail: baldwin@mail.ntu.edu.

New Jersey Institute of Technology, Office of Graduate Studies, Department of Chemical Engineering, Chemistry and Environmental Science, Program in Chemical Engineering, Newark, NJ 07102-1982. Offers MS, PhD, Engineer. *Degree requirements:* For doctorate, dissertation, residency required, foreign language not required, foreign language not required. *Entrance requirements:* For master's, GRE General Test (minimum score of 450 on verbal section, 600 on quantitative, 550 on analytical required); for doctorate, GRE General Test (minimum score of 450 on verbal section, 600 on quantitative, 550 on analytical required), minimum graduate GPA of 3.5.

New Mexico State University, Graduate School, College of Engineering, Department of Chemical Engineering, Las Cruces, NM 88003-8001. Offers MS Ch E, PhD. Part-time programs available. *Faculty:* 4 full-time (1 woman), 1 part-time (0 women). *Students:* 19 full-time (6 women), 12 part-time (4 women); includes 3 minority (all Hispanic Americans), 15 international. Average age 33. 60 applicants, 50% accepted. In 1998, 8 master's, 3 doctorates awarded. *Degree requirements:* For master's, computer language, thesis required (for some programs), foreign language not required; for doctorate, computer language, dissertation required, foreign language not required. *Entrance requirements:* For master's and doctorate, GRE General Test. *Application deadline:* For fall admission, 7/1 (priority date); for spring admission, 11/1. Applications are processed on a rolling basis. Application fee: $15 ($35 for international students). Tuition, state resident: full-time $2,682; part-time $112 per credit. Tuition, nonresident: full-time $8,376; part-time $349 per credit. Tuition and fees vary according to course load. *Financial aid:* Fellowships, research assistantships, teaching assistantships, career-related internships or fieldwork and Federal Work-Study available. Aid available to part-time students. Financial aid application deadline: 3/1. *Faculty research:* Advanced materials, semiconductors, environmental engineering, biochemical engineering, biomedical engineering, food technology, computer-aided design, actinide separations and computer simulation of chemical and thermodynamic properties. *Unit head:* Dr. Stewart Munson-McGee, Interim Head, 505-646-1214, Fax: 505-646-7706, E-mail: stumcgee@nmsu.edu. *Application contact:* Dr. Stewart Munson-McGee, Interim Head, 505-646-1214, Fax: 505-646-7706, E-mail: stumcgee@nmsu.edu.

North Carolina Agricultural and Technical State University, Graduate School, College of Engineering, Program in Engineering, Greensboro, NC 27411. Offers chemical engineering (MSE). Part-time programs available. *Faculty:* 7 full-time (0 women), 1 part-time (0 women). *Students:* 28 full-time (11 women), 12 part-time (8 women); includes 27 minority (20 African Americans, 7 Asian Americans or Pacific Islanders), 7 international. Average age 23. 39 applicants, 41% accepted. In 1998, 7 degrees awarded (60% found work related to degree, 40% continued full-time study). *Degree requirements:* For master's, thesis (for some programs), comprehensive exam required, foreign language not required. *Entrance requirements:* For master's, GRE General Test, GRE Subject Test (recommended), TOEFL. *Average time to degree:* Master's–2 years full-time, 4 years part-time. *Application deadline:* For fall admission, 7/1 (priority date); for spring admission, 1/9. Applications are processed on a rolling basis. Application fee: $35. *Financial aid:* In 1998–99, 22 students received aid, including 2 fellowships with partial tuition reimbursements available (averaging $13,000 per year), 10 research assistantships with partial tuition reimbursements available (averaging $13,000 per year), 10 teaching assistantships with partial tuition reimbursements available (averaging $13,000 per year) Financial aid application deadline: 3/30. *Faculty research:* Thermodynamics, biomediation, membrane reactors, environmental engineering, fuel cells. *Unit head:* Dr. Franklin G. King,

Chemical Engineering

North Carolina Agricultural and Technical State University *(continued)*
Chairperson, 336-334-7564 Ext. 40, Fax: 336-334-7904, E-mail: king@ncat.edu. *Application contact:* Dr. Keith A. Schimmel, Graduate Coordinator, 336-334-7564 Ext. 41, Fax: 336-334-7904, E-mail: schimmel@ncat.edu.

North Carolina State University, Graduate School, College of Engineering, Department of Chemical Engineering, Raleigh, NC 27695. Offers M Ch E, MS, PhD. Part-time programs available. *Faculty:* 23 full-time (2 women), 10 part-time (0 women). *Students:* 77 full-time (29 women), 4 part-time; includes 8 minority (4 African Americans, 4 Asian Americans or Pacific Islanders), 21 international. Average age 26. 145 applicants, 25% accepted. In 1998, 6 master's, 10 doctorates awarded. Terminal master's awarded for partial completion of doctoral program. *Degree requirements:* For master's and doctorate, thesis/dissertation required, foreign language not required. *Entrance requirements:* For master's and doctorate, GRE General Test (average 540 verbal, 670 quantitative, 760 analytical), TOEFL (minimum score of 550 required). *Average time to degree:* Master's–2.5 years full-time; doctorate–5 years full-time. *Application deadline:* For fall admission, 6/25 (priority date); for spring admission, 11/25. Applications are processed on a rolling basis. Application fee: $45. *Financial aid:* In 1998–99, 11 fellowships (averaging $2,366 per year), 59 research assistantships (averaging $4,457 per year), 21 teaching assistantships (averaging $4,987 per year) were awarded. *Faculty research:* Transport phenomena and separation processes, thermodynamics, energy and environmental engineering, biotechnology, electrochemical engineering. Total annual research expenditures: $4.9 million. *Unit head:* Dr. Ruben G. Carbonell, Head, 919-515-2499, Fax: 919-515-3465, E-mail: ruben@ncsu.edu. *Application contact:* Dr. Saad A. Khan, Director of Graduate Programs, 919-515-4519, Fax: 919-515-3465, E-mail: khan@eos.ncsu.edu.

Northeastern University, College of Engineering, Department of Chemical Engineering, Boston, MA 02115-5096. Offers MS, PhD. Part-time programs available. *Faculty:* 4 full-time (1 woman), 3 part-time (0 women). *Students:* 21 full-time (12 women), 12 part-time (4 women), 12 international. Average age 25. 87 applicants, 46% accepted. In 1998, 10 master's awarded. Terminal master's awarded for partial completion of doctoral program. *Degree requirements:* For master's, thesis optional, foreign language not required; for doctorate, dissertation, departmental qualifying exam required, foreign language not required. *Entrance requirements:* For master's and doctorate, GRE General Test. *Average time to degree:* Master's–2.78 years full-time, 3.54 years part-time; doctorate–2.75 years full-time. *Application deadline:* For fall admission, 4/15. Applications are processed on a rolling basis. Application fee: $50. *Financial aid:* In 1998–99, 21 students received aid, including 1 fellowship with tuition reimbursement available (averaging $12,450 per year), 12 research assistantships with full tuition reimbursements available (averaging $12,450 per year), 8 teaching assistantships with full tuition reimbursements available (averaging $12,450 per year); career-related internships or fieldwork, Federal Work-Study, tuition waivers (full), and unspecified assistantships also available. Aid available to part-time students. Financial aid application deadline: 2/15; financial aid applicants required to submit FAFSA. *Faculty research:* Aerogel, catalysts, advanced microgravity materials processing, biomaterials, catalyst development, biochemical reactions. *Unit head:* Dr. Ralph A. Buonopane, Chairman, 617-373-2989, Fax: 617-373-2501. *Application contact:* Stephen L. Gibson, Associate Director, 617-373-2711, Fax: 617-373-2501, E-mail: grad-eng@coe.neu.edu.

Northwestern University, The Graduate School, Robert R. McCormick School of Engineering and Applied Science, Department of Chemical Engineering, Evanston, IL 60208. Offers MS, PhD. Admissions and degrees offered through The Graduate School. Part-time programs available. *Faculty:* 19 full-time (4 women), 1 part-time (0 women). *Students:* 69 full-time (19 women); includes 17 minority (3 African Americans, 11 Asian Americans or Pacific Islanders, 3 Hispanic Americans), 13 international. 187 applicants, 27% accepted. In 1998, 4 master's, 10 doctorates awarded. Terminal master's awarded for partial completion of doctoral program. *Degree requirements:* For master's, thesis optional, foreign language not required; for doctorate, dissertation required, foreign language not required. *Entrance requirements:* For master's and doctorate, GRE General Test, TOEFL (minimum score of 560 required). *Application deadline:* For fall admission, 8/30. Applications are processed on a rolling basis. Application fee: $50 ($55 for international students). *Financial aid:* In 1998–99, 13 fellowships with full tuition reimbursements (averaging $11,673 per year), 24 research assistantships with partial tuition reimbursements (averaging $16,285 per year), 7 teaching assistantships with full tuition reimbursements (averaging $12,042 per year) were awarded.; career-related internships or fieldwork, Federal Work-Study, institutionally-sponsored loans, and scholarships also available. Financial aid application deadline: 1/15; financial aid applicants required to submit FAFSA. *Faculty research:* Materials design and polymers, biotechnology, biomedical engineering, fluid mechanics and transport, catalysis and reaction engineering environmental-bioengineering. Total annual research expenditures: $2.2 million. *Unit head:* Julio M. Ottino, Chair, 847-491-7398, Fax: 847-491-3728. *Application contact:* Gwynne Fox, Admissions Contact, 847-491-2773, Fax: 847-491-2776, E-mail: gfox@nwu.edu.

The Ohio State University, Graduate School, College of Engineering, Department of Chemical Engineering, Columbus, OH 43210. Offers MS, PhD. *Faculty:* 14 full-time, 2 part-time. *Students:* 64 full-time (24 women), 2 part-time (1 woman); includes 8 minority (3 African Americans, 4 Asian Americans or Pacific Islanders, 1 Hispanic American), 45 international. 273 applicants, 15% accepted. In 1998, 14 master's, 7 doctorates awarded. *Degree requirements:* For master's and doctorate, computer language, thesis/dissertation required, foreign language not required. *Entrance requirements:* For master's and doctorate, GRE General Test or minimum GPA of 3.0 (international students). *Application deadline:* For fall admission, 8/15. Applications are processed on a rolling basis. Application fee: $30 ($40 for international students). *Financial aid:* Fellowships, research assistantships, teaching assistantships, career-related internships or fieldwork, Federal Work-Study, institutionally-sponsored loans, and unspecified assistantships available. Aid available to part-time students. *Unit head:* Liang-Shih Fan, Chairman, 614-688-3262, Fax: 614-292-3769, E-mail: fan.1@osu.edu.

Announcement: Research programs include artificial intelligence, process control, image analysis, biochemical engineering, reaction engineering, catalysis, thermodynamics, polymer processing, particle technology, rheology, drag reduction, supercritical fluid separations, and interfacial phenomena. Excellent fellowships and research assistantships available. Students with strong chemical engineering or chemistry backgrounds, especially women and students who are members of minority groups, are encouraged to apply.

Ohio University, Graduate Studies, College of Engineering and Technology, Department of Chemical Engineering, Athens, OH 45701-2979. Offers MS, PhD. Part-time programs available. *Faculty:* 9 full-time (1 woman). *Students:* 25 full-time (7 women), 3 part-time (1 woman), 21 international. Average age 25. 109 applicants, 45% accepted. In 1998, 6 master's awarded (100% found work related to degree); 1 doctorate awarded. Terminal master's awarded for partial completion of doctoral program. *Degree requirements:* For master's, thesis required, foreign language not required; for doctorate, dissertation, comprehensive and qualifying exams required, foreign language not required. *Entrance requirements:* For master's and doctorate, GRE General Test, TOEFL. *Application deadline:* For fall admission, 3/1 (priority date). Applications are processed on a rolling basis. Application fee: $30. Tuition, state resident: full-time $5,754; part-time $238 per credit hour. Tuition, nonresident: full-time $11,055; part-time $457 per credit hour. Tuition and fees vary according to course load, campus/location and program. *Financial aid:* In 1998–99, 3 fellowships, 10 research assistantships, 5 teaching assistantships were awarded.; institutionally-sponsored loans and tuition waivers (full and partial) also available. Financial aid application deadline: 3/15. *Faculty research:* Corrosion and multiphase flow, biochemical engineering, thin film materials, air pollution modelling and control, polymerization. *Unit head:* Dr. Michael E. Prudich, Chairman, 740-593-1501, Fax: 740-593-0873. *Application contact:* Dr. Kendree J. Sampson, Graduate Chairman, 740-593-1503, Fax: 740-593-0873, E-mail: sampson@bobcat.ent.ohiou.edu.

Oklahoma State University, Graduate College, College of Engineering, Architecture and Technology, School of Chemical Engineering, Stillwater, OK 74078. Offers M En, MS, PhD.

Faculty: 11 full-time (1 woman). *Students:* 23 full-time (5 women), 25 part-time (4 women); includes 3 minority (1 Asian American or Pacific Islander, 2 Native Americans), 29 international. Average age 28. In 1998, 7 master's, 4 doctorates awarded. *Degree requirements:* For master's, thesis or alternative required, foreign language not required; for doctorate, dissertation required, foreign language not required. *Entrance requirements:* For master's and doctorate, GRE, TOEFL (minimum score of 550 required). *Application deadline:* For fall admission, 7/1 (priority date). Application fee: $25. *Financial aid:* In 1998–99, 40 students received aid, including 28 research assistantships (averaging $11,798 per year), 12 teaching assistantships (averaging $11,382 per year); career-related internships or fieldwork, Federal Work-Study, and tuition waivers (partial) also available. Aid available to part-time students. Financial aid application deadline: 3/1. *Unit head:* Dr. Russell Rhinehart, Head, 405-744-5280.

Oregon State University, Graduate School, College of Engineering, Department of Chemical Engineering, Corvallis, OR 97331. Offers MAIS, MS, PhD. Part-time programs available. *Faculty:* 8 full-time (1 woman). *Students:* 19 full-time (6 women), 9 part-time (1 woman); includes 2 minority (both Asian Americans or Pacific Islanders), 20 international. Average age 27. In 1998, 6 master's, 1 doctorate awarded. Terminal master's awarded for partial completion of doctoral program. *Degree requirements:* For master's, computer language, thesis or alternative required, foreign language not required; for doctorate, computer language, dissertation required, foreign language not required. *Entrance requirements:* For master's and doctorate, TOEFL (minimum score of 550 required), minimum GPA of 3.0 in last 90 hours. *Application deadline:* For fall admission, 3/1. Applications are processed on a rolling basis. Application fee: $50. *Financial aid:* Research assistantships, teaching assistantships, career-related internships or fieldwork, Federal Work-Study, and institutionally-sponsored loans available. Aid available to part-time students. Financial aid application deadline: 2/1. *Faculty research:* Fluidization, mass transfer, chemical reactor design, combustion and gasification, polymers. *Unit head:* Dr. Shoichi Kimura, Head, 541-737-4791, Fax: 541-737-4600, E-mail: kimuras@che.orst.edu. *Application contact:* Dawn M. Belveal, Secretary, 541-737-4791, Fax: 541-737-2182, E-mail: belveadm@che.orst.edu.

Pennsylvania State University University Park Campus, Graduate School, College of Engineering, Department of Chemical Engineering, State College, University Park, PA 16802-1503. Offers MS, PhD. *Students:* 55 full-time (17 women), 8 part-time (2 women). In 1998, 16 master's, 11 doctorates awarded. *Degree requirements:* For master's and doctorate, thesis/dissertation required, foreign language not required. *Entrance requirements:* For master's and doctorate, GRE General Test. Application fee: $50. *Unit head:* Dr. J. Larry Duda, Head, 814-865-2574.

Announcement: Faculty research includes biotechnology—protein separations, plant biotechnology, bioreactors; biomedical engineering—cardiovascular and respiratory dynamics, physiological transport, drug delivery; catalysis—metal-support effects, hydroisomerization reactions, catalyst characterization; transport phenomena; polymer solution thermodynamics and transport; plasma-catalyst generation modeling; interfacial and colloidal engineering; tribology/lubrication; molecular dynamic and Monte Carlo simulations; turbulent combustion; and optimization.

Polytechnic University, Brooklyn Campus, Department of Chemical Engineering, Chemistry and Materials Science, Major in Chemical Engineering, Brooklyn, NY 11201-2990. Offers MS, PhD. Part-time and evening/weekend programs available. *Faculty:* 9. *Students:* 17 full-time (5 women), 16 part-time (1 woman); includes 3 minority (1 African American, 1 Asian American or Pacific Islander, 1 Hispanic American), 20 international. Average age 33. 47 applicants, 43% accepted. In 1998, 3 master's, 2 doctorates awarded. *Degree requirements:* For master's and doctorate, thesis/dissertation required. *Entrance requirements:* For master's, GRE General Test, BS in chemical engineering; for doctorate, GRE General Test. *Application deadline:* Applications are processed on a rolling basis. Application fee: $45. Electronic applications accepted. *Financial aid:* Fellowships, research assistantships, teaching assistantships, institutionally-sponsored loans available. Aid available to part-time students. Financial aid applicants required to submit FAFSA. *Faculty research:* Plasma polymerization, crystallization of organic compounds, dipolar relaxations in reactive polymers. Total annual research expenditures: $723,453. *Application contact:* John S. Kerge, Dean of Admissions, 718-260-3200, Fax: 718-260-3446, E-mail: admitme@poly.edu.

Polytechnic University, Farmingdale Campus, Graduate Programs, Department of Chemical Engineering, Chemistry and Material Science, Major in Chemical Engineering, Farmingdale, NY 11735-3995. Offers MS, PhD. 1 applicants, 100% accepted. *Degree requirements:* For master's, thesis required (for some programs); for doctorate, dissertation required. *Application fee:* $45. *Unit head:* John S. Kerge, Dean of Admissions, 718-260-3200, Fax: 718-260-3446, E-mail: admitme@poly.edu. *Application contact:* John S. Kerge, Dean of Admissions, 718-260-3200, Fax: 718-260-3446, E-mail: admitme@poly.edu.

Polytechnic University, Westchester Graduate Center, Graduate Programs, Department of Chemical Engineering, Chemistry, and Materials Science, Major in Chemical Engineering, Hawthorne, NY 10532-1507. Offers MS. Average age 33. *Degree requirements:* For master's, computer language required. Application fee: $45. *Unit head:* Dr. Yinlun Huang, Graduate Committee Chair, 313-577-3800. *Application contact:* John S. Kerge, Dean of Admissions, 718-260-3200, Fax: 718-260-3446, E-mail: admitme@poly.edu.

Princeton University, Graduate School, School of Engineering and Applied Science, Department of Chemical Engineering, Princeton, NJ 08544-1019. Offers applied and computational mathematics (PhD); chemical engineering (M Eng, MSE, PhD); plasma science and technology (MSE, PhD); polymer sciences and materials (MSE, PhD). *Degree requirements:* For master's, thesis required; for doctorate, dissertation, general exam required. *Entrance requirements:* For master's and doctorate, GRE General Test, GRE Subject Test, TOEFL.

Announcement: Areas of research include applied mathematics; aerosol physics and chemistry; atmospheric chemistry; biomaterials engineering; biomimetic processing; ceramic materials; colloid, particulate, and interfacial sciences; complex fluids; chemical reactor design and stability; computational chemistry and biology; electrohydrodynamics; environmental applications of catalysis; flow of granular materials; fluid mechanics; gas phase and homogeneous kinetics; heterogeneous catalysis; kinetic theory; materials properties and processing; optimization; polymer morphology, properties, and rheology; process synthesis and control; statistical and quantum mechanics; surface science; supercooled liquids and glasses; supercritical fluids; thermodynamics; transport phenomena.

Purdue University, Graduate School, Schools of Engineering, School of Chemical Engineering, West Lafayette, IN 47907. Offers biomedical engineering (MS Bm E, PhD); chemical engineering (MS Ch E, PhD). *Faculty:* 18 full-time (3 women), 6 part-time (0 women). *Students:* 89 full-time (21 women), 6 part-time (4 women); includes 9 minority (3 African Americans, 4 Asian Americans or Pacific Islanders, 2 Hispanic Americans), 50 international. Average age 22. 130 applicants, 21% accepted. In 1998, 14 master's, 9 doctorates awarded. *Degree requirements:* For master's and doctorate, thesis/dissertation required, foreign language not required. *Entrance requirements:* For master's and doctorate, TOEFL (minimum score of 550 required). *Average time to degree:* Master's–2.5 years full-time; doctorate–5.3 years full-time. *Application deadline:* Applications are processed on a rolling basis. Application fee: $30. Electronic applications accepted. *Financial aid:* In 1998–99, 13 fellowships, 54 research assistantships were awarded.; career-related internships or fieldwork also available. Aid available to part-time students. Financial aid applicants required to submit FAFSA. *Faculty research:* Biochemical and biomedical processes, polymer materials, interfacial and surface phenomena, applied thermodynamics, process systems engineering. *Unit head:* Dr. G. V. Reklaitis, Head, 765-494-4075. *Application contact:* Linda Hawkins, Graduate Administrator, 765-494-4057.

Queen's University at Kingston, School of Graduate Studies and Research, Faculty of Applied Science, Department of Chemical Engineering, Kingston, ON K7L 3N6, Canada.

Offers M Sc Eng, PhD. Part-time programs available. *Students:* 32 full-time (8 women), 11 part-time (1 woman). In 1998, 8 master's, 4 doctorates awarded. *Degree requirements:* For master's, thesis or alternative required, foreign language not required; for doctorate, dissertation, comprehensive exam required, foreign language not required. *Entrance requirements:* For master's and doctorate, TOEFL (minimum score of 580 required). *Application deadline:* For fall admission, 2/28 (priority date). Application fee: $60. Electronic applications accepted. *Financial aid:* Fellowships, research assistantships, teaching assistantships, institutionally-sponsored loans available. Financial aid application deadline: 3/1. *Unit head:* Dr. R. J. Neufeld, Head, 613-533-2785. *Application contact:* Dr. A. S. Daugulis, Graduate Coordinator, 613-533-2784.

Rensselaer Polytechnic Institute, Graduate School, School of Engineering, Howard P. Isermann Department of Chemical Engineering, Troy, NY 12180-3590. Offers M Eng, MS, D Eng, PhD, MBA/M Eng. Part-time programs available. *Faculty:* 11 full-time (0 women). *Students:* 61 full-time (16 women), 7 part-time (2 women); includes 8 minority (5 Asian Americans or Pacific Islanders, 3 Hispanic Americans), 36 international. 187 applicants, 29% accepted. In 1998, 13 master's, 8 doctorates awarded. Terminal master's awarded for partial completion of doctoral program. *Degree requirements:* For master's, thesis required (for some programs), foreign language not required; for doctorate, dissertation required, foreign language not required. *Entrance requirements:* For master's and doctorate, GRE, TOEFL (minimum score of 550 required). *Application deadline:* For fall admission, 2/1 (priority date). Applications are processed on a rolling basis. Application fee: $35. *Financial aid:* In 1998–99, 58 students received aid, including 11 fellowships with full tuition reimbursements available (averaging $16,500 per year), 37 research assistantships (averaging $16,500 per year), 10 teaching assistantships (averaging $16,500 per year); institutionally-sponsored loans and scholarships also available. Financial aid application deadline: 2/1. *Faculty research:* Biocatalysis, bioseparations, polymers, microelectronics, high-temperature kinetics, advanced materials, interfacial phenomena. Total annual research expenditures: $2.2 million. *Unit head:* Dr. Jonathan Dordick, Chair, 518-276-6379. *Application contact:* Dr. E. Bruce Nauman, Admissions Coordinator, 518-276-6929, Fax: 518-276-4030, E-mail: bockea@rpi.edu.

See in-depth description on page 501.

Rice University, Graduate Programs, George R. Brown School of Engineering, Department of Chemical Engineering, Houston, TX 77251-1892. Offers M Ch E, MS, PhD. Part-time programs available. *Faculty:* 12 full-time (1 woman). In 1998, 4 master's, 15 doctorates awarded. *Degree requirements:* For master's, thesis (MS) required; for doctorate, dissertation required, foreign language not required. *Entrance requirements:* For master's and doctorate, GRE General Test, TOEFL (minimum score of 600 required), minimum GPA of 3.0. *Application deadline:* For fall admission, 2/1 (priority date). Applications are processed on a rolling basis. Application fee: $25. *Financial aid:* Fellowships, research assistantships available. *Faculty research:* Thermodynamics, phase equilibria, rheology, fluid mechanics, polymers, biomedical engineering, interfacial phenomena, process control, petroleum engineering, reaction engineering and catalysis, biomaterials, metabolic engineering. *Unit head:* Dr. E. Bruce Nauman, Admissions Coordinator, 518-276-6929, Fax: 518-276-4030, E-mail: bockea@rpi.edu. *Application contact:* Graduate Admissions Committee, 713-527-4720, E-mail: sery.u@rice.edu.

See in-depth description on page 503.

Rose-Hulman Institute of Technology, Faculty of Engineering and Applied Sciences, Department of Chemical Engineering, Terre Haute, IN 47803-3920. Offers MS. Part-time programs available. *Faculty:* 7 full-time (0 women). *Students:* 7 full-time (1 woman); includes 1 minority (African American), 4 international. Average age 22. 6 applicants, 50% accepted. In 1998, 2 degrees awarded. *Degree requirements:* For master's, thesis required, foreign language not required. *Entrance requirements:* For master's, GRE, TOEFL (minimum score of 580 required), minimum GPA of 3.0. *Average time to degree:* Master's–2 years full-time. *Application deadline:* For fall admission, 2/1 (priority date). Applications are processed on a rolling basis. Application fee: $0. *Financial aid:* In 1998–99, 5 students received aid, including 2 fellowships with full and partial tuition reimbursements available (averaging $6,000 per year); research assistantships, teaching assistantships, grants, institutionally-sponsored loans, and tuition waivers (full and partial) also available. Financial aid application deadline: 2/1. *Faculty research:* Process control, transport phenomena, thermodynamics. Total annual research expenditures: $167,433. *Unit head:* Dr. Hossein Hariri, Chairman, 812-877-8292, Fax: 812-877-3198, E-mail: hossein.hariri@rose-hulman.edu. *Application contact:* Dr. Buck F. Brown, Dean for Research and Graduate Studies, 812-877-8403, Fax: 812-877-8102, E-mail: buck.brown@rose-hulman.edu.

Rutgers, The State University of New Jersey, New Brunswick, Graduate School, Program in Chemical and Biochemical Engineering, New Brunswick, NJ 08903. Offers MS, PhD. Part-time and evening/weekend programs available. *Faculty:* 18 full-time (2 women), 3 part-time (1 woman). *Students:* 48 full-time (26 women), 20 part-time (5 women); includes 8 minority (2 African Americans, 6 Asian Americans or Pacific Islanders), 21 international. Average age 25. 163 applicants, 10% accepted. In 1998, 12 master's awarded (33% found work related to degree, 67% continued full-time study); 5 doctorates awarded (40% entered university research/teaching, 60% found other work related to degree). Terminal master's awarded for partial completion of doctoral program. *Degree requirements:* For master's, thesis required (for some programs), foreign language not required; for doctorate, dissertation required, foreign language not required. *Entrance requirements:* For master's, GRE General Test (minimum combined score of 1100 required; average 1320), TOEFL (minimum score of 550 required); for doctorate, GRE General Test (minimum combined score of 1100 required; average 1320). *Application deadline:* For fall admission, 1/15 (priority date); for spring admission, 10/15. Applications are processed on a rolling basis. Application fee: $50. *Financial aid:* In 1998–99, 48 students received aid, including 21 fellowships with full tuition reimbursements available (averaging $16,500 per year), 12 research assistantships with full tuition reimbursements available (averaging $16,500 per year), 10 teaching assistantships with full tuition reimbursements available (averaging $16,500 per year); unspecified assistantships also available. Financial aid application deadline: 1/15; financial aid applicants required to submit FAFSA. *Faculty research:* Biotechnology, environmental engineering, statistical thermodynamics, polymers, pharmaceutical engineering. Total annual research expenditures: $4 million. *Unit head:* Dr. Henrik Pedersen, Graduate Director, 732-445-4950, Fax: 732-445-2421, E-mail: cbemail@sol.rutgers.edu.

San Jose State University, Graduate Studies, College of Engineering, Department of Chemical Engineering and Materials Engineering, Program in Chemical Engineering, San Jose, CA 95192-0001. Offers MS. *Faculty:* 3 full-time (0 women), 2 part-time (0 women). *Students:* 8 full-time (4 women), 17 part-time (6 women); includes 14 minority (13 Asian Americans or Pacific Islanders, 1 Hispanic American), 5 international. Average age 27. 37 applicants, 57% accepted. In 1998, 10 degrees awarded. *Degree requirements:* For master's, thesis or alternative required. *Application deadline:* For fall admission, 6/1. Applications are processed on a rolling basis. Application fee: $59. Tuition, nonresident: part-time $246 per unit. Required fees: $1,939; $1,309 per year. *Unit head:* Dawn M. Belveal, Secretary, 541-737-4791, Fax: 541-737-2182, E-mail: belveadm@che.orst.edu. *Application contact:* Dr. Melanie McNeil, Coordinator, 408-924-3873.

South Dakota School of Mines and Technology, Graduate Division, Department of Chemical Engineering, Rapid City, SD 57701-3995. Offers MS. Part-time programs available. *Faculty:* 5 full-time (0 women). *Students:* 7 full-time (2 women), 3 international. Average age 27. In 1998, 3 degrees awarded. *Degree requirements:* For master's, thesis required, foreign language not required. *Entrance requirements:* For master's, TOEFL (minimum score of 520 required), TWE. *Application deadline:* For fall admission, 6/15 (priority date); for spring admission, 10/15. Applications are processed on a rolling basis. Application fee: $15. Electronic applications accepted. Tuition, state resident: part-time $89 per hour. Tuition, nonresident: part-time $261 per hour. Part-time tuition and fees vary according to program. *Financial aid:* In 1998–99, 5 research assistantships, 3 teaching assistantships were awarded.; fellowships, Federal Work-Study and institutionally-sponsored loans also available. Aid available to part-time students. Financial aid application deadline: 5/15. *Faculty research:* Incineration chemistry, environmental chemistry, polymer surface chemistry. Total annual research expenditures: $79,357. *Unit head:* Dr. Steve McDowell, Chair, 605-394-2421. *Application contact:* Brenda Brown, Secretary, 800-454-8162 Ext. 2493, Fax: 605-394-5360, E-mail: graduate_admissions@silver.sdmt.edu.

South Dakota School of Mines and Technology, Graduate Division, Division of Material Engineering and Science, Doctoral Program in Materials Engineering and Science, Rapid City, SD 57701-3995. Offers chemical engineering (PhD); chemistry (PhD); civil engineering (PhD); electrical engineering (PhD); mechanical engineering (PhD); metallurgical engineering (PhD); physics (PhD). Part-time programs available. *Students:* 14 full-time (2 women), 9 international. *Degree requirements:* For doctorate, dissertation required, foreign language not required. *Entrance requirements:* For doctorate, TOEFL (minimum score of 520 required), TWE, minimum graduate GPA of 3.0. *Application deadline:* For fall admission, 6/15 (priority date); for spring admission, 10/15. Applications are processed on a rolling basis. Application fee: $15. Electronic applications accepted. Tuition, state resident: part-time $89 per hour. Tuition, nonresident: part-time $261 per hour. Part-time tuition and fees vary according to program. *Unit head:* Dr. Chris Jenkins, Coordinator, 605-394-2406. *Application contact:* Brenda Brown, Secretary, 800-454-8162 Ext. 2493, Fax: 605-394-5360, E-mail: graduate_admissions@silver.sdmt.edu.

Stanford University, School of Engineering, Department of Chemical Engineering, Stanford, CA 94305-9991. Offers MS, PhD, Eng. *Faculty:* 10 full-time (1 woman). *Students:* 51 full-time (10 women), 11 part-time (4 women); includes 20 minority (18 Asian Americans or Pacific Islanders, 2 Hispanic Americans), 21 international. Average age 25. 212 applicants, 25% accepted. In 1998, 6 master's, 6 doctorates awarded. Terminal master's awarded for partial completion of doctoral program. *Degree requirements:* For master's, foreign language and thesis not required; for doctorate, dissertation required, foreign language not required; for Eng, thesis required. *Entrance requirements:* For master's, doctorate, and Eng, GRE General Test, TOEFL. *Application deadline:* For fall admission, 1/1. Application fee: $65 ($80 for international students). Electronic applications accepted. Tuition: Full-time $24,588. Required fees: $152. Part-time tuition and fees vary according to course load. *Financial aid:* Fellowships, research assistantships, teaching assistantships, institutionally-sponsored loans available. Financial aid application deadline: 1/1. *Unit head:* Gerald G. Fuller, Chairman, 650-723-9243, Fax: 650-725-7294, E-mail: ggf@chemeng.stanford.edu. *Application contact:* Graduate Admissions Coordinator, 650-723-3143.

State University of New York at Buffalo, Graduate School, School of Engineering and Applied Sciences, Department of Chemical Engineering, Buffalo, NY 14260. Offers M Eng, MS, PhD. Part-time programs available. *Faculty:* 11 full-time (0 women), 1 part-time (0 women). *Students:* 33 full-time (11 women), 25 part-time (2 women); includes 6 minority (3 African Americans, 3 Asian Americans or Pacific Islanders), 34 international. Average age 23. 193 applicants, 10% accepted. In 1998, 12 master's, 3 doctorates awarded. *Degree requirements:* For master's and doctorate, thesis/dissertation required, foreign language not required. *Entrance requirements:* For master's and doctorate, GRE General Test, TOEFL (minimum score of 550 required). *Application deadline:* For fall admission, 2/1 (priority date); for spring admission, 10/1 (priority date). Applications are processed on a rolling basis. Application fee: $35. Tuition, state resident: full-time $5,100; part-time $213 per credit hour. Tuition, nonresident: full-time $8,416; part-time $351 per credit hour. Required fees: $870; $75 per semester. Tuition and fees vary according to course load and program. *Financial aid:* In 1998–99, 28 students received aid, including fellowships with full tuition reimbursements available (averaging $15,000 per year), 14 research assistantships with full tuition reimbursements available (averaging $12,000 per year), 14 teaching assistantships with full tuition reimbursements available (averaging $10,350 per year); career-related internships or fieldwork, Federal Work-Study, institutionally-sponsored loans, tuition waivers (full and partial), and unspecified assistantships also available. Aid available to part-time students. Financial aid application deadline: 2/28; financial aid applicants required to submit FAFSA. *Faculty research:* Transport, polymers, materials, biochemical engineering, catalysis. Total annual research expenditures: $649,704. *Unit head:* Dr. Carl R. F. Lund, Chairman, 716-645-2911 Ext. 2211, Fax: 716-645-3822, E-mail: lund@eng.buffalo.edu. *Application contact:* T. J. Mountziaris, Graduate Committee Chair, 716-645-2911 Ext. 2212, Fax: 716-645-3822, E-mail: tjm@eng.buffalo.edu.

See in-depth description on page 505.

Stevens Institute of Technology, Graduate School, Charles V. Schaefer Jr. School of Engineering, Department of Chemical Engineering, Hoboken, NJ 07030. Offers analysis of polymer processing methods (Certificate); biochemical engineering (M Eng, PhD, Engr); fundamentals of modern chemical engineering (Certificate); polymer engineering (M Eng, PhD, Engr); polymer processing (Certificate); process control (M Eng, PhD, Engr); process engineering (M Eng, PhD, Certificate, Engr). Part-time and evening/weekend programs available. Postbaccalaureate distance learning degree programs offered (no on-campus study). Terminal master's awarded for partial completion of doctoral program. *Degree requirements:* For master's, computer language, thesis or alternative required, foreign language not required; for doctorate, one foreign language, computer language, dissertation required; for other advanced degree, computer language, project or thesis required. *Entrance requirements:* For master's and doctorate, TOEFL (minimum score of 550 required). Electronic applications accepted. *Faculty research:* Biochemical reaction engineering, polymerization engineering, reactor design, biochemical process control and synthesis.

Syracuse University, Graduate School, L. C. Smith College of Engineering and Computer Science, Department of Chemical Engineering and Materials Sciences, Program in Chemical Engineering, Syracuse, NY 13244-0003. Offers MS, PhD. *Students:* 15 full-time (3 women), 15 part-time (4 women); includes 2 minority (1 African American, 1 Asian American or Pacific Islander), 23 international. Average age 27. 143 applicants, 29% accepted. In 1998, 5 master's, 2 doctorates awarded. *Degree requirements:* For master's, thesis required, foreign language not required; for doctorate, computer language, dissertation required, foreign language not required. *Entrance requirements:* For master's and doctorate, GRE General Test, GRE Subject Test. *Application deadline:* Applications are processed on a rolling basis. Application fee: $40. Tuition: Full-time $13,992; part-time $583 per credit hour. *Financial aid:* Fellowships, research assistantships, teaching assistantships, Federal Work-Study and tuition waivers (partial) available. Financial aid application deadline: 3/1. *Faculty research:* Fluid particle technology, water desalination and renovation, membrane technology. *Unit head:* Dr. Philip A. Rice, Head, 315-443-2557.

Tennessee Technological University, Graduate School, College of Engineering, Department of Chemical Engineering, Cookeville, TN 38505. Offers MS, PhD. Part-time programs available. *Faculty:* 8 full-time (0 women). *Students:* 14 full-time (1 woman), 1 part-time; includes 11 minority (all Asian Americans or Pacific Islanders) Average age 26. 68 applicants, 56% accepted. In 1998, 6 master's awarded. *Degree requirements:* For master's, thesis required, foreign language not required; for doctorate, one foreign language (computer language can substitute), dissertation required. *Entrance requirements:* For master's, GRE General Test, TOEFL (minimum score of 525 required); for doctorate, GRE Subject Test, TOEFL (minimum score of 525 required), minimum GPA of 3.5. *Application deadline:* For fall admission, 3/1 (priority date); for spring admission, 8/1. Application fee: $25 ($30 for international students). Tuition, state resident: part-time $137 per hour. Tuition, nonresident: part-time $361 per hour. Required fees: $17 per hour. Tuition and fees vary according to course load. *Financial aid:* In 1998–99, 5 research assistantships (averaging $8,000 per year), 3 teaching assistantships (averaging $8,000 per year) were awarded.; fellowships, career-related internships or fieldwork also available. Financial aid application deadline: 4/1. *Faculty research:* Biochemical conversion, insulation, fuel reprocessing. *Unit head:* Dr. David W. Yarbrough, Chairperson, 931-372-3297, Fax: 931-372-6372, E-mail: dwy1460@tntech.edu. *Application contact:* Dr. Rebecca F. Quattlebaum, Dean of the Graduate School, 931-372-3233, Fax: 931-372-3497, E-mail: rquattlebaum@tntech.edu.

Chemical Engineering

Texas A&M University, College of Engineering, Department of Chemical Engineering, College Station, TX 77843. Offers M Eng, MS, PhD. *Faculty:* 24 full-time (1 woman), 2 part-time (0 women). *Students:* 85 full-time (22 women), 12 part-time (2 women); includes 14 minority (2 African Americans, 6 Asian Americans or Pacific Islanders, 6 Hispanic Americans), 47 international. Average age 27. 265 applicants, 22% accepted. In 1998, 24 master's awarded (50% found work related to degree, 50% continued full-time study); 7 doctorates awarded. Terminal master's awarded for partial completion of doctoral program. *Degree requirements:* For master's, thesis (MS) required; for doctorate, dissertation required, foreign language not required. *Entrance requirements:* For master's and doctorate, GRE General Test, TOEFL. *Application deadline:* For fall admission, 3/31 (priority date); for spring admission, 10/1. Applications are processed on a rolling basis. Application fee: $50 ($75 for international students). *Financial aid:* In 1998–99, 6 fellowships with tuition reimbursements (averaging $12,100 per year), 80 research assistantships with tuition reimbursements (averaging $13,935 per year), 15 teaching assistantships with tuition reimbursements (averaging $18,600 per year) were awarded.; career-related internships or fieldwork and scholarships also available. Financial aid application deadline: 3/31; financial aid applicants required to submit FAFSA. *Faculty research:* Reaction engineering, interface phenomena, environmental applications, biochemical engineering, polymers. Total annual research expenditures: $4.2 million. *Unit head:* Dr. Rayford G. Anthony, Head, 409-845-3361. *Application contact:* Towanna H. Mann, Staff Assistant, 409-895-3364, Fax: 409-895-6446, E-mail: towanna@tamu.edu.

Texas A&M University–Kingsville, College of Graduate Studies, College of Engineering, Department of Chemical Engineering and Natural Gas Engineering, Program in Chemical Engineering, Kingsville, TX 78363. Offers ME, MS. Part-time and evening/weekend programs available. *Faculty:* 4 full-time. *Students:* 20 full-time (2 women), 16 part-time (4 women). *Degree requirements:* For master's, computer language, thesis or alternative, comprehensive exam required, foreign language not required. *Entrance requirements:* For master's, GRE General Test (minimum combined score of 1000 required), TOEFL (minimum score of 525 required), minimum GPA of 3.0. *Application deadline:* For fall admission, 6/1; for spring admission, 11/15. Applications are processed on a rolling basis. Application fee: $15 ($25 for international students). Tuition, state resident: full-time $2,062. Tuition, nonresident: full-time $7,246. *Financial aid:* Fellowships, Federal Work-Study, institutionally-sponsored loans, and tuition waivers (partial) available. Financial aid application deadline: 5/15. *Faculty research:* Process control, error detection and reconciliation, fluid mechanics, handling of solids. *Unit head:* Dr. William Heenan, Coordinator, Department of Chemical Engineering and Natural Gas Engineering, 361-593-2001, Fax: 361-593-2106.

Texas Tech University, Graduate School, College of Engineering, Department of Chemical Engineering, Lubbock, TX 79409. Offers MS Ch E, PhD. Part-time programs available. *Faculty:* 8 full-time (0 women). *Students:* 17 full-time (1 woman), 4 part-time (1 woman); includes 1 minority (Native American), 16 international. Average age 26. 69 applicants, 17% accepted. In 1998, 4 master's, 5 doctorates awarded. *Degree requirements:* For master's and doctorate, computer language, thesis/dissertation required, foreign language not required. *Entrance requirements:* For master's, GRE General Test (minimum combined score of 1000 required; average 1269), minimum GPA of 3.0; for doctorate, GRE General Test (minimum combined score of 1000 required), minimum GPA of 3.0. *Application deadline:* For fall admission, 4/15 (priority date); for spring admission, 11/1 (priority date). Applications are processed on a rolling basis. Application fee: $25 ($50 for international students). Electronic applications accepted. *Financial aid:* In 1998–99, 14 students received aid, including 14 research assistantships (averaging $9,409 per year); fellowships, teaching assistantships, Federal Work-Study and institutionally-sponsored loans also available. Aid available to part-time students. Financial aid application deadline: 5/15; financial aid applicants required to submit FAFSA. *Faculty research:* Cotton fiber use for aquatic crude oil spills, chemical process control, hazardous and toxic waste. Total annual research expenditures: $333,316. *Unit head:* Dr. Alan L. Graham, Chairman, 806-742-3552, Fax: 806-742-3552.

Tufts University, Division of Graduate and Continuing Studies and Research, Graduate School of Arts and Sciences, College of Engineering, Department of Chemical Engineering, Medford, MA 02155. Offers ME, MS, PhD. Part-time programs available. *Faculty:* 11 full-time, 2 part-time. *Students:* 53 (20 women); includes 3 minority (all Asian Americans or Pacific Islanders) 29 international. 45 applicants, 42% accepted. In 1998, 11 master's, 1 doctorate awarded. Terminal master's awarded for partial completion of doctoral program. *Degree requirements:* For master's, thesis required (for some programs), foreign language not required; for doctorate, dissertation required, foreign language not required. *Entrance requirements:* For master's and doctorate, GRE General Test, TOEFL (minimum score of 550 required). *Application deadline:* For fall admission, 2/15; for spring admission, 10/15. Applications are processed on a rolling basis. Application fee: $50. *Financial aid:* Research assistantships with full and partial tuition reimbursements, teaching assistantships with full and partial tuition reimbursements, Federal Work-Study, scholarships, and tuition waivers (partial) available. Financial aid application deadline: 2/15; financial aid applicants required to submit FAFSA. *Unit head:* Jerry Meldon, Chair, 617-627-2580, Fax: 617-627-3991. *Application contact:* Maria Flytzani-Stephanopoulas, 617-627-3900, Fax: 617-627-3991, E-mail: chemstudent@infonet.tufts.edu.

Tulane University, School of Engineering, Department of Chemical Engineering, New Orleans, LA 70118-5669. Offers MS, MSE, PhD, Sc D. MS and PhD offered through the Graduate School. *Students:* 29 full-time (9 women), 1 (woman) part-time; includes 3 minority (2 African Americans, 1 Asian American or Pacific Islander), 12 international. 139 applicants, 13% accepted. In 1998, 3 master's, 7 doctorates awarded. *Degree requirements:* For master's, computer language required (for some programs), thesis optional, foreign language not required; for doctorate, computer language (for some programs), dissertation required, foreign language not required. *Entrance requirements:* For master's and doctorate, GRE General Test, TOEFL, minimum B average in undergraduate course work. *Application deadline:* For fall admission, 7/1; for spring admission, 10/15. Application fee: $35. *Financial aid:* Fellowships, research assistantships, teaching assistantships, tuition waivers (full) available. Financial aid application deadline: 2/1. *Faculty research:* Interfacial phenomena catalysis, electrochemical engineering, environmental science. *Unit head:* Dr. Richard D. Gonzalez, Chairman, 504-865-5772. *Application contact:* Dr. E. Michaelides, Associate Dean, 504-865-5764.

See in-depth description on page 507.

Universidad de las Américas–Puebla, Division of Graduate Studies, School of Engineering, Program in Chemical Engineering, Cholula, 72820, Mexico. Offers chemical engineering (MS); food technology (MS). Part-time and evening/weekend programs available. *Faculty:* 10 full-time (0 women), 2 part-time (0 women). *Students:* 31 full-time (20 women), 20 part-time (1 woman); all minorities (all Hispanic Americans) Average age 29. 40 applicants, 75% accepted. In 1998, 9 degrees awarded. *Degree requirements:* For master's, one foreign language, computer language, thesis required. *Average time to degree:* Master's–2.5 years full-time, 4 years part-time. *Application deadline:* For fall admission, 7/16. Applications are processed on a rolling basis. Application fee: $0. *Financial aid:* In 1998–99, 20 students received aid, including 8 research assistantships Aid available to part-time students. Financial aid application deadline: 5/15. *Faculty research:* Food science, reactors, oil industry, biotechnology. Total annual research expenditures: $75,000. *Unit head:* Dr. Rene Reyes, Coordinator, 22-29-21-26, Fax: 22-29-20-32, E-mail: rreyes@mail.udlap.mx. *Application contact:* Maurlclo Vlllegas, Chair of Admissions Office, 22-29-20-17, Fax: 22-29-20-18, E-mail: admision@mail.udlap.mx.

Université de Sherbrooke, Faculty of Applied Sciences, Department of Chemical Engineering, Sherbrooke, PQ J1K 2R1, Canada. Offers M Sc A, PhD. *Degree requirements:* For master's and doctorate, thesis/dissertation required. *Faculty research:* Conversion processes, high-temperature plasma technologies, system engineering, environmental engineering, textile technologies.

Université Laval, Faculty of Graduate Studies, Faculty of Sciences and Engineering, Department of Chemical Engineering, Sainte-Foy, PQ G1K 7P4, Canada. Offers M Sc, PhD. *Students:* 42 full-time (15 women), 3 part-time (1 woman). 27 applicants, 56% accepted. In 1998, 6 master's, 7 doctorates awarded. *Application deadline:* For fall admission, 3/1. Application fee: $30. *Faculty research:* Biochemical engineering, chemical reaction engineering, polymers, thermodynamics. *Unit head:* Jean-Claude Méthot, Director, 418-656-2131 Ext. 2539, Fax: 418-656-5993, E-mail: jean-claude.methot@gch.ulaval.ca.

The University of Akron, Graduate School, College of Engineering, Department of Chemical Engineering, Akron, OH 44325-0001. Offers MS Ch E, PhD. Part-time and evening/weekend programs available. *Faculty:* 9 full-time, 2 part-time. *Students:* 29 full-time (6 women), 4 part-time (1 woman); includes 2 minority (1 African American, 1 Asian American or Pacific Islander), 26 international. Average age 25. In 1998, 10 master's awarded. *Degree requirements:* For master's, thesis required, foreign language not required; for doctorate, variable foreign language requirement (computer language can substitute for one), dissertation, candidacy exam, qualifying exam required. *Entrance requirements:* For master's, GRE General Test, TOEFL (minimum score of 550 required), minimum GPA of 2.75; for doctorate, GRE General Test, TOEFL (minimum score of 550 required). *Average time to degree:* Master's–2 years full-time, 4 years part-time. *Application deadline:* For fall admission, 3/1. Applications are processed on a rolling basis. Application fee: $25 ($50 for international students). Tuition, state resident: part-time $189 per credit. Tuition, nonresident: part-time $353 per credit. Required fees: $7.3 per credit. *Financial aid:* Fellowships with full tuition reimbursements, research assistantships with full tuition reimbursements, teaching assistantships with full tuition reimbursements, career-related internships or fieldwork and scholarships available. Financial aid application deadline: 3/1. *Faculty research:* Reactor design, catalysis, synthetic fuels, transport phenomena, process engineering. *Unit head:* Dr. Steven S. Chuang, Chair, 330-972-6993, E-mail: schuang@uakron.edu. *Application contact:* Dr. Steven S. Chuang, Chair, 330-972-6993, E-mail: schuang@uakron.edu.

The University of Alabama, Graduate School, College of Engineering, Department of Chemical Engineering, Tuscaloosa, AL 35487. Offers MS Ch E, PhD. Part-time and evening/weekend programs available. *Faculty:* 9 full-time (0 women). *Students:* 19 full-time (2 women), 3 part-time (1 woman), includes 2 minority (both Asian Americans or Pacific Islanders), 13 international. Average age 25. 35 applicants, 31% accepted. In 1998, 4 master's awarded (75% found work related to degree, 25% continued full-time study); 3 doctorates awarded (100% found work related to degree). *Degree requirements:* For master's, thesis or alternative required, foreign language not required; for doctorate, dissertation required, foreign language not required. *Entrance requirements:* For master's, GRE General Test (minimum combined score of 1500 on three sections required) or minimum GPA of 3.0 in last 60 hours; for doctorate, GRE General Test (minimum combined score of 1500 on three sections required) or minimum GPA of 3.0. *Average time to degree:* Master's–2 years full-time; doctorate–5 years full-time. *Application deadline:* For fall admission, 7/1 (priority date); for spring admission, 11/1 (priority date). Applications are processed on a rolling basis. Application fee: $25. Electronic applications accepted. *Financial aid:* In 1998–99, 19 research assistantships with full tuition reimbursements (averaging $15,500 per year), teaching assistantships with full tuition reimbursements (averaging $15,500 per year) were awarded.; fellowships with full tuition reimbursements, Federal Work-Study also available. Financial aid application deadline: 5/15. *Faculty research:* Global environmental change, magnetic tape, process modeling, pollution prevention, polymeric materials. *Unit head:* Dr. Gary C. April, Head, 205-348-6450, Fax: 205-348-7558, E-mail: gcapril@coe.eng.ua.edu. *Application contact:* Dr. John Wiest, 205-348-1727, Fax: 205-348-7558, E-mail: jwiest@coe.eng.ua.edu.

The University of Alabama in Huntsville, School of Graduate Studies, College of Engineering, Department of Chemical and Materials Engineering, Huntsville, AL 35899. Offers MSE. Part-time and evening/weekend programs available. *Faculty:* 5 full-time (0 women). *Students:* 8 full-time (2 women), 4 part-time (1 woman); includes 1 minority (African American), 7 international. Average age 29. 10 applicants, 90% accepted. In 1998, 1 degree awarded. *Degree requirements:* For master's, oral and written exams required, thesis optional, foreign language not required. *Entrance requirements:* For master's, GRE General Test (minimum combined score of 1500 on three sections required), appropriate bachelor's degree, minimum GPA of 3.0. *Application deadline:* For fall admission, 7/24 (priority date); for spring admission, 11/15 (priority date). Applications are processed on a rolling basis. Application fee: $20. Tuition and fees vary according to course load. *Financial aid:* In 1998–99, 9 research assistantships with full and partial tuition reimbursements (averaging $9,060 per year), 4 teaching assistantships with full and partial tuition reimbursements (averaging $8,775 per year) were awarded.; fellowships with full and partial tuition reimbursements, career-related internships or fieldwork, Federal Work-Study, grants, institutionally-sponsored loans, scholarships, and tuition waivers (full and partial) also available. Aid available to part-time students. Financial aid application deadline: 4/1; financial aid applicants required to submit FAFSA. *Faculty research:* Turbulence modeling, computational fluid dynamics, microgravity processing, multiphase transport, blood materials transport. Total annual research expenditures: $126,346. *Unit head:* Dr. Ramon Cerro, Chair, 256-890-6810, Fax: 256-890-6839, E-mail: rlc@eb.uah.edu.

University of Alberta, Faculty of Graduate Studies and Research, Department of Chemical and Materials Engineering, Edmonton, AB T6G 2E1, Canada. Offers chemical engineering (M Eng, M Sc, PhD); materials engineering (M Eng, M Sc, PhD); process control (M Eng, M Sc, PhD); welding (M Eng). Part-time programs available. Postbaccalaureate distance learning degree programs offered (minimal on-campus study). Terminal master's awarded for partial completion of doctoral program. *Degree requirements:* For master's and doctorate, thesis/dissertation required, foreign language not required. *Faculty research:* Advanced materials and polymers, catalytic and reaction engineering, mineral processing, physical metallurgy,fluid mechanics.

The University of Arizona, Graduate College, College of Engineering and Mines, Department of Chemical and Environmental Engineering, Tucson, AZ 85721. Offers chemical engineering (MS, PhD); environmental engineering (MS, PhD). Part-time programs available. *Faculty:* 17. *Students:* 48 full-time (12 women), 11 part-time (1 woman); includes 4 minority (1 Asian American or Pacific Islander, 3 Hispanic Americans), 25 international. Average age 30. 77 applicants, 69% accepted. In 1998, 13 master's, 2 doctorates awarded. *Degree requirements:* For master's, thesis required, foreign language not required; for doctorate, dissertation, comprehensive and departmental qualifying exams required, foreign language not required. *Entrance requirements:* For master's and doctorate, TOEFL (minimum score of 550 required), minimum GPA of 3.0. *Application deadline:* For fall admission, 3/1. Applications are processed on a rolling basis. Application fee: $35. *Financial aid:* Fellowships, research assistantships, teaching assistantships, institutionally-sponsored loans and scholarships available. Financial aid application deadline: 6/1. *Faculty research:* Energy and environment–hazardous waste incineration, fossil fuel combustion, processing high-purity gases and liquids, aerosol reactor theory, pharmacokinetics. *Unit head:* Dr. Thomas Peterson, Head, 520-621-2591. *Application contact:* Wendy Haley, Graduate Secretary, 520-621-6052, Fax: 520-621-6048.

University of Arkansas, Graduate School, College of Engineering, Department of Chemical Engineering, Fayetteville, AR 72701-1201. Offers MS Ch E, MSE, PhD. *Faculty:* 11 full-time (0 women). *Students:* 21 full-time (3 women), 3 part-time; includes 1 minority (Asian American or Pacific Islander), 10 international. 24 applicants, 50% accepted. In 1998, 6 master's, 2 doctorates awarded. *Degree requirements:* For master's, thesis optional, foreign language not required; for doctorate, one foreign language, dissertation required. Application fee: $40 ($50 for international students). Tuition, state resident: full-time $3,186. Tuition, nonresident: full-time $7,560. Required fees: $378. *Financial aid:* In 1998–99, 5 research assistantships, 17 teaching assistantships were awarded.; fellowships, career-related internships or fieldwork and Federal Work-Study also available. Aid available to part-time students. Financial aid application deadline: 4/1; financial aid applicants required to submit FAFSA. *Unit head:* Dr. R. E. Babcock, Chair, 501-575-4951. *Application contact:* Reed Welker, Graduate Coordinator, E-mail: jrw@engr.uark.edu.

University of British Columbia, Faculty of Graduate Studies, Faculty of Applied Science, Department of Chemical Engineering, Vancouver, BC V6T 1Z2, Canada. Offers chemical engineering (M Eng, MA Sc, PhD); fire protection engineering (M Eng); pulp and paper engineering (M Eng). Part-time and evening/weekend programs available. *Degree requirements:* For master's, thesis required (for some programs), foreign language not required; for doctorate, dissertation required, foreign language not required. *Entrance requirements:* For master's and doctorate, TOEFL (minimum score of 550 required). *Faculty research:* Electrochemical engineering, biotechnology, catalysis, polymers.

University of Calgary, Faculty of Graduate Studies, Faculty of Engineering, Department of Chemical and Petroleum Engineering, Calgary, AB T2N 1N4, Canada. Offers M Eng, M Sc, PhD. Part-time programs available. *Faculty:* 19 full-time (2 women), 5 part-time (0 women). *Students:* 46 full-time (15 women), 22 part-time (7 women). Average age 24. 800 applicants, 2% accepted. In 1998, 13 master's awarded (60% found work related to degree, 40% continued full-time study); 9 doctorates awarded. *Degree requirements:* For master's, thesis required (for some programs), foreign language not required; for doctorate, dissertation, candidacy exam required, foreign language not required. *Entrance requirements:* For master's and doctorate, TOEFL (minimum score of 550 required). *Average time to degree:* Master's–2 years full-time, 4 years part-time; doctorate–4 years full-time. *Application deadline:* For fall admission, 5/31 (priority date); for winter admission, 9/30 (priority date); for spring admission, 1/31 (priority date). Applications are processed on a rolling basis. Application fee: $60. *Financial aid:* In 1998–99, 40 fellowships with partial tuition reimbursements, 25 research assistantships with partial tuition reimbursements, 30 teaching assistantships with partial tuition reimbursements were awarded.; grants also available. Financial aid application deadline: 5/31. *Faculty research:* Thermodynamics, transport phenomena, enhanced oil recovery, kinetics and fluidized beds, environmental engineering. Total annual research expenditures: $2 million. *Unit head:* R. G. Moore, Head, 403-220-5750, Fax: 403-284-4852, E-mail: moore@ench.ucalgary.ca. *Application contact:* A. K. Mehrotra, Associate Head, Graduate Studies, 403-220-7406, Fax: 403-284-4852, E-mail: mehrotra@acs.ucalgary.ca.

University of California, Berkeley, Graduate Division, College of Chemistry, Department of Chemical Engineering, Berkeley, CA 94720-1500. Offers MS, PhD. *Faculty:* 17 full-time (2 women), 2 part-time (0 women). *Students:* 96 full-time (28 women); includes 23 minority (3 African Americans, 20 Asian Americans or Pacific Islanders), 8 international. 275 applicants, 23% accepted. In 1998, 2 master's, 17 doctorates awarded. *Degree requirements:* For master's, thesis required; for doctorate, dissertation, qualifying exam required. *Entrance requirements:* For master's and doctorate, GRE General Test, TOEFL, minimum GPA of 3.0. *Application deadline:* For fall admission, 2/10. Application fee: $40. *Faculty research:* Biochemical engineering, electrochemical engineering, electronic materials, heterogeneous catalysis and reaction engineering, complex fluids, molecular theory and simulation, environmental engineering. *Unit head:* Harvey W. Blanch, Chair, 510-643-7610, Fax: 510-642-4778. *Application contact:* Aileen Harris, Graduate Assistant for Admission, 510-642-1533, Fax: 510-6424778, E-mail: aileen@cchem.berkeley.edu.

University of California, Davis, Graduate Studies, College of Engineering, Program in Chemical Engineering and Materials Science, Davis, CA 95616. Offers chemical engineering (MS, PhD); materials science (MS, PhD, Certificate). *Faculty:* 25 full-time (5 women). *Students:* 55 full-time (17 women); includes 12 minority (1 African American, 8 Asian Americans or Pacific Islanders, 2 Hispanic Americans, 1 Native American), 15 international. Average age 26. 120 applicants, 25% accepted. In 1998, 11 master's, 12 doctorates awarded. Terminal master's awarded for partial completion of doctoral program. *Degree requirements:* For master's and doctorate, thesis/dissertation required, foreign language not required. *Entrance requirements:* For master's, GRE General Test (minimum score of 500 on verbal section, 700 on quantitative, 500 on analytical required), minimum GPA of 3.0; for doctorate, GRE General Test (minimum score of 500 on verbal section, 720 on quantitative, 500 on analytical required), TOEFL (minimum score of 600 required), minimum GPA of 3.0. *Application deadline:* For fall admission, 1/15 (priority date). Application fee: $40. Electronic applications accepted. *Financial aid:* In 1998–99, 18 fellowships, 37 research assistantships, 13 teaching assistantships were awarded. Financial aid application deadline: 1/15; financial aid applicants required to submit FAFSA. *Faculty research:* Transport phenomena, colloid science, catalysis, biotechnology, materials. *Unit head:* Subhash Risbud, Chairperson, 530-752-5132.

University of California, Irvine, Office of Research and Graduate Studies, School of Engineering, Department of Chemical and Biochemical Engineering and Materials Science, Irvine, CA 92697. Offers biomedical engineering (MS, PhD), including engineering; chemical and biochemical engineering (MS, PhD). Part-time programs available. *Faculty:* 10 full-time (2 women). *Students:* 30 full-time (9 women), 4 part-time (1 woman); includes 11 minority (10 Asian Americans or Pacific Islanders, 1 Hispanic American), 17 international. 72 applicants, 46% accepted. In 1998, 11 master's, 6 doctorates awarded. Terminal master's awarded for partial completion of doctoral program. *Degree requirements:* For doctorate, dissertation required, foreign language not required, foreign language not required. *Entrance requirements:* For master's, GRE General Test, minimum GPA of 3.0; for doctorate, GRE General Test. *Application deadline:* For fall admission, 1/15 (priority date). Applications are processed on a rolling basis. Application fee: $40. Electronic applications accepted. *Financial aid:* Fellowships, research assistantships, teaching assistantships, institutionally-sponsored loans available. Financial aid application deadline: 3/2; financial aid applicants required to submit FAFSA. *Faculty research:* Bioreactors, recombinant cells, separation operations. *Unit head:* Dr. Enrique J. Lavernia, Chair, 949-824-8277, Fax: 949-824-2541, E-mail: lavernia@uci.edu. *Application contact:* Nancy Carter-Fields, Graduate Coordinator, 949-824-3786, Fax: 949-824-2541, E-mail: nvcarter@uci.edu.

See in-depth description on page 509.

University of California, Los Angeles, Graduate Division, School of Engineering and Applied Science, Department of Chemical Engineering, Los Angeles, CA 90095. Offers MS, PhD. *Faculty:* 11 full-time. *Students:* 56 full-time (8 women); includes 18 minority (1 African American, 14 Asian Americans or Pacific Islanders, 1 Hispanic American, 2 Native Americans), 20 international. 132 applicants, 32% accepted. In 1998, 4 master's, 8 doctorates awarded. *Degree requirements:* For master's, thesis required, foreign language not required; for doctorate, dissertation, qualifying exams required, foreign language not required. *Entrance requirements:* For master's, GRE General Test, minimum GPA of 3.0; for doctorate, GRE General Test, minimum GPA of 3.25. *Application deadline:* For fall admission, 1/15; for spring admission, 12/15. Application fee: $40. Electronic applications accepted. *Financial aid:* In 1998–99, 25 fellowships, 15 research assistantships, 6 teaching assistantships were awarded.; Federal Work-Study, institutionally-sponsored loans, and tuition waivers (full and partial) also available. Financial aid application deadline: 1/15; financial aid applicants required to submit FAFSA. *Unit head:* Dr. Selim Senkan, Chair, 310-206-4106. *Application contact:* Jacqueline Ellis-Trice, Student Affairs Officer, 310-825-9063, Fax: 310-206-4107, E-mail: jacque@ce.ucla.edu.

University of California, Riverside, Graduate Division, College of Engineering, Department of Chemical and Enviromental Engineering, Riverside, CA 92521-0102. Offers MS, PhD. *Entrance requirements:* For master's and doctorate, GRE General Test, TOEFL. Application fee: $40. *Unit head:* Dr. Mark Matsumoto, Chair, 909-787-2423, Fax: 909-787-2425, E-mail: matsumoto@engr.ucr.edu. *Application contact:* 909-787-2423, Fax: 909-787-2425, E-mail: grad-adm@ee.ucr.edu.

See in-depth description on page 511.

University of California, Riverside, Graduate Division, College of Engineering, Department of Chemical and Environmental Engineering, Riverside, CA 92521-0102. Offers MS, PhD. *Faculty:* 8 full-time (0 women), 8 part-time (2 women). *Students:* 2 full-time (0 women), (both international). *Degree requirements:* For doctorate, dissertation required. *Entrance requirements:* For master's and doctorate, GRE General Test (minimum combined score of 1100 required), TOEFL (minimum score of 550 required), minimum GPA of 3.2 (3.5 for fellowships/teaching assistantships). *Application deadline:* For fall admission, 5/1; for winter admission, 9/1; for spring admission, 12/1. Applications are processed on a rolling basis. Application fee: $40. Electronic applications accepted. *Financial aid:* In 1998–99, 2 students received aid, including 2 teaching assistantships with partial tuition reimbursements available (averaging $13,329 per year); fellowships Financial aid application deadline: 1/5; financial aid applicants required to submit FAFSA. *Faculty research:* Bioprocessing, biodegredation, bioremediation, water/wastewater treatment, biosensors and biodetoxification, transportation emissions. Total annual research expenditures: $395,171. *Unit head:* Dr. Mark Matsumoto, Chair, 909-787-5318, Fax: 909-787-2425, E-mail: matsumoto@engr.ucr.edu. *Application contact:* Tracie Burruel, Graduate Student Affairs Assistant, 909-787-2484, Fax: 909-787-2425, E-mail: burruel@ee.urc.edu.

University of California, San Diego, Graduate Studies and Research, Department of Applied Mechanics and Engineering Sciences, Program in Chemical Engineering, La Jolla, CA 92093-5003. Offers MS, PhD. Part-time programs available. *Students:* 15 full-time (2 women); includes 4 minority (all Asian Americans or Pacific Islanders), 1 international. 50 applicants, 22% accepted. In 1998, 3 master's, 7 doctorates awarded. *Degree requirements:* For master's, comprehensive exam or thesis required; for doctorate, dissertation, qualifying exam required. *Entrance requirements:* For master's and doctorate, GRE General Test, TOEFL (minimum score of 550 required), minimum GPA of 3.0. *Application deadline:* For fall admission, 5/31. Application fee: $40. *Financial aid:* In 1998–99, 3 fellowships with full tuition reimbursements (averaging $15,000 per year), 1 research assistantship with full tuition reimbursement (averaging $15,000 per year), teaching assistantships with partial tuition reimbursements (averaging $13,000 per year) were awarded.; scholarships also available. Financial aid application deadline: 1/31; financial aid applicants required to submit FAFSA. *Faculty research:* Semiconductor and composite materials processing, biochemical processing, electrochemistry and catalysis. *Unit head:* Dr. John Wiest, 205-348-1727, Fax: 205-348-7558, E-mail: jwiest@coe.eng.ua.edu. *Application contact:* AMES Graduate Student Affairs, 619-534-4387, Fax: 619-534-1730, E-mail: bwalton@ames.ucsd.edu.

See in-depth description on page 513.

University of California, Santa Barbara, Graduate Division, College of Engineering, Department of Chemical Engineering, Santa Barbara, CA 93106. Offers MS, PhD. *Students:* 61 full-time. 153 applicants, 33% accepted. In 1998, 6 master's, 7 doctorates awarded. Terminal master's awarded for partial completion of doctoral program. *Degree requirements:* For master's, thesis or alternative required, foreign language not required; for doctorate, dissertation required, foreign language not required. *Entrance requirements:* For master's and doctorate, GRE General Test, TOEFL (minimum score of 560 required). *Average time to degree:* Master's–2 years full-time; doctorate–5 years full-time. *Application deadline:* For fall admission, 1/31. Application fee: $40. Electronic applications accepted. *Financial aid:* Fellowships, research assistantships, teaching assistantships, career-related internships or fieldwork, Federal Work-Study, institutionally-sponsored loans, and tuition waivers (partial) available. Financial aid application deadline: 1/15; financial aid applicants required to submit FAFSA. *Faculty research:* Macromolecules, surfaces and catalysis, process control, inorganic materials, transport. Total annual research expenditures: $5 million. *Unit head:* Glenn H. Fredrickson, Chair, 805-893-8308. *Application contact:* Laura Crownover, Graduate Assistant, 805-893-8671, E-mail: laura@engineering.ucsb.edu.

University of Cincinnati, Division of Research and Advanced Studies, College of Engineering, Department of Chemical Engineering, Cincinnati, OH 45221-0091. Offers MS, PhD. Part-time and evening/weekend programs available. *Faculty:* 13 full-time. *Students:* 42 full-time (9 women), 12 part-time (1 woman); includes 6 minority (1 African American, 2 Asian Americans or Pacific Islanders, 3 Hispanic Americans), 35 international. 144 applicants, 40% accepted. In 1998, 13 master's, 5 doctorates awarded. Terminal master's awarded for partial completion of doctoral program. *Degree requirements:* For master's and doctorate, computer language, thesis/dissertation required, foreign language not required. *Entrance requirements:* For master's and doctorate, GRE General Test, TOEFL (minimum score of 550 required). *Average time to degree:* Master's–3.5 years full-time; doctorate–6.3 years full-time. *Application deadline:* For fall admission, 2/1 (priority date). Application fee: $40. *Financial aid:* Fellowships, career-related internships or fieldwork, tuition waivers (full), and unspecified assistantships available. Financial aid application deadline: 2/1. *Faculty research:* Process synthesis, aerosol processes, clean coal technology, membrane technology. Total annual research expenditures: $978,550. *Unit head:* Dr. Joel Fried, Head, 513-556-2768, Fax: 513-556-3473, E-mail: joel.fried@uc.edu. *Application contact:* Neville Pinto, Graduate Program Director, 513-556-3116, Fax: 513-556-3473, E-mail: neville.pinto@uc.edu.

University of Colorado at Boulder, Graduate School, College of Engineering and Applied Science, Department of Chemical Engineering, Boulder, CO 80309. Offers ME, MS, PhD. Part-time programs available. Terminal master's awarded for partial completion of doctoral program. *Degree requirements:* For master's, thesis, comprehensive exam required; for doctorate, dissertation required, foreign language not required. *Entrance requirements:* For master's, minimum undergraduate GPA of 3.0. *Faculty research:* Catalysis, process control, transport phenomena, biotechnology, membranes.

University of Connecticut, Graduate School, School of Engineering, Field of Chemical Engineering, Storrs, CT 06269. Offers MS, PhD. Terminal master's awarded for partial completion of doctoral program. *Degree requirements:* For master's, thesis or alternative required; for doctorate, dissertation required. *Entrance requirements:* For master's and doctorate, GRE General Test. *Faculty research:* Catalysis, electrochemicals, polymers.

University of Dayton, Graduate School, School of Engineering, Department of Chemical Engineering, Dayton, OH 45469-1300. Offers MS Ch E. Part-time and evening/weekend programs available. *Faculty:* 7 full-time (0 women), 1 part-time (0 women). *Students:* 11 full-time (5 women), 10 part-time (1 woman); includes 2 minority (both African Americans), 4 international. Average age 24. In 1998, 8 degrees awarded. *Degree requirements:* For master's, thesis optional, foreign language not required. *Entrance requirements:* For master's, TOEFL (minimum score of 550 required). *Average time to degree:* Master's–1.5 years full-time, 4 years part-time. *Application deadline:* For fall admission, 8/1 (priority date). Applications are processed on a rolling basis. Application fee: $30. *Financial aid:* In 1998–99, 2 fellowships with full tuition reimbursements (averaging $14,000 per year), 8 research assistantships with full tuition reimbursements (averaging $12,000 per year) were awarded.; institutionally-sponsored loans also available. *Faculty research:* Process control, process modeling, expert systems, materials processing, agitation. *Unit head:* Dr. Tony Saliba, Chair, 937-229-2627. *Application contact:* Dr. Donald L. Moon, Associate Dean, 937-229-2241, Fax: 937-229-2471, E-mail: dmoon@engr.udayton.edu.

University of Delaware, College of Engineering, Department of Chemical Engineering, Newark, DE 19716. Offers M Ch E, PhD. Part-time and evening/weekend programs available. Postbaccalaureate distance learning degree programs offered (minimal on-campus study). *Faculty:* 25 full-time (4 women), 24 part-time (1 woman). *Students:* 91 full-time (17 women), 6 part-time (3 women); includes 5 minority (2 Asian Americans or Pacific Islanders, 3 Hispanic Americans), 53 international. Average age 25. 230 applicants, 27% accepted. In 1998, 7 master's awarded (14% entered university research/teaching, 86% found other work related to degree); 15 doctorates awarded (15% entered university research/teaching, 80% found other work related to degree). Terminal master's awarded for partial completion of doctoral program. *Degree requirements:* For master's, thesis required (for some programs), foreign language not required; for doctorate, dissertation required, foreign language not required. *Entrance requirements:* For master's and doctorate, GRE General Test (minimum combined score of 1100 required), TOEFL (minimum score of 600 required). *Average time to degree:* Master's–7 years full-time; doctorate–15 years full-time. *Application deadline:* For fall admission, 3/15 (priority date). Application fee: $45. Electronic applications accepted. *Financial aid:* In 1998–99, 93 students received aid, including 15 fellowships with full tuition reimbursements available (averaging $17,500 per year), 53 research assistantships with full tuition reimbursements available (averaging $17,500 per year), 25 teaching assistantships with full tuition reimbursements available (averaging $17,500 per year); grants and scholarships also available. Financial

Chemical Engineering

University of Delaware (continued)
aid application deadline: 3/15; financial aid applicants required to submit FAFSA. *Faculty research:* Biochemical/biomedical engineer, thermodynamics, polymers/composites, materials, catalysis/reactions, colloid/interfaces, expert systems/process control. Total annual research expenditures: $5.5 million. *Unit head:* Dr. Eric W. Kaler, Chairman, 302-831-8155, Fax: 302-831-8201, E-mail: kaler@che.udel.edu. *Application contact:* Dr. Norman J. Wagner, Associate Professor, 302-831-8079, Fax: 302-831-1048, E-mail: wagner@che.udel.edu.

University of Detroit Mercy, College of Engineering and Science, Department of Chemical Engineering, Program in Chemical Engineering, Detroit, MI 48219-0900. Offers ME, DE. Evening/weekend programs available. *Degree requirements:* For master's, thesis required, foreign language not required; for doctorate, dissertation required. *Entrance requirements:* For master's, minimum GPA of 3.0.

See in-depth description on page 515.

University of Florida, Graduate School, College of Engineering, Department of Chemical Engineering, Gainesville, FL 32611. Offers ME, MS, PhD, Engr. *Faculty:* 17. *Students:* 75 full-time (14 women), 7 part-time (1 woman); includes 10 minority (3 African Americans, 5 Asian Americans or Pacific Islanders, 1 Hispanic American, 1 Native American), 26 international. 162 applicants, 20% accepted. In 1998, 5 master's, 10 doctorates awarded. *Degree requirements:* For master's, thesis optional, foreign language not required; for doctorate, dissertation required, foreign language not required; for Engr, thesis optional. *Entrance requirements:* For master's and doctorate, GRE General Test, TOEFL (minimum score of 550 required), minimum GPA of 3.0; for Engr, GRE General Test. *Application deadline:* For fall admission, 6/1 (priority date). Applications are processed on a rolling basis. Application fee: $20. Electronic applications accepted. *Financial aid:* In 1998–99, 64 students received aid, including 5 fellowships, 59 research assistantships; teaching assistantships Financial aid application deadline: 5/15. *Faculty research:* Chemical engineering systems, polymeric and biochemical materials, applied control theory, electrochemical and surface sciences. *Unit head:* Dr. Timothy J. Anderson, Chairman, 352-392-0882, Fax: 352-392-9513, E-mail: tim@nervm.nerdc.ufl.edu. *Application contact:* Dr. Ranga Narayanan, Graduate Coordinator, 352-392-9103, Fax: 352-392-9513, E-mail: ranga@gibbs.che.ufl.edu.

See in-depth description on page 517.

University of Houston, Cullen College of Engineering, Department of Chemical Engineering, Houston, TX 77004. Offers M Ch E, MS Ch E, PhD. Part-time and evening/weekend programs available. *Faculty:* 12 full-time (0 women), 17 part-time (0 women). *Students:* 60 full-time (15 women), 38 part-time (12 women); includes 25 minority (4 African Americans, 19 Asian Americans or Pacific Islanders, 2 Hispanic Americans), 34 international. Average age 27. 114 applicants, 15% accepted. In 1998, 16 master's, 16 doctorates awarded. Terminal master's awarded for partial completion of doctoral program. *Degree requirements:* For master's, thesis required (for some programs), foreign language not required; for doctorate, dissertation, departmental qualifying exam required, foreign language not required. *Entrance requirements:* For master's and doctorate, GRE General Test, TOEFL. *Application deadline:* For fall admission, 2/15 (priority date). Applications are processed on a rolling basis. Application fee: $25 ($75 for international students). *Financial aid:* In 1998–99, 54 students received aid, including 6 fellowships, 32 research assistantships, 16 teaching assistantships; Federal Work-Study also available. Financial aid application deadline: 2/15. *Faculty research:* Biochemical engineering, electronic materials, chemical reaction engineering, two-phase flow. Total annual research expenditures: $2 million. *Unit head:* Dr. James T. Richardson, Chairman, 713-743-4304, Fax: 713-743-4323, E-mail: jtr@uh.edu. *Application contact:* Rosalind Walker, Graduate Analyst, 713-743-4311, Fax: 713-743-4323.

See in-depth description on page 519.

University of Idaho, College of Graduate Studies, College of Engineering, Department of Chemical Engineering, Moscow, ID 83844-4140. Offers M Engr, MS, PhD. *Faculty:* 8 full-time (1 woman). *Students:* 8 full-time (2 women), 17 part-time (2 women); includes 1 minority (Hispanic American), 6 international. In 1998, 7 master's, 2 doctorates awarded. *Degree requirements:* For master's, thesis required, thesis required; for doctorate, one foreign language (computer language can substitute), dissertation required. *Entrance requirements:* For master's, GRE, minimum GPA of 2.8; for doctorate, GRE, minimum undergraduate GPA of 2.8, 3.0 graduate. *Application deadline:* For fall admission, 8/1; for spring admission, 12/15. Application fee: $35 ($45 for international students). *Financial aid:* In 1998–99, 2 research assistantships (averaging $17,634 per year), 2 teaching assistantships (averaging $11,461 per year) were awarded.; fellowships Financial aid application deadline: 2/15. *Faculty research:* Geothermal energy utilization, alcohol production from agriculture waste material, energy conservation in pulp and paper mills. *Unit head:* Dr. Wudneh Admassu, Chair, 208-885-6793.

University of Illinois at Chicago, Graduate College, College of Engineering, Department of Chemical Engineering, Chicago, IL 60607-7128. Offers MS, PhD. *Faculty:* 9 full-time (0 women). *Students:* 27 full-time (4 women), 7 part-time (1 woman); includes 4 minority (3 Asian Americans or Pacific Islanders, 1 Hispanic American), 21 international. Average age 25. 102 applicants, 28% accepted. In 1998, 6 master's, 3 doctorates awarded. *Degree requirements:* For master's, thesis or project required; for doctorate, dissertation, departmental qualifying exam required, foreign language not required. *Entrance requirements:* For master's and doctorate, GRE General Test, TOEFL (minimum score of 550 required), minimum GPA of 3.75 on a 5.0 scale. *Application deadline:* For fall admission, 7/3; for spring admission, 11/8. Application fee: $40 ($50 for international students). *Financial aid:* In 1998–99, 5 students received aid; fellowships, research assistantships, teaching assistantships, tuition waivers (full) and tuition and service fee waivers available. *Faculty research:* Multiphase flows, interfacial transport, heterogeneous catalysis, coal technology, molecular and static thermodynamics. *Unit head:* Dr. John H. Kiefer, Acting Head, 312-996-3424, Fax: 312-996-0808, E-mail: u18090@uicvm.uic.edu. *Application contact:* John R. Regalbuto, Director of Graduate Studies, 312-996-0288, Fax: 312-996-0808, E-mail: john.r.regalbuto@uicvm.uic.edu.

Announcement: Housed in its newly renovated building less than 1 mile from the Chicago Loop, this expanding department has major research efforts related to energy, materials, and environmental concerns, with sponsorship from all major governmental agencies and industry. Research and teaching assistantships are available for newly admitted students. See in-depth description of College of Engineering in Section 1.

University of Illinois at Urbana–Champaign, Graduate College, College of Liberal Arts and Sciences, School of Chemical Sciences, Department of Chemical Engineering, Urbana, IL 61801. Offers MS, PhD. *Faculty:* 11 full-time (0 women). *Students:* 71 full-time (13 women); includes 8 minority (7 Asian Americans or Pacific Islanders, 1 Hispanic American), 24 international. Average age 23. 172 applicants, 9% accepted. In 1998, 14 master's, 10 doctorates awarded. *Degree requirements:* For master's, thesis required, foreign language not required; for doctorate, dissertation, departmental qualifying exam required, foreign language not required. *Application deadline:* Applications are processed on a rolling basis. Application fee: $40 ($50 for international students). Tuition, state resident: full-time $4,616. Tuition, nonresident: full-time $11,768. Full-time tuition and fees vary according to course load. *Financial aid:* Fellowships, research assistantships, teaching assistantships available. Financial aid application deadline: 2/15. *Unit head:* Charles F. Zukoski, Head, 217-333-3640. *Application contact:* K. Dane Wittrup, Director of Graduate Studies, 217-333-2631, Fax: 217-333-5052, E-mail: wittrup@uiuc.edu.

The University of Iowa, Graduate College, College of Engineering, Department of Chemical and Biochemical Engineering, Iowa City, IA 52242-1316. Offers MS, PhD. *Faculty:* 5 full-time, 1 part-time. *Students:* 16 full-time (6 women), 13 part-time (2 women); includes 2 minority (both African Americans), 16 international. 124 applicants, 12% accepted. In 1998, 4 master's, 4 doctorates awarded. *Degree requirements:* For master's, thesis optional; for doctorate, dissertation, comprehensive exam required. *Entrance requirements:* For master's and doctorate, GRE, TOEFL. *Application deadline:* Applications are processed on a rolling basis. Application fee: $30 ($50 for international students). *Financial aid:* In 1998–99, 3 fellowships, 22 research assistantships, 8 teaching assistantships were awarded. Financial aid applicants required to submit FAFSA. *Faculty research:* Catalysis and reactor design, fine-particle morphology and behavior, air pollution modelling. *Unit head:* Dr. Gregory Carmichael, Interim Chair, 319-335-1414.

University of Kansas, Graduate School, School of Engineering, Department of Chemical and Petroleum Engineering, Lawrence, KS 66045. Offers chemical engineering (MS); chemical/petroleum engineering (PhD); petroleum engineering (MS). Part-time programs available. *Faculty:* 11 full-time. *Students:* 15 full-time (4 women), 25 part-time (2 women); includes 1 minority (Asian American or Pacific Islander), 31 international. Average age 25. In 1998, 9 master's, 3 doctorates awarded. *Degree requirements:* For master's, computer language, thesis (for some programs), exam required, foreign language not required; for doctorate, computer language, dissertation, comprehensive and qualifying exams required, foreign language not required. *Entrance requirements:* For master's and doctorate, GRE General Test (minimum combined score of 2000 on three sections required for international students), TOEFL (minimum score of 600 required, TSE (minimum score of 40 required), minimum GPA of 3.0. *Application deadline:* For fall admission, 7/1 (priority date). Application fee: $30. *Financial aid:* In 1998–99, 1 fellowship, 6 research assistantships, 5 teaching assistantships were awarded.; Federal Work-Study also available. Financial aid application deadline: 1/31. *Faculty research:* Enhanced oil recovery, catalysis and kinetics, electrochemical engineering, biochemical engineering, semiconductor materials processing. *Unit head:* Don W. Green, Chairperson, 785-864-4965.

University of Kentucky, Graduate School, Graduate School Programs from the College of Engineering, Program in Chemical Engineering, Lexington, KY 40506-0032. Offers MS Ch E, PhD. *Degree requirements:* For master's, comprehensive exam required, thesis optional, foreign language not required; for doctorate, dissertation, comprehensive exam required, foreign language not required. *Entrance requirements:* For master's, GRE General Test, minimum undergraduate GPA of 2.5; for doctorate, GRE General Test, minimum graduate GPA of 3.0. *Faculty research:* Aerosol physics and chemistry, biocellular engineering fuel science, poly and membrane science.

University of Louisville, Graduate School, Speed Scientific School, Department of Chemical Engineering, Louisville, KY 40292-0001. Offers M Eng, MS, PhD. *Accreditation:* ABET (one or more programs are accredited). *Faculty:* 10 full-time (1 woman), 1 part-time (0 women). *Students:* 40 full-time (12 women), 17 part-time (8 women); includes 5 minority (2 African Americans, 2 Asian Americans or Pacific Islanders, 1 Hispanic American), 15 international. Average age 27. In 1998, 23 master's, 5 doctorates awarded. *Degree requirements:* For master's and doctorate, thesis/dissertation required, foreign language not required. *Entrance requirements:* For master's and doctorate, GRE General Test (minimum combined score of 1200 required). *Application deadline:* Applications are processed on a rolling basis. Application fee: $25. *Financial aid:* In 1998–99, 2 fellowships with tuition reimbursements (averaging $13,450 per year), 1 research assistantship with tuition reimbursement (averaging $12,465 per year), 4 teaching assistantships with tuition reimbursements (averaging $12,465 per year) were awarded. *Faculty research:* Biotechnology, catalysis and reaction engineering, thermodynamics, polymer science, membrane technology. *Unit head:* Dr. Dermot J. Collins, Acting Chair, 502-852-6347, Fax: 502-852-6355, E-mail: dermot@louisville.edu.

University of Maine, Graduate School, College of Engineering, Department of Chemical Engineering, Orono, ME 04469. Offers MS, PhD. Part-time programs available. *Faculty:* 16 full-time (4 women). *Students:* 25 full-time (13 women), 6 part-time (2 women), 13 international. Average age 23. 96 applicants, 16% accepted. In 1998, 3 master's, 3 doctorates awarded. Terminal master's awarded for partial completion of doctoral program. *Degree requirements:* For master's and doctorate, thesis/dissertation required, foreign language not required. *Entrance requirements:* For master's and doctorate, GRE General Test, TOEFL (minimum score of 550 required). *Application deadline:* For fall admission, 2/1 (priority date); for spring admission, 10/15. Applications are processed on a rolling basis. Application fee: $50. *Financial aid:* In 1998–99, 21 research assistantships with tuition reimbursements (averaging $10,500 per year), 3 teaching assistantships with tuition reimbursements (averaging $10,500 per year) were awarded.; Federal Work-Study and tuition waivers (full and partial) also available. Financial aid application deadline: 3/1. *Faculty research:* Transport phenomena, process modeling, polymer science and engineering, material characterization, unit operations in pulp and paper. *Unit head:* Dr. Douglas Ruthven, Chair, 207-581-2277, Fax: 207-581-2323. *Application contact:* Scott G. Delcourt, Director of the Graduate School, 207-581-3218, Fax: 207-581-3232, E-mail: graduate@maine.edu.

Announcement: The Department of Chemical Engineering at the University of Maine offers MS and PhD degrees. The department has research programs in supercritical fluids, fluid dynamics, reactor kinetics, colloidal phenomena, surface science, bioreactors, and pulp and paper processing. Obtain your advanced degree in an enjoyable setting at the University of Maine.

University of Maryland, Baltimore County, Graduate School, College of Engineering, Department of Chemical and Biochemical Engineering, Baltimore, MD 21250-5398. Offers MS, PhD. *Faculty:* 8 full-time (4 women), 8 part-time (0 women). *Students:* 37 full-time (15 women), 8 part-time (1 woman); includes 4 minority (all Asian Americans or Pacific Islanders), 18 international. 78 applicants, 9% accepted. In 1998, 1 master's, 2 doctorates awarded. *Entrance requirements:* For master's, GRE General Test, minimum GPA of 3.0; for doctorate, GRE General Test, GRE Subject Test, TOEFL, minimum GPA of 3.0. *Application deadline:* For fall admission, 7/1. Applications are processed on a rolling basis. Application fee: $45. *Financial aid:* Fellowships, research assistantships, teaching assistantships available. *Faculty research:* Bioengineering, mammalian cell culture, protein purification, adsorptive separation. *Unit head:* Dr. Douglas D. Frey, Interim Chair, 410-455-3400. *Application contact:* Director, Graduate Program, 410-455-3693.

University of Maryland, College Park, Graduate School, A. James Clark School of Engineering, Department of Chemical Engineering, College Park, MD 20742-5045. Offers M Eng, MS, PhD. Part-time and evening/weekend programs available. *Faculty:* 15 full-time (3 women), 1 part-time (0 women). *Students:* 53 full-time (13 women), 16 part-time (4 women); includes 19 minority (7 African Americans, 8 Asian Americans or Pacific Islanders, 4 Hispanic Americans), 36 international. 210 applicants, 16% accepted. In 1998, 10 master's, 6 doctorates awarded. *Degree requirements:* For master's, thesis or alternative required, foreign language not required; for doctorate, dissertation required. *Entrance requirements:* For master's and doctorate, GRE General Test, TOEFL, minimum GPA of 3.0. *Application deadline:* Applications are processed on a rolling basis. Application fee: $50 ($70 for international students). Tuition, state resident: part-time $272 per credit hour. Tuition, nonresident: part-time $475 per credit hour. Required fees: $632; $379 per year. *Financial aid:* In 1998–99, 4 fellowships with full tuition reimbursements (averaging $11,578 per year), 8 research assistantships with tuition reimbursements (averaging $11,182 per year), 4 teaching assistantships with tuition reimbursements (averaging $10,308 per year) were awarded.; Federal Work-Study, grants, and scholarships also available. Aid available to part-time students. Financial aid applicants required to submit FAFSA. *Faculty research:* Applied polymer science, biochemical engineering, transport phenomena, process systems, thermal properties. Total annual research expenditures: $886,582. *Unit head:* Dr. Jan Sengers, Chairman, 301-405-2983, Fax: 301-405-0523. *Application contact:* Trudy Lindsey, Director, Graduate Admission and Records, 301-405-4198, Fax: 301-314-9305, E-mail: grschool@deans.umd.edu.

University of Maryland, College Park, Graduate School, A. James Clark School of Engineering, Professional Program in Engineering, College Park, MD 20742-5045. Offers aerospace engineering (M Eng); chemical engineering (M Eng); civil engineering (M Eng); electrical engineering (M Eng); fire protection engineering (M Eng); materials science and engineering

(M Eng); mechanical engineering (M Eng); reliability engineering (M Eng); systems engineering (M Eng). Part-time and evening/weekend programs available. Postbaccalaureate distance learning degree programs offered. *Faculty:* 11 part-time (0 women). *Students:* 20 full-time (3 women), 205 part-time (42 women); includes 58 minority (27 African Americans, 25 Asian Americans or Pacific Islanders, 5 Hispanic Americans, 1 Native American), 20 international. *Degree requirements:* For master's, foreign language and thesis not required. *Application deadline:* Applications are processed on a rolling basis. Application fee: $50 ($70 for international students). Tuition, state resident: part-time $272 per credit hour. Tuition, nonresident: part-time $475 per credit hour. Required fees: $632; $379 per year. *Unit head:* Dr. Patrick Cunniff, Associate Dean, 301-405-5256, Fax: 301-314-9477. *Application contact:* Trudy Lindsey, Director, Graduate Admission and Records, 301-405-4198, Fax: 301-314-9305, E-mail: grschool@deans.umd.edu.

University of Massachusetts Amherst, Graduate School, College of Engineering, Department of Chemical Engineering, Amherst, MA 01003. Offers MS, PhD. Part-time programs available. *Faculty:* 16 full-time (1 woman). *Students:* 49 full-time (14 women), 11 part-time (2 women); includes 2 minority (both Hispanic Americans), 36 international. Average age 25. 206 applicants, 21% accepted. In 1998, 14 doctorates awarded. Terminal master's awarded for partial completion of doctoral program. *Degree requirements:* For master's, thesis required, foreign language not required; for doctorate, one foreign language, dissertation required. *Entrance requirements:* For master's and doctorate, GRE General Test. *Application deadline:* For fall admission, 2/1 (priority date). Applications are processed on a rolling basis. Application fee: $40. Tuition, state resident: full-time $2,640; part-time $165 per credit. Tuition, nonresident: full-time $9,756; part-time $407 per credit. Required fees: $1,221 per term. One-time fee: $110. Full-time tuition and fees vary according to course load, campus/location and reciprocity agreements. *Financial aid:* In 1998–99, 58 research assistantships with full tuition reimbursements (averaging $12,955 per year), 4 teaching assistantships with full tuition reimbursements (averaging $4,495 per year) were awarded.; fellowships with full tuition reimbursements, career-related internships or fieldwork, Federal Work-Study, grants, scholarships, traineeships, and unspecified assistantships also available. Aid available to part-time students. Financial aid application deadline: 2/1. *Unit head:* Michael Malone, Head, 413-545-2359, Fax: 413-545-1647, E-mail: mmalone@ecs.umass.edu.

University of Massachusetts Lowell, Graduate School, James B. Francis College of Engineering, Department of Chemical Engineering, Lowell, MA 01854-2881. Offers MS Eng. Part-time programs available. *Faculty:* 9 full-time, 1 part-time. *Students:* 12 full-time (1 woman), 29 part-time (6 women); includes 10 minority (1 African American, 6 Asian Americans or Pacific Islanders, 1 Hispanic American, 2 Native Americans), 2 international. 68 applicants, 51% accepted. In 1998, 7 degrees awarded. *Degree requirements:* Foreign language not required. *Entrance requirements:* For master's, GRE General Test. *Application deadline:* For fall admission, 4/1 (priority date); for spring admission, 10/1. Applications are processed on a rolling basis. Application fee: $20 ($35 for international students). *Financial aid:* In 1998–99, 5 teaching assistantships were awarded.; research assistantships Financial aid application deadline: 4/1. *Unit head:* Dr. Alfred Donatelli, Chairman, 978-934-3156.

University of Michigan, Horace H. Rackham School of Graduate Studies, College of Engineering, Department of Chemical Engineering, Ann Arbor, MI 48109. Offers MSE, PhD, CE. Part-time programs available. Postbaccalaureate distance learning degree programs offered (no on-campus study). *Faculty:* 20 full-time (2 women), 1 (woman) part-time. *Students:* 64 full-time (17 women), 15 part-time (3 women); includes 9 minority (2 African Americans, 2 Asian Americans or Pacific Islanders, 5 Hispanic Americans), 34 international. 262 applicants, 27% accepted. In 1998, 12 master's awarded (8% found work related to degree, 75% continued full-time study); 8 doctorates awarded (12% entered university research/teaching, 88% found other work related to degree). Terminal master's awarded for partial completion of doctoral program. *Degree requirements:* For master's, computer language required, thesis not required; for doctorate, dissertation, oral defense of dissertation, preliminary exams required. *Entrance requirements:* For master's and doctorate, GRE General Test (combined average 1748 on three sections). *Average time to degree:* Master's–2 years full-time, 2.5 years part-time; doctorate–6 years full-time, 8 years part-time. *Application deadline:* For fall admission, 1/15 (priority date); for winter admission, 10/15 (priority date). Application fee: $55. Electronic applications accepted. *Financial aid:* In 1998–99, 13 fellowships with full tuition reimbursements (averaging $17,000 per year), 19 research assistantships with full tuition reimbursements (averaging $17,000 per year), 12 teaching assistantships with full and partial tuition reimbursements (averaging $11,300 per year) were awarded. *Faculty research:* Biochemical, fluid mechanics, polymers, catalysis, reaction engineering. Total annual research expenditures: $2.2 million. *Unit head:* Dr. Ralph T. Yang, Chair, 734-764-2383, Fax: 734-763-0459, E-mail: hamlins@engin.umich.edu. *Application contact:* Teresa Clayton, Department Office, 734-764-2383, Fax: 734-763-0459, E-mail: tereclay@engin.umich.edu.

Announcement: The department is active in the following areas: kinetics, catalysis and surface science, flow in porous media, colloidal and interfacial phenomena, biochemical engineering and the life sciences, environmental engineering, polymer rheology, tissue engineering, educational technology, and process control and simulation. For more information, call collect (734-763-1148) or e-mail Susan Hamlin (hamlins@engin.umich.edu).

University of Minnesota, Twin Cities Campus, Graduate School, Institute of Technology, Department of Chemical Engineering and Materials Science, Program in Chemical Engineering, Minneapolis, MN 55455-0213. Offers M Ch E, MS Ch E, PhD. Part-time programs available. Terminal master's awarded for partial completion of doctoral program. *Degree requirements:* For master's and doctorate, thesis/dissertation required, foreign language not required. *Entrance requirements:* For master's and doctorate, GRE General Test. *Faculty research:* Chemical kinetics, reaction engineering and modeling, gas and membrane separation processes, biochemical engineering, nonequilibrium statistical mechanics.

University of Missouri–Columbia, Graduate School, College of Engineering, Department of Chemical Engineering, Columbia, MO 65211. Offers MS, PhD. *Faculty:* 13 full-time (1 woman). *Students:* 12 full-time (3 women), 8 part-time (1 woman); includes 1 minority (African American), 8 international. 26 applicants, 23% accepted. In 1998, 7 master's, 5 doctorates awarded. *Degree requirements:* For master's and doctorate, thesis/dissertation required, foreign language not required. *Entrance requirements:* For master's and doctorate, GRE General Test, TOEFL, minimum GPA of 3.0. *Application deadline:* Applications are processed on a rolling basis. Application fee: $30 ($50 for international students). *Financial aid:* Research assistantships, teaching assistantships, institutionally-sponsored loans available. *Unit head:* Dr. David Retzloff, Director of Graduate Studies, 573-882-4036.

See in-depth description on page 521.

University of Missouri–Rolla, Graduate School, School of Engineering, Department of Chemical Engineering, Rolla, MO 65409-0910. Offers MS, PhD. Part-time and evening/weekend programs available. *Faculty:* 13 full-time (0 women). *Students:* 33 full-time (5 women); includes 4 minority (1 African American, 2 Asian Americans or Pacific Islanders, 1 Hispanic American), 17 international. Average age 27. 123 applicants, 89% accepted. In 1998, 12 master's awarded. Terminal master's awarded for partial completion of doctoral program. *Degree requirements:* For master's and doctorate, thesis/dissertation required, foreign language not required. *Entrance requirements:* For master's and doctorate, minimum GPA of 3.0. *Application deadline:* For fall admission, 7/1. Applications are processed on a rolling basis. Application fee: $25. *Financial aid:* In 1998–99, 6 fellowships with partial tuition reimbursements (averaging $15,190 per year), 11 research assistantships with partial tuition reimbursements (averaging $4,536 per year), 19 teaching assistantships with partial tuition reimbursements (averaging $6,352 per year) were awarded.; institutionally-sponsored loans also available. Financial aid application deadline: 2/1. *Faculty research:* Polymers, reaction engineering, bioengineering, mixing, physical properties. *Unit head:* Douglas K. Ludlow, Chairman, 573-341-6477, Fax: 573-341-4377. *Application contact:* Parthasakha Neogi, Graduate Coordinator, 573-341-4416.

University of Nebraska–Lincoln, Graduate College, College of Engineering and Technology, Department of Chemical Engineering, Lincoln, NE 68588. Offers chemical engineering (MS); engineering (PhD). *Faculty:* 4 full-time (1 woman), 2 part-time (0 women). *Students:* 10 full-time (1 woman), 2 part-time; includes 2 minority (both Asian Americans or Pacific Islanders), 6 international. Average age 26. 43 applicants, 26% accepted. In 1998, 3 degrees awarded. *Degree requirements:* For master's, thesis required, foreign language not required; for doctorate, dissertation, comprehensive exams required. *Entrance requirements:* For master's and doctorate, GRE General Test, TOEFL (minimum score of 550 required). *Application deadline:* For fall admission, 3/1 (priority date). Applications are processed on a rolling basis. Application fee: $35. Electronic applications accepted. *Financial aid:* In 1998–99, 2 fellowships, 5 teaching assistantships were awarded.; research assistantships, Federal Work-Study also available. Aid available to part-time students. Financial aid application deadline: 2/15. *Faculty research:* Biochemical engineering, surface science, thermodynamics, ellipsometry, polymers and composites. *Unit head:* Dr. James Eakman, Chair, 402-472-2750, Fax: 402-472-6989.

University of Nevada, Reno, Graduate School, College of Engineering, Department of Chemical Engineering, Reno, NV 89557. Offers MS, PhD. *Degree requirements:* For master's, thesis optional, foreign language not required; for doctorate, dissertation required, foreign language not required. *Entrance requirements:* For master's, TOEFL (minimum score of 500 required), minimum GPA of 2.75; for doctorate, TOEFL (minimum score of 500 required), minimum GPA of 3.0.

University of New Brunswick, School of Graduate Studies, Faculty of Engineering, Department of Chemical Engineering, Fredericton, NB E3B 5A3, Canada. Offers M Eng, M Sc E, PhD. Part-time programs available. *Degree requirements:* For master's, thesis required, foreign language not required; for doctorate, dissertation, qualifying exam required, foreign language not required. *Entrance requirements:* For master's and doctorate, TOEFL, TWE, minimum GPA of 3.0.

University of New Hampshire, Graduate School, College of Engineering and Physical Sciences, Programs in Engineering, Department of Chemical Engineering, Durham, NH 03824. Offers MS. *Faculty:* 8 full-time. *Students:* 5 full-time (2 women), 8 part-time, 6 international. Average age 29. 7 applicants, 86% accepted. In 1998, 3 degrees awarded. *Degree requirements:* For master's, thesis required, foreign language not required. *Application deadline:* For fall admission, 4/1 (priority date). Applications are processed on a rolling basis. Application fee: $50. Tuition, area resident: Full-time $5,750; part-time $319 per credit. Tuition, state resident: full-time $8,625. Tuition, nonresident: full-time $14,640; part-time $598 per credit. Required fees: $224 per semester. Tuition and fees vary according to course load, degree level and program. *Financial aid:* In 1998–99, 1 research assistantship, 6 teaching assistantships were awarded.; Federal Work-Study, scholarships, and tuition waivers (full and partial) also available. Aid available to part-time students. Financial aid application deadline: 2/15. *Unit head:* Dr. Stephen S. T. Fan, Chairperson, 603-862-3656.

University of New Hampshire, Graduate School, College of Engineering and Physical Sciences, Programs in Engineering, Doctoral Program in Engineering, Durham, NH 03824. Offers chemical engineering (PhD); civil engineering (PhD); electrical engineering (PhD); mechanical engineering (PhD); systems design engineering (PhD). *Students:* 16 full-time (3 women), 8 part-time (1 woman), 12 international. *Degree requirements:* For doctorate, dissertation required. *Entrance requirements:* For doctorate, GRE (for civil and mechanical engineering options). *Application deadline:* For fall admission, 4/1 (priority date). Applications are processed on a rolling basis. Application fee: $50. Tuition, area resident: Full-time $5,750; part-time $319 per credit. Tuition, state resident: full-time $8,625. Tuition, nonresident: full-time $14,640; part-time $598 per credit. Required fees: $224 per semester. Tuition and fees vary according to course load, degree level and program.

University of New Mexico, Graduate School, School of Engineering, Department of Chemical and Nuclear Engineering, Albuquerque, NM 87131-2039. Offers chemical engineering (MS); engineering (PhD); hazardous waste engineering (ME); nuclear engineering (MS). Part-time programs available. *Faculty:* 22 full-time (3 women), 3 part-time (0 women). *Students:* 43 full-time (11 women), 53 part-time (10 women); includes 14 minority (2 Asian Americans or Pacific Islanders, 12 Hispanic Americans), 28 international. Average age 29. In 1998, 15 master's, 9 doctorates awarded. *Degree requirements:* For doctorate, dissertation required, foreign language not required, foreign language not required. *Entrance requirements:* For master's and doctorate, GRE General Test, minimum GPA of 3.0. *Application deadline:* For fall admission, 7/15; for spring admission, 11/14. Applications are processed on a rolling basis. Application fee: $25. *Financial aid:* In 1998–99, 3 fellowships (averaging $493 per year), 68 research assistantships with tuition reimbursements (averaging $2,603 per year), 1 teaching assistantship with tuition reimbursement (averaging $5,649 per year) were awarded.; career-related internships or fieldwork also available. Financial aid application deadline: 3/15. *Faculty research:* Advanced materials manufacturing, radioactive waste management, nuclear data and measurements, health physics, space nuclear power. Total annual research expenditures: $2 million. *Unit head:* Joseph Cecchi, Chair, 505-277-5431, Fax: 505-277-5433, E-mail: cecchi@unm.edu.

See in-depth description on page 523.

University of North Dakota, Graduate School, School of Engineering and Mines, Department of Chemical Engineering, Grand Forks, ND 58202. Offers M Engr, MS. Part-time programs available. *Faculty:* 4 full-time (0 women). *Students:* 11 full-time (1 woman). 13 applicants, 77% accepted. In 1998, 7 degrees awarded. *Degree requirements:* For master's, thesis or alternative required, foreign language not required. *Entrance requirements:* For master's, GRE General Test, TOEFL (minimum score of 550 required), minimum GPA of 3.0 (MS), 2.5 (M Engr). *Application deadline:* For fall admission, 3/1 (priority date). Applications are processed on a rolling basis. Application fee: $20. *Financial aid:* In 1998–99, 11 students received aid, including 4 research assistantships, 7 teaching assistantships; fellowships, career-related internships or fieldwork, Federal Work-Study, institutionally-sponsored loans, tuition waivers (full and partial), and unspecified assistantships also available. Financial aid application deadline: 3/15. *Faculty research:* Catalysis, fluid flow and heat transfer, application of fractals, modeling and simulation, reaction engineering. *Unit head:* Dr. Tom Owens, Chairperson, 701-777-4244, Fax: 701-777-4838, E-mail: tom_owens@mail.und.nodak.edu.

University of Notre Dame, Graduate School, College of Engineering, Department of Chemical Engineering, Notre Dame, IN 46556. Offers MS, PhD. *Faculty:* 15 full-time (1 woman). *Students:* 53 full-time (15 women), 2 part-time; includes 2 minority (both Asian Americans or Pacific Islanders), 30 international. 110 applicants, 35% accepted. In 1998, 4 master's, 8 doctorates awarded (38% entered university research/teaching, 50% found other work related to degree). *Degree requirements:* For master's and doctorate, thesis/dissertation required, foreign language not required. *Entrance requirements:* For master's, GRE General Test, TOEFL (minimum score of 600 required; 250 for computer-based); for doctorate, GRE General Test, GRE Subject Test (strongly recommended), TOEFL (minimum score of 600 required; 250 for computer-based). *Average time to degree:* Master's–2 years full-time; doctorate–5.9 years full-time. *Application deadline:* For fall admission, 2/1; for spring admission, 11/15. Application fee: $40. *Financial aid:* In 1998–99, 53 students received aid, including 14 fellowships with full tuition reimbursements available (averaging $16,000 per year), 23 research assistantships with full tuition reimbursements available (averaging $13,500 per year), 16 teaching assistantships with full tuition reimbursements available (averaging $13,500 per year); tuition waivers (full) also available. Financial aid application deadline: 2/1. *Faculty research:* Catalysis and reaction engineering, complex fluids and flows, applied molecular thermodynamics and statistical mechanics, science and engineering of advanced materials, process systems, engineering for minimizing environmental impact. Total annual research expenditures: $1.6 million. *Unit head:* Dr. Mark J. McCready, Chairman, 219-631-5580, E-mail: chegdept.1@nd.edu. *Application contact:* Marty Nemeth, 800-528-9487, Fax: 219-631-8366, E-mail: nemeth.1@nd.edu.

Chemical Engineering

University of Oklahoma, Graduate College, College of Engineering, School of Chemical Engineering and Materials Science, Norman, OK 73019-0390. Offers chemical engineering (MS, PhD). *Faculty:* 12 full-time (1 woman), 1 part-time (0 women). *Students:* 18 full-time (7 women), 27 part-time (8 women); includes 5 minority (3 Asian Americans or Pacific Islanders, 1 Hispanic American, 1 Native American), 19 international. Average age 29. 17 applicants, 88% accepted. In 1998, 9 master's, 5 doctorates awarded. *Degree requirements:* For master's, thesis, oral exams required, foreign language not required; for doctorate, dissertation, oral exam, qualifying exams required, foreign language not required. *Entrance requirements:* For master's and doctorate, TOEFL (minimum score of 550 required), minimum GPA of 3.0. *Application deadline:* For fall admission, 6/1 (priority date). Applications are processed on a rolling basis. Application fee: $25. Tuition, state resident: part-time $86 per credit hour. Tuition, nonresident: part-time $275 per credit hour. Tuition and fees vary according to course level, course load and program. *Financial aid:* In 1998–99, 26 research assistantships, 3 teaching assistantships were awarded.; fellowships, tuition waivers (partial) also available. Financial aid application deadline: 8/1. *Faculty research:* Enhanced oil recovery, biochemical and biomedical engineering, kinetics, surface chemistry, thermodynamics. *Unit head:* Dr. Jeffrey Harwell, Director, 405-325-5811. *Application contact:* Dr. Roger Harrison, Graduate Adviser, 405-325-5811.

University of Ottawa, School of Graduate Studies and Research, Faculty of Engineering, Department of Chemical Engineering, Ottawa, ON K1N 6N5, Canada. Offers M Eng, MA Sc, PhD. *Faculty:* 15 full-time, 4 part-time. *Students:* 34 full-time (11 women), 7 part-time (2 women), 5 international. Average age 30. In 1998, 5 master's, 2 doctorates awarded. *Degree requirements:* For master's, thesis or alternative required, foreign language not required; for doctorate, dissertation required, foreign language not required. *Entrance requirements:* For master's, honors degree or equivalent, minimum B average; for doctorate, master's degree, minimum B+ average. *Application deadline:* For fall admission, 3/1. Application fee: $35. *Financial aid:* Fellowships, research assistantships, teaching assistantships, Federal Work-Study available. *Faculty research:* Absorption, applied kinetics and catalysis, biochemical engineering, cryogenics, oil recovery. *Unit head:* Vladimir Hornoff, Chair, 613-562-5800 Ext. 6105, Fax: 613-562-5172. *Application contact:* Nicole Larivière, Administrative Secretary, 613-562-5800 Ext. 6101, Fax: 613-562-5172, E-mail: nlarivie@genie.uottawa.ca.

University of Pennsylvania, School of Engineering and Applied Science, Department of Chemical Engineering, Philadelphia, PA 19104. Offers MSE, PhD, MSE/MBA. Part-time programs available. Terminal master's awarded for partial completion of doctoral program. *Degree requirements:* For master's, foreign language and thesis not required; for doctorate, dissertation required, foreign language not required. *Entrance requirements:* For master's and doctorate, TOEFL (minimum score of 600 required). *Faculty research:* Biochemical engineering, surface and interfacial phenomena, process and design control, zeolites, molecular dynamics.

University of Pittsburgh, School of Engineering, Department of Chemical and Petroleum Engineering, Pittsburgh, PA 15260. Offers chemical engineering (MS Ch E, PhD); petroleum engineering (MSPE). Part-time and evening/weekend programs available. *Faculty:* 16 full-time (3 women), 3 part-time (0 women). *Students:* 60 full-time (19 women), 13 part-time (4 women); includes 4 minority (2 African Americans, 2 Asian Americans or Pacific Islanders), 36 international. 219 applicants, 16% accepted. In 1998, 15 master's, 8 doctorates awarded. *Degree requirements:* For master's, computer language, thesis required, foreign language not required; for doctorate, computer language, dissertation, comprehensive and final oral exams required, foreign language not required. *Entrance requirements:* For master's and doctorate, GRE General Test, TOEFL (minimum score of 550 required), minimum QPA of 3.2. *Average time to degree:* Master's–2 years full-time, 3 years part-time; doctorate–5 years full-time, 7 years part-time. *Application deadline:* For fall admission, 8/1 (priority date); for spring admission, 12/1 (priority date). Applications are processed on a rolling basis. Application fee: $30 ($40 for international students). *Financial aid:* In 1998–99, 1 fellowship (averaging $10,800 per year), 58 research assistantships (averaging $10,704 per year) were awarded.; teaching assistantships, grants, scholarships, traineeships, and tuition waivers (full and partial) also available. Financial aid application deadline: 2/15. *Faculty research:* Biotechnology, polymers, catalysis, energy and environment, computational modeling. Total annual research expenditures: $3.6 million. *Unit head:* Dr. Alan J. Russell, Chairman, 412-624-9630, Fax: 412-624-9639, E-mail: ajrche@vms.cis.pitt.edu. *Application contact:* James G. Goodwin, Graduate Coordinator, 412-624-9641, Fax: 412-624-9639, E-mail: goodwin@engrng.pitt.edu.

University of Puerto Rico, Mayagüez Campus, Graduate Studies, College of Engineering, Department of Chemical Engineering, Mayagüez, PR 00681-5000. Offers M Ch E, MS. Part-time programs available. *Degree requirements:* For master's, thesis, comprehensive exam required, foreign language not required. *Entrance requirements:* For master's, minimum GPA of 2.5, proficiency in English and Spanish. *Faculty research:* Process simulation and optimization, air and water pollution control, mass transport, biochemical engineering.

University of Rhode Island, Graduate School, College of Engineering, Department of Chemical Engineering, Kingston, RI 02881. Offers MS, PhD.

University of Rochester, The College, School of Engineering and Applied Sciences, Department of Chemical Engineering, Rochester, NY 14627-0250. Offers MS, PhD. Part-time programs available. *Faculty:* 8. *Students:* 14 full-time (5 women), 5 part-time (1 woman); includes 2 minority (both Asian Americans or Pacific Islanders), 7 international. 19 applicants, 26% accepted. In 1998, 12 master's, 4 doctorates awarded. Terminal master's awarded for partial completion of doctoral program. *Degree requirements:* For doctorate, dissertation, preliminary and oral exams required, foreign language not required, foreign language not required. *Entrance requirements:* For master's, GRE, TOEFL. *Application deadline:* For fall admission, 2/1. Application fee: $25. *Financial aid:* Fellowships, research assistantships, teaching assistantships, tuition waivers (full and partial) available. Financial aid application deadline: 2/1. *Unit head:* Harvey Palmer, Chair, 716-275-4041. *Application contact:* Donna Porcelli, Graduate Program Secretary, 716-275-4042.

University of Saskatchewan, College of Graduate Studies and Research, College of Engineering, Department of Chemical Engineering, Saskatoon, SK S7N 5A2, Canada. Offers M Eng, M Sc, PhD. *Degree requirements:* For master's and doctorate, thesis/dissertation required. *Entrance requirements:* For master's and doctorate, GRE, TOEFL.

University of South Alabama, Graduate School, College of Engineering, Department of Chemical Engineering, Mobile, AL 36688-0002. Offers MS Ch E. *Faculty:* 5 full-time (0 women). *Students:* 15 full-time (1 woman), 3 part-time; includes 3 minority (2 Asian Americans or Pacific Islanders, 1 Hispanic American), 9 international. 52 applicants, 81% accepted. In 1998, 4 degrees awarded. *Degree requirements:* For master's, project or thesis required. *Entrance requirements:* For master's, GRE General Test (minimum combined score of 1000 required), BS in engineering, minimum GPA of 3.0. *Application deadline:* For fall admission, 9/1 (priority date). Applications are processed on a rolling basis. Application fee: $25. Tuition, state resident: part-time $116 per semester hour. Tuition, nonresident: part-time $230 per semester hour. Required fees: $121 per semester. Part-time tuition and fees vary according to course load and program. *Financial aid:* In 1998–99, 1 research assistantship was awarded.; career-related internships or fieldwork and institutionally-sponsored loans also available. Aid available to part-time students. Financial aid application deadline: 4/1. *Unit head:* Dr. B. Keith Harrison, Chairperson, 334-460-6160. *Application contact:* Dr. Russell M. Hayes, Director of Graduate Studies, 334-460-6117.

University of South Carolina, Graduate School, College of Engineering and Information Technology, Department of Chemical Engineering, Columbia, SC 29208. Offers ME, MS, PhD. Part-time and evening/weekend programs available. Postbaccalaureate distance learning degree programs offered (minimal on-campus study). *Faculty:* 15 full-time (1 woman). *Students:* 59 full-time (15 women), 20 part-time (5 women); includes 3 minority (2 African Americans, 1 Native American), 50 international. Average age 29. In 1998, 8 master's, 4 doctorates awarded. *Degree requirements:* For master's, thesis required (for some programs), foreign language not required; for doctorate, dissertation required, foreign language not required. *Entrance requirements:* For master's, GRE General Test (minimum combined score of 1100 required), TOEFL (minimum score of 500 required); for doctorate, GRE General Test (minimum combined score of 1100 required), TOEFL (minimum score of 550 required). *Application deadline:* For fall admission, 3/1 (priority date); for spring admission, 11/1. Applications are processed on a rolling basis. Application fee: $35. Electronic applications accepted. Tuition, state resident: full-time $4,014; part-time $202 per credit hour. Tuition, nonresident: full-time $8,528; part-time $428 per credit hour. Required fees: $100; $4 per credit hour. Tuition and fees vary according to program. *Financial aid:* In 1998–99, 23 students received aid, including 9 research assistantships with partial tuition reimbursements available (averaging $12,000 per year), 14 teaching assistantships with partial tuition reimbursements available (averaging $12,000 per year); fellowships, career-related internships or fieldwork and institutionally-sponsored loans also available. *Faculty research:* Rheology, liquid and supercritical extractions, electrochemistry, corrosion, heterogeneous and homogeneous catalysis. Total annual research expenditures: $2.7 million. *Unit head:* Dr. R. E. White, Chair, 803-777-4181, Fax: 803-777-8265, E-mail: rew@engr.sc.edu. *Application contact:* Judy Williams, Department Secretary, 803-777-4181.

University of Southern California, Graduate School, School of Engineering, Department of Chemical Engineering, Los Angeles, CA 90089. Offers MS, PhD, Engr. *Faculty:* 7 full-time (1 woman), 1 part-time (0 women). *Students:* 49 full-time (10 women), 15 part-time (4 women); includes 5 minority (4 Asian Americans or Pacific Islanders, 1 Hispanic American), 55 international. Average age 25. 142 applicants, 76% accepted. In 1998, 9 master's, 3 doctorates awarded. *Degree requirements:* For master's, thesis optional; for doctorate, dissertation required. *Entrance requirements:* For master's, doctorate, and Engr, GRE General Test. *Application deadline:* For fall admission, 7/1 (priority date); for spring admission, 12/1. Application fee: $55. Tuition: Part-time $768 per unit. Required fees: $350 per semester. *Financial aid:* In 1998–99, 12 fellowships, 20 research assistantships, 9 teaching assistantships were awarded.; Federal Work-Study, institutionally-sponsored loans, and scholarships also available. Aid available to part-time students. Financial aid application deadline: 2/15; financial aid applicants required to submit FAFSA. *Unit head:* Dr. Katherine Shing, Chairman, 213-740-2227.

University of South Florida, Graduate School, College of Engineering, Department of Chemical Engineering, Tampa, FL 33620-9951. Offers M Ch E, ME, MS Ch E, MSE, PhD. Part-time programs available. *Faculty:* 9 full-time (0 women). *Students:* 47 full-time (19 women), 11 part-time (2 women); includes 13 minority (1 African American, 1 Asian American or Pacific Islander, 11 Hispanic Americans), 22 international. Average age 28. 41 applicants, 90% accepted. In 1998, 12 master's awarded (75% found work related to degree, 25% continued full-time study); 2 doctorates awarded (50% entered university research/teaching, 50% found other work related to degree). Terminal master's awarded for partial completion of doctoral program. *Degree requirements:* For master's, thesis required (for some programs), foreign language not required; for doctorate, dissertation, 2 tools of research as specified by dissertation committee required, foreign language not required. *Entrance requirements:* For master's, GRE General Test (minimum combined score of 1200 required), minimum GPA of 3.0 during previous 2 years; for doctorate, GRE General Test (minimum combined score of 1300 on three sections required). *Average time to degree:* Master's–2.5 years full-time, 4.5 years part-time; doctorate–4 years full-time, 7 years part-time. *Application deadline:* For fall admission, 6/1; for spring admission, 10/15. Application fee: $20. Electronic applications accepted. Tuition, state resident: part-time $148 per credit hour. Tuition, nonresident: part-time $509 per credit hour. *Financial aid:* In 1998–99, 30 students received aid, including 1 fellowship with full tuition reimbursement available (averaging $7,000 per year), 18 research assistantships with full tuition reimbursements available (averaging $11,180 per year), 11 teaching assistantships with full tuition reimbursements available (averaging $8,712 per year); career-related internships or fieldwork, Federal Work-Study, institutionally-sponsored loans, and tuition waivers (partial) also available. Aid available to part-time students. Financial aid applicants required to submit FAFSA. *Faculty research:* Process design and control, sensor development and identification, biomedical engineering, polymer characterization and synthesis, supercritical fluid technology. Total annual research expenditures: $258,782. *Unit head:* Luis Garcia-Rubio, Chairperson, 813-974-3997, Fax: 813-974-3651, E-mail: garcia@eng.usf.edu. *Application contact:* A. K. Sunol, Coordinator, 813-974-3997, Fax: 813-974-3651, E-mail: sunol@eng.usf.edu.

University of Southwestern Louisiana, Graduate School, College of Engineering, Department of Chemical Engineering, Lafayette, LA 70504. Offers MSE. Evening/weekend programs available. *Faculty:* 6 full-time (0 women). *Students:* 26 full-time (6 women), 6 part-time (1 woman), 30 international. 85 applicants, 69% accepted. In 1998, 9 degrees awarded. *Degree requirements:* For master's, computer language, thesis or alternative, comprehensive exam required, foreign language not required. *Entrance requirements:* For master's, GRE General Test, BS in chemical engineering, minimum GPA of 2.85. *Application deadline:* For fall admission, 5/15. Application fee: $5 ($15 for international students). *Financial aid:* In 1998–99, 10 research assistantships with full tuition reimbursements (averaging $4,937 per year) were awarded.; Federal Work-Study and tuition waivers (full and partial) also available. Financial aid application deadline: 5/1. *Faculty research:* Corrosion, transport phenomena and thermodynamics in the oil and gas industry. *Unit head:* Dr. James D. Garber, Head, 318-482-6562. *Application contact:* Dr. James R. Reinhardt, Graduate Coordinator, 318-482-5351.

University of Tennessee, Knoxville, Graduate School, College of Engineering, Department of Chemical Engineering, Knoxville, TN 37996. Offers MS, PhD. *Faculty:* 13 full-time (1 woman), 1 part-time (0 women). *Students:* 27 full-time (5 women), 10 part-time (2 women); includes 5 minority (4 African Americans, 1 Asian American or Pacific Islander), 11 international. 50 applicants, 36% accepted. In 1998, 8 master's, 4 doctorates awarded. *Degree requirements:* For master's, thesis or alternative required, foreign language not required; for doctorate, dissertation required, foreign language not required. *Entrance requirements:* For master's and doctorate, GRE General Test, TOEFL (minimum score of 550 required), minimum GPA of 2.7. *Application deadline:* For fall admission, 2/1 (priority date). Applications are processed on a rolling basis. Application fee: $35. Electronic applications accepted. *Financial aid:* In 1998–99, 4 fellowships, 18 research assistantships, 5 teaching assistantships were awarded.; career-related internships or fieldwork, Federal Work-Study, institutionally-sponsored loans, and unspecified assistantships also available. Financial aid application deadline: 2/1; financial aid applicants required to submit FAFSA. *Unit head:* Dr. Charles F. Moore, Head, 423-974-2421, Fax: 423-974-7076, E-mail: cmoore10@utk.edu. *Application contact:* Dr. David Keffer, Graduate Representative, 423-974-5322, E-mail: dkeffer@utk.edu.

University of Tennessee Space Institute, Graduate Programs, Program in Chemical Engineering, Tullahoma, TN 37388-9700. Offers MS. *Faculty:* 1 full-time (0 women). *Students:* 4 full-time (1 woman); includes 1 minority (African American), 1 international. 11 applicants, 73% accepted. *Degree requirements:* For master's, thesis required, foreign language not required. *Application deadline:* Applications are processed on a rolling basis. Application fee: $35. *Financial aid:* Fellowships, research assistantships, Federal Work-Study available. Financial aid applicants required to submit FAFSA. *Unit head:* Dr. Atul Sheth, Degree Program Chairman, 931-393-7427, Fax: 931-393-7201, E-mail: asheth@utsi.edu. *Application contact:* Dr. Edwin M. Gleason, Assistant Dean for Admissions and Student Affairs, 931-393-7432, Fax: 931-393-7340, E-mail: egleason@utsi.edu.

The University of Texas at Austin, Graduate School, College of Engineering, Department of Chemical Engineering, Austin, TX 78712-1111. Offers MSE, PhD. *Students:* 155 (29 women); includes 10 minority (3 African Americans, 4 Asian Americans or Pacific Islanders, 3 Hispanic Americans) 42 international. 375 applicants, 12% accepted. In 1998, 15 master's, 30 doctorates awarded. *Entrance requirements:* For master's and doctorate, GRE General Test (minimum combined score of 1000 required). Application fee: $50 ($75 for international students). *Financial aid:* Fellowships, research assistantships, teaching assistantships available. Financial aid application deadline: 2/1. *Unit head:* John K. Ekerdt, Chairman, 512-471-5238. *Application contact:* Gary Rochelle, Graduate Adviser, 512-471-6991.

University of Toledo, Graduate School, College of Engineering, Department of Chemical and Environmental Engineering, Toledo, OH 43606-3398. Offers chemical engineering (MS Ch E); engineering sciences (PhD). Part-time and evening/weekend programs available. *Faculty:* 10 full-time (2 women), 4 part-time (0 women). *Students:* 59 full-time (9 women), 10 part-time (1 woman); includes 5 minority (2 African Americans, 2 Asian Americans or Pacific Islanders, 1 Hispanic American), 42 international. Average age 27. 163 applicants, 25% accepted. In 1998, 15 master's, 3 doctorates awarded. *Degree requirements:* For master's, thesis optional, foreign language not required; for doctorate, dissertation required, foreign language not required. *Entrance requirements:* For master's, GRE General Test (minimum combined score of 1100 required), TOEFL (minimum score of 550 required), minimum GPA of 2.7; for doctorate, GRE General Test (minimum combined score of 1100 required), TOEFL (minimum score of 550 required), minimum GPA of 3.3. *Average time to degree:* Master's–2 years full-time; doctorate–4 years full-time. *Application deadline:* For fall admission, 5/31 (priority date). Applications are processed on a rolling basis. Application fee: $30. Electronic applications accepted. *Financial aid:* In 1998–99, 46 students received aid, including 30 research assistantships with full tuition reimbursements available, 12 teaching assistantships with full tuition reimbursements available; fellowships, Federal Work-Study, scholarships, tuition waivers (full), and unspecified assistantships also available. Financial aid application deadline: 4/1. *Faculty research:* Biomedical and environmental chemical engineering, polymers, applied computing, membranes. Total annual research expenditures: $664,013. *Unit head:* Dr. Steven E. LeBlanc, Chairman, 419-530-8080, Fax: 419-530-8086. *Application contact:* Dr. Arun Nadarajah, Graduate Director, 419-530-8031, Fax: 419-530-8086, E-mail: anadaraj@eng.utoledo.edu.

University of Toronto, School of Graduate Studies, Physical Sciences Division, Faculty of Applied Science and Engineering, Department of Chemical Engineering and Applied Chemistry, Toronto, ON M5S 1A1, Canada. Offers M Eng, MA Sc, PhD. Part-time programs available. *Degree requirements:* For master's, thesis required (for some programs); for doctorate, dissertation required.

University of Tulsa, Graduate School, College of Business Administration, Department of Engineering and Technology Management, Tulsa, OK 74104-3189. Offers chemical engineering (METM); computer science (METM); electrical engineering (METM); geological science (METM); mathematics (METM); mechanical engineering (METM); petroleum engineering (METM). Part-time and evening/weekend programs available. *Students:* 3 full-time (1 woman), 1 part-time, 3 international. *Degree requirements:* For master's, foreign language and thesis not required. *Entrance requirements:* For master's, GRE General Test (minimum score of 430 on verbal section, 600 on quantitative required), TOEFL (minimum score of 575 required). *Application deadline:* Applications are processed on a rolling basis. Application fee: $30. Electronic applications accepted. Tuition: Full-time $8,640; part-time $480 per hour. Required fees: $3 per hour. One-time fee: $200 full-time. Tuition and fees vary according to program. *Unit head:* Dr. Richard C. Burgess, Assistant Dean/Director of Graduate Business Studies, 918-631-2242, Fax: 918-631-2142.

University of Tulsa, Graduate School, College of Engineering and Applied Sciences, Department of Chemical Engineering, Tulsa, OK 74104-3189. Offers ME, MSE, PhD. Part-time programs available. *Faculty:* 6 full-time (0 women). *Students:* 16 full-time (1 woman), 3 part-time, 13 international. Average age 31. 54 applicants, 37% accepted. In 1998, 6 master's awarded. *Degree requirements:* For master's, thesis optional, foreign language not required; for doctorate, dissertation required, foreign language not required. *Entrance requirements:* For master's, GRE General Test, TOEFL (minimum score of 550 required); for doctorate, GRE General Test (minimum score of 700 on quantitative section, 1100 combined required), TOEFL (minimum score of 550 required). *Application deadline:* Applications are processed on a rolling basis. Application fee: $30. Electronic applications accepted. Tuition: Full-time $8,640; part-time $480 per hour. Required fees: $3 per hour. One-time fee: $200 full-time. Tuition and fees vary according to program. *Financial aid:* In 1998–99, 11 research assistantships (averaging $6,291 per year), 10 teaching assistantships (averaging $5,846 per year) were awarded.; fellowships, career-related internships or fieldwork, Federal Work-Study, and tuition waivers (partial) also available. Aid available to part-time students. Financial aid application deadline: 2/1; financial aid applicants required to submit FAFSA. *Faculty research:* Refinery design, enhanced hydrocarbon recovery, fluid rheology, phase equilibria, fluid mechanics. *Unit head:* Dr. Keith D. Wisecarver, Chairperson, 918-631-2974. *Application contact:* Dr. Charles Sheppard, Adviser, 918-631-2444, Fax: 918-631-3268, E-mail: charles-sheppard@utulsa.edu.

University of Utah, Graduate School, College of Engineering, Department of Chemical and Fuels Engineering, Salt Lake City, UT 84112-1107. Offers chemical engineering (M Phil, ME, MS, PhD); fuels engineering (ME, MS, PhD). Part-time programs available. *Faculty:* 17 full-time (1 woman), 23 part-time (0 women). *Students:* 32 full-time (7 women), 12 part-time (1 woman), 27 international. Average age 30. In 1998, 11 master's, 7 doctorates awarded. *Degree requirements:* For master's and doctorate, foreign language and thesis not required. *Entrance requirements:* For master's and doctorate, GRE, TOEFL (minimum score of 500 required), minimum GPA of 3.0. *Application deadline:* For fall admission, 7/1. Application fee: $30 ($50 for international students). *Financial aid:* In 1998–99, 4 teaching assistantships were awarded.; fellowships, research assistantships *Faculty research:* Computer-aided process synthesis and design, combustion of solid and liquid fossil fuels, oxygen mass transport in biochemical reactors. *Unit head:* Terry Ring, Chair, 801-585-5705, Fax: 801-581-8692. *Application contact:* Donald A. Dahlstrom, Director of Graduate Studies, 801-581-6934, Fax: 801-581-8692, E-mail: dadahlstrom@cc.utah.edu.

University of Virginia, School of Engineering and Applied Science, Department of Chemical Engineering, Charlottesville, VA 22903. Offers ME, MS, PhD. Postbaccalaureate distance learning degree programs offered (no on-campus study). *Faculty:* 10 full-time (1 woman), 1 part-time (0 women). *Students:* 48 full-time (11 women), 2 part-time; includes 7 minority (4 African Americans, 2 Asian Americans or Pacific Islanders, 1 Hispanic American), 10 international. Average age 25. 53 applicants, 58% accepted. In 1998, 16 master's, 4 doctorates awarded. *Degree requirements:* For master's, thesis required (for some programs), foreign language not required; for doctorate, dissertation, comprehensive exam required, foreign language not required. *Entrance requirements:* For master's and doctorate, GRE General Test. *Application deadline:* For fall admission, 8/1; for spring admission, 12/1. Applications are processed on a rolling basis. Application fee: $60. *Financial aid:* Fellowships available. Financial aid application deadline: 2/1. *Faculty research:* Fluid mechanics, heat and mass transfer, chemical reactor analysis and engineering, biochemical engineering and biotechnology. *Unit head:* Donald J. Kirwan, Chairman, 804-924-7778. *Application contact:* J. Milton Adams, Assistant Dean, 804-924-3897, E-mail: twr2c@virginia.edu.

University of Washington, Graduate School, College of Engineering, Department of Chemical Engineering, Seattle, WA 98195. Offers MS Ch E, PhD. *Faculty:* 15 full-time (1 woman), 3 part-time (1 woman). *Students:* 72 full-time (22 women), 1 (woman) part-time; includes 9 minority (1 African American, 7 Asian Americans or Pacific Islanders, 1 Hispanic American), 19 international. Average age 24. 150 applicants, 19% accepted. In 1998, 3 master's awarded (67% found work related to degree, 33% continued full-time study); 7 doctorates awarded (14% entered university research/teaching, 86% found other work related to degree). *Degree requirements:* For master's, thesis or alternative required, foreign language not required; for doctorate, dissertation required, foreign language not required. *Entrance requirements:* For master's and doctorate, GRE, TOEFL (minimum score of 600 required), minimum GPA of 3.0. *Average time to degree:* Master's–1.4 years full-time; doctorate–4.8 years full-time. *Application deadline:* For fall admission, 1/15 (priority date). Applications are processed on a rolling basis. Application fee: $50. Tuition, state resident: full-time $5,196; part-time $475 per credit. Tuition, nonresident: full-time $13,485; part-time $1,285 per credit. Required fees: $387; $38 per credit. Tuition and fees vary according to course load. *Financial aid:* In 1998–99, 16 fellowships with full tuition reimbursements (averaging $16,860 per year), 42 research assistantships with full tuition reimbursements (averaging $17,400 per year), 13 teaching assistantships with full tuition reimbursements (averaging $17,400 per year) were awarded.; career-related internships or fieldwork and Federal Work-Study also available. Financial aid application deadline: 1/15. *Faculty research:* Materials, biochemical engineering, bioengineering, environmental technology, computers and process control, transport phenomena and physics. Total annual research expenditures: $2.6 million. *Unit head:* Dr. J. W. Rogers, Chair, 206-543-2250. *Application contact:* Graduate Admissions Coordinator, 206-543-2252, Fax: 206-543-3778, E-mail: grad.admissions@cheme.washington.edu.

Announcement: The Department of Chemical Engineering offers a vigorous research program with a nationally recognized faculty, excellent physical facilities, and support for all full-time graduate students. Research programs include de-inking of paper, polymeric composites for airplane parts, particulates and aerosols, remediation of harbors, fiber-filled paper for improved recycling, control and mathematical modeling, biomaterials for implants, genetic engineering for bioremediation, electrochemical deposition of polymers, nanoscale surface science and polymer rheology, fuel-cell chemistry, synthesis of electronic materials, and production of proteins using molecular chaperones. For more information, see the Web page at http://weber.u.washington.edu/~chemeng.

University of Waterloo, Graduate Studies, Faculty of Engineering, Department of Chemical Engineering, Waterloo, ON N2L 3G1, Canada. Offers MA Sc, PhD. Part-time and evening/weekend programs available. *Faculty:* 22 full-time (2 women), 13 part-time (0 women). *Students:* 59 full-time (21 women), 1 part-time. 130 applicants, 25% accepted. In 1998, 14 master's, 8 doctorates awarded. *Degree requirements:* For master's, research paper or thesis required; for doctorate, dissertation, comprehensive exam required. *Entrance requirements:* For master's, TOEFL (minimum score of 550 required), honors degree, minimum B average; for doctorate, TOEFL (minimum score of 550 required), master's degree. *Application deadline:* Applications are processed on a rolling basis. Application fee: $50. *Expenses:* Tuition and fees charges are reported in Canadian dollars. Tuition, state resident: full-time $3,168 Canadian dollars; part-time $792 Canadian dollars per term. Tuition, nonresident: full-time $8,000 Canadian dollars; part-time $2,000 Canadian dollars. Required fees: $45 Canadian dollars per term. Tuition and fees vary according to program. *Financial aid:* In 1998–99, research assistantships (averaging $15,000 per year), teaching assistantships (averaging $4,122 per year) were awarded. *Faculty research:* Biotechnical and environmental engineering, mathematical analysis, statistics and control, polymer science and engineering. Total annual research expenditures: $2 million. *Unit head:* Dr. I. Chatzis, Chair, 519-888-4567 Ext. 2295, Fax: 519-746-4979, E-mail: ichatzis@cape.uwaterloo.ca.. *Application contact:* Dr. R. L. Legge, Graduate Officer, 519-888-4567 Ext. 2484, Fax: 519-746-4979, E-mail: rllegge@engmail.uwaterloo.ca.

University of Wisconsin–Madison, Graduate School, College of Engineering, Department of Chemical Engineering, Madison, WI 53706-1380. Offers MS, PhD. *Faculty:* 15 full-time (1 woman), 3 part-time (0 women). *Students:* 94 full-time (19 women), 3 part-time; includes 10 minority (5 Asian Americans or Pacific Islanders, 5 Hispanic Americans), 36 international. 287 applicants, 30% accepted. In 1998, 6 master's awarded (50% found work related to degree, 33% continued full-time study); 14 doctorates awarded (21% entered university research/teaching, 71% found other work related to degree). Terminal master's awarded for partial completion of doctoral program. *Degree requirements:* For master's, thesis or alternative required; for doctorate, dissertation required. *Entrance requirements:* For master's and doctorate, GRE General Test (combined average 2090 on three sections). *Average time to degree:* Master's–2.5 years full-time; doctorate–5.5 years full-time. *Application deadline:* For fall admission, 1/15; for spring admission, 10/15. Application fee: $45. Electronic applications accepted. *Financial aid:* In 1998–99, 90 students received aid, including fellowships with full tuition reimbursements available (averaging $18,750 per year), research assistantships with full tuition reimbursements available (averaging $17,250 per year), teaching assistantships with full tuition reimbursements available (averaging $17,250 per year) Financial aid application deadline: 1/15. Total annual research expenditures: $7.7 million. *Unit head:* Prof. James A. Dumesic, Chair, 608-262-1092, Fax: 608-262-5434. *Application contact:* Donna M. Gabl, Graduate Coordinator, 608-263-3138, Fax: 608-262-5434, E-mail: gradoffice@che.wisc.edu.

University of Wyoming, Graduate School, College of Engineering, Department of Chemical and Petroleum Engineering, Program in Chemical Engineering, Laramie, WY 82071. Offers MS, PhD. *Faculty:* 6 full-time (0 women). *Students:* 11 full-time (2 women), 5 part-time (1 woman); includes 1 minority (Hispanic American), 8 international. 13 applicants, 31% accepted. In 1998, 1 master's awarded (100% found work related to degree); 1 doctorate awarded. Terminal master's awarded for partial completion of doctoral program. *Degree requirements:* For master's, thesis required, foreign language not required; for doctorate, computer language, dissertation required, foreign language not required. *Entrance requirements:* For master's, GRE General Test, TOEFL (minimum score of 550 required), minimum GPA of 3.0; for doctorate, GRE General Test, TOEFL, minimum GPA of 3.0. *Application deadline:* For fall admission, 4/15 (priority date). Applications are processed on a rolling basis. Application fee: $40. Electronic applications accepted. Tuition, state resident: full-time $2,520; part-time $140 per credit hour. Tuition, nonresident: full-time $7,790; part-time $433 per credit hour. Required fees: $400; $7 per credit hour. Full-time tuition and fees vary according to course load and program. *Financial aid:* In 1998–99, 9 research assistantships, 3 teaching assistantships were awarded.; career-related internships or fieldwork, Federal Work-Study, and institutionally-sponsored loans also available. Financial aid application deadline: 4/15. *Faculty research:* Microwave reactor systems, synthetic fuels, fluidization, coal combustion/gasification, flue-gas cleanup. Total annual research expenditures: $700,000. *Application contact:* Dr. David O. Cooney, Graduate Student Coordinator, 307-766-6464, Fax: 307-766-6777, E-mail: cooney@uwyo.edu.

Vanderbilt University, School of Engineering, Department of Chemical Engineering, Nashville, TN 37240-1001. Offers M Eng, MS, PhD. MS and PhD offered through the Graduate School. Part-time programs available. *Faculty:* 10 full-time (1 woman). *Students:* 26 full-time (8 women), 15 international. Average age 24. 47 applicants, 28% accepted. In 1998, 3 master's awarded (100% found work related to degree); 5 doctorates awarded (100% found work related to degree). *Degree requirements:* For master's and doctorate, thesis/dissertation required, foreign language not required. *Entrance requirements:* For master's and doctorate, GRE General Test, TOEFL. *Average time to degree:* Master's–2 years part-time; doctorate–4.5 years part-time. *Application deadline:* For fall admission, 1/15. Application fee: $40. *Financial aid:* In 1998–99, 1 fellowship with tuition reimbursement (averaging $3,000 per year), 12 research assistantships with tuition reimbursements (averaging $15,000 per year), 9 teaching assistantships with tuition reimbursements (averaging $15,000 per year) were awarded.; Federal Work-Study, institutionally-sponsored loans, and tuition waivers (partial) also available. Aid available to part-time students. Financial aid application deadline: 1/15; financial aid applicants required to submit CSS PROFILE or FAFSA. *Faculty research:* Adsorption and surface chemistry; biochemical engineering and biotechnology; chemical reaction engineering, environment, materials, and process modeling and control; chemical reaction kinetics, thermodynamics. Total annual research expenditures: $860,679. *Unit head:* M. Douglas LeVan, Chair, 615-322-2441, Fax: 615-343-7951, E-mail: mdl@vuse.vanderbilt.edu. *Application contact:* G. Kane Jennings, Director of Graduate Studies, 615-322-2441, Fax: 615-343-7951, E-mail: jenningk@vuse.vanderbilt.edu.

Villanova University, College of Engineering, Department of Chemical Engineering, Villanova, PA 19085-1699. Offers M Ch E. Part-time and evening/weekend programs available. *Faculty:* 8 full-time (1 woman), 1 part-time (0 women). *Students:* 5 full-time (0 women), 35 part-time (12 women); includes 4 minority (1 African American, 2 Asian Americans or Pacific Islanders, 1 Hispanic American), 1 international. Average age 26. 13 applicants, 62% accepted. In 1998, 8 degrees awarded (100% found work related to degree). *Degree requirements:* For master's, comprehensive exam required, thesis optional, foreign language not required. *Entrance requirements:* For master's, GRE General Test (for applicants with degrees from foreign universities), B Ch E, minimum GPA of 3.0. *Average time to degree:* Master's–2 years full-time, 4 years part-time. *Application deadline:* For fall admission, 8/1 (priority date); for spring admission, 12/1 (priority date). Applications are processed on a rolling basis. Application fee: $40. *Financial aid:* In 1998–99, 5 students received aid, including 5 teaching assistantships;

Chemical Engineering–Cross-Discipline Announcements

Villanova University (continued)

research assistantships, Federal Work-Study also available. Financial aid application deadline: 3/15. *Faculty research:* Heat transfer, advanced materials, chemical vapor deposition, pyrolysis and combustion chemistry, industrial waste treatment, model-based control, organic synthesis in supercritical carbon dioxide. *Unit head:* C. Michael Kelly, Chairperson, 610-519-4950, E-mail: cmkelly@email.vill.edu.

Virginia Polytechnic Institute and State University, Graduate School, College of Engineering, Department of Chemical Engineering, Blacksburg, VA 24061. Offers MS, PhD. *Faculty:* 13 full-time (0 women). *Students:* 39 full-time (9 women), 5 part-time (1 woman); includes 4 minority (2 Asian Americans or Pacific Islanders, 1 Hispanic American, 1 Native American), 11 international. 117 applicants, 15% accepted. In 1998, 8 master's, 5 doctorates awarded. Terminal master's awarded for partial completion of doctoral program. *Degree requirements:* For master's and doctorate, computer language, thesis/dissertation required, foreign language not required. *Entrance requirements:* For master's and doctorate, GRE, TOEFL (minimum score of 550 required). *Application deadline:* For fall admission, 12/1 (priority date). Applications are processed on a rolling basis. Application fee: $25. *Financial aid:* In 1998–99, 25 research assistantships, 3 teaching assistantships were awarded.; fellowships, unspecified assistantships also available. Financial aid application deadline: 4/1. *Faculty research:* Polymers, characterization, and rheological properties; synthesis and characterization of zeolite catalysts; affinity separations of biologically-produced materials; high vacuum surface science of single crystal metal oxides; innovative techniques for the removal of and/or deactivation of organic waste materials from soils. *Unit head:* Dr. William L. Conger, Head, 540-231-6631, E-mail: wlconger@vt.edu.

Washington State University, Graduate School, College of Engineering and Architecture, Department of Chemical Engineering, Pullman, WA 99164. Offers MS, PhD. *Faculty:* 8 full-time (1 woman). *Students:* 14 full-time (4 women), 1 part-time; includes 1 minority (Asian American or Pacific Islander), 6 international. In 1998, 11 master's, 4 doctorates awarded. *Degree requirements:* For master's, thesis, oral exam required, foreign language not required; for doctorate, dissertation, oral exam required. *Entrance requirements:* For master's and doctorate, minimum GPA of 3.0. *Average time to degree:* Master's–2 years full-time; doctorate–4 years full-time. *Application deadline:* For fall admission, 3/1 (priority date). Applications are processed on a rolling basis. Application fee: $35. *Financial aid:* In 1998–99, 10 research assistantships, 3 teaching assistantships were awarded.; fellowships, career-related internships or fieldwork, Federal Work-Study, institutionally-sponsored loans, tuition waivers (partial), and teaching associateships also available. Financial aid application deadline: 4/1; financial aid applicants required to submit FAFSA. *Faculty research:* Bioprocessing, kinetics and catalysis, hazardous waste remediation. Total annual research expenditures: $793,948. *Unit head:* Dr. Richard Zollars, Chair, 509-335-4332.

Washington University in St. Louis, School of Engineering and Applied Science, Sever Institute of Technology, Department of Chemical Engineering, St. Louis, MO 63130-4899. Offers chemical engineering (MS, D Sc); environmental engineering (MS, D Sc); materials science and engineering (MS); materials science engineering (D Sc). Part-time programs available. *Faculty:* 10 full-time (0 women), 2 part-time (0 women). *Students:* 36 full-time (8 women), 9 part-time (3 women), 35 international. Average age 25. 143 applicants, 8% accepted. In 1998, 15 master's, 6 doctorates awarded. Terminal master's awarded for partial completion of doctoral program. *Degree requirements:* For master's, thesis optional, foreign language not required; for doctorate, variable foreign language requirement, dissertation, preliminary exam, qualifying exam required. *Entrance requirements:* For master's and doctorate, GRE, minimum B average during final 2 years. *Average time to degree:* Master's–1.8 years full-time; doctorate–4.1 years full-time. *Application deadline:* For fall admission, 2/1 (priority date). Applications are processed on a rolling basis. Application fee: $20. *Financial aid:* In 1998–99, 37 students received aid, including 30 research assistantships; fellowships, career-related internships or fieldwork, Federal Work-Study, and institutionally-sponsored loans also available. Financial aid application deadline: 2/1. *Faculty research:* Reaction engineering, materials processing, catalysis, process control, air pollution control. Total annual research expenditures: $1.2 million. *Unit head:* Dr. M. P. Dudukovic, Chairman, 314-935-6021, Fax: 314-935-7211, E-mail: dudu@wuche3.wustl.edu. *Application contact:* Rose Baxter, Graduate Coordinator, 314-935-6082, Fax: 314-935-7211, E-mail: chedept@wuche3.wustl.edu.

Wayne State University, Graduate School, College of Engineering, Department of Chemical Engineering and Materials Science, Program in Chemical Engineering, Detroit, MI 48202. Offers MS, PhD. *Degree requirements:* For master's, thesis optional, foreign language not required; for doctorate, dissertation required, foreign language not required. *Faculty research:* Environmental management, biochemical engineering, supercritical technology, polymer process catalysis.

Western Michigan University, Graduate College, College of Engineering and Applied Sciences, Department of Paper, Imaging, and Chemical Engineering, Kalamazoo, MI 49008. Offers MS. *Students:* 8 full-time (1 woman), 13 part-time (2 women); includes 1 minority (African American), 14 international. 33 applicants, 70% accepted. In 1998, 6 degrees awarded. *Degree requirements:* For master's, thesis required, foreign language not required. *Entrance requirements:* For master's, minimum GPA of 3.0. *Application deadline:* For fall admission, 2/15 (priority date). Applications are processed on a rolling basis. Application fee: $25. *Financial aid:* Fellowships, research assistantships, teaching assistantships, Federal Work-Study available. Financial aid application deadline: 2/15; financial aid applicants required to submit FAFSA. *Faculty research:* Fiber recycling, paper machine wet end operations, paper coating. *Unit head:* Dr. Thomas Joyce, Chairperson, 616-387-2770. *Application contact:* Paula J. Boodt, Coordinator, Graduate Admissions and Recruitment, 616-387-2000, Fax: 616-387-2355, E-mail: paula.boodt@wmich.edu.

West Virginia University, College of Engineering and Mineral Resources, Department of Chemical Engineering, Morgantown, WV 26506. Offers chemical engineering (MS Ch E); engineering (MSE, PhD). Part-time programs available. Terminal master's awarded for partial completion of doctoral program. *Degree requirements:* For master's, thesis required, foreign language not required; for doctorate, dissertation, comprehensive exam required, foreign

language not required. *Entrance requirements:* For master's and doctorate, TOEFL (minimum score of 550 required), minimum GPA of 3.0. *Faculty research:* Fluidization, bioengineering, catalysis, coal gasification, coal liquefaction, polymer processing, materials engineering.

Widener University, School of Engineering, Program in Chemical Engineering, Chester, PA 19013-5792. Offers ME, ME/MBA. Part-time and evening/weekend programs available. *Faculty:* 5 part-time (0 women). *Students:* 4 full-time (0 women), 5 part-time (1 woman); includes 1 minority (African American), 3 international. 9 applicants, 78% accepted. In 1998, 5 degrees awarded. *Degree requirements:* For master's, thesis optional, foreign language not required. *Entrance requirements:* For master's, GMAT (ME/MBA). *Average time to degree:* Master's–2 years full-time, 4 years part-time. *Application deadline:* For fall admission, 8/1 (priority date); for spring admission, 12/1. Applications are processed on a rolling basis. Application fee: $25 ($300 for international students). *Financial aid:* In 1998–99, 1 teaching assistantship with full tuition reimbursement (averaging $7,500 per year) was awarded.; unspecified assistantships also available. Financial aid application deadline: 3/15. *Faculty research:* Biotechnology, environmental engineering, computational fluid mechanics, reaction kinetics, process design. *Unit head:* Dr. Gennard J. Maffia, Chairman, Department of Chemical Engineering, 610-499-4089, Fax: 610-499-4059, E-mail: gennard.j.maffia@widener.edu. *Application contact:* Dr. David H. T. Chen, Assistant Dean for Graduate Programs and Research, 610-499-4049, Fax: 610-499-4059, E-mail: david.h.chen@widener.edu.

Worcester Polytechnic Institute, Graduate Studies, Department of Chemical Engineering, Worcester, MA 01609-2280. Offers MS, PhD. *Faculty:* 10 full-time (2 women). *Students:* 23 full-time (5 women), 1 part-time, 12 international. 113 applicants, 29% accepted. In 1998, 4 master's awarded. Terminal master's awarded for partial completion of doctoral program. *Degree requirements:* For doctorate, dissertation required, foreign language not required, foreign language not required. *Entrance requirements:* For master's and doctorate, GRE (required for non-native speakers of English; combined average of 2032 on three sections), TOEFL (minimum score of 550 required; average 619). *Average time to degree:* Master's–2 years full-time; doctorate–5 years full-time. *Application deadline:* For fall admission, 2/15 (priority date). For spring admission, 10/15 (priority date). Applications are processed on a rolling basis. Application fee: $50. Electronic applications accepted. *Financial aid:* In 1998–99, 22 students received aid, including 1 fellowship with full tuition reimbursement available (averaging $14,200 per year), 10 research assistantships with full tuition reimbursements available (averaging $15,000 per year), 11 teaching assistantships with full tuition reimbursements available (averaging $11,970 per year); career-related internships or fieldwork, grants, institutionally-sponsored loans, and scholarships also available. Financial aid application deadline: 2/15; financial aid applicants required to submit FAFSA. *Faculty research:* Biochemical engineering, environmental engineering, chemical vapor deposition, advanced materials processing, bioreactor engineering: whole cells, environmental catalysis and reactor designs, bioseparations, Zeolite technology. Total annual research expenditures: $496,586. *Unit head:* Dr. Ravindra Datta, Head, 508-831-5250, Fax: 508-831-5853, E-mail: rdatta@wpi.edu. *Application contact:* Dr. Barbara Wyslouzil, Graduate Coordinator, 508-831-5493, Fax: 508-831-5853, E-mail: barbaraw@wpi.edu.

See in-depth description on page 525.

Yale University, Graduate School of Arts and Sciences, Programs in Engineering and Applied Science, Department of Chemical Engineering, New Haven, CT 06520. Offers MS, PhD. *Faculty:* 18. *Students:* 19 full-time (5 women), 2 part-time; includes 5 minority (4 Asian Americans or Pacific Islanders, 1 Hispanic American), 7 international. 40 applicants, 38% accepted. In 1998, 1 master's, 8 doctorates awarded. Terminal master's awarded for partial completion of doctoral program. *Degree requirements:* For master's, foreign language and thesis not required; for doctorate, dissertation, exam required, foreign language not required. *Entrance requirements:* For master's and doctorate, GRE General Test, TOEFL. *Average time to degree:* Doctorate–5.3 years full-time. *Application deadline:* For fall admission, 1/4. *Application fee:* $65. *Financial aid:* Fellowships, Federal Work-Study and institutionally-sponsored loans available. Aid available to part-time students. *Unit head:* Chair, 203-432-4378. *Application contact:* Admissions Information, 203-432-2770.

Announcement: Actively engaged in research on colloid/interfacial transport phenomena, colloidal forces, rheology of complex fluids, transport of environmental contaminants in groundwater, chromatographic separation of biomaterials, protein interactions, dynamics of adsorption-ion exchange, synthesis of mesoporous materials, catalytically stabilized combustion, spectroscopy of supported catalysts, nucleation and growth of thin films, mechanisms of surface processes, transport phenomena in multiphase chemically reacting systems, fine-particle science and technology, combustion for materials synthesis and processing.

Youngstown State University, Graduate School, William Rayen College of Engineering, Department of Civil, Chemical, and Environmental Engineering, Youngstown, OH 44555-0001. Offers MSE. Part-time and evening/weekend programs available. *Faculty:* 9 full-time (1 woman). *Students:* 13 full-time (2 women), 5 part-time (2 women), 4 international. 4 applicants, 75% accepted. In 1998, 10 degrees awarded. *Degree requirements:* For master's, computer language required, thesis optional, foreign language not required. *Entrance requirements:* For master's, TOEFL (minimum score of 550 required), minimum GPA of 2.75 in field. *Application deadline:* For fall admission, 8/15 (priority date); for winter admission, 11/15 (priority date); for spring admission, 2/15 (priority date). Applications are processed on a rolling basis. Application fee: $30 ($75 for international students). Tuition, state resident: part-time $97 per credit hour. Tuition, nonresident: part-time $219 per credit hour. Required fees: $21 per credit hour. $41 per quarter. *Financial aid:* In 1998–99, 4 students received aid, including 4 research assistantships with full tuition reimbursements available (averaging $7,500 per year); teaching assistantships, Federal Work-Study, institutionally-sponsored loans, and scholarships also available. Aid available to part-time students. Financial aid application deadline: 3/1. *Faculty research:* Structural mechanics, water quality modeling, surface and ground water hydrology, physical and chemical processes in aquatic systems. *Unit head:* Dr. Jack D. Bakos, Chair, 330-742-3027. *Application contact:* Dr. Peter J. Kasvinsky, Dean of Graduate Studies, 330-742-3091, Fax: 330-742-1580, E-mail: amgrad03@ysub.ysu.edu.

Cross-Discipline Announcements

Colorado State University, Graduate School, College of Engineering, Department of Atmospheric Science, Fort Collins, CO 80523-0015.

The Department of Atmospheric Science offers MS and PhD degrees with specialization in atmospheric chemistry. Research areas include modeling, laboratory, and field experiment programs in aerosol physics, water chemistry, cloud physics, climate change, and air pollution. Students with backgrounds in chemistry, engineering, and physics are encouraged to apply.

Institute of Paper Science and Technology, Graduate Programs, Atlanta, GA 30318-5794.

Multidisciplinary program in chemical and mechanical engineering, chemistry, physics, biology, and pulp and paper science. Generous financial assistance provided to all admitted students

meeting eligibility requirements (citizens of the US, Canada, or Mexico or permanent legal residents of US). Admission geared to BS majors in above disciplines. See Book 4, Chemistry section for in-depth description.

Pennsylvania State University University Park Campus, Graduate School, Intercollege Graduate Programs, Intercollege Graduate Program in Integrative Biosciences, State College, University Park, PA 16802-1503.

The Life Sciences Consortium is especially interested in attracting graduate students for the integrative biosciences graduate program who have backgrounds in chemical engineering and plan to study the quantification of biomolecule movement within cells, tissues, and throughout organisms. See in-depth description in Book 3.

Princeton University, Graduate School, Princeton Materials Institute, Princeton, NJ 08540-5211.

The Princeton Materials Institute welcomes students interested in cross-disciplinary research involving any aspect of materials science and engineering. Faculty members from 8 academic departments collaborate in a remarkably broad range of programs, supported by outstanding facilities. Some fellowships are available. See in-depth description in Section 21 of this guide.

University of California, San Francisco, School of Pharmacy, Department of Pharmaceutical Chemistry, San Francisco, CA 94143.

The PhD degree program in pharmaceutical chemistry is directed toward research at the interface between chemistry and biology. The program offers research in areas of bioorganic chemistry, macromolecular structure and function, computer-aided drug design, medicinal chemistry, drug metabolism and biochemical toxicology, drug delivery, molecular parasitology, molecular pharmacology, and pharmacokinetics/pharmacodynamics.

University of Massachusetts Amherst, Graduate School, College of Natural Sciences and Mathematics, Department of Polymer Science and Engineering, Amherst, MA 01003.

The Department of Polymer Science and Engineering offers doctoral polymer studies at the interface between chemistry, materials science, and engineering. See in-depth description in Section 16 of this book. Visit the Web site at http://www.pse.umass.edu.

University of Minnesota, Twin Cities Campus, Medical School, Graduate Programs in Medicine, Program in Microbial Engineering, Minneapolis, MN 55455-0213.

The microbial engineering Master of Science program, coordinated by the Biological Process Technology Institute, involves faculty members and facilities from 12 departments and 6 institutes in cross-disciplinary training and research. The microbial engineering degree fulfills the minor requirement for graduates who choose to pursue a PhD in related fields.

Vanderbilt University, School of Engineering, Department of Biomedical Engineering, Nashville, TN 37240-1001.

The Department of Biomedical Engineering is especially interested in attracting graduate students with backgrounds in chemical engineering who wish to study the applications of their field to biomedical problems. Research is available in cardiopulmonary transport phenomena, cellular bioengineering, cellular exchange and kinetics, and biosensors. See in-depth description in Bioengineering, Biomedical Engineering, and Biotechnology section.

ARIZONA STATE UNIVERSITY

Department of Chemical, Bio and Materials Engineering
Chemical Engineering Program

Programs of Study	The Chemical Engineering Program at Arizona State University (ASU) is in the Department of Chemical, Bio and Materials Engineering and offers graduate programs that lead to the Master of Science and Doctor of Philosophy degrees in chemical engineering. The faculty offers a wide range of course work and research topics, allowing the student an opportunity to develop a program of study to satisfy his or her specific needs. Research is currently being done in the areas of air and water purification, semiconductor materials processing, surface science, control system engineering, environmental remediation, environmentally-benign manufacturing, air pollution modeling, biotechnology, biomass conversion to liquid fuels, and multiphase flow. With their choice of electives and thesis research topics, students are encouraged to take advantage of the synergy that exists between the three degree-granting programs in the department and between the departments within the College of Engineering and Applied Sciences.
	The Master of Science degree program requires a minimum of 30 credit hours, including 6–9 hours of thesis credit. All students are required to enroll in a 1-credit seminar during each semester in residence.
	The Doctor of Philosophy degree program requires a minimum of 84 hours beyond the bachelor's degree. To satisfy the residency requirements, the students must spend a minimum of two semesters of full-time study on campus. All candidates must take a qualifying exam during their first year in the program. The first part of the exam covers material that would be included in an undergraduate degree program. The second part consists of the preparation of a research proposal on a topic selected by the student from a list provided by the faculty. Students are admitted to candidacy after passing a comprehensive exam, which is given near the completion of course work. The exam consists of an oral and written dissertation prospectus.
	Students with a B.S. degree in a field other than chemical engineering are encouraged to apply. The Graduate Committee determines, on an individual basis, which undergraduate courses must be taken to ensure success in the graduate program. These courses are not counted toward the program of study.
Research Facilities	Research facilities within the department are supplemented by the laboratories in the Center for Solid State Electronics Research, the Environmental Fluid Dynamics Program, the Computer-Integrated Manufacturing Systems Research Center, and the Center for Energy Systems Research, all of which provide technical support for students and faculty in the chemical engineering program.
	Engineering Computer Services has a staff of 35 full-time computer specialists who support the wide array of computers and electronic equipment within the college. The college also maintains complete machine, carpentry, electrical, and paint shops to support graduate research.
	The University libraries house more than 2 million volumes, 1.7 million microfilms, and 24,000 journal titles. The Noble Science and Engineering Library, a designated patent depository, houses the entire U.S. patent collection.
Financial Aid	A wide variety of financial support is available for graduate students, including teaching and research assistantships and University and industrial fellowships. Out-of-state tuition is waived for all graduate assistants and fellows. In addition, there are a number of Graduate Tuition Scholarships (GTS) that cover out-of-state tuition and Graduate Academic Scholarships (GAS) tha cover in-state registration fees available on a highly competitive basis.
Cost of Study	In 1999–2000, the registration and tuition fee for 12 hours or more is $1044 per semester for Arizona residents; nonresidents pay an additional $3476 (prorated for fewer than 12 hours).
Living and Housing Costs	Limited on-campus housing is available for unmarried students. For the 1998–99 academic year, room rates ranged from $2180 to $3805. Thirteen different meal plans are available (per week or semester). Moderately priced apartments are available within walking distance of the campus.
Student Group	There are more than 47,000 students at Arizona State, including more than 11,000 graduate students. More than 1,800 of the 6,300 students in the College of Engineering and Applied Sciences are enrolled in graduate programs. In fall 1998, there were 140 graduate students in the department, with 47 of them enrolled in the chemical engineering program.
Student Outcomes	Employment opportunities for graduate degree holders have been excellent because of the environmental, process control, and semiconductor industries that are directly related to the research activities of the students and faculty members. Recent alumni with graduate degrees are currently employed in processing, research, development, and design positions.
Location	Tempe is a suburb bordering the city of Phoenix. The metropolitan area is the population, economic, and industrial center of the state of Arizona. Entertainment centers, art and anthropology museums, and sports arenas are among the wide variety of facilities available in the Valley of the Sun. The area's mild climate also provides for numerous year-round outdoor activities. The urban community surrounding ASU thrives on high-technology industries and is one of the fastest-growing communities in the country.
The University	Arizona State University traces its origin to 1885 and is the largest and oldest institution of higher learning in the state of Arizona. Its present enrollment of 47,000 places it as the fifth-largest central-campus university in the United States. ASU's campus comprises 700 acres and offers outstanding physical facilities to support the University's research and educational mission. Included within the more than 125 buildings are twelve colleges and schools, a University-wide computer system, seven libraries (including the Noble Science and Engineering Library), and more than two dozen specialized centers of research. ASU was recently awarded Research I status as a result of its past and continuing commitment to academic research.
Applying	All students must apply for admission through the Graduate College. For application forms, students should contact Graduate Admissions, Arizona State University, Tempe, Arizona 85287-1003 or call 602-965-6113. Submission of Graduate Record Examinations (GRE) scores and a statement of purpose is required. International students whose native language is not English must submit a Test of English as a Foreign Language (TOEFL) score of 580 or better. Applications are reviewed continuously but, for financial consideration, should be received by February 1 for the following academic year.
Correspondence and Information	Chair of Graduate Committee Chemical, Bio and Materials Engineering Box 876006 Arizona State University Tempe, Arizona 85287-6006 Telephone: 602-965-3313 E-mail: cbmerec@asuvax.eas.asu.edu

Arizona State University

THE FACULTY AND THEIR RESEARCH

Chemical Engineering

Stephen P. Beaudoin, Ph.D., North Carolina State. Semiconductor materials processing and environmentally-benign manufacturing, transport phenomena, surface science, waste minimization.

James R. Beckman, Ph.D., Arizona. Unit operations, applied mathematics, energy-efficient water purification, crystallization control and nucleation, process simulation, data analysis, energy storage, solar cooling.

Neil S. Berman, Ph.D., Texas at Austin. Regional air pollution and global warming physical and numerical models, turbulent mixing experiments and calculations in atmospheric boundary layers and in liquid flows with polymeric additives, experimental measurements using laser-induced fluorescence, transformations and deposition of air toxic materials.

Veronica A. Burrows, Ph.D., Princeton. Surface science; environmental sensors; semiconductor processing; interfacial chemical and physical processes in sensor processing, lubrication, and composite materials.

Antonio A. Garcia, Ph.D., Berkeley. Protein purification, acid-base molecular interactions in separations, solid-liquid interfacial phenomena, scanning probe microscopy, biocolloid chemistry, chromatography, biosensor immobilization.

Gregory B. Raupp, Ph.D., Wisconsin. Gas-solid surface reaction mechanisms and kinetics, interaction between surface reactions and simultaneous transport processes, semiconductor materials processing, thermal and plasma-enhanced chemical vapor deposition (CVD), environmental pollution remediation and control, photocatalytic oxidation.

Daniel E. Rivera, Ph.D., Caltech. Control systems engineering, dynamic modeling via system identification, robust control, computer-aided control system design.

Vernon E. Sater, Ph.D., IIT. Heavy metal removal from wastewater, online process instrumentation.

Robert S. Torrest, Ph.D., Minnesota. Multiphase flow, filtration, polymer solution flow in porous media, in situ processes for energy and mineral recovery, pollution control.

Bioengineering

William J. Dorson (Emeritus), Ph.D., Cincinnati. Biomedical engineering, artificial organ problems and applications, rheology, physicochemical phenomena, transport processes, physiological models and analogs.

Eric J. Guilbeau, Ph.D., Louisiana Tech. Biomedical engineering, biomaterials, skeletal muscle cardiac assist, biological transport phenomena, development of biosensors, physiological systems analysis and simulation, artificial internal organs.

Jiping He, Ph.D., Maryland. Biomechanics, robotics, computational neuroscience, optimal control, system dynamics and control.

Daryl R. Kipke, Ph.D., Michigan. Computational neuroscience, auditory neurophysiology, electrical stimulation, speech recognition, signal processing, biomedical instrumentation, neural modeling using parallel computers and analog VLSI technology.

Vincent B. Pizziconi, Ph.D., Arizona State. Biomedical engineering, molecular and cellular engineering/biological engineering/tissue engineering, artificial organs, biomaterials, biosensors, biotechnology, bioseparations, scanning tunneling microscopy (STM) and atomic force microscopy (AFM) of natural and synthetic biomaterials, fractals/nonlinear dynamical systems.

James D. Sweeney, Ph.D., Case Western Reserve. Biomedical engineering, rehabilitation engineering, applied neural control, neurophysiology, motor physiology, mathematical modeling.

Bruce C. Towe, Ph.D., Penn State. Bioinstrumentation, implantable biochemical sensors, medical ultrasound, bioelectric phenomena, bioimpedance imaging.

Gary T. Yamaguchi, Ph.D., Stanford. Biomechanics and rehabilitation engineering design, including joint mechanics; computer modeling of muscle, tendon, and joints; optimal control and dynamic analysis/simulation of movement, coordination, and functional neuromuscular stimulation.

Materials Science Engineering

Terry L. Alford, Ph.D., Cornell. Microelectronic metallization and reliability, silicide formation, ion-beam modification of materials.

Ray W. Carpenter, Ph.D., Berkeley. Atomic structure and chemistry of interfaces and boundaries in solids, phase transformation mechanisms in metals and ceramics, electron microscopy methods and instrumentation.

Sandwip K. Dey, Ph.D., Alfred, College of Ceramics. Thin-film processing science of electroceramics; characterization of electrical, microstructural, and microchemical properties; high-permittivity dielectrics for ULSI DRAMs and microelectronic packages.

Lester E. Hendrickson (Emeritus), Ph.D., Illinois. Corrosion, fracture and failure analysis, physical and chemical metallurgy.

Dean L. Jacobson, Ph.D., UCLA. Thermionic energy conversion, high-temperature materials, erosion, heat pipes, thermodynamics.

Stephen L. Krause, Ph.D., Michigan. Ordered polymers, composite materials, electronic materials, X-ray diffraction, electron X-ray diffraction, electron microscopy.

James W. Mayer, Ph.D., Purdue. Electronic materials and metallization of integrated circuits; development of new semiconductor materials, such as the ternary alloy SiGeC grown on silicon; development of new metal systems for interconnectors; interdiffusion and reactions in thin films; analysis of paint pigments, art media, and metallic artifacts; ion-beam analysis and Rutherford backscattering analysis.

James T. Stanley (Emeritus), Ph.D., Illinois. Phase transformations, radiation effects on materials, erosion-corrosion.

ILLINOIS INSTITUTE OF TECHNOLOGY

Department of Chemical and Environmental Engineering

Programs of Study

IIT's chemical and environmental engineering department offers programs leading to the degrees of Master of Science, Professional Master, and Doctor of Philosophy in both chemical and environmental engineering. A new master's degree program in food process engineering is also available.

The Master of Science degree requires 32 credit hours, including 6 to 8 credit hours of thesis work. The Professional Master degree does not require a thesis. Students holding a bachelor of science degree in non-engineering fields may enroll by taking up to four additional courses. The candidate for the Doctor of Philosophy degree usually spends at least three years in study and research beyond the bachelor's degree. A minimum of 96 credit hours, which include thesis and research, is required.

Research Facilities

The department administers three interdisciplinary research centers: the Energy + Power Center, the Center of Excellence in Polymer Science and Engineering, and the Center for Electrochemical Science and Engineering. The computational facilities of the department are divided into three groups: the Advanced Computational Laboratory, the Multimedia Personal Computer Lab, and the computer facilities of each research group. The Advanced Computational Laboratory consists of five HP9000/715 workstations, with 100-MHz speed, 64-MB memory, 20-inch color monitors, and a server with dual processors and 256-MB memory. All computers are connected to each other and to the IIT computer network by Ethernet. All Pentium computers in the PC Lab can access the workstations, creating a twenty-seat computational lab for instructional activities at the graduate and undergraduate levels. Both the PCs and workstations access the multimedia system to provide data visualization and high-quality presentations. In addition, each research lab has specialized computer facilities.

Financial Aid

Financial assistance is available through fellowships, teaching assistantships, and research assistantships, which provide varying stipends to help defray living expenses, and through tuition scholarships that range from partial to full support. A limited number of half-tuition scholarships are available for full-time graduate students. Additionally, loans and assistance for students and their spouses in securing employment are available.

Cost of Study

Graduate tuition for 1999–2000 is charged at the rate of $590 per credit hour. International students are required to register for a minimum of 9 credits or the equivalent per semester.

Living and Housing Costs

Housing is available for graduate students in IIT residence halls. The 1999–2000 annual cost for room and board ranges from $5155 to $6880. Unfurnished IIT apartments are available for graduate students and their families at costs ranging from $458 to $927 per month, including utilities. Early application is recommended.

Student Group

IIT's total enrollment in 1998–99 was approximately 5,900. Chemical and environmental engineering graduate students numbered 212. More than 40 percent of these students were full-time, and 22 percent were women.

Location

IIT's Main Campus is located near the heart of Chicago, just 3 miles south of the Loop and central to the greater Chicago area's thriving technological community of business, industry, and research institutions. Internationally known for its architecture, museums, symphony and theater, and beautiful lakefront on the western shore of Lake Michigan and the unusually rich variety of its ethnic communities, Chicago offers a vast array of recreational and cultural opportunities. The Main Campus, designed by Ludwig Mies van der Rohe and regarded internationally as a landmark of twentieth century architecture, occupies fifty buildings on a 120-acre site and includes research institutes, libraries, laboratories, residence halls, a sports center, and other facilities. Among its immediate neighbors are Comiskey Park, home of the Chicago White Sox; two major medical centers; and the McCormick Place Exposition Center. The Downtown Campus is in the Loop near the city's financial trading, banking, and legal centers. The Rice Campus is in suburban Wheaton, convenient to the Interstate 88 research and technology corridor west of the city. The Moffet Campus is in southwest suburban Summit-Argo.

The Institute and The Department

Illinois Institute of Technology was formed in 1940 by the merger of Armour Institute of Technology (founded in 1890) and Lewis Institute (founded in 1896). In the Department of Chemical and Environmental Engineering, 21 full-time faculty members conduct an annual average of approximately $2 million of research. The department is one of the oldest chemical engineering programs in the nation. Courses are offered on the Main Campus, which houses the department's state-of-the-art multimedia classroom and computational lab; at convenient off-site locations in Chicago Loop and Wheaton; and via more than twenty-five interactive TV sites and the Internet.

Applying

Students may be admitted to study at the beginning of any regular semester. Applications and supporting documents should be received by the Office of Graduate Admissions (3300 South Federal Street, Room 301A, Chicago, Illinois 60616) by June 1, November 1, and May 1 for matriculation in the fall, spring, and summer semesters, respectively. Primary consideration for financial aid, which is usually awarded for the academic year beginning in August, is given to applicants whose materials are received before March 1. Application forms and additional information are available on the Web at http://www.grad.iit.edu.

Correspondence and Information

Graduate Admissions Coordinator
Department of Chemical and Environmental Engineering
Illinois Institute of Technology, PH 127
Chicago, Illinois 60616-3793
Telephone: 312-567-3533
E-mail: chee@charlie.cns.iit.edu
World Wide Web: http://www.chee.iit.edu/

Illinois Institute of Technology

THE FACULTY AND THEIR RESEARCH AREAS

The Department of Chemical and Environmental Engineering faculty members are conducting numerous projects in the department's core areas of research competency: air pollution testing and remediation; bioprocessing and bioremediation; chemical process modeling, statistical monitoring, and control; computational multiphase flow; electrochemical science and engineering; energy engineering; environmental chemistry; industrial waste and wastewater treatment; particle technology and fluidization; and polymer science and engineering.

Chemical Engineering Faculty

Hamid Arastoopour, Professor and Chairman; Ph.D. (gas engineering), IIT, 1978. Computational multiphase flow pulverization and agglomeration of particles, fluidization, fluid-particle flow, material processing.

Nader Aderangi, Lecturer of Chemical Engineering and Director of Department Laboratories; Ph.D., IIT, 1978. Unit operations, chemical processes.

Barry Bernstein, Professor of Chemical Engineering and Applied Mathematics; Ph.D. (mathematical mechanics), Indiana, 1956. Computational fluid mechanics, materials properties, polymer rheology.

Ali Cinar, Professor and Associate Chairman of Chemical Engineering; Ph.D., Texas A&M, 1976. Chemical and food process control, nonlinear input-output modeling, statistical process monitoring, intelligent process control, AI applications.

Stuart L. Cooper, Philip Danforth Armour Professor of Engineering and Vice President and Chief Academic Officer; Ph.D., Princeton, 1967. Polymer science and biomaterials.

Dimitri Gidaspow, Professor; Ph.D. (gas technology), IIT, 1962. Hydrodynamics of fluidization using kinetic theory, gas-solid transport, hydrodynamic models for slurry bubble column reactors.

Henry R. Linden, McGraw Professor of Energy and Power Engineering; Ph.D., IIT, 1952; member, National Academy of Engineering. Fossil fuel technologies, energy and resource economics, energy and environmental policy.

Satish Parulekar, Professor; Ph.D., Purdue, 1983. Biochemical engineering, chemical reaction engineering.

Jai Prakash, Associate Professor; Ph.D. (chemical kinetics), Delhi (India), 1985; Ph.D. (electrochemistry), Case Western Reserve, 1990. Electrochemistry, solid-state chemistry, materials synthesis and characterization for energy conversion and energy storage applications.

Jay D. Schieber, Associate Professor; Ph.D., Wisconsin–Madison, 1989. Kinetic theory, polymer rheology predictions, thermal conductivity measurements, transport phenomena, non-Newtonian fluid mechanics.

J. Robert Selman, Professor; Ph.D., Berkeley, 1971. Applied electrochemistry and electrochemical engineering, battery and fuel cell design, electroplating and coatings.

Eugene S. Smotkin, Associate Professor; Ph.D. (chemistry), Texas at Austin, 1989. FTIR spectroscopy of electrode surfaces, electrochemical mass spectroscopy, fuel cells, polymer electrolytes.

Fouad A. Teymour, S.C. Johnson Polymer Associate Professor; Ph.D., Wisconsin–Madison, 1989. Polymer reaction engineering, mathematical modeling, nonlinear dynamics.

David C. Venerus, Associate Professor; Penn State, 1989. Polymer rheology and processing, transport phenomena in polymeric systems.

Darsh T. Wasan, Vice President of International Affairs, and Motorola Chair; Ph.D., Berkeley, 1965. Thin liquid films; foams, emulsions and dispersions; interfacial rheology; environmental technologies.

Environmental Engineering Faculty

Paul R. Anderson, Associate Professor; Ph.D. (civil engineering), Washington (Seattle), 1987. Precipitation kinetics, evaluation of oxide adsorbents for water and wastewater treatment.

H. Ted Chang, Assistant Professor; Ph.D., (civil/environmental engineering), Illinois at Urbana-Champaign, 1985. Biological processes, activated carbon adsorption, hazardous waste remediation, groundwater aquifer remediation, membrane separation.

Nasrin R. Khalili, Assistant Professor; Ph.D. (environmental engineering), IIT, 1992. Synthesis and characterization of adsorbents for waste and water treatment processes, application of waste management strategies to industrial processes.

Demetrios J. Moschandreas, Professor; Ph.D. (physics), Cincinnati, 1972. Ambient and indoor air pollution, statistical methods for multimedia pollution, total exposure and risk analysis.

Kenneth E. Noll, Professor; Ph.D. (civil engineering), Washington, 1975. Air resources engineering, air pollution meteorology, design of control devices, chemistry and physics of atmospheric particles, control of emissions from hazardous waste sites.

Krishna R. Pagilla, Assistant Professor; Ph.D. (civil engineering), Berkeley, 1994. Water and wastewater engineering, environmental microbiology, biological nutrient control, soil remediation, sludge treatment.

Adjunct Faculty

V. M. Balasubramaniam. Food processing.

Michael Caracotsios. Design and optimization of chemical processes.

Ellis Fields. Industrial chemistry, catalysis.

Ted Knowlton. Fluidization, fluid-particle systems.

Harold Lindahl. Process design.

Robert Peters. Hazardous waste treatment.

Hwa-Chi Wang. Aerosol technology.

Research Centers

IIT's Department of Chemical and Environmental Engineering houses three research centers: the IIT Energy + Power Center, the IIT Center of Excellence in Polymer Science and Engineering, and the IIT Center for Electrochemical Science and Engineering.

The IIT Energy + Power Center was established at IIT in 1989 through the endowment of a chaired professorship by the McGraw Foundation. The McGraw Distinguished Professor directs the Center's comprehensive research and education program in sustainable global energy development. Included among the numerous activities of the Center is an interdisciplinary research and education program in Energy/Environment/Economics (E^3). This unique program features efforts in multiphase flow, fluidization, enhanced oil and gas recovery, process monitoring and control, power engineering, pollution prevention, global climate change, energy economics, and energy and environmental policy. The program also offers an E^3 specialization at the undergraduate level and an E^3 concentration at the graduate level.

The Center of Excellence in Polymer Science and Engineering (CEPSE) was established at IIT in 1990 with a grant from the Amoco Foundation to serve as an interdisciplinary research, educational, and technology transfer facility. Research efforts focus on polymer process modeling, polymer rheology, polymer reaction engineering, polymer synthesis, polymer characterization, and polymer and rubber recycling. The CEPSE also offers an undergraduate specialization and a graduate concentration in polymer science and engineering.

The Center for Electrochemical Science and Engineering (CESE) had its origins in the establishment of the Army Research Office Fuel Cell/Battery Manufacturing Research Hub at IIT in 1995. The activities of the Hub program include both fundamental and applied research related to the emerging technologies of low-temperature direct methanol fuel cells and lithium-ion and nickel/metal hydride batteries. Other CESE activities include photoelectrochemistry, photovoltaic energy storage, and high-temperature fuel cells such as solid-oxide and molten carbonate fuel cells. In addition, basic research and development of electrochemical sensors—important in energy, medical, industrial, military, and space applications—is a key focus area of the center. Another objective of the CESE is to make electrochemical science and engineering an interdisciplinary degree specialization by taking advantage of existing graduate and undergraduate course offerings in engineering and basic sciences.

LOUISIANA STATE UNIVERSITY

Gordon A. and Mary Cain Department of Chemical Engineering

Programs of Study

LSU offers the Ph.D. and both thesis and nonthesis M.S. programs in chemical engineering. Part-time students are welcome. For the convenience of those working in nearby industries, key graduate courses are offered in the early evening.

M.S. students are encouraged to elect the thesis option. In this option, 24 semester hours of graduate-level course work are required in addition to at least 6 hours of thesis credit. The 24 hours must include 12 hours from the four core courses: mathematical methods, thermodynamics, transport phenomena, and reactor design.

In the nonthesis M.S. option, 36 semester hours of graduate-level course work are required. The 36 hours must include 15 hours of chemical engineering courses, of which 12 hours must be the core courses. The master's examination for the nonthesis option consists of a written comprehensive examination.

Requirements for the Ph.D. include 30 semester hours of graduate-level work beyond the bachelor's degree. Of these, 18 semester hours must be graduate-level chemical engineering courses and must include the four core courses. The remaining 12 semester hours of graduate-level course work may be a formal minor in another department or may consist of a variety of courses selected from several departments, including chemical engineering. Students already holding a master's degree can transfer up to 24 hours of course work.

Students enrolled in the Ph.D. program take a written qualifying examination within one year of their enrollment. Within one year of successfully passing the qualifying examination, a Ph.D. student forms an examining committee and takes a general examination, which is an oral defense of a written plan for doctoral research. The final examination is an oral defense of the doctoral dissertation.

Research Facilities

Spectroscopic facilities consist of various lasers, a Raman microprobe, two Fourier transform infrared spectrometers, and four quadrupole mass spectrometers. Analytical facilities include an environmental scanning electron microscope, a powder X-ray diffractometer, two thermogravimetric analyzers, a heat-flow calorimeter, a BET surface area apparatus, a mercury porosimeter, several optical microscopes of different types, a total carbon analyzer, a liquid scintillation counter, gel-permeation and supercritical fluid chromatographs, and numerous HPLCs, LCs, and GCs. Incubators, shakers, fermentors, and culture cabinets are available for biochemical research. Properties of melts and solutions are studied on constant-stress capillary and torque rheometers, a dynamic surface tensiometer, and a laser Doppler velocimeter. Polymer processing equipment includes extruders, an injection molder, and a reaction injection molding machine. Fiber processing is studied with a fiber pull adhesion tester, a fiber coating line, and a tube knitter. High-pressure facilities are available for equilibrium thermodynamics and supercritical fluid extraction studies. There are a wet air stripping unit, a distillation column, and a bubble column, all pilot-scale. Extensive computational facilities are available in the department and on campus.

Financial Aid

Assistantships of up to $21,500 per year, as well as Dean's fellowships of $17,000 per year plus tuition exception, are available to qualified students.

Cost of Study

For 1998–99, tuition, the student health service fee, and certain other fees total $3775 per calendar year.

Living and Housing Costs

A student living off campus can expect to spend at least $6925 per academic year for rent, food, clothing, and other living expenses. Married students spend approximately $13,000 per academic year. A limited number of University apartments are available for families and full-time graduate students.

Student Group

The graduate student body of approximately 60 members consists of both American and international students. Of these, about 10 percent are part-time students. Most full-time students are supported financially by the department.

Student Outcomes

Most graduates find employment in the chemical industries, but more than 50 LSU graduates have taken faculty positions around the world. Four have been elected to the National Academy of Engineering, and 2 doctoral and 3 baccalaureate graduates were recently recognized by AIChE as preeminent chemical engineers.

Location

Baton Rouge, adjacent to the Acadiana region, is located in one of the most culturally diverse regions of the United States. The surrounding countryside retains the ambiance of the rural deep South, with plantation homes, salt marshes, lakes, cypress swamps, and pine forests. The city has a Performing Arts Center, an Arts and Science Center, and several cultural institutions, including opera, ballet, symphonic orchestra, and theater companies. New Orleans is located only 80 miles to the southeast, and the Gulf Coast beaches are a 2-hour drive away.

The University and The Department

The University, founded in 1860, has a student population of more than 30,000 and a faculty of more than 1,300. LSU is one of only twenty-five universities nationwide designated as both land-grant and sea-grant and ranks among the top 2 percent of all colleges and universities in the nation by virtue of its Research University I designation. The campus is located on the southern edge of the city and is bordered on the west by the Mississippi River.

Applying

Applicants must have an undergraduate degree in chemical engineering or a related field from an accredited American university or a similar international university. GRE scores are required. Special programs can be developed for those with degrees from other disciplines. Applications for the fall semester should be received by the graduate school by mid-January. Applications received later will be considered but have a reduced chance of acceptance. Students are selected on the basis of undergraduate records and GRE scores.

Correspondence and Information

Graduate Coordinator
Gordon A. and Mary Cain Department of Chemical Engineering
Louisiana State University
Baton Rouge, Louisiana 70803
Telephone: 800-256-2084 (toll-free)
Fax: 225-388-1476
E-mail: gradcoor@che.lsu.edu
World Wide Web: http://svr1.che.lsu.edu/main.htm

Louisiana State University

THE FACULTY AND THEIR RESEARCH

John R. Collier, Ph.D., Case Western Reserve, 1966. Polymer and textiles processing and properties, fluid flow, conversion of agricultural wastes to value-added products.

Armando B. Corripio, Ph.D., LSU, 1970. Application of computers for designing, controlling, and operating chemical processes.

Kerry M. Dooley, Ph.D., Delaware, 1983. Acid and metal oxide catalysis, applications of supercritical fluids in environmental remediation and materials processing.

Gregory L. Griffin, Ph.D., Princeton, 1979. Chemical processing of materials for microelectronics and advanced ceramics.

Douglas P. Harrison, Ph.D., Texas, 1966. Noncatalytic gas-solid reactions, removal of semivolatile contaminants from aqueous solution, combined reaction and separation.

Michael A. Henson, Ph.D., California, Santa Barbara, 1992. Process control and nonlinear estimation, reverse engineering of biological control systems, membrane separation systems, biochemical reactor control.

Martin A. Hjortsø, Ph.D., Houston, 1983. Bioreactor engineering and modeling, hairy root and oscillating yeast cultures.

F. Carl Knopf, Chairman; Ph.D., Purdue, 1980. Extraction by supercritical fluids, transient pump-probe absorption, Raman spectroscopy.

Ralph W. Pike, Ph.D., Georgia Tech, 1962. Fluid dynamics with chemical reactions occurring in the flow, optimization theory.

Elizabeth J. Podlaha, Ph.D., Columbia, 1992. Electrochemical characterization and deposition of advanced materials.

Geoffrey L. Price, Ph.D., Rice, 1979. Zeolites and zeolite catalysis.

Maciej (Mac) Radosz, M. F. Gautreaux/Ethyl Professor; Ph.D., Polytechnic University of Cracow (Poland), 1977. Phase behavior of compressible macromolecular systems and its relationship to molecular interactions.

Danny D. Reible, Director, Hazardous Substance Research Center; Ph.D., Caltech, 1982. Transport phenomena and their applications to environmental mechanics.

Arthur M. Sterling, Ph.D., Washington (Seattle), 1969. Experimental and theoretical investigations of incineration phenomena.

Louis J. Thibodeaux, Ph.D., LSU, 1968. Chemodynamics, the study of the transport and fate of anthropogenic substances in the natural environment.

Karsten E. Thompson, Ph.D., Michigan, 1996. Modeling flow, reaction, and mass transfer in porous materials; application to IOR and environmental transport.

Kalliat T. Valsaraj, Ph.D., Vanderbilt, 1983. Fate and transport of chemicals in the environment, dilute solution separation processes.

David M. Wetzel, F. J. Haydel Jr./Kaiser Aluminum Professor; Ph.D., Delaware, 1975. Modeling and application of crossflow vapor-liquid contacting units, wood-drying processes.

RENSSELAER POLYTECHNIC INSTITUTE

School of Engineering
Howard P. Isermann Department of Chemical Engineering

Programs of Study

The department offers the Doctor of Philosophy (Ph.D.) degree and two degrees at the master's level, Master of Science (M.S.) and Master of Engineering (M.E.). These programs are tailored to fulfill the varying educational needs of each graduate student, as well as reflect a dynamic balance of chemical engineering science and practice.

Ninety credits of graduate-level studies, including a dissertation, are required for a Ph.D.; these must include 12 credits representing four courses treating advanced topics in the core areas of the discipline: thermodynamics, kinetics, and transport phenomena, as well as mathematical analysis. The emphasis is on advanced study in a specialty with major focus on these four areas. The dissertation adviser is selected during the first semester of residence. A doctoral student must pass a comprehensive qualifying examination, prepare a dissertation proposal and the dissertation itself, and present and defend the dissertation. A student may proceed directly toward a Ph.D. without completing a master's degree, or, upon successful completion of the dissertation proposal exam, the student may be awarded an M.S. degree in the Optional Master's Program.

The M.S. degree is a thesis-based master's program, while the M.E. degree is a course-based master's program. Either degree is well-suited for professional entry. Both degrees require 30 credits of graduate-level studies, including the four required courses in the core areas of the discipline constituting 12 credit hours as described above. For the M.S. degree, the credits normally include 6 credits of thesis work and at least 15 credits of 600-level courses. The thesis adviser is selected during the first semester of residence. The M.E. program may be completed in one year of full-time study and is a terminal degree. Formal specialization options, including bioseparations and polymer engineering, are also offered within the context of the M.E. program. Terminal master's degree applicants are not eligible for financial aid.

Research Facilities

The department's research laboratories house myriad analytical and optical instrumentation, minicomputers, and workstations. Major instrumentation, which includes an FTIR spectrometer; a Raman spectrometer with microprobe; an atomic-force microscope; GC/mass spectrometers; a variety of high-power lasers; an X-ray fluorescence analyzer; an ion chromatograph; several HPLC systems; a high-vacuum CVD system; a UV-VIS spectrophotometer; high-temperature furnaces and reactors; a laser zee particle characterization system; several ellipsometers; a surface forces apparatus; polymer blending, compounding, and devolatilization equipment; and complete characterization equipment and extensive membrane filtration testing systems, makes the department's laboratories among the most comprehensively equipped university centers for research in chemical engineering. The computing facilities available at Rensselaer are extensive; a campuswide network provides access to more than 100 Sun SPARC workstations, 500 IBM RISC workstations, and several hundred personal computers. High-speed computation is accomplished by a parallel cluster of IBM RS-6000 workstations.

Financial Aid

Financial aid is available to competitive Ph.D. candidates in the form of Howard P. Isermann Fellowships, research or teaching assistantships, and scholarships. For 1998–99, stipends were a minimum of $15,000. In addition, full tuition is usually granted. Low-interest, deferred-repayment graduate loans are also available to U.S. citizens with demonstrated need. Dean's Topper Fellowships are very competitive awards available to well qualified students.

Cost of Study

Tuition for 1999–2000 is $665 per credit hour. Other fees amount to approximately $535 per semester. Books and supplies cost about $1700 per year.

Living and Housing Costs

The cost of rooms for single students in residence halls or apartments ranges from $3356 to $5298 for the 1999–2000 academic year. Family student housing with monthly rent of $592 to $720 is available.

Student Group

There are about 4,300 undergraduates and 1,750 graduate students representing all fifty states and more than eighty countries at Rensselaer.

Student Outcomes

Starting salaries in 1998 averaged about $55,000 for master's degree recipients and nearly $70,000 for doctoral degree recipients.

Location

Troy, Albany, and Schenectady form an upstate metropolitan area with a population of approximately 750,000. The area is a major center of government, industrial, research, and academic activity. Within easy driving distance are the headquarters of major research centers of some of the world's largest technology-based firms as well as the Wadsworth Laboratories and Axelrod Institute of the New York State Department of Health.

The University

Founded in 1824 and the first American college to award degrees in engineering and science, Rensselaer Polytechnic Institute today is accredited by the Middle States Association of Colleges and Schools and is a private, nonsectarian, coeducational university. Rensselaer has five schools—Architecture, Engineering, Management, Science, and Humanities and Social Sciences. The School of Engineering is ranked among the top twenty engineering schools nationally by the *U.S.News & World Report* survey and is ranked in the top ten by practicing engineers.

Applying

Admission applications and all supporting credentials should be submitted well in advance of the preferred semester of entry to allow sufficient time for departmental review and processing. The application fee is $35. Since the first departmental awards are made in February and March for the next full academic year, applicants requesting financial aid are encouraged to submit all required credentials by February 1 to ensure consideration.

Correspondence and Information

For further information about graduate work:

Graduate Admissions Coordinator
Isermann Department of Chemical Engineering
Rensselaer Polytechnic Institute
110 8th Street
Troy, New York 12180-3590

Telephone: 518-276-6379
Fax: 518-276-4030
E-mail: che_grad_info@rpi.edu
World Wide Web: http://www.rpi.edu/dept/chem-eng

For application forms and admissions information:

Director of Graduate Academic and Enrollment
    Services, Graduate Center
Rensselaer Polytechnic Institute
110 8th Street
Troy, New York 12180-3590

Telephone: 518-276-6789
E-mail: grad-services@rpi.edu
World Wide Web: http://www.rpi.edu

Rensselaer Polytechnic Institute

THE FACULTY AND THEIR RESEARCH

M. M. Abbott, Professor; Ph.D., Rensselaer. Thermodynamics.

E. R. Altwicker, Professor; Ph.D., Ohio State. Air pollution control, heterogeneous combustion and pollution formation.

G. Belfort, Professor; Ph.D., California, Irvine. Membrane separations engineering, biocatalysis, magnetic resonance flow imaging, fluid and mass transport, intermolecular forces.

B. W. Bequette, Associate Professor; Ph.D., Texas at Austin. Chemical process modeling, control, optimization, electronic materials processing, control of drug delivery systems.

H. R. Bungay III, Professor; Ph.D., Syracuse. Water resources, biochemical engineering.

T. S. Cale, Professor; Ph.D., Houston. Semiconductor materials processing, modeling and simulation of transport processes, applied mathematics.

S. M. Cramer, Professor; Ph.D., Yale. Biochemical engineering, chromatographic separations.

J. S. Dordick, Professor and Chairman; Ph.D., MIT. Biochemical engineering, biocatalysis, polymer science, bioseparation.

A. Fontijn, Professor; D.Sc., Amsterdam. Combustion, high-temperature kinetics, gas phase reactions.

W. N. Gill, Professor; Ph.D., Syracuse. Microelectronic materials processing, membrane separation systems, crystal growth phenomena.

M. E. Glicksman, Professor; Ph.D., Rensselaer. Transport phenomena of crystal growth.

R. T. Lahey, Professor; Ph.D., Stanford. Two-phase flow and boiling, heat transfer.

H. Littman, Professor; Ph.D., Yale. Fluidization, fluid-particle systems.

C. Muckenfuss, Professor Emeritus; Ph.D., Wisconsin. Kinetic theory, transport phenomena.

E. B. Nauman, Professor; Ph.D., Leeds (England). Reaction engineering, diffusion phenomena, polymer blends, plastics recycling.

J. L. Plawsky, Associate Professor; Sc.D., MIT. Optical, nonlinear and electrooptic, crystalline, and glassy materials.

H. C. Van Ness, Institute Professor Emeritus; D.Eng., Yale; PE. Thermodynamics.

P. C. Wayner Jr., Professor; Ph.D., Northwestern. Heat transfer, interfacial phenomena, thin dielectric coatings, nanostructured materials, microgravity transport processes.

RESEARCH AREAS

Advanced Materials and Microelectronics: Transport, chemical, and thermodynamic aspects of the production of advanced materials for the optical, electronic, and allied industries; modeling, optimization, and control of CVD of metals and dielectrics, chemical mechanic polishing, and spin-on aerogels for deep submicron devices.

Air Resources: Formation of air pollutants from incineration of heterogeneous wastes, fly ash reactivity and polychlorinated organics, modeling of pollutant formation in combustion.

Biocatalysis: Enzymatic catalysis under extreme conditions, enzymes in polymer synthesis and modification, combinatorial chemistry and biology.

Biochemical Engineering: Bioseparations, fundamental and applied aspects of membrane sorption and separation, displacement chromatography, control theory of biological processes, protein formulation and delivery.

Fluid Mechanics: Mechanics of fluidized beds, spouted beds, and bubbles; low Reynolds number hydrodynamics; kinetic theory; two-phase flow; surfactant behavior in organic-aqueous systems; magnetic resonance imaging of fluid flow.

Heat and Mass Transfer: Free convection stability, forced convection, fluid-to-particle heat transfer in fluidized and spouted beds, heat and mass transfer at interfaces and in porous media, multicomponent diffusion, diffusion and mixing in laminar flow systems, crystal growth phenomena, microgravity transport processes.

High-Temperature Kinetics: Kinetics of combustion reactions of metals, air toxics, hydrocarbons, explosives, and propellants over wide temperature ranges.

Interfacial Phenomena: Interfacial resistance to mass transfer, interaction of surface forces and interfacial convection, condensation and evaporation, distillation from ultrathin films, lubrication, pattern formation in dendritic growth, protein-solid interaction, self-assembled systems flows.

Meso- and Nano-Scale Engineering: Thin films, particles, phases and assemblies wherein the size of the system confers the properties of interest.

Process Control and Design: Impact of process design and scale up on process operability, batch control, nonlinear control, control of drug infusion systems, modeling and control of pharmaceutical reactions.

Thermodynamics: Vapor-liquid equilibria and heats of mixing of liquids, irreversible thermodynamics of multicomponent diffusion, fluid-phase equations of state.

Polymers: Blends and alloys, impact modification, spinodal decomposition, polymer reaction engineering, flash devolatilization, recovery of commingled scrap plastics by selective dissolution of membranes and microelectronics applications.

Separation Engineering: Membrane processes and chromatographic processes.

A gathering of the departmental graduate students on the steps of the Ricketts Building, home of Chemical Engineering.

An image scanning ellipsometer helps this graduate student focus with laser-like intensity on his research.

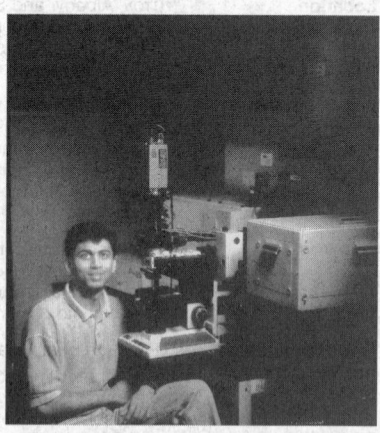

A graduate student with his research tool, a Raman microprobe spectrometer.

RICE UNIVERSITY

George R. Brown School of Engineering
Department of Chemical Engineering

Programs of Study

The Department of Chemical Engineering at Rice offers a program of graduate study leading to the Doctor of Philosophy (Ph.D.) degree. This program prepares students for careers in either the academic or the industrial sector. A nonthesis professional master's degree (M.Ch.E.), involving only course work, is also offered. The first year of the thesis degree program is flexible and allows entering students time to develop a sound basis in advanced areas of chemical engineering and to prepare them for the Ph.D. qualifying examinations. A large portion of the formal course work required for the Ph.D. degree is completed during the first year. Typically, students take eight of the twelve courses required for the Ph.D. degree during their first year at Rice. While there are no required courses, many students find advanced courses in thermodynamics, transport phenomena, applied mathematics, and reaction engineering to be excellent preparation for research and the qualifying examinations. Course selections are based on a student's individual preferences and previous educational background.

Entering graduate students attend a seminar series in which members of the faculty discuss their research programs. After attending these seminars, the students discuss possible Ph.D. thesis projects with individual faculty members with whom they have an interest in conducting research. Entering students normally affiliate with a research group before the beginning of the second semester, at which time work on the Ph.D. project begins. About four to five years of study are normally necessary to complete the Ph.D. degree requirements at Rice University.

Research Facilities

Abercrombie Laboratory houses the research, administrative, and support functions of the chemical engineering department. The Chemical Reaction Laboratory studies dynamic phenomena occurring in non-catalytic gas-solid reactions, using video microscopy and image processing. The Complex Fluids Dynamics Laboratory uses rheology and visualization to study microstructured fluids under flow. The Interfacial Phenomena Laboratory investigates the equilibrium phase behavior of surfactant systems, including emulsions and foams. The Wettability and Spreading Phenomena Laboratory explores the mechanisms for the wetting of fluids on solid substrates and the spreading of liquids on fluid substrates. The Thermodynamics and Transport Properties Laboratory investigates NMR relaxation, diffusion and viscosity of high-pressure hydrocarbons, NMR rock properties, and formation of gas hydrates.

Bioengineering faculty members with joint appointments have their offices and research laboratories in the George R. Brown Hall. The Biomedical Engineering Laboratory is devoted to cellular and tissue engineering projects, while the Biochemical Engineering Laboratory provides bioreactors and other facilities to study microbial, plant, and animal cell cultures.

Financial Aid

Graduate students in the Ph.D. program normally receive full financial support throughout the twelve-month year. The graduate stipend is $17,500 for 1999–2000. In addition, full tuition support is granted. There is no financial aid available for students in the nonthesis M.Ch.E. program.

Cost of Study

Tuition for 1999–2000 is $16,100. In addition, students must pay fees amounting to $255 per year and carry health insurance at an estimated annual cost of $620. Stipends are generally considered as taxable income.

Living and Housing Costs

Rent and utilities in University housing or nearby apartments range from $4200 to $6000 per year. Rice has built a new complex of graduate apartments that is scheduled to open in fall 1999. Information may be found on the World Wide Web at http://housing.rice.edu/rga/. Food and other expenses are estimated at $2800 per year. Living in Houston costs 17 percent less on average than in other large U.S. cities.

Student Group

Rice University has a student body of 1,400 graduate students. The Department of Chemical Engineering recruits highly motivated students with good academic records and the potential to carry out original, independent research. The department has 59 full-time graduate students (54 Ph.D., 5 M.Ch.E.) in 1999–2000. About 40 percent of the Ph.D. students are female, and 48 percent are international students.

Location

Houston is the fourth-largest U.S. city and home to a diverse blend of ethnic groups and cultures. It is the world's energy capital, an international business hub, and a center for the visual and performing arts. Surrounded by beautiful residential neighborhoods, the Rice campus is adjacent to the Texas Medical Center, the museum district, and a large municipal park.

The University

Rice University is a leading American research university—small, private, and highly selective—with outstanding graduate and professional programs. Rice was founded in 1912 and consists of seven major schools: Engineering, Natural Sciences, Humanities, Social Sciences, Music, Architecture, and Business Administration. Its modest size promotes interdisciplinary interactions and close working relationships between students and faculty members. Rice's programs are further enhanced through collaborations with researchers from the Texas Medical Center, the NASA Johnson Space Center, and other academic institutions. Rice has 450 full-time faculty members and 2,700 undergraduate students.

Applying

Selection is based on the student's academic record, teacher and adviser evaluations, and Graduate Record Examinations (GRE) scores. A minimum TOEFL score of 600 is required of international students. The deadline for applications is February 1, but students are advised to apply as early as possible because review of applications begins in early January.

Correspondence and Information

Chair, Graduate Admissions Committee
Department of Chemical Engineering
MS 362
P.O. Box 1892
Rice University
Houston, Texas 77251-1892

Fax: 713-285-5478
E-mail: cega@rice.edu
World Wide Web: http://www.ruf.rice.edu/~che/

Rice University

THE FACULTY AND THEIR RESEARCH

W. W. Akers, Professor Emeritus; Ph.D., Michigan, 1950.

Constantine D. Armeniades, Professor; Ph.D., Case Western Reserve, 1969. Polymer science, biomaterials.

Walter G. Chapman, Associate Professor; Ph.D. Cornell, 1988. Thermodynamics, statistical mechanics, molecular simulation.

Sam H. Davis, Professor; Sc.D., MIT, 1957.

Derek C. Dyson, Professor; Ph.D., London, 1966.

Jacqueline L. Goveas, Assistant Professor; Ph.D., Princeton, 1996. Nonequilibrium statistical mechanics, complex fluids.

J. David Hellums, A. J. Hartsook Professor Emeritus; Ph.D., Michigan, 1961.

Joe W. Hightower, Professor; Ph.D., Johns Hopkins, 1963.

George J. Hirasaki, A. J. Hartsook Professor; Ph.D., Rice, 1967. NMR-measured transport properties of fluids and rocks, wettability and spreading, foams and emulsions, aquifer remediation and enhanced oil recovery.

Riki Kobayashi, Louis Calder Professor Emeritus; Ph.D., Michigan, 1951.

Larry V. McIntire, E. D. Butcher Professor (joint with Bioengineering); Ph.D., Princeton, 1970. Biomedical engineering, rheology and fluid mechanics, tissue engineering.

Antonios G. Mikos, Associate Professor (joint with Bioengineering); Ph.D., Purdue, 1988. Biomaterials, targeted drug delivery, tissue engineering.

Clarence A. Miller, Louis Calder Professor; Ph.D., Minnesota, 1969. Interfacial phenomena, surfactants, foam, emulsions, aquifer remediation.

Matteo Pasquali, Assistant Professor; Ph.D., Minnesota, 1999. Microstructured liquids, free-surface flows, computational modeling of processing flows.

Marc A. Robert, Professor; Ph.D., Swiss Federal Institute of Technology, 1980. Thermodynamics, interfacial phenomena, thin films, random media.

Ka-Yiu San, Professor (joint with Bioengineering); Ph.D., Caltech, 1984. Biochemical engineering, reaction engineering.

Kyriacos Zygourakis, Professor; Ph.D., Minnesota, 1981. Chemical reactor design, cell migration, tissue engineering.

RESEARCH AREAS

The chemical engineering faculty has research interests in a number of areas. In some cases, students may have joint advisers.

Advanced materials: Biomaterials, complex fluids, microstructured liquids, polymers and composites, surfactants, foam, solid thin films, random media.

Bioengineering: Biochemical and metabolic engineering, biomedical engineering, drug delivery, tissue engineering.

Chemical reacting systems: Bioreactors, catalysis, coal combustion, design and optimization of pollution control reactors.

Computational chemical engineering: Computational modeling of processing flows, thin-film hydrodynamics, microstructured liquids, molecular simulation, transport in porous media, reactor design, cell population dynamics, tissue engineering.

Environmental: Surfactant/foam aquifer remediation, clean chemical processing.

Interfacial phenomena: Colloids, detergency, emulsion coalescence, foam, free-surface flows, microemulsions, thin films, wettability and spreading.

Petroleum: NMR rock and fluid properties, wettability and spreading, asphaltene deposition, foam EOR and diversion.

Thermodynamics: Asphaltene deposition, colloids, complex fluids, molecular simulation, statistical mechanics, gas hydrates.

Transport phenomena: Rheology of complex fluids, computational fluid mechanics, foam in porous media, correlation of NMR relaxation with diffusivity and viscosity, estimation of rock pore structure and fluid distribution with NMR.

STATE UNIVERSITY OF NEW YORK AT BUFFALO

Department of Chemical Engineering

Programs of Study

The department offers Ph.D., M.S., and M.Eng. degrees. Graduate research is geared primarily toward the Ph.D. program, which typically requires four to five years of full-time study. Students entering this program usually work directly toward the doctoral degree and do not obtain a master's degree in the process. On their way to completing their doctoral theses, students must take a total of 72 credit hours of graduate courses, pass a qualifying examination immediately following the second semester, and prepare and present a doctoral thesis proposal by the end of the second year. The M.S. degree typically requires eighteen to twenty-four months of full-time study and involves 30 credit hours of graduate course work and the completion of a master's thesis. The M.Eng. degree, a terminal professional degree, is typically completed in twelve to eighteen months and involves 30 credit hours of graduate course work and the completion of an M.Eng. project that is roughly equivalent to half of a master's thesis. Whenever possible, M.Eng. projects are conducted in collaboration with local industry. The course work requirements for the Ph.D. and M.S. degrees include four graduate core courses in the areas of chemical kinetics, thermodynamics, transport phenomena, and mathematical analysis. Chemical engineering electives may be substituted for two of the core courses in the M.Eng. program.

Specialization and research programs are in the frontiers of engineering science in core areas of chemical engineering, in the practical application of engineering science to chemical and environmental processes, and in new emerging areas of technology. Areas of activity include biochemical and biomedical engineering, catalysis and reaction engineering, ceramics, colloids, electronic materials, environmental engineering, interfacial phenomena, kinetics, molecular simulation, polymer processing and rheology, polymer science, process design, optimization and control, separations, statistical physics, surface science, thermodynamics, and transport phenomena.

Research Facilities

Experimental facilities include the Bioengineering Laboratory (blood cell adhesion and migration, vascular engineering, gene therapy, and tissue engineering), the Catalysis and Reaction Engineering Laboratory (catalysts, membrane reactors, kinetics, and adsorption), the Ceramic and Reaction Engineering Laboratory (synthesis of advanced ceramic materials by chemical vapor deposition and other techniques), the Colloid and Surface Science Laboratory (study of polymer solutions, micelles and microemulsions by laser light scattering, fluorescence spectrophotometry, and other techniques), the Electronic and Photonic Materials Laboratory (synthesis and processing of innovative microelectronic materials, diamond films, and nanoparticles by metalorganic vapor phase epitaxy and plasma-assisted chemical vapor deposition), and the Polymer Rheology and Processing Laboratory (processing of polymer solutions and melts by thermoforming and other techniques). Experimental work is pursued parallel to complementary theoretical studies in these laboratories, and independent theoretical programs also exist, including the Molecular Simulation Laboratory (computer experiments to understand how phase behavior correlates with molecular structure), the Process Design Laboratory (process synthesis and optimization, especially for pollution prevention, waste minimization, and the design of environmentally benign chemical processes and products), and the Transport Phenomena Laboratory (microscopic modeling of microstructured materials, biosensor surfaces, and biological pores). Computation is supported by a state-of-the-art computer facility within the department that was established through grants from the National Science Foundation and includes numerous Silicon Graphics and Sun Workstations and a four-processor Origin-class Silicon Graphics minisupercomputer. A University Sun supercomputer and the National Supercomputing Centers are also used.

Financial Aid

Admission to the Ph.D. program generally carries full support in the form of a full tuition scholarship plus a full-time stipend (at least $10,722.60 for the academic year plus additional summer stipend); students act as teaching assistants during the first year and research assistants in subsequent years. In addition, 1 or 2 Ph.D. students are admitted each year with Graduate Assistants in Areas of National Need (GAANN) fellowships that provide full tuition and a twelve-month stipend and are funded by a grant from the Department of Education for work in the area of environmentally benign chemical processes and products. Although the master's programs generally carry no stipend support, full or partial tuition scholarships are offered to highly qualified M.S. and M.Eng. applicants.

Cost of Study

Full-time tuition is $2550 per semester for New York State residents and $4208 per semester for nonresidents. These costs apply only to M.S. or M.Eng. students who are admitted without a tuition scholarship.

Living and Housing Costs

Graduate students generally live off campus, and the Off-Campus Housing Office maintains a file of housing accommodations in various price ranges. Housing and living costs in Buffalo and the surrounding suburbs are well below the national average.

Student Group

A number of student and student-faculty social and scientific activities are organized throughout the year by the Chemical Engineering Graduate Student Association.

Student Outcomes

Master's graduates typically take industrial positions locally and nationally. Ph.D. graduates typically obtain industrial positions locally and nationally as well as postdoctoral and tenure-track teaching positions at leading universities throughout the country and abroad. The department has an excellent track record in placing its advanced-degree recipients.

Location

The department is in the University's North Campus in the suburb of Amherst, New York, which is ranked as one of the most livable communities in the nation and the top community in safety. Cultural and recreational interests are served by world-class museums and art galleries, the Buffalo Philharmonic Orchestra, a host of local theaters, restaurants that offer excellent American and international cuisine, major-league sports teams, and the natural beauties of western New York, including Niagara Falls. The cultural and cosmopolitan center of Toronto is 2 hours away by car.

The University and The Department

The University at Buffalo (UB) is the flagship campus of SUNY, the single largest university system in the world. The Department of Chemical Engineering at UB is world class and holds special distinction as the only such department in the SUNY system. It has consistently been ranked among the nation's top thirty chemical engineering departments in the Gorman and National Research Council Reports. In two recent rankings of all chemical engineering departments in the world, conducted independently by the University of Pittsburgh and the Korean Advanced Institute of Science and Technology, the department was ranked in the top ten worldwide based on research publications per faculty and their impact on the field (as manifested by the quality of the scientific journals in which the papers were published and the number of citations that each paper received).

Applying

Applications and all supporting documents for admission in the fall semester should be received by February 1 to receive full attention; students are notified of the outcomes of their applications by April 15 or sooner. In rare cases, a very few students may be admitted in the spring semester, depending on the availability of positions. Applications for spring admission should be received by October 1.

Correspondence and Information

Graduate Admissions
Department of Chemical Engineering
Furnas Hall
State University of New York at Buffalo
Buffalo, New York 14260

Telephone: 716-645-2911
Fax: 716-645-3822
World Wide Web: http://www.eng.buffalo.edu/Departments/ce

State University of New York at Buffalo

THE FACULTY AND THEIR RESEARCH

Paschalis Alexandridis, Assistant Professor; Ph.D., MIT, 1994. Amphiphilic polymers, structured fluids, self-assembly, interfacial phenomena.

Stelios T. Andreadis, Assistant Professor; Ph.D., Michigan, 1996. Bioengineering, gene therapy, tissue engineering of genetically modified skin for wound healing.

Robert J. Good, Professor Emeritus; Ph.D., Michigan, 1950. Surface chemistry, intermolecular forces, surface tension, adhesion.

Ashish Gupta, Assistant Professor; Ph.D., UCLA, 1995. Process synthesis, simulation and optimization, pollution prevention, separations.

Vladimir Hlavacek, C. C. Furnas Eminent Professor; Ph.D., Institute of Chemical Technology (Prague), 1965. Ceramic and reaction engineering, analysis of complex chemical plants, simulation of separation units, analysis of industrial reactors.

David A. Kofke, Professor and Director of Undergraduate Education; Ph.D., Pennsylvania, 1988. Thermodynamics, statistical physics, molecular simulation.

Carl R. F. Lund, Professor and Chairperson; Ph.D., Wisconsin, 1981. Heterogeneous catalysis, chemical kinetics, reaction engineering.

T. J. Mountziaris, Professor and Director of Graduate Studies; Ph.D., Princeton, 1989. Electronic and photonic materials, biosensors, chemical kinetics, transport phenomena, reactor design, multiphase flows.

Sriram Neelamegham, Assistant Professor; Ph.D., Rice, 1995. Biomedical engineering, cell biomechanics, vascular engineering.

Johannes M. Nitsche, Professor and Director of Graduate Admissions; Ph.D., MIT, 1989. Fluid mechanics and transport phenomena, bioactive surfaces, biological pores, thermodynamics.

Eli Ruckenstein, Distinguished Professor and Member of the National Academy of Engineering; Ph.D., Polytechnic Institute (Bucharest), 1966. Catalysis, surface phenomena, colloids, emulsions, biocompatible surfaces and materials.

Michael E. Ryan, Professor and Associate Dean of Undergraduate Education; Ph.D., McGill, 1978. Polymer and ceramics processing, rheology, non-Newtonian fluid mechanics.

Mark T. Swihart, Assistant Professor; Ph.D., Minnesota, 1997. Chemical kinetics, chemical vapor deposition, reactor modeling, computational chemistry, particle nucleation and growth.

Carel J. van Oss, Adjunct Professor of Chemical Engineering and Geology and Professor of Microbiology; Ph.D., Paris IV (Sorbonne), 1955. Interfacial phenomena and cell interactions, van der Waals and hydrogen-bonding forces, electrokinetic phenomena.

Thomas W. Weber, Professor; Ph.D., Cornell, 1963. Process control, classical thermodynamics, adsorption.

Sol W. Weller, C. C. Furnas Professor Emeritus; Ph.D., Chicago, 1941. Catalysis, catalytic processes, kinetics.

TULANE UNIVERSITY

Department of Chemical Engineering

Program of Study

The Department of Chemical Engineering at Tulane is one of the oldest chemical engineering departments in the country. Graduate programs are available at the M.S. and Ph.D. levels, though the emphasis is on doctoral research. M.S. students receiving financial aid must complete 24 hours of approved graduate course work plus a thesis. Part-time students may select a nonthesis option, which requires 30 credits of approved graduate course work. Ph.D. candidates must complete 48 hours of approved graduate course work, pass a qualifying examination at the end of their first year, complete a thesis prospectus, and defend their thesis. The department offers an interdisciplinary degree program in bioengineering and biotechnology. One of five departments in the School of Engineering, the department has 9 total faculty members involved in teaching and/or research. The department promotes a comfortable environment for research, where one-on-one interactions between faculty members and students are highly encouraged.

Research Facilities

The department is extremely well equipped for research in advanced technologies, with fully equipped laboratories for work in bioengineering, biotechnology, catalysis, electrochemistry, materials engineering, colloid science, molecular simulations, polymers, and rheology. Extensive computational facilities are available in the department and the University. Tulane has also established the centralized Coordinated Instrumentation Facility (CIF) with the intent to operate and maintain sophisticated instrumentation for academic and industrial research. The University fosters collaborative and interdisciplinary research, and chemical engineering faculty members collaborate with faculty members from biology, chemistry, and the School of Medicine. University-operated research centers funded through federal and industrial grants include the Center for Bioenvironmental Research and the Regional Center for Research in Global and Environmental Change. These centers sponsor a number of research projects within the department.

Financial Aid

Financial support is awarded on the basis of academic merit. Assistance is available in the form of tuition waivers and/or stipends. Nearly all full-time graduate students receive a tuition waiver plus stipend. The graduate stipends offered by the chemical engineering department are competitive with those offered by other departments throughout the country. Fellowships are offered by the Graduate Division for highly qualified domestic candidates.

Cost of Study

Students receive scholarships to cover tuition, which was approximately $22,000 per year in 1998–99.

Living and Housing Costs

New Orleans enjoys a rich and diverse architectural history, which lends itself to many unique housing opportunities. This, coupled with a moderate cost of living, means a single student can anticipate an expenditure for the academic year of approximately $10,500, which includes room and board, health costs, and transportation, among other fees.

Student Group

The chemical engineering graduate student body is a diverse mixture of both American and international students. Approximately 25 percent of the students are women.

Location

New Orleans offers a rich variety of cultural and recreational opportunities. The New Orleans Philharmonic, the New Orleans Opera Association, and the New Orleans Jazz and Heritage Festival schedule major events and performing artists throughout the concert season. The city is renowned for its cuisine. Employment opportunities in chemical engineering are enhanced by the proximity to industry.

The University

Tulane is a private, nonsectarian University offering a wide range of undergraduate, professional, and graduate courses for men and women. It dates from 1834 with the formation of the Medical College of Louisiana. While located in a beautiful residential neighborhood adjoining Audubon Park and the Mississippi River, it is only 30 minutes away by streetcar or bus from the central business district. The University enrolls about 7,600 full-time and 1,900 part-time students each year. The University has recently undertaken an ambitious program to enhance its research activities in biotechnology, materials science, and environmental science. The chemical engineering department is expected to play a vital role in realizing these objectives.

Applying

Applicants with a B.S. from a recognized institution may be admitted to study for a graduate degree in chemical engineering if their academic credentials indicate the ability to pursue advanced study successfully. Applicants who have undergraduate degrees in the sciences (e.g., biology or chemistry) or other branches of engineering (e.g., biomedical engineering or mechanical engineering) are encouraged to apply but may be required to make up undergraduate chemical engineering course deficiencies upon admission. Every applicant for admission to the Graduate School is expected to take the GRE General Test. In order to be considered for the full financial aid package, applications for the fall semester should be received by February 1. Students interested in starting in the spring semester should submit their application by October 15.

Correspondence and Information

Graduate Advisor
Department of Chemical Engineering
Tulane University
New Orleans, Louisiana 70118
Telephone: 504-865-5772
Fax: 504-865-6744
E-mail: koc@mailhost.tcs.tulane.edu

Tulane University

THE FACULTY AND THEIR RESEARCH

Daniel C. R. De Kee, Professor; Ph.D., Montreal. Rheology, fluid mechanics, polymers. (e-mail: ddekee@mailhost.tcs.tulane.edu)

Richard D. Gonzalez, Herman and George R. Brown Professor of Chemical Engineering; Ph.D., Johns Hopkins. Synthesis and characterization of supported metal catalysts, fundamental studies in reactor design, in situ spectroscopic methods, reactions in organized media. (e-mail: gonzo@mailhost.tcs.tulane.edu)

Vijay T. John, Professor; D.Eng.Sci., Columbia. Biomimetic and nanostructured materials, interfacial phenomena, polymer-ceramic composites, surfactant science. (e-mail: vj@mailhost.tcs.tulane.edu)

Daniel J. I. Lacks, Associate Professor; Ph.D., Harvard. Molecular simulation, thermodynamics of condensed phases, dynamical processes in solids, physical properties of polymer materials, density functional theory. (e-mail: lacks@mailhost.tcs.tulane.edu)

Victor J. Law, Professor; Ph.D., Tulane. Modeling environmental systems, nonlinear optimization and regression, transport phenomena, numerical methods. (e-mail: law@mailhost.tcs.tulane.edu)

Brian S. Mitchell, Associate Professor; Ph.D., Wisconsin. Fiber technology, materials processing, composites. (e-mail: brian@mailhost.tcs.tulane.edu)

Kim C. O'Connor, Associate Professor; Ph.D., Caltech. Bioengineering, biotechnology, animal-cell technology, organ/tissue regeneration, recombinant protein expression. (e-mail: koc@mailhost.tcs.tulane.edu)

Kyriakos D. Papadopoulas, Professor; D.Eng.Sci., Columbia. Colloid stability, coagulation, transport of multiphase systems through porous media, colloidal interactions. (e-mail: pops@mailhost.tcs.tulane.edu)

Peter N. Pintauro, Professor; Ph.D., UCLA. Electrochemical engineering, membrane separations, electro-organic synthesis, environmental remediation. (e-mail: peter@mailhost.tcs.tulane.edu)

UNIVERSITY OF CALIFORNIA, IRVINE

Department of Chemical and Biochemical Engineering and Materials Science

Programs of Study

The Department of Chemical and Biochemical Engineering and Materials Science offers programs of study leading to the degrees of M.S. and Ph.D. in chemical and biochemical engineering. The M.S. and Ph.D. in engineering, with a concentration in materials science and engineering, are available through an interdepartmental graduate program administered by the Dean's Office in the School of Engineering.

Chemical and biochemical engineering focuses on bioremediation, cellular growth kinetics and regulation, materials science and engineering, optimization and control of reactors, protein engineering, recombinant DNA technology, biomedical engineering, and separations. Materials science focuses on semiconductor materials, processing and modeling, structure of materials, mechanics of solids, biomaterials and polymers, and advanced processing of materials.

About one year is required to complete the requirements for the Master of Science degree, either by course work or thesis option. The Ph.D. requires four to five years of study, during which time the following milestones must be achieved: a satisfactory score on the departmental preliminary examination; satisfactory research preparation; advancement to candidacy; completion of significant research investigation; and the submission and defense of an acceptable dissertation.

Research Facilities

For the program in chemical and biochemical engineering, the University has well-equipped laboratories for separation chemistry, recombinant cell technology, protein engineering, bioremediation, and process modeling, control, and integration. A central research area of highly instrumented and computer-interfaced bioreactor links basic biochemical engineering research to applied biotechnology. The facility contains a minicomputer connected to a microcomputer interfacing with various bioreactors and instruments for online measurements, analyses, and control.

For the program in materials science, the University has excellent research facilities for mechanical testing, spray atomization, sol-gel processing, and semiconductor materials processing and modeling. The facilities also include excellent materials characterization, electron microscopies, X-ray diffraction, and thermal, mechanical, and chemical analysis systems.

Financial Aid

Fellowships and teaching and research assistantships are available on a competitive basis. Except for students on visas, there are opportunities for part-time work in the engineering community of Orange County. With the same exception, financial aid may be obtained from UCI's Financial Aid Office.

Cost of Study

In 1999–2000, student fees are $1726 per quarter for California residents and an additional $3441 per quarter for nonresidents. These fees are subject to change.

Living and Housing Costs

On-campus housing is available. In 1999–2000, monthly apartment rents are from $277 to $725 for single students and from $554 to $1202 for married students and families. Early application is advised for on-campus housing. Privately owned apartments are available close to the campus, and many types of housing can be found in the surrounding communities of Santa Ana, Newport Beach, Costa Mesa, Irvine, Tustin, and Laguna Beach.

Student Group

Current campus enrollment is about 18,209, including 1,263 undergraduate and 309 graduate students in the School of Engineering.

Student Outcomes

Graduates of the degree programs offered within the department hold positions in academia or in the industrial or governmental sectors that involve the development of new technologies for the benefit of society. Many graduates take various professional positions in local, national, and international high-tech companies. Others obtain faculty appointments or research and development positions in their areas of specialization.

Location

The 1,510-acre UCI campus is in Orange County, 40 miles south of Los Angeles. Irvine is one of the nation's fastest-growing residential, industrial, and business areas, yet within view of the campus is a wildlife sanctuary; Pacific Ocean beaches are nearby. Residential areas range from the beach communities of Newport Beach and Laguna Beach to the socially and economically diverse urban centers of Santa Ana, Tustin, and Costa Mesa.

The University

One of the nine campuses in the University of California system, UCI now enrolls 3,638 graduate and professional students. The University offers graduate degrees through the Schools of Biological Sciences, Engineering, Fine Arts, Humanities, Physical Sciences, Social Ecology, and Social Sciences; the Graduate School of Management; the College of Medicine; and the Department of Information and Computer Science. The Department of Education offers courses and training leading to California teaching credentials.

Applying

Application forms and general information may be obtained by writing to the department or by visiting the department's Web site. The deadlines for applications are May 1 for the fall quarter, October 15 for the winter quarter, and January 15 for the spring quarter. Applicants who wish to be considered for fellowships or for teaching or research assistantships should apply by February 1. Applicants must submit official records covering all postsecondary academic work, three letters of recommendation, and official scores on the General Test of the Graduate Record Examinations. International students whose native language is not English must submit the results of the Test of English as a Foreign Language (TOEFL).

Correspondence and Information

Enrique J. Lavernia, Chair
Department of Chemical and Biochemical Engineering
    and Materials Science
916 Engineering Tower
School of Engineering
University of California
Irvine, California 92697-2575

Graduate Admissions Coordinator
916 Engineering Tower
University of California
Irvine, California 92697-2575

Telephone: 949-824-3786
E-mail: nvcarter@uci.edu
World Wide Web: http://www.eng.uci.edu/cbe

University of California, Irvine

THE FACULTY AND THEIR RESEARCH

Ying C. Chang, Assistant Professor of Chemical and Biochemical Engineering and Materials Science; Ph.D., Stanford. Molecular engineering, biophysical chemistry, polymer interfacial chemistry and physics, organic vapor deposition and thin film fabrication.

Nancy DaSilva, Associate Professor of Biochemical Engineering; Ph.D., Caltech. Recombinant cell technology.

James C. Earthman, Associate Professor of Materials Science Engineering; Ph.D., Stanford. High-temperature fracture mechanisms, cavitation processes in superplastic materials, numerical modeling of deformation and damage processes, mechanical behavior and damage mechanisms in biomedical materials, corrosion prevention using regenerative biofilms, novel damage monitoring techniques, automated materials testing and analysis.

Steven C. George, Assistant Professor of Biochemical Engineering; Ph.D., Northwestern. Physiological modeling, tissue engineering, pulmonary gas exchange, pulmonary metabolism and transfer properties of nitric oxide.

G. Wesley Hatfield, Professor of Microbiology and Molecular Genetics, Biological Sciences, and Biochemical Engineering; Ph.D., Purdue. Molecular mechanics of biological control systems.

Juan Hong, Professor of Biochemical Engineering; Ph.D., Purdue. Biochemical reaction processes and separation processes.

Enrique J. Lavernia, Professor of Materials Science Engineering and Chair of the Department; Ph.D., MIT. Manufacturing, composite materials, nanostructured materials, modeling, synthesis by nonequilibrium methods.

Henry C. Lim, Professor of Chemical and Biochemical Engineering; Ph.D., Northwestern. Modeling, optimization and control of bioreactors, cellular regulations and kinetics, bioreaction engineering.

Martha L. Mecartney, Associate Professor of Materials Science Engineering; Ph.D., Stanford. Sol-gel synthesis; analytical electron microscopy; grain boundaries in oxides, crystalline oxide thin films.

Farghalli A. Mohamed, Professor of Materials Science Engineering; Ph.D., Berkeley. Mechanical behavior of engineering materials (metals, composites, ceramics), correlation between behavior and microstructures, creep, superplasticity, strengthening and fracture mechanisms.

Frank G. Shi, Assistant Professor of Chemical and Materials Engineering; Ph.D., Caltech. Semiconductor processing and modeling, interconnecting and packaging polymers, amorphous and nanocrystalline materials, thermodynamics and kinetics of nucleation/crystallization/glass transition processes.

Vasan Venugopalan, Assistant Professor of Chemical and Biochemical Engineering and Materials Science; Ph.D., MIT. Biomedical laser applications; fundamentals of laser-induced thermal, mechanical, and radiative transport phenomena.

UNIVERSITY OF CALIFORNIA, RIVERSIDE

Marlan and Rosemary Bourns College of Engineering
Department of Chemical and Environmental Engineering

Programs of Study

The Marlan and Rosemary Bourns College of Engineering offers Master of Science (M.S.) and Doctor of Philosophy (Ph.D.) degrees in chemical and environmental engineering (CEE). These degree programs emphasize "environmentally friendly" chemical engineering and the application of chemical engineering principles to environmental engineering problems. Because of the complementary skills and expertise of the faculty and the cooperating faculty members in other departments, research emphases are collaborative and interdisciplinary. Degree programs and research are concentrated in four major areas: biochemical engineering, environmental biotechnology, air quality systems engineering, and water quality systems engineering. The intent of the program is to deepen the student's understanding of fundamental principles in chemical and environmental engineering and to increase his or her knowledge of the needs and directions in the chosen field of study. The master's degree program is normally completed in two years, but it can be finished in one year by well-prepared students. Students may either pass a qualifying examination or complete a research thesis. The normal time to complete the Ph.D. degree is three years for students holding an M.S. degree in chemical or environmental engineering and five years for those who enter the program without an M.S. degree. A typical program includes completing course work, passing a preliminary examination, independent research, presenting a dissertation proposal and passing the associated qualifying exam, and completing and presenting the dissertation research.

Research Facilities

The CEE program is housed in the recently opened engineering complex, Bourns Hall. CEE program space consists of 2,000 square feet of teaching lab space, 8,500 square feet of research lab space, and 1,600 square feet of computing lab space.

Research in the Environmental Engineering Laboratory/Particulate Processes Laboratory focuses on solid-solid and water-solid interactions related to colloidal particle kinetics, colloidal particle transport, and adsorption phenomena. The laboratory is also developing process control systems for particulate removal unit processes. The Environmental Processes and Control Laboratory focuses on water and soil remediation, treatment process control, and developing new technologies for contaminant destruction, containment, and control. A 400-square-foot area is reserved for larger bench-scale laboratory studies and pilot plant studies. The Biofiltration Laboratory is devoted to research on the biological treatment of contaminated waste air and biodegradation of xenobiotics. The Biosensors Laboratory supports research on biosensors. A number of sensors are being developed for biotechnological, medical, and environmental monitoring and control applications. The Biochemical Engineering Laboratory is devoted to the study and characterization of proteins, cells, and DNA for possible use in controlled biodegradation of xenobiotic compounds.

The Air Quality Laboratories are associated with the College of Engineering Center for Environmental Research and Technology (CE-CERT). CE-CERT is a center for collaborative research by university, industry, and regulatory agencies on environmental problems. CE-CERT is housed in a 36,000-square-foot office and laboratory complex located two miles from the University of California, Riverside (UCR), campus in an industrial park. The laboratories at CE-CERT have been designed and developed to address air pollution and technology issues. Primary laboratories at CE-CERT include an atmospheric processes laboratory, vehicle emissions research laboratory, advanced vehicle engineering laboratory, environmental modeling laboratory, pollutant analysis laboratory, and stationary source evaluation laboratory. Each of these laboratories is a state-of-the-art test facility, and a number of the labs, especially the vehicle emissions research laboratory, contain equipment that is unique to a university research facility.

Financial Aid

Fellowships are awarded by the Graduate Division on a competitive basis, with stipends up to $15,000 for the nine-month academic year. These awards include payment of all assessed registration fees. Department of Chemical and Environmental Engineering teaching assistantships are also available. A half-time appointment as a teaching assistance carried a stipend of $13,329 for the 1998–99 academic year. Research assistantships on supported programs are also available to well-prepared students.

Cost of Study

For 1998–99, California residents paid $4861 a year in tuition fees. Nonresidents were charged an additional tuition fee of $3128 per quarter. These amounts are subject to change.

Living and Housing Costs

Riverside offers graduate students one of the lowest costs of living of any city with a University of California campus. The cost of room and board in the residence halls averaged $6444 in 1998–99. The University owns 268 houses that are available to married students and single students with children. These houses rent for between $430 and $470 per month. Rents for approximately 150 apartments and 92 suites that are available for single students range from $300 to $775 per month. Off-campus housing is available within walking distance of campus.

Student Group

The chemical and environmental engineering graduate program begins offering classes in winter 1999 and is admitting its first students in the 1999–2000 academic year. The graduate admission committee seeks to expand the current enrollment, admitting up to 50 talented and hardworking students in the next five years.

Student Outcomes

The University of California, Riverside, and the College of Engineering are working with industry in the Inland Empire to establish a strong technical workforce in this part of southern California. The Department of Chemical and Environmental Engineering also seeks to place graduates in top industrial, government, and academic positions worldwide.

Location

Riverside is located between the golden coastline of Los Angeles and Orange Counties (40 miles west), the desert community of Palm Springs (50 miles east), the San Bernardino Mountains (20 miles north), and San Diego (90 miles south). The location offers excellent opportunities for skiing, hiking, and biking. In addition, the climate offers the opportunity to take part in outdoor activities throughout the year. Riverside is a community of 250,000 people, with excellent recreational facilities, a symphony orchestra, an opera association, a community theater, and several free community events.

The College

Founded in 1989, the Marlan and Rosemary Bourns College of Engineering stresses interdisciplinary collaboration. Faculty members and students work together inside Bourns Hall, a three-story, two-building architectural showpiece that opened in 1994.

Applying

Application deadlines for domestic applicants are May 1 for the fall quarter, September 1 for the winter quarter, and December 1 for the spring quarter. Deadlines for international applicants are February 1 for the fall quarter, July 1 for the winter quarter, and October 1 for the spring quarter. To receive full consideration for financial support for fall, all application materials must be received by January 5. Applications postmarked after the published deadline are deferred to the following quarter. The University does not grant waivers of the graduate application fee. Specific requirements for admission can be found on the department's Web site, listed below. The Graduate School Application, an application fee of $40, official transcripts, three letters of recommendation, and official GRE and TOEFL scores (if applicable) should be submitted directly to the department. When determining acceptance, faculty members look for high GRE scores, and high GPAs and prefer a strong research background in students applying for this program.

Correspondence and Information

Chemical and Environmental Engineering
    Graduate Student Affairs
Bourns College of Engineering
University of California, Riverside
Riverside, California 92521

Telephone: 909-787-2423
Fax: 909-787-2425
E-mail: grad-adm@ee.ucr.edu
World Wide Web: http://www.engr.ucr.edu/chemical

University of California, Riverside

THE FACULTY AND THEIR RESEARCH

Wilfred Chen, Assistant Professor; Ph.D., Caltech. Environmental biotechnology, microbial engineering.

Marc Deshusses, Assistant Professor; Ph.D., Swiss Federal Institute of Technology. Biofiltration, bioremediation, mathematical modeling of biochemical systems.

James Lents, Adjunct Professor; Ph.D., Tennessee Tech. Air pollution policy, global air quality.

Mark Matsumoto, Professor and Department Chair; Ph.D., California, Davis. Wastewater treatment, soil remediation, hazardous wastes.

Ashok Mulchandani, Associate Professor; Ph.D., McGill. Biosensors, biochemical engineering.

Joseph Norbeck, Professor, Yeager Families Professor of Environmental Engineering, and Director of the College of Engineering—Center for Environmental Research and Technology; Ph.D., Nebraska. Mobile and stationary source emissions and control, alternative fuel development.

Akula Venkatram, Professor; Ph.D., Purdue. Air quality modeling, atmospheric dispersion.

Anders Wistrom, Assistant Professor; Ph.D., California, Davis. Particulate process mechanisms, water/wastewater treatment systems.

Yushan Yan, Assistant Professor; Ph.D., Caltech. Zeolitic membranes, catalytic systems.

RESEARCH FACULTY

William Carter, Research Chemist; Ph.D., Iowa. Gas-phase atmospheric reactions and mechanisms, secondary ozone formation.

Thomas Durban, Assistant Research Engineer; Ph.D., California, Riverside. Renewable fuels, advanced vehicle design, vehicle emissions.

Dennis Fitz, Assistant Research Engineer; M.S., California, Riverside. Measurement, speciation, and characterization of air pollutants.

COOPERATING FACULTY MEMBERS

Christopher Amrhein, Associate Professor, Environmental Sciences; Ph.D., Utah State. Soil physical chemistry, salt-affected soils, redox processes.

Michael Anderson, Associate Professor, Environmental Sciences; Ph.D., Virginia Tech. Water characterization, fate of inorganic and organic chemicals in soils.

Janet Arey, Professor, Environmental Sciences; Ph.D., Michigan. Atmospheric chemistry, air pollution.

Andrew Chang, Professor, Environmental Sciences; Ph.D., Purdue. Land application of sludge, water reuse, phosphorus soil chemistry, pollutant retention in soils.

David Crohn, Assistant Professor, Environmental Sciences; Ph.D., Cornell. Constructed wetlands treatment, biosolids management, composting.

David Crowley, Associate Professor, Environmental Sciences; Ph.D., Colorado State. Soil microbiology, bioremediation, microbial ecology.

William Frankenberger, Professor, Environmental Sciences; Ph.D., Iowa State. Bioremediation of metals.

William Jury, Professor, Environmental Sciences; Ph.D., Michigan. Soil physics, fate and transport of pollutants in groundwater systems.

Marylynn Yates, Professor, Environmental Sciences; Ph.D., Arizona. Water and wastewater microbiology, pathogen contamination.

Paul Ziemann, Assistant Professor, Environmental Sciences; Ph.D., Penn State. Atmospheric aerosol chemistry.

UNIVERSITY OF CALIFORNIA, SAN DIEGO

Department of Mechanical and Aerospace Engineering
Chemical Engineering Program

Programs of Study	The Chemical Engineering Program within the Department of Mechanical and Aerospace Engineering offers M.S. and Ph.D. degrees in engineering science with a specialization in chemical engineering. The curricula emphasize education in broad principles and fundamentals that provide a common foundation for all engineering subspecialties.

The M.S. program extends and broadens an undergraduate background and equips practicing engineers with fundamental knowledge in their fields. The degree may be terminal or may be obtained on the way to earning the Ph.D. Two plans of study are offered, both requiring successful completion of 48 quarter units of credit. Plan I is a combination of course work and research, culminating in the preparation of a thesis; Plan II involves course work only and requires a comprehensive examination.

The Ph.D. program prepares students for careers in research and/or teaching in their area of specialization. There are no formal course requirements; however, all students, in conjunction with their advisers, develop course programs that prepare them for the departmental Ph.D. qualifying examination, which tests students' capabilities in four areas of specialization and ascertains their potential for independent study, dissertation research, and completion and defense of that research.

Research Facilities
Among the research laboratories and equipment within the department are a cluster of laboratories devoted to biochemical engineering, three low-turbulence wind tunnels and two water tunnels for boundary-layer and low-speed aerodynamics studies, a two-phase flow facility, a stratified flow channel facility, a counterflow spray combustion apparatus, spectroscopic equipment for high-temperature gasdynamics, a Rheometrics fluid rheometer, a molecular beam facility, ESCA/XPS and FTIR surface spectroscopies, a facility to study photoelasticity and micromechanics of granular materials, a facility in the Center for Magnetic Recording Research to study lubrication and chemical materials aspects of magnetic disk and tape storage devices, several high-speed microcinematographic facilities; and two sets of major electron microscope facilities on the campus of the University of California, San Diego (UCSD).

The department maintains several CAD/CAM and computational fluid dynamics laboratories. Campus computing facilities include the CRAY C90 supercomputer at the San Diego Supercomputer Center.

Financial Aid
Financial aid is available in the form of fellowships, traineeships, teaching assistantships, and research assistantships. The department attempts to support all full-time graduate students, especially at the Ph.D. level. Award of financial support is competitive, and stipends range from $5000 to a maximum of $20,000 for the academic year, plus tuition and fees. While several types of support are available, the most common is a half-time research assistantship that provides $16,140 during the twelve-month academic year plus tuition and fees. Funds for support of international students are extremely limited, and the selection process is highly competitive.

Cost of Study
In 1999–2000, full-time students who are California residents pay an estimated $1630 per quarter in registration and incidental fees. Nonresidents pay an estimated total of $5100 per quarter for registration, tuition, and incidental fees. There is a reduced-fee structure for students enrolled on a half-time basis. Costs are subject to change.

Living and Housing Costs
UCSD provides 1,200 apartments for graduate students. Current monthly rates range from $320 for a single student in a shared apartment to $670 for a family. There is also a variety of off-campus housing in the surrounding communities. Prevailing rates range from $400 per month for a room in a private home to $950 or more for a two-bedroom apartment. Information in this regard may be obtained from the UCSD Housing Office.

Student Group
Current campus enrollment is about 18,825; of this number, 2,200 are graduate students.

Location
The 1,200-acre campus spreads from the seashore, where the Scripps Institution of Oceanography is located, across a large wooded portion of the Torrey Pines Mesa overlooking the Pacific Ocean. To the east and north lie mountains, and to the south are Mexico and the almost uninhabited seacoast of Baja California.

The University
One of nine campuses in the University of California System, UCSD comprises the General Campus, the School of Medicine, and the Scripps Institution of Oceanography. Established in La Jolla in 1960, it is one of the newer campuses but in this short time has become one of the major universities in the country, with particular strengths in the physical and biological sciences.

Applying
A minimum GPA of 3.0 (on a 4.0 scale) is required for admission. All applicants are required to take the GRE General Test. International applicants whose native language is not English are required to take the TOEFL and obtain a minimum score of 550. In addition to test scores, applicants must submit a completed Graduate Admission and Award Application, all official transcripts (English translation must accompany official transcripts in other languages), a statement of purpose, and three letters of recommendation. The deadline for filing applications for international applicants and those requesting financial assistance is January 31. The deadline for domestic applicants not requesting financial assistance is May 31.

Correspondence and Information
Department of Mechanical and Aerospace Engineering
Irwin and Joan Jacobs School of Engineering
University of California, San Diego
La Jolla, California 92093-0413
Telephone: 858-534-4387
Fax: 858-534-1730
World Wide Web: http://www-mae.ucsd.edu/

University of California, San Diego

THE FACULTY AND THEIR RESEARCH

P. C. Chau, Professor of Chemical Engineering; Ph.D., Princeton. Catalysis, biochemical engineering.

R. K. Herz, Associate Professor of Chemical Engineering; Ph.D., Berkeley. Heterogeneous catalysis, chemical reaction engineering.

D. R. Miller, Professor of Chemical Engineering; Ph.D., Princeton. Gas-surface interactions, gasdynamics.

K. Seshadri, Professor of Chemical Engineering and Fluid Mechanics; Ph.D., California, San Diego. Combustion, fluid mechanics, applied mathematics.

J. B. Talbot, Professor of Chemical Engineering and Materials Science; Ph.D., Minnesota. Corrosion, electrodeposition, electrophoretic deposition, electrochemical transport phenomena.

Students matriculating for a graduate degree in chemical engineering may select a thesis adviser from among any of the faculty members in the AMES department. Faculty members with interests in areas that are relevant to chemical engineers are listed below.

J. A. Frangos, Associate Professor of Bioengineering; Ph.D., Rice. Mechanisms of shear-induced activation of endothelial membrane function, bone tissue engineering, muscle tissue engineering, signaling by mechanical stimulation in plants.

C. H. Gibson, Professor of Engineering Physics and Oceanography; Ph.D., Stanford. Turbulent flows, mixing, and diffusion in oceanography, astrophysics, combustion, and chemical engineering.

D. A. Gough, Professor of Bioengineering; Ph.D., Utah. Mass transfer in biological systems, biotechnology.

J. C. Lasheras, Professor of Fluid Mechanics; Ph.D., Princeton. Turbulence, two-phase flows combustion.

J. M. McKittrick, Associate Professor of Materials Science; Ph.D., MIT. Experimental materials science, processing of ceramic materials.

B. O. Palsson, Professor of Bioengineering; Ph.D., Wisconsin. Hematopoietic tissue engineering, stem cell technology, bioreactor design, retroviral gene transfer for gene therapy, metabolic dynamics and regulation, whole cell simulators, metabolic engineering.

C. Pozrikidis, Professor of Fluid Mechanics; Ph.D., Illinois at Urbana-Champaign. Fluid mechanics, applied mathematics.

F. A. Williams, Professor of Engineering Physics and Combustion; Ph.D., Caltech. Flame theory, combustion and turbulent flows, fire research, and other areas of combustion.

UNIVERSITY OF DETROIT MERCY

College of Engineering and Science
Department of Chemical Engineering

Programs of Study	The Department of Chemical Engineering offers graduate programs leading to the degrees of Master of Engineering (M.E.) and Doctor of Engineering (D.E.). Students with degrees in related disciplines may become candidates for the M.E. degree upon completion of a number of upper-level courses (usually about eight courses). Thirty credits are required for the M.E. degree, which may include 6 credits of thesis research. Both the thesis and nonthesis options are available. Most graduate courses are offered in the evening to accommodate working professionals. Full-time students generally complete the master's degree program within two years. The D.E. degree requires 51 credits of course work past the baccalaureate; in addition, a minimum of 36 credits of dissertation research is required. The department specializes in polymer engineering and in environmental engineering. Graduate students often conduct research in conjunction with the Polymer Institute, a world-renowned organization within the College of Engineering and Science.
Research Facilities	Research facilities are available within the department, at the Polymer Institute and at the Center of Excellence in Environmental Science and Engineering (all units of the College of Engineering and Science). Most of the research facilities are polymer-oriented and include extruders, reactors, mixers, and reaction-injection molding machines. For polymer testing and characterization, DSC, TGA, FTIR, tensile, flexural, and rheometric capabilities are available. In other areas, air pollution research capabilities and electrochemical engineering facilities are available.
Financial Aid	Most financial aid is derived from research projects and is awarded to full-time students on a competitive basis. Most students receive awards consisting of a stipend in the range of $800 to $900 per month plus tuition. For exceptionally well qualified students, a small number of Frisch Fellowships, which start at $1000 per month, are also awarded. In addition, the Scholarship and Financial Aid Office accepts applications for grants, loans, and work-study assistance. Aid includes the Michigan Tuition Grant (for Michigan residents only), Federal Work-Study, and a variety of loans. The University also accepts third-party payments from employers and government agencies and offers payment plans of its own. For information regarding financial aid programs, students should call 313-993-3350.
Cost of Study	Tuition in 1999–2000 is $545 per credit hour. Registration fees are $100 for full-time students.
Living and Housing Costs	Housing is available on campus. Double-occupancy rates range from $1410 to $3280. Single-occupancy rates range from $2420 to $2720. The University offers a meal plan at a cost of approximately $1085. All rates are for a sixteen-week term. For more information, students should call the Residence Life Office at 313-993-1230.
Student Group	There are approximately 15 chemical engineering graduate students, most of whom are enrolled full-time. Many of the part-time students are professionals working for automotive companies and their suppliers. Most of the full-time students receive tuition and stipend support. Approximately 6,700 students attend classes on three UDM campuses located in northwest and downtown Detroit.
Location	Students enjoy a variety of activities offered on campus and throughout the metropolitan Detroit area, including sports, theater, concerts, and more.
The University and The Department	As Michigan's largest Catholic university, the University of Detroit Mercy has an outstanding tradition of academic excellence firmly rooted in a strong liberal arts curriculum. This tradition dates back to the formation of two Detroit institutions, the University of Detroit, founded in 1877 by the Society of Jesus (Jesuits), and Mercy College of Detroit, founded in 1944 by the Religious Sisters of Mercy of the Americas. In 1990, these schools consolidated to become the University of Detroit Mercy. Today, UDM offers more than 120 majors and programs in nine different schools and colleges and is widely recognized for its programs in engineering, law, dentistry, nursing, and architecture. Faculty members are known for their excellence; more than 90 percent have a Ph.D. or comparable terminal degree. The chemical engineering program has a distinguished history dating from its inception in the 1930s. Many of the department's graduates have careers within the automotive industry concentrated in southeastern Michigan. Because of the importance of polymers in automotive manufacture, the department has strong links with industries engaged in the fabrication of plastics, foams, adhesives, elastomers, coatings, and sealants.
Applying	Applications for admission normally should be completed at least six weeks before the beginning of a term. Applications for financial aid should be submitted by April 1. International students are urged to complete their applications at least three months before classes begin. Admission requirements are a bachelor's degree from an accredited college; a B average in the total undergraduate program and in the proposed field of study; and, normally, an undergraduate major or the equivalent in the proposed field. Official transcripts are required from all colleges attended. Applicants with less than a B average who present other evidence of ability to perform graduate-level work may be admitted as probationary students upon the recommendation of the director of the program.
Correspondence and Information	Department of Chemical Engineering University of Detroit Mercy P.O. Box 19900 Detroit, Michigan 48219 E-mail: engineering@udmercy.edu

University of Detroit Mercy

THE FACULTY AND THEIR RESEARCH

Geoffrey Prentice, Professor and Chair; Ph.D., Berkeley. Electrochemical engineering, current distribution, corrosion, electroorganic synthesis, plating.

Kurt C. Frisch, Professor; Ph.D., Columbia. Synthesis and characterization of polymers, applications of polyurethanes, polymer alloys, IPNs.

Thomas Hamade, Associate Professor; Ph.D., Wayne State. Air pollution, electrostatic air filtration, electret polymers, VOC emissions.

Daniel Klempner, Research Professor; Ph.D., SUNY at Albany. Recycling of thermosetting composites, recycling of commingled thermoplastics, polymer alloys, IPNs.

Jiri Kresta, Research Professor; Ph.D., Czechoslovak Academy of Science. Polymerization kinetics and catalysis, high-performance composites, structure-property relationships in polymers.

Soo-Il Lee, Professor; Ph.D., Osaka City (Japan). Interactive computer interfacing, process control of plastics operations.

Vahid Sendijarevic, Research Professor; Ph.D., Zagreb (Yugoslavia). Polymer encapsulation of hazardous wastes; heat-resistant, isocyanate-based polymers; polymers for biomedical applications.

Sung-Kuk Soh, Associate Professor; Ph.D., New Hampshire. Performance of polymer composites, recycling of thermosetting composites.

UNIVERSITY OF FLORIDA

Department of Chemical Engineering

Programs of Study

The Department of Chemical Engineering offers the Master of Science (M.S.), the Master of Engineering (M.E.), and the Doctor of Philosophy (Ph.D.). The principal requirements for the M.S. degree include 30 graduate-level semester hours of course work distributed as follows: 12 semester hours in the basis of chemical engineering courses, 9 additional hours in the engineering science core area, 1 semester hour of graduate seminar for each academic semester in residence, and individual work on a research thesis for 6 semester hours. The program requires the submission of a thesis approved by the student's Supervisory Committee. The M.E. degree is a nonthesis program intended for students whose B.S. degree is not from an accredited chemical engineering curriculum. Depending upon their prior education, students are generally required to take a set of undergraduate courses in chemical engineering. In addition, the program requires a minimum of 32 graduate-level semester hours. The Ph.D. degree is primarily a research program, the granting of which is based essentially on general proficiency and distinctive attainments in chemical engineering, particularly on the demonstrated ability to conduct an independent investigation as exhibited in a required doctoral dissertation. The formal requirements for the Ph.D. degree include successful completion of a qualifying examination before the end of the second semester in residence, admission to candidacy by means of successful completion of a written examination composed of the candidate's objectives and achievements toward his or her doctoral dissertation plus an oral examination based on the written part and related areas, submission of a written dissertation based on original research, passing a dissertation defense examination, and the completion of 90 graduate-level semester hours of course work, including those required by the M.S. degree.

Research Facilities

The Department of Chemical Engineering is housed in a 51,000-square-foot building, much of which is devoted to research. There are an extensive network of twenty-eight Hewlett-Packard workstations, several DEC workstations, more than twenty Pentium-based PCs, and access to a powerful University-wide computing system (HSR parallel mainframe, n-cube, IBM 3090). The department's research equipment includes a Spire II-VI MOCVD system polaron CV profiler, a Japan oxygen III-V MOCVD system, a Keithley Hall measurement system, four custom CVD systems, a Mellon Bridgman crystal grower, a JY-Ramanor 1000 laser Raman spectrometer, and a scanning electron microscope. Processing and testing equipment for semiconductor device fabrication and testing include a CHA electron beam metal deposition system, a Plasma Therm plasma-enhanced chemical vapor deposition system, a Karl Zuss contact aligner, a JEOL electron beam direct-write system, an HP 4145 semiconductor parameter analyzer, an HP 4284 precision LCR meter, a 50-MHz–40-GHz HP 8722c network analyzer, an HP 4140 PA meter/DC voltage source, and a cascade probe station. The department has Instron, Rheometrics, and Haake instruments for rheological experiments. Also available are a Brabender and a Killion extruder with die attachments for coextrusion, as well as pressure-jump, temperature-jump, and stopped-flow instruments. Some laboratories are equipped with low-angle light scattering, VPO, HTHPLC, ion chromatography, and UV-visible spectroscopy equipment. Two ultrahigh vacuum characterization systems are equipped with XPS, AES, LEED, ELS, ISS, TPD, SIMS, and ESD instrumentation. There is a complete electrochemical laboratory for in situ ellipsometry (Gaertner) and impedance instrumentation (Solartron 1250/1286). Other major instruments include a System 7 DSC/TGA, a Setaram microcalorimeter, a combustion calorimeter, a Brookhaven quasielastic light scattering spectrometer, a Langmuir film balance, and a contact angle goniometer.

Financial Aid

The department offers research assistantships and/or fellowships to most M.S. and Ph.D. students. Graduate assistantships pay very attractive stipends plus tuition and are awarded to students for teaching and/or research duties, depending upon the source of funds available and student desires and qualifications. Provided satisfactory progress is maintained, support is continued for the minimum number of credits to reach the student's degree objective. After completion of the minimum credits, support is normally provided in response to requests by the student and his or her research director. The department awards special fellowships as funds from corporate and governmental sponsors become available. Students can also be nominated by the department for a College of Engineering Fellowship and a University Graduate Minority Fellowship. External fellowships below the department's stipend are usually supplemented up to the level of a graduate assistantship stipend if funds are available.

Cost of Study

The registration fee for most graduate course work is $129.01 per credit hour for Florida residents and $434.40 per credit hour for out-of-state students. Tuition waivers are granted to most students holding graduate assistantships and fellowships.

Living and Housing Costs

Rents for apartments provided by the University for single graduate students and those with families begin at $250 per person per month. There are also many apartment complexes in the area, with rent for one-bedroom apartments starting at approximately $250 per month, excluding utilities.

Student Group

The total enrollment at the University of Florida is more than 40,000. The Department of Chemical Engineering currently has 75 graduate students.

Student Outcomes

The diversity of research activities in the department has produced an equally diverse record of placement of its graduates. Recent alumni with graduate degrees are currently working in academia (New Mexico State University), industry (Intel, Texas Instruments, DuPont, Procter & Gamble, International Paper, Hewlett-Packard, Monsanto), and government laboratories (Argonne National Lab) or attending law and medical schools.

Location

The University of Florida is located in sunny Gainesville, consistently ranked one of the top ten best places to live in the U.S. by *Money* magazine for the past five years. A city of approximately 93,500, it is situated in north-central Florida. Gainesville is served by several airlines and bus lines and is located along I-75, 1 hour south of I-10 and 2 hours north of Orlando and Tampa. The Atlantic Ocean and the Gulf of Mexico are within a 2-hour drive.

The University and The Department

A combined state university and land-grant college, the University of Florida has sixteen upper-division colleges and schools and four professional colleges (dentistry, law, medicine, and veterinary medicine). The College of Engineering consists of twelve degree-granting departments. The chemical engineering department's graduate study and research emphasize chemical engineering science (transport phenomena, thermodynamics, kinetics and catalysis, interfacial phenomena, and materials science), chemical engineering systems (chemical and biochemical reaction engineering, computer-aided design, optimization, and control), and interdisciplinary chemical engineering (electrochemical engineering, microelectronics, biomedical interfaces, and cell and tissue engineering).

Applying

An application package containing the required forms, additional information about the program, and a brochure about the department is available upon request. For fall entrance, the following should be submitted by February 15: admission application and processing fee, official transcripts from all colleges and universities attended, official GRE scores (international students must also submit official TOEFL scores), financial assistance application, and three recommendation letters. The first three items should be sent to the Office of Admissions at the University, and the latter two items should be sent to the Graduate Admissions Coordinator in the Department of Chemical Engineering. Students are encouraged to apply well in advance of the deadline of February 15. Late applications will be considered based on availability of funds.

Correspondence and Information

Graduate Admissions Secretary
Department of Chemical Engineering
University of Florida
P.O. Box 116005
Gainesville, Florida 32611-6005
Telephone: 352-392-4753
E-mail: admissions@che.ufl.edu
World Wide Web: http://www.che.ufl.edu/che/graduate/

University of Florida

THE FACULTY AND THEIR RESEARCH

Timothy J. Anderson, Professor and Chairman; Ph.D., Berkeley, 1979. Electronic materials processing, thermochemistry and phase diagrams, chemical vapor deposition, bulk crystal growth, advanced composite materials.

Novel multilayer process for $CuInSe_2$ thin film formation by rapid thermal annealing. *MRS Symp. Proc.* 485:163–8, 1998. With Chang, Stanbery, Morone, and Davydov.

Solution thermodynamics of electronic materials. *CALPHAD* 21:266–85, 1997. With de Fontaine et al.

Ioannis A. Bitsanis, Associate Professor; Ph.D., Minnesota, 1989. Transport in molecularly confined liquids, molecular structure and dynamics at solid-polymer interfaces, viscoelasticity of nanoscopically thin polymer films.

A lattice Monte Carlo study of long chain conformations at solid-melt interfaces. *J. Chem. Phys.* 99:3100, 1993. With Ten and Brinke.

Structure, conformations and dynamics of polymer chains at solid-melt interfaces. *Macromol. Chem. Macromol. Symp.* 65:211, 1993. With Pan.

Oscar D. Crisalle, Associate Professor; Ph.D., California, Santa Barbara, 1990. Robust control of parametrically uncertain plants, computer-aided design of control systems, adaptive and predictive control strategies, control of microelectronic processing operations, wood-pulp and paper processing systems, emulsion polymerization processing.

The Nyquist robust stability margin—A new metric for robust stability of uncertain systems. *Int. J. Robust Nonlinear Control* 7:211–26, 1997. With Latchman and Basker.

Robust unconstrained predictive control design with guaranteed nominal performance. *AIChE J.* 42(5):1293–303, 1996. With Hrissagis and Sznaier.

Richard B. Dickinson, Assistant Professor; Ph.D., Minnesota, 1992. Cellular bioengineering, cell-biomaterial interactions.

A dynamic model for the attachment of a Brownian particle to a surface mediated by discrete macromolecular bonds. *J. Colloid Interface Sci.* 190:3143–50, 1997.

Quantitative comparison of clumping factor- and coagulase-mediated *Staphylococcus aureus* adhesion to surface-bound fibrinogen under flow. *Infect. Immun.* 63(3):3143, 1995. With Nagel et al.

Arthur L. Fricke, Professor; Ph.D., Wisconsin, 1962. Rheology and processing, thermodynamic properties, polymer characterization, pulp, paper, and polymer processes.

Heat of dilution and enthalpy concentration relations for slash pine Kraft black liquors. *Chem. Eng. Commun.* 155:197–216, 1996. With Zaman.

Steady shear flow properties of high solids softwood Kraft black liquors: Effects of temperature, solids concentrations, lignin molecular weight, and shear rate. *Chem. Eng. Commun.* 139:201–21, 1995. With Zaman.

Gar B. Hoflund, Professor; Ph.D., Berkeley, 1978. Ultrahigh vacuum surface characterization techniques, heterogeneous catalysis, Si and III-V semiconductor, tribological material studies.

Spectroscopic Techniques: X-ray Photoelectron Spectroscopy (XPS), Auger Electron Spectroscopy (AES), and Ion Scattering Spectroscopy (ISS). In *Handbook of Surface and Interface Analysis: Methods in Problem Solving,* eds. J.C. Rivière and S. Myhra. New York: Marcel Dekker, 1998.

Reaction and surface characterization study of higher alcohol synthesis catalysts VII: Cs- and Pd-Promoted 1:1 Zn/Cr spinel. *J. Catal.* 1998. With Epling and Minahan.

Lewis E. Johns Jr., Professor; Ph.D., Carnegie Tech, 1964. Fluid mechanics, solute dispersion.

The construction of dispersion approximations to the solution of the vector convective diffusion equation. *Appl. Sci. Res.* 43:239, 1987.

Infinite sums in the theory of dispersion of chemically reactive solute. *SIAM J. Math. Anal.* 18:473, 1987.

Dale W. Kirmse, Associate Professor; Ph.D., Iowa State, 1964. Process improvement, computer-aided design, expert systems, database management tools.

Creating computer-aided process improvement resources. *Proceedings of the First SUCCEED Annual Conference,* North Carolina State University, Raleigh, N.C., 1994.

The computer aided process improvement laboratory. *The Innovator,* pp. 12–13, Spring, 1994.

Anthony J. C. Ladd, Associate Professor; Ph.D., Cambridge, 1978. Statistical mechanics, dynamics of particle-fluid systems, numerical simulations.

Numerical simulations of particulate suspensions via a discretized Boltmann Equation. Part I. Theoretical foundation. *J. Fluid Mech.* 271:285, 1994. Part II. Numerical results. *J. Fluid Mech.* 271:311, 1994.

New Monte-Carlo method to compute the free-energy of arbitrary solids. Application to the fcc and hcp phases of hard spheres. *J. Chem. Phys.* 81:3188, 1984. With D. A. Frenkel.

Atul Narang, Assistant Professor; Ph.D., Purdue, 1994. Kinetics of microbial growth in mixed-substrate and mixed-culture environments.

The steady states of microbial growth on mixtures of substitutable substrates in a chemostat. *J. Theor. Biol.* 190:241–61, 1998.

The dynamical analogy between microbial growth on mixtures of substrates and population growth of competing species. *Biotechnol. Bioeng.* 59:116–21, 1998.

Ranganathan Narayanan, Professor; Ph.D., IIT, 1978. Interfacial instabilities, bifurcation in transport phenomena, fluid mechanics of crystal growth.

Detection of solutal convection during diffusivity measurements of oxygen in liquid tin. *Met. Trans.* 24 B:91–100, 1993. With Sears, Anderson, and Fripp.

Convection of tin in a Bridgman system I-flow characterization by effective diffusivity measurements. *J. Cryst. Growth* 125(3–4):404–14, 1992. With Sears, Anderson, and Fripp.

Mark E. Orazem, Professor; Ph.D., Berkeley, 1983. Electrochemical engineering: electrochemical impedance spectroscopy, corrosion, behavior and processing of electronic and semiconducting materials, photoelectrochemical processes, new methods for characterization, mathematical modeling.

Application of measurement models for interpretation of impedance spectra for corrosion. *Mater. Sci. Forum* 813:289–92, 1998. With Wojcik, Durbha, Fratuer, and García-Rubio.

Current distribution on a rotating disk electrode below the mass-transfer limited current: correction for finite Schmidt number and determination of surface charge distribution. *J. Electrochem. Soc.* 145:1940–9, 1998. With Durbha.

Chang-Won Park, Associate Professor; Ph.D., Stanford, 1985. Interfacial phenomena in multiphase flows, polymer rheology and processing, structure-property relationships.

Analysis of isothermal two-layer blown film coextrusion. *Polym. Eng. Sci.* 32:1771, 1992. With Yoon.

Influence of soluble surfactants on the motion of a finite bubble in a capillary tube. *Phys. Fluids A* 4:2335, 1992.

Raj Rajagopalan, Professor; Ph.D., Syracuse, 1975. Colloid physics, statistical thermodynamics, particle science.

Interaction forces in charged colloidal dispersions: Inversion of static structure factors. *Phys. Rev. E* 55:4423–32, 1997. With Rao.

Principles of Colloid and Surface Chemistry. New York: Marcel Dekker, 1997. With Hiemenz.

Fan Ren, Associate Professor; Ph.D., Polytechnic, 1991. Plasma etching, device passivation, metallization, GaN-based electronic and photonic devices, GaAs-based MOSFETs.

Demonstration of enhancement-mode p- and n-channel GaAs MOSFETs with $Ga_2O_3(Gd_2O_3)$ as gate oxide. *Solid-State Electron.* 41:1751, 1997. With Hong et al.

Extremely high rate of In-based III-V semiconductor in BCl_3/N_2 based plasma. *J. Electrochem. Soc.* 143:3394, 1997. With Pearton et al.

Dinesh O. Shah, Professor and Director, Center for Surface Science and Engineering; Ph.D., Columbia, 1965. Interfacial phenomena in engineering and biomedical systems, molecular association in micelles, liquid crystals and microemulsions, nanoparticles for superconductors and magnetic materials, enzymatic reactions at interfaces.

Effect of counterions on the interfacial tension and emulsion droplet size in the oil/water/dodecyl sulfate system. *J. Phys. Chem.* 97(2):284–6, 1993. With Oh.

Preparation of YBa2Cu3O7-x superconductor by oxalate coprecipitation. *J. Mater. Sci. Lett.* 12:162–4, 1992. With Kumar and Pillai.

Spyros Svoronos, Professor; Ph.D., Minnesota, 1981. Process modeling, optimization, and control with applications in wastewater treatment, particle separations, and biochemical engineering.

In situ estimation of MOCVD growth rate via a modified Kalman filter. *AIChE J.* 42:1319–25, 1996. With Woo, Sankur, Bajaj, and Irvine.

One-dimensional modeling of secondary clarifiers using a concentration-dependent dispersion coefficient. *Water Res.* 30:2112–24, 1996. With Watts and Koopman.

Jason F. Weaver, Assistant Professor; Ph.D., Stanford, 1998. Dynamics of solid interactions relating to semiconductor processing and heterogeneous catalysis.

Surface corrugation effects: molecular ethane adsorption dynamics on rigid-adsorbate covered surfaces of Pt(111). *Surf. Sci.* 395:148 1998. With Stinnert and Madix.

Coverage dependence of neopentane trapping dynamics on Pt(111). *Surf. Sci.* 400:11, 1998. With Ho, Krzyzowski, and Madix.

UNIVERSITY OF HOUSTON

Department of Chemical Engineering

Programs of Study

The Department of Chemical Engineering at the University of Houston is one of the leading chemical engineering departments in the country. Graduate research programs are available at the M.S. and Ph.D. levels and through the professionally oriented Master of Chemical Engineering (M.Ch.E.) program. Part-time M.S. students may select a nonthesis option. Much of the graduate course work is offered in the evening so working professionals from the Houston area can attend. First-year M.S. and Ph.D. students take a standard set of graduate core courses in thermodynamics, mathematics, and transport phenomena. Research advisers are assigned after the first semester, and M.S. students begin active research in the second semester. Ph.D. candidates take the departmental qualifying examination at the end of the first academic year. The department has firm timelines for completion of graduate degrees; student residence times are typically twenty-four months for M.S. degrees and four years for Ph.D. degrees. Faculty members and students participate in a variety of University-wide and citywide interdisciplinary centers, including the Texas Center for Superconductivity; the Environmental Engineering, Biomedical Engineering, and Materials Engineering Programs; the Keck Center for Computational Biology; the Materials Engineering Research Center; and the Institute for Molecular Design. It is also possible to do research with chemical engineering faculty members through the interdisciplinary graduate programs in biomedical, environmental, and materials engineering.

Research Facilities

The Department of Chemical Engineering has research programs in the areas of biochemical engineering, chemical reaction engineering, chemical vapor deposition, catalysis, colloid science, combinatorial library screening, applied molecular biology, process control, electrochemical systems, interfacial phenomena, numerical simulation, molecular recognition, rheology, fluid flow and phase behavior in porous media, polymer and macromolecular solutions, processing of electronic materials, fuel cells, two-phase flow, solid-fluid separation, reliability theory, superconductivity, thermochemical energy storage, and petroleum engineering. Departmental research equipment includes an X-ray diffractometer with a hot stage, a pulsed excimer-pumped dye laser, a quasielastic laser-light-scattering spectroscopy unit, a computerized axial tomographic scanner (CAT scan) system, rheometers, a fluorescence polarization stopped-flow kinetics apparatus, a titration microcalorimeter, and a large number of computer workstations.

Financial Aid

Financial support is awarded on the basis of academic merit. Nearly all full-time graduate students receive a tuition waiver, stipend for living expenses, and health insurance. Graduate stipend amounts are competitive with those offered by other chemical engineering departments throughout the country, particularly considering the relatively low cost of living in Houston.

Cost of Study

Most full-time students receive scholarships to cover tuition and fees. The current tuition rate is $1260 per semester.

Living and Housing Costs

Houston is a large, cosmopolitan city with a wide variety of housing options. Lodging costs are lower than those of almost any U.S. city of comparable size. The average cost of living for lodging, food, and transportation is approximately $15,200 per year.

Student Group

The chemical engineering graduate student body comprises American and international students drawn from the upper ranks of graduating classes all over the world. Many of the part-time students come from Houston-area chemical industries, which compose one of the largest concentrations of chemical engineers in the world.

Student Outcomes

Graduate alumni of the department find employment with a wide variety of industrial and academic organizations. The proximity of the University of Houston to the local petrochemical industry clearly facilitates the job searches of graduating students. Employers of recent graduates include Shell, Exxon, Mobil, DuPont, Merck, AMGEN, and the University of California.

Location

Houston offers all the cultural and entertainment advantages of a large metropolitan area, with costs and crime rates meaningfully lower than those of most cities of comparable size.

The University

The University of Houston grants doctoral degrees and is the largest university in the University of Houston System, a public system of higher education that includes three other universities: UH–Clear Lake, UH–Downtown, and UH–Victoria. The University was founded in 1927 as Houston Junior College, which was operated by the Houston Independent School District (HISD). In 1945, UH was separated from HISD and operated as a private university; in 1963, the University of Houston became a state-supported institution of higher education.

Applying

Admission is selective, which allows the department to maintain high standards. The process is designed, however, to be as easy and low-cost as possible. Most admitted students have B.S. or M.S. degrees in chemical engineering or a closely related discipline. Applicants who have undergraduate degrees in the sciences or other branches of engineering may be required to make up undergraduate chemical engineering course deficiencies. Applications for the fall semester received by February 1 are most favorably considered, though later applications may also be considered.

Correspondence and Information

Graduate Office
Department of Chemical Engineering
University of Houston
Houston, Texas 77204-4792

Telephone: 713-743-4311
E-mail: grad_che@uh.edu
World Wide Web: http://www.chee.uh.edu/.

University of Houston

THE FACULTY AND THEIR RESEARCH

N. R. Amundson. Chemical reactions, transport, mathematical modeling.

V. Balakotaiah. Chemical reaction engineering, environmental engineering, two-phase flow, dynamics of nonlinear systems, applied mathematics.

M. J. Economides. Petroleum engineering, petroleum production, hydraulic fracture mechanics, well completions, reservoir stimulation, petroleum reservoir exploitation strategies.

D. J. Economou. Plasma-, ion-, and laser-assisted etching and deposition of electronic materials; atomic-layer processing; composites and ceramics.

E. J. Henley. Reliability engineering and risk-assessment, biomedical engineering.

R. Krishnamoorti. Thermodynamic interactions in polymer and model polyolefin blends; rheo-optical techniques to characterize flow and alignment of ordered block copolymers; structure, dynamics, and measurement of bulk and confined polymers.

D. Luss. Chemical reaction engineering, pattern-formation in chemically reacting systems, dynamics of packed-bed reactors, kinetics of solid-solid reactions.

K. K. Mohanty. Fluid flow, interfacial mechanics, porous-media transport, underground contaminants, oil recovery, fabrication of composite materials.

M. Nikolaou. Computer-aided process engineering, process optimization, process simulation, process control.

R. Pollard. Deposition of thin solid films, molecular modeling, electrochemical systems, complex reaction networks, applied kinetics and catalysis, chemical reactor engineering.

J. T. Richardson. Heterogeneous catalysis and catalytic processes, reactor engineering, catalyst preparation and characterization, catalyst design; solar energy, solar-receiver design, solar-related chemical processes; catalytic processes for the destruction of hazardous wastes; high-temperature superconductivity, processing of ceramic superconductors; solid oxide fuel cells.

F. M. Tiller. Fluid/particle separation; filtration, thickening, centrifugation; ceramic processing; moisture-transport in drying solids; CAT scan analysis of solid/liquid systems; separation of biosolids from wastewater sludge; developing agricultural fibers as aids in solid/liquid separation and coalescence of oily waters.

R. C. Willson. Biochemical separations, molecular recognition.

F. L. Worley. Physical modeling of atmospheric dispersion, expert systems applied to air pollution and hazardous waste.

UNIVERSITY OF MISSOURI–COLUMBIA

Department of Chemical Engineering

Programs of Study

The Department of Chemical Engineering offers graduate programs leading to the Master of Science and Doctor of Philosophy degrees. The student's graduate program is designed in consultation with a faculty adviser. Information about current areas of faculty research can be found on the reverse of this page.

Research Facilities

The department has moved into a recently renovated building with excellent laboratory facilities. The department also maintains a comprehensive shop facility for fabricating specialized equipment to support research activity. The department participates in four interdisciplinary research centers in the College of Engineering: Capsule Pipeline Research Center, Particulate Systems Research Center, Waste Management Center, and Surface Science and Plasma Technology Center. These centers have excellent facilities for research in their respective areas and are headed by faculty members with national and international reputations. There is active research in three principal areas: materials, biochemical/environmental engineering, and microelectronics. In the materials initiative, current activity is focused on predicting polymer behavior, applying plasma polymerization techniques for coating and medical applications, polymers at interfaces, ceramics, polymer blends, polymerization and grafting, polymer dynamics, and aerosol dynamics. Research in the biochemical/environmental initiative is directed toward bioprocessing, remediation of contaminated sites and streams, bioreactor dynamics and transport processes, bioremediation, environmental catalysis, C_1-chemistry, supercritical processes, and the electrochemical separation of radioactive elements from spent fuel. In the microelectronics research initiative there is ongoing research on the design of CVD reactors to eliminate crowning and defects during epitaxial growth of silicon, investigation of the dynamics of plasma reactors for applications to next-generation electronic devices, and ceramic packages for VLSI devices. Within these areas the department participates in the college initiatives in the environmental area, the manufacturing area, and the microelectronics area. In each of these areas there is strong interdisciplinary activity. The department anticipates enhancing its strengths in these areas by the addition of several new faculty members over the next three years.

Financial Aid

Teaching and research assistantships are offered at the rate of $10,296 to $20,000 per academic year for a half-time position. Nonresident fee waivers are provided with assistantships, and insurance is available.

Cost of Study

Graduate tuition is waived for students on assistantships taking recommended credit loads. Semester fees for fall 1999 include $167.80 per credit hour for Missouri residents and $504.80 per credit hour for nonresidents. All fees (with the exception of approximately $500 in miscellaneous fees) are waived for students on assistantships taking recommended credit loads.

Living and Housing Costs

The estimated living expenses per year (twelve months) for food, housing, etc. is $5000. For books, supplies, medical insurance (required for international students), and miscellaneous expenses, the cost is estimated at $2000.

Student Group

There are approximately 50 graduate students in the chemical engineering program.

Student Outcomes

Graduates have been employed by a variety of academic and industrial organizations. Recent industrial employers include Rockwell International, International Paper, Kimberly-Clark, Procter and Gamble, DuPont de Nemours, 3M, and Wyeth-Ayerst.

Location

Columbia is a three-college town (University of Missouri, Columbia College, and Stephens College) of approximately 75,000 people. The campus is surrounded by rural farmland yet is just a 2-hour drive from both Kansas City and St. Louis, which are both highly industrialized with major industries such as Monsanto, Anheuser-Busch, Mallinckrodt, MidWest Microelectronics, Black & Veatch, Allied Signal, and Mobay. The state capital, Jefferson City, is 45 minutes south en route to the beautiful Lake of the Ozarks, which offers water sports such as fishing, swimming, skiing, and boating. Other outdoor activities such as hiking, caving, and camping continue throughout the fall.

The University

The University of Missouri–Columbia, the first public university west of the Mississippi River, was founded in 1839 and approved as a land-grant institution. Today it consists of 23,000 students enrolled in fourteen schools and colleges, thus providing significant opportunity for interdisciplinary research within the campus as well as the campuses in St. Louis, Rolla, and Kansas City. The University holds a Carnegie Foundation Research Universities I ranking with the Association of American Universities. The chemical engineering faculty members and graduate students enjoy on-campus use of the Natatorium, intramural programs, Wilderness Adventure programs, and open recreation in the recent multimillion-dollar renovation of the Student Recreation Center, which is adjoined to the Stankowski Athletic field and track.

Applying

Application forms and instructions for submitting required materials may be obtained either directly from the Graduate School or from the department. Study normally begins in August. GRE scores are required. A TOEFL score greater than 500 is required for international students.

Correspondence and Information

Director of Graduate Studies
Department of Chemical Engineering
W2030 EBE
University of Missouri
Columbia, Missouri 65211

Graduate School
210 Jesse Hall
University of Missouri
Columbia, Missouri 65211

University of Missouri–Columbia

THE FACULTY AND THEIR RESEARCH

Rakesh K. Bajpai, Professor; Ph.D., Indian Institute of Technology, 1976. Biochemical engineering, fermentation technology, bioreaction kinetics and reactor analysis, scale-up, mathematical modeling of chemical and biochemical processes, bioremediation.

Paul C.-H. Chan, Associate Professor and Undergraduate Director; Ph.D., Caltech, 1979. Low Reynold's number hydrodynamics, reactor analyses, aerosol mechanics, applied mathematics.

Patricia A. Darcy, Assistant Professor; Ph.D., Iowa, 1998. Biochemical engineering, protein crystallization, protein stability, bioseparations, characterization of soil contamination.

William A. Jacoby, Assistant Professor; Ph.D., Colorado, 1993. Environmental processing, bioremediation.

Sunggyu Lee, C. W. LaPierre Professor and Chairman; Ph.D., Case Western Reserve, 1980. Polymer processing, polymer blends, supercritical fluid technology, C_1-chemistry, catalysis, reaction process engineering.

Stephen J. Lombardo, Assistant Professor; Ph.D., Berkeley, 1990. Ceramic materials, ceramic processing.

Sudarshan K. Loyalka, Curator's Professor and Director, Particulate Systems Research Center; Ph.D., Stanford, 1967. Kinetic theory of gases, neutron transport and mechanics of aerosols, physics and thermal hydraulics of nuclear reactors.

Richard H. Luecke, Professor; Ph.D., Oklahoma, 1966. Process control and optimization, process analysis and modeling, modeling of biological systems.

Thomas R. Marrero, Professor and Associate Director, Capsule Pipeline Research Center; Ph.D., Maryland, 1970. Coal log compaction and hydrodynamics, polymer film properties and processes, pollution control, gaseous diffusion.

David G. Retzloff, Associate Professor, Associate Chairman, and Graduate Director; Ph.D., Pittsburgh, 1967. Mathematical modeling and analysis of dynamical systems, chemical vapor deposition reactors, microelectronics processing.

Truman S. Storvick, Professor Emeritus; Ph.D., Purdue, 1959. Thermodynamics and transport properties of pure gases and gas mixtures, flow in capillaries and porous media, fluid properties correlation and predictions, electrochemistry in molten salts and liquid metals.

Dabir S. Viswanath, James C. Dowell Professor; Ph.D., Rochester, 1962. Thermodynamic properties of gas liquid mixtures, equation of state and law of corresponding states, supercritical fluid extraction, heterogeneous catalysis.

Hirotsugu K. Yasuda, James C. Dowell Professor and Director, Surface Science and Plasma Technology Center; Ph.D., SUNY College of Environmental Science and Forestry, 1961. Adhesion, biomaterials, coating by vacuum deposition polymerization, interfacial phenomena, membranes for separation processes, tribo-electric characteristics of materials and tribology.

UNIVERSITY OF NEW MEXICO

Department of Chemical and Nuclear Engineering

Programs of Study

The Department of Chemical and Nuclear Engineering offers graduate programs leading to the Master of Science (M.S.) in chemical engineering and in nuclear engineering and the Doctor of Philosophy (Ph.D.) in engineering. The master's degree programs for both chemical and nuclear engineering require a minimum of 24 semester hours of graduate course work, with 6 hours for research thesis, for a total of 30 hours. Each discipline requires approximately two years to complete the master's degree. The Ph.D. program requires a minimum of 24 hours beyond the master's degree or 48 hours beyond the bachelor's degree, exclusive of dissertation or master's thesis, and must include at least 24 hours completed at the University of New Mexico. The doctorate normally requires a minimum of three years of graduate study after the M.S. degree and must be completed within ten years after the student has begun the Ph.D. graduate work.

Current research topics include applied artificial intelligence, database management, waste management, health physics, space nuclear power and propulsion systems engineering and design, heat pipes technology, nuclear reactor safety and two-phase flow, reactor thermal-hydraulics, accelerator physics and engineering, interaction of radiation with matter, radiation measurement diagnostics, in situ fractional solidification, application of new techniques in powder and porous materials characterization, microstructure of ceramics, catalysis by metals on model supports, chemical and plasma processing of semiconductors, colloid transport, vapor-phase synthesis of materials, catalytic-waste degradation, inorganic membranes, biomedical sensors and instrumentation, and the use of self-assembled monolayers to study interfacial phenomena. Graduates are recruited by employers, with local and national opportunities that range from operation of high-technology facilities to research and development on the cutting edge.

Research Facilities

Faculty members and students participate in a number of special programs at the Institute for Space Nuclear Power Studies (ISNPS), the Center for Micro-Engineered Materials (CMEM), and the Center for Radioactive Waste Management (CeRaM). Research facilities are located in the Farris Engineering Center and the Nuclear Engineering, Space Technology, Thermal-Hydraulics, and Advanced Materials Laboratories. Collaboration with researchers at Sandia National, Los Alamos National, and Air Force Phillips Laboratories provides access to CRAY supercomputers for carrying on advanced computational physics. The department operates IBM and UNIX workstations and terminals that link the department to the Computer and Information Resources and Technology Center (CIRT), where mainframes are available. The department also maintains its own computer pod for student use of laser printers and Macintosh and IBM-compatible computers. State-of-the-art equipment is available in the research laboratories. The Centennial Science and Engineering Library has approximately 350,000 volumes and about 2,000 current journal subscriptions. Computer-searchable database services are available.

Financial Aid

Research assistantships are awarded to students undertaking graduate research in fields funded by research grants and contracts. Research assistantships pay from $12,000 to $25,000 per year. Assistantships also pay the cost of tuition. Research assistants are selected based upon their ability to perform the research, their interest in the project, and their potential for gaining educational benefits from the research experience. Graduate Studies Fellowships are available and coordinated by the Office of Graduate Studies. Students from underrepresented groups in graduate education are encouraged to apply. The majority of full-time graduate students receive financial aid.

Cost of Study

In 1999–2000, tuition is $1250 per semester for full-time New Mexico residents and $4585.80 per semester for nonresident students carrying 12 or more credit hours. There is also a Graduate Student Association fee of $16 per semester.

Living and Housing Costs

In 1998–99, minimum full-time expenses, excluding tuition, were approximately $7200 on campus and $8700 off campus per semester. This included health and accident insurance, books, supplies, board, room, clothing, laundry, and miscellaneous expenses. University housing is available for both individuals and families.

Student Group

The University of New Mexico has a total undergraduate and graduate enrollment of approximately 23,744. The School of Engineering has about 600 graduate students, of whom 63 percent are master's candidates and 37 percent are doctoral candidates. Women constitute 15 percent of the graduate engineering enrollment. In fall 1998, there were 60 full-time graduate students in the department, with 36 enrolled in chemical engineering and 24 in nuclear engineering. The part-time enrollment is 11 for chemical engineering and 27 for nuclear engineering.

Student Outcomes

Recent recipients of master's and doctoral degrees from chemical and nuclear engineering have found employment in a variety of organizations, including Sandia National and Los Alamos National Laboratories; manufacturing firms, such as Intel, Motorola, and Hoechst Celanese; consulting firms, such as SAIC and Benchmark; government agencies, such as the Nuclear Regulatory Commission and the U.S. Department of Energy; and universities throughout the United States.

Location

The University is situated in Albuquerque, which has a metropolitan population of more than 500,000. The city is a mile above sea level, overlooking the Rio Grande, and it abuts the Sandia Mountains, which reach to 10,678 feet and offer a variety of outdoor opportunities. Although Albuquerque undergoes seasonal changes, the dry, sunny climate rarely exhibits temperature extremes. New Mexico is rich with the traditions of the Native American, Spanish, and Anglo cultures.

The University

The University of New Mexico (UNM) was created by an act of the Territorial Legislature in 1889, twenty-three years before New Mexico became a state. There are five museums and seven libraries among the more than 170 buildings on campus. In addition, there are UNM extension campuses located at Los Alamos, Santa Fe, Taos, Gallup, and Valencia County. Faculty members and students number about 40,000 on all campuses. UNM's mission includes offering comprehensive educational programs, conducting research and participating in other scholarly activities, and contributing to the quality of life in the state.

Applying

Students must have their application, application fee, and transcripts on file in the Office of Graduate Studies. Application deadlines are as follows: fall semester, July 15 and spring semester, November 15. Applications are valid for one semester only. Applicants must hold an accredited bachelor's degree and have above a B average in their last two undergraduate years and in their major field. A complete graduate application file consists of the following: application, transcripts, three letters of recommendation, a letter of intent, and GRE scores.

Correspondence and Information

Ms. Karen Hayes
University of New Mexico
Department of Chemical and Nuclear Engineering
209 Farris Engineering Center
Albuquerque, New Mexico 87131-1341

Telephone: 505-277-5431
Fax: 505-277-5433
E-mail: khayes@unm.edu
World Wide Web: http://www-chne.unm.edu

University of New Mexico

THE FACULTY AND THEIR RESEARCH

Chemical Engineering

Harold M. Anderson, Associate Professor; Ph.D., Wayne State, 1981. Plasma processing of semiconductor materials.

Plamen Atanassov, Research Assistant Professor; Ph.D., Bulgarian Academy of Science, 1992. Biomedical sensors.

C. Jeffrey Brinker, Professor; Ph.D., Rutgers, 1979. Ceramics, sol-gel processing, porous materials, inorganic membranes.

Joseph L. Cecchi, Professor and Chair; Ph.D., Harvard, 1972. Semiconductor manufacturing technology, submicron patterning, plasma etching and deposition.

Abhaya K. Datye, Professor; Ph.D., Michigan, 1984. Catalysis, interfaces, advanced materials.

Sang M. Han, Assistant Professor; Ph.D., California, Santa Barbara, 1998. Semiconductor manufacturing technology, submicron patterning, plasma etching and deposition.

Dmitri M. Ivnitski, Research Professor; Ph.D., Moscow State, 1975. Biomedical engineering.

David Kauffman, Professor; Ph.D., Colorado, 1970. Plant design, environmental engineering.

William J. Kroenke, Research Professor; Ph.D., Case Tech, 1963. Inorganic synthesis and materials science.

Ronald E. Loehman, UNM/NL Distinguished Professor; Ph.D., Purdue, 1969. Glass-metal and ceramic-metal bonding and interfacial reactions.

Gabriel P. López, Assistant Professor; Ph.D., Washington (Seattle), 1991. Chemical sensors, hybrid materials, biotechnology, interfacial phenomena.

Richard W. Mead, Associate Professor; Ph.D., Arizona, 1971. Unit operations, resource extraction.

H. Eric Nuttall, Professor; Ph.D., Arizona, 1971. Environmental science, waste transport management, colloid science.

Victor H. Perez-Luna, Research Assistant Professor; Ph.D., Washington (Seattle), 1995. Surface modification and characterization of biomaterials.

Emmanuil Rabinovich, Research Assistant Professor; Ph.D., Saratov (Russia), 1980. Laser spectroscopy and chemical sensors.

Thomas P. Rieker, Research Associate Professor; Ph.D., Colorado, 1988. Small-angle X-ray and neutron scattering.

David L. Sidebottom, Research Assistant Professor; Ph.D., Kansas State, 1989. Glass-metal and ceramic-metal bonding and interfacial reactions.

Thomas P. Swiler, Research Assistant Professor; Ph.D., Florida, 1994. Material science.

Karel Vanheusden, Research Assistant Professor; Ph.D., Leuven (Belgium), 1993. Electrum para magenetic resonance.

Timothy L. Ward, Associate Professor; Ph.D., Washington (Seattle), 1989. Aerosol materials synthesis, inorganic membranes.

Ebtisam S. Wilkins, Professor; Ph.D., Virginia, 1976. Biomedical sensors and waste treatment.

Nuclear Engineering

Robert D. Busch, Lecturer III; Ph.D., New Mexico, 1976. Nuclear criticality safety, radiation protection.

Gary W. Cooper, Associate Professor; Ph.D., Illinois, 1976. Radiation diagnostics.

Mohamed S. El-Genk, Professor; Ph.D., New Mexico, 1978. Space power, thermal hydraulics.

Werner Lutze, Professor; Dr.rer.nat., Technische Universitäet Berlin (Germany), 1967. Management of radioactive wastes.

Anil K. Prinja, Professor; Ph.D., London, 1980. Radiation transport and effects in materials, fusion physics.

Norman F. Roderick, Professor and Associate Chair; Ph.D., Michigan, 1971. Plasma physics and applications.

Jean-Michel P. Tournier, Research Assistant Professor; Ph.D., New Mexico, 1996. Advanced energy conversion, thermal management of space systems, numerical analysis and heat pipes.

WORCESTER POLYTECHNIC INSTITUTE

Department of Chemical Engineering

Programs of Study

The department offers the Doctor of Philosophy (Ph.D.) and Master of Science (M.S.) degrees. In addition to enhancing their understanding of chemical engineering through advanced course work, students conduct creative work on cutting-edge research projects using state-of-the-art equipment. Faculty research interests can be broadly grouped into the areas of advanced materials processing, catalysis and reaction engineering, environmental engineering, and biochemical engineering.

In order to complete a Ph.D., 90 credits of graduate-level study beyond the bachelor's degree or 60 credits beyond the master's degree are required. A minimum of 30 credits must be dissertation research, while the minimum number of courses depends on the student's background. The dissertation adviser is chosen during the first semester of residence, and each candidate develops a specific program of study in association with his or her adviser. A doctoral student must pass a comprehensive qualifying exam, prepare a dissertation proposal and the dissertation itself, and present and defend the dissertation. The student may proceed directly to the Ph.D. program without completing a master's degree.

The M.S. degree may include a thesis (thesis option) or may consist solely of course work (nonthesis option). In either case, 30 credit hours of graduate-level study beyond the bachelor's degree are required, and both degrees require at least three courses from the core chemical engineering curriculum. The thesis option includes at least 12 credit hours of thesis work, while a maximum of 6 credit hours of independent study under the faculty adviser may be part of the nonthesis program.

Research Facilities

The department is housed in Goddard Hall and contains the following highly advanced laboratories: CVD laboratory, environmental catalysis laboratory, catalyst and reaction engineering laboratory, Center for Inorganic Membrane Studies, zeolite crystallization laboratory, heat and mass transfer laboratory, aerosol laboratory, fuel cell laboratory, and biochemical engineering laboratory. Specialized equipment available includes an X-ray diffractometer, a scanning electron microscope, a surface-area analyzer, various mass spectrometers, a chemiluminescence NO/NOx analyzer, a TEOM pulse mass analyzer, a fuel-cell test station, a UHV chamber equipped with Auger electron spectroscopy, X-ray photoelectron spectroscopy, mass spectrometer and low-energy electron diffraction, an ultraviolet Raman spectrometer, in situ Fourier transform infrared spectrometers, and a complete angle-scattering instrument. In addition to the campuswide computing facilities, the department houses a computer laboratory for graduate students and several workstations for graduate research.

Financial Aid

Financial aid is available to qualified M.S. and Ph.D. students in the form of competitive research and teaching assistantships. For 1999–2000, stipends are a minimum of $16,000. In addition, full tuition is usually granted. The Institute also offers Robert H. Goddard Research Fellowships to outstanding U.S. applicants, as well as several new institutional fellowships to highly qualified Ph.D. candidates whose research falls into the thrust areas of materials science and technology, bioengineering, environmental studies, and computational modeling. Both of these fellowships include a twelve-month stipend and a full tuition waiver. In addition, U.S. citizens and permanent residents may also qualify for student loans.

Cost of Study

Tuition for full-time students is $661 per credit hour in 1999–2000.

Living and Housing Costs

Most graduate students obtain housing in private apartments or houses near the campus. Housing costs are reasonable, and the typical annual rent plus utilities for a student sharing a three-bedroom apartment with two other students is approximately $3200 per year.

Student Group

About 200 undergraduate students and 30 full-time graduate students are enrolled in the chemical engineering department.

Location

Worcester Polytechnic Institute (WPI) is set on an attractive 80-acre hilltop campus located in a residential area of Worcester, Massachusetts, a city of 170,000. Flanked by city parks on two sides, the campus has a feeling of openness, even though it is located only 1 mile from downtown Worcester. The city is well known for its many colleges and its Art Museum, Higgins Armory Museum, and Science Center and Ecoterium and the Worcester Centrum. Music is well represented by several excellent choruses and a symphony orchestra. Concerts are given by distinguished visiting performers in the restored acoustical gem, Mechanics Hall. Several professional and amateur theater companies are located in Worcester. Boston and Cape Cod to the east and the Berkshires to the west are within easy driving distance, and there is good skiing nearby in New Hampshire and Vermont.

The Institute

WPI was founded in 1865 and is the third-oldest college of engineering, science, and management in the United States. Graduate education has been part of the Institute's activity for most of this time, and the first graduate degree was awarded in 1893. Classes and research groups at WPI are small and provide close student-faculty relationships. WPI is one of the ten colleges that form the Worcester Consortium for Higher Education, and graduate students may take courses at any of the colleges, including the University of Massachusetts Medical School. Complete recreational and athletic facilities and a program of concerts and special events are available to graduate students.

Applying

Although applications are accepted at any time, the first offers of admissions and departmental awards are made in February and March. Thus, applicants requesting financial aid are encouraged to submit all required documentation by January 15 to ensure full consideration. All applicants should include GRE General Test scores, as these are required for the Goddard and Institute fellowships. International students must submit TOEFL and GRE General Test scores.

Correspondence and Information

Graduate Coordinator
Chemical Engineering Department
Worcester Polytechnic Institute
100 Institute Road
Worcester, Massachusetts 01609-2280

Telephone: 508-831-5493
Fax: 508-831-5853
E-mail: che-gradinfo@wpi.edu

Worcester Polytechnic Institute

THE FACULTY AND THEIR RESEARCH

W. M. Clark, Associate Professor; Ph.D., Rice. Bioseparations: two-phase electrophoresis, aqueous two-phase extraction, membrane filtration.

R. Datta, Professor and Department Head; Ph.D., California, Santa Barbara. Catalyst and reaction engineering: supported molten metal catalysts, catalytic microkinetics, fuels and chemicals from renewable resources, fuel cells and reformers, transport in porous media and membranes.

D. DiBiasio, Associate Professor; Ph.D., Purdue. Bioreactor engineering: magnetic resonance imaging of bioreactors, mammalian cell culture, hollow fiber reactors, immobilized cell reactors.

A. G. Dixon, Professor; Ph.D., Edinburgh. Reaction engineering: computational fluid dynamics for gas-solid catalytic reactors, dense and porous inorganic membrane reactors, zeolite membrane reactors, heat-transfer problems in fixed-bed membrane and microchannel reactors.

Y. H. Ma, Professor; Sc.D., MIT. Inorganic membranes: palladium membranes for hydrogen separations, perodskite and perovskite-like membranes for air separation, zeolite membranes, membrane reactors, adsorbent development, adsorption and diffusion.

K. M. McNamara, Assistant Professor; Ph.D., MIT. Chemical vapor deposition: CVD growth processes for semiconductors, impurity and defect incorporation, optical and electrical properties, materials for space applications, art and historical objects.

W. R. Moser, Professor; Ph.D., MIT. Nanostructured materials synthesis and catalysis: supercritical fluid catalysis, in situ reaction monitoring, homogeneous catalysis.

F. H. Ribeiro, Assistant Professor; Ph.D., Stanford. Catalysis and surface science: heterogeneous catalysis, kinetics, model catalysts.

R. W. Thompson, Professor; Ph.D., Iowa State. Applied reactor design and particulate systems: zeolite crystallization, polymer degradation, water purification, film formation.

R. E. Wagner, Professor Emeritus; Ph.D., Princeton.

A. H. Weiss, Professor Emeritus; Ph.D., Pennsylvania.

B. E. Wyslouzil, Associate Professor; Ph.D., Caltech. Aerosol science: small-angle neutron scattering from aerosols, multicomponent aerosol formation, condensation in supersonic nozzles, aerosol transport in plant tissue reactors.

RESEARCH AREAS

Advanced Materials Processing: zeolite technology, nanostructured materials synthesis, inorganic membranes for gas separations, chemical vapor deposition, diamond thin films, solid-state materials characterization, materials for space applications, small-angle neutron scattering from aerosols.

Biochemical Engineering: bioreactor engineering for whole cells and plant tissue cultures, magnetic resonance imaging of bioreactors, hollow fiber reactors, immobilized cell reactors, bioseparations, two-phase electrophoresis.

Catalysis and Reactor Engineering: adsorption and transport in porous media, heterogeneous and homogeneous catalysis, supported molten metal catalysis, zeolite catalysis, computational fluid dynamics, catalytic microkinetics.

Environmental Engineering: nucleation and phase transitions, multicomponent aerosol formation, pollution prevention in chemical processes, environmentally benign chemical reactor technology, catalysis of waste hydrocarbon, fuel cells, renewable fuels and chemicals.

Section 7
Civil and Environmental Engineering

This section contains a directory of institutions offering graduate work in civil and environmental engineering, followed by in-depth entries submitted by institutions that chose to prepare detailed program descriptions. Additional information about programs listed in the directory but not augmented by an in-depth entry may be obtained by writing directly to the dean of a graduate school or chair of a department at the address given in the directory.

For programs offering related work, see also in this book Agricultural Engineering; Bioengineering, Biomedical Engineering, and Biotechnology; Engineering and Applied Sciences; Management of Engineering and Technology; and Ocean Engineering. In Book 2, see Public, Regional, and Industrial Affairs (City and Regional Planning and Urban Studies); in Book 3, Ecology, Environmental Biology, and Evolutionary Biology; and in Book 4, Agricultural and Food Sciences, Environmental Sciences and Management, Geosciences, and Marine Sciences and Oceanography.

CONTENTS

Civil Engineering

Arizona State University, Graduate College, College of Engineering and Applied Sciences, Department of Civil and Environmental Engineering, Tempe, AZ 85287. Offers civil engineering (MS, MSE, PhD). *Faculty:* 18 full-time (2 women). *Students:* 60 full-time (18 women), 35 part-time (14 women); includes 6 minority (5 Asian Americans or Pacific Islanders, 1 Hispanic American), 42 international. Average age 29. 129 applicants, 78% accepted. In 1998, 20 master's, 8 doctorates awarded. *Degree requirements:* For master's, thesis or alternative required; for doctorate, dissertation required. *Entrance requirements:* For master's and doctorate, GRE General Test (recommended). Application fee: $45. *Financial aid:* Fellowships, career-related internships or fieldwork available. *Faculty research:* Environmental/sanitary engineering, geotechnical/soil mechanics, structures, transportation, water resources/hydraulics. *Unit head:* Dr. Sandra Houston, Chair, 480-965-3589, E-mail: civil@www.eas.asu.edu. *Application contact:* Graduate Secretary, 480-965-8474.

Auburn University, Graduate School, College of Engineering, Department of Civil Engineering, Auburn, Auburn University, AL 36849-0002. Offers construction engineering and management (MCE, MS, PhD); environmental engineering (MCE, MS, PhD); geotechnical/materials engineering (MCE, MS, PhD); hydraulics/hydrology (MCE, MS, PhD); structural engineering (MCE, MS, PhD); transportation engineering (MCE, MS, PhD). Part-time programs available. *Faculty:* 22 full-time. *Students:* 24 full-time (6 women), 34 part-time (8 women); includes 2 minority (1 African American, 1 Hispanic American), 9 international. 52 applicants, 46% accepted. In 1998, 22 master's, 3 doctorates awarded. *Degree requirements:* For master's, project (MCE), thesis (MS) required; for doctorate, dissertation, comprehensive exam required, foreign language not required. *Entrance requirements:* For master's, GRE General Test; for doctorate, GRE General Test (minimum score of 400 on each section required). *Application deadline:* For fall admission, 9/1; for spring admission, 3/1. Applications are processed on a rolling basis. Application fee: $25 ($50 for international students). Tuition, state resident: full-time $2,760; part-time $76 per credit hour. Tuition, nonresident: full-time $8,280; part-time $228 per credit hour. *Financial aid:* Fellowships, research assistantships, teaching assistantships, Federal Work-Study available. Aid available to part-time students. Financial aid application deadline: 3/15. *Unit head:* Dr. Joseph F. Judkins, Head, 334-844-4320. *Application contact:* Dr. John F. Pritchett, Dean of the Graduate School, 334-844-4700.

Announcement: Department's strength is a faculty of 22 professors committed to excellence in teaching and research. Department housed in modern facility containing 43,000 square feet of laboratory, classroom, and office space. Advanced degrees offered in construction engineering and management, environmental engineering, geotechnical/pavement materials engineering, hydraulics/hydrology, structural engineering, and transportation engineering.

Bradley University, Graduate School, College of Engineering and Technology, Department of Civil Engineering and Construction, Peoria, IL 61625-0002. Offers MSCE. Part-time and evening/weekend programs available. *Degree requirements:* For master's, comprehensive exam required, thesis not required. *Entrance requirements:* For master's, TOEFL (minimum score of 525 required), minimum GPA of 3.0.

Brigham Young University, Graduate Studies, College of Engineering and Technology, Department of Civil and Environmental Engineering, Provo, UT 84602-1001. Offers civil engineering (MS); engineering (PhD). *Faculty:* 17 full-time (0 women), 5 part-time (1 woman). *Students:* 58 full-time (7 women), 15 part-time; includes 1 minority (Hispanic American), 9 international. Average age 27. 28 applicants, 93% accepted. In 1998, 43 master's, 3 doctorates awarded (33% entered university research/teaching, 67% found other work related to degree). *Degree requirements:* For master's, thesis required (for some programs), foreign language not required; for doctorate, dissertation required, foreign language not required. *Entrance requirements:* For master's, GRE General Test (minimum combined score of 1650 on three sections required for international applicants), GRE Subject Test (minimum score of 550 required for international applicants), minimum GPA of 3.0 in last 60 hours for international applicants; for doctorate, GRE General Test (minimum combined score of 1650 on three sections required for international applicants), GRE Subject Test (minimum score of 550 required for international applicants), minimum graduate GPA of 3.3. *Average time to degree:* Master's–1.5 years full-time, 3.5 years part-time; doctorate–3 years full-time, 7 years part-time. *Application deadline:* For fall admission, 5/15; for winter admission, 9/15; for spring admission, 2/20. Applications are processed on a rolling basis. Application fee: $30. Tuition: Full-time $3,330; part-time $185 per credit hour. Tuition and fees vary according to program and student's religious affiliation. *Financial aid:* In 1998–99, 63 students received aid, including 34 research assistantships (averaging $2,400 per year), 18 teaching assistantships (averaging $2,400 per year); fellowships, career-related internships or fieldwork and scholarships also available. Aid available to part-time students. Financial aid application deadline: 3/15. *Faculty research:* Computer graphics, liquefaction, wastewater treatment, hazardous waste management, contaminant transport, transportation. Total annual research expenditures: $1.2 million. *Unit head:* Dr. T. Leslie Youd, Chair, 801-378-2811, Fax: 801-378-4449, E-mail: tyoud@byu.edu. *Application contact:* A. Woodruff Miller, Graduate Coordinator, 801-378-2811, Fax: 801-378-4449, E-mail: wood_miller@byu.edu.

Bucknell University, Graduate Studies, College of Engineering, Department of Civil Engineering, Lewisburg, PA 17837. Offers MS, MSCE. *Faculty:* 7 full-time, 2 part-time. *Students:* 6 (2 women). *Degree requirements:* For master's, thesis required, foreign language not required. *Entrance requirements:* For master's, GRE General Test (minimum combined score of 1000 required), GRE Subject Test, TOEFL (minimum score of 550 required), minimum GPA of 2.8. *Application deadline:* For fall admission, 6/1 (priority date); for spring admission, 12/1 (priority date). Applications are processed on a rolling basis. Application fee: $25. Tuition: Part-time $2,600 per course. Tuition and fees vary according to course load. *Financial aid:* Unspecified assistantships available. Financial aid application deadline: 3/1. *Faculty research:* Pile foundations, rehabilitation of bridges, deep-shaft biological-waste treatment, precast concrete structures. *Unit head:* Dr. Jai B. Kim, Head, 570-577-1112.

California Institute of Technology, Division of Engineering and Applied Science, Option in Civil Engineering, Pasadena, CA 91125-0001. Offers MS, PhD, Engr. *Faculty:* 6 full-time (0 women). *Students:* 10 full-time (2 women), 5 international. 69 applicants, 3% accepted. In 1998, 1 master's, 1 doctorate awarded. *Degree requirements:* For master's, foreign language and thesis not required; for doctorate, dissertation required, foreign language not required. *Application deadline:* For fall admission, 1/15. Application fee: $0. *Faculty research:* Earthquake engineering, soil mechanics, finite-element analysis, hydraulics, coastal engineering. *Unit head:* Dr. John Hall, Executive Officer, 626-395-4160. *Application contact:* Dr. Erik Antonsson, Representative, 626-395-3790.

California Polytechnic State University, San Luis Obispo, College of Engineering, Department of Civil and Environmental Engineering, San Luis Obispo, CA 93407. Offers MS, MCRP/MS. Part-time programs available. *Faculty:* 17 full-time (1 woman), 11 part-time (3 women). *Students:* 10 full-time (4 women), 4 part-time (2 women). 31 applicants, 58% accepted. In 1998, 3 degrees awarded. *Degree requirements:* For master's, thesis or alternative required, foreign language not required. *Entrance requirements:* For master's, GRE General Test, minimum GPA of 3.0 in last 90 quarter units. *Application deadline:* For fall admission, 5/31 (priority date); for spring admission, 12/31. Applications are processed on a rolling basis. Application fee: $55. Tuition, nonresident: part-time $164 per unit. Required fees: $531 per quarter. *Financial aid:* Fellowships, research assistantships, teaching assistantships, career-related internships or fieldwork available. Financial aid application deadline: 3/2; financial aid applicants required to submit FAFSA. *Faculty research:* Soils, structures, transportation, traffic, environmental protection, biomediation. *Unit head:* Dr. Robert Lang, Chair, 805-756-2947, Fax: 805-756-6330, E-mail: rlang@calpoly.edu. *Application contact:* Dr. Nirupam Pal, Professor, 805-756-1355, Fax: 805-756-6330, E-mail: npal@calpoly.edu.

California State University, Fresno, Division of Graduate Studies, School of Engineering, Department of Civil Engineering, Fresno, CA 93740-0057. Offers MS. Part-time and evening/weekend programs available. *Faculty:* 10 full-time (0 women). *Students:* 3 full-time (1 woman), 8 part-time (1 woman); includes 3 minority (all Asian Americans or Pacific Islanders), 3 international. Average age 31. 10 applicants, 100% accepted. In 1998, 3 degrees awarded. *Degree requirements:* For master's, thesis or alternative required, foreign language not required. *Entrance requirements:* For master's, GRE General Test, TOEFL (minimum score of 550 required), minimum GPA of 2.75. *Average time to degree:* Master's–3.5 years full-time. *Application deadline:* For fall admission, 8/1 (priority date); for spring admission, 12/1. Applications are processed on a rolling basis. Application fee: $55. Electronic applications accepted. Tuition, nonresident: part-time $246 per unit. Required fees: $1,906; $620 per semester. *Financial aid:* Career-related internships or fieldwork, Federal Work-Study, and scholarships available. Financial aid application deadline: 3/1; financial aid applicants required to submit FAFSA. *Faculty research:* Surveying, water damage, instrumentation equipment, agricultural drainage, aerial triangulation. *Unit head:* Dr. Mohamad A. Yousef, Chair, 559-278-2889, Fax: 559-278-7071. *Application contact:* Dr. Jesus Larralde-Muro, Graduate Program Coordinator, 559-278-2566, E-mail: jesus_larralde-muro@csufresno.edu.

California State University, Fullerton, Graduate Studies, School of Engineering and Computer Science, Department of Civil Engineering and Engineering Mechanics, Fullerton, CA 92834-9480. Offers MS. *Faculty:* 7 full-time (0 women), 6 part-time. *Students:* 8 full-time (2 women), 25 part-time (4 women); includes 13 minority (1 African American, 11 Asian Americans or Pacific Islanders, 1 Hispanic American), 5 international. Average age 31. 18 applicants, 56% accepted. In 1998, 20 degrees awarded. *Degree requirements:* For master's, computer language, comprehensive exam, project or thesis required. *Entrance requirements:* For master's, minimum undergraduate GPA of 2.5. Application fee: $55. Tuition, nonresident: part-time $264 per unit. Required fees: $1,947; $1,281 per year. *Financial aid:* Career-related internships or fieldwork, Federal Work-Study, grants, and institutionally-sponsored loans available. Aid available to part-time students. Financial aid application deadline: 3/1. *Faculty research:* Soil-structure interaction, finite-element analysis, computer-aided analysis and design. *Unit head:* Dr. Chandra Putcha, Chair, 714-278-3012.

California State University, Long Beach, Graduate Studies, College of Engineering, Department of Civil Engineering, Long Beach, CA 90840. Offers MSCE, MSE, CE. Part-time programs available. *Faculty:* 12 full-time (1 woman), 8 part-time (0 women). *Students:* 17 full-time (3 women), 91 part-time (22 women); includes 54 minority (5 African Americans, 30 Asian Americans or Pacific Islanders, 18 Hispanic Americans, 1 Native American), 12 international. Average age 31. 83 applicants, 63% accepted. In 1998, 32 degrees awarded. *Degree requirements:* For master's, comprehensive exam or thesis required. *Entrance requirements:* For master's, TOEFL (minimum score of 550 required). *Application deadline:* For fall admission, 8/1; for spring admission, 12/1. Application fee: $55. Electronic applications accepted. Tuition, nonresident: part-time $246 per unit. Required fees: $569 per semester. Tuition and fees vary according to course load. *Financial aid:* Career-related internships or fieldwork, Federal Work-Study, grants, institutionally-sponsored loans, and unspecified assistantships available. Financial aid application deadline: 3/2. *Faculty research:* Soils, hydraulics, seismic structures, composite metals, computer-aided manufacturing. *Unit head:* Dr. Steve Tsai, Chairman, 562-985-5118, Fax: 562-985-2380, E-mail: stsai@engr.csulb.edu. *Application contact:* Dr. Peter Cowan, Graduate Adviser, 562-985-5135, Fax: 562-985-2380, E-mail: cowan@engr.csulb.edu.

California State University, Los Angeles, Graduate Studies, School of Engineering and Technology, Department of Civil Engineering, Los Angeles, CA 90032-8530. Offers MS. Part-time and evening/weekend programs available. *Faculty:* 7 full-time, 6 part-time. *Students:* 1 full-time (0 women), 28 part-time (5 women); includes 17 minority (3 African Americans, 11 Asian Americans or Pacific Islanders, 3 Hispanic Americans), 5 international. In 1998, 4 degrees awarded. *Degree requirements:* For master's, computer language, comprehensive exam or thesis required. *Entrance requirements:* For master's, TOEFL (minimum score of 550 required), GRE or minimum GPA of 2.4. *Application deadline:* For fall admission, 6/30; for spring admission, 2/1. Applications are processed on a rolling basis. Application fee: $55. *Financial aid:* In 1998–99, 8 students received aid. Federal Work-Study available. Aid available to part-time students. Financial aid application deadline: 3/1. *Faculty research:* Structure, hydraulics, hydrology, soil mechanics. *Unit head:* Dr. Young Kim, Chair, 323-343-4450.

California State University, Northridge, Graduate Studies, College of Engineering and Computer Science, Department of Civil, Industrial and Applied Mechanics, Northridge, CA 91330. Offers applied mechanics (MSE); civil engineering (MS); engineering (MS); engineering management (MS); industrial engineering (MS); materials engineering (MS); mechanical engineering (MS), including aerospace engineering, applied engineering, machine design, mechanical engineering, structural engineering, thermofluids; mechanics (MS). Part-time and evening/weekend programs available. *Faculty:* 13 full-time, 2 part-time. *Students:* 10 full-time (2 women), 101 part-time (15 women); includes 38 minority (3 African Americans, 22 Asian Americans or Pacific Islanders, 11 Hispanic Americans, 2 Native Americans), 8 international. Average age 32. 58 applicants, 57% accepted. In 1998, 34 degrees awarded. *Degree requirements:* For master's, thesis required, foreign language not required. *Entrance requirements:* For master's, GRE General Test, TOEFL, minimum GPA of 2.5. *Application deadline:* For fall admission, 11/30. Application fee: $55. Tuition, nonresident: part-time $246 per unit. International tuition: $7,874 full-time. Required fees: $1,970. Tuition and fees vary according to course load. *Financial aid:* Teaching assistantships available. Financial aid application deadline: 3/1. *Faculty research:* Composite study. *Unit head:* Dr. Stephen Gadomski, Chair, 818-677-2166. *Application contact:* Dr. Ileana Costa, Graduate Coordinator, 818-677-3299.

California State University, Sacramento, Graduate Studies, School of Engineering and Computer Science, Department of Civil Engineering, Sacramento, CA 95819-6048. Offers MS. Part-time and evening/weekend programs available. *Degree requirements:* For master's, thesis or alternative, writing proficiency exam required, foreign language not required. *Entrance requirements:* For master's, TOEFL (minimum score of 550 required). *Application deadline:* For fall admission, 4/15; for spring admission, 11/1. Application fee: $55. *Financial aid:* Research assistantships, teaching assistantships, career-related internships or fieldwork and Federal Work-Study available. Aid available to part-time students. Financial aid application deadline: 3/1. *Unit head:* Dr. Joan Al Kazily, Chair, 916-278-6982.

Carleton University, Faculty of Graduate Studies, Faculty of Engineering and Design, Department of Civil and Environmental Engineering, Ottawa, ON K1S 5B6, Canada. Offers M Eng, PhD. *Faculty:* 18 full-time (1 woman). *Students:* 55 full-time (17 women), 10 part-time (3 women). Average age 29. In 1998, 14 master's, 2 doctorates awarded. *Degree requirements:* For master's, thesis optional; for doctorate, dissertation required. *Entrance requirements:* For master's, TOEFL (minimum score of 550 required), honors degree; for doctorate, TOEFL (minimum score of 550 required), MA Sc or M Eng. *Average time to degree:* Master's–2.1 years full-time, 3.3 years part-time; doctorate–4.7 years full-time. *Application deadline:* For fall admission, 3/1 (priority date). Applications are processed on a rolling basis. Application fee: $35. *Financial aid:* Application deadline: 3/1. *Faculty research:* Transportation planning and technology, computer applications in civil engineering, structural and continuum mechanics, building design and construction. Total annual research expenditures: $616,000. *Unit head:* Dr. J. L. Humar, Chair, 613-520-2600 Ext. 8389, Fax: 613-520-3951, E-mail: jag_humar@carleton.ca. *Application contact:* Ata M. Khan, Associate Dean of Engineering, 613-520-5659, Fax: 613-520-5682, E-mail: ata_khan@carleton.ca.

Carnegie Mellon University, Carnegie Institute of Technology, Department of Civil and Environmental Engineering, Pittsburgh, PA 15213-3891. Offers civil engineering (MS, PhD); civil engineering and industrial management (MS); civil engineering and robotics (PhD); civil engineering/bioengineering (PhD); civil engineering/engineering and public policy (MS, PhD). Part-time programs available. *Faculty:* 21 full-time (5 women). *Students:* 56 full-time (20 women), 11 part-time (2 women); includes 3 minority (1 African American, 1 Hispanic American, 1 Native American), 44 international. Average age 27. In 1998, 27 master's, 15 doctorates awarded. Terminal master's awarded for partial completion of doctoral program. *Degree requirements:* For master's, thesis required (for some programs), foreign language not required; for doctorate, dissertation, qualifying exam required, foreign language not required. *Entrance requirements:* For master's and doctorate, GRE General Test, TOEFL. *Average time to degree:* Master's–1 year full-time; doctorate–4 years full-time. *Application deadline:* For fall admission, 2/1 (priority date); for spring admission, 10/15. Application fee: $45. *Financial aid:* In 1998–99, 63 students received aid, including 7 fellowships, 36 research assistantships; teaching assistantships, career-related internships or fieldwork, Federal Work-Study, and scholarships also available. Financial aid application deadline: 2/1. *Faculty research:* Computer-aided engineering and management, structured and computational machines, civil systems. Total annual research expenditures: $2.2 million. *Unit head:* Chris Hendrickson, Head, 412-268-2941, Fax: 412-268-7813. *Application contact:* Maxine A. Leffard, Graduate Program Administrator, 412-268-8712, Fax: 412-268-7813, E-mail: ce-addmissions+@andrew.cmu.edu.

See in-depth description on page 569.

Case Western Reserve University, School of Graduate Studies, The Case School of Engineering, Department of Civil Engineering, Cleveland, OH 44106. Offers civil engineering (MS, PhD); engineering mechanics (MS). Part-time programs available. Postbaccalaureate distance learning degree programs offered (minimal on-campus study). *Faculty:* 9 full-time (0 women), 1 part-time (0 women). *Students:* 12 full-time (3 women), 10 part-time (3 women). Average age 24. 89 applicants, 66% accepted. In 1998, 4 master's, 1 doctorate awarded. *Degree requirements:* For master's, thesis required (for some programs), foreign language not required; for doctorate, dissertation required, foreign language not required. *Entrance requirements:* For master's and doctorate, TOEFL (minimum score of 550 required). *Application deadline:* For fall admission, 8/1 (priority date); for spring admission, 1/1. Application fee: $25. Electronic applications accepted. *Financial aid:* In 1998–99, 7 fellowships with full tuition reimbursements (averaging $14,400 per year), 8 research assistantships with full and partial tuition reimbursements (averaging $14,000 per year), 4 teaching assistantships with full tuition reimbursements (averaging $10,800 per year) were awarded.; institutionally-sponsored loans also available. Financial aid application deadline: 8/1. *Faculty research:* Environmental remediation, structural dynamics, damage and fracture mechanics, earthquakes and liquefaction, bifurcation in particulate media. Total annual research expenditures: $559,000. *Unit head:* Robert L. Mullen, Chairman, 216-368-2423, Fax: 216-368-5229, E-mail: rlm@po.cwru.edu. *Application contact:* Kathleen Ballou, Secretary, 216-368-2950, Fax: 216-368-5229, E-mail: kad4@po.cwru.edu.

The Catholic University of America, School of Engineering, Department of Civil Engineering, Washington, DC 20064. Offers civil engineering (MCE, D Engr); construction management (MCE, MS Engr); environmental engineering (MCE, MS Engr); fluid and solid mechanics (PhD); geotechnical engineering (MCE); structures and structural mechanics (MCE, PhD). Part-time and evening/weekend programs available. *Faculty:* 6 full-time (0 women), 18 part-time (0 women). *Students:* 16 full-time (3 women), 29 part-time (4 women); includes 12 minority (9 African Americans, 1 Asian American or Pacific Islander, 2 Hispanic Americans), 17 international. Average age 32. 44 applicants, 61% accepted. In 1998, 20 master's, 2 doctorates awarded. *Degree requirements:* For master's, thesis optional, foreign language not required; for doctorate, dissertation, comprehensive and qualifying exams required, foreign language not required. *Entrance requirements:* For master's, TOEFL (minimum score of 550 required), minimum GPA of 3.0; for doctorate, TOEFL (minimum score of 550 required), minimum GPA of 3.5. *Application deadline:* For fall admission, 8/1 (priority date); for spring admission, 12/1. Applications are processed on a rolling basis. Application fee: $50. *Financial aid:* Research assistantships, teaching assistantships, career-related internships or fieldwork, Federal Work-Study, institutionally-sponsored loans, and tuition waivers (full and partial) available. Aid available to part-time students. Financial aid application deadline: 2/1. *Faculty research:* Wave propagation, geophysical fluid mechanics. Total annual research expenditures: $337,733. *Unit head:* Dr. Timothy W. Kao, Chair, 202-319-5163, Fax: 202-319-4499, E-mail: kao@cua.edu.

Announcement: Doctoral programs in applied mechanics and geotechnical, earthquake, structural, and computer-aided engineering are available. Infrastructure management has been added to existing master's programs in environmental engineering and management, construction management, structural engineering, and systems engineering. Architecture majors are accepted directly into the construction management program.

City College of the City University of New York, Graduate School, School of Engineering, Department of Civil Engineering, New York, NY 10031-9198. Offers ME, MS, PhD. Part-time programs available. *Students:* 8 full-time (1 woman), 97 part-time (26 women). In 1998, 27 degrees awarded. *Degree requirements:* For master's, computer language required, thesis optional, foreign language not required; for doctorate, one foreign language (computer language can substitute), dissertation, comprehensive exams required. *Entrance requirements:* For master's, TOEFL (minimum score of 500 required); for doctorate, GRE General Test, TOEFL. *Application deadline:* Applications are processed on a rolling basis. Application fee: $40. *Faculty research:* Earthquake engineering, transportation systems, groundwater, environmental systems, highway systems. *Unit head:* John Filos, Acting Chairman, 212-650-8010. *Application contact:* Graduate Admissions Office, 212-650-6977.

Clarkson University, Graduate School, School of Engineering, Department of Civil and Environmental Engineering, Potsdam, NY 13699. Offers civil and environmental engineering (PhD); civil engineering (ME, MS). Part-time programs available. *Faculty:* 15 full-time (2 women), 2 part-time (0 women). *Students:* 33 full-time (8 women); includes 2 minority (both Native Americans), 15 international. Average age 27. 82 applicants, 43% accepted. In 1998, 11 master's, 3 doctorates awarded. *Degree requirements:* For master's, thesis required, foreign language not required; for doctorate, dissertation, departmental qualifying exam required, foreign language not required. *Entrance requirements:* For master's, GRE, TOEFL. *Application deadline:* For fall admission, 5/15 (priority date); for spring admission, 10/15 (priority date). Applications are processed on a rolling basis. Application fee: $25 ($35 for international students). Tuition: Part-time $661 per credit hour. Required fees: $215 per semester. *Financial aid:* In 1998–99, 1 fellowship, 17 research assistantships, 13 teaching assistantships were awarded. *Faculty research:* Granular flows, water treatment, environmental systems, geotechnical structural dynamics, fluid dynamics. Total annual research expenditures: $1.2 million. *Unit head:* Dr. Thomas L. Theis, Chairman, 315-268-6529, Fax: 315-268-7985, E-mail: tlto@clarkson.edu. *Application contact:* Dr. Philip K. Hopke, Dean of the Graduate School, 315-268-6447, Fax: 315-268-7994, E-mail: hopkepk@clarkson.edu.

Clemson University, Graduate School, College of Engineering and Science, Department of Civil Engineering, Clemson, SC 29634. Offers M Engr, MS, PhD. Part-time programs available. *Students:* 53 full-time (8 women), 11 part-time (3 women); includes 7 minority (2 African Americans, 2 Asian Americans or Pacific Islanders, 3 Hispanic Americans), 18 international. Average age 24. 73 applicants, 75% accepted. In 1998, 29 master's, 2 doctorates awarded. *Degree requirements:* For master's, thesis or alternative, oral exam, seminar required, foreign language not required; for doctorate, dissertation, oral exam, seminar required, foreign language not required. *Entrance requirements:* For master's and doctorate, GRE General Test, TOEFL, minimum GPA of 3.0. *Application deadline:* For fall admission, 6/1. Application fee: $35. *Financial aid:* Fellowships, research assistantships, teaching assistantships available. Financial aid application deadline: 2/15; financial aid applicants required to submit FAFSA. *Faculty research:* Applied fluid mechanics, construction materials, project management, structural and protechnical engineering. *Unit head:* Dr. Russell H. Brown, Chair, 864-656-3002, Fax: 864-656-2670, E-mail: russel.brown@eng.clemson.edu. *Application contact:* David Rosowsky, Graduate Program Coordinator, 864-656-5942, Fax: 864-656-2670, E-mail: rdavid@eng.clemson.edu.

Cleveland State University, College of Graduate Studies, Fenn College of Engineering, Department of Civil Engineering, Cleveland, OH 44115-2440. Offers MS, D Eng. Part-time programs available. *Faculty:* 9 full-time (0 women). *Students:* 10 full-time (1 woman), 20 part-time (2 women); includes 1 minority (Hispanic American), 15 international. Average age 30. 62 applicants, 73% accepted. In 1998, 28 degrees awarded. *Degree requirements:* For master's, project or thesis required; for doctorate, dissertation, candidacy and qualifying exams required, foreign language not required. *Entrance requirements:* For master's, GRE General Test, GRE Subject Test, TOEFL, minimum GPA of 2.75; for doctorate, GRE General Test, GRE Subject Test, TOEFL, minimum GPA of 3.25. *Application deadline:* For fall admission, 7/15 (priority date). Applications are processed on a rolling basis. Application fee: $25. *Financial aid:* In 1998–99, 3 research assistantships, 6 teaching assistantships were awarded.; career-related internships or fieldwork and unspecified assistantships also available. Financial aid application deadline: 9/1. *Faculty research:* Environmental engineering, solid-waste disposal, composite materials, nonlinear buckling, constitutive modeling. *Unit head:* Dr. Paul Bosela, Chairperson, 216-687-2190, Fax: 216-687-9280.

Columbia University, Fu Foundation School of Engineering and Applied Science, Department of Civil Engineering and Engineering Mechanics, New York, NY 10027. Offers civil engineering (MS, Eng Sc D, PhD, Engr); mechanics (MS, Eng Sc D, PhD, Engr). Part-time programs available. *Faculty:* 11 full-time (0 women), 6 part-time (0 women). *Students:* 31 full-time (8 women), 29 part-time (5 women). 170 applicants, 34% accepted. In 1998, 24 master's, 5 doctorates, 2 other advanced degrees awarded. Terminal master's awarded for partial completion of doctoral program. *Degree requirements:* For master's, foreign language and thesis not required; for doctorate, dissertation, qualifying exam required, foreign language not required. *Entrance requirements:* For master's, doctorate, and Engr, GRE General Test, TOEFL. *Application deadline:* For fall admission, 1/5; for spring admission, 10/1. Application fee: $55. *Financial aid:* In 1998–99, 8 research assistantships, 6 teaching assistantships were awarded.; fellowships, Federal Work-Study and scholarships also available. Financial aid application deadline: 1/5; financial aid applicants required to submit FAFSA. *Faculty research:* Structural deterioration and control structural materials, damage mechanics, geoenvironmental engineering, construction engineering. *Unit head:* Dr. Rene B. Testa, Chairman, 212-854-6283, Fax: 212-854-6267, E-mail: testa@civil.columbia.edu. *Application contact:* Carolyn Waldo, Administrative Assistant, 212-854-3143, Fax: 212-854-6267, E-mail: clw1@columbia.edu.

Concordia University, School of Graduate Studies, Faculty of Engineering and Computer Science, Department of Building, Civil and Environmental Engineering, Montréal, PQ H3G 1M8, Canada. Offers building (M Eng, MA Sc, PhD, Certificate); civil engineering (M Eng, MA Sc, PhD). *Students:* 117 full-time (27 women), 14 part-time (3 women). *Degree requirements:* For master's, computer language, thesis or alternative required; for doctorate, computer language, dissertation, comprehensive exam required. *Application deadline:* For fall admission, 6/1; for spring admission, 10/1. Application fee: $50. *Faculty research:* Structural engineering, geotechnical engineering, water resources and fluid engineering, transportation engineering, systems engineering. *Unit head:* Dr. O. Moselhi, Chair, 514-848-7801, Fax: 514-848-7823. *Application contact:* Dr. F. Haghighat, Director, 514-848-8791, Fax: 514-848-2809.

Cornell University, Graduate School, Graduate Fields of Engineering, Field of Civil and Environmental Engineering, Ithaca, NY 14853-0001. Offers environmental engineering (M Eng, MS, PhD); environmental fluid mechanics and hydrology (M Eng, MS, PhD); environmental systems engineering (M Eng, MS, PhD); geotechnical engineering (M Eng, MS, PhD); remote sensing (M Eng, MS, PhD); structural engineering (M Eng, MS, PhD); transportation engineering (M Eng, MS, PhD); water resource systems (M Eng, MS, PhD). *Faculty:* 29 full-time. *Students:* 143 full-time (32 women); includes 16 minority (11 Asian Americans or Pacific Islanders, 5 Hispanic Americans), 76 international. 365 applicants, 52% accepted. In 1998, 74 master's, 10 doctorates awarded. Terminal master's awarded for partial completion of doctoral program. *Degree requirements:* For master's, thesis (MS) required; for doctorate, dissertation required, foreign language not required. *Entrance requirements:* For master's, TOEFL (minimum score of 600required); for doctorate, GRE General Test, TOEFL (minimum score of 600 required). *Application deadline:* For fall admission, 1/15. Application fee: $65. Electronic applications accepted. *Financial aid:* In 1998–99, 65 students received aid, including 20 fellowships with full tuition reimbursements available, 28 research assistantships with full tuition reimbursements available, 17 teaching assistantships with full tuition reimbursements available; institutionally-sponsored loans, scholarships, tuition waivers (full and partial), and unspecified assistantships also available. Financial aid applicants required to submit FAFSA. *Unit head:* Director of Graduate Studies, 607-255-7560, Fax: 607-255-9004. *Application contact:* Graduate Field Assistant,, 607-255-7560, E-mail: cee_grad@cornell.edu.

See in-depth description on page 573.

Dalhousie University, Faculty of Graduate Studies, DalTech, Faculty of Engineering, Department of Civil Engineering, Halifax, NS B3H 3J5, Canada. Offers M Eng, MA Sc, PhD. *Faculty:* 11 full-time (1 woman), 2 part-time (0 women). *Students:* 38 full-time (8 women), 5 part-time (2 women). Average age 30. 27 applicants, 52% accepted. In 1998, 10 master's, 2 doctorates awarded (100% entered university research/teaching). *Degree requirements:* For master's and doctorate, thesis/dissertation required, foreign language not required. *Entrance requirements:* For master's and doctorate, TOEFL (minimum score of 580 required). *Average time to degree:* Doctorate–4.5 years full-time. *Application deadline:* For fall admission, 6/1 (priority date); for winter admission, 10/1 (priority date); for spring admission, 2/1 (priority date). Applications are processed on a rolling basis. Application fee: $55. *Financial aid:* Fellowships, research assistantships, teaching assistantships, career-related internships or fieldwork and scholarships available. *Faculty research:* Environmental/water resources, bridge engineering, geotechnical engineering, pavement design and management/highway materials, composite materials. *Unit head:* Dr. H. Vaziri, Head, 902-494-3217, Fax: 902-494-3960, E-mail: civil.engineering@dal.ca. *Application contact:* Shelley Parker, Admissions Coordinator, Graduate Studies and Research, 902-494-1288, Fax: 902-494-3149, E-mail: shelley.parker@dal.ca.

Drexel University, Graduate School, College of Engineering, Department of Civil and Architectural Engineering, Program in Civil Engineering, Philadelphia, PA 19104-2875. Offers MS, PhD. Part-time and evening/weekend programs available. *Students:* 15 full-time (3 women), 50 part-time (8 women); includes 2 minority (both Asian Americans or Pacific Islanders), 16 international. Average age 31. 73 applicants, 60% accepted. In 1998, 19 master's, 3 doctorates awarded. *Degree requirements:* For master's, thesis optional, foreign language not required; for doctorate, dissertation required, foreign language not required. *Entrance requirements:* For master's, TOEFL (minimum score of 570 required), minimum GPA of 3.0; for doctorate, TOEFL (minimum score of 570 required), minimum GPA of 3.5, MS in civil engineering. *Application deadline:* For fall admission, 8/21. Applications are processed on a rolling basis. Application fee: $35. Tuition: Full-time $15,795; part-time $585 per credit. Required fees: $375; $67 per term. Tuition and fees vary according to program. *Financial aid:* In 1998–99, 7 research assistantships, 19 teaching assistantships were awarded.; career-related internships or fieldwork, Federal Work-Study, institutionally-sponsored loans, tuition waivers (partial), and unspecified assistantships also available. Financial aid application deadline: 2/1. *Unit head:* Carolyn Waldo, Administrative Assistant, 212-854-3143, Fax: 212-854-6267, E-mail: clw1@columbia.edu. *Application contact:* Kelli Kennedy, Director of Admissions, 215-895-6706, Fax: 215-895-5939, E-mail: crowlka@duvm.ocs.drexel.edu.

Announcement: The Department of Civil and Architectural Engineering at Drexel University offers graduate studies leading to MSCE and PhD degrees in the areas of geoenvironmental, geosynthetics, structural, building systems, construction materials, hydraulics, and coastal engineering. Primary areas of research include geosynthetics, steel structure, stream flow hydraulics, noncorrosive and recycled materials, seismic modeling, building energy, and waste disposal. Visit the Web site at http://www.coe.drexel.edu/cae/cae.dept/cae.home.html.

Duke University, Graduate School, School of Engineering, Department of Civil and Environmental Engineering, Durham, NC 27708-0586. Offers civil and environmental engineer-

Civil Engineering

Duke University (continued)
ing (MS, PhD); environmental engineering (MS, PhD). Part-time programs available. *Faculty:* 18 full-time, 9 part-time. *Students:* 36 full-time (9 women), 18 international. 96 applicants, 40% accepted. In 1998, 9 master's, 3 doctorates awarded. Terminal master's awarded for partial completion of doctoral program. *Degree requirements:* For doctorate, dissertation required, foreign language not required, foreign language not required. *Entrance requirements:* For master's and doctorate, GRE General Test. *Application deadline:* For fall admission, 12/31; for spring admission, 11/1. Application fee: $75. *Financial aid:* Fellowships, research assistantships, Federal Work-Study available. Financial aid application deadline: 12/31. *Unit head:* Tod Laursen, Director of Graduate Studies, 919-660-5200, Fax: 919-660-5219, E-mail: pr@egr.duke.edu.

See in-depth description on page 577.

École Polytechnique de Montréal, Graduate Programs, Department of Civil Engineering, Montréal, PQ H3C 3A7, Canada. Offers environmental engineering (M Eng, M Sc A, PhD); geotechnical engineering (M Eng, M Sc A, PhD); hydraulics engineering (M Eng, M Sc A, PhD); structural engineering (M Eng, M Sc A, PhD); transportation engineering (M Eng, M Sc A, PhD). Part-time and evening/weekend programs available. *Degree requirements:* For master's and doctorate, one foreign language, computer language, thesis/dissertation required. *Entrance requirements:* For master's, minimum GPA of 2.75; for doctorate, minimum GPA of 3.0. *Faculty research:* Water resources management, characteristics of building materials, aging of dams, pollution control.

Florida Agricultural and Mechanical University, Division of Graduate Studies, Research, and Continuing Education, FAMU-FSU College of Engineering, Department of Civil Engineering, Tallahassee, FL 32307-3200. Offers civil engineering (MS, PhD); environmental engineering (MS, PhD). *Students:* 10 (2 women); all minorities (7 African Americans, 2 Asian Americans or Pacific Islanders, 1 Hispanic American) *Entrance requirements:* For master's, GRE General Test (minimum combined score of 1000 required), minimum GPA of 3.0. *Application deadline:* For fall admission, 7/1. Application fee: $20. *Unit head:* Dr. C. J. Chen, Dean, FAMU-FSU College of Engineering, 850-487-6100, Fax: 850-487-6486.

See in-depth description on page 579.

Florida Atlantic University, College of Engineering, Department of Ocean Engineering, Program in Civil Engineering, Boca Raton, FL 33431-0991. Offers MS. Part-time and evening/weekend programs available. *Faculty:* 3 full-time (0 women), 2 part-time (0 women). *Students:* 5 full-time (0 women), 12 part-time (2 women); includes 6 minority (1 African American, 5 Hispanic Americans), 3 international. Average age 34. In 1998, 13 degrees awarded. *Degree requirements:* For master's, thesis required (for some programs), foreign language not required. *Entrance requirements:* For master's, GRE General Test (minimum combined score of 1000 required), TOEFL (minimum score of 550 required), minimum GPA of 3.0. *Application deadline:* For fall admission, 3/1 (priority date); for spring admission, 7/1. Applications are processed on a rolling basis. Application fee: $20. Tuition, state resident: part-time $148 per credit hour. Tuition, nonresident: part-time $509 per credit hour. *Financial aid:* In 1998–99, 5 students received aid, including 5 research assistantships; career-related internships or fieldwork, Federal Work-Study, grants, tuition waivers (full), and unspecified assistantships also available. Financial aid applicants required to submit FAFSA. *Faculty research:* Structures, geotechnical engineering, environmental and water resources engineering, transportation engineering, materials. Total annual research expenditures: $250,000. *Unit head:* Dr. Panagiotis D. Scarlatos, Graduate Coordinator, 561-297-3444, Fax: 561-297-3885, E-mail: scarlatos@oe.fau.edu. *Application contact:* Patricia Capozziello, Graduate Admissions Coordinator, 561-297-2694, Fax: 561-297-2659, E-mail: capozzie@fau.edu.

Florida Institute of Technology, Graduate School, College of Engineering, Division of Engineering Science, Program in Civil Engineering, Melbourne, FL 32901-6975. Offers civil engineering (PhD); geotechnical engineering (MS); structures engineering (MS); water resources (MS). Part-time programs available. *Faculty:* 5 full-time (0 women), 4 part-time (2 women). *Students:* 7 full-time (0 women), 10 part-time (2 women); includes 1 minority (Hispanic American), 9 international. Average age 31. 46 applicants, 59% accepted. In 1998, 10 master's awarded. *Degree requirements:* For master's, thesis optional, foreign language not required; for doctorate, dissertation required, foreign language not required. *Entrance requirements:* For master's, minimum GPA of 3.0; for doctorate, minimum GPA of 3.2. *Application deadline:* Applications are processed on a rolling basis. Application fee: $50. Electronic applications accepted. Tuition: Part-time $575 per credit hour. Required fees: $100. Tuition and fees vary according to campus/location and program. *Financial aid:* In 1998–99, 7 students received aid, including 4 research assistantships with full and partial tuition reimbursements available (averaging $3,460 per year), 2 teaching assistantships with full and partial tuition reimbursements available (averaging $2,850 per year); tuition remissions also available. Financial aid application deadline: 3/1; financial aid applicants required to submit FAFSA. *Faculty research:* Groundwater and surface water modeling, pavements, waste materials, *in situ* soil testing, fiber optic sensors. Total annual research expenditures: $246,690. *Unit head:* Dr. Ashok Pandit, Chair, 407-674-7151, Fax: 407-768-7565, E-mail: apandit@fit.edu. *Application contact:* Carolyn P. Farrior, Associate Dean of Graduate Admissions, 407-674-7118, Fax: 407-723-9468, E-mail: cfarrior@fit.edu.

See in-depth description on page 581.

Florida International University, College of Engineering, Department of Civil and Environmental Engineering, Program in Civil Engineering, Miami, FL 33199. Offers MS, PhD. Part-time and evening/weekend programs available. *Students:* 15 full-time (2 women), 16 part-time (3 women); includes 13 minority (3 African Americans, 1 Asian American or Pacific Islander, 9 Hispanic Americans), 11 international. Average age 32. 46 applicants, 46% accepted. In 1998, 14 degrees awarded. *Degree requirements:* For master's, thesis optional; for doctorate, dissertation required. *Entrance requirements:* For master's, GRE General Test (minimum combined score of 1000 required), TOEFL (minimum score of 500 required), bachelor's degree in related field; for doctorate, GRE General Test (minimum combined score of 1000 required), TOEFL (minimum score of 550 required), minimum graduate GPA of 3.3. *Application deadline:* For fall admission, 4/1 (priority date); for spring admission, 10/1. Applications are processed on a rolling basis. Application fee: $20. Tuition, state resident: part-time $145 per credit hour. Tuition, nonresident: part-time $506 per credit hour. Required fees: $158; $158 per year. *Unit head:* Dr. David Shen, Chairperson, Department of Civil and Environmental Engineering, 305-348-2824, Fax: 305-348-2802, E-mail: shen@eng.fiu.edu.

Florida State University, Graduate Studies, FAMU/FSU College of Engineering, Department of Civil Engineering, Tallahassee, FL 32306. Offers civil engineering (MS, PhD); environmental engineering (MS, PhD). Part-time programs available. *Faculty:* 16 full-time (3 women). *Students:* 21 full-time (4 women), 22 part-time (5 women); includes 21 minority (12 African Americans, 7 Asian Americans or Pacific Islanders, 2 Hispanic Americans) Average age 23. 39 applicants, 56% accepted. In 1998, 10 master's awarded (100% found work related to degree). *Degree requirements:* For master's, thesis optional, foreign language not required; for doctorate, dissertation, preliminary exam, qualifying exam required, foreign language not required. *Entrance requirements:* For master's, GRE General Test (minimum combined score of 1000 required), TOEFL (minimum score of 550 required), BS in engineering or related field, minimum GPA of 3.0; for doctorate, GRE General Test, MS in engineering or related field, minimum GPA of 3.0. *Average time to degree:* Master's–2 years full-time, 3.5 years part-time. *Application deadline:* For fall admission, 7/15; for spring admission, 11/23. Applications are processed on a rolling basis. Application fee: $20. Tuition, state resident: part-time $139 per credit hour. Tuition, nonresident: part-time $482 per credit hour. Tuition and fees vary according to program. *Financial aid:* In 1998–99, 1 fellowship, 15 research assistantships (averaging $13,400 per year), 6 teaching assistantships (averaging $13,400 per year) were awarded.; tuition waivers (full) also available. Financial aid application deadline: 6/15. *Faculty research:* Tidal hydraulics, temperature effects on bridge girders, codes for coastal construction, field performance of pine bridges, river basin management, transportation pavement design, soil dynamics, structural analysis. *Unit head:* Dr. Jerry Wekezer, Chair, 850-410-6143, Fax: 850-410-6142, E-mail: wekezer@eng.fsu.edu. *Application contact:* J. Belinda Morris, Graduate Studies Assistant, 850-487-6319, Fax: 850-487-6142, E-mail: bmorris@eng.fsu.edu.

See in-depth description on page 579.

The George Washington University, School of Engineering and Applied Science, Department of Civil and Environmental Engineering, Washington, DC 20052. Offers MS, D Sc, App Sc, Engr. Part-time and evening/weekend programs available. *Degree requirements:* For master's, thesis optional, foreign language not required; for doctorate, computer language, dissertation, final and qualifying exams required, foreign language not required; for other advanced degree, foreign language and thesis not required. *Entrance requirements:* For master's, TOEFL (minimum score of 550 required; average 580) or George Washington University English as a Foreign Language Test, appropriate bachelor's degree, minimum GPA of 3.0; for doctorate, TOEFL (minimum score of 550 required; average 580) or George Washington University English as a Foreign Language Test, appropriate bachelor's or master's degree, minimum GPA of 3.4, GRE required if highest earned degree is BS; for other advanced degree, TOEFL (minimum score of 550 required; average 580) or George Washington University English as a Foreign Language Test, appropriate master's degree, minimum GPA of 3.0. *Application deadline:* For fall admission, 3/1 (priority date); for spring admission, 10/1. Applications are processed on a rolling basis. Application fee: $55. Tuition: Full-time $17,328; part-time $722 per credit hour. Required fees: $828; $35 per credit hour. Tuition and fees vary according to campus/location and program. *Financial aid:* Fellowships, research assistantships, teaching assistantships, career-related internships or fieldwork, Federal Work-Study, and institutionally-sponsored loans available. Financial aid application deadline: 3/1; financial aid applicants required to submit FAFSA. *Faculty research:* Computer-integrated manufacturing, materials engineering, electronic materials, fatigue and fracture, reliability. *Unit head:* Dr. Theodore Toridis, Chair, 202-994-6749. *Application contact:* Howard M. Davis, Manager, Office of Admissions and Student Records, 202-994-6158, Fax: 202-994-0909, E-mail: data: adms@seas.gwu.edu.

See in-depth description on page 583.

Georgia Institute of Technology, Graduate Studies and Research, College of Engineering, School of Civil and Environmental Engineering, Program in Civil Engineering, Atlanta, GA 30332-0001. Offers biomedical engineering (MS Bio E); civil engineering (MS, MSCE, PhD). Part-time programs available. Terminal master's awarded for partial completion of doctoral program. *Degree requirements:* For doctorate, computer language, dissertation required, foreign language not required, foreign language not required. *Entrance requirements:* For master's, GRE, TOEFL (minimum score of 550 required), minimum GPA of 3.0; for doctorate, GRE, TOEFL (minimum score of 550 required), minimum GPA of 3.2. *Faculty research:* Structural analysis, fluid mechanics, geotechnical engineering, construction management, transportation engineering.

Graduate School and University Center of the City University of New York, Graduate Studies, Program in Engineering, New York, NY 10036-8099. Offers chemical engineering (PhD); civil engineering (PhD); electrical engineering (PhD); mechanical engineering (PhD). *Faculty:* 68 full-time (1 woman). *Students:* 105 full-time (16 women), 11 part-time (2 women); includes 12 African Americans, 5 Asian Americans or Pacific Islanders, 4 Hispanic Americans *Degree requirements:* For doctorate, dissertation required, dissertation required. *Entrance requirements:* For doctorate, GRE General Test. *Application deadline:* For fall admission, 4/15. Application fee: $40. *Unit head:* Dr. Mumtaz Kassir, Acting Executive Officer, 212-650-8030.

Howard University, College of Engineering, Architecture, and Computer Sciences, School of Engineering and Computer Science, Department of Civil Engineering, Washington, DC 20059-0002. Offers M Eng. Offered through the Graduate School of Arts and Sciences. *Faculty:* 7 full-time (2 women). *Students:* 21 full-time (7 women); includes 13 minority (10 African Americans, 3 Asian Americans or Pacific Islanders) Average age 27. 15 applicants, 27% accepted. In 1998, 11 degrees awarded. *Degree requirements:* For master's, thesis optional, foreign language not required. *Entrance requirements:* For master's, GRE General Test, TOEFL, minimum GPA of 3.0, bachelor's degree in engineering or related field. *Average time to degree:* Master's–2.5 years full-time, 4 years part-time. *Application deadline:* For fall admission, 4/1; for spring admission, 11/1. Application fee: $45. Electronic applications accepted. *Financial aid:* In 1998–99, 17 students received aid, including 13 research assistantships with full tuition reimbursements available (averaging $8,000 per year), 4 teaching assistantships with full tuition reimbursements available (averaging $8,000 per year) Financial aid application deadline: 4/1. *Faculty research:* Modeling of concrete, structures, transportation planning, structural analysis, environmental and water resources. Total annual research expenditures: $916,774. *Unit head:* Dr. Lorraine N. Fleming, Chair, 202-806-6570, Fax: 202-806-5271, E-mail: fleming@scs.howard.edu.

See in-depth description on page 585.

Illinois Institute of Technology, Graduate College, Armour College of Engineering and Sciences, Department of Civil and Architectural Engineering, Chicago, IL 60616-3793. Offers M Geoenv E, M Trans E, MCEM, MGE, MPW, MS, MSE, PhD. Part-time and evening/weekend programs available. *Faculty:* 9 full-time (0 women), 11 part-time (0 women). *Students:* 36 full-time (8 women), 60 part-time (6 women); includes 13 minority (2 African Americans, 7 Asian Americans or Pacific Islanders, 4 Hispanic Americans), 55 international. 108 applicants, 63% accepted. In 1998, 23 master's, 4 doctorates awarded. Terminal master's awarded for partial completion of doctoral program. *Degree requirements:* For master's, thesis (for some programs), comprehensive exam required, foreign language not required; for doctorate, dissertation, comprehensive exam required, foreign language not required. *Entrance requirements:* For master's and doctorate, GRE (minimum score of 1200 required), TOEFL (minimum score of 550 required), undergraduate GPA of 3.0 required. *Application deadline:* For fall admission, 7/1; for spring admission, 11/1. Applications are processed on a rolling basis. Application fee: $30. Electronic applications accepted. *Financial aid:* In 1998–99, 1 fellowship, 11 teaching assistantships were awarded.; research assistantships, Federal Work-Study, institutionally-sponsored loans, scholarships, and graduate assistantships also available. Financial aid application deadline: 3/1. *Faculty research:* Structural engineering, construction management, geotechnical engineering, transportation engineering, optimum design of civil structures. Total annual research expenditures: $51,000. *Unit head:* Dr. J. Mohammadi, Chairman, 312-567-3540, Fax: 312-567-3519, E-mail: cemohammadi@minna.cns.iit.edu. *Application contact:* Dr. S. Mohammad Shahidehpour, Dean of Graduate College, 312-567-3024, Fax: 312-567-7517, E-mail: grad@minna.cns.iit.edu.

See in-depth description on page 587.

Instituto Tecnológico y de Estudios Superiores de Monterrey, Campus Monterrey, Graduate and Research Division, Programs in Engineering, Monterrey, 64849, Mexico. Offers applied statistics (M Eng); artificial intelligence (PhD); automation engineering (M Eng); chemical engineering (M Eng); civil engineering (M Eng); electrical engineering (M Eng); electronic engineering (M Eng); environmental engineering (M Eng); industrial engineering (M Eng, PhD); manufacturing engineering (M Eng); mechanical engineering (M Eng); systems and quality engineering (M Eng). M Eng offered jointly with the University of Waterloo; PhD (industrial engineering) offered jointly with Texas A&M University. Part-time and evening/weekend programs available. Terminal master's awarded for partial completion of doctoral program. *Degree requirements:* For master's and doctorate, one foreign language, computer language, thesis/dissertation required. *Entrance requirements:* For master's, PAEG, TOEFL; for doctorate, GRE, TOEFL, master's in related field. *Faculty research:* Flexible manufacturing cells, materials, statistical methods, environmental prevention, control and evaluation.

Iowa State University of Science and Technology, Graduate College, College of Engineering, Department of Civil and Construction Engineering, Ames, IA 50011. Offers civil engineering (MS, PhD), including civil engineering materials, construction engineering and management, environmental engineering, geometronics, geotechnical engineering, structural engineering, transportation engineering. *Faculty:* 36 full-time, 5 part-time. *Students:* 60 full-time (14 women), 51 part-time (12 women). 216 applicants, 28% accepted. In 1998, 6 master's, 39 doctorates awarded. *Degree requirements:* For master's, thesis or alternative required, foreign language not required; for doctorate, dissertation required, foreign language not required. *Entrance requirements:* For master's and doctorate, GRE General Test (foreign students), TOEFL. *Application deadline:* For fall admission, 6/15 (priority date); for spring admission, 11/15 (priority date). Application fee: $20 ($50 for international students). Electronic applications accepted. Tuition, state resident: full-time $3,308. Tuition, nonresident: full-time $9,744. Part-time tuition and fees vary according to course load, campus/location and program. *Financial aid:* In 1998–99, 53 research assistantships with partial tuition reimbursements (averaging $9,639 per year), 15 teaching assistantships with partial tuition reimbursements (averaging $8,775 per year) were awarded.; fellowships, scholarships also available. *Unit head:* Dr. Lowell F. Greimann, Chair, 515-294-2140, E-mail: cceinfo@iastate.edu. *Application contact:* Edward Kannel, Director of Graduate Education, 515-294-2861, E-mail: cceinfo@iastate.edu.

Johns Hopkins University, G. W. C. Whiting School of Engineering, Department of Civil Engineering, Baltimore, MD 21218-2699. Offers MCE, MSE, PhD. Part-time and evening/weekend programs available. *Faculty:* 7 full-time (0 women), 4 part-time (0 women). *Students:* 31 full-time (7 women), 18 international. Average age 28. 73 applicants, 16% accepted. In 1998, 9 master's, 5 doctorates awarded. Terminal master's awarded for partial completion of doctoral program. *Degree requirements:* For master's, thesis required, foreign language not required; for doctorate, dissertation, oral exam required, foreign language not required. *Entrance requirements:* For master's and doctorate, GRE General Test, TOEFL (minimum score of 560 required). *Application deadline:* For fall admission, 2/1. Application fee: $50. Tuition: Full-time $23,660. Tuition and fees vary according to program. *Financial aid:* In 1998–99, 30 students received aid, including 4 fellowships (averaging $12,933 per year), 26 teaching assistantships (averaging $11,195 per year); research assistantships, Federal Work-Study, grants, and institutionally-sponsored loans also available. Financial aid application deadline:2/1. *Faculty research:* Geotechnical engineering, structural dynamics, computer-aided engineering, computational methods, structural reliability. Total annual research expenditures: $599,000. *Unit head:* Poul V. Lade, Chair, 410-516-8680, Fax: 410-516-7473, E-mail: civil@jhu.edu.

Kansas State University, Graduate School, College of Engineering, Department of Civil Engineering, Manhattan, KS 66506. Offers MS, PhD. Postbaccalaureate distance learning degree programs offered (no on-campus study). *Faculty:* 15 full-time (0 women). *Students:* 22 full-time (3 women), 15 part-time (3 women); includes 2 minority (1 Asian American or Pacific Islander, 1 Hispanic American), 24 international. 76 applicants, 72% accepted. In 1998, 8 master's, 9 doctorates awarded. *Degree requirements:* For master's and doctorate, thesis/dissertation required. *Entrance requirements:* For master's and doctorate, GRE General Test (score in 80th percentile or higher on quantitative section required), TOEFL (minimum score of 550 required). *Average time to degree:* Master's–2 years full-time; doctorate–4 years full-time. *Application deadline:* Applications are processed on a rolling basis. Application fee: $0 ($25 for international students). Electronic applications accepted. *Financial aid:* In 1998–99, 20 research assistantships (averaging $9,000 per year) were awarded.; tuition waivers also available. *Faculty research:* Structural testing, transportation systems, soil and water contamination, numerical methods. Total annual research expenditures: $1.3 million. *Unit head:* Dr. Stuart Swartz, Head, 785-532-5862, Fax: 785-532-7717, E-mail: endsley@ce.ksu.edu. *Application contact:* Danita Deters, 785-532-5862, Fax: 785-532-7717, E-mail: grad_ce@engg.ksu.edu.

Lamar University, College of Graduate Studies, College of Engineering, Department of Civil Engineering, Program in Civil Engineering, Beaumont, TX 77710. Offers ME, MES, DE. Part-time programs available. *Faculty:* 5 full-time (0 women). *Students:* 11 full-time (1 woman), 3 part-time; includes 1 minority (Asian American or Pacific Islander), 12 international. Average age 26. In 1998, 3 master's awarded (100% found work related to degree); 1 doctorate awarded (100% found work related to degree). *Degree requirements:* For master's, thesis optional, foreign language not required; for doctorate, computer language, dissertation required, foreign language not required. *Entrance requirements:* For master's, GRE General Test (minimum combined score of 950 required), TOEFL (minimum score of 525 required); for doctorate, GRE General Test, TOEFL (minimum score of 530 required). *Average time to degree:* Master's–1 year full-time, 3 years part-time; doctorate–4 years full-time, 7 years part-time. *Application deadline:* For fall admission, 5/15 (priority date); for spring admission, 10/1 (priority date). Applications are processed on a rolling basis. Application fee: $0. *Financial aid:* In 1998–99, 6 fellowships, 5 research assistantships, 2 teaching assistantships were awarded. Financial aid application deadline: 4/1. *Faculty research:* Coastal engineering, structural composites, construction productivity, geotechnical grids. Total annual research expenditures: $170,000. *Unit head:* Edward Kannel, Director of Graduate Education, 515-294-2861, E-mail: cceinfo@iastate.edu. *Application contact:* Sandy Drane, Coordinator, International Students and Graduate Studies, 409-880-8349, Fax: 409-880-8414, E-mail: dranesl@lub002.lamar.edu.

Lawrence Technological University, College of Engineering, Southfield, MI 48075-1058. Offers automotive engineering (MAE); civil engineering (MCE); manufacturing systems (MEMS). Part-time and evening/weekend programs available. *Faculty:* 8 full-time (1 woman), 6 part-time (0 women). *Degree requirements:* For master's, foreign language and thesis not required. *Application deadline:* For fall admission, 8/1 (priority date); for spring admission, 1/1. Applications are processed on a rolling basis. Application fee: $50. Electronic applications accepted. Tuition: Full-time $5,128; part-time $419 per credit hour. Required fees: $100; $100 per year. $50 per semester. Tuition and fees vary according to course level. *Unit head:* Dr. George Kartsounes, Dean, 248-204-2500, Fax: 248-204-2509, E-mail: kartsounes@ltu.edu. *Application contact:* Lisa Kujawa, Director of Admissions, 248-204-3160, Fax: 248-204-3188, E-mail: admission@hu.edu.

Lehigh University, College of Engineering and Applied Science, Department of Civil and Environmental Engineering, Bethlehem, PA 18015-3094. Offers M Eng, MS, PhD. Part-time programs available. *Faculty:* 18 full-time (2 women), 2 part-time (0 women). *Students:* 70 full-time (7 women), 5 part-time (1 woman); includes 9 minority (6 African Americans, 1 Asian American or Pacific Islander, 2 Hispanic Americans), 31 international. Average age 22. 90 applicants, 74% accepted. In 1998, 24 master's, 3 doctorates awarded. Terminal master's awarded for partial completion of doctoral program. *Degree requirements:* For master's, thesis optional, foreign language not required; for doctorate, dissertation required, foreign language not required. *Entrance requirements:* For master's and doctorate, GRE General Test (score in 75th percentile or higher required), TOEFL (minimum score of 550 required). *Average time to degree:* Master's–2 years full-time; doctorate–4.7 years full-time. *Application deadline:* For fall admission, 7/15; for spring admission, 12/1. Applications are processed on a rolling basis. Application fee: $40. *Financial aid:* In 1998–99, 27 students received aid, including 7 fellowships, 12 research assistantships, 8 teaching assistantships; tuition waivers (partial) also available. Financial aid application deadline: 1/15. *Faculty research:* Structural and geotechnical engineering, water resources and coastal engineering. Total annual research expenditures: $4 million. *Unit head:* Dr. Arup K. SenGupta, Chairman, 610-758-3538, Fax: 610-758-6405, E-mail: aks@lehigh.edu. *Application contact:* Dr. Prisca Vidanage, Graduate Coordinator, 610-758-3530, Fax: 610-758-6405, E-mail: pmv1@lehigh.edu.

Louisiana State University and Agricultural and Mechanical College, Graduate School, College of Engineering, Department of Civil and Environmental Engineering, Baton Rouge, LA 70803. Offers environmental engineering (MSCE, PhD); geotechnical engineering (MSCE, PhD); structural engineering and mechanics (MSCE, PhD); transportation engineering (MSCE, PhD); water resources (MSCE, PhD). Part-time programs available. *Faculty:* 27 full-time (1 woman), 1 part-time (0 women). *Students:* 59 full-time (14 women), 24 part-time (2 women); includes 5 minority (1 African American, 3 Asian Americans or Pacific Islanders, 1 Hispanic American), 48 international. Average age 31. 188 applicants, 31% accepted. In 1998, 13 master's, 5 doctorates awarded. *Degree requirements:* For master's, thesis optional, foreign language not required; for doctorate, one foreign language (computer language can substitute), dissertation required. *Entrance requirements:* For master's and doctorate, GRE General Test, TOEFL, minimum GPA of 3.0. *Application deadline:* For fall admission, 1/25 (priority date). Applications are processed on a rolling basis. Application fee: $25. *Financial aid:* In 1998–99, 4 fellowships, 34 research assistantships with partial tuition reimbursements, 5 teaching assistantships with partial tuition reimbursements were awarded.; career-related internships or fieldwork, institutionally-sponsored loans, and scholarships also available. Financial aid application deadline: 3/1. *Faculty research:* Solid waste management, electrokinetics, composite structures, transportation planning, river mechanics. *Unit head:* Dr. Ronald F. Malone, Acting Chair, 225-388-8666, Fax: 225-388-8652, E-mail: rmalone@unix1.sncc.lsu.edu. *Application contact:* Dr. Vijaya K. A. Gopu, Graduate Coordinator, 225-388-8442, E-mail: cegopu@eng.lsu.edu.

See in-depth description on page 589.

Louisiana Tech University, Graduate School, College of Engineering and Science, Department of Civil Engineering, Ruston, LA 71272. Offers MS, D Eng. Part-time programs available. Terminal master's awarded for partial completion of doctoral program. *Degree requirements:* For master's, thesis or alternative required, foreign language not required; for doctorate, dissertation required, foreign language not required. *Entrance requirements:* For master's, GRE General Test (minimum combined score of 1070 required), TOEFL (minimum score of 550 required), minimum GPA of 3.0 in last 60 hours; for doctorate, TOEFL (minimum score of 550 required), minimum graduate GPA of 3.25 (with MS) or GRE General Test (minimum combined score of 1270 required without MS). *Faculty research:* Environmental engineering, trenchless excavation construction, structural mechanics, transportation materials and planning, water quality modeling.

Loyola Marymount University, Graduate Division, College of Science and Engineering, Department of Civil Engineering and Environmental Science, Programs in Civil Engineering, Los Angeles, CA 90045-8350. Offers MS, MSE. Part-time and evening/weekend programs available. *Faculty:* 5 full-time (0 women), 6 part-time (0 women). *Students:* 3 full-time (2 women), 23 part-time (6 women); includes 11 minority (3 African Americans, 4 Asian Americans or Pacific Islanders, 4 Hispanic Americans) 20 applicants, 70% accepted. In 1998, 15 degrees awarded. *Degree requirements:* For master's, computer language, comprehensive exam required, foreign language and thesis not required. *Entrance requirements:* For master's, TOEFL (minimum score of 550 required). Application fee: $35. Electronic applications accepted. Tuition: Part-time $525 per unit. Required fees: $143; $14 per semester. Tuition and fees vary according to program. *Financial aid:* In 1998–99, 6 students received aid. Grants and laboratory assistantships, available. Aid available to part-time students. Financial aid application deadline: 3/2; financial aid applicants required to submit FAFSA. *Unit head:* Dr. James E. Foxworthy, Professor, 310-338-2828. *Application contact:* Dr. James E. Foxworthy, Professor, 310-338-2828.

Manhattan College, Graduate Division, School of Engineering, Civil Engineering Program, Riverdale, NY 10471. Offers MS. Part-time and evening/weekend programs available. *Degree requirements:* For master's, foreign language and thesis not required. *Entrance requirements:* For master's, GRE, TOEFL, minimum GPA of 3.0. *Faculty research:* Compressible-inclusion function for geofoams used with rigid walls under static loading, validation of sediment criteria.

Marquette University, Graduate School, College of Engineering, Department of Civil and Environmental Engineering, Milwaukee, WI 53201-1881. Offers construction and public works management (MS, PhD); environmental/water resources engineering (MS, PhD); structural/geotechnical engineering (MS, PhD); transportational planning and engineering (MS, PhD). Part-time and evening/weekend programs available. *Faculty:* 12 full-time (0 women), 1 part-time (0 women). *Students:* 17 full-time (7 women), 32 part-time (5 women); includes 5 minority (1 African American, 4 Asian Americans or Pacific Islanders), 12 international. Average age 30. 45 applicants, 51% accepted. In 1998, 16 master's awarded (80% found work related to degree, 20% continued full-time study); 1 doctorate awarded (100% found work related to degree). Terminal master's awarded for partial completion of doctoral program. *Degree requirements:* For master's, thesis or alternative, comprehensive exam required, foreign language not required; for doctorate, dissertation required, foreign language not required. *Entrance requirements:* For master's, TOEFL (minimum score of 550 required); for doctorate, GRE General Test, TOEFL (minimum score of 550 required). *Average time to degree:* Master's–2 years full-time, 5 years part-time; doctorate–5 years full-time. *Application deadline:* For fall admission, 6/1 (priority date). Applications are processed on a rolling basis. Application fee: $40. Tuition: Part-time $510 per credit hour. Tuition and fees vary according to program. *Financial aid:* In 1998–99, 20 students received aid, including 5 research assistantships, 13 teaching assistantships; fellowships, Federal Work-Study, institutionally-sponsored loans, scholarships, and tuition waivers (full and partial) also available. Aid available to part-time students. Financial aid application deadline: 2/15. *Faculty research:* Highway safety, highway performance, and intelligent transportation systems; surface mount technology; watershed management. Total annual research expenditures: $1.2 million. *Unit head:* Dr. Thomas H. Wenzel, Chairman, 414-288-7030, Fax: 414-288-7521, E-mail: wenzelt@vms.csd.mu.edu.

Massachusetts Institute of Technology, School of Engineering, Department of Civil and Environmental Engineering, Cambridge, MA 02139-4307. Offers M Eng, SM, PhD, Sc D, CE, EE. *Faculty:* 36 full-time (6 women). *Students:* 245 full-time (62 women), 3 part-time (1 woman); includes 19 minority (3 African Americans, 8 Asian Americans or Pacific Islanders, 8 Hispanic Americans), 124 international. Average age 27. 503 applicants, 47% accepted. In 1998, 103 master's, 24 doctorates awarded. *Degree requirements:* For master's and other advanced degree, thesis required; for doctorate and other advanced degree, dissertation, comprehensive exams required. *Entrance requirements:* For master's, GRE General Test, TOEFL; for doctorate and other advanced degree, GRE General Test, TOEFL. *Average time to degree:* Master's–2 years full-time; doctorate–5 years full-time. *Application deadline:* For fall admission, 1/15; for spring admission, 11/1. Application fee: $55. *Financial aid:* In 1998–99, 187 students received aid, including 58 fellowships, 145 research assistantships, 36 teaching assistantships; career-related internships or fieldwork, Federal Work-Study, and institutionally-sponsored loans also available. Financial aid application deadline: 1/15; financial aid applicants required to submit FAFSA. *Faculty research:* Environmental chemistry and biology, environmental fluid dynamics and hydrodynamics, geoenvironment and geotechnology, surface and groundwater hydrology, materials and structures. Total annual research expenditures: $12.3 million. *Unit head:* Dr. Rafael L. Bras, Head, 617-253-2117, E-mail: rlbras@mit.edu. *Application contact:* Graduate Admissions Coordinator, 617-253-7119, Fax: 617-258-6775, E-mail: ceed@mit.edu.

McGill University, Faculty of Graduate Studies and Research, Faculty of Engineering, Department of Civil Engineering and Applied Mechanics, Montréal, PQ H3A 2T5, Canada. Offers environmental engineering and water resources management (M Eng, M Sc, PhD); fluid mechanics and hydraulic engineering (M Eng, M Sc, PhD); geotechnical and geoenvironmental engineering (M Eng, M Sc, PhD); structural engineering and construction materials (M Eng, M Sc, PhD). Part-time and evening/weekend programs available. *Degree requirements:* For master's, computer language required, thesis optional, foreign language not required; for doctorate, computer language, dissertation required, foreign language not required. *Entrance requirements:* For master's, TOEFL (minimum score of 550 required), minimum GPA of 3.0; for doctorate, TOEFL (minimum score of 580 required).

McMaster University, School of Graduate Studies, Faculty of Engineering, Department of Civil Engineering, Hamilton, ON L8S 4M2, Canada. Offers M Eng, PhD. *Faculty:* 18 full-time. *Students:* 32 full-time, 1 part-time. *Degree requirements:* For master's, thesis required, foreign language not required; for doctorate, dissertation, comprehensive exam required, foreign language not required. *Application deadline:* For fall admission, 3/1 (priority date). Application fee: $50. *Financial aid:* In 1998–99, teaching assistantships (averaging $7,722 per year); fellowships, research assistantships *Unit head:* Dr. D. Stolle, Chair, 905-525-9140 Ext. 24746.

Civil Engineering

McNeese State University, Graduate School, College of Engineering and Technology, Lake Charles, LA 70609-2495. Offers chemical engineering (M Eng); civil engineering (M Eng); electrical engineering (M Eng); mechanical engineering (M Eng). Part-time and evening/weekend programs available. *Faculty:* 13 full-time (1 woman). *Students:* 5 full-time (0 women), 3 part-time. *Degree requirements:* For master's, computer language, thesis or alternative required, foreign language not required. *Entrance requirements:* For master's, GRE General Test, TOEFL, minimum undergraduate GPA of 3.0. *Application deadline:* For fall admission, 7/15 (priority date). Applications are processed on a rolling basis. Application fee: $10 ($25 for international students). *Unit head:* Dr. O. C. Karkalits, Dean, 318-475-5875.

Memorial University of Newfoundland, School of Graduate Studies, Faculty of Engineering and Applied Science, St. John's, NF A1C 5S7, Canada. Offers civil engineering (M Eng, PhD); electrical engineering (M Eng, PhD); mechanical engineering (M Eng, PhD); ocean engineering (M Eng, PhD). Part-time programs available. *Students:* 75 full-time (11 women), 28 part-time (2 women), 31 international. *Degree requirements:* For master's, thesis optional; for doctorate, dissertation, comprehensive exam required. *Application deadline:* For fall admission, 3/1. Application fee: $40. *Unit head:* Dr. Rangaswany Seshadri, Dean, 709-737-8810, Fax: 709-737-8975, E-mail: sesh@engr.mun.ca. *Application contact:* Dr. J. J. Sharp, Associate Dean, 709-737-8901, Fax: 709-737-3480, E-mail: jsharp@engr.mun.ca.

Michigan State University, Graduate School, College of Engineering, Department of Civil and Environmental Engineering, East Lansing, MI 48824-1020. Offers civil engineering (MS, PhD); civil engineering-environmental toxicology (PhD); civil engineering-urban studies (MS); environmental engineering (MS, PhD), including environmental toxicology (PhD); environmental engineering-environmental toxicology (PhD); environmental engineering-urban studies (MS). Part-time programs available. *Faculty:* 17 full-time (2 women). *Students:* 44 full-time (11 women), 47 part-time (9 women); includes 13 minority (4 African Americans, 5 Asian Americans or Pacific Islanders, 4 Hispanic Americans), 52 international. Average age 28. 147 applicants, 48% accepted. In 1998, 35 master's, 9 doctorates awarded. Terminal master's awarded for partial completion of doctoral program. *Degree requirements:* For master's, foreign language and thesis not required; for doctorate, dissertation required, foreign language not required. *Entrance requirements:* For master's, GRE, TOEFL (minimum score of 570 required), minimum GPA of 3.0; for doctorate, GRE, TOEFL (minimum score of 570 required), minimum GPA of 3.0, MS. *Average time to degree:* Master's–2 years full-time, 3 years part-time; doctorate–4 years full-time, 6 years part-time. *Application deadline:* For fall admission, 5/31; for spring admission, 10/31. Applications are processed on a rolling basis. Application fee: $30 ($40 for international students). *Financial aid:* In 1998–99, 31 research assistantships (averaging $12,647 per year), 9 teaching assistantships with tuition reimbursements (averaging $12,058 per year) were awarded.; fellowships Financial aid application deadline: 2/15; financial aid applicants required to submit FAFSA. *Faculty research:* Highway safety, hazardous waste management, pavement design, concrete materials, waste water treatment. Total annual research expenditures: $1.9 million. *Unit head:* Dr. Ronald S. Harichandran, Chairperson, 517-355-5107, Fax: 517-432-1827, E-mail: harichan@egr.msu.edu. *Application contact:* Dr. Francis X. McKelvey, Graduate Coordinator, 517-355-5107, Fax: 517-432-1827, E-mail: mckelvey@pilot.msu.edu.

See in-depth description on page 593.

Michigan Technological University, Graduate School, College of Engineering, Department of Civil and Environmental Engineering, Program in Civil Engineering, Houghton, MI 49931-1295. Offers MS, PhD. Part-time programs available. *Faculty:* 27 full-time (4 women). *Students:* 41 full-time (6 women), 9 part-time (all women), 17 international. Average age 26. 22 applicants, 82% accepted. In 1998, 14 master's, 2 doctorates awarded. *Degree requirements:* For master's, foreign language and thesis not required; for doctorate, dissertation required, foreign language not required. *Entrance requirements:* For master's, GRE General Test (combined average 1735 on three sections), TOEFL (minimum score of 600 required); for doctorate, GRE General Test, TOEFL (minimum score of 600 required). *Average time to degree:* Master's–1.9 years full-time; doctorate–5.7 years full-time. *Application deadline:* For fall admission, 3/15 (priority date). Applications are processed on a rolling basis. Application fee: $30 ($35 for international students). Tuition, state resident: full-time $4,377. Tuition, nonresident: full-time $9,108. Required fees: $126. Tuition and fees vary according to course load. *Financial aid:* In 1998–99, 3 fellowships (averaging $3,271 per year), 16 research assistantships (averaging $8,294 per year), 13 teaching assistantships (averaging $6,290 per year) were awarded.; Federal Work-Study, institutionally-sponsored loans, and unspecified assistantships also available. Aid available to part-time students. Financial aid application deadline: 4/1; financial aid applicants required to submit FAFSA. *Unit head:* Dr. C. Robert Baillod, Chair, Department of Civil and Environmental Engineering, 906-487-2520, Fax: 906-487-2943, E-mail: baillod@mtu.edu.

Mississippi State University, College of Engineering, Department of Civil Engineering, Mississippi State, MS 39762. Offers MS. Part-time programs available. Postbaccalaureate distance learning degree programs offered (no on-campus study). *Students:* 12 full-time (3 women), 21 part-time (3 women); includes 5 minority (2 African Americans, 3 Asian Americans or Pacific Islanders) Average age 31. 24 applicants, 38% accepted. In 1998, 9 degrees awarded. *Degree requirements:* Foreign language not required. *Entrance requirements:* For master's, GRE General Test, TOEFL (minimum score of 550 required), minimum GPA of 2.75. *Application deadline:* For fall admission, 7/1; for spring admission, 11/1. Applications are processed on a rolling basis. Application fee: $25 for international students. *Financial aid:* Federal Work-Study, institutionally-sponsored loans, and unspecified assistantships available. Financial aid applicants required to submit FAFSA. *Faculty research:* Transportation, environmental structures. Total annual research expenditures: $412,435. *Unit head:* Dr. Adnan Shindala, Interim Head, 662-325-3050, Fax: 662-325-7189, E-mail: shindala@engr.msstate.edu. *Application contact:* Jerry B. Inmon, Director of Admissions, 662-325-2224, Fax: 662-325-7360, E-mail: admit@admissions.msstate.edu.

Montana State University–Bozeman, College of Graduate Studies, College of Engineering, Department of Civil Engineering, Bozeman, MT 59717. Offers civil engineering (MS); construction engineering management (MCEM); engineering (PhD); environmental engineering (MS). Part-time programs available. *Students:* 28 full-time (5 women), 17 part-time (2 women); includes 4 minority (3 Asian Americans or Pacific Islanders, 1 Native American) Average age 28. 27 applicants, 78% accepted. In 1998, 20 master's, 1 doctorate awarded. *Degree requirements:* For master's, thesis or alternative required, foreign language not required; for doctorate, dissertation required, foreign language not required. *Entrance requirements:* For master's and doctorate, GRE General Test, TOEFL (minimum score of 550 required). *Application deadline:* For fall admission, 6/1 (priority date); for spring admission, 11/1. Applications are processed on a rolling basis. Application fee: $50. *Financial aid:* In 1998–99, 24 research assistantships with full tuition reimbursements (averaging $8,910 per year), 1 teaching assistantship with full tuition reimbursement (averaging $8,100 per year) were awarded.; Federal Work-Study, scholarships, and tuition waivers (full and partial) also available. Financial aid application deadline: 3/1; financial aid applicants required to submit FAFSA. *Faculty research:* Geotechnical structures, environment, transportation, hydraulics, snow and ice. Total annual research expenditures: $1.8 million. *Unit head:* Dr. Donald Rabern, Head, 406-994-2111, E-mail: cedept@ce.montana.edu.

New Jersey Institute of Technology, Office of Graduate Studies, Department of Civil and Environmental Engineering, Program in Civil Engineering, Newark, NJ 07102-1982. Offers MS, PhD, Engineer. Part-time and evening/weekend programs available. Terminal master's awarded for partial completion of doctoral program. *Degree requirements:* For doctorate, dissertation, residency required, foreign language not required, foreign language not required. *Entrance requirements:* For master's and doctorate, GRE General Test (minimum score of 450 on verbal section, 600 on quantitative, 550 on analytical required). Electronic applications accepted.

New Mexico State University, Graduate School, College of Engineering, Department of Civil, Agricultural and Geological Engineering, Las Cruces, NM 88003-8001. Offers civil engineering (MSCE, PhD); environmental engineering (MS Env E). *Accreditation:* ABET (one or more programs are accredited). Part-time programs available. *Faculty:* 14 full-time (1 woman). *Students:* 33 full-time (13 women), 22 part-time (5 women); includes 16 minority (14 Hispanic Americans, 2 Native Americans), 10 international. Average age 33. 48 applicants, 44% accepted. In 1998, 14 degrees awarded. *Degree requirements:* For master's, thesis required (for some programs), foreign language not required; for doctorate, dissertation required, dissertation required. *Entrance requirements:* For doctorate, BS in engineering, minimum GPA of 3.0. *Application deadline:* For fall admission, 7/1 (priority date); for spring admission, 11/1. Applications are processed on a rolling basis. Application fee: $15 ($35 for international students). Electronic applications accepted. Tuition, state resident: full-time $2,682; part-time $112 per credit. Tuition, nonresident: full-time $8,376; part-time $349 per credit. Tuition and fees vary according to course load. *Financial aid:* Fellowships, research assistantships, teaching assistantships, career-related internships or fieldwork and Federal Work-Study available. Aid available to part-time students. Financial aid application deadline: 3/1. *Faculty research:* Structural inspection, evaluation and testing, transportation engineering, hydraulics/hydrology, geotechnical engineering. *Unit head:* Dr. Kenneth R. White, Head, 505-646-3801, Fax: 505-646-6049, E-mail: krwhite@nmsu.edu.

North Carolina Agricultural and Technical State University, Graduate School, College of Engineering, Department of Civil Engineering, Greensboro, NC 27411. Offers MSE. *Degree requirements:* Foreign language not required. *Application deadline:* For fall admission, 7/1 (priority date); for spring admission, 1/9. Applications are processed on a rolling basis. *Unit head:* Dr. Miguel Picornell, Chairperson, 336-334-7737.

North Carolina State University, Graduate School, College of Engineering, Department of Civil Engineering, Raleigh, NC 27695. Offers MCE, MS, PhD. Part-time programs available. *Faculty:* 43 full-time (2 women), 20 part-time (0 women). *Students:* 139 full-time (33 women), 47 part-time (7 women); includes 11 minority (4 African Americans, 6 Asian Americans or Pacific Islanders, 1 Hispanic American), 57 international. Average age 29. 238 applicants, 34% accepted. In 1998, 47 master's, 10 doctorates awarded. *Degree requirements:* For master's, thesis (for some programs), oral exams required, foreign language not required; for doctorate, dissertation, oral exams required, foreign language not required. *Entrance requirements:* For master's, GRE General Test (combined average 1850 on three sections), TOEFL (minimum score of 550 required; average 607), minimum B average in major; for doctorate, GRE General Test (combined average 1850 on three sections), TOEFL (minimum score of 550 required; average 607). *Average time to degree:* Master's–2 years full-time; doctorate–4 years full-time. Application fee: $45. *Financial aid:* In 1998–99, 13 fellowships (averaging $2,441 per year), 86 research assistantships (averaging $5,462 per year), 26 teaching assistantships (averaging $5,078 per year) were awarded.; Federal Work-Study also available. Financial aid application deadline: 3/1. *Faculty research:* Construction, environmental systems, transportation, water resources, structural, geotechnical, materials, computer-aided engineering. Total annual research expenditures: $6.7 million. *Unit head:* Dr. E. Downey Brill, Head, 919-515-2331, Fax: 919-515-7908, E-mail: brill@eos.ncsu.edu. *Application contact:* Dr. David W. Johnston, Director of Graduate Programs, 919-515-7412, Fax: 919-515-7908, E-mail: johnston@eos.ncsu.edu.

North Dakota State University, Graduate Studies and Research, College of Engineering and Architecture, Department of Civil Engineering, Fargo, ND 58105. Offers civil engineering (MS); environmental engineering (MS); natural resource management (MS). Part-time programs available. Postbaccalaureate distance learning degree programs offered (minimal on-campus study). *Faculty:* 6 full-time (0 women), 3 part-time (0 women). *Students:* 30 full-time (3 women); includes 1 minority (Asian American or Pacific Islander), 4 international. Average age 22. 38 applicants, 61% accepted. In 1998, 5 degrees awarded (100% found work related to degree). *Degree requirements:* For master's, computer language, thesis or alternative required, foreign language not required. *Entrance requirements:* For master's, TOEFL (minimum score of 525 required). *Application deadline:* For fall admission, 7/1 (priority date). Applications are processed on a rolling basis. Application fee: $25. *Financial aid:* In 1998–99, 1 fellowship with full tuition reimbursement (averaging $24,966 per year), 4 research assistantships with full tuition reimbursements (averaging $6,150 per year), 9 teaching assistantships with full tuition reimbursements (averaging $3,600 per year) were awarded.; career-related internships or fieldwork, Federal Work-Study, and institutionally-sponsored loans also available. Aid available to part-time students. Financial aid application deadline: 4/15. *Faculty research:* Wastewater, hydrology, structures, transportation, solid waste. Total annual research expenditures: $50,000. *Unit head:* Donald A. Andersen, Chair, 701-231-7244, Fax: 701-231-6185, E-mail: danderse@badlands.nodak.edu.

Northeastern University, College of Engineering, Department of Civil and Environmental Engineering, Boston, MA 02115-5096. Offers MS, PhD. Part-time programs available. *Faculty:* 15 full-time (1 woman), 1 part-time (0 women). *Students:* 39 full-time (13 women), 89 part-time (20 women); includes 2 African Americans, 4 Asian Americans or Pacific Islanders, 1 Hispanic American, 31 international. Average age 25. 163 applicants, 70% accepted. In 1998, 23 master's, 1 doctorate awarded. Terminal master's awarded for partial completion of doctoral program. *Degree requirements:* For master's, thesis optional, foreign language not required; for doctorate, dissertation, departmental qualifying exam required, foreign language not required. *Entrance requirements:* For master's and doctorate, GRE General Test. *Average time to degree:* Master's–3 years full-time, 5.25 years part-time; doctorate–4 years full-time. *Application deadline:* For fall admission, 4/15. Applications are processed on a rolling basis. Application fee: $50. *Financial aid:* In 1998–99, 22 students received aid, including 4 research assistantships with full tuition reimbursements available (averaging $12,450 per year), 10 teaching assistantships with full tuition reimbursements available (averaging $12,450 per year); fellowships, career-related internships or fieldwork, Federal Work-Study, tuition waivers (full), and unspecified assistantships also available. Aid available to part-time students. Financial aid application deadline: 2/15; financial aid applicants required to submit FAFSA. *Faculty research:* Earthquake engineering, geotechnical and geoenvironmental engineering, structural engineering, transportation engineering. *Unit head:* Dr. Mishac K. Yegian, Chairman, 617-373-2445, Fax: 617-373-4419. *Application contact:* Stephen L. Gibson, Associate Director, 617-373-2711, Fax: 617-373-2501, E-mail: grad-eng@coe.edu.

Northwestern University, The Graduate School, Robert R. McCormick School of Engineering and Applied Science, Department of Civil Engineering, Evanston, IL 60208. Offers biosolid mechanics (MS, PhD); environmental health engineering (MS, PhD); geotechnical engineering (MS, PhD); health physics/radiological health (MS, PhD); project management (MPM); structural engineering (MS, PhD); structural mechanics (MS, PhD); transportation systems engineering (MS, PhD). MS and PhD admissions and degrees offered through The Graduate School. Part-time programs available. *Faculty:* 26 full-time (2 women), 1 (woman) part-time. *Students:* 110 full-time (38 women), 7 part-time (2 women); includes 5 minority (2 African Americans, 3 Asian Americans or Pacific Islanders), 68 international. 202 applicants, 33% accepted. In 1998, 18 master's, 11 doctorates awarded. Terminal master's awarded for partial completion of doctoral program. *Degree requirements:* For master's, thesis required (for some programs); for doctorate, dissertation required. *Entrance requirements:* For master's and doctorate, GRE General Test, TOEFL (minimum score of 560 required). Application fee: $50 ($55 for international students). *Financial aid:* In 1998–99, 11 fellowships with full tuition reimbursements (averaging $11,673 per year), 35 research assistantships with partial tuition reimbursements (averaging $16,285 per year), 7 teaching assistantships with full tuition reimbursements (averaging $12,042 per year) were awarded.; career-related internships or fieldwork, institutionally-sponsored loans, and scholarships also available. Financial aid application deadline: 1/15; financial aid applicants required to submit FAFSA. *Faculty research:* Environmental health, geotechnics, mechanics of materials and solids, structural engineering and materials, transportation systems analysis and planning. Total annual research expenditures: $6.3 million. *Unit head:* Joseph L. Schofer, Chair, 847-491-3257, Fax: 847-491-4011, E-mail: j-schofer@nwu.edu. *Application contact:* Karm Kerwell, Secretary, 847-491-3176, Fax: 847-491-4011, E-mail: k-kerwell@nwu.edu.

The Ohio State University, Graduate School, College of Engineering, Department of Civil and Environmental Engineering and Geodetic Science, Program in Civil Engineering, Columbus, OH 43210. Offers MS, PhD. *Faculty:* 21 full-time, 8 part-time. *Students:* 51 full-time (16 women), 24 part-time (4 women); includes 9 minority (1 African American, 4 Asian Americans or Pacific Islanders, 4 Hispanic Americans), 41 international. 222 applicants, 30% accepted. In 1998, 27 master's, 9 doctorates awarded. *Degree requirements:* For master's, computer language required, thesis optional, foreign language not required; for doctorate, computer language, dissertation required, foreign language not required. *Entrance requirements:* For master's and doctorate, GRE General Test, TOEFL (minimum score of 525 required). *Application deadline:* For fall admission, 8/15. Applications are processed on a rolling basis. Application fee: $30 ($40 for international students). *Financial aid:* Fellowships, research assistantships, teaching assistantships, career-related internships or fieldwork, Federal Work-Study, institutionally-sponsored loans, and unspecified assistantships available. Aid available to part-time students. *Unit head:* Hojjat Adeli, Graduate Studies Committee Chair, 614-292-7929, Fax: 614-292-3780, E-mail: adeli.1@osu.edu.

Ohio University, Graduate Studies, College of Engineering and Technology, Department of Civil Engineering, Athens, OH 45701-2979. Offers geotechnical and environmental engineering (MS); water resources and structures (MS). *Faculty:* 11 full-time (1 woman). *Students:* 24 full-time (5 women), 2 part-time, 13 international. Average age 28. 58 applicants, 86% accepted. In 1998, 6 degrees awarded. *Degree requirements:* For master's, thesis optional, foreign language not required. *Entrance requirements:* For master's, minimum GPA of 3.0. *Application deadline:* For fall admission, 5/1 (priority date). Applications are processed on a rolling basis. Application fee: $30. Tuition, state resident: full-time $5,754; part-time $238 per credit hour. Tuition, nonresident: full-time $11,055; part-time $457 per credit hour. Tuition and fees vary according to course load, campus/location and program. *Financial aid:* In 1998–99, 1 fellowship (averaging $15,000 per year), 12 research assistantships, 6 teaching assistantships were awarded.; Federal Work-Study, institutionally-sponsored loans, and tuition waivers (full and partial) also available. Financial aid application deadline: 3/15. *Faculty research:* Soil-structure interaction, solid waste management, pipes, pavements, noise pollution, mine reclamation, drought analysis. Total annual research expenditures: $500,000. *Unit head:* Dr. Gayle Mitchell, Chair, 740-593-1465, Fax: 740-593-0625, E-mail: gmitchell@bobcat.ent.ohiou.edu. *Application contact:* Dr. Glenn Hazen, Graduate Chair, 740-593-1469, Fax: 740-593-0625, E-mail: ghazen@bobcat.ent.ohiou.edu.

Oklahoma State University, Graduate College, College of Engineering, Architecture and Technology, School of Civil and Environmental Engineering, Stillwater, OK 74078. Offers civil engineering (M En, MS, PhD); environmental engineering (M En, MS, PhD). *Faculty:* 16 full-time (1 woman). *Students:* 30 full-time (4 women), 68 part-time (11 women); includes 9 minority (2 African Americans, 3 Asian Americans or Pacific Islanders, 1 Hispanic American, 3 Native Americans), 23 international. Average age 32. In 1998, 28 master's, 2 doctorates awarded. *Degree requirements:* For master's, thesis or alternative required, foreign language not required; for doctorate, dissertation required, foreign language not required. *Entrance requirements:* For master's and doctorate, TOEFL (minimum score of 550 required). *Application deadline:* For fall admission, 7/1 (priority date). Application fee: $25. *Financial aid:* In 1998–99, 27 students received aid, including 7 research assistantships (averaging $9,985 per year), 20 teaching assistantships (averaging $8,532 per year); career-related internships or fieldwork, Federal Work-Study, and tuition waivers (partial) also available. Aid available to part-time students. Financial aid application deadline: 3/1. *Unit head:* Robert Hughes, Head, 405-744-5189.

Old Dominion University, College of Engineering and Technology, Department of Civil and Environmental Engineering, Norfolk, VA 23529. Offers civil engineering (ME, MS, PhD); environmental engineering (ME, MS, PhD). Part-time and evening/weekend programs available. Postbaccalaureate distance learning degree programs offered (minimal on-campus study). *Faculty:* 11 full-time. *Students:* 24 full-time (7 women), 45 part-time (10 women); includes 5 minority (3 African Americans, 1 Asian American or Pacific Islander, 1 Hispanic American), 13 international. Average age 33. In 1998, 24 master's, 3 doctorates awarded. *Degree requirements:* For master's, comprehensive exam required; for doctorate, dissertation, candidacy exam required, foreign language not required. *Entrance requirements:* For master's, GRE, TOEFL (minimum score of 550 required), minimum GPA of 3.0; for doctorate, GRE, TOEFL (minimum score of 550 required), minimum GPA of 3.5. *Application deadline:* For fall admission, 7/1; for spring admission, 10/1. Applications are processed on a rolling basis. Application fee: $30. Electronic applications accepted. *Financial aid:* In 1998–99, 32 students received aid, including 12 research assistantships (averaging $6,713 per year), 4 teaching assistantships (averaging $5,200 per year); fellowships, career-related internships or fieldwork, grants, scholarships, and tuition waivers (partial) also available. Aid available to part-time students. Financial aid application deadline: 2/15; financial aid applicants required to submit FAFSA. *Faculty research:* Structural engineering, coastal engineering, geotechnical engineering, water resources. Total annual research expenditures: $626,084. *Unit head:* Dr. Isao Ishibashi, Chair, 757-683-3753, Fax: 757-683-5354, E-mail: iishbas@cee.odu.edu. *Application contact:* Dr. Isao Ishibashi, Chair, 757-683-3753, Fax: 757-683-5354, E-mail: iishbas@cee.odu.edu.

Oregon State University, Graduate School, College of Engineering, Department of Civil Engineering, Corvallis, OR 97331. Offers civil engineering (MAIS, MS, PhD); ocean engineering (M Oc E). Part-time programs available. *Faculty:* 28 full-time (3 women), 1 part-time (0 women). *Students:* 65 full-time (8 women), 12 part-time; includes 5 minority (all Asian Americans or Pacific Islanders), 33 international. Average age 29. In 1998, 32 master's, 5 doctorates awarded. Terminal master's awarded for partial completion of doctoral program. *Degree requirements:* For master's, thesis or alternative required, foreign language not required; for doctorate, one foreign language, computer language, dissertation required. *Entrance requirements:* For master's, GRE General Test (minimum combined score of 1000 required), TOEFL (minimum score of 550 required), minimum GPA of 3.0 in last 90 hours (3.5 for MS); for doctorate, GRE General Test (minimum combined score of 1000 required), TOEFL (minimum score of 550 required), minimum GPA of 3.0 in last 90 hours of undergraduate course work. *Application deadline:* For fall admission, 3/1 (priority date). Applications are processed on a rolling basis. Application fee: $50. *Financial aid:* Fellowships, research assistantships, teaching assistantships, career-related internships or fieldwork and institutionally-sponsored loans available. Aid available to part-time students. Financial aid application deadline: 2/1. *Faculty research:* Hazardous waste management, carbon cycling, wave forces on structures, pavement design, seismic analysis. *Unit head:* Dr. Wayne C. Huber, Head, 541-737-6150, E-mail: wayne.huber@orst.edu. *Application contact:* Linda A. Rowe, Office Manager, 541-737-6149, Fax: 541-737-3052, E-mail: linda.rowe.orst.edu.

Pennsylvania State University University Park Campus, Graduate School, College of Engineering, Department of Civil and Environmental Engineering, Program in Civil Engineering, State College, University Park, PA 16802-1503. Offers M Eng, MS, PhD. *Students:* 56 full-time (15 women), 36 part-time (4 women). In 1998, 38 master's, 5 doctorates awarded. *Degree requirements:* For master's, final paper (M Eng), oral exam and thesis (MS) required; for doctorate, dissertation, comprehensive and oral exams required. *Entrance requirements:* For master's and doctorate, GRE General Test, BS in engineering. Application fee: $50. *Unit head:* Dr. Paul P. Jovanis, Head, Department of Civil and Environmental Engineering, 814-863-3084.

Polytechnic University, Brooklyn Campus, Department of Civil and Environmental Engineering, Major in Civil Engineering, Brooklyn, NY 11201-2990. Offers MS, PhD. Part-time and evening/weekend programs available. *Students:* 9 full-time (2 women), 66 part-time (7 women); includes 14 minority (3 African Americans, 10 Asian Americans or Pacific Islanders, 1 Hispanic American), 18 international. Average age 33. 45 applicants, 49% accepted. In 1998, 15 master's, 4 doctorates awarded. *Degree requirements:* For master's, thesis or alternative required; for doctorate, dissertation required. *Entrance requirements:* For doctorate, qualifying exam, MS in civil engineering. *Application deadline:* Applications are processed on a rolling basis. Application fee: $45. Electronic applications accepted. *Financial aid:* Fellowships, research assistantships, teaching assistantships, institutionally-sponsored loans available. Aid available to part-time students. Financial aid applicants required to submit FAFSA. *Unit head:* 610-648-3242, Fax: 610-889-1334. *Application contact:* John S. Kerge, Dean of Admissions, 718-260-3200, Fax: 718-260-3446, E-mail: admitme@poly.edu.

Polytechnic University, Farmingdale Campus, Graduate Programs, Department of Civil and Environmental Engineering, Major in Civil Engineering, Farmingdale, NY 11735-3995. Offers MS, PhD. Average age 33. 3 applicants, 67% accepted. In 1998, 2 master's awarded. *Degree requirements:* For master's, computer language, thesis required (for some programs); for doctorate, dissertation required. *Application deadline:* Applications are processed on a rolling basis. Application fee: $45. Electronic applications accepted. *Financial aid:* Institutionally-sponsored loans available. Aid available to part-time students. Financial aid applicants required to submit FAFSA. *Unit head:* Kathy Kelly, Director of Admissions, 973-596-3300, Fax: 973-596-3461, E-mail: admissions@njit.edu. *Application contact:* John S. Kerge, Dean of Admissions, 718-260-3200, Fax: 718-260-3446, E-mail: admitme@poly.edu.

Polytechnic University, Westchester Graduate Center, Graduate Programs, Department of Civil and Environmental Engineering, Major in Civil Engineering, Hawthorne, NY 10532-1507. Offers MS, PhD. Average age 33. *Degree requirements:* For master's, computer language, thesis or alternative required; for doctorate, dissertation required. *Application deadline:* Applications are processed on a rolling basis. Application fee: $45. Electronic applications accepted. *Unit head:* John S. Kerge, Dean of Admissions, 718-260-3200, Fax: 718-260-3446, E-mail: admitme@poly.edu. *Application contact:* John S. Kerge, Dean of Admissions, 718-260-3200, Fax: 718-260-3446, E-mail: admitme@poly.edu.

Portland State University, Graduate Studies, School of Engineering and Applied Science, Department of Civil Engineering, Portland, OR 97207-0751. Offers MS, PhD. Part-time and evening/weekend programs available. *Faculty:* 10 full-time (1 woman), 10 part-time (0 women). *Students:* 15 full-time (2 women), 13 part-time (3 women); includes 3 minority (2 Asian Americans or Pacific Islanders, 1 Native American), 5 international. Average age 29. 24 applicants, 75% accepted. In 1998, 14 degrees awarded. *Degree requirements:* For master's, computer language required, foreign language not required; for doctorate, one foreign language, computer language, dissertation, oral and written exams required. *Entrance requirements:* For master's, TOEFL (minimum score of 550 required), minimum GPA of 3.0 in upper-division course work or 2.75 overall, BS in civil engineering or allied field; for doctorate, GRE General Test, GRE Subject Test, minimum GPA of 3.0 in upper-division course work. *Application deadline:* For fall admission, 4/1; for spring admission, 11/1. Applications are processed on a rolling basis. Application fee: $50. *Financial aid:* In 1998–99, 12 research assistantships, 3 teaching assistantships were awarded.; career-related internships or fieldwork, Federal Work-Study, and institutionally-sponsored loans also available. Aid available to part-time students. Financial aid application deadline: 3/1; financial aid applicants required to submit FAFSA. *Faculty research:* Structures, water resources, geotechnical engineering, environmental engineering, transportation. Total annual research expenditures: $379,517. *Unit head:* Dr. Franz Rad, Head, 503-725-4282, Fax: 503-725-4298. *Application contact:* 503-725-4244, Fax: 503-725-4298, E-mail: cedept@eas.pdx.edu.

Portland State University, Graduate Studies, Systems Science Program, Portland, OR 97207-0751. Offers systems science/anthropology (PhD); systems science/business administration (PhD); systems science/civil engineering (PhD); systems science/economics (PhD); systems science/engineering management (PhD); systems science/general (PhD); systems science/mathematical sciences (PhD); systems science/mechanical engineering (PhD); systems science/psychology (PhD); systems science/sociology (PhD). *Faculty:* 3 full-time (0 women), 1 part-time (0 women). *Students:* 45 full-time (17 women), 23 part-time (6 women); includes 5 minority (1 African American, 3 Asian Americans or Pacific Islanders, 1 Hispanic American), 12 international. *Degree requirements:* For doctorate, variable foreign language requirement, computer language, dissertation required. *Entrance requirements:* For doctorate, GMAT (score in 75th percentile or higher required), GRE General Test (score in 75th percentile or higher required), TOEFL (minimum score of 575 required), minimum undergraduate GPA of 3.0. *Application deadline:* For fall admission, 2/1; for spring admission, 11/1. Application fee: $50. *Unit head:* Dr. Nancy Perrin, Director, 503-725-4960, E-mail: perrinn@pdx.edu. *Application contact:* Dawn Kuenle, Coordinator, 503-725-4960, E-mail: dawn@sysc.pdx.edu.

Princeton University, Graduate School, School of Engineering and Applied Science, Department of Civil Engineering and Operations Research, Princeton, NJ 08544-1019. Offers environmental engineering and water resources (PhD); financial engineering (M Eng); statistics and operations research (MSE, PhD); structural engineering (M Eng); structures and mechanics (MSE, PhD); transportation systems (MSE, PhD). *Degree requirements:* For master's and doctorate, thesis/dissertation required. *Entrance requirements:* For master's and doctorate, GRE General Test, GRE Subject Test.

Purdue University, Graduate School, Schools of Engineering, School of Civil Engineering, West Lafayette, IN 47907. Offers MS, MSCE, MSE, PhD. Part-time programs available. *Faculty:* 55 full-time (4 women), 8 part-time (1 woman). *Students:* 205 full-time (37 women), 45 part-time (8 women); includes 18 minority (2 African Americans, 7 Asian Americans or Pacific Islanders, 9 Hispanic Americans), 142 international. Average age 25. 479 applicants, 57% accepted. In 1998, 83 master's, 20 doctorates awarded. Terminal master's awarded for partial completion of doctoral program. *Degree requirements:* For master's, thesis required (for some programs), foreign language not required; for doctorate, dissertation required, foreign language not required. *Entrance requirements:* For master's and doctorate, GRE General Test, TOEFL (minimum score of 575 required). *Application deadline:* For fall admission, 3/1 (priority date); for spring admission, 9/15. Applications are processed on a rolling basis. Application fee: $30. Electronic applications accepted. *Financial aid:* In 1998–99, 160 students received aid, including 12 fellowships, 96 research assistantships, 52 teaching assistantships Aid available to part-time students. Financial aid application deadline: 6/30; financial aid applicants required to submit FAFSA. *Faculty research:* Environmental and hydraulic engineering, geotechnical and materials engineering, structural engineering, construction engineering, transportation and urban engineering. *Unit head:* Dr. V. P. Drnevich, Head, 765-494-2159. *Application contact:* Marcie Duffin, Graduate Secretary, 765-494-2156, Fax: 765-494-0395, E-mail: civlgrad@ecn.purdue.edu.

See in-depth description on page 595.

Queen's University at Kingston, School of Graduate Studies and Research, Faculty of Applied Science, Department of Civil Engineering, Kingston, ON K7L 3N6, Canada. Offers M Sc, M Sc Eng, PhD. Part-time programs available. *Students:* 37 full-time (11 women), 1 (woman) part-time. In 1998, 12 master's, 3 doctorates awarded. *Degree requirements:* For master's, thesis optional, foreign language not required; for doctorate, dissertation, comprehensive exam required. *Entrance requirements:* For master's and doctorate, TOEFL (minimum score of 600 required). *Application deadline:* For fall admission, 2/28 (priority date). Application fee: $60. Electronic applications accepted. *Financial aid:* Fellowships, research assistantships, teaching assistantships, institutionally-sponsored loans available. Financial aid application deadline: 3/1. *Faculty research:* Structural, geotechnical, transportation, hydrotechnical, and environmental engineering. *Unit head:* Dr. D. J. Turcke, Head, 613-533-4228. *Application contact:* Dr. T. I. Campbell, Graduate Coordinator, 613-533-2141.

Rensselaer Polytechnic Institute, Graduate School, School of Engineering, Department of Civil Engineering, Troy, NY 12180-3590. Offers geotechnical engineering (M Eng, MS, D Eng, PhD); mechanics of composite materials and structures (M Eng, MS, D Eng, PhD); structural engineering (M Eng, MS, D Eng, PhD); transportation engineering (M Eng, MS, D Eng, PhD). Part-time programs available. *Faculty:* 11 full-time (0 women), 3 part-time (0 women). *Students:* 31 full-time (4 women), 11 part-time (2 women); includes 6 minority (3 Asian Americans or Pacific Islanders, 3 Hispanic Americans), 18 international. 77 applicants, 45% accepted. In 1998, 10 master's, 5 doctorates awarded. *Degree requirements:* For master's, thesis required (for some programs), foreign language not required; for doctorate, dissertation required,

Civil Engineering

Rensselaer Polytechnic Institute (continued)
foreign language not required. *Entrance requirements:* For master's and doctorate, GRE, TOEFL (minimum score of 550 required). *Application deadline:* For fall admission, 2/1 (priority date). Applications are processed on a rolling basis. Application fee: $35. *Financial aid:* In 1998–99, 22 students received aid, including 6 fellowships, 8 research assistantships, 8 teaching assistantships; career-related internships or fieldwork and institutionally-sponsored loans also available. Financial aid application deadline: 2/1. *Faculty research:* Computational mechanics, earthquake engineering, geo-environmental engineering. Total annual research expenditures: $3.5 million. *Unit head:* Dr. George List, Chair, 518-276-6940, Fax: 518-276-4833, E-mail: gregaja@rpi.edu. *Application contact:* Jo Ann Grega, Admissions Assistant, 518-276-6679, Fax: 518-276-4833, E-mail: gregaj2@rpi.edu.

See in-depth description on page 597.

Rice University, Graduate Programs, George R. Brown School of Engineering, Department of Civil Engineering, Houston, TX 77251-1892. Offers civil engineering (MCE, MS, PhD); structural engineering (MCE, MS, PhD). Part-time programs available. *Degree requirements:* For master's, thesis required (for some programs), foreign language not required; for doctorate, dissertation required, foreign language not required. *Entrance requirements:* For master's and doctorate, GRE General Test, GRE Subject Test, TOEFL (minimum score of 550 required), minimum GPA of 3.0. *Faculty research:* Structural dynamics, probabilistic studies in dynamics, fatigue, reinforced concrete experimental research.

Rose-Hulman Institute of Technology, Faculty of Engineering and Applied Sciences, Department of Civil Engineering, Program in Civil Engineering, Terre Haute, IN 47803-3920. Offers MS. Part-time programs available. 2 applicants, 0% accepted. *Degree requirements:* For master's, thesis required, foreign language not required. *Entrance requirements:* For master's, GRE, TOEFL (minimum score of 580 required), minimum GPA of 3.0. *Application deadline:* For fall admission, 2/1 (priority date). Applications are processed on a rolling basis. Application fee: $0. *Financial aid:* Fellowships, research assistantships, teaching assistantships, grants, institutionally-sponsored loans, and tuition waivers (full and partial) available. Financial aid application deadline: 2/1. *Application contact:* Dr. Buck F. Brown, Dean for Research and Graduate Studies, 812-877-8403, Fax: 812-877-8102, E-mail: buck.brown@rose-hulman.edu.

Rutgers, The State University of New Jersey, New Brunswick, Graduate School, Program in Civil and Environmental Engineering, New Brunswick, NJ 08903. Offers MS, PhD. Part-time and evening/weekend programs available. *Faculty:* 12 full-time (1 woman), 6 part-time (0 women). *Students:* 31 full-time (4 women), 66 part-time (21 women); includes 18 minority (2 African Americans, 13 Asian Americans or Pacific Islanders, 3 Hispanic Americans), 24 international. Average age 24. 96 applicants, 73% accepted. In 1998, 22 master's, 3 doctorates awarded. Terminal master's awarded for partial completion of doctoral program. *Degree requirements:* For master's, thesis optional, foreign language not required; for doctorate, dissertation required, foreign language not required. *Entrance requirements:* For master's and doctorate, GRE General Test. *Average time to degree:* Master's–2.5 years full-time, 4 years part-time; doctorate–3 years full-time, 5 years part-time. *Application deadline:* For fall admission, 6/1 (priority date); for spring admission, 11/1. Applications are processed on a rolling basis. Application fee: $50. *Financial aid:* In 1998–99, 2 fellowships with full and partial tuition reimbursements, 5 research assistantships with full and partial tuition reimbursements, 5 teaching assistantships with full and partial tuition reimbursements were awarded.; Federal Work-Study, grants, scholarships, and tuition waivers (full and partial) also available. Financial aid application deadline: 3/1; financial aid applicants required to submit FAFSA. *Faculty research:* Structural mechanics, soil mechanics, environmental geotechnology, water resources, computational mechanics. *Unit head:* Dr. Nenad Gucunski, Director, 732-445-4413, Fax: 732-445-0577, E-mail: gucunski@dora.rutgers.edu. *Application contact:* Connie Dellamura, Graduate Secretary, 732-445-2232, Fax: 732-445-0577, E-mail: dellamur@rci.rutgers.edu.

San Diego State University, Graduate and Research Affairs, College of Engineering, Department of Civil and Environmental Engineering, San Diego, CA 92182. Offers civil engineering (MS). Part-time and evening/weekend programs available. *Students:* 19 full-time (3 women), 44 part-time (10 women); includes 14 minority (1 African American, 7 Asian Americans or Pacific Islanders, 6 Hispanic Americans), 8 international. Average age 29. 38 applicants, 66% accepted. In 1998, 12 degrees awarded. *Degree requirements:* For master's, thesis optional, foreign language not required. *Entrance requirements:* For master's, GRE General Test (minimum combined score of 950 required), TOEFL (minimum score of 550 required). *Application deadline:* For fall admission, 7/1 (priority date); for spring admission, 12/1. Applications are processed on a rolling basis. Application fee: $55. *Financial aid:* In 1998–99, 2 teaching assistantships were awarded.; career-related internships or fieldwork and Federal Work-Study also available. *Faculty research:* Hydraulics, hydrology, transportation, smart material, concrete material. Total annual research expenditures: $245,000. *Unit head:* Dr. Janasz Supernak, Chair, 619-594-6071, Fax: 619-594-6005, E-mail: supernak@jcsnext.sdsu.edu. *Application contact:* Fang-Hui Chou, Graduate Adviser, 619-594-7007, Fax: 619-594-6005, E-mail: fchou@mail.sdsu.edu.

San Jose State University, Graduate Studies, College of Engineering, Department of Civil Engineering and Applied Mechanics, San Jose, CA 95192-0001. Offers MS. *Faculty:* 15 full-time (2 women), 9 part-time (0 women). *Students:* 29 full-time (6 women), 76 part-time (18 women); includes 39 minority (4 African Americans, 28 Asian Americans or Pacific Islanders, 7 Hispanic Americans), 9 international. Average age 30. 63 applicants, 73% accepted. In 1998, 41 degrees awarded. *Degree requirements:* For master's, thesis or alternative required. *Entrance requirements:* For master's, minimum GPA of 2.7. *Application deadline:* For fall admission, 6/1. Applications are processed on a rolling basis. Application fee: $59. Tuition, nonresident: part-time $246 per unit. Required fees: $1,939; $1,309 per year. *Unit head:* Dr. Thalia Anagnos, Chair, 408-924-3900, Fax: 408-924-4004. *Application contact:* Dr. Rhea Williamson, Graduate Adviser, 408-924-3849.

Santa Clara University, School of Engineering, Department of Civil Engineering, Santa Clara, CA 95053-0001. Offers MSCE. Part-time and evening/weekend programs available. *Students:* 1 (woman) full-time, 2 part-time; includes 2 minority (both Asian Americans or Pacific Islanders), 1 international. Average age 26. 4 applicants, 75% accepted. In 1998, 2 degrees awarded. *Degree requirements:* For master's, thesis or alternative required, foreign language not required. *Entrance requirements:* For master's, GRE General Test, TOEFL (minimum score of 550 required), minimum GPA of 2.75. *Application deadline:* For fall admission, 6/1; for spring admission, 1/1. Applications are processed on a rolling basis. Application fee: $40. *Financial aid:* Research assistantships, teaching assistantships, Federal Work-Study and scholarships available. Aid available to part-time students. Financial aid application deadline: 2/1. *Unit head:* Dr. Sukhmander Singh, Chair, 408-554-4061. *Application contact:* Tina Samms, Assistant Director of Graduate Admissions, 408-554-4313, Fax: 408-554-5474, E-mail: engr-grad@scu.edu.

South Dakota School of Mines and Technology, Graduate Division, Department of Civil and Environmental Engineering, Rapid City, SD 57701-3995. Offers civil engineering (MS). Part-time programs available. *Faculty:* 14 full-time (0 women), 1 (woman) part-time. *Students:* 41 full-time (8 women), 14 international. Average age 29. In 1998, 11 degrees awarded. *Degree requirements:* For master's, foreign language and thesis not required. *Entrance requirements:* For master's, TOEFL (minimum score of 520 required), TWE. *Application deadline:* For fall admission, 6/15 (priority date); for spring admission, 10/15. Applications are processed on a rolling basis. Application fee: $15. Electronic applications accepted. Tuition, state resident: part-time $89 per hour. Tuition, nonresident: part-time $261 per hour. Part-time tuition and fees vary according to program. *Financial aid:* In 1998–99, 7 fellowships, 6 research assistantships, 20 teaching assistantships were awarded.; Federal Work-Study and institutionally-sponsored loans also available. Aid available to part-time students. Financial aid application deadline: 5/15. *Faculty research:* Concrete technology, environmental and sanitation engineering, water resources engineering, composite materials, geotechnical engineering. Total annual research expenditures: $356,502. *Unit head:* Dr. Wendell Hovey, Chair, 605-394-2439. *Application contact:* Brenda Brown, Secretary, 800-454-8162 Ext. 2493, Fax: 605-394-5360, E-mail: graduate_admissions@silver.sdmt.edu.

South Dakota School of Mines and Technology, Graduate Division, Division of Material Engineering and Science, Doctoral Program in Materials Engineering and Science, Rapid City, SD 57701-3995. Offers chemical engineering (PhD); chemistry (PhD); civil engineering (PhD); electrical engineering (PhD); mechanical engineering (PhD); metallurgical engineering (PhD); physics (PhD). Part-time programs available. *Students:* 14 full-time (2 women), 9 international. *Degree requirements:* For doctorate, dissertation required, foreign language not required. *Entrance requirements:* For doctorate, TOEFL (minimum score of 520 required), TWE, minimum graduate GPA of 3.0. *Application deadline:* For fall admission, 6/15 (priority date); for spring admission, 10/15. Applications are processed on a rolling basis. Application fee: $15. Electronic applications accepted. Tuition, state resident: part-time $89 per hour. Tuition, nonresident: part-time $261 per hour. Part-time tuition and fees vary according to program. *Unit head:* Dr. Chris Jenkins, Coordinator, 605-394-2406. *Application contact:* Brenda Brown, Secretary, 800-454-8162 Ext. 2493, Fax: 605-394-5360, E-mail: graduate_admissions@silver.sdmt.edu.

South Dakota State University, Graduate School, College of Engineering, Department of Civil and Environmental Engineering, Brookings, SD 57007. Offers engineering (MS), including civil engineering, environmental engineering. *Degree requirements:* For master's, thesis, oral exam required, foreign language not required. *Entrance requirements:* For master's, TOEFL (minimum score of 520 required for civil engineering; 550 for environmental engineering). *Faculty research:* Groundwater modeling, biological wastewater treatment, corrosion control, highway materials, traffic analysis.

Southern Illinois University Carbondale, Graduate School, College of Engineering, Department of Civil Engineering and Mechanics, Carbondale, IL 62901-6806. Offers MS. *Faculty:* 12 full-time (1 woman), 1 part-time (0 women). *Students:* 21 full-time (1 woman), 4 part-time (3 women). Average age 26. 24 applicants, 63% accepted. In 1998, 9 degrees awarded. *Degree requirements:* For master's, thesis, comprehensive exam required, foreign language not required. *Entrance requirements:* For master's, TOEFL (minimum score of 550 required), minimum GPA of 2.7. *Application deadline:* Applications are processed on a rolling basis. Application fee: $20. *Financial aid:* In 1998–99, 21 students received aid, including 5 research assistantships with full tuition reimbursements available, 9 teaching assistantships with full tuition reimbursements available; fellowships with full tuition reimbursements available, Federal Work-Study, institutionally-sponsored loans, and tuition waivers (full) also available. Aid available to part-time students. Financial aid application deadline: 7/1. *Faculty research:* Composite materials, wastewater treatment, solid waste disposal, slurry transport, geotechnical engineering. Total annual research expenditures: $230,856. *Unit head:* Dr. Buck F. Brown, Dean for Research and Graduate Studies, 812-877-8403, Fax: 812-877-8102, E-mail: buck.brown@rose-hulman.edu.

Southern Illinois University Edwardsville, Graduate Studies and Research, School of Engineering, Program in Civil Engineering, Edwardsville, IL 62026-0001. Offers MS. Part-time programs available. *Students:* 10 full-time (2 women), 27 part-time (2 women); includes 1 minority (African American), 10 international. 59 applicants, 46% accepted. In 1998, 10 degrees awarded. *Degree requirements:* For master's, thesis or research paper, final exam required. *Entrance requirements:* For master's, TOEFL (minimum score of 550 required). *Application deadline:* For fall admission, 7/24. Application fee: $25. *Financial aid:* In 1998–99, 2 research assistantships with full tuition reimbursements, 10 teaching assistantships with full tuition reimbursements were awarded.; fellowships with full tuition reimbursements, career-related internships or fieldwork, Federal Work-Study, institutionally-sponsored loans, traineeships, and unspecified assistantships also available. Aid available to part-time students. *Unit head:* Dr. Chiang Lin, Chair, 618-650-2816, E-mail: clin@siue.edu. *Application contact:* Dr. Susan Morgan, Director, 618-650-5014, E-mail: smorgan@siue.edu.

Stanford University, School of Engineering, Department of Civil and Environmental Engineering, Stanford, CA 94305-9991. Offers MS, PhD, Eng. *Faculty:* 26 full-time (3 women). *Students:* 199 full-time (57 women), 47 part-time (13 women); includes 53 minority (6 African Americans, 26 Asian Americans or Pacific Islanders, 20 Hispanic Americans, 1 Native American), 92 international. Average age 27. 455 applicants, 55% accepted. In 1998, 104 master's, 12 doctorates awarded. Terminal master's awarded for partial completion of doctoral program. *Degree requirements:* For master's, foreign language and thesis not required; for doctorate and Eng, dissertation, qualifying exam required, foreign language not required; for Eng, thesis required, foreign language not required. *Entrance requirements:* For master's, doctorate, and Eng, GRE General Test, TOEFL. *Application deadline:* For fall admission, 1/1. Application fee: $65 ($80 for international students). Electronic applications accepted. Tuition: Full-time $24,588. Required fees: $152. Part-time tuition and fees vary according to course load. *Financial aid:* Fellowships, research assistantships, teaching assistantships, Federal Work-Study, institutionally-sponsored loans, and course assistantships, traineeships available. Financial aid application deadline: 1/1. *Unit head:* Jeffrey Koseff, Chair, 650-725-2385, Fax: 650-725-8662, E-mail: koseff@ce.stanford.edu. *Application contact:* Graduate Admissions Coordinator, 650-725-2387.

See in-depth description on page 599.

State University of New York at Buffalo, Graduate School, School of Engineering and Applied Sciences, Department of Civil, Structural, and Environmental Engineering, Buffalo, NY 14260. Offers computational engineering and mechanics (MS, PhD); construction (M Eng, MS, PhD); geoenvironmental and geotechnical engineering (M Eng, MS, PhD); structural and earthquake engineering (M Eng, MS, PhD); water resources and environmental engineering (M Eng, MS, PhD). Part-time programs available. Postbaccalaureate distance learning degree programs offered (minimal on-campus study). *Faculty:* 25 full-time (0 women), 7 part-time (1 woman). *Students:* 99 full-time (19 women), 71 part-time (13 women); includes 9 minority (2 African Americans, 5 Asian Americans or Pacific Islanders, 1 Hispanic American, 1 Native American), 101 international. Average age 27. 252 applicants, 29% accepted. In 1998, 46 master's, 16 doctorates awarded. Terminal master's awarded for partial completion of doctoral program. *Degree requirements:* For master's, computer language, project or thesis required; for doctorate, computer language, dissertation required, foreign language not required. *Entrance requirements:* For master's and doctorate, GRE General Test (minimum combined score of 1250 required), TOEFL (minimum score of 550 required). *Application deadline:* For fall admission, 1/15 (priority date); for spring admission, 10/1. Applications are processed on a rolling basis. Application fee: $35. Tuition, state resident: full-time $5,100; part-time $213 per credit hour. Tuition, nonresident: full-time $8,416; part-time $351 per credit hour. Required fees: $870; $75 per semester. Tuition and fees vary according to course load and program. *Financial aid:* In 1998–99, 3 fellowships with full tuition reimbursements (averaging $14,700 per year), 79 research assistantships with full tuition reimbursements (averaging $10,700 per year), 30 teaching assistantships with full tuition reimbursements (averaging $10,700 per year) were awarded.; career-related internships or fieldwork, Federal Work-Study, grants, institutionally-sponsored loans, scholarships, tuition waivers (full and partial), and unspecified assistantships also available. Aid available to part-time students. Financial aid application deadline: 1/15; financial aid applicants required to submit FAFSA. *Faculty research:* Earthquake protection, environmental engineering and fluid mechanics, structural dynamics, geomechanics. Total annual research expenditures: $1.5 million. *Unit head:* Dr. Andrei M. Reinhorn, Chairman, 716-645-2114 Ext. 2419, Fax: 716-645-3733, E-mail: reinhorn@civil.eng.buffalo.edu. *Application contact:* Dr. Joe Atkinson, Director of Graduate Admissions, 716-645-2114 Ext. 2326, Fax: 716-645-3667, E-mail: atkinson@acsu.buffalo.edu.

See in-depth description on page 601.

Stevens Institute of Technology, Graduate School, Charles V. Schaefer Jr. School of Engineering, Department of Civil, Environmental, and Ocean Engineering, Program in Civil Engineering, Hoboken, NJ 07030. Offers coastal and ocean engineering (M Eng, PhD, Engr); construction accounting/estimating (Certificate); construction engineering (M Eng, PhD, Certificate, Engr); construction law/disputes (Certificate); construction/quality management (Certificate); geotechnical engineering (Certificate); geotechnical/geoenvironmental engineering (M Eng, PhD, Engr); structures (M Eng, PhD, Engr). *Degree requirements:* For master's, computer language required, thesis optional, foreign language not required; for doctorate, variable foreign language requirement, computer language, dissertation required; for other advanced degree, computer language, project or thesis required. *Entrance requirements:* For master's, TOEFL (minimum score of 500 required); for doctorate, GRE, TOEFL (minimum score of 500 required). Electronic applications accepted.

Syracuse University, Graduate School, L. C. Smith College of Engineering and Computer Science, Department of Civil and Environmental Engineering, Syracuse, NY 13244-0003. Offers civil engineering (MS, PhD); environmental engineering (MS); hydrogeology (MS). *Faculty:* 19. *Students:* 22 full-time (7 women), 15 part-time (4 women), 17 international. Average age 30. 90 applicants, 77% accepted. In 1998, 14 degrees awarded. *Degree requirements:* For doctorate, computer language, dissertation required, foreign language not required, foreign language not required. *Entrance requirements:* For master's and doctorate, GRE General Test, GRE Subject Test. *Application deadline:* Applications are processed on a rolling basis. Application fee: $40. Tuition: Full-time $13,992; part-time $583 per credit hour. *Financial aid:* Fellowships, research assistantships, teaching assistantships, Federal Work-Study and tuition waivers (partial) available. Financial aid application deadline: 3/1. *Unit head:* Dr. Shobha Bhatia, Chair, 315-443-2311.

Temple University, Graduate School, College of Science and Technology, College of Engineering, Program in Civil and Environmental Engineering, Philadelphia, PA 19122-6096. Offers MSE. Part-time programs available. *Faculty:* 6 full-time (0 women). *Students:* 13 (2 women); includes 3 minority (2 African Americans, 1 Asian American or Pacific Islander) 1 international. 11 applicants, 64% accepted. *Degree requirements:* For master's, thesis optional, foreign language not required. *Entrance requirements:* For master's, GRE General Test, TOEFL (minimum score of 575 required). *Application deadline:* For fall admission, 7/1; for spring admission, 11/1. Applications are processed on a rolling basis. Application fee: $40. *Financial aid:* Research assistantships, teaching assistantships, Federal Work-Study available. Financial aid application deadline: 2/15. *Faculty research:* Prestressed masonry structure, recycling processes and products, finite element analysis of highways and runways. Total annual research expenditures: $118,428. *Unit head:* Dr. Philip D. Udo-Inyang, Director, 215-204-7831, Fax: 215-204-6936.

Tennessee Technological University, Graduate School, College of Engineering, Department of Civil Engineering, Cookeville, TN 38505. Offers MS, PhD. Part-time programs available. *Faculty:* 17 full-time (0 women). *Students:* 13 full-time (2 women), 4 part-time (1 woman); includes 8 minority (all Asian Americans or Pacific Islanders) Average age 27. 70 applicants, 63% accepted. In 1998, 11 master's awarded. *Degree requirements:* For master's, thesis required, foreign language not required; for doctorate, one foreign language (computer language can substitute), dissertation required. *Entrance requirements:* For master's, GRE General Test, TOEFL (minimum score of 525 required); for doctorate, GRE Subject Test, TOEFL (minimum score of 525 required), minimum GPA of 3.5. *Application deadline:* For fall admission, 3/1 (priority date); for spring admission, 8/1. Application fee: $25 ($30 for international students). Tuition, state resident: part-time $137 per hour. Tuition, nonresident: part-time $361 per hour. Required fees: $17 per hour. Tuition and fees vary according to course load. *Financial aid:* In 1998–99, 16 students received aid, including 14 research assistantships (averaging $7,200 per year), teaching assistantships (averaging $7,200 per year); career-related internships or fieldwork also available. Financial aid application deadline: 4/1. *Faculty research:* Environmental engineering, transportation, structural engineering, water resources. *Unit head:* Dr. William Paul Bonner, Interim Chairperson, 931-372-3454, Fax: 931-372-6352, E-mail: wpbonner@tntech.edu. *Application contact:* Dr. Rebecca F. Quattlebaum, Dean of the Graduate School, 931-372-3233, Fax: 931-372-3497, E-mail: rquattlebaum@tntech.edu.

Texas A&M University, College of Engineering, Department of Civil Engineering, College Station, TX 77843. Offers construction engineering and project management (M Eng, MS, D Eng, PhD); engineering mechanics (M Eng, MS, PhD); environmental engineering (M Eng, MS, D Eng, PhD); geotechnical engineering (M Eng, MS, D Eng, PhD); hydraulic engineering (M Eng, MS, PhD); hydrology (M Eng, MS, PhD); materials engineering (M Eng, MS, D Eng, PhD); ocean engineering (M Eng, MS, D Eng, PhD); public works engineering and management (M Eng, MS, PhD); structural engineering and structural mechanics (M Eng, MS, D Eng, PhD); transportation engineering (M Eng, MS, D Eng, PhD); water resources engineering (M Eng, MS, D Eng, PhD). Part-time programs available. *Faculty:* 59 full-time (4 women), 7 part-time (2 women). *Students:* 249 full-time, 54 part-time; includes 13 minority (2 African Americans, 4 Asian Americans or Pacific Islanders, 7 Hispanic Americans), 162 international. Average age 29. 235 applicants, 63% accepted. In 1998, 75 master's, 14 doctorates awarded. *Degree requirements:* For master's, thesis (MS) required; for doctorate, dissertation (PhD), internship (D Eng) required. *Entrance requirements:* For master's and doctorate, GRE General Test, TOEFL. *Application deadline:* Applications are processed on a rolling basis. Application fee: $50 ($75 for international students). *Financial aid:* In 1998–99, 196 students received aid, including 27 fellowships (averaging $4,000 per year), 122 research assistantships (averaging $1,000 per year), 47 teaching assistantships (averaging $1,000 per year); career-related internships or fieldwork and institutionally-sponsored loans also available. Financial aid application deadline: 4/1; financial aid applicants required to submit FAFSA. Total annual research expenditures: $7 million. *Unit head:* Dr. John M. Niedzwecki, Head, 409-845-7435, Fax: 409-862-2800, E-mail: ce-grad@tamu.edu. *Application contact:* Dr. Peter B. Keating, Graduate Adviser, 409-845-2498, Fax: 409-862-2800, E-mail: ce-grad@tamu.edu.

Announcement: M Eng, MS, D Eng, and PhD awarded. Specializations: coastal and ocean engineering, construction engineering and project management, engineering mechanics, environmental engineering, geotechnical engineering, hydraulic engineering, hydrology, materials engineering, public works engineering and management, structural engineering and structural mechanics, transportation engineering, water resources engineering. Extensive research laboratories, library, digital computing facilities, field testing facilities. Financial assistance available to outstanding students through research and teaching assistantships, fellowships.

See in-depth description on page 603.

Texas A&M University–Kingsville, College of Graduate Studies, College of Engineering, Department of Civil Engineering, Kingsville, TX 78363. Offers ME, MS. Part-time and evening/weekend programs available. *Faculty:* 4 full-time (0 women), 1 part-time (0 women). *Students:* 1 full-time (0 women), 2 part-time. *Degree requirements:* For master's, computer language, thesis or alternative, comprehensive exam required, foreign language not required. *Entrance requirements:* For master's, GRE General Test (minimum combined score of 1000 required), TOEFL (minimum score of 525 required). *Application deadline:* For fall admission, 6/1; for spring admission, 11/15. Applications are processed on a rolling basis. Application fee: $15 ($25 for international students). Tuition, state resident: full-time $2,062. Tuition, nonresident: full-time $7,246. *Financial aid:* Fellowships, research assistantships, teaching assistantships, career-related internships or fieldwork and institutionally-sponsored loans available. Financial aid application deadline: 5/15. *Faculty research:* Geotechnical engineering, structural mechanics, structural design, transportation engineering. *Unit head:* Dr. John W. Weber, Chairman, 361-593-2003. *Application contact:* Dr. Pat Leelani, Graduate Coordinator, 361-593-2266.

Texas Tech University, Graduate School, College of Engineering, Department of Civil Engineering, Lubbock, TX 79409. Offers civil engineering (MSCE, PhD); environmental engineering (MENVEGR); environmental technology and management (MSETM). Part-time programs available. *Faculty:* 20 full-time (0 women), 1 part-time (0 women). *Students:* 60 full-time (15 women), 22 part-time (6 women); includes 4 minority (all Hispanic Americans), 33 international. Average age 27. 93 applicants, 55% accepted. In 1998, 27 master's, 6 doctorates awarded. *Degree requirements:* For master's and doctorate, computer language, thesis/dissertation required, foreign language not required. *Entrance requirements:* For master's, GRE General Test (minimum combined score of 1000 required; average 1159), minimum GPA of 3.0; for doctorate, GRE General Test (minimum combined score of 1000 required), minimum GPA of 3.0. *Application deadline:* For fall admission, 4/15 (priority date); for spring admission, 11/1 (priority date). Applications are processed on a rolling basis. Application fee: $25 ($50 for international students). Electronic applications accepted. *Financial aid:* In 1998–99, 46 research assistantships (averaging $8,664 per year), 2 teaching assistantships (averaging $8,100 per year) were awarded.; Federal Work-Study and institutionally-sponsored loans also available. Aid available to part-time students. Financial aid application deadline: 5/15; financial aid applicants required to submit FAFSA. *Faculty research:* Wind load/engineering on structures, fluid mechanics, structural dynamics, water resource management. Total annual research expenditures: $2.2 million. *Unit head:* Dr. James R. McDonald, Chairman, 806-742-3523, Fax: 806-742-3488.

See in-depth description on page 607.

Tufts University, Division of Graduate and Continuing Studies and Research, Graduate School of Arts and Sciences, College of Engineering, Department of Civil and Environmental Engineering, Medford, MA 02155. Offers civil engineering (MS, PhD), including geotechnical engineering, structural engineering; environmental engineering (MS, PhD), including environmental engineering and environmental sciences, environmental geotechnology, environmental health, environmental science and management, hazardous materials management, water resources engineering. Part-time programs available. *Faculty:* 13 full-time, 10 part-time. *Students:* 99 (47 women); includes 21 minority (5 African Americans, 7 Asian Americans or Pacific Islanders, 9 Hispanic Americans) 16 international. 124 applicants, 61% accepted. In 1998, 26 master's, 1 doctorate awarded. Terminal master's awarded for partial completion of doctoral program. *Degree requirements:* For master's, thesis or alternative required, foreign language not required; for doctorate, dissertation required, foreign language not required. *Entrance requirements:* For master's and doctorate, GRE General Test, TOEFL (minimum score of 550 required). *Application deadline:* For fall admission, 2/15; for spring admission, 10/15. Applications are processed on a rolling basis. Application fee: $50. *Financial aid:* Research assistantships with full and partial tuition reimbursements, teaching assistantships with full and partial tuition reimbursements, Federal Work-Study, scholarships, and tuition waivers (partial) available. Aid available to part-time students. Financial aid application deadline: 2/15; financial aid applicants required to submit FAFSA. *Unit head:* Dr. Stephen Levine, Chair, 617-627-3211, Fax: 617-627-3994. *Application contact:* Linfield Brown, 617-627-3211, Fax: 617-627-3994.

Tulane University, School of Engineering, Department of Civil and Environmental Engineering, New Orleans, LA 70118-5669. Offers MS, MSE, PhD, Sc D. MS and PhD offered through the Graduate School. Part-time programs available. *Students:* 15 full-time (0 women), 3 part-time (1 woman); includes 3 minority (1 African American, 1 Asian American or Pacific Islander, 1 Hispanic American), 4 international. 29 applicants, 59% accepted. In 1998, 3 master's awarded. *Degree requirements:* For master's, thesis required, foreign language not required; for doctorate, 2 foreign languages (computer language can substitute for one), dissertation required. *Entrance requirements:* For master's and doctorate, GRE General Test, TOEFL, minimum B average in undergraduate course work. *Application deadline:* For fall admission, 7/1. Application fee: $35. *Financial aid:* Fellowships, research assistantships, teaching assistantships available. Financial aid application deadline: 2/1. *Unit head:* Dr. John Niklaus, Chairman, 504-865-5778. *Application contact:* Dr. E. Michaelides, Associate Dean, 504-865-5764.

Université de Moncton, School of Engineering, Program in Civil Engineering, Moncton, NB E1A 3E9, Canada. Offers M Sc A. *Faculty:* 4 full-time (0 women). *Students:* 3 full-time (0 women), 1 part-time, 1 international. Average age 26. 3 applicants, 67% accepted. In 1998, 4 degrees awarded (25% entered university research/teaching, 50% found other work related to degree, 25% continued full-time study). *Degree requirements:* For master's, thesis, proficiency in French required. *Average time to degree:* Master's–3.5 years full-time. *Application deadline:* For fall admission, 6/1 (priority date); for winter admission, 11/15 (priority date). Application fee: $30. *Financial aid:* In 1998–99, 2 students received aid, including fellowships (averaging $17,200 per year), teaching assistantships (averaging $720 per year); research assistantships Financial aid application deadline: 5/31. *Faculty research:* Structures and materials, hydrology and water resources, soil mechanics and statistical analysis, environment, transportation. Total annual research expenditures: $217,800. *Unit head:* Dr. Michel Massiera, Chairman, 506-858-4141, Fax: 506-858-4082, E-mail: massiem@umoncton.ca.

Université de Sherbrooke, Faculty of Applied Sciences, Department of Civil Engineering, Sherbrooke, PQ J1K 2R1, Canada. Offers M Sc A, PhD. *Degree requirements:* For master's and doctorate, thesis/dissertation required. *Faculty research:* High-strength concrete, dynamics of structures, solid mechanics, geotechnical engineering, wastewater treatment.

Université Laval, Faculty of Graduate Studies, Faculty of Sciences and Engineering, Department of Civil Engineering, Sainte-Foy, PQ G1K 7P4, Canada. Offers civil engineering (M Sc, PhD); urban infrastructures engineering (Diploma). *Students:* 65 full-time (18 women), 24 part-time (4 women). 46 applicants, 52% accepted. In 1998, 23 master's, 2 doctorates awarded. *Application deadline:* For fall admission, 3/1. Application fee: $30. *Faculty research:* Structural design and applied mechanics, foundation and soil mechanics, hydraulics, ice mechanics. *Unit head:* André Picard, Director, 418-656-2131 Ext. 2848, Fax: 418-656-2928, E-mail: andre.picard@gci.ulaval.ca.

The University of Akron, Graduate School, College of Engineering, Department of Civil Engineering, Akron, OH 44325-0001. Offers MSCE, PhD. Evening/weekend programs available. *Faculty:* 12 full-time, 4 part-time. *Students:* 27 full-time (9 women), 15 part-time (1 woman); includes 2 minority (1 African American, 1 Asian American or Pacific Islander), 15 international. Average age 29. In 1998, 12 master's awarded. *Degree requirements:* For master's, thesis or alternative required, foreign language not required; for doctorate, dissertation, candidacy exam, qualifying exam required, dissertation, candidacy exam, qualifying exam required. *Entrance requirements:* For master's, TOEFL (minimum score of 550 required), minimum GPA of 2.75; for doctorate, GRE, TOEFL (minimum score of 550 required). *Average time to degree:* Master's–2 years full-time, 4 years part-time. *Application deadline:* For fall admission, 3/1. Applications are processed on a rolling basis. Application fee: $25 ($50 for international students). Tuition, state resident: part-time $189 per credit. Tuition, nonresident: part-time $353 per credit. Required fees: $7.3 per credit. *Financial aid:* In 1998–99, 15 research assistantships with full tuition reimbursements, 15 teaching assistantships with full tuition reimbursements were awarded.; fellowships with full tuition reimbursements, career-related internships or fieldwork and Federal Work-Study also available. Financial aid application deadline: 3/1. *Faculty research:* Development of constitutive relations and numerical analysis of nonlinear problems in structural mechanics, computer modeling of large-scale water supply networks. *Unit head:* Dr. Robert Liang, Chair, 330-972-7228. *Application contact:* Dr. Wieslaw Binienda, Chairman of the Admissions Committee, 330-972-7288, E-mail: wbinienda@uakron.edu.

The University of Alabama, Graduate School, College of Engineering, Department of Civil and Environmental Engineering, Program in Civil Engineering, Tuscaloosa, AL 35487. Offers MSCE, PhD. Part-time programs available. Postbaccalaureate distance learning degree programs offered (no on-campus study). *Faculty:* 10 full-time (1 woman), 3 part-time (1 woman). In 1998, 2 degrees awarded. Terminal master's awarded for partial completion of doctoral program. *Degree requirements:* For master's, thesis or alternative required, foreign language not required; for doctorate, one foreign language (computer language can substitute), dissertation required. *Entrance requirements:* For master's and doctorate, GRE General Test (minimum combined score of 1500 on three sections required), minimum GPA of 3.0 in last 60 hours. *Application deadline:* For fall admission, 7/6. Applications are processed on a rolling basis. Application fee: $25. *Financial aid:* Fellowships, research assistantships, teaching assistantships, Federal Work-Study and institutionally-sponsored loans available. *Unit head:* Dr. Daniel S. Turner, Head, Department of Civil and Environmental Engineering, 205-348-6550, Fax: 205-348-0783.

Civil Engineering

The University of Alabama at Birmingham, Graduate School, School of Engineering, Department of Civil and Environmental Engineering, Birmingham, AL 35294. Offers MSCE, PhD. Evening/weekend programs available. *Students:* 10 full-time (5 women), 12 part-time (2 women), 1 international. 22 applicants, 100% accepted. In 1998, 9 master's awarded. *Degree requirements:* For master's, thesis required (for some programs), foreign language not required; for doctorate, dissertation required. *Entrance requirements:* For master's, GRE General Test (minimum score of 500 on each section required), BS in engineering, physical sciences, life sciences, or mathematics; for doctorate, GRE General Test (minimum score of 600 on each section required), TOEFL (minimum score of 550 required), BS or MS in engineering or related field, minimum undergraduate GPA of 3.0. *Application deadline:* Applications are processed on a rolling basis. Application fee: $30 ($60 for international students). Electronic applications accepted. *Financial aid:* In 1998–99, 2 fellowships with full tuition reimbursements (averaging $9,500 per year), 11 research assistantships (averaging $1,229 per year) were awarded. *Unit head:* Dr. Fouad H. Fouad, Chair, 205-934-8430, Fax: 205-934-9855, E-mail: ffouad@uab.edu.

The University of Alabama in Huntsville, School of Graduate Studies, College of Engineering, Department of Civil and Environmental Engineering, Huntsville, AL 35899. Offers MSE. Part-time and evening/weekend programs available. *Faculty:* 6 full-time (1 woman). *Students:* 2 full-time (0 women), 15 part-time (6 women); includes 3 minority (2 African Americans, 1 Asian American or Pacific Islander), 4 international. Average age 32. 12 applicants, 67% accepted. In 1998, 2 degrees awarded. *Degree requirements:* For master's, oral and written exams required, thesis optional, foreign language not required. *Entrance requirements:* For master's, GRE General Test (minimum combined score of 1500 on three sections required), BSE, minimum GPA of 3.0. *Application deadline:* For fall admission, 7/24 (priority date); for spring admission, 11/15 (priority date). Applications are processed on a rolling basis. Application fee: $20. Tuition and fees vary according to course load. *Financial aid:* In 1998–99, 10 students received aid, including 5 research assistantships with full and partial tuition reimbursements available (averaging $9,330 per year), 4 teaching assistantships with full and partial tuition reimbursements available (averaging $7,362 per year); fellowships with full and partial tuition reimbursements available, career-related internships or fieldwork, Federal Work-Study, grants, institutionally-sponsored loans, scholarships, and tuition waivers (full and partial) also available. Aid available to part-time students. Financial aid application deadline: 4/1; financial aid applicants required to submit FAFSA. *Faculty research:* Hydrologic modeling, orbital debris impact, hydrogeology, environmental engineering, water quality control. Total annual research expenditures: $52,487. *Unit head:* Dr. William Schonberg, Chair, 256-890-6854, Fax: 256-890-6724, E-mail: wschon@eb.uah.edu.

University of Alaska Anchorage, School of Engineering, Program in Civil Engineering, Anchorage, AK 99508-8060. Offers MCE, MS. Part-time and evening/weekend programs available. *Students:* 2 full-time (0 women), 16 part-time (1 woman); includes 4 minority (1 Asian American or Pacific Islander, 3 Native Americans) 8 applicants, 75% accepted. In 1998, 7 degrees awarded. *Degree requirements:* For master's, computer language required, foreign language not required. *Entrance requirements:* For master's, GRE General Test, bachelor's degree in engineering. *Application deadline:* For fall admission, 5/1 (priority date). Applications are processed on a rolling basis. Application fee: $45. *Financial aid:* In 1998–99, 1 research assistantship was awarded.; Federal Work-Study also available. Aid available to part-time students. Financial aid application deadline: 4/1; financial aid applicants required to submit FAFSA. *Faculty research:* Structural engineering, engineering education, astronomical observations related to engineering. *Unit head:* Dr. T. Bartlett Quimby, Chair, 907-786-1046, Fax: 907-786-1079. *Application contact:* Cecile Mitchell, Director for Enrollment Services, 907-786-1558.

University of Alaska Fairbanks, Graduate School, College of Science, Engineering and Mathematics, Department of Civil Engineering, Fairbanks, AK 99775-7480. Offers arctic engineering (MS); civil engineering (MCE, MS); environmental quality engineering (MS); environmental quality science (MS). *Faculty:* 8 full-time (1 woman), 3 part-time (0 women). *Students:* 9 full-time (5 women), 6 part-time (3 women), 2 international. Average age 32. 9 applicants, 100% accepted. In 1998, 6 degrees awarded. *Degree requirements:* For master's, thesis or alternative, comprehensive exam required, foreign language not required. *Entrance requirements:* For master's, GRE General Test, TOEFL (minimum score of 550 required). *Application deadline:* For fall admission, 8/1. Applications are processed on a rolling basis. Application fee: $35. *Financial aid:* Research assistantships, teaching assistantships available. *Faculty research:* Soils, structures, culvert thawing with solar power, pavement drainage. *Unit head:* Dr. Lufti Raad, Head, 907-474-7241.

University of Alberta, Faculty of Graduate Studies and Research, Department of Civil and Environmental Engineering, Edmonton, AB T6G 2E1, Canada. Offers construction engineering and management (M Eng, M Sc, PhD); environmental engineering (M Eng, M Sc, PhD); environmental science (M Sc, PhD); geoenvironmental engineering (M Eng, M Sc, PhD); geotechnical engineering (M Sc); geotechnical engineering (M Eng, PhD); mining engineering (M Eng, M Sc, PhD); petroleum engineering (M Eng, M Sc, PhD); structural engineering (M Eng, M Sc, PhD); water resources (M Eng, M Sc, PhD). Part-time programs available. *Degree requirements:* For master's, thesis required (for some programs), foreign language not required; for doctorate, dissertation required, foreign language not required. *Faculty research:* Mining.

The University of Arizona, Graduate College, College of Engineering and Mines, Department of Civil Engineering and Engineering Mechanics, Program in Civil Engineering, Tucson, AZ 85721. Offers MS, PhD. Part-time programs available. *Students:* 27 full-time (2 women), 11 part-time (2 women); includes 4 minority (1 African American, 2 Asian Americans or Pacific Islanders, 1 Hispanic American), 21 international. Average age 30. 63 applicants, 70% accepted. In 1998, 7 master's, 3 doctorates awarded. *Degree requirements:* For master's, computer language, thesis required, foreign language not required; for doctorate, computer language, dissertation, departmental qualifying exam required, foreign language not required. *Entrance requirements:* For master's, TOEFL (minimum score of 550 required), minimum GPA of 3.0; for doctorate, TOEFL (minimum score of 550 required), minimum GPA of 3.5. *Application deadline:* For fall admission, 8/1. Applications are processed on a rolling basis. Application fee: $35. *Financial aid:* Fellowships, research assistantships, teaching assistantships, institutionally-sponsored loans available. Financial aid application deadline: 4/6. *Faculty research:* Soil-structure interaction, water resources, waste disposal, concrete and steel structures. *Unit head:* Dr. Wieslaw Binienda, Chairman of the Admissions Committee, 330-972-7288, E-mail: wbinienda@uakron.edu. *Application contact:* Mary Jankovsky, Graduate Secretary, 520-621-2266, Fax: 520-621-2550.

University of Arkansas, Graduate School, College of Engineering, Department of Civil Engineering, Program in Civil Engineering, Fayetteville, AR 72701-1201. Offers MSCE, MSE, PhD. *Students:* 19 full-time (5 women), 4 part-time; includes 3 minority (1 African American, 1 Asian American or Pacific Islander, 1 Hispanic American), 10 international. 32 applicants, 63% accepted. In 1998, 14 master's awarded. *Degree requirements:* For master's, thesis optional, foreign language not required; for doctorate, one foreign language, dissertation required. Application fee: $40 ($50 for international students). Tuition, state resident: full-time $3,186. Tuition, nonresident: full-time $7,560. Required fees: $378. *Financial aid:* Research assistantships, teaching assistantships, career-related internships or fieldwork and Federal Work-Study available. Aid available to part-time students. Financial aid application deadline: 4/1; financial aid applicants required to submit FAFSA. *Unit head:* Dr. Robert Elliott, Chair, Department of Civil Engineering, 501-575-4954.

University of British Columbia, Faculty of Graduate Studies, Faculty of Applied Science, Department of Civil Engineering, Vancouver, BC V6T 1Z2, Canada. Offers M Eng, MA Sc, PhD. *Degree requirements:* For master's, thesis or alternative required, foreign language not required; for doctorate, dissertation required, foreign language not required. *Entrance requirements:* For master's and doctorate, TOEFL. *Faculty research:* Geotechnology; structural, water, and environmental engineering; transportation; materials and construction engineering.

University of Calgary, Faculty of Graduate Studies, Faculty of Engineering, Department of Civil Engineering, Calgary, AB T2N 1N4, Canada. Offers M Eng, M Sc, PhD. Part-time and evening/weekend programs available. *Faculty:* 23 full-time (0 women), 3 part-time (0 women). *Students:* 90 full-time (22 women), 42 part-time (7 women). Average age 33. 200 applicants, 10% accepted. In 1998, 10 master's awarded (100% found work related to degree); 4 doctorates awarded (50% entered university research/teaching, 50% found other work related to degree). *Degree requirements:* For master's, thesis required (for some programs), foreign language not required; for doctorate, dissertation, candidacy exam required, foreign language not required. *Entrance requirements:* For master's, TOEFL (minimum score of 570 required), minimum GPA of 3.0; for doctorate, TOEFL (minimum score of 570 required), minimum GPA of 3.5. *Application deadline:* For fall admission, 5/31 (priority date). Applications are processed on a rolling basis. Application fee: $60. *Financial aid:* In 1998–99, 54 students received aid, including 5 fellowships, 21 research assistantships, 19 teaching assistantships Financial aid application deadline: 5/30. *Faculty research:* Structures, including structural materials; transportation; project management and biomechanics; geotechnical engineering; environmental engineering. Total annual research expenditures: $2.7 million. *Unit head:* Dr. Tom Brown, Head, 403-220-5820, Fax: 403-282-7026, E-mail: brownt@ucalgary.ca. *Application contact:* Dr. N. G. Shrive, Graduate Coordinator, 403-220-6630, Fax: 403-282-7026, E-mail: shrive@acs.ucalgary.ca.

University of California, Berkeley, Graduate Division, College of Engineering, Department of Civil and Environmental Engineering, Berkeley, CA 94720-1500. Offers construction engineering and management (M Eng, MS, D Eng, PhD); environmental quality and environmental water resources engineering (M Eng, MS, D Eng, PhD); geotechnical engineering (M Eng, MS, D Eng, PhD); structural engineering, mechanics and materials (M Eng, MS, D Eng, PhD); transportation engineering (M Eng, MS, D Eng, PhD). *Students:* 326 full-time (97 women); includes 59 minority (3 African Americans, 42 Asian Americans or Pacific Islanders, 13 Hispanic Americans, 1 Native American), 113 international. 704 applicants, 45% accepted. In 1998, 161 master's, 4 doctorates awarded. *Degree requirements:* For master's, comprehensive exam or thesis (MS) required; for doctorate, dissertation, qualifying exam required. *Entrance requirements:* For master's, GRE General Test, minimum GPA of 3.0; for doctorate, GRE General Test, minimum GPA of 3.5. *Application deadline:* For fall admission, 2/10. Application fee: $40. *Financial aid:* Fellowships, research assistantships, teaching assistantships available. Financial aid application deadline: 12/15. *Unit head:* Dr. Adib Kanafani, Chair, 510-642-3261. *Application contact:* Mari Cook, Graduate Assistant for Admission, 510-643-8944, E-mail: mcook@ce.berkeley.edu.

See in-depth description on page 609.

University of California, Davis, Graduate Studies, College of Engineering, Program in Civil and Environmental Engineering, Davis, CA 95616. Offers M Engr, MS, D Engr, PhD, Certificate, M Engr/MBA. Part-time programs available. *Faculty:* 28 full-time (3 women), 1 part-time (0 women). *Students:* 143 full-time (46 women), 1 part-time; includes 20 minority (2 African Americans, 13 Asian Americans or Pacific Islanders, 5 Hispanic Americans), 42 international. 226 applicants, 63% accepted. In 1998, 28 master's, 5 doctorates awarded. *Degree requirements:* For master's, foreign language and thesis not required; for doctorate, dissertation required, foreign language not required. *Entrance requirements:* For master's, GRE General Test, minimum GPA of 3.0; for doctorate, GRE, minimum graduate GPA of 3.5. *Average time to degree:* Master's–1 year full-time; doctorate–4 years full-time. *Application deadline:* For fall admission, 1/15 (priority date). Applications are processed on a rolling basis. Application fee: $40. Electronic applications accepted. *Financial aid:* In 1998–99, 21 fellowships with full and partial tuition reimbursements, 69 research assistantships with full and partial tuition reimbursements, 18 teaching assistantships with full and partial tuition reimbursements were awarded.; career-related internships or fieldwork, Federal Work-Study, institutionally-sponsored loans, and tuition waivers (full and partial) also available. Aid available to part-time students. Financial aid application deadline: 1/15; financial aid applicants required to submit FAFSA. *Faculty research:* Environmental water resources, transportation, structural mechanics, structural engineering, geotechnical engineering. *Unit head:* Daniel P. Y. Chang, Chairperson, 530-752-1441, Fax: 530-752-7872. *Application contact:* Donna Douglas, Administrative Assistant, 530-752-1441, Fax: 530-752-7872.

University of California, Irvine, Office of Research and Graduate Studies, School of Engineering, Department of Civil and Environmental Engineering, Irvine, CA 92697. Offers civil engineering (MS, PhD); environmental engineering (MS, PhD), including engineering. Part-time programs available. *Faculty:* 18 full-time (3 women). *Students:* 53 full-time (9 women), 17 part-time (6 women); includes 14 minority (11 Asian Americans or Pacific Islanders, 3 Hispanic Americans), 20 international. 100 applicants, 75% accepted. In 1998, 20 master's, 4 doctorates awarded. Terminal master's awarded for partial completion of doctoral program. *Degree requirements:* For doctorate, dissertation required, foreign language not required, foreign language not required. *Entrance requirements:* For master's, GRE General Test, minimum GPA of 3.0; for doctorate, GRE General Test. *Application deadline:* For fall admission, 1/15 (priority date). Applications are processed on a rolling basis. Application fee: $40. Electronic applications accepted. *Financial aid:* Fellowships, research assistantships, teaching assistantships, institutionally-sponsored loans and tuition waivers (full and partial) available. Financial aid application deadline: 3/2; financial aid applicants required to submit FAFSA. *Faculty research:* Structural mechanics, earthquake and reliability engineering, geotechnical engineering, transportation planning and urban systems. *Unit head:* Dr. Stephen G. Ritchie, Chair, 949-824-4214, Fax: 949-824-2117, E-mail: sritchie@uci.edu. *Application contact:* Eileen Reisert, Administrative Assistant, 949-824-2120, Fax: 949-824-2117, E-mail: efreiser@uci.edu.

See in-depth description on page 613.

University of California, Los Angeles, Graduate Division, School of Engineering and Applied Science, Department of Civil and Environmental Engineering, Los Angeles, CA 90095. Offers environmental engineering (MS, PhD); geotechnical engineering (MS, PhD); structures (MS, PhD), including structural mechanics and earthquake engineering; water resource systems engineering (MS, PhD). *Faculty:* 15 full-time, 11 part-time. *Students:* 116 full-time (29 women); includes 28 minority (3 African Americans, 21 Asian Americans or Pacific Islanders, 4 Hispanic Americans), 37 international. 267 applicants, 44% accepted. In 1998, 57 master's, 15 doctorates awarded. *Degree requirements:* For master's, comprehensive exam or thesis required; for doctorate, dissertation, qualifying exams required, foreign language not required. *Entrance requirements:* For master's, GRE General Test, minimum GPA of 3.0; for doctorate, GRE General Test, minimum GPA of 3.25. *Application deadline:* For fall admission, 1/15; for spring admission, 12/1. Application fee: $40. Electronic applications accepted. *Financial aid:* In 1998–99, 31 fellowships, 26 research assistantships, 11 teaching assistantships were awarded.; Federal Work-Study, institutionally-sponsored loans, and tuition waivers (full and partial) also available. Financial aid application deadline: 1/15; financial aid applicants required to submit FAFSA. *Unit head:* Dr. Michael Stenstrom, Chair, 310-825-1408. *Application contact:* Deeona Columbia, Graduate Affairs Officer, 310-825-1851, Fax: 310-206-2222, E-mail: deeona@ea.ucla.edu.

University of Central Florida, College of Engineering, Department of Civil and Environmental Engineering, Program in Civil Engineering, Orlando, FL 32816. Offers MS, MSCE, PhD, Certificate. Part-time and evening/weekend programs available. *Faculty:* 17 full-time, 16 part-time. *Students:* 32 full-time (7 women), 29 part-time (7 women); includes 9 minority (2 African Americans, 3 Asian Americans or Pacific Islanders, 4 Hispanic Americans), 19 international. Average age 31. 35 applicants, 29% accepted. In 1998, 21 master's, 4 doctorates awarded. *Degree requirements:* For master's, thesis or alternative required, foreign language not required; for doctorate, dissertation, departmental qualifying exam, candidacy exam required, foreign language not required. *Entrance requirements:* For master's, GRE General Test (minimum combined score of 1000 required), TOEFL (minimum score of 550 required; 213 computer-based), minimum GPA of 3.0 in last 60 hours; for doctorate, GRE General Test (minimum combined score of 1100 required), TOEFL (minimum score of 550 required; 213 computer-based), minimum GPA of 3.5 in last 60 hours. *Application deadline:* For fall admission, 7/15; for spring admission, 12/15. Application fee: $20. Tuition, state resident: full-time $2,054; part-time $137 per credit. Tuition, nonresident: full-time $7,207; part-time $480 per credit. Required fees: $47 per term. *Financial aid:* In 1998–99, 37 students received aid, including 17 fellowships with partial tuition reimbursements available (averaging $2,294 per year), 16 teaching assistantships with partial tuition reimbursements available (averaging $2,628 per year); research assistantships with partial tuition reimbursements available, career-related internships or fieldwork, Federal Work-Study, institutionally-sponsored loans, tuition waivers (partial), and unspecified assistantships also available. Financial aid application deadline: 3/1; financial aid applicants required to submit FAFSA. *Unit head:* Dr. A. E. Radwan, Chair, 407-823-2841. *Application contact:* Dr. Roger L. Wayson, Coordinator, 407-823-2841.

University of Cincinnati, Division of Research and Advanced Studies, College of Engineering, Department of Civil and Environmental Engineering, Program in Civil Engineering, Cincinnati, OH 45221-0091. Offers MS, PhD. *Students:* 47 full-time (8 women), 14 part-time (3 women); includes 1 minority (Asian American or Pacific Islander), 38 international. In 1998, 21 master's, 3 doctorates awarded. *Degree requirements:* For master's, project or thesis required; for doctorate, one foreign language (computer language can substitute), dissertation required. *Entrance requirements:* For master's and doctorate, GRE General Test, TOEFL (minimum score of 560 required). *Average time to degree:* Master's–3.9 years full-time; doctorate–6 years full-time. *Application deadline:* For fall admission, 2/1 (priority date). Application fee: $40. *Financial aid:* Fellowships, career-related internships or fieldwork, tuition waivers (full), and unspecified assistantships available. Aid available to part-time students. Financial aid application deadline: 2/1. *Faculty research:* Soil mechanics and foundations, structures, transportation, water resources systems and hydraulics. *Unit head:* Mari Cook, Graduate Assistant for Admission, 510-643-8944, E-mail: mcook@ce.berkeley.edu. *Application contact:* Frank Weisgerber, Graduate Program Director, 513-556-3673, Fax: 513-556-2599, E-mail: fweisger@boss.cee.uc.edu.

University of Colorado at Boulder, Graduate School, College of Engineering and Applied Science, Department of Civil, Environmental, and Architectural Engineering, Boulder, CO 80309. Offers building systems (MS, PhD); construction engineering and management (MS, PhD); environmental engineering (MS, PhD); geoenvironmental engineering (MS, PhD); geotechnical engineering (MS, PhD); structural engineering (MS, PhD); water resource engineering (MS, PhD). *Degree requirements:* For master's, thesis or alternative, comprehensive exam required; for doctorate, dissertation required. *Entrance requirements:* For master's, GRE General Test, minimum undergraduate GPA of 3.0.

See in-depth description on page 619.

University of Colorado at Denver, Graduate School, College of Engineering and Applied Science, Department of Civil Engineering, Denver, CO 80217-3364. Offers MS, PhD. Part-time and evening/weekend programs available. *Faculty:* 13. *Students:* 16 full-time (2 women), 69 part-time (15 women); includes 11 minority (2 African Americans, 4 Asian Americans or Pacific Islanders, 4 Hispanic Americans, 1 Native American), 13 international. Average age 31. 42 applicants, 71% accepted. In 1998, 20 master's, 1 doctorate awarded. *Degree requirements:* For master's, thesis or alternative required; for doctorate, dissertation required. *Entrance requirements:* For master's and doctorate, GRE. *Application deadline:* For fall admission, 6/1; for spring admission, 10/1. Applications are processed on a rolling basis. Application fee: $50 ($60 for international students). Electronic applications accepted. Tuition, state resident: part-time $217 per credit hour. Tuition, nonresident: part-time $783 per credit hour. Required fees: $3 per credit hour. $130 per year. One-time fee: $25 part-time. *Financial aid:* Research assistantships, teaching assistantships, career-related internships or fieldwork and Federal Work-Study available. Financial aid application deadline: 3/1; financial aid applicants required to submit FAFSA. Total annual research expenditures: $255,022. *Unit head:* David Hubly, Chair, 303-556-2871, Fax: 303-556-2368. *Application contact:* Dawn Arge, Program Assistant, 303-556-2871, Fax: 303-556-2368.

University of Connecticut, Graduate School, School of Engineering, Field of Civil Engineering, Storrs, CT 06269. Offers MS, PhD. Terminal master's awarded for partial completion of doctoral program. *Degree requirements:* For master's, thesis or alternative required; for doctorate, dissertation required. *Entrance requirements:* For master's and doctorate, GRE General Test. *Faculty research:* Structures, environmental and transportation engineering.

University of Dayton, Graduate School, School of Engineering, Department of Civil Engineering, Dayton, OH 45469-1300. Offers engineering mechanics (MSEM); environmental engineering (MSCE); soil mechanics (MSCE); structural engineering (MSCE); transport engineering (MSCE). Part-time programs available. *Faculty:* 9 full-time (0 women), 2 part-time (0 women). *Students:* 7 full-time (1 woman), 6 part-time (1 woman); includes 1 minority (African American), 4 international. In 1998, 2 degrees awarded (100% found work related to degree). *Degree requirements:* For master's, thesis or alternative required, foreign language not required. *Entrance requirements:* For master's, TOEFL. *Application deadline:* For fall admission, 8/1. Applications are processed on a rolling basis. Application fee: $30. *Financial aid:* In 1998–99, 1 research assistantship with full tuition reimbursement (averaging $12,000 per year), 1 teaching assistantship (averaging $9,000 per year) were awarded.; institutionally-sponsored loans also available. *Faculty research:* Tire/soil interaction, tilt-up structures, viscoelastic response of restraint systems, composite materials. *Unit head:* Dr. Joseph Saliba, Chairperson, 937-229-3847. *Application contact:* Dr. Donald L. Moon, Associate Dean, 937-229-2241, Fax: 937-229-2471, E-mail: dmoon@engr.udayton.edu.

University of Delaware, College of Engineering, Department of Civil and Environmental Engineering, Newark, DE 19716. Offers environmental engineering (MAS, MCE, PhD); geotechnical engineering (MAS, MCE, PhD); ocean engineering (MAS, MCE, PhD); railroad engineering (MAS, MCE, PhD); structural engineering (MAS, MCE, PhD); transportation engineering (MAS, MCE, PhD); water resource engineering (MAS, MCE, PhD). Terminal master's awarded for partial completion of doctoral program. *Degree requirements:* For master's and doctorate, thesis/dissertation required, foreign language not required. *Entrance requirements:* For master's and doctorate, GRE General Test, TOEFL (minimum score of 550 required).

See in-depth description on page 621.

University of Detroit Mercy, College of Engineering and Science, Department of Civil and Environmental Engineering, Detroit, MI 48219-0900. Offers ME, DE. Evening/weekend programs available. *Degree requirements:* For master's, computer language required, foreign language not required; for doctorate, dissertation required. *Faculty research:* Geotechnical engineering.

See in-depth description on page 623.

University of Florida, Graduate School, College of Engineering, Department of Civil Engineering, Gainesville, FL 32611. Offers MCE, ME, MS, PhD, Engr. Part-time programs available. *Faculty:* 40. *Students:* 134 full-time (24 women), 28 part-time (7 women); includes 28 minority (5 African Americans, 7 Asian Americans or Pacific Islanders, 16 Hispanic Americans), 53 international. 308 applicants, 77% accepted. In 1998, 65 master's, 6 doctorates awarded. *Degree requirements:* For master's, thesis optional, foreign language not required; for doctorate, dissertation required; for Engr, thesis optional. *Entrance requirements:* For master's and doctorate, GRE General Test, TOEFL, minimum GPA of 3.0; for Engr, GRE General Test. *Application deadline:* For fall admission, 6/1 (priority date); for spring admission, 9/1. Applications are processed on a rolling basis. Application fee: $20. Electronic applications accepted. *Financial aid:* In 1998–99, 100 students received aid, including 11 fellowships, 84 research assistantships, 1 teaching assistantship; unspecified assistantships also available. *Faculty research:* Structures, materials, hydrology, public works, surveying and mapping. Total annual research expenditures: $4.2 million. *Unit head:* Dr. Paul Thompson, Chairman, 352-392-9537, Fax: 352-392-3394, E-mail: pthom@ce.ufl.edu. *Application contact:* Dr. Kirk Hatfield, Graduate Coordinator, 352-392-0956, Fax: 352-392-3394, E-mail: khatf@ce.ufl.edu.

University of Hawaii at Manoa, Graduate Division, College of Engineering, Department of Civil Engineering, Honolulu, HI 96822. Offers MS, PhD. Part-time programs available. *Faculty:* 22 full-time (1 woman). *Students:* 22 full-time (9 women), 34 part-time (6 women); includes 25 minority (23 Asian Americans or Pacific Islanders, 2 Hispanic Americans), 24 international. Average age 29. 47 applicants, 66% accepted. In 1998, 16 master's awarded. *Degree requirements:* For master's, exams required, thesis optional, foreign language not required; for doctorate, dissertation, exams required, foreign language not required. *Entrance requirements:* For master's and doctorate, GRE General Test. *Average time to degree:* Master's–2 years full-time, 5 years part-time. *Application deadline:* For fall admission, 3/1; for spring admission, 9/1. Applications are processed on a rolling basis. Application fee: $25 ($50 for international students). *Financial aid:* In 1998–99, 31 students received aid, including 14 research assistantships (averaging $15,533 per year), 4 teaching assistantships (averaging $12,786 per year); career-related internships or fieldwork, Federal Work-Study, and tuition waivers (full and partial) also available. *Faculty research:* Structures, transportation, environmental engineering, geotechnical engineering, construction. Total annual research expenditures: $800,000. *Unit head:* Dr. Edmond Cheng, Chairperson, 808-956-7550. *Application contact:* Panagiotis Prevedouros, Graduate Chairperson, 808-956-7449, E-mail: pap@hawaii.edu.

University of Houston, Cullen College of Engineering, Department of Civil and Environmental Engineering, Houston, TX 77004. Offers MCE, MS Env E, MSCE, PhD. Part-time and evening/weekend programs available. *Faculty:* 15 full-time (1 woman), 3 part-time (0 women). *Students:* 31 full-time (7 women), 27 part-time (7 women); includes 9 minority (1 African American, 3 Asian Americans or Pacific Islanders, 5 Hispanic Americans), 33 international. Average age 29. 65 applicants, 42% accepted. In 1998, 9 master's, 3 doctorates awarded (100% found work related to degree). Terminal master's awarded for partial completion of doctoral program. *Degree requirements:* For master's, thesis required (for some programs), foreign language not required; for doctorate, dissertation, departmental qualifying exam required, foreign language not required. *Entrance requirements:* For master's and doctorate, GRE General Test, TOEFL. *Application deadline:* For fall admission, 7/3 (priority date); for spring admission, 12/4. Applications are processed on a rolling basis. Application fee: $25 ($75 for international students). *Financial aid:* In 1998–99, 26 students received aid, including 17 research assistantships, 9 teaching assistantships; career-related internships or fieldwork and Federal Work-Study also available. Financial aid application deadline: 4/1. *Faculty research:* Structural engineering and construction materials, geotechnical engineering and deep foundation, hydraulic engineering and wave mechanics, water and soil treatment. Total annual research expenditures: $2.1 million. *Unit head:* Dr. Dennis Clifford, Chairman, 713-743-4250, Fax: 713-743-4260, E-mail: daclifford@uh.edu. *Application contact:* Charlene Holliday, Graduate Analyst, 713-743-4254, Fax: 713-743-4260, E-mail: wholliday@uh.edu.

See in-depth description on page 627.

University of Idaho, College of Graduate Studies, College of Engineering, Department of Civil Engineering, Moscow, ID 83844-4140. Offers M Engr, MS, PhD. *Faculty:* 15 full-time (0 women). *Students:* 16 full-time (3 women), 39 part-time (3 women); includes 1 minority (African American), 10 international. In 1998, 11 master's awarded. *Degree requirements:* For master's, thesis required, foreign language not required; for doctorate, dissertation required. *Entrance requirements:* For master's, minimum GPA of 2.8; for doctorate, minimum undergraduate GPA of 2.8, 3.0 graduate. *Application deadline:* For fall admission, 8/1; for spring admission, 12/15. Application fee: $35 ($45 for international students). *Financial aid:* In 1998–99, 2 research assistantships (averaging $12,811 per year), 4 teaching assistantships (averaging $5,957 per year) were awarded.; fellowships, career-related internships or fieldwork also available. Financial aid application deadline: 2/15. *Faculty research:* Water resources, structural engineering, soil mechanics, materials science. *Unit head:* Dr. James Milligan, Chair, 208-885-6782.

University of Illinois at Chicago, Graduate College, College of Engineering, Department of Civil and Materials Engineering, Chicago, IL 60607-7128. Offers MS, PhD. Evening/weekend programs available. *Faculty:* 13 full-time (0 women), 2 part-time (0 women). *Students:* 37 full-time (8 women), 31 part-time (8 women); includes 9 minority (4 African Americans, 4 Asian Americans or Pacific Islanders, 1 Hispanic American), 29 international. Average age 29. 114 applicants, 29% accepted. In 1998, 9 master's, 2 doctorates awarded. *Degree requirements:* For master's, thesis required (for some programs), foreign language not required; for doctorate, dissertation, preliminary and qualifying exams required, foreign language not required. *Entrance requirements:* For master's and doctorate, GRE General Test, TOEFL, minimum GPA of 4.0 on a 5.0 scale. *Application deadline:* For fall admission, 7/3; for spring admission, 11/8. Application fee: $40 ($50 for international students). *Financial aid:* In 1998–99, 14 students received aid; fellowships, research assistantships, teaching assistantships, tuition waivers (full) available. *Faculty research:* Transportation and geotechnical engineering, damage and anisotropic behavior, steel processing. *Unit head:* Dr. Mohsen Issa, Director of Graduate Studies, 312-996-3432.

See in-depth description on page 629.

University of Illinois at Urbana–Champaign, Graduate College, College of Engineering, Department of Civil and Environmental Engineering, Urbana, IL 61801. Offers civil engineering (MS, PhD); environmental engineering and environmental science (MS, PhD), including environmental engineering, environmental science. *Faculty:* 50 full-time (3 women). *Students:* 348 full-time (78 women); includes 31 minority (2 African Americans, 16 Asian Americans or Pacific Islanders, 13 Hispanic Americans), 163 international. 588 applicants, 16% accepted. In 1998, 96 master's, 24 doctorates awarded. *Degree requirements:* For master's, thesis or alternative required, foreign language not required; for doctorate, dissertation required, foreign language not required. *Application deadline:* Applications are processed on a rolling basis. Application fee: $40 ($50 for international students). Tuition, state resident: full-time $4,616. Tuition, nonresident: full-time $11,768. Full-time tuition and fees vary according to course load. *Financial aid:* In 1998–99, 8 fellowships, 182 research assistantships, 34 teaching assistantships were awarded.; tuition waivers (full and partial) also available. Financial aid application deadline: 2/15. *Unit head:* Dr. David E. Daniel, Head, 217-333-1497. *Application contact:* Dr. Frederick V. Lawrence, Director of Graduate Studies, 217-333-6928, Fax: 217-333-9464, E-mail: flawrenc@uiuc.edu.

See in-depth description on page 631.

The University of Iowa, Graduate College, College of Engineering, Department of Civil and Environmental Engineering, Iowa City, IA 52242-1316. Offers MS, PhD. *Faculty:* 23 full-time, 2 part-time. *Students:* 46 full-time (9 women), 29 part-time (10 women); includes 3 minority (2 Asian Americans or Pacific Islanders, 1 Native American), 28 international. 115 applicants, 53% accepted. In 1998, 27 master's, 6 doctorates awarded. *Degree requirements:* For master's, thesis optional; for doctorate, dissertation, comprehensive exam required. *Entrance requirements:* For master's and doctorate, GRE, TOEFL. *Application deadline:* Applications are processed on a rolling basis. Application fee: $30 ($50 for international students). *Financial aid:* In 1998–99, 14 fellowships, 51 research assistantships, 17 teaching assistantships were awarded. Financial aid applicants required to submit FAFSA. *Faculty research:* Environmental quality modeling, engineering hydraulics, structural optimization and design. *Unit head:* Robert Ettema, Chair, 319-335-5224.

University of Kansas, Graduate School, School of Engineering, Department of Civil and Environmental Engineering, Lawrence, KS 66045. Offers civil engineering (MS, DE, PhD); environmental engineering (MS, PhD); environmental science (MS, PhD); water resources engineering (MS); water resources science (MS). *Faculty:* 20 full-time (0 women). *Students:* 23 full-time (6 women), 80 part-time (15 women); includes 9 minority (1 African American, 3 Asian Americans or Pacific Islanders, 4 Hispanic Americans, 1 Native American), 29 international. In 1998, 30 master's, 6 doctorates awarded. *Degree requirements:* For master's, thesis or

Civil Engineering

University of Kansas (continued)
alternative, exam required, foreign language not required; for doctorate, dissertation, comprehensive exam required. *Entrance requirements:* For master's and doctorate, Michigan English Language Assessment Battery, TOEFL, minimum GPA of 3.0. *Application deadline:* For fall admission, 7/1. Application fee: $30 ($45 for international students). *Financial aid:* Fellowships, research assistantships, teaching assistantships, career-related internships or fieldwork available. *Faculty research:* Structures (fracture mechanics), transportation, environmental health. *Unit head:* Steve McCabe, Chair, 785-864-3766. *Application contact:* David Parr, Graduate Director.

See in-depth description on page 633.

University of Kentucky, Graduate School, Graduate School Programs from the College of Engineering, Program in Civil Engineering, Lexington, KY 40506-0032. Offers MCE, MSCE, PhD. *Degree requirements:* For master's, comprehensive exam required, thesis optional, foreign language not required; for doctorate, dissertation, comprehensive exam required. *Entrance requirements:* For master's, GRE General Test, minimum GPA of 2.8; for doctorate, GRE General Test, minimum graduate GPA of 3.0. *Faculty research:* Geotechnical engineering, structures, construction engineering and management, environmental engineering and water resources, transportation and materials.

University of Louisville, Graduate School, Speed Scientific School, Department of Civil and Environmental Engineering, Louisville, KY 40292-0001. Offers M Eng, MS, PhD. *Accreditation:* ABET (one or more programs are accredited). *Faculty:* 14 full-time (0 women), 1 (woman) part-time. *Students:* 17 full-time (5 women), 50 part-time (14 women); includes 9 minority (3 African Americans, 6 Asian Americans or Pacific Islanders), 3 international. Average age 29. In 1998, 17 degrees awarded. *Degree requirements:* For master's and doctorate, thesis/dissertation required, foreign language not required. *Entrance requirements:* For master's and doctorate, GRE General Test (minimum combined score of 1200 required). *Application deadline:* Applications are processed on a rolling basis. Application fee: $25. *Financial aid:* In 1998–99, 3 research assistantships with tuition reimbursements (averaging $10,000 per year), 3 teaching assistantships with tuition reimbursements (averaging $10,000 per year) were awarded. *Unit head:* Dr. Louis F. Cohn, Chair, 502-852-6276, Fax: 502-852-8851, E-mail: cohn@louisville.edu.

University of Maine, Graduate School, College of Engineering, Department of Civil and Environmental Engineering, Orono, ME 04469. Offers civil engineering (MS, PhD), including environmental engineering, geotechnical engineering, structural engineering. *Faculty:* 11 full-time (0 women). *Students:* 19 full-time (3 women), 10 part-time (1 woman), 4 international. 33 applicants, 45% accepted. In 1998, 11 master's, 1 doctorate awarded. *Degree requirements:* For doctorate, dissertation required, foreign language not required, foreign language not required. *Entrance requirements:* For master's and doctorate, GRE General Test, TOEFL (minimum score of 550 required). *Application deadline:* For fall admission, 2/1 (priority date); for spring admission, 10/15. Applications are processed on a rolling basis. Application fee: $50. *Financial aid:* In 1998–99, 15 research assistantships with tuition reimbursements (averaging $9,200 per year), 5 teaching assistantships with tuition reimbursements (averaging $9,900 per year) were awarded.; Federal Work-Study, institutionally-sponsored loans, scholarships, and tuition waivers (full and partial) also available. Financial aid application deadline: 3/1. *Unit head:* Dr. Willem Brutsaert, Chair, 207-581-2170, Fax: 207-581-3888. *Application contact:* Scott G. Delcourt, Director of the Graduate School, 207-581-3218, Fax: 207-581-3232, E-mail: graduate@maine.edu.

University of Maine, Graduate School, College of Engineering, Department of Spatial Information Science and Engineering, Orono, ME 04469. Offers MS, PhD. *Faculty:* 7 full-time (1 woman). *Students:* 35 full-time (8 women), 4 part-time (2 women). Average age 25. 43 applicants, 58% accepted. In 1998, 6 master's, 2 doctorates awarded. *Degree requirements:* For master's, thesis required (for some programs), foreign language not required; for doctorate, dissertation required, foreign language not required. *Entrance requirements:* For master's and doctorate, GRE General Test, TOEFL (minimum score of 550 required). *Application deadline:* For fall admission, 2/1 (priority date); for spring admission, 10/15. Applications are processed on a rolling basis. Application fee: $50. *Financial aid:* In 1998–99, 16 research assistantships (averaging $11,500 per year), 3 teaching assistantships (averaging $8,782 per year) were awarded.; Federal Work-Study, institutionally-sponsored loans, and tuition waivers (full and partial) also available. Financial aid application deadline: 3/1. *Faculty research:* Geographic information systems, analytical photogrammetry, geodesy, global positioning systems, remote sensing. *Unit head:* Dr. Mary Kate Beard, Chair, 207-581-2175, Fax: 207-581-2206. *Application contact:* Scott G. Delcourt, Director of the Graduate School, 207-581-3218, Fax: 207-581-3232, E-mail: graduate@maine.edu.

University of Manitoba, Faculty of Graduate Studies, Faculty of Engineering, Department of Civil Engineering, Winnipeg, MB R3T 2N2, Canada. Offers M Eng, M Sc, PhD. *Degree requirements:* For master's, thesis required. *Unit head:* R. B. Pinkney, Head.

University of Maryland, College Park, Graduate School, A. James Clark School of Engineering, Department of Civil and Environmental Engineering, College Park, MD 20742-5045. Offers M Eng, MS, PhD. Part-time and evening/weekend programs available. Postbaccalaureate distance learning degree programs offered. *Faculty:* 36 full-time (2 women), 8 part-time (4 women). *Students:* 88 full-time (24 women), 84 part-time (14 women); includes 29 minority (14 African Americans, 6 Asian Americans or Pacific Islanders, 8 Hispanic Americans, 1 Native American), 65 international. 189 applicants, 41% accepted. In 1998, 32 master's, 8 doctorates awarded. *Degree requirements:* For master's, thesis or alternative required, foreign language not required; for doctorate, dissertation, qualifying exam required, foreign language not required. *Entrance requirements:* For master's, GRE General Test, minimum GPA of 3.0. *Application deadline:* Applications are processed on a rolling basis. Application fee: $50 ($70 for international students). Tuition, state resident: part-time $272 per credit hour. Tuition, nonresident: part-time $475 per credit hour. Required fees: $632; $379 per year. *Financial aid:* In 1998–99, 13 fellowships with full tuition reimbursements (averaging $14,927 per year), 37 research assistantships with tuition reimbursements (averaging $12,698 per year), 18 teaching assistantships with tuition reimbursements (averaging $10,717 per year) were awarded.; Federal Work-Study, grants, and scholarships also available. Aid available to part-time students. Financial aid applicants required to submit FAFSA. *Faculty research:* Transportation and urban systems, environmental engineering, geotechnical engineering, construction engineering and management, hydraulics, remote sensing, soil mechanics. Total annual research expenditures: $4 million. *Unit head:* Dr. Gregory Baecher, Chairman, 301-405-1977, Fax: 301-405-2585. *Application contact:* Trudy Lindsey, Director, Graduate Admission and Records, 301-405-4198, Fax: 301-314-9305, E-mail: grschool@deans.umd.edu.

See in-depth description on page 635.

University of Maryland, College Park, Graduate School, A. James Clark School of Engineering, Professional Program in Engineering, College Park, MD 20742-5045. Offers aerospace engineering (M Eng); chemical engineering (M Eng), civil engineering (M Eng); electrical engineering (M Eng); fire protection engineering (M Eng); materials science and engineering (M Eng); mechanical engineering (M Eng); reliability engineering (M Eng); systems engineering (M Eng). Part-time and evening/weekend programs available. Postbaccalaureate distance learning degree programs offered. *Faculty:* 11 part-time (0 women). *Students:* 20 full-time (3 women), 205 part-time (42 women); includes 58 minority (27 African Americans, 25 Asian Americans or Pacific Islanders, 5 Hispanic Americans, 1 Native American), 20 international. *Degree requirements:* For master's, foreign language and thesis not required. *Application deadline:* Applications are processed on a rolling basis. Application fee: $50 ($70 for international students). Tuition, state resident: part-time $272 per credit hour. Tuition, nonresident: part-time $475 per credit hour. Required fees: $632; $379 per year. *Unit head:* Dr. Patrick Cunniff, Associate Dean, 301-405-5256, Fax: 301-314-9477. *Application contact:* Trudy Lindsey, Director, Graduate Admission and Records, 301-405-4198, Fax: 301-314-9305, E-mail: grschool@deans.umd.edu.

University of Massachusetts Amherst, Graduate School, College of Engineering, Department of Civil Engineering, Program in Civil Engineering, Amherst, MA 01003. Offers MS, PhD. Part-time programs available. *Faculty:* 24 full-time (4 women). *Students:* 38 full-time (12 women), 16 part-time (4 women), 21 international. Average age 28. 105 applicants, 63% accepted. In 1998, 16 master's, 5 doctorates awarded. Terminal master's awarded for partial completion of doctoral program. *Degree requirements:* For master's, thesis or alternative required, foreign language not required; for doctorate, dissertation required, foreign language not required. *Entrance requirements:* For master's and doctorate, GRE General Test. *Application deadline:* For fall admission, 2/1 (priority date); for spring admission, 10/1. Applications are processed on a rolling basis. Application fee: $40. Tuition, state resident: full-time $2,640; part-time $165 per credit. Tuition, nonresident: full-time $9,756; part-time $407 per credit. Required fees: $1,221 per term. One-time fee: $110. Full-time tuition and fees vary according to course load, campus/location and reciprocity agreements. *Financial aid:* In 1998–99, 16 fellowships with full tuition reimbursements (averaging $7,535 per year), research assistantships with full tuition reimbursements (averaging $10,184 per year), 12 teaching assistantships with full tuition reimbursements (averaging $5,296 per year) were awarded.; career-related internships or fieldwork, Federal Work-Study, grants, scholarships, traineeships, and unspecified assistantships also available. Aid available to part-time students. Financial aid application deadline: 2/1.

University of Massachusetts Lowell, Graduate School, James B. Francis College of Engineering, Department of Civil Engineering, Lowell, MA 01854-2881. Offers civil engineering (MS Eng); environmental studies (MS Eng). Part-time programs available. *Faculty:* 13 full-time (1 woman), 6 part-time (2 women). *Students:* 21 full-time (7 women), 82 part-time (10 women); includes 9 minority (3 African Americans, 5 Asian Americans or Pacific Islanders, 1 Hispanic American), 5 international. 36 applicants, 81% accepted. In 1998, 14 degrees awarded. *Degree requirements:* For master's, thesis optional, foreign language not required. *Entrance requirements:* For master's, GRE General Test. *Application deadline:* For fall admission, 4/1 (priority date); for spring admission, 10/1. Applications are processed on a rolling basis. Application fee: $20 ($35 for international students). *Financial aid:* In 1998–99, 6 research assistantships, 4 teaching assistantships were awarded.; career-related internships or fieldwork also available. Financial aid application deadline: 4/1. *Faculty research:* Bridge design, traffic control, groundwater remediation, pile capacity. *Unit head:* Dr. William B. Moeller, Chairman, 978-934-2287. *Application contact:* Dr. Burton Segall, Graduate Coordinator, 978-934-2288, E-mail: burton_segall@woods.uml.edu.

The University of Memphis, Graduate School, Herff College of Engineering, Department of Civil Engineering, Memphis, TN 38152. Offers civil engineering (PhD); environmental engineering (MS); foundation engineering (MS); structural engineering (MS); transportation engineering (MS); water resources engineering (MS). *Faculty:* 12 full-time (0 women), 1 part-time (0 women). *Students:* 6 full-time (2 women), 22 part-time (1 woman), 7 international. Average age 32. 36 applicants, 56% accepted. In 1998, 5 degrees awarded. *Degree requirements:* For master's, thesis or alternative, comprehensive exam required; for doctorate, dissertation required. *Entrance requirements:* For master's, GRE General Test (minimum combined score of 1000 required) or MAT, BS, minimum undergraduate GPA of 2.5. *Application deadline:* For fall admission, 8/1; for spring admission, 12/1. Application fee: $25 ($50 for international students). Tuition, state resident: full-time $3,410; part-time $178 per credit hour. Tuition, nonresident: full-time $8,670; part-time $408 per credit hour. Tuition and fees vary according to program. *Financial aid:* In 1998–99, 15 research assistantships were awarded.; career-related internships or fieldwork also available. *Faculty research:* Soil structure interaction, groundwater, strength of silt, well design, transitioning to Ada as the primary development language for PC software. *Unit head:* Dr. Martin E. Lipinski, Chairman, 901-678-3279. *Application contact:* Dr. Larry W. Moore, Coordinator of Graduate Studies, 901-678-3278.

University of Miami, Graduate School, College of Engineering, Department of Civil, Architectural, and Environmental Engineering, Coral Gables, FL 33124. Offers architectural engineering (MSAE); civil engineering (MSCE, DA, PhD). Part-time and evening/weekend programs available. *Faculty:* 9 full-time (1 woman), 8 part-time (1 woman). *Students:* 17 full-time (7 women), 2 part-time (1 woman); includes 8 minority (5 Asian Americans or Pacific Islanders, 3 Hispanic Americans) Average age 25. 43 applicants, 74% accepted. In 1998, 4 master's awarded. *Degree requirements:* For master's, thesis required, foreign language not required; for doctorate, dissertation, oral and qualifying exams required, foreign language not required. *Entrance requirements:* For master's and doctorate, GRE General Test (minimum score of 500 on verbal section, 500 on quantitative required), TOEFL (minimum score of 550 required), minimum GPA of 3.0. *Average time to degree:* Master's–1.5 years full-time, 3 years part-time; doctorate–3 years full-time, 4 years part-time. *Application deadline:* For fall admission, 4/1 (priority date); for spring admission, 11/1 (priority date). Applications are processed on a rolling basis. Application fee: $35. Tuition: Full-time $15,336; part-time $852 per credit. Required fees: $174. Tuition and fees vary according to program. *Financial aid:* In 1998–99, 11 students received aid, including 3 research assistantships, 9 teaching assistantships; fellowships, institutionally-sponsored loans and tuition waivers (partial) also available. *Faculty research:* Electron beam, wastewater treatment, water management, structural reliability, structural dynamics, wind engineering. Total annual research expenditures: $2.3 million. *Unit head:* Dr. David A. Chin, Chairman, 305-284-3391, Fax: 305-284-3492. *Application contact:* Dr. Ahmad H. Namini, Adviser, 305-284-3391, Fax: 305-284-3492.

University of Michigan, Horace H. Rackham School of Graduate Studies, College of Engineering, Department of Civil and Environmental Engineering, Ann Arbor, MI 48109. Offers civil engineering (MSE, PhD, CE); construction engineering and management (MSE); environmental engineering (MSE, PhD). *Degree requirements:* For master's, foreign language and thesis not required; for doctorate, computer language, oral defense of dissertation, preliminary and written exams required. *Entrance requirements:* For master's, GRE General Test; for doctorate, GRE General Test, master's degree. Electronic applications accepted. *Faculty research:* Earthquake engineering, environmental and water resources engineering, geotechnical engineering, hydraulics and hydrologic engineering, structural engineering.

See in-depth description on page 637.

University of Michigan, Horace H. Rackham School of Graduate Studies, College of Engineering, Department of Naval Architecture and Marine Engineering, Ann Arbor, MI 48109. Offers concurrent marine design (M Eng); naval architecture and marine engineering (MS, MSE, PhD, Mar Eng, Nav Arch). Part-time programs available. Postbaccalaureate distance learning degree programs offered (minimal on-campus study). *Faculty:* 12 full-time (2 women), 9 part-time (0 women). *Students:* 70 full-time (7 women), 2 part-time; includes 1 minority (Hispanic American), 29 international. Average age 26. 63 applicants, 95% accepted. In 1998, 33 master's, 6 doctorates awarded (100% found work related to degree). Terminal master's awarded for partial completion of doctoral program. *Degree requirements:* For master's, thesis required (for some programs), foreign language not required; for doctorate and other advanced degree, dissertation, oral defense of dissertation, preliminary exams required, foreign language not required; for other advanced degree, thesis, comprehensive written exam, oral defense of dissertation required, foreign language not required. *Entrance requirements:* For master's, GRE General Test (for financial aid), TOEFL (minimum score of 560 required) or Michigan English Language Assessment Battery (minimum score of 80 required); for doctorate, GRE General Test (minimum combined score of 1300 required; average 1352), master's degree; for other advanced degree, GRE General Test (minimum combined score of 1300 required). *Average time to degree:* Master's–1 year full-time; doctorate–4 years full-time. *Application deadline:* For fall admission, 2/1. Applications are processed on a rolling basis. Application fee: $55. Electronic applications accepted. *Financial aid:* In 1998–99, 24 students received aid, including 3 fellowships, 15 research assistantships, 2 teaching assistantships; career-related internships or fieldwork, Federal Work-Study, institutionally-sponsored loans, and scholarships also available. Financial aid application deadline: 2/1. *Faculty research:* Marine mechanics including hydrodynamics, structures, marine environmental engineering, marine design analysis, concurrent marine design, virtual reality. *Unit head:* Dr. Michael M. Bernitsas, Chair, 734-936-0566, Fax: 734-936-8820, E-mail: clv@engin.umich.edu. *Application contact:* Celia A. Eidex, Graduate Program Coordinator, 734-936-0566, Fax: 734-936-8820, E-mail: ceidex@engin.umich.edu.

University of Minnesota, Twin Cities Campus, Graduate School, Institute of Technology, Department of Civil Engineering, Minneapolis, MN 55455-0213. Offers civil engineering (MCE, MS, PhD); geological engineering (M Geo E, MS, PhD). Part-time programs available. *Faculty:* 34 full-time (4 women), 2 part-time (0 women). *Students:* 113 full-time (29 women), 23 part-time (5 women), 31 international. 157 applicants, 44% accepted. In 1998, 36 master's, 15 doctorates awarded. *Degree requirements:* For master's, thesis optional, foreign language not required; for doctorate, dissertation required, foreign language not required. *Entrance requirements:* For master's and doctorate, GRE General Test, TOEFL (minimum score of 550 required). *Average time to degree:* Master's–2.4 years full-time, 4.8 years part-time; doctorate–6 years full-time, 10 years part-time. *Application deadline:* For fall admission, 6/15 (priority date); for spring admission, 10/15. Applications are processed on a rolling basis. Application fee: $50 ($55 for international students). *Financial aid:* In 1998–99, 89 students received aid, including 20 fellowships with tuition reimbursements available (averaging $14,160 per year), 41 research assistantships with tuition reimbursements available (averaging $14,616 per year), 17 teaching assistantships with tuition reimbursements available (averaging $14,616 per year); career-related internships or fieldwork, institutionally-sponsored loans, and unspecified assistantships also available. *Faculty research:* Environmental engineering, rock mechanics, water resources, structural engineering, transportation. Total annual research expenditures: $4 million. *Unit head:* John Gulliver, Head, 612-625-5522, Fax: 612-626-7750. *Application contact:* Roxane McGlade, Student Personnel Worker for Graduate Studies, 612-625-9581, Fax: 612-626-7750, E-mail: gradsec@ce.umn.edu.

University of Missouri–Columbia, Graduate School, College of Engineering, Department of Civil and Environmental Engineering, Columbia, MO 65211. Offers civil engineering (MS, PhD); environmental engineering (MS, PhD); geotechnical engineering (MS, PhD); structural engineering (MS, PhD); transportation and highway engineering (MS); water resources (MS, PhD). *Faculty:* 24 full-time (2 women). *Students:* 19 full-time (2 women), 18 part-time (3 women); includes 1 minority (Hispanic American), 21 international. 74 applicants, 43% accepted. In 1998, 19 master's, 1 doctorate awarded. *Degree requirements:* For master's, report or thesis required; for doctorate, dissertation required. *Entrance requirements:* For master's and doctorate, GRE General Test, TOEFL. *Application deadline:* For fall admission, 7/10. Application fee: $30 ($50 for international students). *Financial aid:* Research assistantships, teaching assistantships, institutionally-sponsored loans available. *Unit head:* Dr. Mark Virkler, Director of Graduate Studies, 573-882-3678.

See in-depth description on page 639.

University of Missouri–Rolla, Graduate School, School of Engineering, Department of Civil Engineering, Program in Civil Engineering, Rolla, MO 65409-0910. Offers MS, PhD. *Faculty:* 12 full-time (0 women). *Students:* 46 full-time (10 women), 18 part-time (6 women); includes 4 minority (1 African American, 2 Asian Americans or Pacific Islanders, 1 Hispanic American), 21 international. Average age 30. 99 applicants, 66% accepted. In 1998, 19 master's, 5 doctorates awarded. *Degree requirements:* For master's, thesis or alternative required, foreign language not required; for doctorate, dissertation required, foreign language not required. *Entrance requirements:* For master's and doctorate, GRE General Test (minimum combined score of 1100 required), TOEFL (minimum score of 550 required), minimum GPA of 3.0. *Application deadline:* For fall admission, 7/1; for spring admission, 12/1. Applications are processed on a rolling basis. Application fee: $25. Electronic applications accepted. *Financial aid:* In 1998–99, 18 fellowships with full tuition reimbursements, 13 research assistantships with partial tuition reimbursements, 14 teaching assistantships with partial tuition reimbursements were awarded.; institutionally-sponsored loans also available. Aid available to part-time students. Financial aid application deadline: 1/1. *Unit head:* Deborah Parker, Administrative Assistant I, 313-593-5582, Fax: 313-593-5386, E-mail: debbie@umdsun2.umd.umich.edu. *Application contact:* Dr. Abdeldjelil Belarbi, Director of Graduate Advising, 573-341-4478, Fax: 573-341-4729, E-mail: ceadvise@novell.civil.umr.edu.

University of Nebraska–Lincoln, Graduate College, College of Engineering and Technology, Department of Civil Engineering, Lincoln, NE 68588. Offers civil engineering (MS); engineering (PhD). *Faculty:* 16 full-time (0 women). *Students:* 16 full-time (2 women), 5 part-time (2 women); includes 1 minority (African American), 5 international. Average age 28. 25 applicants, 56% accepted. In 1998, 9 degrees awarded. *Degree requirements:* For master's, thesis optional, foreign language not required; for doctorate, dissertation, comprehensive exams required. *Entrance requirements:* For master's and doctorate, GRE General Test, TOEFL (minimum score of 550 required). *Application deadline:* For fall admission, 4/15; for spring admission, 10/15. Application fee: $35. Electronic applications accepted. *Financial aid:* In 1998–99, 3 fellowships, 18 research assistantships, 6 teaching assistantships were awarded.; Federal Work-Study also available. Aid available to part-time students. Financial aid application deadline: 2/15. *Faculty research:* Box culverts, crash tests on bridge approaches, biological/chemical removal of nitrates from water, hazardous waste containment systems. *Unit head:* Dr. Raymond Moore, Chair, 402-472-2371, Fax: 402-472-8934.

University of Nevada, Las Vegas, Graduate College, Howard R. Hughes College of Engineering, Department of Civil and Environmental Engineering, Las Vegas, NV 89154-9900. Offers MSE, PhD. *Faculty:* 19 full-time (2 women). *Students:* 21 full-time (3 women), 37 part-time (12 women); includes 10 minority (7 Asian Americans or Pacific Islanders, 3 Hispanic Americans), 11 international. 36 applicants, 64% accepted. In 1998, 12 master's, 1 doctorate awarded. *Degree requirements:* For master's, comprehensive exam required, thesis optional, foreign language not required; for doctorate, dissertation required. *Entrance requirements:* For master's, minimum GPA of 3.0; for doctorate, minimum GPA of 3.5. *Application deadline:* For fall admission, 6/15 (priority date); for spring admission, 11/15. Applications are processed on a rolling basis. Application fee: $40 ($95 for international students). *Financial aid:* In 1998–99, 11 research assistantships with tuition reimbursements (averaging $7,133 per year), 13 teaching assistantships with partial tuition reimbursements (averaging $9,058 per year) were awarded. Financial aid application deadline: 3/1. *Unit head:* Dr. Edward S. Neumann, Chair, 702-895-3701, Fax: 702-895-3936, E-mail: ce-info@ce.unlv.edu.

University of Nevada, Reno, Graduate School, College of Engineering, Department of Civil Engineering, Reno, NV 89557. Offers MS, PhD. Terminal master's awarded for partial completion of doctoral program. *Degree requirements:* For master's, thesis optional, foreign language not required; for doctorate, dissertation required, foreign language not required. *Entrance requirements:* For master's, TOEFL (minimum score of 500 required), minimum GPA of 2.75; for doctorate, TOEFL (minimum score of 500 required), minimum GPA of 3.0.

University of New Brunswick, School of Graduate Studies, Faculty of Engineering, Department of Civil Engineering, Fredericton, NB E3B 5A3, Canada. Offers environmental engineering (M Eng, M Sc E, PhD); geotechnical engineering (M Eng, M Sc E); structures and structural foundations (M Eng, M Sc E, PhD); transportation engineering (M Eng, M Sc E, PhD); water resources and hydrology (M Eng, M Sc E, PhD). Part-time programs available. *Degree requirements:* For master's, thesis required, foreign language not required; for doctorate, dissertation, qualifying exam required, foreign language not required. *Entrance requirements:* For master's and doctorate, TOEFL, TWE, minimum GPA of 3.0.

University of New Hampshire, Graduate School, College of Engineering and Physical Sciences, Programs in Engineering, Department of Civil Engineering, Durham, NH 03824. Offers MS. Part-time programs available. *Faculty:* 14 full-time. *Students:* 12 full-time (5 women), 28 part-time (8 women); includes 1 minority (Asian American or Pacific Islander) Average age 29. 20 applicants, 100% accepted. In 1998, 9 degrees awarded. *Degree requirements:* For master's, thesis or alternative required, foreign language not required. *Entrance requirements:* For master's, GRE. *Application deadline:* For fall admission, 4/1 (priority date). Applications are processed on a rolling basis. Application fee: $50. Tuition, area resident: Full-time $5,750; part-time $319 per credit. Tuition, state resident: full-time $8,625. Tuition, nonresident: full-time $14,640; part-time $598 per credit. Required fees: $224 per semester. Tuition and fees vary according to course load, degree level and program. *Financial aid:* In 1998–99, 8 research assistantships, 10 teaching assistantships were awarded.; fellowships, Federal Work-Study, scholarships, and tuition waivers (full and partial) also available. Aid available to part-time students. Financial aid application deadline: 2/15. *Faculty research:* Environmental, structural materials, geotechnical engineering, water resources, systems analysis. *Unit head:* Dr. Thomas P. Ballestero, Chairperson, 603-862-2405. *Application contact:* Dr. M. Robin Collins, Graduate Coordinator, 603-862-1407.

University of New Hampshire, Graduate School, College of Engineering and Physical Sciences, Programs in Engineering, Doctoral Program in Engineering, Durham, NH 03824. Offers chemical engineering (PhD); civil engineering (PhD); electrical engineering (PhD); mechanical engineering (PhD); systems design engineering (PhD). *Students:* 16 full-time (3 women), 8 part-time (1 woman), 12 international. *Degree requirements:* For doctorate, dissertation required. *Entrance requirements:* For doctorate, GRE (for civil and mechanical engineering options). *Application deadline:* For fall admission, 4/1 (priority date). Applications are processed on a rolling basis. Application fee: $50. Tuition, area resident: Full-time $5,750; part-time $319 per credit. Tuition, state resident: full-time $8,625. Tuition, nonresident: full-time $14,640; part-time $598 per credit. Required fees: $224 per semester. Tuition and fees vary according to course load, degree level and program.

University of New Mexico, Graduate School, School of Engineering, Department of Civil Engineering, Albuquerque, NM 87131-2039. Offers civil engineering (MS); engineering (PhD); hazardous waste engineering (ME). Part-time programs available. *Faculty:* 17 full-time (2 women), 2 part-time (0 women). *Students:* 26 full-time (5 women), 39 part-time (6 women); includes 13 minority (1 African American, 9 Hispanic Americans, 3 Native Americans), 14 international. Average age 33. 21 applicants, 57% accepted. In 1998, 14 master's, 4 doctorates awarded. *Degree requirements:* For master's, thesis required (for some programs), foreign language not required; for doctorate, dissertation required, foreign language not required. *Entrance requirements:* For master's and doctorate, GRE General Test, minimum GPA of 3.0. *Application deadline:* For fall admission, 7/15; for spring admission, 11/14. Applications are processed on a rolling basis. Application fee: $25. *Financial aid:* In 1998–99, 3 fellowships (averaging $3,500 per year), 27 research assistantships with tuition reimbursements (averaging $2,901 per year), 3 teaching assistantships with tuition reimbursements (averaging $6,905 per year) were awarded.; career-related internships or fieldwork and Federal Work-Study also available. Financial aid application deadline: 2/15. *Faculty research:* Construction, environmental engineering, geotechnical engineering, structural engineering, transportation. Total annual research expenditures: $1.2 million. *Unit head:* Dr. Timothy L. Ward, Chair, 505-277-2722, Fax: 505-277-7430, E-mail: tlward@unm.edu. *Application contact:* Richard Heggen, Graduate Adviser, 505-277-5737, Fax: 505-277-7430, E-mail: rheggen@unm.edu.

University of New Orleans, Graduate School, College of Engineering, Concentration in Civil Engineering, New Orleans, LA 70148. Offers MS. Part-time and evening/weekend programs available. *Faculty:* 8 full-time (0 women), 1 part-time (0 women). *Students:* 18 full-time (1 woman), 16 part-time (1 woman); includes 1 minority (African American), 19 international. Average age 30. 67 applicants, 46% accepted. In 1998, 8 degrees awarded. *Degree requirements:* For master's, thesis optional, foreign language not required. *Entrance requirements:* For master's, GRE General Test (minimum combined score of 1200 required), minimum GPA of 3.0. *Application deadline:* For fall admission, 7/1 (priority date). Applications are processed on a rolling basis. Application fee: $20. Tuition, state resident: full-time $2,362. Tuition, nonresident: full-time $7,888. Part-time tuition and fees vary according to course load. *Faculty research:* Dynamic analysis for pile capacity, soil stabilization, groundwater modeling and aquifer remediation, potable water studies, water quality modeling. *Unit head:* Dr. Kenneth L. McManis, Chairman, 504-280-6271, Fax: 504-280-5586, E-mail: klmce@uno.edu. *Application contact:* Dr. Mike Folse, Graduate Coordinator, 504-280-7268, Fax: 504-280-5586, E-mail: mdfce@uno.edu.

University of New Orleans, Graduate School, College of Engineering, Concentration in Naval Architecture and Marine Engineering, New Orleans, LA 70148. Offers MS. Part-time and evening/weekend programs available. *Faculty:* 3 full-time (0 women), 1 part-time (0 women). *Students:* 5 full-time (1 woman), 6 part-time, 5 international. Average age 28. 17 applicants, 53% accepted. *Degree requirements:* For master's, computer language required, thesis optional, foreign language not required. *Entrance requirements:* For master's, GRE General Test (minimum combined score of 1200 required), minimum GPA of 3.0. *Application deadline:* For fall admission, 7/1 (priority date). Applications are processed on a rolling basis. Application fee: $20. Tuition, state resident: full-time $2,362. Tuition, nonresident: full-time $7,888. Part-time tuition and fees vary according to course load. *Financial aid:* Research assistantships, teaching assistantships, institutionally-sponsored loans available. *Faculty research:* Ship structures, hydrodynamics, computer-aided ship design. *Unit head:* Dr. William Vorus, Chairman, 504-280-7180, Fax: 504-280-5542, E-mail: wsvna@uno.edu. *Application contact:* Dr. Jeffrey Falzarano, Graduate Coordinator, 504-280-7184, Fax: 504-280-5542, E-mail: jmfna@uno.edu.

University of North Carolina at Charlotte, Graduate School, The William States Lee College of Engineering, Department of Civil Engineering, Charlotte, NC 28223-0001. Offers MSCE. Part-time and evening/weekend programs available. *Faculty:* 14 full-time (2 women). *Students:* 7 full-time (0 women), 25 part-time (4 women); includes 2 minority (both Asian Americans or Pacific Islanders), 5 international. Average age 28. 23 applicants, 91% accepted. In 1998, 12 degrees awarded. *Degree requirements:* For master's, thesis or project required. *Entrance requirements:* For master's, GRE General Test, minimum GPA of 3.0 in undergraduate major, 2.75 overall. *Application deadline:* For fall admission, 7/15; for spring admission, 11/15. Applications are processed on a rolling basis. Application fee: $35. Electronic applications accepted. *Financial aid:* In 1998–99, 8 research assistantships, 5 teaching assistantships were awarded.; Federal Work-Study also available. Financial aid application deadline: 4/1. *Faculty research:* Energy consumption, management of inorganic materials, geotechnical engineering, water resources, transportation engineering. *Unit head:* Dr. David T. Young, Acting Chair, 704-547-2304, Fax: 704-547-2352, E-mail: dyoung@email.uncc.edu. *Application contact:* Kathy Barringer, Assistant Director of Graduate Admissions, 704-547-3366, Fax: 704-547-3279, E-mail: gradadm@email.uncc.edu.

University of North Dakota, Graduate School, School of Engineering and Mines, Department of Civil Engineering, Grand Forks, ND 58202. Offers civil engineering (M Engr); sanitary engineering (M Engr), including soils and structures engineering, surface mining engineering. Part-time programs available. *Faculty:* 6 full-time (0 women). *Students:* 7 full-time (1 woman). 1 applicants, 100% accepted. In 1998, 4 degrees awarded. *Degree requirements:* For master's, thesis or alternative required, foreign language not required. *Entrance requirements:* For master's, GRE General Test, TOEFL (minimum score of 550 required), minimum GPA of 2.5. *Application deadline:* For fall admission, 3/1 (priority date). Applications are processed on a rolling basis. Application fee: $20. *Financial aid:* In 1998–99, 5 students received aid, including 2 research assistantships, 3 teaching assistantships; fellowships, career-related internships or fieldwork, Federal Work-Study, institutionally-sponsored loans, and tuition waivers (full and partial) also available. Financial aid application deadline: 3/15. *Unit head:* Dr. Ronald Apanian, Chairperson, 701-777-3562, Fax: 701-777-4838, E-mail: ron_apanian@mail.und.nodak.edu.

University of Notre Dame, Graduate School, College of Engineering, Department of Civil Engineering and Geological Sciences, Notre Dame, IN 46556. Offers bioengineering (MS); civil engineering (MS); civil engineering and geological sciences (PhD); environmental engineering (MS); geological sciences (MS). *Faculty:* 13 full-time (1 woman). *Students:* 35 full-time (10 women), 4 part-time; includes 4 minority (2 African Americans, 1 Asian American or Pacific Islander, 1 Hispanic American), 11 international. 86 applicants, 20% accepted. In 1998, 3

Civil Engineering

University of Notre Dame (continued)
master's awarded (100% found work related to degree); 2 doctorates awarded (100% entered university research/teaching). Terminal master's awarded for partial completion of doctoral program. *Degree requirements:* For master's and doctorate, thesis/dissertation required, foreign language not required. *Entrance requirements:* For master's and doctorate, GRE General Test, TOEFL (minimum score of 600 required; 250 for computer-based). *Average time to degree:* Master's–2 years full-time; doctorate–5.5 years full-time. *Application deadline:* For fall admission, 2/1 (priority date); for spring admission, 10/15. Applications are processed on a rolling basis. Application fee: $40. *Financial aid:* In 1998–99, 35 students received aid, including 14 fellowships with full tuition reimbursements available (averaging $16,000 per year), 4 research assistantships with full tuition reimbursements available (averaging $11,500 per year), 17 teaching assistantships with full tuition reimbursements available (averaging $11,500 per year); tuition waivers (full) also available. Financial aid application deadline: 2/1. *Faculty research:* Structural analysis, finite-element methods, environmental modeling, biological-waste treatment, petrology, environmental geology. Total annual research expenditures: $2.1 million. *Unit head:* Dr. Billie F. Spencer, Director of Graduate Studies, 219-631-5381, Fax: 219-631-9236, E-mail: cegeos@nd.edu. *Application contact:* Dr. Terrence J. Akai, Director of Graduate Admissions, 219-631-7706, Fax: 219-631-4183, E-mail: gradad@nd.edu.

University of Oklahoma, Graduate College, College of Engineering, School of Civil Engineering and Environmental Science, Program in Civil Engineering, Norman, OK 73019-0390. Offers civil engineering (PhD); environmental engineering (MS); geotechnical engineering (MS); structures (MS); transportation (MS). *Accreditation:* ABET (one or more programs are accredited). *Students:* 19 full-time (4 women), 36 part-time (6 women); includes 5 minority (3 Asian Americans or Pacific Islanders, 2 Native Americans), 28 international. 48 applicants, 83% accepted. In 1998, 18 master's awarded (6% entered university research/teaching); 6 doctorates awarded. *Degree requirements:* For master's, comprehensive and oral exams required, foreign language and thesis not required; for doctorate, dissertation, oral and qualifying exams required, foreign language not required. *Entrance requirements:* For master's, GRE General Test, TOEFL (minimum score of 575 required), minimum GPA of 3.0; for doctorate, GRE General Test, TOEFL (minimum score of 575 required), minimum graduate GPA of 3.5. *Application deadline:* For fall admission, 4/1 (priority date). Applications are processed on a rolling basis. Application fee: $25. Tuition, state resident: part-time $86 per credit hour. Tuition, nonresident: part-time $275 per credit hour. Tuition and fees vary according to course level, course load and program. *Financial aid:* Fellowships, research assistantships, teaching assistantships, Federal Work-Study and institutionally-sponsored loans available. Financial aid application deadline: 3/1. *Faculty research:* Soils and materials, industrial and hazardous waste. *Unit head:* Robert C. Knox, Graduate Liaison, 405-325-4256, Fax: 405-325-7508. *Application contact:* Robert C. Knox, Graduate Liaison, 405-325-4256, Fax: 405-325-7508.

See in-depth description on page 641.

University of Ottawa, School of Graduate Studies and Research, Faculty of Engineering, Ottawa-Carleton Institute for Civil Engineering, Ottawa, ON K1N 6N5, Canada. Offers M Eng, MA Sc, PhD. *Faculty:* 46 full-time, 7 part-time. *Students:* 95 full-time (24 women), 26 part-time (5 women), 28 international. Average age 32. In 1998, 31 master's, 8 doctorates awarded. *Degree requirements:* For master's, thesis or alternative required, foreign language not required; for doctorate, dissertation required, foreign language not required. *Entrance requirements:* For master's, honors degree or equivalent, minimum B average; for doctorate, master's degree, minimum B+ average. *Application deadline:* For fall admission, 3/1 (priority date). Applications are processed on a rolling basis. Application fee: $35. *Financial aid:* In 1998–99, 1 fellowship, 19 research assistantships, 21 teaching assistantships were awarded. *Faculty research:* Geotechnical structures, environmental and water resources engineering. *Unit head:* Hiroshi Tanaka, Director, 613-562-5800 Ext. 6144, Fax: 613-562-5173. *Application contact:* Bonnie Mavrantzas, Academic Secretary, 613-562-5800 Ext. 6138, Fax: 613-562-5173, E-mail: gradinfo@eng.uottawa.ca.

University of Pittsburgh, School of Engineering, Department of Civil and Environmental Engineering, Pittsburgh, PA 15260. Offers MSCEE, PhD. Part-time and evening/weekend programs available. *Faculty:* 12 full-time (0 women). *Students:* 27 full-time (7 women), 69 part-time (6 women). 64 applicants, 83% accepted. In 1998, 20 master's, 4 doctorates awarded. Terminal master's awarded for partial completion of doctoral program. *Degree requirements:* For master's, computer language required, thesis optional, foreign language not required; for doctorate, computer language, dissertation, comprehensive and final oral exams required, foreign language not required. *Entrance requirements:* For master's and doctorate, TOEFL, minimum QPA of 3.0. *Average time to degree:* Master's–2 years full-time, 4 years part-time; doctorate–5 years full-time, 7 years part-time. *Application deadline:* For fall admission, 8/1 (priority date); for spring admission, 12/1 (priority date). Applications are processed on a rolling basis. Application fee: $30 ($40 for international students). *Financial aid:* In 1998–99, 19 students received aid, including 8 research assistantships with tuition reimbursements available (averaging $8,872 per year), 13 teaching assistantships with tuition reimbursements available (averaging $10,608 per year); fellowships with tuition reimbursements available, grants, scholarships, traineeships, and tuition waivers (full and partial) also available. Financial aid application deadline: 2/15. *Faculty research:* Environmental and water resources, structures and infrastructures, construction management. Total annual research expenditures: $779,722. *Unit head:* Dr. Rafael G. Quimpo, Interim Chairman, 412-624-9870, Fax: 412-624-0135, E-mail: quimpo@civ.pitt.edu. *Application contact:* Attila A. Sooky, Academic Coordinator, 412-624-9869, Fax: 412-624-0135, E-mail: sooky@civ.pitt.edu.

University of Portland, Graduate School, Multnomah School of Engineering, Department of Civil Engineering, Portland, OR 97203-5798. Offers MSCE. Part-time and evening/weekend programs available. 5 applicants, 40% accepted. In 1998, 2 degrees awarded. *Degree requirements:* For master's, computer language required, foreign language and thesis not required. *Entrance requirements:* For master's, GRE General Test, TOEFL (minimum score of 550 required), minimum GPA of 3.0. *Application deadline:* For fall admission, 8/1 (priority date); for spring admission, 12/1. Applications are processed on a rolling basis. Application fee: $40. Tuition: Part-time $563 per semester hour. *Financial aid:* Application deadline: 3/15. *Unit head:* Dr. Khalid Khan, Director, 503-943-7276.

University of Puerto Rico, Mayagüez Campus, Graduate Studies, College of Engineering, Department of Civil Engineering, Mayagüez, PR 00681-5000. Offers MCE, MS, PhD. Part-time programs available. *Degree requirements:* For master's, thesis (MS), comprehensive exam required; for doctorate, one foreign language, dissertation required. *Entrance requirements:* For master's and doctorate, minimum GPA of 2.5, proficiency in English and Spanish. *Faculty research:* Structural design, concrete structure, finite elements, dynamic analysis, transportation, soils.

University of Rhode Island, Graduate School, College of Engineering, Department of Civil and Environmental Engineering, Kingston, RI 02881. Offers environmental engineering (MS, PhD); geotechnical engineering (MS, PhD); structural engineering (MS, PhD); transportation engineering (MS, PhD). *Entrance requirements:* For master's and doctorate, GRE.

University of Saskatchewan, College of Graduate Studies and Research, College of Engineering, Department of Civil Engineering, Saskatoon, SK S7N 5A2, Canada. Offers M Eng, M Sc, PhD. *Degree requirements:* For master's and doctorate, computer language, thesis/dissertation required, foreign language not required. *Entrance requirements:* For master's, TOEFL (minimum score of 500 required), GRE, minimum GPA of 5.0 on an 8.0 scale; for doctorate, GRE, TOEFL. *Faculty research:* Geotechnical engineering, structures, water sciences.

University of South Carolina, Graduate School, College of Engineering and Information Technology, Department of Civil and Environmental Engineering, Columbia, SC 29208. Offers ME, MS, PhD. Part-time and evening/weekend programs available. Postbaccalaureate distance learning degree programs offered (minimal on-campus study). *Faculty:* 13 full-time (2 women). *Students:* 20 full-time (5 women), 53 part-time (15 women); includes 5 minority (2 African Americans, 1 Asian American or Pacific Islander, 1 Hispanic American, 1 Native American), 13 international. Average age 34. In 1998, 28 master's, 1 doctorate awarded. *Degree requirements:* For master's, thesis required (for some programs), foreign language not required; for doctorate, dissertation required, foreign language not required. *Entrance requirements:* For master's and doctorate, GRE General Test (minimum combined score of 1100 required), TOEFL (minimum score of 525 required). *Application deadline:* For fall admission, 3/1 (priority date); for spring admission, 11/1. Applications are processed on a rolling basis. Application fee: $35. Electronic applications accepted. Tuition, state resident: full-time $4,014; part-time $202 per credit hour. Tuition, nonresident: full-time $8,528; part-time $428 per credit hour. Required fees: $100; $4 per credit hour. Tuition and fees vary according to program. *Financial aid:* In 1998–99, research assistantships with partial tuition reimbursements (averaging $12,000 per year), teaching assistantships with partial tuition reimbursements (averaging $12,000 per year) were awarded.; fellowships, career-related internships or fieldwork also available. *Faculty research:* Pavement evaluation, mathematical modeling, cable-guyed towers, *in situ* bioremediation, groundwater hydraulics. Total annual research expenditures: $312,716. *Unit head:* Dr. H. Chaudhry, Chair, 803-777-3652, Fax: 803-777-8265, E-mail: chaudhry@engr.sc.edu. *Application contact:* Jo Wooley, Administrative Assistant, 803-777-3614, Fax: 803-777-8265, E-mail: woolej@engr.sc.edu.

University of Southern California, Graduate School, School of Engineering, Department of Civil Engineering, Program in Civil Engineering, Los Angeles, CA 90089. Offers MS, PhD, Engr. *Faculty:* 16 full-time (1 woman), 4 part-time (0 women). *Students:* 32 full-time (6 women), 24 part-time (2 women); includes 9 minority (7 Asian Americans or Pacific Islanders, 2 Hispanic Americans), 38 international. Average age 31. 218 applicants, 81% accepted. In 1998, 3 master's, 10 doctorates awarded. *Degree requirements:* For master's, thesis optional; for doctorate, dissertation required. *Entrance requirements:* For master's, doctorate, and Engr, GRE General Test. *Application deadline:* For fall admission, 6/1 (priority date); for spring admission, 11/1. Application fee: $55. Tuition: Part-time $768 per unit. Required fees: $350 per semester. *Financial aid:* In 1998–99, 4 fellowships, 26 research assistantships, 10 teaching assistantships were awarded.; Federal Work-Study and institutionally-sponsored loans also available. Aid available to part-time students. Financial aid application deadline: 2/15; financial aid applicants required to submit FAFSA. *Unit head:* Dr. L. Carter Wellford, Chair, Department of Civil Engineering, 213-740-0587.

See in-depth description on page 643.

University of South Florida, Graduate School, College of Engineering, Department of Civil and Environmental Engineering, Tampa, FL 33620-9951. Offers civil engineering (MCE, MSCE, PhD); engineering (ME, MSE); environmental engineering (MEVE, MSEV). Part-time programs available. *Faculty:* 21 full-time (2 women), 3 part-time (0 women). *Students:* 48 full-time (9 women), 84 part-time (15 women); includes 43 minority (2 African Americans, 33 Asian Americans or Pacific Islanders, 8 Hispanic Americans), 36 international. Average age 32. In 1998, 40 master's, 1 doctorate awarded. Terminal master's awarded for partial completion of doctoral program. *Degree requirements:* For master's, thesis required (for some programs), foreign language not required; for doctorate, dissertation, 2 tools of research as specified by dissertation committee required, foreign language not required. *Entrance requirements:* For master's, GRE General Test (minimum score of 400 on each section, 1000 combined required), minimum GPA of 3.0 during previous 2 years; for doctorate, GRE General Test (minimum score of 450 on each section, 1000 combined required). *Application deadline:* For fall admission, 6/1; for spring admission, 10/15. Application fee: $20. Electronic applications accepted. Tuition, state resident: part-time $148 per credit hour. Tuition, nonresident: part-time $509 per credit hour. *Financial aid:* In 1998–99, 2 fellowships, 48 research assistantships, 10 teaching assistantships were awarded.; career-related internships or fieldwork, Federal Work-Study, institutionally-sponsored loans, and tuition waivers (partial) also available. Aid available to part-time students. Financial aid applicants required to submit FAFSA. *Faculty research:* Water resources, structures and materials, transportation, geotechnical engineering. Total annual research expenditures: $7.5 million. *Unit head:* Dr. W. C. Carpenter, Chairperson, 813-974-2275, Fax: 813-974-2957, E-mail: carpente@eng.usf.edu. *Application contact:* Manjriker Gunaratne, Coordinator, 813-974-5818, Fax: 813-974-5927, E-mail: gunaratn@eng.usf.edu.

University of Southwestern Louisiana, Graduate School, College of Engineering, Department of Civil Engineering, Lafayette, LA 70504. Offers MSE. Evening/weekend programs available. *Faculty:* 7 full-time (1 woman). *Students:* 18 full-time (4 women), 3 part-time; includes 1 minority (Native American), 18 international. 66 applicants, 74% accepted. In 1998, 3 degrees awarded. *Degree requirements:* For master's, thesis or alternative, comprehensive exam required, foreign language not required. *Entrance requirements:* For master's, GRE General Test, BS in civil engineering, minimum GPA of 2.85. *Application deadline:* For fall admission, 5/15. Application fee: $5 ($15 for international students). *Financial aid:* In 1998–99, 7 research assistantships with full tuition reimbursements (averaging $5,821 per year) were awarded.; Federal Work-Study and tuition waivers (full and partial) also available. Financial aid application deadline: 5/1. *Faculty research:* Structural mechanics, computer-aided design, environmental engineering. *Unit head:* Dr. Robert Wang, Head, 318-482-6511. *Application contact:* Dr. Xiaoduan Sun, Graduate Coordinator, 318-482-6514.

University of Tennessee, Knoxville, Graduate School, College of Engineering, Department of Civil and Environmental Engineering, Program in Civil Engineering, Knoxville, TN 37996. Offers MS, PhD. Part-time programs available. Postbaccalaureate distance learning degree programs offered (minimal on-campus study). *Students:* 46 full-time (10 women), 33 part-time (6 women); includes 4 minority (2 African Americans, 2 Asian Americans or Pacific Islanders), 18 international. 56 applicants, 64% accepted. In 1998, 21 master's, 3 doctorates awarded. *Degree requirements:* For master's, thesis or alternative required, foreign language not required; for doctorate, dissertation required, foreign language not required. *Entrance requirements:* For master's and doctorate, TOEFL (minimum score of 550 required), minimum GPA of 2.7. *Application deadline:* For fall admission, 2/1 (priority date). Applications are processed on a rolling basis. Application fee: $35. Electronic applications accepted. *Financial aid:* Application deadline: 2/1; *Unit head:* Dr. Gregory D. Reed, Head, Department of Civil and Environmental Engineering, 423-974-2503, Fax: 423-974-2669, E-mail: gdreed@utk.edu.

The University of Texas at Arlington, Graduate School, College of Engineering, Department of Civil and Environmental Engineering, Arlington, TX 76019. Offers M Engr, MS, PhD. *Faculty:* 12 full-time (0 women), 1 part-time (0 women). *Students:* 35 full-time (10 women), 71 part-time (12 women); includes 15 minority (2 African Americans, 8 Asian Americans or Pacific Islanders, 4 Hispanic Americans, 1 Native American), 36 international. 69 applicants, 61% accepted. In 1998, 19 master's, 2 doctorates awarded. *Degree requirements:* For master's, computer language, thesis (for some programs), oral and written exams required, foreign language not required; for doctorate, one foreign language, computer language, oral and written defense of dissertation required. *Entrance requirements:* For master's, GRE General Test (minimum combined score of 1000 required), TOEFL, minimum GPA of 3.0 in last 60 hours of undergraduate course work; for doctorate, GRE General Test (minimum combined score of 1200 required), TOEFL, minimum graduate GPA of 3.5. *Application deadline:* Applications are processed on a rolling basis. Application fee: $25 ($50 for international students). Tuition, state resident: full-time $1,368; part-time $76 per semester hour. Tuition, nonresident: full-time $5,454; part-time $303 per semester hour. Required fees: $66 per semester hour. $86 per term. Tuition and fees vary according to course load. *Financial aid:* Fellowships, research assistantships, teaching assistantships, Federal Work-Study, scholarships, and tuition waivers (partial) available. *Unit head:* Dr. C. E. Parker, Chairman, 817-272-2201, Fax: 817-272-2630, E-mail: parkerce@uta.edu. *Application contact:* Dr. Ernest C. Crosby, Graduate Adviser, 817-272-3500, Fax: 817-272-2630, E-mail: ecrosby@uta.edu.

The University of Texas at Austin, Graduate School, College of Engineering, Department of Civil Engineering, Austin, TX 78712-1111. Offers civil engineering (MSE, PhD); environmental and water resources engineering (MSE). *Accreditation:* ABET (one or more programs are accredited). *Students:* 368 (79 women); includes 33 minority (3 African Americans, 13 Asian

Americans or Pacific Islanders, 17 Hispanic Americans) 151 international. 551 applicants, 41% accepted. In 1998, 120 master's, 32 doctorates awarded. *Degree requirements:* For doctorate, dissertation required, foreign language not required, foreign language not required. *Entrance requirements:* For master's, GRE General Test (minimum combined score of 1000 required); for doctorate, GRE General Test (minimum combined score of 1100 required). *Application deadline:* For fall admission, 1/15 (priority date); for spring admission, 9/1 (priority date). Applications are processed on a rolling basis. Application fee: $50 ($75 for international students). Electronic applications accepted. *Financial aid:* Fellowships, research assistantships, teaching assistantships available. Financial aid application deadline: 2/1. *Unit head:* Dr. James O. Jirsa, Chairman, 512-471-4921, Fax: 512-471-0592, E-mail: jirsa@uts.cc.utexas.edu. *Application contact:* Dr. Howard M. Liljestrand, Graduate Adviser, 512-471-4921, Fax: 512-471-0592, E-mail: liljestrand@mail.utexas.edu.

Announcement: Department of Civil Engineering offers program of graduate study nationally and internationally recognized among leaders. Graduate courses and research opportunities offered in areas of specialization that include architectural engineering; construction engineering and project management; environmental and water resources engineering; and geotechnical, structural, and transportation engineering as well as hazardous-waste treatment, disposal, and management; construction materials; earthquake engineering; and offshore and ocean engineering.

The University of Texas at El Paso, Graduate School, College of Engineering, Department of Civil Engineering, El Paso, TX 79968-0001. Offers MS. Part-time and evening/weekend programs available. *Faculty:* 9 full-time (0 women), 3 part-time (0 women). *Students:* 22 full-time (5 women), 25 part-time (7 women); includes 26 minority (1 Asian American or Pacific Islander, 25 Hispanic Americans), 11 international. Average age 29. 33 applicants, 52% accepted. In 1998, 20 degrees awarded. *Degree requirements:* For master's, thesis optional, foreign language not required. *Entrance requirements:* For master's, GRE General Test, TOEFL (minimum score of 550 required), minimum GPA of 3.0. *Application deadline:* Applications are processed on a rolling basis. Application fee: $15 ($65 for international students). Tuition, state resident: full-time $2,790. Tuition, nonresident: full-time $7,710. *Financial aid:* Fellowships, research assistantships, teaching assistantships, career-related internships or fieldwork, Federal Work-Study, institutionally-sponsored loans, tuition waivers (partial), and stipends available. Financial aid applicants required to submit FAFSA. *Faculty research:* On-site wastewater treatment systems, wastewater reuse, disinfection by-product control, water resources, membrane filtration. Total annual research expenditures: $655,262. *Unit head:* Dr. Carlos Ferregut, Chairperson, 915-747-6921. *Application contact:* Susan Jordan, Director, Graduate Student Services, 915-747-5491, Fax: 915-747-5788, E-mail: sjordan@utep.edu.

The University of Texas at San Antonio, College of Sciences and Engineering, Division of Engineering, San Antonio, TX 78249-0617. Offers civil engineering (MS); electrical engineering (MS); mechanical engineering (MS). Part-time and evening/weekend programs available. *Faculty:* 22 full-time (2 women), 15 part-time (2 women). *Students:* 19 full-time (1 woman), 90 part-time (15 women); includes 38 minority (5 African Americans, 11 Asian Americans or Pacific Islanders, 21 Hispanic Americans, 1 Native American), 21 international. *Degree requirements:* For master's, thesis optional, foreign language not required. *Entrance requirements:* For master's, GRE General Test. *Application deadline:* For fall admission, 7/1; for spring admission, 12/1. Applications are processed on a rolling basis. Application fee: $25. *Unit head:* Dr. Lex Akers, Director, 210-458-4490.

University of Toledo, Graduate School, College of Engineering, Department of Civil Engineering, Toledo, OH 43606-3398. Offers civil engineering (MSCE); engineering sciences (PhD). Part-time programs available. *Faculty:* 13 full-time (1 woman), 2 part-time (0 women). *Students:* 62 full-time (8 women), 26 part-time (1 woman); includes 2 minority (both African Americans), 63 international. Average age 25. 176 applicants, 72% accepted. In 1998, 35 master's, 1 doctorate awarded. Terminal master's awarded for partial completion of doctoral program. *Degree requirements:* For master's, thesis or alternative required, foreign language not required; for doctorate, dissertation required, foreign language not required. *Entrance requirements:* For master's, GRE General Test (minimum combined score of 1000 required), TOEFL (minimum score of 550 required), minimum GPA of 2.7; for doctorate, GRE General Test (minimum combined score of 1000 required), TOEFL (minimum score of 550 required). *Average time to degree:* Master's–2 years full-time; doctorate–4 years full-time. *Application deadline:* For fall admission, 5/31 (priority date). Applications are processed on a rolling basis. Application fee: $30. Electronic applications accepted. *Financial aid:* In 1998–99, 60 students received aid, including 3 research assistantships with full tuition reimbursements available, 16 teaching assistantships with full tuition reimbursements available; fellowships with full tuition reimbursements available, Federal Work-Study, scholarships, tuition waivers (full), and unspecified assistantships also available. Aid available to part-time students. Financial aid application deadline: 4/1. *Faculty research:* Environmental modeling, soil/pavement interaction, structural mechanics, earthquakes, transportation engineering. Total annual research expenditures: $485,987. *Unit head:* Dr. Brian Randolph, Chair, 419-530-8115, Fax: 419-530-8116. *Application contact:* Barbara Laird, Academic Program Coordinator, 419-530-8114, Fax: 419-530-8116, E-mail: blaird@uoft02.utoledo.edu.

University of Toronto, School of Graduate Studies, Physical Sciences Division, Faculty of Applied Science and Engineering, Department of Civil Engineering, Toronto, ON M5S 1A1, Canada. Offers M Eng, MA Sc, PhD. Part-time programs available. *Degree requirements:* For master's, thesis required (for some programs), foreign language not required; for doctorate, dissertation required, foreign language not required.

University of Utah, Graduate School, College of Engineering, Department of Civil Engineering, Salt Lake City, UT 84112-1107. Offers ME, MS, PhD. *Faculty:* 12 full-time (1 woman), 42 part-time (3 women). *Students:* 26 full-time (8 women), 22 part-time (4 women); includes 2 minority (1 Asian American or Pacific Islander, 1 Native American), 15 international. Average age 32. In 1998, 9 master's, 3 doctorates awarded. Terminal master's awarded for partial completion of doctoral program. *Degree requirements:* For master's, project (ME), thesis (MS) required; for doctorate, dissertation, departmental qualifying exam required. *Entrance requirements:* For master's and doctorate, GRE General Test, TOEFL (minimum score of 500 required), minimum GPA of 3.0. *Application deadline:* For fall admission, 7/1. Application fee: $30 ($50 for international students). *Financial aid:* In 1998–99, 7 teaching assistantships were awarded.; research assistantships Aid available to part-time students. *Faculty research:* Wastewater treatment, structural engineering, structural mechanics, hydraulics, composite materials. *Unit head:* Lawrence D. Reaveley, Chair, 801-581-6931, Fax: 801-581-8692. *Application contact:* Dr. Chris Pantelides, Director of Graduate Studies, 801-581-6931.

University of Vermont, Graduate College, College of Engineering and Mathematics, Department of Civil Engineering, Burlington, VT 05405-0160. Offers MS, PhD. *Degree requirements:* For master's, thesis or alternative required, foreign language not required; for doctorate, dissertation required, foreign language not required. *Entrance requirements:* For master's and doctorate, GRE General Test, TOEFL (minimum score of 550 required).

University of Virginia, School of Engineering and Applied Science, Department of Civil Engineering, Charlottesville, VA 22903. Offers applied mechanics (MAM, MS); environmental engineering (ME, MS, PhD), including environmental engineering, water resources; structural engineering (ME, MS, PhD); transportation engineering and management (ME, MS, PhD). Part-time programs available. Postbaccalaureate distance learning degree programs offered (no on-campus study). *Faculty:* 14 full-time (3 women). *Students:* 59 full-time (28 women), 1 part-time; includes 1 minority (Asian American or Pacific Islander), 12 international. Average age 26. 75 applicants, 76% accepted. In 1998, 24 master's awarded (4% entered university research/teaching, 83% found other work related to degree, 13% continued full-time study); 6 doctorates awarded (33% entered university research/teaching, 67% found other work related to degree). Terminal master's awarded for partial completion of doctoral program. *Degree requirements:* For doctorate, dissertation, comprehensive exam required, foreign language not required, foreign language not required. *Entrance requirements:* For master's and doctorate, GRE General Test. *Average time to degree:* Master's–1.7 years full-time; doctorate–3 years full-time. *Application deadline:* For fall admission, 2/1 (priority date). Applications are processed on a rolling basis. Application fee: $60. Electronic applications accepted. *Financial aid:* In 1998–99, 51 students received aid, including 5 fellowships with full tuition reimbursements available (averaging $11,400 per year), 35 research assistantships with full tuition reimbursements available (averaging $11,300 per year), 11 teaching assistantships with full tuition reimbursements available (averaging $10,700 per year) Financial aid application deadline: 2/1. Total annual research expenditures: $2.4 million. *Unit head:* Dr. Nicholas J. Garber, Chairman, 804-924-7464, Fax: 804-982-2951, E-mail: njg@virginia.edu.

University of Washington, Graduate School, College of Engineering, Department of Civil and Environmental Engineering, Seattle, WA 98195. Offers environmental engineering (MS, MS Civ E, MSE, PhD); hydraulic engineering (MS Civ E, MSE, PhD); structural and geotechnical engineering and mechanics (MS, MS Civ E, MSE, PhD); transportation and construction engineering (MS, MS Civ E, MSE, PhD). *Faculty:* 31 full-time (4 women), 26 part-time (1 woman). *Students:* 153 full-time (54 women), 42 part-time (13 women); includes 25 minority (2 African Americans, 19 Asian Americans or Pacific Islanders, 3 Hispanic Americans, 1 Native American), 34 international. Average age 29. 339 applicants, 60% accepted. In 1998, 71 master's, 11 doctorates awarded. Terminal master's awarded for partial completion of doctoral program. *Degree requirements:* For master's, thesis or alternative required, foreign language not required; for doctorate, dissertation required, foreign language not required. *Entrance requirements:* For master's, GRE General Test, TOEFL (minimum score of 580 required), minimum GPA of 3.0; for doctorate, GRE, TOEFL (minimum score of 580 required), minimum GPA of 3.0. *Average time to degree:* Master's–2 years full-time; doctorate–5 years full-time. *Application deadline:* For fall admission, 2/1 (priority date); for winter admission, 11/1; for spring admission, 2/1. Applications are processed on a rolling basis. Application fee: $50. Electronic applications accepted. Tuition, state resident: full-time $5,196; part-time $475 per credit. Tuition, nonresident: full-time $13,485; part-time $1,285 per credit. Required fees: $387; $38 per credit. Tuition and fees vary according to course load. *Financial aid:* In 1998–99, 22 fellowships, 70 research assistantships with tuition reimbursements (averaging $1,160 per year), 22 teaching assistantships with tuition reimbursements (averaging $1,160 per year) were awarded.; scholarships also available. Financial aid application deadline: 2/1. *Faculty research:* Water resources, hydrology. Total annual research expenditures: $3.7 million. *Unit head:* Dr. Fred L. Mannering, Chair, 206-543-2390. *Application contact:* Marcia Buck, Graduate Secretary, 206-543-2574, E-mail: ceginfo@u.washington.edu.

University of Waterloo, Graduate Studies, Faculty of Engineering, Department of Civil Engineering, Waterloo, ON N2L 3G1, Canada. Offers MA Sc, PhD. Part-time programs available. *Faculty:* 25 full-time (2 women), 13 part-time (1 woman). *Students:* 67 full-time (12 women), 10 part-time (3 women). 54 applicants, 35% accepted. In 1998, 20 master's, 13 doctorates awarded. *Degree requirements:* For master's; for doctorate, dissertation, comprehensive exam required. *Entrance requirements:* For master's, TOEFL (minimum score of 550 required), TWE (minimum score of 4.0 required), honors degree, minimum B average; for doctorate, TOEFL (minimum score of 550 required), TWE (minimum score of 4.0 required), master's degree, minimum A average. *Average time to degree:* Master's–2 years full-time, 4 years part-time; doctorate–4 years full-time, 4 years part-time. *Application deadline:* Applications are processed on a rolling basis. Application fee: $50. *Expenses:* Tuition and fees charges are reported in Canadian dollars. Tuition, state resident: full-time $3,168 Canadian dollars; part-time $792 Canadian dollars per term. Tuition, nonresident: full-time $8,000 Canadian dollars; part-time $2,000 Canadian dollars. Required fees: $45 Canadian dollars per term. Tuition and fees vary according to program. *Financial aid:* In 1998–99, 43 research assistantships, 28 teaching assistantships were awarded. *Faculty research:* Water resources, structures, construction management, transportation, geotechnical engineering. *Unit head:* Dr. J. F. Sykes, Chair, 519-888-4567 Ext. 3776, Fax: 519-888-6197, E-mail: sykes@civoffice.uwaterloo.ca. *Application contact:* C. Jones, Graduate Secretary, 519-888-4567 Ext. 2804, Fax: 519-888-6197, E-mail: gradoffice@civil.watstar.uwaterloo.ca.

University of Windsor, College of Graduate Studies and Research, Faculty of Engineering, Department of Civil and Environmental Engineering, Windsor, ON N9B 3P4, Canada. Offers MA Sc, PhD. Part-time programs available. *Degree requirements:* For master's and doctorate, thesis/dissertation required, foreign language not required. *Entrance requirements:* For master's, TOEFL (minimum score of 550 required), minimum B average; for doctorate, TOEFL, master's degree. *Faculty research:* Structures, water resources.

University of Wisconsin–Madison, Graduate School, College of Engineering, Department of Civil and Environmental Engineering, Madison, WI 53706-1380. Offers MS, PhD. Part-time programs available. *Faculty:* 31 full-time (1 woman), 3 part-time (0 women). *Students:* 129 full-time (26 women), 22 part-time (1 woman); includes 3 minority (2 Asian Americans or Pacific Islanders, 1 Hispanic American), 2 international. Average age 25. 173 applicants, 65% accepted. In 1998, 52 master's, 9 doctorates awarded. Terminal master's awarded for partial completion of doctoral program. *Degree requirements:* For master's, thesis or alternative required, foreign language not required; for doctorate, dissertation, preliminary exam and qualifying exams required, foreign language not required. *Entrance requirements:* For master's and doctorate, TOEFL, minimum GPA of 3.0. *Average time to degree:* Master's–2 years full-time, 5 years part-time; doctorate–3.5 years full-time, 7 years part-time. *Application deadline:* For fall admission, 6/30 (priority date); for spring admission, 11/15. Applications are processed on a rolling basis. Application fee: $45. Electronic applications accepted. *Financial aid:* In 1998–99, 95 students received aid, including 9 fellowships with full tuition reimbursements available (averaging $14,600 per year), 53 research assistantships with full tuition reimbursements available (averaging $14,600 per year), 10 teaching assistantships with full tuition reimbursements available (averaging $6,422 per year); Federal Work-Study also available. Aid available to part-time students. Financial aid application deadline: 12/1. *Faculty research:* Environmental geotechnics and soil mechanics, design and analysis of structures, traffic engineering and intelligent transport systems, industrial pollution control, hydrological monitoring. Total annual research expenditures: $4.3 million. *Unit head:* Robert L. Smith, Chair, 608-262-3542, Fax: 608-262-5199, E-mail: cee@engr.wisc.edu. *Application contact:* Lynn Maertz, Student Services Coordinator, 608-262-5198, Fax: 608-262-5199, E-mail: maertz@engr.wisc.edu.

University of Wyoming, Graduate School, College of Engineering, Department of Civil Engineering, Program in Civil Engineering, Laramie, WY 82071. Offers MS, PhD. *Students:* 19 full-time (4 women), 17 part-time (3 women), 1 international. 28 applicants, 64% accepted. In 1998, 12 master's awarded (100% found work related to degree). *Degree requirements:* For master's, computer language, thesis required; for doctorate, variable foreign language requirement, computer language, dissertation required. *Entrance requirements:* For master's and doctorate, GRE General Test, TOEFL, minimum GPA of 3.0. *Application deadline:* For spring admission, 9/1. Applications are processed on a rolling basis. Application fee: $40. Electronic applications accepted. Tuition, state resident: full-time $2,520; part-time $140 per credit hour. Tuition, nonresident: full-time $7,790; part-time $433 per credit hour. Required fees: $400; $7 per credit hour. Full-time tuition and fees vary according to course load and program. *Financial aid:* Application deadline: 3/1. *Application contact:* P. A. Van Houten, Graduate Coordinator, 307-766-5446, Fax: 307-766-2221, E-mail: pvh@uwyo.edu.

Utah State University, School of Graduate Studies, College of Engineering, Department of Civil and Environmental Engineering, Logan, UT 84322. Offers ME, MS, PhD, CE. *Faculty:* 27 full-time (2 women). *Students:* 53 full-time (6 women), 30 part-time (3 women); includes 2 minority (1 Asian American or Pacific Islander, 1 Native American), 23 international. Average age 28. 179 applicants, 55% accepted. In 1998, 23 master's, 6 doctorates awarded. *Degree requirements:* For master's, thesis required (for some programs), foreign language not required; for doctorate, dissertation required, foreign language not required. *Entrance requirements:* For master's and doctorate, GRE General Test (score in 40th percentile or higher required), TOEFL (minimum score of 550 required), minimum GPA of 3.0. *Application*

Civil Engineering–Construction Engineering and Management

Utah State University (continued)

deadline: For fall admission, 6/15 (priority date); for spring admission, 10/15. Applications are processed on a rolling basis. Application fee: $40. Tuition, state resident: full-time $1,492. Tuition, nonresident: full-time $5,232. Required fees: $434. Tuition and fees vary according to course load. *Financial aid:* In 1998–99, 49 research assistantships with partial tuition reimbursements were awarded.; fellowships with partial tuition reimbursements, teaching assistantships with partial tuition reimbursements, career-related internships or fieldwork, Federal Work-Study, and institutionally-sponsored loans also available. Aid available to part-time students. Financial aid application deadline: 3/31. *Faculty research:* Hazardous waste treatment, large space structures, river basin management, earthquake engineering, environmental impact. *Unit head:* Loren R. Anderson, Head, 435-797-2932.

Vanderbilt University, School of Engineering, Department of Civil and Environmental Engineering, Program in Civil Engineering, Nashville, TN 37240-1001. Offers M Eng, MS, PhD. MS and PhD offered through the Graduate School. Part-time programs available. *Faculty:* 8 full-time (0 women), 2 part-time (both women). *Students:* 28 full-time (4 women), 10 part-time (2 women); includes 2 minority (1 Asian American or Pacific Islander, 1 Hispanic American), 20 international. Average age 24. 40 applicants, 38% accepted. In 1998, 3 master's awarded (100% found work related to degree); 2 doctorates awarded (50% entered university research/teaching, 50% found other work related to degree). Terminal master's awarded for partial completion of doctoral program. *Degree requirements:* For master's and doctorate, computer language, thesis/dissertation required, foreign language not required. *Entrance requirements:* For master's, GRE General Test (minimum score of 650 on quantitative section; combined average 1800 on three sections); for doctorate, GRE General Test (minimum score of 650 on quantitative section; combined average 2000 on three sections). *Average time to degree:* Master's–2 years full-time, 4 years part-time; doctorate–4 years full-time, 6 years part-time. *Application deadline:* For fall admission, 1/15 (priority date). Applications are processed on a rolling basis. Application fee: $40. *Financial aid:* In 1998–99, 22 students received aid, including 1 fellowship with full tuition reimbursement available (averaging $5,000 per year), 8 research assistantships with full tuition reimbursements available (averaging $16,000 per year), 5 teaching assistantships with full tuition reimbursements available (averaging $12,000 per year); institutionally-sponsored loans and tuition waivers (full and partial) also available. Financial aid application deadline: 1/15. *Faculty research:* Structural mechanics, finite element analysis, urban transportation, hazardous material transport. Total annual research expenditures: $600,000. *Unit head:* Edward Thackston, Chair, Department of Civil and Environmental Engineering, 615-343-2372.

Villanova University, College of Engineering, Department of Civil and Environmental Engineering, Program in Civil Engineering, Villanova, PA 19085-1699. Offers MCE. Part-time and evening/weekend programs available. *Faculty:* 11 full-time (0 women), 9 part-time (0 women). *Students:* 6 full-time (2 women), 49 part-time (6 women); includes 3 minority (1 African American, 2 Hispanic Americans), 3 international. Average age 25. 26 applicants, 85% accepted. In 1998, 16 degrees awarded. *Degree requirements:* For master's, computer language required, thesis optional, foreign language not required. *Entrance requirements:* For master's, GRE General Test (for applicants with degrees from foreign universities), minimum GPA of 3.0. *Average time to degree:* Master's–1 year full-time, 3.5 years part-time. *Application deadline:* For fall admission, 8/1 (priority date); for spring admission, 12/1. Applications are processed on a rolling basis. Application fee: $40. *Financial aid:* Teaching assistantships, Federal Work-Study, scholarships, and tuition waivers (full and partial) available. Aid available to part-time students. Financial aid application deadline: 4/15. *Faculty research:* Bridge inspection, environment maintenance, economy and risk. *Unit head:* Dr. Lewis J. Mathers, Chairman, Department of Civil and Environmental Engineering, 610-519-4960, Fax: 610-519-6754, E-mail: ldeangel@email.vill.edu.

Virginia Polytechnic Institute and State University, Graduate School, College of Engineering, Department of Civil and Environmental Engineering, Program in Civil Engineering, Blacksburg, VA 24061. Offers M Eng, MS, PhD. *Students:* 159 full-time (31 women), 94 part-time (16 women); includes 28 minority (8 African Americans, 12 Asian Americans or Pacific Islanders, 7 Hispanic Americans, 1 Native American), 53 international. 264 applicants, 62% accepted. In 1998, 83 master's, 7 doctorates awarded. *Degree requirements:* For master's, thesis required (for some programs), foreign language not required; for doctorate, dissertation required. *Entrance requirements:* For master's and doctorate, GRE, TOEFL (minimum score of 570 required). *Application deadline:* For fall admission, 12/1 (priority date). Applications are processed on a rolling basis. Application fee: $25. *Financial aid:* In 1998–99, 26 research assistantships, 19 teaching assistantships were awarded.; fellowships, unspecified assistantships also available. Financial aid application deadline: 4/1. *Unit head:* Dr. Bill Knocke, Head, Department of Civil and Environmental Engineering, 540-231-6637, E-mail: knocke@vt.edu.

See in-depth description on page 649.

Washington State University, Graduate School, College of Engineering and Architecture, Department of Civil and Environmental Engineering, Program in Civil Engineering, Pullman, WA 99164. Offers MS, PhD. *Students:* 45 full-time (10 women), 6 part-time (1 woman); includes 4 minority (1 African American, 2 Asian Americans or Pacific Islanders, 1 Hispanic American), 13 international. In 1998, 4 master's, 4 doctorates awarded. Terminal master's awarded for partial completion of doctoral program. *Degree requirements:* For master's and doctorate, thesis/dissertation, oral exam required, foreign language not required. *Entrance requirements:* For master's and doctorate, GRE General Test, minimum GPA of 3.0. *Average time to degree:* Master's–1.5 years full-time; doctorate–4 years full-time. *Application deadline:* For fall admission, 3/1 (priority date); for spring admission, 10/1. Applications are processed on a rolling basis. Application fee: $35. *Financial aid:* In 1998–99, 11 research assistantships, 9 teaching assistantships were awarded.; fellowships, career-related internships or fieldwork, Federal Work-Study, and institutionally-sponsored loans also available. Financial aid application deadline: 4/1; financial aid applicants required to submit FAFSA. *Application contact:* Maureen Clausen, Graduate Secretary, 509-335-2576, Fax: 509-335-7632, E-mail: mclausen@wsu.edu.

Washington University in St. Louis, School of Engineering and Applied Science, Sever Institute of Technology, Department of Civil Engineering, St. Louis, MO 63130-4899. Offers civil engineering (MSCE); construction engineering (MCE); construction management (MCM); structural engineering (MSE, D Sc); transportation and urban systems engineering (D Sc). Part-time and evening/weekend programs available. Terminal master's awarded for partial completion of doctoral program. *Degree requirements:* For master's, thesis optional, foreign language not required; for doctorate, dissertation, departmental qualifying exam required. *Faculty research:* Composites, earthquake, reinforced concrete steel.

Wayne State University, Graduate School, College of Engineering, Department of Civil and Environmental Engineering, Detroit, MI 48202. Offers MS, PhD. *Degree requirements:* For master's, thesis optional, foreign language not required; for doctorate, dissertation required, foreign language not required. *Faculty research:* Environmental geotechnics, bridge engineering, seismic analysis, construction safety, mass transit.

See in-depth description on page 651.

West Virginia University, College of Engineering and Mineral Resources, Department of Civil and Environmental Engineering, Morgantown, WV 26506. Offers civil engineering (MSCE, MSE); engineering (PhD). Part-time programs available. *Degree requirements:* For master's, thesis required, foreign language not required; for doctorate, dissertation, comprehensive exam required, foreign language not required. *Entrance requirements:* For master's, TOEFL (minimum score of 550 required), minimum GPA of 3.0; for doctorate, TOEFL (minimum score of 550 required), GRE (international students), minimum GPA of 3.0. *Faculty research:* Environmental and hydrotechnical structural composites, bridge innovation and rehabilitation, transport, soil mechanics, geoenvironmental engineering.

Widener University, School of Engineering, Program in Civil Engineering, Chester, PA 19013-5792. Offers ME, ME/MBA. Part-time and evening/weekend programs available. *Faculty:* 6 part-time (1 woman). *Students:* 1 full-time (0 women), 6 part-time (1 woman). 7 applicants, 86% accepted. In 1998, 5 degrees awarded. *Degree requirements:* For master's, thesis optional, foreign language not required. *Entrance requirements:* For master's, GMAT (ME/MBA). *Average time to degree:* Master's–2 years full-time, 4 years part-time. *Application deadline:* For fall admission, 8/1 (priority date); for spring admission, 12/1. Applications are processed on a rolling basis. Application fee: $25 ($300 for international students). *Financial aid:* In 1998–99, 1 teaching assistantship with full tuition reimbursement (averaging $7,500 per year) was awarded.; unspecified assistantships also available. Financial aid application deadline: 3/15. *Faculty research:* Environmental engineering, laws and water supply, structural analysis and design. *Unit head:* Dr. Charles L. Bartholomew, Chairman, Department of Civil Engineering, 610-499-4249, E-mail: cbarth0170@aol.com.

Worcester Polytechnic Institute, Graduate Studies, Department of Civil and Environmental Engineering, Worcester, MA 01609-2280. Offers M Eng, MS, PhD, Advanced Certificate, Certificate. Part-time and evening/weekend programs available. Postbaccalaureate distance learning degree programs offered (minimal on-campus study). *Faculty:* 13 full-time (0 women), 4 part-time (1 woman). *Students:* 25 full-time (6 women), 27 part-time (6 women); includes 5 minority (1 African American, 4 Hispanic Americans), 4 international. 57 applicants, 67% accepted. In 1998, 10 master's, 3 doctorates awarded. *Degree requirements:* For master's, thesis optional, foreign language not required; for doctorate, dissertation required, foreign language not required. *Entrance requirements:* For master's, GRE General Test (combined average 1711 on three sections), TOEFL (minimum score of 550 required average 605); for doctorate, GRE General Test (combined average 1711 on three sections), TOEFL (minimum score of 550 required; average 605). *Application deadline:* For fall admission, 2/15 (priority date); for spring admission, 10/15 (priority date). Applications are processed on a rolling basis. Application fee: $50. Electronic applications accepted. *Financial aid:* In 1998–99, 15 students received aid, including 2 fellowships with full tuition reimbursements available (averaging $13,276 per year), 3 research assistantships with full tuition reimbursements available (averaging $15,000 per year), 10 teaching assistantships with full tuition reimbursements available (averaging $11,970 per year); career-related internships or fieldwork, grants, institutionally-sponsored loans, and scholarships also available. Financial aid application deadline: 2/15; financial aid applicants required to submit FAFSA. *Faculty research:* Water and wastewater treatment, construction management, structural and geotechnical engineering. Total annual research expenditures: $126,039. *Unit head:* Dr. Fred L. Hart, Head, 508-831-5530, Fax: 508-831-5808, E-mail: flhart@wpi.edu. *Application contact:* James O'Shaughnessy, Graduate Coordinator, 508-831-5309, Fax: 508-831-5808, E-mail: jco@wpi.edu.

Youngstown State University, Graduate School, William Rayen College of Engineering, Department of Civil, Chemical, and Environmental Engineering, Youngstown, OH 44555-0001. Offers MSE. Part-time and evening/weekend programs available. *Faculty:* 9 full-time (1 woman). *Students:* 13 full-time (2 women), 5 part-time (2 women), 4 international. 4 applicants, 75% accepted. In 1998, 10 degrees awarded. *Degree requirements:* For master's, computer language required, thesis optional, foreign language not required. *Entrance requirements:* For master's, TOEFL (minimum score of 550 required), minimum GPA of 2.75 in field. *Application deadline:* For fall admission, 8/15 (priority date); for winter admission, 11/15 (priority date); for spring admission, 2/15 (priority date). Applications are processed on a rolling basis. Application fee: $30 ($75 for international students). Tuition, state resident: part-time $97 per credit hour. Tuition, nonresident: part-time $219 per credit hour. Required fees: $21 per credit hour. $41 per quarter. *Financial aid:* In 1998–99, 4 students received aid, including 4 research assistantships with full tuition reimbursements available (averaging $7,500 per year); teaching assistantships, Federal Work-Study, institutionally-sponsored loans, and scholarships also available. Aid available to part-time students. Financial aid application deadline: 3/1. *Faculty research:* Structural mechanics, water quality modeling, surface and ground water hydrology, physical and chemical processes in aquatic systems. *Unit head:* Dr. Jack D. Bakos, Chair, 330-742-3027. *Application contact:* Dr. Peter J. Kasvinsky, Dean of Graduate Studies, 330-742-3091, Fax: 330-742-1580, E-mail: amgrad03@ysub.ysu.edu.

Construction Engineering and Management

Arizona State University, Graduate College, College of Engineering and Applied Sciences, Del E. Webb School of Construction, Tempe, AZ 85287. Offers MS. *Faculty:* 13 full-time (1 woman), 6 part-time (0 women). *Students:* 37 full-time (8 women), 36 part-time (3 women); includes 8 minority (1 African American, 1 Asian American or Pacific Islander, 5 Hispanic Americans, 1 Native American), 29 international. Average age 30. 45 applicants, 80% accepted. In 1998, 10 degrees awarded. *Entrance requirements:* For master's, GRE General Test (recommended), TOEFL (minimum score of 550 required), minimum GPA of 3.0. Application fee: $45. *Financial aid:* Career-related internships or fieldwork available. *Faculty research:* Roof performance, green building, use of waste and recycled materials, international construction alliances, water supply services. *Unit head:* Dr. William W. Badger, Director, 480-965-3615, E-mail: dewsc@asu.edu.

Auburn University, Graduate School, College of Engineering, Department of Civil Engineering, Auburn, Auburn University, AL 36849-0002. Offers construction engineering and management (MCE, MS, PhD); environmental engineering (MCE, MS, PhD); geotechnical/materials engineering (MCE, MS, PhD); hydraulics/hydrology (MCE, MS, PhD); structural engineering (MCE, MS, PhD); transportation engineering (MCE, MS, PhD). Part-time programs available. *Faculty:* 22 full-time. *Students:* 24 full-time (6 women), 34 part-time (8 women); includes 2 minority (1 African American, 1 Hispanic American), 9 international. *Degree requirements:* For master's, project (MCE), thesis (MS) required; for doctorate, dissertation, comprehensive exam required, foreign language not required. *Entrance requirements:* For master's, GRE General Test; for doctorate, GRE General Test (minimum score of 400 on each section required). *Application deadline:* For fall admission, 9/1; for spring admission, 3/1. Applications are processed on a rolling basis. Application fee: $25 ($50 for international students). Tuition, state resident: full-time $2,760; part-time $76 per credit hour. Tuition, nonresident: full-time $8,280; part-time $228 per credit hour. *Unit head:* Dr. Joseph F. Judkins, Head, 334-844-4320. *Application contact:* Dr. John F. Pritchett, Dean of the Graduate School, 334-844-4700.

Bradley University, Graduate School, College of Engineering and Technology, Department of Civil Engineering and Construction, Peoria, IL 61625-0002. Offers MSCE. Part-time and evening/weekend programs available. *Degree requirements:* For master's, comprehensive exam required, thesis not required. *Entrance requirements:* For master's, TOEFL (minimum score of 525 required), minimum GPA of 3.0.

The Catholic University of America, School of Engineering, Department of Civil Engineering, Program in Construction Management, Washington, DC 20064. Offers MCE, MS Engr. *Students:* 3 full-time (0 women), 7 part-time (3 women); includes 3 minority (all African Americans), 4 international. Average age 29. *Degree requirements:* For master's, comprehensive exam required, thesis optional, foreign language not required. *Entrance requirements:* For master's, minimum GPA of 3.0. *Application deadline:* For fall admission, 8/1 (priority date); for spring admission, 12/1. Applications are processed on a rolling basis. Application fee: $50. *Financial aid:* Research assistantships, teaching assistantships, career-related internships or fieldwork, Federal Work-Study, institutionally-sponsored loans, and tuition waivers (full and partial) available. Aid available to part-time students. Financial aid application deadline: 2/1. *Unit head:* Dr. Timothy W. Kao, Chair, Department of Civil Engineering, 202-319-5163, Fax: 202-319-4499, E-mail: kao@cua.edu.

Clemson University, Graduate School, College of Architecture, Arts, and Humanities, School of Design and Building, Department of Construction Science and Management, Clemson, SC 29634. Offers MCSM. Part-time programs available. *Students:* 11 full-time (2 women), 14 part-time (3 women), 6 international. Average age 26. 23 applicants, 52% accepted. In 1998, 14 degrees awarded. *Degree requirements:* For master's, thesis optional, foreign language not required. *Entrance requirements:* For master's, GRE General Test, TOEFL. *Application deadline:* For fall admission, 6/1. Application fee: $35. *Financial aid:* Research assistantships, teaching assistantships available. Financial aid applicants required to submit FAFSA. *Faculty research:* Computer applications, employer incentive programs, artificial intelligence, productivity improvement, financial management. *Unit head:* Roger Liska, Chair, 864-656-3878, Fax: 864-656-0204. *Application contact:* Christine Piper, Graduate Coordinator, 864-656-7581, E-mail: cpiper@clemson.edu.

Colorado State University, Graduate School, College of Applied Human Sciences, Department of Manufacturing Technology and Construction Management, Fort Collins, CO 80523-0015. Offers automotive pollution control (MS); construction management (MS); historic preservation (PhD); industrial technology management (MS); technology education and training (MS); technology of industry (PhD). *Faculty:* 17 full-time (1 woman), 5 part-time (1 woman). *Students:* 14 full-time (1 woman), 8 part-time (3 women); includes 2 minority (both Hispanic Americans), 2 international. *Degree requirements:* For master's, computer language, thesis required (for some programs), foreign language not required; for doctorate, dissertation required. *Entrance requirements:* For master's, GRE General Test (minimum combined score of 1250 required), TOEFL (minimum score of 550 required; 213 for computer-based); for doctorate, GRE General Test, TOEFL. *Application deadline:* For fall admission, 4/1 (priority date). Applications are processed on a rolling basis. Application fee: $30. Electronic applications accepted. *Unit head:* Dr. Larry Grosse, Head, 970-491-7958, Fax: 970-491-2473, E-mail: drfire@aol.com. *Application contact:* Linda Burrous, Secretary, 970-491-7355, Fax: 970-491-2473, E-mail: burrous@cahs.colostate.edu.

Columbia University, Graduate School of Business, MBA Program, New York, NY 10027. Offers accounting (MBA); construction management (MBA); entrepreneurship (MBA); finance and economics (MBA); human resource management (MBA); international business (MBA); management of organizations (MBA); management science (MBA); marketing (MBA); media, entertainment and communications (MBA); production and operations management (MBA); public and nonprofit management (MBA); real estate (MBA). *Faculty:* 114 full-time (20 women), 81 part-time (15 women). *Students:* 1,284 full-time (475 women); includes 285 minority (103 African Americans, 116 Asian Americans or Pacific Islanders, 64 Hispanic Americans, 2 Native Americans), 360 international. *Degree requirements:* For master's, foreign language and thesis not required. *Entrance requirements:* For master's, GMAT, TOEFL (minimum score of 610 required), minimum 2 years of work experience. *Application deadline:* For fall admission, 4/20; for spring admission, 10/1. Applications are processed on a rolling basis. Application fee: $160. Electronic applications accepted. *Unit head:* Prof. Safwan Masri, Vice Dean of Students and the MBA Program, 212-854-8716, Fax: 212-932-0545, E-mail: smm1@columbia.edu. *Application contact:* Linda Meehan, Assistant Dean and Executive Director of Admissions and Financial Aid, 212-854-1961, Fax: 212-662-6754, E-mail: gohermes@claven.gsb.columbia.edu.

Concordia University, School of Graduate Studies, Faculty of Engineering and Computer Science, Department of Building, Civil and Environmental Engineering, Montréal, PQ H3G 1M8, Canada. Offers building (M Eng, MA Sc, PhD, Certificate); civil engineering (M Eng, MA Sc, PhD). *Students:* 117 full-time (27 women), 14 part-time (3 women). *Degree requirements:* For master's, computer language, thesis or alternative required; for doctorate, computer language, dissertation, comprehensive exam required. *Application deadline:* For fall admission, 6/1; for spring admission, 10/1. Application fee: $50. *Faculty research:* Structural engineering, geotechnical engineering, water resources and fluid engineering, transportation engineering, systems engineering. *Unit head:* Dr. O. Moselhi, Chair, 514-848-7801, Fax: 514-848-7823. *Application contact:* Dr. F. Haghighat, Director, 514-848-8791, Fax: 514-848-2809.

Florida International University, College of Engineering, Department of Construction Management, Miami, FL 33199. Offers MS. Part-time and evening/weekend programs available. *Faculty:* 7 full-time (0 women). *Students:* 20 full-time (3 women), 48 part-time (16 women); includes 32 minority (4 African Americans, 2 Asian Americans or Pacific Islanders, 26 Hispanic Americans), 7 international. 37 applicants, 68% accepted. In 1998, 26 degrees awarded. *Degree requirements:* For master's, thesis optional. *Entrance requirements:* For master's, GRE General Test (minimum combined score of 1000 required), TOEFL (minimum score of 500 required). *Application deadline:* For fall admission, 4/1 (priority date); for spring admission, 10/1. Applications are processed on a rolling basis. Application fee: $20. Tuition, state resident: part-time $145 per credit hour. Tuition, nonresident: part-time $506 per credit hour. Required fees: $158; $158 per year. *Unit head:* Dr. Jose D. Mitrani, Chairperson, 305-348-3172, Fax: 305-348-2766, E-mail: mitranij@fiu.edu.

Georgia Institute of Technology, Graduate Studies and Research, College of Engineering, School of Civil and Environmental Engineering, Program in Construction Management, Atlanta, GA 30332-0001. Offers MS, MSCE, PhD. Part-time programs available. Terminal master's awarded for partial completion of doctoral program. *Degree requirements:* For doctorate, dissertation required, foreign language not required, foreign language not required. *Entrance requirements:* For master's, GRE General Test, TOEFL (minimum score of 550 required), minimum GPA of 3.0; for doctorate, GRE General Test, TOEFL (minimum score of 550 required), minimum GPA of 3.2. *Faculty research:* Automation and robotics, risk management, design/construction integration, sustaining technologies, infrastructure rehabilitation.

Iowa State University of Science and Technology, Graduate College, College of Engineering, Department of Civil and Construction Engineering, Ames, IA 50011. Offers civil engineering (MS, PhD), including civil engineering materials, construction engineering and management, environmental engineering, geometronics, geotechnical engineering, structural engineering, transportation engineering. *Faculty:* 36 full-time, 5 part-time. *Students:* 60 full-time (14 women), 51 part-time (12 women). *Degree requirements:* For master's, thesis or alternative required, foreign language not required; for doctorate, dissertation required, foreign language not required. *Entrance requirements:* For master's and doctorate, GRE General Test (foreign students), TOEFL. *Application deadline:* For fall admission, 6/15 (priority date); for spring admission, 11/15 (priority date). Application fee: $20 ($50 for international students). Electronic applications accepted. Tuition, state resident: full-time $3,308. Tuition, nonresident: full-time $9,744. Part-time tuition and fees vary according to course load, campus/location and program. *Unit head:* Dr. Lowell F. Greimann, Chair, 515-294-2140, E-mail: cceinfo@iastate.edu. *Application contact:* Edward Kannel, Director of Graduate Education, 515-294-2861, E-mail: cceinfo@iastate.edu.

Marquette University, Graduate School, College of Engineering, Department of Civil and Environmental Engineering, Milwaukee, WI 53201-1881. Offers construction and public works management (MS, PhD); environmental/water resources engineering (MS, PhD); structural/geotechnical engineering (MS, PhD); transportational planning and engineering (MS, PhD). Part-time and evening/weekend programs available. *Faculty:* 12 full-time (0 women), 1 part-time (0 women). *Students:* 17 full-time (7 women), 32 part-time (5 women); includes 5 minority (1 African American, 4 Asian Americans or Pacific Islanders), 12 international. Terminal master's awarded for partial completion of doctoral program. *Degree requirements:* For master's, thesis or alternative, comprehensive exam required, foreign language not required; for doctorate, dissertation required, foreign language not required. *Entrance requirements:* For master's, TOEFL (minimum score of 550 required); for doctorate, GRE General Test, TOEFL (minimum score of 550 required). *Application deadline:* For fall admission, 6/1 (priority date). Applications are processed on a rolling basis. Application fee: $40. Tuition: Part-time $510 per credit hour. Tuition and fees vary according to program. *Unit head:* Dr. Thomas H. Wenzel, Chairman, 414-288-7030, Fax: 414-288-7521, E-mail: wenzelt@vms.csd.mu.edu.

McGill University, Faculty of Graduate Studies and Research, Faculty of Engineering, Department of Civil Engineering and Applied Mechanics, Program in Structural Engineering and Construction Materials, Montréal, PQ H3A 2T5, Canada. Offers M Eng, M Sc, PhD. Part-time and evening/weekend programs available. *Degree requirements:* For master's, computer language required, thesis optional, foreign language not required; for doctorate, computer language, dissertation required, foreign language not required. *Entrance requirements:* For master's, TOEFL (minimum score of 550 required), minimum GPA of 3.0; for doctorate, TOEFL (minimum score of 580 required). *Faculty research:* Seismic design, risk engineering, composite structures, high strength and prestressed concrete, advanced materials, rehabilitation.

Michigan State University, Graduate School, College of Agriculture and Natural Resources, Department of Agricultural Engineering, Program in Building Construction Management, East Lansing, MI 48824-1020. Offers MS. *Students:* 15; includes 1 minority (African American), 6 international. *Degree requirements:* For master's, foreign language and thesis not required. *Entrance requirements:* For master's, GRE. *Application deadline:* Applications are processed on a rolling basis. Application fee: $30 ($40 for international students). *Financial aid:* Applicants required to submit FAFSA. *Unit head:* Dr. Ajit K. Srivastava, Chairperson, Department of Agricultural Engineering, 517-353-7268, Fax: 517-432-2892, E-mail: srivasta@pilot.msu.edu.

Montana State University–Bozeman, College of Graduate Studies, College of Engineering, Department of Civil Engineering, Bozeman, MT 59717. Offers civil engineering (MS); construction engineering management (MCEM); engineering (PhD); environmental engineering (MS). Part-time programs available. *Students:* 28 full-time (5 women), 17 part-time (2 women); includes 4 minority (3 Asian Americans or Pacific Islanders, 1 Native American) *Degree requirements:* For master's, thesis or alternative required, foreign language not required; for doctorate, dissertation required, foreign language not required. *Entrance requirements:* For master's and doctorate, GRE General Test, TOEFL (minimum score of 550 required). *Application deadline:* For fall admission, 6/1 (priority date); for spring admission, 11/1. Applications are processed on a rolling basis. Application fee: $50. *Unit head:* Dr. Donald Rabern, Head, 406-994-2111, E-mail: cedept@ce.montana.edu.

Southern Polytechnic State University, College of Technology, Department of Construction, Marietta, GA 30060-2896. Offers MS. Part-time and evening/weekend programs available. *Faculty:* 3 full-time (0 women). *Students:* 23 full-time (6 women), 29 part-time (6 women); includes 8 minority (7 African Americans, 1 Hispanic American), 6 international. Average age 35. 19 applicants, 100% accepted. In 1998, 24 degrees awarded (100% found work related to degree). *Degree requirements:* For master's, thesis optional, foreign language not required. *Entrance requirements:* For master's, GMAT or GRE General Test. *Application deadline:* For fall admission, 7/15 (priority date); for spring admission, 12/1. Applications are processed on a rolling basis. Application fee: $20. Tuition, state resident: full-time $2,146; part-time $119 per credit hour. Tuition, nonresident: full-time $7,586; part-time $421 per credit hour. *Financial aid:* In 1998–99, 20 students received aid; teaching assistantships, career-related internships or fieldwork and Federal Work-Study available. Aid available to part-time students. Financial aid application deadline: 5/1; financial aid applicants required to submit FAFSA. *Unit head:* Dr. Arlan Toy, Head, 770-528-7221, Fax: 770-528-4966, E-mail: atoy@spsu.edu.

State University of New York at Buffalo, Graduate School, School of Engineering and Applied Sciences, Department of Civil, Structural, and Environmental Engineering, Buffalo, NY 14260. Offers computational engineering and mechanics (MS, PhD); construction (M Eng, MS, PhD); geoenvironmental and geotechnical engineering (M Eng, MS, PhD); structural and earthquake engineering (M Eng, MS, PhD); water resources and environmental engineering (M Eng, MS, PhD). Part-time programs available. Postbaccalaureate distance learning degree programs offered (minimal on-campus study). *Faculty:* 25 full-time (0 women), 7 part-time (1 woman). *Students:* 99 full-time (19 women), 71 part-time (13 women); includes 9 minority (2 African Americans, 5 Asian Americans or Pacific Islanders, 1 Hispanic American, 1 Native American), 101 international. Terminal master's awarded for partial completion of doctoral program. *Degree requirements:* For master's, computer language, project or thesis required; for doctorate, computer language, dissertation required, foreign language not required. *Entrance requirements:* For master's and doctorate, GRE General Test (minimum combined score of 1250 required), TOEFL (minimum score of 550 required). *Application deadline:* For fall admission, 1/15 (priority date); for spring admission, 10/1. Applications are processed on a rolling basis. Application fee: $35. Tuition, state resident: full-time $5,100; part-time $213 per credit hour. Tuition, nonresident: full-time $8,416; part-time $351 per credit hour. Required fees: $870; $75 per semester. Tuition and fees vary according to course load and program. *Unit head:* Dr. Andrei M. Reinhorn, Chairman, 716-645-2114 Ext. 2419, Fax: 716-645-3733, E-mail: reinhorn@civil.eng.buffalo.edu. *Application contact:* Dr. Joe Atkinson, Director of Graduate Admissions, 716-645-2114 Ext. 2326, Fax: 716-645-3667, E-mail: atkinson@acsu.buffalo.edu.

See in-depth description on page 601.

Stevens Institute of Technology, Graduate School, Wesley J. Howe School of Technology Management, Program in Construction Management, Hoboken, NJ 07030. Offers MS. Offered in cooperation with the Department of Civil, Environmental, and Ocean Engineering. *Degree requirements:* For master's, computer language required, thesis optional, foreign language not required. *Entrance requirements:* For master's, GMAT, GRE, TOEFL. Electronic applications accepted.

Texas A&M University, College of Architecture, Department of Construction Science, College Station, TX 77843. Offers construction management (MS). *Faculty:* 18 full-time (1 woman), 4 part-time (1 woman). Average age 30. 29 applicants, 41% accepted. In 1998, 16 degrees awarded. *Entrance requirements:* For master's, GRE General Test, TOEFL. *Application deadline:* For fall admission, 2/1. Application fee: $50 ($75 for international students). *Financial aid:* Fellowships, research assistantships, teaching assistantships available. Financial aid application deadline: 4/1; financial aid applicants required to submit FAFSA. *Faculty research:* Fire safety, housing foundations, construction project management, quality management. Total annual research expenditures: $300,000. *Unit head:* Dr. James C. Smith, Head, 409-845-1017, Fax: 409-862-1572. *Application contact:* Albert Pedulla, Coordinator, 409-845-7000, Fax: 409-862-1572.

Texas A&M University, College of Engineering, Department of Civil Engineering, Program in Construction Engineering and Project Management, College Station, TX 77843. Offers M Eng, MS, D Eng, PhD. D Eng offered through the College of Engineering. *Students:* 35. *Degree requirements:* For master's, thesis (MS) required; for doctorate, dissertation (PhD), internship (D Eng) required. *Entrance requirements:* For master's and doctorate, GRE General Test, TOEFL. Application fee: $50 ($75 for international students). *Financial aid:* Fellowships, research assistantships, teaching assistantships available. Financial aid application deadline: 4/1; financial aid applicants required to submit FAFSA. *Faculty research:* Engineering management aspects of major engineered construction projects from concept formulation through start-up. *Unit head:* Dr. Paul N. Roschke, Head, Constructed Facilities Division, 409-845-4414, Fax: 409-862-2800, E-mail: ce-grad@tamu.edu. *Application contact:* Dr. Donald A. Maxwell, 409-845-2498, Fax: 409-862-2800, E-mail: ce-grad@tamu.edu.

Construction Engineering and Management

Universidad de las Américas–Puebla, Division of Graduate Studies, School of Engineering, Program in Construction Management, Cholula, 72820, Mexico. Offers M Adm. Part-time and evening/weekend programs available. *Faculty:* 7 full-time (0 women), 3 part-time (0 women). Average age 35. 20 applicants, 60% accepted. In 1998, 1 degree awarded. *Degree requirements:* For master's, one foreign language, computer language, thesis required. *Average time to degree:* Master's–1.5 years full-time, 2.5 years part-time. *Application deadline:* For fall admission, 7/16. Applications are processed on a rolling basis. Application fee: $0. *Financial aid:* In 1998–99, 4 students received aid. Available to part-time students. Application deadline: 5/15. *Faculty research:* Building structures, budget, project management. Total annual research expenditures: $33,000. *Unit head:* Jesus Bravo, Coordinator, 22-29-20-31, Fax: 22-29-20-32, E-mail: jbravo@mail.udlap.mx. *Application contact:* Mauricio Villegas, Chair of Admissions Office, 22-29-20-17, Fax: 22-29-20-18, E-mail: admision@mail.udlap.mx.

University of Alberta, Faculty of Graduate Studies and Research, Department of Civil and Environmental Engineering, Edmonton, AB T6G 2E1, Canada. Offers construction engineering and management (M Eng, M Sc, PhD); environmental engineering (M Eng, M Sc, PhD); environmental science (M Sc, PhD); geoenvironmental engineering (M Eng, M Sc, PhD); geotechnical engineering (M Sc); geotechnical engineering (M Eng, PhD); mining engineering (M Eng, M Sc, PhD); petroleum engineering (M Eng, M Sc, PhD); structural engineering (M Eng, M Sc, PhD); water resources (M Eng, M Sc, PhD). Part time programs available. *Degree requirements:* For master's, thesis required (for some programs), foreign language not required; for doctorate, dissertation required, foreign language not required. *Faculty research:* Mining.

University of California, Berkeley, Graduate Division, College of Engineering, Department of Civil and Environmental Engineering, Berkeley, CA 94720-1500. Offers construction engineering and management (M Eng, MS, D Eng, PhD); environmental quality and environmental water resources engineering (M Eng, MS, D Eng, PhD); geotechnical engineering (M Eng, MS, D Eng, PhD); structural engineering, mechanics and materials (M Eng, MS, D Eng, PhD); transportation engineering (M Eng, MS, D Eng, PhD). *Students:* 326 full-time (97 women); includes 59 minority (3 African Americans, 42 Asian Americans or Pacific Islanders, 13 Hispanic Americans, 1 Native American), 113 international. *Degree requirements:* For master's, comprehensive exam or thesis (MS) required; for doctorate, dissertation, qualifying exam required. *Entrance requirements:* For master's, GRE General Test, minimum GPA of 3.0; for doctorate, GRE General Test, minimum GPA of 3.5. *Application deadline:* For fall admission, 2/10. Application fee: $40. *Unit head:* Dr. Adib Kanafani, Chair, 510-642-3261. *Application contact:* Mari Cook, Graduate Assistant for Admission, 510-643-8944, E-mail: mcook@ce.berkeley.edu.

See in-depth description on page 609.

University of Colorado at Boulder, Graduate School, College of Engineering and Applied Science, Department of Civil, Environmental, and Architectural Engineering, Boulder, CO 80309. Offers building systems (MS, PhD); construction engineering and management (MS, PhD); environmental engineering (MS, PhD); geoenvironmental engineering (MS, PhD); geotechnical engineering (MS, PhD); structural engineering (MS, PhD); water resource engineering (MS, PhD). *Degree requirements:* For master's, thesis or alternative, comprehensive exam required; for doctorate, dissertation required. *Entrance requirements:* For master's, GRE General Test, minimum undergraduate GPA of 3.0.

See in-depth description on page 619.

University of Denver, Daniels College of Business, School of Real Estate and Construction Management, Denver, CO 80208. Offers real estate (MBA); real estate and construction management (MRECM). Part-time programs available. *Faculty:* 4 full-time (0 women). *Students:* 59 (17 women); includes 3 minority (1 Asian American or Pacific Islander, 2 Hispanic Americans) 6 international. Average age 27. 42 applicants, 90% accepted. In 1998, 25 degrees awarded. *Degree requirements:* For master's, foreign language and thesis not required. *Entrance requirements:* For master's, GMAT (average 545). *Application deadline:* For fall admission, 5/1 (priority date); for spring admission, 1/1. Applications are processed on a rolling basis. Application fee: $50. *Financial aid:* In 1998–99, 30 students received aid, including 1 teaching assistantship with full and partial tuition reimbursement available (averaging $2,760 per year); research assistantships with full and partial tuition reimbursements available, career-related internships or fieldwork, Federal Work-Study, grants, institutionally-sponsored loans, and scholarships also available. Aid available to part-time students. Financial aid application deadline: 2/15; financial aid applicants required to submit FAFSA. *Unit head:* Dr. Mark Levine, Director, 303-871-2142. *Application contact:* Jan Johnson, Executive Director, Student Services, 303-871-3416, Fax: 303-871-4466, E-mail: dcb@du.edu.

University of Florida, Graduate School, College of Architecture, M. E. Rinker, Sr. School of Building Construction, Gainesville, FL 32611. Offers MBC, MSBC, PhD. Part-time programs available. *Faculty:* 24. *Students:* 42 full-time (9 women), 8 part-time (3 women); includes 13 minority (4 African Americans, 2 Asian Americans or Pacific Islanders, 7 Hispanic Americans), 6 international. Average age 30. 68 applicants, 53% accepted. In 1998, 21 master's awarded. *Degree requirements:* For master's and doctorate, thesis/dissertation required. *Entrance requirements:* For master's, GRE General Test (minimum combined score of 1000 required), minimum GPA of 3.0; for doctorate, GRE General Test (minimum combined score of 1200 required), minimum GPA of 3.0. *Application deadline:* For fall admission, 3/16 (priority date); for spring admission, 10/15. Applications are processed on a rolling basis. Application fee: $20. Electronic applications accepted. *Financial aid:* In 1998–99, 15 students received aid, including 2 fellowships, 11 research assistantships, 9 teaching assistantships; career-related internships or fieldwork and unspecified assistantships also available. *Faculty research:* Safety, affordable housing, construction management, environmental issues, sustainable construction. Total annual research expenditures: $1.3 million. *Unit head:* Dr. Jimmie Hinze, Director, 352-392-5965, Fax: 352-392-9606. *Application contact:* Dr. Abdol Chini, Graduate Coordinator, 352-392-7510, Fax: 352-392-9606, E-mail: chini@ufl.edu.

University of Houston, College of Technology, Houston, TX 77004. Offers construction management (MT); manufacturing systems (MT); microcomputer systems (MT); occupational technology (MSOT). Part-time and evening/weekend programs available. *Faculty:* 23 full-time (7 women), 3 part-time (0 women). *Students:* 17 full-time (11 women), 75 part-time (41 women); includes 27 minority (15 African Americans, 4 Asian Americans or Pacific Islanders, 8 Hispanic Americans), 6 international. *Degree requirements:* Foreign language not required. *Entrance requirements:* For master's, GMAT, GRE, or MAT (MSOT); GRE (MT), minimum GPA of 3.0 in last 60 hours. *Application deadline:* For fall admission, 7/1; for spring admission, 11/1. Application fee: $35 ($110 for international students). *Unit head:* Bernard McIntyre, Dean, 713-743-4028, Fax: 713-743-4032, E-mail: bmcintyre@uh.edu. *Application contact:* Holly Rosenthal, Graduate Academic Adviser, 713-743-4098, Fax: 713-743-4032, E-mail: hrosenthal@uh.edu.

University of Michigan, Horace H. Rackham School of Graduate Studies, College of Engineering, Department of Civil and Environmental Engineering, Ann Arbor, MI 48109. Offers civil engineering (MSE, PhD, CE); construction engineering and management (MSE); environmental engineering (MSE, PhD). *Degree requirements:* For master's, foreign language and thesis not required; for doctorate, computer language, oral defense of dissertation, preliminary and written exams required. *Entrance requirements:* For master's, GRE General Test; for doctorate, GRE General Test, master's degree. Electronic applications accepted. *Faculty research:* Earthquake engineering, environmental and water resources engineering, geotechnical engineering, hydraulics and hydrologic engineering, structural engineering.

See in-depth description on page 637.

University of Missouri–Rolla, Graduate School, School of Engineering, Department of Civil Engineering, Program in Construction Engineering, Rolla, MO 65409-0910. Offers MS, DE, PhD. *Degree requirements:* For master's, thesis or alternative required, foreign language not required; for doctorate, dissertation required, foreign language not required. *Entrance requirements:* For master's and doctorate, GRE General Test (minimum combined score of 1100 required), TOEFL (minimum score of 550 required), minimum GPA of 3.0.

University of Southern California, Graduate School, School of Engineering, Department of Civil Engineering, Program in Construction Engineering, Los Angeles, CA 90089. Offers MS. *Students:* 8 full-time (0 women), 3 part-time (1 woman); includes 1 minority (Asian American or Pacific Islander), 9 international. Average age 26. 14 applicants, 79% accepted. In 1998, 16 degrees awarded. *Degree requirements:* For master's, thesis optional. *Entrance requirements:* For master's, GRE General Test. *Application deadline:* For fall admission, 6/1 (priority date); for spring admission, 11/1. Application fee: $55. Tuition: Part-time $768 per unit. Required fees: $350 per semester. *Financial aid:* Fellowships, research assistantships, teaching assistantships, Federal Work-Study and institutionally-sponsored loans available. Aid available to part-time students. Financial aid application deadline: 2/15; financial aid applicants required to submit FAFSA. *Unit head:* Dr. L. Carter Wellford, Chair, Department of Civil Engineering, 213-740-0587.

University of Southern California, Graduate School, School of Engineering, Department of Civil Engineering, Program in Construction Management, Los Angeles, CA 90089. Offers MCM. *Students:* 25 full-time (2 women), 3 part-time; includes 6 minority (1 African American, 4 Asian Americans or Pacific Islanders, 1 Hispanic American), 19 international. Average age 26. 60 applicants, 87% accepted. In 1998, 12 degrees awarded. *Degree requirements:* For master's, thesis optional. *Entrance requirements:* For master's, GRE General Test. *Application deadline:* For fall admission, 6/1 (priority date); for spring admission, 11/1. Application fee: $55. Tuition: Part-time $768 per unit. Required fees: $350 per semester. *Financial aid:* In 1998–99, 1 fellowship, 1 research assistantship were awarded.; teaching assistantships, Federal Work-Study and institutionally-sponsored loans also available. Aid available to part-time students. Financial aid application deadline: 2/15; financial aid applicants required to submit FAFSA. *Unit head:* Dr. L. Carter Wellford, Chair, Department of Civil Engineering, 213-740-0587.

See in-depth description on page 645.

University of Washington, Graduate School, College of Architecture and Urban Planning, Department of Construction Management, Seattle, WA 98195. Offers MS. Part-time and evening/weekend programs available. *Faculty:* 5 full-time (0 women), 2 part-time (0 women). *Students:* 9 full-time (3 women), 41 part-time (4 women); includes 4 minority (all African Americans), 8 international. Average age 34. 25 applicants, 44% accepted. In 1998, 12 degrees awarded. *Degree requirements:* For master's, thesis or alternative required, foreign language not required. *Entrance requirements:* For master's, GRE General Test, TOEFL, minimum GPA of 3.0. *Average time to degree:* Master's–1 year full-time, 3 years part-time. *Application deadline:* For fall admission, 7/1; for winter admission, 11/1; for spring admission, 2/1. Applications are processed on a rolling basis. Application fee: $50. Electronic applications accepted. Tuition, state resident: full-time $5,196; part-time $475 per credit. Tuition, nonresident: full-time $13,485; part-time $1,285 per credit. Required fees: $387; $38 per credit. Tuition and fees vary according to course load. *Financial aid:* Application deadline: 2/15. *Faculty research:* Business practices, delivery methods, materials, productivity. *Unit head:* Dr. Saeed Daniali, Chair, 206-685-1764, Fax: 206-685-1976, E-mail: sdaniali@u.washington.edu. *Application contact:* Dr. John E. Schaufelberger, Graduate Coordinator, 206-685-4440, Fax: 206-685-1976, E-mail: jesbcon@u.washington.edu.

University of Washington, Graduate School, College of Engineering, Department of Civil and Environmental Engineering, Seattle, WA 98195. Offers environmental engineering (MS, MS Civ E, MSE, PhD); hydraulic engineering (MS Civ E, MSE, PhD); structural and geotechnical engineering and mechanics (MS, MS Civ E, MSE, PhD); transportation and construction engineering (MS, MS Civ E, MSE, PhD). *Faculty:* 31 full-time (4 women), 26 part-time (1 woman). *Students:* 153 full-time (54 women), 42 part-time (13 women); includes 25 minority (2 African Americans, 19 Asian Americans or Pacific Islanders, 3 Hispanic Americans, 1 Native American), 34 international. Terminal master's awarded for partial completion of doctoral program. *Degree requirements:* For master's, thesis or alternative required, foreign language not required; for doctorate, dissertation required, foreign language not required. *Entrance requirements:* For master's, GRE General Test, TOEFL (minimum score of 580 required), minimum GPA of 3.0; for doctorate, GRE, TOEFL (minimum score of 580 required), minimum GPA of 3.0. *Application deadline:* For fall admission, 2/1 (priority date); for winter admission, 11/1; for spring admission, 2/1. Applications are processed on a rolling basis. Application fee: $50. Electronic applications accepted. Tuition, state resident: full-time $5,196; part-time $475 per credit. Tuition, nonresident: full-time $13,485; part-time $1,285 per credit. Required fees: $387; $38 per credit. Tuition and fees vary according to course load. *Unit head:* Dr. Fred L. Mannering, Chair, 206-543-2390. *Application contact:* Marcia Buck, Graduate Secretary, 206-543-2574, E-mail: ceginfo@u.washington.edu.

Washington University in St. Louis, School of Engineering and Applied Science, Sever Institute of Technology, Department of Civil Engineering, Program in Construction Engineering, St. Louis, MO 63130-4899. Offers MCE. *Degree requirements:* For master's, thesis optional, foreign language not required.

Washington University in St. Louis, School of Engineering and Applied Science, Sever Institute of Technology, Department of Civil Engineering, Program in Construction Management, St. Louis, MO 63130-4899. Offers MCM, M Arch/MCM. *Degree requirements:* For master's, thesis optional, foreign language not required.

Western Michigan University, Graduate College, College of Engineering and Applied Sciences, Department of Construction Engineering, Materials Engineering and Industrial Design, Program in Construction Management, Kalamazoo, MI 49008. Offers MS. *Students:* 1 (woman) full-time, 11 part-time, 11 international. 9 applicants, 67% accepted. *Entrance requirements:* For master's, minimum GPA of 3.0. *Application deadline:* For fall admission, 2/15 (priority date). Applications are processed on a rolling basis. Application fee: $25. *Financial aid:* Application deadline: 2/15. *Unit head:* Maureen Clausen, Graduate Secretary, 509-335-2576, Fax: 509-335-7632, E-mail: mclausen@wsu.edu. *Application contact:* Paula J. Boodt, Coordinator, Graduate Admissions and Recruitment, 616-387-2000, Fax: 616-387-2355, E-mail: paula.boodt@wmich.edu.

Environmental Engineering

Air Force Institute of Technology, School of Engineering, Department of Engineering and Environmental Management, Wright-Patterson AFB, OH 45433-7765. Offers MS. Part-time programs available. *Faculty:* 7 full-time (0 women). *Students:* 35 full-time, 1 part-time. *Degree requirements:* For master's, thesis required, foreign language not required. *Entrance requirements:* For master's, GRE General Test (minimum score of 500 on verbal section, 600 on quantitative required), minimum GPA of 3.0, must be military officer or U.S. citizen. *Average time to degree:* Master's–1.5 years full-time. Application fee: $0. *Faculty research:* Groundwater contaminant modeling/remediation, system dynamics modeling, landfill performance. *Unit head:* Lt. Col. Steven T. Lofgren, Head, 937-255-2998, E-mail: slofgren@afit.af.mil.

Auburn University, Graduate School, College of Engineering, Department of Civil Engineering, Auburn, Auburn University, AL 36849-0002. Offers construction engineering and management (MCE, MS, PhD); environmental engineering (MCE, MS, PhD); geotechnical/materials engineering (MCE, MS, PhD); hydraulics/hydrology (MCE, MS, PhD); structural engineering (MCE, MS, PhD); transportation engineering (MCE, MS, PhD). Part-time programs available. *Faculty:* 22 full-time. *Students:* 24 full-time (6 women), 34 part-time (8 women); includes 2 minority (1 African American, 1 Hispanic American), 9 international. *Degree requirements:* For master's, project (MCE), thesis (MS) required; for doctorate, dissertation, comprehensive exam required, foreign language not required. *Entrance requirements:* For master's, GRE General Test; for doctorate, GRE General Test (minimum score of 400 on each section required). *Application deadline:* For fall admission, 9/1; for spring admission, 3/1. Applications are processed on a rolling basis. Application fee: $25 ($50 for international students). Tuition, state resident: full-time $2,760; part-time $76 per credit hour. Tuition, nonresident: full-time $8,280; part-time $228 per credit hour. *Unit head:* Dr. Joseph F. Judkins, Head, 334-844-4320. *Application contact:* Dr. John F. Pritchett, Dean of the Graduate School, 334-844-4700.

California Institute of Technology, Division of Engineering and Applied Science, Option in Environmental Engineering Science, Pasadena, CA 91125-0001. Offers MS, PhD. *Faculty:* 4 full-time (1 woman). *Students:* 34 full-time (11 women), 12 international. 57 applicants, 5% accepted. In 1998, 7 master's, 8 doctorates awarded. *Degree requirements:* For master's, foreign language and thesis not required; for doctorate, dissertation required, foreign language not required. *Application deadline:* For fall admission, 1/15. Application fee: $0. *Faculty research:* Chemistry of natural waters, physics and chemistry of particulates, fluid mechanics of the natural environment, pollutant formation and control, environmental modeling systems. *Unit head:* Dr. Michael Hoffmann, Executive Officer, 626-395-4391. *Application contact:* Dr. Glen R. Cass, Representative, 626-395-6888.

Announcement: This program is designed to prepare students for careers in specialized research, teaching, and advanced environmental engineering and science. Major areas of specialization include atmospheric chemistry and physics, biotechnology, environmental fluid mechanics, environmental chemistry, and advanced technologies for water treatment. Financial assistance is available and is awarded by Caltech on a competitive basis.

California Polytechnic State University, San Luis Obispo, College of Engineering, Department of Civil and Environmental Engineering, San Luis Obispo, CA 93407. Offers MS, MCRP/MS. Part-time programs available. *Faculty:* 17 full-time (1 woman), 11 part-time (3 women). *Students:* 10 full-time (4 women), 4 part-time (2 women). 31 applicants, 58% accepted. In 1998, 3 degrees awarded. *Degree requirements:* For master's, thesis or alternative required, foreign language not required. *Entrance requirements:* For master's, GRE General Test, minimum GPA of 3.0 in last 90 quarter units. *Application deadline:* For fall admission, 5/31 (priority date); for spring admission, 12/31. Applications are processed on a rolling basis. Application fee: $55. Tuition, nonresident: part-time $164 per unit. Required fees: $531 per quarter. *Financial aid:* Fellowships, research assistantships, teaching assistantships, career-related internships or fieldwork available. Financial aid application deadline: 3/2; financial aid applicants required to submit FAFSA. *Faculty research:* Soils, structures, transportation, traffic, environmental protection, biomediation. *Unit head:* Dr. Robert Lang, Chair, 805-756-2947, Fax: 805-756-6330, E-mail: rlang@calpoly.edu. *Application contact:* Dr. Nirupam Pal, Professor, 805-756-1355, Fax: 805-756-6330, E-mail: npal@calpoly.edu.

Carleton University, Faculty of Graduate Studies, Faculty of Engineering and Design, Department of Civil and Environmental Engineering, Ottawa, ON K1S 5B6, Canada. Offers M Eng, PhD. *Faculty:* 18 full-time (1 woman). *Students:* 55 full-time (17 women), 10 part-time (3 women). Average age 29. In 1998, 14 master's, 2 doctorates awarded. *Degree requirements:* For master's, thesis optional; for doctorate, dissertation required. *Entrance requirements:* For master's, TOEFL (minimum score of 550 required), honors degree; for doctorate, TOEFL (minimum score of 550 required), MA Sc or M Eng. *Average time to degree:* Master's–2.1 years full-time, 3.3 years part-time; doctorate–4.7 years full-time. *Application deadline:* For fall admission, 3/1 (priority date). Applications are processed on a rolling basis. Application fee: $35. *Financial aid:* Application deadline: 3/1. *Faculty research:* Transportation planning and technology, computer applications in civil engineering, structural and continuum mechanics, building design and construction. Total annual research expenditures: $616,000. *Unit head:* Dr. J. L. Humar, Chair, 613-520-2600 Ext. 8389, Fax: 613-520-3951, E-mail: jag_humar@carleton.ca. *Application contact:* Ata M. Khan, Associate Dean of Engineering, 613-520-5659, Fax: 613-520-5682, E-mail: ata_khan@carleton.ca.

Carnegie Mellon University, Carnegie Institute of Technology, Department of Civil and Environmental Engineering, Pittsburgh, PA 15213-3891. Offers civil engineering (MS, PhD); civil engineering and industrial management (MS); civil engineering and robotics (PhD); civil engineering/bioengineering (PhD); civil engineering/engineering and public policy (MS, PhD). Part-time programs available. *Faculty:* 21 full-time (5 women). *Students:* 56 full-time (20 women), 11 part-time (2 women); includes 3 minority (1 African American, 1 Hispanic American, 1 Native American), 44 international. Average age 27. In 1998, 27 master's, 15 doctorates awarded. Terminal master's awarded for partial completion of doctoral program. *Degree requirements:* For master's, thesis required (for some programs), foreign language not required; for doctorate, dissertation, qualifying exam required, foreign language not required. *Entrance requirements:* For master's and doctorate, GRE General Test, TOEFL. *Average time to degree:* Master's–1 year full-time; doctorate–4 years full-time. *Application deadline:* For fall admission, 2/1 (priority date); for spring admission, 10/15. Application fee: $45. *Financial aid:* In 1998–99, 63 students received aid, including 7 fellowships, 36 research assistantships; teaching assistantships, career-related internships or fieldwork, Federal Work-Study, and scholarships also available. Financial aid application deadline: 2/1. *Faculty research:* Computer-aided engineering and management, structured and computational machines, civil systems. Total annual research expenditures: $2.2 million. *Unit head:* Chris Hendrickson, Head, 412-268-2941, Fax: 412-268-7813. *Application contact:* Maxine A. Leffard, Graduate Program Administrator, 412-268-8712, Fax: 412-268-7813, E-mail: ce-addmissions+@andrew.cmu.edu.

See in-depth description on page 569.

Carnegie Mellon University, Graduate School of Industrial Administration, Pittsburgh, PA 15213-3891. Offers accounting (PhD); algorithms, combinatorics, and optimization (PhD); business management and software engineering (MBMSE); civil engineering and industrial management (MS); computational finance (MSCF); economics (PhD); electronic commerce (MS); environmental engineering and management (MEEM); finance (PhD); financial economics (PhD); industrial administration (MSIA), including administration and public management; information science (PhD); manufacturing (MOM); manufacturing and operating systems (PhD), including industrial administration; marketing (PhD); mathematical finance (PhD); operations research (PhD); organizational behavior and theory (PhD); political economy (PhD); public policy and management (MS, MSED); robotics (PhD). Part-time programs available. *Faculty:* 86 full-time (13 women), 13 part-time (2 women). *Students:* 601 full-time (153 women), 236 part-time (43 women); includes 82 minority (8 African Americans, 63 Asian Americans or Pacific Islanders, 6 Hispanic Americans, 5 Native Americans), 383 international. Terminal master's awarded for partial completion of doctoral program. *Degree requirements:* For master's, foreign language and thesis not required; for doctorate, dissertation required, foreign language not required. *Entrance requirements:* For master's, GMAT. Application fee: $50. *Unit head:* Douglas Dunn, Dean, 412-268-2265. *Application contact:* Director of Admissions, 412-268-2272.

The Catholic University of America, School of Engineering, Department of Civil Engineering, Program in Environmental Engineering, Washington, DC 20064. Offers MCE, MS Engr. *Degree requirements:* For master's, thesis optional, foreign language not required. *Entrance requirements:* For master's, minimum GPA of 3.0. *Application deadline:* For fall admission, 8/1 (priority date); for spring admission, 12/1. Applications are processed on a rolling basis. Application fee: $50. *Financial aid:* Application deadline: 2/1. *Unit head:* Dr. Timothy W. Kao, Chair, Department of Civil Engineering, 202-319-5163, Fax: 202-319-4499, E-mail: kao@cua.edu.

Clarkson University, Graduate School, School of Engineering, Department of Civil and Environmental Engineering, Potsdam, NY 13699. Offers civil and environmental engineering (PhD); civil engineering (ME, MS). Part-time programs available. *Faculty:* 15 full-time (2 women), 2 part-time (0 women). *Students:* 33 full-time (8 women); includes 2 minority (both Native Americans), 15 international. Average age 27. 82 applicants, 43% accepted. In 1998, 11 master's, 3 doctorates awarded. *Degree requirements:* For master's, thesis required, foreign language not required; for doctorate, dissertation, departmental qualifying exam required, foreign language not required. *Entrance requirements:* For master's, GRE, TOEFL. *Application deadline:* For fall admission, 5/15 (priority date); for spring admission, 10/15 (priority date). Applications are processed on a rolling basis. Application fee: $25 ($35 for international students). Tuition: Part-time $661 per credit hour. Required fees: $215 per semester. *Financial aid:* In 1998–99, 1 fellowship, 17 research assistantships, 13 teaching assistantships were awarded. *Faculty research:* Granular flows, water treatment, environmental systems, geotechnical structural dynamics, fluid dynamics. Total annual research expenditures: $1.2 million. *Unit head:* Dr. Thomas L. Theis, Chairman, 315-268-6529, Fax: 315-268-7985, E-mail: tlto@clarkson.edu. *Application contact:* Dr. Philip K. Hopke, Dean of the Graduate School, 315-268-6447, Fax: 315-268-7994, E-mail: hopkepk@clarkson.edu.

Clemson University, Graduate School, College of Engineering and Science, School of the Environment, Department of Environmental Engineering and Science, Clemson, SC 29634. Offers M Engr, MS, PhD. *Accreditation:* ABET (one or more programs are accredited). *Students:* 61 full-time (28 women), 18 part-time (4 women); includes 4 minority (3 African Americans, 1 Asian American or Pacific Islander), 20 international. Average age 24. 82 applicants, 48% accepted. In 1998, 19 master's, 2 doctorates awarded. *Degree requirements:* For master's and doctorate, thesis/dissertation required, foreign language not required. *Entrance requirements:* For master's and doctorate, GRE General Test, TOEFL, minimum GPA of 3.0. *Application deadline:* For fall admission, 3/1 (priority date). Application fee: $35. *Financial aid:* Fellowships, research assistantships, teaching assistantships, institutionally-sponsored loans and unspecified assistantships available. Financial aid applicants required to submit FAFSA. *Faculty research:* Water and air pollution control, hazardous waste and environmental management, environmental chemistry and biology, containment transport modeling, risk assessment. *Unit head:* Dr. Alan Elzerman, Interim Chair, 864-656-3276, Fax: 864-656-0672, E-mail: alan.elzerman@ces.clemson.edu. *Application contact:* Pamela S. Fjeld, Student Services, 864-656-1010, Fax: 864-656-0672, E-mail: hpamela@ces.clemson.edu.

See in-depth description on page 571.

Colorado School of Mines, Graduate School, Division of Environmental Science and Engineering, Golden, CO 80401-1887. Offers MS, PhD. Part-time programs available. *Faculty:* 8 full-time (1 woman), 8 part-time (1 woman). *Students:* 50 full-time (13 women), 27 part-time (10 women); includes 2 minority (1 African American, 1 Hispanic American), 17 international. 92 applicants, 72% accepted. In 1998, 33 master's awarded (100% found work related to degree); 2 doctorates awarded (100% found work related to degree). *Degree requirements:* For master's, thesis required (for some programs), foreign language not required; for doctorate, dissertation, comprehensive exam required, foreign language not required. *Entrance requirements:* For master's and doctorate, GRE General Test (combined average 1660 on three sections), minimum GPA of 3.0. *Application deadline:* Applications are processed on a rolling basis. Application fee: $40. Electronic applications accepted. *Financial aid:* In 1998–99, 56 students received aid, including 6 fellowships, 12 research assistantships, 9 teaching assistantships; unspecified assistantships also available. Aid available to part-time students. Financial aid applicants required to submit FAFSA. *Faculty research:* Treatment of water and wastes, environmental law–policy and practice, natural environment systems, hazardous waste management, environmental data analysis. Total annual research expenditures: $505,431. *Unit head:* Dr. Philippe Ross, Head, 303-273-3473, Fax: 303-273-3413, E-mail: pross@mines.edu. *Application contact:* Juanita Chuven, Administrative Assistant, 303-273-3427, Fax: 303-273-3413, E-mail: jchuven@mines.edu.

Colorado State University, Graduate School, College of Engineering, Department of Civil Engineering, Specialization in Environmental Engineering, Fort Collins, CO 80523-0015. Offers MS, PhD. Part-time programs available. *Faculty:* 3 full-time (0 women). In 1998, 15 master's awarded. Terminal master's awarded for partial completion of doctoral program. *Degree requirements:* For master's, thesis or alternative required, foreign language not required; for doctorate, dissertation required, foreign language not required. *Entrance requirements:* For master's and doctorate, GRE General Test, TOEFL (minimum score of 550 required; 213 for computer-based), minimum GPA of 3.0. *Average time to degree:* Master's–2 years full-time, 5 years part-time; doctorate–4 years full-time. *Application deadline:* For fall admission, 3/1 (priority date); for spring admission, 8/1 (priority date). Applications are processed on a rolling basis. Application fee: $30. Electronic applications accepted. *Financial aid:* Fellowships, research assistantships, teaching assistantships available. *Faculty research:* Water treatment, dilute aqueous chemistry, sludge de-watering, water quality monitoring, advanced waste treatment. Total annual research expenditures: $245,000. *Unit head:* Thomas G. Sanders, Leader, 970-491-5448, Fax: 970-491-7727, E-mail: tgs@engr.colostate.edu. *Application contact:* Laurie Howard, Student Adviser, 970-491-5844, Fax: 970-491-7727, E-mail: lhoward@engr.colostate.edu.

Colorado State University, Graduate School, College of Engineering, Department of Mechanical Engineering, Program in Energy and Environmental Engineering, Fort Collins, CO 80523-0015. Offers MS, PhD. Postbaccalaureate distance learning degree programs offered (minimal on-campus study). *Faculty:* 18 full-time (0 women). *Degree requirements:* For doctorate, dissertation required, foreign language not required, foreign language not required. *Entrance requirements:* For master's and doctorate, GRE General Test (minimum combined score of 1850 on three sections required; average 1872), TOEFL (minimum score of 550 required; average 596), minimum GPA of 3.0. *Application deadline:* For fall admission, 2/1 (priority date). Applications are processed on a rolling basis. Application fee: $30. Electronic applications accepted. *Faculty research:* Indoor air quality, solar energy industry energy conservation, building and industrial energy conservation, engine pollution abatement optimal control. *Unit head:* Christine Piper, Graduate Coordinator, 864-656-7581, E-mail: cpiper@clemson.edu. *Application contact:* Dr. Doug Hittle, Graduate Committee Chairman, 970-491-8617, Fax: 970-491-3827, E-mail: hittle@lamar.colostate.edu.

Columbia University, Fu Foundation School of Engineering and Applied Science, Department of Earth and Environmental Engineering, Program in Earth Resources Engineering, New York, NY 10027. Offers earth resources engineering (MS); earth systems engineering (PhD). Part-time programs available. *Faculty:* 7 full-time (0 women), 3 part-time (1 woman). *Students:*

Environmental Engineering

Columbia University (continued)
2 full-time (0 women), 1 (woman) part-time, 2 international. In 1998, 8 master's awarded (50% found work related to degree, 50% continued full-time study); 6 doctorates awarded. *Degree requirements:* For doctorate, dissertation, qualifying exam required, foreign language not required. *Entrance requirements:* For master's and doctorate, GRE General Test, TOEFL. *Average time to degree:* Master's–6 years full-time. *Application deadline:* For fall admission, 2/15; for spring admission, 10/1. Application fee: $55. *Financial aid:* Federal Work-Study available. Financial aid application deadline: 1/5; financial aid applicants required to submit FAFSA. *Faculty research:* Industrial ecology, waste treatment and recycling, water resources, environmental remediation, hazardous waste disposal. *Unit head:* Dr. Barbara Algin, Departmental Administrator, 212-854-2905, Fax: 212-854-7081, E-mail: ba110@columbia.edu. *Application contact:* Dr. Barbara Algin, Departmental Administrator, 212-854-2905, Fax: 212-854-7081, E-mail: ba110@columbia.edu.

Cornell University, Graduate School, Graduate Fields of Engineering, Field of Civil and Environmental Engineering, Ithaca, NY 14853-0001. Offers environmental engineering (M Eng, MS, PhD); environmental fluid mechanics and hydrology (M Eng, MS, PhD); environmental systems engineering (M Eng, MS, PhD); geotechnical engineering (M Eng, MS, PhD); remote sensing (M Eng, MS, PhD); structural engineering (M Eng, MS, PhD); transportation engineering (M Eng, MS, PhD); water resource systems (M Eng, MS, PhD). *Faculty:* 29 full-time. *Students:* 143 full-time (32 women); includes 16 minority (11 Asian Americans or Pacific Islanders, 5 Hispanic Americans), 76 international. 365 applicants, 52% accepted. In 1998, 74 master's, 10 doctorates awarded. Terminal master's awarded for partial completion of doctoral program. *Degree requirements:* For master's, thesis (MS) required; for doctorate, dissertation required, foreign language not required. *Entrance requirements:* For master's, TOEFL (minimum score of 600required); for doctorate, GRE General Test, TOEFL (minimum score of 600 required). *Application deadline:* For fall admission, 1/15. Application fee: $65. Electronic applications accepted. *Financial aid:* In 1998–99, 65 students received aid, including 20 fellowships with full tuition reimbursements available, 28 research assistantships with full tuition reimbursements available, 17 teaching assistantships with full tuition reimbursements available; institutionally-sponsored loans, scholarships, tuition waivers (full and partial), and unspecified assistantships also available. Financial aid applicants required to submit FAFSA. *Unit head:* Director of Graduate Studies, 607-255-7560, Fax: 607-255-9004. *Application contact:* Graduate Field Assistant, 607-255-7560, E-mail: cee_grad@cornell.edu.

See in-depth description on page 573.

Dartmouth College, Thayer School of Engineering, Program in Environmental Engineering, Hanover, NH 03755. Offers MS, PhD. *Degree requirements:* For master's, thesis required; for doctorate, dissertation, candidacy oral exam required. *Entrance requirements:* For master's and doctorate, GRE General Test. *Application deadline:* For fall admission, 1/15 (priority date). Application fee: $20 ($40 for international students). *Financial aid:* Career-related internships or fieldwork, Federal Work-Study, institutionally-sponsored loans, and tuition waivers (full and partial) available. Financial aid application deadline: 1/15. *Faculty research:* Environmental fluid mechanics, large-scale environmental simulation, water resources, physical oceanography, sustainable resource utilization. Total annual research expenditures: $1.1 million. *Unit head:* Shelley Parker, Admissions Coordinator, Graduate Studies and Research, 902-494-1288, Fax: 902-494-3149, E-mail: shelley.parker@dal.ca. *Application contact:* Candace S. Potter, Admissions Coordinator, 603-646-3844, Fax: 603-646-3856, E-mail: candace.potter@dartmouth.edu.

Drexel University, Graduate School, School of Environmental Science, Engineering and Policy, Philadelphia, PA 19104-2875. Offers environmental engineering (MS, PhD); environmental science (MS, PhD). Part-time and evening/weekend programs available. *Faculty:* 7 full-time, 7 part-time. *Students:* 17 full-time (11 women), 50 part-time (20 women); includes 4 minority (all Asian Americans or Pacific Islanders), 12 international. Average age 30. 101 applicants, 60% accepted. In 1998, 29 master's awarded. Terminal master's awarded for partial completion of doctoral program. *Degree requirements:* For master's, thesis optional; for doctorate, dissertation required. *Entrance requirements:* For master's, TOEFL (minimum score of 570 required), minimum GPA of 3.0; for doctorate, TOEFL (minimum score of 570 required), minimum GPA of 3.0, MS. *Application deadline:* For fall admission, 8/21. Applications are processed on a rolling basis. Application fee: $35. Tuition: Full-time $15,795; part-time $585 per credit. Required fees: $375; $67 per term. Tuition and fees vary according to program. *Financial aid:* In 1998–99, 8 research assistantships, 3 teaching assistantships were awarded.; career-related internships or fieldwork and unspecified assistantships also available. Financial aid application deadline: 2/1. *Faculty research:* Environmental health, water quality and resources, hazardous-waste disposal, environmental chemistry. *Unit head:* Dr. Michael Gealt, Director, 215-895-2265. *Application contact:* Director of Graduate Admissions, 215-895-6700, Fax: 215-895-5939.

See in-depth description on page 575.

Duke University, Graduate School, School of Engineering, Department of Civil and Environmental Engineering, Durham, NC 27708-0586. Offers civil and environmental engineering (MS, PhD); environmental engineering (MS, PhD). Part-time programs available. *Faculty:* 18 full-time, 9 part-time. *Students:* 36 full-time (9 women), 18 international. 96 applicants, 40% accepted. In 1998, 9 master's, 3 doctorates awarded. Terminal master's awarded for partial completion of doctoral program. *Degree requirements:* For doctorate, dissertation required, foreign language not required, foreign language not required. *Entrance requirements:* For master's and doctorate, GRE General Test. *Application deadline:* For fall admission, 12/31; for spring admission, 11/1. Application fee: $75. *Financial aid:* Fellowships, research assistantships, Federal Work-Study available. Financial aid application deadline: 12/31. *Unit head:* Tod Laursen, Director of Graduate Studies, 919-660-5200, Fax: 919-660-5219, E-mail: pr@egr.duke.edu.

See in-depth description on page 577.

École Polytechnique de Montréal, Graduate Programs, Department of Civil Engineering, Montréal, PQ H3C 3A7, Canada. Offers environmental engineering (M Eng, M Sc A, PhD); geotechnical engineering (M Eng, M Sc A, PhD); hydraulics engineering (M Eng, M Sc A, PhD); structural engineering (M Eng, M Sc A, PhD); transportation engineering (M Eng, M Sc A, PhD). Part-time and evening/weekend programs available. *Degree requirements:* For master's and doctorate, one foreign language, computer language, thesis/dissertation required. *Entrance requirements:* For master's, minimum GPA of 2.75; for doctorate, minimum GPA of 3.0. *Faculty research:* Water resources management, characteristics of building materials, aging of dams, pollution control.

Florida Agricultural and Mechanical University, Division of Graduate Studies, Research, and Continuing Education, FAMU-FSU College of Engineering, Department of Civil Engineering, Tallahassee, FL 32307-3200. Offers civil engineering (MS, PhD); environmental engineering (MS, PhD). *Students:* 10 (2 women); all minorities (7 African Americans, 2 Asian Americans or Pacific Islanders, 1 Hispanic American) *Entrance requirements:* For master's, GRE General Test (minimum combined score of 1000 required), minimum GPA of 3.0. *Application deadline:* For fall admission, 7/1. Application fee: $20. *Unit head:* Dr. C. J. Chen, Dean, FAMU-FSU College of Engineering, 850-487-6100, Fax: 850-487-6486.

See in-depth description on page 579.

Florida Institute of Technology, Graduate School, College of Engineering, Division of Engineering Science, Program in Environmental Engineering, Melbourne, FL 32901-6975. Offers MS. Part-time programs available. *Students:* 3 full-time (0 women), 3 part-time (1 woman), 5 international. Average age 27. 15 applicants, 53% accepted. In 1998, 2 degrees awarded. *Degree requirements:* For master's, thesis optional, foreign language not required. *Entrance requirements:* For master's, TOEFL (minimum score of 550 required), minimum GPA of 3.0. *Application deadline:* Applications are processed on a rolling basis. Application fee: $50. Electronic applications accepted. Tuition: Part-time $575 per credit hour. Required fees: $100. Tuition and fees vary according to campus/location and program. *Financial aid:* Research assistantships, teaching assistantships, tuition remissions available. Financial aid application deadline: 3/1; financial aid applicants required to submit FAFSA. *Faculty research:* Seepage and salt transport, solid and hazardous waste management, water and wastewater treatment systems. *Unit head:* Dr. Paul A. Jennings, Chair, 407-674-8068, Fax: 407-984-8461, E-mail: jennings@fit.edu. *Application contact:* Carolyn P. Farrior, Associate Dean of Graduate Admissions, 407-674-7118, Fax: 407-723-9468, E-mail: cfarrior@fit.edu.

Florida International University, College of Engineering, Department of Civil and Environmental Engineering, Program in Environmental and Urban Systems, Miami, FL 33199. Offers MS. Part-time and evening/weekend programs available. Average age 38. 5 applicants, 60% accepted. In 1998, 1 degree awarded. *Degree requirements:* For master's, thesis required. *Entrance requirements:* For master's, GRE General Test (minimum combined score of 1000 required), TOEFL (minimum score of 500 required), bachelor's degree in related field. *Application deadline:* For fall admission, 4/1 (priority date); for spring admission, 10/1. Applications are processed on a rolling basis. Application fee: $20. Tuition, state resident: part-time $145 per credit hour. Tuition, nonresident: part-time $506 per credit hour. Required fees: $158; $158 per year. *Faculty research:* Water and wastewater treatment, housing systems. *Unit head:* Dr. David Shen, Chairperson, Department of Civil and Environmental Engineering, 305-348-2824, Fax: 305-348-2802, E-mail: shen@eng.fiu.edu.

Florida International University, College of Engineering, Department of Civil and Environmental Engineering, Program in Environmental Engineering, Miami, FL 33199. Offers MS. Part-time and evening/weekend programs available. *Students:* 7 full-time (3 women), 24 part-time (11 women); includes 15 minority (1 African American, 2 Asian Americans or Pacific Islanders, 12 Hispanic Americans), 8 international. Average age 32. 34 applicants, 41% accepted. In 1998, 6 degrees awarded. *Degree requirements:* For master's, thesis optional. *Entrance requirements:* For master's, GRE General Test (minimum combined score of 1000 required), TOEFL (minimum score of 500 required), bachelor's degree in related field. *Application deadline:* For fall admission, 4/1 (priority date); for spring admission, 10/1. Applications are processed on a rolling basis. Application fee: $20. Tuition, state resident: part-time $145 per credit hour. Tuition, nonresident: part-time $506 per credit hour. Required fees: $158; $158 per year. *Unit head:* Dr. David Shen, Chairperson, Department of Civil and Environmental Engineering, 305-348-2824, Fax: 305-348-2802, E-mail: shen@eng.fiu.edu.

Florida State University, Graduate Studies, FAMU/FSU College of Engineering, Department of Civil Engineering, Tallahassee, FL 32306. Offers civil engineering (MS, PhD); environmental engineering (MS, PhD). Part-time programs available. *Faculty:* 16 full-time (3 women). *Students:* 21 full-time (4 women), 22 part-time (5 women); includes 21 minority (12 African Americans, 7 Asian Americans or Pacific Islanders, 2 Hispanic Americans) *Degree requirements:* For master's, thesis optional, foreign language not required; for doctorate, dissertation, preliminary exam, qualifying exam required, foreign language not required. *Entrance requirements:* For master's, GRE General Test (minimum combined score of 1000 required), TOEFL (minimum score of 550 required), BS in engineering or related field, minimum GPA of 3.0; for doctorate, GRE General Test, MS in engineering or related field, minimum GPA of 3.0. *Application deadline:* For fall admission, 7/15; for spring admission, 11/23. Applications are processed on a rolling basis. Application fee: $20. Tuition, state resident: part-time $139 per credit hour. Tuition, nonresident: part-time $482 per credit hour. Tuition and fees vary according to program. *Unit head:* Dr. Jerry Wekezer, Chair, 850-410-6143, Fax: 850-410-6142, E-mail: wekezer@eng.fsu.edu. *Application contact:* J. Belinda Morris, Graduate Studies Assistant, 850-487-6319, Fax: 850-487-6142, E-mail: bmorris@eng.fsu.edu.

See in-depth description on page 579.

The George Washington University, School of Engineering and Applied Science, Department of Civil and Environmental Engineering, Washington, DC 20052. Offers MS, D Sc, App Sc, Engr. Part-time and evening/weekend programs available. *Degree requirements:* For master's, thesis optional, foreign language not required; for doctorate, computer language, dissertation, final and qualifying exams required, foreign language not required; for other advanced degree, foreign language and thesis not required. *Entrance requirements:* For master's, TOEFL (minimum score of 550 required; average 580) or George Washington University English as a Foreign Language Test, appropriate bachelor's degree, minimum GPA of 3.0; for doctorate, TOEFL (minimum score of 550 required; average 580) or George Washington University English as a Foreign Language Test, appropriate bachelor's or master's degree, minimum GPA of 3.4, GRE required if highest earned degree is BS; for other advanced degree, TOEFL (minimum score of 550 required; average 580) or George Washington University English as a Foreign Language Test, appropriate master's degree, minimum GPA of 3.0. *Application deadline:* For fall admission, 3/1 (priority date); for spring admission, 10/1. Applications are processed on a rolling basis. Application fee: $55. Tuition: Full-time $17,328; part-time $722 per credit hour. Required fees: $828; $35 per credit hour. Tuition and fees vary according to campus/location and program. *Financial aid:* Fellowships, research assistantships, teaching assistantships, career-related internships or fieldwork, Federal Work-Study, and institutionally-sponsored loans available. Financial aid application deadline: 3/1; financial aid applicants required to submit FAFSA. *Faculty research:* Computer-integrated manufacturing, materials engineering, electronic materials, fatigue and fracture, reliability. *Unit head:* Dr. Theodore Toridis, Chair, 202-994-6749. *Application contact:* Howard M. Davis, Manager, Office of Admissions and Student Records, 202-994-6158, Fax: 202-994-0909, E-mail: data: adms@seas.gwu.edu.

See in-depth description on page 583.

Georgia Institute of Technology, Graduate Studies and Research, College of Engineering, School of Civil and Environmental Engineering, Program in Environmental Engineering, Atlanta, GA 30332-0001. Offers MS, MS Env E, PhD. *Accreditation:* ABET (one or more programs are accredited). Part-time programs available. Postbaccalaureate distance learning degree programs offered (no on-campus study). *Degree requirements:* For master's, research report or thesis required; for doctorate, computer language, dissertation required, foreign language not required. *Entrance requirements:* For master's and doctorate, GRE, TOEFL (minimum score of 550 required), minimum GPA of 3.2. *Faculty research:* Advanced microbiology of water and wastes, industrial waste treatment and disposal, air pollution measurements and control.

Harvard University, School of Public Health, Department of Environmental Health, Boston, MA 02115-6096. Offers environmental epidemiology (SM, DPH, SD); environmental health (SM); environmental science and engineering (SM, SD); occupational health (MOH, SM, DPH, SD); physiology (SD). *Accreditation:* CEPH. Part-time programs available. *Faculty:* 23 full-time (5 women), 31 part-time (4 women). *Students:* 75 full-time (38 women), 6 part-time (4 women); includes 8 minority (2 African Americans, 6 Asian Americans or Pacific Islanders), 34 international. *Degree requirements:* For master's, thesis not required; for doctorate, dissertation, qualifying exam required. *Entrance requirements:* For master's, GRE, TOEFL (minimum score of 550 required; 220 for computer-based); for doctorate, GRE, TOEFL (minimum score of 550 required; 220 for computer-based)). *Application deadline:* For fall admission, 1/4. Application fee: $60. *Unit head:* Dr. Joseph D. Brain, Chairman, 617-432-1272. *Application contact:* Margaret R. Watson, Assistant Director of Admissions, 617-432-1031, Fax: 617-432-2009, E-mail: admisofc@hsph.harvard.edu.

Idaho State University, Graduate School, College of Engineering, Pocatello, ID 83209. Offers engineering and applied science (PhD); environmental engineering (MS); hazardous waste management (MS); measurement and control engineering (MS); nuclear science and engineering (MS). MS (hazardous waste management), PhD offered jointly with the University of Idaho. Part-time programs available. *Degree requirements:* For master's, thesis required, foreign language not required; for doctorate, dissertation required. *Entrance requirements:* For master's and doctorate, GRE General Test, TOEFL. *Faculty research:* Isotope separation, control technology,two-phase flow, photosonolysis, criticality calculations.

Illinois Institute of Technology, Graduate College, Armour College of Engineering and Sciences, Department of Chemical and Environmental Engineering, Environmental Engineering Division, Chicago, IL 60616-3793. Offers M Env E, MS, PhD. Part-time and evening/weekend programs available. *Faculty:* 6 full-time (1 woman). *Students:* 27 full-time (11 women), 50 part-time (13 women); includes 11 minority (1 African American, 7 Asian Americans or Pacific Islanders, 3 Hispanic Americans), 28 international. 87 applicants, 75% accepted. In 1998, 33 master's, 6 doctorates awarded. Terminal master's awarded for partial completion of doctoral program. *Degree requirements:* For master's, thesis (for some programs), comprehensive exam required, foreign language not required; for doctorate, dissertation, comprehensive exam required, foreign language not required. *Entrance requirements:* For master's and doctorate, GRE (minimum score of 1200 required), TOEFL (minimum score of 550 required), undergraduate GPA of 3.0 required. *Application deadline:* For fall admission, 7/1; for spring admission, 11/1. Applications are processed on a rolling basis. Application fee: $30. Electronic applications accepted. *Financial aid:* In 1998–99, 6 research assistantships, 8 teaching assistantships were awarded.; fellowships, Federal Work-Study, institutionally-sponsored loans, scholarships, and graduate assistantships also available. Financial aid application deadline: 3/1. *Faculty research:* Bioremediation, industrial wastewater control, water supply and quality, hazardous waste treatment, air pollution control. *Unit head:* Margaret R. Watson, Assistant Director of Admissions, 617-432-1031, Fax: 617-432-2009, E-mail: admisofc@hsph.harvard.edu. *Application contact:* Dr. S. Mohammad Shahidehpour, Dean of Graduate College, 312-567-3024, Fax: 312-567-7517, E-mail: grad@minna.cns.iit.edu.

Announcement: The IIT Department of Chemical and Environmental Engineering offers cutting-edge programs that prepare engineers for the technological challenges of the 21st century. Capitalizing on its interdisciplinary focus, the department provides students with a knowledge of chemical and environmental engineering fundamentals, the capability to design processes that incorporate principles of pollution prevention, and an understanding of economic, environmental, and societal issues that influence intelligent technology choices. Master of Science, Master of Engineering, and PhD degree programs are available in both chemical and environmental engineering. A number of specializations are also offered. See in-depth description in the chemical engineering section.

Instituto Tecnológico y de Estudios Superiores de Monterrey, Campus Estado de México, Graduate Division, Division of Engineering and Architecture, Atizapán de Zaragoza, 52500, Mexico. Offers computer science (MCS); environmental engineering (MEE); industrial engineering (MIE); manufacturing systems (MMS); materials engineering (PhD). *Degree requirements:* For master's; for doctorate, one foreign language, dissertation required. *Entrance requirements:* For master's, interview; for doctorate, research proposal. *Application deadline:* For fall admission, 1/13 (priority date); for spring admission, 4/4. Applications are processed on a rolling basis. Application fee: 750 Mexican pesos. *Unit head:* Juan López Díaz, Headmaster, 5-326-5530, Fax: 5-326-5531, E-mail: jlopez@campus.cem.itesm.mx. *Application contact:* Lourdes Turrubiates, Admissions Officer, 5-326-5776, Fax: 5-326-5788, E-mail: lturrubi@campus.cem.itesm.mx.

Instituto Tecnológico y de Estudios Superiores de Monterrey, Campus Monterrey, Graduate and Research Division, Programs in Engineering, Monterrey, 64849, Mexico. Offers applied statistics (M Eng); artificial intelligence (PhD); automation engineering (M Eng); chemical engineering (M Eng); civil engineering (M Eng); electrical engineering (M Eng); electronic engineering (M Eng); environmental engineering (M Eng); industrial engineering (M Eng, PhD); manufacturing engineering (M Eng); mechanical engineering (M Eng); systems and quality engineering (M Eng). M Eng offered jointly with the University of Waterloo; PhD (industrial engineering) offered jointly with Texas A&M University. Part-time and evening/weekend programs available. Terminal master's awarded for partial completion of doctoral program. *Degree requirements:* For master's and doctorate, one foreign language, computer language, thesis/dissertation required. *Entrance requirements:* For master's, PAEG, TOEFL; for doctorate, GRE, TOEFL, master's in related field. *Faculty research:* Flexible manufacturing cells, materials, statistical methods, environmental prevention, control and evaluation.

Iowa State University of Science and Technology, Graduate College, College of Engineering, Department of Civil and Construction Engineering, Ames, IA 50011. Offers civil engineering (MS, PhD), including civil engineering materials, construction engineering and management, environmental engineering, geometronics, geotechnical engineering, structural engineering, transportation engineering. *Faculty:* 36 full-time, 5 part-time. *Students:* 60 full-time (14 women), 51 part-time (12 women). *Degree requirements:* For master's, thesis or alternative required, foreign language not required; for doctorate, dissertation required, foreign language not required. *Entrance requirements:* For master's and doctorate, GRE General Test (foreign students), TOEFL. *Application deadline:* For fall admission, 6/15 (priority date); for spring admission, 11/15 (priority date). Application fee: $20 ($50 for international students). Electronic applications accepted. Tuition, state resident: full-time $3,308. Tuition, nonresident: full-time $9,744. Part-time tuition and fees vary according to course load, campus/location and program. *Unit head:* Dr. Lowell F. Greimann, Chair, 515-294-2140, E-mail: cceinfo@iastate.edu. *Application contact:* Edward Kannel, Director of Graduate Education, 515-294-2861, E-mail: cceinfo@iastate.edu.

Johns Hopkins University, G. W. C. Whiting School of Engineering, Department of Geography and Environmental Engineering, Baltimore, MD 21218-2699. Offers MA, MS, MSE, PhD. *Faculty:* 16 full-time (3 women), 9 part-time (2 women). *Students:* 67 full-time (31 women); includes 4 minority (1 African American, 3 Asian Americans or Pacific Islanders), 15 international. Average age 28. 170 applicants, 38% accepted. In 1998, 11 master's, 6 doctorates awarded. Terminal master's awarded for partial completion of doctoral program. *Degree requirements:* For master's, thesis required (for some programs), foreign language not required; for doctorate, dissertation required, foreign language not required. *Entrance requirements:* For master's and doctorate, GRE General Test (combined average 1905 on three sections), TOEFL (minimum score of 560 required). *Application deadline:* For fall admission, 1/15 (priority date). Applications are processed on a rolling basis. Application fee: $50. Tuition: Full-time $23,660. Tuition and fees vary according to program. *Financial aid:* In 1998–99, 11 fellowships (averaging $12,420 per year), 17 research assistantships (averaging $11,340 per year), 3 teaching assistantships (averaging $13,230 per year) were awarded.; Federal Work-Study, grants, and institutionally-sponsored loans also available. Financial aid application deadline: 2/1. *Faculty research:* Aquatic chemistry; physical, chemical, and biological processes in aquatic systems; systems analysis and public policy; human geography; ecology; earth surface processor, hydrology. Total annual research expenditures: $1.1 million. *Unit head:* Dr. J. Hugh Ellis, Chair, 410-516-6537, Fax: 410-516-8996. *Application contact:* Dr. Edward Bouwer, 410-516-7437, Fax: 410-516-8996, E-mail: dogee@jhu.edu.

Johns Hopkins University, School of Hygiene and Public Health, Department of Environmental Health Sciences, Division of Environmental Health Engineering, Baltimore, MD 21218-2699. Offers MHS, Sc M, Dr PH, PhD, Sc D. *Degree requirements:* For master's, thesis required (for some programs), foreign language not required; for doctorate, dissertation, 1 year full-time residency, oral and written exams required, foreign language not required. *Entrance requirements:* For master's, GRE General Test, TOEFL (minimum score of 550 required); for doctorate, GRE General Test, TOEFL (minimum score of 580 required). *Application deadline:* For fall admission, 2/1 (priority date). Applications are processed on a rolling basis. Application fee: $60. Electronic applications accepted. Tuition: Full-time $23,660; part-time $493 per unit. Full-time tuition and fees vary according to degree level, campus/location and program. *Financial aid:* Federal Work-Study, institutionally-sponsored loans, and scholarships available. Aid available to part-time students. Financial aid application deadline: 4/15. *Faculty research:* Industrial hygiene and safety, biofluid mechanics, environmental microbiology, aerosol science, microbiological water hazards. *Unit head:* Dr. Patrick Breysse, Director, 410-955-3602, E-mail: pbreysse@jhsph.edu.

Lamar University, College of Graduate Studies, College of Engineering, Department of Civil Engineering, Program in Environmental Engineering, Beaumont, TX 77710. Offers MS. Part-time programs available. *Faculty:* 5 full-time (0 women). *Students:* 9 full-time (2 women), (all international). In 1998, 1 degree awarded (100% found work related to degree). *Degree requirements:* For master's, thesis optional, foreign language not required. *Entrance requirements:* For master's, GRE General Test (minimum combined score of 950 required), TOEFL (minimum score of 525 required). *Average time to degree:* Master's–1.5 years full-time, 3 years part-time. *Application deadline:* For fall admission, 5/15 (priority date); for spring admission, 10/1 (priority date). Applications are processed on a rolling basis. Application fee: $0. *Financial aid:* In 1998–99, 1 fellowship was awarded. Financial aid application deadline:4/1. *Faculty research:* Coastal engineering, lake hydrodynamics, contamination transport. Total annual research expenditures: $20,000. *Application contact:* Sandy Drane, Coordinator, International Students and Graduate Studies, 409-880-8349, Fax: 409-880-8414, E-mail: dranesl@lub002.lamar.edu.

Lehigh University, College of Engineering and Applied Science, Department of Civil and Environmental Engineering, Bethlehem, PA 18015-3094. Offers M Eng, MS, PhD. Part-time programs available. *Faculty:* 18 full-time (2 women), 2 part-time (0 women). *Students:* 70 full-time (7 women), 5 part-time (1 woman); includes 9 minority (6 African Americans, 1 Asian American or Pacific Islander, 2 Hispanic Americans), 31 international. Average age 22. 90 applicants, 74% accepted. In 1998, 24 master's, 3 doctorates awarded. Terminal master's awarded for partial completion of doctoral program. *Degree requirements:* For master's, thesis optional, foreign language not required; for doctorate, dissertation required, foreign language not required. *Entrance requirements:* For master's and doctorate, GRE General Test (score in 75th percentile or higher required), TOEFL (minimum score of 550 required). *Average time to degree:* Master's–2 years full-time; doctorate–4.7 years full-time. *Application deadline:* For fall admission, 7/15; for spring admission, 12/1. Applications are processed on a rolling basis. Application fee: $40. *Financial aid:* In 1998–99, 27 students received aid, including 7 fellowships, 12 research assistantships, 8 teaching assistantships; tuition waivers (partial) also available. Financial aid application deadline: 1/15. *Faculty research:* Structural and geotechnical engineering, water resources and coastal engineering. Total annual research expenditures: $4 million. *Unit head:* Dr. Arup K. SenGupta, Chairman, 610-758-3538, Fax: 610-758-6405, E-mail: aks@lehigh.edu. *Application contact:* Dr. Prisca Vidanage, Graduate Coordinator, 610-758-3530, Fax: 610-758-6405, E-mail: pmv1@lehigh.edu.

Louisiana State University and Agricultural and Mechanical College, Graduate School, College of Engineering, Department of Civil and Environmental Engineering, Baton Rouge, LA 70803. Offers environmental engineering (MSCE, PhD); geotechnical engineering (MSCE, PhD); structural engineering and mechanics (MSCE, PhD); transportation engineering (MSCE, PhD); water resources (MSCE, PhD). Part-time programs available. *Faculty:* 27 full-time (1 woman), 1 part-time (0 women). *Students:* 59 full-time (14 women), 24 part-time (2 women); includes 5 minority (1 African American, 3 Asian Americans or Pacific Islanders, 1 Hispanic American), 48 international. Average age 31. 188 applicants, 31% accepted. In 1998, 13 master's, 5 doctorates awarded. *Degree requirements:* For master's, thesis optional, foreign language not required; for doctorate, one foreign language (computer language can substitute), dissertation required. *Entrance requirements:* For master's and doctorate, GRE General Test, TOEFL, minimum GPA of 3.0. *Application deadline:* For fall admission, 1/25 (priority date). Applications are processed on a rolling basis. Application fee: $25. *Financial aid:* In 1998–99, 4 fellowships, 34 research assistantships with partial tuition reimbursements, 5 teaching assistantships with partial tuition reimbursements were awarded.; career-related internships or fieldwork, institutionally-sponsored loans, and scholarships also available. Financial aid application deadline: 3/1. *Faculty research:* Solid waste management, electrokinetics, composite structures, transportation planning, river mechanics. *Unit head:* Dr. Ronald F. Malone, Acting Chair, 225-388-8666, Fax: 225-388-8652, E-mail: rmalone@unix1.sncc.lsu.edu. *Application contact:* Dr. Vijaya K. A. Gopu, Graduate Coordinator, 225-388-8442, E-mail: cegopu@eng.lsu.edu.

See in-depth description on page 589.

Manhattan College, Graduate Division, School of Engineering, Program in Environmental Engineering, Riverdale, NY 10471. Offers ME, MS. Part-time and evening/weekend programs available. *Degree requirements:* For master's, computer language, thesis or alternative required. *Entrance requirements:* For master's, GRE, TOEFL, minimum GPA of 3.0.

Marquette University, Graduate School, College of Engineering, Department of Civil and Environmental Engineering, Milwaukee, WI 53201-1881. Offers construction and public works management (MS, PhD); environmental/water resources engineering (MS, PhD); structural/geotechnical engineering (MS, PhD); transportational planning and engineering (MS, PhD). Part-time and evening/weekend programs available. *Faculty:* 12 full-time (0 women), 1 part-time (0 women). *Students:* 17 full-time (7 women), 32 part-time (5 women); includes 5 minority (1 African American, 4 Asian Americans or Pacific Islanders), 12 international. Average age 30. 45 applicants, 51% accepted. In 1998, 16 master's awarded (80% found work related to degree, 20% continued full-time study); 1 doctorate awarded (100% found work related to degree). Terminal master's awarded for partial completion of doctoral program. *Degree requirements:* For master's, thesis or alternative, comprehensive exam required, foreign language not required; for doctorate, dissertation required, foreign language not required. *Entrance requirements:* For master's, TOEFL (minimum score of 550 required); for doctorate, GRE General Test, TOEFL (minimum score of 550 required). *Average time to degree:* Master's–2 years full-time, 5 years part-time; doctorate–5 years full-time. *Application deadline:* For fall admission, 6/1 (priority date). Applications are processed on a rolling basis. Application fee: $40. Tuition: Part-time $510 per credit hour. Tuition and fees vary according to program. *Financial aid:* In 1998–99, 20 students received aid, including 5 research assistantships, 13 teaching assistantships; fellowships, Federal Work-Study, institutionally-sponsored loans, scholarships, and tuition waivers (full and partial) also available. Aid available to part-time students. Financial aid application deadline: 2/15. *Faculty research:* Highway safety, highway performance, and intelligent transportation systems; surface mount technology; watershed management. Total annual research expenditures: $1.2 million. *Unit head:* Dr. Thomas H. Wenzel, Chairman, 414-288-7030, Fax: 414-288-7521, E-mail: wenzelt@vms.csd.mu.edu.

Massachusetts Institute of Technology, School of Engineering, Department of Civil and Environmental Engineering, Cambridge, MA 02139-4307. Offers M Eng, SM, PhD, Sc D, CE, EE. *Faculty:* 36 full-time (6 women). *Students:* 245 full-time (62 women), 3 part-time (1 woman); includes 19 minority (3 African Americans, 8 Asian Americans or Pacific Islanders, 8 Hispanic Americans), 124 international. Average age 27. 503 applicants, 47% accepted. In 1998, 103 master's, 24 doctorates awarded. *Degree requirements:* For master's and other advanced degree, thesis required; for doctorate and other advanced degree, dissertation, comprehensive exams required. *Entrance requirements:* For master's, GRE General Test, TOEFL; for doctorate and other advanced degree, GRE General Test, TOEFL. *Average time to degree:* Master's–2 years full-time; doctorate–5 years full-time. *Application deadline:* For fall admission, 1/15; for spring admission, 11/1. Application fee: $55. *Financial aid:* In 1998–99, 187 students received aid, including 58 fellowships, 145 research assistantships, 36 teaching assistantships; career-related internships or fieldwork, Federal Work-Study, and institutionally-sponsored loans also available. Financial aid application deadline: 1/15; financial aid applicants required to submit FAFSA. *Faculty research:* Environmental chemistry and biology, environmental fluid dynamics and hydrodynamics, geoenvironment and geotechnology, surface and groundwater hydrology, materials and structures. Total annual research expenditures: $12.3 million. *Unit head:* Dr. Rafael L. Bras, Head, 617-253-2117, E-mail: rlbras@mit.edu. *Application contact:* Graduate Admissions Coordinator, 617-253-7119, Fax: 617-258-6775, E-mail: ceed@mit.edu.

McGill University, Faculty of Graduate Studies and Research, Faculty of Engineering, Department of Civil Engineering and Applied Mechanics, Program in Environmental Engineering and Water Resources Management, Montréal, PQ H3A 2T5, Canada. Offers M Eng, M Sc, PhD. Part-time and evening/weekend programs available. *Degree requirements:* For master's, computer language required, thesis optional, foreign language not required; for doctorate,

Environmental Engineering

McGill University (continued)
computer language, dissertation required, foreign language not required. *Entrance requirements:* For master's, TOEFL (minimum score of 550 required), minimum GPA of 3.0; for doctorate, TOEFL (minimum score of 580 required). *Faculty research:* Biological/biochemical treatment, physical and chemical treatment, UV disinfection, site remediation, water resource modelling and optimization.

McGill University, Faculty of Graduate Studies and Research, Faculty of Engineering, Department of Civil Engineering and Applied Mechanics, Program in Geotechnical and Geoenvironmental Engineering, Montreal, PQ H3A 2K6, Canada. Offers M Eng, M Sc, PhD. Part-time and evening/weekend programs available. *Degree requirements:* For master's, computer language required, thesis optional, foreign language not required; for doctorate, computer language, dissertation required, foreign language not required. *Entrance requirements:* For master's, TOEFL (minimum score of 550 required), minimum GPA of 3.0; for doctorate, TOEFL (minimum score of 580 required). *Faculty research:* Geomechanics, mechanics of frozen soils, soil-foundation interaction, groundwater contamination and remediation, flow in fractional media, hydrothermomechanical processes in geomaterials.

Memorial University of Newfoundland, School of Graduate Studies, Interdisciplinary Program in Environmental Engineering and Applied Science, St. John's, NF A1C 5S7, Canada. Offers MA Sc. *Students:* 2 full-time (1 woman), 8 part-time. 6 applicants, 0% accepted. In 1998, 20 degrees awarded. *Degree requirements:* For master's, project required, thesis not required. *Entrance requirements:* For master's, honors B Sc or 2nd class B Eng. *Application deadline:* For fall admission, 5/30. Application fee: $40. *Unit head:* Dr. Tahir Husain, Chair, 709-737-8900, E-mail: thusain@engr.mun.ca. *Application contact:* Dr. M. Haddara, Associate Dean, Faculty of Engineering, 709-737-8900, E-mail: mhaddara@engr.mun.ca.

Michigan State University, Graduate School, College of Engineering, Department of Civil and Environmental Engineering, East Lansing, MI 48824-1020. Offers civil engineering (MS, PhD); civil engineering-environmental toxicology (PhD); civil engineering-urban studies (MS); environmental engineering (MS, PhD), including environmental toxicology (PhD); environmental engineering-environmental toxicology (PhD); environmental engineering-urban studies (MS). Part-time programs available. *Faculty:* 17 full-time (2 women). *Students:* 44 full-time (11 women), 47 part-time (9 women); includes 13 minority (4 African Americans, 5 Asian Americans or Pacific Islanders, 4 Hispanic Americans), 52 international. Average age 28. 147 applicants, 48% accepted. In 1998, 35 master's, 9 doctorates awarded. Terminal master's awarded for partial completion of doctoral program. *Degree requirements:* For master's, foreign language and thesis not required; for doctorate, dissertation required, foreign language not required. *Entrance requirements:* For master's, GRE, TOEFL (minimum score of 570 required), minimum GPA of 3.0; for doctorate, GRE, TOEFL (minimum score of 570 required), minimum GPA of 3.0, MS. *Average time to degree:* Master's–2 years full-time, 3 years part-time; doctorate–4 years full-time, 6 years part-time. *Application deadline:* For fall admission, 5/31; for spring admission, 10/31. Applications are processed on a rolling basis. Application fee: $30 ($40 for international students). *Financial aid:* In 1998–99, 31 research assistantships (averaging $12,647 per year), 9 teaching assistantships with tuition reimbursements (averaging $12,058 per year) were awarded.; fellowships Financial aid application deadline: 2/15; financial aid applicants required to submit FAFSA. *Faculty research:* Highway safety, hazardous waste management, pavement design, concrete materials, waste water treatment. Total annual research expenditures: $1.9 million. *Unit head:* Dr. Ronald S. Harichandran, Chairperson, 517-355-5107, Fax: 517-432-1827, E-mail: harichan@egr.msu.edu. *Application contact:* Dr. Francis X. McKelvey, Graduate Coordinator, 517-355-5107, Fax: 517-432-1827, E-mail: mckelvey@pilot.msu.edu.

See in-depth description on page 593.

Michigan Technological University, Graduate School, College of Engineering, Department of Civil and Environmental Engineering, Program in Environmental Engineering, Houghton, MI 49931-1295. Offers MS, PhD. Part-time programs available. *Faculty:* 27 full-time (4 women). *Students:* 20 full-time (12 women); includes 1 minority (Asian American or Pacific Islander), 3 international. Average age 25. 32 applicants, 69% accepted. In 1998, 2 master's awarded. *Degree requirements:* For master's, foreign language and thesis not required; for doctorate, dissertation required, foreign language not required. *Entrance requirements:* For master's, GRE General Test, TOEFL (minimum score of 600 required); for doctorate, GRE General Test (combined average 1848 on three sections), TOEFL (minimum score of 600 required). *Average time to degree:* Master's–3.7 years full-time. *Application deadline:* For fall admission, 3/15 (priority date). Applications are processed on a rolling basis. Application fee: $30 ($35 for international students). Tuition, state resident: full-time $4,377. Tuition, nonresident: full-time $9,108. Required fees: $126. Tuition and fees vary according to course load. *Financial aid:* In 1998–99, 2 fellowships (averaging $1,150 per year), 8 research assistantships (averaging $8,471 per year), 6 teaching assistantships (averaging $5,934 per year) were awarded.; Federal Work-Study, institutionally-sponsored loans, and unspecified assistantships also available. Aid available to part-time students. Financial aid application deadline: 4/1; financial aid applicants required to submit FAFSA. *Unit head:* Dr. C. Robert Baillod, Chair, Department of Civil and Environmental Engineering, 906-487-2520, Fax: 906-487-2943, E-mail: baillod@mtu.edu.

Milwaukee School of Engineering, Department of Architectural Engineering and Building Construction, Program in Environmental Engineering, Milwaukee, WI 53202-3109. Offers MS. Part-time and evening/weekend programs available. *Faculty:* 6 part-time (2 women). *Students:* 13 full-time (8 women), 13 part-time (4 women); includes 1 minority (Hispanic American) Average age 25. *Degree requirements:* For master's, thesis or alternative, design project required, foreign language not required. *Entrance requirements:* For master's, BS in engineering or science. *Application deadline:* For fall admission, 8/15 (priority date); for spring admission, 2/1. Applications are processed on a rolling basis. Application fee: $30. Electronic applications accepted. *Financial aid:* Research assistantships, career-related internships or fieldwork available. Aid available to part-time students. *Unit head:* Dr. Deborah Jackman, Director, 414-277-7472, Fax: 414-277-7479, E-mail: jackman@msoe.edu. *Application contact:* Helen Boomsma, Director, Lifelong Learning Institute, 800-321-6763, Fax: 414-277-7475, E-mail: boomsma@msoe.edu.

Announcement: Milwaukee School of Engineering's MS in environmental engineering degree program develops expertise in environmental systems design and environmental management issues. It is designed for graduate engineers who want to utilize a total systems approach to solve environmental problems. Other MS degrees offered by MSOE include architectural engineering, engineering, perfusion, engineering management, and medical informatics.

Montana State University–Bozeman, College of Graduate Studies, College of Engineering, Department of Chemical Engineering, Bozeman, MT 59717. Offers chemical engineering (MS); engineering (PhD); environmental engineering (MS, PhD); project engineering and management (MPEM). Part-time programs available. *Students:* 17 full-time (6 women), 12 part-time (8 women). *Degree requirements:* For master's, thesis or alternative required, foreign language not required; for doctorate, dissertation required, foreign language not required. *Entrance requirements:* For master's and doctorate, GRE General Test, TOEFL (minimum score of 550 required). *Application deadline:* For fall admission, 6/1 (priority date); for spring admission, 11/1. Applications are processed on a rolling basis. Application fee: $50. *Unit head:* Dr. John T. Sears, Head, 406-994-2221, Fax: 406-994-5308, E-mail: cheme@coe.montana.edu.

Montana State University–Bozeman, College of Graduate Studies, College of Engineering, Department of Civil Engineering, Bozeman, MT 59717. Offers civil engineering (MS); construction engineering management (MCEM); engineering (PhD); environmental engineering (MS). Part-time programs available. *Students:* 28 full-time (5 women), 17 part-time (2 women); includes 4 minority (3 Asian Americans or Pacific Islanders, 1 Native American) *Degree requirements:* For master's, thesis or alternative required, foreign language not required; for doctorate, dissertation required, foreign language not required. *Entrance requirements:* For master's and doctorate, GRE General Test, TOEFL (minimum score of 550 required). *Application deadline:* For fall admission, 6/1 (priority date); for spring admission, 11/1. Applications are processed on a rolling basis. Application fee: $50. *Unit head:* Dr. Donald Rabern, Head, 406-994-2111, E-mail: cedept@ce.montana.edu.

Montana Tech of The University of Montana, Graduate School, Environmental Engineering Program, Butte, MT 59701-8997. Offers MS. Part-time programs available. Postbaccalaureate distance learning degree programs offered (minimal on-campus study). *Faculty:* 6 full-time (1 woman). *Students:* 15 full-time (5 women), 5 part-time; includes 1 minority (Asian American or Pacific Islander) 25 applicants, 72% accepted. In 1998, 9 degrees awarded. *Degree requirements:* For master's, thesis required, foreign language not required. *Entrance requirements:* For master's, GRE General Test, TOEFL (minimum score of 525 required), minimum B average. *Application deadline:* For fall admission, 4/1 (priority date); for spring admission, 10/1 (priority date). Applications are processed on a rolling basis. Application fee: $30. Tuition, state resident: full-time $3,211; part-time $162 per credit hour. Tuition, nonresident: full-time $9,883; part-time $440 per credit hour. International tuition: $15,500 full-time. *Financial aid:* In 1998–99, 13 students received aid, including 11 research assistantships with partial tuition reimbursements available (averaging $6,490 per year), 2 teaching assistantships with partial tuition reimbursements available (averaging $3,400 per year); career-related internships or fieldwork, Federal Work-Study, institutionally-sponsored loans, and tuition waivers (full and partial) also available. Aid available to part-time students. Financial aid application deadline: 4/1; financial aid applicants required to submit FAFSA. *Faculty research:* Air diffusion, modeling, air pollution control, wetlands, water pollution control, bioremediation. Total annual research expenditures: $3.3 million. *Unit head:* Dr. Kumar Ganesan, Department Head, 406-496-4239, Fax: 406-496-4133, E-mail: kganesan@mtech.edu. *Application contact:* Cindy Dunstan, Administrative Assistant, 406-496-4128, Fax: 406-496-4334, E-mail: cdunstan@mtech.edu.

New Jersey Institute of Technology, Office of Graduate Studies, Department of Civil and Environmental Engineering, Program in Environmental Engineering, Newark, NJ 07102-1982. Offers MS, PhD. Part-time and evening/weekend programs available. Terminal master's awarded for partial completion of doctoral program. *Degree requirements:* For master's, thesis or alternative required, foreign language not required; for doctorate, dissertation, residency required, foreign language not required. *Entrance requirements:* For master's, GRE General Test (minimum score of 450 on verbal section, 600 on quantitative, 550 on analytical required); for doctorate, GRE General Test (minimum score of 450 on verbal section, 600 on quantitative, 550 on analytical required), minimum graduate GPA of 3.5. Electronic applications accepted. *Faculty research:* Water resources engineering, solid and hazardous waste management.

New Mexico Institute of Mining and Technology, Graduate Studies, Department of Mineral and Environmental Engineering, Socorro, NM 87801. Offers environmental engineering (MS), including air quality engineering and science, hazardous waste engineering, water quality engineering and science; mineral engineering (MS). *Faculty:* 11 full-time (1 woman). *Students:* 12 full-time (3 women); includes 8 minority (all Hispanic Americans), 4 international. Average age 30. 13 applicants, 46% accepted. In 1998, 2 degrees awarded. *Degree requirements:* For master's, thesis required, foreign language not required. *Entrance requirements:* For master's, GRE General Test, TOEFL (minimum score of 540 required). *Average time to degree:* Master's–3 years full-time. *Application deadline:* For fall admission, 3/1 (priority date); for spring admission, 6/1. Applications are processed on a rolling basis. Application fee: $16. *Financial aid:* In 1998–99, 3 research assistantships (averaging $9,670 per year), 3 teaching assistantships (averaging $9,670 per year) were awarded.; fellowships, Federal Work-Study and institutionally-sponsored loans also available. Financial aid application deadline: 3/1; financial aid applicants required to submit CSS PROFILE or FAFSA. *Faculty research:* Rock mechanics, geological engineering, mining problems, blasting, shock waves. *Unit head:* Dr. Clinton P. Richardson, Chair, 505-835-5346, Fax: 505-835-5252. *Application contact:* Dr. David B. Johnson, Dean of Graduate Studies, 505-835-5513, Fax: 505-835-5476, E-mail: graduate@nmt.edu.

New Mexico State University, Graduate School, College of Engineering, Department of Civil, Agricultural and Geological Engineering, Las Cruces, NM 88003-8001. Offers civil engineering (MSCE, PhD); environmental engineering (MS Env E). *Accreditation:* ABET (one or more programs are accredited). Part-time programs available. *Faculty:* 14 full-time (1 woman). *Students:* 33 full-time (13 women), 22 part-time (5 women); includes 16 minority (14 Hispanic Americans, 2 Native Americans), 10 international. *Degree requirements:* For master's, thesis required (for some programs), foreign language not required; for doctorate, dissertation required. *Entrance requirements:* For doctorate, BS in engineering, minimum GPA of 3.0. *Application deadline:* For fall admission, 7/1 (priority date); for spring admission, 11/1. Applications are processed on a rolling basis. Application fee: $15 ($35 for international students). Electronic applications accepted. Tuition, state resident: full-time $2,682; part-time $112 per credit. Tuition, nonresident: full-time $8,376; part-time $349 per credit. Tuition and fees vary according to course load. *Unit head:* Dr. Kenneth R. White, Head, 505-646-3801, Fax: 505-646-6049, E-mail: krwhite@nmsu.edu.

New York Institute of Technology, Graduate Division, School of Engineering and Technology, Program in Environmental Technology, Old Westbury, NY 11568-8000. Offers MS, Certificate. Part-time and evening/weekend programs available. *Students:* 5 full-time (0 women), 92 part-time (21 women); includes 18 minority (11 African Americans, 6 Asian Americans or Pacific Islanders, 1 Hispanic American), 16 international. Average age 35. 33 applicants, 61% accepted. In 1998, 41 degrees awarded. *Degree requirements:* For master's, thesis or alternative required, foreign language not required; for degree, foreign language not required. *Entrance requirements:* For master's, minimum QPA of 2.85. *Average time to degree:* Master's–3 years part-time. *Application deadline:* For fall admission, 8/1. Applications are processed on a rolling basis. Application fee: $50. Electronic applications accepted. *Financial aid:* Fellowships, research assistantships, career-related internships or fieldwork, institutionally-sponsored loans, tuition waivers (full and partial), and unspecified assistantships available. Aid available to part-time students. *Faculty research:* Develop and test methodology to assess health risks and environmental impacts from separate sanitary sewage; introduction of technology innovation, including GIS. *Unit head:* Stanley Greenwald, Chair, 516-686-7969. *Application contact:* Glenn Berman, Executive Director of Admissions, 516-686-7519, Fax: 516-626-0419, E-mail: gberman@iris.nyit.edu.

North Dakota State University, Graduate Studies and Research, College of Engineering and Architecture, Department of Civil Engineering, Fargo, ND 58105. Offers civil engineering (MS); environmental engineering (MS); natural resource management (MS). Part-time programs available. Postbaccalaureate distance learning degree programs offered (minimal on-campus study). *Faculty:* 6 full-time (0 women), 3 part-time (0 women). *Students:* 30 full-time (3 women); includes 1 minority (Asian American or Pacific Islander), 4 international. *Degree requirements:* For master's, computer language, thesis or alternative required, foreign language not required. *Entrance requirements:* For master's, TOEFL (minimum score of 525 required). *Application deadline:* For fall admission, 7/1 (priority date). Applications are processed on a rolling basis. Application fee: $25. *Unit head:* Donald A. Andersen, Chair, 701-231-7244, Fax: 701-231-6185, E-mail: danderse@badlands.nodak.edu.

Northeastern University, College of Engineering, Department of Civil and Environmental Engineering, Boston, MA 02115-5096. Offers MS, PhD. Part-time programs available. *Faculty:* 15 full-time (1 woman), 1 part-time (0 women). *Students:* 39 full-time (13 women), 89 part-time (20 women); includes 2 African Americans, 4 Asian Americans or Pacific Islanders, 1 Hispanic American, 31 international. Average age 25. 163 applicants, 70% accepted. In 1998, 23 master's, 1 doctorate awarded. Terminal master's awarded for partial completion of doctoral program. *Degree requirements:* For master's, thesis optional, foreign language not required; for doctorate, dissertation, departmental qualifying exam required, foreign language not required. *Entrance requirements:* For master's and doctorate, GRE General Test. *Average* *time to degree:* Master's–3 years full-time, 5.25 years part-time; doctorate–4 years full-time. *Application deadline:* For fall admission, 4/15. Applications are processed on a rolling basis. Application fee: $50. *Financial aid:* In 1998–99, 22 students received aid, including 4 research assistantships with full tuition reimbursements available (averaging $12,450 per year), 10 teaching assistantships with full tuition reimbursements available (averaging $12,450 per year); fellowships, career-related internships or fieldwork, Federal Work-Study, tuition waivers (full), and unspecified assistantships also available. Aid available to part-time students. Financial aid application deadline: 2/15; financial aid applicants required to submit FAFSA. *Faculty research:* Earthquake engineering, geotechnical and geoenvironmental engineering, structural engineering, transportation engineering. *Unit head:* Dr. Mishac K. Yegian, Chairman, 617-373-2445, Fax: 617-373-4419. *Application contact:* Stephen L. Gibson, Associate Director, 617-373-2711, Fax: 617-373-2501, E-mail: grad-eng@coe.edu.

Northwestern University, The Graduate School, Robert R. McCormick School of Engineering and Applied Science, Department of Civil Engineering, Evanston, IL 60208. Offers biosolid mechanics (MS, PhD); environmental health engineering (MS, PhD); geotechnical engineering (MS, PhD); health physics/radiological health (MS, PhD); project management (MPM); structural engineering (MS, PhD); structural mechanics (MS, PhD); transportation systems engineering (MS, PhD). MS and PhD admissions and degrees offered through The Graduate School. Part-time programs available. *Faculty:* 26 full-time (2 women), 1 (woman) part-time. *Students:* 110 full-time (38 women), 7 part-time (2 women); includes 5 minority (2 African Americans, 3 Asian Americans or Pacific Islanders), 68 international. Terminal master's awarded for partial completion of doctoral program. *Degree requirements:* For master's, thesis required (for some programs); for doctorate, dissertation required. *Entrance requirements:* For master's and doctorate, GRE General Test, TOEFL (minimum score of 560 required). Application fee: $50 ($55 for international students). *Unit head:* Joseph L. Schofer, Chair, 847-491-3257, Fax: 847-491-4011, E-mail: j-schofer@nwu.edu. *Application contact:* Karm Kerwell, Secretary, 847-491-3176, Fax: 847-491-4011, E-mail: k-kerwell@nwu.edu.

Ohio University, Graduate Studies, College of Engineering and Technology, Department of Civil Engineering, Athens, OH 45701-2979. Offers geotechnical and environmental engineering (MS); water resources and structures (MS). *Faculty:* 11 full-time (1 woman). *Students:* 24 full-time (5 women), 2 part-time, 13 international. *Degree requirements:* For master's, thesis optional, foreign language not required. *Entrance requirements:* For master's, minimum GPA of 3.0. *Application deadline:* For fall admission, 5/1 (priority date). Applications are processed on a rolling basis. Application fee: $30. Tuition, state resident: full-time $5,754; part-time $238 per credit hour. Tuition, nonresident: full-time $11,055; part-time $457 per credit hour. Tuition and fees vary according to course load, campus/location and program. *Unit head:* Dr. Gayle Mitchell, Chair, 740-593-1465, Fax: 740-593-0625, E-mail: gmitchell@bobcat.ent.ohiou.edu. *Application contact:* Dr. Glenn Hazen, Graduate Chair, 740-593-1469, Fax: 740-593-0625, E-mail: ghazen@bobcat.ent.ohiou.edu.

Ohio University, Graduate Studies, College of Engineering and Technology, Department of Integrated Engineering, Athens, OH 45701-2979. Offers geotechnical and environmental engineering (PhD); intelligent systems (PhD); materials processing (PhD). *Faculty:* 39 full-time (1 woman). *Students:* 9 full-time (0 women), 6 part-time; includes 1 minority (Asian American or Pacific Islander), 12 international. *Degree requirements:* For doctorate, computer language, dissertation required, foreign language not required. *Entrance requirements:* For doctorate, GRE General Test, MS in engineering or related field. *Application deadline:* For fall admission, 3/15. Applications are processed on a rolling basis. Application fee: $30. Tuition, state resident: full-time $5,754; part-time $238 per credit hour. Tuition, nonresident: full-time $11,055; part-time $457 per credit hour. Tuition and fees vary according to course load, campus/location and program. *Unit head:* Dr. Jerrel R. Mitchell, Associate Dean for Research and Graduate Studies, 740-593-1482, E-mail: mitchell@bobcat.ent.ohiou.edu.

Oklahoma State University, Graduate College, College of Engineering, Architecture and Technology, School of Civil and Environmental Engineering, Stillwater, OK 74078. Offers civil engineering (M En, MS, PhD); environmental engineering (M En, MS, PhD). *Faculty:* 16 full-time (1 woman). *Students:* 30 full-time (4 women), 68 part-time (11 women); includes 9 minority (2 African Americans, 3 Asian Americans or Pacific Islanders, 1 Hispanic American, 3 Native Americans), 23 international. Average age 32. In 1998, 28 master's, 2 doctorates awarded. *Degree requirements:* For master's, thesis or alternative required, foreign language not required; for doctorate, dissertation required, foreign language not required. *Entrance requirements:* For master's and doctorate, TOEFL (minimum score of 550 required). *Application deadline:* For fall admission, 7/1 (priority date). Application fee: $25. *Financial aid:* In 1998–99, 27 students received aid, including 7 research assistantships (averaging $9,985 per year), 20 teaching assistantships (averaging $8,532 per year); career-related internships or fieldwork, Federal Work-Study, and tuition waivers (partial) also available. Aid available to part-time students. Financial aid application deadline: 3/1. *Unit head:* Robert Hughes, Head, 405-744-5189.

Old Dominion University, College of Engineering and Technology, Department of Civil and Environmental Engineering, Norfolk, VA 23529. Offers civil engineering (ME, MS, PhD); environmental engineering (ME, MS, PhD). Part-time and evening/weekend programs available. Postbaccalaureate distance learning degree programs offered (minimal on-campus study). *Faculty:* 11 full-time. *Students:* 24 full-time (7 women), 45 part-time (10 women); includes 5 minority (3 African Americans, 1 Asian American or Pacific Islander, 1 Hispanic American), 13 international. Average age 33. In 1998, 24 master's, 3 doctorates awarded. *Degree requirements:* For master's, comprehensive exam required; for doctorate, dissertation, candidacy exam required, foreign language not required. *Entrance requirements:* For master's, GRE, TOEFL (minimum score of 550 required), minimum GPA of 3.0; for doctorate, GRE, TOEFL (minimum score of 550 required), minimum GPA of 3.5. *Application deadline:* For fall admission, 7/1; for spring admission, 10/1. Applications are processed on a rolling basis. Application fee: $30. Electronic applications accepted. *Financial aid:* In 1998–99, 32 students received aid, including 12 research assistantships (averaging $6,713 per year), 4 teaching assistantships (averaging $5,200 per year); fellowships, career-related internships or fieldwork, grants, scholarships, and tuition waivers (partial) also available. Aid available to part-time students. Financial aid application deadline: 2/15; financial aid applicants required to submit FAFSA. *Faculty research:* Structural engineering, coastal engineering, geotechnical engineering, water resources. Total annual research expenditures: $626,084. *Unit head:* Dr. Isao Ishibashi, Chair, 757-683-3753, Fax: 757-683-5354, E-mail: iishbas@cee.odu.edu. *Application contact:* Dr. Isao Ishibashi, Chair, 757-683-3753, Fax: 757-683-5354, E-mail: iishbas@cee.odu.edu.

Oregon Graduate Institute of Science and Technology, Graduate Studies, Department of Environmental Science and Engineering, Portland, OR 97291-1000. Offers ecosystem management and restoration (MS); environmental science (MS, PhD); environmental systems management (MS). Part-time programs available. *Faculty:* 9 full-time (1 woman). *Students:* 20 full-time (11 women), 4 part-time, 1 international. Average age 28. 74 applicants, 18% accepted. In 1998, 13 master's, 3 doctorates awarded. Terminal master's awarded for partial completion of doctoral program. *Degree requirements:* For master's, thesis optional, foreign language not required; for doctorate, comprehensive exam, oral defense of dissertation required. *Entrance requirements:* For master's, GRE General Test (strongly recommended), TOEFL (minimum score of 600 required); for doctorate, GRE General Test, TOEFL (minimum score of 600 required). *Average time to degree:* Master's–1.5 years full-time; doctorate–5 years full-time. Application fee: $50. Electronic applications accepted. *Financial aid:* In 1998–99, 20 students received aid, including 11 research assistantships; fellowships, teaching assistantships, Federal Work-Study and scholarships also available. Financial aid application deadline: 2/15. *Faculty research:* Air and water science, hydrogeology, estuarine and coastal modeling, environmental microbiology, contaminant transport, ecosystems. *Unit head:* Dr. James F. Pankow, Head, 503-690-1196, Fax: 503-690-1273, E-mail: pankow@ese.ogi.edu. *Application contact:* Director of Admissions, 800-685-2423, Fax: 503-690-1285, E-mail: admissions@admin.ogi.edu.

Announcement: The department offers new interdisciplinary educational programs in environmental systems management (an 18-month MS program) and ecosystem management and restoration (a 12-month MS program). For further information, visit the Web site (http://www.ese.ogi.edu/) and see the in-depth description in Book 4 in the Environmental Sciences and Management section.

Oregon State University, Graduate School, College of Forestry, Department of Forest Engineering, Corvallis, OR 97331. Offers MAIS, MF, MS, PhD. *Accreditation:* SAF (one or more programs are accredited). Part-time programs available. *Faculty:* 13 full-time (0 women). *Students:* 14 full-time (4 women). Average age 27. In 1998, 8 master's, 3 doctorates awarded. *Degree requirements:* For master's and doctorate, computer language, thesis/dissertation required, foreign language not required. *Entrance requirements:* For master's and doctorate, GRE General Test, TOEFL (minimum score of 550 required), minimum GPA of 3.0 in last 90 hours. *Application deadline:* For fall admission, 3/1. Applications are processed on a rolling basis. Application fee: $50. *Financial aid:* Fellowships, research assistantships, career-related internships or fieldwork, Federal Work-Study, and institutionally-sponsored loans available. Aid available to part-time students. Financial aid application deadline: 2/1. *Faculty research:* Timber harvesting systems, forest hydrology, slope stability, impacts of harvesting on soil and water, training of logging labor force. *Unit head:* Dr. Steven D. Teschson, Head, 541-737-4952, Fax: 541-737-4316, E-mail: tesch@ccmail.orst.edu. *Application contact:* Rayetta Beall, Office Manager, 541-737-1345, Fax: 541-737-4316, E-mail: beallr@ccmail.orst.edu.

Pennsylvania State University Great Valley School of Graduate Professional Studies, Graduate Studies and Continuing Education, College of Engineering, Program in Environmental Engineering, Malvern, PA 19355-1488. Offers M Eng. In 1998, 6 degrees awarded. Application fee: $50. *Unit head:* Dr. Lily Sehayek, Adviser, 610-648-3243. *Application contact:* 610-648-3242, Fax: 610-889-1334.

Pennsylvania State University Great Valley School of Graduate Professional Studies, Graduate Studies and Continuing Education, Intercollege Graduate Programs, Program in Environmental Pollution Control, Malvern, PA 19355-1488. Offers MEPC. Application fee: $50. *Unit head:* Dr. Lily Sehayek, Adviser, 610-648-3243. *Application contact:* 610-648-3242, Fax: 610-889-1334.

Pennsylvania State University Harrisburg Campus of the Capital College, Graduate Center, School of Science, Engineering and Technology, Program in Environmental Pollution Control, Middletown, PA 17057-4898. Offers M Eng, MEPC, MS. Evening/weekend programs available. *Students:* 8 full-time (4 women), 32 part-time (12 women). Average age 33. In 1998, 14 degrees awarded. *Degree requirements:* For master's, thesis required, foreign language not required. *Entrance requirements:* For master's, GRE General Test, TOEFL (minimum score of 560 required), minimum GPA of 2.75. *Application deadline:* For fall admission, 7/26. Application fee: $50. *Unit head:* Dr. Scott Huebner, Coordinator, 717-948-6127.

Pennsylvania State University University Park Campus, Graduate School, College of Engineering, Department of Civil and Environmental Engineering, Program in Environmental Engineering, State College, University Park, PA 16802-1503. Offers M Eng, MS, PhD. *Students:* 21 full-time (9 women), 16 part-time (4 women). In 1998, 15 master's, 4 doctorates awarded. *Degree requirements:* For master's, final paper (M Eng), oral exam and thesis (MS) required; for doctorate, dissertation, comprehensive and oral exams required. *Entrance requirements:* For master's and doctorate, GRE General Test, BS in engineering. Application fee: $50. *Financial aid:* Fellowships, research assistantships, teaching assistantships available. *Faculty research:* Physical, chemical, and biological treatment processes; reclamation and treatment of hazardous and toxic wastes; subsoil transport of pollutants. *Unit head:* Dr. Paul P. Jovanis, Head, Department of Civil and Environmental Engineering, 814-863-3084.

Polytechnic University, Brooklyn Campus, Department of Civil and Environmental Engineering, Major in Environmental Engineering, Brooklyn, NY 11201-2990. Offers MS. Part-time and evening/weekend programs available. *Students:* 8 full-time (5 women), 18 part-time (4 women); includes 4 minority (3 African Americans, 1 Hispanic American), 7 international. Average age 33. 27 applicants, 30% accepted. In 1998, 6 degrees awarded. *Degree requirements:* For master's, thesis or alternative required. *Application deadline:* Applications are processed on a rolling basis. Application fee: $45. Electronic applications accepted. *Financial aid:* Fellowships, research assistantships, teaching assistantships, institutionally-sponsored loans available. Aid available to part-time students. Financial aid applicants required to submit FAFSA. *Unit head:* Rayetta Beall, Office Manager, 541-737-1345, Fax: 541-737-4316, E-mail: beallr@ccmail.orst.edu. *Application contact:* John S. Kerge, Dean of Admissions, 718-260-3200, Fax: 718-260-3446, E-mail: admitme@poly.edu.

Polytechnic University, Farmingdale Campus, Graduate Programs, Department of Civil and Environmental Engineering, Major in Environmental Engineering, Farmingdale, NY 11735-3995. Offers MS. Average age 33. *Degree requirements:* For master's, computer language required. *Application deadline:* Applications are processed on a rolling basis. Application fee: $45. Electronic applications accepted. *Financial aid:* Institutionally-sponsored loans available. Aid available to part-time students. Financial aid applicants required to submit FAFSA. *Application contact:* John S. Kerge, Dean of Admissions, 718-260-3200, Fax: 718-260-3446, E-mail: admitme@poly.edu.

Polytechnic University, Westchester Graduate Center, Graduate Programs, Department of Civil and Environmental Engineering, Major in Environmental Engineering, Hawthorne, NY 10532-1507. Offers MS. Average age 33. *Degree requirements:* For master's, computer language, thesis required (for some programs). *Application deadline:* Applications are processed on a rolling basis. Application fee: $45. Electronic applications accepted. *Unit head:* John S. Kerge, Dean of Admissions, 718-260-3200, Fax: 718-260-3446, E-mail: admitme@poly.edu. *Application contact:* John S. Kerge, Dean of Admissions, 718-260-3200, Fax: 718-260-3446, E-mail: admitme@poly.edu.

Princeton University, Graduate School, Department of Geosciences, Princeton, NJ 08544-1019. Offers atmospheric and oceanic sciences (PhD); environmental engineering and water resources (PhD); geological and geophysical sciences (PhD). *Degree requirements:* For doctorate, dissertation required. *Entrance requirements:* For doctorate, GRE General Test, GRE Subject Test.

Princeton University, Graduate School, School of Engineering and Applied Science, Department of Civil Engineering and Operations Research, Program in Environmental Engineering and Water Resources, Princeton, NJ 08544-1019. Offers PhD. *Degree requirements:* For doctorate, dissertation required. *Entrance requirements:* For doctorate, GRE General Test, GRE Subject Test.

Rensselaer Polytechnic Institute, Graduate School, School of Engineering, Department of Environmental and Energy Engineering, Program in Environmental Engineering, Troy, NY 12180-3590. Offers M Eng, MS, PhD, MBA/M Eng. Part-time programs available. *Faculty:* 5 full-time (1 woman), 3 part-time (0 women). *Students:* 18 full-time (7 women), 9 part-time; includes 2 minority (both Asian Americans or Pacific Islanders), 7 international. 60 applicants, 38% accepted. In 1998, 13 master's, 2 doctorates awarded. *Degree requirements:* For master's, thesis required (for some programs), foreign language not required; for doctorate, dissertation required, foreign language not required. *Entrance requirements:* For master's and doctorate, GRE, TOEFL (minimum score of 550 required). *Application deadline:* For fall admission, 2/1 (priority date). Applications are processed on a rolling basis. Application fee: $35. *Financial aid:* In 1998–99, 10 students received aid, including 6 teaching assistantships; fellowships, research assistantships, career-related internships or fieldwork, institutionally-sponsored loans, and tuition waivers (full and partial) also available. Financial aid application deadline: 2/1. *Faculty research:* Water treatment, ecosystem modeling, bioremediation, sediment treatment and disposal, waste water treatment. Total annual research expenditures: $250,000. *Unit head:* Dr.

Environmental Engineering

Rensselaer Polytechnic Institute (continued)

Nicholas L. Clesceri, Director, 518-276-6416, Fax: 518-276-2080, E-mail: clescn@rpi.edu. *Application contact:* Pam Zepf, Senior Secretary, 518-276-6402, Fax: 518-276-3055, E-mail: zepf@rpi.edu.

Announcement: This department offers MS, ME, PhD, and Doctor of Engineering degrees in environmental engineering. Research focuses on water quality, including bioremediation and physicochemical techniques. Topics address disinfection byproduct formation, pathogenic protozoan fate in reservoirs, sediment contaminant fate and transport, and mathematical modeling of these and other environmental processes. E-mail: denuej@rpi.edu, WWW: http://www.eng.rpi.edu/dept/neep/public_html/

Rice University, Graduate Programs, George R. Brown School of Engineering, Department of Environmental Science and Engineering, Houston, TX 77251-1892. Offers environmental engineering (MEE, MES, MS, PhD); environmental science (MEE, MES, MS, PhD). Part-time programs available. *Degree requirements:* For master's, thesis required (for some programs), foreign language not required; for doctorate, dissertation required, foreign language not required. *Entrance requirements:* For master's and doctorate, GRE General Test, GRE Subject Test, TOEFL (minimum score of 550 required), minimum GPA of 3.0. *Faculty research:* Biology and chemistry of groundwater, pollutant fate in groundwater systems, water quality monitoring, urban storm water runoff.

Rose-Hulman Institute of Technology, Faculty of Engineering and Applied Sciences, Department of Civil Engineering, Program in Environmental Engineering, Terre Haute, IN 47803-3920. Offers MS. Part-time programs available. *Faculty:* 10 full-time (2 women). *Students:* 6 full-time (2 women), 2 international. Average age 25. 5 applicants, 80% accepted. *Degree requirements:* For master's, thesis required, foreign language not required. *Entrance requirements:* For master's, GRE, TOEFL (minimum score of 580 required), minimum GPA of 3.0. *Average time to degree:* Master's–2 years full-time. *Application deadline:* For fall admission, 2/1 (priority date). Applications are processed on a rolling basis. Application fee: $0. *Financial aid:* In 1998–99, 3 students received aid, including 2 fellowships with full and partial tuition reimbursements available (averaging $6,000 per year); research assistantships, grants, institutionally-sponsored loans, and tuition waivers (full and partial) also available. Financial aid application deadline: 2/1. *Unit head:* Pam Zepf, Senior Secretary, 518-276-6402, Fax: 518-276-3055, E-mail: zepf@rpi.edu. *Application contact:* Dr. Buck F. Brown, Dean for Research and Graduate Studies, 812-877-8403, Fax: 812-877-8102, E-mail: buck.brown@rose-hulman.edu.

Rutgers, The State University of New Jersey, New Brunswick, Graduate School, Program in Civil and Environmental Engineering, New Brunswick, NJ 08903. Offers MS, PhD. Part-time and evening/weekend programs available. *Faculty:* 12 full-time (1 woman), 6 part-time (0 women). *Students:* 31 full-time (4 women), 66 part-time (21 women); includes 18 minority (2 African Americans, 13 Asian Americans or Pacific Islanders, 3 Hispanic Americans), 24 international. Average age 24. 96 applicants, 73% accepted. In 1998, 22 master's, 3 doctorates awarded. Terminal master's awarded for partial completion of doctoral program. *Degree requirements:* For master's, thesis optional, foreign language not required; for doctorate, dissertation required, foreign language not required. *Entrance requirements:* For master's and doctorate, GRE General Test. *Average time to degree:* Master's–2.5 years full-time, 4 years part-time; doctorate–3 years full-time, 5 years part-time. *Application deadline:* For fall admission, 6/1 (priority date); for spring admission, 11/1. Applications are processed on a rolling basis. Application fee: $50. *Financial aid:* In 1998–99, 2 fellowships with full and partial tuition reimbursements, 5 research assistantships with full and partial tuition reimbursements, 5 teaching assistantships with full and partial tuition reimbursements were awarded.; Federal Work-Study, grants, scholarships, and tuition waivers (full and partial) also available. Financial aid application deadline: 3/1; financial aid applicants required to submit FAFSA. *Faculty research:* Structural mechanics, soil mechanics, environmental geotechnology, water resources, computational mechanics. *Unit head:* Dr. Nenad Gucunski, Director, 732-445-4413, Fax: 732-445-0577, E-mail: gucunski@dora.rutgers.edu. *Application contact:* Connie Dellamura, Graduate Secretary, 732-445-2232, Fax: 732-445-0577, E-mail: dellamur@rci.rutgers.edu.

South Dakota State University, Graduate School, College of Engineering, Department of Civil and Environmental Engineering, Brookings, SD 57007. Offers engineering (MS), including civil engineering, environmental engineering. *Degree requirements:* For master's, thesis, oral exam required, foreign language not required. *Entrance requirements:* For master's, TOEFL (minimum score of 520 required for civil engineering; 550 for environmental engineering). *Faculty research:* Groundwater modeling, biological wastewater treatment, corrosion control, highway materials, traffic analysis.

Stanford University, School of Engineering, Department of Civil and Environmental Engineering, Stanford, CA 94305-9991. Offers MS, PhD, Eng. *Faculty:* 26 full-time (3 women). *Students:* 199 full-time (57 women), 47 part-time (13 women); includes 53 minority (6 African Americans, 26 Asian Americans or Pacific Islanders, 20 Hispanic Americans, 1 Native American), 92 international. Average age 27. 455 applicants, 55% accepted. In 1998, 104 master's, 12 doctorates awarded. Terminal master's awarded for partial completion of doctoral program. *Degree requirements:* For master's, foreign language and thesis not required; for doctorate and Eng, dissertation, qualifying exam required, foreign language not required; for Eng, thesis required, foreign language not required. *Entrance requirements:* For master's, doctorate, and Eng, GRE General Test, TOEFL. *Application deadline:* For fall admission, 1/1. Application fee: $65 ($80 for international students). Electronic applications accepted. Tuition: Full-time $24,588. Required fees: $152. Part-time tuition and fees vary according to course load. *Financial aid:* Fellowships, research assistantships, teaching assistantships, Federal Work-Study, institutionally-sponsored loans, and course assistantships, traineeships available. Financial aid application deadline: 1/1. *Unit head:* Jeffrey Koseff, Chair, 650-725-2385, Fax: 650-725-8662, E-mail: koseff@ce.stanford.edu. *Application contact:* Graduate Admissions Coordinator, 650-725-2387.

See in-depth description on page 599.

State University of New York at Buffalo, Graduate School, School of Engineering and Applied Sciences, Department of Civil, Structural, and Environmental Engineering, Buffalo, NY 14260. Offers computational engineering and mechanics (MS, PhD); construction (M Eng, MS, PhD); geoenvironmental and geotechnical engineering (M Eng, MS, PhD); structural and earthquake engineering (M Eng, MS, PhD); water resources and environmental engineering (M Eng, MS, PhD). Part-time programs available. Postbaccalaureate distance learning degree programs offered (minimal on-campus study). *Faculty:* 25 full-time (0 women), 7 part-time (1 woman). *Students:* 99 full-time (19 women), 71 part-time (13 women); includes 9 minority (2 African Americans, 5 Asian Americans or Pacific Islanders, 1 Hispanic American, 1 Native American), 101 international. Average age 27. 252 applicants, 29% accepted. In 1998, 46 master's, 16 doctorates awarded. Terminal master's awarded for partial completion of doctoral program. *Degree requirements:* For master's, computer language, project or thesis required; for doctorate, computer language, dissertation required, foreign language not required. *Entrance requirements:* For master's and doctorate, GRE General Test (minimum combined score of 1250 required), TOEFL (minimum score of 550 required). *Application deadline:* For fall admission, 1/15 (priority date); for spring admission, 10/1. Applications are processed on a rolling basis. Application fee: $35. Tuition, state resident: full-time $5,100; part-time $213 per credit hour. Tuition, nonresident: full-time $8,416; part-time $351 per credit hour. Required fees: $870; $75 per semester. Tuition and fees vary according to course load and program. *Financial aid:* In 1998–99, 3 fellowships with full tuition reimbursements (averaging $14,700 per year), 79 research assistantships with full tuition reimbursements (averaging $10,700 per year), 30 teaching assistantships with full tuition reimbursements (averaging $10,700 per year) were awarded.; career-related internships or fieldwork, Federal Work-Study, grants, institutionally-sponsored loans, scholarships, tuition waivers (full and partial), and unspecified assistantships also available. Aid available to part-time students. Financial aid application deadline: 1/15; financial aid applicants required to submit FAFSA. *Faculty research:* Earthquake protection, environmental engineering and fluid mechanics, structural dynamics, geomechanics. Total annual research expenditures: $1.5 million. *Unit head:* Dr. Andrei M. Reinhorn, Chairman, 716-645-2114 Ext. 2419, Fax: 716-645-3733, E-mail: reinhorn@civil.eng.buffalo.edu. *Application contact:* Dr. Joe Atkinson, Director of Graduate Admissions, 716-645-2114 Ext. 2326, Fax: 716-645-3667, E-mail: atkinson@acsu.buffalo.edu.

See in-depth description on page 601.

State University of New York College of Environmental Science and Forestry, Faculty of Environmental and Resource Engineering, Syracuse, NY 13210-2779. Offers MPS, MS, PhD. Part-time programs available. *Faculty:* 24 full-time (0 women), 1 part-time (0 women). *Students:* 47 full-time (12 women), 52 part-time (10 women); includes 2 minority (1 Asian American or Pacific Islander, 1 Native American), 37 international. Average age 31. 43 applicants, 79% accepted. In 1998, 11 master's, 3 doctorates awarded. Terminal master's awarded for partial completion of doctoral program. *Degree requirements:* For master's, thesis or alternative required, foreign language not required; for doctorate, dissertation required. *Entrance requirements:* For master's and doctorate, GRE General Test (minimum combined score of 1800 on three sections required), minimum GPA of 3.0. *Application deadline:* For fall admission, 4/15 (priority date); for spring admission, 11/15. Applications are processed on a rolling basis. Application fee: $50. *Financial aid:* In 1998–99, 7 fellowships with tuition reimbursements, 19 research assistantships with tuition reimbursements, 12 teaching assistantships with tuition reimbursements were awarded.; Federal Work-Study also available. Aid available to part-time students. *Faculty research:* Forest engineering, paper science and engineering, wood products engineering. Total annual research expenditures: $1.1 million. *Unit head:* Dr. Robert Brock, Chairperson, 315-470-6510, Fax: 315-470-6958, E-mail: rbrock@suadmin.syr.edu. *Application contact:* Dr. Robert H. Frey, Dean, Instruction and Graduate Studies, 315-470-6599, Fax: 315-470-6978, E-mail: esfgrad@esf.edu.

Stevens Institute of Technology, Graduate School, Charles V. Schaefer Jr. School of Engineering, Department of Civil, Environmental, and Ocean Engineering, Program in Environmental Engineering, Hoboken, NJ 07030. Offers environmental compatibility in engineering (Certificate); environmental process (M Eng, PhD, Certificate); groundwater and soil pollution control (M Eng, PhD, Certificate); inland and coastal environmental hydrodynamics (M Eng, PhD, Certificate); water quality (Certificate). *Degree requirements:* For master's, computer language required, thesis optional, foreign language not required; for doctorate, variable foreign language requirement, computer language, dissertation required; for Certificate, computer language, project or thesis required. *Entrance requirements:* For master's, TOEFL (minimum score of 500 required); for doctorate, GRE, TOEFL (minimum score of 500 required). Electronic applications accepted.

Syracuse University, Graduate School, L. C. Smith College of Engineering and Computer Science, Department of Civil and Environmental Engineering, Syracuse, NY 13244-0003. Offers civil engineering (MS, PhD); environmental engineering (MS); hydrogeology (MS). *Faculty:* 19. *Students:* 22 full-time (7 women), 15 part-time (4 women), 17 international. Average age 30. 90 applicants, 77% accepted. In 1998, 14 degrees awarded. *Degree requirements:* For doctorate, computer language, dissertation required, foreign language not required, foreign language not required. *Entrance requirements:* For master's and doctorate, GRE General Test, GRE Subject Test. *Application deadline:* Applications are processed on a rolling basis. Application fee: $40. Tuition: Full-time $13,992; part-time $583 per credit hour. *Financial aid:* Fellowships, research assistantships, teaching assistantships, Federal Work-Study and tuition waivers (partial) available. Financial aid application deadline: 3/1. *Unit head:* Dr. Shobha Bhatia, Chair, 315-443-2311.

Temple University, Graduate School, College of Science and Technology, College of Engineering, Program in Civil and Environmental Engineering, Philadelphia, PA 19122-6096. Offers MSE. Part-time programs available. *Faculty:* 6 full-time (0 women). *Students:* 13 (2 women); includes 3 minority (2 African Americans, 1 Asian American or Pacific Islander) 1 international. 11 applicants, 64% accepted. *Degree requirements:* For master's, thesis optional, foreign language not required. *Entrance requirements:* For master's, GRE General Test, TOEFL (minimum score of 575 required). *Application deadline:* For fall admission, 7/1; for spring admission, 11/1. Applications are processed on a rolling basis. Application fee: $40. *Financial aid:* Research assistantships, teaching assistantships, Federal Work-Study available. Financial aid application deadline: 2/15. *Faculty research:* Prestressed masonry structure, recycling processes and products, finite element analysis of highways and runways. Total annual research expenditures: $118,428. *Unit head:* Dr. Philip D. Udo-Inyang, Director, 215-204-7831, Fax: 215-204-6936.

Texas A&M University, College of Engineering, Department of Civil Engineering, Program in Environmental Engineering, College Station, TX 77843. Offers M Eng, MS, D Eng, PhD. D Eng offered through the College of Engineering. *Students:* 46. *Degree requirements:* For master's, thesis (MS) required; for doctorate, dissertation (PhD), internship (D Eng) required. *Entrance requirements:* For master's and doctorate, GRE General Test, TOEFL. Application fee: $50 ($75 for international students). *Financial aid:* Fellowships, research assistantships, teaching assistantships available. Financial aid application deadline: 4/1; financial aid applicants required to submit FAFSA. *Faculty research:* Prediction and control of environmental consequences, water resources, air resources, liquid and solid waste control technology, public health and sanitation. *Unit head:* Dr. Yavuz Corapcioglu, Head, 409-845-3011, Fax: 409-862-2800, E-mail: ce-grad@tamu.edu. *Application contact:* Dr. Roy W. Hann, 409-845-2498, Fax: 409-862-2800, E-mail: ce-grad@tamu.edu.

Texas A&M University–Kingsville, College of Graduate Studies, College of Engineering, Department of Environmental Engineering, Kingsville, TX 78363. Offers ME, MS. Part-time and evening/weekend programs available. *Faculty:* 4. *Students:* 17 full-time (5 women), 15 part-time (1 woman). *Degree requirements:* For master's, computer language, thesis, comprehensive exam required, foreign language not required. *Entrance requirements:* For master's, GRE General Test (minimum combined score of 1000 required), TOEFL (minimum score of 525 required), bachelor's degree in engineering or physical science, minimum undergraduate GPA of 2.7. *Application deadline:* For fall admission, 6/1; for spring admission, 11/15. Application fee: $15 ($25 for international students). Tuition, state resident: full-time $2,062. Tuition, nonresident: full-time $7,246. *Financial aid:* Fellowships, research assistantships, teaching assistantships, career-related internships or fieldwork, institutionally-sponsored loans, and unspecified assistantships available. Financial aid application deadline: 5/15. *Faculty research:* Biodegradation of hazardous waste, air modeling, toxicology and industrial hygiene, water waste treating. *Unit head:* R. N. Finch, Coordinator, 361-593-3046.

See in-depth description on page 605.

Texas Tech University, Graduate School, College of Engineering, Department of Civil Engineering, Lubbock, TX 79409. Offers civil engineering (MSCE, PhD); environmental engineering (MENVEGR); environmental technology and management (MSETM). Part-time programs available. *Faculty:* 20 full-time (0 women), 1 part-time (0 women). *Students:* 60 full-time (15 women), 22 part-time (6 women); includes 4 minority (all Hispanic Americans), 33 international. *Degree requirements:* For master's and doctorate, computer language, thesis/dissertation required, foreign language not required. *Entrance requirements:* For master's, GRE General Test (minimum combined score of 1000 required; average 1159), minimum GPA of 3.0; for doctorate, GRE General Test (minimum combined score of 1000 required), minimum GPA of 3.0. *Application deadline:* For fall admission, 4/15 (priority date); for spring admission, 11/1 (priority date). Applications are processed on a rolling basis. Application fee: $25 ($50 for international students). Electronic applications accepted. *Unit head:* Dr. James R. McDonald, Chairman, 806-742-3523, Fax: 806-742-3488.

See in-depth description on page 607.

Tufts University, Division of Graduate and Continuing Studies and Research, Graduate School of Arts and Sciences, College of Engineering, Department of Civil and Environmental Engineering, Medford, MA 02155. Offers civil engineering (MS, PhD), including geotechnical engineering, structural engineering; environmental engineering (MS, PhD), including environmental engineering and environmental sciences, environmental geotechnology, environmental health, environmental science and management, hazardous materials management, water resources engineering. Part-time programs available. *Faculty:* 13 full-time, 10 part-time. *Students:* 99 (47 women); includes 21 minority (5 African Americans, 7 Asian Americans or Pacific Islanders, 9 Hispanic Americans) 16 international. 124 applicants, 61% accepted. In 1998, 26 master's, 1 doctorate awarded. Terminal master's awarded for partial completion of doctoral program. *Degree requirements:* For master's, thesis or alternative required, foreign language not required; for doctorate, dissertation required, foreign language not required. *Entrance requirements:* For master's and doctorate, GRE General Test, TOEFL (minimum score of 550 required). *Application deadline:* For fall admission, 2/15; for spring admission, 10/15. Applications are processed on a rolling basis. Application fee: $50. *Financial aid:* Research assistantships with full and partial tuition reimbursements, teaching assistantships with full and partial tuition reimbursements, Federal Work-Study, scholarships, and tuition waivers (partial) available. Aid available to part-time students. Financial aid application deadline: 2/15; financial aid applicants required to submit FAFSA. *Unit head:* Dr. Stephen Levine, Chair, 617-627-3211, Fax: 617-627-3994. *Application contact:* Linfield Brown, 617-627-3211, Fax: 617-627-3994.

Tulane University, School of Engineering, Department of Civil and Environmental Engineering, New Orleans, LA 70118-5669. Offers MS, MSE, PhD, Sc D. MS and PhD offered through the Graduate School. Part-time programs available. *Students:* 15 full-time (0 women), 3 part-time (1 woman); includes 3 minority (1 African American, 1 Asian American or Pacific Islander, 1 Hispanic American), 4 international. 29 applicants, 59% accepted. In 1998, 3 master's awarded. *Degree requirements:* For master's, thesis required, foreign language not required; for doctorate, 2 foreign languages (computer language can substitute for one), dissertation required. *Entrance requirements:* For master's and doctorate, GRE General Test, TOEFL, minimum B average in undergraduate course work. *Application deadline:* For fall admission, 7/1. Application fee: $35. *Financial aid:* Fellowships, research assistantships, teaching assistantships available. Financial aid application deadline: 2/1. *Unit head:* Dr. John Niklaus, Chairman, 504-865-5778. *Application contact:* Dr. E. Michaelides, Associate Dean, 504-865-5764.

Université de Sherbrooke, Faculty of Applied Sciences, Program in the Environment, Sherbrooke, PQ J1K 2R1, Canada. Offers M Env. *Degree requirements:* For master's, thesis required.

The University of Alabama, Graduate School, College of Engineering, Department of Civil and Environmental Engineering, Program in Environmental Engineering, Tuscaloosa, AL 35487. Offers MSE. Part-time programs available. Postbaccalaureate distance learning degree programs offered (no on-campus study). *Faculty:* 10 full-time (1 woman), 3 part-time (1 woman). *Degree requirements:* For master's, thesis or alternative required, foreign language not required. *Entrance requirements:* For master's, GRE General Test (minimum combined score of 1500 on three sections required), minimum GPA of 3.0 in last 60 hours. *Application deadline:* For fall admission, 7/6. Applications are processed on a rolling basis. Application fee: $25. *Financial aid:* Fellowships, research assistantships, teaching assistantships, Federal Work-Study available. *Faculty research:* Heavy metals in groundwater, hydrology, waste treatment, water treatment, water quality. *Unit head:* Dr. Daniel S. Turner, Head, Department of Civil and Environmental Engineering, 205-348-6550, Fax: 205-348-0783.

The University of Alabama at Birmingham, Graduate School, School of Engineering, Department of Civil and Environmental Engineering, Birmingham, AL 35294. Offers MSCE, PhD. Evening/weekend programs available. *Students:* 10 full-time (5 women), 12 part-time (2 women), 1 international. 22 applicants, 100% accepted. In 1998, 9 master's awarded. *Degree requirements:* For master's, thesis required (for some programs), foreign language not required; for doctorate, dissertation required. *Entrance requirements:* For master's, GRE General Test (minimum score of 500 on each section required), BS in engineering, physical sciences, life sciences, or mathematics; for doctorate, GRE General Test (minimum score of 600 on each section required), TOEFL (minimum score of 550 required), BS or MS in engineering or related field, minimum undergraduate GPA of 3.0. *Application deadline:* Applications are processed on a rolling basis. Application fee: $30 ($60 for international students). Electronic applications accepted. *Financial aid:* In 1998–99, 2 fellowships with full tuition reimbursements (averaging $9,500 per year), 11 research assistantships (averaging $1,229 per year) were awarded. *Unit head:* Dr. Fouad H. Fouad, Chair, 205-934-8430, Fax: 205-934-9855, E-mail: ffouad@uab.edu.

The University of Alabama in Huntsville, School of Graduate Studies, College of Engineering, Department of Civil and Environmental Engineering, Huntsville, AL 35899. Offers MSE. Part-time and evening/weekend programs available. *Faculty:* 6 full-time (1 woman). *Students:* 2 full-time (0 women), 15 part-time (6 women); includes 3 minority (2 African Americans, 1 Asian American or Pacific Islander), 4 international. Average age 32. 12 applicants, 67% accepted. In 1998, 2 degrees awarded. *Degree requirements:* For master's, oral and written exams required, thesis optional, foreign language not required. *Entrance requirements:* For master's, GRE General Test (minimum combined score of 1500 on three sections required), BSE, minimum GPA of 3.0. *Application deadline:* For fall admission, 7/24 (priority date); for spring admission, 11/15 (priority date). Applications are processed on a rolling basis. Application fee: $20. Tuition and fees vary according to course load. *Financial aid:* In 1998–99, 10 students received aid, including 5 research assistantships with full and partial tuition reimbursements available (averaging $9,330 per year), 4 teaching assistantships with full and partial tuition reimbursements available (averaging $7,362 per year); fellowships with full and partial tuition reimbursements available, career-related internships or fieldwork, Federal Work-Study, grants, institutionally-sponsored loans, scholarships, and tuition waivers (full and partial) also available. Aid available to part-time students. Financial aid application deadline: 4/1; financial aid applicants required to submit FAFSA. *Faculty research:* Hydrologic modeling, orbital debris impact, hydrogeology, environmental engineering, water quality control. Total annual research expenditures: $52,487. *Unit head:* Dr. William Schonberg, Chair, 256-890-6854, Fax: 256-890-6724, E-mail: wschon@eb.uah.edu.

University of Alaska Fairbanks, Graduate School, College of Science, Engineering and Mathematics, Department of Civil Engineering, Program in Environmental Quality Engineering, Fairbanks, AK 99775-7480. Offers MS. *Faculty:* 2 full-time (0 women). *Students:* 4 full-time (1 woman), 6 part-time (2 women); includes 2 minority (1 Asian American or Pacific Islander, 1 Hispanic American) Average age 29. 3 applicants, 100% accepted. In 1998, 5 degrees awarded. *Degree requirements:* For master's, thesis or alternative, comprehensive exam required, foreign language not required. *Entrance requirements:* For master's, GRE General Test, TOEFL (minimum score of 550 required). *Application deadline:* For fall admission, 8/1. Applications are processed on a rolling basis. Application fee: $35. *Financial aid:* Research assistantships available. *Faculty research:* Waste treatment in arctic, oil spill microbiology. *Unit head:* Dr. Lufti Raad, Head, Department of Civil Engineering, 907-474-7241.

University of Alberta, Faculty of Graduate Studies and Research, Department of Civil and Environmental Engineering, Edmonton, AB T6G 2E1, Canada. Offers construction engineering and management (M Eng, M Sc, PhD); environmental engineering (M Eng, M Sc, PhD); environmental science (M Sc, PhD); geoenvironmental engineering (M Eng, M Sc, PhD); geotechnical engineering (M Sc); geotechnical engineering (M Eng, PhD); mining engineering (M Eng, M Sc, PhD); petroleum engineering (M Eng, M Sc, PhD); structural engineering (M Eng, M Sc, PhD); water resources (M Eng, M Sc, PhD). Part-time programs available. *Degree requirements:* For master's, thesis required (for some programs), foreign language not required; for doctorate, dissertation required, foreign language not required. *Faculty research:* Mining.

The University of Arizona, Graduate College, College of Engineering and Mines, Department of Chemical and Environmental Engineering, Tucson, AZ 85721. Offers chemical engineering (MS, PhD); environmental engineering (MS, PhD). Part-time programs available. *Faculty:* 17. *Students:* 48 full-time (12 women), 11 part-time (1 woman); includes 4 minority (1 Asian American or Pacific Islander, 3 Hispanic Americans), 25 international. Average age 30. 77 applicants, 69% accepted. In 1998, 13 master's, 2 doctorates awarded. *Degree requirements:* For master's, thesis required, foreign language not required; for doctorate, dissertation, comprehensive and departmental qualifying exams required, foreign language not required. *Entrance requirements:* For master's and doctorate, TOEFL (minimum score of 550 required), minimum GPA of 3.0. *Application deadline:* For fall admission, 3/1. Applications are processed on a rolling basis. Application fee: $35. *Financial aid:* Fellowships, research assistantships, teaching assistantships, institutionally-sponsored loans and scholarships available. Financial aid application deadline: 6/1. *Faculty research:* Energy and environment–hazardous waste incineration, fossil fuel combustion, processing high-purity gases and liquids, aerosol reactor theory, pharmacokinetics. *Unit head:* Dr. Thomas Peterson, Head, 520-621-2591. *Application contact:* Wendy Haley, Graduate Secretary, 520-621-6052, Fax: 520-621-6048.

University of Arkansas, Graduate School, College of Engineering, Department of Civil Engineering, Program in Environmental Engineering, Fayetteville, AR 72701-1201. Offers MS En E, MSE. *Students:* 4 full-time (2 women). 12 applicants, 50% accepted. In 1998, 4 degrees awarded. *Degree requirements:* For master's, thesis optional, foreign language not required. Application fee: $40 ($50 for international students). Tuition, state resident: full-time $3,186. Tuition, nonresident: full-time $7,560. Required fees: $378. *Financial aid:* Career-related internships or fieldwork and Federal Work-Study available. Aid available to part-time students. Financial aid application deadline: 4/1; financial aid applicants required to submit FAFSA. *Unit head:* Dr. Robert Elliott, Chair, Department of Civil Engineering, 501-575-4954.

University of California, Berkeley, Graduate Division, College of Engineering, Department of Civil and Environmental Engineering, Berkeley, CA 94720-1500. Offers construction engineering and management (M Eng, MS, D Eng, PhD); environmental quality and environmental water resources engineering (M Eng, MS, D Eng, PhD); geotechnical engineering (M Eng, MS, D Eng, PhD); structural engineering, mechanics and materials (M Eng, MS, D Eng, PhD); transportation engineering (M Eng, MS, D Eng, PhD). *Students:* 326 full-time (97 women); includes 59 minority (3 African Americans, 42 Asian Americans or Pacific Islanders, 13 Hispanic Americans, 1 Native American), 113 international. 704 applicants, 45% accepted. In 1998, 161 master's, 4 doctorates awarded. *Degree requirements:* For master's, comprehensive exam or thesis (MS) required; for doctorate, dissertation, qualifying exam required. *Entrance requirements:* For master's, GRE General Test, minimum GPA of 3.0; for doctorate, GRE General Test, minimum GPA of 3.5. *Application deadline:* For fall admission, 2/10. Application fee: $40. *Financial aid:* Fellowships, research assistantships, teaching assistantships available. Financial aid application deadline: 12/15. *Unit head:* Dr. Adib Kanafani, Chair, 510-642-3261. *Application contact:* Mari Cook, Graduate Assistant for Admission, 510-643-8944, E-mail: mcook@ce.berkeley.edu.

See in-depth description on page 609.

University of California, Davis, Graduate Studies, College of Engineering, Program in Civil and Environmental Engineering, Davis, CA 95616. Offers M Engr, MS, D Engr, PhD, Certificate, M Engr/MBA. Part-time programs available. *Faculty:* 28 full-time (3 women), 1 part-time (0 women). *Students:* 143 full-time (46 women), 1 part-time; includes 20 minority (2 African Americans, 13 Asian Americans or Pacific Islanders, 5 Hispanic Americans), 42 international. 226 applicants, 63% accepted. In 1998, 28 master's, 5 doctorates awarded. *Degree requirements:* For master's, foreign language and thesis not required; for doctorate, dissertation required, foreign language not required. *Entrance requirements:* For master's, GRE General Test, minimum GPA of 3.0; for doctorate, GRE, minimum graduate GPA of 3.5. *Average time to degree:* Master's–1 year full-time; doctorate–4 years full-time. *Application deadline:* For fall admission, 1/15 (priority date). Applications are processed on a rolling basis. Application fee: $40. Electronic applications accepted. *Financial aid:* In 1998–99, 21 fellowships with full and partial tuition reimbursements, 69 research assistantships with full and partial tuition reimbursements, 18 teaching assistantships with full and partial tuition reimbursements were awarded.; career-related internships or fieldwork, Federal Work-Study, institutionally-sponsored loans, and tuition waivers (full and partial) also available. Aid available to part-time students. Financial aid application deadline: 1/15; financial aid applicants required to submit FAFSA. *Faculty research:* Environmental water resources, transportation, structural mechanics, structural engineering, geotechnical engineering. *Unit head:* Daniel P. Y. Chang, Chairperson, 530-752-1441, Fax: 530-752-7872. *Application contact:* Donna Douglas, Administrative Assistant, 530-752-1441, Fax: 530-752-7872.

University of California, Irvine, Office of Research and Graduate Studies, School of Engineering, Department of Civil and Environmental Engineering, Program in Environmental Engineering, Irvine, CA 92697. Offers engineering (MS, PhD). Part-time programs available. *Faculty:* 15 full-time (2 women). *Students:* 10 full-time (4 women), 2 part-time (1 woman); includes 3 minority (1 Asian American or Pacific Islander, 2 Hispanic Americans), 3 international. 26 applicants, 69% accepted. In 1998, 3 master's awarded. Terminal master's awarded for partial completion of doctoral program. *Degree requirements:* For doctorate, dissertation required, foreign language not required, foreign language not required. *Entrance requirements:* For master's, GRE General Test, minimum GPA of 3.0; for doctorate, GRE General Test. *Application deadline:* For fall admission, 1/15 (priority date). Applications are processed on a rolling basis. Application fee: $40. Electronic applications accepted. *Financial aid:* Fellowships, research assistantships, teaching assistantships, institutionally-sponsored loans and tuition waivers (full and partial) available. Financial aid application deadline: 3/2; financial aid applicants required to submit FAFSA. *Faculty research:* Environmental air and water chemistry, environmental microbiology, combustion technologies, aerosol science, transport phenomena. *Unit head:* Dr. Stanley Grant, Director, 949-824-7320, Fax: 949-824-3672, E-mail: sbgrant@uci.edu. *Application contact:* Admissions Assistant, 949-824-3562, Fax: 949-824-3440.

University of California, Los Angeles, Graduate Division, School of Engineering and Applied Science, Department of Civil and Environmental Engineering, Los Angeles, CA 90095. Offers environmental engineering (MS, PhD); geotechnical engineering (MS, PhD); structures (MS, PhD), including structural mechanics and earthquake engineering; water resource systems engineering (MS, PhD). *Faculty:* 15 full-time, 11 part-time. *Students:* 116 full-time (29 women); includes 28 minority (3 African Americans, 21 Asian Americans or Pacific Islanders, 4 Hispanic Americans), 37 international. 267 applicants, 44% accepted. In 1998, 57 master's, 15 doctorates awarded. *Degree requirements:* For master's, comprehensive exam or thesis required; for doctorate, dissertation, qualifying exams required, foreign language not required. *Entrance requirements:* For master's, GRE General Test, minimum GPA of 3.0; for doctorate, GRE General Test, minimum GPA of 3.25. *Application deadline:* For fall admission, 1/15; for spring admission, 12/1. Application fee: $40. Electronic applications accepted. *Financial aid:* In 1998–99, 31 fellowships, 26 research assistantships, 11 teaching assistantships were awarded.; Federal Work-Study, institutionally-sponsored loans, and tuition waivers (full and partial) also available. Financial aid application deadline: 1/15; financial aid applicants required to submit FAFSA. *Unit head:* Dr. Michael Stenstrom, Chair, 310-825-1408. *Application contact:* Deeona Columbia, Graduate Affairs Officer, 310-825-1851, Fax: 310-206-2222, E-mail: deeona@ea.ucla.edu.

University of California, Los Angeles, Graduate Division, School of Public Health, Program in Environmental Science and Engineering, Los Angeles, CA 90095. Offers D Env. *Students:* 37 full-time (16 women); includes 6 minority (1 African American, 4 Asian Americans or Pacific Islanders, 1 Hispanic American), 6 international. 24 applicants, 33% accepted. *Degree requirements:* For doctorate, dissertation, oral and written qualifying exams required, foreign language not required. *Entrance requirements:* For doctorate, GRE General Test (minimum combined score of 1200 required), minimum undergraduate GPA of 3.0, master's degree or equivalent in a natural science, engineering, or public health. *Application deadline:* For fall admission, 12/15. Application fee: $40. Electronic applications accepted. *Financial aid:* In 1998–99, 32 students received aid, including 26 fellowships, 26 research assistantships, 4 teaching assistantships; institutionally-sponsored loans, scholarships, and tuition waivers (full

Environmental Engineering

University of California, Los Angeles (continued)
and partial) also available. Financial aid application deadline: 3/1. *Faculty research:* Toxic and hazardous substances, air and water pollution, risk assessment/management, water resources, marine science. *Unit head:* Dr. Richard F. Ambrose, Direactor, 310-825-9901. *Application contact:* Departmental Office, 310-825-9901, E-mail: app_ese@admin.ph.ucla.edu.

See in-depth description on page 615.

University of California, Riverside, Graduate Division, College of Engineering, Department of Chemical and Enviromental Engineering, Riverside, CA 92521-0102. Offers MS, PhD. *Entrance requirements:* For master's and doctorate, GRE General Test, TOEFL. Application fee: $40. *Unit head:* Dr. Mark Matsumoto, Chair, 909-787-2423, Fax: 909-787-2425, E-mail: matsumoto@engr.ucr.edu. *Application contact:* 909-787-2423, Fax: 909-787-2425, E-mail: grad-adm@ee.ucr.edu.

University of California, Riverside, Graduate Division, College of Engineering, Department of Chemical and Environmental Engineering, Riverside, CA 92521-0102. Offers MS, PhD. *Faculty:* 8 full-time (0 women), 8 part-time (2 women). *Students:* 2 full-time (0 women), (both international). *Degree requirements:* For doctorate, dissertation required. *Entrance requirements:* For master's and doctorate, GRE General Test (minimum combined score of 1100 required), TOEFL (minimum score of 550 required), minimum GPA of 3.2 (3.5 for fellowships/teaching assistantships). *Application deadline:* For fall admission, 5/1; for winter admission, 9/1; for spring admission, 12/1. Applications are processed on a rolling basis. Application fee: $40. Electronic applications accepted. *Financial aid:* In 1998–99, 2 students received aid, including 2 teaching assistantships with partial tuition reimbursements available (averaging $13,329 per year); fellowships Financial aid application deadline: 1/5; financial aid applicants required to submit FAFSA. *Faculty research:* Bioprocessing, biodegredation, bioremediation, water/wastewater treatment, biosensors and biodetoxification, transportation emissions. Total annual research expenditures: $395,171. *Unit head:* Dr. Mark Matsumoto, Chair, 909-787-5318, Fax: 909-787-2425, E-mail: matsumoto@engr.ucr.edu. *Application contact:* Tracie Burruel, Graduate Student Affairs Assistant, 909-787-2484, Fax: 909-787-2425, E-mail: burruel@ee.urc.edu.

University of California, Santa Barbara, Graduate Division, College of Engineering, Department of Mechanical and Environmental Engineering, Santa Barbara, CA 93106. Offers MS, PhD. *Students:* 64 full-time (8 women). 145 applicants, 47% accepted. In 1998, 23 master's, 10 doctorates awarded. *Degree requirements:* For master's, thesis or alternative required, foreign language not required; for doctorate, dissertation required, foreign language not required. *Entrance requirements:* For master's and doctorate, GRE General Test, TOEFL (minimum score of 550 required). *Application deadline:* For fall admission, 6/1. Application fee: $40. *Financial aid:* Fellowships, research assistantships, teaching assistantships, career-related internships or fieldwork, Federal Work-Study, institutionally-sponsored loans, and tuition waivers (full and partial) available. Financial aid application deadline: 1/15; financial aid applicants required to submit FAFSA. *Unit head:* G. Robert Odette, Chair, 805-893-3525. *Application contact:* Linda James, Graduate Program Assistant, 805-893-2239, E-mail: linda@engineering.ucsb.edu.

University of Central Florida, College of Engineering, Department of Civil and Environmental Engineering, Program in Environmental Engineering, Orlando, FL 32816. Offers MS, MS Env E, PhD, Certificate. Part-time and evening/weekend programs available. *Faculty:* 17 full-time, 16 part-time. *Students:* 19 full-time (6 women), 26 part-time (9 women); includes 9 minority (4 Asian Americans or Pacific Islanders, 5 Hispanic Americans), 7 international. Average age 32. 26 applicants, 31% accepted. In 1998, 8 master's, 4 doctorates awarded. *Degree requirements:* For master's, thesis or alternative required, foreign language not required; for doctorate, dissertation, departmental qualifying exam, candidacy exam required, foreign language not required. *Entrance requirements:* For master's, GRE General Test (minimum combined score of 1000 required), TOEFL (minimum score of 550 required; 213 computer-based), minimum GPA of 3.0 in last 60 hours; for doctorate, GRE General Test (minimum combined score of 1100 required), TOEFL (minimum score of 550 required; 213 computer-based), minimum GPA of 3.5 in last 60 hours. *Application deadline:* For fall admission, 7/15; for spring admission, 12/15. Application fee: $20. Tuition, state resident: full-time $2,054; part-time $137 per credit. Tuition, nonresident: full-time $7,207; part-time $480 per credit. Required fees: $47 per term. *Financial aid:* In 1998–99, 26 students received aid, including 5 fellowships with partial tuition reimbursements available (averaging $3,615 per year), 11 teaching assistantships with partial tuition reimbursements available (averaging $2,232 per year); research assistantships with partial tuition reimbursements available, career-related internships or fieldwork, Federal Work-Study, institutionally-sponsored loans, tuition waivers (partial), and unspecified assistantships also available. Financial aid application deadline: 3/1; financial aid applicants required to submit FAFSA. *Unit head:* Dr. A. E. Radwan, Chair, 407-823-2841. *Application contact:* Dr. Roger L. Wayson, Coordinator, 407-823-2841.

University of Cincinnati, Division of Research and Advanced Studies, College of Engineering, Department of Civil and Environmental Engineering, Program in Environmental Engineering, Cincinnati, OH 45221-0091. Offers MS, PhD. *Accreditation:* ABET (one or more programs are accredited). *Students:* 58 full-time (18 women), 20 part-time (4 women); includes 7 minority (1 African American, 6 Asian Americans or Pacific Islanders), 51 international. In 1998, 13 master's, 7 doctorates awarded. *Degree requirements:* For master's, project or thesis required; for doctorate, one foreign language (computer language can substitute), dissertation required. *Entrance requirements:* For master's and doctorate, GRE General Test, TOEFL (minimum score of 540 required). *Average time to degree:* Master's–2.7 years full-time; doctorate–6 years full-time. *Application deadline:* For fall admission, 2/1 (priority date). Application fee: $40. *Financial aid:* Fellowships, career-related internships or fieldwork, tuition waivers (full), and unspecified assistantships available. Financial aid application deadline: 2/1. *Faculty research:* Environmental microbiology, solid-waste management, air pollution control, water pollution control, aerosols. *Unit head:* Dr. Roger L. Wayson, Coordinator, 407-823-2841. *Application contact:* Frank Weisgerber, Graduate Program Director, 513-556-3673, Fax: 513-556-2599, E-mail: fweisger@boss.cee.uc.edu.

University of Cincinnati, Division of Research and Advanced Studies, College of Medicine, Graduate Programs in Medicine, Department of Environmental Health, Cincinnati, OH 45267. Offers environmental and industrial hygiene (MS); environmental and occupational medicine (MS); environmental health (PhD); environmental hygiene science and engineering (MS, PhD); epidemiology and biostatistics (MS); occupational safety (MS); toxicology (MS, PhD). *Faculty:* 20 full-time. *Students:* 69 full-time (34 women), 66 part-time (32 women); includes 29 minority (16 African Americans, 12 Asian Americans or Pacific Islanders, 1 Hispanic American), 31 international. Terminal master's awarded for partial completion of doctoral program. *Degree requirements:* For master's, thesis required, foreign language not required; for doctorate, one foreign language, dissertation, qualifying exam required. *Entrance requirements:* For master's, GRE General Test, TOEFL, bachelor's degree in science; for doctorate, GRE General Test, TOEFL. *Application deadline:* For fall admission, 2/1 (priority date). Applications are processed on a rolling basis. Application fee: $30. *Unit head:* Dr. Marshall W. Anderson, Chairman, 513-558-5701, Fax: 513-558-4397, E-mail: marshall.anderson@uc.edu. *Application contact:* Judy Jarrell, Graduate Program Director, 513-558-1729, Fax: 513-558-4397, E-mail: judy.jarrell@uc.edu.

University of Colorado at Boulder, Graduate School, College of Engineering and Applied Science, Department of Civil, Environmental, and Architectural Engineering, Boulder, CO 80309. Offers building systems (MS, PhD); construction engineering and management (MS, PhD); environmental engineering (MS, PhD); geoenvironmental engineering (MS, PhD); geotechnical engineering (MS, PhD); structural engineering (MS, PhD); water resource engineering (MS, PhD). *Degree requirements:* For master's, thesis or alternative, comprehensive exam required; for doctorate, dissertation required. *Entrance requirements:* For master's, GRE General Test, minimum undergraduate GPA of 3.0.

See in-depth description on page 619.

University of Connecticut, Graduate School, School of Engineering, Field of Environmental Engineering, Storrs, CT 06269. Offers MS, PhD. *Degree requirements:* For doctorate, dissertation required.

University of Dayton, Graduate School, School of Engineering, Department of Civil Engineering, Dayton, OH 45469-1300. Offers engineering mechanics (MSEM); environmental engineering (MSCE); soil mechanics (MSCE); structural engineering (MSCE); transport engineering (MSCE). Part-time programs available. *Faculty:* 9 full-time (0 women), 2 part-time (0 women). *Students:* 7 full-time (1 woman), 6 part-time (1 woman); includes 1 minority (African American), 4 international. *Degree requirements:* For master's, thesis or alternative required, foreign language not required. *Entrance requirements:* For master's, TOEFL. *Application deadline:* For fall admission, 8/1. Applications are processed on a rolling basis. Application fee: $30. *Unit head:* Dr. Joseph Saliba, Chairperson, 937-229-3847. *Application contact:* Dr. Donald L. Moon, Associate Dean, 937-229-2241, Fax: 937-229-2471, E-mail: dmoon@engr.udayton.edu.

University of Delaware, College of Engineering, Department of Civil and Environmental Engineering, Program in Environmental Engineering, Newark, DE 19716. Offers MAS, MCE, PhD. *Degree requirements:* For master's and doctorate, thesis/dissertation required, foreign language not required. *Entrance requirements:* For master's and doctorate, GRE General Test, TOEFL (minimum score of 550 required). *Faculty research:* Transport phenomena, treatment of hazardous wastes, groundwater hydrology, water and wastewater treatment, contaminant removal from water and soil.

University of Detroit Mercy, College of Engineering and Science, Department of Civil and Environmental Engineering, Detroit, MI 48219-0900. Offers ME, DE. Evening/weekend programs available. *Degree requirements:* For master's, computer language required, foreign language not required; for doctorate, dissertation required. *Faculty research:* Geotechnical engineering.

See in-depth description on page 623.

University of Florida, Graduate School, College of Engineering, Department of Environmental Engineering Sciences, Gainesville, FL 32611. Offers ME, MS, PhD, Engr. *Faculty:* 20. *Students:* 94 full-time (32 women), 31 part-time (11 women); includes 15 minority (2 African Americans, 6 Asian Americans or Pacific Islanders, 7 Hispanic Americans), 10 international. 142 applicants, 61% accepted. In 1998, 34 master's, 12 doctorates awarded. Terminal master's awarded for partial completion of doctoral program. *Degree requirements:* For master's, computer language, project or thesis required; for doctorate, computer language, dissertation required, foreign language not required; for Engr, project or thesis required. *Entrance requirements:* For master's and doctorate, GRE General Test (minimum combined score of 1100 required), TOEFL (minimum score of 575 required), minimum GPA of 3.0; for Engr, GRE General Test, TOEFL. *Application deadline:* For fall admission, 6/1 (priority date); for spring admission, 11/1. Applications are processed on a rolling basis. Application fee: $20. Electronic applications accepted. *Financial aid:* In 1998–99, 72 students received aid, including 9 fellowships, 55 research assistantships, 6 teaching assistantships; career-related internships or fieldwork and unspecified assistantships also available. *Faculty research:* Air pollution, potable water supply system, water pollution control, hazardous waste, aquatic ecology and chemistry. *Unit head:* Dr. Joseph J. Delfino, Chair, 352-392-0841, Fax: 352-392-3076, E-mail: jdelf@eng.ufl.edu. *Application contact:* Doris Smithson, Graduate Secretary, 352-392-0842, Fax: 352-392-3076, E-mail: dsmit@eng.ufl.edu.

See in-depth description on page 625.

University of Guelph, Faculty of Graduate Studies, College of Physical and Engineering Science, School of Engineering, Guelph, ON N1G 2W1, Canada. Offers biological engineering (M Sc, PhD); environmental engineering (M Eng, M Sc, PhD); water resources engineering (M Eng, M Sc, PhD). Part-time programs available. *Faculty:* 18 full-time (1 woman), 29 part-time (4 women). *Students:* 54 full-time (15 women), 13 part-time (2 women); includes 22 minority (2 African Americans, 18 Asian Americans or Pacific Islanders, 2 Hispanic Americans), 7 international. *Degree requirements:* For master's, thesis required (for some programs); for doctorate, dissertation required. *Entrance requirements:* For master's, minimum B- average during previous 2 years; for doctorate, minimum B average. *Application deadline:* For fall admission, 8/1 (priority date); for winter admission, 11/1 (priority date); for spring admission, 4/1 (priority date). Applications are processed on a rolling basis. Application fee: $60. *Expenses:* Tuition and fees charges are reported in Canadian dollars. Tuition, area resident: Full-time $4,725 Canadian dollars; part-time $1,055 Canadian dollars per term. International tuition: $6,999 Canadian dollars full-time. Required fees: $295 Canadian dollars per term. *Unit head:* Dr. Lambert Otten, Director, 519-824-4120 Ext. 2043, Fax: 519-836-0227, E-mail: lotten@uoguelph.ca. *Application contact:* Dr. Ramesh P. Rudra, Graduate Coordinator, 519-824-4120 Ext. 2110, Fax: 519-836-0227, E-mail: rrudra@uoguelph.ca.

University of Houston, Cullen College of Engineering, Department of Civil and Environmental Engineering, Houston, TX 77004. Offers MCE, MS Env E, MSCE, PhD. Part-time and evening/weekend programs available. *Faculty:* 15 full-time (1 woman), 3 part-time (0 women). *Students:* 31 full-time (7 women), 27 part-time (7 women); includes 9 minority (1 African American, 3 Asian Americans or Pacific Islanders, 5 Hispanic Americans), 33 international. Average age 29. 65 applicants, 42% accepted. In 1998, 9 master's, 3 doctorates awarded (100% found work related to degree). Terminal master's awarded for partial completion of doctoral program. *Degree requirements:* For master's, thesis required (for some programs), foreign language not required; for doctorate, dissertation, departmental qualifying exam required, foreign language not required. *Entrance requirements:* For master's and doctorate, GRE General Test, TOEFL. *Application deadline:* For fall admission, 7/3 (priority date); for spring admission, 12/4. Applications are processed on a rolling basis. Application fee: $25 ($75 for international students). *Financial aid:* In 1998–99, 26 students received aid, including 17 research assistantships, 9 teaching assistantships; career-related internships or fieldwork and Federal Work-Study also available. Financial aid application deadline: 4/1. *Faculty research:* Structural engineering and construction materials, geotechnical engineering and deep foundation, hydraulic engineering and wave mechanics, water and soil treatment. Total annual research expenditures: $2.1 million. *Unit head:* Dr. Dennis Clifford, Chairman, 713-743-4250, Fax: 713-743-4260, E-mail: daclifford@uh.edu. *Application contact:* Charlene Holliday, Graduate Analyst, 713-743-4254, Fax: 713-743-4260, E-mail: wholliday@uh.edu.

See in-depth description on page 627.

University of Houston, Cullen College of Engineering, Program in Environmental Engineering, Houston, TX 77004. Offers MS, PhD. Part-time and evening/weekend programs available. *Students:* 22 full-time (8 women), 22 part-time (8 women); includes 6 minority (1 African American, 5 Asian Americans or Pacific Islanders), 16 international. Average age 31. 73 applicants, 41% accepted. In 1998, 19 master's awarded. Terminal master's awarded for partial completion of doctoral program. *Degree requirements:* For master's, thesis required (for some programs), foreign language not required; for doctorate, dissertation, departmental qualifying exam required, foreign language not required. *Entrance requirements:* For master's and doctorate, GRE General Test, TOEFL. *Application deadline:* For fall admission, 3/15. Applications are processed on a rolling basis. Application fee: $25 ($75 for international students). *Financial aid:* Research assistantships, teaching assistantships, career-related internships or fieldwork, scholarships, and tuition waivers (partial) available. Financial aid application deadline: 3/15. *Faculty research:* Modeling contaminant transport in soil, advanced treatment processes for removing contaminants from water and soil, disinfection byproducts, bioremediation. Total annual research expenditures: $920,000. *Unit head:* Dr. Theodore G. Cleveland, Director, 713-743-4250, Fax: 713-743-4260. *Application contact:* Charlene Holliday, Graduate Analyst, 713-743-4254, Fax: 713-743-4260, E-mail: wholliday@uh.edu.

University of Illinois at Urbana–Champaign, Graduate College, College of Engineering, Department of Civil and Environmental Engineering, Urbana, IL 61801. Offers civil engineering

(MS, PhD); environmental engineering and environmental science (MS, PhD), including environmental engineering, environmental science. *Faculty:* 50 full-time (3 women). *Students:* 348 full-time (78 women); includes 31 minority (2 African Americans, 16 Asian Americans or Pacific Islanders, 13 Hispanic Americans), 163 international. 588 applicants, 16% accepted. In 1998, 96 master's, 24 doctorates awarded. *Degree requirements:* For master's, thesis or alternative required, foreign language not required; for doctorate, dissertation required, foreign language not required. *Application deadline:* Applications are processed on a rolling basis. Application fee: $40 ($50 for international students). Tuition, state resident: full-time $4,616. Tuition, nonresident: full-time $11,768. Full-time tuition and fees vary according to course load. *Financial aid:* In 1998–99, 8 fellowships, 182 research assistantships, 34 teaching assistantships were awarded.; tuition waivers (full and partial) also available. Financial aid application deadline: 2/15. *Unit head:* Dr. David E. Daniel, Head, 217-333-1497. *Application contact:* Dr. Frederick V. Lawrence, Director of Graduate Studies, 217-333-6928, Fax: 217-333-9464, E-mail: flawrenc@uiuc.edu.

See in-depth description on page 631.

The University of Iowa, Graduate College, College of Engineering, Department of Civil and Environmental Engineering, Iowa City, IA 52242-1316. Offers MS, PhD. *Faculty:* 23 full-time, 2 part-time. *Students:* 46 full-time (9 women), 29 part-time (10 women); includes 3 minority (2 Asian Americans or Pacific Islanders, 1 Native American), 28 international. 115 applicants, 53% accepted. In 1998, 27 master's, 6 doctorates awarded. *Degree requirements:* For master's, thesis optional; for doctorate, dissertation, comprehensive exam required. *Entrance requirements:* For master's and doctorate, GRE, TOEFL. *Application deadline:* Applications are processed on a rolling basis. Application fee: $30 ($50 for international students). *Financial aid:* In 1998–99, 14 fellowships, 51 research assistantships, 17 teaching assistantships were awarded. Financial aid applicants required to submit FAFSA. *Faculty research:* Environmental quality modeling, engineering hydraulics, structural optimization and design. *Unit head:* Robert Ettema, Chair, 319-335-5224.

University of Kansas, Graduate School, School of Engineering, Department of Civil and Environmental Engineering, Lawrence, KS 66045. Offers civil engineering (MS, DE, PhD); environmental engineering (MS, PhD); environmental science (MS, PhD); water resources engineering (MS); water resources science (MS). *Faculty:* 20 full-time (0 women). *Students:* 23 full-time (6 women), 80 part-time (15 women); includes 9 minority (1 African American, 3 Asian Americans or Pacific Islanders, 4 Hispanic Americans, 1 Native American), 29 international. In 1998, 30 master's, 6 doctorates awarded. *Degree requirements:* For master's, thesis or alternative, exam required, foreign language not required; for doctorate, dissertation, comprehensive exam required. *Entrance requirements:* For master's and doctorate, Michigan English Language Assessment Battery, TOEFL, minimum GPA of 3.0. *Application deadline:* For fall admission, 7/1. Application fee: $30 ($45 for international students). *Financial aid:* Fellowships, research assistantships, teaching assistantships, career-related internships or fieldwork available. *Faculty research:* Structures (fracture mechanics), transportation, environmental health. *Unit head:* Steve McCabe, Chair, 785-864-3766. *Application contact:* David Parr, Graduate Director.

See in-depth description on page 633.

University of Louisville, Graduate School, Speed Scientific School, Department of Civil and Environmental Engineering, Louisville, KY 40292-0001. Offers M Eng, MS, PhD. *Accreditation:* ABET (one or more programs are accredited). *Faculty:* 14 full-time (0 women), 1 (woman) part-time. *Students:* 17 full-time (5 women), 50 part-time (14 women); includes 9 minority (3 African Americans, 6 Asian Americans or Pacific Islanders), 3 international. Average age 29. In 1998, 17 degrees awarded. *Degree requirements:* For master's and doctorate, thesis/dissertation required, foreign language not required. *Entrance requirements:* For master's and doctorate, GRE General Test (minimum combined score of 1200 required). *Application deadline:* Applications are processed on a rolling basis. Application fee: $25. *Financial aid:* In 1998–99, 3 research assistantships with tuition reimbursements (averaging $10,000 per year), 3 teaching assistantships with tuition reimbursements (averaging $10,000 per year) were awarded. *Unit head:* Dr. Louis F. Cohn, Chair, 502-852-6276, Fax: 502-852-8851, E-mail: cohn@louisville.edu.

University of Maine, Graduate School, College of Engineering, Department of Civil and Environmental Engineering, Orono, ME 04469. Offers civil engineering (MS, PhD), including environmental engineering, geotechnical engineering, structural engineering. *Faculty:* 11 full-time (0 women). *Students:* 19 full-time (3 women), 10 part-time (1 woman), 4 international. *Degree requirements:* For doctorate, dissertation required, foreign language not required, foreign language not required. *Entrance requirements:* For master's and doctorate, GRE General Test, TOEFL (minimum score of 550 required). *Application deadline:* For fall admission, 2/1 (priority date); for spring admission, 10/15. Applications are processed on a rolling basis. Application fee: $50. *Unit head:* Dr. Willem Brutsaert, Chair, 207-581-2170, Fax: 207-581-3888. *Application contact:* Scott G. Delcourt, Director of the Graduate School, 207-581-3218, Fax: 207-581-3232, E-mail: graduate@maine.edu.

University of Maryland, College Park, Graduate School, A. James Clark School of Engineering, Department of Civil and Environmental Engineering, College Park, MD 20742-5045. Offers M Eng, MS, PhD. Part-time and evening/weekend programs available. Postbaccalaureate distance learning degree programs offered. *Faculty:* 36 full-time (2 women), 8 part-time (4 women). *Students:* 88 full-time (24 women), 84 part-time (14 women); includes 29 minority (14 African Americans, 6 Asian Americans or Pacific Islanders, 8 Hispanic Americans, 1 Native American), 65 international. 189 applicants, 41% accepted. In 1998, 32 master's, 8 doctorates awarded. *Degree requirements:* For master's, thesis or alternative required, foreign language not required; for doctorate, dissertation, qualifying exam required, foreign language not required. *Entrance requirements:* For master's, GRE General Test, minimum GPA of 3.0. *Application deadline:* Applications are processed on a rolling basis. Application fee: $50 ($70 for international students). Tuition, state resident: part-time $272 per credit hour. Tuition, nonresident: part-time $475 per credit hour. Required fees: $632; $379 per year. *Financial aid:* In 1998–99, 13 fellowships with full tuition reimbursements (averaging $14,927 per year), 37 research assistantships with tuition reimbursements (averaging $12,698 per year), 18 teaching assistantships with tuition reimbursements (averaging $10,717 per year) were awarded.; Federal Work-Study, grants, and scholarships also available. Aid available to part-time students. Financial aid applicants required to submit FAFSA. *Faculty research:* Transportation and urban systems, environmental engineering, geotechnical engineering, construction engineering and management, hydraulics, remote sensing, soil mechanics. Total annual research expenditures: $4 million. *Unit head:* Dr. Gregory Baecher, Chairman, 301-405-1977, Fax: 301-405-2585. *Application contact:* Trudy Lindsey, Director, Graduate Admission and Records, 301-405-4198, Fax: 301-314-9305, E-mail: grschool@deans.umd.edu.

See in-depth description on page 635.

University of Maryland, College Park, Graduate School, College of Agriculture and Natural Resources, Department of Biological Resources Engineering, College Park, MD 20742-5045. Offers MS, PhD. *Faculty:* 13 full-time (0 women), 6 part-time (1 woman). *Students:* 17 full-time (10 women), 12 part-time (3 women); includes 7 minority (4 African Americans, 3 Asian Americans or Pacific Islanders), 6 international. 37 applicants, 43% accepted. In 1998, 6 master's, 1 doctorate awarded. *Degree requirements:* For master's, thesis optional, foreign language not required; for doctorate, dissertation required, foreign language not required. *Entrance requirements:* For master's, minimum GPA of 3.0. *Application fee:* $50 ($70 for international students). Tuition, state resident: part-time $272 per credit hour. Tuition, nonresident: part-time $475 per credit hour. Required fees: $632; $379 per year. *Financial aid:* In 1998–99, 17 research assistantships with tuition reimbursements (averaging $13,288 per year), 7 teaching assistantships with tuition reimbursements (averaging $11,818 per year) were awarded.; fellowships with full tuition reimbursements, career-related internships or fieldwork also available. Financial aid applicants required to submit FAFSA. *Faculty research:* Engineering aspects of production; harvesting, processing, and marketing of terrestrial and aquatic food and fiber. Total annual research expenditures: $385,057. *Unit head:* Dr. Frederick Wheaton, Chairman, 301-405-2223, Fax: 301-314-9023. *Application contact:* Trudy Lindsey, Director, Graduate Admission and Records, 301-405-4198, Fax: 301-314-9305, E-mail: grschool@deans.umd.edu.

Announcement: Housed in one of the University's newest buildings, the Department of Biological Resources Engineering provides graduate students with state-of-the-art research facilities, including specialized laboratories for work on instrumentation, cellular engineering, wetland and plant ecology, ecological engineering, geographic information systems, water quality, and bioenvironmental and water resources engineering. The graduate program provides qualified students with the interdisciplinary study and research experience they need to contribute to this exciting field. Under the personal guidance of outstanding faculty members, graduate students design educational programs leading to both Master of Science and Doctor of Philosophy degrees. Programs of study are tailored to meet the individual research interests and career ambitions of each graduate student. For additional information, see the in-depth description in Section 5.

University of Massachusetts Amherst, Graduate School, College of Engineering, Department of Civil Engineering, Program in Environmental Engineering, Amherst, MA 01003. Offers MS. *Accreditation:* ABET. Part-time programs available. *Students:* 15 full-time (9 women), 5 part-time (1 woman); includes 3 minority (2 Hispanic Americans, 1 Native American), 1 international. Average age 26. 63 applicants, 37% accepted. In 1998, 12 degrees awarded. *Degree requirements:* For master's, thesis or alternative required, foreign language not required. *Entrance requirements:* For master's, GRE General Test. *Application deadline:* For fall admission, 2/1 (priority date); for spring admission, 10/1. Applications are processed on a rolling basis. Application fee: $40. Tuition, state resident: full-time $2,640; part-time $165 per credit. Tuition, nonresident: full-time $9,756; part-time $407 per credit. Required fees: $1,221 per term. One-time fee: $110. Full-time tuition and fees vary according to course load, campus/location and reciprocity agreements. *Financial aid:* Fellowships with full tuition reimbursements, research assistantships with full tuition reimbursements, teaching assistantships with full tuition reimbursements, career-related internships or fieldwork, Federal Work-Study, grants, scholarships, traineeships, and unspecified assistantships available. Aid available to part-time students. Financial aid application deadline: 2/1.

University of Massachusetts Lowell, Graduate School, James B. Francis College of Engineering, Department of Civil Engineering, Program in Environmental Studies, Lowell, MA 01854-2881. Offers MS Eng. Part-time programs available. *Faculty:* 13 full-time. *Students:* 6 full-time (1 woman), 59 part-time (20 women); includes 7 minority (all African Americans), 1 international. 71 applicants, 49% accepted. *Degree requirements:* For master's, thesis optional, foreign language not required. *Entrance requirements:* For master's, GRE General Test. *Application deadline:* For fall admission, 4/1 (priority date); for spring admission, 10/1. Applications are processed on a rolling basis. Application fee: $20 ($35 for international students). *Financial aid:* Teaching assistantships, career-related internships or fieldwork available. Financial aid application deadline: 4/1. *Faculty research:* Remote sensing of air pollutants, atmospheric deposition of toxic metals, contaminant transport in groundwater, soil remediation. *Application contact:* Dr. Burton Segall, Graduate Coordinator, 978-934-2288, E-mail: burton_segall@woods.uml.edu.

The University of Memphis, Graduate School, Herff College of Engineering, Department of Civil Engineering, Memphis, TN 38152. Offers civil engineering (PhD); environmental engineering (MS); foundation engineering (MS); structural engineering (MS); transportation engineering (MS); water resources engineering (MS). *Faculty:* 12 full-time (0 women), 1 part-time (0 women). *Students:* 6 full-time (2 women), 22 part-time (1 woman), 7 international. *Degree requirements:* For master's, thesis or alternative, comprehensive exam required; for doctorate, dissertation required. *Entrance requirements:* For master's, GRE General Test (minimum combined score of 1000 required) or MAT, BS, minimum undergraduate GPA of 2.5. *Application deadline:* For fall admission, 8/1; for spring admission, 12/1. Application fee: $25 ($50 for international students). Tuition, state resident: full-time $3,410; part-time $178 per credit hour. Tuition, nonresident: full-time $8,670; part-time $408 per credit hour. Tuition and fees vary according to program. *Unit head:* Dr. Martin E. Lipinski, Chairman, 901-678-3279. *Application contact:* Dr. Larry W. Moore, Coordinator of Graduate Studies, 901-678-3278.

University of Michigan, Horace H. Rackham School of Graduate Studies, College of Engineering, Department of Civil and Environmental Engineering, Ann Arbor, MI 48109. Offers civil engineering (MSE, PhD, CE); construction engineering and management (MSE); environmental engineering (MSE, PhD). *Degree requirements:* For master's, foreign language and thesis not required; for doctorate, computer language, oral defense of dissertation, preliminary and written exams required. *Entrance requirements:* For master's, GRE General Test; for doctorate, GRE General Test, master's degree. Electronic applications accepted. *Faculty research:* Earthquake engineering, environmental and water resources engineering, geotechnical engineering, hydraulics and hydrologic engineering, structural engineering.

See in-depth description on page 637.

University of Missouri–Columbia, Graduate School, College of Engineering, Department of Civil and Environmental Engineering, Columbia, MO 65211. Offers civil engineering (MS, PhD); environmental engineering (MS, PhD); geotechnical engineering (MS, PhD); structural engineering (MS, PhD); transportation and highway engineering (MS); water resources (MS, PhD). *Faculty:* 24 full-time (2 women). *Students:* 19 full-time (2 women), 18 part-time (3 women); includes 1 minority (Hispanic American), 21 international. 74 applicants, 43% accepted. In 1998, 19 master's, 1 doctorate awarded. *Degree requirements:* For master's, report or thesis required; for doctorate, dissertation required. *Entrance requirements:* For master's and doctorate, GRE General Test, TOEFL. *Application deadline:* For fall admission, 7/10. Application fee: $30 ($50 for international students). *Financial aid:* Research assistantships, teaching assistantships, institutionally-sponsored loans available. *Unit head:* Dr. Mark Virkler, Director of Graduate Studies, 573-882-3678.

See in-depth description on page 639.

University of Missouri–Rolla, Graduate School, School of Engineering, Department of Civil Engineering, Program in Environmental Engineering, Rolla, MO 65409-0910. Offers MS. *Faculty:* 4 full-time (0 women). *Students:* 11 full-time (3 women), 2 international. Average age 30. 40 applicants, 53% accepted. In 1998, 8 degrees awarded. *Degree requirements:* For master's, thesis or alternative required, foreign language not required. *Entrance requirements:* For master's, GRE General Test (minimum combined score of 1100 required), TOEFL (minimum score of 550 required), minimum GPA of 3.0. *Application deadline:* For fall admission, 7/1; for spring admission, 12/1. Applications are processed on a rolling basis. Application fee: $25. *Financial aid:* In 1998–99, 3 fellowships with tuition reimbursements (averaging $15,000 per year), 2 research assistantships with partial tuition reimbursements, 2 teaching assistantships with partial tuition reimbursements were awarded.; institutionally-sponsored loans also available. Aid available to part-time students. Financial aid application deadline: 1/1. *Faculty research:* Hazardous waste treatment, groundwater remediation, air pollution control, advanced oxidation, phytoremediation. *Unit head:* Scott G. Delcourt, Director of the Graduate School, 207-581-3218, Fax: 207-581-3232, E-mail: graduate@maine.edu. *Application contact:* Dr. Abdeldjelil Belarbi, Director of Graduate Advising, 573-341-4478, Fax: 573-341-4729, E-mail: ceadvise@novell.civil.umr.edu.

University of Missouri–Rolla, Graduate School, School of Engineering, Department of Civil Engineering, Program in Sanitary Engineering and Environmental Health, Rolla, MO 65409-0910. Offers MS, DE, PhD. *Degree requirements:* For master's, thesis or alternative required, foreign language not required; for doctorate, dissertation required, foreign language not required. *Entrance requirements:* For master's and doctorate, GRE General Test (minimum combined score of 1100 required), TOEFL (minimum score of 550 required), minimum GPA of 3.0.

Environmental Engineering

University of Nebraska–Lincoln, Graduate College, College of Engineering and Technology, Interdepartmental Area of Environmental Engineering, Lincoln, NE 68588. Offers engineering (PhD); environmental engineering (MS). *Students:* 9 full-time (2 women), 9 part-time (3 women); includes 1 minority (Asian American or Pacific Islander), 2 international. Average age 29. 20 applicants, 60% accepted. In 1998, 9 degrees awarded. *Degree requirements:* For master's, thesis optional; for doctorate, dissertation, comprehensive exams required. *Entrance requirements:* For master's and doctorate, GRE General Test, TOEFL (minimum score of 550 required). *Application deadline:* For fall admission, 4/15; for spring admission, 10/15. Application fee: $35. Electronic applications accepted. *Financial aid:* Fellowships available. Financial aid application deadline: 2/15. *Faculty research:* Water supply engineering, wastewater engineering, hazardous waste management, solid waste management, agricultural waste management. *Unit head:* Dr. Mohamed F. Dahab, Graduate Committee Chair, 402-472-5020.

University of Nevada, Las Vegas, Graduate College, Howard R. Hughes College of Engineering, Department of Civil and Environmental Engineering, Las Vegas, NV 89154-9900. Offers MSE, PhD. *Faculty:* 19 full-time (2 women). *Students:* 21 full-time (3 women), 37 part-time (12 women); includes 10 minority (7 Asian Americans or Pacific Islanders, 3 Hispanic Americans), 11 international. 36 applicants, 64% accepted. In 1998, 12 master's, 1 doctorate awarded. *Degree requirements:* For master's, comprehensive exam required, thesis optional, foreign language not required; for doctorate, dissertation required. *Entrance requirements:* For master's, minimum GPA of 3.0; for doctorate, minimum GPA of 3.5. *Application deadline:* For fall admission, 6/15 (priority date); for spring admission, 11/15. Applications are processed on a rolling basis. Application fee: $40 ($95 for international students). *Financial aid:* In 1998–99, 11 research assistantships with full tuition reimbursements (averaging $7,133 per year), 13 teaching assistantships with partial tuition reimbursements (averaging $9,058 per year) were awarded. Financial aid application deadline: 3/1. *Unit head:* Dr. Edward S. Neumann, Chair, 702-895-3701, Fax: 702-895-3936, E-mail: ce-info@ce.unlv.edu.

University of Nevada, Reno, Graduate School, Center for Environmental Sciences and Engineering, Reno, NV 89557. Offers atmospheric sciences (MS, PhD); ecology, evolution and conservation biology (PhD); environmental sciences and health (MS, PhD). *Entrance requirements:* For master's and doctorate, GRE, TOEFL. *Faculty research:* Air, water, and soil pollution; endangered species habitats; restoration ecology; environmental analytical chemistry; control of pollution from mining.

University of New Brunswick, School of Graduate Studies, Faculty of Engineering, Department of Civil Engineering, Fredericton, NB E3B 5A3, Canada. Offers environmental engineering (M Eng, M Sc E, PhD); geotechnical engineering (M Eng, M Sc E); structures and structural foundations (M Eng, M Sc E, PhD); transportation engineering (M Eng, M Sc E, PhD); water resources and hydrology (M Eng, M Sc E, PhD). Part-time programs available. *Degree requirements:* For master's, thesis required, foreign language not required; for doctorate, dissertation, qualifying exam required, foreign language not required. *Entrance requirements:* For master's and doctorate, TOEFL, TWE, minimum GPA of 3.0.

University of New Haven, Graduate School, School of Engineering and Applied Science, Program in Environmental Engineering, West Haven, CT 06516-1916. Offers civil engineering design (Certificate); environmental engineering (MS). Part-time and evening/weekend programs available. *Students:* 7 full-time (0 women), 22 part-time (6 women); includes 3 minority (1 African American, 1 Asian American or Pacific Islander, 1 Native American), 8 international. 19 applicants, 79% accepted. In 1998, 16 degrees awarded. *Degree requirements:* For master's, thesis or alternative required, foreign language not required. *Entrance requirements:* For master's, bachelor's degree in engineering. *Application deadline:* Applications are processed on a rolling basis. Application fee: $50. *Financial aid:* Federal Work-Study available. Aid available to part-time students. Financial aid application deadline: 5/1; financial aid applicants required to submit FAFSA. *Unit head:* Dr. Agamemnon D. Koutsospyros, Coordinator, 203-932-7398.

The University of North Carolina at Chapel Hill, Graduate School, School of Public Health, Department of Environmental Sciences and Engineering, Chapel Hill, NC 27599. Offers environmental engineering (MSEE); environmental sciences and engineering (MS, PhD); public health (MPH, MSPH). *Accreditation:* ABET (one or more programs are accredited). *Faculty:* 36 full-time (5 women), 31 part-time (5 women). *Students:* 99 full-time (45 women), 63 part-time (27 women); includes 13 minority (5 African Americans, 4 Asian Americans or Pacific Islanders, 3 Hispanic Americans, 1 Native American), 31 international. Average age 28. 208 applicants, 53% accepted. In 1998, 20 master's, 15 doctorates awarded. Terminal master's awarded for partial completion of doctoral program. *Degree requirements:* For master's, research paper, comprehensive exam required; for doctorate, dissertation, comprehensive exam required, foreign language not required. *Entrance requirements:* For master's and doctorate, GRE General Test (minimum combined score of 1000 required), minimum GPA of 3.0. *Application deadline:* For fall admission, 1/1 (priority date); for spring admission, 9/15. Applications are processed on a rolling basis. Application fee: $55. *Financial aid:* In 1998–99, 81 students received aid, including 14 fellowships with tuition reimbursements available (averaging $14,000 per year), 48 research assistantships with tuition reimbursements available (averaging $17,000 per year), 6 teaching assistantships with tuition reimbursements available (averaging $12,133 per year); career-related internships or fieldwork, Federal Work-Study, and traineeships also available. Aid available to part-time students. Financial aid application deadline: 1/1. *Faculty research:* Air, radiation and industrial hygiene, aquatic and atmospheric sciences, environmental health sciences, environmental management and policy, water resources engineering. Total annual research expenditures: $6.5 million. *Unit head:* Dr. Donald Fox, Chair, 919-966-1024, Fax: 919-966-7911, E-mail: don_fox@unc.edu. *Application contact:* Nikki Bryant, Assistant Registrar, 919-966-3844, Fax: 919-966-7911.

University of Notre Dame, Graduate School, College of Engineering, Department of Civil Engineering and Geological Sciences, Notre Dame, IN 46556. Offers bioengineering (MS); civil engineering (MS); civil engineering and geological sciences (PhD); environmental engineering (MS); geological sciences (MS). *Faculty:* 13 full-time (1 woman). *Students:* 35 full-time (10 women), 4 part-time; includes 4 minority (2 African Americans, 1 Asian American or Pacific Islander, 1 Hispanic American), 11 international. Terminal master's awarded for partial completion of doctoral program. *Degree requirements:* For master's and doctorate, thesis/dissertation required, foreign language not required. *Entrance requirements:* For master's and doctorate, GRE General Test, TOEFL (minimum score of 600 required; 250 for computer-based). *Application deadline:* For fall admission, 2/1 (priority date); for spring admission, 10/15. Applications are processed on a rolling basis. Application fee: $40. *Unit head:* Dr. Billie F. Spencer, Director of Graduate Studies, 219-631-5381, Fax: 219-631-9236, E-mail: cegeos@nd.edu. *Application contact:* Dr. Terrence J. Akai, Director of Graduate Admissions, 219-631-7706, Fax: 219-631-4183, E-mail: gradad@nd.edu.

University of Oklahoma, Graduate College, College of Engineering, School of Civil Engineering and Environmental Science, Program in Civil Engineering, Norman, OK 73019-0390. Offers civil engineering (PhD); environmental engineering (MS); geotechnical engineering (MS); structures (MS); transportation (MS). *Accreditation:* ABET (one or more programs are accredited). *Students:* 19 full-time (4 women), 36 part-time (6 women); includes 5 minority (3 Asian Americans or Pacific Islanders, 2 Native Americans), 28 international. *Degree requirements:* For master's, comprehensive and oral exams required, foreign language and thesis not required; for doctorate, dissertation, oral and qualifying exams required, foreign language not required. *Entrance requirements:* For master's, GRE General Test, TOEFL (minimum score of 575 required), minimum GPA of 3.0; for doctorate, GRE General Test, TOEFL (minimum score of 575 required), minimum graduate GPA of 3.5. *Application deadline:* For fall admission, 4/1 (priority date). Applications are processed on a rolling basis. Application fee: $25. Tuition, state resident: part-time $86 per credit hour. Tuition, nonresident: part-time $275 per credit hour. Tuition and fees vary according to course level, course load and program. *Unit head:* Robert C. Knox, Graduate Liaison, 405-325-4256, Fax: 405-325-7508. *Application contact:* Robert C. Knox, Graduate Liaison, 405-325-4256, Fax: 405-325-7508.

See in-depth description on page 641.

University of Pennsylvania, School of Engineering and Applied Science, Department of Systems Engineering, Philadelphia, PA 19104. Offers environmental resources engineering (MSE); environmental/resources engineering (PhD); systems engineering (MSE, PhD); technology and public policy (MSE, PhD); transportation (MSE, PhD). Part-time programs available. *Faculty:* 12 full-time (0 women), 9 part-time (2 women). *Students:* 26 full-time (10 women), 16 part-time (2 women); includes 4 minority (2 African Americans, 2 Hispanic Americans), 16 international. Terminal master's awarded for partial completion of doctoral program. *Degree requirements:* For master's, computer language required, foreign language not required; for doctorate, one foreign language, computer language, dissertation required. *Entrance requirements:* For master's and doctorate, TOEFL (minimum score of 600 required). *Application deadline:* For fall admission, 1/2 (priority date). Applications are processed on a rolling basis. Application fee: $65. Electronic applications accepted. *Unit head:* Dr. G. Anandalingam, Chair, 215-898-8790, Fax: 215-898-5020, E-mail: anand@seas.upenn.edu. *Application contact:* Dr. Tony E. Smith, Graduate Group Chair, 215-898-9647, Fax: 215-898-5020, E-mail: tesmith@seas.upenn.edu.

University of Pittsburgh, School of Engineering, Department of Civil and Environmental Engineering, Pittsburgh, PA 15260. Offers MSCEE, PhD. Part-time and evening/weekend programs available. *Faculty:* 12 full-time (0 women). *Students:* 27 full-time (7 women), 69 part-time (6 women). 64 applicants, 83% accepted. In 1998, 20 master's, 4 doctorates awarded. Terminal master's awarded for partial completion of doctoral program. *Degree requirements:* For master's, computer language required, thesis optional, foreign language not required; for doctorate, computer language, dissertation, comprehensive and final oral exams required, foreign language not required. *Entrance requirements:* For master's and doctorate, TOEFL, minimum QPA of 3.0. *Average time to degree:* Master's–2 years full-time, 4 years part-time; doctorate–5 years full-time, 7 years part-time. *Application deadline:* For fall admission, 8/1 (priority date); for spring admission, 12/1 (priority date). Applications are processed on a rolling basis. Application fee: $30 ($40 for international students). *Financial aid:* In 1998–99, 19 students received aid, including 8 research assistantships with tuition reimbursements available (averaging $8,872 per year), 13 teaching assistantships with tuition reimbursements available (averaging $10,608 per year); fellowships with tuition reimbursements available, grants, scholarships, traineeships, and tuition waivers (full and partial) also available. Financial aid application deadline: 2/15. *Faculty research:* Environmental and water resources, structures and infrastructures, construction management. Total annual research expenditures: $779,722. *Unit head:* Dr. Rafael G. Quimpo, Interim Chairman, 412-624-9870, Fax: 412-624-0135, E-mail: quimpo@civ.pitt.edu. *Application contact:* Attila A. Sooky, Academic Coordinator, 412-624-9869, Fax: 412-624-0135, E-mail: sooky@civ.pitt.edu.

University of Regina, Faculty of Graduate Studies and Research, Faculty of Engineering, Program in Regional Environmental Systems Engineering, Regina, SK S4S 0A2, Canada. Offers M Eng, MA Sc, PhD. *Students:* 14 full-time (0 women), 32 part-time (4 women). 33 applicants, 64% accepted. In 1998, 11 master's, 2 doctorates awarded. *Degree requirements:* For master's, thesis required (for some programs), foreign language not required; for doctorate, dissertation required, foreign language not required. *Entrance requirements:* For master's, TOEFL (minimum score of 550 required); for doctorate, TOEFL (minimum score of 550 required), master's degree. *Application deadline:* Applications are processed on a rolling basis. Application fee: $0. *Expenses:* Tuition and fees charges are reported in Canadian dollars. Tuition, state resident: full-time $1,688 Canadian dollars; part-time $94 Canadian dollars per credit hour. International tuition: $3,375 Canadian dollars full-time. Required fees: $65 Canadian dollars per course. Tuition and fees vary according to course load and program. *Financial aid:* In 1998–99, 1 fellowship, 8 research assistantships, 6 teaching assistantships were awarded.; scholarships also available. Financial aid application deadline: 6/15. *Faculty research:* Flood control, waste management and treatment, groundwater flow, traffic system analysis, transportation planning. *Unit head:* Dr. G. Fuller, Head, 306-585-4704, Fax: 306-585-4855, E-mail: fuller@robinhood.engg.uregina.ca. *Application contact:* Dr. Y. C. Jin, Coordinator, 306-585-4567, Fax: 306-585-4855, E-mail: jin@meena.cc.uregina.ca.

University of Rhode Island, Graduate School, College of Engineering, Department of Civil and Environmental Engineering, Kingston, RI 02881. Offers environmental engineering (MS, PhD); geotechnical engineering (MS, PhD); structural engineering (MS, PhD); transportation engineering (MS, PhD). *Entrance requirements:* For master's and doctorate, GRE.

University of Saskatchewan, College of Graduate Studies and Research, College of Engineering, Division of Environmental Engineering, Saskatoon, SK S7N 5A2, Canada. Offers M Eng, M Sc, PhD, Diploma. *Degree requirements:* For master's and doctorate, thesis/dissertation required. *Entrance requirements:* For master's and doctorate, GRE, TOEFL.

University of Southern California, Graduate School, School of Engineering, Department of Civil Engineering, Program in Environmental Engineering, Los Angeles, CA 90089. Offers MS, PhD. *Students:* 23 full-time (7 women), 23 part-time (5 women); includes 9 minority (8 Asian Americans or Pacific Islanders, 1 Hispanic American), 27 international. Average age 28. 42 applicants, 69% accepted. In 1998, 15 master's, 2 doctorates awarded. *Degree requirements:* For master's, thesis optional; for doctorate, dissertation required. *Entrance requirements:* For master's and doctorate, GRE General Test, GRE Subject Test. *Application deadline:* For fall admission, 6/1 (priority date); for spring admission, 11/1. Application fee: $55. Tuition: Part-time $768 per unit. Required fees: $350 per semester. *Financial aid:* In 1998–99, 6 fellowships, 8 research assistantships, 8 teaching assistantships were awarded.; Federal Work-Study, institutionally-sponsored loans, and scholarships also available. Aid available to part-time students. Financial aid application deadline: 2/15; financial aid applicants required to submit FAFSA. *Unit head:* Dr. L. Carter Wellford, Chair, Department of Civil Engineering, 213-740-0587.

See in-depth description on page 647.

University of South Florida, Graduate School, College of Engineering, Department of Civil and Environmental Engineering, Tampa, FL 33620-9951. Offers civil engineering (MCE, MSCE, PhD); engineering (ME, MSE); environmental engineering (MEVE, MSEV). Part-time programs available. *Faculty:* 21 full-time (2 women), 3 part-time (0 women). *Students:* 48 full-time (9 women), 84 part-time (15 women); includes 43 minority (2 African Americans, 33 Asian Americans or Pacific Islanders, 8 Hispanic Americans), 36 international. Average age 32. In 1998, 40 master's, 1 doctorate awarded. Terminal master's awarded for partial completion of doctoral program. *Degree requirements:* For master's, thesis required (for some programs), foreign language not required; for doctorate, dissertation, 2 tools of research as specified by dissertation committee required, foreign language not required. *Entrance requirements:* For master's, GRE General Test (minimum score of 400 on each section, 1000 combined required), minimum GPA of 3.0 during previous 2 years; for doctorate, GRE General Test (minimum score of 450 on each section, 1000 combined required). *Application deadline:* For fall admission, 6/1; for spring admission, 10/15. Application fee: $20. Electronic applications accepted. Tuition, state resident: part-time $148 per credit hour. Tuition, nonresident: part-time $509 per credit hour. *Financial aid:* In 1998–99, 2 fellowships, 48 research assistantships, 10 teaching assistantships were awarded.; career-related internships or fieldwork, Federal Work-Study, institutionally-sponsored loans, and tuition waivers (partial) also available. Aid available to part-time students. Financial aid applicants required to submit FAFSA. *Faculty research:* Water resources, structures and materials, transportation, geotechnical engineering. Total annual research expenditures: $7.5 million. *Unit head:* Dr. W. C. Carpenter, Chairperson, 813-974-2275, Fax: 813-974-2957, E-mail: carpente@eng.usf.edu. *Application contact:* Manjriker Gunaratne, Coordinator, 813-974-5818, Fax: 813-974-5927, E-mail: gunaratn@eng.usf.edu.

University of Tennessee, Knoxville, Graduate School, College of Engineering, Department of Civil and Environmental Engineering, Program in Environmental Engineering, Knoxville, TN 37996. Offers MS. Part-time programs available. Postbaccalaureate distance learning degree programs offered (minimal on-campus study). *Students:* 13 full-time (5 women), 37 part-time (7 women); includes 1 minority (African American), 6 international. 44 applicants, 43% accepted. In 1998, 20 degrees awarded. *Degree requirements:* For master's, thesis or alternative required, foreign language not required. *Entrance requirements:* For master's, TOEFL (minimum score of 550 required), minimum GPA of 2.7. *Application deadline:* For fall admission, 2/1 (priority date). Applications are processed on a rolling basis. Application fee: $35. Electronic applications accepted. *Financial aid:* Application deadline: 2/1; *Unit head:* Dr. Gregory D. Reed, Head, Department of Civil and Environmental Engineering, 423-974-2503, Fax: 423-974-2669, E-mail: gdreed@utk.edu.

The University of Texas at Arlington, Graduate School, College of Engineering, Department of Civil and Environmental Engineering, Arlington, TX 76019. Offers M Engr, MS, PhD. *Faculty:* 12 full-time (0 women), 1 part-time (0 women). *Students:* 35 full-time (10 women), 71 part-time (12 women); includes 15 minority (2 African Americans, 8 Asian Americans or Pacific Islanders, 4 Hispanic Americans, 1 Native American), 36 international. 69 applicants, 61% accepted. In 1998, 19 master's, 2 doctorates awarded. *Degree requirements:* For master's, computer language, thesis (for some programs), oral and written exams required, foreign language not required; for doctorate, one foreign language, computer language, oral and written defense of dissertation required. *Entrance requirements:* For master's, GRE General Test (minimum combined score of 1000 required), TOEFL, minimum GPA of 3.0 in last 60 hours of undergraduate course work; for doctorate, GRE General Test (minimum combined score of 1200 required), TOEFL, minimum graduate GPA of 3.5. *Application deadline:* Applications are processed on a rolling basis. Application fee: $25 ($50 for international students). Tuition, state resident: full-time $1,368; part-time $76 per semester hour. Tuition, nonresident: full-time $5,454; part-time $303 per semester hour. Required fees: $66 per semester hour. $86 per term. Tuition and fees vary according to course load. *Financial aid:* Fellowships, research assistantships, teaching assistantships, Federal Work-Study, scholarships, and tuition waivers (partial) available. *Unit head:* Dr. C. E. Parker, Chairman, 817-272-2201, Fax: 817-272-2630, E-mail: parkerce@uta.edu. *Application contact:* Dr. Ernest C. Crosby, Graduate Adviser, 817-272-3500, Fax: 817-272-2630, E-mail: ecrosby@uta.edu.

The University of Texas at Arlington, Graduate School, Program in Environmental Science and Engineering, Arlington, TX 76019. Offers MS. *Students:* 6 full-time (1 woman), 5 part-time (4 women), 7 international. 24 applicants, 58% accepted. In 1998, 2 degrees awarded. *Application deadline:* Applications are processed on a rolling basis. Application fee: $25 ($50 for international students). Tuition, state resident: full-time $1,368; part-time $76 per semester hour. Tuition, nonresident: full-time $5,454; part-time $303 per semester hour. Required fees: $66 per semester hour. $86 per term. Tuition and fees vary according to course load. *Unit head:* Dr. John S. Wickham, Head, 817-272-2332, Fax: 817-272-2628, E-mail: wickham@uta.edu.

The University of Texas at Austin, Graduate School, College of Engineering, Department of Civil Engineering, Program in Environmental and Water Resources Engineering, Austin, TX 78712-1111. Offers MSE. *Accreditation:* ABET. *Students:* 17 (8 women); includes 2 minority (both Hispanic Americans) 1 international. 34 applicants, 29% accepted. In 1998, 9 degrees awarded. *Degree requirements:* Foreign language not required. *Entrance requirements:* For master's, GRE General Test (minimum combined score of 1000 required). *Application deadline:* For fall admission, 1/15 (priority date); for spring admission, 9/1 (priority date). Applications are processed on a rolling basis. Application fee: $50 ($75 for international students). Electronic applications accepted. *Financial aid:* Fellowships, research assistantships, teaching assistantships available. Financial aid application deadline: 2/1. *Unit head:* Kathy Rose, Graduate Coordinator, 512-471-4921, Fax: 512-471-0592, E-mail: krose@mail.utexas.edu. *Application contact:* Kathy Rose, Graduate Coordinator, 512-471-4921, Fax: 512-471-0592, E-mail: krose@mail.utexas.edu.

The University of Texas at El Paso, Graduate School, Interdisciplinary Program in Environmental Science and Engineering, El Paso, TX 79968-0001. Offers PhD. *Students:* 9 full-time (4 women), 12 part-time (4 women); includes 8 minority (1 Asian American or Pacific Islander, 7 Hispanic Americans), 5 international. Average age 39. 10 applicants, 80% accepted. *Degree requirements:* For doctorate, dissertation required. *Entrance requirements:* For doctorate, GRE General Test, TOEFL (minimum score of 550 required). *Application deadline:* Applications are processed on a rolling basis. Application fee: $15 ($65 for international students). Tuition, state resident: full-time $2,790. Tuition, nonresident: full-time $7,710. *Financial aid:* Fellowships, research assistantships, teaching assistantships, Federal Work-Study, institutionally-sponsored loans, and tuition waivers (partial) available. Financial aid applicants required to submit FAFSA. *Unit head:* Dr. Charles Groat, Director, 915-747-5954. *Application contact:* Susan Jordan, Director, Graduate Student Services, 915-747-5491, Fax: 915-747-5788, E-mail: sjordan@utep.edu.

University of Toronto, School of Graduate Studies, Physical Sciences Division, Faculty of Applied Science and Engineering, Collaborative Program in Environmental Engineering, Toronto, ON M5S 1A1, Canada. Offers M Eng, MA Sc, PhD. Part-time programs available. *Degree requirements:* For master's, thesis required (for some programs), foreign language not required; for doctorate, dissertation required, foreign language not required.

University of Virginia, School of Engineering and Applied Science, Department of Civil Engineering, Program in Environmental Engineering, Charlottesville, VA 22903. Offers environmental engineering (ME, MS, PhD); water resources (ME, MS, PhD). Part-time programs available. *Faculty:* 5 full-time (2 women). *Students:* 36 full-time (21 women), 1 part-time; includes 1 minority (Asian American or Pacific Islander), 8 international. Average age 27. 46 applicants, 74% accepted. In 1998, 13 master's awarded (77% found work related to degree, 23% continued full-time study); 2 doctorates awarded (50% entered university research/teaching, 50% found other work related to degree). Terminal master's awarded for partial completion of doctoral program. *Degree requirements:* For master's, thesis required (for some programs), foreign language not required; for doctorate, dissertation, comprehensive exam required, foreign language not required. *Entrance requirements:* For master's and doctorate, GRE General Test. *Average time to degree:* Master's–1.7 years full-time; doctorate–3 years full-time. *Application deadline:* For fall admission, 2/1 (priority date). Applications are processed on a rolling basis. Application fee: $60. Electronic applications accepted. *Financial aid:* In 1998–99, 28 students received aid, including 3 fellowships with full tuition reimbursements available (averaging $11,000 per year), 16 research assistantships with full tuition reimbursements available (averaging $11,800 per year), 9 teaching assistantships with full tuition reimbursements available (averaging $10,700 per year) Financial aid application deadline: 2/1. *Faculty research:* Stormwater management, nonpoint pollution control, water quality modeling, estuarine and coastal water quality management, groundwater flow and transport. Total annual research expenditures: $900,000. *Unit head:* Dr. Nicholas J. Garber, Chairman, Department of Civil Engineering, 804-924-7464, Fax: 804-982-2951, E-mail: njg@virginia.edu.

University of Washington, Graduate School, College of Engineering, Department of Civil and Environmental Engineering, Seattle, WA 98195. Offers environmental engineering (MS, MS Civ E, MSE, PhD); hydraulic engineering (MS Civ E, MSE, PhD); structural and geotechnical engineering and mechanics (MS, MS Civ E, MSE, PhD); transportation and construction engineering (MS, MS Civ E, MSE, PhD). *Faculty:* 31 full-time (4 women), 26 part-time (1 woman). *Students:* 153 full-time (54 women), 42 part-time (13 women); includes 25 minority (2 African Americans, 19 Asian Americans or Pacific Islanders, 3 Hispanic Americans, 1 Native American), 34 international. Terminal master's awarded for partial completion of doctoral program. *Degree requirements:* For master's, thesis or alternative required, foreign language not required; for doctorate, dissertation required, foreign language not required. *Entrance requirements:* For master's, GRE General Test, TOEFL (minimum score of 580 required), minimum GPA of 3.0; for doctorate, GRE, TOEFL (minimum score of 580 required), minimum GPA of 3.0. *Application deadline:* For fall admission, 2/1 (priority date); for winter admission, 11/1; for spring admission, 2/1. Applications are processed on a rolling basis. Application fee: $50. Electronic applications accepted. Tuition, state resident: full-time $5,196; part-time $475 per credit. Tuition, nonresident: full-time $13,485; part-time $1,285 per credit. Required fees: $387; $38 per credit. Tuition and fees vary according to course load. *Unit head:* Dr. Fred L. Mannering, Chair, 206-543-2390. *Application contact:* Marcia Buck, Graduate Secretary, 206-543-2574, E-mail: ceginfo@u.washington.edu.

University of Windsor, College of Graduate Studies and Research, Faculty of Engineering, Department of Civil and Environmental Engineering, Windsor, ON N9B 3P4, Canada. Offers MA Sc, PhD. Part-time programs available. *Degree requirements:* For master's and doctorate, thesis/dissertation required, foreign language not required. *Entrance requirements:* For master's, TOEFL (minimum score of 550 required), minimum B average; for doctorate, TOEFL, master's degree. *Faculty research:* Structures, water resources.

University of Wisconsin–Madison, Graduate School, College of Engineering, Department of Civil and Environmental Engineering, Madison, WI 53706-1380. Offers MS, PhD. Part-time programs available. *Faculty:* 31 full-time (1 woman), 3 part-time (0 women). *Students:* 129 full-time (26 women), 22 part-time (1 woman); includes 3 minority (2 Asian Americans or Pacific Islanders, 1 Hispanic American), 2 international. Average age 25. 173 applicants, 65% accepted. In 1998, 52 master's, 9 doctorates awarded. Terminal master's awarded for partial completion of doctoral program. *Degree requirements:* For master's, thesis or alternative required, foreign language not required; for doctorate, dissertation, preliminary exam and qualifying exams required, foreign language not required. *Entrance requirements:* For master's and doctorate, TOEFL, minimum GPA of 3.0. *Average time to degree:* Master's–2 years full-time, 5 years part-time; doctorate–3.5 years full-time, 7 years part-time. *Application deadline:* For fall admission, 6/30 (priority date); for spring admission, 11/15. Applications are processed on a rolling basis. Application fee: $45. Electronic applications accepted. *Financial aid:* In 1998–99, 95 students received aid, including 9 fellowships with full tuition reimbursements available (averaging $14,600 per year), 53 research assistantships with full tuition reimbursements available (averaging $14,600 per year), 10 teaching assistantships with full tuition reimbursements available (averaging $6,422 per year); Federal Work-Study also available. Aid available to part-time students. Financial aid application deadline: 12/1. *Faculty research:* Environmental geotechnics and soil mechanics, design and analysis of structures, traffic engineering and intelligent transport systems, industrial pollution control, hydrological monitoring. Total annual research expenditures: $4.3 million. *Unit head:* Robert L. Smith, Chair, 608-262-3542, Fax: 608-262-5199, E-mail: cee@engr.wisc.edu. *Application contact:* Lynn Maertz, Student Services Coordinator, 608-262-5198, Fax: 608-262-5199, E-mail: maertz@engr.wisc.edu.

University of Wyoming, Graduate School, College of Engineering, Department of Chemical and Petroleum Engineering, Laramie, WY 82071. Offers chemical engineering (MS, PhD); environmental engineering (MS); petroleum engineering (MS, PhD). *Faculty:* 11 full-time (0 women). *Students:* 13 full-time (2 women), 8 part-time (1 woman); includes 1 minority (Hispanic American), 12 international. Terminal master's awarded for partial completion of doctoral program. *Degree requirements:* For master's, thesis required, foreign language not required; for doctorate, computer language, dissertation required, foreign language not required. *Entrance requirements:* For master's, GRE General Test, TOEFL (minimum score of 550 required), minimum GPA of 3.0; for doctorate, GRE General Test, TOEFL, minimum GPA of 3.0. *Application deadline:* For fall admission, 4/15 (priority date). Applications are processed on a rolling basis. Application fee: $40. Electronic applications accepted. Tuition, state resident: full-time $2,520; part-time $140 per credit hour. Tuition, nonresident: full-time $7,790; part-time $433 per credit hour. Required fees: $400; $7 per credit hour. Full-time tuition and fees vary according to course load and program. *Unit head:* Dr. Henry W. Haynes, Head, 307-766-4923, Fax: 307-766-6777, E-mail: haynes@uwyo.edu. *Application contact:* Dr. David O. Cooney, Graduate Student Coordinator, 307-766-6464, Fax: 307-766-6777, E-mail: cooney@uwyo.edu.

University of Wyoming, Graduate School, College of Engineering, Department of Civil Engineering, Program in Environmental Engineering, Laramie, WY 82071. Offers MS. *Faculty:* 6. *Students:* 3 full-time (1 woman), 2 part-time (both women). 7 applicants, 43% accepted. In 1998, 6 degrees awarded. *Degree requirements:* For master's, computer language required, thesis optional. *Entrance requirements:* For master's, GRE General Test, TOEFL (minimum score of 550 required), minimum GPA of 3.0. *Application deadline:* For fall admission, 3/1 (priority date); for spring admission, 9/1. Applications are processed on a rolling basis. Application fee: $40. Electronic applications accepted. Tuition, state resident: full-time $2,520; part-time $140 per credit hour. Tuition, nonresident: full-time $7,790; part-time $433 per credit hour. Required fees: $400; $7 per credit hour. Full-time tuition and fees vary according to course load and program. *Financial aid:* In 1998–99, 2 research assistantships, 2 teaching assistantships were awarded.; fellowships, career-related internships or fieldwork, Federal Work-Study, and institutionally-sponsored loans also available. Financial aid application deadline: 3/1. *Faculty research:* Water and waste water, solid and hazardous waste management, air pollution control, flue-gas cleanup. *Unit head:* Dr. M. P. Sharma, Chairman, 307-766-6317, E-mail: sharma@uwyo.edu. *Application contact:* P. A. Van Houten, Graduate Coordinator, 307-766-5446, Fax: 307-766-2221, E-mail: pvh@uwyo.edu.

Utah State University, School of Graduate Studies, College of Engineering, Department of Civil and Environmental Engineering, Logan, UT 84322. Offers ME, MS, PhD, CE. *Faculty:* 27 full-time (2 women). *Students:* 53 full-time (6 women), 30 part-time (3 women); includes 2 minority (1 Asian American or Pacific Islander, 1 Native American), 23 international. Average age 28. 179 applicants, 55% accepted. In 1998, 23 master's, 6 doctorates awarded. *Degree requirements:* For master's, thesis required (for some programs), foreign language not required; for doctorate, dissertation required, foreign language not required. *Entrance requirements:* For master's and doctorate, GRE General Test (score in 40th percentile or higher required), TOEFL (minimum score of 550 required), minimum GPA of 3.0. *Application deadline:* For fall admission, 6/15 (priority date); for spring admission, 10/15. Applications are processed on a rolling basis. Application fee: $40. Tuition, state resident: full-time $1,492. Tuition, nonresident: full-time $5,232. Required fees: $434. Tuition and fees vary according to course load. *Financial aid:* In 1998–99, 49 research assistantships with partial tuition reimbursements were awarded.; fellowships with partial tuition reimbursements, teaching assistantships with partial tuition reimbursements, career-related internships or fieldwork, Federal Work-Study, and institutionally-sponsored loans also available. Aid available to part-time students. Financial aid application deadline: 3/31. *Faculty research:* Hazardous waste treatment, large space structures, river basin management, earthquake engineering, environmental impact. *Unit head:* Loren R. Anderson, Head, 435-797-2932.

Vanderbilt University, School of Engineering, Department of Civil and Environmental Engineering, Program in Environmental Engineering, Nashville, TN 37240-1001. Offers M Eng, MS, PhD. MS and PhD offered through the Graduate School. Part-time programs available. *Faculty:* 8 full-time (1 woman), 1 part-time (0 women). *Students:* 21 full-time (4 women), 9 part-time (1 woman); includes 1 minority (Asian American or Pacific Islander), 15 international. Average age 26. 40 applicants, 25% accepted. In 1998, 6 master's awarded (83% found work related to degree, 17% continued full-time study). Terminal master's awarded for partial completion of doctoral program. *Degree requirements:* For master's, computer language, thesis or alternative required, foreign language not required; for doctorate, computer language, dissertation required, foreign language not required. *Entrance requirements:* For master's, GRE General Test (minimum score of 650 on quantitative section; combined average 1800 on three sections); for doctorate, GRE General Test (minimum score of 650 on quantitative section; combined average 2000 on three sections). *Average time to degree:* Master's–2 years full-time, 4 years part-time; doctorate–4 years full-time, 8 years part-time. *Application deadline:* For fall admission, 1/15 (priority date). Applications are processed on a rolling basis. Application fee: $40. *Financial aid:* In 1998–99, 18 students received aid, including 2 fellowships with full tuition reimbursements available (averaging $5,000 per year), 5 research assistantships with full tuition reimbursements available (averaging $14,000 per year), 3 teaching assistantships with full tuition reimbursements available (averaging $12,000 per year); career-related internships or fieldwork, institutionally-sponsored loans, and tuition waivers (full and partial) also available. Financial aid application deadline: 1/15. *Faculty research:* Waste treatment, hazardous waste management, chemical waste treatment, water quality. Total annual research

Environmental Engineering–Fire Protection Engineering

Vanderbilt University (continued)
expenditures: $300,000. *Unit head:* Edward Thackston, Chair, Department of Civil and Environmental Engineering, 615-343-2372.

Villanova University, College of Engineering, Department of Civil and Environmental Engineering, Program in Water Resources and Environmental Engineering, Villanova, PA 19085-1699. Offers MSWREE. Part-time and evening/weekend programs available. *Faculty:* 5 full-time (0 women), 5 part-time (0 women). *Students:* 5 full-time (2 women), 12 part-time (3 women); includes 2 minority (1 African American, 1 Hispanic American), 2 international. Average age 25. 19 applicants, 89% accepted. In 1998, 12 degrees awarded. *Degree requirements:* For master's, computer language required, thesis optional, foreign language not required. *Entrance requirements:* For master's, GRE General Test (for applicants with degrees from foreign universities), minimum GPA of 3.0, BCE or bachelor's degree in science or related engineering field. *Average time to degree:* Master's–1 year full-time, 3.5 years part-time. *Application deadline:* For fall admission, 8/1 (priority date); for spring admission, 12/1. Applications are processed on a rolling basis. Application fee: $40. *Financial aid:* Federal Work-Study and tuition waivers (full and partial) available. Aid available to part-time students. Financial aid application deadline: 4/15. *Faculty research:* Photocatalytic decontamination and disinfection of water, urban storm water wetlands, economy and risk, removal and destruction of organic acids in water, sludge treatment. *Unit head:* Dr. Lewis J. Mathers, Chairman, Department of Civil and Environmental Engineering, 610-519-4960, Fax: 610-519-6754, E-mail: ldeangel@email.vill.edu.

Virginia Polytechnic Institute and State University, Graduate School, College of Engineering, Department of Civil and Environmental Engineering, Program in Environmental Engineering, Blacksburg, VA 24061. Offers MS. *Accreditation:* ABET. *Students:* 25 full-time (9 women), 21 part-time (9 women); includes 4 minority (1 African American, 2 Asian Americans or Pacific Islanders, 1 Hispanic American), 9 international. 55 applicants, 38% accepted. In 1998, 4 degrees awarded. *Degree requirements:* For master's, thesis required, foreign language not required. *Entrance requirements:* For master's, GRE General Test (minimum score of 400 on each section required), TOEFL (minimum score of 600 required). *Application deadline:* For fall admission, 12/1 (priority date). Applications are processed on a rolling basis. Application fee: $25. *Financial aid:* Fellowships available. Financial aid application deadline: 4/1. *Unit head:* Dr. John T. Novak, Chairman, 540-231-6635.

Virginia Polytechnic Institute and State University, Graduate School, College of Engineering, Department of Civil and Environmental Engineering, Program in Environmental Sciences and Engineering, Blacksburg, VA 24061. Offers MS, PhD. *Students:* 13 full-time (3 women), 17 part-time (6 women); includes 2 minority (both Asian Americans or Pacific Islanders), 4 international. 75 applicants, 24% accepted. In 1998, 10 master's awarded. *Degree requirements:* For master's, thesis required, foreign language not required; for doctorate, dissertation required. *Entrance requirements:* For master's and doctorate, GRE General Test (minimum score of 400 on each section required), TOEFL (minimum score of 600 required). *Application deadline:* For fall admission, 12/1 (priority date). Applications are processed on a rolling basis. Application fee: $25. *Financial aid:* Fellowships, research assistantships, teaching assistantships, unspecified assistantships available. Financial aid application deadline: 4/1. *Unit head:* Dr. John T. Novak, Chairman, 540-231-6635.

Washington State University, Graduate School, College of Engineering and Architecture, Department of Civil and Environmental Engineering, Program in Environmental Engineering, Pullman, WA 99164. Offers MS. *Students:* 12 full-time (5 women), 1 (woman) part-time; includes 3 minority (1 Asian American or Pacific Islander, 2 Hispanic Americans), 2 international. In 1998, 14 degrees awarded. *Degree requirements:* For master's, thesis, oral exam required, foreign language not required. *Entrance requirements:* For master's, GRE General Test, minimum GPA of 3.0. *Average time to degree:* Master's–1.5 years full-time. *Application deadline:* For fall admission, 3/1 (priority date); for spring admission, 10/1. Applications are processed on a rolling basis. Application fee: $35. *Financial aid:* In 1998–99, 7 research assistantships, 8 teaching assistantships were awarded.; fellowships, career-related internships or fieldwork, Federal Work-Study, and institutionally-sponsored loans also available. Financial aid application deadline: 4/1; financial aid applicants required to submit FAFSA. *Application contact:* Maureen Clausen, Graduate Secretary, 509-335-2576, Fax: 509-335-7632, E-mail: mclausen@wsu.edu.

Washington University in St. Louis, School of Engineering and Applied Science, Sever Institute of Technology, Program in Environmental Engineering, St. Louis, MO 63130-4899. Offers MSEE, D Sc. Part-time and evening/weekend programs available. Terminal master's awarded for partial completion of doctoral program. *Degree requirements:* For doctorate, variable foreign language requirement, dissertation, departmental qualifying exam required, foreign language not required. *Faculty research:* Air pollution transport, environmental reaction engineering, environmental informatics, industrial pollution prevention, environmental policy and law.

Wayne State University, Graduate School, College of Engineering, Department of Civil and Environmental Engineering, Detroit, MI 48202. Offers MS, PhD. *Degree requirements:* For master's, thesis optional, foreign language not required; for doctorate, dissertation required, foreign language not required. *Faculty research:* Environmental geotechnics, bridge engineering, seismic analysis, construction safety, mass transit.

See in-depth description on page 651.

West Virginia University, College of Engineering and Mineral Resources, Department of Civil and Environmental Engineering, Morgantown, WV 26506. Offers civil engineering (MSCE, MSE); engineering (PhD). Part-time programs available. *Degree requirements:* For master's, thesis required, foreign language not required; for doctorate, dissertation, comprehensive exam required, foreign language not required. *Entrance requirements:* For master's, TOEFL (minimum score of 550 required), minimum GPA of 3.0; for doctorate, TOEFL (minimum score of 550 required), GRE (international students), minimum GPA of 3.0. *Faculty research:* Environmental and hydrotechnical structural composites, bridge innovation and rehabilitation, transport, soil mechanics, geoenvironmental engineering.

Worcester Polytechnic Institute, Graduate Studies, Department of Civil and Environmental Engineering, Worcester, MA 01609-2280. Offers M Eng, MS, PhD, Advanced Certificate, Certificate. Part-time and evening/weekend programs available. Postbaccalaureate distance learning degree programs offered (minimal on-campus study). *Faculty:* 13 full-time (0 women), 4 part-time (1 woman). *Students:* 25 full-time (6 women), 27 part-time (6 women); includes 5 minority (1 African American, 4 Hispanic Americans), 4 international. 57 applicants, 67% accepted. In 1998, 10 master's, 3 doctorates awarded. *Degree requirements:* For master's, thesis optional, foreign language not required; for doctorate, dissertation required, foreign language not required. *Entrance requirements:* For master's, GRE General Test (combined average 1711 on three sections), TOEFL (minimum score of 550 required average 605); for doctorate, GRE General Test (combined average 1711 on three sections), TOEFL (minimum score of 550 required; average 605). *Application deadline:* For fall admission, 2/15 (priority date); for spring admission, 10/15 (priority date). Applications are processed on a rolling basis. Application fee: $50. Electronic applications accepted. *Financial aid:* In 1998–99, 15 students received aid, including 2 fellowships with full tuition reimbursements available (averaging $13,276 per year), 3 research assistantships with full tuition reimbursements available (averaging $15,000 per year), 10 teaching assistantships with full tuition reimbursements available (averaging $11,970 per year); career-related internships or fieldwork, grants, institutionally-sponsored loans, and scholarships also available. Financial aid application deadline: 2/15; financial aid applicants required to submit FAFSA. *Faculty research:* Water and wastewater treatment, construction management, structural and geotechnical engineering. Total annual research expenditures: $126,039. *Unit head:* Dr. Fred L. Hart, Head, 508-831-5530, Fax: 508-831-5808, E-mail: flhart@wpi.edu. *Application contact:* James O'Shaughnessy, Graduate Coordinator, 508-831-5309, Fax: 508-831-5808, E-mail: jco@wpi.edu.

Youngstown State University, Graduate School, William Rayen College of Engineering, Department of Civil, Chemical, and Environmental Engineering, Youngstown, OH 44555-0001. Offers MSE. Part-time and evening/weekend programs available. *Faculty:* 9 full-time (1 woman). *Students:* 13 full-time (2 women), 5 part-time (2 women), 4 international. 4 applicants, 75% accepted. In 1998, 10 degrees awarded. *Degree requirements:* For master's, computer language required, thesis optional, foreign language not required. *Entrance requirements:* For master's, TOEFL (minimum score of 550 required), minimum GPA of 2.75 in field. *Application deadline:* For fall admission, 8/15 (priority date); for winter admission, 11/15 (priority date); for spring admission, 2/15 (priority date). Applications are processed on a rolling basis. Application fee: $30 ($75 for international students). Tuition, state resident: part-time $97 per credit hour. Tuition, nonresident: part-time $219 per credit hour. Required fees: $21 per credit hour. $41 per quarter. *Financial aid:* In 1998–99, 4 students received aid, including 4 research assistantships with full tuition reimbursements available (averaging $7,500 per year); teaching assistantships, Federal Work-Study, institutionally-sponsored loans, and scholarships also available. Aid available to part-time students. Financial aid application deadline: 3/1. *Faculty research:* Structural mechanics, water quality modeling, surface and ground water hydrology, physical and chemical processes in aquatic systems. *Unit head:* Dr. Jack D. Bakos, Chair, 330-742-3027. *Application contact:* Dr. Peter J. Kasvinsky, Dean of Graduate Studies, 330-742-3091, Fax: 330-742-1580, E-mail: amgrad03@ysub.ysu.edu.

Fire Protection Engineering

University of British Columbia, Faculty of Graduate Studies, Faculty of Applied Science, Department of Chemical Engineering, Vancouver, BC V6T 1Z2, Canada. Offers chemical engineering (M Eng, MA Sc, PhD); fire protection engineering (M Eng); pulp and paper engineering (M Eng). Part-time and evening/weekend programs available. *Degree requirements:* For master's, thesis required (for some programs), foreign language not required; for doctorate, dissertation required, foreign language not required. *Entrance requirements:* For master's and doctorate, TOEFL (minimum score of 550 required). *Faculty research:* Electrochemical engineering, biotechnology, catalysis, polymers.

University of Maryland, College Park, Graduate School, A. James Clark School of Engineering, Department of Fire Protection Engineering, College Park, MD 20742-5045. Offers M Eng, MS. Part-time and evening/weekend programs available. *Faculty:* 6 full-time (0 women), 3 part-time (0 women). *Students:* 11 full-time (2 women), 32 part-time (7 women); includes 5 minority (2 African Americans, 1 Asian American or Pacific Islander, 2 Hispanic Americans), 5 international. Average age 26. 11 applicants, 91% accepted. In 1998, 16 degrees awarded (94% found work related to degree, 6% continued full-time study). *Degree requirements:* For master's, thesis optional, foreign language not required. *Entrance requirements:* For master's, GRE General Test, minimum GPA of 3.0, BS in any engineering or physical science area. *Average time to degree:* Master's–1.5 years full-time, 3 years part-time. *Application deadline:* For fall admission, 8/1; for spring admission, 11/1. Applications are processed on a rolling basis. Application fee: $50 ($70 for international students). Electronic applications accepted. Tuition, state resident: part-time $272 per credit hour. Tuition, nonresident: part-time $475 per credit hour. Required fees: $632; $379 per year. *Financial aid:* In 1998–99, 9 research assistantships with tuition reimbursements (averaging $15,550 per year), 1 teaching assistantship with tuition reimbursement (averaging $15,550 per year) were awarded.; fellowships, career-related internships or fieldwork, Federal Work-Study, institutionally-sponsored loans, and scholarships also available. Financial aid application deadline: 2/1; financial aid applicants required to submit FAFSA. *Faculty research:* Fire and thermal degradation of materials, fire modeling, fire dynamics, smoke detection and management, fire resistance. Total annual research expenditures: $507,942. *Unit head:* Dr. Steven M. Spivak, Chair, 301-405-6651, Fax: 301-405-9383, E-mail: ss60@eng.umd.edu. *Application contact:* Dr. James A. Milke, Graduate Director, 301-405-3995, Fax: 301-405-9383, E-mail: milke@eng.umd.edu.

Announcement: MS degree programs are concerned with fire science, fundamentals of diffusion, flame combustion, mechanics of flame propagation, and prediction of fire development. Other courses concentrate on simulation and risk analysis, quantitative assessment of fire hazard and the probabilities of potential fire incidents. Individual courses of study can be arranged.

University of Maryland, College Park, Graduate School, A. James Clark School of Engineering, Professional Program in Engineering, College Park, MD 20742-5045. Offers aerospace engineering (M Eng); chemical engineering (M Eng); civil engineering (M Eng); electrical engineering (M Eng); fire protection engineering (M Eng); materials science and engineering (M Eng); mechanical engineering (M Eng); reliability engineering (M Eng); systems engineering (M Eng). Part-time and evening/weekend programs available. Postbaccalaureate distance learning degree programs offered. *Faculty:* 11 part-time (0 women). *Students:* 20 full-time (3 women), 205 part-time (42 women); includes 58 minority (27 African Americans, 25 Asian Americans or Pacific Islanders, 5 Hispanic Americans, 1 Native American), 20 international. *Degree requirements:* For master's, foreign language and thesis not required. *Application deadline:* Applications are processed on a rolling basis. Application fee: $50 ($70 for international students). Tuition, state resident: part-time $272 per credit hour. Tuition, nonresident: part-time $475 per credit hour. Required fees: $632; $379 per year. *Unit head:* Dr. Patrick Cunniff, Associate Dean, 301-405-5256, Fax: 301-314-9477. *Application contact:* Trudy Lindsey, Director, Graduate Admission and Records, 301-405-4198, Fax: 301-314-9305, E-mail: grschool@deans.umd.edu.

University of New Haven, Graduate School, School of Public Safety and Professional Studies, Program in Fire Science, West Haven, CT 06516-1916. Offers MS. 8 applicants, 38% accepted. In 1998, 4 degrees awarded. *Degree requirements:* For master's, thesis or alternative required, foreign language not required. *Application deadline:* Applications are processed on a rolling basis. Application fee: $50. *Financial aid:* Career-related internships or fieldwork and Federal Work-Study available. Aid available to part-time students. Financial aid application deadline: 5/1; financial aid applicants required to submit FAFSA. *Unit head:* Robert Sawyer, Director, 203-932-7298.

Worcester Polytechnic Institute, Graduate Studies, Program in Fire Protection Engineering, Worcester, MA 01609-2280. Offers MS, PhD, Advanced Certificate, Certificate. Part-time and evening/weekend programs available. Postbaccalaureate distance learning degree programs offered (no on-campus study). *Faculty:* 3 full-time (0 women), 1 part-time (0 women). *Students:* 39 full-time (5 women), 26 part-time (6 women); includes 6 minority (3 Asian Americans or Pacific Islanders, 3 Hispanic Americans), 7 international. 40 applicants, 88% accepted. In 1998, 26 master's awarded. *Degree requirements:* For master's, foreign language and thesis not required; for doctorate, dissertation required, foreign language not required. *Entrance requirements:* For master's, TOEFL (minimum score of 550 required; average 586), BS in engineering or physical sciences; for doctorate, TOEFL (minimum score of 550 required; average 586). *Average time to degree:* Master's–2 years full-time, 8 years part-time. *Application deadline:* For fall admission, 2/15 (priority date); for spring admission, 10/15 (priority date). Applications are processed on a rolling basis. Application fee: $50. Electronic applica-

tions accepted. *Financial aid:* In 1998–99, 7 students received aid, including 1 fellowship with full tuition reimbursement available (averaging $14,200 per year), 5 research assistantships with full tuition reimbursements available (averaging $15,000 per year), 2 teaching assistantships with full tuition reimbursements available (averaging $11,970 per year); career-related internships or fieldwork, grants, institutionally-sponsored loans, and scholarships also available. Financial aid application deadline: 2/15; financial aid applicants required to submit FAFSA. Total annual research expenditures: $645,070. *Unit head:* David A. Lucht, Director, Center for Firesafety Studies, 508-831-5593, Fax: 508-831-5680, E-mail: dalucht@wpi.edu. *Application contact:* Jonathan Barnett, Graduate Coordinator, 508-831-5113, Fax: 508-831-5680, E-mail: jbarnett@wpi.edu.

Geotechnical Engineering

Auburn University, Graduate School, College of Engineering, Department of Civil Engineering, Auburn, Auburn University, AL 36849-0002. Offers construction engineering and management (MCE, MS, PhD); environmental engineering (MCE, MS, PhD); geotechnical/materials engineering (MCE, MS, PhD); hydraulics/hydrology (MCE, MS, PhD); structural engineering (MCE, MS, PhD); transportation engineering (MCE, MS, PhD). Part-time programs available. *Faculty:* 22 full-time. *Students:* 24 full-time (6 women), 34 part-time (8 women); includes 2 minority (1 African American, 1 Hispanic American), 9 international. *Degree requirements:* For master's, project (MCE), thesis (MS) required; for doctorate, dissertation, comprehensive exam required, foreign language not required. *Entrance requirements:* For master's, GRE General Test; for doctorate, GRE General Test (minimum score of 400 on each section required). *Application deadline:* For fall admission, 9/1; for spring admission, 3/1. Applications are processed on a rolling basis. Application fee: $25 ($50 for international students). Tuition, state resident: full-time $2,760; part-time $76 per credit hour. Tuition, nonresident: full-time $8,280; part-time $228 per credit hour. *Unit head:* Dr. Joseph F. Judkins, Head, 334-844-4320. *Application contact:* Dr. John F. Pritchett, Dean of the Graduate School, 334-844-4700.

The Catholic University of America, School of Engineering, Department of Civil Engineering, Program in Geotechnical Engineering, Washington, DC 20064. Offers MCE. *Degree requirements:* For master's, thesis optional, foreign language not required. *Entrance requirements:* For master's, minimum GPA of 3.0. *Application deadline:* For fall admission, 8/1 (priority date); for spring admission, 12/1. Applications are processed on a rolling basis. Application fee: $50. *Financial aid:* Application deadline: 2/1. *Unit head:* Dr. Timothy W. Kao, Chair, Department of Civil Engineering, 202-319-5163, Fax: 202-319-4499, E-mail: kao@cua.edu.

Colorado State University, Graduate School, College of Engineering, Department of Civil Engineering, Specialization in Structural and Geotechnical Engineering, Fort Collins, CO 80523-0015. Offers MS, PhD. Part-time programs available. Postbaccalaureate distance learning degree programs offered (no on-campus study). *Faculty:* 9 full-time (0 women). In 1998, 13 master's, 2 doctorates awarded. Terminal master's awarded for partial completion of doctoral program. *Degree requirements:* For master's, thesis or alternative required, foreign language not required; for doctorate, dissertation required, foreign language not required. *Entrance requirements:* For master's and doctorate, GRE General Test, TOEFL (minimum score of 550 required; 213 for computer-based), minimum GPA of 3.0. *Average time to degree:* Master's–2 years full-time, 5 years part-time; doctorate–4 years full-time. *Application deadline:* For fall admission, 3/1 (priority date); for spring admission, 8/1 (priority date). Applications are processed on a rolling basis. Application fee: $30. Electronic applications accepted. *Financial aid:* Research assistantships, teaching assistantships, Federal Work-Study and institutionally-sponsored loans available. *Faculty research:* Computational mechanics, solid mechanics, finite elements, timber engineering, concrete structures, expansive soils, blast and earthquake induced liquefaction, export systems. Total annual research expenditures: $700,000. *Unit head:* Erik G. Thompson, Leader, 970-491-6060, Fax: 970-491-7727, E-mail: thompson@engr.colostate.edu. *Application contact:* Laurie Howard, Student Adviser, 970-491-5844, Fax: 970-491-7727, E-mail: lhoward@engr.colostate.edu.

Cornell University, Graduate School, Graduate Fields of Engineering, Field of Civil and Environmental Engineering, Ithaca, NY 14853-0001. Offers environmental engineering (M Eng, MS, PhD); environmental fluid mechanics and hydrology (M Eng, MS, PhD); environmental systems engineering (M Eng, MS, PhD); geotechnical engineering (M Eng, MS, PhD); remote sensing (M Eng, MS, PhD); structural engineering (M Eng, MS, PhD); transportation engineering (M Eng, MS, PhD); water resource systems (M Eng, MS, PhD). *Faculty:* 29 full-time. *Students:* 143 full-time (32 women); includes 16 minority (11 Asian Americans or Pacific Islanders, 5 Hispanic Americans), 76 international. Terminal master's awarded for partial completion of doctoral program. *Degree requirements:* For master's, thesis (MS) required; for doctorate, dissertation required, foreign language not required. *Entrance requirements:* For master's, TOEFL (minimum score of 600required); for doctorate, GRE General Test, TOEFL (minimum score of 600 required). *Application deadline:* For fall admission, 1/15. Application fee: $65. Electronic applications accepted. *Unit head:* Director of Graduate Studies, 607-255-7560, Fax: 607-255-9004. *Application contact:* Graduate Field Assistant, 607-255-7560, E-mail: cee_grad@cornell.edu.

See in-depth description on page 573.

École Polytechnique de Montréal, Graduate Programs, Department of Civil Engineering, Montréal, PQ H3C 3A7, Canada. Offers environmental engineering (M Eng, M Sc A, PhD); geotechnical engineering (M Eng, M Sc A, PhD); hydraulics engineering (M Eng, M Sc A, PhD); structural engineering (M Eng, M Sc A, PhD); transportation engineering (M Eng, M Sc A, PhD). Part-time and evening/weekend programs available. *Degree requirements:* For master's and doctorate, one foreign language, computer language, thesis/dissertation required. *Entrance requirements:* For master's, minimum GPA of 2.75; for doctorate, minimum GPA of 3.0. *Faculty research:* Water resources management, characteristics of building materials, aging of dams, pollution control.

Florida Institute of Technology, Graduate School, College of Engineering, Division of Engineering Science, Program in Civil Engineering, Melbourne, FL 32901-6975. Offers civil engineering (PhD); geotechnical engineering (MS); structures engineering (MS); water resources (MS). Part-time programs available. *Faculty:* 5 full-time (0 women), 4 part-time (2 women). *Students:* 7 full-time (0 women), 10 part-time (2 women); includes 1 minority (Hispanic American), 9 international. *Degree requirements:* For master's, thesis optional, foreign language not required; for doctorate, dissertation required, foreign language not required. *Entrance requirements:* For master's, minimum GPA of 3.0; for doctorate, minimum GPA of 3.2. *Application deadline:* Applications are processed on a rolling basis. Application fee: $50. Electronic applications accepted. Tuition: Part-time $575 per credit hour. Required fees: $100. Tuition and fees vary according to campus/location and program. *Unit head:* Dr. Ashok Pandit, Chair, 407-674-7151, Fax: 407-768-7565, E-mail: apandit@fit.edu. *Application contact:* Carolyn P. Farrior, Associate Dean of Graduate Admissions, 407-674-7118, Fax: 407-723-9468, E-mail: cfarrior@fit.edu.

See in-depth description on page 581.

Iowa State University of Science and Technology, Graduate College, College of Engineering, Department of Civil and Construction Engineering, Ames, IA 50011. Offers civil engineer-

ing (MS, PhD), including civil engineering materials, construction engineering and management, environmental engineering, geometronics, geotechnical engineering, structural engineering, transportation engineering. *Faculty:* 36 full-time, 5 part-time. *Students:* 60 full-time (14 women), 51 part-time (12 women). *Degree requirements:* For master's, thesis or alternative required, foreign language not required; for doctorate, dissertation required, foreign language not required. *Entrance requirements:* For master's and doctorate, GRE General Test (foreign students), TOEFL. *Application deadline:* For fall admission, 6/15 (priority date); for spring admission, 11/15 (priority date). Application fee: $20 ($50 for international students). Electronic applications accepted. Tuition, state resident: full-time $3,308. Tuition, nonresident: full-time $9,744. Part-time tuition and fees vary according to course load, campus/location and program. *Unit head:* Dr. Lowell F. Greimann, Chair, 515-294-2140, E-mail: cceinfo@iastate.edu. *Application contact:* Edward Kannel, Director of Graduate Education, 515-294-2861, E-mail: cceinfo@iastate.edu.

Louisiana State University and Agricultural and Mechanical College, Graduate School, College of Engineering, Department of Civil and Environmental Engineering, Baton Rouge, LA 70803. Offers environmental engineering (MSCE, PhD); geotechnical engineering (MSCE, PhD); structural engineering and mechanics (MSCE, PhD); transportation engineering (MSCE, PhD); water resources (MSCE, PhD). Part-time programs available. *Faculty:* 27 full-time (1 woman), 1 part-time (0 women). *Students:* 59 full-time (14 women), 24 part-time (2 women); includes 5 minority (1 African American, 3 Asian Americans or Pacific Islanders, 1 Hispanic American), 48 international. *Degree requirements:* For master's, thesis optional, foreign language not required; for doctorate, one foreign language (computer language can substitute), dissertation required. *Entrance requirements:* For master's and doctorate, GRE General Test, TOEFL, minimum GPA of 3.0. *Application deadline:* For fall admission, 1/25 (priority date). Applications are processed on a rolling basis. Application fee: $25. *Unit head:* Dr. Ronald F. Malone, Acting Chair, 225-388-8666, Fax: 225-388-8652, E-mail: rmalone@unix1.sncc.lsu.edu. *Application contact:* Dr. Vijaya K. A. Gopu, Graduate Coordinator, 225-388-8442, E-mail: cegopu@eng.lsu.edu.

See in-depth description on page 589.

Marquette University, Graduate School, College of Engineering, Department of Civil and Environmental Engineering, Milwaukee, WI 53201-1881. Offers construction and public works management (MS, PhD); environmental/water resources engineering (MS, PhD); structural/geotechnical engineering (MS, PhD); transportational planning and engineering (MS, PhD). Part-time and evening/weekend programs available. *Faculty:* 12 full-time (0 women), 1 part-time (0 women). *Students:* 17 full-time (7 women), 32 part-time (5 women); includes 5 minority (1 African American, 4 Asian Americans or Pacific Islanders), 12 international. Terminal master's awarded for partial completion of doctoral program. *Degree requirements:* For master's, thesis or alternative, comprehensive exam required, foreign language not required; for doctorate, dissertation required, foreign language not required. *Entrance requirements:* For master's, TOEFL (minimum score of 550 required); for doctorate, GRE General Test, TOEFL (minimum score of 550 required). *Application deadline:* For fall admission, 6/1 (priority date). Applications are processed on a rolling basis. Application fee: $40. Tuition: Part-time $510 per credit hour. Tuition and fees vary according to program. *Unit head:* Dr. Thomas H. Wenzel, Chairman, 414-288-7030, Fax: 414-288-7521, E-mail: wenzelt@vms.csd.mu.edu.

McGill University, Faculty of Graduate Studies and Research, Faculty of Engineering, Department of Civil Engineering and Applied Mechanics, Program in Geotechnical and Geoenvironmental Engineering, Montreal, PQ H3A 2K6, Canada. Offers M Eng, M Sc, PhD. Part-time and evening/weekend programs available. *Degree requirements:* For master's, computer language required, thesis optional, foreign language not required; for doctorate, computer language, dissertation required, foreign language not required. *Entrance requirements:* For master's, TOEFL (minimum score of 550 required), minimum GPA of 3.0; for doctorate, TOEFL (minimum score of 580 required). *Faculty research:* Geomechanics, mechanics of frozen soils, soil-foundation interaction, groundwater contamination and remediation, flow in fractional media, hydrothermomechanical processes in geomaterials.

Michigan Technological University, Graduate School, College of Engineering, Department of Geology, Geophysics and Geological Engineering, Program in Geology, Houghton, MI 49931-1295. Offers geology (MS, PhD); geotechnical engineering (PhD). Part-time programs available. *Faculty:* 8 full-time (1 woman), 1 part-time (0 women). *Students:* 16 full-time (6 women); includes 1 minority (African American), 2 international. *Degree requirements:* For master's and doctorate, thesis/dissertation required, foreign language not required. *Entrance requirements:* For master's, GRE General Test (combined average 1714 on three sections), TOEFL (minimum score of 550 required; average 657); for doctorate, GRE General Test (combined average 1827 on three sections), TOEFL (minimum score of 550 required; average 596). *Application deadline:* For fall admission, 3/15 (priority date). Applications are processed on a rolling basis. Application fee: $30 ($35 for international students). Tuition, state resident: full-time $4,377. Tuition, nonresident: full-time $9,108. Required fees: $126. Tuition and fees vary according to course load. *Unit head:* V. T. V. Nguyen, Chair, Graduate Admissions Committee, 514-398-6870, Fax: 514-398-7361, E-mail: sandy@civil.lan.mcgill.ca. *Application contact:* Dr. Jimmy Diehl, Graduate Coordinator, 906-487-2665, Fax: 906-487-3371, E-mail: jdiehl@mtu.edu.

Michigan Technological University, Graduate School, College of Engineering, Department of Mining Engineering, Houghton, MI 49931-1295. Offers geotechnical engineering (PhD); mining engineering (MS, PhD). Part-time programs available. *Faculty:* 4 full-time (1 woman), 1 part-time (0 women). *Students:* 12 full-time (3 women); includes 2 minority (1 African American, 1 Asian American or Pacific Islander), 5 international. *Degree requirements:* For master's, thesis or alternative required, foreign language not required; for doctorate, dissertation required, foreign language not required. *Entrance requirements:* For master's and doctorate, TOEFL (minimum score of 550 required), BS in engineering or science. *Application deadline:* For fall admission, 3/15 (priority date). Applications are processed on a rolling basis. Application fee: $30 ($35 for international students). Tuition, state resident: full-time $4,377. Tuition, nonresident: full-time $9,108. Required fees: $126. Tuition and fees vary according to course load. *Unit head:* Dr. O. Francis Otuonye, Chair, 906-487-2610, Fax: 906-487-2495, E-mail: frotuony@mtu.edu. *Application contact:* Dr. Jiann-Yang Hwang, Associate Professor, 906-487-2600, Fax: 906-487-2495, E-mail: jhwang@mtu.edu.

Northwestern University, The Graduate School, Robert R. McCormick School of Engineering and Applied Science, Department of Civil Engineering, Evanston, IL 60208. Offers biosolid

Geotechnical Engineering

Northwestern University *(continued)*
mechanics (MS, PhD); environmental health engineering (MS, PhD); geotechnical engineering (MS, PhD); health physics/radiological health (MS, PhD); project management (MPM); structural engineering (MS, PhD); structural mechanics (MS, PhD); transportation systems engineering (MS, PhD). MS and PhD admissions and degrees offered through The Graduate School. Part-time programs available. *Faculty:* 26 full-time (2 women), 1 (woman) part-time. *Students:* 110 full-time (38 women), 7 part-time (2 women); includes 5 minority (2 African Americans, 3 Asian Americans or Pacific Islanders), 68 international. Terminal master's awarded for partial completion of doctoral program. *Degree requirements:* For master's, thesis required (for some programs); for doctorate, dissertation required. *Entrance requirements:* For master's and doctorate, GRE General Test, TOEFL (minimum score of 560 required). Application fee: $50 ($55 for international students). *Unit head:* Joseph L. Schofer, Chair, 847-491-3257, Fax: 847-491-4011, E-mail: j-schofer@nwu.edu. *Application contact:* Karm Kerwell, Secretary, 847-491-3176, Fax: 847-491-4011, E-mail: k-kerwell@nwu.edu.

Ohio University, Graduate Studies, College of Engineering and Technology, Department of Civil Engineering, Athens, OH 45701-2979. Offers geotechnical and environmental engineering (MS); water resources and structures (MS). *Faculty:* 11 full-time (1 woman). *Students:* 24 full-time (5 women), 2 part-time, 13 international. *Degree requirements:* For master's, thesis optional, foreign language not required. *Entrance requirements:* For master's, minimum GPA of 3.0. *Application deadline:* For fall admission, 5/1 (priority date). Applications are processed on a rolling basis. Application fee: $30. Tuition, state resident: full-time $5,754; part-time $238 per credit hour. Tuition, nonresident: full-time $11,055; part-time $457 per credit hour. Tuition and fees vary according to course load, campus/location and program. *Unit head:* Dr. Gayle Mitchell, Chair, 740-593-1465, Fax: 740-593-0625, E-mail: gmitchell@bobcat.ent.ohiou.edu. *Application contact:* Dr. Glenn Hazen, Graduate Chair, 740-593-1469, Fax: 740-593-0625, E-mail: ghazen@bobcat.ent.ohiou.edu.

Ohio University, Graduate Studies, College of Engineering and Technology, Department of Integrated Engineering, Athens, OH 45701-2979. Offers geotechnical and environmental engineering (PhD); intelligent systems (PhD); materials processing (PhD). *Faculty:* 39 full-time (1 woman). *Students:* 9 full-time (0 women), 6 part-time; includes 1 minority (Asian American or Pacific Islander), 12 international. *Degree requirements:* For doctorate, computer language, dissertation required, foreign language not required. *Entrance requirements:* For doctorate, GRE General Test, MS in engineering or related field. *Application deadline:* For fall admission, 3/15. Applications are processed on a rolling basis. Application fee: $30. Tuition, state resident: full-time $5,754; part-time $238 per credit hour. Tuition, nonresident: full-time $11,055; part-time $457 per credit hour. Tuition and fees vary according to course load, campus/location and program. *Unit head:* Dr. Jerrel R. Mitchell, Associate Dean for Research and Graduate Studies, 740-593-1482, E-mail: mitchell@bobcat.ent.ohiou.edu.

Rensselaer Polytechnic Institute, Graduate School, School of Engineering, Department of Civil Engineering, Troy, NY 12180-3590. Offers geotechnical engineering (M Eng, MS, D Eng, PhD); mechanics of composite materials and structures (M Eng, MS, D Eng, PhD); structural engineering (M Eng, MS, D Eng, PhD); transportation engineering (M Eng, MS, D Eng, PhD). Part-time programs available. *Faculty:* 11 full-time (0 women), 3 part-time (0 women). *Students:* 31 full-time (4 women), 11 part-time (2 women); includes 6 minority (3 Asian Americans or Pacific Islanders, 3 Hispanic Americans), 18 international. *Degree requirements:* For master's, thesis required (for some programs), foreign language not required; for doctorate, dissertation required, foreign language not required. *Entrance requirements:* For master's and doctorate, GRE, TOEFL (minimum score of 550 required). *Application deadline:* For fall admission, 2/1 (priority date). Applications are processed on a rolling basis. Application fee: $35. *Unit head:* Dr. George List, Chair, 518-276-6940, Fax: 518-276-4833, E-mail: gregaja@rpi.edu. *Application contact:* Jo Ann Grega, Admissions Assistant, 518-276-6679, Fax: 518-276-4833, E-mail: gregaj2@rpi.edu.

See in-depth description on page 597.

State University of New York at Buffalo, Graduate School, School of Engineering and Applied Sciences, Department of Civil, Structural, and Environmental Engineering, Buffalo, NY 14260. Offers computational engineering and mechanics (MS, PhD); construction (M Eng, MS, PhD); geoenvironmental and geotechnical engineering (M Eng, MS, PhD); structural and earthquake engineering (M Eng, MS, PhD); water resources and environmental engineering (M Eng, MS, PhD). Part-time programs available. Postbaccalaureate distance learning degree programs offered (minimal on-campus study). *Faculty:* 25 full-time (0 women), 7 part-time (1 woman). *Students:* 99 full-time (19 women), 71 part-time (13 women); includes 9 minority (2 African Americans, 5 Asian Americans or Pacific Islanders, 1 Hispanic American, 1 Native American), 101 international. Terminal master's awarded for partial completion of doctoral program. *Degree requirements:* For master's, computer language, project or thesis required; for doctorate, computer language, dissertation required, foreign language not required. *Entrance requirements:* For master's and doctorate, GRE General Test (minimum combined score of 1250 required), TOEFL (minimum score of 550 required). *Application deadline:* For fall admission, 1/15 (priority date); for spring admission, 10/1. Applications are processed on a rolling basis. Application fee: $35. Tuition, state resident: full-time $5,100; part-time $213 per credit hour. Tuition, nonresident: full-time $8,416; part-time $351 per credit hour. Required fees: $870; $75 per semester. Tuition and fees vary according to course load and program. *Unit head:* Dr. Andrei M. Reinhorn, Chairman, 716-645-2114 Ext. 2419, Fax: 716-645-3733, E-mail: reinhorn@civil.eng.buffalo.edu. *Application contact:* Dr. Joe Atkinson, Director of Graduate Admissions, 716-645-2114 Ext. 2326, Fax: 716-645-3667, E-mail: atkinson@acsu.buffalo.edu.

See in-depth description on page 601.

Texas A&M University, College of Engineering, Department of Civil Engineering, Program in Geotechnical Engineering, College Station, TX 77843. Offers M Eng, MS, D Eng, PhD. D Eng offered through the College of Engineering. *Students:* 26. *Degree requirements:* For master's, thesis (MS) required; for doctorate, dissertation (PhD), internship (D Eng) required. *Entrance requirements:* For master's and doctorate, GRE General Test, TOEFL. Application fee: $50 ($75 for international students). *Financial aid:* Fellowships, research assistantships, teaching assistantships available. Financial aid application deadline: 4/1; financial aid applicants required to submit FAFSA. *Faculty research:* Classical geotechnical engineering, marine geotechnical engineering with soil dynamics. *Unit head:* Dr. Paul N. Roschke, Head, Constructed Facilities Division, 409-845-4414, Fax: 409-862-2800, E-mail: ce-grad@tamu.edu. *Application contact:* Dr. Joseph M. Bracci, 409-845-2498, Fax: 409-862-2800, E-mail: ce-grad@tamu.edu.

Tufts University, Division of Graduate and Continuing Studies and Research, Graduate School of Arts and Sciences, College of Engineering, Department of Civil and Environmental Engineering, Medford, MA 02155. Offers civil engineering (MS, PhD), including geotechnical engineering, structural engineering; environmental engineering (MS, PhD), including environmental engineering and environmental sciences, environmental geotechnology, environmental health, environmental science and management, hazardous materials management, water resources engineering. Part-time programs available. *Faculty:* 13 full-time, 10 part-time. *Students:* 99 (47 women); includes 21 minority (5 African Americans, 7 Asian Americans or Pacific Islanders, 9 Hispanic Americans) 16 international. Terminal master's awarded for partial completion of doctoral program. *Degree requirements:* For master's, thesis or alternative required, foreign language not required; for doctorate, dissertation required, foreign language not required. *Entrance requirements:* For master's and doctorate, GRE General Test, TOEFL (minimum score of 550 required). *Application deadline:* For fall admission, 2/15; for spring admission, 10/15. Applications are processed on a rolling basis. Application fee: $50. *Unit head:* Dr. Stephen Levine, Chair, 617-627-3211, Fax: 617-627-3994. *Application contact:* Linfield Brown, 617-627-3211, Fax: 617-627-3994.

University of Alberta, Faculty of Graduate Studies and Research, Department of Civil and Environmental Engineering, Edmonton, AB T6G 2E1, Canada. Offers construction engineering and management (M Eng, M Sc, PhD); environmental engineering (M Eng, M Sc, PhD); environmental science (M Sc, PhD); geoenvironmental engineering (M Eng, M Sc, PhD); geotechnical engineering (M Sc); geotechnical engineering (M Eng, PhD); mining engineering (M Eng, M Sc, PhD); petroleum engineering (M Eng, M Sc, PhD); structural engineering (M Eng, M Sc, PhD); water resources (M Eng, M Sc, PhD). Part-time programs available. *Degree requirements:* For master's, thesis required (for some programs), foreign language not required; for doctorate, dissertation required, foreign language not required. *Faculty research:* Mining.

University of Calgary, Faculty of Graduate Studies, Faculty of Engineering, Department of Geomatics Engineering, Calgary, AB T2N 1N4, Canada. Offers M Eng, M Sc, PhD. Part-time programs available. *Faculty:* 13 full-time (2 women), 3 part-time (0 women). *Students:* 51 full-time (5 women), 10 part-time (2 women). 16 applicants, 56% accepted. In 1998, 5 master's, 5 doctorates awarded (100% found work related to degree). *Degree requirements:* For master's, thesis required (for some programs), foreign language not required; for doctorate, dissertation, candidacy exam required, foreign language not required. *Entrance requirements:* For master's and doctorate, TOEFL (minimum score of 550 required). *Application deadline:* For fall admission, 5/30; for spring admission, 2/15. Applications are processed on a rolling basis. Application fee: $60. *Financial aid:* In 1998–99, 35 students received aid, including research assistantships (averaging $3,920 per year), teaching assistantships (averaging $2,993 per year) *Faculty research:* Gravity and reference systems, positioning and navigation, photogrammetry and remote sensing, precise engineering surveys, spatial information systems. Total annual research expenditures: $287,480. *Unit head:* Dr. Gerard Lachapelle, Head, 403-220-7104, Fax: 403-284-1980, E-mail: lachapel@ensu.ucalgary.ca. *Application contact:* Dr. Michael George Sideris, Associate Head for Research and Graduate Studies, 403-220-4985, Fax: 403-284-1980, E-mail: anne@ensu.ucalgary.ca.

University of California, Berkeley, Graduate Division, College of Engineering, Department of Civil and Environmental Engineering, Berkeley, CA 94720-1500. Offers construction engineering and management (M Eng, MS, D Eng, PhD); environmental quality and environmental water resources engineering (M Eng, MS, D Eng, PhD); geotechnical engineering (M Eng, MS, D Eng, PhD); structural engineering, mechanics and materials (M Eng, MS, D Eng, PhD); transportation engineering (M Eng, MS, D Eng, PhD). *Students:* 326 full-time (97 women); includes 59 minority (3 African Americans, 42 Asian Americans or Pacific Islanders, 13 Hispanic Americans, 1 Native American), 113 international. *Degree requirements:* For master's, comprehensive exam or thesis (MS) required; for doctorate, dissertation, qualifying exam required. *Entrance requirements:* For master's, GRE General Test, minimum GPA of 3.0; for doctorate, GRE General Test, minimum GPA of 3.5. *Application deadline:* For fall admission, 2/10. Application fee: $40. *Unit head:* Dr. Adib Kanafani, Chair, 510-642-3261. *Application contact:* Mari Cook, Graduate Assistant for Admission, 510-643-8944, E-mail: mcook@ce.berkeley.edu.

See in-depth description on page 609.

University of California, Los Angeles, Graduate Division, School of Engineering and Applied Science, Department of Civil and Environmental Engineering, Los Angeles, CA 90095. Offers environmental engineering (MS, PhD); geotechnical engineering (MS, PhD); structures (MS, PhD), including structural mechanics and earthquake engineering; water resource systems engineering (MS, PhD). *Faculty:* 15 full-time, 11 part-time. *Students:* 116 full-time (29 women); includes 28 minority (3 African Americans, 21 Asian Americans or Pacific Islanders, 4 Hispanic Americans), 37 international. *Degree requirements:* For master's, comprehensive exam or thesis required; for doctorate, dissertation, qualifying exams required, foreign language not required. *Entrance requirements:* For master's, GRE General Test, minimum GPA of 3.0; for doctorate, GRE General Test, minimum GPA of 3.25. *Application deadline:* For fall admission, 1/15; for spring admission, 12/1. Application fee: $40. Electronic applications accepted. *Unit head:* Dr. Michael Stenstrom, Chair, 310-825-1408. *Application contact:* Deeona Columbia, Graduate Affairs Officer, 310-825-1851, Fax: 310-206-2222, E-mail: deeona@ea.ucla.edu.

University of Colorado at Boulder, Graduate School, College of Engineering and Applied Science, Department of Civil, Environmental, and Architectural Engineering, Boulder, CO 80309. Offers building systems (MS, PhD); construction engineering and management (MS, PhD); environmental engineering (MS, PhD); geoenvironmental engineering (MS, PhD); geotechnical engineering (MS, PhD); structural engineering (MS, PhD); water resource engineering (MS, PhD). *Degree requirements:* For master's, thesis or alternative, comprehensive exam required; for doctorate, dissertation required. *Entrance requirements:* For master's, GRE General Test, minimum undergraduate GPA of 3.0.

See in-depth description on page 619.

University of Delaware, College of Engineering, Department of Civil and Environmental Engineering, Program in Geotechnical Engineering, Newark, DE 19716. Offers MAS, MCE, PhD. Terminal master's awarded for partial completion of doctoral program. *Degree requirements:* For master's and doctorate, thesis/dissertation required, foreign language not required. *Entrance requirements:* For master's and doctorate, GRE General Test, TOEFL (minimum score of 550 required). *Faculty research:* Computational mechanics, behavior of composite soil structures subjected to seismic loadings, three dimensional stability analysis of inhomogeneous slopes.

University of Illinois at Chicago, Graduate College, College of Liberal Arts and Sciences, Department of Earth and Environmental Sciences, Program in Geotechnical Engineering and Geosciences, Chicago, IL 60607-7128. Offers PhD. *Faculty:* 9 full-time (2 women). *Students:* 7 full-time (2 women), 5 international. Average age 27. 15 applicants, 7% accepted. In 1998, 3 degrees awarded. *Degree requirements:* For doctorate, dissertation required, foreign language not required. *Entrance requirements:* For doctorate, GRE General Test, TOEFL (minimum score of 550 required), minimum GPA of 3.75 on a 5.0 scale. *Application deadline:* For fall admission, 7/3; for spring admission, 11/8. Application fee: $40 ($50 for international students). *Financial aid:* In 1998–99, 7 students received aid; fellowships, research assistantships, teaching assistantships available. *Unit head:* Charlene Holliday, Graduate Analyst, 713-743-4254, Fax: 713-743-4260, E-mail: wholliday@uh.edu. *Application contact:* Martin Flower, Graduate Director, 312-996-9662.

University of Maine, Graduate School, College of Engineering, Department of Civil and Environmental Engineering, Orono, ME 04469. Offers civil engineering (MS, PhD), including environmental engineering, geotechnical engineering, structural engineering. *Faculty:* 11 full-time (0 women). *Students:* 19 full-time (3 women), 10 part-time (1 woman), 4 international. *Degree requirements:* For doctorate, dissertation required, foreign language not required, foreign language not required. *Entrance requirements:* For master's and doctorate, GRE General Test, TOEFL (minimum score of 550 required). *Application deadline:* For fall admission, 2/1 (priority date); for spring admission, 10/15. Applications are processed on a rolling basis. Application fee: $50. *Unit head:* Dr. Willem Brutsaert, Chair, 207-581-2170, Fax: 207-581-3888. *Application contact:* Scott G. Delcourt, Director of the Graduate School, 207-581-3218, Fax: 207-581-3232, E-mail: graduate@maine.edu.

University of Missouri–Columbia, Graduate School, College of Engineering, Department of Civil and Environmental Engineering, Columbia, MO 65211. Offers civil engineering (MS, PhD); environmental engineering (MS, PhD); geotechnical engineering (MS, PhD); structural engineering (MS, PhD); transportation and highway engineering (MS); water resources (MS, PhD). *Faculty:* 24 full-time (2 women). *Students:* 19 full-time (2 women), 18 part-time (3 women); includes 1 minority (Hispanic American), 21 international. *Degree requirements:* For master's, report or thesis required; for doctorate, dissertation required. *Entrance requirements:* For master's and doctorate, GRE General Test, TOEFL. *Application deadline:* For fall admission, 7/10. Application fee: $30 ($50 for international students). *Unit head:* Dr. Mark Virkler, Director of Graduate Studies, 573-882-3678.

See in-depth description on page 639.

University of Missouri–Rolla, Graduate School, School of Engineering, Department of Civil Engineering, Program in Geotechnical Engineering, Rolla, MO 65409-0910. Offers MS, DE, PhD. *Degree requirements:* For master's, thesis or alternative required, foreign language not required; for doctorate, dissertation required, foreign language not required. *Entrance requirements:* For master's and doctorate, GRE General Test (minimum combined score of 1100 required), TOEFL (minimum score of 550 required), minimum GPA of 3.0.

University of New Brunswick, School of Graduate Studies, Faculty of Engineering, Department of Civil Engineering, Fredericton, NB E3B 5A3, Canada. Offers environmental engineering (M Eng, M Sc E, PhD); geotechnical engineering (M Eng, M Sc E); structures and structural foundations (M Eng, M Sc E, PhD); transportation engineering (M Eng, M Sc E, PhD); water resources and hydrology (M Eng, M Sc E, PhD). Part-time programs available. *Degree requirements:* For master's, thesis required, foreign language not required; for doctorate, dissertation, qualifying exam required, foreign language not required. *Entrance requirements:* For master's and doctorate, TOEFL, TWE, minimum GPA of 3.0.

University of Oklahoma, Graduate College, College of Engineering, School of Civil Engineering and Environmental Science, Program in Civil Engineering, Norman, OK 73019-0390. Offers civil engineering (PhD); environmental engineering (MS); geotechnical engineering (MS); structures (MS); transportation (MS). *Accreditation:* ABET (one or more programs are accredited). *Students:* 19 full-time (4 women), 36 part-time (6 women); includes 5 minority (3 Asian Americans or Pacific Islanders, 2 Native Americans), 28 international. *Degree requirements:* For master's, comprehensive and oral exams required, foreign language and thesis not required; for doctorate, dissertation, oral and qualifying exams required, foreign language not required. *Entrance requirements:* For master's, GRE General Test, TOEFL (minimum score of 575 required), minimum GPA of 3.0; for doctorate, GRE General Test, TOEFL (minimum score of 575 required), minimum graduate GPA of 3.5. *Application deadline:* For fall admission, 4/1 (priority date). Applications are processed on a rolling basis. Application fee: $25. Tuition, state resident: part-time $86 per credit hour. Tuition, nonresident: part-time $275 per credit hour. Tuition and fees vary according to course level, course load and program. *Unit head:* Robert C. Knox, Graduate Liaison, 405-325-4256, Fax: 405-325-7508. *Application contact:* Robert C. Knox, Graduate Liaison, 405-325-4256, Fax: 405-325-7508.

See in-depth description on page 641.

University of Rhode Island, Graduate School, College of Engineering, Department of Civil and Environmental Engineering, Program in Geotechnical Engineering, Kingston, RI 02881. Offers MS, PhD.

University of Southern California, Graduate School, School of Engineering, Department of Civil Engineering, Program in Soil Mechanics and Foundations, Los Angeles, CA 90089. Offers MS. *Students:* 3 full-time (2 women), 1 part-time, 2 international. Average age 27. 3 applicants, 33% accepted. In 1998, 3 degrees awarded. *Degree requirements:* For master's, thesis optional. *Entrance requirements:* For master's, GRE General Test. *Application deadline:* For fall admission, 6/1 (priority date); for spring admission, 11/1. Application fee: $55. Tuition: Part-time $768 per unit. Required fees: $350 per semester. *Financial aid:* Fellowships, research assistantships, teaching assistantships, Federal Work-Study and institutionally-sponsored loans available. Aid available to part-time students. Financial aid application deadline: 2/15; financial aid applicants required to submit FAFSA. *Unit head:* Dr. L. Carter Wellford, Chair, Department of Civil Engineering, 213-740-0587.

The University of Texas at Austin, Graduate School, College of Engineering, Department of Petroleum and Geosystems Engineering, Austin, TX 78712-1111. Offers MSE, PhD. *Students:* 97 full-time (17 women), 10 part-time (1 woman); includes 32 minority (1 African American, 30 Asian Americans or Pacific Islanders, 1 Hispanic American) 82 applicants, 49% accepted. In 1998, 14 master's, 9 doctorates awarded. *Entrance requirements:* For master's and doctorate, GRE General Test (minimum combined score of 1000 required). Application fee: $50 ($75 for international students). *Financial aid:* Fellowships, research assistantships, teaching assistantships available. Financial aid application deadline: 2/1. *Unit head:* Dr. Ekwere J. Peters, Chairman, 512-471-3161. *Application contact:* Dr. Mukul Sharma, Graduate Adviser, 512-471-3257.

University of Washington, Graduate School, College of Engineering, Department of Civil and Environmental Engineering, Seattle, WA 98195. Offers environmental engineering (MS, MS Civ E, MSE, PhD); hydraulic engineering (MS Civ E, MSE, PhD); structural and geotechnical engineering and mechanics (MS, MS Civ E, MSE, PhD); transportation and construction engineering (MS, MS Civ E, MSE, PhD). *Faculty:* 31 full-time (4 women), 26 part-time (1 woman). *Students:* 153 full-time (54 women), 42 part-time (13 women); includes 25 minority (2 African Americans, 19 Asian Americans or Pacific Islanders, 3 Hispanic Americans, 1 Native American), 34 international. Terminal master's awarded for partial completion of doctoral program. *Degree requirements:* For master's, thesis or alternative required, foreign language not required; for doctorate, dissertation required, foreign language not required. *Entrance requirements:* For master's, GRE General Test, TOEFL (minimum score of 580 required), minimum GPA of 3.0; for doctorate, GRE, TOEFL (minimum score of 580 required), minimum GPA of 3.0. *Application deadline:* For fall admission, 2/1 (priority date); for winter admission, 11/1; for spring admission, 2/1. Applications are processed on a rolling basis. Application fee: $50. Electronic applications accepted. Tuition, state resident: full-time $5,196; part-time $475 per credit. Tuition, nonresident: full-time $13,485; part-time $1,285 per credit. Required fees: $387; $38 per credit. Tuition and fees vary according to course load. *Unit head:* Dr. Fred L. Mannering, Chair, 206-543-2390. *Application contact:* Marcia Buck, Graduate Secretary, 206-543-2574, E-mail: ceginfo@u.washington.edu.

Hydraulics

Auburn University, Graduate School, College of Engineering, Department of Civil Engineering, Auburn, Auburn University, AL 36849-0002. Offers construction engineering and management (MCE, MS, PhD); environmental engineering (MCE, MS, PhD); geotechnical/materials engineering (MCE, MS, PhD); hydraulics/hydrology (MCE, MS, PhD); structural engineering (MCE, MS, PhD); transportation engineering (MCE, MS, PhD). Part-time programs available. *Faculty:* 22 full-time. *Students:* 24 full-time (6 women), 34 part-time (8 women); includes 2 minority (1 African American, 1 Hispanic American), 9 international. *Degree requirements:* For master's, project (MCE), thesis (MS) required; for doctorate, dissertation, comprehensive exam required, foreign language not required. *Entrance requirements:* For master's, GRE General Test; for doctorate, GRE General Test (minimum score of 400 on each section required). *Application deadline:* For fall admission, 9/1; for spring admission, 3/1. Applications are processed on a rolling basis. Application fee: $25 ($50 for international students). Tuition, state resident: full-time $2,760; part-time $76 per credit hour. Tuition, nonresident: full-time $8,280; part-time $228 per credit hour. *Unit head:* Dr. Joseph F. Judkins, Head, 334-844-4320. *Application contact:* Dr. John F. Pritchett, Dean of the Graduate School, 334-844-4700.

Colorado State University, Graduate School, College of Engineering, Department of Civil Engineering, Specialization in Hydraulics and Wind Engineering, Fort Collins, CO 80523-0015. Offers MS, PhD. Part-time programs available. *Faculty:* 11 full-time (0 women), 1 part-time (0 women). In 1998, 18 master's, 8 doctorates awarded. Terminal master's awarded for partial completion of doctoral program. *Degree requirements:* For master's, thesis or alternative required, foreign language not required; for doctorate, dissertation required, foreign language not required. *Entrance requirements:* For master's and doctorate, GRE General Test, TOEFL (minimum score of 550 required; 213 for computer-based), minimum GPA of 3.0. *Average time to degree:* Master's–2 years full-time, 5 years part-time; doctorate–4 years full-time. *Application deadline:* For fall admission, 3/1 (priority date); for spring admission, 8/1 (priority date). Applications are processed on a rolling basis. Application fee: $30. Electronic applications accepted. *Financial aid:* Fellowships, research assistantships, teaching assistantships, Federal Work-Study and institutionally-sponsored loans available. *Faculty research:* Hydromachinery, hydraulic structures, river mechanics, sedimentation and erosion, physical and numerical modeling, turbulence, transfer processes, wind effects on the environment, building aerodynamics, dispersion and diffusion mode. Total annual research expenditures: $1.9 million. *Unit head:* Pierre Julien, Leader, 970-491-8450, Fax: 970-491-7727, E-mail: pierre@engr.colostate.edu. *Application contact:* Laurie Howard, Student Adviser, 970-491-5844, Fax: 970-491-7727, E-mail: lhoward@engr.colostate.edu.

École Polytechnique de Montréal, Graduate Programs, Department of Civil Engineering, Montréal, PQ H3C 3A7, Canada. Offers environmental engineering (M Eng, M Sc A, PhD); geotechnical engineering (M Eng, M Sc A, PhD); hydraulics engineering (M Eng, M Sc A, PhD); structural engineering (M Eng, M Sc A, PhD); transportation engineering (M Eng, M Sc A, PhD). Part-time and evening/weekend programs available. *Degree requirements:* For master's and doctorate, one foreign language, computer language, thesis/dissertation required. *Entrance requirements:* For master's, minimum GPA of 2.75; for doctorate, minimum GPA of 3.0. *Faculty research:* Water resources management, characteristics of building materials, aging of dams, pollution control.

McGill University, Faculty of Graduate Studies and Research, Faculty of Engineering, Department of Civil Engineering and Applied Mechanics, Program in Fluid Mechanics and Hydraulic Engineering, Montréal, PQ H3A 2T5, Canada. Offers M Eng, M Sc, PhD. Part-time and evening/weekend programs available. *Degree requirements:* For master's, computer language required, thesis optional, foreign language not required; for doctorate, computer language, dissertation required, foreign language not required. *Entrance requirements:* For master's, TOEFL (minimum score of 550 required), minimum GPA of 3.0; for doctorate, TOEFL (minimum score of 580 required). *Faculty research:* Transport processes in inland and coastal waters, experimental and computational methods.

Texas A&M University, College of Engineering, Department of Civil Engineering, College Station, TX 77843. Offers construction engineering and project management (M Eng, MS, D Eng, PhD); engineering mechanics (M Eng, MS, PhD); environmental engineering (M Eng, MS, D Eng, PhD); geotechnical engineering (M Eng, MS, D Eng, PhD); hydraulic engineering (M Eng, MS, PhD); hydrology (M Eng, MS, PhD); materials engineering (M Eng, MS, D Eng, PhD); ocean engineering (M Eng, MS, D Eng, PhD); public works engineering and management (M Eng, MS, PhD); structural engineering and structural mechanics (M Eng, MS, D Eng, PhD); transportation engineering (M Eng, MS, D Eng, PhD); water resources engineering (M Eng, MS, D Eng, PhD). Part-time programs available. *Faculty:* 59 full-time (4 women), 7 part-time (2 women). *Students:* 249 full-time, 54 part-time; includes 13 minority (2 African Americans, 4 Asian Americans or Pacific Islanders, 7 Hispanic Americans), 162 international. *Degree requirements:* For master's, thesis (MS) required; for doctorate, dissertation (PhD), internship (D Eng) required. *Entrance requirements:* For master's and doctorate, GRE General Test, TOEFL. *Application deadline:* Applications are processed on a rolling basis. Application fee: $50 ($75 for international students). *Unit head:* Dr. John M. Niedzwecki, Head, 409-845-7435, Fax: 409-862-2800, E-mail: ce-grad@tamu.edu. *Application contact:* Dr. Peter B. Keating, Graduate Adviser, 409-845-2498, Fax: 409-862-2800, E-mail: ce-grad@tamu.edu.

See in-depth description on page 603.

University of Missouri–Rolla, Graduate School, School of Engineering, Department of Civil Engineering, Program in Hydrology and Hydraulic Engineering, Rolla, MO 65409-0910. Offers MS, DE, PhD. *Degree requirements:* For master's, thesis or alternative required, foreign language not required; for doctorate, dissertation required, foreign language not required. *Entrance requirements:* For master's and doctorate, GRE General Test (minimum combined score of 1100 required), TOEFL (minimum score of 550 required), minimum GPA of 3.0.

University of Washington, Graduate School, College of Engineering, Department of Civil and Environmental Engineering, Seattle, WA 98195. Offers environmental engineering (MS, MS Civ E, MSE, PhD); hydraulic engineering (MS Civ E, MSE, PhD); structural and geotechnical engineering and mechanics (MS, MS Civ E, MSE, PhD); transportation and construction engineering (MS, MS Civ E, MSE, PhD). *Faculty:* 31 full-time (4 women), 26 part-time (1 woman). *Students:* 153 full-time (54 women), 42 part-time (13 women); includes 25 minority (2 African Americans, 19 Asian Americans or Pacific Islanders, 3 Hispanic Americans, 1 Native American), 34 international. Terminal master's awarded for partial completion of doctoral program. *Degree requirements:* For master's, thesis or alternative required, foreign language not required; for doctorate, dissertation required, foreign language not required. *Entrance requirements:* For master's, GRE General Test, TOEFL (minimum score of 580 required), minimum GPA of 3.0; for doctorate, GRE, TOEFL (minimum score of 580 required), minimum GPA of 3.0. *Application deadline:* For fall admission, 2/1 (priority date); for winter admission, 11/1; for spring admission, 2/1. Applications are processed on a rolling basis. Application fee: $50. Electronic applications accepted. Tuition, state resident: full-time $5,196; part-time $475 per credit. Tuition, nonresident: full-time $13,485; part-time $1,285 per credit. Required fees: $387; $38 per credit. Tuition and fees vary according to course load. *Unit head:* Dr. Fred L. Mannering, Chair, 206-543-2390. *Application contact:* Marcia Buck, Graduate Secretary, 206-543-2574, E-mail: ceginfo@u.washington.edu.

Structural Engineering

Auburn University, Graduate School, College of Engineering, Department of Civil Engineering, Auburn, Auburn University, AL 36849-0002. Offers construction engineering and management (MCE, MS, PhD); environmental engineering (MCE, MS, PhD); geotechnical/materials engineering (MCE, MS, PhD); hydraulics/hydrology (MCE, MS, PhD); structural engineering (MCE, MS, PhD); transportation engineering (MCE, MS, PhD). Part-time programs available. *Faculty:* 22 full-time. *Students:* 24 full-time (6 women), 34 part-time (8 women); includes 2 minority (1 African American, 1 Hispanic American), 9 international. *Degree requirements:* For master's, project (MCE), thesis (MS) required; for doctorate, dissertation, comprehensive exam required, foreign language not required. *Entrance requirements:* For master's, GRE General Test; for doctorate, GRE General Test (minimum score of 400 on each section required). *Application deadline:* For fall admission, 9/1; for spring admission, 3/1. Applications are processed on a rolling basis. Application fee: $25 ($50 for international students). Tuition, state resident: full-time $2,760; part-time $76 per credit hour. Tuition, nonresident: full-time $8,280; part-time $228 per credit hour. *Unit head:* Dr. Joseph F. Judkins, Head, 334-844-4320. *Application contact:* Dr. John F. Pritchett, Dean of the Graduate School, 334-844-4700.

California State University, Northridge, Graduate Studies, College of Engineering and Computer Science, Department of Civil, Industrial and Applied Mechanics, Department of Mechanical Engineering, Northridge, CA 91330. Offers aerospace engineering (MS); applied engineering (MS); machine design (MS); mechanical engineering (MS); structural engineering (MS); thermofluids (MS). Part-time and evening/weekend programs available. *Faculty:* 8 full-time, 4 part-time. *Students:* 3 full-time (0 women), 44 part-time (4 women); includes 17 minority (14 Asian Americans or Pacific Islanders, 2 Hispanic Americans, 1 Native American), 3 international. *Degree requirements:* For master's, thesis or alternative required, foreign language not required. *Entrance requirements:* For master's, GRE General Test, TOEFL, minimum GPA of 2.5. *Application deadline:* For fall admission, 11/30. Application fee: $55. Tuition, nonresident: part-time $246 per unit. International tuition: $7,874 full-time. Required fees: $1,970. Tuition and fees vary according to course load. *Unit head:* Dr. William J. Rivers, Chair, 818-677-2187. *Application contact:* Dr. Tom Mincer, Graduate Coordinator, 818-677-2007.

The Catholic University of America, School of Engineering, Department of Civil Engineering, Program in Structures and Structural Mechanics, Washington, DC 20064. Offers MCE, PhD. *Degree requirements:* For master's, thesis optional, foreign language not required; for doctorate, dissertation, comprehensive and oral exams required, foreign language not required. *Entrance requirements:* For master's, minimum GPA of 3.0; for doctorate, minimum GPA of 3.5. *Application deadline:* For fall admission, 8/1 (priority date); for spring admission, 12/1. Applications are processed on a rolling basis. Application fee: $50. *Financial aid:* Application deadline: 2/1. *Unit head:* Dr. Timothy W. Kao, Chair, Department of Civil Engineering, 202-319-5163, Fax: 202-319-4499, E-mail: kao@cua.edu.

Colorado State University, Graduate School, College of Engineering, Department of Civil Engineering, Specialization in Structural and Geotechnical Engineering, Fort Collins, CO 80523-0015. Offers MS, PhD. Part-time programs available. Postbaccalaureate distance learning degree programs offered (no on-campus study). *Faculty:* 9 full-time (0 women). In 1998, 13 master's, 2 doctorates awarded. Terminal master's awarded for partial completion of doctoral program. *Degree requirements:* For master's, thesis or alternative required, foreign language not required; for doctorate, dissertation required, foreign language not required. *Entrance requirements:* For master's and doctorate, GRE General Test, TOEFL (minimum score of 550 required; 213 for computer-based), minimum GPA of 3.0. *Average time to degree:* Master's–2 years full-time, 5 years part-time; doctorate–4 years full-time. *Application deadline:* For fall admission, 3/1 (priority date); for spring admission, 8/1 (priority date). Applications are processed on a rolling basis. Application fee: $30. Electronic applications accepted. *Financial aid:* Research assistantships, teaching assistantships, Federal Work-Study and institutionally-sponsored loans available. *Faculty research:* Computational mechanics, solid mechanics, finite elements, timber engineering, concrete structures, expansive soils, blast and earthquake induced liquefaction, export systems. Total annual research expenditures: $700,000. *Unit head:* Erik G. Thompson, Leader, 970-491-6060, Fax: 970-491-7727, E-mail: thompson@engr.colostate.edu. *Application contact:* Laurie Howard, Student Adviser, 970-491-5844, Fax: 970-491-7727, E-mail: lhoward@engr.colostate.edu.

Cornell University, Graduate School, Graduate Fields of Engineering, Field of Civil and Environmental Engineering, Ithaca, NY 14853-0001. Offers environmental engineering (M Eng, MS, PhD); environmental fluid mechanics and hydrology (M Eng, MS, PhD); environmental systems engineering (M Eng, MS, PhD); geotechnical engineering (M Eng, MS, PhD); remote sensing (M Eng, MS, PhD); structural engineering (M Eng, MS, PhD); transportation engineering (M Eng, MS, PhD); water resource systems (M Eng, MS, PhD). *Faculty:* 29 full-time. *Students:* 143 full-time (32 women); includes 16 minority (11 Asian Americans or Pacific Islanders, 5 Hispanic Americans), 76 international. Terminal master's awarded for partial completion of doctoral program. *Degree requirements:* For master's, thesis (MS) required; for doctorate, dissertation required, foreign language not required. *Entrance requirements:* For master's, TOEFL (minimum score of 600required); for doctorate, GRE General Test, TOEFL (minimum score of 600 required). *Application deadline:* For fall admission, 1/15. Application fee: $65. Electronic applications accepted. *Unit head:* Director of Graduate Studies, 607-255-7560, Fax: 607-255-9004. *Application contact:* Graduate Field Assistant, 607-255-7560, E-mail: cee_grad@cornell.edu.

See in-depth description on page 573.

École Polytechnique de Montréal, Graduate Programs, Department of Civil Engineering, Montréal, PQ H3C 3A7, Canada. Offers environmental engineering (M Eng, M Sc A, PhD); geotechnical engineering (M Eng, M Sc A, PhD); hydraulics engineering (M Eng, M Sc A, PhD); structural engineering (M Eng, M Sc A, PhD); transportation engineering (M Eng, M Sc A, PhD). Part-time and evening/weekend programs available. *Degree requirements:* For master's and doctorate, one foreign language, computer language, thesis/dissertation required. *Entrance requirements:* For master's, minimum GPA of 2.75; for doctorate, minimum GPA of 3.0. *Faculty research:* Water resources management, characteristics of building materials, aging of dams, pollution control.

Florida Institute of Technology, Graduate School, College of Engineering, Division of Engineering Science, Program in Civil Engineering, Melbourne, FL 32901-6975. Offers civil engineering (PhD); geotechnical engineering (MS); structures engineering (MS); water resources (MS). Part-time programs available. *Faculty:* 5 full-time (0 women), 4 part-time (2 women). *Students:* 7 full-time (0 women), 10 part-time (2 women); includes 1 minority (Hispanic American), 9 international. *Degree requirements:* For master's, thesis optional, foreign language not required; for doctorate, dissertation required, foreign language not required. *Entrance requirements:* For master's, minimum GPA of 3.0; for doctorate, minimum GPA of 3.2. *Application deadline:* Applications are processed on a rolling basis. Application fee: $50. Electronic applications accepted. Tuition: Part-time $575 per credit hour. Required fees: $100. Tuition and fees vary according to campus/location and program. *Unit head:* Dr. Ashok Pandit, Chair, 407-674-7151, Fax: 407-768-7565, E-mail: apandit@fit.edu. *Application contact:* Carolyn P. Farrior, Associate Dean of Graduate Admissions, 407-674-7118, Fax: 407-723-9468, E-mail: cfarrior@fit.edu.

See in-depth description on page 581.

Iowa State University of Science and Technology, Graduate College, College of Engineering, Department of Civil and Construction Engineering, Ames, IA 50011. Offers civil engineering (MS, PhD), including civil engineering materials, construction engineering and management, environmental engineering, geometronics, geotechnical engineering, structural engineering, transportation engineering. *Faculty:* 36 full-time, 5 part-time. *Students:* 60 full-time (14 women), 51 part-time (12 women). *Degree requirements:* For master's, thesis or alternative required, foreign language not required; for doctorate, dissertation required, foreign language not required. *Entrance requirements:* For master's and doctorate, GRE General Test (foreign students), TOEFL. *Application deadline:* For fall admission, 6/15 (priority date); for spring admission, 11/15 (priority date). Application fee: $20 ($50 for international students). Electronic applications accepted. Tuition, state resident: full-time $3,308. Tuition, nonresident: full-time $9,744. Part-time tuition and fees vary according to course load, campus/location and program. *Unit head:* Dr. Lowell F. Greimann, Chair, 515-294-2140, E-mail: cceinfo@iastate.edu. *Application contact:* Edward Kannel, Director of Graduate Education, 515-294-2861, E-mail: cceinfo@iastate.edu.

Louisiana State University and Agricultural and Mechanical College, Graduate School, College of Engineering, Department of Civil and Environmental Engineering, Baton Rouge, LA 70803. Offers environmental engineering (MSCE, PhD); geotechnical engineering (MSCE, PhD); structural engineering and mechanics (MSCE, PhD); transportation engineering (MSCE, PhD); water resources (MSCE, PhD). Part-time programs available. *Faculty:* 27 full-time (1 woman), 1 part-time (0 women). *Students:* 59 full-time (14 women), 24 part-time (2 women); includes 5 minority (1 African American, 3 Asian Americans or Pacific Islanders, 1 Hispanic American), 48 international. *Degree requirements:* For master's, thesis optional, foreign language not required; for doctorate, one foreign language (computer language can substitute), dissertation required. *Entrance requirements:* For master's and doctorate, GRE General Test, TOEFL, minimum GPA of 3.0. *Application deadline:* For fall admission, 1/25 (priority date). Applications are processed on a rolling basis. Application fee: $25. *Unit head:* Dr. Ronald F. Malone, Acting Chair, 225-388-8666, Fax: 225-388-8652, E-mail: rmalone@unix1.sncc.lsu.edu. *Application contact:* Dr. Vijaya K. A. Gopu, Graduate Coordinator, 225-388-8442, E-mail: cegopu@eng.lsu.edu.

See in-depth description on page 589.

Marquette University, Graduate School, College of Engineering, Department of Civil and Environmental Engineering, Milwaukee, WI 53201-1881. Offers construction and public works management (MS, PhD); environmental/water resources engineering (MS, PhD); structural/geotechnical engineering (MS, PhD); transportational planning and engineering (MS, PhD). Part-time and evening/weekend programs available. *Faculty:* 12 full-time (0 women), 1 part-time (0 women). *Students:* 17 full-time (7 women), 32 part-time (5 women); includes 5 minority (1 African American, 4 Asian Americans or Pacific Islanders), 12 international. Terminal master's awarded for partial completion of doctoral program. *Degree requirements:* For master's, thesis or alternative, comprehensive exam required, foreign language not required; for doctorate, dissertation required, foreign language not required. *Entrance requirements:* For master's, TOEFL (minimum score of 550 required); for doctorate, GRE General Test, TOEFL (minimum score of 550 required). *Application deadline:* For fall admission, 6/1 (priority date). Applications are processed on a rolling basis. Application fee: $40. Tuition: Part-time $510 per credit hour. Tuition and fees vary according to program. *Unit head:* Dr. Thomas H. Wenzel, Chairman, 414-288-7030, Fax: 414-288-7521, E-mail: wenzelt@vms.csd.mu.edu.

McGill University, Faculty of Graduate Studies and Research, Faculty of Engineering, Department of Civil Engineering and Applied Mechanics, Program in Structural Engineering and Construction Materials, Montréal, PQ H3A 2T5, Canada. Offers M Eng, M Sc, PhD. Part-time and evening/weekend programs available. *Degree requirements:* For master's, computer language required, thesis optional, foreign language not required; for doctorate, computer language, dissertation required, foreign language not required. *Entrance requirements:* For master's, TOEFL (minimum score of 550 required), minimum GPA of 3.0; for doctorate, TOEFL (minimum score of 580 required). *Faculty research:* Seismic design, risk engineering, composite structures, high strength and prestressed concrete, advanced materials, rehabilitation.

Northwestern University, The Graduate School, Robert R. McCormick School of Engineering and Applied Science, Department of Civil Engineering, Evanston, IL 60208. Offers biosolid mechanics (MS, PhD); environmental health engineering (MS, PhD); geotechnical engineering (MS, PhD); health physics/radiological health (MS, PhD); project management (MPM); structural engineering (MS, PhD); structural mechanics (MS, PhD); transportation systems engineering (MS, PhD). MS and PhD admissions and degrees offered through The Graduate School. Part-time programs available. *Faculty:* 26 full-time (2 women), 1 (woman) part-time. *Students:* 110 full-time (38 women), 7 part-time (2 women); includes 5 minority (2 African Americans, 3 Asian Americans or Pacific Islanders), 68 international. Terminal master's awarded for partial completion of doctoral program. *Degree requirements:* For master's, thesis required (for some programs); for doctorate, dissertation required. *Entrance requirements:* For master's and doctorate, GRE General Test, TOEFL (minimum score of 560 required). Application fee: $50 ($55 for international students). *Unit head:* Joseph L. Schofer, Chair, 847-491-3257, Fax: 847-491-4011, E-mail: j-schofer@nwu.edu. *Application contact:* Karm Kerwell, Secretary, 847-491-3176, Fax: 847-491-4011, E-mail: k-kerwell@nwu.edu.

Ohio University, Graduate Studies, College of Engineering and Technology, Department of Civil Engineering, Athens, OH 45701-2979. Offers geotechnical and environmental engineering (MS); water resources and structures (MS). *Faculty:* 11 full-time (1 woman). *Students:* 24 full-time (5 women), 2 part-time, 13 international. *Degree requirements:* For master's, thesis optional, foreign language not required. *Entrance requirements:* For master's, minimum GPA of 3.0. *Application deadline:* For fall admission, 5/1 (priority date). Applications are processed on a rolling basis. Application fee: $30. Tuition, state resident: full-time $5,754; part-time $238 per credit hour. Tuition, nonresident: full-time $11,055; part-time $457 per credit hour. Tuition and fees vary according to course load, campus/location and program. *Unit head:* Dr. Gayle Mitchell, Chair, 740-593-1465, Fax: 740-593-0625, E-mail: gmitchell@bobcat.ent.ohiou.edu. *Application contact:* Dr. Glenn Hazen, Graduate Chair, 740-593-1469, Fax: 740-593-0625, E-mail: ghazen@bobcat.ent.ohiou.edu.

Pennsylvania State University University Park Campus, Graduate School, College of Engineering, Department of Civil and Environmental Engineering, State College, University Park, PA 16802-1503. Offers civil engineering (M Eng, MS, PhD); environmental engineering (M Eng, MS, PhD); structural engineering (M Eng, MS, PhD); transportation and highway engineering (M Eng, MS, PhD); water resources engineering (M Eng, MS, PhD). *Students:* 77 full-time (24 women), 52 part-time (8 women). *Degree requirements:* For master's, final paper (M Eng), oral exam and thesis (MS) required; for doctorate, dissertation, comprehensive and oral exams required. *Entrance requirements:* For master's and doctorate, GRE General Test. Application fee: $50. *Unit head:* Dr. Paul P. Jovanis, Head, 814-863-3084.

Princeton University, Graduate School, School of Engineering and Applied Science, Department of Civil Engineering and Operations Research, Princeton, NJ 08544-1019. Offers environmental engineering and water resources (PhD); financial engineering (M Eng); statistics and operations research (MSE, PhD); structural engineering (M Eng); structures and mechanics (MSE, PhD); transportation systems (MSE, PhD). *Degree requirements:* For master's and doctorate, thesis/dissertation required. *Entrance requirements:* For master's and doctorate, GRE General Test, GRE Subject Test.

Rensselaer Polytechnic Institute, Graduate School, School of Engineering, Department of Civil Engineering, Troy, NY 12180-3590. Offers geotechnical engineering (M Eng, MS, D Eng, PhD); mechanics of composite materials and structures (M Eng, MS, D Eng, PhD); structural engineering (M Eng, MS, D Eng, PhD); transportation engineering (M Eng, MS, D Eng, PhD). Part-time programs available. *Faculty:* 11 full-time (0 women), 3 part-time (0 women). *Students:* 31 full-time (4 women), 11 part-time (2 women); includes 6 minority (3 Asian Americans or Pacific Islanders, 3 Hispanic Americans), 18 international. *Degree requirements:* For master's, thesis required (for some programs), foreign language not required; for doctorate, dissertation required, foreign language not required. *Entrance requirements:* For master's and doctorate,

GRE, TOEFL (minimum score of 550 required). *Application deadline:* For fall admission, 2/1 (priority date). Applications are processed on a rolling basis. Application fee: $35. *Unit head:* Dr. George List, Chair, 518-276-6940, Fax: 518-276-4833, E-mail: gregaja@rpi.edu. *Application contact:* Jo Ann Grega, Admissions Assistant, 518-276-6679, Fax: 518-276-4833, E-mail: gregaj2@rpi.edu.

See in-depth description on page 597.

Rice University, Graduate Programs, George R. Brown School of Engineering, Department of Civil Engineering, Houston, TX 77251-1892. Offers civil engineering (MCE, MS, PhD); structural engineering (MCE, MS, PhD). Part-time programs available. *Degree requirements:* For master's, thesis required (for some programs), foreign language not required; for doctorate, dissertation required, foreign language not required. *Entrance requirements:* For master's and doctorate, GRE General Test, GRE Subject Test, TOEFL (minimum score of 550 required), minimum GPA of 3.0. *Faculty research:* Structural dynamics, probabilistic studies in dynamics, fatigue, reinforced concrete experimental research.

State University of New York at Buffalo, Graduate School, School of Engineering and Applied Sciences, Department of Civil, Structural, and Environmental Engineering, Buffalo, NY 14260. Offers computational engineering and mechanics (MS, PhD); construction (M Eng, MS, PhD); geoenvironmental and geotechnical engineering (M Eng, MS, PhD); structural and earthquake engineering (M Eng, MS, PhD); water resources and environmental engineering (M Eng, MS, PhD). Part-time programs available. Postbaccalaureate distance learning degree programs offered (minimal on-campus study). *Faculty:* 25 full-time (0 women), 7 part-time (1 woman). *Students:* 99 full-time (19 women), 71 part-time (13 women); includes 9 minority (2 African Americans, 5 Asian Americans or Pacific Islanders, 1 Hispanic American, 1 Native American), 101 international. Average age 27. 252 applicants, 29% accepted. In 1998, 46 master's, 16 doctorates awarded. Terminal master's awarded for partial completion of doctoral program. *Degree requirements:* For master's, computer language, project or thesis required; for doctorate, computer language, dissertation required, foreign language not required. *Entrance requirements:* For master's and doctorate, GRE General Test (minimum combined score of 1250 required), TOEFL (minimum score of 550 required). *Application deadline:* For fall admission, 1/15 (priority date); for spring admission, 10/1. Applications are processed on a rolling basis. Application fee: $35. Tuition, state resident: full-time $5,100; part-time $213 per credit hour. Tuition, nonresident: full-time $8,416; part-time $351 per credit hour. Required fees: $870; $75 per semester. Tuition and fees vary according to course load and program. *Financial aid:* In 1998–99, 3 fellowships with full tuition reimbursements (averaging $14,700 per year), 79 research assistantships with full tuition reimbursements (averaging $10,700 per year), 30 teaching assistantships with full tuition reimbursements (averaging $10,700 per year) were awarded.; career-related internships or fieldwork, Federal Work-Study, grants, institutionally-sponsored loans, scholarships, tuition waivers (full and partial), and unspecified assistantships also available. Aid available to part-time students. Financial aid application deadline: 1/15; financial aid applicants required to submit FAFSA. *Faculty research:* Earthquake protection, environmental engineering and fluid mechanics, structural dynamics, geomechanics. Total annual research expenditures: $1.5 million. *Unit head:* Dr. Andrei M. Reinhorn, Chairman, 716-645-2114 Ext. 2419, Fax: 716-645-3733, E-mail: reinhorn@civil.eng.buffalo.edu. *Application contact:* Dr. Joe Atkinson, Director of Graduate Admissions, 716-645-2114 Ext. 2326, Fax: 716-645-3667, E-mail: atkinson@acsu.buffalo.edu.

See in-depth description on page 601.

Texas A&M University, College of Engineering, Department of Civil Engineering, Program in Structural Engineering and Structural Mechanics, College Station, TX 77843. Offers M Eng, MS, D Eng, PhD. D Eng offered through the College of Engineering. *Students:* 49. *Degree requirements:* For master's, thesis (MS) required; for doctorate, dissertation (PhD), internship (D Eng) required. *Entrance requirements:* For master's and doctorate, GRE General Test, TOEFL. Application fee: $50 ($75 for international students). *Financial aid:* Fellowships, research assistantships, teaching assistantships available. Financial aid application deadline: 4/1; financial aid applicants required to submit FAFSA. *Faculty research:* Analysis and design of bridges, buildings, and offshore structures; dynamic loads and structural behavior; structural reliability; computer-aided design. *Unit head:* Dr. Paul N. Roschke, Head, Constructed Facilities Division, 409-845-4414, Fax: 409-862-2800, E-mail: ce-grad@tamu.edu. *Application contact:* Dr. Joseph M. Bracci, 409-845-6554, Fax: 409-862-2800, E-mail: ce-grad@tamu.edu.

Tufts University, Division of Graduate and Continuing Studies and Research, Graduate School of Arts and Sciences, College of Engineering, Department of Civil and Environmental Engineering, Medford, MA 02155. Offers civil engineering (MS, PhD), including geotechnical engineering, structural engineering; environmental engineering (MS, PhD), including environmental engineering and environmental sciences, environmental geotechnology, environmental health, environmental science and management, hazardous materials management, water resources engineering. Part-time programs available. *Faculty:* 13 full-time, 10 part-time. *Students:* 99 (47 women); includes 21 minority (5 African Americans, 7 Asian Americans or Pacific Islanders, 9 Hispanic Americans) 16 international. Terminal master's awarded for partial completion of doctoral program. *Degree requirements:* For master's, thesis or alternative required, foreign language not required; for doctorate, dissertation required, foreign language not required. *Entrance requirements:* For master's and doctorate, GRE General Test, TOEFL (minimum score of 550 required). *Application deadline:* For fall admission, 2/15; for spring admission, 10/15. Applications are processed on a rolling basis. Application fee: $50. *Unit head:* Dr. Stephen Levine, Chair, 617-627-3211, Fax: 617-627-3994. *Application contact:* Linfield Brown, 617-627-3211, Fax: 617-627-3994.

Université du Québec, École de technologie supérieure, Graduate Programs, Program in Building Engineering, Montréal, PQ H3C 1K3, Canada. Offers M Eng, Diploma. *Degree requirements:* For master's and Diploma, thesis not required. *Entrance requirements:* For master's and Diploma, appropriate bachelor's degree, proficiency in French.

University of Alberta, Faculty of Graduate Studies and Research, Department of Civil and Environmental Engineering, Edmonton, AB T6G 2E1, Canada. Offers construction engineering and management (M Eng, M Sc, PhD); environmental engineering (M Eng, M Sc, PhD); environmental science (M Sc, PhD); geoenvironmental engineering (M Eng, M Sc, PhD); geotechnical engineering (M Sc); geotechnical engineering (M Eng, PhD); mining engineering (M Eng, M Sc, PhD); petroleum engineering (M Eng, M Sc, PhD); structural engineering (M Eng, M Sc, PhD); water resources (M Eng, M Sc, PhD). Part-time programs available. *Degree requirements:* For master's, thesis required (for some programs), foreign language not required; for doctorate, dissertation required, foreign language not required. *Faculty research:* Mining.

University of California, Berkeley, Graduate Division, College of Engineering, Department of Civil and Environmental Engineering, Berkeley, CA 94720-1500. Offers construction engineering and management (M Eng, MS, D Eng, PhD); environmental quality and environmental water resources engineering (M Eng, MS, D Eng, PhD); geotechnical engineering (M Eng, MS, D Eng, PhD); structural engineering, mechanics and materials (M Eng, MS, D Eng, PhD); transportation engineering (M Eng, MS, D Eng, PhD). *Students:* 326 full-time (97 women); includes 59 minority (3 African Americans, 42 Asian Americans or Pacific Islanders, 13 Hispanic Americans, 1 Native American), 113 international. *Degree requirements:* For master's, comprehensive exam or thesis (MS) required; for doctorate, dissertation, qualifying exam required. *Entrance requirements:* For master's, GRE General Test, minimum GPA of 3.0; for doctorate, GRE General Test, minimum GPA of 3.5. *Application deadline:* For fall admission, 2/10. Application fee: $40. *Unit head:* Dr. Adib Kanafani, Chair, 510-642-3261. *Application contact:* Mari Cook, Graduate Assistant for Admission, 510-643-8944, E-mail: mcook@ce.berkeley.edu.

See in-depth description on page 609.

University of California, Los Angeles, Graduate Division, School of Engineering and Applied Science, Department of Civil and Environmental Engineering, Los Angeles, CA 90095. Offers environmental engineering (MS, PhD); geotechnical engineering (MS, PhD); structures (MS, PhD), including structural mechanics and earthquake engineering; water resource systems engineering (MS, PhD). *Faculty:* 15 full-time, 11 part-time. *Students:* 116 full-time (29 women); includes 28 minority (3 African Americans, 21 Asian Americans or Pacific Islanders, 4 Hispanic Americans), 37 international. *Degree requirements:* For master's, comprehensive exam or thesis required; for doctorate, dissertation, qualifying exams required, foreign language not required. *Entrance requirements:* For master's, GRE General Test, minimum GPA of 3.0; for doctorate, GRE General Test, minimum GPA of 3.25. *Application deadline:* For fall admission, 1/15; for spring admission, 12/1. Application fee: $40. Electronic applications accepted. *Unit head:* Dr. Michael Stenstrom, Chair, 310-825-1408. *Application contact:* Deeona Columbia, Graduate Affairs Officer, 310-825-1851, Fax: 310-206-2222, E-mail: deeona@ea.ucla.edu.

University of California, San Diego, Graduate Studies and Research, Department of Applied Mechanics and Engineering Sciences, Program in Structural Engineering, La Jolla, CA 92093-5003. Offers MS, PhD. Part-time programs available. *Students:* 37 full-time (7 women); includes 10 minority (4 Asian Americans or Pacific Islanders, 4 Hispanic Americans, 2 Native Americans), 8 international. 67 applicants, 61% accepted. In 1998, 6 master's, 2 doctorates awarded. *Degree requirements:* For master's, comprehensive exam or thesis required; for doctorate, dissertation, qualifying exam required. *Entrance requirements:* For master's and doctorate, GRE General Test, TOEFL (minimum score of 550 required), minimum GPA of 3.0. *Application deadline:* For fall admission, 5/31. Application fee: $40. *Financial aid:* In 1998–99, fellowships with full tuition reimbursements (averaging $15,000 per year), research assistantships with full tuition reimbursements (averaging $15,000 per year), teaching assistantships with partial tuition reimbursements (averaging $13,000 per year) were awarded.; scholarships also available. Financial aid application deadline: 1/31; financial aid applicants required to submit FAFSA. *Faculty research:* Advanced large-scale civil, mechanical, and aerospace structures. *Unit head:* Dr. Frieder Seible, Head, 619-534-4640, Fax: 619-534-1730. *Application contact:* AMES Graduate Student Affairs, 619-822-1421, Fax: 619-534-1730, E-mail: lfloyd@ucsd.edu.

See in-depth description on page 617.

University of Colorado at Boulder, Graduate School, College of Engineering and Applied Science, Department of Civil, Environmental, and Architectural Engineering, Boulder, CO 80309. Offers building systems (MS, PhD); construction engineering and management (MS, PhD); environmental engineering (MS, PhD); geoenvironmental engineering (MS, PhD); geotechnical engineering (MS, PhD); structural engineering (MS, PhD); water resource engineering (MS, PhD). *Degree requirements:* For master's, thesis or alternative, comprehensive exam required; for doctorate, dissertation required. *Entrance requirements:* For master's, GRE General Test, minimum undergraduate GPA of 3.0.

See in-depth description on page 619.

University of Dayton, Graduate School, School of Engineering, Department of Civil Engineering, Dayton, OH 45469-1300. Offers engineering mechanics (MSEM); environmental engineering (MSCE); soil mechanics (MSCE); structural engineering (MSCE); transport engineering (MSCE). Part-time programs available. *Faculty:* 9 full-time (0 women), 2 part-time (0 women). *Students:* 7 full-time (1 woman), 6 part-time (1 woman); includes 1 minority (African American), 4 international. *Degree requirements:* For master's, thesis or alternative required, foreign language not required. *Entrance requirements:* For master's, TOEFL. *Application deadline:* For fall admission, 8/1. Applications are processed on a rolling basis. Application fee: $30. *Unit head:* Dr. Joseph Saliba, Chairperson, 937-229-3847. *Application contact:* Dr. Donald L. Moon, Associate Dean, 937-229-2241, Fax: 937-229-2471, E-mail: dmoon@engr.udayton.edu.

University of Delaware, College of Engineering, Department of Civil and Environmental Engineering, Program in Structural Engineering, Newark, DE 19716. Offers MAS, MCE, PhD. Terminal master's awarded for partial completion of doctoral program. *Degree requirements:* For master's and doctorate, thesis/dissertation required, foreign language not required. *Entrance requirements:* For master's and doctorate, GRE General Test, TOEFL (minimum score of 550 required). *Faculty research:* Structural dynamics, analytical and numerical methods in structural mechanics and geomechanics, analysis of geomaterials, ice mechanics, structural stability.

University of Maine, Graduate School, College of Engineering, Department of Civil and Environmental Engineering, Orono, ME 04469. Offers civil engineering (MS, PhD), including environmental engineering, geotechnical engineering, structural engineering. *Faculty:* 11 full-time (0 women). *Students:* 19 full-time (3 women), 10 part-time (1 woman), 4 international. *Degree requirements:* For doctorate, dissertation required, foreign language not required, foreign language not required. *Entrance requirements:* For master's and doctorate, GRE General Test, TOEFL (minimum score of 550 required). *Application deadline:* For fall admission, 2/1 (priority date); for spring admission, 10/15. Applications are processed on a rolling basis. Application fee: $50. *Unit head:* Dr. Willem Brutsaert, Chair, 207-581-2170, Fax: 207-581-3888. *Application contact:* Scott G. Delcourt, Director of the Graduate School, 207-581-3218, Fax: 207-581-3232, E-mail: graduate@maine.edu.

The University of Memphis, Graduate School, Herff College of Engineering, Department of Civil Engineering, Memphis, TN 38152. Offers civil engineering (PhD); environmental engineering (MS); foundation engineering (MS); structural engineering (MS); transportation engineering (MS); water resources engineering (MS). *Faculty:* 12 full-time (0 women), 1 part-time (0 women). *Students:* 6 full-time (2 women), 22 part-time (1 woman), 7 international. *Degree requirements:* For master's, thesis or alternative, comprehensive exam required; for doctorate, dissertation required. *Entrance requirements:* For master's, GRE General Test (minimum combined score of 1000 required) or MAT, BS, minimum undergraduate GPA of 2.5. *Application deadline:* For fall admission, 8/1; for spring admission, 12/1. Application fee: $25 ($50 for international students). Tuition, state resident: full-time $3,410; part-time $178 per credit hour. Tuition, nonresident: full-time $8,670; part-time $408 per credit hour. Tuition and fees vary according to program. *Unit head:* Dr. Martin E. Lipinski, Chairman, 901-678-3279. *Application contact:* Dr. Larry W. Moore, Coordinator of Graduate Studies, 901-678-3278.

University of Missouri–Columbia, Graduate School, College of Engineering, Department of Civil and Environmental Engineering, Columbia, MO 65211. Offers civil engineering (MS, PhD); environmental engineering (MS, PhD); geotechnical engineering (MS, PhD); structural engineering (MS, PhD); transportation and highway engineering (MS); water resources (MS, PhD). *Faculty:* 24 full-time (2 women). *Students:* 19 full-time (2 women), 18 part-time (3 women); includes 1 minority (Hispanic American), 21 international. *Degree requirements:* For master's, report or thesis required; for doctorate, dissertation required. *Entrance requirements:* For master's and doctorate, GRE General Test, TOEFL. *Application deadline:* For fall admission, 7/10. Application fee: $30 ($50 for international students). *Unit head:* Dr. Mark Virkler, Director of Graduate Studies, 573-882-3678.

See in-depth description on page 639.

University of Missouri–Rolla, Graduate School, School of Engineering, Department of Civil Engineering, Program in Structural Analysis and Design, Rolla, MO 65409-0910. Offers MS, DE, PhD. *Degree requirements:* For master's, thesis or alternative required, foreign language not required; for doctorate, dissertation required, foreign language not required. *Entrance requirements:* For master's and doctorate, GRE General Test (minimum combined score of 1100 required), TOEFL (minimum score of 550 required), minimum GPA of 3.0.

University of Missouri–Rolla, Graduate School, School of Engineering, Department of Civil Engineering, Program in Structural Materials, Rolla, MO 65409-0910. Offers MS. *Degree requirements:* For master's, thesis or alternative required, foreign language not required.

University of Missouri–Rolla (continued)
Entrance requirements: For master's, GRE General Test (minimum combined score of 1100 required), TOEFL (minimum score of 550 required), minimum GPA of 3.0.

University of Missouri–Rolla, Graduate School, School of Engineering, Department of Civil Engineering, Program in Structural Methods, Rolla, MO 65409-0910. Offers DE, PhD. *Degree requirements:* For doctorate, dissertation required, foreign language not required. *Entrance requirements:* For doctorate, GRE General Test (minimum combined score of 1100 required), TOEFL (minimum score of 550 required), minimum GPA of 3.0.

University of New Brunswick, School of Graduate Studies, Faculty of Engineering, Department of Civil Engineering, Fredericton, NB E3B 5A3, Canada. Offers environmental engineering (M Eng, M Sc E, PhD); geotechnical engineering (M Eng, M Sc E); structures and structural foundations (M Eng, M Sc E, PhD); transportation engineering (M Eng, M Sc E, PhD); water resources and hydrology (M Eng, M Sc E, PhD). Part-time programs available. *Degree requirements:* For master's, thesis required, foreign language not required; for doctorate, dissertation, qualifying exam required, foreign language not required. *Entrance requirements:* For master's and doctorate, TOEFL, TWE, minimum GPA of 3.0.

University of North Dakota, Graduate School, School of Engineering and Mines, Department of Civil Engineering, Grand Forks, ND 58202. Offers civil engineering (M Engr); sanitary engineering (M Engr), including soils and structures engineering, surface mining engineering. Part-time programs available. *Faculty:* 6 full-time (0 women). *Students:* 7 full-time (1 woman). *Degree requirements:* For master's, thesis or alternative required, foreign language not required. *Entrance requirements:* For master's, GRE General Test, TOEFL (minimum score of 550 required), minimum GPA of 2.5. *Application deadline:* For fall admission, 3/1 (priority date). Applications are processed on a rolling basis. Application fee: $20. *Unit head:* Dr. Ronald Apanian, Chairperson, 701-777-3562, Fax: 701-777-4838, E-mail: ron_apanian@mail.und.nodak.edu.

University of Oklahoma, Graduate College, College of Engineering, School of Civil Engineering and Environmental Science, Program in Civil Engineering, Norman, OK 73019-0390. Offers civil engineering (PhD); environmental engineering (MS); geotechnical engineering (MS); structures (MS); transportation (MS). *Accreditation:* ABET (one or more programs are accredited). *Students:* 19 full-time (4 women), 36 part-time (6 women); includes 5 minority (3 Asian Americans or Pacific Islanders, 2 Native Americans), 28 international. *Degree requirements:* For master's, comprehensive and oral exams required, foreign language and thesis not required; for doctorate, dissertation, oral and qualifying exams required, foreign language not required. *Entrance requirements:* For master's, GRE General Test, TOEFL (minimum score of 575 required), minimum GPA of 3.0; for doctorate, GRE General Test, TOEFL (minimum score of 575 required), minimum graduate GPA of 3.5. *Application deadline:* For fall admission, 4/1 (priority date). Applications are processed on a rolling basis. Application fee: $25. Tuition, state resident: part-time $86 per credit hour. Tuition, nonresident: part-time $275 per credit hour. Tuition and fees vary according to course level, course load and program. *Unit head:* Robert C. Knox, Graduate Liaison, 405-325-4256, Fax: 405-325-7508. *Application contact:* Robert C. Knox, Graduate Liaison, 405-325-4256, Fax: 405-325-7508.

See in-depth description on page 641.

University of Rhode Island, Graduate School, College of Engineering, Department of Civil and Environmental Engineering, Program in Structural Engineering, Kingston, RI 02881. Offers MS, PhD.

University of Southern California, Graduate School, School of Engineering, Department of Civil Engineering, Program in Structural Engineering, Los Angeles, CA 90089. Offers MS. *Students:* 9 full-time (2 women), 10 part-time (4 women); includes 8 minority (1 African American, 6 Asian Americans or Pacific Islanders, 1 Hispanic American), 7 international. Average age 26. 34 applicants, 88% accepted. In 1998, 11 degrees awarded. *Degree requirements:* For master's, thesis optional. *Entrance requirements:* For master's, GRE General Test. *Application deadline:* For fall admission, 6/1 (priority date); for spring admission, 11/1. Application fee: $55. Tuition: Part-time $768 per unit. Required fees: $350 per semester. *Financial aid:* In 1998–99, 1 fellowship, 2 research assistantships, 2 teaching assistantships were awarded.; Federal Work-Study and institutionally-sponsored loans also available. Aid available to part-time students. Financial aid application deadline: 2/15; financial aid applicants required to submit FAFSA. *Unit head:* Dr. L. Carter Wellford, Chair, Department of Civil Engineering, 213-740-0587.

University of Southern California, Graduate School, School of Engineering, Department of Civil Engineering, Program in Structural Mechanics, Los Angeles, CA 90089. Offers MS. 3 applicants, 100% accepted. *Degree requirements:* For master's, thesis optional. *Entrance requirements:* For master's, GRE General Test. *Application deadline:* For fall admission, 6/1 (priority date); for spring admission, 11/1. Application fee: $55. Tuition: Part-time $768 per unit. Required fees: $350 per semester. *Financial aid:* Fellowships, research assistantships, teaching assistantships, Federal Work-Study and institutionally-sponsored loans available. Aid available to part-time students. Financial aid application deadline: 2/15; financial aid applicants required to submit FAFSA. *Unit head:* Dr. L. Carter Wellford, Chair, Department of Civil Engineering, 213-740-0587.

University of Virginia, School of Engineering and Applied Science, Department of Civil Engineering, Program in Structural Engineering, Charlottesville, VA 22903. Offers ME, MS, PhD. Part-time programs available. Postbaccalaureate distance learning degree programs offered (no on-campus study). *Faculty:* 5 full-time (1 woman). *Students:* 12 full-time (5 women), 2 international. Average age 26. 14 applicants, 79% accepted. In 1998, 6 master's awarded (100% found work related to degree); 1 doctorate awarded (100% found work related to degree). Terminal master's awarded for partial completion of doctoral program. *Degree requirements:* For master's, thesis required (for some programs), foreign language not required; for doctorate, dissertation, comprehensive exam required, foreign language not required. *Entrance requirements:* For master's and doctorate, GRE General Test. *Average time to degree:* Master's–1.7 years full-time; doctorate–3 years full-time. *Application deadline:* For fall admission, 2/1 (priority date). Applications are processed on a rolling basis. Application fee: $60. Electronic applications accepted. *Financial aid:* In 1998–99, 12 students received aid, including 1 fellowship with full tuition reimbursement available (averaging $12,100 per year), 5 research assistantships with full tuition reimbursements available (averaging $10,700 per year), 6 teaching assistantships with full tuition reimbursements available (averaging $10,700 per year) Financial aid application deadline: 2/1. *Faculty research:* Dynamic structural response, computational mechanics, probabilistic methods, mechanics and design of composites, computer-aided design. *Unit head:* Dr. Nicholas J. Garber, Chairman, Department of Civil Engineering, 804-924-7464, Fax: 804-982-2951, E-mail: njg@virginia.edu.

University of Washington, Graduate School, College of Engineering, Department of Civil and Environmental Engineering, Seattle, WA 98195. Offers environmental engineering (MS, MS Civ E, MSE, PhD); hydraulic engineering (MS Civ E, MSE, PhD); structural and geotechnical engineering and mechanics (MS, MS Civ E, MSE, PhD); transportation and construction engineering (MS, MS Civ E, MSE, PhD). *Faculty:* 31 full-time (4 women), 26 part-time (1 woman). *Students:* 153 full-time (54 women), 42 part-time (13 women); includes 25 minority (2 African Americans, 19 Asian Americans or Pacific Islanders, 3 Hispanic Americans, 1 Native American), 34 international. Terminal master's awarded for partial completion of doctoral program. *Degree requirements:* For master's, thesis or alternative required, foreign language not required; for doctorate, dissertation required, foreign language not required. *Entrance requirements:* For master's, GRE General Test, TOEFL (minimum score of 580 required), minimum GPA of 3.0; for doctorate, GRE, TOEFL (minimum score of 580 required), minimum GPA of 3.0. *Application deadline:* For fall admission, 2/1 (priority date); for winter admission, 11/1; for spring admission, 2/1. Applications are processed on a rolling basis. Application fee: $50. Electronic applications accepted. Tuition, state resident: full-time $5,196; part-time $475 per credit. Tuition, nonresident: full-time $13,485; part-time $1,285 per credit. Required fees: $387; $38 per credit. Tuition and fees vary according to course load. *Unit head:* Dr. Fred L. Mannering, Chair, 206-543-2390. *Application contact:* Marcia Buck, Graduate Secretary, 206-543-2574, E-mail: ceginfo@u.washington.edu.

Washington University in St. Louis, School of Engineering and Applied Science, Sever Institute of Technology, Department of Civil Engineering, Program in Structural Engineering, St. Louis, MO 63130-4899. Offers MSE, D Sc. *Degree requirements:* For master's, thesis optional, foreign language not required; for doctorate, variable foreign language requirement, dissertation, departmental qualifying exam required.

Surveying Science and Engineering

The Ohio State University, Graduate School, College of Engineering, Department of Civil and Environmental Engineering and Geodetic Science, Program in Geodetic Science and Surveying, Columbus, OH 43210. Offers MS, PhD. *Faculty:* 10 full-time, 3 part-time. *Students:* 55 full-time (6 women), 3 part-time; includes 6 minority (4 African Americans, 1 Asian American or Pacific Islander, 1 Hispanic American), 42 international. 50 applicants, 54% accepted. In 1998, 18 master's, 5 doctorates awarded. *Degree requirements:* For master's, computer language required, thesis optional, foreign language not required; for doctorate, computer language, dissertation required, foreign language not required. *Application deadline:* For fall admission, 8/15. Applications are processed on a rolling basis. Application fee: $30 ($40 for international students). *Financial aid:* Fellowships, research assistantships, teaching assistantships, Federal Work-Study and institutionally-sponsored loans available. Aid available to part-time students. *Faculty research:* Photogrammetry, cartography, geodesy, land information systems. *Unit head:* Rongxing Li, Graduate Studies Committee Chair, 614-292-6753, Fax: 614-292-3780, E-mail: li.282@osu.edu.

University of New Brunswick, School of Graduate Studies, Faculty of Engineering, Department of Geodesy and Geomatics, Fredericton, NB E3B 5A3, Canada. Offers land information management (Diploma); mapping, charting and geodesy (Diploma); surveying engineering (M Eng, M Sc E, PhD). Part-time programs available. *Degree requirements:* For master's, thesis required, foreign language not required; for doctorate, dissertation, qualifying exam required, foreign language not required; for degree. *Entrance requirements:* For master's and doctorate, TOEFL, TWE, minimum GPA of 3.0; for Diploma, TOEFL, TWE.

Transportation and Highway Engineering

Auburn University, Graduate School, College of Engineering, Department of Civil Engineering, Auburn, Auburn University, AL 36849-0002. Offers construction engineering and management (MCE, MS, PhD); environmental engineering (MCE, MS, PhD); geotechnical/materials engineering (MCE, MS, PhD); hydraulics/hydrology (MCE, MS, PhD); structural engineering (MCE, MS, PhD); transportation engineering (MCE, MS, PhD). Part-time programs available. *Faculty:* 22 full-time. *Students:* 24 full-time (6 women), 34 part-time (8 women); includes 2 minority (1 African American, 1 Hispanic American), 9 international. *Degree requirements:* For master's, project (MCE), thesis (MS) required; for doctorate, dissertation, comprehensive exam required, foreign language not required. *Entrance requirements:* For master's, GRE General Test; for doctorate, GRE General Test (minimum score of 400 on each section required). *Application deadline:* For fall admission, 9/1; for spring admission, 3/1. Applications are processed on a rolling basis. Application fee: $25 ($50 for international students). Tuition, state resident: full-time $2,760; part-time $76 per credit hour. Tuition, nonresident: full-time $8,280; part-time $228 per credit hour. *Unit head:* Dr. Joseph F. Judkins, Head, 334-844-4320. *Application contact:* Dr. John F. Pritchett, Dean of the Graduate School, 334-844-4700.

Central Missouri State University, School of Graduate Studies, College of Applied Sciences and Technology, Department of Power and Transportation, Warrensburg, MO 64093. Offers aviation safety (MS). Part-time programs available. *Faculty:* 7 full-time (0 women), 7 part-time (0 women). *Students:* 24 full-time (3 women), 59 part-time (10 women); includes 5 minority (4 African Americans, 1 Asian American or Pacific Islander), 7 international. In 1998, 52 degrees awarded (85% found work related to degree). *Degree requirements:* For master's, comprehensive exam required, thesis not required. *Entrance requirements:* For master's, minimum GPA of 2.5. *Application deadline:* Applications are processed on a rolling basis. Application fee: $25 ($50 for international students). Tuition, state resident: full-time $3,576; part-time $149 per credit hour. Tuition, nonresident: full-time $7,152; part-time $298 per credit hour. Tuition and fees vary according to course load and campus/location. *Financial aid:* In 1998–99, research assistantships with tuition reimbursements (averaging $3,750 per year), 6 teaching assistantships with tuition reimbursements (averaging $3,750 per year) were awarded.; Federal Work-Study, grants, scholarships, unspecified assistantships, and administrative and laboratory assistantships also available. Aid available to part-time students. Financial aid application deadline: 3/1; financial aid applicants required to submit FAFSA. *Unit head:* Dr.

John Dennison, Interim Chair, 660-543-4975, Fax: 660-543-4979, E-mail: dennison@cmsu1.cmsu.edu.

Cornell University, Graduate School, Graduate Fields of Engineering, Field of Civil and Environmental Engineering, Ithaca, NY 14853-0001. Offers environmental engineering (M Eng, MS, PhD); environmental fluid mechanics and hydrology (M Eng, MS, PhD); environmental systems engineering (M Eng, MS, PhD); geotechnical engineering (M Eng, MS, PhD); remote sensing (M Eng, MS, PhD); structural engineering (M Eng, MS, PhD); transportation engineering (M Eng, MS, PhD); water resource systems (M Eng, MS, PhD). *Faculty:* 29 full-time. *Students:* 143 full-time (32 women); includes 16 minority (11 Asian Americans or Pacific Islanders, 5 Hispanic Americans), 76 international. Terminal master's awarded for partial completion of doctoral program. *Degree requirements:* For master's, thesis (MS) required; for doctorate, dissertation required, foreign language not required. *Entrance requirements:* For master's, TOEFL (minimum score of 600required); for doctorate, GRE General Test, TOEFL (minimum score of 600 required). *Application deadline:* For fall admission, 1/15. Application fee: $65. Electronic applications accepted. *Unit head:* Director of Graduate Studies, 607-255-7560, Fax: 607-255-9004. *Application contact:* Graduate Field Assistant, 607-255-7560, E-mail: cee_grad@cornell.edu.

See in-depth description on page 573.

École Polytechnique de Montréal, Graduate Programs, Department of Civil Engineering, Montréal, PQ H3C 3A7, Canada. Offers environmental engineering (M Eng, M Sc A, PhD); geotechnical engineering (M Eng, M Sc A, PhD); hydraulics engineering (M Eng, M Sc A, PhD); structural engineering (M Eng, M Sc A, PhD); transportation engineering (M Eng, M Sc A, PhD). Part-time and evening/weekend programs available. *Degree requirements:* For master's and doctorate, one foreign language, computer language, thesis/dissertation required. *Entrance requirements:* For master's, minimum GPA of 2.75; for doctorate, minimum GPA of 3.0. *Faculty research:* Water resources management, characteristics of building materials, aging of dams, pollution control.

Iowa State University of Science and Technology, Graduate College, College of Engineering, Department of Civil and Construction Engineering, Ames, IA 50011. Offers civil engineering (MS, PhD), including civil engineering materials, construction engineering and management, environmental engineering, geometronics, geotechnical engineering, structural engineering, transportation engineering. *Faculty:* 36 full-time, 5 part-time. *Students:* 60 full-time (14 women), 51 part-time (12 women). *Degree requirements:* For master's, thesis or alternative required, foreign language not required; for doctorate, dissertation required, foreign language not required. *Entrance requirements:* For master's and doctorate, GRE General Test (foreign students), TOEFL. *Application deadline:* For fall admission, 6/15 (priority date); for spring admission, 11/15 (priority date). Application fee: $20 ($50 for international students). Electronic applications accepted. Tuition, state resident: full-time $3,308. Tuition, nonresident: full-time $9,744. Part-time tuition and fees vary according to course load, campus/location and program. *Unit head:* Dr. Lowell F. Greimann, Chair, 515-294-2140, E-mail: cceinfo@iastate.edu. *Application contact:* Edward Kannel, Director of Graduate Education, 515-294-2861, E-mail: cceinfo@iastate.edu.

Lawrence Technological University, College of Engineering, Southfield, MI 48075-1058. Offers automotive engineering (MAE); civil engineering (MCE); manufacturing systems (MEMS). Part-time and evening/weekend programs available. *Faculty:* 8 full-time (1 woman), 6 part-time (0 women). *Degree requirements:* For master's, foreign language and thesis not required. *Application deadline:* For fall admission, 8/1 (priority date); for spring admission, 1/1. Applications are processed on a rolling basis. Application fee: $50. Electronic applications accepted. Tuition: Full-time $5,128; part-time $419 per credit hour. Required fees: $100; $100 per year. $50 per semester. Tuition and fees vary according to course level. *Unit head:* Dr. George Kartsounes, Dean, 248-204-2500, Fax: 248-204-2509, E-mail: kartsounes@ltu.edu. *Application contact:* Lisa Kujawa, Director of Admissions, 248-204-3160, Fax: 248-204-3188, E-mail: admission@hu.edu.

Louisiana State University and Agricultural and Mechanical College, Graduate School, College of Engineering, Department of Civil and Environmental Engineering, Baton Rouge, LA 70803. Offers environmental engineering (MSCE, PhD); geotechnical engineering (MSCE, PhD); structural engineering and mechanics (MSCE, PhD); transportation engineering (MSCE, PhD); water resources (MSCE, PhD). Part-time programs available. *Faculty:* 27 full-time (1 woman), 1 part-time (0 women). *Students:* 59 full-time (14 women), 24 part-time (2 women); includes 5 minority (1 African American, 3 Asian Americans or Pacific Islanders, 1 Hispanic American), 48 international. *Degree requirements:* For master's, thesis optional, foreign language not required; for doctorate, one foreign language (computer language can substitute), dissertation required. *Entrance requirements:* For master's and doctorate, GRE General Test, TOEFL, minimum GPA of 3.0. *Application deadline:* For fall admission, 1/25 (priority date). Applications are processed on a rolling basis. Application fee: $25. *Unit head:* Dr. Ronald F. Malone, Acting Chair, 225-388-8666, Fax: 225-388-8652, E-mail: rmalone@unix1.sncc.lsu.edu. *Application contact:* Dr. Vijaya K. A. Gopu, Graduate Coordinator, 225-388-8442, E-mail: cegopu@eng.lsu.edu.

See in-depth description on page 589.

Marquette University, Graduate School, College of Engineering, Department of Civil and Environmental Engineering, Milwaukee, WI 53201-1881. Offers construction and public works management (MS, PhD); environmental/water resources engineering (MS, PhD); structural/geotechnical engineering (MS, PhD); transportational planning and engineering (MS, PhD). Part-time and evening/weekend programs available. *Faculty:* 12 full-time (0 women), 1 part-time (0 women). *Students:* 17 full-time (7 women), 32 part-time (5 women); includes 5 minority (1 African American, 4 Asian Americans or Pacific Islanders), 12 international. Terminal master's awarded for partial completion of doctoral program. *Degree requirements:* For master's, thesis or alternative, comprehensive exam required, foreign language not required; for doctorate, dissertation required, foreign language not required. *Entrance requirements:* For master's, TOEFL (minimum score of 550 required); for doctorate, GRE General Test, TOEFL (minimum score of 550 required). *Application deadline:* For fall admission, 6/1 (priority date). Applications are processed on a rolling basis. Application fee: $40. Tuition: Part-time $510 per credit hour. Tuition and fees vary according to program. *Unit head:* Dr. Thomas H. Wenzel, Chairman, 414-288-7030, Fax: 414-288-7521, E-mail: wenzelt@vms.csd.mu.edu.

Massachusetts Institute of Technology, School of Engineering, Center for Transportation Studies, Cambridge, MA 02139-4307. Offers logistics (M Eng); transportation (MST, PhD). In 1998, 20 degrees awarded. *Application deadline:* For fall admission, 1/15; for spring admission, 11/1. Application fee: $55. Total annual research expenditures: $1 million. *Unit head:* Yossi Sheffi, Director, 617-253-5316, Fax: 617-253-4560, E-mail: sheffi@mit.edu. *Application contact:* Sydney Miller, Student Coordinator, 617-253-8069, Fax: 617-253-4560, E-mail: tec@mit.edu.

See in-depth description on page 591.

National Technological University, Programs in Engineering, Fort Collins, CO 80526-1842. Offers chemical engineering (MS); computer engineering (MS); computer science (MS); electrical engineering (MS); engineering management (MS); hazardous waste management (MS); health physics (MS); management of technology (MS); manufacturing systems engineering (MS); materials science and engineering (MS); software engineering (MS); special majors (MS); transportation engineering (MS); transportation systems engineering (MS). Part-time programs available. *Faculty:* 600 part-time (20 women). *Entrance requirements:* For master's, BS in engineering or related field. *Application deadline:* Applications are processed on a rolling basis. Application fee: $50. *Unit head:* Lionel V. Baldwin, President, 970-495-6400, Fax: 970-484-0668, E-mail: baldwin@mail.ntu.edu.

New Jersey Institute of Technology, Office of Graduate Studies, Department of Civil and Environmental Engineering, Interdisciplinary Program in Transportation, Newark, NJ 07102-1982. Offers MS, PhD. Part-time and evening/weekend programs available. Terminal master's awarded for partial completion of doctoral program. *Degree requirements:* For master's, thesis or alternative required, foreign language not required; for doctorate, dissertation, residency required, foreign language not required. *Entrance requirements:* For master's, GRE General Test (minimum score of 450 on verbal section, 600 on quantitative, 550 on analytical required); for doctorate, GRE General Test (minimum score of 450 on verbal section, 600 on quantitative, 550 on analytical required), minimum graduate GPA of 3.5. Electronic applications accepted. *Faculty research:* Transportation planning, administration, and policy; intelligent vehicle highway systems; bridge maintenance.

Northwestern University, The Graduate School, Robert R. McCormick School of Engineering and Applied Science, Department of Civil Engineering, Evanston, IL 60208. Offers biosolid mechanics (MS, PhD); environmental health engineering (MS, PhD); geotechnical engineering (MS, PhD); health physics/radiological health (MS, PhD); project management (MPM); structural engineering (MS, PhD); structural mechanics (MS, PhD); transportation systems engineering (MS, PhD). MS and PhD admissions and degrees offered through The Graduate School. Part-time programs available. *Faculty:* 26 full-time (2 women), 1 (woman) part-time. *Students:* 110 full-time (38 women), 7 part-time (2 women); includes 5 minority (2 African Americans, 3 Asian Americans or Pacific Islanders), 68 international. Terminal master's awarded for partial completion of doctoral program. *Degree requirements:* For master's, thesis required (for some programs); for doctorate, dissertation required. *Entrance requirements:* For master's and doctorate, GRE General Test, TOEFL (minimum score of 560 required). Application fee: $50 ($55 for international students). *Unit head:* Joseph L. Schofer, Chair, 847-491-3257, Fax: 847-491-4011, E-mail: j-schofer@nwu.edu. *Application contact:* Karm Kerwell, Secretary, 847-491-3176, Fax: 847-491-4011, E-mail: k-kerwell@nwu.edu.

Pennsylvania State University University Park Campus, Graduate School, College of Engineering, Department of Civil and Environmental Engineering, State College, University Park, PA 16802-1503. Offers civil engineering (M Eng, MS, PhD); environmental engineering (M Eng, MS, PhD); structural engineering (M Eng, MS, PhD); transportation and highway engineering (M Eng, MS, PhD); water resources engineering (M Eng, MS, PhD). *Students:* 77 full-time (24 women), 52 part-time (8 women). *Degree requirements:* For master's, final paper (M Eng), oral exam and thesis (MS) required; for doctorate, dissertation, comprehensive and oral exams required. *Entrance requirements:* For master's and doctorate, GRE General Test. Application fee: $50. *Unit head:* Dr. Paul P. Jovanis, Head, 814-863-3084.

Polytechnic University, Brooklyn Campus, Department of Civil and Environmental Engineering, Major in Transportation Planning and Engineering, Brooklyn, NY 11201-2990. Offers MS. Part-time and evening/weekend programs available. Average age 33. 9 applicants, 67% accepted. In 1998, 9 degrees awarded. *Degree requirements:* For master's, thesis or alternative required. *Application deadline:* Applications are processed on a rolling basis. Application fee: $45. Electronic applications accepted. *Financial aid:* Fellowships, research assistantships, teaching assistantships, institutionally-sponsored loans available. Aid available to part-time students. Financial aid applicants required to submit FAFSA. *Application contact:* John S. Kerge, Dean of Admissions, 718-260-3200, Fax: 718-260-3446, E-mail: admitme@poly.edu.

Polytechnic University, Farmingdale Campus, Graduate Programs, Department of Civil and Environmental Engineering, Major in Transportation Planning and Engineering, Farmingdale, NY 11735-3995. Offers MS. Average age 33. 2 applicants, 100% accepted. In 1998, 5 degrees awarded. *Degree requirements:* For master's, computer language, thesis required (for some programs). *Application deadline:* Applications are processed on a rolling basis. Application fee: $45. Electronic applications accepted. *Financial aid:* Institutionally-sponsored loans available. Aid available to part-time students. Financial aid applicants required to submit FAFSA. *Unit head:* John S. Kerge, Dean of Admissions, 718-260-3200, Fax: 718-260-3446, E-mail: admitme@poly.edu. *Application contact:* John S. Kerge, Dean of Admissions, 718-260-3200, Fax: 718-260-3446, E-mail: admitme@poly.edu.

Princeton University, Graduate School, School of Engineering and Applied Science, Department of Civil Engineering and Operations Research, Program in Transportation Systems, Princeton, NJ 08544-1019. Offers MSE, PhD. Offered jointly with the Woodrow Wilson School of Public and International Affairs. *Degree requirements:* For master's and doctorate, thesis/dissertation required. *Entrance requirements:* For master's and doctorate, GRE General Test, GRE Subject Test.

Rensselaer Polytechnic Institute, Graduate School, School of Engineering, Department of Civil Engineering, Program in Transportation Engineering, Troy, NY 12180-3590. Offers M Eng, MS, D Eng, PhD, MBA/M Eng. Part-time programs available. *Faculty:* 11 full-time (0 women), 3 part-time (0 women). *Students:* 3 full-time (0 women), 4 part-time (1 woman), 1 international. 10 applicants, 20% accepted. *Degree requirements:* For master's, thesis required (for some programs), foreign language not required; for doctorate, dissertation required, foreign language not required. *Entrance requirements:* For master's and doctorate, GRE, TOEFL (minimum score of 550 required). *Application deadline:* For fall admission, 2/1 (priority date). Applications are processed on a rolling basis. Application fee: $35. *Financial aid:* Fellowships, research assistantships, teaching assistantships, institutionally-sponsored loans available. Financial aid application deadline: 2/1. *Faculty research:* Intelligent transportation systems, routing algorithms, dynamic network management, user behavior. Total annual research expenditures: $331,000. *Application contact:* Jo Ann Grega, Admissions Assistant, 518-276-6679, Fax: 518-276-4833, E-mail: gregaj2@rpi.edu.

Texas A&M University, College of Engineering, Department of Civil Engineering, Program in Transportation Engineering, College Station, TX 77843. Offers M Eng, MS, D Eng, PhD. D Eng offered through the College of Engineering. *Students:* 57. *Degree requirements:* For master's, thesis (MS) required; for doctorate, dissertation (PhD), internship (D Eng) required. *Entrance requirements:* For master's and doctorate, GRE General Test, TOEFL. Application fee: $50 ($75 for international students). *Financial aid:* Fellowships, research assistantships, teaching assistantships available. Financial aid application deadline: 4/1; financial aid applicants required to submit FAFSA. *Faculty research:* Design and operation of transportation facilities and systems, intelligent transportation systems. *Unit head:* Dr. Roger E. Smith, Head, 409-845-9967, Fax: 409-862-2800, E-mail: ce-grad@tamu.edu. *Application contact:* Dr. Daniel B. Fambro, 409-845-2498, Fax: 409-862-2800, E-mail: ce-grad@tamu.edu.

Texas Southern University, Graduate School, School of Technology, Program in Transportation, Houston, TX 77004-4584. Offers MS. *Degree requirements:* For master's, thesis, comprehensive exam required, foreign language not required. *Entrance requirements:* For master's, GRE General Test, TOEFL, minimum GPA of 2.5.

University of Arkansas, Graduate School, College of Engineering, Department of Civil Engineering, Program in Transportation Engineering, Fayetteville, AR 72701-1201. Offers MSE, MSTE. *Accreditation:* ABET. *Students:* 1 full-time (0 women), 1 international. 4 applicants, 100% accepted. In 1998, 4 degrees awarded. *Degree requirements:* For master's, thesis optional, foreign language not required. Application fee: $40 ($50 for international students). Tuition, state resident: full-time $3,186. Tuition, nonresident: full-time $7,560. Required fees: $378. *Financial aid:* Application deadline: 4/1. *Unit head:* Dr. Robert Elliott, Chair, Department of Civil Engineering, 501-575-4954.

University of California, Berkeley, Graduate Division, College of Engineering, Department of Civil and Environmental Engineering, Berkeley, CA 94720-1500. Offers construction engineering and management (M Eng, MS, D Eng, PhD); environmental quality and environmental water resources engineering (M Eng, MS, D Eng, PhD); geotechnical engineering (M Eng, MS, D Eng, PhD); structural engineering, mechanics and materials (M Eng, MS, D Eng, PhD); transportation engineering (M Eng, MS, D Eng, PhD). *Students:* 326 full-time (97 women); includes 59 minority (3 African Americans, 42 Asian Americans or Pacific Islanders, 13 Hispanic

Transportation and Highway Engineering

University of California, Berkeley (continued)

Americans, 1 Native American), 113 international. *Degree requirements:* For master's, comprehensive exam or thesis (MS) required; for doctorate, dissertation, qualifying exam required. *Entrance requirements:* For master's, GRE General Test, minimum GPA of 3.0; for doctorate, GRE General Test, minimum GPA of 3.5. *Application deadline:* For fall admission, 2/10. Application fee: $40. *Unit head:* Dr. Adib Kanafani, Chair, 510-642-3261. *Application contact:* Mari Cook, Graduate Assistant for Admission, 510-643-8944, E-mail: mcook@ce.berkeley.edu.

See in-depth description on page 609.

University of California, Davis, Graduate Studies, Program in Transportation Technology and Policy, Davis, CA 95616. Offers MS, PhD. *Students:* 15 full-time (4 women); includes 3 minority (2 Asian Americans or Pacific Islanders, 1 Hispanic American), 3 international. 23 applicants, 91% accepted. In 1998, 2 master's awarded. *Degree requirements:* For doctorate, dissertation required. *Entrance requirements:* For master's, GRE General Test, minimum GPA of 3.0; for doctorate, GRE General Test, minimum GPA of 3.5. Electronic applications accepted. *Financial aid:* In 1998–99, 10 fellowships with full and partial tuition reimbursements, 10 research assistantships with full and partial tuition reimbursements, 2 teaching assistantships with full and partial tuition reimbursements were awarded. Financial aid application deadline: 1/15; financial aid applicants required to submit FAFSA. *Unit head:* Patricia Makhtarian, Chair, 530-752-0247, E-mail: itsgraduate@ucdavis.edu.

See in-depth description on page 611.

University of Dayton, Graduate School, School of Engineering, Department of Civil Engineering, Dayton, OH 45469-1300. Offers engineering mechanics (MSEM); environmental engineering (MSCE); soil mechanics (MSCE); structural engineering (MSCE); transport engineering (MSCE). Part-time programs available. *Faculty:* 9 full-time (0 women), 2 part-time (0 women). *Students:* 7 full-time (1 woman), 6 part-time (1 woman); includes 1 minority (African American), 4 international. *Degree requirements:* For master's, thesis or alternative required, foreign language not required. *Entrance requirements:* For master's, TOEFL. *Application deadline:* For fall admission, 8/1. Applications are processed on a rolling basis. Application fee: $30. *Unit head:* Dr. Joseph Saliba, Chairperson, 937-229-3847. *Application contact:* Dr. Donald L. Moon, Associate Dean, 937-229-2241, Fax: 937-229-2471, E-mail: dmoon@engr.udayton.edu.

University of Delaware, College of Engineering, Department of Civil and Environmental Engineering, Program in Railroad Engineering, Newark, DE 19716. Offers MAS, MCE, PhD. Terminal master's awarded for partial completion of doctoral program. *Degree requirements:* For master's and doctorate, thesis/dissertation required, foreign language not required. *Entrance requirements:* For master's and doctorate, GRE General Test, TOEFL (minimum score of 550 required). *Faculty research:* Railway analyses.

University of Delaware, College of Engineering, Department of Civil and Environmental Engineering, Program in Transportation Engineering, Newark, DE 19716. Offers MAS, MCE, PhD. Terminal master's awarded for partial completion of doctoral program. *Degree requirements:* For master's and doctorate, thesis/dissertation required, foreign language not required. *Entrance requirements:* For master's and doctorate, GRE General Test, TOEFL (minimum score of 550 required). *Faculty research:* Traffic operations and highway traffic management, public transportation systems, highway maintenance planning, properties of pavement structure, analyses of impact of highway construction on wetlands regions.

The University of Memphis, Graduate School, Herff College of Engineering, Department of Civil Engineering, Memphis, TN 38152. Offers civil engineering (PhD); environmental engineering (MS); foundation engineering (MS); structural engineering (MS); transportation engineering (MS); water resources engineering (MS). *Faculty:* 12 full-time (0 women), 1 part-time (0 women). *Students:* 6 full-time (2 women), 22 part-time (1 woman), 7 international. *Degree requirements:* For master's, thesis or alternative, comprehensive exam required; for doctorate, dissertation required. *Entrance requirements:* For master's, GRE General Test (minimum combined score of 1000 required) or MAT, BS, minimum undergraduate GPA of 2.5. *Application deadline:* For fall admission, 8/1; for spring admission, 12/1. Application fee: $25 ($50 for international students). Tuition, state resident: full-time $3,410; part-time $178 per credit hour. Tuition, nonresident: full-time $8,670; part-time $408 per credit hour. Tuition and fees vary according to program. *Unit head:* Dr. Martin E. Lipinski, Chairman, 901-678-3279. *Application contact:* Dr. Larry W. Moore, Coordinator of Graduate Studies, 901-678-3278.

University of Michigan, Horace H. Rackham School of Graduate Studies, College of Engineering, Department of Electrical Engineering and Computer Science, Ann Arbor, MI 48109. Offers computer science and engineering (MS, MSE, PhD); electrical science and engineering (MS, MSE, PhD, EE), including electrical engineering; systems science and engineering (MS, MSE, PhD), including electrical engineering: systems; transportation studies: intelligent transportation systems (Certificate). *Degree requirements:* For doctorate, dissertation, oral defense of dissertation, preliminary exams required. *Entrance requirements:* For master's, GRE General Test (minimum combined score of 1900 on three sections required; average 2000); for doctorate, GRE General Test (minimum combined score of 1900 on three sections required; average 2000), master's degree; for other advanced degree, GRE General Test.

University of Michigan–Dearborn, College of Engineering and Computer Science, Interdisciplinary Programs, Program in Automotive Systems Engineering, Dearborn, MI 48128-1491. Offers MSE. Part-time and evening/weekend programs available. *Faculty:* 1 full-time (0 women). *Students:* 8 full-time (0 women), 52 part-time (7 women); includes 6 minority (2 African Americans, 2 Asian Americans or Pacific Islanders, 2 Hispanic Americans), 2 international. Average age 29. *Degree requirements:* For master's, computer language required, thesis optional, foreign language not required. *Entrance requirements:* For master's, bachelor's degree in applied mathematics, computer science, engineering, or physical science; minimum GPA of 3.0. *Application deadline:* For fall admission, 8/1 (priority date); for winter admission, 12/1 (priority date); for spring admission, 4/1. Applications are processed on a rolling basis. Application fee: $55. Electronic applications accepted. Tuition, state resident: part-time $259 per credit hour. Tuition, nonresident: part-time $748 per credit hour. Required fees: $80 per course. Tuition and fees vary according to course level, course load and program. *Unit head:* Trudy Lindsey, Director, Graduate Admission and Records, 301-405-4198, Fax: 301-314-9305, E-mail: grschool@deans.umd.edu. *Application contact:* Deborah Parker, Administrative Assistant I, 313-593-5582, Fax: 313-593-5386, E-mail: debbie@umdsun2.umd.umich.edu.

University of Missouri–Columbia, Graduate School, College of Engineering, Department of Civil and Environmental Engineering, Columbia, MO 65211. Offers civil engineering (MS, PhD); environmental engineering (MS, PhD); geotechnical engineering (MS, PhD); structural engineering (MS, PhD); transportation and highway engineering (MS); water resources (MS, PhD). *Faculty:* 24 full-time (2 women). *Students:* 19 full-time (2 women), 18 part-time (3 women); includes 1 minority (Hispanic American), 21 international. *Degree requirements:* For master's, report or thesis required; for doctorate, dissertation required. *Entrance requirements:* For master's and doctorate, GRE General Test, TOEFL. *Application deadline:* For fall admission, 7/10. Application fee: $30 ($50 for international students). *Unit head:* Dr. Mark Virkler, Director of Graduate Studies, 573-882-3678.

See in-depth description on page 639.

University of New Brunswick, School of Graduate Studies, Faculty of Engineering, Department of Civil Engineering, Fredericton, NB E3B 5A3, Canada. Offers environmental engineering (M Eng, M Sc E, PhD); geotechnical engineering (M Eng, M Sc E); structures and structural foundations (M Eng, M Sc E, PhD); transportation engineering (M Eng, M Sc E, PhD); water resources and hydrology (M Eng, M Sc E, PhD). Part-time programs available. *Degree requirements:* For master's, thesis required, foreign language not required; for doctorate, dissertation, qualifying exam required, foreign language not required. *Entrance requirements:* For master's and doctorate, TOEFL, TWE, minimum GPA of 3.0.

University of Oklahoma, Graduate College, College of Engineering, School of Civil Engineering and Environmental Science, Program in Civil Engineering, Norman, OK 73019-0390. Offers civil engineering (PhD); environmental engineering (MS); geotechnical engineering (MS); structures (MS); transportation (MS). *Accreditation:* ABET (one or more programs are accredited). *Students:* 19 full-time (4 women), 36 part-time (6 women); includes 5 minority (3 Asian Americans or Pacific Islanders, 2 Native Americans), 28 international. *Degree requirements:* For master's, comprehensive and oral exams required, foreign language and thesis not required; for doctorate, dissertation, oral and qualifying exams required, foreign language not required. *Entrance requirements:* For master's, GRE General Test, TOEFL (minimum score of 575 required), minimum GPA of 3.0; for doctorate, GRE General Test, TOEFL (minimum score of 575 required), minimum graduate GPA of 3.5. *Application deadline:* For fall admission, 4/1 (priority date). Applications are processed on a rolling basis. Application fee: $25. Tuition, state resident: part-time $86 per credit hour. Tuition, nonresident: part-time $275 per credit hour. Tuition and fees vary according to course level, course load and program. *Unit head:* Robert C. Knox, Graduate Liaison, 405-325-4256, Fax: 405-325-7508. *Application contact:* Robert C. Knox, Graduate Liaison, 405-325-4256, Fax: 405-325-7508.

See in-depth description on page 641.

University of Pennsylvania, School of Engineering and Applied Science, Department of Systems Engineering, Philadelphia, PA 19104. Offers environmental resources engineering (MSE); environmental/resources engineering (PhD); systems engineering (MSE, PhD); technology and public policy (MSE, PhD); transportation (MSE, PhD). Part-time programs available. *Faculty:* 12 full-time (0 women), 9 part-time (2 women). *Students:* 26 full-time (10 women), 16 part-time (2 women); includes 4 minority (2 African Americans, 2 Hispanic Americans), 16 international. Terminal master's awarded for partial completion of doctoral program. *Degree requirements:* For master's, computer language required, foreign language not required; for doctorate, one foreign language, computer language, dissertation required. *Entrance requirements:* For master's and doctorate, TOEFL (minimum score of 600 required). *Application deadline:* For fall admission, 1/2 (priority date). Applications are processed on a rolling basis. Application fee: $65. Electronic applications accepted. *Unit head:* Dr. G. Anandalingam, Chair, 215-898-8790, Fax: 215-898-5020, E-mail: anand@seas.upenn.edu. *Application contact:* Dr. Tony E. Smith, Graduate Group Chair, 215-898-9647, Fax: 215-898-5020, E-mail: tesmith@seas.upenn.edu.

University of Rhode Island, Graduate School, College of Engineering, Department of Civil and Environmental Engineering, Program in Transportation Engineering, Kingston, RI 02881. Offers MS, PhD.

University of Southern California, Graduate School, School of Engineering, Department of Civil Engineering, Program in Transportation Engineering, Los Angeles, CA 90089. Offers MS. *Students:* 4 full-time (1 woman), 1 part-time, (all international). Average age 26. 12 applicants, 83% accepted. In 1998, 2 degrees awarded. *Degree requirements:* For master's, thesis optional. *Entrance requirements:* For master's, GRE General Test. *Application deadline:* For fall admission, 6/1 (priority date); for spring admission, 11/1. Application fee: $55. Tuition: Part-time $768 per unit. Required fees: $350 per semester. *Financial aid:* In 1998–99, 2 teaching assistantships were awarded.; fellowships, research assistantships, Federal Work-Study and institutionally-sponsored loans also available. Aid available to part-time students. Financial aid application deadline: 2/15; financial aid applicants required to submit FAFSA. *Unit head:* Dr. L. Carter Wellford, Chair, Department of Civil Engineering, 213-740-0587.

University of Virginia, School of Engineering and Applied Science, Department of Civil Engineering, Program in Transportation Engineering and Management, Charlottesville, VA 22903. Offers ME, MS, PhD. Part-time programs available. *Faculty:* 4 full-time (0 women). *Students:* 10 full-time (2 women), 2 international. Average age 26. 13 applicants, 85% accepted. In 1998, 5 master's awarded (20% entered university research/teaching, 80% found other work related to degree); 3 doctorates awarded (33% entered university research/teaching, 67% found other work related to degree). Terminal master's awarded for partial completion of doctoral program. *Degree requirements:* For master's, thesis required (for some programs), foreign language not required; for doctorate, dissertation, comprehensive exam required, foreign language not required. *Entrance requirements:* For master's and doctorate, GRE General Test. *Average time to degree:* Master's–1.7 years full-time; doctorate–3 years full-time. *Application deadline:* For fall admission, 2/1 (priority date). Applications are processed on a rolling basis. Application fee: $60. Electronic applications accepted. *Financial aid:* In 1998–99, 10 students received aid, including 10 research assistantships with full tuition reimbursements available (averaging $10,800 per year); fellowships, teaching assistantships Financial aid application deadline: 2/1. *Faculty research:* Intermodal freight planning, highway safety, land use/air quality, intelligent transportation systems, artificial intelligence applications. Total annual research expenditures: $1 million. *Unit head:* Dr. Nicholas J. Garber, Chairman, Department of Civil Engineering, 804-924-7464, Fax: 804-982-2951, E-mail: njg@virginia.edu.

University of Washington, Graduate School, College of Engineering, Department of Civil and Environmental Engineering, Seattle, WA 98195. Offers environmental engineering (MS, MS Civ E, MSE, PhD); hydraulic engineering (MS Civ E, MSE, PhD); structural and geotechnical engineering and mechanics (MS, MS Civ E, MSE, PhD); transportation and construction engineering (MS, MS Civ E, MSE, PhD). *Faculty:* 31 full-time (4 women), 26 part-time (1 woman). *Students:* 153 full-time (54 women), 42 part-time (13 women); includes 25 minority (2 African Americans, 19 Asian Americans or Pacific Islanders, 3 Hispanic Americans, 1 Native American), 34 international. Terminal master's awarded for partial completion of doctoral program. *Degree requirements:* For master's, thesis or alternative required, foreign language not required; for doctorate, dissertation required, foreign language not required. *Entrance requirements:* For master's, GRE General Test, TOEFL (minimum score of 580 required), minimum GPA of 3.0; for doctorate, GRE, TOEFL (minimum score of 580 required), minimum GPA of 3.0. *Application deadline:* For fall admission, 2/1 (priority date); for winter admission, 11/1; for spring admission, 2/1. Applications are processed on a rolling basis. Application fee: $50. Electronic applications accepted. Tuition, state resident: full-time $5,196; part-time $475 per credit. Tuition, nonresident: full-time $13,485, part-time $1,285 per credit. Required fees: $387; $38 per credit. Tuition and fees vary according to course load. *Unit head:* Dr. Fred L. Mannering, Chair, 206-543-2390. *Application contact:* Marcia Buck, Graduate Secretary, 206-543-2574, E-mail: ceginfo@u.washington.edu.

Villanova University, College of Engineering, Department of Civil and Environmental Engineering, Program in Transportation Engineering, Villanova, PA 19085-1699. Offers MSTE. Part-time and evening/weekend programs available. *Faculty:* 2 full-time (0 women), 2 part-time (0 women). *Students:* 2 full-time (1 woman), 5 part-time (2 women); includes 1 minority (Hispanic American), 1 international. Average age 25. 13 applicants, 85% accepted. *Degree requirements:* For master's, computer language required, thesis optional, foreign language not required.

Entrance requirements: For master's, GRE General Test (for applicants with degrees from foreign universities), minimum GPA of 3.0, BCE or bachelor's degree in science, business, or related engineering field. *Application deadline:* For fall admission, 8/1 (priority date); for spring admission, 12/1. Applications are processed on a rolling basis. Application fee: $40. *Financial aid:* Federal Work-Study and tuition waivers (full and partial) available. Aid available to part-time students. Financial aid application deadline: 4/15. *Faculty research:* Simulation of unsignalized intersections, services to the elderly and disabled, recycling of secondary materials into hot mix asphalt concrete pavements. *Unit head:* Dr. Lewis J. Mathers, Chairman, Department of Civil and Environmental Engineering, 610-519-4960, Fax: 610-519-6754, E-mail: ldeangel@email.vill.edu.

Washington University in St. Louis, School of Engineering and Applied Science, Sever Institute of Technology, Department of Civil Engineering, Program in Transportation and Urban Systems Engineering, St. Louis, MO 63130-4899. Offers D Sc. *Degree requirements:* For doctorate, variable foreign language requirement, dissertation, departmental qualifying exam required.

Waste Management

Colorado State University, College of Veterinary Medicine and Biomedical Sciences, Graduate Programs in Veterinary Medicine and Biomedical Sciences, Department of Radiological Health Sciences, Fort Collins, CO 80523-0015. Offers cellular and molecular biology (MS, PhD); health physics (MS, PhD); mammalian radiobiology (MS, PhD); nuclear-waste management (MS); radiobiology (MS); radioecology (MS, PhD); radiology (MS, PhD); veterinary radiology (MS). *Faculty:* 17 full-time (1 woman). *Students:* 13 full-time (4 women), 9 part-time (3 women), 4 international. *Degree requirements:* For master's, thesis required, foreign language not required; for doctorate, dissertation required. *Entrance requirements:* For master's and doctorate, GRE General Test, TOEFL. *Application deadline:* For fall admission, 2/1 (priority date). Applications are processed on a rolling basis. Application fee: $30. Electronic applications accepted. *Unit head:* Dr. F. W. Whicker, Interim Chairman, 970-491-5222, Fax: 970-491-0623, E-mail: wiedeman@cvmbs.colostate.edu.

Idaho State University, Graduate School, College of Engineering, Pocatello, ID 83209. Offers engineering and applied science (PhD); environmental engineering (MS); hazardous waste management (MS); measurement and control engineering (MS); nuclear science and engineering (MS). MS (hazardous waste management), PhD offered jointly with the University of Idaho. Part-time programs available. *Degree requirements:* For master's, thesis required, foreign language not required; for doctorate, dissertation required. *Entrance requirements:* For master's and doctorate, GRE General Test, TOEFL. *Faculty research:* Isotope separation, control technology,two-phase flow, photosonolysis, criticality calculations.

Idaho State University, Graduate School, Department of Interdisciplinary Studies, Pocatello, ID 83209. Offers biology (MNS); chemistry (MNS); general interdisciplinary (M Ed, MA); geology (MNS); hazardous waste management (MS), including general interdisciplinary; mathematics (MNS); physics (MNS). *Degree requirements:* Foreign language not required. *Entrance requirements:* For master's, GRE General Test.

National Technological University, Programs in Engineering, Fort Collins, CO 80526-1842. Offers chemical engineering (MS); computer engineering (MS); computer science (MS); electrical engineering (MS); engineering management (MS); hazardous waste management (MS); health physics (MS); management of technology (MS); manufacturing systems engineering (MS); materials science and engineering (MS); software engineering (MS); special majors (MS); transportation engineering (MS); transportation systems engineering (MS). Part-time programs available. *Faculty:* 600 part-time (20 women). *Entrance requirements:* For master's, BS in engineering or related field. *Application deadline:* Applications are processed on a rolling basis. Application fee: $50. *Unit head:* Lionel V. Baldwin, President, 970-495-6400, Fax: 970-484-0668, E-mail: baldwin@mail.ntu.edu.

New Mexico Institute of Mining and Technology, Graduate Studies, Department of Mineral and Environmental Engineering, Socorro, NM 87801. Offers environmental engineering (MS), including air quality engineering and science, hazardous waste engineering, water quality engineering and science; mineral engineering (MS). *Faculty:* 11 full-time (1 woman). *Students:* 12 full-time (3 women); includes 8 minority (all Hispanic Americans), 4 international. *Degree requirements:* For master's, thesis required, foreign language not required. *Entrance requirements:* For master's, GRE General Test, TOEFL (minimum score of 540 required). *Application deadline:* For fall admission, 3/1 (priority date); for spring admission, 6/1. Applications are processed on a rolling basis. Application fee: $16. *Unit head:* Dr. Clinton P. Richardson, Chair, 505-835-5346, Fax: 505-835-5252. *Application contact:* Dr. David B. Johnson, Dean of Graduate Studies, 505-835-5513, Fax: 505-835-5476, E-mail: graduate@nmt.edu.

Rutgers, The State University of New Jersey, New Brunswick, Graduate School, Program in Environmental Sciences, New Brunswick, NJ 08903. Offers air resources (MS, PhD); aquatic biology (MS, PhD); aquatic chemistry (MS, PhD); chemistry and physics of aerosol and hydrosol systems (MS, PhD); environmental chemistry (MS, PhD); environmental microbiology (MS, PhD); environmental toxicology (MS, PhD); exposure assessment (PhD); water and wastewater treatment (MS, PhD); water resources (MS, PhD). Part-time and evening/weekend programs available. *Faculty:* 33 full-time (7 women), 36 part-time (6 women). *Students:* 42 full-time (16 women), 92 part-time (30 women); includes 22 minority (17 Asian Americans or Pacific Islanders, 5 Hispanic Americans), 31 international. Terminal master's awarded for partial completion of doctoral program. *Degree requirements:* For master's, thesis or alternative, oral final exam required, foreign language not required; for doctorate, dissertation, thesis defense, qualifying exam required, foreign language not required. *Entrance requirements:* For master's and doctorate, GRE General Test (minimum score of 500 on verbal section, 600 on quantitative required), TOEFL (minimum score of 590 required). *Application deadline:* For fall admission, 3/1; for spring admission, 11/1. Applications are processed on a rolling basis. Application fee: $50. *Unit head:* Dr. Peter F. Strom, Director, 732-932-8078, Fax: 732-932-8644, E-mail: strom@aesop.rutgers.edu. *Application contact:* Paul J. Lioy, Graduate Admissions Committee, 732-932-0150, Fax: 732-445-0116, E-mail: plioy@eohsi.rutgers.edu.

Southern Methodist University, School of Engineering and Applied Science, Center for Special Studies, Dallas, TX 75275. Offers applied science (MS, PhD); hazardous and waste materials management (MS); material science and engineering (MS); systems engineering (MS). *Faculty:* 12 part-time (0 women). *Students:* 10 full-time (2 women), 112 part-time (32 women); includes 22 minority (5 African Americans, 7 Asian Americans or Pacific Islanders, 10 Hispanic Americans), 7 international. *Degree requirements:* For master's, thesis optional, foreign language not required; for doctorate, dissertation, oral and written qualifying exams, oral final exam required. *Entrance requirements:* For master's, GRE General Test (minimum score of 650 on quantitative section required), TOEFL (minimum score of 550 required), minimum GPA of 3.0 in last 2 years, bachelor's degree in related field; for doctorate, preliminary counseling exam, minimum GPA of 3.0, bachelor's degree in related field. *Application deadline:* For fall admission, 8/1 (priority date); for spring admission, 12/15. Applications are processed on a rolling basis. Application fee: $25. Tuition: Full-time $9,216; part-time $512 per credit hour.

Required fees: $88 per credit hour. Part-time tuition and fees vary according to course load and campus/location. *Unit head:* Dr. Buck F. Brown, Dean for Research and Graduate Studies, 812-877-8403, Fax: 812-877-8102, E-mail: buck.brown@rose-hulman.edu. *Application contact:* Dr. Zeynep Celik-Butler, Assistant Dean for Graduate Studies and Research, 214-768-3979, Fax: 214-768-3845, E-mail: zcb@seas.smu.edu.

State University of New York at Stony Brook, School of Professional Development and Continuing Studies, Stony Brook, NY 11794. Offers chemistry-grade 7-12 (MAT); coaching (Certificate); earth science-grade 7-12 (MAT); educational computing (Certificate); English-grade 7-12 (MAT); environmental/occupational health and safety (Certificate); French-grade 7-12 (MAT); German-grade 7-12 (MAT); human resource management (Certificate); information systems management (Certificate); Italian-grade 7-12 (MAT); liberal studies (MA); Long Island regional studies (Certificate); physics-grade 7-12 (MAT); Russian-grade 7-12 (MAT); school administration and supervision (Certificate); school district administration (Certificate); social science and the professions (MPS), including labor management, public affairs, waste management; social studies (MAT); waste management (Certificate); women's studies (Certificate). Part-time and evening/weekend programs available. *Faculty:* 1 full-time, 101 part-time. *Students:* 224 full-time (122 women), 910 part-time (606 women); includes 92 minority (41 African Americans, 14 Asian Americans or Pacific Islanders, 34 Hispanic Americans, 3 Native Americans), 2 international. *Degree requirements:* For master's, one foreign language, thesis or alternative, teacher certification courses required. *Application deadline:* For fall admission, 1/15. Application fee: $50. *Unit head:* Dr. Paul J. Edelson, Dean, 516-632-7052, Fax: 516-632-9046, E-mail: paul.edelson@sunysb.edu. *Application contact:* Sandra Romansky, Director of Admissions and Advisement, 516-632-7050, Fax: 516-632-9046, E-mail: sandra.romansky@sunysb.edu.

University of Idaho, College of Graduate Studies, College of Engineering, Program in Waste Management, Moscow, ID 83844-4140. Offers MS. In 1998, 6 degrees awarded. *Degree requirements:* For master's, thesis required, foreign language not required. *Entrance requirements:* For master's, GRE, minimum GPA of 2.8. *Application deadline:* For fall admission, 8/1; for spring admission, 12/15. Application fee: $35 ($45 for international students). *Financial aid:* Application deadline: 2/15. *Unit head:* Dr. David E. Thompson, Dean, College of Engineering, 208-885-6479.

University of New Mexico, Graduate School, School of Engineering, Program in Hazardous Waste Engineering, Albuquerque, NM 87131-2039. Offers ME. Part-time programs available. Average age 44. 2 applicants, 50% accepted. In 1998, 4 degrees awarded. *Degree requirements:* Foreign language not required. *Entrance requirements:* For master's, GRE General Test, minimum GPA of 3.0. *Application deadline:* For fall admission, 7/15; for spring admission, 11/14. Applications are processed on a rolling basis. Application fee: $25. *Faculty research:* Radioactive waste management, groundwater contamination. *Unit head:* Richard W. Mead, Head, 505-277-3221.

University of Oklahoma, Graduate College, College of Engineering, School of Civil Engineering and Environmental Science, Program in Environmental Science, Norman, OK 73019-0390. Offers air (M Env Sc); environmental science (PhD); groundwater management (M Env Sc); hazardous solid waste (M Env Sc); occupational safety and health (M Env Sc); process design (M Env Sc); water quality resources (M Env Sc). Part-time and evening/weekend programs available. *Students:* 10 full-time (3 women), 24 part-time (14 women); includes 3 minority (1 African American, 1 Hispanic American, 1 Native American), 6 international. Terminal master's awarded for partial completion of doctoral program. *Degree requirements:* For master's, comprehensive and oral exams required, foreign language and thesis not required; for doctorate, dissertation, comprehensive, oral, and qualifying exams required. *Entrance requirements:* For master's, GRE General Test, TOEFL (minimum score of 575 required), minimum GPA of 3.0; for doctorate, GRE General Test, TOEFL (minimum score of 575 required), minimum graduate GPA of 3.5. *Application deadline:* For fall admission, 4/1 (priority date). Applications are processed on a rolling basis. Application fee: $25. Tuition, state resident: part-time $86 per credit hour. Tuition, nonresident: part-time $275 per credit hour. Tuition and fees vary according to course level, course load and program. *Unit head:* Dr. Jeffrey Falzarano, Graduate Coordinator, 504-280-7184, Fax: 504-280-5542, E-mail: jmfna@uno.edu. *Application contact:* Robert C. Knox, Graduate Liaison, 405-325-4256, Fax: 405-325-7508.

University of South Carolina, Graduate School, School of Public Health, Department of Environmental Health Sciences, Program in Hazardous Materials Management, Columbia, SC 29208. Offers MPH, MSPH, PhD. *Degree requirements:* For master's, thesis, practicum (MPH) required, foreign language not required; for doctorate, dissertation required. *Entrance requirements:* For master's and doctorate, GRE. *Application deadline:* Applications are processed on a rolling basis. Application fee: $35. Electronic applications accepted. Tuition, state resident: full-time $4,014; part-time $202 per credit hour. Tuition, nonresident: full-time $8,528; part-time $428 per credit hour. Required fees: $100; $4 per credit hour. Tuition and fees vary according to program. *Faculty research:* Environmental/human health protection; use and disposal of hazardous materials; site safety; exposure assessment; migration, fate and transformation of materials. *Unit head:* Tina Samms, Assistant Director of Graduate Admissions, 408-554-4313, Fax: 408-554-5474, E-mail: engr-grad@scu.edu. *Application contact:* Dr. Edward Oswald, Graduate Director, 803-777-6994, Fax: 803-777-3391, E-mail: eoswald@sph.sc.edu.

Wayne State University, Graduate School, College of Engineering, Department of Chemical Engineering and Materials Science, Programs in Hazardous Waste, Detroit, MI 48202. Offers environmental auditing (Certificate); hazardous materials management on public lands (Certificate); hazardous waste control (Certificate); hazardous waste management (MS). Part-time programs available. *Degree requirements:* For master's, thesis optional, foreign language not required. *Faculty research:* Environmental management.

Water Resources Engineering

California Polytechnic State University, San Luis Obispo, College of Engineering, Program in Engineering, San Luis Obispo, CA 93407. Offers biochemical engineering (MS); industrial engineering (MS); integrated technology management (MS); materials engineering (MS); mechanical engineering (MS); water engineering (MS). *Faculty:* 98 full-time (8 women), 82 part-time (14 women). *Students:* 25 full-time (3 women), 6 part-time. *Degree requirements:* Foreign language not required. *Entrance requirements:* For master's, GRE General Test, minimum GPA of 2.5 in last 90 quarter units. Application fee: $55. Tuition, nonresident: part-time $164 per unit. Required fees: $531 per quarter. *Unit head:* Dr. Paul E. Rainey, Associate Dean, 805-756-2131, Fax: 805-756-6503, E-mail: prainey@calpoly.edu. *Application contact:* Dr. Paul E. Rainey, Associate Dean, 805-756-2131, Fax: 805-756-6503, E-mail: prainey@calpoly.edu.

Cornell University, Graduate School, Graduate Fields of Engineering, Field of Civil and Environmental Engineering, Ithaca, NY 14853-0001. Offers environmental engineering (M Eng, MS, PhD); environmental fluid mechanics and hydrology (M Eng, MS, PhD); environmental systems engineering (M Eng, MS, PhD); geotechnical engineering (M Eng, MS, PhD); remote sensing (M Eng, MS, PhD); structural engineering (M Eng, MS, PhD); transportation engineering (M Eng, MS, PhD); water resource systems (M Eng, MS, PhD). *Faculty:* 29 full-time. *Students:* 143 full-time (32 women); includes 16 minority (11 Asian Americans or Pacific Islanders, 5 Hispanic Americans), 76 international. Terminal master's awarded for partial completion of doctoral program. *Degree requirements:* For master's, thesis (MS) required; for doctorate, dissertation required, foreign language not required. *Entrance requirements:* For master's, TOEFL (minimum score of 600required); for doctorate, GRE General Test, TOEFL (minimum score of 600 required). *Application deadline:* For fall admission, 1/15. Application fee: $65. Electronic applications accepted. *Unit head:* Director of Graduate Studies, 607-255-7560, Fax: 607-255-9004. *Application contact:* Graduate Field Assistant, 607-255-7560, E-mail: cee_grad@cornell.edu.

See in-depth description on page 573.

Florida Institute of Technology, Graduate School, College of Engineering, Division of Engineering Science, Program in Civil Engineering, Melbourne, FL 32901-6975. Offers civil engineering (PhD); geotechnical engineering (MS); structures engineering (MS); water resources (MS). Part-time programs available. *Faculty:* 5 full-time (0 women), 4 part-time (2 women). *Students:* 7 full-time (0 women), 10 part-time (2 women); includes 1 minority (Hispanic American), 9 international. *Degree requirements:* For master's, thesis optional, foreign language not required; for doctorate, dissertation required, foreign language not required. *Entrance requirements:* For master's, minimum GPA of 3.0; for doctorate, minimum GPA of 3.2. *Application deadline:* Applications are processed on a rolling basis. Application fee: $50. Electronic applications accepted. Tuition: Part-time $575 per credit hour. Required fees: $100. Tuition and fees vary according to campus/location and program. *Unit head:* Dr. Ashok Pandit, Chair, 407-674-7151, Fax: 407-768-7565, E-mail: apandit@fit.edu. *Application contact:* Carolyn P. Farrior, Associate Dean of Graduate Admissions, 407-674-7118, Fax: 407-723-9468, E-mail: cfarrior@fit.edu.

See in-depth description on page 581.

Louisiana State University and Agricultural and Mechanical College, Graduate School, College of Engineering, Department of Civil and Environmental Engineering, Baton Rouge, LA 70803. Offers environmental engineering (MSCE, PhD); geotechnical engineering (MSCE, PhD); structural engineering and mechanics (MSCE, PhD); transportation engineering (MSCE, PhD); water resources (MSCE, PhD). Part-time programs available. *Faculty:* 27 full-time (1 woman), 1 part-time (0 women). *Students:* 59 full-time (14 women), 24 part-time (2 women); includes 5 minority (1 African American, 3 Asian Americans or Pacific Islanders, 1 Hispanic American), 48 international. *Degree requirements:* For master's, thesis optional, foreign language not required; for doctorate, one foreign language (computer language can substitute), dissertation required. *Entrance requirements:* For master's and doctorate, GRE General Test, TOEFL, minimum GPA of 3.0. *Application deadline:* For fall admission, 1/25 (priority date). Applications are processed on a rolling basis. Application fee: $25. *Unit head:* Dr. Ronald F. Malone, Acting Chair, 225-388-8666, Fax: 225-388-8652, E-mail: rmalone@unix1.sncc.lsu.edu. *Application contact:* Dr. Vijaya K. A. Gopu, Graduate Coordinator, 225-388-8442, E-mail: cegopu@eng.lsu.edu.

See in-depth description on page 589.

Marquette University, Graduate School, College of Engineering, Department of Civil and Environmental Engineering, Milwaukee, WI 53201-1881. Offers construction and public works management (MS, PhD); environmental/water resources engineering (MS, PhD); structural/geotechnical engineering (MS, PhD); transportational planning and engineering (MS, PhD). Part-time and evening/weekend programs available. *Faculty:* 12 full-time (0 women), 1 part-time (0 women). *Students:* 17 full-time (7 women), 32 part-time (5 women); includes 5 minority (1 African American, 4 Asian Americans or Pacific Islanders), 12 international. Terminal master's awarded for partial completion of doctoral program. *Degree requirements:* For master's, thesis or alternative, comprehensive exam required, foreign language not required; for doctorate, dissertation required, foreign language not required. *Entrance requirements:* For master's, TOEFL (minimum score of 550 required); for doctorate, GRE General Test, TOEFL (minimum score of 550 required). *Application deadline:* For fall admission, 6/1 (priority date). Applications are processed on a rolling basis. Application fee: $40. Tuition: Part-time $510 per credit hour. Tuition and fees vary according to program. *Unit head:* Dr. Thomas H. Wenzel, Chairman, 414-288-7030, Fax: 414-288-7521, E-mail: wenzelt@vms.csd.mu.edu.

McGill University, Faculty of Graduate Studies and Research, Faculty of Engineering, Department of Civil Engineering and Applied Mechanics, Program in Environmental Engineering and Water Resources Management, Montréal, PQ H3A 2T5, Canada. Offers M Eng, M Sc, PhD. Part-time and evening/weekend programs available. *Degree requirements:* For master's, computer language required, thesis optional, foreign language not required; for doctorate, computer language, dissertation required, foreign language not required. *Entrance requirements:* For master's, TOEFL (minimum score of 550 required), minimum GPA of 3.0; for doctorate, TOEFL (minimum score of 580 required). *Faculty research:* Biological/biochemical treatment, physical and chemical treatment, UV disinfection, site remediation, water resource modelling and optimization.

New Mexico Institute of Mining and Technology, Graduate Studies, Department of Mineral and Environmental Engineering, Socorro, NM 87801. Offers environmental engineering (MS), including air quality engineering and science, hazardous waste engineering, water quality engineering and science; mineral engineering (MS). *Faculty:* 11 full-time (1 woman). *Students:* 12 full-time (3 women); includes 8 minority (all Hispanic Americans), 4 international. *Degree requirements:* For master's, thesis required, foreign language not required. *Entrance requirements:* For master's, GRE General Test, TOEFL (minimum score of 540 required). *Application deadline:* For fall admission, 3/1 (priority date); for spring admission, 6/1. Applications are processed on a rolling basis. Application fee: $16. *Unit head:* Dr. Clinton P. Richardson, Chair, 505-835-5346, Fax: 505-835-5252. *Application contact:* Dr. David B. Johnson, Dean of Graduate Studies, 505-835-5513, Fax: 505-835-5476, E-mail: graduate@nmt.edu.

Ohio University, Graduate Studies, College of Engineering and Technology, Department of Civil Engineering, Athens, OH 45701-2979. Offers geotechnical and environmental engineering (MS); water resources and structures (MS). *Faculty:* 11 full-time (1 woman). *Students:* 24 full-time (5 women), 2 part-time, 13 international. *Degree requirements:* For master's, thesis optional, foreign language not required. *Entrance requirements:* For master's, minimum GPA of 3.0. *Application deadline:* For fall admission, 5/1 (priority date). Applications are processed on a rolling basis. Application fee: $30. Tuition, state resident: full-time $5,754; part-time $238 per credit hour. Tuition, nonresident: full-time $11,055; part-time $457 per credit hour. Tuition

and fees vary according to course load, campus/location and program. *Unit head:* Dr. Gayle Mitchell, Chair, 740-593-1465, Fax: 740-593-0625, E-mail: gmitchell@bobcat.ent.ohiou.edu. *Application contact:* Dr. Glenn Hazen, Graduate Chair, 740-593-1469, Fax: 740-593-0625, E-mail: ghazen@bobcat.ent.ohiou.edu.

Oregon State University, Graduate School, College of Engineering, Department of Bioresource Engineering, Corvallis, OR 97331. Offers M Agr, MAIS, MS, PhD. *Faculty:* 13 full-time (2 women), 7 part-time (2 women). Average age 26. In 1998, 2 degrees awarded (100% found work related to degree). Terminal master's awarded for partial completion of doctoral program. *Degree requirements:* For master's, thesis or alternative required, foreign language not required; for doctorate, dissertation required, foreign language not required. *Entrance requirements:* For master's and doctorate, TOEFL (minimum score of 550 required), minimum GPA of 3.0 in last 90 hours. *Application deadline:* For fall admission, 3/1. Applications are processed on a rolling basis. Application fee: $50. *Financial aid:* Fellowships, research assistantships, teaching assistantships, Federal Work-Study and institutionally-sponsored loans available. Aid available to part-time students. Financial aid application deadline: 2/1. *Faculty research:* Bioengineering, water resources engineering, food engineering, cell culture and fermentation, vadose zone transport, regional hydrology modeling, bioseparations, post-harvest processing, biomedical engineering, nonpoint pollution abatement, drug formulation and delivery, waste management, stochastic hydrology, biological modeling. *Unit head:* Dr. James A. Moore, Head, 541-737-2041, Fax: 541-737-2082, E-mail: info@pandora.bre.orst.edu.

Pennsylvania State University University Park Campus, Graduate School, College of Engineering, Department of Civil and Environmental Engineering, State College, University Park, PA 16802-1503. Offers civil engineering (M Eng, MS, PhD); environmental engineering (M Eng, MS, PhD); structural engineering (M Eng, MS, PhD); transportation and highway engineering (M Eng, MS, PhD); water resources engineering (M Eng, MS, PhD). *Students:* 77 full-time (24 women), 52 part-time (8 women). *Degree requirements:* For master's, final paper (M Eng), oral exam and thesis (MS) required; for doctorate, dissertation, comprehensive and oral exams required. *Entrance requirements:* For master's and doctorate, GRE General Test. Application fee: $50. *Unit head:* Dr. Paul P. Jovanis, Head, 814-863-3084.

Princeton University, Graduate School, Department of Geosciences, Princeton, NJ 08544-1019. Offers atmospheric and oceanic sciences (PhD); environmental engineering and water resources (PhD); geological and geophysical sciences (PhD). *Degree requirements:* For doctorate, dissertation required. *Entrance requirements:* For doctorate, GRE General Test, GRE Subject Test.

Princeton University, Graduate School, School of Engineering and Applied Science, Department of Civil Engineering and Operations Research, Program in Environmental Engineering and Water Resources, Princeton, NJ 08544-1019. Offers PhD. *Degree requirements:* For doctorate, dissertation required. *Entrance requirements:* For doctorate, GRE General Test, GRE Subject Test.

State University of New York at Buffalo, Graduate School, School of Engineering and Applied Sciences, Department of Civil, Structural, and Environmental Engineering, Buffalo, NY 14260. Offers computational engineering and mechanics (MS, PhD); construction (M Eng, MS, PhD); geoenvironmental and geotechnical engineering (M Eng, MS, PhD); structural and earthquake engineering (M Eng, MS, PhD); water resources and environmental engineering (M Eng, MS, PhD). Part-time programs available. Postbaccalaureate distance learning degree programs offered (minimal on-campus study). *Faculty:* 25 full-time (0 women), 7 part-time (1 woman). *Students:* 99 full-time (19 women), 71 part-time (13 women); includes 9 minority (2 African Americans, 5 Asian Americans or Pacific Islanders, 1 Hispanic American, 1 Native American), 101 international. Terminal master's awarded for partial completion of doctoral program. *Degree requirements:* For master's, computer language, project or thesis required; for doctorate, computer language, dissertation required, foreign language not required. *Entrance requirements:* For master's and doctorate, GRE General Test (minimum combined score of 1250 required), TOEFL (minimum score of 550 required). *Application deadline:* For fall admission, 1/15 (priority date); for spring admission, 10/1. Applications are processed on a rolling basis. Application fee: $35. Tuition, state resident: full-time $5,100; part-time $213 per credit hour. Tuition, nonresident: full-time $8,416; part-time $351 per credit hour. Required fees: $870; $75 per semester. Tuition and fees vary according to course load and program. *Unit head:* Dr. Andrei M. Reinhorn, Chairman, 716-645-2114 Ext. 2419, Fax: 716-645-3733, E-mail: reinhorn@civil.eng.buffalo.edu. *Application contact:* Dr. Joe Atkinson, Director of Graduate Admissions, 716-645-2114 Ext. 2326, Fax: 716-645-3667, E-mail: atkinson@acsu.buffalo.edu.

See in-depth description on page 601.

Texas A&M University, College of Engineering, Department of Civil Engineering, Program in Water Resources Engineering, College Station, TX 77843. Offers M Eng, MS, D Eng, PhD. D Eng offered through the College of Engineering. *Students:* 27. *Degree requirements:* For master's, thesis (MS) required; for doctorate, dissertation (PhD), internship (D Eng) required. *Entrance requirements:* For master's and doctorate, GRE General Test, TOEFL. Application fee: $50 ($75 for international students). *Financial aid:* Fellowships, research assistantships, teaching assistantships available. Financial aid application deadline: 4/1; financial aid applicants required to submit FAFSA. *Faculty research:* Water resources development, planning, and management; water resources system engineering, hydrology, and hydraulics; groundwater systems analysis. *Unit head:* Dr. M. Yavuz Corapcioglu, Head, 409-845-3011, Fax: 409-862-2800, E-mail: ce-grad@tamu.edu. *Application contact:* Dr. Thomas M. Over, Graduate Adviser, 409-845-2498, Fax: 409-862-2800, E-mail: ce-grad@tamu.edu.

Tufts University, Division of Graduate and Continuing Studies and Research, Graduate School of Arts and Sciences, College of Engineering, Department of Civil and Environmental Engineering, Medford, MA 02155. Offers civil engineering (MS, PhD), including geotechnical engineering, structural engineering; environmental engineering (MS, PhD), including environmental engineering and environmental sciences, environmental geotechnology, environmental health, environmental science and management, hazardous materials management, water resources engineering. Part-time programs available. *Faculty:* 13 full-time, 10 part-time. *Students:* 99 (47 women); includes 21 minority (5 African Americans, 7 Asian Americans or Pacific Islanders, 9 Hispanic Americans) 16 international. Terminal master's awarded for partial completion of doctoral program. *Degree requirements:* For master's, thesis or alternative required, foreign language not required; for doctorate, dissertation required, foreign language not required. *Entrance requirements:* For master's and doctorate, GRE General Test, TOEFL (minimum score of 550 required). *Application deadline:* For fall admission, 2/15; for spring admission, 10/15. Applications are processed on a rolling basis. Application fee: $50. *Unit head:* Dr. Stephen Levine, Chair, 617-627-3211, Fax: 617-627-3994. *Application contact:* Linfield Brown, 617-627-3211, Fax: 617-627-3994.

University of Alberta, Faculty of Graduate Studies and Research, Department of Civil and Environmental Engineering, Edmonton, AB T6G 2E1, Canada. Offers construction engineering and management (M Eng, M Sc, PhD); environmental engineering (M Eng, M Sc, PhD); environmental science (M Sc, PhD); geoenvironmental engineering (M Eng, M Sc, PhD); geotechnical engineering (M Sc); geotechnical engineering (M Eng, PhD); mining engineering (M Eng, M Sc, PhD); petroleum engineering (M Eng, M Sc, PhD); structural engineering (M Eng, M Sc, PhD); water resources (M Eng, M Sc, PhD). Part-time programs available. *Degree requirements:* For master's, thesis required (for some programs), foreign language not required; for doctorate, dissertation required, foreign language not required. *Faculty research:* Mining.

University of California, Berkeley, Graduate Division, College of Engineering, Department of Civil and Environmental Engineering, Berkeley, CA 94720-1500. Offers construction engineering and management (M Eng, MS, D Eng, PhD); environmental quality and environmental water resources engineering (M Eng, MS, D Eng, PhD); geotechnical engineering (M Eng, MS, D Eng, PhD); structural engineering, mechanics and materials (M Eng, MS, D Eng, PhD); transportation engineering (M Eng, MS, D Eng, PhD). *Students:* 326 full-time (97 women); includes 59 minority (3 African Americans, 42 Asian Americans or Pacific Islanders, 13 Hispanic Americans, 1 Native American), 113 international. *Degree requirements:* For master's, comprehensive exam or thesis (MS) required; for doctorate, dissertation, qualifying exam required. *Entrance requirements:* For master's, GRE General Test, minimum GPA of 3.0; for doctorate, GRE General Test, minimum GPA of 3.5. *Application deadline:* For fall admission, 2/10. Application fee: $40. *Unit head:* Dr. Adib Kanafani, Chair, 510-642-3261. *Application contact:* Mari Cook, Graduate Assistant for Admission, 510-643-8944, E-mail: mcook@ce.berkeley.edu.

See in-depth description on page 609.

University of California, Los Angeles, Graduate Division, School of Engineering and Applied Science, Department of Civil and Environmental Engineering, Los Angeles, CA 90095. Offers environmental engineering (MS, PhD); geotechnical engineering (MS, PhD); structures (MS, PhD), including structural mechanics and earthquake engineering; water resource systems engineering (MS, PhD). *Faculty:* 15 full-time, 11 part-time. *Students:* 116 full-time (29 women); includes 28 minority (3 African Americans, 21 Asian Americans or Pacific Islanders, 4 Hispanic Americans), 37 international. *Degree requirements:* For master's, comprehensive exam or thesis required; for doctorate, dissertation, qualifying exams required, foreign language not required. *Entrance requirements:* For master's, GRE General Test, minimum GPA of 3.0; for doctorate, GRE General Test, minimum GPA of 3.25. *Application deadline:* For fall admission, 1/15; for spring admission, 12/1. Application fee: $40. Electronic applications accepted. *Unit head:* Dr. Michael Stenstrom, Chair, 310-825-1408. *Application contact:* Deeona Columbia, Graduate Affairs Officer, 310-825-1851, Fax: 310-206-2222, E-mail: deeona@ea.ucla.edu.

University of Colorado at Boulder, Graduate School, College of Engineering and Applied Science, Department of Civil, Environmental, and Architectural Engineering, Boulder, CO 80309. Offers building systems (MS, PhD); construction engineering and management (MS, PhD); environmental engineering (MS, PhD); geoenvironmental engineering (MS, PhD); geotechnical engineering (MS, PhD); structural engineering (MS, PhD); water resource engineering (MS, PhD). *Degree requirements:* For master's, thesis or alternative, comprehensive exam required; for doctorate, dissertation required. *Entrance requirements:* For master's, GRE General Test, minimum undergraduate GPA of 3.0.

See in-depth description on page 619.

University of Delaware, College of Engineering, Department of Civil and Environmental Engineering, Program in Water Resource Engineering, Newark, DE 19716. Offers MAS, MCE, PhD. Terminal master's awarded for partial completion of doctoral program. *Degree requirements:* For master's and doctorate, thesis/dissertation required, foreign language not required. *Entrance requirements:* For master's and doctorate, GRE General Test, TOEFL (minimum score of 550 required). *Faculty research:* Computer program for groundwater modeling, *in situ* stress determination by microhydraulic fracturing, stochastic groundwater flow modeling, poroelasticity in rock mechanics.

University of Guelph, Faculty of Graduate Studies, College of Physical and Engineering Science, School of Engineering, Guelph, ON N1G 2W1, Canada. Offers biological engineering (M Sc, PhD); environmental engineering (M Eng, M Sc, PhD); water resources engineering (M Eng, M Sc, PhD). Part-time programs available. *Faculty:* 18 full-time (1 woman), 29 part-time (4 women). *Students:* 54 full-time (15 women), 13 part-time (2 women); includes 22 minority (2 African Americans, 18 Asian Americans or Pacific Islanders, 2 Hispanic Americans), 7 international. *Degree requirements:* For master's, thesis required (for some programs); for doctorate, dissertation required. *Entrance requirements:* For master's, minimum B- average during previous 2 years; for doctorate, minimum B average. *Application deadline:* For fall admission, 8/1 (priority date); for winter admission, 11/1 (priority date); for spring admission, 4/1 (priority date). Applications are processed on a rolling basis. Application fee: $60. *Expenses:* Tuition and fees charges are reported in Canadian dollars. Tuition, area resident: Full-time $4,725 Canadian dollars; part-time $1,055 Canadian dollars per term. International tuition: $6,999 Canadian dollars full-time. Required fees: $295 Canadian dollars per term. *Unit head:* Dr. Lambert Otten, Director, 519-824-4120 Ext. 2043, Fax: 519-836-0227, E-mail: lotten@uoguelph.ca. *Application contact:* Dr. Ramesh P. Rudra, Graduate Coordinator, 519-824-4120 Ext. 2110, Fax: 519-836-0227, E-mail: rrudra@uoguelph.ca.

University of Kansas, Graduate School, School of Engineering, Department of Civil and Environmental Engineering, Lawrence, KS 66045. Offers civil engineering (MS, DE, PhD); environmental engineering (MS, PhD); environmental science (MS, PhD); water resources engineering (MS); water resources science (MS). *Faculty:* 20 full-time (0 women). *Students:* 23 full-time (6 women), 80 part-time (15 women); includes 9 minority (1 African American, 3 Asian Americans or Pacific Islanders, 4 Hispanic Americans, 1 Native American), 29 international. *Degree requirements:* For master's, thesis or alternative, exam required, foreign language not required; for doctorate, dissertation, comprehensive exam required. *Entrance requirements:* For master's and doctorate, Michigan English Language Assessment Battery, TOEFL, minimum GPA of 3.0. *Application deadline:* For fall admission, 7/1. Application fee: $30 ($45 for international students). *Unit head:* Steve McCabe, Chair, 785-864-3766. *Application contact:* David Parr, Graduate Director.

See in-depth description on page 633.

University of Maryland, College Park, Graduate School, College of Agriculture and Natural Resources, Department of Biological Resources Engineering, College Park, MD 20742-5045. Offers MS, PhD. *Faculty:* 13 full-time (0 women), 6 part-time (1 woman). *Students:* 17 full-time (10 women), 12 part-time (3 women); includes 7 minority (4 African Americans, 3 Asian Americans or Pacific Islanders), 6 international. 37 applicants, 43% accepted. In 1998, 6 master's, 1 doctorate awarded. *Degree requirements:* For master's, thesis optional, foreign language not required; for doctorate, dissertation required, foreign language not required. *Entrance requirements:* For master's, minimum GPA of 3.0. Application fee: $50 ($70 for international students). Tuition, state resident: part-time $272 per credit hour. Tuition, nonresident: part-time $475 per credit hour. Required fees: $632; $379 per year. *Financial aid:* In 1998–99, 17 research assistantships with tuition reimbursements (averaging $13,288 per year), 7 teaching assistantships with tuition reimbursements (averaging $11,818 per year) were awarded.; fellowships with full tuition reimbursements, career-related internships or fieldwork also available. Financial aid applicants required to submit FAFSA. *Faculty research:* Engineering aspects of production; harvesting, processing, and marketing of terrestrial and aquatic food and fiber. Total annual research expenditures: $385,057. *Unit head:* Dr. Frederick Wheaton, Chairman, 301-405-2223, Fax: 301-314-9023. *Application contact:* Trudy Lindsey, Director, Graduate Admission and Records, 301-405-4198, Fax: 301-314-9305, E-mail: grschool@deans.umd.edu.

The University of Memphis, Graduate School, Herff College of Engineering, Department of Civil Engineering, Memphis, TN 38152. Offers civil engineering (PhD); environmental engineering (MS); foundation engineering (MS); structural engineering (MS); transportation engineering (MS); water resources engineering (MS). *Faculty:* 12 full-time (0 women), 1 part-time (0 women). *Students:* 6 full-time (2 women), 22 part-time (1 woman), 7 international. *Degree requirements:* For master's, thesis or alternative, comprehensive exam required; for doctorate, dissertation required. *Entrance requirements:* For master's, GRE General Test (minimum combined score of 1000 required) or MAT, BS, minimum undergraduate GPA of 2.5. *Application deadline:* For fall admission, 8/1; for spring admission, 12/1. Application fee: $25 ($50 for international students). Tuition, state resident: full-time $3,410; part-time $178 per credit hour. Tuition, nonresident: full-time $8,670; part-time $408 per credit hour. Tuition and fees vary according to program. *Unit head:* Dr. Martin E. Lipinski, Chairman, 901-678-3279. *Application contact:* Dr. Larry W. Moore, Coordinator of Graduate Studies, 901-678-3278.

University of Missouri–Columbia, Graduate School, College of Engineering, Department of Civil and Environmental Engineering, Columbia, MO 65211. Offers civil engineering (MS, PhD); environmental engineering (MS, PhD); geotechnical engineering (MS, PhD); structural engineering (MS, PhD); transportation and highway engineering (MS); water resources (MS, PhD). *Faculty:* 24 full-time (2 women). *Students:* 19 full-time (2 women), 18 part-time (3 women); includes 1 minority (Hispanic American), 21 international. *Degree requirements:* For master's, report or thesis required; for doctorate, dissertation required. *Entrance requirements:* For master's and doctorate, GRE General Test, TOEFL. *Application deadline:* For fall admission, 7/10. Application fee: $30 ($50 for international students). *Unit head:* Dr. Mark Virkler, Director of Graduate Studies, 573-882-3678.

See in-depth description on page 639.

University of Southern California, Graduate School, School of Engineering, Department of Civil Engineering, Program in Water Resources, Los Angeles, CA 90089. Offers MS. Average age 25. 5 applicants, 100% accepted. In 1998, 1 degree awarded. *Degree requirements:* For master's, thesis optional. *Entrance requirements:* For master's, GRE General Test. *Application deadline:* For fall admission, 6/1 (priority date); for spring admission, 11/1. Application fee: $55. Tuition: Part-time $768 per unit. Required fees: $350 per semester. *Financial aid:* Fellowships, research assistantships, teaching assistantships, Federal Work-Study and institutionally-sponsored loans available. Aid available to part-time students. Financial aid application deadline: 2/15; financial aid applicants required to submit FAFSA. *Unit head:* Dr. L. Carter Wellford, Chair, Department of Civil Engineering, 213-740-0587.

The University of Texas at Austin, Graduate School, College of Engineering, Department of Civil Engineering, Program in Environmental and Water Resources Engineering, Austin, TX 78712-1111. Offers MSE. *Accreditation:* ABET. *Students:* 17 (8 women); includes 2 minority (both Hispanic Americans) 1 international. 34 applicants, 29% accepted. In 1998, 9 degrees awarded. *Degree requirements:* Foreign language not required. *Entrance requirements:* For master's, GRE General Test (minimum combined score of 1000 required). *Application deadline:* For fall admission, 1/15 (priority date); for spring admission, 9/1 (priority date). Applications are processed on a rolling basis. Application fee: $50 ($75 for international students). Electronic applications accepted. *Financial aid:* Fellowships, research assistantships, teaching assistantships available. Financial aid application deadline: 2/1. *Unit head:* Kathy Rose, Graduate Coordinator, 512-471-4921, Fax: 512-471-0592, E-mail: krose@mail.utexas.edu. *Application contact:* Kathy Rose, Graduate Coordinator, 512-471-4921, Fax: 512-471-0592, E-mail: krose@mail.utexas.edu.

University of Virginia, School of Engineering and Applied Science, Department of Civil Engineering, Program in Environmental Engineering, Charlottesville, VA 22903. Offers environmental engineering (ME, MS, PhD); water resources (ME, MS, PhD). Part-time programs available. *Faculty:* 5 full-time (2 women). *Students:* 36 full-time (21 women), 1 part-time; includes 1 minority (Asian American or Pacific Islander), 8 international. Terminal master's awarded for partial completion of doctoral program. *Degree requirements:* For master's, thesis required (for some programs), foreign language not required; for doctorate, dissertation, comprehensive exam required, foreign language not required. *Entrance requirements:* For master's and doctorate, GRE General Test. *Application deadline:* For fall admission, 2/1 (priority date). Applications are processed on a rolling basis. Application fee: $60. Electronic applications accepted. *Unit head:* Dr. Nicholas J. Garber, Chairman, Department of Civil Engineering, 804-924-7464, Fax: 804-982-2951, E-mail: njg@virginia.edu.

Utah State University, School of Graduate Studies, College of Engineering, Department of Biological and Irrigation Engineering, Logan, UT 84322. Offers biological and agricultural engineering (MS, PhD); irrigation engineering (MS, PhD). Part-time programs available. *Faculty:* 10 full-time (0 women). *Students:* 26 full-time (4 women), 11 part-time (4 women); includes 1 minority (Hispanic American), 27 international. Average age 28. 21 applicants, 76% accepted. In 1998, 7 master's, 8 doctorates awarded. Terminal master's awarded for partial completion of doctoral program. *Degree requirements:* For master's, computer language, thesis required (for some programs), foreign language not required; for doctorate, computer language, dissertation required, foreign language not required. *Entrance requirements:* For master's and doctorate, GRE General Test (score in 40th percentile or higher required), TOEFL (minimum score of 550 required), minimum GPA of 3.0. *Application deadline:* For fall admission, 6/15 (priority date); for spring admission, 10/15. Applications are processed on a rolling basis. Application fee: $40. Tuition, state resident: full-time $1,492. Tuition, nonresident: full-time $5,232. Required fees: $434. Tuition and fees vary according to course load. *Financial aid:* In 1998–99, 15 fellowships with partial tuition reimbursements, 15 research assistantships with partial tuition reimbursements, 2 teaching assistantships with partial tuition reimbursements were awarded.; tuition waivers (partial) also available. Aid available to part-time students. *Faculty research:* Surge flow, on-farm water management, crop-water yield modeling, drainage, groundwater modeling. *Unit head:* Wynn Walker, Head, 435-797-2788, Fax: 435-797-1248, E-mail: wynnwalk@cc.usu.edu. *Application contact:* Lyman Willardson, Graduate Adviser, 435-797-2789, Fax: 435-797-1248, E-mail: fath8@cc.usu.edu.

Villanova University, College of Engineering, Department of Civil and Environmental Engineering, Program in Water Resources and Environmental Engineering, Villanova, PA 19085-1699. Offers MSWREE. Part-time and evening/weekend programs available. *Faculty:* 5 full-time (0 women), 5 part-time (0 women). *Students:* 5 full-time (2 women), 12 part-time (3 women); includes 2 minority (1 African American, 1 Hispanic American), 2 international. Average age 25. 19 applicants, 89% accepted. In 1998, 12 degrees awarded. *Degree requirements:* For master's, computer language required, thesis optional, foreign language not required. *Entrance requirements:* For master's, GRE General Test (for applicants with degrees from foreign universities), minimum GPA of 3.0, BCE or bachelor's degree in science or related engineering field. *Average time to degree:* Master's–1 year full-time, 3.5 years part-time. *Application deadline:* For fall admission, 8/1 (priority date); for spring admission, 12/1. Applications are processed on a rolling basis. Application fee: $40. *Financial aid:* Federal Work-Study and tuition waivers (full and partial) available. Aid available to part-time students. Financial aid application deadline: 4/15. *Faculty research:* Photocatalytic decontamination and disinfection of water, urban storm water wetlands, economy and risk, removal and destruction of organic acids in water, sludge treatment. *Unit head:* Dr. Lewis J. Mathers, Chairman, Department of Civil and Environmental Engineering, 610-519-4960, Fax: 610-519-6754, E-mail: ldeangel@email.vill.edu.

Cross-Discipline Announcements

Pennsylvania State University University Park Campus, Graduate School, College of Engineering, Department of Architectural Engineering, State College, University Park, PA 16802-1503.

Two of the major teaching and research groups in the Department of Architectural Engineering are Building Structures, whose research focuses on masonry, steel, and reinforced-concrete structural design and analysis for buildings, loadings, and performance of the building envelope; and Building Construction, whose research focuses on project delivery systems, process and information modeling, domestic and international construction project organization and control, advanced technology projects, and integration of the design, build, and operate processes.

Princeton University, Graduate School, Princeton Materials Institute, Princeton, NJ 08540-5211.

The Princeton Materials Institute welcomes students interested in cross-disciplinary research involving any aspect of materials science and engineering. Faculty members from 8 academic departments collaborate in a remarkably broad range of programs, supported by outstanding facilities. Some fellowships are available. See in-depth description in Section 21 of this guide.

Rensselaer Polytechnic Institute, Graduate School, Lally School of Management and Technology, Program in Environmental Management and Policy, Troy, NY 12180-3590.

Environmental Management and Policy (EMP) is a Master of Science program that integrates technical course work with managerial and policy studies to develop management capability for decision-making positions in industry, environmental consulting firms, government, public interest groups, and research institutions. The practice-oriented EMP program trains candidates to integrate better, cheaper, cleaner solutions to environmental challenges both in the US and overseas.

University of Southern California, Graduate School, School of Architecture, Program in Building Science, Los Angeles, CA 90089.

The Master of Building Science program emphasizes independent research in energy-conscious design, natural forces and systems, structural analysis, and research design methods suited to the complexity of building. The requirement for the degree, typically a 2-year program, is a first accredited degree in architecture, engineering, computer science, or physical science.

Carnegie Mellon

CARNEGIE MELLON UNIVERSITY

Department of Civil and Environmental Engineering

Programs of Study	Graduate study in this highly ranked department is designed to provide opportunities for the development of professional engineering competence and scholarly achievement. The program emphasizes fundamental knowledge in civil and environmental engineering methodology and in related fields of computer science, engineering, physical and social science, and management. The study programs are individually planned with faculty consultation; each student designs his or her program, drawing on the available strengths in the department and the University. The department has graduate programs in the following areas: environmental engineering, including programs in green design, solid and hazardous waste, pollution prevention, and air, water, and soil quality engineering; structural computational mechanics, including applications in earthquake engineering, biomechanics, materials processing, solid and fluid mechanics, fracture mechanics, and optimal design; and computer-aided engineering and civil systems, including computer-aided design, robotics, transportation systems, environmental systems, and project management. Dual-degree programs (e.g., with Architecture or the Graduate School of Industrial Administration) and programs aimed at practitioners who cannot pursue regular full-time study are available. In addition to the traditional M.S. in Civil Engineering, the following degrees are offered: M.S. in Civil Engineering/Engineering and Public Policy, M.S. in Environmental Science, M.S. in Environmental Management, and M.S. in Civil Engineering and Management. The M.S. in Civil Engineering can be taken either as a course-only option or as a thesis option. The professional course-only option can usually be completed in two semesters of courses. The requirements for the thesis option can usually be completed in two or three semesters plus a summer. After completion of the master's degree at CMU or elsewhere, students of superior ability are eligible for admission as candidates for the Ph.D. degree. Candidates normally complete at least one additional year of course work and two years of research.
Research Facilities	The Civil and Environmental Engineering Laboratories, located in a four-story wing adjacent to the department, encompass more than 12,000 square feet. Included are individual laboratories for research in water chemistry, air pollution, materials, soil mechanics, and structural mechanics. The environmental engineering facilities encompass 6,000 square feet and were completely renovated in 1994. They include a water chemistry research laboratory, a trace metals clean laboratory, a fully equipped geoenvironmental laboratory, and a radioactive material laboratory. Analytical facilities are available for characterization of samples for both major and trace constituents in air, water, and soil samples. The laboratories also have a complement of structural testing equipment, including a closed-loop 100,000-pound MTS system. All graduate students are assigned an office with a desk, filing space, and some computing equipment.
	Computing resources within the Department of Civil and Environmental Engineering include more than eighty general-purpose multimedia PCs, various systems dedicated to research, and numerous privately owned PCs and laptops. Departmental machines are shared throughout the faculty, staff, and student offices, along with an undergraduate computer cluster. The computing infrastructure includes printers distributed throughout the department, high-speed networks, and various specialized equipment and peripherals. General computing services include file, print, e-mail, and Web servers. All equipment in the department is connected to the campus network and campus file systems, providing connections to other University computing facilities (including Computing Services, the School of Computer Science, and the Pittsburgh Supercomputing Center) and the Internet.
Financial Aid	Many graduate students are financially supported by various appointments, which are normally made for nine months, from the first of September to the end of May. Separate appointments for research and academic programs in the summer are often made for graduate students in residence. A few students enter research programs in the summer before beginning graduate study in September. In addition to taking a full program of graduate courses, a graduate assistant is usually expected to devote 15 to 20 hours per week in combined research and teaching activity for the department. This activity is considered to be an integral part of graduate education. A limited number of partial tuition scholarships are also available. Professional course-only M.S. students are only eligible for partial-tuition scholarships.
Cost of Study	The cost of tuition is $22,100 for the academic year 1999–2000. The cost of books and supplies is typically $1450 for the academic year.
Living and Housing Costs	Graduate students live off campus in nearby apartments, most often with 1 or 2 other students. An Off-Campus Housing Referral System is available through the CMU Housing Office. In 1998–99, a nine-meal-per-week food-service contract cost approximately $825 per semester.
Student Group	The graduate student body numbers 78. Of these, about 60 are full-time students and most receive some form of financial aid.
Location	Pittsburgh is a large metropolitan area of more than 2 million people and is the headquarters for many of the nation's largest corporations. There is an unusually large concentration of research laboratories and software firms in the area. Carnegie Mellon is located in the Oakland section of Pittsburgh, on a tract of 90 acres adjacent to Schenley Park. Oakland is also home to world-renowned medical and educational institutions, as well as the Carnegie Museum and Library. The campus is close to the many cultural and sports activities of the city and is only 4 miles from the downtown business district.
The University	Carnegie Mellon was first established in 1900 as the Carnegie Technical School through a gift from Andrew Carnegie. In 1912, the name of the school was changed to Carnegie Institute of Technology. Mellon Institute, founded in 1913 by A. W. and R. B. Mellon merged with Carnegie Institute of Technology in 1967 to become Carnegie Mellon University. The University has an enrollment of about 7,500, approximately 2,700 of whom are engaged in graduate study. The teaching and research faculty numbers more than 1,000 members.
Applying	Applications and complete credentials should be received by February 1.
Correspondence and Information	Chairman of the Graduate Committee Department of Civil and Environmental Engineering Carnegie Mellon University Pittsburgh, Pennsylvania 15213-3890 E-mail: ce-admissions@cmu.edu World Wide Web: http://www.ce.cmu.edu/

Carnegie Mellon University

THE FACULTY AND THEIR RESEARCH

Burku H. Akinci, Assistant Professor; Ph.D. candidate, Stanford. Computer-aided engineering, computer-integrated construction, automation of production analysis prior to construction, integration of design and construction, standardization efforts (IFC), symbolic modeling of lean construction priciples.

John Bares, Senior Research Scientist; Ph.D., Carnegie Mellon, 1991. Conception to testing of intelligent machines for hazardous environments, construction, and heavy industry; automated robot design and optimization; technology transfer to industry.

Jacobo Bielak, Professor; Ph.D., Caltech, 1971. Applied and computational mechanics, earthquake engineering and engineering seismology.

Paul P. Christiano, Professor and Provost; Ph.D., Carnegie Mellon, 1967. Stability and vibrations of structures, dynamic soil-structure interactions, expert systems for site characterization, soil mechanics and foundation engineering.

Jared Cohon, Professor and President, Carnegie Mellon University; Ph.D., MIT, 1973. Multiobjective decision making, water resources.

Cliff I. Davidson, Professor and Director of the Environmental Institute; Ph.D., Caltech, 1977. Transport of anthropogenic aerosols from source to sink, pollutants in remote areas, use of the glacial record to understand historical air pollution trends, indoor air pollution modeling and measurement, atmospheric deposition.

David A. Dzombak, Professor; Ph.D., MIT, 1986. Environmental engineering, geochemistry and groundwater pollution, physicochemical processes in natural aquatic systems and in water/wastewater treatment.

Susan Finger, Associate Professor; Ph.D., MIT, 1981. Computer-aided design, representation languages for designs, integration of design and manufacturing concerns.

James Garrett, Professor; Ph.D., Carnegie Mellon, 1986. Computer-aided engineering, wearable/mobile computing applications in civil engineering, knowledge-based methods for engineering design, object-oriented methods for standards representation and usage.

Omar Ghattas, Associate Professor; Ph.D., Duke, 1988. Computational and applied mechanics, optimal design and control, inverse problems, parallel scientific computing, biomechanics, applied numerical algorithms.

Chris T. Hendrickson, Duquesne Light Company Professor and Head; Ph.D., MIT, 1978. Civil systems, project management and finance, product and process design for the environment, computer applications in engineering planning and management.

Richard G. Luthy, Thomas A. Lord Professor; Ph.D., Berkeley, 1976. Environmental engineering—water quality, physicochemical processes, chemistry and aqueous systems, problems in soil and groundwater contamination, industrial waste reduction.

Francis C. McMichael, Professor of Civil Engineering/Engineering and Public Policy and Walter J. Blenko, Sr., Professor of Environmental Engineering; Ph.D., Caltech, 1963. Environmental regulatory processes and regulatory risk analysis, statistical and time-series analysis, product and process design for the environment, solid-waste management: source reduction and recycling.

Sue McNeil, Professor; Ph.D., Carnegie Mellon, 1983. Analytical methods for the management of transportation infrastructure systems, application of new technologies to infrastructure management and infrastructure financing, brownfields development.

Irving J. Oppenheim, Professor of Civil Engineering and Architecture; Ph.D., Cambridge, 1972. Robotics, structural mechanics, building design.

Marina Pantazidou, Assistant Professor; Ph.D., Berkeley, 1991. Geoenvironmental engineering, experimental and numerical modeling of multiphase fluid flow, environmental decision making, engineering ethics.

Daniel R. Rehak, Professor; Ph.D., Illinois, 1981. Computer applications in civil engineering, design of computer-aided engineering systems, alternative hardware and software environments for computer applications, educational systems.

Sunil Saigal, Professor; Ph.D., Purdue, 1985. Computational mechanics, finite-element and boundary-element methods, biomechanics, MEMS in civil infrastructure systems, cohesive zone modeling of fracture.

Mitchell J. Small, Professor of Civil Engineering/Engineering and Public Policy and Associate Head of Engineering and Public Policy; Ph.D., Michigan, 1982. Environmental engineering; mathematical modeling of surface water, groundwater, and air quality; probabilistic methods and uncertainty.

Jeanne Marie VanBriesen, Assistant Professor; Ph.D., Northwestern, 1998. Environmental engineering, environmental biotechnology, environmental risk assessment and biogeochemical processes.

CURRENT RESEARCH INTERESTS

Civil Systems. Civil systems encompass the physical and built environment that supports a large number of social, industrial, and service activities. Systematic study of civil systems is the basis for the planning, design, construction, and management of facilities comprising the built environment. Due to the complexity of these facilities and of the interactions that take place within them, the study of civil systems is highly interdisciplinary. The fundamentals for such study include basic concepts from management science and economics, as well as basic tools and methods from computational and information technologies. Furthermore, the study of civil systems is invariably tied to the study of a specific segment of the physical and built environment, such as environmental systems.

Computer-Aided Engineering. Computer-aided engineering (CAE) encompasses the intersection between the traditional disciplines of civil engineering and the enabling technologies provided by computer science and computer-based information technologies. The objective of CAE is to understand the process of civil engineering so as to propose improvements, both in the process of engineering and in the resultant products and facilities, through the application of computer-based tools and techniques. CAE emphasizes the fundamental methodological components occurring throughout civil and environmental engineering: information modeling, representation, and management and computer support for calculation, display, communication, and collaboration. Domain-specific models, methods, and knowledge are superimposed on and integrated with the fundamental components. Study in CAE begins with the fundamentals of computer-based problem solving and software development. Upon this base, relevant concepts from computer science and information technologies are added: artificial intelligence, databases, sensing, multimedia, and networking are examples of the ever-changing toolkit of CAE. The emphasis is on evaluating, refining, and extending these concepts and technologies to meet engineering needs and improve the practice of civil and environmental engineering.

Environmental Engineering. Development and application of fundamental scientific knowledge and engineering methodology to problems of contaminated water; transport and reactions of species in groundwater systems; air pollution monitoring and evaluation; pollution prevention; and mathematical modeling of surface water, groundwater, and air resource systems. A joint graduate program with the Department of Engineering and Public Policy addresses interdisciplinary issues, with particular emphasis on methodologies for evaluating the economic and public decision-making implications of engineering and scientific research. A joint M.S. program with the Graduate School of Industrial Administration provides an opportunity to study environmental engineering and management in a business context. Current research interests include transport and remediation of chemicals in groundwater systems, stochastic modeling of contaminant transport in environmental systems, and site remediation. Active air pollution research addresses the characterization and fate of airborne particulate matter in both indoor and global environments and the assessment of national air pollution control strategies, especially in relation to fine particles. Additional research areas include product and process design for the environment, waste minimization, natural amelioration of acid mine drainage, causes and impact of global climate change, and the source and fate of air toxics in ambient and indoor environments.

Structural and Computational Mechanics. Computational mechanics involves a coordinated blend of insightful modeling of mechanical phenomena with the development of appropriate computational methods. With mechanics, mathematics, and scientific computing as its foundation, it addresses the numerical simulation of a wide variety of physical phenomena with a view toward the analysis and optimum design of engineering systems. Researchers also seek to gain a better understanding of complex physical phenomena, which are difficult, if not impossible, to study by alternate approaches. Computational mechanics efforts in the department focus on selected challenging problems that pose difficulty due to their size, geometric or physical complexity, nonlinearity, and coupled or multiscale nature. Currently, research is being conducted in biomechanics, earthquake engineering and engineering seismology, fluid-structure interaction, materials processing, and micromechanics. Because large-scale problems are often intractable on conventional computers, the program also focuses on algorithms for high-performance computing, in particular parallel supercomputers.

CLEMSON UNIVERSITY

School of the Environment
Department of Environmental Engineering and Science

Programs of Study

Clemson University offers the Master of Engineering (M.Engr.), the Master of Science (M.S.), and the Ph.D. degree in environmental engineering and science. Course work for the M.S. thesis or M.Engr. degree program can be completed in three semesters (usually sixteen to twenty-four months). The nonthesis M.S. program can be completed in one calendar year by taking courses during both regular semesters and during the summer. A core curriculum of courses is required of all master's students. In addition, students pursuing the M.Engr. degree are required to take four additional courses, including one from a select list of eight. The remaining course work is taken to meet career specialization and is not restricted to environmental engineering and science (EE&S) courses. A special project or thesis defense is required.

The Ph.D. program of study is flexible, and each student's specific program is uniquely tailored by the student and his or her committee. The Ph.D. program can be completed in three to five years. All doctoral candidates are required to take a written qualifying examination at the end of the first year of study. Comprehensive examinations, written and oral, are scheduled after completion of course work and after formulation of the dissertation research proposal, respectively. A dissertation defense is required prior to the approval and acceptance of the completed work.

The aim of the degree programs in environmental engineering and science is to prepare professionals capable of leading and participating in interdisciplinary teams that develop informed, practical, and technically sound solutions to challenging environmental protection and restoration problems in the air, water, and soil environments. The program does this within the scope of five specialty areas: process engineering, nuclear environmental (including risk management), environmental chemistry, environmental fate and transport, and environmental waste management.

Research Facilities

The 42,000-square-foot Rich Environmental Research Laboratory is located in the Clemson Research Park, 8 miles from the main campus. State-of-the-art analytical instrumentation in EE&S laboratories includes atomic absorption spectrophotometers, computer-based gas chromatographs, a GC/MS, two high-pressure liquid chromatographs, an ion chromatograph, electroanalytical instruments, two TOC analyzers, an electrobalance, a fluorescence spectrophotometer, and a UV-Vis spectrophotometer. Other EE&S laboratories include a trace organics analysis sample preparation facility, a clean room, sediment and particle-size analysis equipment, and extensive air pollution sampling equipment. A radiation laboratory with a complete array of gaseous and solid-state radiation detectors contains a high-resolution gamma ray spectrometry system, including an intrinsic germanium detector with an online computer for data analysis; an alpha spectrometry system; a thermoluminescent dosimetry system; and two liquid scintillation counters. Sources and instrumentation for conducting health physics applications are available. Operations and processes experiments are conducted in a two-story laboratory equipped for mixed-media filtration, slurry thickening, clarification, reverse osmosis, ozonation, and pilot-scale research. Two electrolytic respirometers, one N-con respirometer, three BioFlo reactors, and five magnetically coupled Microferm reactors are available for biological process research.

Financial Aid

Financial aid is available on a competitive basis in the form of full and partial fellowships. Also available are teaching and research assistantships, which require up to 20 hours of work per week and provide stipends from $12,780 to $15,280 annually.

Cost of Study

In-state academic fees for full-time (minimum 12 hours) students are $1577 per semester; part-time (maximum 11 hours) students pay $130 per hour. Out-of-state tuition is $3226 per semester for full-time students or $264 per hour for part-time students. Graduate assistants pay $493 per semester and $165 per summer for an unlimited number of credit hours. All full-time students pay a $95 medical fee each semester.

Living and Housing Costs

Most graduate students prefer to live off campus in apartments or houses, where the costs range from $250 to $375 per month, depending on the number of residents. Detailed information may be obtained by writing the Housing Office, 200 Mell Hall, Clemson University, Clemson, South Carolina 29634-4075.

Student Group

The Department of Environmental Engineering and Science had 30 women and 46 men enrolled in the fall of 1998 from all over the U.S. (typically less than 20 percent of the students are international). Almost all students have some type of funding either by the department, the University, national fellowships (NSF and DOE), or their home governments.

Location

Clemson is a typical college town, with a resident population of approximately 12,000. It is midway between Atlanta, Georgia, and Charlotte, North Carolina. Both of these cities can be reached in about 2 hours' driving time, and an international airport is only 45 minutes away. Clemson lies in the foothills of the Blue Ridge Mountains. The climate is very moderate yet there are four distinct seasons. A wide variety of outdoor recreation opportunities are close at hand. Water sports such as swimming, fishing, boating, and sailing can be enjoyed at University recreation areas and on lakes Hartwell, Keowee, and Jocassee. White-water canoeing, trout fishing, hiking, and camping are available on or along several mountain streams within 45 minutes of campus, including the well-known Chattooga River, now designated a Wild and Scenic River by the U.S. Forest Service. Nearby national forests and national and state parks complement other recreational opportunities. The historic city of Charleston, Myrtle Beach on South Carolina's renowned Grand Strand, the barrier islands such as Hilton Head and Kiawah, and Alpine skiing at various Georgia and North Carolina mountain resorts are all within a few hours' driving time from campus.

The University

Founded in 1889, Clemson is located on the 1,000-mile shoreline of Lake Hartwell in South Carolina's lake and mountain region. Clemson's 1,400-acre main campus is surrounded by 17,000 acres of University farms and woodlands with an additional 12,390 acres of engineering research lands throughout the state. Clemson offers 112 graduate programs in seventy-one areas of study. Approximately 16,000 students, including 4,000 graduate students, are enrolled. A member of the Atlantic Coast Conference, Clemson competes in eleven intercollegiate sports for men and six for women.

Applying

In considering applicants for admission, EE&S faculty members look for the overall academic performance of the student and the quality of the undergraduate institution. Letters of recommendation and comments are carefully read to determine if the student is serious about graduate school, and an essay is required. General admission requirements are a GPA of 3.0 out of 4.0 and a minimum GRE of 1600 (minimum 450 verbal). Students required to take the Test of English as a Foreign Language (TOEFL) in addition to the GRE must attain a score of 600. Students can find further information on the department's Web page below. There is no deadline for applications. Applications are available on the World Wide Web at http://www.grad.clemson.edu or by e-mail at grdapp@clemson.edu.

Correspondence and Information

Pamela Fjeld
Environmental Engineering and Science
Clemson University
Clemson, South Carolina 29634-0919

Telephone: 864-656-1010
Fax: 864-656-0672
E-mail: ees-admissions@clemson.edu
World Wide Web: http://www.eng.clemson.edu/ees

Clemson University

THE FACULTY AND THEIR RESEARCH

The EE&S faculty is a dynamic group—diverse yet complementary—with a broad spectrum of teaching specialties and research interests. Faculty members have won numerous awards and honors, published articles in leading scholarly journals, chaired sessions at symposia in the U.S. and abroad, made technical presentations at major professional meetings, served as officers of national and international professional societies, and sponsored research activities, which are consistent with the $1–$2 million in active external research funding. Faculty members have continued to attract significant support for the department since it was organized as a separate graduate department in 1968.

Christos Christoforou, Assistant Professor; Ph.D. (mechanical engineering), Caltech, 1995. Air pollution, particle sampling and characterization.

Tim DeVol, Assistant Professor; Ph.D. (nuclear engineering), Michigan, 1993. Environmental monitoring of ionizing radiation, the usage of ionizing radiation for elemental analysis.

Alan W. Elzerman, Chair and Director of the Environmental Science and Policy Program; Ph.D. (water chemistry), Wisconsin–Madison, 1976. Environmental chemistry and analytical chemistry.

Robert A. Fjeld, Dempsey Professor of Waste Management; Ph.D. (nuclear engineering), Penn State, 1976. Environmental risk assessment, radioactive waste management, environmental transport of radioactivity.

David L. Freedman, Assistant Professor; Ph.D. (environmental engineering), Cornell, 1990. Hazardous waste management, water and wastewater treatment, biodegradation/bioremediation of recalcitrant organic compounds.

C. P. Leslie Grady Jr., R. A. Bowen Professor; Ph.D. (bioenvironmental engineering), Oklahoma State, 1969. Environmental biotechnology, modeling of wastewater treatment systems, the application of process engineering techniques to the design of treatment systems for water, wastewater, and aquifer restoration.

Tanju Karanfil, Assistant Professor; Ph.D. (environmental engineering), Michigan, 1995. Physicochemical processes in natural and engineered environmental systems.

Cindy M. Lee, Associate Professor; Ph.D. (geochemistry), Colorado School of Mines, 1990. Chemistry of environmentally significant organic compounds, analytical chemistry, especially chromatographic techniques.

Fred J. Molz III, Westinghouse Distinguished Professor; Ph.D. (hydrology), Stanford, 1970. Hydraulics and transport process in the groundwater-soil-plant-atmosphere system.

Thomas J. Overcamp, Professor; Ph.D. (mechanical engineering), MIT, 1973. Hazardous waste treatment and air pollution control.

CORNELL UNIVERSITY

School of Civil and Environmental Engineering

Programs of Study	The School of Civil and Environmental Engineering offers two distinct graduate degree programs. The first is a research-oriented program that leads to the degrees of Master of Science and Doctor of Philosophy. It is intended primarily for students who seek research or academic careers, although a number of degree recipients enter professional practice. The second program is a one-year academic professional course of study that leads to the degree of Master of Engineering (civil engineering). It is intended primarily for those who plan to enter professional engineering practice, but it can also be used as preparation for the Ph.D. program.
	The M.S. and Ph.D. degree programs are offered through the Graduate School. Major subject areas are environmental engineering, environmental fluid mechanics and hydrology, environmental systems engineering, geotechnical engineering, remote sensing, structural engineering, transportation systems engineering, and water resource systems. Each student plans an individualized course of study with the assistance of a special committee made up of faculty members from the major and minor areas of study. The typical M.S. program takes 1½ to 2 years to complete, and the Ph.D. degree usually requires at least 2 additional years. Both degrees require a thesis and a final oral examination, and the Ph.D. program also involves qualifying and comprehensive examinations.
	The M.Eng. degree, offered through the College of Engineering, is normally completed in nine months of intensive study. Completion of a design project and course work in professional practice is required. A combined M.Eng./M.B.A. program is also available. Students may pursue an M.Eng. degree program with a design or engineering management option.
Research Facilities	The School's research facilities, housed in Hollister Hall and in adjoining Thurston Hall, include separate laboratories for research in environmental engineering, hydraulic engineering, geotechnical engineering, remote sensing, materials of construction, structural modeling, and static and dynamic loading of large structural components and systems. The University library is one of the largest in the United States and has an excellent collection in civil and environmental engineering. Computer facilities include the School's workstation cluster, the University's central system, equipment for digital image processing, and several walk-in microcomputer facilities.
Financial Aid	Research and teaching assistantships and fellowships are available on a competitive basis. The assistantships require 20 hours of work per week and carried a stipend of $12,450 for the 1998–99 academic year plus full tuition. Support for M.Eng. candidates is provided mainly by the College of Engineering in the form of part-time teaching assistantships and partial scholarships. All financial aid is based on merit and is awarded on a competitive basis.
Cost of Study	Tuition for the 1999–2000 nine-month academic year is $23,760.
Living and Housing Costs	For the 1998–99 twelve-month academic year, living expenses (cost of books and supplies, housing and dining, personal expenses, and medical insurance) for a single graduate student were between $10,200 and $13,800. The additional expenses for a spouse were about $7000. Tuition and travel expenses should be added to estimate the cost of attendance for the academic year.
Student Group	The School has had about 160 graduate students in residence during each of the past few years, approximately half of whom are M.Eng. students. Students come from all parts of the United States and many other countries.
Student Outcomes	Employment opportunities in professional practice and in research and teaching continue to be very good for holders of graduate degrees in civil and environmental engineering. More than half of the School's graduates find employment in professional practice, approximately 20 percent go on to other professional degrees, and the remainder are employed in academic or research positions.
Location	Ithaca, a city of about 40,000, is located on Cayuga Lake in the Finger Lakes region of upper New York State. The city is one of the country's great educational communities, offering cultural advantages that rival those of many large cities. Many recreational facilities, including ski areas, museums, marinas, wineries, and state parks, are located nearby.
The University and The School	At Cornell, graduate study in engineering is conducted within the context of a large and diverse university with an international reputation, and the resources of the entire University are available to each graduate student. The fourteen colleges and schools at Cornell have about 4,700 graduate students and 12,500 undergraduates. Civil engineering has been an important area of study for more than a century; the first class of engineers, all civil engineers, graduated in 1871.
	Cornell University is an Equal Opportunity/Affirmative Action educator and employer.
Applying	For fall admission, M.S. and Ph.D. applicants who request financial support should file applications before January 15; M.Eng. applications are due by February 1. Applicants must hold a baccalaureate degree or the equivalent from a recognized university, have adequate preparation for graduate study in civil and environmental engineering, be fluent in the English language, and present evidence of promise for high achievement in engineering. The Graduate Record Examinations (GRE) are strongly recommended, especially for any applicant who requests financial aid and/or admission to the Ph.D. program.
Correspondence and Information	Graduate Program Coordinator School of Civil and Environmental Engineering Hollister Hall Cornell University Ithaca, New York 14853-3501 Telephone: 607-255-7560 Fax: 607-255-9004 E-mail: cee_grad@cornell.edu World Wide Web: http://www.cee.cornell.edu/

Cornell University

THE FACULTY AND THEIR RESEARCH

John F. Abel, Professor; Ph.D., Berkeley; PE. Structural analysis and mechanics, computer-aided design, numerical methods, structural dynamics.

Sarah L. Billington, Assistant Professor; Ph.D., Texas. Behavior and design of concrete structures, repair and rehabilitation of concrete structures.

James J. Bisogni Jr., Associate Professor; Ph.D., Cornell. Environmental engineering, aquatic chemistry.

Wilfried H. Brutsaert, Professor; Ph.D., California, Davis. Hydrology, hydraulics, groundwater flow.

Edwin A. Cowen, Assistant Professor; Ph.D., Stanford. Environmental fluid mechanics; mixing, transport, and air-sea processes; field and laboratory measurement technologies.

Richard I. Dick, Joseph P. Ripley Professor of Engineering; Ph.D., Illinois; PE. Environmental engineering, sludge treatment.

James M. Gossett, Professor; Ph.D., Stanford. Environmental engineering, biological treatment processes.

Mircea D. Grigoriu, Professor; Ph.D., MIT. Structural engineering, structural reliability, random vibration, structural dynamics.

Douglas A. Haith, Professor; Ph.D., Cornell. Water resource systems, nonpoint-source pollution.

Kenneth C. Hover, Professor; Ph.D., Cornell; PE. Civil engineering materials, construction techniques, durability.

Anthony R. Ingraffea, Professor; Ph.D., Colorado; PE. Structural mechanics, fracture mechanics, numerical methods.

Fred H. Kulhawy, Professor; Ph.D., Berkeley; PE, GE. Geotechnical engineering, soil-structure interaction, numerical analysis.

Leonard W. Lion, Professor; Ph.D., Stanford. Environmental engineering, reactions to pollutants in aqueous systems.

Philip L.-F. Liu, Professor; Sc.D., MIT. Fluid mechanics, wave hydrodynamics, coastal engineering.

Daniel P. Loucks, Professor; Ph.D., Cornell. Water resource and environmental management systems, interactive simulating and design support.

Walter R. Lynn, Professor Emeritus; Ph.D., Northwestern; PE. Environmental systems analysis, public health, natural hazard reduction.

Arnim H. Meyburg, Professor and Director of the School; Ph.D., Northwestern. Transportation engineering and planning.

Linda K. Nozick, Associate Professor; Ph.D., Pennsylvania. Engineering management, transportation systems analysis, systems engineering.

Thomas D. O'Rourke, Professor; Ph.D., Illinois. Geotechnical engineering, underground technologies, soil-structure interaction.

Teoman Peköz, Professor; Ph.D., Cornell. Steel and aluminum structures.

William D. Philpot, Associate Professor; Ph.D., Delaware. Remote sensing, hydrologic optics, image processing.

Mary J. Sansalone, Professor; Ph.D., Cornell. Structural analysis and mechanics, nondestructive testing, stress wave propagation.

Richard E. Schuler, Professor and Director, Institute for Public Affairs; Ph.D., Brown; PE. Urban, spatial, and regional economics; public finance; utility regulation.

Christine A. Shoemaker, Professor; Ph.D., USC. Water resource and water quality systems, groundwater contamination, pest management, optimization, algorithms.

Jery R. Stedinger, Professor; Ph.D., Harvard. Stochastic hydrology, water resource systems, risk analysis.

Harry E. Stewart, Associate Professor; Ph.D., Massachusetts; PE. Geotechnical engineering, dynamic behavior of soils.

Mark A. Turnquist, Professor; Ph.D., MIT. Transportation systems planning, analysis, and design; manufacturing logistics; engineering management.

Richard N. White, James A. Friend Family Distinguished Professor of Engineering; Ph.D., Wisconsin; PE. Behavior and design of concrete structures, earthquake engineering.

RESEARCH AREAS

Environmental Engineering. The phenomena, concepts, and technology essential for maintaining the quality of the air-land-water environment are the concerns of this area. Specialization in environmental engineering requires an understanding of the biological, chemical, and physical phenomena that affect the planning, design, and operation of the engineering facilities needed to ensure effective control of environmental quality. Faculty members concentrate their efforts on contaminant fate in the environment as well as the various aspects of water quality control engineering, with emphasis on water and wastewater treatment, the disposal of treated effluents in water and on land, the treatment of sludges and solid residuals generated by these processes, and the mechanisms of pollutant transformation in nature.

Environmental Fluid Mechanics and Hydrology. The primary objective of the research program is to achieve a better understanding of the physical processes of the motion of water, other fluids, and air in the environment. Equal emphasis is placed on experimental studies (centered in the DeFrees Hydraulics Laboratory) and on the development of engineering solutions and advanced numerical computing methods. Research includes wave hydrodynamics and coastline processes, the interaction of man-made structures or pollutant discharges and the environment, groundwater studies, hydrologic processes, gas exchange at fluid surfaces, and sediment transport.

Environmental Systems Engineering and Water Resource Systems. Research and instruction address development and application of scientific principles, economic theory, and mathematical techniques to the management and planning of public infrastructure and environmental and water resource systems. Within this area students may select degree programs in environmental systems engineering or water resource systems. Research projects include evaluation of engineering projects, groundwater contaminant modeling and remediation optimization, statistical analysis of hydrologic processes, hydropower system optimization, water supply systems management, water quality planning, risk analysis, river basin and groundwater systems planning and operation, ecological system management, sustainable development, and computer graphics–oriented decision support systems.

Geotechnical Engineering. Emphasis is placed on developing an understanding of soil and rock as engineering materials and how they behave as construction materials, supporting media for structures, host media for structures such as tunnels, and structures in themselves such as dams. Research activities may be grouped into the areas of soil and rock behavior; soil/rock-structure interaction; analytical, finite-element, and probabilistic modeling; marine and coastal geotechnique; soil dynamics and geomechanics.

Remote Sensing. Instruction and research opportunities are provided in remote sensing, airphoto interpretation, evaluation of physical environment, digital image processing, and geographic information systems. Cornell pioneered in the development of engineering interpretation of aerial photographs and now maintains an extensive collection of aircraft and satellite images, along with manual and digital image analysis equipment. The scope of research activities encompasses methods for collecting, analyzing, and interpreting remotely sensed data, and applications in engineering, agriculture, oceanography, geology, and planning in major geographic regions of the world.

Structural Engineering and Structural Materials. Advanced study and research are offered in many aspects of structural engineering, especially (but not exclusively) related to infrastructural problems and encompassing the broad areas of structural behavior, analysis, and design; fundamental and engineering properties of structural materials; and computational aspects of structural engineering and mechanics. Current research is focused on finite-element analysis, computer-aided design, earthquake engineering, fracture mechanics, random vibration, retrofit of buildings for seismic effects, composite structures, structural reliability, progressive collapse, behavior of steel members and connections, high-strength concrete, concrete production, concrete construction methods, concrete durability, cold-formed steel construction, active control, dynamics of structures in space, nondestructive testing of materials using transient stress wave propagation, and nonlinear steel-frame and composite-steel/concrete analysis and design.

Transportation Systems Engineering. Research and instruction focus on the planning, design, operation, and management of multimodal transportation systems. Special emphasis is given to transportation of freight. Areas of current research include urban goods movement, risk analysis for transporting hazardous materials, heavy truck regulation, and information support for intermodal operations.

DREXEL UNIVERSITY

School of Environmental Science, Engineering and Policy

Program of Study

The School of Environmental Science, Engineering and Policy (SESEP) offers the M.S. in Environmental Engineering (M.S.E.E.), the M.S. in Environmental Science (M.S.E.S.), and the Ph.D. A minimum of 45 quarter credits is required for the M.S. degree with a thesis option, 48 quarter credits without a thesis. Part-time study is available for the M.S. programs. Course requirements for all doctoral students are determined on an individual basis; students must have a plan of study approved by the program director at the end of their first term of course work.

The SESEP is a special unit within Drexel bringing together faculty and students who traditionally would be in separate and distinct disciplines. The School creates an atmosphere that enables students to broaden their knowledge and enhance their problem-solving abilities. Because of its multidisciplinary nature, the School is able to tailor students' course work and research to meet their personal interests in the following areas of specialization: air pollution, environmental assessment, environmental biotechnology, environmental chemistry, environmental health, hazardous and solid waste, subsurface contaminant hydrology, water resources, and water and wastewater treatment. Curriculums may be further specialized through course work from other departments such as biotechnology and bioscience; chemical, civil, or mechanical engineering; chemistry; management; or economics. All full-time research students are expected to gain hands-on laboratory and field experience in their areas of concentration through participation in ongoing research projects.

Research Facilities

The School maintains its own research facilities, including more than 22,000 square feet of laboratory space. Among the equipment available for student use are two atomic absorption spectrophotometers with graphite furnace and autosampler; a UV-visible spectrophotometer; an ion analyzer with autoburette and automatic titrator; gas chromatographs; liquid chromatographs with UV, fluorescence, and radiolabel detectors; a liquid scintillation counter; a total organic carbon analyzer with sludge/sediment sampler; high-speed centrifuges; an ion chromatograph; ozone generators; and a UV photochemical reactor. Fully equipped research laboratories in microbiology and hydrology are also available. In addition, many SESEP faculty members maintain laboratories in chemistry, physics, biotechnology, and engineering in various locations throughout the University.

Financial Aid

Research and teaching graduate assistantships are available on a very limited basis. However, many of the School's faculty have research grants that support the employment of graduate students. Information on general need-based loan programs can be secured from the Financial Aid Office, Drexel University, Philadelphia, Pennsylvania 19104.

Cost of Study

Tuition for master's programs in the SESEP is $585 per credit hour in 1999–2000. For Ph.D. candidates, tuition is charged at this same rate for credit hours taken in course work, research, or thesis preparation. The general University fee is $125 per term for full-time students and $67 per term for part-time students.

Living and Housing Costs

Accommodations for single students are available in University residence halls. Ample housing is also available in the neighborhood bordering campus. For the nine-month academic year, transportation and living expenses for a single student are estimated at $11,450.

Student Group

Drexel University has a total enrollment of about 9,590 students, including 2,785 at the graduate level. The School maintains a student population of approximately 125 students. Many are employed full-time and attend the School on a part-time basis in order to enter or advance in the environmental field. (Most graduate courses are scheduled in the late afternoon or evening to accommodate part-time students.)

Location

As a part of the University City area of west Philadelphia, Drexel is conveniently located within minutes of downtown Philadelphia, a great cultural, educational, and industrial center. Amtrak's 30th Street Station, a hub for national and local transportation, is located within three blocks of the University.

The University

Founded in 1891, Drexel University is a private, coeducational institution offering undergraduate and graduate programs in arts and sciences, business and administration, design arts, engineering, and information studies. The University features a nationally recognized cooperative education program, which is facilitated by an academic calendar of four terms per year.

Applying

Graduate students may apply with the intention of enrolling in any of Drexel's four terms. (These begin in January, March, June, and September; application deadlines vary accordingly.) Transcripts and letters of recommendation are required. Because most assistantships begin in September, students interested in applying for assistantships must submit their applications by the preceding February 1.

Correspondence and Information

Office of Graduate Admissions
Drexel University
3141 Chestnut Street
Philadelphia, Pennsylvania 19104
Telephone: 215-895-6700
E-mail: admissions-grad@post.drexel.edu

School of Environmental Science, Engineering and Policy
Drexel University
Philadelphia, Pennsylvania 19104
Telephone: 215-895-2266
World Wide Web: http://www.coas.drexel.edu/environ/

Drexel University

THE FACULTY AND THEIR RESEARCH

Arthur Baehr, Associate Research Professor of Civil Engineering; Ph.D., Delaware. Fate and transport of organic contaminants in groundwater.

R. Christopher Barry, Assistant Professor of Environmental Engineering; Ph.D., Iowa. Microbiological and chemical processes in natural systems, solid waste, water and wastewater treatment.

Nicholas P. Cernansky, Frederic O. Hess Chair Professor of Combustion, Mechanical Engineering; Ph.D., Berkeley. Pollution formation and control in combustion systems, combustion and fuels chemistry, incineration of solid and hazardous wastes.

James P. Friend, Robert S. Hanson Professor of Atmospheric Chemistry; Ph.D., Columbia. Chemical mechanisms of aerosol formation, global sulfur cycle.

Michael A. Gealt, Professor and Director, SESEP; Ph.D., Rutgers. Microbial ecology and environmental microbiology; regulation of bacterial genes involved in DNA transfer; effect of environmental factors on bacterial physiology, including gene transfer frequency, replication rates, and capability for biodegradation.

Elihu D. Grossmann, Professor of Chemical Engineering; Ph.D., Pennsylvania. Thermodynamics of mixtures, process analysis–energy conversion, process safety and reliability analysis, process design, fluid mechanics of accelerated flows, environmental process engineering.

Mirat D. Gurol, Professor of Chemical Engineering; Ph.D., North Carolina at Chapel Hill. Chemical and biological oxidation processes for treatment of hazardous chemicals in water, air, and soil; kinetics and mechanisms of chemical oxidation reactions and water chemistry.

Charles N. Haas, L. D. Betz Chair Professor of Environmental Engineering; Ph.D., Illinois at Urbana-Champaign. Water, wastewater and hazardous-waste treatment; risk assessment; environmental data analysis.

A. Philip Handel, Assistant Professor of Food Science; Ph.D., Massachusetts. Food science, especially lipid chemistry; food composition and functionality; valuation and analysis of frying fats and fried foods.

Jack Garvin Kay, Professor of Chemistry; Ph.D., Kansas. Inorganic and physical chemistry: spectroscopy of diatomic molecules, flash heating and kinetic spectroscopy, chemical effects of nuclear transformation, flash photolysis and radiation of solids, annealing and reactions of defects in crystals, solar furnaces, chemistry at surfaces, radon and radon daughters in the atmosphere.

Susan S. Kilham, Professor of Bioscience and Biotechnology; Ph.D., Duke. Aquatic ecology: phytoplankton; physiological ecology, especially of diatoms in freshwater and marine systems; large lakes; food webs; biogeochemistry.

Robert M. Koerner, Harry L. Bowman Professor of Civil Engineering and Director of the Geosynthetic Research Institute; Ph.D., Duke. Geosynthetics, waste containment systems, liners, covers, leachate collection, recycling and disposal.

Carl W. Kreitzberg, Professor of Physics and Atmospheric Science and Director of the Data Processing and Modeling Center; Ph.D., Washington (Seattle). Atmospheric science: mesoscale computational weather prediction, dynamical data assimilation, air-quality modeling on scales of 10 km to 1,000 km.

Michael P. O'Connor, Assistant Professor of Bioscience and Biotechnology; M.D., Johns Hopkins; Ph.D., Colorado State. Biophysical ecology, physiological ecology, thermoregulation of vertebrates, ecological modeling.

Wesley O. Pipes, Professor of Civil Engineering; Ph.D., Northwestern. Microbiology of water distribution systems, aerobic biological wastewater treatment, sludge treatment and disposal.

Stanley Segall, Professor of Nutrition and Food Science; Ph.D., MIT. Food science, metabolic and nutritional chemistry, food preservation processes, nutritional impact of food processing.

James R. Spotila, L. D. Betz Chair Professor of Environmental Science; Ph.D., Arkansas. Physiological and biophysical ecology, thermoregulation of aquatic vertebrates, biology of sea turtles.

J. Richard Weggel, Professor of Civil Engineering; Ph.D., Illinois at Urbana-Champaign. Ocean and coastal engineering, surface-water hydrology, hydraulic engineering.

Claire Welty, Associate Professor of Civil and Environmental Engineering and Associate Director, SESEP; Ph.D., MIT. Groundwater flow and pollutant transportation, water resources and hydrology.

DUKE UNIVERSITY

Department of Civil and Environmental Engineering

Programs of Study	The Department of Civil and Environmental Engineering offers graduate study leading to the degrees of Master of Science and Doctor of Philosophy. Programs of advanced course work and research are available in five major areas: engineering mechanics, environmental engineering, geomechanics and geophysics, structural engineering, and water resources. Interdisciplinary programs combining other areas of engineering, including environmental science; mathematics; the biological, chemical, and physical sciences; computational science; and business administration are also available. Under the Reciprocal Agreement with Neighboring Universities, a student may include some courses offered at the University of North Carolina at Chapel Hill and North Carolina State University in Raleigh. The faculty members and their research interests are specified on the reverse of this page.
Research Facilities	Research and teaching facilities in engineering mechanics, structural engineering, fracture mechanics, and geomechanics include four independent closed-loop electrohydraulic dynamic loading systems (MTS) with a frequency range up to 100 Hz and load capacities to 220,000 pounds. A 6,000-pound actuator can develop a constant crosshead speed up to 50,000 inches per minute. Equipment is available for fabricating specimens and testing fiber-reinforced polymer composites. An environmental chamber has a temperature range of -100°F to +350°F; equipment for spectral and modal dynamic analysis is available.
	Research and teaching facilities in environmental engineering include wet and dry laboratories equipped to study a range of physical, chemical, and biological processes. Calorimetry for the measurement of heat values of secondary fuels and air classifiers interfaced with computer monitors, as well as indoor and outdoor water resources monitoring devices, including flumes, venturi meters, and digital computation hardware, are available. The biotechnology and physical-chemical laboratories are equipped with autoclaves, a media preparation room, walk-in environmental rooms, numerous fume hoods, a biohazard containment facility for cultivation of genetically engineered microorganisms, fully instrumented bioreactors with online control, and various analytical instrumentation, including instrumentation for liquid scintillation counting; autoradiography; atomic absorption spectroscopy; and total carbon analysis to ppb levels; gas chromatographs equipped with ECO, FID, and TCD detectors; HPLCs; computer-assisted image analysis microscopes; and a recently acquired Fourier-transform infrared spectrometer facility.
	Extensive computer resources networked with the School of Engineering's Ethernet backbone are easily accessible from within the department and throughout the campus. Depending on the specific application, students can investigate problems in computational fluid mechanics, structural dynamics, and mathematical optimization, as well as in transportation and environmental systems engineering. If additional computing capabilities are needed, access to the Microelectronics Center of North Carolina's CRAY Y-MP vector processing supercomputer is available. Numerous software packages are available to students through the existing Computational Resource Center.
Financial Aid	Financial support is available to a select number of students. Fellowships valued at up to $33,000 for the calendar year and research and teaching assistantships of $30,000 for the academic year are awarded on an annual basis and can be maintained until a student receives a degree. Summer research assistantship appointments are also available. Fellowships and assistantships typically include all tuition and fees regularly required of graduate students in addition to monthly stipends.
Cost of Study	For 1999–2000, tuition is $8760 per semester (12 units at $730 per unit). In addition to tuition, a registration fee of $1250 and a health fee of $240 are required each semester.
Living and Housing Costs	Duke University has two residential apartment facilities available through an application process. These apartments are available for continuous occupancy throughout the calendar year. All of the apartments are completely furnished. The fee for Town House Apartments, including utilities, is $2700 per occupant for the fall and spring semester, on the basis of 2 students to a two-bedroom apartment. Rates in Central Campus Apartments range from $2300 per occupant for 3 students in a three-bedroom apartment to $3463 for an efficiency apartment. Several large apartment complexes are within walking distance of the campus.
Student Group	There are approximately 236 graduate students enrolled in the School of Engineering; of whom 34 are in the Department of Civil and Environmental Engineering.
Location	Located in the rolling central Piedmont area of North Carolina, the campus of Duke University is widely regarded as one of the most beautiful in the country. The four-season climate is mild, but good winter skiing is available in the North Carolina mountains a few hours drive to the west, and ocean recreation is a similar drive to the east. Duke is readily accessible by interstate highways 85 and 40 and from Raleigh-Durham International Airport, which is about a 20-minute drive from the campus via Interstate 40 and the Durham Expressway.
The University	Trinity College, founded in 1859, was selected by James B. Duke as the major recipient of a 1924 endowment that enabled a university to be organized around the college and to be named for Washington B. Duke, the family patriarch. A Department of Engineering was established at Trinity in 1910, and, following the establishment of Duke University, the Departments of Civil and of Electrical Engineering were formed in 1927. Duke remains a privately supported university, with more than 11,000 students enrolled in degree programs.
Applying	Admission to the department is based on a review of previous education and experience, campus or telephone interviews, the applicant's statement of intent, letters of evaluation, standardized test scores (GRE and TOEFL), and grade point average. Applications for admission may be submitted at any time. Applications for financial awards, however, should generally be submitted along with all supporting documents before December 31. While offers of financial support may be made at any time, most are awarded in the winter and early spring for the following academic year.
Correspondence and Information	Director of Graduate Studies Department of Civil and Environmental Engineering School of Engineering Box 90287 Duke University Durham, North Carolina 27708-0287 Telephone: 919-660-5200 E-mail: dgs@cee.egr.duke.edu

Duke University

THE FACULTY AND THEIR RESEARCH

The department's graduate programs require active participation in research. Current areas of interest include adaptive structures; real-time analysis and optimization of servo-elastic systems; fracture mechanics and failure analysis; constitutive properties, failure criteria, and damage models of modern construction materials; structural stability; modal techniques for evaluation of structural and material integrity; computer-aided design and optimization of structures; theory of plasticity; geomechanics of nuclear-waste isolation in sedimentary rocks; thermoplasticity of clays and deep rocks; chemo-plasticity of soils subjected to environmental loadings; physical and chemical processes of water quality control; resource recovery and solid-waste management; wastewater treatment process: design and optimization; wastewater solids management; groundwater simulation and control; pollutant fate and transport; hazardous-waste management; environmental and engineering geophysics; transportation planning and optimization; and travel behavior analysis.

John F. Ahearne, Adjunct Professor; Ph.D., Princeton, 1966. Nuclear reactor safety, radioactive waste disposal, risk analysis.
Dianne Ahmann, Assistant Professor; Ph.D., MIT, 1996. Microbial ecology, biogeochemical cycling, bioremediation of inorganic contaminants.
* Fred K. Boadu, Assistant Professor; Ph.D., Georgia Tech, 1994. Environmental and engineering geophysics fractal theory, wave propagation in poroelastic media, neural network applications, environment and engineering properties of soils from geophysical measurements.
* Henri P. Gavin, Assistant Professor; Ph.D., Michigan, 1994. Earthquake hazard mitigation, vibration control, controllable materials, semi-active control devices.
Peter K. Haff, Professor; Ph.D., Virginia, 1970. Development of new modeling tools for computer simulation of complex nonlinear mechanical systems, sediment transport.
* Tomasz A. Hueckel, Associate Professor; Ph.D., Polish Academy of Sciences, 1974; Sc.D., Grenoble Polytechnique, 1985. Theoretical soil and rock mechanics, theory of plasticity, environmental mechanics.
* Zbigniew J. Kabala, Associate Professor; Ph.D., Princeton, 1988. Deterministic and stochastic water and contaminant transport in heterogeneous porous media, theory of related measurements.
Gabriel G. Katul, Assistant Professor; Ph.D., California, Davis, 1993. Coupled mass, energy, and momentum exchange processes near the land-atmosphere interface: theory, models, and measurement.
* Tod A. Laursen, Associate Professor; Ph.D., Stanford, 1992. Structural and solid mechanics, inelastic material modeling, large deformation kinematics, finite-element concepts.
Peter E. Malin, Associate Professor; Ph.D., Princeton, 1978. Seismic wave propagation in planetary crusts, earthquake processes.
Ross E. McKinney Sr., Adjunct Professor; Sc.D., MIT, 1951. Environmental engineering, biological wastewater treatment systems design and operations, and solid wastes.
* Miguel A. Medina, Professor; Ph.D., Florida, 1976. Water resources, hydrologic and water quality mathematical modeling, integration of surface and groundwater contaminant transport models within decision-analysis frameworks for risk assessment.
* Joseph C. Nadeau, Assistant Professor; Ph.D., Berkeley, 1996. Applied mechanics, mechanics of composite and microstructured materials.
* J. Jeffrey Peirce, Associate Professor; Ph.D., Wisconsin–Madison, 1977. Environmental engineering; physical, chemical, and microbiological aspects of particle-fluid interactions; nitrogen flux from soil to air; hazardous waste processing and long-term storage.
* Henry Petroski, Aleksandar S. Vesic Professor and Chairman; Ph.D., Illinois at Urbana-Champaign, 1968. Failure analysis, design theory, case histories.
Warren T. Piver, Adjunct Associate Professor; Ph.D., North Carolina State, 1972. Transport and transformation of contaminant chemicals in subsurface, waste site remediation and risk assessment.
Kenneth H. Reckhow, Professor; Ph.D., Harvard, 1977. Water quality simulation, monitoring design and trend detection, environmental decision making under uncertainty.
Stuart Rojstaczer, Associate Professor; Ph.D., Stanford, 1988. Hydrology of natural and human-influenced groundwater flow systems, engineering geology with emphasis on failure in the shallow subsurface.
Ralph R. Rumer, Visiting Professor; Sc.D., MIT, 1962. Environmental fluid mechanics, hydraulic engineering, hydrology, and water resources.
* Senol Utku, Professor; Sc.D., MIT, 1960. Adaptive structures and their control, structural and continuum mechanics, computational analysis.
Dharni Vasudevan, Assistant Professor; Ph.D., Johns Hopkins, 1996. Environmental chemistry, fate of organic contaminants, chemistry at the solid/water interface.
* P. Aarne Vesilind, Professor; Ph.D., North Carolina at Chapel Hill, 1968. Environmental engineering, resource recovery, dewatering of industrial and wastewater sludges, management of municipal solid waste, professional ethics.
Lawrence N. Virgin, Associate Professor; Ph.D., London, 1986. Nonlinear dynamics and chaos, buckling, ship capsize, aeroelastic flutter, experimental and theoretical mechanics.

RECENT PUBLICATIONS

Ahearne, J.F. How safe is safe enough? *Reliability Engineering and Systems Safety* 62:5-7, 1998.
Ahmann, D., et al. Microbial mobilization of arsenic from sediments of the Aberjona Watershed. *Environ. Sci. Tech.* 31:2923–30, 1997.
Boadu, F. K. Inversion of fracture density from field seismic velocities using artificial neural networks. *Geophysics* 63:534-45, 1998.
Gavin, H. P., R. D. Hanson, and F. E. Filisko. Electrorheological devices I: Analysis and design. *J. Appl. Mech.* 63(3):669–82, 1996.
Haff, P. K. Discrete mechanics. In *Granular Mechanics*, pp. 141–60, ed. A. Mehta. Springer-Verlag, 1994.
Kaczmarek, M. and **T. Hueckel** Chemomechanical consolidation of clays: Analytical solutions for a linearized one-dimensional problem. *Transport in Porous Media* 32(1)49-74, 1998.
Kabala, Z. J., and G. Sposito. Statistical moments of reactive solute concentration in a heterogeneous aquifer. *Water Resources Res.* 30:759–68, 1994.
Katul, G.G. and C. R. Chu. A theoretical and experimental investigation of energy-containing scales in the dynamic sublayer of boundary-layer flows. *Boundary Layer Meteorology* 86:279-312, 1998.
Laursen, T. A., and J. C. Simo. A continuum-based finite element formulation for the implicit solution of multibody, large deformation frictional contact problems. *Int. J. Numer. Methods Eng.* 36:3451–85, 1993.
Malin, P. E., and M. G. Alvarez. Stress diffusion along the San Andreas fault at Parkfield, CA. *Science* 256:1005–7, 1992.
McKinney, R. E., Sr. Wastewater treatment, municipal. In *Encyclopedia of Microbiology*, pp. 363–75, ed. Joshua Lederberg. San Diego, California: Academic Press, 1992.
Cassiani, G. and **M. A. Medina, Jr.** Incorporating auxillary geophysical data into groundwater flow parameter estimation. *Ground Water* 35(1):79-91, 1997.
Nadeau, J. C., and M. Ferrari. Invarient tensor-to-matrix mappings for evaluation of tensorial expressions. *J. Elasticity*, 1999.
Yum, K. J., and **J. J. Peirce.** Biodegration of cholorophenols in immobilized-cell reactors using white-rot fungus on wood chips. *Water Environ. Res.* 70(2):121–9, 1998.
Petroski, H. Case histories and the study of structural failures. *Struct. Eng. Int.* 5:250–5, 1995.
Piver, W. T., and F. T. Lindstrom. Numerical methods for describing chemical transport in the unsaturated zone of the subsurface. *J. Contam. Hydrol.* 8:243–62, 1991.
Reckhow, K. H. Water quality simulation modeling and uncertainty analysis for risk assessment and decision making. *Ecol. Model.* 72:1–20, 1994.
Rojstaczer, S. Gone for Good: Tales of University Life After the Golden Age. New York: Oxford University Press, 1999.
Atkinson, J. F., J. V. DePinto, and **R. Rumer.** Linking hydrodynamic and water quality models with different scales. *J. Environ. Eng.* 5:399-407, 1998.
Utku, S. Theory of adaptive structures. CRC Press, 1998.
Vasudevan, D. and A. T. Stone. Absorption of 4-Nitrocatechol, 4-Niro-2-Aminophenol, and 4-Nitro-1, 2-phenylenediamine at the metal (hydr)oxide/water interface: Effect of metal (hydr)oxide properties. *J. Colloid and Interface Science* 202:1-19, 1998.
Vesilind, P. A., and T. Ramsey. Effect of temperature on the fuel value of wastewater sludge. *Waste Manage. Res.* 12:9, 1995.
Murphy, K. D., **L. N. Virgin,** and S. A. Rizzi. Characterizing the response of a thermally loaded, acoustically excited plate. *J. Sound Vib.* 196:635–58, 1996.

---

* Denotes faculty members with primary appointments in the Department of Civil and Environmental Engineering.

FLORIDA A&M UNIVERSITY–
FLORIDA STATE UNIVERSITY

College of Engineering
Department of Civil Engineering

Program of Study

The Florida A&M University–Florida State University (FAMU–FSU) College of Engineering is a joint program between two universities within the State University System of Florida. The Department of Civil Engineering offers Master of Science (M.S.) and Doctor of Philosophy (Ph.D.) programs in civil and environmental engineering with concentrations in structural, geotechnical, transportation, water resources, and environmental engineering. Special areas of emphasis include, but are not limited to, bridge design and rehabilitation, coastal construction and storm effects, computer-aided design, decision support systems, geotextiles, hazardous and solid wastes, transportation networks and multimodal systems, pavements, soil dynamics, structural stability, and storm water. The department currently has thesis and nonthesis M.S. degree programs. The thesis option is strongly recommended due to its research contribution to the graduate student experience.

Research Facilities

Located in Innovation Park and equidistant between the two campuses of FAMU and FSU, the College of Engineering building features classroom and research facilities. Specific laboratories for the Department of Civil Engineering are geotechnical, environmental, hydraulics, pavement, construction materials, and structures. Geotechnical laboratory facilities include equipment for soil classification, compaction, hydraulic conductivity, slurry evaluation, shear strength, and compressibility of soils. Electronic data acquisition systems, sampling devices, and a machine shop are also available for student use. The environmental laboratories include equipment and instrumentation needed for physical and chemical analysis of water quality, sampling and filtering devices, and space for bench-scale experiments. The hydraulics laboratory is used to reinforce the concepts of hydraulic equipment and instrumentation and the basics of data collection and analysis. Students can perform experiments of hydrostatic pressure, hydrostatic forces on submerged bodies, flow measurement, friction in pipe flow, pump power, open channel flow, hydraulic jump, and wave mechanics. Pavement laboratory facilities include equipment for resilient modulus characterization of highway materials (such as an MTS Load System, a TestStar Control Unit, a Triaxial Testing System, and a compaction set). Electronic data acquisition systems, personal computers, and pavement engineering software systems are available. Construction materials laboratory facilities include equipment for compression strength testing, a concrete mixer, an MTS shock tester, an L.A. abrasion test machine, and an MTS test system. The structures laboratory, which is more than two stories high with a reinforced concrete reaction slab, is equipped with 100 kips anchorage pods spaced at four-foot intervals. When fully equipped, this facility will provide students with applied instruction on specialized testing of materials and structures, support for research in developing new construction materials, applications for existing materials, and innovative structural systems.

Students have access to a large number and variety of computer systems. A network of nearly 700 computing devices is available for the academic and research efforts of the College. The College computers are connected to a high-speed, switched, fiber-optic LAN and to the Internet via the FSU connection to the NSFvBNS network. Desktop computers are supported by a cluster of Sun, DEC, and SGI servers. Other nearby resources include the Supercomputer Computations Research Institute (SCRI), FSU Academic Computing and NetworkServices (ACNS), and FAMU Computing Services.

Financial Aid

Students may be supported through research or teaching assistantships on a competitive basis. Most civil engineering graduate students currently hold half-time assistantships equivalent to 20 hours per week. Inquiries about research assistantships should be made to the professor directing individual research projects. Students can visit the departmental Web site to learn more about individual faculty research at http://www.eng.fsu.edu. The department chairman should be contacted about prospects of teaching assistantships. For other financial and scholarship opportunities, students can contact FAMU (telephone: 850-599-3730; e-mail: aid@ famu.edu) or FSU (telephone: 850-644-5716; e-mail: finaid @admin.fsu.edu).

Cost of Study

The 1998–99 cost for Florida residents was $131.07 per credit hour, while non–Florida residents paid $436.46 per credit hour for graduate courses. The non-Florida portion of the tuition fees is waived for fellowship holders, and some portion is waived for graduate teaching and research assistants. The current estimated annual cost (based on two 12-semester-hour terms), including tuition, fees, housing, food, books, and supplies, is $10,660 for Florida residents and $18,905 for out-of-state students.

Living and Housing Costs

University-owned apartments are available to single and married students and students with children.

Student Group

During the 1998–99 academic year, the Department of Civil Engineering had an enrollment of 50 at the graduate level; members of minority groups and women comprised 54 percent of students.

Location

Located in Tallahassee, the state capital of Florida, the metropolitan area enjoys a stable economy, comparatively low unemployment rates, and modest living expenses and is close to beautiful beaches.

The Universities

Florida A&M University and Florida State University are located in Tallahassee, the state capital. Although Tallahassee is among the state's fastest-growing cities, its natural beauty has been preserved. Five large lakes surrounding Tallahassee and the nearby Gulf of Mexico offer numerous recreational opportunities, including canoeing, fishing, waterskiing, boating, camping, hunting, and cycling. Among Florida cities, Tallahassee is distinguished by its gently rolling hills and roads canopied by majestic oaks.

Applying

Students enroll in the college through either FAMU or FSU. Admission requirements for the M.S. program include a baccalaureate degree in civil engineering (or an allied academic discipline) from an accredited college or university, good standing in the academic institution last attended, a score of 1000 on the GRE, and a GPA of 3.0 on a 4.0 scale as an upper-division student. International students must have a degree from a recognized academic institution and have a score of 550 (paper-based) or 213 (computer-based) on the TOEFL.

Admission requirements for the Ph.D. degree are similar to the M.S. requirements, except that applicants must have a prior M.S. degree in civil or environmental engineering (or a closely related field) and a score of 1100 on the GRE.

Correspondence and Information

Graduate Admissions
Florida A & M University
Foote-Hilyer Administration Building
Tallahassee, Florida 32307

Graduate Admissions
Florida State University
A2500 University Center
Tallahassee, Florida 32306-2400

Florida A&M University–Florida State University

THE FACULTY AND THEIR RESEARCH

Makola Abdullah, Assistant Professor; Ph.D., Northwestern, 1994. Structural engineering, structural dynamics, active control.

Andrew A. Dzurik, Professor; Ph.D., Cornell, 1969; PE. Water resources planning and management, stormwater runoff, constructed wetlands for water treatment.

Gary J. Foose, Assistant Professor; Ph.D., Wisconsin, 1997. Environmental geotechnics, geotechnical engineering, performance of waste containment systems, groundwater remediation systems, beneficial reuse of waste materials, flow and contaminant transport in porous media.

Millard W. Hall, Professor; Ph.D., Illinois, 1968. Water quality, water resources policy, environmental engineering.

Wenrui Huang, Assistant Professor; Ph.D., Rhode Island, 1993. Large-scale natural flow systems, modeling pollutant transport, coastal engineering analysis, time-series data analysis.

Danuta Leszczynska, Associate Professor; Ph.D., Technical University (Poland), 1978. Environmental engineering and chemistry, water pollution control issues, removing organic pollutants from water matrix.

Primus V. Mtenga, Assistant Professor; Ph.D., Wisconsin–Madison, 1991; PE. Structural systems: behavior modeling and analysis, wood and wood-based structural components and structures, nondestructive evaluation (NDE) of structures, biocomposites, structural mechanics.

Renatus Mussa, Assistant Professor; Ph.D., Arizona State, 1996; PE. Incident detection and management systems, traffic operations and control, highway safety analysis and remedial measures, intelligent transportation systems (ITS).

Soronnadi Nnaji, Professor; Ph.D., Arizona, 1981. Water resources systems, hydrology, hydraulics, engineering decision support systems.

Wei-Chou Virgil Ping, Associate Professor; Ph.D., Texas at Austin, 1989; PE. Transportation design and materials, pavement design and management, geotechnical engineering.

John Sobanjo, Assistant Professor; Ph.D., Texas A&M, 1991; PE. Transportation engineering, infrastructure engineering and management, construction engineering, computer applications.

Lisa Spainhour, Assistant Professor; Ph.D., North Carolina State, 1994. Computer applications in civil engineering, engineering data management, computer-aided analysis and design, composite materials, structural analysis.

Kamal Tawfiq, Associate Professor and Associate Chairman; Ph.D., Maryland, 1987; PE. Geotechnical engineering, soil-structure interaction, dynamic/nondestructive testing, numerical modeling, geotextiles.

Jerry Wekezer, Professor and Chairman; Ph.D., Technical Gdansk, (Poland), 1974; PE. Solid mechanics, finite elements, structural analysis, transportation safety.

Nur Yazdani, Professor; Ph.D., Maryland, 1984; PE. Prestressing systems for highway bridges, general bridge design and rehabilitation, timber bridge design, coastal construction and building codes, wind effect on structures.

FLORIDA INSTITUTE OF TECHNOLOGY

College of Engineering
Graduate Programs in Civil Engineering

Programs of Study

Florida Institute of Technology offers programs of study leading to the Master of Science and Doctor of Philosophy in civil engineering. These programs are designed to provide opportunities for students' development of professional engineering competence and scholarly achievement.

The Master of Science in civil engineering allows for specialization in construction, geotechnical, structures, and water resources. The degree is conferred upon students who have successfully completed a minimum of 30 credit hours in either a thesis or a nonthesis program consisting of required and elective course work. Students in the thesis program must successfully defend their thesis, while students in the nonthesis program are required to pass a comprehensive examination.

The Doctor of Philosophy degree program in civil engineering is offered for those students who wish to conduct advanced research in environmental/water resources and structural/geotechnical engineering. The program consists of advanced studies to prepare the student for scientific engineering research and completion of a research project that leads to a significant contribution to the knowledge of a particular problem. A minimum of 24 credit hours of course work and 24 credit hours of dissertation beyond a master's degree are required.

Research Facilities

Laboratories for research and instructional activities are available in the areas of materials, structures, soil mechanics, solid waste, water quality, and interactive graphics. Other campus laboratories may be used by students conducting graduate research. Analytical capabilities are extended by means of cooperative projects with the academic programs in biology, chemical engineering, chemistry, ocean engineering, and oceanography and with the Life Sciences Research Complex.

Financial Aid

Graduate teaching and research assistantships are available to qualified students. For 1999–2000, typical stipends range upward from $7800 for twelve months for approximately half-time duties. Teaching assistants who are required to be in the thesis option are granted partial or full tuition remission.

Cost of Study

For 1999–2000, tuition is $575 per semester credit hour for all students. As noted above, tuition is waived for some graduate assistants.

Living and Housing Costs

Room and board on campus cost approximately $2200 per semester in 1999–2000. On-campus housing (dormitories and apartments) is available for full-time single and married graduate students, but priority for dormitory rooms is given to undergraduate students. Many apartment complexes and rental houses are available near the campus.

Student Group

The College of Engineering has 878 undergraduate and 492 graduate students. Approximately 20 students are enrolled in graduate programs in civil engineering.

Graduates of these programs have found careers with industry, consulting engineering firms, and state, regional, and local agencies, such as water management districts and highway and environmental departments. About 25 percent of graduates have continued on to the Ph.D. degree.

Student Outcomes

Graduates of the Graduate Programs in Civil Engineering are employed by numerous counties including Brevard, Collier, and Orange, and by state agencies such as the U.S. Army Corps of Engineers, the Florida Department of Environmental Protection, and the Florida Department of Transportation. Graduates also find employment with numerous consulting firms such as Camp, Presser & McKee, Inc.; CHZM Hill; Berryman & Henigan; Agnoli, Barber & Brundage, Inc.; GeoSyntec Consultants; and Technical Solutions.

Location

The campus is located in Melbourne, on Florida's east coast. It is an area, located 3 miles from Atlantic Ocean beaches, with a year-round subtropical climate. The area's economy is supported by a well-balanced mix of industries in electronics, aviation, light manufacturing, optics, communications, agriculture, and tourism. Many industries support activities at the Kennedy Space Center.

The Institute

Florida Institute of Technology, founded in 1958, has developed into a distinctive independent university that provides undergraduate and graduate education in engineering and sciences for students from throughout the United States and many other countries. Florida Tech is supported by local industry and is the recipient of many research grants and contracts.

Applying

Applicants for graduate study in civil engineering should have an undergraduate degree in civil engineering. An applicant whose degree is in another field of engineering or in the applied sciences will be reviewed; however, undergraduate course work may be required prior to starting the Master of Science program.

Forms and instructions for applying for admission and assistantships are sent upon request. Applicants are asked to submit two letters of recommendation from academic references and a statement of purpose giving their reason for graduate study. Although the GRE is not required, it is considered for students with marginal undergraduate academic performance. International students applying for assistantships must have a TOEFL score greater than 600. International students without English proficiency and with a TOEFL score less than 550 are required to enroll in English studies before beginning their graduate studies. Separate application for an assistantship should be made on forms available from the Graduate School.

Correspondence and Information

Graduate Admissions Office
Florida Institute of Technology
150 West University Boulevard
Melbourne, Florida 32901
Telephone: 407-674-8027
       800-944-4348 (toll-free)
Fax: 407-723-9468

Dr. Ashok Pandit, Civil Engineering
Florida Institute of Technology
150 West University Boulevard
Melbourne, Florida 32901
Telephone: 407-674-8048
Fax: 407-674-7565
E-mail: pandit@fit.edu

Florida Institute of Technology

THE FACULTY AND THEIR RESEARCH

Paul J. Cosentino, Associate Professor; Ph.D., Texas A&M; PE. Pavement design and evaluation; transportation planning; containment of hazardous wastes; geotechnical engineering, with emphasis on in situ testing of soils, instrumentation evaluation of pavements, and slope stability. Current funded research projects are the application of fiber-optic sensors for traffic monitoring, the development of a fiber-optic weigh-in-motion sensor, and the use of reclaimed asphalt pavement as a highway fill material.

Howell H. Heck, Associate Professor; Ph.D., Arkansas; PE. Solid-wastes management, degradable materials, determining the ultimate fate of chemicals in disposal facilities, volatile chemical emissions from treatment facilities, movement of chemicals across the interface boundaries of land, air, and water. Current funded research projects are the anaerobic decomposition of waste materials, the effects of phosphogypsum on the biological degradation process in a landfill, and the measurement of groundwater exchange with surface water.

Edward H. Kalajian, Professor; Ph.D., Massachusetts; PE. Geotechnical engineering, foundations, stabilization of waste materials for beneficial uses. Current funded research projects are the use of reclaimed asphalt pavement as a highway material.

Ashok Pandit, Professor; Ph.D., Clemson; PE. Groundwater hydraulics and hydrology, numerical groundwater modeling, stormwater management. Current research projects are the estimation of removal efficiency of sediment traps using experimental procedures, modeling of saltwater intrusion in aquifers, groundwater mounding below retention/detention basins, assessment of water quality from citrus groves, continuous simulation runoff models, and rainfall/runoff relationships.

Jean-Paul Pinelli, Assistant Professor; Ph.D., Georgia Tech; PE. Structural dynamics, earthquake, and wind engineering; modeling and optimization of nonlinear mechanical systems; computer-aided design techniques in structural engineering. Current research projects are the redesign of break-away highway-sign connections subject to windloading, the effect of tornadic and hurricane winds and structures, the use of energy dissipative devices to dampen vibrations, and the use of structural physical models as teaching aids.

Jack W. Schwalbe, Associate Professor; M.S., Connecticut; PE. Structural analysis and design, math modeling techniques, structural dynamics, experimental stress analysis.

THE GEORGE WASHINGTON UNIVERSITY

School of Engineering and Applied Science
Department of Civil and Environmental Engineering

Programs of Study	Programs are offered leading to the Master of Science degree (M.S.), the professional degrees of Applied Scientist (App.Sc.) and Engineer (Engr.), and the Doctor of Science degree (D.Sc.). In the field of civil and environmental engineering, areas of concentration include engineering mechanics, environmental engineering, structural engineering, geotechnical engineering, and water resources engineering. Each student's program is individually planned with the adviser. The M.S. programs require satisfactory completion of 30 semester hours of graduate-level courses with a thesis (equivalent to 6 semester hours), or 33 semester hours without a thesis. Students who select the thesis option are required to defend their thesis. Professional degree programs consist of a minimum of 30 semester hours of approved graduate courses beyond a master's degree and a design project. The doctoral program prepares students for careers of creative scholarship. The program requires a minimum of 30 semester hours of formal study beyond the master's degree or a minimum of 54 credit hours of approved graduate work for students whose highest earned degree is a baccalaureate, prior to the qualifying examination. The research phase culminates in the presentation and oral defense of a dissertation. Work on the dissertation is equivalent to a minimum of 24 semester hours.
Research Facilities	The department has a wide variety of laboratory facilities that include a fully instrumented low-speed wind tunnel, an open-channel water flume, a water pollution laboratory, and a soil and foundation engineering laboratory. There are three MTS random-cycle testing systems and other materials-testing machines; a scanning electron microscope; a rooftop solar energy facility; and laboratories for materials science and engineering, thin-film development, surface analysis, fluid mechanics and propulsion, and combustion diagnostics. The School provides UNIX, PC, and Macintosh resources, featuring a Sun Microsystems S1000 multiprocessor server and Ultra 3D workstations, Dell Pentium Pro II computers, and Apple Power Macintosh G3 computers. Systems are equipped with licensed software from leading manufacturers, such as CADENCE, MathWorks, Synopsys, AdaCore Technologies, Sun Microsystems, Autodesk, Asymetrix, Adobe, and Microsoft. In addition, the University provides extensive computing resources with a Sun Microsystems SPARC Center 2000E for academic information system services and a Sun Microsystems Enterprise 4000, with four processors, dedicated to research computing support. A high-performance ATM network integrates these resources, and Internet connectivity is provided for research support. The University libraries contain more than 1.7 million volumes, and students have access to the Library of Congress and the libraries of six other local universities.
Financial Aid	Teaching assistantships provide remission of tuition for 9 hours per semester and $1000 to $3500 for each course taught in 1999–2000. Research assistants receive $8000 to $15,000 for the calendar year. School Graduate Fellow, Dean's Fellow, and Department Fellow awards range from $7500 to $15,000 for full-time students. Full-time students who are U.S. citizens or permanent residents may be eligible for Graduate Engineering Honors Fellowships.
Cost of Study	In 1998–99, tuition was charged at the rate of $680 per semester hour, payable on a course-by-course basis.
Living and Housing Costs	Apartments for students attending the George Washington University are available in the area at a wide range of costs, starting at about $600 a month.
Student Group	Students in the School of Engineering and Applied Science include graduates of most colleges and universities in the United States and a number of other countries. Approximately 1,000 students are working on master's degrees, 20 on professional degrees, and 440 on the D.Sc. The department has approximately 40 master's degree students and 24 D.Sc. students.
Location	The Washington, D.C., area has the second-largest concentration of research and development activity in the United States. The department offers programs at both the main campus in the Foggy Bottom historic district and in the branch campuses in Virginia and Maryland.
The University and The School	The School was organized in 1884. It operates on a two-semester academic year. Limited course work in engineering, engineering administration, operations research, physical science, mathematics, economics, and statistics is available during the University's summer sessions.
Applying	Admission to a master's degree program generally requires an appropriate bachelor's degree from a recognized institution and evidence of a capacity for productive work in the field selected. Admission to professional degree programs requires an appropriate master's degree from a recognized institution and evidence of capacity for productive work in the field selected, as indicated by prior scholarship and/or professional experience. Applicants for doctoral study must have adequate preparation for advanced study, including demonstrated capacity for original scholarship and, if the highest degree earned is the baccalaureate, course work pertinent to the field, or else a master's degree or the equivalent in engineering or a related field. March 1 and October 1 are the respective fall and spring priority application deadlines.

Correspondence and Information

Professor Theodore G. Toridis, Chair
Department of Civil and Environmental Engineering
School of Engineering and Applied Science
The George Washington University
Washington, D.C. 20052
Telephone: 202-994-6749
Fax: 202-994-0238
E-mail: cmee@seas.gwu.edu

Interim Dean Thomas A. Mazzuchi
School of Engineering and Applied Science
The George Washington University
Washington, D.C. 20052
Telephone: 202-994-3096
    800-537-7327 (toll-free)
Fax: 202-994-4522

The George Washington University

THE FACULTY AND THEIR RESEARCH

Nabih E. Bedewi, Associate Professor of Engineering and Applied Science; Ph.D., Maryland, 1986. Vehicle and highway safety engineering, payload/spacecraft dynamic launch load interaction analysis, stability, stability analysis of a time-varying contact model.

Muhammad I. Haque, Professor of Engineering and Applied Science; Ph.D., Colorado State, 1973. Finite-element methods, computational methods in fluid and solid mechanics, hydraulics.

Khalid Mahmood, Professor of Engineering; Ph.D., Colorado State, 1971. Hydraulics, water resources, river mechanics.

Majid T. Manzari, Assistant Professor of Civil Engineering; Ph.D., California, Davis, 1994. Soil mechanics, physical and numerical modeling of soil structures, constitutive modeling, large deformation in porous media.

Rumana Riffat, Assistant Professor of Civil and Environmental Engineering; Ph.D., Iowa State, 1994. Anaerobic biotechnology for wastewater treatment, process kinetics and microbiology.

Irving H. Shames, Professor of Engineering and Applied Science; Ph.D., Maryland, 1953. Applied mechanics, aerospace and nuclear engineering.

Theodore G. Toridis, Professor of Engineering and Applied Science and Chairman of the Department; Ph.D., Michigan State, 1964. Structural mechanics, analysis of structures subjected to earthquake excitation, numerical analysis and computer-aided design.

The George Washington University is an equal opportunity/affirmative action institution.

HOWARD UNIVERSITY

College of Engineering, Architecture, and Computer Sciences
Department of Civil Engineering

Program of Study	The Department of Civil Engineering offers study leading to the degree of Master of Engineering. The degree program requires the completion of 31 credit hours, which may include 6 credit hours for the thesis, a final oral examination, or a written comprehensive examination if no thesis is written. Major areas of study are environmental and water resources engineering, geotechnical engineering, structural engineering and mechanics, and transportation systems engineering. Current areas of research include mutagenicity studies of nonionic detergents; incineration and biodegradation of hazardous wastes; beneficial aspects of wastewater sludges; effects of cyclic temperature variation on soil permeability; mechanical behavior of large, repetitive lattice structures; completion risks analysis in construction projects; turbulent open-channel flow studies; and transportation subsidization trends.
Research Facilities	In the Department of Civil Engineering there are currently five major environmental laboratories and an air sampling station. These laboratories contain the analytical instrumentation that supports the teaching and research efforts for both air- and water-related areas. The laboratory facilities also include the basic equipment needed for research in the areas of environmental microbiology and toxicology and the apparatus necessary for performing the classical analytical techniques. Of particular interest is a laboratory-scale incinerator that supports research efforts in hazardous-waste treatment and sludge incineration. Experimental research in the structural engineering and mechanics program is conducted in the structures and materials laboratory, which contains a wide range of equipment, including a 600,000-pound BLH Universal Testing Machine and load measuring and recording devices with ranges from 1,000 pounds to 240,000 pounds. Research in geotechnical engineering is conducted in a well-equipped soil mechanics laboratory. All laboratories, as well as faculty and administrative offices, are housed in the L. K. Downing Building.
	The School maintains excellent computing facilities that provide students with the resources needed to perform word processing, engineering problem solving, general graphics, computer-aided design, and computer programming. These facilities include the Computer Learning and Design Center (CLDC), which provides access to the University's mainframe and World Wide Web and houses several minicomputers. Additional personal computer laboratories are also operated within the School of Engineering and Computer Science.
Financial Aid	Graduate student assistantships for instruction or research are available for well-qualified students. In 1999–2000, teaching assistantships provide a stipend of $8000 (half-time) plus tuition. Research assistantships normally include summer support. A limited number of fellowships are also available.
Cost of Study	Tuition and fees vary depending on the credit hour load. The 1999–2000 tuition for graduate students enrolled for 9 or more credit hours is $5250 per semester. For students carrying fewer than 9 credit hours, the charge is prorated on the basis of the full-time rate.
Living and Housing Costs	The rent for a University-owned apartment ranges from $724 to $1500 per month. In addition, a large number of privately owned apartments are available in the metropolitan Washington, D.C., area.
Student Group	The Department of Civil Engineering typically enrolls 10 new graduate students each year and has a total enrollment of 30. There are approximately 850 undergraduate and 200 graduate students in the School of Engineering and Computer Science. Nearly 11,000 students are currently enrolled in all schools and colleges of Howard University.
Location	The University is in the heart of Washington, D.C., a city that offers students unsurpassed opportunities for study and research.
The University	Founded in 1867, Howard University is a privately governed institution. The University has twelve schools and colleges, and its main campus occupies more than 89 acres. West and East campuses house the School of Divinity and the Law School, respectively, and a 108-acre site will be developed for the Center for Advanced Research in the Life and Physical Sciences. Howard is a member of the National Consortium for Graduate Degrees for Minorities in Engineering, Inc.
Applying	Information and application forms are obtainable on request. To be assured of consideration, applicants should provide credentials before April 1 for August admission and before November 1 for January admission. Applicants requiring financial assistance should provide credentials as soon as possible before the April 1 and November 1 deadlines. All new applicants are required to submit scores on the General Test of Graduate Record Examinations. International students must report scores from the Test of English as a Foreign Language (TOEFL).
Correspondence and Information	Chair Department of Civil Engineering College of Engineering, Architecture, and Computer Sciences Howard University Washington, D.C. 20059 Telephone: 202-806-6570 Fax: 202-806-5271 E-mail: plw@scs.howard.edu

Howard University

THE FACULTY AND THEIR RESEARCH

Taft H. Broome Jr., Professor, Structures and Mechanics; Sc.D., George Washington. Structural analysis of large space structures, nuclear plants, naval deep submersibles, philosophy of engineering, engineering ethics.

Robert E. Efimba, Associate Professor, Structures and Mechanics; Sc.D., MIT; PE. Structural mechanics, computer-aided structural engineering, finite-element models in biomechanics and equilibrium, stability and dynamic analysis of building frame shearwall systems, stiffened and/or folded plates and shells.

Lorraine N. Fleming, Professor and Chair, Geotechnical Engineering; Ph.D., Berkeley; PE. Experimental soil mechanics, cyclic temperature effects on soil permeability, modeling of fluid transport through soil, modeling of soil-structure interaction.

James H. Johnson Jr., Professor, Environmental and Water Resources, and Dean, College of Engineering, Architecture, and Computer Sciences; Ph.D., Delaware; PE. Solid-liquid separation techniques, sludge treatment, biodegradation of hazardous wastes, beneficial aspects of wastewater sludge. (On leave)

Kimberly Jones, Assistant Professor, Environmental Engineering (Environmental and Water Resources); Ph.D., Johns Hopkins. Drinking water quality, membrane processes, physical-chemical processes.

Edward J. Martin, Professor, Environmental and Water Resources, and Samuel P. Massey Chair of Excellence in Environmental Engineering; Ph.D., RPI; PE. Applications of risk analysis techniques to the assessment and management of environmental problems; incineration of hazardous wastes and hazardous materials.

Errol C. Noel, Professor, Transportation; Ph.D., Maryland; PE. Highway traffic safety and control, highway and traffic signal design, transportation planning, computer-aided analysis and transportation subsidization trends.

Kwamena Ocran; Assistant Professor, General Medicine and Biophysics and Nuclear Medicine; M.D., Ph.D., Karlova (Prague). Health effects of environmental degradation, environmental science, environmental health, occupational environment, environmental epidemiology, chemical risk assessment and risk management.

Gajanan M. Sabnis, Professor, Structures and Mechanics; Ph.D., Cornell; PE. Modeling of concrete structures, risk analysis in construction projects.

Bruce Schimming, Professor, Geotechnical Engineering; Ph.D., Northwestern; PE. Engineering design.

Ahlam I. Shalaby, Assistant Professor, Environmental and Water Resources; Ph.D., Maryland. Estimating probable maximum flood in hydraulic design, turbulent open channel flow studies.

Shahram E. Zanganeh, Lecturer, Structures and Mechanics; M.E., Howard. Structural mechanics, nonlinear finite-element models of structures subjected to abnormal loads, computer and mathematical methods.

ILLINOIS INSTITUTE OF TECHNOLOGY

Department of Civil and Architectural Engineering

Programs of Study

The Department of Civil and Architectural Engineering offers programs leading to the degrees of Master of Science (M.S.) and Doctor of Philosophy (Ph.D.) in the general areas of structural engineering, geotechnical engineering, geoenvironmental engineering, transportation engineering, and construction engineering and management. In addition, professional master's degrees are offered, including Master of Construction Engineering and Management (M.C.M.), Master of Geotechnical Engineering (M.Geotech.E.), Master of Geoenvironmental Engineering (M.Geonv.E.), Master of Structural Engineering (M.Struct.E.), and Master of Transportation Engineering and Planning (M.Trans.E.). Through cooperation with the public administration program, the department also offers graduate study in infrastructure engineering and management leading to the Master of Public Works (M.P.W.) degree.

The M.S. program requires 32 credit hours, of which 6 to 8 are for thesis research. The professional master's programs require 32 credit hours of course work. The Master of Science program typically takes three semesters to complete on a full-time basis, while the professional master's program can be completed in one calendar year.

The program for the Ph.D. requires a course of study in one major field and in distinct but supporting minor fields. Formal requirements include a qualifying examination taken upon completion of course work in both the major and minor fields, a comprehensive examination in which the student presents a proposal of the dissertation research and his or her engineering background is examined, and a defense in which the dissertation is presented to the examining committee. The Ph.D. program can be completed in three to four years after the award of the M.S. degree.

Research Facilities

The Department of Civil and Architectural Engineering offers state-of-the-art computing and laboratory equipment. Research facilities include a Concrete Laboratory, a Metal Testing Laboratory, a Materials Laboratory, a Geotechnical Engineering Laboratory, a Transportation Engineering Laboratory, and a Construction Engineering and Management Laboratory. In addition, researchers have access to the extensive on-campus facilities of the IIT Research Institute and the nearby Argonne National Laboratory.

The IIT Paul V. Galvin Library is well stocked with engineering monographs and periodicals. It has online access to a multitude of libraries nationwide. Interlibrary loan services are fast and efficient. Online and CD-ROM search services are available.

The central computing facilities are accessible to civil and architectural engineering students. In addition, the department has its own computer lab exclusively for civil engineering students, and a computer-aided engineering and design lab equipped with state-of-the-art hardware and software. The Cray supercomputer at the University of Illinois at Urbana–Champaign can be conveniently accessed via the Internet for research purposes.

Financial Aid

Merit scholarships covering half of the cost of tuition for the first year of study are awarded to a limited number of outstanding full-time master's candidates. Teaching and research assistantships are offered to students on a competitive basis, generally after the first semester. Several industrial employers in the metropolitan Chicago area give selected employees company fellowships for graduate study in the department; these include time off for study and payment of fees. Loans may be arranged through the Financial Aid Office. Assistance for students and their spouses in securing employment on campus or in the Chicago area is available through the Career Advancement office.

Cost of Study

Tuition for graduate students for 1999–2000 is $590 per credit hour.

Living and Housing Costs

Housing is available for unmarried graduate students in IIT residence halls. The 1999–2000 cost for room and board ranges from $5155 to $6880. IIT apartments are available for families. Rents range from $458 to $927 per month, including utilities. Early application for apartments is recommended. Several off-campus apartment complexes are located within a mile of the Institute. Living in the suburbs and commuting by car or by public transportation is also possible.

Student Group

There are approximately 5,900 students enrolled at IIT; more than half are graduate students. There are nearly 100 students enrolled in graduate degree programs in Civil and Architectural Engineering; 70 percent are international students and 11 percent are women.

Student Outcomes

IIT graduates go on to academia as well as work as engineers, consultants, and contractors for private firms and government agencies.

Location

IIT's Main Campus is located near the heart of Chicago, just 3 miles south of the Loop. The Main Campus's immediate neighbors include Comiskey Park, home of the Chicago White Sox; two major medical centers; two independent professional colleges; a major technical high school; McCormick Place exposition center; and three privately developed residential complexes. The Chicago metropolitan area, home to approximately 7 million people, offers an abundance of employment opportunities and an unusually rich variety of ethnic communities and cultural activities. The Main Campus, designed by Ludwig Mies van der Rohe and regarded internationally as a landmark of twentieth century architecture, occupies fifty buildings on a 120-acre site and includes research institutes, libraries, classrooms, laboratories, residence halls, and a sports center. Students can easily take advantage of Chicago's many museums, sporting events, theaters, and shops. Nearby Lake Michigan and the surrounding parks provide opportunities for biking, swimming, jogging, and other recreational activities.

The Institute

Illinois Institute of Technology was established in 1940 with the merger of Armour Institute of Technology (founded in 1890) and Lewis Institute (founded in 1896). IIT offers programs in engineering and the sciences, architecture, design, psychology, technical communication, business, and law. IIT is a member of the Association of Independent Technological Universities (AITU).

Applying

Applications and supporting documents, which include transcripts, test scores, and letters of recommendation, should be received by the Office of Graduate Admissions (3300 South Federal, Room 301A, Chicago, Illinois 60616) no later than June 1 for fall matriculation and November 1 for spring matriculation. The application deadline for the summer session is May 1. Applications that include requests to be considered for financial aid should be submitted by March 1. Application forms and additional information are available on line at http://www.grad.iit.edu.

Correspondence and information

Dr. Jamshid Mohammadi, Chairman
Department of Civil and Architectural Engineering
Alumni Hall, Room 228
Illinois Institute of Technology
3201 South Dearborn Street
Chicago, Illinois 60616

Telephone: 312-567-3540
Fax: 312-567-3519
E-mail: katyb@alpha1.ais.iit.edu
World Wide Web: http://www.iit.edu/~ce

Illinois Institute of Technology

THE FACULTY AND THEIR EXPERTISE

D. Arditi, Professor (Construction Engineering and Management); Ph.D., Loughborough (England).
S. A. Guralnick, Perlstein Distinguished Professor (Structures of Plastic Design and Experimental Analysis); Ph.D., Cornell; PE, SE.
C. J. Khisty, Professor (Transportation Systems and Traffic Engineering); Ph.D., Ohio State; PE.
J. Mohammadi, Professor (Structural Reliability and Bridge Engineering); Ph.D., Illinois at Urbana-Champaign; PE, SE.
J. Budiman, Associate Professor (Geotechnical Engineering and Advanced Soils Testing); Ph.D., Colorado at Boulder; PE.
J. R. O'Leary, Associate Professor (Solid Mechanics and Computational Methods); Ph.D., Texas at Austin.
A. Mokhtar, Assistant Professor (Architectural Engineering and Applications of Information Technology in Construction); Ph.D., Concordia.
E. De Santiago, Assistant Professor (Nonlinear Finite Element Analysis and Computational Mechanics); Ph.D., Stanford.
J.-H. Shen, Assistant Professor (Structural Engineering and Seismic Design); Ph.D., Berkeley; PE, SE.
D. Carreira, Adjunct Professor (Concrete Structures); Ph.D., IIT; PE, SE.
S. Gill, Adjunct Professor (Geotechnical Engineering); Ph.D., Northwestern.
A. Paintal, Adjunct Associate Professor (Hydraulics and Hydrology); Ph.D., Minnesota.
A. W. Domel, Adjunct Assistant Professor (Project Administration and Legal Issues); Ph.D., Illinois at Chicago; J.D., Loyola; PE, SE.
J. Fazio, Adjunct Assistant Professor (Traffic and Transportation Engineering); Ph.D., Illinois at Chicago.
A. Frano, Adjunct Assistant Professor (Construction Cost Accounting); J.D., IIT.
J. Jahedi, Adjunct Assistant Professor (Structural Engineering); Ph.D., IIT; PE, SE, FAIA.
R. Lemming, Adjunct Assistant Professor (Construction Cost Estimating and Scheduling); M.B.A., Central Michigan; J.D., IIT.
S. Pinjarkar, Adjunct Assistant Professor (Structural Engineering); Ph.D., IIT; PE, SE.

RESEARCH AREAS

Architectural Engineering and Building Science. Research in database design, management of information systems, and computer-aided design, as applied to construction management.

Construction Engineering and Management. Research in construction engineering and management includes studies in construction techniques; site productivity; contracts and specifications; planning, scheduling, and control of construction activities; quantity takeoff and estimating; economic decision analysis; construction equipment; systems analysis; construction administration; and computerized applications in scheduling, estimating, resource planning, and cost control. Faculty research interests cover such topics as optimization techniques, construction productivity, construction safety, repetitive scheduling techniques and associated software, expert/knowledge-based systems in forensic scheduling, and software for scheduling, estimating, and cost control.

Fire Protection and Safety Engineering. Research areas include prediction of structural collapse of buildings during fire, fire-fighting personnel locator in buildings during fire, fire prevention and protection during construction, fire load environment, and interior gas distribution systems.

Geotechnical and Geoenvironmental Engineering. Research in geotechnical engineering includes studies in soil mechanics, rock mechanics, engineering geology, earthquake engineering, soil structure, and soil-water interactions. The faculty research interests cover such topics as soil dynamics, foundation design, practical problems in soils engineering, rock mechanics and loads on underground structures, soil-structure interaction and earthquake engineering, experimental and numerical methods in geomechanics, groundwater hydrology and subsidence due to withdrawal of fluids, engineering geology, geoenvironmental engineering, and offshore structure foundations.

Public Works (Infrastructure Engineering and Management). Faculty research interests in public works cover such topics as public policy evaluation; management of engineering operations; maintenance, repair, and rehabilitation of bridges; and rehabilitation and renovation of the existing infrastructure.

Structural Engineering. Research in structural engineering includes studies in structural mechanics, analysis methods, and design in concrete and steel. Faculty research interests cover such topics as elastoplastic and viscoelastic behavior; static or dynamic response of bridges; design and analysis of plates; shell structures; cable-supported and inflatable structures; modern computational methods of analysis; finite element methods of analysis; experimental analysis of structures; seismic and wind loadings on structures; approximate and variational methods of linear and nonlinear analysis; probabilistic analysis and reliability design of structures; behavior of materials; design of reinforced and prestressed concrete structures; fatigue analysis, behavior, and design of metal structures; and plastic methods of design.

Transportation Engineering. Research in transportation engineering includes traffic engineering, urban transportation planning, traffic flow theory, public transport, railroad engineering, and transportation systems management. Faculty research interests cover such topics as waterborne transportation, crowd dynamics, transportation modal systems, socioeconomic issues connected with transportation infrastructure, and railroad engineering. A vigorous program of research is conducted by faculty members in cooperation with the Illinois Universities Transportation Research Consortium.

LOUISIANA STATE UNIVERSITY

Department of Civil and Environmental Engineering

Programs of Study	The department offers graduate programs leading to either an M.S. or a Ph.D. degree. Five major areas in civil engineering are offered: environmental engineering, geotechnical engineering, structural engineering and mechanics, transportation engineering, and water resources. Course work is also available in remote sensing.
	Research in geotechnical engineering is focused on constitutive modeling of geomaterials, numerical methods in geomechanics, pile foundations, testing of soils, soil behavior, environmental geotechnics, soil reinforcement, electrokinetic remediation, and geotechnics of recyclable materials.
	Recent research projects in the structures area include experimental studies of heat straightening of damaged steel bridge members, structural identification, design and analysis of articulated glulam timber members, and residual stresses in timber structural elements. Current research in the mechanics area is in refined theories of plates, damage mechanics, constitutive modeling of finite strain plasticity of metals, computational mechanics, constitutive modeling of composites, and timber mechanics.
	Current research in the environmental engineering area emphasizes water quality, remote sensing, and solid-waste management and includes lake restoration, aquaculture, hazardous wastes, stochastic water quality modeling, urban runoff, and coastal zone impact studies.
	Research in water resources stresses free-boundary problems, mathematical modeling of hydrologic systems response, surface and groundwater conjunctive use, experimental and theoretical studies of transport processes, and ocean measurements and modeling.
	Research in transportation engineering is focused on pavement design and management, materials characterization, maintenance and rehabilitation, systems analysis, economics, and urban planning.
	Remote-sensing projects include the detection and mapping of hazardous waste, hydrologic systems, land use, coastal processes, geologic features, and geographic database development.
	Civil engineering students are also involved in projects with the Institute for Recyclable Materials that encompass fundamental materials, leachate, geotechnical applications, and highway pavements and materials studies. New research funding received during the 1997–98 fiscal year for all departmental program areas exceeded $3.8 million.
	There are two M.S. options. The research option requires 25 hours of approved course work and an acceptable thesis. The course work option requires 34 hours of approved course work and an acceptable master's report. A student becomes a doctoral candidate by passing a qualifying examination, satisfying residence requirements, fulfilling 54 credit hours of course work beyond the B.S., and passing the general examination. The Ph.D. is conferred after successful completion of the dissertation.
Research Facilities	The department is housed in the Center for Engineering and Business Administration (CEBA) building. Eleven laboratories covering 23,600 square feet of floor area are available. Among the laboratories used for graduate studies and research are geotechnical, environmental, structural, mechanics of materials, water resources, and computer laboratories. The TEM/SEM microscopy mineralogical analysis laboratory of the mechanical engineering department and the remote sensing and image processing facilities of the Civil and Environmental Engineering Research Infrastructure Center (CERIC) are also used by graduate civil engineering students. Laboratories for state-of-the-art materials testing are available for use at the nearby Louisiana Transportation Research Center (LTRC). The department shares the Geosynthetic Engineering Research Laboratory with LTRC. The department maintains an array of more than thirty microcomputers for word processing, data acquisition, plotting, and research functions. A general use I/O room with terminals, VAXstations, and PCs is also located in CEBA. University Computing Services maintains a high performance UNIX cluster with thirteen IBM RS/600 servers for research needs.
Financial Aid	Graduate assistantships carry stipends of $10,000 to $21,000 for approximately half-time duties. Fiscal-year assistantships carry proportionately higher stipends. Fellowships, some of which cover fees, are also available; they require no duties. Support is provided by the University, the Louisiana Water Resources Research Institute, the Hazardous Waste Research Center, the Louisiana Transportation Research Center, the Institute for Recyclable Materials, federal and state agencies, and industry. Loan funds are available. Excellent jobs can be found on campus and in Baton Rouge for students' spouses.
Cost of Study	Resident tuition and fees in 1998–99 for full-time students were $5094 per year; nonresidents paid $10,044 per year. Diploma and thesis-binding fees were $60 for the master's degree and $105 for the doctorate.
Living and Housing Costs	Residence halls are open to all graduate students and range from $965 to $1345 per person per semester. One-bedroom apartments rent for $355 per month; two-bedroom apartments range from $305 to $350 per month; and three-bedroom apartments range from $380 to $405 per month.
Student Group	Enrollment at the Baton Rouge campus is about 28,000; graduate and professional enrollment exceeds 5,000. More than half of the graduate students are married. Graduate enrollment in the department as of January 1998 was 154, with 34 working toward the Ph.D. degree. LSU students come from every state and more than sixty countries. In 1997–98, the department awarded sixteen Ph.D. and eighteen M.S. degrees.
Location	Baton Rouge, the state capital, is a growing industrial metropolis of 300,000. It is centered on the oil and petrochemical industries and is a deepwater port on the Mississippi River. The semitropical climate makes activities such as golf and tennis popular the year round, and the area is known for its fishing, boating, and hunting. New Orleans is just 80 miles to the southeast.
The University	LSU, a land- and sea-grant institution, was founded in 1860. It records its first graduate degree as a "Civil Engineer" degree awarded in 1869. The principal buildings, grouped on a beautifully landscaped 300-acre tract, exhibit a pleasing blend of contemporary design and the older Italian style, with its tile roofs and colonnaded passageways. Growth and improvements in buildings, facilities, and staff have been steady.
Applying	Applications are considered the year round but should be made as early as possible. Where financial assistance is desired, applications should be completed by March 1 for fall and November 1 for spring. GRE General Test scores are required. International students whose native language is not English must pass the TOEFL with a minimum score of 550 and obtain appropriate scores on the TSE; a TOEFL score of 575 is required to be eligible for financial assistance.
Correspondence and Information	Chairman, Department of Civil and Environmental Engineering CEBA Building (3502) Louisiana State University Baton Rouge, Louisiana 70803 Telephone: 504-388-8442

Louisiana State University

THE FACULTY AND THEIR RESEARCH

Donald D. Adrian, Professor; Ph.D., Stanford; PE. Environmental engineering, pollutant transport and attenuation in aquifers, land application of wastes.

Mohamed Alawady, Associate Professor; Ph.D., Oklahoma State; PE. Water resources, hydraulics, river mechanics.

R. Richard Avent, Professor; Ph.D., North Carolina State; PE. Structural engineering and mechanics, epoxy repair of damaged structures, latticed structures, structural failures.

W. David Constant, Professor, Director of the Civil and Environmental Infrastructure Research Center, and Assistant Director of the Hazardous Substance Research Center/South and Southwest; Ph.D., LSU; PE. Fate and transport of hazardous substances, water resources management, waste site remediation technology.

Vijaya K. A. Gopu, Professor and Program Director of Large Structural and Building Systems, National Science Foundation; Ph.D., Colorado State; PE. Design and behavior of heavy engineered timber structures, analysis of composite and concrete structures.

George M. Hammitt, Associate Professor and Chair of the Department of Construction Management; Ph.D., Texas A&M; PE. Pavements, materials-airports.

Michael D. Heiler, Assistant Professor; Ph.D., Carnegie Mellon. Transportation engineering, remote sensing, application of advanced technology to transportation infrastructure assessment.

Ernest Heymsfield, Assistant Professor; Ph.D., CUNY, City College; PE. Structures and design of steel, evaluation of existing highway bridges, design methodology for new bridges.

Gary G. Kelly, Instructor and Coordinator, Surveying/Photogrammetry/Remote Sensing; M.S., Ohio State. Application of geodesy, photogrammetry, and remote sensing to problems in engineering; earth sciences.

Marc L. Levitan, Associate Professor; Ph.D., Texas Tech. Structural engineering, wind engineering, wind loads on industrial/petrochemical structures, hurricane sheltering and evacuation, design of steel structures.

Ronald Malone, Chevron Professor; Ph.D., Utah State; PE. Environmental engineering, water quality, mathematical models, recirculating aquaculture.

Emir J. Macari, Bingham C. Stewart Distinguished Professor of Engineering and Department Chairman; Ph.D., Colorado at Boulder; White House/NSF Presidential Faculty Fellow, 1992–97. Experimental and computational geomechanics.

John B. Metcalf, Freeport-McMohan Professor; Ph.D., Leeds (England). Highway materials and pavements, low-cost roads, use of nonstandard pavement materials, pavement design, accelerated testing, construction quality control.

William M. Moe, Assistant Professor; Ph.D., Notre Dame. Environmental engineering, biological treatment of gas-phase contaminants, treatment of domestic and industrial wastewaters, periodic processes in environmental systems, sequencing batch reactors, biofilms.

Louay N. Mohammad, Research Assistant Professor; Ph.D., LSU. Highway construction materials, pavement design and analysis, pavement maintenance and rehabilitation.

David J. Mukai, Assistant Professor; Ph.D., Washington (Seattle). Reinforced concrete (flow characteristics, computer modeling, fracture mechanics), heat straightening of damaged steel girders, repair of damaged timber piles and poles, theoretical fracture mechanics, engineering education.

John H. Pardue, Associate Professor and E. H. Stewart Engineering Leader Chair; Ph.D., LSU; PE. Environmental engineering, bioremediation, fate and transport of contaminants, wetlands, wastewater treatment.

Amitava Roy, Research Assistant Professor; Ph.D., LSU. Materials science of cement chemistry, utilization of industrial by-products, hazardous-waste treatment.

Kelly A. Rusch, Assistant Professor; Ph.D., LSU. Environmental engineering, water quality, aquacultural engineering, process control.

John J. Sansalone, Assistant Professor; Ph.D., Cincinnati; PE. Unit operations for stormwater nonpoint pollution, experimental methods in geoenvironmental and urban hydrology, drainage infrastructure, geoenvironmental engineering, wastewater treatment.

Roger K. Seals, Professor; Ph.D., North Carolina State; PE. Geotechnical engineering properties and utilization of residual materials, earthwork design.

Vijay P. Singh, Professor; Ph.D., Colorado State; D.Sc., Witwatersrand (Johannesburg); PE, PH. Surface and groundwater hydrology, irrigation hydraulics, mathematical and stochastic modeling, environmental pollution.

Peter R. Stopher, Professor; Ph.D., London. Traffic studies, travel demand forecasting methods, urban transportation planning models.

Joseph Suhayda, Associate Professor and Director of Louisiana Water Resources Research Institute; Ph.D., California, San Diego. Coastal engineering/offshore technology: environmental design criteria, wave forces, sediment transport, stability.

Mehmet T. Tumay, Professor and Associate Dean of Engineering for Research, Louisiana Transportation Research Center; Ph.D., Istanbul Technical; PE. In situ soil properties/testing, fiber soil reinforcement, centrifuge modeling, flow through porous media, statistical modeling.

George Z. Voyiadjis, Boyd Professor and Coordinator of Graduate Programs; D.Sc., Columbia. Micromechanical modeling of the behavior of metals and composites, finite strain plasticity, damage mechanics, computational mechanics, refined theory of plates and shells.

Clinton S. Willson, Assistant Professor; Ph.D., Texas at Austin. Environmental fluid mechanics, multiphase flow, and transport in porous media; in situ remediation and assessment techniques.

Chester G. Wilmot, Research Associate Professor; Ph.D., Northwestern. Transportation planning, prioritization, transferability of travel demand models.

Brian Wolshon, Assistant Professor; Ph.D., Michigan State; PE, PTOE. Transportation engineering, geometric highway design and safety analysis, traffic operations, development/application/evaluation of intelligent transportation systems.

MASSACHUSETTS INSTITUTE OF TECHNOLOGY

Center for Transportation Studies

Programs of Study

Transportation and logistics education at Massachusetts Institute of Technology (MIT) involves graduate programs at both the master's and doctoral levels. Approximately 40 faculty members at MIT are actively involved in teaching transportation, logistics, and related subjects and in conducting research in these areas.

The Master of Science in Transportation (M.S.T.) is designed for students with a broad range of undergraduate education and interests and produces graduates with strong analytic and problem-solving skills. Students select from a wide range of transportation and related subjects to structure a program tailored to their specific interests. The 72-unit M.S.T. program takes 1½ to 2 years to complete and is structured with five half-semester core classes, a three-subject program area, and a research-based thesis.

The interdisciplinary Ph.D. in transportation is for students who already have a thesis-based master's degree in transportation and who wish to continue their studies at the doctoral level. Students interested in the 150-unit Ph.D. program generally take their General Exams in the third semester at MIT, having demonstrated their ability to conduct research. The General Exam encompasses a written exam on three pairs of subjects selected by the student in both methodological and transportation areas, followed by an oral exam.

The Master of Engineering (M.Eng.) in logistics is designed for logistics professionals or for those interested in working in logistics and related areas. The intensive nine-month, 90-unit program includes several required courses, with electives in business, transportation, or other areas, according to the student's interests. In addition to their course work, students complete a thesis project and write a short thesis.

Research Facilities

Center for Transportation Studies (CTS) students have full access to MIT's extensive computer facilities as well as to their own new, state-of-the-art computer lab. In addition, CTS students are assigned office space, with the specialized facilities, equipment, hardware, and software necessary for their research project or study group.

Financial Aid

CTS has several fellowships and research and teaching assistantships for transportation students. Eighty percent of MIT's M.S.T. and Ph.D. students receive full funding to cover tuition and are provided with a monthly living stipend. M.Eng. students are eligible for several tuition fellowships that are provided through CTS.

Cost of Study

Graduate student tuition for the 1999–2000 academic year is $25,000. Estimated medical insurance is $650 for a single student and $2250 for a married student.

Living and Housing Costs

MIT currently estimates that monthly living expenses are $1400 for a single student and $1800 for a married student during the 1999–2000 academic year. The Institute has several housing options on campus for single and married students, with dorms and apartments available for single students and for families. In addition, many MIT students live off campus in apartments in Cambridge, Boston, and the surrounding areas.

Student Group

MIT's total enrollment is approximately 10,000, with somewhat fewer undergraduates than graduates. Students represent all fifty states and the District of Columbia, three territories, and 104 other countries. CTS students come from a variety of different academic disciplines, with a wide range of interests in transportation and logistics. The center has a close academic affiliation with the Departments of Civil and Environmental Engineering, Urban Studies and Design, Aeronautics and Astronautics, and Ocean Engineering and the Sloan School of Business Management.

Location

MIT is set on the Cambridge bank of the Charles River, facing south to the city of Boston, which is one of the nation's leading centers of art, education, and medicine. MIT is one of more than fifty schools located within the Boston area, including Harvard, Boston University, Northeastern, Boston College, Brandeis, Tufts, Simmons, Wellesley, and many specialized professional art and music schools. The concentration of academic, cultural, and intellectual activities in the Boston area is one of the densest in the country.

The Institute and The Center

The Institute is an independent, coeducational university that is privately endowed. It has five academic schools (Architecture and Planning, Engineering, Humanities and Social Science, Management, and Science) and departmental laboratories, centers, and divisions, which extend beyond the traditional department boundaries. The MIT faculty numbers approximately 1,100, with a total teaching staff of more than 2,000.

The Center for Transportation Studies was established in 1973 to provide a focal point at MIT for transportation education. CTS currently facilitates transportation and logistics research, conducts an outreach program to the transportation and logistics industries, and encourages a sense of common purpose among the many departments, centers, and laboratories involved in transportation and logistics at MIT.

Applying

The application deadline is January 15 for fall admission to the M.S.T., M.Eng., and Ph.D. programs, and Ph.D. applicants may also apply by November 1 for the spring semester. Later applications are considered but may put students at a disadvantage for funding and on-campus housing. There is no preapplication process or interview required for these programs, but CTS welcomes visits from prospective students.

Correspondence and Information

Center for Transportation Studies
Massachusetts Institute of Technology
77 Massachusetts Avenue, Room 1-235
Cambridge, Massachusetts 02139
Telephone: 617-253-8069
Fax: 617-253-4560
E-mail: tec@mit.edu
World Wide Web: http://web.mit.edu/cts/www/

Massachusetts Institute of Technology

THE FACULTY AND THEIR RESEARCH

More than 50 faculty and staff members are involved in various aspects of transportation and logistics at MIT through participation in courses, research projects, industry affiliate programs, and interaction with students. Provided below is an abbreviated list of faculty and research staff members who are directly involved in the center's educational programs through teaching of transportation and logistics subjects or supervising students performing research.

Arnold Barnett, Professor: aviation safety, statistics. Cynthia Barnhart, Associate Professor and Co-Director, Operations Research Center (ORC): distribution, logistics, large-scale network optimization. Peter Belobaba, Principal Research Scientist: air transportation economics, airline marketing and management, applied operations research, aerospace industry analysis. Moshe Ben-Akiva, Professor: transportation systems analysis, transportation demand forecasting, transportation and urban models, behavioral models and econometric methods. Erik Brynjolfsson, Associate Professor: information technology, organizational transformation and the Internet. Didier Burton, Research Associate: real-time dynamic traffic control. Oral Buyukozturk, Professor: rail transportation. Jonathan Byrnes, Senior Lecturer: supply chain management. Ismail Chabini, Assistant Professor: information technology in transportation analysis, models and algorithms for intelligent transportation systems, high-performance computing and optimization methods. Alan Chachich, Research Affiliate: intelligent transportation systems. Wai Kong Cheng, Professor: motor vehicle engine technology. John-Paul Clarke, Assistant Professor: air traffic management, intelligent application of advanced flight guidance technology, environmental impact of aircraft operations. Joseph Coughlin, Director, University Transportation Center (UTC): public policy, strategic management, transport and the environment, transportation needs of an aging population. Richard De Neufville, Professor and Chairman, Technology and Policy Program (TPP): airport systems planning, transportation technology and policy, geographic databases. Herbert Einstein, Professor: manufacturing underground space. Joseph Ferreira, Professor: geographic information systems, urban planning. Charles Fine, Affiliate, Leaders for Manufacturing (LFM) Program: international motor vehicle industry. Ernst Frankel, Professor: port development and planning, systems analysis, ocean systems design. Ralph Gakenheimer, Professor: urban transportation planning, transfer of methodology to developing countries, transportation infrastructure. Ahmed Ghoniem, Professor: motor vehicle engine technology. Stephen Graves, Professor and Co-Director, LFM: operations management, optimization models, inventory analysis. Amar Gupta, Senior Research Scientist: information management systems, air traffic control systems, financial services. John Hansman, Professor: air transportation. Wesley Harris, Professor: air transportation. John Heywood, Professor: automotive engines and fuels, use of ceramic materials in internal combustion engines, development and application of new analysis and experimental techniques. Sandy Jap, Assistant Professor: supply chain and distribution channel management. Mithilesh Jha, Research Associate: intelligent transportation systems modeling, driver behavior modeling, transportation network analysis, applications of distributed computing in transportation. Jack Kerrebrock, Professor: space transportation. Judith Kildow, Associate Professor: technology risk management for ships, particularly oil tankers. Thomas Kochan, Professor: labor management partnership. Jin Au Kong, Professor and Director, Research Laboratory of Electronics/Center for Electromagnetic Theory and Applications (RLE/CETA): air landing systems. Kenneth Kruckemeyer, Research Associate: public transport system development. James Kuchar, Assistant Professor: air transportation. Richard Larson, Professor and Director, Center for Advanced Educational Services (CAES): operations research and analysis of public systems, evaluation of technology applied to urban systems, postal services, urban transportation. Paul Levy, Adjunct Professor: financing infrastructure. Nancy Lynch, Professor: advanced vehicle control systems. Ronald MacNeil, Principal Research Associate: railroad information displays. Stuart Madnick, Professor: information systems engineering. Thomas Magnanti, Dean: transportation planning, including vehicle fleet planning, personnel scheduling, distribution system design, and urban traffic management; network analysis; mathematical programming; combinatorial theory. Thomas Malone, Professor: use of computers and communications technology in groups and organizations. Henry Marcus, Professor: transportation management, ocean transportation, ocean systems management, public policy. David Marks, Professor: environmental impacts of transportation. Carl Martland, Senior Research Associate: railroad operations and management. Ichiro Masaki, Principal Research Associate: intelligent vehicle systems. James Masters, Executive Director, Master of Engineering in Logistics Program: analytic techniques for logistics decision making in inventory deployment and logistics network design. Dennis Mathaisel, Research Associate: air transportation. Peter Metz, Deputy Director, Center for Transportation Studies (CTS): logistics systems design and strategy, supply chain management. John Miller, Associate Professor: procurement strategies for long-term delivery and finance of transportation infrastructure, integration of transportation infrastructure facilities with long-term economic activity, optimal choice of project delivery and project finance methods for transportation infrastructure facilities. Fred Moavenzadeh, Director, Center for Technology, Policy, and Industrial Development (CTPID): construction, transportation in developing countries, transfer and adaptation of technology to developing countries. Amedeo Odoni, Professor: operations research, airport and air traffic control problems, analysis of urban service systems, project evaluation. Carl Much, Lab Group Leader: intelligent transportation systems. Earll Murman, Professor: air transportation. Charles Oman, Senior Research Engineer: human performance in flight. James Orlin, Professor and Co-Director, ORC: mathematical programming, combinatorial and network optimization, design and analysis of heuristics. Karen Polenske, Professor: regional economic analysis, with emphasis on transportation, employment, energy, and environmental policies. James Rice, Director, Integrated Supply Chain Management (ISCM) Program: integrated supply chain management. Daniel Roos, Associate Dean, Engineering Systems Division (ESD): information systems, transportation systems, policy issues in transportation, automotive industry. Nancy Rose, Professor: economic regulation and deregulation, airline pricing behavior, airline safety. Donald Rosenfield, Senior Lecturer: supply chain management, global production and distribution strategies. Jeanne Ross, Principal Research Associate: information technology management practices, design and management of global information technology infrastructures. Jerome Rothenberg, Professor Emeritus: urban transportation systems, relationship between transportation and urban forms and industrial and residual location, urban transportation and environment. Ted Rybeck, MIT Affiliate: increasing shareholder wealth through value chain network strategy, benchmarking the ROI and software enablers for Internet-enabled demand and supply chain convergence. Fred Salvucci, Senior Lecturer: transportation planning, government policy. Harvey Sapolsky, Professor: public policy and organization. Yossi Sheffi, Professor and Director, CTS: transportation network analysis and design, business logistics, freight transportation, rail and truck systems, decision support systems. Qing Shen, Associate Professor: economic development, land use and transportation policy. Thomas Sheridan, Professor: human factors in transportation. Michael Shiffer, Lecturer and Principal Research Scientist: transportation planning presentation aids. Joseph Sussman, Professor: transportation systems management and operations, systems analysis, simulation methods, rail systems, intelligent transportation systems. Richard Thornton, Professor: magnetic levitation transportation. Lester Thurow, Professor: economic and transport development. Harold Tuller, Professor and Director, Crystal Physics and Electroceramics (CPE): development of motor vehicle fuels. Malcolm Weiss, Senior Research Staff: environmental impacts of transportation. Tomasz Wierzbicki, Professor: collision protection and crashworthiness of land, air, and sea vehicles. Shi-Chang Wooh, Assistant Professor: infrastructure assessment. Nigel Wilson, Professor: urban transport, public transport operations, planning and management, transport systems analysis

MICHIGAN STATE UNIVERSITY

College of Engineering
Department of Civil and Environmental Engineering

Programs of Study
The department offers six degree programs, including the M.S. and Ph.D. in both civil engineering and environmental engineering. The faculty members are active in teaching and research, and their specialties encompass civil engineering materials, environmental engineering, fluid mechanics, geotechnical engineering, pavement engineering, structural engineering, and transportation engineering. For some students, their program focuses on one of the traditional areas of study while others are more integrative—for example, combining geotechnical-oriented courses with selections in structural or environmental engineering. Similarly, some students may take courses in other disciplines (e.g., a transportation engineering student may take courses in statistics and urban planning), while others may concentrate on department-based offerings. Examples of areas of research and a listing of faculty interests are given below. Of seven College-wide research thrust areas, the department has a key role in three: environmental engineering, materials science and structural engineering, and transportation engineering. The department has three Michigan Department of Transportation–sponsored Centers of Excellence: one in pavement research, one in traffic safety and operations research, and one in transit research. It is also a member of the Great Lakes and Mid-Atlantic Hazardous Substance Research Center, sponsored by the U.S. Environmental Protection Agency and the Michigan Department of Environmental Quality.

Research Facilities
The department has more than 20,000 square feet of laboratory space in two buildings. A 10,000-square-foot Infrastructure Laboratory is planned to be constructed by early 2000. These buildings include facilities for teaching and research. Available equipment ranges from that required to mix, cure, and test concrete materials to up-to-date PC-based computer laboratories. Facilities also include a transportation laboratory, two controlled environment rooms, a hazardous waste analysis laboratory, and instrumentation capabilities for gas chromatography, mass spectroscopy, atomic absorption, ion chromatography, total organic carbon, and several light spectroscopy techniques. Faculty members and graduate students have access to the Composite Materials and Structures Center, which houses a wide variety of equipment for composite fabrication and characterization. The Division of Engineering Computing Services provides an advanced computing environment for faculty members and students through an extensive array of hardware, software, and support services. Mainframe computing facilities are also available.

Financial Aid
There are numerous sources of financial aid available from the department, the College, the University, and special programs such as Affirmative Action. There are fellowships as well as some teaching and research assistantships that are administered by the department, although most research assistantships are awarded by individual professors. Department-based fellowships generally range from $500 to $6000 per year. The most common assistantship is for one-quarter-time (10 hours of work required per week in support of a course or on a research project). This translates to a monthly stipend of about $630 (for 1998–99) and also covers 6 credits of tuition per semester and health insurance for the student. There are a very limited number of fellowship and/or assistantship opportunities that range as high as $15,000 per year. Currently, approximately 45 percent of the graduate students in civil and environmental engineering programs have assistantships. There are no special applications for awards made through the department, though students should indicate their desire for financial aid when applying.

Cost of Study
Based on 1998–99 rates, graduate tuition for Michigan residents was $223 per semester credit and $450 per semester credit for nonresidents. Including registration and related fees, a Michigan resident paid about $2540, while a nonresident paid about $4600 for 9 credits (three courses) per semester.

Living and Housing Costs
There is a variety of housing available on campus and in the adjacent community. Typical on-campus housing rates for 1998–99 for an apartment were $390 per month for a one-bedroom and $432 per month for a two-bedroom apartment. Graduate student dormitory rates for 1998–99 were $1649 per semester for a double room and $1929 per semester for a single room. Additional details and current rates can be obtained from the University Housing Office.

Student Group
The total number of students at Michigan State is approximately 43,200, including about 7,700 involved in graduate studies. In 1998–99, there were approximately 90 graduate students in civil and environmental engineering programs, of whom more than 30 were in the department's Ph.D. programs. Of the total number of graduate students, 80 to 90 percent are full-time. There are several student organizations, including ASCE and specialized groups in transportation and environmental engineering.

Student Outcomes
The vast majority of M.S. graduates gain employment in engineering companies and federal, state, and local government organizations. Some M.S. graduates proceed on to additional graduate study for the Ph.D. Most Ph.D. graduates are employed by universities and research organizations, though some also are employed by engineering companies.

Location
The University is located in East Lansing in south-central Michigan, about 80 miles from Detroit and 210 miles from Chicago. The Lansing area has more than 300,000 residents and includes the state capitol and the main offices of state government, the MSU campus, and major manufacturing facilities (e.g., General Motors). Recreational opportunities within a few hours' drive range from Great Lakes beaches to near-wilderness areas in Michigan's Upper Peninsula. Locally, there are libraries, museums, and numerous outdoor recreation opportunities and extensive on-campus opportunities ranging from plays and galleries to intramural and spectator sports.

The University
The University was founded in 1855 and is a land-grant and AAU institution offering more than 200 different programs of study by more than 4,000 faculty members in fourteen colleges. The campus has more than 5,000 acres, of which about 40 percent are dedicated to about 150 campus buildings and their immediate environs, with the remainder being used for experimental farms, outlying research centers, and natural areas.

Applying
Application forms and instructions for submitting transcripts, a statement of purpose, letters of recommendation, and related materials may be obtained from the department or from the Web site listed below. In civil engineering, study may begin in any term while for environmental engineering, the fall semester is more appropriate (although there are exceptions for students who have done some work in the area or who need to do collateral work). TOEFL scores are required from all international students. A minimum TOEFL score of 570 is required for admission. The GRE is recommended for all students.

Correspondence and Information
Correspondence should be addressed to:

Graduate Program and Admissions Coordinator—GPAPG
Department of Civil and Environmental Engineering
Michigan State University
East Lansing, Michigan 48824-1226

Telephone: 517-355-5107
Fax: 517-432-1827
E-mail: ceegrad@egr.msu.edu
World Wide Web: http://www.egr.msu.edu/cee

Michigan State University

THE FACULTY AND THEIR RESEARCH

Gilbert Y. Baladi, Professor; Ph.D., Purdue. Pavements, geotechnical, and materials: fatigue and plastic properties of soils, unbounded materials and asphaltic mixtures, pavement performance models, engineering analysis of existing pavements, pavement management, pavement evaluation and rehabilitation.

Neeraj J. Buch, Assistant Professor; Ph.D., Texas A&M. Pavements and materials: performance modeling of rigid pavements, pavement rehabilitation and management, nondestructive testing, mechanical properties of concrete and recycled materials in Portland cement concrete.

Karim Chatti, Assistant Professor; Ph.D., Berkeley. Pavements, geotechnical, and materials: pavement modeling, dynamics of pavements, truck-pavement interaction, mechanical properties of asphalt-concrete mixtures, nondestructive testing of pavements, pavement performance.

Mackenzie L. Davis, Associate Professor; Ph.D., Illinois. Environmental: risk assessment as a tool in pollution prevention.

Ronald S. Harichandran, Professor and Chairperson; Ph.D., MIT. Structures: random vibrations and probabilistic methods, earthquake engineering, structural dynamics, applications of finite-element method, fiber-reinforced polymers for infrastructure.

Frank J. Hatfield, Professor; Ph.D., Illinois at Urbana–Champaign. Structures: design in steel and wood.

Richard W. Lyles, Professor and Associate Chairperson; Ph.D., Carnegie Mellon. Transportation: evaluation of traffic control devices, highway work-zone safety, general highway and traffic safety, mobility and safety of older persons, public transportation, transportation planning.

Thomas L. Maleck, Associate Professor and University Traffic Engineer; Ph.D., Michigan State. Transportation: geometric design, traffic flow and control, traffic simulation, highway traffic safety.

Susan J. Masten, Associate Professor; Ph.D., Harvard. Environmental: use of chemical oxidants for remediation of soils, water, and leachates contaminated with hazardous organic chemicals; effect of chemical treatments on the biodegradability and toxicity of recalcitrant organics; formation of disinfection by-products.

Francis X. McKelvey, Professor; Ph.D., Penn State. Transportation: airport planning and design, highway traffic safety associated with older drivers, highway work-zone safety, transportation planning.

William E. Saul, Professor; Ph.D., Northwestern. Structures: structural dynamics, including wind and earthquake engineering; human loads; simulation; computer methods.

Virginia P. Sisiopiku, Assistant Professor; Ph.D., Illinois at Chicago. Transportation: traffic flow theory and traffic operations, computer-based simulation modeling and intelligent transportation systems applications.

Parviz Soroushian, Professor; Ph.D., Cornell. Structures and materials: fiber-reinforced concrete composites, durability and repair of concrete materials and structures, recycling in concrete.

William C. Taylor, Professor; Ph.D., Ohio State. Transportation: remediation of highway congestion, improved traffic safety computer modeling, intelligent transportation systems.

Thomas C. Voice, Professor; Ph.D., Michigan. Environmental: mass-transfer of chemical contaminants in systems of environmental interest, with an emphasis on the interactions between pollutants and soils, sediments, and suspended solids.

Roger B. Wallace, Associate Professor; Ph.D., Michigan. Hydrology and environmental: mechanics of fluids in porous media, contaminant transport and remediation in groundwater systems, with emphasis on the role of nonaqueous phase liquids.

David C. Wiggert, Professor; Ph.D., Michigan. Fluids and environmental: environmental fluid mechanics associated with groundwater remediation, fluid transients associated with free-surface flows and pipeline systems.

Thomas F. Wolff, Associate Professor; Ph.D., Purdue. Geotechnical: geotechnical reliability, probabilistic methods in geotechnical and pavement engineering, engineering properties of athletic field soils.

PURDUE UNIVERSITY

School of Civil Engineering

Programs of Study	Tremendous research and study opportunities exist in all areas of civil engineering, including environmental and hydraulic engineering (wastewater and water process systems, environmental chemistry and microbiology, pollution prevention and control, and surface and subsurface hydrology); environmental fluid mechanics; sediment transport; computational hydraulics, hydrology, and fluid mechanics; geotechnical engineering (soil and rock mechanics, foundations, dams, levees, highways and airfields, and geoenvironmental projects); materials engineering (nonmetallic construction materials, including concrete, cement, aggregates, bituminous materials, and other materials); structural engineering (design in concrete and/or steel, classical and computer analysis of structures, and solid mechanics); geomatic engineering (surveying, geodesy, photogrammetry, remote sensing, cartography, and geographic information systems); transportation and infrastructure systems engineering (transportation systems planning and management, traffic engineering, public infrastructure, intelligent transportation systems, and airport planning and design); and construction and engineering management (technical and managerial responsibilities of the constructor, scheduling, utilization of construction equipment, and management of fiscal and human resources).
Research Facilities	Excellent research facilities with cutting-edge equipment are available in all civil engineering disciplines and include internationally recognized endowed experimental laboratories and a broad range of computational facilities. The environmental and hydraulics engineering area has approximately 3,400 square meters of working space available for research. Geotechnical engineering's Bechtel Laboratory has conventional and automated laboratory equipment. The Pankow Concrete Materials Laboratory is used for research in cement and concrete. The Bituminous Laboratory and an accelerated pavement-testing facility offer extensive research opportunities. The Kettelhut Structural Engineering Laboratory is a 500-square-meter space designed to test full-scale building and bridge components. Modern loading equipment instrumentation, including a number of multichannel computer-controlled data acquisition systems, is available for measurement and data collection. Geomatics engineering has four laboratory facilities available for instruction and research. Transportation and infrastructure engineering facilities include an interactive traffic signal system laboratory with real-time traffic systems interface.
Financial Aid	Numerous fellowships, scholarships, and traineeships awarded by Purdue University, various state and federal agencies, and industrial sponsors are available each year. Teaching and research assistantships are available, with stipends that ranged from $950 to $1300 per month in 1998–99. The stipends also carry exemptions from University fees and tuition, except for approximately $330 per semester.
Cost of Study	In 1998–99, Indiana residents paid $1750 per semester for tuition and fees, and nonresidents paid $5860 per semester. Summer session charges are half of the semester rates.
Living and Housing Costs	University-supervised graduate residences cost from $285 to $575 per month in 1998–99. The University operates more than 1,300 married student apartments that rent at a reasonable rate. A variety of off-campus housing is available in the Lafayette and West Lafayette communities.
Location	Purdue's West Lafayette campus is situated across the Wabash River from Lafayette. The population of the combined metropolitan area (excluding students) is approximately 130,000. The area offers an exceptional variety of cultural activities, historic landmarks, and recreational attractions. The campus is located 60 miles north of Indianapolis and 130 miles south of Chicago.
The University and the School	Purdue is a comprehensive research university with an international reputation. Established as a land-grant institution in 1869, Purdue has grown from one campus with 39 students and 6 instructors to four campuses with a total enrollment of more than 65,000 students and more than 3,500 faculty members. The graduate programs in the Schools of Engineering have long been recognized as leaders in their disciplines based on the scope and quality of the curriculum, faculty, facilities, and student body.
Applying	Applications and supporting materials should be submitted at least six months before the beginning of the semester for which admission is sought. International students whose native language is not English are required to take the Test of English as a Foreign Language and obtain a score of 575 or better.
Correspondence and Information	Graduate Studies Office School of Civil Engineering Purdue University 1284 Civil Engineering Building West Lafayette, Indiana 47907-1284 E-mail: cegrad@ecn.purdue.edu World Wide Web: http://CE.www.ecn.purdue.edu/CE/

Purdue University

THE FACULTY AND THEIR RESEARCH

Vincent P. Drnevich, Professor and Head; Ph.D., Michigan. Earthquake engineering; soil structural dynamics; engineering properties of soils; nondestructive testing of soils, pavements, and structures; experimental methods.

V. James Meyers, Associate Professor and Assistant Head; Ph.D., Purdue. Computer analysis of structures.

Julio A. Ramirez, Professor and Director of Graduate Programs; Ph.D., Texas. Reinforced and prestressed concrete, experimental methods, prestressed concrete, structural analysis and design.

Dulcy M. Abraham, Associate Professor; Ph.D., Maryland. Construction automation and robotics, international construction, management information systems, environmental issues in construction, infrastructure assessment, infrastructure rehabilitation, resource leveling and allocation.

James E. Alleman, Professor; Ph.D., Notre Dame. Biological wastewater treatment, residuals management, environmental engineering history.

Adolph G. Altschaeffl, Professor; Ph.D., Purdue. Soils (engineering behavior), foundation engineering, dam engineering.

Graham C. Archer, Assistant Professor; Ph.D., Berkeley. Structural analysis, finite element analysis, philosophy of object-oriented design and programming, advanced research in nonlinear dynamic systems.

Katherine Banks, Associate Professor; Ph.D., Duke. Phytoremediation, bioremediation, wastewater treatment, environmental microbiology.

James S. Bethel, Associate Professor; Ph.D., Purdue. Photogrammetry/remote sensing, data adjustment, digital image processing.

Ernest R. Blatchley III, Professor; Ph.D., Berkeley. Water treatment, wastewater treatment, environmental chemistry.

Antonio Bobet, Assistant Professor; Ph.D., MIT. Numerical and experimental modeling, engineering geology, rock mechanics, soil mechanics.

P. L. Bourdeau, Associate Professor; Dr.Eng.Sc., Swiss Federal Institute of Technology. Soil mechanics, geosynthetics, reliability, probability and stochastic processes, computer modeling.

Mark D. Bowman, Professor; Ph.D., Illinois. Structural steel design, structural analysis and design, structural fatigue, structural models and experimental methods, codes and standards.

Dan D. Budny, Associate Professor; Ph.D., Michigan State. Hydraulics, structures, innovative teaching, engineering education.

Darcy Bullock, Associate Professor; Ph.D., Carnegie Mellon. Real-time traffic signal systems, traffic engineering.

Luh-Maan Chang, Associate Professor; Ph.D., Texas. Construction productivity/quality management, autoidentification systems in construction, construction materials.

Wai-Fah Chen, George E. Goodwin Distinguished Professor; Ph.D., Brown. Structural mechanics, behavior of structures, mechanics of materials.

Menashi D. Cohen, Professor; Ph.D., Stanford. Cement, concrete, composites.

Sidney Diamond, Professor; Ph.D., Purdue. Chemistry and physics of cement and concrete, microstructure evaluation by scanning electron microscopy, admixtures, alkali-aggregate reactions and other durability problems, clay mineralogy and soil chemistry.

Patrick J. Fox, Associate Professor; Ph.D., Wisconsin. Soil behavior, geoenvironmental engineering, computer modeling, experimental methods.

Jon D. Fricker, Professor; Ph.D., Carnegie Mellon. Transportation planning, network analysis, transportation systems analysis, public transportation.

Robert J. Frosch, Assistant Professor; Ph.D., Texas. Behavior and design of structural concrete, earthquake engineering, repair and rehabilitation of structures, autoadaptive structures.

John T. Gaunt, Associate Professor; Ph.D., Purdue. Structural steel design, architectural engineering, structural analysis and design, structural stability.

Rao S. Govindaraju, Associate Professor; Ph.D., California, Davis. Modeling of surface and subsurface flows, contaminant transport, and surface erosion.

D. W. Halpin, Professor and Head, Division of Construction Engineering and Management; Ph.D., Illinois. Construction methods, simulation of operations, automation and robotics in construction, international competition in engineering and construction, emerging construction technology.

David J. Harmelink, Assistant Professor; Ph.D., Iowa State. Computer applications in construction, linear scheduling, computer-aided design.

Adam J. Hand, Assistant Professor; Ph.D., Nevada. Transportation materials, pavement design, pavement rehabilitation and maintenance, pavement management systems, performance-related specifics, geotechnical engineering.

Midhat Hondzo, Assistant Professor; Ph.D., Minnesota. Environmental fluid dynamics, water quality modeling.

I. Hua, Assistant Professor; Ph.D., Caltech. Water quality, aquatic and environmental chemistry, ultrasonic irradiation, remediation processes.

Robert B. Jacko, Professor; Ph.D., Purdue. Air pollution management and control, transportation noise problems, environmental occupational safety.

Chad T. Jafvert, Professor; Ph.D., Iowa. Environmental chemistry, water quality modeling, environmental remediation.

Garrett D. Jeong, Associate Professor; Ph.D., Caltech. Structural mechanics, dynamics and vibrations, earthquake engineering, computer applications.

Steven D. Johnson, Associate Professor; Ph.D., Wisconsin. Land surveying, engineering surveys and analysis, remote sensing/photogrammetry.

Dennis A. Lyn, Associate Professor; Ph.D., Caltech. Experimental, environmental, and computational fluid mechanics; turbulent flows; sediment transport.

Edward M. Mikhail, Professor; Ph.D., Cornell. Photogrammetry/remote sensing, image exploitation and analysis, automated/digital mapping.

Loring F. Nies, Assistant Professor; Ph.D., Michigan. Aerobic/anaerobic biotransformation of organic pollutants, abiotic organometallic and microbially mediated reductive dehalogenation, in-situ bioremediation, pollutants in groundwater and sediments.

Jan Olek, Associate Professor and Director, North Central Superpave Center; Ph.D., Purdue. Durability of portland cement concrete, corrosion of reinforcement, high-performance concrete, mineral admixtures in highway applications, asphalt emulsions, SUPERPAVE technology.

Srinivas Peeta, Assistant Professor; Ph.D., Texas. Transportation systems analysis, traffic flow theory, operations research, simulation modeling, dynamic analysis of transportation systems usage, network analysis.

A. Ramachandra Rao, Professor; Ph.D., Illinois. Hydrology, water resources, hydraulic engineering, sediment transportation, surface runoff, hydromechanics, computational hydraulics and hydrology, stochastic hydrology.

Rodrigo Salgado, Associate Professor; Ph.D., Berkeley. Foundation engineering, geotechnical analysis, soil dynamics and earthquake engineering.

Gilbert T. Satterly, Professor; Ph.D., Northwestern. Transportation planning, highway design, public transportation, traffic engineering.

Charles F. Scholer, Professor; Ph.D., Purdue. Portland cement concrete, aggregates, properties of concrete for construction, mechanical/durability properties of hardened concrete, curing and finishing, mineral aggregates for construction, low-volume road maintenance and construction.

Jie Shan, Assistant Professor; Ph.D., Wuhan Technical (China). Space mapping, GIS imagery integration, geocoding, feature extraction.

Kumares C. Sinha, Olson Distinguished Professor and Director, Joint Transportation Research Program; Ph.D., Connecticut. Infrastructure planning and management, transportation economics and evaluation, transportation safety, urban and regional planning.

M. J. Skibniewski, Professor and Assistant Executive Vice President for Academic Affairs, Purdue University; Ph.D., Carnegie Mellon. Construction automation and robotics, computer-aided engineering decision support systems, management and legal aspects of engineering.

Elisa D. Sotelino, Associate Professor; Ph.D., Brown. Computational mechanics, high-performance computing, structural dynamics.

M. A. Sozen, Kettelhut Distinguished Professor; Ph.D., Illinois. Earthquake and blast response of structures, reinforced and prestressed concrete, concrete dams.

C. Douglas Sutton, Associate Professor; Ph.D., Purdue. Reinforced and prestressed concrete structures, experimental analysis and physical models, masonry structures, structural analysis.

Andrzej Tarko, Assistant Professor; Ph.D., Cracow University of Technology (Poland), Ph.D., Illinois. Traffic flow theory, simulation modeling, highway safety.

Robert K. Tener, Assistant Professor and Industry-Student Internship Program Liaison; Ph.D., Iowa State. Construction organization management, construction program management, design-build, life-cycle engineering, quality management, educational outcomes assessment.

Boudewijn H. W. van Gelder, Associate Professor; Ph.D., Ohio State. Geodesy and satellite geodesy, surveying and data adjustment, deformation analysis.

Timothy M. Whalen, Assistant Professor; Ph.D., Cornell. Engineering mechanics, linear and nonlinear dynamics, structural dynamics, wind engineering, stochastic and random dynamics, continuum mechanics.

Thomas D. White, Professor; Ph.D., Purdue. Asphalt and asphalt mixtures; highway and airfield pavement design; construction, evaluation, and performance; airfield design; prototype and laboratory accelerated testing; material and pavement structural modeling.

Robert K. Whitford, Professor; Ph.D., Purdue. Airport design, freight transportation, air transportation systems, urban planning.

Jeff R. Wright, Professor; Director, Water Resources Research Center; and Assistant Dean, Schools of Engineering; Ph.D., Johns Hopkins. Civil and environmental systems engineering and infrastructure management.

Ronald F. Wukasch, Professor; Ph.D., Purdue. Environmental engineering, environmental management, solid and hazardous waste.

RENSSELAER POLYTECHNIC INSTITUTE

Department of Civil Engineering

Programs of Study

Rensselaer offers Doctor of Philosophy, Doctor of Engineering, Master of Science, and Master of Engineering degrees in civil engineering and transportation engineering. Graduate degree specializations offered include computational mechanics, geotechnical engineering, infrastructure engineering, mechanics of composite materials and structures, structural engineering, and transportation engineering. The M.Eng. program is a nonthesis degree intended for professional practice. A student with an accredited B.S. or its equivalent can typically complete this degree in one year.

Degree requirements for the Doctor of Philosophy and Doctor of Engineering are determined on an individual basis according to the background and interests of the candidate and faculty adviser suggestions. Ninety credit hours are required beyond the bachelor's degree (60 beyond the master's degree) plus successful completion of a preliminary examination, a candidacy examination, and a thesis. Master's programs consist of 30 credits of required and elective courses, including a thesis or project. Students select an area of specialization in consultation with advisers.

Research Facilities

Graduate students are an integral part of all research programs at Rensselaer. They participate in research on a wide range of projects. The mechanics of composite materials and structures program at Rensselaer has the benefit of two dedicated composite materials laboratories. The scientific computation research center operates a parallel computer and a UNIX workstation environment with common file servers. A state-of-the-art infrastructure engineering research facility is housed in the Jonsson Engineering Center. The cyclic soils laboratory has conducted significant research for more than fifteen years in the areas of earthquake engineering and soil dynamics. The Rensselaer Geotechnical Centrifuge facility started operating in 1989. The centrifuge is a 100 g-ton, state-of-the-art machine that can test up to a 1-ton payload at centrifugal acceleration of 100 g or 0.5 tons at 200 g. The arm has a nominal radius of 2.7 meters to the center of the soil or soil structure model. Structural testing facilities include 150 k and 300 k universal testing machines as well as a water flume for evaluating snow drift loads on model building roofs. The transportation laboratory consists of a large number of PCs, principally 486/66 MHz machines equipped with laser printers.

Financial Aid

Financial aid is available to full-time students in the form of fellowships, research or teaching assistantships, and scholarships. The stipend ranged between $10,000 and $11,000 for the nine-month academic year in 1998–99. In addition, full tuition is usually granted. Additional compensation for study during the summer months may also be available. Outstanding students may qualify for University-supported Rensselaer Scholar Fellowships ($15,000 plus full tuition and fees scholarships). Low-interest, deferred repayment graduate loans are also available to U.S. citizens with demonstrated need.

Cost of Study

Tuition for 1999–2000 is $665 per credit hour. Other fees amount to approximately $535 per semester. Books and supplies cost about $1700 per year.

Living and Housing Costs

The cost of rooms for single students in residence halls or apartments ranges from $3356 to $5298 for the 1999–2000 academic year. Family student housing, with a monthly rent of $592 to $720, is available.

Student Group

There are about 4,300 undergraduates and 1,750 graduate students representing all fifty states and more than eighty countries at Rensselaer.

Student Outcomes

Eighty-eight percent of Rensselaer's 1998 graduating students were hired after graduation with starting salaries that averaged $56,259 for master's degree recipients and $57,000–$75,000 for doctoral degree recipients.

Location

Rensselaer is situated on a scenic 260-acre hillside campus in Troy, New York, across the Hudson River from the state capital of Albany. Troy's central Northeast location provides students with a supportive, active, medium-sized community in which to live; an easy commute to Boston, New York, and Montreal; and some of the country's finest outdoor recreation sites, including Lake George, Lake Placid, and the Adirondack, Catskill, Berkshire, and Green Mountains. The Capital Region has one of the largest concentrations of academic institutions in the United States. Sixty thousand students attend fourteen area colleges and benefit from shared activities and courses.

The University

Founded in 1824 and the first U.S. college to award degrees in engineering and science, Rensselaer Polytechnic Institute today is accredited by the Middle States Association of Colleges and Schools and is a private, nonsectarian, coeducational university. All engineering undergraduate degree programs are also accredited by the Accreditation Board for Engineering and Technology (ABET). Rensselaer has five schools—Architecture, Engineering, Management, Science, and Humanities and Social Sciences. The School of Engineering is nationally ranked among the top twenty engineering schools by a *U.S.News & World Report* survey and is ranked in the top ten by practicing engineers.

Applying

Admission applications and all supporting credentials should be submitted well in advance of the preferred semester of entry to allow sufficient time for departmental review and processing. The application fee is $35. Since the first departmental awards are made in February and March for the next full academic year, applicants requesting financial aid are encouraged to submit all required credentials by February 1 to ensure consideration.

Correspondence and Information

For written information about graduate study:
Department of Civil Engineering
Rensselaer Polytechnic Institute
110 8th Street
Troy, New York 12180-3590
Telephone: 518-276-6360
World Wide Web: http://www.rpi.edu/dept/civil/html/

For applications and admissions information:
Director of Graduate Academic and Enrollment
    Services, Graduate Center
Rensselaer Polytechnic Institute
110 8th Street
Troy, New York 12180-3590
Telephone: 518-276-6789
E-mail: grad-services@rpi.edu
World Wide Web: http://www.rpi.edu

Rensselaer Polytechnic Institute

THE FACULTY AND THEIR RESEARCH

George F. List, Professor and Chair; Ph.D., Pennsylvania; PE. Intelligent transportation systems, real-time control of system operation, especially for networks of signalized intersections; hazardous materials transportation; capacity investment decision making, especially for highway and railroad networks; operations planning, routing, scheduling, and fleet sizing.

Professors

Nicholas L. Clesceri, Ph.D., Wisconsin; PE. Heavy metals recovery from industrial wastes, site remediation, physical-chemical soil containment, pretreatment for enhancing biodegradation, unsaturated soil organic contaminant modeling for clean-up kinetics, Environmentally Sound Manufacturing (ESM) planning and development.

Ricardo Dobry, D.Sc., MIT. Earthquake site response and building/bridge code development, liquefaction failure and permanent deformations of natural soil deposits and earth embankments, earthquake geotechnical centrifuge model testing, seismic design and retrofitting of highway structures, characterization of soil properties for cyclic loading evaluation, constitutive stress-strain relations, micromechanical computer simulations of granular soils.

George J. Dvorak, William Howard Hart Professor of Mechanics; Ph.D., Brown. Mechanics and mechanical behavior of heterogeneous media, composite materials and structures, fracture and fatigue.

Larry J. Feeser, Ph.D., Carnegie Mellon; PE. Structures, computer applications and computer-aided design, structural optimization.

Jacob Fish, Ph.D., Northwestern. Mathematical modeling, finite element and boundary element methods, adaptive technique, multiscale and hierarchical computational methods, micromechanics of composite materials, postprocessing techniques.

Dimitri A. Grivas, Ph.D., Purdue. Infrastructure engineering, geostochastics, risk assessment and management, design of geotechnical structures under static and seismic loads, damage analysis, design of experiments and data analysis, application of fuzzy sets and expert systems in civil and transportation engineering.

Michael J. O'Rourke, Ph.D., Northwestern; PE. Snow loads on building, particularly drift loads on multilevel roofs; design and behavior of buried pipelines subject to earthquakes.

Mark S. Shephard, Ph.D., Cornell. Automated, adaptive finite element modeling techniques; parallel processing; material process analysis; computational biomechanics; analysis procedures for unsteady aerodynamics; analysis of advanced composites.

Thomas F. Zimmie, Ph.D., Connecticut; PE. Landfill siting and design, groundwater hydrology, groundwater contamination, centrifuge modeling of geoenvironmental problems, physical-chemical phenomena in soils, subsurface drainage, geosynthetics, experimental soil dynamics, solid and hazardous waste disposal, sediment transport in rivers, problems on the geotechnical-environmental interface.

Assistant Professors

Jeffrey L. Adler, Ph.D., California, Irvine. Intelligent vehicle/highway systems: advanced traveler information systems and advanced traffic management systems, interactive computer simulation, pre-trip and en-route driver behavioral choice, expert systems and artificial intelligence applications for transportation systems modeling and analysis, adaptive signal control and network optimization.

James Kilduff, Ph.D., Michigan. Physicochemical processes, separations and recovery processes in water and wastewater treatment, effects of adsorption and mass transfer on pollutant fate and transport in natural systems, membrane processes for water quality control.

Simeon J. Komisar, Ph.D., Washington. Biological wastewater treatment, fate and effect of pollutants, treatment of hazardous wastes.

J. Russell Manson, Ph.D., Glasgow (Scotland). Mathematical modeling of flow, fate and transport in fluvial and lacustrine systems, environmental hydrology fluid mechanics, applied mathematics, episodic pollution.

Mourad Zeghal, Ph.D., Princeton. Soil dynamics and geotechnical earthquake engineering, computational geomechanics, geotechnical system identification and seismic response monitoring, damage diagnosis and nondestructive evaluation, seismic risk analyses.

Research Areas

Computational Mechanics: Automated, adaptive finite element modeling techniques; parallel processing; multiscale modeling; nonlinear analysis of composite materials; analysis procedures for unsteady aerodynamics problems; materials process modeling; computational biomechanics; automatic mesh generation.

Geotechnical Engineering: Static and dynamic behavior of soils and foundations (experimental and analytical), deterministic and probabilistic evaluation of the performance of earth structures and soil-foundation systems, centrifuge modeling of soil and soil-structure systems for earthquake engineering and environmental-geotechnical purposes, bridge foundation and abutment seismic retrofitting and design, water flow through soils, system identification and computational modeling of seismic soil response, geotechnical environmental engineering.

Infrastructure Engineering: Decision methodologies for pavement and bridge management, network optimization, probabilistic bridge scour analysis, computerized image processing for condition assessment, reliability-based maintenance, decision support and expert systems, database design for infrastructure engineering, risk analysis and management.

Mechanics of Composite Materials and Structures: Development of micromechanical material models and overall constitutive equations, elastic-plastic and thermoviscoplastic behavior, damage mechanics, development of finite element programs, structural response in plate and shell structures, modeling of electromechanical response of piezocomposites and smart materials, modeling of Functionally Gradient Material (FGM) systems.

Structural Engineering: Uses of new materials, development of model analysis techniques, structural frames, analysis and design, computer graphics, environmental loads on structures, structural optimization.

Transportation Engineering: Intelligent transportation systems, real-time management of transport networks, goods movement, hazardous materials transport, freight service planning, emergency response team siting and planning, advanced public transportation systems.

STANFORD UNIVERSITY

School of Engineering
Department of Civil and Environmental Engineering

Programs of Study

Programs of study in construction engineering and management, design/construction integration, environmental engineering and science, environmental fluid mechanics and hydrology, and structural engineering and geomechanics lead to the M.S., Engineer, and Ph.D. degrees.

Students with a bachelor's degree in civil engineering from an accredited curriculum can satisfy the requirements for the M.S. by completing a minimum of 45 units of study. No thesis is required. The M.S. can be completed in three quarters. Students without a civil engineering background may be required to complete undergraduate courses that are prerequisites to required graduate courses.

Students with an M.S. in civil engineering may satisfy the requirements for the engineer degree by completing a further 45 units of study and an acceptable thesis. The engineer degree can be completed in six quarters.

Requirements for the Ph.D. include a minimum of 90 units of study, a general qualifying exam, a University dissertation oral examination, and an acceptable dissertation. All Ph.D. candidates are required to serve as teaching or course assistants for one quarter in order to gain instructional experience. The Ph.D. can be completed in three to four years after the M.S.

Research Facilities

The John A. Blume Earthquake Engineering Center provides facilities for theoretical and experimental research in earthquake engineering. Within the center are the structural testing, model testing, and advanced technologies laboratories. The center contains equipment for static and dynamic experimentation, including two shaking tables, structural dynamics analyzer equipment, and a test bed for investigating structural components and systems. Its data processing and numerical analysis laboratory contains computers and peripherals.

The Center for Integrated Facility Engineering encompasses a computing lab with graphical workstations and advanced artificial intelligence and three-dimensional CAD software for the purpose of research on the use of information technology for improved integration and automation of the facility development (A/E/C) process.

The Project-Based Learning Laboratory provides a flexible computing facility for teaching and learning multidisciplinary, collaborative, geographically distributed teamwork. The computational infrastructure is equipped with emerging hardware and software information technologies that include videoconferencing and desktop sharing, Internet-mediated collaboration tools, ED CAD and 4D CAD, and World Wide Web workspaces for teamwork.

The Water Quality Control Research Lab is equipped for studies of trace organic and inorganic contaminants, treatment, colloid, surface, and radiochemical research. Analytical facilities include GC/MS/MS, GC/MS, 9GS HPLC, IC, AA, ICP-MS, and instruments for solids characterization.

The Environmental Microbiology Laboratory is equipped to conduct isolations of novel aerobic and strict anaerobic bacteria and to investigate bacterial metabolism quantitatively with analytical instruments of the Water Quality Control Research Lab. Microbial ecology and cell movement are currently studied with state-of-the-art epifluorescence microscopes and a digital image processing unit. Instruments integral in conducting experiments in molecular microbiology are PCR amplifier, gene pulser, DNA sequencing and DNA separation equipment, and pulsed field gel apparatus.

The Western Region Hazardous Substance Research Center is an EPA-sponsored center that conducts cooperative research between Stanford and Oregon State University and the respective Schools of Engineering, Earth Sciences, and Medicine to develop advanced physical, chemical, and biological processes for restoration of quality of water, soils, and groundwater contaminated with hazardous chemicals.

The Air Quality Research Lab is specially equipped for field sampling and analysis of organic and inorganic particulate matter. In addition to the analytical equipment listed under the Water Quality Control Research Lab, facilities include microbalances, a tandem differential mobility analyzer, a monodisperse aerosol generator, and a termooptical organic and elemental carbon analyzer.

The Environmental Fluid Mechanics Laboratory has extensive facilities for experimental and computational research. The laboratory has flumes for studying wavy and boundary-layer flows, a rotating table for geophysical fluid dynamics experiments, and tanks for studying thermally stratified and salt-stratified flows. Instrumentation for these facilities includes several laser and acoustic Doppler velocimeters and a variety of high-performance cameras and laser-light sources for quantitative flow visualization techniques like laser-induced fluorescence and digital particle image velocimetry. Computational resources include a Cray J90 minisupercomputer, a Silicon Graphics Origin 2000 workstation, and a network of approximately twenty workstations and personal computers.

The department Intranet centers on two Novell NetWare file and print servers and one Macintosh e-mail and Web server. The 10-Mbps Ethernet is employed for Intranet, campus backbone, and Internet connections. Department computing resources include Macintosh and Windows computers as well as DEC, Sun, and SGI workstations/hosts for advanced projects.

The Engineering Library, one of eight science libraries on campus, contains approximately 1,800 active serial titles, 50,000 monographs, 5,500 Stanford theses, and thousands of technical reports. All the holdings of the library are available via Socrates, the online catalog. For further information, students should refer to the World Wide Web (http://www-sul.stanford.edu/depts/eng).

Financial Aid

The department maintains a large and continuing program of financial aid for graduate students. Fellowship or scholarship awards for the academic year typically range from $1000 to $30,000.

Cost of Study

Tuition in the School of Engineering is $8196 per quarter. Books are approximately $380 per quarter.

Living and Housing Costs

Several types of graduate housing are available on campus. Students should visit the home page for housing services at http://www.stanford.edu/dept/hds/ for current rates.

Student Group

There are 177 men and 72 women in the department. In the 1998–99 academic year, there were 92 international students representing thirty-four other countries. Sixty-four percent of the students received financial aid.

Location

The campus extends from the wooded area surrounding Palo Alto to the foothills of the Coast Range. San Francisco is 35 miles to the north. Boating is available on nearby San Francisco Bay and on Lagunita, the campus lake. Pacific beaches and the redwoods are a 45-minute drive to the west. The Sierra Nevada snow country, the wine-producing areas of the state, the Gold Rush country, the Monterey-Carmel area, and Big Sur are within easy reach of the campus.

The School

Enrollment at Stanford University is approximately 7,000 graduate students and 6,500 undergraduates. The School of Engineering is the second largest on campus, with an enrollment of 4,173.

Applying

Applications become available August 1. The deadline for applications for applicants who request financial aid is January 1. The application deadline for admission only is March 1. Students who are seeking an M.S. degree must start in Autumn Quarter. Requests for applications should be directed to the Student Services Office, Department of Civil and Environmental Engineering.

Correspondence and Information

Student Services Office
Department of Civil and Environmental Engineering
Stanford University
Stanford, California 94305-4020

Telephone: 650-725-2387
Fax: 650-725-8662
E-mail: lane@ce.stanford.edu
World Wide Web: http://www-ce.stanford.edu/

Stanford University

THE FACULTY AND THEIR RESEARCH

Construction Engineering and Management. The CEM program prepares technically qualified students for responsible management roles in all phases of the development of major constructed facilities. Emphasis is placed on management techniques that are useful in planning, coordinating, and controlling the activities of diverse specialists—designers, contractors, subcontractors, and client representatives—within the unique project environment of the construction industry.

Hans Björnsson, Professor; Ph.D., Chalmers University of Technology (Sweden), 1974. Computer-supported collaboration, modeling design and construction processes, evaluation of IT investments.

Martin Fischer, Assistant Professor; Ph.D., Stanford, 1991. Symbolic product and process models, integrating models to facilitate data exchange among project participants and phases, computer-aided construction planning, automating constructability feedback.

Raymond E. Levitt, Professor; Ph.D., Stanford, 1975. Systematic (re)engineering of organizations engaged in project-oriented work processes, computational models of project-oriented work processes.

Boyd Paulson, Professor; Ph.D., Stanford, 1971. Computer applications in construction, operations analysis and design, project management, affordable housing.

Clyde B. Tatum, Professor; Ph.D., Stanford, 1983. Technological innovation in construction, mechanisms and strategies for innovation in construction and design/construction integration.

Environmental Engineering and Science. This program covers areas such as water quality and hazardous substance control; air and land pollution; and physicochemical, biological, and engineering aspects of water quality and water pollution control, including groundwater remediation and hazardous chemical treatment. Opportunities also exist for specializations in atmospheric chemistry, physics, and pollutant transport.

Craig S. Criddle, Associate Professor; Ph.D., Stanford, 1990. Environmental biotechnology, bioaugmentation, applied microbial ecology, bioremediation, fate of persistent contaminants.

Lynn M. Hildemann, Associate Professor; Ph.D., Caltech, 1989. Air pollution engineering, atmospheric chemistry, air pollution sources and modeling, control strategy design, aerosol dynamics, sources and fates of organic aerosols in the atmosphere.

James O. Leckie, Professor; Ph.D., Harvard, 1970. Chemical pollutant behavior in natural aquatic systems and engineered processes, environmental aspects of surface and colloid chemistry and low temperature aqueous geochemistry of trace elements.

Gilbert M. Masters, Professor; Ph.D., Stanford, 1966. Environmental quality and energy consumption, renewable energy systems and energy conservation, impact of pollution prevention technology on reducing global climate change and acid rain, improving urban air quality.

Martin Reinhard, Professor; Dr. Tech.Sc., Eidgenossische Tech (Switzerland), 1977. Environmental fate of organic substances, methods for instrumental analysis of organic trace contaminants in environmental matrices, processes that affect transformation and attenuation of organic chemicals in soils, natural waters, and treatment systems.

Paul V. Roberts, Professor; Ph.D., Cornell, 1965. Physical and chemical processes affecting water quality in natural systems, water treatment, trace organic behavior, groundwater contaminants, water reuse, mass transfer.

Alfred M. Spormann, Assistant Professor; Dr. Rer. Nat., Phillips (Germany), 1989. Bacterial physiology and microbial ecology, metabolism and biochemistry of aromatic hydrocarbon degradation by anaerobic bacteria, molecular mechanism of bacterial cell movement on solid surfaces.

Environmental Fluid Mechanics and Hydrology. The EFMH program focuses on developing an understanding of the physical processes that control the movement of mass, energy, and momentum in water and atmosphere environments. Courses address fluid and sediment transport and mixing processes, turbulence and its modeling, fluid mechanics of stratified flows, natural flows in coastal waters, estuaries, lakes, and open channels, and experimental methods. Hydrology courses consider stochastic methods in both surface and subsurface hydrology, watershed hydrology and modeling, and flow and transport in porous media. Atmospheric modeling and environmental policy courses are also given.

David L. Freyberg, Associate Professor; Ph.D., Stanford, 1981. Uncertainty in predicting hydrologic phenomena, contaminant transport in spatially variable aquifers, numerical analysis techniques applied to simulation of subsurface flow and transport processes.

Mark Z. Jacobson, Assistant Professor; Ph.D., UCLA, 1994. Computer modeling of atmospheric pollution, numerical algorithms for simulating gas and aerosol chemical and microphysical processes.

Peter K. Kitanidis, Professor; Ph.D., MIT, 1978. Physically based stochastic hydrology to estimate spatial processes and validate and calibrate groundwater models, dilution and mixing of soluble substances in heterogeneous geologic formations, in-situ decay of pollutants in the ground.

Jeffrey R. Koseff, Professor; Ph.D., Stanford, 1983. Estaurine and coastal systems hydrodynamics, nature of turbulence in stratified environments, numerical techniques for simulating estuarine and geophysical flows, hydrodynamics of benthic grazing by bivalve feeders, hydrodynamics of chemical sensing in marine environments.

Stephen G. Monismith, Associate Professor; Ph.D., Berkeley, 1983. Application of fluid mechanics principals to the analysis of flow processes operating in rivers, lakes, estuaries, and the oceans.

Leonard Ortolano, Professor; Ph.D., Harvard, 1969. Implementation of environmental policies and programs in the U.S. and developing countries.

Robert L. Street, Professor; Ph.D., Stanford, 1963. Geophysical and sediment transport fluid motions, numerical simulations of ocean and atmospheric flows, stratified flows in lakes and coastal seas, modeling of turbulence in fluid flows.

Structural Engineering and Geomechanics. The program is designed to prepare students for careers in the consulting profession, industry, government, and academia. Instructional programs and research opportunities are available in the areas of structural analysis and design, earthquake engineering, structural dynamics, risk and reliability analysis, computational mechanics, geomechanics, and computer-aided engineering.

Ronaldo I. Borja, Associate Professor; Ph.D., Stanford, 1984. Geomechanics, geotechnical engineering, computational mechanics, developing constitutive and strain localization models for soils, modeling the dynamic nonlinear response of soil-structure interaction systems.

C. Allin Cornell, Professor; Ph.D., Stanford, 1964. Application of risk-based engineering criteria and probabilistic models to the analysis of buildings and offshore structures.

Gregory G. Deierlein, Associate Professor; Ph.D., Texas at Austin, 1988. Nonlinear structural analysis, design and behavior of steel and concrete structures, performance-based earthquake engineering, fracture mechanics applications to steel structures.

Anne S. Kiremidjian, Professor; Ph.D., Stanford, 1976. Development of spatial and temporal stochastic models for earthquake occurrences and seismic ground motion; structural component and systems reliability methods; structural damage evaluation models; regional damage.

Helmut Krawinkler, Professor; Ph.D., Berkeley, 1971. Nonlinear dynamic behavior of steel and reinforced concrete structures, knowledge-based systems for conceptual structural design, seismic code developments.

Kincho H. Law, Professor; Ph.D., Carnegie-Mellon, 1981. Engineering informatics; computational algorithm and mechanics; software platform for engineering analysis and design; computational mechanics; engineering information management; parallel, distributed, and Internet computing.

Laura N. Lowes, Assistant Professor; Ph.D., Berkeley, 1999. Development of analytical models for investigating and predicting the response of structures subjected to dynamic loading: constitutive theory, finite element methods, nonlinear analysis of structural systems, experimental investigation of structural component behavior.

H. Allison Smith, Assistant Professor; Ph.D., Duke, 1989. Development of finite-element modeling techniques for dynamic analyses, modeling and analysis of actively controlled structures, use of adaptive finite element modeling techniques.

STATE UNIVERSITY OF NEW YORK AT BUFFALO

Department of Civil, Structural and Environmental Engineering

Programs of Study

The Department of Civil, Structural and Environmental Engineering (CSEE) offers programs of study leading to M.Eng., M.S., and Ph.D. degrees. Areas of excellence in the department include high-performance structures and systems, earthquake engineering and advanced structural control technologies, environmental engineering and science, and advanced computational engineering mechanics and boundary-element techniques. In addition, construction management and engineering, geomechanics, geoenvironmental and foundation engineering, water resources engineering, and environmental fluid mechanics are major areas of study in the department. Master's degrees require completion of 30 credits of graduate course work, usually including from 3 to 6 credits for a project (M.Eng.) or thesis (M.S.). The Ph.D. degree requires completion of 72 credits of graduate course work, of which 30 credits may be counted from a master's program and of which 12 to 24 credits may be used for dissertation research. The M.Eng. program is a design- and practice-oriented program suitable for students planning to pursue a professional career in consulting, industry, or government service. Full-time students complete an M.Eng. degree within ten months. The M.S. program provides an intensive education in both fundamental and applied aspects of civil engineering. General degree requirements include 24 credits of approved graduate course work and 6 credits of thesis or 30 credits of course work with a comprehensive exam. Core course requirements must be satisfied, depending on the specific program of study chosen. The department also administers a degree program leading to an M.S. in environmental science. Core course requirements must be satisfied by students in the Ph.D. program, who must also pass a qualifying exam and defend a dissertation.

Research Facilities

The department houses structural and geotechnical laboratories, including one of the most advanced structural engineering and earthquake simulation testing facilities in the country; state-of-the-art computational hardware and software; and extensive environmental engineering and environmental hydraulics laboratories. The earthquake simulator, a platform of 12 feet by 12 feet, allows testing with five degrees of freedom that can be individually programmed. Structural models of up to 50 metric tons can be shaken with a maximum acceleration of .625g in the horizontal direction and 1.05g in the vertical direction. The structural laboratories offer material testing equipment for static and dynamic loading, for structural models, or for full-scale components testing. Two geotechnical laboratories have facilities for carrying out soil-structure interaction studies under both static and dynamic loading conditions. The primary capabilities of the Environmental Engineering Research laboratories include development and testing of biochemical and physicochemical processes for environmental management, modeling fate and transport of contaminants in natural and engineered systems, and development of advanced techniques for pollutant analysis. The Environmental Fluid Mechanics Laboratory contains a recirculating water tunnel, a 20-meter tilting open channel flume with sediment transport capacity, and a 3.5 meter by 5 meter rotating laboratory for the study of geophysical fluid flow.

Financial Aid

Financial support is available on a competitive basis, either through tuition scholarships, research and teaching assistantships, or University Scholarships and Lectureships. Students in the M.Eng. program are eligible for tuition scholarships only. The current stipend for assistantships ranges from $10,700 to $15,000 for the academic year (summer support may also be available), usually in addition to a full tuition scholarship. Research assistantships are funded through extramural grants and are awarded by individual faculty members. University Scholarships and Lectureships are awarded through campuswide competition.

Cost of Study

Full-time tuition for the 1998–99 academic year was $2550 per semester for New York residents and $4208 per semester for out-of-state residents. Fees, books, and other miscellaneous expenses are an additional $500–$600 per semester.

Living and Housing Costs

Graduate students generally live off campus, where different housing accommodations are available in various price ranges. The Off-Campus Housing Office maintains a file of housing availabilities. The basic cost of living for a single student in the Buffalo area is about $9000 to $10,000 per year. Moderate entertainment and travel expenses are additional.

Student Group

There are currently about 160 students in the department, 140 of whom are full-time. Of the full-time students, approximately two thirds are international and 20 percent are women. About one quarter of graduate students are working toward their Ph.D. More than 80 percent of full-time students receive financial aid. The average undergraduate grade point average for entering students is above 3.2, and average GRE scores (combined verbal and quantitative) are above 1200.

Student Outcomes

Most graduates with master's degrees find employment in private industry, consulting, or government agencies, while those with doctoral degrees generally work in research labs or in academic positions. Over the past twenty years, nearly half of the Ph.D. graduates from the department have gone on to take positions in academia.

Location

Buffalo is located in western New York along the shores of Lake Erie and the Niagara River. It has a population of about 350,000, with the county total more than 1 million. Scenic, recreational, and cultural opportunities include museums, a music hall, professional sports teams, a downtown theater district, Niagara Falls 15 miles to the north, and nearby winter skiing resorts.

The University and The Department

The University was founded in 1846 as a medical school. In 1962 it was incorporated into the rapidly expanding State University of New York System and became one of its four main University centers. Expansion in the late 1960s led to development of the north campus on a 1,600-acre site in Amherst, where the School of Engineering is presently located. The CSEE department was established in 1958 and currently has 25 full-time faculty members.

Applying

Applications for admission and financial aid should be submitted before January 15 for September admission. All international students must submit TOEFL scores. GRE scores (General Test) are required for financial aid consideration. Applications for admission only should be submitted at least several months before the start of the semester of entrance (earlier deadlines may apply for applicants from certain countries). First-round decisions for September admission are mailed by the end of February. Application forms and updated instructions are available from the Web site listed below.

Correspondence and Information

Department of Civil, Structural and Environmental Engineering
212 Ketter Hall
State University of New York at Buffalo
Buffalo, New York 14260

Telephone: 716-645-2114 Ext. 2333
Fax: 716-645-3667
E-mail: dianeh@civil.eng.buffalo.edu
World Wide Web: http://www.civil.buffalo.edu

State University of New York at Buffalo

THE FACULTY AND THEIR RESEARCH

Department Chair: Andrei M. Reinhorn

Director of Graduate Studies: Joseph F. Atkinson

Shahid Ahmad, Ph.D., SUNY at Buffalo. Foundation dynamics, boundary-element method, wave propagation, vibration isolation.

Amjad Aref, Ph.D., Illinois. Computational mechanisms, composite materials, earthquake engineering.

Joseph F. Atkinson, Ph.D., MIT. Environmental fluid mechanics, sediment transport, water quality modeling, geophysical fluid dynamics.

Prasanta K. Banerjee, Ph.D., Southampton (U.K.). Soil dynamics, constitutive relationships, boundary-element methods, soil-structure interaction.

Cemal Basaran, Ph.D., Arizona. Damage mechanics, constitutive modeling, finite-element method, dynamic analysis.

Michel Bruneau, Ph.D., Berkeley. Seismic evaluation and retrofit of steel bridges, steel and masonry buildings.

Stuart S. Chen, Ph.D., Lehigh. Expert systems, metal structures, bridge engineering.

Michael C. Constantinou, Ph.D., Rensselaer. Earthquake engineering, aseismic base isolation.

Gary Dargush, Ph.D., SUNY at Buffalo. Finite-element methods, boundary-element methods, structural dynamics.

Joseph V. DePinto, Ph.D., Notre Dame. Water quality modeling, fate and transport of particle-associated contaminants in natural systems, applied limnological studies.

Michael P. Gaus, Ph.D., Illinois. Natural hazard engineering, structural analysis and design, construction engineering and management, infrastructure assessment, repair and retrofit.

James N. Jensen, Ph.D., North Carolina. Environmental engineering, environmental chemistry, chemistry of drinking water, wastewater treatment.

George C. Lee, Ph.D., Lehigh. Structural analysis, nonlinear mechanics, biomechanics.

John B. Mander, Ph.D., Canterbury (New Zealand). Structural design, earthquake engineering, masonry, reinforced and prestressed concrete.

Dale D. Meredith, Ph.D., Illinois. Hydrology, hydraulics, water resources, urban water systems.

Satish Mohan, Ph.D., Purdue. Construction management, expert systems, transportation.

Apostoles Papageorgiou, Ph.D., MIT. Structural engineering.

Alan Rabideau, Ph.D., North Carolina. Groundwater modeling, subsurface remediation, numerical methods.

Andrei M. Reinhorn, D.Sc., Technion (Israel). Reinforced concrete, seismic behavior of structures, experimental dynamics, materials for structural repair.

Rowland R. Richards Jr., Ph.D., Princeton. Buried structures, shape mechanics, experimental stress analysis, optimum design, seismic behavior of soil structures.

Eddie Rojas, Ph.D., Colorado. Construction engineering and management, information technologies, infrastructure systems, economics.

Tsu-Teh Soong, Ph.D., Purdue. Active control of structures, reliability analysis, stochastic methods, structural dynamics.

S. Thevanayagam, Ph.D., Purdue. Soil liquefaction, seismic stability, ground improvement, geoenvironmental engineering, site characterizations, soil remediation.

John E. Van Benshoten, Ph.D., Massachusetts. Environmental engineering, unit operations, speciation of metals in water, remediation of contaminated soils.

A. Scott Weber, Ph.D., California, Davis. Environmental engineering, biological process microbiology, kinetics and modeling.

TEXAS A&M UNIVERSITY

Civil Engineering Department

Programs of Study

The Civil Engineering Department offers graduate programs with specialization in the following areas: coastal and ocean engineering (leading to a degree in ocean engineering), construction engineering and project management, engineering mechanics, environmental engineering, geotechnical engineering, hydraulic engineering, hydrology, materials engineering, public works engineering and management, structural engineering and structural mechanics, transportation engineering, and water resources engineering.

The department offers the Master of Science, Master of Engineering, and Doctor of Philosophy degrees in civil engineering and in ocean engineering. The College of Engineering offers the Doctor of Engineering degree, with emphasis in civil engineering or ocean engineering.

The Master of Science degree program requires a minimum of two full semesters of approved courses and research (32 semester hours). This requirement is ordinarily met by completing at least 24 hours of course work and 8 hours of research. An acceptable thesis embodying original research is required. The student must pass a final examination covering his or her graduate program. The examination is administered by the student's graduate committee and may be either written or oral or both.

The Master of Engineering degree requires a minimum of 36 semester hours, of which one third must be taken in fields other than the major field. A thesis is not required for this degree, but work in the major field includes one or two written reports. With these exceptions, the requirements are the same as those for the Master of Science degree.

Students in the Ph.D. program must spend two academic years in resident study. A minimum of 96 credit hours beyond the baccalaureate degree is normally required. The student must pass a preliminary examination, given by his or her graduate committee, and present a final defense of the dissertation.

Research Facilities

Facilities are available for extensive research in most areas of civil engineering. These facilities include the Texas A&M Research Center; the Computing Services Center, which houses, among other units, a Cray supercomputer and two Amdahl 470V/6's; an IBM 3090 microcomputing laboratory; the Texas Transportation Institute; the Texas Water Resources Institute; a hydromechanics laboratory; the Center for Dredging Studies; the Center for Marine Geotechnical Engineering; the Offshore Technology Research Center; environmental, geotechnical, and structural engineering laboratories; a coastal engineering laboratory; an advanced computational laboratory; and a library housing more than a million volumes.

Financial Aid

The Civil Engineering Department administers graduate teaching and research assistantships as well as other grants-in-aid. Assistantships qualify out-of-state students for in-state tuition. The department also has several fellowships sponsored by individuals, industrial corporations, and private foundations. Inquiries concerning financial aid should be made as early as possible.

Cost of Study

Tuition and fees in 1999–2000 for Texas residents for a regular semester are approximately $1350 plus laboratory fees. For nonresidents, the cost is approximately $3300 plus laboratory fees.

Living and Housing Costs

Room and board for single students were available off campus for approximately $400–$800 per month in 1998–99. Numerous apartments for single and married students are available in areas surrounding the campus and throughout the Bryan–College Station community. For married students, a limited number of University-owned apartments, both furnished and unfurnished, are available. Rents range from $230 to $360 per month, excluding utilities. Further information regarding types of accessible housing and general living costs can be obtained through the Housing Office (409-845-2261).

Student Group

Texas A&M, the oldest public institution of higher learning in the state of Texas, has on its main campus an approximate undergraduate and graduate enrollment of 35,000 and 8,000, respectively. Students are drawn from every state and more than sixty countries. The Civil Engineering Department has more than 300 graduate students in its programs.

Location

The Bryan–College Station area is a progressive community with a population of about 100,000. Located centrally in relation to the metropolitan areas of Houston, Dallas, Fort Worth, and Austin, the area offers many recreational and cultural activities. The climate is relatively mild.

The University

The University was founded as a land-grant college in 1876. It is now a dynamic force in the advanced educational program of the state of Texas. The University has an excellent athletic program, and cultural activities are provided by the Town Hall series, the Aggie Players, the Singing Cadets, the Artists Showcase, the Opera and Performing Arts Society, and many other groups and programs. The academic environment challenges and stimulates both faculty members and students to achieve their educational goals.

Applying

All communications concerning admission and registration should be addressed to the director of admissions. Applications for admission, including transcripts and scores on the Graduate Record Examinations, should be received at least four weeks before the beginning of the semester in which a student wishes to enroll.

Correspondence and Information

Dr. John M. Niedzwecki, Head
Civil Engineering Department
Texas A&M University
College Station, Texas 77843
Telephone: 409-845-2438

Dr. Peter B. Keating, Graduate Advisor
Civil Engineering Department
Texas A&M University
College Station, Texas 77843
Telephone: 409-845-2498

Texas A&M University

FACULTY HEADS AND RESEARCH AREAS

Listed below are the major areas of study in the Civil Engineering Department of Texas A&M University, the professor in charge of each division, and a general description of some of the research being conducted in the division. Graduate studies may lead to specialization in one or more of these divisions.

Coastal and Ocean Engineering
Professor Billy L. Edge, Ph.D., Georgia Tech; PE. Applied hydrodynamics, coastal structures, dynamic coastal processes, mathematical modeling of natural systems, marine pollution control, physical modeling of hydraulic phenomena, sediment transport and estuarine analysis.

Construction Engineering and Project Management
Associate Professor Paul N. Roschke, Ph.D., Purdue; PE. The management of engineered construction projects, from conception to startup; roles of the owner, engineer, and contractor; construction risk, productivity, and methods; project management systems; cost, schedule, materials, and quality.

Engineering Mechanics
Associate Professor Roger E. Smith, Ph.D., Illinois; PE. Traffic engineering and operations, accident studies and investigation, traffic signals, system design, highway design, traffic safety.

Environmental Engineering
Professor M. Yavuz Corapcioglu, Ph.D., Cornell; PE. Estuarine water resources investigations; groundwater contaminant transport; waste containment; hydrocarbon contamination of groundwater; mobility of viruses, colloids, and metals in soils and groundwater; solid waste management; water and wastewater treatment and disinfection; particle-size analysis and chemical description of fossil fuel power plant emissions; virus transport in groundwater; advanced remediation technology.

Geotechnical Engineering
Associate Professor Paul N. Roschke, Ph.D., Purdue; PE. Soil-structure interaction: soil-pile interaction under both dynamic (impact) and static loads, long-term instrumentation, soil dynamics, stress-strain-strength relationships of soils, geotechnical properties of marine soils, soil stabilization, basic studies on expansive and collapsing soils, deep foundations for near-shore and offshore structures.

Hydraulic Engineering
Professor M. Yavuz Corapcioglu, Ph.D., Cornell; PE. Mechanical behavior of water in physical systems and processes; steady and unsteady flow in pipelines, pipe networks, and open channels; erosion and sediment transport; potential- and viscous-flow theory; physical and numerical modeling; design of various hydraulic structures and facilities.

Materials Engineering
Associate Professor Roger E. Smith, Ph.D., Illinois; PE. Pavement management, infrastructure management, transportation materials, modeling pavement performance, pavement maintenance and rehabilitation, pavement design, pavement evaluation, implementation of infrastructure and pavement management systems.

Public Works Engineering and Management
Professor Edward J. Rhomberg, Ph.D., Iowa State; PE. Public transportation; paving, airports, and environmental control (including water, wastewater, drainage, and pollution); space utilization; city management and records; laws, codes, and regulations.

Structural Engineering and Structural Mechanics
Associate Professor Paul N. Roschke, Ph.D., Purdue; PE. Structural analysis and design, properties of structural materials as they influence structural design criteria, dynamic loads and dynamic behavior of structures, probabilistic structural mechanics, structural reliability, offshore and coastal structures, intelligent structural systems.

Transportation Engineering
Associate Professor Roger E. Smith, Ph.D., Illinois; PE. Traffic engineering and operations, accident studies and investigation, traffic signals, system design, highway design, traffic safety.

Water Resources Engineering
Professor M. Yavuz Corapcioglu, Ph.D., Cornell; PE. Hydrology, river basin hydrology and geomorphology, groundwater flow and contaminant transport, hydrometeorology, water resources systems analysis and planning, modeling of two-phase fluid flow, stochastic hydrology.

TEXAS A&M UNIVERSITY–KINGSVILLE

Department of Environmental Engineering

Program of Study

The Environmental Engineering (EVEN) Program at Texas A&M University–Kingsville was implemented in spring 1990, with the first course offered in the fall of the same year. The EVEN program is exclusively graduate level and offers the only M.S. in environmental engineering in the state of Texas. The EVEN program is interdisciplinary in nature, offering courses in conjunction with most of the physical sciences and engineering departments in addition to courses offered within the program itself. Students with both undergraduate engineering and physical science degrees are accepted into the program; preparatory courses are offered to allow the students to meet minimum ABET basic engineering requirements, as stipulated by the American Academy of Environmental Engineering (AAEE). The program is seeking ABET accreditation in spring 1999.

Current enrollment is 31 graduate students who are majoring in environmental engineering; an additional 22 students have selected environmental engineering as a minor. This enrollment is more than double what the program had initially anticipated. Expansion that will enable the department to offer the Ph.D. degree is envisioned for the near future.

Research Facilities

The Environmental Engineering Laboratory is well equipped and includes nearly 2,200 square feet of research space. In addition to having the capacity to perform traditional water quality analyses, the laboratory contains an atomic absorption spectrophotometer, a total organic carbon analyzer, gas chromatography equipment, a computer-driven titrator, a UV/VIS spectrophotometer, and several computer-controlled respirometers. Because of the interdisciplinary nature of the program, equipment and laboratory space in other departments of the University are available for graduate student research. Students also have access to a wide range of advanced multipurpose computing facilities at the University. The EVEN program supports a full-time laboratory technician who is charged with operations and maintenance of the EVEN laboratories.

Financial Aid

Financial support is available through research and graduate fellowships. Research assistantships are funded through extramural grants and are awarded by individual faculty members. International students who receive competitive academic scholarships and part-time teaching assistantships are eligible for resident tuition rates.

Cost of Study

Tuition and fees for 1998–99 were approximately $1940, based on a 13-credit-hour course load for a Texas resident. Costs are subject to change.

Living and Housing Costs

The cost of living is moderate to low, making it financially affordable for most students. The University has some dormitory facilities for single students and apartments for married students and single students with dependents. Many students live in adjacent off-campus areas. Additional information about residential facilities in Kingsville can be obtained through the Chamber of Commerce, 635 East King, Kingsville, Texas 78363.

Student Group

In fall 1998, there were more than 850 students in the College of Engineering and 31 graduate students in the Environmental Engineering Program.

Location

Texas A&M University–Kingsville is located approximately 40 miles southwest of Corpus Christi, 153 miles southeast of San Antonio, and 120 miles north of Mexico. Texas A&M University–Kingsville serves an area that comprises the citrus region of the Rio Grande Valley, extensive ranch and farm land, productive oil and gas regions, and the expanding industrial area along the Gulf Coast. Kingsville, the county seat of Kleberg County, is a city of approximately 30,000 residents.

The University and The Department

Texas A&M University–Kingsville is a comprehensive university and the only predominantly residential university in south Texas. Texas A&M University–Kingsville offers a large inventory of academic programs at the bachelor's and master's levels and a doctorate in education. The College of Engineering offers EAC/ABET–accredited Bachelor of Science degrees and Master of Science degrees in civil, mechanical, electrical, chemical, and natural gas engineering, plus Master of Science degrees in environmental engineering, computer science, industrial technology, and, most recently, industrial engineering.

Applying

Students who wish to be admitted should contact the Department of Environmental Engineering. In addition to general requirements for graduate courses, applicants should have an undergraduate grade point average of 3.0 or higher and a GRE General Test score of 1100 or higher for unconditional acceptance.

Correspondence and Information

Andrew N. Ernest, Department Chairman
Department of Environmental Engineering
Campus Box 213
Texas A&M University–Kingsville
Kingsville, Texas 78363
Telephone: 512-593-3046
Fax: 512-593-2069
E-mail: allen@tamuk.edu

Kuruvilla John, Graduate Coordinator
Department of Environmental Engineering
Campus Box 213
Texas A&M University–Kingsville
Kingsville, Texas 78363

Texas A&M University–Kingsville

THE FACULTY AND THEIR RESEARCH

Andrew N. Ernest, Associate Professor and Chairman, Department of Environmental Engineering; Ph.D., Texas A&M. Surface and subsurface water quality modeling, particle and particle-mediated contaminant transport, sediment–water column interactions, parameter estimation and process optimization, in situ bioremediation.

Kuruvilla John, Assistant Professor, Department of Environmental Engineering; Ph.D., Iowa. Air quality modeling, air pollution monitoring, emissions inventory, regional/urban tropospheric ozone, fine particulate matter, environmental risk and impact assessment.

Chen-Yu Cheng, Visiting Professor, Department of Environmental Engineering; Ph.D., SUNY at Buffalo. Solid-waste management, environmental fluid mechanics, pollutant transport and particle mechanics.

Mauro E. Castro, Assistant Professor of Chemistry; Ph.D., Texas A&M. Analytical techniques in monitoring water, air, and soil contaminants by mass spectrometry.

Paul H. Cox, Associate Professor of Physics; Ph.D., Harvard. Meteorology, solar energy, undergraduate physics lab improvement.

Duane T. Gardiner, Assistant Professor of Soil Sciences; Ph.D., Oregon State. Soil fertility and environmental quality.

Jerry W. Hedrick, Associate Professor of Industry and Technology; M.S., Texas A&M. Industrial occupational safety.

Blake Kidd, Assistant Professor of Chemistry; Ph.D., Texas at Austin. Chemical and biological methods for removal of chlorinated solvents and metal contaminants from groundwater and soil.

Thomas L. McGahee, Assistant Professor of Geosciences; Ph.D., Texas at Dallas. Hydrogeology and groundwater modeling, low-temperature geochemistry, fate of elements in the subsurface.

James C. Pierce, Assistant Professor of Biology; Ph.D., Texas at Austin. Molecular biology of rattlesnake venoms and antivenoms, DNA damage and repair, microbial iron acquisition.

Joseph O. Sai, Assistant Professor of Civil Engineering; Ph.D., Texas A&M. Water resources, wastewater treatment and disposal, groundwater hydrology.

Robert W. Serth, Professor of Chemical and Natural Gas Engineering; Ph.D., SUNY at Buffalo. Mathematical modeling and air pollution control, optimization.

TEXAS TECH UNIVERSITY

College of Engineering
Department of Civil Engineering

Programs of Study

The Department of Civil Engineering offers programs leading to the Master of Science in Civil Engineering, Master of Environmental Engineering, Master of Science in Environmental Technology and Management, and Doctor of Philosophy degrees. Research is conducted within the traditional core areas of structural, geotechnical, transportation, environmental, and water resources engineering. However, graduate students have opportunities for more specialized degree paths in such areas as wind engineering, glass research, disaster mitigation, and geoenvironmental engineering. These programs are designed to produce graduates who are attuned to the needs of society, who possess professional competence and judgment, and who have the personal attributes that will help them to become leaders in their career fields.

Two general plans of study are available for the Master of Science degree: a 30-semester-hour plan that includes 6 hours of credit for a thesis or a 36-hour plan that includes 3 hours of credit for a master's report. The student and faculty adviser confer to select the most appropriate plan. Most full-time master's students complete the program in 1½ to 2 years.

Students pursuing the master's degree in environmental technology and management may choose from six areas of specialization: air quality, environmental technology management, hazardous and toxic waste, land quality, solid waste, and water quality. Certain leveling courses may be required, depending on the baccalaureate credits presented. As a prerequisite, all graduate students must possess the computer and computational skills needed to perform analytical work associated with graduate-level engineering courses. Master's candidates are required to pass a comprehensive examination and a thesis or report defense before a faculty committee.

Doctoral studies consist of at least 60 semester hours of graduate work plus independent research culminating in a dissertation. Each student's degree plan is formulated through consultation with a faculty advisory committee. Nearly all doctoral candidates have opportunities for teaching and substantive research experience. Time needed to fulfill Ph.D. degree requirements varies considerably among individuals, but 2½ to 3½ years beyond the master's level is typical. A preliminary exam, a qualifying exam, and a dissertation defense are the principal milestones in completing Ph.D. degree requirements.

Research Facilities

Laboratory and research facilities in the department are modern and well equipped. The recently renovated environmental sciences laboratory contains nearly 4,000 square feet of space and features mostly state-of-the-art equipment for biological studies and measurements of standard water quality parameters. The geotechnical engineering laboratory occupies 2,300 square feet with state-of-the-art testing and data acquisition equipment for soil characterization, shear, and consolidation tests. Structural and materials laboratories occupy 5,000 square feet and 2,200 square feet, respectively. The structures lab features a 30-foot by 60-foot test deck and a 10-ton overhead rolling crane. In addition to the usual static and dynamic testing machines, a tornado cannon is employed to test the resistance of walls, cladding, and structural glazing to the impact of wind-borne missiles. A wind engineering field research facility features a 30-foot by 45-foot highly instrumented movable metal building that may be rotated to any orientation relative to ambient winds. Researchers employing computationally intensive analysis procedures have access to IBM and VAX mainframe systems as well as to a Cray computer.

Financial Aid

Some U.S. citizens qualify for student loans from the government or from private lenders; the Financial Aid Office can assist students in applying for these loans. Scholarships and assistantships are available on a competitive basis. Receipt of a scholarship or appointment as a teaching or research assistant may qualify the recipient for in-state tuition fees for the duration of the appointment. The sooner an application for financial aid is made, the better the chances for a favorable outcome. Salaries for graduate students with appointments in the department range from $700 to $1600 monthly.

Cost of Study

The current tuition fee is $72 per semester hour for Texas residents and qualified non-Texas residents; $90 per semester hour for New Mexico and Oklahoma residents; and $252 per semester hour for other non-Texas residents. Other additional fees (e.g., student health) are assessed. Total semester fees for a full-time resident student can be as low as $1200; a nonresident may require up to $7500 for a long semester. Scholarship or appointed students must register for 12 hours each semester.

Living and Housing Costs

Moderately priced living quarters ($300 to $400 per month) are available on campus for single students. All students can purchase meals at on-campus dining facilities. Off-campus living in Lubbock is relatively inexpensive, with a wide range of choices available within walking distance of the campus.

Student Group

The department has approximately 400 full-time undergraduate students, of whom approximately 18 percent are women and fewer than 3 percent are international students. There are 70 active graduate students, of whom 14 percent are women and 57 percent are international students.

Location

The 1,839-acre main campus is bounded by retail outlets and residential neighborhoods on all sides. The host city, Lubbock, is situated high on the south plains of the Texas panhandle at an elevation of 3,300 feet and has an approximate population of 200,000. Lubbock citizens are extremely friendly and are highly supportive of University activities and students. The climate is moderate, typically with more than 260 days of sunshine per year; seldom is it excessively hot or cold. The low humidity helps prevent discomfort on all but the hottest days of summer. Local cultural activities include concerts, a symphonic orchestra, a civic ballet, a civic chorale, and an arts festival.

The University

Founded in 1923, Texas Tech University and Health Sciences Center is one of the four major state-supported comprehensive institutions of higher education in Texas. With approximately 110 approved master's degree programs and approximately 65 approved Ph.D. programs, Texas Tech has unusual diversity in graduate degree opportunities. Nominal enrollment in 1998–99 was 24,000 undergraduate and 3,600 graduate students. The library is excellent, with more than 1.2 million volumes, 7,300 periodical subscriptions, and several unique archival collections. Texas Tech University and Texas Tech Health Sciences Center are composed of seven colleges, a museum, a graduate school, and schools of law, medicine, nursing, and allied health, which combine to make Texas Tech a complete university.

Applying

Candidates for graduate school should apply at least three months prior to the date of intended enrollment to the Office of Graduate Admissions, Texas Tech University, P.O. Box 41030, Lubbock, Texas 79409-1030 (telephone: 806-742-2787; e-mail: aqgrd@ttuvmi.ttu.edu). Admission is determined by a review of the applicant's transcripts, Graduate Record Examinations (GRE) scores, an evaluation by the Director of Graduate Admissions, and consent of the Civil Engineering graduate faculty. The minimum GRE combined verbal and quantitative score is 1000. International students must have an above-average academic record, an official Test of English as a Foreign Language score of at least 550, and documented evidence of financial support. International students not residing in the U.S. should commence the application process a year in advance of intended enrollment.

Correspondence and Information

Graduate Advisor
Department of Civil Engineering
Texas Tech University
Lubbock, Texas 79409-1023
Telephone: 806-742-3523

Texas Tech University

THE FACULTY AND THEIR RESEARCH

James R. McDonald, Professor and Chair; Ph.D., Purdue; PE. Engineering for extreme winds, natural phenomena hazards, simulations of tornado-generated missiles, building codes and standards, structural steel design.

John Borrelli, Professor; Ph.D., Penn State; PE. Irrigation engineering, economics, water resources, hydrology.

William R. Burkett, Assistant Professor; Ph.D., Texas at Austin; PE. Reinforced and prestressed concrete structures, structural analysis and dynamics, finite element analysis, concrete materials and constituents.

Clifford B. Fedler, Associate Professor; Ph.D., Illinois at Urbana-Champaign; PE. Movement of chemicals through the soil profile to groundwater, environmental characterizations of specific geographical areas, livestock waste and agricultural runoff management.

James M. Gregory, Professor and Associate Dean for Undergraduate Studies; Ph.D., Iowa State; PE. Erosion and sediment pollution control, desert aeolian processes, soil cover using crop residue, irrigation scheduling.

W. Andrew Jackson, Assistant Professor; Ph.D., Louisiana State. Hazardous waste remediation.

Priyantha W. Jayawickrama, Assistant Professor; Ph.D., Texas A&M. Geotechnical aspects of hazardous waste disposal, hydraulic conductivity, mechanistic design/rehabilitation of flexible pavements, unsaturated soils.

Ernst W. Kiesling, Professor; Ph.D., Michigan State; PE. Energy conservation and earth-sheltered housing, site characterization studies, professional development and leadership, structural mechanics.

Kishor C. Mehta, P. W. Horn Professor and Director, Wind Engineering Research Center; Ph.D., Texas at Austin; PE. Wind loads on structures, characterization of wind profile, development of technical codes and standards for wind engineering, concrete design.

Tony Mollhagen, Assistant Professor and Director, Environmental Science Laboratory; Ph.D., Texas Tech. Wetlands ecology, surface water and groundwater quality, bioassays and index species, environmental contaminants.

H. Scott Norville, Professor and Director, Glass Research and Testing Laboratory; Ph.D., Purdue; PE. Design and characterization of window glass as a structural component; effects of blast and missile impacts on glass; dynamic failure prediction models; strength of laminated, heat strengthened, and fully tempered glass.

Kenneth A. Rainwater, Associate Professor; Ph.D., Texas at Austin; PE. Subsurface contaminant transport, assessment, and remediation; groundwater hydrology; fluid mechanics; water resources systems.

R. Heyward Ramsey, Associate Professor; Ph.D., Oklahoma; PE. Solid waste management, geologic and hydrologic characterization of industrial sites, assessment of nonpoint source contamination of playa basins, characterization of basin sediments.

Partha P. Sarkar, Assistant Professor; Ph.D., Johns Hopkins. Wind tunnel studies of structural vibrations, aeroelastic parameters of flexible bridges, flutter derivatives and stability of long span bridges, structural mechanics.

Sanjaya Senadheera, Assistant Professor; Ph.D., Texas A&M; PE. Geotechnical engineering, pavement materials, and transportation.

Douglas A. Smith, Assistant Professor; Ph.D., Texas Tech; PE. Wind-structure interaction, full-scale measurement of wind effects on low-rise buildings, structural analysis, design.

Jimmy H. Smith, Professor and Director, Murdough Center for Engineering Professionalism; Ph.D., Arizona; PE. Development of course materials and teaching methods regarding engineering ethics and professionalism, dynamics of complex structures, manufacturing systems.

Robert M. Sweazy, Professor and Vice Provost for Research; Ph.D., Oklahoma; PE. Urban runoff, wastewater reuse, effluent irrigation, water strategies for arid and semiarid regions.

David B. Thompson, Assistant Professor; Ph.D., Missouri–Rolla; PE. Modeling of surface water and groundwater hydrologic processes associated with transport of waterborne constituents through the land based part of the hydrologic cycle.

Lloyd V. Urban, Professor and Director, Water Resources Center; Ph.D., Texas at Austin; PE. Environmental impact analyses, water quality and resources, environmental engineering, biomass conversion, feasibility studies.

C. V. Girija Vallabhan, Professor; Ph.D., Texas at Austin; PE. Continuum mechanics, finite element and finite difference modeling, probability and statistics in engineering analyses, impact and blast dynamics, soil mechanics.

W. Pennington Vann, Associate Professor; Ph.D., Rice; PE. Free vibration analysis of insulating glass units, engineering ethics, flow-induced vibrations in power plant coal silos, dynamic soil-structure interaction, wind effects on tied-arch bridges.

Current Research Interests

The stated objective of the Department of Civil Engineering is to address the needs of society. Research conducted by department faculty members is directed toward these needs. Faculty members hold leadership roles and participate in various research centers and institutes that carry out the research mission of the department. These organizations are multidisciplinary and in some cases engage in cooperative research with other universities. The Wind Engineering Research Center (WERC), consisting of faculty members, postdoctoral research associates, graduate students, undergraduate students, and staff members from at least several departments and four colleges, engages in a variety of wind engineering research topics. Origin of WERC is traced to May 11, 1970, when a massive tornado devastated the city of Lubbock. In the ensuing years, the group has studied wind-structure interaction; wind hazards; effects of tornadoes, hurricanes, and thunderstorms on the built environment; protection of people; and mitigation of wind damage. Field and laboratory facilities, including a full-size test building, wind tunnel, tow tank, and tornado missile cannon, are available for current research. Current areas of activity include wind damage documentation and assessment, debris impact, physical testing, retrofit of existing buildings, economic effects of wind damage, and wind erosion. The WERC is uniquely equipped to further the understanding of wind through its long-term record of achievements and its interdisciplinary team of researchers. The Center is directed by one of the foremost researchers in the field, Dr. Kishor C. Mehta. Dr. Mehta chaired the Wind Load Task Committee of American Society of Civil Engineering (ASCE 7) for many years. He has served as President of the American Association for Wind Engineers and chaired the Committee on Natural Disaster of the National Academy of Sciences. Another center that reflects the interest and expertise of the faculty is the Water Resources Center (WRC). This center, in cooperation with the Environmental Science Laboratory, addresses hydrological and agricultural aspects of land and water use, contaminated soil remediation, contaminant transport, and all aspects of environmental characterization. An outgrowth of the wind damage investigation was recognition of the enormous damage to window glass in severe windstorms and hailstorms. This led to the creation of the Glass Research and Testing Laboratory. The objective of the lab is to study and test both new and existing window glass and make recommendations for improving its resistance to wind pressures and windstorm debris impacts. In recent years, blast effects on glass have been of concern as threats of urban terrorism increase.

Recently the Center of Multidisciplinary Research in Transportation (TechMRT) was approved by the Texas Tech Board of Regents. The mission of the center is to create the necessary research and educational environment and promote multidisciplinary studies of highway construction. Faculty members and students from civil and chemical engineering and engineering technology are currently working on projects that relate to highway materials, lighting, and recycling.

Another area of growing interest among the faculty is ethics and professionalism. Assisted by faculty members and professional leaders from off campus, the Director of the Murdough Center for Ethics and Professionalism has generated much interest in the role of ethics in professional practice. Courses are available and being developed for correspondence, distance learning, continuing education, and classroom presentations. The Center Director is recognized as a national and international leader in the ethics field. Although there is no geotechnical engineering center, a dedicated group of faculty members focuses on research in areas of expansive and unsaturated soils, geoenvironmental problems, and numerical modeling of soil-structures interaction and slab-on-ground foundation.

UNIVERSITY OF CALIFORNIA, BERKELEY

Department of Civil and Environmental Engineering

Programs of Study

The Department of Civil and Environmental Engineering in the College of Engineering offers excellent opportunities for graduate study to qualified students. Teaching and research are carried out in five major groups: Construction Engineering and Management; Environmental Engineering; Geotechnical Engineering; Structural Engineering, Mechanics and Materials; and Transportation Engineering.

The department offers two degrees at the master's level (Master of Science and Master of Engineering) and two degrees at the doctoral level (Doctor of Philosophy and Doctor of Engineering). The Master of Science program is based upon a program of study in the major field and may include research or individual study as an option. The Master of Science Plan I program requires a minimum of 20 semester units plus a thesis. The Master of Science Plan II program requires a minimum of 24 units and a comprehensive final examination. The Master of Engineering degree is a professionally oriented program of study in the major field combined with study in two additional fields to provide related technical and nontechnical breadth. This degree requires a minimum of 40 to 44 semester units, including 4 units of individual study or applied research culminating in a written report. The doctoral programs are directed toward research and teaching careers or the highest levels of professional practice. There is no strict course or unit requirement for the doctoral degrees, but a minimum of eight to twelve semester courses taken while in graduate standing (at least 33 semester units) is recommended. Two related minor fields of study, a qualifying examination, and a dissertation are required. Students can complete the M.S. in two semesters, the M.Eng. in four semesters, and the doctoral programs in three to four years after the M.S.

Research Facilities

Excellent modern facilities for advanced study and research are located on the Berkeley campus and at the Richmond Field Station. They include Structures and Materials laboratories—facilities ranging from miniaturized precision equipment to a 4-million-pound-capacity testing machine and physical-chemical and fire-testing facilities; the Hydraulic and Coastal Engineering Laboratory—a model basin, wind-wave channels, and large flumes for estuary studies; the Environmental Engineering and Health Sciences Laboratory—field and pilot installations; the Institute of Transportation Studies—transportation-related computer activities, including video and hard-copy terminals, and an extensive library of programs and data files; the Soil Mechanics and Bituminous Materials Laboratory—in situ measuring devices and a high-chamber-pressure triaxial cell; the Surveying Laboratory—geodetic instruments, including GPS receivers and total station systems; and the Earthquake Engineering Research Center—earthquake simulation using a 20-foot by 20-foot shaking table. Departmental computer facilities include local area networks of graphics workstations and personal computers, all of which are connected to the main campus network and the Internet. Extensive additional mainframe, workstation, and personal computer facilities are also available to students and are located throughout the campus.

Financial Aid

Various types of graduate appointments and aid are available, including teaching and research assistantships and fellowships.

Cost of Study

For fall 1999, residents of California pay fees of about $2200 per semester. Nonresidents pay tuition and fees of approximately $6700 per semester.

Living and Housing Costs

In 1998–99, most accommodations in University-supervised residences cost approximately $7450 per year. For students with families, the University operates 1,022 apartments, which are located in Berkeley and in neighboring Albany, 4 miles from campus. In 1998–99, they rented for approximately $320–$620 per month. Because of the eight-to-twelve-month waiting list, prospective students are encouraged to apply for student apartments at the same time that they apply to the graduate program. A variety of other housing facilities exists in the Berkeley area, with cost depending on proximity to the campus.

Student Group

There are approximately 350 graduate students in the department. The graduate population is widely varied in interests and includes students from all over the world. A student chapter of the American Society of Civil Engineers provides opportunities for contact with professionals as well as participation in student-faculty social events.

Location

The University is located at the base of the Berkeley hills, directly across the bay from San Francisco. The San Francisco Bay Area has a tremendous variety of cultural and entertainment activities to suit all tastes and interests. Students have ready access to the Pacific Coast and beaches, and excellent skiing areas are located 3½ hours to the east in the Sierra Nevadas. The climate is cool in the summer, with little or no rainfall. Winters are mild, with intermittent rainy and sunny weather. It is an excellent working climate.

The University

The Berkeley campus of the University of California has an enrollment of 30,000, of whom 9,000 are graduate students. The campus is noted for the academic distinction of its faculty, the high quality and wide scope of its research activities, and the variety and vitality of student activities. It is generally ranked by its academic peers as one of the best graduate institutions in the United States.

Applying

The deadline for fall fellowship and scholarship applications is December 15. The application deadline is February 10 for the fall semester.

Applicants for graduate study should have satisfied the substantial equivalent of a B.S. in an engineering field. Further details concerning each program may be obtained by writing to the Academic Affairs Office, Department of Civil and Environmental Engineering.

Correspondence and Information

Academic Affairs Office
Department of Civil and Environmental Engineering
750 Davis Hall #1714
University of California
Berkeley, California 94720-1714

Telephone: 510-642-6464
Fax: 510-643-5264
E-mail: aao@ce.berkeley.edu
World Wide Web: http://www.ce.berkeley.edu

University of California, Berkeley

THE FACULTY AND THEIR RESEARCH

L. Alvarez-Cohen, Ph.D., Associate Professor. Groundwater remediation and hazardous waste treatment technologies.

F. Armero, Ph.D., Associate Professor. Computational solid and fluid mechanics, nonlinear continuum mechanics.

A. Astaneh-Asl, Ph.D., Professor. Experimental and analytical behavior of steel structures.

R. G. Bea, M.S., Professor. Research and development in design, requalification, and construction of offshore structures and pipelines.

J. D. Bray, Ph.D., Professor. Foundation engineering, numerical and physical modeling, earthquake engineering, environmental geotechnics.

M. J. Cassidy, Ph.D., Associate Professor. Transportation operations.

A. K. Chopra, Ph.D., Professor. Dynamics of structures, earthquake engineering.

C. F. Daganzo, Ph.D., Professor. Transportation theory, mathematical analysis.

A. Der Kiureghian, Ph.D., Professor. Structural risk and reliability analysis, earthquake engineering.

G. Fenves, Ph.D., Professor. Structural dynamics and computer-aided engineering.

F. Filippou, Ph.D., Professor. Analysis and design of structures.

M. A. Foda, Sc.D., Professor. Coastal and offshore engineering.

S. Glaser, Ph.D., Assistant Professor. Rock mechanics, acoustics, system identification, tunneling.

S. Govindjee, Ph.D., Associate Professor. Theoretical and computational solid mechanics, constitutive theory, micromechanics.

M. M. Hansen, Ph.D., Associate Professor. Transportation planning, policy, and economics.

R. A. Harley, Ph.D., Associate Professor. Atmospheric chemistry and physics, urban-scale air quality modeling.

S. W. Hermanowicz, Ph.D., Associate Professor. Biological wastewater treatment and biological activities in drinking water, biological fixed-film reactors, dynamics and modeling of water and wastewater treatment processes.

A. J. Horne, Ph.D., Professor. Ecology and management of aquatic systems.

J. R. Hunt, Ph.D., Professor. Contaminant transport processes.

C. W. Ibbs, Ph.D., Professor. Computer applications to construction management.

D. Jenkins, Ph.D., Lawrence E. Peirano Professor of Civil and Environmental Engineering and Professor of the Graduate School. Water, wastewater chemistry, biological waste treatment.

A. Kanafani, Ph.D., Professor. Transportation engineering, air transport engineering.

J. M. Kelly, Ph.D., Professor. Structural mechanics.

X. Liang, Ph.D., Assistant Professor; Surface-water hydrology, hydrologic processes, land-atmosphere interactions, hydrometeorology.

S. Madanat, Ph.D., Associate Professor. Transportation systems analysis, transportation infracture management.

S. A. Mahin, Ph.D., Professor. Design and behavior of structures, earthquake engineering.

N. Makris, Ph.D., Associate Professor. Seismic protection of structures, structural mechanics, earthquake engineering.

J. P. Moehle, Ph.D., Professor. Design and inelastic behavior of reinforced concrete structures.

C. L. Monismith, M.S., Robert Horonjeff Professor of Civil Engineering and Professor of the Graduate School. Transportation engineering, pavement design, pavement materials.

P. J. M. Monteiro, Ph.D., Professor. Concrete behavior, structural materials.

K. M. Mosalam, Ph.D., Assistant Professor. Behavior of reinforced concrete and masonry structures, fracture and damage mechanics, earthquake engineering.

W. W. Nazaroff, Ph.D., Professor. Air quality control engineering, emphasizing fundamental chemical and physical processes governing the concentrations and fates of air pollutants.

C. P. Ostertag, Ph.D., Assistant Professor. Reinforced concrete, toughening mechanisms.

J. M. Pestana-Nascimento, Sc.D., Assistant Professor. Constitutive modelling of soils, soil property characterization, environmental geotechnics.

E. Popov, Ph.D., Professor; Analytical and experimental research on structural systems, seismic structural engineering.

Y. Rubin, Ph.D., Professor. Groundwater hydrology, transport phenomena, stochastic processes, geostatistics.

D. L. Sedlak, Ph.D., Assistant Professor. Environmental chemistry.

R. B. Seed, Ph.D., Professor. Soil mechanics, numerical analysis and physical testing, earthquake engineering.

N. Sitar, Ph.D., Professor. Engineering geology and behavior of recent sediments.

R. J. Sobey, Ph.D., Professor. Physical oceanology, coastal and ocean engineering.

G. Sposito, Ph.D., Professor. Environmental geochemistry, mass transport in porous media.

R. L. Taylor, Ph.D., Professor of the Graduate School. Mechanics of solids, computational mechanics.

I. D. Tommelein, Ph.D., Associate Professor. Construction materials management, site layout, operations analysis for productivity improvement, decision support systems.

K. S. Udell, Ph.D., Professor. Soil and groundwater remediation, subsurface heat and mass transfer.

M. Wachs, Ph.D., Professor. Transportation planning, transportation economics and finance.

W. C. Webster, Ph.D., Professor. Nonlinear coupled motions of offshore structures, operations research, shallow-water wave mechanics.

R. B. Williamson, Ph.D., Professor. Fire research, materials engineering.

UNIVERSITY OF CALIFORNIA, DAVIS

Institute of Transportation Studies
Graduate Group in Transportation Technology and Policy

Programs of Study

In the past, the principal transportation challenge was to build more roads and other infrastructure. That is no longer the case. The Graduate Group in Transportation Technology and Policy provides students with the mix of skills and insights needed to solve the pressing economic, social, and environmental problems facing today's metropolitan transportation systems. The program emphasizes energy and environmental aspects of transportation, design and evaluation of advanced transportation technologies, and metropolitan transportation planning, policy, and management. Davis faculty members are internationally recognized for their leadership in these areas of research.

M.S. and Ph.D. degrees are awarded for students pursuing a technology or a planning/policy track. The technology track is for students trained in engineering and the physical sciences and interested in advanced transportation technologies (especially "intelligent" and propulsion technologies). The planning/policy track is aimed at students from a broader range of disciplines. The curriculum includes courses in civil, mechanical, and environmental engineering; economics; policy sciences; statistics; travel behavior; management; technology assessment; and environmental studies.

All students take between 17 and 19 units from a common set of core courses. Master's degrees require a minimum of 36 quarter units and doctoral degrees a minimum of 54 units (of which 21 units must be in the chosen track and 9 units in the other track). Full-time students can expect to complete the master's program in one to two years and the doctoral program in four years.

The Institute of Transportation Studies is the administrative home of the Graduate Group and the center of transportation research activities on the campus. A total of 106 graduate students and 30 faculty members from a broad array of disciplines are affiliated with the Institute.

Research Facilities

Well-equipped laboratories for electric vehicle power technologies (including fuel cells), travel behavior analysis, geographic information systems, driving simulation, and highway maintenance robotics are available to students for teaching and research. The campus library has a large collection of transportation books and journals and supports several information-retrieval services.

Financial Aid

A number of research assistantships, fellowships, and dissertation grants are available, as are a few teaching assistantships. Fellowships and dissertation awards are available through the Institute of Transportation Studies, and research assistantships are available directly from faculty advisers.

Cost of Study

In 1999–2000, California residents pay $1478 per quarter in registration and incidental fees. Nonresidents pay an additional fee of $3441 per quarter. Books and supplies cost an average of $325 per quarter. The academic year consists of three quarters (registration in the spring quarter carries through the summer quarter).

Living and Housing Costs

Campus housing (dormitory rooms with board contracts) is available for rent at an average cost of $3277 per quarter. Off-campus housing is available in attractive residential areas within bicycling distance of the University at an average cost of $3041 per quarter.

Student Group

The University of California, Davis, has a total enrollment of 24,299, including 5,167 graduate and professional students. In coordination with the Graduate Group, the campus offers strong established transportation graduate programs in civil and environmental engineering and ecology. Students in those other programs participate together with students in the Graduate Group on research projects, take many of the same courses, and share office space. Students have considerable flexibility to tailor their course work to their transportation interests.

Location

Davis is a pleasant university town (population 53,000) with the largest network of bicycle paths of any U.S. city its size and is a national leader in the development of solar power. It is on Interstate 80, about 70 miles from San Francisco and 15 miles west of Sacramento, the state capital. In addition to the many concerts, plays, and special lectures held in Davis, recreational activities include an extensive intramural sports program, nearby sailing, skiing in the Sierra Nevada (a 2-hour drive), and a wide variety of entertainment in the San Francisco Bay Area. Lake Tahoe and Yosemite National Park are just a few hours' drive from the campus.

The University and The Program

The University opened its doors to students in 1908 as a University of California branch campus. Davis became a general campus offering its own degrees in 1959. The Graduate Group is affiliated with the flourishing transportation research program centered in the Institute of Transportation Studies and has transportation faculty members in management, engineering, environmental studies, and the social sciences.

Applying

Students are generally admitted for the fall quarter only. Applications for fellowships and research assistantships are required by January 15. Applications for admission are required by February 1 for international students and March 1 for domestic students. Early application is encouraged. All applicants must submit scores on the General Test of the Graduate Record Examinations (GRE) and three letters of recommendation. Master's degree applicants must have earned a grade point average of at least 3.0 in the junior and senior years of college, and doctoral students must have a minimum grade point average of 3.5.

Correspondence and Information

Graduate Program Assistant
Institute of Transportation Studies
One Shields Avenue
University of California, Davis
Davis, California 95616
Telephone: 530-752-0247
Fax: 530-752-6572
E-mail: itsgraduate@ucdavis.edu
World Wide Web: http://www.engr.ucdavis.edu/~its

University of California, Davis

THE FACULTY AND THEIR RESEARCH

Ralph C. Aldredge III, Associate Professor of Mechanical and Aeronautical Engineering; Ph.D., Princeton, 1990. Combustion propulsion and emissions.

Rahman Azari, Lecturer of Statistics; Ph.D., George Washington, 1975. Air pollution and transportation demand models.

Andrew F. Burke, Research Engineer, ITS–Davis; Ph.D., Princeton, 1967. Electric and hybrid vehicle design, batteries, ultracapacitors, fuel cells, vehicle energy simulation models.

Mark A. Delucchi, Research Ecologist, ITS–Davis; Ph.D., California, Davis, 1990. Social costs of motor vehicle use; fuel cycle analyses of air pollution, energy, and greenhouse gases; costs of electric-drive vehicles.

Mark Francis, Professor of Landscape Architecture/Environmental Design; M.L.A., Harvard, 1975. Community and urban design, sustainable communities, open space and transportation, community participation, new urbanism.

Joanna R. Groza, Professor of Chemical Engineering and Materials Science; Ph.D., Polytechnic Institute (Bucharest), 1973. Materials characterization and processing.

Britt A. Holmén, Research Geochemist, Crocker Nuclear Laboratory; Ph.D., MIT, 1995. Air quality, automobile emissions and transport, transformation of organic chemicals in the environment.

Robert A. Johnston, Professor of Environmental Science and Policy; M.S., Nevada, Reno, 1975. Transportation planning and policy analysis, including the improvement and use of regional travel demand forecasting models to evaluate policy alternatives.

Ian Kennedy, Vice Chair and Professor of Mechanical and Aeronautical Engineering; Ph.D., Sydney (Australia), 1980. Engine emissions and related health effects.

Ryuichi Kitamura, Research Engineer, ITS–Davis, and Professor, Kyoto University; Ph.D., Michigan, 1978. Advanced travel demand modeling, lifestyle and travel behavior.

Kenneth S. Kurani, Research Engineer, ITS–Davis; Ph.D., California, Davis, 1992. Travel behavior, consumer/user response to new transportation and information technology, research methodology.

Marshall Miller, Research Associate, ITS–Davis; Ph.D., Pennsylvania, 1988. Design and testing of electric power systems for vehicles, fuel cells, hybrid vehicles.

Patricia L. Mokhtarian, Associate Professor of Civil and Environmental Engineering and Chair of the Graduate Group; Ph.D., Northwestern, 1981. Travel behavior modeling; travel demand forecasting; impacts of telecommunications on transportation, land use, and the environment; transportation–land use interactions.

Janet Momsen, Professor of Human and Community Development; Ph.D., London, 1969. Gender and development; gender and land rights; small-scale agriculture; regional geography of the Caribbean, Brazil, Central America, and China.

Robert Moore, Research Associate, ITS–Davis, and Director, Advanced Vehicle Modeling Program; Ph.D., George Washington, 1966. Future vehicle technology, electric hybrid vehicles, fuel cell applications, alternative fuels for fuel cells in transportation and stationary applications.

Debbie A. Niemeier, Associate Professor of Civil and Environmental Engineering; Ph.D., Washington (Seattle), 1994. Travel behavior modeling, transportation air quality, land use–transportation relationships, infrastructure prioritization.

Ahmet Palazoglu, Professor of Chemical and Materials Science Engineering; Ph.D., Rensselaer, 1984. Dynamic modeling and control of chemical process systems.

G. Tayhas R. Palmore, Assistant Professor of Chemistry; Ph.D., MIT, 1992. Fuel cell catalysts.

Bahram Ravani, Professor and Chair of Mechanical and Aeronautical Engineering; Ph.D., Stanford, 1982. CAD/CAM, robotics and kinematics, design and manufacturing, automated highway technology.

David M. Rocke, Professor, Graduate School of Management; Ph.D., Illinois at Chicago, 1972. Statistical analysis of emissions and fuel composition.

Paul Sabatier, Professor of Environmental Science and Policy; Ph.D., Chicago, 1974. Policy implementation, bureaucratic decision making, role of science in policymaking, air pollution policy.

Daniel Sperling, Professor of Civil and Environmental Engineering and Environmental Science and Policy and Director, ITS–Davis; Ph.D., Berkeley, 1982. Alternative fuels and electric-drive vehicles, technology policy, energy and air quality impacts of transportation, developing countries.

Tom Turrentine, Research Anthropologist, ITS–Davis; Ph.D., California, Davis, 1994. Lifestyle and travel behavior, survey design methodology, alternative fuels, electric vehicles.

Steven Velinsky, Professor of Mechanical and Aeronautical Engineering; Ph.D., Illinois, 1981. Mechanical design, vehicle design and dynamics, solid mechanics, automated highway technology.

Michael Zhang, Assistant Professor of Civil and Environmental Engineering; Ph.D., California, Irvine, 1995. Mathematical modeling of transportation systems, system operations and control, analysis and design of urban transportation networks.

UNIVERSITY OF CALIFORNIA, IRVINE

Department of Civil and Environmental Engineering

Programs of Study	The Department of Civil and Environmental Engineering at University of California, Irvine (UCI) offers programs leading to the degrees of Master of Science (M.S.) and Doctor of Philosophy (Ph.D.) in civil engineering. The programs are designed for those who plan to engage in the public or private sectors or in teaching and research.

The Structures Program emphasizes the application of analytical and experimental approaches to the investigation of the effects of earthquakes and other extreme hazards on constructed facilities. Areas of specific interest include reliability of engineering systems; random vibration; passive, active, and hybrid control of structural vibration; elastomeric and sliding base isolation systems; dynamic behavior of liquid storage tanks; seismic response of equipment and other secondary systems; liquefaction; fragility of lifelines; the retrofitting of buildings and bridges; and stochastic fatigue, fracture, and maintenance of structures.

The Water Resources and Environmental Engineering Graduate Program focuses on hydraulics and modeling, contaminant fate and transport, pollution control technologies, and microbial diagnostics and chemical processes in natural waters. Particular research emphasis is placed on contaminant fate in saturated and unsaturated subsurface formations; coastal, river, and estuarine surface waters; and atmospheric droplets. Innovative treatment technologies are being developed for drinking water, hazardous and toxic waste, and for water reclamation and reuse. The objective of the program is to prepare graduates for a career in private engineering firms and public agencies, in research and development, and in academic positions.

Among leading centers for transportation research, the department offers a graduate program that is distinguished by its interdisciplinary approach to the study of contemporary urban transportation issues and by its unique relationship with the UC Irvine Institute of Transportation Studies. The program focuses on the planning, design, operation, and management of modern urban transportation systems. Emphasis is on the development of fundamental skills and knowledge in engineering, systems analysis, modeling, and planning, combined with advanced computational techniques to address transportation problems affecting urban travelers and the movement of goods.

Research Facilities Well-equipped laboratories provide opportunities for analytic and experimental research in the areas mentioned above. Excellent computer-based modeling facilities are available, including a departmental graduate research computer laboratory, a cluster of VAX minicomputers and other campus computers, such as a Hewlett-Packard/Convex Exemplar SPP2000 minisupercomputer with 16 CPUs and 2 gigabytes of memory, and network connections to National Science Foundation supercomputer centers. The structural test hall and structural dynamics laboratories include a strong floor and wall, a large MTS feedback system, sophisticated shaking equipment, accelerometers, seismometers, signal processors, a Fourier analyzer, and online computers. The Institute of Transportation Studies–Irvine provides excellent support facilities for transportation research, including a state-of-the-art computational lab. Water resources laboratories include a hydraulics laboratory with several flumes up to 50 feet in length, a laboratory of environmental transport phenomena and physicochemical hydrodynamics, and a modern computational laboratory. Environmental water quality laboratories are instrumented for a wide range of chemical analyses, biotechnology research, bench-scale reactor studies, and field investigations.

Financial Aid Fellowships and teaching and research assistantships are available on a competitive basis. Except for students on visas, there are opportunities for part-time work in the engineering community of Orange County. With the same exception, financial aid may be obtained from UCI's Financial Aid Office.

Cost of Study In 1999–2000, student fees are $1726 per quarter for California residents and an additional $3441 per quarter for nonresidents. These fees are subject to change.

Housing On-campus housing is available. Early application is advised for on-campus housing.

Student Group Current campus enrollment is 18,209, including 1,263 undergraduate and 309 graduate students in the School of Engineering.

Student Outcomes Graduates of the degree programs offered within the department hold positions in academia or in the industrial or governmental sectors that involve the development of new technologies for the benefit of society. Many graduates take various professional positions in local, national, and international high-tech companies. Others obtain faculty appointments or research and development positions in their areas of specialization.

Location The 1,510-acre UCI campus is in Orange County, 40 miles south of Los Angeles. Irvine is one of the nation's fastest-growing residential, industrial, and business areas, yet within view of the campus is a wildlife sanctuary. Pacific Ocean beaches are nearby.

The University One of the nine campuses in the University of California system, UCI enrolls 3,638 graduate and professional students. Some of the graduate degrees offered include biological sciences, engineering, physical sciences, and information and computer science.

Applying Application forms and general information may be obtained by writing to the department. The deadlines for applications are May 15 for the fall quarter, October 15 for the winter quarter, and January 15 for the spring quarter. Applicants who wish to be considered for fellowships or for teaching or research assistantships should apply by February 1. The General Test of the Graduate Record Examinations is required, as is the Test of English as a Foreign Language (TOEFL) for international students whose native language is not English. The minimum TOEFL score accepted is 550.

Correspondence and Information For applications and information about the department:

Department of Civil and Environmental Engineering
University of California
Irvine, California 92697-2175
Telephone: 949-824-5333
E-mail: ceeinfo@eng.uci.edu
World Wide Web: http://www.eng.uci.edu/civil/graduate

University of California, Irvine

THE FACULTY AND THEIR RESEARCH

Alfredo H.-S. Ang, Professor Emeritus of Civil Engineering; Ph.D., Illinois at Urbana-Champaign; PE. Structural and earthquake engineering, risk and reliability engineering.

Constantinos V. Chrysikopoulos, Associate Professor of Civil and Environmental Engineering; Ph.D., Stanford. Subsurface solute transport, nonaqueous phase liquid dissolution in porous media, mathematical modeling.

Maria Q. Feng, Associate Professor of Civil Engineering; Ph.D., Tokyo. Structural engineering and intelligent control of structural systems.

Stanley B. Grant, Associate Professor of Environmental Engineering; Ph.D., Caltech. Environmental engineering, coagulation and filtration of colloidal contaminants, environmental microbiology.

Gary L. Guymon, Professor Emeritus of Civil Engineering; Ph.D., California, Davis; PE. Water resources, groundwater, modeling uncertainty.

Medhat A. Haroun, Professor of Civil Engineering; Ph.D., Caltech; PE. Numerical and experimental modeling of the seismic behavior of structural systems for design, retrofit, and repair: systems evaluated are liquid storage tanks, bridge-supporting columns and piers, and concrete/masonry/composite buildings.

R. Jayakrishnan, Associate Professor of Civil Engineering; Ph.D., Texas at Austin. Transportation systems analysis.

Michael G. McNally, Associate Professor of Civil Engineering; Ph.D., California, Irvine. Travel behavior, activity-based approaches, transportation planning and modeling, transportation and land use, computer applications in transportation.

Terese M. Olson, Associate Professor of Civil and Environmental Engineering; Ph.D., Caltech. Environmental aquatic chemistry, colloidal processes, pollutant transformation processes in natural and water treatment systems.

Gerard C. Pardoen, Professor of Civil Engineering; Ph.D., Stanford; PE. Structural dynamics, including ambient and forced vibration testing of large- and small-scale structures; static and dynamic testing of timber, steel, and concrete components, with applications to earthquake engineering; nonlinear finite-element analysis of lightweight, composite material structures.

Wilfred W. Recker, Professor of Civil Engineering and Director, Institute of Transportation Studies; Ph.D., Carnegie Mellon. Transportation engineering, demand analysis, intermodal transfer analysis, urban planning and transportation interaction.

Amelia C. Regan, Assistant Professor of Civil Engineering; Ph.D., Texas at Austin. Logistics, freight and fleet management, intermodal transportation systems.

Stephen G. Ritchie, Professor of Civil Engineering and Department Chair; Ph.D., Cornell. Transportation systems engineering, advanced traffic management and control systems, development and application of emerging technologies in transportation.

Brett F. Sanders, Assistant Professor of Civil and Environmental Engineering; Ph.D., Michigan. Environmental and computational fluid dynamics, flood mitigation, urban runoff, adjoint methods.

Jan Scherfig, Professor Emeritus of Civil Engineering; Ph.D., Berkeley; PE. Water resources and water quality, effect of low temperature on waste treatment processes, modeling, water reclamation and reuse.

Robin Shepherd, Professor Emeritus of Civil Engineering; Ph.D., Canterbury (New Zealand); D.Sc., Leeds (England); PE. Structural engineering, structural dynamic analysis of buildings, bridges, and aerospace structures.

Roberto Villaverde, Professor of Civil Engineering; Ph.D., Illinois at Urbana-Champaign; PE. Earthquake engineering, nonlinear structural mechanics, passive structural control, seismic response of nonstructural components, earthquake ground motion characterization.

Jann N. Yang, Professor of Civil Engineering; D.Sc., Columbia; PE. Fatigue, reliability, maintainability, and control of structures.

UNIVERSITY OF CALIFORNIA, LOS ANGELES

Environmental Science and Engineering Program

Program of Study	The Environmental Science and Engineering Program is an interdepartmental graduate degree curriculum administered through the School of Public Health that culminates in the award of the Doctor of Environmental Science and Engineering (D.Env.) degree. This professional degree was established in 1973 with the conviction that, to resolve complex environmental problems, individuals are needed who are not specialists in the narrow traditional sense but who have a broad understanding of the environment as well as the technical and managerial skills for environmental problem solving. The purpose of the program is to supply this much-needed kind of professional. A graduate of the D.Env. program has an area of specialization (represented by the student's master's degree), a background that includes several disciplinary areas, experience gained through working with experts in a variety of fields, and an understanding of how a particular discipline interacts with others. More than 180 students have graduated from UCLA with the D.Env. degree, and they are much sought after by government, industry, and private consulting firms.
	Applicants must qualify for admission to the UCLA Graduate Division; hold a master's degree or the equivalent in one of the natural sciences, engineering, or public health; have a good background in basic science and mathematics; and have strong communication skills. Following admission, a student is required to take a program of courses to broaden his or her education in environmental problem areas. Usually in the second year, the student enrolls in three quarters of environmental problems courses—projects that provide intensive exposure to multidisciplinary professional work. Recent problems course topics have emphasized groundwater pollution, air pollution, water quality, toxic substances, hazardous-waste control, and habitat restoration, and they often focus on the interaction between policy and technology. The student advances to candidacy after passing written and oral qualifying examinations. There is no language requirement. An approved internship of 1½ to 2 years with government, industry, or consulting firms follows, during which time the student completes a dissertation on a topic related to the internship experience. The candidate is required to present a written prospectus, including an outline of the dissertation, and defend it before the doctoral committee not later than nine months after advancement to candidacy and the beginning of the internship. The student then returns to UCLA to defend the dissertation. Completion of the program normally requires four years.
Research Facilities	UCLA has some of the finest library resources and computer facilities in the nation. Several laboratories are available to support workshops and field studies, although the program is not primarily a laboratory research one. Campus organizations formally affiliated with the program are the School of Engineering and Applied Science, the School of Public Health, and the Departments of Atmospheric Sciences, Chemistry, Biology, Earth and Space Sciences, Urban Planning, and Geography. Other campus groups, including those involved in the social sciences and management, frequently participate when the subject matter warrants it. Students, therefore, have the opportunity to take advantage of the full spectrum of campus resources.
Financial Aid	In the second year, the program offers graduate research assistantships that paid $2400 per quarter for 1998–99. A limited number of fellowships are available as well. Students may also be eligible for aid from funds administered through the Graduate Division. Currently, 100 percent of the students entering the program receive some form of financial aid.
Cost of Study	In 1998–99, fees for California residents were $1519 per quarter. Nonresident fees were an additional $3128 per quarter.
Living and Housing Costs	The University provides housing for single and married graduate students. In 1998–99, monthly rates for married student housing ranged from $556 to $875. Single graduate students were housed in a coed graduate hall at a cost of about $1633 per quarter. Early enrollment for housing is advised. There is adequate housing in the west Los Angeles area, within bicycling distance of UCLA, or on bus routes leading directly to UCLA.
Student Group	There are about 24,000 undergraduate and 10,000 graduate students enrolled at UCLA. Each year the Environmental Science and Engineering Program enrolls 8–10 doctoral students who come from many schools and hold master's degrees in science or engineering disciplines.
Student Outcomes	In the last several years, environmental science and engineering students have been successfully placed in a wide range of professional positions, such as with the U.S. EPA, California EPA, U.S. Army Corps of Engineers, California Air Resources Board, Lawrence Berkeley Laboratory, the State of Washington Department of Ecology, and regional water quality control boards. Graduates also find employment with environmental consulting companies and in industry. Many graduates have risen to positions of leadership in government and the private sector.
Location	UCLA is located on the west side of Los Angeles, 5 miles from the Pacific Ocean and 12 miles from downtown Los Angeles. The many diverse cultural and recreational opportunities in the region are within easy reach, and the University itself is a vigorous community center.
The University	UCLA, established in 1919, is academically ranked among the leading universities in the United States and has attracted distinguished scholars and researchers from all over the world. Undergraduate and graduate programs offered in the colleges and schools cover the academic spectrum. UCLA has also developed research programs and curricula outside the usual departmental structures. Interdisciplinary research facilities include institutes, centers, projects, bureaus, nondepartmental laboratories, stations, and museums. There are also many interdisciplinary programs of study, one of which is the Environmental Science and Engineering Program. UCLA's library is the largest in the Southwest. The University's Center for the Health Sciences contains one of the nation's leading hospitals and several nationally known institutes. UCLA's performing arts program of music, dance, theater, film, and lectures is one of the largest and most diverse offered by any university in the country.
Applying	Application forms for admission and financial aid may be obtained from the Graduate Admissions Office. The GRE General Test is required. TOEFL scores are required for international applicants whose native language is not English. The application deadline is December 15 for fall quarter admission.
Correspondence and Information	Dr. Richard F. Ambrose, Director Environmental Science and Engineering Program School of Public Health, Rm. 46-081 CHS University of California Los Angeles, California 90095-1772 Telephone: 310-825-9901 E-mail: app-ese@admin.ph.ucla.edu World Wide Web: http://www.ph.ucla.edu/ese

University of California, Los Angeles

THE FACULTY

Environmental Science and Engineering is an interdepartmental program drawing faculty members from participating campus departments and organizations. This unusual structure precludes identifying faculty permanently associated with Environmental Science and Engineering except for 4 core faculty members. Any faculty member of the nine participating departments may be associated with the program, depending upon the theme of the problems courses being offered. He or she may also serve as a member of a student's doctoral committee or contribute to regularly scheduled candidacy examinations.

The program is administered through the School of Public Health by an Interdepartmental Committee appointed by the Dean of the Graduate Division. The Interdepartmental Committee determines administrative and academic policy within the program and ensures interdepartmental participation. Members of the current Interdepartmental Committee and ESE-affiliated faculty are presented below.

Birgitte Ahring, Professor of Civil and Environmental Engineering; Ph.D., Copenhagen, 1986.
Richard F. Ambrose, Associate Professor of Environmental Health Sciences and Director, ESE Program; Ph.D., UCLA, 1982.
Mario Baur, Professor of Chemistry and Biochemistry; Ph.D., MIT, 1959.
Richard A. Berk, Professor of Sociology; Ph.D., Johns Hopkins, 1970.
Trudy Cameron, Professor of Economics; Ph.D., Princeton, 1982.
Yoram Cohen, Professor of Chemical Engineering; Ph.D., Delaware, 1981.
Michael Collins, Associate Professor of Environmental Health Sciences; Ph.D., Missouri, 1982.
William Cumberland, Professor of Biostatistics; Ph.D., Johns Hopkins, 1975.
Climis Davos, Professor of Environmental Health Sciences; Ph.D., Michigan, 1974.
John A. Dracup, Professor of Engineering and Applied Science; Ph.D., Berkeley, 1966.
L. Donald Duke, Assistant Professor of Environmental Health Sciences; Ph.D., Stanford, 1991.
Peggy Fong, Assistant Professor of Biology; Ph.D., California, Davis; San Diego State, 1991.
Jody Freeman, Acting Professor, School of Law; S.J.D., Harvard, 1995.
John Froines, Professor of Environmental Health Sciences; Ph.D., Yale, 1967.
Malcolm S. Gordon, Professor of Biology; Ph.D., Yale, 1958.
Thomas Harmon, Assistant Professor of Civil and Environmental Engineering; Ph.D., Stanford, 1992.
William Hinds, Professor of Environmental Health Sciences; Sc.D., Harvard, 1972.
Raymond Ingersoll, Professor of Earth and Space Sciences; Ph.D., Stanford, 1976.
Vasilios Manousiothakis, Professor of Chemical Engineering; Ph.D., Rensselaer, 1986.
Antony Orme, Professor of Geography; Ph.D., Birmingham (England), 1961.
Suzanne Paulson, Assistant Professor of Atmospheric Sciences, Chemistry Department; Ph.D., Caltech, 1991.
Richard L. Perrine, Professor Emeritus of Civil and Environmental Engineering; Ph.D., Stanford, 1953.
Theodore Porter, Professor of History; Ph.D., Princeton, 1991.
Shane Que Hee, Professor of Environmental Health Sciences; Ph.D., Saskatchewan, 1976.
Walter Reed, Associate Professor of Earth and Space Sciences; Ph.D., Berkeley, 1972.
Michael K. Stenstrom, Professor of Civil and Environmental Engineering; Ph.D., Clemson, 1976.
Irwin H. Suffet, Professor of Environmental Health Sciences; Ph.D., Rutgers, 1968.
Stanley W. Trimble, Professor of Geography; Ph.D., Georgia, 1973.
Richard P. Turco, Professor of Atmospheric Sciences; Ph.D., Illinois, 1971.
Arthur M. Winer, Professor of Environmental Health Sciences; Ph.D., Ohio State, 1969.
William Yeh, Professor of Civil and Environmental Engineering; Ph.D., Stanford, 1967.

Sampling for fish using beach seines at Malibu Lagoon.

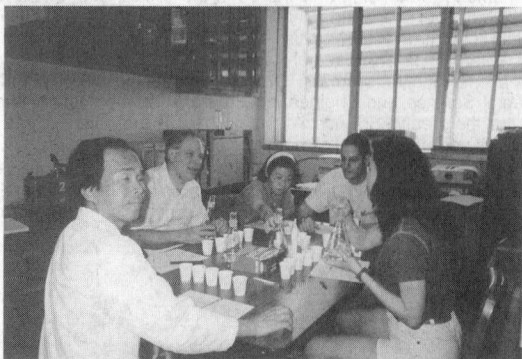

Dr. Mel Suffet and his Flavor Profile Analysis Panel test Los Angeles's water supply.

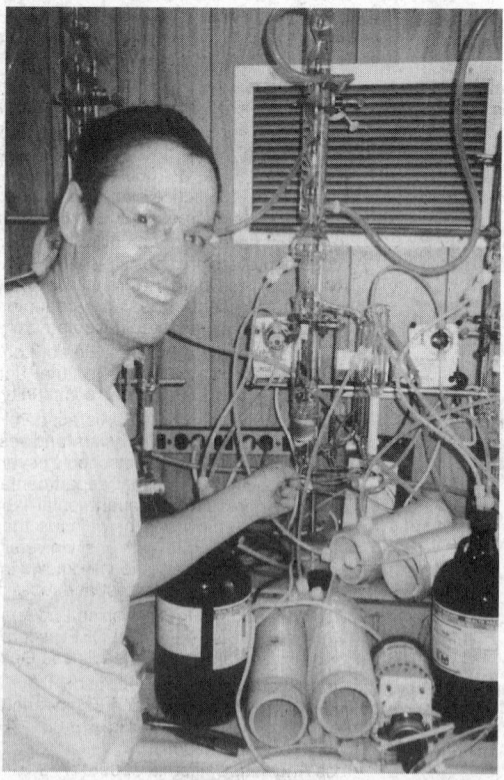

A continuous liquid extractor collects 500-liter extracts at a water reuse project at West Basin in El Segundo, California.

UNIVERSITY OF CALIFORNIA, SAN DIEGO

Department of Structural Engineering

Programs of Study	The Department of Structural Engineering Program offers M.S. and Ph.D. degrees in structural engineering (civil or aerospace) and aerospace engineering. The program encompasses analysis and design of civil, mechanical, aerospace, and marine/offshore structures, using both traditional and advanced materials such as composites. Graduate courses provide students with a strong background in structural analysis and design based on a thorough knowledge of linear and nonlinear behavior of solids, structural dynamics, composite materials, and fluid mechanics as they relate to structural response, applied mathematics for the generation of theoretical structural models, numerical analysis and computer methods for structural response quantification, and large-scale experimental testing for structural response verification.

The M.S. program extends and broadens an undergraduate structural background and equips practicing engineers with specialized knowledge concerning structural systems behavior in their fields. Students may complete their M.S. degree in one year of full-time study or two years of part-time study. The Ph.D. program is intended to prepare students for a variety of careers in research and teaching. Students admitted to the Ph.D. program may obtain their master's degrees while they pursue doctorates. |
Research Facilities	Excellent experimental and computer facilities at the University of California, San Diego (UCSD) provide students with unique opportunities to conduct research. The Powell Structural Research Laboratories are among the world's leading laboratories for full-scale testing of structural systems up to five stories tall and 120 feet in length. In the Structural Systems Laboratory, full-scale tests on bridge and building systems under simulated seismic loads are conducted, while individual structural members are tested to failure in the Structural Components Laboratory. Experimental research on composite materials is conducted in the Advanced Composite Materials Fabrication and Testing Laboratories, which include state-of-the-art equipment for processing and testing of materials and structures. Also included in the Powell Structural Research Laboratories are the Geotechnical Laboratory, Advanced Composite Manufacturing Laboratories, and a Seismic Response Modification Device Test Laboratory. The experimental facilities' many workstations and minicomputers, as well as the San Diego Supercomputer Center on the UCSD campus, greatly enhance graduate research opportunities.
Financial Aid	Financial aid is available in the form of fellowships, teaching assistantships, and research assistantships. The department attempts to support all full-time graduate students, especially at the Ph.D. level. Award of financial support is competitive, and stipends range from $9000 to a maximum of $15,000 for the academic year, plus tuition and fees. While several types of support are available, the most common is a half-time research assistantship, which provides about $12,300 during the academic year plus tuition and fees. Funds for support of international students are extremely limited, and the selection process is highly competitive.
Cost of Study	In 1999–2000, full-time students who are California residents pay an estimated $1630 per quarter in registration and incidental fees. Nonresidents pay an estimated total of $5100 per quarter for registration, tuition, and incidental fees. There is a reduced-fee structure for students enrolled on a half-time basis. Costs are subject to change.
Living and Housing Costs	UCSD provides 1,200 apartments for graduate students. Current monthly rates range from $320 for a single student in a shared apartment to $670 for a family. There is also a variety of off-campus housing in the surrounding communities. Prevailing rates range from $400 per month for a room in a private home to $950 or more for a two-bedroom apartment. Information in this regard may be obtained from the UCSD Housing Office.
Student Group	Current campus enrollment is about 18,825; of this number, 2,200 are graduate students. The Department of Structural Engineering has an undergraduate enrollment of about 130 and a graduate enrollment of about 45.
Location	The 1,200-acre campus spreads from the seashore, where the Scripps Institution of Oceanography is located, across a large wooded portion of the Torrey Pines Mesa overlooking the Pacific Ocean. To the east and north lie mountains, and to the south are Mexico and the almost uninhabited seacoast of Baja, California.
The University	One of nine campuses in the University of California System, UCSD comprises the general campus, the School of Medicine, and the Scripps Institution of Oceanography. Established in La Jolla in 1960, it has become one of the major universities in the country, with particular strengths in the physical and biological sciences.
Applying	A minimum GPA of 3.0 (on a 4.0 scale) is required for admission. All applicants are required to take the GRE General Test. International applicants whose native language is not English are required to take the TOEFL and obtain a minimum score of 550. In addition to test scores, applicants submit a completed Graduate Admission and Award Application, all official transcripts (English translation must accompany official transcripts in other languages), a statement of purpose, and three letters of recommendation. The deadline for filing applications for international applicants and those requesting financial assistance is January 31. The deadline for domestic applicants not requesting financial assistance is May 31.
Correspondence and Information	Department of Structural Engineering, 0085 Irwin and Joan Jacobs School of Engineering University of California, San Diego La Jolla, California 92093-0085 Telephone: 619-822-1421 World Wide Web: http://www.structures.ucsd.edu

University of California, San Diego

THE FACULTY AND THEIR RESEARCH

R. J. Asaro, Professor of Materials Science and Applied Mechanics; Ph.D., Stanford. Experimental and computational studies of nonlinear material behavior.

S. A. Ashford, Assistant Professor of Geotechnical Engineering; Ph.D., Berkeley. Geotechnical earthquake engineering, soil dynamics, foundation engineering, soil-structure interaction, slope stability, landfill liner design.

A.-W. Elgamal, Professor of Geotechnical Engineering; Ph.D., Princeton. Retrofit for mitigation of earthquake hazards in soil systems, interpretation of recorded seismic response, earthquake engineering, nondestructive evaluation techniques, environmental geotechnology, geomechanics.

A. Filiatrault, Professor of Structural Engineering; Ph.D., British Columbia. Earthquake engineering, structural dynamics, seismic design and analysis of structures, shake table testing, passive energy dissipation systems.

G. A. Hegemier, Professor of Applied Mechanics and Structural Engineering; Ph.D., Caltech. Mechanics of composite materials and structures, with applications to civil, aerospace, and marine systems.

V. M. Karbhari, Associate Professor of Structural Engineering; Ph.D., Delaware. Infrastructure renewal, mechanics of composite structures, manufacturing/processing science of polymers and composites, durability of polymers and composites.

J. B. Kosmatka, Associate Professor of Aerospace Structures; Ph.D., UCLA. Design, computational analysis, and experimental testing of light-weight composite structures; linear and nonlinear structural dynamics, stability, and aeroelasticity of fixed- and rotary-wing aerospace structures; development of damped composite materials and structural systems; damage detection and nondestructive evaluation.

J. E. Luco, Professor of Structural Engineering, Ph.D., UCLA. Soil-structure interaction, strong-motion seismology, earthquake engineering, active and passive control.

M. J. N. Priestley, Professor of Structural Engineering, Ph.D., Canterbury (New Zealand). Reinforced and prestressed concrete design, seismic design philosophy, seismic retrofitting, seismic assessment, design of buildings and bridges.

F. Seible, Professor of Structural Engineering, Ph.D., Berkeley. Design of bridges and structural evaluation, rehabilitation, reinforced and prestressed concrete design and analysis, large-scale structural testing and polymer matrix composite structural systems.

C.-M. Uang, Associate Professor of Structural Engineering; Ph.D., Berkeley. Seismic behavior of steel structures, seismic design philosophy.

UNIVERSITY OF COLORADO AT BOULDER

Department of Civil, Environmental, and Architectural Engineering

Programs of Study

The Department of Civil, Environmental, and Architectural Engineering offers the Master of Science and Doctor of Philosophy degrees. Areas of concentration are construction engineering and management, building systems, environmental engineering, geoenvironmental engineering, geotechnical engineering, structural engineering, and water resource engineering. Emphasis is placed on professional training at the master's level and on research and mechanics at the doctoral level. The department has earned national recognition for the breadth and depth of its research, which covers a wide field of applications in civil, environmental, and architectural engineering. A broad range of computers, from PCs to high-performance workstations, are used extensively. A state-of-the-art CAD workstation laboratory is also available.

The Master of Science degree can be earned in one calendar year, although most students take 1½ years to complete all requirements. It is offered under two plans—one with thesis and one without thesis (30 semester hours minimum). A final-comprehensive examination is given at the end of the master's program. The doctoral program includes approximately two years of course work and one or more years for completion of the dissertation. A preliminary examination is administered in the early part of the course program, and a comprehensive examination is given when all course work is completed. The dissertation defense completes the examination schedule. There is no foreign language requirement.

Each graduate student is assigned an adviser, and students doing a thesis or dissertation work closely with a research advisory committee. Employment opportunities exist with consulting engineering firms, government agencies, industry, and universities.

Research Facilities

The department has a 400 g-ton geotechnical centrifuge and 1-million-pound and 110-kip servo-controlled testing machines, dynamic structural test equipment, Sun and other workstations, and specialized equipment for research in building HVAC and lighting systems, water quality, groundwater, hazardous-materials modeling, and a state-of-the-art microscopy and imaging facility. The department includes the Center for Advanced Decision Support in Water and Environmental Systems, the Joint Center for Energy Management, and the Center for Drinking Water Optimization. Nearby laboratories include those of the National Bureau of Standards, National Center for Atmospheric Research, U.S. Bureau of Reclamation, and U.S. Geological Survey.

Financial Aid

Research and teaching assistantships are available to especially well qualified applicants. These awards include tuition and fees. Graduate School fellowships are available on a competitive basis to doctoral students, to women (as underrepresented in engineering), and to members of minority groups. Scores on the General Test of the Graduate Record Examinations are required for fellowship applicants. Stipends for 1998–99 were approximately $1200 per month for the academic year for a half-time teaching or research assistantship. International students are expected to be self-supporting during their first year in residence, after which they are considered for research assistantships on a competitive basis with other applicants for aid.

Cost of Study

In 1998–99, graduate tuition for Colorado residents was $1225 for 6 credit hours and $1827 for 9–18 credit hours per semester. Nonresident graduate tuition was $4974 for 6 credit hours and $7461 for 9–18 credit hours per semester. Student fees were approximately $350.

Living and Housing Costs

On-campus housing was $2842 per semester for a single room with twenty-one meals a week in 1998–99. Information about single-student housing may be obtained from the Supervisor of Reservations, Campus Box 154, University of Colorado at Boulder, Boulder, Colorado 80310-0154. Family housing information may be obtained from the Family Housing Office, University of Colorado at Boulder, 1350 20th Street, Boulder, Colorado 80302. A separate housing application is required.

Student Group

The enrollment at the University of Colorado at Boulder is more than 25,500. The department has approximately 250 graduate students, nearly equally divided among the six areas of concentration.

Location

The Boulder campus is located along the Front Range of the Rocky Mountains. Outdoor recreation is nearby, offering celebrated skiing, backpacking, fishing, mountain climbing, and cycling in a health and fitness-oriented area. Boulder is a community of 96,000 people; metropolitan Denver, 30 miles away, offers all the cultural amenities of a large city and is easily accessible from Boulder by public transportation.

The University and The Department

The University awarded its first advanced degree in 1893. Today, the graduate programs have an international reputation; about 5,000 graduate students are enrolled annually. The University is the flagship research institution in the region. The College of Engineering and Applied Science and the department have greatly increased their scope of research and academic excellence in the last decade. The civil engineering graduate program ranks among the top twenty in the United States in its quality of research and education.

Applying

The department welcomes applications from those students with an engineering background. A minimum grade point average of 3.0 (A = 4.0) or its equivalent is required for admission consideration. Applicants with nonengineering backgrounds are invited to inquire about possible eligibility for graduate programs. Application deadline for fall semester is April 30; deadline for spring semester is October 30. Financial awards are usually made in early spring for the following academic year.

Correspondence and Information

Graduate Coordinator
Department of Civil, Environmental,
  and Architectural Engineering
Campus Box 428
University of Colorado at Boulder
Boulder, Colorado 80309-0428

Telephone: 303-492-7316
E-mail: cvengrad@spot.colorado.edu
World Wide Web: http://bechtel.colorado.edu

University of Colorado at Boulder

THE FACULTY AND THEIR RESEARCH

Bernard Amadei, Professor; Ph.D., Berkeley. Rock mechanics, geological engineering, mechanics of anisotropic rocks and rock masses, in situ stress measurements, rock-structure interaction, stability analysis of concrete dams. (303-492-7734)

Gary Amy, Professor; Ph.D., Berkeley. Aquatic chemistry, potable water treatment, hazardous-waste management. (303-492-6274)

Angela Bielefeldt, Assistant Professor; Ph.D., Washington (Seattle). Environmental engineering, bioremediation of hazardous wastes, in situ and ex situ bioprocesses for water, air, and soil treatment. (303-492-8433)

Michael J. Brandemuehl, Associate Professor; Ph.D., Wisconsin. HVAC systems, simulation and testing of energy systems, adaptive control for building systems, indoor air quality, desiccant cooling. (303-492-8594)

Hyman Brown, Senior Instructor; B.S., CUNY City College. Project administration, construction management, business development. (303-492-7994)

Steven C. Chapra, Professor; Ph.D., Michigan. Mathematical modeling of environmental systems, water quality analysis, systems analysis, numerical methods, computer applications. (303-492-7573)

Ross Corotis, Professor; Ph.D., MIT. Structural reliability, probabilistic code formulation, system reliability and optimization, natural hazard evaluation, structural analysis and design, random vibrations, life-cycle costs. (303-492-7006)

Robert G. Davis, Senior Instructor; M.S., Penn State. Lighting design, human response to lighting, lighting technology, energy use in buildings. (303-492-7614)

James E. Diekmann, Professor; Ph.D., Washington (Seattle). Project control systems, construction contracts and claims, expert systems, risk analysis. (303-492-7642)

David L. DiLaura, Senior Instructor; B.A., Wayne State. Illumination engineering, lighting energy analysis, daylighting. (303-492-4798)

John O. Dow, Associate Professor; Ph.D., Colorado. Structural dynamics, computational mechanics, a posteriori evaluation of finite-element results, equivalent continuum analysis of discrete structural models. (303-492-8561)

Dan M. Frangopol, Professor; Ph.D., Liège (Belgium). Safety and reliability in structural engineering, optimal design of civil infrastructure systems, bridge engineering and life-cycle safety management, deterioration modeling of materials and structures, natural hazard mitigation. (303-492-7165)

Kurt H. Gerstle, Professor Emeritus; Ph.D., Colorado. Elastic and inelastic analysis; theory, behavior, and design of steel and concrete structures; constitutive modeling of plain concrete. (303-492-6383)

Vijay Gupta, Professor; Ph.D., Arizona. Global hydrology, scaling theory of hydrology, hydraulics and geometry of river networks. (303-492-3696)

Milan F. Halek, Senior Instructor; M.S., Czech Technical. Surveying and photogrammetry. (303-492-7007)

James P. Heaney, Professor; Ph.D., Northwestern. Water resource engineering, stormwater infrastructure, water management. (303-492-3276)

George Hearn, Associate Professor; D.E.S., Columbia. Nondestructive evaluation, modal analysis, cable-supported structures, long-span bridges, nonlinear dynamic response of structures, stochastic processes in structural vibration. (303-492-6381)

Mark Hernandez, Assistant Professor; Ph.D., Berkeley. Environmental engineering, environmental microbiology, biological waste treatment processes, bioaerosols and indoor air quality, microbially induced corrosion. (303-492-5991)

Hon-Yim Ko, Professor and Chairman; Ph.D., Caltech. Soil and rock mechanics, constitutive relationships, multiaxial testing and constitutive modeling of geomaterials, centrifuge modeling. (303-492-6716)

Moncef Krarti, Associate Professor; Ph.D., Colorado. Energy conservation, ground-coupled heat transfer, mathematical modeling of energy systems, HVAC-system simulation, load prediction using neural networks. (303-492-3389)

Jan F. Kreider, Professor; Ph.D., Colorado. Solar engineering, energy conservation, expert systems, HVAC system testing, energy monitoring and prediction for buildings, fluid dynamics, chemical and mechanical engineering. (303-492-7603)

Diane McKnight, Associate Professor; Ph.D., MIT. Organic geochemistry, reactive solute transport in streams, lake and stream ecology, surface water hydrology. (303-492-4687)

Ralph Muehleisen, Assistant Professor; Ph.D., Penn State. Architectural and engineering acoustics, thermoacoustic refrigeration, active and passive noise and vibration control, acoustic measurements and instrumentation. (303-492-5736)

Ronald Pak, Professor; Ph.D., Caltech. Soil-structure interaction, soil dynamics, elasticity and continuum mechanics, wave propagation and earthquake engineering. (303-492-8613)

Russell J. Qualls, Assistant Professor; Ph.D., Cornell. Hydrology, land-atmosphere interactions, evapotranspiration, remote sensing of environmental processes. (303-492-5968)

Harihar Rajaram, Assistant Professor; Sc.D., MIT. Fluid mechanics, hydrology, numerical and stochastic modeling of environmental transport processes. (303-492-6604)

Joseph N. Ryan, Assistant Professor; Ph.D., MIT. Environmental engineering, aqueous geochemistry, surface chemistry, colloid transport. (303-492-0772)

Victor Saouma, Professor; Ph.D., Cornell. Fracture mechanics and computer-aided design. (303-492-1622)

Benson Shing, Professor; Ph.D., Berkeley. Structural dynamics, earthquake engineering, online computer-controlled seismic testing, inelastic behavior of structures, development of analytical models, performance of rc bridge structures. (303-492-8026)

JoAnn Silverstein, Associate Professor; Ph.D., California, Davis. Biological wastewater-treatment process, sludge separation and disposal, biodegradation of organic compounds in soils. (303-492-7211)

Anthony D. Songer, Assistant Professor; Ph.D., Berkeley. Computer-integrated civil systems, construction engineering and management, alternative delivery mechanisms and project control systems. (303-492-2627)

Enrico Spacone, Assistant Professor; Ph.D., Berkeley. Inelastic behavior of structures, computational modeling. (303-492-7607)

Kenneth Strzepek, Associate Professor; Ph.D., MIT. Water resources planning and management, hydrology, river basin planning, agricultural and environmental systems, artificial intelligence, advanced decision support and computer applications. (303-492-7111)

Stein Sture, Professor; Ph.D., Colorado. Soil and rock mechanics, constitutive modeling, numerical methods in structural mechanics, laboratory modeling, finite-element analysis. (303-492-7651)

Luis H. Summers, Professor; Ph.D., Notre Dame. Interface and design integration of building systems, HVAC system design and simulation, building energy management, cogeneration, radiant comfort. (303-492-1094)

R. Scott Summers, Professor; Ph.D., Stanford. Drinking water quality and treatment, disinfection by-products, mass transfer, natural organic matter.

Kaspar J. Willam, Professor; Dr.-Ing. habil., Stuttgart; Ph.D., Berkeley. Computational mechanics, finite-element analysis of transient and nonlinear problems, constitutive modeling of nonlinear and inelastic material behavior. (303-492-7011)

Yunping Xi, Assistant Professor; Ph.D., Northwestern. Concrete durability, solid waste conversion and applications of smart materials. (303-492-8991)

Dobroslav Znidarcic, Associate Professor; Ph.D., Colorado. Soil mechanics, earth structures, slope stability, groundwater and seepage, theories of consolidation, laboratory testing, centrifuge modeling, partially saturated soil behavior. (303-492-7577)

Jorge G. Zornberg, Assistant Professor; Ph.D., Berkeley. Geotechnical and geoenvironmental engineering, soil reinforcement and improvement, geosynthetics, waste containment facilities, numerical and centrifuge modeling of geotechnical and geoenvironmental systems. (303-492-4699)

UNIVERSITY OF DELAWARE

Department of Civil and Environmental Engineering

Programs of Study

The Department of Civil and Environmental Engineering offers graduate programs leading to three degrees: Master of Civil Engineering (M.C.E.), Master of Applied Sciences (M.A.S.), and Doctor of Philosophy (Ph.D.) in civil engineering. The M.C.E. degree normally includes a thesis. However, a nonthesis option is available.

Major subject areas in the M.C.E., M.A.S., and Ph.D. programs are environmental and water resources engineering, soil mechanics and geotechnical engineering, structural engineering, materials and engineering mechanics, coastal and ocean engineering, transportation engineering, and railroad engineering. Each student develops an individualized plan of study and course work in cooperation with a chosen supervisor. An interdepartmental doctoral committee consisting of faculty members involved in the student's major and minor areas of study is formed for each Ph.D. candidate.

The M.C.E. and M.A.S. degrees require 24 semester credit hours of course work plus 6 credits of thesis or additional course work and can normally be completed in 1½ to 2 years by full-time students. There is also a nonthesis option that requires 30 credit hours. The Ph.D. usually requires a total of 48 semester credit hours of course work above the B.S. level or 24 credit hours above the M.S. degree. A final comprehensive examination and a thesis are required as well. A Ph.D. degree usually requires at least two years of study beyond the master's degree.

Research Facilities

The department has large research facilities and separate laboratories for structural engineering and materials, geotechnical engineering, environmental engineering, and hydraulic and ocean engineering. Constant upgrading of equipment through University and external research support renders the laboratories well suited for state-of-the-art experimental research. The ocean laboratory is one of the largest and best equipped in the country.

The departmental computer system consists of more than thirty Sun Workstations networked to a central file server and to University mainframes. The University library contains more than 2 million volumes and more than 20,000 journal subscriptions and has an excellent collection in engineering and science disciplines. Its card catalogs can be accessed electronically from all the Sun Workstations and networked microcomputers.

Financial Aid

The department has substantial funding available to support its M.S. and Ph.D. students as teaching assistants, research fellows, and research assistants. In 1999–2000, these positions carry an average stipend of $1200 per month, plus tuition, for first-year students. A number of Davis and University fellowships are awarded each year to incoming students. Many student spouses find employment at the University or in the community.

Cost of Study

Full-time tuition in 1999–2000 is approximately $2125 per semester for Delaware residents and $6125 per semester for nonresidents. Part-time tuition is approximately $236 and $681 per credit hour for residents and nonresidents, respectively.

Living and Housing Costs

A large number of apartments are available within comfortable distance from the University. Rents range from $450 to $600 per month depending on unit size. A number of University-owned apartments are available for married students. Generally, the cost of living is substantially less than in neighboring large cities.

Student Group

Over 20,000 students are enrolled at the University. The department currently has about 250 undergraduate and 95 graduate students, of whom 29 are part-time. Approximately half the graduate students come from various parts of the United States, and about half come from other countries. Women constitute 22 percent of the undergraduate and 24 percent of the graduate student group.

Location

The suburban community of Newark, Delaware, is the site of the University's main campus, where the department is located. Newark is within an hour's drive of Philadelphia, Baltimore, and the Chesapeake Bay. Atlantic beaches and Washington, D.C., can be reached in less than two hours and New York in less than three.

The University

The University was founded in 1743 as a small liberal arts school and became a land-grant college in 1867. It now ranks among the finest of the nation's medium-sized universities. The campus, which exhibits a classical architectural style, is situated in an area rich in technical talent and interests. The cultural and technical interaction between students and faculty and the surrounding community provides a stimulating environment.

Applying

Applications for admission and supporting credentials should be received by the Office of Graduate Studies before July 1 for the fall semester, December 1 for the spring semester, and April 1 for the summer session. Applications with requests for financial aid should be received by February 1 for the fall semester. The General Test of the Graduate Record Examinations is required of all applicants. International students must also pass the TOEFL.

Correspondence and Information

Coordinator of Graduate Studies
Department of Civil and Environmental Engineering
137 P. S. DuPont Hall
University of Delaware
Newark, Delaware 19716
Telephone: 302-831-6570
Fax: 302-831-3640
E-mail: mary.horstman@mvs.udel.edu

University of Delaware

THE FACULTY AND THEIR RESEARCH

Chin-Pao Huang, Distinguished Professor and Chairman (joint appointment with the College of Marine Studies); Ph.D., Harvard; PE. Industrial-waste management, environmental surface chemistry, in situ soil and groundwater remediation technology, aquatic chemistry.

Herbert E. Allen, Professor; Ph.D., Michigan. Environmental engineering, fate and transport of environmental pollutants, sediments soil and water quality standards, reactions of inorganic and charged organic compounds in environmental systems.

Daniel K. Cha, Assistant Professor; Ph.D., Berkeley. Process control of biological wastewater treatment facilities, biotransformation of toxic pollutants, microbially mediated systems for the remediation of hazardous waste problems, bioenvironmental engineering.

Michael J. Chajes, Associate Professor; Ph.D., California, Davis; PE. Structural design and analysis, infrastructure evaluation and rehabilitation, composite material applications, structural dynamics.

Alexander H.-D. Cheng, Professor; Ph.D., Cornell. Water resources, groundwater, multiaquifer theory, boundary-element technique, poroelasticity, poroelastodynamics, geomechanics, borehole stability, chaotic dynamics, hydraulic fracture.

Pei-Chun Chiu, Assistant Professor; Ph.D., Stanford. Transformation and treatment of hazardous organic contaminants, environmental redox chemistry.

Robert A. Dalrymple, Edward C. Davis Professor and Director of Center for Applied Coastal Research (joint appointment with the College of Marine Studies); Ph.D., Florida; PE. Coastal engineering, wave mechanics, sediment transport.

Steven K. Dentel, Professor; Ph.D., Cornell; PE. Environmental engineering, aquatic and colloid chemistry applications in water and wastewater treatment, chemical dose optimization and sludge treatment.

Ardeshir Faghri, Associate Professor; Ph.D., Virginia. Transportation systems engineering, geographic information systems, artificial intelligence, applied probability theory.

Kevin J. Folliard, Assistant Professor; Ph.D., Berkeley. Concrete materials, high-performance concrete, durability of construction materials.

Paul T. Imhoff, Assistant Professor; Ph.D., Princeton; PE. Groundwater hydrology, storm water pollution, contaminant transport, mass transfer processes.

Victor N. Kaliakin, Associate Professor; Ph.D., California, Davis. Computational mechanics, geomechanics, structural mechanics.

Arnold D. Kerr, Professor; Ph.D., Northwestern. Stress and vibration analysis of structures, stability of structures, analysis of continuously supported structures, railroad engineering, analysis of concrete pavements, mechanics of floating ice plates, engineering mathematics.

Shinya Kikuchi, Professor (joint appointment with the Operations Research Program); Ph.D., Pennsylvania; PE. Urban transportation systems and operation, application of operations research to transport problems, analysis of uncertainty.

James T. Kirby, Professor (joint appointment with the College of Marine Studies); Ph.D., Delaware. Ocean and coastal engineering, wave spectra, nonlinear wave mechanics.

Nobuhisa Kobayashi, Professor (joint appointment with the College of Marine Studies); Ph.D., MIT. Hydrodynamics, coastal engineering, coastal structures, sediment transport mechanics.

Dov Leshchinsky, Professor; Ph.D., Illinois at Chicago. Geotechnical engineering: foundation engineering, retaining structures, slope stability, geosynthetics.

Dennis Mertz, Associate Professor; Ph.D., Lehigh. Structural engineering, bridges, fatigue and fracture, design methodologies.

Harry W. Shenton III, Assistant Professor; Ph.D., Johns Hopkins. Structural dynamics, earthquake and wind engineering, civil structural control, experimental methods.

Ib A. Svendsen, Distinguished Professor (joint appointment with the College of Marine Studies); Ph.D., Denmark Technical; PE. Coastal and ocean engineering, wave breaking and the dynamics of coastal surf zones, wave generation of coastal currents, sediment transport.

W. David Teter, Assistant Professor; M.S.M.E., West Virginia; PE. Computer-aided drawing and design, engineering graphics and analysis.

Associated Faculty

The following faculty members have joint appointments with the Department of Civil and Environmental Engineering.

Mohsen Badiey, Associate Professor, College of Marine Studies; Ph.D., Miami (Florida). Field and laboratory measurements of water waves, acoustic waves and seismic waves, acoustic wave propagation, underwater and environmental noise measurement and analysis.

John W. Gillespie Jr., Associate Professor and Technical Director, Center for Composite Materials; Ph.D., Delaware. Use of composite materials in infrastructure repair, rehabilitation and construction.

George Luther, Professor, College of Marine Studies; Ph.D., Pittsburgh. Marine chemistry, trace element cycling and speciation in the environment, redox processes in sediments and waters.

William Ritter, Chair and Professor, Agricultural Engineering; Ph.D. Iowa; PE.

Donald L. Sparks, Professor and Chairman, Plant and Soil Science; Ph.D., Virginia Tech. Kinetics of soil chemical reactions and use of surface spectroscopic and microscopic techniques to elucidate contaminant reaction mechanisms and speciation in soils and natural systems.

Robert M. Stark, Professor and Chairman, Department of Mathematical Sciences; Ph.D., Delaware. Operations research, systems analysis, random processes, technological management, reliability of structures.

RESEARCH ACTIVITIES

Coastal and Ocean Engineering. Water-wave mechanics, hydrodynamics, nearshore circulation, littoral processes, sediment transport mechanics, coastal structures, Arctic coastal engineering. The Center for Applied Coastal Research supports engineers and scientists conducting research on the coastal and nearshore problems and trains students for industry, government, and academia.

Environmental Engineering. Biological, physical, and chemical aspects of environmental systems; sludge handling and disposal; groundwater modeling; industrial wastewater treatment; chemistry of heavy metals; nonpoint sources of pollution; engineering limnology; physics and chemistry of waterborne and airborne particulates; response of biological treatment processes to changes in environmental conditions; mathematical modeling of microbial growth and substrate removal; soil remediation; resource recovery; advanced oxidation processes.

Geotechnical Engineering. Soil mechanics, foundation engineering, geosynthetics, slope stability, retaining walls, groundwater hydrology, construction materials, computational geomechanics, geoenvironmental engineering.

Structural Engineering and Engineering Mechanics. Structural design, evaluation, and rehabilitation; bridge engineering; structural dynamics; soil-structure interaction; stability of structures; plates and shells; applications of advanced composite materials; analyses of floating ice covers; railroad track analysis; analyses of failures in structures; experimental stress analysis; fracture mechanics; nonlinear analysis of structural frameworks; computational mechanics; computer-assisted design and analysis of structures; rehabilitation of civil structures using composite materials; structural control; passive energy dissipation systems; seismic design; railway engineering.

Transportation Engineering. Transportation policy and planning; urban public transportation design and operations; traffic engineering and control; marine, air, and railroad transportation; application of operations research, expert systems, fuzzy set theory, and neural networks to transport problems; intelligent transportation systems.

Water Resources. Groundwater hydraulics, contaminant transport, numerical modeling, poroelastic consolidation, multiaquifer theory, hydraulic fracture, chaotic dynamics.

UNIVERSITY OF DETROIT MERCY

College of Engineering and Science
Department of Civil and Environmental Engineering

Programs of Study

The Department of Civil and Environmental Engineering offers courses leading to the degrees of Master of Engineering, Master of Engineering Management, and Doctor of Engineering. The department emphasizes three areas of concentration: environmental engineering, geotechnical engineering, and structural engineering. Environmental engineering includes water and wastewater engineering, solid waste management, environmental chemistry and biology, and physical, chemical, and biological unit operations. Geotechnical engineering includes soil mechanics, soil dynamics, tunneling, earth dams, and pavement design. Structural engineering emphasizes finite element methods, computer-aided structural analyses and design, prestressed concrete, and plastic analyses. The graduate program prepares engineers to apply their analytical and experimental skills to engineering design, development, research, and management. Candidates focus their research on industrial and consulting challenges that require the development and application of new engineering knowledge, analytical methods, and experimental systems and tools. Local engineering activities and facilities, including research and development, industrial operations, and construction, offer students considerable opportunities for study and potential future employment. Many students are practicing engineers who take the program on a part-time basis; to accommodate these students, all graduate classes are offered in the evenings. The Master of Engineering degree requires a minimum of 30 semester hours and can be completed with or without a thesis. If a thesis is elected, the research is initiated and completed during the second half of the study period. The average duration for the Master of Engineering degree is four semesters. The Doctor of Engineering degree at UDM is an interdisciplinary program that requires postbaccalaureate completion of four engineering/mathematics core courses, 30 semester hours in the general area of specialization, 9 semester hours in approved technical electives, and 36 hours of dissertation research. The first qualifying examination is completed during the first year after admission to the program. The final qualifying examination must be taken within three years following admission to the program. At least 30 semester hours of dissertation research must be taken following successful completion of the final qualifying examination. The final examination/defense for the doctorate must be completed within five years of admission to the program. Dissertation work can involve any combination of analytical and experimental research, as guided by a faculty adviser in the area of interest to the student. The Master of Engineering Management program is designed for experienced engineers who desire additional technical depth and the ability to manage technical projects and teams. Further information about the program can be obtained from Richard Schneider, Master of Engineering Management Director, University of Detroit Mercy, 4001 West McNichols Road, Detroit, Michigan 48219-0900.

Research Facilities

The department has laboratory facilities that are fully equipped for analytical and experimental research in the areas mentioned above. In addition, environmental engineering has facilities and equipment housed in the U.S. EPA-sponsored Center of Excellence in Environmental Engineering and Science. Available equipment for environmental engineering includes an atomic absorption spectrometer, gas chromatograph, spectrophotometer, and water and wastewater quality analysis apparatuses. The construction materials laboratory includes SUPERPAVE test facilities and a universal testing machine. The geotechnical engineering laboratory equipment includes a triaxial test setup with online data acquisition for measurement of stress drain for water pressure and back pressure and a motorized residual direct shear test apparatus for testing geotextile interfaces against soils and residual soil properties. The structures laboratory has a universal testing machine with a graphic recorder, an extensometer, a signal condition module, and data acquisition equipment. The College of Engineering and Science has an advanced computing laboratory that supports the computational component of the research needs. The computing laboratory has advanced computer workstations (IBM RS-6000 Model 530's) and appropriate software for geometric, solid, and finite-element modeling. The University's library facilities include about 500,000 volumes. The library participates in a cooperative network with other Detroit-area universities and hospital libraries.

Financial Aid

A variety of teaching and research fellowships/assistantships are available in the College of Engineering and Science. This financial aid is competitively awarded on a yearly basis. Full-time graduate students are eligible usually after their first semester of study. In addition, the Scholarship and Financial Aid Office accepts applications for grants, loans, and work-study assistance. Aid includes the Michigan Tuition Grant (for Michigan residents only), Federal Work-Study, and a variety of loans. The University also accepts third-party payments from employers and government agencies and offers payment plans of its own. For information regarding financial aid programs, students should call 313-993-3350.

Cost of Study

Tuition in 1999–2000 is $545 per credit hour. Registration fees are $100 for full-time students.

Living and Housing Costs

Housing is available on campus. Double-occupancy rates range from $1410 to $3280. Single-occupancy rates range from $2420 to $2720. The University offers meal plans at about $1085. All rates are for a sixteen-week term. For more information, students should call the Residence Life Office at 313-993-1230.

Student Group

The graduate students in the civil and environmental engineering department represent many groups, including men, women, and minority and international students. Many students are pursuing their degree part-time in the evening while working full-time in industry. Approximately 6,700 students attend classes on three UDM campuses located in northwest and downtown Detroit.

Location

Students enjoy a variety of activities offered on campus and throughout the metropolitan Detroit area, including sports, theater, concerts, and more.

The University

As Michigan's largest Catholic university, the University of Detroit Mercy has an outstanding tradition of academic excellence firmly rooted in a strong liberal arts curriculum. This tradition dates back to the formation of two Detroit institutions, the University of Detroit, founded in 1877 by the Society of Jesus (Jesuits), and Mercy College of Detroit, founded in 1944 by the Sisters of Mercy of the Americas. In 1990, these schools consolidated to become the University of Detroit Mercy. Today, UDM offers more than 120 majors and programs in nine different schools and colleges and is widely recognized for its programs in engineering, law, dentistry, nursing, and architecture. Faculty members are known for their personal attention to students and teaching excellence; more than 87 percent have a Ph.D. or comparable terminal degree.

Applying

Applications for admission normally should be completed at least six weeks before the beginning of a term. Applications for financial aid should be submitted by April 1. International students are urged to complete their applications at least three months before classes begin. Admission requirements are a bachelor's degree from an accredited college; a B average in the total undergraduate program and in the proposed field of study; and, normally, an undergraduate major or the equivalent in the proposed field. Official transcripts are required from all colleges attended. Applicants with less than a B average who present other evidence of ability to perform graduate-level work may be admitted as probationary students upon the recommendation of the director of the program.

Correspondence and Information

Records Office
College of Engineering and Science
University of Detroit Mercy
4001 West McNichols Road
P.O. Box 19900
Detroit, Michigan 48219-0900

Telephone: 313-993-3335
Fax: 313-993-1187

Professor U. Dutta, Chairman
Department of Civil and Environmental Engineering
University of Detroit Mercy
4001 West McNichols Road
P.O. Box 19900
Detroit, Michigan 48219-0900

Telephone: 313-993-1040
Fax: 313-993-1187
E-mail: duttau@udmercy.edu

University of Detroit Mercy

THE FACULTY AND THEIR RESEARCH

Utpal Dutta, Associate Professor; Ph.D., Oklahoma, 1985; PE. Transportation, utilization of waste tires and automotive shredder residue in road materials, construction materials.

Chitta R. Gangopadhyay, Professor; Ph.D., Illinois at Urbana-Champaign, 1962; PE. Geotechnical engineering, pollution containment, hydrogeology, groundwater modeling, geotechnical failure.

Alan Hoback, Assistant Professor; Ph.D., Washington (St. Louis), 1993. Structural design, structural optimization, computer-aided engineering.

Albert Ku, Professor; Ph.D., Ohio State, 1965; PE. Finite element methods with nonlocking plate elements, internal constraints, local stress fills.

Kevin P. Olmstead, Assistant Professor; Ph.D., Michigan, 1989; PE. Chemical and biological environmental processes, metal removal from wastewater, activated carbon adsorption, remediation processes, pollution prevention, Geographic Information System (GIS).

UNIVERSITY OF FLORIDA

Department of Environmental Engineering Sciences

Programs of Study

The department offers programs leading to the degree of Master of Engineering, Master of Science, and Doctor of Philosophy in environmental engineering sciences. Major concentrations include air pollution control, drinking water treatment systems, environmental biology, microbiology and toxicology, groundwater protection and remediation, industrial hygiene, hazardous waste, solid waste, radiological health and health physics, systems ecology and energy analysis, water chemistry, water resources management, water pollution control, and wetlands ecology. A joint program with the law school is also available.

Master's degree candidates may choose a thesis or nonthesis option. The programs require 30 semester credit hours (thesis) or 34 semester credit hours (nonthesis). Students who pursue the nonthesis option for the Master of Science are required to pass a comprehensive written examination. The student who chooses the thesis option must write a thesis and defend it during an oral final examination. The Master of Engineering degree is earned by students with a baccalaureate degree in engineering, while the Master of Science degree is earned by students who have a bachelor's degree in a discipline other than engineering, such as chemistry, geology, physics, or biology. Students with a degree in a nontechnical field may also be admitted into this program upon the completion of articulation courses.

The Ph.D. student is required to pass both oral and written doctoral qualifying examinations. The student is also expected to complete a dissertation that reflects independent investigation and to defend it during an oral final examination. Ninety graduate semester credit hours, which can include some or most of the master's degree course work, are required for the Ph.D. degree.

Research Facilities

The Department of Environmental Engineering Sciences is housed in its own four-story building, A. P. Black Hall, which contains numerous laboratories, classrooms, and computers for teaching and research. Additional facilities include the Center for Wetlands and extensive space in the new engineering building.

Financial Aid

Graduate teaching and research assistantships are available from department faculty members and are subject to funding. These assistantships are awarded on the basis of GRE scores, college transcripts, and letters of recommendation. A limited number of fellowships are also available.

Cost of Study

For 1999–2000, the tuition for most graduate course work is approximately $160 per semester credit hour for Florida residents and $520 per semester credit hour for out-of-state students. These costs are subject to annual increases by the Florida legislature and are announced each summer. Tuition may be waived for those who hold assistantships. Graduate students are required to register for 12 credit hours per semester unless they have received an assistantship, which would allow a reduced credit hour load.

Living and Housing Costs

Rent for apartments provided by the University of Florida for single graduate and professional students begins at $900 per person per semester (double occupancy). The University also operates five apartment villages for families; these rent from $300 to $400 per month, including utilities. In addition to on-campus housing, there are many apartment complexes in the area, with rent starting at $400 per month, not including utilities.

Student Group

The total enrollment at the University of Florida is approximately 43,000 students. There are approximately 120 graduate students enrolled in the Department of Environmental Engineering Sciences. Many graduate students are from out of state. Several are pursuing an advanced degree after having worked in the profession for some time.

Student Outcomes

Students who have earned graduate degrees from the master's and doctoral programs at the University of Florida have found employment in academe (Minnesota, Ohio State, Maryland, Rice, West Point, Toledo, Mercer, and the Air Force Institute of Technology), industry (International Paper, Cargill, and Merck), agencies (USEPA, Florida DEP, Florida Water Management Districts, and the Metropolitan District of Southern California), and consulting firms (CDM, CH2MHILL, Law Engineering, Post Buckley, and Montgomery Watson). The employment market is often influenced by national and state environmental regulations and funding.

Location

The beautiful campus of the University of Florida is located in Gainesville, an urban area of approximately 150,000 inhabitants in north central Florida. Gainesville is a 1-hour drive from the Gulf of Mexico and 2 hours from the Atlantic Ocean, and there are abundant facilities close by for recreational activities. Bike paths run alongside most streets in Gainesville. The city also has many cultural events, including professional theater, art exhibits, open-air festivals, and concerts.

The University

The University of Florida is a member of the prestigious Association of American Universities. It is one of the nation's twenty-five largest universities and ranks among the top three universities, public and private, in the number of academic programs offered. The College of Engineering's graduate programs are often rated among the top thirty in the United States.

Applying

There are three ways to apply to the department: obtain the appropriate forms from the department, apply on line, or print out the forms from the department's Web site. Admission into the graduate program requires a Bachelor of Science degree in engineering or science from an accredited college and an upper-division grade point average of 3.0 or better (on a 4.0 scale). Students must also submit acceptable scores on the General Test of the Graduate Record Examinations (500 or better in both the verbal and quantitative sections and an overall score greater than 1100 for M.S. and M.E. candidates; higher scores are required for students who plan to pursue the Ph.D. degree). Students from non-English-speaking countries must submit a score of 575 or better on the TOEFL. Applicants must also submit three letters of recommendation, preferably from faculty members with whom they have studied, and a statement of interest in which the applicant's goals and desired area of concentration are explained. No action is taken on applications until all required information has been received. Students can be admitted in fall or spring semesters, but applications, transcripts, and other required information must be received at least three months in advance of the desired semester of entrance.

Correspondence and Information

Graduate Academic Office
Department of Environmental Engineering Sciences
216 Black Hall
P.O. Box 116450
University of Florida
Gainesville, Florida 32611-6450

Telephone: 352-392-8450
Fax: 352-392-3076
E-mail: dsmit@eng.ufl.edu
World Wide Web: http://www.enveng.ufl.edu

University of Florida

THE FACULTY AND THEIR RESEARCH

Jean M. Andino, Assistant Professor; Ph.D., Caltech, 1996; EIT. Air pollution.
  Mechanism of atmospheric photoxidation of aromatics: A theoretical study. *J. Phys. Chem.* 100:10967–80, 1996. With Smith et al.
  Tropospheric chemistry. *Adv. Chem. Eng.* 19:325–407, 1994. With Seinfeld et al.
  Atmospheric oxidation of biogenic hydrocarbons: Reaction of ozone with -pinene, d-limonene, and trans-caryophyllene. *Env. Sci. Tech.* 27:2754, 1993. With Grosjean et al.
Michael D. Annable, Associate Professor; Ph.D., Michigan State, 1991; PE. Hydrology.
  Field implementation of a Winsor type I surfactant/alcohol mixture for in situ solubilization of a complex LNAPL as a single-phase microemulsion. *Environ. Sci. Tech.* 32:523–30, 1998. With Jawitz, Rao, and Rhue.
  Field evaluation of interfacial and partitioning tracers for characterization of effective NAPL-water contact areas. *Groundwater* 36:459–502, 1998. With Jawitz et al.
Gabriel Bitton, Professor; Ph.D., Hebrew (Jerusalem), 1973. Microbiology.
  Short-term toxicity test based on algal uptake by *Ceriodaphnia dubia. Water Environ. Res.* 69:1207–10, 1997. With Lee et al.
  Validity of fluorochrome-stained bacteria as tracers of short-term microbial transport through porous media. *J. Cont. Hydrol.* 31:349–57, 1998. With Kucukcolak, Koopman, and Farrah.
W. Emmett Bolch, Professor and Associate Chairman; Ph.D., Berkeley, 1967; PE. Radiological health.
  A methodology for uncertainty analysis: Demonstration with a simple pathway equation. *Radiat. Protection Manage.* 12:55–65, 1995. With Stanford, Birky, and Huston.
Matthew M. Booth, Assistant in Engineering; Ph.D., Florida, 1996. Mass spectrometry.
  Analysis of human skin emanations by gas chromatography/mass spectrometry. Thermal desorption of attractants for the yellow fever mosquito *(Aedes aegypti)* from handled glass beads. *Anal. Chem.* 71:1–7, 1999. With Bernier and Yost.
  Practical ion trap technology: GC/MS and GC/MS/MS in practical aspects of ion trap mass spectrometry. In *Chemical, Biomedical, and Environmental Applications,* vol. 2. Florida: CRC Press, 1993. With Yates, Stephenson, and Yost.
Mark T. Brown, Assistant Professor; Ph.D., Florida, 1980. Systems ecology.
  Energy based indicies and ratios to evaluate sustainability: Monitoring technology and economies toward environmentally sound innovation. *Ecological Eng.* 9:51–69, 1997. With Ulgiati.
  Embodied energy analysis and energy analysis: A comparative view. *Ecological Econ.* 19:219–35, 1996. With Herendeen.
  Energy analysis of Thailand and Mekong River dam proposals. *Ecological Modeling* 91:105–30, 1996. With McClanahan.
Paul A. Chadik, Assistant Professor; Ph.D., Arizona, 1985; PE. Water treatment.
  Effect of bromide ion on haloacetic acid formation during chlorination of Biscayne Aquifer water. *J. Environ. Eng. (ASCE)* 124:932–8, 1998. With Wu.
  Disinfection byproduct formation from the preparation of instant tea. *J. Agric. Food Chem.* 46:3272–9, 1998. With Wu et al.
  Sulfide-oxidizing bacteria: Their role during air-stripping. *J. Am. Water Works Assoc.* 90:107–15, 1998. With Dell'Orco, Bitton, and Neumann.
Thomas L. Crisman, Professor; Ph.D., Indiana, 1976. Aquatic ecology.
  Food web structure in the pelagic and littoral regions of a subtropical lake ecosystem. *Oikos* 75:20–32, 1996. With Havens et al.
  Chironomidae (diptera) and vegetation in a created wetland and implications for sampling. *Wetlands* 15:285–9, 1996. With Streever, Evans, and Kenner.
  Zooplankton seasonality and trophic state relationships in Lake Okeechobee, Florida. *Arch. Hydrobiol.* 45:213–32, 1995. With Phlips and Beaver.
Joseph J. Delfino, Professor and Chairman; Ph.D., Wisconsin, 1968. Water chemistry, water quality.
  Ecological-economic evaluations of wetland management alternatives. *Ecological Eng.* 11:291–302, 1998. With Ton and Odum.
  Comparison of mercury accumulation in Lake Eden and the Savannas Marsh wetland, Florida, U.S.A. *Verh. Int. Verein. Limnol.* 26:1365–9, 1998. With Rood and Gottgens.
  Uncertainty in paleoecological studies of mercury in sediment cores. *Water, Air, Soil Pollut.* 110:313–33, 1999. With Gottgens, Rood, and Summers.
Ben Koopman, Professor; Ph.D., Berkeley, 1981; PE. Biological wastewater treatment.
  Full-scale test of methanol addition for enhanced nitrogen removal in a Ludzack-Ettinger process. *Water Environ. Res.* 70:376–81, 1998. With Regan, Svoronos, and Lee.
  Long-term evaluation of aluminum hydroxide–coated sand for removal of bacteria from wastewater. *Water Res.* 32:2171–9, 1998. With Chen et al.
  Nitrogen removal in a partial nitrification/complete denitrification process. *Water Environ. Res.* 70:334–42, 1998. With Potter and Tseng.
Angela S. Lidner, Ph.D., Michigan, 1998. Bioremediation, pollution prevention.
  Microbial ecology of PCB transformations in the environment: A niche for methanotrophs? In *Microbial Growth on C1 Compounds.* Dortrecht, the Netherlands: Kluwer Academic Publishers, 1995. With Adriaens.
Clay L. Montague, Associate Professor; Ph.D., Georgia, 1980. Systems ecology.
  The future of coastal Florida. *J. Publ. Interest Environ. Conf.* 1:51–9, 1997.
  Estuaries. In *Water Resources Handbook,* pp. 12.1–114, ed. L.W. Mays. New York: McGraw-Hill, 1996. With Ward.
  Influence of the energy relationships of trophic levels and of elements of bioaccumulation. *Ecotoxicol. Environ. Safety* 30:203–18, 1995. With Genoni.
Howard T. Odum, Graduate Research Professor Emeritus; Ph.D., Yale, 1951. Systems ecology.
  Environment and Society in Florida. Boca Raton: St. Lucie Press, 1998. With E. C. Odum and M. T. Brown.
  Environmental Accounting: EMERGY and Decision Making. New York: John Wiley, 1995.
William S. Properzio, Associate Professor; Ph.D., Florida, 1975. Health physics.
  A review of experience with diagnostic X-ray quality assurance in the U.S.A. *Br. J. Radiol. Suppl.* 18:75–8, 1985. With Burkhard.
Timothy G. Townsend, Assistant Professor; Ph.D., Florida, 1994; EIT. Solid waste.
  Leachate recycle using horizontal injection. *Adv. Environ. Res.* 2:129–38, 1998. With Miller.
  Management of solid waste from abrasive blasting. *Pract. Periodical Hazardous Toxic Radiol. Waste Manage. (ASCE)* 2:72–7, 1998. With Carlson.
  Management of discarded pharmaceuticals. *Pract. Periodical Hazardous Toxic Radiol. Waste Manage. (ASCE)* 2:89–92, 1998. With Musson.
Warren Viessman Jr., Professor and Associate Dean; D.Eng., Johns Hopkins, 1961; PE. Water resources policy.
  Introduction to Hydrology, 4th edition. New York: Harper Collins, 1995. With Lewis.
  Water Supply and Pollution Control, 6th edition. New York: Harper Collins, 1998. With Hammer.
William R. Wise, Associate Professor; Ph.D., Texas, 1989; PE. Hydrology.
  NAPL characterization via partitioning tracer tests: Quantifying effects of partitioning nonlinearities. *J. Cont. Hydrology* 36:167–83, 1999.
  NAPL characterization via partitioning tracer tests: A modified langmuir relation to describe partitioning nonlinearities. *J. Cont. Hydrology* 36:153–65, 1999. With Dai et al.
Chang-Yu Wu, Assistant Professor; Ph.D., Cincinnati, 1996; EIT. Aerosols.
  Control of toxic metal emissions from combustors using sorbents: A review. *J. Air Waste Manage. Assoc.* 48:113–27, 1998.
  Capture of mercury combustion systems by in situ-generated titania particles with UV irradiation. *Environ. Eng. Sci.* 15:137–48, 1998. With Lee et al.
John Zoltek Jr., Professor; Ph.D., Caltech, 1973; PE. Wastewater treatment.
  Bench scale treatability of leachate from an abandoned phenolic waste site. *J. Water Pollut. Control Fed.* 58:1057–65, 1986. With Drinkwater and Delfino.
  Organic priority pollutants in wetland-treated leachates at a landfill in central Florida. *Chemosphere* 31:3455–65, 1995. With Chen.

UNIVERSITY OF HOUSTON

Cullen College of Engineering
Department of Civil and Environmental Engineering

Programs of Study	The Department of Civil and Environmental Engineering offers graduate study in environmental, geotechnical, hydraulic/water resources, offshore, and structural engineering. Programs of study include the Master of Science in Civil Engineering (M.S.C.E.), the Master of Civil Engineering (M.C.E.), the Master of Science in Environmental Engineering (M.S.Env.E.), and the Doctor of Philosophy (Ph.D.). The M.S.C.E. and M.S.Env.E. programs have both thesis and nonthesis options. The M.S.C.E. thesis option requires a minimum of 21 credit hours in addition to a thesis; the M.S.Env.E. thesis option requires 24 credit hours. Both programs require a satisfactory thesis defense before a thesis committee. The nonthesis option of both programs requires a minimum of 36 credit hours, including a 3-credit-hour project in the M.S.C.E. nonthesis option. The M.C.E. program is intended to prepare students for professional practice in civil engineering and requires a minimum of 30 credit hours. The M.S.C.E. and M.C.E. programs may be completed in three semesters of full-time study or in two to three years of part-time study. The M.S.Env.E. program may be completed in three semesters and a summer. The Ph.D. programs require a minimum of 24 credit hours of approved graduate study beyond the master's degree, plus at least 30 credit hours of research and dissertation.
Research Facilities	The department has five modern research laboratories. The structural laboratory has more than 2,500 square feet of strong floor with a 2.5-million-pound MTS testing system, a giant panel tester with forty 100-ton jacks, and a biaxial fatigue tester. The materials testing laboratory is equipped with a 2-foot-deep strong floor and a 400,000-pound Tinnis-Olsen universal testing machine. The geotechnical facility includes a laboratory with triaxial testing capability, sophisticated pile test chambers, fully equipped grouting and soil mechanics laboratories, and a National Geotechnical Experimentation Site. The environmental engineering labs are among the best in the world and contain new GC-LC-MS and ICP-MS instruments. The hydraulic laboratory includes a 120-foot-long wave tank and equipment for fluid mechanics testing of pipe and open channel flow. The University, the College, and the department maintain a wide variety of state-of-the-art computer facilities that are dedicated to teaching and academic research.
Financial Aid	All accepted graduate students who are U.S. citizens and who have a record of outstanding undergraduate academic performance (a high GPA) at an accredited university in the United States and high GRE scores are eligible for research or teaching assistantships. Assistantships range from $10,800 to $15,000 per year, plus tuition. Financial assistance is also available to exceptional international students.
Cost of Study	In 1999–2000, tuition is $192 per 3-credit-hour course for Texas residents and $762 per 3-credit-hour course for nonresidents. Additional fees are also collected at registration. Tuition and fees for 12 credit hours total $1358.50 per semester for Texas residents and $3638.50 per semester for nonresidents. Students receiving financial assistance qualify for in-state tuition.
Living and Housing Costs	On-campus room and board cost about $8890 per academic year. Off-campus apartments begin at approximately $450 per month. The cost of books is approximately $500 per semester.
Student Group	The University of Houston has a diverse student population of approximately 32,300 enrolled in fourteen colleges and schools. In the fall semester of 1998, the Cullen College of Engineering enrolled 2,172 students with a wide variety of academic and cultural backgrounds. The Department of Civil and Environmental Engineering had 300 students, of whom 96 were graduate students, 28 were students with postbaccalaureate status, and 28 were in the doctoral programs.
Location	The University's 545-acre campus is located in the nation's fourth-largest city, just a short drive from downtown. Also within driving distance are the internationally acclaimed Texas Medical Center, NASA's Johnson Space Center, Rice University, many corporate research laboratories, and petrochemical and industrial complexes. These institutions and Houston's many engineering firms offer numerous interdisciplinary research activities and job opportunities. The Alley Theater, the Museum of Fine Arts, the Contemporary Arts Museum, the Houston Ballet, the Houston Symphony, the Houston Grand Opera, and other cultural resources offer the finest in the classical and contemporary arts.
The University and The Department	The University of Houston is the largest campus of the four-campus, state-supported University of Houston System and serves as a strong research and intellectual base for the city of Houston, the state of Texas, and the United States. A large majority of students commute, and graduate courses are offered in the evening to accommodate students' scheduling needs. The Cullen College of Engineering is one of fourteen colleges at the University. The department is one of five departments in the College. Excellent faculty members and research facilities result in exceptional civil and environmental engineering programs that offer very personalized instruction and attention.
Applying	Applications for admission, official transcripts, test scores, and other required materials must be submitted to the appropriate program director as early as possible. The civil engineering program admits students in the fall, spring, and summer semesters; the environmental engineering program admits students only in the fall. All required credentials must be submitted to the civil engineering program director no later than May 1 for fall admission (June 1 for domestic students). Credentials for admission to the environmental engineering program must be submitted by March 15. To be considered for admission, students must have good scores on the Graduate Record Examinations, with a minimum score of 650 on the quantitative portion. In addition, the University requires that all international students obtain a score of 550 or more on the TOEFL and pay a nonrefundable $75 application processing fee. U.S. students must pay a nonrefundable $25 application processing fee. The Ph.D. programs require a GPA of at least 3.5 on work done at the master's level.
Correspondence and Information	Director, Civil Engineering Program (for civil engineering applicants only) Director, Environmental Engineering Program (for environmental engineering applicants only) Department of Civil and Environmental Engineering University of Houston Houston, Texas 77204-4791

University of Houston

THE FACULTY AND THEIR RESEARCH

Theodore G. Cleveland, Associate Professor; Ph.D., UCLA, 1989. Groundwater modeling, water distribution modeling, fate and transport of contaminants.

Dennis A. Clifford, Professor; Ph.D., Michigan, 1976. Water chemistry, physical-chemical treatment processes for water and soil treatment.

Osman I. Ghazzaly, Professor; Ph.D., Texas at Austin, 1965. Geotechnical engineering, foundations, geoenvironmental engineering.

Kye J. Han, Associate Professor; D.Sc., Washington (St. Louis), 1981. Engineering mechanics, tubular structures, finite elements.

Todd A. Helwig, Assistant Professor; Ph.D., Texas at Austin, 1994. Structural stability, bracing of steel structures, steel design, bridge design.

Robert D. Hill, Adjunct Associate Professor; Ph.D., Rice, 1985. Design, operation, and computer control of biological treatment processes.

Thomas T. C. Hsu, Professor; Ph.D., Cornell, 1962. Concrete, structural mechanics, engineering materials.

Michael W. O'Neill, Professor; Ph.D., Texas at Austin, 1970. Deep foundations, pile-group behavior, soil mechanics, soil dynamics.

Hanadi Rifai, Assistant Professor; Ph.D., Rice, 1989. Surface water and groundwater hydrology, risk assessment, geographical information systems modeling.

William G. Rixey, Associate Professor; Ph.D., Berkeley, 1987. Solid- and hazardous-waste treatment processes, contaminant fate and transport.

Deborah J. Roberts, Associate Professor; Ph.D., Alberta, 1990. Microbial degradation of toxic compounds, microbial corrosion of concrete.

Jerry R. Rogers, Associate Professor; Ph.D., Northwestern, 1970. Water resources, stormwater runoff, urban development, engineering history.

Sami W. Tabsh, Assistant Professor; Ph.D., Michigan, 1990. Bridge design, reliability methods, finite-element analysis.

Cumaraswamy Vipulanandan, Professor; Ph.D., Northwestern, 1984. Polymer composites, soil properties, constitutive modeling, fracture mechanics, contaminated soils, coating materials, grouting.

Keh-Han Wang, Associate Professor; Ph.D., Iowa, 1985. Coastal/estuary hydrodynamics, wave-structure interactions, hydraulics.

Anthony N. Williams, Professor; Ph.D., Reading (England), 1983. Hydrodynamics, wave-structure interactions, applied mathematics.

CURRENT RESEARCH INTERESTS

Environmental Engineering: Ion exchange and biological processes for removing perchlorate from groundwater, strategies for improving water quality in nondesignated streams, phosphate stabilization of lead-contaminated battery waste site soils, analysis and removal of arsenic from water at the sub-ppb level, environmentally acceptable endpoints for hydrocarbon-contaminated soils, removal of lead from soil by extraction with concentrated salt solutions, mass transfer characteristics of residually trapped NAPLs, natural attenuation of contaminants in the environment, risk-based corrective action, biosurfactant production and characterization, biological corrosion of concrete, biodegradation of munitions compounds from contaminated soils.

Geotechnical Engineering: Response of piles and pile groups to high-amplitude transient and steady-state loading, response of laterally loaded piles to high-amplitude cyclic loads, behavior of soil composites, interactions of foundations with expansive soils, geostatistical site characterization, vibratory installation of piles, design methods for deep foundations, treatment of contaminated soils, recycled materials, constitutive modeling of geomaterials, backfill materials for flexible pipes, grouting materials and behavior under dynamic/chemical environments.

Offshore and Coastal Engineering: Wave-structure interactions, springing and ringing loads on offshore platforms, nonlinear wave generation and propagation, coastal processes, hydrodynamic modeling in estuaries, compliant and floating breakwaters, global response of TLPs.

Structural Engineering: Biaxial testing of concrete panels, cyclic loading of shear walls, unified theory of reinforced concrete, behavior of high-strength concrete, shear and torsional behavior of reinforced and prestressed concrete, behavior of fiber-reinforced concrete, properties of polymer concrete, system identification and damage assessment from earthquake records, buckling of steel tubular members, ultimate strength of steel tubular members, flexibility of tubular joints, steel–polymer concrete sandwich elements, structural bracing, trapezoidal box girder systems, analysis of irregular bridges, reliability evaluation of bridges, seismic analysis of bridges, coating materials.

Hydraulic and Water-Resource Engineering: Watershed urbanization and stormwater management and modeling, sediment control and stability of earthen drainage channels, open-channel hydraulics and overland flow modeling, physical and numerical modeling of surcharged sewer systems, energy management for water distribution system.

Professor Thomas T. C. Hsu (foreground) directs a setting for a concrete panel test in shear; on the platform, students Jeffrey Pollack and Gabriel Garza set the test specimen.

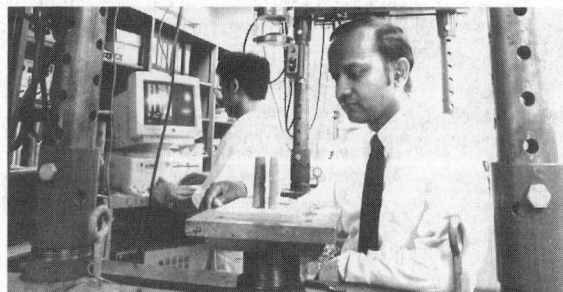

Professor Cumaraswamy Vipulanandan inspects the specimen after a compression test; in the background, graduate assistant Yaogen Ge works on a computer terminal.

Professor Neil Williams discusses the boundary element mesh that is used for the analysis of wave effects on a tension-leg platform with Ph.D. candidate Mahmoud Darwiche.

UNIVERSITY OF ILLINOIS AT CHICAGO

Department of Civil and Materials Engineering

Programs of Study

The Department of Civil and Materials Engineering offers programs of graduate instruction leading to the Master of Science and Doctor of Philosophy degrees in civil engineering and materials engineering. The areas of specialization for civil engineering are environmental and water resources engineering, geotechnical and geoenvironmental engineering, structural mechanics and engineering, and transportation engineering. For the degrees in materials engineering, the principal specializations are in mechanics of materials and in materials science and engineering.

Two program options are available leading to an M.S. degree in civil engineering or materials engineering. The nonthesis option involves 36 semester hours of course work only. The thesis option may include up to 12 semester hours of thesis credit within the total of 36 semester hours.

The Ph.D. program requires 112 semester hours of credit beyond the baccalaureate. A maximum transfer of 32 hours of M.S. credits is allowed, and a minimum of 36 semester hours of course work beyond the M.S. (or the equivalent) is required. In addition, a minimum of 44 semester hours of thesis credit is mandatory. A qualifying examination must be taken no later than one semester after admission to the program. The examination may be retaken once if so recommended by the examining committee. In addition, toward the end of their course work, students are required to pass a preliminary examination administered by a faculty committee.

M.S. and Ph.D. theses must be defended before an examination committee appointed by the Graduate College. The departmental *Graduate Study Brochure* has further details.

Research Facilities

The department has laboratories for geotechnical studies, structural and mechanical testing, experimental stress analysis, fracture mechanics and failure analysis, processing and solidification, corrosion, welding, and scanning, transmission, electron, and optical microscopy, X-ray diffraction, and transportation research. Automatic electronic data measuring systems are provided.

Departmental computer facilities include a microcomputer lab containing both IBM- and Macintosh-based platforms and a Sun Workstation lab dedicated to more advanced computational research.

Financial Aid

Financial aid is offered on a competitive basis and is available in the form of University fellowships, teaching and research assistantships, and/or tuition and fee waivers.

Cost of Study

In 1999–2000, tuition and fees per semester for full-time study (12 or more semester hours) are $2820 for Illinois residents and $6239 for nonresidents. Illinois resident students taking 6 to 11 semester hours are charged $2129 per semester; nonresidents taking the same number of hours pay $4409. Illinois residents taking 1 to 5 semester hours pay $1353 per semester; nonresidents taking 1 to 5 semester hours pay $2493.

Living and Housing Costs

On-campus housing for engineering graduate students is available. In addition, most students are able to find off-campus housing in nearby areas. Costs are comparable to those found in other large metropolitan areas.

Student Group

The University's total enrollment is about 25,500, and the College of Engineering has a total enrollment of 3,200. The department has about 60 graduate students, of whom 37 are full-time.

Location

The campus is located just southwest of downtown Chicago. The lakefront and parks and many recreational facilities are close by. The campus is easily reached by bus, subway, and car.

The University

The University of Illinois at Chicago acquired its name in 1982 when the Chicago Circle Campus (established in 1965) and the Medical Center Campus (established in 1884) of the University of Illinois merged. The campus architecture is modern and dramatic. The department is housed in one of the most recently completed buildings, the Engineering Research Facility.

Applying

Applications for admission should be sent directly to the Office of Admissions. Applications for financial aid should be sent to the department. All applicants must submit scores from the GRE General Test. The Test of English as a Foreign Language (TOEFL) is required of all applicants from abroad.

Correspondence and Information

Director of Graduate Studies
Department of Civil and Materials Engineering (M/C 246)
University of Illinois at Chicago
2095 Engineering Research Facility
842 West Taylor Street
Chicago, Illinois 60607-7023
Telephone: 312-996-3428
Fax: 312-996-2426
E-mail: missa@uic.edu

University of Illinois at Chicago

THE FACULTY AND THEIR RESEARCH

Farhad Ansari, Professor; Ph.D., Illinois at Chicago, 1983. Fiber-optic sensors for condition monitoring of civil structures, fiber-reinforced polymer composites for construction and repair, nondestructive testing, mechanics of concrete.

David E. Boyce, Professor; Ph.D., Pennsylvania, 1965. Urban and regional transportation planning methods, intelligent transportation systems.

Alexander Chudnovsky, Professor; Ph.D., Leningrad Civil Engineering Institute, 1971. Theoretical and experimental fracture mechanics, crack-damage interaction, critical phenomena, probability and statistics in engineering applications, reliability, failure analysis.

Stephen V. Harren, Associate Professor; Ph.D., Brown, 1988. Plasticity, mechanics of materials, computational mechanics.

J. Ernesto Indacochea, Associate Professor; Ph.D., Colorado School of Mines, 1981. Solidification, phase stability, and transformations in metals; welding, joining, and coating of metals and ceramics; molten carbonate fuel cells: electrotransport, degradation, and diffusion in fuel cell components.

Mohsen A. Issa, Associate Professor; Ph.D., Texas at Arlington, 1986. Reinforced and prestressed concrete structures, steel structures, high-performance concrete, advanced composites for infrastructure, theoretical and experimental stress analysis, seismic design, dynamic resistance of structures, fracture mechanics of concrete.

Donald G. Lemke, Associate Professor; Ph.D., Pennsylvania, 1968. Rotor dynamics, structural dynamics and vibrations.

Arif Masud, Assistant Professor; Ph.D., Stanford, 1993. Computational mechanics and finite element analysis, constitutive modeling of nonsimple materials, computational fluid dynamics, fluid-structure interaction.

Michael J. McNallan, Professor; Ph.D., MIT, 1977. High-temperature materials chemistry focusing on processing and corrosion of metals and ceramics.

Krishna R. Reddy, Assistant Professor; Ph.D., IIT, 1990. Environmental remediation engineering, waste containment systems, waste and recycled material applications, groundwater and contaminant hydrology, foundation engineering, geotechnical earthquake engineering.

Thomas C. T. Ting, Professor; Ph.D., Brown, 1962. Wave propagation, viscoelasticity, anisotropic elasticity, composite materials.

Ming L. Wang, Professor; Ph.D., New Mexico, 1983. Experimental mechanics, fiber-reinforced composites, construction materials, recycled waste materials, structural dynamics, earthquake engineering, structural monitoring, structural damage assessment, random vibrations.

Chien H. Wu, Professor; Ph.D., Minnesota, 1965. Linear and nonlinear elasticity, fracture mechanics and interfacial phenomena, nonlinear vibrations, asymptotic analysis and perturbation methods.

MOST RECENT PUBLICATIONS

Li., Q. B., and **F. Ansari.** Mechanics of damage and constitutive relationships for high strength concrete in triaxial compression. *ASCE J. Eng. Mech.,* Vol. 125, 1999.

Ansari, F., and Y. Libo. Mechanics of bond and interface shear transfer in optical fiber sensors. *ASCE J. Eng. Mech.,* Vol. 124, 1998.

Boyce, D. E., and M. Daskin. Urban transportation. In *Design and Operation of Civil and Environmental Engineering Systems,* eds. C. ReVelle and A. McGarity. New York: John Wiley & Sons, 1997.

Boyce, D. E., and B. Ran. Modeling dynamic transportation networks. Spring-Verlag, Berlin, 1996.

Chudnovsky, A., W. Zhou, C. P. Bosnyak, and K. Sehanobish. The time dependency of the necking process in polyethylene. *Proc. SPE/ANTEC 1999,* Vol. 3, No. 57, 1999.

Chudnovsky, A., and S. K. Kanaun. A model of quasibrittle fracture of solid. *Int. J. Damage Mech.,* Vol. 8, 1999.

Harren, S. V., and J. Botsis. On the constitutive response of 63/37 Sn/Pb eutectic solder. *ASME J. Eng. Mater. Technol.,* Vol. 45, 1997.

Harren, S. V. A yield surface and flow rule for orientationally hardening polymers subjected to arbitrary deformations. *J. Mech. Phys. Solids,* Vol. 45, 1997.

Indacochea, J. E., J. L. Smith, K. R. Litko, and E. J. Karrel. Corrosion performance of ferrous and refractory metals in molten salts under reducing conditions. *J. Mater. Res.,* Vol. 14, 1999.

Indacochea, J. E., I. Bloom, M. Krumpelt, and T. Benjamin. A comparison of two aluminizing methods for corrosion protection in the wet seal of molten carbonate fuel cells. *J. Mater. Res.,* Vol. 13, 1998.

Issa, M. A. Investigation of cracking of concrete bridge decks at early ages. *ASCE J. Bridge Eng.,* Vol. 4, 1999.

Issa, M. A., and A. B. Shafiq. Fatigue characteristics of aligned fiber reinforced mortar. *ASCE J. Eng. Mech.,* Vol. 125, 1999.

Masud, A., Z. Zhang, and J. Botsis. Strengths of composites with long aligned fibers: Fiber-fiber and fiber-crack interaction. *Composites Part B,* Vol. 29B, 1998.

Masud, A., and T. J. R. Hughes. Galerkin/least-squares finite element formulation of the Navier-Stokes equations for moving domain problems. *Comput. Methods Appl. Mech. Eng.,* Vol. 146, 1997.

Gogotsi, Y. G., I. Jeon, and **M. J. McNallan.** Carbon coatings on silicon carbide by reaction with chlorine-containing gases. *J. Mater. Chem.,* Vol. 17, 1997.

Park, Y., **M. J. McNallan,** and D. P. Butt. Endothermic reactions between mullite and silicon carbide in an argon plasma environment. *J. Am. Ceramic Soc.,* Vol. 81, 1998.

Reddy, K. R., and U. S. Parupudi. Removal of chromium, nickel and cadmium from clays by in-situ electrokinetic remediation. *J. Soil Contam.,* Vol. 6, 1997.

Reddy, K. R., and J. A. Adams. System effects on benzene removal from saturated soils and groundwater using air sparging. *J. Environ. Eng.,* Vol. 124, 1998.

Ting, T. C. T. A modified Lekhnitskii formalism a la Stroh for anisotropic elasticity and classifications of the 6X6 matrix N. *Proc. Roy. Soc. London,* Vol. A455, 1999.

Ting, T. C. T. The remarkable nature of cylindrically orthotropic elastic materials under plane strain deformations. *Q. J. Mech. Appl. Math.,* Vol. 52, No. 3, 1999.

Wang, M. L., G. Heo, and D. Sapathi. A health monitoring system for large structural systems. *J. Smart Mater. Struct.,* Vol. 7, 1998.

Rutland, C. A., and **M. L. Wang.** The effects of confinement of the failure orientation in cementitious materials I. *J. Cement Concrete Composites,* Vol. 19, 1998.

Wu, C. H., J. H. Hsu, and C. H. Chen. The effect of surface stress on the stability of surfaces of stressed solids. *Acta Materialia,* Vol. 46, 1998.

Wu, C. H. The chemical potential for stress-driven surface diffusion. *J. Mech. Phys. Solids,* Vol. 44, 1996.

ILLINOIS

UNIVERSITY OF ILLINOIS AT URBANA–CHAMPAIGN

Department of Civil and Environmental Engineering

Programs of Study

The Department of Civil and Environmental Engineering offers programs of graduate instruction leading to the degrees of Master of Science and Doctor of Philosophy in civil engineering, environmental engineering, and environmental science. Areas of study and research include air quality; aquatic biology and ecology; computer-aided engineering systems—artificial intelligence, expert systems, and neural networks; construction engineering and management; earthquake engineering; engineering risk, decision, and reliability analysis; environmental chemistry; environmental engineering and environmental systems analysis; geotechnical engineering—rock mechanics, soil mechanics, and foundation engineering; hazardous-waste management; hydrosystems engineering—hydrology, water resources, and hydraulic engineering; materials engineering; nondestructive diagnostics; ocean engineering; railway engineering; solid-waste management; stochastic structural dynamics and random vibrations; structural and computational mechanics; structures—analysis, design, and behavior; traffic engineering; transportation—planning, systems design, and operations; and water quality and treatment.

The master's program can be completed in one academic year of full-time study. Nine units without a thesis, or 8 units of course work including a thesis, are required. A unit of study is equivalent to 4 semester hours.

The doctoral program, consisting primarily of research, includes three stages: the completion of the M.S. degree or its equivalent; the completion of a minimum of 8 additional units of work and any special course requirements; and research with a minimum of 8 units of thesis credit, preparation of a dissertation, and the final examination. The three stages may be completed in a minimum of three years of full-time study.

Research Facilities

Newmark Civil Engineering Laboratory's research facilities include static and dynamic loading frames, a large shake table, and fully equipped laboratories for highway materials and geotechnical and environmental engineering. The department maintains a computer laboratory with HP workstations, microcomputers, and access via the 80-megabit Engineering College network to other campus computer facilities. In addition, the researchers in the department make use of the CRAY X-MP in the nearby National Supercomputing Center.

Financial Aid

Research assistantships with a nine-month, half-time appointment are the most common form of support for graduate students. Federal, state, and industrial fellowships are available. Fellowships also may be supplemented by quarter-time assistantships. Assistantship and fellowship awards include payment of, or exemption from, tuition and some fees.

Cost of Study

For a full-time program, tuition and fees were $2819 per semester for the 1998–99 academic year (nine months) for residents of Illinois and $6257 per semester for nonresidents.

Living and Housing Costs

Living costs for a single student, including books and incidentals but not the cost of study, are approximately $1200 per month. Housing for single graduate students is available in two dormitories; meal contracts are available for a dining facility located within easy walking distance. There are also numerous apartments. Bus service is excellent. Apartment rents start at approximately $425 per month.

Student Group

There are both graduate and undergraduate students enrolled in the Department of Civil and Environmental Engineering. Graduate students come from approximately thirty states and thirty countries. Almost half of the graduate students receive some type of financial assistance administered through the University.

Location

The campus is located 130 miles south of Chicago in the twin cities of Urbana and Champaign (population 100,000). The area is primarily a university community, with excellent public schools, park systems, and modern shopping facilities.

The University and The Department

The Department of Civil and Environmental Engineering, with 63 faculty members, has been rated consistently among the top two or three in the nation in civil engineering and environmental engineering. The College of Engineering, founded in 1868, is an internationally known center of engineering research and education.

Leisure activities are enhanced by the Krannert Center for the Performing Arts, the 16,000-seat Assembly Hall, Big Ten sports events, and the 1,500-acre Allerton Park.

Applying

The University evaluates admission and all requests for financial assistance through the use of one combined application form, which may be obtained by writing to the department. Admission requires a B.S. degree or the equivalent and a scholastic average of at least 3.0 (A = 4.0) for the last 60 semester hours or 90 quarter hours. GRE General Test scores are required. International students must have a TOEFL score above 550. Applicants must pay a nonrefundable application fee ($40 for U.S. citizens and $50 for international students).

Correspondence and Information

Graduate Admissions Officer
1110 Newmark Civil Engineering Laboratory
University of Illinois at Urbana–Champaign
205 North Mathews Avenue
Urbana, Illinois 61801

Telephone: 217-333-8038
Fax: 217-333-9464
E-mail: civil@uiuc.edu
World Wide Web: http://www.cee.ce.uiuc.edu

University of Illinois at Urbana–Champaign

THE FACULTY AND THEIR RESEARCH

Daniel P. Abrams, Professor; Ph.D., Illinois. Reinforced concrete, masonry, earthquake engineering.

Mark A. Aschheim, Assistant Professor; Ph.D., Berkeley. Reinforced concrete, seismic design, structural dynamics and earthquake engineering.

Ernest J. Barenberg, Professor; Ph.D., Illinois. Transportation facilities, pavements, railroads, materials.

Rahim F. Benekohal, Associate Professor; Ph.D., Ohio State. Modeling and simulation of transportation systems, traffic control and operations, accident analysis, traffic engineering.

L. T. Boyer, Professor; Ph.D., Minnesota. Construction management, construction cost analysis, estimates.

William G. Buttlar, Assistant Professor; Ph.D., Penn State. Transportation materials, pavements, asphalt technology, viscoelasticity/fracture.

Samuel H. Carpenter, Professor; Ph.D., Texas A&M. Bituminous materials, pavement evaluation, climatic factors.

Mark M. Clark, Associate Professor; Ph.D., Johns Hopkins. Flocculation and mixing processes, membrane science, fluid mechanics.

Fred Coleman III, Assistant Professor; Ph.D., Michigan State. Transportation planning, traffic engineering, accident analysis, public transportation systems analysis.

Edward J. Cording, Professor; Ph.D., Illinois. Foundation engineering, rock mechanics, underground construction.

David E. Daniel, Professor and Head of the Department; Ph.D., Texas at Austin. Waste containment and disposal, landfills, geosynthetic clay liners, remediation, soil vapor extraction.

Barry J. Dempsey, Professor; Ph.D., Illinois. Climatic effects on pavements and materials, drainage, geosynthetic materials.

Robert H. Dodds, Professor; Ph.D., Illinois. Nonlinear fracture mechanics, computational mechanics, software engineering.

J. Wayland Eheart, Professor; Ph.D., Wisconsin–Madison. Environmental systems analysis and management, water quality modeling.

Douglas A. Foutch, Professor; Ph.D., Caltech. Dynamic characteristics of full-scale structures, analysis and design of earthquake-resistant bridges and steel buildings, wind effects.

William L. Gamble, Professor; Ph.D., Illinois. Structural concrete, prestressed concrete bridges, creep and shrinkage of concrete, slabs, fire-resistant structures.

Marcelo H. Garcia, Associate Professor; Ph.D., Minnesota. Environmental hydraulics, stratified flows, river mechanics, sediment transport.

Jamshid Ghaboussi, Professor; Ph.D., Berkeley. Structural mechanics and geomechanics, computational mechanics, structural control, applications of artificial intelligence to civil engineering problems.

German R. Gurfinkel, Professor; Ph.D., Illinois. Structural design, failure investigations, retrofitting of wood, reinforced concrete, prestressed concrete and steel structures including silos, tanks, and tall buildings.

William J. Hall, Professor; Ph.D., Illinois. Structural engineering, dynamics, earthquake engineering.

Youssef Hashash, Assistant Professor; Ph.D., MIT. Numerical modeling and static and dynamic soil-structure interaction analysis for open-cut excavations and tunnels.

Neil M. Hawkins, Professor; Ph.D., Illinois. Reinforced, prestressed, and composite steel and concrete structures; fracture mechanics.

Edwin E. Herricks, Professor; Ph.D., Virginia Tech. Aquatic ecology, ecosystem management, water quality management standards and legislation.

Keith D. Hjelmstad, Professor; Ph.D., Berkeley. Structural mechanics, structural engineering, earthquake engineering.

Daniel A. Kuchma, Assistant Professor; Ph.D., Toronto. Design and behavior of reinforced concrete structures: experimental testing and development of numerical models.

Praveen Kumar, Assistant Professor; Ph.D., Minnesota. Large-scale hydrogeologic processes, hydroclimatology and hydrometeorology, multiscale structure of precipitation.

David A. Lange, Assistant Professor; Ph.D., Northwestern. Construction materials, microstructure and properties of cement and concrete.

Susan M. Larson, Associate Professor; Ph.D., Caltech. Air quality monitoring and modeling, aerosol physics.

Frederick V. Lawrence, Professor and Associate Head of the Department; Sc.D., MIT. Materials, metal engineering properties, welding/joining, fatigue/fracture.

Jon C. Liebman, Professor; Ph.D., Cornell. Water resource and water quality systems, solid-waste management and disposal.

Liang-Y. Liu, Visiting Adjunct Professor; Ph.D., Michigan. Construction schedule and cost analysis, project controls, management information systems, construction modeling and simulation.

James H. Long, Associate Professor; Ph.D., Texas at Austin. Soil-structure interaction, foundation engineering.

Leonard A. Lopez, Professor; Ph.D., Illinois. Software architecture for parallel processing.

Benito J. Mariñas, Associate Professor; Ph.D., Berkeley. Physicochemical water and wastewater treatment processes, water and wastewater chemistry.

W. H. C. Maxwell, Professor; Ph.D., Minnesota. Hydromechanics, hydraulic engineering.

John W. Melin, Professor; Ph.D., Illinois. Construction management, construction planning and control.

Gholamreza Mesri, Professor; Ph.D., Illinois. Soil mechanics, foundation engineering, creep behavior of clays and shales.

Roger A. Minear, Professor; Ph.D., Washington (Seattle). Chemistry of drinking water, wastewater and natural water, environmental analytical chemistry.

Barbara S. Minsker, Assistant Professor; Ph.D., Cornell. Environmental systems analysis and management, groundwater remediation design using optimization and simulation models, bioremediation modeling.

Joseph P. Murtha, Professor; Ph.D., Illinois. Construction technology, nondestructive diagnostics, structural dynamics.

I. Dennis Parsons, Associate Professor; Ph.D., Caltech. Structural analysis, computational mechanics, composite structures.

Stanley L. Paul, Associate Professor; Ph.D., Illinois. Experimental and analytical studies of reinforced concrete.

Glaucio H. Paulino, Assistant Professor; Ph.D., Cornell. Structural analysis and computational mechanics.

David A. Pecknold, Professor; Ph.D., Illinois. Structural engineering and dynamics, shell structures, composite materials.

John T. Pfeffer, Professor; Ph.D., Florida. Biological wastewater treatment processes, processing and disposal of solid wastes.

Lutgarde Raskin, Assistant Professor; Ph.D., Illinois. Biological treatment, molecular microbial ecology, environmental microbiology.

Chris R. Rehmann, Assistant Professor; Ph.D., Stanford. Environmental fluid mechanics, turbulence in stratified fluids, modeling mixing in natural flows.

Mark J. Rood, Professor; Ph.D., Washington (Seattle). Aerosol chemistry, air quality engineering, aerosol sampling, instrumentation.

Robert A. Sanford, Assistant Professor; Ph.D., Michigan State. Reductive dechlorination in aerobic systems, microbial characterization.

Stephen P. Schneider, Associate Professor; Ph.D., Washington (Seattle). Steel and structural steel/concrete composite construction, structural dynamics and earthquake engineering.

Vernon L. Snoeyink, Professor; Ph.D., Michigan. Environmental chemistry, drinking water treatment.

Lucio Soibelman, Assistant Professor; Ph.D., MIT. Construction management, design rationale, civil engineering information systems, distributed artificial intelligence, data mining and KDD.

Timothy D. Stark, Associate Professor; Ph.D., Virginia Tech. Stability of natural and man-made slopes, soil liquefaction, foundation engineering.

Leslie J. Struble, Associate Professor; Ph.D., Purdue. Materials; chemical, microstructural, and physical properties of cement and concrete.

Marshall R. Thompson, Professor; Ph.D., Illinois. Flexible pavements, transportation, railroad track structures, soil stabilization.

Erol Tutumluer, Assistant Professor; Ph.D., Georgia Tech. Flexible pavement, pavement materials, aggregates and soils, analysis of layered system, artificial neural networks.

Albert J. Valocchi, Professor; Ph.D., Stanford. Transport processes in porous media, groundwater contamination, numerical methods.

William H. Walker, Professor and Associate Head of the Department; Ph.D., Illinois. Structural mechanics, structural dynamics, highway bridge dynamics, fatigue/fracture.

Yi-Kwei Wen, Professor; D.E.S., Columbia. Random vibration, structural reliability, earthquake and offshore engineering.

Charles J. Werth, Assistant Professor; Ph.D., Stanford. Subsurface transport and fate of organic chemicals, sorption, mass transfer, soil and sediment characterization.

Kam W. Wong, Professor; Ph.D., Cornell. Surveying, photogrammetry, computer vision metrology.

Ben C. Yen, Professor; Ph.D., Iowa. Surface water hydrology, river mechanics, water resources systems, urban drainage.

J. Francis Young, Professor; Ph.D., London. Advanced cement-based materials, chemical and physical properties of cement and concrete.

UNIVERSITY OF KANSAS

Department of Civil and Environmental Engineering

Programs of Study

The Department of Civil and Environmental Engineering at the University of Kansas offers Doctor of Philosophy and Doctor of Engineering degrees in civil engineering and Doctor of Philosophy degrees in environmental engineering and environmental science. Master of Science degrees are offered in civil engineering, environmental engineering, environmental science, water resources engineering, and water resources science. A Master of Civil Engineering degree is also available. Areas of specialization for the civil engineering Master of Science degree include structures, transportation, pavement and structural materials, environmental engineering, engineering mechanics, water resources, and geotechnical engineering. A baccalaureate engineering degree is required for admission to all the programs above except the environmental science and water resources science degrees, which are intended for qualified students with nonengineering undergraduate degrees.

General requirements for the doctoral degree programs include a written qualifying examination, course work, a research skills requirement, a comprehensive oral examination, a dissertation, and a final oral examination. Specific requirements are set on an individual basis by the student's major professor and dissertation committee. The Master of Science degrees require 30 semester hours of credit, which must include either a 6-semester-hour thesis or a 3- or 4-semester-hour special problem investigation. Both options require a final examination. The Master of Civil Engineering degree requires 34 semester hours of course work, including a minimum of 7 hours of engineering management courses. There are no thesis or special project report requirements. A final examination is required in the speciality area unless the Professional Engineering (PE) exam has been passed.

Research Facilities

The Department of Civil and Environmental Engineering has many experimental laboratories and has access to excellent computing facilities. The following experimental laboratories are available for graduate student research at the master's and doctoral levels: a climate-controlled strength-of-materials lab, a structural testing lab, a bituminous materials lab, a structural materials lab, a SEM microanalysis lab, a hydraulics lab, a geotechnical lab, a microbiology lab, and environmental engineering and science labs.

Financial Aid

Financial aid is available for full-time graduate students in the form of fellowships, research assistantships, teaching assistantships, and scholarships. The stipends range from approximately $9000 to $18,000 per academic year. A tuition waiver is included for half-time teaching assistants and is often covered for research appointments. Additional compensation for the summer session is typically available for qualified students.

Cost of Study

Tuition and fees for 1998–99 were $145 per semester hour for in-state engineering students. Nonresident students paid $373 per semester hour. Nonresident students holding half-time appointments are charged tuition at the in-state rate. An additional fee of $39 per semester hour is assessed for graduate courses taken at the Regents Center in Kansas City, Kansas.

Living and Housing Costs

The academic-year costs for single students are $1736–$6640 (graduate student apartments without a meal plan) to $3736–$4548 (graduate student dorm rooms with a meal plan). Married student housing ranges from $1998 to $3906 (plus utilities) per academic year.

Student Group

There are about 19,000 undergraduate students and 8,000 graduate students at the University of Kansas.

Location

The 1,000-acre main campus is located in Lawrence, Kansas, on the Mount Oread ridge. Lawrence, a city with a population of 68,000, is located about 40 miles west of Kansas City, Kansas. The community and the University provide many cultural and recreational opportunities for students. Graduate courses in civil engineering are also taught at the Edwards Campus in Overland Park, Kansas.

The University

The University of Kansas was founded in 1866 and is the only Kansas Regents University to be among the fifty-eight public and private universities in the country that hold membership in the prestigious Association of American Universities. The University is committed to excellence and strives to provide the best educational experience possible for both undergraduate and graduate students. High-level research is strongly emphasized.

Applying

Admission applications with all supporting documents should be submitted well before the semester of entry if financial support is sought. For fall semester admission, applications should be completed by March 1 of the prior spring semester. While applications received later are processed and evaluated by the same admission criteria, the likelihood of receiving financial aid is diminished.

Correspondence and Information

Graduate Studies
Department of Civil and Environmental Engineering
University of Kansas
Lawrence, Kansas 66045
Telephone: 785-864-3826
Fax: 785-864-3199
E-mail: cwiley@ukans.edu

University of Kansas

THE FACULTY AND THEIR RESEARCH

Professors

Steven L. McCabe, Department Chair; Ph.D., Illinois at Urbana-Champaign, 1987. Structural dynamics, structural analysis and design, earthquake engineering, structural mechanics.

Ernest Angino, Ph.D., Kansas, 1961. Rock-water interaction, hazardous waste materials, energy materials.

Carl E. Burkhead, Ph.D., Kansas, 1966. Biological-waste treatment, biomonitoring and toxicity studies of industrial wastes, hazardous wastes and sludge management.

David Darwin, Deane E. Ackers Distinguished Professor; Ph.D., Illinois at Urbana-Champaign, 1974. Structural engineering, engineering materials, reinforced concrete design and analysis, plain concrete, composite construction, finite-element analysis, earthquake engineering, experimental studies.

Carl E. Kurt, Ph.D., Oklahoma State, 1969. Inspection, analysis, and rating of highway bridges; structural steel design; fatigue of steel connections; lateral-torsional buckling of crane girders; microcomputer applications; GIS development, modeling, and simulation.

Dennis D. Lane, N. T. Veatch Distinguished Professor; Ph.D., Illinois at Urbana-Champaign, 1976. Air pollution aerosol science, acid rain, pesticide runoff from agricultural lands.

Joe Lee, Ph.D., Ohio State, 1971. Traffic flow theory, traffic control, transportation planning, system analysis, transportation design.

Glen A. Marotz, Ph.D., Illinois at Urbana-Champaign, 1971. Air quality, air quality modeling, toxic air pollutants, hydrometeorology.

Bruce M. McEnroe, Ph.D., Kansas, 1983. Hydraulics, hydrology, water resources engineering, stormwater management, groundwater.

Thomas E. Mulinazzi, Ph.D., Purdue, 1973. Traffic engineering, highway engineering, surveying, airport and railroad planning and design.

A. David Parr, Ph.D., Iowa, 1976. Hydraulics, geohydrology, surface water hydrology, dispersion processes, aquifer thermal energy storage.

Stephen J. Randtke, Ph.D., Stanford, 1977. Environmental engineering, water and wastewater treatment, removal of soluble organic contaminants, organic chlorine formation.

Stanley T. Rolfe, A. P. Learned Distinguished Professor; Ph.D., Illinois at Urbana-Champaign, 1962. Fracture mechanics, fatigue fracture control plans for structures, structural mechanics.

Francis M. Thomas, Ph.D., Illinois at Urbana-Champaign, 1969. Structural mechanics, stress analysis, vibrational analysis and structural dynamics.

Associate Professors

Stephen A. Cross, Ph.D., Auburn, 1992. Construction materials, pavement design and construction, pavement maintenance and rehabilitation.

David W. Graham, Ph.D., Arizona, 1992. Biological wastewater treatment bioremediation, applied microbiology, biotechnology.

W. M. Kim Roddis, Ph.D., MIT, 1989. Expert systems, artificial intelligence, structural design, infrastructure maintenance, bridge fatigue, engineering ethics.

Assistant Professors

JoAnn Browning, Ph.D., Purdue, 1998. Structural dynamics, earthquake engineering.

Eric Meyer, Ph.D., Vanderbilt, 1995. Risk analysis; hazardous waste transportation; GIS development, modeling, and simulation.

Robert Parsons, Ph.D., Georgia Tech, 1998. Geotechnical, GIS, landfill design.

William Ramirez, Ph.D., Texas at Austin, 1999. Structural analysis and design, composites.

UNIVERSITY OF MARYLAND, COLLEGE PARK

College of Engineering
Department of Civil and Environmental Engineering

Programs of Study

The Department of Civil and Environmental Engineering offers programs of study leading to the degrees of Master of Science and Doctor of Philosophy. Areas of specialization include engineering project management, environmental engineering, geotechnical and materials engineering, structural engineering, transportation engineering, remote sensing, and water resource systems. Each student plans an individualized course of study with the assistance of a faculty committee.

The Master of Science degree is awarded after completion of 30 semester credits of approved work if no thesis is written or after 24 semester credits, including thesis work. Research assistants are required to write a thesis. Nonthesis students are required to prepare a scholarly paper. A final comprehensive examination is required in either case. There is no foreign language requirement. The average duration of stay for research assistants is 1½ years.

There is no minimum number of semester credits required for the Doctor of Philosophy degree. Upon substantial completion of a program of study approved by a 5-member faculty committee, a comprehensive examination is required for admission to candidacy. At this time, candidates are also expected to submit a research plan for faculty review. All Ph.D. students must pass an oral exam, administered by a faculty committee, in defense of their dissertation.

Graduate research topics vary and may be either experimental or analytical. The principal requirement for doctoral research is original scholarly work. Research guidance is available from the student's faculty adviser.

Research Facilities

Department research facilities are housed in the Engineering Laboratory Building and the Engineering Classroom Building. The classroom building contains departmental and faculty offices, the Environmental Engineering Laboratories, and the Remote Sensing Systems Laboratory. Research facilities in geotechnical engineering, structural engineering, transportation engineering, and water resources are provided in the laboratory building. Departmental computational facilities include an undergraduate and an advanced computer computational facility and the Remote Sensing Systems Laboratory.

The Engineering Classroom Building is located adjacent to the Engineering Library. The University libraries include, in all, more than 1.5 million volumes, approximately 1.5 million microfilm units, and approximately 19,600 current periodicals and newspapers.

Financial Aid

Graduate student assistantships for instruction and for research are available to well-qualified students. Assistantship stipends vary, depending on the status of the student (i.e., first-year graduate student, Ph.D. candidate, etc.). In 1998–99, the minimum stipend was $16,000 per academic year, in addition to tuition remission. Graduate assistants are limited to 10 credits per semester. Research assistants are normally expected to continue research work on a full-time basis during the summer. The minimum summer stipend was $5600. A limited number of fellowships are also available.

Cost of Study

In 1998–99, tuition was $272 per credit hour for Maryland residents and $400 per credit hour for nonresidents. Other costs included mandatory fees of $150 per semester, depending on the number of credits taken.

Living and Housing Costs

There are a very limited number of University-owned apartments within walking distance of the campus that are available for graduate students. A few efficiency apartments for single students are available, with rates from about $450 to $550 per month. One- and two-bedroom apartment rents for married graduate students range from about $500 to $600 per month. Off-campus housing is available, with costs of approximately $300 per month for rooms in private homes and $600–$850 per month for a two-bedroom apartment.

Student Group

Approximately 32,000 students are enrolled at the College Park campus; nearly 5,000 are enrolled in the College of Engineering. Graduate enrollment in the Department of Civil and Environmental Engineering is 327, with degree candidates divided as follows: M.S., 247, and Ph.D., 80. The graduate population consists of approximately 20 percent women, 15 percent members of minority groups, and 40 percent international students.

Location

The University campus is located in a suburban area about 10 miles north of Washington, D.C. There is a vast range of cultural, social, recreational, and academic activities available in the Washington-Baltimore area. The Appalachian Mountains to the west and the Chesapeake Bay and Atlantic Ocean to the east provide a complete spectrum of outdoor activities. There are many major federal scientific laboratories and the headquarters of many federal agencies in the area.

The University and The College

The present form of the University of Maryland dates from a 1920 act of the Maryland state legislature, which created the University of Maryland at College Park and the University of Maryland at Baltimore. Since then, the University has added three other campuses, including the worldwide University College, headquartered at College Park. The College of Engineering consists of the Departments of Aerospace Engineering, Chemical Engineering, Civil and Environmental Engineering, Electrical Engineering, Fire Protection Engineering, and Materials and Nuclear Engineering and laboratories.

Applying

Applications for admission to the Graduate School must be filed with the Dean for Graduate Studies by May 1 for fall semester and by November 1 for spring semester admission. International students must apply at least seven months in advance of the intended entrance time. A nonrefundable $50 application fee is required for domestic students, and $70 is required for international applicants.

Correspondence and Information

Dr. Gregory Baecher, Professor and Chairman
Department of Civil and Environmental Engineering
University of Maryland, College Park
College Park, Maryland 20742
Telephone: 301-405-1974
World Wide Web: http://www.ence.umd.edu

Al Santos, Academic Coordinator
Department of Civil and Environmental Engineering
University of Maryland, College Park
College Park, Maryland 20742
E-mail: asantos@eng.umd.edu

University of Maryland, College Park

THE FACULTY AND THEIR RESEARCH

Mohamed S. Aggour, Professor; Ph.D., Washington (Seattle), 1972; PE. Earthquake engineering, foundation engineering.

Pedro Albrecht, Professor; Ph.D., Lehigh, 1972; PE. Fracture mechanics, steel design.

Amde M. Amde, Professor; Ph.D., SUNY at Buffalo, 1976; PE. Nonlinear systems, structural analysis.

Mark Austin, Associate Professor; Ph.D., Berkeley, 1985. Earthquake-resistant design, computer-aided design.

Bilal Ayyub, Professor; Ph.D., Georgia Tech, 1983; PE. Risk analysis, structural analysis and design.

Gregory Baecher, Professor and Chairman; Ph.D., MIT, 1972; PE. Geotechnical engineering, rock mechanics, geotechnical risk and reliability.

Francis B. Birkner, Professor; Ph.D., Florida, 1965. Aquatic chemistry, environmental engineering.

Kaye Brubaker, Assistant Professor; Ph.D., MIT, 1995. Water resources, hydroclimatology, remote sensing.

Everett C. Carter, Professor; Ph.D., Northwestern, 1969. Transportation planning, traffic engineering.

Gang-Len Chang, Professor; Ph.D., Texas at Austin, 1985. Transportation systems analysis, traffic systems control.

Peter C. Chang, Associate Professor; Ph.D., Illinois, 1982. Dynamics of structures, structural analysis.

James Colville, Professor; Ph.D., Texas at Austin, 1970; PE. Masonry design, structural analysis.

Allen Davis, Associate Professor; Ph.D., Delaware, 1989; PE. Environmental engineering.

Bruce Donaldson, Professor; Ph.D., Illinois, 1968. Numerical analysis, structural dynamics.

Deborah J. Goodings, Associate Professor; Ph.D., Cambridge, 1979. Centrifuge modeling, frozen soils, cratering, reinforced soil.

Ali Haghani, Associate Professor; Ph.D., Northwestern, 1981. Traffic systems analysis, transportation logistics.

Oliver J. Hao, Professor; Ph.D., Berkeley, 1982; PE. Biological treatment of waste, waste reuse.

David Lovell, Assistant Professor; Ph.D., Berkeley, 1997. Transportation engineering.

Richard H. McCuen, Professor; Ph.D., Georgia Tech, 1970. Urban hydrology, stormwater management.

Glenn Moglen, Assistant Professor; Ph.D., MIT, 1995. Hydrology, river mechanics, geomorphology.

Robert M. Ragan, Professor; Ph.D., Cornell, 1965; PE. Water resources, remote sensing.

David R. Schelling, Professor; Ph.D., Maryland, 1968; PE. Computer-aided design, bridge design.

Paul M. Schonfeld, Professor; Ph.D., Berkeley, 1978; PE. Transportation planning, public transportation system.

Charles W. Schwartz, Associate Professor; Ph.D., MIT, 1979. Rock mechanics, numerical methods for geotechnical engineering.

Eric Seagren, Assistant Professor; Ph.D., Illinois, 1994. Bioremediation and biodegration of contaminants.

M. William Sermons, Assistant Professor; Ph.D., Northwestern, 1998. Metropolitan transportation planning, travel demand analysis and transportation planning.

Jayanta K. Sircar, Affiliate Associate Professor; Ph.D., Maryland, 1986. Information systems.

Yaron M. Sternberg, Professor; Ph.D., California, Davis, 1965. Groundwater hydrology, land-treatment processes.

Alba Torrents, Associate Professor; Ph.D., Johns Hopkins, 1992. Environmental engineering.

Chung-Li Tseng, Assistant Professor; Ph.D., Berkeley, 1996. Financial engineering, project management, optimization.

Donald W. Vannoy, Professor; Ph.D., Virginia, 1975; PE. Building design, fatigue.

Matthew W. Witczak, Professor; Ph.D., Purdue, 1969. Pavement design, geotechnical engineering.

Engineering classroom building at the University of Maryland, College Park.

UNIVERSITY OF MICHIGAN

Department of Civil and Environmental Engineering

Programs of Study

The Department of Civil and Environmental Engineering offers three Master of Science in Engineering (M.S.E.) degree programs: the M.S.E. in civil engineering, M.S.E. in construction engineering and management, and M.S.E. in environmental engineering. The M.S.E. degree programs require 30 credit hours of graduate work (typically ten courses) and do not require a thesis or other major research project. At least two courses, one of which must be mathematically oriented, must be taken in departments other than the Department of Civil and Environmental Engineering. The M.S.E. in civil engineering program includes areas of specialization in construction engineering and management and in geotechnical, hydraulic and hydrologic, materials and highway, and structural engineering. The department also offers a Master of Engineering (M.Eng.) degree (established in 1994) in construction engineering and management. The M.Eng. degree is for students concentrating on state-of-the-art construction professional practice. It also requires 30 credit hours of course work.

The department also offers the Doctor of Philosophy (Ph.D.) in two disciplines: the Ph.D. in civil engineering and Ph.D. in environmental engineering. The Ph.D. programs usually include 50 to 60 hours of graduate course work beyond the bachelor's degree level. The focus of doctoral studies is the student's dissertation research, which must make a significant contribution to professional knowledge in the field. A written and oral qualifying examination is required. The Ph.D. in civil engineering program includes areas of specialization in construction engineering and management and in geotechnical, hydraulic and hydrologic, materials and highway, and structural engineering. Areas of specialization for the Ph.D. in environmental engineering include environmental chemistry and microbiology, fate and transport of surface and groundwater contaminants, solid- and hazardous-waste treatment and management, water quality engineering, and environmental policy and economics.

Research Facilities

The department maintains outstanding experimental and digital computational facilities. Many of the laboratories are recently constructed, and each contains state-of-the-art equipment. These include the Advanced Civil Engineering Materials, Construction Engineering, Environmental and Water Resources Engineering, W. S. Housel Materials, Hydraulics, F. E. Richart Soil Dynamics, Soil Mechanics, Solid Wastes, and Structural Engineering and Dynamics laboratories. Environmental contamination remediation technology research facilities are also available to faculty members and students. Externally funded research in the department amounts to $6 million annually.

The University has an outstanding library system with one of the nation's largest book collections. An excellent research facility, the Engineering and Transportation Library has a collection of 350,000 volumes and 3,000 technical periodicals and has computerized bibliographical services. The Computer Aided Engineering Network (CAEN) is composed of more than 1,300 engineering-class workstations and more than 1,600 microcomputers and has become one of the largest integrated networks in the world.

Financial Aid

Financial aid is available in the form of fellowships or assistantships in research or teaching for students with outstanding records. In the case of international students, financial aid is generally not provided until a satisfactory performance record has been established in residence at Michigan. In addition to research and teaching assistantships, a number of fellowships are available in the department for exceptionally well-qualified students. These awards include Bailey, Borchardt, Bottum, Alumni Friends, Rackham, Regents, Riggs, and Streeter fellowships and scholarships. An NSF Graduate Research Traineeship on Infrastructure Facilities has been funded from 1993 to 1999 to support Ph.D.-bound women and underrepresented minorities with specific interests in repair, rehabilitation, and maintenance of public and private infrastructure facilities. Research and teaching assistantships provide a complete tuition waiver and health insurance benefits.

Cost of Study

In 1998–99, tuition was $5749 per term for Michigan residents and $10,848 for nonresidents. Tuition for the M.Eng. degree for Michigan residents was $6042. Other mandatory student fees total about $90. Books and supplies typically cost about $600 per academic year.

Living and Housing Costs

Living expenses for an academic year (eight months) are estimated to be $6800. This figure is based on a periodic student budget survey that includes costs for room and board, transportation, and personal needs for a single student with no dependents. There are two campus coeducational dormitories for single graduate students and two apartment facilities for family housing. Students should apply for this housing several months in advance of arrival.

Student Group

There are approximately 165 graduate students from all parts of the United States and many other countries in the department. The University of Michigan chapters of the American Society of Civil Engineers, Chi Epsilon, the American Society for Engineering Education, and the Earthquake Engineering Research Institute provide opportunities for students to participate in student-faculty social events and to be in contact with professional engineers.

Location

Predominantly a university town, Ann Arbor is also the home of several major government and industrial laboratories. The active cultural life includes the Summer Street Art Fair and several ethnic festivals. There are excellent shops and restaurants, a large farmer's market, and many public amenities, such as parks and recreational facilities.

The University

One of the oldest public universities in the country, the University of Michigan was founded in 1817. Total enrollment at the Ann Arbor campus is currently about 36,800; roughly 40 percent of these students are enrolled in the graduate and professional schools. The College of Engineering was established more than 140 years ago and has long been one of the country's major engineering schools. Current enrollment is about 4,700 students in fourteen undergraduate programs and 2,000 students in more than twenty-five graduate programs.

Applying

The application deadline for the fall term is February 8. Scores on the Graduate Record Examinations (GRE) are required for all applicants. International students must also take either the Test of English as a Foreign Language (TOEFL) or the Michigan English Language Assessment Battery (MELAB). Minimum scores required for admission are 560 for the TOEFL and 80 for the MELAB. International students are also required to certify funds to cover academic expenses for the entire period of proposed study at the University. Students are also asked to submit letters of recommendation with their application. A nonrefundable application fee of $55 is required for all students. Inquiries should be addressed to either the specific program adviser or to the address below.

Correspondence and Information

Richard D. Woods, Chair
Department of Civil and Environmental Engineering
2340 G. G. Brown Building
University of Michigan
Ann Arbor, Michigan 48109-2125
Telephone: 734-764-8495
Fax: 734-764-4292

University of Michigan

THE FACULTY AND THEIR RESEARCH

Linda M. Abriola, Professor; Ph.D., Princeton, 1983. Environmental and water resources engineering, groundwater flow/contaminant transport modeling, aquifer remediation, mathematical modeling and laboratory experimentation.

Peter Adriaens, Associate Professor; Ph.D., California, Riverside, 1989. Environmental and water resources engineering, pollution microbiology and biotreatment, biotransformation and mineralization of organic compounds, hazardous-waste biodegradation.

Rajendra Aggarwala, Lecturer; M.S., Michigan, 1975. Surveying, photogrammetry, cartography, remote sensing, GIS systems.

Michael J. Barcelona, Professor; Ph.D., Puerto Rico, 1977. Subsurface contamination investigative techniques, field sampling and hydrogeochemical analysis, organic geochemistry of organic contaminants, microbial and redox processes, geostatistical approaches to contamination site remediation.

Jonathan W. Bulkley, Professor; Ph.D., MIT, 1966; PE. Environmental and water resources engineering, environmental policy/risk analysis.

Robert I. Carr, Professor; Ph.D., Stanford, 1971; PE. Construction engineering and management, cost engineering with uncertainty, decision support systems, building construction.

Kevin R. Collins, Assistant Professor; Ph.D., Illinois at Urbana-Champaign, 1995. Structural engineering, earthquake engineering and structural dynamics, structural mechanics, structural reliability.

Avery H. Demond, Associate Professor; Ph.D., Stanford, 1988. Environmental and water resources engineering, fate and transport of subsurface contaminants, aquifer remediation, laboratory experimentation and field studies.

Subhash C. Goel, Professor; Ph.D., Michigan, 1968; PE. Structural engineering, steel design, seismic behavior of steel and hybrid steel-R/C or masonry structures.

Pierre Goovaerts, Assistant Professor; Ph.D., Louvain-la-Neuve (Belgium), 1992. Environmental and water resources engineering, geostatistical modeling of the space-time variability of environmental variables, spatial interpolation, risk assessment, incorporation of uncertainty in decision making.

Donald H. Gray, Professor; Ph.D., Berkeley, 1966. Geotechnical engineering, physicochemical properties, slope stability, soil bioengineering, global change.

Will Hansen, Associate Professor; Ph.D., Illinois at Urbana-Champaign, 1983. Materials, physicochemical properties, cement-based composites.

Robert D. Hanson, Professor; Ph.D., Caltech, 1965; PE. Structural engineering, structural dynamics, supplemental mechanical damping devices for buildings, pseudodynamic testing.

Kim F. Hayes, Associate Professor; Ph.D., Stanford, 1987. Environmental and water resources engineering, environmental chemistry/interfacial processes, transport and transformation processes of environmental contaminants.

Roman D. Hryciw, Professor; Ph.D., Northwestern, 1986. Geotechnical engineering, physical properties, in situ testing, soil dynamics, wave propagation, rock mechanics, liquefaction.

Charles J. Hurbis, Adjunct Professor; J.D., Detroit, 1965. Engineering law.

Photios G. Ioannou, Associate Professor; Ph.D., MIT, 1984. Construction engineering and management, process performance and process-interaction simulation systems, intelligent database systems, risk-sensitive dynamic decision processes.

Nikolaos D. Katopodes, Professor; Ph.D., California, Davis, 1977. Environmental and water resources engineering, free surface flows/hydrologic modeling, computational fluid mechanics.

Gerald J. Keeler, Associate Professor; Ph.D., Michigan, 1987. Atmospheric processes, atmospheric chemistry, transport and sources of hazardous air pollutants, air-water exchange of environmental pollutants, global change issues related to environmental health.

Victor C. Li, Professor; Ph.D., Brown, 1981. Civil engineering materials, high-performance fiber-reinforced cement-based composites, structural element design, civil infrastructure, recycled materials, micromechanics and fracture mechanics, fiber composite design, property optimization.

Antoine E. Naaman, Professor; Ph.D., MIT, 1972. Structural engineering, composite materials, partially prestressed concrete, high-performance fiber-reinforced cement-based composites, nondestructive testing.

Andrzej S. Nowak, Professor; Ph.D., Warsaw Technical, 1975. Structural engineering, structural reliability, bridge structures, diagnostic procedures for bridges, modeling human error.

Jeremy D. Semrau, Assistant Professor; Ph.D., Caltech, 1995. Microbial physiology and diversity, in situ biodegradation, enumeration and characterization of natural microbial populations, bacterial and contaminant transport in the subsurface.

Božidar Stojadinovič, Assistant Professor; Ph.D., Berkeley, 1995. Earthquake-resistant design of structures, computer software for nonlinear dynamic analysis, experimental investigation and modeling of composite steel-R/C structures.

Walter J. Weber Jr., Professor; Ph.D., Harvard, 1962; PE. Environmental and water resources engineering, physicochemical processes/contaminant fate and transport, water and waste treatment and water reclamation.

James K. Wight, Professor; Ph.D., Illinois at Urbana-Champaign, 1973; PE. Structural engineering, reinforced concrete, earthquake-resistant design of concrete structures, seismic behavior of hybrid steel-R/C structures.

Richard D. Woods, Professor and Chair; Ph.D., Michigan, 1967; PE. Geotechnical engineering, soil dynamics, spectral analysis of surface waves, vibration instrumentation, soil sampling and evaluation.

Steven J. Wright, Professor; Ph.D., Caltech, 1977; PE. Environmental and water resources engineering, environmental fluid mechanics/contaminant transport, turbulent mixing, subsurface contaminant transport, density current propagation, laboratory experimentation.

UNIVERSITY OF MISSOURI–COLUMBIA

Department of Civil and Environmental Engineering

Programs of Study

The Department of Civil and Environmental Engineering offers programs of study leading to the M.S. and Ph.D. degrees. The master's degree program requires a minimum of 30 credit hours; a minimum of 15 hours of this credit must be in courses at the 400 level. The candidate must submit a thesis or a formal report to an examining committee. Students who receive research appointments are required to submit a thesis. A final oral examination is required of all M.S. candidates. The doctoral degree program requires a written and oral qualifying examination for formal admission during the first semester of post-master's work. One year of credit is usually given for the M.S. degree. The candidate must pass a comprehensive examination and submit and defend a dissertation at a final oral examination.

Major subject areas include structural mechanics, structural engineering and materials, geotechnical engineering, sanitary and environmental engineering, hydraulic engineering, hydrology, water resources engineering, and transportation engineering.

Research Facilities

The structural and materials engineering laboratories are equipped with several conventional universal testing machines and machines for torsion, fatigue, and pendulum-impact testing. The high-bay laboratories also have several servo-controlled hydraulic actuators and testing frames and an instrumented drop-weight impact machine. Support instrumentation includes high-speed digital storage oscilloscopes, analog and digital plotters, strain gauge instrumentation, laser holographic equipment, amplifiers/conditioners, load and displacement transducers, and several personal-computer-based data acquisition and test control systems. A 100-foot by 20-foot high-bay structural floor with anchoring capability on a 4-foot by 4-foot grid allows testing of full-size structures and structural components. A 5-foot by 5-foot 5-kip capacity shake-table with support instrumentation for dynamic testing is available for earthquake studies on model structures.

The geotechnical engineering laboratories include advanced triaxial testing facilities; unconfined compression testing; manual and automated direct shear equipment; rigid-wall hydraulic conductivity equipment; large- and small-diameter flexible-wall hydraulic conductivity test equipment; automated-, backpressure-, and constant gradient-consolidation testing capabilities; and porosimetry testing equipment, along with complete facilities for soil characterization. In addition, a separate geosynthetics laboratory includes a permittivity device, two transmissivity devices, a large-scale tilt table, eight creep loading frames for geosynthetics, two creep frames for polymeric structural members, and two environmental chambers for accelerated creep testing.

The environmental engineering laboratories contain analytical equipment for the complete physical, chemical, and microbiological analysis of environmental samples, such as a high-performance liquid chromatograph, a gas chromatograph, an AA spectrophotometer, a bioluminescence meter, photodegradation apparatus, a fluorescence meter, and a respirometer.

The hydraulics laboratory has a high bay and a floor area of 3,000 square feet. It contains a 35-foot constant-head tank, a large underground sump, three large pumps, two open-channel flumes, and three pipeline test loops. A wind tunnel lab houses a wind tunnel of 3 feet by 3 feet in cross section for wind engineering research and education.

Financial Aid

Approximately thirty graduate research and teaching assistantships are available each year. Half-time appointments pay $1080 to $1485 per month in addition to a waiver of tuition fees.

Cost of Study

Estimated expenses for 1999–2000 are $167.80 per credit hour for residents or $489.10 for nonresidents, $117 per semester for the activity fee, $8.30 per credit hour for the computer fee, and $60 per semester for the student health fee. The engineering course fee is $36.80 per credit hour.

Living and Housing Costs

Residence halls (dormitories including meals) are available at an average cost of $6000 per year. Total living expenses (fees, housing, books and supplies, and personal expenses) approximate $17,550 for residents and $18,900 for nonresidents.

Student Group

The Department of Civil and Environmental Engineering enrolls approximately 50 graduate students and 250 undergraduate students. The total engineering enrollment is approximately 2,000.

Student Outcomes

Most graduates are placed prior to graduation. During the last five years, approximately 18 percent have gone on to higher education, 3 percent have taken teaching positions, 7 percent are in state and federal government positions, 58 percent have taken jobs with private industry, and 14 percent are employed overseas.

Location

The University is located in Columbia, Missouri, halfway between St. Louis and Kansas City. The city population is approximately 77,000. Also located in Columbia are two small, private colleges: Stephens College and Columbia College. The city has consistently ranked high on *Money* magazine's list of most livable cities (as high as second in 1992).

The University

The University of Missouri–Columbia (MU), established in 1839, is the oldest state university west of the Mississippi River. MU is the largest of the four campuses of the Missouri system. Other campuses are located in Kansas City, Rolla, and St. Louis. MU enrolls a total of approximately 22,700 students, including 5,000 graduate students in ninety-seven graduate degree programs. Also located on the MU campus are the School of Law, School of Medicine, College of Veterinary Medicine, and School of Journalism. With its diverse programs, MU is one of the five most comprehensive universities in the nation.

Applying

Further information and application forms may be obtained from the department's Director of Graduate Studies. GRE General Test scores, transcripts from schools attended, and three letters of recommendation from faculty members are required of all applicants. A minimum TOEFL score of 550 is required of international applicants. Application deadlines are May 30 for the fall semester and October 15 for the winter semester.

Correspondence and Information

Director of Graduate Studies
Department of Civil and Environmental Engineering
Room E2509-A Engineering Building East
University of Missouri–Columbia
Columbia, Missouri 65211-2200
Telephone: 573-882-6269
Fax: 573-882-4784
E-mail: civilgrad@missouri.edu
World Wide Web: http://www.missouri.edu/~civilwww/

University of Missouri–Columbia

THE FACULTY AND THEIR RESEARCH

Shankha K. Banerji, Professor; Ph.D., Illinois at Urbana-Champaign; PE. Water and wastewater treatment processes, sludge stability, solid- and hazardous-waste treatment, water corrosion prevention, waste stabilization ponds and other natural treatment processes.

Michael G. Barker, Associate Professor; Ph.D., Minnesota; PE. Postelastic response of structures, bridges, reliability.

John J. Bowders Jr., Associate Professor; Ph.D., Texas at Austin; PE. Environmental geotechnics, performance assessment of containment barriers, geosynthetics, geophysical methods in geoenvironmental engineering applications.

Zhen Chen, Associate Professor; Ph.D., New Mexico. Applied mechanics, constitutive modeling using continuous and discrete approaches, computer simulation of structural failure subjected to long-term and impact loading, theoretical and computational aspects of particle mechanics.

Thomas E. Clevenger, Associate Professor; Ph.D., Missouri–Columbia. Fate and movement of chemicals in the environment, chemical speciation, recycling of wastes and wastewater sludges.

Aderbal C. Correa, Research Associate Professor; Ph.D., Stanford. Integration of remote sensing and geographic information systems for multidisciplinary applications, remote sensing in forested tropical areas, monitoring environmental conditions in coastal and fluvial areas.

V. S. Gopalaratnam, Professor ; Ph.D., Northwestern; PE. Fracture mechanics and strain-rate behavior of concrete and related composites, interface mechanics, ceramic and metal matrix composites, residual stresses in composites, high-performance hybrid cement composites.

Brett W. Gunnink, Associate Professor; Ph.D., Iowa State; PE. Microstructure of porous materials, conductometric phase transition porosimetry, pollutant transport through soil, concrete and aggregate durability.

Sam A. Kiger, Professor and Chairman; Ph.D., Illinois at Urbana-Champaign; PE. Explosion-resistant design, structural dynamics, earthquake engineering, soil-structure interaction, construction materials.

Charles Lenau, Professor; Ph.D., Stanford; PE. Flood routing and dispersion in open-channel flows; seepage through stratified soils, water hammer in pipes.

Henry Liu, Professor and Director of Capsule Pipeline Research Center; Ph.D., Colorado State; PE. Pipeline engineering, hydropower and other hydraulic engineering problems.

J. Erik Loehr, Assistant Professor; Ph.D., Texas at Austin. Stability and performance of earth slopes and waste fills, foundations and earth retaining structures, soil improvement techniques, computer methods and numerical modeling.

Jay B. McGarraugh, Professor and Assistant Dean; Ph.D., Purdue; PE. Reinforced concrete, concrete-on-steel composite construction.

Lee Peyton, Associate Professor and Director of Center for Environmental Technology; Ph.D., Colorado State; PE. Contaminant transport in soils, groundwater modeling, modeling of leachate generation and movement through landfills, flood plain hydrology and hydraulics.

Hani A. Salim, Assistant Professor; Ph.D., West Virginia. Advanced composite materials, wood composites, thin-walled structures, bridge engineering, mechanics, finite element modeling, testing of structures.

Kristen L. Sanford, Assistant Professor; Ph.D., Carnegie Mellon. Computer modeling of infrastructure management, analysis of transportation networks, decision support for infrastructure operation, intelligent transportation systems.

Robert L. Segar Jr., Assistant Professor; Ph.D., Texas at Austin; PE. Biological treatment processes, biofilm reactor technology and modeling, groundwater pollution control.

Mark R. Virkler, Associate Professor and Director of Graduate Studies; Ph.D., Virginia; PE. Traffic engineering, intersection operations, automatic vehicle detection, pedestrian characteristics, transportation planning.

RESEARCH AREAS

Structural mechanics, structural engineering, and materials

Emphasis areas within these programs include fracture and failure of composites, constitutive modeling, computational mechanics, impact mechanics, inelastic response of materials and structural systems, bridge engineering, computer-aided analysis and design, computer simulation of structural failure, active control, base isolation, structural dynamics, soil-structure interaction, earthquake engineering, finite element analysis, microstructure of porous materials, conductometric phase transition porosimetry, and concrete and aggregate durability.

Geotechnical and Geoenvironmental Engineering

This group performs research and technology development in environmental geotechnics and geotechnical infrastructure. Geoenvironmental emphases include: in situ remediation of contaminated soil and groundwater and in situ containment of contamination and pollutant transport through soil. Geotechnical infrastructure emphases include: stability and performance of earth slopes and waste fills, stabilization and maintenance of earth slopes, ground improvement techniques, waste utilization, engineering behavior of wind-blown soils, and field performance of geosynthetics.

Sanitary and environmental engineering

The principal emphases are on water pollution control, water purification, wastewater treatment, disposal of residues from these processes, hazardous- and solid-waste management, bioremediation, and groundwater contamination. Concentration is also on the application of physical, chemical, and microbiological principles to design for water supply and pollution control facilities, biological treatment processes, biofilm reactor technology, and groundwater pollution evaluation and control.

Transportation engineering

The emphasis is on course work ranging from highway design to traffic operations and the development of advanced transportation systems for urban areas. Current research emphases include intelligent transportation systems, infrastructure management, signalized intersection operations, highway flow characteristics, highway safety, and pedestrian flow characteristics.

Hydraulic engineering, hydrology, and water resources engineering

This area concentrates on the fundamentals of fluid flow and design in hydraulic and hydrologic systems. Current research emphases are on pipeline engineering, groundwater modeling, water and chemical transport in soils, landfill hydrology, soil macropore flow, and characteristics of organic chemical residuals in groundwater systems. The Capsule Pipeline Research Center conducts research on the development of large freight pipelines.

Capsule Pipeline Research Center

The center is a state-industry-University cooperative interdisciplinary research effort supported by the National Science Foundation. The only pipeline research center at a U.S. university, the center's research efforts are focused on new pipeline technologies for transporting coal, agricultural products, and solid waste.

Center for Environmental Technology

This is an interdisciplinary center for research and applications, technology development and transfer, and education and training in the areas of environmental and waste management. The focus of the center is on control and treatment, waste minimization, clean production, pollution prevention, monitoring and assessment, recovery and reuse, remediation, and ultimate disposal.

UNIVERSITY OF OKLAHOMA

School of Civil Engineering and Environmental Science

Programs of Study

The School of Civil Engineering and Environmental Science (CEES) offers graduate programs with areas of specialization in environmental science/engineering (subsurface remediation technologies, hydrologic applications of GIS, hazardous and solid waste management, risk assessment, industrial hygiene, air quality management, environmental chemistry, environmental microbiology, and wetlands ecology), structural engineering (high performance concrete materials, prestressed and reinforced concrete structures, and steel structures), and geotechnical engineering (behavior of unsaturated soils, earthquake engineering, finite element and constitutive modeling, geoenvironmental engineering, in situ testing, pavement materials and systems, rock mechanics and soil dynamics, and geomechanics).

The Doctor of Philosophy (Ph.D.) degree program is concerned with expansion of professional knowledge in the fundamental concepts of environmental science, environmental engineering, and civil engineering. CEES offers two master's degrees: the Master of Environmental Science (M.E.S.) and the Master of Science in Civil Engineering (M.S.C.E.). The M.E.S. degree is open to students with non-engineering undergraduate degrees in the physical or life sciences. Environmental or civil engineering doctoral students and M.S.C.E. students must have an undergraduate engineering degree or must complete remedial course work equivalent to an undergraduate degree. Environmental science doctoral students must have a master's degree in the physical or life sciences. CEES offers both a thesis and a nonthesis option for the master's degree.

Research Facilities

The environmental laboratory facilities are devoted to studying physical, chemical, and biological processes that affect the transport and fate of chemicals in the environment, and are equipped with state-of-the-art analytical equipment for measuring chemical concentrations in various media. Remediation laboratories are studying advanced processes for aquifer restoration and soil remediation. Specialized computer facilities support research in GIS, environmental process modeling, and visualization. Geotechnical laboratories include a triaxial device that controls air and pore pressure for testing unsaturated soils. The Constitutive Modeling Laboratory is equipped with two high-capacity polyaxial frames for testing of geologic materials under 3-D loading. The Fears Structural Engineering Laboratory has 8,400 square feet of laboratory space and an 1,800-square-foot reaction floor that can handle 320,000 pounds and accommodate configurations up to 22 feet high.

Computing facilities in CEES, including the Environmental Computing Applications System (ECAS), are excellent. ECAS consists of an 8 CPU Cray J90 supercomputer and a research and teaching lab composed of twelve IBM RS/6000 42T workstations. There are also eight IBM RS/6000 42T workstations distributed across campus (two in CEES).

Financial Aid

Research or teaching assistantships and tuition waivers are available for full-time graduate students. Nine hours of nonresident tuition and four hours of resident tuition per semester are waived for full-time graduate students with teaching or research assistantships. CEES traditionally provides additional tuition waivers for in-state and nonresident graduate students who are in good standing.

CEES has been awarded ten doctoral fellowships in environmental science and engineering from the Department of Education's Graduate Assistantships in Areas of National Need (GAANN) Program. Qualified recipients receive a stipend of up to $15,000 per year (based on financial need), full tuition waivers, and an educational allowance that can be used for educational and research expenses (books, laboratory and field supplies, computer hardware and software, and travel). Detailed information, including program objectives, application procedures, and deadlines, can be found on the departmental Web site.

Cost of Study

Tuition for 1998–99 was $80 per semester hour for in-state students. Nonresident students paid an additional $174.50 per semester hour (a total of $254.50 per semester hour). Additional fees are added to all graduate courses. Estimated total tuition and fees are $2256.90 for residents and $6028.50 for nonresidents.

Living and Housing Costs

Estimated costs for the 1998–99 academic year for a graduate student living off-campus were books and supplies ($766) and room and board ($3926). The University of Oklahoma (OU) offers single and married student housing in proximity to the campus. University housing and many private housing complexes provide convenient shuttle service to the main campus.

Student Group

The University's more than 25,400 students are enrolled in nineteen colleges that are located on the Norman campus and at the Health Sciences Center in Oklahoma City. Current graduate enrollment includes approximately 60 master's students and 30 doctoral students. CEES is the leading school in the College of Engineering in recruiting students from groups that are underrepresented in the sciences and engineering. One quarter of the students in CEES are women or members of minority groups (excluding nonresident aliens).

Location

The main campus for OU is located in Norman, a city of 80,000 residents. Norman is an independent community, with extensive parks and recreation programs, a 10,000-acre lake and park area, a community theater, an art center, and other amenities of a university town. Within 20 miles of Norman is Oklahoma City, the state capitol, with the National Cowboy Hall of Fame, the Firefighters Museum, and the Oklahoma City Zoo.

The University

OU was established in 1890, 17 years before Oklahoma became a state. Today, the University is a major national research university that serves the educational, cultural, and economic needs of the state, region, and nation. Scholars are attracted by the outstanding research facilities and unique resources of the University, but they also appreciate the mild Oklahoma climate, the varied cultural environment, and the friendly, informal atmosphere. Civil engineering, started in 1902, was the first engineering program at OU.

Applying

For admission, students must have a GPA of 3.0 or higher for their last 60 hours of course work, current (less than 5 years old) GRE scores, and two letters of reference if applying for financial aid (three for Ph.D. applicants). International applicants must have TOEFL scores of 575 or higher. Deadlines for applications by semester are July 15 for the fall semester, December 1 for the spring semester, and April 1 for the summer semester. International student application deadlines are April 1 for the fall semester, September 1 for the spring semester, and February 1 for the summer semester.

Correspondence and Information

For additional information about CEES, students may visit the Web page listed below. Interested applicants should send a request for application materials to:

Dr. Robert C. Knox
Professor and Graduate Liaison
202 West Boyd, Room 334
University of Oklahoma
Norman, Oklahoma 73019
Telephone: 405-325-5911
Fax: 405-325-4217
E-mail: knox@mailhost.ecn.ou.edu
World Wide Web: http://www.ou.edu/cees

University of Oklahoma

THE FACULTY AND THEIR RESEARCH

Professors

Larry W. Canter, George Lynn Cross Research Professor and Sun Company Chair; Ph.D., Texas; PE. Methods for environmental impact assessment, ground water protection and remediation, market-based approaches for air quality management.

Robert C. Knox, Samuel Roberts Noble Foundation Presidential Professor; Ph.D., Oklahoma; PE. Subsurface transport and fate processes, subsurface remediation technologies, environmental impacts of oil and gas exploration and production activities.

Anant Kukreti, Samuel Roberts Noble Foundation Presidential Professor; Ph.D., Colorado; PE. Analysis of linear and nonlinear structural systems, steel frame connections, constitutive modeling of anisotropic materials, coupled flow-deformation modeling.

David A. Sabatini, Ph.D., Iowa State; PE. Subsurface transport and fate processes, advanced subsurface remediation processes, innovative water treatment processes.

Ronald L. Sack, Professor and Director; Ph.D., Minnesota; PE. Environmental loads, dynamic loads produced by occupants, semiactive control for buildings and bridges.

Musharraf Zaman, Kerr McGee Presidential Professor; Ph.D., Arizona; PE. Rock mechanics, geomechanics, constitutive modeling, soil-structure interaction, pavement materials and systems, neural network modeling, behavior of soft sediments.

Associate Professors

Mark Meo, Director, Science and Public Policy Program; Ph.D., California, Davis. Environmental policy analysis, global change, corporate environmental management, energy technology policy.

Deborah Imel Nelson, Ph.D., Oklahoma Health Sciences Center; CIH. Human health risk assessment, establishment of risk-based cleanup levels, industrial hygiene, risk-based occupational exposure limits.

Bruce Russell, Ph.D., Texas at Austin; PE. High performance and high strength concrete materials, structural concrete, design philosophy and development of buildings codes, fatigue and fracture of offshore structural piping, structural stability and earthquake engineering.

Baxter E. Vieux, Ph.D., Michigan State; PE. Hydrologic and environmental process modeling and visualization, integrating numerical methods with Geographic Information Systems analysis and simulation of spatially distributed processes.

Ben Wallace, Ph.D., Stanford; PE. Structural systems, design and behavior of complete structural systems (particularly steel structures), novel framing systems for seismic resistance, serviceability and constructability of structural steel connections, small scale models of structural systems.

Assistant Professors

Tom Bush, Ph.D., Texas at Austin; PE. Design, behavior, and testing of structural components and systems (particularly prestressed and reinforced concrete structures), concrete materials, earthquake engineering.

Randall L. Kolar, Ph. D., Notre Dame; PE. Deterministic modeling of hydrologic processes in both the surface and groundwater regime, numerical methods for differential equations, high technology and alternative learning in the classroom.

Gerald A. Miller, Ph.D., Massachusetts Amherst; PE. Laboratory soil behavior, soil stabilization, unsaturated soil mechanics, swelling and dispersive clays, foundations, in situ testing, field prototype testing of geotechnical systems.

Mike Mooney, Ph.D., Northwestern. Soil dynamics, vibratory and dynamic compaction, geotechnical construction technologies, soil improvement, localization of strain in geomaterials, pavement rehabilitation.

Kanthasamy Muraleethran, Ph.D., California, Davis; PE. Soil dynamics, specifically behavior of unsaturated soils under earthquake loading conditions; pavement drainage; neural network modeling; centrifuge modeling; pollutant transport through soils.

Robert Nairn, Ph.D., Ohio State. Ecosystem biogeochemistry and ecology, wetlands science, ecosystem structure and function, pollutant removal capabilities of wetlands and passive treatment systems.

Mark A. Nanny, Ph.D. Illinois at Urbana-Champaign. Aquatic, sediment, and soil chemistry; nutrient cycling; pollutant-natural organic matter interactions; environmental applications of nuclear magnetic resonance (NMR) spectroscopy.

Keith Strevett, Ph.D., Connecticut. Association between microbial surface properties, interfacial forces, genetic alteration, microbial physiology, and microbial biophysical chemistry.

UNIVERSITY OF SOUTHERN CALIFORNIA

School of Engineering
Department of Civil Engineering

Programs of Study

The Graduate Program in Civil Engineering at USC offers three professional degrees: the Master of Science, the Engineer degree, and the Doctor of Philosophy. The graduate program has two parallel objectives: to train engineers who will enter the profession in critical fields and to prepare students for fundamental research in both basic and emerging areas of science and technology. Students seeking the M.S. degree in civil engineering may specialize in construction engineering and management, coastal and ocean engineering, earthquake engineering, environmental engineering, geotechnical engineering, structural engineering, structural mechanics, or water resources engineering. Students also have the option of earning a Master of Construction Management, a degree program that is open to students with a B.Sc. or B.A. degree, or a Master of Engineering in computer-aided engineering. Those pursuing studies in environmental engineering may specialize in sanitary engineering design, air pollution control, water quality control, hazardous and solid waste management, or energy and the environment. Students seeking the Ph.D. degree may specialize in any of these fields. In addition, they may pursue other related areas of study as determined in consultation with their faculty adviser.

Research Facilities

The department is fully equipped to carry out state-of-the-art analytical and laboratory research. A wide range of computer facilities are available to support analytical research. The University Computing Services maintain mini supercomputers, which are networked to department workstations and personal computers. All systems offer a broad range of language compilers, engineering packages, word processing, and e-mail. The structural engineering laboratory has two reaction frames for testing large-scale building components under computer-controlled static and cyclic loading. A 1-million-pound compression machine and a 600,000-pound universal machine are used for testing specimens under tension and compression loads. The concrete laboratory contains a 4.5-cubic-foot-high intensity mixer and equipment for aggregate testing, creep testing, and concrete curing. The engineering dynamics laboratory is equipped with two large synchronized shake tables for simulating earthquake loading. The earthquake data processing laboratory maintains equipment to both process and archive strong motion records and maintains the Southern California strong motion array. The Center for Research and Construction Engineering links research and applications by developing procedures for improved engineering design and construction and by experimenting on new earthquake resistant materials and structural systems. The Research Center for Computational Geomechanics maintains an array of data workstations and personal computers. It has soil testing equipment with computerized acquisition. The wave mechanics laboratory is equipped with a short-wave channel and a programmable wave plate and associated hardware for digital control of all experiments. The environmental engineering laboratories are equipped with a wide array of standard instrumentation, including a substantial capability for environmental microbiology utilizing incubators, sterilizers, microscopes, and support hardware. Professors in the department working on civil engineering research programs have made major contributions to the fields of earthquake-resistant design, behavior of high-strength concrete, earthquake ground motion studies and liquefaction evaluation methods, numerical analysis of soil media, finite element analyses of structural systems, structural control systems, wave mechanics, and analysis of tsunamis. Professors in the Environmental Engineering Program have made significant research contributions on the effects of air pollution on visibility, advanced methods for determining sources of air pollution, new methods of treating difficult industrial wastes, releases of toxic vapors from asphalt operations, biological systems for removing sulfur from coal, and microbiological decontamination of water and air.

Financial Aid

The department provides financial assistance to qualified students through teaching and research assistantships; fellowships are available from the Graduate School. The deadline for financial aid applications is January 31. Applications for Graduate School fellowships and teaching and research assistantships are available from the department.

Cost of Study

The tuition cost for the 1999–2000 academic year is $768 per unit per semester, not including mandatory fees. Full-time students typically take 9–12 units per semester.

Living and Housing Costs

University housing averages $2000 per semester. Privately owned off-campus apartments can be rented in the immediate vicinity of campus for $500 to $800 per month. Housing is available in the greater Los Angeles area at all levels of cost.

Student Group

The University of Southern California has a total enrollment of approximately 27,000 students, 42 percent of whom are women. The Department of Civil Engineering has a steady enrollment of about 200 graduate students and 275 undergraduate students. About 35 percent of the civil engineering students are women.

Location

USC is located at University Park, near downtown Los Angeles, across from the Museum of Natural History and the Museum of Science and Industry.

The southern California area is a world center for many engineering and construction companies, and the California economy is the sixth-largest economy in the industrialized world. The Los Angeles landscape has a striking combination of beaches, rolling hills, and mountains. Ski slopes are within a 90-minute drive of the campus. The San Gabriel Mountains, the Santa Monica Mountains, and Venice beach are all less than a half an hour's drive from USC.

The University

USC was founded in 1880. It is the oldest and largest private university on the West Coast. The University has a central library with 2.4 million volumes, as well as twelve specialized libraries, and subscribes to 18,000 periodicals. The University offers 183 different degree programs and has twenty professional schools. The School of Engineering ranks fifth in the nation in terms of government-supported research. USC sponsors twelve varsity sports, five intramural programs, and forty on-campus and fifteen off-campus sports clubs. An estimated 6,000 students participate in the sports program. The civil engineering department has a strong tradition of participation in USC sports, particularly football and swimming.

Applying

Deadlines for applications are June 1 for fall and November 1 for spring. Applicants must have a B.Sc. in science or engineering. The GRE General Test is required for civil engineering applicants and the GRE General and Subject tests are required for environmental engineering applicants. Students applying for a Master of Construction Management can have either a B.Sc. or a B.A. They can choose to take either the GRE General Test or the GMAT. All programs require three letters of recommendation and a statement of purpose. All applications are to be sent directly to admissions.

Correspondence and Information

Irene Soloff
Student Advisor in Civil Engineering
University of Southern California–2531
Los Angeles, California 90089-2531
Telephone: 213-740-0587
E-mail: civileng@mizar.usc.edu
World Wide Web: http://www.usc.edu/dept/civil_eng/dept/

University of Southern California

THE FACULTY AND THEIR RESEARCH

Ahmed M. Abdel-Ghaffar, Professor of Civil Engineering; Ph.D., Caltech, 1976. Earthquake engineering, dynamics of bridges, dams, and other civil engineering structures.

James C. Anderson, Professor of Civil Engineering; Ph.D., Berkeley, 1969. Computer-aided design of structures, earthquake-resistant design, nonlinear dynamic response analysis, modeling of structures, design of steel and concrete structures.

Jean-Pierre Bardet, Professor of Civil Engineering; Ph.D., Caltech, 1984. Experimental and theoretical soil mechanics and numerical methods in geotechnical engineering.

George V. Chillingarian, Professor of Civil and Environmental Engineering; Ph.D., USC, 1956. Environmental aspects of oil and gas production, petrophysical properties of rocks, drilling fluids, surface and subsurface operations in petroleum production, subsidence due to the fluid withdrawal, testing and storage of petroleum products.

Joseph S. Devinny, Professor of Civil and Environmental Engineering; Ph.D., Caltech, 1975. Environmental impact of human activity on natural ecosystems and the problems of hazardous wastes.

Ronald C. Henry, Associate Professor of Civil and Environmental Engineering; Ph.D., Oregon Graduate Center, 1977. Environmental engineering, specifically, air pollution.

Jiin-Jen Lee, Professor of Civil and Environmental Engineering and Director of the Foundation of Cross-Connection Control; Ph.D., Caltech, 1970. Hydraulics, water resource engineering, coastal engineering.

Vincent W. Lee, Associate Professor of Civil Engineering; Ph.D., USC, 1979. Seismic data processing, risk analysis, wave propagation and numerical methods.

Geoffrey R. Martin, Professor of Civil Engineering; Ph.D., Berkeley, 1965. Geotechnical aspects of earthquake engineering, particularly soil behavior under cyclic loading; liquefaction, response of earth structures and seismic design of foundations.

Sami F. Masri, Professor of Civil Engineering; Ph.D., Caltech, 1965. Analysis, control, and modeling of nonlinear dynamic systems.

Najmedin Meshkati, Associate Professor of Civil Engineering; Ph.D., USC, 1983. Environmental health and safety assessment, risk and reliability of complex technological systems.

James Moore, Associate Professor of Civil Engineering and of Urban and Regional Planning; Ph.D., Stanford, 1986. Transportation systems analysis, discrete facility location, local area traffic control, optimal control of congestable transportation networks.

Robert L. Nigbor, Research Assistant Professor of Civil Engineering; Ph.D., USC, 1989. Earthquake engineering and structural vibrations.

Massoud Pirbazari, Professor of Civil and Environmental Engineering and Director of Environmental Engineering Program; Ph.D., Michigan, 1980. Water quality control, bioremediation of toxic and hazardous pollutants, adsorption and membrane processes.

Masanobu Shinozuka, Fred Champion Professor of Civil Engineering; Ph.D., Columbia, 1960. Random vibrations, reliability of structural systems, structural dynamics, structural control, inelasticity, infrastructure systems, lifeline systems.

Constantinos Sioutas, Assistant Professor of Civil and Environmental Engineering; Sc.D., Harvard, 1994. Environmental health aspects of air pollution, air pollution monitoring, and control.

Costas Synolakis, Professor of Civil and Aerospace Engineering; Ph.D., Caltech, 1985. Breaking waves, wave run-up, hydrodynamic pressures on dams, two-phase flow.

Craig Taylor, Research Assistant Professor of Civil Engineering; Ph.D., Illinois, 1974. Risk assessment and uncertainty analysis.

Maria I. Todorovska, Research Associate Professor of Civil Engineering; Ph.D., USC, 1988. Elastic wave propagation in soils and structures, soil-structure interaction, passive isolation of structural vibrations and control of structural response, probabilistic assessment of losses from earthquakes, seismic risk analysis, earthquake engineering.

Mihailo D. Trifunac, Professor of Civil Engineering; Ph.D., Caltech, 1969. Strong motion seismology, earthquake engineering, structural dynamics, wave propagation, random vibrations, and instrumentation and measurement.

Landon C. Wellford, Professor of Civil Engineering and Chairman; Ph.D., Alabama, 1975. Numerical methods in engineering, finite element methods for linear and nonlinear structural analysis.

Hung Leung Wong, Professor of Civil Engineering; Ph.D., Caltech, 1975. Numerical methods in engineering, wave propagation in solids, active control of high-rise buildings and earthquake engineering.

Yan Xiao, Assistant Professor of Civil Engineering; Ph.D., Kyushu (Japan), 1989. Earthquake-resistant design of buildings and bridges, structural concrete, steel-concrete composite structures, properties of structural materials.

Teh Fu Yen, Professor of Civil and Environmental Engineering; Ph.D., Virginia Tech, 1956. Environmental chemistry, fossil fuels, alternative resource conversion processes, solid-waste management, asphalt chemistry, organic geochemistry.

Structural test facilities.

UNIVERSITY OF SOUTHERN CALIFORNIA

Department of Civil Engineering and
School of Urban Planning and Development
Master of Construction Management Program

Program of Study

The Master of Construction Management (M.C.M.) interdisciplinary degree program is administered by the Department of Civil Engineering and is associated with the School of Urban Planning and Development Lusk Center for Real Estate Development, the School of Architecture's Building Science Program, the Graduate School of Business Administration's Finance and Business Economics Program in Real Estate, and other participating schools such as the Law Center and the Leonard Davis School of Gerontology. The program is innovative and unique in its approach to educating and training multidisciplinary professionals to understand and execute the broad array of technical and nontechnical activities associated with construction management. This array of skills draws on a variety of academic disciplines in which USC enjoys prominence, reflecting the University's blend of professional and liberal education. The Master of Construction Management degree program includes leading theories and practices and prepares students for careers as industrial leaders in the real estate/construction industry while providing special attention to the function of construction in real estate development. Potential employers include general contractors, real estate developers, and related enterprises.

Research Facilities

The University is fully equipped to carry out state-of-the-art analytical and laboratory research. A wide range of computer facilities are available to support analytical research. The University Computing Services maintain minisupercomputers, which are networked to department workstations and personal computers. Systems offer a range of language compilers, engineering packages, word processing, and electronic mail and are connected to the National Computer Internet.

Dedicated workstations and microcomputer equipment are available through the Department of Civil Engineering and the School of Urban Planning and Development microcomputer laboratories. The USC libraries house approximately 2.5 million volumes and receive more than 16,000 scholarly journals. USC has noteworthy collections in the areas of cinema, international and public affairs, American literature, regional history, marine science, philosophy, Latin American studies, and Korean studies. In addition, through an agreement with other libraries and library associations, the USC libraries are able to extend their reach beyond the campus to resources at other institutions. USCInfo, the library's campus information service, provides access to books, periodicals, articles, and other information through networked computers at all library locations and for remote users with authorized passwords.

The Lusk Center Research Institute is the research branch of the School of Urban Planning and Development. The center facilitates collaborative research on all aspects of real estate development and urban planning. The Lusk Center provides academic support to government agencies, consulting firms, nonprofit organizations, and other private groups seeking effective approaches to complex planning and real estate development problems. The School of Urban Planning and Development has developed expertise in a broad range of areas, including urban planning and design, international planning and development, enterprise planning, transportation planning, regional economics, and community development.

The Center for Advanced Transportation Technologies is an interdisciplinary center for transportation research that brings together USC scholars in civil engineering, urban planning and development, electrical engineering, industrial and systems engineering, human factors, safety, and other areas to study advanced research questions relating to the development and implementation of new technologies in transportation.

The structural engineering laboratory has two reaction frames for testing large-scale building components under computer-controlled static and cyclic loading. A 1-million-pound compression machine and a 600-pound universal machine are used for testing specimens under tension and compression loads. The concrete laboratory contains a 4.5-cubic-foot high-intensity mixer and equipment for aggregate testing, creep testing, and concrete curing. The engineering dynamics laboratory is equipped with two large synchronized shake tables for simulating earthquake loading. The earthquake data processing laboratory maintains equipment to process strong motion records and maintains the southern California strong motion array. The Center for Research and Construction Engineering links research and applications by developing procedures for improved engineering design and construction and by experimenting on new earthquake-resistant materials and structural systems. The Research Center for Computational Geomechanics maintains an array of workstations and personal computers and has available standard soil testing hardware. USC civil engineering professors have made major contributions to the fields of earthquake-resistant design, behavior of high-strength concrete, earthquake ground motion studies and liquefaction evaluations methods, numerical analysis of soil media, finite-element analyses of structural systems, and structural control systems.

Financial Aid

The M.C.M. program's sponsoring units offer financial assistance to qualified students through teaching assistantships. The deadline for applications is January 31. Applications for teaching assistantships are available from the Department of Civil Engineering or the School of Urban Planning and Development.

Cost of Study

The tuition cost for the 1999–2000 academic year is $768 per unit per semester, not including mandatory fees. Full-time students typically take 9–15 units per semester. A minimum of 33 units is required.

Living and Housing Costs

University housing averages $2000 per semester. Privately owned off-campus apartments can be rented in the immediate vicinity of campus for $500–$800 per month. Housing is available in the greater Los Angeles area at all levels of cost.

Student Group

The University of Southern California has a total enrollment of approximately 27,000, 42 percent of whom are women. The student body includes 14,000 undergraduate, 13,000 graduate, and 4,000 international students.

Location

USC is located at University Park, near downtown Los Angeles, across from the Museum of Natural History and the Museum of Science and Industry.

The southern California area is a world center for many engineering and construction companies, and the California economy is the sixth-largest economy in the industrialized world. The Los Angeles landscape has a striking combination of beaches, rolling hills, and mountains. World-class ski slopes are within a 90-minute drive of the campus. The San Gabriel Mountains, the Santa Monica Mountains, and Venice Beach are all less than half an hour's drive from USC.

The University

USC was founded in 1880. It is the oldest and largest private university on the West Coast. The University has a central library with 2.4 million volumes, as well as twelve specialized libraries, and subscribes to 18,000 periodicals. The University offers 183 different degree programs and has twenty professional schools. The School of Engineering ranks fifth in the nation in terms of government-supported research. USC sponsors twelve varsity sports, five intramural programs, and forty on-campus and fifteen off-campus sports clubs. An estimated 6,000 students participate in the sports program. Graduates have a high success rate in finding jobs with the aid of the strong USC alumni network.

Applying

Women and men interested in pursuing a career in construction management are encouraged to apply. Applicants must have a B.Sc. or B.A. from an accredited school. The GRE General and Subject tests or the GMAT are required. Application deadlines are June 1 for fall and November 1 for spring. Three letters of recommendation and a statement of purpose are required. All applications are to be sent directly to Admissions.

Correspondence and Information

Irene Soloff
Student Advisor in Civil Engineering
University of Southern California
Los Angeles, California 90089-2531

Telephone: 213-740-0587
Fax: 213-744-1426
E-mail: civileng@mizar.usc.edu
World Wide Web: http://www.usc.edu/dept/civil_eng/dept/

University of Southern California

THE FACULTY

Marc Angelil, Associate Professor of Architecture and Building Sciences.
Edward J. Blakely, Dean of the School of Urban Planning and Development and Lusk Professor of Planning and Development.
David Dale-Johnson, Associate Professor of Finance and Business Economics and Director, Real Estate Program and Finance and Business Economics Program.
Michael D'Artuono, Lecturer in Civil Engineering.
Stuart Gabriel, Associate Professor of Business Administration and of Finance and Business Economics.
Peter Gordon, Professor of Urban Planning and Development.
Susan Kamei, Lecturer in Urban Planning and Development.
Michael Keston, Adjunct Professor of Urban Planning and Development.
Henry M. Koffman, Director, Construction Engineering and Management Program and Master of Construction Management Program.
Allan Kotin, Adjunct Professor of Urban Planning and Development.
John Kuprenas, Research Assistant Professor of Civil Engineering.
Brian Jackson, Lecturer in Urban Planning and Development.
Vincent Lee, Associate Professor of Civil Engineering.
Michael Markus, Lecturer in Civil Engineering.
Geoffrey R. Martin, Professor, Department of Civil Engineering.
Leonard Marvin, Lecturer in Civil Engineering.
Steve Minassian, Lecturer in Civil Engineering.
James E. Moore II, Associate Professor of Urban Planning and Development and Civil Engineering and Director, Transportation Engineering Program.
Richard Peiser, Associate Professor of Urban Planning and Development and Director, Lusk Center for Real Estate Development.
Dominick Pescarolo, Lecturer in Civil Engineering.
F. Edward Reynolds Jr., Lecturer in Civil Engineering.
Mark Schiler, Associate Professor of Architecture and Building Sciences.
Gregory Schwann, Assistant Professor of Urban Planning and Development.
Dana Sherman, Lecturer in Civil Engineering, The Law Center.
Goetz Shierle, Professor of Architecture and Building Sciences.
Rena Sivitanidou, Assistant Professor of Urban Planning and Development.
Richard Smith, Adjunct Professor of Urban Planning and Development.
Jon Sommers, Lecturer in Civil Engineering.
Kevin Starr, Professor of Urban Planning and Development.
George Stumpf, Lecturer in Civil Engineering.
Don Valachi, Clinical Associate Professor of Business Administration and of Finance and Business Economics.
Johannes Van Tilburg, Lecturer in Urban Planning and Development.

A construction site on the University of Southern California campus.

UNIVERSITY OF SOUTHERN CALIFORNIA

School of Engineering
Department of Civil Engineering
Environmental Engineering Program

Programs of Study

The Environmental Engineering Program of the Department of Civil Engineering offers programs leading to the degrees of Master of Science (M.S.) and Doctor of Philosophy (Ph.D.) in environmental engineering. These professional degrees were established in 1970 for those who plan to pursue a career in public or private sectors or in research and teaching. The M.S. program is available to students with B.S. degrees in science and engineering and requires 27 units of course work beyond the bachelor's degree. The 27 units must include 12 units of designated core courses and 9 units minimum from a selected specialty. Major specialty areas in the M.S. program include air pollution control, water and wastewater engineering, solid- and hazardous-waste management, energy and environment, water quality control, groundwater pollution control, and general environmental engineering. The doctoral program requires a minimum of 60 units of course work beyond the bachelor's degree and completion of a doctoral dissertation. Full-time students can complete the master's program in two to three semesters and the doctoral program in three to four years.

Research Facilities

The environmental engineering laboratories are equipped with state-of-the-art analytical and experimental research instruments and are involved in a broad spectrum of research activities. There are four major areas of research, including water quality engineering and control, air pollution control and atmospheric research, energy and environment, and environmental microbiology and biotechnology. There is also access to excellent University-wide research facilities through collaborative and interdisciplinary research projects. A wide range of computer facilities is available to support analytical research. The University Computing Services maintain minisupercomputers, which are networked to department workstations and personal computers. The USC computing network is telelinked with Cray systems of Supercomputer Centers in San Diego and other locations. The on-campus library facilities have been considerably upgraded to create an ideal environment for sophisticated research. The USC library system provides accessibility to nearly all the important bibliographic databanks available today. The establishment of a computerized worldwide network through the new Leavey Teaching Library has further enhanced the overall capability of the library system.

Financial Aid

The department provides financial assistance to qualified students through teaching and research assistantships. The deadline for applications is February 1. Fellowships are available from the graduate school; the deadline for applications is February 1. Applications for graduate school fellowships and teaching and research assistantships are available from the department.

Cost of Study

The tuition cost for the 1999–2000 academic year is $768 per unit per semester, not including mandatory fees. Full-time students typically take 9–12 units per semester.

Living and Housing Costs

University housing averages $2000 per semester. Various University meal plans can be purchased for approximately $1500 per semester. Privately owned off-campus apartments can be rented in the immediate vicinity of campus for $500 to $800 per month. Housing is available in the greater Los Angeles area at all levels of cost.

Student Group

The University of Southern California has a total enrollment of approximately 27,000 students, 42 percent of whom are women. The Department of Civil Engineering has a steady enrollment of about 200 graduate students and 275 undergraduate students. About 35 percent of the civil engineering students are women.

Location

USC is located at University Park, near downtown Los Angeles, across from the Museum of Natural History and the Museum of Science and Industry.

The southern California area is a world center for many engineering and construction companies, and the California economy is the sixth-largest economy in the industrialized world. The Los Angeles landscape has a striking combination of beaches, rolling hills, and mountains. Ski slopes are within a 90-minute drive of the campus. The San Gabriel Mountains, the Santa Monica Mountains, and Venice beach are all less than a half an hour's drive from USC.

The University

USC was founded in 1880. It is the oldest and largest private university on the West Coast. The University has a central library with 2.4 million volumes, as well as twelve specialized libraries, and subscribes to 18,000 periodicals. The University offers 183 different degree programs and has twenty professional schools. The School of Engineering ranks fifth in the nation in terms of government-supported research. USC sponsors twelve varsity sports, five intramural programs, and forty on-campus and fifteen off-campus sports clubs. An estimated 6,000 students participate in the sports program. The civil engineering department has a strong tradition of participation in USC sports, particularly football and swimming.

Applying

Deadlines for applications are June 1 for fall and November 1 for spring. Applicants must have a B.Sc. in science or engineering. The GRE General and Subject tests are required for all environmental engineering applicants. Three letters of recommendation and a statement of purpose are required. All applications are to be sent directly to admissions.

Correspondence and Information

Irene Soloff
Student Advisor in Civil Engineering
University of Southern California–2531
Los Angeles, California 90089-2531
Telephone: 213-740-0587
E-mail: civileng@mizar.usc.edu
World Wide Web: http://www.usc.edu/dept/civil_eng/dept/

University of Southern California

THE FACULTY AND THEIR RESEARCH

Ahmed M. Abdel-Ghaffar, Professor of Civil Engineering; Ph.D., Caltech, 1976. Earthquake engineering, dynamics of bridges, dams, and other civil engineering structures.

James C. Anderson, Professor of Civil Engineering; Ph.D., Berkeley, 1969. Computer-aided design of structures, earthquake-resistant design, nonlinear dynamic response analysis, modeling of structures, design of steel and concrete structures.

Jean-Pierre Bardet, Professor of Civil Engineering; Ph.D., Caltech, 1984. Experimental and theoretical soil mechanics and numerical methods in geotechnical engineering.

George V. Chillingarian, Professor of Civil and Environmental Engineering; Ph.D., USC, 1956. Environmental aspects of oil and gas production, petrophysical properties of rocks, drilling fluids, surface and subsurface operations in petroleum production, subsidence due to the fluid withdrawal, testing and storage of petroleum products.

Joseph S. Devinny, Professor of Civil and Environmental Engineering; Ph.D., Caltech, 1975. Environmental impact of human activity on natural ecosystems and the problems of hazardous wastes.

Ronald T. Eguchi, Research Associate Professor of Civil Engineering; M.S., UCLA, 1975. Risk analysis, earthquake engineering, natural hazards engineering.

Ronald C. Henry, Associate Professor of Civil and Environmental Engineering; Ph.D., Oregon Graduate Center, 1977. Environmental engineering, specifically, air pollution.

Jiin-Jen Lee, Professor of Civil and Environmental Engineering and Director of the Foundation of Cross-Connection Control; Ph.D., Caltech, 1970. Hydraulics, water resource engineering, coastal engineering.

Vincent W. Lee, Associate Professor of Civil Engineering; Ph.D., USC, 1979. Seismic data processing, risk analysis, wave propagation and numerical methods.

Geoffrey R. Martin, Professor of Civil Engineering and Chairman; Ph.D., Berkeley, 1965. Geotechnical aspects of earthquake engineering, particularly soil behavior under cyclic loading; liquefaction, response of earth structures and seismic design of foundations.

Sami F. Masri, Professor of Civil Engineering; Ph.D., Caltech, 1965. Analysis, control, and modeling of nonlinear dynamic systems.

Najmedin Meshkati, Associate Professor of Civil Engineering; Ph.D., USC, 1983. Environmental health and safety assessment, risk and reliability of complex technological systems.

James Moore, Associate Professor of Civil Engineering and of Urban and Regional Planning; Ph.D., Stanford, 1986. Transportation systems analysis, discrete facility location, local area traffic control, optimal control of congestable transportation networks.

Massoud Pirbazari, Professor of Civil and Environmental Engineering and Director of Environmental Engineering Program; Ph.D., Michigan, 1980. Water quality control, bioremediation of toxic and hazardous pollutants, adsorption and membrane processes.

Masanobu Shinozuka, Fred Champion Professor of Civil Engineering; Ph.D., Columbia, 1960. Random vibrations, reliability of structural systems, structural dynamics, structural control, inelasticity, infrastructure systems, lifeline systems.

Constantinos Sioutas, Assistant Professor of Civil and Environmental Engineering; Sc.D., Harvard, 1994. Environmental health aspects of air pollution, air pollution monitoring, and control.

Costas Synolakis, Professor of Civil and Aerospace Engineering; Ph.D., Caltech, 1985. Breaking waves, wave run-up, hydrodynamic pressures on dams, two-phase flow.

Maria I. Todorovska, Research Associate Professor of Civil Engineering; Ph.D., USC, 1988. Elastic wave propagation in soils and structures, soil-structure interaction, passive isolation of structural vibrations and control of structural response, probabilistic assessment of losses from earthquakes, seismic risk analysis, earthquake engineering.

Mihailo D. Trifunac, Professor of Civil Engineering; Ph.D., Caltech, 1969. Strong motion seismology, earthquake engineering, structural dynamics, wave propagation, random vibrations, and instrumentation and measurement.

Landon C. Wellford, Professor of Civil Engineering and Associate Chairman; Ph.D., Alabama, 1975. Numerical methods in engineering, finite element methods for linear and nonlinear structural analysis.

Hung Leung Wong, Professor of Civil Engineering; Ph.D., Caltech, 1975. Numerical methods in engineering, wave propagation in solids, active control of high-rise buildings and earthquake engineering.

Yan Xiao, Assistant Professor of Civil Engineering; Ph.D., Kyushu (Japan), 1989. Earthquake-resistant design of buildings and bridges, structural concrete, steel-concrete composite structures, properties of structural materials.

Teh Fu Yen, Professor of Civil and Environmental Engineering; Ph.D., Virginia Tech, 1956. Environmental chemistry, fossil fuels, alternative resource conversion processes, solid-waste management, asphalt chemistry, organic geochemistry.

Ray Ruichong Zhang, Research Assistant Professor of Civil Engineering; Ph.D., Florida Atlantic, 1992. Engineering seismology, earthquake engineering, structural dynamics, reliability analysis, stochastic process, random vibrations.

VIRGINIA POLYTECHNIC INSTITUTE AND STATE UNIVERSITY

Via Department of Civil and Environmental Engineering

Programs of Study	The department awards the Master of Science in civil engineering and environmental engineering, the Master of Engineering in civil engineering, and the Doctor of Philosophy in civil engineering. Master of Science and Ph.D. degrees in environmental science and engineering are also administered by the College of Engineering.
	For the civil engineering degree, major emphasis may be placed on construction engineering, geodetic engineering, geotechnical engineering, hydraulic engineering, materials engineering, structural engineering, transportation engineering, water resources engineering, air conservation, or environmental statistics. A minor may be taken in one or more of the alternate branches of civil engineering or in an allied field, such as mathematics, engineering mechanics, geology, or urban and regional studies. Studies in hydraulic engineering and water resources engineering range from fundamental analytical and experimental efforts in hydraulics and hydrology to the management and design of broad water resource systems, with a wide variety of options and possible emphases. The transportation engineering program includes all aspects of the planning, operation, and design of transportation facilities. Studies in structural engineering generally emphasize computer techniques in analysis and design and are closely supported by offerings in the engineering mechanics department. The environmental engineering program is concerned primarily with the areas of water quality management (including strong emphasis on treatment processes), air resources engineering, and solid-waste disposal. The environmental statistics program is an interdisciplinary program offered in cooperation with the Department of Statistics. The environmental science and engineering program is a broadly based environmental studies program designed for students with undergraduate degrees in one of the natural sciences as well as for some with degrees in other branches of engineering. For degree and residence requirements, see the page describing the College of Engineering.
Research Facilities	Separate laboratories are available for research in hydraulics, soil mechanics, environmental engineering, air pollution, structural testing, materials, computer-aided design, and geodetic engineering. Recent equipment acquisitions include advanced computer workstations and apparatus for laboratory and in situ soil testing, electron capture gas chromatography, atomic absorption spectrophotometry, and large scale dynamic and static structural testing. Laboratories have been completed recently for computer-aided engineering and structural materials research. An off-campus laboratory is available for study of runoff pollution effects, reservoir eutrophication control, and watershed modeling and management.
Financial Aid	A variety of fellowships, assistantships, and traineeships are available that pay up to approximately $16,000 a year with incremental raises each year. A $6-million endowment allows the department to offer a significant number of fellowships for highly qualified students. It also provides an independent source of funds for the enhancement of faculty salaries and student activities. It continues to grow through interest accrual. Information on application for these awards may be obtained from the head of the Via Department of Civil and Environmental Engineering. Graduate students may receive long-term or short-term loans from a number of funds administered by the University. Teaching and research assistantships paid $1290–$1350 per month for half-time appointments during the 1998–99 year. Students on an assistantship may take a maximum of 12 and a minimum of 9 credits per semester. Candidates for M.Eng. degrees are generally not eligible for research assistantships.
Cost of Study	Fees for full-time study in 1999–2000 are $2475 (in-state) and $3879 (out-of-state) per semester.
Living and Housing Costs	Private housing, including both rooms and apartments, is available in Blacksburg and the surrounding area. Apartment rents range from $300 (for one or two bedrooms) to $500 (for two or three bedrooms) per month; normally these rates include all utilities except electricity. Single students often share apartments to reduce their expenses.
Student Group	The Via Department of Civil and Environmental Engineering has more than 270 graduate students registered in the various degree programs offered.
Location	The Virginia Tech campus is located in Blacksburg, a town of almost 30,000 people. Blacksburg is located in the Appalachian Mountains, west of the Blue Ridge Mountains, in southwest Virginia. Forty miles to the east is the Roanoke-Salem metropolitan area, which has a population of approximately 200,000. The high student population density in Blacksburg provides an ideal atmosphere for student life. Nearby Claytor Lake and Hungry Mother state parks provide boating, swimming, camping, and hiking opportunities. Over the past few years Blacksburg has undergone rapid growth, with the addition of several restaurants, shopping centers, and apartment complexes.
The University	Virginia Tech, Virginia's land-grant university, was founded in 1872 and awarded its first graduate degree in 1892. Currently, the total enrollment of approximately 24,000 includes 4,500 graduate students. About one fourth of the total enrollment is in engineering. For many years, Virginia Tech has ranked among the ten largest institutions in the United States in terms of the number of engineering bachelor's degrees awarded each year.
	The University provides facilities for outdoor and indoor sports, including an eighteen-hole golf course, an indoor swimming pool, and indoor tennis courts. There is a very active faculty–graduate student intramural program. Extensive cultural programs, including musical and dramatic events, lectures, and discussions, are held on campus.
Applying	Each applicant must submit a completed application form, two official transcripts of his or her undergraduate and graduate records to date, and three letters of recommendation from former professors. Application should be made as early as possible before the opening of the term for which admission is sought. GRE General Test scores are required of applicants who graduated from a non-ABET-accredited institution and are strongly recommended for all other applicants.
Correspondence and Information	William R. Knocke, Head Via Department of Civil and Environmental Engineering Virginia Polytechnic Institute and State University Blacksburg, Virginia 24061-0105

Virginia Polytechnic Institute and State University

THE FACULTY AND THEIR RESEARCH

Construction Engineering and Management

Jesus M. de la Garza, Associate Professor and Program Coordinator; Ph.D., Illinois at Urbana-Champaign; PE. Financial management, expert systems, scheduling, cost control, infrastructure management, computer applications in construction.

Julio C. Martinez, Assistant Professor; Ph.D., Michigan. Simulation of construction processes.

W. Eric Showalter, Assistant Professor; Ph.D., Purdue; PE. Green construction, geoenvironmental and project management applications with a focus on cost-optimized selection of soil remediation technologies.

Michael C. Vorster, David H. Burrows Professor of Construction Engineering and Management; Ph.D., Stellenbosch (South Africa). Equipment management, equipment productivity, project management techniques, contract administration.

Environmental Engineering

Gregory D. Boardman, Associate Professor; Ph.D., Maine; PE. Hazardous-waste management, environmental toxicology, industrial-waste treatment.

Andrea M. Dietrich, Associate Professor; Ph.D., North Carolina at Chapel Hill. Toxicology, trace organics analysis, water treatment.

Marc A. Edwards, Associate Professor; Ph.D., Washington (Seattle); EIT. Corrosion of plumbing materials in water supplies; physical and chemical drinking water processes; areas of applied aquatic chemistry, including arsenic geochemistry.

Daniel L. Gallagher, Associate Professor; Ph.D., North Carolina at Chapel Hill; PE. Water quality and hydrologic models.

Thomas J. Grizzard, Associate Professor and Director, Occoquan Monitoring Research Laboratory, Manassas, Virginia; Ph.D., Virginia Tech. Nonpoint pollution control, water quality monitoring, eutrophication assessment and management, sediment-water interactions.

John M. Hughes, Associate Professor; Ph.D., Illinois at Urbana-Champaign; PE. Air pollution impact assessment and modeling, air pollution control systems, instrumental concepts and methodology.

William R. Knocke, W. Curtis English Professor of Civil Engineering and Department Head; Ph.D., Missouri–Columbia; PE. Water treatment processes, industrial-waste treatment, and sludge thickening/dewatering.

John C. Little, Assistant Professor; Ph.D., Berkeley; PE. Environmental mass transfer, indoor air pollution, physical/chemical treatment processes.

Nancy G. Love, Assistant Professor; Ph.D., Clemson. Biodegradation of xenobiotic compounds, biological processes.

John T. Novak, Nick Prillaman Professor of Environmental Engineering and Program Coordinator; Ph.D., Washington (Seattle); PE. Sludge treatment, treatment of industrial wastes, subsurface biodegradation.

Clifford W. Randall, Charles P. Lunsford Professor of Civil Engineering; Ph.D., Texas at Austin. Biological process design, industrial-waste treatment, biological nutrient removal, nonpoint source pollution control, sludge conditioning and dewatering.

Geotechnical Engineering

Thomas L. Brandon, Associate Professor; Ph.D., Berkeley. Laboratory testing and instrumentation, shear strength and stress/strain behavior of soils, soil property evaluation.

J. Michael Duncan, University Distinguished Professor; Ph.D., Berkeley; PE. Slope stability, foundation design, seepage, earth dams.

George M. Filz, Associate Professor and Program Coordinator; Ph.D., Virginia Tech. Geoenvironmental engineering, foundation engineering, soil-structure interaction, finite-element analyses.

Thangavelu Kuppusamy, Professor; Ph.D., Indian Institute of Technology (Kanpur). Constitutive laws, hazardous wastes, soil dynamics, soil-structure interaction, finite-element application.

James R. Martin II, Associate Professor; Ph.D., Virginia Tech. Earthquake engineering, in situ testing, groundwater pollution.

Hydrosystems Engineering

William E. Cox, Professor and Assistant Department Head; Ph.D., Virginia Tech. Water resources planning and management, water law and policy.

Panayiotis Diplas, Associate Professor and Program Coordinator; Ph.D., Minnesota. Sediment transport and hydrodynamics.

Randel L. Dymond, Associate Professor; Ph.D., Penn State. Geographical information systems (GIS), computer applications.

David F. Kibler, Professor; Ph.D., Colorado State; PE. Stormwater management, urban hydrology, hydraulics and water resources management.

G. V. Loganathan, Associate Professor; Ph.D., Purdue. Hydrology, systems analysis of water resources planning management.

Mark A. Widdowson, Associate Professor; Ph.D., Auburn; PE. Computer modeling, groundwater hydraulics, contaminant transport, bioremediation.

Structural Engineering

Richard M. Barker, Professor; Ph.D., Minnesota; PE. Reinforced concrete, bridge design specifications, soil-structure interaction.

Thomas E. Cousins, Associate Professor; Ph.D., North Carolina State; PE. Prestressed and reinforced concrete, bridge engineering behavior.

W. Samuel Easterling, Associate Professor and Program Coordinator; Ph.D., Iowa State. Composite floor systems, cold-formed steel structures, structural steel design.

Siegfried M. Holzer, Alumni Distinguished Professor; Ph.D., Illinois at Urbana-Champaign. Nonlinear analysis of structures systems, structural stability and dynamics, application of finite-element method, behavior of glulam domes.

Thomas M. Murray, Montague-Betts Professor of Structural Steel; Ph.D., Kansas; PE. Steel structures, structural connections, serviceability of floor systems, experimental research, expert systems.

Raymond H. Plaut, Dan H. Pletta Professor; Ph.D., Berkeley. Dynamic stability of structures, nonlinear oscillations, optimal structural design, dynamic behavior of inflatable dams.

Kamal B. Rojiani, Associate Professor; Ph.D., Illinois at Urbana-Champaign; PE. Computer applications, expert systems, structural safety and reliability, wind engineering.

Richard E. Weyers, Professor; Ph.D., Penn State; PE. Physical and environmental behavior of Portland cement, the economics of rehabilitation/replacement strategies.

Surveying and Geodetics

Richard G. Greene, Assistant Professor; Ph.D., LSU; PS. Surveying, Geographic Information System (GIS), remote sensing, hydrology.

Transportation Engineering

Imad L. Al-Qadi, Associate Professor and Program Coordinator; Ph.D., Penn State. Asphaltic materials and mixtures, pavement rehabilitation and maintenance, nondestructive pavement evaluation.

John Collura, Professor; Ph.D., North Carolina State; PE. Advanced transportation systems.

Gerardo W. Flintsch, Assistant Professor; Ph.D., Arizona State. Bituminous materials and mixtures, pavement management systems, artificial neural networks applications.

Antoine G. Hobeika, Professor; Ph.D., Purdue. Mass transit, airport planning, transportation economics and finance.

Wei H. Lin, Assistant Professor; Ph.D., Berkeley. Dynamic traffic assignment, intelligent transportation systems, network optimization.

Antonio A. Trani, Associate Professor; Ph.D., Virginia Tech. Aviation systems planning, air transportation, airport engineering, transport systems engineering.

Michel W. VanAerde, Professor; Ph.D., Waterloo; PE. Traffic operations, simulation and modeling, intelligent transportation systems.

WAYNE STATE UNIVERSITY

Department of Civil and Environmental Engineering

Programs of Study

The Department of Civil and Environmental Engineering offers M.S. and Ph.D. degree programs, with specialization in the following areas: structures, geotechnical engineering, environmental engineering, transportation, and construction. The M.S. program accommodates the needs of full-time, on-campus students as well as part-time students who are concurrently employed. Most full-time students participate in research with the faculty while pursuing their graduate courses. All graduate classes are held in the evening.

For admission to the M.S. program, the student must have an undergraduate engineering degree from an institution that is accredited by the Accrediting Board for Engineering and Technology (ABET) or from a comparable international institution. Students from other backgrounds may be required to complete prerequisite undergraduate courses before credit may be accrued. An overall undergraduate honor point average (HPA) of at least 3.0 (on a 4.0 scale) is required for regular admission. Qualified admission may be granted to students who have practical experience but a lower HPA. The M.S. program requires 32 graduate-level credits, with or without a thesis option.

For admission to the Ph.D. program, the student's overall honor point average must be 3.2 or better (3.4 in the last two years as an undergraduate student). Students who do not satisfy these minimum standards are not considered for admission to the program until they have completed an M.S. degree program and have earned a graduate honor point average that is not less than 3.5. Candidates for the doctoral degree must complete 90 credits beyond the baccalaureate, including 30 credits of dissertation direction and 60 credits of course work and directed study.

Research Facilities

The department maintains state-of-the-art experimental and computer facilities for teaching and research. These include laboratories in the areas of structural and materials engineering, environmental engineering, hydraulic engineering, geotechnical engineering, geoenvironmental engineering, and transportation engineering. Two new laboratories for civil infrastructure information systems and civil infrastructure materials are currently being developed.

The Wayne State University (WSU) library system includes the new Adamany Undergraduate Library, which contains state-of-the-art technology and information resources, and the Walter P. Reuther Library of Labor and Urban Affairs, the largest archive of its kind. In addition, Arthur Neef Law Library, Purdy/Kresge Library, Science and Engineering Library, and Vera Shiffman Medical Library serve the campus community. Wayne State's academic and administrative computing environment includes open-access computer workstations, e-mail and directory services for all students, central mainframes and servers, off-campus dial-in access, data and voice networks, high-speed Internet connections, and a variety of consulting, training, and documentation services.

Financial Aid

Financial aid is available in the form of scholarships and research and teaching assistantships for students with outstanding records. There are a limited number of scholarships. Most research assistantships are awarded by individual professors. In the case of international students, financial aid is generally not provided until a satisfactory performance record has been established in residence at the University. Currently, approximately 20 graduate students have assistantships. Research and teaching assistantships provide a complete tuition waiver and health insurance benefits. There are no formal applications for awards made through the department, although students should indicate their desire for financial aid when applying.

Cost of Study

For 1998–99, graduate tuition was $163 per semester credit for Michigan residents and $355 per semester credit for nonresidents. Including the registration fee, Michigan residents paid $1755, while nonresidents paid $3483, for 9 credits (typically three courses) per semester, excluding books.

Living and Housing Costs

There is a variety of housing options available on campus and in the adjacent community. Typical on-campus rates were $500 per month for a one-bedroom apartment and $750 per month for a two-bedroom apartment in 1998–99. Additional details and current rates can be obtained from the University Housing Office (telephone: 313-577-2116; World Wide Web: http://www.wsuhousing.com/info.html).

Student Group

The total number of students at Wayne State University is approximately 30,000, including about 13,000 who are involved in graduate studies. In 1999, there are approximately 150 students in the civil and environmental engineering graduate program, of whom 25 are in the Ph.D. program. Of the total number of graduate students, about 50 percent are part-time. There are several student organizations, including ASCE and Chi Epsilon, that provide opportunities for students to participate in student-faculty social events and to be in contact with professional engineers.

Location

WSU is located in the heart of the cultural center of Detroit. The Detroit Institute of the Arts, Detroit Historical Museum, Detroit Science Center, Museum of African American History, and the main branch of the Detroit Public Library are all within easy walking distance. Detroit is one of the largest industrial areas of the world, with numerous engineering and construction project activities.

The University

Founded in 1868, Wayne State University is now a comprehensive university. There are fifteen colleges and schools. The University offers 355 major subject areas and 128 bachelor's, 136 master's, and sixty-one different doctoral degrees plus thirty different certificate, specialist, and professional programs. In 1994, Wayne State University joined a select group of eighty-seven other universities nationwide that are classified as Research I universities by the Carnegie Foundation in recognition of its broad range of baccalaureate programs, commitment to graduate education, and strong emphasis on research.

Applying

Application forms and instructions for submitting transcripts, a statement of purpose, letters of recommendation, and related materials may be obtained from the department. Study may begin in any semester. TOEFL scores are required from all international students. A minimum TOEFL score of 550 is required for admission. The GRE is recommended for all students. International students are also required to certify funds to cover academic expenses for at least the first year of proposed study at the University.

Correspondence and Information

Correspondence should be directed to:
Mumtaz Usmen, Chair
   or
Takaaki Kagawa, Graduate Program Officer
Department of Civil and Environmental Engineering
Wayne State University
Detroit, Michigan 48202
Fax: 313-577-3881
E-mail: musmen@ce.eng.wayne.edu (Mumtaz Usmen)
     tkagawa@ce.eng.wayne.edu (Takaaki Kagawa)
World Wide Web: http://www.ce.eng.wayne.edu/

Wayne State University

THE FACULTY AND THEIR RESEARCH

The faculty of the Department of Civil and Environmental Engineering has been active in a variety of research projects sponsored by federal, state, and local government agencies and industrial firms and organizations. Representative research efforts performed by the faculty are listed below.

Haluk M. Aktan, Professor; Ph.D., Michigan; PE. Nondestructive methods of structural evaluation, structural materials, structural dynamics and design, finite-element analysis, structural analysis and design.

Tapan K. Datta, Professor; Ph.D., Michigan State; PE. Transportation systems analysis and economics, highway safety and risk management, traffic operation, geographic information systems (GIS) applications in civil engineering.

Gongkang Fu, Associate Professor; Ph.D., Case Western Reserve; PE. Bridge engineering, infrastructure engineering, engineering reliability and stochastic analysis, construction materials, structural and materials testing, structural dynamics and earthquake engineering.

Thomas Heidtke, Associate Professor; Ph.D., Michigan. Surface water hydrology, Great Lake water quality, watershed management, environmental systems analysis.

Takaaki Kagawa, Associate Professor; Ph.D., Berkeley; PE. Earthquake responses of geotechnical systems, numerical simulation of liquefaction-related problems, performance of pile foundations in liquefying sand, dynamic responses of pile foundations, earthquake damage prevention research.

Snehamay Khasnabis, Professor; Ph.D., North Carolina State; PE. Transportation planning and system analysis, mass transit and urban transportation systems, numerical and physical simulations, transit land use interactions, automated highway systems.

Carol J. Miller, Associate Professor; Ph.D., Michigan; PE. Groundwater contaminant transport, numeric modeling of geoenvironmental systems, physical models of geoenvironmental systems, leachate characterization, waste containment systems and materials.

Mumtaz A. Usmen, Professor; Ph.D., West Virginia; PE. Stabilization of geomaterials, construction safety and quality management, construction workforce development, facilities management.

Hwai-Chung Wu, Assistant Professor; Ph.D., MIT. Structural mechanics, civil infrastructure materials/rehabilitation, advanced composites, micromechanics of composites.

Nazli Yesiller, Assistant Professor; Ph.D., Wisconsin–Madison. Geotechnical/geoenvironmental engineering, geosynthetics, characterization of earthen materials using nondestructive testing.

Section 8
Computer Science and Information Technology

This section contains a directory of institutions offering graduate work in computer science and information technology, followed by in-depth entries submitted by institutions that chose to prepare detailed program descriptions. Additional information about programs listed in the directory but not augmented by an in-depth entry may be obtained by writing directly to the dean of a graduate school or chair of a department at the address given in the directory.

For programs offering related work, see also in this book Electrical and Computer Engineering, Engineering and Applied Sciences, and Industrial Engineering. In Book 2, see Communication and Media; in Book 4, Mathematical Sciences; and in Book 6, Business Administration and Management and Library and Information Studies.

CONTENTS

CONTENTS

Artificial Intelligence/Robotics

Carnegie Mellon University, Carnegie Institute of Technology, Department of Civil and Environmental Engineering, Pittsburgh, PA 15213-3891. Offers civil engineering (MS, PhD); civil engineering and industrial management (MS); civil engineering and robotics (PhD); civil engineering/bioengineering (PhD); civil engineering/engineering and public policy (MS, PhD). Part-time programs available. *Faculty:* 21 full-time (5 women). *Students:* 56 full-time (20 women), 11 part-time (2 women); includes 3 minority (1 African American, 1 Hispanic American, 1 Native American), 44 international. Terminal master's awarded for partial completion of doctoral program. *Degree requirements:* For master's, thesis required (for some programs), foreign language not required; for doctorate, dissertation, qualifying exam required, foreign language not required. *Entrance requirements:* For master's and doctorate, GRE General Test, TOEFL. *Application deadline:* For fall admission, 2/1 (priority date); for spring admission, 10/15. Application fee: $45. *Unit head:* Chris Hendrickson, Head, 412-268-2941, Fax: 412-268-7813. *Application contact:* Maxine A. Leffard, Graduate Program Administrator, 412-268-8712, Fax: 412-268-7813, E-mail: ce-addmissions+@andrew.cmu.edu.

Carnegie Mellon University, School of Computer Science, Robotics Institute, Pittsburgh, PA 15213-3891. Offers PhD. *Faculty:* 67 full-time (4 women), 1 part-time (0 women). *Students:* 67 full-time (15 women), 11 part-time (2 women); includes 1 minority (Asian American or Pacific Islander), 34 international. Average age 27. In 1998, 13 doctorates awarded. *Degree requirements:* For doctorate, dissertation required, foreign language not required. *Entrance requirements:* For doctorate, GRE General Test, GRE Subject Test, TOEFL. *Application deadline:* For fall admission, 2/1. Application fee: $50. *Financial aid:* Grants available. *Faculty research:* Perception, cognition, manipulation, robot systems, manufacturing. Total annual research expenditures: $27.4 million. *Unit head:* Takeo Kanade, Director, 412-268-3016, Fax: 412-268-5571. *Application contact:* Takeo Kanade, Director, 412-268-3016, Fax: 412-268-5571.

See in-depth description on page 715.

The Catholic University of America, School of Engineering, Department of Mechanical Engineering, Program in Design and Robotics, Washington, DC 20064. Offers MME, D Engr, PhD. Part-time and evening/weekend programs available. *Degree requirements:* For master's, comprehensive exam required, thesis optional, foreign language not required; for doctorate, dissertation, comprehensive and oral exams required, foreign language not required. *Entrance requirements:* For master's, minimum GPA of 3.0; for doctorate, minimum GPA of 3.5. *Application deadline:* For fall admission, 8/1 (priority date); for spring admission, 12/1. Applications are processed on a rolling basis. Application fee: $50. *Financial aid:* Research assistantships, teaching assistantships, career-related internships or fieldwork, Federal Work-Study, institutionally-sponsored loans, and tuition waivers (full and partial) available. Aid available to part-time students. Financial aid application deadline: 2/1. *Faculty research:* Active constrained damping, smart traversing beams. Total annual research expenditures: $86,000. *Unit head:* Dr. Amr Baz, Director, 202-319-5170.

Cornell University, Graduate School, Graduate Fields of Engineering, Field of Computer Science, Ithaca, NY 14853-0001. Offers algorithms (M Eng, PhD); applied logic and automated reasoning (M Eng, PhD); artificial intelligence (M Eng, PhD); computer graphics (M Eng, PhD); computer science (M Eng, PhD); computer vision (M Eng, PhD); concurrency and distributed computing (M Eng, PhD); information organization and retrieval (M Eng, PhD); operating systems (M Eng, PhD); parallel computing (M Eng, PhD); programming environments (M Eng, PhD); programming languages and methodology (M Eng, PhD); robotics (M Eng, PhD); scientific computing (M Eng, PhD); theory of computation (M Eng, PhD). *Faculty:* 34 full-time. *Students:* 141 full-time (20 women); includes 20 minority (19 Asian Americans or Pacific Islanders, 1 Hispanic American), 64 international. Terminal master's awarded for partial completion of doctoral program. *Degree requirements:* For doctorate, dissertation required, foreign language not required, foreign language not required. *Entrance requirements:* For master's, GRE General Test, GRE Subject Test, TOEFL (minimum score of 550 required); for doctorate, GRE General Test, GRE Subject Test (computer science), TOEFL (minimum score of 550 required). *Application deadline:* For fall admission, 1/1. Application fee: $65. Electronic applications accepted. *Unit head:* Director of Graduate Studies, 607-255-8593. *Application contact:* Graduate Field Assistant, 607-255-8593, E-mail: phd@cs.cornell.edu.

See in-depth description on page 729.

Howard University, College of Engineering, Architecture, and Computer Sciences, School of Engineering and Computer Science, Department of Mechanical Engineering, Washington, DC 20059-0002. Offers aerospace engineering/dynamics and controls (M Eng, PhD); applied mechanics (M Eng, PhD); CAD/CAM and robotics (M Eng, PhD); fluid and thermal sciences (M Eng, PhD). Part-time programs available. *Faculty:* 9 full-time (1 woman). *Students:* 17 full-time (7 women), 2 part-time; includes 7 African Americans, 1 Asian American or Pacific Islander, 7 international. Terminal master's awarded for partial completion of doctoral program. *Degree requirements:* For master's, computer language, comprehensive exam required; for doctorate, one foreign language, computer language, dissertation, 2 terms of residency required. *Entrance requirements:* For master's and doctorate, GRE General Test, TOEFL, minimum GPA of 3.0. *Application deadline:* For fall admission, 4/1 (priority date); for spring admission, 11/1. Applications are processed on a rolling basis. Application fee: $45. Electronic applications accepted. *Unit head:* Dr. Lewis Thigpen, Chair, 202-806-6600, Fax: 202-806-5258, E-mail: lthigpen@scs.howard.edu. *Application contact:* Dr. Sonya Smith, Graduate Director, 202-806-4837.

Instituto Tecnológico y de Estudios Superiores de Monterrey, Campus Monterrey, Graduate and Research Division, Program in Computer Science, Monterrey, 64849, Mexico. Offers artificial intelligence (PhD); computer science (MS); information systems (MS); information technology (MS). Part-time programs available. *Degree requirements:* For master's and doctorate, thesis/dissertation required. *Entrance requirements:* For master's, PAEG, TOEFL; for doctorate, TOEFL, master's in related field. *Faculty research:* Distributed systems, software engineering, decision support systems.

Instituto Tecnológico y de Estudios Superiores de Monterrey, Campus Monterrey, Graduate and Research Division, Programs in Engineering, Monterrey, 64849, Mexico. Offers applied statistics (M Eng); artificial intelligence (PhD); automation engineering (M Eng); chemical engineering (M Eng); civil engineering (M Eng); electrical engineering (M Eng); electronic engineering (M Eng); environmental engineering (M Eng); industrial engineering (M Eng, PhD); manufacturing engineering (M Eng); mechanical engineering (M Eng); systems and quality engineering (M Eng). M Eng offered jointly with the University of Waterloo; PhD (industrial engineering) offered jointly with Texas A&M University. Part-time and evening/weekend programs available. Terminal master's awarded for partial completion of doctoral program. *Degree requirements:* For master's and doctorate, one foreign language, computer language, thesis/dissertation required. *Entrance requirements:* For master's, PAEG, TOEFL; for doctorate, GRE, TOEFL, master's in related field. *Faculty research:* Flexible manufacturing cells, materials, statistical methods, environmental prevention, control and evaluation.

New Hampshire College, Graduate School of Business, Program in Business Administration, Manchester, NH 03106-1045. Offers accounting (Certificate); artificial intelligence (Certificate); business administration (MBA); finance (Certificate); government administration (Certificate); health administration (Certificate); human resource management (Certificate); international business (Certificate); operations management (Certificate); taxation (Certificate); training and development (Certificate). Part-time and evening/weekend programs available. *Degree requirements:* For master's, thesis or alternative required, foreign language not required. *Entrance requirements:* For master's, minimum GPA of 2.7 during previous 2 years, 2.5 overall.

Ohio University, Graduate Studies, College of Engineering and Technology, Department of Integrated Engineering, Athens, OH 45701-2979. Offers geotechnical and environmental engineering (PhD); intelligent systems (PhD); materials processing (PhD). *Faculty:* 39 full-time (1 woman). *Students:* 9 full-time (0 women), 6 part-time; includes 1 minority (Asian American or Pacific Islander), 12 international. *Degree requirements:* For doctorate, computer language, dissertation required, foreign language not required. *Entrance requirements:* For doctorate, GRE General Test, MS in engineering or related field. *Application deadline:* For fall admission, 3/15. Applications are processed on a rolling basis. Application fee: $30. Tuition, state resident: full-time $5,754; part-time $238 per credit hour. Tuition, nonresident: full-time $11,055; part-time $457 per credit hour. Tuition and fees vary according to course load, campus/location and program. *Unit head:* Dr. Jerrel R. Mitchell, Associate Dean for Research and Graduate Studies, 740-593-1482, E-mail: mitchell@bobcat.ent.ohiou.edu.

San Jose State University, Graduate Studies, College of Engineering, Department of Computer, Information and Systems Engineering, Program in Computer Engineering, San Jose, CA 95192-0001. Offers computer engineering (MS); computer software (MS); computerized robots and computer applications (MS); microprocessors and microcomputers (MS). *Faculty:* 5 full-time (0 women), 12 part-time (1 woman). *Students:* 57 full-time (28 women), 90 part-time (19 women); includes 106 minority (1 African American, 103 Asian Americans or Pacific Islanders, 2 Hispanic Americans), 27 international. *Degree requirements:* For master's, computer language, thesis, comprehensive exam required, foreign language not required. *Entrance requirements:* For master's, GRE General Test (minimum combined score of 1500 on three sections required), BS in computer science or 24 credits in related area. *Application deadline:* For fall admission, 6/1. Applications are processed on a rolling basis. Application fee: $59. Tuition, nonresident: part-time $246 per unit. Required fees: $1,939; $1,309 per year. *Unit head:* Dr. Haluk Ozemek, Coordinator, 408-924-4100. *Application contact:* Dr. Haluk Ozemek, Coordinator, 408-924-4100.

University of California, San Diego, Graduate Studies and Research, Department of Electrical and Computer Engineering, La Jolla, CA 92093-5003. Offers applied ocean science (MS, PhD); applied physics (MS, PhD); communication theory and systems (MS, PhD); computer engineering (MS, PhD); electrical engineering (M Eng, MS, PhD); electronic circuits and systems (MS, PhD); intelligent systems, robotics and control (MS, PhD); photonics (MS, PhD); signal and image processing (MS, PhD). *Faculty:* 35. *Students:* 251 (24 women). *Entrance requirements:* For master's and doctorate, GRE General Test. Application fee: $40. *Unit head:* William Coles, Chair. *Application contact:* Graduate Coordinator, 619-534-6606.

University of Georgia, Graduate School, College of Arts and Sciences, Program in Artificial Intelligence, Athens, GA 30602. Offers MS. *Faculty:* 1 full-time (0 women). *Students:* 20 full-time (4 women), 8 part-time (1 woman), 18 international. 24 applicants, 54% accepted. In 1998, 6 degrees awarded. *Degree requirements:* For master's, thesis required, foreign language not required. *Entrance requirements:* For master's, GRE General Test. *Application deadline:* For fall admission, 7/1 (priority date); for spring admission, 11/15. Application fee: $30. Electronic applications accepted. *Financial aid:* Unspecified assistantships available. *Unit head:* Dr. Walter D. Potter, Graduate Coordinator, 706-542-0358, Fax: 706-542-0349, E-mail: potter@ai.uga.edu.

University of Southern California, Graduate School, School of Engineering, Department of Computer Science, Program in Robotics and Automation, Los Angeles, CA 90089. Offers MS. *Students:* 3 full-time (1 woman); includes 1 minority (Asian American or Pacific Islander), 2 international. Average age 24. 17 applicants, 47% accepted. *Entrance requirements:* For master's, GRE General Test. *Application deadline:* For fall admission, 6/1 (priority date); for spring admission, 10/1. Applications are processed on a rolling basis. Application fee: $55. Tuition: Part-time $768 per unit. Required fees: $350 per semester. *Financial aid:* In 1998–99, 1 research assistantship was awarded.; fellowships, teaching assistantships, Federal Work-Study, institutionally-sponsored loans, and scholarships also available. Aid available to part-time students. Financial aid application deadline: 2/15; financial aid applicants required to submit FAFSA. *Unit head:* Dr. Ellis Horowitz, Chairman, Department of Computer Science, 213-740-4494.

University of Tennessee, Knoxville, Graduate School, College of Engineering, Department of Mechanical and Aerospace Engineering and Engineering Science, Program in Engineering Science, Knoxville, TN 37996. Offers applied artificial intelligence (MS); biomedical engineering (MS, PhD); composite materials (MS, PhD); computational mechanics (MS, PhD); engineering science (MS, PhD); fluid mechanics (MS, PhD); industrial engineering (MS, PhD); optical engineering (MS, PhD); solid mechanics (MS, PhD). Part-time programs available. *Students:* 33 full-time (9 women), 16 part-time (1 woman); includes 4 minority (1 African American, 2 Asian Americans or Pacific Islanders, 1 Native American), 10 international. *Degree requirements:* For master's, thesis or alternative required, foreign language not required; for doctorate, dissertation required, foreign language not required. *Entrance requirements:* For master's and doctorate, TOEFL (minimum score of 550 required), minimum GPA of 2.7. *Application deadline:* For fall admission, 2/1 (priority date). Applications are processed on a rolling basis. Application fee: $35. Electronic applications accepted. *Application contact:* Dr. Allen Yu, Graduate Representative, 923-974-4159, E-mail: nyu@utk.edu.

Computer Science

Acadia University, Faculty of Pure and Applied Science, School of Computer Science, Wolfville, NS B0P 1X0, Canada. Offers M Sc. *Faculty:* 10 full-time (0 women). *Students:* 15 full-time (1 woman), 7 international. Average age 22. In 1998, 1 degree awarded. *Degree requirements:* For master's, thesis required, foreign language not required. *Entrance requirements:* For master's, honors degree in computer science. *Average time to degree:* Master's–2 years full-time. *Application deadline:* For fall admission, 2/1. Application fee: $25. *Expenses:* Tuition and fees charges are reported in Canadian dollars. Tuition, state resident: full-time $4,361 Canadian dollars. International tuition: $8,722 Canadian dollars full-time. Required fees: $147 Canadian dollars. Full-time tuition and fees vary according to program and student level. *Financial aid:* Teaching assistantships, career-related internships or fieldwork available. Financial aid application deadline: 2/1. *Faculty research:* Visual and object-oriented programming, concurrency, artificial intelligence, hypertext and multimedia, algorithm analysis. *Unit head:* Dr. A. Trudel, Director, 902-585-1331, Fax: 902-585-1067, E-mail: cs@acadiau.ca. *Application contact:* Secretary, 902-585-1585, Fax: 902-585-1067, E-mail: cs@acadiau.ca.

Air Force Institute of Technology, School of Engineering, Department of Electrical and Computer Engineering, Program in Computer Systems/Science, Wright-Patterson AFB,

Computer Science

Air Force Institute of Technology (continued)
OH 45433-7765. Offers MS, PhD. Part-time programs available. *Faculty:* 9 full-time (0 women). *Students:* 57 full-time, 7 part-time. In 1998, 3 master's awarded (100% found work related to degree). *Degree requirements:* For master's and doctorate, computer language, thesis/dissertation required, foreign language not required. *Entrance requirements:* For master's, GRE General Test (minimum score of 500 on verbal section, 600 on quantitative required), minimum GPA of 3.0, must be military officer or U.S. citizen; for doctorate, GRE General Test (minimum score of 550 on verbal section, 650 on quantitative required), minimum GPA of 3.0, must be military officer or U.S. citizen. Application fee: $0. *Faculty research:* Artificial intelligence, formal methods of software engineering, information survivability, database systems. *Unit head:* Maj. Rick Raines, Chief, 937-255-3636 Ext. 4527, Fax: 937-656-4055, E-mail: rraines@afit.af.mil.

Alabama Agricultural and Mechanical University, School of Graduate Studies, School of Arts and Sciences, Department of Mathematics and Computer Science, Normal, AL 35762-1357. Offers computer science (MS). Evening/weekend programs available. *Faculty:* 9 full-time (0 women), 2 part-time (0 women). *Students:* 40; includes 10 minority (8 African Americans, 2 Asian Americans or Pacific Islanders), 25 international. Average age 30. In 1998, 23 degrees awarded. *Degree requirements:* For master's, computer language, comprehensive exam required, thesis optional, foreign language not required. *Entrance requirements:* For master's, GRE General Test (minimum score of 500 on each section required), TOEFL (minimum score of 500 required). *Application deadline:* For fall admission, 5/1. Application fee: $15 ($20 for international students). Tuition, state resident: full-time $1,932. Tuition, nonresident: full-time $3,864. Tuition and fees vary according to course load. *Financial aid:* In 1998–99, 5 research assistantships with tuition reimbursements (averaging $4,900 per year) were awarded.; career-related internships or fieldwork also available. Financial aid application deadline: 4/1. *Faculty research:* Computer-assisted instruction, database management, software engineering, operating systems, neural networks. *Unit head:* Dr. Imao Chen, Chair, 256-851-5570.

Alcorn State University, School of Graduate Studies, School of Arts and Sciences, Department of Mathematical Sciences, Lorman, MS 39096-9402. Offers computer and information sciences (MS).

American University, College of Arts and Sciences, Department of Computer Science and Information Systems, Program in Computer Science, Washington, DC 20016-8001. Offers MS. Part-time and evening/weekend programs available. *Faculty:* 16 full-time (5 women), 16 part-time (1 woman). *Students:* 22 full-time (6 women), 17 part-time (7 women); includes 6 minority (4 African Americans, 2 Asian Americans or Pacific Islanders), 25 international. 65 applicants, 83% accepted. In 1998, 21 degrees awarded. *Degree requirements:* For master's, computer language, thesis or alternative, comprehensive exam required. *Entrance requirements:* For master's, minimum GPA of 3.0. *Average time to degree:* Master's–2 years full-time, 4 years part-time. *Application deadline:* For fall admission, 2/1 (priority date); for spring admission, 10/1. Applications are processed on a rolling basis. Application fee: $50. *Financial aid:* In 1998–99, 8 fellowships with full tuition reimbursements were awarded.; career-related internships or fieldwork, Federal Work-Study, institutionally-sponsored loans, tuition waivers (full and partial), and unspecified assistantships also available. Financial aid application deadline: 2/1. *Faculty research:* Artificial intelligence, database systems, software engineering, expert systems. Total annual research expenditures: $153,000. *Application contact:* Sonia McKie, Staff Assistant, 202-885-1486, Fax: 202-885-1479, E-mail: smckie@american.edu.

See in-depth description on page 697.

American University, College of Arts and Sciences, Department of Mathematics and Statistics, Program in Statistical Computing, Washington, DC 20016-8001. Offers MS. Part-time and evening/weekend programs available. *Faculty:* 19 full-time (6 women), 5 part-time (3 women). In 1998, 1 degree awarded. *Degree requirements:* For master's, one foreign language required, (computer language can substitute), thesis optional. *Entrance requirements:* For master's, BA in mathematics. *Application deadline:* For fall admission, 2/1; for spring admission, 10/1. Application fee: $50. *Financial aid:* Fellowships, teaching assistantships, career-related internships or fieldwork, Federal Work-Study, and institutionally-sponsored loans available. Aid available to part-time students. Financial aid application deadline: 2/1. *Faculty research:* Data analysis; random processes; environmental, meteorological, and biological applications. *Unit head:* Dr. Virginia Stallings, Chair, Department of Mathematics and Statistics, 202-885-3166, Fax: 202-885-3155.

American University in Cairo, Graduate Studies and Research, School of Sciences and Engineering, Department of Computer Science, Cairo, 11511, Egypt. Offers MS. *Students:* 21. *Degree requirements:* For master's, thesis required. *Entrance requirements:* For master's, English entrance exam and/or TOEFL. *Application deadline:* For fall admission, 3/31 (priority date); for spring admission, 1/10 (priority date). Application fee: $45. *Financial aid:* Fellowships, unspecified assistantships available. *Faculty research:* Software engineering, artificial intelligence, robotics, data and knowledge bases. *Unit head:* Dr. A. Goneid, Chair, 202-357-5308, Fax: 202-355-7565, E-mail: goneid@aucegypt.edu. *Application contact:* Mary Davidson, Coordinator of Student Affairs, 212-730-8800, Fax: 212-730-1600, E-mail: davidson@aucnyo.edu.

Angelo State University, Graduate School, College of Professional Studies, Department of Computer Science, San Angelo, TX 76909. Offers MBA. Part-time and evening/weekend programs available. *Faculty:* 4 full-time (1 woman). *Degree requirements:* For master's, computer language, comprehensive exam required, thesis optional, foreign language not required. *Entrance requirements:* For master's, GMAT, GRE General Test, minimum GPA of 2.5. *Application deadline:* For fall admission, 8/7 (priority date); for spring admission, 1/2. Applications are processed on a rolling basis. Application fee: $25 ($50 for international students). Tuition, state resident: part-time $38 per semester hour. Tuition, nonresident: part-time $249 per semester hour. Required fees: $40 per semester hour. $71 per semester. Tuition and fees vary according to degree level. *Financial aid:* Fellowships, Federal Work-Study and tuition waivers (partial) available. Aid available to part-time students. Financial aid application deadline: 8/1. *Unit head:* Dr. Fred Homeyer, Head, 915-942-2101.

Appalachian State University, Cratis D. Williams Graduate School, College of Arts and Sciences, Department of Computer Science, Boone, NC 28608. Offers MS. *Faculty:* 8 full-time (3 women). *Students:* 5 full-time (1 woman), 3 part-time, 1 international. 6 applicants, 17% accepted. In 1998, 1 degree awarded (100% found work related to degree). *Degree requirements:* For master's, one foreign language (computer language can substitute), comprehensive exam required, thesis not required. *Entrance requirements:* For master's, GRE General Test. *Average time to degree:* Master's–2 years full-time, 3 years part-time. *Application deadline:* For fall admission, 7/15 (priority date). Application fee: $35. *Financial aid:* In 1998–99, fellowships (averaging $2,000 per year), 6 research assistantships (averaging $6,750 per year) were awarded.; teaching assistantships, scholarships and unspecified assistantships also available. Financial aid application deadline: 7/15; financial aid applicants required to submit FAFSA. *Unit head:* Dr. Ray Russell, Chairperson, 828-262-2612. *Application contact:* Dr. James Wilkes, Adviser, 828-262-3050, Fax: 828-265-8617, E-mail: wilkesjt@appstate.edu.

Arizona State University, Graduate College, College of Engineering and Applied Sciences, Department of Computer Science and Engineering, Tempe, AZ 85287. Offers computer science (MCS, MS, PhD). *Faculty:* 35 full-time (5 women), 1 part-time (0 women). *Students:* 185 full-time (44 women), 119 part-time (30 women); includes 22 minority (4 African Americans, 14 Asian Americans or Pacific Islanders, 3 Hispanic Americans, 1 Native American), 183 international. Average age 28. 564 applicants, 48% accepted. In 1998, 106 master's, 6 doctorates awarded. *Degree requirements:* For master's, thesis or alternative required; for doctorate, dissertation required. *Entrance requirements:* For master's and doctorate, GRE General Test (recommended). Application fee: $45. *Financial aid:* Fellowships available. *Faculty research:* Software engineering, graphics, computer-aided geometric design microprocessor applications, digital system design. *Unit head:* Dr. Stephen Yau, Chair, 480-965-3190, E-mail: cse.graduate.office@asu.edu. *Application contact:* Dr. Ben Huey, Assistant Chair, 480-965-3190.

See in-depth description on page 699.

Arkansas State University, Graduate School, College of Arts and Sciences, Department of Computer Science and Mathematics, Jonesboro, State University, AR 72467. Offers computer science (MS); mathematics (MS, MSE). Part-time programs available. *Faculty:* 14 full-time (2 women). *Students:* 11 full-time (1 woman), 12 part-time (6 women), 6 international. Average age 26. In 1998, 12 degrees awarded. *Degree requirements:* For master's, thesis or alternative, comprehensive exam required. *Entrance requirements:* For master's, GRE General Test or MAT, appropriate bachelor's degree. *Application deadline:* For fall admission, 7/1 (priority date); for spring admission, 11/15 (priority date). Applications are processed on a rolling basis. Application fee: $15 ($25 for international students). *Financial aid:* Teaching assistantships available. Aid available to part-time students. Financial aid application deadline: 7/1; financial aid applicants required to submit FAFSA. *Unit head:* Dr. Roger Abernathy, Chair, 870-972-3090, Fax: 870-972-3950, E-mail: raber@caddo.astate.edu.

Auburn University, Graduate School, College of Engineering, Department of Computer Science and Engineering, Auburn, Auburn University, AL 36849-0002. Offers MCSE, MS, PhD. Part-time programs available. *Faculty:* 12 full-time (2 women). *Students:* 38 full-time (11 women), 29 part-time (11 women); includes 12 minority (8 African Americans, 3 Asian Americans or Pacific Islanders, 1 Hispanic American), 24 international. 61 applicants, 39% accepted. In 1998, 25 master's, 5 doctorates awarded. *Degree requirements:* For master's, thesis (MS) required; for doctorate, dissertation required, foreign language not required. *Entrance requirements:* For master's, GRE General Test, GRE Subject Test; for doctorate, GRE General Test (minimum score of 400 on each section required), GRE Subject Test. *Application deadline:* For fall admission, 9/1; for spring admission, 3/1. Applications are processed on a rolling basis. Application fee: $25 ($50 for international students). Tuition, state resident: full-time $2,760; part-time $76 per credit hour. Tuition, nonresident: full-time $8,280; part-time $228 per credit hour. *Financial aid:* Research assistantships, teaching assistantships, Federal Work-Study available. Aid available to part-time students. Financial aid application deadline: 3/15. *Faculty research:* Parallelizable, scalable software translations; graphical representations of algorithms, structures, and processes; graph drawing. Total annual research expenditures: $400,000. *Unit head:* Dr. James Cross, Head, 334-844-4330. *Application contact:* Dr. John F. Pritchett, Dean of the Graduate School, 334-844-4700.

Azusa Pacific University, Graduate Studies, College of Liberal Arts and Sciences, Department of Computer Science, Azusa, CA 91702-7000. Offers applied computer science and technology (MS), including client/server technology, computer information systems, end-user support, inter-emphasis, software engineering, technical programming, telecommunications; client/server technology (Certificate); computer information systems (Certificate); computer science (Certificate); end-user training and support (Certificate); software engineering (MSE, Certificate); technical programming (Certificate); telecommunications (Certificate). Part-time and evening/weekend programs available. *Faculty:* 9 full-time (1 woman), 10 part-time (2 women). *Students:* 187. 126 applicants, 94% accepted. In 1998, 62 degrees awarded. *Degree requirements:* For master's, computer language, thesis or alternative, project required, foreign language not required. *Entrance requirements:* For master's, minimum GPA of 3.0; proficiency in one programming language, college-level algebra, and applied calculus. *Application deadline:* For fall admission, 9/1 (priority date). Applications are processed on a rolling basis. Application fee: $45 ($65 for international students). *Financial aid:* Teaching assistantships, career-related internships or fieldwork available. Aid available to part-time students. *Faculty research:* Applied artificial intelligence, programming languages, engineering, database systems. *Unit head:* Dr. Samuel E. Sambasivam, Acting Chairman, 626-815-5476, Fax: 626-815-5323. *Application contact:* Dr. Samuel E. Sambasivam, Acting Chairman, 626-815-5476, Fax: 626-815-5323.

See in-depth description on page 701.

Ball State University, Graduate School, College of Sciences and Humanities, Department of Computer Science, Muncie, IN 47306-1099. Offers MA, MS. *Faculty:* 10. *Students:* 46 full-time (11 women), 33 part-time (9 women). Average age 30. 62 applicants, 85% accepted. In 1998, 45 degrees awarded. *Degree requirements:* Foreign language not required. Application fee: $15 ($25 for international students). *Financial aid:* Teaching assistantships available. *Faculty research:* Numerical methods, programmer productivity, graphics. *Unit head:* Dr. Clinton Fuelling, Chairman, 765-285-8641, E-mail: cfuelling@bsu.edu.

Baylor University, Graduate School, School of Engineering and Computer Science, Waco, TX 76798. Offers computer science (MS). Part-time programs available. *Students:* 13 full-time (4 women), 2 part-time; includes 2 minority (1 Asian American or Pacific Islander, 1 Hispanic American), 8 international. In 1998, 3 degrees awarded. *Degree requirements:* For master's, computer language required, thesis optional, foreign language not required. *Entrance requirements:* For master's, GRE General Test (minimum combined score of 1050 required), minimum GPA of 3.0. *Application deadline:* For fall admission, 8/1; for spring admission, 12/1. Applications are processed on a rolling basis. Application fee: $25. *Financial aid:* Teaching assistantships available. Financial aid application deadline: 3/15. *Faculty research:* Database systems, advanced architecture, operations research. *Unit head:* Dr. Greg Speegle, Director of Graduate Studies, 254-710-3871, Fax: 254-710-3839, E-mail: greg_speegle@baylor.edu. *Application contact:* Suzanne Keener, Administrative Assistant, 254-710-3588, Fax: 254-710-3870, E-mail: suzanne_keener@baylor.edu.

Boise State University, Graduate College, College of Arts and Sciences, Program in Computer Science, Boise, ID 83725-0399. Offers MS. Part-time programs available. *Faculty:* 27 full-time (5 women). *Students:* 2 full-time (0 women), 7 part-time (2 women); includes 1 minority (Hispanic American), 2 international. Average age 36. 6 applicants, 100% accepted. *Degree requirements:* For master's, computer language, thesis, written comprehensive exam required. *Entrance requirements:* For master's, GRE General Test, minimum GPA of 3.0. *Application deadline:* For fall admission, 7/23 (priority date); for spring admission, 11/24. Applications are processed on a rolling basis. Application fee: $20 ($30 for international students). Electronic applications accepted. *Financial aid:* Career-related internships or fieldwork, Federal Work-Study, and institutionally-sponsored loans available. Aid available to part-time students. Financial aid application deadline: 3/1. *Unit head:* Dr. Alex Feldman, Coordinator, 208-426-3374. *Application contact:* Dr. Alex Feldman, Coordinator, 208-426-3374.

Boston University, Graduate School of Arts and Sciences, Department of Computer Science, Boston, MA 02215. Offers MA, PhD. Part-time programs available. *Faculty:* 11 full-time (1 woman), 2 part-time (0 women). *Students:* 39 full-time (7 women), 7 part-time (1 woman); includes 1 minority (Asian American or Pacific Islander), 28 international. Average age 28. 170 applicants, 56% accepted. In 1998, 164 master's awarded (72% found work related to degree, 28% continued full-time study); 5 doctorates awarded (100% found work related to degree). *Degree requirements:* For master's, one foreign language, project required, thesis optional; for doctorate, one foreign language, dissertation, oral and written qualifying exams required. *Entrance requirements:* For master's and doctorate, GRE General Test, TOEFL (minimum score of 550 required). *Average time to degree:* Master's–1.75 years full-time; doctorate 6 years full-time. *Application deadline:* For fall admission, 6/1 (priority date); for spring admission, 10/15. Applications are processed on a rolling basis. Application fee: $50. Tuition: Full-time $23,770; part-time $743 per credit. Required fees: $220. Tuition and fees vary according to class time, course level, campus/location and program. *Financial aid:* In 1998–99, 24 students received aid, including 9 fellowships, 8 research assistantships; teaching assistantships, Federal Work-Study and scholarships also available. Aid available to part-time students. Financial aid application deadline: 1/15; financial aid applicants required to submit FAFSA. *Faculty research:* Networking, real time systems, computational complexity, database systems, mathematical logic, algorithmic information theory. *Unit head:* Wayne Snyder, Acting Chairman, 617-353-3840, Fax: 617-353-6457, E-mail: snyder@bu.edu.

Boston University, Metropolitan College, Program in Computer Science, Boston, MA 02215. Offers computer information systems (MS); computer science (MS); telecommunications (MS). Part-time and evening/weekend programs available. *Faculty:* 8 full-time (2 women), 64 part-time. *Students:* 11 full-time (0 women), 406 part-time (84 women); includes 73 minority (11 African Americans, 59 Asian Americans or Pacific Islanders, 3 Hispanic Americans), 21 international. Average age 33. 118 applicants, 95% accepted. In 1998, 75 degrees awarded. *Degree requirements:* For master's, computer language required, foreign language and thesis not required. *Average time to degree:* Master's–1 year full-time, 2 years part-time. *Application deadline:* Applications are processed on a rolling basis. Application fee: $50. Tuition: Part-time $508 per credit. Required fees: $40 per semester. Part-time tuition and fees vary according to class time. *Financial aid:* In 1998–99, 12 students received aid, including 2 research assistantships; career-related internships or fieldwork, Federal Work-Study, and tuition waivers (full and partial) also available. Aid available to part-time students. *Faculty research:* Software engineering, information systems architecture, process control, operating systems, parallel processing, object-oriented methods. *Unit head:* Dr. Tanya Zlateva, Chairman, 617-353-2566. *Application contact:* Administrative Secretary, 617-353-2566, Fax: 617-353-2367, E-mail: csinfo@bu.edu.

Bowie State University, Graduate Programs, Program in Computer Science, Bowie, MD 20715-9465. Offers MS. Part-time and evening/weekend programs available. *Degree requirements:* For master's, research paper, written comprehensive exam required, thesis optional. *Faculty research:* Holographics, launch vehicle ground truth ephemeras.

Bowling Green State University, Graduate College, College of Arts and Sciences, Department of Computer Science, Bowling Green, OH 43403. Offers MS. Part-time and evening/weekend programs available. *Faculty:* 10 full-time (3 women). *Students:* 28 full-time (9 women), 7 part-time (5 women); includes 2 minority (both Asian Americans or Pacific Islanders), 18 international. 91 applicants, 34% accepted. In 1998, 23 degrees awarded. *Degree requirements:* For master's, thesis or alternative required, foreign language not required. *Entrance requirements:* For master's, GRE General Test, TOEFL (minimum score of 550 required). Application fee: $30. Electronic applications accepted. *Financial aid:* Research assistantships with full tuition reimbursements, teaching assistantships with full tuition reimbursements, career-related internships or fieldwork, tuition waivers (full and partial), and unspecified assistantships available. Financial aid applicants required to submit FAFSA. *Faculty research:* Artificial intelligence, real time and concurrent programming languages, behavioral aspects of computing, network protocols. *Unit head:* Dr. Ron Lancaster, Chair, 419-372-2337. *Application contact:* Dr. Sub Ramakrishnan, Graduate Coordinator, 419-372-8783.

Bradley University, Graduate School, College of Liberal Arts and Sciences, Department of Computer Science, Peoria, IL 61625-0002. Offers computer information systems (MS); computer science (MS). Part-time and evening/weekend programs available. *Degree requirements:* For master's, thesis or alternative, comprehensive exam required, foreign language not required. *Entrance requirements:* For master's, TOEFL (minimum score of 500 required).

Brandeis University, Graduate School of Arts and Sciences, Michtom School of Computer Science, Waltham, MA 02454-9110. Offers MA, PhD. Part-time programs available. *Faculty:* 12 full-time (1 woman), 1 part-time. *Students:* 41 full-time (9 women), 4 part-time; includes 9 minority (all Asian Americans or Pacific Islanders), 11 international. 57 applicants, 68% accepted. In 1998, 9 master's, 1 doctorate awarded. *Degree requirements:* For doctorate, dissertation, general exam required. *Entrance requirements:* For master's and doctorate, GRE. *Application deadline:* For fall admission, 3/1. Application fee: $60. *Financial aid:* In 1998–99, 33 students received aid; fellowships, research assistantships, teaching assistantships, institutionally-sponsored loans, scholarships, and tuition waivers (full and partial) available. Aid available to part-time students. Financial aid application deadline: 4/15; financial aid applicants required to submit CSS PROFILE or FAFSA. *Faculty research:* Artificial intelligence, programming languages, parallel computing, computer linguistics, data compression. *Unit head:* Dr. James Pustejovsky, Director of Graduate Studies, 781-736-2709, Fax: 781-736-2741. *Application contact:* Myrna Fox, Department Administrator, 781-736-2701.

Announcement: Faculty and graduate student research in computer science at Brandeis University includes artificial intelligence, machine learning, logic programming, parallel computing, data compression, information theory, algorithms design, operating systems, and the semantics of programming languages. A typical PhD program consists of 2 years of graduate courses, the general examination, and 2 or 3 years of dissertation research. Master's students typically complete a 2-year program of 10 courses and a master's project. Brandeis University is located in the Cambridge/Boston area and provides a rich environment for academic, cultural, and recreational activities. For more information, call 781-736-2701 or visit the Web site at http://www.cs.brandeis.edu.

Bridgewater State College, Graduate School, School of Arts and Sciences, Department of Mathematics and Computer Science, Bridgewater, MA 02325-0001. Offers computer science (MS); mathematics (MAT). *Accreditation:* NCATE (one or more programs are accredited). Part-time and evening/weekend programs available. *Entrance requirements:* For master's, GRE General Test. *Application deadline:* For fall admission, 3/1; for spring admission, 10/1. Application fee: $25. *Unit head:* John Guest, Director, 305-899-3300. *Application contact:* Graduate School, 508-697-1300.

Brigham Young University, Graduate Studies, College of Physical and Mathematical Sciences, Department of Computer Science, Provo, UT 84602-1001. Offers MS, PhD. *Faculty:* 24 full-time (0 women). *Students:* 42 full-time (11 women), 67 part-time (3 women); includes 5 minority (3 Asian Americans or Pacific Islanders, 2 Hispanic Americans), 37 international. Average age 26. 62 applicants, 50% accepted. In 1998, 32 master's awarded (85% found work related to degree, 15% continued full-time study); 2 doctorates awarded (100% found work related to degree). Terminal master's awarded for partial completion of doctoral program. *Degree requirements:* For master's and doctorate, thesis/dissertation required, foreign language not required. *Entrance requirements:* For master's, GRE General Test, minimum GPA of 3.0 in last 60 hours; for doctorate, GRE General Test, GRE Subject Test. *Average time to degree:* Master's–3 years full-time; doctorate–5 years full-time. *Application deadline:* For fall admission, 2/20; for winter admission, 5/15; for spring admission, 9/15. Application fee: $30. Tuition: Full-time $3,330; part-time $185 per credit hour. Tuition and fees vary according to program and student's religious affiliation. *Financial aid:* In 1998–99, 8 teaching assistantships with full tuition reimbursements (averaging $13,500 per year) were awarded.; research assistantships, scholarships and tuition waivers (full and partial) also available. Financial aid application deadline: 2/20. *Faculty research:* Software development, graphics, image processing, neural networks and machine learning. Total annual research expenditures: $350,000. *Unit head:* Dr. William A. Barrett, Chair, 801-378-3027, Fax: 801-378-7775, E-mail: barrett@cs.byu.edu. *Application contact:* Scott Woodfield, Graduate Coordinator, 801-378-2915, Fax: 801-378-7775, E-mail: gradinfo@cs.byu.edu.

Brooklyn College of the City University of New York, Division of Graduate Studies, Department of Computer and Information Science, Brooklyn, NY 11210-2889. Offers computer and information science (MA, PhD); computer science and health science (MS); economics and computer and information science (MPS); information systems (MS). Part-time and evening/weekend programs available. *Faculty:* 10 full-time (2 women), 11 part-time (1 woman). *Students:* 23 full-time (10 women), 184 part-time (62 women); includes 93 minority (33 African Americans, 56 Asian Americans or Pacific Islanders, 4 Hispanic Americans), 37 international. Average age 27. In 1998, 42 degrees awarded. *Degree requirements:* For master's, computer language, comprehensive exam or thesis required. *Entrance requirements:* For master's, TOEFL (minimum score of 525 required), previous course work in computer science. *Application deadline:* For fall admission, 3/1; for spring admission, 11/1. Application fee: $40. *Financial aid:* Fellowships, research assistantships, teaching assistantships, career-related internships or fieldwork, Federal Work-Study, institutionally-sponsored loans, scholarships, and tuition waivers (partial) available. Aid available to part-time students. Financial aid application deadline: 5/1; financial aid applicants

required to submit FAFSA. *Faculty research:* Networks and distributed systems, programming languages, modeling and computer applications, algorithms, artificial intelligence, theoretical computer science. *Unit head:* Dr. Aaron H. Tenenbaum, Chairperson, 718-951-5657. *Application contact:* Gerald Weiss, Graduate Counselor, 718-951-5217, Fax: 718-951-4842, E-mail: weiss@sci.brooklyn.cuny.edu.

Brown University, Graduate School, Department of Computer Science, Providence, RI 02912. Offers Sc M, PhD. *Degree requirements:* For master's, thesis or alternative required, foreign language not required; for doctorate, dissertation, comprehensive exam required. *Entrance requirements:* For master's and doctorate, GRE General Test, GRE Subject Test.

California Institute of Technology, Division of Engineering and Applied Science, Option in Computer Science, Pasadena, CA 91125-0001. Offers MS, PhD. *Faculty:* 5 full-time (0 women). *Students:* 23 full-time (5 women), 8 international. 286 applicants, 1% accepted. In 1998, 8 master's, 8 doctorates awarded. *Degree requirements:* For master's and doctorate, thesis/ dissertation required, foreign language not required. *Application deadline:* For fall admission, 1/ 15. Application fee: $0. *Faculty research:* VLSI systems, concurrent computation, high-level programming languages, signal and image processing, graphics. *Unit head:* Dr. K. Mani Chandy, Executive Officer, 626-395-6559.

See in-depth description on page 703.

California Polytechnic State University, San Luis Obispo, College of Engineering, Department of Computer Science, San Luis Obispo, CA 93407. Offers MSCS. Part-time programs available. *Faculty:* 19 full-time (2 women), 22 part-time (3 women). *Students:* 17 full-time (2 women), 19 part-time (4 women); includes 6 Asian Americans or Pacific Islanders, 6 international. 49 applicants, 24% accepted. In 1998, 14 degrees awarded. *Degree requirements:* For master's, thesis required, foreign language not required. *Entrance requirements:* For master's, GRE General Test (minimum combined score of 1650 on three sections required, 400 on verbal), TOEFL (minimum score of 550 required), TWE (minimum score of 5 required), minimum GPA of 3.0 in last 90 quarter units. *Average time to degree:* Master's–2.5 years full-time, 4.5 years part-time. *Application deadline:* For fall admission, 5/31 (priority date); for spring admission, 2/28. Applications are processed on a rolling basis. Application fee: $55. Electronic applications accepted. Tuition, nonresident: part-time $164 per unit. Required fees: $531 per quarter. *Financial aid:* In 1998–99, 20 teaching assistantships were awarded.; career-related internships or fieldwork, Federal Work-Study, and institutionally-sponsored loans also available. Financial aid application deadline: 3/2; financial aid applicants required to submit FAFSA. *Faculty research:* Computer systems, software, graphics, hardware design, expert systems, software engineering and computer networks. *Unit head:* Dr. James Beug, Chair, 805-756-2824, Fax: 805-756-2956, E-mail: jlbeug@calpoly.edu.

California State Polytechnic University, Pomona, Graduate Studies, College of Science, Program in Computer Science, Pomona, CA 91768-2557. Offers MS. Part-time and evening/ weekend programs available. *Faculty:* 13. *Students:* 7 full-time (2 women), 27 part-time (11 women); includes 20 minority (19 Asian Americans or Pacific Islanders, 1 Hispanic American), 9 international. Average age 31. 64 applicants, 20% accepted. In 1998, 5 degrees awarded. *Degree requirements:* For master's, thesis required. *Entrance requirements:* For master's, GRE General Test. *Application deadline:* Applications are processed on a rolling basis. Application fee: $55. Tuition, nonresident: part-time $164 per unit. *Financial aid:* In 1998–99, 4 students received aid. Career-related internships or fieldwork, Federal Work-Study, and institutionally-sponsored loans available. Aid available to part-time students. Financial aid application deadline: 3/2; financial aid applicants required to submit FAFSA. *Unit head:* Dr. Norton Riley, Coordinator, 909-869-3444, E-mail: hnriley@csupomona.edu.

California State University, Chico, Graduate School, College of Engineering, Computer Science, and Technology, Department of Computer Science, Chico, CA 95929-0722. Offers MS. *Faculty:* 15 full-time (3 women), 5 part-time (1 woman). *Students:* 50 full-time (12 women), 20 part-time (7 women); includes 14 minority (10 Asian Americans or Pacific Islanders, 4 Hispanic Americans), 29 international. Average age 31. In 1998, 65 degrees awarded. *Degree requirements:* For master's, computer language (for some programs), thesis or alternative, oral exam required, foreign language not required. *Entrance requirements:* For master's, GRE General Test. *Application deadline:* For fall admission, 4/1. Applications are processed on a rolling basis. Application fee: $55. *Financial aid:* Fellowships, research assistantships, teaching assistantships, career-related internships or fieldwork available. *Unit head:* John Zenor, Chair, 530-898-6442. *Application contact:* Ralph Huntsinger, Graduate Adviser, 530-898-5740.

California State University, Fresno, Division of Graduate Studies, School of Engineering, Department of Computer Science, Fresno, CA 93740-0057. Offers MS. Part-time and evening/ weekend programs available. *Faculty:* 8 full-time (1 woman). *Students:* 21 full-time (3 women), 30 part-time (18 women); includes 10 minority (2 African Americans, 6 Asian Americans or Pacific Islanders, 2 Hispanic Americans), 34 international. Average age 31. 79 applicants, 78% accepted. In 1998, 11 degrees awarded. *Degree requirements:* For master's, thesis or alternative required, foreign language not required. *Entrance requirements:* For master's, GRE General Test, TOEFL (minimum score of 550 required), minimum GPA of 3.0. *Average time to degree:* Master's–3.5 years full-time. *Application deadline:* For fall admission, 8/1 (priority date); for spring admission, 12/1. Applications are processed on a rolling basis. Application fee: $55. Electronic applications accepted. Tuition, nonresident: part-time $246 per unit. Required fees: $1,906; $620 per semester. *Financial aid:* In 1998–99, 8 teaching assistantships were awarded.; fellowships, research assistantships, career-related internships or fieldwork, Federal Work-Study, scholarships, and unspecified assistantships also available. Financial aid application deadline: 3/1; financial aid applicants required to submit FAFSA. *Faculty research:* Software design, parallel processing, computer engineering. *Unit head:* Dr. Brent Auernheimer, Chair, 559-278-4373, Fax: 559-278-4197. *Application contact:* Lan Jin, Coordinator, 559-278-4373, Fax: 559-278-4197, E-mail: lan_jin@csufresno.edu.

California State University, Fullerton, Graduate Studies, School of Engineering and Computer Science, Department of Computer Science, Fullerton, CA 92834-9480. Offers applications administrative information systems (MS); applications mathematical methods (MS); computer science (MS); information processing systems (MS). Part-time programs available. *Faculty:* 11 full-time, 32 part-time. *Students:* 10 full-time (4 women), 98 part-time (27 women); includes 47 minority (2 African Americans, 42 Asian Americans or Pacific Islanders, 3 Hispanic Americans), 24 international. Average age 32. 115 applicants, 38% accepted. In 1998, 24 degrees awarded. *Degree requirements:* For master's, computer language, comprehensive exam, project or thesis required. *Entrance requirements:* For master's, GRE General Test (minimum combined score of 1100 required), minimum undergraduate GPA of 2.5. Application fee: $55. Tuition, nonresident: part-time $264 per unit. Required fees: $1,947; $1,281 per year. *Financial aid:* Career-related internships or fieldwork, Federal Work-Study, grants, and institutionally-sponsored loans available. Aid available to part-time students. Financial aid application deadline: 3/1. *Faculty research:* Software engineering, development of computer networks. *Unit head:* Dr. Nick Mousouris, Chair, 714-278-3700. *Application contact:* Dr. Susamma Barua, Adviser, 714-278-3700.

California State University, Hayward, Graduate Programs, School of Science, Department of Mathematics and Computer Science, Computer Science Program, Hayward, CA 94542-3000. Offers MS. *Faculty:* 17 full-time (2 women). *Students:* 81 full-time (52 women), 85 part-time (41 women); includes 137 minority (1 African American, 134 Asian Americans or Pacific Islanders, 2 Hispanic Americans), 2 international. 63 applicants, 70% accepted. In 1998, 24 degrees awarded. *Degree requirements:* For master's, computer language, comprehensive exam or thesis required. *Entrance requirements:* For master's, minimum GPA of 3.0 in field, 2.75 overall. Application fee: $55. Tuition, nonresident: part-time $164 per unit. Required fees: $587 per quarter. *Financial aid:* Career-related internships or fieldwork, Federal Work-Study, and institutionally-sponsored loans available. Aid available to part-time students. Financial aid application deadline: 3/1. *Unit head:* Donald L. Wolitzer, Coordinator, 510-885-

Computer Science

California State University, Hayward (continued)
3467. *Application contact:* Jennifer Rice, Graduate Program Assistant, 510-885-3286, Fax: 510-885-4795, E-mail: gradprograms@csuhayward.edu.

California State University, Long Beach, Graduate Studies, College of Engineering, Department of Computer Engineering and Computer Science, Long Beach, CA 90840. Offers computer engineering (MS); computer science (MS). Part-time programs available. *Faculty:* 17 full-time (3 women), 8 part-time (0 women). *Students:* 127 full-time (35 women), 189 part-time (51 women); includes 137 minority (5 African Americans, 124 Asian Americans or Pacific Islanders, 8 Hispanic Americans), 101 international. Average age 31. 386 applicants, 55% accepted. In 1998, 48 degrees awarded. *Degree requirements:* For master's, computer language, thesis or alternative required, foreign language not required. *Entrance requirements:* For master's, TOEFL (minimum score of 550 required). *Application deadline:* For fall admission, 8/1; for spring admission, 12/1. Application fee: $55. Electronic applications accepted. Tuition, nonresident: part-time $246 per unit. Required fees: $569 per semester. Tuition and fees vary according to course load. *Financial aid:* Teaching assistantships, Federal Work-Study, grants, institutionally-sponsored loans, and unspecified assistantships available. Financial aid application deadline: 3/2. *Faculty research:* Artificial intelligence, software engineering, computer simulation and modeling, user-interface design, networking. *Unit head:* Dr. Sandra Cynar, Chair, 562-985-4285, Fax: 562-985-7561, E-mail: cynar@csulb.edu. *Application contact:* Dr. Dar-Biau Liu, Graduate Adviser, 562-985-1594, Fax: 562-985-7561, E-mail: liu@csulb.edu.

California State University, Northridge, Graduate Studies, College of Engineering and Computer Science, Department of Computer Science, Northridge, CA 91330. Offers MS. Part-time and evening/weekend programs available. *Faculty:* 15 full-time, 25 part-time. *Students:* 33 full-time (16 women), 62 part-time (21 women); includes 22 minority (1 African American, 16 Asian Americans or Pacific Islanders, 5 Hispanic Americans), 26 international. Average age 31. 105 applicants, 61% accepted. In 1998, 14 degrees awarded. *Degree requirements:* For master's, computer language, thesis required, foreign language not required. *Entrance requirements:* For master's, GRE General Test, TOEFL, minimum GPA of 2.5. *Application deadline:* For fall admission, 11/30. Application fee: $55. Tuition, nonresident: part-time $246 per unit. International tuition: $7,874 full-time. Required fees: $1,970. Tuition and fees vary according to course load. *Financial aid:* Application deadline: 3/1. *Faculty research:* Radar data processing. *Unit head:* Dr. Diane Schwartz, Chair, 818-677-3398.

California State University, Sacramento, Graduate Studies, School of Engineering and Computer Science, Department of Computer Science, Sacramento, CA 95819-6048. Offers computer systems (MS); software engineering (MS). Part-time and evening/weekend programs available. *Degree requirements:* For master's, computer language, thesis or alternative, writing proficiency exam required, foreign language not required. *Entrance requirements:* For master's, TOEFL (minimum score of 550 required). *Application deadline:* For fall admission, 4/15; for spring admission, 11/1. Application fee: $55. *Financial aid:* Research assistantships, teaching assistantships, career-related internships or fieldwork and Federal Work-Study available. Aid available to part-time students. Financial aid application deadline: 3/1. *Unit head:* Dr. Ann Louis Radimsky, Chair, 916-278-5843. *Application contact:* Dr. Fred Blackwell, Coordinator, 916-278-6834.

California State University, San Bernardino, Graduate Studies, School of Natural Sciences, Department of Computer Science, San Bernardino, CA 92407-2397. Offers MS. *Degree requirements:* Foreign language not required.

California State University, San Marcos, Program in Computer Science, San Marcos, CA 92096-0001. Offers MS. *Students:* 4 full-time (1 woman), 8 part-time (2 women). *Degree requirements:* For master's, computer language, thesis required (for some programs). *Entrance requirements:* For master's, GRE General Test, GRE Subject Test (recommended), TOEFL (minimum score of 550 required). *Average time to degree:* Master's–2 years full-time. *Application deadline:* For fall admission, 6/30; for spring admission, 11/30. Application fee: $55. *Financial aid:* In 1998–99, 2 teaching assistantships (averaging $4,400 per year) were awarded. *Faculty research:* Networks, multimedia, parallel algorithms, software engineering, artificial intelligence. *Unit head:* Rochelle L. Boehning, Chair, 760-750-4118, Fax: 760-750-3439, E-mail: chelle@csusm.edu. *Application contact:* JoAnn Espinosa, Program Support, 760-750-4118, Fax: 760-750-3439, E-mail: jespinoz@csusm.edu.

Carleton University, Faculty of Graduate Studies, Faculty of Science, School of Computer Science, Ottawa, ON K1S 5B6, Canada. Offers MCS, PhD. Part-time programs available. *Faculty:* 20 full-time (0 women). *Students:* 53 full-time (5 women), 39 part-time (10 women). Average age 31. In 1998, 18 master's, 3 doctorates awarded. *Degree requirements:* For master's, project required, thesis optional; for doctorate, dissertation, comprehensive exam required. *Entrance requirements:* For master's, TOEFL (minimum score of 550 required), honors degree; for doctorate, TOEFL (minimum score of 550 required). *Average time to degree:* Master's–2.1 years full-time, 2.9 years part-time; doctorate–5.7 years full-time. *Application deadline:* For fall admission, 2/1 (priority date). Applications are processed on a rolling basis. Application fee: $35. *Financial aid:* Application deadline: 3/1. *Faculty research:* Programming systems, theory of computing, computer applications, computer systems. Total annual research expenditures: $516,000. *Unit head:* E. Kranakis, Director, 613-520-2600 Ext. 4358, Fax: 613-520-4334, E-mail: scs@carleton.ca.

Carnegie Mellon University, Graduate School of Industrial Administration, Program in Electronic Commerce, Pittsburgh, PA 15213-3891. Offers MS. *Entrance requirements:* For master's, GRE General Test or GMAT, TOEFL. Application fee: $60. *Unit head:* Dr. Tridas Mukhopadhyay, Director. *Application contact:* Vickie Motz, Program Manager, 412-268-1322, Fax: 412-268-6837, E-mail: vmotz@andrew.cmu.edu.

See in-depth description on page 707.

Carnegie Mellon University, School of Computer Science, Department of Computer Science, Pittsburgh, PA 15213-3891. Offers algorithms, combinatorics, and optimization (PhD); computer science (PhD); pure and applied logic (PhD). *Faculty:* 81 full-time (13 women), 5 part-time (0 women). *Students:* 133 full-time (15 women), 9 part-time (1 woman); includes 8 minority (7 Asian Americans or Pacific Islanders, 1 Hispanic American), 62 international. Average age 27. In 1998, 19 doctorates awarded. *Degree requirements:* For doctorate, dissertation required, foreign language not required. *Entrance requirements:* For doctorate, GRE General Test, GRE Subject Test, TOEFL, BS in computer science or equivalent. *Application deadline:* For fall admission, 1/1. Application fee: $65. *Financial aid:* Fellowships, research assistantships, teaching assistantships, outside fellowships available. *Faculty research:* Software systems, theory of computations, artificial intelligence, computer systems, programming languages. Total annual research expenditures: $18.2 million. *Unit head:* Jim Morris, Head, 412-268-2574. *Application contact:* Martha Clarke, Admissions Coordinator, 412-268-3863, Fax: 412-681-5739, E-mail: grad_admin@cs.cmu.edu.

See in-depth description on page 705.

Carnegie Mellon University, School of Computer Science, Department of Human-Computer Interaction, Pittsburgh, PA 15213-3891. Offers MHCI. *Faculty:* 19 full-time (6 women), 1 part-time (0 women). *Students:* 20 full-time (9 women), 13 part-time (7 women); includes 3 minority (1 African American, 2 Asian Americans or Pacific Islanders), 8 international. Average age 27. In 1998, 11 degrees awarded. *Degree requirements:* For master's, foreign language and thesis not required. *Entrance requirements:* For master's, GRE General Test, GRE Subject Test. *Application deadline:* For fall admission, 2/1. Application fee: $50. Total annual research expenditures: $5.8 million. *Unit head:* Dan R. Olsen, Director, 412-268-2980. *Application contact:* Admissions Coordinator, 412-268-6493, E-mail: hcii-masters@cs.cmu.edu.

See in-depth description on page 709.

Carnegie Mellon University, School of Computer Science, Language Technologies Institute, Pittsburgh, PA 15213-3891. Offers MS, PhD. *Faculty:* 14 full-time (5 women). *Students:* 21 full-time (6 women), 4 part-time (1 woman); includes 2 minority (1 Asian American or Pacific Islander, 1 Hispanic American), 13 international. 54 applicants, 54% accepted. In 1998, 3 master's awarded (67% found work related to degree, 33% continued full-time study); 2 doctorates awarded (50% entered university research/teaching, 50% found other work related to degree). Terminal master's awarded for partial completion of doctoral program. *Degree requirements:* For master's, foreign language and thesis not required; for doctorate, dissertation required, foreign language not required. *Entrance requirements:* For master's and doctorate, GRE General Test, GRE Subject Test, TOEFL. *Average time to degree:* Master's–2 years full-time. *Application deadline:* For fall admission, 1/5. Application fee: $50. *Financial aid:* In 1998–99, research assistantships with tuition reimbursements (averaging $16,800 per year) *Faculty research:* Machine translation, natural language processing, speech and information retrieval, literacy. Total annual research expenditures: $2.2 million. *Unit head:* Jaime G. Carbonell, Director, 412-268-7279, Fax: 412-268-6298, E-mail: jgc@cs.cmu.edu. *Application contact:* Catherine Morrow, Admissions Coordinator, 412-268-6591, E-mail: ltp@cs.cmu.edu.

See in-depth description on page 713.

Case Western Reserve University, School of Graduate Studies, The Case School of Engineering, Department of Electrical, Systems, Computer Engineering and Science, Cleveland, OH 44106. Offers computer engineering and science (MS, PhD), including computer engineering, computing and information science; electrical engineering (MS, PhD); systems and control engineering (MS, PhD). Part-time and evening/weekend programs available. Postbaccalaureate distance learning degree programs offered (minimal on-campus study). *Faculty:* 28 full-time (2 women). *Students:* 73 full-time (13 women), 101 part-time (17 women). Average age 25. 484 applicants, 38% accepted. In 1998, 68 master's, 20 doctorates awarded. Terminal master's awarded for partial completion of doctoral program. *Degree requirements:* For master's and doctorate, thesis/dissertation required, foreign language not required. *Entrance requirements:* For master's and doctorate, GRE General Test, TOEFL (minimum score of 550 required). *Average time to degree:* Master's–2 years full-time, 3 years part-time; doctorate–4 years full-time, 5 years part-time. *Application deadline:* For fall admission, 3/1; for spring admission, 11/1. Applications are processed on a rolling basis. Application fee: $25. Electronic applications accepted. *Financial aid:* In 1998–99, 96 students received aid, including 70 research assistantships with full and partial tuition reimbursements available (averaging $15,600 per year), 33 teaching assistantships with full and partial tuition reimbursements available (averaging $10,170 per year); fellowships, career-related internships or fieldwork, Federal Work-Study, and institutionally-sponsored loans also available. Aid available to part-time students. Financial aid application deadline: 3/1. *Faculty research:* Microelectromechanical systems, control, artificial intelligence, mixed signals. Total annual research expenditures: $5.8 million. *Unit head:* Dr. Robert V. Edwards, Acting Chairman, 216-368-2800, Fax: 216-368-6888, E-mail: rve2@po.cwru.edu. *Application contact:* Elizabethanne M. Fuller, Department Assistant, 216-368-4080, Fax: 216-368-2668, E-mail: emf4@po.cwru.edu.

The Catholic University of America, School of Engineering, Department of Electrical Engineering and Computer Science, Washington, DC 20064. Offers MEE, MS Engr, D Engr, PhD. Part-time and evening/weekend programs available. *Faculty:* 7 full-time (0 women), 4 part-time (0 women). *Students:* 11 full-time (1 woman), 20 part-time (4 women); includes 10 minority (4 African Americans, 4 Asian Americans or Pacific Islanders, 2 Hispanic Americans), 10 international. Average age 30. 40 applicants, 50% accepted. In 1998, 11 master's, 1 doctorate awarded. *Degree requirements:* For master's, thesis optional, foreign language not required; for doctorate, dissertation, comprehensive and oral exams required, foreign language not required. *Entrance requirements:* For master's, TOEFL (minimum score of 550 required), minimum GPA of 3.0; for doctorate, TOEFL (minimum score of 550 required), minimum GPA of 3.4. *Application deadline:* For fall admission, 8/1 (priority date); for spring admission, 12/1. Applications are processed on a rolling basis. Application fee: $50. *Financial aid:* Research assistantships, career-related internships or fieldwork, Federal Work-Study, institutionally-sponsored loans, tuition waivers (full and partial), and unspecified assistantships available. Aid available to part-time students. Financial aid application deadline: 2/1. *Faculty research:* Signal and image processing, computer communications, robotics, intelligent controls, bioelectromagnetics, properties of materials. Total annual research expenditures: $320,000. *Unit head:* Dr. Charles Nguyen, Chair, 202-319-5193.

Central Michigan University, College of Graduate Studies, College of Science and Technology, Department of Computer Science, Mount Pleasant, MI 48859. Offers MS. *Faculty:* 16 full-time (1 woman). *Students:* 55 full-time (13 women), 14 part-time (5 women); includes 6 minority (all Asian Americans or Pacific Islanders), 53 international. Average age 30. In 1998, 20 degrees awarded. *Degree requirements:* For master's, computer language, thesis or alternative required, foreign language not required. *Entrance requirements:* For master's, TOEFL (minimum score of 550 required), minimum GPA of 2.5 in last 2 undergraduate years. *Application deadline:* For fall admission, 3/15 (priority date). Applications are processed on a rolling basis. Application fee: $30. Tuition, state resident: part-time $144 per credit hour. Tuition, nonresident: part-time $285 per credit hour. Required fees: $240 per semester. Tuition and fees vary according to degree level and program. *Financial aid:* In 1998–99, 2 research assistantships with tuition reimbursements, 16 teaching assistantships with tuition reimbursements were awarded.; fellowships with tuition reimbursements, career-related internships or fieldwork and Federal Work-Study also available. Financial aid application deadline: 3/7. *Faculty research:* Compiler construction, artificial intelligence, database theory, software engineering, operating systems. *Unit head:* Dr. Gongzhu Hu, Chairperson, 517-774-3774, Fax: 517-774-6652, E-mail: hu@cps.cmich.edu.

Christopher Newport University, Graduate Studies, Department of Physics, Computer Science, and Engineering, Newport News, VA 23606-2998. Offers applied physics and computer science (MS). Part-time and evening/weekend programs available. *Faculty:* 14 full-time (2 women), 1 part-time (0 women). *Students:* 2 full-time (1 woman), 25 part-time (3 women); includes 5 minority (3 African Americans, 1 Asian American or Pacific Islander, 1 Hispanic American) Average age 36. In 1998, 3 degrees awarded (100% found work related to degree). *Degree requirements:* For master's, computer language, comprehensive exam required, thesis optional, foreign language not required. *Entrance requirements:* For master's, GRE, minimum GPA of 3.0. *Average time to degree:* Master's–2 years full-time, 3 years part-time. *Application deadline:* For fall admission, 7/1 (priority date); for spring admission, 12/15. Applications are processed on a rolling basis. Application fee: $40. Electronic applications accepted. Tuition, state resident: part-time $145 per credit hour. Tuition, nonresident: part-time $351 per credit hour. Required fees: $20 per year. *Financial aid:* In 1998–99, 1 research assistantship with full tuition reimbursement (averaging $5,000 per year) was awarded.; career-related internships or fieldwork and Federal Work-Study also available. Aid available to part-time students. Financial aid application deadline: 3/1; financial aid applicants required to submit FAFSA. *Faculty research:* Advanced programming methodologies, experimental nuclear physics, computer architecture, semiconductor nanophysics, laser and optical fiber sensors. *Unit head:* Dr. David Hibler, Coordinator, 757-594-7360, Fax: 757-594-7919, E-mail: dhibler@pcs.cnu.edu. *Application contact:* Gary Clark, Graduate Admissions, 757-594-7993, Fax: 757-594-7333, E-mail: admit@cnu.edu.

City College of the City University of New York, Graduate School, School of Engineering, Department of Computer Sciences, New York, NY 10031-9198. Offers MS, PhD. *Students:* 75 full-time (36 women), 159 part-time (56 women). In 1998, 69 degrees awarded. *Degree requirements:* For master's, computer language required, thesis optional, foreign language not required; for doctorate, dissertation, comprehensive exams required. *Entrance requirements:* For master's, TOEFL (minimum score of 500 required); for doctorate, GRE General Test, TOEFL. *Application deadline:* Applications are processed on a rolling basis. Application fee: $40. *Financial aid:* Fellowships, teaching assistantships, Federal Work-Study and tuition waivers (partial) available. Aid available to part-time students. Financial aid application deadline: 6/1. *Faculty research:* Complexities of algebraic research, human issues in computer science, scientific computing, supercompilers, parallel algorithms. *Unit head:* Steven Lucci, Chairman, 212-650-6152. *Application contact:* Graduate Admissions Office, 212-650-6977.

Clark Atlanta University, School of Arts and Sciences, Department of Computer and Information Science, Atlanta, GA 30314. Offers MS. *Students:* 30 full-time (12 women), 10 part-time (4 women); includes 19 African Americans, 12 Asian Americans or Pacific Islanders, 9 international. In 1998, 23 degrees awarded. *Degree requirements:* For master's, one foreign language (computer language can substitute), thesis required. *Entrance requirements:* For master's, GRE General Test, minimum GPA of 2.5. *Application deadline:* For fall admission, 4/1; for spring admission, 11/1. Applications are processed on a rolling basis. Application fee: $40. *Financial aid:* Fellowships, research assistantships available. Financial aid application deadline: 4/30. *Unit head:* Dr. Kenneth Perry, Chairperson, 404-880-6951. *Application contact:* Michelle Clark-Davis, Graduate Program Assistant, 404-880-8709.

Clark Atlanta University, School of Arts and Sciences, Department of Mathematical Sciences, Atlanta, GA 30314. Offers applied mathematics (MS); computer science (MS). Part-time programs available. *Students:* 11 full-time (2 women), 4 part-time (1 woman); all minorities (all African Americans) *Degree requirements:* For master's, one foreign language (computer language can substitute), thesis required. *Entrance requirements:* For master's, GRE General Test, minimum GPA of 2.5. *Application deadline:* For fall admission, 4/1; for spring admission, 11/1. Applications are processed on a rolling basis. Application fee: $40. *Unit head:* Dr. Michael Bleicher, Chairperson, 404-880-8272. *Application contact:* Michelle Clark-Davis, Graduate Program Assistant, 404-880-8709.

Clarkson University, Graduate School, School of Science, Department of Mathematics and Computer Science, Potsdam, NY 13699. Offers computer science (MS); mathematics (MS, PhD). *Faculty:* 15 full-time (2 women), 1 part-time (0 women). *Students:* 17 full-time (9 women); includes 1 minority (Asian American or Pacific Islander), 9 international. Average age 27. 49 applicants, 45% accepted. In 1998, 3 master's, 1 doctorate awarded. Terminal master's awarded for partial completion of doctoral program. *Degree requirements:* For master's, foreign language and thesis not required; for doctorate, dissertation, departmental qualifying exam required, foreign language not required. *Entrance requirements:* For master's, GRE, TOEFL. *Application deadline:* For fall admission, 5/15 (priority date); for spring admission, 10/15 (priority date). Applications are processed on a rolling basis. Application fee: $25 ($35 for international students). Tuition: Part-time $661 per credit hour. Required fees: $215 per semester. *Financial aid:* In 1998–99, 2 research assistantships, 10 teaching assistantships were awarded.; fellowships *Faculty research:* Fiber optics, hydrodynamics, inverse scattering, nonlinear optics, nonlinear waves. Total annual research expenditures: $136,542. *Unit head:* Dr. David L. Powers, Chair, 315-268-2369, Fax: 315-268-6670, E-mail: dpowers@clarkson.edu.

Announcement: A program leading to the Master of Science in computer science is offered jointly by the Departments of Electrical and Computer Engineering and Mathematics and Computer Science at Clarkson University. With strengths in engineering and science, Clarkson offers students an exceptional opportunity to study computer science in an interdisciplinary environment. There is a wide variety of courses, ranging from theoretical topics in computer science to design and layout of VLSI components. Students with appropriate undergraduate backgrounds in computer science/engineering are invited to apply. For more information, contact Tess Casler, Clarkson University, Box 5802, Potsdam, NY 13699-5802 (e-mail: caslertc@clarkson.edu).

Clemson University, Graduate School, College of Engineering and Science, Department of Computer Science, Clemson, SC 29634. Offers MS, PhD. *Students:* 93 full-time (17 women), 21 part-time (7 women); includes 2 minority (1 African American, 1 Asian American or Pacific Islander), 69 international. 208 applicants, 65% accepted. In 1998, 29 master's, 2 doctorates awarded. Terminal master's awarded for partial completion of doctoral program. *Degree requirements:* For master's, thesis optional, foreign language not required; for doctorate, dissertation required, foreign language not required. *Entrance requirements:* For master's and doctorate, GRE General Test, TOEFL. *Application deadline:* For fall admission, 5/1; for spring admission, 10/1. Applications are processed on a rolling basis. Application fee: $35. *Financial aid:* Fellowships, research assistantships, teaching assistantships, institutionally-sponsored loans available. Financial aid application deadline: 3/1; financial aid applicants required to submit FAFSA. *Faculty research:* Parallel computation, performance modeling, operating systems, software engineering, design and analysis of algorithms. Total annual research expenditures: $537,017. *Unit head:* Dr. Stephen T. Hedetniemi, Chair, 864-656-5858, Fax: 864-656-0145, E-mail: hedet@cs.clemson.edu. *Application contact:* Dr. James Westall, Graduate Coordinator, 864-656-3444, Fax: 864-656-0145, E-mail: westall@clemson.edu.

See in-depth description on page 721.

The College of Saint Rose, Graduate Studies, School of Mathematics and Sciences, Program in Computer Information Systems, Albany, NY 12203-1419. Offers MS. Part-time and evening/weekend programs available. *Faculty:* 1 full-time (0 women). 9 applicants, 78% accepted. *Degree requirements:* Foreign language not required. *Application deadline:* For fall admission, 7/15 (priority date); for spring admission, 12/1. Applications are processed on a rolling basis. Application fee: $30. *Financial aid:* Research assistantships, career-related internships or fieldwork and tuition waivers (partial) available. Aid available to part-time students. Financial aid application deadline: 3/1; financial aid applicants required to submit FAFSA.

College of Staten Island of the City University of New York, Graduate Programs, Program in Computer Science, Staten Island, NY 10314-6600. Offers MS, PhD. Part-time and evening/weekend programs available. *Faculty:* 10 full-time (3 women), 4 part-time (1 woman). Average age 30. *Degree requirements:* For master's, thesis optional; for doctorate, dissertation required. *Entrance requirements:* For master's, previous undergraduate course work in computer science. *Application deadline:* For fall admission, 6/1 (priority date); for spring admission, 12/1. Applications are processed on a rolling basis. Application fee: $40. Tuition, state resident: full-time $4,350; part-time $185 per credit. Tuition, nonresident: full-time $7,600; part-time $320 per credit. Required fees: $53; $27 per term. *Financial aid:* Research assistantships, teaching assistantships available. *Faculty research:* Scientific visualization, knowledge engineering, image processing, performance evaluation, database networks, neural computing, distributed operating systems. *Unit head:* Dr. Miriam Tausner, Graduate Coordinator, 718-982-2845, E-mail: tausner@postbox.csi.cuny.edu. *Application contact:* Earl Teasley, Director of Admissions, 718-982-2010, Fax: 718-982-2500.

College of William and Mary, Faculty of Arts and Sciences, Department of Computer Science, Williamsburg, VA 23187-8795. Offers computational operations research (MS); computational science (PhD); computer science (MS, PhD). Part-time programs available. *Faculty:* 16 full-time (4 women), 1 part-time (0 women). *Students:* 44 full-time (17 women), 25 part-time (5 women); includes 8 minority (3 African Americans, 4 Asian Americans or Pacific Islanders, 1 Hispanic American), 12 international. Average age 30. 175 applicants, 33% accepted. In 1998, 13 master's, 1 doctorate awarded (100% entered university research/teaching). Terminal master's awarded for partial completion of doctoral program. *Degree requirements:* For master's, computer language, research project required, thesis optional; for doctorate, computer language, dissertation, oral exam required. *Entrance requirements:* For master's, GRE General Test, minimum GPA of 2.5; for doctorate, GRE General Test (minimum combined score of 1700 on three sections required; average 2010), minimum GPA of 3.0. *Average time to degree:* Master's–2 years full-time, 3 years part-time. *Application deadline:* For fall admission, 3/1 (priority date); for spring admission, 11/1. Applications are processed on a rolling basis. Application fee: $30. *Financial aid:* In 1998–99, 32 students received aid, including 2 fellowships (averaging $17,000 per year), 8 research assistantships (averaging $14,000 per year), 22 teaching assistantships (averaging $14,000 per year) Financial aid application deadline: 3/1; financial aid applicants required to submit FAFSA. *Faculty research:* Distributed and parallel systems, simulation, stochastic modeling, computer architecture, scientific computing. Total annual research expenditures: $264,225. *Unit head:* Dr. Stephen K. Park, Chair, 757-221-3455, Fax: 757-221-1717, E-mail: chair@cs.wm.edu. *Application contact:* Vanessa Godwin, Administrative Director, 757-221-3455, Fax: 757-221-1717, E-mail: gradinfo@cs.wm.edu.

Announcement: Areas of faculty research include parallel and distributed systems, databases, image processing and graphics, simulation, petri nets, networks, programming languages and compilers, algorithms, theory of computation, and artificial intelligence. Computing equipment includes network of UNIX workstations, multiprocessor SGI Onyx, and a MasPar MP2. Visit the Web site at http://www.cs.wm.edu

Colorado School of Mines, Graduate School, Department of Mathematical and Computer Sciences, Golden, CO 80401-1887. Offers MS, PhD. Part-time programs available. *Faculty:* 18 full-time (2 women), 11 part-time (5 women). *Students:* 12 full-time (3 women), 24 part-time (5 women); includes 2 minority (1 African American, 1 Asian American or Pacific Islander), 16 international. 40 applicants, 68% accepted. In 1998, 13 master's awarded (100% found work related to degree); 3 doctorates awarded (100% found work related to degree). *Degree requirements:* For master's, thesis required, foreign language not required; for doctorate, dissertation, written comprehensive exams required, foreign language not required. *Entrance requirements:* For master's and doctorate, GRE General Test (combined average 1680 on three sections), minimum GPA of 3.0. *Application deadline:* Applications are processed on a rolling basis. Application fee: $40. Electronic applications accepted. *Financial aid:* In 1998–99, 28 students received aid, including 1 fellowship, 9 research assistantships, 10 teaching assistantships; unspecified assistantships also available. Aid available to part-time students. Financial aid applicants required to submit FAFSA. *Faculty research:* Applied statistics, numerical computation, artificial intelligence, linear optimization. Total annual research expenditures: $410,734. *Unit head:* Dr. Graeme Fairweather, Head, 303-273-3860, E-mail: gfairwea@mines.edu. *Application contact:* Willy Hereman, Associate Professor, 303-273-3881, Fax: 303-273-3875, E-mail: whereman@mines.edu.

Colorado State University, Graduate School, College of Natural Sciences, Department of Computer Science, Fort Collins, CO 80523-0015. Offers MS, PhD. *Faculty:* 15 full-time (2 women). *Students:* 45 full-time (12 women), 31 part-time (6 women); includes 1 minority (Asian American or Pacific Islander), 41 international. Average age 30. 228 applicants, 69% accepted. In 1998, 14 master's, 1 doctorate awarded. Terminal master's awarded for partial completion of doctoral program. *Degree requirements:* For master's, thesis or alternative required, foreign language not required; for doctorate, dissertation, qualifying, preliminary, and final exams required, foreign language not required. *Entrance requirements:* For master's, GRE General Test, GRE Subject Test, TOEFL (minimum score of 580 required), minimum GPA of 3.2, computer science background; for doctorate, GRE General Test, GRE Subject Test, TOEFL (minimum score of 580 required), minimum GPA of 3.2, BSC in computer science. *Application deadline:* For fall admission, 2/1 (priority date); for spring admission, 9/1 (priority date). Applications are processed on a rolling basis. Application fee: $30. Electronic applications accepted. *Financial aid:* In 1998–99, 2 fellowships, 28 research assistantships, 30 teaching assistantships were awarded.; career-related internships or fieldwork and Federal Work-Study also available. Financial aid application deadline: 2/10. *Faculty research:* Architecture, artificial intelligence, parallel and distributed computing, software engineering, computer vision/graphics. Total annual research expenditures: $1.1 million. *Unit head:* Dr. Stephen B. Seidman, Chairman, 970-491-5792, Fax: 970-491-2466, E-mail: seidman@cs.colostate.edu. *Application contact:* Graduate Coordinator, 970-491-5792, Fax: 970-491-2466, E-mail: gradinfo@cs.colostate.edu.

See in-depth description on page 723.

Colorado Technical University, Graduate Studies, Program in Computer Science, Colorado Springs, CO 80907-3896. Offers MSCS, DCS. Part-time and evening/weekend programs available. *Degree requirements:* For master's, computer language, thesis or alternative required, foreign language not required; for doctorate, computer language, dissertation required, foreign language not required. *Entrance requirements:* For master's, minimum undergraduate GPA of 3.0; for doctorate, minimum graduate GPA of 3.0, 5 years of related work experience. *Faculty research:* Software engineering, systems engineering.

Colorado Technical University Denver Campus, Program in Computer Science, Greenwood Village, CO 80111. Offers MSCS, DCS.

Columbia University, Fu Foundation School of Engineering and Applied Science, Department of Computer Science, New York, NY 10027. Offers MS, PhD, CSE. PhD offered through the Graduate School of Arts and Sciences. Part-time programs available. Postbaccalaureate distance learning degree programs offered (no on-campus study). *Faculty:* 25 full-time (2 women), 11 part-time (2 women). *Students:* 89 full-time (13 women), 59 part-time (14 women); includes 29 minority (2 African Americans, 24 Asian Americans or Pacific Islanders, 3 Hispanic Americans), 64 international. 327 applicants, 35% accepted. In 1998, 45 master's, 8 doctorates awarded. Terminal master's awarded for partial completion of doctoral program. *Degree requirements:* For master's, computer language required, thesis optional, foreign language not required; for doctorate, computer language, dissertation, candidacy exam, qualifying exam required, foreign language not required; for CSE, computer language required, foreign language not required. *Entrance requirements:* For master's, doctorate, and CSE, GRE General Test, GRE Subject Test, TOEFL. *Application deadline:* For fall admission, 1/5 (priority date); for spring admission, 10/1 (priority date). Application fee: $55. *Financial aid:* In 1998–99, 82 students received aid, including 2 fellowships, 70 research assistantships, 10 teaching assistantships; Federal Work-Study and outside fellowships also available. Financial aid application deadline: 1/5; financial aid applicants required to submit FAFSA. *Faculty research:* Algorithms and complexity, robotics, software systems, parallel processing, artificial intelligence. Total annual research expenditures: $4.7 million. *Unit head:* Dr. Kathleen McKeown, Chairman, 212-939-7000, Fax: 212-666-0140, E-mail: kathy@cs.columbia.edu. *Application contact:* Martha Ruth Zadok, Graduate Program Office, 212-939-7000, Fax: 212-666-0140, E-mail: gradinfo@cs.columbia.edu.

See in-depth description on page 725.

Columbus State University, Graduate Studies, College of Science, Department of Computer Science, Columbus, GA 31907-5645. Offers applied computer science (MS). *Faculty:* 4 full-time (0 women). *Students:* 5 full-time (1 woman), 20 part-time (4 women); includes 2 minority (1 African American, 1 Asian American or Pacific Islander), 2 international. 25 applicants, 32% accepted. In 1998, 3 degrees awarded. *Degree requirements:* For master's, computer language required, foreign language not required. *Entrance requirements:* For master's, GRE General Test (minimum combined score of 800 required), MAT (minimum score of 44 required). *Average time to degree:* Master's–1.5 years full-time, 3 years part-time. *Application deadline:* For fall admission, 8/4 (priority date); for spring admission, 12/17. Application fee: $20. *Financial aid:* Application deadline: 6/4. *Unit head:* Dr. Leary Bell, Chair, 706-68-2410, Fax: 706-565-3529, E-mail: bell_leary@colstate.edu. *Application contact:* Katie Thornton, Graduate Admissions, 706-568-2279, Fax: 706-568-2462, E-mail: thornton_katie@colstate.edu.

Concordia University, School of Graduate Studies, Faculty of Engineering and Computer Science, Department of Computer Science, Montréal, PQ H3G 1M8, Canada. Offers computer science (MCS, PhD, Diploma); software engineering (MCS). *Students:* 335 full-time (113 women), 84 part-time (16 women). *Degree requirements:* For master's, one foreign language, computer language required, thesis optional; for doctorate, one foreign language, computer language, dissertation, comprehensive exam required. *Application deadline:* For fall admission, 6/1; for spring admission, 10/1. Application fee: $50. *Faculty research:* Computer systems and applications, mathematics of computation, pattern recognition, artificial intelligence and robotics. *Unit head:* Dr. H. F. Li, Director, 514-848-3043, Fax: 514-848-2830. *Application contact:* Dr. T. D. Bui, Director, 514-848-3043, Fax: 514-848-2830.

Cornell University, Graduate School, Graduate Fields of Engineering, Field of Computer Science, Ithaca, NY 14853-0001. Offers algorithms (M Eng, PhD); applied logic and automated

Computer Science

Cornell University (continued)

reasoning (M Eng, PhD); artificial intelligence (M Eng, PhD); computer graphics (M Eng, PhD); computer science (M Eng, PhD); computer vision (M Eng, PhD); concurrency and distributed computing (M Eng, PhD); information organization and retrieval (M Eng, PhD); operating systems (M Eng, PhD); parallel computing (M Eng, PhD); programming environments (M Eng, PhD); programming languages and methodology (M Eng, PhD); robotics (M Eng, PhD); scientific computing (M Eng, PhD); theory of computation (M Eng, PhD). *Faculty:* 34 full-time. *Students:* 141 full-time (20 women); includes 20 minority (19 Asian Americans or Pacific Islanders, 1 Hispanic American), 64 international. 414 applicants, 19% accepted. In 1998, 68 master's, 13 doctorates awarded. Terminal master's awarded for partial completion of doctoral program. *Degree requirements:* For doctorate, dissertation required, foreign language not required, foreign language not required. *Entrance requirements:* For master's, GRE General Test, GRE Subject Test, TOEFL (minimum score of 550 required); for doctorate, GRE General Test, GRE Subject Test (computer science), TOEFL (minimum score of 550 required). *Application deadline:* For fall admission, 1/1. Application fee: $65. Electronic applications accepted. *Financial aid:* In 1998–99, 90 students received aid, including 25 fellowships with full tuition reimbursements available, 39 research assistantships with full tuition reimbursements available, 26 teaching assistantships with full tuition reimbursements available; institutionally-sponsored loans, scholarships, tuition waivers (full and partial), and unspecified assistantships also available. Financial aid applicants required to submit FAFSA. *Faculty research:* Numerical analysis, distributed systems, multimedia. *Unit head:* Director of Graduate Studies, 607-255-8593. *Application contact:* Graduate Field Assistant, 607-255-8593, E-mail: phd@cs.cornell.edu.

See in-depth description on page 729.

Creighton University, Graduate School, College of Arts and Sciences, Department of Mathematics, Statistics, and Computer Science, Program in Computer Sciences, Omaha, NE 68178-0001. Offers MCS. *Students:* 4 full-time (0 women), 3 part-time (1 woman). In 1998, 12 degrees awarded. *Entrance requirements:* For master's, GRE General Test, TOEFL (minimum score of 550 required). *Application deadline:* For fall admission, 3/1. Applications are processed on a rolling basis. Application fee: $30. *Unit head:* Dr. Mark Wierman, Director. *Application contact:* Dr. Barbara J. Braden, Dean, Graduate School, 402-280-2870, Fax: 402-280-5762.

Dalhousie University, Faculty of Graduate Studies, DalTech, Faculty of Computer Science, Halifax, NS B3H 3J5, Canada. Offers computer science (MC Sc, PhD); electronic commerce (MEC). *Faculty:* 20 full-time (1 woman), 5 part-time (0 women). *Students:* 38 full-time (8 women), 10 part-time (3 women). 66 applicants, 44% accepted. In 1998, 15 master's awarded. *Degree requirements:* For master's and doctorate, thesis/dissertation required, foreign language not required. *Entrance requirements:* For master's and doctorate, TOEFL (minimum score of 580 required). *Application deadline:* For fall admission, 6/1 (priority date); for winter admission, 10/1 (priority date); for spring admission, 2/1 (priority date). Applications are processed on a rolling basis. Application fee: $55. *Financial aid:* Fellowships, research assistantships, teaching assistantships, career-related internships or fieldwork, scholarships, and unspecified assistantships available. *Unit head:* Dr. Jacob Slonim, Dean, 902-494-2093, Fax: 902-492-1517, E-mail: computer.science@dal.ca. *Application contact:* Shelley Parker, Admissions Coordinator, Graduate Studies and Research, 902-494-1288, Fax: 902-494-3149, E-mail: shelley.parker@dal.ca.

Dartmouth College, School of Arts and Sciences, Department of Computer Science, Hanover, NH 03755. Offers MS, PhD. *Faculty:* 14 full-time (2 women), 1 part-time (0 women). *Students:* 34 full-time (5 women); includes 2 minority (1 African American, 1 Native American), 13 international. 79 applicants, 41% accepted. In 1998, 3 master's awarded (33% found work related to degree, 67% continued full-time study); 3 doctorates awarded (100% found work related to degree). Terminal master's awarded for partial completion of doctoral program. *Degree requirements:* For master's and doctorate, thesis/dissertation required, foreign language not required. *Entrance requirements:* For master's and doctorate, GRE General Test, GRE Subject Test. *Average time to degree:* Doctorate–5 years full-time. *Application deadline:* For fall admission, 2/1 (priority date). Application fee: $30. *Financial aid:* In 1998–99, 11 fellowships with full tuition reimbursements, 23 research assistantships with full tuition reimbursements were awarded.; career-related internships or fieldwork, grants, institutionally-sponsored loans, scholarships, and tuition waivers (full and partial) also available. Aid available to part-time students. Financial aid application deadline: 2/1. *Unit head:* Dr. Scot Drysdale, Chair, 603-646-2101. *Application contact:* Phyllis Bellmore, Administrative Assistant, 603-646-2206.

See in-depth description on page 731.

DePaul University, School of Computer Science, Telecommunications, and Information Systems, Program in Computer Science, Chicago, IL 60604-2287. Offers MS, PhD. Part-time and evening/weekend programs available. *Students:* 223 full-time (81 women), 268 part-time (67 women); includes 192 minority (42 African Americans, 129 Asian Americans or Pacific Islanders, 21 Hispanic Americans), 56 international. Average age 32. 293 applicants, 68% accepted. In 1998, 68 master's, 1 doctorate awarded. *Degree requirements:* For master's, computer language, comprehensive exam required, foreign language and thesis not required; for doctorate, computer language, dissertation, comprehensive exam required, foreign language not required. *Entrance requirements:* For master's, passing grade on the department's Graduate Assessment Examination; for doctorate, GRE, master's degree in computer science. *Application deadline:* For fall admission, 8/1 (priority date); for winter admission, 11/5 (priority date); for spring admission, 5/1 (priority date). Applications are processed on a rolling basis. Application fee: $25. *Financial aid:* Fellowships, research assistantships, teaching assistantships, Federal Work-Study, tuition waivers (partial), and unspecified assistantships available. Aid available to part-time students. Financial aid application deadline: 3/20. *Unit head:* Dr. Richard Johnsonbaugh, Director, 312-362-8728. *Application contact:* Anne B. Morley, Director of Student Services, 312-362-8714, Fax: 312-362-6116.

See in-depth description on page 733.

DePaul University, School of Computer Science, Telecommunications, and Information Systems, Program in Human-Computer Interaction, Chicago, IL 60604-2287. Offers MS. *Students:* 26 full-time (18 women), 22 part-time (17 women); includes 16 minority (9 African Americans, 6 Asian Americans or Pacific Islanders, 1 Hispanic American), 1 international. Average age 32. 25 applicants, 72% accepted. *Degree requirements:* For master's, computer language, comprehensive exam required. *Entrance requirements:* For master's, passing grade on the department's Graduate Assessment Examination. *Application deadline:* For fall admission, 8/1; for winter admission, 11/15 (priority date); for spring admission, 5/1 (priority date). Applications are processed on a rolling basis. Application fee: $25. *Unit head:* Dr. Rosalee Wolfe, Director, 312-362-8381, Fax: 312-362-6116. *Application contact:* Anne B. Morley, Director of Student Services, 312-362-8714, Fax: 312-362-6116.

Drexel University, Graduate School, College of Arts and Sciences, Department of Mathematics and Computer Science, Program in Computer Science, Philadelphia, PA 19104-2875. Offers MS. *Students:* 21 full-time (4 women), 51 part-time (10 women); includes 7 minority (5 Asian Americans or Pacific Islanders, 2 Native Americans), 29 international. Average age 32. 135 applicants, 41% accepted. In 1998, 34 degrees awarded. *Degree requirements:* For master's, thesis not required. *Entrance requirements:* For master's, GRE, TOEFL (minimum score of 570 required), TSE (for teaching assistants). *Application deadline:* For fall admission, 8/21. Applications are processed on a rolling basis. Application fee: $35. Tuition: Full-time $15,795; part-time $585 per credit. Required fees: $375; $67 per term. Tuition and fees vary according to program. *Financial aid:* In 1998–99, 2 research assistantships, 8 teaching assistantships were awarded.; unspecified assistantships also available. Financial aid application deadline: 2/1. *Unit head:* Anne B. Morley, Director of Student Services, 312-362-8714, Fax: 312-362-6116. *Application contact:* Director of Graduate Admissions, 215-895-6700, Fax: 215-895-5939.

See in-depth description on page 735.

Duke University, Graduate School, Department of Computer Science, Durham, NC 27708-0586. Offers MS, PhD. *Faculty:* 26 full-time, 7 part-time. *Students:* 68 full-time, 2 part-time; includes 3 minority (2 Asian Americans or Pacific Islanders, 1 Hispanic American), 50 international. 203 applicants, 36% accepted. In 1998, 10 master's, 4 doctorates awarded. *Degree requirements:* For doctorate, dissertation required. *Entrance requirements:* For master's, GRE General Test; for doctorate, GRE General Test, GRE Subject Test (recommended). *Application deadline:* For fall admission, 12/31. Application fee: $75. *Financial aid:* Fellowships, research assistantships, teaching assistantships, Federal Work-Study available. Financial aid application deadline: 12/31. *Unit head:* Robert A. Wagner, Director of Graduate Studies, 919-660-6538, Fax: 919-660-6519, E-mail: driggs@cs.duke.edu.

See in-depth description on page 739.

East Carolina University, Graduate School, College of Arts and Sciences, Department of Mathematics, Program in Computer Science, Greenville, NC 27858-4353. Offers MS. *Students:* 3 full-time (1 woman), 6 part-time (1 woman); includes 1 minority (Asian American or Pacific Islander), 2 international. Average age 35. 4 applicants, 50% accepted. In 1998, 2 degrees awarded. *Degree requirements:* For master's, comprehensive exams required, thesis optional, foreign language not required. *Entrance requirements:* For master's, GRE General Test, TOEFL. *Application deadline:* For fall admission, 6/1 (priority date); for spring admission, 10/15. Applications are processed on a rolling basis. Application fee: $40. Tuition, state resident: full-time $1,012. Tuition, nonresident: full-time $8,578. Required fees: $1,006. Part-time tuition and fees vary according to course load. *Financial aid:* Research assistantships with partial tuition reimbursements, teaching assistantships with partial tuition reimbursements available. Financial aid application deadline: 6/1. *Application contact:* Dr. Ronnie Smith, Senior Director of Graduate Studies, 252-328-6461, Fax: 252-328-6414, E-mail: smithron@mail.ecu.edu.

Eastern Michigan University, Graduate School, College of Arts and Sciences, Department of Computer Science, Ypsilanti, MI 48197. Offers MS. *Students:* 24 full-time, 58 part-time. *Degree requirements:* For master's, computer language, thesis or alternative required, foreign language not required. *Entrance requirements:* For master's, TOEFL. *Application deadline:* For fall admission, 5/15. Application fee: $30. *Financial aid:* Application deadline: 3/15. *Unit head:* Dr. George Haynam, Head, 734-487-1063.

Eastern Washington University, Graduate School, College of Science, Mathematics and Technology, Department of Computer Science, Cheney, WA 99004-2431. Offers M Ed, MS. *Accreditation:* NCATE. Part-time programs available. *Faculty:* 10 full-time (2 women). *Students:* 3 full-time (2 women), 5 part-time (1 woman); includes 1 minority (Asian American or Pacific Islander), 1 international. In 1998, 5 degrees awarded. *Degree requirements:* For master's, thesis or alternative, comprehensive oral exam required. *Entrance requirements:* For master's, minimum GPA of 3.0. *Application deadline:* For fall admission, 4/1 (priority date); for spring admission, 1/15. Applications are processed on a rolling basis. Application fee: $35. Tuition, state resident: full-time $4,368. Tuition, nonresident: full-time $13,284. *Financial aid:* Research assistantships, teaching assistantships, Federal Work-Study and institutionally-sponsored loans available. Financial aid application deadline: 2/1. *Unit head:* Dr. Ray Hamel, Chairman, 509-359-6260. *Application contact:* Dr. Steve Simmons, Adviser, 509-359-6064.

East Stroudsburg University of Pennsylvania, Graduate School, School of Arts and Sciences, Department of Computer Science, East Stroudsburg, PA 18301-2999. Offers MS. Part-time and evening/weekend programs available. *Faculty:* 5 full-time (1 woman). *Students:* 6 full-time (3 women), 4 part-time; includes 1 minority (Asian American or Pacific Islander), 4 international. Average age 29. In 1998, 12 degrees awarded. *Degree requirements:* For master's, computer language, thesis or alternative, comprehensive exam required. *Entrance requirements:* For master's, bachelor's degree in computer science or related field. *Application deadline:* For fall admission, 7/31 (priority date); for spring admission, 11/30. Applications are processed on a rolling basis. Application fee: $15. Tuition, state resident: full-time $3,780; part-time $210 per credit. Tuition, nonresident: full-time $6,610; part-time $367 per credit. Required fees: $724; $40 per credit. *Financial aid:* In 1998–99, 5 students received aid, including 5 research assistantships with full tuition reimbursements available (averaging $5,000 per year); career-related internships or fieldwork, Federal Work-Study, and institutionally-sponsored loans also available. Financial aid application deadline: 3/1; financial aid applicants required to submit FAFSA. *Unit head:* Dr. Richard Prince, Graduate Coordinator, 570-422-3772, Fax: 570-422-3777, E-mail: rprince@esu.edu.

East Tennessee State University, School of Graduate Studies, College of Applied Science and Technology, Department of Computer and Information Sciences, Johnson City, TN 37614-0734. Offers computer science (MS); information sciences (MS). Part-time and evening/weekend programs available. *Degree requirements:* For master's, computer language, thesis, written comprehensive exam required, foreign language not required. *Entrance requirements:* For master's, GRE General Test (minimum combined score of 1050 required), TOEFL (minimum score of 550 required), minimum GPA of 2.5. *Faculty research:* Operating systems, database design, artificial intelligence, simulation, parallel algorithms.

École Polytechnique de Montréal, Graduate Programs, Department of Electrical and Computer Engineering, Montréal, PQ H3C 3A7, Canada. Offers automation (M Eng, M Sc A, PhD); computer science (M Eng, M Sc A, PhD); electrotechnology (M Eng, M Sc A, PhD); microelectronics (M Eng, M Sc A, PhD); microwave technology (M Eng, M Sc A, PhD). Part-time and evening/weekend programs available. *Degree requirements:* For master's and doctorate, one foreign language, computer language, thesis/dissertation required. *Entrance requirements:* For master's, minimum GPA of 2.75; for doctorate, minimum GPA of 3.0. *Faculty research:* Microwaves, telecommunications, software engineering.

Emory University, Graduate School of Arts and Sciences, Department of Mathematics and Computer Science, Atlanta, GA 30322-1100. Offers mathematics (PhD); mathematics/computer science (MS). *Faculty:* 22 full-time (4 women), 4 part-time (0 women). *Students:* 35 full-time (17 women); includes 4 minority (2 African Americans, 1 Asian American or Pacific Islander, 1 Hispanic American), 16 international. 105 applicants, 28% accepted. In 1998, 13 master's, 4 doctorates awarded. Terminal master's awarded for partial completion of doctoral program. *Degree requirements:* For master's, thesis required; for doctorate, dissertation, comprehensive exams required. *Entrance requirements:* For master's, GRE General Test (combined average 1600 on three sections), TOEFL; for doctorate, GRE General Test (combined average 1800 on three sections), TOEFL. *Application deadline:* For fall admission, 1/20. Application fee: $45. *Financial aid:* Fellowships, teaching assistantships, scholarships available. Financial aid application deadline: 1/20. Total annual research expenditures: $1.1 million. *Unit head:* Dr. Dwight Duffus, Chairman, 404-727-7580, Fax: 404-727-5611. *Application contact:* Ron Gould, Director of Graduate Studies, 404-727-7580, Fax: 404-727-5611, E-mail: dgs@mathcs.emory.edu.

Announcement: The department offers an MS in math/computer science. Computer science research areas include algorithms, networking, parallel processing, distributed systems, operating systems, software engineering, fault tolerance, and theory of computation. Full tuition and funding are available in both the math/computer science MS and math PhD programs. See also the math announcement in Book 4 of this series.

Fairleigh Dickinson University, Florham-Madison Campus, Maxwell Becton College of Arts and Sciences, Department of Mathematics, Computer Science and Physics, Madison, NJ 07940-1099. Offers computer science (MS); mathematics (MS). *Faculty:* 5 full-time (0 women), 1 part-time (0 women). *Degree requirements:* For master's, computer language required, foreign language and thesis not required. *Entrance requirements:* For master's, GRE General Test. *Application deadline:* Applications are processed on a rolling basis. Application fee: $35. Tuition: Full-time $9,396; part-time $522 per credit. Required fees: $69 per semester. *Financial aid:* Fellowships, research assistantships, teaching assistantships available. *Unit head:* Dr. Richard Wagner, Chairperson, 973-443-8691.

Fairleigh Dickinson University, Teaneck–Hackensack Campus, University College: Arts, Sciences, and Professional Studies, School of Computer Science and Information Systems, Program in Computer Science, Teaneck, NJ 07666-1914. Offers MS. *Degree requirements:* For master's, computer language required, foreign language and thesis not required. *Entrance requirements:* For master's, GRE General Test. *Faculty research:* Real time computer systems, software design modeling and simulation, parallel processing, pattern recognition, image processing.

Fitchburg State College, Division of Graduate and Continuing Education, Program in Computer Science, Fitchburg, MA 01420-2697. Offers MS. Part-time and evening/weekend programs available. *Students:* 25 full-time (11 women), 19 part-time (11 women); includes 7 minority (2 African Americans, 5 Asian Americans or Pacific Islanders), 23 international. In 1998, 13 degrees awarded. *Degree requirements:* For master's, computer language required, foreign language and thesis not required. *Entrance requirements:* For master's, GRE General Test or MAT, appropriate bachelor's degree, interview. *Application deadline:* Applications are processed on a rolling basis. Application fee: $10. Tuition, state resident: part-time $140 per credit. Tuition, nonresident: part-time $140 per credit. *Financial aid:* In 1998–99, research assistantships with partial tuition reimbursements (averaging $5,500 per year), teaching assistantships with partial tuition reimbursements (averaging $5,500 per year) were awarded.; Federal Work-Study and unspecified assistantships also available. Aid available to part-time students. Financial aid application deadline: 3/1; financial aid applicants required to submit FAFSA. *Unit head:* Dr. Robert McGuire, Chair, 978-665-3305, Fax: 978-665-3658, E-mail: dgce@fsc.edu. *Application contact:* James DuPont, Director of Admissions, 978-665-3144, Fax: 978-665-4540, E-mail: admissions@fsc.edu.

Florida Atlantic University, College of Engineering, Department of Computer Science and Engineering, Program in Computer Science, Boca Raton, FL 33431-0991. Offers MS, PhD. Part-time and evening/weekend programs available. *Faculty:* 11 full-time (1 woman). *Students:* 36 full-time (14 women), 52 part-time (18 women); includes 25 minority (2 African Americans, 18 Asian Americans or Pacific Islanders, 5 Hispanic Americans), 42 international. Average age 32. In 1998, 29 master's awarded. *Degree requirements:* For master's, computer language required, thesis optional, foreign language not required; for doctorate, computer language, dissertation, qualifying exam required, foreign language not required. *Entrance requirements:* For master's, GRE General Test (minimum combined score of 1000 required), TOEFL (minimum score of 550 required), minimum GPA of 3.0 in last 60 hours of undergraduate work; for doctorate, TOEFL (minimum score of 550 required), GRE General Test (minimum combined score of 1100 required) and minimum GPA of 3.0 or GRE General Test (minimum combined score of 1000 required) and minimum GPA of 3.5. *Application deadline:* For fall admission, 4/10 (priority date); for spring admission, 10/1. Applications are processed on a rolling basis. Application fee: $20. Tuition, state resident: part-time $148 per credit hour. Tuition, nonresident: part-time $509 per credit hour. *Financial aid:* Fellowships, research assistantships, teaching assistantships, career-related internships or fieldwork, Federal Work-Study, and unspecified assistantships available. Aid available to part-time students. Financial aid application deadline: 4/1; financial aid applicants required to submit FAFSA. *Faculty research:* Software engineering, artificial intelligence, performance evaluation, queuing theory. *Application contact:* Patricia Capozziello, Graduate Admissions Coordinator, 561-297-2694, Fax: 561-297-2659, E-mail: capozzie@fau.edu.

Florida Institute of Technology, Graduate School, College of Engineering, Division of Electrical and Computer Science and Engineering, Program in Computer Science, Melbourne, FL 32901-6975. Offers MS, PhD. Part-time and evening/weekend programs available. *Faculty:* 10 full-time (2 women), 3 part-time (1 woman). *Students:* 15 full-time (1 woman), 61 part-time (11 women); includes 6 minority (2 African Americans, 1 Asian American or Pacific Islander, 2 Hispanic Americans, 1 Native American), 31 international. Average age 32. 121 applicants, 69% accepted. In 1998, 10 master's, 4 doctorates awarded. Terminal master's awarded for partial completion of doctoral program. *Degree requirements:* For master's, computer language, comprehensive exam required, thesis optional, foreign language not required; for doctorate, computer language, dissertation, comprehensive exam required, foreign language not required. *Entrance requirements:* For master's, minimum GPA of 3.0; for doctorate, GRE General Test, GRE Subject Test (computer science), minimum GPA of 3.2. *Application deadline:* Applications are processed on a rolling basis. Application fee: $50. Electronic applications accepted. Tuition: Part-time $575 per credit hour. Required fees: $100. Tuition and fees vary according to campus/location and program. *Financial aid:* In 1998–99, 24 students received aid, including 14 research assistantships (averaging $4,113 per year), 9 teaching assistantships (averaging $4,032 per year); tuition remissions also available. Financial aid application deadline: 3/1; financial aid applicants required to submit FAFSA. *Faculty research:* Computer graphics, artificial intelligence, software engineering, parallel processing, programming languages. Total annual research expenditures: $596,490. *Unit head:* Dr. William D. Shoaff, Chair, 407-674-8066, Fax: 407-676-8192, E-mail: wds@cs.fit.edu. *Application contact:* Carolyn P. Farrior, Associate Dean of Graduate Admissions, 407-674-7118, Fax: 407-723-9468, E-mail: cfarrior@fit.edu.

Florida Institute of Technology, Graduate School, School of Extended Graduate Studies, Program in Computer Science, Melbourne, FL 32901-6975. Offers MS. Part-time and evening/weekend programs available. Average age 34. 5 applicants, 40% accepted. In 1998, 7 degrees awarded (100% found work related to degree). *Degree requirements:* For master's, computer language required, thesis optional, foreign language not required. *Entrance requirements:* For master's, minimum GPA of 3.0. *Average time to degree:* Master's–1 year full-time, 3 years part-time. *Application deadline:* Applications are processed on a rolling basis. Application fee: $50. Electronic applications accepted. Tuition: Part-time $270 per credit hour. Part-time tuition and fees vary according to campus/location. *Financial aid:* Application deadline: 3/1; *Unit head:* Dr. James Pleasant, Graduate Coordinator, 423-439-6962, Fax: 423-439-7119. *Application contact:* Carolyn P. Farrior, Associate Dean of Graduate Admissions, 407-674-7118, Fax: 407-723-9468, E-mail: cfarrior@fit.edu.

See in-depth description on page 743.

Florida International University, College of Arts and Sciences, School of Computer Science, Miami, FL 33199. Offers MS, PhD. Part-time and evening/weekend programs available. *Faculty:* 24 full-time (3 women), 5 part-time (0 women). *Students:* 45 full-time (10 women), 29 part-time (7 women); includes 20 minority (3 African Americans, 4 Asian Americans or Pacific Islanders, 13 Hispanic Americans), 42 international. Average age 30. 175 applicants, 34% accepted. In 1998, 16 master's, 2 doctorates awarded. *Degree requirements:* For master's, computer language required, thesis optional, foreign language not required; for doctorate, computer language, dissertation required. *Entrance requirements:* For master's and doctorate, GRE General Test (minimum combined score of 1650 on three sections required), TOEFL (minimum score of 550 required). *Application deadline:* For fall admission, 4/1 (priority date); for spring admission, 10/1. Applications are processed on a rolling basis. Application fee: $20. Tuition, state resident: part-time $145 per credit hour. Tuition, nonresident: part-time $506 per credit hour. Required fees: $158; $158 per year. *Financial aid:* Application deadline: 4/1. *Faculty research:* Computer graphics, database management systems, simulation. *Unit head:* Dr. Samuel S. Shapiro, Acting Director, 305-348-2744, Fax: 305-348-3549, E-mail: shapiro@cs.fiu.edu.

See in-depth description on page 745.

Florida State University, Graduate Studies, College of Arts and Sciences, Department of Computer Science, Tallahassee, FL 32306. Offers computer and network system administration (MA, MS); computer science (MA, MS, PhD); software engineering (MA, MS). Part-time programs available. *Faculty:* 13 full-time (1 woman), 6 part-time (0 women). *Students:* 91 full-time (30 women), 3 part-time. Average age 24. 200 applicants, 20% accepted. In 1998, 20 master's, 3 doctorates awarded (100% entered university research/teaching). *Degree requirements:* For master's, computer language, thesis or alternative required, foreign language not required; for doctorate, computer language, dissertation required, foreign language not required. *Entrance requirements:* For master's, GRE General Test (minimum score of 650 on quantitative section, 1100 combined required), minimum undergraduate GPA of 3.0; for doctorate, GRE General Test (minimum score of 650 on quantitative section, 1100 combined required), minimum GPA of 3.0. *Average time to degree:* Master's–3 years full-time; doctorate–5 years full-time. *Application deadline:* For fall admission, 3/3 (priority date); for spring admission, 7/1 (priority date). Applications are processed on a rolling basis. Application fee: $20. Electronic applications accepted. Tuition, state resident: part-time $139 per credit hour. Tuition, nonresident: part-time $482 per credit hour. Tuition and fees vary according to program. *Financial aid:* In 1998–99, 48 students received aid, including 9 fellowships with full tuition reimbursements available (averaging $10,000 per year), 8 research assistantships with full tuition reimbursements available, 38 teaching assistantships with full tuition reimbursements available (averaging $13,000 per year); career-related internships or fieldwork, Federal Work-Study, and institutionally-sponsored loans also available. Financial aid application deadline: 3/3; financial aid applicants required to submit FAFSA. *Faculty research:* Expert systems, compiler design, artificial intelligence, neural networks, database theory and design, real time systems, logic, networking. *Unit head:* Theodore P. Baker, Chairman, 850-644-4029, Fax: 850-644-0058, E-mail: baker@cs.fsu.edu. *Application contact:* David Gaitros, Graduate Admissions, 850-644-4055, Fax: 850-644-0058.

Fordham University, Graduate School of Arts and Sciences, Department of Computer Science and Information Systems, New York, NY 10458. Offers computer science (MS). Part-time and evening/weekend programs available. *Faculty:* 11 full-time (1 woman). *Students:* 2 full-time (1 woman), 25 part-time (13 women); includes 5 minority (1 African American, 3 Asian Americans or Pacific Islanders, 1 Hispanic American), 7 international. 45 applicants, 64% accepted. In 1998, 10 degrees awarded. *Degree requirements:* For master's, computer language, comprehensive exam required, thesis not required. *Entrance requirements:* For master's, GRE General Test. *Application deadline:* For fall admission, 1/15 (priority date); for spring admission, 12/1. Applications are processed on a rolling basis. Application fee: $50. *Financial aid:* In 1998–99, 5 students received aid, including 1 teaching assistantship; research assistantships, career-related internships or fieldwork, institutionally-sponsored loans, tuition waivers (full and partial), and unspecified assistantships also available. Financial aid application deadline: 1/15. *Unit head:* Chair, 718-817-4480, Fax: 718-817-4488. *Application contact:* Dr. Craig W. Pilant, Assistant Dean, 718-817-4420, Fax: 718-817-3566, E-mail: ada_pilant@lars.fordham.edu.

George Mason University, School of Information Technology and Engineering, Department of Computer Science, Fairfax, VA 22030-4444. Offers MS. Part-time and evening/weekend programs available. *Faculty:* 25 full-time (5 women), 13 part-time (2 women). *Students:* 49 full-time (19 women), 299 part-time (79 women); includes 103 minority (12 African Americans, 84 Asian Americans or Pacific Islanders, 7 Hispanic Americans), 96 international. Average age 31. 276 applicants, 63% accepted. In 1998, 102 degrees awarded. *Degree requirements:* For master's, thesis optional, foreign language not required. *Entrance requirements:* For master's, GRE General Test, TOEFL (minimum score of 575 required), minimum GPA of 3.0 in last 60 hours. *Application deadline:* For fall admission, 5/1; for spring admission, 11/1. Application fee: $30. Electronic applications accepted. Tuition, state resident: full-time $4,416; part-time $184 per credit hour. Tuition, nonresident: full-time $12,516; part-time $522 per credit hour. Tuition and fees vary according to program. *Financial aid:* Fellowships, research assistantships, teaching assistantships, career-related internships or fieldwork and Federal Work-Study available. Aid available to part-time students. Financial aid application deadline: 3/1; financial aid applicants required to submit FAFSA. *Faculty research:* Artificial intelligence, image processing/graphics, parallel/distributed systems, software engineering systems. Total annual research expenditures: $1.8 million. *Unit head:* Dr. Henry Hamburger, Chairman, 703-993-1530, Fax: 703-993-1710, E-mail: hhamburger@gmu.edu. *Application contact:* Graduate Coordinator, 703-993-1530, Fax: 703-993-3729, E-mail: csinfo@cs.gmu.edu.

The George Washington University, School of Engineering and Applied Science, Department of Computer Science, Washington, DC 20052. Offers MS, D Sc, App Sc, Engr. Part-time and evening/weekend programs available. *Degree requirements:* For master's, thesis optional, foreign language not required; for doctorate and other advanced degree, dissertation defense, qualifying exam required; for other advanced degree, foreign language and thesis not required. *Entrance requirements:* For master's, TOEFL (minimum score of 550 required; average 580) or George Washington University English as a Foreign Language Test, appropriate bachelor's degree, minimum GPA of 3.0; for doctorate, TOEFL (minimum score of 550 required; average 580) or George Washington University English as a Foreign Language Test, appropriate bachelor's or master's degree, minimum GPA of 3.3, GRE required if highest earned degree is BS; for other advanced degree, TOEFL (minimum score of 550 required; average 580) or George Washington University English as a Foreign Language Test, appropriate master's degree, minimum GPA of 3.4. *Application deadline:* For fall admission, 3/1 (priority date); for spring admission, 10/1. Applications are processed on a rolling basis. Application fee: $55. Tuition: Full-time $17,328; part-time $722 per credit hour. Required fees: $828; $35 per credit hour. Tuition and fees vary according to campus/location and program. *Financial aid:* Fellowships, research assistantships, teaching assistantships, career-related internships or fieldwork and institutionally-sponsored loans available. Financial aid application deadline: 3/1; financial aid applicants required to submit FAFSA. *Faculty research:* Computer graphics, multimedia, VLSI, parallel processing. *Unit head:* Dr. Bhagirath Narahari, Chair, 202-994-6083, Fax: 202-994-0227, E-mail: cs@seas.gwu.edu. *Application contact:* Howard M. Davis, Manager, Office of Admissions and Student Records, 202-994-6158, Fax: 202-994-0909, E-mail: data: adms@seas.gwu.edu.

See in-depth description on page 747.

Georgia Institute of Technology, Graduate Studies and Research, College of Computing, Atlanta, GA 30332-0001. Offers algorithms, combinatorics, and optimization (PhD); computer science (MS, MSCS, PhD); human computer interaction (MSHCI). Part-time programs available. Terminal master's awarded for partial completion of doctoral program. *Degree requirements:* For master's, thesis optional, foreign language not required; for doctorate, dissertation, comprehensive exam required, foreign language not required. *Entrance requirements:* For master's, GRE General Test, TOEFL (minimum score of 600 required), GRE Subject Test, minimum GPA of 3.0; for doctorate, GRE General Test, GRE Subject Test, TOEFL (minimum score of 600 required), minimum GPA of 3.3. *Faculty research:* Computer systems, graphics, intelligent systems and artificial intelligence, networks and telecommunications, software engineering.

See in-depth description on page 749.

Georgia Institute of Technology, Graduate Studies and Research, Ivan Allen College of Policy and International Affairs, Multidisciplinary Program in Human Computer Interaction, Atlanta, GA 30332-0001. Offers MSHCI. Offered jointly through the College of Computing, the Program in Information Design and Technology, and the School of Psychology. *Entrance requirements:* For master's, TOEFL.

Georgia Southwestern State University, Graduate Studies, School of Computer and Information Science, Americus, GA 31709-4693. Offers computer information systems (MS); computer science (MS). Part-time programs available. *Faculty:* 5 full-time (0 women). *Students:* 24 full-time (10 women), 13 part-time (6 women); includes 15 minority (3 African Americans, 12 Asian Americans or Pacific Islanders), 12 international. Average age 22. 20 applicants, 25% accepted. In 1998, 10 degrees awarded (100% found work related to degree). *Degree requirements:* For master's, computer language, thesis required (for some programs), foreign language not required. *Entrance requirements:* For master's, GRE General Test (minimum combined score of 900 required), minimum GPA of 3.0. *Average time to degree:* Master's–2 years full-time, 3 years part-time. *Application deadline:* For fall admission, 8/1; for spring admission, 12/15. Applications are processed on a rolling basis. Application fee: $20. Tuition, state resident: full-time $2,000; part-time $83 per hour. Tuition, nonresident: full-time $5,000; part-time $333 per hour. Required fees: $235. *Financial aid:* In 1998–99, 10 students received aid, including fellowships (averaging $6,000 per year), 10 teaching assistantships with full

Computer Science

Georgia Southwestern State University (continued)
tuition reimbursements available (averaging $4,000 per year); grants also available. Financial aid application deadline: 9/1. *Faculty research:* Database, Internet technologies, computational complexity, encryption. Total annual research expenditures: $30,000. *Unit head:* Dr. John J. Stroyls, Acting Chairman, 912-931-2263, Fax: 912-931-2270, E-mail: mdaniels@canes.gsw.edu. *Application contact:* Chris Laney, Graduate Admissions Specialist, 912-931-2027, Fax: 912-931-2983, E-mail: claney@canes.gsw.edu.

Georgia State University, College of Arts and Sciences, Department of Computer Science, Atlanta, GA 30303-3083. Offers MS. Part-time and evening/weekend programs available. *Faculty:* 36 full-time (12 women), 1 part-time (0 women). *Students:* 21 full-time (8 women), 22 part-time (8 women); includes 12 minority (4 African Americans, 8 Asian Americans or Pacific Islanders), 17 international. Average age 29. 93 applicants, 26% accepted. In 1998, 36 degrees awarded. *Degree requirements:* For master's, one foreign language (computer language can substitute), thesis or alternative required. *Entrance requirements:* For master's, GRE General Test, TOEFL (minimum score of 550 required). Application fee: $25. Tuition, state resident: full-time $2,896; part-time $121 per credit hour. Tuition, nonresident: full-time $11,584; part-time $483 per credit hour. Required fees: $468. Tuition and fees vary according to program. *Financial aid:* Research assistantships, teaching assistantships, career-related internships or fieldwork, Federal Work-Study, institutionally-sponsored loans, and tuition waivers (full) available. Aid available to part-time students. *Faculty research:* Linear algebra and graph theory, numerical and functional analysis, computer graphics and software engineering, applied statistics and probability, mathematics education. *Unit head:* Dr. Martin D. Fraser, Chair, 404-651-2245, Fax: 404-651-2246. *Application contact:* Dr. Sushil Prasad, Director of Graduate Studies, 404-651-2253, Fax: 404-651-2246, E-mail: sprasad@cs.gsu.edu.

See in-depth description on page 751.

Governors State University, College of Arts and Sciences, Division of Science, Program in Computer Science, University Park, IL 60466. Offers MS. Part-time and evening/weekend programs available. *Faculty:* 8 full-time (1 woman), 8 part-time (1 woman). *Students:* 16 full-time, 41 part-time; includes 10 African Americans, 12 Asian Americans or Pacific Islanders, 1 Hispanic American In 1998, 7 degrees awarded. *Degree requirements:* For master's, computer language, thesis or alternative required, foreign language not required. *Entrance requirements:* For master's, minimum GPA of 2.75. *Application deadline:* For fall admission, 7/15 (priority date); for spring admission, 11/10. Applications are processed on a rolling basis. Application fee: $0. *Financial aid:* Research assistantships, career-related internships or fieldwork, Federal Work-Study, institutionally-sponsored loans, and scholarships available. Aid available to part-time students. Financial aid application deadline: 5/1. *Unit head:* Dr. Edwin Cehelnik, Chairperson, Division of Science, 708-534-4520.

Graduate School and University Center of the City University of New York, Graduate Studies, Program in Computer Science, New York, NY 10036-8099. Offers PhD. *Faculty:* 56 full-time (12 women). *Students:* 74 full-time (21 women), 2 part-time; includes 6 African Americans, 9 Asian Americans or Pacific Islanders, 3 Hispanic Americans Average age 36. 70 applicants, 44% accepted. In 1998, 4 degrees awarded. *Degree requirements:* For doctorate, one foreign language (computer language can substitute), dissertation required. *Entrance requirements:* For doctorate, GRE General Test. *Application deadline:* For fall admission, 4/15. Application fee: $40. *Financial aid:* In 1998–99, 36 students received aid, including 26 fellowships, 3 teaching assistantships; research assistantships Financial aid application deadline: 2/1; financial aid applicants required to submit FAFSA. *Unit head:* Dr. Stanley Habib, Executive Officer, 212-642-2201.

Hampton University, Graduate College, Department of Computer Science, Hampton, VA 23668. Offers MS. Part-time and evening/weekend programs available. *Faculty:* 8 full-time (3 women). *Degree requirements:* For master's, computer language, thesis or alternative required, foreign language not required. *Entrance requirements:* For master's, GRE General Test (minimum score of 450 on verbal section required). *Application deadline:* For fall admission, 6/1 (priority date); for spring admission, 11/1. Applications are processed on a rolling basis. Application fee: $25. Tuition: Full-time $9,490; part-time $230 per semester hour. Required fees: $60; $35 per semester. Tuition and fees vary according to course load. *Financial aid:* Research assistantships, career-related internships or fieldwork, Federal Work-Study, and scholarships available. Aid available to part-time students. Financial aid application deadline: 5/1; financial aid applicants required to submit FAFSA. *Faculty research:* Software testing, neural networks, parallel processing, computer graphics, natural language processing. *Unit head:* Dr. Edward Hill, Interim Chair, 757-727-5552. *Application contact:* Erika Henderson, Director, Graduate Programs, 757-727-5454, Fax: 757-727-5084.

Harvard University, Graduate School of Arts and Sciences, Division of Engineering and Applied Sciences, Center for Research in Computing Technology, Cambridge, MA 02138. Offers PhD. *Degree requirements:* For doctorate, dissertation required, foreign language not required. *Entrance requirements:* For doctorate, GRE General Test, GRE Subject Test, TOEFL (minimum score of 550 required). *Application deadline:* For fall admission, 12/15. Application fee: $60. *Financial aid:* Career-related internships or fieldwork, Federal Work-Study, and institutionally-sponsored loans available. Financial aid application deadline: 12/30. *Unit head:* Thomas E. Cheatham, Director, 617-495-3989. *Application contact:* Office of Admissions and Financial Aid, 617-495-5315.

Hofstra University, College of Liberal Arts and Sciences, Division of Natural Sciences, Mathematics, Engineering, and Computer Science, Department of Computer Science, Hempstead, NY 11549. Offers MA, MS. Part-time and evening/weekend programs available. *Faculty:* 6 full-time (1 woman), 4 part-time (0 women). *Students:* 2 full-time (0 women), 17 part-time (2 women), 3 international. Average age 32. In 1998, 11 degrees awarded. *Degree requirements:* For master's, computer language, thesis, projects (MA) required, foreign language not required. *Entrance requirements:* For master's, GRE General Test (minimum combined score of 1650 on three sections required), minimum GPA of 3.0. *Application deadline:* Applications are processed on a rolling basis. Application fee: $40 ($75 for international students). *Financial aid:* Fellowships, institutionally-sponsored loans available. Aid available to part-time students. Financial aid applicants required to submit FAFSA. *Faculty research:* Graphics, medical imaging, modeling graphics, artificial intelligence. *Unit head:* Dr. Paul Nagin, Chairperson, 516-463-5558, Fax: 516-463-5790, E-mail: cscpan@magic.hofstra.edu. *Application contact:* Mary Beth Carey, Vice President of Enrollment Services, 516-463-6700, Fax: 516-560-7660, E-mail: hofstra@hofstra.edu.

Hollins University, Graduate Programs, Program in Liberal Studies, Roanoke, VA 24020-1688. Offers computer science (MALS); general studies (MALS); humanities (MALS); liberal studies (CAS); social studies (MALS). Part-time and evening/weekend programs available. *Faculty:* 10 full-time (6 women), 5 part-time (all women). *Students:* 26 full-time (22 women), 140 part-time (108 women); includes 14 minority (13 African Americans, 1 Native American), 3 international. *Degree requirements:* For master's, thesis required, foreign language not required; for degree, foreign language not required. *Entrance requirements:* For master's, interview. *Application deadline:* For fall admission, 8/14 (priority date); for spring admission, 1/10. Applications are processed on a rolling basis. Application fee: $25. *Application contact:* Cathy S. Koon, Administrative Assistant, 540-362-6575, Fax: 540-362-6288, E-mail: ckoon@hollins.edu.

Hood College, Graduate School, Programs in Computer and Information Sciences, Frederick, MD 21701-8575. Offers MS. *Students:* 16 full-time (12 women), 150 part-time (66 women); includes 34 minority (5 African Americans, 28 Asian Americans or Pacific Islanders, 1 Hispanic American), 10 international. Average age 33. In 1998, 37 degrees awarded. *Degree requirements:* For master's, thesis or alternative required, foreign language not required. *Entrance requirements:* For master's, minimum GPA of 2.5. *Application deadline:* Applications are processed on a rolling basis. Application fee: $30. *Financial aid:* Career-related internships or fieldwork, institutionally-sponsored loans, and tuition waivers (partial) available. Aid available to part-time students. Financial aid applicants required to submit FAFSA. *Unit head:* John Boon, Chairperson, 301-696-3763, Fax: 301-696-3597, E-mail: boon@hood.edu. *Application contact:* Graduate School Office, 301-696-3600, Fax: 301-696-3597, E-mail: hoodgrad@hood.edu.

Howard University, College of Engineering, Architecture, and Computer Sciences, School of Engineering and Computer Science, Department of Systems and Computer Science, Washington, DC 20059-0002. Offers MCS. Offered through the Graduate School of Arts and Sciences. Part-time and evening/weekend programs available. *Faculty:* 9 full-time (1 woman), 5 part-time (0 women). *Students:* 12 full-time (4 women), 9 part-time (4 women); includes 18 minority (15 African Americans, 3 Asian Americans or Pacific Islanders) 50 applicants, 40% accepted. In 1998, 9 degrees awarded (100% found work related to degree). *Degree requirements:* For master's, computer language required, thesis optional, foreign language not required. *Entrance requirements:* For master's, GRE General Test, TOEFL, minimum GPA of 3.0. *Average time to degree:* Master's–1.88 years full-time, 3.5 years part-time. *Application deadline:* For fall admission, 4/1 (priority date); for spring admission, 11/1. Applications are processed on a rolling basis. Application fee: $45. *Financial aid:* In 1998–99, 3 students received aid, including 5 research assistantships with full tuition reimbursements available (averaging $8,000 per year), 5 teaching assistantships with full tuition reimbursements available (averaging $8,000 per year); fellowships, career-related internships or fieldwork, grants, and institutionally-sponsored loans also available. Financial aid application deadline: 4/1; financial aid applicants required to submit FAFSA. *Faculty research:* Software engineering, software fault-tolerance, software reliability, artificial intelligence. Total annual research expenditures: $549,722. *Unit head:* Dr. Ronald J. Leach, Acting Chair, 202-806-6595, Fax: 202-806-4531, E-mail: rjl@scs.howard.edu. *Application contact:* Dr. John Trimble, Director of Graduate Studies, 202-806-4822, Fax: 202-806-4531, E-mail: trimble@scs.howard.edu.

Hunter College of the City University of New York, Graduate School, School of Arts and Sciences, Department of Computer Science, New York, NY 10021-5085. Offers PhD. Part-time and evening/weekend programs available. *Faculty:* 10 full-time (3 women), 7 part-time (3 women). *Degree requirements:* For doctorate, one foreign language, computer language, dissertation required. Application fee: $40. Tuition, state resident: full-time $4,350; part-time $185 per credit. Tuition, nonresident: full-time $7,600; part-time $320 per credit. Required fees: $8 per term. *Faculty research:* Artificial intelligence, software engineering, graph theory, combinatorics. *Unit head:* Dr. Howard A. Rubin, Chair, 212-772-5213, Fax: 212-772-5219, E-mail: howard.rubin@hunter.cuny.edu.

Illinois Institute of Technology, Graduate College, Armour College of Engineering and Sciences, Department of Computer Science, Chicago, IL 60616-3793. Offers computer science (MS, PhD); teaching (MST); telecommunications and software engineering (MTSE). Part-time and evening/weekend programs available. *Faculty:* 16 full-time (3 women), 15 part-time (3 women). *Students:* 161 full-time (34 women), 574 part-time (112 women); includes 230 minority (34 African Americans, 185 Asian Americans or Pacific Islanders, 11 Hispanic Americans), 239 international. 605 applicants, 50% accepted. In 1998, 116 master's, 14 doctorates awarded. Terminal master's awarded for partial completion of doctoral program. *Degree requirements:* For master's, computer language, thesis (for some programs), comprehensive exam required, foreign language not required; for doctorate, computer language, dissertation, comprehensive exam required, foreign language not required. *Entrance requirements:* For master's and doctorate, GRE (minimum score of 1200 required), TOEFL (minimum score of 550 required), undergraduate GPA of 3.0 required. *Application deadline:* For fall admission, 7/1; for spring admission, 11/1. Applications are processed on a rolling basis. Application fee: $30. Electronic applications accepted. *Financial aid:* In 1998–99, 3 fellowships, 7 research assistantships, 33 teaching assistantships were awarded.; Federal Work-Study, institutionally-sponsored loans, scholarships, and graduate assistantships also available. Financial aid application deadline: 3/1. *Faculty research:* Networking, computer graphics, database, parallel computing, software engineering. Total annual research expenditures: $280,084. *Unit head:* Dr. Bodgen Korel, Interim Chairman, 312-567-5150, Fax: 312-567-5067, E-mail: korel@charlie.cns.iit.edu. *Application contact:* Dr. S. Mohammad Shahidehpour, Dean of Graduate College, 312-567-3024, Fax: 312-567-7517, E-mail: grad@minna.cns.iit.edu.

See in-depth description on page 753.

Illinois State University, Graduate School, College of Applied Science and Technology, Department of Applied Computer Science, Normal, IL 61790-2200. Offers MS. *Faculty:* 15 full-time (3 women), 1 part-time (0 women). *Students:* 31 full-time (11 women), 38 part-time (13 women); includes 5 minority (2 African Americans, 3 Asian Americans or Pacific Islanders), 17 international. 54 applicants, 30% accepted. In 1998, 30 degrees awarded. *Degree requirements:* For master's, computer language, thesis required (for some programs). *Entrance requirements:* For master's, GRE General Test (minimum score of 400 on verbal section, 1000 combined required), minimum GPA of 3.0 in last 60 hours; proficiency in COBOL, FORTRAN, Pascal, or P12. *Application deadline:* Applications are processed on a rolling basis. Application fee: $0. Tuition, state resident: full-time $2,526; part-time $105 per credit hour. Tuition, nonresident: full-time $7,578; part-time $316 per credit hour. Required fees: $1,082; $38 per credit hour. Tuition and fees vary according to course load and program. *Financial aid:* In 1998–99, 1 research assistantship, 12 teaching assistantships were awarded.; fellowships, tuition waivers (full) and unspecified assistantships also available. Financial aid application deadline: 4/1. *Faculty research:* Technology training for state employees, foundation computing. Total annual research expenditures: $1.8 million. *Unit head:* Dr. Robert Zant, Chairperson, 309-438-8338.

Indiana University Bloomington, Graduate School, College of Arts and Sciences, Department of Computer Science, Bloomington, IN 47405. Offers computer science (MS, PhD); computer science/cognitive science (PhD); computer science/logic (PhD). PhD offered through the University Graduate School. *Faculty:* 20 full-time (1 woman). *Students:* 82 full-time (28 women), 32 part-time (7 women); includes 7 minority (2 African Americans, 3 Asian Americans or Pacific Islanders, 2 Hispanic Americans), 69 international. In 1998, 40 master's, 10 doctorates awarded. Terminal master's awarded for partial completion of doctoral program. *Degree requirements:* For master's, computer language required, thesis optional, foreign language not required; for doctorate, computer language, dissertation, oral and written exams required, foreign language not required. *Entrance requirements:* For master's and doctorate, GRE General Test (minimum combined score of 1800 on three sections required), TOEFL (minimum score of 600 required). *Average time to degree:* Master's–2 years full-time. *Application deadline:* For fall admission, 1/15 (priority date); for spring admission, 9/1 (priority date). Applications are processed on a rolling basis. Application fee: $40. Electronic applications accepted. Tuition, state resident: part-time $161 per credit hour. Tuition, nonresident: part-time $468 per credit hour. Required fees: $360 per year. Tuition and fees vary according to course load and program. *Financial aid:* In 1998–99, 8 fellowships with full tuition reimbursements (averaging $15,000 per year), 22 research assistantships with full tuition reimbursements (averaging $11,400 per year), 54 teaching assistantships with full tuition reimbursements (averaging $11,400 per year) were awarded.; Federal Work-Study and traineeships also available. Financial aid application deadline: 2/1. *Faculty research:* Hardware/VLSI, parallel programming, graphics/visualization, programming language, cognitive science/artificial intelligence. *Unit head:* Dr. Daniel Friedman, Chairman, 812-855-6488, Fax: 812-855-4829. *Application contact:* Pam Larson, Admissions Secretary, 812-855-6487, Fax: 812-855-4829, E-mail: admissions@cs.indiana.edu.

See in-depth description on page 755.

Indiana University–Purdue University Fort Wayne, School of Engineering, Technology, and Computer Science, Department of Computer Science, Fort Wayne, IN 46805-1499. Offers applied computer science (MS). Part-time and evening/weekend programs available. *Faculty:* 6 full-time (1 woman). *Students:* 1 full-time (0 women), 15 part-time (2 women). Average age 36. *Degree requirements:* For master's, computer language required, foreign language and thesis not required. *Entrance requirements:* For master's, GRE, TOEFL, minimum GPA of 3.0. *Application deadline:* For fall admission, 2/15 (priority date); for spring admission, 9/1. Applications are processed on a rolling basis. Application fee: $30. *Financial aid:* Career-related internships or fieldwork available. Financial aid application deadline: 3/1; financial aid applicants required to submit FAFSA. *Unit head:* James L. Silver, Chair, 219-481-6803, Fax: 219-481-5734.

Indiana University–Purdue University Indianapolis, School of Science, Department of Computer and Information Science, Indianapolis, IN 46202-5132. Offers computer science (MS). Part-time and evening/weekend programs available. *Students:* 2 full-time (1 woman), 35 part-time (11 women); includes 7 minority (all Asian Americans or Pacific Islanders), 17 international. Average age 30. In 1998, 10 degrees awarded (90% entered university research/teaching, 10% found other work related to degree). *Degree requirements:* For master's, computer language required, thesis optional, foreign language not required. *Entrance requirements:* For master's, GRE, BS or equivalent in computer science. *Average time to degree:* Master's–2 years full-time, 3 years part-time. *Application deadline:* For fall admission, 1/15 (priority date); for spring admission, 9/15. Applications are processed on a rolling basis. Application fee: $30 ($50 for international students). Electronic applications accepted. Tuition, state resident: part-time $158 per credit hour. Tuition, nonresident: part-time $455 per credit hour. Required fees: $121 per year. Tuition and fees vary according to course load and degree level. *Financial aid:* In 1998–99, 8 students received aid, including 1 fellowship, 5 research assistantships with tuition reimbursements available; teaching assistantships with tuition reimbursements available, career-related internships or fieldwork, institutionally-sponsored loans, and tuition waivers (full and partial) also available. Aid available to part-time students. Financial aid application deadline: 1/15; financial aid applicants required to submit FAFSA. *Faculty research:* Artificial intelligence, graphics and visualization, computational geometry, database systems, distributed computing. *Unit head:* Mathew J. Palakal, Chair, 317-274-9727, Fax: 317-274-9742, E-mail: grad_advisor@cs.iupui.edu. *Application contact:* 317-274-9727, Fax: 317-274-9742, E-mail: admissions@cs.iupui.edu.

Instituto Tecnológico y de Estudios Superiores de Monterrey, Campus Estado de México, Graduate Division, Division of Engineering and Architecture, Atizapán de Zaragoza, 52500, Mexico. Offers computer science (MCS); environmental engineering (MEE); industrial engineering (MIE); manufacturing systems (MMS); materials engineering (PhD). *Degree requirements:* For master's; for doctorate, one foreign language, dissertation required. *Entrance requirements:* For master's, interview; for doctorate, research proposal. *Application deadline:* For fall admission, 1/13 (priority date); for spring admission, 4/4. Applications are processed on a rolling basis. Application fee: 750 Mexican pesos. *Unit head:* Juan López Díaz, Headmaster, 5-326-5530, Fax: 5-326-5531, E-mail: jlopez@campus.cem.itesm.mx. *Application contact:* Lourdes Turrubiates, Admissions Officer, 5-326-5776, Fax: 5-326-5788, E-mail: lturrubi@campus.cem.itesm.mx.

Instituto Tecnológico y de Estudios Superiores de Monterrey, Campus Monterrey, Graduate and Research Division, Program in Computer Science, Monterrey, 64849, Mexico. Offers artificial intelligence (PhD); computer science (MS); information systems (MS); information technology (MS). Part-time programs available. *Degree requirements:* For master's and doctorate, thesis/dissertation required. *Entrance requirements:* For master's, PAEG, TOEFL; for doctorate, TOEFL, master's in related field. *Faculty research:* Distributed systems, software engineering, decision support systems.

Instituto Tecnológico y de Estudios Superiores de Monterrey, Campus Morelos, Programs in Information Science, Cuernavaca, 62000, Mexico. Offers administration of information technology (MATI); computer science (MCC, DCC); information technology (MTI).

Iona College, School of Arts and Science, Program in Computer Science, New Rochelle, NY 10801-1890. Offers MS. Part-time and evening/weekend programs available. *Faculty:* 7 full-time (2 women), 6 part-time (1 woman). Average age 33. In 1998, 14 degrees awarded. *Degree requirements:* For master's, thesis or alternative required. *Entrance requirements:* For master's, minimum GPA of 3.0. *Application deadline:* Applications are processed on a rolling basis. Application fee: $25. *Financial aid:* Tuition waivers (partial) and unspecified assistantships available. Aid available to part-time students. *Faculty research:* Telecommunications, computer graphics, expert systems, database design, compiler design. *Unit head:* Dr. John Mallozzi, Chair, 914-633-2578. *Application contact:* Arlene Melillo, Director of Graduate Recruitment, 914-633-2328, Fax: 914-633-2023.

Iowa State University of Science and Technology, Graduate College, College of Liberal Arts and Sciences, Department of Computer Science, Ames, IA 50011. Offers MS, PhD. *Faculty:* 20 full-time. *Students:* 66 full-time (15 women), 28 part-time (8 women); includes 2 minority (1 Asian American or Pacific Islander, 1 Hispanic American), 74 international. 432 applicants, 16% accepted. In 1998, 32 master's, 4 doctorates awarded. *Degree requirements:* For master's, thesis or alternative required; for doctorate, dissertation required. *Entrance requirements:* For master's and doctorate, GRE General Test, TOEFL (minimum score of 550 required). *Application deadline:* For fall admission, 6/15 (priority date); for spring admission, 11/15 (priority date). Application fee: $20 ($50 for international students). Electronic applications accepted. Tuition, state resident: full-time $3,308. Tuition, nonresident: full-time $9,744. Part-time tuition and fees vary according to course load, campus/location and program. *Financial aid:* In 1998–99, 37 research assistantships with partial tuition reimbursements (averaging $12,104 per year), 46 teaching assistantships with partial tuition reimbursements (averaging $11,802 per year) were awarded.; fellowships, scholarships also available. *Unit head:* Dr. Leslie L. Miller, Interim Chair, 515-294-4377, Fax: 515-294-0258, E-mail: grad_adm@cs.iastate.edu.

See in-depth description on page 757.

Jackson State University, Graduate School, School of Science and Technology, Department of Computer Science, Jackson, MS 39217. Offers MS. Part-time and evening/weekend programs available. *Faculty:* 12 full-time (3 women). *Students:* 43 full-time (16 women), 35 part-time (9 women); includes 38 minority (25 African Americans, 13 Asian Americans or Pacific Islanders), 37 international. In 1998, 20 degrees awarded. *Degree requirements:* For master's, computer language, thesis, comprehensive exam required. *Entrance requirements:* For master's, GRE General Test (minimum combined score of 1000 required), TOEFL (minimum score of 550 required). *Application deadline:* For fall admission, 3/1 (priority date); for spring admission, 10/1. Applications are processed on a rolling basis. Application fee: $20. *Financial aid:* In 1998–99, 12 students received aid. Career-related internships or fieldwork, Federal Work-Study, scholarships, and unspecified assistantships available. Aid available to part-time students. Financial aid application deadline: 3/1; financial aid applicants required to submit FAFSA. *Unit head:* Dr. Loretta Moore, Chair. *Application contact:* Curtis Gore, Admissions Coordinator, 601-974-5841, Fax: 601-974-6196, E-mail: cgore@ccaix.jsums.edu.

James Madison University, Graduate School, College of Integrated Science and Technology, Department of Computer Science, Harrisonburg, VA 22807. Offers MS. Postbaccalaureate distance learning degree programs offered (no on-campus study). *Faculty:* 7 full-time (1 woman), 3 part-time (0 women). *Students:* 25 full-time (9 women), 91 part-time (15 women); includes 13 minority (4 African Americans, 8 Asian Americans or Pacific Islanders, 1 Hispanic American), 12 international. Average age 30. In 1998, 9 degrees awarded. *Degree requirements:* For master's, thesis or alternative required, foreign language not required. *Entrance requirements:* For master's, GRE General Test. *Application deadline:* For fall admission, 7/1 (priority date). Applications are processed on a rolling basis. Application fee: $50. Tuition, state resident: full-time $3,240; part-time $135 per credit hour. Tuition, nonresident: full-time $9,960; part-time $415 per credit hour. *Financial aid:* Fellowships, teaching assistantships, Federal Work-Study and unspecified assistantships available. Financial aid application deadline: 2/15; financial aid applicants required to submit FAFSA. *Unit head:* Dr. Charles W. Reynolds, Director, 540-568-2770.

Johns Hopkins University, G. W. C. Whiting School of Engineering, Department of Computer Science, Baltimore, MD 21218-2699. Offers MSE, PhD. *Faculty:* 13 full-time (0 women), 6 part-time (0 women). *Students:* 59 full-time (16 women), 3 part-time (2 women); includes 1 minority (Asian American or Pacific Islander), 36 international. Average age 26. 156 applicants, 38% accepted. In 1998, 21 master's, 3 doctorates awarded. Terminal master's awarded for partial completion of doctoral program. *Degree requirements:* For master's, thesis optional, foreign language not required; for doctorate, dissertation required, foreign language not required. *Entrance requirements:* For master's, GRE General Test, TOEFL (minimum score of 560 required); for doctorate, GRE General Test (average 654 verbal, 793 quantitative, 767 analytical), GRE Subject Test (recommended), TOEFL (minimum score of 560 required). *Application deadline:* For fall admission, 2/1. Application fee: $50. Tuition: Full-time $23,660. Tuition and fees vary according to program. *Financial aid:* In 1998–99, 1 fellowship (averaging $13,230 per year), 24 research assistantships (averaging $13,230 per year), 9 teaching assistantships (averaging $13,230 per year) were awarded.; Federal Work-Study, grants, and institutionally-sponsored loans also available. Financial aid application deadline: 2/1. *Faculty research:* Artificial intelligence, networking parallel algorithms, programming languages, fault-tolerant computing, geometric computing. Total annual research expenditures: $2.2 million. *Unit head:* Dr. Gerald M. Masson, Chair, 410-516-7013, Fax: 410-516-6134, E-mail: masson@cs.jhu.edu. *Application contact:* Linda Rorke, Admissions Coordinator, 410-516-8775, Fax: 410-516-6134.

See in-depth description on page 759.

Kansas State University, Graduate School, College of Engineering, Department of Computing and Information Sciences, Manhattan, KS 66506. Offers computer science (MS, PhD); software engineering (MSE). Part-time programs available. Postbaccalaureate distance learning degree programs offered (minimal on-campus study). *Faculty:* 16 full-time (2 women). *Students:* 70 full-time (22 women), 50 part-time (13 women); includes 10 minority (1 African American, 8 Asian Americans or Pacific Islanders, 1 Hispanic American), 76 international. 312 applicants, 27% accepted. In 1998, 47 master's, 3 doctorates awarded (100% entered university research/teaching). Terminal master's awarded for partial completion of doctoral program. *Degree requirements:* For master's, computer language required, thesis optional, foreign language not required; for doctorate, computer language, dissertation required, foreign language not required. *Entrance requirements:* For master's, TOEFL (minimum score of 575 required); for doctorate, GRE General Test (minimum score of 400 on verbal section, 650 on quantitative, 600 on analytical required), TOEFL (minimum score of 575 required). *Average time to degree:* Master's–2 years full-time, 3 years part-time; doctorate–5 years full-time. *Application deadline:* For fall admission, 4/1 (priority date); for spring admission, 11/1. Applications are processed on a rolling basis. Application fee: $0 ($25 for international students). Electronic applications accepted. *Financial aid:* In 1998–99, 10 research assistantships (averaging $9,000 per year), 29 teaching assistantships (averaging $9,000 per year) were awarded.; career-related internships or fieldwork and institutionally-sponsored loans also available. Financial aid application deadline: 2/15. *Faculty research:* Programming language semantics, distributed systems, real time systems and algorithms, database management systems. Total annual research expenditures: $511,000. *Unit head:* Dr. Virgil E. Wallentine, Head, 785-532-6350, Fax: 785-532-7353, E-mail: virg@cis.ksu.edu. *Application contact:* Dr. David Gustafson, Graduate Coordinator, 785-532-6350, Fax: 785-532-7353, E-mail: dag@cis.ksu.edu.

See in-depth description on page 761.

Kent State University, College of Arts and Sciences, Department of Mathematics and Computer Science, Kent, OH 44242-0001. Offers applied mathematics (MA, MS, PhD); computer science (MA, MS, PhD); pure mathematics (MA, MS, PhD). *Faculty:* 41 full-time. *Students:* 69 full-time (19 women), 60 part-time (23 women); includes 2 minority (both Asian Americans or Pacific Islanders), 78 international. 135 applicants, 76% accepted. In 1998, 14 master's, 9 doctorates awarded. *Degree requirements:* For master's, thesis optional, foreign language not required; for doctorate, dissertation required. *Entrance requirements:* For master's, GRE, minimum GPA of 2.75; for doctorate, GRE, minimum GPA of 3.0. *Application deadline:* For fall admission, 7/12; for spring admission, 11/29. Applications are processed on a rolling basis. Application fee: $30. *Financial aid:* Fellowships, research assistantships, teaching assistantships, Federal Work-Study, institutionally-sponsored loans, and tuition waivers (full) available. Financial aid application deadline: 2/1. *Unit head:* Dr. Austin C. Melton, Chairman, 330-672-2430, Fax: 330-672-7824.

Knowledge Systems Institute, Program in Computer and Information Sciences, Skokie, IL 60076. Offers MS. *Faculty:* 4 full-time (0 women), 23 part-time (3 women). *Students:* 83 full-time (25 women), 61 part-time (21 women). Application fee: $40. Electronic applications accepted. Tuition: Full-time $6,730; part-time $275 per credit hour. Required fees: $5 per course. $30 per term. *Unit head:* Judy Pan, Executive Director, 847-679-3135, Fax: 847-679-3166, E-mail: judy@ksi.edu. *Application contact:* Margaret Price, Admissions Officer, 847-679-3135, Fax: 847-679-3166, E-mail: mprice@ksi.edu.

Announcement: Knowledge Systems Institute Graduate School is NCA accredited and offers master's degrees in computer and information sciences with 7 concentrations. Programs include certificate, job training, and distance learning. Evening and weekend classes. All PhD full-time faculty members. Campus near Northwestern University. Financial aid and job placement available. Contact the School at 3420 Main Street, Skokie, IL 60076; phone: 847-679-3135; fax: 847-679-3166; e-mail: office@ksi.edu; WWW: http://www.ksi.edu.

Kutztown University of Pennsylvania, College of Graduate Studies and Extended Learning, College of Liberal Arts and Sciences, Program in Computer and Information Science, Kutztown, PA 19530-0730. Offers MS. *Faculty:* 10 full-time (3 women). *Students:* 2 full-time (1 woman), 11 part-time (4 women); includes 5 minority (all Asian Americans or Pacific Islanders) Average age 33. In 1998, 5 degrees awarded. *Degree requirements:* For master's, computer language, comprehensive exam or thesis required. *Entrance requirements:* For master's, GRE General Test, TOEFL, TSE. *Application deadline:* For fall admission, 3/1; for spring admission, 8/1. Application fee: $25. *Financial aid:* Career-related internships or fieldwork, Federal Work-Study, tuition waivers (partial), and unspecified assistantships available. Financial aid application deadline: 3/15; financial aid applicants required to submit FAFSA. *Faculty research:* Artificial intelligence, expert systems, neural networks. *Unit head:* William Bateman, Chairperson, 610-683-4410, E-mail: bateman@kutztown.edu.

Lakehead University, Graduate Studies and Research, Faculty of Arts and Science, School of Mathematical Sciences, Thunder Bay, ON P7B 5E1, Canada. Offers computer science (M Sc, MA); mathematics and statistics (M Sc, MA). Part-time and evening/weekend programs available. *Degree requirements:* For master's, thesis or alternative required, foreign language not required. *Entrance requirements:* For master's, TOEFL (minimum score of 550 required), minimum B average. *Faculty research:* Numerical analysis, classical analysis, theoretical computer science, abstract harmonic analysis, functional analysis.

Lamar University, College of Graduate Studies, College of Engineering, Department of Computer Science, Beaumont, TX 77710. Offers MS. Part-time programs available. *Faculty:* 5 full-time (1 woman). *Students:* 58 full-time (13 women), 26 part-time (5 women); includes 5 minority (all Asian Americans or Pacific Islanders), 72 international. Average age 31. 50 applicants, 90% accepted. In 1998, 21 degrees awarded. *Degree requirements:* For master's, computer language, comprehensive exams and project or thesis required. *Entrance requirements:* For master's, GRE General Test (minimum combined score of 1050 required), TOEFL (minimum score of 500 required), minimum GPA of 3.0 or 3.3 in last 60 hours of undergraduate course work. *Average time to degree:* Master's–2 years full-time, 4 years part-time. *Application deadline:* For fall admission, 5/15 (priority date); for spring admission, 10/1 (priority date). Applications are processed on a rolling basis. Application fee: $0. *Financial aid:* In 1998–99, 29 students received aid, including 2 research assistantships with tuition reimbursements available (averaging $6,000 per year), 11 teaching assistantships with tuition reimbursements available (averaging $6,000 per year); institutionally-sponsored loans, scholarships, and tuition waivers (partial) also available. Financial aid application deadline: 4/1. *Faculty research:* Artificial intelligence, complexity, databases, networks, distributed systems. *Unit head:* Dr. Lawrence J. Osborne, Chair, 409-880-8775, Fax: 409-880-2364, E-mail: osborne@

Computer Science

Lamar University (continued)
hal.lamar.edu. *Application contact:* Sandy Drane, Coordinator, International Students and Graduate Studies, 409-880-8349, Fax: 409-880-8414, E-mail: dranesl@lub002.lamar.edu.

Announcement: Lamar University Department of Computer Science offers an MS degree in computer science. The curriculum is designed to prepare those who aspire to careers related to networking, database design, and intelligent systems. It is an excellent program for those who have a background in computer science, but it is designed so that those without such a background can still complete the requirements for the degree within 2 years.

La Salle University, School of Arts and Sciences, Program in Computer Information Science, Philadelphia, PA 19141-1199. Offers MS. Part-time and evening/weekend programs available. *Faculty:* 7 full-time (3 women), 3 part-time (0 women). *Students:* 5 full-time (1 woman), 72 part-time (11 women); includes 7 minority (4 African Americans, 3 Asian Americans or Pacific Islanders), 1 international. Average age 35. In 1998, 10 degrees awarded. *Degree requirements:* For master's, computer language required, foreign language not required. *Entrance requirements:* For master's, GRE or MAT, 18 undergraduate credits in computer science, professional experience. *Average time to degree:* Master's–2 years full-time, 4 years part-time. *Application deadline:* Applications are processed on a rolling basis. Application fee: $30. *Financial aid:* Grants and institutionally-sponsored loans available. Aid available to part-time students. Financial aid application deadline: 7/15; financial aid applicants required to submit FAFSA. *Faculty research:* Human-computer interaction, networks, technology trends, databases, groupware. *Unit head:* Dr. Margaret McManus, Director, 215-951-1222, Fax: 215-951-1805, E-mail: macis@lasalle.edu.

Lehigh University, College of Engineering and Applied Science, Department of Electrical Engineering, Computer Science and Computer Engineering, Program in Computer Science, Bethlehem, PA 18015-3094. Offers MS, PhD. Part-time programs available. *Faculty:* 6 full-time (1 woman). *Students:* 36 full-time (6 women), 14 part-time (3 women); includes 9 minority (3 African Americans, 5 Asian Americans or Pacific Islanders, 1 Hispanic American), 13 international. Average age 24. 164 applicants, 4% accepted. In 1998, 7 master's, 4 doctorates awarded. *Degree requirements:* For master's, oral presentation of thesis required; for doctorate, dissertation, qualifying, general, and oral exams required. *Entrance requirements:* For master's, GRE General Test (minimum combined score of 1600 on three sections required), TOEFL (minimum score of 550 required), minimum GPA of 3.0; for doctorate, GRE General Test (minimum combined score of 1600 on three sections required), TOEFL (minimum score of 550 required), MS, minimum GPA of 3.25. *Application deadline:* For fall admission, 7/15; for spring admission, 12/1. Applications are processed on a rolling basis. Application fee: $40. Electronic applications accepted. *Financial aid:* Fellowships, research assistantships, teaching assistantships available. Financial aid application deadline: 1/15. *Application contact:* Anne Nierer, Graduate Coordinator, 610-758-4072, Fax: 610-758-6279, E-mail: aln3@lehigh.edu.

See in-depth description on page 763.

Lehman College of the City University of New York, Division of Natural and Social Sciences, Department of Mathematics and Computer Science, Program in Computer Science, Bronx, NY 10468-1589. Offers MS. *Students:* 10 full-time (2 women), 21 part-time (6 women). *Degree requirements:* For master's, one foreign language, computer language, thesis or alternative required. *Application deadline:* For fall admission, 4/1; for spring admission, 11/1. Applications are processed on a rolling basis. Application fee: $40. Tuition, state resident: full-time $4,350; part-time $185 per credit. Tuition, nonresident: full-time $7,600; part-time $320 per credit. *Financial aid:* Federal Work-Study and tuition waivers (full and partial) available. Aid available to part-time students. Financial aid application deadline: 5/15; financial aid applicants required to submit FAFSA. *Unit head:* Charles Berger, Adviser, 718-960-8117, Fax: 718-960-8969.

Long Island University, Brooklyn Campus, School of Business and Public Administration, Department of Computer Science, Brooklyn, NY 11201-8423. Offers MS. *Faculty:* 5 full-time (0 women). *Students:* 34 full-time (16 women), 38 part-time (12 women); includes 60 minority (17 African Americans, 37 Asian Americans or Pacific Islanders, 5 Hispanic Americans, 1 Native American) 104 applicants, 66% accepted. In 1998, 20 degrees awarded. *Entrance requirements:* For master's, GMAT or GRE. *Application deadline:* Applications are processed on a rolling basis. Application fee: $30. Electronic applications accepted. *Financial aid:* Scholarships and unspecified assistantships available. Aid available to part-time students. *Unit head:* Dr. Walter Vasilaky, Chair, 718-488-1073. *Application contact:* Bernard W. Sullivan, Associate Director of Admissions, 718-488-1011, Fax: 718-797-2399, E-mail: attend@liu.edu.

Long Island University, Rockland Graduate Campus, Graduate School, Program in Computer Science, Orangeburg, NY 10962. Offers MS. *Degree requirements:* For master's, foreign language and thesis not required. *Entrance requirements:* For master's, GRE General Test (minimum combined score of 1100 required). *Application deadline:* Applications are processed on a rolling basis. *Financial aid:* Scholarships available. Aid available to part-time students. *Unit head:* Elena M. Quiroz, Assistant Dean, 518-276-6142, E-mail: bioinformatics@rpi.edu. *Application contact:* Irene R. Delgado, Director of Admissions/Marketing, 914-359-7200, Fax: 914-359-7248, E-mail: delgado@liu.edu.

Louisiana State University and Agricultural and Mechanical College, Graduate School, College of Basic Sciences, Department of Computer Science, Program in Computer Science, Baton Rouge, LA 70803. Offers MSSS, PhD. Part-time programs available. *Students:* 25 full-time (5 women), 22 part-time (4 women); includes 3 minority (1 African American, 2 Asian Americans or Pacific Islanders), 26 international. 53 applicants, 40% accepted. In 1998, 10 doctorates awarded. Terminal master's awarded for partial completion of doctoral program. *Degree requirements:* For master's and doctorate, computer language, thesis/dissertation required, foreign language not required. *Entrance requirements:* For master's, GRE General Test (minimum combined score of 1000 required), minimum GPA of 3.0; for doctorate, GRE General Test (minimum combined score of 1200 required), minimum GPA of 3.0. *Application deadline:* Applications are processed on a rolling basis. Application fee: $25. *Financial aid:* In 1998–99, 5 fellowships, 2 research assistantships with partial tuition reimbursements, 18 teaching assistantships with partial tuition reimbursements were awarded.; unspecified assistantships also available. *Application contact:* Dr. Doris Carver, Graduate Coordinator, 225-388-1495, Fax: 225-388-1465, E-mail: carver@bit.csc.lsu.edu.

Louisiana Tech University, Graduate School, College of Engineering and Science, Department of Computer Science, Ruston, LA 71272. Offers MS. Part-time programs available. *Degree requirements:* For master's, computer language, thesis or alternative required, foreign language not required. *Entrance requirements:* For master's, GRE General Test (minimum combined score of 1070 required), TOEFL (minimum score of 550 required), minimum GPA of 3.0 in last 60 hours. *Faculty research:* Computer systems organization, artificial intelligence, expert systems, graphics, program language.

Loyola Marymount University, Graduate Division, College of Science and Engineering, Department of Electrical Engineering and Computer Science, Program in Computer Science, Los Angeles, CA 90045-8350. Offers MS. Part-time and evening/weekend programs available. *Faculty:* 4 part-time (0 women). *Students:* 11 full-time (2 women), 5 part-time; includes 8 minority (1 African American, 6 Asian Americans or Pacific Islanders, 1 Hispanic American), 2 international. 42 applicants, 31% accepted. In 1998, 6 degrees awarded. *Degree requirements:* For master's, computer language, research seminar required, foreign language and thesis not required. *Entrance requirements:* For master's, TOEFL (minimum score of 550 required). Application fee: $35. Electronic applications accepted. Tuition: Part-time $525 per unit. Required fees: $143; $14 per semester. Tuition and fees vary according to program. *Financial aid:* In 1998–99, 1 student received aid. Grants available. Aid available to part-time students. Financial aid application deadline: 3/2; financial aid applicants required to submit FAFSA. *Unit head:* Dr. Paul A. Rude, Graduate Director, 310-338-5101. *Application contact:* Dr. Paul A. Rude, Graduate Director, 310-338-5101.

Loyola University Chicago, Graduate School, Department of Mathematical and Computer Sciences, Chicago, IL 60611-2196. Offers computer science (MS); mathematical science (MS). Part-time and evening/weekend programs available. *Degree requirements:* For master's, oral and written comprehensive exams required. *Entrance requirements:* For master's, GRE General Test or TOEFL (minimum score of 550 required), minimum B average. *Faculty research:* Parallel computing, programming language, analysis of algorithms, logic.

See in-depth description on page 765.

Loyola University New Orleans, College of Arts and Sciences, Department of Mathematics and Computer Science, New Orleans, LA 70118-6195. Offers MS. Part-time and evening/weekend programs available. *Faculty:* 6 full-time (1 woman). Average age 37. 2 applicants, 100% accepted. In 1998, 4 degrees awarded. *Degree requirements:* For master's, foreign language and thesis not required. *Entrance requirements:* For master's, documentation of previous work, courses currently being taught, minimum GPA of 3.0 in last 30 hours. *Application deadline:* For fall admission, 8/1 (priority date); for spring admission, 12/1 (priority date). Applications are processed on a rolling basis. Application fee: $20. Electronic applications accepted. *Financial aid:* Tuition waivers (partial) available. Aid available to part-time students. Financial aid application deadline: 5/1; financial aid applicants required to submit FAFSA. *Unit head:* Dr. Bogdan Czejdo, Chair, 504-865-3340, Fax: 504-865-2051, E-mail: czejdo@loyno.edu. *Application contact:* Dr. Antonio Lopez, Adviser, 504-865-3340, Fax: 504-865-2051, E-mail: tlopez@loyno.edu.

Maharishi University of Management, Graduate Studies, Program in Computer Science, Fairfield, IA 52557. Offers MS. *Degree requirements:* For master's, computer language, thesis or alternative required, foreign language not required. *Entrance requirements:* For master's, GRE General Test, TOEFL (minimum score of 550 required), minimum GPA of 3.0. *Faculty research:* Parallel processing, computer systems in architecture.

Marist College, Graduate Programs, School of Computer Science and Mathematics, Poughkeepsie, NY 12601-1387. Offers computer science (MS), including information systems, software development. Part-time and evening/weekend programs available. *Faculty:* 12 full-time (2 women), 3 part-time (0 women). *Students:* 50 full-time (9 women), 79 part-time (19 women). Average age 33. 280 applicants, 81% accepted. In 1998, 30 degrees awarded. *Degree requirements:* For master's, thesis optional, foreign language not required. *Application deadline:* For fall admission, 8/1 (priority date); for spring admission, 12/15. Applications are processed on a rolling basis. Application fee: $30. *Financial aid:* Federal Work-Study and tuition waivers (partial) available. Aid available to part-time students. Financial aid application deadline: 8/15; financial aid applicants required to submit FAFSA. *Unit head:* Dr. Jose Torres, Dean, 914-575-3000 Ext. 2610, E-mail: jose.torres@marist.edu. *Application contact:* Dr. H. Griffin Walling, Dean of Graduate and Continuing Education, 914-575-3530, Fax: 914-575-3640, E-mail: griffin.walling@marist.edu.

Marycrest International University, Graduate Studies, Department of Computer Science, Davenport, IA 52804-4096. Offers MS. Part-time and evening/weekend programs available. *Faculty:* 2 full-time (0 women). *Students:* 13 full-time (4 women), 11 part-time (3 women); includes 1 minority (Asian American or Pacific Islander), 8 international. Average age 30. 14 applicants, 71% accepted. In 1998, 9 degrees awarded (100% found work related to degree). *Degree requirements:* For master's, computer language, comprehensive exams required, foreign language and thesis not required. *Entrance requirements:* For master's, minimum undergraduate GPA of 2.8. *Application deadline:* Applications are processed on a rolling basis. Application fee: $25. Tuition: Part-time $430 per semester hour. Required fees: $12 per semester hour. One-time fee: $30 part-time. Part-time tuition and fees vary according to degree level. *Financial aid:* In 1998–99, 10 students received aid, including 10 research assistantships; Federal Work-Study also available. Financial aid application deadline: 3/1. *Faculty research:* Distributed database, network management, distributed fault tolerance, object-oriented systems. *Unit head:* Jeffrey Dickerson, Coordinator, 319-326-9273, Fax: 319-327-9615, E-mail: jdickerson@mcrest.edu. *Application contact:* Meg Farber, Admissions Director, 319-327-9609, Fax: 319-327-9620, E-mail: mfarber@mcrest.edu.

Marymount University, School of Arts and Sciences, Program in Computer Science, Arlington, VA 22207-4299. Offers MS. *Degree requirements:* For master's, thesis optional, foreign language not required. *Entrance requirements:* For master's, GRE, interview.

Massachusetts Institute of Technology, School of Engineering, Department of Electrical Engineering and Computer Science, Cambridge, MA 02139-4307. Offers computer science (EE); electrical engineering (EE); electrical engineering and computer science (M Eng, SM, PhD, Sc D). *Faculty:* 101 full-time (6 women), 2 part-time (0 women). *Students:* 771 full-time (153 women), 25 part-time (4 women); includes 189 minority (18 African Americans, 149 Asian Americans or Pacific Islanders, 21 Hispanic Americans, 1 Native American), 179 international. Average age 26. 2185 applicants, 23% accepted. In 1998, 306 master's, 68 doctorates awarded. Terminal master's awarded for partial completion of doctoral program. *Degree requirements:* For master's and EE, thesis required; for doctorate and EE, dissertation, comprehensive exams required. *Application deadline:* For fall admission, 1/15; for spring admission, 11/1. Application fee: $55. *Financial aid:* In 1998–99, 724 students received aid, including 148 fellowships, 511 research assistantships, 117 teaching assistantships; career-related internships or fieldwork, Federal Work-Study, and institutionally-sponsored loans also available. Financial aid applicants required to submit FAFSA. *Faculty research:* Modem control and system theory, radio astronomy, knowledge-based application systems, artificial intelligence, electrohydrodynamics. Total annual research expenditures: $49.3 million. *Unit head:* Dr. John V. Guttag, Head, 617-253-6022, E-mail: guttag@eecs.mit.edu. *Application contact:* Peggy Carney, Administrator, 617-253-4603, E-mail: peggy@eecs.mit.edu.

McGill University, Faculty of Graduate Studies and Research, School of Computer Science, Montréal, PQ H3A 2T5, Canada. Offers M Sc, PhD. *Faculty:* 17 full-time (2 women), 3 part-time (0 women). *Students:* 104 full-time (22 women), 20 international. 212 applicants, 24% accepted. In 1998, 29 master's awarded (90% found work related to degree, 10% continued full-time study); 3 doctorates awarded (33% entered university research/teaching, 67% found other work related to degree). Terminal master's awarded for partial completion of doctoral program. *Degree requirements:* For master's, project or thesis required; for doctorate, dissertation, comprehensive exam required. *Entrance requirements:* For master's, GRE General Test (minimum combined score of 1900 on three sections required), TOEFL (minimum score of 580 required); for doctorate, Gre General Test (minimum combined score of 1900 on three sections required), TOEFL (minimum score of 580 required). *Average time to degree:* Master's–3 years full-time; doctorate–4.5 years full-time. *Application deadline:* For fall admission, 2/1 (priority date); for winter admission, 10/1. Applications are processed on a rolling basis. Application fee: $60. *Financial aid:* In 1998–99, 18 research assistantships (averaging $15,000 per year) were awarded.; fellowships, teaching assistantships, tuition waivers (full) also available. Financial aid application deadline: 2/1. *Faculty research:* Computational geometry and robotics, software engineering, database systems and compilers, concurrency cryptography and numerical computation. Total annual research expenditures: $1 million. *Unit head:* D. Thérien, Director, 514-398-7072 Ext. 3744, Fax: 514-398-3883, E-mail: grad@cs.mcgill.ca. *Application contact:* Franca Cianci, Graduate Program Secretary, 514-398-7071 Ext. 3744, Fax: 514-398-3883, E-mail: grad-sec@cs.mcgill.ca.

McMaster University, School of Graduate Studies, Faculty of Science, Department of Computing and Software, Hamilton, ON L8S 4M2, Canada. Offers computer science (M Sc). Part-time programs available. *Degree requirements:* For master's, thesis required, foreign language not required. *Faculty research:* Software engineering; theory of non-sequential systems; parallel and distributed computing; artificial intelligence; complexity, design, and analysis of algorithms; combinatorial computing, especially applications to molecular biology.

McNeese State University, Graduate School, College of Science, Department of Mathematics, Computer Science, and Statistics, Lake Charles, LA 70609-2495. Offers computer science (MS); mathematics (MS); statistics (MS). Evening/weekend programs available. *Faculty:* 14 full-time (3 women). *Students:* 4 full-time (3 women), 5 part-time (2 women). In 1998, 10 degrees awarded. *Degree requirements:* For master's, computer language, thesis or alternative, written exam required, foreign language not required. *Entrance requirements:* For master's, GRE General Test. *Application deadline:* For fall admission, 7/15 (priority date). Applications are processed on a rolling basis. Application fee: $10 ($25 for international students). *Financial aid:* Teaching assistantships available. Financial aid application deadline: 5/1. *Unit head:* Sid Bradley, Head, 318-475-5788.

Memorial University of Newfoundland, School of Graduate Studies, Department of Computer Science, St. John's, NF A1C 5S7, Canada. Offers M Sc, PhD. *Students:* 13 full-time (6 women), 4 part-time (1 woman), 11 international. 47 applicants, 4% accepted. In 1998, 3 degrees awarded. *Degree requirements:* For master's, thesis required; for doctorate, dissertation, comprehensive exam required. *Entrance requirements:* For master's, TOEFL (minimum score of 600 required), GRE, honors degree in computer science or related field; for doctorate, GRE, master's degree in computer science. *Application deadline:* Applications are processed on a rolling basis. Application fee: $40. *Financial aid:* Fellowships, research assistantships, teaching assistantships available. *Unit head:* Dr. Paul Gillard, Head, 709-737-8652, Fax: 709-737-2009, E-mail: bartha@cs.mun.ca. *Application contact:* Elaine Boone, 709-737-8627, Fax: 709-737-2009, E-mail: elaine@cs.mun.ca.

Miami University, Graduate School, School of Applied Science, Department of Systems Analysis, Oxford, OH 45056. Offers MS. *Faculty:* 12. *Students:* 22 full-time (8 women), 6 part-time (2 women); includes 3 minority (1 African American, 2 Asian Americans or Pacific Islanders), 10 international. 37 applicants, 84% accepted. In 1998, 9 degrees awarded. *Degree requirements:* For master's, computer language, thesis, final exam required, foreign language not required. *Entrance requirements:* For master's, GRE, minimum undergraduate GPA of 3.0 during previous 2 years or 2.75 overall. *Application deadline:* For fall admission, 3/1 (priority date). Applications are processed on a rolling basis. Application fee: $35. *Financial aid:* In 1998–99, 3 research assistantships, 5 teaching assistantships were awarded.; Federal Work-Study and tuition waivers (full) also available. Financial aid application deadline: 3/1. *Unit head:* Dr. Douglas Troy, Director of Graduate Studies, 513-529-5928.

Michigan State University, Graduate School, College of Engineering, Department of Computer Science and Engineering, East Lansing, MI 48824-1020. Offers computer science (MS, PhD). *Faculty:* 23. *Students:* 66 full-time (10 women), 64 part-time (17 women); includes 10 minority (7 African Americans, 2 Asian Americans or Pacific Islanders, 1 Hispanic American), 85 international. Average age 27. 286 applicants, 22% accepted. In 1998, 50 master's, 9 doctorates awarded. *Degree requirements:* For master's, written exam or substantial design project required, foreign language and thesis not required; for doctorate, dissertation, comprehensive and qualifying exams required. *Entrance requirements:* For master's, GRE General Test, GRE Subject Test, TOEFL; for doctorate, GRE General Test, GRE Subject Test, TOEFL, sample of published work. *Application deadline:* Applications are processed on a rolling basis. Application fee: $30 ($40 for international students). *Financial aid:* In 1998–99, 66 teaching assistantships with tuition reimbursements (averaging $12,413 per year) were awarded.; fellowships, research assistantships Financial aid applicants required to submit FAFSA. Total annual research expenditures: $1.8 million. *Unit head:* Dr. Anil K. Jain, Chairperson, 517-353-3148, Fax: 517-432-1061, E-mail: jain@egr.msu.edu. *Application contact:* Dr. George C. Stockman, Graduate Coordinator, 517-353-1679.

See in-depth description on page 767.

Michigan Technological University, Graduate School, College of Sciences and Arts, Department of Computer Science, Houghton, MI 49931-1295. Offers computer science (MS); engineering-computational science (PhD). Part-time programs available. *Faculty:* 9 full-time (2 women). *Students:* 5 full-time (all women), 41 part-time (11 women); includes 1 minority (Asian American or Pacific Islander), 33 international. Average age 26. 87 applicants, 62% accepted. In 1998, 11 master's awarded. *Degree requirements:* For master's, computer language required, foreign language not required; for doctorate, computer language, dissertation required, foreign language not required. *Entrance requirements:* For master's, GRE General Test (minimum combined score of 1780 on three sections required; average 1998), TOEFL (minimum score of 600 required; average 631); for doctorate, GRE General Test (combined average 2145 on three sections), TOEFL (minimum score of 575 required; average 588). *Average time to degree:* Master's–2.8 years full-time. *Application deadline:* For fall admission, 3/15 (priority date). Applications are processed on a rolling basis. Application fee: $30 ($35 for international students). Tuition, state resident: full-time $4,377. Tuition, nonresident: full-time $9,108. Required fees: $126. Tuition and fees vary according to course load. *Financial aid:* In 1998–99, 3 fellowships (averaging $3,217 per year), 11 research assistantships (averaging $6,505 per year), 16 teaching assistantships (averaging $6,495 per year) were awarded.; Federal Work-Study, institutionally-sponsored loans, and unspecified assistantships also available. Aid available to part-time students. Financial aid application deadline: 3/15; financial aid applicants required to submit FAFSA. *Faculty research:* Software engineering, parallel algorithms, graphics and computational biology, geometric modeling/graphics, instruction level parallelism. Total annual research expenditures: $190,527. *Unit head:* Dr. Linda Ott, Chair, 906-487-2209, Fax: 906-487-2283, E-mail: linda@mtu.edu. *Application contact:* Dr. Steve Carr, Assistant Professor, 906-487-2950, Fax: 906-487-2283, E-mail: carr@mtu.edu.

See in-depth description on page 769.

Middle Tennessee State University, College of Graduate Studies, College of Basic and Applied Sciences, Department of Computer Science, Murfreesboro, TN 37132. Offers MS. Part-time programs available. *Faculty:* 12 full-time (4 women). *Students:* 19 full-time (8 women), 13 part-time (3 women); includes 6 minority (1 African American, 4 Asian Americans or Pacific Islanders, 1 Hispanic American), 9 international. Average age 30. 38 applicants, 42% accepted. In 1998, 19 degrees awarded. *Degree requirements:* For master's, one foreign language, computer language, thesis, comprehensive exams required. *Entrance requirements:* For master's, GRE or MAT (minimum score of 30 required). *Application deadline:* For fall admission, 8/1 (priority date). Applications are processed on a rolling basis. Application fee: $25. Electronic applications accepted. *Financial aid:* Teaching assistantships, institutionally-sponsored loans available. Aid available to part-time students. Financial aid application deadline: 5/1; financial aid applicants required to submit FAFSA. Total annual research expenditures: $1,969. *Unit head:* Dr. Richard Detmer, Chair, 615-898-2397, Fax: 615-898-5567, E-mail: rdetmer@mtsu.edu.

Midwestern State University, Graduate Studies, Division of Mathematical Sciences, Computer Science Program, Wichita Falls, TX 76308-2096. Offers MS. Part-time and evening/weekend programs available. *Faculty:* 3 full-time (1 woman). *Students:* 30 full-time (2 women), 18 part-time (2 women). Average age 35. 40 applicants, 95% accepted. In 1998, 14 degrees awarded (100% found work related to degree). *Degree requirements:* For master's, computer language, thesis or alternative required, foreign language not required. *Entrance requirements:* For master's, GRE General Test, TOEFL (minimum score of 550 required). *Application deadline:* For fall admission, 8/7; for spring admission, 12/15. Application fee: $0 ($50 for international students). Tuition, state resident: full-time $1,542; part-time $46 per hour. Tuition, nonresident: full-time $5,376; part-time $304 per hour. Tuition and fees vary according to course load. *Financial aid:* In 1998–99, 19 research assistantships were awarded.; teaching assistantships, Federal Work-Study, institutionally-sponsored loans, tuition waivers (partial), and unspecified assistantships also available. Aid available to part-time students. *Unit head:* Dr. Stewart Carpenter, Graduate Adviser, 940-397-4279.

Mills College, Graduate Studies, New Horizons Program in Mathematics and Computer Science, Oakland, CA 94613-1000. Offers computer science (Certificate). Part-time programs available. *Faculty:* 7 full-time (5 women). Average age 27. 5 applicants, 100% accepted. *Degree requirements:* For Certificate, computer language required, thesis not required. *Entrance requirements:* For degree, TOEFL (minimum score of 600 required). *Average time to degree:* 1 year full-time, 2 years part-time. *Application deadline:* For fall admission, 2/1 (priority date); for spring admission, 11/1. Applications are processed on a rolling basis. Application fee: $50. Electronic applications accepted. Tuition: Full-time $11,130; part-time $2,690 per course. One-time fee: $977. Tuition and fees vary according to course load and program. *Financial aid:* In 1998–99, 5 teaching assistantships (averaging $5,300 per year) were awarded.; career-related internships or fieldwork, institutionally-sponsored loans, scholarships, and residence awards also available. Aid available to part-time students. Financial aid application deadline: 2/1; financial aid applicants required to submit FAFSA. *Faculty research:* Dynamical systems, human interface, parallel computation, compiling techniques, fault tolerance, operating systems. *Unit head:* Matt Merzbacher, Director, 510-430-2201, Fax: 510-430-3314, E-mail: matthew@mills.edu. *Application contact:* La Vonna S. Brown, Coordinator of Graduate Studies, 510-430-3309, Fax: 510-430-2159, E-mail: lavonna@mills.edu.

Mills College, Graduate Studies, Program in Interdisciplinary Computer Science, Oakland, CA 94613-1000. Offers MA. Part-time programs available. *Faculty:* 7 full-time (5 women). *Students:* 8 full-time (6 women); includes 2 minority (1 Asian American or Pacific Islander, 1 Hispanic American) Average age 24. 8 applicants, 50% accepted. In 1998, 2 degrees awarded (100% found work related to degree). *Degree requirements:* For master's, computer language, thesis required. *Entrance requirements:* For master's, TOEFL (minimum score of 600 required). *Average time to degree:* Master's–2 years full-time, 4 years part-time. *Application deadline:* For fall admission, 2/1 (priority date); for spring admission, 11/1. Applications are processed on a rolling basis. Application fee: $50. Electronic applications accepted. Tuition: Full-time $11,130; part-time $2,690 per course. One-time fee: $977. Tuition and fees vary according to course load and program. *Financial aid:* In 1998–99, 8 teaching assistantships (averaging $2,981 per year) were awarded.; career-related internships or fieldwork, institutionally-sponsored loans, scholarships, and residence awards also available. Aid available to part-time students. Financial aid application deadline: 2/1; financial aid applicants required to submit CSS PROFILE or FAFSA. *Faculty research:* Dynamical systems, linear programming, theory of computer viruses, interface design, intelligent tutoring systems. *Unit head:* Matthew Merzbacher, Director, 510-430-2201, Fax: 510-430-3314, E-mail: matthew@mills.edu. *Application contact:* La Vonna S. Brown, Coordinator of Graduate Studies, 510-430-3309, Fax: 510-430-2159, E-mail: lavonna@mills.edu.

Minnesota State University, Mankato, College of Graduate Studies, College of Science, Engineering and Technology, Department of Computer Science, Mankato, MN 56002-8400. Offers MS. *Faculty:* 11 full-time (2 women). *Students:* 6 full-time (0 women), 10 part-time (3 women); includes 1 minority (Asian American or Pacific Islander) Average age 31. In 1998, 6 degrees awarded. *Degree requirements:* For master's, thesis or alternative, comprehensive exam required. *Entrance requirements:* For master's, GRE General Test (minimum combined score of 1500 on three sections required), minimum GPA of 3.0 during previous 2 years. *Application deadline:* For fall admission, 7/9 (priority date); for spring admission, 11/27. Applications are processed on a rolling basis. Application fee: $20. *Financial aid:* Research assistantships with partial tuition reimbursements, teaching assistantships with partial tuition reimbursements available. Financial aid application deadline: 3/15; financial aid applicants required to submit FAFSA. *Unit head:* Dr. Leon Tietz, Chairperson, 507-389-5319. *Application contact:* Joni Roberts, Admissions Coordinator, 507-389-2321, Fax: 507-389-5974, E-mail: grad@mankato.msus.edu.

Minnesota State University, Mankato, College of Graduate Studies, College of Science, Engineering and Technology, Department of Mathematics and Statistics, Program in Computers, Mankato, MN 56002-8400. Offers mathematics: computer science (MS). *Students:* 5 full-time (0 women). Average age 32. *Degree requirements:* For master's, one foreign language, computer language, thesis or alternative, comprehensive exam required. *Entrance requirements:* For master's, GRE General Test, GRE Subject Test, minimum GPA of 3.0 during previous 2 years. *Application deadline:* For fall admission, 7/9 (priority date); for spring admission, 11/27. Applications are processed on a rolling basis. Application fee: $20. *Financial aid:* Fellowships with partial tuition reimbursements, research assistantships with partial tuition reimbursements, teaching assistantships with partial tuition reimbursements, Federal Work-Study and institutionally-sponsored loans available. Aid available to part-time students. Financial aid application deadline: 3/15; financial aid applicants required to submit FAFSA. *Unit head:* Dr. Lee Cornell, Chairperson, 507-389-2968. *Application contact:* Joni Roberts, Admissions Coordinator, 507-389-2321, Fax: 507-389-5974, E-mail: grad@mankato.msus.edu.

Mississippi College, Graduate School, College of Arts and Sciences, Department of Mathematics and Computer Science, Clinton, MS 39058. Offers computer science (MS); mathematics (MS). *Faculty:* 12 full-time (6 women), 7 part-time (4 women). *Degree requirements:* For master's. *Entrance requirements:* For master's, minimum GPA of 2.5. *Application deadline:* For fall admission, 8/15 (priority date). Applications are processed on a rolling basis. Application fee: $25 ($75 for international students). *Financial aid:* Application deadline: 4/1. *Unit head:* Dr. Thomas Leavelle, Head, 601-925-3463.

Mississippi State University, College of Engineering, Department of Computer Science, Mississippi State, MS 39762. Offers MS, PhD. Part-time programs available. Postbaccalaureate distance learning degree programs offered (minimal on-campus study). *Students:* 66 full-time (16 women), 31 part-time (8 women); includes 40 minority (2 African Americans, 37 Asian Americans or Pacific Islanders, 1 Hispanic American), 12 international. Average age 29. 157 applicants, 20% accepted. In 1998, 10 master's, 2 doctorates awarded. *Degree requirements:* For master's, comprehensive oral or written exam required, thesis optional, foreign language not required; for doctorate, dissertation, comprehensive oral or written exam required, foreign language not required. *Entrance requirements:* For master's, GRE General Test, TOEFL (minimum score of 550 required), minimum GPA of 2.75; for doctorate, GRE, TOEFL (minimum score of 550 required). *Application deadline:* For fall admission, 7/1; for spring admission, 11/1. Applications are processed on a rolling basis. Application fee: $25 for international students. *Financial aid:* Federal Work-Study, institutionally-sponsored loans, and unspecified assistantships available. Financial aid applicants required to submit FAFSA. *Faculty research:* Artificial intelligence, software engineering, visualization, high performance computing. Total annual research expenditures: $603,564. *Unit head:* Dr. Julia E. Hodges, Head, 662-325-2756, Fax: 662-325-8997, E-mail: hodges@cs.msstate.edu. *Application contact:* Jerry B. Inmon, Director of Admissions, 662-325-2224, Fax: 662-325-7360, E-mail: admit@admissions.msstate.edu.

See in-depth description on page 771.

Monmouth University, Graduate School, Department of Computer Science, West Long Branch, NJ 07764-1898. Offers MS. *Faculty:* 5 full-time (1 woman), 2 part-time (0 women). *Students:* 42 full-time (22 women), 54 part-time (20 women); includes 14 minority (2 African Americans, 12 Asian Americans or Pacific Islanders), 46 international. Average age 31. In 1998, 18 degrees awarded. *Degree requirements:* For master's, computer language required, thesis optional, foreign language not required. *Entrance requirements:* For master's, minimum GPA of 3.0 in major, 2.5 overall. *Application deadline:* For fall admission, 8/15 (priority date); for spring admission, 12/15 (priority date). Applications are processed on a rolling basis. Application fee: $35 ($40 for international students). Electronic applications accepted. *Financial aid:* In 1998–99, 23 students received aid. Federal Work-Study and unspecified assistantships available. Aid available to part-time students. Financial aid application deadline: 3/1; financial aid applicants required to submit FAFSA. *Faculty research:* Databases, natural language processing, protocols, performance analysis, communications networks (systems), telecommunications. *Unit head:* Dr. Wlodek Dobosiewicz, Chair, 732-571-3441, Fax: 732-571-4415. *Application contact:* 732-571-3452, Fax: 732-571-5123, E-mail: gradadm@monmouth.edu.

Montana State University–Bozeman, College of Graduate Studies, College of Engineering, Department of Computer Science, Bozeman, MT 59717. Offers computer science (MS); engineering (PhD). Part-time programs available. *Students:* 28 full-time (7 women), 14 part-

Computer Science

Montana State University–Bozeman (continued)
time (4 women); includes 1 minority (Asian American or Pacific Islander) Average age 29. 21 applicants, 71% accepted. In 1998, 11 master's awarded. *Degree requirements:* For master's, thesis or alternative required, foreign language not required; for doctorate, dissertation required, foreign language not required. *Entrance requirements:* For master's, GRE General Test (minimum combined score of 1200 required), TOEFL (minimum score of 580 required), minimum GPA of 3.0; for doctorate, GRE General Test (minimum combined score of 1200 required), TOEFL (minimum score of 580 required). *Application deadline:* For fall admission, 6/1 (priority date); for spring admission, 11/1. Applications are processed on a rolling basis. Application fee: $50. *Financial aid:* In 1998–99, 1 research assistantship with full tuition reimbursement (averaging $6,500 per year), 1 teaching assistantship with full tuition reimbursement (averaging $6,500 per year) were awarded.; fellowships, Federal Work-Study and scholarships also available. Financial aid application deadline: 3/1; financial aid applicants required to submit FAFSA. *Faculty research:* Graphics, artificial intelligence, networks, algorithms, computer science education. Total annual research expenditures: $178,229. *Unit head:* Dr. J. Denbigh Starkey, Head, 406-994-4780, E-mail: gradappl@cs.montana.edu.

Montclair State University, Office of Graduate Studies, College of Science and Mathematics, Department of Mathematics and Computer Science, Upper Montclair, NJ 07043-1624. Offers computer science (MS), including applied mathematics, applied statistics; mathematics (MS), including computer science, mathematics education, pure and applied mathematics, statistics. Part-time and evening/weekend programs available. *Degree requirements:* For master's, written comprehensive exam required, foreign language and thesis not required. *Entrance requirements:* For master's, GRE General Test, minimum GPA of 2.67.

Montclair State University, Office of Graduate Studies, College of Science and Mathematics, Department of Mathematics and Computer Science, Programs in Mathematics, Concentration in Computer Science, Upper Montclair, NJ 07043-1624. Offers MS. *Degree requirements:* For master's, written comprehensive exam required, foreign language and thesis not required. *Entrance requirements:* For master's, GRE General Test, minimum GPA of 2.67.

National Technological University, Programs in Engineering, Fort Collins, CO 80526-1842. Offers chemical engineering (MS); computer engineering (MS); computer science (MS); electrical engineering (MS); engineering management (MS); hazardous waste management (MS); health physics (MS); management of technology (MS); manufacturing systems engineering (MS); materials science and engineering (MS); software engineering (MS); special majors (MS); transportation engineering (MS); transportation systems engineering (MS). Part-time programs available. *Faculty:* 600 part-time (20 women). *Entrance requirements:* For master's, BS in engineering or related field. *Application deadline:* Applications are processed on a rolling basis. Application fee: $50. *Unit head:* Lionel V. Baldwin, President, 970-495-6400, Fax: 970-484-0668, E-mail: baldwin@mail.ntu.edu.

Naval Postgraduate School, Graduate Programs, Department of Computer Science, Monterey, CA 93943. Offers MS, PhD. Program only open to commissioned officers of the United States and friendly nations and selected United States federal civilian employees. Part-time programs available. Postbaccalaureate distance learning degree programs offered (minimal on-campus study). *Students:* 118 full-time, 33 international. In 1998, 26 master's, 1 doctorate awarded. *Degree requirements:* For master's, computer language, thesis required, foreign language not required; for doctorate, one foreign language, computer language, dissertation required. *Unit head:* Dr. Dan C. Boger, Chairman, 831-656-2449. *Application contact:* Theodore H. Calhoun, Director of Admissions, 831-656-3093, Fax: 831-656-2891, E-mail: tcalhoon@nps.navy.mil.

New Jersey Institute of Technology, Office of Graduate Studies, Department of Computer and Information Science, Newark, NJ 07102-1982. Offers bioinformatics (MS, PhD); computer and information science (PhD); computer science (MS); information systems (MS). MS and PhD (bioinformatics) offered jointly with the University of Medicine and Dentistry of New Jersey. Part-time and evening/weekend programs available. Terminal master's awarded for partial completion of doctoral program. *Degree requirements:* For master's, computer language, thesis required, foreign language not required; for doctorate, computer language, dissertation, residency required, foreign language not required. *Entrance requirements:* For master's, GRE General Test (minimum score of 450 on verbal section, 600 on quantitative, 550 on analytical required); for doctorate, GRE General Test (minimum score of 450 on verbal section, 600 on quantitative, 550 on analytical required), minimum graduate GPA of 3.5. Electronic applications accepted. *Faculty research:* Computer systems, communications and networking, artificial intelligence, database engineering, systems analysis.

New Mexico Highlands University, Graduate Office, College of Arts and Sciences, Department of Communication and Fine Arts, Las Vegas, NM 87701. Offers media arts and computer science (MA), including design studies, digital audio and video production. *Faculty:* 4 full-time (2 women). *Entrance requirements:* For master's, minimum undergraduate GPA of 3.0. *Application deadline:* For fall admission, 8/1 (priority date). Applications are processed on a rolling basis. Application fee: $15. Tuition, state resident: full-time $1,988; part-time $83 per credit hour. Tuition, nonresident: full-time $8,034; part-time $83 per credit hour. Tuition and fees vary according to course load. *Unit head:* Dr. Margaret Mertz, Chair, 505-454-3359, Fax: 505-454-3068, E-mail: mertz_m@merlin.nmhu.edu. *Application contact:* Dr. Glen W. Davidson, Provost, 505-454-3311, Fax: 505-454-3558, E-mail: glendavidson@venus.nmhu.edu.

New Mexico Highlands University, Graduate Office, College of Arts and Sciences, Department of Mathematics and Computer Science, Las Vegas, NM 87701. Offers media arts and computer science (MS), including cognitive science, computer graphics, networking technologies. *Faculty:* 2 full-time (0 women). *Degree requirements:* For master's, thesis or alternative required, foreign language not required. *Entrance requirements:* For master's, minimum undergraduate GPA of 3.0. *Application deadline:* For fall admission, 8/1 (priority date). Applications are processed on a rolling basis. Application fee: $15. Tuition, state resident: full-time $1,988; part-time $83 per credit hour. Tuition, nonresident: full-time $8,034; part-time $83 per credit hour. Tuition and fees vary according to course load. *Unit head:* Dr. Wayne Summers, Chair, 505-454-3295, Fax: 505-545-3103, E-mail: wsummer@cs.nmhu.edu. *Application contact:* Dr. Glen W. Davidson, Provost, 505-454-3311, Fax: 505-454-3558, E-mail: glendavidson@venus.nmhu.edu.

New Mexico Institute of Mining and Technology, Graduate Studies, Department of Computer Science, Socorro, NM 87801. Offers MS, PhD. *Faculty:* 6 full-time (0 women). *Students:* 20 full-time (2 women), 12 part-time (7 women); includes 1 minority (Hispanic American), 15 international. Average age 30. 113 applicants, 79% accepted. In 1998, 5 master's awarded. *Degree requirements:* For master's, computer language required, thesis optional, foreign language not required; for doctorate, computer language, dissertation required, foreign language not required. *Entrance requirements:* For master's, GRE General Test, TOEFL (minimum score of 540 required); for doctorate, GRE General Test, GRE Subject Test, TOEFL (minimum score of 540 required). *Average time to degree:* Master's–4 years full-time; doctorate–7 years full-time. *Application deadline:* For fall admission, 3/1 (priority date); for spring admission, 6/1. Applications are processed on a rolling basis. Application fee: $16. *Financial aid:* In 1998–99, 7 research assistantships (averaging $9,670 per year), 9 teaching assistantships (averaging $9,670 per year) were awarded.; fellowships, Federal Work-Study and institutionally-sponsored loans also available. Financial aid application deadline: 3/1; financial aid applicants required to submit CSS PROFILE or FAFSA. *Unit head:* Dr. Victor Yodaiken, Chairman, 505-835-5126, Fax: 505-835-5587, E-mail: yodaiken@.nmt.edu. *Application contact:* Dr. David B. Johnson, Dean of Graduate Studies, 505-835-5513, Fax: 505-835-5476, E-mail: graduate@nmt.edu.

New Mexico State University, Graduate School, College of Arts and Sciences, Department of Computer Science, Las Cruces, NM 88003-8001. Offers MS, PhD. Part-time programs available. *Faculty:* 15 full-time (2 women). *Students:* 46 full-time (13 women), 9 part-time (2 women); includes 5 minority (1 African American, 4 Hispanic Americans), 37 international. Average age 31. 143 applicants, 76% accepted. In 1998, 18 master's, 4 doctorates awarded. *Degree requirements:* For master's, computer language, thesis or alternative required, foreign language not required; for doctorate, one foreign language, computer language, dissertation required. *Entrance requirements:* For master's and doctorate, GRE General Test. *Application deadline:* For fall admission, 7/1 (priority date); for spring admission, 11/1. Applications are processed on a rolling basis. Application fee: $15 ($35 for international students). Electronic applications accepted. Tuition, state resident: full-time $2,682; part-time $112 per credit. Tuition, nonresident: full-time $8,376; part-time $349 per credit. Tuition and fees vary according to course load. *Financial aid:* Research assistantships, teaching assistantships, career-related internships or fieldwork and Federal Work-Study available. Aid available to part-time students. Financial aid application deadline: 3/1. *Faculty research:* Programming languages, artificial intelligence, databases, operating systems, computer networks. *Unit head:* Dr. Arthur Karshmer, Head, 505-646-3723, Fax: 505-646-1002, E-mail: arthur@nmsu.edu. *Application contact:* Dr. Joseph J. Pfeiffer, Chair, Graduate Committee, 505-646-1605, Fax: 505-646-1002, E-mail: pfeiffer@nmsu.edu.

Announcement: Faculty research interests focus on parallel processing and artificial intelligence and include computational science, computer music, databases, natural language, programming languages, software engineering, visual programming, and theory of computation. The department also enjoys close ties with NMSU's Computing Research Laboratory. Computing facilities include Sun OS, Solaris, Linux, Windows 95, and Macintosh.

New York Institute of Technology, Graduate Division, School of Engineering and Technology, Program in Computer Science, Old Westbury, NY 11568-8000. Offers MS. Part-time and evening/weekend programs available. *Students:* 160 full-time (54 women), 176 part-time (47 women); includes 91 minority (12 African Americans, 71 Asian Americans or Pacific Islanders, 6 Hispanic Americans, 2 Native Americans), 187 international. Average age 31. 278 applicants, 64% accepted. In 1998, 82 degrees awarded. *Degree requirements:* For master's, computer language, project required, foreign language and thesis not required. *Entrance requirements:* For master's, GRE General Test, TOEFL, minimum QPA of 2.85, BS in computer science or related field. *Average time to degree:* Master's–2 years full-time, 3 years part-time. *Application deadline:* For fall admission, 8/1. Applications are processed on a rolling basis. Application fee: $50. Electronic applications accepted. *Financial aid:* Fellowships, research assistantships, teaching assistantships, institutionally-sponsored loans, tuition waivers (full and partial), and unspecified assistantships available. Aid available to part-time students. *Faculty research:* Image processing, multimedia CD-Rom, fuzzy logic controller design. *Unit head:* Dr. Ayat Jafari, Chair, 516-686-7523. *Application contact:* Glenn Berman, Executive Director of Admissions, 516-686-7519, Fax: 516-626-0419, E-mail: gberman@iris.nyit.edu.

New York University, Graduate School of Arts and Science, Courant Institute of Mathematical Sciences, Department of Computer Science, New York, NY 10012-1019. Offers computer science (MS, PhD); information systems (MS); scientific computing (MS). Part-time and evening/weekend programs available. *Faculty:* 30 full-time (1 woman). *Students:* 96 full-time (19 women), 254 part-time (50 women); includes 35 minority (2 African Americans, 26 Asian Americans or Pacific Islanders, 7 Hispanic Americans), 168 international. Average age 27. 420 applicants, 39% accepted. In 1998, 105 master's, 13 doctorates awarded. *Degree requirements:* For master's, computer language required, foreign language and thesis not required; for doctorate, computer language, dissertation, oral and written exams required, foreign language not required. *Entrance requirements:* For master's and doctorate, GRE General Test, GRE Subject Test, TOEFL. *Application deadline:* For fall admission, 1/4; for spring admission, 11/1. Application fee: $60. Tuition: Full-time $17,880; part-time $745 per credit. Required fees: $1,140; $35 per credit. Tuition and fees vary according to course load and program. *Financial aid:* Fellowships, research assistantships, teaching assistantships, Federal Work-Study and tuition waivers (full and partial) available. Financial aid application deadline: 1/4; financial aid applicants required to submit FAFSA. *Faculty research:* Distributed and parallel computing, multimedia and graphics, algorithmics and theory of computation, programming languages, artificial intelligence. *Unit head:* Richard Cole, Chair, 212-998-3011, Fax: 212-995-4124. *Application contact:* Ralph Grishman, Director of Graduate Studies, 212-998-3011, Fax: 212-995-4124, E-mail: admissions@cs.nyu.edu.

See in-depth description on page 773.

North Carolina Agricultural and Technical State University, Graduate School, College of Engineering, Department of Computer Science, Greensboro, NC 27411. Offers MSCS. Part-time programs available. *Faculty:* 9 full-time (1 woman), 1 part-time (0 women). *Students:* 49 full-time (27 women), 30 part-time (15 women); includes 60 minority (38 African Americans, 22 Asian Americans or Pacific Islanders) Average age 25. 51 applicants, 47% accepted. In 1998, 18 degrees awarded. *Degree requirements:* For master's, thesis (for some programs), comprehensive exam required, foreign language not required. *Average time to degree:* Master's–1.6 years full-time, 4 years part-time. *Application deadline:* For fall admission, 7/1 (priority date); for spring admission, 1/9. Applications are processed on a rolling basis. Application fee: $35. *Financial aid:* In 1998–99, 24 students received aid, including 16 research assistantships, 4 teaching assistantships; fellowships, career-related internships or fieldwork also available. Financial aid application deadline: 3/30. *Faculty research:* Object-oriented analysis, artificial intelligence, distributed computing, societal implications of computing, testing. Total annual research expenditures: $1.2 million. *Unit head:* Dr. Joseph Monroe, Chairperson, 336-334-7245, Fax: 336-334-7244, E-mail: monroe@ncat.edu. *Application contact:* Dr. David Bellin, Graduate Coordinator, 336-334-7245, Fax: 336-334-7244, E-mail: dbellin@ncat.edu.

North Carolina State University, Graduate School, College of Engineering, Department of Computer Science, Raleigh, NC 27695. Offers MC Sc, MS, PhD. Part-time programs available. *Faculty:* 33 full-time (3 women), 14 part-time (2 women). *Students:* 87 full-time (18 women), 88 part-time (18 women); includes 36 minority (10 African Americans, 23 Asian Americans or Pacific Islanders, 3 Hispanic Americans), 47 international. Average age 31. 246 applicants, 31% accepted. In 1998, 37 master's, 8 doctorates awarded. *Degree requirements:* For master's, computer language, thesis required (for some programs), foreign language not required; for doctorate, computer language, dissertation required, foreign language not required. *Entrance requirements:* For master's, GRE General Test, GRE Subject Test, TOEFL, minimum GPA of 3.0; for doctorate, GRE General Test, GRE Subject Test, TOEFL, minimum GPA of 3.5. *Application deadline:* For fall admission, 4/1 (priority date); for spring admission, 10/1. Application fee: $45. *Financial aid:* In 1998–99, 24 fellowships (averaging $4,425 per year), 123 research assistantships (averaging $4,308 per year), 78 teaching assistantships (averaging $5,608 per year) were awarded.; career-related internships or fieldwork and institutionally-sponsored loans also available. Financial aid application deadline: 2/1. *Faculty research:* Software systems, networking and performance analysis, theory and algorithms, architecture, multimedia systems. Total annual research expenditures: $3.4 million. *Unit head:* Dr. Alan L. Tharp, Head, 919-515-7435, Fax: 919-515-7896, E-mail: tharp@adm.csc.ncsu.edu. *Application contact:* Dr. Edward Davis, Interim Director of Graduate Programs, 919-515-7045, Fax: 919-515-7896, E-mail: davis@csc.ncsu.edu.

North Carolina State University, Graduate School, College of Management, Program in Management, Raleigh, NC 27695. Offers biotechnology (MS); computer science (MS); engineering (MS); forest resources management (MS); general business (MS); management information systems (MS); operations research (MS); statistics (MS); telecommunications systems engineering (MS); textile management (MS); total quality management (MS). Part-time programs available. *Faculty:* 40 full-time (9 women), 4 part-time (0 women). *Students:* 48 full-time (15 women), 156 part-time (43 women); includes 33 minority (16 African Americans, 15 Asian Americans or Pacific Islanders, 1 Hispanic American, 1 Native American), 4 international. *Degree requirements:* For master's, computer language required, foreign language and thesis not required. *Entrance requirements:* For master's, GRE or GMAT, TOEFL (minimum score of 550 required), minimum undergraduate GPA of 3.0. *Application deadline:* For fall admission, 6/25; for spring admission, 11/25. Applications are processed on a rolling basis.

Application fee: $45. *Unit head:* Dr. Jack W. Wilson, Director of Graduate Programs, 919-515-4327, Fax: 919-515-6943, E-mail: jack_wilson@ncsu.edu. *Application contact:* Dr. Steven G. Allen, Director of Graduate Programs, 919-515-6941, Fax: 919-515-5073, E-mail: steve_allen@ncsu.edu.

North Central College, Graduate Programs, Department of Computer Science, Naperville, IL 60566-7063. Offers MS. Part-time and evening/weekend programs available. *Faculty:* 7 full-time (1 woman), 11 part-time (1 woman). *Students:* 31. Average age 28. In 1998, 13 degrees awarded. *Degree requirements:* For master's, computer language, project required, thesis not required. *Entrance requirements:* For master's, interview. *Application deadline:* For fall admission, 8/15. Applications are processed on a rolling basis. Application fee: $25. Tuition: Part-time $1,464 per course. Tuition and fees vary according to program. *Financial aid:* Available to part-time students. *Faculty research:* Experimental broadband network. *Unit head:* Dr. Judy Walters, Coordinator, 630-637-5840, Fax: 630-637-5844. *Application contact:* Frank Johnson, Director of Graduate Programs, 630-637-5840, Fax: 630-637-5844, E-mail: frjohnson@noctrl.edu.

North Dakota State University, Graduate Studies and Research, College of Science and Mathematics, Department of Computer Science, Fargo, ND 58105. Offers computer science (MS, PhD); operations research (MS). Part-time programs available. *Faculty:* 12 full-time (1 woman), 2 part-time (0 women). *Students:* 73 full-time (11 women), 19 part-time (7 women); includes 4 minority (2 African Americans, 2 Asian Americans or Pacific Islanders), 72 international. Average age 24. 39 applicants, 100% accepted. In 1998, 44 master's, 1 doctorate awarded (100% entered university research/teaching). *Degree requirements:* For master's, computer language, comprehensive exam required, thesis optional, foreign language not required; for doctorate, computer language, dissertation, qualifying exam required, foreign language not required. *Entrance requirements:* For master's, TOEFL (minimum score of 600 required), minimum GPA of 3.0, BS in computer science or related field; for doctorate, TOEFL (minimum score of 600 required), minimum GPA of 3.25, MS in computer science or related field. *Application deadline:* For fall admission, 8/15 (priority date); for spring admission, 12/15 (priority date). Application fee: $25. *Financial aid:* In 1998–99, 22 research assistantships (averaging $10,000 per year), 12 teaching assistantships (averaging $8,000 per year) were awarded.; career-related internships or fieldwork, Federal Work-Study, and institutionally-sponsored loans also available. Financial aid application deadline: 4/15. *Faculty research:* Networking, software engineering, artificial intelligence, database, programming languages. *Unit head:* Dr. Kendall E. Nygard, Chair, 701-231-8562, Fax: 701-231-8255, E-mail: nygard@plains.nodak.edu.

Northeastern Illinois University, Graduate College, College of Arts and Sciences, Department of Computer Science, Program in Computer Science, Chicago, IL 60625-4699. Offers MS. Part-time and evening/weekend programs available. *Faculty:* 12 full-time (3 women), 5 part-time (2 women). *Students:* 28 full-time (5 women), 45 part-time (10 women); includes 47 minority (7 African Americans, 37 Asian Americans or Pacific Islanders, 3 Hispanic Americans), 6 international. Average age 33. 38 applicants, 58% accepted. In 1998, 17 degrees awarded. *Degree requirements:* For master's, computer language, comprehensive exam, research project, or thesis required. *Entrance requirements:* For master's, minimum GPA of 2.75, proficiency in 2 higher-level computer languages, 1 course in discrete mathematics. *Application deadline:* For fall admission, 3/31 (priority date); for spring admission, 9/30. Applications are processed on a rolling basis. Application fee: $0. *Financial aid:* In 1998–99, 22 students received aid, including 10 research assistantships; career-related internships or fieldwork, Federal Work-Study, institutionally-sponsored loans, and tuition waivers (full and partial) also available. Aid available to part-time students. Financial aid applicants required to submit FAFSA. *Faculty research:* Telecommunications, database inference problems, decision making under uncertainty, belief networks, analysis of algorithms. *Unit head:* Dr. Rich Neopolitan, Graduate Adviser, 773-794-2973. *Application contact:* Dr. Mohan K. Sood, Dean of Graduate College, 773-583-4050 Ext. 6143, Fax: 773-794-6670, E-mail: m-sood@neiu.edu.

Northeastern University, College of Computer Science, Boston, MA 02115-5096. Offers MS, PhD. Part-time and evening/weekend programs available. *Faculty:* 18 full-time (5 women), 3 part-time (0 women). *Students:* 85 full-time (26 women), 67 part-time (14 women). Average age 26. 351 applicants, 66% accepted. In 1998, 51 master's, 4 doctorates awarded. Terminal master's awarded for partial completion of doctoral program. *Degree requirements:* For master's, computer language required, thesis optional; for doctorate, computer language, dissertation required. *Entrance requirements:* For master's and doctorate, GRE General Test, TOEFL. *Application deadline:* For fall admission, 8/15; for winter admission, 11/1; for spring admission, 2/1. Applications are processed on a rolling basis. Application fee: $50. *Financial aid:* In 1998–99, 32 students received aid, including 1 fellowship, 5 research assistantships with full tuition reimbursements available (averaging $12,000 per year), 8 teaching assistantships with full tuition reimbursements available (averaging $12,000 per year); career-related internships or fieldwork, Federal Work-Study, institutionally-sponsored loans, and industrial research assistantships also available. Financial aid application deadline: 2/15. *Faculty research:* Database theory, parallel computing, programming languages and software development, artifical intelligence, network and cryptography. Total annual research expenditures: $1.2 million. *Unit head:* Dr. Larry A. Finkelstein, Dean, 617-373-2462, Fax: 617-373-5121. *Application contact:* Dr. Agnes Chan, Associate Dean and Director of Graduate Program, 617-373-2462, Fax: 617-373-5121.

See in-depth description on page 775.

Northern Illinois University, Graduate School, College of Liberal Arts and Sciences, Department of Computer Science, De Kalb, IL 60115-2854. Offers MS. Part-time and evening/weekend programs available. *Faculty:* 11 full-time (1 woman). *Students:* 78 full-time (34 women), 56 part-time (16 women); includes 26 minority (25 Asian Americans or Pacific Islanders, 1 Hispanic American), 77 international. Average age 29. 126 applicants, 44% accepted. In 1998, 60 degrees awarded. *Degree requirements:* For master's, computer language, comprehensive exam required, foreign language and thesis not required. *Entrance requirements:* For master's, GRE General Test, TOEFL (minimum score of 550 required; 213 for computer-based), minimum GPA of 2.75. *Application deadline:* For fall admission, 6/1; for spring admission, 11/1. Applications are processed on a rolling basis. Application fee: $30. *Financial aid:* In 1998–99, 3 research assistantships, 18 teaching assistantships were awarded.; fellowships, career-related internships or fieldwork, Federal Work-Study, tuition waivers (full), and unspecified assistantships also available. Aid available to part-time students. *Faculty research:* Databases, theorem proving. *Unit head:* Dr. Rodney Angotti, Chair, 815-753-0378.

Northwestern Polytechnic University, School of Engineering, Fremont, CA 94539-7482. Offers computer science (MS); computer systems engineering (MS); electrical engineering (MS). Part-time and evening/weekend programs available. *Faculty:* 8 full-time, 52 part-time. *Students:* 262. *Degree requirements:* For master's, computer language, thesis required, foreign language not required. *Entrance requirements:* For master's, TOEFL. *Application deadline:* For fall admission, 8/15; for winter admission, 12/15; for spring admission, 7/15. Applications are processed on a rolling basis. Application fee: $50 ($75 for international students). Tuition: Full-time $6,750; part-time $375 per unit. Required fees: $135 per term. Tuition and fees vary according to course load and program. *Unit head:* Dr. Pochang Hsu, Dean, 510-657-5911, Fax: 510-657-8975, E-mail: npuadm@npu0.npu.edu. *Application contact:* Dr. Fred Kuttner, Dean of Academic Affairs and Admissions, 510-657-5911, Fax: 510-657-8975, E-mail: npuadm@npu0.npu.edu.

Northwestern University, The Graduate School, Robert R. McCormick School of Engineering and Applied Science, Department of Computer Science, Evanston, IL 60208. Offers PhD. Admissions and degrees offered through The Graduate School. *Faculty:* 9 full-time (0 women), 1 part-time (0 women). *Students:* 28 full-time (4 women); includes 5 minority (1 African American, 4 Asian Americans or Pacific Islanders), 7 international. Average age 25. 168 applicants, 13% accepted. In 1998, 8 doctorates awarded. *Degree requirements:* For doctorate, dissertation required. *Entrance requirements:* For doctorate, GRE General Test (minimum combined score of 1500 on three sections required; average 2000), TOEFL (minimum score of 560 required). *Application deadline:* For fall admission, 8/30. Application fee: $50 ($55 for international students). *Financial aid:* In 1998–99, 6 fellowships with full tuition reimbursements (averaging $11,673 per year), 4 research assistantships with partial tuition reimbursements (averaging $16,285 per year), 7 teaching assistantships with full tuition reimbursements (averaging $12,042 per year) were awarded.; Federal Work-Study, institutionally-sponsored loans, and scholarships also available. Financial aid application deadline: 1/15; financial aid applicants required to submit FAFSA. *Faculty research:* Autonomous mobile robots, education and technology, intelligent information, qualitative reasoning, systems research. Total annual research expenditures: $900,000. *Unit head:* Lawrence Birnbaum, Chair, 847-491-2500, E-mail: birnbaum@ils.nwu.edu. *Application contact:* Christopher Riesbeck, Admission Officer, 847-491-7279, E-mail: riesbeck@ils.nwu.edu.

See in-depth description on page 777.

Northwest Missouri State University, Graduate School, College of Professional and Applied Studies, Department of Computer Science and Information Systems, Maryville, MO 64468-6001. Offers school computer studies (MS). Part-time programs available. *Faculty:* 9 full-time (3 women). *Students:* 2 full-time (1 woman), 3 part-time (1 woman), 2 international. 2 applicants, 100% accepted. In 1998, 3 degrees awarded. *Degree requirements:* For master's, comprehensive exam required, foreign language and thesis not required. *Entrance requirements:* For master's, GRE General Test (minimum combined score of 700 required), TOEFL (minimum score of 550 required), minimum GPA of 3.0. *Application deadline:* Applications are processed on a rolling basis. Application fee: $0 ($50 for international students). *Financial aid:* In 1998–99, 1 research assistantship, 1 teaching assistantship were awarded. Financial aid application deadline: 3/1. *Unit head:* Dr. Phillip Heeler, Chairperson, 660-562-1200. *Application contact:* Dr. Frances Shipley, Dean of Graduate School, 660-562-1145, E-mail: gradsch@mail.nwmissouri.edu.

Nova Southeastern University, School of Computer and Information Sciences, Fort Lauderdale, FL 33314-7721. Offers computer information systems (MS, PhD); computer science (MS, PhD); computing technology in education (MS, Ed D, PhD); information science (PhD); information systems (PhD); management information systems (MS). Part-time and evening/weekend programs available. Postbaccalaureate distance learning degree programs offered. *Students:* 149 full-time (49 women), 382 part-time (145 women); includes 116 minority (45 African Americans, 31 Asian Americans or Pacific Islanders, 39 Hispanic Americans, 1 Native American), 34 international. Average age 41. In 1998, 90 master's, 34 doctorates awarded. Terminal master's awarded for partial completion of doctoral program. *Degree requirements:* For master's, computer language required, thesis optional; for doctorate, computer language, dissertation required, foreign language not required. *Entrance requirements:* For master's, GRE or portfolio; for doctorate, GRE, master's degree, or portfolio. *Average time to degree:* Doctorate–4 years full-time. *Application deadline:* For fall admission, 6/1 (priority date); for spring admission, 1/1. Applications are processed on a rolling basis. Application fee: $50. Tuition: Part-time $370 per credit hour. Required fees: $50 per semester. Tuition and fees vary according to degree level and program. *Financial aid:* In 1998–99, 3 teaching assistantships were awarded.; Federal Work-Study also available. Aid available to part-time students. Financial aid application deadline: 5/1. *Faculty research:* Artificial intelligence, database management, human-computer interaction, distance education, computer education. *Unit head:* Dr. Edward Lieblein, Dean. *Application contact:* Karen Shoemaker, Administrative Assistant, 800-986-2247 Ext. 2000, Fax: 954-262-3915, E-mail: scisinfo@scis.nova.edu.

See in-depth description on page 779.

Oakland University, Graduate Studies, School of Engineering and Computer Science, Program in Computer Science and Engineering, Rochester, MI 48309-4401. Offers computer science (MS); software engineering (MS). Part-time and evening/weekend programs available. *Faculty:* 12 full-time, 3 part-time. *Students:* 48 full-time (22 women), 118 part-time (30 women); includes 21 minority (3 African Americans, 18 Asian Americans or Pacific Islanders), 58 international. Average age 31. 113 applicants, 71% accepted. In 1998, 43 degrees awarded. *Degree requirements:* For master's, foreign language and thesis not required. *Entrance requirements:* For master's, minimum GPA of 3.0 for unconditional admission. *Application deadline:* For fall admission, 7/15; for spring admission, 3/15. Application fee: $30. Tuition, state resident: part-time $221 per credit hour. Tuition, nonresident: part-time $488 per credit hour. Required fees: $214 per semester. Part-time tuition and fees vary according to program. *Financial aid:* Federal Work-Study, institutionally-sponsored loans, and tuition waivers (full) available. Financial aid application deadline: 3/1; financial aid applicants required to submit FAFSA. *Unit head:* Dr. Christian Wagner, Chair, 248-370-2200.

The Ohio State University, Graduate School, College of Engineering, Department of Computer and Information Science, Columbus, OH 43210. Offers MS, PhD. *Faculty:* 33. *Students:* 149 full-time (30 women), 20 part-time (6 women); includes 2 African Americans, 4 Hispanic Americans, 126 international. 429 applicants, 25% accepted. In 1998, 64 master's, 11 doctorates awarded. *Degree requirements:* For master's, computer language required, thesis optional, foreign language not required; for doctorate, computer language, dissertation required, foreign language not required. *Entrance requirements:* For master's and doctorate, GRE General Test. *Application deadline:* For fall admission, 7/1. Applications are processed on a rolling basis. Application fee: $30 ($40 for international students). *Financial aid:* Fellowships, teaching assistantships, career-related internships or fieldwork, Federal Work-Study, institutionally-sponsored loans, and administrative assistantships available. Aid available to part-time students. Financial aid application deadline: 1/15. *Unit head:* Stuart H. Zweben, Chairman, 614-292-5813, Fax: 614-292-2911, E-mail: zweben@cis.ohio-state.edu.

See in-depth description on page 781.

Oklahoma City University, Petree College of Arts and Sciences, Division of Mathematics and Science, Oklahoma City, OK 73106-1402. Offers computer science (MS). Part-time and evening/weekend programs available. *Faculty:* 7 full-time (2 women), 2 part-time (0 women). *Students:* 180 full-time (31 women), 35 part-time (6 women); includes 9 minority (all Asian Americans or Pacific Islanders), 198 international. Average age 27. 234 applicants, 64% accepted. In 1998, 56 degrees awarded. *Degree requirements:* For master's, computer language, comprehensive exam required, foreign language and thesis not required. *Entrance requirements:* For master's, minimum GPA of 3.0. *Application deadline:* For fall admission, 8/25 (priority date); for spring admission, 1/15. Applications are processed on a rolling basis. Application fee: $35 ($70 for international students). *Financial aid:* Fellowships with partial tuition reimbursements, career-related internships or fieldwork, Federal Work-Study, institutionally-sponsored loans, and tuition waivers (partial) available. Aid available to part-time students. Financial aid application deadline: 8/1; financial aid applicants required to submit FAFSA. *Unit head:* Dr. Robert Trail, Chairperson, 405-521-5137. *Application contact:* Laura L. Rahhal, Director of Graduate Admissions, 800-633-7242 Ext. 4, Fax: 405-521-5356, E-mail: gadmissions@frodo.okcu.edu.

Oklahoma State University, Graduate College, College of Arts and Sciences, Department of Computer Science, Stillwater, OK 74078. Offers computer education (Ed D); computer science (MS, PhD). *Faculty:* 8 full-time (2 women). *Students:* 99 full-time (28 women), 106 part-time (30 women); includes 30 minority (6 African Americans, 22 Asian Americans or Pacific Islanders, 2 Hispanic Americans), 148 international. Average age 32. In 1998, 35 master's, 2 doctorates awarded. *Degree requirements:* For master's and doctorate, thesis/dissertation required. *Entrance requirements:* For master's, GRE General Test (minimum score of 700 on quantitative section required), TOEFL (minimum score of 550 required); for doctorate, GRE General Test (minimum score of 700 on quantitative section required), GRE Subject Test (score in 50th percentile or higher required), TOEFL (minimum score of 550 required). *Application deadline:* For fall admission, 7/1 (priority date). Application fee: $25. *Financial aid:* In 1998–99, 37 students received aid, including 7 research assistantships (averaging $12,326 per year), 30 teaching assistantships (averaging $10,481 per year); career-related internships or fieldwork, Federal Work-Study, and tuition waivers (partial) also available.

Computer Science

Oklahoma State University (continued)
Aid available to part-time students. Financial aid application deadline: 3/1. *Unit head:* Dr. Blaine Mayfield, Head, 405-744-5668, Fax: 405-774-9097.

See in-depth description on page 783.

Old Dominion University, College of Sciences, Department of Computer Science, Norfolk, VA 23529. Offers MS, PhD. Part-time programs available. *Faculty:* 19 full-time (1 woman). *Students:* 59 full-time (15 women), 84 part-time (29 women); includes 11 minority (3 African Americans, 5 Asian Americans or Pacific Islanders, 3 Hispanic Americans), 82 international. Average age 31. 120 applicants, 81% accepted. In 1998, 27 master's, 2 doctorates awarded. Terminal master's awarded for partial completion of doctoral program. *Degree requirements:* For master's, computer language, comprehensive diagnostic exam required, thesis optional, foreign language not required; for doctorate, computer language, dissertation, comprehensive exam required, foreign language not required. *Entrance requirements:* For master's, GRE General Test, GRE Subject Test, TOEFL (minimum score of 550 required), minimum GPA of 3.0 in major, 2.5 overall; for doctorate, GRE General Test (minimum combined score of 1000 required), GRE Subject Test (minimum score of 600 required), TOEFL. *Application deadline:* For fall admission, 7/1. Applications are processed on a rolling basis. Application fee: $30. *Financial aid:* In 1998–99, 98 students received aid, including 44 research assistantships (averaging $11,661 per year); fellowships, teaching assistantships, career-related internships or fieldwork, grants, and tuition waivers (partial) also available. Aid available to part-time students. Financial aid application deadline: 2/15; financial aid applicants required to submit FAFSA. *Faculty research:* Software engineering, artificial intelligence, foundations, high-performance computing, networking. Total annual research expenditures: $1.4 million. *Unit head:* Dr. Shunichi Toida, Chair, 757-683-4817, E-mail: toida@cs.odu.edu. *Application contact:* Dr. Shunichi Toida, Chair, 757-683-4817, E-mail: toida@cs.odu.edu.

Oregon Graduate Institute of Science and Technology, Graduate Studies, Department of Computer Science and Engineering, Portland, OR 97291-1000. Offers computational finance (MS, Certificate); computer science and engineering (MS, PhD). Part-time and evening/weekend programs available. *Faculty:* 16 full-time (2 women), 31 part-time (6 women). *Students:* 70 full-time (16 women), 61 part-time (20 women); includes 5 Asian Americans or Pacific Islanders, 1 Native American, 54 international. Average age 31. 257 applicants, 43% accepted. In 1998, 29 master's, 2 doctorates awarded. Terminal master's awarded for partial completion of doctoral program. *Degree requirements:* For master's, computer language required, thesis optional, foreign language not required; for doctorate, computer language, comprehensive exam, oral defense of dissertation required. *Entrance requirements:* For master's and doctorate, GRE General Test, TOEFL (minimum score of 600 required). *Average time to degree:* Master's–1.8 years full-time, 3.9 years part-time; doctorate–5.6 years full-time. *Application deadline:* For fall admission, 3/1 (priority date). Applications are processed on a rolling basis. Application fee: $50. Electronic applications accepted. *Financial aid:* In 1998–99, 41 students received aid, including 38 research assistantships, 3 teaching assistantships; fellowships, grants and scholarships also available. Financial aid application deadline: 3/1. *Faculty research:* Computer systems architecture, intelligent and interactive systems, programming models and systems, theory of computation. *Unit head:* Dr. Andrew P. Black, Head, 503-690-1250, E-mail: black@cse.ogi.edu. *Application contact:* Shirley Kapsch, Enrollment Manager, 503-690-1255, Fax: 503-690-1285, E-mail: kapsch@cse.ogi.edu.

See in-depth description on page 785.

Oregon Graduate Institute of Science and Technology, Graduate Studies, Department of Electrical and Computer Engineering, Portland, OR 97291-1000. Offers computational finance (Certificate); computer engineering (MS, PhD); electrical engineering (MS, PhD). Part-time programs available. *Faculty:* 13 full-time (0 women), 34 part-time (1 woman). *Students:* 48 full-time (15 women), 67 part-time (14 women); includes 16 minority (all Asian Americans or Pacific Islanders) Terminal master's awarded for partial completion of doctoral program. *Degree requirements:* For master's, thesis optional, foreign language not required; for doctorate, comprehensive exam, oral defense of dissertation required. *Entrance requirements:* For master's, TOEFL (minimum score of 550 required); for doctorate, GRE General Test, GRE Subject Test, TOEFL (minimum score of 550 required). *Application deadline:* For fall admission, 3/1 (priority date). Applications are processed on a rolling basis. Application fee: $50. Electronic applications accepted. *Unit head:* Dr. Dan Hammerstrom, Head, 503-690-4037, Fax: 503-690-1406. *Application contact:* Don Johansen, Enrollment Manager, 503-690-1315, E-mail: johansen@ece.ogi.edu.

Oregon Graduate Institute of Science and Technology, Graduate Studies, Department of Management in Science and Technology, Portland, OR 97291-1000. Offers computational finance (Certificate); management in science and technology (MS). Part-time and evening/weekend programs available. *Faculty:* 3 full-time (0 women), 9 part-time (1 woman). *Students:* 4 full-time (1 woman), 68 part-time (16 women); includes 5 minority (all Asian Americans or Pacific Islanders), 4 international. *Degree requirements:* For master's, foreign language and thesis not required. *Entrance requirements:* For master's, TOEFL (minimum score of 650 required). *Application deadline:* Applications are processed on a rolling basis. Application fee: $50. Electronic applications accepted. *Unit head:* Dr. Fred Young Phillips, Head, 503-690-1353, Fax: 503-690-1268, E-mail: fphillips@admin.ogi.edu. *Application contact:* Victoria Tyler, Enrollment Manager, 503-690-1335, Fax: 503-690-1285, E-mail: vtyler@admin.ogi.edu.

Oregon State University, Graduate School, College of Engineering, Department of Computer Science, Corvallis, OR 97331. Offers MA, MAIS, MS, PhD. Part-time programs available. *Faculty:* 14 full-time (2 women). *Students:* 48 full-time (14 women), 8 part-time (1 woman); includes 3 minority (all Asian Americans or Pacific Islanders), 29 international. Average age 30. In 1998, 18 master's, 4 doctorates awarded (100% entered university research/teaching). Terminal master's awarded for partial completion of doctoral program. *Degree requirements:* For master's, thesis or alternative required, foreign language not required; for doctorate, dissertation required, foreign language not required. *Entrance requirements:* For master's and doctorate, GRE General Test, TOEFL (minimum score of 550 required), minimum GPA of 3.0 in last 90 hours. *Application deadline:* For fall admission, 3/1. Applications are processed on a rolling basis. Application fee: $50. *Financial aid:* Fellowships, research assistantships, teaching assistantships, career-related internships or fieldwork, Federal Work-Study, and institutionally-sponsored loans available. Aid available to part-time students. Financial aid application deadline: 2/1. *Faculty research:* Artificial intelligence, software systems, theory and algorithms, parallel computing. *Unit head:* Dr. Michael J. Quinn, Head, 541-737-3273, Fax: 541-737-3014, E-mail: quinn@cs.orst.edu. *Application contact:* Bernadette Feyerherm, Student Coordinator, 541-737-3273, Fax: 541-737-3014, E-mail: bernie@cs.orst.edu.

See in-depth description on page 789.

Pace University, School of Computer Science and Information Systems, New York, NY 10038. Offers computer communications and networks (Certificate); computer science (MS); computing studies (DPS); information systems (MS); New 9/1/1999 (DPS); object-oriented programming (Certificate); telecommunications (MS, Certificate). Part-time and evening/weekend programs available. *Faculty:* 36 full-time, 44 part-time. *Students:* 93 full-time (35 women), 468 part-time (170 women); includes 101 minority (40 African Americans, 41 Asian Americans or Pacific Islanders, 20 Hispanic Americans), 80 international. Average age 31. 339 applicants, 84% accepted. In 1998, 154 master's, 14 other advanced degrees awarded. *Degree requirements:* For master's, computer language required, foreign language and thesis not required. *Entrance requirements:* For master's, GRE General Test. *Application deadline:* For fall admission, 7/31 (priority date); for spring admission, 11/30. Applications are processed on a rolling basis. Application fee: $60. Electronic applications accepted. *Financial aid:* Research assistantships, career-related internships or fieldwork available. Aid available to part-time students. Financial aid applicants required to submit FAFSA. *Unit head:* Dr. Susan Merritt, Dean, 914-422-4375. *Application contact:* Lois Rich, Associate Director, 914-422-4283, Fax: 914-422-4287, E-mail: gradwp@ny2.pace.edu.

See in-depth description on page 791.

Pacific States University, College of Electrical Engineering and Computer Science, Los Angeles, CA 90006. Offers MSCS. *Students:* 13. *Entrance requirements:* For master's, TOEFL (minimum score of 450 required), bachelor's degree in physics, engineering, computer science, or applied mathematics; minimum undergraduate GPA of 2.5 during last 90 hours. *Application deadline:* For fall admission, 6/15; for spring admission, 12/15. Application fee: $100. *Unit head:* Meyer Pollack, Dean, 888-200-0383, Fax: 323-731-7276, E-mail: admission@psuca.edu. *Application contact:* Woo Yeol Lee, Admissions Officer, 888-200-0383, Fax: 323-731-7276, E-mail: admission@psuca.edu.

Pennsylvania State University Harrisburg Campus of the Capital College, Graduate Center, School of Science, Engineering and Technology, Program in Computer Science, Middletown, PA 17057-4898. Offers MS. *Students:* 7 full-time (2 women), 15 part-time (6 women). *Degree requirements:* For master's, thesis required, foreign language not required. *Entrance requirements:* For master's, TOEFL (minimum score of 550 required). *Application deadline:* For fall admission, 7/26. Application fee: $50. *Unit head:* Dr. Jeff Hartzler, Chair, 717-948-6091.

Pennsylvania State University University Park Campus, Graduate School, College of Engineering, Department of Computer Science and Engineering, State College, University Park, PA 16802-1503. Offers M Eng, MS, PhD. *Students:* 91 full-time (20 women), 32 part-time (9 women). In 1998, 32 master's, 6 doctorates awarded. *Degree requirements:* For doctorate, dissertation required, foreign language not required. *Entrance requirements:* For master's and doctorate, GRE General Test. Application fee: $50. *Unit head:* Dr. Dale A. Miller, Head, 814-865-9505.

See in-depth description on page 793.

Polytechnic University, Brooklyn Campus, Department of Computer and Information Science, Major in Computer Science, Brooklyn, NY 11201-2990. Offers MS, PhD. Part-time and evening/weekend programs available. *Students:* 41 full-time (11 women), 155 part-time (40 women); includes 34 minority (4 African Americans, 25 Asian Americans or Pacific Islanders, 5 Hispanic Americans), 33 international. Average age 33. 185 applicants, 52% accepted. In 1998, 49 master's, 5 doctorates awarded. *Degree requirements:* For master's, computer language required, thesis not required; for doctorate, computer language, dissertation required. *Entrance requirements:* For master's, BA or BS in computer science, mathematics, science, or engineering; working knowledge of a high-level program; for doctorate, GRE General Test, GRE Subject Test, qualifying exam, BA or BS in science, engineering, or management; MS or 1 year of graduate work. *Application deadline:* Applications are processed on a rolling basis. Application fee: $45. Electronic applications accepted. *Financial aid:* Research assistantships, teaching assistantships, institutionally-sponsored loans available. Aid available to part-time students. Financial aid applicants required to submit FAFSA. *Unit head:* Victoria Tyler, Enrollment Manager, 503-690-1335, Fax: 503-690-1285, E-mail: vtyler@admin.ogi.edu. *Application contact:* John S. Kerge, Dean of Admissions, 718-260-3200, Fax: 718-260-3446, E-mail: admitme@poly.edu.

See in-depth description on page 797.

Polytechnic University, Farmingdale Campus, Graduate Programs, Department of Computer and Information Science, Major in Computer Science, Farmingdale, NY 11735-3995. Offers MS, PhD. *Students:* 1 full-time (0 women), 83 part-time (14 women); includes 3 minority (all Asian Americans or Pacific Islanders), 4 international. Average age 33. 23 applicants, 65% accepted. In 1998, 11 degrees awarded. *Degree requirements:* For master's, computer language, thesis required (for some programs). *Application deadline:* Applications are processed on a rolling basis. Application fee: $45. Electronic applications accepted. *Financial aid:* Institutionally-sponsored loans available. Aid available to part-time students. Financial aid applicants required to submit FAFSA. *Unit head:* 610-648-3242, Fax: 610-889-1334. *Application contact:* John S. Kerge, Dean of Admissions, 718-260-3200, Fax: 718-260-3446, E-mail: admitme@poly.edu.

Polytechnic University, Westchester Graduate Center, Graduate Programs, Department of Computer and Information Science, Major in Computer Science, Hawthorne, NY 10532-1507. Offers MS, PhD. *Students:* 2 full-time (1 woman), 73 part-time (20 women); includes 8 minority (1 African American, 7 Asian Americans or Pacific Islanders), 5 international. Average age 33. 26 applicants, 65% accepted. In 1998, 15 master's awarded. *Degree requirements:* For master's, computer language, thesis required (for some programs); for doctorate, dissertation required. *Application deadline:* Applications are processed on a rolling basis. Application fee: $45. Electronic applications accepted. *Unit head:* John S. Kerge, Dean of Admissions, 718-260-3200, Fax: 718-260-3446, E-mail: admitme@poly.edu. *Application contact:* John S. Kerge, Dean of Admissions, 718-260-3200, Fax: 718-260-3446, E-mail: admitme@poly.edu.

Portland State University, Graduate Studies, School of Engineering and Applied Science, Department of Computer Science, Portland, OR 97207-0751. Offers MS. Part-time programs available. *Faculty:* 14 full-time (2 women), 11 part-time (2 women). *Students:* 24 full-time (12 women), 20 part-time (8 women); includes 6 minority (5 Asian Americans or Pacific Islanders, 1 Native American), 32 international. Average age 28. 82 applicants, 70% accepted. In 1998, 19 degrees awarded. *Degree requirements:* For master's, thesis optional, foreign language not required. *Entrance requirements:* For master's, GRE General Test, TOEFL (minimum score of 550 required), minimum GPA of 3.0 in upper-division course work or 2.75 overall. *Application deadline:* For fall admission, 3/15 (priority date). Applications are processed on a rolling basis. Application fee: $50. *Financial aid:* In 1998–99, 6 research assistantships, 13 teaching assistantships were awarded.; career-related internships or fieldwork and Federal Work-Study also available. Aid available to part-time students. Financial aid application deadline: 3/1; financial aid applicants required to submit FAFSA. *Faculty research:* Formal methods, database systems, parallel programming environments, computer security, software tools. Total annual research expenditures: $344,148. *Unit head:* Cynthia Brown, Head, 503-725-4036, Fax: 503-725-3211, E-mail: brown@cs.pdx.edu. *Application contact:* Beth Phelps, Office Coordinator, 503-725-4036, Fax: 503-725-3211, E-mail: phelps@cs.pdx.edu.

Princeton University, Graduate School, School of Engineering and Applied Science, Department of Computer Science, Princeton, NJ 08544-1019. Offers M Eng, MSE, PhD. *Degree requirements:* For master's, computer language required, thesis not required; for doctorate, computer language, dissertation required. *Entrance requirements:* For master's and doctorate, GRE General Test, GRE Subject Test. *Faculty research:* Algorithms, complexity, systems, VLSI.

See in-depth description on page 799.

Purdue University, Graduate School, School of Science, Department of Computer Sciences, West Lafayette, IN 47907. Offers MS, PhD. Part-time programs available. *Faculty:* 30 full-time (1 woman), 3 part-time (1 woman). *Students:* 89 full-time (17 women), 23 part-time (9 women); includes 8 minority (all Asian Americans or Pacific Islanders), 104 international. Average age 28. 434 applicants, 39% accepted. In 1998, 57 master's, 6 doctorates awarded (17% entered university research/teaching, 83% found other work related to degree). Terminal master's awarded for partial completion of doctoral program. *Degree requirements:* For master's, thesis optional, foreign language not required; for doctorate, dissertation required, foreign language not required. *Entrance requirements:* For master's and doctorate, GRE General Test, TOEFL (minimum score of 600 required), TWE (minimum score of 5.0 required), minimum GPA of 3.5. *Average time to degree:* Master's–1.6 years full-time; doctorate–6 years full-time. *Application deadline:* For fall admission, 12/15. Application fee: $30. Electronic applications accepted. *Financial aid:* In 1998–99, 100 students received aid, including 4 fellowships (averaging $15,830 per year), 36 research assistantships (averaging $12,000 per year), 60 teaching assistantships (averaging $11,270 per year) Financial aid application deadline: 12/15. *Faculty research:* Computer systems, geometric modeling, information security, scientific computing, software systems, theory and algorithms. Total annual research expenditures: $3.7 million. *Unit head:* Prof. Ahmed H. Sameh, Head, 765-494-6003. *Application contact:* Dr. William J.

Gorman, Assistant to the Head, 765-494-6004, Fax: 765-494-0739, E-mail: gradinfo-p@cs.purdue.edu.

See in-depth description on page 801.

Purdue University, Graduate School, School of Science, Department of Statistics, West Lafayette, IN 47907. Offers applied statistics (MS); statistics (PhD); statistics and computer science (MS); theoretical statistics (MS). *Faculty:* 24 full-time (4 women). *Students:* 28 full-time (7 women), 25 part-time (8 women); includes 4 minority (1 African American, 3 Asian Americans or Pacific Islanders), 39 international. *Degree requirements:* For master's, foreign language and thesis not required; for doctorate, dissertation required, foreign language not required. *Entrance requirements:* For master's and doctorate, GRE, TOEFL (minimum score of 600 required). *Application deadline:* For fall admission, 2/1. Application fee: $30. Electronic applications accepted. *Unit head:* Dr. M. E. Bock, Head, 765-494-3141, Fax: 765-494-0558. *Application contact:* Angie Murphy, Graduate Secretary, 765-494-5794, Fax: 765-494-0558, E-mail: graduate@stat.purdue.edu.

Queens College of the City University of New York, Division of Graduate Studies, Mathematics and Natural Sciences Division, Department of Computer Science, Flushing, NY 11367-1597. Offers MA. Part-time and evening/weekend programs available. *Faculty:* 18 full-time (2 women). *Students:* 51 full-time (25 women), 196 part-time (76 women); includes 149 minority (1 African American, 148 Asian Americans or Pacific Islanders), 81 international. 220 applicants, 87% accepted. In 1998, 38 degrees awarded. *Degree requirements:* For master's, computer language, comprehensive exam required, thesis optional, foreign language not required. *Entrance requirements:* For master's, GRE, TOEFL (minimum score of 550 required), minimum GPA of 3.0. *Application deadline:* For fall admission, 4/1; for spring admission, 11/1. Applications are processed on a rolling basis. Application fee: $40. Tuition, state resident: full-time $4,350; part-time $185 per credit. Tuition, nonresident: full-time $7,600; part-time $320 per credit. Required fees: $114; $57 per semester. Tuition and fees vary according to course load and program. *Financial aid:* Career-related internships or fieldwork, Federal Work-Study, institutionally-sponsored loans, tuition waivers (partial), unspecified assistantships, and adjunct lectureships available. Aid available to part-time students. Financial aid application deadline: 4/1; financial aid applicants required to submit FAFSA. *Faculty research:* Fifth-generation computing, hardware/software development, analysis of algorithms and theoretical computer science. *Unit head:* Dr. Theodore Brown, Chairperson, 718-997-3500. *Application contact:* Dr. K. Yukawa, Graduate Adviser, 718-997-3500.

Queen's University at Kingston, School of Graduate Studies and Research, Faculty of Arts and Sciences, Department of Computing and Information Science, Kingston, ON K7L 3N6, Canada. Offers M Sc, PhD. Part-time programs available. *Students:* 45 full-time (7 women), 14 part-time (5 women). In 1998, 19 master's, 1 doctorate awarded. *Degree requirements:* For master's, thesis optional, foreign language not required; for doctorate, dissertation, comprehensive exam required. *Entrance requirements:* For master's and doctorate, TOEFL (minimum score of 600 required). *Application deadline:* For fall admission, 2/28 (priority date). Application fee: $60. Electronic applications accepted. *Financial aid:* Fellowships, research assistantships, teaching assistantships, institutionally-sponsored loans available. Financial aid application deadline: 3/1. *Unit head:* Dr. J. I. Glasgow, Head, 613-533-6058. *Application contact:* Dr. R. A. Browse, Graduate Coordinator, 613-533-6069.

Rensselaer at Hartford, Lally School of Management and Technology, Program in Computer and Information Science, Hartford, CT 06120-2991. Offers MS. Part-time and evening/weekend programs available. *Degree requirements:* For master's, computer language, seminar required, thesis optional, foreign language not required. *Entrance requirements:* For master's, TOEFL (minimum score of 570 required).

Rensselaer Polytechnic Institute, Graduate School, School of Science, Department of Computer Science, Troy, NY 12180-3590. Offers MS, PhD. Part-time programs available. Postbaccalaureate distance learning degree programs offered (no on-campus study). *Faculty:* 14 full-time (2 women), 7 part-time (0 women). *Students:* 76 full-time (19 women), 30 part-time (7 women); includes 12 minority (4 African Americans, 8 Asian Americans or Pacific Islanders), 42 international. 347 applicants, 34% accepted. In 1998, 42 master's, 12 doctorates awarded. *Degree requirements:* For master's, thesis or alternative required; for doctorate, dissertation required, foreign language not required. *Entrance requirements:* For master's and doctorate, GRE General Test, TOEFL (minimum score of 550 required). *Application deadline:* For fall admission, 2/1 (priority date). Applications are processed on a rolling basis. Application fee: $35. *Financial aid:* In 1998–99, 50 students received aid, including 23 research assistantships, 30 teaching assistantships; fellowships, career-related internships or fieldwork and institutionally-sponsored loans also available. Financial aid application deadline: 2/1. *Faculty research:* Computer vision, engineering databases, parallel computing, scientific computation, human-computer interaction. *Unit head:* Dr. Franklin Luk, Chair, 518-276-8326, Fax: 518-276-4033, E-mail: luk@cs.rpi.edu. *Application contact:* Terry Hayden, Coordinator of Graduate Admissions, 518-276-8419, Fax: 518-276-4033, E-mail: grad-adm@cs.rpi.edu.

See in-depth description on page 803.

Rice University, Graduate Programs, George R. Brown School of Engineering, Department of Computer Science, Houston, TX 77251-1892. Offers MCS, MS, PhD. Part-time programs available. Terminal master's awarded for partial completion of doctoral program. *Degree requirements:* For master's, thesis required (for some programs), foreign language not required; for doctorate, dissertation required, foreign language not required. *Entrance requirements:* For master's and doctorate, GRE General Test, GRE Subject Test, TOEFL (minimum score of 550 required), minimum GPA of 3.0. *Faculty research:* Operating systems, distributed systems, programming languages, algorithms, automatic program testing.

See in-depth description on page 807.

Rivier College, School of Professional Studies, Department of Computer Science and Mathematics, Nashua, NH 03060-5086. Offers computer science (MS); information science (MS). Part-time and evening/weekend programs available. *Faculty:* 2 full-time (1 woman), 13 part-time (2 women). *Students:* 31 full-time (12 women), 53 part-time (19 women); includes 4 minority (all Asian Americans or Pacific Islanders), 36 international. Average age 36. 56 applicants, 52% accepted. In 1998, 16 degrees awarded. *Degree requirements:* For master's, thesis not required. *Entrance requirements:* For master's, GRE Subject Test. *Average time to degree:* Master's–3 years part-time. *Application deadline:* Applications are processed on a rolling basis. Application fee: $25. Tuition: Part-time $309 per credit. Required fees: $75 per year. *Financial aid:* Available to part-time students. Application deadline: 2/1; *Unit head:* Dr. Stephen Cooper, Director, 603-888-1311, E-mail: scooper@rivier.edu. *Application contact:* Paula Bailly-Burton, Director of Graduate and Evening Admissions, 603-888-1311, Fax: 603-888-9124, E-mail: geaadmit@rivier.edu.

Rochester Institute of Technology, Part-time and Graduate Admissions, College of Applied Science and Technology, Department of Computer Science and Information Technology, Program in Applied Computer Studies, Rochester, NY 14623-5604. Offers AC. *Students:* 1 (woman) full-time; minority (Asian American or Pacific Islander) 1 applicants, 100% accepted. In 1998, 1 degree awarded. *Entrance requirements:* For degree, GRE, TOEFL (minimum score of 550 required), minimum GPA of 3.0. *Application deadline:* For fall admission, 3/1 (priority date). Applications are processed on a rolling basis. Application fee: $40. *Financial aid:* Unspecified assistantships available. *Unit head:* Dr. Rayno Niemi, Graduate Coordinator, 716-475-2202, E-mail: rdn@it.rit.edu.

Rochester Institute of Technology, Part-time and Graduate Admissions, College of Applied Science and Technology, Department of Computer Science and Information Technology, Program in Computer Science, Rochester, NY 14623-5604. Offers MS. *Students:* 62 full-time (17 women), 95 part-time (25 women); includes 16 minority (1 African American, 11 Asian Americans or Pacific Islanders, 2 Hispanic Americans, 2 Native Americans), 71 international. 181 applicants,

75% accepted. In 1998, 42 degrees awarded. *Degree requirements:* For master's, computer language, thesis required. *Entrance requirements:* For master's, GRE General Test, TOEFL, minimum GPA of 3.0. *Application deadline:* For fall admission, 3/1 (priority date). Applications are processed on a rolling basis. Application fee: $40. *Financial aid:* Research assistantships, teaching assistantships, scholarships available. *Unit head:* Dr. Peter Anderson, Graduate Coordinator, 716-475-2979, E-mail: pga@cs.rit.edu.

See in-depth description on page 809.

Roosevelt University, Graduate Division, College of Arts and Sciences, School of Computer Science and Telecommunications, Program in Computer Science, Chicago, IL 60605-1394. Offers MSC. Part-time and evening/weekend programs available. *Degree requirements:* For master's, computer language required, foreign language and thesis not required. *Application deadline:* For fall admission, 6/1 (priority date). Applications are processed on a rolling basis. Application fee: $25 ($35 for international students). *Financial aid:* Application deadline: 2/15. *Faculty research:* Artificial intelligence, software engineering, distributed databases, parallel processing. Total annual research expenditures: $40,000. *Unit head:* Ken W. Mehavics, Director, 312-341-3685, Fax: 312-341-3860, E-mail: kmihavic@acfsysv.roosevelt.edu. *Application contact:* Joanne Canyon-Heller, Coordinator of Graduate Admissions, 312-341-3612, Fax: 312-341-3523, E-mail: applyru@roosevelt.edu.

Rutgers, The State University of New Jersey, New Brunswick, Graduate School, Program in Computer Science, New Brunswick, NJ 08903. Offers MS, PhD. Part-time programs available. *Faculty:* 34 full-time (2 women), 4 part-time (0 women). *Students:* 92 full-time (11 women), 63 part-time (13 women); includes 8 minority (2 African Americans, 6 Hispanic Americans), 97 international. Average age 28. 330 applicants, 28% accepted. In 1998, 28 master's, 7 doctorates awarded. Terminal master's awarded for partial completion of doctoral program. *Degree requirements:* For master's, foreign language and thesis not required; for doctorate, dissertation required, foreign language not required. *Entrance requirements:* For master's and doctorate, GRE General Test (score in the 80th percentile or higher required), GRE Subject Test (score in the 80th percentile or higher required). *Average time to degree:* Master's–2 years full-time, 2.75 years part-time; doctorate–5.5 years full-time. *Application deadline:* For fall admission, 2/1; for spring admission, 11/1. Applications are processed on a rolling basis. Application fee: $50. Electronic applications accepted. *Financial aid:* In 1998–99, 84 students received aid, including fellowships with full tuition reimbursements available (averaging $13,000 per year), 32 research assistantships with full tuition reimbursements available (averaging $12,336 per year), 52 teaching assistantships with full tuition reimbursements available (averaging $12,336 per year); Federal Work-Study also available. Financial aid application deadline: 2/1; financial aid applicants required to submit FAFSA. *Faculty research:* Theoretical computer science, distributed/networked/wireless systems, artificial intelligence, compiling. *Unit head:* Vasek Chvatal, Graduate Director, 732-445-3908, Fax: 732-445-0537, E-mail: chvatal@cs.rutgers.edu. *Application contact:* Valentine Rolfe, Secretary, 732-445-3547, Fax: 732-445-0537, E-mail: rolfe@cs.rutgers.edu.

See in-depth description on page 811.

Sacred Heart University, Graduate Studies, College of Arts and Sciences, Faculty of Computer and Information Science, Fairfield, CT 06432-1000. Offers MS. *Faculty:* 5 full-time (1 woman), 7 part-time (2 women). Average age 35. 22 applicants, 73% accepted. In 1998, 3 degrees awarded. *Degree requirements:* For master's, thesis optional. *Application deadline:* Applications are processed on a rolling basis. Application fee: $40 ($100 for international students). Tuition: Part-time $375 per credit. Required fees: $83 per term. Tuition and fees vary according to campus/location and program. *Financial aid:* Applicants required to submit FAFSA. *Unit head:* Domenick Pinto, Acting Director, 203-371-7799, Fax: 203-371-0506. *Application contact:* Mike Kennedy, Graduate Admissions Counselor, 203-365-7619, Fax: 203-365-4732, E-mail: gradstudies@sacredheart.edu.

Announcement: Located just 55 miles northeast of New York City, the MSCIS program at Sacred Heart University offers course work on a trimester basis, which allows for earlier degree completion than traditional tracked programs. Students also have the option to concentrate in computer or information technology, with courses such as Visual Basic, Java, Web Design, Windows NT, and Multimedia Design.

St. Cloud State University, School of Graduate Studies, College of Science and Engineering, Department of Computer Science, St. Cloud, MN 56301-4498. Offers MS. *Faculty:* 10 full-time (3 women). *Degree requirements:* For master's, thesis or alternative required, foreign language not required. *Entrance requirements:* For master's, GRE General Test, minimum GPA of 2.75. *Application fee:* $20. *Financial aid:* Unspecified assistantships available. Financial aid application deadline: 3/1. *Unit head:* Dr. Larry Grover, Chairperson, 320-255-4966. *Application contact:* Ann Anderson, Graduate Studies Office, 320-255-2113, Fax: 320-654-5371, E-mail: aeanderson@stcloudstate.edu.

St. John's University, Graduate School of Arts and Sciences, Department of Mathematics and Computer Science, Jamaica, NY 11439. Offers algebra (MA); analysis (MA); applied mathematics (MA); computer science (MA); geometry-topology (MA); logic and foundations (MA); probability and statistics (MA). Part-time and evening/weekend programs available. *Faculty:* 16 full-time (3 women), 5 part-time (2 women). *Students:* 7 full-time (3 women), 5 part-time (3 women); includes 5 minority (2 African Americans, 1 Asian American or Pacific Islander, 2 Hispanic Americans), 2 international. Average age 24. 13 applicants, 100% accepted. In 1998, 1 degree awarded. *Degree requirements:* For master's, comprehensive exam required, thesis optional, comprehensive exam required, thesis optional. *Entrance requirements:* For master's, minimum GPA of 3.0. *Application deadline:* Applications are processed on a rolling basis. Application fee: $40. Tuition: Full-time $13,200; part-time $500 per credit. Required fees: $150; $75 per term. Tuition and fees vary according to degree level and program. *Financial aid:* In 1998–99, 2 research assistantships were awarded.; scholarships also available. Aid available to part-time students. Financial aid application deadline: 3/1; financial aid applicants required to submit FAFSA. *Faculty research:* Development of a computerized metabolicmap. *Unit head:* Dr. Charles Traina, Chair, 718-990-6166, E-mail: trainac@stjohns.edu. *Application contact:* Patricia G. Armstrong, Director, Office of Admission, 718-990-2028, Fax: 718-990-2096, E-mail: armstrop@stjohns.edu.

Saint Joseph's University, College of Arts and Sciences, Program in Computer Science, Philadelphia, PA 19131-1395. Offers MS. Part-time and evening/weekend programs available. *Students:* 91 (51 women); includes 71 minority (4 African Americans, 67 Asian Americans or Pacific Islanders) 12 international. Average age 32. In 1998, 48 degrees awarded. *Degree requirements:* For master's, computer language required. *Entrance requirements:* For master's, GRE General Test, TOEFL. *Application deadline:* For fall admission, 7/15. Application fee: $30. *Unit head:* Dr. Gary Laison, Director, 610-660-1571.

St. Mary's University of San Antonio, Graduate School, Program in Computer Information Systems, San Antonio, TX 78228-8507. Offers MS. *Faculty:* 6 full-time (0 women), 3 part-time (1 woman). *Students:* 14 full-time (3 women), 64 part-time (12 women); includes 28 minority (6 African Americans, 3 Asian Americans or Pacific Islanders, 19 Hispanic Americans), 10 international. Average age 30. In 1998, 14 degrees awarded (100% found work related to degree). *Degree requirements:* For master's, computer language, comprehensive exams required, thesis optional, foreign language not required. *Entrance requirements:* For master's, GMAT or GRE General Test. *Application deadline:* For fall admission, 8/1. Application fee: $15. *Financial aid:* Research assistantships, career-related internships or fieldwork and institutionally-sponsored loans available. *Faculty research:* Artificial intelligence, database/knowledge base, software engineering, expert systems. *Unit head:* Dr. Douglas Hall, Adviser, 210-436-3317.

Sam Houston State University, College of Arts and Sciences, Division of Mathematical and Information Sciences, Program in Computing Science, Huntsville, TX 77341. Offers M Ed, MS. Part-time programs available. *Students:* 23 full-time (7 women), 6 part-time (2 women); includes

Computer Science

Sam Houston State University (continued)

8 minority (7 Asian Americans or Pacific Islanders, 1 Hispanic American), 15 international. Average age 30. In 1998, 4 degrees awarded. *Degree requirements:* For master's, foreign language and thesis not required. *Entrance requirements:* For master's, GRE General Test (minimum combined score of 1000 required), TOEFL (minimum score of 550 required). *Application deadline:* Applications are processed on a rolling basis. Application fee: $15. *Financial aid:* Teaching assistantships available. *Faculty research:* Language design, networks, database, operating systems, multimedia. *Unit head:* Dr. David Burris, Graduate Adviser, 409-294-1568, Fax: 409-294-1882, E-mail: csc_dsb@shsu.edu. *Application contact:* Dr. David Burris, Graduate Adviser, 409-294-1568, Fax: 409-294-1882, E-mail: csc_dsb@shsu.edu.

San Diego State University, Graduate and Research Affairs, College of Sciences, Department of Mathematical Sciences, Program in Computer Science, San Diego, CA 92182. Offers MS. *Students:* 77 full-time (26 women), 125 part-time (33 women); includes 62 minority (2 African Americans, 57 Asian Americans or Pacific Islanders, 2 Hispanic Americans, 1 Native American), 59 international. Average age 28. In 1998, 13 degrees awarded. *Degree requirements:* For master's, computer language, comprehensive exam or thesis required. *Entrance requirements:* For master's, GRE General Test (minimum combined score of 950 required), TOEFL (minimum score of 550 required). *Application deadline:* For fall admission, 7/1 (priority date); for spring admission, 12/1. Applications are processed on a rolling basis. Application fee: $55. *Unit head:* Mike Kennedy, Graduate Admissions Counselor, 203-365-7619, Fax: 203-365-4732, E-mail: gradstudies@sacredheart.edu. *Application contact:* Carl Eckberg, Graduate Adviser, 619-594-6834, Fax: 619-594-6746, E-mail: eckberg@saturn.sdsu.edu.

San Francisco State University, Graduate Division, College of Science and Engineering, Department of Computer Science, San Francisco, CA 94132-1722. Offers MS. Part-time programs available. *Faculty:* 9 full-time (1 woman), 4 part-time (0 women). *Students:* 59 (12 women). Average age 26. In 1998, 16 degrees awarded. *Degree requirements:* For master's, thesis or alternative required. *Entrance requirements:* For master's, GRE, minimum GPA of 2.5 in last 60 units. *Application deadline:* For fall admission, 11/30 (priority date). Applications are processed on a rolling basis. Application fee: $55. *Financial aid:* In 1998–99, 2 research assistantships, 4 teaching assistantships were awarded.; unspecified assistantships also available. Financial aid application deadline: 3/1. *Faculty research:* Parallel computing, real time systems, database systems, neural networks, computer graphics. *Unit head:* Dr. Jozo Dujmovic, Chair, 415-338-1008, Fax: 415-338-6136, E-mail: cs@sfsu.edu. *Application contact:* Dr. Barry Levine, Graduate Coordinator, 415-338-1661, E-mail: levine@sfsu.edu.

San Jose State University, Graduate Studies, College of Engineering, Department of Computer, Information and Systems Engineering, Program in Computer Engineering, San Jose, CA 95192-0001. Offers computer engineering (MS); computer software (MS); computerized robots and computer applications (MS); microprocessors and microcomputers (MS). *Faculty:* 5 full-time (0 women), 12 part-time (1 woman). *Students:* 57 full-time (28 women), 90 part-time (19 women); includes 106 minority (1 African American, 103 Asian Americans or Pacific Islanders, 2 Hispanic Americans), 27 international. *Degree requirements:* For master's, computer language, thesis, comprehensive exam required, foreign language not required. *Entrance requirements:* For master's, GRE General Test (minimum combined score of 1500 on three sections required), BS in computer science or 24 credits in related area. *Application deadline:* For fall admission, 6/1. Applications are processed on a rolling basis. Application fee: $59. Tuition, nonresident: part-time $246 per unit. Required fees: $1,939; $1,309 per year. *Unit head:* Dr. Haluk Ozemek, Coordinator, 408-924-4100. *Application contact:* Dr. Haluk Ozemek, Coordinator, 408-924-4100.

San Jose State University, Graduate Studies, College of Science, Department of Mathematics and Computer Science, San Jose, CA 95192-0001. Offers computer science (MS); mathematics (MA, MS). Part-time and evening/weekend programs available. *Faculty:* 51 full-time (5 women), 3 part-time (0 women). *Students:* 46 full-time (27 women), 101 part-time (46 women); includes 79 minority (1 African American, 75 Asian Americans or Pacific Islanders, 3 Hispanic Americans), 23 international. Average age 31. 365 applicants, 22% accepted. In 1998, 12 degrees awarded. *Degree requirements:* For master's, thesis (for some programs), comprehensive exam required, foreign language not required. *Entrance requirements:* For master's, GRE Subject Test. *Application deadline:* For fall admission, 6/1. Applications are processed on a rolling basis. Application fee: $59. Tuition, nonresident: part-time $246 per unit. Required fees: $1,939; $1,309 per year. *Financial aid:* In 1998–99, 20 teaching assistantships were awarded.; career-related internships or fieldwork and Federal Work-Study also available. Aid available to part-time students. *Faculty research:* Artificial intelligence, algorithms, numerical analysis, software database, number theory. *Unit head:* Dr. Michael Burke, Chair, 408-924-5100, Fax: 408-924-5080. *Application contact:* Dr. John Mitchem, Graduate Adviser, 408-924-5135.

Santa Clara University, School of Engineering, Department of Computer Science and Engineering, Santa Clara, CA 95053-0001. Offers computer science and engineering (MSCSE, PhD); high performance computing (Certificate); software engineering (Certificate). Part-time and evening/weekend programs available. *Students:* 141 full-time (76 women), 229 part-time (86 women); includes 91 minority (88 Asian Americans or Pacific Islanders, 3 Hispanic Americans), 222 international. Average age 34. 264 applicants, 58% accepted. In 1998, 173 master's, 1 doctorate awarded. *Degree requirements:* For master's, computer language, thesis or alternative required, foreign language not required; for doctorate and Certificate, computer language, dissertation required, foreign language not required. *Entrance requirements:* For master's, GRE General Test, TOEFL (minimum score of 550 required), minimum GPA of 2.75; for doctorate, GRE General Test, GRE Subject Test, TOEFL (minimum score of 550 required), master's degree or equivalent; for Certificate, master's degree, published paper. *Application deadline:* For fall admission, 6/1; for spring admission, 1/1. Applications are processed on a rolling basis. Application fee: $40. *Financial aid:* Fellowships, research assistantships, teaching assistantships, Federal Work-Study available. Aid available to part-time students. Financial aid application deadline: 2/1; financial aid applicants required to submit CSS PROFILE or FAFSA. *Unit head:* Dr. Daniel W. Lewis, Chair, 408-554-5281. *Application contact:* Tina Samms, Assistant Director of Graduate Admissions, 408-554-4313, Fax: 408-554-5474, E-mail: engrgrad@scu.edu.

Shippensburg University of Pennsylvania, School of Graduate Studies and Research, College of Arts and Sciences, Department of Mathematics and Computer Science, Shippensburg, PA 17257-2299. Offers computer science (MS); information systems (MS); mathematics (M Ed, MS). Part-time and evening/weekend programs available. *Faculty:* 19 full-time (3 women). *Students:* 19 full-time (7 women), 64 part-time (15 women); includes 3 minority (2 African Americans, 1 Native American), 11 international. Average age 33. In 1998, 45 degrees awarded. *Degree requirements:* For master's, foreign language and thesis not required. *Entrance requirements:* For master's, TOEFL (minimum score of 237 required for computer-based), GRE General Test or minimum GPA of 2.75. *Application deadline:* Applications are processed on a rolling basis. Application fee: $25. Electronic applications accepted. *Financial aid:* Career-related internships or fieldwork, Federal Work-Study, institutionally-sponsored loans, and unspecified assistantships available. Aid available to part-time students. Financial aid application deadline: 3/1; financial aid applicants required to submit FAFSA. *Unit head:* Dr. Fred Nordai, Chairperson, 717-532-1431, Fax: 717-530-4009, E-mail: flnord@ship.edu. *Application contact:* Renee Payne, Assistant Dean of Graduate Studies, 717-532-1213, Fax: 717-530-4038, E-mail: rmpayn@ship.edu.

Simon Fraser University, Graduate Studies, Faculty of Applied Science, School of Computing Science, Burnaby, BC V5A 1S6, Canada. Offers M Sc, PhD. *Faculty:* 33 full-time (4 women). *Students:* 90 full-time (27 women), 10 part-time (3 women). Average age 30. In 1998, 10 master's, 3 doctorates awarded. *Degree requirements:* For master's, thesis required; for doctorate, dissertation, qualifying exams required. *Entrance requirements:* For master's, GRE General Test, GRE Subject Test (minimum score of 570 required), TOEFL, TWE (minimum score of 5 required), or International English Language Test (minimum score of 7.5 required), minimum GPA of 3.0; for doctorate, GRE General Test, GRE Subject Test (minimum score of 570 required), TOEFL, TWE (minimum score of 5 required), or International English Language Test (minimum score of 7.5 required), minimum GPA of 3.5. *Application deadline:* For fall admission, 3/31; for spring admission, 7/31. Application fee: $55. *Financial aid:* In 1998–99, 29 fellowships were awarded.; research assistantships, teaching assistantships *Faculty research:* Artificial intelligence, computer graphics, computer hardware, computer systems, database systems. *Unit head:* W. S. Luk, Director, 604-291-4277, Fax: 604-291-3045. *Application contact:* Graduate Secretary, 604-291-4842, Fax: 604-291-3045, E-mail: gradpgm@cs.sfu.ca.

South Dakota School of Mines and Technology, Graduate Division, Department of Computer Science, Rapid City, SD 57701-3995. Offers MS. Part-time programs available. *Faculty:* 6 full-time (1 woman). *Students:* 15 full-time (5 women), 6 international. Average age 30. In 1998, 2 degrees awarded. *Degree requirements:* For master's, foreign language and thesis not required. *Entrance requirements:* For master's, TOEFL (minimum score of 520 required), TWE. *Application deadline:* For fall admission, 6/15 (priority date); for spring admission, 10/15. Applications are processed on a rolling basis. Application fee: $15. Electronic applications accepted. Tuition, state resident: part-time $89 per hour. Tuition, nonresident: part-time $261 per hour. Part-time tuition and fees vary according to program. *Financial aid:* In 1998–99, 2 research assistantships, 7 teaching assistantships were awarded.; fellowships, Federal Work-Study and institutionally-sponsored loans also available. Aid available to part-time students. Financial aid application deadline: 5/15. *Faculty research:* Database systems, remote sensing, numerical modeling, artificial intelligence, neural networks. *Unit head:* Dr. Donald Teets, Chair, 605-394-2471. *Application contact:* Brenda Brown, Secretary, 800-454-8162 Ext. 2493, Fax: 605-394-5360, E-mail: graduate_admissions@silver.sdmt.edu.

South Dakota State University, Graduate School, College of Engineering, Department of Computer Science, Brookings, SD 57007. Offers engineering (MS), including computer science. *Degree requirements:* For master's, thesis, oral exam required, foreign language not required. *Entrance requirements:* For master's, TOEFL (minimum score of 525 required).

Southeastern University, College of Graduate Studies, Program in Computer Science, Washington, DC 20024-2788. Offers MS. Part-time and evening/weekend programs available. *Faculty:* 2 full-time (0 women), 28 part-time (4 women). *Students:* 88 full-time (50 women), 37 part-time (24 women); includes 4 minority (all African Americans), 120 international. Average age 34. In 1998, 53 degrees awarded. *Degree requirements:* For master's, computer language required, foreign language and thesis not required. *Entrance requirements:* For master's, GRE General Test, TOEFL. *Application deadline:* Applications are processed on a rolling basis. Application fee: $45. Tuition: Full-time $6,156; part-time $228 per credit hour. Required fees: $100 per term. Tuition and fees vary according to course load and degree level. *Financial aid:* In 1998–99, 7 students received aid. Federal Work-Study available. Aid available to part-time students. Financial aid application deadline: 8/21; financial aid applicants required to submit CSS PROFILE or FAFSA. *Unit head:* Dr. Abe Eftekari, Head, 202-488-8162 Ext. 254, Fax: 202-488-8093. *Application contact:* Jack Flinter, Director of Admissions, 202-265-5343, Fax: 202-488-8093.

Southern Illinois University Carbondale, Graduate School, College of Science, Department of Computer Science, Carbondale, IL 62901-6806. Offers MS. *Faculty:* 10 full-time (0 women). *Students:* 37 full-time (12 women), 8 part-time (1 woman). 26 applicants, 69% accepted. In 1998, 31 degrees awarded. *Degree requirements:* For master's, computer language, thesis required, foreign language not required. *Entrance requirements:* For master's, TOEFL (minimum score of 550 required), previous undergraduate course work in computer science, minimum GPA of 2.7. *Application deadline:* Applications are processed on a rolling basis. Application fee: $20. *Financial aid:* In 1998–99, 32 students received aid, including 3 research assistantships with full tuition reimbursements available, 22 teaching assistantships with full tuition reimbursements available; fellowships with full tuition reimbursements available, Federal Work-Study, institutionally-sponsored loans, and tuition waivers (full) also available. Aid available to part-time students. Financial aid application deadline: 3/1. *Faculty research:* Analysis of algorithms, VLSI testing, database systems, artificial intelligence, computer architecture. *Unit head:* Dr. Mehdi Zargham, Chairperson, 618-536-2327. *Application contact:* Georgia L. Marine, Graduate Program Secretary, 618-536-2327, Fax: 618-453-6044, E-mail: csinfo@cs.siu.edu.

Southern Illinois University Edwardsville, Graduate Studies and Research, School of Engineering, Department of Computer Information Systems, Edwardsville, IL 62026-0001. Offers MS. *Students:* 36 full-time (13 women), 31 part-time (9 women); includes 10 minority (all Asian Americans or Pacific Islanders), 40 international. 81 applicants, 25% accepted. In 1998, 7 degrees awarded. *Degree requirements:* For master's, thesis or research paper, final exam required. *Entrance requirements:* For master's, TOEFL (minimum score of 550 required). *Application deadline:* For fall admission, 7/24. Application fee: $25. *Financial aid:* In 1998–99, 1 research assistantship with full tuition reimbursement, 12 teaching assistantships with full tuition reimbursements were awarded.; fellowships with full tuition reimbursements, career-related internships or fieldwork, Federal Work-Study, institutionally-sponsored loans, scholarships, traineeships, and unspecified assistantships also available. Aid available to part-time students. *Unit head:* Bernard Waxman, Chairperson, 618-650-2386, E-mail: bwaxman@siue.edu. *Application contact:* Dr. John Schrage, Director, 618-650-2433, E-mail: jschrag@siue.edu.

Southern Methodist University, School of Engineering and Applied Science, Department of Computer Science and Engineering, Dallas, TX 75275. Offers computer engineering (MS Cp E, PhD); computer science (MS, PhD); engineering management (MSEM, DE); operations research (MS, PhD); software engineering (MS). Part-time programs available. Postbaccalaureate distance learning degree programs offered (no on-campus study). *Faculty:* 13 full-time (2 women), 12 part-time (1 woman). *Students:* 57 full-time (22 women), 294 part-time (60 women); includes 85 minority (24 African Americans, 44 Asian Americans or Pacific Islanders, 16 Hispanic Americans, 1 Native American), 69 international. Average age 32. 236 applicants, 44% accepted. In 1998, 87 master's, 7 doctorates awarded. *Degree requirements:* For master's, thesis optional, foreign language not required; for doctorate, dissertation, oral and written qualifying exams, oral final exam (PhD) required. *Entrance requirements:* For master's, GRE General Test (minimum score of 650 on quantitative section required), TOEFL (minimum score of 550 required), minimum GPA of 3.0 in last 2 years; bachelor's degree in engineering, mathematics, or sciences; for doctorate, preliminary counseling exam (PhD), minimum GPA of 3.0, bachelor's degree in related field, MA (DE). *Application deadline:* For fall admission, 8/1 (priority date); for spring admission, 12/15. Applications are processed on a rolling basis. Application fee: $25. Tuition: Full-time $9,216; part-time $512 per credit hour. Required fees: $88 per credit hour. Part-time tuition and fees vary according to course load and campus/location. *Financial aid:* Fellowships, research assistantships, teaching assistantships available. Financial aid applicants required to submit FAFSA. *Faculty research:* Computer arithmetic, distributed and fault-tolerant computing, main memory databse systems, natural language processing. *Unit head:* Dr. Richard V. Helgason, Interim Chair, 214-768-3278, E-mail: helgason@seas.smu.edu. *Application contact:* Dr. Zeynep Celik-Butler, Assistant Dean for Graduate Studies and Research, 214-768-3979, Fax: 214-768-3845, E-mail: zcb@seas.smu.edu.

See in-depth description on page 813.

Southern Oregon University, Graduate Office, School of Sciences, Ashland, OR 97520. Offers environmental education (MA, MS); mathematics/computer science (MA, MS); science (MA, MS). *Degree requirements:* For master's, comprehensive exam (MA) required, thesis optional. *Entrance requirements:* For master's, GRE General Test, minimum GPA of 3.0.

Southern Polytechnic State University, College of Arts and Sciences, Department of Computer Science, Marietta, GA 30060-2896. Offers computer science (MS); software engineering (MSSE). Part-time and evening/weekend programs available. Postbaccalaureate distance learning degree programs offered. *Faculty:* 12 full-time (0 women). *Students:* 87 full-time, 169 part-time. Average age 33. 79 applicants, 99% accepted. In 1998, 77 degrees awarded (100% found work related to degree). *Degree requirements:* For master's, thesis optional, foreign language not required. *Entrance requirements:* For master's, GRE General Test. *Application deadline:* For fall admission, 7/1 (priority date); for spring admission, 11/1. Applications are processed on a rolling basis. Application fee: $20. Tuition, state resident: full-time $2,146; part-time $119 per credit hour. Tuition, nonresident: full-time $7,586; part-time $421 per credit hour. *Financial aid:* In 1998–99, 100 students received aid; teaching assistantships, career-related internships or fieldwork and Federal Work-Study available. Aid available to part-time students. Financial aid application deadline: 5/1; financial aid applicants required to submit FAFSA. *Unit head:* Dr. Jorge Diaz-Herrerea, Head, 770-528-5558, Fax: 770-528-5501, E-mail: jdiaz@spsu.edu.

Southern University and Agricultural and Mechanical College, Graduate School, College of Sciences, Department of Computer Science, Baton Rouge, LA 70813. Offers information systems (MS); micro/minicomputer architecture (MS); operating systems (MS). *Faculty:* 9 full-time (1 woman). *Students:* 26 full-time (11 women), 36 part-time (19 women); includes 36 minority (32 African Americans, 2 Asian Americans or Pacific Islanders, 2 Hispanic Americans), 23 international. In 1998, 9 degrees awarded. *Degree requirements:* For master's, computer language, thesis, comprehensive exam required, foreign language not required. *Entrance requirements:* For master's, GRE General Test (minimum combined score of 1000 required), TOEFL, minimum GPA of 2.7, bachelor's degree in computer science or related field. *Average time to degree:* Master's–2 years full-time, 4 years part-time. *Application deadline:* For fall admission, 6/1 (priority date); for spring admission, 11/1. Applications are processed on a rolling basis. Application fee: $5. *Financial aid:* In 1998–99, 3 research assistantships (averaging $7,000 per year), 5 teaching assistantships were awarded. Financial aid application deadline: 4/15. *Faculty research:* Network theory, computational complexity, high speed computing, neural networking, data warehousing/mining. *Unit head:* Dr. Erold W. Hinds, Chair, 225-771-2060, Fax: 225-771-4223, E-mail: ewhinds@aol.com. *Application contact:* Dr. John A. Dyer, Professor, 225-771-2060, Fax: 225-771-4223, E-mail: johna@mail.cmps.subr.edu.

Southwest Texas State University, Graduate School, School of Science, Department of Computer Science, San Marcos, TX 78666. Offers MA, MS. Part-time programs available. *Faculty:* 7 full-time (1 woman), 1 part-time (0 women). *Students:* 48 full-time (24 women), 66 part-time (27 women); includes 10 minority (1 African American, 8 Asian Americans or Pacific Islanders, 1 Hispanic American), 78 international. Average age 31. In 1998, 32 degrees awarded. *Degree requirements:* For master's, computer language, thesis (for some programs), comprehensive exam required, foreign language not required. *Entrance requirements:* For master's, GRE General Test (minimum combined score of 1000 required), TOEFL (minimum score of 550 required), minimum GPA of 2.75 in last 60 hours. *Application deadline:* For fall admission, 6/15 (priority date); for spring admission, 10/15 (priority date). Applications are processed on a rolling basis. Application fee: $25 ($50 for international students). Tuition, state resident: full-time $684; part-time $38 per semester hour. Tuition, nonresident: full-time $4,572; part-time $254 per semester hour. *Financial aid:* In 1998–99, 15 teaching assistantships were awarded.; career-related internships or fieldwork, Federal Work-Study, and institutionally-sponsored loans also available. Aid available to part-time students. Financial aid application deadline: 4/1; financial aid applicants required to submit FAFSA. *Faculty research:* Software engineering, artificial intelligence, multimedia, distributed/parallel computing, database systems, operating systems. *Unit head:* Dr. Moonis Ali, Chair, 512-245-3409, Fax: 512-245-8750, E-mail: ma04@swt.edu. *Application contact:* Dr. Tom McCabe, Graduate Adviser, 512-245-3409, Fax: 512-245-8750, E-mail: tm03@swt.edu.

Stanford University, School of Engineering, Department of Computer Science, Stanford, CA 94305-9991. Offers MS, PhD. *Faculty:* 33 full-time (4 women). *Students:* 349 full-time (69 women), 77 part-time (10 women); includes 98 minority (3 African Americans, 86 Asian Americans or Pacific Islanders, 8 Hispanic Americans, 1 Native American), 160 international. Average age 26. 476 applicants, 35% accepted. In 1998, 147 master's, 22 doctorates awarded. Terminal master's awarded for partial completion of doctoral program. *Degree requirements:* For master's, computer language required, foreign language and thesis not required; for doctorate, computer language, dissertation required, foreign language not required. *Entrance requirements:* For master's, GRE General Test, TOEFL; for doctorate, GRE General Test, GRE Subject Test, TOEFL. *Application deadline:* For fall admission, 12/15. Application fee: $65 ($80 for international students). Electronic applications accepted. Tuition: Full-time $24,588. Required fees: $152. Part-time tuition and fees vary according to course load. *Financial aid:* Fellowships, research assistantships, teaching assistantships, Federal Work-Study and institutionally-sponsored loans available. Financial aid application deadline: 1/1. *Unit head:* Jean-Claude La Tombe, Chair, 650-723-0350, Fax: 650-725-0748, E-mail: latombe@cs.stanford.edu. *Application contact:* Graduate Administrator, 650-723-1519.

See in-depth description on page 815.

Stanford University, School of Engineering, Program in Scientific Computing and Computational Mathematics, Stanford, CA 94305-9991. Offers MS, PhD. *Students:* 33 full-time (3 women), 7 part-time (1 woman); includes 11 minority (9 Asian Americans or Pacific Islanders, 2 Hispanic Americans), 17 international. Average age 27. 37 applicants, 38% accepted. In 1998, 10 master's, 7 doctorates awarded. *Degree requirements:* For doctorate, dissertation required. *Entrance requirements:* For master's, GRE General Test, TOEFL; for doctorate, GRE General Test, GRE Subject Test, TOEFL. Application fee: $65 ($80 for international students). Electronic applications accepted. Tuition: Full-time $24,588. Required fees: $152. Part-time tuition and fees vary according to course load. *Financial aid:* Fellowships, research assistantships, institutionally-sponsored loans available. Financial aid application deadline: 2/15. *Unit head:* Dr. Andrew M. Stuart, Director, 650-723-8142, Fax: 650-723-1778, E-mail: stuart@sccm.stanford.edu. *Application contact:* Admissions Coordinator, 650-723-0572.

See in-depth description on page 819.

State University of New York at Albany, College of Arts and Sciences, Department of Computer Science, Albany, NY 12222-0001. Offers MS, PhD. *Faculty:* 13 full-time (2 women), 1 part-time (0 women). *Students:* 52 full-time (16 women), 20 part-time (6 women); includes 2 minority (both Asian Americans or Pacific Islanders), 47 international. Average age 25. 139 applicants, 45% accepted. In 1998, 36 master's awarded. *Degree requirements:* For master's, comprehensive exam, project or thesis required; for doctorate, dissertation, area and comprehensive exams required, foreign language not required. *Entrance requirements:* For master's and doctorate, GRE General Test, TOEFL. Application fee: $50. Tuition, state resident: full-time $5,100; part-time $213 per credit. Tuition, nonresident: full-time $8,416; part-time $351 per credit. Required fees: $31 per credit. *Financial aid:* Fellowships, research assistantships, teaching assistantships, career-related internships or fieldwork and Federal Work-Study available. *Faculty research:* Algorithm design and analysis, artificial intelligence, computational logic, databases, numerical analysis. *Unit head:* Daniel J. Rosenkrantz, Chair, 518-442-4270. *Application contact:* Neil V. Murray, 518-442-3393.

See in-depth description on page 821.

State University of New York at Binghamton, Graduate School, School of Arts and Sciences, Department of Mathematical Sciences, Binghamton, NY 13902-6000. Offers computer science (MA, PhD); probability and statistics (MA, PhD). *Faculty:* 23 full-time, 9 part-time. *Students:* 44 full-time (16 women), 6 part-time (2 women); includes 1 minority (Asian American or Pacific Islander), 14 international. Terminal master's awarded for partial completion of doctoral program. *Degree requirements:* For master's, thesis or alternative required; for doctorate, dissertation required. *Entrance requirements:* For master's and doctorate, GRE General Test, GRE Subject Test, TOEFL. *Application deadline:* For fall admission, 4/15 (priority date); for spring admission, 11/1. Applications are processed on a rolling basis. Application fee: $50. Electronic applications accepted. Tuition, state resident: full-time $5,100; part-time $213 per credit. Tuition, nonresident: full-time $8,416; part-time $351 per credit. Required fees: $77 per credit. Part-time tuition and fees vary according to course load. *Unit head:* Dr. David Hanson, Chairperson, 607-777-2147.

State University of New York at Binghamton, Graduate School, Thomas J. Watson School of Engineering and Applied Science, Department of Computer Science, Binghamton, NY 13902-6000. Offers MS, PhD. *Faculty:* 13 full-time, 14 part-time. *Students:* 79 full-time (25 women), 54 part-time (16 women); includes 13 minority (2 African Americans, 7 Asian Americans or Pacific Islanders, 3 Hispanic Americans, 1 Native American), 74 international. Average age 29. 228 applicants, 56% accepted. In 1998, 48 master's, 6 doctorates awarded. *Degree requirements:* For master's, thesis or alternative required, foreign language not required; for doctorate, dissertation required, foreign language not required. *Entrance requirements:* For master's and doctorate, GRE General Test, GRE Subject Test, TOEFL (minimum score of 550 required). *Application deadline:* For fall admission, 4/15 (priority date); for spring admission, 11/1. Applications are processed on a rolling basis. Application fee: $50. Electronic applications accepted. Tuition, state resident: full-time $5,100; part-time $213 per credit. Tuition, nonresident: full-time $8,416; part-time $351 per credit. Required fees: $77 per credit. Part-time tuition and fees vary according to course load. *Financial aid:* In 1998–99, 57 students received aid, including 1 fellowship with full tuition reimbursement available (averaging $8,100 per year), 19 research assistantships with full tuition reimbursements available (averaging $8,517 per year), 32 teaching assistantships with full tuition reimbursements available (averaging $8,452 per year); unspecified assistantships also available. Financial aid application deadline: 2/15. *Unit head:* Dr. Sudhir Aggarwal, Chair, 607-777-4802.

See in-depth description on page 825.

State University of New York at Buffalo, Graduate School, College of Arts and Sciences, Department of Computer Science and Engineering, Buffalo, NY 14260. Offers MS, PhD. *Faculty:* 25 full-time (6 women), 8 part-time (2 women). *Students:* 80 full-time (14 women), 22 part-time (3 women); includes 3 minority (all Asian Americans or Pacific Islanders), 74 international. Average age 25. 373 applicants, 48% accepted. In 1998, 29 master's, 4 doctorates awarded. Terminal master's awarded for partial completion of doctoral program. *Degree requirements:* For master's, computer language, thesis or alternative required, foreign language not required; for doctorate, computer language, dissertation, comprehensive qualifying exam required, foreign language not required. *Entrance requirements:* For master's and doctorate, GRE General Test, GRE Subject Test (computer science), TOEFL (minimum score of 550 required). *Application deadline:* For fall admission, 12/31. Application fee: $35. Electronic applications accepted. Tuition, state resident: full-time $5,100; part-time $213 per credit hour. Tuition, nonresident: full-time $8,416; part-time $351 per credit hour. Required fees: $870; $75 per semester. Tuition and fees vary according to course load and program. *Financial aid:* In 1998–99, 100 students received aid, including 35 research assistantships with tuition reimbursements available (averaging $10,350 per year), 65 teaching assistantships with tuition reimbursements available (averaging $10,350 per year); fellowships with tuition reimbursements available, Federal Work-Study, institutionally-sponsored loans, and unspecified assistantships also available. Financial aid application deadline: 12/31; financial aid applicants required to submit FAFSA. *Faculty research:* Artificial intelligence, computer vision, theoretical computer science, parallel architecture, operating systems. Total annual research expenditures: $1.4 million. *Unit head:* Dr. Stuart C. Shapiro, Chairman, 716-645-3180 Ext. 125, Fax: 716-645-3464, E-mail: cse-chair@cse.buffalo.edu. *Application contact:* Dr. Raj Acharya, Director of Graduate Studies, 716-645-3180 Ext. 141, Fax: 716-645-3464, E-mail: cse-dgs@cse.buffalo.edu.

See in-depth description on page 827.

State University of New York at New Paltz, Graduate School, Faculty of Liberal Arts and Sciences, Department of Mathematics and Computer Science, Program in Computer Science, New Paltz, NY 12561. Offers MS. *Students:* 22 full-time (11 women), 33 part-time (16 women); includes 11 minority (all Asian Americans or Pacific Islanders), 33 international. In 1998, 19 degrees awarded. *Degree requirements:* For master's, computer language, thesis (for some programs), comprehensive exam required, foreign language not required. *Entrance requirements:* For master's, GRE General Test, minimum GPA of 3.0, proficiency in program assembly. *Application deadline:* For fall admission, 3/15 (priority date). Applications are processed on a rolling basis. Application fee: $50. *Financial aid:* Teaching assistantships, tuition waivers (full) available. *Application contact:* Keqin Li, Graduate Adviser, 914-257-3535.

State University of New York at Stony Brook, Graduate School, College of Engineering and Applied Sciences, Department of Computer Science, Stony Brook, NY 11794. Offers computer science (MS, PhD); software engineering (Certificate). *Faculty:* 26 full-time (1 woman), 2 part-time (both women). *Students:* 129 full-time (39 women), 44 part-time (11 women); includes 25 minority (2 African Americans, 20 Asian Americans or Pacific Islanders, 3 Hispanic Americans), 123 international. 347 applicants, 45% accepted. In 1998, 58 master's, 10 doctorates, 20 other advanced degrees awarded. *Degree requirements:* For master's, computer language, thesis or alternative required, foreign language not required; for doctorate, computer language, dissertation, comprehensive exams required, foreign language not required. *Entrance requirements:* For master's and doctorate, GRE General Test, TOEFL. *Application deadline:* For fall admission, 1/15. Application fee: $50. *Financial aid:* In 1998–99, 6 fellowships, 30 research assistantships, 46 teaching assistantships were awarded. *Faculty research:* Artificial intelligence, computer architecture, database management systems, VLSI, operating systems. Total annual research expenditures: $2.6 million. *Unit head:* Dr. David Warren, Chairman, 516-632-8470. *Application contact:* Dr. Michael Kiefer, Director, 516-632-8443, Fax: 516-632-8334, E-mail: mkiefer@notes.cc.sunysb.edu.

State University of New York Institute of Technology at Utica/Rome, School of Information Systems and Engineering Technology, Program in Computer and Information Science, Utica, NY 13504-3050. Offers MS. Part-time and evening/weekend programs available. *Faculty:* 10 full-time (2 women). *Students:* 13 full-time (3 women), 43 part-time (11 women); includes 5 minority (4 Asian Americans or Pacific Islanders, 1 Hispanic American), 4 international. Average age 35. 34 applicants, 76% accepted. In 1998, 10 degrees awarded (100% found work related to degree). *Degree requirements:* For master's, computer language, comprehensive exam required, thesis optional, foreign language not required. *Entrance requirements:* For master's, GRE General Test, TOEFL (minimum score of 550 required), minimum GPA of 3.0. *Average time to degree:* Master's–2 years full-time, 4 years part-time. *Application deadline:* For fall admission, 6/15 (priority date). Applications are processed on a rolling basis. Application fee: $50. *Financial aid:* In 1998–99, 20 students received aid, including 2 fellowships with full tuition reimbursements available (averaging $4,900 per year), 3 research assistantships with full tuition reimbursements available; career-related internships or fieldwork and Federal Work-Study also available. Aid available to part-time students. Financial aid applicants required to submit FAFSA. *Faculty research:* Parallel processing, databases and data fusion, networks, artificial intelligence. *Unit head:* Dr. Michael Pittarelli, Chair, 315-792-7234, Fax: 315-792-7222, E-mail: mike@sunyit.edu. *Application contact:* Marybeth Lyons, Director of Admissions, 315-792-7500, Fax: 315-792-7837, E-mail: smbl@sunyit.edu.

Stephen F. Austin State University, Graduate School, College of Business, Department of Computer Science, Nacogdoches, TX 75962. Offers MS. Part-time programs available. *Faculty:* 6 full-time (1 woman), 1 part-time (0 women). *Students:* 3 full-time (0 women), 2 part-time, 2 international. 5 applicants, 60% accepted. In 1998, 3 degrees awarded. *Degree requirements:* For master's, computer language, comprehensive exam required, foreign language and thesis not required. *Entrance requirements:* For master's, GRE General Test (minimum combined score of 1000 required), TOEFL. *Application deadline:* For fall admission, 8/1 (priority date); for spring admission, 12/15. Applications are processed on a rolling basis. Application fee: $0 ($50 for international students). Tuition, state resident: full-time $1,792. Tuition, nonresident: full-time $6,880. *Financial aid:* In 1998–99, teaching assistantships (averaging $6,400 per year); research assistantships, Federal Work-Study also available. Financial aid application deadline: 3/1. *Unit head:* Dr. Craig A. Wood, Chair, 409-468-2508.

Stevens Institute of Technology, Graduate School, School of Applied Sciences and Liberal Arts, Department of Computer Science, Hoboken, NJ 07030. Offers advanced programming: theory, design and verification (Certificate); artificial intelligence and robotics (MS, PhD); computer and information systems (MS, PhD); computer architecture and digital system

Computer Science

Stevens Institute of Technology (continued)

design (MS, PhD); database systems (Certificate); elements of computer science (Certificate); information systems (MS, Certificate); network and graph theory (Certificate); software design (MS, PhD); software engineering (Certificate); theoretical computer science (MS, PhD, Certificate); wireless communications (Certificate). MS and Certificate offered in cooperation with the Program in Information Systems. Part-time and evening/weekend programs available. Terminal master's awarded for partial completion of doctoral program. *Degree requirements:* For master's, computer language required, thesis optional, foreign language not required; for doctorate, variable foreign language requirement, computer language, dissertation required; for Certificate, computer language required, foreign language not required. *Entrance requirements:* For master's and doctorate, GRE, TOEFL. Electronic applications accepted. *Faculty research:* Semantics, reliability theory.

See in-depth description on page 829.

Suffolk University, College of Arts and Sciences, Department of Mathematics and Computer Science, Boston, MA 02108-2770. Offers computer science (MS). Part-time and evening/weekend programs available. *Faculty:* 5 full-time (0 women). *Students:* 10 full-time (5 women), 13 part-time (7 women), 16 international. Average age 30. 41 applicants, 83% accepted. In 1998, 1 degree awarded. *Degree requirements:* For master's, computer language required, thesis optional, foreign language not required. *Entrance requirements:* For master's, GRE General Test (average 500 on each section) or MAT (average 50). *Application deadline:* For fall admission, 6/15 (priority date); for spring admission, 11/15. Applications are processed on a rolling basis. Application fee: $50. Tuition: Full-time $11,770; part-time $1,953 per course. Required fees: $10 per semester. *Financial aid:* In 1998–99, 14 students received aid, including 9 fellowships with partial tuition reimbursements available (averaging $2,889 per year); career-related internships or fieldwork, Federal Work-Study, and institutionally-sponsored loans also available. Financial aid application deadline: 3/15; financial aid applicants required to submit FAFSA. *Unit head:* Dan Stefanescu, Head, 617-573-8679, Fax: 617-573-8591, E-mail: dstefane@acad.suffolk.edu. *Application contact:* Judith Reynolds, Director of Graduate Admissions, 617-573-8302, Fax: 617-523-0116, E-mail: grad.admission@suffolk.edu.

See in-depth description on page 831.

Syracuse University, Graduate School, L. C. Smith College of Engineering and Computer Science, Department of Electrical Engineering and Computer Science, Program in Computer and Information Science, Syracuse, NY 13244-0003. Offers computer and information science (PhD); computer science (MS). *Students:* 182 full-time (24 women), 40 part-time (8 women), 65 international. Average age 28. 300 applicants, 79% accepted. In 1998, 26 degrees awarded. *Degree requirements:* For master's, thesis required, foreign language not required; for doctorate, computer language, dissertation required, foreign language not required. *Entrance requirements:* For master's and doctorate, GRE General Test, GRE Subject Test. *Application deadline:* Applications are processed on a rolling basis. Application fee: $40. Tuition: Full-time $13,992; part-time $583 per credit hour. *Financial aid:* Fellowships, research assistantships, teaching assistantships, Federal Work-Study and tuition waivers (partial) available. Financial aid application deadline: 3/1. *Unit head:* Kishan Mehrotra, Graduate Director.

Temple University, Graduate School, College of Science and Technology, Department of Computer and Information Sciences, Philadelphia, PA 19122-6096. Offers MS, PhD. Part-time programs available. *Faculty:* 11 full-time (1 woman). *Students:* 54 (12 women); includes 36 minority (2 African Americans, 34 Asian Americans or Pacific Islanders) 4 international. 238 applicants, 58% accepted. In 1998, 30 master's, 2 doctorates awarded. Terminal master's awarded for partial completion of doctoral program. *Degree requirements:* For master's, computer language required, foreign language and thesis not required; for doctorate, computer language, dissertation required, foreign language not required. *Entrance requirements:* For master's and doctorate, GRE General Test, TOEFL, minimum GPA of 2.8. *Application deadline:* For fall admission, 2/1; for spring admission, 9/30. Applications are processed on a rolling basis. Application fee: $40. Electronic applications accepted. *Financial aid:* Fellowships, research assistantships, teaching assistantships, unspecified assistantships available. Financial aid application deadline: 2/1. *Faculty research:* Artificial intelligence, information systems, software engineering, network-distributed systems. *Unit head:* Dr. Frank Friedman, Chairperson, 215-204-8450, Fax: 215-204-5082, E-mail: friedman@cis.temple.edu. *Application contact:* Dr. James Korsh, Graduate Chair, 215-204-8199, Fax: 215-204-5082, E-mail: korsh@cis.temple.eou.

Texas A&M University, College of Engineering, Department of Computer Science, College Station, TX 77843. Offers computer engineering (MCE, MS, PhD); computer science (MCS, MS, PhD). Part-time programs available. *Faculty:* 26 full-time (3 women), 1 part-time (0 women). *Students:* 162 full-time (24 women), 68 part-time (14 women); includes 8 minority (5 African Americans, 3 Hispanic Americans), 155 international. Average age 28. 517 applicants, 40% accepted. In 1998, 66 master's, 12 doctorates awarded. *Degree requirements:* For master's, computer language, thesis (MS) required; for doctorate, computer language, dissertation required, foreign language not required. *Entrance requirements:* For master's and doctorate, GRE General Test, TOEFL. *Application deadline:* For fall admission, 5/1 (priority date). Application fee: $50 ($75 for international students). *Financial aid:* In 1998–99, 8 fellowships with full tuition reimbursements (averaging $11,875 per year), 63 research assistantships (averaging $8,772 per year), 43 teaching assistantships (averaging $8,117 per year) were awarded. Financial aid application deadline: 3/1. *Faculty research:* Software development, numerical applications and controls, data structures. Total annual research expenditures: $1.9 million. *Unit head:* Dr. Wei Zhao, Head, 409-845-5534, Fax: 409-847-8578, E-mail: csdept@cs.tamu.edu. *Application contact:* S. Bart Childs, Graduate Adviser, 409-845-8981, E-mail: csdept@cs.tamu.edu.

See in-depth description on page 833.

Texas A&M University–Commerce, Graduate School, College of Arts and Sciences, Department of Computer Science and Information Systems, Commerce, TX 75429-3011. Offers computer science (MS). Part-time programs available. *Faculty:* 6 full-time (1 woman), 4 part-time (1 woman). *Students:* 89 full-time (27 women), 51 part-time (21 women); includes 24 minority (1 African American, 23 Asian Americans or Pacific Islanders), 70 international. Average age 36. In 1998, 36 degrees awarded. *Degree requirements:* For master's, thesis (for some programs), comprehensive exam required. *Entrance requirements:* For master's, GMAT (minimum score of 450 required) or GRE General Test (minimum combined score of 850 required). *Average time to degree:* Master's–2 years full-time, 2.75 years part-time. *Application deadline:* For fall admission, 6/1 (priority date); for spring admission, 11/1 (priority date). Applications are processed on a rolling basis. Application fee: $0 ($25 for international students). Electronic applications accepted. *Financial aid:* In 1998–99, research assistantships (averaging $7,750 per year), teaching assistantships (averaging $7,750 per year) were awarded.; Federal Work-Study and institutionally-sponsored loans also available. Financial aid application deadline: 5/1; financial aid applicants required to submit FAFSA. Total annual research expenditures: $6,620. *Unit head:* Dr. Sam Saffer, Head, 903-886-5409. *Application contact:* Betty Hunt, Graduate Admissions Adviser, 903-886-5167, Fax: 903-886-5165, E-mail: betty_hunt@tamu_commerce.edu.

Texas A&M University–Corpus Christi, Graduate Programs, College of Science and Technology, Program in Computing and Mathematical Sciences, Corpus Christi, TX 78412-5503. Offers computer science (MS); mathematics (MS). Part-time and evening/weekend programs available. *Degree requirements:* For master's, computer language, thesis required, foreign language not required. *Entrance requirements:* For master's, GRE General Test.

Texas A&M University–Kingsville, College of Graduate Studies, College of Engineering, Department of Electrical Engineering and Computer Science, Program in Computer Science, Kingsville, TX 78363. Offers MS. *Faculty:* 3 full-time, 1 part-time. *Students:* 12 full-time (1 woman), 12 part-time. *Degree requirements:* For master's, computer language, thesis or alternative, comprehensive exam required, foreign language not required. *Entrance requirements:* For master's, GRE General Test (minimum combined score of 1000 required), TOEFL (minimum score of 525 required), minimum GPA of 3.0. *Application deadline:* For fall admission, 6/1; for spring admission, 11/15. Applications are processed on a rolling basis. Application fee: $15 ($25 for international students). Tuition, state resident: full-time $2,062. Tuition, nonresident: full-time $7,246. *Financial aid:* Research assistantships available. Financial aid application deadline: 5/15. *Faculty research:* Operating systems, programming languages, database systems, computer architecture, artificial intelligence. *Unit head:* Dr. Rajab Challoo, Coordinator, 361-593-2001. *Application contact:* H. D. Gorakhpurwalla, Graduate Coordinator, 361-593-2004.

Texas Tech University, Graduate School, College of Engineering, Department of Computer Science, Lubbock, TX 79409. Offers MS, PhD. Part-time programs available. *Faculty:* 9 full-time (3 women), 1 part-time (0 women). *Students:* 39 full-time (10 women), 10 part-time (1 woman); includes 1 minority (Asian American or Pacific Islander), 34 international. Average age 29. 132 applicants, 24% accepted. In 1998, 54 master's, 2 doctorates awarded. *Degree requirements:* For master's and doctorate, computer language, thesis/dissertation required, foreign language not required. *Entrance requirements:* For master's, GRE General Test (minimum combined score of 1000 required; average 1202), minimum GPA of 3.0; for doctorate, GRE General Test (minimum combined score of 1000 required), minimum GPA of 3.0. *Application deadline:* For fall admission, 4/15 (priority date); for spring admission, 11/1 (priority date). Applications are processed on a rolling basis. Application fee: $25 ($50 for international students). Electronic applications accepted. *Financial aid:* In 1998–99, 33 students received aid, including 7 research assistantships (averaging $9,129 per year), 6 teaching assistantships (averaging $7,800 per year); fellowships, Federal Work-Study and institutionally-sponsored loans also available. Aid available to part-time students. Financial aid application deadline: 5/15; financial aid applicants required to submit FAFSA. *Faculty research:* Generic controller software development, neural networks/speech recognition, neural-type network for solving 2-point boundary value. Total annual research expenditures: $364,914. *Unit head:* Dr. Daniel Earl Cooke, Director, 806-742-3527, Fax: 806-742-3519.

Towson University, Graduate School, Program in Computer Science, Towson, MD 21252-0001. Offers MS. Part-time and evening/weekend programs available. *Faculty:* 4 full-time (0 women). *Students:* 89 full-time, 93 part-time. In 1998, 32 degrees awarded. *Degree requirements:* For master's, exam required, thesis optional. *Application deadline:* For fall admission, 3/1 (priority date); for spring admission, 10/1. Applications are processed on a rolling basis. Application fee: $40. *Financial aid:* Federal Work-Study and unspecified assistantships available. Aid available to part-time students. Financial aid application deadline: 4/1; financial aid applicants required to submit FAFSA. *Faculty research:* Deductive databases, neural nets, software engineering, data communications and networks. *Unit head:* Dr. Ramesh Karne, Director, 410-830-3955, Fax: 410-830-3868, E-mail: rkarne@towson.edu. *Application contact:* Bob Baer, Assistant Director of Graduate School, 410-830-2501, Fax: 410-830-4675, E-mail: petgrad@towson.edu.

Trent University, Graduate Studies, Program in Applications of Modelling in the Natural and Social Sciences, Department of Computer Studies, Peterborough, ON K9J 7B8, Canada. Offers M Sc. *Degree requirements:* For master's, computer language, thesis required, foreign language not required. *Entrance requirements:* For master's, honours degree. *Application deadline:* For fall admission, 2/1 (priority date). Applications are processed on a rolling basis. Application fee: $45. *Unit head:* Dr. R. Hurley, Chair, 705-748-1542, E-mail: rhurley@trentu.ca. *Application contact:* Graduate Studies Officer, 705-748-1245, Fax: 705-748-1587, E-mail: gradstudies@trentu.ca.

Tufts University, Division of Graduate and Continuing Studies and Research, Graduate School of Arts and Sciences, College of Engineering, Department of Electrical Engineering and Computer Science, Medford, MA 02155. Offers computer science (MS, PhD); electrical engineering (MS, PhD). Part-time programs available. *Faculty:* 17 full-time, 5 part-time. *Students:* 99 (26 women); includes 9 minority (2 African Americans, 7 Asian Americans or Pacific Islanders) 32 international. 147 applicants, 41% accepted. In 1998, 19 master's, 6 doctorates awarded. Terminal master's awarded for partial completion of doctoral program. *Degree requirements:* For master's, thesis or alternative required, foreign language not required; for doctorate, dissertation required, foreign language not required. *Entrance requirements:* For master's and doctorate, GRE General Test, TOEFL (minimum score of 550 required). *Application deadline:* For fall admission, 3/1; for spring admission, 10/15. Applications are processed on a rolling basis. Application fee: $50. *Financial aid:* Research assistantships with full and partial tuition reimbursements, teaching assistantships with full and partial tuition reimbursements, Federal Work-Study, scholarships, and tuition waivers (partial) available. Financial aid application deadline: 2/15; financial aid applicants required to submit FAFSA. *Unit head:* Robert Gonsalves, Chair, 617-627-3217, Fax: 617-627-3220. *Application contact:* Anselm Blumer, 617-623-3217, Fax: 617-627-3220, E-mail: webmaster@eecs.tufts.edu.

Tufts University, Division of Graduate and Continuing Studies and Research, Professional and Continuing Studies, Human-Computer Interaction Program, Medford, MA 02155. Offers Certificate. Part-time and evening/weekend programs available. Average age 44. 2 applicants, 100% accepted. *Average time to degree:* 1 year part-time. *Application deadline:* For fall admission, 8/15 (priority date); for spring admission, 12/12. Applications are processed on a rolling basis. Application fee: $40. *Financial aid:* Available to part-time students. Application deadline: 5/1; *Unit head:* H. D. Gorakhpurwalla, Graduate Coordinator, 361-593-2004. *Application contact:* H. D. Gorakhpurwalla, Graduate Coordinator, 361-593-2004.

Tulane University, School of Engineering, Department of Computer Science, New Orleans, LA 70118-5669. Offers MS, MSCS, PhD, Sc D. MS and PhD offered through the Graduate School. Part-time programs available. *Students:* 16 full-time (6 women), 3 part-time (1 woman); includes 7 minority (3 African Americans, 2 Asian Americans or Pacific Islanders, 2 Hispanic Americans), 4 international. 103 applicants, 13% accepted. In 1998, 13 master's, 3 doctorates awarded. Terminal master's awarded for partial completion of doctoral program. *Degree requirements:* For master's, thesis or alternative required, foreign language not required; for doctorate, computer language, dissertation required, foreign language not required. *Entrance requirements:* For master's and doctorate, GRE General Test, TOEFL, minimum B average in undergraduate course work. *Average time to degree:* Master's–2 years full-time; doctorate–5 years full-time. *Application deadline:* For fall admission, 7/1 (priority date); for spring admission, 11/1. Applications are processed on a rolling basis. Application fee: $35. *Financial aid:* In 1998–99, 3 fellowships were awarded.; research assistantships, teaching assistantships, career-related internships or fieldwork, institutionally-sponsored loans, and tuition waivers (full and partial) also available. Financial aid application deadline: 2/1. *Faculty research:* Software engineering, robotics, artificial intelligence, fuzzy sets, programming languages, neural nets. *Unit head:* Dr. Enrique Barbieri, Chairperson, 504-865-5840. *Application contact:* Dr. E. Michaelides, Associate Dean, 504-865-5764.

Union College, Graduate and Continuing Studies, Division of Engineering and Computer Science, Department of Electrical Engineering and Computer Science, Program in Computer Science, Schenectady, NY 12308-2311. Offers MS. 8 applicants, 100% accepted. In 1998, 10 degrees awarded. *Degree requirements:* For master's, computer language, comprehensive exam, project, or thesis required. *Entrance requirements:* For master's, minimum GPA of 3.0. *Application deadline:* Applications are processed on a rolling basis. Application fee: $50. Tuition: Part-time $1,786 per course. *Faculty research:* Microprocessor applications. *Unit head:* Dr. Robert Hemmendinger, Chair, 518-388-6319.

Universidad de las Américas–Puebla, Division of Graduate Studies, School of Engineering, Program in Computer Engineering, Cholula, 72820, Mexico. Offers computer science (MS). Part-time and evening/weekend programs available. *Faculty:* 9 full-time (1 woman). *Students:* 31 full-time (11 women), 4 part-time (2 women); all minorities (all Hispanic Americans) Average age 25. 30 applicants, 37% accepted. In 1998, 18 degrees awarded. *Degree requirements:* For master's, one foreign language, computer language, thesis required. *Average time to* *degree:* Master's–2.5 years full-time, 3.5 years part-time. *Application deadline:* For fall admission, 7/16. Applications are processed on a rolling basis. Application fee: $0. *Financial aid:* In 1998–99, 25 students received aid, including 7 research assistantships Aid available to part-time students. Financial aid application deadline: 5/15. *Faculty research:* Computers in education, robotics, artificial intelligence. Total annual research expenditures: $33,000. *Unit head:* Dr. Juan Manuel Ahuactzin, Coordinator, 22-29-20-29, Fax: 22-29-20-32, E-mail: jmal@mail.udlap.mx. *Application contact:* Mauricio Villegas, Chair of Admissions Office, 22-29-20-17, Fax: 22-29-20-18, E-mail: admision@mail.udlap.mx.

Université de Montréal, Faculty of Graduate Studies, Faculty of Arts and Sciences, Department of Computer Science and Operations Research, Montréal, PQ H3C 3J7, Canada. Offers computer systems (M Sc, PhD). Part-time programs available. *Faculty:* 40 full-time (5 women), 1 part-time (0 women). *Students:* 143 full-time (27 women), 9 part-time (2 women). 87 applicants, 34% accepted. In 1998, 28 master's, 15 doctorates awarded. Terminal master's awarded for partial completion of doctoral program. *Degree requirements:* For master's, one foreign language, computer language, thesis required; for doctorate, one foreign language, computer language, dissertation, general exam required. *Entrance requirements:* For master's, B Sc in related field; for doctorate, MA or M Sc in related field. *Application deadline:* For fall admission, 2/1 (priority date). Application fee: $30. *Financial aid:* Available to part-time students. Application deadline: 10/31. *Faculty research:* Optimization statistics, programming languages, telecommunications, theoretical computer science, artificial intelligence. *Unit head:* Sang Nguyen, Chairman, 514-343-7090. *Application contact:* Jean-Yves Potvin, Chairman, 514-343-7093.

Université du Québec à Trois-Rivières, Graduate Programs, Program in Mathematics and Computer Science, Trois-Rivières, PQ G9A 5H7, Canada. Offers M Sc. *Students:* 9 full-time (2 women). 22 applicants, 77% accepted. *Application deadline:* For fall admission, 2/1. Application fee: $30. *Faculty research:* Probability, statistics, scientific calculation. *Unit head:* Dr. Robert La Barre, Director, 819-376-5125 Ext. 3817, Fax: 819-376-5012, E-mail: robert_labarre@uqtr.uquebec.ca. *Application contact:* Suzanne Camirand, Admissions Officer, 819-376-5045 Ext. 2591, Fax: 819-376-5210, E-mail: suzanne_camirand@uqtr.uquebec.ca.

Université Laval, Faculty of Graduate Studies, Faculty of Sciences and Engineering, Department of Computer Science, Sainte-Foy, PQ G1K 7P4, Canada. Offers M Sc, PhD. *Students:* 43 full-time (10 women), 30 part-time (4 women). 122 applicants, 38% accepted. In 1998, 19 master's, 2 doctorates awarded. *Application deadline:* For fall admission, 3/1. Application fee: $30. *Faculty research:* Software and information systems design, networking, management-oriented computer science and office automation. *Unit head:* Nadir Belkhiter, Director, 418-656-2131 Ext. 5929, Fax: 418-656-2324, E-mail: nadir.belkhiter@ift.ulaval.ca.

The University of Akron, Graduate School, Buchtel College of Arts and Sciences, Department of Mathematics and Computer Science, Program in Computer Science, Akron, OH 44325-0001. Offers MS. *Students:* 32 full-time (14 women), 21 part-time (5 women); includes 5 minority (1 African American, 4 Asian Americans or Pacific Islanders), 27 international. Average age 30. In 1998, 17 degrees awarded. *Degree requirements:* For master's, thesis optional, foreign language not required. *Entrance requirements:* For master's, minimum GPA of 2.75. *Average time to degree:* Master's–2 years full-time, 4 years part-time. *Application deadline:* For fall admission, 3/1. Applications are processed on a rolling basis. Application fee: $25 ($50 for international students). Tuition, state resident: part-time $189 per credit. Tuition, nonresident: part-time $353 per credit. Required fees: $7.3 per credit. *Financial aid:* Application deadline: 3/1. *Unit head:* Dr. Wolfgang Pelz, Coordinator, 330-972-8019, E-mail: wolfgangpelz@uakron.edu.

The University of Alabama, Graduate School, College of Engineering, Department of Computer Science, Tuscaloosa, AL 35487. Offers MSCS, PhD. *Faculty:* 10 full-time (2 women). *Students:* 36 full-time (13 women), 6 part-time; includes 1 minority (Hispanic American), 32 international. Average age 27. 75 applicants, 87% accepted. In 1998, 11 master's awarded (100% found work related to degree); 2 doctorates awarded (100% entered university research/teaching). *Degree requirements:* For master's, thesis or alternative required, foreign language not required; for doctorate, dissertation required, foreign language not required. *Entrance requirements:* For master's, GRE General Test (minimum combined score of 1600 on three sections required), minimum GPA of 3.0 in last 60 hours; for doctorate, GRE General Test (minimum combined score of 1800 on three sections required), minimum GPA of 3.0. *Application deadline:* For fall admission, 7/6. Applications are processed on a rolling basis. Application fee: $25. *Financial aid:* In 1998–99, 23 students received aid, including 1 fellowship with full tuition reimbursement available, 8 research assistantships with full tuition reimbursements available (averaging $8,500 per year), 14 teaching assistantships with full tuition reimbursements available (averaging $8,500 per year); Federal Work-Study also available. *Faculty research:* Software engineering, artificial intelligence, database management, algorithms, human-computer interaction. Total annual research expenditures: $650,000. *Unit head:* Dr. David Cordes, Head, 205-348-6363, Fax: 205-348-0219, E-mail: cs@cs.us.edu. *Application contact:* Dr. Hui-Chuan Chen, Professor, 205-348-6363, Fax: 205-348-0219.

See in-depth description on page 835.

The University of Alabama at Birmingham, Graduate School, School of Natural Sciences and Mathematics, Department of Computer and Information Sciences, Birmingham, AL 35294. Offers MS, PhD. *Students:* 45 full-time (15 women), 13 part-time (3 women); includes 10 minority (4 African Americans, 6 Asian Americans or Pacific Islanders), 40 international. 142 applicants, 84% accepted. In 1998, 20 master's, 4 doctorates awarded. *Degree requirements:* For master's, thesis optional, foreign language not required; for doctorate, dissertation required, foreign language not required. *Entrance requirements:* For master's, GRE General Test (minimum combined score of 1100 required); for doctorate, GRE General Test (minimum combined score of 1300 required). *Application deadline:* Applications are processed on a rolling basis. Application fee: $30 ($60 for international students). *Financial aid:* In 1998–99, 7 fellowships with full tuition reimbursements (averaging $13,200 per year), 2 research assistantships with full tuition reimbursements (averaging $13,200 per year), 9 teaching assistantships with full tuition reimbursements (averaging $13,200 per year) were awarded.; career-related internships or fieldwork, Federal Work-Study, institutionally-sponsored loans, tuition waivers (full and partial), and unspecified assistantships also available. Aid available to part-time students. Financial aid application deadline: 3/10. *Faculty research:* Theory and software systems, intelligent systems, systems architecture. *Unit head:* Dr. Warren T. Jones, Chairman, 205-934-2213, Fax: 205-934-5473, E-mail: wjones@uab.edu.

See in-depth description on page 837.

The University of Alabama in Huntsville, School of Graduate Studies, College of Science, Department of Computer Science, Huntsville, AL 35899. Offers MS, PhD. Part-time and evening/weekend programs available. *Faculty:* 13 full-time (3 women), 3 part-time (0 women). *Students:* 39 full-time (12 women), 47 part-time (15 women); includes 17 minority (3 African Americans, 11 Asian Americans or Pacific Islanders, 2 Hispanic Americans, 1 Native American), 37 international. Average age 31. 66 applicants, 83% accepted. In 1998, 26 master's, 4 doctorates awarded. *Degree requirements:* For master's, computer language, oral and written exams required, thesis optional, foreign language not required; for doctorate, computer language, dissertation, oral and written exams required, foreign language not required. *Entrance requirements:* For master's and doctorate, GRE General Test (minimum combined score of 1500 on three sections required), GRE Subject Test, minimum GPA of 3.0. *Application deadline:* For fall admission, 7/24 (priority date); for spring admission, 11/15 (priority date). Applications are processed on a rolling basis. Application fee: $20. Tuition and fees vary according to course load. *Financial aid:* In 1998–99, 37 students received aid, including 19 research assistantships with full and partial tuition reimbursements available (averaging $7,675 per year), 18 teaching assistantships with full and partial tuition reimbursements available (averaging $6,865 per year); fellowships with full and partial tuition reimbursements available, career-related internships or fieldwork, Federal Work-Study, grants, institutionally-sponsored loans, scholarships, and tuition waivers (full and partial) also available. Aid available to part-time students. Financial aid application deadline: 4/1; financial aid applicants required to submit FAFSA. *Faculty research:* Numerical analysis, programming languages, software systems, artificial intelligence, visualization systems. Total annual research expenditures: $852,992. *Unit head:* Dr. Carl Davis, Chair, 256-890-6088, Fax: 256-890-6239, E-mail: cdavis@cs.uah.edu.

University of Alaska Fairbanks, Graduate School, College of Science, Engineering and Mathematics, Department of Mathematical Sciences, Fairbanks, AK 99775-7480. Offers computer science (MS); mathematics (MAT, MS, PhD). Part-time programs available. *Faculty:* 23 full-time (3 women), 1 part-time (0 women). *Students:* 8 full-time (3 women), 10 part-time (4 women); includes 2 minority (both Asian Americans or Pacific Islanders), 2 international. Terminal master's awarded for partial completion of doctoral program. *Degree requirements:* For master's, comprehensive exam, project required; for doctorate, one foreign language (computer language can substitute), dissertation, comprehensive exam required. *Entrance requirements:* For master's and doctorate, GRE General Test, GRE Subject Test, TOEFL (minimum score of 550 required). *Application deadline:* For fall admission, 8/1 (priority date). Application fee: $35. *Unit head:* Dr. Clifton Lando, Head, 907-474-7332.

University of Alberta, Faculty of Graduate Studies and Research, Department of Computing Science, Edmonton, AB T6G 2E1, Canada. Offers M Sc, PhD. Part-time programs available. Terminal master's awarded for partial completion of doctoral program. *Degree requirements:* For master's and doctorate, thesis/dissertation, oral exam, seminar required. *Entrance requirements:* For master's and doctorate, TOEFL (minimum score of 600 required), GRE General Test. *Faculty research:* Artificial intelligence, multimedia, distributed computing, theory, software engineering.

See in-depth description on page 839.

The University of Arizona, Graduate College, College of Science, Department of Computer Science, Tucson, AZ 85721. Offers MS, PhD. Part-time programs available. *Faculty:* 13 full-time (1 woman). *Students:* 67 full-time (5 women), 1 (woman) part-time; includes 5 minority (1 African American, 4 Asian Americans or Pacific Islanders), 49 international. Average age 27. 248 applicants, 38% accepted. In 1998, 21 master's awarded (95% found work related to degree, 5% continued full-time study); 3 doctorates awarded (100% found work related to degree). Terminal master's awarded for partial completion of doctoral program. *Degree requirements:* For master's, computer language required, thesis optional, foreign language not required; for doctorate, computer language, dissertation required, foreign language not required. *Entrance requirements:* For master's, GRE General Test, TOEFL, minimum GPA of 3.2; for doctorate, GRE General Test, GRE Subject Test, TOEFL, minimum undergraduate GPA of 3.5. *Average time to degree:* Master's–2 years full-time; doctorate–6 years full-time. *Application deadline:* For fall admission, 1/15; for spring admission, 8/1. Applications are processed on a rolling basis. Application fee: $45. *Financial aid:* Fellowships, research assistantships, teaching assistantships, career-related internships or fieldwork, institutionally-sponsored loans, scholarships, and tuition waivers (full and partial) available. Financial aid application deadline: 1/15. *Faculty research:* Operating systems, theory of computation, programming languages, databases, algorithms, networks, web searching, parallel and distributed systems. Total annual research expenditures: $2.5 million. *Unit head:* Dr. Peter J. Downey, Head, 520-621-2207, Fax: 520-621-4246. *Application contact:* Sonia A. Economou, Graduate Coordinator, 520-621-4049, Fax: 520-621-4246, E-mail: soniae@cs.arizona.edu.

University of Arkansas, Graduate School, J. William Fulbright College of Arts and Sciences, Department of Computer Science, Fayetteville, AR 72701-1201. Offers MS, PhD. *Faculty:* 5 full-time (0 women). *Students:* 20 full-time (3 women); includes 2 minority (1 Asian American or Pacific Islander, 1 Native American), 12 international. 49 applicants, 55% accepted. In 1998, 7 master's awarded. *Degree requirements:* For master's, computer language required, foreign language and thesis not required; for doctorate, dissertation required. Application fee: $40 ($50 for international students). Tuition, state resident: full-time $3,186. Tuition, nonresident: full-time $7,560. Required fees: $378. *Financial aid:* In 1998–99, 11 teaching assistantships were awarded.; career-related internships or fieldwork and Federal Work-Study also available. Aid available to part-time students. Financial aid application deadline: 4/1; financial aid applicants required to submit FAFSA. *Unit head:* Dr. Dennis Brewer, Chair, 501-575-6427.

University of Arkansas at Little Rock, Graduate School, College of Sciences and Engineering Technology, Department of Computer and Information Science, Little Rock, AR 72204-1099. Offers MS. Part-time and evening/weekend programs available. *Degree requirements:* For master's, computer language required, thesis optional, foreign language not required. *Entrance requirements:* For master's, GRE General Test, minimum GPA of 3.0, bachelor's degree in computer science, mathematics, or appropriate alternative.

University of Bridgeport, College of Graduate and Undergraduate Studies, School of Science, Engineering, and Technology, Department of Computer Science and Engineering, Bridgeport, CT 06601. Offers computer engineering (MS); computer science (MS). *Faculty:* 8 full-time (0 women), 5 part-time (0 women). *Students:* 87 full-time (15 women), 148 part-time (57 women); includes 20 minority (19 Asian Americans or Pacific Islanders, 1 Hispanic American), 212 international. Average age 31. 236 applicants, 64% accepted. In 1998, 23 degrees awarded. *Degree requirements:* For master's, thesis optional, foreign language not required. *Entrance requirements:* For master's, TOEFL. *Application deadline:* Applications are processed on a rolling basis. Application fee: $35 ($50 for international students). *Financial aid:* In 1998–99, 58 students received aid; research assistantships, teaching assistantships, career-related internships or fieldwork, Federal Work-Study, institutionally-sponsored loans, and tuition waivers (partial) available. Aid available to part-time students. Financial aid application deadline: 6/1; financial aid applicants required to submit FAFSA. *Unit head:* Dr. Stephen F. Grodzinsky, Chairman, 203-576-4145.

See in-depth description on page 841.

University of British Columbia, Faculty of Graduate Studies, Faculty of Science, Department of Computer Science, Vancouver, BC V6T 1Z2, Canada. Offers M Sc, PhD. Part-time programs available. *Degree requirements:* For master's, computer language required, foreign language and thesis not required; for doctorate, computer language, dissertation required, foreign language not required. *Entrance requirements:* For master's and doctorate, GRE, TOEFL (minimum score of 600 required). *Faculty research:* Artificial intelligence, databases, robotics, graphics, systems.

University of Calgary, Faculty of Graduate Studies, Faculty of Science, Department of Computer Science, Calgary, AB T2N 1N4, Canada. Offers M Sc, PhD. *Faculty:* 21 full-time (2 women), 4 part-time (0 women). *Students:* 46 full-time (10 women), 5 part-time (3 women). Average age 28. 54 applicants, 50% accepted. In 1998, 11 master's awarded (73% found work related to degree, 27% continued full-time study); 3 doctorates awarded (33% entered university research/teaching, 67% found other work related to degree). *Degree requirements:* For master's, computer language, thesis required (for some programs), foreign language not required; for doctorate, computer language, dissertation, candidacy exam required, foreign language not required. *Entrance requirements:* For master's and doctorate, TOEFL (minimum score of 600 required), GRE General Test (requested for overseas applicants). *Average time to degree:* Master's–2.5 years full-time; doctorate–3.7 years full-time. *Application deadline:* For fall admission, 3/15; for spring admission, 8/15. Applications are processed on a rolling basis. Application fee: $60. *Financial aid:* In 1998–99, 31 students received aid, including 14 research assistantships, 17 teaching assistantships (averaging $10,700 per year); fellowships, grants and scholarships also available. Financial aid application deadline: 3/15. *Faculty research:* Algorithm design and analysis, computer graphics and animation, distributed computer systems, human-computer interaction, software engineering. *Unit head:* Dr. John Kendall, Graduate Director, 403-220-6319, Fax: 403-284-4707, E-mail: jkendall@cpsc.ucalgary.ca. *Application contact:* Lorraine Storey, Graduate Secretary, 403-220-3528, Fax: 403-284-4707, E-mail: storey1@cpsc.ucalgary.ca.

Computer Science

University of California, Berkeley, Graduate Division, College of Engineering, Department of Electrical Engineering and Computer Sciences, Computer Science Division, Berkeley, CA 94720-1500. Offers MS, PhD. In 1998, 28 master's, 24 doctorates awarded. *Degree requirements:* For master's, comprehensive exam or thesis required; for doctorate, dissertation, qualifying exam required, foreign language not required. *Entrance requirements:* For master's and doctorate, GRE General Test, GRE Subject Test, TOEFL (minimum score of 570 required), minimum GPA of 3.0. *Application deadline:* For fall admission, 12/15. Application fee: $40. *Financial aid:* Fellowships, research assistantships, teaching assistantships, scholarships available. Financial aid application deadline: 12/15. *Application contact:* Admission Assistant, 510-642-3068, E-mail: gradadm@eecs.berkeley.edu.

University of California, Davis, Graduate Studies, Program in Computer Science, Davis, CA 95616. Offers MS, PhD. *Faculty:* 35 full-time (0 women), 8 part-time (1 woman). *Students:* 65 full-time (11 women); includes 9 minority (6 Asian Americans or Pacific Islanders, 3 Hispanic Americans), 9 international. 123 applicants, 66% accepted. In 1998, 7 master's, 4 doctorates awarded. *Degree requirements:* For master's, thesis optional, foreign language not required; for doctorate, dissertation required, foreign language not required. *Entrance requirements:* For master's and doctorate, GRE General Test, GRE Subject Test, minimum GPA of 3.0. *Application deadline:* For fall admission, 1/15 (priority date). Applications are processed on a rolling basis. Application fee: $40. Electronic applications accepted. *Financial aid:* In 1998–99, 18 fellowships with full and partial tuition reimbursements, 23 research assistantships with full and partial tuition reimbursements, 28 teaching assistantships with full and partial tuition reimbursements were awarded.; Federal Work-Study, grants, scholarships, traineeships, and readerships also available. Financial aid application deadline: 1/15; financial aid applicants required to submit FAFSA. *Faculty research:* Intrusion detection, malicious code detection, next generation light wave computer networks, biological algorithms, parallel processing. *Unit head:* Charles Martel, Chair, 530-752-7004, Fax: 530-752-4767, E-mail: martel@cs.ucdavis.edu. *Application contact:* Graduate Adviser, 530-752-7004, Fax: 530-752-4767, E-mail: gradinfo@cs.ucdavis.edu.

University of California, Irvine, Office of Research and Graduate Studies, Department of Information and Computer Science, Irvine, CA 92697. Offers MS, PhD. *Faculty:* 30 full-time (5 women), 6 part-time (0 women). *Students:* 128 full-time (30 women); includes 25 minority (22 Asian Americans or Pacific Islanders, 3 Hispanic Americans), 58 international. Average age 31. 198 applicants, 46% accepted. In 1998, 15 master's, 12 doctorates awarded. Terminal master's awarded for partial completion of doctoral program. *Degree requirements:* For master's and doctorate, thesis/dissertation required, foreign language not required. *Entrance requirements:* For master's and doctorate, GRE General Test, GRE Subject Test. *Application deadline:* For fall admission, 1/15 (priority date). Applications are processed on a rolling basis. Application fee: $40. Electronic applications accepted. *Financial aid:* Fellowships, research assistantships, teaching assistantships, institutionally-sponsored loans and tuition waivers (full and partial) available. Financial aid application deadline: 3/2; financial aid applicants required to submit FAFSA. *Faculty research:* Artificial intelligence, computer system design, software, biomedical computing, theory, computing policy and society, data mining, imbedded systems. *Unit head:* Michael Pazzani, Chair, 949-824-7405. *Application contact:* Kris Domiccio, 949-824-2277.

See in-depth description on page 843.

University of California, Los Angeles, Graduate Division, School of Engineering and Applied Science, Department of Computer Science, Los Angeles, CA 90095. Offers MS, PhD, MBA/MS. *Faculty:* 20 full-time, 19 part-time. *Students:* 218 full-time (26 women); includes 67 minority (1 African American, 64 Asian Americans or Pacific Islanders, 1 Hispanic American, 1 Native American), 77 international. 399 applicants, 30% accepted. In 1998, 54 master's, 16 doctorates awarded. *Degree requirements:* For master's, comprehensive exam or thesis required; for doctorate, dissertation, qualifying exams required, foreign language not required. *Entrance requirements:* For master's, GRE General Test, GRE Subject Test, minimum GPA of 3.0; for doctorate, GRE General Test, GRE Subject Test, minimum GPA of 3.25. *Application deadline:* For fall admission, 1/15; for winter admission, 8/15. Application fee: $40. Electronic applications accepted. *Financial aid:* In 1998–99, 13 fellowships, 130 research assistantships, 30 teaching assistantships were awarded.; Federal Work-Study, institutionally-sponsored loans, and tuition waivers (full and partial) also available. Financial aid application deadline: 1/15; financial aid applicants required to submit FAFSA. *Unit head:* Dr. Richard Muntz, Chair, 310-825-3546. *Application contact:* Verra Morgan, Student Affairs Officer, 310-825-6830, Fax: 310-UCLA-CSD, E-mail: verra@cs.ucla.edu.

University of California, Riverside, Graduate Division, College of Engineering, Department of Computer Science, Riverside, CA 92521-0102. Offers MS, PhD. Part-time programs available. *Degree requirements:* For master's, comprehensive exams required, foreign language and thesis not required; for doctorate, dissertation, qualifying exams required, foreign language not required. *Entrance requirements:* For master's and doctorate, GRE General Test (minimum combined score of 1100 required), TOEFL (minimum score of 550 required). *Application deadline:* For fall admission, 5/1; for spring admission, 12/1. Applications are processed on a rolling basis. Application fee: $40. *Financial aid:* Fellowships, research assistantships, teaching assistantships, career-related internships or fieldwork, Federal Work-Study, institutionally-sponsored loans, and tuition waivers (full and partial) available. Financial aid application deadline: 2/1; financial aid applicants required to submit FAFSA. *Faculty research:* Compiler construction, operating systems, theory of computation, computer architecture, computer networks, design automation. *Unit head:* Dr. Tom Payne, Chair, 909-787-3119, Fax: 909-787-4643, E-mail: thp@cs.ucr.edu. *Application contact:* Graduate Student Affairs, 909-787-5639, Fax: 909-787-4643, E-mail: gradadmissions@cs.ucr.edu.

See in-depth description on page 845.

University of California, San Diego, Graduate Studies and Research, Department of Computer Science and Engineering, La Jolla, CA 92093-5003. Offers computer engineering (MS, PhD); computer science (MS, PhD). *Faculty:* 18. *Students:* 144 (30 women). 428 applicants, 30% accepted. In 1998, 19 master's, 9 doctorates awarded. *Degree requirements:* For master's, foreign language and thesis not required; for doctorate, dissertation required. *Entrance requirements:* For master's and doctorate, GRE General Test. *Application deadline:* For fall admission, 1/15. Application fee: $40. *Faculty research:* Analysis of algorithms, combinatorial algorithms, discrete optimization. *Unit head:* Jeanne Ferrante, Chair. *Application contact:* Graduate Coordinator, 619-534-6005.

See in-depth description on page 847.

University of California, San Diego, Graduate Studies and Research, Interdisciplinary Program in Cognitive Science, La Jolla, CA 92093-5003. Offers cognitive science/anthropology (PhD); cognitive science/communication (PhD); cognitive science/computer science and engineering (PhD); cognitive science/linguistics (PhD); cognitive science/neuroscience (PhD); cognitive science/philosophy (PhD); cognitive science/psychology (PhD); cognitive science/sociology (PhD). Admissions through affiliated departments. *Faculty:* 51 full-time (6 women). *Students:* 10 full-time (3 women). *Degree requirements:* For doctorate, dissertation required. *Entrance requirements:* For doctorate, GRE General Test. *Application deadline:* Applications are processed on a rolling basis. Application fee: $40. *Unit head:* Walter J. Savitch, Director, 619-534-7141, Fax: 619-534-1128, E-mail: wsavitch@ucsd.edu. *Application contact:* Gris Arellano-Ramirez, Graduate Coordinator, 619-534-7141, Fax: 619-534-1128, E-mail: gradinfo@cogsci.ucsd.edu.

University of California, Santa Barbara, Graduate Division, College of Engineering, Department of Computer Science, Santa Barbara, CA 93106. Offers MS, PhD. *Students:* 106 full-time (18 women). 382 applicants, 51% accepted. In 1998, 16 master's, 2 doctorates awarded. *Degree requirements:* For master's, thesis or alternative required, foreign language not required; for doctorate, dissertation required, foreign language not required. *Entrance requirements:* For master's and doctorate, GRE, TOEFL (minimum score of 600 required). *Application deadline:* For fall admission, 5/1. Application fee: $40. *Financial aid:* Fellowships, research assistantships, teaching assistantships, career-related internships or fieldwork, Federal Work-Study, institutionally-sponsored loans, and tuition waivers (full and partial) available. Financial aid application deadline: 1/1; financial aid applicants required to submit FAFSA. *Unit head:* Oscar H. Ibarra, Chair, 805-893-2207. *Application contact:* Amy Mills, Graduate Program Assistant, 805-893-4323, E-mail: grad-advisor@cs.ucsb.edu.

University of California, Santa Cruz, Graduate Division, Division of Natural Sciences, Department of Computer Science, Santa Cruz, CA 95064. Offers MS, PhD. *Faculty:* 12 full-time. *Students:* 62 full-time (14 women). 105 applicants, 48% accepted. In 1998, 12 master's, 5 doctorates awarded. *Degree requirements:* For master's, thesis required; for doctorate, dissertation, qualifying exam required. *Entrance requirements:* For master's and doctorate, GRE General Test, GRE Subject Test. *Application deadline:* For fall admission, 2/1. Application fee: $40. *Financial aid:* Fellowships, research assistantships, teaching assistantships, Federal Work-Study and institutionally-sponsored loans available. Financial aid application deadline:2/1. *Faculty research:* Algorithm analysis, artificial intelligence, computer graphics, information and communication theory, problem-solving techniques. *Unit head:* Dr. Phokion Kolaitis, Chairperson, 831-459-4768. *Application contact:* Graduate Admissions, 831-459-2301.

See in-depth description on page 851.

University of Central Florida, College of Arts and Sciences, Program in Computer Science, Orlando, FL 32816. Offers MS, PhD. Part-time and evening/weekend programs available. *Faculty:* 19 full-time, 10 part-time. *Students:* 69 full-time (14 women), 66 part-time (16 women); includes 24 minority (6 African Americans, 15 Asian Americans or Pacific Islanders, 3 Hispanic Americans), 52 international. Average age 31. 75 applicants, 32% accepted. In 1998, 23 master's, 1 doctorate awarded. *Degree requirements:* For master's, computer language, thesis or alternative required, foreign language not required; for doctorate, computer language, dissertation, candidacy exam, qualifying exam required, foreign language not required. *Entrance requirements:* For master's, GRE General Test, GRE Subject Test, TOEFL (minimum score of 550 required; 213 computer-based), minimum GPA of 3.0 in last 60 hours; for doctorate, GRE Subject Test, TOEFL (minimum score of 550 required; 213 computer-based), minimum GPA of 3.0 in last 60 hours. *Application deadline:* For fall admission, 3/1 (priority date); for spring admission, 12/15. Application fee: $20. Tuition, state resident: full-time $2,054; part-time $137 per credit. Tuition, nonresident: full-time $7,207; part-time $480 per credit. Required fees: $47 per term. *Financial aid:* In 1998–99, 13 fellowships with partial tuition reimbursements (averaging $2,903 per year), 107 research assistantships with partial tuition reimbursements (averaging $3,228 per year), 49 teaching assistantships with partial tuition reimbursements (averaging $4,073 per year) were awarded.; career-related internships or fieldwork, Federal Work-Study, institutionally-sponsored loans, tuition waivers (partial), and unspecified assistantships also available. Financial aid application deadline: 3/1; financial aid applicants required to submit FAFSA. *Faculty research:* Parallel processing, databases, algorithms, virtual reality. *Unit head:* Dr. T. Frederick, Chair, 407-823-2209, Fax: 407-823-5419, E-mail: fred@cs.ucf.edu. *Application contact:* Dr. Ronald Dutton, Coordinator, 407-823-2341, Fax: 407-823-5419, E-mail: dutton@cs.ucf.edu.

University of Central Oklahoma, Graduate College, College of Mathematics and Science, Department of Mathematics and Statistics, Edmond, OK 73034-5209. Offers applied mathematical sciences (MS), including computer science, mathematics, mathematics/computer science teaching, statistics. *Accreditation:* NCATE. Part-time programs available. *Faculty:* 11 full-time (3 women), 1 part-time (0 women). *Degree requirements:* For master's, computer language, thesis required, foreign language not required. *Application deadline:* For fall admission, 8/23 (priority date). Applications are processed on a rolling basis. Application fee: $15. *Unit head:* Dr. David Bridge, Chairperson, 405-974-5697. *Application contact:* Dr. James Yates, Adviser, 405-974-5386, Fax: 405-974-3824, E-mail: jyates@aix1.ucok.edu.

University of Chicago, Division of the Physical Sciences, Conversion Master's Program in Computer Science, Chicago, IL 60637-1513. Offers SM. Program offered by the Department of Computer Science. Part-time and evening/weekend programs available. *Degree requirements:* For master's, foreign language and thesis not required. *Entrance requirements:* For master's, GRE.

University of Chicago, Division of the Physical Sciences, Department of Computer Science, Chicago, IL 60637-1513. Offers SM, PhD. Terminal master's awarded for partial completion of doctoral program. *Degree requirements:* For master's, computer language, thesis required, foreign language not required; for doctorate, one foreign language, computer language, dissertation required. *Entrance requirements:* For doctorate, GRE Subject Test. *Faculty research:* Theory of computing, artificial intelligence, programming languages, robotics, computational geometry.

University of Cincinnati, Division of Research and Advanced Studies, College of Engineering, Department of Electrical and Computer Engineering and Computer Science, Program in Computer Science, Cincinnati, OH 45221-0091. Offers MS. *Students:* 29 full-time (6 women), 38 part-time (11 women); includes 9 minority (2 African Americans, 6 Asian Americans or Pacific Islanders, 1 Hispanic American), 45 international. In 1998, 32 degrees awarded. *Degree requirements:* For master's, thesis or alternative required, foreign language not required. *Entrance requirements:* For master's, GRE General Test, TOEFL (minimum score of 520 required), BS in electrical engineering or related field. *Average time to degree:* Master's–2.9 years full-time. *Application deadline:* For fall admission, 2/1 (priority date). Application fee: $40. *Financial aid:* Fellowships, tuition waivers (full) and unspecified assistantships available. Financial aid application deadline: 2/1. *Unit head:* Barbara J. Paschke, Coordinator, 415-502-7788, Fax: 415-476-0688, E-mail: mis@cgl.ucsf.edu. *Application contact:* Dieter Schmidt, Graduate Program Director, 513-556-1816, Fax: 513-556-7326, E-mail: dieter.schmidt@uc.edu.

University of Cincinnati, Division of Research and Advanced Studies, College of Engineering, Department of Electrical and Computer Engineering and Computer Science, Program in Computer Science and Engineering, Cincinnati, OH 45221-0091. Offers PhD. *Students:* 26 full-time (4 women), 3 part-time; includes 1 African American, 21 international. In 1998, 4 degrees awarded. *Degree requirements:* For doctorate, dissertation required, foreign language not required. *Entrance requirements:* For doctorate, GRE General Test, TOEFL. *Average time to degree:* Doctorate–4.9 years full-time. *Application deadline:* For fall admission, 2/1 (priority date). Application fee: $40. *Financial aid:* Fellowships, tuition waivers (full) and unspecified assistantships available. Financial aid application deadline: 2/1. *Unit head:* Graduate Student Affairs, 909-787-5639, Fax: 909-787-4643, E-mail: gradadmissions@cs.ucr.edu. *Application contact:* Dieter Schmidt, Graduate Program Director, 513-556-1816, Fax: 513-556-7326, E-mail: dieter.schmidt@uc.edu.

University of Colorado at Boulder, Graduate School, College of Engineering and Applied Science, Department of Computer Science, Boulder, CO 80309. Offers ME, MS, PhD. *Degree requirements:* For master's, thesis or alternative, comprehensive exam required; for doctorate, dissertation required. *Entrance requirements:* For master's, minimum undergraduate GPA of 3.0. *Faculty research:* Parallel and numerical computation, human computer interaction, neural networks, software systems, theory.

See in-depth description on page 853.

University of Colorado at Colorado Springs, Graduate School, College of Engineering and Applied Science, Department of Computer Science, Colorado Springs, CO 80933-7150. Offers MS. Part-time programs available. *Faculty:* 11 full-time (2 women), 8 part-time (2 women). *Students:* 35 full-time (12 women), 25 part-time (7 women); includes 10 minority (7 Asian Americans or Pacific Islanders, 3 Hispanic Americans), 8 international. Average age 29. 23 applicants, 74% accepted. In 1998, 27 degrees awarded (100% found work related to degree). *Degree requirements:* For master's, oral final exam required, thesis optional, foreign language not required. *Entrance requirements:* For master's, GRE General Test (minimum combined score of 1200 required), TOEFL (minimum score of 550 required), minimum GPA of 3.0. *Application deadline:* For fall admission, 7/1 (priority date); for spring admission, 12/1.

Applications are processed on a rolling basis. Application fee: $40 ($50 for international students). Tuition, state resident: full-time $2,768; part-time $118 per credit. Tuition, nonresident: full-time $10,392; part-time $425 per credit. Required fees: $265; $7.5 per credit. One-time fee: $28. Tuition and fees vary according to program and student level. *Financial aid:* In 1998–99, 9 students received aid; teaching assistantships available. Financial aid application deadline:5/1. *Faculty research:* Analytical intelligence, software engineering, networks, database systems, graphics. Total annual research expenditures: $170,000. *Unit head:* Dr. Robert W. Sebesta, Chairman, 719-262-3325, Fax: 719-262-3369, E-mail: rws@sneffels.uccs.edu. *Application contact:* Marijke Augusteijn, Academic Adviser, 719-262-3325, Fax: 719-262-3369, E-mail: mfa@antero.uccs.edu.

University of Colorado at Denver, Graduate School, College of Engineering and Applied Science, Department of Computer Science, Denver, CO 80217-3364. Offers MS. Part-time and evening/weekend programs available. *Faculty:* 8. *Students:* 19 full-time (9 women), 88 part-time (20 women); includes 26 minority (2 African Americans, 23 Asian Americans or Pacific Islanders, 1 Hispanic American), 21 international. Average age 28. 62 applicants, 76% accepted. In 1998, 23 degrees awarded. *Degree requirements:* For master's, thesis or alternative required. *Entrance requirements:* For master's, GRE. *Application deadline:* For fall admission, 2/1; for spring admission, 10/1. Applications are processed on a rolling basis. Application fee: $50 ($60 for international students). Electronic applications accepted. Tuition, state resident: part-time $217 per credit hour. Tuition, nonresident: part-time $783 per credit hour. Required fees: $3 per credit hour. $130 per year. One-time fee: $25 part-time. *Financial aid:* Research assistantships, teaching assistantships, career-related internships or fieldwork and Federal Work-Study available. Financial aid application deadline: 3/1; financial aid applicants required to submit FAFSA. Total annual research expenditures: $18,706. *Unit head:* Gita Alaghband, Chair, 303-556-4314, Fax: 303-556-8369. *Application contact:* Mary Stephens, Program Assistant, 303-556-4314, Fax: 303-556-8369.

University of Connecticut, Graduate School, School of Engineering, Field of Computer Science and Engineering, Storrs, CT 06269. Offers artificial intelligence (MS, PhD); computer architecture (MS, PhD); computer science (MS, PhD); operating systems (MS, PhD); robotics (MS, PhD); software engineering (MS, PhD). Terminal master's awarded for partial completion of doctoral program. *Degree requirements:* For master's, thesis or alternative required; for doctorate, dissertation required. *Entrance requirements:* For master's and doctorate, GRE General Test.

University of Dayton, Graduate School, College of Arts and Sciences, Department of Computer Science, Dayton, OH 45469-1300. Offers MCS. Part-time and evening/weekend programs available. *Faculty:* 9 full-time (2 women), 8 part-time (2 women). *Students:* 4 full-time (2 women), 36 part-time (6 women); includes 6 minority (all Asian Americans or Pacific Islanders), 18 international. Average age 23. 34 applicants, 74% accepted. In 1998, 15 degrees awarded (100% found work related to degree). *Degree requirements:* For master's, computer language, project required, foreign language and thesis not required. *Entrance requirements:* For master's, GRE General Test (minimum combined score of 1450 on three sections required), 4 undergraduate courses in computer science, minimum undergraduate GPA of 3.0. *Average time to degree:* Master's–2 years full-time, 3.5 years part-time. *Application deadline:* For fall admission, 8/1. Applications are processed on a rolling basis. Application fee: $30. Electronic applications accepted. *Financial aid:* In 1998–99, 4 students received aid, including 4 teaching assistantships (averaging $9,000 per year) *Faculty research:* Software engineering, networking, databases. Total annual research expenditures: $50,000. *Unit head:* Dr. Barbara Smith, Chair, 937-229-3831, Fax: 937-229-4000, E-mail: smithb@cps.udayton.edu. *Application contact:* Dr. James P. Buckley, Graduate Director, 937-229-3808, Fax: 937-229-4000, E-mail: grad@udcps.cps.udayton.edu.

University of Delaware, College of Arts and Science, Department of Computer and Information Sciences, Newark, DE 19716. Offers MS, PhD. Part-time programs available. *Faculty:* 18 full-time (3 women), 3 part-time (0 women). *Students:* 68 full-time (14 women), 10 part-time (2 women); includes 29 minority (27 Asian Americans or Pacific Islanders, 2 Hispanic Americans), 17 international. In 1998, 32 master's, 4 doctorates awarded. Terminal master's awarded for partial completion of doctoral program. *Degree requirements:* For master's, thesis optional, foreign language not required; for doctorate, dissertation required, foreign language not required. *Entrance requirements:* For master's and doctorate, GRE General Test (minimum combined score of 1750 on three sections required), TOEFL (average 600). *Average time to degree:* Master's–2 years full-time; doctorate–5 years full-time. *Application deadline:* For fall admission, 7/1; for spring admission, 12/1. Applications are processed on a rolling basis. Application fee: $45. Electronic applications accepted. *Financial aid:* In 1998–99, 56 students received aid, including 6 fellowships, 26 research assistantships, 16 teaching assistantships; tuition waivers (full) and unspecified assistantships also available. Financial aid application deadline: 3/1. *Faculty research:* Computer networks and distributed systems, natural language processing and artificial intelligence, parallel computing, theory, computer algebra, graphics. Total annual research expenditures: $1.2 million. *Unit head:* Prof. Errol Lloyd, Chair, 302-831-2711, Fax: 302-831-8458, E-mail: elloyd@cis.udel.edu. *Application contact:* Patricia L. Beazley, Graduate Secretary, 302-831-2713, Fax: 302-831-8458, E-mail: beazley@cis.udel.edu.

See in-depth description on page 855.

University of Denver, Graduate Studies, Faculty of Natural Sciences, Mathematics and Engineering, Department of Engineering, Denver, CO 80208. Offers computer science and engineering (MS); electrical engineering (MS); management and general engineering (MSMGEN); materials science (PhD); mechanical engineering (MS). Part-time and evening/weekend programs available. *Faculty:* 15. *Students:* 23 (9 women) 8 international. Terminal master's awarded for partial completion of doctoral program. *Degree requirements:* For master's, thesis required (for some programs), foreign language not required; for doctorate, dissertation required, foreign language not required. *Entrance requirements:* For master's and doctorate, GRE General Test, TOEFL (minimum score of 570 required), TSE (minimum score of 230 required). *Application deadline:* Applications are processed on a rolling basis. Application fee: $40 ($45 for international students). *Unit head:* Dr. Albert J. Rosa, Chair, 303-871-2102. *Application contact:* Louise Carlson, Assistant to Chair, 303-871-2107.

University of Denver, Graduate Studies, Faculty of Natural Sciences, Mathematics and Engineering, Department of Mathematics and Computer Science, Denver, CO 80208. Offers applied mathematics (MA, MS); computer science (MS); mathematics and computer science (PhD). Part-time programs available. *Faculty:* 14. *Students:* 56 (16 women); includes 8 minority (1 African American, 6 Asian Americans or Pacific Islanders, 1 Hispanic American) 34 international. 104 applicants, 99% accepted. In 1998, 27 master's, 3 doctorates awarded. Terminal master's awarded for partial completion of doctoral program. *Degree requirements:* For master's, computer language, foreign language, or laboratory experience required, thesis not required; for doctorate, one foreign language (computer language can substitute), dissertation, oral and written exams required. *Entrance requirements:* For master's and doctorate, GRE General Test, TOEFL (minimum score of 550 required). *Application deadline:* Applications are processed on a rolling basis. Application fee: $40 ($45 for international students). *Financial aid:* In 1998–99, 23 students received aid, including 6 fellowships with full and partial tuition reimbursements available (averaging $12,000 per year), 3 research assistantships with full and partial tuition reimbursements available (averaging $11,724 per year), 14 teaching assistantships with full and partial tuition reimbursements available (averaging $11,316 per year); career-related internships or fieldwork, Federal Work-Study, institutionally-sponsored loans, and scholarships also available. Aid available to part-time students. Financial aid application deadline: 3/1; financial aid applicants required to submit FAFSA. *Faculty research:* Real-time software, convex bodies, multidimensional data, parallel computer clusters. Total annual research expenditures: $163,312. *Unit head:* Dr. Joel Cohen, Chairperson, 303-871-3292. *Application contact:* Rick Ball, Graduate Adviser, 303-871-2453.

University of Denver, University College, Denver, CO 80208. Offers applied communication (MSS); computer information systems (MCIS); environmental policy and management (MEPM); healthcare systems (MHS); liberal studies (MLS); library and information services (MLIS); public health (MPH); technology management (MoTM); telecommunications (MTEL). Part-time and evening/weekend programs available. Postbaccalaureate distance learning degree programs offered (no on-campus study). *Faculty:* 1 (woman) full-time, 553 part-time (181 women). *Students:* 1,529 (804 women); includes 189 minority (67 African Americans, 33 Asian Americans or Pacific Islanders, 75 Hispanic Americans, 14 Native Americans) 61 international. *Entrance requirements:* For master's, minimum undergraduate GPA of 3.0. *Application deadline:* For fall admission, 8/10; for spring admission, 2/22. Applications are processed on a rolling basis. Application fee: $25. *Unit head:* Peter Warren, Dean, 303-871-3268, Fax: 303-871-4047, E-mail: pwarren@du.edu. *Application contact:* Bryan Ehrlich, Admission Coordinator, 303-871-3969, Fax: 303-871-3303, E-mail: behrlich@du.edu.

University of Detroit Mercy, College of Engineering and Science, Department of Mathematics and Computer Science, Program in Computer Science, Detroit, MI 48219-0900. Offers MSCS. Evening/weekend programs available. *Degree requirements:* For master's, computer language required, foreign language not required. *Entrance requirements:* For master's, minimum GPA of 3.0.

See in-depth description on page 857.

University of Florida, Graduate School, College of Engineering, Department of Computer and Information Science and Engineering, Gainesville, FL 32611. Offers computer and information science and engineering (ME); computer organization (MS, PhD, Engr); information systems (MS, PhD, Engr); manufacturing systems engineering (Certificate); software systems (MS, PhD, Engr). *Faculty:* 38. *Students:* 124 full-time (25 women), 42 part-time (10 women); includes 21 minority (2 African Americans, 14 Asian Americans or Pacific Islanders, 4 Hispanic Americans, 1 Native American), 82 international. 658 applicants, 35% accepted. In 1998, 45 master's, 14 doctorates awarded. *Degree requirements:* For master's, computer language, thesis required (for some programs), foreign language not required; for doctorate, computer language, dissertation required, foreign language not required. *Entrance requirements:* For master's and doctorate, GRE General Test (minimum combined score of 1100 required), minimum GPA of 3.0; for other advanced degree, GRE General Test. *Application deadline:* For fall admission, 6/1 (priority date); for spring admission, 11/1. Applications are processed on a rolling basis. Application fee: $20. Electronic applications accepted. *Financial aid:* In 1998–99, 90 students received aid, including 3 fellowships, 83 research assistantships, 20 teaching assistantships; unspecified assistantships also available. Financial aid application deadline:6/1. *Faculty research:* Artificial intelligence, networks security, distributed computing, parallel processing system, vision and visualization, database systems. *Unit head:* Dr. Gerhard Ritter, Chair, 352-392-1212, Fax: 352-392-1220, E-mail: ritter@cise.ufl.edu. *Application contact:* Dr. Doug Dankel, Graduate Coordinator, 352-392-1387, Fax: 352-392-1220, E-mail: ddd@cise.ufl.edu.

See in-depth description on page 859.

University of Georgia, Graduate School, College of Arts and Sciences, Department of Computer Science, Athens, GA 30602. Offers applied mathematical science (MAMS); computer science (MS, PhD). *Faculty:* 16 full-time (1 woman). *Students:* 56 full-time (12 women), 14 part-time (3 women). 163 applicants, 26% accepted. In 1998, 21 master's awarded. *Degree requirements:* For doctorate, dissertation required, foreign language not required. *Entrance requirements:* For master's and doctorate, GRE General Test. *Application deadline:* For fall admission, 7/1 (priority date); for spring admission, 11/15. Application fee: $30. Electronic applications accepted. *Financial aid:* Fellowships, research assistantships, teaching assistantships, unspecified assistantships available. *Unit head:* Dr. John A. Miller, Graduate Coordinator, 706-542-2911, Fax: 706-542-2966, E-mail: jam@cs.uga.edu.

University of Guelph, Faculty of Graduate Studies, College of Physical and Engineering Science, Department of Computing and Information Science, Guelph, ON N1G 2W1, Canada. Offers applied computer science (M Sc); computer science (M Sc). Part-time programs available. *Faculty:* 16 full-time (3 women). *Students:* 28 full-time (6 women), 2 part-time. 79 applicants, 9% accepted. In 1998, 6 degrees awarded (80% found work related to degree, 20% continued full-time study). *Degree requirements:* For master's, thesis required. *Entrance requirements:* For master's, minimum B- average during previous 2 years. *Application deadline:* For fall admission, 5/1 (priority date); for winter admission, 10/1 (priority date). Applications are processed on a rolling basis. Application fee: $60. *Expenses:* Tuition and fees charges are reported in Canadian dollars. Tuition, area resident: Full-time $4,725 Canadian dollars; part-time $1,055 Canadian dollars per term. International tuition: $6,999 Canadian dollars full-time. Required fees: $295 Canadian dollars per term. *Financial aid:* In 1998–99, 11 students received aid; fellowships, research assistantships, teaching assistantships, grants, institutionally-sponsored loans, and scholarships available. *Faculty research:* Interactive systems, distributed systems, information management, knowledge-based systems, VLSI-(A). Total annual research expenditures: $476,000. *Unit head:* Dr. J. G. Linders, Chair, 519-824-4120 Ext. 2250. *Application contact:* Prof. D. Banerji, Graduate Coordinator, 519-824-4120 Ext. 3005, Fax: 519-837-0323, E-mail: gradinfo@snowhite.cis.uoguelph.ca.

University of Hawaii at Manoa, Graduate Division, College of Arts and Sciences, College of Natural Sciences, Department of Information and Computer Sciences, Honolulu, HI 96822. Offers communication and information science (PhD); computer science (PhD); information and computer sciences (MS); library and information science (MLI Sc, PhD, Certificate, JD/MLI Sc, MLI Sc/MA, MLI Sc/MS), including advanced library and information science (Certificate), communication and information science (PhD), library and information science (MLI Sc). Part-time programs available. *Faculty:* 73 full-time (13 women). *Students:* 21 full-time (5 women), 16 part-time (6 women); includes 8 minority (6 Asian Americans or Pacific Islanders, 2 Hispanic Americans), 20 international. 59 applicants, 47% accepted. In 1998, 24 degrees awarded. *Application deadline:* For fall admission, 10/15; for spring admission, 8/10. Application fee: $25 ($50 for international students). *Financial aid:* In 1998–99, 23 students received aid, including 6 research assistantships (averaging $15,260 per year), 10 teaching assistantships (averaging $12,890 per year); tuition waivers (full and partial) also available. *Faculty research:* Software engineering, telecommunications, artificial intelligence, multimedia. *Unit head:* Dr. Stephen Y. Itoga, Chair, 808-956-7420, E-mail: itoga@hawaii.edu. *Application contact:* Dr. Philip Johnson, Graduate Field Chairperson, 808-956-8249, Fax: 808-956-3548, E-mail: johnson@ics.hawaii.edu.

University of Houston, College of Natural Sciences and Mathematics, Department of Computer Science, Houston, TX 77004. Offers MS, PhD. Part-time programs available. Postbaccalaureate distance learning degree programs offered. *Faculty:* 18 full-time (0 women), 6 part-time (0 women). *Students:* 179 full-time (55 women), 121 part-time (45 women); includes 38 minority (3 African Americans, 31 Asian Americans or Pacific Islanders, 3 Hispanic Americans, 1 Native American), 213 international. Average age 30. 372 applicants, 58% accepted. In 1998, 38 master's, 1 doctorate awarded. Terminal master's awarded for partial completion of doctoral program. *Degree requirements:* For master's and doctorate, computer language, thesis/dissertation required. *Entrance requirements:* For master's and doctorate, GRE General Test, TOEFL (minimum score of 550 required). *Application deadline:* For fall admission, 7/3 (priority date); for spring admission, 12/4. Applications are processed on a rolling basis. Application fee: $0 ($75 for international students). *Financial aid:* In 1998–99, 17 research assistantships, 33 teaching assistantships were awarded.; Federal Work-Study and institutionally-sponsored loans also available. Aid available to part-time students. Financial aid application deadline: 3/1; financial aid applicants required to submit FAFSA. *Faculty research:* Parallel and distributed systems, software engineering, numerical analysis, databases, graphics and virtual reality. *Unit head:* Dr. Lennart Johnsson, Chairman, 713-743-3350. *Application contact:* Amanda Vaughan, Graduate Academic Advising Assistant, 713-743-3364, Fax: 713-743-3335, E-mail: vaughan@cs.uh.edu.

See in-depth description on page 861.

Computer Science

University of Houston, College of Technology, Houston, TX 77004. Offers construction management (MT); manufacturing systems (MT); microcomputer systems (MT); occupational technology (MSOT). Part-time and evening/weekend programs available. *Faculty:* 23 full-time (7 women), 3 part-time (0 women). *Students:* 17 full-time (11 women), 75 part-time (41 women); includes 27 minority (15 African Americans, 4 Asian Americans or Pacific Islanders, 8 Hispanic Americans), 6 international. *Degree requirements:* Foreign language not required. *Entrance requirements:* For master's, GMAT, GRE, or MAT (MSOT); GRE (MT), minimum GPA of 3.0 in last 60 hours. *Application deadline:* For fall admission, 7/1; for spring admission, 11/1. Application fee: $35 ($110 for international students). *Unit head:* Bernard McIntyre, Dean, 713-743-4028, Fax: 713-743-4032, E-mail: bmcintyre@uh.edu. *Application contact:* Holly Rosenthal, Graduate Academic Adviser, 713-743-4098, Fax: 713-743-4032, E-mail: hrosenthal@uh.edu.

University of Houston–Clear Lake, School of Natural and Applied Sciences, Program in Computer Science, Houston, TX 77058-1098. Offers MS. *Faculty:* 8 full-time (2 women), 4 part-time (0 women). *Students:* 110 full-time (43 women), 75 part-time (21 women); includes 59 minority (3 African Americans, 52 Asian Americans or Pacific Islanders, 3 Hispanic Americans, 1 Native American), 98 international. Average age 32. *Degree requirements:* Foreign language not required. *Entrance requirements:* For master's, GRE General Test. *Application deadline:* Applications are processed on a rolling basis. Application fee: $30 ($70 for international students). *Financial aid:* Research assistantships, teaching assistantships available. Financial aid application deadline: 5/1. *Unit head:* Dr. Sadegh Davari, Chair, 281-283-3850, Fax: 281-283-3707. *Application contact:* Dr. Robert Ferebee, Associate Dean, 281-283-3700, Fax: 281-283-3707, E-mail: ferebee@uhcl4.cl.uh.edu.

University of Idaho, College of Graduate Studies, College of Engineering, Department of Computer Science, Moscow, ID 83844-4140. Offers MS, PhD. *Faculty:* 14 full-time (2 women), 1 part-time (0 women). *Students:* 12 full-time (3 women), 37 part-time (5 women); includes 2 minority (1 African American, 1 Asian American or Pacific Islander), 9 international. In 1998, 6 master's, 2 doctorates awarded. *Degree requirements:* For master's, thesis required, foreign language not required; for doctorate, dissertation required. *Entrance requirements:* For master's, GRE General Test, TOEFL (minimum score of 550 required), minimum GPA of 3.0; for doctorate, minimum undergraduate GPA of 2.8, 3.0 graduate. *Application deadline:* For fall admission, 8/1; for spring admission, 12/15. Application fee: $35 ($45 for international students). *Financial aid:* In 1998–99, 9 research assistantships (averaging $8,354 per year), 6 teaching assistantships (averaging $9,732 per year) were awarded.; career-related internships or fieldwork also available. Financial aid application deadline: 2/15. *Faculty research:* Artificial intelligence, theory of computation, software engineering. *Unit head:* Dr. Paul Oman, Chair, 208-885-6589.

University of Illinois at Chicago, Graduate College, College of Engineering, Department of Electrical Engineering and Computer Science, Program in Computer Science and Engineering, Chicago, IL 60607-7128. Offers MS, PhD. Evening/weekend programs available. *Degree requirements:* For master's, computer language, thesis or alternative required, foreign language not required; for doctorate, computer language, dissertation, departmental qualifying exam required, foreign language not required. *Entrance requirements:* For master's, TOEFL (minimum score of 550 required), minimum GPA of 3.75 on a 5.0 scale, BS in related field; for doctorate, GRE General Test, TOEFL (minimum score of 550 required), minimum GPA of 3.75 on a 5.0 scale, MS in related field. *Application deadline:* For fall admission, 6/7; for spring admission, 11/1. Application fee: $40 ($50 for international students). *Financial aid:* Fellowships, research assistantships, teaching assistantships available. *Unit head:* Gyan Agarwal, Director of Graduate Studies, 312-996-8679.

See in-depth description on page 863.

University of Illinois at Chicago, Graduate College, College of Liberal Arts and Sciences, Department of Mathematics, Statistics, and Computer Science, Chicago, IL 60607-7128. Offers applied mathematics (MS, DA, PhD); computer science (MS, DA, PhD); probability and statistics (MS, DA, PhD); pure mathematics (MS, DA, PhD); teaching of mathematics (MST). *Faculty:* 69 full-time (4 women). *Students:* 119 full-time (47 women), 30 part-time (12 women). *Degree requirements:* For master's, comprehensive exam required, foreign language and thesis not required; for doctorate, one foreign language, dissertation required. *Entrance requirements:* For master's and doctorate, GRE General Test, TOEFL (minimum score of 550 required), minimum GPA of 3.75 on a 5.0 scale. *Application deadline:* For fall admission, 7/3; for spring admission, 11/8. Application fee: $40 ($50 for international students). *Unit head:* Henri Gillet, Head, 312-996-3044. *Application contact:* David Marker, Director of Graduate Studies, 312-996-3041.

University of Illinois at Springfield, Graduate Programs, College of Liberal Arts and Sciences, Program in Computer Science, Springfield, IL 62794-9243. Offers MA. *Faculty:* 4 full-time (0 women), 8 part-time (2 women). *Students:* 26 full-time (9 women), 75 part-time (27 women); includes 15 minority (4 African Americans, 11 Asian Americans or Pacific Islanders), 25 international. Average age 34. 37 applicants, 78% accepted. In 1998, 6 degrees awarded. *Degree requirements:* Foreign language not required. *Entrance requirements:* For master's, GRE General Test, TOEFL (minimum score of 550 required), minimum undergraduate GPA of 2.7. *Financial aid:* In 1998–99, 20 students received aid, including 3 research assistantships (averaging $1,492 per year); grants also available. Financial aid applicants required to submit FAFSA. *Unit head:* Ted Mims, Convener, 217-206-7326.

University of Illinois at Urbana–Champaign, Graduate College, College of Engineering, Department of Computer Science, Urbana, IL 61801. Offers MCS, MS, MST, PhD. *Faculty:* 45 full-time (4 women). *Students:* 363 full-time (71 women); includes 30 minority (1 African American, 27 Asian Americans or Pacific Islanders, 2 Hispanic Americans), 232 international. 812 applicants, 10% accepted. In 1998, 104 master's, 37 doctorates awarded. *Degree requirements:* For master's and doctorate, thesis/dissertation required, foreign language not required. *Entrance requirements:* For master's and doctorate, GRE General Test. *Application deadline:* For fall admission, 1/15. Applications are processed on a rolling basis. Application fee: $40 ($50 for international students). Tuition, state resident: full-time $4,616. Tuition, nonresident: full-time $11,768. Full-time tuition and fees vary according to course load. *Financial aid:* In 1998–99, 14 fellowships, 254 research assistantships, 53 teaching assistantships were awarded.; tuition waivers (full and partial) also available. Financial aid application deadline: 1/15. *Unit head:* Daniel A. Reed, Head, 217-333-3373. *Application contact:* Michael Faiman, Director of Graduate Studies, 217-333-6952, Fax: 217-333-3501, E-mail: faiman@cs.uiuc.edu.

See in-depth description on page 865.

The University of Iowa, Graduate College, College of Liberal Arts, Department of Computer Science, Iowa City, IA 52242-1316. Offers MCS, MS, PhD. *Faculty:* 16 full-time, 6 part-time. *Students:* 63 full-time (19 women), 19 part-time (2 women). 290 applicants, 15% accepted. In 1998, 45 master's, 4 doctorates awarded. *Degree requirements:* For master's, thesis optional; for doctorate, dissertation, comprehensive exam required. *Entrance requirements:* For master's and doctorate, GRE General Test, GRE Subject Test, TOEFL (minimum score of 600 required). *Application deadline:* For fall admission, 3/1; for spring admission, 10/1. Application fee: $30 ($50 for international students). *Financial aid:* In 1998–99, 10 research assistantships, 45 teaching assistantships were awarded. Financial aid applicants required to submit FAFSA. *Unit head:* Steven C. Bruell, Chair, 319-335-0713, Fax: 319-335-3624, E-mail: cs_info@cs.uiowa.edu.

Announcement: The University of Iowa Department of Computer Science is an energetic, young, focused, and relatively small department for a major university. Class enrollment is small and students work closely with members of the faculty. Graduate degrees offered include the Master of Computer Science (MCS), Master of Science (MS), and Doctor of Philosophy (PhD) in computer science. Research emphases include artificial intelligence (vision, robotics, automatic theorem proving, cognitive science), databases, distributed and concurrent systems and networks, programming languages (including parallel programming and compilers), simulation

(discrete and continuous), and software engineering. Visit the Web site at http://www.cs.uiowa.edu for more information about graduate programs.

University of Kansas, Graduate School, School of Engineering, Department of Electrical Engineering and Computer Science, Program in Computer Science, Lawrence, KS 66045. Offers MS, PhD. *Students:* 15 full-time (3 women), 37 part-time (7 women), 31 international. Terminal master's awarded for partial completion of doctoral program. *Degree requirements:* For master's, exam required, thesis optional, foreign language not required; for doctorate, one foreign language (computer language can substitute), dissertation, comprehensive and qualifying exams required. *Entrance requirements:* For master's and doctorate, GRE, TOEFL (minimum score of 600 required), minimum GPA of 3.0. *Application deadline:* For fall admission, 4/15 (priority date); for spring admission, 10/15. Applications are processed on a rolling basis. Application fee: $30. *Financial aid:* Fellowships, research assistantships, teaching assistantships, career-related internships or fieldwork available. *Application contact:* Victor Wallace, Graduate Director, 785-864-4487, Fax: 785-864-3226, E-mail: grad.admissions@eecs.ukans.edu.

See in-depth description on page 867.

University of Kentucky, Graduate School, Graduate School Programs from the College of Engineering, Program in Computer Science, Lexington, KY 40506-0032. Offers MS, PhD. *Degree requirements:* For master's, comprehensive exam required, thesis optional, foreign language not required; for doctorate, dissertation, comprehensive exam required. *Entrance requirements:* For master's, GRE General Test, minimum undergraduate GPA of 2.5; for doctorate, GRE General Test, minimum graduate GPA of 3.0. *Faculty research:* Artificial intelligence and databases, communication networks and operating systems, graphics and vision, numerical analysis, theory.

See in-depth description on page 869.

University of Louisville, Graduate School, Speed Scientific School, Department of Engineering Mathematics and Computer Science, Program in Computer Science and Engineering, Louisville, KY 40292-0001. Offers PhD. *Students:* 105 full-time (18 women), 39 part-time (11 women); includes 14 minority (1 African American, 13 Asian Americans or Pacific Islanders), 105 international. Average age 29. In 1998, 16 degrees awarded. *Degree requirements:* For doctorate, dissertation required, foreign language not required. *Entrance requirements:* For doctorate, GRE General Test (minimum combined score of 1200 required). *Application deadline:* Applications are processed on a rolling basis. Application fee: $25. *Unit head:* Dr. Peter Aronhime, Director.

University of Louisville, Graduate School, Speed Scientific School, Department of Engineering Mathematics and Computer Science, Program in Engineering Mathematics and Computer Science, Louisville, KY 40292-0001. Offers M Eng, MS. *Accreditation:* ABET (one or more programs are accredited). *Students:* 10 full-time (1 woman), 29 part-time (5 women); includes 4 minority (1 African American, 2 Asian Americans or Pacific Islanders, 1 Hispanic American) Average age 28. In 1998, 11 degrees awarded. *Degree requirements:* For master's, thesis required, foreign language not required. *Entrance requirements:* For master's, GRE General Test (minimum combined score of 1200 required). *Application deadline:* Applications are processed on a rolling basis. Application fee: $25. *Unit head:* Dr. Khaled A. Kamel, Chair, Department of Engineering Mathematics and Computer Science, 502-852-6304.

University of Maine, Graduate School, College of Liberal Arts and Sciences, Department of Computer Science, Orono, ME 04469. Offers MS, PhD. Part-time programs available. *Faculty:* 5 full-time (0 women), 1 part-time (0 women). *Students:* 11 full-time (0 women), 2 part-time (both women). 19 applicants, 58% accepted. In 1998, 7 degrees awarded. *Degree requirements:* For master's, thesis optional; for doctorate, dissertation required. *Entrance requirements:* For master's, GRE General Test, GRE Subject Test, TOEFL (minimum score of 550 required); for doctorate, GRE General Test, TOEFL (minimum score of 550 required). *Application deadline:* For fall admission, 2/1 (priority date); for spring admission, 10/15. Applications are processed on a rolling basis. Application fee: $50. *Financial aid:* In 1998–99, 1 research assistantship with tuition reimbursement (averaging $9,470 per year), 7 teaching assistantships with tuition reimbursements (averaging $9,388 per year) were awarded.; career-related internships or fieldwork, Federal Work-Study, institutionally-sponsored loans, and tuition waivers (full) also available. Financial aid application deadline: 3/1. *Faculty research:* Theory, software engineering, graphics, applications, artificial intelligence. *Unit head:* Dr. George Markowsky, Chair, 207-581-3941, Fax: 207-581-4977. *Application contact:* Scott G. Delcourt, Director of the Graduate School, 207-581-3218, Fax: 207-581-3232, E-mail: graduate@maine.edu.

University of Manitoba, Faculty of Graduate Studies, Faculty of Science, Department of Computer Science, Winnipeg, MB R3T 2N2, Canada. Offers M Sc, PhD. *Degree requirements:* For master's, thesis or alternative required; for doctorate, dissertation required. *Unit head:* P. R. King, Head.

University of Maryland, Baltimore County, Graduate School, College of Engineering, Department of Computer Science and Electrical Engineering, Program in Computer Science, Baltimore, MD 21250-5398. Offers MS, PhD. *Faculty:* 16 full-time (1 woman), 33 part-time (3 women). *Students:* 66 full-time (17 women), 46 part-time (13 women); includes 13 minority (5 African Americans, 7 Asian Americans or Pacific Islanders, 1 Hispanic American), 49 international. 254 applicants, 29% accepted. In 1998, 17 master's, 4 doctorates awarded. *Entrance requirements:* For master's and doctorate, GRE General Test, GRE Subject Test, minimum GPA of 3.2. *Application deadline:* For fall admission, 7/1. Applications are processed on a rolling basis. Application fee: $45. *Financial aid:* Fellowships, research assistantships, teaching assistantships available. *Faculty research:* Artificial intelligence, quantum computation, computer and communication security, electronic commerce cryptology, computer graphics and animation. *Application contact:* Dr. Yun Peng, Director, 410-455-3788.

See in-depth description on page 871.

University of Maryland, College Park, Graduate School, College of Computer, Mathematical and Physical Sciences, Department of Computer Science, College Park, MD 20742-5045. Offers MS, PhD. *Faculty:* 54 full-time (7 women), 6 part-time (2 women). *Students:* 120 full-time (13 women), 94 part-time (17 women); includes 30 minority (8 African Americans, 16 Asian Americans or Pacific Islanders, 6 Hispanic Americans), 114 international. 475 applicants, 24% accepted. In 1998, 43 master's, 11 doctorates awarded. *Degree requirements:* For master's, thesis or alternative required, foreign language not required; for doctorate, dissertation required. *Entrance requirements:* For master's, GRE General Test, GRE Subject Test, minimum GPA of 3.0; for doctorate, GRE General Test, GRE Subject Test. *Application deadline:* Applications are processed on a rolling basis. Application fee: $50 ($70 for international students). Tuition, state resident: part-time $272 per credit hour. Tuition, nonresident: part-time $475 per credit hour. Required fees: $632; $379 per year. *Financial aid:* In 1998–99, 2 fellowships with full tuition reimbursements (averaging $10,969 per year), 64 research assistantships with tuition reimbursements (averaging $12,871 per year), 75 teaching assistantships with tuition reimbursements (averaging $11,857 per year) were awarded.; career-related internships or fieldwork, Federal Work-Study, grants, and scholarships also available. Aid available to part-time students. Financial aid applicants required to submit FAFSA. *Faculty research:* Artificial intelligence, computer applications, information processing. Total annual research expenditures: $3.9 million. *Unit head:* Dr. John Gannon, Chairman, 301-405-2662, Fax: 301-405-6707. *Application contact:* Trudy Lindsey, Director, Graduate Admission and Records, 301-405-4198, Fax: 301-314-9305, E-mail: grschool@deans.umd.edu.

University of Massachusetts Amherst, Graduate School, College of Natural Sciences and Mathematics, Department of Computer Science, Amherst, MA 01003. Offers MS, PhD. Part-time programs available. *Faculty:* 34 full-time (7 women). *Students:* 58 full-time (9 women), 92 part-time (18 women); includes 3 minority (2 Asian Americans or Pacific Islanders, 1 Hispanic American), 85 international. Average age 28. 705 applicants, 11% accepted. In 1998, 25 master's, 8 doctorates awarded. Terminal master's awarded for partial completion of doctoral program. *Degree requirements:* For master's, foreign language and thesis not required; for doctorate, dissertation required, foreign language not required. *Entrance requirements:* For master's and doctorate, GRE General Test, TOEFL, TWE. *Application deadline:* For fall admission, 1/15 (priority date); for spring admission, 10/1. Applications are processed on a rolling basis. Application fee: $40. Tuition, state resident: full-time $2,640; part-time $165 per credit. Tuition, nonresident: full-time $9,756; part-time $407 per credit. Required fees: $1,221 per term. One-time fee: $110. Full-time tuition and fees vary according to course load, campus/location and reciprocity agreements. *Financial aid:* In 1998–99, 8 fellowships with full tuition reimbursements (averaging $6,933 per year), 127 research assistantships with full tuition reimbursements (averaging $10,946 per year), 47 teaching assistantships with full tuition reimbursements (averaging $6,357 per year) were awarded.; career-related internships or fieldwork, Federal Work-Study, grants, scholarships, traineeships, and unspecified assistantships also available. Aid available to part-time students. Financial aid application deadline: 1/15. *Faculty research:* Artificial intelligence, systems, theory, robotics. *Unit head:* Dr. Roderic Grupen, Director, 413-545-3640, Fax: 413-545-1249, E-mail: csinfo@cs.umass.edu. *Application contact:* Chair, Admissions Committee, 413-545-3640, E-mail: gradinfo@dpc.umassp.edu.

Announcement: The Computer Science Department has research programs in architecture, artificial intelligence, compilers, databases, distributed systems, education/tutoring, empirical methods, graph theory and combinatorics, information retrieval, machine learning, multimedia systems, networks, operating systems, parallel computation, performance evaluation, programming languages, robotics and computer vision, software engineering, and theoretical computer science.

See in-depth description on page 875.

University of Massachusetts Boston, Graduate Studies, College of Arts and Sciences, Faculty of Sciences, Program in Computer Science, Boston, MA 02125-3393. Offers MS, PhD. *Degree requirements:* For master's, comprehensive exams required, thesis optional, foreign language not required; for doctorate, dissertation, comprehensive exams required, foreign language not required. *Entrance requirements:* For master's and doctorate, GRE General Test, minimum GPA of 2.75.

University of Massachusetts Dartmouth, Graduate School, College of Engineering, Program in Computer Science, North Dartmouth, MA 02747-2300. Offers MS. Part-time programs available. *Faculty:* 7 full-time (0 women), 1 (woman) part-time. *Students:* 49 full-time (13 women), 31 part-time (2 women); includes 4 minority (2 African Americans, 1 Asian American or Pacific Islander, 1 Hispanic American), 54 international. Average age 32. 50 applicants, 98% accepted. In 1998, 11 degrees awarded. *Degree requirements:* For master's, thesis or alternative required, foreign language not required. *Entrance requirements:* For master's, GRE General Test, GRE Subject Test, TOEFL. *Application deadline:* For fall admission, 4/20 (priority date); for spring admission, 11/15 (priority date). Applications are processed on a rolling basis. Application fee: $40 for international students. Tuition, area resident: Full-time $3,107; part-time $129 per credit. Tuition, state resident: full-time $2,071; part-time $86 per credit. Tuition, nonresident: full-time $7,845; part-time $327 per credit. Required fees: $2,888. Full-time tuition and fees vary according to program and reciprocity agreements. Part-time tuition and fees vary according to course load and reciprocity agreements. *Financial aid:* In 1998–99, 6 teaching assistantships (averaging $4,667 per year) were awarded.; research assistantships, Federal Work-Study and unspecified assistantships also available. Aid available to part-time students. Financial aid application deadline: 3/15; financial aid applicants required to submit FAFSA. *Faculty research:* Learning software design, computer architecture, parallel architecture. Total annual research expenditures: $55,000. *Unit head:* Dr. Edmund Staples, Director, 508-999-8294, Fax: 508-999-9144, E-mail: estaples@umassd.edu. *Application contact:* Carol A. Novo, Graduate Admissions Office, 508-999-8026, Fax: 508-999-8183, E-mail: graduate@umassd.edu.

University of Massachusetts Lowell, Graduate School, College of Arts and Sciences, Department of Computer Science, Lowell, MA 01854-2881. Offers MS, PhD, Sc D. Part-time programs available. *Faculty:* 18 full-time (0 women), 4 part-time (0 women). *Students:* 61 full-time (18 women), 174 part-time (48 women); includes 46 minority (1 African American, 37 Asian Americans or Pacific Islanders, 6 Hispanic Americans, 2 Native Americans), 108 international. 122 applicants, 71% accepted. In 1998, 50 master's, 8 doctorates awarded. *Degree requirements:* For master's, thesis optional, foreign language not required; for doctorate, computer language, dissertation required. *Entrance requirements:* For master's and doctorate, GRE General Test. *Application deadline:* For fall admission, 4/1 (priority date); for spring admission, 10/1. Applications are processed on a rolling basis. Application fee: $20 ($35 for international students). *Financial aid:* In 1998–99, 2 research assistantships, 12 teaching assistantships were awarded.; fellowships, career-related internships or fieldwork and Federal Work-Study also available. Financial aid application deadline: 4/1. *Faculty research:* Networks, multimedia systems, human-computer interaction, graphics and visualization databases. *Unit head:* Dr. Tom Costello, Chair, 978-934-3633. *Application contact:* Dr. Stuart Smith, Coordinator, 978-934-3616, E-mail: stuart_smith@woods.uml.edu.

The University of Memphis, Graduate School, College of Arts and Sciences, Department of Mathematical Sciences, Memphis, TN 38152. Offers applied mathematics (MS); applied statistics (PhD); computer science (PhD); computer sciences (MS); mathematics (MS, PhD); statistics (MS, PhD). *Faculty:* 35 full-time (3 women), 2 part-time (1 woman). *Students:* 73 full-time (23 women), 37 part-time (9 women); includes 10 minority (5 African Americans, 5 Asian Americans or Pacific Islanders), 77 international. Terminal master's awarded for partial completion of doctoral program. *Degree requirements:* For master's, comprehensive exams required, thesis not required; for doctorate, dissertation, oral exams required. *Entrance requirements:* For master's, GRE General Test, MAT, TOEFL (minimum score of 550 required), minimum GPA of 2.5; for doctorate, GRE General Test, TOEFL (minimum score of 550 required). *Application deadline:* For fall admission, 8/1; for spring admission, 12/1. Applications are processed on a rolling basis. Application fee: $25 ($50 for international students). Tuition, state resident: full-time $3,410; part-time $178 per credit hour. Tuition, nonresident: full-time $8,670; part-time $408 per credit hour. Tuition and fees vary according to program. *Unit head:* Dr. Jerome A. Goldstein, Chairman, 901-678-2482, Fax: 901-678-2480, E-mail: goldstej@msci.memphis.edu. *Application contact:* Dr. Fernanda M. Botelho, Coordinator of Graduate Studies, 901-678-2482, Fax: 901-678-2480, E-mail: lisa@msci.memphis.edu.

University of Miami, Graduate School, College of Arts and Sciences, Department of Mathematics and Computer Science, Coral Gables, FL 33124. Offers computer science (MS); mathematics (MS, DA, PhD). Part-time and evening/weekend programs available. *Faculty:* 27. *Students:* 25 full-time (10 women), 18 part-time (4 women); includes 13 minority (3 African Americans, 1 Asian American or Pacific Islander, 9 Hispanic Americans), 14 international. Average age 30. 71 applicants, 75% accepted. In 1998, 7 master's awarded (100% found work related to degree); 1 doctorate awarded (100% entered university research/teaching). Terminal master's awarded for partial completion of doctoral program. *Degree requirements:* For master's, comprehensive exam or project required, foreign language and thesis not required; for doctorate, one foreign language, dissertation, qualifying exams required. *Entrance requirements:* For master's and doctorate, GRE General Test (minimum combined score of 1000 required), TOEFL (minimum score of 550 required), minimum GPA of 3.0. *Average time to degree:* Master's–2 years full-time, 4 years part-time; doctorate–8 years full-time. *Application deadline:* For fall admission, 7/1. Applications are processed on a rolling basis. Application fee: $35. Tuition: Full-time $15,336; part-time $852 per credit. Required fees: $174. Tuition and fees vary according to program. *Financial aid:* In 1998–99, 27 students received aid, including 1 fellowship with tuition reimbursement available, 25 teaching assistantships with tuition reimbursements available; career-related internships or fieldwork and institutionally-sponsored loans also available. Aid available to part-time students. Financial aid application deadline: 3/1. *Unit head:* Dr. Alan Zame, Chairman, 305-284-2348. *Application contact:* Dr. Marvin Mielke, Graduate Adviser, 305-284-2348.

University of Michigan, Horace H. Rackham School of Graduate Studies, College of Engineering, Department of Electrical Engineering and Computer Science, Division of Computer Science and Engineering, Ann Arbor, MI 48109. Offers MS, MSE, PhD. Terminal master's awarded for partial completion of doctoral program. *Degree requirements:* For master's, thesis not required; for doctorate, dissertation, oral defense of dissertation, preliminary exams required. *Entrance requirements:* For master's, GRE General Test (minimum combined score of 1900 on three sections required; average 2062); for doctorate, GRE General Test (minimum combined score of 1900 on three sections required; average 2062), master's degree.

University of Michigan, Horace H. Rackham School of Graduate Studies, School of Information, Ann Arbor, MI 48109. Offers archives and records management (MS); human-computer interaction (MS); information (PhD); information economics, management and policy (MS); library and information services (MS). *Accreditation:* ALA (one or more programs are accredited). Part-time programs available. *Degree requirements:* For master's, thesis not required; for doctorate, dissertation, oral defense of dissertation, preliminary exam required. *Entrance requirements:* For master's and doctorate, GRE General Test.

University of Michigan–Dearborn, College of Engineering and Computer Science, Department of Computer and Information Science, Dearborn, MI 48128-1491. Offers computer and information science (MS); software engineering (MS). Part-time and evening/weekend programs available. *Faculty:* 11 full-time (1 woman), 2 part-time (0 women). *Students:* 7 full-time (0 women), 111 part-time (24 women); includes 23 minority (4 African Americans, 17 Asian Americans or Pacific Islanders, 2 Hispanic Americans), 7 international. Average age 29. In 1998, 13 degrees awarded. *Degree requirements:* For master's, computer language required, thesis optional, foreign language not required. *Entrance requirements:* For master's, bachelor's degree in mathematics, computer science, or engineering. *Application deadline:* For fall admission, 6/15; for spring admission, 2/15. Applications are processed on a rolling basis. Application fee: $55. Electronic applications accepted. Tuition, state resident: part-time $259 per credit hour. Tuition, nonresident: part-time $748 per credit hour. Required fees: $80 per course. Tuition and fees vary according to course level, course load and program. *Financial aid:* In 1998–99, 1 research assistantship, 1 teaching assistantship were awarded. *Faculty research:* Information systems, geometric modelling, networks, databases. *Unit head:* Dr. Kenneth Modesitt, Chair, 313-593-5680, Fax: 313-593-9967, E-mail: modesitt@umdsun2.umd.umich.edu. *Application contact:* Mary Tamsen, Graduate Secretary, 313-436-9145, Fax: 313-593-9967, E-mail: mtamsen@umdsun2.umd.umich.edu.

University of Minnesota, Duluth, Graduate School, College of Science and Engineering, Department of Computer Science, Duluth, MN 55812-2496. Offers MS. Part-time programs available. *Faculty:* 8 full-time (2 women). *Students:* 16 full-time (2 women). Average age 27. 56 applicants, 59% accepted. In 1998, 7 degrees awarded (100% found work related to degree). *Degree requirements:* For master's, computer language, thesis required (for some programs), foreign language not required. *Entrance requirements:* For master's, GRE General Test (average 2050), TOEFL, minimum GPA of 3.0. *Average time to degree:* Master's–2 years full-time. *Application deadline:* For fall admission, 7/15; for spring admission, 11/15. Applications are processed on a rolling basis. Application fee: $50 ($55 for international students). *Financial aid:* In 1998–99, 12 students received aid, including 1 fellowship with partial tuition reimbursement available, 2 research assistantships with full and partial tuition reimbursements available (averaging $9,641 per year), 9 teaching assistantships with full and partial tuition reimbursements available (averaging $9,641 per year); Federal Work-Study, institutionally-sponsored loans, and tuition waivers (partial) also available. Financial aid application deadline: 3/15. *Faculty research:* Information retrieval, user-system interfaces, artificial intelligence, machine learning, parallel/distributed computing. *Unit head:* Dr. Carolyn J. Crouch, Director of Graduate Studies, 218-726-7607, Fax: 218-726-8240, E-mail: cs@d.umn.edu.

University of Minnesota, Twin Cities Campus, Graduate School, Institute of Technology, Department of Computer Science and Engineering, Minneapolis, MN 55455-0213. Offers computer and information sciences (MCIS, MS, PhD). Part-time programs available. *Faculty:* 25 full-time (2 women), 10 part-time (0 women). *Students:* 153 full-time (34 women), 76 part-time (12 women); includes 8 minority (1 African American, 6 Asian Americans or Pacific Islanders, 1 Hispanic American), 135 international. 500 applicants, 45% accepted. In 1998, 70 master's, 20 doctorates awarded. Terminal master's awarded for partial completion of doctoral program. *Degree requirements:* For master's, foreign language and thesis not required; for doctorate, dissertation required, foreign language not required. *Entrance requirements:* For master's and doctorate, GRE General Test. *Average time to degree:* Master's–2 years full-time, 4 years part-time; doctorate–5 years full-time, 8 years part-time. *Application deadline:* For fall admission, 5/31. Applications are processed on a rolling basis. Application fee: $50 ($55 for international students). *Financial aid:* In 1998–99, 1 fellowship with tuition reimbursement (averaging $12,000 per year), 95 research assistantships with tuition reimbursements (averaging $14,184 per year), 55 teaching assistantships with tuition reimbursements (averaging $14,184 per year) were awarded.; career-related internships or fieldwork, Federal Work-Study, and institutionally-sponsored loans also available. Financial aid application deadline:1/2. *Faculty research:* Software systems, numerical analysis, theory, artificial intelligence. Total annual research expenditures: $4.2 million. *Unit head:* Yousef Saad, Head, 612-625-0726, Fax: 612-625-0572, E-mail: saad@cs.umn.edu. *Application contact:* Haesun Park, Director of Graduate Studies, 612-625-4002, Fax: 612-625-0572.

See in-depth description on page 877.

University of Minnesota, Twin Cities Campus, Graduate School, Scientific Computation Program, Minneapolis, MN 55455-0213. Offers MS, PhD. Part-time programs available. *Faculty:* 35 full-time (3 women). *Students:* 6 full-time (2 women), 5 international. 10 applicants, 30% accepted. *Degree requirements:* For master's and doctorate, thesis/dissertation required, foreign language not required. *Entrance requirements:* For doctorate, GRE. *Application deadline:* For fall admission, 1/2 (priority date). Applications are processed on a rolling basis. Application fee: $50 ($55 for international students). *Financial aid:* In 1998–99, 6 students received aid; fellowships, research assistantships, teaching assistantships, career-related internships or fieldwork and Federal Work-Study available. *Faculty research:* Parallel computations, quantum mechanical dynamics, computational materials science, computational fluid dynamics. *Unit head:* Vipin Kumar, Director of Graduate Studies, 612-625-4002, E-mail: kumar@cs.umn.edu. *Application contact:* Georganne E. Tolaas, Graduate Secretary, 612-625-1592, Fax: 612-625-0572, E-mail: scic@cs.umn.edu.

See in-depth description on page 879.

University of Missouri–Columbia, Graduate School, College of Engineering, Department of Computer Engineering and Computer Science, Columbia, MO 65211. Offers MS, PhD. Part-time programs available. *Faculty:* 15 full-time (4 women). *Students:* 25 full-time (5 women), 34 part-time (7 women); includes 7 minority (1 African American, 3 Asian Americans or Pacific Islanders, 2 Hispanic Americans, 1 Native American), 33 international. 32 applicants, 47% accepted. In 1998, 22 master's, 3 doctorates awarded. *Degree requirements:* For doctorate, dissertation required. *Entrance requirements:* For master's, GRE General Test, minimum GPA of 3.0; for doctorate, GRE General Test, TOEFL. *Application deadline:* For fall admission, 7/1 (priority date). Applications are processed on a rolling basis. Application fee: $30 ($50 for international students). *Financial aid:* Research assistantships, teaching assistantships, institutionally-sponsored loans available. *Unit head:* Dr. Gordon Soringer, Director of Graduate Studies, 573-882-7422.

University of Missouri–Kansas City, Program in Computer Science Telecommunications, Kansas City, MO 64110-2499. Offers computer networking (MS, PhD); software engineering (MS); telecommunications networking (MS, PhD). PhD offered through the School of Graduate Studies. Part-time programs available. *Faculty:* 15 full-time (1 woman). *Students:* 48 full-time (20 women), 74 part-time (21 women); includes 8 minority (2 African Americans, 6 Asian Americans or Pacific Islanders), 85 international. Average age 29. In 1998, 44 degrees awarded. *Degree requirements:* For master's, computer language required, foreign

Computer Science

University of Missouri–Kansas City (continued)

language not required; for doctorate, computer language, dissertation required, foreign language not required. *Entrance requirements:* For master's, GRE General Test (score in 75th percentile or higher on quantitative section required, 50th percentile or higher on verbal), minimum GPA of 3.0; for doctorate, GRE General Test (score in 85th percentile or higher on quantitative section required, 50th percentile or higher on verbal), minimum GPA of 3.5. *Application deadline:* For fall admission, 3/1 (priority date); for spring admission, 10/1. Applications are processed on a rolling basis. Application fee: $25. *Financial aid:* In 1998–99, 15 research assistantships, 15 teaching assistantships were awarded.; career-related internships or fieldwork, Federal Work-Study, institutionally-sponsored loans, and tuition waivers (partial) also available. Aid available to part-time students. Financial aid application deadline: 3/1. *Faculty research:* Multimedia networking, distributed systems/databases, data/network security. *Unit head:* Dr. Richard Hetherington, Director, 816-235-1193, Fax: 816-235-5159, E-mail: info@cstp.umkc.edu.

University of Missouri–Rolla, Graduate School, College of Arts and Sciences, Department of Computer Science, Rolla, MO 65409-0910. Offers MS, PhD. Part-time programs available. *Faculty:* 10 full-time (0 women). *Students:* 46 full-time (9 women), 32 part-time (3 women); includes 1 minority (Asian American or Pacific Islander), 33 international. Average age 30. 178 applicants, 65% accepted. In 1998, 36 master's, 1 doctorate awarded. Terminal master's awarded for partial completion of doctoral program. *Degree requirements:* For master's, foreign language and thesis not required; for doctorate, dissertation, departmental qualifying exam required, foreign language not required. *Entrance requirements:* For master's, GRE General Test (minimum combined score of 1200 on quantitative and analytical sections required); for doctorate, GRE Subject Test. *Application deadline:* For fall admission, 7/1. Applications are processed on a rolling basis. Application fee: $25. Electronic applications accepted. *Financial aid:* In 1998–99, 8 research assistantships with partial tuition reimbursements (averaging $13,000 per year), 11 teaching assistantships with partial tuition reimbursements (averaging $13,000 per year) were awarded.; fellowships, institutionally-sponsored loans also available. *Faculty research:* Intelligent systems, software engineering, distributed systems, database systems, computer systems. Total annual research expenditures: $300,000. *Unit head:* Dr. George Zobrist, Chairman, 573-341-4491, Fax: 573-341-4501, E-mail: zobrist@umr.edu.

University of Missouri–St. Louis, Graduate School, College of Arts and Sciences, Department of Mathematical Sciences, St. Louis, MO 63121-4499. Offers applied mathematics (MA, PhD); computer science (MS). Part-time and evening/weekend programs available. *Faculty:* 20. *Students:* 1 (woman) full-time, 21 part-time (11 women); includes 4 minority (3 African Americans, 1 Asian American or Pacific Islander), 1 international. *Degree requirements:* For master's, thesis optional, foreign language not required; for doctorate, dissertation required, foreign language not required. *Entrance requirements:* For master's, GRE if no BS in computer science; for doctorate, GRE General Test. *Application deadline:* For fall admission, 5/1 (priority date); for spring admission, 12/1. Applications are processed on a rolling basis. Application fee: $25 ($40 for international students). Electronic applications accepted. *Unit head:* Dr. Grant Welland, Director of Graduate Studies, 314-516-5741, Fax: 314-516-5400, E-mail: welland@eads.umsl.edu. *Application contact:* Graduate Admissions, 314-516-5458, Fax: 314-516-6759, E-mail: gradadm@umsl.edu.

The University of Montana–Missoula, Graduate School, College of Arts and Sciences, Department of Computer Science, Missoula, MT 59812-0002. Offers MS. Part-time programs available. *Faculty:* 5 full-time (0 women), 1 part-time (0 women). *Students:* 9 full-time (0 women), 10 part-time (5 women); includes 9 minority (all Asian Americans or Pacific Islanders) Average age 33. 3 applicants, 0% accepted. In 1998, 8 degrees awarded. *Degree requirements:* For master's, computer language, project or thesis required. *Entrance requirements:* For master's, GRE General Test, TOEFL, bachelor's degree in technical field. *Application deadline:* For fall admission, 3/15 (priority date). Applications are processed on a rolling basis. Application fee: $45. *Financial aid:* Research assistantships, teaching assistantships, Federal Work-Study available. Financial aid application deadline: 3/1. *Faculty research:* Parallel and distributed systems, neural networks, genetic algorithms, machine learning, data visualization, artificial intelligence. *Unit head:* Dr. Jerry Esmay, Chair, 406-243-2883. *Application contact:* Kathy Lockridge, Graduate Secretary, 406-243-2883.

University of Nebraska at Omaha, Graduate Studies and Research, College of Information Science and Technology, Department of Computer Science, Omaha, NE 68182. Offers MA, MS. Part-time programs available. *Faculty:* 8 full-time (1 woman), 1 part-time (0 women). *Students:* 24 full-time (10 women), 61 part-time (13 women); includes 22 minority (2 African Americans, 18 Asian Americans or Pacific Islanders, 2 Hispanic Americans) Average age 34. 49 applicants, 55% accepted. In 1998, 5 degrees awarded. *Degree requirements:* For master's, thesis (for some programs), comprehensive exams required, foreign language not required. *Entrance requirements:* For master's, GRE General Test, minimum GPA of 3.0, previous course work in computer science. *Application deadline:* For fall admission, 7/1; for spring admission, 12/1. Applications are processed on a rolling basis. Application fee: $35. Tuition, state resident: part-time $100 per credit hour. Tuition, nonresident: part-time $239 per credit hour. Required fees: $12 per credit hour. $91 per semester. Tuition and fees vary according to course load. *Financial aid:* In 1998–99, 17 students received aid; research assistantships, teaching assistantships, institutionally-sponsored loans and tuition waivers (full) available. Aid available to part-time students. Financial aid application deadline: 3/1; financial aid applicants required to submit FAFSA. *Unit head:* Dr. Peter Ng, Chairperson, 402-554-2834.

University of Nebraska–Lincoln, Graduate College, College of Arts and Sciences, Department of Computer Science and Engineering, Lincoln, NE 68588. Offers MS, PhD. *Faculty:* 13 full-time (1 woman), 3 part-time (0 women). *Students:* 72 full-time (15 women), 28 part-time (8 women); includes 7 minority (1 African American, 6 Asian Americans or Pacific Islanders), 71 international. Average age 31. 202 applicants, 39% accepted. In 1998, 24 master's, 1 doctorate awarded. *Degree requirements:* For master's, thesis optional, foreign language not required; for doctorate, dissertation, comprehensive exams required. *Entrance requirements:* For master's and doctorate, GRE General Test, TOEFL (minimum score of 550 required). *Average time to degree:* Doctorate–6.9 years full-time. *Application deadline:* For fall admission, 3/1; for spring admission, 10/1. Application fee: $35. Electronic applications accepted. *Financial aid:* In 1998–99, 2 fellowships, 25 research assistantships, 22 teaching assistantships were awarded.; Federal Work-Study also available. Aid available to part-time students. Financial aid application deadline: 1/15. *Faculty research:* Communications theory and foundations, distributed systems, human-centered software design, visual information processing, enterprise software engineering and information systems. *Unit head:* Dr. Stephen Reichenbach, Chair, 402-472-2401, Fax: 402-472-7767. *Application contact:* Dr. Ashok Samal, Graduate Committee Chair, E-mail: samal@cse.unl.edu.

Announcement: The J. D. Edwards Honors Program in Computer Science and Management at the University of Nebraska–Lincoln offers two innovative master's programs. The M Eng in software engineering combines 48 credit hours of computer science and business courses with studio design projects. The MBA in Information and software systems combines 48 credit hours of business and computer science courses. Each program produces well-rounded, highly-skilled graduates who understand the software development process for solving business problems. Students work in groups with faculty members on focused projects. Students selected to participate receive a $20,000-per-year assistantship with tuition paid.

See in-depth description on page 881.

University of Nevada, Las Vegas, Graduate College, Howard R. Hughes College of Engineering, Department of Computer Science, Las Vegas, NV 89154-9900. Offers MS, PhD. Part-time programs available. *Faculty:* 10 full-time (0 women). *Students:* 13 full-time (5 women), 13 part-time (2 women); includes 2 minority (1 Asian American or Pacific Islander, 1 Hispanic American), 12 international. 43 applicants, 26% accepted. In 1998, 4 master's awarded.

Degree requirements: For master's, comprehensive exam required, thesis optional, foreign language not required; for doctorate, dissertation required. *Entrance requirements:* For master's, GRE General Test, GRE Subject Test, minimum GPA of 3.0; for doctorate, minimum GPA of 3.5. *Application deadline:* For fall admission, 6/15 (priority date); for spring admission, 11/15. Applications are processed on a rolling basis. Application fee: $40 ($95 for international students). *Financial aid:* In 1998–99, 12 teaching assistantships with partial tuition reimbursements (averaging $8,815 per year) were awarded.; research assistantships Financial aid application deadline: 3/1. *Unit head:* Dr. Walter Vodrazka, Interim Chair, 702-895-3681. *Application contact:* Graduate College Admissions Evaluator, 702-895-3320.

University of Nevada, Reno, Graduate School, College of Engineering, Program in Computer Science, Reno, NV 89557. Offers MS. *Degree requirements:* For master's, thesis optional, foreign language not required. *Entrance requirements:* For master's, GRE, TOEFL (minimum score of 500 required), minimum GPA of 2.75.

University of New Brunswick, School of Graduate Studies, Faculty of Engineering, School of Computer Science, Fredericton, NB E3B 5A3, Canada. Offers M Sc CS, PhD. Part-time programs available. *Degree requirements:* For master's, thesis required, foreign language not required; for doctorate, dissertation, qualifying exam required, foreign language not required. *Entrance requirements:* For master's, TOEFL, TWE, minimum GPA of 3.0; for doctorate, TOEFL, TWE.

University of New Hampshire, Graduate School, College of Engineering and Physical Sciences, Department of Computer Science, Durham, NH 03824. Offers MS, PhD. Part-time and evening/weekend programs available. *Faculty:* 12 full-time. *Students:* 16 full-time (4 women), 36 part-time (6 women), 19 international. Average age 33. 31 applicants, 81% accepted. In 1998, 11 master's, 2 doctorates awarded. *Degree requirements:* For master's, computer language, thesis or alternative required, foreign language not required; for doctorate, computer language, dissertation required, foreign language not required. *Entrance requirements:* For master's and doctorate, GRE General Test, GRE Subject Test. *Application deadline:* For fall admission, 4/1 (priority date). Applications are processed on a rolling basis. Application fee: $50. Tuition, area resident: Full-time $5,750; part-time $319 per credit. Tuition, state resident: full-time $8,625. Tuition, nonresident: full-time $14,640; part-time $598 per credit. Required fees: $224 per semester. Tuition and fees vary according to course load, degree level and program. *Financial aid:* In 1998–99, 4 research assistantships, 13 teaching assistantships were awarded.; fellowships, career-related internships or fieldwork, Federal Work-Study, scholarships, and tuition waivers (full and partial) also available. Aid available to part-time students. Financial aid application deadline: 2/15. *Faculty research:* Programming languages, compiler design, parallel algorithms, computer graphics, artificial intelligence. *Unit head:* Philip Hatcher, Chairperson, 603-862-2678. *Application contact:* Pilar de la Torre, Graduate Coordinator, 603-862-2682.

University of New Haven, Graduate School, School of Engineering and Applied Science, Program in Computer and Information Science, West Haven, CT 06516-1916. Offers applications software (MS); management information systems (MS); systems software (MS). Part-time and evening/weekend programs available. *Students:* 40 full-time (17 women), 157 part-time (41 women); includes 18 minority (4 African Americans, 14 Asian Americans or Pacific Islanders), 68 international. 76 applicants, 66% accepted. In 1998, 62 degrees awarded. *Degree requirements:* For master's, thesis or alternative required, foreign language not required. *Application deadline:* Applications are processed on a rolling basis. Application fee: $50. *Financial aid:* Federal Work-Study available. Aid available to part-time students. Financial aid application deadline: 5/1; financial aid applicants required to submit FAFSA. *Unit head:* Dr. Tahany Fergany, Coordinator, 203-932-7067.

University of New Mexico, Graduate School, School of Engineering, Department of Computer Science, Albuquerque, NM 87131-2039. Offers MS, PhD. Part-time programs available. *Faculty:* 21 full-time (2 women), 8 part-time (0 women). *Students:* 42 full-time (8 women), 50 part-time (9 women); includes 6 minority (4 Asian Americans or Pacific Islanders, 2 Hispanic Americans), 24 international. Average age 33. 138 applicants, 26% accepted. In 1998, 22 master's, 1 doctorate awarded. *Degree requirements:* For master's, computer language required, foreign language not required; for doctorate, computer language, dissertation required, foreign language not required. *Entrance requirements:* For master's and doctorate, GRE General Test, minimum GPA of 3.0. *Application deadline:* For fall admission, 7/15; for spring admission, 11/14. Applications are processed on a rolling basis. Application fee: $25. *Financial aid:* In 1998–99, 60 students received aid, including 9 fellowships (averaging $788 per year), 40 research assistantships with tuition reimbursements available (averaging $4,137 per year), 17 teaching assistantships with tuition reimbursements available (averaging $12,124 per year); career-related internships or fieldwork and Federal Work-Study also available. Financial aid application deadline: 4/1. *Faculty research:* Artificial life, genetic algorithms, database systems, complexity theory, interactive computer graphics. Total annual research expenditures: $1.9 million. *Unit head:* Dr. Deepak Kapur, Chair, 505-277-3112, Fax: 505-277-6927, E-mail: kapur@cs.unm.edu.

University of New Orleans, Graduate School, College of Sciences, Department of Computer Science, New Orleans, LA 70148. Offers MS. *Students:* 24 full-time (5 women), 29 part-time (6 women); includes 7 minority (6 Asian Americans or Pacific Islanders, 1 Hispanic American), 29 international. Average age 31. 99 applicants, 56% accepted. In 1998, 16 degrees awarded. *Entrance requirements:* For master's, GRE General Test. *Application deadline:* For fall admission, 7/1 (priority date). Applications are processed on a rolling basis. Application fee: $20. Tuition, state resident: full-time $2,362. Tuition, nonresident: full-time $7,888. Part-time tuition and fees vary according to course load. *Unit head:* Dr. Mahdi Abdelguerfi, Chairman, 504-280-7076, Fax: 504-280-7228, E-mail: macs@uno.edu. *Application contact:* Dr. Mahdi Abdelguerfi, Chairman, 504-280-7076, Fax: 504-280-7228, E-mail: macs@uno.edu.

The University of North Carolina at Chapel Hill, Graduate School, College of Arts and Sciences, Department of Computer Science, Chapel Hill, NC 27599. Offers MS, PhD. *Degree requirements:* For master's, comprehensive exam required; for doctorate, dissertation, comprehensive exam required, foreign language not required. *Entrance requirements:* For master's and doctorate, GRE General Test, minimum GPA of 3.0.

See in-depth description on page 883.

University of North Carolina at Charlotte, Graduate School, The William States Lee College of Engineering, Department of Computer Science, Charlotte, NC 28223-0001. Offers MS. *Faculty:* 23 full-time (2 women). *Students:* 28 full-time (14 women), 59 part-time (15 women); includes 22 minority (3 African Americans, 18 Asian Americans or Pacific Islanders, 1 Hispanic American), 34 international. Average age 30. 76 applicants, 89% accepted. In 1998, 44 degrees awarded. *Degree requirements:* For master's, thesis optional, foreign language not required. *Entrance requirements:* For master's, GRE General Test, minimum GPA of 3.0 during previous 2 years, 2.8 overall. *Application deadline:* For fall admission, 7/15; for spring admission, 11/15. Applications are processed on a rolling basis. Application fee: $35. Electronic applications accepted. *Financial aid:* In 1998–99, 16 research assistantships, 22 teaching assistantships were awarded. Financial aid application deadline: 4/1. *Faculty research:* Computer programming, data retrieval and processing, robotics. *Unit head:* Dr. Mirsad Hadzikadic, Chair, 704-547-4880, Fax: 704-547-3516, E-mail: mirsad@email.uncc.edu. *Application contact:* Kathy Barringer, Assistant Director of Graduate Admissions, 704-547-3366, Fax: 704-547-3279, E-mail: gradadm@email.uncc.edu.

University of North Dakota, Graduate School, Center for Aerospace Studies, Department of Computer Science, Grand Forks, ND 58202. Offers MS. Part-time programs available. *Faculty:* 8 full-time (1 woman). *Students:* 15 full-time (5 women), 2 part-time (both women). 15 applicants, 80% accepted. In 1998, 4 degrees awarded. *Degree requirements:* For master's, comprehensive exam required, thesis optional, foreign language not required. *Entrance requirements:* For master's, GRE General Test, TOEFL (minimum score of 550 required), minimum GPA of 3.0.

Application deadline: For fall admission, 3/1 (priority date). Applications are processed on a rolling basis. Application fee: $20. *Financial aid:* In 1998–99, 12 students received aid, including 3 research assistantships, 9 teaching assistantships; fellowships, Federal Work-Study, institutionally-sponsored loans, and tuition waivers (full and partial) also available. Financial aid application deadline: 3/15. *Faculty research:* Operating systems, simulation, parallelcomputation, hypermedia, graph theory. *Unit head:* Dr. Brajendra Panda, Director, 701-777-4107, Fax: 701-777-3330, E-mail: panda@cs.und.edu.

University of Northern Iowa, Graduate College, College of Natural Sciences, Department of Computer Science, Cedar Falls, IA 50614. Offers MS. *Students:* 13 full-time (9 women), 7 part-time (2 women); includes 2 minority (1 African American, 1 Asian American or Pacific Islander), 4 international. Average age 33. 19 applicants, 63% accepted. In 1998, 1 degree awarded. *Degree requirements:* For master's, thesis or alternative required, foreign language not required. *Application deadline:* For fall admission, 8/1 (priority date). Applications are processed on a rolling basis. Application fee: $20 ($30 for international students). Tuition, state resident: full-time $3,308; part-time $184 per hour. Tuition, nonresident: full-time $8,156; part-time $454 per hour. Required fees: $202; $88 per semester. Tuition and fees vary according to course load. *Financial aid:* Application deadline: 3/1. *Unit head:* Dr. John McCormick, Head, 319-273-2618, Fax: 319-273-7123, E-mail: mccormick@cs.uni.edu.

University of North Florida, College of Arts and Sciences, Department of Mathematics and Statistics, Jacksonville, FL 32224-2645. Offers computer science (MS); mathematical sciences (MS); statistics (MS). Part-time and evening/weekend programs available. *Faculty:* 17 full-time (4 women). *Students:* 7 full-time (3 women), 11 part-time (6 women); includes 2 minority (1 African American, 1 Hispanic American), 3 international. *Degree requirements:* For master's, comprehensive exam required, thesis optional, foreign language not required. *Entrance requirements:* For master's, GRE Subject Test, TOEFL (minimum score of 500 required), GRe General Test (minimum combined score of 1000 required) or minimum GPA of 3.0 in last 60 hours. *Application deadline:* For fall admission, 12/31 (priority date). Applications are processed on a rolling basis. Application fee: $20. Electronic applications accepted. *Unit head:* Dr. William Caldwell, Chair, 904-620-2653, E-mail: wcaldwell@unf.edu. *Application contact:* Dr. Leonard Lipkin, Coordinator, 904-620-2468, E-mail: llipkin@unf.edu.

University of North Florida, College of Computer Sciences and Engineering, Jacksonville, FL 32224-2645. Offers computer and information sciences (MS). Part-time programs available. *Faculty:* 15 full-time (3 women). *Students:* 8 full-time (5 women), 44 part-time (18 women); includes 11 minority (all Asian Americans or Pacific Islanders), 5 international. Average age 33. 15 applicants, 73% accepted. In 1998, 11 degrees awarded. *Degree requirements:* For master's, computer language, comprehensive exam required, thesis optional, foreign language not required. *Entrance requirements:* For master's, GRE General Test (minimum combined score of 1000 required), minimum GPA of 3.0 in last 60 hours. *Application deadline:* For fall admission, 12/31 (priority date). Applications are processed on a rolling basis. Application fee: $20. Electronic applications accepted. *Financial aid:* In 1998–99, 11 students received aid, including 1 teaching assistantship (averaging $1,106 per year); Federal Work-Study and tuition waivers (partial) also available. Aid available to part-time students. Financial aid application deadline: 4/1; financial aid applicants required to submit FAFSA. *Faculty research:* Parallel processing, software engineering, artificial intelligence, human factors, human-machine interfacing. *Unit head:* Dr. Neal Coulter, Dean, 904-620-2985, E-mail: ncoulter@unf.edu. *Application contact:* Dr. Krissten N. Cooper, Director of Graduate Studies, 904-620-2985, E-mail: kcooper@unf.edu.

University of North Texas, Robert B. Toulouse School of Graduate Studies, College of Arts and Sciences, Department of Computer Sciences, Denton, TX 76203. Offers MA, MS, PhD. *Faculty:* 14 full-time (1 woman). *Students:* 111 full-time (22 women), 57 part-time (13 women); includes 13 minority (4 African Americans, 8 Asian Americans or Pacific Islanders, 1 Hispanic American), 114 international. In 1998, 45 master's, 2 doctorates awarded. Terminal master's awarded for partial completion of doctoral program. *Degree requirements:* For master's, thesis (for some programs), comprehensive exam required, foreign language not required; for doctorate, dissertation, comprehensive exam required, foreign language not required. *Entrance requirements:* For master's, GRE General Test (minimum score of 650 on quantitative section, 1050 combined required), minimum undergraduate GPA of 3.0; for doctorate, GRE General Test (minimum score of 700 on quantitative section, 1150 combined required), minimum GPA of 3.5. *Application deadline:* For fall admission, 7/17. Application fee: $25 ($50 for international students). *Financial aid:* Fellowships, research assistantships, teaching assistantships, career-related internships or fieldwork, Federal Work-Study, and institutionally-sponsored loans available. Financial aid application deadline: 4/1. *Faculty research:* Parallel algorithms, artificial intelligence, operating systems, software engineering, databases. *Unit head:* Dr. Roy T. Jacob, Chair, 940-565-2767, Fax: 940-565-2799, E-mail: jacob@cs.unt.edu. *Application contact:* Dr. Steve R. Tate, Graduate Adviser, 940-565-2767, Fax: 940-565-2799, E-mail: srt@cs.unt.edu.

See in-depth description on page 887.

University of Notre Dame, Graduate School, College of Engineering, Department of Computer Science and Engineering, Notre Dame, IN 46556. Offers MS, PhD. Part-time programs available. *Faculty:* 13 full-time (1 woman), 2 part-time (0 women). *Students:* 35 full-time (9 women), 5 part-time (1 woman); includes 3 minority (2 African Americans, 1 Native American), 23 international. 143 applicants, 15% accepted. In 1998, 13 master's, 5 doctorates awarded (40% entered university research/teaching, 60% found other work related to degree). Terminal master's awarded for partial completion of doctoral program. *Degree requirements:* For master's and doctorate, computer language, thesis/dissertation required, foreign language not required. *Entrance requirements:* For master's and doctorate, GRE General Test, TOEFL (minimum score of 600 required; 250 for computer-based). *Average time to degree:* Doctorate–6 years full-time. *Application deadline:* For fall admission, 2/1 (priority date); for spring admission, 10/1. Applications are processed on a rolling basis. Application fee: $40. *Financial aid:* In 1998–99, 35 students received aid, including 6 fellowships with full tuition reimbursements available (averaging $16,000 per year), 13 research assistantships with full tuition reimbursements available (averaging $11,500 per year), 13 teaching assistantships with full tuition reimbursements available (averaging $11,500 per year); tuition waivers (full) also available. Financial aid application deadline: 2/1. *Faculty research:* Parallel architectures, VLSI design, VLSI CAD, parallel and distributed computing, operating systems. Total annual research expenditures: $1.3 million. *Unit head:* Dr. Edwin H.-M. Sha, Chair, 219-631-8320, Fax: 219-631-9260, E-mail: cse@cse.nd.edu. *Application contact:* Dr. Terrence J. Akai, Director of Graduate Admissions, 219-631-7706, Fax: 219-631-4183, E-mail: gradad@nd.edu.

See in-depth description on page 889.

University of Oklahoma, Graduate College, College of Engineering, School of Computer Science, Norman, OK 73019-0390. Offers MS, PhD. Part-time programs available. *Faculty:* 9 full-time (2 women). *Students:* 52 full-time (9 women), 44 part-time (10 women); includes 4 minority (1 African American, 2 Asian Americans or Pacific Islanders, 1 Native American), 71 international. Average age 30. 80 applicants, 80% accepted. In 1998, 27 master's, 3 doctorates awarded. *Degree requirements:* For master's, oral exams, qualifying exam required, thesis optional, foreign language not required; for doctorate, dissertation, general exam, qualifying exam required, foreign language not required. *Entrance requirements:* For master's and doctorate, GRE General Test (minimum combined score of 1150 required), TOEFL (minimum score of 550 required). *Application deadline:* For fall admission, 4/1 (priority date); for spring admission, 9/1. Applications are processed on a rolling basis. Application fee: $25. Tuition, state resident: part-time $86 per credit hour. Tuition, nonresident: part-time $275 per credit hour. Tuition and fees vary according to course level, course load and program. *Financial aid:* In 1998–99, 16 research assistantships, 11 teaching assistantships were awarded.; fellowships, tuition waivers (partial) also available. Financial aid application deadline: 4/15. *Faculty research:* Artificial intelligence, database, parallel processing and distributed computation, computer architecture. *Unit head:* Sudarshan K. Dhall, Interim Director, 405-325-4397, Fax: 405-325-4044. *Application contact:* Dr. Sridhar Rahhakrishan, Graduate Liaison, 405-325-4042.

University of Oregon, Graduate School, College of Arts and Sciences, Department of Computer and Information Science, Eugene, OR 97403. Offers MA, MS, PhD. Part-time programs available. *Faculty:* 17 full-time (4 women), 5 part-time (1 woman). *Students:* 63 full-time (18 women), 8 part-time; includes 5 minority (all Asian Americans or Pacific Islanders), 32 international. 121 applicants, 43% accepted. In 1998, 14 master's awarded (100% found work related to degree); 1 doctorate awarded (100% entered university research/teaching). Terminal master's awarded for partial completion of doctoral program. *Degree requirements:* For master's, computer language required, foreign language and thesis not required; for doctorate, computer language, dissertation required, foreign language not required. *Entrance requirements:* For master's and doctorate, GRE General Test (minimum score of 480 on verbal section, 640 on quantitative, 600 on analytical required), TOEFL, TSE (for teaching assistants). *Average time to degree:* Master's–2 years full-time; doctorate–7 years full-time. *Application deadline:* For fall admission, 2/1 (priority date). Application fee: $50. *Financial aid:* Fellowships, research assistantships, teaching assistantships, Federal Work-Study and institutionally-sponsored loans available. Financial aid application deadline: 2/1. *Faculty research:* Artificial intelligence, graphics, natural-language processing, expert systems, operating systems. *Unit head:* Zary Segall, Head, 541-346-4408. *Application contact:* Jan Saunders, Graduate Secretary, 541-346-4408, Fax: 541-346-5373.

See in-depth description on page 891.

University of Ottawa, School of Graduate Studies and Research, Faculty of Engineering, Ottawa-Carleton Institute for Computer Science, Ottawa, ON K1N 6N5, Canada. Offers MCS, PhD. *Faculty:* 42 full-time, 2 part-time. *Students:* 85 full-time (13 women), 55 part-time (10 women), 17 international. Average age 30. In 1998, 19 master's, 2 doctorates awarded. *Degree requirements:* For master's, computer language, thesis or alternative required, foreign language not required; for doctorate, computer language, dissertation, computer exam required, foreign language not required. *Entrance requirements:* For master's, honors degree or equivalent, minimum B average; for doctorate, minimum B+ average. *Application deadline:* For fall admission, 3/1 (priority date). Applications are processed on a rolling basis. Application fee: $35. *Financial aid:* Fellowships, research assistantships, teaching assistantships available. Financial aid application deadline: 2/15. *Faculty research:* Algorithms and complexity, artificial intelligence, simulation software engineering. *Unit head:* Stan Matwin, Director, 613-562-5800 Ext. 6679, Fax: 613-562-5187. *Application contact:* Johanne Forgues, Academic Assistant, 613-562-5800 Ext. 6700, Fax: 613-562-5187, E-mail: johanne@site.uottawa.ca.

University of Pennsylvania, School of Engineering and Applied Science, Department of Computer and Information Science, Philadelphia, PA 19104. Offers MSE, PhD, MSE/MBA. Part-time programs available. Terminal master's awarded for partial completion of doctoral program. *Degree requirements:* For master's, computer language required, thesis optional, foreign language not required; for doctorate, computer language, dissertation required, foreign language not required. *Entrance requirements:* For master's and doctorate, GRE General Test, TOEFL (minimum score of 600 required). *Faculty research:* Robotics, graphics, theory, ai, networks and distributed systems, databases, computational biology, natural language processing.

See in-depth description on page 893.

University of Phoenix, Graduate Programs, Computer Science and Information Technology Program, Phoenix, AZ 85072-2069. Offers MSCIS. Programs offered at campuses in Colorado, New Mexico, Northern California, Tucson, Utah, and on-line. Evening/weekend programs available. Postbaccalaureate distance learning degree programs offered (no on-campus study). *Students:* 618 full-time (167 women); includes 99 minority (74 African Americans, 25 Hispanic Americans) Average age 36. *Degree requirements:* For master's, thesis or alternative required. *Entrance requirements:* For master's, TOEFL (minimum score of 580 required), minimum GPA of 2.5, 3 years of work experience, comprehensive cognitive assessment (COCA). *Application deadline:* Applications are processed on a rolling basis. Application fee: $50. *Financial aid:* Applicants required to submit FAFSA. *Unit head:* Hugh McBride, Dean, 602-966-9577. *Application contact:* Campus Information Center, 602-966-9577.

University of Pittsburgh, Faculty of Arts and Sciences, Department of Computer Science, Pittsburgh, PA 15260. Offers MS, PhD. Part-time programs available. *Faculty:* 22 full-time (3 women), 7 part-time (0 women). *Students:* 62 full-time (13 women), 21 part-time (6 women); includes 4 minority (1 African American, 2 Asian Americans or Pacific Islanders, 1 Hispanic American), 43 international. 189 applicants, 24% accepted. In 1998, 18 master's, 7 doctorates awarded. Terminal master's awarded for partial completion of doctoral program. *Degree requirements:* For master's, computer language, thesis or alternative required, foreign language not required; for doctorate, computer language, dissertation, comprehensive and preliminary exams required, foreign language not required. *Entrance requirements:* For master's and doctorate, GRE General Test, TOEFL. *Average time to degree:* Master's–2 years full-time, 4 years part-time; doctorate–5 years full-time, 8 years part-time. *Application deadline:* For fall admission, 3/1; for spring admission, 10/1. Applications are processed on a rolling basis. Application fee: $30 ($40 for international students). *Financial aid:* In 1998–99, 60 students received aid, including 1 fellowship (averaging $13,650 per year), 25 research assistantships (averaging $10,752 per year), 34 teaching assistantships (averaging $10,600 per year); career-related internships or fieldwork, Federal Work-Study, scholarships, and tuition waivers (partial) also available. Financial aid application deadline: 2/1. *Faculty research:* Algorithms and theory, artificial intelligence, parallel and distributed systems, software systems and interfaces. Total annual research expenditures: $1.7 million. *Unit head:* Dr. Siegfried Treu, Chairman, 412-624-8493, Fax: 412-624-8854, E-mail: treu@cs.pitt.edu. *Application contact:* Loretta Shabatura, Graduate Secretary, 412-624-8495, Fax: 412-624-8854, E-mail: loretta@cs.pitt.edu.

University of Regina, Faculty of Graduate Studies and Research, Faculty of Science, Department of Computer Science, Regina, SK S4S 0A2, Canada. Offers M Sc, PhD. *Faculty:* 15 full-time (2 women), 3 part-time (1 woman). *Students:* 28 full-time (5 women), 34 part-time (9 women). 84 applicants, 36% accepted. In 1998, 13 master's, 2 doctorates awarded. *Degree requirements:* For master's and doctorate, computer language, thesis/dissertation required, foreign language not required. *Entrance requirements:* For master's, TOEFL (minimum score of 550 required); for doctorate, TOEFL. *Application deadline:* Applications are processed on a rolling basis. Application fee: $0. *Expenses:* Tuition and fees charges are reported in Canadian dollars. Tuition, state resident: full-time $1,688 Canadian dollars; part-time $94 Canadian dollars per credit hour. International tuition: $3,375 Canadian dollars full-time. Required fees: $65 Canadian dollars per course. Tuition and fees vary according to course load and program. *Financial aid:* In 1998–99, 13 research assistantships, 10 teaching assistantships were awarded.; fellowships, career-related internships or fieldwork and scholarships also available. Financial aid application deadline: 6/15. *Faculty research:* Expert systems, image processing, artificial intelligence, parallel computing data and knowledge bases. Total annual research expenditures: $387,000. *Unit head:* Dr. L. Saxton, Head, 306-585-4632, Fax: 306-585-4745, E-mail: saxton@cs.uregina.ca.

University of Rhode Island, Graduate School, College of Arts and Sciences, Department of Computer Science and Statistics, Kingston, RI 02881. Offers MS, PhD. *Degree requirements:* For master's, thesis optional; for doctorate, dissertation required. *Entrance requirements:* For master's, GRE Subject Test.

University of Rochester, The College, Arts and Sciences, Department of Computer Science, Rochester, NY 14627-0250. Offers MS, PhD. *Faculty:* 11. *Students:* 36 full-time (10 women), 2 part-time; includes 4 minority (2 Asian Americans or Pacific Islanders, 1 Hispanic American, 1 Native American), 18 international. 297 applicants, 4% accepted. In 1998, 8 master's, 9 doctorates awarded. *Degree requirements:* For doctorate, dissertation, qualifying exam required, foreign language not required. *Entrance requirements:* For master's, GRE General Test; for doctorate, GRE General Test, TOEFL. *Application deadline:* For fall admission, 2/1 (priority date). Application fee: $25. *Financial aid:* Fellowships, research assistantships, teaching

Computer Science

University of Rochester (continued)

assistantships, tuition waivers (full and partial) available. Financial aid application deadline:2/1. *Unit head:* Michael Scott, Chair, 716-275-5671. *Application contact:* Peggy Meeker, Graduate Program Secretary, 716-275-7737.

See in-depth description on page 897.

University of San Francisco, College of Arts and Sciences, Department of Computer Science, San Francisco, CA 94117-1080. Offers MS. Part-time programs available. *Faculty:* 7 full-time (0 women), 4 part-time (0 women). *Students:* 20 full-time (5 women), 5 part-time (2 women); includes 4 minority (1 African American, 2 Asian Americans or Pacific Islanders, 1 Hispanic American), 18 international. Average age 26. 47 applicants, 70% accepted. In 1998, 9 degrees awarded. *Degree requirements:* For master's, computer language required, thesis optional, foreign language not required. *Entrance requirements:* For master's, GRE General Test, GRE Subject Test, TOEFL, BS in computer science or related field. *Application deadline:* For fall admission, 7/1 (priority date); for spring admission, 12/1. Applications are processed on a rolling basis. Application fee: $40 ($50 for international students). Tuition: Full-time $12,618; part-time $701 per unit. Tuition and fees vary according to course load, degree level, campus/location and program. *Financial aid:* In 1998–99, 16 students received aid; fellowships, teaching assistantships, career-related internships or fieldwork and Federal Work-Study available. Financial aid application deadline: 3/2. *Faculty research:* Software engineering, computer graphics, computer networks. *Unit head:* Dr. Peter Pacheco, Chairman, 415-422-6630. *Application contact:* Dr. Benjamin Wells, Graduate Adviser, 415-422-6530, E-mail: wells@usfca.edu.

University of South Alabama, Graduate School, Division of Computer and Information Sciences, Mobile, AL 36688-0002. Offers computer science (MS); information science (MS). Part-time and evening/weekend programs available. *Faculty:* 10 full-time (1 woman). *Students:* 56 full-time (15 women), 22 part-time (7 women); includes 6 minority (2 African Americans, 3 Asian Americans or Pacific Islanders, 1 Native American), 31 international. 153 applicants, 60% accepted. In 1998, 23 degrees awarded. *Degree requirements:* For master's, computer language, project required, thesis optional, foreign language not required. *Entrance requirements:* For master's, GRE General Test (minimum combined score of 1000 required), minimum GPA of 2.5. *Application deadline:* For fall admission, 9/1 (priority date). Applications are processed on a rolling basis. Application fee: $25. Tuition, state resident: part-time $116 per semester hour. Tuition, nonresident: part-time $230 per semester hour. Required fees: $121 per semester. Part-time tuition and fees vary according to course load and program. *Financial aid:* In 1998–99, 4 research assistantships were awarded.; career-related internships or fieldwork and institutionally-sponsored loans also available. Aid available to part-time students. Financial aid application deadline: 4/1. *Faculty research:* Numerical analysis, artificial intelligence, simulation, medical applications, software engineering. *Unit head:* Dr. David Feinstein, Chairman, 334-460-6390.

University of South Carolina, Graduate School, College of Science and Mathematics, Department of Computer Science, Columbia, SC 29208. Offers MS, PhD. *Faculty:* 11 full-time (1 woman). *Students:* 100 full-time (28 women), 53 part-time (11 women); includes 15 minority (4 African Americans, 11 Asian Americans or Pacific Islanders), 93 international. Average age 30. 157 applicants, 58% accepted. In 1998, 25 master's, 4 doctorates awarded. *Degree requirements:* For master's and doctorate, thesis/dissertation required, foreign language not required. *Entrance requirements:* For master's and doctorate, GRE General Test. *Application deadline:* For fall admission, 3/1 (priority date); for spring admission, 11/1. Applications are processed on a rolling basis. Application fee: $35. Electronic applications accepted. Tuition, state resident: full-time $4,014; part-time $202 per credit hour. Tuition, nonresident: full-time $8,528; part-time $428 per credit hour. Required fees: $100; $4 per credit hour. Tuition and fees vary according to program. *Financial aid:* In 1998–99, fellowships with partial tuition reimbursements (averaging $10,000 per year), 26 research assistantships with partial tuition reimbursements (averaging $15,000 per year), 31 teaching assistantships with partial tuition reimbursements (averaging $13,000 per year) were awarded.; Federal Work-Study also available. Financial aid application deadline: 3/1. *Faculty research:* Computer vision, pattern recognition, artificial intelligence, database management. Total annual research expenditures: $1.7 million. *Unit head:* Dr. Robert L. Oakman, Chair, 803-777-2880.

See in-depth description on page 901.

University of South Dakota, Graduate School, College of Arts and Sciences, Department of Computer Science, Vermillion, SD 57069-2390. Offers MA. *Faculty:* 5 full-time (0 women), 1 part-time (0 women). *Students:* 11 full-time (3 women), 3 part-time (1 woman), 4 international. 25 applicants, 32% accepted. In 1998, 6 degrees awarded. *Degree requirements:* For master's, computer language, thesis required, foreign language not required. *Entrance requirements:* For master's, GRE General Test. Application fee: $15. *Financial aid:* Teaching assistantships available. Aid available to part-time students. *Unit head:* John Lushbough, Chair, 605-677-5388. *Application contact:* Dr. Rich McBride, Graduate Adviser, 605-677-5388.

University of Southern California, Graduate School, School of Engineering, Department of Computer Science, Program in Computer Science, Los Angeles, CA 90089. Offers MS, PhD. *Students:* 183 full-time (33 women), 120 part-time (24 women); includes 35 minority (2 African Americans, 31 Asian Americans or Pacific Islanders, 2 Hispanic Americans), 227 international. Average age 28. 401 applicants, 60% accepted. In 1998, 77 master's, 21 doctorates awarded. *Degree requirements:* For doctorate, dissertation required. *Entrance requirements:* For master's and doctorate, GRE General Test. *Application deadline:* For fall admission, 1/1 (priority date); for spring admission, 10/1. Applications are processed on a rolling basis. Application fee: $55. Tuition: Part-time $768 per unit. Required fees: $350 per semester. *Financial aid:* In 1998–99, 11 fellowships, 144 research assistantships, 34 teaching assistantships were awarded.; Federal Work-Study, institutionally-sponsored loans, and scholarships also available. Aid available to part-time students. Financial aid application deadline: 2/15; financial aid applicants required to submit FAFSA. *Unit head:* Dr. Ellis Horowitz, Chairman, Department of Computer Science, 213-740-4494.

Announcement: Master's and PhD degree programs in computer science. Teaching and research assistantships available to qualified PhD students. Research: multimedia technology, neural networks, languages, databases, theory, AI, robotics, networks, software engineering, systems, vision. Sun Workstations, IBM PC, Macintosh, SGI, COMVEX. Graduate admissions: 3.0–3.5 GPA, suitable GRE scores, references. Contact Graduate Admission, 213-740-4496.

See in-depth description on page 903.

University of Southern Maine, School of Applied Science, Department of Computer Science, Portland, ME 04104-9300. Offers MS. Part-time programs available. *Faculty:* 6 full-time (0 women). *Students:* 8 full-time (4 women), 8 part-time. Average age 30. 11 applicants, 73% accepted. In 1998, 3 degrees awarded. *Degree requirements:* For master's, computer language, thesis required, foreign language not required. *Entrance requirements:* For master's, GRE Subject Test, minimum GPA of 3.0. *Average time to degree:* Master's–2 years full-time, 5 years part-time. *Application deadline:* For fall admission, 3/1 (priority date); for spring admission, 10/1. Application fee: $25. *Financial aid:* In 1998–99, 4 students received aid, including 1 research assistantship, 3 teaching assistantships; Federal Work-Study also available. Aid available to part-time students. Financial aid application deadline: 4/1; financial aid applicants required to submit FAFSA. *Faculty research:* Computer networks, database systems, software engineering, theory of computability, human factors. Total annual research expenditures:$47,863. *Unit head:* David A. Briggs, Chair, 207-780-4723, Fax: 207-780-4933, E-mail: briggs@usm.maine.edu. *Application contact:* Mary Sloan, Assistant Director of Graduate Studies, 207-780-4386, Fax: 207-780-4969, E-mail: msloan@usm.maine.edu.

University of Southern Mississippi, Graduate School, College of Science and Technology, School of Mathematical Sciences, Department of Computer Science, Hattiesburg, MS 39406-5106. Offers MS. *Faculty:* 14 full-time (1 woman), 2 part-time (1 woman). *Students:* 16 full-time (7 women), 6 part-time (1 woman); includes 12 minority (all Asian Americans or Pacific Islanders) Average age 30. 50 applicants, 44% accepted. In 1998, 15 degrees awarded. *Degree requirements:* For master's, thesis or alternative, oral/written comprehensive exam required, foreign language not required. *Entrance requirements:* For master's, GRE General Test (minimum combined score of 1000 required), TOEFL (minimum score of 580 required), minimum GPA of 2.75. *Application deadline:* For fall admission, 8/6 (priority date). Applications are processed on a rolling basis. Application fee: $0 ($25 for international students). Tuition, state resident: full-time $2,250; part-time $137 per semester hour. Tuition, nonresident: full-time $3,102; part-time $172 per semester hour. Required fees: $602. *Financial aid:* Research assistantships, teaching assistantships, Federal Work-Study and institutionally-sponsored loans available. Financial aid application deadline: 3/15. *Faculty research:* Satellite telecommunications, advanced life-support systems, artificial intelligence. *Unit head:* Dr. Frank Nagurney, Chair, 601-266-4949.

Announcement: The University of Southern Mississippi offers the MS degree in computer science and the MS and PhD degrees in scientific computing. While the MS degree in computer science follows the standard computer science curriculum, the degrees in scientific computing offer a unique combination of computer science, mathematics, and physics. The areas of faculty and graduate student computer science–oriented research include parallel and distributed processing, neural networks, nested/coupled domain decomposition algorithms, databases, software engineering, graphics, artificial intelligence, knowledge-based systems, telecommunications, human-computer interaction, formal languages, computer security, visualization software, operating systems, operations research, and others. For further information, contact Dr. Adel Ali, Director of Graduate Studies in Computer Science, Box 5106, Hattiesburg, MS 39406-5106; e-mail: adel.ali@usm.edu

University of Southern Mississippi, Graduate School, College of Science and Technology, School of Mathematical Sciences, Program in Scientific Computing, Hattiesburg, MS 39406-5167. Offers PhD. Part-time programs available. *Faculty:* 2 part-time (0 women). Average age 35. 24 applicants, 54% accepted. In 1998, 1 doctorate awarded. *Degree requirements:* For doctorate, 2 foreign languages (computer language can substitute for one), dissertation, written comprehensive exam required. *Entrance requirements:* For doctorate, GRE General Test (minimum combined score of 1000 required), TOEFL, minimum GPA of 3.5. *Application deadline:* For fall admission, 8/6 (priority date). Applications are processed on a rolling basis. Application fee: $0 ($25 for international students). Tuition, state resident: full-time $2,250; part-time $137 per semester hour. Tuition, nonresident: full-time $3,102; part-time $172 per semester hour. Required fees: $602. *Financial aid:* Teaching assistantships, Federal Work-Study and institutionally-sponsored loans available. Financial aid application deadline: 3/15. *Unit head:* Dr. Grayson Rayborn, Director, School of Mathematical Sciences, 601-266-4739.

University of South Florida, Graduate School, College of Engineering, Department of Computer Science and Engineering, Tampa, FL 33620-9951. Offers computer engineering (M Cp E, MS Cp E); computer science (MCS, MSCS); computer science and engineering (PhD). Part-time programs available. *Faculty:* 17 full-time (2 women). *Students:* 95 full-time (20 women), 72 part-time (14 women); includes 27 minority (5 African Americans, 17 Asian Americans or Pacific Islanders, 5 Hispanic Americans), 85 international. Average age 30. 198 applicants, 78% accepted. In 1998, 40 master's awarded (80% found work related to degree, 20% continued full-time study); 6 doctorates awarded (50% entered university research/teaching, 50% found other work related to degree). Terminal master's awarded for partial completion of doctoral program. *Degree requirements:* For master's, computer language required, foreign language not required; for doctorate, dissertation, 2 tools of research as specified by dissertation committee required, foreign language not required. *Entrance requirements:* For master's, GRE General Test (minimum combined score of 1200 required), minimum GPA of 3.0 during previous 2 years; for doctorate, GRE General Test (minimum combined score of 1200 required, 500 on verbal section). *Average time to degree:* Master's–2.5 years full-time, 5 years part-time; doctorate–4.5 years full-time, 7 years part-time. *Application deadline:* For fall admission, 6/1; for spring admission, 10/15. Application fee: $20. Electronic applications accepted. Tuition, state resident: part-time $148 per credit hour. Tuition, nonresident: part-time $509 per credit hour. *Financial aid:* In 1998–99, 2 fellowships with full tuition reimbursements (averaging $9,500 per year), 60 research assistantships with full tuition reimbursements (averaging $11,993 per year), 22 teaching assistantships with full tuition reimbursements (averaging $11,746 per year) were awarded.; career-related internships or fieldwork, Federal Work-Study, institutionally-sponsored loans, and tuition waivers (partial) also available. Aid available to part-time students. Financial aid applicants required to submit FAFSA. *Faculty research:* Computer vision, databases, VLSI design and test, networks, artificial intelligence. Total annual research expenditures: $1.1 million. *Unit head:* Dr. Abe Kandel, Chairperson, 813-974-3652, Fax: 813-974-5456, E-mail: kandel@csee.usf.edu. *Application contact:* Dr. Dmitry B. Goldgof, Graduate Director, 813-974-3033, Fax: 813-974-5456, E-mail: msphd@csee.usf.edu.

See in-depth description on page 905.

University of Southwestern Louisiana, Graduate School, College of Engineering, Center for Advanced Computer Studies, Lafayette, LA 70504. Offers computer engineering (MS, PhD); computer science (MS, PhD). Part-time programs available. *Faculty:* 28 full-time (5 women). *Students:* 106 full-time (26 women), 18 part-time (3 women); includes 4 minority (1 African American, 2 Asian Americans or Pacific Islanders, 1 Hispanic American), 97 international. 401 applicants, 64% accepted. In 1998, 53 master's, 9 doctorates awarded. Terminal master's awarded for partial completion of doctoral program. *Degree requirements:* For master's, computer language, thesis or alternative required, foreign language not required; for doctorate, computer language, dissertation, final oral exam required, foreign language not required. *Entrance requirements:* For master's, GRE General Test, TOEFL, minimum GPA of 2.75; for doctorate, GRE General Test, TOEFL, minimum GPA of 3.0. *Application deadline:* For fall admission, 5/15. Application fee: $5 ($15 for international students). *Financial aid:* In 1998–99, 10 fellowships (averaging $15,700 per year), 74 research assistantships with full tuition reimbursements (averaging $6,824 per year) were awarded.; teaching assistantships, Federal Work-Study and tuition waivers (full) also available. Financial aid application deadline: 3/1. *Unit head:* Dr. Magdy A. Bayoumi, Chair, 318-482-6147. *Application contact:* Dr. William Edwards, Graduate Coordinator, 318-482-6284.

See in-depth description on page 907.

University of Southwestern Louisiana, Graduate School, College of Sciences, Department of Computer Science, Lafayette, LA 70504. Offers MS. *Degree requirements:* Foreign language not required. *Entrance requirements:* For master's, GRE General Test, minimum GPA of 2.75. *Application deadline:* For fall admission, 5/15. Application fee: $5 ($15 for international students). *Unit head:* Ursula Jackson, Acting Head, 318-482-6768.

University of Tennessee at Chattanooga, Graduate Division, School of Engineering, Department of Computer Science, Chattanooga, TN 37403-2598. Offers MS. Part-time and evening/weekend programs available. *Faculty:* 4 full-time (1 woman). *Students:* 9 full-time (0 women), 14 part-time (2 women); includes 3 minority (1 African American, 2 Asian Americans or Pacific Islanders), 8 international. Average age 28. 38 applicants, 53% accepted. In 1998, 4 degrees awarded. *Degree requirements:* For master's, computer language, thesis required, foreign language not required. *Entrance requirements:* For master's, GRE General Test (combined average 1593 on three sections). *Application deadline:* Applications are processed on a rolling basis. Application fee: $25. *Financial aid:* Fellowships, research assistantships, Federal Work-Study and institutionally-sponsored loans available. Aid available to part-time students. Financial aid application deadline: 4/1. *Unit head:* Peggy Meeker, Graduate Program Secretary, 716-275-7737. *Application contact:* Dr. Deborah E. Arfken, Assistant Provost for Graduate Studies, 423-755-4667, Fax: 423-755-4478, E-mail: deborah-arfken@utc.edu.

University of Tennessee, Knoxville, Graduate School, College of Arts and Sciences, Department of Computer Science, Knoxville, TN 37996. Offers MS, PhD. Part-time programs available. *Faculty:* 16 full-time (0 women). *Students:* 40 full-time (7 women), 86 part-time (26 women); includes 6 minority (2 African Americans, 2 Asian Americans or Pacific Islanders, 1 Hispanic American, 1 Native American), 33 international. 119 applicants, 49% accepted. In 1998, 25 master's, 4 doctorates awarded. *Degree requirements:* For master's, computer language, thesis or alternative required, foreign language not required; for doctorate, computer language, dissertation required, foreign language not required. *Entrance requirements:* For master's and doctorate, GRE General Test, TOEFL (minimum score of 550 required), minimum GPA of 2.7. *Application deadline:* For fall admission, 2/1 (priority date). Applications are processed on a rolling basis. Application fee: $35. Electronic applications accepted. *Financial aid:* In 1998–99, 2 fellowships, 22 research assistantships, 38 teaching assistantships were awarded.; Federal Work-Study, institutionally-sponsored loans, and unspecified assistantships also available. Financial aid application deadline: 2/1; financial aid applicants required to submit FAFSA. *Unit head:* Dr. Robert Ward, Head, 423-974-5067, Fax: 423-974-4404, E-mail: ward@cs.utk.edu. *Application contact:* Dr. David Straight, Graduate Representative, E-mail: straight@cs.utk.edu.

See in-depth description on page 909.

University of Tennessee Space Institute, Graduate Programs, Program in Computer Science, Tullahoma, TN 37388-9700. Offers MS. *Faculty:* 2 full-time (0 women), 1 part-time (0 women). *Students:* 5 full-time (3 women), 9 part-time (3 women); includes 4 minority (2 African Americans, 1 Asian American or Pacific Islander, 1 Hispanic American), 1 international. 7 applicants, 71% accepted. In 1998, 1 degree awarded. *Degree requirements:* For master's, thesis required (for some programs). *Entrance requirements:* For master's, GRE General Test. *Application deadline:* Applications are processed on a rolling basis. Application fee: $35. *Financial aid:* Fellowships, research assistantships, Federal Work-Study available. Financial aid applicants required to submit FAFSA. *Unit head:* Dr. Bruce Whitehead, Degree Program Chairman, 931-393-7296, Fax: 931-454-2271, E-mail: bwhitehe@utsi.edu. *Application contact:* Dr. Edwin M. Gleason, Assistant Dean for Admissions and Student Affairs, 931-393-7432, Fax: 931-393-7346, E-mail: egleason@utsi.edu.

The University of Texas at Arlington, Graduate School, College of Engineering, Department of Computer Science and Engineering, Arlington, TX 76019. Offers M Engr, M Sw En, MCS, MS, PhD. *Faculty:* 16 full-time (2 women). *Students:* 218 full-time (44 women), 138 part-time (18 women); includes 35 minority (1 African American, 30 Asian Americans or Pacific Islanders, 4 Hispanic Americans), 252 international. 434 applicants, 36% accepted. In 1998, 121 master's, 3 doctorates awarded. *Degree requirements:* For master's, computer language, thesis required (for some programs), foreign language not required; for doctorate, computer language, dissertation required, foreign language not required. *Entrance requirements:* For master's, GRE General Test (minimum combined score of 1100 required), TOEFL (minimum score of 560 required); for doctorate, GRE General Test (minimum combined score of 1250 required), TOEFL (minimum score of 560 required). *Application deadline:* Applications are processed on a rolling basis. Application fee: $25 ($50 for international students). Tuition, state resident: full-time $1,368; part-time $76 per semester hour. Tuition, nonresident: full-time $5,454; part-time $303 per semester hour. Required fees: $86 per semester hour. $86 per term. Tuition and fees vary according to course load. *Financial aid:* Research assistantships, teaching assistantships, career-related internships or fieldwork and tuition waivers (partial) available. *Unit head:* Dr. Bill D. Carroll, Chairman, 817-272-3785, Fax: 817-272-3784, E-mail: carroll@cse.uta.edu. *Application contact:* Dr. Bob P. Weems, Graduate Adviser, 817-272-3785, Fax: 817-272-3784, E-mail: weems@cse.uta.edu.

See in-depth description on page 911.

The University of Texas at Austin, Graduate School, College of Natural Sciences, Department of Computer Sciences, Austin, TX 78712-1111. Offers MA, MSCS, PhD. *Students:* 225 (44 women); includes 31 minority (2 African Americans, 19 Asian Americans or Pacific Islanders, 10 Hispanic Americans) 113 international. 247 applicants, 54% accepted. In 1998, 53 master's, 17 doctorates awarded. *Degree requirements:* For master's, thesis optional; for doctorate, dissertation, qualifying exam required. *Entrance requirements:* For master's and doctorate, GRE General Test, GRE Subject Test. *Application deadline:* For fall admission, 1/2. Application fee: $50 ($75 for international students). *Financial aid:* Fellowships, research assistantships, teaching assistantships, institutionally-sponsored loans available. Financial aid application deadline: 1/2. *Unit head:* Dr. Benjamin J. Kuipers, Chairman. *Application contact:* Dr. Bruce Porter, Graduate Adviser, 512-471-9503, E-mail: csadmis@cs.utexas.edu.

See in-depth description on page 913.

The University of Texas at Dallas, Erik Jonsson School of Engineering and Computer Science, Program in Computer Science, Richardson, TX 75083-0688. Offers MS, PhD. Part-time and evening/weekend programs available. *Students:* 301 full-time (110 women), 193 part-time (53 women); includes 110 minority (3 African Americans, 102 Asian Americans or Pacific Islanders, 5 Hispanic Americans), 290 international. Average age 30. In 1998, 237 master's, 3 doctorates awarded. *Degree requirements:* For master's, minimum GPA of 3.0 required, thesis optional, foreign language not required; for doctorate, dissertation, minimum grade of B in core courses required, foreign language not required. *Entrance requirements:* For master's, GRE General Test (minimum combined score of 1100 required), TOEFL (minimum score of 550 required), minimum GPA of 3.0 in undergraduate course work, 3.3 in quantitative course work; for doctorate, GRE General Test (minimum combined score of 1100 required with master's degree, 1300 with bachelor's degree), TOEFL (minimum score of 550 required), minimum GPA of 3.5. *Application deadline:* For fall admission, 7/15; for spring admission, 11/15. Applications are processed on a rolling basis. Application fee: $25 ($75 for international students). *Financial aid:* Fellowships, research assistantships, teaching assistantships, career-related internships or fieldwork, Federal Work-Study, grants, institutionally-sponsored loans, and scholarships available. Aid available to part-time students. Financial aid application deadline: 4/30; financial aid applicants required to submit FAFSA. *Faculty research:* Telecommunication networks, parallel processing, analysis of algorithms, artificial intelligence, software engineering. *Unit head:* Dr. Dung T. Huynh, Head, 972-883-2169, Fax: 972-883-2349, E-mail: huynh@utdallas.edu. *Application contact:* Deborah Chen, Graduate Secretary, 972-883-2185, Fax: 972-883-2349, E-mail: cs-grad-info@utdallas.edu.

See in-depth description on page 915.

The University of Texas at El Paso, Graduate School, College of Engineering, Department of Computer Science, El Paso, TX 79968-0001. Offers MS. Part-time and evening/weekend programs available. *Faculty:* 7 full-time (1 woman), 3 part-time (0 women). *Students:* 16 full-time (2 women), 15 part-time (5 women); includes 7 minority (all Hispanic Americans), 17 international. Average age 27. 45 applicants, 49% accepted. In 1998, 10 degrees awarded. *Degree requirements:* For master's, thesis optional, foreign language not required. *Entrance requirements:* For master's, GRE General Test, TOEFL (minimum score of 550 required), minimum GPA of 3.0. *Application deadline:* Applications are processed on a rolling basis. Application fee: $15 ($65 for international students). Electronic applications accepted. Tuition, state resident: full-time $2,790. Tuition, nonresident: full-time $7,710. *Financial aid:* Research assistantships, teaching assistantships, Federal Work-Study, institutionally-sponsored loans, and tuition waivers (partial) available. Financial aid applicants required to submit FAFSA. Total annual research expenditures: $395,129. *Unit head:* Dr. Daniel Cooke, Chairperson, 915-747-5480. *Application contact:* Susan Jordan, Director, Graduate Student Services, 915-747-5491, Fax: 915-747-5788, E-mail: sjordan@utep.edu.

The University of Texas at San Antonio, College of Sciences and Engineering, Division of Computer Science, San Antonio, TX 78249-0617. Offers MS, PhD. *Faculty:* 13 full-time (1 woman), 10 part-time (3 women). *Students:* 28 full-time (7 women), 38 part-time (9 women); includes 11 minority (1 African American, 6 Asian Americans or Pacific Islanders, 4 Hispanic Americans), 25 international. Average age 34. 54 applicants, 17% accepted. In 1998, 9 master's awarded. *Degree requirements:* For doctorate, dissertation, comprehensive exam required. *Entrance requirements:* For master's, GRE General Test; for doctorate, GRE General Test (minimum combined score of 1000 required), TOEFL (minimum score of 550 required), minimum GPA of 3.0. *Application deadline:* For fall admission, 7/1. Applications are processed on a rolling basis. Application fee: $25. *Unit head:* Dr. Richard F. Sincovec, Director, 210-458-4453.

The University of Texas at Tyler, Graduate Studies, College of Sciences and Mathematics, Department of Computer Science, Tyler, TX 75799-0001. Offers computer science (MS); interdisciplinary studies (MA, MS). *Faculty:* 6 full-time (1 woman), 1 part-time (0 women). *Students:* 6 full-time (2 women), 21 part-time (7 women). Average age 28. In 1998, 39 degrees awarded. *Degree requirements:* For master's, computer language, comprehensive exam, project or thesis required, thesis optional, foreign language not required. *Entrance requirements:* For master's, GRE General Test (minimum combined score of 1000 required), previous course work in data structures and computer organization, 6 hours of calculus and statistics required. *Application deadline:* Applications are processed on a rolling basis. Application fee: $0. *Financial aid:* In 1998–99, 3 research assistantships with tuition reimbursements (averaging $6,000 per year) were awarded. Financial aid application deadline: 7/1. *Faculty research:* Artificial neural systems, artificial intelligence, image processing, computer graphics, protein identification using the computer, data mining, database design, distributed objects. *Unit head:* Dr. Ron King, Chair, 903-566-7097, Fax: 903-566-7189, E-mail: rking@mail.uttyl.edu. *Application contact:* Martha D. Wheat, Director of Admissions and Student Records, 903-566-7201, Fax: 903-566-7068.

The University of Texas–Pan American, College of Science and Engineering, Department of Computer Science, Edinburg, TX 78539-2999. Offers MS. *Degree requirements:* For master's, final written exam, project required. *Entrance requirements:* For master's, GRE General Test, TOEFL, minimum GPA of 3.0 in last 60 hours.

University of Toledo, Graduate School, College of Engineering, Department of Electrical Engineering and Computer Science, Toledo, OH 43606-3398. Offers computer science (MSES); electrical engineering (MSEE); engineering sciences (PhD). Part-time and evening/weekend programs available. *Faculty:* 21 full-time (3 women), 1 part-time (0 women). *Students:* 123 full-time (25 women), 34 part-time (4 women); includes 5 minority (1 African American, 3 Asian Americans or Pacific Islanders, 1 Native American), 128 international. Average age 25. 528 applicants, 52% accepted. In 1998, 78 master's, 1 doctorate awarded. *Degree requirements:* For master's, thesis or alternative required, foreign language not required; for doctorate, dissertation required, foreign language not required. *Entrance requirements:* For master's, GRE General Test (minimum combined score of 1700 on three sections required), TOEFL (minimum score of 550 required), minimum GPA of 2.7; for doctorate, GRE General Test (minimum combined score of 1700 on three sections required), TOEFL (minimum score of 550 required). *Average time to degree:* Master's–2 years full-time; doctorate–4 years full-time. *Application deadline:* For fall admission, 5/31 (priority date). Applications are processed on a rolling basis. Application fee: $30. Electronic applications accepted. *Financial aid:* In 1998–99, 146 students received aid, including 1 fellowship with full tuition reimbursement available, 7 research assistantships with full tuition reimbursements available, 12 teaching assistantships with full tuition reimbursements available; Federal Work-Study, scholarships, and tuition waivers (full) also available. Aid available to part-time students. Financial aid application deadline: 4/1. *Faculty research:* Power electronics, digital television, satellite communications, computer networks, fault-tolerant computing, weather and intelligent transportation. Total annual research expenditures: $539,300. *Unit head:* Dr. Adel Ghandakly, Chairman, 419-530-8146, E-mail: aghanda2@uoft02.utoledo.edu. *Application contact:* Sylvia Pinkerman, Academic Program Coordinator, 419-530-8144, Fax: 419-530-8146, E-mail: spinkerm@eng.utoledo.edu.

University of Toronto, School of Graduate Studies, Physical Sciences Division, Department of Computer Science, Toronto, ON M5S 1A1, Canada. Offers M Sc, PhD. Part-time programs available. *Degree requirements:* For master's and doctorate, thesis/dissertation required.

University of Tulsa, Graduate School, College of Business Administration, Department of Engineering and Technology Management, Tulsa, OK 74104-3189. Offers chemical engineering (METM); computer science (METM); electrical engineering (METM); geological science (METM); mathematics (METM); mechanical engineering (METM); petroleum engineering (METM). Part-time and evening/weekend programs available. *Students:* 3 full-time (1 woman), 1 part-time, 3 international. *Degree requirements:* For master's, foreign language and thesis not required. *Entrance requirements:* For master's, GRE General Test (minimum score of 430 on verbal section, 600 on quantitative required), TOEFL (minimum score of 575 required). *Application deadline:* Applications are processed on a rolling basis. Application fee: $30. Electronic applications accepted. Tuition: Full-time $8,640; part-time $480 per hour. Required fees: $3 per hour. One-time fee: $200 full-time. Tuition and fees vary according to program. *Unit head:* Dr. Richard C. Burgess, Assistant Dean/Director of Graduate Business Studies, 918-631-2242, Fax: 918-631-2142.

University of Tulsa, Graduate School, College of Engineering and Applied Sciences, Department of Mathematical and Computer Sciences, Program in Computer Science, Tulsa, OK 74104-3189. Offers MS, PhD. Part-time programs available. *Students:* 23 full-time (9 women), 5 part-time (1 woman), 22 international. Average age 28. 36 applicants, 94% accepted. In 1998, 13 master's, 2 doctorates awarded. *Degree requirements:* For master's, computer language required, thesis optional, foreign language not required; for doctorate, computer language, dissertation, comprehensive exams required, foreign language not required. *Entrance requirements:* For master's and doctorate, GRE General Test, TOEFL (minimum score of 550 required). *Application deadline:* Applications are processed on a rolling basis. Application fee: $30. Electronic applications accepted. Tuition: Full-time $8,640; part-time $480 per hour. Required fees: $3 per hour. One-time fee: $200 full-time. Tuition and fees vary according to program. *Financial aid:* In 1998–99, 24 students received aid, including 14 research assistantships (averaging $3,125 per year), 10 teaching assistantships (averaging $5,453 per year); fellowships, Federal Work-Study and tuition waivers (partial) also available. Aid available to part-time students. Financial aid application deadline: 2/1; financial aid applicants required to submit FAFSA. *Faculty research:* Genetic algorithms, medical imaging, parallel and scientific computation, database security, fuzzy control. *Unit head:* Dr. Roger L. Wainwright, Adviser, 918-631-3143, Fax: 918-631-3077.

University of Utah, Graduate School, College of Engineering, Department of Computer Science, Salt Lake City, UT 84112-1107. Offers M Phil, ME, MS, PhD. *Faculty:* 19 full-time (2 women), 14 part-time (0 women). *Students:* 52 full-time (9 women), 31 part-time (1 woman). Average age 28. In 1998, 12 master's, 5 doctorates awarded. *Degree requirements:* For master's, thesis (MS) required; for doctorate, dissertation required, foreign language not required. *Entrance requirements:* For master's and doctorate, GRE General Test, GRE Subject Test, TOEFL (minimum score of 500 required), minimum GPA of 3.0. *Application deadline:* For fall admission, 7/1. Application fee: $30 ($50 for international students). *Financial aid:* In 1998–99, 25 teaching assistantships were awarded.; fellowships, research assistantships *Faculty research:* Computer-aided graphic design, VLSI, information retrieval, portable artificial intelligence systems, functional programming. *Unit head:* Robert R. Kessler, Chair, 801-581-8224, Fax: 801-581-5843, E-mail: kessler@cs.utah.edu. *Application contact:* Gary Linstrom, Director of Graduate Admissions, 801-581-5586.

See in-depth description on page 917.

University of Vermont, Graduate College, College of Engineering and Mathematics, Department of Computer Science and Electrical Engineering, Program in Computer Science, Burlington, VT 05405-0160. Offers MS. *Degree requirements:* For master's, thesis or alternative required, foreign language not required. *Entrance requirements:* For master's, GRE General Test, TOEFL (minimum score of 550 required).

University of Victoria, Faculty of Graduate Studies, Faculty of Engineering, Department of Computer Science, Victoria, BC V8W 2Y2, Canada. Offers M Sc, MA, PhD. Part-time programs available. *Faculty:* 23 full-time (4 women), 6 part-time (0 women). *Students:* 74

University of Victoria (continued)
full-time (19 women), 4 part-time (1 woman), 29 international. Average age 28. 190 applicants, 9% accepted. In 1998, 8 master's, 1 doctorate awarded. *Degree requirements:* For master's and doctorate, thesis/dissertation required. *Entrance requirements:* For master's, TOEFL (minimum score of 550 required), BS in computer science (recommended); for doctorate, TOEFL (minimum score of 575 required), MS in computer science (recommended). *Average time to degree:* Master's–2.3 years full-time; doctorate–4.47 years full-time. *Application deadline:* For fall admission, 5/1 (priority date); for spring admission, 10/1. Applications are processed on a rolling basis. Application fee: $50. *Financial aid:* In 1998–99, 50 students received aid, including 42 research assistantships, 28 teaching assistantships; fellowships, career-related internships or fieldwork, institutionally-sponsored loans, and awards also available. Financial aid application deadline: 2/15. *Faculty research:* Functional and logic programming, numerical analysis, parallel and distributed computing, software systems, theoretical computer science, VLSI design and testing. Total annual research expenditures: $1.1 million. *Unit head:* Dr. R. Nigel Horspool, Chair, 250-721-7220, Fax: 250-721-7292, E-mail: nigelh@csr.uvic.ca. *Application contact:* Dr. Wendy Myrvold, Graduate Admissions Officer, 250-721-7224, Fax: 250-721-7792, E-mail: wendym@csc.uvic.ca.

University of Virginia, School of Engineering and Applied Science, Department of Computer Science, Charlottesville, VA 22903. Offers MCS, MS, PhD. *Faculty:* 24 full-time (4 women), 1 part-time (0 women). *Students:* 69 full-time (13 women), 2 part-time; includes 9 minority (5 African Americans, 3 Asian Americans or Pacific Islanders, 1 Hispanic American), 22 international. Average age 27. 106 applicants, 49% accepted. In 1998, 15 master's, 5 doctorates awarded. *Degree requirements:* For master's, thesis required (for some programs), foreign language not required; for doctorate, dissertation, comprehensive exam required, foreign language not required. *Entrance requirements:* For master's and doctorate, GRE General Test. *Application deadline:* For fall admission, 8/1; for spring admission, 12/1. Applications are processed on a rolling basis. Application fee: $60. *Financial aid:* Fellowships available. Financial aid application deadline: 2/1. *Faculty research:* Systems programming, operating systems, analysis of programs and computation theory, programming languages, software engineering. *Unit head:* John A. Stankovic, Chairman, 804-924-7605. *Application contact:* J. Milton Adams, Assistant Dean, 804-924-3897, E-mail: twr2c@virginia.edu.

See in-depth description on page 919.

University of Washington, Graduate School, College of Engineering, Department of Computer Science and Engineering, Seattle, WA 98195. Offers computer science (MS, PhD). Part-time programs available. *Faculty:* 32 full-time (4 women). *Students:* 116 full-time (16 women), 7 part-time; includes 19 minority (5 African Americans, 13 Asian Americans or Pacific Islanders, 1 Hispanic American), 42 international. Average age 28. 609 applicants, 18% accepted. In 1998, 19 master's awarded (16% found work related to degree, 84% continued full-time study); 7 doctorates awarded (100% found work related to degree). Terminal master's awarded for partial completion of doctoral program. *Degree requirements:* For master's, thesis or alternative required, foreign language not required; for doctorate, dissertation, comprehensive exam, depth exam required, foreign language not required. *Entrance requirements:* For master's and doctorate, GRE General Test, TOEFL (minimum score of 600 required), minimum GPA of 3.0. *Average time to degree:* Master's–2 years full-time, 4 years part-time; doctorate–5.5 years full-time. *Application deadline:* For fall admission, 1/1; for winter admission, 11/1; for spring admission, 2/1. Application fee: $50. Electronic applications accepted. Tuition, state resident: full-time $5,196; part-time $475 per credit. Tuition, nonresident: full-time $13,485; part-time $1,285 per credit. Required fees: $387; $38 per credit. Tuition and fees vary according to course load. *Financial aid:* In 1998–99, 26 fellowships with full tuition reimbursements (averaging $12,600 per year), research assistantships with partial tuition reimbursements (averaging $12,340 per year), 35 teaching assistantships with partial tuition reimbursements (averaging $11,530 per year) were awarded. Financial aid application deadline: 2/1. *Faculty research:* Theory, systems, artificial intelligence, graphics, databases. Total annual research expenditures: $3.8 million. *Unit head:* Dr. Edward D. Lazowska, Chair, 206-543-1695. *Application contact:* 206-543-1695, E-mail: grad_admissions@cs.washington.edu.

See in-depth description on page 921.

University of Waterloo, Graduate Studies, Faculty of Mathematics, Department of Computer Science, Waterloo, ON N2L 3G1, Canada. Offers computer science (M Math, PhD); computer science (software engineering) (M Math); computer science (statistics-computing) (M Math). Part-time programs available. *Faculty:* 47 full-time (7 women), 31 part-time (3 women). *Students:* 121 full-time (26 women), 25 part-time (5 women). 194 applicants, 44% accepted. In 1998, 27 master's, 7 doctorates awarded. *Degree requirements:* For master's, computer language, research paper or thesis required; for doctorate, computer language, dissertation required, foreign language not required. *Entrance requirements:* For master's, TOEFL (minimum score of 580 required), honors degree in field, minimum B+ average; for doctorate, TOEFL (minimum score of 580 required), master's degree. *Average time to degree:* Master's–2 years full-time; doctorate–5 years full-time. *Application deadline:* For fall admission, 2/28 (priority date); for winter admission, 9/30; for spring admission, 1/31. Applications are processed on a rolling basis. Application fee: $50. *Expenses:* Tuition and fees charges are reported in Canadian dollars. Tuition, state resident: full-time $3,168 Canadian dollars; part-time $792 Canadian dollars per term. Tuition, nonresident: full-time $8,000 Canadian dollars; part-time $2,000 Canadian dollars. Required fees: $45 Canadian dollars per term. Tuition and fees vary according to program. *Financial aid:* In 1998–99, 110 students received aid, including research assistantships (averaging $8,500 per year), teaching assistantships (averaging $10,000 per year); scholarships also available. *Faculty research:* Computer graphics, data structures, artificial intelligence, symbolic computation, theory of computing. Total annual research expenditures: $2.2 million. *Unit head:* Dr. N. J. Cecone, Chair, 519-888-4567 Ext. 3292, E-mail: csgrad@jeeves.uwaterloo.ca. *Application contact:* Dr. G. Labahn, Graduate Officer, 519-888-4567 Ext. 4439, E-mail: csgrad@math.uwaterloo.ca.

The University of Western Ontario, Faculty of Graduate Studies, Physical Sciences Division, Department of Computer Science, London, ON N6A 5B8, Canada. Offers M Sc, PhD. Part-time programs available. *Degree requirements:* For master's, thesis, project, or course work required; for doctorate, dissertation required, foreign language not required. *Entrance requirements:* For master's, TOEFL (minimum score of 580 required), B Sc in computer science or comparable academic qualifications; for doctorate, TOEFL (minimum score of 580 required), M Sc in computer science or comparable academic qualifications. *Faculty research:* Artificial intelligence and logic programming, graphics and image processing, software and systems, theory of computing.

University of West Florida, College of Science and Technology, Department of Computer Science, Pensacola, FL 32514-5750. Offers computer science (MS); systems and control engineering (MS). Part-time and evening/weekend programs available. *Students:* 37 full-time (14 women), 98 part-time (19 women); includes 21 minority (7 African Americans, 9 Asian Americans or Pacific Islanders, 5 Hispanic Americans), 10 international. Average age 35. 65 applicants, 97% accepted. In 1998, 25 degrees awarded. *Degree requirements:* For master's, computer language required, thesis optional, foreign language not required. *Entrance requirements:* For master's, GRE General Test (minimum combined score of 1000 required). *Application deadline:* For fall admission, 7/1; for spring admission, 11/1. Applications are processed on a rolling basis. Application fee: $20. Tuition, state resident: full-time $3,582; part-time $149 per credit hour. Tuition, nonresident: full-time $12,240; part-time $510 per credit hour. *Financial aid:* Fellowships, research assistantships available. *Unit head:* Dr. Ed Rodgers, Chairperson, 850-474-2542.

University of Windsor, College of Graduate Studies and Research, Faculty of Science, School of Computer Science, Windsor, ON N9B 3P4, Canada. Offers M Sc. Part-time programs available. *Degree requirements:* For master's, thesis optional. *Entrance requirements:* For master's, GRE, TOEFL (minimum score of 550 required), minimum B average. *Faculty research:* Database management, information retrieval systems, programming languages, computer graphics, artificial intelligence.

University of Wisconsin–Madison, Graduate School, College of Letters and Science, Department of Computer Sciences, Madison, WI 53706-1380. Offers MS, PhD. Part-time programs available. *Faculty:* 35 full-time (5 women). *Students:* 219 full-time (42 women), 1 part-time; includes 13 minority (1 African American, 11 Asian Americans or Pacific Islanders, 1 Hispanic American), 90 international. 425 applicants, 38% accepted. In 1998, 63 master's, 12 doctorates awarded. Terminal master's awarded for partial completion of doctoral program. *Degree requirements:* For master's, foreign language and thesis not required; for doctorate, dissertation required, foreign language not required. *Entrance requirements:* For master's, GRE General Test, GRE Subject Test; for doctorate, GRE General Test, GRE Subject Test (minimum score of 600 required). *Average time to degree:* Master's–1.5 years full-time, 3 years part-time; doctorate–6 years full-time. *Application deadline:* For fall admission, 12/31. Application fee: $45. Electronic applications accepted. *Financial aid:* In 1998–99, 19 fellowships (averaging $9,540 per year), 68 research assistantships with tuition reimbursements (averaging $11,946 per year), 92 teaching assistantships with tuition reimbursements (averaging $9,634 per year) were awarded. *Unit head:* James Goodman, Chair, 608-262-1204, Fax: 608-262-9777, E-mail: chair@cs.wisc.edu. *Application contact:* Lorene Webber, Graduate Coordinator, 608-262-7967, Fax: 608-262-9777, E-mail: admissions@cs.wisc.edu.

See in-depth description on page 923.

University of Wisconsin–Milwaukee, Graduate School, College of Engineering and Applied Science, Department of Electrical Engineering and Computer Science, Milwaukee, WI 53201-0413. Offers computer science (MS, PhD). Part-time programs available. *Faculty:* 21 full-time (1 woman). *Students:* 19 full-time (6 women), 56 part-time (14 women); includes 6 minority (all Asian Americans or Pacific Islanders), 28 international. 64 applicants, 48% accepted. In 1998, 14 master's awarded. *Degree requirements:* For master's, thesis or alternative required, foreign language not required; for doctorate, dissertation, internship required, foreign language not required. *Entrance requirements:* For master's, minimum GPA of 2.75; for doctorate, minimum GPA of 3.5. *Application deadline:* For fall admission, 1/1 (priority date); for spring admission, 9/1. Applications are processed on a rolling basis. Application fee: $45 ($75 for international students). *Financial aid:* In 1998–99, 3 research assistantships, 10 teaching assistantships were awarded.; fellowships, career-related internships or fieldwork and unspecified assistantships also available. Aid available to part-time students. Financial aid application deadline: 4/15. *Unit head:* K. Vairavan, Co-Chair, 414-229-4677.

University of Wyoming, Graduate School, College of Arts and Sciences, Department of Computer Science, Laramie, WY 82071. Offers computer science (MS, PhD); mathematics-computer science (PhD). Part-time programs available. *Faculty:* 10 full-time (0 women), 1 (woman) part-time. *Students:* 23 full-time (6 women), 14 part-time (3 women); includes 1 minority (Asian American or Pacific Islander), 20 international. 83 applicants, 23% accepted. In 1998, 6 master's, 1 doctorate awarded. Terminal master's awarded for partial completion of doctoral program. *Degree requirements:* For master's, thesis required; for doctorate, dissertation required. *Entrance requirements:* For master's, GRE General Test (minimum combined score of 900 required), minimum GPA of 3.0; for doctorate, GRE General Test (minimum combined score of 1000 required), minimum GPA of 3.0. *Application deadline:* For fall admission, 3/1; for spring admission, 10/1. Applications are processed on a rolling basis. Application fee: $40. Electronic applications accepted. Tuition, state resident: full-time $2,520; part-time $140 per credit hour. Tuition, nonresident: full-time $7,790; part-time $433 per credit hour. Required fees: $400; $7 per credit hour. Full-time tuition and fees vary according to course load and program. *Financial aid:* Research assistantships, teaching assistantships, career-related internships or fieldwork, Federal Work-Study, and tuition waivers (partial) available. Financial aid application deadline: 3/1. *Faculty research:* Fault-tolerant computing, distributed systems, knowledge representation, case-based reasoning, automated reasoning, parallel database access, parallel compilers. Total annual research expenditures: $169,701. *Unit head:* Dr. Henry Bauer, Chairman, 307-766-5190, Fax: 307-766-4036. *Application contact:* Graduate Coordinator, 307-766-5190, Fax: 307-766-4036, E-mail: cosc@uwyo.edu.

Utah State University, School of Graduate Studies, College of Science, Department of Computer Science, Logan, UT 84322. Offers MS. Part-time and evening/weekend programs available. *Faculty:* 12 full-time (1 woman). *Students:* 55 full-time (17 women), 39 part-time (5 women); includes 3 minority (2 Asian Americans or Pacific Islanders, 1 Native American), 57 international. Average age 26. 64 applicants, 78% accepted. In 1998, 25 degrees awarded. *Degree requirements:* For master's, computer language, thesis (for some programs), research project required, foreign language not required. *Entrance requirements:* For master's, GRE General Test (score in 40th percentile or higher required), GRE Subject Test, TOEFL (minimum score of 550 required), minimum GPA of 3.0. *Application deadline:* For fall admission, 6/15 (priority date); for spring admission, 10/15. Applications are processed on a rolling basis. Application fee: $40. Tuition, state resident: full-time $1,492. Tuition, nonresident: full-time $5,232. Required fees: $434. Tuition and fees vary according to course load. *Financial aid:* In 1998–99, 3 fellowships with partial tuition reimbursements, 19 research assistantships with partial tuition reimbursements, 38 teaching assistantships with partial tuition reimbursements were awarded.; career-related internships or fieldwork, Federal Work-Study, institutionally-sponsored loans, and tuition waivers (partial) also available. Aid available to part-time students. Financial aid application deadline: 2/16. *Faculty research:* Artificial intelligence, software engineering, parallelism. Total annual research expenditures: $275,000. *Unit head:* Donald H. Cooley, Head, 435-797-2451, Fax: 435-797-3265, E-mail: usucs@cc.usu.edu. *Application contact:* Greg Jones, Graduate Adviser, 435-797-3267, Fax: 435-797-3265, E-mail: jones@greg.cs.usu.edu.

Vanderbilt University, School of Engineering, Department of Computer Science, Nashville, TN 37240-1001. Offers M Eng, MS, PhD. MS and PhD offered through the Graduate School. Part-time programs available. *Faculty:* 9 full-time (1 woman), 1 (woman) part-time. *Students:* 43 full-time (9 women), 9 part-time (3 women); includes 4 minority (2 African Americans, 1 Asian American or Pacific Islander, 1 Hispanic American), 26 international. Average age 26. 84 applicants, 42% accepted. In 1998, 10 master's awarded (100% found work related to degree); 3 doctorates awarded (100% found work related to degree). *Degree requirements:* For master's, thesis required (for some programs), foreign language not required; for doctorate, dissertation required, foreign language not required. *Entrance requirements:* For master's, GRE General Test (minimum combined score of 1700 on three sections required), TOEFL (minimum score of 600 required); for doctorate, GRE General Test (minimum combined score of 1800 on three sections required), TOEFL (minimum score of 600 required). *Application deadline:* For fall admission, 1/15. Application fee: $40. *Financial aid:* In 1998–99, 33 students received aid, including 2 fellowships with tuition reimbursements available (averaging $3,000 per year), 6 research assistantships with tuition reimbursements available (averaging $15,000 per year), 18 teaching assistantships with tuition reimbursements available (averaging $11,250 per year); institutionally-sponsored loans and tuition waivers (full and partial) also available. Financial aid application deadline: 1/15. *Faculty research:* Artificial intelligence, performance evaluation, databases, software engineering, computational science, image processing, graph algorithms, learning theory. Total annual research expenditures: $448,815. *Unit head:* Edward J. White, Acting Chair, 615-322-2796. *Application contact:* Douglas Fisher, Director of Graduate Studies, 615-322-2976, Fax: 615-343-8006, E-mail: csdgs@vuse.vanderbilt.edu.

Villanova University, Graduate School of Liberal Arts and Sciences, Department of Computing Sciences, Villanova, PA 19085-1699. Offers MS. Part-time and evening/weekend programs available. *Students:* 63 full-time (25 women), 122 part-time (44 women); includes 17 minority (2 African Americans, 13 Asian Americans or Pacific Islanders, 2 Hispanic Americans), 103 international. Average age 30. 115 applicants, 84% accepted. In 1998, 73 degrees awarded. *Degree requirements:* For master's, computer language, independent study project required, thesis optional, foreign language not required. *Entrance requirements:* For master's, minimum GPA of 3.0. *Application deadline:* For fall admission, 8/1 (priority date); for spring admission, 12/1. Application fee: $40. *Financial aid:* Research assistantships, Federal Work-Study and scholarships available. Financial aid application deadline: 4/1; financial aid applicants required to submit FAFSA. *Unit head:* Dr. Robert Beck, Director, 610-519-7310.

Virginia Commonwealth University, School of Graduate Studies, College of Humanities and Sciences, Department of Mathematical Sciences, Program in Computer Science, Richmond, VA 23284-9005. Offers MS. *Students:* 8 full-time (1 woman), 28 part-time (7 women); includes 16 minority (2 African Americans, 12 Asian Americans or Pacific Islanders, 1 Hispanic American, 1 Native American) In 1998, 12 degrees awarded. *Degree requirements:* Foreign language not required. *Entrance requirements:* For master's, GRE General Test, GRE Subject Test, TOEFL. *Application deadline:* For fall admission, 7/1; for spring admission, 11/15. Application fee: $30. Tuition, state resident: full-time $4,031; part-time $224 per credit hour. Tuition, nonresident: full-time $11,946; part-time $664 per credit hour. Required fees: $1,081; $40 per credit hour. Tuition and fees vary according to campus/location and program. *Unit head:* Gary Linstrom, Director of Graduate Admissions, 801-581-5586. *Application contact:* Dr. James A. Wood, Director of Graduate Studies, 804-828-1301, Fax: 804-828-8785, E-mail: jawood@vcu.edu.

Virginia Polytechnic Institute and State University, Graduate School, College of Arts and Sciences, Department of Computer Science, Program in Computer Science, Blacksburg, VA 24061. Offers MS, PhD. *Students:* 128 full-time (36 women), 94 part-time (24 women); includes 29 minority (4 African Americans, 20 Asian Americans or Pacific Islanders, 4 Hispanic Americans, 1 Native American), 118 international. 433 applicants, 31% accepted. In 1998, 49 master's, 5 doctorates awarded. *Degree requirements:* For master's, computer language, thesis required (for some programs); for doctorate, computer language required. *Entrance requirements:* For master's and doctorate, GRE General Test (minimum combined score of 1200 required), TOEFL. *Application deadline:* For fall admission, 12/1 (priority date). Applications are processed on a rolling basis. Application fee: $25. *Financial aid:* Application deadline: 4/1. *Application contact:* Dr. Verna Schuetz, Assistant Head, 540-231-6931.

See in-depth description on page 925.

Wake Forest University, Graduate School, Department of Computer Science, Winston-Salem, NC 27109. Offers MS. Part-time programs available. *Faculty:* 4 full-time (0 women). *Students:* 7 full-time (3 women), 3 part-time (1 woman); includes 2 minority (1 Asian American or Pacific Islander, 1 Hispanic American) Average age 27. 24 applicants, 38% accepted. In 1998, 5 degrees awarded. *Degree requirements:* For master's, one foreign language required, (computer language can substitute), thesis optional. *Entrance requirements:* For master's, GRE General Test, GRE Subject Test. *Application deadline:* For fall admission, 2/15. Application fee: $25. *Financial aid:* In 1998–99, 10 students received aid, including 1 fellowship, 9 teaching assistantships; scholarships also available. Aid available to part-time students. Financial aid application deadline: 2/15; financial aid applicants required to submit FAFSA. *Unit head:* Dr. Stan Thomas, Director, 336-758-5354, E-mail: sjt@wfu.edu.

Washington State University, Graduate School, College of Engineering and Architecture, School of Electrical Engineering and Computer Science, Program in Computer Science, Pullman, WA 99164. Offers MS, PhD. *Faculty:* 10 full-time (0 women). *Students:* 31 full-time (9 women), 10 part-time (2 women); includes 3 minority (all Asian Americans or Pacific Islanders), 29 international. In 1998, 19 master's, 2 doctorates awarded. *Degree requirements:* For master's, oral exam required, thesis optional, foreign language not required; for doctorate, dissertation, oral exam, qualifying exam required, foreign language not required. *Entrance requirements:* For master's and doctorate, GRE General Test, GRE Subject Test, minimum GPA of 3.0. *Average time to degree:* Master's–2 years full-time; doctorate–4 years full-time. *Application deadline:* For fall admission, 3/1 (priority date). Applications are processed on a rolling basis. Application fee: $35. *Financial aid:* In 1998–99, 5 research assistantships, 18 teaching assistantships were awarded.; career-related internships or fieldwork, Federal Work-Study, institutionally-sponsored loans, tuition waivers (partial), and teaching associateships also available. Financial aid application deadline: 4/1; financial aid applicants required to submit FAFSA. *Application contact:* Dr. Zoran Obradovic, Graduate Coordinator, 509-335-6601, Fax: 509-335-3818, E-mail: zoran@eecs.wsu.edu.

Washington University in St. Louis, School of Engineering and Applied Science, Sever Institute of Technology, Department of Computer Science, St. Louis, MO 63130-4899. Offers MS, D Sc. Part-time programs available. *Faculty:* 14 full-time (1 woman), 11 part-time (1 woman). *Students:* 69 full-time (11 women), 23 part-time (2 women); includes 9 minority (1 African American, 7 Asian Americans or Pacific Islanders, 1 Hispanic American), 45 international. 239 applicants, 27% accepted. In 1998, 23 master's, 1 doctorate awarded (100% found work related to degree). Terminal master's awarded for partial completion of doctoral program. *Degree requirements:* For master's, thesis optional, foreign language not required; for doctorate, dissertation required, foreign language not required. *Entrance requirements:* For master's, GRE General Test, GRE Subject Test, TOEFL (minimum score of 575 required), TWE (minimum score of 4.5 required), minimum undergraduate GPA of 2.75; for doctorate, GRE General Test, GRE Subject Test, TOEFL (minimum score of 575 required), TWE (minimum score of 4.5 required). *Average time to degree:* Doctorate–3 years full-time. *Application deadline:* For fall admission, 5/1; for spring admission, 9/15. Application fee: $20. *Financial aid:* In 1998–99, 44 research assistantships with tuition reimbursements (averaging $16,620 per year), 4 teaching assistantships with tuition reimbursements (averaging $16,620 per year) were awarded.; fellowships, career-related internships or fieldwork and Federal Work-Study also available. Financial aid application deadline: 1/15. *Faculty research:* Multimedia networking, distributed systems, artificial intelligence, computational science, multimedia-user interfaces, computer engineering, compilers. Total annual research expenditures: $6.1 million. *Unit head:* Dr. Gruia-Catalin Roman, Chairman, 314-935-6132, Fax: 314-935-7302, E-mail: roman@cs.wustl.edu. *Application contact:* Jean Grothe, Graduate Admissions Coordinator, 314-935-6160, Fax: 314-935-7302, E-mail: admissions@cs.wustl.edu.

See in-depth description on page 927.

Wayne State University, Graduate School, College of Science, Department of Computer Science, Detroit, MI 48202. Offers computer science (MA, MS, PhD); electronics and computer control systems (MS). *Degree requirements:* For master's, computer language, thesis required (for some programs), foreign language not required; for doctorate, computer language, dissertation required, foreign language not required. *Entrance requirements:* For master's, TOEFL (minimum score of 550 required), GRE General Test (minimum combined score of 1700 on three sections required); for doctorate, TOEFL (minimum score of 550 required), GRE General Test (minimum combined score of 1800 on three sections required). *Faculty research:* Neural computation, artificial intelligence, software engineering, distributed systems, databases.

See in-depth description on page 931.

Webster University, School of Business and Technology, Department of Mathematics and Computer Science, St. Louis, MO 63119-3194. Offers computer distributed systems (Certificate); computer science (MS). *Faculty:* 3 full-time (1 woman). *Students:* 121 full-time (29 women), 52 part-time (15 women); includes 49 minority (22 African Americans, 19 Asian Americans or Pacific Islanders, 8 Hispanic Americans), 19 international. In 1998, 31 degrees awarded. *Degree requirements:* For master's, computer language required, foreign language and thesis not required. *Entrance requirements:* For master's, 36 hours of graduate course work. *Application deadline:* Applications are processed on a rolling basis. Application fee: $25 ($50 for international students). *Financial aid:* Federal Work-Study available. Aid available to part-time students. Financial aid application deadline: 4/1; financial aid applicants required to submit FAFSA. *Faculty research:* Databases, computer information systems networks, operating systems, computer architecture. *Unit head:* Anna Barbara Sakurai, Department Chair, 314-968-7027, Fax: 314-963-6050, E-mail: sakuraab@webster.edu. *Application contact:* Dr. Beth Russell, Director of Graduate Admissions, 314-968-7089, Fax: 314-968-7166, E-mail: russelmb@webster.edu.

West Chester University of Pennsylvania, Graduate Studies, College of Arts and Sciences, Department of Computer Science, West Chester, PA 19383. Offers MS. Part-time and evening/weekend programs available. *Students:* 38. Average age 34. *Degree requirements:* For master's, computer language, comprehensive exam required, foreign language and thesis not required. *Entrance requirements:* For master's, GRE General Test, interview. *Application deadline:* For fall admission, 4/15 (priority date); for spring admission, 10/15. Applications are processed on a rolling basis. Application fee: $25. Tuition, state resident: full-time $3,780; part-time $210 per credit. Tuition, nonresident: full-time $6,610; part-time $367 per credit. Required fees: $684; $39 per credit. Tuition and fees vary according to course load. *Financial aid:* In 1998–99, research assistantships with full tuition reimbursements (averaging $5,000 per year) Aid available to part-time students. Financial aid application deadline: 2/15. *Unit head:* Dr. John Weaver, Chair, 610-436-3228. *Application contact:* Dr. Elaine Milito, Graduate Coordinator, 610-436-2690, E-mail: emilito@wcupa.edu.

Western Carolina University, Graduate School, College of Arts and Sciences, Department of Mathematics and Computer Science, Cullowhee, NC 28723. Offers MA Ed, MAT, MS. Part-time and evening/weekend programs available. *Faculty:* 13. *Students:* 10 full-time (4 women), 7 part-time (4 women). 17 applicants, 71% accepted. In 1998, 4 degrees awarded. *Degree requirements:* For master's, comprehensive exam required, thesis optional, foreign language not required. *Entrance requirements:* For master's, GRE General Test, GRE Subject Test (applied mathematics applicants). *Application deadline:* For fall admission, 5/1 (priority date); for spring admission, 10/1 (priority date). Applications are processed on a rolling basis. Application fee: $35. Tuition, state resident: full-time $918. Tuition, nonresident: full-time $8,188. Required fees: $881. *Financial aid:* In 1998–99, 7 students received aid, including 7 research assistantships with full and partial tuition reimbursements available (averaging $5,679 per year); fellowships, teaching assistantships, Federal Work-Study, grants, and institutionally-sponsored loans also available. Financial aid application deadline: 3/15; financial aid applicants required to submit FAFSA. *Unit head:* Harold Williford, Head, 828-227-7245. *Application contact:* Kathleen Owen, Assistant to the Dean, 828-227-7398, Fax: 828-227-7480, E-mail: kowen@wcu.edu.

Western Connecticut State University, Division of Graduate Studies, School of Arts and Sciences, Department of Mathematics and Computer Science, Danbury, CT 06810-6885. Offers mathematics and computer science (MA); theoretical mathematics (MA). Part-time and evening/weekend programs available. In 1998, 3 degrees awarded. *Degree requirements:* For master's, comprehensive exam, thesis, or research project required. *Entrance requirements:* For master's, minimum GPA of 2.5. *Application deadline:* For fall admission, 8/1 (priority date). Applications are processed on a rolling basis. Application fee: $40. *Financial aid:* Fellowships, career-related internships or fieldwork and Federal Work-Study available. Aid available to part-time students. Financial aid application deadline: 5/1; financial aid applicants required to submit FAFSA. *Unit head:* Dr. Josephine Hamer, Chair, 203-837-9347. *Application contact:* Chris Shankle, Associate Director of Graduate Admissions, 203-837-8244, Fax: 203-837-8338, E-mail: shanklec@wcsu.edu.

Western Illinois University, School of Graduate Studies, College of Business and Technology, Department of Computer Science, Macomb, IL 61455-1390. Offers MS. Part-time programs available. *Faculty:* 12 full-time (2 women). *Students:* 46 full-time (13 women), 16 part-time (4 women); includes 3 minority (1 African American, 2 Asian Americans or Pacific Islanders), 38 international. Average age 28. 48 applicants, 85% accepted. In 1998, 9 degrees awarded. *Degree requirements:* For master's, thesis or alternative required, foreign language not required. *Entrance requirements:* For master's, proficiency in Pascal. *Application deadline:* Applications are processed on a rolling basis. Application fee: $0 ($25 for international students). *Financial aid:* In 1998–99, 20 students received aid, including 11 research assistantships with full tuition reimbursements available (averaging $4,880 per year), 9 teaching assistantships with full tuition reimbursements available (averaging $6,376 per year) Financial aid applicants required to submit FAFSA. *Faculty research:* Space-based life support, public Internet services, artificial intelligence, artificial languages, algorithmic fluency. *Unit head:* Dr. James Bradford, Chairperson, 309-298-1452. *Application contact:* Barbara Baily, Director of Graduate Studies, 309-298-1806, Fax: 309-298-2245, E-mail: barb_baily@ccmail.wiu.edu.

Western Kentucky University, Graduate Studies, Ogden College of Science, Technology, and Health, Department of Computer Science, Bowling Green, KY 42101-3576. Offers MS. Part-time programs available. *Faculty:* 9 full-time (3 women). *Students:* 9 full-time (4 women), 6 part-time; includes 1 minority (Native American), 9 international. Average age 29. 54 applicants, 61% accepted. In 1998, 7 degrees awarded. *Degree requirements:* For master's, computer language required (for some programs), comprehensive exam required, foreign language and thesis not required. *Entrance requirements:* For master's, GRE General Test. *Application deadline:* For fall admission, 8/1 (priority date); for spring admission, 12/1. Applications are processed on a rolling basis. Application fee: $30. Tuition, state resident: full-time $2,590; part-time $140 per hour. Tuition, nonresident: full-time $6,430; part-time $387 per hour. Required fees: $370. *Financial aid:* In 1998–99, 6 research assistantships with partial tuition reimbursements (averaging $4,400 per year) were awarded.; Federal Work-Study, institutionally-sponsored loans, and service awards also available. Aid available to part-time students. Financial aid application deadline: 4/1; financial aid applicants required to submit FAFSA. *Faculty research:* Artificial intelligence. *Unit head:* Dr. Arthur Shindhelm, Head, 270-745-6247.

Western Michigan University, Graduate College, College of Arts and Sciences, Department of Computer Science, Kalamazoo, MI 49008. Offers MS, PhD. *Students:* 74 full-time (10 women), 49 part-time (8 women); includes 3 minority (1 African American, 2 Asian Americans or Pacific Islanders), 105 international. 184 applicants, 50% accepted. In 1998, 49 master's awarded. *Degree requirements:* For master's, oral exams required, thesis not required; for doctorate, dissertation required. *Entrance requirements:* For master's and doctorate, GRE General Test. *Application deadline:* For fall admission, 2/15 (priority date). Applications are processed on a rolling basis. Application fee: $25. *Financial aid:* Fellowships, research assistantships, teaching assistantships, career-related internships or fieldwork and institutionally-sponsored loans available. Financial aid application deadline: 2/15; financial aid applicants required to submit FAFSA. *Unit head:* Dr. Ajay Gupta, Chairperson, 616-387-5645. *Application contact:* Paula J. Boodt, Coordinator, Graduate Admissions and Recruitment, 616-387-2000, Fax: 616-387-2355, E-mail: paula.boodt@wmich.edu.

Western Michigan University, Graduate College, College of Arts and Sciences, Department of Mathematics and Statistics, Kalamazoo, MI 49008. Offers applied mathematics (MS); biostatistics (MS); computational mathematics (MS); graph theory and computer science (PhD); mathematics (MA, PhD), including mathematics (MA), mathematics education; statistics (MS, PhD). *Students:* 28 full-time (11 women), 67 part-time (36 women); includes 6 minority (5 Asian Americans or Pacific Islanders, 1 Hispanic American), 18 international. *Degree requirements:* For master's, thesis not required; for doctorate, dissertation required. *Entrance requirements:* For doctorate, GRE General Test. *Application deadline:* For fall admission, 2/15 (priority date). Applications are processed on a rolling basis. Application fee: $25. *Unit head:* Dr. John Petro, Chairperson, 616-387-4513. *Application contact:* Paula J. Boodt, Coordinator, Graduate Admissions and Recruitment, 616-387-2000, Fax: 616-387-2355, E-mail: paula.boodt@wmich.edu.

Western Washington University, Graduate School, College of Arts and Sciences, Department of Computer Science, Bellingham, WA 98225-5996. Offers MS. Part-time programs available. *Faculty:* 11. *Students:* 5 full-time (1 woman), 3 part-time (1 woman). 8 applicants, 100% accepted. In 1998, 4 degrees awarded. *Degree requirements:* For master's, computer language, project required, thesis optional, foreign language not required. *Entrance requirements:* For master's, GRE General Test, TOEFL, minimum GPA of 3.0 in last 60 semester hours or last 90 quarter hours. *Application deadline:* For fall admission, 5/1 (priority date); for winter admission, 10/1; for spring admission, 2/1. Applications are processed on a rolling basis. Application fee: $35. Tuition, state resident: full-time $3,247; part-time $146 per credit hour.

Computer Science–Information Science

Western Washington University (continued)
Tuition, nonresident: full-time $13,364; part-time $445 per credit hour. Required fees: $254; $85 per quarter. *Financial aid:* In 1998–99, 1 research assistantship with partial tuition reimbursement (averaging $7,563 per year), 5 teaching assistantships with partial tuition reimbursements (averaging $7,563 per year) were awarded.; Federal Work-Study, institutionally-sponsored loans, scholarships, and tuition waivers (partial) also available. Aid available to part-time students. Financial aid application deadline: 2/15; financial aid applicants required to submit FAFSA. *Unit head:* Dr. Debra Jusak, Chair, 360-650-3805. *Application contact:* Dr. Michael Meehan, Graduate Adviser, 360-650-3795.

West Virginia University, College of Engineering and Mineral Resources, Department of Computer Science and Electrical Engineering, Program in Computer Science, Morgantown, WV 26506. Offers MS, PhD. *Degree requirements:* For master's, computer language, thesis required, foreign language not required; for doctorate, one foreign language, computer language, dissertation, comprehensive exam required. *Entrance requirements:* For master's, GRE General Test (score in 50th percentile or higher required), TOEFL (minimum score of 550 required), minimum GPA of 3.0; for doctorate, GRE General Test (score in 50th percentile or higher required), GRE Subject Test, TOEFL (minimum score of 550 required). *Faculty research:* Artificial intelligence, knowledge-based simulation, data communications, mathematical computations, software engineering.

Wichita State University, Graduate School, Fairmount College of Liberal Arts and Sciences, Department of Computer Science, Wichita, KS 67260. Offers MS. *Faculty:* 6 full-time (0 women), 8 part-time (0 women). *Students:* 70 full-time (11 women), 36 part-time (11 women); includes 6 minority (1 African American, 4 Asian Americans or Pacific Islanders, 1 Hispanic American), 80 international. Average age 28. 176 applicants, 56% accepted. In 1998, 16 degrees awarded. *Degree requirements:* For master's, project required, thesis optional. *Entrance requirements:* For master's, GRE, TOEFL (minimum score of 550 required), minimum GPA of 2.75. *Application deadline:* For spring admission, 1/1. Applications are processed on a rolling basis. Application fee: $25 ($40 for international students). Electronic applications accepted. *Financial aid:* In 1998–99, 2 research assistantships (averaging $5,000 per year), 12 teaching assistantships with full tuition reimbursements (averaging $3,000 per year) were awarded.; unspecified assistantships also available. Financial aid application deadline: 4/1. *Faculty research:* Software engineering, database systems. Total annual research expenditures: $217,628. *Unit head:* Dr. Prakash Ramanan, Interim Chair, 316-978-3920, Fax: 316-978-3984, E-mail: ramanan@cs.twsu.edu. *Application contact:* Dr. Rajiv Bagai, Coordinator, 316-978-3923, Fax: 316-978-3984, E-mail: bagai@cs.twsu.edu.

Worcester Polytechnic Institute, Graduate Studies, Department of Computer Science, Worcester, MA 01609-2280. Offers MS, PhD, Advanced Certificate, Certificate. Part-time and evening/weekend programs available. *Faculty:* 18 full-time (4 women), 7 part-time (0 women). *Students:* 70 full-time (19 women), 70 part-time (11 women); includes 17 minority (3 African Americans, 13 Asian Americans or Pacific Islanders, 1 Hispanic American), 51 international. 331 applicants, 61% accepted. In 1998, 28 master's awarded. Terminal master's awarded for partial completion of doctoral program. *Degree requirements:* For master's, computer language required, thesis optional, foreign language not required; for doctorate, computer language, dissertation required, foreign language not required. *Entrance requirements:* For master's and doctorate, GRE General Test (combined average 2009 on three sections), TOEFL (minimum score of 550 required; average 618). *Application deadline:* For fall admission, 2/15 (priority date); for spring admission, 10/15 (priority date). Applications are processed on a rolling basis. Application fee: $50. Electronic applications accepted. *Financial aid:* In 1998–99, 31 students received aid, including 3 fellowships with full tuition reimbursements available (averaging $12,000 per year), 12 research assistantships with full tuition reimbursements available (averaging $15,000 per year), 16 teaching assistantships with full tuition reimbursements available (averaging $11,970 per year); career-related internships or fieldwork, grants, institutionally-sponsored loans, and scholarships also available. Financial aid application deadline: 2/15; financial aid applicants required to submit FAFSA. *Faculty research:* Artificial intelligence, network performance and analysis, language processor systems, system design simulation. Total annual research expenditures: $305,995. *Unit head:* Dr. Micha Hofri, Head, 508-831-5670, Fax: 508-831-5776, E-mail: hofri@cs.wpi.edu. *Application contact:* Elke Rundensteiner, Graduate Coordinator, 508-831-5815, Fax: 508-831-5776, E-mail: rundenst@cs.wpi.edu.

See in-depth description on page 933.

Wright State University, School of Graduate Studies, College of Engineering and Computer Science, Department of Computer Science and Engineering, Computer Science Program, Dayton, OH 45435. Offers MSCS. *Students:* 73 full-time (18 women), 21 part-time (9 women); includes 10 minority (all Asian Americans or Pacific Islanders), 65 international. 173 applicants, 45% accepted. In 1998, 45 degrees awarded. *Degree requirements:* For master's, thesis optional, foreign language not required. *Entrance requirements:* For master's, GRE General Test, TOEFL (minimum score of 550 required), minimum GPA of 3.0 in major, 2.7 overall. Application fee: $25. *Financial aid:* Fellowships, research assistantships, teaching assistantships, unspecified assistantships available. Aid available to part-time students. Financial aid application deadline: 3/31; financial aid applicants required to submit FAFSA. *Faculty research:* Artificial intelligence, human-computer interaction, graphics, software engineering, logic and symbolic programming. *Unit head:* Dr. David H. T. Chen, Assistant Dean for Graduate Programs and Research, 610-499-4049, Fax: 610-499-4059, E-mail: david.h.chen@widener.edu. *Application contact:* Dr. Jay E. Dejongh, Graduate Adviser, 937-775-5136, Fax: 937-775-5133.

See in-depth description on page 935.

Wright State University, School of Graduate Studies, College of Engineering and Computer Science, Department of Computer Science and Engineering, Program in Computer Science and Engineering, Dayton, OH 45435. Offers PhD. *Students:* 20 full-time (2 women), 5 part-time; includes 1 minority (Asian American or Pacific Islander), 13 international. 67 applicants, 60% accepted. In 1998, 1 degree awarded. *Degree requirements:* For doctorate, dissertation, candidacy and general exams required, foreign language not required. *Entrance requirements:* For doctorate, GRE General Test, TOEFL (minimum score of 550 required), minimum GPA of 3.3. Application fee: $25. *Financial aid:* Application deadline: 3/31. *Unit head:* Dr. P. Bruce Berra, Director, 937-775-5138, Fax: 937-775-5133.

Yale University, Graduate School of Arts and Sciences, Department of Computer Science, New Haven, CT 06520. Offers PhD. *Students:* 30 full-time (4 women), 2 part-time; includes 2 minority (both Asian Americans or Pacific Islanders), 20 international. 124 applicants, 23% accepted. In 1998, 7 doctorates awarded. *Degree requirements:* For doctorate, dissertation required, foreign language not required. *Entrance requirements:* For doctorate, GRE General Test, GRE Subject Test. *Average time to degree:* Doctorate–6.2 years full-time. *Application deadline:* For fall admission, 1/4. Application fee: $65. *Financial aid:* Fellowships, research assistantships, teaching assistantships, Federal Work-Study and institutionally-sponsored loans available. Aid available to part-time students. *Unit head:* Chair, 203-432-1997. *Application contact:* Admissions Information, 203-432-2770.

See in-depth description on page 937.

York University, Faculty of Graduate Studies, Faculty of Science, Program in Computer Science, Toronto, ON M3J 1P3, Canada. Offers M Sc. *Degree requirements:* For master's, thesis required, foreign language not required.

Information Science

Alcorn State University, School of Graduate Studies, School of Arts and Sciences, Department of Mathematical Sciences, Lorman, MS 39096-9402. Offers computer and information sciences (MS).

Allentown College of St. Francis de Sales, Graduate Division, Program in Information Systems, Center Valley, PA 18034-9568. Offers MSIS. Part-time and evening/weekend programs available. Average age 31. 14 applicants, 100% accepted. In 1998, 8 degrees awarded. *Degree requirements:* For master's, computer language, comprehensive exam required, thesis optional, foreign language not required. *Average time to degree:* Master's–3 years part-time. *Application deadline:* For fall admission, 8/30 (priority date). Applications are processed on a rolling basis. Application fee: $35. *Faculty research:* Digital communication, numerical analysis, database design. *Unit head:* Dr. Julius G. Bede, Director, 610-282-1100 Ext. 1280, E-mail: bede@accnov.allencol.edu.

American University, College of Arts and Sciences, Department of Computer Science and Information Systems, Program in Information Systems, Washington, DC 20016-8001. Offers MS, Certificate. Part-time and evening/weekend programs available. *Faculty:* 16 full-time (5 women), 16 part-time (1 woman). *Students:* 25 full-time (9 women), 81 part-time (33 women); includes 32 minority (21 African Americans, 6 Asian Americans or Pacific Islanders, 5 Hispanic Americans), 31 international. 102 applicants, 73% accepted. In 1998, 62 degrees awarded. *Degree requirements:* For master's, computer language, thesis or alternative, comprehensive exam required, foreign language not required. *Entrance requirements:* For master's, minimum GPA of 3.0. *Average time to degree:* Master's–2 years full-time, 4 years part-time. *Application deadline:* For fall admission, 2/1 (priority date); for spring admission, 10/1. Applications are processed on a rolling basis. Application fee: $50. *Financial aid:* In 1998–99, 8 fellowships with full tuition reimbursements were awarded.; teaching assistantships, career-related internships or fieldwork, Federal Work-Study, institutionally-sponsored loans, and unspecified assistantships also available. Financial aid application deadline: 2/1. *Faculty research:* Artificial intelligence, database systems, software engineering, expert systems. Total annual research expenditures: $153,000. *Unit head:* Paula J. Boodt, Coordinator, Graduate Admissions and Recruitment, 616-387-2000, Fax: 616-387-2355, E-mail: paula.boodt@wmich.edu. *Application contact:* Sonia McKie, Staff Assistant, 202-885-1486, Fax: 202-885-1479, E-mail: smckie@american.edu.

Arizona State University East, College of Technology and Applied Sciences, Department of Information and Management Technology, Mesa, AZ 85212. Offers MS. Part-time and evening/weekend programs available. *Faculty:* 7 full-time (1 woman), 1 part-time (0 women). *Students:* 6 full-time (0 women), 3 part-time; includes 2 minority (1 Asian American or Pacific Islander, 1 Hispanic American), 1 international. Average age 36. 4 applicants, 75% accepted. *Degree requirements:* For master's, thesis or applied project and oral defense required. *Entrance requirements:* For master's, minimum GPA of 3.0. *Application deadline:* Applications are processed on a rolling basis. Application fee: $45. *Financial aid:* In 1998–99, 2 students received aid; research assistantships, teaching assistantships, career-related internships or fieldwork, Federal Work-Study, grants, scholarships, tuition waivers (full and partial), and unspecified assistantships available. Aid available to part-time students. Financial aid application deadline: 3/1; financial aid applicants required to submit FAFSA. *Faculty research:* Environmental technology management, digital imaging and publications, web development, interactive multimedia, high performance project management. Total annual research expenditures: $68,242. *Unit head:* Dr. Thomas Schildgen, Chair, 602-727-1005, Fax: 602-727-1684, E-mail: ts@asu.edu. *Application contact:* Dr. Thomas Schildgen, Chair, 602-727-1005, Fax: 602-727-1684, E-mail: ts@asu.edu.

Ball State University, Graduate School, College of Communication, Information, and Media, Center for Information and Communication Sciences, Muncie, IN 47306-1099. Offers MS. *Faculty:* 8. *Students:* 94 full-time (26 women), 54 part-time (15 women). Average age 28. 114 applicants, 84% accepted. In 1998, 79 degrees awarded. Application fee: $15 ($25 for international students). *Financial aid:* Teaching assistantships available. *Unit head:* Dr. Rayford Steele, Director, 765-285-1889, E-mail: rsteele@bsu.edu.

Barry University, School of Adult and Continuing Education, Program in Information Technology, Miami Shores, FL 33161-6695. Offers MS. *Unit head:* Dr. Ben Huey, Assistant Chair, 480-965-3190. *Application contact:* John Guest, Director, 305-899-3300.

Bradley University, Graduate School, College of Liberal Arts and Sciences, Department of Computer Science, Peoria, IL 61625-0002. Offers computer information systems (MS); computer science (MS). Part-time and evening/weekend programs available. *Degree requirements:* For master's, thesis or alternative, comprehensive exam required, foreign language not required. *Entrance requirements:* For master's, TOEFL (minimum score of 500 required).

Brooklyn College of the City University of New York, Division of Graduate Studies, Department of Computer and Information Science, Brooklyn, NY 11210-2889. Offers computer and information science (MA, PhD); computer science and health science (MS); economics and computer and information science (MPS); information systems (MS). Part-time and evening/weekend programs available. *Faculty:* 10 full-time (2 women), 11 part-time (1 woman). *Students:* 23 full-time (10 women), 184 part-time (62 women); includes 93 minority (33 African Americans, 56 Asian Americans or Pacific Islanders, 4 Hispanic Americans), 37 international. Average age 27. In 1998, 42 degrees awarded. *Degree requirements:* For master's, computer language, comprehensive exam or thesis required. *Entrance requirements:* For master's, TOEFL (minimum score of 525 required), previous course work in computer science. *Application deadline:* For fall admission, 3/1; for spring admission, 11/1. Application fee: $40. *Financial aid:* Fellowships, research assistantships, teaching assistantships, career-related internships or fieldwork, Federal Work-Study, institutionally-sponsored loans, scholarships, and tuition waivers (partial) available. Aid available to part-time students. Financial aid application deadline: 5/1; financial aid applicants required to submit FAFSA. *Faculty research:* Networks and distributed systems, programming languages, modeling and computer applications, algorithms, artificial intelligence, theoretical computer science. *Unit head:* Dr. Aaron H. Tenenbaum, Chairperson, 718-951-5657. *Application contact:* Gerald Weiss, Graduate Counselor, 718-951-5217, Fax: 718-951-4842, E-mail: weiss@sci.brooklyn.cuny.edu.

California State University, Fullerton, Graduate Studies, School of Engineering and Computer Science, Department of Computer Science, Fullerton, CA 92834-9480. Offers applications administrative information systems (MS); applications mathematical methods (MS); computer science (MS); information processing systems (MS). Part-time programs available. *Faculty:* 11 full-time, 32 part-time. *Students:* 10 full-time (4 women), 98 part-time (27 women); includes 47 minority (2 African Americans, 42 Asian Americans or Pacific Islanders, 3 Hispanic Americans), 24 international. *Degree requirements:* For master's, computer language, comprehensive exam, project or thesis required. *Entrance requirements:* For master's, GRE General Test (minimum combined score of 1100 required), minimum undergraduate GPA of 2.5. Application fee: $55. Tuition, nonresident: part-time $264 per unit. Required fees: $1,947; $1,281 per year. *Unit head:* Dr. Nick Mousouris, Chair, 714-278-3700. *Application contact:* Dr. Susamma Barua, Adviser, 714-278-3700.

Capitol College, Graduate School, Laurel, MD 20708-9759. Offers electronic commerce management (MS); information and telecommunications systems management (MS); systems management (MS). Part-time and evening/weekend programs available. *Faculty:* 1 full-time (0 women), 24 part-time (4 women). *Students:* 6 full-time (3 women), 145 part-time (39 women); includes 23 minority (9 African Americans, 7 Asian Americans or Pacific Islanders, 6 Hispanic Americans, 1 Native American), 2 international. *Degree requirements:* For master's, foreign language and thesis not required. *Entrance requirements:* For master's, GRE General Test (minimum score of 500 on each section required), minimum GPA of 3.0. *Application deadline:* For fall admission, 7/1 (priority date); for winter admission, 12/1 (priority date); for spring admission, 3/1 (priority date). Applications are processed on a rolling basis. Application fee: $25 ($100 for international students). Electronic applications accepted. Tuition: Full-time $5,328; part-time $888 per course. Tuition and fees vary according to campus/location and program. *Unit head:* Dr. Joe Goldsmith, Dean of Graduate Studies, 301-369-2800, Fax: 301-953-3876. *Application contact:* Sandy Perriello, Coordinator of Graduate Administration, 703-998-5503, Fax: 703-379-8239, E-mail: gradschool@capitol-college.edu.

Carleton University, Faculty of Graduate Studies, Faculty of Engineering and Design, Ottawa-Carleton Institute for Electrical and Computer Engineering, Department of Systems and Computer Engineering, Ottawa, ON K1S 5B6, Canada. Offers electrical engineering (M Eng, PhD); information and systems science (M Sc); telecommunications technology management (M Eng). *Faculty:* 23 full-time (3 women). *Students:* 64 full-time (14 women), 51 part-time (9 women). *Degree requirements:* For master's, thesis optional; for doctorate, dissertation, comprehensive exam required. *Entrance requirements:* For master's, TOEFL (minimum score of 550 required), honors degree; for doctorate, TOEFL (minimum score of 550 required), MA Sc or M Eng. *Application deadline:* For fall admission, 3/1. Applications are processed on a rolling basis. Application fee: $35. *Unit head:* Rafik Goubran, Chair, 613-520-5740, Fax: 613-520-5742, E-mail: goubran@sce.carleton.ca. *Application contact:* D. G. Swartz, Supervisor of Graduate Studies, 613-520-5740, Fax: 613-520-5727, E-mail: swartz@sce.carleton.ca.

Carleton University, Faculty of Graduate Studies, Faculty of Science, Information and Systems Science Committee, Ottawa, ON K1S 5B6, Canada. Offers M Sc. *Students:* 61 full-time (23 women), 31 part-time (7 women). Average age 33. In 1998, 18 degrees awarded. *Degree requirements:* For master's, computer language required, thesis optional. *Entrance requirements:* For master's, TOEFL (minimum score of 550 required), honors degree. *Average time to degree:* Master's–1.8 years full-time, 3.5 years part-time. *Application deadline:* For fall admission, 3/1 (priority date). Applications are processed on a rolling basis. Application fee: $35. *Financial aid:* Application deadline: 3/1. *Faculty research:* Software engineering, real-time and microprocessor programming, computer communications. *Unit head:* Mike J. Moore, Coordinator, 613-520-2600 Ext. 2160, Fax: 613-520-5733.

Carnegie Mellon University, Graduate School of Industrial Administration, Program in Information Science, Pittsburgh, PA 15213-3891. Offers PhD. *Faculty:* 6 full-time (0 women). *Degree requirements:* For doctorate, dissertation required, foreign language not required. *Entrance requirements:* For doctorate, GRE General Test. *Application deadline:* For fall admission, 2/1. Application fee: $50. *Financial aid:* Fellowships available. Financial aid application deadline:5/1. *Unit head:* Sandy Perriello, Coordinator of Graduate Administration, 703-998-5503, Fax: 703-379-8239, E-mail: gradschool@capitol-college.edu. *Application contact:* Jackie Cavendish, Administrative Assistant, 412-268-2301.

Carnegie Mellon University, Information Networking Institute, Pittsburgh, PA 15213-3891. Offers MS. *Students:* 34 full-time (5 women), 4 part-time (1 woman); includes 2 Asian Americans or Pacific Islanders, 1 Hispanic American, 24 international. Average age 25. In 1998, 30 degrees awarded. *Degree requirements:* For master's, computer language, thesis required, foreign language not required. *Entrance requirements:* For master's, GRE General Test, previous course work in computer science, computer engineering, or electrical engineering. *Average time to degree:* Master's–1.5 years full-time, 2 years part-time. *Application deadline:* For fall admission, 2/1 (priority date). Applications are processed on a rolling basis. Application fee: $30. *Financial aid:* Federal Work-Study and scholarships available. Financial aid application deadline: 2/1. *Faculty research:* Wireless, including protocols, architecture, innovative platforms, compression research, human factors, high speed networks, and prototype for industrial clients. *Unit head:* Bernard Bennington, Director, 412-268-7195, E-mail: ini-director@ini.cmu.edu. *Application contact:* Susan Jones, Graduate Coordinator, 412-268-5721, Fax: 412-268-7196, E-mail: sj1q@andrew.cmu.edu.

Announcement: The MS in Information Networking is a cooperative endeavor of the Schools of Engineering, Computer Science, and Business, providing an alternative to the conventional one-year computer science or electrical engineering graduate program by integrating both and adding some features of an MBA program. Now in its eleventh year, the program carefully selects 35 people from the engineering and computer science disciplines to form each new class. The program provides technical electives aimed at several areas of specialization including (a) the telecommunications and computing industries, (b) wireless and mobile computing, (c) the financial services industry, and (d) systems integrating and consulting.

See in-depth description on page 711.

Carnegie Mellon University, School of Computer Science, Language Technologies Institute, Pittsburgh, PA 15213-3891. Offers MS, PhD. *Faculty:* 14 full-time (5 women). *Students:* 21 full-time (6 women), 4 part-time (1 woman); includes 2 minority (1 Asian American or Pacific Islander, 1 Hispanic American), 13 international. 54 applicants, 54% accepted. In 1998, 3 master's awarded (67% found work related to degree, 33% continued full-time study); 2 doctorates awarded (50% entered university research/teaching, 50% found other work related to degree). Terminal master's awarded for partial completion of doctoral program. *Degree requirements:* For master's, foreign language and thesis not required; for doctorate, dissertation required, foreign language not required. *Entrance requirements:* For master's and doctorate, GRE General Test, GRE Subject Test, TOEFL. *Average time to degree:* Master's–2 years full-time. *Application deadline:* For fall admission, 1/5. Application fee: $50. *Financial aid:* In 1998–99, research assistantships with tuition reimbursements (averaging $16,800 per year) *Faculty research:* Machine translation, natural language processing, speech and information retrieval, literacy. Total annual research expenditures: $2.2 million. *Unit head:* Jaime G. Carbonell, Director, 412-268-7279, Fax: 412-268-6298, E-mail: jgc@cs.cmu.edu. *Application contact:* Catherine Morrow, Admissions Coordinator, 412-268-6591, E-mail: ltp@cs.cmu.edu.

See in-depth description on page 713.

Case Western Reserve University, School of Graduate Studies, The Case School of Engineering, Department of Electrical, Systems, Computer Engineering and Science, Cleveland, OH 44106. Offers computer engineering and science (MS, PhD), including computer engineering, computing and information science; electrical engineering (MS, PhD); systems and control engineering (MS, PhD). Part-time and evening/weekend programs available. Postbaccalaureate distance learning degree programs offered (minimal on-campus study). *Faculty:* 28 full-time (2 women). *Students:* 73 full-time (13 women), 101 part-time (17 women). Terminal master's awarded for partial completion of doctoral program. *Degree requirements:* For master's and doctorate, thesis/dissertation required, foreign language not required. *Entrance requirements:* For master's and doctorate, GRE General Test, TOEFL (minimum score of 550 required). *Application deadline:* For fall admission, 3/1; for spring admission, 11/1. Applications are processed on a rolling basis. Application fee: $25. Electronic applications accepted. *Unit head:* Dr. Robert V. Edwards, Acting Chairman, 216-368-2800, Fax: 216-368-6888, E-mail: rve2@po.cwru.edu. *Application contact:* Elizabethanne M. Fuller, Department Assistant, 216-368-4080, Fax: 216-368-2668, E-mail: emf4@po.cwru.edu.

Claremont Graduate University, Graduate Programs, School of Information Sciences, Claremont, CA 91711-6163. Offers information systems (MIS); management of information systems (MSMIS, PhD). Part-time programs available. *Faculty:* 4 full-time (0 women), 3 part-time (0 women). *Students:* 24 full-time (11 women), 79 part-time (34 women); includes 28 minority (4 African Americans, 19 Asian Americans or Pacific Islanders, 5 Hispanic Americans), 38 international. Average age 36. 97 applicants, 49% accepted. In 1998, 31 master's, 2 doctorates awarded. Terminal master's awarded for partial completion of doctoral program. *Degree requirements:* For master's, computer language required, thesis not required; for doctorate, computer language, dissertation required. *Entrance requirements:* For master's and doctorate, GMAT, GRE General Test. *Application deadline:* For fall admission, 2/15 (priority date). Applications are processed on a rolling basis. Application fee: $40. Electronic applications accepted. Tuition: Full-time $20,950; part-time $913 per unit. Required fees: $65 per semester. Tuition and fees vary according to program. *Financial aid:* Fellowships, research assistantships, teaching assistantships, Federal Work-Study and institutionally-sponsored loans available. Aid available to part-time students. Financial aid application deadline: 2/15; financial aid applicants required to submit FAFSA. *Faculty research:* GPSS, man-machine interaction, organizational aspects of computing, implementation of information systems, information systems practice. *Unit head:* Lorne Olfman, Chair, 909-621-8209, Fax: 909-621-8390, E-mail: lorne.olfman@cgu.edu. *Application contact:* Nancy Back, Program Coordinator, 909-621-8209, Fax: 909-621-8390, E-mail: infosci@cgu.edu.

See in-depth description on page 719.

Clark Atlanta University, School of Arts and Sciences, Department of Computer and Information Science, Atlanta, GA 30314. Offers MS. *Students:* 30 full-time (12 women), 10 part-time (4 women); includes 19 African Americans, 12 Asian Americans or Pacific Islanders, 9 international. In 1998, 23 degrees awarded. *Degree requirements:* For master's, one foreign language (computer language can substitute), thesis required. *Entrance requirements:* For master's, GRE General Test, minimum GPA of 2.5. *Application deadline:* For fall admission, 4/1; for spring admission, 11/1. Applications are processed on a rolling basis. Application fee: $40. *Financial aid:* Fellowships, research assistantships available. Financial aid application deadline: 4/30. *Unit head:* Dr. Kenneth Perry, Chairperson, 404-880-6951. *Application contact:* Michelle Clark-Davis, Graduate Program Assistant, 404-880-8709.

Coleman College, Graduate Program in Information Systems, La Mesa, CA 91942-1532. Offers MS. Evening/weekend programs available. *Degree requirements:* For master's, computer language required, thesis optional, foreign language not required.

The College of Saint Rose, Graduate Studies, School of Mathematics and Sciences, Program in Computer Information Systems, Albany, NY 12203-1419. Offers MS. Part-time and evening/weekend programs available. *Faculty:* 1 full-time (0 women). 9 applicants, 78% accepted. *Degree requirements:* Foreign language not required. *Application deadline:* For fall admission, 7/15 (priority date); for spring admission, 12/1. Applications are processed on a rolling basis. Application fee: $30. *Financial aid:* Research assistantships, career-related internships or fieldwork and tuition waivers (partial) available. Aid available to part-time students. Financial aid application deadline: 3/1; financial aid applicants required to submit FAFSA.

DePaul University, School of Computer Science, Telecommunications, and Information Systems, Program in Information Systems, Chicago, IL 60604-2287. Offers MS. Part-time and evening/weekend programs available. *Students:* 128 full-time (54 women), 125 part-time (52 women); includes 99 minority (21 African Americans, 70 Asian Americans or Pacific Islanders, 8 Hispanic Americans), 15 international. Average age 30. 110 applicants, 76% accepted. In 1998, 46 degrees awarded. *Degree requirements:* For master's, computer language, comprehensive exam required, foreign language and thesis not required. *Application deadline:* For fall admission, 8/1 (priority date); for winter admission, 11/15 (priority date); for spring admission, 5/1 (priority date). Applications are processed on a rolling basis. Application fee: $25. *Financial aid:* Research assistantships, teaching assistantships, Federal Work-Study, tuition waivers (partial), and unspecified assistantships available. Financial aid application deadline: 4/1. *Unit head:* Dr. Susy Chan, Director, 312-362-5247, E-mail: schan@wppost.depaul.edu. *Application contact:* Anne B. Morley, Director of Student Services, 312-362-8714, Fax: 312-362-6116.

Drexel University, Graduate School, College of Information Science and Technology, Philadelphia, PA 19104-2875. Offers information studies (PhD, CAS); information systems (MSIS); library and information science (MS). *Accreditation:* ALA (one or more programs are accredited). Part-time and evening/weekend programs available. *Faculty:* 17 full-time (7 women), 17 part-time (6 women). *Students:* 67 full-time (49 women), 454 part-time (217 women); includes 68 minority (33 African Americans, 30 Asian Americans or Pacific Islanders, 4 Hispanic Americans, 1 Native American), 40 international. Average age 37. 260 applicants, 65% accepted. In 1998, 174 master's, 3 doctorates awarded. *Degree requirements:* For master's, foreign language and thesis not required; for doctorate, dissertation required, foreign language not required. *Entrance requirements:* For master's, GRE General Test, TOEFL (minimum score of 600 required); for doctorate, GRE General Test, TOEFL (minimum score of 600 required), master's degree. *Application deadline:* For fall admission, 8/21. Applications are processed on a rolling basis. Application fee: $35. Tuition: Part-time $452 per credit. Required fees: $375; $67 per term. *Financial aid:* In 1998–99, 2 research assistantships, 5 teaching assistantships were awarded.; career-related internships or fieldwork, Federal Work-Study, institutionally-sponsored loans, tuition waivers (partial), and unspecified assistantships also available. Aid available to part-time students. Financial aid application deadline: 2/1. *Faculty research:* Bibliometric analysis, information management, scientific communication, expert systems, man-machine interfaces in information transfer. *Unit head:* Dr. Tom Childers, Interim Dean, 215-895-2474. *Application contact:* Anne B. Tanner, Associate Dean, 215-895-2474.

See in-depth description on page 737.

East Tennessee State University, School of Graduate Studies, College of Applied Science and Technology, Department of Computer and Information Sciences, Johnson City, TN 37614-0734. Offers computer science (MS); information sciences (MS). Part-time and evening/weekend programs available. *Degree requirements:* For master's, computer language, thesis, written comprehensive exam required, foreign language not required. *Entrance requirements:* For master's, GRE General Test (minimum combined score of 1050 required), TOEFL (minimum score of 550 required), minimum GPA of 2.5. *Faculty research:* Operating systems, database design, artificial intelligence, simulation, parallel algorithms.

Florida Institute of Technology, Graduate School, College of Engineering, Division of Electrical and Computer Science and Engineering, Program in Computer Information Systems, Melbourne, FL 32901-6975. Offers MS. Part-time and evening/weekend programs available. *Students:* 31 full-time (13 women), 43 part-time (14 women); includes 9 minority (3 African Americans, 4 Asian Americans or Pacific Islanders, 2 Hispanic Americans), 41 international. Average age 32. 111 applicants, 72% accepted. In 1998, 23 degrees awarded. *Degree requirements:* For master's, computer language, comprehensive exam required, foreign language and thesis not required. *Entrance requirements:* For master's, minimum GPA of 3.0. *Application deadline:* Applications are processed on a rolling basis. Application fee: $50. Electronic applications accepted. Tuition: Part-time $575 per credit hour. Required fees: $100. Tuition and fees vary according to campus/location and program. *Financial aid:* In 1998–99, 7 students received aid, including 4 research assistantships with full and partial tuition reimbursements available (averaging $2,705 per year), 2 teaching assistantships with full and partial tuition reimbursements available (averaging $3,610 per year); tuition remission also available. Financial aid application deadline: 3/1; financial aid applicants required to submit FAFSA. *Faculty research:* Artificial intelligence, software engineering, parallel processing, computer graphics, programming languages. *Unit head:* Dr. William D. Shoaff, Chair, 407-674-8066, Fax: 407-676-8192, E-mail: wds@cs.fit.edu. *Application contact:* Carolyn P. Farrior, Associate Dean of Graduate Admissions, 407-674-7118, Fax: 407-723-9468, E-mail: cfarrior@fit.edu.

Information Science

Florida Institute of Technology, Graduate School, School of Extended Graduate Studies, Program in Computer Information Systems, Melbourne, FL 32901-6975. Offers MS. Part-time and evening/weekend programs available. Postbaccalaureate distance learning degree programs offered (no on-campus study). *Students:* 3 full-time (2 women), 84 part-time (23 women); includes 15 minority (5 African Americans, 9 Asian Americans or Pacific Islanders, 1 Hispanic American), 2 international. Average age 36. 38 applicants, 55% accepted. In 1998, 16 degrees awarded (100% found work related to degree). *Degree requirements:* For master's, computer language, comprehensive exam required, foreign language and thesis not required. *Entrance requirements:* For master's, minimum GPA of 3.0. *Average time to degree:* Master's–1 year full-time, 3 years part-time. *Application deadline:* Applications are processed on a rolling basis. Application fee: $50. Electronic applications accepted. Tuition: Part-time $270 per credit hour. Part-time tuition and fees vary according to campus/location. *Financial aid:* Application deadline: 3/1; *Unit head:* Lu Uber, Administrative Assistant, 518-276-2660, E-mail: uberm@rpi.edi. *Application contact:* Carolyn P. Farrior, Associate Dean of Graduate Admissions, 407-674-7118, Fax: 407-723-9468, E-mail: cfarrior@fit.edu.

George Mason University, School of Information Technology and Engineering, Department of Information and Software Engineering, Fairfax, VA 22030-4444. Offers information systems (MS); software systems engineering (MS). Part-time and evening/weekend programs available. *Faculty:* 16 full-time (1 woman), 9 part-time (1 woman). *Students:* 59 full-time (29 women), 469 part-time (174 women); includes 156 minority (29 African Americans, 115 Asian Americans or Pacific Islanders, 11 Hispanic Americans, 1 Native American), 77 international. Average age 33. 287 applicants, 70% accepted. In 1998, 159 degrees awarded. *Degree requirements:* For master's, computer language required, thesis optional, foreign language not required. *Entrance requirements:* For master's, GMAT or GRE General Test, TOEFL (minimum score of 575 required), minimum GPA of 3.0 in last 60 hours. *Application deadline:* For fall admission, 5/1; for spring admission, 11/1. Application fee: $30. Electronic applications accepted. Tuition, state resident: full-time $4,416; part-time $184 per credit hour. Tuition, nonresident: full-time $12,516; part-time $522 per credit hour. Tuition and fees vary according to program. *Financial aid:* Fellowships, research assistantships, teaching assistantships available. Aid available to part-time students. Financial aid application deadline: 3/1; financial aid applicants required to submit FAFSA. *Faculty research:* Security, database management, real time systems, software quality. Total annual research expenditures: $380,638. *Unit head:* Dr. Sushil Jajodia, Chairperson, 703-993-1653, Fax: 703-993-1638, E-mail: sjajodia@gmu.edu. *Application contact:* Sandy Mayo, Student Adviser, 703-993-1640, Fax: 703-993-1638.

George Mason University, School of Information Technology and Engineering, Interdisciplinary Program in Information Technology, Fairfax, VA 22030-4444. Offers PhD. Part-time and evening/weekend programs available. *Faculty:* 3 full-time (1 woman). *Students:* 43 full-time (11 women), 307 part-time (62 women); includes 59 minority (19 African Americans, 36 Asian Americans or Pacific Islanders, 2 Hispanic Americans, 2 Native Americans), 102 international. Average age 38. 167 applicants, 64% accepted. In 1998, 29 degrees awarded. *Degree requirements:* For doctorate, dissertation, comprehensive oral and written exams required, foreign language not required. *Entrance requirements:* For doctorate, GRE General Test, TOEFL (minimum score of 575 required), minimum graduate GPA of 3.5. *Application deadline:* For fall admission, 5/1; for spring admission, 11/1. Application fee: $30. Electronic applications accepted. Tuition, state resident: full-time $4,416; part-time $184 per credit hour. Tuition, nonresident: full-time $12,516; part-time $522 per credit hour. Tuition and fees vary according to program. *Financial aid:* Fellowships, research assistantships, teaching assistantships, Federal Work-Study and institutionally-sponsored loans available. Aid available to part-time students. Financial aid application deadline: 3/1; financial aid applicants required to submit FAFSA. *Unit head:* Dr. W. Murray Black, Associate Dean, 703-993-1500, Fax: 703-993-1734, E-mail: mblack@gmu.edu. *Application contact:* Student Services, 703-993-1499, Fax: 703-993-1497, E-mail: sitegrad@gmu.edu.

Georgia Southwestern State University, Graduate Studies, School of Computer and Information Science, Americus, GA 31709-4693. Offers computer information systems (MS); computer science (MS). Part-time programs available. *Faculty:* 5 full-time (0 women). *Students:* 24 full-time (10 women), 13 part-time (6 women); includes 15 minority (3 African Americans, 12 Asian Americans or Pacific Islanders), 12 international. Average age 22. 20 applicants, 25% accepted. In 1998, 10 degrees awarded (100% found work related to degree). *Degree requirements:* For master's, computer language, thesis required (for some programs), foreign language not required. *Entrance requirements:* For master's, GRE General Test (minimum combined score of 900 required), minimum GPA of 3.0. *Average time to degree:* Master's–2 years full-time, 3 years part-time. *Application deadline:* For fall admission, 8/1; for spring admission, 12/15. Applications are processed on a rolling basis. Application fee: $20. Tuition, state resident: full-time $2,000; part-time $83 per hour. Tuition, nonresident: full-time $5,000; part-time $333 per hour. Required fees: $235. *Financial aid:* In 1998–99, 10 students received aid, including fellowships (averaging $6,000 per year), 10 teaching assistantships with full tuition reimbursements available (averaging $4,000 per year); grants also available. Financial aid application deadline: 9/1. *Faculty research:* Database, Internet technologies, computational complexity, encryption. Total annual research expenditures: $30,000. *Unit head:* Dr. John J. Stroyls, Acting Chairman, 912-931-2263, Fax: 912-931-2270, E-mail: mdaniels@canes.gsw.edu. *Application contact:* Chris Laney, Graduate Admissions Specialist, 912-931-2027, Fax: 912-931-2983, E-mail: claney@canes.gsw.edu.

Grand Valley State University, Science and Mathematics Division, Department of Computer Science and Information Systems, Allendale, MI 49401-9403. Offers information systems (MS); software engineering (MS). Part-time and evening/weekend programs available. *Faculty:* 7 full-time (0 women), 1 part-time (0 women). *Students:* 15 full-time (5 women), 55 part-time (17 women); includes 10 minority (2 African Americans, 8 Asian Americans or Pacific Islanders), 16 international. Average age 34. 25 applicants, 80% accepted. In 1998, 4 degrees awarded (100% found work related to degree). *Degree requirements:* For master's, computer language, thesis or alternative required, foreign language not required. *Entrance requirements:* For master's, GMAT or GRE General Test. *Average time to degree:* Master's–2 years full-time, 3.5 years part-time. *Application deadline:* For fall admission, 2/1. Application fee: $20. *Faculty research:* Object technology, distributed computing, information systems management. *Unit head:* Bruce J. Klein, Associate Professor, 616-895-2048, Fax: 616-895-2106, E-mail: kleinb@gvsu.edu.

Harvard University, Extension School, Cambridge, MA 02138-3722. Offers applied sciences (CAS); English for graduate and professional studies (DGP); information technology (ALM); liberal arts (ALM); museum studies (CMS); premedical studies (Diploma); public health (CPH); publication and communication (CPC); special studies in administration and management (CSS). Part-time and evening/weekend programs available. *Faculty:* 450 part-time. *Degree requirements:* For master's, thesis required, foreign language not required; for other advanced degree, computer language required, foreign language and thesis not required. *Entrance requirements:* For master's and other advanced degree, TOEFL (minimum score of 600 required), TWE (minimum score of 5 required). *Application deadline:* Applications are processed on a rolling basis. Application fee: $75. *Unit head:* Michael Shinagel, Dean. *Application contact:* Program Director, 617-495-4024, Fax: 617-495-9176.

Hood College, Graduate School, Programs in Computer and Information Sciences, Frederick, MD 21701-8575. Offers MS. *Students:* 16 full-time (12 women), 150 part-time (66 women); includes 34 minority (5 African Americans, 28 Asian Americans or Pacific Islanders, 1 Hispanic American), 10 international. Average age 33. In 1998, 37 degrees awarded. *Degree requirements:* For master's, thesis or alternative required, foreign language not required. *Entrance requirements:* For master's, minimum GPA of 2.5. *Application deadline:* Applications are processed on a rolling basis. Application fee: $30. *Financial aid:* Career-related internships or fieldwork, institutionally-sponsored loans, and tuition waivers (partial) available. Aid available to part-time students. Financial aid applicants required to submit FAFSA. *Unit head:* John Boon, Chairperson, 301-696-3763, Fax: 301-696-3597, E-mail: boon@hood.edu. *Application contact:* Graduate School Office, 301-696-3600, Fax: 301-696-3597, E-mail: hoodgrad@hood.edu.

Instituto Tecnológico y de Estudios Superiores de Monterrey, Campus Monterrey, Graduate and Research Division, Program in Computer Science, Monterrey, 64849, Mexico. Offers artificial intelligence (PhD); computer science (MS); information systems (MS); information technology (MS). Part-time programs available. *Degree requirements:* For master's and doctorate, thesis/dissertation required. *Entrance requirements:* For master's, PAEG, TOEFL; for doctorate, TOEFL, master's in related field. *Faculty research:* Distributed systems, software engineering, decision support systems.

Instituto Tecnológico y de Estudios Superiores de Monterrey, Campus Monterrey, Graduate and Research Division, Program in Informatics, Monterrey, 64849, Mexico. Offers PhD. Part-time programs available. *Degree requirements:* For doctorate, dissertation, technological project, arbitrated publication of articles required. *Entrance requirements:* For doctorate, GRE General Test, GRE Subject Test, TOEFL, master's in related field. *Faculty research:* Artificial intelligence, distributed systems, software engineering, decision support systems.

Instituto Tecnológico y de Estudios Superiores de Monterrey, Campus Morelos, Programs in Information Science, Cuernavaca, 62000, Mexico. Offers administration of information technology (MATI); computer science (MCC, DCC); information technology (MTI).

Instituto Tecnológico y de Estudios Superiores de Monterrey, Campus Sonora Norte, Program in Technological Information Management, Hermosillo, 83000, Mexico. Offers MA.

Kansas State University, Graduate School, College of Engineering, Department of Computing and Information Sciences, Manhattan, KS 66506. Offers computer science (MS, PhD); software engineering (MSE). Part-time programs available. Postbaccalaureate distance learning degree programs offered (minimal on-campus study). *Faculty:* 16 full-time (2 women). *Students:* 70 full-time (22 women), 50 part-time (13 women); includes 10 minority (1 African American, 8 Asian Americans or Pacific Islanders, 1 Hispanic American), 76 international. 312 applicants, 27% accepted. In 1998, 47 master's, 3 doctorates awarded (100% entered university research/teaching). Terminal master's awarded for partial completion of doctoral program. *Degree requirements:* For master's, computer language required, thesis optional, foreign language not required; for doctorate, computer language, dissertation required, foreign language not required. *Entrance requirements:* For master's, TOEFL (minimum score of 575 required); for doctorate, GRE General Test (minimum score of 400 on verbal section, 650 on quantitative, 600 on analytical required), TOEFL (minimum score of 575 required). *Average time to degree:* Master's–2 years full-time, 3 years part-time; doctorate–5 years full-time. *Application deadline:* For fall admission, 4/1 (priority date); for spring admission, 11/1. Applications are processed on a rolling basis. Application fee: $0 ($25 for international students). Electronic applications accepted. *Financial aid:* In 1998–99, 10 research assistantships (averaging $9,000 per year), 29 teaching assistantships (averaging $9,000 per year) were awarded.; career-related internships or fieldwork and institutionally-sponsored loans also available. Financial aid application deadline: 2/15. *Faculty research:* Programming language semantics, distributed systems, real time systems and algorithms, database management systems. Total annual research expenditures: $511,000. *Unit head:* Dr. Virgil E. Wallentine, Head, 785-532-6350, Fax: 785-532-7353, E-mail: virg@cis.ksu.edu. *Application contact:* Dr. David Gustafson, Graduate Coordinator, 785-532-6350, Fax: 785-532-7353, E-mail: dag@cis.ksu.edu.

See in-depth description on page 761.

Kennesaw State University, College of Science and Mathematics, Program in Information Systems, Kennesaw, GA 30144-5591. Offers MSIS. *Entrance requirements:* For master's, GMAT (minimum score of 500 required) or GRE General Test (minimum combined score of 1350 on three sections required), minimum GPA of 2.75. *Application deadline:* For fall admission, 7/16; for spring admission, 9/19. *Unit head:* Dr. Meryl S. King, Director, 770-423-6354, Fax: 770-423-6731, E-mail: mking@kennesaw.edu. *Application contact:* Dr. Thomas M. Hughes, Acting Associate Director of Graduate Admissions, 770-499-3008, Fax: 770-423-6541, E-mail: thughes@kennesaw.edu.

Knowledge Systems Institute, Program in Computer and Information Sciences, Skokie, IL 60076. Offers MS. *Faculty:* 4 full-time (0 women), 23 part-time (3 women). *Students:* 83 full-time (25 women), 61 part-time (21 women). Application fee: $40. Electronic applications accepted. Tuition: Full-time $6,730; part-time $275 per credit hour. Required fees: $5 per course. $30 per term. *Unit head:* Judy Pan, Executive Director, 847-679-3135, Fax: 847-679-3166, E-mail: judy@ksi.edu. *Application contact:* Margaret Price, Admissions Officer, 847-679-3135, Fax: 847-679-3166, E-mail: mprice@ksi.edu.

Kutztown University of Pennsylvania, College of Graduate Studies and Extended Learning, College of Liberal Arts and Sciences, Program in Computer and Information Science, Kutztown, PA 19530-0730. Offers MS. *Faculty:* 10 full-time (3 women). *Students:* 2 full-time (1 woman), 11 part-time (4 women); includes 5 minority (all Asian Americans or Pacific Islanders) Average age 33. In 1998, 5 degrees awarded. *Degree requirements:* For master's, computer language, comprehensive exam or thesis required. *Entrance requirements:* For master's, GRE General Test, TOEFL, TSE. *Application deadline:* For fall admission, 3/1; for spring admission, 8/1. Application fee: $25. *Financial aid:* Career-related internships or fieldwork, Federal Work-Study, tuition waivers (partial), and unspecified assistantships available. Financial aid application deadline: 3/15; financial aid applicants required to submit FAFSA. *Faculty research:* Artificial intelligence, expert systems, neural networks. *Unit head:* William Bateman, Chairperson, 610-683-4410, E-mail: bateman@kutztown.edu.

Long Island University, C.W. Post Campus, College of Liberal Arts and Sciences, Department of Computer Sciences, Brookville, NY 11548-1300. Offers computer science education (MS); information systems (MS); management engineering (MS). Part-time and evening/weekend programs available. *Faculty:* 9 full-time (3 women), 9 part-time (0 women). *Students:* 26 full-time (9 women), 104 part-time (34 women); includes 45 minority (10 African Americans, 15 Asian Americans or Pacific Islanders, 20 Hispanic Americans), 25 international. *Degree requirements:* For master's, computer language, thesis or alternative, comprehensive exam required, foreign language not required. *Entrance requirements:* For master's, bachelor's degree in science, mathematics, or engineering. *Application deadline:* Applications are processed on a rolling basis. Application fee: $30. Electronic applications accepted. *Unit head:* Dr. Susan Dorchak, Chair, 516-299-2293, E-mail: dorchak@homet.liunet.edu. *Application contact:* John Keane, Graduate Adviser, 516-299-2293.

Marist College, Graduate Programs, School of Computer Science and Mathematics, Poughkeepsie, NY 12601-1387. Offers computer science (MS), including information systems, software development. Part-time and evening/weekend programs available. *Faculty:* 12 full-time (2 women), 3 part-time (0 women). *Students:* 50 full-time (9 women), 79 part-time (19 women). *Degree requirements:* For master's, thesis optional, foreign language not required. *Application deadline:* For fall admission, 8/1 (priority date); for spring admission, 12/15. Applications are processed on a rolling basis. Application fee: $30. *Unit head:* Dr. Jose Torres, Dean, 914-575-3000 Ext. 2610, E-mail: jose.torres@marist.edu. *Application contact:* Dr. H. Griffin Walling, Dean of Graduate and Continuing Education, 914-575-3530, Fax: 914-575-3640, E-mail: griffin.walling@marist.edu.

Marshall University, Graduate College, Graduate School of Information, Technology and Engineering, Program in Information Systems, Huntington, WV 25755-2020. Offers MS. Part-time and evening/weekend programs available. *Students:* 11 full-time (5 women), 50 part-time (12 women); includes 5 minority (1 African American, 4 Asian Americans or Pacific Islanders) Average age 32. In 1998, 7 degrees awarded. *Degree requirements:* For master's, computer language, final project, oral exam required. *Entrance requirements:* For master's, GRE General Test, minimum undergraduate GPA of 2.5. *Financial aid:* Tuition waivers (full) available. Aid available to part-time students. Financial aid application deadline: 8/1; financial aid applicants required to submit FAFSA. *Unit head:* Dr. Tom Hankins, Director, 304-746-2044, E-mail: thankins@marshall.edu. *Application contact:* Ken O'Neal, Assistant Vice President, Adult Student Services, 304-746-2500 Ext. 1907, Fax: 304-746-1902, E-mail: oneal@marshall.edu.

Naval Postgraduate School, Graduate Programs, Program in Information Systems, Monterey, CA 93943. Offers MS. Program open only to commissioned officers of the United States and friendly nations and selected United States federal civilian employees. Part-time programs available. *Students:* 115 full-time, 15 international. In 1998, 47 degrees awarded. *Degree requirements:* For master's, computer language, thesis required, foreign language not required. *Unit head:* Dr. Dan C. Boger, Academic Group Chairman, 831-656-3671.

New Jersey Institute of Technology, Office of Graduate Studies, Department of Computer and Information Science, Newark, NJ 07102-1982. Offers bioinformatics (MS, PhD); computer and information science (PhD); computer science (MS); information systems (MS). MS and PhD (bioinformatics) offered jointly with the University of Medicine and Dentistry of New Jersey. Part-time and evening/weekend programs available. Terminal master's awarded for partial completion of doctoral program. *Degree requirements:* For master's, computer language, thesis required, foreign language not required; for doctorate, computer language, dissertation, residency required, foreign language not required. *Entrance requirements:* For master's, GRE General Test (minimum score of 450 on verbal section, 600 on quantitative, 550 on analytical required); for doctorate, GRE General Test (minimum score of 450 on verbal section, 600 on quantitative, 550 on analytical required), minimum graduate GPA of 3.5. Electronic applications accepted. *Faculty research:* Computer systems, communications and networking, artificial intelligence, database engineering, systems analysis.

New York University, Graduate School of Arts and Science, Courant Institute of Mathematical Sciences, Department of Computer Science, New York, NY 10012-1019. Offers computer science (MS, PhD); information systems (MS); scientific computing (MS). Part-time and evening/weekend programs available. *Faculty:* 30 full-time (1 woman). *Students:* 96 full-time (19 women), 254 part-time (50 women); includes 35 minority (2 African Americans, 26 Asian Americans or Pacific Islanders, 7 Hispanic Americans), 168 international. *Degree requirements:* For master's, computer language required, foreign language and thesis not required; for doctorate, computer language, dissertation, oral and written exams required, foreign language not required. *Entrance requirements:* For master's and doctorate, GRE General Test, GRE Subject Test, TOEFL. *Application deadline:* For fall admission, 1/4; for spring admission, 11/1. Application fee: $60. Tuition: Full-time $17,880; part-time $745 per credit. Required fees: $1,140; $35 per credit. Tuition and fees vary according to course load and program. *Unit head:* Richard Cole, Chair, 212-998-3011, Fax: 212-995-4124. *Application contact:* Ralph Grishman, Director of Graduate Studies, 212-998-3011, Fax: 212-995-4124, E-mail: admissions@cs.nyu.edu.

See in-depth description on page 773.

Northeastern University, College of Engineering, Information Systems Program, Boston, MA 02115-5096. Offers MS. Part-time programs available. *Students:* 111 full-time (74 women), 83 part-time (38 women); includes 7 minority (1 African American, 6 Asian Americans or Pacific Islanders), 96 international. Average age 26. 164 applicants, 73% accepted. In 1998, 91 degrees awarded. *Degree requirements:* For master's, computer language required, thesis optional, foreign language not required. *Entrance requirements:* For master's, GRE General Test. *Average time to degree:* Master's–2.45 years full-time, 3.86 years part-time. *Application deadline:* For fall admission, 4/15. Applications are processed on a rolling basis. Application fee: $50. *Financial aid:* In 1998–99, 28 students received aid, including 4 research assistantships with full tuition reimbursements available (averaging $12,450 per year), teaching assistantships with full tuition reimbursements available (averaging $12,450 per year); fellowships, career-related internships or fieldwork, Federal Work-Study, tuition waivers (full), and unspecified assistantships also available. Aid available to part-time students. Financial aid application deadline: 2/15; financial aid applicants required to submit FAFSA. *Faculty research:* Simulation analysis. *Unit head:* Dr. Agnes Chan, Associate Dean and Director of Graduate Program, 617-373-2462, Fax: 617-373-5121. *Application contact:* Stephen L. Gibson, Associate Director, 617-373-2711, Fax: 617-373-2501, E-mail: grad-eng@coe.neu.edu.

Northwestern University, The Graduate School, Robert R. McCormick School of Engineering and Applied Science, Department of Electrical and Computer Engineering, Program in Information Technology, Evanston, IL 60208. Offers MIT. Average age 32. *Entrance requirements:* For master's, GRE General Test, TOEFL (minimum score of 560 required), 2 years of professional experience. *Application deadline:* For fall admission, 6/30. Application fee: $50 ($55 for international students). *Unit head:* Abraham Haddad, Director, 847-491-5931, Fax: 847-467-3550, E-mail: ahaddad@ece.nwu.edu.

Nova Southeastern University, School of Computer and Information Sciences, Fort Lauderdale, FL 33314-7721. Offers computer information systems (MS, PhD); computer science (MS, PhD); computing technology in education (MS, Ed D, PhD); information science (PhD); information systems (PhD); management information systems (MS). Part-time and evening/weekend programs available. Postbaccalaureate distance learning degree programs offered. *Students:* 149 full-time (49 women), 382 part-time (145 women); includes 116 minority (45 African Americans, 31 Asian Americans or Pacific Islanders, 39 Hispanic Americans, 1 Native American), 34 international. Average age 41. In 1998, 90 master's, 34 doctorates awarded. Terminal master's awarded for partial completion of doctoral program. *Degree requirements:* For master's, computer language required, thesis optional; for doctorate, computer language, dissertation required, foreign language not required. *Entrance requirements:* For master's, GRE or portfolio; for doctorate, GRE, master's degree, or portfolio. *Average time to degree:* Doctorate–4 years full-time. *Application deadline:* For fall admission, 6/1 (priority date); for spring admission, 1/1. Applications are processed on a rolling basis. Application fee: $50. Tuition: Part-time $370 per credit hour. Required fees: $50 per semester. Tuition and fees vary according to degree level and program. *Financial aid:* In 1998–99, 3 teaching assistantships were awarded.; Federal Work-Study also available. Aid available to part-time students. Financial aid application deadline: 5/1. *Faculty research:* Artificial intelligence, database management, human-computer interaction, distance education, computer education. *Unit head:* Dr. Edward Lieblein, Dean. *Application contact:* Karen Shoemaker, Administrative Assistant, 800-986-2247 Ext. 2000, Fax: 954-262-3915, E-mail: scisinfo@scis.nova.edu.

See in-depth description on page 779.

The Ohio State University, Graduate School, College of Engineering, Department of Computer and Information Science, Columbus, OH 43210. Offers MS, PhD. *Faculty:* 33. *Students:* 149 full-time (30 women), 20 part-time (6 women); includes 2 African Americans, 4 Hispanic Americans, 126 international. 429 applicants, 25% accepted. In 1998, 64 master's, 11 doctorates awarded. *Degree requirements:* For master's, computer language required, thesis optional, foreign language not required; for doctorate, computer language, dissertation required, foreign language not required. *Entrance requirements:* For master's and doctorate, GRE General Test. *Application deadline:* For fall admission, 7/1. Applications are processed on a rolling basis. Application fee: $30 ($40 for international students). *Financial aid:* Fellowships, teaching assistantships, career-related internships or fieldwork, Federal Work-Study, institutionally-sponsored loans, and administrative assistantships available. Aid available to part-time students. Financial aid application deadline: 1/15. *Unit head:* Stuart H. Zweben, Chairman, 614-292-5813, Fax: 614-292-2911, E-mail: zweben@cis.ohio-state.edu.

See in-depth description on page 781.

Pace University, School of Computer Science and Information Systems, New York, NY 10038. Offers computer communications and networks (Certificate); computer science (MS); computing studies (DPS); information systems (MS); New 9/1/1999 (DPS); object-oriented programming (Certificate); telecommunications (MS, Certificate). Part-time and evening/weekend programs available. *Faculty:* 36 full-time, 44 part-time. *Students:* 93 full-time (35 women), 468 part-time (170 women); includes 101 minority (40 African Americans, 41 Asian Americans or Pacific Islanders, 20 Hispanic Americans), 80 international. Average age 31. 339 applicants, 84% accepted. In 1998, 154 master's, 14 other advanced degrees awarded. *Degree requirements:* For master's, computer language required, foreign language and thesis not required. *Entrance requirements:* For master's, GRE General Test. *Application deadline:*

For fall admission, 7/31 (priority date); for spring admission, 11/30. Applications are processed on a rolling basis. Application fee: $60. Electronic applications accepted. *Financial aid:* Research assistantships, career-related internships or fieldwork available. Aid available to part-time students. Financial aid applicants required to submit FAFSA. *Unit head:* Dr. Susan Merritt, Dean, 914-422-4375. *Application contact:* Lois Rich, Associate Director, 914-422-4283, Fax: 914-422-4287, E-mail: gradwp@ny2.pace.edu.

See in-depth description on page 791.

Pennsylvania State University Great Valley School of Graduate Professional Studies, Graduate Studies and Continuing Education, School of Graduate Professional Studies, Department of Engineering and Information Science, Program in Information Science, Malvern, PA 19355-1488. Offers MS. *Students:* 15 full-time (8 women), 293 part-time (89 women). Average age 34. Application fee: $50. *Unit head:* Lois Rich, Associate Director, 914-422-4283, Fax: 914-422-4287, E-mail: gradwp@ny2.pace.edu. *Application contact:* 610-648-3242, Fax: 610-889-1334.

See in-depth description on page 795.

Polytechnic University, Brooklyn Campus, Department of Computer and Information Science, Major in Information Systems Engineering, Brooklyn, NY 11201-2990. Offers MS. Part-time and evening/weekend programs available. Average age 33. 4 applicants, 50% accepted. *Degree requirements:* For master's, computer language required, thesis not required. *Entrance requirements:* For master's, BA or BS in computer science, mathematics, science, or engineering; working knowledge of a high-level program. *Application deadline:* Applications are processed on a rolling basis. Application fee: $45. Electronic applications accepted. *Financial aid:* Research assistantships, teaching assistantships, institutionally-sponsored loans available. Aid available to part-time students. Financial aid applicants required to submit FAFSA. *Unit head:* Shirley Kapsch, Enrollment Manager, 503-690-1255, Fax: 503-690-1285, E-mail: kapsch@cse.ogi.edu. *Application contact:* John S. Kerge, Dean of Admissions, 718-260-3200, Fax: 718-260-3446, E-mail: admitme@poly.edu.

Polytechnic University, Farmingdale Campus, Graduate Programs, Department of Computer and Information Science, Major in Information Systems Engineering, Farmingdale, NY 11735-3995. Offers MS. *Degree requirements:* For master's, computer language, thesis required (for some programs). *Application deadline:* Applications are processed on a rolling basis. Application fee: $45. Electronic applications accepted. *Financial aid:* Institutionally-sponsored loans available. Aid available to part-time students. Financial aid applicants required to submit FAFSA. *Unit head:* John S. Kerge, Dean of Admissions, 718-260-3200, Fax: 718-260-3446, E-mail: admitme@poly.edu. *Application contact:* John S. Kerge, Dean of Admissions, 718-260-3200, Fax: 718-260-3446, E-mail: admitme@poly.edu.

Polytechnic University, Westchester Graduate Center, Graduate Programs, Department of Computer and Information Science, Major in Information Systems Engineering, Hawthorne, NY 10532-1507. Offers MS. *Students:* 1 (woman) full-time, 45 part-time (10 women); includes 7 minority (3 African Americans, 4 Asian Americans or Pacific Islanders), 1 international. Average age 33. 23 applicants, 100% accepted. In 1998, 6 degrees awarded. *Degree requirements:* For master's, computer language, thesis required (for some programs). *Application deadline:* Applications are processed on a rolling basis. Application fee: $45. Electronic applications accepted. *Application contact:* John S. Kerge, Dean of Admissions, 718-260-3200, Fax: 718-260-3446, E-mail: admitme@poly.edu.

Princeton University, Graduate School, School of Engineering and Applied Science, Department of Electrical Engineering, Princeton, NJ 08544-1019. Offers computer engineering (MSE, PhD); electrical engineering (M Eng); electronic materials and devices (MSE, PhD); information sciences and systems (MSE, PhD); optoelectronics (MSE, PhD). Part-time programs available. *Faculty:* 27 full-time (3 women). *Students:* 135 full-time (18 women), 10 part-time (1 woman). *Degree requirements:* For master's, thesis optional; for doctorate, dissertation required. *Entrance requirements:* For master's and doctorate, GRE General Test, TOEFL. *Application deadline:* For fall admission, 1/3. Electronic applications accepted. *Unit head:* Prof. Wayne Wolf, Director of Graduate Studies, 609-258-3335, Fax: 609-258-3745, E-mail: dgs@ee.princeton.edu. *Application contact:* Prof. Wayne Wolf, Director of Graduate Studies, 609-258-3335, Fax: 609-258-3745, E-mail: dgs@ee.princeton.edu.

Queen's University at Kingston, School of Graduate Studies and Research, Faculty of Arts and Sciences, Department of Computing and Information Science, Kingston, ON K7L 3N6, Canada. Offers M Sc, PhD. Part-time programs available. *Students:* 45 full-time (7 women), 14 part-time (5 women). In 1998, 19 master's, 1 doctorate awarded. *Degree requirements:* For master's, thesis optional, foreign language not required; for doctorate, dissertation, comprehensive exam required. *Entrance requirements:* For master's and doctorate, TOEFL (minimum score of 600 required). *Application deadline:* For fall admission, 2/28 (priority date). Application fee: $60. Electronic applications accepted. *Financial aid:* Fellowships, research assistantships, teaching assistantships, institutionally-sponsored loans available. Financial aid application deadline: 3/1. *Unit head:* Dr. J. I. Glasgow, Head, 613-533-6058. *Application contact:* Dr. R. A. Browse, Graduate Coordinator, 613-533-6069.

Regis University, School for Professional Studies, Program in Computer Information Systems, Denver, CO 80221-1099. Offers data base technology (MSCIS); management of technology (Certificate); multimedia technology (Certificate); networking technology (Certificate); object oriented technology (Certificate). Offered at Boulder Campus, Northwest Denver Campus, Southeast Denver Campus, Fort Collins Campus, and Colorado Springs Campus. Part-time and evening/weekend programs available. *Students:* 587. Average age 36. In 1998, 40 degrees awarded. *Degree requirements:* For master's, computer language, final research project required, foreign language and thesis not required. *Entrance requirements:* For master's, TOEFL (minimum score of 550 required), 3 years of related experience, interview. *Average time to degree:* Master's–2 years full-time, 3 years part-time. *Application deadline:* Applications are processed on a rolling basis. Application fee: $75. Tuition: Part-time $290 per credit hour. *Financial aid:* Federal Work-Study available. Aid available to part-time students. Financial aid applicants required to submit FAFSA. *Faculty research:* Application of computer techniques to solving organizational problems. *Unit head:* Don Archer, Chair, 303-458-4302. *Application contact:* 800-677-9270, Fax: 303-964-5538, E-mail: admarg@regis.edu.

Rensselaer at Hartford, Lally School of Management and Technology, Program in Computer and Information Science, Hartford, CT 06120-2991. Offers MS. Part-time and evening/weekend programs available. *Degree requirements:* For master's, computer language, seminar required, thesis optional, foreign language not required. *Entrance requirements:* For master's, TOEFL (minimum score of 570 required).

Rensselaer Polytechnic Institute, Graduate School, Interdisciplinary Program in Information Technology, Troy, NY 12180-3590. Offers MS. Part-time and evening/weekend programs available. Postbaccalaureate distance learning degree programs offered (no on-campus study). *Faculty:* 105 full-time (5 women), 8 part-time (0 women). *Degree requirements:* For master's, thesis not required. *Entrance requirements:* For master's, TOEFL (minimum score of 500 required). *Application deadline:* For fall admission, 2/1 (priority date). Application fee: $35. *Faculty research:* Database systems, telecommunications, software design, human-computer interaction, management of technology. *Unit head:* Gregory N. Hughes, Vice Provost, 518-276-2590, E-mail: hughesg@rpi.edu. *Application contact:* Lu Uber, Administrative Assistant, 518-276-2660, E-mail: uberm@rpi.edi.

See in-depth description on page 805.

Rivier College, School of Professional Studies, Department of Computer Science and Mathematics, Nashua, NH 03060-5086. Offers computer science (MS); information science (MS). Part-time and evening/weekend programs available. *Faculty:* 2 full-time (1 woman), 13 part-time (2 women). *Students:* 31 full-time (12 women), 53 part-time (19 women); includes 4 minority (all Asian Americans or Pacific Islanders), 36 international. *Degree requirements:*

Information Science

Rivier College (continued)

For master's, thesis not required. *Entrance requirements:* For master's, GRE Subject Test. *Application deadline:* Applications are processed on a rolling basis. Application fee: $25. Tuition: Part-time $309 per credit. Required fees: $75 per year. *Unit head:* Dr. Stephen Cooper, Director, 603-888-1311, E-mail: scooper@rivier.edu. *Application contact:* Paula Bailly-Burton, Director of Graduate and Evening Admissions, 603-888-1311, Fax: 603-888-9124, E-mail: geaadmit@rivier.edu.

Rochester Institute of Technology, Part-time and Graduate Admissions, College of Applied Science and Technology, Department of Computer Science and Information Technology, Program in Information Technology, Rochester, NY 14623-5604. Offers MS. *Students:* 68 full-time (26 women), 125 part-time (38 women); includes 13 minority (4 African Americans, 7 Asian Americans or Pacific Islanders, 1 Hispanic American, 1 Native American), 46 international. 133 applicants, 83% accepted. In 1998, 6 degrees awarded. *Entrance requirements:* For master's, minimum GPA of 3.0. *Application deadline:* For fall admission, 3/1 (priority date). Applications are processed on a rolling basis. Application fee: $40. *Unit head:* Dr. Rayno Niemi, Graduate Coordinator, 716-475-2202, E-mail: rdn@it.rit.edu.

Sacred Heart University, Graduate Studies, College of Arts and Sciences, Faculty of Computer and Information Science, Fairfield, CT 06432-1000. Offers MS. *Faculty:* 5 full-time (1 woman), 7 part-time (2 women). Average age 35. 22 applicants, 73% accepted. In 1998, 3 degrees awarded. *Degree requirements:* For master's, thesis optional. *Application deadline:* Applications are processed on a rolling basis. Application fee: $40 ($100 for international students). Tuition: Part-time $375 per credit. Required fees: $83 per term. Tuition and fees vary according to campus/location and program. *Financial aid:* Applicants required to submit FAFSA. *Unit head:* Domenick Pinto, Acting Director, 203-371-7799, Fax: 203-371-0506. *Application contact:* Mike Kennedy, Graduate Admissions Counselor, 203-365-7619, Fax: 203-365-4732, E-mail: gradstudies@sacredheart.edu.

St. Mary's University of San Antonio, Graduate School, Program in Computer Information Systems, San Antonio, TX 78228-8507. Offers MS. *Faculty:* 6 full-time (0 women), 3 part-time (1 woman). *Students:* 14 full-time (3 women), 64 part-time (12 women); includes 28 minority (6 African Americans, 3 Asian Americans or Pacific Islanders, 19 Hispanic Americans), 10 international. Average age 30. In 1998, 14 degrees awarded (100% found work related to degree). *Degree requirements:* For master's, computer language, comprehensive exams required, thesis optional, foreign language not required. *Entrance requirements:* For master's, GMAT or GRE General Test. *Application deadline:* For fall admission, 8/1. Application fee: $15. *Financial aid:* Research assistantships, career-related internships or fieldwork and institutionally-sponsored loans available. *Faculty research:* Artificial intelligence, database/knowledge base, software engineering, expert systems. *Unit head:* Dr. Douglas Hall, Adviser, 210-436-3317.

San Jose State University, Graduate Studies, College of Engineering, Department of Computer, Information and Systems Engineering, Program in Information and Systems Engineering, San Jose, CA 95192-0001. Offers MS. Part-time programs available. *Faculty:* 5 full-time (0 women), 9 part-time (1 woman). *Students:* 14 full-time (5 women), 27 part-time (4 women); includes 19 minority (3 African Americans, 13 Asian Americans or Pacific Islanders, 3 Hispanic Americans), 7 international. Average age 32. 30 applicants, 60% accepted. In 1998, 9 degrees awarded. *Degree requirements:* For master's, comprehensive exam required. *Entrance requirements:* For master's, minimum GPA of 3.0. *Application deadline:* For fall admission, 6/1. Applications are processed on a rolling basis. Application fee: $59. Tuition, nonresident: part-time $246 per unit. Required fees: $1,939; $1,309 per year. *Financial aid:* Federal Work-Study available. *Unit head:* 800-677-9270, Fax: 303-964-5538, E-mail: admarg@regis.edu. *Application contact:* Dr. Louis Freund, Graduate Coordinator, 408-924-3890.

Shippensburg University of Pennsylvania, School of Graduate Studies and Research, College of Arts and Sciences, Department of Mathematics and Computer Science, Shippensburg, PA 17257-2299. Offers computer science (MS); information systems (MS); mathematics (M Ed, MS). Part-time and evening/weekend programs available. *Faculty:* 19 full-time (3 women). *Students:* 19 full-time (7 women), 64 part-time (15 women); includes 3 minority (2 African Americans, 1 Native American), 11 international. *Degree requirements:* For master's, foreign language and thesis not required. *Entrance requirements:* For master's, TOEFL (minimum score of 237 required for computer-based), GRE General Test or minimum GPA of 2.75. *Application deadline:* Applications are processed on a rolling basis. Application fee: $25. Electronic applications accepted. *Unit head:* Dr. Fred Nordai, Chairperson, 717-532-1431, Fax: 717-530-4009, E-mail: flnord@ship.edu. *Application contact:* Renee Payne, Assistant Dean of Graduate Studies, 717-532-1213, Fax: 717-530-4038, E-mail: rmpayn@ship.edu.

State University of New York at Albany, Nelson A. Rockefeller College of Public Affairs and Policy, Information Science Program, Albany, NY 12222-0001. Offers MS, PhD. *Students:* 96 full-time (70 women), 125 part-time (90 women); includes 7 minority (4 African Americans, 2 Asian Americans or Pacific Islanders, 1 Hispanic American), 13 international. 132 applicants, 77% accepted. In 1998, 103 master's, 1 doctorate awarded. *Degree requirements:* For doctorate, dissertation required. *Entrance requirements:* For doctorate, GRE General Test. Application fee: $50. Tuition, state resident: full-time $5,100; part-time $213 per credit. Tuition, nonresident: full-time $8,416; part-time $351 per credit. Required fees: $31 per credit. *Unit head:* Dr. Thomas J. Galvin, Director, 518-442-3306.

See in-depth description on page 823.

State University of New York Institute of Technology at Utica/Rome, School of Information Systems and Engineering Technology, Program in Computer and Information Science, Utica, NY 13504-3050. Offers MS. Part-time and evening/weekend programs available. *Faculty:* 10 full-time (2 women). *Students:* 13 full-time (3 women), 43 part-time (11 women); includes 5 minority (4 Asian Americans or Pacific Islanders, 1 Hispanic American), 4 international. Average age 35. 34 applicants, 76% accepted. In 1998, 10 degrees awarded (100% found work related to degree). *Degree requirements:* For master's, computer language, comprehensive exam required, thesis optional, foreign language not required. *Entrance requirements:* For master's, GRE General Test, TOEFL (minimum score of 550 required), minimum GPA of 3.0. *Average time to degree:* Master's–2 years full-time, 4 years part-time. *Application deadline:* For fall admission, 6/15 (priority date). Applications are processed on a rolling basis. Application fee: $50. *Financial aid:* In 1998–99, 20 students received aid, including 2 fellowships with full tuition reimbursements available (averaging $4,900 per year), 3 research assistantships with full tuition reimbursements available; career-related internships or fieldwork and Federal Work-Study also available. Aid available to part-time students. Financial aid applicants required to submit FAFSA. *Faculty research:* Parallel processing, databases and data fusion, networks, artificial intelligence. *Unit head:* Dr. Michael Pittarelli, Chair, 315-792-7234, Fax: 315-792-7222, E-mail: mike@sunyit.edu. *Application contact:* Marybeth Lyons, Director of Admissions, 315-792-7500, Fax: 315-792-7837, E-mail: smbl@sunyit.edu.

Stevens Institute of Technology, Graduate School, Wesley J. Howe School of Technology Management, Program in Information Systems, Hoboken, NJ 07030. Offers MS, Certificate. Offered in cooperation with the Department of Computer Science. *Degree requirements:* For master's, computer language required, thesis optional, foreign language not required; for Certificate, computer language required, foreign language not required. *Entrance requirements:* For master's, GMAT, GRE, TOEFL. Electronic applications accepted.

Syracuse University, Graduate School, L. C. Smith College of Engineering and Computer Science, Department of Electrical Engineering and Computer Science, Program in Computer and Information Science, Syracuse, NY 13244-0003. Offers computer and information science (PhD); computer science (MS). *Students:* 182 full-time (24 women), 40 part-time (8 women), 65 international. Average age 28. 300 applicants, 79% accepted. In 1998, 26 degrees awarded. *Degree requirements:* For master's, thesis required, foreign language not required; for doctorate, computer language, dissertation required, foreign language not required. *Entrance requirements:* For master's and doctorate, GRE General Test, GRE Subject Test. *Application deadline:* Applications are processed on a rolling basis. Application fee: $40. Tuition: Full-time $13,992; part-time $583 per credit hour. *Financial aid:* Fellowships, research assistantships, teaching assistantships, Federal Work-Study and tuition waivers (partial) available. Financial aid application deadline: 3/1. *Unit head:* Kishan Mehrotra, Graduate Director.

Syracuse University, Graduate School, L. C. Smith College of Engineering and Computer Science, Department of Electrical Engineering and Computer Science, Program in Systems and Information Science, Syracuse, NY 13244-0003. Offers MS, PhD. *Students:* 64 full-time (14 women), 12 part-time (1 woman), 64 international. Average age 33. 86 applicants, 74% accepted. In 1998, 1 master's, 11 doctorates awarded. *Degree requirements:* For doctorate, computer language, dissertation required, foreign language not required, foreign language not required. *Entrance requirements:* For master's and doctorate, GRE General Test, GRE Subject Test. *Application deadline:* Applications are processed on a rolling basis. Application fee: $40. Tuition: Full-time $13,992; part-time $583 per credit hour. *Financial aid:* Fellowships, research assistantships, teaching assistantships, Federal Work-Study and tuition waivers (partial) available. Financial aid application deadline: 3/1. *Application contact:* James Fawcett, Contact, 315-443-2655.

Temple University, Graduate School, College of Science and Technology, Department of Computer and Information Sciences, Philadelphia, PA 19122-6096. Offers MS, PhD. Part-time programs available. *Faculty:* 11 full-time (1 woman). *Students:* 54 (12 women); includes 36 minority (2 African Americans, 34 Asian Americans or Pacific Islanders) 4 international. 238 applicants, 58% accepted. In 1998, 30 master's, 2 doctorates awarded. Terminal master's awarded for partial completion of doctoral program. *Degree requirements:* For master's, computer language required, foreign language and thesis not required; for doctorate, computer language, dissertation required, foreign language not required. *Entrance requirements:* For master's and doctorate, GRE General Test, TOEFL, minimum GPA of 2.8. *Application deadline:* For fall admission, 2/1; for spring admission, 9/30. Applications are processed on a rolling basis. Application fee: $40. Electronic applications accepted. *Financial aid:* Fellowships, research assistantships, teaching assistantships, unspecified assistantships available. Financial aid application deadline: 2/1. *Faculty research:* Artificial intelligence, information systems, software engineering, network-distributed systems. *Unit head:* Dr. Frank Friedman, Chairperson, 215-204-8450, Fax: 215-204-5082, E-mail: friedman@cis.temple.edu. *Application contact:* Dr. James Korsh, Graduate Chair, 215-204-8199, Fax: 215-204-5082, E-mail: korsh@cis.temple.eou.

Université du Québec, Institut national de la recherche scientifique, Graduate Programs, Research Center—Telecommunications, Ste-Foy, PQ G1V 4C7, Canada. Offers information technology (Diploma); software engineering (M Sc); telecommunications (M Sc, PhD). Part-time programs available. *Degree requirements:* For master's and doctorate, thesis/dissertation required; for Diploma, thesis not required. *Entrance requirements:* For master's and Diploma, appropriate bachelor's degree, proficiency in French; for doctorate, appropriate master's degree, proficiency in French. *Faculty research:* Visual and verbal systems, communications networks.

The University of Alabama at Birmingham, Graduate School, School of Natural Sciences and Mathematics, Department of Computer and Information Sciences, Birmingham, AL 35294. Offers MS, PhD. *Students:* 45 full-time (15 women), 13 part-time (3 women); includes 10 minority (4 African Americans, 6 Asian Americans or Pacific Islanders), 40 international. 142 applicants, 84% accepted. In 1998, 20 master's, 4 doctorates awarded. *Degree requirements:* For master's, thesis optional, foreign language not required; for doctorate, dissertation required, foreign language not required. *Entrance requirements:* For master's, GRE General Test (minimum combined score of 1100 required); for doctorate, GRE General Test (minimum combined score of 1300 required). *Application deadline:* Applications are processed on a rolling basis. Application fee: $30 ($60 for international students). *Financial aid:* In 1998–99, 7 fellowships with full tuition reimbursements (averaging $13,200 per year), 2 research assistantships with full tuition reimbursements (averaging $13,200 per year), 9 teaching assistantships with full tuition reimbursements (averaging $13,200 per year) were awarded.; career-related internships or fieldwork, Federal Work-Study, institutionally-sponsored loans, tuition waivers (full and partial), and unspecified assistantships also available. Aid available to part-time students. Financial aid application deadline: 3/10. *Faculty research:* Theory and software systems, intelligent systems, systems architecture. *Unit head:* Dr. Warren T. Jones, Chairman, 205-934-2213, Fax: 205-934-5473, E-mail: wjones@uab.edu.

See in-depth description on page 837.

University of Arkansas at Little Rock, Graduate School, College of Sciences and Engineering Technology, Department of Computer and Information Science, Little Rock, AR 72204-1099. Offers MS. Part-time and evening/weekend programs available. *Degree requirements:* For master's, computer language required, thesis optional, foreign language not required. *Entrance requirements:* For master's, GRE General Test, minimum GPA of 3.0, bachelor's degree in computer science, mathematics, or appropriate alternative.

University of California, Irvine, Office of Research and Graduate Studies, Department of Information and Computer Science, Irvine, CA 92697. Offers MS, PhD. *Faculty:* 30 full-time (5 women), 6 part-time (0 women). *Students:* 128 full-time (30 women); includes 25 minority (22 Asian Americans or Pacific Islanders, 3 Hispanic Americans), 58 international. Average age 31. 198 applicants, 46% accepted. In 1998, 15 master's, 12 doctorates awarded. Terminal master's awarded for partial completion of doctoral program. *Degree requirements:* For master's and doctorate, thesis/dissertation required, foreign language not required. *Entrance requirements:* For master's and doctorate, GRE General Test, GRE Subject Test. *Application deadline:* For fall admission, 1/15 (priority date). Applications are processed on a rolling basis. Application fee: $40. Electronic applications accepted. *Financial aid:* Fellowships, research assistantships, teaching assistantships, institutionally-sponsored loans and tuition waivers (full and partial) available. Financial aid application deadline: 3/2; financial aid applicants required to submit FAFSA. *Faculty research:* Artificial intelligence, computer system design, software, biomedical computing, theory, computing policy and society, data mining, imbedded systems. *Unit head:* Michael Pazzani, Chair, 949-824-7405. *Application contact:* Kris Domiccio, 949-824-2277.

See in-depth description on page 843.

University of Delaware, College of Arts and Science, Department of Computer and Information Sciences, Newark, DE 19716. Offers MS, PhD. Part-time programs available. *Faculty:* 18 full-time (3 women), 3 part-time (0 women). *Students:* 68 full-time (14 women), 10 part-time (2 women); includes 29 minority (27 Asian Americans or Pacific Islanders, 2 Hispanic Americans), 17 international. In 1998, 32 master's, 4 doctorates awarded. Terminal master's awarded for partial completion of doctoral program. *Degree requirements:* For master's, thesis optional, foreign language not required; for doctorate, dissertation required, foreign language not required. *Entrance requirements:* For master's and doctorate, GRE General Test (minimum combined score of 1750 on three sections required), TOEFL (average 600). *Average time to degree:* Master's–2 years full-time; doctorate–5 years full-time. *Application deadline:* For fall admission, 7/1; for spring admission, 12/1. Applications are processed on a rolling basis. Application fee: $45. Electronic applications accepted. *Financial aid:* In 1998–99, 56 students received aid, including 6 fellowships, 26 research assistantships, 16 teaching assistantships; tuition waivers (full) and unspecified assistantships also available. Financial aid application deadline: 3/1. *Faculty research:* Computer networks and distributed systems, natural language processing and artificial intelligence, parallel computing, theory, computer algebra, graphics. Total annual research expenditures: $1.2 million. *Unit head:* Prof. Errol Lloyd, Chair, 302-831-2711, Fax: 302-831-8458, E-mail: elloyd@cis.udel.edu. *Application contact:* Patricia L. Beazley, Graduate Secretary, 302-831-2713, Fax: 302-831-8458, E-mail: beazley@cis.udel.edu.

See in-depth description on page 855.

University of Florida, Graduate School, College of Engineering, Department of Computer and Information Science and Engineering, Gainesville, FL 32611. Offers computer and information science and engineering (ME); computer organization (MS, PhD, Engr); information systems (MS, PhD, Engr); manufacturing systems engineering (Certificate); software systems (MS, PhD, Engr). *Faculty:* 38. *Students:* 124 full-time (25 women), 42 part-time (10 women); includes 21 minority (2 African Americans, 14 Asian Americans or Pacific Islanders, 4 Hispanic Americans, 1 Native American), 82 international. 658 applicants, 35% accepted. In 1998, 45 master's, 14 doctorates awarded. *Degree requirements:* For master's, computer language, thesis required (for some programs), foreign language not required; for doctorate, computer language, dissertation required, foreign language not required. *Entrance requirements:* For master's and doctorate, GRE General Test (minimum combined score of 1100 required), minimum GPA of 3.0; for other advanced degree, GRE General Test. *Application deadline:* For fall admission, 6/1 (priority date); for spring admission, 11/1. Applications are processed on a rolling basis. Application fee: $20. Electronic applications accepted. *Financial aid:* In 1998–99, 90 students received aid, including 3 fellowships, 83 research assistantships, 20 teaching assistantships; unspecified assistantships also available. Financial aid application deadline:6/1. *Faculty research:* Artificial intelligence, networks security, distributed computing, parallel processing system, vision and visualization, database systems. *Unit head:* Dr. Gerhard Ritter, Chair, 352-392-1212, Fax: 352-392-1220, E-mail: ritter@cise.ufl.edu. *Application contact:* Dr. Doug Dankel, Graduate Coordinator, 352-392-1387, Fax: 352-392-1220, E-mail: ddd@cise.ufl.edu.

See in-depth description on page 859.

University of Great Falls, Graduate Studies Division, Programs in Information Systems, Great Falls, MT 59405. Offers MIS. Part-time and evening/weekend programs available. Postbaccalaureate distance learning degree programs offered (minimal on-campus study). *Degree requirements:* For master's, computer language required, foreign language not required.

University of Hawaii at Manoa, Graduate Division, College of Arts and Sciences, College of Natural Sciences, Department of Information and Computer Sciences, Program in Communication and Information Science, Honolulu, HI 96822. Offers PhD. Part-time programs available. *Faculty:* 57 full-time (12 women). *Students:* 26; includes 5 minority (1 African American, 4 Asian Americans or Pacific Islanders), 15 international. Average age 30. 24 applicants, 38% accepted. In 1998, 3 degrees awarded. *Degree requirements:* For doctorate, computer language, dissertation required. *Entrance requirements:* For doctorate, GMAT or GRE, TOEFL (minimum score of 760 required), master's in closely related field, knowledge of computer programming. *Average time to degree:* Doctorate–5 years full-time. *Application deadline:* For fall admission, 1/15. Application fee: $25 ($50 for international students). *Financial aid:* In 1998–99, 5 research assistantships (averaging $16,667 per year), 7 teaching assistantships (averaging $14,067 per year) were awarded.; tuition waivers (full and partial) also available. Financial aid application deadline: 2/1. *Faculty research:* Data communications, organizational communications, communication policies, information systems, computer software systems, human-computer interaction. *Unit head:* Dr. Rebecca Knuth, Chair, 808-956-7321, Fax: 808-956-5835, E-mail: knuth@hawaii.edu.

University of Houston–Clear Lake, School of Natural and Applied Sciences, Program in Computer Information Systems, Houston, TX 77058-1098. Offers MA. *Faculty:* 5 full-time (1 woman), 2 part-time (0 women). *Students:* 24 full-time (13 women), 25 part-time (14 women); includes 11 minority (3 African Americans, 6 Asian Americans or Pacific Islanders, 2 Hispanic Americans), 27 international. Average age 32. *Degree requirements:* Foreign language not required. *Entrance requirements:* For master's, GRE General Test. *Application deadline:* Applications are processed on a rolling basis. Application fee: $30 ($70 for international students). *Financial aid:* Research assistantships, teaching assistantships available. Financial aid application deadline: 5/1. *Unit head:* Dr. Kwok-Bun Yue, Chair, 281-283-3850, Fax: 281-283-3707. *Application contact:* Dr. Robert Ferebee, Associate Dean, 281-283-3700, Fax: 281-283-3707, E-mail: ferebee@uhcl4.cl.uh.edu.

University of Maryland, Baltimore County, Graduate School, Department of Information Systems, Baltimore, MD 21250-5398. Offers operations analysis (MS, PhD). *Faculty:* 16 full-time (6 women), 22 part-time (5 women). *Students:* 36 full-time (13 women), 83 part-time (35 women); includes 24 minority (9 African Americans, 15 Asian Americans or Pacific Islanders), 26 international. 63 applicants, 49% accepted. In 1998, 19 master's awarded. *Entrance requirements:* For master's and doctorate, GRE General Test, minimum GPA of 3.0. *Application deadline:* For fall admission, 5/1. Applications are processed on a rolling basis. Application fee: $45. *Financial aid:* Fellowships, research assistantships, teaching assistantships available. *Faculty research:* Human-computer interaction, medical informatics, security, networking. *Unit head:* Dr. Jennifer Preece, Chairman, 410-455-3206. *Application contact:* Dr. Patricia Fletcher, Director, 410-455-3154.

See in-depth description on page 873.

University of Michigan–Dearborn, College of Engineering and Computer Science, Department of Computer and Information Science, Dearborn, MI 48128-1491. Offers computer and information science (MS); software engineering (MS). Part-time and evening/weekend programs available. *Faculty:* 11 full-time (1 woman), 2 part-time (0 women). *Students:* 7 full-time (0 women), 111 part-time (24 women); includes 23 minority (4 African Americans, 17 Asian Americans or Pacific Islanders, 2 Hispanic Americans), 7 international. Average age 29. In 1998, 13 degrees awarded. *Degree requirements:* For master's, computer language required, thesis optional, foreign language not required. *Entrance requirements:* For master's, bachelor's degree in mathematics, computer science, or engineering. *Application deadline:* For fall admission, 6/15; for spring admission, 2/15. Applications are processed on a rolling basis. Application fee: $55. Electronic applications accepted. Tuition, state resident: part-time $259 per credit hour. Tuition, nonresident: part-time $748 per credit hour. Required fees: $80 per course. Tuition and fees vary according to course level, course load and program. *Financial aid:* In 1998–99, 1 research assistantship, 1 teaching assistantship were awarded. *Faculty research:* Information systems, geometric modelling, networks, databases. *Unit head:* Dr. Kenneth Modesitt, Chair, 313-593-5680, Fax: 313-593-9967, E-mail: modesitt@umdsun2.umd.umich.edu. *Application contact:* Mary Tamsen, Graduate Secretary, 313-436-9145, Fax: 313-593-9967, E-mail: mtamsen@umdsun2.umd.umich.edu.

University of Minnesota, Twin Cities Campus, Graduate School, Institute of Technology, Department of Computer Science and Engineering, Minneapolis, MN 55455-0213. Offers computer and information sciences (MCIS, MS, PhD). Part-time programs available. *Faculty:* 25 full-time (2 women), 10 part-time (0 women). *Students:* 153 full-time (34 women), 76 part-time (12 women); includes 8 minority (1 African American, 6 Asian Americans or Pacific Islanders, 1 Hispanic American), 135 international. 500 applicants, 45% accepted. In 1998, 70 master's, 20 doctorates awarded. Terminal master's awarded for partial completion of doctoral program. *Degree requirements:* For master's, foreign language and thesis not required; for doctorate, dissertation required, foreign language not required. *Entrance requirements:* For master's and doctorate, GRE General Test. *Average time to degree:* Master's–2 years full-time, 4 years part-time; doctorate–5 years full-time, 8 years part-time. *Application deadline:* For fall admission, 5/31. Applications are processed on a rolling basis. Application fee: $50 ($55 for international students). *Financial aid:* In 1998–99, 1 fellowship with tuition reimbursement (averaging $12,000 per year), 95 research assistantships with tuition reimbursements (averaging $14,184 per year), 55 teaching assistantships with tuition reimbursements (averaging $14,184 per year) were awarded.; career-related internships or fieldwork, Federal Work-Study, and institutionally-sponsored loans also available. Financial aid application deadline:1/2. *Faculty research:* Software systems, numerical analysis, theory, artificial intelligence. Total annual research expenditures: $4.2 million. *Unit head:* Yousef Saad, Head, 612-625-0726, Fax: 612-625-0572, E-mail: saad@cs.umn.edu. *Application contact:* Haesun Park, Director of Graduate Studies, 612-625-4002, Fax: 612-625-0572.

See in-depth description on page 877.

University of New Haven, Graduate School, School of Engineering and Applied Science, Program in Computer and Information Science, West Haven, CT 06516-1916. Offers applications software (MS); management information systems (MS); systems software (MS). Part-time and evening/weekend programs available. *Students:* 40 full-time (17 women), 157 part-time (41 women); includes 18 minority (4 African Americans, 14 Asian Americans or Pacific Islanders), 68 international. 76 applicants, 66% accepted. In 1998, 62 degrees awarded. *Degree requirements:* For master's, thesis or alternative required, foreign language not required. *Application deadline:* Applications are processed on a rolling basis. Application fee: $50. *Financial aid:* Federal Work-Study available. Aid available to part-time students. Financial aid application deadline: 5/1; financial aid applicants required to submit FAFSA. *Unit head:* Dr. Tahany Fergany, Coordinator, 203-932-7067.

University of North Carolina at Charlotte, Graduate School, The William States Lee College of Engineering, School of Information Technology, Charlotte, NC 28223-0001. Offers PhD. *Students:* 2 full-time (0 women), 8 part-time (3 women), 4 international. Average age 35. 19 applicants, 63% accepted. *Degree requirements:* For doctorate, dissertation required. *Entrance requirements:* For doctorate, GRE General Test or GMAT, TOEFL. *Application deadline:* For fall admission, 7/15; for spring admission, 11/15. Applications are processed on a rolling basis. Application fee: $35. Electronic applications accepted. *Financial aid:* In 1998–99, 7 teaching assistantships were awarded.; research assistantships Financial aid application deadline: 4/1. *Unit head:* Dr. Joanna R. Baker, Director, 704-547-3124, E-mail: jrbaker@email.uncc.edu. *Application contact:* Kathy Barringer, Assistant Director of Graduate Admissions, 704-547-3366, Fax: 704-547-3279, E-mail: gradadm@email.uncc.edu.

See in-depth description on page 885.

University of North Florida, College of Computer Sciences and Engineering, Jacksonville, FL 32224-2645. Offers computer and information sciences (MS). Part-time programs available. *Faculty:* 15 full-time (3 women). *Students:* 8 full-time (5 women), 44 part-time (18 women); includes 11 minority (all Asian Americans or Pacific Islanders), 5 international. Average age 33. 15 applicants, 73% accepted. In 1998, 11 degrees awarded. *Degree requirements:* For master's, computer language, comprehensive exam required, thesis optional, foreign language not required. *Entrance requirements:* For master's, GRE General Test (minimum combined score of 1000 required), minimum GPA of 3.0 in last 60 hours. *Application deadline:* For fall admission, 12/31 (priority date). Applications are processed on a rolling basis. Application fee: $20. Electronic applications accepted. *Financial aid:* In 1998–99, 11 students received aid, including 1 teaching assistantship (averaging $1,106 per year); Federal Work-Study and tuition waivers (partial) also available. Aid available to part-time students. Financial aid application deadline: 4/1; financial aid applicants required to submit FAFSA. *Faculty research:* Parallel processing, software engineering, artificial intelligence, human factors, human-machine interfacing. *Unit head:* Dr. Neal Coulter, Dean, 904-620-2985, E-mail: ncoulter@unf.edu. *Application contact:* Dr. Krissten N. Cooper, Director of Graduate Studies, 904-620-2985, E-mail: kcooper@unf.edu.

University of North Texas, Robert B. Toulouse School of Graduate Studies, Interdisciplinary Studies, Denton, TX 76203. Offers information science (PhD); interdisciplinary studies (MA, MS). Part-time programs available. *Students:* 18 full-time (12 women), 38 part-time (26 women); includes 6 minority (4 African Americans, 2 Hispanic Americans), 1 international. *Degree requirements:* For master's, thesis optional, foreign language not required; for doctorate, one foreign language (computer language can substitute), dissertation required. *Entrance requirements:* For master's, GRE General Test (minimum score of 400 on each section, 1000 combined required), minimum GPA of 2.8. *Application deadline:* For fall admission, 7/17. Application fee: $25 ($50 for international students). *Unit head:* Dr. David Bellin, Graduate Coordinator, 336-334-7245, Fax: 336-334-7244, E-mail: dbellin@ncat.edu. *Application contact:* Dr. Sandra L. Terrell, Associate Dean, 940-565-2383, Fax: 940-565-2141.

University of Oregon, Graduate School, College of Arts and Sciences, Department of Computer and Information Science, Eugene, OR 97403. Offers MA, MS, PhD. Part-time programs available. *Faculty:* 17 full-time (4 women), 5 part-time (1 woman). *Students:* 63 full-time (18 women), 8 part-time; includes 5 minority (all Asian Americans or Pacific Islanders), 32 international. 121 applicants, 43% accepted. In 1998, 14 master's awarded (100% found work related to degree); 1 doctorate awarded (100% entered university research/teaching). Terminal master's awarded for partial completion of doctoral program. *Degree requirements:* For master's, computer language required, foreign language and thesis not required; for doctorate, computer language, dissertation required, foreign language not required. *Entrance requirements:* For master's and doctorate, GRE General Test (minimum score of 480 on verbal section, 640 on quantitative, 600 on analytical required), TOEFL, TSE (for teaching assistants). *Average time to degree:* Master's–2 years full-time; doctorate–7 years full-time. *Application deadline:* For fall admission, 2/1 (priority date). Application fee: $50. *Financial aid:* Fellowships, research assistantships, teaching assistantships, Federal Work-Study and institutionally-sponsored loans available. Financial aid application deadline: 2/1. *Faculty research:* Artificial intelligence, graphics, natural-language processing, expert systems, operating systems. *Unit head:* Zary Segall, Head, 541-346-4408. *Application contact:* Jan Saunders, Graduate Secretary, 541-346-4408, Fax: 541-346-5373.

See in-depth description on page 891.

University of Pennsylvania, School of Engineering and Applied Science, Department of Computer and Information Science, Philadelphia, PA 19104. Offers MSE, PhD, MSE/MBA. Part-time programs available. Terminal master's awarded for partial completion of doctoral program. *Degree requirements:* For master's, computer language required, thesis optional, foreign language not required; for doctorate, computer language, dissertation required, foreign language not required. *Entrance requirements:* For master's and doctorate, GRE General Test, TOEFL (minimum score of 600 required). *Faculty research:* Robotics, graphics, theory, ai, networks and distributed systems, databases, computational biology, natural language processing.

See in-depth description on page 893.

University of Phoenix, Graduate Programs, Computer Science and Information Technology Program, Phoenix, AZ 85072-2069. Offers MSCIS. Programs offered at campuses in Colorado, New Mexico, Northern California, Tucson, Utah, and on-line. Evening/weekend programs available. Postbaccalaureate distance learning degree programs offered (no on-campus study). *Students:* 618 full-time (167 women); includes 99 minority (74 African Americans, 25 Hispanic Americans) Average age 36. *Degree requirements:* For master's, thesis or alternative required. *Entrance requirements:* For master's, TOEFL (minimum score of 580 required), minimum GPA of 2.5, 3 years of work experience, comprehensive cognitive assessment (COCA). *Application deadline:* Applications are processed on a rolling basis. Application fee: $50. *Financial aid:* Applicants required to submit FAFSA. *Unit head:* Hugh McBride, Dean, 602-966-9577. *Application contact:* Campus Information Center, 602-966-9577.

University of Pittsburgh, Faculty of Arts and Sciences, Program in Intelligent Systems, Pittsburgh, PA 15260. Offers MS, PhD. *Students:* 17 full-time (7 women), 6 part-time; includes 2 minority (both Asian Americans or Pacific Islanders), 13 international. 28 applicants, 11% accepted. In 1998, 2 master's, 5 doctorates awarded. Terminal master's awarded for partial completion of doctoral program. *Degree requirements:* For master's, computer language required, foreign language and thesis not required; for doctorate, computer language, dissertation required, foreign language not required. *Entrance requirements:* For doctorate, GRE General Test, TOEFL (average 653). *Average time to degree:* Master's–3 years full-time; doctorate–5 years full-time. *Application deadline:* For fall admission, 2/1 (priority date). Applications are processed on a rolling basis. Application fee: $30 ($40 for international students). Electronic applications accepted. *Financial aid:* In 1998–99, 22 students received aid. Federal Work-Study, grants, institutionally-sponsored loans, scholarships, traineeships, and unspecified assistantships available. Financial aid application deadline: 2/1. *Faculty research:* Medical artificial intelligence, expert systems, clinical decision support, plan generation and recognition, special cognition. *Unit head:* Martha E. Pollack, Director, 412-624-5775, Fax: 412-624-6089, E-mail: pollack@cs.pitt.edu. *Application contact:* Kelly Lloyd, Administrator, 412-624-5755, Fax: 412-624-6089, E-mail: secry@isp.pitt.edu.

Information Science–Medical Informatics

University of Pittsburgh, School of Information Sciences, Department of Information Science and Telecommunications, Program in Information Science, Pittsburgh, PA 15260. Offers MSIS, PhD, Certificate, MSIS/MPA, MSIS/MPIA. Part-time and evening/weekend programs available. *Faculty:* 11 full-time (1 woman), 5 part-time (1 woman). *Students:* 122 full-time (53 women), 119 part-time (52 women); includes 37 minority (6 African Americans, 29 Asian Americans or Pacific Islanders, 2 Hispanic Americans), 83 international. 142 applicants, 88% accepted. In 1998, 80 master's, 4 doctorates awarded. *Degree requirements:* For master's, computer language required, thesis optional, foreign language not required; for doctorate, computer language, dissertation required, foreign language not required. *Entrance requirements:* For master's, GRE General Test, previous course work in statistics and mathematics; for doctorate, GRE General Test, master's degree; minimum QPA of 3.0; previous course work in statistics, mathematics, and algebra. *Average time to degree:* Master's–1 year full-time, 3 years part-time; doctorate–4 years full-time, 8 years part-time. Application fee: $30 ($40 for international students). *Financial aid:* In 1998–99, 7 fellowships, 80 research assistantships, 71 teaching assistantships were awarded.; career-related internships or fieldwork, grants, scholarships, and tuition waivers (full and partial) also available. Aid available to part-time students. Financial aid application deadline: 1/15; financial aid applicants required to submit FAFSA. *Faculty research:* Visualization, information storage and retrieval, systems analysis and design, telecommunications and networking, cognitive science. Total annual research expenditures: $1.6 million. *Unit head:* Loretta Shabatura, Graduate Secretary, 412-624-8495, Fax: 412-624-8854, E-mail: loretta@cs.pitt.edu. *Application contact:* Ninette Kay, Admissions Coordinator, 412-624-5146, Fax: 412-624-5231, E-mail: nk@sis.pitt.edu.

See in-depth description on page 895.

University of South Alabama, Graduate School, Division of Computer and Information Sciences, Mobile, AL 36688-0002. Offers computer science (MS); information science (MS). Part-time and evening/weekend programs available. *Faculty:* 10 full-time (1 woman). *Students:* 56 full-time (15 women), 22 part-time (7 women); includes 6 minority (2 African Americans, 3 Asian Americans or Pacific Islanders, 1 Native American), 31 international. 153 applicants, 60% accepted. In 1998, 23 degrees awarded. *Degree requirements:* For master's, computer language, project required, thesis optional, foreign language not required. *Entrance requirements:* For master's, GRE General Test (minimum combined score of 1000 required), minimum GPA of 2.5. *Application deadline:* For fall admission, 9/1 (priority date). Applications are processed on a rolling basis. Application fee: $25. Tuition, state resident: part-time $116 per semester hour. Tuition, nonresident: part-time $230 per semester hour. Required fees: $121 per semester. Part-time tuition and fees vary according to course load and program. *Financial aid:* In 1998–99, 4 research assistantships were awarded.; career-related internships or fieldwork and institutionally-sponsored loans also available. Aid available to part-time students. Financial aid application deadline: 4/1. *Faculty research:* Numerical analysis, artificial intelligence, simulation, medical applications, software engineering. *Unit head:* Dr. David Feinstein, Chairman, 334-460-6390.

University of Tennessee, Knoxville, Graduate School, College of Communications, Knoxville, TN 37996. Offers advertising (MS, PhD); broadcasting (MS, PhD); communications (MS, PhD); information sciences (PhD); journalism (MS, PhD); public relations (MS, PhD); speech communication (MS, PhD). *Accreditation:* ACEJMC (one or more programs are accredited). Part-time and evening/weekend programs available. Postbaccalaureate distance learning degree programs offered (no on-campus study). *Faculty:* 25 full-time (8 women). *Students:* 64 full-time (32 women), 61 part-time (35 women); includes 9 minority (7 African Americans, 2 Hispanic Americans), 14 international. *Degree requirements:* For master's, thesis or alternative required, foreign language not required; for doctorate, dissertation required, foreign language not required. *Entrance requirements:* For master's and doctorate, GRE General Test, TOEFL (minimum score of 550 required), minimum GPA of 2.7. *Application deadline:* For fall admission, 2/1 (priority date). Applications are processed on a rolling basis. Application fee: $35. Electronic applications accepted. *Unit head:* Dr. Dwight Teeter, Dean, 423-974-3031, Fax: 423-974-3896. *Application contact:* Dr. Herbert Howard, Program Head, 423-974-6651, Fax: 423-974-3896, E-mail: hhoward@utk.edu.

Virginia Polytechnic Institute and State University, Graduate School, College of Arts and Sciences, Department of Computer Science, Program in Information Systems, Blacksburg, VA 24061. Offers MIS. Part-time programs available. *Faculty:* 22 full-time (2 women). *Students:* 7 full-time (5 women), 49 part-time (16 women); includes 20 minority (18 Asian Americans or Pacific Islanders, 2 Hispanic Americans), 4 international. Average age 24. 34 applicants, 26% accepted. In 1998, 14 degrees awarded. *Degree requirements:* For master's, computer language required. *Entrance requirements:* For master's, GRE General Test (minimum combined score of 1200 required), TOEFL. *Application deadline:* For fall admission, 12/1 (priority date). Applications are processed on a rolling basis. Application fee: $25. *Financial aid:* Fellowships, research assistantships, teaching assistantships, Federal Work-Study available. Financial aid application deadline: 4/1. *Faculty research:* Software engineering, operating systems, simulation, artificial intelligence. *Application contact:* Dr. Verna Schuetz, Assistant Head, 540-231-6931.

Medical Informatics

Case Western Reserve University, School of Graduate Studies, The Case School of Engineering, Department of Civil Engineering, Cleveland, OH 44106. Offers civil engineering (MS, PhD); engineering mechanics (MS). Part-time programs available. Postbaccalaureate distance learning degree programs offered (minimal on-campus study). *Faculty:* 9 full-time (0 women), 1 part-time (0 women). *Students:* 12 full-time (3 women), 10 part-time (3 women). *Degree requirements:* For master's, thesis required (for some programs), foreign language not required; for doctorate, dissertation required, foreign language not required. *Entrance requirements:* For master's and doctorate, TOEFL (minimum score of 550 required). *Application deadline:* For fall admission, 8/1 (priority date); for spring admission, 1/1. Application fee: $25. Electronic applications accepted. *Unit head:* Robert L. Mullen, Chairman, 216-368-2423, Fax: 216-368-5229, E-mail: rlm@po.cwru.edu. *Application contact:* Kathleen Ballou, Secretary, 216-368-2950, Fax: 216-368-5229, E-mail: kad4@po.cwru.edu.

College of St. Scholastica, Graduate Studies, Program in Health Information Management, Duluth, MN 55811-4199. Offers MA. Part-time programs available. Postbaccalaureate distance learning degree programs offered (minimal on-campus study). *Faculty:* 2 full-time (both women), 6 part-time (3 women). Average age 30. 5 applicants, 100% accepted. *Degree requirements:* For master's, thesis required. *Entrance requirements:* For master's, interview, minimum GPA of 3.0. *Application deadline:* Applications are processed on a rolling basis. Application fee: $50. Tuition: Part-time $512 per credit. Tuition and fees vary according to course load and program. *Financial aid:* In 1998–99, 2 students received aid. Available to part-time students. Applicants required to submit FAFSA. *Unit head:* Kathy LaTour, Director, 218-723-6011, Fax: 218-733-2239, E-mail: klatour@css.edu. *Application contact:* Shirley Eichenwald, Admissions Office, 218-723-6448, Fax: 218-733-2239, E-mail: seichenw@css.edu.

Columbia University, College of Physicians and Surgeons, Graduate School of Arts and Sciences at the College of Physicians and Surgeons, Department of Medical Informatics, New York, NY 10032. Offers M Phil, MA, PhD, MD/PhD. *Degree requirements:* For master's, foreign language and thesis not required; for doctorate, dissertation required. *Entrance requirements:* For master's, GRE General Test, TOEFL (minimum score of 550 required).

See in-depth description on page 727.

Duke University, School of Nursing, Durham, NC 27708-0586. Offers adult acute care (Certificate); adult cardiovascular (Certificate); adult oncology/HIV (Certificate); adult primary care (Certificate); clinical nurse specialist (MSN), including adult oncology/HIV (MSN, MSN), gerontology (MSN, MSN), neonatal (MSN, MSN), pediatric (MSN, MSN), pediatric acute care (MSN, MSN); clinical trials management (MSN, Certificate); family (Certificate); gerontology (Certificate); health systems leadership and outcomes (MSN, Certificate); neonatal (Certificate); nurse practitioner (MSN), including adult acute care, adult cardiovascular, adult oncology/HIV (MSN, MSN), adult primary care, family, gerontology (MSN, MSN), neonatal (MSN, MSN), pediatric (MSN, MSN), pediatric acute care (MSN, MSN); nursing informatics (Certificate); pediatric (Certificate); pediatric acute care (Certificate). *Accreditation:* NLN (one or more programs are accredited). Part-time programs available. Postbaccalaureate distance learning degree programs offered (minimal on-campus study). *Faculty:* 24 full-time (22 women), 4 part-time (all women). *Students:* 55 full-time (52 women), 175 part-time (168 women); includes 35 minority (23 African Americans, 5 Asian Americans or Pacific Islanders, 2 Hispanic Americans, 5 Native Americans) *Degree requirements:* For master's, computer language required, thesis optional, foreign language not required; for Certificate, computer language required, foreign language and thesis not required. *Entrance requirements:* For master's, GRE General Test or MAT, BSN, minimum GPA of 3.0, previous course work in statistics, 1 year of nursing experience; for Certificate, MSN. *Application deadline:* For fall admission, 4/1 (priority date); for spring admission, 10/1 (priority date). Applications are processed on a rolling basis. Application fee: $50. *Unit head:* Dr. Mary T. Champagne, Dean, 919-684-3786, Fax: 919-681-8899, E-mail: champ001@mc.duke.edu. *Application contact:* Liz Kelly, Director of Admissions, 919-684-4248, Fax: 919-681-8899, E-mail: kelly043@mc.duke.edu.

George Mason University, College of Arts and Sciences, Department of Biology, Master's Program in Biology, Fairfax, VA 22030-4444. Offers bioinformatics (MS); ecology, systematics and evolution (MS); environmental science and public policy (MS); interpretive biology (MS); molecular, microbial, and cellular biology (MS); organismal biology (MS). Part-time programs available. *Faculty:* 32 full-time (11 women), 26 part-time (16 women). *Students:* 1 full-time (0 women), 63 part-time (34 women); includes 6 minority (4 African Americans, 1 Asian American or Pacific Islander, 1 Hispanic American), 2 international. *Degree requirements:* For master's, thesis or alternative required, foreign language not required. *Entrance requirements:* For master's, GRE General Test (minimum combined score of 1100 required), GRE Subject Test, bachelor's degree in biology or equivalent. *Application deadline:* For fall admission, 5/1; for spring admission, 11/1. Application fee: $30. Electronic applications accepted. Tuition, state resident: full-time $4,416; part-time $184 per credit hour. Tuition, nonresident: full-time $12,516; part-time $522 per credit hour. Tuition and fees vary according to program. *Unit head:* Dr. George E. Andrykovitch, Director, 703-993-1027, Fax: 703-993-1046.

Kirksville College of Osteopathic Medicine, Arizona School of Health Sciences, Phoenix, AZ 85017. Offers medical informatics (MS); occupational therapy (MS); physical therapy (MS); physician assistant (MS); physician assistant studies (MS); sports health care (MS). *Accreditation:* AOTA. *Faculty:* 27 full-time (18 women), 139 part-time (57 women). *Students:* 351 full-time (225 women); includes 30 minority (18 Asian Americans or Pacific Islanders, 9 Hispanic Americans, 3 Native Americans), 1 international. *Degree requirements:* For master's, thesis required, foreign language not required. *Entrance requirements:* For master's, GRE General Test. *Application deadline:* For fall admission, 2/1. Applications are processed on a rolling basis. Application fee: $50. Tuition: Full-time $16,600. *Unit head:* Dr. Craig Phelps, Provost, 602-841-4077, Fax: 602-841-4092, E-mail: phelpsc@az.swc.kcom.edu. *Application contact:* Stephanie Seyer, Assistant Director of Admissions, 660-626-2237, Fax: 660-626-2969.

Massachusetts Institute of Technology, Whitaker College of Health Sciences and Technology, Division of Health Sciences and Technology, Program in Medical Informatics, Cambridge, MA 02139-4307. Offers SM. Offered jointly with Harvard University. *Degree requirements:* For master's, computer language, thesis required, foreign language not required. *Entrance requirements:* For master's, MD or current enrollment in an MD program.

Medical College of Georgia, School of Graduate Studies, Programs in Allied Health Sciences, Department of Health Information Management, Augusta, GA 30912. Offers MHE, MS. Part-time programs available. *Faculty:* 1 (woman) full-time. 1 applicants, 100% accepted. *Degree requirements:* For master's, thesis required (for some programs), foreign language not required. *Entrance requirements:* For master's, GRE General Test (minimum combined score of 1000 required), TOEFL (minimum score of 600 required; 250 for computer-based). *Application deadline:* For fall admission, 6/30 (priority date); for spring admission, 11/1 (priority date). Applications are processed on a rolling basis. Application fee: $25. *Financial aid:* Federal Work-Study and institutionally-sponsored loans available. Financial aid application deadline: 3/31; financial aid applicants required to submit FAFSA. *Unit head:* Dr. Carol Campbell, Acting Chair, 706-721-3436, E-mail: cacampbe@mail.mcg.edu.

Medical College of Wisconsin, Graduate School of Biomedical Sciences, Program in Medical Informatics, Milwaukee, WI 53226-0509. Offers MS. Offered jointly with Milwaukee School of Engineering. Part-time and evening/weekend programs available. *Degree requirements:* For master's, computer language, thesis or alternative required, foreign language not required. *Entrance requirements:* For master's, GMAT or GRE, TOEFL (minimum score of 580 required;average 600). *Faculty research:* Computer science.

Medical University of South Carolina, College of Health Professions, Department of Health Administration and Policy, Program in Health Information Administration, Charleston, SC 29425-0002. Offers MHS, MHA/MHS. Part-time and evening/weekend programs available. Postbaccalaureate distance learning degree programs offered. *Faculty:* 10 full-time (4 women), 2 part-time (1 woman). In 1998, 9 degrees awarded (100% found work related to degree). *Degree requirements:* For master's, minimum GPA of 3.0 in each course, 30 to 40 hours of community service required, foreign language and thesis not required. *Entrance requirements:* For master's, GRE General Test (minimum combined score of 1000 required), MAT (minimum score of 42 required) (MHS), minimum GPA of 3.0. *Average time to degree:* Master's–2 years full-time, 4 years part-time. Application fee: $55. *Financial aid:* Fellowships, research assistantships, Federal Work-Study available. Aid available to part-time students. Financial aid application deadline: 4/1; financial aid applicants required to submit FAFSA. *Faculty research:* Computer-based patient records, Internet use in health care, health information networks, continuous quality improvement, organizational behavior. Total annual research expenditures: $21,294. *Unit head:* Karen A. Wager, Director, 843-792-4491, Fax: 843-792-3327, E-mail: wagerka@musc.edu. *Application contact:* Kelly Long, Student Services Coordinator, 843-792-8510, Fax: 843-792-3327, E-mail: longk@musc.edu.

Milwaukee School of Engineering, Department of Electrical Engineering and Computer Science, Program in Medical Informatics, Milwaukee, WI 53202-3109. Offers MS. *Faculty:* 5. *Students:* 22 full-time (13 women), 17 part-time (7 women); includes 4 minority (3 Asian Americans or Pacific Islanders, 1 Hispanic American), 3 international. Average age 25. *Degree requirements:* Foreign language not required. *Entrance requirements:* For master's, GRE General Test. *Application deadline:* For fall admission, 8/15. Applications are processed on a rolling basis. Application fee: $30. Electronic applications accepted. *Unit head:* Dr. Carol

Mannino, Director, 414-277-7105, Fax: 414-277-7479. *Application contact:* Helen Boomsma, Director, Lifelong Learning Institute, 800-321-6763, Fax: 414-277-7475, E-mail: boomsma@msoe.edu.

Announcement: MSOE offers a joint master's degree in medical informatics with the Medical College of Wisconsin. Medical informatics combines medical science with several technologies and disciplines in the information and computer sciences. This program educates professionals who design, implement, and manage information systems used in health-care organizations. Full- or part-time study is available.

New Jersey Institute of Technology, Office of Graduate Studies, Department of Computer and Information Science, Newark, NJ 07102-1982. Offers bioinformatics (MS, PhD); computer and information science (PhD); computer science (MS); information systems (MS). MS and PhD (bioinformatics) offered jointly with the University of Medicine and Dentistry of New Jersey. Part-time and evening/weekend programs available. Terminal master's awarded for partial completion of doctoral program. *Degree requirements:* For master's, computer language, thesis required, foreign language not required; for doctorate, computer language, dissertation, residency required, foreign language not required. *Entrance requirements:* For master's, GRE General Test (minimum score of 450 on verbal section, 600 on quantitative, 550 on analytical required); for doctorate, GRE General Test (minimum score of 450 on verbal section, 600 on quantitative, 550 on analytical required), minimum graduate GPA of 3.5. Electronic applications accepted. *Faculty research:* Computer systems, communications and networking, artificial intelligence, database engineering, systems analysis.

New York Medical College, Graduate School of Health Sciences, Program in Health Informatics, Valhalla, NY 10595-1691. Offers MPH. *Degree requirements:* For master's, thesis required, foreign language not required. *Entrance requirements:* For master's, TOEFL.

Oregon Health Sciences University, School of Medicine, Graduate Programs in Medicine, Division of Medical Informatics and Outcomes Research, Portland, OR 97201-3098. Offers medical informatics (MS). Part-time programs available. *Degree requirements:* For master's, computer language, thesis, thesis required, foreign language not required. *Entrance requirements:* For master's, GRE General Test (combined average 1600 on three sections), MCAT (average 8). *Application deadline:* For fall admission, 2/1 (priority date). Applications are processed on a rolling basis. Application fee: $60. *Financial aid:* Research assistantships, Federal Work-Study, institutionally-sponsored loans, and scholarships available. Financial aid application deadline: 3/1; financial aid applicants required to submit FAFSA. *Faculty research:* Information retrieval, outcomes research, telemedicine, consumer health informatics, information needs assessment. Total annual research expenditures: $3 million. *Unit head:* Dr. William Hersh, Associate Professor, 503-494-4563, Fax: 503-494-4551, E-mail: hersh@ohsu.edu.

See in-depth description on page 787.

Rensselaer Polytechnic Institute, Graduate School, School of Science, Department of Bioinformatics, Troy, NY 12180-3590. Offers MS. *Unit head:* Vickie Motz, Program Manager, 412-268-1322, Fax: 412-268-6837, E-mail: vmotz@andrew.cmu.edu. *Application contact:* Elena M. Quiroz, Assistant Dean, 518-276-6142, E-mail: bioinformatics@rpi.edu.

Announcement: Rensselaer offers an interdisciplinary program in bioinformatics drawing from biology, chemistry, and computer science. This powerful curriculum prepares science professionals in research laboratories, hospitals, and life science firms who will use computer tools to extract vital information for research and product development from immense genetic databases. For additional information, e-mail bioinformatics@rpi.edu.

Stanford University, School of Medicine, Graduate Programs in Medicine, Medical Information Sciences Program, Stanford, CA 94305-9991. Offers medical computer science (MS, PhD); medical decision science (MS, PhD). *Students:* 18 full-time (3 women), 10 part-time (4 women); includes 10 minority (1 African American, 6 Asian Americans or Pacific Islanders, 3 Hispanic Americans), 6 international. Average age 30. 45 applicants, 16% accepted. In 1998, 2 doctorates awarded. Terminal master's awarded for partial completion of doctoral program. *Degree requirements:* For master's and doctorate, computer language, thesis/dissertation required. *Entrance requirements:* For master's, GRE or MCAT; for doctorate, GRE or MCAT, TOEFL. *Application deadline:* For fall admission, 1/1. Application fee: $65 ($80 for international students). Electronic applications accepted. Tuition: Full-time $23,058. Required fees: $152. Part-time tuition and fees vary according to course load. *Financial aid:* Research assistantships available. Financial aid application deadline: 1/1. *Unit head:* Dr. Tom McCabe, Graduate Adviser, 512-245-3409, Fax: 512-245-8750, E-mail: tm03@swt.edu. *Application contact:* Darlene Vian, Administrator, 650-725-3388, Fax: 650-725-7944, E-mail: vian@smi.stanford.edu.

See in-depth description on page 817.

The University of Alabama at Birmingham, Graduate School, School of Health Related Professions, Department of Health Services Administration, Program in Health Informatics, Birmingham, AL 35294. Offers MS. *Students:* 46 full-time (21 women), 12 part-time (8 women); includes 9 minority (7 African Americans, 2 Asian Americans or Pacific Islanders), 2 international. 39 applicants, 97% accepted. In 1998, 12 degrees awarded. *Degree requirements:* For master's, computer language, thesis or alternative required. *Entrance requirements:* For master's, GRE General Test (minimum combined score of 1500 required), GMAT (minimum score of 480 required), MAT (minimum score of 50 required), minimum GPA of 3.0, previous course work in computing fundamentals and programming. Application fee: $30 ($60 for international students). Electronic applications accepted. *Financial aid:* Career-related internships or fieldwork and Federal Work-Study available. *Faculty research:* Healthcare/medical informatics, natural language processing, application of expert systems, graphical user interface design. *Unit head:* Dr. Helmuth F. Orthner, Director, 205-934-3509, Fax: 205-975-6608, E-mail: horthner@uab.edu.

University of California, San Francisco, Graduate Division, Program in Medical Information Science, San Francisco, CA 94143. Offers MS, PhD. *Faculty:* 54 part-time (13 women). *Students:* 9 full-time (3 women); includes 4 minority (3 Asian Americans or Pacific Islanders, 1 Hispanic American), 4 international. *Degree requirements:* For master's, research project required, foreign language and thesis not required; for doctorate, dissertation, cumulative qualifying exams, proposal defense required, foreign language not required. *Entrance requirements:* For master's and doctorate, GRE General Test. *Application deadline:* For winter admission, 2/15. Application fee: $40. *Financial aid:* In 1998–99, 8 students received aid, including 6 fellowships with full tuition reimbursements available (averaging $17,600 per year), 2 research assistantships with full tuition reimbursements available *Faculty research:* Bioinformatics, biomedical computing, decision science and engineering, imaging informatics, knowledge management/telehealth/health services research. *Unit head:* Thomas E. Ferrin, Director, 415-476-2299, Fax: 415-502-1755, E-mail: tef@cgl.ucsf.edu. *Application contact:* Barbara J. Paschke, Coordinator, 415-502-7788, Fax: 415-476-0688, E-mail: mis@cgl.ucsf.edu.

See in-depth description on page 849.

University of Medicine and Dentistry of New Jersey, School of Health Related Professions, Newark, NJ 07107-3001. Offers biomedical informatics (MS, PhD); clinical nutrition (MS); dietetic internship (Certificate); health care informatics (Certificate); health professions education (MA); health science (MS); health sciences (PhD); nurse midwifery (Certificate); physical

therapy (MPT, MS); physician assistant (MS); psychiatric rehabilitation (MS). *Accreditation:* ACNM/DOA (one or more programs are accredited); APTA (one or more programs are accredited). Part-time programs available. *Faculty:* 44 full-time (36 women), 15 part-time (10 women). *Students:* 394 full-time (271 women), 177 part-time (118 women); includes 174 minority (49 African Americans, 85 Asian Americans or Pacific Islanders, 34 Hispanic Americans, 6 Native Americans) *Degree requirements:* For master's, thesis required (for some programs), foreign language not required; for Certificate, foreign language and thesis not required. *Entrance requirements:* For master's, GRE; for Certificate, RN license (nurse midwifery). *Application deadline:* Applications are processed on a rolling basis. Application fee: $35. *Unit head:* Dr. David M. Gibson, Dean, 973-972-4276, Fax: 973-972-7028. *Application contact:* Dr. Laura Nelson, Associate Dean of Academic and Student Services, 973-972-5453, Fax: 973-972-7028, E-mail: shrp.adm@umdnj.edu.

University of Medicine and Dentistry of New Jersey, School of Nursing, Newark, NJ 07107-3001. Offers nursing (MSN, PMC); nursing informatics (MSN). *Accreditation:* NLN (one or more programs are accredited). Part-time programs available. *Faculty:* 30 full-time (28 women), 6 part-time (all women). *Students:* 19 full-time (all women), 147 part-time (131 women); includes 60 minority (25 African Americans, 30 Asian Americans or Pacific Islanders, 5 Hispanic Americans) *Degree requirements:* For master's, foreign language and thesis not required. *Entrance requirements:* For master's, GRE, TOEFL, RN license; basic life support, statistics, and health assessment experience. *Application deadline:* For fall admission, 5/15; for spring admission, 10/15. Applications are processed on a rolling basis. Application fee: $30. *Unit head:* Dr. Frances W. Quinless, Dean, 973-972-4322, Fax: 973-972-3225. *Application contact:* Joan Z. Shields, Manager, Enrollment and Student Services, 973-972-5447, Fax: 973-972-7453.

University of Minnesota, Twin Cities Campus, Graduate School, Program in Health Informatics, Minneapolis, MN 55455-0213. Offers MS, PhD. Part-time programs available. *Faculty:* 15 full-time (5 women), 8 part-time (0 women). *Students:* 26 full-time (12 women), 7 part-time (4 women), 16 international. Average age 35. 24 applicants, 67% accepted. In 1998, 6 master's awarded (100% found work related to degree); 2 doctorates awarded (100% found work related to degree). *Degree requirements:* For master's, thesis or alternative, project paper required; for doctorate, dissertation required. *Entrance requirements:* For master's, GRE General Test, previous course work in calculus, linear algebra, life sciences, programming, and biology; for doctorate, GRE General Test, previous course work in life sciences, programming, and differential equations. *Average time to degree:* Master's–2.5 years full-time; doctorate–7 years part-time. *Application deadline:* For fall admission, 6/30. Applications are processed on a rolling basis. Application fee: $50 ($55 for international students). *Financial aid:* In 1998–99, 20 students received aid, including 12 research assistantships with full and partial tuition reimbursements available (averaging $11,762 per year), 2 teaching assistantships with full and partial tuition reimbursements available (averaging $11,762 per year); fellowships with full tuition reimbursements available, Federal Work-Study, grants, traineeships, and tuition waivers (full and partial) also available. Financial aid application deadline: 1/15. *Faculty research:* Medical decision making, physiological control systems, population studies, clinical information systems, telemedicine. Total annual research expenditures: $1.4 million. *Unit head:* Dr. Stuart Speedie, Director, 612-625-8440, Fax: 612-625-7166, E-mail: sspeedie@mailbox.mail.umn.edu. *Application contact:* Doreen Gruebele, Principal Secretary, 612-625-8440, Fax: 612-625-7166, E-mail: doreen@email.labmed.umn.edu.

Announcement: Interdisciplinary program applies computer and information sciences to the quantitative and decision needs of the health and life sciences. Method development and applied research in health services, patient information systems, population genetics, expert systems, physiological/epidemiological simulation, clinical decision support, control systems, biomedical engineering, and bioinformatics. Research and employment opportunities in academic, clinical, industrial, and government settings.

University of Puerto Rico, Medical Sciences Campus, College of Health Related Professions, Program in Health Information Management, San Juan, PR 00936-5067. Offers MS. Part-time programs available. *Faculty:* 3 full-time (all women), 4 part-time (all women). *Students:* 18 full-time (16 women), 4 part-time (all women); all minorities (all Hispanic Americans) Average age 27. 18 applicants, 56% accepted. In 1998, 4 degrees awarded. *Degree requirements:* For master's, one foreign language, computer language, thesis, internship required. *Entrance requirements:* For master's, PAEG (minimum score of 295 required; average 521), interview. *Average time to degree:* Master's–2 years full-time, 4 years part-time. *Application deadline:* For fall admission, 2/15 (priority date). Applications are processed on a rolling basis. Application fee: $25. *Financial aid:* In 1998–99, 13 students received aid, including 2 research assistantships with full tuition reimbursements available (averaging $7,000 per year), 2 teaching assistantships with full tuition reimbursements available (averaging $7,000 per year); career-related internships or fieldwork, Federal Work-Study, institutionally-sponsored loans, and tuition waivers (partial) also available. Financial aid application deadline: 4/30. *Faculty research:* Quality of medical records. *Unit head:* Ana Orabona, Director, 787-758-2525 Ext. 4507. *Application contact:* Genoveva Ruiz, Student Affairs Office Director, 787-758-2525 Ext. 4000.

University of Utah, School of Medicine, Graduate Programs in Medicine, Department of Medical Informatics, Salt Lake City, UT 84112-1107. Offers MS, PhD. Part-time programs available. *Faculty:* 12 full-time (4 women), 10 part-time (1 woman). *Students:* 43 full-time (8 women), 3 part-time (2 women); includes 6 minority (5 Asian Americans or Pacific Islanders, 1 Hispanic American), 5 international. Average age 32. 57 applicants, 25% accepted. In 1998, 3 master's, 2 doctorates awarded. *Degree requirements:* For master's and doctorate, computer language, thesis/dissertation required, foreign language not required. *Entrance requirements:* For master's, GRE General Test, TOEFL (minimum score of 600 required), minimum GPA of 3.3; for doctorate, GRE, TOEFL (minimum score of 600 required), minimum GPA of 3.3. *Average time to degree:* Master's–2.5 years full-time; doctorate–4 years full-time. *Application deadline:* For fall admission, 2/1 (priority date). Applications are processed on a rolling basis. Application fee: $40 ($60 for international students). *Financial aid:* In 1998–99, 22 students received aid, including 5 fellowships, 7 research assistantships; career-related internships or fieldwork and traineeships also available. Financial aid application deadline: 2/1. *Faculty research:* Health information systems, expert systems, genetic epidemiology, medical imaging. Total annual research expenditures: $6.4 million. *Unit head:* Dr. Reed M. Gardner, Chairman, 801-585-9428, Fax: 801-581-4297, E-mail: reed.gardner@hsc.utah.edu. *Application contact:* J. Lynn Ford, Graduate Student Adviser, 801-581-3121, Fax: 801-581-4297, E-mail: lynn.ford@hsc.utah.edu.

University of Virginia, College and Graduate School of Arts and Sciences, Department of Health Evaluation Sciences, Charlottesville, VA 22903. Offers clinical investigation (MS); epidemiology (MS); health care informatics (MS); health care resource management (MS); health services research and outcomes evaluation (MS). Part-time programs available. *Faculty:* 18 full-time (4 women), 1 (woman) part-time. *Students:* 19 full-time (13 women), 7 part-time (4 women); includes 4 minority (1 African American, 2 Asian Americans or Pacific Islanders, 1 Hispanic American), 1 international. *Degree requirements:* For master's, thesis required (for some programs), foreign language not required. *Entrance requirements:* For master's, GRE or MCAT. *Application deadline:* For fall admission, 3/1 (priority date). Application fee: $60. *Unit head:* Dr. Paige Hornsby, Director, 804-924-0496, Fax: 804-924-8437, E-mail: pph8c@virginia.edu. *Application contact:* Robyn Kells, Coordinator, 804-924-8646, Fax: 804-924-8437, E-mail: ms-hes@virginia.edu.

Software Engineering

Azusa Pacific University, Graduate Studies, College of Liberal Arts and Sciences, Department of Computer Science, Azusa, CA 91702-7000. Offers applied computer science and technology (MS), including client/server technology, computer information systems, end-user support, inter-emphasis, software engineering, technical programming, telecommunications; client/server technology (Certificate); computer information systems (Certificate); computer science (Certificate); end-user training and support (Certificate); software engineering (MSE, Certificate); technical programming (Certificate); telecommunications (Certificate). Part-time and evening/weekend programs available. *Faculty:* 9 full-time (1 woman), 10 part-time (2 women). *Students:* 187. *Degree requirements:* For master's, computer language, thesis or alternative, project required, foreign language not required. *Entrance requirements:* For master's, minimum GPA of 3.0; proficiency in one programming language, college-level algebra, and applied calculus. *Application deadline:* For fall admission, 9/1 (priority date). Applications are processed on a rolling basis. Application fee: $45 ($65 for international students). *Unit head:* Dr. Samuel E. Sambasivam, Acting Chairman, 626-815-5476, Fax: 626-815-5323. *Application contact:* Dr. Samuel E. Sambasivam, Acting Chairman, 626-815-5476, Fax: 626-815-5323.

See in-depth description on page 701.

California State University, Sacramento, Graduate Studies, School of Engineering and Computer Science, Department of Computer Science, Sacramento, CA 95819-6048. Offers computer systems (MS); software engineering (MS). Part-time and evening/weekend programs available. *Degree requirements:* For master's, computer language, thesis or alternative, writing proficiency exam required, foreign language not required. *Entrance requirements:* For master's, TOEFL (minimum score of 550 required). *Application deadline:* For fall admission, 4/15; for spring admission, 11/1. Application fee: $55. *Unit head:* Dr. Ann Louis Radimsky, Chair, 916-278-5843. *Application contact:* Dr. Fred Blackwell, Coordinator, 916-278-6834.

Carnegie Mellon University, Graduate School of Industrial Administration, Pittsburgh, PA 15213-3891. Offers accounting (PhD); algorithms, combinatorics, and optimization (PhD); business management and software engineering (MBMSE); civil engineering and industrial management (MS); computational finance (MSCF); economics (PhD); electronic commerce (MS); environmental engineering and management (MEEM); finance (PhD); financial economics (PhD); industrial administration (MSIA), including industrial administration and public management; information science (PhD); manufacturing (MOM); manufacturing and operating systems (PhD), including industrial administration; marketing (PhD); mathematical finance (PhD); operations research (PhD); organizational behavior and theory (PhD); political economy (PhD); public policy and management (MS, MSED); robotics (PhD). Part-time programs available. *Faculty:* 86 full-time (13 women), 13 part-time (2 women). *Students:* 601 full-time (153 women), 236 part-time (43 women); includes 82 minority (8 African Americans, 63 Asian Americans or Pacific Islanders, 6 Hispanic Americans, 5 Native Americans), 383 international. Terminal master's awarded for partial completion of doctoral program. *Degree requirements:* For master's, foreign language and thesis not required; for doctorate, dissertation required, foreign language not required. *Entrance requirements:* For master's, GMAT. Application fee: $50. *Unit head:* Douglas Dunn, Dean, 412-268-2265. *Application contact:* Director of Admissions, 412-268-2272.

Carnegie Mellon University, School of Computer Science, Software Engineering Program, Pittsburgh, PA 15213-3891. Offers MSE. *Students:* 65 full-time (10 women), 11 part-time (2 women); includes 7 minority (2 African Americans, 4 Asian Americans or Pacific Islanders, 1 Native American), 25 international. Average age 32. In 1998, 19 degrees awarded (100% found work related to degree). *Degree requirements:* For master's, foreign language and thesis not required. *Entrance requirements:* For master's, GRE General Test, GRE Subject Test (computer science), 2 years of experience in large-scale software development project. *Application deadline:* For fall admission, 3/1. Application fee: $50. *Financial aid:* Teaching assistantships available. *Unit head:* Dr. Steve Cross, Director, 412-268-7740, Fax: 412-681-5758. *Application contact:* Coordinator, 412-268-7713, Fax: 412-681-5739, E-mail: mse-info@cs.cmu.edu.

See in-depth description on page 717.

Central Michigan University, College of Extended Learning, Program in Administration, Mount Pleasant, MI 48859. Offers general administration (MSA); health services administration (MSA, Certificate); hospitality and tourism (MSA, Certificate); human resources administration (MSA, Certificate); information resource management (MSA, Certificate); international administration (MSA, Certificate); public administration (MSA, Certificate); software engineering administration (MSA, Certificate). Part-time and evening/weekend programs available. Postbaccalaureate distance learning degree programs offered. *Faculty:* 1,157 part-time. *Entrance requirements:* For master's, minimum GPA of 2.5 in major. Application fee: $50. Tuition, state resident: part-time $220 per credit. Part-time tuition and fees vary according to campus/location. *Unit head:* Dr. Susan Smith, Director, 517-774-4373. *Application contact:* 800-950-1144, Fax: 517-774-2461, E-mail: celinfo@mail.cel.cmich.edu.

Concordia University, School of Graduate Studies, Faculty of Engineering and Computer Science, Department of Computer Science, Montréal, PQ H3G 1M8, Canada. Offers computer science (MCS, PhD, Diploma); software engineering (MCS). *Students:* 335 full-time (113 women), 84 part-time (16 women). *Degree requirements:* For master's, one foreign language, computer language required, thesis optional; for doctorate, one foreign language, computer language, dissertation, comprehensive exam required. *Application deadline:* For fall admission, 6/1; for spring admission, 10/1. Application fee: $50. *Unit head:* Dr. H. F. Li, Director, 514-848-3043, Fax: 514-848-2830. *Application contact:* Dr. T. D. Bui, Director, 514-848-3043, Fax: 514-848-2830.

DePaul University, School of Computer Science, Telecommunications, and Information Systems, Program in Software Engineering, Chicago, IL 60604-2287. Offers MS. Part-time and evening/weekend programs available. *Students:* 23 full-time (4 women), 56 part-time (12 women); includes 23 minority (6 African Americans, 13 Asian Americans or Pacific Islanders, 4 Hispanic Americans), 4 international. Average age 32. 39 applicants, 79% accepted. In 1998, 2 degrees awarded. *Degree requirements:* For master's, computer language, thesis, comprehensive exam required, foreign language not required. *Application deadline:* For fall admission, 8/1 (priority date); for winter admission, 11/15 (priority date); for spring admission, 5/1 (priority date). Applications are processed on a rolling basis. Application fee: $25. *Financial aid:* Fellowships, Federal Work-Study and tuition waivers (partial) available. Financial aid application deadline: 4/1. *Faculty research:* Formal methods, object oriented technolgy, measurement of human-computer interaction, architecture. *Unit head:* Dr. George Knafl, Director, 312-362-8715. *Application contact:* Anne B. Morley, Director of Student Services, 312-362-8714, Fax: 312-362-6116.

Drexel University, Graduate School, College of Arts and Sciences, Department of Mathematics and Computer Science, Program in Software Engineering, Philadelphia, PA 19104-2875. Offers MS. *Students:* 1 full-time (0 women), 12 part-time (6 women); includes 1 minority (Hispanic American), 3 international. 20 applicants, 43% accepted. *Degree requirements:* For master's, thesis not required. *Entrance requirements:* For master's, GRE, TOEFL (minimum score of 570 required), TSE (for teaching assistants). *Application deadline:* For fall admission, 8/21. Applications are processed on a rolling basis. Application fee: $35. Tuition: Full-time $15,795; part-time $585 per credit. Required fees: $375; $67 per term. Tuition and fees vary according to program. *Financial aid:* Application deadline: 2/1. *Unit head:* Anne B. Morley, Director of Student Services, 312-362-8714, Fax: 312-362-6116. *Application contact:* Director of Graduate Admissions, 215-895-6700, Fax: 215-895-5939.

Embry-Riddle Aeronautical University, Daytona Beach Campus Graduate Program, Department of Computing and Mathematics, Daytona Beach, FL 32114-3900. Offers software engineering (MSE). Part-time and evening/weekend programs available. *Faculty:* 4 full-time (0 women). *Students:* 27 full-time (6 women), 27 part-time (4 women); includes 9 minority (4 African Americans, 4 Asian Americans or Pacific Islanders, 1 Hispanic American), 21 international. Average age 27. 41 applicants, 73% accepted. In 1998, 13 degrees awarded. *Degree requirements:* For master's, computer language, thesis or alternative required, foreign language not required. *Entrance requirements:* For master's, TOEFL (minimum score of 550 required), minimum GPA of 3.0 in senior year, 2.5 overall; previous course work in computer science. *Application deadline:* Applications are processed on a rolling basis. Application fee: $30 ($50 for international students). Tuition: Full-time $8,820; part-time $490 per credit. Required fees: $105 per semester. Tuition and fees vary according to program. *Financial aid:* In 1998–99, 8 research assistantships with tuition reimbursements (averaging $8,640 per year), 13 teaching assistantships with tuition reimbursements (averaging $8,640 per year) were awarded.; fellowships, career-related internships or fieldwork, Federal Work-Study, and unspecified assistantships also available. Financial aid application deadline: 4/15; financial aid applicants required to submit FAFSA. *Faculty research:* Software processes, software process improvement models, software testing, real-time software systems. Total annual research expenditures: $800,000. *Unit head:* Dr. Soheil Khajenoori, Program Chair, 904-226-7036, Fax: 904-226-6678, E-mail: khajenos@cts.db.erau.edu. *Application contact:* Ginny Tait, Graduate Admissions Specialist, 904-226-6115, Fax: 904-226-6299, E-mail: taitg@cts.db.erau.edu.

See in-depth description on page 741.

Fairfield University, School of Engineering, Fairfield, CT 06430-5195. Offers management of technology (MS); software engineering (MS). Part-time and evening/weekend programs available. *Faculty:* 44 part-time (4 women). *Students:* 1 full-time (0 women), 39 part-time (4 women); includes 2 minority (1 African American, 1 Hispanic American), 2 international. *Degree requirements:* For master's, thesis, final exam required. *Entrance requirements:* For master's, interview, minimum GPA of 2.8. *Application deadline:* For fall admission, 6/30 (priority date). Applications are processed on a rolling basis. Application fee: $40. *Unit head:* Dr. Evangelos Hadjimichael, Dean, 203-254-4000 Ext. 4147, Fax: 203-254-4013, E-mail: hadjm@fair1.fairfield.edu.

Florida State University, Graduate Studies, College of Arts and Sciences, Department of Computer Science, Tallahassee, FL 32306. Offers computer and network system administration (MA, MS); computer science (MA, MS, PhD); software engineering (MA, MS). Part-time programs available. *Faculty:* 13 full-time (1 woman), 6 part-time (0 women). *Students:* 91 full-time (30 women), 3 part-time. *Degree requirements:* For master's, computer language, thesis or alternative required, foreign language not required; for doctorate, computer language, dissertation required, foreign language not required. *Entrance requirements:* For master's, GRE General Test (minimum score of 650 on quantitative section, 1100 combined required), minimum undergraduate GPA of 3.0; for doctorate, GRE General Test (minimum score of 650 on quantitative section, 1100 combined required), minimum GPA of 3.0. *Application deadline:* For fall admission, 3/3 (priority date); for spring admission, 7/1 (priority date). Applications are processed on a rolling basis. Application fee: $20. Electronic applications accepted. Tuition, state resident: part-time $139 per credit hour. Tuition, nonresident: part-time $482 per credit hour. Tuition and fees vary according to program. *Unit head:* Theodore P. Baker, Chairman, 850-644-4029, Fax: 850-644-0058, E-mail: baker@cs.fsu.edu. *Application contact:* David Gaitros, Graduate Admissions, 850-644-4055, Fax: 850-644-0058.

Gannon University, School of Graduate Studies, College of Sciences, Engineering, and Health Sciences, School of Sciences and Engineering, Program in Engineering, Erie, PA 16541-0001. Offers electrical engineering (MS); embedded software engineering (MS); mechanical engineering (MS). Part-time and evening/weekend programs available. *Students:* 23 full-time (4 women), 29 part-time (5 women); includes 1 minority (Asian American or Pacific Islander), 13 international. *Degree requirements:* For master's, thesis or alternative, comprehensive exam required. *Entrance requirements:* For master's, GRE Subject Test, bachelor's degree in engineering, minimum QPA of 2.5. *Application deadline:* Applications are processed on a rolling basis. Application fee: $25. *Unit head:* Dr. Mehmet Cultu, Co-Director, 814-871-7624. *Application contact:* Beth Nemenz, Director of Admissions, 814-871-7240, Fax: 814-871-5803, E-mail: admissions@gannon.edu.

George Mason University, School of Information Technology and Engineering, Department of Information and Software Engineering, Fairfax, VA 22030-4444. Offers information systems (MS); software systems engineering (MS). Part-time and evening/weekend programs available. *Faculty:* 16 full-time (1 woman), 9 part-time (1 woman). *Students:* 59 full-time (29 women), 469 part-time (174 women); includes 156 minority (29 African Americans, 115 Asian Americans or Pacific Islanders, 11 Hispanic Americans, 1 Native American), 77 international. Average age 33. 287 applicants, 70% accepted. In 1998, 159 degrees awarded. *Degree requirements:* For master's, computer language required, thesis optional, foreign language not required. *Entrance requirements:* For master's, GMAT or GRE General Test, TOEFL (minimum score of 575 required), minimum GPA of 3.0 in last 60 hours. *Application deadline:* For fall admission, 5/1; for spring admission, 11/1. Application fee: $30. Electronic applications accepted. Tuition, state resident: full-time $4,416; part-time $184 per credit hour. Tuition, nonresident: full-time $12,516; part-time $522 per credit hour. Tuition and fees vary according to program. *Financial aid:* Fellowships, research assistantships, teaching assistantships available. Aid available to part-time students. Financial aid application deadline: 3/1; financial aid applicants required to submit FAFSA. *Faculty research:* Security, database management, real time systems, software quality. Total annual research expenditures: $380,638. *Unit head:* Dr. Sushil Jajodia, Chairperson, 703-993-1653, Fax: 703-993-1638, E-mail: sjajodia@gmu.edu. *Application contact:* Sandy Mayo, Student Adviser, 703-993-1640, Fax: 703-993-1638.

Grand Valley State University, Science and Mathematics Division, Department of Computer Science and Information Systems, Allendale, MI 49401-9403. Offers information systems (MS); software engineering (MS). Part-time and evening/weekend programs available. *Faculty:* 7 full-time (0 women), 1 part-time (0 women). *Students:* 15 full-time (5 women), 55 part-time (17 women); includes 10 minority (2 African Americans, 8 Asian Americans or Pacific Islanders), 16 international. Average age 34. 25 applicants, 80% accepted. In 1998, 4 degrees awarded (100% found work related to degree). *Degree requirements:* For master's, computer language, thesis or alternative required, foreign language not required. *Entrance requirements:* For master's, GMAT or GRE General Test. *Average time to degree:* Master's–2 years full-time, 3.5 years part-time. *Application deadline:* For fall admission, 2/1. Application fee: $20. *Faculty research:* Object technology, distributed computing, information systems management. *Unit head:* Bruce J. Klein, Associate Professor, 616-895-2048, Fax: 616-895-2106, E-mail: kleinb@gvsu.edu.

Illinois Institute of Technology, Graduate College, Armour College of Engineering and Sciences, Department of Computer Science, Chicago, IL 60616-3793. Offers computer science (MS, PhD); teaching (MST); telecommunications and software engineering (MTSE). Part-time and evening/weekend programs available. *Faculty:* 16 full-time (3 women), 15 part-time (3 women). *Students:* 161 full-time (34 women), 574 part-time (112 women); includes 230 minority (34 African Americans, 185 Asian Americans or Pacific Islanders, 11 Hispanic Americans), 239 international. Terminal master's awarded for partial completion of doctoral program. *Degree requirements:* For master's, computer language, thesis (for some programs), comprehensive exam required, foreign language not required; for doctorate, computer language, dissertation, comprehensive exam required, foreign language not required. *Entrance requirements:* For master's and doctorate, GRE (minimum score of 1200 required), TOEFL (minimum score of 550 required), undergraduate GPA of 3.0 required. *Application deadline:* For fall admission, 7/1; for spring admission, 11/1. Applications are processed on a rolling basis. Application fee: $30. Electronic applications accepted. *Unit head:* Dr. Bodgen Korel, Interim Chairman, 312-567-5150, Fax: 312-567-5067, E-mail: korel@charlie.cns.iit.edu. *Applica-*

tion contact: Dr. S. Mohammad Shahidehpour, Dean of Graduate College, 312-567-3024, Fax: 312-567-7517, E-mail: grad@minna.cns.iit.edu.

See in-depth description on page 753.

Kansas State University, Graduate School, College of Engineering, Department of Computing and Information Sciences, Program in Software Engineering, Manhattan, KS 66506. Offers MSE. Part-time programs available. Postbaccalaureate distance learning degree programs offered (no on-campus study). *Degree requirements:* For master's, computer language required, thesis optional, foreign language not required. *Entrance requirements:* For master's, TOEFL. *Faculty research:* Distributed systems, database systems.

Mercer University, School of Engineering, Macon, GA 31207-0003. Offers biomedical engineering (MSE); electrical engineering (MSE); engineering management (MSE); mechanical engineering (MSE); software engineering (MSE); software systems (MS); technical management (MS). Part-time and evening/weekend programs available. *Faculty:* 23 full-time (1 woman), 6 part-time (0 women). *Degree requirements:* For master's, computer language, thesis or alternative required, foreign language not required. *Entrance requirements:* For master's, GRE, minimum undergraduate GPA of 3.0. *Application deadline:* For fall admission, 7/1; for spring admission, 11/15. Applications are processed on a rolling basis. Application fee: $35 ($50 for international students). *Unit head:* Dr. Benjamin S. Kelley, Dean, 912-752-2459, Fax: 912-752-5593, E-mail: kelley_bs@mercer.edu. *Application contact:* Kathy Olivier, Coordinator, Special Programs, 912-752-2196, E-mail: oliver_kh@mercer.edu.

Mercer University, Cecil B. Day Campus, School of Engineering, Atlanta, GA 30341-4155. Offers electrical engineering (MSE); engineering management (MSE); software engineering (MSE); software systems (MS); technical communication management (MS). Part-time and evening/weekend programs available. Postbaccalaureate distance learning degree programs offered (no on-campus study). *Faculty:* 5 full-time (1 woman), 1 part-time (0 women). *Degree requirements:* For master's, computer language, thesis or alternative required, foreign language not required. *Entrance requirements:* For master's, GRE, minimum GPA of 3.0 in major. *Application deadline:* For fall admission, 7/1; for spring admission, 11/15. Applications are processed on a rolling basis. Application fee: $35 ($50 for international students). *Unit head:* Dr. Benjamin S. Kelley, Acting Dean, 912-752-2459, E-mail: kelley_bs@mercer.edu. *Application contact:* Dr. David Leonard, Director of Admissions, 770-986-3203.

Monmouth University, Graduate School, Department of Software Engineering, West Long Branch, NJ 07764-1898. Offers MS. Part-time and evening/weekend programs available. *Faculty:* 5 full-time (1 woman), 3 part-time (1 woman). *Students:* 31 full-time (6 women), 63 part-time (19 women); includes 24 minority (3 African Americans, 17 Asian Americans or Pacific Islanders, 4 Hispanic Americans), 5 international. Average age 35. In 1998, 63 degrees awarded. *Degree requirements:* For master's, computer language required, thesis optional, foreign language not required. *Entrance requirements:* For master's, bachelor's degree in computer science, engineering, mathematics, or physics; minimum GPA of 3.0; 1 year of software development experience. *Application deadline:* For fall admission, 8/15 (priority date); for spring admission, 12/15 (priority date). Applications are processed on a rolling basis. Application fee: $35 ($40 for international students). Electronic applications accepted. *Financial aid:* In 1998–99, 11 students received aid. Career-related internships or fieldwork, Federal Work-Study, tuition waivers (partial), and unspecified assistantships available. Aid available to part-time students. Financial aid application deadline: 3/1; financial aid applicants required to submit FAFSA. *Faculty research:* Formal protocol modeling with abstract data types and finite state machines, network computing, object orientation, distributed object base, artificial intelligence, real time systems. *Unit head:* Dr. Jorge Diaz-Herrera, Chairperson, 732-571-7501, Fax: 732-728-5253. *Application contact:* 732-571-3452, Fax: 732-571-5123, E-mail: gradadm@monmouth.edu.

Announcement: Master's program in software engineering, offered by the Software Engineering Department. The program focus is on team-oriented development of software products. Program combines elements of computer science, engineering management, and process improvement. Includes modern techniques in requirements analysis, software design, software verification, formal techniques, software project management, network-based systems, security, and reuse.

National Technological University, Programs in Engineering, Fort Collins, CO 80526-1842. Offers chemical engineering (MS); computer engineering (MS); computer science (MS); electrical engineering (MS); engineering management (MS); hazardous waste management (MS); health physics (MS); management of technology (MS); manufacturing systems engineering (MS); materials science and engineering (MS); software engineering (MS); special majors (MS); transportation engineering (MS); transportation systems engineering (MS). Part-time programs available. *Faculty:* 600 part-time (20 women). *Entrance requirements:* For master's, BS in engineering or related field. *Application deadline:* Applications are processed on a rolling basis. Application fee: $50. *Unit head:* Lionel V. Baldwin, President, 970-495-6400, Fax: 970-484-0668, E-mail: baldwin@mail.ntu.edu.

National University, Graduate Studies, School of Business and Technology, Department of Technology, La Jolla, CA 92037-1011. Offers e-commerce (MBA, MS); electronic engineering (MS); engineering management (MS); environmental management (MBA, MS); industrial engineering management (MS); software engineering (MS); technology management (MBA, MS); telecommunication systems management (MS). Part-time and evening/weekend programs available. Postbaccalaureate distance learning degree programs offered (minimal on-campus study). *Faculty:* 12 full-time, 125 part-time. *Students:* 305 (79 women); includes 122 minority (34 African Americans, 69 Asian Americans or Pacific Islanders, 17 Hispanic Americans, 2 Native Americans) 53 international. *Degree requirements:* For master's, foreign language and thesis not required. *Entrance requirements:* For master's, interview, minimum GPA of 2.5. *Application deadline:* Applications are processed on a rolling basis. Application fee: $60 ($100 for international students). Tuition: Full-time $7,830; part-time $870 per course. One-time fee: $60. Tuition and fees vary according to campus/location. *Unit head:* Dr. Leonid Preiser, Chair, 858-642-8425, Fax: 858-642-8716, E-mail: lpreiser@nu.edu. *Application contact:* Nancy Rohland, Director of Enrollment Management, 858-642-8180, Fax: 858-642-8709, E-mail: nrohland@nu.edu.

Oakland University, Graduate Studies, School of Engineering and Computer Science, Program in Computer Science and Engineering, Rochester, MI 48309-4401. Offers computer science (MS); software engineering (MS). Part-time and evening/weekend programs available. *Faculty:* 12 full-time, 3 part-time. *Students:* 48 full-time (22 women), 118 part-time (30 women); includes 21 minority (3 African Americans, 18 Asian Americans or Pacific Islanders), 58 international. *Degree requirements:* For master's, foreign language and thesis not required. *Entrance requirements:* For master's, minimum GPA of 3.0 for unconditional admission. *Application deadline:* For fall admission, 7/15; for spring admission, 3/15. Application fee: $30. Tuition, state resident: part-time $221 per credit hour. Tuition, nonresident: part-time $488 per credit hour. Required fees: $214 per semester. Part-time tuition and fees vary according to program. *Unit head:* Dr. Christian Wagner, Chair, 248-370-2200.

Pennsylvania State University Great Valley School of Graduate Professional Studies, Graduate Studies and Continuing Education, School of Graduate Professional Studies, Department of Engineering and Information Science, Program in Software Engineering, Malvern, PA 19355-1488. Offers MSE. Application fee: $50. *Unit head:* Dr. Thomas M. Hughes, Acting Associate Director of Graduate Admissions, 770-499-3008, Fax: 770-423-6541, E-mail: thughes@kennesaw.edu. *Application contact:* 610-648-3242, Fax: 610-889-1334.

Rochester Institute of Technology, Part-time and Graduate Admissions, College of Applied Science and Technology, Department of Computer Science and Information Technology, Program in Software Development and Management, Rochester, NY 14623-5604. Offers MS. *Students:* 10 full-time (3 women), 70 part-time (24 women); includes 14 minority (3 African Americans, 6 Asian Americans or Pacific Islanders, 5 Hispanic Americans), 8 international. 51 applicants, 63% accepted. In 1998, 33 degrees awarded. *Degree requirements:* For master's, computer language, thesis required. *Entrance requirements:* For master's, GRE General Test, TOEFL, minimum GPA of 3.0. *Application deadline:* For fall admission, 3/1 (priority date). Applications are processed on a rolling basis. Application fee: $40. *Financial aid:* Scholarships and unspecified assistantships available. *Unit head:* Dr. Rayno Niemi, Graduate Coordinator, 716-475-2202, E-mail: rdn@it.rit.edu.

Rochester Institute of Technology, Part-time and Graduate Admissions, College of Applied Science and Technology, Department of Computer Science and Information Technology, Program in Telecommunications Software Technology, Rochester, NY 14623-5604. Offers MS. 1 applicants, 0% accepted. In 1998, 2 degrees awarded. *Degree requirements:* For master's, computer language, thesis required. *Entrance requirements:* For master's, minimum GPA of 3.0. *Application deadline:* For fall admission, 3/1 (priority date). Applications are processed on a rolling basis. Application fee: $40. *Financial aid:* Unspecified assistantships available. *Unit head:* Dr. Rayno Niemi, Graduate Coordinator, 716-475-2202, E-mail: rdn@it.rit.edu.

San Jose State University, Graduate Studies, College of Engineering, Department of Computer, Information and Systems Engineering, Program in Computer Engineering, San Jose, CA 95192-0001. Offers computer engineering (MS); computer software (MS); computerized robots and computer applications (MS); microprocessors and microcomputers (MS). *Faculty:* 5 full-time (0 women), 12 part-time (1 woman). *Students:* 57 full-time (28 women), 90 part-time (19 women); includes 106 minority (1 African American, 103 Asian Americans or Pacific Islanders, 2 Hispanic Americans), 27 international. *Degree requirements:* For master's, computer language, thesis, comprehensive exam required, foreign language not required. *Entrance requirements:* For master's, GRE General Test (minimum combined score of 1500 on three sections required), BS in computer science or 24 credits in related area. *Application deadline:* For fall admission, 6/1. Applications are processed on a rolling basis. Application fee: $59. Tuition, nonresident: part-time $246 per unit. Required fees: $1,939; $1,309 per year. *Unit head:* Dr. Haluk Ozemek, Coordinator, 408-924-4100. *Application contact:* Dr. Haluk Ozemek, Coordinator, 408-924-4100.

Santa Clara University, School of Engineering, Department of Computer Science and Engineering, Santa Clara, CA 95053-0001. Offers computer science and engineering (MSCSE, PhD); high performance computing (Certificate); software engineering (Certificate). Part-time and evening/weekend programs available. *Students:* 141 full-time (76 women), 229 part-time (86 women); includes 91 minority (88 Asian Americans or Pacific Islanders, 3 Hispanic Americans), 222 international. *Degree requirements:* For master's, computer language, thesis or alternative required, foreign language not required; for doctorate and Certificate, computer language, dissertation required, foreign language not required. *Entrance requirements:* For master's, GRE General Test, TOEFL (minimum score of 550 required), minimum GPA of 2.75; for doctorate, GRE General Test, GRE Subject Test, TOEFL (minimum score of 550 required), master's degree or equivalent; for Certificate, master's degree, published paper. *Application deadline:* For fall admission, 6/1; for spring admission, 1/1. Applications are processed on a rolling basis. Application fee: $40. *Unit head:* Dr. Daniel W. Lewis, Chair, 408-554-5281. *Application contact:* Tina Samms, Assistant Director of Graduate Admissions, 408-554-4313, Fax: 408-554-5474, E-mail: engr-grad@scu.edu.

Seattle University, School of Science and Engineering, Program in Software Engineering, Seattle, WA 98122. Offers MSE. Part-time and evening/weekend programs available. *Faculty:* 8 full-time (2 women), 4 part-time (1 woman). *Students:* 9 full-time (4 women), 65 part-time (13 women); includes 19 minority (5 African Americans, 11 Asian Americans or Pacific Islanders, 1 Hispanic American, 2 Native Americans), 10 international. Average age 34. 37 applicants, 95% accepted. In 1998, 28 degrees awarded. *Degree requirements:* For master's, computer language, thesis or alternative required, foreign language not required. *Entrance requirements:* For master's, GRE General Test, 2 years of related work experience. *Average time to degree:* Master's–2 years full-time, 3 years part-time. *Application deadline:* For fall admission, 5/1 (priority date). Application fee: $55. *Financial aid:* Career-related internships or fieldwork and Federal Work-Study available. Aid available to part-time students. Financial aid applicants required to submit FAFSA. *Unit head:* Dr. David Umphress, Director, 206-296-5510, Fax: 206-296-2071, E-mail: umphress@seattleu.edu.

Southern Adventist University, School of Computing, Collegedale, TN 37315-0370. Offers MSE. *Application deadline:* Applications are processed on a rolling basis. Application fee: $25. Tuition: Full-time $5,040; part-time $280 per credit hour. *Unit head:* Dr. Jared Bruckner, Associate Dean, 423-238-2935, Fax: 423-238-2234.

Southern Methodist University, School of Engineering and Applied Science, Department of Computer Science and Engineering, Dallas, TX 75275. Offers computer engineering (MS Cp E, PhD); computer science (MS, PhD); engineering management (MSEM, DE); operations research (MS, PhD); software engineering (MS). Part-time programs available. Postbaccalaureate distance learning degree programs offered (no on-campus study). *Faculty:* 13 full-time (2 women), 12 part-time (1 woman). *Students:* 57 full-time (22 women), 294 part-time (60 women); includes 85 minority (24 African Americans, 44 Asian Americans or Pacific Islanders, 16 Hispanic Americans, 1 Native American), 69 international. *Degree requirements:* For master's, thesis optional, foreign language not required; for doctorate, dissertation, oral and written qualifying exams, oral final exam (PhD) required. *Entrance requirements:* For master's, GRE General Test (minimum score of 650 on quantitative section required), TOEFL (minimum score of 550 required), minimum GPA of 3.0 in last 2 years; bachelor's degree in engineering, mathematics, or sciences; for doctorate, preliminary counseling exam (PhD), minimum GPA of 3.0, bachelor's degree in related field, MA (DE). *Application deadline:* For fall admission, 8/1 (priority date); for spring admission, 12/15. Applications are processed on a rolling basis. Application fee: $25. Tuition: Full-time $9,216; part-time $512 per credit hour. Required fees: $88 per credit hour. Part-time tuition and fees vary according to course load and campus/location. *Unit head:* Dr. Richard V. Helgason, Interim Chair, 214-768-3278, E-mail: helgason@seas.smu.edu. *Application contact:* Dr. Zeynep Celik-Butler, Assistant Dean for Graduate Studies and Research, 214-768-3979, Fax: 214-768-3845, E-mail: zcb@seas.smu.edu.

See in-depth description on page 813.

Southern Polytechnic State University, College of Arts and Sciences, Department of Computer Science, Marietta, GA 30060-2896. Offers computer science (MS); software engineering (MSSE). Part-time and evening/weekend programs available. Postbaccalaureate distance learning degree programs offered. *Faculty:* 12 full-time (0 women). *Students:* 87 full-time, 169 part-time. *Degree requirements:* For master's, thesis optional, foreign language not required. *Entrance requirements:* For master's, GRE General Test. *Application deadline:* For fall admission, 7/1 (priority date); for spring admission, 11/1. Applications are processed on a rolling basis. Application fee: $20. Tuition, state resident: full-time $2,146; part-time $119 per credit hour. Tuition, nonresident: full-time $7,586; part-time $421 per credit hour. *Unit head:* Dr. Jorge Diaz-Herrerea, Head, 770-528-5558, Fax: 770-528-5501, E-mail: jdiaz@spsu.edu.

State University of New York at Stony Brook, Graduate School, College of Engineering and Applied Sciences, Department of Computer Science, Stony Brook, NY 11794. Offers computer science (MS, PhD); software engineering (Certificate). *Faculty:* 26 full-time (1 woman), 2 part-time (both women). *Students:* 129 full-time (39 women), 44 part-time (11 women); includes 25 minority (2 African Americans, 20 Asian Americans or Pacific Islanders, 3 Hispanic Americans), 123 international. *Degree requirements:* For master's, computer language, thesis or alternative required, foreign language not required; for doctorate, computer language, dissertation, comprehensive exams required, foreign language not required. *Entrance requirements:* For master's and doctorate, GRE General Test, TOEFL. *Application deadline:* For fall admission, 1/15. Application fee: $50. *Unit head:* Dr. David Warren, Chairman, 516-632-8470. *Application contact:* Dr. Michael Kiefer, Director, 516-632-8443, Fax: 516-632-8334, E-mail: mkiefer@notes.cc.sunysb.edu.

Software Engineering

Texas Christian University, Add Ran College of Arts and Sciences, Department of Computer Science, Fort Worth, TX 76129-0002. Offers software engineering (MSE). Part-time and evening/weekend programs available. *Students:* 10. In 1998, 4 degrees awarded. *Degree requirements:* For master's, foreign language and thesis not required. *Entrance requirements:* For master's, GRE General Test (minimum combined score of 1000 required), TOEFL (minimum score of 550 required). *Application deadline:* For fall admission, 3/1; for spring admission, 12/1. Applications are processed on a rolling basis. Application fee: $0. *Financial aid:* Application deadline: 3/1. *Unit head:* Dr. Dick Rinewalt, Chairperson, 817-257-7166.

Université du Québec, Institut national de la recherche scientifique, Graduate Programs, Research Center—Telecommunications, Ste-Foy, PQ G1V 4C7, Canada. Offers information technology (Diploma); software engineering (M Sc); telecommunications (M Sc, PhD). Part-time programs available. *Degree requirements:* For master's and doctorate, thesis/dissertation required; for Diploma, thesis not required. *Entrance requirements:* For master's and Diploma, appropriate bachelor's degree, proficiency in French; for doctorate, appropriate master's degree, proficiency in French. *Faculty research:* Visual and verbal systems, communications networks.

University of Connecticut, Graduate School, School of Engineering, Field of Computer Science and Engineering, Storrs, CT 06269. Offers artificial intelligence (MS, PhD); computer architecture (MS, PhD); computer science (MS, PhD); operating systems (MS, PhD); robotics (MS, PhD); software engineering (MS, PhD). Terminal master's awarded for partial completion of doctoral program. *Degree requirements:* For master's, thesis or alternative required; for doctorate, dissertation required. *Entrance requirements:* For master's and doctorate, GRE General Test.

University of Houston–Clear Lake, School of Natural and Applied Sciences, Program in Software Engineering, Houston, TX 77058-1098. Offers MS. *Faculty:* 4 full-time (1 woman), 2 part-time (0 women). *Students:* 2 full-time (1 woman), 47 part-time (14 women); includes 13 minority (2 African Americans, 7 Asian Americans or Pacific Islanders, 4 Hispanic Americans), 5 international. *Degree requirements:* Foreign language not required. *Entrance requirements:* For master's, GRE General Test. *Application deadline:* Applications are processed on a rolling basis. Application fee: $30 ($70 for international students). *Financial aid:* Research assistantships, teaching assistantships available. Financial aid application deadline: 5/1. *Unit head:* Dr. Sharon White, Chair, 281-283-3850, Fax: 281-283-3707, E-mail: white5@uhcl4.cl.uh.edu. *Application contact:* Dr. Robert Ferebee, Associate Dean, 281-283-3700, Fax: 281-283-3707, E-mail: ferebee@uhcl4.cl.uh.edu.

University of Maryland, College Park, Graduate School, College of Computer, Mathematical and Physical Sciences, Software Engineering Program, College Park, MD 20742-5045. Offers MSWE. Part-time programs available. 41 applicants, 78% accepted. *Degree requirements:* For master's, computer language required. *Entrance requirements:* For master's, minimum GPA of 3.0, previous experience in software design. *Application deadline:* Applications are processed on a rolling basis. Application fee: $50 ($70 for international students). Tuition, state resident: part-time $272 per credit hour. Tuition, nonresident: part-time $475 per credit hour. Required fees: $632; $379 per year. *Financial aid:* Federal Work-Study, grants, and scholarships available. Aid available to part-time students. Financial aid applicants required to submit FAFSA. *Unit head:* Dr. Jamel Portilo, Director, 301-405-2706. *Application contact:* Admissions Representative, University College, 301-985-7155.

University of Maryland University College, Graduate School of Management and Technology, Program in Software Engineering, College Park, MD 20742-1600. Offers M Sw E. Offered evenings and weekends only. Part-time and evening/weekend programs available. *Students:* 2 full-time (both women), 72 part-time (20 women); includes 23 minority (6 African Americans, 14 Asian Americans or Pacific Islanders, 3 Hispanic Americans), 16 international. 20 applicants, 100% accepted. *Degree requirements:* For master's, foreign language and thesis not required. *Entrance requirements:* For master's, programming language, software development experience, previous course work in discrete mathematics. *Application deadline:* Applications are processed on a rolling basis. Application fee: $50. Electronic applications accepted. Tuition, state resident: full-time $5,058; part-time $281 per credit. Tuition, nonresident: full-time $6,876; part-time $382 per credit. Tuition and fees vary according to program. *Financial aid:* Federal Work-Study, grants, and scholarships available. Aid available to part-time students. Financial aid application deadline: 6/1; financial aid applicants required to submit FAFSA. *Unit head:* John Richardson, Director, 301-985-7200, Fax: 301-985-4611, E-mail: jrichard@ucsfs1.umd.edu. *Application contact:* Coordinator, Graduate Admissions, 301-985-7155, Fax: 301-985-7175, E-mail: gradinfo@nova.umuc.edu.

University of Michigan–Dearborn, College of Engineering and Computer Science, Department of Computer and Information Science, Dearborn, MI 48128-1491. Offers computer and information science (MS); software engineering (MS). Part-time and evening/weekend programs available. *Faculty:* 11 full-time (1 woman), 2 part-time (0 women). *Students:* 7 full-time (0 women), 111 part-time (24 women); includes 23 minority (4 African Americans, 17 Asian Americans or Pacific Islanders, 2 Hispanic Americans), 7 international. *Degree requirements:* For master's, computer language required, thesis optional, foreign language not required. *Entrance requirements:* For master's, bachelor's degree in mathematics, computer science, or engineering. *Application deadline:* For fall admission, 6/15; for spring admission, 2/15. Applications are processed on a rolling basis. Application fee: $55. Electronic applications accepted. Tuition, state resident: part-time $259 per credit hour. Tuition, nonresident: part-time $748 per credit hour. Required fees: $80 per course. Tuition and fees vary according to course level, course load and program. *Unit head:* Dr. Kenneth Modesitt, Chair, 313-593-5680, Fax: 313-593-9967, E-mail: modesitt@umdsun2.umd.umich.edu. *Application contact:* Mary Tamsen, Graduate Secretary, 313-436-9145, Fax: 313-593-9967, E-mail: mtamsen@umdsun2.umd.umich.edu.

University of Minnesota, Twin Cities Campus, Graduate School, Institute of Technology, Center for the Development of Technological Leadership, Program in Software Engineering, Minneapolis, MN 55455-0213. Offers MS. Part-time and evening/weekend programs available. *Faculty:* 10 part-time (0 women). *Students:* 62 (16 women); includes 15 minority (all Asian Americans or Pacific Islanders) Average age 27. 65 applicants, 58% accepted. *Degree requirements:* For master's, thesis, capstone project required, foreign language not required. *Entrance requirements:* For master's, 1 year of work experience in software field, preferably in Twin Cities area; minimum undergraduate GPA of 3.0. *Application deadline:* For fall admission, 7/15 (priority date). Applications are processed on a rolling basis. Application fee: $50 ($55 for international students). Electronic applications accepted. *Financial aid:* Institutionally-sponsored loans available. Aid available to part-time students. Financial aid applicants required to submit FAFSA. *Faculty research:* Database systems, human-computer interaction, software development, high performance neural systems, data mining. *Unit head:* Haesun Park, Director of Graduate Studies, 612-625-4002, Fax: 612-625-0572. *Application contact:* Toni Limon, Admissions.

University of Missouri–Kansas City, Program in Computer Science Telecommunications, Kansas City, MO 64110-2499. Offers computer networking (MS, PhD); software engineering (MS); telecommunications networking (MS, PhD). PhD offered through the School of Graduate Studies. Part-time programs available. *Faculty:* 15 full-time (1 woman). *Students:* 48 full-time (20 women), 74 part-time (21 women); includes 8 minority (2 African Americans, 6 Asian Americans or Pacific Islanders), 85 international. *Degree requirements:* For master's, computer language required, foreign language not required; for doctorate, computer language, dissertation required, foreign language not required. *Entrance requirements:* For master's, GRE General Test (score in 75th percentile or higher on quantitative section required, 50th percentile or higher on verbal), minimum GPA of 3.0; for doctorate, GRE General Test (score in 85th percentile or higher on quantitative section required, 50th percentile or higher on verbal), minimum GPA of 3.5. *Application deadline:* For fall admission, 3/1 (priority date); for spring admission, 10/1. Applications are processed on a rolling basis. Application fee: $25. *Unit head:* Dr. Richard Hetherington, Director, 816-235-1193, Fax: 816-235-5159, E-mail: info@cstp.umkc.edu.

University of New Haven, Graduate School, School of Engineering and Applied Science, Program in Computer and Information Science, West Haven, CT 06516-1916. Offers applications software (MS); management information systems (MS); systems software (MS). Part-time and evening/weekend programs available. *Students:* 40 full-time (17 women), 157 part-time (41 women); includes 18 minority (4 African Americans, 14 Asian Americans or Pacific Islanders), 68 international. *Degree requirements:* For master's, thesis or alternative required, foreign language not required. *Application deadline:* Applications are processed on a rolling basis. Application fee: $50. *Unit head:* Dr. Tahany Fergany, Coordinator, 203-932-7067.

University of St. Thomas, Graduate Studies, Graduate School of Applied Science and Engineering, Program in Software, St. Paul, MN 55105-1096. Offers MS, MSDD, MSS, Certificate. Part-time and evening/weekend programs available. *Faculty:* 6 full-time (2 women), 26 part-time (2 women). *Students:* 39 full-time (21 women), 536 part-time (168 women); includes 102 minority (25 African Americans, 73 Asian Americans or Pacific Islanders, 3 Hispanic Americans, 1 Native American), 137 international. Average age 33. 256 applicants, 99% accepted. In 1998, 78 master's, 20 other advanced degrees awarded. *Degree requirements:* For master's, thesis optional, foreign language not required; for Certificate, foreign language and thesis not required. *Entrance requirements:* For master's, TOEFL. *Average time to degree:* Master's–3.5 years full-time, 7 years part-time; Certificate–1.5 years full-time, 2.5 years part-time. *Application deadline:* For fall admission, 8/1 (priority date); for spring admission, 1/1 (priority date). Applications are processed on a rolling basis. Application fee: $30. Tuition: Part-time $437 per credit. Tuition and fees vary according to degree level, program and student level. *Financial aid:* In 1998–99, 66 students received aid; fellowships, research assistantships, grants and institutionally-sponsored loans available. Aid available to part-time students. Financial aid application deadline: 4/1; financial aid applicants required to submit FAFSA. *Faculty research:* Distributed databases, fault tolerant computing, expert systems, object-oriented software. *Unit head:* Dr. Bernice Folz, Director, 651-962-5501, E-mail: gradsoftware@stthomas.edu.

See in-depth description on page 899.

The University of Scranton, Graduate School, Program in Software Engineering, Scranton, PA 18510. Offers MS. Part-time and evening/weekend programs available. *Faculty:* 9 full-time (1 woman). *Students:* 10 full-time (3 women), 9 part-time (4 women); includes 2 minority (both Asian Americans or Pacific Islanders), 8 international. Average age 29. 9 applicants, 100% accepted. In 1998, 8 degrees awarded. *Degree requirements:* For master's, computer language, thesis, capstone experience required, foreign language not required. *Entrance requirements:* For master's, GMAT or GRE, TOEFL (minimum score of 550 required), minimum GPA of 3.0. *Application deadline:* For fall admission, 3/1 (priority date). Application fee: $35. Tuition: Part-time $490 per credit. Required fees: $25 per semester. Tuition and fees vary according to program. *Financial aid:* In 1998–99, 7 students received aid, including 7 teaching assistantships; career-related internships or fieldwork, Federal Work-Study, and teaching fellowships also available. Aid available to part-time students. Financial aid application deadline: 3/1. *Faculty research:* Database, parallel and distributed systems, computer network, real time systems. *Unit head:* Dr. Yaodong Bi, Director, 570-941-6108, Fax: 570-941-4250, E-mail: biy1@uofs.edu.

University of Southern California, Graduate School, School of Engineering, Department of Computer Science, Program in Software Engineering, Los Angeles, CA 90089. Offers MS. *Students:* 20 full-time (4 women), 18 part-time (5 women); includes 9 minority (all Asian Americans or Pacific Islanders), 24 international. Average age 28. 83 applicants, 57% accepted. In 1998, 9 degrees awarded. *Entrance requirements:* For master's, GRE General Test. *Application deadline:* For fall admission, 6/1 (priority date); for spring admission, 10/1. Applications are processed on a rolling basis. Application fee: $55. Tuition: Part-time $768 per unit. Required fees: $350 per semester. *Financial aid:* In 1998–99, 1 fellowship, 6 research assistantships, 1 teaching assistantship were awarded.; Federal Work-Study, institutionally-sponsored loans, and scholarships also available. Aid available to part-time students. Financial aid application deadline: 2/15; financial aid applicants required to submit FAFSA. *Unit head:* Dr. Ellis Horowitz, Chairman, Department of Computer Science, 213-740-4494.

The University of Texas at Arlington, Graduate School, College of Engineering, Department of Computer Science and Engineering, Arlington, TX 76019. Offers M Engr, M Sw En, MCS, MS, PhD. *Faculty:* 16 full-time (2 women). *Students:* 218 full-time (44 women), 138 part-time (18 women); includes 35 minority (1 African American, 30 Asian Americans or Pacific Islanders, 4 Hispanic Americans), 252 international. 434 applicants, 36% accepted. In 1998, 121 master's, 3 doctorates awarded. *Degree requirements:* For master's, computer language, thesis required (for some programs), foreign language not required; for doctorate, computer language, dissertation required, foreign language not required. *Entrance requirements:* For master's, GRE General Test (minimum combined score of 1100 required), TOEFL (minimum score of 560 required); for doctorate, GRE General Test (minimum combined score of 1250 required), TOEFL (minimum score of 560 required). *Application deadline:* Applications are processed on a rolling basis. Application fee: $25 ($50 for international students). Tuition, state resident: full-time $1,368; part-time $76 per semester hour. Tuition, nonresident: full-time $5,454; part-time $303 per semester hour. Required fees: $66 per semester hour. $86 per term. Tuition and fees vary according to course load. *Financial aid:* Research assistantships, teaching assistantships, career-related internships or fieldwork and tuition waivers (partial) available. *Unit head:* Dr. Bill D. Carroll, Chairman, 817-272-3785, Fax: 817-272-3784, E-mail: carroll@cse.uta.edu. *Application contact:* Dr. Bob P. Weems, Graduate Adviser, 817-272-3785, Fax: 817-272-3784, E-mail: weems@cse.uta.edu.

See in-depth description on page 911.

University of Toronto, School of Graduate Studies, Physical Sciences Division, Collaborative Program in Software Engineering, Toronto, ON M5S 1A1, Canada. Offers M Eng, M Sc. Offered jointly with Carleton University, University of Ottawa, Queen's University at Kingston, University of Waterloo, and The University of Western Ontario. *Degree requirements:* For master's, thesis required (for some programs).

University of Waterloo, Graduate Studies, Faculty of Engineering, Department of Electrical and Computer Engineering, Waterloo, ON N2L 3G1, Canada. Offers electrical engineering (MA Sc, PhD); electrical engineering (software engineering) (MA Sc). Part-time programs available. *Faculty:* 45 full-time (3 women), 10 part-time (0 women). *Students:* 159 full-time (34 women), 14 part-time (1 woman). *Degree requirements:* For master's, thesis (for some programs), research paper or thesis required, foreign language not required; for doctorate, dissertation, comprehensive exam required, foreign language not required. *Entrance requirements:* For master's, TOEFL (minimum score of 550 required), honors degree, minimum B+ average; for doctorate, TOEFL (minimum score of 550 required), master's degree. *Application deadline:* For fall admission, 2/1; for winter admission, 6/1; for spring admission, 10/1. Application fee: $50. *Expenses:* Tuition and fees charges are reported in Canadian dollars. Tuition, state resident: full-time $3,168 Canadian dollars; part-time $792 Canadian dollars per term. Tuition, nonresident: full-time $8,000 Canadian dollars; part-time $2,000 Canadian dollars. Required fees: $45 Canadian dollars per term. Tuition and fees vary according to program. *Unit head:* Dr. A. Vanelli, Chair, 519-888-4016 Ext. 4016, Fax: 519-746-3077, E-mail: vanelli@cheetah.vlsi.uwaterloo.ca. *Application contact:* Dr. D. W. L. Wang, Graduate Officer, 519-888-4567 Ext. 3330, Fax: 519-746-3077, E-mail: dwang@kingcong.uwaterloo.ca.

University of Waterloo, Graduate Studies, Faculty of Mathematics, Department of Computer Science, Waterloo, ON N2L 3G1, Canada. Offers computer science (M Math, PhD); computer science (software engineering) (M Math); computer science (statistics-computing) (M Math). Part-time programs available. *Faculty:* 47 full-time (7 women), 31 part-time (3 women). *Students:* 121 full-time (26 women), 25 part-time (5 women). *Degree requirements:* For master's, computer language, research paper or thesis required; for doctorate, computer language, dissertation required, foreign language not required. *Entrance requirements:* For master's, TOEFL (minimum score of 580 required), honors degree in field, minimum B+ average; for doctorate, TOEFL (minimum score of 580 required), master's degree. *Application deadline:* For fall admission, 2/28 (priority date); for winter admission, 9/30; for spring admission, 1/31. Applications are processed on a rolling basis. Application fee: $50. *Expenses:* Tuition and fees charges are reported in Canadian dollars. Tuition, state resident: full-time $3,168 Canadian dollars; part-time $792 Canadian dollars per term. Tuition, nonresident: full-time $8,000 Canadian dollars; part-time $2,000 Canadian dollars. Required fees: $45 Canadian dollars per term. Tuition and fees vary according to program. *Unit head:* Dr. N. J. Cecone, Chair, 519-888-4567 Ext. 3292, E-mail: csgrad@jeeves.uwaterloo.ca. *Application contact:* Dr. G. Labahn, Graduate Officer, 519-888-4567 Ext. 4439, E-mail: csgrad@math.uwaterloo.ca.

Wayne State University, Graduate School, College of Science, Department of Computer Science, Detroit, MI 48202. Offers computer science (MA, MS, PhD); electronics and computer control systems (MS). *Degree requirements:* For master's, computer language, thesis required (for some programs), foreign language not required; for doctorate, computer language, dissertation required, foreign language not required. *Entrance requirements:* For master's, TOEFL (minimum score of 550 required), GRE General Test (minimum combined score of 1700 on three sections required); for doctorate, TOEFL (minimum score of 550 required), GRE General Test (minimum combined score of 1800 on three sections required). *Faculty research:* Neural computation, artificial intelligence, software engineering, distributed systems, databases.

See in-depth description on page 931.

West Virginia University, College of Engineering and Mineral Resources, Department of Computer Science and Electrical Engineering, Program in Software Engineering, Morgantown, WV 26506. Offers MS.

Widener University, School of Engineering, Program in Computer and Software Engineering, Chester, PA 19013-5792. Offers ME, ME/MBA. Part-time and evening/weekend programs available. *Faculty:* 5 part-time (1 woman). *Students:* 8 full-time (2 women), 29 part-time (7 women); includes 6 minority (2 African Americans, 4 Asian Americans or Pacific Islanders), 15 international. 28 applicants, 93% accepted. In 1998, 16 degrees awarded. *Degree requirements:* For master's, thesis optional, foreign language not required. *Entrance requirements:* For master's, GMAT (ME/MBA). *Average time to degree:* Master's–2 years full-time, 4 years part-time. *Application deadline:* For fall admission, 8/1 (priority date); for spring admission, 12/1. Applications are processed on a rolling basis. Application fee: $25 ($300 for international students). *Financial aid:* In 1998–99, 1 research assistantship with full tuition reimbursement (averaging $7,500 per year) was awarded.; unspecified assistantships also available. Financial aid application deadline: 3/15. *Faculty research:* Computer and software engineering, computer network fault-tolerant computing, and optical computing. *Unit head:* Dr. Alfred T. Johnson, Chairman, Department of Electrical/Telecommunication Engineering, 610-499-4053, Fax: 610-499-4059, E-mail: alfred.t.johnson@widener.edu. *Application contact:* Dr. David H. T. Chen, Assistant Dean for Graduate Programs and Research, 610-499-4049, Fax: 610-499-4059, E-mail: david.h.chen@widener.edu.

Systems Science

Carleton University, Faculty of Graduate Studies, Faculty of Science, Information and Systems Science Committee, Ottawa, ON K1S 5B6, Canada. Offers M Sc. *Students:* 61 full-time (23 women), 31 part-time (7 women). Average age 33. In 1998, 18 degrees awarded. *Degree requirements:* For master's, computer language required, thesis optional. *Entrance requirements:* For master's, TOEFL (minimum score of 550 required), honors degree. *Average time to degree:* Master's–1.8 years full-time, 3.5 years part-time. *Application deadline:* For fall admission, 3/1 (priority date). Applications are processed on a rolling basis. Application fee: $35. *Financial aid:* Application deadline: 3/1. *Faculty research:* Software engineering, real-time and microprocessor programming, computer communications. *Unit head:* Mike J. Moore, Coordinator, 613-520-2600 Ext. 2160, Fax: 613-520-5733.

Fairleigh Dickinson University, Teaneck–Hackensack Campus, University College: Arts, Sciences, and Professional Studies, Department of Systems Science, Teaneck, NJ 07666-1914. Offers computer engineering (MS); environmental studies (MS). *Degree requirements:* For master's, foreign language and thesis not required. *Entrance requirements:* For master's, GRE General Test.

Louisiana State University and Agricultural and Mechanical College, Graduate School, College of Basic Sciences, Department of Computer Science, Program in Systems Science, Baton Rouge, LA 70803. Offers MSSS. *Faculty:* 1 full-time (0 women). *Students:* 42 full-time (10 women), 20 part-time (3 women); includes 2 minority (both Asian Americans or Pacific Islanders), 56 international. Average age 29. 106 applicants, 44% accepted. In 1998, 78 degrees awarded. *Degree requirements:* For master's, computer language, thesis required, foreign language not required. *Entrance requirements:* For master's, GRE General Test (minimum combined score of 1200 required), minimum GPA of 3.0. *Application deadline:* For fall admission, 5/15 (priority date); for spring admission, 10/15. Applications are processed on a rolling basis. Application fee: $25. *Financial aid:* In 1998–99, 12 research assistantships with partial tuition reimbursements, 14 teaching assistantships with partial tuition reimbursements were awarded.; fellowships Financial aid application deadline: 2/1. *Faculty research:* Robotics, artificial intelligence, algorithms, database software engineering, high-performance computing. Total annual research expenditures: $391,000. *Application contact:* Dr. Doris Carver, Graduate Coordinator, 225-388-1495, Fax: 225-388-1465, E-mail: carver@bit.csc.lsu.edu.

Louisiana State University in Shreveport, College of Sciences, Shreveport, LA 71115-2399. Offers systems technology (MST). Part-time and evening/weekend programs available. *Faculty:* 22 full-time (4 women). Average age 30. 12 applicants, 100% accepted. In 1998, 14 degrees awarded (100% found work related to degree). *Degree requirements:* For master's, computer language, comprehensive exam required, foreign language and thesis not required. *Entrance requirements:* For master's, GRE General Test (minimum combined score of 1400 on three sections required). *Average time to degree:* Master's–2.5 years part-time. *Application deadline:* For fall admission, 8/5 (priority date); for spring admission, 12/15. Applications are processed on a rolling basis. Application fee: $10. *Financial aid:* Teaching assistantships with partial tuition reimbursements available. *Faculty research:* Graphics, software quality, programming languages, tutoring systems. *Unit head:* Dr. Alfred McKinney, Dean, 318-797-5231, Fax: 318-797-5230, E-mail: amckinne@pilot.lsus.edu.

Portland State University, Graduate Studies, Systems Science Program, Portland, OR 97207-0751. Offers systems science/anthropology (PhD); systems science/business administration (PhD); systems science/civil engineering (PhD); systems science/economics (PhD); systems science/engineering management (PhD); systems science/general (PhD); systems science/mathematical sciences (PhD); systems science/mechanical engineering (PhD); systems science/psychology (PhD); systems science/sociology (PhD). *Faculty:* 3 full-time (0 women), 1 part-time (0 women). *Students:* 45 full-time (17 women), 23 part-time (6 women); includes 5 minority (1 African American, 3 Asian Americans or Pacific Islanders, 1 Hispanic American), 12 international. Average age 35. 98 applicants, 31% accepted. In 1998, 11 degrees awarded. *Degree requirements:* For doctorate, variable foreign language requirement, computer language, dissertation required. *Entrance requirements:* For doctorate, GMAT (score in 75th percentile or higher required), GRE General Test (score in 75th percentile or higher required), TOEFL (minimum score of 575 required), minimum undergraduate GPA of 3.0. *Application deadline:* For fall admission, 2/1; for spring admission, 11/1. Application fee: $50. *Financial aid:* In 1998–99, 8 research assistantships, 1 teaching assistantship were awarded.; career-related internships or fieldwork, Federal Work-Study, and institutionally-sponsored loans also available. Aid available to part-time students. Financial aid application deadline: 3/1; financial aid applicants required to submit FAFSA. *Faculty research:* Systems theory and methodology, artificial intelligence neural networks, information theory, nonlinear dynamics/chaos, modeling and simulation. Total annual research expenditures: $13,250. *Unit head:* Dr. Nancy Perrin, Director, 503-725-4960, E-mail: perrinn@pdx.edu. *Application contact:* Dawn Kuenle, Coordinator, 503-725-4960, E-mail: dawn@sysc.pdx.edu.

Salve Regina University, Graduate School, Program in Information Systems Science, Newport, RI 02840-4192. Offers MS. Part-time and evening/weekend programs available. *Faculty:* 3 part-time (0 women). *Students:* 3 full-time (1 woman), 23 part-time (9 women). Average age 37. 6 applicants, 67% accepted. In 1998, 10 degrees awarded. *Degree requirements:* For master's, computer language required, foreign language and thesis not required. *Entrance requirements:* For master's, GMAT, GRE General Test, or MAT. *Average time to degree:* Master's–2 years full-time, 3 years part-time. *Application deadline:* Applications are processed on a rolling basis. Application fee: $35. *Financial aid:* Career-related internships or fieldwork and Federal Work-Study available. Aid available to part-time students. Financial aid application deadline: 3/1. *Unit head:* Frederick E. Lupone, Director, 401-847-6650 Ext. 3135, Fax: 401-847-0372. *Application contact:* Laura E. McPhie-Oliveira, Dean of Enrollment Services, 401-847-6650 Ext. 2908, Fax: 401-848-2823, E-mail: sruadmis@salve.edu.

State University of New York at Binghamton, Graduate School, Thomas J. Watson School of Engineering and Applied Science, Department of Systems Science and Industrial Engineering, Binghamton, NY 13902-6000. Offers M Eng, MS, MSAT, PhD. Part-time and evening/weekend programs available. *Faculty:* 6 full-time, 3 part-time. *Students:* 57 full-time (5 women), 36 part-time (8 women); includes 8 minority (5 African Americans, 1 Asian American or Pacific Islander, 2 Hispanic Americans), 46 international. Average age 31. 82 applicants, 73% accepted. In 1998, 20 master's, 5 doctorates awarded. Terminal master's awarded for partial completion of doctoral program. *Degree requirements:* For master's, thesis or alternative required, foreign language not required; for doctorate, dissertation required, foreign language not required. *Entrance requirements:* For master's and doctorate, GRE General Test, GRE Subject Test, TOEFL (minimum score of 550 required). *Application deadline:* For fall admission, 4/15 (priority date); for spring admission, 11/1. Applications are processed on a rolling basis. Application fee: $50. Electronic applications accepted. Tuition, state resident: full-time $5,100; part-time $213 per credit. Tuition, nonresident: full-time $8,416; part-time $351 per credit. Required fees: $77 per credit. Part-time tuition and fees vary according to course load. *Financial aid:* In 1998–99, 49 students received aid, including 27 research assistantships with full tuition reimbursements available (averaging $8,833 per year), 8 teaching assistantships with full tuition reimbursements available (averaging $8,536 per year); fellowships, career-related internships or fieldwork, Federal Work-Study, institutionally-sponsored loans, and unspecified assistantships also available. Aid available to part-time students. Financial aid application deadline: 2/15. *Faculty research:* Problem restructuring, protein modeling. *Unit head:* Dr. Robert Emerson, Chair, 607-777-6509.

Syracuse University, Graduate School, L. C. Smith College of Engineering and Computer Science, Department of Electrical Engineering and Computer Science, Program in Systems and Information Science, Syracuse, NY 13244-0003. Offers MS, PhD. *Students:* 64 full-time (14 women), 12 part-time (1 woman), 64 international. Average age 33. 86 applicants, 74% accepted. In 1998, 1 master's, 11 doctorates awarded. *Degree requirements:* For doctorate, computer language, dissertation required, foreign language not required, foreign language not required. *Entrance requirements:* For master's and doctorate, GRE General Test, GRE Subject Test. *Application deadline:* Applications are processed on a rolling basis. Application fee: $40. Tuition: Full-time $13,992; part-time $583 per credit hour. *Financial aid:* Fellowships, research assistantships, teaching assistantships, Federal Work-Study and tuition waivers (partial) available. Financial aid application deadline: 3/1. *Application contact:* James Fawcett, Contact, 315-443-2655.

University of Ottawa, School of Graduate Studies and Research, Faculty of Administration, Systems Science Program, Ottawa, ON K1N 6N5, Canada. Offers M Sc. Part-time programs available. *Faculty:* 19 full-time. *Students:* 29 full-time (10 women), 25 part-time (7 women), 9 international. Average age 31. In 1998, 11 degrees awarded. *Degree requirements:* For master's, computer language required, thesis optional, foreign language not required. *Entrance requirements:* For master's, honors degree or equivalent, minimum B average. *Application deadline:* For fall admission, 2/1 (priority date). Applications are processed on a rolling basis. Application fee: $35. *Financial aid:* Fellowships, research assistantships, teaching assistantships, career-related internships or fieldwork and Federal Work-Study available. *Faculty research:* Deterministic and probabilistic modelling, optimization, computer science, information systems. *Unit head:* Jean-Michel Thizy, Chair, 613-562-5800 Ext. 4787, Fax: 613-562-5164. *Application contact:* Diane Sarrazin, Administrator, 613-562-5884 Ext. 4713, Fax: 613-562-5164, E-mail: infomba@admin.uottawa.ca.

Washington University in St. Louis, School of Engineering and Applied Science, Sever Institute of Technology, Department of Systems Science and Mathematics, St. Louis, MO 63130-4899. Offers control engineering (MCE); systems science and mathematics (MS, D Sc); systems science, mathematics, and economics (D Sc). Part-time programs available. *Faculty:* 8 full-time (0 women), 7 part-time (0 women). *Students:* 34 full-time (2 women), 9 part-time; includes 20 minority (19 Asian Americans or Pacific Islanders, 1 Hispanic American), 7 international. Average age 27. 56 applicants, 41% accepted. In 1998, 10 master's, 3 doctorates awarded. Terminal master's awarded for partial completion of doctoral program. *Degree requirements:* For master's, thesis optional, foreign language not required; for doctorate, dissertation, departmental qualifying exam required, foreign language not required. *Entrance requirements:* For master's and doctorate, TOEFL. *Application deadline:* For fall admission, 2/15 (priority date); for spring admission, 10/14. Application fee: $20. *Financial aid:* In 1998–99, 27 students received aid, including 27 research assistantships (averaging $10,000 per year); fellowships, teaching assistantships, career-related internships or fieldwork, Federal Work-Study, and institutionally-sponsored loans also available. Financial aid application deadline: 2/15. *Faculty research:* Linear and nonlinear control systems, robotics and automation, scheduling and transportation systems, computer vision, discrete event dynamical systems. *Unit head:* Dr. I. Norman Katz, Chairman, 314-935-6001, Fax: 314-935-6121, E-mail: katz@zach.wustl.edu. *Application contact:* Sandra Devereaux, Administrative Secretary, 314-935-6001, Fax: 314-935-6121, E-mail: sandra@zach.wustl.edu.

See in-depth description on page 929.

Cross-Discipline Announcements

Baylor College of Medicine, Graduate School of Biomedical Sciences, Division of Neuroscience, Houston, TX 77030-3498.

The Division of Neuroscience offers a course of study leading to the PhD in neuroscience (and the MD/PhD in conjunction with the Medical School) for students interested in advanced study of the nervous system. Faculty research programs include molecular neurobiology, neuroanatomy, neurophysiology, neural systems analysis, neurobiology of disease, biophysics, imaging, behavioral neuroscience, and computer-assisted neural modeling. See in-depth description in the Neuroscience section.

Bentley College, Graduate School of Business, Program in Computer Information Systems, Waltham, MA 02452-4705.

Bentley College offers the Master of Science in Computer Information Systems. The MSCIS is designed to develop the expertise needed to plan for, capitalize on, and deploy information technology in commercial, government, and not-for-profit sectors. The program includes in-depth courses on IS development, IS management, and IS corporate policy. Contact Holly Chase (phone: 781-891-2108; e-mail: gradadm@bentley.edu). Visit the Web site at http://www.bentley.edu/graduate. See in-depth description in section 1.

Boston University, College of Engineering, Department of Aerospace and Mechanical Engineering, Boston, MA 02215.

The department offers graduate programs leading to the PhD and MS degrees. Research is focused in waves and acoustics; dynamics, control, and robotics; fluid mechanics; and precision engineering. Areas of specialization include aerodynamics, biomechanics, medical acoustics, multiphase flow, noise and vibrational control, structural dynamics, thermal processes, theoretical fluid dynamics, and turbulence.

Carnegie Mellon University, H. John Heinz III School of Public Policy and Management, Pittsburgh, PA 15213-3891.

The School offers 2-year master's programs in public policy and management and arts management and a 1-year master's program in health care policy and management. A doctoral program in policy analysis is also offered. Course work is interdisciplinary and analytical, with a strong emphasis on quantitative analysis, team-based problem solving, and real-world applications. The School awards $2 million in financial aid annually.

Georgia Institute of Technology, Graduate Studies and Research, Ivan Allen College of Policy and International Affairs, Sam Nunn School of International Affairs, Atlanta, GA 30332-0001.

The Sam Nunn School of International Affairs at Georgia Tech offers an 18-month master's degree in international affairs that enables graduates to assume professional positions within business, government, and international organizations. The program is built around a core of 6 courses that provide strong theoretical and methodological skills and an understanding of the major issues in international security and international political economy. Students also have the opportunity to tailor the program to their individual interests through elective offerings within the School and interdisciplinary work in economics, management, public policy, computer science, engineering, and other fields.

Indiana University Bloomington, School of Library and Information Science, Bloomington, IN 47405.

The School's information science curriculum (MIS and PhD) focuses on analysis and design of information and systems, emphasizing the social and behavioral dimensions of information technology. The programs appeal to students who are interested in careers in information management, systems analysis and design, online searching and information brokerage, competitive intelligence and research analysis, Internet and network management, Web design, and database development and marketing.

Louisiana State University and Agricultural and Mechanical College, Graduate School, College of Engineering, Department of Electrical and Computer Engineering, Baton Rouge, LA 70803.

A modern program encompassing the areas of parallel and distributed computing, interconnection networks, parallel algorithms, parallelizing compilers, image processing, neural networks, computer networks and VLSI systems, and theory of computing. Opportunities for graduate assistants to participate on several funded research projects at the cutting edge of the above areas.

Northwestern University, The Graduate School, School of Education and Social Policy, Program in Education and Social Policy-Learning Sciences, Evanston, IL 60208.

Northwestern University's interdisciplinary Learning Sciences PhD Program prepares professionals to advance the scientific understanding and practice of learning and education. A major focus of the program concerns theory, design, construction, and assessment of effective technologies and learning environments for schools and workplaces. The program integrates 3 research foci in its core course work, methodological foundations, and research apprenticeships: (1) social context, the social, organizational, and cultural dynamics of learning, (2) cognition, the structures and processes by which domain knowledge and skills are acquired, and (3) design, the theory-guided design and use of curricula, multimedia computing and communications technologies for supporting learning and teaching. Faculty members in the program are drawn from the School of Education and Social Policy, the Department of Computer Science, and the Department of Psychology.

The Ohio State University, Graduate School, College of Engineering, Department of Civil and Environmental Engineering and Geodetic Science, Program in Geodetic Science and Surveying, Columbus, OH 43210.

Today's information-oriented society has created a need for specialists in spatial information science with a foundation in computer mapping. MS and PhD programs offer this unique combination, including spatial information and analysis, image understanding, computer vision/photogrammetry, geodesy, and earth and environmental remote sensing. Center for Mapping on campus. WWW: http://www-ceg.eng.ohio-state.edu.

Pace University, Lubin School of Business, New York, NY 10038.

A recognized leader in business education, the Lubin School of Business of Pace University offers dynamic and innovative programs on campuses in New York City and White Plains, New York.

Polytechnic University, Brooklyn Campus, Department of Electrical Engineering, Brooklyn, NY 11201-2990.

The Department of Electrical Engineering offers graduate programs leading to the degrees of Master of Science and Doctor of Philosophy in the following areas: telecommunications, high-speed networking, optical and wireless communications, wave propagation and scattering, microwaves, antennas, signal processing, image and video processing, multimedia, robotics and control, VLSI design, plasmas, electric power, and power electronics. WWW: http://www.poly.edu

University of Michigan, Horace H. Rackham School of Graduate Studies, School of Information, Ann Arbor, MI 48109.

Students may explore a human-centered approach to information access, management, and systems. Students utilize their scientific and computing backgrounds in a curriculum that includes specializations in human-computer interaction and information economics, management, and policy. The School of Information offers interdisciplinary education, cutting-edge research opportunities, and a state-of-the-art technology environment.

AMERICAN UNIVERSITY

Department of Computer Science and Information Systems

Programs of Study	The Department of Computer Science and Information Systems offers programs leading to the M.S. degree in computer science or the M.S. degree in information systems, both in a broad array of formats. Both programs balance the practical and theoretical aspects of computer science and information systems. The 30-credit-hour M.S. in computer science provides a thorough background in computer science and its applications. The 36-credit-hour M.S. in information systems is a professionally oriented program covering all aspects of the analysis, design, development, and maintenance of computerized information systems. Formats include traditional day and evening classes as well as cohort-based weekend classes that meet in twelve sequenced courses, each course running for six weekend sessions on Friday and Saturday. The Master of Science in statistical computing program, offered jointly with the Department of Mathematics and Statistics, emphasizes computer-oriented data analysis, including the design of algorithms, data structures, and numerical and graphical procedures for statistical applications. The department also offers professional advancement courses through its graduate certificate programs in information systems, information resources management, and systems and project management. Graduate students may also integrate cooperative education experiences into their curriculum, earning credit while working in paid positions in the Washington, D.C., region. Graduate degree programs are designed to prepare students for professional careers or for further doctoral study. Department emphasis is on software, with concentrations in intelligent systems, database management, and software development.
Research Facilities	The campus is fully networked with Windows-based, UNIX-based, and Macintosh workstations, and all are accessible through EagleNet, the University's campuswide fiber-optic network. Network-based software for personal computers and workstations is provided on EagleNet, which also serves as the University's gateway to the Internet. The department supports five specialized laboratories, including an intelligent systems laboratory, an Oracle development laboratory, and a multimedia laboratory. Labs are used for instruction and research and are maintained by department personnel to ensure smooth operation. Graduate students have ample opportunities to develop a variety of experience and skills. The Bender Library and Learning Resources Center houses more than 700,000 volumes and 3,000 periodicals as well as extensive microform collections and a nonprint media center. In addition, more than fourteen indexes in compact disk format are searchable using library microcomputers. Students have unlimited book check-out privileges and access to online bibliographical search services. As a member of the Washington Research Library Consortium, AU graduate students have borrowing privileges at six college and university libraries in the Washington, D.C., area. Dozens of other private and governmental collections, including the Library of Congress, are easily accessible.
Financial Aid	Fellowships, scholarships, and graduate assistantships are available to full-time students. There are special awards for members of minority groups and international students. Part-time work is also available, as are loans and deferred-payment programs. Duties of graduate fellows and assistants usually include helping students in the department's computer center and assisting faculty members in their research.
Cost of Study	Graduate tuition for 1999–2000 is $721 per semester hour.
Living and Housing Costs	Although many graduate students live off campus, the University provides graduate dormitory rooms and apartments. The Off-Campus Housing Office maintains a referral file of rooms and apartments. Housing costs in Washington, D.C., are comparable to those in most other major metropolitan areas.
Student Group	The department has approximately 200 graduate students, with a number of these in part-time status. The University's favorable student-faculty ratio of 14:1 allows ample opportunity for one-on-one interaction among faculty members and fellow students.
Student Outcomes	Graduates from the master's programs have begun their own start-up ventures or are employed in networking, databases, and system development and administration with government organizations such as the Census Bureau, the Department of Defense, and Fannie Mae and with private companies such as Oracle Corporation, America Online, Discovery Corporation, and Hughes Network Systems.
Location	The national capital area offers students access to a variety of educational, governmental, and cultural resources that enrich the student's degree program with opportunities for practical applications of theoretical studies. In recent years, students have completed internships at organizations such as IBM, AT&T, MCI, the American Council on Education, the Internal Revenue Service, the Department of Treasury, the General Accounting Office, and the National Security Agency.
The University	American University was founded as a Methodist institution, chartered by Congress in 1893, and intended originally for graduate study only. The University is located on an 84-acre site in a residential area of northwest Washington. As a member of the Consortium of Universities of the Washington Metropolitan Area, AU offers its degree candidates the option of taking courses at other consortium universities for residence credit.
Applying	Applications for admission should be submitted prior to February 1 if the student is also applying for financial aid. Deadlines vary for different fields of study, but early application is always encouraged. The application fee is $50.

Correspondence and Information	To contact faculty members and for specific program information:	For an application and University catalog:
	Department of Computer Science and Information Systems	Graduate Admissions Office
	College of Arts and Sciences	American University
	American University	4400 Massachusetts Avenue, NW
	4400 Massachusetts Avenue, NW	Washington, D.C. 20016-8001
	Washington, D.C. 20016-8116	Telephone: 202-885-6000
	Telephone: 202-885-1470	E-mail: afa@american.edu
	E-mail: csis@american.edu	
	WWW: http://www.american.edu/csis.html	

American University

THE FACULTY AND THEIR RESEARCH

Judith Barlow, Assistant Professor; Ph.D., Colorado. Databases, decision support systems, simulation.

Shirley Becker, Associate Professor; Ph.D., Maryland. Computer-assisted software engineering, expert systems, database and information systems.

Thomas J. Bergin, Professor; Ph.D., American. History of computing, information resources management, ethical issues in computing.

Frank W. Connolly, Professor; Ph.D., American. Legal and ethical issues of technology, educational computing, intellectual property.

Richard Gibson, Assistant Professor; Ph.D., Maryland. Global information technology and software process improvement.

Parke Godfrey, Instructor; M.S., Georgia Tech. Database and knowledge base systems, artificial intelligence and logic programming, algorithms and complexity.

Michael Gray, Associate Professor; Ph.D., Penn State. Artificial intelligence (AI), distributed AI, intelligent systems architecture.

Maliha Haddad, Assistant Professor; Ph.D., George Washington. Information engineering.

Reza Khorramshahgol, Associate Professor; Ph.D., George Washington. Data communications, quantitative methods, software engineering.

Anita J. LaSalle, Associate Professor; Ph.D., Stevens. Software engineering, expert systems, communications, multimedia design and development.

Jack Ligon, Assistant Professor; Ph.D., Maryland. Information resources management and business process re-engineering.

Charles Linville, Assistant Professor; Ph.D., Carnegie Mellon. Environmental computing, geographic information systems, simulation, decision analysis.

Gene McGuire, Associate Professor; Ph.D., American. Organizational and behavior aspects of information systems, software quality, hypermedia computing, expert systems, educational computing.

Larry Medsker, Professor; Ph.D., Indiana. Hybrid intelligent systems, expert systems, fuzzy logic, neural networks, database systems.

Vincent Ribiere, Assistant Professor; M.S., Marseilles (France). Information systems quality, multimedia, knowledge management.

Mehdi Owrang, Associate Professor; Ph.D., Oklahoma. Database systems, expert systems, knowledge discovery in databases.

Angela Wu, Professor and Chair; Ph.D., Maryland. Computer vision systems and computational geometry.

ARIZONA STATE UNIVERSITY

College of Engineering and Applied Sciences
Department of Computer Science and Engineering

Programs of Study

The Department of Computer Science and Engineering at Arizona State University offers three graduate degree programs: the M.C.S., M.S., and Ph.D. The M.C.S. primarily involves course work but does include a project at the master's level. The M.S. combines course work with research at the master's level, and the Ph.D. is a research-intensive program. Highly motivated and research-oriented candidates are encouraged to apply for the Ph.D. program. While professionally inclined applicants may apply for the M.C.S. program, those desiring a balance between course work and research should choose the M.S. program.

Research Facilities

The department maintains various instructional laboratories with UNIX workstations (Sun, Silicon Graphics, etc.) and Pentium PCs. These laboratories support special applications required for computer science and engineering courses not available elsewhere on the ASU campus. The department has three laboratories with equipment and software specifically designed for instructions at the microprocessor level. They support Motorola and Intel processors and VLSI design.

The department has various research laboratories ranging from personal computers to UNIX and graphics workstations to parallel computers such as the IBM RISC workstation cluster and the Silicon Graphics Power Challenge Supercomputer located at ASU Information Technology (IT) in the Computing Commons.

All computers in the department are networked, with some research laboratories using high-speed ATM switches, Myrinet, mobile ATM, and 100 mbps LAN Hubs. The College of Engineering and Applied Sciences provides various servers to support client/server applications and development in the department. All computers in the department are connected through networking to ASU Information Technology. IT maintains the general-purpose computer laboratories throughout the campus.

Financial Aid

Approximately 100 graduate teaching or research assistantships are available, including sponsored research support. Students should write to the Graduate Office for further information. Positions are highly competitive, and only exceptional new students are offered financial aid upon admission to the Ph.D. program. Arrangements are made for numerous graduate students who are supported around the campus.

Cost of Study

In 1998–99, registration and tuition for 12 hours or more were $1094 per semester for Arizona residents; nonresidents paid $4670 per semester (prorated for fewer than 12 hours).

Living and Housing Costs

Limited on-campus housing is available for unmarried students. In 1998–99, room and board ranged from $4500 to $6725 per academic year. Twelve different meal plans are available (per week or per semester).

Student Group

There are more than 49,000 students at Arizona State, including more than 10,500 graduate students. More than 1,766 of the 5,188 students in the College of Engineering and Applied Sciences are enrolled in graduate programs. There are 409 doctoral students enrolled in the College. More than 300 of the 1,200 students in the Department of Computer Science and Engineering are enrolled in the graduate programs. There are about 70 students enrolled in the Ph.D. program within the department.

Student Outcomes

Excellent opportunities exist for graduates of the program, either in the local high-technology industry or in the major software/hardware organizations. The department fosters contacts with industry during graduate studies through colloquia, assistantships, and contacts with part-time graduate students who are full-time employees.

Location

Arizona is well known for its scenic attractions, which range from desert to mountain woodlands, lakes, and streams. The urban community immediately surrounding Arizona State University thrives on high-technology industries that have large research and development staffs involved with computers, airborne electronics, semiconductors, turbines, energy production and conservation, food processing, and health services. Approximately 11,000 scientists and engineers are employed in the immediate vicinity of the University.

The College

The rapid growth of the College began in 1956 with the authorization of the B.S.E. degree. The faculty now numbers about 230. Enrollment in the graduate program in engineering is the largest among engineering colleges in the Rocky Mountain region. The College has several research centers targeted toward specific research areas.

Applying

Application forms may be obtained by writing to the Admissions Office of the Graduate College or by accessing the Web site http://www.asu.edu/forms/adm.html. At the beginning of the semester prior to the student's enrollment (by August 15 for spring semester enrollment or by January 15 for fall semester enrollment), the Graduate College should have received the application for admission and two copies of transcripts of all undergraduate and graduate work. If the applicant is an international student, then a TOEFL score is required before evaluation can begin. It is required that scores on the General Test of the Graduate Record Examinations also be submitted. Regular admission requires a grade point average of at least 3.0 (on a scale of 4.0) in the last two years of course work leading to the bachelor's degree.

Graduate admission to the M.C.S., M.S., and Ph.D. programs in the Department of Computer Science and Engineering is based on a competitive evaluation of the applicant's background. The evaluation includes scholastic background, experience, and performance on the GRE and TOEFL (for international applicants). The equivalent of a bachelor's degree in computer science in previous academic experience is expected. However, highly accomplished candidates from related disciplines are also considered. In addition, applicants to the Ph.D. program are expected to provide three letters of recommendation, a statement of purpose, GRE General Test scores, and GRE Subject Test scores (recommended).

The department looks for demonstrable academic capabilities in the prior degree program. Most of the admitted applicants have GRE scores above 50 percent in verbal, 90 percent in quantitative, and 70 percent in analytical. Admitted applicants to the M.S. and M.C.S. programs typically have a grade point average of 3.25 out of 4.0. Admissible applicants to the Ph.D. program normally have a grade point average of 3.5 out of 4.0. In addition, all successful international applicants score the Graduate College standard of 550 or higher on the TOEFL.

Correspondence and Information

Graduate Office
Department of Computer Science and Engineering
Arizona State University, Box 875406
Tempe, Arizona 85287-5406

Telephone: 602-965-3190
Fax: 602-965-2751
E-mail: cse.graduate.office@asu.edu
World Wide Web: http://www.eas.asu.edu/~csedept

Arizona State University

THE FACULTY AND THEIR RESEARCH

Stephen S. Yau, Professor and Chair; Ph.D., Illinois at Urbana-Champaign, 1961. Software engineering, parallel processing and distributed computing systems, fault-tolerant computing systems.

Edward A. Ashcroft, Professor; Ph.D., Imperial College (London), 1970. Program verification, declarative language, intensional programming, high-level parallel programming language, language Lucid.

Chitta Baral, Associate Professor; Ph.D., Maryland, 1991. Artificial intelligence, multimedia, visualization of databases.

Rida A. Bazzi, Assistant Professor; Ph.D., Georgia Tech, 1994. Distributed computing, software engineering for distributed systems, fault-tolerance algorithms, computer vision.

Sourav Bhattacharya, Associate Professor; Ph.D., Minnesota, 1993. Networked and parallel computing, dependable communication, ATM networks.

K. Selcuk Candan, Assistant Professor; Ph.D., Maryland, 1997. Distributed multimedia authoring systems, video servers for video-on-demand systems, content-based video indexing, query/retrieval of multimedia data, security, query processing.

Prasad Chalasani, Assistant Professor; Ph.D., Carnegie Mellon, 1994. Design, analysis, and implementation of algorithms in finance, machine learning, software agents, and operations research.

James S. Collofello, Professor; Ph.D., Northwestern, 1978. Software engineering, project management.

Partha Dasgupta, Associate Professor; Ph.D., SUNY at Stony Brook, 1984. Distributed operating systems, system software.

Suzanne W. Dietrich, Associate Professor; Ph.D., SUNY at Stony Brook, 1987. Databases, knowledge management, object management, active features.

Leonard Faltz, Associate Professor; Ph.D., Berkeley, 1977. Formal linguistics, computational linguistics.

Gerald E. Farin, Professor; Ph.D., Braunschweig (Germany), 1979. Computer-aided geometric design, NURBS.

Nicholas V. Findler, Emeritus Professor; Ph.D., Budapest Technical, 1956. Artificial intelligence, heuristic programming, expert systems, pattern recognition, information retrieval.

Gerald C. Gannod, Assistant Professor; Ph.D., Michigan State, 1998. Software engineering, formal methods for software development, reverse engineering, object-oriented analysis and design.

Sumit Ghosh, Associate Professor and Associate Chair of Research and Graduate Programs; Ph.D., Stanford, 1984. Networking, network security, asynchronous distributed algorithms, hardware design languages, modeling and distributed simulation.

Forouzan Golshani, Professor; Ph.D., Warwick (England), 1982. Multimedia information system, digital video processing, advanced databases, intelligent systems.

Ben M. Huey, Associate Professor and Associate Chair for Undergraduate Program; Ph.D., Arizona, 1975. Language-based models for architecture, silicon compilation, design verification, automatic test generation.

Subbarao Kambhampati, Associate Professor; Ph.D., Maryland, 1989. Artificial intelligence, automated planning, machine learning.

William E. Lewis, Professor and Vice Provost for Information Technology; Ph.D., Northwestern, 1966. Analytical modeling, information systems.

Timothy E. Lindquist, Associate Professor; Ph.D., Iowa State, 1979. Programming language, software engineering, human-computer interaction, software engineering environments.

Donald S. Miller, Associate Professor; Ph.D., USC, 1972. Address space operating systems, distributed and multiprocessor operating systems, computer architecture, local area networks.

Gregory M. Nielson, Professor; Ph.D. Utah, 1970. Interactive design of curves and surfaces, multivariate data fitting, computer-aided geometric design, computer graphics, visualization of scientific computing.

Pearse O'Grady, Associate Professor; Ph.D., Arizona, 1969. Parallel processing, computer architecture, continuous system simulation.

Sethuraman Panchanathan, Associate Professor; Ph.D., Ottawa, 1989. Multimedia computing and communications, multimedia hardware architectures, VLSI architectures for real-time video processing, indexing/storage/browsing/retrieval of images and video.

David C. Pheanis, Associate Professor; Ph.D., Arizona State, 1974. Software/hardware interface in embedded microprocessor system, real-time system.

Andrea W. Richa, Assistant Professor; Ph.D., Carnegie Mellon, 1998. Design and analysis of algorithms for distributed networks, graph algorithms, combinatorial optimization, parallel computation.

Alyn P. Rockwood, Associate Professor; Ph.D., Cambridge, 1988. Mechanical CAD, computer graphics.

Arunabha Sen, Associate Professor; Ph.D., South Carolina, 1987. Parallel computing, computer interconnection networks, combinatorial optimization.

Wei-Tek Tsai, Professor; Ph.D., Berkeley, 1986. Software engineering, Internet, parallel and distributed processing.

Joseph E. Urban, Professor; Ph.D., Southwestern Louisiana, 1977. CASE, computer languages, data engineering, distributed computing, executable specification languages, software prototyping.

Susan D. Urban, Associate Professor; Ph.D., Southwestern Louisiana, 1987. Active database systems, heterogeneous database systems, object-oriented database systems.

Michael G. Wagner, Assistant Professor; Ph.D., Technical University of Vienna, 1994. Computer-aided geometric design, geometric modeling and processing, computer animation, theoretical kinematics, robotics and robot dynamics.

Marvin C. Woodfill, Professor; Ph.D., Iowa State, 1964. Digital logic, microcomputers.

RESEARCH CONCENTRATIONS

Artificial Intelligence
Internationally recognized in distributed planning systems, incremental planning, and applications. Funded by the NSF and the U.S. Coast Guard.

Computer-Aided Geometric Design/Graphics
Internationally recognized. The new IEEE *Transactions on Visualization and Computer Graphics* and the *CAGD Journal* edited at ASU. Funding from NSF, DOE, and NASA. Recent work in biomedical application (with University of Arizona Medical Center, Good Samaritan) is funded by the Flinn Foundation. A new project on multiresolution flow visualization is being funded by NASA/Ames.

Database/Multimedia Systems
There is a strong foundation in advanced database research and multimedia information systems. Current research is targeted toward commercial and manufacturing applications. Funded by NSF and industrial sources. Emerging national leadership.

Distributed Computing and Networking
Internationally recognized group working on distributed systems and networking, including distributed algorithms, fault-tolerant applications, software development, protocols, and security. Government and industrial support.

Microprocessors
Extensive research collaboration with industries such as AT&T, En. Gen. Inc., Enhanced Software Inc., Inter-Tel, Motorola, Municipal Services & Software, and Unizone, Inc.

Software Engineering
This internationally recognized group spans software life cycle and covers central as well as distributed and parallel systems. It has strong DoD and industrial support.

AZUSA PACIFIC UNIVERSITY

Department of Computer Science
Graduate Programs

Programs of Study

The Department offers the Master of Software Engineering (M.S.E.) and the Master of Science (M.S.) in applied computer science and technology. The department also offers three undergraduate programs. The Master of Science in applied computer science and technology is a practical and applied degree with seven program emphases. A minimum of 40 semester units is required to complete the Master of Science program with a capstone option. Without the capstone, 46 units are needed. Each emphasis requires core course work totaling 24 units, plus 16 units in the emphasis (including a capstone project) or 22 units in the emphasis without a capstone project. These emphases are client/server, computer information systems, end-user support, inter-emphasis, software engineering, technical programming, and telecommunications.

The department also offers graduate certificate programs in several emphasis areas. Each certificate comprises 18 units of graduate course work that may also be applied to the appropriate master's degree. Graduate certificate programs include client/server, which provides for a career in networked client/server database applications; computer information systems, which provides a foundation for a career in computer information systems; end-user training and support, which provides knowledge and skills needed for a career in end-user support; software engineering, which addresses current software engineering knowledge, research, and practice; technical programming, which focuses on skills and knowledge for employment as a technical programmer; and telecommunications, which teaches practical and theoretical networking and telecommunications.

The Master of Software Engineering (M.S.E.) degree is designed for those engaged in the development and maintenance of large-scale software products. The M.S.E. program addresses the three key components of software development: people, process, and technology. The M.S.E. program requires 30 units but has higher prerequisites than the 46-unit M.S. program. The M.S.E. program is not offered in the 1999-2000 academic year.

Graduate computer science programs at Azusa Pacific University (APU) are offered in a nine-week half-semester format with five terms per year. Because the graduate program is geared toward working professionals, classes usually meet one evening per week from 5:30 p.m. to 9:45 p.m., with some classes scheduled on Saturdays. Six units per term is considered a full load, but students may take more or less than that.

Research Facilities

The computer science department operates six computer science laboratories (including the telecommunications lab, the multimedia lab, the advanced technologies lab, the computer engineering lab, two PC labs, and one Macintosh lab) and also shares access through the University's T1 line as well as access to a variety of software available on the various department servers. Most workstations are Pentium class machines running multiple operating systems, including Windows 95, OS/2, DOS, and UNIX. The University library system includes extensive book collections, numerous electronic search facilities, and other research support resources.

Financial Aid

Students at the University are eligible for financial aid in the form of employment, California Graduate Fellowships, and Federal Stafford, Federal Perkins, and CLAS loans. For more information, students should contact the Office of Student Financial Services.

Cost of Study

Tuition for the M.S. and the M.A. programs is $390 per unit in 1999–2000 and $500 per unit for most M.S.E. courses.

Living and Housing Costs

There is no on-campus housing offered for graduate students attending Azusa Pacific University, but private housing is widely available in the area at various prices.

Student Group

Approximately 2,600 graduate students were enrolled in the various master's and doctoral programs at the University in fall 1998. Approximately 50 percent were women. More than 200 students were enrolled in graduate computer science programs.

Student Outcomes

More than 200 graduate students are enrolled in computer science department graduate programs. The student population is divided between full-time students and working professionals. Working professionals frequently obtain promotions during the course of their studies, generally avoiding company downsizing and layoffs. Most students find excellent employment in their professions soon after graduation. Graduates are working in every aspect of computer science and technology, with positions as network administrators, software engineers, technical programmers, software developers, consultants, client/server specialists, and in many other areas.

Location

Azusa Pacific University is located 30 miles east of downtown Los Angeles. The area offers students a wide variety of cultural, recreational, intellectual, and athletic opportunities. The climate is moderate.

The University

Azusa Pacific University, founded in 1899, is a coeducational, independent, nondenominational Christian university. APU remains committed to the goal of fostering each student's personal, spiritual, physical, and academic growth.

Applying

Applicants for master's programs must hold a bachelor's degree from an accredited university or college with a minimum GPA of 3.0. Applicants must file a university and department application, pay a nonrefundable application fee of $45, and have all materials sent to the Graduate Admissions Office. The University operates under a policy of continuous admission. Students should call or write for an information packet with applications.

Correspondence and Information

Graduate Admissions Office
Azusa Pacific University
901 East Alosta
P.O. Box 7000
Azusa, California 91702-7000
Telephone: 800-TALK-APU (toll-free)

Dr. Samuel E. Sambasivam, Chairman
Department of Computer Science
Azusa Pacific University
901 East Alosta
P.O. Box 7000
Azusa, California 91702-7000
Telephone 626-815-5310
E-mail: cs@apu.edu
World Wide Web: http://www.apu.edu/~cs

Azusa Pacific University

THE FACULTY AND THEIR RESEARCH

Samuel E. Sambasivam, Professor and Chairman; Ph.D. (mathematics), Moscow State (USSR), 1986; M.S. (computer science), Western Michigan, 1990. Dr. Sambasivam has done extensive research, written for publications, and given presentations in both computer science and mathematics. He has taught computer science courses for fourteen years at three universities in the U.S. and has taught mathematics courses for eighteen years at five universities in both the U.S. and India. Dr. Sambasivam teaches database management systems, information structures and algorithm design, microcomputer programming with C++, discrete structures, client/server applications, advanced database applications, applied artificial intelligence, Java programming, and other courses and has developed and introduced several new courses for computer science majors during the past nine years. He was the director of a regional Association for Computing Machinery (ACM) Programming Contest for six years. Professor Sambasivam is a voting member of the ACM and is a member of the Institute of Electrical and Electronics Engineers (IEEE), the International WHO's WHO of Professionals, Kappa Mu Epsilon, and Ypsilon Pi Epsilon. He is named in Marquis's *Who's Who in the World, Who's Who in America,* and *Who's Who in the Midwest.*

Gerald Boerner, Associate Professor; M.A. (experimental psychology), Claremont, 1972. Professor Boerner has more than twenty years of significant experience in various aspects of computing and has taught at APU since 1986. He is widely recognized for his expertise in end-user support, including applications software on many computer platforms. His experience includes educational evaluation, research and testing, data processing, telecommunications and networking, printing, desktop publishing, customized training programs, and extensive university teaching. Professor Boerner coordinates the End-User Support emphasis for the department.

Yufeng F. Chen, Assistant Professor; Ph.D. (computer engineering), South Carolina, 1993. Dr. Chen's teaching and research area interests center on software engineering and artificial intelligence. His current research and publications focus on the topics of large-scale software reuse and agent-based software engineering. Dr. Chen began working at Clark Atlanta University in Georgia in 1993 as an assistant professor and research scientist and also acted as a chairperson of the CIS department's undergraduate curriculum committee. In addition, Dr. Chen has ten years of industrial experience and is a voting member of the ACM and the American Association for Artificial Intelligence (AAAI). Professor Chen currently sponsors a Christian Fellowship for Chinese students at APU.

Richard Eckhart, Associate Professor; Ph.D. (chemical engineering), Penn State, 1964. Dr. Eckhart's professional experience includes more than thirty years of teaching, scientific programming, software engineering, computer simulation and modeling, and business applications programming. He has taught software engineering, programming logic, numerical analysis, computer-aided design, management methods, computer simulation and modeling, mathematics, physics, and engineering at five different universities and colleges. In addition, he has a number of publications to his credit and frequently works as a consultant developing software products for scientific applications and computer simulation. He coordinates the interemphasis track for the department.

Donald Johnson, Associate Professor; Ph.D. (chemistry), Michigan State, 1970; Ph.D. (computer science), Minnesota, 1997. Professor Johnson's specialities include low-level software and hardware, such as operating systems and computer architecture. His recent publications deal with new hardware for reducing the interaction latencies and synchronization times in multiprocessor systems. Of particular interest to him is to make a laboratory of PCs or workstations applicable to fine-grained problems with near-supercomputer performance. He has worked for many years as a senior research scientist in laboratory automation and embedded systems in the medical and scientific instrument field, and he served as president and technical expert in an independent computer consulting firm for twelve years. He taught computer science courses for more than sixteen years at universities in Wisconsin, Minnesota, and Europe before joining APU in September 1998. He coordinates the technical programming emphasis.

Wendel Scarbrough, Associate Professor; M.A. (educational administration), New Mexico Highlands, 1952. Professor Scarbrough has more than forty years of varied teaching and computing experience, including analysis, design, and implementation of computer systems; scientific programming; information systems management; and computer consulting. He has more than twenty years of instructional design, including development, implementation, and teaching of several major programs with more than fifty different computer courses. He has also taught a wide variety of mathematics and computer courses at six other universities, including the Universities of Texas, New Mexico, and Oklahoma. After joining the faculty of APU in 1983, Professor Scarbrough developed the undergraduate computer science program, the Master of Educational Computing, and the Master of Science in applied computer science and technology. He has received numerous awards and fellowships, including several National Science Fellowships, the Walter Anderson Fellowship, and the Chase Sawtell Inspirational Teacher Award (1991). Professor Scarbrough is a member of the ACM and the IEEE.

Carol Stoker, Associate Professor; Ph.D. (education), USC, 1992. Among Professor Stoker's many contributions to APU, she developed numerous courses, was the leader in the redesign of the undergraduate computer science curriculum in 1990, and helped develop the computer information systems major. During her more than thirteen years of teaching experience, she has taught a variety of computer science courses and other courses. Professor Stoker has coauthored a textbook on Java programming and is a member of the ACM.

Peter Yoon, Assistant Professor; Ph.D. (computer science), Penn State, 1996. Professor Yoon has done extensive research, publications, and presentations in the mathematics and computer fields. His research interests include parallel algorithms, high-performance computing, numerical linear algebra, and signal processing. He has published a number of papers in various international journals on these subjects. Professor Yoon taught mathematics at Purdue University for three years; spent three years as a systems programmer for IBM Corporation in Research Triangle Park, North Carolina; and was a research assistant at the Applied Research Laboratory at University Park, Pennsylvania. He is a member of the Society for Industrial and Applied Mathematics (SIAM) and the ACM. He coordinates the technical programming emphasis at APU.

CALIFORNIA INSTITUTE OF TECHNOLOGY

Program in Computer Science

Program of Study

Graduate study leading to M.S. and Ph.D. degrees in computer science is oriented principally toward doctoral research. Students join a research group and participate in research from the very beginning. The first year emphasizes course work and master's thesis research. Students in their second year complete their master's thesis preliminary to the candidacy examination. After passing the candidacy exam, students devote their time to research on the doctoral dissertation. The program requires a minimum of three academic years of residence.

The small-college atmosphere of Caltech facilitates close working relationships between faculty members and students and encourages novel interdisciplinary research. Research in the department stresses the integration between theory and implementation; hardware and software; and computing systems and their application to all fields. Students are encouraged to tailor their programs of courses and research to fit their specific needs.

Students have unusual opportunities for research in the following areas: VLSI systems, including asynchronous and analog VLSI; computer graphics and human-computer interaction; neural networks and learning systems; concurrent (parallel) systems; mathematics of program construction; fault-tolerant systems; distributed object systems; and computer vision. Research frequently involves work in several of these areas, as well as connections with fields such as electrical engineering, environmental engineering, control and dynamical systems, mathematics, biology, physics, and chemistry.

Research Facilities

The Computer Science Department and its research groups have extensive computing facilities. The department has sizable laboratories of Intel, SGI, and HP workstations, including multiprocessor systems. In addition, each graduate student has a workstation. The computer graphics laboratory is a site of the NSF Science and Technology Center for Computer Graphics and Scientific Visualization. The lab has specialized equipment for fast rendering of models and for generating computer animation sequences. The graphics group is also developing two responsive workbenches for three-dimensional visualization. The VLSI labs are equipped with complete facilities for constructing and testing VLSI systems. Caltech has state-of-the-art parallel supercomputers, and students have access to equipment provided by the NSF Science and Technology Center in Parallel Computing. The Sherman Fairchild Library houses an extensive collection and superb computing facilities.

Financial Aid

Almost all graduate students are supported by fellowships, scholarships, or research assistantships throughout their program at Caltech. This support includes tuition. Some students also elect to serve as teaching assistants. Students are encouraged to apply for outside fellowships such as NSF and DoD fellowships, which are supplemented by the department if necessary. Many students get fellowships from corporations, including IBM, Intel, and Microsoft. In 1999–2000, full-time research assistantships range from $17,040 to $19,476 and also covered tuition.

Cost of Study

Tuition, health plan coverage, and other graduate student fees for the academic year 1999–2000 are $19,260. A typical figure for books and supplies is $1026 a year.

Living and Housing Costs

Single-students find on-campus housing in dormitory- or apartment-style accommodations, which range in price from approximately $347 to $466 per month. Married students find on-campus housing in apartment-style accommodations, which range in price from approximately $725 to $840 per month. More detailed information is available on the Web at http://www.caltech.edu/~cabs/housing/index.html.

Student Group

Currently enrolled at Caltech are 901 undergraduate students, including 256 women, and 957 graduate students, including 226 women. There are 25 students currently enrolled in the graduate program in computer science.

Location

Pasadena, a city of approximately 125,000, is located about 10 miles northeast of Los Angeles. The Institute, although located in the center of a residential district, is within a few blocks of shopping facilities. Pasadena and the Metropolitan Los Angeles area provide abundant cultural and recreational opportunities.

The Institute

Caltech is an independent, privately supported institution. With its off-campus facilities (such as the Jet Propulsion Laboratory), it constitutes one of the world's major research centers. Because of the Institute's relatively small size (810 faculty members and 1,885 students), close interaction exists between students and research staff members.

Caltech scientists have achieved wide recognition for their distinguished achievements. Twenty-seven Nobel Prizes have been awarded to Caltech alumni and faculty, and 44 alumni and faculty members have received the National Medal of Science for outstanding contributions to the development of science and engineering.

Applying

Applications for September admission should be received by January 1. GRE General Test scores are required. Admission to graduate study in computer science is highly competitive; only about 6 new students are admitted each year. The department seeks applicants with exceptional promise for scientific research.

Application materials may be obtained from the Office of the Dean of Graduate Studies. Inquiries for further information should be sent to the computer science department. Caltech is an Affirmative Action institution and encourages women and members of minority groups to apply.

Correspondence and Information

Dean of Graduate Studies, 02-31
California Institute of Technology
Pasadena, California 91125
Telephone: 626-395-6346
Fax: 626-577-9246
World Wide Web: http://www.cs.caltech.edu/

California Institute of Technology

THE FACULTY AND THEIR RESEARCH

Yaser S. Abu-Mostafa, Professor of Electrical Engineering and Computer Science; Ph.D., Caltech, 1983. Learning theory, learning systems, computational finance, neural networks, pattern recognition.

James R. Arvo, Associate Professor; Ph.D., Yale, 1995. Mathematical foundations of computer graphics, human-computer interaction, intelligent interfaces.

Alan H. Barr, Professor; Ph.D., Rensselaer, 1983. Three-dimensional mathematical and computational modeling techniques, image synthesis and computer graphics, biomechanics and simulation of natural phenomena.

Jehoshua Bruck, Professor of Computation and Neural Systems and Electrical Engineering; Ph.D., Stanford, 1989. Parallel and distributed computing, computation theory and neural systems, fault-tolerant computing and error-correcting codes.

K. Mani Chandy, Simon Ramo Professor; Ph.D., MIT, 1969. Concurrent computation: formal methods, performance modeling, software engineering.

Alain J. Martin, Professor; Engineer in Applied Mathematics, Grenoble Polytechnic (France), 1969. Asynchronous VLSI design, concurrent and distributed computations, highly parallel computers, computer architecture, verification.

Carver A. Mead, Gordon and Betty Moore Professor; Ph.D., Caltech, 1960. VLSI design, ultraconcurrent systems, physics of computation, real-time vision and hearing.

Peter Schröder, Associate Professor; Ph.D., Princeton, 1994. Illumination computations, wavelets, massively parallel graphics, scientific visualization, scientific computation.

Frederick Burtis Thompson, Professor of Applied Philosophy and Computer Science, Emeritus; Ph.D., Berkeley, 1951. Philosophy of information, computational linguistics, user interface, dynamics of information.

The following faculty members and professionals at Caltech actively contribute to the computer science research and curriculum:

Alan H. Bond, Lecturer in Computer Science and Senior Research Scientist in Electrical Engineering; Ph.D., London, 1966. Brain modeling, multiagent systems.

John C. Doyle, Professor of Electrical Engineering; Ph.D., Berkeley, 1984. Integrating modeling, ID, analysis and design of uncertain nonlinear systems.

Joel N. Franklin, Professor of Applied Mathematics; Ph.D., Stanford, 1953. Mathematical programming, numerical analysis, computer algorithms.

Glen George, Lecturer in Electrical Engineering and Computer Science; M.S., UCLA, 1984. Embedded systems, evolvable hardware, bioinformatics.

Alexander S. Kechris, Professor of Mathematics; Ph.D., UCLA, 1972. Mathematical logic, computability theory.

Herbert B. Keller, Professor of Applied Mathematics; Ph.D., NYU (Courant), 1954. Numerical analysis, bifurcation theory, large-scale scientific computing.

Robert J. McEliece, Professor of Electrical Engineering; Ph.D., Caltech, 1967. Information theory, error-correcting codes, applied mathematics.

Daniel I. Meiron, Professor of Applied Mathematics; Sc.D., MIT, 1981. Computational physics, large-scale scientific computing.

Paul C. Messina, Assistant Vice President for Scientific Computing, Faculty Associate in Scientific Computing, and Director of the Center for Advanced Computing Research; Ph.D., Cincinnati, 1972. Parallel computing architectures, data-intensive computing, large-scale scientific computing.

Pietro Perona, Professor of Electrical Engineering; Ph.D., Berkeley, 1990. Computational foundations of vision; psychophysics and modeling of the human visual system; applications of computer vision to vehicle navigation, human-machine interfaces, multimedia, and automated image analysis.

Philip G. Saffman, Theodore von Kármán Professor of Applied Mathematics and Aeronautics, Emeritus; Ph.D., Cambridge, 1956. Computational fluid mechanics.

Carnegie Mellon

CARNEGIE MELLON UNIVERSITY

School of Computer Science
Doctoral Program in Computer Science

Program of Study

The Department of Computer Science offers the degree of Doctor of Philosophy. Requirements for the Ph.D. are successful completion of eight core course requirements and submission of a thesis describing original, independent research. An initial acculturation program (the "Immigration Course") involves students in all the activities of the Department. Participation in one or more of the ongoing research projects is a key factor in a student's education. Visitors come to Carnegie Mellon in a steady stream from universities and laboratories throughout the world for various lengths of time, joining the faculty members and students in an active colloquium program.

Research Facilities

The School of Computer Science (SCS) research facility provides numerous and diverse computers for faculty and graduate student use—currently more than 3,000 machines. Nearly every person in the School has the individual use of a workstation. These machines include UNIX platforms, primarily from Sun, DEC, SGI, HP, and IBM; a growing population of PCs that run both UNIX and NT; and Macintosh systems; all have transparent access to the Andrew File System, a 625-Gbyte, shared filespace, and to one another through the Network File System protocol. SCS maintains several terabytes of secondary storage.

Beyond these resources, the University provides various independent facilities for general use. Computationally intensive applications also use PSC computers, including Cray T3E, C90-16/512, and J90 supercomputers.

Financial Aid

Most students in the department are supported by graduate research fellowships during the academic year. In 1999–2000, each student receives full tuition plus a stipend of $1525 per month for the academic year. Dependency allowances are available; students who receive external fellowships may be given supplementary stipends.

Summer support is normally available for many students, particularly those working on the dissertation. However, since the University believes that it is also good for students to gain experience in industry for one or two summers during their careers at Carnegie Mellon, faculty and staff members are able to provide valuable help in finding suitable summer employment.

Cost of Study

Tuition and fees for full-time graduate students in 1999–2000 are $22,230 for the academic year. This figure is subject to change.

Living and Housing Costs

The University does not provide housing for graduate students, but accommodations are available in the community at a variety of costs.

Student Group

Carnegie Mellon has a total enrollment of approximately 7,500 students. About 150 full-time students are enrolled in the Doctoral Program in Computer Science, which makes the student-faculty member ratio about 2:1. Admission to the program is highly competitive; about 25 students are admitted each year.

Student Outcomes

Carnegie Mellon's Computer Science doctoral program aims to produce well-educated researchers and future leaders in computer science. Approximately 25 students graduate each year, with slightly more than half accepting positions in industrial research laboratories such as DEC, IBM, Intel, Lucent Labs, and Xerox; those preferring academic careers accept both tenure- and research-track positions at many of the top universities in the country, including the University of California at Berkeley, Cornell University, the University of Pennsylvania, and Stanford University.

Location

Carnegie Mellon is located in Oakland, the cultural center of Pittsburgh, on a 90-acre campus adjacent to Schenley Park, the largest city park. The campus is close to the many cultural and sports activities of the city and is only 4 miles from the downtown business district. Pittsburgh is the headquarters for many of the nation's biggest corporations. There is a large concentration of research laboratories in the area.

The University

Founded in 1900 by Andrew Carnegie, the Carnegie Institute of Technology joined with Mellon Institute (now the Carnegie Mellon Research Institute) in 1967 to become Carnegie Mellon University. With this merger, one of the leading research and education institutions in the country was established.

Applying

The application, Graduate Record Examinations scores, and letters of reference must be received by January 1. Notification of acceptance is made by March 15. Minimum preparation normally includes an undergraduate program in mathematics, physics, electrical engineering, or computer science and some experience in computer programming. Excellence and promise may balance a lack of formal preparation. No applications are considered unless accompanied by the GRE scores (the General Test and Subject Test in mathematics, computer science, physics, or engineering).

Correspondence and Information

Graduate Admissions
Department of Computer Science
Carnegie Mellon University
5000 Forbes Avenue
Pittsburgh, Pennsylvania 15213
Telephone: 412-268-3863
E-mail: grad-adm@cs.cmu.edu
World Wide Web: http://www.cs.cmu.edu/csd/

Carnegie Mellon University

THE FACULTY AND THEIR RESEARCH

J. Anderson, Professor. Cognitive psychology, artificial intelligence, human computer interaction.

G. Blelloch, Associate Professor. Compilers, parallel architectures, parallel languages, parallel algorithms.

A. Blum, Associate Professor. Machine-learning theory, online algorithms, approximation algorithms.

P. Braam, Senior Systems Scientist. Distributed file systems and storage architectures, particularly on Linux.

S. Brookes, Associate Professor. Mathematical theory of computation, theory and semantics of programming languages.

R. E. Bryant, President's Professor. Formal verification of digital systems, data structures and algorithms for representing and reasoning about discrete functions.

J. Cagan, Professor. Computational design.

J. Carbonell, Professor. Artificial intelligence, natural-language processing, machine learning, machine translation.

M. Christel, Senior Systems Scientist. Digital video interfaces, information visualization, digital libraries.

E. Clarke, FORE Professor. Hardware and software verification, automatic theorem proving, symbolic computation, parallel algorithms, programming.

S. Cochran, Senior Systems Scientist. Image understanding, aerial photo interpretation, stereo vision.

A. Corbett, Senior Research Scientist. Application of AI in education, use of cognitive models in computer-based learning environments.

R. Dannenberg, Senior Research Scientist. Computer music, interactive real-time systems.

M. Erdmann, Associate Professor. Robotics: planning under uncertainty, mechanics of manipulation of cooperating robots; computational biology: protein folding, automatic assignment, homology.

S. Fahlman, Principal Research Scientist. Artificial intelligence, machine learning, artificial neural networks, software development environments.

C. Faloutsos, Associate Professor. Multimedia databases, indexing, data mining, database performance.

A. Fisher, Principal Systems Scientist and Associate Dean. Networks, parallel computing.

M. Furst, Professor. Algorithm design, computational complexity, learning theory.

D. Garlan, Associate Professor. Software engineering, software architecture, formal methods, programming environments.

G. Gibson, Associate Professor. Computer systems, computer architecture, operating systems, file systems, storage systems, networking.

S. Goldstein, Assistant Professor. Compilers and architectures for parallel and reconfigurable systems.

T. Gross, Associate Professor. Computer systems, parallel computing, compilers, debugging, software construction.

R. W. Harper, Associate Professor. Type theory, formal logics, semantics, compilers, functional programming, module systems.

A. Hauptmann, Senior Systems Scientist. Multimedia digital libraries, information retrieval from speech and video.

P. Heckbert, Associate Professor. Computer graphics, image processing, scientific computing.

S. Hudson, Associate Professor. User interface software and technology.

B. John, Associate Professor. Human-computer interaction, engineering models of human performance.

D. Johnson, Associate Professor. Wireless and mobile networking, network protocols, operating systems, distributed systems, fault tolerance.

T. Kanade, Whitaker University Professor. Computer vision, virtualized reality, autonomous mobile robots, medical robotics, sensors.

K. Koedinger, Senior Research Scientist. Cognitive modeling, problem solving and learning, intelligent tutoring systems, educational technology, enhancing system learnability.

R. Kraut, Professor. Computer-mediated communication, social impact of technology.

J. Lafferty, Associate Professor. Probability and information theory, natural-language processing, stochastic modeling.

P. Lee, Associate Professor. Compilers for advanced programming languages, functional programming, formal semantics.

T. Lee, Assistant Professor. Medical engineering, computational neuroscience, computer vision.

J. Fain Lehman, Senior Research Scientist. Natural-language understanding, generation, and acquisition; machine learning; cognitive modeling.

B. Maggs, Associate Professor. Networks for parallel and distributed computing systems.

M. Mason, Professor. Robotics, mobile manipulators, mechanics of manipulation, manufacturing automation.

R. Maxion, Senior Systems Scientist. Human and machine diagnosis of complex systems, discovery of structure, fault tolerance.

C. McGlone, Senior Systems Scientist. Photogrammetry and image understanding.

D. McKeown, Principal Research Scientist. Image understanding, remote sensing, cartography, spatial databases.

G. Miller, Professor. Parallel computation; sparse matrix, graph and number-theoretic algorithms.

T. Mitchell, Professor. Machine learning, intelligent Web agents, data mining, artificial intelligence.

A. Moore, Associate Professor. Reinforcement learning, machine learning, AI in manufacturing, scheduling, Markov decision processes.

J. Morris, Simon Professor and Department Head. Distributed personal computer systems, functional programming, user interfaces.

T. Mowry, Associate Professor. Computer systems and architecture, compiler algorithms for prefetching in uniprocessors and shared-memory multiprocessors.

D. Nagle, Assistant Professor. Computer architecture, embedded systems, I/O.

C. Neuwirth, Associate Professor. Computer support for cooperative work, human-computer interaction.

D. O'Hallaron, Associate Professor. Architectures, compilers, and applications for parallel computer systems.

R. Pausch, Associate Professor and Codirector, Entertainment Technology Center. Human-computer interaction, virtual reality, interactive 3-D computer graphics, theme park and other entertainment technologies, design of consumer market devices.

F. Pfenning, Senior Research Scientist. Logic and computation, type theory, logic programming, functional programming, automated deduction.

D. Plaut, Associate Professor. Neural network modeling of language acquisition and language impairments following brain damage.

R. Rajkumar, Senior Systems Scientist. Real-time and multimedia operating systems, scheduling theory, networking and distributed systems.

M. Ravishankar, Systems Scientist. Speech decoding algorithms, acoustic modeling, speech applications, interface design.

R. Reddy, University Professor and Dean. AI, speech recognition and understanding, integrated manufacturing systems.

J. Reynolds, Professor. Design of programming languages, mathematical semantics, program specification.

J. Roberts, Principal Lecturer and Freshman Advisor.

R. Rosenfeld, Associate Professor. Human language technologies, statistical language modeling, speech recognition, stochastic modeling.

S. Roth, Senior Research Scientist. Information visualization and exploration, user interface design.

S. Rudich, Associate Professor. Complexity theory, cryptography, combinatorics, probability.

A. Rudnicky, Senior Systems Scientist. Speech recognition, spoken language interaction, interface design.

M. Satyanarayanan, Professor. Mobile computing, distributed file systems, measurement and evaluation, security.

W. Scherlis, Senior Research Scientist. Program manipulation and restructuring, collaboration information management, software engineering.

D. Scott, Hillman University Professor. Semantics of computer languages, computer algebra.

S. Seitz, Assistant Professor. Computer vision, computer graphics, image-based rendering, motion analysis and synthesis.

M. Shaw, Perlis Professor and Associate Dean. Software architecture, software engineering, programming systems and methodologies.

D. Siewiorek, Buhl Professor. Computer architecture, fault-tolerance, design automation, parallel processing, mobile computing, rapid prototyping.

R. Simmons, Senior Research Scientist. Artificial intelligence, autonomous mobile robots, autonomous spacecraft, task-level control architectures, planning, model-based reasoning, reasoning under uncertainty.

H. Simon, Mellon Professor. Computer simulation of cognitive processes, artificial intelligence, management science.

D. Sleator, Professor. Data structures, graph algorithms, online algorithms, parsing natural languages.

P. Steenkiste, Senior Research Scientist. High-performance networking, quality of service in networking, distributed systems, operating systems.

S. Thrun, Assistant Professor. Artificial intelligence, machine learning, and probabilistic reasoning, application to robotics and intelligent control.

J. Tomayko, Principal Lecturer. Reliability of embedded real-time systems, software development management, history of computing.

D. Touretzky, Senior Research Scientist. Computational neuroscience, spatial representations in the brain, animal and robot learning.

R. Valdes-Perez, Senior Research Scientist. AI-computational science, interactive discovery tools for biology, chemistry, physics, and other sciences.

M. Veloso, Associate Professor. Artificial intelligence; planning, execution, and learning in autonomous agents; multiagent systems; robotic soccer.

A. Waibel, Principal Research Scientist. Perceptual user interfaces, speech, language, multimodal human-computer interaction, face/body tracking and interpretation.

W. Ward, Senior Research Scientist. Artificial intelligence, speech understanding, stochastic modeling.

J. Wing, Professor. Formal methods, concurrent and distributed systems, security protocols, programming language design and implementation.

Y. Yang, Senior Research Scientist. Natural language analysis systems, machine translation, text retrieval and categorization.

H. Zhang, Assistant Professor. Computer networks, Internet, quality of service, distributed systems.

Carnegie Mellon

CARNEGIE MELLON UNIVERSITY

Graduate School of Industrial Administration
and School of Computer Science
Program in Electronic Commerce

Program of Study

Carnegie Mellon's new one-year Master of Science in Electronic Commerce (M.S.E.C.) program is designed to educate a new generation of system developers, planners, analysts, programmers, managers, and executives in both the technology and business aspects of e-commerce. The curriculum starts with interdisciplinary courses from Carnegie Mellon's top-ranked schools of business and computer science. Establishing an electronic business demands a variety of skills that span Internet marketing, software development, Web page design, networks and security, and the ability to manage these elements to spawn a successful enterprise. Students acquire experience in all of these topics as well as in electronic payments systems, Internet technology, business-to-business transactions, the law of the electronic marketplace, and the accounting components needed to support auditable electronic transactions.

Research Facilities

Computers are used heavily in course work. There are 106 PC terminals, linked by a campuswide computer network, available for use by students in the program. They are located in classrooms, the library, the computer center, and computer labs. Computer resources and online services include training, access to the Internet, e-mail, a videoconferencing center, a multimedia center, and the Dow Jones News/Retrieval Service.

The main library contains 852,241 bound volumes, 756,985 titles on microform, 3,889 periodical subscriptions, 21,340 records/tapes/CDs, and sixteen CD-ROMs for students to use.

Financial Aid

There are no institutional grants or scholarships designated for this program. However, for those individuals who are eligible, federal and private student loans may be processed through the Graduate School of Industrial Administration (GSIA) financial aid office.

Cost of Study

Tuition for the full-year program beginning in 1999 is $38,000.

Living and Housing Costs

Graduate students are responsible for their own housing needs. Housing is available in the area within a range of costs.

Student Group

In the first year of the M.S.E.C. program (1999–2000), 37 students enrolled.

Location

Pittsburgh has been called the "City of Bridges." Located at the confluence of the Allegheny and Monongahela Rivers, where the Ohio River begins, it is a city of downtown skyscrapers, scenic rivers, and steep hillsides. Picturesque neighborhoods scattered throughout the city create a friendly, small-town atmosphere with parks, tree-lined streets, convenient shopping areas, and active community centers. The Oakland section of the city, where Carnegie Mellon is located, houses the Carnegie Library, Museum of Art, and Museum of Natural History as well as several major concert halls.

The University

Carnegie Mellon University was founded by steel magnate and philanthropist Andrew Carnegie in 1900 and has become one of America's leading universities. Some 3,400 of its 49,000 alumni are chief executive officers, company chairmen, presidents, or vice presidents of their respective companies.

Applying

GSIA has a self-managed application procedure; students are fully responsible for completing and submitting their applications on time and in accordance with set guidelines. Application must be made in writing or electronically. One official transcript is required from each college or university attended at both the undergraduate and graduate levels, even if a degree was not earned. Two evaluation forms are to be completed by individuals who are able to provide specific information regarding the applicant's intellectual ability, performance, initiative, motivation, and mathematical and programming skills. Scores from the Graduate Record Examinations (GRE) or the Graduate Management Admission Test (GMAT) are required of all applicants. The Test of English as a Foreign Language (TOEFL) is required of applicants whose native language is not English. There is a nonrefundable application fee of $60. Evaluative interviews are strongly recommended of all applicants. Students are encouraged to refer to the program's catalog or Web page, listed below, for applicable deadlines.

Correspondence and Information

Vickie Motz, Program Manager
Master of Science in Electronic Commerce
Graduate School of Industrial Administration
Carnegie Mellon University
Pittsburgh, Pennsylvania 15213-3890
Telephone: 412-268-1322
       877-544-ECOM (toll-free)
Fax: 412-268-6837
E-mail: vmotz@andrew.cmu.edu
World Wide Web: http://www.ecom.cmu.edu

Carnegie Mellon University

THE FACULTY AND THEIR RESEARCH

Jaime G. Carbonell, Allen Newell Professor of Computer Science and Director, Languages Technologies Institute; Ph.D., Yale, 1978. His research interests span several areas of artificial intelligence, from machine translation to multimedia digital libraries.

Scott E. Fahlman, Principal Research Scientist; Ph.D., MIT, 1977. His principal area of research is artificial intelligence, especially artificial neural nets and their practical applications.

Tridas Mukhopadhyay, Professor and Director, Master of Science in Electronic Commerce Program; P.G.D.M., Indian Institute of Management, 1981; Ph.D., Michigan, 1987. His latest research project is HomeNet, which studies patterns of computer use in the home.

Raj Reddy, Herbert A. Simon Professor of Computer Science and Robotics and Dean, School of Computer Science; Ph.D., Stanford, 1966. Professor Reddy was named a member of the President's Advisory Committee on Information Technology in 1997. His current research includes speech recognition and understanding systems, gigabit networks, universal digital libraries, and electronic commerce.

William L. Scherlis, Principal Research Scientist and Director, Information Technology Center; Ph.D., Stanford, 1980. He is currently a principal investigator for the HomeNet Project, which is examining how families use the Internet.

Michael Ian Shamos, Principal Systems Scientist and Codirector of the Master of Science in Electronic Commerce Program; Ph.D., Yale, 1978; J.D., Duquesne, 1981. He is the founder of two Pittsburgh computer software companies. His research interests lie primarily in the analysis of algorithms, computational mathematics, the law of electronic commerce, and digital libraries.

Mary Shaw, Alan J. Perlis Professor of Computer Science; Ph.D., Carnegie-Mellon, 1972. Her current research interests are in the areas of programming systems—particularly software architecture and reliable software development.

Stephen Spear, Professor of Economics and Chair, Ph.D. Program in Economics; Ph.D., Pennsylvania, 1982. His research interests include microeconomic theory, dynamic general equilibrium theory, mathematical economics, and strategy.

Kannan Srinivasan, H. J. Heinz II Professor of Management, Marketing and Information Systems; Ph.D., UCLA, 1986. The majority of his research centers on how companies hope to operate in real time as they instantly adjust product mix, inventory levels, cash reserves, marketing programs, or other factors that change business conditions.

Robert H. Thibadeau, Senior Research Scientist and Director, Imaging Systems Laboratory; Ph.D., Virginia, 1976. The architect of a number of systems, he developed the dollar bill inspection device at the U.S. Bureau of Engraving and the General Motors CAD/CAM headlamp/taillamp design system.

Michael A. Tick, Associate Professor and President, Carnegie Bosch Institute; Ph.D., Georgia Tech, 1987. His current research includes applications of operations, computational combinatorial optimization, and applications in scheduling and distribution. His biggest research project in recent years was a three-year contract to schedule all the men's and women's basketball games for the nine teams of the Atlantic Coast Conference.

CARNEGIE MELLON UNIVERSITY

Human-Computer Interaction Institute
Master of Human-Computer Interaction

Program of Study

The computer industry is spending an increasing proportion of its development funds in human-computer interaction (HCI), but there is a shortage of personnel with the breadth of training or experience to effectively work on the multidisciplinary teams that produce software with a significant HCI component. With instructors from the School of Computer Science, the Graduate School of Industrial Administration, the School of Humanities and Social Sciences, the College of Fine Arts, the Robotics Institute, and the Software Engineering Institute, the Human Computer Interaction Institute at Carnegie Mellon is one of the few institutes in the country with the breadth of expertise to offer such a program.

The program prepares students to participate in the design and implementation of software systems that can be used easily, effectively, and enjoyably. Students are expected to have a strong undergraduate degree or comparable work experience in either computer science, a behavioral science (psychology, sociology, anthropology, or organizational behavior), or visual or information design. All students are expected to have had at least one prerequisite course in statistics, design, and elementary programming.

During the first two semesters of the twelve-month program, students complete courses to give them a broad background in computer science, human behavior, evaluation and assessment, and design. The curriculum consists of four core courses and six elective courses. In the spring and summer, students participate in the extensive team-oriented Human-Computer Interaction Project, in which they apply their classroom knowledge and develop skills by working on a multidisciplinary team.

Research Facilities

The joint computing facility of the HCI program and the School of Computer Science has a large number and a wide variety of computers available to faculty members and graduate students, ranging from Macintosh and Intel-based personal computers to large multiprocessors from DEC (8800) and Encore (Multimax-16). All personal computers provide transparent access to the Andrew File System.

The User Studies Labs are University-wide facilities for research on human-computer interaction. The three laboratory facilities have space for studying individuals and groups. Equipment includes facilities for video recording and analysis of video data.

Financial Aid

The Human-Computer Interaction Program does not provide students with financial support. Financial aid is available through the University's financial aid office (telephone: 412-268-1946). The University also accepts third-party payments from employers and government agencies.

A limited number of research and teaching assistant positions may be available to qualified students. Accepting such a position extends the length of the program.

Cost of Study

Tuition for full-time graduate students accepted in 1999–2000 is $11,050 per semester (The program length is three semesters). This figure is subject to change.

Living and Housing Costs

Graduate students are responsible for their own housing needs. There is no on-campus graduate housing; however, housing is available in the area within a range of costs. The Office of International Education suggests living costs of $9100 for single students and $12,100 for married students.

Student Group

Carnegie Mellon has an enrollment of approximately 7,500 students, about 2,700 of whom are graduate students. Within the HCI program, there are 37 students, including 16 women and 8 international students.

Student Outcomes

Graduates of the Master of Human-Computer Interaction Program are in high demand. Recent graduates have accepted positions at such companies as Microsoft, IBM, Bellcore, and Trilogy.

Location

Carnegie Mellon University is located in Oakland, the educational center of Pittsburgh. The 90-acre campus is adjacent to Schenley Park, the city's largest park. The campus is close to many cultural and sports activities in the city and is only 4 miles from the downtown business district. There is a large concentration of research laboratories in the area.

The University

Founded in 1900 by Andrew Carnegie, the Carnegie Institute of Technology joined with Mellon Institute (now the Carnegie Mellon Research Institute) in 1967 to become Carnegie Mellon University. With this merger, one of the leading research and education institutions in the country was established.

Applying

The application, Graduate Record Examinations scores, a statement of purpose, an up-to-date resume, letters of reference, and the application fee must be received by February 1. A strong undergraduate degree or comparable work in computer science, a behavioral science (psychology, sociology, anthropology, or organizational behavior), or visual or information design is required, as is at least one course in statistics, design, and elementary programming.

Correspondence and Information

Admissions Coordinator
Human-Computer Interaction Program
School of Computer Science
Carnegie Mellon University
Pittsburgh, Pennsylvania 15213
Telephone: 412-268-6493
E-mail: hcii-masters@cs.cmu.edu
World Wide Web: http://www.cs.cmu.edu/~hcii

Carnegie Mellon University

THE FACULTY

Members of the faculty are drawn from the membership of the Human-Computer Interaction Institute (HCII) of Carnegie Mellon.

Human-Computer Interaction Institute Faculty

John Anderson, Professor of Psychology and Computer Science. Cognitive psychology, artificial intelligence, human-computer interaction.

Daniel Boyarski, Professor of Design. How words, pictures, sound, and motion may be combined for effective communication.

Albert T. Corbett, Senior Research Scientist, HCII. Development and evaluation of computer-based learning environments.

Nancy Green, Systems Scientist, HCII. Human-computer interaction, natural language processing, multimedia.

Scott Hudson, Associate Professor of HCI and Computer Science. User interface software and technology, computer-supported cooperative work, tangible interfaces, user interface development tools.

Bonnie E. John, Associate Professor of Computer Science and Psychology. Usability evaluation methods, engineering models of human performance.

Sara Kiesler, Professor of Social and Decision Sciences. Social and behavioral aspects of computers and computer-based communication technologies.

Roberta L. Klatzky, Professor and Head of Department, Psychology. Perceptual/motor interactions and their implications for interface design.

Kenneth R. Koedinger, Senior Research Scientist, HCII. Cognitive modeling of problem solving, intelligent tutoring systems.

Robert Kraut, Professor of Social Psychology and Human-Computer Interaction, Computer Science, Social and Decision Sciences, and Graduate School of Industrial Administration. Use of emerging computer and telecommunication technologies for communication by groups and individuals.

F. Javier Lerch, Associate Professor, Graduate School of Industrial Administration. Working memory, problem representation.

James H. Morris, Professor and Dean, School of Computer Science. Computer-mediated communication.

Brad A. Myers, Senior Research Computer Scientist, Computer Science Department. User interface design, user interface software, demonstrational interfaces, programming environments.

Christine M. Neuwirth, Associate Professor of English, HCI, and Computer Science. Computer-supported cooperative work, collaborative writing.

Randy Pausch, Associate Professor of Human-Computer Interaction, Computer Science and Design and Codirector of the Entertainment Technology Center. Virtual reality, interactive 3-D graphics, theme park and entertainment technologies, design of consumer market devices.

Steven Roth, Senior Research Scientist, Robotics Institute. Presentation of information, visualization, multimedia presentation.

Richard Scheines, Associate Professor of Philosophy. Causal reasoning with statistical data, interactive intelligent Web-based courseware.

Jane Siegel, Senior Systems Scientist, HCII. Usability evaluation methods, behavioral aspects of computer-supported cooperative work systems.

Dan Siewiorek, Professor of Electrical and Computer Engineering and Computer Science and Director of HCII. Computer architecture, parallel processing, mobile computers, fault-tolerant computing, design automation, rapid prototyping.

Alex Waibel, Senior Research Scientist, Computer Science, Center for Machine Translation, Robotics, and Computational Linguistics. Speech understanding and translation, multimodal interfaces, neural networks, machine learning.

Jie Yang, Research Computer Scientist. Multimodal human-computer interaction, computer vision, pattern recognition.

Human-Computer Interaction Institute Members

Joseph M. Ballay, Professor of Design. Design for human-computer interaction, nonverbal communication.

Len Bass, Senior Member, Technical Staff, Software Engineering Institute. Evaluating user interface tools, user interfaces for hands-free operation.

Peter Brusilovsky, Adjunct Research Scientist. Intelligent user interfaces, user and student modeling, hypermedia, adaptive Web-based systems, intelligent tutoring systems.

Richard Buchanan, Professor and Head of Department, Design. Rhetorical theory and its application in graphic and communications design, industrial design, strategic planning, design management.

Kathleen M. Carley, Professor of Sociology and Organizations. Computational organization and social science; social and knowledge networks; computer simulation of groups, organizations, and technology; multiagent models; adaptive and emergent systems.

Michael Christel, Senior Systems Scientist. Multimedia interfaces, digital video computer systems.

Stephen E. Cross, Director, Software Engineering Institute. Broad adoption of innovative software development methods, transition of advanced technology, application of intelligent systems to strategic decision making, models of technology transition.

Roger B. Dannenberg, Senior Research Computer Scientist, School of Computer Science. Computer languages, computer music.

Mark Derthick, Project Scientist, Robotics Institute. Visual database query languages, interactive data visualization.

Maxine Eskenazi, Visiting Research Scholar, Robotics Institute. Phonetics, automatic speech processing, assessment of speech recognition systems.

Scott E. Fahlman, Principal Research Scientist, School of Computer Science. Artificial intelligence, artificial neural networks, software development environments, biomedical image processing.

Ulrich Flemming, Professor of Architecture. Computers as a novel tool and medium for building design.

Nick V. Flor, Assistant Professor, Graduate School of Industrial Administration. Distributed cognition frameworks, distributed collaboration tools.

Jolene Galegher, Associate Professor of English. Communication in small, collaborative workgroups, collaborative writing.

David Garlan, Assistant Professor of Computer Science. Software engineering, software architecture, formal methods, programming environments.

William E. Hefley, Lecturer of Social and Decision Sciences. Intelligent human-computer interaction, training systems.

Suguru Ishizaki, Assistant Professor of Communication Design. Kinetic typography, facilitation of dynamic design solutions.

Takeo Kanade, Professor of Computer Science and Robotics and Director of the Robotics Institute. Computer vision and sensors, development of self-mobile space manipulators.

David Kaufer, Professor and Head of Department, English. Computer interfaces for supporting individual writing and collaborative writing.

Stephan Kerpedjiev, Project Scientist, Robotics Institute. Information visualization, task-centered visualization, multimedia generation, text and graphics coordination.

Jill Fain Lehman, Research Computer Scientist, Computer Science Department. Natural language understanding, language acquisition, machine learning, cognitive modeling.

Don Marinelli, Professor of Drama and Codirector, Entertainment Technology Center. Interactive drama, dramatic structure, location-based entertainment, immersive environments, role-playing/role-playing games, performance art, synthetic interviews.

Richard L. Martin, Senior Systems Scientist, Robotics Institute. Systems engineering and development of software-intensive systems, evolutionary development models.

Roy Maxion, Systems Scientist, Computer Science Department. Human and machine diagnosis of complex systems, discovery of structure in high-dimensional knowledge bases, distributed systems.

John F. McClusky, Assistant Professor of Industrial Design. User-centered product design, the flow and utilization of computer data throughout the product design and development process.

Philip Miller, Principal Lecturer and Director, Introductory Programming, School of Computer Science. Computer support for education.

Tom M. Mitchell, Professor of Computer Science and Robotics. Artificial intelligence, machine learning, robotics, interfaces that learn.

Jack Mostow, Principal Research Scientist, Robotics Institute, HCII, and Language Technologies Institute. Applying speech recognition and AI to literacy.

Dan R. Olsen Jr., Professor of Human-Computer Interaction. Personal computing systems that integrate people with information; individuals, organizations, and services that are needed to accomplish this work.

Raj Reddy, University Professor of Robotics and Computer Science. AI, speech recognition and understanding, integrated manufacturing.

Lynne M. Reder, Professor of Psychology. Human memory, cognitive skill learning, transfer of training.

Alexander I. Rudnicky, Senior Research Scientist, School of Computer Science. Speech recognition, spoken language interaction, interface design.

William L. Scherlis, Principal Research Scientist and Director, Information Technology Center. Collaboration information management, program manipulation and restructuring, analysis of home Internet use.

Mary Shaw, Professor and Associate Dean for Professional Programs, School of Computer Science. Software architecture, software engineering, programming systems and methodologies.

Mel Siegel, Senior Research Scientist, School of Computer Science, and Director, Measurement and Control Lab, Robotics Institute. Measurement, diagnosis, and control; perception system invention and modeling; stereoscopic computer and TV systems.

Herbert A. Simon, Richard King Mellon University Professor of Computer Science and Psychology. Computer simulation of cognitive processes, artificial intelligence, management science.

Scott M. Stevens, Senior Member, Technical Staff, Software Engineering Institute. Multimedia, digital video.

Susan G. Straus, Assistant Professor, Graduate School of Industrial Administration. Work groups and technology, communication media.

Katia P. Sycara, Senior Research Scientist, Robotics Institute. Case-based and analogical reasoning and learning, group problem solving, negotiation, intelligent agents.

CARNEGIE MELLON UNIVERSITY

Information Networking Institute

Program of Study

The M.S. in information networking is a cooperative endeavor of the Schools of Engineering, Computer Science, and Business, providing an alternative to the conventional one-year computer science or electrical engineering graduate program by integrating both programs and adding some features of an M.B.A. program. Now in its eleventh year, the program carefully selects 35 people from the engineering and computer science disciplines to form each new class. The program provides technical electives aimed at several areas of specialization, including the telecommunications and computing industries, wireless and mobile computing, the financial services industry, and systems integrating and consulting.

The information networking degree has been designed as a sixteen-month program with three major elements: technology, business applications, and an intensive design project. The first two elements constitute a twelve-month program of study, equivalent to three semesters. In each semester, students carry five courses and associated lab work. The summer semester is devoted to a thesis or a design project on which students and faculty members collaborate. Building on the Institute's substantial research program, many of these efforts have made significant contributions to the Institute.

The technology courses cover the fields of computer science and electrical engineering, including packet switching, operating systems, distributed systems design, wireless and mobile computing, software engineering, and broadband networks. In addition, students have the opportunity to pursue a wide range of technical electives in which they have a special interest.

In the business applications section, specially designed courses show how businesses use and process information, how to analyze information flows in the firm and identify opportunities to use information networking, how telecommunications and information regulation both constrain and create opportunities, and how to use information networking as an element of competitive strategy. This group of courses also provides an introduction to fundamental principles of management and economics.

For the research element of the program, students may choose either an interdisciplinary thesis or an intensive design project that is equivalent in rigor to a thesis in current graduate programs at Carnegie Mellon and other leading universities.

Research Facilities

The Information Networking Institute (INI) has a computer cluster exclusively for use by its students. The cluster contains more than two dozen high-end desktop machines (Sun Ultra Workstations, Pentium PCs, G3 Macintoshes) hosting a variety of operating systems (Solaris, Windows NT, Linux, FreeBSD, Windows 95/98, MacOS, and others). In addition, cluster equipment also includes a variety of portable computers, video cameras, scanners, CD recorders, RAID storage, tape libraries, and wireless access devices. The Institute continuously updates its software and machines. In addition, the University has exceptional computer and library facilities that include extensive online access to citation and full-text databases. The campus is served by a high-speed wireless network, which was built by the Institute as part of its wireless research program. In addition, research facilities are available through the School of Computer Science and the Department of Electrical and Computer Engineering.

Financial Aid

The Information Networking Institute provides financial aid for approximately 30 percent of its graduate students. There are usually full scholarships (some with competitive stipends) for exceptional applicants who are U.S. citizens or permanent residents.

Cost of Study

The Institute has successfully kept its tuition unchanged for the last two years. Tuition is $39,500 for the sixteen-month program during the 1999–2000 academic year. Additional fees may be required for students wishing to extend the program to two years.

Living and Housing Costs

All graduate students attending Carnegie Mellon University live off campus in nearby areas. The estimated living expense for 1999–2000, including room and board, insurance, transportation, and miscellaneous expenses, is $16,000.

Student Group

The graduate enrollment at Carnegie Mellon University totals more than 2,600 students, who come from colleges and universities throughout the United States and from many countries. The M.S. in information networking program has 35 students who use their own computer cluster, with individual desks and common conference spaces, for the duration of the program.

Student Outcomes

Information Networking Institute alumni currently work in more than 100 corporations worldwide. These include the telecommunications and computer industries and financial and systems consulting firms. More than 150 companies aggressively recruit the University's graduates each year.

Location

Carnegie Mellon is located in Oakland, the cultural center of Pittsburgh, on a 90-acre campus adjacent to Schenley Park, the city's largest park. The campus is conveniently located for access to many cultural and sporting events and is only 4 miles from the downtown business district. Pittsburgh is the headquarters for many of the nation's largest corporations and is also home to numerous research laboratories. Many recreational facilities, including ski areas and state parks, are located nearby.

The University

Founded by Andrew Carnegie, the Carnegie Institute of Technology joined the Mellon Institute in 1967 to become Carnegie Mellon University. With this merger, one of the leading research institutions in the country was established. The Information Networking Institute is a joint endeavor of the College of Engineering, the School of Computer Science, and the Graduate School of Industrial Administration.

Applying

The application, Graduate Record Examinations General Test scores, and letters of reference must be received by February 1. Early application is encouraged. Notification of acceptance and financial aid awards is completed by April 15. Minimum preparation normally includes an undergraduate program in electrical engineering, computer engineering, or computer science. Application materials are available on line at the Web site listed below.

Correspondence and Information

Admissions Coordinator
Information Networking Institute
Carnegie Mellon University
Pittsburgh, Pennsylvania 15213-3890

Telephone: 412-268-7195
Fax: 412-268-7196
E-mail: ini-admissions+@andrew.cmu.edu
World Wide Web: http://www.ini.cmu.edu

Carnegie Mellon University

THE FACULTY AND THEIR RESEARCH

Linda Argote, David M. and Barbara A. Kirr Professor of Industrial Administration; Ph.D., Michigan. Organizational learning, productivity, technology transfer, organizational structure, group decision making and performance.

Tsuhan Chen, Associate Professor of Electrical and Computer Engineering; Ph.D., Caltech. Multimedia processing and communication, image/video coding, audiovisual interaction, and multirate signal processing.

Alex Hills, Distinguished Service Professor of Engineering and Public Policy; Ph.D., Carnegie Mellon. Telecommunications policy, wireless telecommunications technology, remote and rural telecommunications systems.

David Johnson, Professor of Computer Science; Ph.D., Rice. Distributed systems, operating systems, and network protocols, particularly the interaction between these areas in the field of mobile computing. His teams were the developers of Mobile IP.

Hyong Kim, Associate Professor of Electrical and Computer Engineering; Ph.D., Toronto. Computer networks, high-speed multimedia networks, advanced switching system, BISDN.

Charles H. Kriebel, Professor of Industrial Administration; Ph.D., MIT. Computers and information systems, information economics, management science, operations management, robotics, applied economics, productivity, manufacturing control systems, information resources management.

B. V. K. Vijaya Kumar, Professor of Electrical and Computer Engineering; Ph.D., Carnegie Mellon. Information theory, statistical signal processing, pattern recognition, optical data processing, magnetic recording.

F. Javier Lerch, Senior Lecturer of Information Systems; Ph.D., Michigan. Human-computer interaction, including detection and correction of human errors in computer-based managerial tasks, design and evaluation of user interfaces for computer decision support and office automation, trust in machine advice, and cognitive modeling of organizations.

José Moura, Professor of Electrical and Computer Engineering; Sc.D., MIT. Digital telecommunications, wireless communications, video coding and compression, content-based video analysis, video and broadband networks, parameter identification, statistical image processing.

Tridas Mukhopadhyay, Professor of Industrial Administration; Ph.D., Michigan. Business value of information technology, electronic data interchange, software cost estimation, cost analysis.

Jon M. Peha, Associate Professor of Electrical and Computer Engineering and of Engineering and Public Policy; Ph.D., Stanford. Computer and telecommunications networks (e.g., integrated services/multimedia networks, wireless networks, telecom policy, telecom for developing countries).

Mahadev Satyanarayanan, Carnegie Group Professor of Computer Science; Ph.D., Carnegie Mellon. Mobile computing, Coda, Odyssey, distributed file systems, performance, and security.

Daniel Siewiorek, Buhl Professor of Electrical and Computer Engineering and Computer Science; Ph.D. Stanford. Modular design of reliable computing structures, with emphasis on design automation and wearable computers.

Marvin Sirbu, Professor of Engineering and Public Policy, Industrial Administration, and Electrical and Computer Engineering; Sc.D., MIT. Telecommunications technology, policy, and management; regulation and industrial structure of computer and communication technologies; communications networks and standards; economics of information and networks; electronic commerce.

Sandra Slaughter, Assistant Professor of Information Systems; Ph.D., Minnesota. Productivity and quality in software development and maintenance, software project management, information systems employment strategies.

Daniel D. Stancil, Professor and Associate Dean of the College of Engineering; Ph.D., MIT. Wireless communication, integrated optics, optical data storage.

Richard Stern, Professor of Electrical Engineering, Computer Science, and Biomedical Engineering; Ph.D., MIT. Automatic speech recognition, particularly signal processing for robust speech recognition and dynamic adaptation; auditory perception, particularly models of binaural perception based on optimal detection and estimation theory.

Hui Zhang, Assistant Professor of Computer Science; Ph.D., Berkeley. Theory, design, and implementation of scalable and efficient integrated services computer networks. Current research includes development of protocols that support real-time delivery of bursty compressed video streams, scalable multicast protocols, and integration of ATM and Internet protocols.

INFORMATION NETWORKING INSTITUTE MAJOR RESEARCH INITIATIVES

The Institute has taken a leadership role in several areas of research. These have formed opportunities for M.S. in information networking students to undertake research and make significant contributions. Among these are:

The Wireless Research Program: Since 1993, the INI has taken a leadership role in wireless and mobile computing research. The Wireless Research Program includes systems research, innovative computer platforms for mobile use, antennas and propagation, compression research, and human factors in mobile usage. In addition, the Institute was the prime initiator and implementer of Wireless Andrew, the largest high-speed wireless network in the world, covering most of the campus buildings. The Institute has also been a leader in developing mobile applications. These include emergency response, health care, vehicle maintenance systems, and the construction and mining industry. *The IEEE Personal Communications* February 1996 Issue (Mobile Computing at Carnegie Mellon), was devoted to papers from this initiative. Additional information is available on the World Wide Web (http://www.ini.cmu.edu/WIRELESS/Wireless.html).

Voice over IP: The Institute recently began a major research initiative into the technology, policy, and economic considerations of moving voice services from traditional networks to Internetworks.

NetBill: This project is an Internet billing system developed at the Information Networking Institute that supports electronic commerce. NetBill functions as an intermediary between a merchant and a consumer who wish to conduct business. It provides the mechanisms for consumer-merchant pairs to securely transfer goods and perform financial transactions. More information is available on the World Wide Web (http://www.ini.cmu.edu/netbill).

CARNEGIE MELLON UNIVERSITY

School of Computer Science
Language Technologies Institute
Graduate Programs in Language Technologies

Programs of Study

The Language Technologies Institute (LTI) offers a Ph.D. in language and information technologies and a Master's in Language Technologies (M.L.T.). These language technologies graduate programs are unique in North America and build on Carnegie Mellon University's (CMU) strengths in machine translation, information management, speech understanding, and computational linguistics. These fields of study have recently grown rapidly and are poised for further breakthroughs as they take advantage of emerging technological infrastructures, such as the World Wide Web, mobile computing, and multimedia interfaces. Directed research in these areas under the guidance of a faculty adviser is an integral part of both graduate programs.

The M.L.T. program leads to a professional degree with both regular two-year and accelerated twelve-month tracks. Students choose an individualized curriculum from a flexible set of courses and self-paced laboratory modules. These cover linguistic and statistical approaches, basic computer science, and applied areas such as machine translation, information retrieval, and speech recognition. No thesis is required for the M.L.T. degree.

The Ph.D. in language and information technologies draws on the same set of courses and laboratories as the M.L.T. degree; Ph.D. students normally complete their courses in their first two years. Additional breadth requirements ensure that LTI Ph.D. students receive a broad and deep education. Ph.D. students must also demonstrate proficiency in writing, presentation, programming, and teaching as well as complete and defend a Ph.D. thesis that contains significant original research. The entire program is expected to require approximately five years to complete.

Research Facilities

As part of the School of Computer Science (SCS), the faculty members and graduate students of the LTI have access to a wide variety of more than 2,200 machines. All new students in LTI's graduate programs are assigned personal workstations. The LTI computing environment has personal computers that run Windows and LINUX, Sun workstations that run SunOS, and several Macintoshes and other architectures. All personal computers in SCS have transparent access to the Andrew File System. The Carnegie Mellon Internet is a fully interconnected, multimedia, multiprotocol infrastructure that spans more than 100 segments. These segments are attached to an inverted backbone, which enables access between all systems on the campus. Network protocols utilized are IP/TCP, AppleTalk, and IPX. Carnegie Mellon is connected to the Internet with a T3 (45 Mbit/sec) link.

Financial Aid

Most students in the LTI are supported by graduate research fellowships, which typically include summer support. In 1999–2000, each Ph.D. student receives full tuition plus a stipend of $1525 per month. M.L.T. students also usually receive tuition and year-round stipend support through research fellowships, although this is not guaranteed. Dependency allowances are available; students who receive external fellowships may be given supplementary stipends.

Cost of Study

Tuition and fees for full-time graduate students in 1999–2000 are $22,100 for the academic year. This figure is subject to change.

Living and Housing Costs

The University does not provide housing for graduate students, but accommodations are available in the community at a variety of costs. Pittsburgh has a low cost of living and a good quality of life.

Student Group

Carnegie Mellon has a total enrollment of approximately 7,500 students. Twenty-three full-time students are enrolled in the Ph.D. program, and 12 are enrolled in the master's program, which makes the student-faculty ratio about 2:1. This includes 9 women, 2 members of minority groups, and 20 international students. Admission to these programs is highly competitive; 5 Ph.D. students and 7 M.L.T. students were admitted for the 1998–99 academic year (this number is expected to increase each year).

Location

Carnegie Mellon is located in Oakland, the cultural center of Pittsburgh, on a 90-acre campus adjacent to the city's largest park, Schenley Park. The campus is close to several other universities and the many cultural and sports activities of the city and is only 4 miles from the downtown business district. The mountains of western Pennsylvania and West Virginia are a 1-hour drive away, providing numerous outdoor recreation opportunities. Pittsburgh is the headquarters for many of the nation's biggest corporations.

The University

Founded in 1900 by Andrew Carnegie, the Carnegie Institute of Technology joined with Mellon Institute (now the Carnegie Mellon Research Institute) in 1967 to become Carnegie Mellon University, establishing itself as one of the leading research and educational institutions in the country.

Applying

The application, official transcripts, scores on the Graduate Record Examinations, TOEFL scores (for students whose native language is not English), and three letters of recommendation must be received January 5. Notification of acceptance and financial support is made by February 28. Students may apply directly to the Ph.D. program; a previous M.S. degree is not required.

Correspondence and Information

Admissions Coordinator
Language Technologies Institute
School of Computer Science
Carnegie Mellon University
5000 Forbes Avenue
Pittsburgh, Pennsylvania 15213

Telephone: 412-268-2623
E-mail: ltp@cs.cmu.edu
World Wide Web: http://www.lti.cs.cmu.edu/

Carnegie Mellon University

THE FACULTY AND THEIR RESEARCH

R. Brown, Systems Scientist. Natural-language processing, corpus-based machine translation, topic detection and tracking.

J. Callan, Associate Professor. Intelligent information filtering/retrieval, machine learning, data mining, K–12 electronic information literacy.

J. Carbonell, Professor and Director, Language Technologies Institute. Artificial intelligence, natural-language processing, machine learning, machine translation.

M. Eskenazi, Systems Scientist. Speech recognition and synthesis, speech database creation, foreign language teaching systems.

D. Evans, Adjunct Faculty, ClariTech. Natural-language processing, text mining/analysis/summarization, information retrieval and filtering.

R. Frederking, Senior Systems Scientist. Machine translation, natural-language processing, information retrieval, human factors design.

N. Green, Systems Scientist, HCII. Human–computer interaction, intelligent multimedia, conversational agents, discourse, argumentation and pragmatics.

A. Hauptmann, Systems Scientist. Integrating speech recognition, language processing, and information retrieval research.

J. Lafferty, Senior Research Scientist. Probability and information theory, natural-language processing, stochastic modeling.

A. Lavie, Research Scientist. Machine translation, parsing algorithms for spoken language analysis.

J. Fain Lehman, Senior Research Scientist. Natural-language understanding, generation, and acquisition; machine learning; cognitive modeling.

L. Levin, Senior Research Scientist. Machine translation, spoken dialogue understanding, computerized language instruction, linguistic theory.

M. Mauldin, Adjunct Faculty, Lycos and Virtual Personalities. Information retrieval, natural-language processing, virtual personalities.

T. Mitamura, Research Scientist. Machine translation, efficient knowledge base construction and text extraction.

T. Mitchell, Professor. Machine learning, intelligent Web agents, data mining, artificial intelligence.

V. Mittal, Adjunct Faculty, JustResearch. Natural-language processing, multimedia generation, interactive interfaces, rehabilitation research.

J. Mostow, Senior Research Scientist, Robotics. AI, machine learning, using automated speech recognition to help children learn to read.

E. Nyberg, Research Scientist. Machine translation, parsing and generation, knowledge representation, information retrieval, Web-based content delivery.

M. Ravishankar, Systems Scientist. Speech decoding algorithms, acoustic modeling, speech applications, interface design.

R. Reddy, University Professor and Dean. AI, speech recognition and understanding, integrated manufacturing systems.

R. Rosenfeld, Assistant Professor. Human language technologies, statistical language modeling, speech recognition, stochastic modeling.

A. Rudnicky, Senior Systems Scientist. Speech recognition, spoken language interaction, interface design.

M. Shamos, Principal Research Scientist. Universal libraries, e-commerce, intellectual property law.

K. Sycara, Research Scientist, Robotics. AI, planning and scheduling, case-based reasoning, coordination of autonomous agents.

D. Waeltermann, Senior Systems Scientist. Germanic linguistics, psycholinguistics, applied linguistics, translation studies.

A. Waibel, Principal Research Scientist. Perceptual user interfaces, speech, language, multimodal human-computer interactions, face/body tracking.

W. Ward, Senior Research Scientist. Artificial intelligence, speech understanding, stochastic modeling.

Y. Yang, Senior Research Scientist. Natural-language analysis systems, machine translation, text retrieval and categorization.

CARNEGIE MELLON UNIVERSITY

Programs in Robotics

Programs of Study	The robotics doctoral and master's programs bring together areas of robotics research that would otherwise be spread across different departments or separate universities, preparing students to take a leading role in the research and development of future generations of integrated robotics technologies and systems. Requirements for the degree include course work and a research qualifier. The Ph.D. also requires the submission of a thesis that describes original, independent research.

Master's students can typically complete the program in twelve months of full-time, self-funded study or, if financial assistance is available and needed, two years of study while working as a research assistant. Doctoral students are expected to complete their program in four to six years. All students are expected to split their time between research and course work during their first years in the program. Doctoral students, after completion of their course work, concentrate entirely on research. Students are involved in every aspect of research, from initial problem formulation to the final publication of results. Research is conducted in the laboratories of the Robotics Institute under the supervision of faculty advisers and in collaboration with student colleagues.

Research Facilities Students in the robotics programs work in the Robotics Institute's various research laboratories, including Advanced Mechatronics, Intelligent Modeling, Intelligent Sensors, Learning Hand-Eye Robot, Manipulation, Manufacturing Logistics, Manufacturing Systems Architecture, Medical Robotics, Mobile Robot, Field Robotics, Intelligent Measurement and Control, Production Planning, Rapid Manufacturing, Shape Deposition, Concurrent Engineering, Production Control, and High Definition Systems.

The Institute's research laboratories hold and maintain a multitude of unique and general research equipment. In the area of mobile robots, Carnegie Mellon is known for the Navlab vehicles, Ambler, Dante, Demeter, Heli, and Uranus. Robotic manipulators include the Troikabot system, Pumas, Adepts, Robot World, IMB-RS, and others. Institute laboratories house more than a dozen vision systems, a stereolithographer, Silicon Graphics Reality Engines, a shape deposition facility, and special sensors, such as fast-range finders and olfactory and photometric sensors. Computer-controlled moving platforms, high-precision calibration equipment, and solid modeling systems are also used in the research. The Institute's electronic labs, machine shops, and fabrication facilities support all of these activities.

The joint computing facility of the Robotics Institute and the School of Computer Science has more than 2,200 machines of a wide variety available for faculty members and graduate students. All new students in the robotics programs are assigned personal workstations.

Financial Aid All students in the Ph.D. program are supported by graduate fellowships during the academic year. Graduate fellowships may also be available for some students in the master's program. This support is provided on the basis of the student's participation in one or more of the ongoing robotics research projects. This participation is an integral part of the student's education. In 1999–2000, each funded student receives full tuition and fees plus a stipend of $1525 per month for the nine-month academic year. Additional allowances for dependents are available. Students holding outside fellowships may be given supplementary stipends. In the summer, students may accept research support from the University or seek employment in industry.

Cost of Study Tuition and fees for 1999–2000 are $22,230. Fellowships are available for all doctoral students and may also be available for master's students.

Living and Housing Costs The University does not provide housing for graduate students. Accommodations in the community are available at a variety of costs.

Student Group Carnegie Mellon has a total enrollment of approximately 7,500 students, of whom about 2,700 are graduate students. Within the robotics programs there are 52 faculty members and approximately 90 students.

Student Outcomes The goal of the robotics graduate programs is to prepare students to conduct independent research and become the future leaders formulating the ideas and building the systems that determine the basic understanding of robots and purposeful behavior in general. Graduates of the programs are in positions at top universities, research groups, and government research laboratories all over the world.

Location Carnegie Mellon is located in Pittsburgh. The 90-acre campus is adjacent to Schenley Park, the largest city park. The campus is close to the many cultural and sports activities of the city and is only 4 miles from the downtown business district.

The University and The Programs Founded in 1900 by Andrew Carnegie, the Carnegie Institute of Technology joined with the Mellon Institute (now the Carnegie Mellon Research Institute) in 1967 to become Carnegie Mellon University. Through this merger one of the nation's leading research and education institutions was established.

The robotics programs are administered by the Robotics Institute, which is part of the School of Computer Science.

Applying Application materials are available by writing to the address below. The application, official transcripts, Graduate Record Examinations General Test scores, and three letters of recommendation must be received by February 1. The TOEFL is required for students whose native language is not English. While formal admission requirements are flexible, minimum preparation normally includes an undergraduate program in science or engineering and some experience in computer programming. Excellence and promise may balance a lack of formal preparation.

Correspondence and Information
Robotics Graduate Programs
Carnegie Mellon University
5000 Forbes Avenue
Pittsburgh, Pennsylvania 15213-3890

Telephone: 412-268-3733
E-mail: robotics.admissions@ri.cmu.edu
World Wide Web: http://www.ri.cmu.edu

Carnegie Mellon University

THE FACULTY AND THEIR RESEARCH

O. Amidi, Systems Scientist, Robotics; Ph.D., Carnegie Mellon, 1996. Aerial robotics, vision-based robot navigation, high-speed industrial inspection.

D. Apostolopoulos, Systems Scientist, Robotics; Ph.D. Carnegie Mellon, 1998. Robots for extreme environments, planetary work systems, science autonomy robots, terramechanics for robotic locomotion.

J. Bares, Senior Research Scientist, Robotics; Ph.D., Carnegie Mellon, 1991. Conception to testing of intelligent machines for hazardous environments, construction, and heavy industry; automated robot design and optimization; technology transfer to industry.

D. A. Bourne, Senior Systems Scientist, Robotics; Ph.D. candidate, Pennsylvania. Expert systems, automated manufacturing, computer vision.

V. M. Brajovic, Research Scientist, Robotics; Ph.D., Carnegie Mellon, 1996. Analog and mixed-signal VLSI design, VLSI computational sensors for vision, machine vision and biologically motivated robot perception, image sensors, range sensors, real-time embedded vision systems.

H. Choset, Assistant Professor, Mechanical Engineering and Robotics; Ph.D., Caltech, 1996. Sensor-based exploration, motion planning, serpentine robots, mobile robots, robotic de-mining, material transport, robot painting.

R. Collins, Research Scientist, Robotics; Ph.D., Massachusetts, 1993. Video understanding, multi-image stereo, site modeling, camera calibration, projective geometry.

S. Cross, Senior Research Scientist, Robotics and Software Engineering Institute (SEI), and Director, SEI; Ph.D., Illinois at Urbana-Champaign, 1983. Artificial intelligence, large multimedia information systems, knowledge banks for medicine, articulate agents, associate systems.

A. DiGioia, Senior Research Scientist, Robotics, and Orthopedic Surgeon; M.D., Harvard, 1986. Clinical applications of medical robotics, image guide surgical navigation and clinical outcomes database development.

J. Dolan, Systems Scientist, Robotics; Ph.D., Carnegie Mellon, 1991. Distributed agent-based control, sensor-based real-time control, industrial automation, man-machine interaction.

M. Erdmann, Associate Professor, Computer Science and Robotics; Ph.D., MIT, 1989. Motion planning, planning under uncertainty, probabilistic strategies, sensing strategies, mechanics of manipulation, parts assembly, cooperating robots, computational biology.

G. K. Fedder, Associate Professor, Electrical and Computer Engineering and Robotics; Ph.D., Berkeley, 1994. Microelectromechanical systems (MEMS), microrobotics.

K. J. Gabriel, Professor, Electrical and Computer Engineering and Robotics; Sc.D., MIT, 1983. MEMS, human-machine interfaces, biomimetic systems.

R. Grace, Senior Systems Scientist, Robotics; Ph.D., Carnegie Mellon, 1985. Human factors, drowsiness measurement, truck driving simulator.

M. Hebert, Principal Research Scientist, Robotics; Ph.D., Paris IX, 1983. Range data analysis, 3-D perception for unstructured environments, underwater perception.

P. Heckbert, Associate Professor, Computer Science; Ph.D., Berkeley, 1991. Computer graphics, scientific computing, image processing.

R. Hollis, Principal Research Scientist, Robotics; Ph.D., Colorado, 1975. Mechatronic design, MEMS, magnetic actuators, scaled teleoperation, intelligent mobile robots.

B. Jaramaz, Research Scientist, Robotics; Ph.D., Carnegie Mellon, 1992. Surgical simulation, computational mechanics, medical robotics and computer-assisted surgery.

A. Jordan, Keithley University Professor Emeritus of Electrical and Computer Engineering and Robotics; Ph.D., Carnegie Tech, 1959. High-definition systems, digital television, studies of computer industries.

T. Kanade, U. A. and Helen Whitaker University Professor of Computer Science and Robotics and Director of the Robotics Institute; Ph.D., Kyoto, 1974. Computer vision, autonomous robots, medical robotics, sensors, virtualized reality.

A. Kelly, Research Scientist, Robotics; Ph.D., Carnegie Mellon, 1997. Perception, planning, control, simulation, operator interfaces for mobile robots.

P. K. Khosla, Philip and Marsha Dowd Professor of Engineering and Robotics and Director, Institute for Complex Engineered Systems; Ph.D., Carnegie Mellon, 1986. Distributed and reconfigurable robots, composable simulations, reconfigurable and adaptive embedded software, agent-based control, agent-based distributed information systems, intelligent assembly modeling.

A. Lipton, Systems Scientist, Robotics; Ph.D., Monash (Australia), 1996. Video surveillance and monitoring.

R. Martin, Principal Systems Scientist, Robotics; Ph.D. candidate, California, San Diego. Engineering of software-intensive systems using evolutionary processes, methods, and tools.

M. T. Mason, Professor of Computer Science and Robotics and Chair, Robotics Doctoral Program; Ph.D., MIT, 1982. Manipulation, automatic planning of robot manipulator programs, mechanics of manipulation.

T. M. Mitchell, Professor, Computer Science; Ph.D., Stanford, 1979. AI, process optimization, information agents, data mining, machine learning.

A. Moore, Associate Professor, Computer Science and Robotics; Ph.D., Cambridge, 1991. Reinforcement and machine learning, AI in manufacturing, Markov decision processes.

H. P. Moravec, Principal Research Scientist, Robotics; Ph.D., Stanford, 1980. Mobile robots, computer vision, 3-D modeling, robot manipulators, space applications.

J. Mostow, Principal Research Scientist, Robotics, Language Technologies, and Human-Computer Interaction; Ph.D., Carnegie Mellon, 1981. AI, machine learning, using automated speech recognition to help children learn to read.

I. R. Nourbakhsh, Assistant Professor, Robotics; Ph.D., Stanford, 1996. Robot architecture, robot communication and cooperation, nonprehensile manipulation, robot learning.

S. Penny, Associate Professor, Art and Robotics. Robotic art, simulation, electronic media.

R. Reddy, Simon University Professor of Computer Science and Robotics; Ph.D., Stanford, 1966. Spoken language systems, multimedia/human-computer interaction, learning from example.

A. A. Rizzi, Research Scientist, Robotics; Ph.D., Yale, 1994. Distributed and hybrid control of robot systems, dynamically dexterous mechanisms for locomotion and manipulation.

S. F. Roth, Senior Research Scientist, Robotics; Ph.D., Pittsburgh, 1981. HCI, information visualization and exploration, AI applied to visualization design and multimedia explanation generation.

H. Schempf, Senior Systems Scientist, Robotics; Ph.D., MIT, 1990. Hazardous waste robots, long-reach manipulators for space and infrastructure inspection.

J. Schneider, Research Scientist, Robotics; Ph.D., Rochester, 1994. Machine learning, reinforcement learning, optimization, scheduling.

S. M. Seitz, Assistant Professor, Robotics; Ph.D., Wisconsin, 1997. Computer vision, computer graphics, image-based rendering, video analysis and synthesis.

M. W. Siegel, Senior Research Scientist, Robotics; Ph.D., Colorado at Boulder, 1970. Sensors, mobile robots for measurement in difficult environments, 3-D stereoscopic optics.

R. Simmons, Senior Research Scientist, Computer Science and Robotics; Ph.D., MIT, 1987. AI, mobile robots, task-level control architectures, planning, reasoning under uncertainty, multi-robot coordination, indoor and outdoor robot navigation, formal verification of robot control programs.

S. Singh, Systems Scientist, Robotics; Ph.D., Carnegie Mellon, 1995. Forceful robot-world interaction, motion planning, machine learning, automated earthmoving, mine detection.

S. F. Smith, Senior Research Scientist, Robotics; Ph.D., Pittsburgh, 1980. AI, planning and scheduling, distributed and dynamic problem solving, evolutionary computation, knowledge-based production management.

A. X. Stentz, Senior Research Scientist, Robotics; Ph.D., Carnegie Mellon, 1989. Robotic excavation, mining, and farming; mobile robots; AI; robot architectures; computer vision.

K. P. Sycara, Senior Research Scientist, Robotics; Ph.D., MIT, 1987. AI, planning and scheduling, case-based reasoning, multiagent systems.

R. H. Thibadeau, Principal Research Scientist, Robotics; Ph.D., Virginia, 1976. Automated optical inspection devices, intelligent visual and verbal perception.

C. Thorpe, Principal Research Scientist, Robotics, and Chair, Robotics Master's Program; Ph.D., Carnegie Mellon, 1984. Vision, navigation, and systems for mobile robots; the Automated Highway System.

S. Thrun, Assistant Professor, Computer Science; Ph.D., Bonn (Germany), 1995. Artificial intelligence, robotics, machine learning, probabilistic reasoning, intelligent control, human-robot interaction.

D. S. Touretzky, Senior Research Scientist, Computer Science and Robotics, and Center for the Neural Basis of Cognition; Ph.D., Carnegie Mellon, 1984. Computational neuroscience, representations of space in the rodent brain, incremental learning/operant conditioning in animals and robots.

L. E. Weiss, Principal Research Scientist, Robotics; Ph.D., Carnegie Mellon, 1984. Rapid prototyping, micromechanisms, human tissue engineering.

W. L. Whittaker, Fredkin Research Professor, Robotics; Ph.D., Carnegie Mellon, 1979. Robots for unstructured environments, navigation and manipulation, integration of robot systems.

CARNEGIE MELLON UNIVERSITY

Master of Software Engineering

Program of Study

Software engineering is that form of computer science that applies the principles of computer science, mathematics, engineering, and management to achieve cost-effective solutions to software problems. Carnegie Mellon University, in response to industry's growing demand for skilled software professionals, offers a one-year master's degree program in software engineering aimed at practitioners from industry. The Master of Software Engineering (MSE) program is sponsored jointly by the University's School of Computer Science (SCS) and the Software Engineering Institute (SEI).

A set of core courses develops skills in the fundamentals of software engineering, with an emphasis on design, analysis, and management of large-scale software systems. A rich collection of elective courses, selected not only from the offerings of the School of Computer Science but also from other schools within the University, allows students to develop deeper technical expertise in one of several specialties, such as real-time systems, human-computer interaction, business, or the economic and organizational environment of software systems.

The capstone of the program is the software development studio component that runs throughout the entire program. The studio provides an opportunity for students to apply the knowledge and skills gained in other courses. Students work in teams to analyze a problem, plan a software development project, and find a solution. Recent studio projects include the movement software for a robot that services the Space Shuttle tiles, a robot for lunar exploration studies, a tool for integrating architectural information, and a computer-aided device that assists with hip replacement procedures.

Research Facilities

The School of Computer Science research facility has both a large number and a wide variety of computers available for faculty and graduate student use. Personal computing employs various UNIX workstations primarily from Sun, DEC, HP, IBM, and SGI. The MSE facility has personal computers running Windows NT, MS/DOS, and LINUX; Suns running SUNOS; and several Macintoshes. All personal computers have transparent access to the Andrew File System. The Carnegie Mellon Internet is a fully interconnected, multimedia, multiprotocol infrastructure spanning more than 100 segments. These segments are attached to an inverted backbone, enabling access between all systems on the campus. Various media such as fiber optics, shielded and unshielded twisted pair, support Fast, Ethernet, ATM, HIPPPI, Token Ring, and LocalTalk. Network protocols utilized are IP/TCP, AppleTalk, and IPX.

Financial Aid

Students in the MSE program are generally supported by their companies. Students may be able to find employment to complement the program. University-wide work-study opportunities are available to graduate students. Students may apply for Stafford loans.

Cost of Study

Tuition for the full-time graduate students accepted for the 1999–2000 academic year is $11,050 per semester. This figure is subject to change.

Living and Housing Costs

Graduate students are responsible for their own housing needs. Housing is available in the area within a range of costs.

Student Group

The number of students admitted to the MSE program is dependent on the number of qualified applicants. The present class has 22 students from the United States and four other countries, all with at least two years' experience. Previous classes have averaged 15 students, with an average of six years of work experience.

Student Outcomes

The program's graduates do exceptionally well in obtaining important professional positions after graduation. Some graduates have received starting salaries as high as $80,000 (salaries in some areas of the country are higher than others.) Graduates advance rapidly in their careers, and many have achieved influential management and policy positions within a few years after graduating.

Location

Located 5 miles east of the city in Oakland, the cultural center of Pittsburgh, the University is adjacent to Schenley Park, the city's largest, and close to many cultural and sports activities.

The University

Founded in 1900 by Andrew Carnegie, the Carnegie Institute of Technology joined with Mellon Institute (now the Carnegie Mellon Research Institute) in 1967 to become Carnegie Mellon University. With this merger, one of the leading research and education institutions in the country was established.

Applying

The application, Graduate Record Examinations scores (General Test and Subject Test in computer science), statement of purpose, up-to-date resume, letters of reference, and application fee must be received by January 31. Minimum preparation normally includes at least two years' experience working in a sizable software development project; foundation in discrete mathematics and programming-in-the-small; competence in using an imperative block-structured language; an undergraduate degree in computer science, engineering, mathematics, or physics (this requirement is flexible); and practical knowledge of programming methods, computer organization, data structures, compiling techniques, comparative programming languages, operating systems, and database systems.

Correspondence and Information

Admissions Coordinator
Master of Software Engineering Program
School of Computer Science
Carnegie Mellon University
Pittsburgh, Pennsylvania 15213
Telephone: 412-268-6493
E-mail: mse-info@cs.cmu.edu

Carnegie Mellon University

THE FACULTY AND THEIR RESEARCH

Members of the MSE faculty are drawn from the SEI staff as well as the School of Computer Science. In addition, members of SEI technical projects often act as advisers to student teams in the MSE studio course.

David Garlan, Assistant Professor, School of Computer Science; Ph.D., Carnegie Mellon. Formal methods in software engineering, programming environments, domain-specific software architectures.

Nancy R. Mead, Senior Member of the Technical Staff, Software Engineering Institute; Ph.D., Polytechnic of New York. Study of survivable systems architectures, development of professional infrastructure for software engineers, and real-time systems.

Mary M. Shaw, Professor and Associate Dean for Professional Programs, School of Computer Science; Ph.D., Carnegie Mellon. Programming systems and software engineering, particularly software architecture; programming languages, specifications, and abstraction techniques.

James E. Tomayko, Director of the MSE Program and Coordinator of the MSE Program's Software Development Studio; D.A., Carnegie Mellon. Fault tolerance of distributed embedded systems and the history of software development practices and techniques.

Associated Faculty in the School of Computer Science

Bernd O. Bruegge, Assistant Professor. Programming environments, distributed debugging, high-speed networking, software engineering.

Roger B. Dannenberg, Senior Research Scientist. Computer languages, computer music, human-computer interaction.

Allan L. Fisher, Systems Scientist and Associate Dean. Theory, design, and implementation of efficient, high-performance computing systems.

Merrick L. Furst, Professor and Associate Dean. Algorithm design, computational complexity, learning theory.

Bonnie E. John, Assistant Professor and Director, Master of Human-Computer Interaction Program. Creating engineering models of human performance to be used in the design of human-computer interaction.

James H. Morris, Professor and Department Head. Developing distributed computer systems, software engineering, functional programming, user interfaces.

Brad A. Myers, Senior Research Scientist. User interface design and management systems, programming environments.

Haj Reddy, University Professor and Dean. Artificial intelligence, speech recognition and understanding, integrated management systems.

M. Satyanarayanan, Associate Professor. Large-scale distributed systems, file systems, measurement and evaluation, security.

William Scherlis, Director, Information Technology Center. Program manipulation, information structures, HomeNet research, information infrastructure.

Jeannette M. Wing, Associate Professor. Formal specification, concurrent and distributed systems, object management, language design and implementation.

Associated Technical Staff in the Software Engineering Institute

Mario R. Barbacci. Software architecture description languages.

Mary Beth Chrissis. Software process.

Stephen E. Cross, Director, Software Engineering Institute. Broad adoption of innovative software development methods, transition of advanced technology, application of intelligent systems to strategic decision-making models of technology transition.

David Gluch. Dependable system upgrade and reliable real-time systems.

Will Hayes. Software measurement and analysis.

Clifford H. Huff. Software visualization, information, repositories, Web technologies.

Watts Humphrey, Institute Fellow. Software process research.

Rick Kazman. Software architecture, architecture tradeoff analysis method (ATAM).

Mark Klein. Rate monotonic analysis for real-time systems.

Richard Linger. Survivable systems, cleanroom software engineering.

Tom Longstaff. Computer and information security, research on survivable systems.

Linda Northrop. Object-oriented development, system design, databases, minicomputers in business environments, micrographics.

Mark Paulk. Capability Maturity Model (CMM®), software process, SPICE, software engineering standards.

Linda Hutz Pesante. Technical communications.

Ray Williams. Software risk management.

Associated Faculty from Industry

James Rozum, Mellon Bank. Software measurement methods and software process improvement methods and evaluation techniques.

CLAREMONT GRADUATE UNIVERSITY

Program in Information Science

Programs of Study	Claremont Graduate University (CGU) offers four academic programs in information science (IS): the Master of Science in Electronic Commerce (M.S.E.C.), the Master of Information Systems (M.I.S.), the Master of Science in the Management of Information Systems (M.S.M.I.S.), and the Ph.D. in the management of information systems.
	The 32-unit M.S.E.C. degree is designed to provide a professional education for persons who are working in or planning to work in the field of electronic commerce. The 32-unit M.I.S. degree program is designed for professionals in information systems and information systems users who want to improve both their professional and management skills and do not plan to continue beyond the master's level. Courses focus on information science issues. The 56-unit M.S.M.I.S. degree program is designed for professionals who want a master's degree that leads to management or senior staff positions in information systems organizations or for students who plan to continue on to a doctoral program. Specially designed courses allow students in each program to be exposed to the areas emphasized in all three programs.
	The 76-unit Ph.D. program in the management of information systems is designed to prepare graduates to make advanced contributions in either university or applied organizational settings. Since its inception in 1987, the Ph.D. program has produced 41 graduates, more than half of whom hold faculty appointments at universities in the United States, Korea, and Israel.
	The comprehensive combination of the technical, organizational, and systems elements in the curricula trains future managers to interact effectively with the technical specialists within their organization and trains technical specialists to be more sensitive to the management aspects of information systems. It is this integrated, two-culture approach that distinguishes the CGU vision of information science from a more modest data processing emphasis or from the highly theoretical training of traditional computer science programs. The programs in information science are affiliated with the Peter F. Drucker Graduate School of Management. Students can take management courses as part of their studies in the M.S.M.I.S. and Ph.D. degree programs.
	As part of its staffing concept in applied fields, the Graduate University appoints adjunct faculty members from industry to teach courses in specialized areas. Because of the University's location in the Los Angeles basin, a large pool of qualified professionals with relevant experience and superior teaching skills is readily available.
Research Facilities	The Claremont Information and Technology Institute (CITI) provides leadership in research, policy, and business strategy development for decision makers in the information economy. The Academic Computing Building houses the Graduate University's computers, the School of Information Science, and the Digital Laboratory, which supports groupware, geographic information systems, and data warehousing applications; a wide-band local area network; and personal computer laboratories. The library system, which includes the central Honnold Library, has extensive holdings of more than 1.9 million volumes, 6,000 periodicals and other serials, and a large collection of materials in microtext format.
Financial Aid	Financial aid packages are available through the information science department and the Graduate University's Financial Aid Office. Institutional fellowships are awarded by the program faculty. Federal and state loans and grants are also available. Information about need-based aid is available from the Financial Aid Office (telephone: 909-621-8337).
Cost of Study	For 1999–2000, full-time tuition at CGU is $10,145 per semester (12 units). Part-time tuition is $913 per unit of credit.
Living and Housing Costs	Most students make their own residential arrangements, living off campus in the city of Claremont or in one of the surrounding communities. Limited on-campus opportunities are available. Further information on housing may be obtained from the Housing Office, 1263 North Dartmouth Street, Claremont, California 91711 (telephone: 909-607-2609).
Student Group	The program's current student population consists of 60 part-time and 25 full-time students, with plans to grow by 50 percent within five years. The student population is split between U.S. citizens (60 percent) and international students (40 percent). Students come from a diversity of backgrounds, including science/computer science/ engineering, business, and social science/humanities. The student population is also diverse in terms of gender and race.
Location	Claremont Graduate University is in Claremont, a residential community of 36,000, which is located 35 miles east of Los Angeles. Situated in the foothills of the San Gabriel Mountains, Claremont is ideally located for those who enjoy skiing, hiking, swimming, and camping. Mountains, deserts, and beaches are all within an hour's drive. The climate is sunny most of the year.
The University	Founded in 1925, Claremont Graduate University (formerly the Claremont Graduate School) is affiliated with a group of five distinguished undergraduate colleges (Pomona, Scripps, Claremont McKenna, Harvey Mudd, and Pitzer) and the Keck Graduate Institute of Applied Life Sciences, which together form the Claremont Colleges. The small-college atmosphere at CGU offers students close relationships with faculty members and ample opportunity for individual development. At the same time, the Graduate University cooperates with the other colleges in Claremont to provide many facilities and services that are characteristic of a large university. CGU awards master's and doctoral degrees in both traditional academic programs and professional programs with an interdisciplinary emphasis. The regular CGU faculty of more than 80 members is supplemented by approximately 550 faculty members from the undergraduate colleges in Claremont and other institutions.
Applying	New students are admitted on a rolling basis in the fall, spring, and summer. Applicants should submit a completed application form, transcripts of all college-level work showing the satisfactory completion of a bachelor's degree, GRE scores, at least three letters of recommendation, and a statement of purpose. The TOEFL is required of applicants whose native language is not English and who do not hold a degree from a U.S. college or university. Students requesting institutional fellowship awards must complete their application by February 15 to receive priority consideration.
Correspondence and Information	Nancy M. Back, Program Coordinator School of Information Science Claremont Graduate University 130 East Ninth Street Claremont, California 91711-6190 Telephone: 909-621-8209 Fax: 909-621-8564 World Wide Web: http://www.cgu.edu/is

Claremont Graduate University

THE FACULTY AND THEIR RESEARCH

Paul Gray, Ph.D., Stanford. Professor Gray, professor of information science, was the founding chair of the Claremont Graduate University's Program in Information Science. He specializes in decision support systems and is the originator of the Graduate University's Decision Laboratory. He worked for eighteen years in research and development organizations, including nine years at SRI. Before coming to Claremont in 1983, he was a faculty member at Stanford University, the Georgia Institute of Technology, the University of Southern California, and Southern Methodist University. He has held numerous offices in national professional societies. He was president of the Institute of Management Sciences for 1992–93 and was formerly president-elect, vice president, and secretary of the institute. He is on the editorial board of several journals. He is the author of twelve books and more than seventy-five journal articles.

M. Lynne Markus, Ph.D., Case Western Reserve. Professor Markus, professor of information science and management in the Drucker School, joined CGU in 1992 after teaching at the University of California, Los Angeles, and the Massachusetts Institute of Technology. Her current research interests include "reengineering" the organization through the use of information technology, the information systems professional as "change agent," and the management of information technology from the perspective of the general manager. Among her publications is the book *Systems in Organizations: Bugs and Features*.

Magid Igbaria, Ph.D., Tel-Aviv. Professor Igbaria, professor of information science, joined CGU in 1995. Before he joined the faculty, he taught at the University of Hawaii and at Drexel University. He has published numerous articles in journals such as the *Communications of ACM, MIS Quarterly, Journal of MIS, Omega,* and *Information and Management*. His current research interests focus on electronic commerce, virtual workplace, technology impacts, IS personnel, computer technology acceptance, and management of IS.

Lorne Olfman, Ph.D., Indiana. Professor Olfman, professor and director of the School of Information Science, joined the faculty in 1987. His teaching specialties are system analysis design and planning, end-user computing, and research methodology. He is actively involved in research projects on the design and adoption of group work technologies, end-user training, and the impact of information systems on organizational memory. He served as chair of the 1996 conference of the Association for Computing Machinery Computer Personnel Research and Management Information Systems. He has been an economist and a systems analyst in transportation and telecommunications organizations and has previously worked as a programmer.

The Program in Information Science benefits enormously from a unique arrangement between CGU and Israel's Tel-Aviv University, which is widely recognized for the strength of its IS program. Each year, a Tel-Aviv University faculty member serves as a visiting professor at CGU; the following Tel-Aviv University professors have taught at CGU:

Niv Ahituv is a professor of information systems at Tel-Aviv University and former dean of its faculty of management.

Gad Ariav, professor on the faculty of management at Tel-Aviv University, specializes in database management and decision support systems.

Israel Borovits is a professor of computers and information systems on the faculty of management at Tel-Aviv University, where he served for a number of years as the chair of its CIS program.

Phillip Ein-Dor is a professor on the faculty of management at Tel-Aviv University and has served as chairperson for the Program in Information Science.

Seev Neumann was highly influential in the establishment of the Ph.D. program at the Claremont Graduate University's Program in Information Science and has served as chair of the program.

Eli Segev is a professor on the faculty of management at Tel-Aviv University.

Israel Spiegler is chair of the Computers and Information Systems Department of the faculty of management at Tel-Aviv University.

Moshe Zviran is a professor of information systems on the faculty of management at Tel-Aviv University.

The School of Information Science is in the process of hiring two new faculty members, who will join the School by the 2000–01 academic year.

CLEMSON UNIVERSITY

College of Engineering and Science
Department of Computer Science

Programs of Study

The Department of Computer Science offers programs leading to both Master of Science and Doctor of Philosophy degrees in computer science.

Requirements for a Ph.D. degree include passing a written qualification examination on core computer science subjects, a comprehensive examination in the student's specialization, and an oral defense of the dissertation. The areas of possible research span computer science. Interdisciplinary work is possible.

Programs leading to the M.S. degree are tailored to the student's interest, while ensuring a strong base in core subjects. Typical requirements include ten courses and a scholarly paper or a written examination.

Research Facilities

The Department of Computer Science provides a wide variety of computing facilities for student use. The departmental server is a Sun Enterprise 450 with dual 300-Mhz UltraSparc IIs, 256MB RAM, and 40GB of user disk space. More than eighty workstations running Sun's Solaris 2.6 are provided in four departmental general-purpose labs. Dedicated labs equipped with Linux PCs and Silicon Graphics workstations support graduate instruction in operating systems and graphics, respectively. Departmental labs support research in graphics with O2 and Octane class SGI machines and research in networking with IBM ATM equipment. Graduate students working in graphics also have unconstrained access to both an SGI Onyx and an SGI Onyx2 with Infinite Reality Graphics.

Financial Aid

Financial support for graduate students is available through limited research assistantships associated with grants and contracts, teaching assistantships, and graduate fellowships. Students with assistantships are normally expected to work an average of 20 hours each week while taking three courses per semester (two courses for Ph.D. students). Both twelve-month and nine-month assistantships are available. Alternative employment opportunities for qualified applicants are sometimes available.

Cost of Study

In-state tuition for 1999–2000 was $1735 per semester for graduate students without assistantships and $612 per semester for graduate students with assistantships. Out-of-state tuition was $3561 per semester for graduate students without assistantships and $612 per semester for graduate students with assistantships.

Living and Housing Costs

In 1999–2000, University housing—including dormitory rooms, apartments, and duplex units—ranges in cost from $835 to $1320 per semester. The cost of utilities varies. Many privately owned apartments are available; costs vary considerably.

Student Group

Clemson University's on-campus enrollment of more than 17,000 students includes approximately 4,000 graduate students. The Department of Computer Science enrolls about 110 full-time graduate students.

Student Outcomes

Many M.S. graduates find employment in the computer industry, with salaries typically starting at more than $50,000. Recent graduates have taken positions with AT&T, Bell Northern Research, Data General, IBM, Intel, Microsoft, Sun Microsystems, and Transarc.

Location

Clemson, South Carolina, is a small college town that enjoys the beauty and water sports of Lakes Hartwell (61,350 acres), Keowee (18,500 acres), and Jocassee (7,500 acres); the scenery of the South Carolina, Georgia, and North Carolina mountains (40-minute drive); and the challenge of many wild and scenic rivers such as the nearby Chattooga. Opportunities to participate in and enjoy plays, concerts, lectures, films, and sports events are provided by many University and community groups. While shopping in Clemson is limited to specialty stores and shopping centers, Seneca, Greenville, and Anderson are only a few minutes away and offer more extensive shopping and entertainment.

The University

Clemson University is a fully accredited, state-supported, coeducational, land-grant university founded in 1889. The main campus is situated on a 1,400-acre site, part of which was once the John C. Calhoun plantation. The campus is surrounded by 21,000 acres of agricultural research land and bordered by Lake Hartwell. Clemson offers sixty-eight undergraduate and ninety-five graduate curricula.

Applying

Applications for admission are considered at any time of the year. Applications for financial assistance, however, should be submitted before March 1 for full consideration.

Correspondence and Information

Student Services Specialist
Department of Computer Science
Clemson University
Box 341906
Clemson, South Carolina 29634-1906
Telephone: 864-656-5853

Clemson University

THE FACULTY AND THEIR RESEARCH

Andrew T. Duchowski, Assistant Professor; Ph.D., Texas A&M. Visual perception, HCI, graphics and VR, computer vision, image and video processing.

Robert M. Geist III, Professor; Ph.D., Notre Dame. Systems modeling, performance evaluation, reliability modeling, graphics.

Harold C. Grossman, Associate Professor; Ph.D., Michigan State. Programming language theory, design, and implementation; software development methodology.

Eleanor O'M. Hare, Associate Professor; Ph.D., Clemson. Graph algorithms, graph representation.

Sandra M. Hedetniemi, Professor; Ph.D., Virginia. Data structures, analysis of algorithms.

Stephen T. Hedetniemi, Professor and Head; Ph.D., Michigan. Design and analysis of algorithms, parallel algorithms, computational complexity and combinatorial optimization.

David P. Jacobs, Associate Professor; Ph.D., Missouri. Algorithms, algebraic computation.

Alan Kaplan, Assistant Professor; Ph.D., Massachusetts. Programming languages, object-oriented databases, software engineering, bioinformatics.

A. Wayne Madison, Associate Professor; Ph.D., Virginia. Operating systems, performance measurement and evaluation.

Brian A. Malloy, Associate Professor; Ph.D., Pittsburgh. Languages, compilers, parallel processing, software maintenance and testing, simulation modeling.

John D. McGregor, Associate Professor; Ph.D., Vanderbilt. Software engineering, graphical systems, object-oriented development.

Edward W. Page III, Professor; Ph.D., Duke. Neural computation, computer architecture, telecommunications.

Roy P. Pargas, Associate Professor; Ph.D., North Carolina. Parallel computation, genetic algorithms.

John C. Peck, Professor; Ph.D., Southwestern Louisiana. Operating systems, database systems, real-time manufacturing systems.

Mark K. Smotherman, Associate Professor; Ph.D., North Carolina. Computer architecture, superscalar processors, reliability.

D. E. Stevenson, Associate Professor; Ph.D., Clemson. Computational science, numerical analysis, computation theory.

Albert J. Turner, Professor; Ph.D., Maryland. Software engineering, computer science education.

James M. Westall, Professor; Ph.D., North Carolina. Systems software, performance measurement and evaluation, character recognition, neural networks.

COLORADO STATE UNIVERSITY

Computer Science Department

Programs of Study	The Computer Science Department at Colorado State University seeks outstanding students for study leading to the M.S. and Ph.D. degrees. The M.S. program provides maximum flexibility so participants may prepare for professional employment or advanced graduate study. The Ph.D. program is meant for well-qualified individuals who desire to prepare for careers in academic or industrial research. While students may work in many areas, active research programs in algorithms, applicative languages, architecture, artificial intelligence, distributed systems, fault tolerance, genetic algorithms, graphics, languages and compilers, neural networks, operating systems, parallel processing, performance evaluation, and software engineering provide special opportunities for thesis research topics and assistantship support.

The department retains an applied orientation, with industrial adjunct faculty supplying insight into practical issues and industrial approaches. Weekly seminars with speakers from many academic and industrial organizations provide students with a broad introduction to many research areas.

The M.S. curriculum requires at least 39 credits. M.S. students pass an oral final examination on their thesis or project work. Highly successful M.S. students may transfer to the Ph.D. program with no loss of efficiency. The Ph.D. curriculum requires at least 72 credits in total, comprising both course work and research. Required examinations are a written qualifying examination, an oral preliminary examination, and a final examination in defense of the dissertation. |
| **Research Facilities** | The department's computing facilities include 124 UNIX-based systems (Sun, HP, IBM, Linux), fifty X-terminals (NCD), 108 Microsoft NT–based PCs and five Macintoshes. Of these systems, thirty-seven UNIX workstations, thirty-eight X-terminals, eight NT PCs, and three Macintoshes are devoted solely to faculty and graduate student work, while the remainder are housed in general-purpose open laboratories or are available for general department use as computation or file servers on the network. The department maintains an Adaptive Solutions CNAPS 128-processor SIMD machine. All departmental machines are connected to the Internet via a subnet of the campus backbone.

In addition to the department's resources, the University provides a dial-up modem-pool with approximately 400 lines, campuswide general-purpose computing with large servers (IBM RS/6000), a computer visualization lab (largely SGI), and some general-purpose computing laboratories (Microsoft PCs and Apple Macintoshes). Department researchers have used large parallel systems at LANL, LLNL, Sandia Labs, NCAR, and NEPAC. |
Financial Aid	The University has grants, loans, and work-study opportunities available to graduate students. The department offers teaching and research assistantships that require duties of 20 hours per week. Assistantships pay a competitive stipend and tuition; supported students pay only fees.
Cost of Study	In 1999–2000, full-time graduate tuition is $1394 per semester for Colorado residents and $5330 for nonresidents. Fees are $361.10 per semester.
Living and Housing Costs	On-campus apartments for single students rent for $332 to $497 per month. Family housing ranges from $498 to $724 per month (utilities included). Off-campus, rent for a two-bedroom apartment averages $650 per month (plus utilities).
Student Group	The University enrolled about 22,000 regular, on-campus students in 1998–99, including 3,361 graduate students. Thirty percent of approximately 80 computer science graduate students are noncitizens. More than 50 percent receive financial aid from the department, and more than 20 percent work in the computer industry while pursuing graduate work.
Location	Fort Collins is a community of more than 100,000 located along the foothills of the Rocky Mountains, 60 miles north of Denver. The climate is moderate—15 inches of precipitation and 290 days of sunshine annually. Cultural offerings include a museum, library, symphony, chorale, and community center, all with many activities and performances. The spectrum of cultural and outdoor activities, the climate, and the mountain setting contribute to making Fort Collins an attractive community.
The University	Colorado State was designated Colorado's land-grant college in 1879; it was named Colorado State University in 1957. Today the University has 8 colleges, 56 departments, and more than 100 academic programs. The central 666-acre campus includes nearly 100 academic buildings, research facilities, dormitories, married students' housing, and the Veterinary Teaching Hospital. Other campuses support instruction and research in agriculture, engineering, natural resources, and biological sciences.
Applying	Applications are considered throughout the year. Applicants who desire financial assistance should apply six months prior to their first semester. Applicants should have an undergraduate GPA of at least 3.2 on a 4.0 scale. All applicants must take the GRE General Test. International applicants must also take the TOEFL; a score of at least 580 is strictly required. A nonrefundable application fee of $30 is required from all applicants. An online, Web-based application form is also available.
Correspondence and Information	Graduate Programs Coordinator Department of Computer Science Colorado State University Fort Collins, Colorado 80523 Telephone: 970-491-5792 Fax: 970-491-2466 E-mail: csgradinfo@cs.colostate.edu

Colorado State University

THE FACULTY AND THEIR RESEARCH

Charles W. Anderson, Associate Professor; Ph.D., Massachusetts. Neural networks, machine learning, artificial intelligence, pattern classification, adaptive control, graphics.

J. Ross Beveridge, Associate Professor; Ph.D., Massachusetts. Computer vision, robot navigation, model matching.

James Bieman, Associate Professor; Ph.D., Southwestern Louisiana. Software engineering, executable specifications, software analysis and testing, programming languages.

A. P. Wim Böhm, Professor; Ph.D., Utrecht (Netherlands). Parallel and dataflow architectures, software, and languages.

Bruce Draper, Assistant Professor; Ph.D., Massachusetts. Computer vision, machine learning, image understanding.

Robert B. France, Associate Professor; Ph.D., Massey (New Zealand). Software engineering, formal techniques, object-oriented modeling, software architecture, domain-specific software engineering environments.

Dale H. Grit, Associate Professor; Ph.D., Minnesota. Parallel functional languages and architectures, operating systems.

Sandeep Gupta, Assistant Professor; Ph.D., Ohio State. Parallel and distributed systems, scientific computing, compilers.

Adele Howe, Associate Professor; Ph.D., Massachusetts. Artificial intelligence, planning, agent architectures, failure recovery.

Robert Kelman, Professor Emeritus; Ph.D., Berkeley. Computational methods, mathematical software.

Yashwant Malaiya, Professor; Ph.D., Utah State. Fault-tolerant computing, architecture, VLSI design, hardware and software reliability evaluation.

Walid Najjar, Associate Professor; Ph.D., USC. Parallel processing, dataflow architectures, performance and reliability evaluation, parallel simulation.

Rodney Oldehoeft, Professor; Ph.D., Purdue. Parallel processing software and systems, functional programming, operating systems.

Stephen Seidman, Professor and Chair; Ph.D., Michigan. Parallel computation, formal methods in software engineering.

Pradip Srimani, Professor; Ph.D., Calcutta. Parallel and distributed computing, operating systems, graph theory applications.

Anneliese von Mayrhauser, Professor; Ph.D., Duke. Software engineering, maintenance, metrics, testing, reliability, performance evaluation.

Darrell Whitley, Professor; Ph.D., Southern Illinois. Artificial intelligence, machine learning, genetic algorithms, neural networks.

Affiliate Faculty
Anura Jayasumana, Ph.D., Michigan State. Computer networks, VLSI.

Selected Research Areas

Artificial Intelligence: Learning algorithms for neural networks, neural networks for control and signal processing, genetic algorithms and applications to scheduling problems, planning, agent architectures.

Software Engineering: Evaluation of testing criteria, executable specifications, foundations of software measurement, automatic analysis from algebraic specifications, static evaluation of sequencing constraints, software maintenance toolkit, software reliability simulation.

Parallel and Distributed Computing: Applicative language features, dataflow parallelisms versus locality, SISAL language and implementations, representation of parallel programs, distributed algorithms.

Graphics and Computer Vision: Detail enhancement in volume images, ray tracing of fractal terrain, image understanding.

Architecture and Networks: Reliability management through self-testing; reliability, performability, and scalability of large-scale distributed systems; network reliability and topology.

COLUMBIA UNIVERSITY

Department of Computer Science

Programs of Study	The doctoral program of the Department of Computer Science is geared toward the exceptional student. The faculty believes that the best way to learn how to do research is by doing it. Starting in their first semester, students do joint research with the faculty. They also prepare themselves for the Ph.D. comprehensive examinations, which test breadth in computer science. The primary educational goal is to prepare students for research and teaching careers either in universities or in industry.
	The department enjoys a low doctoral student–faculty ratio (about 4:1). Ph.D. students are viewed as colleagues by the faculty.
	Current research areas include algorithmic analysis, artificial intelligence, collaborative work, computational complexity, computer-aided design of digital systems, databases, distributed computing, graphics, logic synthesis, network management, parallel processing, software development environments, vision, robotics, user interfaces, and virtual environments.
	The department also offers the Master of Science degree in computer science. This program can be completed within one academic year of full-time classwork. However, completing the optional thesis generally stretches the program to two years. The M.S. degree can also be earned through part-time study.
Research Facilities	The department has well-equipped lab areas for research in collaborative work, computer graphics, computer-aided digital design, computer vision, databases and digital libraries, data mining and knowledge discovery, distributed systems, mobile computing, natural language processing, networking, robotics, user interfaces, and real-time multimedia.
	The computer facilities include a shared infrastructure of Sun multiprocessor file servers, a cluster of high-speed multiprocessor CPU servers, a student interactive teaching and research lab of high-end multimedia workstations (with audio/video capabilities), a department-wide switched Fast-Ethernet network, high-speed dial-up and ISDN lines, FDDI rings connecting the main servers together, and an ATM connection to Internet. The research infrastructure includes numerous Sun and SGI Workstations, a network of high-end X86 Solaris, Windows NT and Linux PCs, and an Onyx 2 server with InfiniteReality graphics. In addition, high-speed, high-resolution color and black and white printers and scanners are available for use, as well as portable projection and audio/video equipment. The vast amounts of disk space are being backed up by a 2.1TB DLT7000 and 600-gigabyte Exabyte robotic tape units.
	Research labs contain Unimate Puma 500 and IBM robotic arms; a UTAH-MIT dextrous hand; a DataCube image processor; an Adept-1 robot; three mobile research robots; a real-time defocus range sensor; numerous multimedia PCs; SGI, Sun, and HP PC-based interactive 3-D graphics Workstations with true 3-D input (via a VPL DataGlove hand-tracking system and assorted 3-D position and orientation trackers); wall-sized stereo projection systems; see-through head-mounted displays; experimental packet-radio-based workstations; and additional network gateways.
	The research facility is supported by a full-time staff of professional systems administrators, programmers, and engineers, aided by many part-time student technicians.
Financial Aid	Most doctoral students receive graduate research assistantships. The stipend for 1999–2000 is $1664 per month for the academic year. In addition, graduate research assistants receive full tuition exemption. A limited number of teaching assistantships are available to master's students meeting selection criteria.
Cost of Study	Tuition and fees total approximately $26,040 for the 1999–2000 academic year.
Living and Housing Costs	In 1998–99, dormitory rooms were available for approximately $4445 for the academic year. Apartments in University-owned buildings cost $750 per month and up. Rooms are also available at International House; these cost approximately $4135 per academic year.
Student Group	There are approximately 70 Ph.D. students in the department. A large proportion of Columbia University's student body is at the graduate level: of approximately 18,800 students, 10,500 are in the graduate or professional schools.
Location	New York City is the intellectual, artistic, cultural, gastronomic, corporate, financial, and media center of the United States, and perhaps of the world. The city is renowned for its theaters, museums, libraries, restaurants, opera, and music. Inexpensive student tickets for cultural and sporting events are frequently available, and the museums are open to students at very modest cost or are free. The ethnic variety of the city adds to its appeal.
	The city is bordered by uncongested areas of great beauty that provide varied types of recreation, such as hiking, camping, skiing, and ocean and lake swimming. There are superb beaches on Long Island and in New Jersey, while to the north lie the Catskill, Green, Berkshire, and Adirondack mountains. Close at hand is the beautiful Hudson River valley.
The University	Columbia University was established as King's College in 1754. Today it consists of sixteen schools and faculties and is one of the leading universities in the world. The University draws students from many countries. The high caliber of the students and faculty makes it an intellectually stimulating place to be. Columbia University is located on Morningside Heights, close to Lincoln Center for the Performing Arts, Greenwich Village, Central Park, and midtown Manhattan. Columbia athletic teams compete in the Ivy League.
Applying	For maximum consideration for admission to the doctoral program, students should submit the following before January 5: official applications for the fall term, transcripts, two recommendation letters, and a $55 application fee. The General and Subject tests of the Graduate Record Examinations are required for all computer science graduate students and should be taken by October for doctoral admission or December for master's admission. The deadlines for applications to the master's program are February 15 for fall admission and October 1 for spring admission.

Correspondence and Information

For information about the programs:

Admissions Committee
Department of Computer Science
450 Computer Science Building, MC 0401
Columbia University
New York, New York 10027
Telephone: 212-939-7000
E-mail: gradinfo@cs.columbia.edu
World Wide Web: http://www.cs.columbia.edu

For applications:

Office of Engineering Admissions
524 S. W. Mudd Building
Columbia University
New York, New York 10027
Telephone: 212-854-6442

Columbia University

THE FACULTY AND THEIR RESEARCH

Kathleen R. McKeown, Professor and Chair. Artificial intelligence, natural-language processing, language generation, multimedia explanation, text summarization, user interfaces, user modeling, digital libraries.

Alfred V. Aho, Professor. Evolvable information systems and networks, software production, algorithms for information retrieval.

Peter K. Allen, Associate Professor. Sensor-based robotics, computer vision, 3-D modeling.

Steven K. Feiner, Associate Professor. Computer graphics, knowledge-based picture generation, user interfaces, animation, virtual worlds, augmented reality, visual languages, hypermedia.

Zvi Galil, Morris A. and Alma Schapiro Professor of Engineering and Julian Clarence Levi Professor of Computer Science. Analysis of algorithms, computational complexity, cryptography, parallel processing, theory of computation.

Luis Gravano, Assistant Professor. Databases, Internet information sources, distributed indexing and searching, digital libraries.

Jonathan L. Gross, Professor of Computer Science, Mathematics, and Mathematical Statistics. Combinatorial and probabilistic models, parallel architectural models, topological graph theory and low-dimensional algebraic topology.

William N. Grundy, Assistant Professor. Bioinformatics, artificial intelligence, Bayesian inference, artificial life.

Gail E. Kaiser, Professor. Software development environments, software process, object-oriented databases, collaborative work, workflow applications, cooperative transactions, 3-D virtual environments.

John R. Kender, Associate Professor. Computer vision, robotic navigation, artificial intelligence.

Andrew P. Kosoresow, Lecturer. Artificial intelligence, analysis of algorithms, computational biology and multiagent systems.

Shree K. Nayar, Professor. Computer vision, physical models, vision sensors, realistic rendering, robotics.

Jason Nieh, Assistant Professor. Parallel and distributed operating systems, network and multiprocessor performance evaluation, interactive real-time multimedia systems, network computing.

Steven M. Nowick, Associate Professor. Asynchronous circuits, computer-aided digital design, logic synthesis, low-power and high-performance systems, formal hardware verification.

Christopher Okasaki, Assistant Professor. Programming languages, functional programming, algorithms and data structures.

Daniel N. Port, Assistant Professor. Software engineering, component and object-oriented architecture, distributed collaboration, software patterns, causal event posets.

Kenneth A. Ross, Associate Professor. Databases, query optimization, declarative languages for database systems, logic programming.

Henning Schulzrinne, Associate Professor. Internet real-time and multimedia services and protocols, modeling and analysis of computer-communication networks, operating systems, network security.

Eric Siegel, Lecturer. Natural language processing, machine learning, genetic algorithms.

Salvatore J. Stolfo, Professor. Artificial intelligence, distributed data mining, knowledge discovery in databases.

Joseph F. Traub, Edwin Howard Armstrong Professor and Professor of Mathematics. Computational complexity, information-based complexity, financial computations, limits to scientific knowledge.

Luca Trevisan, Assistant Professor. Computational complexity, randomness in computation, foundations of cryptography.

Stephen H. Unger, Professor of Computer Science and Electrical Engineering. Logic circuits theory, digital systems, self-timed systems, parallel processing, technology-society interface, engineering ethics.

Henryk Woźniakowski, Professor. Computational complexity, information-based complexity, algorithmic analysis, numerical mathematics.

Yechiam Yemini, Professor. Algorithms and protocols for computer networks, network management, high-speed networks, organization of distributed systems, modeling and analysis of computer-communication networks.

Associated Faculty

Theodore R. Bashkow, Professor Emeritus. Computer architecture, data communications.

Danilo Florissi, Associate Research Scientist. Computer networks, network management, high-speed networks.

Vasileios Hatzvivassiloglou, Associate Research Scientist. Natural language processing, digital libraries, statistical methods, evaluation methodology, genomics applications.

Judith Klavans, Research Scientist and Adjunct Associate Professor. Text databases, natural language processing, digital library, computational lexicons, information retrieval, multilingual information access.

Jaron Lanier, Visiting Scholar. Visualization of extremely complex systems, "slow machine"—an exotic hypothetical computation device, programming language design and virtual reality systems.

Leora Morgenstern, Adjunct Associate Professor. Artificial intelligence, knowledge representation, non-nomotonic reasoning.

Soumitra Sengupta, Assistant Professor, Department of Medical Informatics. Distributed systems, deductive and object-oriented databases, concurrency control, medical informatics, application system development.

Athanasios M. Tsantilas, Adjunct Lecturer. Parallel computing, randomized algorithms, computational complexity.

Nina Wacholder, Associate Research Scientist. Computational linguistics, syntactive processing, medical applications.

Arthur G. Werschulz, Adjunct Senior Reseach Scientist. Information-based complexity, especially that of partial differential equations, integral equations, and ill-posed problems.

George Wolberg, Adjunct Lecturer. Image processing, computer graphics, computer vision, and image warping, reconstruction, and enhancement.

Moti Yung, Adjunct Lecturer. Cryptography.

COLUMBIA UNIVERSITY

Graduate School of Arts and Sciences
College of Physicians and Surgeons
Department of Medical Informatics
Training Programs in Medical Informatics

Programs of Study

Medical informatics studies the flow of information in health care, medical education, and medical research. The field explores techniques for assessing current information practices, determining the information needs of health-care providers and patients, developing interventions using computer technology, and evaluating the impact of those interventions. This research seeks to optimize the use of information in order to improve the quality of health care, reduce cost, provide better education for providers and patients, and conduct medical research more effectively. Columbia University offers two programs in medical informatics: an advanced degree program leading to the M.A., M.Phil., or Ph.D. degree and a postdoctoral training program.

The degree program focuses on the theory and application of information science in the domain of medicine. The program trains students for academic careers as researchers and teachers as well as for professional positions in health-care information processing. The curriculum incorporates course work from computer science, public health, and biostatistics, with core courses and projects in medical informatics serving to integrate approaches and illustrate practical applications. The predoctoral program requires three years of full-time study to complete the M.Phil. and typically two additional years to conduct research and write the dissertation. The M.A. degree can be completed in two years of part-time study.

The postdoctoral training program trains medical informatics scientists for careers as productive researchers and teachers. The fellowship is funded by the National Library of Medicine, an agency of the National Institutes of Health, and is open to United States citizens or permanent residents with an M.D. or Ph.D. degree. Support for each fellow is for three years in most cases. The program is individualized to address the needs and directions of each fellow. Each fellow is expected to develop, conduct, and report on an original research project.

Research Facilities

Computing resources for medical informatics research at Columbia include the Integrated Academic Information Management System (IAIMS), a sophisticated, heterogeneous distributed information system for patient care, administration, research, and scholarship. This environment provides an unparalleled living laboratory for informatics projects. The Health Sciences Library provides access to medical informatics journals, online materials, and a microcomputer classroom. Additional computer laboratories and modern classroom facilities are available in a newly renovated Student Learning Center located in Presbyterian Hospital.

Financial Aid

All students accepted into the predoctoral program are awarded support that fully provides for the tuition and the medical insurance fees required by the University. Students also receive a stipend toward living expenses that commences with registration and normally continues throughout graduate study. This stipend was $20,000 for the 1998–99 academic year. International students as well as U.S. citizens are eligible for these fellowships. The postdoctoral training program includes a stipend and tuition support for course work.

Cost of Study

Information about fees can be obtained on the World Wide Web at http://www.columbia.edu/cu/gsas or by telephone at 212-305-5780.

Living and Housing Costs

Housing is available on the Health Sciences Campus and on the Morningside Heights Campus of Columbia University. Accommodations include University residence halls, which consist of furnished 2- or 4-person suites, and institutional Real Estate Apartments, which include studios and one-, two-, and three-bedroom apartments. Membership at the Athletic Club is free to students. The club contains a swimming pool, squash courts, a gymnasium, a sauna, and exercise equipment and is accessible to the handicapped.

Student Group

The degree program has support for up to 20 Ph.D. students and 15 master's students. The postdoctoral training program can support 9 fellows. About one third of students and fellows are women, and slightly fewer than half are international.

Location

The Department of Medical Informatics at the College of Physicians and Surgeons of Columbia University is located on the Health Sciences Campus at 165th Street and Fort Washington Avenue in upper Manhattan. The complex includes the Columbia-Presbyterian Medical Center and its subdivisions and the New York Psychiatric Institute. New York's world-renowned cultural activities are easily accessible by public transportation, as are sporting events and other recreational opportunities.

The University

Columbia University, a privately supported institution, is one of the world's leading educational, scholarly, and research centers. Founded by charter as King's College in 1754, it is one of the oldest universities in the country. Medical informatics programs enjoy close collaborative relationships with the Columbia University Health Sciences Library, basic and clinical science departments on the Health Sciences Campus, and the computer science department at the Morningside Campus.

Applying

The basic requirement for admission as an M.A. student in the medical informatics degree program is a bachelor's degree. Applicants should indicate clearly on the application whether they are applying for the part-time M.A. degree program or the full-time Ph.D. program.

The postdoctoral training program is open to U.S. citizens and permanent residents with an M.D. or Ph.D. degree. Applicants should send a letter of purpose, transcripts (from college and from medical or graduate school), a curriculum vitae, and three letters of reference.

Correspondence and Information

For advanced degree programs:
Office of Graduate Affairs, Room 406
College of Physicians and Surgeons
Columbia University
701 West 168th Street
New York, New York 10032
Telephone: 212-305-8058
Fax: 212-305-1031
E-mail: midegree@cucis.cpmc.columbia.edu
World Wide Web: http://www.cpmc.columbia.edu

For the postdoctoral training program:
James J. Cimino, M.D.
Atchley Pavilion 1310
161 Fort Washington Avenue
New York, New York 10032
Telephone: 212-305-8127
Fax: 212-305-3302
E-mail: ciminoj@cucis.cis.columbia.edu

Columbia University

THE FACULTY AND THEIR RESEARCH

George Hripcsak, M.D., Acting Chair. Data mining, evaluation of systems, mobile computing.
Stephen B. Johnson, Ph.D., Degree Program Director. Data modeling, database design, language interfaces.
James J. Cimino, M.D., Postdoctoral Program Director. Medical concept representation, linking medical records to knowledge sources.

Barry A. Allen, Ph.D. Advanced user interfaces, genome informatics.
Randolph C. Barrows Jr., M.D. Outpatient applications, user interface design, cancer information systems.
Bruce H. Forman, M.D. Methods and standards for data interchange.
Carol Friedman, Ph.D. Natural language processing.
Robert A. Jenders, M.D. Expert systems, electronic medical records, medical vocabulary.
Desmond Jordan, M.D. Intensive care information systems, critical paths, care plans.
Pat Molholt, Ph.D. Electronic medical curricula.
Andrey Rzhetsky, Ph.D. Computational genomics.
Soumitra Sengupta, Ph.D. Networking, security, systems architecture.
Justin B. Starren, M.D., Ph.D. User interface design, telemedicine.
John L. Zimmerman, D.D.S. Dental informatics, education and computer-based records.

CORNELL UNIVERSITY

Department of Computer Science

Programs of Study	The Department of Computer Science offers one of the country's oldest and top-ranked doctoral (M.S./Ph.D.) and M.Eng. programs. The emphasis of the doctoral program is on research, and most graduates of the program find employment in research positions. The department offers programs of study in artificial intelligence, distributed computing and concurrency, computer systems (operating systems and databases), programming languages and methodology, logic and reasoning about uncertainty, theory and analysis of computing and algorithms, multimedia and video processing, graphics, and scientific computing (numerical analysis). Recently, the department has expanded into networks and mobile computing, computational biology and genomics, natural language processing, and digital libraries; many of these programs are considered to be among the top one or two worldwide in their area.
	The dominant characteristic of the department, and its principal advantage for a graduate student, is the unusual opportunity for close interaction between faculty members and students. All faculty members are easily accessible to graduate students and actively engaged in research, their teaching load is relatively light, and the student-faculty ratio is quite favorable. One indication of the degree of interaction is the large number of papers coauthored by faculty members and students.
	A typical Ph.D. program requires about five years of study. Students often become involved in research very soon after entering the program. Most complete their course work within one to two years and identify a thesis topic after about two years. A written examination is taken early in the first year of studies, and an oral exam is needed for approval of the thesis topic.
Research Facilities	The department's research computing facility consists of more than 300 high-performance PCs and engineering workstations from Intel, Sun, Hewlett-Packard, and other sources. The department is unusual in viewing Microsoft NT as an important research platform. The Duffield Systems Laboratory includes state-of-the-art technology dedicated to research in systems and multimedia, including a programmable telephone switch with the capacity to handle half of the voice telephony traffic for the whole campus. A separate laboratory is dedicated to use by the database research group, and another to the AI group. The graphics laboratory has a tremendous variety of graphics computing systems. Researchers also have access to the IBM SP2 supercomputer in the Cornell Theory Center. Departmental machines are on a local network with gateways to University-wide, national, and international networks. An off-campus ADSL network provides megabit connectivity to the homes of many researchers. Cornell has many other computing facilities that are also available to graduate researchers and views information technology as one of the top three University priorities for the coming decades, along with computational genomics (biology) and advanced material sciences.
	The Cornell University library, one of the largest in the country, maintains an excellent collection in computer science. A major new digital library research effort is underway.
Financial Aid	All Ph.D. students receive financial aid in the form of a fellowship, research assistantship, or teaching assistantship. For 1998–99, fellowship stipends ranged from $17,000 to $19,000 for nine months, and assistantships paid $14,000. The department pays tuition in addition to the amounts cited and helps students find summer employment in the form of internships with major industry research labs, teaching summer courses, or summer research assistantships.
Cost of Study	For 1999–2000, tuition for the two-semester academic year is $23,760.
Living and Housing Costs	For 1998–99, graduate dormitory accommodations costed about $7000 per academic year. University-operated married student apartments rented for about $800 per month. In addition, a considerable range of privately owned accommodations can be found within commuting distance.
Student Group	There are 150 resident graduate students (90 Ph.D. and 60 M.Eng.) in the department; all are full-time. Each year, about 20 to 25 new Ph.D. students are enrolled.
Student Outcomes	The department's graduates are a good source of information about the Cornell program—they hold computer science positions at Arizona, Berkeley, Brown, Carnegie Mellon, Colorado, Colorado State, CUNY, Dartmouth, Florida State, Harvard, Illinois, Johns Hopkins, Kansas, Maryland, Massachusetts, Minnesota, Montreal, North Carolina, Ohio, Penn State, Pennsylvania, Princeton, Purdue, Queens, Rhode Island, Rice, Stanford, SUNY at Albany, SUNY at Stony Brook, Toronto, Washington, and Wisconsin. Many graduates have gone into research positions in industry (Hewlett-Packard, IBM, ITT, Bell Labs, Control Data, Ford Aerospace, Mitre, Softech, Microsoft, and Xerox) and government, and many have formed their own companies.
Location	Ithaca, New York, is a small town in the heart of the Finger Lakes region. It offers the cultural activities of a large university and the pleasures of a rural environment. Facilities for skiing, sailing, camping, hiking, soaring, and other sports are close at hand.
The University and The Department	Cornell is a prominent research university, one of the largest producers of Ph.D.'s in the country, and it stands out as a major contributor of women scholars. Computer Science interacts most closely with Cornell's distinguished Departments of Mathematics, Operations Research, Mechanical Engineering, Linguistics, Electrical Engineering, and Architecture, as well as the interdisciplinary Center for Applied Mathematics.
	The computer science department, which was organized in 1965, is one of the oldest departments of its kind in the country. It has a full-time faculty of 29, associated faculty members from other departments, and numerous visitors each year.
Applying	To be admitted to the Graduate School, an applicant must have a B.S. or equivalent degree. In addition, applicants are expected to have had significant programming experience, a solid background in mathematics (at least calculus and linear algebra and preferably other subjects such as logic, statistics, or analysis), and, depending on the specialization chosen, an appropriate background that would permit immediate enrollment in graduate-level courses in that specialization.
Correspondence and Information	Graduate Field Office Department of Computer Science 4126 Upson Hall Cornell University Ithaca, New York 14853-7501 World Wide Web: http://www.cs.cornell.edu/gradstudies.html

Cornell University

THE FACULTY AND THEIR RESEARCH

Kenneth P. Birman, Professor; Ph.D., Berkeley, 1981. Distributed computing, fault-tolerant network systems, distributed systems security, large-scale network applications.

Claire Cardie, Assistant Professor; Ph.D., Massachusetts, 1994. Natural language processing, machine learning, artificial intelligence.

Tom Coleman, Professor and Director of the Cornell Theory Center; Ph.D., Waterloo, 1979. Numerical analysis, scientific computing, sparse optimization, algorithms.

Robert L. Constable, Professor and Chairman; Ph.D., Wisconsin, 1968. Computational complexity, formal semantics, applied logic, automated reasoning.

Ronald Elber, Professor; Ph.D., Hebrew (Jerusalem), 1984. Computational biology, genomics.

Donald P. Greenberg, Jacob Gould Schurman Professor of Computer Graphics; Ph.D., Cornell, 1968. Realistic image synthesis, modeling, scientific visualization, computer-aided design, image processing.

David Gries, Professor; Ph.D., Munich, 1966. Programming methodology, programming languages, compiler construction.

Joseph Halpern, Professor; Ph.D., Harvard, 1981. Logics, artificial intelligence, distributed computing, reasoning about uncertainty.

Juris Hartmanis, Walter R. Read Professor of Engineering; Ph.D., Caltech, 1955. Theory of computation.

Sheila S. Hemami, Assistant Professor; Ph.D., Stanford, 1995. Image processing, networks.

Daniel P. Huttenlocher, Associate Professor; Ph.D., MIT, 1988. Computer vision.

S. Keshav, Associate Professor; Ph.D., Berkeley, 1991. Networks, quality of service, multimedia; telephony and telecommunications networks.

Jon Kleinberg, Assistant Professor; Ph.D., MIT, 1996. Theory of computation, optimization, computational biology.

Dexter Kozen, Professor; Ph.D., Cornell, 1977. Theory of computation, proof-carrying code.

Lillian Lee, Assistant Professor; Ph.D., Harvard, 1997. Natural language processing.

Andrew Myers, Assistant Professor; Ph.D., MIT, 1998. Programming languages, security, mobile code.

Gregory Morrisett, Assistant Professor; Ph.D., Carnegie Mellon, 1996. Programming languages, compilers.

Anil Nerode, Professor; Ph.D., Chicago, 1956. Logic, applied mathematics.

Keshav Pingali, Associate Professor; Ph.D., MIT, 1986. Compilers, parallel computing, programming languages.

Ronitt Rubinfeld, Associate Professor; Ph.D., Berkeley, 1990. Theory of computation, randomized algorithms, computational complexity.

Fred B. Schneider, Professor; Ph.D., SUNY at Stony Brook, 1978. Distributed systems security and fault tolerance, mobile code, concurrent programming, operating systems.

Bart Selman, Assistant Professor; Ph.D., Toronto, 1991. Artificial intelligence, theory of learning and decision making.

Praveen Seshadri, Assistant Professor; Ph.D., Wisconsin, 1996. Database systems.

Rosen Sharma, Assistant Professor; Ph.D., Cornell, 1998. Network management and protocols, quality of service, multimedia.

David Shmoys, Associate Professor; Ph.D., Berkeley, 1984. Scheduling, computational complexity.

Brian Smith, Assistant Professor; Ph.D., Berkeley, 1994. Multimedia systems.

Eva Tardos, Associate Professor; Ph.D., Eotvos Lorand (Budapest), 1981. Combinatorics, complexity theory, communication networks, quality of service and data flow.

Ray (Tim) Teitelbaum, Associate Professor; Ph.D., Carnegie-Mellon, 1975. Programming languages and systems.

Sam Toueg, Professor; Ph.D., Princeton, 1979. Distributed computing, fault tolerance.

Charles Van Loan, Professor; Ph.D., Michigan, 1973. Matrix computations.

Stephen Vavasis, Associate Professor; Ph.D., Stanford, 1989. Numerical analysis, optimization.

Thorsten von Eicken, Assistant Professor; Ph.D., Berkeley, 1993. Parallel and distributed systems.

Ramin Zabih, Assistant Professor; Ph.D., Stanford, 1993. Multimedia, computer vision.

RESEARCH AREAS

Algorithms. Work in this traditionally strong area continues but with a more applied flavor. Some of the problems under investigation in the computational biology and genomics area fall under the heading of analysis of algorithms. In addition, several faculty members are studying various combinatorial algorithms that are used to solve sparse matrix problems arising in numerical analysis.

Artificial Intelligence. Faculty interests include logic and reasoning about uncertainty, automated theorem proving, machine learning, and natural language processing.

Computational Genomics and Biology. This is a new area for the department and is expected to grow rapidly. Research includes computational models of protein-enzyme bonding, protein folding, and relating gene sequences to function across species.

Computing Theory. This area is concerned with fundamental mathematical problems of computer science. Specific topics include computational complexity, analysis of algorithms, formal languages and automata, semantics, and program verification. Computational complexity has always been a research interest at Cornell, and current work is concerned with the intrinsic difficulty of computing problems and the relationships among various measures of computational and structural complexity, such as run time, space, and program size. Recently, there has been extensive work on the classification of problems and algorithms.

Concurrency and Distributed Computing. Several researchers are investigating theoretical issues in concurrent programming, with particular attention focused on distributed systems security, decentralized control (i.e., distributed systems), and fault tolerance. Cornell-developed software from the Isis Project (now Ensemble) is in use worldwide, including in the Swiss and New York Stock Exchanges and the PHIDEAS component of the new French Air Traffic Control system.

Database Systems. The Predator project is exploring multimedia databases, including retrieval of nonstandard datatypes and efficient representation of multimedia databases.

Graphics. The graphics laboratory is well known for work on rendering and illumination and has focused on architectural and scientific visualization problems.

Multimedia and Vision. Several researchers are exploring issues that include automated analysis of video and other images, computer vision, object recognition, shape comparison, structure from motion, motion tracking, and automated processing of video and other multimedia information. Ramin Zabih's work on automated testing of HTML rendering systems has revolutionized the approach to testing Web software in industry.

Networks and Operating Systems. Research focuses on protocols for the Next Generation Internet, including mobile IP, network management for collaboration and group computing, quality of service and other aspects of reliability, network modeling and simulation, and wireless communication. Other work explores high-performance communication in operating systems, new operating system support for mobile code and Java, cluster-style computing systems, fault-tolerance, and security.

Numerical Analysis. Research is conducted in the field of matrix computations, sparse optimization, and differential equations. Topics include parallel-matrix algorithms, generalized and inverse eigenvalue problems, matrix problems in statistics, signal processing, graph methods for handling matrix sparsity, and grid generation.

Programming Environments. The Synthesizer Generator, a system for creating interactive, language-based environments from formal specification, continues to be a focal point of much research. It is being used to create interactive environments for theorem proving as well as for programming and has been licensed to more than 290 sites worldwide.

Programming Languages and Methodology. Mobile code correctness and proof-carrying code are important current topics. Proofs of program correctness and their influence on program development are being studied for concurrent and sequential programs. The Nuprl system permits interactive development and verification of programs specified in a top-down fashion, using constructive mathematical proofs, and is being used to prove properties of the Ensemble distributed computing system. There are significant efforts in developing semantic theories of concurrency and of types, as well as in extracting parallelism from sequential programs for various parallel machines.

DARTMOUTH COLLEGE

Department of Computer Science

Programs of Study	The Department of Computer Science offers programs leading to the M.S. and Ph.D. degrees in computer science. The M.S. and Ph.D. program faculty includes all members of the Department of Computer Science and adjunct members from the Department of Mathematics and the Thayer School of Engineering.

Programs of Study

The Department of Computer Science offers programs leading to the M.S. and Ph.D. degrees in computer science. The M.S. and Ph.D. program faculty includes all members of the Department of Computer Science and adjunct members from the Department of Mathematics and the Thayer School of Engineering.

Active research areas include algorithms, combinatorial optimization, computational biology, computational geometry, discrete-event simulation, distributed systems, graphics, information retrieval, image processing, microelectromechanical systems, multimedia, operating systems, parallel computing, performance modeling, physical geometric modeling, robotics, and signal processing.

The requirements for the Ph.D. degree include the following: six core courses and four advanced topics courses; passing five written qualifying exams (algorithms, architecture, operating systems, programming languages, and one of theory, artificial intelligence, or numerical analysis) by the end of the second year; participation in teaching; an oral thesis proposal; six terms in residence at Dartmouth; and acceptance and public defense of a thesis. Most students who complete the Ph.D. program do so in four to five years.

The requirements for the M.S. degree include the following: six courses, at least five of which are graduate courses; three advanced topics courses; six course equivalents of research; an oral thesis proposal; and the preparation and public defense of a master's thesis.

The Department of Computer Science is housed in the Sudikoff Laboratory for Computer Science, a modern building with offices for all faculty and graduate students. Each office has Ethernet and Appletalk connections. A weekly colloquium series and a weekly graduate student seminar series add to a lively environment for the exchange of ideas.

Research Facilities

Graduate students have access to many computer laboratories in Sudikoff Laboratory with networks of DEC, Silicon Graphics, Linux, Windows NT, and Macintosh computers. All graduate student offices have computing facilities.

Other research facilities in the department include a Silicon Graphics Origin 2000 with eight processors, 512 megabytes of memory, and 72 gigabytes of disk storage; an agents lab with thirty Linux machines and fast Ethernet; seventeen Linux laptops on a wireless network and a collection of PAs; the FLEET Laboratory for file system research; a mobile robotics laboratory with two RWI B12 autonomous mobile robots with on-board vision and communication, a pioneer robot, a wheel-based inchworm, and self-reconfigurable robots; the Dartmouth Experimental Visualization Laboratory, which conducts research in multimedia technology; a laboratory for the study of physical geometrical algorithms, containing six high-end SGI and Wintel graphics workstations, Phantom haptic control devices, hardware benches for MEMS control and testing, 100Mbit CDDI Ethernet, small robots, and digital cameras.

Financial Aid

All Ph.D. students accepted into the Department of Computer Science are granted a tuition scholarship. Most Ph.D. students are supported either by Dartmouth Fellowships or by faculty member grants; the stipend is $1278 per month for nine months in 1999–2000. Summer support is often available for Ph.D. students working with faculty members.

Cost of Study

Tuition for 1999–2000 is $24,624 for nine months, but all full-time Ph.D. students receive full tuition scholarships. A health plan fee of approximately $935 for a single student is required unless the student can verify equivalent coverage from another plan. Stipends are subject to federal income tax.

Living and Housing Costs

A single graduate student should expect to spend up to $5500 per year on rent and utilities, $4000 per year on food, and $4000 per year on transportation and personal items. Many Dartmouth graduate students are able to spend much less. College housing is available for single and married students on a priority basis. Rents in Hanover are relatively high for New Hampshire, but rents are considerably lower in many nearby towns.

Student Group

The Department of Computer Science currently has 40 graduate students, 35 of whom are full-time. Approximately one third are international students. Almost all the full-time Ph.D. students have fellowship or grant support.

Student Outcomes

Graduates have gone into both academia (e.g., tenure-track positions at Wellesley, Puget Sound, Hartford, Vermont, Michigan State, and Southern Illinois; a postdoctoral position at Carnegie Mellon; and a position at the University of Minnesota Supercomputing Center) and industry (e.g., BBN, Transarc, Thinking Machines, NEC Research Labs, and IBM Almaden Research Center).

Location

Hanover lies along the eastern bank of the Connecticut River, between the Green Mountains of Vermont and the White Mountains of New Hampshire. It is a small town of about 9,200 residents. Hanover has tried hard to preserve traditional New England small town qualities. Outdoor activities—especially hiking, skiing, canoeing, and bicycling—are popular. Hanover also offers a modest but active cultural life. The Hopkins Center at Dartmouth is a magnet for the arts in New Hampshire and Vermont; it sponsors concerts by visiting and local musicians, film series, art shows, and theater. Dartmouth provides a full assortment of athletic facilities.

Because Hanover is near the junction of Interstates 89 and 91, students can drive to Boston in 2½ hours, Montréal in 3½ hours, and New York in 5 hours. The airport in Lebanon, New Hampshire, is about 10 minutes from campus and has direct flights to Boston, New York, and Philadelphia.

The College and The Department

Founded in 1769, Dartmouth College combines the advantages of a liberal arts college and a research university. The approximately 4,300 undergraduates benefit from a faculty renowned for its teaching excellence, and the approximately 300 graduate students in the arts and sciences and 800 graduate students in the engineering, business, and medical schools enjoy close working relationships with faculty members actively involved in research.

Dartmouth has a long tradition of leadership in computing, pioneering time-sharing availability to all students and the BASIC programming language in the 1960s. The undergraduate major in computer science was created in 1979, the Ph.D. program in computer science started in 1986, the Department of Computer Science was formed in 1994, and the M.S. program in computer science began in 1998.

Applying

Applications for the Ph.D. program are on the World Wide Web and are mailed upon request. Completed applications are due by February 1 and require college transcripts, a statement, three letters of reference, GRE scores (copy is accepted), and, if English is not the student's native language, TOEFL scores (copy is accepted). There is no application fee and GRE Subject Tests are not required; however, if the applicant has no academic degree in computer science, the GRE Subject Test in computer science is recommended. Admissions are announced by March 15. Applicants have until April 15 to accept or decline. There is no admissions fee.

Applications for the M.S. program are on the World Wide Web and are mailed upon request. Completed applications must be accompanied by college transcripts, three letters of reference, GRE General Test and Subject Test scores, a $40 application fee, and, if English is not the student's native language, TOEFL scores. Completed applications are due by February 15, with admission announcements by March 15. Applicants have until April 15 to accept or decline.

Correspondence and Information

Ph.D. Program/M.S. Program
Department of Computer Science
Dartmouth College
6211 Sudikoff Laboratory
Hanover, New Hampshire 03755-3510

Telephone: 603-646-2206
E-mail: phd@cs.dartmouth.edu
      ms@cs.dartmouth.edu
World Wide Web: http://www.cs.dartmouth.edu/

Dartmouth College

THE FACULTY AND THEIR RESEARCH

Javed A. Aslam, Assistant Professor; Ph.D., MIT, 1994. Machine learning, design and analysis of algorithms, computational biology, computational geometry.

Thomas H. Cormen, Associate Professor; Ph.D., MIT, 1992. Parallel disk systems, languages, analysis of algorithms.

George Cybenko, Dorothy and Walter Gramm Professor of Engineering Sciences (Adjunct, Thayer); Ph.D., Princeton, 1978. Signal processing, networked information systems, parallel computing.

Bruce Randall Donald, Professor; Ph.D., MIT, 1987. Physical geometric algorithms, microelectromechanical systems, robotics, graphics, and computational biology.

Robert L. (Scot) Drysdale III, Professor and Chair; Ph.D., Stanford, 1979. Algorithms, computational geometry, Voronoi diagrams, triangulations.

Hany Farid, Assistant Professor; Ph.D., Pennsylvania, 1997. Image processing, computational and human aspects of perception.

Dennis M. Healy Jr., Associate Professor; Ph.D., California, San Diego, 1986. Nonabelian harmonic analysis, architectures for distributed statistical pattern recognition.

Prasad Jayanti, Associate Professor; Ph.D., Cornell, 1994. Asynchronous concurrent systems, synchronization, fault tolerance.

David Kotz, Associate Professor; Ph.D., Duke, 1991. Operating systems, mobile agents, parallel architectures and systems.

Fillia S. Makedon, Professor; Ph.D., Northwestern, 1982. Multimedia systems and analysis, information retrieval of multimedia data, electronic publishing, multimedia data access.

M. Douglas McIlroy, Adjunct Professor; Ph.D., MIT, 1958. Bell Laboratories, 1958–1997; Text processing, security, maps.

David M. Nicol, Professor; Ph.D., Virginia, 1985. Performance modeling, analysis, optimization, and tools, parallel processing; discrete-event simulation; reliability modeling and tools.

Daniel Rockmore, Associate Professor (Joint, Mathematics); Ph.D., Harvard, 1989. Algorithms related to representation theory of finite groups with applications in statistics.

Daniela Rus, Assistant Professor; Ph.D., Cornell, 1992. Multimedia information capture and access, electronic libraries, applications of geometric algorithms, robotics.

Clifford Stein, Associate Professor; Ph.D., MIT, 1992. Design and analysis of algorithms, combinatorial optimization, network algorithms, parallel algorithms, scheduling, computational biology.

Linda F. Wilson, Assistant Professor (Adjunct, Thayer); Ph.D., Texas at Austin, 1994. Parallel and distributed computing, parallel discrete-event simulation, computer performance analysis, computer architecture.

Neal Young, Assistant Professor; Ph.D., Princeton, 1991. Design and analysis of approximation algorithms for combinatorial optimization, network design, and online problems.

Sudikoff Laboratory for Computer Science.

Dartmouth Hall.

Baker Library.

DEPAUL UNIVERSITY

School of Computer Science, Telecommunications and Information Systems

Programs of Study

The School of Computer Science, Telecommunications and Information Systems (CTI) offers programs leading to the M.S. degree in computer science, distributed systems, e-commerce technology, human-computer interaction, information systems, management information systems, software engineering, and telecommunication systems and to the Ph.D. degree in computer science. Separate master's degree concentrations allow specialization in the areas of artificial intelligence, computer science telecommunications, data analysis, database, data communications, information systems, quality management, software engineering, standard telecommunications systems, systems foundations, and visual computing.

All master's programs are organized into three phases: Prerequisite, Core Knowledge, and Advanced. Most degree candidates are admitted conditionally into the Prerequisite Phase and become fully admitted to the graduate program by achieving a passing grade on the department's Graduate Assessment Examination or by proving equivalent competence in undergraduate course work or work experience. In order to obtain the M.S. degree, students must complete three Core Knowledge Phase courses and ten Advanced Phase courses. A comprehensive examination covering material from Core Knowledge Phase courses is required to enter the Advanced Phase.

The M.S. curriculum is flexible and appropriate for students seeking professional development through academic study. Each student, in consultation with a faculty adviser, plans a phase-by-phase program suited to the student's interests and goals. The Ph.D. program prepares computer scientists for positions in industry or academia. A Ph.D. candidate must complete fifteen courses beyond the master's degree, pass a doctoral candidacy examination, pass an oral examination on the dissertation area, present a dissertation proposal, and prepare and successfully defend the dissertation.

Research Facilities

DePaul's Information Services (IS) division houses a large network of computers and allows students access to a rich computing environment. The configuration includes several Sun SPARC centers for student use. In addition, students have access to PC laboratories at the Loop and Lincoln Park campuses. There are numerous dial-up phone numbers available for off-campus work. DePaul's suburban campuses, in the Lake County, Naperville, O'Hare, and South areas, also offer excellent student laboratory facilities. Permanent student Internet access accounts are available, through a service called DePaul Online, along with dial-in connections.

The School itself operates specialized laboratories for artificial intelligence, computer vision and graphics, database, group support systems, multimedia, software engineering, telecommunications, local area networks, and computer telephony. One laboratory allows students to explore specialized software. The laboratories include both PCs and UNIX workstations. The School also operates an IBM ES 9000/9221. Telecommunications laboratory equipment available for student use includes Cisco routers, LAN switches, and ATM switches as well as a Definity PBX system from Lucent Technologies. In addition, a multimedia laboratory provides hardware and software for students to support high-bandwidth synchronized audio, video, and control applications among multiple high-performance Sun and NT workstations. Eight new laboratories open in September 1999: the computer-supported collaborative learning lab, the database teaching lab, the database research lab, the student project teamwork collaborative learning lab, the distributed systems lab, the human-computer interaction lab, the software engineering lab, and a second multimedia lab.

There are a number of ways that DePaul CTI is incorporating technology into the classroom experience, including the introduction of Internet-based courses, videotaped course offerings, distance learning courses, Web-supported classes, and testing on line.

Financial Aid

The School provides a number of full and partial graduate assistantships that carry stipends and tuition waivers. Application should be made directly to the School. There is one filing period for each academic year, from February 2 to April 1.

The University's Financial Aid Office assists interested students in applying to the Federal Perkins Loan and Federal Stafford Student Loan programs. Students are encouraged to apply before May 1 to receive maximum consideration.

Cost of Study

Graduate tuition for the 1999–2000 academic year is $420 per quarter hour. Full-time graduate study generally consists of 8 to 12 hours per quarter.

Living and Housing Costs

There is no on-campus housing for graduate students, but the Housing Office has listings of apartments and residential hotels near the Lincoln Park campus. Apartments in the Chicago area are available at various rents.

Student Group

CTI has approximately 1,200 students in the graduate program; nearly half attend full-time. This sizable enrollment allows the School to offer a wide variety of courses each quarter. Many students are already employed in a computer or computer-related profession and enroll in the program for career enhancement.

Student Outcomes

Graduates of CTI have experienced a high degree of success in locating employment. Many graduates have reported starting salaries that exceed national averages. The Career Development Center assists students with job searches through career development seminars, job fairs, networking programs, interviewing skills workshops, and on-campus recruiting.

Location

The city of Chicago offers DePaul students a wide range of cultural and recreational opportunities. The Loop campus is minutes from the Art Institute; Orchestra Hall; museums of art, natural history, and science; and the LaSalle Street business district. The 25-acre Lincoln Park campus is less than 1 mile from Lake Michigan beaches, the lakefront bicycle and running paths, the zoo, and other public recreational facilities. The stores, theaters, musical groups, and other attractions of the Lincoln Park community reflect the broad interests of people who live and work there. The Naperville, O'Hare, and South campuses serve students in the outlying suburbs.

The University

DePaul University was founded in 1898 by the Vincentian Fathers and is now one of the largest Catholic universities in the world. Urban in style, the University today still strives to maintain the heritage of St. Vincent DePaul.

Numerous student organizations offer considerable opportunities for participation in both community and University activities. There are music performance groups, theater groups, student publications, sports, and honor and service societies. Athletic facilities include two gymnasiums, a swimming pool, racquetball courts, and extensive physical education equipment.

Applying

Master's degree students may begin their course work in any academic quarter; Ph.D. degree students are admitted twice a year. Application materials for the graduate programs in computer science, distributed systems, information systems, software engineering, telecommunication systems, or management information systems may be obtained by sending a request to the address given below. The application fee is $25.

The CTI Web site (listed below) offers information and resources for current and prospective students. Students can order admission applications, view the class schedule, and visit faculty member and student home pages.

Correspondence and Information

School of Computer Science, Telecommunications and Information Systems
Graduate Programs
DePaul University
243 South Wabash
Chicago, Illinois 60604-2302
World Wide Web: http://www.cs.depaul.edu

DePaul University

THE FACULTY AND THEIR RESEARCH

L. Edward Allemand, Professor Emeritus; Ph.D., Louvain (Belgium), 1970. Information systems, human-computer interaction.

Ehab S. Al-Shaer, Assistant Professor; Ph.D., Old Dominion, 1998. Management of distributed systems, multimedia networks, multicast protocols.

Gary Andrus, Associate Professor; Ph.D., Wayne State, 1977. Formal language theory, compiler design.

Karen Bernstein, Assistant Professor; Ph.D., SUNY at Stony Brook, 1996. Programming environments, programming languages, software engineering and concurrent systems.

André Berthiaume, Assistant Professor; Ph.D., Montreal, 1995. Quantum computation and quantum information processing.

Gregory Brewster, Assistant Professor; Ph.D., Wisconsin–Madison, 1994. Telecommunication systems, data communications, computer networks, performance analysis of communication systems.

Susy Chan, Associate Professor; Ph.D., Syracuse, 1979. IT management, planning, and strategies; systems analysis and design; electronic commerce.

I-Ping Chu, Associate Professor; Ph.D., SUNY at Stony Brook, 1981. Data communication, computer networks, combinatorial algorithms, database, distributed database.

Anthony Wai Man Chung, Associate Professor; Ph.D., Maryland, 1992. Communication networks, distributed systems, automated tools for software development, operating systems, programming languages.

Kamal Dahbur, Instructor; M.S., DePaul, 1993. Artificial intelligence, information systems.

Charles Earl, Visiting Assistant Professor; Ph.D., Chicago, 1998. Transformational planning and scheduling, mobile autonomous agents and machine learning for autonomous software agents, artificial intelligence, computational modeling.

Clark Elliott, Associate Professor; Ph.D., Northwestern, 1992. Artificial intelligence.

Helmut Epp, Associate Professor and Dean; Ph.D., Northwestern, 1966. Expert systems, artificial intelligence, computer security, hardware description languages.

Xiaowen Fang, Assistant Professor; Ph.D., Purdue, 1999. Human-computer interaction, information systems, Web applications.

Robert Fisher, Associate Professor; Ph.D., Harvard, 1975. Graphics, operating systems.

Jacob D. Furst, Assistant Professor; Ph.D., North Carolina, 1999. Image processing.

Gerald Gordon, Associate Professor; Ph.D., Berkeley, 1968. Computer vision.

Henry Harr, Associate Professor; Ph.D., IIT, 1988. Parallel processing, operating systems.

Alan Jeffrey, Associate Professor; Ph.D., Oxford, 1992. Semantics of programming languages.

Xiaoping Jia, Associate Professor; Ph.D., Northwestern, 1989. Software engineering, formal methods, object-oriented software development.

Richard Johnsonbaugh, Professor; Ph.D., Oregon, 1969. Combinatorial algorithms, pattern recognition.

Steve Jost, Associate Professor; Ph.D., Northwestern, 1985. Statistics, pattern recognition, image processing.

Martin Kalin, Professor; Ph.D., Northwestern, 1969. Programming languages, architecture, distributed systems.

George Knafl, Professor; Ph.D., Northwestern, 1978. Software quality and reliability, data mining, statistical computing.

Linda V. Knight, Assistant Professor; Ph.D., DePaul, 1998. Information systems, artificial intelligence.

Vladimir Kulyukin, Assistant Professor; Ph.D., Chicago, 1998. Distributed information retrieval, artificial intelligence.

Glenn Lancaster, Associate Professor; Ph.D., California, Irvine, 1972. Compiler design.

Chengwen Liu, Associate Professor and Assistant Dean; Ph.D., Illinois, 1991. Database management systems, information retrieval, data compression.

King-Lup Liu, Visiting Assistant Professor; Ph.D., Illinois at Chicago, 1999. Database systems, information retrieval, multimedia retrieval.

Steve Lytinen, Associate Professor; Ph.D., Yale, 1984. Artificial intelligence, natural language processing.

Will Marrero, Assistant Professor; Ph.D., Carnegie Mellon, 1999. Computer security, formal methods.

John McDonald, Assistant Professor; Ph.D., Northwestern, 1996. Computational geometry, mathematical modeling and simulation.

Craig S. Miller, Assistant Professor; Ph.D., Michigan, 1993. Human cognition, user modeling, machine learning.

David Miller, Associate Professor and Associate Dean; Ph.D., Chicago, 1981. Artificial intelligence, computation theory.

Daniel Mittleman, Assistant Professor; Ph.D., Arizona, 1995. Information systems, groupware, GSS, design of technology-supported physical environments.

Bamshad Mobasher, Assistant Professor; Ph.D., Iowa State, 1994. Data mining, autonomous software agents, knowledge-based systems.

Thomas Muscarello, Assistant Professor; Ph.D., Illinois, 1993. Hospital/medical informatics, police/fraud, information systems, software systems upgrades and maintenance, IT workforce issues.

Makoto Nakayama, Assistant Professor; Ph.D., UCLA, 1999. Strategic use of information systems, electronic commerce.

Corin Pitcher, Assistant Professor; Ph.D., Oxford, 1999. Semantics of programming languages, mobile code, security.

James W. Riley, Assistant Professor; Ph.D., North Carolina, 1999. Distributed computing, programming languages and computer design, software engineering, theoretical computer science.

John Rogers, Assistant Professor; Ph.D., Chicago, 1995. Computational complexity, mathematical logic, quantum computation.

LoriLee Sadler, Visiting Assistant Professor; Ph.D. Indiana, 1999. High-speed data networking, human-computer interaction, human cognition.

Marcus Schaefer, Assistant Professor; Ph.D., Chicago, 1999. Computational complexity theory, computability theory.

Eric J. Schwabe, Associate Professor; Ph.D., MIT, 1991. Algorithms, parallel computation.

Eric Sedgwick, Visiting Assistant Professor; Ph.D., Texas at Austin, 1997. Computational geometry and topology.

Amber Settle, Assistant Professor; Ph.D., Chicago, 1999. Distributed algorithms, cellular automata.

Paul A. Sisul, C.M., Instructor; M.Div., DeAndreis Institute of Theology, 1975. Language and applications.

Norma G. Sutcliffe, Visiting Assistant Professor; Ph.D., UCLA, 1997. Information systems, IT management and consulting.

Ian Sutherland, Assistant Professor; Ph.D., Northwestern, 1996. Fault tolerance, computer security, mathematical logic, semantics of real-number computation.

George K. Thiruvathukal, Assistant Professor; Ph.D., IIT, 1995. Concurrent, parallel, and distributed computing; object-oriented methods and programming; programming language design and implementation; software systems; computational science.

Noriko Tomuro, Instructor; Ph.D., DePaul, 1999. Artificial intelligence, natural language processing.

Gary B. Weinstein, Instructor; M.S., Ohio, 1982. Database, operating systems.

Curt M. White, Visiting Associate Professor; Ph.D., Wayne State, 1986. Computer science education research, data communications, computer networks, genetic algorithms.

Nathaniel Whitmal, Assistant Professor; Ph.D., Northwestern, 1997. Signal and image processing, digital communications.

Rosalee Wolfe, Associate Professor; Ph.D., Indiana, 1987. Computer graphics, human-computer interaction.

DREXEL UNIVERSITY

College of Arts and Sciences
Program in Computer Science

Programs of Study	Drexel University prepares students through its comprehensive graduate programs in mathematics and computer science leading to the Master of Science (M.S.) and Doctor of Philosophy (Ph.D.) degrees.
	In the M.S. program in computer science, areas of emphasis include artificial intelligence, computer graphics, compiler design, software engineering, parallel and distributed computing, computer algebra systems, and scientific computation. The program is intended to prepare students for employment as computing professionals in business, industry, or government. Studies are offered on a full- or part-time basis; full-time students normally complete the program within two years. A minimum of 45 quarter credits is required. The Ph.D. program emphasizes computer algebra systems and parallel and distributed computing. Details on this program can be obtained from the Department of Mathematics and Computer Science.
	Students with strong backgrounds in both mathematics and computer science are encouraged to pursue the dual M.S. in mathematics and computer science. Typically, this requires an additional one-half year of study beyond the time required for an M.S. degree in one field. The degrees are awarded simultaneously upon completion of the program.
Research Facilities	A department-wide local area network, MCSNET, is the cornerstone for the computing facilities in the Department of Mathematics and Computer Science. MCSNET directly supports the administrative, instructional, and research activities of the department and provides access to centralized resources operated by Drexel's Office of Computing Services; resources operated by other Drexel departments, making possible joint instructional and research efforts; remote resources via the Internet; and the Pennsylvania Education Network (PrepNET) via a T3 connection. The departmental research computers have a connection to the campus backbone at 100 megabits per second and are on the vBNS via a campus OCS ATM connection.
	Student-oriented departmental facilities include an E3000 computer server with an E150 file server, twenty-five lab machines (Suns, PC/NTs, and Macs), and a tutorial area with PC/NTs and two Macs. Research-specific facilities include various Sun workstations, an SGI workstation, NT and linus workstations and servers, and X-terminals. General faculty/departmental facilities include Sun MP630 SPARC-10 OSs (Solaris 2.x, Windows NT, Windows 95, MacOS, and Iris), Linus Languages (C, C++, Pascal, Fortran, Lisp, Java, and others), and databases (Oracle Packages, Nag Fortran library, Matlab, Maple, and Network Linda).
	The Laboratory for Geometric and Intelligent Computation includes a heterogeneous network of Intel/Windows NT-based CAD/CAE graphics workstations (five P-IIs 300 and 400 MHz) and Unix (Solaris SPARC, Solaris x86, and Linix) graphics workstations and servers (one Ultra 30, two Ultra 2s, and one P-II 266), as well as a Bridgeport VMC 600 vertical machining center (four-axis).
	The Drexel Hagerty Library, which has more than 480,000 volumes, has approximately 120,000 in the science and technology section.
Financial Aid	A significant number of teaching and research assistantships are available. A teaching assistant position includes a teaching stipend and a tuition stipend. For 1998–99, teaching stipends averaged $10,000 per academic year. The tuition stipend includes tuition remission of up to 27 credits for the academic year.
Cost of Study	Tuition varies with the program of study. For 1999–2000, tuition for computer science students is $585 per credit hour. The general University fee is $125 per term for full-time students and $67 per term for part-time students.
Living and Housing Costs	Accommodations for single students are available in University residence halls. Ample housing is also available in the neighborhood bordering campus. For the nine-month academic year, transportation and living expenses for a single student are estimated at $11,450.
Student Group	The University has a total enrollment of 9,590 students, including 2,785 at the graduate level. Approximately 100 graduate students are enrolled in the computer science program. Evening course offerings are sufficient to offer a robust degree program for part-time evening students.
	The Philadelphia/Delaware Valley area is part of a technological corridor with a wealth of companies that hire Drexel graduates as full-time computing professionals. In most cases, students get jobs involving some aspect of software development. The graduate student body is a diverse mixture of part-time students with full-time jobs in the computing industry, students with backgrounds other than computing science, students from other countries spanning the globe, and full-time students with interests in advanced graduate study. Job opportunities exist not only in the Delaware Valley but all over the country and the world.
Location	As a part of the University City area of west Philadelphia, Drexel is conveniently located within minutes of downtown Philadelphia, a great cultural, educational, and industrial center. From campus, New York City and Washington, D.C., are easily reached by train, bus, or car. Amtrak's 30th Street Station, a hub for national and local transportation, is located within three blocks of the University.
The University	Founded in 1891, Drexel University is a private institution offering undergraduate and graduate programs in arts and sciences, business and administration, design arts, engineering, and information studies. The University operates on an academic calendar of four terms per year.
Applying	Graduate students may apply with the intention of enrolling in any of Drexel's four terms (these begin in January, March, June, and September; application deadlines vary accordingly). Transcripts and letters of recommendation are required. The GRE General Test (aptitude portion) is recommended for the program in computer science. For assistantship consideration, students must submit their application by February 1.
Correspondence and Information	For further information and an application form, students should contact: Office of Graduate Admissions, Box P Drexel University Philadelphia, Pennsylvania 19104 Telephone: 215-895-6700 E-mail: admissions-grad@post.drexel.edu

Drexel University

THE FACULTY AND THEIR RESEARCH

Loren N. Argabright, Professor; Ph.D., Washington (Seattle). Functional analysis, wavelets, abstract harmonic analysis and the theory of group representations.

Robert P. Boyer, Professor and Associate Department Head; Ph.D., Pennsylvania. Functional analysis, C* algebras and the theory of group representations.

Robert C. Busby, Professor; Ph.D., Pennsylvania. Functional analysis, C* algebras and group representations, computer science.

Bruce W. Char, Professor; Ph.D., Berkeley. Symbolic mathematical computation, algorithms and systems for computer algebra, automatic scientific programming, parallel and distributed computation.

William M. Y. Goh, Associate Professor; Ph.D., Ohio State. Number theory, approximation theory and special functions, combinatorial enumeration, asymptotic analysis.

Herman Gollwitzer, Associate Professor; Ph.D., Minnesota. Applied mathematics, differential equations, data analysis, user interface design, visualization and scientific computing.

William J. Gordon, Professor; Ph.D., Brown. Numerical analysis, multivariate interpolation and approximation, numerical solution of partial differential equations, computer graphics.

Lloyd G. Greenwald, Assistant Professor; Ph.D., Brown. Time-critical planning and scheduling, robotics, resource-bounded reasoning, sequential decision making, reinforcement learning, medical informatics.

Nira Herrmann, Associate Professor and Department Head; Ph.D., Stanford. Mathematical and applied statistics, early decision problems, expert systems in statistics, computer science, computer science education, multivariate analysis, biostatistics.

Thomas T. Hewett, Professor; Ph.D., Illinois at Urbana-Champaign. Applied cognitive psychology, human-computer interaction, scientific problem-solving environments, networked engineering design, instructional computing.

Jeremy R. Johnson, Associate Professor; Ph.D., Ohio State. Computer algebra, parallel computation, algebraic algorithms, scientific computing.

Bernard Kolman, Professor; Ph.D., Pennsylvania. Lie algebras; theory, applications, and computational techniques; operations research.

Yagati N. Lakshman, Assistant Professor; Ph.D., RPI. Computational algebra, design and analysis of algorithms, symbolic computation systems.

Spiros Mancoridis, Assistant Professor; Ph.D., Toronto. Software engineering.

Charles J. Mode, Professor; Ph.D., California, Davis. Probability and statistics, biostatistics, epidemiology, mathematical demography, data analysis, computer-intensive methods.

Ljubomir Perkovic, Assistant Professor; Ph.D., Carnegie Mellon. Graph theory and algorithms, combinatorial optimization, probabilistic methods, theoretical computer science.

Ronald K. Perline, Associate Professor; Ph.D., Berkeley. Applied mathematics, numerical analysis, symbolic computation, differential geometry, mathematical physics.

Marci A. Perlstadt, Associate Professor; Ph.D., Berkeley. Applied mathematics, special functions, numerical analysis of function reconstruction, signal processing, combinatorics.

Jeffrey L. Popyack, Associate Professor; Ph.D., Virginia. Operations research, stochastic optimization, computational methods for Markov decision processes, artificial intelligence, computer science education.

William Regli, Assistant Professor; Ph.D., Maryland. Solid modeling, intelligent design and manufacturing.

Chris Rorres, Professor; Ph.D., NYU (Courant). Applied mathematics, scattering theory, mathematical modeling in biological sciences, dynamical systems.

Eric Schmutz, Associate Professor; Ph.D., Pennsylvania. Probability, algorithms, discrete mathematics.

Li Sheng, Assistant Professor; Ph.D., Rutgers. Discrete optimization, probabilistic methods in combinatorics, operations research, graph theory and its application in molecular biology, social sciences and communication networks, biostatistics.

Justin Smith, Professor; Ph.D., NYU (Courant). Computer science; parallel algorithms, artificial intelligence, and computer vision.

Chunguang Sun, Assistant Professor; Ph.D., Penn State. Parallel and distributed computing, graph algorithms, numerical linear algebra, design and implementation of programming languages.

Jet Wimp, Professor; Ph.D., Edinburgh. Applied mathematics; special functions, approximation theory, numerical techniques, and asymptotic analysis.

Stanley Zietz, Associate Professor; Ph.D., Berkeley. Population dynamics, applied mathematics, mathematical biology, biophysics, image analysis.

DREXEL UNIVERSITY

College of Information Science and Technology

Programs of Study

Drexel University's College of Information Science and Technology prepares practitioners and researchers for the information systems professions. The College's systems-related graduate degrees are Master of Science in Information Systems (M.S.I.S.), offered both on line and on campus, Master of Science in Software Engineering, and Ph.D. The Certificate of Advanced Study (C.A.S.) program enrolls professionals who already hold a master's degree in information systems or a related field.

The M.S.I.S. normally requires 60 credits and may be completed on a full- or part-time basis. The Drexel University calendar includes four terms per calendar year; most information science and technology courses carry 4 credits. The program includes required courses, distribution courses, and free electives as well as a strong focus on the design, implementation, and evaluation of software-intensive information systems. Subjects taught include systems analysis, artificial intelligence, knowledge base systems, cognition and information retrieval, language processing, and software engineering. Workshops and other special offerings allow students to pursue studies in various programming languages and research or professionally oriented topics. M.S.I.S. applicants are asked to demonstrate competencies in basic statistics, the use of basic software packages (word processors, spreadsheets, and database management systems), and basic computer programming in a third-generation programming language or object-oriented language. Students lacking one or more of these competencies may enroll in foundation courses that do not count toward the degree. The Ph.D. program comprises an approved plan of study, candidacy examinations, and a dissertation. One year of full-time residency is required; otherwise, doctoral students may pursue studies on either a full- or a part-time basis. The doctoral program offers two tracks of study, computer information systems and information and library science; one track must be chosen for program planning and examinations. The Ph.D. normally requires 60 credits beyond the master's, or 90 credits beyond the bachelor's if no applicable master's is held. The C.A.S. requires 32 credits and offers specialization beyond the master's. This program is regarded as continuing professional education.

Research Facilities

The College's Computing Resource Center supports students and faculty members with such features as microcomputer hardware and software, access to the University's mainframes, a networked computer training room, online and CD-ROM information resources, a collection of 60 periodicals, and audio and video equipment. The College is also home to the Alfred P. Sloan Center for Asynchronous Learning and Training, which conducts research and development for Internet-based distance learning.

The University's library holds extensive collections of materials for all major areas of library and information science, computer science, systems engineering, and information systems. Students also have access to libraries on the adjacent University of Pennsylvania campus as well as to other libraries and information centers in the Philadelphia area.

Financial Aid

A number of library, graduate, research, and teaching assistantships are awarded each year to incoming students, as are partial tuition scholarships. Assistants receive tuition remission and stipends in return for fulfilling specific work requirements in the College or the University's library. Teaching assistantships are available only for Ph.D. or advanced master's students. Enrolled students may also borrow limited funds through a College-administered loan program. Information on federal and state loan programs is available from Drexel's Graduate Financial Aid Office (Room 241, Randell Hall). No financial aid is available for international students.

Cost of Study

In 1999–2000, tuition is $452 per credit. Each term, students are also charged a general University fee, based on full-time ($125) or part-time ($67) status.

Living and Housing Costs

Ample housing is available in the neighborhood bordering campus.

Student Group

The students represent diverse academic and professional backgrounds and hold varied career expectations. In addition to the M.S.I.S and Ph.D. degrees, graduate students can pursue the M.S., Library and Information Science, and undergraduates can pursue the B.S. in Information Systems. About 75 percent of the 531 graduate students come from the mid-Atlantic region, with other regions and countries also represented.

The College of Information Science and Technology maintains its own placement office with a full-time director. The director helps students find preprofessional positions and internships and assists graduates in locating professional employment. The placement rate for graduates is high. Recent master's graduates have accepted systems positions in corporate, government, and academic settings.

The business, industry, and government resources of Philadelphia also provide ample opportunities for students to pursue preprofessional employment, internships, and permanent employment.

Location

The campus is located in the University City section of Philadelphia. As one of the nation's oldest and largest cities, Philadelphia is rich in cultural, historical, and academic institutions and is home to an extraordinary variety of general and special libraries.

The University

Drexel is a private institution with an approximate enrollment of 2,800 graduate and 6,800 undergraduate students. In addition to the information science and technology curricula, degree programs are offered in the arts and sciences, business and administration, design arts, and engineering. These varied programs feature a strong professional orientation.

With College approval, graduate students may include courses from other Drexel departments in their program of study; related curricula include computer science, management and other business specializations, neuropsychology, and technical and science communication.

Applying

Graduate students may apply for admission in any term. Those seeking assistantships and scholarships should apply by February 1 for admission in the following September. An application and fee, transcripts, letters of recommendation, and a personal statement are required. Scores for the GRE General Test are required for some master's and all Ph.D. applicants; scores for the GRE are required for all applicants seeking assistantships.

Correspondence and Information

Anne B. Tanner, Associate Dean
College of Information Science and Technology-P
Drexel University
Philadelphia, Pennsylvania 19104
Telephone: 215-895-2474
World Wide Web: http://www.cis.drexel.edu

Drexel University

THE FACULTY AND THEIR RESEARCH

Michael E. Atwood, Professor; Ph.D., Colorado. Human-computer interaction, computer-supported cooperative work, organizational memory.

Thomas A. Childers, Alice B. Kroeger Professor and Associate Dean; Ph.D., Rutgers. Management and evaluation of information organizations and services, foundations of information work, the quality of information services, effectiveness of information organizations.

M. Carl Drott, Associate Professor; Ph.D., Michigan. Computer programming for information processing, search strategy techniques for information retrieval and dissemination, use of systems analysis techniques for dealing with problems in large organizations such as libraries.

William Evanco, Associate Professor; Ph.D., Cornell. Software engineering and systems analysis, software and process measurement, software performance.

David E. Fenske, Isaac L. Auerbach Professor of Information Science and Dean of the College; Ph.D., Wisconsin–Madison. Digital libraries, knowledge management and information technologies.

Abby Goodrum, Assistant Professor; Ph.D., North Texas. Visual information retrieval, electronic publishing and digital libraries, social impact of information technology and new media.

Belver C. Griffith, Research Professor and Professor Emeritus; Ph.D., Connecticut. Research methods; design, planning, and evaluation of information services and products; functions of information in technical and scientific work; communications.

John B. Hall, Associate Professor; Ph.D., Florida State. Academic library service, library administration, organization of materials, technical processes, social aspects of information systems, academic library management information systems and their relation to management decision making, academic library services and use, collective bargaining, cooperation among libraries, application of simulation and gaming techniques to library education.

Lewis Hassell, Instructor; Ph.D., Drexel. Systems analysis techniques, database management systems, computer-supported cooperative work (CSCW), use of computers to support collaboration, applications of linguistics to CSCW, object-oriented analysis and design.

Gregory W. Hislop, Assistant Professor; Ph.D., Drexel. Software development and modification, software evaluation and reuse, and organizing and staffing information systems groups; systems management, software product development, and technology support for large organizations.

Maxwell Hughes, Research Professor; Ph.D., Cambridge. Management of information systems, generally in large organizations, specifically in the health care industry.

Sandra M. Hughes, Assistant Professor; Ph.D., North Carolina at Chapel Hill. Information resources and services for children and young adults, school library media programs, social impact of information technology, library services to the disadvantaged.

Lee Leitner, Instructor; Ph.D., Nova Southeastern. Networks and distributed computing, systems analysis, software engineering, programming languages.

Xia Lin, Assistant Professor; Ph.D., Maryland. Information seeking in digital environments, digital libraries, information visualization, information technologies, human-computer interactions, visual interface design.

Cynthia L. Lopata, Instructor; Ph.D., Drexel. Technology and organizational change processes, technology impact assessment, qualitative research methodologies.

Jacqueline C. Mancall, Professor; Ph.D., Drexel. Collection development, management, delivery of information services to children and adolescents, instructional role of the information specialist, design of library collections and services for user groups based on their communication behavior, application of survey methodology and statistical analysis to collection planning and management.

Katherine W. McCain, Professor; Ph.D., Drexel. Resources in science and technology, serial literature, abstracting and indexing, bibliometric studies of scholarly literatures, information transfer in the biomedical sciences.

Carol Hansen Montgomery, Professor; Ph.D., Drexel. Library automation, evaluation of information systems, user interface design, medical informatics.

Scott P. Overmyer, Associate Professor; Ph.D., George Mason. System and software requirements analysis, rapid prototyping, human-computer interaction.

Katherine M. Shelfer, Assistant Professor; Ph.D., Florida State. Business information resources, competitive intelligence, information services to organizations, and the design, marketing, and evaluation of information products and services.

Il-Yeol Song, Associate Professor; Ph.D., LSU. Database management systems and systems analysis and design, database modeling and design, object-oriented analysis and design, object-oriented database systems, client-server systems, data warehousing, digital libraries.

Marilyn Mantei Tremaine, Professor; Ph.D., USC. Cognitive modeling and design of user interfaces, management and incorporation of user interface activities in software development, design and evaluation of collaborative work systems.

June M. Verner, Professor; Ph.D., Massey (New Zealand). Software project management, software metrics, software process improvement, software development tools and techniques.

Howard D. White, Professor; Ph.D., Berkeley. Issues surrounding information work with well-defined clienteles concentrating on services that involve resources of the social sciences, improvement of statistics for use by library management, foundations of information work, expert systems in reference service, library collection evaluation, co-citation mapping of subject specialties, and American attitudes toward censorship.

Associate Professor John Hall chats with prospective students during the College's annual Open House.

Students at work in the College's Computing Resource Center.

DUKE UNIVERSITY

Department of Computer Science

Program of Study	The Department of Computer Science has many exciting research efforts in the areas of systems and architecture, algorithms and complexity, scientific computing, and artificial intelligence. Scalable parallel computing is a unifying theme in much of the department's research. The department offers instruction leading to the M.S. and Ph.D. degrees. There are strong ties and collaborations with researchers in chemistry, engineering, mathematics, physics, and medicine through faculty members who have joint appointments in these areas. The Ph.D. program emphasizes early research experience in the first two years. The M.S. degree program is designed to allow completion in twelve months and provides practical training with a strong theoretical base. The department especially encourages M.S. candidates in the areas of operating systems, parallel computation, scientific computing, and VLSI design.
Research Facilities	The Department of Computer Science maintains state-of-the-art computing facilities within the Levine Science Research Center to satisfy a variety of research and teaching needs, in part due to a $1-million NSF academic instrumentation award. Duke's networking and next-generation collaborative computing and research environments use state-of-the-art Myrinet gigabit networking to connect many of the high-speed Alpha servers. An SGI PowerChallenge L 4-processor machine provides a powerful computational research platform, while an SGI Octane dual processor supports graphics research. A $1.6-million equipment grant from Intel provides Pentium II PCs for systems and architecture research and education. Students and faculty members have ready access to a 4-processor CRAY T3E at the nearby North Carolina Supercomputing Center.
Financial Aid	Full-time Ph.D. students are supported by fellowships, teaching assistantships, or research assistantships during their first two years. By the third year, students are expected to work with an adviser as a research assistant. Financial support provides tuition, most fees, and a stipend of $13,500 for the academic year. Additional support for the summer can be provided by research assistantships or internships. Financial support is typically not available for M.S. candidates. No separate application for financial aid is required when applying. U.S. applicants are encouraged to apply for NSF Graduate Fellowships and other external sources of funding.
Cost of Study	For 1999–2000, tuition, registration, and fees are $20,550 for a nine-month year of full-time study toward the M.S. or Ph.D. degree.
Living and Housing Costs	The cost of living in Durham for a single student is estimated at $10,500 per academic year. Comfortable and affordable student housing is available on or close to campus.
Student Group	All full-time students are provided with office space within the Department of Computer Science. In the last few years, about half of the graduate students have been from other countries, and 30 percent have been women. The department is quite successful in placing graduates in strong positions in industry, consulting, and academics.
Location	Duke is located on a beautiful 1,635-acre campus in Durham, North Carolina, a city of approximately 137,000 inhabitants. Durham sits at the apex of North Carolina's famous Research Triangle, which includes the nearby University of North Carolina at Chapel Hill and North Carolina State University in Raleigh. The adjacent Research Triangle Park, situated in 6,750 acres of rolling woodland, is home to many sophisticated research and manufacturing facilities. Duke is centrally located for a variety of cultural activities, and the mild climate makes the area a sports paradise. There are also outstanding recreational opportunities in the Outer Banks (eastern coast of the state) and in the Blue Ridge mountain range (western part of the state).
The University	Duke University was founded by James Buchanan Duke in 1924 and is now recognized as one of the top educational and research schools in the country. The University currently enrolls more than 9,000 students annually.
Applying	Applicants should have a strong background in mathematics, preferably three semesters of calculus and one semester of linear algebra. They should also have some knowledge of data structures, assembly language, and a higher-level computer programming language, especially C++. Most students accepted into the Ph.D. program have a grade point average of 3.5 or higher (on a 4.0 scale) and GRE scores in the 90th percentile or above. The GRE Subject Test in computer science is recommended but not required. International students must score well in the TOEFL test. Excellent references and transcripts, prior research experience, and a good essay explaining the applicant's research interests and plan of study play important roles in the admission process.

Correspondence and Information

For an application:
Graduate School Admissions Office
127 Allen Building
Duke University
Durham, North Carolina 27706
Telephone: 919-684-3913
E-mail: grad-admissions@acpub.duke.edu
World Wide Web: http://www.gradschool.duke.
    edu/admissions

For department information:
Director of Graduate Studies
Department of Computer Science
Duke University
Box 90129
Durham, North Carolina 27708
Telephone: 919-660-6538
E-mail: dgs@cs.duke.edu
World Wide Web: http://www.cs.duke.edu

Duke University

THE FACULTY AND THEIR RESEARCH

Full-Time Faculty

Pankaj K. Agarwal, Professor; Ph.D., NYU (Courant), 1989. Analysis of algorithms, computational and combinatorial geometry, robotics, data structures. (E-mail: pankaj@cs.duke.edu; telephone: 919-660-6540)

Lars Arge, Assistant Professor; Ph.D., Aarhus (Denmark), 1996. Design and analysis of algorithms and data structures, I/O efficient computation, computational geometry. (E-mail: large@cs.duke.edu; telephone: 919-660-6557)

Owen L. Astrachan, Associate Professor of the Practice and Co-Director of Undergraduate Studies for Teaching and Learning; Ph.D., Duke, 1992. Automated theorem proving, parallel and distributed computing, computer science education. (E-mail: ola@cs.duke.edu; telephone: 919-660-6522)

Alan W. Biermann, Professor; Ph.D., Berkeley, 1968. Artificial intelligence, automatic program synthesis, learning and inference theory, natural language processing. (E-mail: awb@cs.duke.edu; telephone: 919-660-6539)

Jeffrey S. Chase, Assistant Professor; Ph.D., Washington (Seattle), 1995. Operating systems, distributed systems, storage systems, parallel programming. (E-mail: chase@cs.duke.edu; telephone: 919-660-6559)

Robert Duvall, Lecturer; M.S., Brown, 1997. Computer science education. (E-mail: rcd@cs.duke.edu; telephone: 919-660-6567)

Herbert Edelsbrunner, Professor; Ph.D., Graz Technical (Austria), 1982. Theoretical computing, algorithms, data structures, combinatorics, discrete geometry, computational geometry. (E-mail: edels@cs.duke.edu; telephone: 919-660-6509)

Carla S. Ellis, Professor; Ph.D., Washington (Seattle), 1979. Operating systems, parallel systems, distributed data structures. (E-mail: carla@cs.duke.edu; telephone: 919-660-6523)

Gershon Kedem, Associate Professor of Computer Science and of Electrical and Computer Engineering; Ph.D., Wisconsin–Madison, 1978. Parallel architecture and VLSI design algorithms. (E-mail: kedem@cs.duke.edu; telephone: 919-660-6555)

Alvin R. Lebeck, Assistant Professor of Computer Science and of Electrical and Computer Engineering; Ph.D., Wisconsin–Madison, 1995. Computer architecture, memory systems, parallel and distributed systems, performance analysis. (E-mail: alvy@cs.duke.edu; telephone: 919-660-6551)

Michael L. Littman, Assistant Professor; Ph.D., Brown, 1996. Artificial intelligence, machine learning, planning under uncertainty, statistical natural language processing, algorithms and complexity, user interfaces. (E-mail: mlittman@cs.duke.edu; telephone: 919-660-6537)

Donald W. Loveland, Professor; Ph.D., NYU, 1964. Automated theorem proving, logic programming, test and treatment problem, knowledge evaluation. (E-mail: dwl@cs.duke.edu; telephone: 919-660-6542)

Richard A. Lucic, Associate Professor of the Practice, Associate Chair, and Director of External Relations; M.S., Stanford, 1971. Tools for computer science education, management of research, leadership training, industrial relations, technology management and transfer. (E-mail: lucic@cs.duke.edu; telephone: 919-660-6524)

Dietolf Ramm, Associate Professor of the Practice and Co-Director of Undergraduate Studies; Ph.D., Duke, 1969. Communications, applications of personal computers to education. (E-mail: dr@cs.duke.edu; telephone: 919-660-6532)

John H. Reif, Professor; Ph.D., Harvard, 1977. Theoretical computer science, efficient algorithms, parallel computation, robotics. (E-mail: reif@cs.duke.edu; telephone: 919-660-6568)

Susan H. Rodger, Associate Professor of the Practice and Co-Director of Undergraduate Studies for Teaching and Learning; Ph.D., Purdue, 1989. Interactive and visual tools for theoretical computer science, computer science education, algorithm animation, analysis of algorithms, parallel algorithms, data structures, computational geometry. (E-mail: rodger@cs.duke.edu; telephone: 919-660-6595)

Donald J. Rose, Professor; Ph.D., Harvard, 1970. Numerical solution of PDEs, numerical algebra, numerical methods for semiconductor device and circuit simulation. (E-mail: djr@cs.duke.edu; telephone: 919-660-6510)

Xiaobai Sun, Assistant Professor; Ph.D., Maryland College Park, 1991. Successive band reduction and banded approaches to eigenvalue problems, block householder transformations, numerical libraries for high-performance architectures, numerical methods for the solutions of Markov chains. (E-mail: xiaobai@cs.duke.edu; telephone: 919-660-6518)

Amin M. Vahdat, Assistant Professor; Ph.D., Berkeley, 1998. Operating systems, wide-area distributed systems, networks, architecture. (E-mail: vahdat@cs.duke.edu; telephone: 919-660-6566)

Jeffrey S. Vitter, Gilbert, Louis, and Edward Lehrman Professor and Chair; Ph.D., Stanford, 1980. Design and analysis of algorithms, large-scale computation (including I/O efficiency, parallel computation, incremental algorithms), computational geometry, data compression, machine learning, order statistics. (E-mail: jsv@cs.duke.edu; telephone: 919-660-6548)

Robert A. Wagner, Associate Professor and Director of Graduate Studies; Ph.D., Carnegie Mellon, 1969. Experimental VLSI architectures, applications of dynamic programming to algorithm design and systems design, design of optimal software and hardware systems, time-cost tradeoffs in abstract parallel computer models. (E-mail: raw@cs.duke.edu; telephone: 919-660-6536)

Adjunct, Associate, and Emeritus Faculty

Robert P. Behringer, James B. Duke Professor of Physics, of Mechanical Engineering and Material Science, and of Computer Science; Ph.D., Duke. Experiment and computation in nonlinear dynamics, Rayleigh-Bénard convection, flow in porous media, dynamics of granular materials, low-temperature physics.

John A. Board Jr., Anne T. and Robert M. Bass Associate Professor of Electrical and Computer Engineering and of Computer Science; D.Phil., Oxford. Application of high-performance computing technology to large scientific and industrial computing problems.

Siddhartha Chatterjee, Assistant Professor, North Carolina at Chapel Hill and Adjunct Assistant Professor of Computer Science; Ph.D., Carnegie Mellon, 1991. Programming systems, mathematical modeling and high-performance algorithms.

William W. Coughran, Adjunct Professor of Computer Science and Head, Scientific Computing Research, AT&T Bell; Ph.D., Stanford. Differential equations, distributed computing, and scientific programming environments.

Thomas M. Gallie, Professor Emeritus; Ph.D., Rice. Complex variable theory, numerical analysis, compiler design, computer graphics, real-time computing, computing services for education and cardiology.

Henry S. Greenside, Associate Professor of Physics and of Computer Science; Ph.D., Princeton, 1981. Nonlinear dynamics, computational fluid dynamics, vector and parallel scientific computing.

Peter N. Marinos, Professor of Electrical and Computer Engineering and of Computer Science; Ph.D., North Carolina State. Design of digital systems, fault diagnosis, applied automata theory.

Tassos Markas, Adjunct Assistant Professor and Research Engineer at Research Triangle Institute; Ph.D, Duke, 1993. Multimedia, data compression, ASIC and system design. (E-mail: am@cs.duke.edu; telephone: 919-462-6541)

Thomas Narten, Adjunct Assistant Professor and Research Staff Member at IBM-RTP; Ph.D., Purdue, 1988. Operating systems, networking and software systems. (E-mail: narten@cs.duke.edu; telephone: 919-660-6500)

Richard G. Palmer, Professor of Physics, of Psychology-Experimental, and of Computer Science; Ph.D., Cambridge. Theory and modeling of complex systems including glasses, neural networks, genetic algorithms, and economic markets.

Merrell L. Patrick, Professor Emeritus; Ph.D., Carnegie Mellon (at National Science Foundation). Research administration, computer and information science and engineering.

Mark Alan Peot, Scholar in Residence; Ph.D., Stanford, 1998. Probability and Bayesain reasoning, decision analysis, AI expert systems, belief networks.

C. Franklin Starmer Jr., Professor Emeritus; Ph.D., North Carolina at Chapel Hill, 1968. Cellular communication, biological modeling, medical research databases.

Kishor S. Trivedi, Professor of Electrical and Computer Engineering and of Computer Science; Ph.D., Illinois. Computer architecture, performance evaluation, systems modeling and fault-tolerant computing.

Senol Utku, Professor of Civil and Environmental Engineering and of Computer Science; Ph.D., MIT. Finite element methods and parallel processing.

Mazin Yousif, Adjunct Assistant Professor and Performance Architect, IBM Corporation; Ph.D., Penn State, 1991. Computer architecture, multiprocessors, clusters, I/O subsytem, performance evaluation.

EMBRY-RIDDLE AERONAUTICAL UNIVERSITY

Office of Graduate Programs and Research
Department of Computing and Mathematics
Program of Software Engineering

Program of Study

Embry-Riddle Aeronautical University's Master of Software Engineering (M.S.E.) degree is designed to give recent college graduates, or college graduates who have had several years of professional life, an opportunity to enhance their careers and work on the cutting edge of modern software development. Students learn technical tools and techniques in combination with skills in communication, group interaction, management, and planning. The program provides graduates with in-depth understanding and ability in the areas of software process engineering, software project planning and management, software analysis and design, communications, and teamwork. In addition, the M.S.E. curriculum takes full notice of the Software Engineering Institute's Capability Maturity Model (CMM) by incorporating the key practices throughout the course work.

The curriculum is structured into two groups of courses: core (15 credits) and specified electives (12 credits). In addition, each student is required to complete a graduate research project (3 credits). Students may elect to take an M.S.E. elective in lieu of the graduate research project. Courses available as specified electives include metrics and statistical methods for software engineering, performance analysis of software systems, concurrent and distributed systems, software safety, and formal methods for software engineering.

Research Facilities

A range of laboratories is available to software engineering students. The Real-Time Software Project Laboratory is equipped with industry-strength, real-time operating systems and interfacing hardware for real-time applications. The Team Software Development Laboratory and the Personal Software Development Laboratory enable students to explore, in teams and individually, the latest software development methods. The Guidant Software Center Laboratory is a multifaceted facility where students gain experience developing and testing safety-critical software. All laboratories are equipped with top-of-the-line hardware and software.

Financial Aid

Embry-Riddle makes every effort, within the limitations of the financial resources available, to ensure that no qualified student is denied the opportunity to obtain an education because of inadequate funds. However, the primary responsibility for financing an education must be assumed by the student. A number of graduate assistantships, providing a stipend and a tuition waiver, are available on a competitive basis each year. Opportunities for graduate research and teaching assistantships are excellent, since most University functions employ M.S.E. students to assist them with their information technology needs. Other financial aid programs include Federal Stafford Student Loans, the Embry-Riddle Student Employment Program, Embry-Riddle short-term loans, and scholarship and fellowship programs. All graduate programs are approved for Veterans Administration education benefits.

Cost of Study

In 1999–2000, tuition costs are $490 per semester hour. Books and supplies cost approximately $300 per semester.

Living and Housing Costs

Some on-campus housing is available to graduate students. The cost for a standard double-occupancy room is $1400 per semester. Off-campus housing is reasonably priced. Single students who are sharing rental and utility expenses with someone can expect off-campus room and board yearly expenses of $4000. Married students can expect higher average yearly expenses.

Student Group

The graduate programs currently enroll 250 students on the Daytona Beach campus. The College of Career Education enrolls more than 3,000 students in graduate degree programs off campus. On the Daytona Beach campus, 40 percent are from other countries, 41 percent are women, and 42 percent are members of minority groups. More than 10 percent of the campus-based graduate students are employed full-time.

Student Outcomes

Employment opportunities are excellent for graduates of the M.S.E. degree program. In an era of information technology, software engineers are in high demand. Recent M.S.E. graduates were recruited by companies such as Motorola, Andersen Consulting, Lockheed Martin, Boeing, and Sikorsky.

Location

The Daytona Beach campus is adjacent to the Daytona Beach International Airport and is 10 minutes from the Daytona beaches. Within an hour's drive are Disney World and EPCOT, the Kennedy Space Center, Sea World, and St. Augustine.

The University

The University comprises the main campus at Daytona Beach; a western campus in Prescott, Arizona; and the College of Career Education. Within the field of aviation, Embry-Riddle Aeronautical University has built a reputation for the high quality of instruction in its programs since its founding in 1926.

Applying

The minimum desired undergraduate cumulative GPA is 2.5 out of a possible 4.0, with a 3.0 in the senior year. Applications from U.S. citizens and permanent residents should be received at least thirty days prior to the first day of the term in which they plan to enroll. International students should submit all of their documents at least ninety days prior to the first day of the term in which they plan to enroll.

Correspondence and Information

Graduate Admissions
Embry-Riddle Aeronautical University
600 South Clyde Morris Boulevard
Daytona Beach, Florida 32114-3900
Telephone: 904-226-6115
          800-388-3728 (toll-free)
Fax: 904-226-7050
E-mail: admit@db.erau.edu
World Wide Web: http://erau.db.erau.edu/curriculum/mseprog.html

Embry-Riddle Aeronautical University

THE FACULTY AND THEIR AREAS OF SPECIALIZATION

The following are faculty members at the Daytona Beach Campus.

Thomas Hilburn, Professor; Ph.D., Louisiana Tech. Software engineering, fuzzy clustering, Ada programming.

Iraj Hirmanpour, Professor; Ed.D., Florida Atlantic. Software engineering, information modeling.

Soheil Khajenoori, Professor; Ph.D., Central Florida. Software development methodologies, CASE, software metrics and software process engineering.

Andrew Kornecki, Professor; Ph.D., University of Mining and Metallurgy, Krakow (Poland). Computer simulation, object-oriented programming, real-time systems, microprocessor applications, AI, applications, air traffic control automation.

Rodney O. Rogers, Associate Professor; Ph.D., Central Florida. Computer graphics, design and analysis of algorithms, parallel computation, theory of computation, programming languages.

Massood Towhidnejad, Associate Professor; Ph.D., Central Florida. AI, intelligent CBT systems, multimedia, computer networking and architecture.

Andres G. Zellweger, Professor; Ph.D., Harvard. System safety.

Florida Tech

FLORIDA INSTITUTE OF TECHNOLOGY

College of Engineering
Computer Science Program

Programs of Study

The Computer Science Program offers programs of graduate study leading to the degrees of Master of Science in Computer Information Systems, Master of Science in Computer Science, and Doctor of Philosophy. Major areas of study include software engineering, artificial intelligence/expert systems, computer graphics, programming languages, and database systems.

The master's degree in computer information systems is for students who do not have an undergraduate degree in computer science but who wish to obtain advanced training in this field. The course work required for this degree provides a broad background in the major areas of computer science. All students must pass a final program examination during their last semester.

The master's degree in computer science offers the student the opportunity to pursue advanced studies in various areas of computer science. The program is designed for students with baccalaureate degrees in computer science and provides a solid preparation for those who may pursue a doctorate. All students must complete and defend a thesis or pass a final program examination during their last semester.

The doctoral program is designed to provide research in the disciplines of computer science. The program requires understanding the fundamentals of computer science, mastery of a specialized subject, and the creativity to produce a dissertation based on original research.

Research Facilities

The Computer Science Program occupies approximately 6,000 square feet of laboratory and office space. Computer laboratories support research programs in artificial intelligence, graphics, software engineering, and systems development. Computer resources include Silicon Graphics and Sun workstations, X-terminals, and PC networks.

The Academic and Research Computing Services (ARCS) provides graduate students with a wide range of computing resources for course work and research. These resources include a Sun Enterprise 3000 and several Sun SPARC Workstations. These machines are connected internally as part of the campus network and externally to the Internet. Many programs and departments have their own computing resources, which are also connected to the campus network. Access to these computing resources is available in computer labs and academic departments and through dial-up lines. The programming languages that are supported include C, Pascal, ADA, FORTRAN, and C++. A staff of professionals is available to assist users with consultation and documentation. In addition to these resources, ARCS maintains a large microcomputer center located in the Library Pavilion.

Financial Aid

Graduate teaching and research assistantships are available to qualified students. For 1999–2000, stipends range from $10,830 to $13,680 for twelve months. All assistantships include tuition remission. Computer-based information on scholarships, loan funds, and other student assistance may be obtained from the Financial Aid Office. A limited number of assistantships providing tuition remission only or stipend only are also available.

Cost of Study

In 1999–2000, tuition is $575 per semester credit hour for all graduate students. As noted above, however, tuition is remitted for some graduate assistants.

Living and Housing Costs

Room and board on campus cost approximately $2400 per semester. On-campus housing (dormitories and apartments) is available for full-time single and married graduate students, but priority for dormitory rooms is given to undergraduate students. Many apartment complexes and rental houses are available near the campus.

Student Group

The program currently has an enrollment of 200 graduate students from colleges throughout the United States. Approximately 19 percent of the graduate students are women, and 30 percent are international students.

Student Outcomes

Graduates of the College of Engineering have found employment with such firms as IBM, Microsoft, Texas Instruments, NASA, Harris Corp., AT&T, General Electric, Keane Inc., Matrox, Northrop Grumman, Lockheed Martin, McDonnell Douglas, DBA Systems, Rockwell International, Advanced Micro Devices, USF&G, United Technologies, Honeywell, Computer Sciences Raytheon, ITT Aerospace, U.S. Patent Office, CIA, KPMG Audit, Los Alamos National Laboratory, Hewlett-Packard, Intel, Naval Air Systems Command, Naval Undersea Warfare Center, Macintosh Software Development, and Computer Task Group.

Location

Florida Tech's main campus is located in Melbourne, a residential community on Florida's Space Coast. Melbourne is the key city in south Brevard County, which also encompasses nine other smaller communities on the mainland and beachside. The Kennedy Space Center and Disney World are within a 90-minute drive of the Institute. The area's economy is a well-balanced mix of electronics, aviation, light manufacturing, opticals, communications, agriculture, and tourism.

The Institute and The Program

Florida Tech was founded in 1958 and has developed rapidly into a university that provides both undergraduate and graduate education in the sciences and engineering for selected students from throughout the United States and many countries. Current enrollment on the Melbourne campus is about 4,000. In addition to computer science, Florida Tech offers graduate programs in aerospace engineering, airport development management, applied mathematics, aquaculture, aviation science, biotechnology, business administration, cell and molecular biology, chemical engineering, chemistry, civil engineering, computer information systems, computer education, ecology, electrical engineering, environmental engineering, environmental management, environmental resource management, environmental science, industrial/organizational psychology, managerial communication, marine biology, mathematics education, mechanical engineering, ocean engineering, oceanography, operations research, physics, science education, space sciences, systems engineering, and technical and professional communication.

Applying

Further information and application forms for admission may be obtained from the Graduate Admissions Office. Students are encouraged to take the GRE General Test and Subject Test in computer science and submit scores for program consideration. Separate application for financial aid must be made on forms available from the Graduate School and must be submitted to the department by March 1.

Correspondence and Information

Graduate Admissions Office
Florida Institute of Technology
150 West University Boulevard
Melbourne, Florida 32901-6975
Telephone: 407-674-8027
Fax: 407-723-9468
World Wide Web: http://www.fit.edu

Dr. W. D. Shoaff, Chair
Computer Science Program
Florida Institute of Technology
150 West University Boulevard
Melbourne, Florida 32901-6975
Telephone: 407-674-8060
E-mail: wds@cs.fit.edu

Florida Institute of Technology

THE FACULTY AND THEIR RESEARCH

Shirley Ann Becker, Associate Professor; Ph.D., Maryland, 1990. Software engineering, statistical testing of software, software reliability, software management. (E-mail: sab@cs.fit.edu)

Philip Bernhard, Associate Professor; Ph.D., SUNY at Albany, 1988. Databases, database performance tuning and optimization, software engineering, object-oriented analysis and design. (E-mail: pbernhar@cs.fit.edu)

Frederick B. Buoni, Professor and Associate Dean; Ph.D., Ohio State, 1971. Decision analysis, artificial intelligence, systems simulations modeling, reliability and risk assessment, operations research applications, engineering quality control, engineering reliability. Introducing "quality" into an undergraduate computer science curriculum. *Fifty-Fourth Annual Meeting of the Florida Academy of Sciences.* Melbourne, Florida, March 23, 1990 (with Hadjilogiou and Newman). Comparison of performance for fuzzy expert system shell implementations in Pascal and in ADA. *Proceedings of AIDA-89, Fifth Annual Conference on Artificial Intelligence and Ada.* George Mason University, November 16–17, 1989 (with Schneider and Cornett). (E-mail: fbb@cs.fit.edu)

Philip K. Chan, Assistant Professor; Ph.D., Columbia, 1996. Scalable and adaptive systems, machine learning, data mining, parallel and distributed computing. Systems for knowledge discovery in databases. *IEEE Trans. Knowledge Data Eng.*, 1993 (with Matheus and Piatetsky-Shapiro). On the accuracy of meta-learning for scalable data mining. *J. Intelligent Information Syst.*, 1997 (with Stolfo). (E-mail: pkc@cs.fit.edu)

David W. Clay, Assistant Professor; M.S., Florida Tech, 1984. Advanced data structures, object-oriented design, concurrency in operating systems, parallel processing systems, computer science education. Internet: Features and resources—an unexplored infoscape. *Proceedings of the Third National Conference on College Teaching and Learning.* Jacksonville, Florida, 1992. *Information Structures—Implementing Imagination.* St. Paul, Minn.: West Publishing, 1985. (E-mail: dclay@cs.fit.edu)

Lina Khatib, Assistant Professor; Ph.D., Florida Tech, 1994. Artificial intelligence, temporal reasoning, constraint satisfaction, analysis of algorithms. *Path Consistency in a Network of Non-Convex Intervals,* IJCAI, 1993.

Robert A. Morris, Professor; Ph.D., Indiana, 1984. Temporal reasoning theory and applications, distributed artificial intelligence, model-based diagnostic reasoning, knowledge representation. An interval-based temporal relational calculus for events with gaps. *J. Exp. Theoret. Artific. Intell.* 3:87–107, 1991 (with Al-Khatib). Distributed intelligent monitoring and control for space applications. *Proceedings of the Fourth International Conference on Industrial/Engineering Applications of Artificial Intelligence and Expert Systems (AEI/AIE-91).* Kauai, Hawaii, June 2–5, 1991 (with Gonzalez and Raval). (E-mail: morris@cs.fit.edu)

J. Richard Newman, Professor and Associate Dean; Ph.D., Southwestern Louisiana, 1976. Software engineering, information systems management, CASE tools for clean room software engineering, legal issues, program specification tools. An undergraduate curriculum in software engineering. *Proceedings of the Fourth Annual Conference on Software Engineering Education, SEI.* April 1990 (with Mills and Engle). Performance issues for an expert system written in Ada. *Fifty-Fourth Annual Meeting of the Florida Academy of Sciences.* Melbourne, Florida, March 23, 1990 (with Buoni and Baggs). (E-mail: newman@cs.fit.edu)

William D. Shoaff, Associate Professor; Ph.D., Southern Illinois, 1981. Mathematical programming, parallel algorithms, parallel processing, supercomputers, computer modeling in genetics, computer graphics. The recognition of imperfect strings generated by fuzzy context sensitive grammars. *Int. J. Fuzzy Sets Syst.* 62:21–29, 1994 (with Inui, Fausett, and Schneider). Supercomputer simulation of chromosomes. In *Proceedings of the First International Conference on Electrophoresis, Supercomputing and the Human Genome,* eds. H. A. Lim and C. R. Cantor. Teaneck, N.J.: Supercomputer Computations Research Institute and Florida State University, 1990 (with Newman, Nuttall, and Hozier). A parallel algorithm for the singular value decomposition of rectangular matrices. In *Parallel Processing for Scientific Computing,* ed. Garry Rodrique. SIAM, 1989. (E-mail: wds@cs.fit.edu)

Ryan Stansifer, Associate Professor; Ph.D., Cornell, 1985. Programming languages, compilers, information systems, internationalization. *The Study of Programming Languages,* Prentice Hall, 1994; *M. L. Primer,* Prentice Hall, 1992; *The Foundation of Program Verification,* Wiley, 1987. (E-mail: ryan@cs.fit.edu)

James Whittaker, Assistant Professor; Ph.D., Tennessee, 1992. Statistical testing of software, software reliability engineering, software metrics, clean room software engineering. *Clean Room Systems Engineering Practices,* IDEA Publishing, 1996. (E-mail: jw@cs.fit.edu)

FLORIDA INTERNATIONAL UNIVERSITY

School of Computer Science

Programs of Study

The School offers programs leading to Master of Science (M.S.) and Doctor of Philosophy (Ph.D.) degrees in computer science.

The master's degree in computer science is a terminal professional degree; the degree program offers the student course work in the most current concepts and theory of computer science. The program consists of 30 semester hours of course work, including a thesis. The intent of this degree program is to prepare students to assume professional leadership positions in industry, government, and education.

The doctoral program in computer science consists initially of nine required graduate courses, followed by a qualifying examination. The student then participates in a number of advanced seminars that culminate in the Ph.D. candidacy examination. Upon being admitted to candidacy, the student prepares and presents a dissertation for review by a committee, after which the degree may be awarded.

Research Facilities

The library has more than 8,000 volumes in the mathematical and computer sciences and receives more than 125 periodicals.

The School's research is performed on an integrated ATM/Ethernet network of Sun and SGI enterprise servers, workstations, and personal computers. Researchers typically collaborate on distributed databases, parallel processing, and software engineering interest with NASA, DOD, NSF, and other government and commercial settings.

Financial Aid

There are several Presidential Doctoral Fellowships available for exceptional students, carrying an initial stipend of $24,750 per year (plus tuition and fee waiver) and eventually rising to $29,354. Regular research assistantships are also awarded to excellent students. These carry an initial stipend of $15,350 (plus tuition and fee waiver), eventually rising to $18,400.

Cost of Study

In 1998–99, tuition for in-state graduate students was $138.08 per semester hour for regular course work, dissertation, or thesis credit. In addition, there was a $46 assessment each semester for the use of the campus health center. Tuition for out-of-state graduate students was $481.64 per semester hour for regular course work, dissertation, or thesis credit. (These figures are subject to change without notice in 1999–2000.)

Living and Housing Costs

Recently built, privately owned off-campus apartments were available for approximately $1000 to $2000 per semester in 1998–99. (These figures are subject to change without notice in 1999–2000.)

Student Group

There are 30,000 students on campus, with more than 120 graduate students in computer science. Graduate students come from many areas of the United States and from a number of other countries.

Location

Florida International University is located at the gateway to the Everglades in suburban Dade County, 10 miles from downtown Miami. There are ample recreational activities available, and the mild weather allows these activities to be enjoyed the year round.

The University

Florida International University, a public institution, is a member of the State University System of Florida. It is composed of a College of Arts and Sciences, a College of Engineering and Design, and Schools of Computer Science, Business Administration, Hospitality Management, Public Administration, Education, Health Sciences, and Nursing.

Applying

Graduate study may begin in any semester, but assistantships are usually awarded for the fall semester. The application deadline for assistantships is January 15. Applicants must submit up-to-date transcripts and current Graduate Record Examinations scores to the graduate program director as early as possible. The GRE General Test is required of all applicants, and the minimum acceptable score is a combined total score (verbal, quantitative, and analytical) of 1650; the scores must have been earned within the last five years. The department requires a grade average of at least B in all upper-division undergraduate class work and a background in mathematics that includes calculus and statistics.

Correspondence and Information

Graduate Program Coordinator
School of Computer Science
Florida International University
Miami, Florida 33199-2788

Telephone: 305-348-1038
World Wide Web: http://www.cs.fiu.edu.

Florida International University

THE FACULTY AND THEIR RESEARCH

Walid Akache, Instructor; M.S., Miami, 1984. Computer science.

Paul Attie, Assistant Professor; Ph.D., Texas, 1995. Temporal logic, distributed computing, verification of programs.

David Barton, Professor; Ph.D., Cambridge, 1966. Distributed systems and data communications.

Toby S. Berk, Professor; Ph.D., Purdue, 1972. Computer graphics and operating systems.

Shu-Ching Chen, Assistant Professor; Ph.D., Purdue, 1998. Electrical and computer engineering.

Yi Deng, Assistant Professor; Ph.D., Pittsburgh, 1992. Software engineering and knowledge engineering, distributed systems, multimedia systems.

Timothy Downey, Instructor; M.S., SUNY at Albany, 1986. Computer science.

Raimund Ege, Associate Professor and Graduate Program Director; Ph.D., Oregon Graduate Center, 1987. Object-oriented programming.

Michael Evangelist, Professor and Director of the School; Ph.D., Northwestern, 1978. Distributed computing and software engineering.

Mbola Fanomezantsoa, Instructor; M.S., SUNY Institute of Technology, 1994. Computer science.

William Feild, Instructor; M.S., Florida International, 1989. Computer science.

Xudong He, Assistant Professor; Ph.D., Virginia Tech, 1989. Computer science.

Bill Kraynek, Associate Professor and Associate Director; Ph.D., Carnegie Mellon, 1968. Programming languages and computer science education.

Masoud Milani, Associate Professor; Ph.D., Central Florida, 1986. Programming language environments.

Jainendra Navlakha, Professor; Ph.D., Case Western Reserve, 1977. Analysis of algorithms, program verification, software metrics.

Ana Pasztor, Professor; D.R.N., Darmstadt (Germany), 1979. Cognitive sciences and program verification.

Alexander Pelin, Associate Professor; Ph.D., Pennsylvania, 1977. Automated reasoning.

Norman Pestaina, Instructor; M.S., Penn State, 1979. Computer science.

Nagarajan Prabhakaran, Associate Professor; Ph.D., Queensland (Australia), 1985. Database systems, graphics.

Naphtali Rishe, Professor; Ph.D., Tel-Aviv, 1984. Database management and systems.

Joslyn Smith, Instructor; M.S., Canada, 1994. Computer science.

Wei Sun, Associate Professor; Ph.D., Illinois at Chicago, 1990. Database systems and knowledge-based systems.

Mark A. Weiss, Professor; Ph.D., Princeton, 1987. Data structures and algorithm analysis.

THE GEORGE WASHINGTON UNIVERSITY

School of Engineering and Applied Science
Department of Computer Science

Programs of Study

The department offers the M.S., Engineer (Engr.), and Applied Scientist (App.Sc.) professional degrees and the D.Sc. in computer science. Computer science concentrations include algorithms and theory, artificial intelligence and computer vision, computer and communications security, computer engineering and architecture, computer graphics and multimedia systems, parallel and distributed computing, and software engineering and systems. M.S. programs require a minimum of 24 semester hours of graduate-level courses and 6 semester hours of thesis or a minimum of 30 semester hours for the nonthesis option. Engr. and App.Sc. professional degree programs emphasize applied subject material. At least two years of professional experience beyond a master's degree are required for admission to these programs, which require at least 30 semester hours of approved post-master's courses. The D.Sc. program normally consists of one major and two minor areas of concentration, totaling a minimum of 30 course semester hours beyond the master's level or a minimum of 54 credit hours of approved graduate work for students whose highest earned degree is a baccalaureate, and 24 dissertation research semester hours. Each doctoral student is assigned an adviser whose research specialty is in the student's major area of study; together they determine the major and minor areas of study.

Research Facilities

The School provides UNIX, PC, and Macintosh resources, featuring a Sun Microsystems S1000 multiprocessor server and Ultra 3D workstations, Dell Pentium Pro II computers, and Apple Power Macintosh G3 computers. Systems are equipped with licensed software from leading manufacturers, such as AdaCore Technologies, Sun Microsystems, Autodesk, Asymetrix, Adobe, and Microsoft. The facility's labs support engineering design, artificial intelligence, multimedia, software development, Web technology, graphics, analysis, and project management and include high-quality laser and color printing and scanning equipment. In addition, the University provides extensive computing resources, with a Sun Microsystems SPARCcenter 2000E for academic information system services and a Sun Microsystems Enterprise 4000 with four processors dedicated to research computing support. A high-performance ATM network integrates these resources, and Internet connectivity is provided for research support. University libraries contain 1.7 million volumes. Their computer system provides online access to the shared catalog of all seven member libraries of the Washington Research Library Consortium, as well as to periodical and newspaper index databases. Gelman Library provides additional databases on CD-ROM, including several specific to engineering, and offers research consultation services. Students have ready access to the Library of Congress.

Financial Aid

Teaching assistantships provide remission of tuition for 9 hours per semester and a salary of $12,000 per academic year. Research assistants receive a salary of $8000 to $16,000 for the calendar year. School Graduate Fellow, Dean's Fellow, and Department Fellow awards range from $7500 to $15,000 for eligible full-time students. Full-time students who are U.S. citizens or permanent residents may be eligible for Graduate Engineering Honors Fellowships.

Cost of Study

For the 1998–99 academic year, tuition was charged at the rate of $680 per semester hour, payable on a course-by-course basis.

Living and Housing Costs

Apartments for students enrolled in the School of Engineering and Applied Science are available in the surrounding area at a wide range of costs. These costs start at about $600 a month.

Location

The Washington area has the second-largest concentration of research and development activity in the nation. Library facilities are unsurpassed in scope. The campus is in the Foggy Bottom historic district of Washington, D.C.

The University and The School

The School, organized in 1884, offers limited course work in engineering, engineering management and systems, physical science, mathematics, economics, and statistics during summer sessions.

Applying

Admission to the M.S. program requires an appropriate bachelor's degree from a recognized institution and evidence of a capacity for productive work in the field. Admission to professional degree programs requires an appropriate master's degree from a recognized institution and evidence of capacity for productive work in the field. Applicants for doctoral study must have adequate preparation for advanced study, including a bachelor's or a master's degree or the equivalent and a capacity for creative scholarship. GRE scores are required. March 1 and October 1 are the respective priority deadlines for applications for the fall and spring; applications are accepted on a space-available basis thereafter.

Correspondence and Information

Professor Bhagirath Narahari, Chair
Department of Computer Science
School of Engineering and Applied Science
The George Washington University
Washington, D.C. 20052
Telephone: 202-994-6083
Fax: 202-994-0227
E-mail: cs@seas.gwu.edu
World Wide Web: http://www.cs.seas.gwu.edu

Interim Dean Thomas A. Mazzuchi
School of Engineering and Applied Science
The George Washington University
Washington, D.C. 20052
Telephone: 202-994-3096
          800-537-7327(toll-free)
Fax: 202-994-4522

The George Washington University

THE FACULTY AND THEIR RESEARCH

Simon Y. Berkovich, Professor; Ph.D., Institute of Precise Mechanics and Computer Technology (Moscow), 1964. Information systems, data structures, associative memories and processors, computer organization.

Peter Bock, Professor; M.S., Purdue, 1964. Computer systems, artificial intelligence, robotics, simulation, programming languages, microprocessor systems.

Hyeong-Ah Choi, Professor; Ph.D., Northwestern, 1986. Graph theory, design and analysis of algorithms, computational complexity.

Michael B. Feldman, Professor; Ph.D., Pennsylvania, 1973. Computer science, programming languages, data structures, software engineering, concurrency and parallelism.

James K. Hahn, Associate Professor; Ph.D., Ohio State, 1989. Computer graphics, computer animation, display algorithms.

Rachelle S. Heller, Professor; Ph.D., Maryland, 1986. Computer literacy, computers in education, programming languages.

Lance J. Hoffman, Professor; Ph.D., Stanford, 1970. Computer science, security and privacy, viruses, social implications.

C. Dianne Martin, Professor; Ed.D., George Washington, 1987. Computer literacy, computers in education, computers and society, computer ethics.

W. Douglas Maurer, Professor; Ph.D., Berkeley, 1965. Computer science, correctness of programs, analysis of algorithms, semantics of programming languages.

Arnold C. Meltzer, Professor; D.Sc., George Washington, 1967. Computer architecture, design of computer systems, database systems, multiprocessor systems, information storage and retrieval.

Forest K. Musgrave, Assistant Professor; Ph.D., Yale, 1993. Computer graphics, modeling of natural phenomena.

Bhagirath Narahari, Associate Professor; Ph.D., Pennsylvania, 1987. Parallel processing, algorithms and interconnection networks, reconfigurable parallel computer architectures, special-purpose computing.

Shmuel Rotenstreich, Associate Professor; Ph.D., California, San Diego, 1983. Software engineering, operation systems.

John L. Sibert, Professor; Ph.D., Michigan, 1974. Computer graphics, human-computer interaction.

Abdou Youssef, Associate Professor; Ph.D., Princeton, 1988. Interconnection networks, parallel computer architecture, parallel algorithms, algorithms and data structures, theory of computing.

The George Washington University is an equal opportunity/affirmative action institution.

GEORGIA INSTITUTE OF TECHNOLOGY
A Unit of the University System of Georgia

College of Computing

Programs of Study

The College of Computing currently offers programs of study leading to the degrees of Doctor of Philosophy and Master of Science in computer science. The College also offers an interdisciplinary program leading to the degree of Doctor of Philosophy in algorithms, combinatorics, and optimization as well as an interdisciplinary master's program in human-computer interaction (HCI). In addition, the College awards an interdisciplinary Ph.D. certificate in cognitive science. The principal requirements for the Ph.D. are successful completion of certain comprehensive examinations and submission of a dissertation that describes original independent research. The degree of Doctor of Philosophy is awarded in recognition of high achievement in research and is intended for people who intend to pursue research careers. A student may earn the M.S. degree by completing one of the following: 30 semester hours of approved course work, 27 hours of approved course work and a master's project, or 24 hours of approved course work and a thesis. The master's degree is intended for people who have a variety of career objectives, including technical and managerial positions in industry and government. A wide variety of graduate courses and research seminars are offered, and several active colloquium series featuring outside speakers are held.

Research Facilities

The College maintains a variety of computer systems for general support of academic and research activities. These include more than fifty Sun, Silicon Graphics, and Intel systems used as file and computer servers (ten of which are quad-processor machines); and more than 600 workstation-class machines from Sun, Silicon Graphics, Intel, and Apple. A number of specialized facilities augment the College's general-purpose computing capabilities. The Graphics, Visualization, and Usability (GVU) Center houses a variety of graphics and multimedia equipment, including high-performance systems from Silicon Graphics, Sun, IBM, and Apple. The affiliated Multimedia, Computer Animation, Audio/Video Production, Usability/Human-Computer Interface, Virtual Reality/Environments, Biomedical Imaging, Educational Technology, and Future Computing Environments laboratories provide shared facilities that target specific research areas. A Scientific Visualization Laboratory with additional equipment from Silicon Graphics is jointly operated by the GVU Center and the Institute's Office of Information Technology (OIT). The High-Performance and Parallel Computation Experimentation Laboratory (HPPCEL), another joint operation between the College and OIT, serves as a focus for interdisciplinary research involving high-performance computer systems. Shared facilities, all linked by a dedicated network that utilizes OC12C ATM (622 Mbps) and Gigabit Ethernet (1000 Mbps), include two Silicon Graphics Origin 2000 systems, with twenty-eight and sixteen R10000 superscalar RISC processors, respectively; an 8-node IBM SP-2; an 8-processor IBM RS/6000 Model R50; a cluster of sixteen Intel Pentium quad processors that utilize Myrinet (full-duplex 1280 Mbps) and switched FastEthernet (100 Mbps) interconnects; a cluster of sixteen Sun UltraSPARC processors and five UltraSPARC dual processors that utilize Myrinet (sixteen systems), Dolphin (four systems), switched 100BaseT (all systems), and ATM interconnects (all systems); a cluster of forty-eight Intel Pentium dual-processors that utilize a switched FastEthernet interconnect; an SGI Origin 200 quad-processor video server; and a laboratory of Silicon Graphics, Sun, Intel, and IBM workstations. The Networking and Telecommunications group and the related Broadband Telecommunications Center (BTC) house the Hybrid Fiber/Coax, ATM, Protocols, Wireless Technologies, and Video Sources labs, which are equipped with leading-edge computing, communications, and test equipment. Other specialized laboratories support research in databases, open systems, software engineering, robotics, and intelligent systems. All of the College's facilities are linked via local area networks that can provide a choice of communications capabilities from 10 to 1,000 mbps in most locations, including offices, labs, and classrooms. The College's network employs a high-performance OC12C ATM and Gigabit Ethernet backbone with connectivity to the campus ATM network via an OC12C ATM link and a redundant OC3C ATM (155 Mbps) link for fail-over. The primary campus Internet connection is provided by a direct FDDI (100 mbps) link to the service provider's switching center, which is augmented by an OC3C ATM connection to the NSF vBNS (very high performance Backbone Network Service) research network. The Institute is also leading efforts to establish a southern regional gigabit network as part of Internet2. Additional computing facilities are provided to the Georgia Tech campus by OIT, including five public-access clusters of Apple, Dell, and Sun workstations; a Sun SPARCcenter 2000 with twelve superscalar RISC processors; and various mainframes.

Financial Aid

Most doctoral students and some master's students are supported throughout the year by assistantships. Research assistants participate in the College's ongoing research projects. The range of teaching assistants' duties includes grading, instructional laboratory support, and some teaching. Assistantships are usually awarded on a ⅜ time basis and begin at $1365 per month for 1999–2000, including a tuition waiver. A limited number of fellowships are available, as is an industry-college cooperative program. In addition, Georgia Tech President's Fellowships at $5500 per calendar year are available as supplements to normal assistantships or fellowship awards given to chosen students.

Cost of Study

Georgia Tech converts to a semester system in fall 1999. For 1999–2000, estimated fees are $1847 per semester for Georgia residents and $6173 per semester for nonresidents. Resident and nonresident students with at least ⅓-time research or teaching assistantships pay an estimated $430 per semester. Costs are subject to change.

Living and Housing Costs

The Housing Office supervises 6,267 single spaces and 300 married student apartments. In 1999–2000, room rent costs $1400 to $1951 per semester. Total expenses for a single student living on campus, excluding tuition and fees, are estimated at $4797 per semester.

Student Group

The Georgia Institute of Technology has a total enrollment of about 14,000 students, of whom approximately 3,600 are graduate students. The College's graduate enrollment is approximately 230.

Student Outcomes

Recent Ph.D. graduates have taken faculty positions at Washington (St. Louis), Penn State, University of Virginia, Michigan State, and Rensselaer Polytechnic Institute and research positions at Xerox PARC, Sun Microsystems, Silicon Graphics, Hewlett-Packard, and Intel.

Location

Situated close to the center of Atlanta on a 330-acre campus, Georgia Tech is near the many cultural and sports activities of the city. Atlanta, the site of the 1996 Olympic Games, is the headquarters of many large corporations and is the financial hub of the rapidly growing southeastern United States.

The Institute

The Georgia Institute of Technology was founded in 1885. It offers graduate and undergraduate degrees in computing, the sciences, engineering, architecture, management, the social sciences, and the humanities. As the site of the Olympic Village for the 1996 Olympic Games, the Georgia Tech campus was improved by a significant building program that included a coliseum, a natatorium, and dormitories.

Applying

Students with a variety of backgrounds are encouraged to apply, provided they have strong computer science and mathematics preparation. All applicants must submit GRE scores on the General Test and one of the Subject Tests, preferably in computer science. The M.S. program requires that the Subject Test be in computer science. All international students from countries where English is not the native language must submit their TOEFL results. For Ph.D. applicants, completed applications for fall quarter should be received by January 5; for M.S. applicants, by May 1.

Correspondence and Information

College of Computing
Georgia Institute of Technology
Atlanta, Georgia 30332-0280
Telephone: 404-894-3152
World Wide Web: http://www.cc.gatech.edu

Georgia Institute of Technology

THE FACULTY AND THEIR RESEARCH

G. Abowd, Assistant Professor; D.Phil., Oxford. Software engineering, formal methods, human-computer interaction, software architecture, requirements engineering.

M. Ahamad, Professor; Ph.D., SUNY at Stony Brook. Distributed operating systems, distributed algorithms, software fault tolerance, networks.

M. H. Ammar, Professor; Ph.D., Waterloo. Computer networks, communication protocols, performance evaluation, distributed database systems, distributed computing systems.

R. C. Arkin, Professor; Ph.D., Massachusetts. Artificial intelligence, computer vision and mobile robotics.

C. Atkeson, Associate Professor; Ph.D., MIT. Intelligent systems and robotics.

A. N. Badre, Professor; Ph.D., Michigan. Human factors in computer systems, software engineering.

A. Bruckman, Assistant Professor; Ph.D., MIT. Virtual communities, technology and education.

B. Burnham, Principal Research Scientist; Ph.D., Arizona State. Information security.

A. Chervenak, Assistant Professor; Ph.D., Berkeley. Computer systems, computer architecture, performance modeling and evaluation.

D. Colestock, Lecturer; M.S., Georgia Tech. Computer network engineering, telecommunications.

R. Das, Assistant Professor; Ph.D., William and Mary. High-performance computing systems, programming languages.

C. Eastman, Professor; M.S., Berkeley. Computer-based design environments, geometric modeling, integrated databases.

K. P. Eiselt, Assistant Dean; Ph.D., California, Irvine. Artificial intelligence, cognitive science, natural-language understanding, models of human sentence processing, computational psycholinguistics.

P. H. Enslow Jr., Professor; Ph.D., Stanford. Computer networks, telecommunication systems, data communications, distributed processing, operating systems, computer systems.

I. Essa, Assistant Professor; Ph.D., MIT. Computational perception, computer vision, computer graphics, perception and cognition.

N. F. Ezquerra, Associate Professor; Ph.D., Florida State. Medical informatics, artificial intelligence, computer vision.

P. A. Freeman, Professor and Dean; Ph.D., Carnegie Mellon. Software engineering, design processes, science and technology policy.

R. M. Fujimoto, Professor; Ph.D., Berkeley. Computer architecture, parallel processing, simulation.

J. J. Goda Jr., Assistant Professor; M.S., Georgia Tech. Computer programming, programming languages.

A. K. Goel, Associate Professor; Ph.D., Ohio State. Artificial intelligence, planning and design, problem solving, qualitative modeling and model-based reasoning, case-based reasoning and learning.

J. K. Greenlee, Lecturer; M.S., Georgia Tech. Computer architecture.

M. Guzdial, Associate Professor; Ph.D., Michigan. Human-computer interactions, design support environments, interactive technologies for learning.

L. F. Hodges, Associate Professor; Ph.D., North Carolina State. Computer graphics, scientific visualization, 3-D display technology, virtual environments.

J. Hodgins, Associate Professor; Ph.D., Carnegie Mellon. Computer graphics and artificial intelligence, with emphasis on robotics.

H. Karloff, Professor; Ph.D., Berkeley. Theoretical computer science, randomized and parallel algorithms.

S. Koënig, Assistant Professor; Ph.D., Carnegie Mellon. Artificial intelligence.

J. L. Kolodner, Professor; Ph.D., Yale. Artificial intelligence, cognitive science, learning and problem solving, case-based reasoning.

R. J. LeBlanc Jr., Professor and Associate Dean; Ph.D., Wisconsin–Madison. Programming languages and environments, compilers, distributed processing, software engineering.

J. Limb, Professor; Ph.D., Western Australia. High-speed networks, multimedia telecommunications, interactive video systems, video coding, video teleconferencing.

B. MacIntyre, Assistant Professor; Ph.D., Columbia. HCI, computer graphics, augmented reality, virtual environments.

K. MacKenzie, Assistant Professor; Ph.D., MIT. Computer architecture and systems.

L. Mark, Associate Professor; Ph.D., Aarhus (Denmark). Database system architecture, data models, database design, Metadata management, data dictionary systems, information exchange.

W. M. McCracken, Principal Research Scientist; M.S., Georgia Tech. Design cognition, learning theory, software engineering.

M. Mikhail, Associate Professor; Ph.D., Harvard. Networks and technology transfer, theory.

E. Mynatt, Assistant Professor; Ph.D., Georgia Tech. Everyday and ubiquitous computing, HCI, audio interfaces, augmented reality.

S. B. Navathe, Professor; Ph.D., Michigan. Database modeling, database design, manufacturing systems, CAD/CAM and office systems, information systems analysis and design, distributed and heterogeneous databases.

N. Nersessian, Professor; Ph.D., Case Western Reserve. Philosophy and history of science, conceptual change, creativity, learning.

E. R. Omiecinski, Associate Professor; Ph.D., Northwestern. Database systems, file management, parallel algorithms.

C. Potts, Associate Professor; Ph.D., Sheffield (England). Software requirements analysis and design methods with related interodes in hypertext software, documentation and computer support for design decision making.

J. Preston, Lecturer; M.S., Georgia Tech. Software engineering, educational software, graphics, human-computer interaction.

A. Ram, Associate Professor; Ph.D., Yale. Artificial intelligence, story understanding, memory, learning.

U. Ramachandran, Associate Professor; Ph.D., Wisconsin–Madison. Computer architecture and VLSI, distributed operating systems, networking.

D. Randall, Assistant Professor; Ph.D., Berkeley. Algorithms, computational complexity, discrete mathematics.

W. Ribarsky, Senior Research Scientist; Ph.D., Cincinnati. Visualization and virtual environments.

J. Rossignac, Professor and Director of Graphics, Visualization and Usability; Ph.D., Rochester. 3-D modeling, compression, graphics acceleration, user interface technologies and system architecture.

J. S. Rugaber, Senior Research Scientist; Ph.D., Yale. Software engineering, program understanding.

L. Schulman, Assistant Professor; Ph.D., MIT. Theory of computation and discrete mathematics.

K. Schwan, Professor; Ph.D., Carnegie Mellon. Operating and programming systems for parallel and distributed computers, real-time systems.

R. Shackelford, Academic Professional; Ph.D., Georgia Tech. Director of Lower Division Studies.

T. Starner, Instructor; S.M., MIT. Wearable computing, enabled intelligent agents.

J. T. Stasko, Associate Professor; Ph.D., Brown. Software and information visualization, human-computer interaction, programming environments.

A. Tew, Lecturer; M.S., Georgia Tech. Educational technology, software engineering, human-computer interaction.

C. Tovey, Professor; Ph.D., Stanford. Industrial and systems engineering.

G. Turk, Assistant Professor; Ph.D., North Carolina at Chapel Hill. Computer graphics, scientific visualization, geometric modeling, image processing.

V. Vazirani, Professor; Ph.D., Berkeley. Design and analysis of algorithms, complexity theory.

H. Venkateswaran, Associate Professor; Ph.D., Washington (Seattle). Theoretical computer science, computational complexity, parallel computation, algorithms.

E. W. Zegura, Assistant Professor; Ph.D., Washington (St. Louis). Advanced communication systems, analysis of switching networks, parallel and distributed algorithms.

GEORGIA STATE UNIVERSITY

College of Arts and Sciences
Department of Computer Science

Programs of Study	The Department of Computer Science offers a program leading to the degree of Master of Science (M.S.). The M.S. degree in computer science provides students with advanced training in the fundamental principles and processes of computation. The program focuses on the technical aspects of both software and hardware. An M.S. degree requires 31 semester hours of study, which includes eight courses, a thesis, and a seminar. A variety of courses are available to students, and several colloquiums featuring outside speakers are held.
Research Facilities	Departmental computing facilities for research and instruction include more than 130 networked computers, a sixteen-processor Origin-2000 high-performance computer, and six laboratories, including one with ATM switches for network and distributed computing research and another for hypermedia and visualization research. The departmental local area network consists of thirty-three UNIX machines, eighty-five Pentium PCs, fifteen Pentium II PCs, and twenty Macintosh machines. Georgia State University (GSU) libraries, along with online access and borrowing privileges at a number of Atlanta area libraries and other research universities and institutions, provide students and faculty members with excellent facilities. University-wide, a four-processor Silicon Graphics Power Challenge L provides support for research and instructional use. More than 100 network fileservers provide access to centrally supplied software, support e-mail (GroupWise), and provide services to more than 4,000 microcomputer workstations, including more than 450 workstations in open and instructional labs. The University operates a Digital Arts and Entertainment Laboratory, with state-of-the-art Kodak equipment and a four-processor Onyx machine. The University is one of a few selected to participate in the Internet2 project.
Financial Aid	Several graduate laboratory, research, and teaching assistantships are available to graduate students. A limited number of nonresident fee waivers are also available. Students working on specific research programs are often supported by extramural funds.
Cost of Study	For current tuition figures, students should visit the University's Web site at http://www.gsu.edu. Graduate students receiving assistantships are charged a fee of $37.50 per semester, plus the activity, athletics, health, recreation, and transportation fees. Nonresident fees are also waived for students receiving assistantships.
Living and Housing Costs	Georgia State University has a nonresidential campus located in downtown Atlanta at the center of a network of highways and rapid-transit services that extend throughout the greater metropolitan area. This transportation network makes it possible to live anywhere in the metropolitan area and get downtown easily. The cost of living in Atlanta is moderate compared with that in other urban centers in the United States. Dormitory housing is available at the Georgia State Village, a short distance from Georgia State's downtown campus.
Student Group	Georgia State University is a public institution with more than 24,000 students. The graduate student population of more than 7,000 is one of the largest in the Southeast. The average age of graduate students is 33. Students from 113 countries and all fifty states attend the University.
Location	The University is located in the heart of Atlanta's central business district. The city is a rapidly growing metropolitan area characterized by a spectacular skyline and a culturally diverse population. Atlanta's Hartsfield International Airport is the world's largest and busiest, making the city easily accessible from anywhere in the world. The climate is moderate, with a mean July temperature of 73.4°F (23°C) and a mean January temperature of 50°F (10°C). Atlanta is located in the foothills of the southern Appalachian mountain range and is close to both the Great Smoky Mountains and the Atlantic and Gulf coasts.
The University	Georgia State University is responsive to students' career goals and provides educational and research programs that are relevant to the practical needs of both the students and the community. The University offers nearly fifty undergraduate and graduate degree programs that cover some 200 fields of study through its five colleges—Arts and Sciences, Business Administration, Education, Health and Human Sciences, and Law—and its School of Policy Studies.
Applying	Application materials may be obtained from the department or from the Office of Graduate Studies of the College of Arts and Sciences. Applicants must submit the Application for Graduate Study and the University Information forms, a $25 application fee, official copies of transcripts from each institution attended, General Test scores on the Graduate Record Examinations, a statement of background and goals, and three letters of recommendation. Application materials can also be downloaded from the department's Web site (listed below). Applicants for graduate assistantships must apply by February 15 for fall semester. Applicants may obtain additional information about the Department of Computer Science by contacting the Director of Graduate Studies or by viewing the Web page.
Correspondence and Information	Director of Graduate Studies Department of Computer Sciences Georgia State University University Plaza Atlanta, Georgia 30303-3083 Telephone: 404-651-2253 Fax: 404-651-2246 E-mail: info@cs.gsu.edu World Wide Web: http://www.cs.gsu.edu/

Georgia State University

THE FACULTY AND THEIR RESEARCH

Krishnan Balakrishnan, Assistant Professor; Ph.D., Clemson, 1996. Network design, performance analysis, simulation.

Martin D. Fraser, Professor; Ph.D., St. Louis, 1972. Software engineering, simulation.

Ross A. Gagliano, Associate Professor; Ph.D., Georgia Tech, 1976. Modeling and simulation, software engineering.

K. N. King, Associate Professor; Ph.D., Berkeley, 1980. Programming languages.

G. Scott Owen, Professor; Ph.D., Washington (Seattle), 1970. Hypermedia, computer graphics, visualization, digital libraries.

Sushil K. Prasad, Associate Professor; Ph.D., Central Florida, 1990. Parallel and distributed algorithms and data structures, network-based computing, parallel and distributed simulation.

Raj Sunderraman, Associate Professor; Ph.D., Iowa State, 1988. Deductive databases, logic programming, incomplete information in databases, object-oriented databases.

Alexander Zelikovsky, Assistant Professor; Ph.D., Belarus State, 1989. Discrete algorithms, VLSI CAD, combinatorial optimization, computational geometry, computational biology, graph theory.

Yanqing Zhang, Assistant Professor; Ph.D., South Florida, 1997. Artificial intelligence, computational intelligence, fuzzy logic, neural networks, genetic algorithms, data mining and knowledge discovery, parallel and distributed processing.

RESEARCH ACTIVITIES

Artificial intelligence and neural networks: Fuzzy logic, neural networks, genetic algorithms, data mining and knowledge discovery, hybrid intelligent systems, distributed intelligent agents.

Combinatorial optimization: Discrete optimization, computational geometry, complexity theory, graph algorithms, applications.

Computer architecture: VLSI CAD, applications to VLSI manufacturing, computer interconnection networks, parallel logic simulation, layout algorithms, placements and routing.

Computer networks: Network design, performance analysis, survivability, reliability.

Databases: Relational databases, object-oriented databases, logic-based databases.

Digital libraries: Structure, query, navigation.

Graphics and visualization: Hypermedia, multimedia, computer-supported cooperative work, user interface design, computer-based training, multimedia training systems, visual representations of concepts and processes in mathematics, applications of fractals in mathematical visualization, visualization methods for data analysis and teaching.

Parallel and distributed computing: Experimental parallel and distributed computing, parallel algorithms and data structures, network-based computing, parallel discrete event simulation, parallelizing compilers, parallel graph algorithms, interconnection network topology, computational biology.

Programming languages: Language evaluation and implementation, languages for the World Wide Web.

Simulation: Simulation of artificial neural networks, modeling methodologies, discrete element simulation, distributed control systems, simulation of computer and telecommunication networks, parallel simulation.

Software engineering: Formal specification methods, safety critical software, software reusability, object orientation.

ILLINOIS INSTITUTE OF TECHNOLOGY

Department of Computer Science

Programs of Study

Graduate programs lead to the Master of Science (M.S.) and Doctor of Philosophy (Ph.D.) degrees in computer science and the Master of Science in Teaching (M.S.T.) for teachers/trainers of computer science. The graduate programs provide both a practical preparation for those seeking careers in the computer industry and a solid foundation for those aiming at careers in research and education. The programs offer the necessary background in the core areas and exposure to cutting-edge computer technologies that have high impact in the academic and industrial worlds. Courses are offered on IIT's Main Campus in Chicago, at the Daniel F. and Ada L. Rice Campus in suburban Wheaton, and via IITV, the University's interactive television network, at twenty-six public and corporate sites in the greater Chicago area.

The M.S. degree program prepares the research-oriented student for candidacy into the Ph.D. program. Faculty members supervise Ph.D. research in computer algorithms, computer networks and data communication, compilers and operating systems, distributed systems, graphics and image processing, intelligent tutoring/information systems, natural language processing, neural networks, performance analysis and evaluation, programming languages, relational and object-oriented information systems, robotics, software engineering, and software verification and testing. The M.S. degree program also includes several program options designed for working professionals. Popular among these are concentrated programs in software engineering, computer networking and telecommunications, intelligent information systems, and distributed system software.

The M.S.T. program is designed for experienced certified teachers and corporate trainers to strengthen their knowledge of computer science and the effective utilization of educational technologies in computer instruction.

Research Facilities

Research computing facilities include Sun SPARC and Silicon Graphics UNIX workstations and Windows-based PCs.

Financial Aid

Financial assistance is available for new and continuing students through fellowships, teaching assistantships, and research assistantships that include both varying stipends and full or partial tuition and through scholarships that provide all or some portion of tuition. Loans for eligible students may be arranged through the Financial Aid Office. Primary consideration for financial aid is given to applications received before March 1.

Cost of Study

For 1999–2000, tuition is $590 per credit hour. International students are required to register for a minimum of 9 credit hours or the equivalent per semester.

Living and Housing Costs

Housing is available for graduate students in IIT residence halls; the 1999–2000 cost of room and board ranges from $5155 to $6880. Unfurnished IIT apartments are available for graduate students at costs ranging from $458 to $927 per month, including utilities. Early application for apartments is recommended. Several off-campus apartment complexes are located within a mile of the Main Campus.

Student Group

IIT's total enrollment in 1998–99 was approximately 5,900; of this number, 946 were enrolled as full-time graduate students and 1,979 as part-time graduate students. Total undergraduate enrollment was 1,718. The remainder were enrolled in the Chicago-Kent College of Law. More than 800 students were enrolled in computer science graduate programs in 1998–99.

Location

IIT's Main Campus is located near the heart of Chicago, just 3 miles south of the Loop and central to the greater Chicago area's thriving technological community of business, industry, and research institutions. Internationally known for its architecture, museums, symphony, and theater; its beautiful lakefront on the western shore of Lake Michigan; and the unusually rich variety of its ethnic communities, Chicago offers a vast array of recreational and cultural opportunities. The Main Campus, designed by Ludwig Mies van der Rohe and regarded internationally as a landmark of twentieth-century architecture, occupies fifty buildings on a 120-acre site and includes research institutes, libraries, laboratories, residence halls, a sports center, and other facilities. Among its immediate neighbors are Comiskey Park (home of the Chicago White Sox), two major medical centers, and the McCormick Place Exposition Center. The Downtown Campus is in the Loop near the city's financial trading, banking, and legal centers. The Rice Campus is in suburban Wheaton, convenient to the Interstate 88 research and technology corridor west of the city. The Moffett Campus is in southwest suburban Summit-Argo.

The Institute

Illinois Institute of Technology was formed in 1940 by the merger of Armour Institute of Technology (founded in 1890) and Lewis Institute (founded in 1896). IIT offers programs of study in engineering and the sciences, architecture, design, public administration, technical communications and information design, psychology, business, and law. IIT is a member of the prestigious Association of Independent Technological Universities (AITU).

Applying

Applicants must submit an application form, official transcripts (or certified copies) of all college-level work, letters of recommendation, and required GRE and TOEFL scores. Application forms and additional information may be found at the Graduate College's Web site (http://www.iit.edu/colleges/grad). Applications should be received no later than June 1 for fall matriculation and November 1 for spring matriculation.

Correspondence and Information

Department of Computer Science
Illinois Institute of Technology
10 West 31st Street
Chicago, Illinois 60616-3793

Telephone: 312-567-5150
Fax: 312-567-5067
E-mail: info@cs.iit.edu
World Wide Web: http://www.cs.iit.edu/

Illinois Institute of Technology

THE FACULTY AND THEIR RESEARCH

Ilene Burnstein, Associate Professor; Ph.D., IIT. Software engineering, knowledge-based testing and debugging tools, test management.

Graham Campbell, Professor; Ph.D., Penn State. Computer networking, wideband communication protocols, protocol verification, computer performance analysis.

C. Robert Carlson, Professor; Ph.D., Iowa. Relational databases, database design tools and methodologies, information modeling techniques and tools, software design patterns and information architecture.

Morris Chang, Assistant Professor; Ph.D., North Carolina State. Computer architecture, object-oriented co-design.

Thomas Christopher, Associate Professor; Ph.D., IIT. Message driven computing, distributed and parallel computing, programming languages and compilers.

Phillip Dickens, Assistant Professor; Ph.D., Virginia. Parallel and distributed computation, parallel simulation, network-based distributed computation, thread-based computation.

Tzilla Elrad, Associate Professor; Ph.D., Technion (Israel). Concurrent programming, formal verification, embedded real-time systems and ADA standards.

Martha Evens, Professor; Ph.D., Northwestern. Natural language processing, expert systems and intelligent tutoring/information systems.

Ophir Frieder, IITRI Chair Professor; Ph.D., Michigan. Parallel and distributed information retrieval systems, communication systems, high-performance database systems, biological and medical data processing architectures.

Peter Greene, Associate Professor; Ph.D., Chicago. Neural networks, feeling-based reasoning, artificial intelligence and robotics.

Cynthia Hood, Assistant Professor; Ph.D., Rensselaer. Network management, statistical signal processing.

Bogdan Korel, Associate Professor and Interim Chair; Ph.D., Oakland. Software engineering, program testing and operating systems.

Ratko Orlandic, Assistant Professor; Ph.D., Virginia. Database systems, software architecture, multimedia databases.

James Roberge, Associate Professor; Ph.D., Northwestern. Computer graphics, medical imaging, image processing, educational technology, pedagogy.

Peng-Jun Wan, Assistant Professor; Ph.D., Minnesota. Optical computer networks, distributed computing.

INDIANA UNIVERSITY BLOOMINGTON

Department of Computer Science

Programs of Study

The department offers several options for graduate study. The master's and doctoral programs lead to the degrees of Master of Science and Doctor of Philosophy. The professional master's program is designed to lead to both the Bachelor of Science and Master of Science degrees in five years of study.

The Ph.D. program prepares students to attain research expertise in one field of study, combined with a broad background in computer science in general. The requirements include written qualifying examinations, an oral area exam, a thesis proposal, and a thesis defense.

The department's master's program began in 1973 and has awarded more than 1,000 degrees to date. Requirements include 30 credit hours of course work, of which 18 must be in computer science, and evidence of creativity, such as a research project or a master's thesis.

All incoming graduate students who are not native English speakers must take an English proficiency test.

Research Facilities

The department's modern facilities in Lindley Hall serve as a base for research and education in computer science. The department's computing environment is based upon UNIX, Ethernet, Fast Ethernet, and high-speed ATM networking and is fully connected to the Internet and vBNS. Sun, SGI, and Intel-based UNIX workstations along with Windows and Macintosh personal computers provide desktop computing for faculty members, staff members, and students. Silicon Graphics workstations provide powerful graphics capabilities. There are more than 250 computing systems in the department with access to more than 200 GB of online and near-line disk storage. The department is on the standard national and international research networks. A 10-processor SGI Power Challenge with a total capacity of 3 GFLOPS and 2 GB of main memory provides support for various research groups in the department. There is also an SGI Origin 2000 on campus for parallel computing research support and a room-sized CAVE virtual reality research environment, which is maintained in Lindley Hall by the University's Information Technology Services. For input and output, there are numerous laser printers, three color printers, a large format plotter, multiple scanners, video frame-grabbers, and CD-ROM players and recorders. There is a digital hardware lab with logic design and PC-board fabrication facilities. The computing facilities are complemented by a research, instructional, and administrative support staff.

Financial Aid

Most Ph.D. and some M.S. students receive financial aid in the form of research or teaching assistantships. For teaching awards, all students who are not native speakers of English must pass an English proficiency exam designed for teachers. The best Ph.D. students receive full or partial fellowships. Most awards include a stipend of at least $11,856 for ten months plus tuition support for all fees except an unremittable portion of $21 per credit hour. About twenty summer teaching assistantships are available in addition to research assistantships. Many computer-related jobs are also available throughout the University and the Bloomington area.

Cost of Study

Tuition and fees were $153 per credit hour for in-state students and $445 per credit hour for out-of-state students for the 1998–99 academic year. Mandatory fees and student health fees total about $400 per year.

Living and Housing Costs

The cost of a single room and board in the graduate dorm is about $4700 for the academic year. Average University family housing costs range from $400 to $730 per month, including utilities. There are a large number of apartment complexes, condominiums, and houses for rent, as well as studios, efficiencies, and single rooms with shared kitchen and bathroom privileges. The standard lease is twelve months and begins in August.

Student Group

Indiana University has about 25,000 undergraduates and 6,000 graduate students. The Department of Computer Science has 149 graduate and 250 undergraduate students.

Location

The 1,800-acre main campus of Indiana University is located in Bloomington, a city of 60,000 (excluding students), 45 miles southwest of Indianapolis in the rolling hills of southern Indiana. Bloomington combines small-town charm with a wide variety of cultural activities, including Indiana University's School of Music (ranked in the top five music schools in the country), the theater and drama department, and the School of Fine Arts. With the Hoosier National Forest, three state parks, and Lake Monroe all less than an hour away, there are ample opportunities for outdoor recreation.

The University and The Department

Founded in 1820, Indiana University is a liberal arts school with professional programs in medicine, law, nursing, and optometry. The computer science department has grown from an interdisciplinary program in 1967 to a robust research and teaching department. The department's faculty and students conduct research in a wide variety of areas, including interdisciplinary work with other departments and nationwide collaboration. With the Cognitive Science Program, the Indiana University Logic Group, and the Center for Innovative Computer Applications, there is strong interaction with the linguistics, philosophy, mathematics, and psychology departments. Students should contact the department at the address below to receive the most recent computer science annual report and to arrange to visit the department.

Applying

All applicants must have an accredited baccalaureate degree or its equivalent. A strong computer science background is not required, but maturity in mathematics and writing is expected. Three letters of recommendation, official transcripts from all previously attended institutions, current GRE scores, TOEFL scores (for international students), and a 300–500-word statement of purpose must accompany the official graduate school application. Admission decisions are based on a careful review of these documents. Decisions are not made until all application materials have arrived. The application priority date for the fall semester is January 15 for admission with aid and April 1 for admission only. International students are advised to submit application materials to the Office of International Admissions by December 15 to allow ample time for processing.

Correspondence and Information

Admissions Secretary
Department of Computer Science
Lindley Hall 215
Indiana University
Bloomington, Indiana 47405
E-mail: admissions@cs.indiana.edu
World Wide Web: http://www.cs.indiana.edu

Indiana University Bloomington

THE FACULTY AND THEIR RESEARCH

K. Jon Barwise (1990), University Professor of Philosophy, Mathematics, and Logic Philosophy and Adjunct Professor of Computer Science; Ph.D. (mathematics), Stanford, 1967. Logic, information-theoretic approaches to semantics, heterogeneous inference. (E-mail: barwise@cs.indiana.edu)

Randall Bramley (1992), Associate Professor of Computer Science; Ph.D., (computer science), Illinois at Urbana-Champaign, 1989. Scientific computation, parallel numerical algorithms, computational optimization, numerical linear algebra. (E-mail: bramley@cs.indiana.edu)

J. Michael Dunn (1987), Professor of Computer Science, Oscar R. Ewing Professor of Philosophy, and Associate Dean, College of Arts and Sciences; Ph.D. (philosophy), Pittsburgh, 1966. Algebraic logic, proof theory, nonstandard logics (especially relevance logic), relations between logic and computer science. (E-mail: dunn@cs.indiana.edu)

R. Kent Dybvig (1985), Associate Professor of Computer Science; Ph.D. (computer science), North Carolina at Chapel Hill, 1987. Programming language design and implementation, parallel architectures and languages, compiler design, code optimization. (E-mail: dyb@cs.indiana.edu)

Daniel P. Friedman (1973), Professor of Computer Science; Ph.D. (computer science), Texas at Austin, 1973. Programming languages. (E-mail: dfried@cs.indiana.edu)

Dennis Gannon (1985), Professor of Computer Science and Chair; Ph.D. (mathematics), California, Davis, 1974; Ph.D. (computer science), Illinois, 1980. Parallel computation, programming systems, network and distributed computation. (E-mail: gannon@cs.indiana.edu)

Michael E. Gasser (1988), Associate Professor of Computer Science; Ph.D. (applied linguistics), UCLA, 1988. Natural language processing, language acquisition, machine translation, speech acts, and the lexicon; knowledge representation; connectionist models of cognition. (E-mail: gasser@cs.indiana.edu)

Andrew J. Hanson (1989), Professor of Computer Science; Ph.D. (physics), MIT, 1971. Artificial intelligence, machine vision, computer graphics, interactive human interfaces, scientific visualization, virtual reality methods. (E-mail: hanson@cs.indiana.edu)

Christopher T. Haynes (1982), Associate Professor of Computer Science; Ph.D. (computer science), Iowa, 1982. Programming languages. (E-mail: chaynes@cs.indiana.edu)

Douglas R. Hofstadter (1988), University Professor of Cognitive Science and Computer Science; Adjunct Professor of Psychology, Philosophy, History and Philosophy of Science, and Comparative Literature; and Director, Center for Research on Concepts and Cognition; Ph.D. (physics), Oregon, 1975. Artificial intelligence, philosophy of mind, cognitive science. (E-mail: dughof@cs.indiana.edu)

Steven D. Johnson (1982), Associate Professor of Computer Science; Ph.D. (computer science), Indiana, 1983. Formal design methods, program verification and synthesis, hardware verification and synthesis, functional programming, parallel computation. (E-mail: sjohnson@cs.indiana.edu)

David Leake (1990), Associate Professor of Computer Science; Ph.D. (computer science), Yale, 1990. Artificial intelligence and cognitive science, especially case-based reasoning, explanation-based learning, natural language processing, story understanding, and memory. (E-mail: leake@cs.indiana.edu)

Daniel Leivant (1991), Professor of Computer Science and Adjunct Professor of Philosophy; Ph.D. (mathematics), Amsterdam, 1975. Theory of computing, theory of programming languages, mathematical logic and foundations of mathematics. (E-mail: leivant@cs.indiana.edu)

Annie E. Liu (1997), Assistant Professor of Computer Science; Ph.D. (computer science), Cornell, 1996. Programming languages, compilers, and software systems. (E-mail: liu@cs.indiana.edu)

Jonathan W. Mills (1988), Associate Professor of Computer Science; Ph.D. (computer science), Arizona State, 1988. Computer architecture, analog VLSI circuits, robotics, computational sensors, neural networks, continuous-valued logic. (E-mail: jwmills@cs.indiana.edu)

Lawrence S. Moss (1990), Associate Professor of Mathematics and Adjunct Assistant Professor of Computer Science; Ph.D. (mathematics), UCLA, 1984. Logic, theory and semantics of computation, interaction of logic, linguistics, and computer science. (E-mail: moss@cs.indiana.edu)

Benjamin C. Pierce (1997), Assistant Professor of Computer Science; Ph.D. (computer science), Carnegie Mellon, 1991. Programming languages design and implementation, object-oriented programming, static type systems, subtyping, concurrency, process calculi, and distributed programming. (E-mail: pierce@cs.indiana.edu)

Robert F. Port (1986), Professor of Computer Science and Linguistics; Ph.D. (linguistics), Connecticut, 1976. Artificial intelligence, speech recognition, cognitive science, natural language processing. (E-mail: port@cs.indiana.edu)

Franklin Prosser (1969), Professor of Computer Science; Ph.D. (physical chemistry), Penn State, 1961. Digital hardware, computer science education. (E-mail: fpp@cs.indiana.edu)

Paul W. Purdom (1971), Professor of Computer Science; Ph.D. (physics), Caltech, 1966. Analysis of algorithms, rewriting systems, compilers, game playing. (E-mail: pwp@cs.indiana.edu)

Gregory J. E. Rawlins (1987), Associate Professor of Computer Science; Ph.D. (computer science), Waterloo, 1987. Combinatorial algorithms, computational geometry, shape editors and computer animation, models of machine learning. (E-mail: rawlins@cs.indiana.edu)

Edward L. Robertson (1978), Professor of Computer Science; Ph.D. (computer science), Wisconsin, 1970. Database systems, theory of computation, computational complexity, software engineering. (E-mail: edrbtsn@cs.indiana.edu)

Brian Cantwell Smith (1997), Professor of Computer Science and Cognitive Science; Ph.D. (computer science), MIT, 1982. Conceptual foundations of computing, use of computational metaphors in other fields, computational reflection, meta-level architectures, programming languages, knowledge representation. (E-mail: bcsmith@cs.indiana.edu)

George Springer (1986), Emeritus Professor of Computer Science and Mathematics; Ph.D. (mathematics), Harvard, 1949. Programming languages, numerical analysis, complex analysis. (E-mail: springer@cs.indiana.edu)

Scott D. Stoller (1997), Assistant Professor of Computer Science; Ph.D. (computer science), Cornell, 1997. Techniques and supporting tools for design, optimization, and validation of distributed systems. (E-mail: stoller@cs.indiana.edu)

Dirk Van Gucht (1985), Associate Professor of Computer Science; Ph.D. (computer science), Vanderbilt, 1985. Database theory and systems, machine learning. (E-mail: vgucht@cs.indiana.edu)

David E. Winkel (1983), Emeritus Professor of Computer Science; Ph.D. (chemistry), Iowa State, 1957. Digital design, applicative architectures. (E-mail: winkel@cs.indiana.edu)

David S. Wise (1972), Professor of Computer Science; Ph.D. (computer science), Wisconsin, 1971. Applicative programming, multiprocessing architectures and algorithms. (E-mail: dswise@cs.indiana.edu)

IOWA STATE UNIVERSITY

Department of Computer Science

Programs of Study	The Department of Computer Science at Iowa State University offers programs of study leading to either the M.S. or the Ph.D. degree, with areas of concentration in AI databases, programming languages, complexity theory, algorithms, computer architecture, software engineering, operating systems, distributed and parallel computing, VLSI systems, motion planning, vision systems, and robotics.
	For the M.S. degree, students must complete 31 credits and either a master's thesis or a formal paper. Ph.D. candidates must complete 72 credits, fulfill the proficiency requirement in three areas, pass a preliminary exam, and complete and defend a dissertation.
Research Facilities	Faculty and graduate student research is supported by local area networks with workstations, personal computers, and X-terminals. Access is available to machines throughout campus and to the Internet. Parallel processing is provided by the University Scalable Computing Laboratory.
	Computer science graduate students are involved in research projects within the department and in numerous research centers and laboratories across the campus.
	The University library collections total more than 5.6 million items, including more than 2 million books and bound serials and 2.7 million microforms. The library receives more than 21,000 journals and other serial publications.
Financial Aid	Many qualified graduate students in computer science receive financial support in the form of teaching or research assistantships. Assistantship stipends start at $1300 per month in 1999–2000. Eligible graduate students receive a Graduate College tuition scholarship, which covers a portion of the tuition and fee assessment. The tuition scholarship for 1999–2000 is $827 per semester for students on at least half-time appointment and $413 per semester for students on at least quarter-time but less than half-time appointment. Appointments are normally made on a half-time basis, committing the student to 20 hours of work per week. Ordinarily, the student is eligible for reappointment if academic progress and the performance of duties of the previous appointment have been satisfactory. The Graduate College permits a student holding a half-time graduate assistantship to register for no more than 12 semester credits.
	Applicants who have outstanding undergraduate records are eligible for nomination for a Premium for Academic Excellence (PACE) Award. The amount of the PACE Award is equal to one half of the resident tuition. A nomination for the PACE Award is initiated by the department and must be awarded prior to enrollment.
Cost of Study	Fall 1999 fees for full-time study are $1654 per semester for Iowa residents and $4872 per semester for nonresidents. Fees and tuition are subject to change without notice. Students on assistantships are assessed Iowa resident tuition.
Living and Housing Costs	The University provides graduate housing facilities for approximately 336 students and 920 families. There is also apartment space for 174 single graduate students. Private rooms and apartments are available in Ames and other nearby communities.
	Buchanan (air-conditioned) is the graduate residence hall. In 1999–2000, the proposed Buchanan rates are $256 per month for double occupancy and $352 for single occupancy. Rates include electricity, phone, and cable television. A full meal plan costs $1904.
Student Group	In spring 1999, there were 103 M.S. students and 10 Ph.D. students in the department.
Student Outcomes	Graduates of the program are heavily recruited by major computer companies, energy corporations, software producers, and application developers. The types of employment include both research and development positions and range from internal system design to application development. During the course of their studies, some students accept summer internships with corporations. Many Ph.D. graduates have elected to assume research and teaching positions in academic institutions.
Location	Ames, a city with a population of 50,000, is located near the geographical center of Iowa and is 35 miles north of the state capital, Des Moines. Ames is 300 miles west of Chicago, 200 miles south of Minneapolis–St. Paul, and 200 miles north of Kansas City. Students new to the area are generally pleased with the range of available cultural, social, and recreational activities.
The University	Iowa State University was founded in 1858 as one of the first land-grant institutions in the United States. The campus occupies more than 1,000 acres of land on the west side of Ames. Currently, nearly 25,000 students are enrolled in the University and of those, 4,500 are graduate students. The computer science department is part of the College of Liberal Arts and Sciences.
Applying	Applications for fall semester admission with financial aid should be received by February 1 and by September 15 for the spring semester. The department requires submission of GRE General Test scores and recommends submission of scores on the GRE Subject Test in computer science. International students are required to submit TOEFL results, with a minimum score of 550. Applicants are expected to have been in the upper quartile of their undergraduate class and must normally have a strong and comprehensive background in computer science. Provisional admission is sometimes granted to promising students whose prior training is outside the discipline.
Correspondence and Information	Inquiries about the graduate program or requests for information can be made by regular mail, telephone, or fax. Students can also send e-mail to "grad__adm" or to "almanac"; the former generates a package of information that is returned by e-mail and a secretary will request the Graduate Admissions Office to send the necessary application forms. Information about the graduate program (without application forms) may be electronically obtained by sending e-mail directly to "almanac" as described below or by accessing either the departmental Internet Gopher (gopher.cs.iastate.edu) or World Wide Web (www.cs.iastate.edu) servers.

Traditional correspondence:

Graduate Admissions
Department of Computer Science
Iowa State University
Ames, Iowa 50011-1040

Telephone: 515-294-8361
Fax: 515-294-0258
E-mail: grad__adm@cs.iastate.edu

Almanac Mail Server:

Send e-mail to:

almanac@cs.iastate.edu
with the following lines in the message
send gi pamphlet
send gi faculty__interests
send gi evaluation
send gi catalog

Iowa State University

THE FACULTY AND THEIR RESEARCH

Albert L. Baker, Associate Professor; Ph.D., Ohio State. Software engineering, specification languages and CASE tools, natural language text analysis.

J. Peter Boysen, Adjunct Assistant Professor; Ph.D., Iowa State. Computer-assisted instruction, programming languages, object-oriented software development.

Harrington C. Brearley, Professor Emeritus; Ph.D., Illinois at Urbana-Champaign. Computer architecture, switching theory, fault detection.

Soma Chaudhuri, Associate Professor; Ph.D., Washington (Seattle). Theory of distributed computing, parallel computation and complexity, theory of computation.

Hui-Hsien Chou, Assistant Professor; Ph.D., Maryland, College Park. Bioinformatics, computational biology, cellular automata, self-organization phenomena.

Carolina Cruz-Neira, Litton Assistant Professor; Ph.D., Illinois at Chicago. Virtual reality, high-speed networks.

David Fernandez-Baca, Associate Professor; Ph.D., California, Davis. Design and analysis of algorithms, combinatorial optimization.

Shashi K. Gadia, Associate Professor; Ph.D., Illinois at Urbana-Champaign. Temporal databases, databases, geographical information systems.

Don E. Heller, Adjunct Associate Professor; Ph.D., Carnegie Mellon. Programming and debugging support tools for distributed memory systems, performance evaluation and analysis, parallel algorithms.

Vasant Honavar, Associate Professor; Ph.D., Wisconsin. Artificial intelligence; neural networks; artificial life; cognitive science; computational neuroscience; neural, parallel, and distributed algorithms for AI; machine learning; machine perception; evolutionary computation; intelligent multimedia information systems; computer networks.

Jan-Bin Jia, Assistant Professor; Ph.D., Carnegie Mellon. Robotics, artificial intelligence, computational geometry.

Suresh C. Kothari, Professor; Ph.D., Purdue. Computer architecture, parallel and distributed computing, performance analysis, neural networks, computational science.

Steven M. LaValle, Assistant Professor; Ph.D., Illinois at Urbana-Champaign. Robotics, computer vision, artificial intelligence, geometric planning, visibility analysis, and graphical animation.

Gary T. Leavens, Associate Professor; Ph.D., MIT. Programming language design and semantics, programming methodology, specification, verification, distributed systems, object-oriented programming.

Jack H. Lutz, Professor; Ph.D., Caltech. Computational complexity, algorithmic information, randomness and pseudorandomness.

Robyn R. Lutz, Affiliate Assistant Professor; Ph.D., Kansas. High-integrity systems, software safety, real-time embedded software, spacecraft.

Leslie L. Miller, Professor; Ph.D., SMU. Database design, file organization and parallel searching.

Arthur E. Oldehoeft, Professor Emeritus; Ph.D., Purdue. Operating systems, parallel processing, distributed processing, computer security.

Wayne Ostendorf, Associate Professor; B.S., Iowa State. Applied systems technology, computer-based information systems, data center management, large databases, interactive systems.

G. M. Prabhu, Associate Professor; Ph.D., Washington State. Parallel processing, computer architecture and information technology.

Giora Slutzki, Professor; Ph.D., Tel-Aviv. Algorithms, computational complexity, formal languages, automata theory, relational database theory.

Robert M. Stewart, Professor Emeritus; Ph.D., Iowa State. Computer architecture.

George O. Strawn, Associate Professor; Ph.D., Iowa State. Expert systems, optimizing compilers, programming language translation theory, data structures.

Akhilesh Tyagi, Associate Professor; Ph.D., Washington (Seattle). VLSI: complexity theory, design, and architectures; computer architecture; parallel computers.

Johnny S. K. Wong, Associate Professor; Ph.D., Sydney. Computer networks, operating systems, performance evaluation of communication protocols, distributed computing, object-oriented database, multimedia and hypermedia systems.

JOHNS HOPKINS UNIVERSITY

G. W. C. Whiting School of Engineering
Department of Computer Science

Program of Study

Graduate study is oriented toward the Ph.D. degree. Faculty research interests are concentrated in the following areas: algorithms, distributed and fault-tolerant computing and networks, biomedical applications of computer science, object-oriented programming languages and methodologies, robotics and computer vision, geometric computing and computer graphics, concurrent and parallel computer systems, machine learning, computational biology, and natural language processing. The department encourages students to become involved in research-oriented studies guided by a faculty research adviser shortly after enrolling in the program. Ph.D. students must qualify for the Ph.D. by satisfactorily completing both course and project requirements by the end of the second academic year of graduate study. A mutually agreeable selection of a faculty research adviser is then made. The student and research adviser together plan the remainder of the graduate program. Ph.D. students must also pass a Graduate Board oral examination, prepare and defend a preliminary research proposal, and present a department seminar. Once the thesis is completed, there is a dissertation defense open to the public.

The department also offers both full-time and part-time Master of Science in Engineering programs, which require successful completion of a minimum of 8 one-semester courses at an acceptable level, plus satisfactory completion of either a master's thesis/project or 2 additional one-semester courses.

Research Facilities

The department maintains extensive facilities for research and teaching in the New Engineering Building on the Homewood Campus of Johns Hopkins. Department laboratories include a Center for Geometric Computing, a Center for Networks and Distributed Systems, a Computed Integrated Surgery Lab, a Computer Vision Lab, a Machine Learning and Computational Biology Lab, and a Natural Language Processing Lab. The department's general computing facilities include numerous Sun and SGI workstations, Macs and PCs, X-terminals, scanners, and laser printers.

Financial Aid

Financial assistance is available through tuition scholarships, teaching assistantships, research assistantships, and fellowships, including the Abel Wolman Graduate Fellowship, as well as fellowships from the Center for Geometric Computing, the Networks and Distributed Systems Center, and the Center for Language and Speech Processing. Financial assistance should be requested when applying for admission. The standard assistantship stipend is $13,635. Normally, financial aid is not available for students working toward only an M.S. degree.

Cost of Study

Tuition for 1999–2000 is $23,660.

Living and Housing Costs

University-owned apartments are available for single and married students. Private off-campus housing in Baltimore is available in a wide range of prices.

Student Group

The department currently has approximately 80 full-time graduate students.

Location

The campus is in a residential neighborhood of both single-family homes and apartments, located 4 miles from downtown Baltimore. There are churches, restaurants, drugstores, grocery stores, and other shops nearby. The 140-acre tree-lined Homewood campus offers a wide variety of areas for gatherings and recreation. Generally, graduate students find that their academic and social lives tend to center on their departments. The three most widely used buildings are the Milton S. Eisenhower Library; the Newton H. White, Jr. Athletic Center; and the Hopkins Union (Levering Hall), which is the University student center.

The University

Privately endowed, the Johns Hopkins University was founded in 1876 as the first American educational institution committed to the university idea: giving students and faculty the freedom of choice and opportunity necessary for learning and creativity. It remains committed to this idea today. Johns Hopkins is a small coeducational university. To preserve close intellectual association, the University community and the student-faculty ratio are intentionally small. Approximately 3,400 undergraduates, 1,400 graduates, and 140 postdoctoral students are currently enrolled. There are more than 350 faculty members.

Applying

Admissions materials should be requested between August 1 and December 1; completed applications must be received by February 15. Students may also apply by electronic mail by sending to the address below. Acceptance letters are sent by mid-March. Students should take the Graduate Record Examinations (both the General Test and a Subject Test in computer science, mathematics, or engineering) no later than December and should specify the Department of Computer Science when having scores sent. Students without a degree from an American college or university must also take the Test of English as a Foreign Language (TOEFL) and have the score sent to the Department of Computer Science.

Qualified individuals are encouraged to visit the campus to discuss their plans for graduate study with the faculty. Midyear entrance into the graduate program is difficult and not recommended.

Correspondence and Information

Chair, Graduate Admissions
Department of Computer Science
Johns Hopkins University
3400 North Charles Street
Baltimore, Maryland 21218-2694
Telephone: 410-516-8775
Fax: 410-516-6134
E-mail: admissions@cs.jhu.edu
World Wide Web: http://www.cs.jhu.edu/

Johns Hopkins University

THE FACULTY AND THEIR RESEARCH

Yair Amir, Assistant Professor; Ph.D., Hebrew (Jerusalem), 1995. Distributed systems, communication protocol, conferencing replication.

Giuseppe Ateniese, Assistant Professor; Ph.D., Genova (Italy), 1999. Applied cryptography, network security, electronic commerce.

Baruch Awerbuch, Professor; D.Sc., Technion (Israel), 1984. Algorithmic theory of communications, online and distributed computing.

Eric Brill, Assistant Professor; Ph.D., Pennsylvania, 1993. Natural language and speech processing, machine learning and artificial intelligence.

Jagdish Chandra, Adjunct Professor; Ph.D., Rensselaer, 1965. Intelligent systems, computational methods.

Jonathan D. Cohen, Assistant Professor; Ph.D., North Carolina, 1998. Interactive 3-D computer graphics, virtual environments, modeling of 3-D objects.

Michael T. Goodrich, Professor; Ph.D., Purdue, 1987. Design and analysis of algorithms, parallel algorithms, computational geometry, computer graphics.

Gregory D. Hager, Professor; Ph.D., Pennsylvania, 1988. Computer vision, robotics, software systems, sensor fusion.

S. Rao Kosaraju, Edward J. Schaefer Professor; Ph.D., Pennsylvania, 1969. Design of algorithms, parallel computation, pattern matching computation geometry, computational biology.

Subodh Kumar, Assistant Professor; Ph.D., North Carolina, 1996. Interactive 3-D computer graphics, virtual environments, computational geometry.

F. Thomson Leighton, Professor MIT, Adjunct Professor; Ph.D., MIT, 1981. Algorithms, parallel computation, cryptography.

Gerald M. Masson, Professor and Chair; Ph.D., Northwestern, 1971. Computer and communication networking, fault-tolerant computing, interconnection structures.

Scott F. Smith, Professor; Ph.D., Cornell, 1988. Programming languages, semantics.

Roberto Tamassia, Adjunct Associate Professor; Ph.D., Illinois at Urbana-Champaign, 1988. Design and analysis of algorithms, graph drawing, computational geometry.

Russell Taylor, Professor; Ph.D., Stanford, 1976. Medical robotics and computer-assisted surgery.

Lawrence B. Wolff, Associate Professor; Ph.D., Columbia, 1990. Computer vision.

David Yarowsky, Assistant Professor; Ph.D., Pennsylvania, 1995. Speech and natural language processing, machine translation, information retrieval and machine learning.

Research Faculty

Arthur L. Delcher, Research Scientist; Ph.D., Johns Hopkins, 1989. Artificial intelligence, parallel algorithms, parallel programming.

Stacey Franklin Jones, Research Scholar; D.Sc., George Washington, 1997. Analog/digital convergence, interactive/integrated systems.

Raghu Raghavan, Research Faculty; Ph.D., Wisconsin, 1976.

Steven L. Salzberg, Associate Professor; Ph.D., Harvard, 1989. Artificial intelligence, machine learning, computational biology.

Gregory Sullivan, Research Scholar; Ph.D., Yale, 1986. Design of algorithms and fault-tolerant computing.

Joint and Visiting Appointments

Andreas Andreou, Associate Professor (joint appointment with Electrical and Computer Engineering); Ph.D., Johns Hopkins, 1986. Electron devices, analog VLSI, sensor micropower electronics.

Yossi Azar, Visiting Research Scholar; Ph.D., Tel Aviv, 1989. Online algorithms.

Amnon Barak, Visiting Professor; Ph.D., Illinois at Urbana-Champaign, 1971. Distributed operating systems.

Michael R. Brent, Assistant Professor (joint appointment with Cognitive Science); Ph.D., MIT, 1991. Computational linguistics and algorithms for learning natural languages.

Greg Chirikjian, Assistant Professor (joint appointment with Mechanical Engineering); Ph.D., Caltech, 1992. Design, kinematics, dynamics, and control of mechanisms, robotics, and controls.

Lenore Cowen, Assistant Professor (joint appointment with Mathematical Sciences); Ph.D., MIT, 1993. Combinatorics graph theory routing and scheduling, probabilistic method.

Frederick Jelinek, Professor (joint appointment with Electrical and Computer Engineering) and Director of Center for Speech Processing; Ph.D., MIT, 1962. Speech recognition, statistical methods of natural language processing, information theory.

Simon Kasif, Visiting Associate Professor; Ph.D., Maryland, 1985. Artificial intelligence, logic programming, parallel computation.

Edward R. Scheinerman, Professor (joint appointment with Mathematical Sciences); Ph.D., Princeton, 1984. Graph theory, partially ordered sets and random graphs.

Raimond L. Winslow, Associate Professor (joint appointment with Biomedical Engineering); Ph.D., Johns Hopkins, 1985. Modeling of biological systems, large-scale computation, visualization and nonlinear dynamical systems theory.

Yaacov Yesha, Visiting Professor; Ph.D., 1979. Weizmann (Israel). Algorithmic theory of communications and distributed computing.

Yelena Yesha, Visiting Research Scholar; Ph.D., Ohio State, 1989. Electronic commerce and databases.

Amy E. Zwarico, Visiting Assistant Professor; Ph.D., Pennsylvania, 1988. Programming languages, distributed and real-time processing.

Undergraduate Lecturers

Lewis L. Beach, M.S., West Virginia, 1989. General computer science.

Mark E. Giuliano, Ph.D., Maryland, 1990. Parallel languages, artificial intelligence, logic programming.

Joanne Houlahan, Ph.D., Johns Hopkins, 1996. Fault-tolerant computing.

Harold Lehmann, Ph.D., Stanford, 1992. Medical informatics, Bayesian reasoning, statistical software.

Robert Massof, Ph.D., Indiana, 1975. Ophthalmology, virtual reality and vision.

Fernando Pineda, Ph.D., Maryland, 1986. Neural computation, machine learning.

John Sadowsky, Ph.D., Maryland, 1980. Sensory engineering, virtual reality, applied number theory, signal and image processing.

Dwight Wilson, Ph.D., Johns Hopkins, 1996. Fault-tolerant computing.

KANSAS STATE UNIVERSITY

Department of Computing and Information Sciences

Programs of Study

The Department of Computing and Information Sciences at Kansas State University offers graduate programs leading to the Master of Software Engineering, Master of Science, and Doctor of Philosophy degrees.

The program of study for the M.S.E. program consists of 33 credits. Each student specializes in an application area and does a project related to that application area. Each student produces and presents a software portfolio that contains a collection of documents related to the software development activity.

The M.S. degree requires a minimum of 30 credit hours of graduate-level course work. The degree program can take one of three forms: a nonthesis-report option requiring 33 hours, a report option requiring 30 hours, or a thesis option requiring 30 hours. An oral presentation is required for each option, and further original research is required for the thesis option.

The Ph.D. degree requirements include 90 semester hours of graduate-level credit. General requirements include passing a preliminary examination, writing a research proposal about the dissertation research, and writing and successfully defending the dissertation in an open forum.

Kansas State University maintains research programs in programming languages, distributed and real-time systems, software engineering, and database systems.

Research Facilities

The Department of Computing and Information Sciences maintains a large network of servers, workstations, and graphics display terminals for graduate study and faculty research. Servers include Sun SPARC and Intel-based symmetrical multiprocessor systems. Access to these servers is available in offices and laboratories equipped with more than 120 workstations, including Sun SPARCstations, X window system terminals, and NT systems. Direct access to the Internet (vbns) permits communication with computer science researchers worldwide. Programming languages include Ada, C, C++, Concurrent C, Java, FORTRAN-77, Common LISP, Miranda, ML, OBJ3, Parlog, Pascal, PROLOG, and Scheme. Many other software packages are available, including CASE tools, relational database management systems, simulation, expert systems, interface builders, and document publishing. Additional campuswide computer facilities are provided by central Computing and Network Services. These facilities include an IBM 3090 mainframe, Sun servers, Sun Workstations, a Hewlett-Packard (convex) multiprocessor(64), and several labs throughout campus with PCs.

Hale Library subscribes to the standard computer science publications and is able to obtain nearly any title through a nationwide interlibrary loan system.

Financial Aid

Graduate teaching assistantships carry stipends for the academic year and are normally offered on a half-time basis. Resident staff tuition is available for teaching assistants, with a substantial additional fee reduction based on the time of the appointment. A few graduate research assistantships are available and are in effect for the full year. University fellowships are available to qualified U.S. citizens and permanent residents who are interested in pursuing doctoral degrees.

Cost of Study

For 1999–2000, basic fees for all students are $64 for 1 credit hour, then $17 per credit hour for the next 11 credit hours. In addition, graduate tuition for state residents is $103.40 per credit hour; nonresident graduate tuition is $337.65 per credit hour. Tuition is waived for graduate teaching assistants. Tuition is subject to change.

Living and Housing Costs

The cost of a dormitory room and board for single students is $2410 per semester in 1999–2000, with the contract taken on a yearly basis. There is a $25 application fee for housing. Married student housing on campus ranges in cost from $266 to $326 per month plus electricity. Off-campus one-bedroom apartments rent for $250 per month and up. Overall living costs are among the lowest in the nation.

Student Group

The total enrollment at Kansas State University in 1998–99 was 20,885; of this figure, 2,958 were graduate students. Forty-eight percent of the graduate students were women. Students are drawn from every state in the Union and from seventy-one countries. The graduate enrollment in computer science was about 75 students; 36 of these students received some form of financial support.

Location

Manhattan, Kansas, a pleasant, tree-shaded city of 42,000 in the northern Flint Hills, is the trade and cultural center of a five-county area. The area enjoys a stable economy based on agriculture and ranching, trade, light industry, education, and government installations. Manhattan offers most of the advantages of a big city, yet it retains the wholesome flavor and attractions of small-town living. Tuttle Creek Lake, with 165 miles of shoreline and nearly 16,000 surface acres, offers some of the finest recreational areas in Kansas. Manhattan is located about 100 miles west of Kansas City via Interstate 70.

The University

Kansas State University was founded in 1863 as a land-grant institution under the Morrill Act. The University was first located on the grounds of the old Bluemont Central College, which was chartered in 1858, but in 1875 the University was moved to its present site. The 315-acre campus is in northern Manhattan, convenient to both business and residential sections. Most of the buildings on campus are constructed of native limestone.

Applying

Applications for admission are considered throughout the year but should be completed as early as possible. Application for financial aid should be made by February 15 for the following academic year. Applicants are required to take the GRE General Test; international students must take the TOEFL, and they must score at least 50 out of 60 on the TSE to be considered for financial assistance.

Correspondence and Information

Dr. Virgil E. Wallentine, Head
Department of Computing and Information Sciences
Nichols Hall
Kansas State University
Manhattan, Kansas 66506-2302
Telephone: 913-532-6350
Fax: 913-532-7353
E-mail: gradinfo@cis.ksu.edu
World Wide Web: http://www.cis.ksu.edu/

Kansas State University

THE FACULTY AND THEIR RESEARCH

Professors
Virgil E. Wallentine, Head of the Department; Ph.D., Iowa State. Operating systems, computer networks, parallel simulation, concurrent programming systems, knowledge engineering.
David A. Gustafson, Ph.D., Wisconsin–Madison. Software engineering, AI techniques in software development, software measures, expert systems, software testing.
William J. Hankley, Ph.D., Ohio State. Formal specification of software systems, software engineering, graphical user interfaces.
David A. Schmidt, Ph.D., Kansas State. Programming language semantics, abstract interpretation.
Elizabeth A. Unger, Ph.D., Kansas. Data and knowledge based systems, security, integrity of data, office automation systems.
Maarten van Swaay (Emeritus), Ph.D., Leiden (Netherlands). Laboratory instrumentation, social and ethical issues in computing.

Associate Professors
Myron A. Calhoun (Emeritus), Ph.D., Arizona State. Digital systems design, computer applications.
Matthew Dwyer, Ph.D., Massachusetts Amherst. Analysis of concurrent systems, software validation and verification, parallel software architectures.
Rodney Howell, Ph.D., Texas at Austin. Self-stabilizing systems, design and analysis of algorithms.
Masaaki Mizuno, Ph.D., Iowa State. Distributed systems, operating systems.
Gurdip Singh, Ph.D., SUNY at Stony Brook. Concurrent and distributed systems, network management protocols, modular verification, specification languages, database concurrency control.
Stefan Sokolowski (visiting), Ph.D., Polish Academy of Sciences. Program semantics, program verification, program specification, type theories, algebraic topology in studies of concurrency.
Allen Stoughton, Ph.D., Edinburgh (Scotland). Programming language semantics, intuitionistic logic.
Maria Zamfir-Bleyberg, Ph.D., UCLA. Database and knowledge-base systems, data mining, data warehousing.

Assistant Professors
Daniel A. Andresen, Ph.D., California, Santa Barbara. Parallel and distributed systems, digital libraries/World Wide Web applications.
John Hatcliff, Ph.D., Kansas State. Partial evaluation and program transformations, formal methods in software engineering, software verification, semantics of programming languages, static analyses of programs, compiler construction, logics and type theory.
William H. Hsu, Ph.D., Illinois at Urbana-Champaign. Artificial intelligence and knowledge-based software engineering.
Michael R. A. Huth, Ph.D., Tulane. Quantitative performance analysis, model checking, computation and logic, programming language semantics, computers and society.
Mitchell L. Neilsen, Ph.D., Kansas State. Operating systems, computer networks, distributed systems.

REPRESENTATIVE PUBLICATIONS
McCune, T., and **D. Andresen.** Towards a hierarchical scheduling system for distributed WWW server clusters. *Proc. of the Seventh IEEE Int. Symp. on High Performance Distributed Computing (HPDC98), Chicago, Illinois.* 301–9, 1998.
Andresen, D., T. Yang, O. Ibarra, and O. Egecioglu. Adaptive partitioning and scheduling for enhancing WWW application performance. *J. Parallel Distributed Computing* 49(1):57–85, 1998.
Dwyer, M. B., J. Hatcliff, and M. Nanda. Using partial evaluation to enable verification of concurrent software. In *ACM Computing Surveys dedicated to 1998 Symp. on Partial Evaluation,* September 1998.
Dwyer, M. B., G. S. Avrunin and J. C. Corbett. Property specification patterns for finite-state verification. In *Proceedings of the 2nd Workshop on Formal Methods in Software Practice,* March 1998.
Gustafson, D. A. Profile of a winner: Kansas State University. *AI Magazine* p. 27, 1998.
Gustafson, D. A. Software testing and inspection. In *McGraw-Hill Encyclopedia of Science and Technology,* 1999.
Gustafson, D. A., and **W. Hankley.** Software measurement. In *McGraw-Hill Encyclopedia of Science and Technology,* 1999.
Barthe, G., **J. Hatcliff,** and M. H. Soerensen. CPS transformations: The lambda-cube and beyond. *J. Higher-Order Symbolic Computing,* in press.
Hatcliff, J., M. Dwyer, and S. Laubach. Staging static analysis using abstraction-based program specialization. *Proc. of the 1998 Int. Conf. on Programming Languages, Implementation, Logics, and Programming, Pisa, Italy, September 1998.* Lecture Notes in Computer Science, vol. 1490.
Howell, R., M. Nesterenko, and **M. Mizuno.** Finite-state self-stabilizing protocols in message-passing systems. *Proc. of the Fourth Workshop on Self-Stabilizing Systems (WSS '99), June 1999,* in press.
Cherkosova, L., **R. Howell,** and L. Rosier. Bounded self-stabilizing petri nets. *Acta Informatica* 32:189–207, 1995.
Ray, S. R., and **W. H. Hsu.** Self-organized-expert modular network for classification of spatiotemporal sequences. *J. Intelligent Data Anal.,* in press.
Hsu, W. H., S. R. Ray, and D. C. Wilkins. A new multistrategy approach to classifier learning from time series. *Machine Learning,* under review, 1998.
Huth, M. The interval domain: A matchmaker for aCTL and aPCTL. US–Brazil Joint Workshops on the Formal Specifications of Software Systems. *Electronic Notes in Theoretical Computer Science* 14, 1999. (http://www.elsevier.nl/locate/entcs/volume14.html).
Huth, M., and M. Kwiatkowska. Comparing CTL and PCTL on labeled Markov chains. In *Proc. of Programming Concepts and Methods 1998,* pp. 244–62, IFIP, Chapman & Hall, June 2–6, Shelter Island, NY, 1998.
Mizuno, M. A structured approach for developing concurrent programs in Java. *Information Processing Lett.,* in press.
Mizuno, M., M. Hurfin, M. Raynal, and M. Singhal. Efficient distributed detection of conjunction of local predicates. *IEEE Transactions on Software Engineering* 24(8):664–77, 1998.
Neilsen, M. A dynamic probe strategy for quorum systems. In *Proceedings IEEE 17th Int. Conference on Distributed Computing Syst.,* pp. 95–99, 1997.
Neilsen, M. Properties of nondominated k-coteries. *J. Systems Software* 36(4):91–96, 1997.
Schmidt, D. A. Trace-based abstract interpretation of operational semantics. *J. Lisp Symbolic Computing* 10:237–71, 1998.
Schmidt, D. A., and B. Steffen. Data flow analysis as model checking of abstract interpretations. *Proc. 5th Static Analysis Symp.,* Springer LNCS 1503, 1998.
Singh, G. Constraint based structuring of network protocols. *Distributed Computing* 12(1), 1999.
Singh, G. Leader election in compete networks. *SIAM J. Computing* 26(3), 1997.
Pawlowski, W., P. Paczkowski, and **S. Sokolowski.** Specifying and verifying parametric processes. *Math. Foundations of CS 1996, 21st Int. Symp., Cracow, Poland, September 1996, Proceedings,* eds. W. Penczek and A. Szalas, 1113:469–81, 1996.
Sokolowski, S. Investigation of concurrent processes by means of homotopy functors. Work in progress.
Stoughton, A. An operational semantics framework supporting the incremental construction of derivation trees. *Second Workshop on Higher-Order Operational Techniques in Semantics, Electronic Notes in Theor. Computer Sci.* Vol. 10, Elsevier Science B.V., 1998.
Stoughton, A. Fully abstract models of programming languages. *Research Notes in Theoretical Computer Science.* New York: Pitman/Wiley, 1988.
McNulty, S. K., and **E. A. Unger.** A mediator model for inferential security. *J. Official Stat.,* 1998.
Slack, J. M. and **E. A. Unger.** Integrity in object-oriented database systems. *Comp. Sec. J.,* 1994.
van Swaay, M. The value and protection of privacy. *Computer Networks and ISDN Systems* 26(4):149, 1995.
van Swaay, M. Magic or mischief: The illusions of cyberspace as a "technological fix". *National Computer Ethics Conference,* Washington, D.C., pp. 27–8, April 1995.
Wallentine, V., et al. Multiprocessing. *Advan. Comp.,* ed. Marshall C. Yovits, vol. 33. Academic Press, 1994.
Wallentine, V., and J. Butler. A performance study of the RPE mechanism for PDES. *Proceedings of the 1994 International Workshop on Modeling, Analysis and Simulation of Computer and Telecommunications Systems,* eds. Madisetti, Gelenbe, and Walrand. IEEE Computer Society Press, January 1994.
Zamfir-Bleyberg, M., and D. Ben-Arleh. Real-time distributed shopfloor control using data mining. In *IEEE International Conference on Systems, Man, and Cybernetics,* 1998.
Zamfir-Bleyberg, M. Inference rules for text data mining, CIS TR-5-99.

LEHIGH UNIVERSITY

Department of Electrical Engineering, Computer Science and Computer Engineering

Programs of Study

The department offers programs of study leading to the Master of Science (M.S.), Master of Engineering (M.E.), and Ph.D. degrees in electrical engineering; to the M.S. degree in computer engineering; and to the M.S. and Ph.D. degrees in computer science. The department programs span the spectrum from the mathematical and physical aspects of electrical engineering through computer engineering to the natural language processing and artificial intelligence of computer science. The M.S. degree requires 30 credit hours of work, which may include a 6-credit-hour thesis for the electrical engineering degree program and a 3-credit-hour thesis for the computer science degree program. The M.S. in computer science has 9 credit hours of specific course requirements. The M.E. degree requires 30 credit hours of work, which include design-oriented courses and an engineering project. An oral presentation of the thesis or project is required. The Ph.D. degree requires completing 42 credit hours of work (including the dissertation) beyond the master's degree (48 credit hours if the master's degree is not from Lehigh), passing a departmental qualifying examination within one year after completion of the master's degree and a general examination in the area of specialization, fulfilling the University's residence requirement, and writing and defending a dissertation. Competence in a foreign language is not required. The master's thesis is optional for Ph.D. candidates. Major areas of emphasis in electrical engineering and computer engineering include VLSI design, semiconductor device physics, monolithic microwave circuit design and characterization, fiber optics and optical computing, CMOS/VLSI chip technology, digital signal processing, image processing, special audio chips, signal processing algorithms for HDTV, pattern recognition, error-control coding, nonlinear optics and electromagnetics, logic design and verification, computer architectures, and computer vision. Major areas of emphasis in computer engineering include computer architectures, operating systems, software engineering, VLSI design and design automation, parallel and distributed computing, digital signal processing algorithms and architectures, communication systems, network architectures, and optical communications. Major areas of emphasis in computer science include computer vision, pattern recognition, expert systems, databased systems, large-scale software, programming languages, parallel processing and program semantics, artificial intelligence, natural language interfacing, and computational linguistics.

Research Facilities

The department's research laboratories include the computer architectures and arithmetic laboratory, VLSI design automation laboratory, electron device physics laboratory, microelectronics fabrication laboratory, microwave measurements laboratory, microwave monolithic circuits laboratory, multimedia laboratory, parallel and distributed processing laboratory (PDPL), optical computing and communications laboratory, signal processing and communications research laboratory, and vision and software technology laboratory (VAST). The department facilities include Sun SPARCstations, a variety of personal computers, and a complete CMOS integrated-circuit processing laboratory. Facilities available for graduate research include the University Computer Center's workstations in public sites. Collaborative research is ongoing with the facilities of the Sherman Fairchild Laboratory for Solid State Studies, the Energy Research Center, the Materials Research Center, and the Engineering Research Center for Advanced Technology for Large Structural Systems.

Financial Aid

Approximately 50 of the department's graduate students receive financial aid in the form of fellowships and scholarships from Lehigh University and departmental research and teaching assistantships. Beginning teaching assistants in 1999–2000 receive $1200 per month and a 20 hours' tuition award for the academic year and are obligated for 20 hours per week of teaching service.

Cost of Study

Tuition for the 1999–2000 academic year is $860 per credit hour.

Living and Housing Costs

The University's Saucon Valley apartments for married and graduate students provide one- to three-bedroom garden-style efficiency apartments in a rural setting near the athletic complex, with rates that range from $375 to $510 per month in 1999–2000. For further information, students should contact the Office of Residence Operations, Rathbone Hall 63, Lehigh University. Many graduate students live in rooms or apartments in the neighboring community. Living costs are below the national average.

Student Group

Lehigh has more than 4,000 undergraduates and 2,000 graduate students. The Department of Electrical Engineering, Computer Science and Computer Engineering has 169 graduate students; 114 are full-time and 84 are Ph.D. students.

Location

Lehigh University, located on the north slope of South Mountain, overlooks the Lehigh Valley and its cities of Allentown, Bethlehem, and Easton. The Lehigh and Saucon Valleys provide a choice of urban or rural living with ready access to the Poconos, the Appalachian Trail, the Delaware and Lehigh River systems, and the Pennsylvania Dutch region. The Jersey shore is a 2- to 3-hour drive east, and buses run frequently to Atlantic City. Downtown Philadelphia and New York can be reached in 1 and 2 hours, respectively, by car or bus. Lehigh is readily accessible by air and by bus or car on Interstate 78.

The University

Lehigh is a private, coeducational university, founded in 1865 by Asa Packer to "provide students with a scientific education tempered by humanistic study." This innovative concept of offering contiguous technical and nontechnical courses of study continues to provide Lehigh's unusual and successful flavor today. Approximately 50 percent of the students are enrolled in the College of Engineering and Physical Sciences. Research is conducted in all major departments and in the thirteen interdepartmental and interdisciplinary research centers and institutes.

Applying

Application forms for admission and financial aid may be obtained from the department. The admission decision is based on the applicant's transcripts, letters of recommendation, and research and professional goals. GRE scores are required of all applicants. Scores on the Test of English as a Foreign Language (TOEFL) are required of all applicants whose native language is not English, and the Test of Spoken English (TSE) is required of aid applicants whose native language is not English. Applicants who wish to be considered for financial aid must apply by January 15.

Correspondence and Information

Graduate Coordinator
Department of Electrical Engineering, Computer Science and Computer Engineering
Packard Laboratory
Lehigh University
19 Memorial Drive, West
Bethlehem, Pennsylvania 18015-3084

Telephone: 610-758-4072
Fax: 610-758-6279
E-mail: gradinfo@eecs.lehigh.edu
World Wide Web: http://www.eecs.lehigh.edu

Lehigh University

THE FACULTY AND THEIR RESEARCH

Glenn D. Blank, Associate Professor; Ph.D. (cognitive science), Wisconsin, 1984. Computational linguistics, artificial intelligence, cognitive science.

Rick S. Blum, Assistant Professor; Ph.D. (electrical engineering), Pennsylvania, 1991. Signal processing, communications. Signal Processing and Communications Research Laboratory.

Terrance Boult, Associate Professor; Ph.D. (computer science), Columbia, 1986. Computer vision, software systems. Vision and Software Technology Laboratory (VAST).

Dragana Brzakovic, Associate Professor; Ph.D. (electrical engineering), Florida, 1984. Image processing, pattern recognition.

Demetrios Christodoulides, Associate Professor; Ph.D. (electrical engineering), Johns Hopkins, 1986. Light-wave technology, nonlinear optics, solitons.

D. Richard Decker, Professor; Ph.D. (electrical engineering), Lehigh, 1970. Microwave integrated circuit design, packaging design.

Douglas R. Frey, Associate Professor and Graduate Officer; Ph.D. (electrical engineering), Lehigh, 1977. Nonlinear circuit analysis.

Bruce D. Fritchman, Professor and Chair; Ph.D. (electrical engineering), Lehigh, 1967. Image processing.

Samuel L. Gulden, Professor; M.A. (mathematics), Princeton, 1950. Distributed operating systems, program verification.

Miltiadis Hatalis, Professor; Ph.D. (electrical engineering), Carnegie Mellon, 1987. Microelectronic device structures and fabrication processes for active matrix displays.

Frank H. Hielscher, Professor and Associate Chair; Ph.D. (electrical engineering), Illinois, 1966. VLSI design and testing.

Donald J. Hillman, Professor; Ph.D. (mathematics, logic), Cambridge, 1961. Database systems, expert systems.

Carl S. Holzinger, Professor; Ph.D. (electrical engineering), Lehigh, 1963. Microprocessor system design.

James C. M. Hwang, Professor; Ph.D. (material science), Cornell, 1976. Compound semiconductor materials and devices.

Edwin J. Kay, Associate Professor; Ph.D. (mathematics), 1968, Ph.D. (psychology), 1971, Lehigh. Computer applications in psychology, cognitive science.

Weiping Li, Associate Professor; Ph.D. (electrical engineering), Stanford, 1987. Signal processing and computer algorithms, processor architecture, ASIC design. Digital Signal Processing Laboratory (DSP).

Alastair D. McAulay, Chandler-Weaver Professor; Ph.D. (electrical engineering), Carnegie-Mellon, 1974. Optical information processing, communications.

Roger Nagel, Harvey E. Wagner Professor of Manufacturing Systems Engineering; Ph.D. (computer science), Maryland, 1976. Robotics, computer-aided manufacturing.

Karl H. Norian, Associate Professor; Ph.D. (electrical materials), London, 1977. Electronic materials, electrical properties of thin films and biomembranes.

Peggy A. Ota, Vice Provost for Academic Administration; Ph.D. (computer and information sciences), Pennsylvania, 1972. Software testing, pattern recognition.

Eunice E. Santos, Assistant Professor; Ph.D. (computer science), Berkeley, 1995. Parallel and distributed computing, algorithm design and implementation. Parallel and Distributed Processing Laboratory (PDPL).

Michael Schulte, Assistant Professor; Ph.D. (electrical engineering), Texas, 1992. Computer architecture, computer arithmetic, VLSI design. VLSI Measurement, Computer Architecture and Arithmetic Laboratory.

Kenneth K. Tzeng, Professor; Ph.D. (electrical engineering), Illinois, 1969. Coding theory computer networks.

Meghanad D. Wagh, Associate Professor; Ph.D. (electrical engineering), Indian Institute of Technology (Bombay), 1977. Computer architecture, signal processing algorithms.

Marvin H. White, Sherman Fairchild Professor; Ph.D. (electrical engineering), Ohio State, 1969. Submicron device physics, solid-state device modeling, VLSI design.

LOYOLA UNIVERSITY CHICAGO

Department of Mathematical and Computer Sciences
Computer Science Program

Programs of Study

The M.S. program at Loyola University Chicago offers a rigorous course of study to introduce students to new modes of inquiry and to deepen their awareness and understanding of recent progress in computer science. This program appeals to the computer science professional seeking an advanced degree, to professionals in other disciplines contemplating a career change, and to students who wish to prepare for entering a Ph.D. program in computer science. Entering students who do not have an undergraduate degree in computer science will be required to take a minimal number of undergraduate prerequisites in computer science.

The program consists of ten 3-credit graduate courses: five are core courses in algorithms, architecture, operating systems, software engineering, and object-oriented programming; five are elective courses. Students may choose electives in various areas, including artificial intelligence, database design, programming languages, computer networks, parallel processing, and computer graphics. These course offerings parallel faculty research interests. A student may choose, as an elective course, individual study and research or a programming project directed by a faculty member.

A well-prepared student may finish the program in 1 to 1½ years. A student attending part-time may require a minimum of two years to complete the degree.

Research Facilities

The department has approximately thirty Sun Workstations and seventy Pentium PCs running Windows NT. These are all networked with the University computer system and the Internet and have access to Loyola's larger machines, which include an IBM 3081D mainframe and an IBM RS/6000. Dial-up and SLIP access is also available. The department runs its own Gopher and World Wide Web servers offering information about the department, courses offered, and faculty research. A variety of programming languages is available, including Ada, C, C++, Haskell, Java, LISP, ML, Pascal, Perl, PROLOG, S+, Scheme, and Tcl/Tk. Installed UNIX software includes FrameMaker, Gopher, IslandWrite, Mathematica, Mosaic, Netscape, SoftWindows, Tex/LaTex, XEmacs, Xess (a spreadsheet), and others.

Loyola's Sullivan Science Library contains an excellent collection of books in computer science, mathematics, and statistics and receives most major national and international journals in these areas.

Financial Aid

The department offers teaching assistantships that pay full tuition and a stipend of $10,600 for nine months. Also available are research assistantships within the department and the data center.

Cost of Study

Tuition in 1999–2000 for a 3-hour graduate course is $1500, and there is a University services and programs fee of $58. Students holding graduate assistantships are provided with tuition scholarships and stipends.

Living and Housing Costs

There are two campus residence halls that are exclusively for graduate students. Rooms and apartments are available in the campus neighborhood as well as throughout the Chicago area at widely varying costs.

Student Group

There are about 155 graduate students in the department, with 120 studying full-time. It is an international group, with students coming from the United States, Canada, Europe, and Asia. Approximately 80 students graduate each year, and the majority of these take positions in software companies, research institutes, high schools, and colleges. Other graduates continue their studies toward a Ph.D. degree.

Location

The department is located at the Lake Shore Campus of the University, which is on the north side of Chicago on the shore of Lake Michigan. The University has an active theater program and a museum of medieval art and sponsors many cultural events. The Chicago area has a large concentration of universities and colleges, and there are many world-renowned museums such as the Art Institute, the Museum of Science and Industry, and the Field Museum of Natural History. Chicago also has an outstanding array of musical organizations, including the Chicago Symphony Orchestra and the Lyric Opera.

The University and The Department

Loyola University Chicago was founded by the Society of Jesus in 1870 and is committed to the Jesuit tradition of education. Loyola is a Carnegie Doctoral I institution and enrolls 14,000 students. The University offers bachelor's degrees in forty-two fields, master's degrees in forty, and doctoral degrees in thirty-two. In fall 1998, there were 5,200 students enrolled in graduate and professional programs at the University. The department offers undergraduate and graduate degrees in computer science, mathematics, and statistics.

Applying

Applicants must take the GRE General Test and have the equivalent of a four-year undergraduate degree with a major in computer science or a related area. The Graduate School requires a minimum undergraduate GPA of 3.0 for admission, and international applicants must submit a TOEFL score of at least 550 on the paper-based test or 213 on the computer-based test. Three letters of recommendation are required. Applicants who wish to be considered for an assistantship must submit applications by February 1 for the following academic year. Students may apply to be admitted for fall, spring, or summer sessions. Loyola University is an equal opportunity educator and employer.

Correspondence and Information

Graduate Program Director
Department of Mathematical and Computer Sciences
Loyola University Chicago
6525 North Sheridan Road
Chicago, Illinois 60626
Telephone: 773-508-3322
E-mail: info@math.luc.edu
World Wide Web: http://www.math.luc.edu/

Loyola University Chicago

THE FACULTY AND THEIR RESEARCH

Emmanuel Nicholas Barron, Professor; Ph.D., Northwestern.
Martin Buntinas, Professor; Ph.D., IIT.
Christopher Colby, Assistant Professor; Ph.D., Carnegie Mellon.
John Del Greco, Associate Professor; Ph.D., Purdue.
Peter Lars Dordal, Associate Professor; Ph.D., Harvard.
Stephen Doty, Professor; Ph.D., Notre Dame.
Gerald Funk, Associate Professor; Ph.D., Michigan State.
Anthony Giaquinto, Associate Professor; Ph.D., Pennsylvania.
Ronald Greenberg, Associate Professor; Ph.D., MIT.
Michael Handel, Visiting Professor; Ph.D., Berkeley.
Andrew N. Harrington, Associate Professor; Ph.D., Stanford.
Christine Haught, Associate Professor; Ph.D., Cornell.
William Cary Huffman, Professor; Ph.D., Caltech.
Anne Peters Hupert, Associate Professor; Ph.D., Chicago.
Radha Jagadeesan, Associate Professor; Ph.D., Cornell.
Robert Jensen, Professor; Ph.D., Northwestern.
Konstantin Läufer, Assistant Professor; Ph.D., NYU.
Satya Lokam, Assistant Professor; Ph.D., Chicago.
Richard J. Lucas, Professor; Ph.D., Illinois at Chicago.
Richard J. Maher, Associate Professor; Ph.D., Northwestern.
Joseph H. Mayne, Associate Professor; Ph.D., IIT.
Anne Leggett McDonald, Associate Professor; Ph.D., Yale.
Gerard McDonald, Associate Professor; Ph.D., SUNY at Stony Brook.
Timothy O'Brien, Assistant Professor; Ph.D., North Carolina State.
Alan Saleski, Associate Professor; Ph.D., Berkeley.
Chandra Sekharan, Associate Professor; Ph.D., Clemson.
J. Richard VandeVelde, S.J., Associate Professor; Ph.D., Chicago.
Changyou Wang, Assistant Professor; Ph.D., Rice.

Faculty research interests include programming languages, algorithms, software engineering, parallel processing, architecture, database design, artificial intelligence, graphics, system performance evaluation, combinatorics, graph theory, logic and automata, and algebraic coding theory.

MICHIGAN STATE UNIVERSITY

Department of Computer Science and Engineering

Programs of Study
The Department of Computer Science and Engineering offers graduate study leading to the Master of Science and Doctor of Philosophy degrees. Advanced study is available in the areas of computer architecture, high-performance computing, design automation, distributed systems, computer networks, artificial intelligence, knowledge-based systems, database systems, parallel systems and algorithms, pattern recognition, image processing, computer vision, software engineering, and theory of computing. Interdisciplinary work with other departments is encouraged. There are three options for the M.S. program: thesis, project, and course work. Ph.D. students must pass a written qualifying examination within the first two years. A comprehensive examination with both written and oral parts is administered by the guidance committee prior to the dissertation work.

Research Facilities
The department maintains state-of-the-art instructional and research facilities that are constantly upgraded. Instructional facilities include more than 200 color workstations, including Sun, Silicon Graphics, Pentium machines running NT and Macintosh platforms, and servers. Facilities are networked campuswide and are available for students 24 hours a day, seven days a week. Individual research groups in the department have their own laboratories consisting of specialized equipment. These include the Advanced Computing Systems Lab, the High-Speed Networking and Performance Lab, the Intelligent Systems Lab, the Image Processing Lab, and the Software Engineering and Network Systems Lab.

All computer science and engineering graduate students have a permanent account and electronic mail address. Students can keep in contact with colleagues, research groups, databases, and agencies throughout the world.

Financial Aid
The department supports a total of about 100 graduate teaching and research assistants. Stipends range from about $1200 to $1400 per month for half-time appointments. A health plan is included with all assistantships. Assistants qualify for in-state tuition rates and receive tuition grants for 6 credits of study per term. Additional fellowship opportunities are available. Many students avail themselves of the opportunity to work as teaching or research assistants for other campus units.

Cost of Study
In 1998–99, graduate tuition was $222.50 per semester credit for Michigan residents and $450 per semester credit for nonresidents. In addition, matriculation fees of $288 for students enrolled for more than 4 credits or $238 for students enrolled for 4 or fewer credits are assessed each semester. Matriculation fees include the student information technology fee and the infrastructure/technology support fee. An engineering program fee of $237 for students enrolled for more than 4 credits or $131 for students enrolled for 4 or fewer credits is assessed each semester. There is also a $3 FM radio station tax, and students who carry 6 or more credits are charged a $4 tax for the student newspaper and a $6.75 Council of Graduate Students (COGS) tax.

Living and Housing Costs
Owen Graduate Residence Hall offers comfortable living in an atmosphere that is conducive to advanced study and the exchange of ideas. The 1998–99 rates were $1928 per single-room occupancy per semester and $1649 per double-room occupancy per semester, including residence hall tax. Furnished University apartments are available for married students. The 1998–99 family monthly rates were $420 for a one-bedroom apartment and $465 for a two-bedroom.

Student Group
Currently, there are 94 M.S. and 59 Ph.D. students. Seventy-four students are supported as teaching assistants, 51 are supported as research assistants by the department, and approximately 20 are supported by other departments. Fifty-six M.S. and 10 Ph.D. students graduated in 1998.

Location
Located near the center of lower Michigan, the East Lansing campus is famous for its beauty. The campus consists of 5,263 acres crossed by the Red Cedar River. There are more than 400 buildings and 8,000 trees. The metropolitan area has a population of about 400,000. Nearby Lansing is the capital of Michigan. There are cultural events to suit nearly every taste. The Wharton Center for the Performing Arts and the Breslin Student Events Center bring world-famous entertainment to campus. The cost of living is moderate and the secondary schools excellent. The area has many of the amenities of a large city without the congestion or pollution.

The Department
The Department of Computer Science and Engineering was formed in 1968 and continues to expand its graduate and research programs. Michigan State has been a pioneer in computing, providing computer courses for more than thirty years. The faculty members have varied research interests covering the areas mentioned under Program of Study. Related courses are offered in the Departments of Mathematics, Statistics, Electrical Engineering, Business, Psychology, and Linguistics and in the School of Communication.

Applying
In addition to other University and College of Engineering requirements, applicants must provide scores from the GRE General Test and from a Subject Test in computer science or in a closely related area. Applicants for the M.S. program must score in at least the 50th percentile on the Subject Test of the GRE. Ph.D. applicants must score in at least the 85th percentile on the GRE Subject Test in computer science; a related field may be accepted for students with exceptional records. Ph.D. applicants are encouraged to contact faculty members in their area of research interest. International students must submit a minimum score of 600 on the TOEFL.

Correspondence and Information
Graduate Program Director
Department of Computer Science and Engineering
Michigan State University
East Lansing, Michigan 48824-1226
Telephone: 517-353-1679
Fax: 517-432-1061
E-mail: graddir@cse.msu.edu
World Wide Web: http://www.cse.msu.edu

Michigan State University

THE FACULTY AND THEIR RESEARCH

Betty H. C. Cheng, Associate Professor; Ph.D., Illinois. Software tools, synthesis of procedural and data abstractions from formal specifications.

Moon Jung Chung, Professor; Ph.D., Northwestern. Theoretical computer science, algorithms, design automation.

Laura K. Dillon, Professor; Ph.D., Massachusetts at Amherst. Specification and analysis of real-time systems, temporal test oracles, formal methods.

Richard Enbody, Associate Professor; Ph.D., Minnesota. Design automation for digital systems, CAD, parallel algorithms, computer architecture.

Abdol H. Esfahanian, Associate Professor; Ph.D., Northwestern. Applied graph theory, fault-tolerant computing, analysis of algorithms.

John J. Forsyth, Associate Professor; Ph.D., Michigan State. Database theory.

Lewis H. Greenberg, Professor; Ph.D., Michigan State. Operating systems, programming languages, networks.

Herman D. Hughes, Professor; Ph.D., Southwestern Louisiana. Performance measurement and evaluation, programming languages, simulation.

Anil K. Jain, Professor and Chairperson; Ph.D., Ohio State. Pattern recognition, image processing, artificial intelligence.

Sandeep Kulkarni, Assistant Professor; Ph.D., Ohio State. Distributed systems, operating systems, networks, fault tolerance, software engineering, security, reliability.

Sridhar Mahadevan, Assistant Professor; Ph.D., Rutgers. Artificial intelligence, autonomous agents, machine learning, robotics.

Philip K. McKinley, Associate Professor; Ph.D., Illinois. Computer networks, distributed systems, parallel processing, fault-tolerant systems.

Prasant Mohapatra, Associate Professor; Ph.D., Penn State. Multimedia systems, storage architectures, parallel and distributed systems, computer architecture, Internet and World Wide Web issues.

Matthew W. Mutka, Associate Professor; Ph.D., Wisconsin. Resource management in distributed and parallel systems, operating systems, modeling and simulation.

Lionel M. Ni, Professor; Ph.D., Purdue. Parallel and distributed processing, networks, operating systems, computer architecture.

Charles B. Owen, Assistant Professor; Ph.D., Dartmouth. Multimedia information retrieval, multimedia authoring, computer graphics.

Sakti Pramanik, Professor; Ph.D., Yale. Database systems, distributed database systems, parallel processing, optical computing, parallel architectures.

William F. Punch, Associate Professor; Ph.D., Ohio State. Artificial intelligence, expert systems, diagnostic reasoning, genetic algorithms.

Jon Sticklen, Associate Professor; Ph.D., Ohio State. Artificial intelligence, expert systems, knowledge engineering.

R. E. Kurt Stirewalt, Assistant Professor; Ph.D., Georgia Tech. Formal methods, user interfaces, software engineering.

George C. Stockman, Professor; Ph.D., Maryland. Computer vision, artificial intelligence.

Eric K. Torng, Assistant Professor; Ph.D., Stanford. Online algorithms, scheduling, computational complexity, computational biology.

Donald J. Weinshank, Professor; Ph.D., Wisconsin. Computer simulations for science education, textual analysis via computer.

John J. Weng, Associate Professor; Ph.D., Illinois. Computer vision, parallel architectures for real-time vision systems, neural networks, robotics.

Anthony S. Wojcik, Professor; Ph.D., Illinois. Design automation for digital systems, computer architecture, software engineering, automated reasoning.

MICHIGAN TECHNOLOGICAL UNIVERSITY

Department of Computer Science

Program of Study

The Department of Computer Science offers a Master of Science in computer science and, through the College of Engineering, the department offers an interdisciplinary Ph.D. in computational science and engineering (CS&E).

The M.S. program requires the completion of 45 quarter credits. Breadth is achieved through course work in operating systems, software engineering, design and analysis of algorithms, theory of computation, programming languages, and computer architecture. Students are expected to do additional course work in computer science and/or related areas as well as select an area in which they will achieve greater depth. This depth is generally achieved through the preparation of a thesis, though students can also fulfill this requirement with a project or additional course work. Students with a half-time assistantship usually take two years to complete the M.S. program.

The computational science and engineering Ph.D. program is open to students who have completed an M.S. degree in the sciences or engineering. It is an interdisciplinary program that allows students either to focus on large-scale, computational problems in the sciences and engineering or to pursue traditional computer science research programs at the doctoral level. The program also supports research activities that produce environments and methods for solving computational problems in the sciences and engineering. Students must complete a minimum of 24 credits of course work, pass a comprehensive examination, and successfully defend a dissertation. After completing the first two of these requirements, students spend the majority of their efforts contributing to a research project with a faculty member or research group. Students with a half-time assistantship are expected to take approximately four years to complete the Ph.D. program.

Research Facilities

Departmental research equipment available to faculty members and graduate students includes high-end Sun workstations as well as special-purpose DEC Alpha-based and Silicon Graphics workstations. Access to supercomputer centers off campus is also available. Computational resources for the computational engineering program include a symmetric multiprocessor funded by the National Science Foundation and MTU. Visit the web site at http://www.cs.mtu.edu/grad/PHD-CSE.html for the equipment acquired through this project.

Financial Aid

A number of graduate teaching assistantships are available within the department. The 1998–99 stipends were $2680 for M.S. students and $3110 for Ph.D. students, plus tuition for each of the three quarters of the academic year. Graduate research assistantships are available on a limited basis.

Cost of Study

The 1998–99 tuition costs were $1292 for full-time graduate students residing in Michigan and $3359 for nonresident students.

Living and Housing Costs

Apartments and dormitory rooms for students are available in University-owned buildings; room and board costs started at $4488 (double occupancy) for the 1998–99 academic year. Housing for married students was $338 per month for a one-bedroom apartment and $375 for a two-bedroom apartment. Utilities are included, and the apartments are partially furnished. Reasonably priced off-campus housing is located nearby.

Student Group

Approximately 40 students are currently enrolled in the M.S. program. Most of these students are supported by graduate teaching assistantships, graduate research assistantships, or other forms of University-funded support. The CS&E Ph.D. program started in 1994 and currently has 7 students enrolled. Michigan Technological University has more than 6,000 students enrolled in its undergraduate and graduate programs. The department has approximately 400 undergraduate majors.

Student Outcomes

Graduates of the program have found employment in the commercial sector and in government laboratories. Some have gone on to earn Ph.D. degrees at other institutions, such as Clemson University, Georgia Institute of Technology, Michigan State University, and Rice University, and several have started their own companies. Commercial sector employers include AT&T, EDS, Dow Chemical, Ford, General Dynamics, GE Aerospace, Hewlett Packard, Honeywell, IBM, Motorola, Sprint, Sun Microsystems, Texas Instruments, and Unisys. Government laboratory employers include Goddard Spaceflight Center, NASA Ames Research Center, and Los Alamos National Laboratory.

Location

Michigan Technological University is located in Houghton, Michigan, a town of 7,000 on the Keeweenaw Peninsula, on the northwestern side of the Upper Peninsula of Michigan. In the lee of Lake Superior, summers are mild and winters are temperate, with an average of 250 inches of snow. Autumn and spring are cool and colorful. The remoteness imposed by geography is mitigated by the beauty of the countryside, which is unspoiled and free of pollution. There are abundant opportunities for outdoor recreational activities, such as hiking, fishing, boating, and skiing.

The University

Michigan Technological University has an excellent reputation in engineering and science education. By national standards, the University has superior laboratories and equipment and excels in several spheres of science and technology. Michigan Tech was founded in 1885 as a school of mining and metallurgical engineering, and, although it maintains leadership in these areas, it has expanded its curriculum to encompass all areas of science and technology.

There are approximately thirty buildings on the main campus. They house laboratories, classrooms, lecture halls, gymnasiums, the student union, and the library. The University also operates an indoor ice arena, an eighteen-hole golf course, 8 kilometers of Nordic ski trails, its own ski hill with a chair lift, and an indoor tennis center. A major athletic-recreational complex contains a ½-mile indoor track, two pools, handball courts, volleyball courts, basketball courts, dance rooms, and a rifle range. A diverse selection of cultural and entertainment activities, including many of national and international reputation, are sponsored by the University.

Applying

Application forms and instructions for submitting the appropriate materials can be obtained from the department and the department's Web site. Applicants for the M.S. program should have a B.S. or B.A. degree in computer science. Students with a background in another area are considered for admission if there is evidence in their backgrounds to indicate likelihood of success in the M.S. program. Evaluations are based on GRE scores, previous academic performance, experience in computer science, and letters of recommendation. Students may also submit papers or projects in the area of computer science for consideration by the admission committee.

Applicants for the CS&E Ph.D. program should have an M.S. degree in some field of science, engineering, or mathematics. Applicants backgrounds are matched with on-going research projects across the University.

GRE General Test scores are required for all applicants whose degrees are not from a U.S. institution. Although not required, the GRE Subject Test in computer science is also strongly recommended for M.S. applicants. TOEFL scores are required for all applicants whose native language is not English.

Applications for the fall term received by February 15 are generally evaluated by March 30. Most financial aid decisions are also made at this time. Applications received later, as well as applications for other terms, are evaluated on an individual basis.

Correspondence and Information

Graduate Admissions Committee
Department of Computer Science
Michigan Technological University
Houghton, Michigan 49931-1295

Telephone: 906-487-2209
Fax: 906-487-2283
E-mail: csdept@mtu.edu
World Wide Web: http://www.cs.mtu.edu

Michigan Technological University

THE FACULTY AND THEIR RESEARCH

S. Carr, Assistant Professor; Ph.D. (computer science), Rice, 1993. Compiler optimization, multithreaded programming.

X. Huang, Associate Professor; Ph.D. (computer science), Penn State, 1990. Sequence comparison algorithms, parallel applications in computational biology.

J. Lowther, Associate Professor; Ph.D. (computer science), Iowa, 1975. Artificial intelligence, computer graphics.

J. Mayo, Assistant Professor; Ph.D. (computer science), William and Mary, 1997. Distributed systems, clock synchronization, and operating systems.

L. Ott, Chair and Associate Professor; Ph.D. (computer science), Purdue, 1978. Software metrics.

D. Poplawski, Associate Professor; Ph.D. (computer science), Purdue, 1978. Parallel computer architectures, parallel processing, performance evaluation.

A. Sandu, Assistant Professor; Ph.D. (applied mathematical and computational sciences), Iowa, 1997. Scientific computing, mathematical software development, numerical methods for Stiff, ODE, DAE.

S. Seidel, Associate Professor; Ph.D. (computer science), Iowa, 1979. Interprocessor communication algorithms, massively parallel computers, interconnection networks.

C. Shene, Assistant Professor; Ph.D. (computer science), Johns Hopkins, 1992. Geometric/solid modeling, computer-aided design, computer graphics, computational geometry, multithreaded programming.

P. Sweany, Assistant Professor; Ph.D. (computer science), Colorado State, 1992. Compiler optimization and parallel architectures.

RESEARCH AREAS

Artificial Intelligence. Work in artificial intelligence emphasizes issues in knowledge representation. This research concerns the acquisition, use, verification, inference, and/or revision of knowledge that is critical in the operation of problem solvers. To study these issues in knowledge representation, various representation techniques and representation languages are used to develop specific problem solvers.

Computer Architecture and Computer Science Education. Current work in architecture is focused on instruction-level parallelism, in particular highly efficient speculative execution. Efficient speculative execution requires fetching and executing instructions before it is known whether the instructions are supposed to execute, and current research is examining ways to fetch a lot of information quickly in the presence of conditional branch instructions.

Research in computer science education primarily involves finding new ways of motivating students to achieve, including the expanded use of computers in the classroom and interactive and group learning.

Compiler Optimization. Current work is in developing techniques to generate excellent code for instruction-level parallel (ILP) architectures such as VLIW and superscalar computers and embedded processors. Efficient code generation for ILP architecture requires careful ordering of the operations to be performed during execution. This ordering, called instruction scheduling, is the driving force behind ILP code generation. Within that larger context, on-going projects include developing and testing novel techniques to perform register assignment for ILP architectures, improving existing loop scheduling ("software pipelining") methods, and making effective use of an ILP architecture's memory hierarchy, including cache, and partitioning data on processors with partitioned register banks.

Computational Biology. Research in this area is a combination of computer science and molecular biology. Current work is focused on the development of algorithms and software for computational problems in genomic DNA sequencing and analysis. The DNA sequence assembly problem is to reconstruct the original sequence from overlapping pieces. The gene-finding problem is to identify the exact structures of genes in genomic DNA sequences. Some of the programs created by Michigan Tech researchers are currently used in genome sequencing centers to assemble DNA sequences and find genes.

Computational Science. Research in this area is interdisciplinary in nature, combining current computing practices and applied mathematics with applications from the sciences and engineering. Topics pursued are determined on the basis of the combined expertise of the students and faculty members and often involves faculty members from other departments of the University.

Communication Algorithms. Interprocessor communication is often an expensive part of applications designed for parallel distributed memory computers. Recent work has centered on the development of a communication model for a network of Sun workstations. Algorithms have been studied for frequently encountered communication problems, such as the broadcast, scatter/gather, and complete exchange problems. Models of message passing protocols are also being developed in order to discover ways in which the latency of message passing can be reduced.

Geometric Modeling. Geometric modeling is the technique used to describe the shape of an object and to simulate dynamic processes. It is a primary ingredient in computer-aided design and computer-aided manufacturing (CAD/CAM), computer graphics, computer art, animation, simulation, computer vision, visualization, and robotics. Recent research has been aimed at investigating the possible use of lower-degree algebraic surfaces in geometric modeling. This includes robust algorithms for detecting and calculating the intersection of two geometric models, simple geometric representations for lower degree (i.e., three or four) algebraic surfaces, and surface approximation, interpolation, and reconstruction. Another important focus is in developing techniques for visualizing and interrogating surfaces with animation.

Distributed Systems. Recent work includes the development of methods for the efficient evaluation of the global state of distributed computations based on the use of roughly synchronized clocks. Changes in technology for both hardware and software clock synchronization are providing ever tighter clock skews at lower costs. Research in this area will consider the design of new distributed applications based on an assumption of the availability of a global time base.

Numerical Computing. Current research in this area focuses on numerical methods for ordinary and algebraic differential equations, transport equations, sparse linear algebra, automatic differentiation, sensitivity analysis, parallel computations, and applications in air pollution modeling and molecular dynamics.

Software Measurement. A significant difficulty in improving the software development process and the quality of software produced is in identifying when actual improvements have occurred. One technique for identifying such improvements is through measurement. Previous research has focused on measuring the functional cohesion of software developed using an imperative paradigm and exploring the existence of relationships between functional cohesion and other software quality and software process attributes in that paradigm. Current research is focused on developing a clearer understanding of what are good practices in the object-oriented paradigm so that appropriate measurement techniques can be developed.

MISSISSIPPI STATE UNIVERSITY

College of Engineering
Programs in Computational Engineering,
Computer Science, and Computer Engineering

Programs of Study

The College of Engineering offers graduate programs of study and research leading to the M.S. and Ph.D. degrees in three computing fields: computer science, computer engineering, and computational engineering. The computer science program is administered by the Department of Computer Science. The computer engineering program is jointly administered by the Department of Electrical and Computer Engineering and the Department of Computer Science. The computational engineering (CME) graduate program is interdisciplinary and includes faculty members from most engineering disciplines, computer science, and mathematics.

Areas of research and study supported by the three fields are artificial intelligence (CS), graphics and visualization (CS, CPE, CME), software engineering (CS), digital integrated circuit design (CPE), high-performance computing (CS, CPE, CME), signal and information processing (CPE), embedded microprocessor systems (CPE), communications (CPE, CS), domain-specific computational technologies (CME), computational mathematics (CME), and parallel and distributed computing (CS, CPE, CME). The computational engineering program requires students to complete a program of study with adequate work in a specific computational engineering application area, high-performance computing, and computational mathematics.

All three M.S. programs require 30 to 35 credit hours and offer both thesis and nonthesis options. The Ph.D. programs consist of course work, examinations, a dissertation that reports original scholarly research, and an oral presentation and defense of the dissertation.

Research Facilities

Research facilities are housed in the Simrall Electrical Engineering Building, a modern 95,000-square-foot facility; the Butler Computer Science Building, a newly renovated 35,000-square-foot facility; and the Engineering Research Center for Computational Field Simulation, a 44,000-square-foot building designed for cross-disciplinary research. Both Simrall and Butler house faculty offices, modern classrooms, research laboratories, and workstation laboratories for graduate students. The Engineering Research Center Building has an electronic classroom, a state-of-the-art workstation laboratory/classroom, and a general-purpose workstation laboratory supplementing the extensive desktop infrastructure. Computer equipment includes one Sun Ultra HPC10000 supercomputer, three Silicon Graphics Onyx RealityEngine2 servers, two Silicon Graphics Power Challenge XL servers, one Power Challenge L compute server, one Sun Ultra Enterprise 5000 NFS file server, one Sun Ultra Enterprise 5000 compute server, a dedicated Archive/FTP/WWW server, a 32-processor SuperMSPARC (designed and constructed at the ERC), 140 Sun SPARCstation class workstations, eighty Silicon Graphics workstations, two Virtual Reality booms, and assorted other personal computers, printers, and peripherals. The center heavily uses both serial and parallel supercomputers at installations around the country via a T3 connection to BBNPlanet and the Internet.

Financial Aid

Research and teaching assistantships are available for highly qualified applicants. Stipends for assistantships vary. The College of Engineering offers Barrier and Honda Graduate Fellowships for both M.S. and Ph.D. students. Mississippi State University (MSU) is a member of the National Consortium for Graduate Degrees for Minorities in Engineering and Science (GEM), which supports students from minority groups in advanced study in engineering. Approximately 90 percent of graduate students are on assistantship or fellowship.

Cost of Study

In 1998–99, tuition and fees were $1508.50 per semester for Mississippi residents and $3059.50 per semester for nonresidents. Assistantships and fellowships include a waiver of both resident and nonresident tuition.

Living and Housing Costs

Critz Hall is an on-campus dormitory for graduate students that costs $250 to $300 per month, which includes all utilities, including cable television. Each unit contains a refrigerator-freezer-microwave combination. Housing for married students is available in Aiken Village, where the cost of unfurnished one- and two-bedroom apartments is $200 to $225, including all utilities except electricity. Off-campus apartments rent for $300 to $600 per month.

Student Group

Approximately 160 graduate students are enrolled on campus in the three programs. Because of the nature of the programs and the location of the University, almost all of the students are full-time. More than one third are enrolled in the doctoral programs. Additional students are enrolled in programs at the U.S. Corps of Engineers Waterways Experiment Station and at the Stennis Space Flight Center.

Student Outcomes

Students who are awarded graduate degrees in computing fields at Mississippi State University work for a wide variety of high-tech companies, including Microsoft, Intel, Federal Express, Intergraph, Texas Instruments, Sun Microsystems, and many others. Graduates are also placed in a variety of government positions and at universities involved in both teaching and research.

Location

Mississippi State University is located in a rural area of east Mississippi and adjoins the small town of Starkville (population 18,500). Mississippi State is within easy driving distance of Jackson, Mississippi (120 miles); Birmingham, Alabama (140 miles); Memphis, Tennessee (179 miles); and New Orleans, Louisiana (307 miles). Air service is available through the Golden Triangle Regional Airport, located 16 miles east of the campus.

The University and The College

Mississippi State was established as a land-grant college in 1880 and has grown to be the largest university in the state of Mississippi. It is the only institution in the state that is typically listed among the nation's top 100 research institutions by the National Science Foundation and designated as a Doctoral I university by the Southern Regional Educational Board. The College of Engineering consists of ten operating departments and units. Since the first freshman class enrolled in engineering in 1892, more than 15,000 engineers have gone on to become respected leaders in many diverse fields across the state, nation, and world.

Applying

Applications for graduate study for the 2000–2001 academic year are accepted at any time. Preference for awarding assistantships is given to applications received by February 1 for summer or fall semester admission or October 1 for spring semester admission. All applicants are required to submit scores for the general test of the GRE, and the TOEFL is required of international students whose native language is not English.

Correspondence and Information

Computer Science Graduate
 Program
Graduate Studies Committee
Box 9637
Mississippi State, Mississippi
 39762
Telephone: 601-325-2756
Fax: 601-325-8997
E-mail:grad-coord@cs.msstate.
 edu
WWW: http://www.cs.msstate.edu/

Computer Engineering Graduate
 Program
Electrical and Computer
 Engineering
Box 9571
Mississippi State, Mississippi
 39762
Telephone: 601-325-3667
Fax: 601-325-2298
E-mail: harden@ece.msstate.edu
WWW: http://www.ece.msstate.edu/

Computational Engineering
 Graduate Program
Coordinator of Graduate Studies
NSF Engineering Research Center
Box 9627
Mississippi State, Mississippi
 39762
Telephone: 601-325-8278
Fax: 601-325-7692
E-mail: grad-coord@erc.msstate.
 edu
WWW: http://www.erc.msstate.edu/

Mississippi State University

THE FACULTY AND THEIR RESEARCH

Artificial Intelligence, Information and Signal Processing
G. Boggess, Assistant Professor; Ph.D., Illinois. Cognitive science, neural networks, genetic algorithms, computational linguistics.
L. Boggess, Professor; Ph.D., Illinois. Artificial intelligence, natural language processing, speech recognition.
S. Bridges, Associate Professor; Ph.D., Alabama in Huntsville. Expert systems, knowledge discovery in databases.
D. Dearholt, Professor; Ph.D., Washington (Seattle). Human-computer interaction, associative graphs.
E. Hansen, Assistant Professor; Ph.D., Massachusetts. Artificial intelligence, planning and reasoning under uncertainty, resource-bounded computing.
J. Hodges, Professor; Ph.D., Southwestern Louisiana. Knowledge representation, knowledge discovery in databases, document understanding.
H. Jamil, Assistant Professor; Ph.D., Concordia (Montreal). Deductive databases, uncertainty management.
R. King, Professor; Ph.D., Wales. Neural networks, knowledge-based expert systems.
N. Miller, Associate Professor; Ph.D., Iowa State. Curriculum, database, CAI, intelligent tutoring.
J. Picone, Associate Professor; Ph.D., IIT. Signal processing, speech processing.

Graphics and Visualization
D. Banks, Assistant Professor; Ph.D., North Carolina. Graphics, flow visualization, mathematical visualization.
R. Machiraju, Assistant Professor; Ph.D., Ohio State. Graphics, visualization, image analysis.
R. Moorhead, Associate Professor; Ph.D., North Carolina State. Scientific visualization, digital image processing.

High Performance, Distributed, and Parallel Computing
I. Banicescu, Assistant Professor; Ph.D., Polytechnic. Parallel algorithms, scientific computing.
J. Harden, Professor; Ph.D., Texas A&M. Performance monitoring, memory-hierarchy design, real-time embedded systems.
S. Howard, Assistant Professor; Ph.D., North Carolina State. Algorithms, parallel computing, signal processing.
R. Little, Associate Professor; Ph.D., Louisiana Tech. Data compression, instruction set architecture.
D. Reese, Associate Professor; Ph.D., Texas A&M. Distributed computing, computer architecture, object-oriented programming.
S. Russ, Assistant Professor; Ph.D., Georgia Tech. Distributed computing, resource allocation, computer architecture.
A. Skjellum, Associate Professor; Ph.D., Caltech. Parallel algorithms, parallel and distributed software, scientific computing.

Software Engineering
B. Carter, Professor; Ph.D., Arkansas. Software engineering, software metrics.
T. Philip, Professor; Ph.D., Mississippi State. Software engineering, real-time systems, software design and testing.
R. Vaughn, Associate Professor; Ph.D., Kansas State. Software security, software engineering.

Domain Specific Computational Applications
J. Beggs, Assistant Professor; Ph.D., Penn State. Computational electromagnetics, solution algorithms, time-domain electromagnetics.
R. Briley, Professor; Ph.D., Texas. Computational fluid dynamics, parallel computing.
P. Cinnella, Associate Professor; Ph.D., Virginia Tech. Computational fluid dynamics; hypersonic, nonequilibrium, and reactive flows.
B. Gatlin, Associate Professor; Ph.D., Mississippi State. Computational biofluid mechanics, computational particle dynamics.
D. Huddleston, Associate Professor; Ph.D., Tennessee. Computational methods in water resources, environmental quality modeling.
M. Janus, Associate Professor; Ph.D., Mississippi State. Computational fluids, algorithms, unsteady flows, dynamic grids.
J. Newman, Assistant Professor; Ph.D., Virginia Tech. Computational fluid-structure interaction, design optimization.
D. Whitfield, Distinguished Professor; Ph.D., Tennessee. Computational fluid dynamics.

Grid Technology and Computational Mathematics
D. Marcum, Professor; Ph.D., Purdue. Fluid mechanics, computational fluid dynamics, unstructured grid generation, numerical methods.
B. Soni, Professor; Ph.D., Texas. Computational fluid dynamics, numerical grid generation, computer-aided geometry design.
J. Thompson, Distinguished Professor; Ph.D., Georgia Tech. Grid generation computational fluid dynamics, high-performance computing.
J. Zhu, Associate Professor; Ph.D., SUNY at Stony Brook. Numerical analysis, parallel computing, mathematical modeling.

Microelectronics
B. Blalock, Assistant Professor; Ph.D., Georgia Tech. Circuit design, CMOS design, monolithic sensors.
D. Linder, Assistant Professor; Ph.D., Mississippi State. Computer architecture, CAD tools.
R. Reese, Associate Professor; Ph.D., Texas A&M. CAD for integrated circuits, digital systems.
D. Trotter, Professor; Ph.D., Texas. Microelectronics design, VLSI.
R. Winton, Professor; Ph.D., Duke. Microelectronics, optoelectronics.

Research Centers
The National Science Foundation Engineering Research Center (ERC) at Mississippi State University has as its mission the enhancement of global competitiveness of United States industry by reducing the time and cost necessary for complex field simulations for engineering analysis and design. The ERC for CFS is an interdisciplinary research center within the College of Engineering, with faculty members and students from most engineering programs, computer science, mathematics, and physics. The Center conducts coordinated cross-disciplinary research (approximately $9 million per year), with industrial affiliate interactions (twenty-eight government laboratories and agencies and sixteen industrial companies).
The Diagnostic Instrumentation and Analysis Laboratory (DIAL) is an interdisciplinary research department in the College of Engineering. DIAL is supported primarily through funding from the Office of Technology Development within the Office of Environmental Management in the U.S. Department of Energy (DOE). DIAL's mission is to improve effectiveness and competitiveness by employing modern diagnostic techniques to monitor, control, and optimize processes, thereby improving process understanding while minimizing environmental impact. Measurements are made in extremely hot, highly corrosive atmospheres in which conventional measurement devices are ineffective.
The Institute for Signal and Information Processing (ISIP) is a multidisciplinary program to develop next generation information-processing techniques. Research at ISIP is centered on intelligent information processing. ISIP draws upon a wide range of research experience in areas such as signal processing, communications, natural language, database query, intelligent systems, and discrete controls. Its vision is to develop systems capable of intelligent interactions with users by the integration of multiple interface technologies, including speech, natural language, database query, and imaging.
The High Performance Computing Laboratory (HPC Lab) is located in the Engineering Research Center and the Butler Computer Science Building. Addressing high-performance system software, including parallel and distributed middleware for scientific and real-time settings, the HPC Lab research includes leading standards-based efforts involving the Message-Passing Interface, IETF PacketWay, and high-speed networking protocol software. A well-equipped laboratory includes ATM and Myrinet networking, connecting modern multiprocessor PC's and UltraSPARC systems, and several small-scale parallel machines. Research is supported by DARPA, NSF, DOE, and industry.
The Microsystems Prototyping Laboratory (MPL) is located at the Engineering Research Center and represents a growing presence in the government, industrial, and academic research communities. By aligning its capabilities with existing and projected Department of Defense projects relating to a variety of microsystem topics, the MPL provides a state-of-the-art research environment in the following areas: VLSI design and test; microsystem specification, design, and test; and VHDL (VHSIC hardware description language) modeling, simulation, and test.
The Computational Fluid Dynamics Laboratory (CFD Lab), located at the Engineering Research Center, has the objective of advancing the state of the art in the computational solution of real-world problems involving complex geometry and complex physics. Strengths of the ERC are brought to bear on computational problems dealing with complex flows, which include results from various grid generation projects; research in parallel and vector computational methods; and data visualization. Experiences with these state-of-the-art computational challenges shape the details of a research program involved in numerical solution algorithms, with emphasis on three-dimensional time-dependent compressible and incompressible viscous flows.
The Visualization, Analysis and Imaging Laboratory (VAIL), located at the Engineering Research Center, specializes in interactive analysis and visualization of very-large-scale data sets; immersive visualization; distributed and parallel visualization; multiresolutional analysis, feature detection, extraction, and classification; and exploration of new interaction paradigms. Lab equipment includes a CAVE.

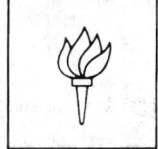

NEW YORK UNIVERSITY

Courant Institute of Mathematical Sciences
Department of Computer Science

Programs of Study

The Department of Computer Science offers courses leading to the M.S. and Ph.D. degrees. The program offers instruction in the fundamental principles, design, and applications of computer systems and computer technologies. In addition to its computer science degrees, the department offers an M.S. in information systems, in collaboration with the Stern School of Business, and the M.S. in scientific computing, jointly established with the mathematics department.

The M.S. degree in computer science qualifies students for significant development work in the computer industry or important application areas. The emphasis in the M.S. in information systems program is on the management of computer systems in business. The M.S. in scientific computing is designed to provide broad training in areas related to scientific computing using modern computing technology and mathematical modeling arising in various applications. Doctoral degree recipients are in a position to hold faculty appointments and do research and development work at the forefront of this rapidly changing and expanding field.

Research Facilities

The primary facility for graduate educational and research computing is a network of workstations, including several Sun servers. In addition, individual research groups have various other machines, including UNIX workstations (SGIs, IBMs, DECs, and HPs), Macintoshes, and PCs. Access to the Internet is provided through a T3 connection. All supported doctoral students have access to their own dedicated UNIX workstation. Many other research machines provide for abundant access to a variety of computer architectures. The Multimedia Center for Advanced Technology and Media Research Laboratory has an extensive range of state-of-the-art graphics equipment, a sound studio, and access to related facilities in the Tisch School of the Arts.

Financial Aid

Fellowships and assistantships are awarded exclusively to students who study full-time for the Ph.D. degree. They cover tuition and, in 1998–99, provided a stipend of $16,600 for the nine-month academic year. Additional summer support may be available. Students who perform well have their awards renewed for a period of four to five years. Low-interest loans (available to students who qualify on the basis of need) are another source of support for graduate study in the department.

Cost of Study

In 1999–2000, tuition and fees are calculated at $750 per point. A full-time program normally consists of four 3-point courses per term (24 points for the year). A limited deferred-tuition plan is in effect.

Living and Housing Costs

University housing for graduate students is limited. It consists mainly of shared studio apartments in University apartment buildings and shared suites with private bedrooms in residence halls. University housing rents in 1998–99 ranged from approximately $6310 to $10,020 for the nine-month academic year.

Student Group

The department has a substantial number of both full- and part-time students. Most of the part-time students are computer professionals employed by a wide range of corporations in the metropolitan area, including IBM and Bell Laboratories, as well as many banking and finance corporations. The student body numbers approximately 380.

Location

New York University is located at Washington Square in Greenwich Village in a residential neighborhood consisting of apartments, art galleries, theaters, restaurants, and shops.

The University and The Institute

New York University, founded in 1831, enrolls 46,000 students and is the largest private university in the country. Its various schools offer a wide range of undergraduate, graduate, and professional degrees. Among its units of international stature is the Courant Institute of Mathematical Sciences. The Institute combines research of mathematics and computer science with advanced training at the graduate and postdoctoral levels. Its activities are supported by the University, government, industry, private foundations, and individuals.

Applying

Students seeking departmental financial aid, including assistantships, must apply by January 4. Such awards are open only to Ph.D. applicants. The deadlines for summer and fall admission to the M.S. and Ph.D. programs in computer science are April 1 for international applicants and April 15 for others. For spring admission, the deadline is November 1. M.S. in information systems applications are due March 1 for the summer and fall terms and November 1 for spring. GRE general test scores are required for admission to each of the M.S. programs; the Ph.D. program application requires both general and computer science subject test scores. Applicants whose native language is not English must submit TOEFL scores.

Correspondence and Information

For program and financial aid information:
Computer Science Department
Courant Institute of Mathematical Sciences
New York University
251 Mercer Street
New York, New York 10012-1185
Telephone: 212-998-3063
E-mail: admissions@cs.nyu.edu
World Wide Web: http://cs.nyu.edu

For applications and a bulletin:
Office of Admissions and Financial Aid
Graduate School of Arts and Science
New York University
P.O. Box 907, Cooper Station
New York, New York 10276-0907
Telephone: 212-998-8050
E-mail: gsas.admissions@nyu.edu
World Wide Web: http://nyu.edu/gsas/degree/admission

New York University

THE FACULTY AND THEIR RESEARCH

Professors

Marsha J. Berger. Computational fluid dynamics, adaptive methods for partial differential equations, parallel computing. berger@cs.nyu.edu

Richard J. Cole. Algorithmics, pattern matching, amortized complexity, communication. cole@cs.nyu.edu

Martin D. Davis (Emeritus). Mathematical logic, theory of computation, diophantine decision problems, history of logic. davism@cs.nyu.edu

Robert B. K. Dewar. Programming languages, compilers, operating systems, microprocessor architectures. dewar@cs.nyu.edu

Allan Gottlieb. Parallel computing/supercomputing, computer architecture, digital libraries, operating systems. gottlieb@cs.nyu.edu

Ralph Grishman. Computational linguistics (natural language processing). grishman@cs.nyu.edu

Zvi M. Kedem. Metacomputing and web-based computing, parallel and distributed processing. kedem@cs.nyu.edu

Bhubaneswar Mishra. Robotics, mathematical and theoretical computer science, computational biology and computational finance. mishra@cs.nyu.edu

Michael L. Overton. Numerical optimization, numerical linear algebra, mathematical programming, linear, semidefinite and convex programming. overton@cs.nyu.edu

Robert Paige. Program transformations, compilers and programming languages, software environments, algorithms. paige@cs.nyu.edu

Edmond Schonberg. Programming languages, compiler construction, software engineering, software prototyping, parallel programming, Ada 95. schonberg@cs.nyu.edu

Jacob T. Schwartz. Design of algorithms and systems for computational logic, interactive multimedia systems and their applications to education. schwartz@cs.nyu.edu

Dennis E. Shasha. Biological computing, information navigation, data mining, database tuning. shasha@cs.nyu.edu

Joel H. Spencer. Discrete mathematics, theoretical computer science, probabilistic methods, random graphs. spencer@cs.nyu.edu

Olof B. Widlund. Numerical analysis, parallel computing, partial differential equations, continuum mechanics. widlund@cs.nyu.edu

Chee K. Yap. Computational geometry, computer algebra, visualization, robotics, complexity theory. yap@cs.nyu.edu

Research Professors

Naomi Sager. Natural language processing, science information structures, medical informatics, speech recognition. sager@cs.nyu.edu

Micha Sharir. Robotics, computational geometry, analysis of algorithms, combinatorial geometry. sharir@cs.nyu.edu

Associate Professors

Ravi Boppana. Theoretical computer science, computational complexity, probabilistic methods, analysis of algorithms. boppana@cs.nyu.edu

Ernest Davis. Knowledge representation, commonsense reasoning, physical and spatial reasoning. davise@cs.nyu.edu

Benjamin F. Goldberg. Program analysis and optimization, programming language design and implementation, functional programming languages, languages for parallel computation. goldberg@cs.nyu.edu

Robert A. Hummel. Computer vision, evidential reasoning, automatic target recognition, medical image processing. hummel@cs.nyu.edu

Krishna Palem. Compilers and optimization, programming tools for embedded and adaptive computing, parallel computing, string and pattern matching. palem@cs.nyu.edu

Kenneth Perlin. Computer graphics, computer/human interfaces, multimedia, simulation. perlin@cs.nyu.edu

Alan R. Siegel. Probabilistic computation, parallel computation, graphics design systems. siegel@cs.nyu.edu

Research Associate Professor

Stephane Mallat. Computer vision, signal processing, harmonic analysis. mallat@cs.nyu.edu

Assistant Professors

Thomas Anantharaman. Large-scale software development, statistical techniques for genomics, computer chess. tsa@cs.nyu.edu

Davi Geiger. Computer vision, memory, learning and their applications. geiger@cs.nyu.edu

Vijay Karamcheti. Computer architecture, parallel and distributed computation, experimental computer systems. vijayk@cs.nyu.edu

Dennis Zorin. Graphics, surface representation, perception of computer-generated images. zorin@cs.nyu.edu

Associated Faculty

Richard Pollack, Professor of Mathematics and Computer Science. Discrete geometry, computational geometry and algorithmic real algebraic geometry. pollack@cims.nyu.edu

Tamar Schlick, Associate Professor of Chemistry, Mathematics and Computer Science. Mathematical biology, numerical analysis, computational chemistry.

David Schwartz, Associate Professor of Chemistry and Computer Science. Genomics, DNA, physical chemistry, microscopy.

MAJOR RESEARCH AREAS

Algorithmics and computational complexity: computational geometry, probabilistic methods, string and pattern matching, computer vision and biology, online algorithms and real-time computation, parallel and distributed computation, symbolic computing.

Artificial intelligence and natural language processing: natural language processing, neural networks, reasoning with uncertainty, commonsense reasoning, learning, database modeling.

Image processing and computer vision: analysis and synthesis of digital images, human-machine interfaces, object recognition and image data representation.

Multimedia: computer graphics, human/computer interface, authoring tools, digital audio and video, high-level visual languages, multiuser simulation worlds, new input and output technologies, visualization.

Parallel and distributed systems: reliable distributed computing, parallel computer architectures, systems software, databases.

Programming languages and compilers: programming language design and implementation, program development methodology (including automatic derivation of programs from high-level specifications), languages for parallel computation, instruction level scheduling.

Scientific computing: computations in specific application domains, numerical analysis, optimization and control.

NORTHEASTERN UNIVERSITY

College of Computer Science

Programs of Study	The College of Computer Science offers programs leading to the M.S. and Ph.D. degrees. The M.S. program is designed for those who are seeking to prepare themselves for organizations that design, develop, market, or utilize computing systems. Forty-eight quarter hours of study are required. Areas of concentration include artificial intelligence, communications and networks, databases, graphics and imaging, operating systems, programming languages and compilers, software engineering, and theory. Admission normally requires a B.S. in computer science. College graduates with equivalent industrial or technical experience may also apply.
	The goal of the Ph.D. program is to equip its graduates to conduct state-of-the-art research in computer science, either in academia or in industry. The curriculum aims to fulfill this goal by providing the student with a broad background in the fundamentals of computer science, advanced courses in the dissertation area, and an intensive research experience, culminating in the writing of a dissertation.
	The Graduate School of the College of Computer Science offers courses on a quarter system. Most full-time students complete the M.S. program in 1½ to 2 years. Part-time students usually elect one or two courses per academic quarter and can complete the M.S. degree in two or three years. Most of the graduate courses are offered in the late afternoon and early evening, which enables many students to pursue their graduate degrees while continuing with their daytime employment.
Research Facilities	The main computing facility in the College of Computer Science is a network of about seventy-five UNIX workstations connected via a mixed 10/100 switched Ethernet network to a variety of special-purpose equipment and servers. The workstations are primarily Sun UltraSPARCs or SPARCs and DEC Alphas. The main file server is an Auspex with more than 38 gigabytes of storage. Microcomputers include more than 100 personal computers, including Macintoshes and Windows NT Pentium/PII machines. These machines are available in four laboratories in Cullinane Hall. Dial-in access to UNIX workstations is provided via high-speed modems. SLIP/PPP access is available to computer science majors, graduate students, and faculty members. A ring of eight Alpha workstations connected over a high-speed CDDI network is available for research purposes. The College also has two research ATM networks for experiments in network and distributed computing. Graduate students in the College of Computer Science also have access to the University-wide facilities, including microcomputer laboratories, a computer mail and conferencing system, and an array of specialized computing equipment. A high-speed data network links users and facilities on the central campus and three satellite campuses. The campus network is also connected via the Internet to computing resources around the world.
	The University Libraries system contains more than 852,000 volumes, 2 million microforms, 171,000 government documents, 8,200 serial subscriptions, and 18,900 audiotape, videotape, and software titles. A central library contains technologically sophisticated services, including an online catalog and circulation systems, a gateway to external networked information resources, and a network of CD-ROM optical disk databases. Students also have access to major research collections through the Boston Library Consortium.
Financial Aid	Northeastern University awards need-based financial aid to graduate students through the Federal Perkins Loan, Federal Work-Study, and Federal Stafford Student Loan programs. The University offers a limited number of fellowships and Martin Luther King Jr. Scholarships to students from minority groups. The Graduate School of the College of Computer Science also provides financial assistance through teaching, research, and administrative assistantship awards that include tuition remission and a stipend, typically ranging from $11,900 to $13,400. These assistantships require a maximum of 20 hours of work per week. A limited number of cooperative education positions are also available. Co-op is an integration of classroom work and professional experience in an organized program under which qualified graduate students have an opportunity to combine their classroom activities with employment in industry, business, and government.
Cost of Study	Tuition for 1998–99 was $475 per quarter hour of credit. Where applicable, special tuition charges are made for theses, dissertations, teaching, practicums, and fieldwork. Other charges include the Student Center fee and the health and accident insurance fee required of all full-time students.
Living and Housing Costs	On-campus living expenses are estimated at $900 per month, with on-campus housing available on a limited basis to newly accepted students. Off-campus living expenses are estimated at $1000 per month. A public transportation system services the greater Boston area, and there are convenient subway and bus services.
Student Group	Approximately 30,600 students are enrolled at Northeastern University, representing a wide variety of academic, professional, geographic, and cultural backgrounds. The Graduate School of the College of Computer Science has 130 students, 64 percent of whom attend on a full-time basis.
Location	Boston, the capital city of Massachusetts, offers students extraordinary academic, cultural, and recreational opportunities. In addition to the abundant resources available within Northeastern, there are those of the other educational and cultural institutions of greater Boston. Boston is a mixture of Colonial tradition and modern America, and it is home to people of every intellectual, political, economic, racial, ethnic, and religious background. It is a place where the past is appreciated, the present enjoyed, and the future anticipated.
The University and The College	Founded in 1898, Northeastern is a privately endowed nonsectarian institution of higher learning and is among the largest private universities in the country. Today, Northeastern has nine undergraduate schools and colleges, ten graduate and professional schools, two undergraduate divisions offering part-time study, a number of continuing and special education programs and institutes, several suburban campuses, and a large research division. The College of Computer Science is a fully accredited, degree-granting academic unit in the United States, now offering the M.S. and Ph.D. in computer science, dedicated to computer science under the cooperative education program. Northeastern's proximity to leading computer companies aids it in promoting exchanges of ideas and improved research and teaching.
Applying	An applicant must submit an application form, a nonrefundable application fee, complete official transcripts indicating the award of a bachelor's degree from a recognized institution, a typed 250- to 300-word personal statement, an official copy of scores on the GRE General Test, and three letters of recommendation. Applicants whose native language is not English must submit official TOEFL scores. Acceptance into the College of Computer Science is granted to an applicant upon recommendation of the College's Graduate Committee after a review of the completed application. Applicants may begin their study in any quarter.
Correspondence and Information	Director of Graduate Studies College of Computer Science 161 Cullinane Hall Northeastern University Boston, Massachusetts 02115 Telephone: 617-373-2464 Fax: 617-373-5121 E-mail: csgradinfo@ccs.neu.edu

Northeastern University

THE FACULTY AND THEIR RESEARCH

Larry A. Finkelstein, Dean of the College; Ph.D., Birmingham (England). Symbolic problems in algebra, group theory algorithms and applications, fast algorithms for signal processing.

Agnes H. Chan, Associate Dean and Director of the Graduate Program; Ph.D., Ohio State. Coding theory, cryptography and computer security, algorithms.

Richard A. Rasala, Associate Dean and Director of the Undergraduate Program; Ph.D., Harvard. Computer-aided instruction, algorithm animation, graphics, multimedia, software engineering.

Professors

Gene Cooperman, Ph.D., Brown. Symbolic algebra, computational algebra, search and enumeration in large data sets, distributed systems.

Jill Crisman, Ph.D., Carnegie Mellon (Joint with Department of Electrical and Computer Engineering). Intelligent robotic systems and active computer vision.

Harriet J. Fell, Ph.D., MIT. Interactive graphics systems, raster graphics algorithms, digital typography, cryptography.

Karl J. Lieberherr, Ph.D., ETH (Switzerland). Software development methodology and tools.

Viera K. Proulx, Ph.D., Columbia. Computer science education, object-oriented design, curriculum development and design.

Betty J. Salzberg, Ph.D., Michigan. Database access methods, online reorganization, concurrency and recovery.

Raoul N. Smith, Ph.D., Brown. Intelligent interfaces, knowledge representation, natural-language processing, expert systems.

Mitchell Wand, Ph.D., MIT. The semantics of programming languages, program verification and construction, algebra and logic.

Patrick S. P. Wang, Ph.D., Oregon State. Artificial intelligence, pattern recognition, programming languages, automata.

Associate Professors

Kenneth P. Baclawski, Ph.D., Harvard. Distributed and object-oriented database systems, high-performance concurrency control methods, data semantics and view integration.

John Casey, B.A., Boston College. Exploring the possibilities of cooperation among numbers of computers on common tasks.

William Clinger, Ph.D., MIT. Semantics and implementation of programming languages.

Robert P. Futrelle, Ph.D., MIT. Artificial intelligence and the construction of an intelligent "scientists' assistant," natural-language and diagram understanding, representation and reasoning about biological knowledge.

Carole D. Hafner, Ph.D., Michigan. Artificial intelligence, knowledge representation, natural-language processing.

Ronald J. Williams, Ph.D., California, San Diego. Machine learning, reinforcement learning, neural networks.

Bryant W. York, Ph.D., Massachusetts Amherst. Parallel and distributed computing, computational science, applications of advanced technology to education.

Assistant Professors

David Lorenz, Ph.D., Technion-Israel. Software engineering, machine learning, genetic algorithms.

Ibrahim Matta, Ph.D., Maryland. Integrated-services networks, routing protocols, modeling and performance evaluation.

Rajmohan Rajaraman, Ph.D., Texas. Algorithms, distributed systems, combinatorial optimization.

Adjunct Professors

John Makhoul, Ph.D., MIT. Image processing.

Eytan Modiano, Ph.D., Maryland. Communication networks.

Homer Pien, Ph.D., Northeastern. Computer vision, image processing.

NORTHWESTERN UNIVERSITY

McCormick School of Engineering and Applied Science
Department of Computer Science

Program of Study

The department offers a comprehensive program of course work and research leading to the degree of Doctor of Philosophy. A Master of Science degree may be obtained upon completion of a two-year program of courses; however, students are normally admitted only if they intend to complete the Ph.D. program and exhibit the ability to successfully do so. A candidate for the Ph.D. degree is required to complete nine quarters of full-time registration (normally three years, but credits may be transferred from other graduate institutions), pass the departmental qualifying examination, prepare a thesis that presents the results of original research, and pass an oral examination based on the thesis.

Most course work is accomplished in the first two years. Involvement in research typically commences as early as the second quarter of the first year. At the close of their second year, students take their Ph.D. qualifying examinations. After admission to candidacy, students work primarily on research.

Research Facilities

The department maintains an extensive network of more than 150 high-end personal computers and workstations, with eighteen assorted servers. The internal network is an FDDI Ring with switched 10T hubs linked to Northwestern's backbone network by ATM fiber. The entire infrastructure is maintained by a professional support staff. Specific services include a 24-gigabyte real-time video server, an Oracle server, and a code manager server. Multimedia production is supported by a complete graphics and video production facility with ten dedicated high-end Macintosh computers. The Autonomous Mobile Robot Laboratory has two high-performance robot systems. The robots are equipped with real-time color vision, simple object grippers, radio modems, and high-end speech synthesizers. The lab also has workstations, video equipment, hardware assembly, and test equipment.

Financial Aid

All applicants for admission are automatically considered for financial aid. Students receiving teaching assistantships are given a nine-month stipend ($12,078 in 1999–2000) plus full tuition. Teaching assistants spend 7 to 8 contact hours per week in laboratory teaching or other equivalent duties and approximately another 3 to 4 hours per week in preparation. Fellowships are also available; fellows may receive up to $20,000 over twelve months, with no teaching duties. Students normally receive financial support during the summer.

Cost of Study

Tuition in 1999–2000 is $21,798 for the three-quarter academic year. (Full tuition remission is offered with teaching assistantships, as noted above.) Books and supplies vary in cost, but a typical figure is $700 per year.

Living and Housing Costs

In 1998 rental rates in the Evanston area ranged from $650 to $950 per month for a one-bedroom apartment and $1000 to $2000 per month for a three-bedroom apartment, depending upon location and amenities. Single rooms in shared houses ranged from $300 to $600 per month. Campus dining facilities are available.

Student Group

There are currently 25 graduate students in computer science and most receive full financial support. The undergraduate enrollment in the University is 7,300 with approximately equal numbers of men and women. The graduate enrollment on the Evanston campus of Northwestern is 4,000.

Student Outcomes

Recent graduates have accepted tenure-track positions at Columbia University and University of Michigan. They have also accepted industrial and business positions in companies such as Xerox, Learning Sciences Corporation, and Andersen Consulting as well as becoming independent consultants.

Location

The Evanston campus of Northwestern University stretches for a mile along the western shore of Lake Michigan. Evanston is the first suburb north of Chicago and is one of the most pleasant residential towns of the area. It has an excellent shopping district within walking distance of the campus and four lakefront parks with sandy beaches and picnic areas. Chicago, with its wide variety of shopping facilities, cultural activities, and entertainment, is easily reached by the elevated railroad running close to the campus. A wide variety of activities in the form of sports events, plays, concerts, and public lectures are an integral part of life at Northwestern University.

The University

Northwestern's geographical location enables the student to profit from the cultural advantages of a large city, the recreational opportunities of the local environment, and the quieter pace of town life when on campus. There is housing within a 20-minute walk of the campus.

Applying

Except in special cases, students are admitted only in September. Completed applications for the forthcoming year should be received by February 1. The General Test of the Graduate Record Examinations is required and students are encouraged to take a Subject Test. These tests should be taken early enough so that test scores are available by February 1. TOEFL scores are required of candidates from non-English-speaking countries. Completed applications and Graduate Record Examinations scores must be submitted by February 1 to be considered for financial support.

Correspondence and Information

Graduate Admission Coordinator
Computer Science Department
Northwestern University
1890 Maple Avenue
Evanston, Illinois 60201
Telephone: 847-467-1174
E-mail: compsci@cs.nwu.edu

Northwestern University

THE FACULTY AND THEIR RESEARCH

Bradley Adelberg, Assistant Professor; Ph.D., Stanford, 1997. Databases, real-time systems, computer architecture.

Lawrence A. Birnbaum, Associate Professor; Ph.D., Yale, 1986. Semantic information processing, educational software design, natural language processing, memory and learning, interface design/HCI, computer vision.

Brian Dennis, Assistant Professor; Ph.D., Berkeley, 1998. Programming language design and implementation, multimedia systems and Internet-based services.

Daniel Edelson, Assistant Professor; Ph.D., Northwestern, 1993. Computer-based learning environments, case-based teaching, computer-supported collaborative learning (CSCL), scientific visualization tools for open-ended inquiry.

Kenneth Forbus, Professor; Ph.D., MIT, 1984. Qualitative physics, analogical reasoning and learning, cognitive simulation.

Louis M. Gomez, Associate Professor; Ph.D., Berkeley, 1979. Design of collaborative learning environments, computer-supported collaborative work, human-computer interaction.

Kristian J. Hammond, Professor; Ph.D., Yale, 1986. Artificial intelligence, intelligent information systems, information retrieval, electronic commerce, integrated knowledge management systems.

Ian Horswill, Assistant Professor; Ph.D., MIT, 1993. Autonomous agents, robotics, and computer vision; cognitive architecture and situated agency and biological modeling.

Christopher K. Riesbeck, Associate Professor; Ph.D., Stanford, 1974. Natural language understanding, case-based reasoning, intelligent tutoring systems, intelligent interfaces for knowledge acquisition and teaching, authoring tools.

Roger C. Schank, John Evans Professor and Director of the Institute for the Learning Sciences; Ph.D., Texas, 1969. Artificial intelligence, cognitive science, natural language processing, learning, models of human reasoning and human memory, computers and education.

Jennifer M. Schopf, Assistant Professor; Ph.D., California, San Diego, 1998. Parallel distributed computation, metacomputing, performance prediction and modeling, scheduling and application characterization.

RESEARCH FACULTY

Alex Kass, Research Associate Professor; Ph.D., Yale, 1990. Simulation-based training and case presentation, theory-rich authoring tools for learning environments, case-based reasoning, story understanding, hypothesis formation, machine learning, computational creativity.

RESEARCH ACTIVITIES

Northwestern's Department of Computer Science currently emphasizes research in the following five areas:

Autonomous Mobile Robot Group

Robot control and perceptual systems combining logical inference, color vision, and natural language instruction.

Education and Technology

Authoring tools for intelligent learning environments, scaffolded scientific visualization environments, learning technology for urban schools.

Intelligent Information Laboratory

Computation for communication, query-less information access, task modeling for intelligent performance support systems, interface design, and knowledge management.

Qualitative Reasoning Group

Articulate virtual labs, self-explanatory simulators, cognitive simulations of analogy and similarity.

Systems Research Group

Databases: technology to extract data from text and other semistructured sources, new information management systems blurring the line between traditional file systems and semistructured data stores, improved performance of main memory database systems.

Programming environments: programming language design and implementation, programming environments, multimedia systems, mobile computing and wireless networking scripting languages for transforming structured documents with an eye toward building Web-based personal information systems and agents.

MAJOR RESEARCH EFFORTS

Computer-based education and training.

Design of intelligent learning-by-doing environments, construction and evaluation of education and training systems, development of theory-rich authoring tools to speed high-quality system construction.

Intelligent task support.

Generation and use of task models, multimedia corporate memory, semantically appropriate interfaces.

Cognitive simulation of analogical processing.

Structure-mapping engine, MAC/FAC models similarity-based retrieval, MARS models analogical problem solving, problem of solving and reasoning technology.

Qualitative physics.

Qualitative process theory, compositional modeling, qualitative spatial reasoning, self-explanatory simulation, articulate virtual laboratories for science and engineering education.

Autonomous robots.

Development of autonomous robots that seamlessly integrate sensory-motor activities such as navigation and object manipulation with cognitive activities such as following a set of instructions, techniques for real-time visual obstacle avoidance and navigation, object tracking and real-time parallel reasoning systems that interface cleanly with active vision systems.

NOVA SOUTHEASTERN UNIVERSITY

The School of Computer and Information Sciences

Programs of Study

A major force in educational innovation, the School of Computer and Information Sciences (SCIS) is distinguished by its ability to offer both traditional and nontraditional choices in educational programs and formats that enable professionals to pursue advanced degrees without career interruption. SCIS offers programs leading to the Master of Science, Doctor of Philosophy, and Doctor of Education in several disciplines. It has more than 1,000 graduate students from across the United States and other countries, and has been awarding graduate degrees since 1980. The School offers programs leading to the M.S. in computer information systems, computer science, computing technology in education, and management information systems; the Ph.D. in computer information systems, computer science, information systems, and information science; and the Ph.D. and Ed.D. in computing technology in education. A combined master's/doctoral degree program is available. The School offers master's degree programs in the evening on campus or online via the Internet. Master's programs require 36 credit hours for graduation and may be completed in eighteen months. To earn the degree in eighteen months, the student must enroll in two courses per term. Terms are twelve weeks long, and there are four terms each year. Terms begin in September, January, April, and July. Doctoral students, depending on the program, may take one of two formats: cluster or institute. Cluster students attend four cluster meetings per year, held quarterly over an extended weekend (Friday, Saturday, and half-day Sunday) at the University. Cluster terms start in March and September. Cluster weekends take place in March, June, September, and December. Institute students attend weeklong institutes in January and July at the University at the start of each five-month term. Clusters and institutes bring together students and faculty for participation in courses, workshops, seminars, and dissertation counseling. Between meetings, students work on assignments and projects and participate in online activities. The School of Computer and Information Sciences pioneered online graduate education and has been offering programs with an online component since 1983. Online activities require use of a computer and modem from home or office or while traveling. Students may participate in online activities or online courses from anywhere in the U.S. or outside the U.S. where Internet access is available. Online interactive learning methods, used throughout the instructional sequence, facilitate frequent interaction with faculty, classmates, and colleagues.Online instruction and interaction include a wide variety of sophisticated techniques such as the real-time electronic classroom; online forums; online submission of assignments for review by faculty; e-mail; the electronic library; World Wide Web pages to access course material, announcements, etc.; and use of the Internet and World Wide Web for research. The Ph.D. programs require 64 credit hours, including eight courses, four projects, and the dissertation. They may be completed in three years.

Research Facilities

Computing facilities include ten Sun servers, nine of which are SPARCserver 1000E's and one of which is a SPARCserver 20 running Solaris; a DEC 5910 running Ultrix; and a DEC VAX 6610 and DEC VAX 4500 running VMS. The Sun servers are connected to two Sun storage arrays that provide a total of 30 gigabytes of mirrored critical data using RAID 0+1. The two VAX machines are connected to a DEC storage works box with 24 gigabytes of disk space. There are also fourteen Sun workstations used by faculty members and staff. Students have access to four Sun workstations, 320 IBM personal computers, and fifty Apple Macintoshes in fourteen microcomputer laboratories. The PCs are connected to six servers with an application storage area of 16 gigabytes. The network includes a T-1 Internet connection, capacity for more than 200 dial-up connections via a local modem pool and a national switched public data network, five WAN connections to satellite campus locations in the area, FDDI connections between all major buildings, and support for 5,700 connections at all the campus locations.

Financial Aid

The Office of Student Financial Aid administers the University's financial aid programs of grants, loans, scholarships, and student employment and provides professional financial advisers to help students plan for the most efficient use of their financial resources for education. To qualify for financial aid, a student must be admitted into a University program, must be a U.S. citizen or a U.S. immigrant, and must plan on registering for a minimum of 6 credit hours per term. A prospective student who requires financial assistance should apply for financial aid while a candidate for admission. For financial aid information or application forms, students should call 800-522-3243.

Cost of Study

For the 1999–2000 academic year, tuition is $370 per credit for master's students; for doctoral students, semiannual tuition is $4150.

Living and Housing Costs

There are many furnished apartments available for lease on an annual basis to graduate students and married students without children. These apartments are located in several buildings on the Main Campus. Application for housing for the fall term should be submitted prior to May 31. For additional information, students should call the Office of Student Housing at 800-541-6682 Ext. 5654.

Student Group

The School of Computer and Information Sciences has more than 1,000 graduate students from across the U.S. and other countries and has been awarding graduate degrees since 1983.

Location

The School is located on NSU's East Campus in Fort Lauderdale, Florida. In addition to the Main Campus and East Campus, NSU has facilities in downtown Fort Lauderdale, Coral Springs, Port Everglades, and North Miami Beach.

The University

Located on a beautiful 232-acre campus in Fort Lauderdale, Nova Southeastern University (NSU) has approximately 16,000 students and is the largest private, independent institution of higher education in Florida. It ranks twenty-fifth in the size of its postbaccalaureate programs among the 1,560 public and private universities in the U.S. with graduate and professional programs, and tenth among private universities. In addition to the School of Computer and Information Sciences, NSU has an undergraduate college and graduate schools of medicine, dentistry, pharmacy, allied health, optometry, law, psychology, education, business, oceanography, and social and systemic studies. To date, the institution has produced approximately 50,000 alumni. NSU has enjoyed full accreditation by the Commission on Colleges of the Southern Association of Colleges and Schools (SACS) since 1971. SACS is recognized by the U.S. Department of Education as the regional accrediting body for this region of the United States.

Applying

Applications, including transcripts and recommendations, should be submitted at least three months before the anticipated starting term. Students who wish to matriculate in a shorter amount of time must contact the SCIS admissions office by telephone to begin the process. Copies of transcripts are acceptable for unofficial early review. Students applying late may be granted provisional acceptance pending completion of the application process. Master's terms start in September, January, April, and July. Doctoral cluster terms start in September and March. Doctoral institute terms start in January and July.

Correspondence and Information

The School of Computer and Information Sciences
Nova Southeastern University
3100 Southwest 9th Avenue
Fort Lauderdale, Florida 33315
Telephone: 954-262-2000
 800-986-2247 Ext. 2000 (toll-free)
E-mail: scisinfo@scis.nova.edu
World Wide Web: http://www.scis.nova.edu

Nova Southeastern University

THE FACULTY AND THEIR RESEARCH AREAS

Gertrude W. Abramson, Ed.D., Columbia. Computer-supported education, hypermedia/multimedia, instructional systems design and development, distance learning.

Maxine S. Cohen, Ph.D., SUNY at Binghamton. Human-computer interaction, multimedia, usability engineering, database systems, distance learning.

Laurie P. Dringus, Ph.D., Nova Southeastern. Human-computer interaction, group support systems, usability engineering, learning theory, instructional delivery systems, distance learning.

George K. Fornshell, Ph.D., Nova Southeastern. Instructional systems development, multimedia, authoring systems, human factors, distance education.

William L. Hafner, M.S.E.E., Pennsylvania. Human-computer interaction, data warehousing, information storage and retrieval, computer security, artificial intelligence.

William M. Hartman, Ph.D., Nova Southeastern. Software engineering, data communications, computer networks, decision support systems, mathematics in computing.

Michael J. Laszlo, Ph.D., Princeton. Data structures and algorithms, software engineering, programming, computer graphics.

Jacques Levin, Ph.D., Grenoble (France). Database management, modeling, distance education, decision support systems, numerical analysis.

Edward Lieblein, Ph.D., Pennsylvania. Software engineering, object-oriented design, programming languages, automata theory.

Marlyn Kemper Littman, Ph.D., Nova Southeastern. Computer networks, ATM, wirefree and wire-based communications, network security, distance learning.

Frank Mitropoulos, M.S., Nova Southeastern. Programming languages, data structures, software engineering, object-oriented design, C, C++.

Sumitra Mukherjee, Ph.D., Carnegie Mellon. Database, decision support systems, information systems, network security, artificial intelligence, telecommunications.

John Scigliano, Ed.D., Florida. Online information systems, information systems management, distance education.

Greg Simco, Ph.D., Nova Southeastern. Operating systems, data communications, computer networks, client-server computing, online learning environments, C++, Java.

Junping Sun, Ph.D., Wayne State. Database management systems, object-oriented database systems, artificial neural networks.

Raisa Szabo, M.S., Budapest Technical. Computer architecture, artificial intelligence, neural networks, robotics, operations research, concurrent languages.

Steven R. Terrell, Ed.D., Florida International. Research methodology and statistics, learning theory, distance education, computer-managed instruction.

Visiting and Adjunct Faculty

Susan Dorchak, Ph.D.
Andres Folleco, Ph.D.
Rollins Guild, Ph.D.
Lee Leitner, Ph.D.
Robert Lipton, Ph.D.
Richard Manning, Ph.D.
Ronald McFarland, Ph.D.
Terry McQueen, D.B.A.
David Metcalf II, Ph.D.
Michael Moody, Ph.D.
Elena Schultz, M.S.
Steven Zink, Ph.D.

The Mailman Hollywood building houses the computer link to students throughout the world.

Students attending class in NSU's state-of-the-art computer facilities.

NSU is located on a 315-acre campus in Ft. Lauderdale, Florida.

THE OHIO STATE UNIVERSITY

Department of Computer and Information Science

Programs of Study	The department offers graduate programs leading to the M.S. and Ph.D. degrees. It also offers joint M.S. degree programs with the Departments of Mathematics and Biomedical Engineering and with the Center for Mapping. Research areas include artificial intelligence, combinatorial algorithms, computational geometry, concurrent programming, database systems, data translation, graphics, image analysis, multimedia, networking, neural networks, object-oriented applications, parallel and distributed computing, parallel computer architecture, performance evaluation, reusable software, robotics, scientific computing, software engineering, theoretical computer science, and visualization.
	Requirements for the Ph.D. degree include course work and a preliminary examination based on initial and proposed thesis research. The M.S. degree, whether completed with a thesis option or a comprehensive examination option, usually takes two years.
Research Facilities	Computing facilities used for instruction and research consist of Unix, NT Operating Systems, and NCD X-terminals. The Unix platforms are Sun Solaris and Hewlett Packard HP-UX. The NT 3.51 Operating System is in the process of being upgraded to NT 4.0.
Financial Aid	Financial aid is available to students in the form of teaching assistantships, research assistantships, and fellowships. These provide tuition and fee waivers and a significant stipend. Special fellowships for members of minority groups are also available.
Cost of Study	In-state tuition for 1999–2000 is $1944 per quarter, and out-of-state tuition is $4844 per quarter.
Living and Housing Costs	In addition to dormitories, University-operated one- and two-bedroom apartments are available for married students; the cost is $365 to $465 per month. There is also a substantial amount of off-campus housing in the immediate area with similar costs.
Student Group	The department has 170 graduate students, with approximately 50 new students entering each year. Approximately thirteen Ph.D. degrees and fifty M.S. degrees are awarded each year.
Student Outcomes	Of the Ph.D. graduates, approximately 70 percent take jobs in industry and 30 percent in academia.
Location	The Ohio State University is located about 3 miles north of downtown Columbus, the capital of Ohio. The city has a metropolitan population of approximately 1.4 million and is one of the fastest-growing urban areas in the United States. Columbus is a global center for high technology (especially in information services), is home to an active arts community, and serves as the corporate and divisional headquarters for a number of major corporations.
The University and The Department	The University was founded as a land-grant institution in 1870. It has one of the most comprehensive academic programs of any university in the world, many of which are ranked among the top twenty-five in the country. *U.S. News & World Report* recently ranked The Ohio State University fifteenth among public universities in the United States for overall academic reputation.
	The Department of Computer and Information Science, formed in 1968, has a tenure-track faculty of 33 members, all of whom hold a Ph.D., as well as several auxiliary and part-time faculty members. All regular faculty members lead active research programs, and a high-quality graduate student population contributes to the intellectual vitality of the department. A 1996 survey published in *Communications of the ACM* ranked the department fourteenth among academic computer science departments in the number of research publications appearing in prestigious IEEE and ACM transactions.
Applying	For fall quarter admission, all application materials must be received by August 15 (July 1 for those applying from abroad). All assistantship and fellowship applications must be received in the admissions office by January 15. Applicants must take the GRE General Test, and if their undergraduate degree is not in computer science, it is strongly recommended that they submit scores from the GRE Subject Test in computer science. If undergraduate course work is not in computer science, significant related course work is expected. The TOEFL is required and the TSE highly recommended for applicants whose native language is not English.

Correspondence and Information

Graduate Admissions Secretary
Department of Computer and Information Science
The Ohio State University
2015 Neil Avenue
Columbus, Ohio 43210-1277
Telephone: 614-292-7084
Fax: 614-292-2911
E-mail: oneill@cis.ohio-state.edu
World Wide Web: http://www.cis.ohio-state.edu/
~grad-adm/

Office of Admissions
The Ohio State University
3rd Floor, Lincoln Tower
1800 Cannon Drive
Columbus, Ohio 43210-1277
Telephone: 614-292-3980
World Wide Web: http://www.afa.adm.ohio-state.edu/

The Ohio State University

THE FACULTY AND THEIR RESEARCH

Anish Arora, Associate Professor; Ph.D., Texas at Austin, 1992. Distributed and concurrent systems, fault-tolerant computing, software engineering, formal methods, communication protocols, computer networks.

Gerald Baumgartner, Assistant Professor; Ph.D., Purdue, 1996. Programming languages, object-oriented programming.

Wayne Carlson, Associate Professor; Ph.D., Ohio State, 1982. Graphics, human-computer interaction, mapping.

Roger Crawfis, Assistant Professor; Ph.D., California, Davis, 1995. Graphics, visualization.

Tamal Dey, Associate Professor; Ph.D., Purdue, 1991. Computational geometry, computer graphics.

Wu-Chi Feng, Assistant Professor; Ph.D., Michigan, 1996. Networking, multimedia computing.

Eitan M. Gurari, Associate Professor; Ph.D., Minnesota, 1978. Programs and automata theory, computational complexity, software engineering.

Mary Jean Harrold, Associate Professor; Ph.D., Pittsburgh, 1988. Software analysis and testing, software engineering.

Raj Jain, Professor; Ph.D., Harvard, 1978. Performance analysis, computer networking,

Douglas S. Kerr, Associate Professor; Ph.D., Purdue, 1967. Database systems, software engineering.

Ten-Hwang Lai, Professor; Ph.D., Minnesota, 1982. Design and analysis of algorithms, parallel algorithms, computational aspects of VLSI.

Richard Lewis, Assistant Professor; Ph.D., Carnegie Mellon, 1993. Cognitive modeling, computational psycholinguistics, artificial intelligence.

Ming-Tsan Liu, Professor; Ph.D., Pennsylvania, 1964. Computer architecture and organization, computer networking, parallel and distributed processing/minicomputer/microcomputer systems, fault-tolerant computing systems.

Timothy J. Long, Associate Professor; Ph.D., Purdue, 1978. Complexity theory, software engineering education.

Sandra A. Mamrak, Professor; Ph.D., Illinois, 1975. Distributed processing, operating systems, computer network architecture, heterogeneous networks.

Renee Miller, Assistant Professor; Ph.D., Wisconsin, 1994. Database systems.

Ramon E. Moore, Professor; Ph.D., Stanford, 1963. Numerical computations.

Mervin E. Muller, Professor; Ph.D., UCLA, 1954. Management systems; design and developmental systems, including performance analysis and systems engineering, statistical computing, simulation, distributed systems, databases, financial systems.

William F. Ogden, Associate Professor; Ph.D., Stanford, 1969. Software engineering, program verification.

Dhabaleswar K. Panda, Associate Professor; Ph.D., USC, 1991. Parallel computer architecture, interprocessor communication, high-performance computing, wormhole routing, ATM switching, clustered systems, networks of workstations.

Richard Parent, Associate Professor; Ph.D., Ohio State, 1977. Computer graphics, artificial intelligence and animation.

Ponnuswamy Sadayappan, Professor; Ph.D., SUNY at Stony Brook, 1983. Parallel computing, scientific computing.

Mukesh Singhal, Associate Professor; Ph.D., Maryland, 1986. Distributed systems, distributed databases, systems modeling and performance evaluation.

Paul A. G. Sivilotti, Assistant Professor; Ph.D., Caltech, 1997. Distributed computing, software engineering.

Neelamegam Soundarajan, Associate Professor; Ph.D., Bombay, 1978. Theory of computation, semantics of programming languages, semantics of parallel processing.

Kenneth J. Supowit, Associate Professor; Ph.D., Illinois, 1981. Combinatorial algorithms, design automation.

DeLiang Wang, Associate Professor; Ph.D., USC, 1991. Temporal sequence processing, neural mechanisms of visuomotor coordination, visual pattern perception, neural engineering.

Bruce W. Weide, Professor; Ph.D., Carnegie-Mellon, 1978. Analysis of algorithms, data structures, combinatorics, computer architecture, parallel and distributed computing, reusable software.

Rephael Wenger, Associate Professor; Ph.D., McGill, 1988. Theory of algorithms and computational complexity.

Feng Zhao, Associate Professor; Ph.D., MIT, 1992. Artificial intelligence, numerical analysis, parallel processing, machine recognition.

Song Chun Zhu, Assistant Professor; Ph.D., Harvard, 1996. Artificial intelligence, computer vision.

Stuart H. Zweben, Professor and Chairman; Ph.D., Purdue, 1974. Software engineering, programming methodology, analysis of algorithms, data structures.

OKLAHOMA STATE UNIVERSITY

Computer Science Department

Programs of Study

The Computer Science Department of Oklahoma State University offers graduate programs leading to the Master of Science, Doctor of Philosophy, and Doctor of Education degrees.

The M.S. degree requires 30 hours of course work, including a thesis. For all M.S. students, the department requires courses in programming languages, operating systems, advanced data structures, and computer architecture. Graduates of the M.S. program are well prepared to serve as industry leaders in software development or to move into a doctoral program.

The Ph.D. program requires at least 30 semester hours of course work beyond the master's degree and an additional 30 semester hours of research. The presentation and defense of a dissertation describing original research are required. Ph.D. graduates have a sound background for original research in educational and industrial environments.

The Ed.D. program requires at least 30 hours of course work beyond the master's degree and an additional component consisting of 30 hours of research and advanced course work. The presentation and defense of an expository dissertation are required. Graduates of the Ed.D. program are prepared to teach in colleges and universities.

The Computer Science Department is also a participant in the cross-disciplinary Master's of Science in Telecommunications Management (M.S.T.M.) degree, representing the College of Arts and Sciences. This program combines the expertise of the Computer Science Department with that of the College of Engineering, Architecture and Technology and the College of Business Administration. Students in this program may choose computer science as their home department.

Research Facilities

Graduate students have access to departmental and University computers as well as a dedicated research computer running the UNIX operating system. There is an on-campus network linking research, departmental, and University computers. OSU is an active participant in national networks and the ONENET regional network.

The Edmond Low Library subscribes to numerous publications in the computing area and can obtain additional publications through the interlibrary loan system.

Financial Aid

Graduate assistantships require one-quarter-time or one-half-time teaching or research duties and carry a stipend for an academic year. Several such assistantships are available each year. Most Ph.D. students who apply for a teaching assistantship receive one. Students on assistantships pay Oklahoma resident tuition. A limited number of fellowships are available for Ph.D.-level study. Some students may qualify for jobs as computer programmers in any of several departments on campus.

Cost of Study

In 1998–99, state resident tuition and fees totaled approximately $1300 per semester; nonresident tuition and fees were approximately $2500 per semester. Costs are subject to change.

Living and Housing Costs

In 1998–99, the cost of dormitory room and board for single students ranged from $2000 to $3100 per academic year. Married student housing ranged upward from $294 to $600 per month. Off-campus, one-bedroom apartments rented for approximately $500 or higher per month, including utilities.

Student Group

In 1998–99, student enrollment on the Stillwater campus was approximately 20,000, of whom about 7,500 were pursuing graduate degrees. Graduate enrollment in the Computer Science Department was 200 full-time students.

The majority of graduate students in the department earned their undergraduate degrees in the sciences, but students with undergraduate degrees in other areas have completed a graduate program successfully. The majority of graduate students are studying for the M.S. and spend 1½–2 years on campus. Ed.D. and Ph.D. students usually require 3 to 4 years beyond the master's degree. Most graduate students in the department are active in the local chapter of the Association for Computing Machinery.

Student Outcomes

Graduates of the M.S. program have been extremely successful in finding software development and research positions all over the world. Most of these positions are in the fields of scientific computing, business applications, and telecommunications.

The majority of recent doctoral graduates have taken university positions. Many others have accepted industrial research positions.

Location

Stillwater is a small, attractive university city of about 38,000, located on the prairie in north-central Oklahoma. The city is 65 miles west of Tulsa and 65 miles north of Oklahoma City. There are numerous cultural activities to be found in the Stillwater community and many more within a 2-hour drive of Stillwater.

The University

Oklahoma State University was founded in 1890 as Oklahoma Agricultural and Mechanical College. The name was changed to reflect its university status in 1957. Proud of its land-grant heritage, the University takes seriously the commitment to promote liberal and practical education on the campus, throughout the state of Oklahoma, and in those areas of the nation and world where its special talents can be put to use.

The OSU campus is one of exceptional beauty, with modified Georgian architecture in all buildings. University property includes the main campus of 415 acres at Stillwater and lands and farms totaling 5,300 acres. In addition, the University holds title to the Lake Carl Blackwell area, which contains 19,364 acres and a lake covering 3,380 acres that provides recreational and experimental facilities as well as the University's water supply.

Applying

Applications for admission are considered throughout the year, but they should be submitted early enough to allow six weeks for processing prior to initial enrollment. Application for financial aid should be made by March 1 for the following academic year. All applicants are urged to take the GRE General Test, and Ph.D. students are required to submit scores on the Subject Test in computer science. International students whose native language is not English must take the TOEFL, and international students seeking financial aid must take the TSE.

Correspondence and Information

John P. Chandler, Ph.D.
Director of Graduate Programs
Computer Science Department
Math Sciences Building 219
Oklahoma State University
Stillwater, Oklahoma 74078-1053

Telephone: 405-744-5668
E-mail: grad_info@cs.okstate.edu
World Wide Web: http://www.cs.okstate.edu/

Oklahoma State University

THE FACULTY AND THEIR RESEARCH

John P. Chandler, Professor; Ph.D., Indiana. Numerical methods, data structures, statistical computation.
H. K. Dai, Assistant Professor; Ph.D., Washington (Seattle). Computing networks, theoretical computer science.
Judy Edgmand, Adjunct Assistant Professor; Ed.D., Oklahoma State. Statistical computing, social issues of computing.
Olac Fuentes, Visiting Assistant Professor; Ph.D., Rochester. Databases, data structures.
K. M. George, Professor; Ph.D., SUNY at Stony Brook. Mathematical foundations, programming language theory.
G. E. Hedrick, Regents' Service Professor; Ph.D., Iowa State. Programming languages, compiler implementation, scientific applications.
Rajgopal Kannan, Assistant Professor; Ph.D., Denver. Architecture, networks.
Jacques LaFrance, Visiting Associate Professor; Ph.D., Illinois. Artificial intelligence, programming languages.
Huizhu Lu, Professor; Ph.D., Oklahoma. Computer architecture, automata theory, database theory.
Richard W. Matzen, Visiting Assistant Professor; Ph.D., Oklahoma State. Text processing, markup languages, programming languages.
Blayne Mayfield, Associate Professor and Department Head; Ph.D., Missouri–Rolla. Artificial intelligence, data structures.
J. Terry Nutter, Assistant Professor; Ph.D., SUNY at Buffalo. Artificial intelligence, computational linguistics.
Mansur H. Samadzadeh, Associate Professor; Ph.D., Southwestern Louisiana. Software engineering, systems measurement.

Emeritus Faculty
Donald D. Fisher, Regents' Service Professor; Ph.D., Stanford. Computer organization, data and information, numerical analysis.
Donald W. Grace, Professor; Ph.D., Stanford. Optimization, combinatorics, numerical analysis.

Supporting Faculty
Herman G. Burchard (mathematics), Ph.D., Purdue. Numerical analysis.
Louis G. Johnson (computer engineering), Ph.D., MIT. Computer architecture, software engineering.
Marilyn G. Kletke (management), Ph.D., Oklahoma State. Management information systems, business applications.
J. Scott Turner (management), Ph.D., SMU. Optimization.

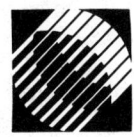

OREGON GRADUATE INSTITUTE OF SCIENCE AND TECHNOLOGY

Department of Computer Science and Engineering

Programs of Study

The goal of Oregon Graduate Institute's (OGI) Department of Computer Science and Engineering (CSE) is to give its students the knowledge and intellectual discipline they need to become innovators and leaders within their professional communities. CSE places special emphasis on preparing students for the technological evolution that they will experience throughout their careers.

CSE awards four degrees: an M.S. in computer science and engineering, with a thesis or nonthesis option; a Ph.D. in computer science and engineering; an M.S. in computational finance; and the Oregon Master of Software Engineering (O.M.S.E.).

The M.S. in computer science and engineering program prepares students for professional careers in business, industry, or government. All M.S./CSE students take the same core courses, covering at an advanced level concepts that are common to most disciplines of computer science. Each student selects an area of interest for in-depth study, choosing the M.S. thesis option or one of seven nonthesis areas of emphasis: adaptive systems, computational finance, data-intensive systems, human-computer interfaces, software engineering, spoken language systems, or systems engineering.

The Ph.D. in computer science and engineering program is oriented toward preparation for research. Each student works closely with a faculty research adviser throughout his or her residency at OGI. The program of study for each Ph.D. student is tailored to meet individual needs and interests. The doctoral dissertation documents a significant, original research contribution and must be of publishable quality, both in content and presentation.

The nonthesis M.S. in computational finance is a twelve-month intensive program that stresses applications skills in computer science, applied mathematics, and management science. The computational finance M.S. provides a solid foundation in finance at an advanced quantitative level. Unlike a standard M.B.A., OGI's computational finance curriculum is designed for scientists, engineers, and technically oriented finance professionals.

OGI is a partner with Portland State University, Oregon State University, and the University of Oregon in the Oregon Master of Software Engineering program for software professionals. O.M.S.E. focuses on principles, methods, and tools used to create high-quality products that serve the needs of customers. The curriculum emphasizes the technical leadership aspects of software engineering, teamwork, and communication skills and the business aspect of developing industrial-strength software.

The department has 36 faculty members; 22 are full-time and 14 are either part-time or jointly appointed. Local computer science professionals serve as adjunct faculty members, teaching in their areas of specialty.

Departmental research strengths include active databases, computer security, database query processing, distributed operating systems, domain-specific languages, formal specifications and designs, functional programming, meta-programming, mobile computing, natural language processing, network data management, neural networks, object-oriented databases, parallel processing, pattern recognition, sensor fusion, software reusability, speech recognition, spoken language identification, statistical computing, superimposed information models, system support for audio and video, the architecture/os interface, type systems for programming languages, and wide-area distributed systems.

Research Facilities

CSE maintains a computing infrastructure that is capable of supporting a high degree of heterogeneity, as required for high-quality research. Central services such as mail, news, dial-up access, file and printer sharing, and World Wide Web access are distributed across Sun computers and a Network Appliance file server, which are connected at up to 100 Mbits/second through a Cisco 5500 Ethernet switch. In addition to Sun computers, CSE's research network includes Intel (NT and Linux), HP, DEC, and other systems. The department accesses the Internet via Verio Northwest and participates in NERO, an ATM-based network of educational and research institutions in Oregon.

Financial Aid

The department offers research assistantships to Ph.D. students on a competitive basis. OGI's Office of Academic and Student Services provides assistance for students who seek to secure externally funded fellowships or student loans.

Cost of Study

Tuition for 1999–2000 is $4465 per academic quarter or $495 per credit. Computational finance courses are $765 per credit.

Living and Housing Costs

In general, the cost of living in Portland is less than in most major metropolitan areas on the West Coast. Students live off campus; most share housing with other students. Monthly rent for off-campus apartments ranges from $550 to $850.

Student Group

The department has approximately 100 matriculated students; 28 percent are women, and about 55 percent are international. In addition, more than 200 students, primarily engineers and scientists from nearby high-technology firms, take CSE courses to continue their professional education.

Location

OGI is 12 miles west of downtown Portland and 60 miles from the Oregon Coast, the Mount Hood ski areas, and the Columbia River Gorge. National and state parks, scenic areas, and recreational areas are at distances ranging from a few hours to a day's drive away.

The Institute

Founded in 1963, Oregon Graduate Institute combines the vigorous research emphasis and instrumentation of a large university with the personal interaction and collaboration that are characteristic of a small research institution. In addition to computer science and engineering, OGI offers M.S. and Ph.D. degrees in environmental science and engineering, electrical engineering, materials science and engineering, and biochemistry and molecular biology as well as the M.S. in management in science and technology.

Applying

M.S. applications are accepted any time; Ph.D. applications are due March 1. Applicants should hold a bachelor's degree in computer science, mathematics, engineering, a biological or physical science, or one of the quantitative social sciences. Candidates with a degree in a field other than computer science must have completed courses in data structures, discrete mathematics, logic design and computer organization, and calculus or other college-level mathematics and have an introduction to programming in a high-level language.

Required admissions materials include official college or university transcripts with a minimum GPA of 3.5 (on a 4.0 scale), GRE General Test scores (a Subject Test in computer science is not required; OGI's institutional GRE code is 4592), three letters of recommendation, and a statement of purpose that describes goals for graduate study and the field of specialization chosen for graduate work. A minimum score of 650 on the TOEFL is recommended for applicants whose native language is not English, unless the applicant earned an undergraduate degree in the United States.

Correspondence and Information

Admissions
Office of Academic and Student Services
Oregon Graduate Institute
P.O. Box 91000
Portland, Oregon 97291-1000
Telephone: 503-748-1027
        800-685-2423 (toll-free)
E-mail: admissions@admin.ogi.edu
      csedept@cse.ogi.edu
World Wide Web: http://www.ogi.edu/
      http://www.cse.ogi.edu/

Oregon Graduate Institute of Science and Technology

THE FACULTY AND THEIR RESEARCH

Andrew P. Black, Department Chair; D. Phil., Oxford. Programming languages; distributed systems; wide-area networking, particularly the World Wide Web; object-oriented languages and systems; types for objects; and the ways in which these areas interrelate.

Thomas Bundt, Associate Professor; Ph.D. (economics), Michigan State, 1985. International transmission of monetary and fiscal policies, empirical testing of derivative pricing models, applied corporate and international finance, financial cooperatives, foreign exchange markets.

Phil Cohen, Professor and Co-Director of the Center for Human Computer Communication; Ph.D. (computer science), Toronto, 1978. Multimodal interfaces, human-computer interaction, natural language processing, dialog, delegation technology, cooperating agents, communicative action, applications to mobile computing, information management, network management, manufacturing.

Crispin Cowan, Assistant Research Professor; Ph.D. (computer science), Western Ontario, 1995. Operating systems, distributed and parallel systems, computer architecture, programming languages, optimism.

Lois Delcambre, Professor and Director of DISC: A Center for Systems Software Research. Database system data models, data models for loosely structured data such as documents, object-oriented models (for requirements, analysis, and design), scientific data management.

Richard Fairley, Professor and Director of Software Engineering; Ph.D. (computer science), UCLA, 1971. Software systems engineering, software process improvement, software requirements engineering, software quality engineering, software metrics, software project management, software risk management.

Peter A. Heeman, Assistant Professor; Ph.D. (computer science), Rochester, 1997. Spoken dialog understanding, spontaneous speech recognition, intonation, modeling disfluencies, natural language processing, discourse, collaboration, statistical learning, spoken dialog systems.

James Hook, Associate Professor and Director of the Pacific Software Research Center; Ph.D. (computer science), Cornell, 1988. Type theory, programming language semantics, program verification, software engineering.

Michael Johnston, Assistant Research Professor; Ph.D. (linguistics), California, Santa Cruz, 1994. Natural language processing; human-computer interaction; multimodal interfaces; spoken dialog systems; syntax, semantics, and pragmatics of human language; linguistic processing of text for content characterization and information extraction and navigation; computational models of phonology, morphology, and the lexicon; natural language understanding and computational semantics.

Mark P. Jones, Associate Professor; D.Phil. (computation), Oxford, 1992. Programming language design and implementation, programming paradigms, module and component systems, type theory, semantics, program transformation and analysis.

Richard Kieburtz, Professor; Ph.D. (electrical engineering), Washington (Seattle), 1961. Functional programming, program transformation, software specification, deriving programs from specifications, semantics of programming languages.

John Launchbury, Associate Professor; Ph.D. (computing science), Glasgow, 1990. Functional programming languages, semantics-based program analysis, program transformation, partial evaluation.

Todd K. Leen, Associate Professor; Ph.D. (physics), Wisconsin, 1982. Neural learning algorithms, architecture and theory, dynamics, noise, model complexity and pruning, applications to signal processing.

Ling Liu, Assistant Research Professor; Ph.D. (computer science), Tilburg (The Netherlands), 1993. Information dissemination on the Internet, distributed object management, distributed database systems, data mining and data warehousing technology and applications, fundamentals of object-oriented systems.

David Maier, Professor; Ph.D. (electrical engineering and computer science), Princeton, 1978. Database systems (including object-oriented database management systems, query processing, and scientific information management), scientific computing, object-oriented and logic programming languages, algorithms, survivability of information systems, health information technology.

Dylan McNamee, Assistant Professor; Ph.D., Washington (Seattle), 1996. Operating systems, application/operating system interactions, parallel and distributed systems.

John Moody, Professor; Ph.D. (theoretical physics), Princeton, 1984. Computational finance, time-series analysis, and statistical learning theory and algorithms; foundations of neural networks, machine learning, and nonparametric statistics and the application of these methods to problems in finance, economics, and time-series analysis.

Sharon L. Oviatt, Associate Professor and Co-Director of the Center for Human Computer Communication; Ph.D. (experimental psychology), Toronto, 1979. Human language technology and multimodal systems, modality effects in communication, communication models, telecommunications and technology-mediated communication, interactive systems, human-computer interaction, empirically based design and evaluation of human-computer interfaces, cognitive science, research methodology.

Calton Pu, Professor; Ph.D. (computer science), Washington (Seattle), 1986. Transaction processing, distributed databases, scientific databases, parallel and distributed operating systems.

Tim Sheard, Associate Professor; Ph.D. (computer and information science), Massachusetts Amherst, 1985. Functional programming, software specification, program generation, reflection, automatic theorem proving, partial evaluation.

David Steere, Assistant Professor; Ph.D. (computer science), Carnegie Mellon, 1997. Operating systems, mobile computing, distributed information systems.

Jonathan Walpole, Professor; Ph.D. (computer science), Lancaster (England), 1987. Operating systems, distributed systems, multimedia computing.

JOINT APPOINTMENTS

Ronald A. Cole, Professor, Department of Electrical and Computer Engineering, Oregon Graduate Institute.

Dan Hammerstrom, Professor and Department Head, Department of Electrical and Computer Engineering, Oregon Graduate Institute.

Hynek Hermansky, Professor, Department of Electrical and Computer Engineering, Oregon Graduate Institute.

Michael W. Macon, Assistant Professor, Department of Electrical and Computer Engineering, Oregon Graduate Institute.

Misha Pavel, Professor, Department of Electrical and Computer Engineering, Oregon Graduate Institute.

Leonard Shapiro, Professor, Department of Computer Science, Portland State University.

Andrew Tolmach, Assistant Professor, Department of Computer Science, Portland State University.

Eric A. Wan, Associate Professor, Department of Electrical and Computer Engineering, Oregon Graduate Institute.

Yonghong Yan, Associate Professor, Department of Electrical and Computer Engineering, Oregon Graduate Institute.

The OGI campus and Science Park share a beautiful natural setting.

The Cooley Science Center.

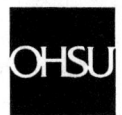

OREGON HEALTH SCIENCES UNIVERSITY

School of Medicine
Division of Medical Informatics and Outcomes Research

Programs of Study

The field of medical informatics is concerned with the development, dissemination, and evaluation of information technology in the health-care field. Oregon Health Sciences University (OHSU) is a national leader in this area, with one of the largest externally funded research and educational programs in the country. OHSU offers two educational programs in medical informatics: a Master of Science degree program and a postdoctoral training program.

The major goal of the degree program is to educate future developers and managers of health-care information systems. The curriculum includes core courses from medical informatics, health and medicine, computer science, and quantitative methods. Students have the opportunity to work on a variety of research projects with individual faculty members or to complete a practicum that allows them to gain experience in a work setting, such as a hospital information service center. The curriculum is divided between 48 hours of course work and 12 hours of thesis work. Full-time students should be able to complete the program in two years. Part-time enrollment is allowed.

The postdoctoral training program prepares medical informatics scientists for careers in academia or industry. The fellowship is funded by the National Library of Medicine (NLM) and the Department of Veterans Affairs. It is open to United States citizens who have an M.D. or Ph.D. Fellows survey the field broadly during their two- to three-year fellowship, and the program can be tailored to individual needs. They are expected to develop, conduct, and report on an original research project. Fellows can optionally enroll in the degree program.

Research Facilities

The Division of Medical Informatics and Outcomes Research is housed in the Biomedical Information Communication Center, which contains offices, classrooms, the OHSU Health Sciences Library, and other information technology resources. A state-of-the-art computer laboratory available to students enrolled in the informatics program contains twelve high-end Pentium and Macintosh workstations, a Windows NT Server, and a color laser printer. All computers at OHSU are on a single campuswide network, which is built around a 622-megabit Asynchronous Transfer Mode (ATM) backbone and is directly connected to the Internet. There are also a variety of other resources and technologies available for student projects.

Financial Aid

The major form of financial aid for degree students is the federal student loan program. A limited amount of funding may be available to provide research stipend support for students in the program. Postdoctoral fellowship trainees receive a stipend and tuition waiver consistent with NLM rules. For more information, interested students should write to: Financial Aid Director, Oregon Health Sciences University, 3181 Southwest Sam Jackson Park Road, Portland, Oregon 97201.

Cost of Study

In 1998–99, tuition and fees for a full-time student were $1944 per quarter for Oregon residents and $2970 per quarter for nonresidents. Tuition for part-time students in 1998–99 was $168 per credit for Oregon residents and $282 per credit for nonresidents.

Living and Housing Costs

There are several apartments near the Oregon Health Sciences University campus. Rent for a one-bedroom apartment downtown is between $500 and $800 per month. The surrounding suburbs tend to be slightly cheaper.

Housing is also available in the OHSU Residence Hall. For more information on the residence hall, interested students should call 503-494-8255.

Student Group

There are currently 11 part-time and 10 full-time students in the master's program. There are 4 postdoctoral fellows and 2 library fellows. Students in the program have a variety of backgrounds, including medicine, nursing, veterinary medicine, computer science, psychology, and law. In the admissions process, demonstration of self-motivation and ability to complete projects is of paramount importance. The program also values intellectual diversity, aiming to achieve a complementary mix of students with varied backgrounds.

Location

OHSU is located in the hills of southwest Portland, a metropolitan area of 1.5 million people, and offers excellent views of downtown, Mount St. Helens, and Mount Hood. The surrounding areas provide many opportunities for recreation, including hiking, camping, and skiing. Portland provides many cultural offerings, including the Oregon Symphony Orchestra, the Portland Opera, and the Portland Center for Performing Arts.

The University

OHSU is one of America's top academic health science centers and includes the Schools of Medicine, Nursing, and Dentistry. It has a national reputation for its expertise, leadership, and commitment to service and education and is widely recognized for its distinguished faculty and innovative research endeavors. OHSU is in the top 3 percent of institutions nationally in receiving NIH research funds. Medical informatics students collaborate with basic and clinical science departments, OHSU's Information Technology Group, and the computer science department at Portland State University.

Applying

The requirements for admission to the Master of Science degree program are a bachelor's degree in any area, an introductory computer programming course, an introductory course in anatomy and psychology, and GRE scores (for those without a doctoral degree from an American university). Priority is given to applications received by February 1.

The postdoctoral training program is open to U.S. citizens with an M.D. or Ph.D. in biological science or other relevant area. Positions are filled on a flexible schedule and applications should be received at least six months prior to the proposed start date. Applications should include a letter of interest, a curriculum vitae, and three letters of recommendation.

Correspondence and Information

For the Master's Degree Program:
Program Administrator
Division of Medical Informatics and Outcomes
  Research
Oregon Health Sciences University
3181 Southwest Sam Jackson Park Road, Mailcode
  BICC
Portland, Oregon 97201
Telephone: 503-494-4502
Fax: 503-494-4551
E-mail: informat@ohsu.edu
World Wide Web: http://www.ohsu.edu/
  bicc-informatics/ms

For the Postdoctoral Training Program:
Kent Spackman, M.D., Ph.D.
Division of Medical Informatics and Outcomes
  Research
Oregon Health Sciences University
3181 Southwest Sam Jackson Park Road, Mailcode
  BICC
Portland, Oregon 97201
Telephone: 503-494-6161
Fax: 503-494-4551
E-mail: spackman@ohsu.edu
World Wide Web: http://www.ohsu.edu/
  bicc-informatics/fellowship

Oregon Health Sciences University

THE FACULTY AND THEIR RESEARCH

Joint appointments are listed in parentheses.

William Hersh, M.D., Associate Professor and Chief (Internal Medicine). Information retrieval and digital libraries, electronic medical records, evidence-based medicine.

Richard Appleyard, Ph.D., Assistant Professor (Information Technology Group).

Joan Ash, Ph.D., Assistant Professor (Associate Professor, Library). Organizational aspects of informatics and the application of diffusion of innovations theory to information technology in health care; use of oral history as a qualitative research methodology.

Dana Braner, M.D., Assistant Professor (Pediatrics). Multimedia methods, including imaging, in health care.

Nancy Carney, Ph.D., Assistant Professor (Emergency Medicine). Rehabilitation of traumatic brain-injury victims.

Katherine Caton, Ph.D., Assistant Professor (Nursing). Usability evaluation for nursing information systems in long-term care, development and evaluation of an expert system prototype for community resource allocation in the care of AIDS clients.

Homer Chin, M.D., M.S., Clinical Assistant Professor (Kaiser-Permanente Northwest). Institutional electronic medical record systems.

Christopher Dubay, Ph.D., Assistant Professor (Medical Genetics). Complex genetic diseases and the dissection of the genetic basis using biostatistical techniques of linkage analysis in model systems and humans.

Karen Eden, Ph.D., Assistant Professor. Health outcomes research and evidence-based diagnostic technology assessment related to various screening tests.

Alan Ertle, M.D., M.P.H., Clinical Assistant Professor (Corvallis Clinic).

Linda Felver, Ph.D., R.N., Assistant Professor (Nursing). Pathophysiology.

Richard Gibson, M.D., Ph.D., Clinical Assistant Professor (Providence Health System). Institutional electronic medical record systems.

Paul Gorman, M.D., Assistant Professor (Internal Medicine). Assessment of information needs of primary-care practitioners.

Jeanne-Marie Guise, M.D., Assistant Professor (Obstetrics and Gynecology). Evidence-based medicine.

Warren Harrison, Ph.D., Adjunct Professor. Software quality assurance, software measurement, empirical studies of software engineering, use of the World Wide Web as an information delivery vehicle within the context of education and training.

Brian Hazelhurst, Ph.D., Clinical Assistant Professor (Sapient Health Network). Intelligent Query Engine, knowledge-based systems, neural networks, information retrieval, visual/interactive interface design.

Mark Helfand, M.D., M.P.H., Associate Professor (Internal Medicine). Health outcomes research and evidence-based diagnostic technology assessment related to various screening tests.

David Hickam, M.D., M.P.H., Professor (Internal Medicine). Health services research.

Holly Jimison, Ph.D., Assistant Professor (Public Health). Computer decision models to tailor consumer health information, evaluation of self-care and shared decision-making interventions.

Sandra Joos, Ph.D., Assistant Professor. Health services research.

Dale Kraemer, Ph.D., Assistant Professor. Statistical analyses and cost-effectiveness analyses of medical interventions, including pharmaceutical therapies.

Judith Logan, M.D., Assistant Professor. Usability evaluation of electronic medical records.

Richard Lowensohn, M.D., Associate Professor (Obstetrics and Gynecology). Clinical information management systems for obstetrics and gynecology.

Phillip Marshall, M.D., M.P.H., Clinical Assistant Professor.

Fred Masarie, M.D., Clinical Assistant Professor (Medicalogic, Inc.). Electronic medical record systems.

Hugo Maynard, Ph.D., Adjunct Professor. Evidence-based medicine.

Blackford Middleton, M.D., M.S., Clinical Assistant Professor (Medicalogic, Inc.). Electronic medical record systems.

Cynthia Morris, Ph.D., M.P.H., Associate Professor (Internal Medicine). Population-based studies to evaluate long-term natural history of congenital heart defects, assessment of the effects of dietary interventions on cardiovascular risk factors.

Heidi Nelson, M.D., M.P.H., Assistant Professor (Internal Medicine). Clinical epidemiology and women's health, assessment of the effects of moderate alcohol on health and functional outcomes.

Hanyu Ni, Ph.D., M.P.H., Assistant Professor (Cardiology). Outcomes research to evaluate cost effectiveness and factors that can be modified to improve heart failure care.

Patricia Patterson, Ph.D., Assistant Professor (Nursing). Development and testing of information systems for geriatric and chronically ill populations.

Douglas Perednia, M.D., Associate Professor (Dermatology). Telemedicine and the applications of computer imaging in medicine.

Susan Price, M.D., Assistant Professor. Use of the Internet to provide and exchange medical information; evaluating the quality of health information on the World Wide Web; linking patient-specific information retrieval with electronic medical records; security, privacy, and confidentiality issues related to electronic medical information.

Craig Redfern, D.O., Clinical Assistant Professor (Providence Health System). Evidence-based medicine.

Paul Sher, M.D., Adjunct Professor (Pathology). Workstation technology, graphic representation of laboratory data, and mathematical algorithms in computer-aided medical decision making; quality assurance and the use of computers to present quality information to health-care consumers.

Kent Spackman, M.D., Ph.D., Associate Professor (Pathology). Standardization of clinical terminology for computer-based patient records.

Andrew Zechnich, M.D., Assistant Professor (Emergency Medicine). Automated protocols for emergency medicine.

Melanie Zimmer-Gembeck, Assistant Professor. Data analysis in outcomes research.

OREGON STATE UNIVERSITY

College of Engineering
Department of Computer Science

Programs of Study	Graduate degree programs leading to the M.A., M.A.I.S., M.S., and Ph.D. degrees in computer science are offered in the following areas of concentration: parallel computing, artificial intelligence, programming languages, theory of computation, software engineering, information-based systems, numerical analysis (with the Department of Mathematics), and computer architecture (with the Department of Electrical and Computer Engineering).
	OSU operates on the quarter system. There are three quarters in the academic year. The master's degree programs require 45 hours of graduate-level courses, a final oral examination, and a thesis or research paper. The doctoral program requires approximately 120 graduate credit hours, demonstration of both breadth and depth of knowledge, and completion of significant independent research.
Research Facilities	The department has an exceptional computing environment. Parallel computing facilities include a 17-node Meiko CS-2 multicomputer, an 8-node Sun Enterprise 4000, a cluster of Sun SPARC20 quad-processors, a cluster of thirty-two Pentium II computers, and a 128-node Adaptive Systems CNAPS neural net builder. The department also maintains dozens of Hewlett-Packard, Sun, and Apple, as well as Windows, workstations distributed to faculty members, research assistants, and teaching assistants.
	Open laboratories on campus contain dozens of state-of-the-art workstations and X-terminals and three multimedia laboratories contain dozens of PowerPC Macintosh computers and PC-compatible computers.
	The Valley Library contains more than 1 million volumes and subscribes to most computer science journals.
Financial Aid	Financial support is available for qualified graduate students in the form of teaching and research assistantships. New graduate students with teaching assistantships receive a package paying $12,500 over the academic year. Stipends for half-time research assistants usually range between $16,000 and $18,000 for a twelve-month appointment.
Cost of Study	Estimated resident tuition and fees for 1999–2000 total $2175 for 9 to 16 term hours; nonresidents pay tuition and fees of $3699 per term. These figures are reduced to $300 per term for students with assistantships. (Figures are subject to change without notice.)
Living and Housing Costs	For 1999–2000, the rate for double occupancy in University dormitories and election of a full meal plan is $5102 for the academic year. A few University apartments are available for married students and rent for $360 to $420 per month. Off-campus one-bedroom apartments rent for approximately $450 to $550 per month, excluding electricity.
Student Group	The total enrollment for 1998–99 was 14,618, including 2,620 graduate students. About 47 percent of the graduate students were women. There were 75 on-campus computer science graduate majors in 1998–99.
Location	Corvallis, a city of 46,000, is located in the Willamette Valley. Portland is 80 miles to the north, the Oregon coast is 50 miles to the west, and the Cascade mountain range is 60 miles to the east. Skiing, camping, hiking, and climbing opportunities are all within easy driving distance. The climate is mild (ranging from the low 30s in the winter to 85 in the summer).
	The University and the community offer concerts, plays, movies, and art exhibits.
The University and The Department	Oregon State University is a land-grant and sea-grant university and dates back to 1868. Graduate work and research are carried out in many disciplines, including science, engineering, agriculture, and oceanography. The campus is a visual delight, beautifully landscaped with trees, flowering shrubs, and flowing lawns—all of which contribute to a relaxed atmosphere.
	The Department of Computer Science was formed in 1972 and now has 13 graduate faculty members. The department maintains close ties with neighboring institutions and high-technology industry.
Applying	A $50 fee is required of all applicants. Application for graduate admission should be made to the Office of Admissions, Oregon State University. Applications for financial assistance should be submitted to the department by March 1 for the following academic year. All graduate applicants are required to submit scores on the General Test (quantitative and verbal sections) of the GRE. International students must pass the TOEFL to be considered for University admission. Oregon State University supports equal educational opportunity without regard to sex, race, handicap, national origin, marital status, or religion.

Correspondence and Information

For additional information:
Graduate Advisor
Department of Computer Science
Dearborn Hall 303
Oregon State University
Corvallis, Oregon 97331
Telephone: 541-737-3273
Fax: 541-737-3014
World Wide Web: http://www.cs.orst.edu

For application forms:
Director of Admissions
Oregon State University
Corvallis, Oregon 97331
Telephone: 541-737-4411
E-mail: osuadmit@ccmail.orst.edu
World Wide Web: http://osu.orst.edu

Oregon State University

THE FACULTY AND THEIR RESEARCH

Bella Bose, Professor; Ph.D., SMU, 1980. Error-control codes, VLSI testing, parallel processing.

William S. Bregar, Adjunct Associate Professor (at Hewlett-Packard, Vancouver, Washington); Ph.D., Wisconsin–Madison, 1974. Artificial intelligence—problem solving, intelligent-computer-assisted instruction, representation of knowledge.

Timothy A. Budd, Associate Professor; Ph.D., Yale, 1980. Programming languages, programming environments, program testing.

Margaret M. Burnett, Associate Professor; Ph.D., Kansas, 1991. Visual programming languages; alternative programming paradigms, especially object-oriented and functional.

Curtis R. Cook, Professor; Ph.D., Iowa, 1970. Software quality, software complexity measures, program comprehension.

Paul Cull, Professor; Ph.D., Chicago, 1970. Analysis of algorithms, mathematical biology.

Bruce D. D'Ambrosio, Associate Professor; Ph.D., Berkeley, 1986. Artificial intelligence, management of uncertainty, real-time problem solving, qualitative reasoning.

Thomas G. Dietterich, Professor; Ph.D., Stanford, 1984. Machine learning.

Toshimi Minoura, Associate Professor; Ph.D., Stanford; 1980. Actionbase system for manufacturing control.

Cherri M. Pancake, Professor; Ph.D., Auburn, 1986. Software support for parallel computing, graphical user interface design, parallel languages.

Michael J. Quinn, Professor and Interim Department Head; Ph.D., Washington State, 1983. Parallel algorithms, parallel programming languages and environments.

Gregg Rothermel, Assistant Professor; Ph.D., Clemson, 1996. Software engineering.

Walter G. Rudd, Professor; Ph.D., Rice, 1969. Computer architecture, parallel computing, computational science.

Prasad Tadepalli, Associate Professor, Ph.D., Rutgers, 1989. Artificial intelligence, machine learning, problem solving, computer-integrated manufacturing.

PACE UNIVERSITY

School of Computer Science and Information Systems

Programs of Study	Graduate degrees offered by the School of Computer Science and Information Systems at both the White Plains and New York City campuses are the Master of Science (M.S.) in computer science, the Master of Science (M.S.) in information systems, and the Master of Science (M.S.) in telecommunications. The M.S. program in computer science emphasizes a hands-on approach to software systems engineering. The 36-credit curriculum balances theory and practice in software design and development, culminating in the implementation of a large software system. The program usually requires three semesters of full-time study or five semesters of part-time study.
	The M.S. in information systems provides the technological-organizational balance necessary for effectively dealing with information systems in all organizations. The program integrates information systems technology, information systems concepts and processes, and organizational functions and management into a unified information systems approach, making the relationship between information systems technology and managerial effectiveness more apparent. The 36-credit curriculum can generally be completed in three semesters of full-time study or five semesters of part-time study.
	The M.S. in telecommunications is designed to integrate the technology, management, and policy disciplines underlying telecommunications with up-to-date coverage of the issues and trends. Case studies and the use of computer-based tools and simulations integrate the concepts and disciplines learned in the program by providing a context for complex problem analysis in the planning and management of telecommunications networks. The 36-credit curriculum can generally be completed in three semesters of full-time study or five semesters of part-time study.
	A 48-credit post-master's Doctor of Professional Studies (D.P.S.) in computing sciences if offered. The D.P.S. is structured for working information technology professionals. The program can be completed through part-time asynchronous study with limited residency in three years.
	Certificate programs are available in computer communications and networks, information systems, telecommunications, object-oriented programming, and computing for teachers.
Research Facilities	Pace University's totally integrated online library system holds approximately 825,000 volumes and subscribes to nearly 4,000 serial publications. Electronic access to internal and external information and knowledge sources, including locally mounted CD-ROM databases, online retrieval systems, and the Internet, is available. The Pace libraries annually contract with Dialog, BRS, LEXIS-NEXIS, and Dow Jones News Retrieval to access statistical, bibliographic, directory, and full-text databases that cover all major subjects. The University computing network provides access to a range of both mainframe and microcomputing hardware and software. Currently, more than 250 computers are located in academic computing facilities. Pace University's wide area network (Pace Net) can be accessed from labs, dormitory rooms, and offices.
Financial Aid	A number of graduate scholarships and assistantships are available. Grants are made on the basis of outstanding academic performance as indicated by the applicant's previous college record, test scores, and demonstrated financial need. Research assistantships are available for full- and part-time students and are awarded on the basis of previous academic achievement. Graduate assistants receive stipends of up to $5100 for 1999–2000 and 24 credits in tuition remission. Half assistantships are also available. For further information, students should contact the Financial Aid Office, Pace University, 78 North Broadway, White Plains, New York 10603, 914-422-4050, or Pace Plaza, New York, New York 10038, 212-346-1300.
Cost of Study	Tuition for graduate courses is $575 per credit in 1999–2000; the registration and library fees vary according to the number of credits taken.
Living and Housing Costs	Dormitory rooms for graduate students at Pace University cost $4720 for a double and $5350 for a single in the 1999–2000 academic year. A variety of housing is available in the area.
Student Group	Current enrollment in the computer science and information systems graduate programs is about 650 students. One third are professionals in the field, one third are recent graduates, and the other third are preparing for a career change. Most students attend classes part-time in the evening.
Location	The Civic Center campus and Midtown branch are located in Manhattan, the cultural and financial center of New York City. The nearby city of White Plains is a major retail center and home to national and international corporate headquarters. It is surrounded by towns and villages that have resident artisans, local musical and theatrical groups, and rural museums. New York City is easily accessible from White Plains by bus, train, or car.
The University and The School	Founded in 1906, Pace University is a private, nonsectarian, coeducational institution. It has multiple campuses—one in Manhattan and two in Westchester County. In 1948 Pace Institute became Pace College; in 1973 the New York Board of Regents approved a charter change to designate Pace a university. The School of Computer Science and Information Systems was established in 1983.
Applying	Admission to the M.S. in computer science, M.S. in information systems, and M.S. in telecommunications programs requires satisfactory completion of a baccalaureate degree at an accredited institution. A GPA of at least 3.0 is preferred. In addition to the application, two letters of reference, an essay, and official college transcripts must be submitted. International students must also submit official TOEFL score reports and professional translations of their university transcripts. Official copies of transcripts in the original language must also be submitted. For admission information, students should write to Graduate Admission, Graduate Center, Pace University, 1 Martine Avenue, White Plains, New York 10606 (telephone: 914-422-4283; fax: 914-422-4287; e-mail: gradwp@pace.edu) or Graduate Admission, Pace University, Pace Plaza, New York, New York 10038 (telephone: 212-346-1531; fax: 212-346-1585; e-mail: gradnyc@pace.edu). Applications should be received by August 1 for the fall semester, by December 1 for the spring semester, and by May 1 for the summer semesters. Applications for admission are accepted throughout the year and are evaluated on a rolling basis.
Correspondence and Information	School of Computer Science and Information Systems School of Computer Science and Information Systems

School of Computer Science
   and Information Systems
Pace University
1 Martine Avenue
White Plains, New York 10606

Telephone: 914-422-4191
E-mail: lkleinbaum@mail.pace.edu

School of Computer Science
   and Information Systems
Pace University
Pace Plaza
New York, New York 10038

Telephone: 212-346-1687
E-mail: knorz@mail.pace.edu

Pace University

THE FACULTY

Mehdi Badii, Professor; Ph.D., Loughborough (England), 1987.
D. Paul Benjamin, Associate Professor; Ph.D., NYU, 1997.
Joseph Bergin, Professor; Ph.D., Michigan State, 1989.
Howard Blum, Professor; Ph.D., Polytechnic of Brooklyn, 1973.
Don M. M. Booker, Associate Professor; D.B.A., Nova, 1985.
Frederick B. Bunt, Professor Emeritus-in-Residence; Ed.D., Columbia, 1966.
Linda Jo Calloway, Associate Professor; Ph.D., NYU, 1994.
Mary F. Courtney, Associate Professor; Ed.D., Columbia, 1991.
Edgar G. DuCasse, Professor; Ph.D., Illinois at Urbana-Champaign, 1973.
Samuel S. Epelbaum, Lecturer; Ph.D., Institute of Telecommunications (Russia), 1972.
Daniel J. Farkas, Assistant Professor; M.A., NYU, 1977.
Dietrich Fischer, Professor; Ph.D., NYU, 1976.
Michael L. Gargano, Professor; Ph.D., CUNY Graduate Center, 1980.
Fred Grossman, Professor; Ph.D., NYU, 1973.
Frances Gustavson, Professor; Ph.D., Polytechnic of Brooklyn, 1971.
Constance Knapp, Associate Professor; Ph.D., CUNY Graduate Center, 1995.
Babette Kronstadt, Instructor; Ed.S., Michigan, 1989.
Frank Lo Sacco, Professor Emeritus-in-Residence; Ph.D., Columbia, 1960.
Joseph Malerba, Associate Professor; Ph.D., Yale, 1977.
Francis T. Marchese, Professor; Ph.D., Cincinnati, 1983.
Susan M. Merritt, Professor; Ph.D., NYU, 1982.
Jeanine Meyer, Associate Professor; Ph.D., NYU, 1994.
John C. Molluzzo, Professor; Ph.D., Yeshiva, 1972.
Narayan Murthy, Professor; Ph.D., Rhode Island, 1981.
Richard M. Nemes, Associate Professor; Ph.D., CUNY Graduate Center, 1983.
Nihal Nounou, Assistant Professor; Ph.D., Columbia, 1997.
Bel G. Raggad, Associate Professor; Ph.D., Penn State, 1996.
David A. Sachs, Professor; Ed.D., Columbia, 1978.
Allen Stix, Associate Professor; Ph.D., Pittsburgh, 1973.
Nanda C. Surendra, Assistant Professor; Ph.D., Cincinatti, 1997.
Jennifer Thomas, Associate Professor; Ph.D., Concordia (Montreal), 1997.
Sylvester Tuohy, Professor; Ph.D., Stevens, 1978.
Stuart Varden, Professor; Ed.D., Columbia, 1974.
Gerald Wohl, Professor; M.B.A., CCNY, 1961.
Carol E. Wolf, Professor; Ph.D., Cornell, 1964.
Charles T. Zahn, Professor; M.A., Wisconsin–Madison, 1983.

PENNSYLVANIA STATE UNIVERSITY

College of Engineering
Department of Computer Science and Engineering

Programs of Study

The Department of Computer Science and Engineering offers Master of Engineering (30 credits), Master of Science (30 credits), and doctoral degree (48 credits) programs in computer science and engineering (CSE) at the University Park Campus. Doctoral candidates must take a 1-credit CSE research experience course within the first two semesters, and a written, two-part candidacy examination must be completed during the first three regular semesters after entering the Ph.D. program. Students must pass a comprehensive examination and English competency/communication requirements. A thesis must be defended. Detailed regulations are in the department's graduate brochure, which is available through the department or on line at the Web site listed below.

Research Facilities

The Department of Computer Science and Engineering maintains computer system laboratories at Pond Laboratory. The department currently supports 1,600 user accounts on 240 UNIX workstations and servers. A number of computer vendors are represented in the department's collection of systems, including Sun Microsystems, Silicon Graphics, IBM, and others. These computer systems are connected to one or more of the twenty currently running subnets. A connection to the campus ATM backbone allows any user to easily communicate with other research facilities around the world. The University has connections to the VbNS and commercial ISPs for access to other sites. Programming languages available to the users in the department include C, C++, Pascal, FORTRAN, Scheme, PROLOG, ML, and Common LISP. TeX, troff, and Framemaker are available for typesetting and document preparation. There is a large collection of VLSI/CAD tools, including Berkeley OCTTOOLS, and layout software from Cadence and Synopsys. The MathWorks's Matlab package and a number of its toolboxes are also available.

Financial Aid

Graduate students can be supported by a variety of fellowships and research or teaching assistantships. In 1998–99, stipends for half-time research or teaching assistantships ranged from $11,205 to $12,015. Grants-in-aid that cover tuition accompany these stipends. About one half of the graduate students in the department now receive financial assistance. A half-time assistant is expected to work 20 hours per week and must schedule 8–11 credits per semester. To be considered for financial aid, international applicants must present a score from the Test of Spoken English (TSE). Fall applicants who wish to be considered for financial aid should have completed applications on file before February 1.

Cost of Study

Tuition in 1998–99 was $3267 for Pennsylvania residents and $6730 for nonresidents. Tuition fees; computer ($90), activity ($36), and surcharge ($233) fees; and 80 percent of the mandatory health insurance coverage are included with a teaching/research assistantship. The above fees are for 9 or more credits.

Living and Housing Costs

University and private housing are available. The University's facilities range from dormitory rooms for single students to two-bedroom units for families. In 1998–99, dormitory rooms cost $1255 or more per semester, room and full board were approximately $2300 per semester, shared apartments were $1270 or more per semester, and family housing started at $345 per month. Privately owned apartments are available and range in price from $200 per month for a shared apartment to $700 per month for single occupancy.

Student Group

Overall Penn State enrollment includes approximately 73,000 students, with 40,000 students at University Park (the main campus). There are more than 6,600 graduate students enrolled at University Park, approximately 1,500 of whom are in the College of Engineering. Currently, there are 125 graduate students enrolled in the Department of Computer Science and Engineering. Students come from all parts of the United States. Penn State also has a diverse international population.

Student Outcomes

The demand for computer science and engineering graduates with master's and doctoral degrees is at an all-time high. Starting salaries for doctoral graduates average $80,000 per year in industrial positions. Recruiting for academic positions that begin in 1999–2000 is strong, and candidates can expect annual salary offers in the range of $60,000. Almost all employers in today's economy utilize the expertise of computer science and engineering professionals. This demand for CSE graduates will continue for the foreseeable future.

Location

The University Park campus is located at the center of the state in the borough of State College. The town and its surrounding area, with a population of about 75,000, are located in low, rolling-mountain country and offer a variety of recreational activities. The community and the University present a wide array of cultural and athletic events.

The University and The Department

Penn State, founded in 1855, is a land-grant institution that offers undergraduate and graduate programs. The Department of Computer Science and Engineering was formed in 1993 and combines the strengths of the former Department of Computer Science and the Computer Engineering Program. The department at Penn State consists of 30 faculty members, 25 of whom are graduate research faculty members. Research labs that are set up by both individuals and faculty groups enhance graduate education in the department and provide opportunities for real-world projects that prepare students for future employment.

Applying

Applicants should hold a baccalaureate degree in computer science/engineering or a related field, with a minimum junior/senior GPA of 3.0. The graduate application, recommendation forms, and assistantship application are obtained from the department; electronic versions can be accessed on the World Wide Web (http://www.cse.edu/grad.admissions.html). Applicants must provide a one-page statement of purpose and GRE General Test scores. The GRE Subject Test in computer science or engineering is also recommended. International students whose native language is not English must submit TOEFL scores unless they have received a degree from a U.S. institution. International students seeking financial aid must submit TSE scores.

Correspondence and Information

CSE Graduate Office
Department of Computer Science and Engineering
Pennsylvania State University
201A Pond Lab
University Park, Pennsylvania 16802-6103

Telephone: 814-865-9505
Fax: 814-865-3176
E-mail: vicki@cse.psu.edu
World Wide Web: http://www.cse.psu.edu/

Pennsylvania State University

THE FACULTY AND THEIR RESEARCH

Jesse Barlow, Professor; Ph.D., Northwestern, 1981. Numerical linear algebra, parallel scientific computing, special computer arithmetic.

Piotr Berman, Associate Professor; Ph.D., MIT, 1985. Computational complexity, approximation algorithms, theory distributed systems, computational molecular biology.

Lee D. Coraor, Associate Professor; Ph.D., Iowa, 1978. Digital systems, field-programmable gate arrays, microprocessor systems, computer architecture.

Chita R. Das, Professor; Ph.D., Southwestern Louisiana, 1986. Computer architecture, parallel and distributed computing, performance evaluation, fault-tolerant computing, multimedia, NOW, parallel I/O.

Tse-yun Feng, Binder Professor; Ph.D., Michigan, 1967. Computer architecture, parallel processors and processing, interconnection networks, parallel algorithms.

Martin Furer, Associate Professor; Ph.D., ETH-Zurich (Switzerland), 1978. Graph algorithms, approximation algorithms, computational molecular biology, computational complexity, graph isomorphism problem.

Jonathan Goldstine, Associate Professor; Ph.D., Berkeley, 1970. Automata and formal languages.

Dima Grigoriev, Professor; Ph.D., USSR Academy of Sciences, 1979. Complexity in computer algebra.

John Hannan, Associate Professor; Ph.D., Pennsylvania, 1990. Programming language semantics and implementation, logic and computation, type theory, functional programming, verification of compiler algorithms, software engineering.

Ali R. Hurson, Professor; Ph.D., Central Florida, 1980. Conventional and unconventional concurrent and parallel systems, multidatabases, object-oriented databases, dataflow architecture, global information processing in mobile/wireless environments.

Mary Jane Irwin, Professor; Ph.D., Illinois at Urbana-Champaign, 1977. Computer architecture, application-specific processor design, VLSI system design and electron design automation, computer arithmetic, low-power IC design.

Rangachar Kasturi, Professor; Ph.D., Texas Tech, 1982. Computer vision, document image analysis, visual information systems, image processing, pattern recognition.

Thomas F. Keefe, Associate Professor; Ph.D., Minnesota, 1990. Computer security, database management systems, transaction processing.

Joseph M. Lambert, Associate Professor; Ph.D., Purdue, 1970. Numerical analysis, approximation theory, operations research, software engineering.

John J. Metzner, Professor; Eng.Sc.D., NYU, 1958. Data communications, error-correcting codes, wireless data networks, computer networks, replicated file comparison.

Dale A. Miller, Professor and Department Head; Ph.D., Carnegie Mellon, 1983. Theory, design, and implementation of functional and logic programming languages; concurrency; proof theory.

Webb C. Miller, Professor; Ph.D., Washington (Seattle), 1969. Algorithms and software for molecular biology.

Vijaykrishnan Narayanan, Assistant Professor; Ph.D., South Florida, 1998. Java implementation and performance issues, VLSI design, computer architecture, parallel processing.

Simin Pakzad, Associate Professor; Ph.D., Oklahoma, 1986. Parallel computer systems, database systems, artificial neural networks.

Catusia Palamidessi, Professor; Ph.D., Pisa (Italy), 1988. Theoretical computer science: principles of programming languages, logic and functional programming, theory of concurrency.

Paul Plassmann, Assistant Professor; Ph.D., Cornell, 1990. Development of algorithms and software to enable the solution of large-scale scientific and engineering applications on high-performance parallel computers.

Rajeev Sharma, Assistant Professor; Ph.D., Maryland, 1993. Computer vision, robotics, advanced human-computer interfaces.

Anand Sivasubramaniam, Assistant Professor; Ph.D., Georgia Tech, 1995. Operating systems, computer architecture, performance evaluation, high-performance computing, multimedia.

Hongyuan Zha, Associate Professor; Ph.D., Stanford, 1993. Scientific computing; matrix computations, with applications in statistics and signal processing; parallel computing.

Research plays an important role in graduate education at Penn State. Below is a listing of the research activities of the CSE faculty and students.

Architecture: CAD tools, computer architecture, computer arithmetic, dataflow architecture, database machines, digital system design, fault-tolerant computing, interconnection networks in highly parallel machines, parallel processing, real-time control systems, supercomputers, switching theory, text-retrieval machines, and VLSI design automation.

Computational Molecular Biology: Design and construction of algorithms and software systems for analysis of biological sequence data, including genomic DNA sequences, expressed DNA sequences, and protein sequences.

Computational Theory: Algorithmic design, artificial neural networks, automata and formal languages, combinatorial pattern matching, computational complexity, cryptography, numerical analysis, numerical computing, numerical linear algebra, operations research, parallel algorithms, and symbolic computing.

Computer Applications: Artificial intelligence, computer vision, digital image processing, document image analysis, medical image analysis, pattern recognition, and visual information management systems.

Software Systems: Computer networks; computer security; data communication; databases; multidatabases; distributed systems; error-correcting codes; logic and functional programming; programming language design, semantics, and implementation; object-oriented databases; real-time control systems; software engineering; and transaction processing.

PENNSYLVANIA STATE UNIVERSITY GREAT VALLEY SCHOOL OF GRADUATE PROFESSIONAL STUDIES

Graduate Studies in Information Science and Software Engineering

Programs of Study

The Master of Science in Information Science (M.S.I.S.) and the Master of Software Engineering (M.S.E.) degrees are interdisciplinary programs that blend course work from engineering and management disciplines. These programs prepare graduates for positions in a wide variety of organizations in business, industry, government, and education that use computer-based information systems or will develop the next generation of software products and services.

The programs offer a balance of information systems and management theories and emphasize technical competence, leadership skills, and business expertise. The master's degree in information science requires completion of 39 credits of course work or 30 credits plus a professional paper. The Master of Software Engineering requires 36 credits of course work. Course work focuses on information engineering, system design and integration, software engineering methodologies, information ethics and law, information management, computer hardware and software, process reengineering, project management, managerial communication skills, and the Internet. The program is designed for working professionals with varied levels of experience and academic backgrounds.

Research Facilities

The Computer Center provides laboratories and classroom networks of more than 150 Intel-based microcomputer workstations for student use. These workstations are connected to the campus's Local Area Network (LAN) and to the University's Wide Area Network (WAN). The latter allows student access to the Internet and the World Wide Web. Students are provided with e-mail accounts and dial-up access to facilitate remote use of University-wide computer resources, libraries, and the World Wide Web.

The research library houses more than 24,000 books; 360 current professional, trade, and popular periodicals; and a collection of government publications, microfiche, CD-ROM, and books on audiotape. Drawing on the resources of the entire University, the library at Great Valley is part of Penn State's University Libraries system, one of the leading academic research library organizations in the nation. Students have access to more than 4 million cataloged volumes, 1.4 million government publications, and 32,000 current journals and serials, plus a number of informational materials in various formats, from maps to microforms. Other accessible resources include materials at all Big Ten university libraries, other national research centers, and the Tri-State College Library Cooperative, an organization that provides members with access to the library resources of more than thirty colleges in the Philadelphia area.

Financial Aid

Financial assistance at Penn State Great Valley exists in the form of scholarships, grants, Federal Stafford Student Loans, graduate research assistantships, a minority fellowship, and Federal Work-Study.

Cost of Study

Part-time tuition for 1998–99 was $343 per graduate semester credit for Pennsylvania residents. Tuition for non-Pennsylvania residents was $611 per graduate semester credit.

Living and Housing Costs

Most graduate students are enrolled at Penn State Great Valley on a part-time basis and take evening or Saturday courses. They live and are employed in the greater Philadelphia region. Penn State Great Valley does not offer on-campus housing.

Student Group

More than 500 part-time graduate students are enrolled at Great Valley in graduate programs in information science and software engineering. They are a diverse group of students employed full-time in business, computer science, engineering, health and human services, information, and scientific fields. These students enroll to keep up with the latest technological challenges, build academic credentials to advance their careers, or redirect their career paths into emerging, growing fields.

Location

The campus is situated in the Great Valley Corporate Center, along Route 202, the Philadelphia region's high-technology corridor. The campus is the nation's first university facility permanently housed in a corporate park, alongside world-class companies.

The University and The School

The University is dedicated to preparing students for the information age and is emerging as a national leader in the integration of information sciences and technology into the curriculum.

Penn State Great Valley is designed specifically for adult learners. To meet the needs of working adults, most courses meet two nights a week in seven-week sessions, allowing students to take one course at a time and complete two courses each semester. With six sessions offered each year (in the fall, spring, and summer semesters), students may complete as many as six graduate courses (18 credits) in one year and earn their degree in about three years on a part-time basis. Evening or Saturday classes enable students to participate in the program while maintaining full-time professional positions.

In addition to the master's programs in information science and software engineering, the campus offers an M.B.A. as well as graduate programs in corporate training, education, engineering, health policy, and instructional systems.

Applying

To receive admission consideration, applicants must hold a bachelor's degree from a regionally accredited U.S. institution or a comparable degree from a recognized college or university outside the United States.

New students are admitted to the master's degree program in information science in the beginning of the fall or spring semester. For further information and applications, interested students should contact the Admissions Office.

Correspondence and Information

Admissions Office
Pennsylvania State University Great Valley School of Graduate Professional Studies
30 East Swedesford Road
Malvern, Pennsylvania 19355
Telephone: 610-648-3243
Fax: 610-889-1334
E-mail: gvengr@psu.edu
World Wide Web: http://www.gv.psu.edu

Pennsylvania State University Great Valley School of Graduate Professional Studies

THE FACULTY AND THEIR RESEARCH

Penn State Great Valley faculty members have established national and international reputations as scholars, consultants, practitioners, and researchers, and they bring a broad range of backgrounds and experiences to the classroom. They publish in premier journals and refereed conference proceedings. Faculty members present their research at national and international conferences and often thread their consulting experience into classroom instruction.

Adding to this vital resource of full-time faculty members is a cadre of affiliate and part-time faculty members, as well as adjunct professors, most of whom hold full-time corporate positions or manage their own businesses. As working professionals, they draw on limitless practical wisdom, showing students how fundamental theories relate to real-world issues. As with the full- and part-time faculty members, most adjunct faculty members hold a terminal degree in their field.

James Alpigini, Assistant Professor of Electrical Engineering; Ph.D., Wales. System modeling, simulation, and visualization; control of ill-defined and nonlinear systems; artificial intelligence in real-time control; chaos theory and dynamics.

Eugene Kozik, Associate Professor of Industrial and Manufacturing Engineering; Ph.D., Pittsburgh. Systems engineering, database technology, information systems design, computer-integrated manufacturing, data communications technology, computer network security.

John S. Mullin, Lecturer in Software Engineering; M.Eng., Penn State. Software engineering, project management, information systems, computer performance analysis, database and expert systems, human-computer interaction.

Colin Neill, Assistant Professor of Software Engineering; Ph.D., Wales. Real-time software engineering, software process assessment, communications systems, bioelectronics, manufacturing systems.

Effy Oz, Associate Professor of Management Science and Information Systems; D.B.A., Boston University. Information technology and ethics, World Wide Web, information technology in decision making, strategic planning, electronic monitoring, information technology productivity, codes of conduct for information technology professionals, strategic information systems.

Michael J. Piovoso, Associate Professor of Electrical Engineering; Ph.D., Delaware. AI-based systems, digital signal processing, control engineering, neural nets, fuzzy logic, expert systems, data mining.

Hindupur Ramakrishna, Associate Professor of Management Science and Information Systems; Ph.D., Georgia State. Management of information systems/technology, problem solving and decision making, applied quantitative analysis, MIS curriculum, information technology productivity, information systems success and failure, selection and training issues for MIS.

David Russell, Professor of Electrical Engineering and Academic Division Head; Ph.D., Council for National Academic Awards (Liverpool Polytechnic, Manchester University, and the United Kingdom Atomic Energy Authority, U.K.). Application of artificial intelligence in the control of real-time, ill-defined systems; factory information systems; the philosophy of machine intelligence; software engineering.

Eric Stein, Associate Professor of Management Science and Information Systems; Ph.D., Pennsylvania (Wharton). Information systems theory, management and organization theory, organizational communication, social network analysis, strategic planning, organizational memory and information systems support for it, methods to identify experts in organizational settings, characteristics of human experts, social network communication, organizational learning.

James Weisbecker, Associate Professor of Computer Science; Ph.D., Temple. Parallel computing, programming languages, system development methods.

POLYTECHNIC UNIVERSITY

Computer and Information Science Department

Programs of Study

Polytechnic's Computer and Information Science Department offers programs of study leading to the M.S. and Ph.D. degrees as well as a graduate-level certificate in software engineering. The M.S. program is intended to develop competence in a broad range of fundamental areas, including data structures and algorithms, programming languages, compilers, architecture, operating systems, and artificial intelligence. Graduates are prepared for challenging careers in software and hardware development in such industries as telecommunications, financial services, and computer manufacturing. Degree requirements include completion of 36 credits, including six required courses, two 2-semester course sequences, and additional electives. The M.S. program is offered on all three of Polytechnic's campuses in the New York metropolitan area (Brooklyn, Long Island, and Westchester). Evening classes are an option. A sequence of preparatory courses is available for students whose undergraduate background is not in computer science.

The Ph.D. program is intended to develop competence in a broad range of areas and expertise in one or more specific areas, as well as critical thinking ability and the ability to conduct independent research. Graduates are prepared for state-of-the-art industrial research and for academic research and teaching careers. The main requirement of the Ph.D. program is completion of a thesis embodying significant original research. Preliminary conditions include the completion of the M.S. requirements, the passing of written qualifying exams, and the writing of a survey paper. The relatively small size of the program affords students the opportunity for close collaboration with faculty members working in such research areas as combinatorial optimization, computational biology, computational geometry, databases, image processing and understanding, pattern matching, parallel and distributed architectures and systems, parallel algorithms, network design and management, randomized algorithms, and software engineering, reliability, and testing. The Ph.D. program is centered at Polytechnic's main campus in Brooklyn.

Research Facilities

Polytechnic's research facilities include a network of workstations (Sun SPARC, IBM RS-6000, HP 9000/425), a Convex C120 superminicomputer, several Sun SPARC-10s serving a network of X-terminals, a Silicon Graphics Indigo, a Tandem fault-tolerant computer, a DataCube Image Processing system, and other special purpose equipment for image analysis.

Financial Aid

A limited number of teaching and research assistantships are available to exceptionally qualified Ph.D. students. In 1998–99, these awards covered full tuition and carried a stipend of $1250 per month for junior graduate students (before passing the qualifying exam) and of $1330 per month for advanced students. These awards are renewed annually for up to five years, provided the student is making satisfactory progress. The National Science Foundation awards several graduate research traineeships for Ph.D. students studying at Polytechnic University who are U.S. citizens or permanent residents. Other forms of financial support include grants and loans from state and national programs.

Cost of Study

In 1999–2000, tuition is $695 per credit for full- and part-time study.

Living and Housing Costs

A limited number of dormitory rooms are available at Pratt Institute and Long Island University (near Polytechnic's Brooklyn campus) and at Polytechnic's Farmingdale, Long Island, campus. Privately owned apartments are also available near campus.

Student Group

Polytechnic's computer science program draws students from across the United States and around the world. The M.S. program has about 300 students and the Ph.D. program about 50 students. Approximately 20 percent of the students are women. Most of the M.S. students and some of the Ph.D. students pursue their degrees on a part-time basis.

Location

Polytechnic is located on three campuses in the New York City metropolitan area. The main campus is in downtown Brooklyn, adjacent to Brooklyn Heights, one of New York's more desirable residential communities. The department is located in a University building that is part of the 16-acre Metrotech Center for academic, research, and commercial activities. As a result of the University's favorable location, faculty members and students enjoy close interactions with major companies in the financial, telecommunications, and computer industries. Two suburban campuses are located in Farmingdale, Long Island, and in Hawthorne, Westchester County. New York City's many and diverse cultural attractions are easily accessible from all three campuses.

The University and The Department

Founded in 1854 as the Polytechnic Institute of Brooklyn ("Brooklyn Poly"), Polytechnic University is one of the major technological universities in the Greater New York area. Programs of full-time and part-time study are offered during the day and evening at all three campuses. The evening graduate program of Polytechnic allows students unusual latitude in adjusting their programs to the dictates of employment. The Computer Science and Information Department enjoys close relationships with the Departments of Electrical Engineering and Mathematics and with Polytechnic's Center for Advanced Technology in Telecommunications and Center for Applied Large Scale Computing.

Applying

Applications to the Master of Science program in computer science are accepted until one month prior to the start of each semester. The deadline for Ph.D. program and fellowship applications is March 15. The GRE General Test is required for Ph.D. applicants. TOEFL scores (minimum 550) are required for all students whose native language is not English. The application fee is $45.

Correspondence and Information

For application material and general information:
Office of Graduate Admissions
Polytechnic University
Six Metrotech Center
Brooklyn, New York 11201
Telephone: 718-260-3200
      800-POLYTECH (toll-free)

For specific questions about the Computer Science Program:
Ms. Jeanette M. Magee, Coordinator of Academic Advising
Computer Science and Information Department
Polytechnic University
Six Metrotech Center
Brooklyn, New York 11201
Telephone: 718-260-3210

Polytechnic University

THE FACULTY AND THEIR RESEARCH

Stuart Steele, Industry Professor and Department Head; Ph.D., Penn State, 1965. Software engineering and management, programming languages, real-time systems.

Boris Aronov, Associate Professor; Ph.D., NYU (Courant), 1989. Algorithms, computational and combinatorial geometry.

Ifay Chang, Industry Professor; Ph.D., Rhode Island. Information system development and applications, Internet technologies, multimedia, distance teaching methodology.

Yi-Jen Chiang, Assistant Professor; Ph.D., Brown, 1995. Computer graphics, computer algorithms.

Alex Delis, Assistant Professor; Ph.D., Maryland College Park, 1993. Database management systems, analysis of systems, and software engineering.

David R. Doucette, Industry Professor; Ph.D., Polytechnic of Brooklyn, 1974. Systems integration, software engineering, operating systems.

Robert J. Flynn, Industry Professor; Ph.D., Polytechnic of Brooklyn, 1966. Computer architecture, operating systems.

Phyllis G. Frankl, Associate Professor; Ph.D., NYU (Courant), 1987. Software testing and analysis.

Ivan T. Frisch, Professor and Provost; Ph.D., Columbia, 1962. Information systems, computer networks and network control.

Linda Ann Grieco, Industry Associate Professor; Ph.D., Rutgers, 1976. Programming methodology, computer software.

Haldun Hadimioglu, Industry Associate Professor; Ph.D., Polytechnic, 1991. Computer architecture, high-performance analysis.

Lisa Hellerstein, Associate Professor; Ph.D., Berkeley. Computational learning, complexity theory.

Barry Jones, Industry Assistant Professor; M.S., Marist. Electromechanical systems, real-time computer systems.

Aaron Kershenbaum, Professor; Ph.D., Polytechnic of New York, 1976. Algorithms, telecommunications network design.

Gad M. Landau, Research Professor; Ph.D., Tel-Aviv (Israel), 1986. Algorithms, string matching, computational biology, pattern recognition, communication networks.

Nasir Memon, Associate Professor; Ph.D., Nebraska, 1992. Pulse compression, computer security, image processing.

Paul F. Pickel, Professor; Ph.D., Rice. Mathematical programming, computer graphics, artificial intelligence

Henry Ruston, Professor Emeritus; Ph.D., Michigan. Software engineering, programming languages.

Martin L. Shooman, Professor; D.E.E., Polytechnic of Brooklyn, 1961. Software engineering; software, hardware, and system reliability and safety.

Fred Strauss, Industry Associate Professor; M.S., Polytechnic. Operating systems, software engineering, computer architecture.

Torsten Suel, Assistant Professor; Ph.D., Texas, 1994. Database, parallel computation, algorithms.

Richard Van Slyke, Professor; Ph.D., Berkeley, 1965. Combinatorial optimization, information network design, algorithms.

Joel Wein, Associate Professor; Ph.D., MIT, 1991. Parallel and distributed computation, combinatorial optimization, scheduling theory, algorithms, network optimization.

Edward K. Wong, Associate Professor; Ph.D., Purdue, 1986. Computer vision, image analysis, pattern recognition, computer graphics.

The Computer and Information Science Department is located in the Dibner Building, part of the Metrotech academic/industrial complex in downtown Brooklyn.

PRINCETON UNIVERSITY

Department of Computer Science

Programs of Study	The Department of Computer Science offers programs of study leading to the Master of Engineering (M.E.) and Doctor of Philosophy (Ph.D.) degrees. The requirements for the M.E. can normally be completed in one academic year of full-time graduate study. Requirements for the Ph.D. degree are to (1) complete at least one academic year of full-time residence as a degree candidate; (2) present a research seminar and sustain the Ph.D. general examination; (3) submit a doctoral dissertation and have it accepted by the faculty; and (4) satisfactorily sustain a final public oral examination, which includes a presentation and a defense of the dissertation.
	Dissertation topics are selected through an informal matching of students' interests and faculty expertise. Graduate courses, seminars, and informal work groups have a major influence on this process. External speakers in a regular seminar series also play a role, and student interactions are important as well.
Research Facilities	The Department of Computer Science is located in a recently constructed four-story building. The department has a number of computer systems to support instruction and research in architecture, graphics, networking, and systems, including a High End and DEC mainframes, and a variety of personal workstations and special equipment. These machines are networked and connected via a broadband cable to various University and national networks.
	Princeton's libraries include the Mathematics Library and the Engineering Library, both of which are nearby and are well-stocked with primary reference materials in computer science.
Financial Aid	Financial assistance, equivalent to tuition plus $14,500 to $16,400 for the 1999–2000 academic year, is normally available to all doctoral students in the form of fellowships or assistantships in instruction or research. The Wu Fellowship in Engineering, a prestigious four-year fellowship, will be offered to top entering students by the department. The assistantships are so arranged as to be an integral part of the student's training and to permit a full program of study. Additional compensation in the summer is available. No financial aid is provided to M.E. students.
Cost of Study	Tuition is $25,050 for the 1999–2000 academic year; this includes use of University facilities and comprehensive health and accident insurance coverage.
Living and Housing Costs	The Graduate College provides rooms for about 500 single students at costs of $2312 to $4027 for the 1999–2000 academic year. Some rooms are available in the Princeton area for about $800 a month, and students may take meals at the Graduate College for the board rate, which varies from $2529 to $3636. University housing at $480 to $1251 a month plus utilities is available for 400 married students. Jobs are available at the University or nearby for graduate students' spouses.
Student Group	Of the 6,250 students at Princeton University, approximately 1,800 are graduate students. The Department of Computer Science has 60 graduate students and 90 undergraduates. The student population represents many universities and countries.
Location	The University is located in the historic town of Princeton, about 50 miles from both New York City and Philadelphia. There are a variety of athletic and cultural activities in the area. The University has many athletic facilities for such sports as golf, tennis, sailing, swimming, and squash, plus two large gymnasiums. The Pocono Mountains, about 85 miles away, offer skiing and camping. The beaches of the Jersey shore are an hour's drive away. In addition, McCarter Theatre and Richardson Auditorium present an outstanding variety of concerts and plays.
The University and The Department	Founded in 1746 by charter of King George II, Princeton is the country's fourth-oldest university. The University annually enrolls about 6,250 students (4,450 undergraduate students and 1,800 graduate students). The faculty, numbering about 800, engages in teaching at both the undergraduate and graduate levels and in a diversity of research and scholarly activities.
	Research in computer science at Princeton has a distinguished history that dates back to von Neumann and Turing. The primary center of research activity has shifted from mathematics in the forties and fifties to electrical engineering in the sixties and seventies to an independent Department of Computer Science since 1985. The department has intellectual ties throughout the University, ranging from mathematics to mechanical engineering to molecular biology to music.
Applying	Completed applications for admission received before January 3 receive first consideration; fellowship and assistantship appointments are made from these applications. Applicants are notified of the results around March 15. Applicants should include the results of the Graduate Record Examinations, including scores on the General Test and a Subject Test, as part of their application for admission. Students who are in doubt as to which Subject Test to take usually should register for the one covering their field of undergraduate concentration. The TOEFL is required if English was not the primary language of undergraduate instruction.
Correspondence and Information	Director of Graduate Studies Department of Computer Science 35 Olden Street Princeton University Princeton, New Jersey 08544 Telephone: 609-258-5387 E-mail: gradinfo@cs.princeton.edu World Wide Web: http://www.cs.princeton.edu/gradpgm

Princeton University

THE FACULTY AND THEIR RESEARCH

Andrew Appel, Ph.D., Carnegie Mellon. Programming languages.
Sanjeev Arora, Ph.D., Berkeley. Complexity theory.
Bernard Chazelle, Ph.D., Yale. Computational geometry, data structures.
Douglas Clark, Ph.D., Carnegie Mellon. Architecture.
Perry Cook, Ph.D., Stanford. Computer music, human-computer interface.
David Dobkin, Ph.D., Harvard. Graphics, analysis of algorithms, geometry.
Edward Felten, Ph.D., Washington (Seattle). Computer security, distributed computing.
Adam Finkelstein, Ph.D., Washington (Seattle). Computer graphics.
Thomas Funkhouser, Ph.D., Berkeley. Interactive computer graphics, human-computer interface, collaborative systems.
Hisashi Kobayashi, Ph.D., Princeton. Performance analysis.
Andrea LaPaugh, Ph.D., MIT. Digital libraries, applied algorithms.
Kai Li, Ph.D., Yale. Operating systems, architecture.
Richard Lipton, Ph.D., Carnegie Mellon. Architecture, complexity, DNA computing.
Larry Peterson, Ph.D., Purdue. Networks.
Robert Sedgewick, Ph.D., Stanford. Analysis of algorithms, algorithm visualization.
J. P. Singh, Ph.D., Stanford. Scientific applications of high-performance computers.
Kenneth Steiglitz, Ph.D., NYU. Highly parallel computation, combinatorial optimization, computer music.
Robert Tarjan, Ph.D., Stanford. Data structures, graph algorithms, complexity.
Randolph Wang, Ph.D., Berkeley. Operating systems, I/O systems.
Andrew Yao, Ph.D., Illinois. Computational complexity, analysis of algorithms.

Areas of Graduate Study and Research

Architecture for high-performance systems; complexity; computational biology; computational geometry; computer graphics; computer music; computer networks and operating systems; computer security; data structures and combinatorial algorithms; digital libraries; distributed computer systems; human-computer interface; information and scientific visualization; multimedia; nonstandard computation; parallel computation; parallel and distributed applications; programming languages and environments; software for high-performance systems.

PURDUE UNIVERSITY

Department of Computer Sciences

Program of Study	The Department of Computer Sciences offers challenging programs leading to the Master of Science and Doctor of Philosophy degrees. The areas of the department include analysis of algorithms and theory of computation, compilers and programming languages, databases, geometric modeling and scientific visualization, information security, networking and operating systems, scientific computing, and software engineering.

The requirements for the Ph.D. include a qualifying process involving four examinations, seven specified courses (21 semester hours), a master's thesis or some other demonstration of research potential, seven elective courses, a preliminary examination, and a thesis, the novel results of which merit publication in a refereed journal. For students starting after a bachelor's degree, the qualifying process takes three or four semesters. Courses taken as part of a master's program in computer science at another university can often be counted toward the Ph.D. course requirements. The qualifying process is likely to be changed.

For the M.S., students must complete at least ten courses (30 semester hours) or eight courses and a thesis. The department offers a terminal M.S. degree (33 semester hours) jointly with the Department of Statistics and master's and doctoral specializations in computational science and engineering. |
| **Research Facilities** | The department, committed to experimental as well as theoretical research, has established more than a dozen laboratories. The department has state-of-the-art UNIX and Windows NT systems and direct access to research computing facilities in the Purdue University Computing Center (PUCC). Departmental facilities include a 32-processor Origin2000, two quad-processor Sun E3500 servers, two quad SPARCserver 1000s, a 16-cpu PC cluster, two Sun 4/470 multiprocessors, four Intel-based multiprocessor servers, and about 500 Sun, Intel, and Silicon Graphics workstations. These machines are connected by twenty-eight Ethernet networks, two ATM networks, and several gateways to each other and to the campus network, which includes a 23-megabit-per-second link to the Internet. Additional facilities include 400 gigabytes of disk storage, sixty laser printers, and a fully equipped video production facility.

The PUCC facilities include a 23-node IBM SP2, a 144-node Intel Paragon supercomputer, a 16-node Intel iPSC/860 hypercube, clusters of IBM RISC System/6000 systems, and more than fifty instructional laboratories. |
| **Financial Aid** | Prospective doctoral students may apply for a first-year fellowship, which may be extended for an additional year. United States citizens and permanent residents who are members of certain ethnic minority groups, including African American, Native American, Native Alaskan (Eskimo or Aleut), Native Pacific Islander (Polynesian or Micronesian), and Hispanic, may be eligible for special master's and doctoral fellowships.

The usual form of support is the graduate assistantship. Duties may include teaching classes or recitation sections, grading papers, assisting students with their programs, or working on a research project. A few summer assistantships are usually available. Teaching assistantships require high proficiency in spoken English. The normal stipend for graduate assistants in 1999–2000 is in the range of $1100 to $1375 per month for ten months. |
Cost of Study	In 1999–2000, all graduate assistants and fellows pay fees of $330 per semester. Tuition and fees are $1862 per semester for Indiana residents and $6174 per semester for others. Summer rates are half of those for fall and spring semesters.
Living and Housing Costs	In 1999–2000, single accommodations in the Graduate Houses cost $280–$480 per month. University-operated apartments for married students cost $402 to $505 per month. The rent for most two-bedroom unfurnished apartments in West Lafayette and Lafayette ranges from $540 to $650 per month.
Student Group	In the Department of Computer Sciences, there are about 980 undergraduate students and 155 graduate students, of whom almost all are full-time. Of the graduate students, 60 percent have listed the Ph.D. as a degree objective, 80 percent have assistantships or fellowships, 15 percent are women, and 80 percent are international students.
Location	West Lafayette is a university community across the Wabash River from Lafayette, 65 miles northwest of Indianapolis, the state capital, and 126 miles southeast of Chicago. The population of the two cities exceeds 70,000, not including Purdue's student population. Cultural and athletic events are available throughout the academic year. Convenient transportation is provided by air (Purdue University Airport), bus, and rail (Amtrak) carriers. U.S. 52 passes through both cities, and Interstate 65 is nearby.
The University	Purdue is a public land-grant university with principal emphases in engineering, science, and agriculture. The first regular classes began in September 1874, and the first degree was awarded in June 1875. The West Lafayette Campus has almost 2,100 full-time faculty members who teach and engage in scholarly activities and research within more than 145 principal buildings located on 650 acres. Instructional work is organized in the Schools of Agriculture, Consumer and Family Sciences, Education, Engineering, Health Sciences, Liberal Arts, Management, Nursing, Pharmacy and Pharmacal Sciences, Science, Technology, and Veterinary Medicine.

There are about 31,000 undergraduate students and 6,000 graduate students at the West Lafayette Campus. |
| **Applying** | Applications and all supporting documents should be submitted by December 15 for the following fall semester, which begins about the third week of August. Supporting documents should include official reports of the GRE General Test and Subject Test in computer science scores. Students whose native language is not English normally must submit TOEFL and TWE reports. If applying for a teaching assistantship, they should submit a TSE report.

Applicants should have completed or be close to completing a bachelor's or equivalent degree with a major or the near equivalent of a major in computer science and have general and major GPAs of at least 3.5 on a 4.0 scale. |
| **Correspondence and Information** | Graduate Admissions
Department of Computer Sciences
1398 Computer Science Building
Purdue University
West Lafayette, Indiana 47907-1398
Telephone: 765-494-6004
Fax: 765-494-0739
E-mail: gradinfo-p@cs.purdue.edu
World Wide Web: http://www.cs.purdue.edu |

Purdue University

THE FACULTY AND THEIR RESEARCH

S. S. Abhyankar, Marshall Distinguished Professor; Ph.D. (mathematics), Harvard, 1955. Algebraic geometry, commutative algebra, local algebra, theory of functions of several complex variables, circuit theory, combinatorics, computer-aided design, robotics.

D. C. Anderson, Professor; Ph.D. (mechanical engineering), Purdue, 1974. Computer-aided design, computer graphics, mechanical engineering design.

A. Apostolico, Professor; Dr.Eng. (electronics engineering), Naples, 1973. Algorithmic analysis and design, parallel computation, computational molecular biology.

W. G. Aref, Associate Professor; Ph.D. (computer science), Maryland, 1993. Database systems, spatial and multimedia data indexing, video servers, network-attached storage devices, data mining, algorithms and data structures, computer graphics.

M. J. Atallah, Professor; Ph.D. (computer science), Johns Hopkins, 1982. Algorithms in computer security, geometry, and parallel computation.

B. Bhargava, Professor; Ph.D. (electrical engineering), Purdue, 1974. Transaction processing in distributed systems, videoconferencing, multimedia and communication software, experiments in distributed systems and networks.

D. E. Comer, Professor; Ph.D. (computer science), Penn State, 1976. Design of computer operating systems and network protocols.

S. D. Conte, Professor Emeritus; Ph.D. (mathematics), Michigan, 1950. Numerical analysis and mathematical software; software engineering, particularly software metrics.

H. E. Dunsmore, Associate Professor; Ph.D. (computer science), Maryland, 1978. The Internet, the World Wide Web, Web browsers, Web site design and implementation, object-oriented design and programming, object-oriented implementation using C++, software engineering, information systems.

W. R. Dyksen, Associate Professor; Ph.D. (applied mathematics), Purdue, 1982. Numerical analysis, user interface design.

A. K. Elmagarmid, Professor; Ph.D. (computer and information science), Ohio State, 1985. Consistency aspects of distributed databases; heterogeneous, federated, and multidatabases; transaction management for advanced database applications.

S. Fahmy, Assistant Professor; Ph.D. (computer and information science), Ohio State, 1999. Network traffic management, broadband network architectures, multipoint communication, high-speed internetworking technologies, network support for multimedia applications, distributed systems, performance analysis.

G. N. Frederickson, Professor; Ph.D. (computer science), Maryland, 1977. Analysis of algorithms, data structures, graph and network algorithms.

W. Gautschi, Professor; Ph.D. (mathematics), Basel (Switzerland), 1953. Numerical analysis.

A. Y. Grama, Assistant Professor; Ph.D. (computer science), Minnesota, 1996. Parallel algorithms, architectures, and applications; scientific computation.

C. Guerra, Associate Professor; Dr.Sc.Math. (mathematics), Naples, 1972. Computer vision, image processing.

S. E. Hambrusch, Professor; Ph.D. (computer science), Penn State, 1982. Parallel and distributed computation, analysis of algorithms.

C. M. Hoffmann, Professor; Ph.D. (computer science), Wisconsin–Madison, 1974. Geometric and solid modeling.

A. L. Hosking, Assistant Professor; Ph.D. (computer science), Massachusetts, 1994. Programming language design and implementation; compilation, interpretation, and optimization; run-time systems; object-oriented database systems; database and persistent programming languages and systems; empirical performance evaluation of prototype systems.

E. N. Houstis, Professor; Ph.D. (mathematics), Purdue, 1974. Numerical analysis; parallel, neural, and mobile computing; performance evaluation and modeling; expert systems for scientific computing; problem-solving environments.

Z. Li, Associate Professor; Ph.D. (computer science), Illinois at Urbana-Champaign, 1989. Optimizing compilers for advanced processors and computer systems, interface between compilers and operating systems, static analysis of large programs, performance evaluation of concurrent systems, object-oriented compiler design.

B. J. Lucier, Professor; Ph.D. (applied mathematics), Chicago, 1981. Wavelets, image processing, numerical analysis.

R. E. Lynch, Professor Emeritus; Ph.D. (applied mathematics), Harvard, 1963. Differential equations, linear algebra, software for solving elliptic partial differential equations, computational biology.

D. C. Marinescu, Professor; Ph.D. (EECS), Polytechnic Institute (Bucharest), 1975. Real-time systems, computer networks, performance evaluation of computer and communication systems, parallel and distributed systems, scientific computing.

A. P. Mathur, Professor and Associate Head, Department of Computer Sciences; Ph.D. (computer science), BITS, 1977. Software testing and reliability.

J. Palsberg, Associate Professor; Ph.D. (computer science), Aarhus (Denmark), 1992. Programming languages.

K. Park, Assistant Professor; Ph.D. (computer science), Boston University, 1996. QoS in high-speed networks; design/control of distributed operating systems; cellular automata and fault-tolerant distributed computing.

J. Peters, Associate Professor; Ph.D. (computer science), Wisconsin–Madison, 1990. Representation and analysis of geometry on the computer.

S. Prabhakar, Assistant Professor; Ph.D. (computer science), California, Santa Barbara, 1998. Digital libraries, large-scale storage, parallel I/O, multimedia databases, I/O scheduling, tertiary storage, content-based retrieval and browsing.

V. J. Rego, Professor; Ph.D. (computer science), Michigan State, 1985. Software systems for parallel and distributed computing, threads systems, network protocols, distributed systems, distributed simulation, probabilistic performance and software engineering.

J. R. Rice, W. Brooks Fortune Distinguished Professor; Ph.D. (mathematics), Caltech, 1959. Analysis of numerical methods and problem-solving environments for scientific computing.

E. P. Sacks, Associate Professor; Ph.D. (computer science), MIT, 1988. Scientific and engineering problem solving, geometric reasoning, artificial intelligence.

A. H. Sameh, Samuel D. Conte Professor and Head, Department of Computer Sciences; Ph.D. (civil engineering), Illinois at Urbana-Champaign, 1968. Parallel algorithms in numerical linear algebra.

C. Shields, Assistant Professor; Ph.D. (computer engineering), California, Santa Cruz, 1999. Networking, security.

E. H. Spafford, Professor; Ph.D. (information and computer science), Georgia Tech, 1986. Security, software engineering and computing systems.

J. M. Steele, Associate Professor and Director, Purdue University Computing Center; M.S. (electrical engineering), Purdue, 1965. Computer data communications, computer circuits and systems.

W. Szpankowski, Professor; Ph.D. (electrical engineering and computer science), Gdansk Technical, 1980. Performance evaluation, analysis and design of algorithms, computational pattern matching, data compression, computational combinatorics, queuing theory.

S. S. Wagstaff Jr., Professor; Ph.D. (mathematics), Cornell, 1970. Cryptography, parallel computation, analysis of algorithms.

J. Vitek, Assistant Professor; Ph.D. (computer science), Geneva, 1999. Programming languages, object-oriented software engineering, formal statement and analysis of software security properties.

D. K. Y. Yau, Assistant Professor; Ph.D. (computer science), Texas at Austin, 1997. Networking with quality of service guarantees, operating systems, distributed multimedia.

RENSSELAER POLYTECHNIC INSTITUTE

Department of Computer Science

Programs of Study

The Department of Computer Science offers Master of Science and Doctor of Philosophy degrees.

Master's degree students take courses in software, hardware, theory of computation, and applications. They choose elective courses based on their individual interests. Each master's student completes a Masters Project. This is a substantial piece of software that is expected to have some practical significance. The master's program requires the equivalent of three semesters of work beyond the baccalaureate degree.

Students in the Ph.D. program conduct original research under the guidance of a faculty adviser. Areas of research include computational science and engineering, computer vision, engineering databases, generic programming, human-computer interaction, parallel and distributed computing, and theory and algorithms.

Research Facilities

The department operates more than 100 networked computers. Most of these machines are Sun Workstations or Intel-based PC's, but several other platforms are also represented. The department's more specialized equipment includes a 2000-processor Maspar and a 12-processor SGI Origin 2000.

In addition, all students have access to the Rensselaer Computer System, a campuswide network of several hundred centrally managed workstations. There are many other research-oriented computing resources on campus, including an IBM SP parallel computer and a lab of high-performance graphics computers used for scientific visualization.

Financial Aid

Rensselaer Scholar Fellowships, competitive across the university, provide stipends of $15,000, together with full remission of tuition and fees. Graduate assistantships that involve teaching, research, or a combination of these provide stipends of $11,000 and tuition remission.

Cost of Study

Tuition for 1999–2000 is $665 per credit hour. Other fees amount to approximately $535 per semester. Books and supplies cost about $1700 per year.

Living and Housing Costs

The cost of rooms for single students in residence halls or apartments ranges from $3356 to $5298 for the 1999–2000 academic year. Family student housing, with a monthly rent of $592 to $720, is available.

Student Group

There are about 4,300 undergraduates and 1,750 graduate students representing all fifty states and more than eighty countries at Rensselaer.

Student Outcomes

Eighty-eight percent of Rensselaer's 1998 graduating students were hired after graduation with starting salaries that averaged $56,259 for master's degree recipients and $57,000–$75,000 for doctoral degree recipients.

Location

Rensselaer is situated on a scenic 260-acre hillside campus in Troy, New York, across the Hudson River from the state capital of Albany. Troy's central Northeast location provides students with a supportive, active, medium-sized community in which to live; an easy commute to Boston, New York, and Montreal; and some of the country's finest outdoor recreation sites, including Lake George, Lake Placid, and the Adirondack, Catskill, Berkshire, and Green Mountains. The Capital Region has one of the largest concentrations of academic institutions in the United States. Sixty thousand students attend fourteen area colleges and benefit from shared activities and courses.

The University

Founded in 1824 and the first American college to award degrees in engineering and science, Rensselaer Polytechnic Institute today is accredited by the Middle States Association of Colleges and Schools and is a private, nonsectarian, coeducational university. Rensselaer has five schools—Architecture, Engineering, Management, Science, and Humanities and Social Sciences—that offer a total of ninety-eight graduate degrees in forty-seven fields.

Applying

Students with varied backgrounds are encouraged to apply. Minimum requirements include a knowledge of several higher-level programming languages, assembly language programming, and data structures. In addition, an applicant should have taken at least two years of college-level mathematics, including calculus, discrete mathematics, and linear algebra. GRE General Test scores are required.

Admissions applications and all supporting credentials should be submitted well in advance of the preferred semester of entry to allow sufficient time for departmental review and processing. The application fee is $35. Since the first departmental awards are made in February for the next full academic year, applicants requesting financial aid are encouraged to submit all required credentials by February 1 to ensure consideration.

Correspondence and Information

For written information about graduate study:
Graduate Admissions
Department of Computer Science
Rensselaer Polytechnic Institute
110 8th Street
Troy, New York 12180-3590
Telephone: 518-276-8326
World Wide Web: http://www.cs.rpi.edu

For application and admissions information:
Director of Graduate Academic and Enrollment
  Services, Graduate Center
Rensselaer Polytechnic Institute
110 8th Street
Troy, New York 12180-3590
Telephone: 518-276-6789
E-mail: grad-services@rpi.edu
World Wide Web: http://www.rpi.edu

Rensselaer Polytechnic Institute

THE FACULTY AND THEIR RESEARCH

Sibel Adali, Assistant Professor; Ph.D., Maryland, 1996. Heterogeneous distributed databases, semantic interoperability, query optimization, multimedia information systems.

Christopher D. Carothers, Assistant Professor; Ph.D., Georgia Tech, 1997. Experimental distributed systems, simulation, wireless networks, computer architecture.

Michael Danchak, Clinical Professor; Ph.D., Rensselaer, 1974. Human-computer interaction, usability, information visualization, techniques for distance learning and human learning models.

Arthur G. Duncan, Adjunct Professor; Ph.D., California, Irvine, 1976. Programming languages, functional programming, software engineering, computer security.

Joseph E. Flaherty, Professor; Ph.D., Polytechnic of Brooklyn, 1969. Numerical analysis, scientific computation, adaptive and parallel methods for partial differential equations.

W. Randolph Franklin, Associate Professor; Ph.D., Harvard, 1978. Computer cartography, computational geometry, computer graphics, geographic information systems, computer security. (Joint appointment with Electrical, Computer, and Systems Engineering)

Lester A. Gerhardt, Professor; Ph.D., SUNY at Buffalo, 1969. Communication systems, digital voice and image processing, integrated manufacturing, sensing technologies. (Joint appointment with Electrical, Computer, and Systems Engineering)

Ephraim P. Glinert, Professor; Ph.D., Washington (Seattle), 1985. Computer-based assistive technology for people with disabilities, multimedia information visualization, groupware for collaborative and distance learning, smart/multi-interface multimodal computing environments, visual programming.

Mark K. Goldberg, Professor; Ph.D., Institute of Mathematics (Russia), 1968. Algorithm design, combinatorial optimization, graph theory and combinatorics, software development for symbolic set manipulation.

Martin Hardwick, Professor; Ph.D., Bristol (England), 1982. Database systems for engineering applications.

Wes Huang, Assistant Professor; Ph.D., Carnegie Mellon, 1997. Robotic manipulation, mobile robotics.

Robert P. Ingalls, Executive Officer; Ph.D., Connecticut, 1972. Systems programming.

David Isaacson, Professor; Ph.D., NYU, 1976. Mathematical problems in medical imaging and physics. (Joint appointment with Mathematical Sciences)

Shivkumar Kalyanaraman, Assistant Professor; Ph.D., Ohio State, 1997. Networking architectural issues such as traffic management, quality-of-service support, pricing, multicasting, network management, internetworking, and interoperability with a keen view on performance and specialization in TCP/IP and ATM networks. (Joint appointment with Electrical, Computer, and Systems Engineering)

Mukkai S. Krishnamoorthy, Associate Professor; Ph.D., Indian Institute of Technology (Kanpur), 1976. Programming environments, design and analysis of combinatorial and algebraic algorithms, compiler design.

Franklin T. Luk, Professor and Chair; Ph.D., Stanford, 1978. Numerical linear algebra, parallel computing, signal and image processing.

Paul A. McGloin, Professor Emeritus; Ph.D., Rensselaer, 1968. Data processing, graph theory.

Harry W. McLaughlin, Professor; Ph.D., Maryland, 1966. Applied geometry, computational geometry. (Joint appointment with Mathematical Sciences)

Robert McNaughton, Professor Emeritus; Ph.D., Harvard, 1951. Automata theory, formal languages, combinatorics on words.

Alok Kumar Mehta, Adjunct Professor; M.S., Rensselaer, 1990. Database management systems, object-oriented programming languages, laboratory information systems.

Joseph L. Mundy, Adjunct Professor; Ph.D., Rensselaer, 1969. Artificial intelligence, computer vision.

David R. Musser, Professor; Ph.D., Wisconsin, 1971. Programming methodology, generic software libraries, formal methods of specification and verification, automated theorem proving.

Edwin H. Rogers, Professor; Ph.D., Carnegie-Mellon, 1962. Collaborative design processes and environments, systems for software and hardware design.

Sibylle Schupp, Assistant Professor; Ph.D., Tubingen (Germany), 1996. Generic programming, programming languages, symbolic computation.

Mark S. Shephard, Professor; Ph.D., Cornell, 1979. Scientific computation, mesh generation, adaptive and parallel finite element methods. (Joint appointment with Civil Engineering)

Richard Shuey, Adjunct Professor; Ph.D., Berkeley, 1950. Distributed computer and information systems architecture, assistive systems for the hard of hearing, audiology.

Michael M. Skolnick, Associate Professor; Ph.D., Michigan, 1984. Computer vision, image processing, learning algorithms.

David L. Spooner, Professor; Ph.D., Penn State, 1981. Engineering database systems, object-oriented systems, database security.

Charles V. Stewart, Associate Professor; Ph.D., Wisconsin, 1988. Computer vision, medical applications, computational geometry.

Boleslaw K. Szymanski, Professor; Ph.D., Polish Academy of Sciences, 1976. Languages and compilers for parallel computation, distributed and parallel algorithms, network computing, computational biology.

John D. Valois, Adjunct Professor; Ph.D., Rensselaer, 1995. Distributed computing, data structures.

Mohammed J. Zaki, Assistant Professor; Ph.D., Rochester, 1998. Data mining and knowledge discovery in databases, parallel and distributed computing, parallel algorithms.

Research Areas

Algorithms Design. The research involves the design and analysis of efficient algorithms for combinatorial optimization problems. The emphasis of the research is on the development of learning algorithms. The experimental approach to the algorithm design problem includes work on the development of software systems for set and graph manipulations. Several faculty members participated in the development of two large software systems—GraphPack and LINK.

Computational Science and Engineering. Students and faculty members are working on the application of computational methods to the solution of problems in natural sciences and engineering. Specific research includes adaptive methods for solving partial differential equations, the development of finite-element software, the development of fast and stable matrix algorithms, and algorithms for high-performance computation.

Computer Vision. This research focuses on algorithms for reliably estimating object surface properties from range data and for combining geometric and photometric information for automatic object recognition. A number of industrial and medical applications are being investigated. The research also involves parallel cellular array transformations to obtain measurements from biomedical, scientific, and industrial images and visual-learning algorithms that employ neural network and genetic algorithms.

Engineering Database Development. This research includes databases for design, manufacturing, and other engineering applications. The research uses object-oriented data abstraction methods to develop databases for graphical and other nontextual information. Current research also focuses on distributed collaborative environments for design and manufacturing.

Generic Programming. Research in this area focuses on new ways to design and implement computational methods so that they can adapt easily to different problems, providing solutions that are as efficient as programs written especially for each problem. Research interests span many diverse issues: behavioral concept specifications, language design, compiler optimizations, efficient memory handling, and constructing and using libraries of multipurpose, high-performance software components. Results include the C++ Standard Template Library, which is based on joint research with colleagues in industry; new generic sorting and searching algorithms; and a new memory manager for C++ programs.

Human-Computer Interfaces. Regardless of how embedded a computer may be in a system, the ultimate end user is a human being. Humans are the most complex of all information processors, so the design of human-computer interactions is extraordinarily important and challenging. HCI research at Rensselaer spans human cognition to media hardware, with emphasis on those with handicaps and interfaces for teams of people collaborating on complex assignments.

Parallel Computation. Students and faculty members are involved in all phases of parallel computation, from algorithm design to the design of new languages to run-time support for load balance and computation optimization. To support these efforts, the department has a number of parallel (multiprocessor) computers.

RENSSELAER POLYTECHNIC INSTITUTE

Master of Science in Information Technology

Program of Study	Rensselaer is launching a new professional Master of Science degree program in information technology (IT); New York State approval is expected in fall 1999. The IT program is highly interdisciplinary and is administered by the Faculty of Information Technology, which is composed of faculty members from Rensselaer's five academic schools (Architecture, Engineering, Humanities and Social Sciences, Management and Technology, and Science).
	IT at Rensselaer is understood in the broadest sense to incorporate the computing and communications industries; the information, communications, and entertainment services; and the research and application of IT in and to all fields. The primary intent of the Master of Science program in information technology is to prepare graduates for professional practice in information technology. A secondary intent is to provide a point of entry into Ph.D. programs. Graduates receive a breadth of experience in a variety of IT core areas.
	The degree program requires 30 credits of course work. Students must take one course in each of the following core areas: database systems, telecommunications, software design, management of technology, and human-computer interaction. Students also select an application area that builds from one of these core areas. Applications include, but are not limited to, networking, electronic commerce, human-computer interaction, software engineering, digital publishing, information systems engineering, artificial intelligence, computer hardware design, management information systems, technological entrepreneurship, marketing, psychology, ecoinformatics, computational cybernetics, and graphics and data visualization. Students must also complete a culminating experience, such as a Masters Project or a focused design/studio course.
Research Facilities	The information technology program at Rensselaer has designed an IT Home that fuses the critical components of the program into an interactive forum for analysis, synthesis, design, and production that facilitates a total learning and research experience for IT students. The IT Home features three high-technology design studio spaces; laboratory space for interactive lectures, seminars, demonstrations, and research activities; and resource/office space. A high-speed network links the IT Home to specialized research laboratories around the campus. Students in the IT program have full access to Rensselaer's extensive computer and studio facilities, both in the IT Home and around the campus.
Financial Aid	The student or employer pays most of the costs of this professional degree. Outstanding students may qualify for university-supported Rensselaer Scholar Fellowships, which carry a stipend of $15,000 plus 30 hours of tuition credit, including fees. Low-interest, deferred-repayment graduate loans are also available to U.S. citizens with demonstrated need.
Cost of Study	Tuition for 1999–2000 is $665 per credit hour. Other fees amount to approximately $535 per semester. Books and supplies cost about $1700 per year.
Living and Housing Costs	The cost of rooms for single students in residence halls or apartments ranges from $3356 to $5298 for the 1999–2000 academic year. Family student housing, with monthly rents of $592 to $720, is available.
Student Group	Rensselaer's student body is composed of 4,300 undergraduates and 1,750 graduate students, who represent all fifty states and include approximately 1,000 international students from more than seventy-five countries.
Student Outcomes	Eighty-five percent of Rensselaer's 1998 graduate students were hired after graduation, with starting salaries that averaged $56,259 for master's degree recipients and $57,000 to $75,000 for doctoral degree recipients.
Location	Rensselaer is situated on a scenic 260-acre hillside campus in Troy, New York, across the Hudson River from the state capital of Albany. Troy's central northeast location provides students with a supportive, active, medium-sized community in which to live and an easy commute to Boston, New York, Montreal, and some of the country's finest outdoor recreation, including Lake George, Lake Placid, and the Adirondack, Catskill, Berkshire, and Green Mountains.
	The capital region has one of the largest concentrations of academic institutions in the United States. Fourteen area colleges serve 60,000 students, who benefit from shared activities and courses.
The Institute	Founded in 1824 and the first U.S. college to award degrees in engineering and science, Rensselaer Polytechnic Institute is accredited by the Middle States Association of Colleges and Schools and is a private, nonsectarian, coeducational university. Rensselaer has five schools—Architecture, Engineering, Management and Technology, Science, and Humanities and Social Sciences—and offers a total of ninety-four graduate degrees in forty-seven fields.
Applying	Admissions applications and all supporting credentials should be submitted well in advance of the preferred semester of entry to allow sufficient time for departmental review and processing. The application fee is $35. Since the first departmental awards are made in February and March for the next full academic year, applicants requesting financial aid are encouraged to submit all required credentials by February 1 to ensure consideration.
Correspondence and Information	For written information about graduate study:

For written information about graduate study:
Director of IT Graduate Studies
Lally Building, Room 205
Rensselaer Polytechnic Institute
110 8th Street
Troy, New York 12180-3590
Telephone: 518-276-2660
Fax: 518-276-6687
E-mail:uberm@rpi.edu
World Wide Web: http://www.rpi.edu/IT/

For application and admissions information:
Director of Graduate Academic
    and Enrollment Services, Graduate Center
Rensselaer Polytechnic Institute
110 8th Street
Troy, New York 12180-3590
Telephone: 518-276-6789
E-mail: grad-services@rpi.edu
World Wide Web: http://www.rpi.edu

Rensselaer Polytechnic Institute

THE FACULTY AND THEIR RESEARCH

The Faculty of Information Technology is interdisciplinary and draws its members from all five academic schools at Rensselaer. Currently, there are 118 members of the Faculty of Information Technology. Research activities of the faculty include discipline-specific research as well as interdisciplinary activities related to information technology.

School of Architecture
Alan Balfour, Dean
Three faculty members, with research interests in the areas of architectural design and electronic media.

School of Engineering
James Tien, Acting Dean
Thirty-five faculty members, with affiliations and research interests in the areas of biomedical engineering; chemical engineering; civil engineering; decision sciences and engineering systems; electric power engineering; electrical, computer, and systems engineering; environmental and energy engineering; materials science and engineering; and mechanical engineering, aeronautical engineering, and mechanics.

School of Humanities and Social Sciences
Faye Duchin, Dean
Eighteen faculty members, with affiliations and research interests in the areas of electronic media, arts, and communication; economics; language, literature, and communication; philosophy, psychology, and cognitive science; and science and technology studies.

School of Management and Technology
Joseph G. Ecker, Dean
Seven faculty members, with research interests in the areas of technological entrepreneurship, new product development, financial technology, and environmental management and policy.

School of Science
G. Doyle Daves Jr., Dean
Fifty-five faculty members, with affiliations and research interests in the areas of biochemistry and biophysics, biology, chemistry, computer science, mathematical sciences, and physics, applied physics, and astronomy.

Information Technology Program
Gregory N. Hughes, Vice-Provost
Interdisciplinary research initiatives in the areas of computer networks, electronic commerce, biotechnology, simulation-based science and engineering, communications and optoelectronics, human-computer interaction, visualization, entrepreneurship in information technology, electronic arts and entertainment, and software engineering.

RICE UNIVERSITY

Department of Computer Science

Programs of Study	Rice University offers graduate study programs leading to Master of Computer Science, Master of Science, and Doctor of Philosophy degrees. The M.C.S. is a professional degree that usually takes one to two years to complete. It prepares students for careers as practitioners rather than researchers and does not require a thesis or oral examination. The M.S. program requires a thesis in addition to course work. M.S. recipients usually go on to pursue a Ph.D. The Ph.D. is a research degree that prepares its recipients for careers in independent research, teaching, and advanced development. Four to six years are normally needed to earn a Ph.D. The first year of the program provides the student with a sound basis for independent research. Doctoral students must take comprehensive and qualifying examinations during their first three years. The comprehensive exam covers eight areas: algorithms, architecture, automata, compilers, graphics, logic, operating systems, and programming languages, five of which must be passed during a student's first academic year in the program. The qualifying exam covers one area of computer science, presumably that in which the student wishes to pursue thesis research. Students typically take it after their second year. Ph.D. students are encouraged to become involved in research as quickly as possible after entering the program. While researching with a faculty member, the student finds a problem of interest and develops it into a thesis proposal. Eventually, the student presents the proposal in a departmental colloquium, solves the problem raised by the proposal, and writes a dissertation on the work. Teaching is an integral part of graduate education at Rice. Ph.D. students are required to work as teaching assistants for four semesters. The amount of work varies by course, but is not expected to exceed 10 hours a week averaged over a semester.
Research Facilities	Department research is supported by a shared computing research facility available to all Ph.D. students. The resources are in public laboratories on a local area network of heterogeneous machines running variations of UNIX operating systems and various networking software. Ph.D. student offices are equipped with at least one X-terminal. Students working on projects with the Computer and Information Technology Institute at Rice have access to an 8-way cluster of 4-way high-performance multiprocessors. Rice is the lead institution in the Center for Research on Parallel Computation, a consortium that uses the parallel computing laboratories at seven participating institutions for physically distributed, shared computing resources. The Fondren Library's collection of 1.7 million volumes, 2.2 million microforms, and 14,000 periodical and serial titles is an excellent resource for students.
Financial Aid	In their first academic year, Ph.D. students in the department normally receive financial support in the form of Rice University fellowships, which carry a stipend of between $1555 and $1777 per month, plus tuition. In subsequent years, nearly all students receive financial support, including stipend and tuition, through research and teaching assistantships supported by various grants. M.S. students are eligible for research assistantships. M.C.S. students are not eligible for financial aid.
Cost of Study	The tuition for full-time graduate study in 1998–99 was $15,300 per year, but was waived for most Ph.D. graduate students. Health insurance and other miscellaneous fees are about $230 annually. The cost of books varies by course, and the 1998–99 thesis fee was $93.
Living and Housing Costs	The cost of living in Houston is low compared to other large cities in the U.S. One-bedroom apartments or shared larger apartments can cost $350–$500 in middle-class neighborhoods. The Graduate Apartments, located near the campus, offers private or shared rooms, group kitchens, a commons room, and free transportation to academic buildings. Rooms are reserved as space is available and cost $338–$588 per month, single occupancy.
Student Group	The number of Ph.D. students in computer science at Rice is about 40, and 15 students are in the M.C.S. program. Most Ph.D. students are full-time with full fellowships or research assistantships. Acceptance into the programs is based on scholastic record as reflected by the courses chosen and quality of performance, evaluation of former teachers and advisers, GRE scores, and TOEFL scores (if applicable). M.S. and Ph.D. students are also evaluated on their to ability to conduct independent research. M.C.S. students are evaluated by their ability to take advanced courses.
Student Outcomes	Students who have completed the Ph.D. program in computer science at Rice can be found in the nation's premier universities and corporate research labs. Graduates are teaching or have taught at the California Institute of Technology, Stanford University, Carnegie Mellon University, University of Maryland–College Park, University of Victoria, and Rutgers University, among others. They are working or have worked as researchers for IBM, Tera Computers, Lucent Technologies, Motorola, Daimler Benz, and Hewlett-Packard.
Location	Rice University is located 3 miles from downtown Houston, Texas, the nation's fourth-largest city. Houston's diverse population is reflected in the city's restaurants and cultural events. The city has a symphony, ballet, opera, and theater, as well as professional men's and women's basketball and baseball and hockey teams. Houston's seaport, the nation's third largest, is linked to the Gulf of Mexico by a 50-mile channel.
The University	Rice University is a private, coeducational, nondenominational university founded in 1891 from the estate of William Marsh Rice. It has faculties of liberal arts, science, and engineering and about 2,500 undergraduate and 1,300 graduate students. Rice has a 9:1 student-faculty ratio and is regarded as a top teaching and research university. Students work closely with faculty members and have many opportunities for participation in research activities. The graduate program in computer science is ranked in the top twenty among American universities.
Applying	Typical GRE scores for admitted applicants are in the 90–99th percentile for the quantitative, analytical, and Subject Tests. The minimum acceptable TOEFL score is 600. Completed applications should be sent directly to the Department of Computer Science and must be received by February 1. Applications may be downloaded from http://www.cs.rice.edu/Applications/
Correspondence and Information	Chairman Department of Computer Science Mail Stop 132 Rice University P.O. Box 1892 Houston, Texas 77251-1892 Telephone: 713-527-4834 Fax: 713-285-5930 E-mail: cs@rice.edu World Wide Web: http://www.cs.rice.edu/

Rice University

THE FACULTY AND THEIR RESEARCH

Robert S. Cartwright Jr., Professor; Ph.D., Stanford, 1977. Programming languages and methodology, with an emphasis on type inference to verify program invariants and eliminate run-time checks, a technique dubbed "soft typing;" codirector with M. Felleisen of the Rice Education Infrastructure Project, the goal of which is to create a new core computer science curriculum that integrates the algebraic model, embodied in advanced languages such as Scheme and ML, and the physical model, embodied in languages such as C and C++.

Keith D. Cooper, Associate Professor; Ph.D., Rice, 1983. Code optimization for modern microprocessors, interprocedural analysis and optimization, code generation issues such as register allocation and scheduling, rethinking classical optimizations, practical and effective techniques for interprocedural optimizations, classical optimization, compiler management of latency, issues that arise in the design and use of deep memory hierarchies.

Alan L. Cox, Associate Professor; Ph.D., Rochester, 1992. Parallel computing, operating systems for distributed and multiprocessor systems, computer architecture; member of the Rice team that developed TreadMarks, a state-of-the-art distributed shared-memory system that enables parallel programs written using the shared-memory paradigm to run efficiently on networks of workstations; recipient of the National Science Foundation Young Investigator Award (1994).

Peter Druschel, Assistant Professor; Ph.D., Arizona, 1994. Operating systems, networks, computer architecture, providing operating system support for high-speed networking; recipient of an NSF CAREER grant for early career development as an educator and researcher (1995).

Matthias Felleisen, Professor; Ph.D., Indiana, 1987. High-level programming languages, including semantics, implementation, analysis tools, pragmatics, and teaching; developer of syntax-based methods for describing the behavior of almost all Scheme programs, which are used to study implementation strategies of advanced programming languages; codirector with R. Cartwright of the Rice Education Infrastructure Project.

Ron Goldman, Professor; Ph.D., Johns Hopkins, 1973. Mathematical representation, manipulation, and analysis of shape using computers; algorithms for polynomial and piece-wise polynomial curves and surfaces; parametrically and implicitly represented geometry; computer-aided geometric design, solid modeling, computer graphics, and splines.

G. Anthony Gorry, Professor; Ph.D., MIT, 1967. Impact of information technology on organizations and society, application of artificial intelligence in medicine and the development of decision support systems; director of Rice's Center for Technology in Teaching and Learning, which is developing computing and telecommunications for sharing knowledge in schools, universities, the workplace, and the home; member of the Institute of Medicine of the National Academy of Sciences; Fellow of the American College of Medical Informatics.

Lydia E. Kavraki, Assistant Professor; Ph.D, Stanford, 1995. Physical algorithms with applications to robotics (path planning, assembly sequencing, mechanical part orientation), medicine (noninvasive radiosurgery), and computational chemistry (computer-aided pharmaceutical drug design); recipient of the NSF CAREER Award for early career development as researcher and educator (1996).

Ken Kennedy, Ann and John Doerr Professor of Computational Engineering; Ph.D., NYU, 1971. Parallel computing in science and engineering, scientific programming environments, optimization of compiled code, computer architecture, performance analysis, graph algorithms, extending the techniques of program analysis and parallelization to provide language support for highly parallel supercomputers and advanced programming environments for scientific and engineering professionals; director of the Center for Research on Parallel Computation, a consortium that uses the parallel computing laboratories at seven participating institutions for physically distributed, shared computing resources; recipient of the IEEE Computer Society W. Wallace McDowell Award (1995); Fellow of the ACM, IEEE, and the American Association for the Advancement of Science; member of the National Academy of Engineering; cochair, President's Advisory Committee on High-Performance Computing and Communications, Information Technology, and the Next Generation Internet (1997).

John Mellor-Crummey, Senior Faculty Fellow; Ph.D., Rochester, 1989. Topics in large-scale parallel computation, including architectures, operating systems, programming environments, and algorithms; compiler and programming environment support for data-parallel languages.

Devika Subramanian, Associate Professor; Ph.D., Stanford, 1989. Artificial intelligence aimed at the design and analysis of adaptive, discrete, limited-resource agents that perform tasks in dynamic environments; recipient of the George Forsythe Memorial Award for Excellence in Teaching, Stanford University (1986), and the Outstanding Educator Award, Merrill Presidential Scholar Program, Cornell University (1991, 1993).

Dan Wallach, Assistant Professor; Ph.D., Princeton, 1999. Computer security and the security of mobile code environments, including Java; issues of authentication, access control, and resource management in mobile and distributed computing.

Moshe Y. Vardi, Noah Harding Professor of Computer Science and Chair, Department of Computer Science; Ph.D., Hebrew (Jerusalem), 1981. Applications of logic to computer science, specifically to databases, finite-model theory, knowledge theory, and program specification and verification; connection between finite-model theory and areas of computer science such as complexity theory and database theory; developer of a theory of knowledge-based agents that will have applications to the design and analysis of multiagent systems, such as distributed computer systems or teams of cooperating robots; recipient of the IBM Research Outstanding Innovation Award (1987, 1989, 1992).

Joe Warren, Associate Professor; Ph.D., Cornell, 1986. Application of computers to geometric problems centered around the general problem of representing geometric shapes, geometric modeling (the construction and manipulation of data structures for representing geometric objects), computational geometry (using algorithms to solve geometric problems), algorithms for solving and manipulating systems of polynomial equations.

Willy E. Zwaenepoel, Noah Harding Professor of Computer Science and Professor of Electrical and Computer Engineering; Ph.D., Stanford, 1984. Distributed parallel systems, fault tolerance, operating systems, the extent to which a network of workstations can be used to construct a loosely coupled multicomputer; director of Rice's NSF Research Infrastructure Award, which provides funds to build a state-of-the-art cluster of shared-memory multiprocessors; director of Computer Information Technology Institute; Fellow of the IEEE.

R·I·T

ROCHESTER INSTITUTE OF TECHNOLOGY

College of Applied Science and Technology
Programs in Computer Science, Information Technology,
and Software Development and Management

Programs of Study

The College of Applied Science and Technology offers Master of Science degree programs in computer science, information technology, and software development and management. The computer science program is composed of the computer science graduate core, electives, advanced electives, and a thesis paper or project, for a total of 45 quarter credit hours. Two degree tracks are available: the thesis track and the project track. The information technology and software development and management programs both consist of 48 quarter credit hours of study and a thesis or master's project. Both the information technology and software development and management programs may also be completed via a distance learning format.

Research Facilities

The computer science department provides extensive facilities for students and faculty members, including a graduate lab with eighteen SGI Indy color workstations; seventy-five Sun Ultra1 color workstations; twelve SGI Indy, Indy Modeler, and Indigo 2 color workstations; four SPARC 10 and 20 file servers with more than 20 GB of disk available; a 64-processor Transputter-based parallel processing platform; and a network/distributed systems lab with twelve SGI indys and its own internal network as well as a digital logic laboratory and access to the information technology labs. The information technology and software development and management departments have access to labs containing seventy Power PC Macintosh systems and 135 Pentium PCs.

Financial Aid

A variety of scholarships and assistantships are available through the Department of Computer Science and Information Technology. Federal, state, and institutional aid are also available to those who qualify. Applicants seeking financial aid must have all documents submitted to the Office of Financial Aid by February 15 to be considered for entry with support the following September.

Cost of Study

In 1999–2000, tuition fees are $546 per credit hour (1 to 11 credits) and $6487 per quarter (12 to 18 credits). Fees for internships in the Master of Engineering degree programs are $290 per credit hour. An estimated cost of books and supplies for full-time students ranges from approximately $500 to $2500. Depending on the number of courses, part-time students' books and supplies may cost approximately $300 to $450. All full-time graduate students are required to pay a student activities fee of $48 per quarter. All fees are subject to change.

Living and Housing Costs

Room and board for full-time students for 1999–2000 is $2284 per quarter for a standard meal plan and double room occupancy. A variety of residence hall and apartment housing options and meal plans are available, and costs vary according to the options selected. Housing within the surrounding community is plentiful and moderately priced.

Student Group

Current enrollment for the Institute is approximately 13,000, including 8,000 full-time and 2,700 part-time undergraduates and 2,200 graduate students. The College of Applied Science and Technology has approximately 4,000 undergraduates and 770 graduate students, including 1,400 women.

Location

The campus occupies 1,300 acres in suburban Rochester, the third-largest city in New York State. Rochester has a thriving arts community that includes the International Museum of Photography, the Memorial Art Gallery, the Rochester Philharmonic Orchestra, the Eastman School of Music, the Rochester Museum, and the Science Center and Planetarium. Also close by are the vineyards and wine-tasting region of the Finger Lakes and Lake Ontario. The population of the metropolitan area is just under 1 million, with industries such as Eastman Kodak, Xerox, and Bausch & Lomb providing the area's economic base.

The University

Founded in 1829 and emphasizing career education, the Institute is a privately endowed, coeducational university consisting of seven colleges. RIT is the fourth-oldest and one of the largest cooperative education institutions in the world, annually placing 2,600 students in co-op positions with approximately 1,300 employers. Enrolled students represent all fifty states and more than eighty other countries. The National Technical Institute for the Deaf has a current enrollment of approximately 1,200 students.

Applying

Admission to the computer science degree program is granted to qualified holders of a bachelor's degree from a regionally accredited university or college who have acceptable mathematics credits and experience with a modern high-level language as well as exposure to a variety of programming languages. Those not meeting the mathematics requirement are admitted at the discretion of the department. To be considered for admission to any of the three departments, it is necessary to have a minimum grade point average of 3.0 out of 4.0 and to submit an application for admission to graduate study accompanied by the appropriate undergraduate and graduate transcripts and two letters of recommendation. Applications are accepted on a rolling basis and must be accompanied by a $40 application fee. In addition, TOEFL scores may be required of students whose native language is not English.

Correspondence and Information

Office of Part-time and Graduate Enrollment Services
Rochester Institute of Technology
58 Lomb Memorial Drive
Rochester, New York 14623-5604
Telephone: 716-475-2229
E-mail: opes@rit.edu
World Wide Web: http://www.rit.ed

Rochester Institute of Technology

THE FACULTY

Department of Computer Science

Walter A. Wolf, Associate Professor and Department Chair; Ph.D., Brandeis.

Peter G. Anderson, Professor and Graduate Program Coordinator; Ph.D., MIT.

Warren Carithers, Associate Professor; M.S., Kansas.

Lawrence Coon, Associate Professor; Ph.D., Ohio State.

Henry Etlinger, Associate Professor and Undergraduate Program Coordinator; M.S., Syracuse.

James Heliotis, Associate Professor; Ph.D., Rochester.

Fereydoun Kazemian, Associate Professor; Ph.D., Kansas State.

Andrew Kitchen, Professor; Ph.D., Rochester.

Michael J. Lutz, Professor; M.S., SUNY at Buffalo.

Fernando Naveda, Associate Professor; Ph.D., Minnesota.

Stanislaw Radziszowski, Professor; Ph.D., Warsaw.

Kenneth Reek, Professor; M.S., RIT.

Nan Schaller, Professor; M.S., Union (New York).

Paul Tymann, Assistant Professor; M.S., Syracuse.

James R. Vallino, Assistant Professor; M.S., Rochester.

Department of Information Technology

Edith Lawson, Associate Professor and Department Chair; M.S., RIT.

A'isha Ajayi, Assistant Professor; M.S., Syracuse.

Kumiko Aoki, Assistant Professor; Ph.D., Hawaii.

John A. Biles, Professor and Undergraduate Program Coordinator; M.S., Kansas.

Gordon Goodman, Associate Professor; M.S., RIT.

Stephen Jacobs, Assistant Professor; M.A., New School.

Daryl Johnson, Assistant Professor; M.S., RIT.

Stephen Kurtz, Associate Professor; M.S., RIT.

Jeffrey Lasky, Professor; M.S., Minnesota.

Peter Lutz, Professor; Ph.D., SUNY at Buffalo.

Ronald Perry, Associate Professor and Graduate Program Coordinator; M.S., RIT.

Evelyn P. Rozanski, Professor; Ph.D., SUNY at Buffalo.

William Stratton, Associate Professor; Ph.D., SUNY at Buffalo.

Timothy Wells, Assistant Professor; M.B.A., California State, Bakersfield.

Michael A. Yacci, Associate Professor; Ph.D., Syracuse.

RUTGERS, THE STATE UNIVERSITY OF NEW JERSEY, NEW BRUNSWICK

Graduate School
Department of Computer Science

Program of Study

The Department of Computer Science at Rutgers University, New Brunswick, offers a comprehensive program of study in most areas of computer science leading to either an M.S. or a Ph.D. degree. The department's program includes these areas: algorithms, artificial intelligence, combinatorics, complexity theory, computational biology, computational geometry, computational linguistics, databases, data structures, distributed systems, expert systems, graphics, human-computer interaction, knowledge representation, machine learning, mathematical programming, numerical analysis, optimization, parallel computation, programming languages, software engineering, and vision.

For the M.S. degree, a student must complete 24 credits of course work, a master's thesis of 30 credits, and an expository essay and pass a comprehensive examination of courses. A candidate for the Ph.D. degree must complete 48 credits of course work and pass a written and oral qualifying examination. A student who enters with a master's degree may apply to transfer up to 24 credits toward the 48 required. The candidate must also complete a research project and successfully defend a dissertation written about the project. Normally, a one-year residence in the department is required.

Research Facilities

Three coupled computing environments supporting faculty, graduates, and undergraduates are accessible from a variety of desktop workstations (Sun, Digital, NCD, Dell, Apple, among others). These environments support shared printers, modems, and Internet connections and provide cycle and file service using multi-user servers such as Sun and SGI over high-speed networks. All faculty and graduate student offices are equipped with workstations connected over a 100 Mb Ethernet. Computer-intensive work is supported by a DEC alpha with three processors and 1 GB of memory. Dedicated research and instructional laboratories composed of multiprocessors connected over a low latency/high bandwidth network are also available.

These systems run a variety of operating systems, such as Windows NT 4/5, Windows 95/98, Linux, IRIX, SunOS, Solaris, HP-UX, and Digital Unix, and programming languages, such as Java, Prolog, Lisp, Eiffel, C, C++, and Fortran. Document preparation tools, such as Scribe, TeX, and Publisher, and various specialized tools for specific research areas, such as Matlab and Maple, are also supported.

Financial Aid

A large number of full-time graduate students receive financial support, which includes University fellowships, teaching assistantships, and graduate assistantships. In 1998–99, stipends ranged from $12,000 to $14,000 for nine months. Tuition is remitted for all assistants. Summer support is available for graduate assistants on some research projects.

Cost of Study

Full-time graduate tuition during the 1998–99 academic year was $6492 per year (24 credits) for residents of New Jersey and $9820 for nonresidents, and there was a student fee of $650. For part-time students, resident tuition was $267 per credit, nonresident tuition was $395 per credit, and the student fee was about $176 per year. (These fees are likely to rise.) Books cost approximately $800 per semester.

Living and Housing Costs

Graduate student apartments cost about $4242 to $4344 per year. Graduate dormitory rooms cost from $3960 to $4648 per year with various possible meal plans. Married student housing is available at a cost of $563 per month for a one-bedroom apartment and $716 per month for a two-bedroom apartment. (These fees are likely to rise.)

Student Group

In fall 1998, there were 88 full-time and 61 part-time graduate students in the department. Between January 1990 and January 1998, seventy-eight Ph.D.'s were awarded.

Location

New Brunswick is located on a main railroad line, 33 miles from New York City and 60 miles from Philadelphia. Mountains and shore areas are close by and are easily reached by car. There is an active program of concerts, art exhibits, lectures, and recreational activities at the University, and the unsurpassed cultural advantages of New York and Philadelphia are within easy reach.

The University

Rutgers was founded before the American Revolution; it was the eighth college to be established in this country. In the nineteenth century the state of New Jersey began to provide support for certain programs, and in 1945 Rutgers as a whole became the state university. Currently, about 49,000 students are enrolled in the three principal divisions of the University—at New Brunswick, Newark, and Camden. In New Brunswick there are four resident undergraduate colleges: Rutgers, Douglass, Livingston, and Cook. These colleges are spread over four areas in and around New Brunswick. The computer science department is part of the Faculty of Arts and Sciences, and it is located in the Hill Center for Mathematical Sciences on the Busch campus. An outstanding gymnasium is a block away. This suburban area is also the home of the chemistry, mathematics, geology, psychology, and physics departments; the Engineering School; the College of Pharmacy; Robert Wood Johnson School of Medicine of the University of Medicine and Dentistry of New Jersey; and the New Jersey Mental Health Center.

Applying

Applications for September admission with financial aid must be received by February 1. Ph.D. applications received after February 1 will not be considered. Applicants to the master's program who are not seeking financial aid may apply until April 1; for January admission, until November 1. The application fee is $50. The applicant's academic record should exhibit at least a B+ average and must show distinction in computer science, mathematics, and related fields. Results of the Graduate Record Examinations (General Test and Subject Test in computer science) are required in addition to letters of recommendation and all transcripts. International applicants are required to submit TOEFL and TWE results.

Correspondence and Information

For information on the program:
Department of Computer Science, PG
Hill Center for Mathematical Sciences
Rutgers, The State University of New Jersey
Piscataway, New Jersey 08854

For applications:
Graduate Admissions Office, PG
Rutgers, The State University of New Jersey
New Brunswick, New Jersey 08903

Rutgers, The State University of New Jersey, New Brunswick

THE FACULTY AND THEIR RESEARCH

Eric Allender, Professor; Ph.D., Georgia Tech. Complexity theory, parallel and probabilistic computation.

Saul Amarel, Turing Professor of Computer Science; D.Eng.Sc., Columbia. Artificial intelligence: representation, theory formation, computational design.

B. R. Badrinath, Associate Professor; Ph.D., Massachusetts. Distributed systems and databases, mobile wireless computing.

Sandeep Bhatt, Professor; Ph.D., MIT. Parallel computing, algorithms and architectures.

Alexander T. Borgida, Professor; Ph.D., Toronto. Artificial intelligence in the design of information systems.

Vaclav Chvátal, Professor and Graduate Director; Ph.D., Waterloo. Algorithms, combinatorics, graph theory, operations research.

Doug DeCarlo, Assistant Professor; Ph.D., Pennsylvania. Computer graphics, computer vision, human-computer interaction.

Sven Dickinson, Assistant Professor; Ph.D., Maryland. Computer vision, object modeling, artificial intelligence.

Thomas Ellman, Assistant Professor; Ph.D., Columbia. Artificial intelligence, machine learning, knowledge compilation, qualitative physics.

Martin Farach-Colton, Associate Professor; M.D., Johns Hopkins; Ph.D., Maryland. Computational biology, design and analysis of algorithms.

Michael L. Fredman, Professor; Ph.D., Stanford. Algorithms, data structures, computational complexity.

Apostolos Gerasoulis, Professor; Ph.D., SUNY at Stony Brook. Parallel processing, algorithms, numerical analysis.

Michael D. Grigoriadis, Professor; Ph.D., Wisconsin–Madison. Algorithms for network optimization.

Haym Hirsh, Associate Professor; Ph.D., Stanford. Machine learning and artificial intelligence.

Liviu Iftode, Assistant Professor; Ph.D., Princeton. Distributed and parallel systems, operating systems.

Tomasz Imielinski, Professor and Chairman; Ph.D., Polish Academy of Sciences. Logical foundations of databases, mobile wireless computing.

Bahman Kalantari, Associate Professor; Ph.D., Minnesota. Mathematical programming, global and discrete optimization.

Kenneth R. Kaplan, Associate Professor and Associate Chairman; Ph.D., Polytechnic of Brooklyn. Algorithms, queuing theory, modeling, discrete simulation.

Leonid Khachiyan, Professor; Ph.D., D.Sc., USSR Academy of Sciences. Mathematical programming, computational complexity, discrete optimization.

Ulrich Kremer, Assistant Professor; Ph.D., Rice. Computation techniques and interactive programming environments for distributed-memory and shared-memory multiprocessor.

Casimir Kulikowski, Board of Governors Professor; Ph.D., Hawaii. Artificial intelligence, pattern recognition, imaging, bioinformatics.

Saul Levy, Associate Professor; Ph.D., Yeshiva. Massively parallel architectures, algorithms, environments.

Richard Martin, Assistant Professor; Ph.D., Berkeley. High performance, network design and evaluation, parallel architectures and languages, high throughput I/O systems.

L. Thorne McCarty, Professor of Computer Science and Law; J.D., Harvard. Artificial intelligence, legal reasoning, logic programming.

Naftaly Minsky, Professor; Ph.D., Hebrew (Jerusalem). Software engineering, programming languages, distributed systems.

Craig Nevill-Manning, Assistant Professor; Ph.D., Waikato (New Zealand). Bioinformatics, digital libraries, machine learning, data compression.

Thu Nguyen, Assistant Professor; Ph.D., Washington (Seattle). Operating systems, distributed and parallel systems, networking, architecture, fault-tolerance, system security.

Marvin C. Paull, Professor; B.S., Clarkson. Design and analysis of algorithms—principles and practice.

Gerard Richter, Professor; Ph.D., Harvard. Numerical solution of differential and integral equations.

Barbara Ryder, Professor; Ph.D., Rutgers. Programming languages, software engineering, parallel computation.

Donald Smith, Director of Laboratory for Computer Science Research; Ph.D., Rutgers. Massively parallel architectures, VLSI, artificial intelligence.

Diane L. Souvaine, Associate Professor; Ph.D., Princeton. Development and analysis of geometric and graph-theoretic algorithms, applications of these algorithms to practical problems, parallel algorithms.

William Steiger, Professor; Ph.D., Australian National. Computational geometry, parallel computation, probabilistic algorithms.

Louis Steinberg, Associate Professor; Ph.D., Stanford. Artificial intelligence, computer-aided design with applications to the design of thermodynamic and aerodynamic structures.

Suzanne Stevenson, Assistant Professor; Ph.D., Maryland. Computational linguistics, cognitive modeling.

Matthew Stone, Assistant Professor; Ph.D., Pennsylvania. Computational linguistics, knowledge representation, human-computer interaction.

Endre Szemerédi, State of New Jersey Professor of Computer Science; Ph.D., Moscow. Number theory, external graph theory, parallel algorithms, theoretical computer science.

Robert Vichnevetsky, Professor; Ph.D., Brussels. Numerical analysis, computer methods for partial differential equations, optimization theory, modeling and simulation systems, environmental systems, computational fluid dynamics.

Brett Vickers, Assistant Professor; Ph.D., California, Irvine. High-speed networks, multimedia systems and networking, network traffic control.

Affiliated Faculty

Stanley Dunn, Associate Professor of Biomedical Engineering; Ph.D., Maryland. Computer vision, image understanding, pattern recognition, software engineering for vision, signal-processing applications.

Herbert Freeman, Professor of Computer Engineering; Dr.Eng.Sc., Columbia. Computer graphics, pattern recognition, image processing, computer vision.

Peter L. Hammer, Professor and Director of the Center for Operations Research; Ph.D., Bucharest. Boolean methods in operations research and related areas, theory of graphs and networks.

Jeffry Kahn, Professor of Mathematics; Ph.D., Ohio State. Combinatorics.

János Komlós, Professor of Mathematics; Ph.D., Eotvos Lorand (Budapest). Combinatorics, probability, theoretical computer science.

Evangelina Micheli-Tzanakou, Professor of Biomedical Engineering; Ph.D., Syracuse. Pattern recognition, computer vision, brain information processing, neural networks.

Michael Saks, Professor of Mathematics; Ph.D., MIT. Combinatorics, complexity theory, distributed computing, online algorithms.

Charles Schmidt, Professor of Psychology; Ph.D., Iowa. Human and machine planning and plan recognition, human and machine problem-solving and learning, human-computer interaction.

Eduardo D. Sontag, Professor of Mathematics; Ph.D., Florida. Nonlinear control, neural nets, learning theory.

SOUTHERN METHODIST UNIVERSITY

Department of Computer Science and Engineering

Programs of Study

The department offers the following degrees: Master of Science in Computer Engineering, Master of Science (with majors in computer science, operations research, and software engineering), Master of Science in Engineering Management, Ph.D. (with majors in computer engineering, computer science, and operations research), and D.Eng. (with a major in engineering management). SMU's CSE department emphasizes the following major areas of interest: algorithms engineering, artificial intelligence, computer architecture, computer networks, data and knowledge engineering, mathematical programming, natural-language processing, parallel and distributed processing, and software engineering and systems.

Research Facilities

Students in the Department of Computer Science and Engineering have access to a wide range of facilities and equipment. The department's computing environment has evolved into an Ethernet-based network of minicomputers and workstations. It now includes workstations from Sun Microsystems and Digital Equipment Corporation, including four fast (250–300 megahertz) Alpha Server 2100s. These are networked with a Sequent Symmetry S81 configured with twenty CPUs.

Financial Aid

The Graduate Admissions Committee awards a limited number of merit-based research and teaching assistantships to incoming students, which pay up to $1600 per month and cover tuition. Separate tuition assistantships are also available.

Cost of Study

Tuition and fees for graduate study in 1999–2000 are $600 per semester hour.

Living and Housing Costs

Dormitory housing charges each semester are approximately $1950 per person for double occupancy. Board, including tax, costs $1510 per semester. Furnished efficiency and one- and two-bedroom apartments are available on campus, some with paid utilities, at costs ranging from $2500 to $3000 per semester. The Dallas area offers inexpensive housing options within driving distance of the campus, with apartments starting at $500 per month.

Student Group

The department has about 60 doctoral students and 300 master's degree students. Approximately 10 to 15 doctorates and 50 to 70 master's degrees are awarded each year.

Student Outcomes

Recent graduates have obtained appointments at the University of Oklahoma, the University of Arkansas, and the College of the Ozarks, as well as industrial research positions at Texas Instruments, Sabre Decision Technologies (American Airlines), DSC Communications, MCI, Sun Microsystems, Bell (AT&T), Cyrix, Alcatel, and Science Applications International Corporation.

Location

Dallas is the center of an attractive metropolitan area of 2 million people. It has fine parks, lakes, museums, theaters, orchestras, libraries, and places of worship. Clean and progressive, the city continues to grow as a center of business and light industry. The manufacturing, microelectronics, and telecommunications industries provide many opportunities for employment.

The University and The School

Southern Methodist University is a private, nonprofit, coeducational institution located in suburban University Park, an incorporated residential district of Dallas, Texas. The School of Engineering and Applied Science (SEAS) traces its roots to 1925, when the Technical Club of Dallas, a professional organization of practicing engineers, petitioned SMU to fulfill the need for an engineering school in the Southwest.

Applying

Students may apply for admission at any time. However, initial review for admission in a given semester depends upon receipt by the Graduate Division of all requisite application materials no later than July 1 for fall admission, November 15 for spring admission, or April 15 for summer admission. All international students must use the following dates: May 15 for fall admission, September 1 for spring admission, and February 1 for summer admission. GRE General Test scores are required.

Correspondence and Information

Graduate Admissions
School of Engineering and Applied Science
Southern Methodist University
Dallas, Texas 75275-0335
Telephone: 214-768-3900
World Wide Web: http://www.seas.smu.edu/

Southern Methodist University

THE FACULTY AND THEIR RESEARCH

The following is a partial list of faculty members affiliated with the Department of Computer Science and Engineering.

Professors

Jeffery L. Kennington, Ph.D., Georgia Tech, 1973; PE. Network optimization, mathematical programming, telecommunications networks.

David W. Matula, Ph.D., Berkeley, 1966. Computer arithmetic, network and graph algorithms, algorithm engineering.

Dan I. Moldovan, Ph.D., Columbia, 1978. Computer architecture, parallel and distributed processing, artificial intelligence, natural-language processing.

Associate Professors

Richard S. Barr, Ph.D., Texas at Austin, 1978. Network optimization, data mining, system modeling.

Weidong Chen, Ph.D., SUNY at Stony Brook, 1990. Databases, mobile computing, computer networking and communication.

Margaret H. Dunham, Ph.D., SMU, 1984. Database recovery, data mining, mobile computing, temporal databases.

Richard V. Helgason, Interim Chair; Ph.D., SMU, 1980. Network optimization, mathematical programming, computational geometry.

Sukumaran Nair, Ph.D., Illinois at Urbana-Champaign, 1990. Fault-tolerant computing, computer networks, VLSI systems.

Assistant Professors

Sanda Harabagiu, Ph.D., USC, 1997. Artificial intelligence, natural language processing, parallel processing.

Jacob Kornerup, Ph.D., Texas at Austin, 1998. Parallel computing, powerlists, parlists, plists.

Jeff Tian, Ph.D., Maryland, 1992. Software testing techniques and tools, measurement and analysis of software products and processes, software reliability and safety, software engineering.

Lecturers

Frank Coyle, Ph.D., SMU, 1992. Software engineering.

Mary Alyo Lillard, M.S., SMU, 1983. Computer education, telecommunications.

STANFORD UNIVERSITY

Computer Science Department

Programs of Study	Founded in 1965, the Computer Science Department is a center for research and education at the graduate level. Strong research groups exist in the areas of analysis of algorithms and theory of computation, artificial intelligence, scientific computing, robotics, and systems. Basic research in computer science is the main goal of these groups, but there is also a strong emphasis on interdisciplinary work and on applications. Fields in which interdisciplinary work has been undertaken include chemistry, genetics, linguistics, physics, engineering, and medicine. Close ties are maintained with researchers in the Departments of Electrical Engineering, Mathematics, Statistics, Operations Research, and others with similar interests. In addition, both faculty and students commonly work with investigators at nearby research or industrial institutions. The main educational goal is to prepare students for research and teaching careers, either in universities or in industry.
	Students admitted to the Ph.D. program usually combine course work and participation in a research group during their first year and devote themselves entirely to research thereafter. Students must pass comprehensive examinations that test the breadth of their computer science knowledge and a qualifying examination in their specialty area.
	The department also offers the Master of Science in Computer Science (M.S.C.S.). The M.S.C.S. program chiefly involves course work and usually takes from four to six quarters to complete.
Research Facilities	In addition to equipment provided by the University, the department provides a wide variety of machines used by students and researchers, such as Xenon, a Sun-4/670 with 4 CPUs, the primary student machine; Radon, an HP 9000-755 compute server for student use; various medium to large UNIX machines; and hundreds of workstations and X terminals from Sun Microsystems, Digital Equipment Corporation, Hewlett-Packard, Silicon Graphics, NeXT, IBM, and NCD.
Financial Aid	All incoming Ph.D. students are supported by a departmental assistantship or by a fellowship. Assistantship holders receive a stipend plus a 9-unit tuition credit each quarter. If there is an insufficient number of Ph.D. students to staff teaching and research assistantships, then such positions are offered to qualified students in the M.S.C.S. program.
Cost of Study	For 1999–2000, tuition charges for all students not holding assistantships are approximately $8000 per quarter or $24,000 per academic year.
Living and Housing Costs	On-campus housing for single students costs approximately $2800 for the 1999–2000 academic year. Housing for married students is available and averages about $600 per month. Off-campus housing tends to be more expensive. The cost of living in the area is relatively high.
Student Group	The department has approximately 175 doctoral students and 350 master's students; they come from all parts of the nation and the world. Approximately 30 Ph.D. degrees and 150 master's degrees are awarded each year. Career choices are divided about evenly between academic and industrial positions.
Location	Stanford University is located on a spacious campus on the San Francisco peninsula, 30 miles south of the city of San Francisco. A wide variety of natural attractions are located within a short drive, including the Pacific Ocean and the Sierra Nevada. The climate is mild the year round. The San Francisco Bay Area is a major and diverse cultural center and is also the site of many industrial corporations and research centers in computer and other technologies.
The University	Stanford is a private, nonsectarian, coeducational university with an international reputation as an outstanding educational institution. It operates on the quarter system with a shortened summer session. Enrollment is approximately 14,000, including 7,000 graduate students. Among the approximately 1,300 faculty members are 10 Nobel laureates and many others who have achieved wide academic distinction. The University provides superb academic and athletic facilities.
Applying	Applications and all supporting documentation for admission to the Ph.D. and the M.S.C.S. programs must be received before December 15. Exceptions are made for applicants to the M.S.C.S. program who are either Honors Co-op applicants or already students at Stanford. Information on these deadlines is available from the department. Absolutely no exceptions will be made for the Ph.D. program. Application forms and information packets may be obtained from the Office of Graduate Admissions or, for non-U.S. citizens, the Office of Foreign Graduate Admissions. Financial aid information for Ph.D. students will be made available upon the students' acceptance.
Correspondence and Information	Admissions Central Computer Science Department Gates 1B, Room 196 Stanford University Stanford, California 94305-9015 World Wide Web: http://www.stanford.edu

Stanford University

THE FACULTY AND THEIR RESEARCH

The following is a partial list of faculty members affiliated with the Computer Science Department.

Charles Bigelow, Associate Professor of Art and Computer Science; B.A., Reed, 1967. Digital typographic design.

Thomas O. Binford, Professor of Research in Computer Science; Ph.D., Wisconsin, 1965. Computer vision, robotics, artificial intelligence, computer-aided design, manufacturing.

Dan Boneh, Assistant Professor of Computer Science and Electrical Engineering; Ph.D., Princeton, 1996. Cryptography and network security.

Christoph Bregler, Assistant Professor; Ph.D., Berkeley, 1998. Computer vision, graphics, AI and HCI.

David R. Cheriton, Associate Professor; Ph.D., Waterloo, 1978. Computer operating system design, distributed systems, computer communications, multiprocessor architectures, parallel computation.

William J. Dally, Professor; Ph.D., Caltech, 1986. Computer architecture and the implementation of multiprocessors.

George B. Dantzig, Professor Emeritus of Operations Research and Computer Science; Ph.D., Berkeley, 1946. Modeling and optimization of large-scale energy systems, combinatorial mathematics, mathematical programming.

David L. Dill, Assistant Professor; Ph.D., Carnegie Mellon, 1987. Concurrency, hardware verification, asynchronous circuits, compilers, LISP.

Dawson Engler, Assistant Professor; Ph.D., MIT, 1998. Operating systems.

Edward A. Feigenbaum, Professor; Ph.D., Carnegie Tech, 1960. Knowledge engineering, expert systems, artificial intelligence, large-scale knowledge-based systems.

Richard E. Fikes, Professor of Research and Co-Scientific Director of the HPP in the Knowledge Systems Laboratory; Ph.D., Carnegie Mellon, 1968. Knowledge-based systems technology, declarative knowledge representation.

Robert W. Floyd, Professor; B.A., 1953, B.S., 1958, Chicago. Design and analysis of algorithms.

Armando Fox, Assistant Professor; Ph.D., Berkeley, 1998. Internet services and systems, wireless and mobile computing.

Hector Garcia-Molina, Professor; Ph.D., Stanford, 1979. Database systems, distributed computing.

Michael R. Genesereth, Associate Professor; Ph.D., Harvard, 1978. Artificial intelligence, logic, automated reasoning, and agent architecture with applications in engineering and robotics.

Gene H. Golub, Professor; Ph.D., Illinois, 1959. Numerical analysis, scientific computing, mathematical programming, statistical computing.

Leonidas J. Guibas, Professor; Ph.D., Stanford, 1976. Computational geometry, computer graphics, VLSI algorithms and design aids, analysis of algorithms and data structures, complexity theory, programming techniques, personal computing.

Anoop Gupta, Assistant Professor; Ph.D., Carnegie Mellon, 1985. Highly parallel computer architectures, programming languages and operating systems for such machines, parallel applications studies.

Pat Hanrahan, Professor; Ph.D., Wisconsin, 1986. Computer graphics, rendering algorithms and high-performance graphics systems.

John Hennessy, Willard and Inez Bell Professor of Electrical Engineering and Computer Science; Ph.D., SUNY at Stony Brook, 1977. Computer architecture and optimizing compilers, especially the interaction between compiler technology and architecture; multiprocessors; parallel computing.

John G. Herriot, Professor Emeritus; Ph.D., Brown, 1941. Numerical analysis.

Oussama Khatib, Associate Professor; Ph.D., Toulouse (France), 1980. Robotics, control architectures, strategies, sensing, design.

Donald E. Knuth, Professor of the Art of Computer Programming; Ph.D., Caltech, 1963. Analysis of algorithms, programming languages, mathematical typography, combinatorial mathematics, history of computer science.

Daphne Koller, Assistant Professor of Computer Science; Ph.D., Stanford, 1993. AI: creating systems that reason and act under uncertainty.

Monica S. Lam, Assistant Professor; Ph.D., Carnegie Mellon, 1987. Parallel computer systems, programming languages, optimizing compilers, computer architectures.

Jean-Claude Latombe, Associate Professor; Thèse d'état, Grenoble (France), 1977. Robotics, artificial intelligence, geometrical reasoning, planning.

Marc Levoy, Assistant Professor of Computer Science and Electrical Engineering; Ph.D., North Carolina at Chapel Hill, 1989. Computer graphics, scientific visualization, interactive techniques.

Zohar Manna, Professor; Ph.D., Carnegie Mellon, 1968. Mathematical theory of computation, logic of programs, automated deduction, logic programming, concurrent programming, artificial intelligence.

John McCarthy, Charles Piggot Professor of Computer Science; Ph.D., Princeton, 1951. Artificial intelligence, computing with symbolic expressions, time sharing, formalizing common sense, nonmonotonic logic.

Edward J. McCluskey, Professor of Electrical Engineering and Computer Science; Sc.D., MIT, 1956. Fault-tolerant computing, computer reliability, diagnosis, and testing; organization of computer systems; switching theory and logic design.

Nick McKeown, Assistant Professor of Computer Science; Ph.D., Berkeley, 1995. Architectures for high-speed switched, scheduling algorithms, multicast support, traffic management.

William F. Miller, Professor of Computer Science and Public and Private Management, Graduate School of Business; Ph.D., Purdue, 1956. Computer systems design, software systems, strategic planning and management, economic technological development.

John C. Mitchell, Associate Professor; Ph.D., MIT, 1984. Programming language theory, functional programming, object-oriented programming, applications of classical and nonclassical logic to computational problems.

Rajeev Motwani, Assistant Professor; Ph.D., Berkeley, 1988. Design and analysis of algorithms, data structures, complexity theory.

Nils J. Nilsson, Professor; Ph.D., Stanford, 1958. Artificial intelligence, knowledge representation, reasoning systems.

Joseph Oliger, Professor; Ph.D., Uppsala (Sweden), 1973. Numerical analysis, numerical methods for partial differential equations, applications in meteorology, oceanography, and geophysics.

Serge A. Plotkin, Assistant Professor; Ph.D., MIT, 1988. Parallel and distributed computation, analysis of algorithms, combinatorial optimization.

Vaughan Pratt, Professor; Ph.D., Stanford, 1972. Process specification languages, models of concurrency, applications of algebraic geometry and category theory, digital typography.

Eric Roberts, Associate Professor and Assistant Chair for Educational Affairs; Ph.D., Harvard, 1980. Computer science education, social implications of computing, programming languages, programming environments.

Mendel Rosenblum, Assistant Professor of Computer Science; Ph.D., Berkeley, 1992. Operating systems, computer architecture.

Arthur L. Samuel, Professor Emeritus of Research in Computer Science; S.M., MIT, 1926. Artificial intelligence.

Yoav Shoham, Assistant Professor; Ph.D., Yale, 1987. Artificial intelligence, spatiotemporal reasoning, formalizing common sense.

Andrew M. Stuart, Assistant Professor of Computer Science and Mechanical Engineering; D.Phil., Oxford, 1986. Dynamical systems.

Jeffrey D. Ullman, Professor; Ph.D., Princeton, 1966. Database systems, logic programming, parallel computation.

Gio Wiederhold, Professor of Research in Medicine and Computer Science; Ph.D., California, San Francisco, 1976. Databases, knowledge bases, information systems, parallel problem solving.

Terry Winograd, Professor; Ph.D., MIT, 1970. Artificial intelligence, human-computer interaction, work-centered system design.

STANFORD UNIVERSITY

Medical Information Sciences Program

Program of Study

The Medical Information Sciences (MIS) Program is an interdepartmental program offering instruction and research opportunities leading to an M.S. or a Ph.D. in medical information sciences (medical informatics). The program is administered by the School of Medicine, but its curriculum and degree requirements are coordinated with the Office of Graduate Studies of the University. The program is designed to train researchers and educators in the field of medical informatics. Emphasis is placed on providing innovations of relevance to clinical medicine or biomedical research. Although Stanford researchers have expertise in a cross section of MIS activities, the University is recognized in particular for its studies of clinical decision making and of the interface between developing technology and the decision sciences. These topics are accordingly emphasized in the program's curriculum. Required courses come from five major topic areas: medical informatics, computer science, decision science and statistics, biomedical domain knowledge, and social and ethical issues.

Research Facilities

Five large computing servers and allied student computing clusters are located around campus and are available to all students in the University for instruction, unsponsored research, e-mail, and World Wide Web services. In addition, a number of systems are available to students of computer science, to which MIS trainees may have access for relevant work. Stanford Medical Informatics resources include file servers and a variety of UNIX, Macintosh, and PC client workstations for instruction and research. Essentially all computing and information resources at Stanford are linked together by a high-speed Internet communications network. Students can consult Stanford's World Wide Web page (listed below) for more information.

Financial Aid

A limited amount of funding is available to provide research stipend support for students in the program. Most students become associated with projects that provide them with support soon after their arrival at Stanford.

Cost of Study

Tuition in 1999–2000 is $7686 per quarter. This figure is expected to increase annually.

Living and Housing Costs

Housing is available for both on-campus and off-campus living. On-campus housing for graduate students is limited. The cost of living in the area is relatively high.

Student Group

Stanford University has a faculty of about 1,300 members and a total student enrollment of more than 13,000, of whom nearly half are graduate students. The MIS Program, now in its sixteenth year, has 28 graduate students.

Student Outcomes

There is a high demand for individuals with formal training in medical informatics. Among the program's 56 graduates, 22 are in academic positions, 23 work in industry, 1 works for a hospital, 3 work for the federal government, 4 are in clinical practice, and 3 are completing residency training.

Location

Stanford is located next to Palo Alto, a community of 60,000, about 35 miles south of San Francisco.

The University

Stanford (founded in 1885) is a private, nonsectarian, coeducational university with an international reputation as an outstanding educational and research institution. It operates on the quarter system with a shortened summer session. Among the almost 1,350 faculty members are 12 Nobel laureates and many others who have achieved wide academic distinction. The Medical Information Sciences Program is administered from the offices and laboratory of Stanford Medical Informatics, located in the Stanford Medical Center complex. The University is committed to the principles of affirmative action in the admission of students and in the employment of faculty and staff, and does not use any racial, religious, ethnic, geographic, or sex-related quotas.

Applying

Applications for the master's and Ph.D. programs are normally considered for admission in the autumn quarter only and must be submitted with supporting documents no later than the preceding January 1. Applicants must report GRE scores. M.D. applicants applying for the M.S. degree may submit MCAT scores.

Correspondence and Information

Darlene Vian, Program Administrator
Stanford Medical Center, MSOB X-215
251 Campus Drive
Stanford University
Stanford, California 94305-5479

Telephone: 650-725-3388
Fax: 650-498-4162
E-mail: vian@smi.stanford.edu
World Wide Web: http://www.smi.stanford.edu

For application forms:

Office of Graduate Admissions
Old Union 141
Stanford University
Stanford, California 94305-3005
World Wide Web: http://www.stanford.edu/dept/
  registrar/admissions/applyinfo.html

Stanford University

THE FACULTY AND THEIR RESEARCH

Edward H. Shortliffe, M.D., Ph.D., Program Director; Department of Medicine (Stanford Medical Informatics) and, by courtesy, Computer Science. Computer-based medical consultation systems, with emphasis on integrating decision analytic and artificial intelligence techniques.

Lawrence M. Fagan, M.D., Ph.D., Program Co-Director; Department of Medicine (Medical Informatics). Techniques for relating qualitative and quantitative physiological models, design of speech and graphic interfaces for expert systems.

John R. Adler, M.D., Department of Neurosurgery. Development of computerized tools for surgical guidance.

Russ B. Altman, M.D., Ph.D., Department of Medicine. Application of computing technologies to basic molecular biological problems, particularly the analysis of protein structure and function.

Thomas O. Binford, Ph.D., Department of Computer Science. Sensing, machine perception and computer vision, robotics, geometric modeling, reasoning with geometry, evidential reasoning.

Terrence Blaschke, M.D., Department of Medicine (Clinical Pharmacology). Automated monitoring of therapeutic decisions, pharmacokinetic modeling of drug distribution.

Byron W. Brown Jr., Ph.D., Departments of Statistics and Health Research and Policy. Methodologies for the design and analysis of experiments in all phases of medical research, using mathematical work and computer simulation.

Douglas Brutlag, Ph.D., Department of Biochemistry. Knowledge representation and reasoning, pattern recognition and sequence classification in biological-sequence databases.

Robert W. Carlson, M.D., Department of Medicine (Oncology). Computer-based physician consultation systems for treating patients with cancer and AIDS.

Stanley N. Cohen, M.D., Departments of Genetics and Medicine. Molecular genetics and use of computers for monitoring drug use to assist in therapeutic decisions.

Parvati Dev, Ph.D., SUMMIT. Computers in medical education, hypermedia, 3-D medical imaging, simulation of clinical encounters, virtual reality, cognitive modeling of students.

Alain C. Enthoven, Ph.D., Graduate School of Business. Use of decision analysis techniques in reforming the financing and delivery of health care in the United States.

Edward A. Feigenbaum, Ph.D., Department of Computer Science. Mechanization of scientific reasoning, formalization of scientific knowledge, design of computers for artificial intelligence applications.

James F. Fries, M.D., Department of Medicine (Immunology). National chronic-disease computer databank systems, formal approaches to clinical decision making.

Victor R. Fuchs, Ph.D., Department of Economics. Socioeconomic determinants of health, functioning of health-care markets, relation between health and postindustrial society.

David M. Gaba, M.D., Department of Anesthesia. Human error in anesthesia-related accidents, intelligent decision support for decision making in anesthesia.

Alan M. Garber, M.D., Ph.D., Department of Medicine (General Internal Medicine) and, by courtesy, Economics. Health policy and health economics, particularly methods to forecast the utilization of and expenditures for long-term care of the elderly.

Gary H. Glover, Ph.D., Department of Radiology. MR imaging and spectroscopy, particularly chemical-shift resolved tissue characterization.

Mark A. Hlatky, M.D., Departments of Health Research and Policy and Medicine. Costs and outcomes of cardiovascular care, technology assessment, physician decision making, clinical epidemiology.

Samuel Holtzman, Ph.D., Department of Engineering–Economic Systems. Computer-based decision aids for industry and medicine, economics and ethics of medical decision making, modeling of medical preferences.

Ronald A. Howard, Ph.D., Department of Engineering–Economic Systems. Development of systematic, logical procedures for decision making in uncertain, complex, and dynamic settings.

Daphne Koller, Ph.D., Department of Computer Science. Artificial intelligence and theoretical computer science.

John Koza, Ph.D., Department of Computer Science. Genetic algorithms.

Emmet J. Lamb, M.D., Department of Obstetrics and Gynecology. Clinical infertility, reproductive endocrinology, clinical decision analysis.

Marc Levoy, Ph.D., Departments of Computer Science and Electrical Engineering. Computer graphics for scientific data visualization.

Albert Macovski, Ph.D., Department of Electrical Engineering. Systems approach to the imaging of the internal structures of the body.

Mark A. Musen, M.D., Ph.D., Department of Medicine (Stanford Medical Informatics) and, by courtesy, Computer Science. Computer-based tools for describing the knowledge of how to conduct clinical trials, knowledge-based systems for clinical-trial design and administration.

Sandy A. Napel, Ph.D., Department of Radiology. Visualization of flow from magnetic resonance images; segmentation of tumors from 3-D medical imaging data; multidimensional, multimodality image correlation and display.

Richard A. Olshen, Ph.D., Department of Health Research and Policy. Statistics and mathematics as applied to problems in medicine.

Douglas K. Owens, M.D., Department of Medicine. Technology assessment and the application of decision theory to clinical and health policy problems.

Norbert J. Pelc, Sc.D., Department of Diagnostic Radiology and Nuclear Medicine. Medical imaging modalities, medical imaging, magnetic resonance.

Thomas C. Rindfleisch, Department of Medicine (Medical Informatics), Director, Lane Medical Library. Symbolic systems, integrated workstations for biomedicine, information retrieval from optical databases.

Geoffrey D. Rubin, M.D., Department of Radiology, Co-director of the Radiology 3-D Laboratory. Volumetric imaging, display, and analysis.

Ross D. Schachter, Ph.D., Department of Engineering–Economic Systems and of Operations Research. Modeling of uncertain processes over time, decision making under uncertainty.

Yuval Shahar, M.D., Ph.D., Department of Medicine. Planning, temporal reasoning, knowledge representation, problem-solving methods, medical decision analysis.

Ramin Shahidi, Ph.D., Department of Neurosurgery. Preoperative surgical planning and intraoperative volumetric image navigation techniques and apparatuses.

Lee S. Shulman, Ph.D., Graduate School of Education. Clinical judgment and problem solving in medicine and teaching, assessment of professional and clinical expertise in teaching and medicine.

Howard H. Sussman, M.D., Department of Pathology. Laboratory information systems, networking for data sharing.

Gio Wiederhold, Ph.D., Departments of Computer Science, Medicine, and Electrical Engineering. Effective information systems, acquisition of real-time data.

Terry Winograd, Ph.D., Department of Computer Science. Development of conceptual models and interactive structures.

Lei Xing, Ph.D., Department of Radiation Oncology. Computer optimization of treatment plans in conformal radiation therapy and stereotactic radiosurgery.

STANFORD UNIVERSITY

Scientific Computing and Computational Mathematics Program

Programs of Study

The Scientific Computing and Computational Mathematics (SCCM) Program was founded in 1988 to provide interdisciplinary graduate training and to foster research in areas of the applied sciences and engineering that require interactions among modeling, mathematical and numerical analysis, and scientific computing. The primary educational goal is to prepare students for research or teaching positions within universities or in industry. Both M.S. and Ph.D. degrees are offered.

Successful research in scientific computing involves formulation of a mathematical model of a phenomenon; mathematical analysis of the model; reduction to a finite dimensional form appropriate for numerical simulation, together with attendant numerical analysis; and computer implementation, including exploitation of appropriate computer architecture. The SCCM program provides training that recognizes the broad interrelation of these four areas but that is also sufficiently focused to provide intellectual rigor and challenge. Mathematical and numerical analysis form the core of the course work offered in the SCCM Program; however, it is important to recognize that such analysis can and should be significantly influenced by an understanding of the application areas of interest on the one hand and developments in computer science and scientific computing on the other. Hence a thorough study of an application area, together with fundamental knowledge of computer science, is also required. A training of the kind outlined is not available through traditional departments; the SCCM program provides unification and coherence to doctoral training involving many disciplines. This is reflected in the truly interdisciplinary nature of the group of faculty members involved in the program.

For Ph.D. students, the majority of the first year is spent taking core courses in mathematical and numerical analysis that form the basis of a written comprehensive exam, usually taken twelve months after arrival. During the first eighteen months, students also typically associate themselves with a research group; thereafter, the majority of time is spent in research, leading to a thesis defense and thesis in this area. Additional course work, augmenting the core material, giving a firm foundation in an application area, and covering important issues in computer science, is also required.

Studying for the M.S. degree primarily involves course work of the same type taken for the Ph.D. degree and usually takes between four and six quarters to complete.

Research Facilities

A variety of computer systems are available to all students in the University. In addition, the SCCM group has a cluster of computers: SGI, SUN, and IBM workstations. Parallel computers are also available through specific research groups. All computers are linked by a University-wide Ethernet system.

Financial Aid

Incoming Ph.D. students are generally supported by a departmental assistantship or by a fellowship. Assistantship holders receive a stipend plus a 9-unit tuition credit each quarter. Master's students are not guaranteed aid. If additional research assistantships become available, they are offered to M.S. students.

Cost of Study

For 1999–2000, tuition charges for all students not holding assistantships are $8196 per quarter.

Living and Housing Costs

Housing is available both on campus and off campus. On-campus housing is available for married graduate students. The cost of living in the area is relatively high.

Student Group

Stanford University has a total student enrollment of 14,000, of whom more than half are graduate students. The SCCM Program averages between 30 and 40 graduate students a year.

Location

Stanford has a spacious campus located on the San Francisco peninsula. The San Francisco Bay area is a major and diverse cultural center and is also the site of many industrial corporations and research centers.

The University

Stanford is a private, nonsectarian, coeducational university with an international reputation as an outstanding educational and research institution. It operates on the quarter system with a shortened summer session. There are more than 1,300 faculty members, many of whom have achieved wide academic distinction.

Applying

Applications and all supporting documentation for admission to the Ph.D. program must be received before February 15. Applications for the M.S. program are accepted until April 15, though it is beneficial to apply before March 1 to be eligible for the initial on-campus housing pool. Applicants must submit GRE scores; Ph.D. applicants must also take a GRE Subject Test in math, CS, engineering, or physics.

Correspondence and Information

Admissions
SCCM Program
Gates 2B, Room 291
Stanford University
Stanford, California 94305-9025

Telephone: 650-723-0572
Fax: 650-723-2411
E-mail: admissions@sccm.stanford.edu

Stanford University

THE FACULTY AND THEIR RESEARCH

Program Director
Gene H. Golub.

Associate Program Director
Walter Murray.

Core Faculty
Robert Dutton, Ph.D., Department of Electrical Engineering. Computational electronics for the design and manufacturing of integrated circuits.
Gene H. Golub, Ph.D., Department of Computer Science. Design and analysis of algorithms arising in linear algebra, matrix methods in signal processing, large sparse systems of equations.
Joseph Keller, Ph.D., Department of Mathematics. Applied mathematics, wave propagation, asymptotic analysis, electromagnetic theory, optics and acoustics.
Walter Murray, Ph.D., Department of Operations Research. Linear and nonlinear programming, sparse matrix methods, linear algebra, theoretical and practical optimization problems.
Joseph Oliger, Ph.D., Department of Computer Science. Numerical analysis, numerical methods for partial differential equations, computer simulation or analysis of physical processes.
George Papanicolaou, Ph.D., Department of Mathematics. Stochastic equations and random media, seismic signal processing and wavelets, singularities in nonlinear waves.
Andrew Stuart, Ph.D., Department of Computer Science and Mechanical Engineering. Numerical analysis of evolution equations, dynamical systems, applied and computational mathematics.

Associate Faculty
Khalid Aziz, Ph.D., Department of Petroleum Engineering. Multiphase flow of oil/gas mixtures and steam in pipes and wells, multiphase flow in porous media, reservoir simulation.
Joel Ferziger, Ph.D., Department of Mechanical Engineering. Computational fluid dynamics, simulation of turbulent flows, flow phenomena, numerical methods for solving the equations of fluid mechanics.
Thomas J. Hughes, Ph.D., Department of Mechanical Engineering. Computational methods to problems in solid and fluid mechanics, including finite element methods for nonlinear plate and shell response.
Thomas Kailath, Ph.D., Department of Electrical Engineering. Signal processing algorithms for various applications, emphasizing speed, numerical robustness, and ease of design and implementation in current technology.
Tai-Ping Liu, Ph.D., Department of Mathematics. Hyperbolic equations and conservation laws.

STATE UNIVERSITY OF NEW YORK AT ALBANY

Department of Computer Science

Programs of Study

The Department of Computer Science at Albany offers programs of study leading to the degrees of Doctor of Philosophy and Master of Science. In addition, several members of the department participate in the University's interdisciplinary Ph.D. program in information science. Instruction covers a wide range of areas. The current areas of research include data structures and algorithms, automated reasoning and theorem proving, theory of computation, artificial intelligence, high-performance computing, operating systems and distributed systems, natural language processing and robotics, artificial neural networks, computational biology, knowledge representation, hardware and software specification and verification, computer algebra, VLSI circuit testing and analysis, fault-tolerant computing, database systems, information management, combinatorics, software engineering, and compiler design.

Doctoral students are required to pass examinations in several areas of computer science and an oral examination in their field of research interest and they must submit a dissertation that describes original research. The Ph.D. is awarded in recognition of high achievement in research. The doctoral program is intended for students with career interests in academia, industrial research and development, or government research agencies.

For the M.S. degree, students must complete at least 32 credits of approved course work, complete a programming project of significant scope, and pass a comprehensive exam in computer science. The programming project requirement may be waived for students with appropriate work experience.

Research Facilities

The Departmental Laboratory supports research in High-Performance Parallel Computation C and FORTRAN (HPF) compilers, operating systems, and networks with an ATM switch that joins two Sun SPARC20s, a multiprocessor SPARC1000E, and a Silicon Graphics Indigo2. It also includes an Ethernet subnetwork of other Sun, Silicon Graphics, and (Linux) PC-type workstations and a Pioneer I mobile robot.

All of these systems are accessible from the University's high-speed (T3) Internet connection, modem pool, and dormitory/classroom network.

Financial Aid

In 1999–2000, teaching and research assistantships that provide tuition waivers and stipends of $8000 to $12,000 for the nine-month academic year are available. In some instances, summer appointments are also made. Applications for assistantships beginning in the fall session must be submitted by February 15, while applications for assistantships beginning in the spring session must be submitted by October 15.

Presidential fellowships with a stipend of $15,000 and an appropriate tuition waiver are open to students who have been admitted to the doctoral program. Applications for these awards must be submitted by February 15.

Cost of Study

Graduate tuition for New York State residents is $5100 per year for 12 or more credits and $213 per credit for fewer than 12 credits for 1999–2000. Tuition for out-of-state residents is $8416 for 12 or more credits and $351 per credit for fewer than 12 credits. Other fees total approximately $705 for full-time students.

Living and Housing Costs

On-campus accommodations start at $5472 for room and board for the 1999–2000 academic year. Off-campus apartments average $250–$300 per person per month. Total costs for a full year of study, including tuition, fees, books, room, board, and modest entertainment, are approximately $16,000. International students must have at least $19,925 available to meet all expenses for a calendar year.

Student Group

The graduate student body of the department numbers 72 students, 21 of whom are Ph.D. students.

Student Outcomes

Most M.S. graduates find employment as software professionals, including positions as analysts and developers. Recent doctoral graduates and some M.S. graduates have positions in government, industrial, or academic research. A concentration of class program, individual project, and research participation can lead to the career specialty in which the student is most interested. Graduates have located to large, as well as small, metropolitan areas.

Location

Albany is the capital of New York State, and the Albany-Schenectady-Troy capital region metropolitan area has a population of more than 200,000. The Albany Medical Center, various professional schools, other colleges and universities, and technology parks, as well as museums, theaters, the Pepsi Arena, world-renowned classical and popular music, and nearby summer stock performing arts, all lend a cosmopolitan tone to the city.

The University

Albany, one of four university centers in the State University of New York (SUNY) system, was founded in 1814 and has about 5,250 graduate and 11,617 undergraduate students.

Applying

Degree students may be admitted for the fall or spring terms, although the curriculum is oriented toward fall admission. Application deadlines are flexible, but financial aid applications for the fall semester are due February 15. Admission requirements for the M.S. and Ph.D. programs are a bachelor's degree, superior undergraduate achievement as indicated by transcripts and overall grade point average, three recommendations, and acceptable scores on the General Test of the Graduate Record Examinations. Ph.D. applicants are encouraged to take the GRE Subject Test in either mathematics, physics, computer science, or engineering.

Correspondence and Information

For application forms and admissions information:
Graduate Studies
AD-152
State University of New York at Albany
Albany, New York 12222
Telephone: 518-442-3980
E-mail: graduate@cnsibm.albany.edu

For further information about graduate work:
Professor Daniel J. Rosenkrantz, Chair
Department of Computer Science, LI-67A
State University of New York at Albany
Albany, New York 12222
Telephone: 518-442-4270
E-mail: info@cs.albany.edu
World Wide Web: http://www.cs.albany.edu

State University of New York at Albany

THE FACULTY AND THEIR RESEARCH

George Berg, Associate Professor; Ph.D., Northwestern. Artificial intelligence, especially machine learning. Current projects are in computational molecular biology and autonomous robots. The machine learning laboratory is investigating how proteins fold into their three-dimensional shapes and how those shapes can be characterized. In robotics, the goal is for robots to learn about the features of their environment they encounter using vision, sonar, and others. Additional interests include natural language processing and high-performance computing. (E-mail: berg@cs.albany.edu)

Peter A. Bloniarz, Associate Professor and Research Director of the Center for Technology in Government; Ph.D., MIT. Software engineering, particularly data modeling and object-oriented databases; information management and the use of technology in the public and private sectors. Current research involves cost-benefit analysis of networked hypertext access to documents. The Center for Technology in Government is a research laboratory whose mission is to investigate and facilitate the use of technology in the public sector. (E-mail: pb@cs.albany.edu)

Seth Chaiken, Associate Professor; Ph.D., MIT. Combinatorics, electrical networks, analysis of algorithms, computer architecture. Current work involves mathematical foundations of electrical network theory, including graph and matroid theory and discrete models of the analog behavior of electronic circuits important in digital system design specification, analysis, and verification. Other research is in average system performance studies, environments for graph algorithm study, and window systems. (E-mail: sdc@cs.albany.edu)

Mei-Hwa Chen, Assistant Professor; Ph.D., Purdue. Process modeling, software design, verification and validation, software reliability engineering and GUI. Current research focuses on object-oriented software testing and reliability estimation. Projects involve building a C++ testing and coverage analysis tool and developing white box reliability models for both traditional and object-oriented software. (E-mail: mhc@cs.albany.edu)

Andrew Haas, Associate Professor; Ph.D., Rochester. Artificial intelligence, natural-language syntax and parsing, semantics, propositional attitudes, planning. Current research involves the construction of a simulated robot which carries out commands given in English. The robot can understand indexicals, demonstratives, and descriptions of objects that it has not yet seen. It plans not only to rearrange the physical world but also to acquire knowledge by perception. (E-mail: haas@cs.albany.edu)

Harry B. Hunt III, Professor; Ph.D., Cornell. Theory of computation, analysis of algorithms, combinatorial optimization, parallel and distributed computation, computation science. Current work involves the study of problems on various algebraic structures to obtain a precise characterization of the complexities of the decision, optimization, and counting versions of these problems. Applications in the areas of VLSI, database systems, recursive, hierarchical, or dynamic specifications have been demonstrated. Other topics under investigation include computational issues in distributed computing and model checking and high-performance parallel scientific computing. (E-mail: hunt@cs.albany.edu)

Lenore M. R. Mullin, Associate Professor and 1993 NSF Presidential Faculty Fellow; Ph.D., Syracuse. Psi calculus, advanced scalable computing and communications and their relationship to whole array operations in scientific computing. The mechanization of psi reductions in a preprocessing compiler and general distribution schemes for languages such as HPF. Other interests include OS's tuned for scientific computing, scientific polyalgorithms, and hardware coprocessors to assist OS support of regular memory access patterns. (E-mail: lenore@cs.albany.edu)

Neil V. Murray, Professor; Ph.D., Syracuse. Automated reasoning in classical and multiple-valued logics; logic programming; deduction methods that permit the avoidance of so-called clause form, including analytic tableaux, dissolution, and nonclausal resolution. Current research includes application of these ideas to annotated logic programming, diagnosis, and computation of prime implicants/implicates. (E-mail: nvm@cs.albany.edu)

Paliath Narendran, Associate Professor; Ph.D., Rensselaer. Formal hardware specification and verification, automated reasoning, term rewriting systems, unification. Current interests include reasoning and specification methods suited for hardware, algebraic and computational aspects of term rewriting and unification, and applications of unification and pattern-matching in knowledge representation. (E-mail: dran@cs.albany.edu)

S. S. Ravi, Associate Professor; Ph.D., Pittsburgh. Design and analysis of algorithms, operations research, fault-tolerant computing, VLSI. Current research centers on the design and analysis of algorithms and/or heuristics for optimization problems that arise in areas such as network design, transportation, fault tolerant computing, and VLSI. (E-mail: ravi@cs.albany.edu)

Daniel J. Rosenkrantz, Professor and Chair, ACM Fellow; Ph.D., Columbia. Database systems, algorithms, compiler design, parallel computation, fault-tolerant computing. Recent research includes distributed database transaction processing, data mining, distributed algorithms, algorithms for transportation problems, and network design. (E-mail: djr@cs.albany.edu)

Richard E. Stearns, Professor, ACM Fellow, and 1993 Turing Award Winner; Ph.D., Princeton. General research interests include computational complexity, automata theory, analysis of algorithms, and game theory. One current study involves the use of algebraic formulas with generalized quantifiers to describe computational problems. This approach allows one to attribute structure to problem instances and to develop genetic algorithms that solve structured problem instances much faster than brute-force methods. These techniques are being used to study approximation algorithms and extensions to succinct representations. (E-mail: res@cs.albany.edu)

Dan E. Willard, Professor; Ph.D., Harvard. Database systems, data structures, analysis of algorithms, computational geometry, and proof theory from mathematical logic. Most recent work has included the development of self-verifying axiom systems that can prove their own consistency and the study of faster than LogN methods for searching and faster than NlogN methods for sorting. Other work includes the design and analysis of algorithms for database retrieval, geometric search problems, range queries, and dynamic set manipulation. (E-mail: dew@cs.albany.edu)

Adjunct Faculty

Theodor J. Borys, Data Center Director, New York State Office of Mental Health; M.S., SUNY at Albany. Very large and complex databases, transaction processing, client-server models, local area networks.

Thomas Irvin, Chief of Computer Services, New York State Executive Chamber; M.S., SUNY at Albany. Database design, full text storage and retrieval, large-scale optical storage and retrieval systems, computer-related workflow analysis.

Affiliated Faculty

Jacquelyn Fetrow, Associate Professor, The Scripps Research Institute; Ph.D., Pennsylvania State University College of Medicine. Primary research interests include predicting protein structure and function and understanding protein structure and function relationships. Computational techniques are used to study these problems.

Joachim Frank, Professor, Biomedical Sciences; Ph.D., Technical University of Munich. Methods of three-dimensional reconstruction of biological macromolecules are being developed based on electron microscopic images of single particles. To this end, cross-correlation, multivariate statistical analysis, classification, weighted back-projection and SIRT (Simultaneous Iterative Reconstruction Technique) are utilized. These methods are being applied to ribosomes, hemocyanin, and calcium-release channels.

David Goodall, Director of Fiscal and Human Resource, New York State Department of Motor Vehicles; Ph.D., SUNY at Albany. Social impact of computers; privacy, security, and risks of computing; Internet problems and opportunities; computer ethics and professional responsibilities.

Pawel A. Penczek, Research Scientist, Wadsworth Center; Ph.D., Warsaw. Current research focuses on the three-dimensional (3D) reconstruction methods in application to electron microscopy. Several 3D reconstruction algorithms have been developed, most notably iterative techniques to deal with extremely noisy and inconsistent data. Research interests include signal processing and pattern recognition techniques. These techniques are employed in the studies of 3D structures of biological macromolecules. Some problems, such as transfer function correction, can be expressed as inverse problems and solved with the use of a priori information. Currently, work involves the improvement of the resolution of results. This can be achieved by collecting and processing very large amounts of data. To be able to process such a wealth of data in a reasonable time, attention is turned toward parallel algorithms and distributed processing.

Giri Kumar Tayi, Associate Professor, Department of Management Science and Information Systems; Ph.D., Carnegie Mellon. Current research is in the areas of data quality, data warehousing, data mining, mathematical models of manufacturing and service operations, and design of algorithms and heuristics for problems in telecommunication networks.

STATE UNIVERSITY OF NEW YORK AT ALBANY

Information Science Doctoral Program

Program of Study	The State University of New York at Albany offers an interdisciplinary program leading to the degree of Doctor of Philosophy in information science administered by the Nelson A. Rockefeller College of Public Affairs and Policy. It is a collaborative program of the School of Business; the Departments of Communication, Geography and Planning, and Computer Science of the College of Arts and Sciences; the School of Information Science and Policy; and the Department of Public Administration and Policy of the Graduate School of Public Affairs.
	The program centers on advanced study and applied research in the nature of information as a phenomenon and in the character of the information transfer process. It prepares graduates both for academic and research careers in information science and for senior information management and policy positions in government or the private sector. The program requires a minimum of three years of full-time postbaccalaureate study or its part-time equivalent. A minimum of two terms of full-time resident study are required. The major components are four required interdisciplinary core proseminars, research tool and information technology competencies, course work in primary and secondary areas of specialization, and a doctoral dissertation. Areas of specialization are expert systems, geographic information systems, group decision support modeling, information decision systems, organization of knowledge records, and public information policy and a secondary specialization in organizational studies.
Research Facilities	The Center for Technology in Government seeks practical solutions to problems of information technology and management in government settings. Other research programs in which doctoral students are encouraged to participate are currently under way in research library and archival management and electronic records management.
	The University libraries house more than 1.3 million volumes and maintain some 6,700 current periodical subscriptions. There are forty major libraries in the Albany area, including the New York State Library. Cooperative relationships with these and other research libraries provide access to collections nationwide.
	The University provides access to mainframe computing, microcomputers, graphics equipment, laser printing, and regional and worldwide computer networks, as well as instruction in the use of these facilities. Public computer rooms are available at several campus locations.
Financial Aid	Full-time students may apply for University fellowships and assistantships carrying annual stipends of up to $13,000 plus full or partial tuition scholarships in return for service in teaching, research, or administration. Similar research assistantships are available through the Center for Technology in Government. Information about other state and federal financial aid programs for which doctoral students may be eligible is available from the University's Office of Financial Aid. For 1998–99, most full-time students were supported through fellowships or assistantships.
Cost of Study	In 1999–2000, estimated full-time graduate tuition is $5100 per year for residents of New York and $8416 per year for out-of-state students. Tuition for part-time students is $213 per credit hour for residents and $351 per credit hour for out-of-state students. Special fees, books, and supplies cost approximately $1000 per year.
Living and Housing Costs	University graduate student apartment rents are $2096 per term. Rents for off-campus studio and single-bedroom apartments begin at $350 and shared apartments at $300 per person per month. Total annual costs, including tuition, fees, books, room, board, and incidentals, are approximately $21,000 to $27,000. International students must have at least $21,000 available to meet all expenses for each calendar year.
Student Group	Enrollment for fall 1999 is 40 doctoral students, one third of whom are full-time. Approximately half are women, and 10 percent are international students. Most are experienced information professionals holding master's degrees in information-related disciplines.
Student Outcomes	Graduates hold university faculty positions in library and information science or computer information departments or senior information management positions in government or the private sector.
Location	Albany is headquarters for the state legislature and many major state agencies. The Empire State Plaza, with its Cultural Education Center, museum, and theater, lends a cosmopolitan tone to the city. Classic and popular music concerts are held during the winter season. The Saratoga Raceway, Saratoga Performing Arts Festival, Tanglewood (summer home of the Boston Symphony Orchestra), Jacob's Pillow, and Woodstock all come alive each summer with horse racing, theater, music, arts, and crafts.
The University and The Program	Founded in 1844, the State University of New York at Albany is one of four university centers in the sixty-four-member SUNY System. Albany currently enrolls about 16,000 students, 4,500 of whom are graduate students. The interdisciplinary doctoral program in information science enrolled its initial group of students in 1990–91. Information science is one of the State University of New York at Albany's eight major themes under the state's Graduate Education and Research Initiative.
Applying	New doctoral students are admitted only for the fall semester. Applications for admission and for financial aid for domestic students must be submitted to the Graduate Admissions Office, Nelson A. Rockefeller College of Public Affairs and Policy, Draper Hall, Room 112 by April 1. International applicants are expected to hold a degree from a U.S. university and must submit applications to the Office of Graduate Admissions, AD 112, State University of New York at Albany, New York, 12222. Candidates must have a substantial background of high-quality previous academic work, preferably at the graduate level, in a discipline concerned with perception, evaluation, and manipulation of information and should possess appropriate technical and analytical skills. Admission is highly selective, based on grade point average, scores on the General Test of the Graduate Record Examinations, and academic and professional references. Prospective students are strongly encouraged to arrange a personal interview.
Correspondence and Information	Program Director Information Science Doctoral Program School of Information Science and Policy Draper Hall, Room 113 State University of New York at Albany Albany, New York 12222
	Telephone: 518-442-5115 Fax: 518-442-5367 E-mail: infosci@cnsvax.albany.edu World Wide Web: http://www.albany.edu/rcinf

State University of New York at Albany

THE FACULTY AND THEIR RESEARCH

Senior Program Faculty

David F. Andersen, Professor, Public Administration and Policy, Graduate School of Public Affairs; Ph.D., MIT, 1977. Public management, simulation and decision support systems in public policy, government information management. Coauthor *Government Information Management*, 1991.

Donald P. Ballou, Associate Professor, Management Science and Information Systems, School of Business; Ph.D., Michigan, 1969. Information systems, quantitative methods, impact of information quality on decision making, enhancing data quality.

Salvatore Belardo, Associate Professor, Management Science and Information Systems, School of Business; Ph.D., RPI, 1981. Information systems, management science, statistics, operations research, behavioral analysis.

Peter A. Bloniarz, Associate Professor, Computer Science, College of Arts and Sciences and Research Director, Center for Technology in Government; Ph.D., MIT, 1977. Software engineering, operating systems, computational complexity.

Anthony M. Cresswell, Associate Professor, Educational Administration and Policy Studies, School of Education; Ed.D., Columbia, 1970. Collective bargaining, public school finance, management and computer systems. International information systems development research projects for USAID in Haiti, Indonesia, Yemen. *Navigating the Networks*, 1995; author *Rights in Conflict: Issues in Information and Public Policy*, in preparation.

Philip B. Eppard, Associate Professor and Dean, School of Information Science and Policy; Ph.D., Brown, 1979. Management of archives and manuscripts, rare books, preservation management, electronic records management. Editor *The American Archivist*, 1996–present.

Jagdish S. Gangolly, Associate Professor, Accounting and Law, School of Business; Ph.D, Pittsburgh, 1977. Knowledge representation issues in the accounting domain. Author "Some Thoughts on the Engineering of Financial Accounting Standards," *Artificial Intelligence in Accounting and Auditing*, 1993.

Floyd M. Henderson, Professor, Geography and Planning, College of Arts and Sciences; Ph.D., Kansas, 1973. Remote sensing, digital image analysis, geographic information systems applications. Symposium paper "An Analysis of Settlement Detectability in Central Europe Using SIR-B Radar Imagery," International Society of Photogrammetry and Remote Sensing, 1990.

William K. Holstein, Distinguished Professor, Management Science and Information Systems, School of Business; Ph.D., Purdue, 1964. Management of information systems, production management. Strategic issues in information systems development and implementation, information management in international settings.

Bruce Kingma, Associate Professor, Economics, College of Arts and Sciences and School of Information Science and Policy; Ph.D., Rochester, 1989. Information management and policy, nonprofit management and fund raising. Author *Economics of Information*, 1996; *Economics of Access Versus Ownership*, 1996.

Lakshmi Mohan, Associate Professor, Management Science and Information Systems, School of Business; Ph.D., Columbia, 1960. Decision support systems, executive information systems. Coauthor "Market Decision Support Systems in Transition," *The Information Revolution in Marketing*.

Jeryl L. Mumpower, Associate Professor, Public Administration and Policy, Graduate School of Public Affairs; Ph.D., Colorado, 1976. Director, Center for Policy Research. Social and quantitative psychology. Analysis of formal structure of negotiations, scientific disagreement about policy issues.

Neil V. Murray, Professor, Computer Science, College of Arts and Sciences; Ph.D., Syracuse, 1979. Methods of logical deduction for formulas not necessarily in conjunctive or disjunctive normal form, multivalued logics. National Science Foundation research project "Implementation and Analysis of Proof Techniques Employing Negation Normal Form."

John S. Pipkin, Professor, Geography and Planning, College of Arts and Sciences; Associate Vice President; and Dean of Undergraduate Studies; Ph.D., Northwestern, 1974. Analytical and urban geography.

George P. Richardson, Professor, Public Administration and Policy, Graduate School of Public Affairs; Ph.D., MIT, 1985. Policy-oriented research and computer simulation concerning significant dynamic problems. Author *Feedback Thought in Social Science and Systems Theory*, 1991.

Roger W. Stump, Associate Professor, Geography and Planning, College of Arts and Sciences; Ph.D., Kansas, 1981. Spatial analysis, cultural geography, social applications of geographic information systems.

Affiliated Faculty

John Carlo Bertot, Associate Professor, School of Information Science and Policy; Ph.D., Syracuse, 1996. Public information policy, telecommunications policy, decision support systems, information technology. Coauthor *World Libraries on the Information Superhighway*, in press.

Indushobha N. Chengalur-Smith, Associate Professor, Management Science and Information Systems, School of Business; Ph.D., Virginia Tech, 1989. Statistics, statistical decision systems, operations research, database management, data quality. Coauthor "Surviving Client Server: Some Management Pointers" and "The Impact of Data Quality Information on Decision Making," *IEEE Trans.*, in press.

Sharon S. Dawes, Lecturer in Information Science and Executive Director, Center for Technology in Government; Ph.D., SUNY at Albany, 1991. Information management and policy in the public sector. Coauthor *Government Information Management*, 1991.

Edward J. DeFranco, Public Service Professor of Information Science and Assistant Director for Management Information and Analysis, New York State Division of Substance Abuse Services; Ph.D., NYU, 1967. Information systems. Quality, efficiency, and effectiveness of information infrastructure in the public sector. Research studies on public data systems.

Peter J. Duchessi, Associate Professor, Management Science and Information Systems, School of Business; Ph.D., Union (New York), 1982. Management science, information systems, knowledge-based systems. Coauthor "A Research Perspective: Artificial Intelligence, Management and Organizations," *International Journal of Intelligent Systems in Accounting, Finance and Management*, 1993.

Saurav K. Dutta, Associate Professor, Accounting and Law, School of Business; Ph.D., Kansas, 1991. Accounting and information systems, auditing decision support systems, Bayesian functions. Author "A Decision Support System for Efficient Audit Planning," 1995; "A Bayesian Network Model for Business Viability," 1999.

Sue R. Faerman, Associate Professor, Public Administration and Policy, Graduate School of Public Affairs; D.P.A., SUNY at Albany. Research design, organizational behavior, managerial and leadership effectiveness. Coauthor "Productivity and the Personnel Process," *Handbook of Public Productivity*, 1991; *Electronic Information Access Technologies: A Faculty Needs Assessment*, 1993.

Richard Hall, Distinguished Service Professor, Sociology, College of Arts and Sciences, and Director, Organizational Studies Ph.D. Program; Ph.D., Ohio State, 1961. Organizational theory, sociology of organizations, sociology of information. Author *Sociology of Work*, 1994; *Organizations: Structures, Processes, and Outcomes*, 1993.

Hemalata Iyer, Associate Professor, School of Information Science and Policy; Ph.D., Mysore (India), 1984. Classification theory, information organization and retrieval, natural language representation. Author *Classificatory Structures 1995*.

Paul Miesing, Associate Professor, Management, School of Business; Ph.D., Colorado, 1977. Strategic management, organizational change, technology transfer. Coauthor "Size and Scope of Strategic Planning in State Agencies," *American Review of Public Administration*, 1991; "Market Forces or Technological Rate of Change," *Technology-Based Entrepreneurship*, in press.

James Mower, Associate Professor, Geography and Planning, College of Arts and Sciences; Ph.D., SUNY at Buffalo, 1989. Automated cartography, geographic information systems. Research in applications of parallel computing to automated projection of maps.

Giri Kumar Tayi, Associate Professor, Management Science and Information Systems, School of Business; Ph.D., Carnegie Mellon, 1982. Data communications and communication networks, information economics and policy, quantitative models for policy analysis. Coauthor "Heuristics and Special Case Algorithms for Dispersion Problems" and "Determining Priorities for Data Management," both in preparation.

STATE UNIVERSITY OF NEW YORK AT BINGHAMTON

Thomas J. Watson School of Engineering and Applied Science
Department of Computer Science

Programs of Study

The computer science department in the Thomas J. Watson School of Engineering and Applied Science offers graduate degrees leading to the M.S. and Ph.D. degrees in computer science. The department offers a wide variety of courses. Recent and current faculty research areas include computer architecture, computer networks, database systems, distributed systems, information retrieval, fault testing and diagnosis, operating systems, parallel processing, real-time systems, software specification and verification, and VLSI systems.

Doctoral students are required to have a minimum of 24 credit hours in residence. Students have to pass two qualifying exams: a general comprehensive exam and a specialization exam covering the intended area of research. The general comprehensive exam covers the following five areas: algorithms, architecture, operating systems, programming languages, and one of the following: AI, compilers, database, automata theory and computability, or networks. The doctoral candidate is also required to present and defend a prospectus that describes the intended research topic. Finally, the Ph.D. dissertation has to be successfully defended.

To fulfill the requirements for the M.S. degree, students must (1) complete one course in each of the three core areas of architecture and operating systems, programming languages and compilers, and theoretical computer science; (2) complete seven additional courses and pass a comprehensive examination, complete five additional courses and write and defend a thesis, or complete six additional courses and develop and present a project; and (3) maintain a B average in all course work.

Research Facilities

In addition to the facilities available through the University Computing Center and the Watson School of Engineering, the department operates several research laboratories. These laboratories are equipped with workstations, high-end PCs, multimedia equipment, and multiprocessor systems. The department is also currently developing an experimental high-bandwidth testbed for addressing research issues in a network of tightly coupled workstations.

The department has research collaborations with local industry such as IBM, Loral, Lockheed Martin, Hughes Training, and Universal. The department also has strong ties with Watson School research centers such as the Center for Computing Technologies and the Integrated Electronics Engineering Center.

Financial Aid

Many students hold fellowships, traineeships, or assistantships (graduate, research, or teaching). Most awards include a full waiver of tuition and medical benefits. Other sources of financial aid include the New York State Tuition Assistance Program, the Federal Stafford Student Loan Program, the graduate and professional school College Work-Study Program, and campus jobs. The department awards approximately twenty teaching assistantships and a number of research assistantships each year. Requests for assistantships can be indicated on the Graduate Admissions application form; no separate form is required.

Cost of Study

For full-time matriculated graduate students, tuition in 1998–99 cost $2550 per semester for state residents and $4208 per semester for nonresidents.

Living and Housing Costs

A recently completed apartment complex, the Graduate Community, has 3- and 4-person apartments, with living room, dining area, kitchen, and bath. Based on a 1998–99 academic-year lease, the semester rate for a single bedroom was $2050, and for a double bedroom, $1775 per person and $3085 per couple. The cost of meal plans per semester is as follows: basic, $787; standard, $1022; and ultra, $1097. Assistance in locating off-campus housing is provided by the listing services of Off-Campus College.

Student Group

Of the 12,259 students enrolled at Binghamton University, 2,700 are graduate students. In the Watson School, there are 876 undergraduates and 368 graduate students. Many obtain jobs in local high-technology enterprises during their enrollment at the School and after graduation.

Location

The University's 606-acre campus is in a suburban setting just west of Binghamton. More than 300,000 people live within commuting distance of the campus. Cultural offerings in the community include the museum and programs of the Roberson Center for the Arts and Sciences as well as performances by the Binghamton Symphony, Tri-Cities Opera, Civic Theater, and other groups. The University's Art Gallery has a permanent collection representing all periods and also displays works from special loan exhibitions. The annual concert series of the Anderson Center brings a wide variety of performing artists to campus. The Department of Theater stages more than twenty-five productions each year.

The University and The School

The State University of New York at Binghamton is one of four university centers in the State University of New York System. The faculty numbers about 700. Graduate programs were initiated in 1961 with the establishment of Master of Arts programs in English and mathematics.

The Watson School was created in 1983 by combining the established graduate programs in computer science and systems science from the School of Advanced Technology with new programs in electrical, industrial, and mechanical engineering.

Applying

Applicants should have a baccalaureate degree in computer science or a closely related field and must submit official transcripts, GRE scores, two letters of recommendation, and a statement of personal academic goals. For international students, TOEFL results and a statement of financial means are also required. Requests for application forms can be sent via e-mail to the Office of Graduate Admissions at gradad@binghamton.edu. To ensure consideration for assistantships, admission credentials should be received by February 15. Online application is available at http://www.gradschool.binghamton.edu.

Correspondence and Information

Director of Graduate Studies
Department of Computer Science
Thomas J. Watson School of Engineering and Applied Science
State University of New York at Binghamton
P.O. Box 6000
Binghamton, New York 13902-6000
World Wide Web: http://www.cs.binghamton.edu

State University of New York at Binghamton

THE FACULTY AND THEIR RESEARCH

Nael B. Abu-Ghazaleh, Assistant Professor; Ph.D., Cincinnati, 1997. Parallel and distributed processing, computer architecture, parallel discrete event simulation.

Sudhir Aggarwal, Professor and Department Chair; Ph.D., Michigan, 1975. Computer networks, distributed systems, protocols, information retrieval from the Web, simulation, networks, real-time systems.

Michal Cutler, Associate Professor; Ph.D., Weizmann (Israel), 1979. Information retrieval, expert systems, software reliability.

Richard Eckert, Associate Professor; Ph.D., Kansas, 1971. Computer graphics, human-computer interaction, computer architecture, microprocessor-based systems, computer science education.

Dennis Foreman, Lecturer; M.S., SUNY at Binghamton, 1974. Design and development of operating systems and computers.

Kanad Ghose, Associate Professor; Ph.D., Iowa State, 1988. Parallel processing, computer architecture, VLSI design, high-performance networking.

Margaret E. Iwobi, Lecturer; M.S., SUNY at Binghamton, 1975. Software engineering principles, programming languages.

Walker Land, Lecturer; M.S., George Washington, 1964. Neural networks, evolutionary computing, fuzzy sets and approximate reasoning, object-oriented design, system design and applications.

Leslie Lander, Associate Professor; Ph.D., Liverpool, 1973. Formal aspects of software engineering, programming languages and paradigms.

Michael J. Lewis, Assistant Professor; Ph.D., Virginia, 1999. Distributed computing, metasystems, parallel computing, object-orientation, component-based software development, operating systems.

Patrick H. Madden, Assistant Professor; Ph.D., UCLA, 1998. VLSI computer-aided design, computational geometry, optimization for NP-hard problems.

Weiyi Meng, Associate Professor; Ph.D., Illinois at Chicago, 1992. Internet-based information retrieval, heterogeneous database systems, database query processing and optimization.

Walter G. Piotrowski, Associate Professor; Ph.D., SUNY at Binghamton, 1990. Operating systems, networks, distance learning.

Stephen Y. H. Su, Professor; Ph.D., Wisconsin–Madison, 1967. Fault-tolerant computing, design automation, computer architecture, digital logic and system design.

William L. Ziegler, Associate Professor; M.S., Syracuse, 1982. Programming languages and paradigms, logic design and digital systems, university-industry collaboration.

STATE UNIVERSITY OF NEW YORK AT BUFFALO

Department of Computer Science and Engineering

Programs of Study	The Department of Computer Science and Engineering at Buffalo offers programs of study leading to the degrees of Doctor of Philosophy and Master of Science. Courses cover a wide range of interests, with particular research emphasis on the areas of artificial intelligence, expert systems, parallel systems, pattern recognition, parallel computations, natural-language processing, computer vision, analysis of algorithms, theory of computation, VLSI algorithms, performance evaluation, and numerical analysis.
	Doctoral students are required to qualify in each of three general areas and submit a dissertation that describes original independent research. The Ph.D. is awarded in recognition of high achievement in research, and the program is intended for persons interested in research careers.
	For the M.S. degree, students must complete at least 30 credits of course work. These 30 credits must include either a thesis (usually 6 credits) or a project (usually 3 credits), or the student must achieve a Ph.D. pass in the qualifying process.
Research Facilities	The department's research facilities include more than fifty Sun Workstations as well as X-terminals and SPARCserver systems, including an Ultra60/2360, an Ultra2/1300, an Ultra1/140, two SS20/712s, and an SS10/514, totaling 150 gigabytes of disk space. A Sun Ultra Enterprise 4000, with eight CPUs and 1 gigabyte of memory, handles computer-intensive processing. Power Macintoshes, Windows NT machines, and laser printers are readily available. The department also has access to computing facilities provided by the University and Science and Engineering Node Services (SENS), including a Sun time-sharing system and workstations. Also associated with the department is the Center for Computational Research, a supercomputing center, with its SGI Origin 2000 and IBM SP2. The department's systems are connected to the University backbone by a 100-Mbps Ethernet port directly on an FDDI router. The department's internal networking is a mix of 100BaseT, switched 100BaseT, and 10BaseT Ethernet. Through NYSERNet, the University has an FDDI link to a SprintNet T3 connection to the Internet. By summer 1999, the department will be connected to Internet2. Dial-up access, including PPP, is available.
Financial Aid	A variety of assistantships and fellowships, available for qualified graduate students, carry support levels of up to $13,970 plus tuition remission for the nine-month 1999–2000 academic year. Teaching assistants are assigned part-time duties in the instructional program or in support of departmental laboratories. Research assistants are supported part-time on faculty-supervised research projects.
	Opportunities exist for summer support in both research and teaching.
Cost of Study	For 1999–2000, tuition for full-time study (12 or more credits) is $5952 per academic year for New York State residents and $9268 per academic year for out-of-state students. (For supported students, 9 credits is considered full-time.) Other University fees and required health insurance total approximately $500 in the first year.
Living and Housing Costs	Graduate students live off campus or in university housing. In 1999–2000, rooms in the residence halls cost $3330 for a single and $2710 for a double per academic year, plus approximately $2100 per year for board. New apartment-style on-campus housing is available.
Student Group	There are 137 full-time graduate students in the department. Most students are recent graduates, but some acquired professional experience before returning to graduate school.
Student Outcomes	Recent Ph.D. employment is evenly divided between academics and industry. Many hold tenure-track, teaching, or postdoctoral positions in American universities, and others are research scientists, software engineers, and senior programmer/analysts at companies that include Xerox, AT&T Bell Labs, Hewlett-Packard, and Disney.
Location	The University is located in western New York State just outside Buffalo and near Niagara Falls. The area's recreational activities include swimming, boating, and fishing on the Great Lakes in summer and skiing and sledding in winter. Hiking and camping are available in nearby parks. Buffalo is home to the Buffalo Philharmonic Orchestra, the Albright-Knox Art Gallery, and an active theater district. Many professional sports teams are based in Buffalo, including the Bills (football), Sabres (hockey), and Bisons (baseball). Toronto is an easy drive north. The weather, tempered by Lake Erie to the west, is cool in summer and mild in winter.
The University	With a student enrollment of 24,500, the State University of New York at Buffalo is the largest of the four University Centers of the State University of New York. It offers comprehensive study in the arts and sciences and in the schools of engineering, management, law, medicine, dentistry, and health-related professions.
Applying	Applications for the fall semester should be submitted by December 31. For further information and application forms, prospective applicants should write to the address given below.
Correspondence and Information	Director of Graduate Studies Department of Computer Science and Engineering 226 Bell Hall State University of New York at Buffalo Buffalo, New York 14260-2000 Telephone: 716-645-3180 Fax: 716-645-3464 E-mail: cse-gradinfo@cse.buffalo.edu World Wide Web: http://www.cse.buffalo.edu/

State University of New York at Buffalo

THE FACULTY AND THEIR RESEARCH

Raj Acharya, Professor; Ph.D., Minnesota. Multimedia computing, image processing/vision, medical imaging.

Carl Alphonce, Lecturer; M.S., British Columbia. Computational linguistics, parsing, natural-language syntax.

Debra Burhans, Lecturer; M.S., SUNY at Buffalo. Knowledge representation, natural-language understanding.

Jin-Yi Cai, Professor; Ph.D., Cornell. Theory of computation.

Alistair Campbell, Lecturer; M.S., SUNY at Buffalo. Artificial intelligence, knowledge representation.

Ashim Garg, Assistant Professor; Ph.D., Brown. Information visualization, graph drawing, computational geometry.

Xin He, Associate Professor; Ph.D., Ohio State. Parallel algorithms, data structures, computational complexity, combinatorics.

Bharadwaj Jayaraman, Professor; Ph.D., Utah. Programming languages.

Helene Kershner, Lecturer and Assistant Chair; M.S.E., Pennsylvania. Computer literacy, software engineering.

Nihar Mahapatra, Assistant Professor; Ph.D., Minnesota. Parallel and distributed computing, computer architecture, VLSI.

Russ Miller, Professor; Ph.D., SUNY at Binghamton. Parallel algorithms, image processing, computational geometry, computational crystallography, parallel processing education.

Chunming Qiao, Assistant Professor; Ph.D., Pittsburgh. Computer communication networks, parallel and distributed processing, optical communications.

Bina Ramamurthy, Teaching Assistant Professor; Ph.D., SUNY at Buffalo. Computer architecture, parallel processing.

William J. Rapaport, Associate Professor; Ph.D., Indiana. Artificial intelligence, computational linguistics, cognitive science, philosophical issues of computer science.

Kenneth W. Regan, Associate Professor; Ph.D., Oxford. Theoretical computer science.

Peter D. Scott, Associate Professor; Ph.D., Cornell. Controls, signals, and systems.

Alan L. Selman, Professor; Ph.D., Penn State. Complexity theory.

Stuart C. Shapiro, Professor; Ph.D., Wisconsin–Madison. Artificial intelligence, computational linguistics, cognitive science, knowledge representation, reasoning, natural-language understanding and generation.

Ramalingam Sridhar, Associate Professor; Ph.D., Washington State. Computer architecture, VLSI systems.

Rohini Srihari, Associate Professor; Ph.D., SUNY at Buffalo. Multimedia information retrieval, multimodal interfaces, computational linguistics, context-based vision.

Sargur N. Srihari, Distinguished Professor; Ph.D., Ohio State. Artificial intelligence, spatial knowledge representation and reasoning, computer vision.

Shambhu Upadhyaya, Associate Professor; Ph.D., Newcastle (Australia). Fault-tolerant computing, VLSI testing, knowledge engineering, diagnostic reasoning.

Deborah K. W. Walters, Associate Professor; Ph.D., Birmingham (England). Computational vision, cognitive science, neural networks, visual perception, parallel processing.

Aidong Zhang, Assistant Professor; Ph.D., Purdue. The distributed database system, multimedia database systems, digital libraries.

Adjunct Faculty

Laurence Boxer, Professor, Ph.D., Illinois at Urbana-Champaign. Parallel algorithms.

Herbert Hauptman, Professor, Department of Biophysical Sciences; Ph.D., Maryland. X-ray crystallography.

Surya Mantha, Assistant Professor; Ph.D., Utah.

David Udin, Professor; Ph.D., Harvard. Software engineering.

Research Faculty

Venugopal Govindaraju, Research Assistant Professor; Ph.D., SUNY at Buffalo. Pattern recognition.

Jeannette Neal, Research Associate Professor; Ph.D., SUNY at Buffalo. Natural-language understanding.

Wei Shu, Research Associate Professor; Ph.D., Illinois at Urbana-Champaign. Operating systems, compiler optimization.

Zhongfei Zhang, Research Assistant Professor; Ph.D., Massachusetts. Pattern recognition.

STEVENS INSTITUTE OF TECHNOLOGY

Department of Computer Science

Programs of Study

The Department of Computer Science offers a Master of Science and a Doctor of Philosophy in computer science.

The M.S. degree program in computer science has both a thesis and a nonthesis option. In the nonthesis option, students are required to complete 30 credits (twelve courses) of course work. In the thesis option, students are required to complete 30 credits, including 5 credits dedicated to thesis research. Students in the master's program in computer science take courses and conduct research in the following areas: artificial intelligence, computer communications, database systems, programming languages, software engineering, and theory of computing.

The Ph.D. degree in computer science requires a total of 90 credits beyond the B.S. degree. Generally, a master's degree is required for admission into the Ph.D. program. Students admitted into the Ph.D. program must complete at least 30 credits of thesis work and at least 20 credits of course work at Stevens beyond the master's degree. In addition, the candidate must pass a qualifying examination in a maximum of two attempts.

Research Facilities

There are numerous laboratories and research facilities in the Department of Computer Science. The laboratories provide the latest equipment and computational facilities and are heavily involved in both undergraduate and graduate education and research. The major facilities operated and maintained by the department include the laboratories for information networks, multimedia systems, computer interfacing, and advanced microcomputer development. In addition, the Computer Center of the Institute, a nationally recognized facility, provides high-quality, state-of-the-art computing and networking services for the school.

Financial Aid

Graduate research assistantships and teaching assistantships are available for qualified students. These assistantships pay tuition and a stipend. In 1999–2000, a stipend for a nine-month teaching assistantship pays from $11,000 to 13,000, depending on the student's qualifications.

Cost of Study

In 1999–2000, tuition is $695 per credit. Each graduate course is 2.5 credits.

Living and Housing Costs

Single students may obtain housing in the school's dormitories, several of which are reserved for graduate students. In 1998–99, rates for rooms varied from $1715 to $2950 per semester. Furnished married student apartments are available. Privately owned housing near the school is available at moderate rents.

Student Group

The Department of Computer Science has 230 students (180 men, 50 women) in its master's program and 30 students (21 men, 9 women) in its doctoral program. There are 94 international students enrolled in the programs.

Location

Overlooking the Hudson River and midtown Manhattan from Castle Point in Hoboken, New Jersey, the 55-acre campus encompasses more than thirty buildings, including classroom, departmental, administration, and research facilities and residence halls.

The Institute

Stevens Institute is a private, independent college situated on the banks of the Hudson River in Hoboken, New Jersey, 10 minutes from New York City. The Institute is at the forefront of applied engineering research and has an excellent reputation as a leader in transferring technology to industry. Established in 1870, Stevens offers baccalaureate, master's, engineer, and doctoral degrees in engineering, science, computer science, and management and a baccalaureate degree in humanities.

Applying

Applications for admission to the Graduate School may be submitted at any time. However, students should allow several months for processing the application and transcripts (transcripts are required from all previous colleges attended). Generally, admission into any computer science graduate program requires a B.S. degree in computer science or computer engineering with a minimum GPA of 3.0 on a 4.0 scale. Students with an outstanding academic record in areas such as mathematics, physics, or engineering are encouraged to apply. It is strongly recommended that applicants for the M.S. take the GRE General Test. Applicants who are from outside the U.S. or are seeking financial support must take the GRE General Test. Ph.D. applicants are required to take the GRE General Test and the computer science Subject Test. The application deadlines for international students are May 15 for fall admission and October 15 for spring admission.

Correspondence and Information

For application forms:
Dean of the Graduate School
Stevens Institute of Technology
Castle Point on the Hudson
Hoboken, New Jersey 07030

For additional information:
Dr. Stephen L. Bloom, Director
Department of Computer Science
Stevens Institute of Technology
Castle Point on the Hudson
Hoboken, New Jersey 07030

Stevens Institute of Technology

THE FACULTY AND THEIR RESEARCH

Ellie Angelopoulou, Assistant Professor. Computer vision.
Stephen L. Bloom, Professor and Director. Algebraic analyses of computational structures.
Anindya Banerjee, Assistant Professor. Programming languages and semantics.
Damianos Chatziantoniou, Assistant Professor. Databases.
Adriana Compagnoni, Assistant Professor. Programming languages.
Dominic Duggan, Associate Professor. Development of language tools.
David Klappholz, Associate Professor. Compilers and programming languages.
Patricia Morreale, Associate Professor. Multimedia.
David Naumann, Assistant Professor. Concurrent object-oriented languages and algebraic programming calculi.
A. Satyanarayana, Professor. Graph theory, algorithms and network reliability.

SUFFOLK UNIVERSITY

Department of Mathematics and Computer Science
Program in Computer Science

Programs of Study

The Department of Mathematics and Computer Science offers two programs in computer science: a Master of Science in Computer Science (M.S.C.S.) and a Certificate of Computer Science Studies. The M.S.C.S. program has a strong applied component in software engineering and databases and also provides a thorough grounding in the fundamental concepts of computer science. It is designed to both enable computer professionals to advance to a higher level of professional activity and provide a practical way for entry into this exciting field.

Depending on their background in math and computer science, candidates for the M.S.C.S. degree must complete courses totaling 30 to 54 credits, with a cumulative grade point average of at least 3.0. The courses fall into four categories: foundational mathematics, foundational computer science, required core computer science, and required applied computer science. The required core computer science courses are designed to provide the necessary depth of understanding in key areas of computer science. The required applied courses are intended to provide students with advanced knowledge, skills, and techniques in various critical areas; hence, these courses change to reflect current industrial needs and trends. For more details, students should visit the departmental Web site listed below.

Any student who enters the M.S.C.S. program without an undergraduate degree in computer science may be eligible to receive a Certificate of Computer Science Studies after completing the foundational courses mentioned above with satisfactory grades.

Research Facilities

The department has substantial computer and network facilities, which are constantly upgraded. Among the computers available to students are numerous networked SUN and Pentium workstations. Suffolk also has an IBM RS6000 R24 UNIX server, which is dedicated solely to academic computing. All of the UNIX systems are on the Internet and can be accessed anytime from either on- or off-campus.

Financial Aid

Graduate students are eligible for financial assistance in the form of partial tuition graduate fellowships, Federal Work-Study, and loans. International students are eligible for institutional fellowships. M.S.C.S. students are encouraged to explore various co-op opportunities after their first year of study. Qualified graduate students can apply for teaching fellowships, Web Resource Center Fellowships, and other work-study assignments. Applications for financial aid should be submitted by March 15 for the fall semester and November 1 for the spring semester. Applications received after these deadlines may be considered if funds are still available.

Cost of Study

Tuition for the M.S.C.S. program for the 1999–2000 year is $11,770 per year for full-time study or $1953 per course for part-time study.

Living and Housing Costs

Suffolk University is an urban university, and many students choose to live in apartments near the campus. Additional information on housing can be obtained from the Office of Enrollment Management.

Student Group

The M.S.C.S. program admitted its first students in the fall 1996 semester.

Location

The Suffolk University campus is located on the top of historic Beacon Hill, in the heart of Boston. It adjoins the state house and is easily accessible by subway, commuter rail, bus service, or car. Boston has an outstanding variety of academic, social, cultural, and athletic activities throughout the year. The Suffolk University Web page has links to other Web sites that give a more detailed picture of these activities.

The University

Founded to overcome barriers of income and discrimination, Suffolk University has a proud history of enabling its students to become honored members of the academic community, business world, and professions. Suffolk began as an evening law school in 1906. The College of Liberal Arts and Sciences (home to the Department of Mathematics and Computer Science) was established in 1934 as one of the first institutions of higher learning in New England at which a student could earn a Bachelor of Arts degree entirely through evening study. Suffolk currently has more than 2,000 full-time and 1,000 part-time undergraduates and 3,000 graduate students in the College of Liberal Arts and Sciences, the Frank Sawyer School of Management, and the Suffolk University Law School.

Applying

To apply, an applicant must hold a bachelor's degree from an accredited college or university and submit an application for admission, a current resume, a statement of professional goals, two letters of reference, and official copies of transcripts of all prior academic work. International students must also submit TOEFL scores and a statement of financial resources. Optimally, a candidate admitted to this program should have a strong undergraduate background in computer science and mathematics. Without such preparation, applicants can be considered for admission provided they have the necessary background to take the foundational courses mentioned above. The prerequisite for these foundational courses is a good working knowledge of calculus (at the level of Suffolk's MATH 161-162) and a rigorous course in C programming (equivalent to Suffolk's CMPSC 131). Applications are accepted on a rolling basis for the fall and spring semesters.

Correspondence and Information

For information about the program:
Graduate Program Director
Department of Mathematics and Computer Science
Suffolk University
41 Temple Street
Boston, Massachusetts 02114
Telephone: 617-305-1917
Fax: 617-573-8591
E-mail: gradinfo@clas.suffolk.edu
World Wide Web: http://www.cs.suffolk.edu

Applications and credentials must be sent to:
Office of Graduate Admission
Suffolk University
8 Ashburton Place
Boston, Massachusetts 02108
Telephone: 617-573-8302
Fax: 617-523-0116
E-mail: grad.admission@admin.suffolk.edu

Suffolk University

THE FACULTY

Donald L. Cohn, Professor; Ph.D., Harvard.
Paul N. Ezust, Professor and Chair; Ph.D., Tufts.
Eric R. Myrvaagnes, Professor; Ph.D., Tufts.
Fei Shi, Assistant Professor; Ph.D., ETH Zurich.
Dan C. Stefanescu, Professor and Program Director; Ph.D., Harvard.

TEXAS A&M UNIVERSITY

Department of Computer Science

Programs of Study

The department offers graduate studies in computer science and computer engineering leading to the Master of Science, Master of Computer Science, Master of Engineering, and Doctor of Philosophy degrees. Computer engineering programs are offered jointly with the Department of Electrical Engineering. The fields of computer science and computer engineering are rapidly changing and expanding. This has generated a need for highly trained individuals. Graduate areas of specialization in computer science include architecture, artificial intelligence and cognitive modeling, computational mathematics, computer systems and networks, computer vision, data structures, distributed systems, fault-tolerant computing, graphics, hypertext/hypermedia, neural networks, real-time systems, robotics, simulation, software engineering, software systems, theoretical computer science, and VLSI design automation and simulation. The diverse research interests of the faculty provide opportunities for work in many areas, with curricula suited to individual interests.

Research Facilities

The Department of Computer Science has an installed base of more than $5 million in computing equipment. Research laboratories are equipped with state-of-the-art Sun, Silicon Graphics, Hewlett-Packard (including an HP 16-processor V class), and IBM workstations. For general faculty and student use, the department has an instructional network of more than 400 workstations (UNIX, Pentium, and Macintosh) coupled to several Sun servers. Specialized teaching labs exist for microcomputer architecture, real-time and robotics, and senior design. Workstations, laser printers, and plotters are distributed throughout the department in fifteen power-conditioned laboratories.

The department used a small-scale Infrastructure Grant from the National Science Foundation to provide a distributed systems lab, upgrades to the parallel machines, department-wide FDDI and ATM networks, and an SGI Onyx.

Campus facilities include a CRAY Jedi and an SGI Power Challenge in the Supercomputing Center and several large mainframes with a wide assortment of supported software.

Financial Aid

Graduate assistantships, research assistantships, and fellowships are available. In 1998–99, stipends were approximately $13,000 for the nine-month academic year, depending upon the student's academic attainment and experience. Assistantships require 20 hours per week and permit the holder to carry a full academic program of graduate work. Students with financial support qualify for resident tuition fees.

Cost of Study

For tuition and fees, refer to http://vpfn.tamu.edu/sfs/ or call the Student Financial Aid office at 409-845-3236.

Living and Housing Costs

There is a waiting list for University-owned apartments; the current waiting time is from 1½ to 2 years. Town houses, patio homes, and apartments are available off campus. For information on University-owned apartments, students should contact: Student Apartments Office, Texas A&M University, College Station, Texas 77843. For off-campus housing information, students should contact: Off-Campus Center, Department of Student Affairs, Texas A&M University, College Station, Texas 77843-1257.

Student Group

There are more than 200 graduate students in the computer science department. The department has student chapters of the ACM, IEEE Computer Society, Upsilon Pi Epsilon, and Graduate Student Association.

Student Outcomes

Graduates of the programs in the Department of Computer Science can network with a large number of graduates across the country. University graduates are found in every area, with a significant activity in the computing profession. The segments that are most common include telecommunications, software, systems, electronics, and applications to a wide array of industries.

Location

Texas A&M University is located in the Bryan/College Station area (population approximately 124,000), 90 miles northwest of Houston and 170 miles south of Dallas. A growing industrial base, excellent housing, strong public school systems, and many recreational and entertainment activities characterize the area. The park systems include numerous parks, swimming pools, golf courses, and tennis courts. The area also has two lakes available for recreational activities. The newest attraction is the George Bush Library and Museum.

The University

Texas A&M University, a land-grant university, was established in 1876 as Texas's first public institution of higher education. The spacious campus is within easy driving distance of the four largest cities in Texas. Texas A&M University is noted for its accomplishments in the areas of teaching, public service, and research. It is one of the few universities with space-grant, land-grant, and sea-grant titles. The high quality of Texas A&M's programs is based on the talented men and women who constitute the more than 2,400 members of the faculty. In 1998, the enrollment at Texas A&M University was 41,461. The Department of Computer Science is located in the College of Engineering, which enrolls more than 1,700 graduate students.

Applying

Inquiries regarding admission to the Graduate College should be addressed to the Office of Admissions and Records.

Admission to Texas A&M University and any of its sponsored programs is open to qualified individuals regardless of race, color, religion, sex, age, national origin, or educationally unrelated handicaps.

Correspondence and Information

Graduate Advisor
Department of Computer Science
Texas A&M University
College Station, Texas 77843-3112

Telephone: 409-845-8981
E-mail: csdept@cs.tamu.edu
World Wide Web: http://www.cs.tamu.edu/gradinfo/

Texas A&M University

THE FACULTY AND THEIR RESEARCH

Wei Zhao, Professor and Department Head; Ph.D., Massachusetts. Real-time computing, distributed operating systems, computer networks.

S. Bart Childs, Professor and Associate Department Head for Industrial and Sister School Relations; Ph.D., Oklahoma State. Computational engineering and science, literate programming, programming environments, documentation.

Donald K. Friesen, Professor and Associate Department Head; Ph.D., Illinois at Urbana-Champaign. Algorithm analysis, parallel algorithms, artificial intelligence, neural networks.

Professors

Laxmi N. Bhuyan, Ph.D., Wayne State. Computer architecture, parallel and distributed computing, performance evaluation, fault-tolerant computing.

John Leggett, Ph.D., Texas A&M. Hypertext/hypermedia systems, hyperbase systems, computer-supported collaborative systems, computer-human interaction, programming/operating systems, distributed and object-oriented computation.

William M. Lively, Ph.D., SMU. Software engineering, AI/KB software engineering, computer-human interaction.

Bruce H. McCormick, Ph.D., Harvard. Scientific visualization and modeling, computer vision, neural networks, brain mapping.

Paul Nelson, Ph.D., New Mexico. Mathematical software, numerical analysis, parallel numerical analysis.

Udo W. Pooch, E-Systems Professor of Computer Science; Ph.D., Notre Dame. Operating systems, system architecture, computer networking, fault-tolerant systems, real-time computing.

Dhiraj K. Pradhan, Endowed Chair in Computer Science; Ph.D., Iowa. Fault-tolerant computing, VLSI testing, computer architecture.

Sallie Sheppard, Professor Emeritus; Ph.D., Pittsburgh. Simulation, software engineering, high-level languages.

Dick B. Simmons, Ph.D., Pennsylvania. Software engineering, expert systems, software models and metrics, software and hardware automation, productivity improvement techniques, computer architecture, computer networks.

Richard A. Volz, Ph.D., Northwestern. Real-time embedded computing, robotics and manufacturing, distributed program languages, task planning for robots.

Glen N. Williams, Ph.D., Texas A&M. Computer graphics, scientific and engineering applications, computational mathematics.

John Yen, Ph.D., Berkeley. Fuzzy logic, intelligent control, real-time AI, artificial intelligence.

Associate Professors

Jianer Chen, Ph.D., NYU. Algorithms, complexity theory, combinatorics, parallel processing.

Daniel Colunga, Ph.D., Texas at Austin. Software engineering, database systems design, management information systems.

Richard Furuta, Ph.D., Washington (Seattle). Hypertext, hypermedia, modeling structured documents, virtual reality, software engineering, computer-human interaction.

Jyh-Charn (Steve) Liu, Ph.D., Michigan. Parallel processing systems, fault-tolerant computing, real-time distributed computing systems, congestion control, Intelligent Vehicle and Highway Systems (IVHS).

Jeffrey Trinkle, Ph.D., Pennsylvania. Robotic motion planning, assembly planning, control theory, analytical mechanics.

Duncan Walker, Ph.D., Carnegie Mellon. Computer-aided design and manufacturing of integrated circuits, computer architecture.

Jennifer L. Welch, Ph.D., MIT. Theory of distributed computing, algorithm analysis, distributed systems.

Assistant Professors

Nancy Amato, Ph.D., Illinois at Urbana-Champaign. Parallel algorithms, computational geometry, robotics.

Riccardo Bettati, Ph.D., Illinois at Urbana-Champaign. Real-time communications, resource allocation in very large distributed systems, design of very large distributed systems.

Thomas Ioerger, Ph.D., Illinois at Urbana-Champaign. Artificial intelligence, machine learning.

Lawrence Rauchwerger, Ph.D., Illinois at Urbana-Champaign. Parallelizing compilers, runtime detection and exploitation of coarse-grained parallelism, architectures for parallel computers.

Frank Shipman, Ph.D., Colorado at Denver. Computer-human interaction, computer-supported cooperative work, hypermedia, intelligent user interfaces.

Nitin H. Vaidya, Ph.D., Massachusetts. Fault-tolerant computing, distributed systems.

THE UNIVERSITY OF ALABAMA

Department of Computer Science

Programs of Study

Programs of study lead to the M.S. and Ph.D. degrees in computer science. Major areas of concentration include software engineering, artificial intelligence, databases, algorithms, systems, and human-computer interaction.

M.S. students are required to complete 30 hours, including a thesis or 36 hours with a research project.

Requirements for the Ph.D. include a minimum of 72 hours of courses, seminars, and research. Ph.D. students are strongly encouraged to become actively involved in research during the first year. Students must pass a written qualifying exam upon completion of their course work.

Research Facilities

Computer science graduate students have access to a wide range of computing equipment and other resources within the department, in the College of Engineering and the University, and externally through the Internet. The department's configuration is based on networked UNIX-based graphics workstations that are available for faculty and graduate student research. There are about 75 UNIX-based workstations and approximately 180 PCs available for student use. Each UNIX-based workstation has a megapel color display, and all workstations share a common networked file system. The workstations are networked to all other computing resources on campus and are also connected to the Internet.

The campus network allows access to the University's Sun Ultra/6000 SPARCstation as well as the library online card catalog computer and the Alabama Supercomputer Center, which houses a CRAY X-MP/24 and an NCUBE 128-node machine. Dial-up access to the campus network is available.

Financial Aid

Financial aid is available in the form of teaching and research assistantships and various University fellowships up to $10,000. The vast majority of the unconditionally admitted full-time students receive some form of assistance. Current stipend rates (nine-month appointment) are normally $8500 for master's-level students and $10,000 for doctoral students. Half-time assistantships carry a full tuition waiver, and assistants are allowed to carry up to 10 hours of graduate courses. Graduate assistants on half-time appointments are expected to work 20 hours each week.

Cost of Study

Tuition and fees for Alabama residents who were full-time students were $1297 per semester, while nonresidents paid $3404 per semester for the 1998–99 academic year.

Living and Housing Costs

University housing includes dorm rooms at $940 per semester and more than 200 apartments ranging from unfurnished one-bedroom apartments at $290 to furnished three-bedroom apartments at $750 per month. The Off-Campus Housing Association assists students in selecting from a large number of off-campus apartments, many within walking distance of the campus. Meal plans are available on campus for all students. A single student sharing an apartment should budget at least $6500 for living expenses for an academic year.

Student Group

The University has a total enrollment of 17,400 students, of whom 3,400 are graduate students. The Department of Computer Science has about 225 undergraduate students, 35 master's students, and 12 doctoral students. The College of Engineering ranks in the top five nationally with respect to enrollment of African-American students in non-historically black institutions.

Student Outcomes

Ph.D. students can expect to find positions in academia and research. Most M.S. graduates find industrial jobs.

Location

The University is located in Tuscaloosa, a city of about 100,000. The city is 50 miles southwest of Birmingham by interstate highway and not far from Atlanta and New Orleans. Several large lakes, as well as the Great Smoky Mountains and the Gulf of Mexico, offer recreational opportunities. Tuscaloosa is serviced by a local airport and Interstate I-59/I-20.

The University

Chartered in 1820, the University of Alabama, a comprehensive research institution, is one of the oldest state universities in the nation. As the "capstone" of the educational system within Alabama, the University of Alabama offers graduate degree programs in many areas, ranging from the sciences and engineering to business administration, education, and the fine arts. Library holdings are computerized with access through an online catalog system to more than 300 specialized databases. The University offers a wealth of cultural events, including music, theater, dance, and an impressive art collection.

Applying

Applications for admission are accepted from students with undergraduate degrees in computer science and, subject to remedial course work, from students with degrees in other areas. Applicants should have at least a 3.0 grade point average and present a GRE score of at least 1600 for the M.S. program and 1800 for the Ph.D. program. Students graduating from an accredited undergraduate program may omit the GRE, but the department still strongly recommends taking the exam. International students must score at least 550 on the TOEFL. A $25 application fee is required.

Correspondence and Information

Additional information may be obtained by sending e-mail to cs@cs.ua.edu or by accessing the World Wide Web (http://cs.ua.edu).

Graduate Admissions
Department of Computer Science
Box 870290
The University of Alabama
Tuscaloosa, Alabama 35487

Telephone: 205-348-6363
World Wide Web: http://cs.ua.edu

For an application package:

Office of the Graduate School
102 Rose Administration Building
Box 870118
The University of Alabama
Tuscaloosa, Alabama 35487-0118

The University of Alabama

THE FACULTY AND THEIR RESEARCH

David W. Cordes, Associate Professor and Interim Department Head; Ph.D., LSU, 1988. Software engineering: object-based development and testing; component engineering.

Richard B. Borie, Associate Professor; Ph.D., Georgia Tech, 1988. Algorithm design and analysis, graph theory, parallel computation, computational complexity, data structures, combinatorial optimization, discrete mathematics, compilers.

David B. Brown, Professor; Ph.D., Texas Tech, 1969. Information mining, large database systems, software testing, traffic safety applications.

Marcus E. Brown, Associate Professor; Ph.D., Texas A&M, 1988. Human-computer interface: hypertext, virtual reality, ethics and social impact.

Hui-Chuan Chen, Professor; Ph.D., SUNY at Buffalo, 1972. Artificial intelligence, expert systems, neural networks and fuzzy set theory.

Brandon Dixon, Assistant Professor; Ph.D., Princeton, 1993. Parallel and sequential algorithms, concurrent data structures, efficient implementations of parallel algorithms.

Allen S. Parrish, Associate Professor; Ph.D., Ohio State, 1990. Software engineering, software testing and verification, software reuse, programming languages, software component engineering.

Ron Sun, Associate Professor; Ph.D., Brandeis, 1991. Artificial intelligence and cognitive science, especially reasoning, machine and human learning, and connectionist models/hybrid systems.

Susan V. Vrbsky, Associate Professor; Ph.D., Illinois, 1993. Databases; object-oriented, temporal, and real-time federated database systems; imprecise computation; database security.

RESEARCH AREAS

Software Engineering. Research includes automated testing of object-oriented software and the investigation of new paradigms for component-based software. Current research in software components includes development of a taxonomy of component architectures, as well as the investigation into new approaches to component interconnection.

Algorithms. This area concerns the design and analysis of efficient sequential and parallel algorithms for discrete optimization problems. Recent research has included solving problems on various classes of graphs such as series-parallel, tree-decomposable, chordal comparability, and perfect graphs.

Artificial Intelligence. Research includes both theoretical and practical work. The theoretical work is mainly concerned with connectionism and cognitive modeling with connectionist models. The theoretical foundation of neural networks is examined as an alternative paradigm for AI. The emphasis is on unified models of cognitive processes involving both neural and symbolic processing. Current research investigates the uses of connectionist networks for addressing hard problems in AI, such as common-sense reasoning, analogical reasoning, skill acquisition, and the issue of consciousness, which are important for both theoretical advances and practical applications. The practical work centers primarily on expert systems, including fuzzy expert systems. Expert system applications have included systems for mineral identification, oil and gas exploration, and interpretation of clastic sediments.

Database Research. This area includes theoretical and practical work. One research project focuses on an approximate query processor that produces approximate results when real-time constraints will not allow completion of a query. This query processor works with a standard relational algebra framework and is currently being extended to include temporal data. Other projects center on information mining of large databases such as traffic accident records for several Southeastern states. In view of the obvious implications of this application, work is proceeding to automate this process using object-oriented technology and an inference engine.

Human-Computer Interaction. This area concerns adapting the computer to better assist the human user. The research has several different aspects. Literate programming concentrates on programs written for programmer comprehension, often presented in a hypertext interface. Educational applications of hypermedia and virtual reality are also being explored. Another area of research in human-computer interaction has led to a user authentication technique based on typing characteristics alone, with application in computer system security.

UNIVERSITY OF ALABAMA AT BIRMINGHAM

Department of Computer and Information Sciences

Programs of Study

The department offers programs of study and research leading to the M.S. and Ph.D. degrees in computer and information sciences. Fields of specialization that are available include computer graphics, artificial intelligence, data mining and knowledge discovery, object-oriented technology, computer chess, and distributed systems. One of the features of the graduate program is the opportunity to participate in interdisciplinary research that includes researchers from the Academic Health Center, a world-class biomedical research center.

The M.S. program requires 36 semester hours of study and can be completed in six quarters by students entering with no undergraduate computer science background deficiencies. Both thesis and nonthesis options are available.

The Ph.D. program consists of three phases: preparation for the qualifying examination by taking a specified set of core courses, additional course work and development of a research proposal, and dissertation research and a final defense. There is no foreign language requirement.

Research Facilities

Graduate students have access to Sun and Silicon Graphics workstations, two 4-processor parallel machines, an SGI-based Beowulf machine, and PC systems. All systems are networked within the department and to the campus network. The department has direct access to the Alabama Supercomputer Network Authority CRAY C90 supercomputer for research and instruction. Dial-up lines to the department networks are available for off-campus access. UAB is an Internet 2 site.

The University of Alabama at Birmingham (UAB) library is an institutional member of the Association for Computing Machinery (ACM), the Institute of Electrical and Electronics Engineers (IEEE), and the Society for Industrial and Applied Mathematics (SIAM). These memberships provide complete and extensive collections of journals and other resources in the field of computer science.

Financial Aid

Financial aid is available in the form of teaching and research assistantships. Laboratory assistant positions are also available. The number of research positions varies from year to year since many are related to faculty grant and contract activities. Most students are successful in obtaining assistantships or on-campus employment during their first year of study. Assistantship stipends vary from $9500 to $15,000 (twelve-month appointments), with half tuition for M.S. students and full tuition for Ph.D. students.

Cost of Study

Tuition is $99 per semester hour for Alabama residents and $198 per semester hour for nonresidents. Since UAB operates on a quarter calendar, which compresses a normal fifteen-week semester of work into a ten-week quarter timeframe, 6 semester hours per quarter is considered full-time. Other fees may be required, depending on the student's situation and courses involved, such as a student health fee and laboratory fees.

Living and Housing Costs

University housing includes furnished efficiency apartments, which range from $401 to $488 per month. These rates include utilities, with the exception of telephone and cable. For families, unfurnished apartments that rent for $360 per month are available. Water is the only utility included in these rates. Off-campus housing information is available in the Student Housing and Residential Life Office.

Student Group

Total UAB student enrollment is approximately 16,000, including more than 3,500 graduate students. The undergraduate computer science enrollment is about 150 students. There are about 40 M.S. and 25 Ph.D. students enrolled in the graduate programs. Graduate students are actively involved in the local student chapter of the Association for Computing Machinery (ACM) and the campus Graduate Student Association.

Student Outcomes

Most recent graduates at both the M.S. and Ph.D. levels have accepted positions in the computer industry with such companies as Amdahl, American Airlines, AT&T Bell Labs, Bell Northern Research, BellSouth, Boeing, Borland, Cisco, DHL, GE Laboratories, IBM, Intel, Intergraph, Lockheed, LSI Logic, Mentor Graphics, Microsoft, NYNEX, Ricoh, Unisys, Xerox, and other national and regional corporations.

Location

Birmingham is a dynamic, progressive urban center of great natural beauty. Almost a million people live in the metropolitan area, ranking it in the top fifty-eight nationwide. Birmingham is the cultural and entertainment center of the state and offers beautiful residential neighborhoods and parks, a thriving business climate, and a relatively low cost of living. Birmingham's high quality of life has been recognized nationally for many years, most recently by the U.S. Conference of Mayors, which awarded Birmingham its "Most Livable City" designation.

The University

UAB is a comprehensive urban institution in Alabama's major city and is a nationally and internationally respected center for educational, research, and service programs. The University is composed of twelve schools, as well as hospitals and clinics that house internationally renowned patient-care programs. The Department of Computer and Information Sciences is located in the School of Natural Sciences and Mathematics. UAB also includes the Schools of Arts and Humanities, Business, Dentistry, Education, Engineering, Health-Related Professions, Medicine, Natural Sciences and Mathematics, Nursing, Optometry, Public Health, and Social and Behavioral Sciences. The campus encompasses a seventy-block area on Birmingham's Southside, offering all of the advantages of a university within a city.

Applying

Application forms for admission can be obtained from the Graduate School. Scores on the Graduate Record Examinations General Test are required. Admission is competitive. 1100 and 1350 are typical GRE scores (verbal and quantitative) for admission to the M.S. and Ph.D. programs, respectively. Ph.D. applicants should include a personal statement of specific graduate study and research interests and objectives. International applicants are required to submit TOEFL scores of 600 or higher to be considered for admission.

Applicants who are interested in being considered for financial aid for fall 2000 should apply by March 1, 2000.

Correspondence and Information

Graduate Program Director
Department of Computer and Information Sciences
Room 114 Campbell Hall
University of Alabama at Birmingham
Birmingham, Alabama 35294-1170

Telephone: 205-934-2213
Fax: 205-934-5473
E-mail: gradinfo@cis.uab.edu
World Wide Web: http://www.cis.uab.edu

University of Alabama at Birmingham

THE FACULTY AND THEIR RESEARCH

Anthony C. L. Barnard, Professor; Ph.D., Birmingham (England), 1957. Software and hardware architecture, microprocessor systems.

Barrett R. Bryant, Associate Professor; Ph.D., Northwestern, 1983. Programming languages, compiler design, object-oriented technology.

*Gary Grimes, Professor and Bunn Chair of Telecommunications; Ph.D., Colorado, 1973. Virtual reality, telecommunications systems.

Robert M. Hyatt, Associate Professor; Ph.D., Alabama at Birmingham, 1988. Parallel processing, parallel search, distributed processing using "Tuple Space."

John K. Johnstone, Associate Professor; Ph.D., Cornell, 1987. Computer graphics, geometric modeling, biomedical visualization medical informatics.

Warren T. Jones, Professor and Chair; Ph.D., Georgia Tech, 1973. Machine learning, data mining and knowledge discovery, information filtering, medical informatics.

*Charles R. Katholi, Associate Professor of Biostatistics and Biomathematics; Ph.D., Adelphi, 1970. Numerical analysis, parallel processing.

Kevin D. Reilly, Professor; Ph.D., Chicago, 1966. Simulation, artificial intelligence, software engineering.

Kenneth R. Sloan, Associate Professor; Ph.D., Pennsylvania, 1977. Computer graphics, vision, image processing, parallel search, medical informatics.

Alan P. Sprague, Associate Professor; Ph.D., Ohio State, 1988. Parallel algorithms, data mining and knowledge discovery, medical informatics.

*Ernest M. Stokely, Professor of Biomedical Engineering and Associate Dean of Engineering; Ph.D., Southern Methodist, 1972. Medical imaging.

*Amy E. Zwarico, Adjunct Associate Professor and Manager with BellSouth Telecommunications; Ph.D., Pennsylvania, 1988. Programming languages, concurrency, object-oriented technologies.

*Faculty members whose primary appointment or position is outside the department.

RESEARCH AREAS

Computer Graphics and Geometric Modeling. Research directions include (1) reconstruction of three-dimensional surfaces, (2) geometric modeling of solids using rational Bezier representations of swept surfaces, and (3) visualization and modeling. This research involves collaboration with such biomedical research groups as Ophthalmology, Cardiac Rhythm Measurement Laboratory, Orthopaedics, and Biomedical Engineering.

Data Mining and Knowledge Discovery. This research involves data-driven extraction of information from large repositories of data. It is the process of automated presentation of patterns, rules, or functions to a knowledgeable user for review and examination. Various machine learning models are used. Research collaborators include the Department of Pathology and the Medical Informatics Section of the Department of Medicine.

Computer Chess. Research involves the development of a chess-playing program named Crafty, which is the successor of Cray Blitz, the program that was world computer champion from 1983 to 1989. This is a joint project with Lawrence Livermore National Laboratory and Cray Research. Strategies include (1) using parallel machines and search depth, (2) improving chess knowledge, and (3) selective search depth as a function of positions.

Distributed Systems. This research is involved with the study of various parallel machine architectures and how they can best be used to improve the speed of software applications. A recent result of this research is a distributed system called "Tuple Space," which simplifies the programming effort required to distribute an application over a heterogeneous network of machines.

Object-Oriented Technology. Ongoing research is concentrated in the areas of (1) automated generation of compilers for object-oriented programming languages from denotational semantics and (2) software reuse at the specification level. One recent result of this research is an object-oriented language called SmallC++, which generates parallel code in the form of Tuple Space primitives on a network of processors.

Artificial Intelligence. Current research activities include (1) combining artificial neural network and expert systems into mixed problem–solving systems, (2) use of combined discrete and continuous simulators with artificial neural network capabilities, (3) derivation strategies for artificial neural networks from fuzzy expert systems and vice versa, and (4) distributed artificial intelligence using Tcl-Dp, PVM, and Tuple Space and Java. Research collaborators are the Civitan International Research Center, the Department of Psychology, and the Cognitive Science Research Group.

Simulation Environments. A simulation environment includes elements that help in (1) building models, (2) executing (or exercising) them, (3) analyzing and interpreting modeling results, and (4) storing the knowledge representing all phases to include previous knowledge upon which building occurs, etc. Research involves defining SEs in general purpose frameworks and investigating key problems, e.g., the role of formal logic, automated tools in specifying systems, prototyping, model abstraction, distributing models, and animation.

UNIVERSITY OF ALBERTA

Department of Computing Science

Programs of Study

The department offers graduate programs leading to the M.Sc., Master of Software Technology (M.O.S.T.), and Ph.D. degrees. Instruction and opportunities for research exist in the areas of algorithmics, artificial intelligence, computer vision, database systems, graphics, networks, parallel/distributed computing, programming languages, robotics, and software engineering.

The M.Sc. requirements include four graduate courses, two seminar sessions, a thesis, a department presentation, and an oral examination. The Ph.D. requirements include a minimum of four graduate courses beyond those for the M.Sc. degree, a thesis, a department presentation, and an oral examination. The Master of Software Technology program is a part-time degree program intended for computer professionals.

Research Facilities

The department has its own research computing facility and maintains a pool of central resources, including staff for software, hardware, and daily operations, to support a wide diversity of computing science research by about 35 faculty members and 100 graduate students. Local networks link this "backbone facility" to eight research labs containing specialized equipment. Researchers also have access to University facilities that include a farm of twenty-four IBM RS6000s, an eight-processor IBM SP-2, and a forty-processor SGI Origin 2000 computer.

In addition to general facilities that include hundreds of workstations, general and specialized print servers, scanners, and data archival/retrieval facilities, several labs have specialized equipment. The graphics laboratory includes virtual reality and animation facilities powered by a variety of SGI/DEC/HP workstations. Attached to these are a TV camera, a three-space digitizer, a Data-glove, and Eyephone stereo goggles. The lab also includes a CAVE, the only one in Canada. The parallel systems laboratory includes an eight-processor SGI Challenge and a sixty-four-processor Myrias SPS-2. The network laboratory deploys an experimental wireless network. The robotics laboratory has several robotics platforms, including a Puma 260 arm, a TRC mobile platform, a three-fingered hand, and a submarine. These are controlled by real-time processors, including a DataCube MV-20 image analysis processor. Image understanding and compression for multimedia are supported with several cameras and digitizers. The Research Institute for Multimedia Systems (RIMS) provides access to a number of multimedia input and output devices.

Financial Aid

Normal stipends for a four-month term range from approximately $5070 to $5800 and require 12 hours of service per week. Summer support may be available.

Cost of Study

The full-time instruction fee for 1999–2000 is about $2500, plus additional general fees of about $600. Student visitors (visa students) are assessed a differential fee of 100 percent of the graduate full-time instruction fee.

Living and Housing Costs

A single graduate student needs a total of approximately $14,500 per year for living expenses (including tuition) in the Edmonton, Alberta, area.

Student Group

The University of Alberta has an enrollment of about 29,800 students, of whom about 4,500 are pursuing graduate studies. Currently, the Department of Computing Science has 93 full-time graduate students in residence.

Student Outcomes

Recent graduates of the Ph.D. program have found academic positions at Canadian and international universities, along with research and development positions in the computing industry.

The graduates of the M.Sc. program either find positions in industry or continue their studies at the Ph.D. level.

Location

Edmonton, the capital of the province, is an attractive modern city of more than half a million people. The city is within a few hours' drive of Banff and Jasper national parks in the Canadian Rockies, which offer year-round recreational facilities. Edmonton not only is the home of the largest university of the province but also supports a variety of cultural activities, such as theater, opera, and symphony performances. Although predominantly English-speaking, the local community supports active French-, German-, Ukrainian-, Polish-, and Chinese-speaking populations.

The University

In 1906, the Alberta legislature authorized the establishment of the University of Alberta. It experienced its greatest growth during the years 1945–69, during which time its enrollment increased from 5,000 to 17,000 students. It has continued to grow and now consists of fifteen faculties.

Applying

Canadian applicants may submit applications until June 1. Applications from international students and from all students requesting financial support must be received by March 1. Application forms are available from the department. Students from other countries should be aware that it takes three to six months to obtain a student visa.

Correspondence and Information

Graduate Program Coordinator
Department of Computing Science
University of Alberta
Edmonton, Alberta T6G 2H1
Canada

Telephone: 780-492-4194
Fax: 780-492-1071
E-mail: gradinfo@cs.ualberta.ca
World Wide Web: http://web.cs.ualberta.ca

University of Alberta

THE FACULTY AND THEIR RESEARCH

William W. Armstrong, Professor; Ph.D., British Columbia, 1966. Knowledge discovery in data, adaptive logic networks.

Anup Basu, Professor; Ph.D., Maryland, 1990. Computer vision, multimedia communications.

Stanley Cabay, Professor; Ph.D., Toronto, 1971. Numerical analysis, symbolic and algebraic computation.

Joseph Culberson, Associate Professor; Ph.D., Waterloo, 1986. Data structures, algorithms and genetic algorithms.

Wayne A. Davis, Professor Emeritus; Ph.D., Ottawa, 1967. Interactive graphics, computer vision, spatial databases.

Renée Elio, Professor; Ph.D., Carnegie-Mellon, 1981. Computational models of human learning and reasoning, belief revision, agent communication for problem solving.

Ehab S. Elmallah, Associate Professor; Ph.D., Waterloo, 1987. Combinatorial parallel and distributed algorithms, design and analysis of interconnection networks, fault tolerant computing.

Pawel Gburzynski, Professor; Ph.D., Warsaw, 1982. Telecommunication, simulation, systems programming.

Randy G. Goebel, Professor and Chair; Ph.D., British Columbia, 1985. Artificial intelligence, knowledge representation, logic programming, learning, scheduling.

Mark Green, Professor; Ph.D., Toronto, 1984. Computer graphics, human-computer interaction, virtual reality, computer animation.

Russell Greiner, Professor; Ph.D., Stanford, 1985. Artificial intelligence, machine learning, learnability, knowledge representation, probabilistic models.

Janelle J. Harms, Associate Professor; Ph.D., Waterloo, 1992. Computer networks, performance evaluation, high-speed networks.

H. James Hoover, Associate Professor; Ph.D., Toronto, 1987. Software engineering, theory of computation, programming methodology.

Xiaobo Li, Professor; Ph.D., Michigan State, 1984. Pattern recognition, image processing, computer vision, parallel processing.

Paul Lu, Assistant Professor; Ph.D. (candidate), Toronto. Systems software, parallel and distributed computing, operating systems, system area networks.

T. Anthony Marsland, Professor; Ph.D., Washington (Seattle), 1967. Distributed systems studies, tree search algorithms.

Ursula M. Maydell, Associate Professor; M.Sc., Alberta, 1963. Analysis and design of communication networks, computer system and network performance (modeling, measurement, and evaluation), medical imaging on high-speed local area networks, high-speed circuit switched networks, integrated services, multicast access and routing.

Ioanis Nikolaidis, Assistant Professor; Ph.D., Georgia Tech, 1994. Computer networks, performance evaluation, parallel and distributed simulation.

M. Tamer Özsu, Professor; Ph.D., Ohio State, 1983. Database systems, distributed databases, object-oriented databases, multimedia information systems, electronic commerce.

Francis Jeffry Pelletier, Professor; Ph.D., UCLA, 1971. Formal semantics, knowledge representation, cognitive science, computational linguistics, automated theorem proving, artificial intelligence, philosophy of language, logic. (Joint appointment with Philosophy)

Piotr Rudnicki, Associate Professor; Ph.D., Warsaw, 1979. Program correctness and verification, proof checkers, logic, programming languages.

Jonathan Schaeffer, Professor; Ph.D., Waterloo, 1986. Distributed systems, parallel processing, search algorithms, heuristics, computer games.

Keith W. Smillie, Professor Emeritus; Ph.D., Toronto, 1952. Programming languages, history of computing.

Paul G. Sorenson, Professor; Ph.D., Toronto, 1974. Software engineering, software quality.

Lorna K. Stewart, Associate Professor; Ph.D., Toronto, 1985. Graph theory, design and analysis of algorithms.

Eleni Stroulia, Assistant Professor; Ph.D., Georgia Tech, 1994. Model-based reasoning and learning, knowledge-based systems, design, software architecture, legacy-systems reengineering.

Duane Szafron, Professor; Ph.D., Waterloo, 1978. Object-oriented computing (programming languages, design and object bases), distributed computing, multimedia.

John Tartar, Professor Emeritus; Ph.D., Arizona State, 1967. Computer architectures, multiprocessor systems, graphics, local area networks.

Peter van Beek, Professor; Ph.D., Waterloo, 1990. Artificial intelligence, temporal reasoning, constraint programming.

Benjamin Watson, Assistant Professor, Ph.D., Georgia Tech, 1997. Computer graphics, virtual reality, 3-D and novel human-computer interfaces.

Jia-Huai You, Professor; Ph.D., Utah, 1985. Logic programming, artificial intelligence.

Li Yan Yuan, Professor; Ph.D., Case Western Reserve, 1986. Database management systems, logic programming and deductive databases, artificial intelligence.

Hong Zhang, Associate Professor; Ph.D., Purdue, 1986. Dextrous robot manipulation, tactile sensing, telerobotics and collective robotics.

UNIVERSITY OF BRIDGEPORT

School of Engineering and Design
Program in Computer Science

Program of Study

The Computer Science and Engineering Department offers the M.S. in computer science, with an emphasis on professional applications. The core curriculum includes advanced C programming, advanced algorithms and data structures, operating systems, database design, and theory of computation. Among the electives offered are artificial intelligence, computer networking, robotics, parallel processing, computer music, computer animation, multimedia computing, computer vision, C++, and Java programming. Courses are available at both the main Bridgeport campus and the University's Stamford Center. The program may be pursued on a full-time, part-time, or weekend basis. Co-op and internship opportunities are frequently available, and graduates of the program are held in high demand by prospective employers.

The program in computer science requires the completion of 33 credit hours of study. At least one of the following is also required: a comprehensive examination, a thesis based on independent research, or the completion of an appropriate special project. A weekend option is also available and can be completed in fewer than eighteen months through the University of Bridgeport–Stamford (UB-Stamford) campus.

In addition, the department also offers the M.S. in the allied field of computer engineering. The program in computer engineering differs in its greater emphasis on computer hardware. Courses not found in the normal computer science curriculum include computer architecture, digital signal processing, digital system processing, VLSI design, logic synthesis with VHDL, and image processing.

Research Facilities

The computing facilities at the University of Bridgeport are among the best available. SPARC, digital design, digital signal processing, mixed-signal, image sequence, and microprocessor laboratories are among those available to students in the programs. Hardware platforms include Apple Power PCs, SPARC workstations from Sun Microsystems, and Pentium PCs. Software tools include those from Oracle, Macromedia, Microsoft, Altera, Xilinx, and Exemplar Logic. The University's Wahlstrom Library contains approximately 275,000 bound volumes, including bound journals and indexes, and more than 1 million microforms and subscribes to more than 1,700 periodicals and other serials. Online databases available from off campus via the library's Internet Web site include more than sixty databases in OCLC's FirstSearch that cover all subject areas, EBSCOhost's Academic Search FullTEXT Elite, UMI's ProQuest Direct ABI Global, and MANTIS. Online databases available throughout campus from the library's Web site include LEXIS-NEXIS Academic Universe, Moody's Company Data Direct, Britannica Online, and STAT-USA. CD-ROM databases in the library include ERIC, MEDLINE, ALT HealthWatch, Allied & Alternative Medicine, and Index to Chiropractic Literature. All students have access to e-mail, Netscape, and word processing. Residence halls are wired for individual computer hookups. An extension library is maintained at the UB-Stamford campus, with more than 1,000 volumes, more than forty-five periodicals, and extensive electronic access.

Financial Aid

Financial aid is available in the form of the Federal Work-Study Program, Federal Perkins Loans, Federal Stafford Student Loans, graduate assistantships, and internships. The University also hires graduate students as residence hall directors and assistant hall directors. Additional information can be obtained from the Financial Aid Office at 203-576-4568. The University also has a long-standing partnership with local corporations, which provide employees with excellent educational opportunities that lead to degrees and career advancement.

Cost of Study

In 1999–2000, tuition is $380 per credit hour in the School of Engineering and Design for students taking up to 12 credit hours per semester.

Living and Housing Costs

Graduate students may reside either in the University's on-campus residence halls or in private, off-campus apartments or rooms. The cost of off-campus living varies widely. Additional information related to on-campus residence may be obtained from the Office of Residential Life at 203-576-4395.

Student Group

As of the fall 1998 semester, there were 220 students enrolled in the computer science program among approximately 1,400 graduate students enrolled at the University. Of the total University graduate population, approximately 51 percent are women, 33 percent are international, and 15 percent are members of minority groups.

Location

The University of Bridgeport's 86-acre campus is situated on Long Island Sound. Both the Bridgeport and Stamford campuses are easily accessible from Westchester County, New York City, and northern New Jersey. Sixty-five percent of Connecticut's largest corporations are located in Fairfield County; these companies provide students with excellent opportunities for jobs both before and after graduation. University faculty members maintain close relationships with area corporations, school systems, and agencies.

The University

Founded in 1927, the University of Bridgeport is a private, nonsectarian, urban, comprehensive university. Professional accreditations include those from the ADA, ABA, ACBSP, NASAD, CCE, and ABET. The University's campus is composed of ninety-one buildings of diverse architectural styles. The Bernhard Arts and Humanities Center is a cultural hub, and the Wheeler Recreation Center is a complete recreation and physical fitness facility.

Applying

Students are encouraged to apply well in advance of the term they expect to enter but no later than thirty days before the beginning of the semester. Applications are accepted for fall, spring, and summer semesters. Electronic applications are welcome through the University's Web site (address below).

Correspondence and Information

For general inquiries:
Office of Admissions
University of Bridgeport
126 Park Avenue
Bridgeport, Connecticut 06601
Telephone: 203-576-4552
          800-EXCEL-UB (toll-free)
Fax: 203-576-4941
E-mail: admit@bridgeport.edu
World Wide Web: http://www.bridgeport.edu

For additional information:
Computer Science and Engineering Department
School of Engineering and Design
Charles A. Dana Hall of Science
169 University Avenue
University of Bridgeport
Bridgeport, Connecticut 06601
Telephone: 203-576-4702
Fax: 203-576-4766
E-mail: deptcse@bridgeport.edu

University of Bridgeport

THE FACULTY AND THEIR RESEARCH

Julius Dichter, Assistant Professor of Computer Science and Engineering; Ph.D., Connecticut. Neural networks, parallel processing.

Stephen Grodzinsky, Professor of Computer Science and Engineering (Chair); Ph.D., Illinois. Digital design, logic synthesis, VLSI design, microelectronics.

Gonhsin Liu, Associate Professor of Computer Science and Engineering; Ph.D., SUNY at Buffalo. Signal processing, image processing, computer vision, UNIX programming.

Douglas Lyon, Assistant Professor of Computer Science and Engineering; Ph.D., Rensselaer. Computer-generated music, diffraction rangefinding, image sequencing, signal processing.

Ausif Mahmood, Associate Professor of Computer Science and Engineering; Ph.D., Washington State. Algorithms, computer architecture, parallel VLSI simulation, compiler design, numerical methods.

Valluru Rao, Professor of Computer Science; D.Sc., Washington (St. Louis). Fuzzy logic, neural networks, programming languages.

Tarek Sobh, Associate Professor of Computer Science and Engineering; Ph.D., Pennsylvania. Control and simulation of electromechanical systems, parallel architecture, reverse engineering, robotics.

UNIVERSITY OF CALIFORNIA, IRVINE

Department of Information and Computer Science

Program of Study	Graduate students may work toward a general M.S. or Ph.D. degree or specialize in one of the following research areas: algorithms and data structures, artificial intelligence; computational biology; computer systems design; computer systems and networks; computing, organizations, policy, and society (CORPS); embedded systems; information access; and software. In addition, there are two focused M.S. degrees in embedded systems and knowledge discovery in databases.
Research Facilities	The University of California is the state's primary research institution. Scholarly research in ICS is supported through a variety of grants from public and private institutions such as the State of California, the Department of Education (DOE), DARPA, ONR, the Air Force, NSF, and several companies, including Intel and Motorola.
Financial Aid	University financial support for graduate students is available through three channels: teaching and research assistantships and fellowships. All prospective students are encouraged to apply for fellowship programs that are sponsored by federal agencies, the University of California, the California Student Aid Commission, foundations, and other private organizations. Financial assistance may be available through teaching assistantships and corporate internships for M.S. students. Research assistantships are also available in most ICS areas.
Cost of Study	Graduate fees for the academic year 1999–2000 are projected to be $5252.50 for California residents. Out-of-state and international students should expect to pay an additional $15,675.50 per year. All graduate students are assessed $380 per quarter for health insurance. All fees are subject to change.
Living and Housing Costs	Approximately 50 percent of graduate and health sciences students currently live on campus. Three different types of housing options are available: Verano Place ($245 to $640) or Palo Verde Apartments ($386 to $1115); Quenya residence hall (approximately $6400 with meals); or a recreational vehicle park (space rental is $130 per month). Off-campus housing is also available.
Student Group	Thirty M.S. and 160 Ph.D. graduate students are currently enrolled full-time. Of the current students, there are 40 women and 70 international students. Average GRE scores for fall 1999 accepted students are 582 (verbal), 768 (quantitative), and 712 (analytical). Recent graduates have pursued both academic and professional careers. Faculty members are interested in students with experience in academic or industrial research and who share academic interests with faculty research interests.
Student Outcomes	Recent Ph.D. graduates have found academic or research positions at University of Washington, University of Pennsylvania, and Cambridge University. Many entrepreneurial alumni have made the transition to the private sector by founding their own companies or working in such companies as IBM, Intel, Synopsys, Hewlett-Packard, Hughes, and Qualcom.
Location	Forty miles south of Los Angeles, 5 miles from the Pacific Ocean, and only a few hours away from deserts and mountains, University of California, Irvine (UCI) lies amid rapidly growing residential communities and a dynamic multinational business and industrial complex that affords many research and employment opportunities.
The University and The Department	Established in 1965, UCI is one of the nine campuses of the University of California. Graduate and undergraduate programs leading to the bachelor's, master's, M.D., and Ph.D. degrees are offered in a wide range of disciplines.
Applying	Completed applications should be submitted by January 15. Electronic applications are available on the World Wide Web (http://www.rgs.uci.edu/). Applicants must take the GRE General Test. If an applicant's undergraduate degree is not in computer science or a related field, it is strongly recommended to take the computer science Subject Test. International applicants must also take the TOEFL (minimum acceptable score is 550), and if they anticipate a need for financial assistance, they must take the Test of Spoken English (TSE) and receive a score of 50 or higher.
Correspondence and Information	Graduate Counselor Information and Computer Science University of California, Irvine Irvine, California 92697-3425 Telephone: 949-824-2277 or 5156 E-mail: gcounsel@ics.uci.edu World Wide Web: http://www.ics.uci.edu/~gcounsel

University of California, Irvine

THE FACULTY AND THEIR RESEARCH

Mark S. Ackerman, Associate Professor; Ph.D., MIT. Computer-supported cooperative work, human-computer interaction, information retrieval, sociology of computing.

Lubomir Bic, Professor; Ph.D., California, Irvine. Parallel processing, multiprocessor architectures, semantic and object-oriented database systems.

Rina Dechter, Professor; Ph.D., UCLA. Complexity of automated reasoning models: constraint-based reasoning, distributed connectionist models, causal models, probabilistic reasoning.

Michael Dillencourt, Associate Professor; Ph.D., Maryland. Computational geometry, graph theory, analysis of algorithms and data structures.

Nikil Dutt, Professor; Ph.D., Illinois. Design modeling, languages and synthesis, CAD tools, computer architecture.

Magda El Zarki, Professor; Ph.D., Columbia. Computer networks, wireless.

David Eppstein, Professor; Ph.D., Columbia. Analysis of algorithms, computational geometry, graph theory.

Michael Franz, Assistant Professor; Dr.Sci., Swiss Federal Institute of Technology. Programming languages and their implementation, extensible systems, software architectures, componentware and portable software that migrates across computer networks.

Daniel D. Gajski, Professor; Ph.D., Pennsylvania. Computer and information systems; software/hardware codesign, algorithms, and methodologies for embedded systems; CAD environments; science of design.

John H. Gennari, Adjunct Assistant Professor; Ph.D., California, Irvine. Biomedical informatics, knowledge-based systems, software reuse.

Richard H. Granger, Associate Professor; Ph.D., Yale. Computational and cognitive neuroscience, neural modeling, learning and memory.

Jonathan T. Grudin, Professor; Ph.D., California, San Diego. Computer-supported cooperative work, human-computer interaction, development of interactive systems.

Rajesh Gupta, Associate Professor; Ph.D., Stanford. System-level design and CAD for embedded and portable systems, VLSI design, computer systems architecture and organization.

Daniel Hirschberg, Professor; Ph.D., Princeton. Analysis of algorithms, concrete complexity, data structures, models of computation.

Sandra S. Irani, Associate Professor; Ph.D., Berkeley. Analysis of algorithms, online algorithms, graph theory and combinatorics.

Dennis F. Kibler, Professor; Ph.D., California, Irvine; Ph.D., Rochester. Machine learning, genomic analysis.

John L. King, Professor; Ph.D., California, Irvine. Economics of computing, policies for computer management and use in organizations, public policy and social aspects of computer use.

Richard H. Lathrop, Associate Professor; Ph.D., MIT. Modeling structure and function, intelligent systems and molecular biology, protein structure-function prediction, machine learning.

George S. Lueker, Professor; Ph.D., Princeton. Computational complexity, probabilistic analysis of algorithms and data structures.

Sharad Mehrotra, Assistant Professor; Ph.D., Illinois. Multimedia information systems, multidimensional databases, uncertainty processing in databases, data structures, information retrieval, distributed databases, workflow automation.

Alexandru Nicolau, Professor; Ph.D., Yale. Architecture, parallel computation, programming languages, compilers.

Michael J. Pazzani, Professor and Chair; Ph.D., UCLA. Human and machine learning, natural language understanding, cognitive science.

Wanda Pratt, Assistant Professor; Ph.D., Stanford. Knowledge-based systems, information organization, interfaces, evaluation, integration of multiple sources, knowledge-based reuse, information-gathering agents.

David F. Redmiles, Assistant Professor; Ph.D., Colorado at Boulder. Design environments, human-computer interaction, usability engineering, knowledge-based support.

Debra J. Richardson, Associate Professor; Ph.D., Massachusetts Amherst. Software engineering, program testing, lifecycle validation, software environments.

David S. Rosenblum, Associate Professor; Ph.D., Stanford. Software engineering, software testing, formal specification of software systems, software system evaluation, distributed object technology.

Isaac Scherson, Professor; Ph.D., Weizmann (Israel). Parallel computing architectures, massively parallel systems, parallel algorithms, complexity, orthogonal multiprocessing systems.

Padhraic Smyth, Associate Professor; Ph.D., Caltech. Statistical pattern recognition, automated analysis of large data sets, applications of probability and statistics to problems in artificial intelligence.

Tatsuya Suda, Professor; Ph.D., Kyoto. Computer networks, distributed systems, performance evaluation.

Richard Taylor, Professor; Ph.D., Colorado. Software engineering, user interfaces, environments, team support.

Nalini Venkatasubramanian, Assistant Professor; Ph.D., Illinois. Parallel and distributed systems, multimedia servers and applications, internetworking, high-performance architectures, resource management.

Emeritus Faculty

Alfred M. Bork, Professor; Ph.D., Brown. Computer-based learning, production systems for computer-based learning, screen design, simulation, computer graphics.

Julian Feldman, Professor; Ph.D., Carnegie Tech. Management of computing resources; problems involved in managing the computer resources of an organization, including resources allocation and financing organizations; teaching of programming; development of techniques that will facilitate the learning of programming.

Thomas A. Standish, Professor; Ph.D., Carnegie Tech. Algorithms and data structures.

UNIVERSITY OF CALIFORNIA, RIVERSIDE

College of Engineering
Department of Computer Science and Engineering
Program in Computer Science

Program of Study

The Department of Computer Science and Engineering offers programs leading to the M.S. and Ph.D. degrees.

Instruction and opportunity for directed study exist in a variety of areas, including artificial intelligence, compiler design, complexity of computation, computer architecture, database systems, design and analysis of algorithms, modeling and simulation, operating systems, software engineering, VLSI design, and multimedia technologies and programming.

The master's degree program is usually completed within two years, although it is possible for a well-prepared student to earn a master's degree in one year. A minimum of 36 quarter units of course work is required. The student must also pass qualifying examinations in theory of computation and computer systems or finish a thesis research.

Normative time to the Ph.D. is five years. After fulfilling the course requirements, a student must pass three comprehensive field examinations and the qualifying examination and produce a dissertation based on original research.

Research Facilities

The main campus library and the newly completed science library maintain extensive holdings of mathematics and computer science books and journals, including back issues. Publications not available locally may be obtained through interlibrary loan from all other campuses of the University of California.

The Department of Computer Science and Engineering provides four networked public labs, with a total of 130 machines that run a combination of Windows NT and UNIX. The department is served by three Windows NT servers, two UltraSPARCs, and ten Linux servers. A variety of operating systems and hardware is supported in six research labs.

Financial Aid

All applicants for admission are automatically considered for financial aid. Fellowships are awarded by the Graduate Division on a competitive basis, with stipends ranging from $10,000 to $14,000 for the nine-month academic year. These awards include payment of all assessed registration fees. The department offers teaching and research assistantships. A half-time appointment as a teaching assistant carries an award of $16,281 for the 1999–2000 academic year. Research assistantships carry an award of $13,871.

Cost of Study

For 1999–2000, California residents pay approximately $4861 a year in fees. Nonresidents are charged an additional tuition fee of $3441 per quarter. These amounts are subject to change.

Living and Housing Costs

Riverside offers graduate students one of the lowest costs of living of any city with a UC campus. Room and board in residence halls cost from $6309 to $6579 for the 1998–99 academic year. The University owns 268 houses that are available to married students and single students with children and that rented from approximately $430 to $470 per month. Rents for the 150 apartments and 92 suites available for single students ranged from $300 to $775 per month. Off-campus housing is available within walking distance of the campus.

Student Group

The department currently enrolls about 50 graduate students, who come from all sections of the United States and from abroad. Most are supported by teaching assistantships, research assistantships, or fellowships. The campus has about 10,000 students, of whom more than 1,300 are graduate students.

Location

The Riverside campus, consisting of more than 1,000 acres, is located 3 miles east of the center of Riverside in the shelter of the Box Springs Mountains. A community of more than 250,000 people, Riverside has excellent recreational facilities, a symphony orchestra, an opera association, a community theater, an art center, and several other colleges. Within a 60-mile radius are the mountains, the desert, the ocean, and metropolitan Los Angeles. The average year-round maximum temperature is 79 degrees. The region is semiarid, with relatively low rainfall; consequently, students can spend much of their leisure time out of doors.

The University and The Department

The Riverside campus of the University of California began as a Citrus Experimental Station in 1907. In 1954 the College of Letters and Science opened for classes, and in 1959 Riverside became a general campus. The department began the graduate program in computer science in 1982.

Applying

Students may begin graduate study in the fall, spring, or winter quarters. Application materials may be obtained from the Student Affairs Office at the address below. Applicants are required to submit directly to the department scores from the GRE General Test, a statement outlining interests and professional goals, and a list of courses in progress or planned that do not appear on official transcripts. The Graduate School Application for Admission, the application fee, official transcripts, three letters of recommendation, and a TOEFL score (if applicable) should be submitted directly to the Graduate Admissions Office or to the Department of Computer Science and Engineering.

To receive full consideration for financial support, applications, together with a $40 application fee, should be received by January 5. Later applications are considered if any support is still available.

Correspondence and Information

Graduate Student Affairs
Department of Computer Science and Engineering
Bourns Hall A242
University of California, Riverside
Riverside, California 92521-0304

Telephone: 909-787-2903
Fax: 909-787-4643
E-mail: gradadmission@cs.ucr.edu
World Wide Web: http://www.cs.ucr.edu

University of California, Riverside

THE FACULTY AND THEIR RESEARCH

Marek Chrobak, Professor; Ph.D., Warsaw (Poland). Theory of computation, algorithms, data structures, combinatorics, graph theory.

Michael Faloutsos, Assistant Professor; Ph.D., Toronto. Multicast routing, Internet protocols, QoS and ATM networks, competitive analysis, distributed algorithms.

Brett D. Fleisch, Associate Professor; Ph.D., UCLA. Operating systems, distributed systems, parallel computer systems, shared memory.

Dimitrios Gunopulos, Assistant Professor; Ph.D., Princeton. Data mining, databases, computational geometry, algorithms.

Yang-Chang Hong, Associate Professor; Ph.D., Florida. Computer architecture, database systems.

Yu-Chin Hsu, Associate Professor; Ph.D., Illinois at Urbana-Champaign. VLSI, design automation, computer architecture.

Lawrence L. Larmore, Professor Emeritus; Ph.D., California, Irvine. Algorithms, data structures.

Mart Molle, Professor; Ph.D., UCLA. Computer networking, performance evaluation, distributed algorithms.

Thomas H. Payne, Associate Professor; Ph.D., Notre Dame. Theory, computational logic, architecture.

Teodor C. Przymusinski, Professor; Ph.D., Polish Academy of Sciences. Computational logic, artificial intelligence, logic programming, deductive databases.

Scott Tilley, Assistant Professor; Ph.D., Victoria. Software engineering, program understanding, net-centric computing.

Satish Tripathi, Professor; Ph.D., Toronto. Computer networks, mobile computing, multimedia, performance modeling.

Vassilis J. Tsotras, Associate Professor; Ph.D., Columbia. Database management, access methods, temporal databases, parallel databases.

Frank Vahid, Assistant Professor; Ph.D., California, Irvine. Embedded computing systems, hardware/software codesign, synthesis, IP.

UNIVERSITY OF CALIFORNIA, SAN DIEGO

Department of Computer Science and Engineering

Programs of Study	The Department of Computer Science and Engineering offers programs leading to the M.S. and Ph.D. degrees in computer science and computer engineering.
	Graduate instruction takes place in a wide range of research areas, including algorithms and computation theory (computational complexity, algorithms, data structures, distributed algorithms, circuit complexity, operations research, probabilistic proof systems, approximation algorithms), computational geometry (mathematical foundations for the design and analysis of algorithms), cryptography and security (secure protocol design, provable security, electrical commerce), artificial intelligence (neural nets, automated reasoning, machine learning, cognitive modeling, genetic algorithms, vision, pattern recognition, expert systems, data mining, and natural language processing), computer architecture (computer architecture principles, high-performance processors, and parallel and concurrent machines), computer systems (operating distributed fault-tolerant systems, file systems, distributed high-performance communication and computing systems, highly dependable systems, real-time systems, performance evaluation, load balancing, and mass storage), communication networks (high-speed multimedia networks, ATM, wireless, Internet, protocols, performance analysis), multimedia systems (digital video and audio on-demand servers, media synchronization, and multimedia communication and collaboration), databases (relational database theory; complexity-tailored query language design; content-based retrieval; and deductive, object-oriented, temporal, multimedia, integration of heterogeneous data, semistructured data, and active databases), parallel computation (models of parallel algorithms and architectures, abstraction mechanism, compilers and programming languages, parallel programming environment and software tools, cluster and heterogeneous computing, and efficient large-scale computation), programming languages (language design and implementation, optimization for performance, and complexity and correctness of language), scientific computation (run-time support for concurrent execution, parallel algorithms, and efficient implementation techniques), software engineering (modular hierarchical system design techniques, programming tools, software maintenance, software specification and documentation, program testing and validation, functional program testing, and analysis of distributed and real-time protocols and systems), and VLSI/CAD (combinatorial and graph algorithms to solve circuit layout problems, high-level synthesis of VLSI circuits, hardware-software codesign, self-testable VLSI systems, online test, fault-resilient IC synthesis, and hardware-software issues of rapid system prototyping using field-programmable devices).
Research Facilities	The computer science and engineering department's computing facility provides research, instructional, and administrative resources. The computer hardware includes dozens of Macs, PCs, and X terminals; hundreds of workstations from a variety of vendors; and numerous fileservers as well as direct access to the vector and parallel supercomputers as well as a state-of-the-art scientific visualization laboratory at the San Diego Supercomputer Center (SDSC). There are more than a dozen separate networks in the department's computing facility with access to resources in all departments at UCSD, SDSC, Scripps Institution of Oceanography, and the UCSD Medical Center as well as to Internet and other networks.
Financial Aid	Research assistantships are projected to provide $1512 per month during the 1999–2000 academic year. Teaching assistantships are projected to pay $1510 per month for nine months. Various kinds of fellowships and scholarships for U.S. citizens are available and provide $9000–$15,000 for nine to twelve months, plus tuition and/or fees. A limited number of nonresident tuition or fee scholarships are sometimes available to Ph.D. students.
Cost of Study	Fees for the academic year 1999–2000 are projected to be $4886.50 for California residents. Out-of-state tuition is projected to be $10,324. California residents are exempted from paying tuition. United States citizens may become state residents after living in California for one year and fulfilling other requirements. Students supported by teaching or research assistantships usually also receive tuition and/or fees.
Living and Housing Costs	University Housing Services operates more than 1,300 apartments for couples, families, and single graduate students. They consist of studios and one-, two-, and three-bedroom apartments. Most of the two- and three-bedroom apartments are reserved for students with children. Studio apartments are offered only to single graduate students, and one-bedroom units are reserved for married couples without children. Rents range from $550 to $1000 per month. The rent for some of the two- and three-bedroom apartments excludes utilities. Because University apartments cost much less than comparable private housing, there are long waiting lists for married couples without children and for single graduate students. For further information, students should contact the Residential Apartments Office, UC San Diego, 9224 B Regents Road, La Jolla, California 92093 (World Wide Web: http://hdsu.ucsd.edu/hsgaffil/affhome.htm).
Student Group	The total UCSD enrollment in the fall of 1998 was 19,370; approximately 50 percent of these students were women. There were 143 graduate students in the department; all full-time Ph.D. students received financial aid.
Location	The University is in La Jolla, a suburban seaside resort north of San Diego. There are excellent beaches for swimming, sailing, surfing, and skin diving. La Jolla and San Diego offer a wide variety of cultural events.
The University	Established in 1964, UCSD is one of the nine campuses of the University of California System. Graduate and undergraduate programs leading to the bachelor's, master's, M.D., and Ph.D. degrees are offered in a wide range of disciplines. The campus structure is based on five undergraduate colleges, the Scripps Institution of Oceanography, the School of Medicine, the School of Engineering, and the Graduate School of International Relations and Pacific Studies, which accepted its first students in fall 1992.
Applying	A completed application should be submitted by January 15. Candidates must hold a bachelor's degree or its equivalent in computer science, mathematics, physics, electrical engineering, or a related area from an institution of acceptable standing. In special circumstances, alternative undergraduate preparation will be considered (e.g., a biology major may be appropriate for a student interested in the application of information and computer science to biological problems). Scores on the General Test of the Graduate Record Examinations should be sent to the department. It is recommended that applicants take a GRE Subject Test on a subject of their choice (not necessarily computer science). Students can apply for the fall quarter only. The University of California, San Diego, is an equal opportunity employer. The University encourages applications from men and women (including qualified handicapped students) of all racial, religious, and (within the limits imposed by University regulations) age groups.
Correspondence and Information	Julie Conner Assistant to the Graduate Chair Department of Computer Science and Engineering, 0123 University of California, San Diego La Jolla, California 92093-0123 E-mail: grad_apps@cs.ucsd.edu World Wide Web: http://www-cse.ucsd.edu

University of California, San Diego

THE FACULTY AND THEIR RESEARCH

Donald W. Anderson, Professor Emeritus of Computer Science; Ph.D. (mathematics), Berkeley. Computer graphics and applications of computers to education.

Scott B. Baden, Associate Professor of Computer Science; Ph.D. (computer science), Berkeley. Scientific computation, parallel processing, software design, adaptive irregular representations.

Michael J. Bailey, Associate Adjunct Professor of Computer Science and Senior Principal Scientist, San Diego Supercomputer Center; Ph.D. (engineering), Purdue. Computer graphics, scientific visualization, computational geometry, rapid prototyping.

Richard K. Belew, Associate Professor of Computer Science; Ph.D. (computer science), Michigan. Adaptive representations, including neural networks and genetic algorithms, computational biology and information retrieval (e.g., WWW) applications.

Mihir Bellare, Associate Professor of Computer Science; Ph.D. (computer science), MIT. Cryptography and security, complexity theory, probabilistic proof systems, approximation algorithms.

Francine Berman, Professor of Computer Science; Ph.D. (mathematics), Washington (Seattle). Heterogeneous computing, parallel programming environments, mapping and scheduling of distributed high-performance systems.

Kenneth L. Bowles, Professor Emeritus; Ph.D., Cornell. Computer networks, intelligent terminals, computer-based instruction.

Walter A. Burkhard, Professor of Computer Science; Ph.D. (electrical engineering and computer science), Berkeley. Distributed systems, database systems, programming languages, data structures, storage subsystems.

Samuel R. Buss, Adjunct Professor of Computer Science; Ph.D. (mathematics), Princeton. Mathematical logic, complexity theory, proof theory.

Brad Calder, Assistant Professor of Computer Science and Engineering; Ph.D. (computer science), Colorado at Boulder. Computer architecture, compiler optimizations, instruction-level parallelism.

J. Lawrence Carter, Professor of Computer Science; Ph.D. (mathematics), Berkeley. Scientific computation, performance programming, parallel computation, machine and system architecture for high performance.

Chung-Kuan Cheng, Professor of Computer Science and Engineering; Ph.D. (electrical engineering and computer science), Berkeley. Computer-aided design, VLSI layout automation, circuit partitioning, network flow optimization, physical design of multichip modules for hybrid package.

Andrew A. Chien, SAIC Chair Professor of Computer Science and Engineering; Ph.D. (computer science and electrical engineering), MIT. Networks, distributed objects, operating systems, compilers and runtimes, object-oriented languages, computer architecture.

Garrison W. Cottrell, Professor of Computer Science; Ph.D. (computer science), Rochester. Connectionist models of cognitive processes, simple biological circuits, pattern recognition, dynamical systems, computational philosophy.

Flaviu Cristian, Professor of Computer Science; Ph.D. (computer science), Grenoble (France). Programming methodology, modular hierarchical software design techniques, distributed fault-tolerant systems, operating and communication systems, distributed algorithms, real-time systems.

Charles Elkan, Associate Professor of Computer Science; Ph.D. (computer science), Cornell. Automated reasoning, machine learning, database systems, expert systems, computational biology, data mining.

Jeanne Ferrante, Professor and Chair of Computer Science; Ph.D. (mathematics), MIT. Compiling for high performance, automatic detection and exploitation of parallelism, optimizing data movement and resource usage.

Joseph A. Goguen, Professor of Computer Science; Ph.D. (mathematics), Berkeley. Software engineering, requirements, algebraic semantics of computation, user interface design, semiotics, social issues in computing.

Fan Chung Graham, Professor of Computer Science; Ph.D. (mathematics), Pennsylvania. Graph theory, algorithms, combinatorics, communications networks.

Ronald Graham, Irwin and Joan Jacobs Endowed Chair of Computer and Information Sciences; Ph.D. (mathematics), Berkeley. Algorithms, combinatorics, number theory.

William G. Griswold, Associate Professor of Computer Science; Ph.D. (computer science), Washington (Seattle). Software engineering, programming tools, software design, software maintenance, programming languages, parallel systems, and compilers.

William E. Howden, Professor of Computer Science; Ph.D. (computer science), California, Irvine. Software engineering, system design, software testing and validation, functional program testing, analysis of real-time systems.

T. C. Hu, Professor of Computer Science; Ph.D. (applied mathematics), Brown. Combinatorial algorithms, communications networks, computer-aided design, distributed computing, operations research.

Russell Impagliazzo, Associate Professor of Computer Science; Ph.D. (mathematics), Berkeley. Computational complexity, cryptography, circuit complexity, computational randomness.

Ramesh C. Jain, Professor of Computer Science and of Electrical Engineering; Ph.D. (computer and control engineering), Indian Institute of Technology. Multimedia databases, computer vision, artificial intelligence.

Sidney Karin, Professor and Director of the San Diego Supercomputer Center (SDSC) and the National Partnership for Advanced Computational Infrastructure (NPACI); Ph.D. (nuclear engineering), Michigan. High-performance computing, computational science and engineering, distributed heterogeneous computing, scientific visualization, networking and communications, operating systems, data-intensive computing, integration of high-performance computing resources.

Walter H. Ku, Adjunct Professor of Computer Science, Professor of Electrical and Computer Engineering, and Director, NSF Center for Ultra-High Speed Integrated Circuits and Systems (ICAS); Ph.D. (electrical engineering and electrophysics), Polytechnic Institute of Brooklyn. Computer-aided design, VLSI chip design, VLSI algorithms and architectures.

Keith Marzullo, Associate Professor of Computer Science; Ph.D. (electrical engineering), Stanford. Fault-tolerance and high availability, distributed computing, group-based programming, responsive systems, application management.

Eric Mjolsness, Research Scientist; Ph.D. (physics and computer science), Caltech. Mathematical methods for neural networks; computer vision, pattern recognition, parallel optimization, biological modeling.

Alex Orailoglu, Professor of Computer Science; Ph.D. (computer science), Illinois at Urbana-Champaign. Computer-aided design, synthesis of testable ICs, DSP test, high-level synthesis of fault-tolerant ASICs, microprocessor test, hardware/software codesign.

Alon Orlitsky, Professor of Computer Science and Electrical Engineering; Ph.D. (electrical engineering), Stanford. Information theory, data compression, computer learning, speech recognition.

Philip M. Papadopoulos, Research Scientist; Ph.D. (electrical engineering), California, Santa Barbara. High-performance clustering, heterogeneous distributed computing, parallel computing.

Yannis G. Papakonstantinou, Assistant Professor of Computer Science and Engineering; Ph.D. (computer science), Stanford. Databases, integration of heterogeneous sources, multimedia information systems.

Joseph Pasquale, Beyster Professor of Computer Science; Ph.D. (computer science), Berkeley. Operating systems, networks, multimedia, agent-based computing, mobile computing.

Ramamohan Paturi, Professor of Computer Science; Ph.D. (computer science), Penn State. Complexity theory, circuit complexity, neural networks, learning theory, parallel computation, optical computing.

George Polyzos, Associate Professor of Computer Science; Ph.D. (computer science), Toronto. Communication networks and protocols, wireless mobile communications and computing, multiaccess channels, multimedia distributed systems, systems performance evaluation.

P. Venkat Rangan, Professor of Computer Science; Ph.D. (computer science), Berkeley. Multimedia (digital video and audio) systems, multimedia networking, visual asset management systems, multimedia content-based retrieval, database systems.

Jeffrey B. Remmel, Adjunct Professor of Computer Science and Professor of Mathematics; Ph.D. (mathematics), Cornell. Nonmonotonic logic, logic programming, knowledge representation, program verification, hybrid control.

J. Ben Rosen, Adjunct Professor of Computer Science; Ph.D. (applied mathematics), Columbia. Large-scale numerical optimization algorithms, global optimization with application to molecular structure and docking, structure-preserving approximation algorithms.

Walter J. Savitch, Professor of Computer Science; Ph.D. (mathematics), Berkeley. Computational linguistics, formal language theory, complexity theory.

Terrence J. Sejnowski, Adjunct Professor of Computer Science; Ph.D. (physics), Princeton. Computational neuroscience, neural computation, massively parallel architectures.

Dean Tullsen, Assistant Professor of Computer Science and Engineering; Ph.D. (computer science), Washington (Seattle). Computer architecture, multithreading architectures, instruction-level parallelism, compiling for high-performance processors.

Victor Vianu, Professor of Computer Science; Ph.D. (computer science), USC. Data and knowledge base systems.

S. Gill Williamson, Professor of Computer Science; Ph.D. (mathematics), California, Santa Barbara. Algorithms, combinatorial mathematics.

Bennet S. Yee, Assistant Professor of Computer Science; Ph.D. (computer science), Carnegie Mellon. Computer security, electronic commerce, cryptography, distributed systems, operating systems.

UNIVERSITY OF CALIFORNIA, SAN FRANCISCO

Graduate Program in Medical Information Science

Programs of Study

Medical information science (MIS) encompasses data, information and knowledge acquisition, representation, modeling, integration, communication, interpretation, and transformation that ranges across the basic sciences and engineering through clinical practice and policy. The MIS program at the University of California, San Francisco (UCSF), offers graduate training that leads to both the M.S. and the Ph.D. degrees. Application for the master's degree is open to students who have or will soon have a health sciences–related professional degree, such as an M.D., D.D.S., or Pharm.D. Students accepted into the master's program may later be admitted to the Ph.D. program upon demonstration of superior performance and completion of the comprehensive examination.

The MIS program offers a core curriculum and three focus areas: bioinformatics and biocomputing, health-care knowledge management, and imaging informatics. Students generally take courses over five quarters. In the first year, students do three rotations in faculty laboratories to familiarize themselves with ongoing research and to select an adviser to guide their research. In the fifth quarter, all students take a written comprehensive exam. After successful completion of this exam, master's students work on a research project to complete their degree, while Ph.D. students work on preparing for their oral qualifying exam. Upon successful completion of the oral exam, students advance to candidacy and begin work on their dissertation. It generally takes two years to complete the master's degree, and it takes about five years to complete the Ph.D. degree. A variety of opportunities in various areas of research exist, ranging from the basic sciences to clinical decision making and support.

Research Facilities

Research facilities are decentralized and vary according to the needs of individual principal investigators and research projects. For example, the Computer Graphics Laboratory (CGL) and the Laboratory for Radiological Informatics (LRI) focus on applications involving interactive molecular graphics and medical imaging, respectively, and both maintain state-of-the-art computer facilities with high-performance servers, workstations, and network infrastructure. The UCSF campus network provides high-bandwidth, low-latency access to the Internet and to CalREN-2, a high-performance network serving the needs of higher education in California. Access is also available to remote computer facilities such as the NSF-sponsored supercomputer centers (e.g., the San Diego Supercomputer Center).

Financial Aid

Doctoral students receive funding as research assistants, trainees, or fellows at a standardized level ($19,600 in 1999–2000). Master's degree students are not funded by the program.

Cost of Study

The graduate program covers annual fees (estimated at $4650 for 1999–2000) for its students. Nonresident tuition is estimated at $10,322 per year for 1999–2000 in addition to fees. Tuition is paid for the first year only for U.S. citizens and permanent residents, after which time students are expected to have established California residency. Tuition support for noncitizens is limited.

Living and Housing Costs

The University has a limited amount of off-campus student housing, including some units for students who are married and/or have full-time legal custody of a dependent child. The Campus Housing Office offers assistance in finding housing. The cost of living is relatively high; a conservative estimate of monthly expenses (presuming shared housing) is about $1400 a month.

Student Group

The first class of Ph.D. students was admitted in 1998. As of fall 1999, 12 Ph.D. students are enrolled in the program. All Ph.D. students receive financial aid. In 1999–2000, 34 percent of the students are women and approximately 41 percent are international students.

Location

UCSF is located in a residential neighborhood, close to the center of San Francisco and to Golden Gate Park. The campus community enjoys the social and cultural advantages of a cosmopolitan, metropolitan area as well as easy access to beaches, redwood forests, and mountains. Available activities range from outdoor recreation (hiking, mountain biking, camping, surfing, and skiing) to top-caliber museums and the performing arts (theater, opera, symphony, dance).

The University

One of nine campuses of the University of California, UCSF is a world-class health and biomedical sciences campus for graduate and professional students only. It includes the Schools of Medicine, Pharmacy, Nursing, and Dentistry; graduate programs in basic and behavioral sciences; two health policy institutes; three hospitals; an ambulatory care center; and a student union. A second San Francisco campus is currently under construction. UCSF's mission is to attract and educate the nation's most promising students to future careers in the health sciences and health-care professions, with continuing emphasis on open access and diversity.

Applying

The minimum requirement for admission is a baccalaureate degree. Although there are no specific prerequisites, the core curriculum assumes that students are computer literate and knowledgeable of the basic life sciences, introductory calculus, matrix manipulation, and elementary statistics. Required materials include official transcripts from all previous colleges attended (minimum 3.0 undergraduate GPA), GRE General Test scores, letters of recommendation, a statement of purpose, the Graduate Division application form, and a $40 application fee. Applications are considered for the fall quarter only. The deadline for applications is January 15.

Correspondence and Information

Program Coordinator
Graduate Program in Medical Information Science
University of California, San Francisco
513 Parnassus Avenue
San Francisco, California 94143-0446
Telephone: 415-502-7788
Fax: 415-514-0502
E-mail: mis@cgl.ucsf.edu
World Wide Web: http://www.mis.ucsf.edu (Program)
     http://www.ucsf.edu (University)

University of California, San Francisco

THE FACULTY AND THEIR RESEARCH

The MIS program is organized into three focus areas. Participating faculty members are listed below in their primary focus areas.

Bioinformatics and Biocomputing: Biotechnology; gene mapping and genome structure; sequence analysis and database searching; macromolecular structure; protein sequence, structure, and function relationships; genome information representation; computational biology, chemistry, and medicine; database science and engineering; high-performance computing; computer graphics; biostatistics and epidemiology; modeling and simulation; molecular design and systems simulation.

Patricia C. Babbitt, Ph.D., Assistant Professor in Residence, Biopharmaceutical Sciences and Pharmaceutical Chemistry.
Sally Blower, Ph.D., Associate Professor in Residence, Center for AIDS Prevention Studies.
A. L. Burlingame, Ph.D., Professor, Chemistry and Pharmaceutical Chemistry.
Fred E. Cohen, M.D., Ph.D., Professor, Medicine and Cellular and Molecular Pharmacology.
Ken A. Dill, Ph.D., Professor, Pharmaceutical Chemistry and Biopharmaceutical Sciences.
John Featherstone, Ph.D., Professor, Restorative Dentistry and Stomatology.
Thomas E. Ferrin, Ph.D., Professor in Residence, Pharmaceutical Chemistry, and Director, Computer Graphics Lab.
Conrad Huang, Ph.D., Assistant Adjunct Professor, Pharmaceutical Chemistry.
C. Anthony Hunt, Ph.D., Associate Professor, Biopharmaceutical Sciences and Pharmaceutical Chemistry.
Teri E. Klein, Ph.D., Associate Adjunct Professor, Pharmaceutical Chemistry.
Peter A. Kollman, Ph.D., Professor, Chemistry and Pharmaceutical Chemistry.
Irwin D. Kuntz, Ph.D., Professor, Pharmaceutical Chemistry, and Director, Molecular Design Institute.
Thomas Lietman, M.D., Assistant Clinical Professor, Ophthalmology.
Kenneth Miller, Ph.D., Assistant Professor, Physiology and Otolaryngology.
Jorge Oksenberg, Ph.D., Assistant Professor in Residence, Neurology.
Wolfgang Sadee, Ph.D., Professor, Biopharmaceutical Sciences and Pharmaceutical Chemistry.
Lewis B. Sheiner, M.D., Professor, Laboratory Medicine and Pharmaceutical Chemistry.
Stephen Shiboski, Ph.D., Assistant Adjunct Professor, Epidemiology and Biostatistics.
Davide Verotta, Ph.D., Associate Professor in Residence, Biopharmaceutical Sciences, Pharmaceutical Chemistry, and Biostatistics.

Health-Care Knowledge Management: Clinical informatics; expert systems; decision support systems; human-computer interactions; outcomes research and analysis; policy and economic analysis; public health informatics; quality assessment and improvement; information and knowledge acquisition, retrieval, and representation; computer-aided education; distance learning, practice, and research; health-care records; unified language systems and thesauri.

Lisa Bero, Ph.D., Associate Professor, Clinical Pharmacy and Institute for Health Policy Studies.
Jon Bowersox, M.D., Ph.D., Associate Professor in Residence, Vascular Surgery.
Keith Campbell, M.D., Ph.D., Assistant Adjunct Professor, Biopharmaceutical Sciences.
Maurice E. Cohen, Ph.D., Adjunct Professor, Radiology, and Professor of Mathematics, California State University, Fresno.
Christopher Cullander, Ph.D., Assistant Professor in Residence, Biopharmaceutical Sciences and Pharmaceutical Chemistry.
R. Adams Dudley, M.D., Assistant Adjunct Professor, Institute for Health Policy Studies.
Laura J. Esserman, M.D., Assistant Professor in Residence, Surgery and Institute for Health Policy Studies.
Eric Goldman, M.D., Assistant Clinical Professor, Laboratory Medicine.
Charlene Harrington, Ph.D., Professor, Social and Behavioral Sciences.
Suzanne Henry, D.N.Sc., Associate Professor, Community Health Systems.
William L. Holzemer, Ph.D., Professor and Chair, Community Health Systems.
Donna L. Hudson, Ph.D., Professor, Computer Science, California State University, Fresno.
Patricia Katz, Ph.D., Assistant Adjunct Professor, Medicine and Institute for Health Policy Studies.
Laleh Khonsari, Ph.D., Assistant Adjunct Professor, Community Health Systems.
Helene Lipton, Ph.D., Professor, Clinical Pharmacy and Institute for Health Policy Studies.
Harold Luft, Ph.D., Professor in Residence, Medicine, and Director, Institute for Health Policy Studies.
Gary M. McCart, Pharm.D., Professor, Clinical Pharmacy.
Charles Mead, Ph.D., Assistant Adjunct Professor, Community Health Systems.
Robert Miller, Ph.D., Associate Professor in Residence, Institute for Health and Aging and Institute for Health Policy Studies.
Robert J. Newcomer, Ph.D., Professor, Social and Behavioral Sciences.
Thomas Newman, M.D., Professor, Epidemiology and Biostatistics, Pediatrics.
Drummond Rennie, Ph.D., Adjunct Professor, Medicine and Institute for Health Policy Studies.
Jonathan Showstack, Ph.D., Adjunct Professor, Institute for Health Policy Studies.
Ida Sim, M.D., Ph.D., Assistant Professor in Residence, Medicine.
Robert Slaughter, Ph.D., Lecturer, Physiological Nursing.
Daniel P. Stites, M.D., Professor and Chair, Laboratory Medicine.
Leslie Wilson, Ph.D., Assistant Adjunct Professor, Clinical Pharmacy.

Imaging Informatics: Medical imaging; image processing, analysis, storage, retrieval, and display; signal conditioning, compression, and enhancement; high-performance communication.

Katherine P. Andriole, Ph.D., Assistant Adjunct Professor, Radiology.
Ronald L. Arenson, M.D., Professor and Chair, Radiology.
Thomas Budinger, M.D., Ph.D., Professor in Residence, Radiology.
Gary R. Caputo, M.D., Associate Professor in Residence, Radiology.
Bruce H. Hasegawa, Ph.D., Associate Professor in Residence, Radiology.
H. K. Huang, D.Sc., Professor, Radiology, and Director, Laboratory for Radiological Informatics.
Andrew Lou, Ph.D., Assistant Professor in Residence, Radiology.
Stephen Wong, Ph.D., Assistant Adjunct Professor, Radiology.

UNIVERSITY OF CALIFORNIA, SANTA CRUZ

Graduate Program in Computer Science

Program of Study

The Department of Computer Science (CS) offers M.S. and Ph.D. degrees and conducts research in theoretical computer science, including analysis of algorithms, parallel and distributed computation, and computational learning theory and logic; AI, including machine learning, pattern recognition and retrieval, heuristic search, computational biology and chemistry, nonmonotonic reasoning and theorem proving, and natural language understanding; programming languages and environments, including compilers, object-oriented programming, and parallel and logic programming; computer graphics and image processing, including scientific visualization, physical simulation, computer modeling, image synthesis and animation, signal processing, and image storage, retrieval, and transmission; computer architecture and operation systems, including parallel computers and distributed computing systems; and computer systems design and applications, including real-time systems, embedded systems, special-purpose processors, digital networks, data compression, design of high-speed adders, high-speed packet switching, and performance prediction, evaluation, and optimization. CS works closely with the Department of Computer Engineering, whose curriculum and research encompass some of those listed above as well as communications and networks and CAD of digital systems. Faculty members carry out joint research projects, supervise students, and teach courses for both departments. Students are provided with a general education in computer science and are then provided with the opportunity to begin research projects. M.S. degrees are usually completed in two years, although it is possible to complete the program in one year. A student must also write a master's thesis. Ph.D. degrees are usually completed in six years, although it is possible to complete in less time. After fulfilling the course requirements, a student must pass an oral qualifying exam in his or her research area and defend a dissertation.

Research Facilities

The Jack Baskin School of Engineering operates several computer laboratories and facilities in support of research and graduate instruction in computer science, computer engineering, and electrical engineering. The School of Engineering supports a network of more than 250 UNIX and NT systems, network printers, X-terminals, and various graphics devices. The School's network environment includes shared and switched 10Mb/sec and 100Mb/sec Ethernet. In support of general purpose computing, the School of Engineering operates three Sun servers that provide file service to all School of Engineering systems, several Sun servers that provide for general and specific computing needs (e.g., compute server, Web server, and print server), a general access DEC (Compaq) server, a general access SGI server, an NT domain server, and several graduate student computer labs that consist of a combination of Sun, SGI, DEC, and NT workstations, X-terminals, and printers. In addition to these facilities, the School of Engineering operates the following research labs with the indicated equipment: Machine Learning and Computational Biology Labs; Scientific Visualization, Computer Graphics, and Image Processing Labs; a Computer-Aided VLSI Design Lab; a Concurrent Systems Lab; and Computer Communications and High-Speed Networking Labs. In addition to the facilities provided by the Jack Baskin School of Engineering, students have access to the campus computing facilities of the UCSC Computer Center, which provides computing to more than 2000 users campuswide via a network of distributed UNIX workstations, an IBM ES9000, and a Micom data switch with 100 dial-in modems. The distributed UNIX computing facilities consist of a laboratory of twenty Sun workstations, which are supported by several Sun SPARC Station compute servers and file servers. The UNIX facilities are based on MIT's Project Athena distributed computing environment. The Computer Center maintains the University's network connection to the global Internet computer network.

Financial Aid

A limited number of fellowships provide $9999 plus payment of all University fees except nonresident tuition to first-year students. A number of nonresident tuition waivers are awarded to students who are not residents of California. Half-time teaching assistantships provide a salary of $5233 per quarter, half-time research assistantships provide a salary of $6188 per quarter, and both allow time for taking two lecture courses per quarter.

Cost of Study

Fees for the 1999–2000 academic year are approximately $5205. Students who are not California residents must pay the additional nonresident tuition fee of $10,320 per year.

Living and Housing Costs

Housing for single students living on the University campus is $5406 for the academic year. For married students and single parents, housing costs $5080. For students living off campus, the amount is about $6000, not including board.

Student Group

Total enrollment at UC Santa Cruz is approximately 10,000 students, of whom about 960 are graduate students. The number of graduate students currently in computer science is 82.

Location

Santa Cruz is one of America's most beautiful campuses. Overlooking Monterey Bay, it occupies 2,000 acres in protected redwood forests and meadows above the city of Santa Cruz. Nearby Santa Clara Valley, one of the world's most important centers for the computer industry, is a significant resource.

The University

Founded in 1965, Santa Cruz is a small collegiate university devoted to excellence in undergraduate education and enriched by a select group of graduate programs and the presence of major research units.

Applying

Students must begin the program in the fall quarter. Completed applications must be received by February 1, 2000. Files of applicants for 2000–2001 are reviewed in February 2000. Requests for applications should be directed to the Graduate Division, 150 Social Sciences II. The application fee is $40 and subject to increase without notification. Each applicant must take the GRE General Test and the Subject Test in either engineering or computer science and have the scores sent to the UCSC Graduate Division as part of the application file. The graduate brochure is available on line.

Correspondence and Information

Graduate Representative, CS Department
University of California
Santa Cruz, California 95064
Telephone: 831-459-2576
E-mail: soegradadm@cse.ucsc.edu
World Wide Web: http://www.cse.ucsc.edu

University of California, Santa Cruz

THE FACULTY AND THEIR RESEARCH

Computer Engineering

Alexandre Brandwajn, Professor; Ph.D. (computer science), Paris, 1975. Computer architecture, performance modeling, queuing network models of computer systems and operating systems.

Pak K. Chan, Associate Professor; Ph.D. (computer science), UCLA, 1987. Placement and routing, accelerators, partitioning, system prototyping using FPGAs.

Wayne Wei-ming Dai, Associate Professor; Ph.D. (electrical engineering), Berkeley, 1988. Computer-aided design of VLSI circuits, multichip modules, graph theory, computational geometry.

Joel Ferguson, Associate Professor; Ph.D. (computer engineering), Carnegie Mellon, 1987. Fault modeling, test generation and design-for-test of digital circuits and systems, fault-tolerant computing, VLSI design, computer-aided manufacturing for VLSI.

J. J. Garcia-Luna-Aceves, Professor; Ph.D. (electrical engineering), Hawaii at Manoa, 1983. Computer networks and multimedia information systems.

Richard Hughey, Associate Professor; Ph.D. (computer science), Brown, 1991. Computer architecture, parallel processing, parallel programming languages and environments, computer applications in biology, programmable systolic arrays.

Kevin Karplus, Associate Professor; Ph.D. (computer science), Stanford, 1983. Bioinformatics (computational biology): applying information theory and stochastic modeling to biological sequence analysis, hidden Markov models, regularizers, mutual information, protein threading, multipurpose computer architecture for sequence analysis.

Harwood G. Kolsky, Adjunct Professor; Ph.D. (physics), Harvard, 1950. Scientific computing, compilers, image processing.

Glen G. Langdon Jr., Professor; Ph.D. (electrical engineering), Syracuse, 1968. Data compression, image and video coding, image segmentation, adaptive estimation of probability distributions.

Tracy Larrabee, Assistant Professor; Ph.D. (computer science), Stanford, 1990. Test pattern generation, design verification, logic synthesis.

Patrick E. Mantey, Professor and Dean, School of Engineering; Ph.D. (electrical engineering), Stanford, 1965. Image storage and retrieval; electronic libraries and multimedia; educational applications of computer technology; image and signal processing; graphics and workstation hardware; system architecture, design, and performance; simulation and modeling of complex systems; real-time data acquisition and control systems; graphics and database applications, including geographic information systems; and user-machine interaction.

Martine D. F. Schlag, Professor; Ph.D. (computer science), UCLA, 1986. VLSI design tools and algorithms; VLSI theory; formal specifications of VLSI circuits; Field-Programmable Gate Arrays.

Anujan Varma, Associate Professor; Ph.D. (computer engineering), USC, 1986. Computer networking, computer systems architecture, parallel processing.

Computer Science

Neil Balmforth, Associate Professor; Ph.D. (applied math and theoretical physics), Cambridge, 1990. Astrophysics, dynamical systems, fluid dynamics, mathematical biology, non-Newtonian fluids, plasma physics.

Scott Brandt, Assistant Professor; Ph.D. (computer science), Colorado, 1999. Operating systems, parallel and distributed systems, experimental systems, software systems, soft real-time processing, real-time and embedded systems, networking and Internet technology, asynchronous computer architectures, computer vision and robotics, image processing, virtual reality.

David Haussler, Professor; Ph.D. (computer science), Colorado, 1982. Machine learning, computational biology, neural networks, statistical decision theory, algorithms and complexity.

David Helmbold, Associate Professor; Ph.D. (computer science), Stanford, 1987. Machine learning, theoretical computer science, analysis of algorithms, intelligent systems.

David A. Huffman, Professor Emeritus; Sc.D. (electrical engineering), MIT, 1953. Information theory and coding, scene analysis, graph theory, signal design and processing, discrete systems, sequential circuits.

Phokion Kolaitis, Professor; Ph.D. (mathematics), UCLA, 1978. Logic in computer science, database theory, logic programming, complexity theory.

Robert A. Levinson, Associate Professor; Ph.D. (computer science), Texas at Austin, 1985. Artificial intelligence, machine learning, heuristic search, hierarchical reinforcement learning, associate pattern retrieval, computer chess.

Suresh K. Lodha, Associate Professor; Ph.D. (computer science), Rice, 1992. Computer graphics, scientific visualization, computer-aided geometric design, computer animation, image processing, scientific sonification.

Darrell Long, Associate Professor; Ph.D. (computer science), California, San Diego, 1988. Distributed computing systems, operating systems, performance evaluation, data management.

Tara Madhyastha, Assistant Professor; Ph.D. (computer science), Illinois, 1997. Parallel input/output, adaptive and predictive resource management policies, performance visualization and prediction, input/output characterization.

Charles E. McDowell, Professor; Ph.D. (computer science), California, San Diego, 1983. Computer architecture, parallel computing, microprogramming, compilers, operating systems.

Alex Pang, Associate Professor; Ph.D. (computer science), UCLA, 1990. Visualization (scientific, environmental, and uncertainty), computer graphics, virtual reality interfaces, collaborative software.

Ira Pohl, Professor; Ph.D. (computer science), Stanford, 1969. Artificial intelligence, programming languages, heuristic methods, educational and social issues, combinatorial algorithms.

R. Michael Tanner, Professor; Ph.D. (electrical engineering), Stanford, 1971. Information theory, error-correcting codes, complexity, VLSI systems, fault tolerance.

Allen Van Gelder, Associate Professor; Ph.D. (computer science), Stanford, 1986. Logic programming algorithms, parallel algorithms, complexity, programming languages, automated theorem proving, scientific visualization.

Hongyun Wang, Assistant Professor; Ph.D. (applied math and statistics), Berkeley, 1996. Molecular modeling, fluid mechanics, numerical analysis, instability in physical systems, molecular structure analysis and computer visualization.

Manfred K. Warmuth, Professor; Ph.D. (computer science), Colorado, 1981. Machine learning, neural networks, parallel and distributed algorithms, online learning algorithms, complexity theory.

Jane P. Wilhelms, Associate Professor; Ph.D. (computer science), Berkeley, 1985. Computer graphics, computer animation, scientific visualization, modeling articulated bodies, physical simulation, behavioral animation.

Electrical Engineering

Jiayuan Fang, Associate Professor; Ph.D. (electrical engineering), Berkeley, 1989. Computational electromagnetics and microwaves; numerical techniques in computational electromagnetics; electromagnetic modeling of VLSI interconnections; simulation of electrical performance of electronic packaging; signal integrity and power supply noise analysis in electronic packaging; experimental characterization of integrated-circuit packages and printed circuit boards.

Claire Gu, Associate Professor; Ph.D. (electrical engineering), Caltech, 1989. Optical storage, optical fiber communications, optical information processing.

Peyman Milanfar, Assistant Professor; Ph.D. (electrical engineering), MIT, 1993. Statistical signal and image processing, inverse problems, detection and estimation, scientific computing, applied mathematics.

Ali Shakouri, Assistant Professor; Ph.D. (electrical engineering), Caltech, 1995. Integrated cooling of electronic components, electron and photon transport in semiconductor superlattices, photonic switching, fiber optic communication systems.

John Vesecky, Professor; Ph.D. (electrical engineering), Stanford, 1967. Remote sensing of the ocean surface; ocean current measuring radar for coastal ecology and oceanography; radar and radar systems, especially synthetic aperture radar (SAR) and wave scattering; remote sensing and public health; global change.

UNIVERSITY OF COLORADO AT BOULDER

College of Engineering and Applied Science
Department of Computer Science

Programs of Study

The Department of Computer Science offers programs for the Master of Science (M.S.), Master of Engineering (M.E.), and Doctor of Philosophy (Ph.D.) degrees. Courses are offered in nine areas: artificial intelligence, database systems, numerical computation, operating systems, parallel computing, programming languages, software engineering, computer graphics, and theory. A master's program student may select either a thesis or nonthesis option. All M.S. program students must have a minimum of 30 credit hours, satisfy the breadth requirement, and maintain a 3.0 GPA in all courses taken as a graduate student. An M.E. student must maintain a 3.0 GPA while completing the required 30 credit hours. A written report on a creative investigation is the final program requirement. The program is offered both on campus and through the Center for Advanced Training in Engineering and Computer Science (CATECS). A Ph.D. student must have a minimum of 30 credit hours of graduate courses, 30 credit hours of thesis work, pass three area exams, satisfy the breadth requirement, take a comprehensive exam (thesis proposal), and defend a thesis.

Research Facilities

The Department of Computer Science in the College of Engineering provides its graduate students with state-of-the-art computing resources to complement their academic studies with research interests. For the last several years, the department has been the recipient of a National Science Foundation Research Infrastructure Grant. These funds have enabled the department to build a high-speed internal network, a parallel processor cluster from alpha-based PCs, instructional and collaborative audio/video equipment, and a broad variety of other equipment. Other research funding has augmented this grant, providing additional resources, such as an IBM SP2, ATM switches and workstations connected to the vBNS, Myrinet high-speed message-passing equipment, gigabit Ethernet, alpha releases of the new workstations from Digital Equipment Corporation, and software that utilizes this equipment to the utmost. Specifically, the department has research laboratories for research groups in databases, software engineering, groupware, operating systems and designs, lifelong learning, parallel processing, and artificial intelligence. The department has chosen to support a variety of platforms and operating systems. The laboratories are filled with x-terminals, PCs (running Solaris, Linus, and NT), Macintoshes, HP workstations, Sun workstations, DEC workstations, and SGI workstations. This equipment is internally networked in the department with Cat5 cabling, which allows bandwidth expansion in the future. The department has a firewall and several routers, connecting the department with the campus FDDI backbone. In addition, the department has a connection to the campus ATM network, which is directly connected to the vBNS. The Department of Computer Science, through research funding, provides the campus with connectivity to the department's Internet service provider with an OC3 fiber link. This link will be upgraded during summer 1999 to OC12. There are also plans for an Internet2 connection in the near future.

Financial Aid

Financial aid is available to Ph.D. students in the form of teaching and research assistantships, fellowships, and GPTIs. Currently, 25 students are supported by fellowships, 43 are supported by research assistantships, 23 are supported by teaching assistantships, and 45 are supported by other types of financial aid. Aid is only rarely available for master's program students. Grading positions are sometimes available to master's program students on a per-semester basis.

Cost of Study

In 1998–99, graduate tuition for Colorado residents was $1242 for 6 credit hours and $1855 for 9 to 18 credit hours per semester; nonresident graduate tuition was $4890 for 6 credit hours and $7335 for 9 to 18 credit hours per semester. Student fees were approximately $350.

Living and Housing Costs

On-campus housing was $2644 per semester for a single room with twenty-one meals per week in 1998–99. Information about single-student housing may be obtained from the Supervisor of Reservations, Campus Box 154, University of Colorado at Boulder, Boulder, Colorado 80310-0154. Family housing is available for qualified applicants. Rent ranges from $345 to $608 per month. A separate housing application is required. Information may be obtained from the Family Housing Office, University of Colorado at Boulder, 1350 20th Street, Boulder, Colorado 80302 (telephone: 303-492-6384). Average rent for an off-campus apartment ranges from $500 to $850 per month, plus utilities and living expenses. Information about off-campus housing may be obtained from the Off-Campus Student Services Office, Campus Box 206, University of Colorado at Boulder, Boulder, Colorado 80309-0206 (telephone: 303-492-7053).

Student Group

Enrollment in the computer science programs is about 410, including 183 graduate students. There are 145 full-time graduate students enrolled, of whom 32 are women and 113 are men. Of the part-time graduate students, 8 are women and 30 are men. There are 48 international graduate students.

Student Outcomes

Graduates of the department are readily placed in academia, government laboratories, and a vast array of computer and information technology companies. Graduate students usually have several job offers before graduation. Recent graduates have been employed by Hewlett-Packard, Microsoft, Quark, Lucent Technologies, Lockheed Martin, IBM, Nortel, VR-I, Sun Microsystems, Qualcomm, Motorola, and U.S. West.

Location

Boulder, a community of 96,000 people, is located along the Front Range of the Rocky Mountains, 5,400 feet above sea level. The area enjoys about 250 days of sunshine each year. Outdoor recreation includes cycling, rock climbing, backpacking, and world-class skiing. Professional teams in nearby Denver include baseball, football, hockey, and men's and women's basketball. Musical events, art, dance, and theater are prevalent in the area.

The University and The Department

The University of Colorado was founded in Boulder in 1876 with one building, 1 instructor, and 44 students. The Boulder campus is the flagship of four campuses in the University System. It occupies 565 acres and offers its more than 25,000 students approximately 150 areas of study on the baccalaureate and graduate levels. It is widely recognized as one of the nation's top research universities. The Department of Computer Science was founded in 1970 in the College of Arts and Sciences and found its permanent home in the College of Engineering and Applied Science in 1980. It is the youngest department in the College.

Applying

Requests for admission to the Graduate School from U.S. citizens and permanent residents should be sent to the Graduate Advisor in care of the Department of Computer Science at the address below. An application packet is then sent in return. For doctoral program applicants, there is a January 2 deadline for fall semester. For master's program applicants, there is a February 28 deadline for fall semester and an October 15 deadline for spring semester. International doctoral program applicants have a December 1 deadline for the fall semester. International master's program applicants have a December 1 deadline for the fall semester and a September 15 deadline for the spring semester. A Ph.D. student's application for financial support should be included with the application for admission.

Correspondence and Information

Vicki Kunz, Graduate Program Advisor
Department of Computer Science
College of Engineering and Applied Science
Campus Box 430
University of Colorado at Boulder
Boulder, Colorado 80309-0430

Telephone: 303-492-6361
Fax: 303-492-2844
E-mail: csgradinfo@cs.colorado.edu
World Wide Web: http://www.cs.colorado.edu/current/grad/grad.html

University of Colorado at Boulder

THE FACULTY AND THEIR RESEARCH

Ken Anderson, Assistant Professor; Ph.D., California, Irvine. Software engineering, hypermedia, human-computer interaction.

Elizabeth Bradley, Assistant Professor (joint with Electrical and Computer Engineering); Ph.D., MIT. Scientific computation and artificial intelligence, nonlinear dynamics and chaos, classical mechanics, network theory and circuit design.

Richard H. Byrd, Professor; Ph.D., Rice. Numerical computation, nonlinear optimization.

Xiao-Chuan Cai, Associate Professor; Ph.D., NYU (Courant). Numerical analysis and scientific computing.

Andrzej Ehrenfeucht, Professor; Ph.D., Mathematical Institute (Poland). Theory of computation, artificial intelligence.

Michael Eisenberg, Associate Professor; Ph.D., MIT. Educational computing, mathematics and science education, learnability of programming languages, scientific computation.

Clarence (Skip) Ellis, Professor; Ph.D., Illinois at Urbana-Champaign. Groupware, computer-supported cooperative work, visual programming and languages, concurrency control, office information systems, operating systems.

Gerhard Fischer, Professor; Ph.D., Hamburg (Germany). Artificial intelligence, human-computer interaction.

Lloyd D. Fosdick, Professor Emeritus; Ph.D., Purdue. Scientific computation, numerical methods, history of computation.

Harold N. Gabow, Professor; Ph.D., Stanford. Design and analysis of algorithms: graphs and networks, matroids, combinatorial optimization, data structures.

Dirk Grunwald, Associate Professor; Ph.D., Illinois. Parallel computation, compilers, systems.

Dennis Heimbigner, Research Associate Professor; Ph.D., USC. Configuration management and in paradigms for the engineering of distributed software systems.

Elizabeth Jessup, Assistant Professor; Ph.D., Yale. Parallel and scientific computation.

Harry F. Jordan, Professor (joint with Electrical and Computer Engineering); Ph.D., Illinois. Parallel architectures and algorithms.

Roger (Buzz) King, Professor; Ph.D., USC. Interoperability of heterogeneous systems, database systems, graphical interfaces to databases, software engineering.

Clayton Lewis, Professor; Ph.D., Michigan. Human-computer interaction, artificial intelligence, programming languages.

Michael G. Main, Associate Professor; Ph.D., Washington State. Theory of computation.

James Martin, Associate Professor; Ph.D., Berkeley. Natural language processing.

Oliver McBryan, Professor; Ph.D., Harvard. Numerical and parallel computation.

Michael Mozer, Associate Professor; Ph.D., California, San Diego. Cognitive science, neural networks, intelligent control.

Evi Nemeth, Associate Professor; Ph.D., Waterloo. Computer networks, combinatorial mathematics, encryption.

Gary J. Nutt, Professor; Ph.D., Washington (Seattle). Parallel and distributed systems, operating systems, performance, computer modeling and simulation, collaboration technology.

Rick Osborne, Assistant Research Professor; Ph.D., Michigan State. Heterogenous distributed database systems.

Leysia Palen, Assistant Research Professor; Ph.D., California, Irvine. Human-computer interaction, computer-supported cooperative work, social analysis of information technologies, technology adoption, situated cognition.

Alexander Repenning, Research Assistant Professor; Ph.D., Colorado. Visual programming, interactive simulation, computers in education, agents.

Robert B. Schnabel, Professor; Ph.D., Cornell. Numerical computation, parallel computation.

Satinder Singh (Baveja), Assistant Professor; Ph.D., Massachusetts Amherst. Machine learning, reinforcement learning, automatic decision making.

Tamara Sumner, Assistant Professor; Ph.D., Colorado at Boulder. Interactive publishing, educational technology, sociotechnical design, human-computer interaction.

William M. Waite, Professor (joint with Electrical and Computer Engineering); Ph.D., Columbia. Compiler construction, programming languages.

Karl Winklmann, Associate Professor and Chair; Ph.D., Purdue. Theory of computation.

Alexander L. Wolf, Associate Professor; Ph.D., Massachusetts. Software engineering, object database systems, computer languages.

Benjamin Zorn, Associate Professor; Ph.D., Berkeley. Programming languages, compilers, object database systems, visual programming languages, storage management systems.

UNIVERSITY OF DELAWARE

Department of Computer and Information Sciences

Programs of Study	The Department of Computer and Information Sciences offers programs leading to the Master of Science and Doctor of Philosophy degrees. The M.S. program (normally completed in four semesters of full-time study) prepares students for doctoral studies or for professional employment. The doctoral program consists of additional course work and supervised research leading to a dissertation. There is no foreign language requirement.
	Departmental research areas include algorithms: design and analysis, approximation algorithms, fully dynamic algorithms; artificial intelligence and human-computer interfaces: natural language generation and understanding, grammatical formalisms and parsing, multimodal interfaces, augmentative communication devices for people with disabilities, speech processing and synthesis, planning, rehabilitation robotics, neural networks, intelligent tutoring systems; computational theory: computational learning theory, recursive function theory; languages and compilers for parallel computing: optimizing and parallelizing compilers, compiler phase integration, language and compiler support for cluster computing, optimizing explicitly parallel programs, parallel compilers; networking: protocol specification and testing, network management, flow and congestion control, multiaccess protocols for packet radio and wireless environments, high-speed metropolitan area networks, internetworking virtual private networks, network security, distributed architectures, performance modeling and analysis; and symbolic mathematical computation: algebraic algorithms, analytic algorithms, parallelization.
	Students normally enter with undergraduate preparation in mathematics and computer science. However, well-qualified students with varied backgrounds are encouraged to apply; minor deficiencies can be made up after matriculation.
	The University of Delaware operates on a two-semester system, with additional summer sessions and a one-month winter session. No graduate courses are offered during the summer and winter sessions; however, graduate students often have opportunities to teach and do research during these sessions.
Research Facilities	The department operates a joint research lab with the Department of Electrical Engineering that contains an 8-process DEC Alpha system, several Sun file and print servers, and gateways to the NSFNET national network via SURAnet, and a high-performance computing software lab. A variety of single-user machines, including more than 300 Sun Workstations and four SGI Workstations, are available. All of the above machines are connected via fast Ethernet and a campus backbone network.
	The University computing center operates a Cray Research J90, an IBM 3090-600E running MVS, several large Sun SPARCcenter time-sharing systems running Solaris, and an SGI Power Challenge computer. Numerous single-user machines are also available at more than twenty sites on campus, including Sun Workstations, IBM PCs, and Apple Macintoshes.
	The recently expanded University library system contains more than 2.3 million bound volumes and is a government depository library, housing more than 400,000 government publications, including U.S. patents. The library subscribes to more than 20,000 periodicals, including a wide variety of computer science publications. Materials can be conveniently located using the DELCAT online computer catalog, which includes circulation status. DELCAT Plus provides access to a variety of online publication indices and summaries of journal contents.
Financial Aid	Fellowships and teaching assistantships are available. For 1998–99, fellowship stipends ranged from $9940 upward, and teaching assistantships ranged from $9940 to $11,020; both include waiver of tuition. More than 50 percent of the full-time computer and information sciences graduate students receive fellowships or assistantships. Fellowships and traineeships are also available under a number of federal programs. Advanced students may be supported as research assistants. Some summer stipends are available.
Cost of Study	For 1999–2000, course fees for full-time students are $4250 per academic year for residents of Delaware and $12,250 per academic year for nonresidents. Fees for the summer sessions and for part-time students are $177 per credit for Delaware residents and $681 per credit for nonresidents.
Living and Housing Costs	The University has a limited number of one- and two-bedroom apartments for married students and graduate students who are enrolled in full-time programs of study. Off-campus housing prices vary widely; typical monthly rents are $450 for an efficiency, $550 for one bedroom, and $650 for two bedrooms. The Off-Campus Housing Office maintains a listing of available accommodations near the University. For 1999–2000, meals in the campus dining halls cost $1126 per semester for nineteen meals per week.
Student Group	There are currently 78 graduate students in the department. Approximately 70 percent are full-time students, and most of these are supported by assistantships, fellowships, or external business organizations. The total campus enrollment is about 20,000, including 3,200 graduate students.
Location	Newark (pronounced New Ark), Delaware, is a pleasant university community of 26,000 people. Located midway between Philadelphia and Baltimore, it offers the advantages of a small community but is still within easy traveling distance of New York and Washington, D.C. Newark is also close to the recreational areas along the Chesapeake Bay and Atlantic Ocean.
The University	The University of Delaware developed from a small private academy founded in 1743 and is today a state-assisted, privately controlled, coeducational land-grant and sea-grant university. The beautifully landscaped Newark campus consists of 1,100 acres with nearly 400 buildings in a predominately Georgian architectural style.
Applying	The general application deadlines are July 1 and December 1 for the fall and spring semesters, respectively. Notification of the admissions decision is provided promptly upon receipt of credentials. In addition to the completed application form and the application fee of $45, applicants must forward official transcripts of their previous academic records, including at least three letters of recommendation, GRE General Test scores, and a TOEFL score if English is not their first language and they have not received a degree from a U.S. institution. Applications for fellowships and assistantships, including three letters of recommendation, are due by March 1 for the fall semester and November 1 for the spring semester; late applications are considered if positions exist. Interested students are invited to write to individual faculty members or the Chairperson of the department, at the address below.
Correspondence and Information	Professor B. F. Caviness Chair, Graduate Committee Department of Computer and Information Sciences University of Delaware Newark, Delaware 19716 Telephone: 302-831-2713 Fax: 302-831-8458 E-mail: gradprgm@cis.udel.edu World Wide Web: http://www.udel.edu (U. of Delaware home page) http://www.cis.udel.edu (CIS home page index)

University of Delaware

THE FACULTY AND THEIR RESEARCH

Gagan Agrawal, Assistant Professor; Ph.D., Maryland, 1996. Compiler optimizations, programming languages, parallel and distributed systems.

Paul D. Amer, Professor; Ph.D., Ohio State, 1979. Computer networks, transport protocols, formal specification and testing (Estelle).

M. Sandra Carberry, Professor and Chair; Ph.D., Delaware, 1985. Artificial intelligence, natural-language processing, planning systems.

John Case, Professor; Ph.D., Illinois, 1969. Computational learning theory, recursive function theory, algorithms and architecture design for massive parallelism.

B. F. Caviness, Professor; Ph.D., Carnegie Mellon, 1968. Computer algebra, analysis of algorithms.

Daniel L. Chester, Associate Professor; Ph.D., Berkeley, 1973. Artificial intelligence, natural-language processing, theorem proving, knowledge representation.

Keith S. Decker, Assistant Professor; Ph.D., Massachusetts, 1995. Distributed problem solving, multi-agent systems, real-time problem solving, computational organization design, concurrent engineering, parallel and distributed planning and scheduling, distributed information gathering.

Chandra Kambhamettu, Assistant Professor; Ph.D., South Florida, 1994. Computer vision, image processing, computer graphics and multimedia.

Errol L. Lloyd, Professor; Ph.D., MIT, 1980. Design and analysis of algorithms.

Kathleen F. McCoy, Associate Professor; Ph.D., Pennsylvania, 1985. Artificial intelligence, natural-language generation and understanding, knowledge representation.

Lori L. Pollock, Associate Professor; Ph.D., Pittsburgh, 1986. Compiler construction, code optimization, compilation for parallel machines, programming environments.

B. David Saunders, Associate Professor; Ph.D., Wisconsin, 1975. Computer algebra, analysis of algorithms, parallel computation.

Tuncay Saydam, Professor; Ph.D., Istanbul Technical, 1964. Computer networks, network management, telecommunications software engineering, object-oriented software technologies.

Adarshpal S. Sethi, Associate Professor; Ph.D., Indian Institute of Technology (Kanpur), 1978. Computer networks, network management, distributed systems, protocol modeling and testing.

Chien-Chung Shen, Assistant Professor; Ph.D., UCLA, 1992. Broadband and PCS networks, network and service management, network-centric multimedia applications, distributed object and mobile computing, distributed interactive simulation.

K. Vijayashanker, Associate Professor; Ph.D., Pennsylvania, 1987. Artificial intelligence, natural-language processing, unification-based grammatical systems, knowledge-representation languages.

Joint, Adjunct, and Research Faculty

Charles G. Boncelet, Professor (joint with Electrical and Computer Engineering); Ph.D., Princeton, 1984. Signal processing, algorithms, networking.

H. Timothy Bunnell, Research Associate Professor (A. I. DuPont Institute); Ph.D., Penn State, 1983. Speech perception, computer enhancement of speech.

George E. Collins, Professor; Ph.D., Cornell, 1955. Computer algebra.

Guang Gao, Professor (joint with Electrical and Computer Engineering); Ph.D., MIT, 1996. Computer architecture and systems, parallel and distributed systems.

Robert P. Gilbert, Unidel Professor (joint with Mathematical Sciences); Ph.D., Carnegie Tech, 1958. Hybrid computation, mathematical modeling.

Ashfaq Khokhar, Assistant Professor (joint with Electrical and Computer Engineering); Ph.D., USC, 1993. Parallel computation and software systems, computational aspects of computer vision and multimedia.

Renate Scheidler, Associate Professor (joint with Mathematical Sciences); Ph.D., Manitoba (Canada), 1993. Computational number theory.

John K. Scoggin Jr., Adjunct Associate Professor. Computer networks.

Richard L. Venezky, Unidel Professor (joint with Educational Studies); Ph.D., Stanford, 1965. Intelligent tutoring systems, natural-language processing, lexicography.

David Wood, Research Associate Professor; Ph.D., Rhode Island, 1972. DNA computing, computer algebra, analysis of algorithms.

Professional Staff

Richard A. Albright, Associate to the Chair; Ph.D., Delaware, 1971. Numerical methods.

UNIVERSITY OF DETROIT MERCY

College of Engineering and Science
Department of Mathematics and Computer Science

Program of Study	The Master of Science in Computer Science degree is earned upon completion of a 33-credit-hour program of study. It is designed to prepare students for doctoral study or positions in industry or government service in computer-related areas. In recognition of various student interests, courses may be selected from areas related to computer science, primarily quantitative business decisions, software and hardware engineering, and mathematics appropriate for the theory of computers or computer applications. The strong blend of computer mastery with mathematics makes the program unique. This allows the faculty to lead students in both independent and team-based explorations. This paradigm of exploration and knowledge seeking leads to the development of a much stronger graduate. Most students complete the program within a two-year time frame. Some prerequisite courses are required if areas such as programming language or knowledge of calculus have not been previously completed. Students complete a research paper in their chosen area of computer science as a means of meeting one of the requisites in the program.
Research Facilities	Several computer labs are available on the campus. Each professor has a state-of-the-art computer available in his or her office to work with the students. Each computer has access to the Internet.
Financial Aid	The University's Scholarship and Financial Aid Office accepts applications for grants, loans, and work-study assistance. Aid includes the Michigan Tuition Grant (for Michigan residents only), Federal Work-Study, and a variety of loans. The University also accepts third-party payments from employers and government agencies and offers payment plans of its own. For information regarding financial aid programs, students should call 313-993-3350.
Cost of Study	Tuition in 1999–2000 is $545 per credit hour. Registration fees are $100 for full-time students.
Living and Housing Costs	Housing is available on campus. Double-occupancy rates range from $1410 to $3280. Single-occupancy rates range from $2420 to $2720. The University offers meal plans at about $1085. All rates are for a sixteen-week term. For more information, students should call the Residence Life Office at 313-993-1230.
Student Group	The average student population in the program is 30 students; currently, 60 percent are men, 40 percent are women. Both minority and international students are enrolled. Students in the program develop a great sense of camaraderie through working closely together on their projects. Approximately 6,700 students attend classes on three UDM campuses located in northwest and downtown Detroit.
Location	Students enjoy a variety of activities offered on campus and throughout the metropolitan Detroit area, including sports, theater, concerts, and more.
The University and The Department	As Michigan's largest Catholic university, the University of Detroit Mercy has an outstanding tradition of academic excellence firmly rooted in a strong liberal arts curriculum. This tradition dates back to the formation of two Detroit institutions, the University of Detroit, founded in 1877 by the Society of Jesus (Jesuits), and Mercy College of Detroit, founded in 1944 by the Sisters of Mercy of the Americas. In 1990, these schools consolidated to become the University of Detroit Mercy. Today, UDM offers more than 120 majors and programs in nine different schools and colleges and is widely recognized for its programs in engineering, law, dentistry, nursing, and architecture. Faculty members are known for their excellence; more than 87 percent have a Ph.D. or comparable terminal degree. Since its beginning more than 25 years ago, the department has served many students. These graduates have found rewarding employment in business and industry.
Applying	Applications for admission normally should be completed at least six weeks before the beginning of a term. Applications for financial aid should be submitted by April 1. International students are urged to complete their applications at least three months before classes begin. Admission requirements are a bachelor's degree from an accredited college; a B average in the total undergraduate program and in the proposed field of study; and, normally, an undergraduate major or the equivalent in the proposed field. Official transcripts are required from all colleges attended. Applicants with less than a B average who present other evidence of ability to perform graduate-level work may be admitted as probationary students upon the recommendation of the director of the program.
Correspondence and Information	Professor R. M. Canjar, Chair Department of Mathematics and Computer Science University of Detroit Mercy P.O. Box 19900 Detroit, Michigan 48219-0900 Telephone: 313-993-1209 E-mail: canjarrm@udmercy.edu

University of Detroit Mercy

THE FACULTY AND THEIR RESEARCH

R. Michael Canjar, Professor and Chairman; Ph.D., Michigan. Object-oriented programming, mathematical logic.

D. Kevin Daimi, Associate Professor; Ph.D., Cranfield Institute of Technology (England). Artificial intelligence, expert systems, natural language processing.

John Dwyer, Associate Professor; Ph.D, Texas A&M. Statistics, computer science.

Nancy Dwyer, Assistant Professor; Ph.D., Toledo. Math education.

Robert Kane, Associate Professor; M.A., Detroit. Geometry and numerical analysis.

John O'Neill, S.J., Professor; Ph.D., Wayne State. Algebra, groups and rings.

Michael Skaff, Professor; Ph.D., UCLA. Model theory, operating systems.

Katherine Snyder, Instructor; M.S., Detroit. Artificial neural networks, evolutionary programming, cultural algorithms.

Raymond Travis, Associate Professor; M.A., Wayne State. Celestial mechanics and astrophysics.

Kathy Zhong, Assistant Professor; Ph.D., Wayne State. Complex analysis, wavelet theory.

UNIVERSITY OF FLORIDA

Computer and Information Science and Engineering Department

Programs of Study

The Computer and Information Science and Engineering (CISE) Department offers the Master of Science degree through the Colleges of Engineering and Liberal Arts and Sciences. The Master of Engineering, Engineer and the Ph.D. degrees are offered through the College of Engineering only.

There are five broad areas of specialization in the department: computer systems and communications, which includes computer architecture, distributed systems, networks and communication, operating systems, simulation, and performance evaluation; database systems, which includes database management systems, database design, database theory and implementation, database machines, distributed databases, and information retrieval; high-performance computing, which includes parallel processing, parallel algorithms, and parallel software systems; intelligent systems, which includes computer vision and visualization, pattern recognition, image processing, computer graphics, robotics, expert systems, machine learning, and artificial intelligence; and software engineering, which includes large-scale software development and maintenance paradigms and processes for various computing systems (including parallel and distributed computing systems), software quality assurance, and programming environments and languages.

Research Facilities

The department has a variety of generalized and specialized computer systems, a six-node IBM SP-2, a high-availability cluster of Sun 3500s, and numerous Sun, HP, Silicon Graphics, and IBM workstations and PCs. Along with the general campus computing facilities, a broad base of computing resources is provided. The departmental computers provide full connectivity to the Internet. The department is connected to the rest of the campus via a fiber-optic link, providing both video and digital data communications.

Financial Aid

A number of research and teaching assistantships are available for CISE graduate students both within the department and within many other academic departments and research institutes on the campus. Based on one-half-time employment, assistantships paid $10,920 to $18,500 for nine months in 1998–99. Appointments range from one-quarter time (requiring 10 hours of service per week) to one-half time (requiring 20 hours of service per week). Tuition payments may be granted for students holding assistantships.

Cost of Study

In 1998–99, Florida students paid $129.01 per graduate credit hour, and non-Florida students paid $434.40. A 10 percent rate increase is expected for 1999–2000.

Living and Housing Costs

University and off-campus housing is available for single and married graduate students. Single students should allow $7620 to $13,000 per academic year for housing, food, and normal living expenses. International students must certify that they have a minimum of $21,365 in financial support per year for the first two years ($42,730 total).

Student Group

The total enrollment at the University is 42,000, including approximately 7,000 graduate students. The Department of Computer and Information Science and Engineering has approximately 165 graduate students, of whom 55 are enrolled in the Ph.D. program. Most of these are enrolled through the College of Engineering. The department has approximately 1,000 undergraduate students, and about one half of these are enrolled through the College of Engineering.

Location

The University of Florida is located in Gainesville, a city with a metropolitan area population of approximately 220,000 and situated in north-central Florida, midway between the Atlantic Ocean and the Gulf of Mexico. The city has a strong research focus, with comprehensive University and medical research programs.

The University

The University of Florida is the senior institution in the State University System of Florida. A land-grant institution, it is one of the leading universities in the southeastern United States and is generally regarded as a major national university. There are thirteen upper-division colleges and four professional colleges. The University offers graduate programs in more than ninety-five fields at the master's level and in fifty-eight fields at the doctoral level.

Applying

Students are strongly encouraged to send all their application materials to the Admissions Office at least six months prior to the desired date of enrollment. The general admission requirements include a baccalaureate degree from an accredited college or university with an upper-division GPA of at least 3.0 and a combined score of not less than 1100 on the verbal and quantitative portions of the GRE General Test. International students are required to submit a TOEFL score greater than 600.

Correspondence and Information

The detailed description of the graduate program is available via the World Wide Web (http://www.cise.ufl.edu/). Students can also obtain this information by typing *ftp ftp.cise.ufl.edu* and using the login *anonymous*. Students should use their login ID as the password and then type *cd /cise/grad*. They should specify *binary* and use *get* or *ls* to view available files.

Information can also be obtained by contacting:

Graduate Secretary
Computer and Information Science and Engineering Department
405 Computer Science and Engineering Building
University of Florida
P.O. Box 116120
Gainesville, Florida 32611-6120

Telephone: 352-392-1090 or 1220
Fax: 352-392-6889
E-mail: gradsec@cise.ufl.edu
World Wide Web: http://www.cise.ufl.edu/

University of Florida

THE FACULTY AND THEIR RESEARCH

Gerhard X. Ritter, Professor and Chairman; Ph.D., Wisconsin, 1971. Computer vision, pattern recognition. (E-mail: ritter@cise.ufl.edu)

Manuel E. Bermudez, Associate Professor; Ph.D., California, Santa Cruz, 1984. Programming languages, automata theory, compilers. (E-mail: manuel@cise.ufl.edu)

Sharma Chakravarthy, Associate Professor; Ph.D., Maryland, 1985. Query/rule optimization; active, heterogeneous, temporal, and deductive databases; logic programming. (E-mail: sharma@cise.ufl.edu)

Y. C. Chow, Professor; Ph.D., Massachusetts, 1977. Parallel processors, computer networks, performance evaluation. (E-mail: chow@cise.ufl.edu)

Douglas D. Dankel II, Assistant Professor and Graduate Coordinator; Ph.D., Illinois, 1979. Artificial intelligence, expert systems, software development environments. (E-mail: ddd@cise.ufl.edu)

Timothy A. Davis, Associate Professor; Ph.D., Illinois, 1989. Parallel algorithms and architectures, parallel sparse matrix algorithms. (E-mail: davis@cise.ufl.edu)

Keith L. Doty, Professor; Ph.D., Berkeley, 1967. Microcomputers, robotics, concurrent processing. (E-mail: doty@milcise.ufl.edu)

Paul A. Fishwick, Associate Professor; Ph.D., Pennsylvania, 1986. Simulation, artificial intelligence, systems science, computer animation and visualization. (E-mail: fishwick@cise.ufl.edu)

Li-Min Fu, Associate Professor; Ph.D., Stanford, 1985. Artificial intelligence, expert systems, neural networks. (E-mail: fu@cise.ufl.edu)

Joachim Hammer, Associate Professor; Ph.D., USC, 1994. Knowledge-based and information systems, information system interoperation, data warehousing. (E-mail: jhammer@cise.ufl.edu)

Eric N. Hanson, Assistant Professor; Ph.D., Berkeley, 1987. Database management systems. (E-mail: hanson@cise.ufl.edu)

Abdelsalam Ali Helal, Associate Professor; Ph.D., Purdue, 1991. Database systems, mobile computing, collaborative computing, Internet applications, data warehousing. (E-mail: helal@cise.ufl.edu)

Yann-Hang Lee, Associate Professor; Ph.D., Michigan, 1985. Distributed computing and parallel processing, database systems, performance evaluation. (E-mail: yhlee@cise.ufl.edu)

Panos E. Livadas, Assistant Professor; Ph.D., Florida, 1980. Computer graphics, data and file structures, software engineering. (E-mail: pel@cise.ufl.edu)

Richard Newman, Assistant Professor; Ph.D., Rochester, 1986. Distributed systems and networks, distributed conferencing and groupware, automata theory and complexity, parallel algorithms and architectures. (E-mail: nemo@cise.ufl.edu)

Jih-Kwon Peir, Associate Professor; Ph.D., Illinois, 1985. Parallel system architecture and algorithms. (E-mail: peir@cise.ufl.edu)

Jorg Peters, Associate Professor; Ph.D., Wisconsin, 1990. Computer graphics, geometric design and visualization. (E-mail: jorg@cise.ufl.edu)

Sanjay Ranka, Associate Professor; Ph.D., Minnesota, 1988. Compilers and software environments for parallel machines, high-performance computing, design and analysis of parallel algorithms, models of parallel computation and neural networks. (E-mail: ranka@cise.ufl.edu)

Sartaj Sahni, Professor; Ph.D., Cornell, 1973. Design and analysis of algorithms, parallel computing, VLSI, CAD. (E-mail: sahni@cise.ufl.edu)

Rajasekaran Sanguthevar, Associate Professor; Ph.D., Harvard, 1988. Algorithms and parallel computation. (E-mail: raj@cise.ufl.edu)

Beverly Sanders, Associate Professor; Ph.D., Harvard, 1985. Parallel computation, formal method, software specification and verification, parallel and distributed systems. (E-mail: sanders@cise.ufl.edu)

Ralph G. Selfridge, Professor; Ph.D., Oregon, 1953. Language development and implementation, numerical analysis, computation theory, graphics. (E-mail: selfridg@cise.ufl.edu)

Meera Sitharam, Associate Professor; Ph.D., Wisconsin, 1990. Discrete modeling and algorithmics, geometric constraints, networking economics, complexity. (E-mail: sitharam@cise.ufl.edu)

Stanley Y. W. Su, Professor of Computer and Information Sciences and Electrical Engineering; Ph.D., Wisconsin, 1968. Database management, software systems, computer architecture for nonnumeric processing. (E-mail: su@cise.ufl.edu)

Fred Taylor, Professor of Computer and Information Sciences and Electrical Engineering; Ph.D., Colorado, 1969. Digital systems, signal processing. (E-mail: ft@gamma.ee@cise.ufl.edu)

Stephen M. Thebaut, Assistant Professor; Ph.D., Purdue, 1983. Software engineering, requirements elicitation, cost estimation. (E-mail: smt@cise.ufl.edu)

Baba C. Vemuri, Associate Professor; Ph.D., Texas, 1987. Computer vision, image processing, geometric modeling, artificial intelligence. (E-mail: vemuri@cise.ufl.edu)

Joseph Wilson, Assistant Professor and Associate Chair; Ph.D., Virginia, 1984. Programming languages, computer vision, image processing. (E-mail: jnw@cise.ufl.edu)

RESEARCH CENTERS

Center for Computer Vision and Visualization (Director: G. X. Ritter). This center is an interdisciplinary research center focusing on all aspects of computer vision technology. The center provides coordination, direction, and focus in the area of computer vision at the University of Florida as well as providing a structure for interaction between University faculty and graduate students and industrial and DoD laboratories.

Database Systems Research and Development Center (Director: S. Y. W. Su). This center deals with the following three categories of research and development activities: the database management aspects of information processing, the hardware aspects of information system design and development, and the behavioral aspects of information transfer.

Software Engineering and Research Center (Site Director: Stephen M. Thebaut). The Software Engineering and Research Center is a cooperative research program with Purdue University, the University of Oregon, and West Virginia University funded by the National Science Foundation and fifteen industrial sponsors. It is involved in developing software tools, environments, and metrics to assist the development and maintenance of reliable, efficient, reusable, and easily maintained software systems.

Space Communication Technology Center (Associate Director: R. E. Newman). The Space Communication Technology Center is a NASA-sponsored center for the commercial development of space (CCDS) with a primary focus on high-definition television (HDTV). Research is conducted in the area of digital signal handling and satellite networking, specifically in error control coding, signal modulation, terminal interconnect equipment, the Broadband Integrated Services Digital Network, protocols, and satellite network design, analysis, and management.

UNIVERSITY OF HOUSTON

Department of Computer Science
M.S. and Ph.D. Degree Programs

Programs of Study

The Department of Computer Science at the University of Houston (UH) offers programs leading to the Master of Science (M.S.) and Doctor of Philosophy (Ph.D.) degrees in computer science. Fields of specialization include artificial intelligence (AI), computer vision, computer graphics, databases, parallel and distributed computing, software engineering, virtual reality, and theory.

The M.S. degree requires 24 hours of course work and a thesis. Requirements for the Ph.D. degree include a dissertation and 48 hours of course work after the B.S. Candidates must also meet specific requirements with respect to both breadth and depth in their academic background and may be assigned remedial work if these are not satisfied.

Research Facilities

The department maintains UNIX servers for Solaris, SunOS, and DEC Ultrix. Students have access to around 100 workstations and X-terminals in several open areas within the department as well as 40 high-performance SGI O-2 graphics workstations. Remote access via dial-up, nongraphics connections is provided free by the University, utilizing more than 800 telephone lines. PPP graphics access is also available upon request, as are special University rates for Internet service providers. Access to University supercomputers such as a sixty-four-node IBM SP2, a sixteen-pipe SX-3, a thirty-two-node Cenju-3, and others is also provided. The University of Houston is on the Internet backbone and is part of the experimental Internet II.

Financial Aid

Graduate students with teaching or research assistantships are provided a stipend of between $850 and $1200 per month, depending on the student's qualifications. Application for the assistantship should be made directly to the department. The deadlines are March 1 (for fall) and October 1 (for spring). Approximately 35 teaching assistants/fellows and 15 research assistants are currently employed by the department.

Cost of Study

Tuition and fees for Texas residents are approximately $1300 per semester in 1999–2000; nonresidents, including international students, pay about $4000 per semester. Students receiving financial assistance qualify for resident status.

Living and Housing Costs

On-campus room and board cost approximately $2300 per semester. For students in off-campus housing, apartments begin at about $500 per month.

Student Group

There are approximately 400 graduate students in the department, 60 of whom are Ph.D. students. These students come from all over the world, including Asia, Europe, the Middle East, and the Americas. The department maintains exchange student agreements with universities in Germany, Mexico, and Sweden.

Student Outcomes

The Ph.D. graduates find employment in both academic institutions and industry, such as GTE labs, nationwide. The demand for graduates who earn a master's degree is great and most M.S. students find positions prior to graduation. Local employers include NASA/Johnson Space Center, Texas Medical Center, Shell Oil, and Compaq.

Location

Houston is the fourth-largest city in the U.S. and maintains a high rate of economic growth. UH's 500-acre campus is within a short drive from downtown Houston and the Texas Medical Center. Drawing upon the University's strong relationship with Houston business and research communities (e.g., Compaq, Exxon, Shell, NASA/JSC, Texas Medical Center), students may obtain numerous research and employment opportunities.

The University and The Department

The University of Houston is the only doctoral degree–granting component and the largest campus of the state-supported UH System and serves as a strong research and intellectual base for the city of Houston, the state of Texas, and the United States. Serving 33,000 students in fourteen colleges, UH ranks among the top eighty research universities in the country. Research grants and awards to UH exceeded $40 million in 1998.

The department, one of six departments in the College of Natural Sciences and Mathematics, maintains strong ties with several research institutes on campus, including the High Performance Computer Center, the Virtual Environment Technology Laboratory, and the Texas Center for Advanced Molecular Design.

Applying

Applicants must have a bachelor's degree, and should have a minimum GPA of 3.0 over the last 60 hours, a combined GRE (verbal and quantitative) score of 1150 or better, six or more hours of mathematics beyond Calculus II, and at least two programming courses. Students whose native language is not English must obtain a TOEFL score of 550 or above; for these students, a minimum GRE score of 1000 with at least 600 on the quantitative part is sufficient. Students with degrees in fields other than computer science may be admitted; they may have to complete additional course work.

Correspondence and Information

Dr. Ernst L. Leiss
Director of Graduate Studies
Department of Computer Science
University of Houston
Houston, Texas 77204-3475

Telephone: 713-743-3350
Fax: 713-743-3335
E-mail: gradinfo@cs.uh.edu
World Wide Web: http://www.cs.uh.edu

University of Houston

THE FACULTY AND THEIR RESEARCH

Robert B. Anderson, Associate Professor and Director of Undergraduate Studies; Ph.D., Texas at Austin. Theory of computation, artificial intelligence.

Barbara Chapman, Associate Professor; Ph.D., Trinity (Dublin). High-performance computing.

Albert M. K. Cheng, Associate Professor and Director of Real-Time Systems Laboratory; Ph.D., Texas at Austin. Real-time systems, distributed systems, AI, software engineering, computer security, computer networks.

Kam-Hoi Cheng, Associate Professor; Ph.D., Minnesota. Artificial intelligence, object-oriented analysis and design, VLSI, parallel and distributed processing, networks, architecture, algorithm and complexity.

Christoph F. Eick, Associate Professor; Ph.D., Karlsruhe. Artificial intelligence, rule-based programming, expert systems, genetic algorithms, software reliability.

J. C. Huang, Professor; Ph.D., Pennsylvania. Software engineering, real-time computer systems, program analysis and testing.

S.-H. Stephen Huang, Professor; Ph.D., Texas at Austin. Data structures, design and analysis of algorithms, parallel and distributed processing, data management.

Olin G. Johnson, Professor; Ph.D., Berkeley. Numerical analysis, high-performance computing.

S. Lennart Johnsson, Cullen Professor of Computer Science and Chairman; Ph.D., Chalmers Institute of Technology (Sweden). Parallel computing, scientific computation.

Ioannis A. Kakadiaris, Assistant Professor; Ph.D., Pennsylvania. Computer vision, medical imaging, computer graphics, physics-based modeling, estimation and synthesis of the shape and motion of articulated objects .

Willis K. King, Associate Professor; Ph.D., Pennsylvania. Computer architecture, distributed systems.

Ernst L. Leiss, Professor and Director of Graduate Studies; Dr. Techn., TU Wien (Vienna). Vector and parallel computing, data security, formal and programming languages, geophysical data processing.

H. Bowen Loftin, Professor of Computer Science and Director of Virtual Environment Technology Laboratory; Ph.D., Rice. Computer graphics, data visualization, artificial intelligence, simulation.

Jehan-Francois Paris, Associate Professor; Ph.D., Berkeley. Distributed systems, file systems, fault-tolerant computing, performance evaluation.

B. Montgomery Pettitt, Cullen Professor of Chemistry, Computer Science, and Biochemical and Biophysical Sciences and Director of Institute for Molecular Design; Ph.D., Houston. High-performance computing, parallel processing, numerical analysis, visualization.

D. Sivakumar, Assistant Professor; Ph.D., SUNY at Buffalo. Computational complexity theory, randomized computation, probabilistic self-testing and self-correcting of programs.

Jaspal Subhlok, Associate Professor; Ph.D., Rice. Parallel and distributed systems, adaptive and network-aware computing, compilers.

Rakesh M. Verma, Associate Professor; Ph.D., SUNY at Stony Brook. Symbolic computation, declarative programming languages, automated deduction, parallel computing, temporal and spatial databases.

CURRENT RESEARCH PROJECTS

Input/output management in high-performance computing, including parallel I/O.
Data security in object-oriented and in multimedia systems.
Language equations.
Appropriateness evaluation of drug prescriptions.
Learning Bayesian rule-sets for classification tasks.
Knowledge discovery in databases.
Distributed environment for interactive data and document review.
Smaran: an efficient congruence-closure-based system for equational systems.
Techniques for analyzing recursive algorithms.
Efficient sequential and parallel storage and access methods for temporal and spatial databases.
Virtual environments for training.
Assessing the potential of virtual reality in science education.
Vision-based human motion estimation.
Segmentation and visualization of MRI human brain data.
Fluid flow visualization.
Data visualization science and engineering.
Rigorous reliability assessment of embedded safety-critical systems.
Program transformation for high-performance parallel computing.
Self-stabilizing safety-critical applications over wide area networks.
OMNIBASE: a multidatabase management system.
NARADA: interoperability in a heterogeneous computing environment.
High-performance robust parallel and distributed multi-agent systems.
Software reliability for payloads in space.
Parallel and distributed algorithms.
Fundamental properties of rule-based systems.
A learning program.

UIC

UNIVERSITY OF ILLINOIS AT CHICAGO

Department of Electrical Engineering and Computer Science
Program in Computer Science and Engineering

Programs of Study	The Department of Electrical Engineering and Computer Science offers a broad range of programs in computer science and engineering and electrical engineering leading to the M.S. and Ph.D. degrees. The M.S. degree requires 36 semester hours of study, including an optional thesis. The Ph.D. degree requires an additional 76 semester hours of credit. Ph.D. students are required to pass a written qualifying examination, an oral thesis proposal examination, and a final examination defending the thesis at the conclusion of the research.
	While the department offers a comprehensive range of courses in computer science and engineering, it has special strengths in the areas of computer graphics, software engineering, database systems, parallel and distributed systems, human-computer interaction, theory, computer architecture, programming languages and environments, computer vision, and artificial intelligence.
Research Facilities	The department maintains two large, modern instructional computing facilities, including ten UNIX file servers managing a total of more than ninety-five GB of disk space, four UltraSPARC compute servers, and 130 student-accessible Sun Workstations for software development, database programming, VLSI design, HTML development, and numerical computing. There are also four Silicon Graphics workstations equipped with 24-bit graphics and thirty-four Macintosh computers used for assembly language programming and HTML development. All computers are networked via ten megabit switched Ethernet.
	In addition to the modern instructional computing facility, the department contains several specialized research laboratories, most of which are housed in the $30-million Engineering Research Facility. The Electronic Visualization Laboratory (EVL), the Interactive Computing Environments Laboratory, the Concurrent Software Systems Laboratory, the Intelligent Vehicle Highway Systems and Artificial Intelligence Laboratory, the Software Engineering Laboratory, the Knowledge and Database Systems Laboratory, the Distributed Real-Time Intelligent Systems Laboratory, the Parallel Systems Laboratory, the Laboratory for Advanced Computing, the Vision Interface and Systems Laboratory, and the Signal and Image Research Laboratory contain more than 100 additional workstations and servers as well as an extensive array of computer-based multimedia equipment. The Microfabrication Applications Laboratory, the Communications Laboratory, the Electromagnetics and Optics Laboratory, the Power Electronics Laboratory, the Visual/Motor Laboratory, and the Biomedical Functional Imaging and Computational Laboratory contain a wide array of specialized equipment for advanced research in electrical engineering.
	The departmental computing facilities are networked to general University computing resources and national networks, permitting high-speed access to specialized computing facilities, such as Connection Machine, Power Challenge Array, and Convex supercomputers at the National Center for Supercomputing Applications (NCSA) at the University of Illinois at Urbana-Champaign and to the IBM SP-2 at Argonne National Laboratory.
Financial Aid	Financial aid is available in the form of teaching and research assistantships as well as various University fellowships, and opportunities are excellent for research-oriented students with solid undergraduate backgrounds in computer science and engineering. Half-time assistantships for new graduate students require 19 hours per week and pay $11,300 for the academic year ($13,816 for the calendar year). In 1998–99, the electrical engineering and computer science department supported more than 140 graduate students, about one third of whom were teaching assistants. Approximately 75 graduate students were supported by external research grants totaling more than $5 million.
Cost of Study	For 1998–99, Illinois residents paid $2746 and nonresidents paid $6064 per semester if registered for 12 or more hours, with a sliding scale for registration of fewer than 12 hours. Tuition and fees are waived for recipients of teaching and research assistantships, tuition and fee awards, and some University fellowships. Fees are subject to change.
Living and Housing Costs	In addition to UIC residence halls, off-campus apartments and rooms are available. Meals may be taken in campus dining facilities or a variety of nearby commercial establishments. Food, housing, transportation, medical care, personal items, clothing, and incidentals are estimated to cost $11,000 per calendar year.
Student Group	The department has an undergraduate student body of more than 1,100 students and more than 500 graduate students. The graduate student body includes about 250 full-time (average of 12 or more hours) graduate students, of whom more than 100 are pursuing the Ph.D. degree. Most graduate students are affiliated with one of the research laboratories listed above.
Location	Chicago is a vibrant, friendly, beautiful city where there is never a dull moment. The city has countless restaurants, signature blues clubs, fantastic museums, and 12 miles of lakefront beach. The University, a 5-minute train ride from the center-city Loop, is situated on the original site of Jane Addam's Hull House, the first social settlement house in the United States. The University is bordered by Greektown to the north and Little Italy to the west. Chinatown is a 10-minute drive away.
The University	The University of Illinois at Chicago (UIC), with approximately 25,000 students, is the largest institution of higher learning in the Chicago area. UIC is listed among the Research-I universities in the United States in external research funding. The University offers master's degrees in eighty-seven fields and doctorates in fifty-four areas.
Applying	Application materials for admission and financial aid may be obtained by writing to the address below. Requests for applications for teaching assistantships and fellowships (including tuition and fee waivers) should also be directed to the address below. Each student seeking a research assistantship is advised to write directly to faculty members in his or her area of interest. All international applicants are required to take the Graduate Record Examinations for admission. The GRE is recommended for all applicants for financial aid. Applications for teaching and research assistantships should be received no later than March 1. To be considered for a fellowship, students should submit applications before the January 1 deadline. For fall 2000, the application for admission deadlines are March 15 for international students and June 1 for all other students. Applicants from non-English-speaking countries must take the TOEFL and score at least 570 to be considered for admission.
Correspondence and Information	Director of Graduate Admissions (M/C 154) Department of Electrical Engineering and Computer Science University of Illinois at Chicago 851 South Morgan Street (1120 SEO) Chicago, Illinois 60607-7053 Telephone: 312-996-2290 312-413-2291 Fax: 312-413-0024 E-mail: grad-info@eecs.uic.edu World Wide Web: http://www.eecs.uic.edu

University of Illinois at Chicago

THE FACULTY AND THEIR RESEARCH

Artificial Intelligence and Computer Vision

Barbara Di Eugenio, Assistant Professor; Ph.D., Pennsylvania, 1993. Artificial intelligence and natural languages processing; knowledge representation, planning, dialogue interfaces; NLP for human computer interaction and educational technology.

Simon Kasif, Associate Professor; Ph.D., Maryland, 1985. High-performance intelligent systems machine learning, data mining, computational biology, computational neuroscience, parallel computation.

Peter C. Nelson, Associate Professor; Ph.D., Northwestern, 1988. AI research in heuristic search and applied AI research in the areas of transportation, optimizing manufacturing scheduling problems (genetic algorithms and rule-based systems), and computational biology.

Francis K. H. Quek, Associate Professor; Ph.D., Michigan, 1990. Human-computer interaction, computer vision, robot navigation.

Boaz J. Super, Assistant Professor; Ph.D., Texas at Austin, 1992. Computer and biological vision, image processing, pattern recognition.

Computer Architecture and VLSI Systems

Wai-Kai Chen, Professor; Ph.D., Illinois at Urbana-Champaign, 1964. Broadband matching, applied graph theory, networks, systems, filters, VLSI placement, routing, and layout.

Shantanu Dutt, Associate Professor; Ph.D., Michigan, 1990. Parallel and distributed computing, VLSI CAD, fault-tolerant computing and computer architecture.

Joseph Hummel, Assistant Professor; Ph.D., California, Irvine, 1997. Compilers, programming languages, parallelism.

Jon A. Solworth, Associate Professor; Ph.D., NYU, 1987. Computer architecture, programming language design and compilation techniques for parallel processors, I/O, object stores and file systems.

Database Systems

Jorge Lobo, Associate Professor; Ph.D., Maryland, 1990. Knowledge representation, logic and databases, logic programming, nonmonotonic reasoning.

Ouri Wolfson, Associate Professor; Ph.D., NYU, 1984. Database systems, distributed systems, rule processing, mobile computing.

Clement T. Yu, Professor; Ph.D., Cornell, 1973. Database management, information retrieval and knowledge-base management, multimedia retrieval.

Electromagnetics and Power Electronics

Wolfgang-M. Boerner, Professor; Ph.D., Pennsylvania, 1967. Electromagnetics, inverse scattering, modern optics, geoelectromagnetism, electromagnetic imaging, remote sensing, wideband radar, optical polarimetry.

Sharad R. Laxpati, Associate Professor; Ph.D., Illinois at Urbana-Champaign, 1965. Antennas, electromagnetic theory, computational electromagnetic scattering, microwaves, wave propagation and communication.

James C. Lin, Professor; Ph.D., Washington (Seattle), 1971. Electromagnetics in biology and medicine, biomedical instrumentation, telemedicine, wireless communications.

Korada R. Umashankar, Professor; Ph.D., Mississippi, 1974. Electromagnetic field theory and applications, microwave and millimeter waves, analytical and numerical techniques for electromagnetic scattering and interaction by complex materials, time-domain inverse-scattering methods.

Piergiorgio L. E. Uslenghi, Professor; Ph.D., Michigan, 1967. Electromagnetics, scattering theory, modern optics, solid-state, applied mathematics.

Hung-Yu Yang, Associate Professor; Ph.D., UCLA, 1988. Applied electromagnetics in ferromagnetic integrated circuits and antennas, novel microstrip antennas and arrays, planar integrated transformers for power electronics, advanced computational techniques.

Graphics and Interactive Systems

Richard Beigel, Associate Professor; Ph.D., Stanford, 1988. Human-computer interfaces, distributed data mining, algorithm design, complexity theory, fault diagnosis, molecular computing.

Thomas A. DeFanti, Professor; Ph.D., Ohio State, 1973. Computer graphics and video animation, virtual reality, electronic art, educational technology, operating systems, scientific and volume visualization, networks, parallel computers, supercomputers.

Andrew Johnson, Assistant Professor; Ph.D., Wayne State, 1995. Collaborative virtual reality, human-computer interaction, educational applications.

Robert V. Kenyon, Associate Professor; Ph.D., Berkeley, 1978. Human visual and motor systems, human spatial orientation, computer graphics and display technology, virtual environments, flight simulation, spaceflight and human adaptation.

Thomas G. Moher, Associate Professor; Ph.D., Minnesota, 1983. Human-computer interaction, programming environments, cognitive models.

Signal and Image Processing

Gyan C. Agarwal, Professor; Ph.D., Purdue, 1965. Automatic control, systems analysis, bioengineering, neuroscience, neural networks.

Rashid Ansari, Associate Professor; Ph.D., Princeton, 1981. Image/video processing, coding, transmission, layered video coding schemes for ATM networks, packet video transmissions, theoretical work on general framework for multirate processing of two-dimensional signals.

Jezekiel Ben-Arie, Associate Professor; Ph.D., Technion (Israel), 1986. Object and target recognition, image understanding and processing, shape and signal representation, wavelets and nonorthogonal expansions, neural networks, biomedical engineering.

Daniel Graupe, Professor; Ph.D., Liverpool, 1963. Control systems, time-series analysis, signal processing, neural networks, wavelets and electrical stimulation.

Bin He, Assistant Professor; Ph.D., Tokyo Institute of Technology, 1988. Imaging systems, signal and image processing, bioelectric phenomena.

Arye Nehorai, Professor; Ph.D., Stanford, 1983. Signal and image processing, biomedicine and communications.

William D. O'Neill, Professor; Ph.D., Notre Dame, 1965. Signal and image processing, bioengineering, and information theory.

Roland Priemer, Associate Professor; Ph.D., IIT, 1969. Optimal and adaptive digital signal processing, digital filters, control systems, microprocessor-based system design.

Dan Schonfeld, Associate Professor; Ph.D., Johns Hopkins, 1990. Signal and image processing, pattern recognition, and computer vision.

Oliver Yu, Assistant Professor; Ph.D., British Columbia, 1997. Wireless packet data network, mobile ATM and IP, telecommunications network architectures and protocols, internetworking among fixed and mobile networks.

Software Engineering

Ugo A. Buy, Associate Professor; Ph.D., Massachusetts, 1990. Software engineering, concurrency and real-time analysis.

Carl K. Chang, Associate Professor; Ph.D., Northwestern, 1982. Software engineering: specification, verification, and validation; testing: real-time systems, distributed computer systems, telecommunication software, and object-oriented methods.

Tadao Murata, Professor; Ph.D., Illinois at Urbana-Champaign, 1966. Petri net modeling and analysis of concurrent computer systems.

Sol M. Shatz, Associate Professor; Ph.D., Northwestern, 1983. Distributed computing systems, software engineering, operating systems.

Jeffrey J. P. Tsai, Professor; Ph.D., Northwestern, 1986. Knowledge-based software systems, artificial intelligence, expert systems, requirements specification language, distributed real-time software testing, debugging, and software metrics.

Solid State and Microfabrication

Davorin Babic, Assistant Professor; Ph.D., Pennsylvania, 1988. Physical principles of nanoscale Si particles, their modeling and application to optical modulator, physics and chemistry of silicon interfaces and surfaces.

Alan D. Feinerman, Associate Professor; Ph.D., Northwestern, 1987. Miniaturization of the scanning electron microscope, linear accelerator/undulator and other analytical instruments, 3-D fabrication techniques.

Gary D. Friedman, Associate Professor; Ph.D., Maryland, 1989. Electromagnetics, magnetic and dielectric materials, microfabrication.

Peter J. Hesketh, Associate Professor; Ph.D., Pennsylvania, 1987. Solid-state sensors and actuators, biosensors and microfabrication.

David L. Naylor, Associate Professor; Ph.D., USC, 1988. Development and application of microfabricated photonic materials and devices, diffractive optics: design, fabrication, and systems applications.

Krishna Shenai, Associate Professor; Ph.D., Stanford, 1986. Solid-state and microelectronics, power electronics, semiconductor manufacturing.

Theory and Algorithms

Ajay Kshemkalyani, Assistant Professor; Ph.D., Ohio State, 1991. Computer networks, distributed algorithms, theory of distributed computing, distributed systems, operating systems.

John Lillis, Assistant Professor; Ph.D., California, San Diego, 1996. Computer-aided design for VLSI circuits and combinatorial optimization.

A. Prasad Sistla, Associate Professor; Ph.D., Harvard, 1989. Distributed systems, semantics and verification of concurrent systems and database management systems.

Robert H. Sloan, Associate Professor; Ph.D., MIT, 1989. Design and analysis of algorithms, computational learning theory, machine learning, cryptography, software engineering, analysis of real-time programs, automatic program verification.

UNIVERSITY OF ILLINOIS
AT URBANA-CHAMPAIGN

Department of Computer Science

Programs of Study

The Department of Computer Science offers graduate programs leading to the degrees of Ph.D., M.S., and M.C.S. (a professional option). The Ph.D. requires 24 units of course work and research credit, which may include an M.S., and a thesis describing original research. The M.S. requires 8 units of course work, including a 1-unit thesis (1 unit equals 4 semester hours). The M.C.S., a terminal degree, requires 9 units of course work. There are also programs that enable students to combine an M.C.S. degree with an M.Arch. or M.B.A. degree and a Medical Scholars program to combine the Ph.D. and M.D. degrees. The M.C.S. can also be earned off campus through the Internet Master of Computer Science (IMCS) program. M.S. and Ph.D. programs include a specialization in the interdisciplinary field of computer science and engineering (CSE). Ph.D. students who enter with a bachelor's degree are encouraged, though not required, to earn an M.S. on their way to the doctorate. Within their first three semesters they must demonstrate core knowledge in specified areas of computer science by taking specified courses. The qualifying examination, taken in the fourth semester, focuses on the student's individual research interests. The other examinations, which are required of all Ph.D. candidates at the University, are the preliminary examination of the thesis proposal and initial research results and the final examination, in which the student defends the thesis at the conclusion of his or her research. The time to complete a degree depends on many factors, but typical ranges are one to three years for the M.S. and the M.C.S. and four to seven years for the Ph.D., including an M.S.

Research Facilities

The department's computing facilities are organized around a backbone network, with a number of subnets serving the various research groups and instructional laboratories. The central facilities consist of multiple servers (mail/ftp, software distribution, computation, and NSF) that provide services to more than 400 workstations, color and monochrome printing, copying, backup/archive, and Internet access. Workstation clients and terminals are all distributed. Some research groups operate their own computing facilities. Graduate students have ready access to UNIX workstations either in their offices or in nearby laboratories. The department has installed ATM networking. Large instructional laboratories offer high-end graphics and multimedia support. In addition to these departmental facilities, the services of the Computing and Communications Services Office (CCSO) and the National Center for Supercomputing Applications (NCSA) are available to faculty members and students. CCSO provides campus-wide research and instructional computing and network services. It operates UIUCnet, which is centered on a fiber-optic backbone, and provides various computation, server, personal computer, and UNIX services. NCSA operates a variety of supercomputers and provides service to the national research community, the local campus community, and industrial partners. It also has an active program in the visualization of scientific data. The Grainger Engineering Library and Information Center is located conveniently across the street from the department. It is the center of one of the national Digital Library Initiatives and is currently working with a variety of publishers to put their material totally in electronic form and accessible via the network. The Grainger library is part of the University library (one of the largest in the country) and, together with libraries at twenty-eight other academic institutions in the state of Illinois, is accessible from any terminal via a computerized catalog.

Financial Aid

Fellowships, research assistantships, and teaching assistantships are available competitively. In recent years, most graduate students in the department have received some form of financial aid from the department or from other academic units of the campus. Assistantships include a nine-month salary, which ranged from $1320 to $1532 per month in 1998-99, and a tuition and service fee waiver. Some assistantships may be continued into the summer for two additional months. The best applicants are offered three-year Illiac Scholarships, which provide a $3000 annual supplement over a regular assistantship. Computer-related employment is also available elsewhere on campus and in the community at large. In addition, the University runs a Financial Aid Office with attendant counseling services.

Cost of Study

Tuition for the 1999–2000 academic year is expected to be $4460 for Illinois residents and $10,760 for nonresidents. In addition, there is a service fee of up to $590 per semester, including health insurance, that is not covered by the tuition and service fee waiver. The figures for the optional summer session vary according to the amount of credit registered.

Living and Housing Costs

The University's graduate student housing ranges from approximately $2300 to $2900 for the academic year. Optional board contracts are available from about $2800 to $3500. There are also many privately owned apartments and houses of all sizes for rent, both close to and away from campus, with monthly rents starting at about $400.

Student Group

The department has about 360 graduate students. About 40 percent are Ph.D. candidates, with the remainder divided between the M.S. and M.C.S. programs. Approximately half of the graduate students are international.

Location

The Urbana-Champaign campus of the University of Illinois is 130 miles south of Chicago. It is served by several major and minor airlines at Willard Airport, 8 miles south of campus; by Amtrak; and by Interstate Highways 57, 72, and 74. The area surrounding the twin cities is predominantly agricultural. Chicago is about 2½ hours away by train, bus, or car.

The University and The Department

The University has an international reputation as a research institution. Research facilities abound in most departments, and there is ample scope for interdisciplinary studies. It is also a cultural center, offering many of the amenities of a large city, such as the Krannert Center for Performing Arts with its year-round program of concerts, theater, and dance, given by both local and visiting artists. The Department of Computer Science is one of the oldest such programs in the world. As the Digital Computer Laboratory, it was the home for the design and construction of the first two ILLIACs. It evolved into a department in 1964 and went on to create the ILLIAC III and IV computers. Since then, it has blossomed into one of the largest departments within the College of Engineering, with about 40 full-time, regular faculty members; three graduate and two undergraduate programs that involve about 1,400 students; and a broad range of research activities that range from theory to practice and from microprocessors to supercomputers. The department has built a major addition to integrate these facilities. This new building is part of a College of Engineering Expansion program, which also includes the new Beckman Institute for Advanced Science and Technology, the Microelectronics Center, and the Center for Reliable and High-Performance Computing.

Applying

Applicants should visit the department's Web site (listed below) for application forms and information. They may also request some of these materials by e-mail, mail, or telephone. Those interested in the CSE option should visit the Web site (http://www.cse.uiuc.edu). For fall and spring admission, the deadlines for receipt of applications are January 15 and September 1, respectively. Applications must include official transcripts of previous degree work, GRE General Test scores, and three letters of reference. Applicants to the Ph.D. program should include the score from a GRE Advanced Test, whether computer science (preferred), mathematics, or engineering. Admission is competitive: the most recent averages for entering graduate students have been about 2100 for the GRE General Test, about 800 for the computer science Subject Test, and about 3.6 out of 4.0 grade point average. International applicants whose native language is not English and who have attended a university in an English-speaking country for less than two years must submit a Test of English as a Foreign Language (TOEFL) score of at least 570. State law also requires that nonnative English speakers who wish to be considered for a teaching assistantship submit a Test of Spoken English (TSE) score of 50 or more.

Correspondence and Information

Director of Graduate Programs
Department of Computer Science
University of Illinois
1304 West Springfield Avenue
Urbana, Illinois 61801

Telephone: 217-333-4428
Fax: 217-244-6073
E-mail: academic@cs.uiuc.edu
World Wide Web: http://www.cs.uiuc.edu

University of Illinois at Urbana-Champaign

THE FACULTY AND THEIR RESEARCH

Sarita Adve, Associate Professor. Computer Architecture.

Vikram Adve, Assistant Professor. High-performance computing.

Gul Agha, Associate Professor. Open distributed systems: mobile computing, agents, software architectures, concurrent programming languages, semantics.

Geneva G. Belford, Professor. Distributed processing, distributed database management systems, emphasis on reliability and performance issues.

Roy H. Campbell, Professor. Software engineering, operating systems, security, streamed multimedia.

Gerald F. DeJong, Professor. Explanation-based machine learning, intelligent control, nonlinear dynamical systems, learning and adaptive human interfaces.

Eric DeStrurler, Assistant Professor. Computational science.

Jeff Erickson, Assistant Professor. Algorithms and data structures, computational geometry.

Michael Faiman, Professor Emeritus and Director of Graduate Programs. Computer architecture, computer networks.

H. George Friedman Jr., Associate Professor Emeritus and Director of Undergraduate Programs. Operating systems, computer-aided instruction.

Michael Garland, Assistant Professor. Computer graphics.

Mehdi T. Harandi, Associate Professor. Artificial intelligence, software engineering.

Michael T. Heath, Professor. Numerical linear algebra and optimization, parallel computing.

H. Z. Jagadish, Professor. Database systems.

Ralph E. Johnson, Adjunct Assistant Professor. Object-oriented programming languages, object-oriented design of software systems, distributed systems and user interfaces.

Laxmikant V. Kale, Associate Professor. Parallel programming tools and techniques, parallel applications in science and engineering, parallel symbolic computations.

Samuel N. Kamin, Associate Professor. Program development and verification, programming language semantics, functional programming.

Thomas Kerkhoven, Professor. Numerical simulation of microelectronic devices, numerical analysis, scientific computing.

David J. Kriegman, Associate Professor. Computer vision; robot motion, navigation, and assembly planning; human object recognition.

William J. Kubitz, Professor and Associate Head. Object-oriented graphics and multimedia.

Duncan H. Lawrie, Professor Emeritus. Computer organization and software.

C. L. Liu, Professor Emeritus. Computer-aided design of VLSI circuits, analysis of algorithms.

Jane W. -S. Liu, Professor. Real-time systems, communication networks, distributed systems.

M. Dennis Mickunas, Associate Professor. Operating systems, compiler construction.

Saburo Muroga, Professor. Design automation of VLSI: automation of logic design, switching theory, VLSI hardware.

Klara Nahrstedt, Assistant Professor. Multimedia systems: high-speed networks, operating systems, quality of service, system software, resource management, multimedia security.

David A. Padua, Professor. Compilers, problem-solving environments, parallel computing, performance models.

Leonard Pitt, Associate Professor. Theoretical computer science, computational learning theory, artificial intelligence, machine learning.

Jean A. Ponce, Professor. Computer vision, robotics, geometric modeling, computer graphics.

Sylvian R. Ray, Professor. Artificial intelligence: applications to biomedical signal interpretation.

Uday S. Reddy, Associate Professor. Programming language semantics, logics of programs, functional programming, logic programming, object-oriented programming, automated deduction.

Daniel A. Reed, Professor and Head of the Department. Parallel computation, computer systems modeling.

Edward M. Reingold, Professor. Analysis of algorithms, data structures.

Larry A. Rendell, Associate Professor. Artificial intelligence, inductive and genetic systems, machine learning.

Dan Roth, Assistant Professor. Artificial intelligence: computational learning theory, machine learning; learning in natural language; knowledge representation; learning to reason.

Paul E. Saylor, Professor. Numerical linear algebra, scientific computing.

Lui Sha, Professor. Embedded systems, real-time systems.

Robert D. Skeel, Professor. Numerical analysis and scientific computing: computational methods for biomolecular modeling.

Shang-Hua Teng, Associate Professor. Parallel computing, numerical and scientific computing.

Josep Torrellas, Associate Professor. Scalable multiprocessors: architecture design and implementation, operating systems support, compilation, memory hierarchy design, programming.

Pravin Vaidya, Associate Professor. Analysis of algorithms, computational geometry.

Marianne S. Winslett, Associate Professor. Databases, parallel I/O, security.

UNIVERSITY OF KANSAS

Department of Electrical Engineering and Computer Science
Programs in Electrical Engineering and Computer Science

Programs of Study	The Department of Electrical Engineering and Computer Science at the University of Kansas offers graduate programs of study and research leading to the degrees of M.S., Ph.D., and D.E. in electrical engineering and the M.S. and Ph.D. in computer science.
	Specializations include algorithms, artificial intelligence, computer-aided design, computer systems, digital signal processing, distributed and parallel computing, electromagnetics, expert systems, graphics, high-speed networking, image processing, information retrieval, information systems, language processing, lightwave systems, performance modeling and analysis, programming languages, radar remote sensing, radar systems, semiconductor processing, telecommunications, theory of computing, and wireless communications.
	The M.S. programs require 30 semester hours of graduate work with thesis and nonthesis options. The Ph.D. programs require 24 hours of course work beyond the M.S. and 18 hours of dissertation. The D.E. program requires 30 hours of course work beyond the M.S. and 12 to 18 hours of industrial internship.
Research Facilities	Almost all of the graduate research is conducted using facilities of four major laboratories: the Telecommunications and Information Sciences Laboratory (TISL), the Radar Systems and Remote Sensing Laboratory (RSL), the Center for Excellence in Computer-Aided Systems Engineering (CECASE), and the Design Technologies Laboratory (DesignLab).
	All research is supported by an extensive network of workstations. Special facilities include a multigigabit experimental telecommunications network called "MAGIC," radar systems ranging from VHF through 140 GHz and lightwaves, wide-screen wall-panel graphics, and a new scalable multiprocessor computer for research in computer-aided design and graphics computing.
Financial Aid	Fellowships, research assistantships, teaching assistantships, graduate assistantships, and scholarships are available to the best qualified applicants. Teaching assistants receive a tuition waiver, and research assistants receive resident tuition. The department offers a competitive package of financial support. More than 135 of the graduate students receive some support.
Cost of Study	During the 1998–99 academic year, tuition and fees were approximately $94 per credit hour for residents and $309 per credit hour for nonresidents. An equipment fee of $15 per credit hour is assessed on all School of Engineering students and is used solely for student-accessible laboratory equipment.
Living and Housing Costs	Housing is available both on and off campus. Lawrence is a relatively inexpensive place to live compared to the national average.
Student Group	There are approximately 160 graduate students in the department, including 40 in doctoral programs. The student body is composed of both traditional and nontraditional students, who may also be engaged in industry or the military. The student population is diverse, coming from all parts of the world and all parts of the United States.
Location	Kansas is located in the heart of the United States. The city of Lawrence is in the eastern part of the state, and has a cosmopolitan population of 78,000 people.
The University and The Department	The University of Kansas is a major educational and research institution with nearly 29,000 students and 2,100 faculty members. The university includes the main campus in Lawrence and several satellite campuses throughout the state.
	The Department of Electrical Engineering began in 1887. It has continued to grow and develop along with the state-of-the-art, incorporating programs in computer science, begun in 1967, and computer engineering, begun in 1987. The department has cutting-edge facilities and 31 faculty members who conduct research with funding exceeding $3 million per year.
Applying	To quality for admission, the applicant must hold a four-year baccalaureate degree or its equivalent from an accredited institution and meet other criteria demonstrating potential for success. Application forms, transcripts, GRE General Test scores, three letters of recommendation, a statement of objectives, and an application fee of $30 are required. In addition, for persons whose native language is not English, a minimum TOEFL score of 600 is required. Deadlines and other details are explained in the application materials.
Correspondence and Information	Department of Electrical Engineering and Computer Science Graduate Admissions 415 Snow Hall University of Kansas Lawrence, Kansas 66045 Telephone: 913-864-4487 Fax: 913-864-3226 E-mail: grad_admissions@eecs.ukans.edu

University of Kansas

THE FACULTY AND THEIR RESEARCH

W. Perry Alexander, Associate Professor; Ph.D., Kansas, 1993. Formal methods, software engineering, formal synthesis, architectures, software reuse.

Christopher T. Allen, Associate Professor; Ph.D., Kansas, 1984. Radar systems, high-speed digital circuits.

Allen L. Ambler, Professor; Ph.D., Wisconsin, 1973. Programming paradigms and languages; program design principles, approaches, and tools; visual languages; end-user programming systems; scientific visualization; functionally distributed programming.

Frank M. Brown, Associate Professor; Ph.D., Edinburgh, 1978. Automatic deductions, artificial intelligence.

Swapan Chakrabarti, Associate Professor; Ph.D., Nebraska, 1986. Neural networks and fuzzy systems, pattern classification, signal processing.

Don G. Daugherty, Professor; Ph.D., Wisconsin, 1964. Electronic circuits for communications and control.

Raymond H. Dean, Professor Emeritus; Ph.D., Princeton, 1966. Digital control, systems modeling, simulation, alternative energy sources.

Kenneth R. Demarest, Professor; Ph.D., Ohio State, 1980. Electromagnetic theory and computational techniques, antennae, electromagnetic interference and compatibility.

Harvey H. Doemland, Associate Professor Emeritus; Ph.D., Illinois, 1963. Electronic circuits, pulse circuits, biomedical engineering.

Joseph B. Evans, Professor; Ph.D., Princeton, 1988. Networking, communications, signal processing, VLSI design.

Victor S. Frost, Dan F. Servey Distinguished Professor; Ph.D., Kansas, 1982. Telecommunications, communications network control, modeling and analysis.

John M. Gauch, Associate Professor and Acting Chair; Ph.D., North Carolina, 1989. Image processing, computer vision, computer graphics.

Susan E. Gauch, Associate Professor; Ph.D., North Carolina, 1990. Information retrieval, corpus linguistics, multimedia databases.

Siva Prasad Gogineni, Dean E. Ackers Distinguished Professor; Ph.D., Kansas, 1984. Radar systems, microwave engineering.

Jerzy W. Grzymala-Busse, Professor; Ph.D., Technical University of Poznan (Poland), 1969. Expert systems, reasoning under uncertainty, knowledge acquisition, machine learning, rough set theory.

Jeremiah W. James, Assistant Professor; Ph.D., California, Santa Barbara, 1999. Distributed systems, concurrent objects, data consistency, fault tolerance.

Nancy G. Kinnersley, Associate Professor; Ph.D., Washington State, 1989. Design and analysis of algorithms, graph algorithms, discrete mathematics, tree automata, computational complexity.

Man C. Kong, Associate Professor; Ph.D., Nebraska, 1986. Design and analysis of algorithms, graph and network algorithms, parallel algorithms, combinatorial optimization, computational complexity, operations research.

Stephen P. Lohmeier, Assistant Professor; Ph.D., Massachusetts, 1996. Remote sensing, radar, microwave engineering, electromagnetics, signal processing, communications.

James R. Miller, Associate Professor; Ph.D., Purdue, 1979. Geometric modeling and computer-aided design, computer graphics, scientific visualization, object-oriented programming.

Gary J. Minden, Professor; Ph.D., Kansas, 1982. Digital systems, microprocessors, artificial intelligence.

Richard K. Moore, Distinguished Professor Emeritus; Ph.D., Cornell, 1951. Radar remote sensing, radio wave propagation, communications systems, antennas, radar imaging and backscatter systems.

Douglas Niehaus, Associate Professor; Ph.D., Massachusetts, 1994. Operating systems, real-time and distributed systems, programming environments.

David W. Petr, Associate Professor; Ph.D., Kansas, 1990. Communications, networking, digital signal processing.

Glenn E. Prescott, Associate Professor; Ph.D., Georgia Tech, 1984. Communications theory, digital signal processing.

James A. Roberts Jr., Professor and Vice-Chancellor; Ph.D., Santa Clara, 1979. Telecommunications, information theory, wireless communications.

James R. Rowland, Professor; Ph.D., Purdue, 1966. Stochastic systems modeling and analysis, control systems, radar detection and estimation.

Dale I. Rummer, Professor Emeritus; Ph.D., Kansas, 1963. Design of digital systems, design of microprocessor-based systems, computer-aided design tools.

Earl J. Schweppe, Professor; Ph.D., Illinois, 1955. Computer architecture, computer organization, computer history, personal computers, computer communications, interactive systems, concurrent processes, computer science curriculum.

K. Sam Shanmugan, Southwestern Bell Distinguished Professor; Ph.D., Oklahoma State, 1970. Telecommunications, pattern recognition, general systems theory, statistical communications theory, image processing, digital signal processing techniques.

William P. Smith, Professor Emeritus; Ph.D., Texas, 1950. Electrical power systems and alternate energy sources.

James M. Stiles, Assistant Professor; Ph.D., Michigan, 1995. Radar remote-sensing, propagation and scattering in random media, electromagnetic theory.

Harry E. Talley, Professor Emeritus; Ph.D., Kansas, 1954. Semiconductor devices, solid-state physics and electronics.

Costas Tsatsoulis, Professor; Ph.D., Purdue, 1987. Artificial intelligence, expert systems.

Hillel Unz, Professor Emeritus; Ph.D., Berkeley, 1957. Electromagnetic theory, antenna arrays, plasma propagation, acoustic waves, applied mathematics.

Victor L. Wallace, Professor; Ph.D., Michigan, 1969. Operating systems, computer graphics, user-machine interaction, computer networks, distributed systems, queuing theory, numerical analysis, performance evaluation.

UNIVERSITY OF KENTUCKY

Department of Computer Science

Programs of Study

The Department of Computer Science offers the Master of Science and Doctor of Philosophy degrees. Two years of study are usually required for the M.S. degree and four or more years for the Ph.D.

Two master's options are available. Students may either complete 24 hours of course work and write a thesis, which often involves a major implementation of existing algorithms but can also contain original research, or they may complete 30 hours of course work and complete a project. The project often involves software development. Three credit hours of course work may be devoted to the project. Both options require a final thesis/project defense directed by a committee chaired by the thesis or project director.

The Ph.D. program includes a qualifying examination with both oral and written parts, typically completed by the end of the third year of study. The written part includes two general examinations in approved areas and one that covers the major area in detail. The final step toward the degree is preparing and defending a dissertation that displays independent research.

Research Facilities

The Department of Computer Science has an excellent research computing facility composed mostly of SPARCstations running UNIX (Solaris 2.5.1). These machines include, in decreasing order of performance, four DEC Alpha/533s that run Linux and NT, an SGI Power Challenge R10000/200 with eight CPUs, two SGI O2s, an UltraSPARC 1/167 with 256MB, two SGI O2/180s, thirty HyperSPARC 20/125s with 64MB, five SPARC 5/170s with 64MB, two SGI Indigos, four SGI Indys, five SPARCstation 5/70s with 32MB, and eighteen PC clones. The department is purchasing a cluster of thirty-two AMD K6/233 machines. Once a year, the department upgrades to the most recent stable versions of the operating systems on the research machines, except when older versions are needed for research. The University provides high-end computing on the Exemplar computer; the department's research generally does not use this machine.

The computer science department owns and operates the MultiLab instructional facility, which is composed of twenty-nine PC clones, including twenty-two Pentium 100 machines with 32MB and seven Pentium 200 machines with 1.6GB. One acts as a file server; the others can be rebooted by students to run Windows NT, OS/2 Warp, or Linux. A recent grant has purchased about twenty laptop PCs with radio ether cards for use in classroom instruction. Students often use the SunLab, which has about thirty SPARCstations (built with a grant to the department and matched and run by UKCC). Graduate students are advised to get a permanent account in the Engineering Computer Center (ECC), which has several Sun and HP machines running UNIX.

The department has exceptional networking facilities. Fast Ethernet (100 Mbps) serves as the primary network connection for most machines. In addition, an ATM network consisting of seven FORE and CISCO ATM switches connects the machines in three labs. A CISCO 7000 router connects the research labs and the department and also ties directly into the campus backbone FDDI network. The research networks are also connected to a high-speed ATM network called the SEPSCoR network, which connects the flagship universities of six southeastern states. A recent NSF award will also connect the department's research networks to the NSF vBNS background via a LAN connection.

Financial Aid

Teaching assistantships and research assistantships are available through the department. In 1998–99, these had stipends of about $11,450 to $11,700 on a ten-month basis for 20 hours of work per week. Some opportunities for summer support are also available. Graduate assistants are nominated for partial- or full-tuition scholarships offered by the Graduate School. Teaching assistants usually teach introductory courses, and research assistants lend support to research projects. Half assistantships, which provide half the stipend, are also granted. Competitive fellowships are available through the department, the Graduate School, and the University. Fellowships usually include payment of tuition. Part-time employment opportunities are abundant on campus.

Cost of Study

Full-time tuition, including all fees, is $3276 for in-state students and $9156 for out-of-state students for the 1999–2000 academic year (two semesters).

Living and Housing Costs

There is extensive on-campus graduate housing, both for single students and for families. On-campus graduate student apartment housing ranges from $324 per month for a single/efficiency to $471 per month for two bedrooms. Numerous apartments are also available within walking distance of campus.

Student Group

The department has 431 undergraduates, 46 master's degree students, and 27 Ph.D. students. About 15 percent of the graduate students are women; approximately 50 percent are American. The largest international groups are Indian (20 percent) and Chinese (25 percent). Twenty-six students have teaching assistantships, 24 have research assistantships, and 4 have competitive fellowships.

Location

The University is in the heart of the beautiful Bluegrass region. The Appalachian mountains are several hours to the east, and many scenic parks are within an easy drive of the city. Lexington, with a population of about 230,000, is a thriving community whose major industries are the University and Lexmark. No heavy industry exists in the area, and the city is becoming more cosmopolitan yearly. There is fairly good theater, a ballet company, and an orchestra. Two tracks (harness and flat) and the Kentucky Horse Park (steeplechase and polo) provide equine attractions, but the major spectator sport is the University's nationally ranked basketball team. Both Louisville and Cincinnati are within a 2 hours' drive.

The University

The University of Kentucky, with a total enrollment of more than 23,000 students (of whom 20 percent are graduate students), is the flagship educational institution in Kentucky. It places a high priority on educational training at the postdoctoral, graduate, and undergraduate levels. The Carnegie Foundation classified it as a Research I University in 1987, placing it among the top forty-five public research institutions in the nation.

Applying

Application are processed throughout the year, and students may enter at the beginning of any semester. Applications for fall semester are due February 1 for international students and for those wishing to be considered for fellowships. Applications from students applying for assistantships are due by March 1. All application materials (two official transcripts, GRE scores, and various forms) must be on file by those dates. International applicants whose native language is not English must satisfactorily complete the TOEFL. Additional information can be received by sending e-mail to csgradpro@cs.engr.uky.edu with the subject line INFORMATION REQUEST.

Correspondence and Information

For application forms:
The Graduate School
351 Patterson Office Tower
University of Kentucky
Lexington, Kentucky 40506-0027
Telephone: 606-257-4613

For Computer Science information:
DGS, Department of Computer Science
773 Anderson Hall
University of Kentucky
Lexington, Kentucky 40506-0046
Telephone: 606-257-4997
E-mail: csgradpro@cs.engr.uky.edu (subject: Info Request)
World Wide Web: http://www.cs.engr.uky.edu

University of Kentucky

THE FACULTY AND THEIR RESEARCH

Anthony Q. Baxter, Associate Professor and Associate Chair; Ph.D., Virginia, 1973. Programming and systems, performance monitoring and evaluation, database systems. (e-mail: tony@cs.engr.uky.edu)

Kenneth L. Calvert, Associate Professor; Ph.D., Texas at Austin, 1991. Computer network protocols, network security. (e-mail: calvert@cs.engr.uky.edu)

Fuhua Cheng, Professor; Ph.D., Ohio State, 1982. Computer-aided geometric design, computer graphics, numerical analysis. (e-mail: cheng@cs.engr.uky.edu)

Raphael A. Finkel, Professor; Ph.D., Stanford, 1976. Operating systems, distributed algorithms, programming languages. (e-mail: raphael@cs.engr.uky.edu)

Judy Goldsmith, Associate Professor; Ph.D., Wisconsin, 1988. Structural complexity. (e-mail: goldsmit@cs.engr.uky.edu)

James Griffioen, Associate Professor; Ph.D., Purdue, 1991. Computer networks, operating systems. (e-mail: griff@cs.engr.uky.edu)

J. Robert Heath, Associate Professor; Ph.D., Auburn, 1973. Computer engineering, digital signal processing, software engineering. (Joint appointment with Electrical Engineering) (e-mail: ele185@ukcc.uky.edu)

Jerzy W. Jaromczyk, Associate Professor; Ph.D., Warsaw, 1984. Computational geometry, algorithms and applications. (e-mail: jurek@cs.engr.uky.edu)

Christopher Jaynes, Assistant Professor; Ph.D., Massachusetts, 1998. Computer vision, image processing, artificial intelligence and perception. (e-mail: jaynes@cs.engr.uky.edu)

Andrew Klapper, Associate Professor; Ph.D., Brown, 1982. Cryptography. (e-mail: klapper@cs.engr.uky.edu)

Kenneth K. Kubota, Professor; Ph.D., Facultés des Sciences de Paris, 1969. Number theory, operating systems. (Joint appointment with Mathematics) (e-mail: ken@ms.uky.edu)

Forbes D. Lewis, Professor; Ph.D., Cornell, 1970. Computational complexity, CAD algorithms for VLSI systems. (e-mail: lewis@cs.engr.uky.edu)

D. Manivannan, Assistant Professor; Ph.D., Ohio State, 1997. Distributed computing systems, database systems, operating systems. (e-mail: manivann@cs.engr.uky.edu)

Victor W. Marek, Professor; Ph.D., 1968, D.Sc., 1972, Warsaw. Logical foundations of AI, theory of databases, logic programming. (e-mail: marek@cs.engr.uky.edu)

A.C.R. Newbery, Professor Emeritus; Ph.D., London 1962. Numerical analysis, interpolation. (e-mail: ode@cs.engr.uky.edu)

Brent Seales, Associate Professor; Ph.D., Wisconsin, 1991. Image processing, graphics. (e-mail: seales@cs.engr.uky.edu)

Robert Tannenbaum, Adjunct Professor; Ed.D., Columbia, 1968. Graphics, fractals, CAI. (Joint appointment with Computing Center) (e-mail: rst@pop.uky.edu)

Mirek Truszczynski, Professor and Chair; Ph.D., Warsaw Technical, 1980. Artificial intelligence, knowledge representation, logic programming. (e-mail: mirek@cs.engr.uky.edu)

Grzegorz W. Wasilkowski, Professor and Director of Graduate Studies; Ph.D., Warsaw, 1980. Computational complexity, numerical analysis. (e-mail: greg@cs.engr.uky.edu)

Jun Zhang, Assistant Professor; Ph.D., George Washington, 1997. Scientific and parallel computing, computational sciences, numerical algorithms. (e-mail: jzhang@cs.engr.uky.edu)

A research lab for distributed computing and ATM networking.

Anderson Hall, which houses the College of Engineering and the Department of Computer Science.

UNIVERSITY OF MARYLAND, BALTIMORE COUNTY

Department of Computer Science and Electrical Engineering

Programs of Study

The Department of Computer Science (CS) and Electrical Engineering (EE) administers graduate programs that lead to the M.S. and Ph.D. degrees and provide opportunities for studies in a broad selection of six research areas in computer science and electrical engineering. These research areas are algorithms, theory, and scientific computation; communications and signal processing; computer systems and networks; databases, information, and knowledge management; graphics, animation, and visualization; and photonics and microelectronics.

The M.S. program consists of 30 to 33 credit hours of course work with the option of either writing a scholarly paper (3 additional credit hours) and taking a comprehensive examination or writing a research thesis and taking an oral examination based on that thesis. The Ph.D. program consists of course work, comprehensive and preliminary examinations, a dissertation based on original research, and an oral examination on the dissertation.

Research Facilities

The department is housed in a new building devoted to engineering and computer science. The department maintains an extensive research computing facility that includes a large network of more than 50 UNIX workstations (SGI, Sun, and IBM), several large computer servers (SGI, Sun), and numerous other machines. The department is part of the University of Maryland Institute of Advanced Computer Science (UMIACS) and has access to its research facilities, including a Connection Machine CM-5. The University is a member of the San Diego Supercomputing Center Consortium through which it has access to a number of large supercomputers. The department has a number of well-equipped laboratories that support the research activities of the faculty. They include a CAIBE facility; the Advanced Information Technology Lab; a MOCVD lab; a nonlinear EM theory computations lab; the Computer Graphics, Animation, and Visualization Lab; the Crypto Lab; the Center For Telecommunications Research; the Communications and Signal Processing Lab; the DIODE Laser Lab; the Information Technology Lab; the Parallel Processing Lab; and the Remote Sensing, Signal, and Image Processing Lab.

The University's computing system for instructional and research use consists of UNIX-based and VAX/VMS multiuse systems. These include a new 20-processor SGI Challenge XL UNIX system augmented by two SGI Crimsons, seventy-three Indigo graphic workstations, and a new VAX 4000 model 500. The College of Engineering has also recently acquired a Cray System model YMP-EL. In addition, University of Maryland, Baltimore County (UMBC), is a component of the larger University of Maryland Instructional and Research Computer Network, which provides access to computers at other University of Maryland campuses. From UMBC, users may communicate with other researchers on a national and international basis via the Internet.

UMBC's library ranks among the nation's leading university libraries in providing automated-enhanced systems. There is an integrated on-line catalog and circulation system that also provides access to the holdings at other University of Maryland libraries, the Uncover database of journal articles, and other databases. There are 550,000 monographs and bound periodicals as well as a current standing subscription order for 4,000 journals. Library patrons can request interlibrary loans from other Maryland campuses through the on-line catalog. The library staff also places requests for books and articles nationally, through the OCLC system.

Financial Aid

Research and teaching assistantships are available for well-qualified applicants. In 1998–99, stipends ranged from $11,102 to $11,964, plus tuition, for ten months.

Cost of Study

In 1998–99, tuition was $260 per credit hour for Maryland residents and $468 per credit hour for nonresidents. Fees of $39 per credit hour are also levied.

Living and Housing Costs

UMBC housing and board in campus apartments for graduate students cost approximately $1734, plus utilities, per semester for the 1998–99 academic year. Off-campus housing is available at somewhat higher rates.

Student Group

The department has 170 graduate students (115 computer science and 55 electrical engineering), including 80 Ph.D. students. UMBC has a graduate enrollment of 1,500 students.

Location

UMBC is located on a 488-acre site adjacent to the Baltimore Beltway at Catonsville, in suburban Baltimore County, south of Baltimore city. Exit 47 off Interstate 95 leads directly to the campus. UMBC is conveniently located near both Baltimore and Washington, D.C., and is able to benefit from the immense concentration of academic, government, cultural, and recreational facilities in these urban centers.

The University and The Department

UMBC was founded in 1966 and was designed to offer the greater Baltimore area a major public research university. Currently, it has approximately 10,700 students enrolled in undergraduate and graduate programs. The department's full-time faculty of 34 has substantial industrial experience. A large number of research scientists and engineers from local industry and government laboratories serve as visiting and part-time faculty members. Several international scientists serve as visiting and research faculty members.

Applying

In addition to the completed application form and official college transcripts, applicants should have three letters of recommendation submitted on their behalf. A score on the Graduate Record Examinations General Test (school code 5835) is also required. Applicants whose native language is not English must take the TOEFL (Test of English as a Foreign Language). Application deadlines are specified by the Graduate School, but the review process begins by December 15 for admission in the fall semester and by June 15 for admission in the following spring. The application fee is $40. Potential candidates are encouraged to obtain more details via the World Wide Web and to fill out the free online preapplication.

Correspondence and Information

For information on programs or assistantships:

Director, Graduate Program
Department of Computer Science and
   Electrical Engineering
University of Maryland, Baltimore County
1000 Hilltop Circle
Baltimore, Maryland 21250
Voice mail: 410-455-3500
Fax: 410-455-3000 or 1048
E-mail: gradsecretary@csee.umbc.edu
World Wide Web: http://www.csee.umbc.edu/graduate

For application forms and general information:

Dean of the Graduate School and Vice Provost for
   Research
University of Maryland, Baltimore County
1000 Hilltop Circle
Baltimore, Maryland 21250
Voice mail: 410-455-2537
Fax: 410-455-1130

University of Maryland, Baltimore County

THE FACULTY AND THEIR RESEARCH

Computer Science Faculty

Richard Chang, Assistant Professor; Ph.D., Cornell. Computational complexity theory, structural complexity, analysis of algorithms.

David Ebert, Assistant Professor; Ph.D., Ohio State. Computer graphics and animation; realistic image generation; interactive scientific, medical, and information visualization; procedural modeling and animation; modeling natural phenomena (gases, clouds, water, and fire); advanced rendering and animation techniques; volumetric rendering; accuracy of medical visualization techniques; volumetric displays.

Tim Finin, Professor; Ph.D., Illinois at Urbana-Champaign. Artificial intelligence, knowledge representation and reasoning, knowledge and database systems, natural language processing, intelligent agents.

Anupam Joshi, Assistant Professor; Ph.D., Purdue. Networked/distributed and mobile computing, data/web mining, multimedia databases, computational intelligence and multiagent systems, scientific computing.

Konstantinos Kalpakis, Assistant Professor; Ph.D., Maryland Baltimore County. Digital libraries, electronic commerce, databases, multimedia, parallel and distributed computing, combinatorial optimization.

Samuel Lomonaco, Professor; Ph.D., Princeton. Quantum computation, algebraic coding theory, cryptography, numerical and symbolic computation, analysis of algorithms, applications of topology to physics, knot theory and 3-manifolds, algebraic and differential topology, differential geometry.

Ethan Miller, Assistant Professor; Ph.D., Berkeley. Mass storage systems, parallel file systems, information retrieval, parallel computing, computer systems.

Charles Nicholas, Associate Professor; Ph.D., Ohio State. Electronic document processing, software engineering, and intelligent information systems.

Yun Peng, Associate Professor; Ph.D., Maryland College Park. Artificial intelligence, neural networks, artificial life, and intelligent software agents.

James Plusquellic, Assistant Professor; Ph.D., Pittsburgh. VLSI device testing, optoelectronic integrated circuits.

Penny Rheingans. Assistant Professor; Ph.D., North Carolina at Chapel Hill. Visualization of data with potential uncertainty, multivariate visualization, dynamic interaction, computer graphics and animation, application of perceptual principles to computer graphics and visualization.

Alan T. Sherman, Associate Professor; Ph.D., MIT. Discrete algorithms, cryptology, VLSI layout algorithms.

Deepinder Sidhu, Professor; Ph.D., SUNY at Stony Brook. Computer networks, distributed systems, distributed and heterogeneous databases, parallel and distributed algorithms, computer and communication security, distributed artificial intelligence, high-performance computing.

Brooke Stephens, Associate Professor; Ph.D., Maryland College Park. Numerical analysis, combinatorics, resource allocation, optimization.

Yaacov Yesha, Professor; Ph.D., Weizmann (Israel). Parallel computing, computational complexity, algorithms, source coding, speech and image compression.

Yelena Yesha, Professor; Ph.D., Ohio State. Distributed systems, database systems, digital libraries, electronic commerce, performance modeling, design tools for optimizing availability in replicated database systems, efficient and highly fault tolerant mutual exclusion algorithms, and analytical performance models for distributed and parallel systems.

Electrical Engineering Faculty

Tulay Adali, Assistant Professor; Ph.D., North Carolina State. Statistical signal processing, neural computation, adaptive signal processing and their applications in channel equalization, biomedical image analysis, system identification, time series prediction, image and speech coding.

Gary M. Carter, Professor; Ph.D., MIT. Optoelectronics, diode lasers, nonlinear optics, coherent optical communications.

Chein-I Chang, Associate Professor; Ph.D., Maryland College Park. Information theory and coding, signal detection and estimation, image processing, medical imaging, remote sensing, neural networks.

Jyh-Chia Chen, Associate Professor; Ph.D., SUNY at Buffalo. Optoelectronic materials/devices, thin-film technology.

Yung-Jui Chen, Professor; Ph.D., Pennsylvania. Integrated optics and optoelectronics, optical and electronic properties of materials, ultra-short optical pulse spectroscopy.

Fow-Sen Choa, Associate Professor; Ph.D., SUNY at Buffalo. Semiconductor lasers, optoelectronic integrated circuits.

Curtis R. Menyuk, Professor; Ph.D., UCLA. Light propagation, optical fibers, nonlinear phenomena.

Joel M. Morris, Professor; Ph.D., Johns Hopkins. Communications and signal processing, signal detection and estimation, information theory, joint time-frequency/time-scale representations and analysis techniques.

John Pinkston, Professor and Chair; Ph.D., MIT. Coding theory, information security, electronic commerce, antennas.

Andrew Veronis, Professor (Visiting), Ph.D., Manchester. Computer architecture, microprocessors, digital and logic design, parallel processing, digital signal processing.

Li Yan, Associate Professor; Ph.D., Maryland College Park. Quantum electronics, ultrashort pulse formation, ultrafast nonlinear optics, general aspects of laser physics.

The Engineering/Computer Science Building was dedicated in 1992.

UNIVERSITY OF MARYLAND, BALTIMORE COUNTY

Department of Information Systems

Programs of Study

The Department of Information Systems offers programs that lead to the M.S. and Ph.D. degrees. The principal objective of the information systems graduate program is to provide professional, graduate-level training in computer-based information systems. The course work encompasses the theoretical foundations upon which information systems are constructed as well as the current range of information systems applications. Completion of the program provides career opportunities in every organization in the nation that uses computer-based information services. Information systems program instruction provides the competence needed to become systems analysts, database designers, software designers, decision support system administrators, documentation and technical manual designers, interfacers between engineering and management personnel, expert system designers, administrators of databases, and information systems scholars for higher education institutions.

The graduate program is an acknowledgment of the need for trained information specialists and scholars to continue the exploration of the synergism evolving among computers, information systems, and humans. The graduate program endeavors to meet these needs by providing trained professionals and contributing scholars. Areas of specialization include health sciences information, management information, operations research, public services information, and expert systems. The M.S. typically takes two to three years to complete, and the Ph.D. takes three to five.

Research Facilities

The department maintains extensive hardware and software facilities. These include UNIX-based workstations from Sun Microsystems and Silicon Graphics, PowerPC Macintosh computers, and numerous Pentium II class machines. The department has multiple network laser and ink-jet printers with graphics and color capabilities. The department has a laboratory for personal applications with machines capable of producing object-oriented simulations, multimedia demonstrations, and other graphic adventures. All equipment is networked via switched Ethernet, allowing access to all hardware resources from any one location and to additional computing resources, including a VAX/VMS multiuser server, various high-end Silicon Graphics servers, and other departmental servers across the Internet that allow networking services such as SMTP e-mail, Usenet news, and the World Wide Web.

The department maintains four dedicated laboratories. One lab focuses on Systems Design (SD) and Human Computer Interaction (HCI) research issues. Another lab supporting psychophysiological research focuses on stress effects of VDT-based performance and countermeasures to those effects includes blood pressure, heart rate, and EMG monitoring equipment with online graphical displays. The department also maintains the Health Informatics Lab containing a variety of hardware platforms and software systems for studying the delivery of health care through advanced computing technology. Another lab focuses on Knowledge Management. The Laboratory for Knowledge Management concentrates on developing, applying, and evaluating knowledge management systems, methodologies, tools, techniques, processes, and instruments for performing knowledge management. The department maintains the latest in graphical interfaces and working environments and is one of a few that integrate PC, Macintosh, and UNIX computers into a fully operational working environment. The department houses many applications that facilitate everyday operations such as word processing, spreadsheet, and graphical presentation software; it also has many applications specific to its undergraduate and graduate courses.

Financial Aid

The graduate program offers a limited number of research and teaching assistantships. These awards include full remission of tuition fees and an annual stipend. Assistantships are awarded for one year only, but they can be renewed from year to year. Inquiries about availability should be made to the department.

Cost of Study

Tuition for graduate study during 1999–2000 is $268 per credit hour for residents and $474 per credit hour for nonresidents. Books and other supplies average $750 to $950 per academic year.

Living and Housing Costs

Single students should anticipate off-campus living costs of between $4000 and $8000 per year. The campus is surrounded by numerous apartment and town-house communities.

Student Group

Current enrollment reflects 90 percent master's students and 10 percent doctoral students. Degree-seeking students have prior degrees in every field from arts and humanities to business, representing the broad range of computer-based information systems applications.

Student Outcomes

To date, the department has produced 20 Ph.D.'s, of whom 16 have gone into academia and the rest into high-level research. Virtually all M.S. graduates have progressed and advanced in their professional careers.

Location

The campus is located close to Baltimore and Washington, D.C., with convenient interstate highway access to both. The area abounds with opportunities for practical experience, research, and employment in business and government.

The University and The Department

The University of Maryland, Baltimore County, a public institution founded in 1968, is one of the four campuses of the University of Maryland System. It has grown to be identified with high-technology research and application programs, with an approximate day and evening enrollment of 10,000 students. The department has the largest undergraduate major on the campus, enrolling more than 1,000 students yearly. The IFSM graduate program includes faculty members whose expertise crosses all areas of computer-based information systems applications.

Applying

Students should review course prerequisites or their equivalents required for graduate admission. These prerequisites are never waived. Students who do not satisfy these prerequisites are not admitted. If they do not meet the prerequisites, applicants should apply for admission as a non–degree-seeking undergraduate special student. When prerequisites are satisfied, the student should apply for admission to the appropriate degree program. Students who wish to be considered for an assistantship should send a letter that describes any teaching, research, or other experiences directly to the department. Deadlines for assistantships are the same as admissions deadlines. International students should submit completed applications one year in advance of expected enrollment.

Correspondence and Information

Director of Graduate Programs
Department of Information Systems
University of Maryland, Baltimore County
1000 Hilltop Circle
Baltimore, Maryland 21250
Telephone: 410-455-3688
Fax: 410-455-1073
E-mail: ifsm-gradinfo@umbc.edu

Administration Building
University of Maryland, Baltimore County
1000 Hilltop Circle
Baltimore, Maryland 21250

University of Maryland, Baltimore County

THE FACULTY

Monica Adya, Assistant Professor; Ph.D. (information systems), Case Western Reserve, 1996.
Marion Ball, Affiliate Professor; Ph.D. (continuing medical education), Pennsylvania, 1978.
Douglas Bradham, Affiliate Associate Professor; Dr.P.H. (medical outcomes and policy analysis), North Carolina at Chapel Hill, 1981.
Gerald Canfield, Associate Professor; Ph.D. (medical informatics), Utah, 1990.
Henry H. Emurian, Associate Professor; Ph.D. (psychology), American, 1975.
Patricia Fletcher, Associate Professor; Ph.D. (information studies), Syracuse, 1991.
Guisseppi A. Forgionne, Professor; Ph.D. (management science), California, Riverside, 1973.
Aryya Gangopadhyay, Assistant Professor; Ph.D. (information systems), Rutgers, 1993.
Jay Liebowitz, Professor; D.Sc. (knowledge management; information technology management); George Washington, 1984.
Peng Liu, Assistant Professor; Ph.D. (information systems), George Mason, 1999.
Ross Malaga, Assistant Professor; Ph.D. (information systems), George Mason, 1998.
David Millis, Affiliate Assistant Professor; M.D., Howard, 1983.
Anthony F. Norcio, Professor; Ph.D. (psychology), Catholic University, 1978.
Jennifer J. Preece, Professor; Ph.D. (human-computer interaction), Open (London), 1985.
Bonnie Rubenstein, Assistant Professor; Ph.D. (systems engineering), Pennsylvania, 1997.
Carolyn Seaman, Assistant Professor; Ph.D. (computer science), Maryland College Park, 1996.
Andrew L. Sears, Associate Professor; Ph.D. (computer science), Maryland College Park, 1993.
James Smith, Adjunct Professor; Ph.D. (physics); Michigan, 1970.
James Sorace, Affiliate Professor; M.D. (pathology), Virginia, 1982.
Henry H. Walbesser, Professor Emeritus; Ph.D. (mathematics and education), Maryland.

UNIVERSITY OF MASSACHUSETTS AMHERST

Computer Science Department

Programs of Study	The Computer Science Department at the University of Massachusetts Amherst has a broad, flexible program that combines mastery across the breadth of computer science, with depth in those areas most relevant to a student's career objectives. The M.S. and Ph.D. programs provide foundations and traditions in computing as well as opportunities to participate in advanced research. Students receive a solid technical background and develop professional skills in preparation for lifelong contributions in industrial, research, and teaching careers. The core graduate program is a blend of course work in computer systems, theoretical computer science, and artificial intelligence, and emphasizes programs of study that combine depth and synthesis across those areas.
	Courses and research span a wide range of interests, including computer networks, distributed and real-time systems, parallel processing, computer architecture, software engineering, algorithms and theory of computation, information retrieval, machine learning, computer vision and robotics, and knowledge engineering. Graduate seminars expose students to advanced topics in the faculty's research areas and encourage close interactions between faculty members and students.
Research Facilities	The department has approximately 500 computer systems available, all of which are on a state-of-the-art, multiswitched ethernet with a FDDI backbone. The main external Internet connection is a 4.5 Mbit/second link to Nearnet via BBN Cambridge. Multiuser resources include an IBM SP-2 supercomputer, several multi-CPU DEC Alpha servers, a twelve-processor SGI Challenge, and two Sun UltraServers in addition to ninety Alpha workstations, ninety-six DECStations, forty-four Sun SparcStations and UltraSparcs, sixteen SGIs, and assorted HP workstations, TI Lisp machines, and Linux systems. For personal computing, there are more than 100 Macintosh computers and nearly 50 PCs running under various operating systems. Printing needs are accommodated by more than eighty laser printers, including two high-speed, high-density duplex print servers as well as three full-color printers.
	The Educational Laboratory (EdLab) is reserved for undergraduate and graduate course work. The EdLab has twenty DEC AlphaStations, twenty DECStation 5000 systems, and five Pentium/133 Dell computers running Windows NT 4.0, all served by two Alpha AXP fileservers and two DECStation servers.
	The Computer Science Department moved to the first wing of the Engineering and Computer Science Research Center in early 1999. The layout for the three-floor, 78,500-square-foot building was developed collaboratively to yield not only the best use of the space, but also to provide a superior educational and research environment. The second phase of the plan is the construction of an engineering wing and a classroom and auditorium to connect the two structures.
Financial Aid	Financial aid is available in the form of teaching and research assistantships as well as various University fellowships. Stipends vary with the type of work involved but are normally $12,400 for the academic year or $17,000 for the calendar year for a 20-hour-per-week assignment. Fellowships and assistantships usually include a waiver of tuition and fees. An assistant is permitted to carry up to 20 semester hours of courses. Students who need additional financial aid can apply through the University's Financial Aid Office, 243 Whitmore Administration Building, University of Massachusetts Amherst, Amherst, Massachusetts 01003.
Cost of Study	For 1999–2000, full-time tuition is expected to be $2640 for Massachusetts residents, $3960 for the New England Regional Student Program, and $9756 for out-of-state students. Additional fees of $3000 are charged to all full-time graduate students. Tuition and most fees are waived with most assistantships.
Living and Housing Costs	The minimum living cost (room, board, books, and incidentals) is estimated at $6500 per year in 1999-2000. Some University housing is available for single and married graduate students. Rooms and apartments are available in the surrounding area. The University Housing Office assists students in finding suitable housing on or off campus.
Student Group	In 1998–99, approximately 6,000 students were pursuing graduate degrees. There were 158 graduate students in the Computer Science Department representing all regions of the U.S. as well as fifteen other countries.
Location	The University of Massachusetts Amherst is situated in one of the most picturesque sections of New England. The Five Colleges, which include UMass Amherst and its neighbors—Amherst, Hampshire, Smith, and Mount Holyoke colleges—provide a rich array of educational and cultural opportunities within the beautiful Connecticut River Valley region. These other colleges are conveniently accessible via a free bus system. The area is rural in character, with good local skiing and hiking. Two hours' drive away are the educational and cultural attractions of Boston, Providence, and New Haven as well as the New England mountain and coastal areas. New York City is roughly 3 hours' drive away.
The University	One of today's leading centers of public higher education in the Northeast, the University of Massachusetts Amherst was established in 1863 under the Morrill Land Grant Act as an agricultural college. It became Massachusetts State College in 1931 and the University of Massachusetts in 1947. In 1991, the Amherst campus became the flagship of the Commonwealth's university system, and has grown to be the largest state university in New England. Nine colleges and schools are housed in more than 150 buildings on approximately 1,400 acres of land. There are about 23,900 students enrolled at the University of Massachusetts Amherst, of whom approximately 5,800 are pursuing graduate degrees.
	The department staff comprises 40 faculty members, 15 postdoctoral research associates, 50 technical professionals, and 20 administrative personnel. The department granted its first M.S. in 1967 and its first Ph.D. in 1974. A new, state-of-the-art facility is under construction to house the department in the new millennium.
Applying	Admission to the graduate program is granted for the fall term. Complete applications must be submitted by January 15, 2000 for September enrollment. An undergraduate cumulative grade point average of at least 3.0 is required for an application to be considered. The graduate school requires all applicants to take the GRE General Test, and international students must submit TOEFL and TWE scores. Students should exhibit a strong background in computer science and mathematics or related engineering disciplines, or show other evidence of ability to succeed in the graduate program. Admission and financial support are awarded on a competitive basis. The University provides equal educational opportunities without regard to race, gender, or religion.
Correspondence and Information	Graduate School Goodell Building University of Massachusetts Amherst, Massachusetts 01003 E-mail: gradinfo@dpc.umassp.edu	Computer Science Department Lederle Graduate Research Center University of Massachusetts Amherst, Massachusetts 01003-4610 Telephone: 413-545-2744 Fax: 413-545-1249 E-mail: csinfo@cs.umass.edu World Wide Web: http://www.cs.umass.edu/

University of Massachusetts Amherst

THE FACULTY AND THEIR RESEARCH

W. Richards Adrion, Professor; Ph.D., Texas at Austin, 1971. Concurrency, software development environments, authoring tools, testing and analysis.

James Allan, Research Assistant Professor; Ph.D., Cornell, 1995. Information retrieval, organization, and visualization; topic detection and tracking; textual data mining.

David A. Barrington, Associate Professor; Ph.D., MIT, 1986. Algorithms and complexity, theory of finite automata.

Andrew G. Barto, Professor; Ph.D., Michigan, 1975. Computational neuroscience, reinforcement learning, adaptive motor control, artificial neural networks, adaptive and learning control, motor development.

James Callan, Research Assistant Professor; Ph.D., Massachusetts, 1993. Intelligent information filtering and retrieval, data mining from text, electronic information literacy in K–12 education.

Lori A. Clarke, Professor; Ph.D., Colorado, 1976. Concurrency, software development environments, testing and analysis.

Paul R. Cohen, Professor; Ph.D., Stanford, 1983. Planning, simulation, natural language, agent-based systems, intelligent data analysis, intelligent user interfaces.

W. Bruce Croft, Professor; Ph.D., Cambridge, 1979. Database systems and information retrieval, intelligent user interfaces, office automation.

Andrew H. Fagg, Research Assistant Professor; Ph.D., USC, 1996. Computational neuroscience, biological and robot motor control, hybrid supervised and reinforcement learning systems, wearable computing.

Robert M. Graham, Professor Emeritus; M.A., Michigan, 1957. Operating systems.

Roderic A. Grupen, Associate Professor; Ph.D., Utah, 1988. Autonomous sensorimotor systems, embedded control, developmental dynamics, multifingered hands, walking machines, animate vision.

Allen R. Hanson, Professor; Ph.D., Cornell, 1969. Vision, mobile robotics, aerial image analysis, motion, stereo, 3-D reconstruction, sensor calibration.

Neil Immerman, Professor; Ph.D., Cornell, 1980. Descriptive complexity, database theory.

David Jensen, Research Assistant Professor; D.Sc., Washington (St. Louis), 1992. Knowledge discovery and data mining, data analysis, computing policy.

James F. Kurose, Professor and Chair; Ph.D., Columbia, 1984. Real time, multimedia communication, operating systems support.

Wendy G. Lehnert, Professor; Ph.D., Yale, 1977. Natural language processing.

Barbara S. Lerner, Research Assistant Professor; Ph.D., Carnegie Mellon, 1989. Software design, processes, object management, agent coordination in multi-agent systems.

Victor R. Lesser, Professor; Ph.D., Stanford, 1972. Multi-agent systems, cooperative distributed problem solving, intelligent user interfaces.

Raghavan Manmatha, Research Assistant Professor; Ph.D., Massachusetts Amherst, 1997. Multimedia indexing and retrieval, image and video retrieval, document image analysis, digital libraries, computer vision, information retrieval.

Kathryn S. McKinley, Assistant Professor; Ph.D., Rice, 1992. Compiler algorithms for memory hierarchies and heterogeneous architectures.

Robert N. Moll, Associate Professor; Ph.D., MIT, 1973. Knowledge-based systems, combinatorial optimization.

J. Eliot Moss, Associate Professor; Ph.D., MIT, 1981. Object-oriented languages, performance measurement and optimization, garbage collection, persistent object stores.

Leon J. Osterweil, Professor; Ph.D., Maryland, 1971. Process modeling and process programs, analysis of concurrency, software architectures.

Robin J. Popplestone, Professor; B.Sc., Manchester (England), 1960. Functional languages, robot assembly planning, intelligent user interfaces.

Krithivasan Ramamritham, Professor; Ph.D., Utah, 1981. Real-time systems, transaction processing in real-time databases, data management on the Web.

Edward M. Riseman, Professor; Ph.D., Cornell, 1969. Vision, mobile robotics, aerial image analysis, motion, stereo, 3-D reconstruction, environmental monitoring applications.

Edwina Rissland, Professor; Ph.D., MIT, 1977. Case-based reasoning, knowledge-based systems, knowledge acquisition, user interfaces.

Arnold L. Rosenberg, Distinguished University Professor; Ph.D., Harvard, 1966. Theoretical and algorithmic aspects of parallel architectures and communication networks.

Howard Schultz, Research Assistant Professor; Ph.D., Michigan, 1984. Sensor calibration.

Prashant Shenoy, Assistant Professor; Ph.D., Texas at Austin, 1998. Multimedia systems, operating systems, computer networks, distributed systems.

Ramesh K. Sitaraman, Assistant Professor; Ph.D., Princeton, 1993. Parallel and distributed systems, communication networks, performance analysis, theoretical computer science.

D. N. Spinelli, Professor; M.D., Milan (Italy), 1958. Computational neuroscience, biological vision.

David W. Stemple, Professor Emeritus; Ph.D., Massachusetts, 1977. Distributed software systems and databases.

Donald F. Towsley, Distinguished University Professor; Ph.D., Texas at Austin, 1975. Networks, performance analysis.

Paul E. Utgoff, Associate Professor; Ph.D., Rutgers, 1984. Incremental decision trees, knowledge acquisition and knowledge-based control.

Charles C. Weems, Associate Professor; Ph.D., Massachusetts, 1984. Computer architecture, parallel processing, heterogeneous and configurable processor architectures and software support.

Jack C. Wileden, Professor; Ph.D., Michigan, 1978. Programming languages, convergent computing systems, interoperability, methods for name management.

Conrad A. Wogrin, Professor Emeritus; D.Eng., Yale, 1955. Computer vision and image understanding.

Beverly P. Woolf, Research Associate Professor; Ph.D., Massachusetts, 1984. Intelligent tutors, computational strategies in learning and education.

Shlomo Zilberstein, Assistant Professor; Ph.D., Berkeley, 1993. Autonomous agents, decision theory, intelligent information gathering, real-time planning, resource-bounded reasoning.

RESEARCH ACTIVITIES

Architecture (Rosenberg, Weems).

Artificial intelligence (Barto, Cohen, Fagg, Grupen, Hanson, Jensen, Lehnert, Lesser, Moll, Popplestone, Riseman, Rissland, Utgoff, Woolf, Zilberstein).

Compilers (McKinley, Moss).

Databases (Ramamritham, Stemple, Moss, Wileden).

Distributed Systems (Kurose, McKinley, Ramamritham, Shenoy, Sitaraman, Towsley, Wileden).

Education/Tutoring (Adrion, Kurose, Woolf).

Empirical methods (Jensen, Cohen).

Graph theory and combinatorics (Moll, Sitaraman).

Information retrieval (Allan, Callan, Croft, Lehnert, Manmatha).

Machine learning (Barto, Callan, Cohen, Grupen, Jensen, Rissland, Utgoff).

Multimedia systems (Adrion, Allan, Kurose, Shenoy, Woolf).

Networks (Adrion, Kurose, Ramamritham, Shenoy, Sitaraman, Towsley).

Operating systems (Kurose, Ramamritham, Shenoy, Towsley).

Parallel computation (McKinley, Rosenberg, Sitaraman, Weems).

Performance evaluation (Kurose, Sitaraman, Towsley).

Programming languages (McKinley, Moss, Popplestone, Weems, Wileden).

Robotics and computer vision (Cohen, Fagg, Grupen, Hanson, Manmatha, Popplestone, Riseman, Spinelli, Wogrin).

Software engineering (Adrion, Clarke, Osterweil, Lerner).

Theoretical computer science (Barrington, Immerman, Rosenberg, Sitaraman).

UNIVERSITY OF MINNESOTA

Institute of Technology
Department of Computer Science and Engineering

Program of Study	The department offers a program of study leading to the M.S. and Ph.D. degrees in computer and information science. Applicants with related U.S. industrial experience can apply for a terminal M.S. course-work-only program. Students can select a program of study and research in all core areas of computer science, or they can structure a more interdisciplinary program to match their own special interests. The faculty conduct instruction and research in software engineering, parallel processing, numerical analysis, artificial intelligence, computer graphics, computer-aided design, theory of computation, computer networks, distributed systems, fault-tolerant computing, computer vision, robotics, neural networks, computer security, and computer architecture. Many faculty members actively participate in the graduate minors in scientific computation and in cognitive science. The objective of the graduate program is to prepare the student to carry out advanced research and development in the rapidly expanding areas of computer and information science. M.S. students may elect a course work or a thesis degree program. In addition to fulfilling course requirements for the M.S. degree, a candidate for the Ph.D. degree must pass a written preliminary examination. A typical Ph.D. program requires four years of graduate study, including the completion of an original Ph.D. dissertation with one of the department's faculty members. Lectures on significant current work in computer and information science are presented in a weekly colloquium series by visiting and local faculty members. In addition, seminars on special research topics are offered.
Research Facilities	The department has specialized laboratories in the areas of artificial intelligence, distributed computing, computer-aided design, robotics, vision, and database. The department's current computing resources include more than 400 workstations and personal computers from vendors such as SGI, Sun, and Apple and various Pentium-based PCs in a mixture of research, instructional, and administrative roles. Research programs have access to departmental computing resources, such as four 4-processor SGI Challenge computers connected via HIPPI and Fibre Channel, eight IBM RS6000/590 computers and an IBM SP2 computer with twenty nodes and forty-eight processors connected via ATM, and eight dual-processor Pentium II computers connected by MyriNet switch. Workstation access is also provided by the Institute of Technology public labs, which have more than 400 Sun, SGI, and Macintosh- and Pentium-based computers. Dial-in SLIP service is provided up to 36.6 KBaud. The Minnesota Supercomputer Institute supports research carried out using the supercomputers, which include a 160-processor IBM SP WinterHawk supercomputer, a 68-processor IBM SP NightHawk supercomputer, and a 128-processor SGI Origin 2000 R10000 supercomputer. Research access to supercomputers is also provided by the Army High Performance Computing Research Center, which provides both Web-based (Teraweb) and standard UNIX access to distributed large-memory HPC systems. Library facilities are excellent and have online access to the catalog.
Financial Aid	Fellowships and half-time assistantships are available. Fellowships may be awarded to outstanding graduate students in their first year of study. Teaching and research assistants receive a stipend and full or partial tuition waivers. Summer support is also available. Typically, a half-time teaching assistant is expected to teach laboratory sections for one of the introductory courses; this involves up to 20 hours per week including preparation, consulting, and grading. Research assistants are normally expected to devote an equivalent amount of time to funded research work. Because of the concentration of computer and computer-based industry in the Twin Cities area, graduate students often can find a variety of interesting part-time and summer job opportunities.
Cost of Study	Tuition is based on the number of credits, with a plateau from 6 to 14 credits. For 1999–2000, resident tuition for graduate students is $2520 per semester for 6 to 14 credits. Tuition above 14 credits is $420 per credit. Nonresident tuition is $4950 per semester for 6 to 14 credits. Tuition above 14 credits is $825 per credit. Teaching assistants receive full or partial tuition waivers and health insurance but are responsible for certain fees each quarter.
Living and Housing Costs	Housing options include University residence halls, married student housing cooperatives, and rental properties throughout the Twin Cities. Detailed housing information can be obtained from Housing Services in Comstock Hall.
Student Group	There are approximately 250 full- and part-time M.S. and Ph.D. students in the department. At least twenty Ph.D. degrees have been awarded in each of the past three years. Many of the students who have received a Ph.D. from the department now hold faculty positions in other computer science departments offering doctoral study.
Location	The Twin Cities area of Minneapolis and St. Paul justly deserves its reputation for providing a high quality of life. Recreational opportunities abound, ranging from sailing, canoeing, and camping in the summer to skiing and skating in the winter. The internationally known Tyrone Guthrie Theatre, the Walker Art Center, Orchestra Hall, Fitzgerald Theatre, and Ordway Music Center are but a few of the cultural attractions to be found. The Twin Cities area is also a center for computer research and design, and contacts are encouraged between the University and local industry.
The University and The Department	The computer science and engineering department is a part of the Institute of Technology, which also includes all of the engineering departments, mathematics, and the physical sciences. This broad spectrum of disciplines under a single administrative umbrella is an unusual arrangement and provides students with an excellent opportunity for study in a wide variety of related fields. The University is also well-known for its medical and health sciences program, the Carlson School of Management, and the College of Biological Sciences. The computer science and engineering department maintains close ties with many of these other areas. In particular, various faculty members from electrical and computer engineering, mathematics, health sciences, and the management information systems program are also part of the computer and information science graduate faculty. The structure of the University's Graduate School encourages multidisciplinary research.
Applying	To be admitted to the Graduate School, an applicant must have a bachelor's degree from a U.S. institution or a comparable degree from a recognized college or university outside the United States. Minimum preparation normally includes an undergraduate degree in computer science or in a related field with some computational experience. All applicants for the M.S. and Ph.D. must submit scores on the GRE General Test. GRE Subject Test scores are highly recommended, especially for those applicants seeking financial assistance. Application for admission should be made directly to the Graduate School. Admission applications received by December 15 for the following fall term will be considered for financial aid.
Correspondence and Information	Admission and Awards Committee Department of Computer Science and Engineering University of Minnesota 200 Union Street, SE Minneapolis, Minnesota 55455 Telephone: 612-625-4002

University of Minnesota

THE FACULTY AND THEIR RESEARCH

Professors

Daniel Boley, Ph.D., Stanford. Numerical analysis, linear algebra, control theory.

David Hung-Chuang Du, Ph.D., Washington (Seattle). Computer-aided design for VLSI, computer networking, database design, parallel and distributed architectures and processing.

Ding-Zhu Du, Ph.D., California, Santa Barbara. Complexity theory, theory of computation, combinatorial optimization.

David W. Fox, Ph.D., Maryland. Applied mathematics, eigenvalue problems.

Maria Gini, Doctor of Physics, Milan. Artificial intelligence, robotics.

Vipin Kumar, Ph.D., Maryland. Parallel processing, artificial intelligence.

Arthur Norberg, Ph.D., Wisconsin. History of science and technology.

Yousef Saad, Department Head; Doctorat, Grenoble (France). Sparse matrix computations, parallel computation, nonlinear equations, control theory, partial differential equations.

Eugene Shragowitz, Ph.D., Moscow. Combinatorial optimization, CAD of VLSI and computers, parallel and learning algorithms, learning automata, nonlinear networks.

James R. Slagle, Ph.D., MIT. Artificial intelligence.

Wei-Tek Tsai, Ph.D., Berkeley. Software engineering, parallel and distributed processing, computer security, artificial intelligence.

Pen-Chung Yew, Ph.D., Illinois at Urbana-Champaign. Computer architecture, optimizing compiler, parallel systems, performance evaluation.

Associate Professors

John Carlis, Ph.D., Minnesota. Database systems, management information systems, systems analysis and design.

Ravi Janardan, Ph.D., Purdue. Computational geometry, graph algorithms, data structures, distributed computation.

Joseph Konstan, Ph.D., Berkeley. Human-computer interaction, collaborative filtering, multimedia systems, hypermedia education and applications.

Nikolaos Papanikolopoulos, Ph.D., Carnegie Mellon. Computer vision, robotics, computer engineering, computer integrated manufacturing.

Haesun Park, Ph.D., Cornell. Numerical analysis, parallel computing, signal processing algorithms.

John Riedl, Ph.D., Purdue. Collaborative systems, database systems, fault tolerance, computer networks, object-oriented systems.

Shashi Shekhar, Ph.D., Berkeley. Neural networks, software engineering databases, geographic information systems, intelligent transportation systems.

Jaideep Srivastava, Ph.D., Berkeley. Databases, distributed and parallel processing.

Anand Tripathi, Ph.D., Texas at Austin. Architecture, operating systems, distributed systems, parallel computing.

Assistant Professors

Mats Per Erik Heimdahl, Ph.D., California, Irvine. Software engineering, formal methods, requirements specification, embedded systems, safety-critical systems.

Victoria Interrante, Ph.D., North Carolina at Chapel Hill. Visualization, visual perception, computer graphics, image processing, virtual reality, human-computer interaction.

Richard M. Voyles, Ph.D., Carnegie Mellon. Robotics, computer vision, mechatronics, MEMS, manufacturing methods, entrepreneurial engineering, computer engineering.

Zhi-Li Zhang, Ph.D., Massachusetts Amherst. Computer networks, real-time, distributed, multimedia systems.

UNIVERSITY OF MINNESOTA

Scientific Computation Program

Programs of Study

The graduate degree program in scientific computation encompasses course work and research on the fundamental principles necessary to use intensive computation to support research in the physical, biological, and social sciences and engineering. There is a special emphasis on research issues, state-of-the-art methods, and the application of these methods to outstanding problems in science, engineering, and other fields that use scientific computation, numerical analysis and algorithm development, symbolic and logical analysis, high-performance computing tools, supercomputing and heterogeneous networks, and visualization.

The Scientific Computation Program provides opportunities and options for graduate training in the problem-solving process and for integrating advances in all segments of the process into a coordinated program of course and thesis study. It offers a new interdisciplinary path of formal course and examination requirements toward a Ph.D. or M.S. degree, and this path should also be augmented by a strong participation in one of the more traditional disciplinary departments.

Research Facilities

The Scientific Computation Program has access to the outstanding computational facilities of the University of Minnesota, including the Supercomputer Institute, the Army High Performance Computing Research Center, and the University of Minnesota IBM Shared Research Project. The Minnesota Supercomputer Institute (MSI) supports research carried out using supercomputers, including a 160-processor IBM SP WinterHawk supercomputer, a 68-processor IBM SP NightHawk, and a 128-processor SGI Origin 2000 R10000 supercomputer. MSI provides many other resources; organizes and hosts symposia, workshops, and seminars; and coordinates other activities to create a favorable high-performance computing environment, increase University-industry collaboration, and promote technology transfer. In addition, under the auspices of the IBM Shared University Research (SUR) program, IBM has provided nine IBM RS/6000s; a 16-node SP supercomputer; a 4-node, 32-processor SMP (200 MHz) supercomputer; and a 4-node, 16-processor SMP (332 MHz) supercomputer. These computers are linked by high-speed networks that constitute the SUR cluster. Research access to supercomputers is also provided by the Army High Performance Computing Research Center, which provides both Web-based (Teraweb) and standard UNIX access to distributed large-memory HPC systems.

Financial Aid

Teaching assistantships, research assistantships, and fellowships are available to qualified students. Students may also apply to individual graduate faculty members for research assistantships or for suggestions of fellowship programs in individual fields of scientific computation. Students are also urged to apply for federal and other external fellowships.

Cost of Study

Tuition is based on the number of credits, with a plateau from 6 to 14 credits. For 1999–2000, resident tuition for graduate students is $2520 per semester for 6 to 14 credits. Tuition above 14 credits is $420 per credit. Nonresident tuition is $4950 per semester for 6 to 14 credits. Tuition above 14 credits for nonresidents is $825 per credit. Teaching assistants receive full or partial tuition waivers and health insurance but are responsible for certain fees each quarter.

Living and Housing Costs

Housing options include University residence halls, married student housing cooperatives, and rental properties throughout the Twin Cities. Detailed housing information can be obtained from Housing Services in Comstock Hall.

Student Group

There are currently 7 Ph.D. candidates, 1 master's candidate, and 8 minor students within the Scientific Computation Program. Of the currently enrolled students, 30 percent are women. All Ph.D. students are full-time students; 4 are international students. All students are eligible for some form of financial aid. Faculty members look for students who are self-motivated and who are interested in working in an interdisciplinary environment.

Location

The Twin Cities area, with a population of about 2 million people, has all the advantages of a major metropolitan area but few of the usual urban drawbacks. It offers two flourishing downtowns, sophisticated educational and cultural institutions, countless recreational opportunities, and citizens who are known for their friendliness. Within the city limits of both Minneapolis and St. Paul are numerous large lakes, parks, bicycle and walking trails, and parkways. The Twin Cities are home to professional baseball, basketball, and football teams, in addition to the numerous University teams. The Guthrie Theater's reputation as a top regional theater is firmly established throughout the country. The Twin Cities are also the home of two internationally respected musical ensembles: the St. Paul Chamber Orchestra, which performs in the recently completed Ordway Center, and the Minnesota Orchestra, which performs in Orchestra Hall downtown.

The University

The University of Minnesota was chartered in 1851, seven years before the Minnesota Territory became a state. Today, the University has 4,500 full-time faculty members and 58,000 students enrolled in day classes, with tens of thousands more students in evening, continuing education, and noncredit courses. As one of the largest public institutions of higher learning in the United States, the University offers a rich variety of highly respected programs leading to baccalaureate, graduate, and professional degrees.

Applying

Interested students should complete both an application for admission to the Gradate School and an application for financial aid. These forms may be obtained from the department at the address listed below and should be returned, accompanied by the necessary supporting documents, as soon as possible. Applications returned by January 2 will be eligible for more sources of financial support. Transcripts from all colleges or universities attended and three letters of recommendation are required. Scores on the TOEFL and GRE General Test are required of all international applicants.

Correspondence and Information

Director of Graduate Studies in Scientific Computation
Computer Science Department
4–192 EE/CSci Building
200 Union Street Southeast
Minneapolis, Minnesota 55455
E-mail: scic@cs.umn.edu
World Wide Web: http://www.cs.umn.edu/grad-info/SCP/scp.html

University of Minnesota

THE FACULTY AND THEIR RESEARCH

Norma Allewell, Professor of Biochemistry; Ph.D., Yale. Physical chemistry of proteins, protein structure, electrostatic effects in macromolecules, mechanisms of allosteric regulation.

Ronald Anderson, Professor of Sociology; Ph.D., Stanford. Methodology, sociology of technology and sociology of education, gender roles, organizations.

Phillip Barry, Assistant Professor of Computer Science; Ph.D., Utah. Computer-aided geometric design, computer graphics.

Daniel Boley, Ph.D., Stanford. Numerical analysis, linear algebra, control theory.

Graham Candler, Assistant Professor of Aerospace Engineering and Mechanics; Ph.D., Stanford. Hypersonic aerodynamics, computational fluid dynamics, high-temperature gas physics.

John V. Carlis, Associate Professor of Computer Science and Engineering; Ph.D., Minnesota. Database systems.

James R. Chelikowsky, Professor of Chemical Engineering and Materials Science; Ph.D., Berkeley. Structural and electronic properties of solids.

Bernardo Cockburn, Associate Professor of Mathematics; Ph.D., Chicago. Numerical methods convection, Euler equations for gas dynamics and hydrodynamic equations.

Christopher Cramer, Ph.D., Illinois. Theoretical organic chemistry.

Jeffrey J. Derby, Associate Professor of Chemical Engineering and Materials Science; Ph.D., MIT. Process modeling, materials processing, high-performance computing.

Timothy J. Ebner, Professor of Neurosurgery and Physiology; M.D., Ph.D., Minnesota. Neurophysiology of cerebellum and motor cortex.

Lynne K. Edwards, Associate Professor of Educational Psychology; Ph.D., Washington (Seattle). Computational methods for statistical observations, partial differential equations.

David Ferguson, Assistant Professor of Medicinal Chemistry; Ph.D., Southern Florida. Computational biophysics, molecular dynamics, thermodynamic pertubation theory.

Efi Foufoula-Georgiou, Associate Professor of Civil Engineering; Ph.D., Florida. Stochastic modeling, surface and subsurface hydrologic processes, optimal operation of water resources.

Avner Friedman, Professor of Mathematics; Ph.D., Hebrew University. Partial differential equations, free boundaries, coating flows.

Daniel Kersten, Associate Professor of Psychology; Ph.D., Minnesota. Visual perception, computer vision, visual system models.

Vipin Kumar, Associate Professor of Computer Science; Ph.D., Maryland. Parallel processing, artificial intelligence.

Theodore Labuza, Ph.D., MIT. Physical chemistry of foods, with emphasis on reaction kinetics, water activity, and the glass/rubber state.

David J. Lilja, Associate Professor of Electrical and Computer Engineering; Ph.D., Illinois at Urbana-Champaign. High-performance computer architecture, parallel processing, hardware-software interactions, computer systems performance analysis.

John Lowengrub, Ph.D., NYU (Courant). Numerical analysis.

Mitchell Luskin, Professor of Mathematics; Ph.D., Chicago. Computational methods, nonlinear partial differential equations, numerical analysis.

John L. Nieber, Associate Professor of Agricultural Engineering; Ph.D., Cornell. Deterministic and stochastic equations for fluid flow, heat transport; groundwater contamination.

Nikolaos P. Papanikolopoulos, Associate Professor of Computer Science and Engineering; Ph.D., Carnegie Mellon. Computer vision, robotics, sensors for transportation applications.

Haesun Park, Associate Professor of Computer Science; Ph.D., Cornell. Numerical analysis, parallel computing, signal processing algorithms.

Suhas V. Patankar, Professor of Mechanical Engineering; Ph.D., London. Numerical prediction techniques for practical problems involving heat transfer and fluid flow.

Youcef Saad, Professor of Computer Science; Doctorat d'Etat, Grenoble (France). Sparse matrix computations, parallel computation, nonlinear equations, control theory.

L. E. Scriven, Regents Professor of Chemical Engineering and Materials Science; Ph.D., Delaware. Flow processing solidification, porous media, microstructured liquids.

George R. Sell, Professor of Mathematics; Ph.D., Michigan. Approximation theories for dissipative dynamical systems, fluid flow, and finite-dimensional systems.

J. Ilja Siepmann, Assistant Professor of Chemistry; Ph.D., Cambridge. Physical chemistry, assembly.

Charles C. S. Song, Professor of Civil Engineering; Ph.D., Minnesota. Mathematical modeling, numerical analysis, calculation of three-dimensional turbulent flow.

Jaideep Srivastava, Associate Professor of Computer Science and Engineering; Ph.D., Berkeley. Databases, multimedia systems, distributed computing and data mining.

Harlan W. Stech, Professor of Mathematics and Statistics; Ph.D., Michigan State. Computational methods, dynamical systems modeling, time-delayed effects, computer-aided analysis.

Michael R. Taaffe, Associate Professor of Operations and Management Science; Ph.D., Ohio State. Computational probability, stochastic networks, stochastic models of manufacturing systems.

A. H. Tewfik, Associate Professor of Electrical Engineering; Sc.D., MIT. Statistical signal processing, speech and image processing, computer vision.

David D. Thomas, Professor of Biochemistry (Medical School); Ph.D., Stanford. Spectroscopic and computational techniques, macromolecular structure, physiological processes.

Luke Tierney, Professor of Statistics; Ph.D., Cornell. Numerical techniques, statistical analysis, computational methods for statistical investigation.

Norman Troullier, Ph.D., Minnesota. Computational mathematics.

Donald G. Truhlar, Institute of Technology Professor of Chemistry; Ph.D., Caltech. Computational chemistry, chemical dynamics, POLYRATE, MORATE, AMSOL, RMPROP.

Vaughan R. Voller, Associate Professor of Civil Engineering; Ph.D., Sunderland (England). Numerical techniques for thermal-fluid problems, phase-change boundaries, analysis of metallurgical solidification.

George L. Wilcox, Professor of Pharmacology; Ph.D., Colorado at Boulder. Spinal nociception and analgesia, with emphasis on tolerance, synergy, allodynia, and excitatory amino acids.

Paul R. Woodward, Professor of Astronomy; Ph.D., Berkeley. Numerical astrophysics, hydrodynamic simulations for astrophysical problems.

David A. Yuen, Professor of Geology and Geophysics; Ph.D., UCLA. Large-scale numerical simulation, geophysical fluid dynamics, scientific visualization of complex flow processes.

UNIVERSITY OF NEBRASKA–LINCOLN

Computer Science and Engineering Department

Programs of Study

The Computer Science and Engineering (CSE) Department offers Doctor of Philosophy degree programs in both computer science and computer engineering and Master of Science degree programs with both thesis and nonthesis project options. Research focus areas are communications and networking, including cryptography and optical communications; distributed systems, including real-time and dependable performance, distributed processing, and design and testing; human-centered software design, including software engineering, software architecture, and testing; and visual information processing, including geospatial decision systems, image processing, and computer vision. The Ph.D. in computer engineering is offered through the unified engineering Ph.D. program. The CSE Department also offers a cooperative Ph.D. program with the Department of Mathematics and Statistics. The University also offers the J.D. Edwards Honors Program in Computer Science and Management, with graduate education focusing on enterprise software engineering and information systems.

Research Facilities

The Computer Science and Engineering Department has extensive computing facilities for research. These include a Silicon Graphics Origin 2000, Challenge L, other servers, more than fifty workstations, and a large number of PCs. All graduate offices are furnished with state-of-the-art computing facilities. Advanced laboratories in cryptology, human factors, multimedia, vision and image processing, and VLSI are available for specialized research. A high-speed distributed computing system interconnected by an ATM network is also available for research in networking and distributed computing. All the machines are connected by a campuswide high-performance network utilizing a fiber-optic backbone. The University of Nebraska–Lincoln is a member of the University Corporation for Advanced Internet Development (UCAID) and is a participant in the Internet2 Abilene advanced network project.

Financial Aid

Financial support is available in the form of teaching assistantships, research assistantships, tuition waivers, and fellowships to highly qualified candidates. These carry academic-year stipends of $10,500 to $12,500 plus tuition remission. Graduate students may also apply for loans through the Office of Scholarships and Financial Aid, 16 Administration Building. In the J. D. Edwards Program, fellowships and scholarships provide substantial student support. There are numerous part-time employment opportunities both on and off campus for experienced programmers, but these normally require application in person.

Cost of Study

Graduate tuition in 1999–2000 is $115.50 per credit hour for Nebraska residents and approximately $285 per credit hour for nonresidents. Fees for all full-time students are approximately $250 per semester. Teaching and research assistants pay at the same rate as Nebraska residents, except that their first 9 credit hours each semester are free.

Living and Housing Costs

Room and board charges for students living on campus are approximately $3930 to $4870 for the 1999–2000 academic year. Off-campus housing costs and living expenses in Lincoln are affordable.

Student Group

The University of Nebraska–Lincoln has a total enrollment of 22,408 students. The Computer Science and Engineering Department has 69 M.S. and 23 Ph.D. candidates and grants about 50 B.S. degrees per year. Students actively participate in student and state chapters of the Association for Computing Machinery (ACM) and the Institute of Electrical and Electronics Engineers (IEEE).

Location

Lincoln, a city of 200,000, is an educational center and the state capital. It is situated in a gently rolling terrain and has attractive wide streets, fine parks, and excellent public transportation and bicycle routes. The city has a strong, progressive public school system. The average educational level of Lincoln's citizens is one of the highest in the country. The University, two other colleges, and the community together offer a wide variety of recreational and cultural activities.

The University

The University of Nebraska, founded in 1869, has an enrollment of 45,130 students on its four campuses. It was the first institution west of the Mississippi River to offer degree work beyond the baccalaureate level. The University is a member of the Association of American Universities. It is among the top twenty-five American universities in terms of the number of its graduates listed in *Who's Who*. The Carnegie Foundation for Advancement of Teaching has designated the University of Nebraska–Lincoln a Research I University.

Applying

An application form and official transcripts of all college work must be filed with the Graduate College. At the same time, three letters of reference, a statement of purpose, and a brief resume should be filed with the department. For admission in the fall semester the application deadline is March 1, and for admission in the spring semester the application deadline is October 1. Admission to full graduate standing in the M.S. program requires the equivalent of the undergraduate major in computer science or computer engineering, as determined by the department. Students applying to the Ph.D. program must generally have an M.S. degree in computer science or the equivalent. Students seeking financial assistance should complete the application by January 15. The General Test of the GRE is required, and the Subject Test in computer science is recommended. A TOEFL score of 550 or higher is required for applicants whose native language is not English. High scores on these tests may increase the student's chances of getting an assistantship. Nebraska law requires tests of spoken English before nonnative speakers are allowed to lecture.

Correspondence and Information

Computer Science and Engineering Department
Ferguson Hall
University of Nebraska
Lincoln, Nebraska 68588-0115

Telephone: 402-472-2401
E-mail: gradinfo@cse.unl.edu (information inquiries only)
World Wide Web: http://www.cse.unl.edu

University of Nebraska–Lincoln

THE FACULTY AND THEIR RESEARCH

Prabir Bhattacharya, Professor; D.Phil., Oxford. Computer vision, image processing, soft computing, parallel algorithms. E-mail: prabir@cse.unl.edu

Jean-Camille Birget, Adjunct Associate Professor; Ph.D., Berkeley. Algorithmic problems in algebra, complexity, automata theory. E-mail: birget@cse.unl.edu

Berthe Y. Choueiry, Assistant Professor; Ph.D., Swiss Federal Institute of Technology. Artificial intelligence; constraint satisfaction; abstraction and reformulation; scheduling and resource allocation; interactive, collaborative, and distributed problem solving. E-mail: choueiry@cse.unl.edu

Cecilia R. Daly, Assistant Professor Emeritia; M.S., Nebraska. Applications, education. E-mail: daly@cse.unl.edu

Jitender S. Deogun, Professor; Ph.D., Illinois at Urbana-Champaign. Design and analysis of algorithms, VLSI design automation, graph algorithms, information retrieval, combinatorics. E-mail: deogun@cse.unl.edu

Sebastian Elbaum, Assistant Professor; Ph.D., Idaho. Software engineering, measurement, testing, reliability and intrusion detection. E-mail: elbaum@cse.unl.edu

Mohamed E. Fayad, Associate Professor; Ph.D., Minnesota. Object-oriented (OO) systems and software engineering, OO-distributed computing, design patterns, application and enterprise frameworks, project management, software economics, real-time systems modeling and applications, software stability, software engineering in-the-small, software modeling and architectures, software process improvement. E-mail: fayad@cse.unl.edu

Steve Goddard, Assistant Professor; Ph.D., North Carolina at Chapel Hill. Real-time systems, software engineering, computer networks, distributed systems, multimedia systems, scheduling theory. E-mail: goddard@cse.unl.edu

Scott Henninger, Associate Professor; Ph.D., Colorado at Boulder. Software engineering, multimedia design, human-computer interaction, cognitive science. E-mail: scotth@cse.unl.edu

Hong Jiang, Associate Professor; Ph.D., Texas A&M. Computer architecture, interconnection networks, parallel/distributed processing, supercomputing, performance evaluation, computer storage technology. E-mail: jiang@cse.unl.edu

Roy F. Keller, Professor Emeritus; Ph.D., Missouri–Columbia. Language design and implementation, program verification, projection methods for solving systems of algebraic equations. E-mail: keller@cse.unl.edu

Roger Kieckhafer, Associate Professor; Ph.D., Cornell. Computer architecture, fault tolerance, scheduling, real-time systems, reliability and performance modeling. E-mail: rogerk@cse.unl.edu

Spyros Magliveras, Henson Professor; Ph.D., Birmingham (England). Data encryption, coding theory, data compression, symbolic and algebraic computation, combinatorics, computational group theory. E-mail: spyros@cse.unl.edu

Stuart Margolis, Adjunct Professor; Ph.D., Berkeley. Automata and formal languages; relationships among algebra, automata, and logic; complexity theory; theory of computation. E-mail: margolis@cse.unl.edu

Sarit Mukherjee, Adjunct Assistant Professor; Ph.D., Maryland. Computer network architecture and protocol, multimedia, distributed and real-time systems. E-mail: sarit@cse.unl.edu

Don J. Nelson, Professor; Ph.D., Stanford. Simulation and error analysis, especially power production in electrical utilities and communication networks, software engineering. E-mail: system2@engvms.unl.edu

Byrav Ramamurthy, Assistant Professor; Ph.D., California, Davis. Computer networks, telecommunications, distributed computing, system optimization, performance evaluation, fault tolerance. E-mail: byrav@cse.unl.edu

Stephen E. Reichenbach, Associate Professor and Chair; Ph.D., William and Mary. Digital image processing, computer vision, multimedia computing. E-mail: reich@cse.unl.edu

Peter Revesz, Associate Professor; Ph.D., Brown. Database systems, constraint programming, constraint databases, geographic information systems, multimedia, information retrieval. E-mail: revesz@cse.unl.edu

Charles Riedesel, Assistant Professor; Ph.D., Nebraska. Algorithms, graph theory, computer science education. E-mail: riedesel@cse.unl.edu

Ashok Samal, Associate Professor; Ph.D., Utah. Computer vision, document analysis, parallel and distributed computing. E-mail: samal@cse.unl.edu

Stephen D. Scott, Assistant Professor; D.Sc., Washington (St. Louis). Computational learning theory, machine learning, online algorithms, design and analysis of algorithms, genetic algorithms, reconfigurable computing. E-mail: sscott@cse.unl.edu

Sharad C. Seth, Professor; Ph.D., Illinois at Urbana-Champaign. VLSI design and testing, document image analysis, geographic information systems. E-mail: seth@cse.unl.edu

Alvin Surkan, Professor; Ph.D., Western Ontario. Machine intelligence by simulation and methods of soft computing, pattern learning by neural networks, genetic algorithms and fuzzy logic in database mining. E-mail: surkan@cse.unl.edu

Susan Wiedenbeck, Adjunct Associate Professor; Ph.D., Pittsburgh. Cognition of programming, human-computer interaction, intelligent tutoring. E-mail: susan@cse.unl.edu

UNIVERSITY OF NORTH CAROLINA AT CHAPEL HILL

Department of Computer Science

Programs of Study

The department offers the Ph.D. and a professional M.S. degree. Study for the M.S. degree includes algorithms, programming languages, and hardware as well as important areas of application. The Ph.D. program includes courses in specialized areas and preparation for teaching and advanced research. Students pursue particular areas of their choice and are actively involved in research. The curricula emphasize the design and application of real computer systems and that portion of theory that guides and supports practice. The department's orientation is experimental, with clusters of research in computer architectures, computer graphics and image analysis, computer-supported cooperative work, distributed systems, geometric modeling and computation, hardware systems and design, human-machine interaction, hypertext, the Monte Carlo method, multimedia systems, networking, parallel computing, programming language design and implementation, real-time systems, software engineering and environments, and theorem proving and term rewriting. Students holding an assistantship can typically expect to earn the M.S. degree in two academic years and the Ph.D. in four or five years.

Research Facilities

All of the department's computing facilities are housed in a four-story computer science building that features specialized research laboratories for graphics and image processing; computer building and design; and distributed, parallel, and collaborative systems. The labs, offices, conference areas, and classrooms are bound together by the department's fully integrated distributed computing environment, which includes more than 500 computers that range in performance from 12 MIPS to more than 25 BIPS. These systems are integrated by high-speed networks and by software that is consistent at the user level over the many architectural platforms. The department's research labs contain specialized equipment and facilities. Each student is assigned a computer. The parallel computing facilities include both the department's own designs, such as the PixelFlow graphics supercomputer, and commercial machines, including several parallel SGI Onyx and Power Onyx machines, several Sun multiprocessor systems, and the SGI Reality Monster. The nearby Brauer Library has extensive holdings in mathematics, physics, statistics, operations research, and computer science.

Financial Aid

During the academic year, most students are supported by assistantships and fellowships. The stipend for research and teaching assistantships for the nine-month academic year in 1999–2000 is $12,750 (20 hours per week). Full-time summer employment on a research project is normally available to students who would like to receive support. The rate for summer 1999 is $650 (40 hours per week) for ten to twelve weeks. This produces a combined annual financial package for graduate assistants of approximately $20,000. Students with assistantships qualify for a Graduate Student Tuition Grant and pay no tuition; they are responsible for paying student fees of approximately $400 per semester. At no additional cost, students are also covered by a comprehensive major medical insurance program, underwritten by Blue Cross/Blue Shield of North Carolina. Each semester, the department provides a $500 educational fund to any student who receives a fellowship that is not funded by University of North Carolina at Chapel Hill (UNC–Chapel Hill). The fund may be used for education-related expenses, including books, journals, travel, computer supplies and accessories, and professional memberships. The department also awards a $1500 supplement each semester to nonservice fellowship holders who join a research team. To apply for an assistantship, the applicant should check the appropriate item on the admission application form. Applicants for assistantships are automatically considered for all available fellowships. Students can expect continued support, contingent upon satisfactory work performance and academic progress. Students are encouraged to gain professional experience through summer internships with companies in the Research Triangle area or in other parts of the country.

Cost of Study

For the 1999–2000 academic year, tuition and fees for those attending the University of North Carolina at Chapel Hill are $2252 for state residents and $11,418 for nonresidents.

Living and Housing Costs

Annual living costs for single graduate students in the Chapel Hill area are estimated by University staff to be $9000 or higher. On-campus housing is available for both married and single students attending the University.

Student Group

The Department of Computer Science enrolls up to 140 graduate students, most of whom attend full-time.

Student Outcomes

A majority of the department's master's graduates work in industry, in companies ranging from small start-up operations to government research labs and large research and development corporations. Ph.D. graduates work in both academia and industry. Academic employment ranges from positions in four-year colleges, where teaching is the primary focus, to positions at major research universities. Some graduates take postdoctoral positions at research laboratories prior to continuing in industry or joining academia.

Location

Chapel Hill (population 44,000) is a scenic college town located in the heart of North Carolina, where small-town charm mixes with a cosmopolitan atmosphere to provide students with a rich and varied living experience. The town and the surrounding area offer many cultural advantages, including excellent theater and music, museums, and a planetarium. There are also many opportunities to watch and to participate in sports. The Carolina beaches, the Outer Banks, Great Smoky Mountains National Park, and the Blue Ridge Mountains are only a few hours' drive away.

The Research Triangle of North Carolina is formed by the University of North Carolina at Chapel Hill, Duke University in Durham, and North Carolina State University in Raleigh. The universities have a combined enrollment of more than 63,000 students, have libraries with more than 8 million volumes with interconnected catalogs, and have national prominence in a variety of disciplines.

The University and The Department

The 689-acre central campus of UNC–Chapel Hill is among the most beautiful in the country. Of the approximately 24,200 students enrolled, nearly 9,000 are graduate and professional students. The department's primary missions are graduate teaching and research. It offers the M.S. and Ph.D. degrees and participates in the computer science option of the B.S. curriculum in mathematical sciences. The Computer Science Students' Association sponsors both professional and social events and represents the students in departmental matters. Its president is a voting member at faculty meetings. There is much interaction between students and faculty, and students contribute to nearly every aspect of the department's operation.

Applying

Applications for fall admission, complete with a personal statement, all transcripts, and recommendations, should be received by the Graduate School no later than January 1. To ensure meeting that deadline, students should take the GRE no later than December. Early submission of applications is encouraged. A few assistantships are sometimes available for those who wish to begin in the spring semester. To be considered for these, students should submit completed applications by October 15 and take the GRE no later than June. International applicants should have their applications completed earlier to allow time for processing visa documents. Applicants whose native language is not English must submit TOEFL scores.

Correspondence and Information

For written information about graduate study:

Admissions and Graduate Studies
Department of Computer Science
Campus Box 3175, Sitterson Hall
University of North Carolina
Chapel Hill, North Carolina 27599-3175

Telephone: 919-962-1900
Fax: 919-962-1799
E-mail: admit@cs.unc.edu
World Wide Web: http://www.cs.unc.edu

For applications and admissions information:

The Graduate School
Campus Box 4010, 200 Bynum Hall
University of North Carolina
Chapel Hill, North Carolina 27599-4010

Telephone: 919-966-2611
World Wide Web: http://www.gradschool.unc.edu/

University of North Carolina at Chapel Hill

THE FACULTY AND THEIR RESEARCH

James Anderson, Associate Professor; Ph.D., Texas at Austin, 1990. Distributed and concurrent algorithms, real-time systems, fault-tolerant computing, formal methods.

Sanjoy K. Baruah, Associate Professor; Ph.D., Texas at Austin, 1993. Real-time systems, scheduling theory, computer networks.

Gary Bishop, Associate Professor; Ph.D., North Carolina at Chapel Hill, 1984. Hardware and software for man-machine interaction, 3-D interactive computer graphics.

Frederick P. Brooks Jr., Kenan Professor; Ph.D., Harvard, 1956. 3-D interactive computer graphics, human-computer interaction, virtual worlds, molecular graphics, computer architecture.

Peter Calingaert, Professor Emeritus; Ph.D., Harvard, 1955.

Siddhartha Chatterjee, Assistant Professor; Ph.D., Carnegie Mellon, 1991. High-level programming languages, compilation for highly parallel machines, object-oriented programming, parallel algorithms and architectures.

James M. Coggins, Associate Professor; Ph.D., Michigan State, 1983. Artificial visual systems, pattern recognition, computer graphics, human-computer interaction, object-oriented design and programming, distributed systems.

Prasun Dewan, Professor; Ph.D., Wisconsin–Madison, 1986. User interfaces, distributed collaboration, software engineering environments, object-oriented databases.

Nick England, Research Professor; E.E., North Carolina State, 1974. Systems architectures for graphics and imaging, scientific visualization, volume rendering, interactive surface modeling.

John G. Eyles, Research Associate Professor; Ph.D., North Carolina at Chapel Hill, 1982. Graphics architectures, rapid system prototyping, virtual environments, VLSI-based system design.

Henry Fuchs, Federico Gil Professor; Ph.D., Utah, 1975. High-performance graphics hardware, 3-D medical imaging, head-mounted display, virtual environments.

Guido Gerig, Taylor Grandy Professor of Computer Science and Psychiatry; Ph.D., Swiss Federal Institute of Technology, 1987. Image analysis, shape-based object recognition, 3-D object representation and quantitative analysis, medical image processing.

John H. Halton, Professor; D.Phil., Oxford, 1960. Applications of combinatorial and probabilistic methods and of scientific and mathematical analysis to computational, scientific, and engineering problems.

Kye S. Hedlund, Associate Professor; Ph.D., Purdue, 1982. Computer-aided design, computer architecture, algorithm design and analysis, parallel processing.

Doug L. Hoffman, Research Assistant Professor; Ph.D., North Carolina at Chapel Hill, 1996. Parallel algorithms, parallel architecture, distributed systems, bioinformatics, computer-aided protein science.

Kevin Jeffay, Associate Professor; Ph.D., Washington (Seattle), 1989. Real-time systems, operating systems, distributed systems, multimedia networking, computer-supported cooperative work, performance evaluation.

Anselmo A. Lastra, Research Associate Professor; Ph.D., Duke, 1988. Computer graphics, parallel computing.

Ming C. Lin, Assistant Professor; Ph.D., Berkeley, 1993. Physically based and geometric modeling, applied computational geometry, robotics, distributed interactive simulation, virtual environments, algorithm analysis.

Gyula A. Magó, Professor; Ph.D., Cambridge, 1970. Parallel computation, computer architecture, programming languages.

Dinesh Manocha, Associate Professor; Ph.D., Berkeley, 1992. Geometric and solid modeling, physically based modeling, computer graphics, simulation-based design, symbolic and scientific computation, computational geometry.

Lars S. Nyland, Research Associate Professor, Ph.D., Duke, 1991. High-performance computing, parallel algorithms, parallel computer architecture and hardware systems, programming languages, program transformation and optimization techniques, scientific computing.

Stephen M. Pizer, Kenan Professor; Ph.D., Harvard, 1967. Image analysis and display, human and computer vision, graphics, numerical computing, medical imaging.

David A. Plaisted, Professor; Ph.D., Stanford, 1976. Mechanical theorem proving, term rewriting systems, logic programming, algorithms.

John Poulton, Research Professor; Ph.D., North Carolina at Chapel Hill, 1980. Graphics architectures, VLSI-based system design, design tools, rapid system prototyping.

Jan F. Prins, Associate Professor and Director of Graduate Studies; Ph.D., Cornell, 1987. Parallel algorithms, languages and architectures, high-level programming languages, compilers, formal techniques in program development.

Timothy L. Quigg, Lecturer and Associate Chairman for Administration and Finance; M.P.A., North Carolina State, 1979. Intellectual property rights, industrial relations, contract management, resource allocation.

F. Donelson Smith, Research Professor; Ph.D., North Carolina at Chapel Hill, 1978. Computer networks, operating systems, distributed systems, multimedia, computer-supported cooperative work.

John B. Smith, Professor; Ph.D., North Carolina at Chapel Hill, 1970. Computer-supported cooperative work, hypermedia systems, World Wide Web architecture and programming, Java object storage and access.

Donald F. Stanat, Professor Emeritus; Ph.D., Michigan, 1966.

David Stotts, Associate Professor and Associate Chairman for Academic Affairs; Ph.D., Virginia, 1985. Computer-supported cooperative work, hypermedia, software engineering and formal methods, programming languages and concurrency, interoperable distributed systems.

Russell M. Taylor II, Research Assistant Professor; Ph.D., North Carolina at Chapel Hill, 1994. 3-D interactive computer graphics, virtual worlds, distributed computing, scientific visualization, human-computer interaction.

Leandra Vicci, Lecturer and Director of the Microelectronic Systems Laboratory; B.S., Antioch (Ohio), 1964. Information processing hardware: novel architectures and systems; electrical signals and clock distribution; physical layer technologies: electronic, optical, mechanical, and acoustic; physical first-principles models of computation.

Jeannie M. Walsh, Lecturer; M.S., Oklahoma State, 1984. Computer education; social, legal, and ethical issues in computing.

Stephen F. Weiss, Professor and Chairman; Ph.D., Cornell, 1970. Information storage and retrieval, natural language processing, communications and distributed systems, computer-supported cooperative work.

Gregory F. Welch, Research Assistant Professor; Ph.D., North Carolina at Chapel Hill, 1997. Human-machine interaction, 3-D interactive computer graphics, virtual/augmented environment tracking systems, shared virtual environments and telecollaboration.

Mary C. Whitton, Research Assistant Professor and Project Manager for Virtual Environments Research; M.S., North Carolina State, 1984. Virtual and augmented reality systems for data visualization, computer graphics system architectures.

William V. Wright, Research Professor Emeritus; Ph.D., North Carolina at Chapel Hill, 1972. Interactive systems for supporting scientific research, molecular graphics, architecture and implementation of computing systems.

Adjunct Faculty

Hussein Abdel-Wahab, Adjunct Professor; Ph.D., Waterloo, 1976. Computer-supported cooperative work, multimedia systems and communications, distance learning, distributed systems, operating systems, networking.

Stephen R. Aylward, Adjunct Assistant Professor; Ph.D., North Carolina at Chapel Hill, 1997. Statistical pattern recognition, shape-based object representation, image processing, neural networks.

Bert Dempsey, Adjunct Assistant Professor; Ph.D., Virginia, 1994. Computer supported cooperative work, computer networks, multimedia communications, digital library systems.

Steven E. Molnar, Adjunct Assistant Professor; Ph.D., North Carolina at Chapel Hill, 1991. Architectures for real-time computer graphics, VLSI-based system design, parallel rendering algorithms.

Julian Rosenman, Adjunct Professor; Ph.D., Texas at Austin, 1971; M.D., Texas Health Science Center at Dallas, 1977. Computer graphics for treatment of cancer patients, contrast enhancement of poor quality X rays.

Raj K. Singh, Adjunct Associate Professor; Ph.D., SUNY at Albany, 1986. High-performance systems design and integration, CAD tools, parallel architectures and algorithms, VLSI design and fabrication, networking, computational biology, scientific computing.

Turner Whitted, Adjunct Professor; Ph.D., North Carolina State, 1978. Computer graphics.

UNIVERSITY OF NORTH CAROLINA AT CHARLOTTE

Doctor of Philosophy in Information Technology

Program of Study

The University of North Carolina offers multidisciplinary programs that lead to the Ph.D. degree in information technology. The Ph.D. program offers students opportunities to conduct research in a diverse range of areas. To accommodate the interdisciplinary nature of the program, students are encouraged to work with faculty advisers to produce flexible programs of study that are tailored to fit each student's career goals.

Students who seek employment in industry, commerce, or government agencies and laboratories have opportunities to participate in high-quality research projects, with heavy participation from industry and government agencies. Students whose goal is to teach at comprehensive universities or four-year colleges have opportunities to familiarize themselves with recent advances in educational technology and are able to select a broadly based program of advanced course work. Students who aspire to positions in research universities that support doctoral programs benefit from contact with a strong research faculty of international stature and the exposure to practical applications of their specialties.

Students in the Ph.D. program are expected to pass a preliminary written exam in the Core of Information Technology. This exam ensures that all students master a common set of basic knowledge in information technology. Students are also required to pass a written preliminary exam in an area that is closely related to their area of research. By the end of the third year in the program, Ph.D. students are expected to pass a qualifying exam for advancement to candidacy, based on advanced topics in an area of specialization. Finally, Ph.D. students are expected to complete, under the direction of their thesis adviser, a Ph.D. dissertation comprising a substantial and original contribution to their area of study.

Areas of research include computer networks and communication; decision support systems; data visualization; electronic commerce; enterprise integration; evolutionary computing; geographical information systems; human, social, cultural, and ethical issues in IS; information systems quality management; intelligent information systems; knowledge discovery and data mining; management of IS resources; measurement of IS effectiveness and productivity; parallel programming; supply-chain management; and systems development management. More information on the program can be found on the Web at http://www.sit.uncc.edu.

Research Facilities

The School of Information Technology has a number of laboratories that are equipped with state-of-the-art hardware and software. All computer systems are linked through a 100Mbit fiber-optic campus network. This enables students and faculty members to share computing resources across lab boundaries. The distributed computing laboratory has a large network of heterogeneous computing platforms. Students can benefit from a wide variety of software, including many programming languages and tools, an Oracle database management system, distributed object brokering and management systems, modeling and analysis, and decision support tools such as optimization, statistical analysis, and neural networks. Students also have the opportunity to experiment with different system configurations. The graphics and visualization lab offers special hardware and software for 3-D graphics and visualization. The computer engineering lab offers specialized tools for the design and implementation of embedded systems as well as for large-scale parallel programming. The robotics and machine vision lab offers facilities for building advanced robotics and control systems. The distance learning and collaborative computing systems labs have facilities to conduct interactive two-way audio and video communications. In addition, students and faculty members with very computing-intensive applications have access to the Cray Y-MP at the North Carolina Super Computing Program (NCSP).

Financial Aid

Financial assistance for qualified students is available on a competitive basis in the form of graduate teaching and research assistantships. A limited number of out-of-state and in-state tuition grants are also competitively awarded.

Cost of Study

In 1998–99, tuition and academic fees for a full course load totaled $919 per semester for North Carolina residents; for nonresidents, the total was $4554. Out-of-state students who have been recruited for special talents and who perform substantial academic duties may, in some cases, be eligible for resident tuition. Tuition and fees are subject to change.

Living and Housing Costs

Room and board are available on campus for unmarried students. The cost in 1998–99 ranged from $2860 to $4794 per academic year. Private housing adjacent to the campus is readily available for married and single students. City transit is available to the campus for students who choose to live elsewhere in the community.

Student Group

The total University enrollment for fall 1998 was 16,670 students, more than 2,600 of whom were graduate students. Approximately 55 percent of the students were women. African-American students comprised nearly 16 percent, out-of-state students 12 percent, and international students 3 percent of the total.

Student Outcomes

The program's objective is to enable students to master a significant body of knowledge in information creation, access, analysis, dissemination, and management; relate this knowledge to areas that employ information technology; and carry on fundamental research in information technology at a nationally and internationally competitive level.

Location

A population of 465,895 makes Charlotte one of the country's largest cities. A metropolitan population of 1.3 million makes it the thirty-fourth largest metropolitan area. It is the nation's second-largest banking center, and *Fortune* magazine has ranked its business climate first in the country. It is also noted for trees, cleanliness, quality of life, air service, and a thriving research park.

The University and The School

UNC Charlotte is North Carolina's fourth-largest university. The School of Information Technology, established in 1998, is distinguished by the participation of a strong research faculty with members from a variety of academic disciplines. Many of the faculty members have strong ties to both industry and government agencies.

Applying

Applicants should have a baccalaureate degree in a related field and should submit a completed application, official transcripts from all colleges and universities attended, official GRE or GMAT scores, TOEFL scores (if applicable), a statement of purpose, three letters of recommendation, and an application fee of $35. Students should contact the Graduate Admissions Office for any specific deadlines or admission requirements. Application deadlines are subject to change.

Correspondence and Information

Graduate School Admissions
University of North Carolina at Charlotte
332 Kennedy Building
9201 University City Boulevard
Charlotte, North Carolina 28223

Telephone: 704-547-3366
Fax: 704-547-3279
E-mail: gradadm@email.uncc.edu
World Wide Web: http://www.uncc.edu

University of North Carolina at Charlotte

THE FACULTY AND THEIR RESEARCH

C. Michael Allen, Ph.D., SUNY at Buffalo, 1968. Computer engineering, medical imaging, embedded controllers.

Haldun Aytug, Ph.D., Florida, 1993. Artificial intelligence, machine learning, genetic algorithms, database design.

Joanna Baker, Ph.D., Clemson, 1983. Decision support systems, public-sector decision making, applied mathematical programming.

Maureen Brown, Ph.D., Georgia, 1995. Information technology.

Keh-Hsun Chen, Ph.D., Duke, 1976. Artificial intelligence, knowledge-based systems, heuristic search, computer Go, theory of computing.

Bei-Tseng Chu, Ph.D., Maryland, College Park, 1988. Enterprise integration, electronic commerce, software engineering.

W. Douglas Cooper, Ph.D., North Carolina State, 1968. Operations management, inventory control and scheduling.

John Gretes, Ed.D., Virginia, 1982. Information technology and education.

Mirsad Hadzikadic, Ph.D., SMU, 1987. Medical informatics, artificial intelligence, cognitive science, machine learning, knowledge acquisition, expert systems.

Moutaz Khouja, Ph.D., Kent State, 1991. Technology selection, inventory management, aggregate planning.

Ram L. Kumar, Ph.D., Maryland, College Park, 1993. Management of investments in information and manufacturing technologies, management support systems.

Junsheng Long, Ph.D., Illinois at Urbana-Champaign, 1992. Object-oriented programming, system programming, security and networking, distributed and parallel processing, computer architecture.

Zbigniew Michalewicz, Ph.D., Polish Academy of Sciences, 1981. Genetic algorithms, evolutionary computation, computer algorithms, search techniques.

Taghi Mostafavi, Ph.D., Oklahoma State, 1986. Computer engineering, computer and VLSI architecture, VLSI design, CAD and test, computer vision.

Joseph Quinn, Ph.D., Michigan State, 1970. Probability, stochastic processes.

Zbigniew Ras, Ph.D., Warsaw, 1973. Cooperative and collaborative systems, knowledge discovery, query answering systems, distributed agents, logic for artificial intelligence.

Hassan Razzavi, Ph.D., West Virginia, 1978. Threshold logic, computer arithmetic circuits, computer organization and architecture, fault-tolerant computing.

Gyorgy Revesz, Ph.D., Eotvos Lorand (Budapest), 1968. Formal languages and automata theory, compiler design, semantics of programming languages, functional programming, symbolic computing, parallel programming.

Stephanie S. Robbins, Ph.D., LSU, 1982.

Cem Saydam, Ph.D., Clemson, 1985. Applied mathematical modeling, location analysis, decision support systems.

Anthony Stylianou, Ph.D., Kent State, 1987. TQM, electronic commerce, strategic use of information systems.

Kalpathi Subramanian, Ph.D., Texas at Austin, 1990. Scientific data visualization, medical imaging, biomedical application/data analysis, global illumination algorithms, ray tracing, representation of geometric models.

William Tolone, Ph.D., Illinois at Urbana-Champaign, 1966. Computer-supported cooperative work, group-ware, human-centered systems, object-oriented design agent technologies, meta-level architectures.

Robert Wilhelm, Ph.D., Illinois at Urbana-Champaign, 1992. Computer-integrated manufacturing, enterprise integration.

Anthony Barry Wilkinson, Ph.D., Manchester, 1974. Parallel processing, multiprocessor systems design, multiprocessor interconnection networks.

Wei-Ning Xiang, Ph.D., Berkeley, 1989. Geographical Information Systems, decision analysis and planning support systems.

Jing Xiao, Ph.D., Michigan, 1990. Robotics, computational intelligence, artificial intelligence, computer-integrated manufacturing.

Jan Zytkow, Ph.D. Warsaw, 1968. Artificial intelligence, automated discovery of knowledge in databases, autonomous robot discoverer, intelligent information systems.

UNIVERSITY OF NORTH TEXAS

Department of Computer Sciences

Programs of Study

The Department of Computer Sciences offers programs leading to the M.S. and Ph.D. degrees. A wide range of courses and research areas are available to graduate students. These include algorithm analysis, artificial intelligence, database systems, distributed computing, image processing, networking, neural networks, numerical analysis, operating systems, parallel computing, pattern recognition, programming languages, and simulation and modeling. Through its Center for Research in Parallel and Distributed Computing, the department offers special emphases in all areas of parallel and distributed computing.

Research Facilities

The Department of Computer Sciences provides students and faculty with a broad assortment of hardware and software. The facilities and equipment include a Sun Ultra Enterprise 4000; a networking laboratory consisting of Pentium-Pro workstations connected with gigabit networking; multiple UNIX workstations; and laboratories with X-window terminals, Macintoshes, and IBM PC compatibles, all connected via the departmental Ethernet. In addition to the graphics lab and several open-use labs, the University Computing Center's equipment includes an IBM 9672/R51 mainframe and two UNIX superminicomputers. Dialup text and PPP access are available with telephone service points in Denton, Dallas, and Fort Worth.

Financial Aid

Approximately thirty-five teaching assistantships and research fellowships with stipends beginning at $3800 per semester are available to qualified graduate students. Remission of out-of-state tuition is also included.

Cost of Study

In 1998–99, graduate tuition and fees were $325 per semester credit hour for out-of-state students and $112 per semester credit hour for in-state students (rates are for students enrolled in a full-time course load).

Living and Housing Costs

The University housing system offers ample facilities for single graduate students and a limited number of residences for married students. Room and board at the residence halls cost about $4000 for the academic year. A wide variety of off-campus accommodations is available. The UNT Student Association's annual apartment survey lists apartments beginning at approximately $250 per month.

Student Group

Nearly 25 percent of the University's more than 27,000 students are in the Graduate School. The department has approximately 25 doctoral students and 115 master's students. In the past four years, a total of ten Ph.D. degrees have been awarded; approximately fifty master's degrees are conferred yearly.

Location

The University of North Texas is located in Denton, Texas. Denton, located 35 miles north of the Dallas–Fort Worth metroplex, combines a small-town atmosphere with the advantages of a major metropolitan area. With a population of more than 4 million, the metropolitan area is the largest in Texas and the seventh-largest in the United States. There is a wide range of employment, cultural, and recreational opportunities.

The University

Founded in 1890, the University of North Texas is a comprehensive state-supported institution that combines education and research with public service. The Governor's Select Committee recommended in 1986 that the University be designated as one of Texas's five major research and graduate institutions. UNT is the most comprehensive graduate and research university in the region and the fourth-largest in the state of Texas. There are 131 buildings on the 456-acre campus.

Applying

Students applying to the master's degree program must hold the equivalent of a bachelor's degree and submit GRE scores (a minimum of 650 on the quantitative portion of the GRE and 1050 on the combined quantitative and verbal portions) and transcripts (a minimum grade point average of 3.0 on a 4.0 scale). Students applying to the Ph.D. program must submit three letters of recommendation, GRE scores (700 quantitative, 1150 combined quantitative and verbal), and transcripts (3.5 minimum grade point average).

Applications for admission must be received by the School of Graduate Studies by June 1 for the fall semester and October 1 for the spring semester. Students requesting financial aid should send their completed application to the Department of Computer Sciences by March 1 for the fall semester and September 15 for the spring semester. Students whose native language is not English must submit minimum TOEFL scores 580.

Application forms and information may be obtained by writing the Office of Graduate Admissions or the department. The application fee is $25 for U.S. citizens, $50 for all others.

Correspondence and Information

Department of Computer Sciences
University of North Texas
P.O. Box 311366
Denton, Texas 76203-1366
Fax: 940-565-2799
E-mail: gradinfo@cs.unt.edu
World Wide Web: http://www.cs.unt.edu/

School of Graduate Studies
University of North Texas
Denton, Texas 76203-5459

University of North Texas

THE FACULTY AND THEIR RESEARCH

Azzedine Boukerche, Assistant Professor; Ph.D., McGill, 1995. Distributed systems, parallel simulation, wireless and mobile computing, performance evaluation, distributed interactive simulation, distributed databases.

Robert Brazile, Associate Professor; Ph.D., Texas at Dallas, 1985. Databases.

Sajal Das, Professor; Ph.D., Central Florida, 1988. Wireless mobile computing, cellular wireless networks, parallel computing, multiprocessor interconnection networks.

Paul Fisher, Professor; Ph.D., Arizona State, 1969. Image processing and multimedia.

Tom Irby, Assistant Professor; Ph.D., SMU, 1976. Data structures.

Tom Jacob, Associate Professor; Ph.D., Emory, 1974. Distributed computing.

Robert Kallman, Professor; Ph.D., MIT, 1968. Signal processing, image processing, engineering optimization.

Armin R. Mikler, Assistant Professor; Ph.D., Iowa State, 1995. Computer networks, traffic management, intelligent agents, distributed and collaborative systems, systems security.

Ian Parberry, Professor; Ph.D., Warwick (England), 1984. Computational complexity, parallel computing, neural networks, computer games.

Robert Renka, Associate Professor; Ph.D., Texas at Austin, 1981. Numerical analysis, mathematical software, curve and surface fitting.

Don Retzlaff, Lecturer; M.S., North Texas State, 1979. Software engineering.

Farhad Shahrokhi, Associate Professor; Ph.D., Western Michigan, 1987. Design and analysis of algorithms, combinatorial optimization, graph theory, theory of VLSI.

Weiping Shi, Associate Professor; Ph.D., Illinois at Urbana-Champaign, 1992. Defect and fault tolerance, computer-aided design, VLSI.

Kathleen Swigger, Professor; Ph.D., Iowa, 1977. Artificial intelligence, human factors.

Paul Tarau, Associate Professor; Ph.D., Montreal, 1990. Internet programming, intelligent mobile agents, compilers and abstract machines, distributed logic programming.

Steve Tate, Assistant Professor; Ph.D., Duke, 1992. Algorithms, computational complexity, data compression.

Chao-Chih Yang, Professor; Ph.D., Northwestern, 1966. Artificial intelligence, databases.

UNIVERSITY OF NOTRE DAME

College of Engineering
Department of Computer Science and Engineering

Programs of Study

The department offers programs of study and research leading to the M.S. and Ph.D. degrees. Research emphases within the department are VLSI, high-performance parallel and distributed computing, new parallel computing architectures (especially those that map well into VLSI implementations), and parallel computing algorithms. Other research efforts are also under way in hardware/software codesign and parallel compilers.

All new graduate students are admitted to the master's program unless they hold an equivalent degree. This program requires a minimum of 24 semester hours of course work credit beyond the bachelor's degree, plus a thesis. These requirements can be completed by a full-time student in three regular academic semesters plus the summer, although many students take four semesters. The student must, upon the acceptance of the thesis, successfully pass an oral thesis defense examination. Students who complete the master's program may apply for admission to the doctoral program during their final semester of master's work. Doctoral students are normally required to accumulate a minimum of 42 hours of satisfactory course credit beyond the bachelor's degree plus a dissertation. Additional requirements for the Ph.D. include three years in resident study and passing of qualifying and candidacy examinations and the final examination.

Research Facilities

Notre Dame's College of Engineering maintains a cluster of ninety-nine Sun MicroSystems, Inc., UltraSPARC 140E workstations with 3-D graphics display capability. The cluster also contains fifteen PowerMac 7300s, several Dell Optiplex GXPRO 180 workstations, six Hewlett-Packard 5SiMX laser printers, and a color printer; all are available to students and researchers.

The University's Computing Center supports AFS file service with ten UltraSPARC Enterprise fileservers. These fileservers provide more than 1 terabyte of RAID (0+1) mirrored/striped file storage space for the campus community. The Computing Center also supports a cluster of IBM RS/6000s, a sixteen-processor IBM SP-1, an eight-processor IBM SP-2 array, and two Silicon Graphics compute servers. The campus is currently connected to the VbNS Internet-II backbone via a 45-million bit-per-second connection. This connection was upgraded to 155-million bits per second during the summer of 1998.

The Department of Computer Science and Engineering recently obtained a two-node/four-processor IBM SP-2 array, a parallel multinode Sun SPARCstation array, and two Compaq NT/Netware fileservers. In addition, the department provides twenty Sun UltraSPARC workstations, a Sun SPARCstation 10, four Sun SPARCstation 5s, an Asynchronous Transfer Mode (ATM) teaching network, an ATM research network, a Myrinet Gigabit research network, thirty assorted other workstations (Sun SPARCstation 2, IBM RS/6000-320H, Macintosh, Sun 4/630 fileserver, Pentium-Pro workstations, and an IBM RS/6000-590), and twenty-one printer-plotters.

The department's Design Automation Laboratory is equipped with three Sun SPARCstation workstations and several PC graphics stations. A full suite of design software tools, including Mentor Graphics and Xilinx Field Programmable Gate array development tools, is available on the departmental workstations. The department also maintains a printed circuit board prototyping facility, which includes capabilities for laminating, drilling, and through-hole plating.

A specialized College of Engineering research library holds more than 50,000 volumes. The Engineering Library augments the University's Theodore M. Hesburgh Library, which contains more than 1.8 million volumes and receives 625 journals related to engineering. The Hesburgh Library also provides database searches and bibliographic instruction.

Financial Aid

Teaching assistantships provide stipends as well as remission of academic-year tuition. Research assistantships, with remuneration commensurate with an applicant's qualifications, are available in numbers that depend upon the scope of the research program in progress. Fellowships and traineeships supported by federal agencies, industry, and the University are also available.

Cost of Study

Tuition was $10,400 per semester for full-time study in 1998–99. For part-time study, the cost was $1155 per credit hour.

Living and Housing Costs

A limited number of on-campus living accommodations are available for graduate students. Rooms in private homes adjacent to the South Bend campus rent for $75 to $100 per week. Apartment rentals range from $275 to $500 per month, depending upon occupancy and size.

Student Group

Fifty graduate students are enrolled in the department's programs; 48 are full-time, and 2 are part-time. There are 40 men and 10 women. Twenty-four of the students are U.S. nationals.

Student Outcomes

Graduates of the department's programs find challenging employment as professors in universities, researchers in government/academic laboratories, research and development team leaders in software and hardware firms, and project managers working within technical consulting organizations.

Location

The University is a cultural center of the northern Indiana–southwestern Michigan area and offers cultural, social, athletic, and political events throughout the year. Saint Mary's College, a liberal arts college for women, augments Notre Dame's offerings. The Morris Civic Auditorium hosts road shows of Broadway plays and is the home of a symphony orchestra. Notre Dame is 2 hours by automobile from Chicago.

The University and The College

The University was founded in 1842 by the Reverend Edward Frederick Sorin and 6 brothers of the Congregation of Holy Cross. It was chartered as a university by a special act of the Indiana legislature in 1844, and engineering studies were begun in 1873. The University's 1,250-acre campus is situated immediately north of the city of South Bend, an industrial center of about 130,000 people, approximately 90 miles east of Chicago. Its twin lakes and many wooded areas provide a setting of natural beauty for more than seventy University buildings, many of which have been erected in the past thirty years.

Applying

Applicants should arrange for GRE General Test scores, two original transcripts showing previous academic credits and degrees earned, and letters of recommendation from 3 or 4 college teachers to be sent to the dean of the Graduate School at least six months prior to the beginning of the academic session in which enrollment is sought. The GRE should be taken no later than December preceding the academic year of enrollment.

Correspondence and Information

Director of Graduate Studies
Department of Computer Science and Engineering
University of Notre Dame
Notre Dame, Indiana 46556
Telephone: 219-631-8320
Fax: 219-631-9260
E-mail: cse@cse.nd.edu

University of Notre Dame

THE FACULTY AND THEIR RESEARCH

Steven C. Bass, Professor and Schubmehl-Prein Chair; Ph.D., Purdue, 1971. VLSI and parallel computing architectures.

Jay B. Brockman, Assistant Professor; Ph.D., Carnegie Mellon, 1992. VLSI, computer-aided design frameworks.

Ziyi D. Chen, Associate Professor; Ph.D., Purdue, 1992. Parallel algorithms in computational geometry, theoretical computer science, robot motion and navigation.

Nikos P. Chrisochoides, Assistant Professor; Ph.D., Purdue, 1992. Parallel compilers and problem-solving environments for scientific computing.

Vincent W. Freeh, Assistant Professor; Ph.D., Arizona, 1996. Programming languages, compilers, and operating systems in distributed and parallel computing.

J. Curt Freeland, Assistant Professional Specialist; B.S.E., Purdue, 1985. System administration and network management.

Eugene W. Henry, Professor; Ph.D., Stanford, 1960. Electronic design automation, VLSI design, computer simulation.

Sharon Hu, Assistant Professor; Ph.D., Purdue, 1989. Algorithm design and analysis in VLSI, hardware-software codesign, real-time embedded systems.

Peter M. Kogge, McCourtney Chaired Professor; Ph.D., Stanford, 1972. High-performance computing, parallel computing, computer architectures.

Andrew Lumsdaine, Associate Professor; Ph.D., MIT, 1992. Parallel processing, scientific computing, numerical analysis, VLSI circuit and semiconductor device simulation, mathematical software.

Edwin Sha, Associate Professor; Ph.D., Princeton, 1992. VLSI processor arrays, parallel computer architectures, software for parallel systems.

John J. Uhran Jr., Professor and Associate Dean; Ph.D., Purdue, 1967. Electronic design automation, neural networks, robot motion, computer vision.

Adjunct and Visiting Appointment

Ramzi Bualuan, Lecturer; M.S., Notre Dame, 1986. Database techniques, knowledge engineering.

Matthias Scheutz, Visiting Assistant Professor; Ph.D., Vienna, 1995. Neural networks, artificial intelligence.

David R. Surma, Visiting Assistant Professor; Ph.D., Notre Dame, 1998. Parallel systems.

Professor Andrew Lumsdaine teaches in one of his research specialties, software engineering.

The Fitzpatrick/Cushing Hall of Engineering.

Sunrise at the University of Notre Dame as seen from the opposite shore of Lake St. Mary's.

UNIVERSITY OF OREGON

College of Arts and Sciences
Department of Computer and Information Science

Programs of Study

The Department of Computer and Information Science (CIS) offers programs leading to the degrees of Master of Arts (M.A.), Master of Science (M.S.), and Doctor of Philosophy (Ph.D.). The primary research areas are software systems, software engineering, computational science, human-computer interaction, visualization and graphics, networking, wearable computers, programming languages, artificial intelligence, and theoretical computer science. The University's interdisciplinary program in cognitive science provides for joint study between CIS and the Departments of Psychology and Linguistics in the areas of human vision and natural language. Interdisciplinary research with the Department of Mathematics includes ongoing work in algebraic algorithms stressing both computational complexity and symbolic computation.

The M.S. program usually consists of six quarters of full-time graduate course work. The degree requirements are 54 credits, with a minimum of 42 credit hours in computer science. Twelve credit hours may be taken in a related minor area. The M.S. program has a thesis option, which is recommended for those continuing to the Ph.D. program. Students in the Ph.D. program must take six courses in the core area of the master's program. Courses, along with a directed research project, are completed in the first or second year. Students then choose their dissertation research area, survey current research, and learn research methodology. Advancement to candidacy follows the passing of an oral comprehensive examination and a research proposal exam. A research dissertation is required. Students in both the M.S. and Ph.D. programs receive close guidance from faculty advisers.

Research Facilities

The departmental computing environment is a mix of UNIX, Apple Macintosh, and PCs. The main servers are two Sun SPARCservers that provide NFS file service and support World Wide Web, ftp, e-mail, USENET News, and other network services. Graduate students use a Sun workstation lab. Research labs operate a variety of UNIX workstations and PCs: the Interactive Systems Lab is equipped with Sun workstations, Macintoshes, and PCs; the Computer Graphics Lab includes color scanners, color film recorders, color-calibrated monitors, and HP 755 series workstations with CRX 48Z graphics; and two Silicon Graphics Power Challenge systems with ten and twelve R8000 CPUs and Indigo2 High Impact graphics, a Power Onyx with eight R10000 CPU's, Reality Engine graphics, Sirius Video, and a 4096 processor Maspar SIMD machine support research in high-performance computing in the Computational Science Institute (CSI). The Network Research Lab contains a model of a wide-area network that is used to develop and prototype new Internet applications. any of the computers may also be used for simulations and as general-purpose workstations. The network lab is comprised of custom-built PCs running the FreeBSD operating system.

Financial Aid

Teaching and research assistantships are available for qualified graduate students. In 1999–2000, stipends range from $9284 to $10,754 for the nine-month academic year, with the amount of the award depending on the entry level of the student. The stipends also include a waiver of tuition. The department attempts to support all of its Ph.D. candidates through teaching and research assistantships. A few teaching fellowships are available from the summer school. Also, the Computing Center regularly employs students as programmers, consultants, and computer operators. Various scholarships and work-study opportunities are available through the University's Office of Student Financial Aid.

Cost of Study

Graduate tuition for 1999–2000 is $2230 per term for Oregon residents; nonresident tuition is $3780 per term. Students with graduate assistantships pay only a fee of $221.50 per term.

Living and Housing Costs

The rate for single occupancy in University dormitories with a complete meal plan is $6962 for the academic year 1999–2000. Married student housing is available, with rent between $285 and $865 per month. Off-campus one-bedroom apartments rent for approximately $650 per month excluding electricity.

Student Group

The current student enrollment at the University is more than 16,000 students (more than 3,100 graduate students). The department has more than 60 graduate students.

Location

The University is located in Eugene, a city of approximately 120,000 at the south end of the Willamette Valley. Parks with bike and running paths are built around the confluence of the city's two rivers, the Willamette and the McKenzie. Adjacent to the University are shops, restaurants, parks, and jogging and biking trails along the Willamette River. Within 60 miles of the campus are the Pacific coast to the west and the Cascade mountain range to the east. Thus there are extensive opportunities for camping, hiking, boating, white-water rafting, and cross-country and downhill skiing. Downtown Eugene is the site of a major center for the performing arts (home of the city's symphony orchestra, ballet, opera, and ensembles). The city itself is a compact convention center with major hotels, yet it is within a short distance of wilderness areas.

The University and The Department

The University of Oregon, which dates from 1876, has 750 full-time faculty members engaged in teaching and research. The University operates on the quarter system with a shortened summer session. The campus is richly landscaped with broad lawns and more than 400 varieties of trees and flowering plants. It includes extensive athletic facilities.

The Department of Computer and Information Science is housed in its own building. This three-story, 27,000-square-foot science facility has extensive laboratory space for research and instruction.

Applying

A $50 fee is required of all applicants. Applications for graduate admission should be submitted to the department. All applications should be received by February 1 for the following academic year. All graduate applicants must submit GRE General Test scores, and Ph.D. program applicants should submit scores on the computer science Subject Test of the GRE. International students must pass the TOEFL for admission to the University (the department requires a score of at least 610) and the Test of Spoken English (TSE) for teaching award consideration.

Correspondence and Information

Department of Computer and Information Science
1202 University of Oregon
Eugene, Oregon 97403-1202

Telephone: 541-346-4408
E-mail: info@cs.uoregon.edu
World Wide Web: http://www.cs.uoregon.edu/

University of Oregon

THE FACULTY AND THEIR RESEARCH

Zena M. Ariola, Associate Professor; Ph.D., Harvard, 1992. Programming languages.

John S. Conery, Professor; Ph.D., California, Irvine, 1983. Architecture, parallel processing, computational science.

Jan Cuny, Associate Professor; Ph.D., Michigan, 1981. Parallel processing, programming environments, computational science.

Sarah A. Douglas, Professor; Ph.D., Stanford, 1983. Human-computer interaction.

Arthur M. Farley, Professor; Ph.D., Carnegie-Mellon, 1974. Artificial intelligence, applied graph theory.

Stephen F. Fickas, Professor; Ph.D., California, Irvine, 1982. Requirements engineering, software engineering and the World Wide Web.

Anthony J. Hornof, Assistant Professor; Ph.D., Michigan, 1999. Human-computer interaction, cognitive modeling, visual search.

Virginia M. Lo, Associate Professor; Ph.D., Illinois at Urbana-Champaign, 1983. Parallel and distributed computing.

Eugene M. Luks, Professor; Ph.D., MIT, 1966. Computational complexity, algebraic algorithms, symbolic computation.

Allen D. Malony, Associate Professor; Ph.D., Illinois at Urbana-Champaign, 1990. Performance evaluation of parallel and supercomputing systems, parallel processing.

Gary Meyer, Associate Professor; Ph.D., Cornell, 1986. Computer graphics, color synthesis and reproduction.

Andrzej Proskurowski, Professor; Ph.D., Royal Institute of Technology (Stockholm), 1974. Algorithmic graph theory, models of communication networks.

Amr Sabry, Assistant Professor; Ph.D., Rice, 1994. Programming languages, semantics, compilers.

Zary Segall, Professor and Department Head; D.Sc., Technion (Israel), 1979. Building computer systems with guaranteed properties, software systems, wearable information systems.

Kent A. Stevens, Professor; Ph.D., MIT, 1979. Three-dimensional visualization, computational biomechanics.

Christopher B. Wilson, Associate Professor; Ph.D., Toronto, 1984. Computational complexity, models of computation.

Michal Young, Associate Professor; Ph.D., California, Irvine, 1989. Software engineering, software test and analysis.

Daniel Zappala, Assistant Professor; Ph.D., USC, 1997. Networks.

Deschutes Hall.

UNIVERSITY OF PENNSYLVANIA

School of Engineering and Applied Science
Department of Computer and Information Science

Programs of Study

Research and teaching at the Department of Computer and Information Science covers a wide range of topics in theory and applications, including algorithms, architecture, programming languages, compilers, operating systems, logic and computation, software engineering, databases, parallel and distributed systems, real-time systems, high-speed networks, graphics, computational biology, natural-language processing, artificial intelligence, machine vision, and robotics. Much of this work involves multidisciplinary collaborations with other departments, including the Departments of Electrical Engineering, Systems Science and Engineering, Decision Sciences, Mechanical Engineering and Applied Mechanics, Chemical Engineering, Mathematics, Linguistics, Philosophy, Psychology, Bioengineering, and Neuroscience. The department also has a number of ongoing research collaborations with national and international organizations and laboratories. The main educational goal is to prepare students for research and teaching careers in either academic institutions or industry.

The Ph.D. program combines both course work and research in one of the major computer science areas. Students must pass a written qualifying examination testing breadth in computer science and an oral examination testing depth in the general area of research.

The Master of Science in Engineering program provides course work and research training for students who generally have experience in computer science. A master's degree in virtual environments is also offered.

Research Facilities

The primary educational equipment is a collection of Sun Workstations and servers, running Sun's version of UNIX. In addition to the vast array of utilities provided with manufacturer's UNIX, the department provides various software packages and languages, including C++, C, LISP, PROLOG, Standard ML, FORTRAN 77, most GNU utilities, the MIT/X Consortium release of the X-window system, LaTex, and elm. Each CIS computer is connected to an Ethernet network and supports the TCP/IP protocol. The Ethernet is gatewayed to the Penn Campus Network and to the Internet. All faculty members and graduate students have workstations or terminals in their offices capable of accessing any system in the department. The CIS Computing Facility has five groups of computers configured for specific applications.

The General Robotics and Active Sensory Perception (GRASP) Laboratory includes one Sun 4/280, several Sun SPARCstations, and numerous smaller Sun and MicroVAX machines, all running the manufacturer's version of UNIX. They are used mainly for research in robotics and manipulator control and vision. Users are CIS researchers in robotics and vision.

The Language Information and Computation (LINC) Laboratory includes a Sun 4/490, a Sun 4/280, two Sun 4/110s, and several HP 9000s, all running the manufacturer's version of UNIX, plus numerous Symbolics LISP Machines and X-terminals. The primary applications are for research in artificial intelligence. Users are researchers in artificial intelligence.

The Graphics Laboratory includes about twenty Silicon Graphics workstations running UNIX. The primary applications are for research in computer graphics and animation. The user community consists of CIS researchers in graphics and non-CIS researchers sponsored by CIS faculty members.

The Distributed Systems Laboratory (DSL) includes several Sun 3 Workstations, an IBM PC/RT running the manufacturer's version of UNIX, and numerous microcomputer systems. Some of the computer systems include special-purpose subsystems for distributed or parallel operations. The primary applications are for research in digital design and distributed systems. The user community for this lab includes graduate and undergraduate students in electrical engineering and computer science taking classes assigned to the lab and CIS researchers in distributed systems.

The General Research Computing Facilities of CIS include a SPARCserver 2, several desktop Sun Workstations, and numerous HP 9000s, all running the manufacturer's version of UNIX. In addition, numerous X-terminals are used to connect to server machines. The primary applications used are general programming and research and administrative support (text formatting, e-mail, and others). The user community includes CIS faculty members, staff, graduate students, and non-CIS researchers as sponsored by CIS faculty members. The CIS department also has a Thinking Machines Corporation Connection Machine (CM2a) that is shared by several of the labs. The CM2a is used for research in massively parallel applications such as image and language processing.

Financial Aid

A limited number of fellowships and scholarships are available for Ph.D. candidates; competition for this funding is very intense. No funding is available for master's candidates. The University of Pennsylvania's Office of Student Financial Services for Graduate and Professional Students (http://www.upenn.edu/sfs/) has information on funding sources.

Cost of Study

Tuition and fees for full-time study for the 1999–2000 academic year are $25,674. Tuition and fees for one course are $3243.

Living and Housing Costs

The estimated cost for lodging, board, and books for a twelve-month period is $18,780. Information about on-campus housing can be found at http://www.upenn.edu/resliv/; information about off-campus living can be found at http://www.upenn.edu/resliv/ocl/.

Student Group

In 1998–99, the department had 126 doctoral students and 99 master's students.

Location

The University is located in west Philadelphia, just a few blocks from the heart of the city. Philadelphia is a twentieth-century city with seventeenth-century origins. Renowned museums, concert halls, theaters, and sports arenas provide cultural outlets for students. Fairmount Park extends through large sections of Philadelphia, occupying both banks of the Schuylkill River. The Jersey shore is not far to the east, Pennsylvania Dutch country to the west, and the Poconos to the north. Equidistant from New York City and Washington, D.C., the city of Philadelphia is a patchwork of distinctive neighborhoods that range from Colonial Society Hill to Chinatown. Students can obtain more information about the University of Pennsylvania campus and the city of Philadelphia via the World Wide Web at http://www.upenn.edu/fm/map.html and http://www.libertynet.org.mmp/ respectively.

The School

The School of Engineering and Applied Science has a distinguished reputation for the quality of its programs. Its alumni have achieved international distinction in research, management, industrial development, government service, and engineering education. Its faculty leads a research program that is at the forefront of modern technology and has made major contributions in a wide variety of fields. The School is in fact the birthplace of the modern computer, for it was at its Moore School of Electrical Engineering that ENIAC, the world's first electronic large-scale, general-purpose digital computer, was created.

Applying

Candidates must have a bachelor's degree. A candidate's academic record, letters of recommendation, and test scores are all evaluated. Scores on the GRE General Test are required; the Subject Test is not required. The Test of English as a Foreign Language (TOEFL) is required of all candidates whose native language is not English; the minimum score accepted is 600. Applicants are encouraged to use the online application, which can be found at the Web site listed below.

Correspondence and Information

Graduate Admissions
Department of Computer and Information Science
University of Pennsylvania
Philadelphia, Pennsylvania 19104-6389

Telephone: 215-898-8560
E-mail: cis-grad-admin@central.cis.upenn.edu
World Wide Web: http://www.cis.upenn.edu/

University of Pennsylvania

THE FACULTY AND THEIR RESEARCH

Rajeev Alur, Associate Professor; Ph.D., Stanford. Design tools and formal techniques for building reliable hardware and software systems, high-level design languages, requirements analysis, hardware verification, model checking, hybrid systems, design and control of multiagent control systems.

Norman Badler, Cecilia Fitler Moore Professor; Ph.D., Toronto. Computer graphics, human movement simulation, connections between language and action, three-dimensional modeling and interaction techniques.

Ruzena Bajcsy, Professor; Ph.D., Stanford. Computer vision, robotics, language and vision, biomedical imaging.

Peter Buneman, Professor; Ph.D., Warwick. Databases: database languages, information integration, digital libraries, semistructured data, Web data, and partial information; logic and computation: functional and object-oriented programming languages and types; general: cognitive science, bioinformatics, and mathematical phylogeny.

Kostas Daniilidis, Assistant Professor; Ph.D., Karlsruhe (Germany). Computer vision: visual motion analysis, 3-D reconstruction, and omnidirectional sensing; Multimedia: teleimmersion and augmented reality; Robotics: mathematics of calibration.

Susan Davidson, Professor and Co-Director of the Center for Bioinformatics; Ph.D., Princeton. Databases: database models and languages, update languages, warehouse creation and maintenance, information integration, digital libraries, semistructured data, and Web data; general: bioinformatics and real-time systems.

David Farber, Alfred S. Fitler Moore Professor of Telecommunications Systems; M.S., Stevens. High-speed computer networks, distributed computer systems, distributed collaboration and software productivity.

Peter Freyd, Professor of Computer and Information Science (secondary appointment) and of Mathematics; Ph.D., Princeton. Programming semantics, categorical algebra and geometric logic, topology and knot theory.

Jean Gallier, Professor and Graduate Chair of the Department; Ph.D., UCLA. Geometry of curves and surfaces, CAGD, motion planning, classical mechanics, robotics, computer vision.

Vijay Gehlot, Lecturer; Ph.D., Pennsylvania. Formal description techniques, real-time systems, concurrency, software specification/verification, programming languages, software engineering.

Michael Greenwald, Assistant Professor; Ph.D., Stanford. Computer networking, operating systems, software engineering.

Roch Guerin, Professor of Computer and Information Science (secondary appointment) and of Electrical Engineering; Ph.D., Caltech. Networking, quality of service, and the intersection of the two.

Carl Gunter, Associate Professor; Ph.D., Wisconsin–Madison. Programming languages, semantics, logic, software specification.

Aravind Joshi, Henry Salvatori Professor of Computer and Cognitive Science; Ph.D., Pennsylvania. Natural-language processing: computational models of syntax, semantics, and discourse; mathematical linguistics: grammar, machines, parsing, and statistical language processing; logic and computation: application to language processing; applications: machine translation, natural-language interfaces, information extraction, and summarization; general: cognitive science and artificial intelligence.

Sampath Kannan, Associate Professor; Ph.D., Berkeley. Monitoring and checking real-time systems: randomized algorithms and program checking, process algebras and temporal logic, and code instrumentation; computational biology: algorithms for phylogeny construction, and maximum likelihood estimation; contention resolution: analysis of protocol stability for age-based protocols and backoff protocol.

Vijay Kumar, Professor of Computer and Information Science (secondary appointment) and of Mechanical Engineering and Applied Mechanics; Ph.D., Ohio. Robotics, control, manipulation and locomotion, multiagent coordination.

Insup Lee, Professor; Ph.D., Wisconsin. Real-time systems: specifications and analysis, end-to-end design support, object-oriented real-time systems, runtime monitoring and checking, operating systems, and schedulability analysis; formal methods: real-time process algebra, probabilistic modeling, hybrid systems, equivalence and model checking, abstraction and refinement, runtime assurance, and tools and methodologies; mobile computing: wireless communication, infosystems, and mobile computing; adaptive wireless and mobile systems; applications: embedded systems, medical information systems, e-commerce, robotics, and distributed simulation; general: distributed systems, operating systems, and software engineering.

Mark Y. Liberman, Trustee Professor of Phonetics, Linguistics, and Computer and Information Science (secondary appointment); Ph.D., MIT. Computational linguistics, speech processing, corpus linguistics, phonetics, phonology.

Mitch Marcus, RCA Professor of Artificial Intelligence and Chair; Ph.D., MIT. Statistical natural-language processing, computational theories of grammar, cognitive science.

Dimitri Metaxas, Associate Professor; Ph.D., Toronto. Computer vision: deformable model-based shape and motion estimation, segmentation methods, augmented reality, shape from shading, sign-language recognition, facial and body tracking, and visual cue integration; medical image analysis: modeling of the heart and lungs from MRI-SPAMM, segmentation methods, minimally invasive breast surgery, joint modeling, modeling and correction of abnormal gait, medical education, and modeling the anatomy and physiology of internal organs; computer graphics: recursive dynamics and optimal control methods, gait modeling, flexible body simulation, gas and fluid simulation, and behavior modeling.

Max Mintz, Associate Professor and Undergraduate Chair; Ph.D., Cornell. Decision making under uncertainty in machine perception and robotics: robust multisensor fusion, pose estimation and map building, distributed multiagent coordination and control, statistical decision theory and applications, minimax estimation theory, robust inference techniques, set-valued estimation procedures, game theory and applications, and dynamic stochastic games.

Martha Palmer, Associate Professor; Ph.D., Edinburgh. Computational linguistics and artificial intelligence: machine translation, multilingual information retrieval and extraction, computational lexical semantics, word sense disambiguation, computational sense inventories, and empirical methods; computer graphics, cognitive science, human-computer interaction, and linguistics.

Jim Ostrowski, Assistant Professor of Computer and Information Science (secondary appointment) and of Mechanical Engineering and Applied Mechanics; Ph.D., Caltech. Robotic locomotion and medical applications of robots.

Chris Overton, Associate Professor of Computer and Information Science and of Genetics; Ph.D., Johns Hopkins. Control of gene expression during development and differentiation, bioinformatics, computational genomics, knowledge representation in biology, database evolution, transformation and integration.

Richard Paul, Professor; Ph.D., Stanford. Robotics, teleoperation, real-time numerical methods.

Benjamin C. Pierce, Associate Professor; Ph.D., Carnegie Mellon. Programming languages: type systems, subtyping, foundations of object-oriented languages, and database programming languages; distributed systems: file synchronization, process calculi, and mobile agents.

Ellen Prince, Professor of Computer and Information Science (secondary appointment) and of Linguistics; Ph.D., Pennsylvania. Discourse functions of syntax and related phenomena, language contact, code switching and dialect shift, Yiddish.

Noah Prywes, Professor Emeritus; Ph.D., Harvard. Concurrency/real-time analysis and software understanding, active networks, startups.

Andre Scedrov, Professor of Computer and Information Science (secondary appointment) and of Mathematics; Ph.D., SUNY at Buffalo. Mathematical logic, programming structures, computer security, randomized protocols.

Barry Silverman, Professor of Computer and Information Science (secondary appointment) and of Systems Engineering; Ph.D., Pennsylvania. Knowledge-based systems: decision-support methodology, life-cycle knowledge management workbenches; visual software programming environments and open software standards; human-computer interaction and cognitive engineering; medical informatics; computer-aided decision support environment.

Jonathan Smith, Professor; Ph.D., Columbia. Distributed systems, operating systems, multimedia communications systems, applications of randomness, computer security, cryptology.

Val Tannen, Associate Professor; Ph.D., MIT. Programming languages, databases, logic in computer science.

Camillo J. Taylor, Assistant Professor; Ph.D., Yale. Computer vision: 3-D reconstruction, object recognition, and vision-based human-computer interfaces; robotics: vision-based control of mobile robots, human-robot interfaces.

Lyle Ungar, Associate Professor; Ph.D., MIT. Machine learning and database mining: text mining, clustering methods, data cleaning, statistical methods for machine learning; market-based systems: auction design, intelligent agent design, and distributed optimization.

Scott Weinstein, Professor of Computer and Information Science (secondary appointment) and of Philosophy; Ph.D., Rockefeller. Logic in computer science, finite model theory and its connections with complexity theory and database theory.

Shiyou Zhou, Assistant Professor; Ph.D., Rutgers. Randomness and computation, specifically combinatorial constructions and their applications to achieving pseudorandomness; data structure and algorithm design; protocol design for adaptive wireless mobile communication networks.

UNIVERSITY OF PITTSBURGH

Department of Information Science and Telecommunications
Graduate Program in Information Science

Programs of Study

Programs of study lead to the Master of Science in Information Science (M.S.I.S.), the Ph.D. in information science, and a post-master's Certificate of Advanced Study. Graduate degrees in telecommunications are described in the telecommunications listing. The M.S.I.S. requires 36 credit hours of course work and can be completed in four terms of full-time study. Students may pursue such specialties as cognitive science (including artificial intelligence, neural networks, natural language processing, and human cognition), human-computer interface and visualization, information retrieval, networks and telecommunications, and systems analysis and design. Two joint-degree programs are offered, one with the Graduate School of Public and International Affairs and the other with the Center for Biomedical Informatics. The M.S.T. degree, designed for both beginning and experienced telecommunications professionals, is described in a separate listing. The Ph.D. programs provide research-oriented study and professional specialization in the sciences of information and telecommunications. Candidates must give evidence of superior scholarship, mastery of a specialized field of knowledge, and the ability to do significant and relevant research. The Ph.D. in information science requires 36 course or seminar credits beyond the master's degree, successful completion of the preliminary and comprehensive examinations, three terms of full-time academic study on campus, 6 credits of linguistics, at least 18 dissertation credits, and submission and defense of a dissertation.

Research Facilities

Departmental computing and networking labs are housed in a modern 5,000-square-foot area. The school's computational facilities include a Sun Enterprise 4000 SMP ten-processor system with associated Sun ULTRAsparc 2 systems that collectively are arranged via ATM interconnects to supply both load-sharing and course-grain distributed computing facilities in order to support research activities. The above cluster storage needs are addressed by two RAID arrays that facilitate large-scale data gathering and processing. A HYPERsparc upgraded Sun 670 serves as a general student access system. Workstations in the labs include Sun SPARC 5s, IPXs, and IPCs. Two labs are configured as classrooms with Pentium systems. The internal labs' network environment is being upgraded to employ a mixture of both ATM and fast Ethernet technology. Laser printing is provided throughout the labs. The telecommunications and network labs are built around a heterogeneous collection of UNIX workstations and microcomputer systems. Ethernet, token ring, and ATM test network environments are maintained for research as well as to supplement classroom instruction. These labs also house training, diagnostic, and testing equipment for AT&T phone systems, T1 and T3 connections, and M13, FT3C, and TASI transmission equipment. All workstations and PCs in the labs are linked via a local area network to general-purpose University UNIX and VMS systems. Additional University computer facilities include an advanced technology and computer graphics lab and other labs located throughout the campus. All students have access to national and international electronic mail and assorted other network services.

Financial Aid

In 1998–99, 33 percent of full-time graduate students received financial aid, mainly in the form of full or half graduate assistantships. Full graduate student assistants (GSAs) earned $4325 per term plus remission of tuition and assisted a faculty member for 20 hours per week. Half GSAs earned half the stipend and remission of half the tuition for 10 hours of work per week. Financial aid awards are granted on the basis of academic achievement and financial need. Assistantships are normally awarded for the fall and spring terms. Budget permitting, they may also be offered for the summer term.

Cost of Study

The tuition per term (four months) in 1998–99 for full-time study (9–15 credits) was $8680. Because Pitt is state-related, the University received funding from the state that enabled it to reduce tuition for Pennsylvania residents to $4216 per term. Tuition for part-time students was $720 per credit for out-of-state students and $348 per credit for residents of Pennsylvania.

Living and Housing Costs

Pittsburgh, ranked by Rand McNally among the most livable cities in the United States, is noted for its low cost of living. Average monthly rent is $400 for a one-bedroom apartment and $575 for a two-bedroom apartment; it is estimated that students require at least $2200 per term to cover living expenses exclusive of tuition. Comfortable and affordable housing in attractive residential neighborhoods is readily available within walking distance of the University.

Student Group

The current enrollment in information science graduate programs is 247 graduate students, of whom approximately 49 percent are full-time. Thirty-six percent are international students and 43 percent are women. Members of minority groups represent approximately 18 percent of the U.S. enrollment.

Location

The University is located in the heart of the city's educational center, with museums of art and natural history, music and lecture halls, Carnegie Mellon University and two smaller colleges, restaurants and shops, and a 450-acre park adjacent to the campus. The downtown corporate and cultural center is just a 10-minute bus ride away.

The University

The University of Pittsburgh, a privately organized state-related institution, enrolls approximately 32,000 students. The School of Information Sciences has an enrollment of 733 students in four programs (three graduate and one undergraduate).

Applying

Applicants for all programs must submit a recent score (within three years) from the GRE General Test. Requirements for admission to the M.S.I.S. degree program are a degree from an accredited college or university with a 3.0 GPA and a 3-credit course in a structured programming language (preferably C), math, statistics, and cognitive science. Ph.D. applicants must have a master's degree from an accredited program with a QPA of 3.3 or better, the same prerequisites as the M.S.I.S. degree program, and additional courses in mathematics. Provisional acceptance into any of the degree programs may be granted to students lacking some of the prerequisites, with the condition that deficiencies be made up during the first two terms. A $30 application fee is required ($40 for international applicants). Deadlines for receipt of application materials are July 1 for fall admission, November 1 for spring admission, and March 1 for summer admission. Applications for financial aid should be submitted by January 15 for fall term and October 1 for spring term. Candidates are usually notified of acceptance within six weeks of receipt of all application materials.

Correspondence and Information

Admissions Coordinator
505 SLIS Building
University of Pittsburgh
Pittsburgh, Pennsylvania 15260

Telephone: 412-624-5146
Fax: 412-624-2788
E-mail: isadmit@sis.pitt.edu
World Wide Web: http://www.sis.pitt.edu/~dist

University of Pittsburgh

THE FACULTY AND THEIR RESEARCH

Toni Carbo, Professor and Dean, School of Information Sciences; Ph.D., Drexel. National and international information policies, measurement and use of scientific and technical information, role of information in the economy, education for the information professions.

Stephen C. Hirtle, Associate Professor and Chair; Ph.D., Michigan. Spatial information classification, mathematical psychology, cognitive science, geographic information systems, hypertext and multimedia systems, visualization, neural networks.

Sujata Banerjee, Assistant Professor; Ph.D., USC. Design and analysis of high-speed networking protocols, traffic modeling, network reliability, concurrency control, failure recovery of distributed database systems.

Stefan Brass, Assistant Professor; Ph.D., Hannover (Germany). Databases, World Wide Web query languages and search agents, logic programming, and knowledge representation.

Marek Druzdzel, Assistant Professor; Ph.D., Carnegie Mellon. Decision support systems, strategic business planning, decision making under uncertainty, decision-theoretic methods in intelligent information systems.

Ida M. Flynn, Assistant Professor and Director of the Undergraduate Program; Ph.D., Pittsburgh. Information science, computer science, mathematics, education, information retrieval systems for young users, multimedia systems.

Roger R. Flynn, Associate Professor; Ph.D., Pittsburgh. Education in information science, knowledge representation and inference, database design, artificial intelligence, systems analysis and design, data structures, human-computer interaction, database management systems.

Charles P. Friedman, Professor and Director, Center for Biomedical Informatics; Ph.D., North Carolina at Chapel Hill. Biomedical informatics, clinical reasoning, program evaluation.

Wesley Jamison, Associate Professor (Greensburg Campus); Ph.D., Penn State. Human-computer interaction, computer-supported cooperative work, information technology, human factors, telecommunications.

Michael Lewis, Associate Professor; Ph.D., Georgia Tech. Operator modeling in human-machine systems, ecological models of visualization, virtual realities.

Douglas Metzler, Associate Professor; Ph.D., California, Davis. Artificial intelligence, cognitive science, knowledge representation, natural language processing, expert systems, information storage and retrieval, cognitive modeling, intelligent tutoring systems, education systems, research methods and statistics.

Paul Munro, Associate Professor; Ph.D., Brown. Connectionist systems, neural information processing, image processing, modeling and simulation, cognitive science, models of learning, visualization, genetic algorithms and artificial life.

Kai A. Olsen, Professor; Molde College (Norway); M.S., Norwegian Institute of Technology. Visualization, visual languages, information retrieval, programming languages, operating systems, computers in society.

Edie M. Rasmussen, Associate Professor and Chair, Department of Library and Information Science; Ph.D., Sheffield (England). Information storage and retrieval, applications of parallel processors to information retrieval, geographic information systems, indexing systems and software, microcomputer applications.

Kenneth M. Sochats, Assistant Professor; M.S.E.E., Pittsburgh. Information networks, simulation, databases, artificial intelligence, management information systems (MIS), systems analysis and design, software engineering, network design, microcomputer applications, graphics.

Michael B. Spring, Associate Professor; Ph.D., Pittsburgh. Collaborative authoring, document processing and office automation, client server systems, interactive system design, standards and standardization.

Richard A. Thompson, Professor and Codirector, Telecommunications Program; Ph.D., Connecticut. Communications switching systems, especially photonic switching; intelligent networks; terminals, user services, and the human interface; fault tolerance and cellular automata; probabilistic formal languages.

David W. Tipper, Associate Professor; Ph.D., Arizona. Design and performance analysis of computer and telecommunication networks, control of communication networks, simulation methodology, queuing theory with emphasis on nonstationary/transient behavior, network survivability, application of control theory to communication networks and queuing systems.

Martin B. H. Weiss, Associate Professor and Codirector, Telecommunications Program; Ph.D., Carnegie Mellon. Telecommunications policy, technical standards, information system capacity management, network management and control.

James G. Williams, Professor; Ph.D., Pittsburgh. Information systems, networks, systems design, software engineering, simulation, system architecture, client server computing, database management.

Taieb Znati, Associate Professor; Ph.D., Michigan State. Real-time communication, networks and protocols to support multimedia environments, multimedia synchronization and presentation, design and analysis of medium access control protocols to support distributed real-time systems, network performance.

UNIVERSITY OF ROCHESTER

Department of Computer Science

Program of Study

The Department of Computer Science at the University of Rochester offers an intensive research-oriented program leading to the degree of Doctor of Philosophy. Emphasis is currently being placed on the areas of natural language understanding, robotics and machine perception, systems software for parallel and distributed computing, and the theory of computation. A number of joint faculty appointments and programs with other departments (including the Departments of Linguistics, Mathematics, Philosophy, Electrical and Computer Enginering, Psychology, Cognitive Science, Dermatology, and Neuroscience) add breadth to the program. Additional enrichment is gained from an extensive program of seminar presentations and the participation of visiting professors.

Milestones in the doctoral program include a broad comprehensive exam at the end of the first year, a deeper area exam in the student's chosen subfield at the end of the second year (at which point the master's degree is generally awarded), and a thesis proposal at the end of the third year. Completion of a doctoral dissertation typically requires one to two more years, with formal feedback from the student's thesis committee twice a year.

The department is entering its twenty-fifth year of operation and has a young, energetic faculty, with a student-faculty ratio of 3.5:1. Students receive individual attention in the shaping of their graduate programs and have an active role in the design of laboratory facilities and software.

Research Facilities

The department is well equipped. Its local area network connects sixty Sun SPARCstations and Ultra Workstations, miscellaneous other workstations and file servers, a twelve-node Silicon Graphics Challenge multiprocessor, an eight-processor Sun Ultra Enterprise 4000, a four-processor Sun Ultra Enterprise 3000, and a thirty-two-processor, 600-MHz DEC Alpha cluster connected by DEC's Memory Channel backbone. The Computer Vision and Robotics Laboratory is equipped with two Unimation Puma robot arms, a Utah/MIT four-fingered anthropomorphic hand, a custom-designed "robot head," and numerous DataCube real-time image processing boards. The lab is equipped with a "Quickframe" digital video editing/recording device and a PHANTOM haptic feedback device used for "augmented reality" research. In the Virtual Reality Laboratory, there are eye trackers, a head-mounted stereo display, high-performance Silicon Graphics workstations, and a specially designed "dual" PHANTOM haptic feedback device.

Financial Aid

The department works to ensure that all doctoral students are fully supported by graduate assistantships or fellowships. In return, students are expected to participate in the department's research activities from the outset of their graduate career and to assist with teaching duties during two or three semesters over the course of the graduate program. The appointment in 1999–2000 provides an assistantship stipend of $1430 per month as well as full tuition remission. Many students accept summer research support from the department (a twelve-month total of $17,160); most others take summer jobs in industry.

Cost of Study

Full-time graduate tuition for the 1999–2000 academic year is $22,304 for 32 credit hours but is waived for supported students. Mandatory health fees ($338 for 1999–2000) are the responsibility of the student; required medical insurance is available from the University for $636.

Living and Housing Costs

University-owned housing facilities include more than 800 apartments. Rents range from $365 per month for a single furnished sleeping room to $645 per month for a furnished two-bedroom apartment. The Housing Office maintains a listing of accommodations near the University. A comprehensive board plan is available.

Student Group

The current full-time population at the University of Rochester is about 7,300 students, including more than 2,500 graduate students. There are 40 full-time graduate students in the Department of Computer Science, all of whom are supported by assistantships, fellowships, or tuition grants.

Student Outcomes

Many of the department's graduates have gone on to academic appointments at top schools, including Carnegie Mellon, Stanford, Maryland, Illinois, Pennsylvania, North Carolina, Wisconsin, Northwestern, Rice, Boston, Virginia, Princeton, Chicago, MIT, and Oregon. Those accepting nonacademic positions have gone to top research labs, including Bell Labs, IBM, Xerox, Olivetti Research Center, Philips Laboratories, Lockheed, Matsushita, the International Computer Science Institute, Microsoft, Martin Marietta, David Sarnoff Research Center, Siemens, Digital Equipment Corporation, and Silicon Graphics.

Location

Located on the south shore of Lake Ontario, a short drive from the Finger Lakes region, Rochester is a cultural center of upstate New York and has a metropolitan area population of just over a million. Opportunities for cultural activities are offered the year round by the Strasenburgh Planetarium, the University's Memorial Art Gallery, the International Museum of Photography, the Rochester Museum and Science Center, and the Eastman Theatre. Rochester and the surrounding area have many lovely parks, and in the winter there are many ski areas within an hour's drive. Known as the photographic and optical capital of the world, Rochester is the home of Eastman Kodak, the Xerox Corporation, Bausch & Lomb, and many other high-tech companies.

The University

The University of Rochester is an independent university that offers more than forty-five doctoral programs and some ninety master's degree programs in the following schools and colleges: the College that is made up of Arts and Sciences and the School of Engineering and Applied Sciences, the Eastman School of Music, the School of Medicine and Dentistry, the School of Nursing, the William E. Simon Graduate School of Business Administration, and the Margaret Warner Graduate School of Education and Human Development. The River Campus, where the Department of Computer Science is located, is situated on the east bank of the Genesee River, about 2 miles south of downtown Rochester.

Applying

Applicants are urged to apply electronically (see addresses below) or write directly to the department for application forms and additional information. For maximum consideration for the fall term, the application, transcripts, recommendations, and scores on the Graduate Record Examinations (GRE) must be submitted no later than February 1. A $25 application fee is waived if Part I is postmarked by December 15. The GRE should be taken no later than December. The General Test is required, and the Subject Test in computer science, math, or physics is highly recommended. International students should also submit their scores on the Test of English as a Foreign Language (TOEFL).

Correspondence and Information

Admissions Committee
Department of Computer Science
University of Rochester
Rochester, New York 14627-0226

Telephone: 716-275-5478
E-mail: admissions@cs.rochester.edu
World Wide Web: http://www.cs.rochester.edu

University of Rochester

THE FACULTY AND THEIR RESEARCH

David H. Albonese, Assistant Professor of Electrical and Computer Engineering and of Computer Science; Ph.D., Massachusetts Amherst, 1996. Computer architecture, microprocessor design, adaptive architectures, multiprocessor systems, performance evaluation.

James F. Allen, John H. Dessauer Professor of Computer Science; Ph.D., Toronto, 1979. Artificial intelligence; natural language processing; dialog systems; planning; representation of plans, goals, time, and action.

Dana H. Ballard, Professor of Computer Science; Ph.D., California, Irvine, 1974. Computer vision, artificial intelligence, computational neuroscience.

Christopher M. Brown, Professor of Computer Science; Ph.D., Chicago, 1972. Artificial intelligence, computer vision, graphics, robotics.

Sandhya Dwarkadas, Assistant Professor of Computer Science; Ph.D., Rice, 1992. Parallel and distributed computing, compiler and run-time support for parallelism, computer architecture, networks, simulation methodology, performance evaluation, parallel applications research.

George Ferguson, Scientist, Computer Science; Ph.D., Rochester, 1995. Intelligent systems, planning, knowledge representation, agents and architectures, collaboration.

Lane A. Hemaspaandra, Associate Professor of Computer Science; Ph.D., Cornell, 1987. Computational complexity theory, algorithms from complexity, probabilistic and unambiguous computation, approximate computation, fault-tolerant computation, semi-feasible algorithms, cryptography, complexity-theoretic aspects of voting systems.

Robert A. Jacobs, Associate Professor of Brain and Cognitive Sciences and of Computer Science; Ph.D., Massachusetts, 1990. Machine learning, reasoning under uncertainty, artificial intelligence, computational cognitive science.

Kyros Kutalakos, Assistant Professor of Dermatology and of Computer Science; Ph.D., Wisconsin, 1994. Computer vision, augmented reality, interactive 3-D graphics, human-computer interaction, robotics.

Henry Kyburg, Gideon Webster Burbank Professor of Moral and Intellectual Philosophy and Professor of Computer Science; Ph.D., Columbia, 1954. Uncertain inference, nonmonotonic logic, logical foundations of probability and statistical inference, measurement theory, cognitive science and artificial intelligence.

Thomas J. LeBlanc, Professor of Computer Science and College Dean of the Faculty of Arts, Sciences, and Engineering; Ph.D., Wisconsin, 1982. Parallel programming environments, multiprocessor systems, parallel program debugging and performance tuning.

Randal C. Nelson, Associate Professor of Computer Science; Ph.D., Maryland, 1988. Artificial intelligence, computer vision with an emphasis on the use of visual information for control of systems in real-world environments, robotics.

Mitsunori Ogihara, Assistant Professor of Computer Science; Ph.D., Tokyo Institute of Technology, 1993. Computational complexity theory, recursive function theory, number-theoretic algorithms.

Thaddeus F. Pawlicki, Instructor/Undergraduate Advisor; Ph.D., Buffalo, 1989. Computer vision, medical image processing.

Lenhart K. Schubert, Professor of Computer Science; Ph.D., Toronto, 1970. Knowledge representation and organization, general and specialized inference methods, natural language understanding, planning and acting.

Michael L. Scott, Associate Professor and Chair of Computer Science; Ph.D., Wisconsin, 1985. Parallel and distributed systems software, operating systems, programming languages, program development tools.

Joel I. Seiferas, Associate Professor of Computer Science; Ph.D., MIT, 1974. Computational, descriptional, and combinatorial complexity; lower bound techniques.

UNIVERSITY OF ST. THOMAS

Graduate Programs in Software

Programs of Study

Graduate Programs in Software (GPS), created in 1985 at the University of St.Thomas, has more than 550 active graduate students and is the largest program of its kind in the U.S. Three master degrees and three graduate-level certificates are offered. Classes are available evenings and weekends. These graduate programs are accredited by the North Central Association of Colleges and Schools and are approved by the Minnesota Higher Education Board/Services.

Course content focuses on improving the software development environment by providing knowledge, skills, and experience that apply classroom theory to the workplace. Faculty members have extensive industrial software development experience.

The Master of Science (M.S.) in software engineering degree emphasizes the technical nature of systems software development; the Master of Software Design and Development (M.S.D.D.) degree focuses on applications software development. The Master of Software Systems (M.S.S.) degree provides an opportunity for recent college graduates, career changers, and professionals in education, medicine, business, science, and other functional areas to pursue software development. The M.S. and M.S.D.D. programs consist of fourteen graduate courses. The M.S.S. program consists of ten graduate courses. Concentrations are available in object-oriented, computer communications, embedded systems, knowledge-based systems, databases, and information retrieval. The Certificate in Information Systems provides an opportunity to pursue an M.S. or M.S.D.D. degree without software development experience. The Certificate of Software Design and Development is designed for those interested in upgrading their software skills without completing a master's degree. The Certificate of Advanced Study is for software professionals who have a master's degree in software engineering, computer science, or a closely related field. Nondegree professional development seminars are available for individuals at the Univeristy as well as at a company's site.

Research Facilities

The Graduate Programs in Software's full-time faculty members engage in applied research with local industry. Areas of research collaboration include parallel processing, artificial intelligence, data warehouse, software quality assurance, relational and object databases, architecture performance, microprocessors, computer architecture, objects, and systems architecture.

Financial Aid

Financial aid is available through a variety of private, institutional, and federal programs on both a need and merit basis. For information on these programs, students should contact Student Financial Services at 651-962-6550. In addition, many employers pay or subsidize part or all of their employee's tuition expenses.

Cost of Study

Tuition for 1998–99 was $410 per graduate credit hour for master's and graduate certificate programs. Book expenses vary by program and course.

Living and Housing Costs

Graduate students may apply for housing through Residence Life at 651-962-6470.

Student Group

Most students in the graduate programs are employed full-time during the day and attend classes evenings and Saturdays. Approximately 33 percent of GPS students are international students.

Local corporations employ international students (20 hours per week during the academic year and full-time during the summer) who have taken the basic software development courses. Assistance in finding employment opportunities is available to all GPS students.

Student Outcomes

Most students in the graduate programs work full-time with local industry. For students seeking employment, placement assistance is available through GPS. The program's students and alumni are highly respected for their software development knowledge, skills, and expertise.

Location

The University of St. Thomas is located 5 miles west of St. Paul and 3 miles east of Minneapolis along the Mississippi River.

The University

The University of St. Thomas was founded in 1885 by Archbishop John Ireland. What began as the St. Thomas Aquinas Seminary—with 62 students and a faculty of 5—has grown to be Minnesota's largest independent university, with four campuses and more than 10,000 students, half of whom are working adults who take graduate courses in the late afternoon, early evening, and weekends. Built near a river bluff, the University's main campus is midway between the downtowns of St. Paul and Minneapolis.

Applying

Students entering the M.S., M.S.D.D., M.S.S., or Certificate in Information Systems programs are required to have a bachelor's degree from an accredited educational institution and a GPA of 2.7 or better. International applicants must submit either a minimum TOEFL score of 550 or a minimum MELAB score of 85. Students desiring to enter the M.S. or M.S.D.D. program are required to have software development experience. Students must submit an application, official transcripts, an essay, a list of computer courses taken, a resume, and a $30 nonrefundable application fee.

Correspondence and Information

Graduate Programs in Software
University of St. Thomas
2115 Summit Avenue, OSS301
St. Paul, Minnesota 55105-1079

Telephone: 651-962-5500
Fax: 651-962-5543
E-mail: gradsoftware@stthomas.edu
World Wide Web: http://www.GPS.stthomas.edu

University of St. Thomas

THE FACULTY

Bernice M. Folz, Professor and Director, Graduate Programs in Software; Ph.D., Minnesota.
Michael Allen, Adjunct Lecturer; Ph.D., Ohio State. CEO of Allen Interactions, primary architect of Authorware Professional.
Bonnie Holte Bennett, Associate Professor; Ph.D., Minnesota.
Gary Berosik, Adjunct Lecturer; B.A., St. Thomas (Minnesota). Systems Architect/Educator, West Group.
Keith D. Cramer, Adjunct Lecturer; M.B.A., M.S., Minnesota. Staff Programmer, IBM Rochester.
William D. Darling, Adjunct Lecturer; M.S.S.E., St. Thomas (Minnesota).
Susan M. Dray, Adjunct Lecturer; Ph.D., UCLA. President, Dray & Associates.
Patrick S. Gonia, Adjunct Lecturer; M.S.E.E., Minnesota. Program Manager, Honeywell.
Dmitry L. Gringauz, Adjunct Lecturer; B.S., Notre Dame (California). Tech Lead, ADC Telecommunications.
James K. Habinek, Adjunct Lecturer; Ph.D., Loyola Chicago. Human Factors Engineer, IBM Rochester.
Magdy S. Hanna, Associate Professor; Ph.D., Minnesota.
David R. Hoska, Adjunct Lecturer; Design Technician, St. Paul Technical Institute; Certified Manufacturing Engineer: Special Machine Design and Robotics; Consultant, D & D Engineering, Inc.
David R. Jones, Adjunct Lecturer; M.S., St. Thomas (Minnesota). Systems Design Engineer, Unisys.
Byung Suk Lee, Assistant Professor; Ph.D., Stanford.
Eric V. Level, Adjunct Lecturer; M.S., Minnesota. 47 Software.
Dennis R. McClelland, Adjunct Lecturer; Ph.D., Iowa. Technical Writer, Guidant/CPI.
Bhabani S. Misra, Associate Professor; Ph.D., North Dakota State.
Nasiruddin Mohammed, Adjunct Lecturer; M.S., LSU. Principal Research Engineer, Honeywell.
Ted Mongeon, Adjunct Lecturer; M.C.I.S., Minnesota. Software Engineer, IBM Rochester.
Ernest L. Owens Jr., Instructor in Management, M.B.A., St. Thomas (Minnesota).
Robert L. Raymond, Associate Professor; Ph.D. Minnesota.
Carolyn Sandberg, Adjunct Lecturer; J.D., Hamline. Attorney, Honeywell.
David Spoor, Adjunct Lecturer; M.S., Illinois. Senior Principal Research Scientist, Honeywell.
Paul D. Stachour, Adjunct Lecturer; Ph.D., Waterloo. Software Engineer and Methodologist.
Arthur Sumner, Adjunct Lecturer; M.I.M., M.B.A., St. Thomas (Minnesota). Computer Analyst, LimeLight, Inc.
Paul Thompson, Adjunct Lecturer; Ph.D., Berkeley. Senior Research Scientist, West Publishing.
Nancy E. Thornton, Adjunct Lecturer; Ph.D., Arizona State. Multimedia Developer/Trainer, Allen Interactions.
Jamshid A. Vayghan, Adjunct Lecturer; M.S., Minnesota. Senior Software Architect, IBM Rochester.
Robert G. Waite, Adjunct Lecturer; Ph.D., Kentucky. Information Developer, IBM.
Gerald R. Werth, Adjunct Lecturer; M.D., Wisconsin–Madison; Ph.D., Minnesota. Principal Consultant, Clinician Computing, Inc.
David M. West, Associate Professor; Ph.D., Wisconsin–Madison.
Stephen Zvolner, Adjunct Lecturer; M.S., Northeastern Illinois. Zvolner Software.

UNIVERSITY OF SOUTH CAROLINA

College of Science and Mathematics
Department of Computer Science

Programs of Study

The Department of Computer Science offers programs leading to Master of Science (M.S.) and Doctor of Philosophy (Ph.D.) degrees.

The requirements for the M.S. degree can be satisfied in two ways: with a project thesis and eleven courses or with a research thesis and nine courses. About one third of the M.S. program course work is in the core areas of algorithms, architecture, and compiler construction. This leaves two thirds of the course work as electives that can be focused as the student desires. This program generally requires two years, but frequently has been completed in less time.

Requirements for the Ph.D. degree include a written qualifying examination and a public defense of the dissertation. There are no required courses for the Ph.D., which allows students to focus their course work in support of their research. Strong research groups in computer networks, image compaction, object-oriented methodologies, information systems, and artificial intelligence welcome new students and provide substantial opportunities for research support.

Research Facilities

The department has laboratories for research in image processing, mobile computing, image compaction, artificial intelligence, scientific modeling and visualization, database, and networks. The department has approximately 80 UNIX workstations, including a lab of DEC alpha machines with dual 160 MHz processors. These machines and several PC labs are connected to each other and the Internet via Ethernet and an FDDI ring.

Financial Aid

Financial aid is available to students in the form of research assistantships, teaching assistantships, and fellowships. The Metro Information Services Fellowship awards $20,000 per year and is renewable. There are also a number of other supplemental fellowships. Four departmental research groups, along with other researchers in the College, recently received a $9.3-million grant from the Department of Defense, providing a large number of $15,000 research assistantships. The typical starting stipend for teaching assistants is $11,000 for nine months, with additional support available for the summer. There are also many opportunities for internships and research assistantships in local industries.

Cost of Study

Tuition and fees for 9 graduate hours per semester currently cost $760 for graduate assistants, $1947 for nonassistant residents of South Carolina, and $4057 for nonassistant nonresidents.

Living and Housing Costs

For 1998–99, University housing provided single dorm rooms for $375 per month and two-bedroom apartments for $622 per month. These prices included all utilities, local phone, and cable service.

Student Group

The department has 130 graduate students, 25 of whom are in the Ph.D. program. At the university level, more than 25,000 students are enrolled on the Columbia campus, of whom 8,000 are enrolled in graduate programs.

Student Outcomes

Many of the recent M.S. graduates have worked on internships and co-op opportunities in local industry, which led to full-time employment at the completion of their degrees. Examples of such companies include AT&T, NCR, IBM, and Summus. Many graduates have found employment in the NC Research Triangle Park and Atlanta areas.

Location

The University is located in Columbia, the capital of South Carolina. Columbia has many attractive features and combines the advantages of a metropolitan area with the pace of smaller cities. There are nearby lakes for recreation, a very nice zoological park, and numerous cultural activities. Columbia is only a short distance from the South Carolina beaches and the Blue Ridge Mountains.

The University

The University was founded in 1801 and has grown to an eight campus system with 38,000 students. The graduate school offers master's degrees in 167 areas, the Ph.D. in sixty-eight, and professional doctorates in law, medicine, pharmacy, and public health. About 1,800 master's and 250 Ph.D. degrees are awarded annually.

Applying

Applications for admission are considered throughout the year, but must be received by February 15 to receive full consideration for financial aid. Applicants are required to supply GRE General Test scores, official copies of transcripts of prior academic work, and two letters of recommendation. International applicants are also required to take the TOEFL and score at least 570.

Correspondence and Information

Director of Graduate Studies
Department of Computer Science
University of South Carolina
Columbia, South Carolina 29208
Telephone: 803-777-7849
E-mail: graduate@cs.sc.edu
World Wide Web: http://www.cs.sc.edu

University of South Carolina

THE FACULTY AND THEIR RESEARCH

Carter Bays, Professor; Ph.D., Oklahoma, 1974. Algorithms, simulation, programming languages, functional programming, operating systems.

Robert L. Cannon, Professor; Ph.D., North Carolina, 1973. Digital image processing, expert systems, text collation.

Karel Culik II, Professor; Ph.D., Czechoslovak Academy of Sciences, 1966; RNDr., Charles University, 1967. Formal languages and automata theory, theory of programming languages, systolic systems, cellular automata.

Caroline M. Eastman, Professor; Ph.D., North Carolina, 1977. Information retrieval, database management systems, file organizations, programming and natural languages.

Stephen A. Fenner, Associate Professor; Ph.D., Chicago, 1991. Computational complexity theory, programming languages, semantics.

Toshiro Kubota, Assistant Professor; Ph.D., Georgia Tech, 1995. Computer vision, image processing, neural networks.

Manton M. Matthews, Associate Professor; Ph.D., South Carolina, 1980. Artificial intelligence, parallel processing, user interfaces, graph theory.

Robert L. Oakman, Professor; Ph.D., Indiana University, 1971. Computer literacy, computational linguistics, natural language processing, humanities applications.

John R. Rose, Assistant Professor; Ph.D., SUNY at Stony Brook, 1991. Machine learning, knowledge discovery in databases, computational chemistry knowledge-based systems.

Abhijit Sengupta, Professor; Ph.D., Calcutta (India), 1976. Fault-tolerant computing, multiple-valued logic, parallel architectures.

M. A. Sridhar, Associate Professor; Ph.D., Wisconsin, 1986. Analysis of algorithms, automata theory, compiler design, complexity theory.

Marco Valtorta, Associate Professor; Ph.D., Duke, 1987. Artificial intelligence, expert systems.

UNIVERSITY OF SOUTHERN CALIFORNIA

Computer Science Department

Programs of Study

The department offers the degrees of Master of Science and Doctor of Philosophy. In addition, the department offers an M.S. degree with specialization in computer networks, an M.S. degree with specialization in software engineering, an M.S. degree with specialization in multimedia and creative technologies, and an M.S. degree with specialization in robotics and automation. The M.S. degree requires completion of 27 units of course work with a minimum grade point average of 3.0. No thesis is required for the M.S. degree. Full- or part-time study is possible, and one year is the expected time for completion of the M.S. program for full-time students with adequate backgrounds.

The doctoral program in computer science requires 60 units of course work, with a minimum grade point average of 3.5, and at least 4 units of dissertation research. All course work beyond the M.S. must be taken in residence at USC. Ph.D. students must pass through a screening process about a year into the program. Students must then pass the qualifying examination to be admitted to candidacy and later complete the dissertation to receive the Ph.D. A well-qualified full-time student should be able to finish the program within four to five years of entering graduate school.

The USC Interactive Instructional Television Network broadcasts regular University courses to companies in the Los Angeles area. Students employed at these companies may take some of their degree courses through this system without having to commute frequently to campus.

Research Facilities

The research facilities of the Computer Science Department are divided into the following laboratories: Programmable Automation, Robotics, Computer Vision, Molecular Robotics, Brain Simulation, Computational and Biological Vision, Molecular Science, Computer Networks and Distributed Systems, Database, Software Engineering, Computer Graphics and Creative Technologies, and Computer Animation. Each lab is run by 1 or more faculty members and each has a wide variety of equipment. Most laboratories have Sun-3 and Sun-4 Workstations, some have IBM RISC System 600 workstations and servers, and others have Symbolics Lisp Machines. In addition to the individual laboratory facilities, the department and graduate students have access to a Sun-4/490 with 64 MB of memory that is used for homework, word processing, and general departmental administration. The graduate students have a private laboratory that contains a variety of Sun-4, IBM, and DEC workstations plus personal computers, including Apple Macintosh and NeXT machines. Laser printers are available for output. All machines are connected to the local network and to the Internet, providing access to researchers all over the world.

Financial Aid

Teaching and research assistantships are available for qualified Ph.D. students. These awards cover 12 units of tuition each semester and during the summer, in addition to providing a nine-month stipend of $1336 to $1481 per month in 1999–2000. In addition, the Graduate School awards certain fellowships and scholarships. Interested individuals should contact the Graduate School directly. Information on other types of state and federal programs can be obtained from the Financial Aid Office.

Cost of Study

In 1999–2000, tuition is $10,630 per semester for full-time students and $758 per unit for students taking fewer than 15 units. Student fees are $350, and the optional parking fee is $306.

Living and Housing Costs

For students living on campus, the cost of room and board ranges from $3500 to $4000 per semester in 1999–2000. Comparable housing is available in the immediate vicinity of the campus for about 25 to 40 percent more. Housing can be found at all price ranges in the Greater Los Angeles area.

Student Group

USC has a total enrollment of approximately 27,000 students, 36 percent of whom are women. The School of Engineering has about 3,500 students, of whom 54 percent are graduate students.

The Computer Science Department has a total of about 360 graduate students; roughly 220 are M.S. students, and the remaining 140 are Ph.D. students. The student body is very diverse, with students from all over the United States and many other countries.

Location

Only 5 minutes by freeway from the center of Los Angeles, the University is secluded on an extensive, landscaped campus with a quiet academic atmosphere. Students may take advantage of the broad cultural offerings of a major metropolis. USC is situated midway between the mountains and the sea—less than an hour from each—offering students a choice of many outdoor diversions, such as surfing, boating, hiking, and skiing.

The University

The University of Southern California is private, nonsectarian, and coeducational. It is the oldest major university in the West. USC's diversity of programs has marked it as one of the most distinguished universities in the world. Its membership in the Association of American Universities attests to its educational excellence in general and graduate and research excellence in particular. Attracted by the variety and high quality of the University's offerings, faculty members and students come from every corner of the globe to engage in teaching, learning, and research. At USC, freedom and responsibility characterize the atmosphere—for the freshman as well as the postdoctoral student.

Applying

Admission to the Computer Science Department is highly competitive. Applicants are required to have a bachelor's degree from an accredited college or university and must have taken the GRE General Test. The GRE Subject Test in computer science, mathematics, or engineering is recommended. Three letters of recommendation from professors and an essay on specific research goals must be submitted to the department. International students whose native language is not English must take the Test of English as a Foreign Language and submit their scores to the department. Applicants with a background in computer science or programming are given particular consideration. The application fee is $55.

Correspondence and Information

Graduate Admissions
Computer Science Department
Room 300
Henry Salvatori Computer Science Center
University of Southern California
Los Angeles, California 90089-0781
Telephone: 213-740-4496
E-mail: csdept@pollux.usc.edu
World Wide Web: http://www.usc.edu/dept/cs

University of Southern California

THE FACULTY AND THEIR RESEARCH

Leonard Adleman, Henry Salvatori Professor; Ph.D., Berkeley, 1976. Complexity theory, public key cryptosystems, number theory.

Cengiz Alaettinoglu, Research Assistant Professor; Ph.D., Maryland, 1994. Computer networks, operating systems, distributed algorithms.

Michael Arbib, Professor; Ph.D., MIT, 1963. Neural networks, visuomotor coordination, dextrous hands, AI and cognition, schema theory.

Robert Balzer, Research Professor; Ph.D., Carnegie Mellon, 1966. Automatic programming, artificial intelligence, program specification.

Anindo Banerjea, Research Assistant Professor; Ph.D., Berkeley, 1994. Computer networks, network protocols multimedia systems.

George Bekey, Gordon Marshall Professor; Ph.D., UCLA, 1962. Robotics, artificial intelligence, biological systems.

Irving Biederman, William Keck Professor; Ph.D., Michigan, 1966. Computational, neural, and biological models of shape recognition.

Edward Blum, Professor; Ph.D., Columbia, 1952. Numerical analysis and computer programming theory, neural modeling.

Barry Boehm, TRW Professor; Ph.D., UCLA, 1964. Software engineering, software processes, metrics and architectures.

Melvin Breuer, Charles Lee Powell Professor; Ph.D., Berkeley, 1965. Testing, design-for-testability, computer-aided design of VLSI systems.

Steve Chien, Adjunct Assistant Professor; Ph.D., Illinois, 1991. AI: planning and scheduling, machine learning, intelligent scientific data systems.

Peter Danzig, Associate Professor; Ph.D., Berkeley, 1989. Performance analysis of operating systems, computer networks, distributed systems.

Alvin Despain, Charles Lee Powell Professor; Ph.D., Utah, 1966. AI: architecture, machine organization, and design automation.

Pedro Diniz, Research Assistant Professor; Ph.D., California, Santa Barbara, 1997. Parallel and distributed computing.

Deborah Estrin, Associate Professor; Ph.D., MIT, 1985. Computer networks, security and organizational context of computing.

Robert Felderman, Adjunct Assistant Professor; Ph.D., UCLA, 1991. Parallel and distributed systems, high-speed LAN, distributed simulation.

Martin Frank, Research Assistant Professor; Ph.D., Georgia Tech, 1995. End-user programming, high-quality user interfaces, programming tools.

Shahram Ghandeharizadeh, Associate Professor; Ph.D., Wisconsin–Madison, 1990. Database management systems; parallel hypermedia.

Yolanda Gil, Research Assistant Professor; Ph.D., Carnegie Mellon, 1992. AI: machine learning, planning, problem solving, robotics.

Seymour Ginsburg, Fletcher Jones Professor (Emeritus); Ph.D., Michigan, 1952. Databases theory, automata, formal languages.

Ashish Goel, Assistant Professor; Ph.D., Stanford, 1999. Algorithms and strategies for routing, multicasting, network design and packet scheduling, graph algorithms, online and approximation algorithms.

Ramesh Govindan, Research Assistant Professor; Ph.D., Berkeley, 1992. Network OS, media software, next-generation Internet protocol.

Jonathan Gratch, Research Assistant Professor; Ph.D., Illinois, 1995; AI: machine learning, planning, uncertainty reasoning, decision theory.

Mary Hall, Research Assistant Professor; Ph.D., Rice, 1991. Compilers, program analysis, automatic parallelization.

John Heidemann, Research Assistant Professor; Ph.D., UCLA, 1995. Operating systems, file systems and networking.

Randall W. Hill Jr., Research Assistant Professor; Ph.D., USC, 1993. AI: perception, agent modeling, plan recognition, cognitive modeling.

Ellis Horowitz, Professor; Ph.D., Wisconsin–Madison, 1970. Software engineering, programming environments, computer-based instruction.

Eduard Hovy, Research Assistant Professor; Ph.D., Yale, 1987. AI: natural language processing, computational linguistics.

Ming-Deh Huang, Associate Professor; Ph.D., Princeton, 1984. Parallel computation, computational complexity, number-theoretic algorithms.

Kai Hwang, Professor; Ph.D., Berkeley, 1972. Computer architecture, parallel processing.

Douglas Ierardi, Adjunct Assistant Professor; Ph.D., Cornell, 1989. Algorithms in algebra and number theory, computational complexity.

Lewis Johnson, Research Associate Professor; Ph.D., Yale, 1985. Software specification and design.

Carl Kesselman, Research Associate Professor; Ph.D., UCLA, 1991. High-performance distributed computing, parallel programming processing.

Kevin Knight, Research Assistant Professor; Ph.D., Carnegie Mellon, 1991. AI: natural language processing and machine translation.

Craig Knoblock, Research Associate Professor; Ph.D., Carnegie Mellon, 1991. AI: planning, problem solving, machine learning and abstraction.

Sukhan Lee, Adjunct Associate Professor; Ph.D., Purdue, 1982. Robotics, AI: planning, problem solving and reformulation, machine learning.

Stephen C. Y. Lu, David Packard Chair Professor; Ph.D., Carnegie Mellon, 1984. Development of information technologies to support engineering decision making as collaborative negotiation.

Raymond J. Madachy, Adjunct Assistant Professor; Ph.D., USC, 1994. Software engineering: software metrics, cost estimation, risk management, process modeling and improvement.

Daniel Marcu, Research Assistant Professor; Ph.D., Toronto, 1997. Computational linguistics, text/discourse theories, natural language generation.

Maja Mataric, Assistant Professor; Ph.D, MIT, 1994. Multiagent and multirobot control and learning, imitation learning.

Larry Mathies, Adjunct Assistant Professor; Ph.D., Carnegie Mellon, 1989. Computer vision, 3-D shape and motion estimation, mobile robotics.

Dennis McLeod, Professor; Ph.D., MIT, 1978. Database systems, knowledge management systems.

Gerard Medioni, Associate Professor; Ph.D., USC, 1983. Computer vision, artificial intelligence.

Nenad Medvidovic, Assistant Professor; Ph.D., California, Irvine, 1998. Software architecture, software reuse, software evolution.

Steve Minton, Research Associate Professor; Ph.D., Carnegie Mellon, 1988. Machine learning, planning, scheduling, constraint-based reasoning, and program synthesis.

David Moriarty, Research Assistant Professor; Ph.D., Texas at Austin, 1997. AI, machine learning, knowledge representation.

Robert Neches, Research Associate Professor; Ph.D., Carnegie Mellon, 1981. Intelligent interfaces, very large knowledge bases, computer supported cooperative work.

Clifford Neuman, Research Assistant Professor; Ph.D., Washington (Seattle), 1992. Parallel and distributed systems, operating systems, computer security and pervasive computing.

Ulrich Neumann, Assistant Professor; Ph.D., North Carolina, 1993. High-performance interactive computer graphics, virtual environments, parallel systems, volume visualization, medical imaging, multimedia.

Ramakant Nevatia, Professor; Ph.D., Stanford, 1975. Machine vision, robotics, artificial intelligence.

Katia Obraczka, Research Assistant Professor; Ph.D., USC, 1994. Computer networks, distributed systems, operating systems.

Pavel A. Pevzner, Professor; Ph.D., Moscow Institute of Physics and Technology, 1988. Computational molecular biology, combinatorical and theoretical computer science.

Daniel Port, Research Assistant Professor; Ph.D., MIT, 1993. Component- and object-oriented software engineering, distributed multimedia systems, integration of software architectures.

Keith Price, Research Associate Professor; Ph.D., Carnegie Mellon, 1977. Computer vision, artificial intelligence.

Irving Reed, Charles Lee Powell Professor (Emeritus); Ph.D., Caltech, 1949. Computer architecture, VLSI design, neural networks, coding theory.

Aristides Requicha, Professor; Ph.D., Rochester, 1970. Geometric modeling, AI in computer-aided design and manufacturing.

Jeff Rickel, Research Assistant Professor; Ph.D., Texas, 1995. Knowledge representation and reasoning, knowledge acquisition, machine learning, explanation generation, intelligent tutoring.

Paul Rosenbloom, Associate Professor; Ph.D., Carnegie Mellon, 1983. AI: machine learning, integrated intelligent architectures, cognitive science.

Stefan Schaal, Assistant Professor; Ph.D., Munich Technical, 1991. AI, neural networks, statistical learning, computational neuroscience.

Herbert Schorr, Research Professor; Ph.D., Princeton, 1963. Expert systems, design automation, system architecture, artificial intelligence.

Mark Seidenberg, Professor; Ph.D., Columbia, 1980. Computational models of normal and distorted language.

Cyrus Shahabi, Research Assistant Professor; Ph.D., USC. Multimedia database management systems, continuous media services.

Wei Min Shen, Research Assistant Professor; Ph.D., Carnegie Mellon, 1989. Machine learning, autonomous agents and robots, data mining.

Scott Shenker, Adjunct Associate Professor; Ph.D., Chicago, 1983. Internet architecture.

Gaurav Sukhatme, Research Assistant Professor; Ph.D., USC, 1997. Robotics, modeling and simulation intelligent systems.

William Swartout, Research Associate Professor; Ph.D., MIT, 1981. AI: expert systems and knowledge acquisition.

Pedro Szekely, Research Assistant Professor; Ph.D., Carnegie Mellon, 1987. User interfaces.

Milind Tambe, Research Assistant Professor; Ph.D., Carnegie Mellon, 1991. Intelligent agents, multiagent systems, rule-based systems.

Joseph Touch, Research Assistant Professor; Ph.D., Pennsylvania, 1992. Protocols, networks, distributed systems.

Gene Tsudik, Research Associate Professor; Ph.D., USC, 1991. Computer security, distributed systems, computer networks, operating systems.

Christoph von der Malsburg, Professor; Ph.D., Heidelberg (Germany), 1970. Brain theory, neural computer vision.

Michael Waterman, Professor; Ph.D., Michigan State, 1969. Computational biology.

Richard Weinberg, Research Assistant Professor; Ph.D., Minnesota, 1982. Computer graphics and animation.

David Wile, Research Professor; Ph.D., Carnegie Mellon, 1974. Programming languages and environment.

Wayne W. Zhang, Research Assistant Professor; Ph.D., UCLA, 1994. Heuristic search, combinatorial optimization problems, distributed multiagent systems, parallel and distributed algorithms, robotics.

UNIVERSITY OF SOUTH FLORIDA

Department of Computer Science and Engineering

Programs of Study

The Department of Computer Science and Engineering offers the degrees of Master of Science in Computer Science and Computer Engineering and Doctor of Philosophy in Computer Science and Engineering. The major areas of research are image processing, computer vision, robotics, graphics, artificial intelligence, expert systems, coding theory, databases, computer networks, VLSI design, computer architecture, and fault-tolerant computing.

The master's degree program requires a total of 30 to 33 semester credit hours of work beyond the baccalaureate. Courses in four core areas are required; other courses are chosen to serve the individual student's interest.

The doctoral degree is attained by completing course work as advised by the doctoral committee and a minimum of 20 hours of dissertation beyond the master's degree. Other requirements include passing a written screening and a written preliminary examination, achieving candidacy by means of thesis proposal and oral examination, and presenting and defending thesis research.

Research Facilities

The Department of Computer Science and Engineering operates a research-oriented local area network consisting of several DEC/IBM machines, several SG workstations, more than seventy Sun SPARC workstations, image-processing workstations, and an Intel Hypercube of 16 nodes. The department's facilities also include a microprocessor laboratory, a hardware/architecture laboratory, and a PC-compatible laboratory for instructional purposes. College of Engineering facilities available to the department include a second network of Sun Workstations. In addition, the University operates a large IBM mainframe, which is available for the department's instructional and research purposes.

Financial Aid

Teaching and research assistantships are available to qualified students, with estimated stipends of $10,000 to $12,000 for nine months for M.S. degree students and $10,500 to $16,000 for Ph.D. students. A thesis is required on the subject covered by the research assistantship. Duties normally require 20 hours per week. Students holding assistantships are eligible for partial in-state and out-of-state tuition waivers. Fellowship appointments are available with stipends ranging upwards from $7000 through the graduate school.

Cost of Study

Fall semester 1999 in-state tuition was $137.04 per semester credit hour; out-of-state students paid $480.60 per semester credit hour.

Living and Housing Costs

The cost of living in the Tampa Bay area compares favorably with that of most other parts of the United States. Limited facilities for unmarried students are available on campus. A wide range of off-campus housing is available immediately adjacent to the University. Meal plans are available in the University's dining halls. Excluding tuition, books, and transportation, minimum living costs are estimated at $2000 per semester.

Student Group

There are more than 36,000 students at the University of South Florida, with 6,200 enrolled in graduate programs. Students come from virtually all states of the Union and many other countries. The College of Engineering has an enrollment of about 3,000, including more than 500 graduate students. The department has more than 90 full-time graduate students with a total enrollment of 180.

Location

The Tampa Bay area, with a population of more than 1.5 million, is rich in recreational, cultural, and athletic activities. The University is located approximately 10 miles from downtown Tampa, 35 miles from Clearwater and St. Petersburg. Symphonies, operas, professional theater, chamber music, and professional sports events are regularly available. Fine beaches and parks make the area an elite resort and vacation spot. Tampa Bay is undergoing rapid industrial growth and houses divisions of many internationally known high-technology industries.

The University

The University of South Florida opened its doors in 1960 and now has more than 35,000 students on five campuses. The Tampa campus is the largest, with more than 28,148 students. It comprises the Colleges of Engineering, Arts and Sciences, Business Administration, Education, Fine Arts, Medicine, Nursing, and Social and Behavioral Sciences and the School of Public Health. The campus is modern, airy, and attractive and provides a wide range of recreational facilities, including a golf course. More than $100 million in sponsored research is conducted annually.

The College of Engineering maintains close contact with local industry, where many of its students find part-time or full-time employment. Several of the department facility's research projects are funded by various federal and state agencies and local industries. The college is housed in two modern, 100,000-square-foot, four-story facilities.

Applying

Applicants for the master's program must have a bachelor's degree in computer science, computer engineering, or a closely related area. Students must have either a grade point average of 3.3 or higher during the last two years of their undergraduate work and a combined verbal and quantitative score of 1200 or better on the GRE General Test. Applicants for the doctoral program are expected to have an academic record that exceeds the foregoing minimum requirements. Applications for admissions should be received by the University's Office of Graduate Admissions at least three months prior to first expected enrollment.

Correspondence and Information

Graduate Program Coordinator
Department of Computer Science and Engineering
University of South Florida
Tampa, Florida 33620
Telephone: 813-974-3033
E-mail: msphd@csee.usf.edu
World Wide Web: http//www.csee.usf.edu

University of South Florida

THE FACULTY AND THEIR RESEARCH

Sami Al-Arian, Associate Professor; Ph.D., North Carolina State, 1985. Digital systems and microcomputer-based design, fault-tolerant computing, VLSI/WSI system testing, architecture.

Kevin Bowyer, Professor; Ph.D., Duke, 1980. Image processing and computer vision, pattern recognition, computer science education.

Ken Christensen, Assistant Professor; Ph.D., North Carolina State, 1991. Computer networking, high-speed LANs, performance evaluation, simulation modeling, queueing systems.

Martha Escobar-Molano, Assistant Professor; Ph.D., Southern California, 1996. Conceptual modeling, querying languages, and storage management for multimedia databases.

Eugene Fink, Assistant Professor; Ph.D., Carnegie Mellon, 1999. Machine learning search, automatic problem reformulation, theoretical foundations of AI, computational geometry, algorithm theory, cognitive science.

Dmitry Goldgof, Associate Professor; Ph.D., Illinois, 1989. Computer vision, image processing, biomedical applications, pattern recognition, parallel algorithms for computer vision, visualization techniques.

Lawrence Hall, Professor; Ph.D., Florida State, 1986. Pattern recognition, intelligent systems development, fuzzy set theory, connectionist/symbolic learning.

Abraham Kandel, Professor and Chair; Ph.D., New Mexico, 1977. Intelligent autonomous systems, expert and hybrid systems, neural networks, fuzzy set theory, robotics.

Srinivas Katkoori, Assistant Professor; Ph.D., Cincinnati, 1996. VLSI design, CAD.

Wai-Kei Mak, Assistant Professor; Ph.D., Texas at Austin, 1998. Computer-aided design of VLSI circuits, FPGA design and algorithms for FPGA, design and analysis of discrete optimization algorithms.

Peter Maurer, Associate Professor; Ph.D., Iowa State, 1982. Computer architecture, VLSI, VLSI logic simulation, VLSI layout verification, multiprocessors and parallel programming.

Robin Murphy, Associate Professor; Ph.D., Georgia Tech, 1992. Robotics, artificial intelligence, sensor fusion.

Rafael Perez, Professor; Ph.D., Pittsburgh, 1973. Genetic algorithms, neural networks and expert systems.

Les Piegl, Professor; Ph.D., Eotvos Lorand (Budapest), 1982. Geometric modeling, computer graphics, computer-aided geometric design, data structures, engineering and applied computing.

Dimitrios Plexousakis, Assistant Professor; Ph.D., Toronto, 1996. Databases, knowledge-based systems.

N. Ranganathan, Professor; Ph.D., Central Florida, 1988. VLSI design, hardware algorithms, computer architecture, parallel algorithms and architecture, VLSI for vision, image processing and pattern recognition.

Dewey Rundus, Associate Professor; Ph.D., Stanford, 1971. User interface design, software engineering, computer graphics.

Sudeep Sarkar, Associate Professor; Ph.D., Ohio State, 1993. Computer vision in the area of signals as opposed to symbols, perceptual organization, and image processing.

Murali Varanasi, Professor; Ph.D., Maryland, 1973. Coding theory, computer arithmetic, implementation of communication and signal processing algorithms, fault-tolerant computing.

UNIVERSITY OF SOUTHWESTERN LOUISIANA

Center for Advanced Computer Studies

Programs of Study

The primary mission of the Center for Advanced Computer Studies (CACS) is to conduct research and provide graduate-level education in computer engineering and computer science. The Center offers four graduate degrees: Ph.D. in computer engineering and in computer science, M.S.C.E., and M.S.C.S. Areas of specialization are artificial intelligence and cognitive science, parallel and distributed computing, VLSI, database systems, information retrieval, software systems and engineering, formal aspects of computing, robotics and automation, visual and image computing, neural networks, information technology, multimedia systems architectures, cryptography and error-correcting codes, mobile and wireless communication, and computer communication and networks.

Research Facilities

The CACS research computing facilities consist mainly of more than 100 UNIX workstations, primarily from Sun Microsystems but also including SGI, DEC, IBM, and HP machines. The department also maintains many Windows-based workstations. All of these machines are connected by an extensive Ethernet network and are supported by file and application servers. There are numerous peripherals supported on the network, including monochrome and color laser printers, plotters, and scanners. A wide variety of software is available and ranges from artificial intelligence to VLSI design. CACS houses state-of-the-art research laboratories that specialize in a number of fields, including artificial reasoning, computer vision and pattern recognition, database design and implementation, Internet computing, multimedia systems, remote sensing and mapping, robotic systems, software engineering, and VLSI design.

Financial Aid

A number of fellowships valued at up to $18,000 for up to four years are available for entering Ph.D. students with a superior academic record and exceptional GRE scores. Fellowships valued at up to $10,000 are available for entering M.S. students with superior academic records and strong GRE scores. All fellowship application materials must be received by February 15 for consideration. The Center has a large number of teaching and research assistantships and scholarships. The value of financial support ranges from about $6000 to more than $10,000 for the academic year. Figures quoted include a waiver of tuition and most fees.

Cost of Study

In spring 1999, Louisiana resident tuition and fees totaled $990.75 per semester. Nonresident students paid approximately $3700 per semester. Tuition and fee amounts are subject to change without notice.

Living and Housing Costs

The cost of room and board (fifteen meals per week) for single students living in residence halls in 1998–99 was $1496 per semester. A limited number of rooms in the University Conference Center cost $893 per semester. Married student apartments are available at $300 per month. All costs are subject to change. The community of Lafayette provides a wide selection of privately owned housing convenient to the campus.

Student Group

Total University enrollment is approximately 17,000 students. There are 166 students enrolled in the Center's graduate programs, including 65 pursuing the Ph.D. degree. The University of Southwestern Louisiana (USL) is known to attract international graduate students in the fields of computer science and engineering at both the master's and Ph.D. levels. The master's degree program is currently serving 88 computer science majors and 13 computer engineering majors; the Ph.D. program is currently serving 45 computer science majors and 20 computer engineering majors.

Student Outcomes

Among the universities employing some of the Center's recent graduates are Texas at Austin, Arkansas, South Carolina, Houston, Stevens Institute of Technology, South Alabama, Tulane, Louisiana Tech, Grambling State, Northern Arizona, Mississippi State, Haceteppe (Turkey), National Institute of Saudi Arabia, and Chungbuk National University (Korea). Some Ph.D. graduates have accepted employment in industry with IBM (Durham, Boca Raton), MIT, Schlumberger (Austin), Intel (San Jose, Portland), LSI Logic (Milpitas, California), Centigram Communication (San Jose), and Stock Exchange of Thailand.

Location

The University of Southwestern Louisiana is located in Lafayette, the central city of the geographic area known as Acadiana. The more than 500,000 inhabitants of this locale are mainly descendants of the exiled Acadians of Nova Scotia. Culturally, the region is characterized by a joie de vivre that has given it an international reputation. Lafayette is located approximately 52 miles from the state capital of Baton Rouge and 129 miles from New Orleans.

The University and The Department

The University has an impressive physical plant that is steadily being enlarged on all parts of the campus. It includes the administrative complex; Dupre Library; French House; academic buildings; athletic facilities; housing for men, women, and married students; an art museum; and a Student Union Complex situated on Cypress Lake. Located on the agricultural extension of the campus are Blackham Coliseum, Cajun Field, and the Cajun Dome, which seats approximately 12,000. Since its inception in 1984, the Center for Advanced Computer Studies has demonstrated a strong contribution to both the quality of education and the quality of research in the fields of computer science and engineering. The University and CACS have created an environment unique in the nation. The Center is one of the first to merge the overlapping, yet disjointed, disciplines of computer science and computer engineering into a successful graduate program. CACS has 18 faculty and 7 support staff members.

Applying

Applications for admission for the fall semester must be submitted to the Graduate School thirty days before classes begin. Applications for graduate assistantships for the fall semester must be submitted to the Graduate School by March 1 and for the spring semester by November 1. Students will be notified by April 1 and December 1, respectively, of their acceptance. Requirements for admission include a baccalaureate degree from an accredited institution, an excellent GPA, GRE and TOEFL scores (if the degree is earned outside the United States), three letters of recommendation, and a fluent command of English. Application fees (nonrefundable) are $5 for U.S. citizens and $15 for students who are non-U.S. citizens. Applications, inquiries, letters of recommendation, transcripts, and GRE and TOEFL scores should be sent to the address below.

Correspondence and Information

The Graduate School
University of Southwestern Louisiana
P.O. Box 44610
Lafayette, Louisiana 70504-4610

Telephone: 318-482-6965
Fax: 318-482-6195
World Wide Web: http://www.usl.edu

University of Southwestern Louisiana

THE FACULTY AND THEIR RESEARCH

Professors

Magdy A. Bayoumi, Ph.D., Windsor, 1984. VLSI design, image and video signal processing, parallel processing, neural networks, wide-band, multimedia network architectures.

Subrata Dasgupta, Ph.D., Alberta, 1976. Theory of design, cognitive aspects of creativity in science and technology, history of science and technology, computational models of creativity.

Vijay V. Raghavan, Ph.D., Alberta, 1978. Information storage and retrieval, rough sets and knowledge discovery in databases, content-based image retrieval, data warehousing and online analytical processing (OLAP).

T. R. N. Rao, Ph.D., Michigan, 1964. Fault-tolerant computers, information theory and coding, cryptology and data security.

Nian-Feng Tzeng, Ph.D., Illinois, 1986. Distributed and parallel systems, computer communication and networks, high-performance computer systems.

Kimon P. Valavanis, Ph.D., RPI, 1986. Control systems, robotics, intelligent systems and machines, automated manufacturing systems.

Associate Professors

Chee-Hung Henry Chu, Ph.D., Purdue, 1988. Computer vision, signal and image processing, data compression.

William R. Edwards Jr., Ph.D., Kansas, 1973. Theory of computation, software engineering, modeling and measurement of conceptual complexity.

Kemal Efe, Ph.D., Leeds (England), 1985. Parallel and distributed computer systems, data parallel algorithms.

Gui-Liang Feng, Ph.D., Lehigh, 1990. Error-correcting codes, data compression, fault tolerance, cryptography, computational biology.

Jung Kim, Ph.D., Iowa, 1987. Mobile communication and neural networks.

Miroslav Kubat, Ph.D., Brno Technical, 1990. Machine learning and neural networks.

Arun Lakhotia, Ph.D., Case Western Reserve, 1989. Program understanding, software reengineering, programming flow analysis.

Rasiah Loganantharaj, Ph.D., Colorado State, 1985. Intelligent planning and scheduling, temporal reasoning, knowledge discovery, and application of KB systems.

Niki Pissinou, Ph.D., USC, 1991. Databases; distributed systems and computer networks, including mobile and wireless communications and information systems.

Gunasekaran Seetharaman, Ph.D., Miami (Florida), 1988. Computer vision, image and signal processing, algorithm analysis, data compression, 3-D displays.

Assistant Professors

William H. Bares, Ph.D., North Carolina State, 1998. Computer graphics, multimedia.

Anthony S. Maida, Ph.D., SUNY at Buffalo, 1980. Artificial intelligence, cognitive science.

Adjunct Faculty

István S. N. Berkeley, Ph.D., Alberto, 1997. Artificial neural networks for the study of human thinking.

Adrienne Broadwater, Ph.D., Southwestern Louisiana, 1997. Parallel processing: algorithms, interconnection networks, mapping, scheduling.

Claude G. Čech, Ph.D., Illinois, 1981. Models of mental comparisons, discourse processes, categories and attention.

Fahmida N. Chawdhury, Ph.D., LSU, 1988. Control, neural networks.

Sherri L. Condon, Ph.D., Texas at Austin, 1983. Discourse analysis, computational linguistics, Louisiana languages.

Steve Giambrone, Ph.D., Australian National, 1984. Logic language and automata communication.

Denis Gračanin, Ph.D., Southwestern Louisiana, 1994. Petri nets, virtual reality, computer and communication networks.

Robert R. Henry, Ph.D., New Mexico State, 1976. Computer design and architecture, telecommunications.

Mohammad R. Madani, Ph.D., LSU, 1990. VLSI circuit processing and manufacturing, ion implantation, VLSI circuit characterization and testing.

Bill Manaris, Ph.D., Southwestern Louisiana, 1990. Artificial intelligence, natural-language processing, human-computer interaction, software engineering.

Renee McCauley, Ph.D., Southwestern Louisiana, 1992. Computer science education, software measurement, theory of computation.

Daniel J. Povinelli, Ph.D., Yale, 1991. Origin of self-recognition systems, attributions of intention and belief in young children, chimpanzees and other primates.

Mark G. Radle, Ph.D., Southwestern Louisiana, 1997. Applied artificial intelligence, cognitive science.

UNIVERSITY OF TENNESSEE, KNOXVILLE

Department of Computer Science

Programs of Study

The Department of Computer Science at the University of Tennessee, Knoxville (UTK) offers programs of study leading to the degrees of Doctor of Philosophy and Master of Science. The principal requirements for the Ph.D. are successful completion of certain comprehensive examinations and submission of a dissertation that describes original independent research. The degree of Doctor of Philosophy is awarded in recognition of high achievement in research. The M.S. degree may be earned by completing 30 semester hours of approved course work or course work plus a thesis or a project. A comprehensive examination is required in lieu of a thesis.

Research Facilities

The department operates research laboratories with more than 300 machines, including Sun UltraSPARC workstations, Linux workstations, SGI workstations, Wintel PCs, and a 4096-processor MasPar MP-2. University resources include a 48-processor IBM SP2 with 100 GB of disk and 10 GB of memory. Faculty members collaborate with scientists at the Oak Ridge National Laboratory (ORNL) and have access to their facilities, including three Intel Paragon XP/S supercomputers. The Paragons have 128, 512, and 1024 nodes, respectively. Local area networking technologies include a 10/100Mbps Ethernet and a 155Mbps ATM. The Knoxville campus has six T1 connections and one DS3 connection to the Internet. A 155Mbps ATM link connects UTK to ORNL and through ORNL to ESNet. UTK is a charter member of Internet2, a national collaboration of more than 130 research universities. UTK recently installed a 45Mbps link to the regional GigaPOP at Georgia Tech, connecting UTK to both the SURA network and to the national vBNS network.

Financial Aid

Financial aid is available in the form of graduate teaching assistantships and research assistantships. In 1999–2000, assistantships pay a stipend of $9100 to $9900 for nine months and offer a waiver of tuition and fees. Additional supplements are available under the state-funded Science Alliance and faculty research grants.

Graduate fellowships, awarded on the basis of scholastic record, are available through the Graduate School and directly from the department. For departmental fellowships, students can consult the Web site or contact the Computer Science Office.

Cost of Study

Tuition for 1999–2000 for full-time study is about $1750 per semester for Tennessee residents and $4200 per semester for out-of-state students, plus nominal activities and technology fees. All computer science students are encouraged to own a personal computer; to facilitate this, discount programs are available at the University bookstore.

Living and Housing Costs

Living accommodations of all types, ranging from residence halls to privately owned apartments, are available within walking distance of the campus. Single students may live in the residence halls for approximately $2050 per semester in 1999–2000, including room and board.

Student Group

The department has about 65 undergraduates and 130 graduate students majoring in computer science.

Student Outcomes

Graduates of the department are highly sought by prospective employers such as AT&T, Cray Research, Texas Instruments, Intel, Federal Express, Hewlett-Packard, IBM, and Oak Ridge National Laboratory, as well as a number of universities and colleges.

Location

UTK is located in Knoxville, which was previously listed in a national survey as one of the most livable cities in the nation. Knoxville provides a variety of athletic, cultural, and entertainment activities. The area has several lakes that offer excellent fishing, boating, and swimming. Knoxville is located 40 miles from the Great Smoky Mountains, where numerous outdoor activities, including skiing and white-water canoeing, are available. The Oak Ridge National Laboratory, 30 miles from the campus, is an important part of the intellectual community. The area has a relatively mild climate with a moderate range of temperatures.

The University

The University of Tennessee, with a statewide enrollment of about 43,000, has all the intellectual and social characteristics associated with a state university system. Students may specialize in a great number of professional and occupational fields. There is a distinguished group of faculty members, nationally recognized for their professional accomplishments. The Knoxville campus, with an enrollment of 26,000, is the main campus of the multicampus system and is the center for advanced graduate studies.

Applying

Admission applications are processed throughout the year, and students may enter at the beginning of any semester. Applications for assistantships beginning in August should be submitted to the department by March 1. Consideration may be given to applications received after that date. Fellowship applications must be received in the Office of Graduate Admissions and Records by February 15. The Graduate Admissions Office will supply specific deadlines for international applicants.

Correspondence and Information

Department of Computer Science
107 Ayres Hall
The University of Tennessee, Knoxville
Knoxville, Tennessee 37996-1301

Telephone: 423-974-5067
Fax: 423-974-4404
E-mail: straight@cs.utk.edu
World Wide Web: http://www.cs.utk.edu

For UTK fellowship applications:
Assistant Director
Office of Graduate Admissions and Records
The University of Tennessee, Knoxville
Knoxville, Tennessee 37996-0220

University of Tennessee, Knoxville

THE FACULTY AND THEIR RESEARCH

Michael W. Berry, Associate Professor; Ph.D., Illinois. Scientific computing, parallel numerical algorithms, information retrieval, computational science and performance evaluation.

Jack Dongarra, Distinguished Professor and Oak Ridge National Laboratory Distinguished Scientist; Ph.D., New Mexico. Scientific computing, numerical linear algebra, parallel processing, software tools, mathematical software and software repositories.

Jens Gregor, Associate Professor; Ph.D., Aalborg (Denmark). Pattern and image analysis, computed imaging.

Michael A. Langston, Professor; Ph.D., Texas A&M. Analysis of algorithms, graph theory, operations research, parallel computing, VLSI design.

Bruce J. MacLennan, Associate Professor; Ph.D., Purdue. Neural networks and connectionism, theory of knowledge, massively parallel analog computation, emergent computation.

J. Wallace Mayo, Instructor; M.S., Tennessee. Responsible for teaching and developing core courses, supervising graduate teaching assistants, and student advising.

James S. Plank, Associate Professor; Ph.D., Princeton. Fault-tolerance and checkpointing, operating systems, parallel programming, architecture.

Jesse H. Poore, Professor and Ericsson-Harlan D. Mills Chair in Software Engineering; Ph.D., Georgia Tech. Economical production of high-quality software, federal software policy.

Padma Raghavan, Assistant Professor; Ph.D., Penn State. Parallel processing, graph algorithms, sparse matrices, linear algebra, scientific computing.

Gordon R. Sherman, Professor Emeritus; Ph.D., Purdue. Probability and statistics, discrete optimization, high-performance computing resources, administration of computer services.

David W. Straight, Assistant Professor; Ph.D., Texas. LANS architecture, parallel processing.

Michael G. Thomason, Professor; Ph.D., Duke. Image and pattern analysis, stochastic processes, parallel algorithms.

Bradley T. VanderZanden, Associate Professor; Ph.D., Cornell. Graphical programming environments, programming languages, constraint solving.

Michael D. Vose, Associate Professor; Ph.D., Texas. Genetic algorithms.

Robert C. Ward, Professor and Department Head; Ph.D., Virginia. Numerical linear algebra, scientific computing.

Richard M. Wolski, Assistant Professor; Ph.D., California, Davis. Metacomputing, performance forecasting, distributed systems, parallel computing.

Adjunct Faculty

Stephen G. Batsell, Assistant Professor and Network Research Group Leader, Oak Ridge National Laboratory; Ph.D., Texas Tech. Internetworking, optical networks, mobile computing.

Micah Beck, Associate Professor and Research Associate Professor; Ph.D., Cornell. High-performance parallel and distributed computing, next-generation internetworking.

Heather Booth, Assistant Professor; Ph.D., Princeton. Data structures, graph algorithms, computational geometry.

Shirley Browne, Assistant Professor and Research Associate, Computer Science Department; Ph.D., Purdue. Software reuse in high-performance computing, wide-area information retrieval, security aspects of safe execution environments.

Jeffrey Case, Associate Professor and President, SNMP Research Inc.; Ph.D., Illinois. Networking, network protocols, network management.

June M. Donato, Assistant Professor, Senior Programmer/Analyst at Scientific Applications Incorporated Corporation (SAIC), and 1991 Householder Fellow at ORNL; Ph.D., UCLA. Software development, numerical methods and algorithms for data mining.

Thomas H. Dunigan, Associate Professor and Staff Research Scientist at Oak Ridge National Laboratory; Ph.D., North Carolina at Chapel Hill. Operating systems, networks, parallel programming.

Sara R. Jordan, Associate Professor and Project Manager at Lockheed Martin Energy Systems, Inc.; Ph.D., Wisconsin. Advanced information technologies, information architectures.

Reinhold C. Mann, Associate Professor and Director, Life Sciences Division at Oak Ridge National Laboratory; Dipl.Math., Dr.rer.nat., Mainz (Germany). Computer vision, pattern recognition, computational biology, bioinformatics.

Thomas E. Potok, Assistant Professor and Research Staff Member, Oak Ridge National Laboratory; Ph.D., North Carolina State. Software engineering.

Stacy J. Prowell, Research Associate Professor; Ph.D., Tennessee. Software engineering, formal methods for software specification and testing.

Charles H. Romine, Associate Professor and Research Staff Member at Oak Ridge National Laboratory; Ph.D., Virginia. Numerical linear algebra, parallel computing, analysis of parallel numerical algorithms, development of software environments for parallel computing.

Erich Strohmaier, Assistant Professor and Research Associate; Dr.rer.nat., Heidelberg (Germany). Performance evaluation and prediction, parallel computing, computer architecture, HPC market and technology analysis.

Carmen J. Trammell, Assistant Professor and Software Quality Manager, CTI-PET Systems, Inc.; Ph.D., Tennessee. Software engineering processes, methods, environments, and standards.

UNIVERSITY OF TEXAS AT ARLINGTON

College of Engineering
Department of Computer Science and Engineering

Programs of Study	The Department of Computer Science and Engineering offers graduate programs leading to Master of Science, Master of Computer Science, Master of Engineering, Master of Software Engineering (M.Sw.E.), and Doctor of Philosophy degrees. Studies may be pursued over a spectrum of areas covering computer architecture, database systems, distributed processing, fault-tolerant computing, image processing, intelligent systems, operating systems, parallel processing, real-time systems, and software engineering. The M.S. is a thesis degree that requires a minimum of 31 semester hours, of which at least 24 hours must be in approved course work and 6 hours in thesis. The M.Comp.Sci. and M.Engr. require at least 37 semester hours of approved course work, which may include 6 semester hours of a master's project. The M.Comp.Sci. requires 38 semester credit hours for the structured option. The M.Sw.E. is a practice-oriented program requiring a minimum of 37 semester hours of course work that includes an 18-hour core, 6 hours of design studio courses, and the remainder in approved electives. The Ph.D. is a research-oriented degree that has no specific course or credit-hour requirements and usually takes three years of full-time study after the master's degree.
Research Facilities	The University operates a wide spectrum of computing systems, which include Sun, Compaq, and Silicon Graphics servers; Sun and Compaq workstations; and Macintosh and Windows personal computers. The department operates several UNIX, Linux, and Windows servers and clusters of UNIX workstations and personal computers. Most computers on campus are connected to a local area network and are accessible from both on and off campus. The department also operates laboratories supporting the design and development of microcomputer-based and/or special-purpose digital systems. The UTA Automation and Robotics Research Institute offers networks of workstations, robots, and other facilities for use by department students and faculty members with interests in robotics and automated manufacturing.
Financial Aid	Teaching assistantships and research assistantships are available in limited numbers to qualified students. Unconditional admittance to the program is required for a student to be eligible for an assistantship. Stipends range from $4500 to $9000 for nine months. Assistant instructorships are sometimes available for qualified students in the final stages of a Ph.D. program. Part-time employment is often available for full-time students who are not receiving assistantships or other financial support from the University.
Cost of Study	For 1999–2000, tuition and fees for a 12-semester-hour load are $1631 per semester for Texas residents and $4307 per semester for nonresidents. Half-time teaching or research assistants qualify for Texas resident rates. One-time fees include a general property deposit of $10, a diploma fee of $15, a library fee of $48, a thesis or dissertation binding fee of $12.50, and a thesis or dissertation microfilming fee of $40 or $50, respectively. Tuition and fees are subject to change by legislative or administrative action.
Living and Housing Costs	University apartments are available at rates ranging from $321 to $557 per month. Dormitory rooms range from $777 to $865 per semester. Numerous private apartments are available. Rates are subject to change without notice.
Student Group	Enrollment at the University is 22,480, with students coming from forty-five states and sixty-five countries. College of Engineering enrollment is 2,860, including approximately 350 undergraduates and 350 graduate students in computer science and engineering. Two hundred three of the graduate students are full-time, with about 75 of them receiving financial support from the department. Most of the part-time students are practicing computer scientists or engineers pursuing advanced degrees.
Student Outcomes	Alumni are employed in attractive academic positions and exciting jobs in leading companies throughout the state and the nation, including AT&T, Ericsson, Fujitsu, IBM, Intel, Lockheed-Martin, Motorola, NASA–Johnson Space Center, Nortel (Bell Northern Research), Raytheon, SABRE, and Texas Instruments.
Location	Arlington has a population of more than 300,000 and is within a few minutes' driving time of either Dallas or Fort Worth. The metropolitan area offers a variety of cultural and recreational opportunities.
The University and The Department	The University is located on a modern 347-acre campus in the center of the Dallas–Fort Worth metropolitan area. The University was founded in 1895 as Arlington College, a small private liberal arts school. The college changed with the times and its surroundings, undergoing a succession of names and affiliations until 1967, when it became the University of Texas at Arlington (UTA). Currently, UTA is the fifth-largest institution of higher learning in the state of Texas. The College of Engineering consists of six departments, including Computer Science and Engineering. The graduate degree programs in computer science and computer engineering were started in 1973 and were administered by the Department of Industrial Engineering until formation of the Department of Computer Science and Engineering in 1979.
Applying	Applications for admission to the program should be submitted to the Graduate School at least two months (U.S. students) or four months (international students) prior to the start of new student registration for the semester in which the student plans to enroll. Students seeking financial support should submit applications for admission and for support by March 1 for fall or by October 1 for spring enrollment. Applications for financial support should be submitted to the CSE department. General requirements for acceptance to the master's programs include a minimum combined quantitative and verbal GRE score of 1100, a minimum quantitative GRE score of 700, and a minimum GPA of 3.0 out of 4.0 for the last 60 hours of undergraduate course work. The Ph.D. program requires a minimum combined quantitative and verbal GRE score of 1250 and a GPA of 3.5 out of 4.0 for master's-level course work. All international students must submit a TOEFL score, with a minimum score of 560. International students who do not meet the above GRE combined score requirements may be accepted if their quantitative scores exceed 750 for the master's programs or 780 for the Ph.D. program. Students whose primary language is not English must have scored 250 or higher on the TSE-A or SPEAK in order to be eligible for a teaching assistantship.
Correspondence and Information	For catalogs and application forms: Graduate School University of Texas at Arlington P.O. Box 19088, UTA Station Arlington, Texas 76019 Telephone: 817-272-2688 E-mail: graduate.school@uta.edu	For departmental information: Graduate Advisor Department of Computer Science and Engineering University of Texas at Arlington P.O. Box 19015, UTA Station Arlington, Texas 76019 Telephone: 817-272-3785 E-mail: csegrad@cse.uta.edu World Wide Web: http://www-cse.uta.edu

University of Texas at Arlington

THE FACULTY AND THEIR RESEARCH

Linda S. Barasch, Visiting Associate Professor; Ph.D., Oklahoma, 1988. Programming languages, compilers, database systems.

Carl T. Bruggeman, Assistant Professor; Ph.D., Indiana, 1995. Programming languages, compilers, computer architecture.

Bill D. Carroll, Professor; Ph.D., Texas at Austin, 1969. Fault-tolerant computing, computer architecture, distributed computing.

Sharma Chakravarthy, Professor; Ph.D., Maryland, 1985. Database and information systems, distributed systems, real-time database systems.

Diane J. Cook, Associate Professor; Ph.D., Illinois at Urbana-Champaign, 1990. Machine learning, planning, parallel algorithms, robotics.

Sajal K. Das, Professor; Ph.D., Central Florida, 1988. Wireless networks and mobile computing, communication networks and protocols, performance modeling and simulation.

Ramez A. Elmasri, Professor; Ph.D., Stanford, 1980. Databases, temporal databases, distributed and multidatabase systems.

Leonidas Fegaras, Assistant Professor; Ph.D., Massachusetts at Amherst, 1993. Databases, programming languages.

Piotr J. Gmytrasiewicz, Assistant Professor; Ph.D., Michigan, 1992. Artificial intelligence, multiagent coordination and communication, planning.

Lawrence B. Holder, Associate Professor; Ph.D., Illinois at Urbana-Champaign, 1991. Artificial intelligence, machine learning, data mining, parallel and distributed computing.

Pei Hsia, Professor; Ph.D., Texas at Austin, 1972. Software engineering, requirements engineering, scenario-based prototyping, incremental delivery, object-oriented software testing, scenario management.

Manfred Huber, Assistant Professor; Ph.D., Massachusetts, 1999. Artificial intelligence, robotics, machine learning, neural networks.

Farhad A. Kamangar, Associate Professor; Ph.D., Texas at Arlington, 1980. Artificial intelligence, computer graphics, digital signal/image processing, neural networks, time series analysis.

David C. Kung, Professor; Ph.D., Norwegian Institute of Technology, 1984. Object-oriented software testing, object-oriented real-time systems, modeling and verification.

David E. Levine, Visiting Assistant Professor; M.S.C.S., Texas at Arlington, 1975. Software engineering, operating systems, computer networks.

Lynn L. Peterson, Professor and Associate Dean; Ph.D., Texas Health Science Center at Dallas, 1978. Artificial intelligence, computer-based instructional systems, medical computer science.

Arthur A. Reyes, Assistant Professor; Ph.D., California at Irvine, 1999. Software engineering, software specification and testing, avionics software.

Behrooz Shirazi, Professor and Chairman; Ph.D., Oklahoma, 1985. Software development tools, parallel and distributed processing, task allocation and load balancing, distributed real-time systems, resource management in distributed systems.

L. David Umbaugh, Senior Lecturer; Ph.D., Ohio State, 1983. Computer networks and data communications, systems programming.

Roger S. Walker, Professor; Ph.D., Texas at Austin, 1972. Digital signal processing, microcomputer applications.

Bob P. Weems, Associate Professor; Ph.D., Northwestern, 1984. Parallel algorithms, parallel processing, automated deduction, relational dependency theory.

Ramesh Yerraballi, Assistant Professor, Ph.D., Old Dominion, 1996. Real-time systems, operating systems, mobile computing.

Hee Yong Youn, Associate Professor; Ph.D., Massachusetts at Amherst, 1988. Parallel and distributed computing, computer architecture, fault-tolerant computing, multimedia systems.

UNIVERSITY OF TEXAS AT AUSTIN

Department of Computer Sciences

Programs of Study

The Department of Computer Sciences offers programs leading to the M.S.C.S., M.A., and Ph.D. degrees. Programs provide students with a broad education in the various areas of computer sciences and allow specialization through a thesis. The master's program with thesis requires 30 semester hours of course work. The M.S.C.S. is a nonthesis option and requires 36 hours of course work. The Ph.D. requires 18 semester hours of general course work and about 15 more in the area of specialization. This is followed by an examination (based on the student's preliminary dissertation proposal) and by a dissertation.

Research Facilities

Many different computer systems are available for research use by faculty members and graduate students in the department. Machines available for parallel processing research include two 14-processor Sun Enterprise 5500s, a four-processor IBM F50, an eleven-processor IBM SP2, and a four-processor IBM SP2. The department has two Silicon Graphics Onyx 2 Infinite Reality machines, an SGI Indigo 2, an SGI Indy, fourteen SGI O2s, and an immersive theater for graphics and visualization research. More than sixty Pentium-based machines, including twenty-eight dual-processor Xeons, as well as dual- and quad-processor servers, have been donated by Intel for multimedia research. Other research workstations include four dual-processor Sun Ultra II 3-Ds, two dual-processor Silicon Graphics Origin 200s, more than fifty Sun Ultra workstations, and numerous Sun SPARC workstations. In addition, there are 100 Dell Pentium Gx Pros on graduate desks, donated by Dell in summer 1997. In fall 1996, the department was awarded an NSF CISE grant with a total of $1.6 million over five years to be devoted to the research infrastructure of the department. All departmental computers are networked together using Ethernet, with the majority of the network converting to 100 Mbps and gigabit subnets. There are also several ATM research subnets, and one subnet that uses Myrinet. Network servers include a Sun Enterprise 4000 with more than 600 GB of disk space that is used for home directory service, a Web server that is a Sun SPARC 1000e, many file servers, print servers, and communications servers. The department continues to expand these existing departmental computing facilities, both through donations of equipment by manufacturers and through funds provided by the University.

The equipment mentioned above is used for research; there is substantial additional equipment for educational computing.

Financial Aid

The Doctoral Fellows program guarantees a minimum stipend of $1400 per month for four years. This support consists of a fellowship the first year, a teaching assistantship the second year, and either a teaching or a research assistantship during the third and fourth years. Tuition and required fees are covered during the fellowship period.

The department also employs students as research and teaching assistants at a minimum starting salary of $1389 per month. The computation center has opportunities for qualified students as systems programmers and consultants, and local industries such as MCC, IBM, TI, Schlumberger, and Motorola employ students on a part-time basis. Low-interest loans are also available.

Cost of Study

In 1998–99, tuition and fees for 9 credit hours were approximately $1542 per semester for Texas residents and $3459 per semester for nonresidents.

Living and Housing Costs

The cost of a dormitory room averages $4516 for nine months (including board). University housing for married students ranges in cost from $367 to $568 per month. Private housing is available in all price ranges.

Student Group

There are 202 graduate students in computer sciences; more than half of them are Ph.D. candidates. The department confers forty to fifty master's degrees and fifteen to twenty Ph.D. degrees annually.

Location

Austin, a metropolitan area of approximately 485,000 people, is set in the scenic Hill Country of central Texas. The Colorado River, which flows through the city, has been dammed to form the chain of Highland Lakes, which provide opportunities for swimming, boating, and fishing. There are many fine parks and playgrounds. Austin also has five theater groups, several ballet troupes, an excellent symphony orchestra, and about fifty art galleries.

The University and The Department

The University of Texas, founded in 1883, is one of the largest universities in the country, with an enrollment of 48,906 on the Austin campus alone. There are other branches of the University in Houston, Galveston, San Antonio, and Dallas.

The graduate program in computer sciences was initiated in 1966. The undergraduate program was started in 1974 and has grown to include more than 2,000 computer sciences majors.

Applying

The department deadline for fall applications is January 2. Although the University deadline for graduate students' applications is February 1, all applicants are urged to submit materials by January 2. Financial aid applications should also be submitted by January 2. Applications are accepted for the spring semester, and the deadline for spring admission is September 1 for all applicants. The GRE General Test and Subject Test in computer science are required of all applicants. International applicants are strongly urged to take the TOEFL. Admission standards are high. Normally, a student admitted to the Ph.D. program has a bachelor's degree in computer science, at least a 3.5 GPA, a combined verbal and quantitative GRE score of at least 1400, and a minimum score of 85 percent on the computer science Subject Test.

All admission and application information can be found on the Web site listed below. Applications are available electronically through the Graduate and International Admissions Center's (GIAC) Web page, which is accessible through the department's Web page.

Students who do not have Web access many contact the department by phone for information and the paper application. Students should specify the type of application that they need (U.S. or international).

Correspondence and Information

For application forms and further information:

Graduate Admissions
Department of Computer Sciences
University of Texas at Austin
Austin, Texas 78712-1188

Telephone: 512-471-9503
E-mail: csadmis@cs.utexas.edu
World Wide Web: http://www.cs.utexas.edu

University of Texas at Austin

THE FACULTY AND THEIR RESEARCH

The following list is limited to those faculty members whose primary appointment is in the Department of Computer Sciences.

Lorenzo Alvisi. Distributed computing and fault tolerance in distributed systems.
Nina Amenta. Computational geometry, algorithms, computer graphics.
Chandrajit L. Bajaj. Computational algebra, geometry, computer graphics, geometric design, scientific data visualization.
Don Batory. Software architectures, extensible and object-oriented databases, domain modeling, software system generators.
Robert Blumofe. Parallel computation, combinatorial approximation and optimization algorithms, communication networks, and operating systems.
Robert S. Boyer. Program verification, automatic theorem proving, artificial intelligence.
James C. Browne. Parallel computation, with the major focus on parallel programming, high-level specification languages, and integration of computer science with application areas.
Jeffrey A. Brumfield. Performance analysis, distributed systems, operating systems.
Douglas Burger. Computer architecture, VLSI design, embedded applications.
Alan K. Cline. Mathematical software and numerical analysis.
Michael D. Dahlin. Operating systems, distributed systems, networking.
Chris C. Edmondson-Yurkanan. Computer networks, computer science education, managing large software projects, mobile networking, database design.
E. Allen Emerson. Formal methods, logics and semantics of programs, concurrent and distributed computing.
Donald S. Fussell. Computer architecture, computer graphics, VLSI systems design, database concurrency control.
Anna Gal. Computational complexity, lower bounds for complexity of Boolean functions, fault-tolerant computing.
Suzy C. Gallagher. Computer science education, library and information processing.
Mohamed Gawdat Gouda. Distributed and concurrent computing, fault-tolerant computing, computer networks, network protocols.
Roy M. Jenevein Jr. Interconnection networks and parallel processing in computer architecture.
Stephen W. Keckler. Computer architecture, microprocessor and VLSI design, billion-transistor chips, parallel computer systems.
David R. Kincaid. Mathematical software, high-performance computers, numerical analysis.
Benjamin J. Kuipers. Artificial intelligence, robotics, qualitative reasoning.
Simon S. Lam. Network protocols, performance models, formal verification methods, security.
Vladimir Lifschitz. Mathematical logic, logic programming, knowledge representation.
Calvin Lin. Compilers and languages for parallel computing, parallel performance analysis, and scientific computing.
Risto Miikkulainen. Neural networks, natural language processing, cognitive modeling.
Daniel P. Miranker. Parallel computer architecture, active/expert database system, high-performance artificial intelligence systems.
Jayadev Misra. Parallel programming.
Aloysius K. Mok. Fault-tolerant hard-real-time systems, system architecture, computer-aided system design tools, software engineering.
Raymond J. Mooney. Artificial intelligence, machine learning, natural language understanding.
J Strother Moore. Mechanical theorum proving, formal methods of computer architecture, programming languages.
Gordon S. Novak Jr. Artificial intelligence, automatic programming, physics problem solving, expert systems, compilers.
C. Greg Plaxton. Parallel computation, analysis of algorithms, lower bounds, randomization.
Bruce W. Porter. Artificial intelligence, machine learning, knowledge-based systems.
Vijaya Ramachandran. Design and analysis of algorithms, parallel computation, computational complexity.
Hamilton Richards Jr. Functional programming, concurrent processing, object-oriented programming.
Robert A. van de Geijn. Numerical analysis, high-performance parallel computing.
Harrick M. Vin. Multimedia systems, high-speed networking, databases, mobile computing, distributed systems.
Paul R. Wilson. Design and implementation of programming languages, operating systems, and debuggers; memory hierarchies.
Martin D. F. Wong. Computer-aided design of VLSI, design and analysis of algorithms.
David I. Zuckerman. Role of randomness in computation, complexity theory, design and analysis of algorithms.

THE UNIVERSITY OF TEXAS AT DALLAS

Engineering and Computer Science School
Computer Science Program

Programs of Study

The Computer Science Program offers a broad range of computer science– and engineering-related courses leading to the M.S. and Ph.D. degrees. There are three separate tracks leading to an M.S. degree: traditional computer science, networks/telecommunications, and software engineering.

A total of 33 credit hours is required for the M.S. Normally, a student takes one year of full-time work or two years of part-time work to complete the M.S. degree requirements. A total of 90 credit hours and a Ph.D. dissertation are required for the Ph.D. Typically, a student takes four years of full-time work to complete the Ph.D. degree requirements.

Research Facilities

The Information Resources Department maintains a Sun Ultra Enterprise 6000 with eighteen processors, two Sun Enterprise 3000 Servers, an IBM 4381, many Sun SPARC workstations, and PCs with Windows 95, Windows-NT, and Macintosh systems. The Computer Science Program has eight research labs equipped with high-performance workstations and high-end PCs. Computers on campus are connected via the Ethernet and have access to the Internet. The University library contains a large collection of computer science journals and other publications.

Financial Aid

Financial aid is available in the form of scholarships, research and teaching assistantships, and work-study arrangements with local industry. In 1998–99, stipends for research and teaching assistantships were $1100 per month. Additional compensation is generally available in the summertime, as a substantial schedule of classes is usually offered each summer. There are several additional government-, University-, and industry-backed sources for the support of graduate students. Interested prospective students may write for specific information to the graduate secretary of the department, EC 3.1, University of Texas at Dallas, or view the department's home page on the World Wide Web (http://www.utdallas.edu/dept/cs).

Cost of Study

In 1998–99, tuition and fees for Texas residents were $2098 for 15 semester hours; for nonresident and international students, tuition and fees were $5293 for 15 semester hours.

Living and Housing Costs

The cost of an apartment varies from $400 per month to more than $700 per month. The University has a substantial amount of housing. Some students find housing in neighboring apartment complexes.

Student Group

There are more than 4,000 graduate students and about 4,000 undergraduate students at UTD, representing most states in the United States and many other countries. Freshman and sophomore students were admitted for the first time in fall 1990. A large number of the undergraduate and part-time graduate students work in industry in the surrounding high-technology corridor.

Location

UTD is located on the north side of Richardson, a suburb about 17 miles north of downtown Dallas. The campus is surrounded by University-owned land that is often used for soccer, jogging, bicycle riding, and other recreational activities. There are several lakes in the Dallas and surrounding north Texas area. Dallas has museums, concert halls, a zoo, and other facilities that offer a rich cultural life to students and their families. The area is the home of many computer, electronics, and communication companies that are large supporters of UTD engineering and computer science programs.

The University and The Program

The University of Texas at Dallas was created in 1969 by an act of the Texas legislature that enabled the transfer of the Southwest Center for Advanced Studies to the state of Texas. The University began as a graduate school. The enrollment of undergraduate juniors and seniors began in 1975. Electrical engineering programs were approved in 1985. The Computer Science Program was part of the general program in mathematical sciences until 1982, when it became an independent program in the School of Natural Sciences and Mathematics. In 1986, the program became part of the new Engineering and Computer Science School. In 1992, the Computer Science Program moved into the Engineering and Computer Science Building, which is equipped with offices, laboratories, and working areas.

Applying

M.S. students should have a bachelor's degree that includes a full calculus sequence and the following: (1) a GPA of at least 3.0, (2) a combined GRE General Test score of at least 1100 or a quantitative GRE score of at least 750 and a TOEFL score of at least 550, and (3) a GPA in computer science, math, engineering, and related courses of at least 3.3.

Students seeking direct admission to the Ph.D. program (and not entering the M.S. program first) must satisfy the requirements stated above for M.S. students, plus either (1) have an M.S. degree with a GPA of at least 3.5, or (2) have a GPA of at least 3.5 in upperclass undergraduate and graduate (if any) courses and either a combined GRE General Test score of at least 1300 or a quantitative GRE score of at least 750 and a TOEFL score of at least 550.

Applications for admission and all supporting documents should be submitted well in advance of the preferred semester of entry. Since department awards for teaching assistantships, research assistantships, and other scholarships are usually made in the spring for the next full academic year, applicants requesting financial aid should submit all required documents as early as possible to ensure consideration. U.S. citizens and permanent residents pay a $25 application fee and international students pay a $75 application fee.

Correspondence and Information

Dr. D. T. Huynh, Head
Computer Science Program
EC 3.1
The University of Texas at Dallas
Richardson, Texas 75083-0688
Telephone: 972-883-6810
Fax: 972-883-2349

The University of Texas at Dallas

THE FACULTY AND THEIR RESEARCH

Farokh Bastani, Professor and Director, Center of Application-Specific Systems and Software Engineering; Ph.D., Berkeley, 1980. Safety-critical systems, high-assurance systems, hardware/software reliability assessment, fault-tolerant software, software reliability engineering, parallel and distributed programming, self-stabilizing systems.

Biao Chen, Assistant Professor; Ph.D., Texas A&M, 1996. Communications networks, real-time computing systems, distributed systems, fault-tolerant systems.

Imrich Chlamtac, Distinguished Chair Professor in Telecommunications; Ph.D., Minnesota, 1979. Design and analysis of network architectures, optical and wireless telecommunication systems, protocol design and performance analysis.

Lawrence Chung, Assistant Professor; Ph.D., Toronto, 1993. Software engineering, requirements engineering, nonfunctional requirements, information systems (Re-) engineering, software architectures, knowledge-based software engineering, computer-aided software engineering, software processes.

Jorge Cobb, Assistant Professor; Ph.D., Texas at Austin, 1996. Computer networks and distributed computing, quality of service in computer networks, wireless networks.

G. R. Dattatreya, Associate Professor; Ph.D., Indian Institute of Science, 1981. Stochastic modeling, parameter estimation and performance, optimization in communication, signal and image processing, computer network systems.

Michael Durbin, Senior Lecturer; M.S., Texas at Dallas, 1983. Discrete mathematics, object-oriented technology, modern operating systems, programmer productivity.

Don Evans, Senior Lecturer; D.M.A., North Texas, 1987. 8086-based assembler, C++, object-oriented design methodologies, architecture, visual basic.

Andras Farago, Professor; Ph.D., Budapest Technical (Hungary), 1981. Telecommunication network design and analysis, wireless network protocols, modeling, optimization, algorithms, traffic management in high-speed networks.

Dung T. Huynh, Professor and Program Head; Ph.D., Saarlandes (Germany), 1978. Computational complexity theory, automata and formal languages, concurrency theory, communications networks and protocols, parallel computation, software metrics.

Rym Mili, Assistant Professor; Ph.D., Tunis (Tunisia), 1991; Ph.D., Ottawa, 1996. Software engineering, software reuse, software metrics, software reengineering, clean-room software engineering, formal specifications.

Simeon Ntafos, Professor and Associate Program Head; Ph.D., Northwestern, 1979. Program testing, software reliability estimation, testing of distributed and concurrent software, computational geometry, robotics.

Ivor Page, Associate Professor, Associate Dean of Undergraduate Programs, and College Master; Ph.D., Brunel (England), 1979. Distributed algorithms, telecommunication networks, packet radio, resource allocation in distributed systems, computer graphics.

William J. Pervin, Professor; Ph.D., Pittsburgh, 1957. Software engineering, tools for program testing, program verification, processes in distributed and parallel systems, programming languages, applications to hearing disabilities and to management.

Ravi Prakash, Assistant Professor; Ph.D., Ohio State, 1996. Mobile computing, wireless networks, distributed systems, operating systems.

Balaji Raghavachari, Associate Professor; Ph.D., Penn State, 1992. Design and analysis of algorithms, graphs, telecommunication networks, topological network design, combinatorial optimization, approximation algorithms, sequencing problems.

L. Tissa Samaratunga, Senior Lecturer; Ph.D., Wayne State, 1994. Software engineering, object-oriented programming, analysis and design.

Haim Schweitzer, Associate Professor; Ph.D., Hebrew (Jerusalem) 1986. Artificial intelligence, computer vision, machine learning, multimedia.

David W. Storer, Senior Lecturer; M.S., Texas at Dallas, 1997. Beginning and advanced C++ programming, object-oriented design, assembler language.

Hal Sudborough, Founders Professor; Ph.D., Penn State, 1971. Telecommunication networks, parallel computation networks, efficient parallel (and sequential) algorithms, structure of complexity classes, picture processing, automata and formal languages, graph/network algorithms (especially embedding and layout problems), combinatorial problems (especially sorting by prefix reversals), computational biology.

Violet R. Syrotiuk, Senior Lecturer; Ph.D., Waterloo, 1992. Distributed algorithms, distributed systems, networks and telecommunications, with a special interest in wireless networks.

Ioannis G. Tollis, Professor; Ph.D., Illinois at Urbana-Champaign, 1987. Graph drawing and visualization, computer-aided design, telecommunication networks, VLSI layout, graph layout, computational geometry, algorithms and applications.

Klaus Truemper, Professor; Ph.D., Case Western Reserve, 1973. Computational logic and intelligent computer systems; ongoing applications projects: natural language processing, handwriting interpretation, traffic control, and expert systems providing optimal decisions.

Nancy Van Ness, Senior Lecturer; M.S., Stanford, 1966. C++, discrete mathematics, automata.

S. Venkatesan, Associate Professor; Ph.D., Pittsburgh, 1988. Distributed systems, fault tolerance, telecommunication networks, mobile/nomadic computing.

I-Ling Yen, Associate Professor; Ph.D., Houston, 1992. Parallel and distributed operating systems, fault tolerance for parallel and distributed systems, networking, object-oriented concurrent programming, multimedia software systems, embedded systems, self-stabilizing programs.

S. Q. Zheng, Professor; Ph.D., California, Santa Barbara, 1987. Algorithms, architectures, combinatorial optimization, optical interconnects, parallel and distributed processing, telecommunication and networks, VLSI.

UNIVERSITY OF UTAH

Department of Computer Science

Programs of Study	The Department of Computer Science offers programs leading to the M.S. and Ph.D. degrees. The graduate programs are open to computer science majors and also to students whose preparation is outside of computer science. Most of a doctoral student's time is devoted to course work and research, including participation in the research and teaching environment of the department on a day-to-day basis. The Ph.D. normally requires five years of graduate study, assuming that students undertake some teaching obligations during that time. A full-time student working on an M.S. program normally completes the degree requirements, including thesis, within two calendar years. The Department of Computer Science has an active, highly visible faculty engaged in a variety of research areas, including asynchronous digital systems, compilers, computer-aided geometric design, computer graphics, computer vision, educational computing, formal VLSI design methods, high-speed GaAs circuits, information-based complexity, information retrieval, natural-language processing, numerical analysis, operating systems, parallel and distributed computing, programming languages, rendering, robotics, scientific computing, scientific visualization, software engineering, VLSI design, and virtual reality and teleoperation.Some of the graduate courses offered by the department include computer architecture, computer vision, robotics, scientific computation, computer graphics, software engineering, programming languages, algorithm and data structure design, object-oriented software engineering, theoretical computer science, operating systems, compiler principles and techniques, data communication/networks, parallel programming, scientific visualization, artificial intelligence, integrated circuit design and VLSI theory/architectures, digital signal processing, and formal languages.
Research Facilities	The major research computing facilities are composed of six laboratories: Computer-Aided Design and Graphics, Computer Systems Laboratory, Asynchronous Digital Systems and VLSI, Robotics and Vision, Scientific Computing and Imaging, and Information Retrieval and Natural Language Processing. The department is divided into two computing environments; one is dedicated to research computing, and the other is for general/instructional computing.
	Both facilities share a common network infrastructure that is based on an ATM fabric running at OC-12 (622 Mbps). The departmental network connects the campus's OC-48 ATM mesh at OC-12 rates that connect to the vBNS and the Internet via dual OC-3 pipes. The base level of service to the desktop is Fast Ethernet (100 Mbps), which has the capability to accept OC-3, OC-12, or Gigabit Ethernet connections where necessary. The department's network fabric will be fully switched by fall 1999.
	In addition to the shared network infrastructure, the two facilities share centralized servers that provide firewall, gateway, ftp, news, Web/cgi, interactive, file, printing, dial-in, dns, ntp, calendar, and integrated UNIX and NT user accounts. The services are supported on a range of Solaris-based hosts, ranging from Ultra1's to an Enterprise 5000. Backups are automatically performed nightly via a dedicated network to a DTL library.
	The Research Computing Facility is a heterogeneous mix of 238 machines from SGI, HP, IBM, Sun, and Intel that include specialized equipment such as a 64-node Origin 2000 with eight infinite reality pipes, an Evans and Sutherland Harmony system, and a 14 cpu Power Challenge. In addition to the systems, there are specialized labs for multisource nonlinear video editing, real-time signal processing, haptic (robotic arm-based force feedback), image analysis, and various types of custom hardware design.
	The General Computing Facility supports both UNIX- and NT-based operating systems that total 131 machines. Three major labs make up the bulk of the facility. The first consists of IBM PowerPC systems, which run AIX; the second contains SGI machines, which range from Indys to O2s; and the third is an instructional NT lab, which houses Pentium-based PCs. The Department of Computer Science also has access to the College of Engineering Workstation Laboratory, which consists of five servers, 100 Sun workstations, and twenty-five HP workstations. The machines are divided into two separate rooms and are used for undergraduate and graduate instruction.
Financial Aid	Teaching and research assistantships are available to all full-time graduate students. Stipends are $12,150 for M.S. and Ph.D. students (preproposal) and $13,050 for Ph.D. students (postproposal) for the nine-month 1999–2000 academic year. Research support is also available for the summer.
Cost of Study	For the 1999–2000 academic year, resident tuition is $1495.70 per semester ($2991.40 per academic year) for 12 credit hours. Nonresident tuition is $3897.40 per semester ($7794.80 per academic year) for 12 credit hours. Tuition waivers are available for all supported teaching and research assistants.
Living and Housing Costs	In 1999–2000, dormitory room and board costs are approximately $5600 for the academic year. Apartments in University apartment communities rent for $340 to $680 per month; utilities are included. Off-campus housing is moderately priced. The cost of living in the area is close to or slightly less than the national average.
Student Group	The department has 95 full-time graduate students.
Location	Thirty minutes from campus are some of the world's finest ski and recreational areas. Hiking, fishing, mountaineering, river running, and desert solitude are only hours away, as are more than ten national parks and numerous national monuments and wilderness areas. Cultural activities include the Utah Symphony, Ballet West, Repertory Dance Theatre, and Utah Opera Company. Utah is home to professional basketball, hockey, and baseball teams. Computer-related industries include Evans & Sutherland, Unisys, Signetics, Intersil, Hercules, National Semiconductor, Terratek, and Novell.
The University and The Department	The more than 1,400-acre University of Utah campus, located in the foothills of the Wasatch Mountains, is the oldest state university west of the Missouri River. The University operates on the semester system with a shortened summer session. With more than 3,600 regular and auxiliary faculty members, who are among the nation's most prolific researchers, Utah ranks consistently among the top thirty American colleges and universities in funded research. The University offers its more than 25,000 students excellent recreational facilities, including a nine-hole golf course, racquetball and squash courts, an indoor jogging track, indoor swimming pools, and comprehensive physical education facilities. The University is also home to nationally ranked athletic programs, in particular Coach Rick Majerus's Running Utes basketball team and Coach Greg Marsden's Lady Utes gymnastics team. The Department of Computer Science is located in the Merrill Engineering Building at the northern edge of the campus.
Applying	Forms for admission and financial aid, advice concerning application procedures and official deadlines, and copies of the department handbook and research brochure may be obtained from the address below. Applications for admission, official transcripts, letters of recommendation, GRE scores on the General Test and Subject Test in computer science, and a statement addressing specific research goals should be sent to the Department of Computer Science. Official TOEFL scores are required of all international students whose native language is not English and should be sent to the University of Utah Admissions Office; a photocopy should be sent directly to the Department of Computer Science. Admission decisions are made by the department's Graduate Admissions Committee after careful review of each complete application.Applications for admission can be found on the World Wide Web (http://www.cs.utah.edu/admissions-webform.html).
Correspondence and Information	Graduate Coordinator Department of Computer Science University of Utah 50 South Central Campus Drive, Room 3190 Salt Lake City, Utah 84112-9205 Telephone: 801-581-8224 E-mail: grad-coordinator@cs.utah.edu World Wide Web: http://www.cs.utah.edu/

University of Utah

THE FACULTY AND THEIR RESEARCH

Erik L. Brunvand, Associate Professor; Ph.D., Carnegie Mellon, 1991. Computer architecture and VLSI systems, self-timed and asynchronous systems.

John B. Carter, Assistant Professor; Ph.D., Rice, 1992. Multiprocessor computer architecture, operating systems, distributed computing, computer networks.

Elaine Cohen, Professor; Ph.D., Syracuse, 1974. Geometric modeling, graphics, scientific visualization, physically based modeling, CAD/CAM, process planning, CAE.

Al Davis, Professor; Ph.D., Utah, 1972. Convergence parallel processing system architectures, VLSI, VLSI CAD, high-performance communication, asynchronous circuits.

Samuel H. Drake, Research Associate Professor; Sc.D., MIT, 1977. Mechanical design, industrial code.

Ganesh Gopalakrishnan, Associate Professor; Ph.D., SUNY at Stony Brook, 1986. Asynchronous circuit design and formal verification.

David Hanscom, Clinical Professor; Ph.D., Case Western Reserve, 1970. Undergraduate education.

Charles Hansen, Associate Professor; Ph.D., Utah, 1987. Scientific visualization, computer graphics, computer vision.

Thomas C. Henderson, Professor; Ph.D., Texas at Austin, 1979. Artificial intelligence, computer vision, robotics.

Lee A. Hollaar, Professor; Ph.D., Illinois at Urbana-Champaign, 1975. Legal issues regarding computers, intellectual property protection of software and information.

John M. Hollerbach, Professor; Ph.D., MIT, 1978. Robotics, teleoperation, virtual reality, human motor control.

Wilson C. Hsieh, Assistant Professor; Ph.D., MIT, 1995. Compilers, programming languages, systems.

Christopher R. Johnson, Associate Professor; Ph.D., Utah, 1989. Scientific computing, scientific visualization.

Robert R. Johnson, Emeritus Professor; Ph.D., Caltech, 1956. Computer architecture, system design, graphical programming, information theory.

Robert R. Kessler, Professor; Ph.D., Utah, 1981. Programming languages, parallel programming.

Jay Lepreau, Research Assistant Professor; B.S., Utah, 1983. Operating systems, information security, programming and domain-specific languages, compilers, networks.

Gary E. Lindstrom, Professor; Ph.D., Carnegie-Mellon, 1971. Programming languages, databases, parallel and distributed computing.

Sally McKee, Research Assistant Professor; Ph.D., Virginia, 1995. Processor and memory systems architecture, hardware design, compilers, operating systems.

Chris Myers, Research Assistant Professor; Ph.D., Stanford, 1995. Digital VLSI systems, computer architectures.

Richard F. Riesenfeld, Professor; Ph.D., Syracuse, 1973. Computer graphics, animation, computer-aided geometric design, CAD/CAM.

Ellen M. Riloff, Assistant Professor; Ph.D., Massachusetts Amherst, 1994. Natural-language processing, information retrieval, artificial intelligence.

Peter Shirley, Associate Professor; Ph.D., Illinois, 1990. Computer graphics, virtual reality.

Kris Sikorski, Associate Professor; Ph.D., Utah, 1982. Computational complexity, information-based complexity, numerical analysis.

Kent F. Smith, Professor; Ph.D., Utah, 1982. Integrated-circuit design.

Brian Smits, Research Assistant Professor; Ph.D., Cornell, 1993. Computer graphics, global illumination, reflection models.

Frank Stenger, Professor; Ph.D., Alberta, 1965. Numerical analysis, geometric modeling.

William B. Thompson, Professor; Ph.D., USC, 1975. Computer vision, artificial intelligence.

Joseph L. Zachary, Clinical Professor; Ph.D., MIT, 1987. Software engineering, programming and specification languages.

UNIVERSITY OF VIRGINIA

Department of Computer Science

Programs of Study

The Department of Computer Science offers programs leading to the degrees of Master of Computer Science, Master of Science in computer science, and Ph.D. in computer science. Students at all levels have the opportunity to begin research in their first year on algorithms analysis, computation theory, computer networks, operating systems, real-time computing, multimedia, distributed computing, global computing, computer architecture, databases, graphics, human-computer interaction, parallel computers and systems, programming environments, programming languages/compilers, software engineering, and VLSI design. The department has a strong research orientation and believes in close interaction between faculty members and students. It has an excellent student-faculty ratio, and the faculty members encourage students to publish their research.

Students typically complete the master's degree programs in eighteen months to two years and write a scholarly thesis on their research. The Ph.D. degree requires approximately five years, with qualifying examinations taken during the first or second year, depending on the status of the student. The department requires that students complete a core curriculum of computer science courses to ensure breadth of knowledge in the field.

Research Facilities

The department's computer facilities are primarily UNIX- and Windows NT–based. The department has many centralized services connected by a 100-MB switched Ethernet infrastructure. The central file servers offer a total of 200 GB of storage, much of which is RAID storage. For large, computational jobs, the department runs many multiprocessor Sun UltraSPARC servers with 512 MB of memory and 2 GB of virtual memory. Support for parallel computation and research is provided by 130 500-MHz DEC Alpha workstations and 130 fast multiprocessor Pentium-class machines connected by a 1.3-GB bidirectional Myrinet. Several Silicon Graphics systems are available for virtual reality and graphics research. The department maintains an IBM 8260 and six LightStream 2020 ATM switches. The department runs a 100 MB switched Ethernet to each desktop and provides a 155-MB (OC3) connection to the outside world.

Financial Aid

Financial aid is available through research assistantships, teaching assistantships, and graduate fellowships. In-state students are eligible for tuition fellowships, and tuition waivers are available to out-of-state students. Assistants are normally supported through the summer.

Cost of Study

For 1999–2000, in-state tuition for the academic year is $5000; out-of-state tuition is approximately $15,000.

Living and Housing Costs

The cost of living for the 1999–2000 year is estimated to be $12,000.

Student Group

The department has approximately 75 graduate students, half of whom are in the Ph.D. program. Most students attend full-time and receive financial aid, including tuition and fees.

Student Outcomes

Of the graduate students receiving master's degrees in 1998, 60 percent sought employment; the average starting salary was $54,000. Twenty-five percent continued on to Ph.D. programs.

Location

Charlottesville, Virginia, was settled before the Revolutionary War. The Charlottesville metropolitan area is composed of approximately 100,000 residents and is located within 2 hours of several major metropolitan areas. The area around the town is still primarily agricultural, and the nearby Shenandoah National Park provides excellent recreational opportunities. The University is located on the edge of Charlottesville, with shopping and cultural events accessible via the town's public transportation system.

The University and The Department

The University was founded by Thomas Jefferson in 1819, and his spirit is still in evidence today. The original University buildings are considered an outstanding example of American architecture. To continue Jefferson's tradition of togetherness, the University has remained small compared to other major state institutions, with an enrollment fixed at 17,500. The student-run honor system fosters an environment of community trust.

Computer science was instituted as part of the Department of Applied Mathematics and Computer Science in 1965. In 1984, computer science became a separate department. It is part of the School of Engineering and Applied Science, and its faculty encourages the cross-disciplinary nature of the subject.

Applying

Applications for both fall and spring semesters, with or without financial aid, are considered, but preference is given to fall admissions. Admission to the graduate school requires a bachelor's degree (or the equivalent), scores on the GRE General and Subject Tests, relevant transcripts, three letters of recommendation, and TOEFL scores from applicants whose native language is not English. Applications for the fall semester must be received by February 1 for financial aid consideration.

Correspondence and Information

Graduate Admissions
Department of Computer Science
Thornton Hall
University of Virginia
Charlottesville, Virginia 22903
Telephone: 804-924-7605

University of Virginia

THE FACULTY AND THEIR RESEARCH

Alan Batson, Professor; Ph.D., Birmingham, 1956. Computer systems, modeling and performance evaluation.

Stephen J. Chapin, Research Assistant Professor; Ph.D., Purdue, 1993. Operating and distributed systems.

James P. Cohoon, Associate Professor; Ph.D., Minnesota, 1982. Algorithms, electronic design automation, computational geometry, FPGAs.

Jack W. Davidson, Professor; Ph.D., Arizona, 1981. Compilers, computer architecture, systems software.

James C. French, Research Associate Professor; Ph.D., Virginia, 1982. Information retrieval, digital library and scientific databases.

Andrew Grimshaw, Associate Professor; Ph.D., Illinois, 1988. Parallel and distributed systems, object-oriented systems, computer architecture.

Anita K. Jones, University Professor; Ph.D., Carnegie-Mellon, 1973. Survivable systems, interactive simulation, programmed systems, scientific databases, U.S. science and technology policy.

John C. Knight, Professor; Ph.D., Newcastle, 1973. Safety critical systems, software dependability, software engineering.

Jörg Liebeherr, Assistant Professor; Ph.D., Georgia Tech, 1991. Computer networks, data communications, distributed systems.

David P. Luebke, Assistant Professor; Ph.D., North Carolina, 1998. Computer graphics.

Lois Mansfield, Professor; Ph.D., Utah, 1969. Computational mechanics, finite element methods and parallel algorithms for scientific computing.

Worthy N. Martin, Associate Professor; Ph.D., Texas at Austin, 1981. Computer vision, image databases, artificial intelligence, genetic algorithms.

James M. Ortega, Charles Henderson Professor; Ph.D., Stanford, 1962. Numerical algorithms for parallel architectures.

John L. Pfaltz, Professor; Ph.D., Maryland, 1968. Databases, parallel processes, graph theory.

Norman Ramsey, Research Assistant Professor; Ph.D., Princeton, 1993. Programming languages, environments.

Paul F. Reynolds Jr., Associate Professor; Ph.D., Texas at Austin, 1979. Parallel and distributed systems, concurrency control, simulation.

Gabriel Robins, Walter Munster Associate Professor; Ph.D., UCLA, 1992. Combinatorial optimization, computational geometry, algorithms, VLSI CAD, computational biology.

Kathy Ryall, Assistant Professor; Ph.D., Harvard, 1997. Computer graphics, user interfaces, human-computer interaction.

Sang H. Son, Associate Professor; Ph.D., Maryland, 1986. Databases, distributed systems, real-time systems, information security.

John A. Stankovic, BP America Professor and Chair; Ph.D., Brown, 1979. Real-time systems, operating systems, distributed computing, distributed multimedia.

Kevin Sullivan, Assistant Professor; Ph.D., Washington (Seattle), 1994. Software engineering, design, integration, evolution, economics.

Alfred C. Weaver, Professor; Ph.D., Illinois, 1976. Computer networks, electronic commerce, telemedicine.

William Wulf, University and AT&T Professor of Engineering; Ph.D., Virginia, 1968. Architecture, compilers.

UNIVERSITY OF WASHINGTON

Department of Computer Science and Engineering

Programs of Study	The Department of Computer Science and Engineering offers M.S. and Ph.D. degrees. The M.S. degree program typically takes two years. The Ph.D. degree program typically takes five years and includes both a comprehensive evaluation and a depth exam in addition to the dissertation.

The department has significant strengths in most aspects of the field, with particular emphasis on VLSI, embedded systems, and CAD; architecture; operating systems, networks, and communication; programming systems; information retrieval, database systems, and softbots; software engineering, safety, and human-computer interaction; computer graphics and computer vision; artificial intelligence; theory of computation; and computational biology. With a small undergraduate program, a moderate-sized graduate program, and an open, active faculty, the department feels that graduate student/faculty interaction, both within and across specialties, is one of the strengths of the programs. |
Research Facilities	The department is well equipped to support advanced education and research in computer science. General research equipment includes approximately 500 UNIX and Windows-based workstations and servers. All graduate student desks are equipped with workstations; new graduate student desks have state-of-the-art desktop equipment. A wide variety of special-purpose equipment supports research in graphics, computer vision, VLSI, parallel computing, and other areas.
Financial Aid	Most full-time graduate students receive some financial aid from the department, such as a teaching or research assistantship. For 1999–2000, these pay $1281 to $1473 per month, plus a waiver of most tuition and fees (including nonresident tuition).
Cost of Study	Tuition and fees for 1999–2000 are $1811 per quarter for Washington State residents and $4493 per quarter for nonresidents. (U.S. citizens are often able to establish state residency after twelve months.)
Living and Housing Costs	The University provides some low-cost housing for both married and unmarried students. Otherwise, Seattle is moderately priced for a major metropolitan area.
Student Group	There are approximately 150 graduate students in the program, with 30 new students entering each year. About 80 percent of the entering students are seeking a Ph.D. Most are full-time. Students come from all over the United States and from about twenty other countries. Graduates receive job offers from major universities and industrial labs throughout the world.
Location	Seattle is a cosmopolitan city of approximately 500,000 people, situated between Puget Sound on the west and Lake Washington on the east. Consistently rated as one of the most livable cities in the United States, Seattle has lively regional theater, music, dance, and opera; a wide selection of movies; and other cultural activities. Opportunities for outdoor recreation abound, including boating, hiking, biking, camping, fishing, rock climbing, mountain climbing, and skiing, both downhill and cross-country. Thirty miles to the east, the Cascade Mountains offer alpine lakes, trails, blueberry picking in the fall, and skiing during the winter. Mount Rainier, height 14,408 feet, is visible from campus on a clear day and is about 2 hours away by car. Across Puget Sound by ferry is the Olympic Peninsula, with Olympic National Park, the rain forest, and isolated ocean beaches. Winters are mild, but it does rain—about 34 inches per year, much of it drizzle.
The University and The Department	The University of Washington was founded in 1861. It is one of the major research institutions in the United States, averaging twelfth nationally over a variety of disciplines in a recent assessment, routinely ranking among the top five nationally in total federal awards for research and development, and among the top ten in industrial support of research and development. The enrollment is about 34,000, with approximately 8,000 in professional and graduate programs.

The Department of Computer Science and Engineering was formed in 1967 and has conferred 207 Ph.D. and 592 M.S. degrees. In the 1995 National Research Council study, the department was again ranked among the top ten in the nation both for excellence of faculty and for effectiveness of graduate program. Current faculty members have received fifteen prestigious Presidential/NSF Young Investigator awards from the National Science Foundation, two Guggenheim Fellowships, three ONR Young Investigator Awards, three Presidential Faculty Fellow Awards, two Sloan Research Fellowship, eight Fulbright Research Scholarships, ten ACM Fellowships, and six IEEE Fellowships, among other honors. Essentially all faculty members have research support from federal agencies, such as DARPA, NSF, and ONR. |
| **Applying** | Applicants must have a baccalaureate or equivalent degree. A solid background in computer science is the norm, but lack of formal training can be offset by strong evidence of potential. Transcripts, letters of recommendation, and GRE General Test scores are required; a GRE Subject Test score in computer science or another area is strongly recommended. Admission is very competitive. The department receives many more applications than it can accept. The typical undergraduate GPA of successful applicants is above 3.5, and typical GRE scores are above the 90th percentile. All application materials from international students should reach the department by December 1. All application materials from U.S. residents are due by January 10. |
| **Correspondence and Information** | Graduate Admissions
Department of Computer Science and Engineering, Box 352350
University of Washington
Seattle, Washington 98195-2350
Telephone: 206-543-1695
E-mail: grad-admissions@cs.washington.edu
World Wide Web: http://www.cs.washington.edu/ |

University of Washington

THE FACULTY AND THEIR RESEARCH

Richard Anderson, Associate Professor; Ph.D., Stanford, 1985. Parallel algorithms. Recipient of NSF Presidential Young Investigator award, 1987.

Tom Anderson, Associate Professor; Ph.D., Washington (Seattle), 1991. Local and wide area distributed systems, operating systems, computer architecture.

Jean-Loup Baer, Professor; Diplome d'Ingenieur, 1960, Doctorat 3e cycle, 1963, Grenoble; Ph.D., UCLA, 1968. Parallel and distributed processing, computer architecture.

Paul Beame, Associate Professor; Ph.D., Toronto, 1987. Sequential and parallel computational complexity theory. Recipient of NSF Presidential Young Investigator award, 1988.

Brian Bershad, Associate Professor; Ph.D., Washington (Seattle), 1990. Operating systems, architecture, distributed systems, parallel systems. Recipient of NSF Presidential Young Investigator award, 1991.

Alan Borning, Professor; Ph.D., Stanford, 1979. Computer languages, constraint systems, object-oriented languages.

Gaetano Borriello, Associate Professor; Ph.D., Berkeley, 1988. CAD for VLSI design and system integration, user interfaces, expert systems applications in CAD, VLSI processor and controller architecture. Recipient of NSF Presidential Young Investigator award, 1988.

Craig Chambers, Assistant Professor; Ph.D., Stanford, 1992. Object-oriented language design and implementation.

Brian Curless, Assistant Professor; Ph.D., Stanford, 1997. Computer graphics, active machine vision.

Martin Dickey, Lecturer; Ph.D., Arizona State, 1992. Computational linguistics, computer science education.

Chris Diorio, Assistant Professor; Ph.D., Caltech, 1997. Neurally inspired silicon-learning chips, neural networks and learning algorithms, ultra-high-speed integrated-circuit design.

Pedro Domingos, Assistant Professor; Ph.D., California, Irvine, 1997. Artificial intelligence, machine learning, data mining.

Carl Ebeling, Associate Professor; Ph.D., Carnegie-Mellon, 1986. Special-purpose hardware, VLSI, computer-aided design of complex systems. Recipient of NSF Presidential Young Investigator award, 1987.

Susan Eggers, Associate Professor; Ph.D., Berkeley, 1989. Computer architecture, memory system design, trace-driven methodology. Recipient of NSF Presidential Young Investigator award, 1990.

Oren Etzioni, Associate Professor; Ph.D., Carnegie Mellon, 1990. Artificial intelligence: machine learning, integrated architectures, planning. Recipient of NSF Young Investigator award, 1993.

Anna R. Karlin, Associate Professor; Ph.D., Stanford, 1987. Online algorithms, probabilistic algorithms, and probabilistic analysis.

Richard E. Ladner, Professor; Ph.D., Berkeley, 1971. Distributed computing theory, specification and analysis of distributed protocols, computational complexity theory, design/analysis of algorithms, learning theory, applications to aid deaf/deaf-blind people.

Edward D. Lazowska, Professor; Ph.D., Toronto, 1977. Distributed and parallel computer systems and system performance analysis, using queuing network models.

Alon Levy, Assistant Professor; Ph.D., Stanford, 1993. Database systems, artificial intelligence, query optimization, data integration, knowledge representation.

Henry M. Levy, Professor; M.S., Washington (Seattle), 1981. Operating systems, architecture, distributed and parallel systems.

David Notkin, Professor; Ph.D., Carnegie-Mellon, 1984. Extendable software systems, heterogeneous computer systems, environments for parallel programming. Recipient of the NSF Presidential Young Investigator award, 1988.

Hal Perkins, Lecturer; M.S., Cornell, 1982. Programming languages and compilers.

Walter L. Ruzzo, Professor; Ph.D., Berkeley, 1978. Computational complexity and parallel computation.

David Salesin, Associate Professor; Ph.D., Stanford, 1991. Computer graphics, user interfaces, computational geometry. Recipient of NSF Young Investigator award, 1993.

Linda Shapiro, Professor (joint appointment with Electrical Engineering); Ph.D., Iowa, 1974. Computer vision, AI, robotics.

Alan Shaw, Professor; Ph.D., Stanford, 1968. Operating systems, real-time systems, software specification methods.

Lawrence Snyder, Professor; Ph.D., Carnegie-Mellon, 1973. Parallel computation and VLSI.

Steven Tanimoto, Professor; Ph.D., Princeton, 1975. Computer analysis of images, computer graphics, artificial intelligence.

Martin Tompa, Professor; Ph.D., Toronto, 1978. Computational complexity. Recipient of NSF Presidential Young Investigator award, 1984.

Daniel Weld, Associate Professor; Ph.D., MIT, 1988. Qualitative reasoning, qualitative physics, artificial intelligence. Recipient of NSF Presidential Young Investigator award, 1989.

David Wetherall, Assistant Professor; Ph.D., MIT, 1998. Networks, distributed systems, operating systems.

John Zahorjan, Professor; Ph.D., Toronto, 1980. Computer systems, analytic modeling. Recipient of NSF Presidential Young Investigator award, 1984.

Adjunct, Affiliate, and Emeritus Appointments

Loyce Adams, Associate Professor of Applied Mathematics. Parallel processing, numerical linear algebra.

Les Atlas, Associate Professor of Electrical Engineering. Neural networks, digital signal processing, speech processing.

Philip Bernstein, Affiliate Professor; Microsoft. Databases.

Karl Bohringer, Assistant Professor of Electrical Engineering. Microelectromechanical systems.

James Brinkley, Research Assistant Professor of Biological Structure. Systems, biomedical applications of computers.

Steve Burns, Affiliate Professor; Intel. VLSI.

David Callahan, Affiliate Assistant Professor and Corporate Scientist, Tera Computers, Seattle. Parallel programming, compilers.

Michael Cohen, Affiliate Associate Professor; Microsoft.

Steve Corbato, Affiliate Associate Professor. UW networks and distributed computing.

David Cutler, Affiliate Professor of Computer Science; Microsoft. Implementation of extremely large hardware and software systems.

David B. Dekker, Associate Professor Emeritus of Mathematics and Computer Science. Differential geometry and numerical analysis.

Tony DeRose, Affiliate Professor; Pixar. Computer graphics.

Tom Duchamp, Professor of Mathematics. Differential geometry, applications to graphics.

Hellmut Golde, Professor Emeritus of Computer Science; Ph.D., Stanford, 1959. Computer networks, compilers.

Terence Gray, Affiliate Professor and Director of University Network and Distributed Computing. Operating systems, networks.

Philip Green, Associate Professor of Molecular Biotechnology. Genome analysis.

Robert M. Haralick, Professor of Electrical Engineering. Computer vision, artificial intelligence, image processing, pattern recognition.

Leroy Hood, Professor of Molecular Biotechnology. Molecular immunology and evolution, large-scale DNA mapping.

Earl Hunt, Professor of Psychology. Human and artificial intelligence, computer applications in teaching.

Ira J. Kalet, Associate Professor of Radiation Oncology. Medical applications of artificial intelligence, computer graphics, interface design, process control systems, distributed computing applications.

Gretchen Kalonji, Professor of Materials Science and Engineering. Computer simulation techniques in materials science.

Theodore Kehl, Professor Emeritus of Computer Science. Hardware systems.

Yongmin Kim, Professor of Electrical Engineering. Image processing and microprocessing.

Janusz Kowalik, Affiliate Professor of Computer Science and Manager of Technology Transfer, Boeing Computer Services. Parallel processing systems.

Paul Leach, Affiliate Professor; Microsoft. Distributed and object-oriented systems.

John Lewis, Associate Technical Fellow; Boeing Computer Services. Numerical mathematics and scientific computing.

Jerre D. Noe, Professor Emeritus of Computer Science. Distributed computer systems, operating systems, performance evaluation.

Maynard Olson, Professor of Molecular Biotechnology. DNA sequencing.

Eve Riskin, Assistant Professor of Electrical Engineering. Data compression, image processing, signal processing, information theory.

Burton Smith, Affiliate Professor and Chairman and Chief Scientist, Tera Computers, Seattle. Computer architecture.

Werner Stuetzle, Professor of Statistics. Computational and graphical methods in multivariate analysis, computer vision.

Rick Szeliski, Affiliate Associate Professor; Microsoft.

Paul Young, Professor Emeritus of Computer Science. Theory.

Gregory L. Zick, Professor of Electrical Engineering. Computer engineering, sorting, I/O subsystems, image databases.

UNIVERSITY OF WISCONSIN–MADISON

Department of Computer Sciences

Programs of Study	The Department of Computer Sciences offers M.S. and Ph.D. degrees in computer science. Research areas include artificial intelligence, computer architecture, computer networks, computer vision, database systems, distributed systems, graphics, mathematical programming, modeling and analysis of computer systems, numerical analysis, operating systems, performance evaluation, programming languages, software development environments, and theory of computation. Ongoing collaborative research with other departments includes projects in computational biology, computational chemistry, robotics, space sciences, and medical diagnosis.
	Requirements for the Ph.D. degree include course work and a preliminary examination based on initial and proposed thesis research. The M.S. degree usually takes two years to complete.
Research Facilities	There are more than 600 departmental machines, mostly high-performance Intel, Sun, and HP workstations, and all supported graduate students have their own workstations. Also, there are extensive parallel computing facilities.
Financial Aid	Almost all full-time graduate students are supported by a departmental assistantship or fellowship. Several students are offered fellowships each year, and most of the other new students receive teaching assistantships, with a multi-year guarantee of support. Most graduate students who have passed the Ph.D. preliminary examination are supported as research assistants.
Cost of Study	Starting in 1998, tuition is included in most assistantships. For others, in-state tuition in 1998–99 was $2463.95 per semester, and the nonresident tuition was $7594.95 per semester.
Living and Housing Costs	In addition to dormitories, University-operated one- and two-bedroom apartments are available for married students; these cost $367 to $544 per month. Off-campus housing is abundant and relatively inexpensive.
Student Group	The department has about 200 graduate students, with about 65 new students entering each year. Approximately fifteen Ph.D. degrees and sixty M.S. degrees are awarded each year.
Student Outcomes	M.S. students are in high demand at a range of major computer companies. Half of Ph.D. students accept academic positions and half accept positions in industry. Many major corporations regularly visit the department to describe their work and to recruit graduating students.
Location	Built on an isthmus between two large lakes, Madison—the state capital of Wisconsin—has a population of about 200,000 people. Madison offers a rich cultural life and is regularly rated as one of America's most livable cities. The area is a four-season display of beauty, encouraging recreational activities such as sailing, bicycling, hiking, and skiing. The campus extends 1½ miles along the shores of Lake Mendota, covering more than 1,000 acres. The University has more than 100 sailboats at its docks next to the student union.
The University and The Department	Founded in 1849 as a public, land-grant institution, the University of Wisconsin–Madison is one of America's top research universities; a recent National Academy of Sciences study ranked 16 UW–Madison research-doctorate programs in the top ten and 35 programs in the top twenty-five. The campus ranks third nationally in total research spending. There are 130 departments, making it one of the most comprehensive universities in the nation.
	The Department of Computer Sciences, formed in 1963, is consistently ranked as one of the top ten computer science departments in the country. Research funding for 1998–99 exceeded $7 million, supporting about seventy grants and contracts.
Applying	Applications for admission and all supporting materials must be received by December 31. No distinction is made at admission time between M.S. and Ph.D. applicants. Applicants must take both the GRE General Test and a GRE Subject Test in any area. The TSE is strongly recommended for applicants whose native language is not English and whose primary medium of instruction is not English. Although applicants need not have pursued an undergraduate major in computer science, significant related course work is expected.
Correspondence and Information	Graduate Admissions Department of Computer Sciences 1210 West Dayton Street University of Wisconsin Madison, Wisconsin 53706 Telephone: 608-262-1204 Fax: 608-262-9777 E-mail: admissions@cs.wisc.edu World Wide Web: http://www.cs.wisc.edu/

University of Wisconsin–Madison

THE FACULTY AND THEIR RESEARCH

Eric Bach, Professor; Ph.D., Berkeley, 1984. Theoretical computer science, computational number theory, algebraic algorithms, complexity theory, cryptography, six-string automata.

Carl de Boor, Steenbock Professor of Mathematical Sciences; Ph.D., Michigan, 1966. Approximation theory, numerical analysis.

Pei Cao, Assistant Professor; Ph.D., Princeton, 1995. Operating systems, storage management, parallel and distributed systems.

Anne Condon, Associate Professor; Ph.D., Washington (Seattle), 1987. Complexity theory, randomized complexity classes, theory of parallel computation, interactive proof systems, DNA computing.

Edouard J. Desautels, Professor; Ph.D., Purdue, 1969. Systems programming, personal computer systems and applications.

David J. DeWitt, Professor and Romnes Fellow; Ph.D., Michigan, 1976. Object-oriented database systems, parallel database systems, database benchmarking, geographic information systems.

Charles R. Dyer, Professor of Computer Sciences and of Biostatistics and Medical Informatics; Ph.D., Maryland College Park, 1979. Computer vision, view synthesis, shape representation, motion analysis, visual exploration.

Michael C. Ferris, Professor of Computer Sciences and Industrial Engineering and Vilas Associate; Ph.D., Cambridge, 1989. Large-scale optimization: theory, algorithms, and applications.

Charles N. Fischer, Professor; Ph.D., Cornell, 1974. Compiler theory and design, interactive program development environments, automatic register allocation and code generation, optimization.

Michael L. Gleicher, Assistant Professor of Computer Sciences; Ph.D., Carnegie Mellon, 1994. Computer graphics, computer animation, user interfaces, computer vision, digital video.

James R. Goodman, Professor of Computer Sciences and of Electrical and Computer Engineering and Chair; Ph.D., Berkeley, 1980. Computer architecture, large-scale computing, parallel computing, shared-memory multiprocessors.

Mark D. Hill, Professor of Computer Sciences and of Electrical and Computer Engineering and Romnes Fellow; Ph.D., Berkeley, 1987. Computer architecture, parallel computing, memory systems, performance evaluation.

Susan Horwitz, Professor; Ph.D., Cornell, 1985. Software development environments, language-based tools, static analysis of programs, program slicing, differencing, merging.

Yannis E. Ioannidis, Professor; Ph.D., Berkeley, 1986. Database management systems, complex query optimization, scientific databases, user interfaces, information visualization.

Deborah A. Joseph, Associate Professor of Computer Sciences and Mathematics; Ph.D., Purdue, 1981. Structural and applied complexity theory, mathematical logic, computational biology, computational geometry.

Sheldon Klein, Professor of Computer Sciences and Linguistics; Ph.D., Berkeley, 1963. Archaeology of cognition, simulation of language transmission and language change, language understanding and generation in the context of knowledge structures.

Lawrence H. Landweber, John P. Morgridge Professor; Ph.D., Purdue, 1967. Computer networks and protocols, high-speed networks.

James R. Larus, Associate Professor; Ph.D., Berkeley, 1989. Programming languages, parallel languages, compilers, program measurement.

Miron Livny, Professor; Ph.D., Weizmann (Israel), 1984. Resource management algorithms, performance modeling and analysis, discrete-event simulation.

Olvi L. Mangasarian, John von Neumann Professor of Mathematics and Computer Sciences; Ph.D., Harvard, 1959. Mathematical programming, machine learning, data mining.

Robert R. Meyer, Professor of Computer Sciences and Industrial Engineering; Ph.D., Wisconsin, 1968. Linear and nonlinear network optimization, parallel algorithms for large-scale optimization.

Barton P. Miller, Professor; Ph.D., Berkeley, 1984. Parallel and distributed debugging, parallel program performance tools, network management and name services, user interface design, extensible operating systems.

Jeffrey F. Naughton, Professor; Ph.D., Stanford, 1987. Parallel object-relational database systems, multidimensional database systems.

Raghu Ramakrishnan, Professor and Vilas Associate; Ph.D., Texas, 1987. Database query languages, including query by image content and spatiotemporal queries; data integration; interactive information visualization.

Thomas Reps, Professor; Ph.D., Cornell, 1982. Language-based programming environments; semantics-based program manipulation; program slicing; dataflow analysis and abstract interpretation; alias-analysis, pointer-analysis, and shape-analysis; incremental algorithms.

Stephen M. Robinson, Professor of Computer Sciences and Industrial Engineering; Ph.D., Wisconsin, 1971. Operations research, management science.

Amos Ron, Professor; Ph.D., Tel Aviv, 1987. Multivariate splines, wavelets, radial basis function approximation, polynomial interpolation, windowed Fourier transform, approximation to scattered data.

Jude W. Shavlik, Professor of Computer Sciences and of Biostatistics and Medical Informatics and Vilas Associate; Ph.D., Illinois, 1988. Machine learning, neural networks, artificial intelligence, computational biology, software agents, information retrieval.

Gurindar S. Sohi, Professor of Computer Sciences and of Electrical and Computer Engineering; Ph.D., Illinois, 1985. Instruction-level parallel processing, compiling for parallel architectures, memory systems.

Marvin H. Solomon, Professor; Ph.D., Cornell, 1977. Object-oriented database systems, distributed operating systems, computer networks, program development systems, programming languages.

John C. Strikwerda, Professor; Ph.D., Stanford, 1976. Numerical analysis, scientific computing, applied mathematics.

Mary K. Vernon, Professor of Computer Sciences and Industrial Engineering; Ph.D., UCLA, 1983. Techniques and applications for computer systems performance analysis, performance of parallel systems, parallel architectures and operating systems, multimedia storage servers.

David A. Wood, Associate Professor of Computer Sciences and of Electrical and Computer Engineering and Romnes Fellow; Ph.D., Berkeley, 1990. Computer architecture, performance evaluation, parallel processing, VLSI design.

VIRGINIA POLYTECHNIC INSTITUTE AND STATE UNIVERSITY

College of Arts and Sciences
Department of Computer Science

Programs of Study

The computer science faculty seeks talented graduate students to participate in study and research leading to the degrees of Master of Science and Doctor of Philosophy in computer science. These programs offer research opportunities in a wide range of areas, including algorithms, artificial intelligence, computer-aided education, digital libraries, human-computer interaction, information storage and retrieval, numerical analysis, parallel processing, real-time computation, simulation, and software engineering.

Master's degree candidates in computer science may choose to write a thesis or they may complete a course work–only degree. All students take a final examination: an oral defense for thesis students and a written exam for nonthesis students. Completion of the program takes 2 years.

Each Ph.D. student must take 90 semester hours beyond the baccalaureate. The dissertation usually represents 45–60 of the 90 hours. Up to 30 hours may be transferred from another institution. Students working toward the Ph.D. degree must pass three examinations: qualifying (tests background knowledge), preliminary (tests the student's ability to undertake the dissertation research), and final (an oral defense of the dissertation).

At Virginia Tech's Graduate Center in northern Virginia, the department offers both the M.S. and Ph.D. in computer science as well as a Master of Information Systems. This interdisciplinary program requires a minimum of 33 credits, including courses in five specified areas and a student-selected area of specialization.

Research Facilities

The Department of Computer Science facilities house computers and related equipment for both instruction and research. The department shares a 119-node Intel Paragon parallel computer and a state-of-the-art digital video editing system for multimedia production and jointly operates a 2-terabyte digital library server with the University's Computing Center. The department's computational environment also includes more than 200 DEC RISC- and Alpha-based, Intel Pentium-based, Sun, SGI, and Macintosh workstations, all of which are connected via a high-speed network. There are specially equipped laboratories offering opportunity for work in human-computer interaction, software engineering, simulation, CD-ROM/multimedia, and high-performance computing.

Financial Aid

Both graduate teaching and research assistantships are available. Teaching assistantships are awarded on a competitive basis. Research assistantships generally go to students who have already been at Virginia Tech at least one semester. The stipend associated with assistantships is based on the student's academic level. In 1999–2000, stipends range from $1335 to $1400 per month for nine months. Students on assistantships are exempt from tuition. A small number of highly qualified students receive offers of multiple-year assistantships as Computer Science Scholars; several multiple-year Graduate Research Traineeships are available for Ph.D. applicants in human-computer interaction.

Cost of Study

In 1999–2000, tuition is $2061 per semester for full-time in-state students and $3465 per semester for full-time out-of-state students. All students also pay a comprehensive fee of $396 per semester and a technology fee of $18 per semester.

Living and Housing Costs

Graduate student housing is available on campus, and there are many modestly priced private apartment complexes nearby. Information about housing may be obtained from the Office of Housing and Residence Life.

Student Group

Students in the department's on-campus graduate programs come from a variety of academic backgrounds and geographic areas. Of the 120 students enrolled in 1998–99, approximately 70 percent were in the master's program and 30 percent were working on the Ph.D.

Student Outcomes

Master's students readily find jobs in industry all over the United States, ranging from North Carolina, Virginia, and Maryland on the East Coast (e.g., Northern Telecom, TRW, Hughes Network Systems) to California, Oregon, and Washington on the West Coast (e.g., Qualcomm, Intel, Microsoft). Ph.D. graduates divide about equally between those going to research-oriented jobs in industry (e.g., Lucent Technologies, Batelle Lab, Mitre) and those going into teaching (e.g., Arizona State, James Madison, Villanova).

Location

Blacksburg is a small university community of 35,000 people in the Appalachian Mountains. This area of southwest Virginia is noted for its scenic beauty and outdoor recreational opportunities. The Blue Ridge Parkway and the Appalachian Trail are within an hour's drive of the campus. Roanoke, Virginia, with a population of 100,000, is 45 miles northeast of Blacksburg.

The University and The Department

Virginia Tech, Virginia's senior land-grant university, is the largest university in the state. Of the 25,000 students enrolled, approximately 4,000 are graduate students. More than 1,250 faculty members participate in the University's 118 graduate degree programs.

Applying

The program accepts both students with undergraduate degrees in computer science and students with degrees in such technical areas as the mathematical sciences, the physical sciences, and engineering who have had at least one course or equivalent training in each of the following areas: discrete mathematics, calculus, calculus-based statistics, computer organization, data structures, and operating systems. Deficiencies can be made up by taking undergraduate courses. Applicants who have completed significant research work (e.g., a thesis, conference paper, or journal article) are invited to submit abstracts or copies of this work. The GRE General Test is required of all applicants. The deadline for applications both for admission and for support is February 1 for the fall semester and August 15 for the spring semester.

Correspondence and Information

Computer Science Department
660 McBryde Hall, 0106
Virginia Polytechnic Institute and State University
Blacksburg, Virginia 24061
Telephone: 540-231-6932
E-mail: gradprog@cs.vt.edu
World Wide Web: http://www.cs.vt.edu/

Virginia Polytechnic Institute and State University

THE FACULTY AND THEIR RESEARCH

Marc Abrams, Associate Professor; Ph.D., Maryland. Visualization and performance modeling of communication networks and parallel and distributed programs, human-computer interaction.

Donald C. S. Allison, Professor; Ph.D., Queen's (Belfast). Computational geometry, analysis of algorithms, mathematical software, parallel computation.

James D. Arthur, Associate Professor; Ph.D., Purdue. Software engineering, verification and validation, parallel and distributed computation, user support environments, programming languages and translation.

Osman Balci, Professor; Ph.D., Syracuse. Software engineering, simulation and modeling, World Wide Web.

**John M. Carroll, Professor; Ph.D., Columbia. Human-computer interaction, evaluation methods, scenario-based system development, education and design applications, cooperative work.

Ing-Ray Chen, Associate Professor; Ph.D., Houston. Software engineering, performance and reliability modeling, evaluation of designs of distributed database/mobile/multimedia systems.

Stephen H. Edwards, Assistant Professor; Ph.D., Ohio State. Software reusability, software engineering, languages, object-oriented programming, formal methods.

Csaba J. Egyhazy, Associate Professor; Ph.D., Case Western Reserve. Operations research; management information systems; data, text, and knowledge processing.

*Roger W. Ehrich, Professor; Ph.D., Northwestern. Digital picture processing, automatic visual inspection, human-computer interface design and evaluation.

Edward A. Fox, Professor; Ph.D., Cornell. Multimedia information storage and retrieval, digital libraries, hypertext/hypermedia, electronic publishing and text processing, educational technology, library automation, network/WWW information and modeling.

William B. Frakes, Associate Professor and Director, Northern Virginia Program; Ph.D., Syracuse. Software reusability, software engineering, experimental methods, information storage and retrieval.

Sanjay Gupta, Assistant Professor; Ph.D., Ohio State. Theoretical computer science, design and analysis of algorithms, neural networks, quantum computing.

H. Rex Hartson, Professor; Ph.D., Ohio State. Human-computer interaction, usability methods, human factors in computing, interactive system development, software engineering.

Lenwood S. Heath, Associate Professor; Ph.D., North Carolina. Algorithms, graph theory, computational biology, symbolic computation, computational geometry, theoretical computer science, combinatorics, topology, quantum computing.

Sallie M. Henry, Associate Professor; Ph.D., Iowa State. Software engineering, software metrics, human factors, object-oriented paradigm, empirical studies, operating systems.

Deborah Hix, Assistant Professor; Ph.D., Virginia Tech. Human-computer interaction, development, and evaluation; interactive multimedia systems, virtual environments.

Dennis G. Kafura, Professor and Head; Ph.D., Purdue. Operating systems, software engineering, object-oriented systems.

John A. N. Lee, Professor; Ph.D., Nottingham (England). Programming languages, compiler design, industry standards, software engineering, history of computing, computer ethics.

Richard E. Nance, RADM John Adolphus Dahlgren Professor; Ph.D., Purdue. Computer simulation, distributed systems, software engineering, performance evaluation.

Naren Ramakrishnan, Assistant Professor; Ph.D., Purdue. Recommender systems, problem-solving environments, computational science and data mining.

† Calvin J. Ribbens, Associate Professor; Ph.D., Purdue. Parallel computation, mathematical software, numerical analysis, scientific computing.

*John W. Roach, Associate Professor; Ph.D., Texas. Artificial intelligence, agents, logic programming, natural language, robot problem solving, scene analysis.

Mary Beth Rosson, Associate Professor; Ph.D., Texas. Human-computer interaction, psychology of programming, object-oriented paradigm, community networks.

Clifford A. Shaffer, Associate Professor; Ph.D., Maryland. Data structures and algorithms, computer-aided education, CSCW, data visualization, computer graphics.

† Layne T. Watson, Professor; Ph.D., Michigan. Numerical analysis, nonlinear programming, mathematical software, fluid mechanics, solid mechanics, image processing, parallel computation.

Courtesy Appointments in the Department of Computer Science

Christopher A. Beattie, Professor; Ph.D., Johns Hopkins. Numerical analysis, computational linear algebra, spectral approximation of linear operators, scientific computing.

F. Gail Gray, Professor; Ph.D., Michigan. Fault-tolerant computing, switching and automata theory, computer architecture, algebraic coding theory, modeling and design with hardware description languages.

Charles E. Nunnally, Associate Professor; Ph.D., Virginia. Microprocessor/microcontroller systems, sensor-based real-time systems, embedded systems.

Joseph G. Tront, Professor; Ph.D., SUNY at Buffalo. VLSI design and testing, parallel processing, multimedia design, microprocessor systems, fault-tolerant computing.

Robert C. Williges, Professor; Ph.D., Ohio State. Human factors engineering, human-computer interface design, multimedia information presentation, computer-based training, usability testing, experimental design, human factors research methodology.

---

*Courtesy appointment in the Department of Electrical and Computer Engineering.
† Courtesy appointment in the Department of Mathematics.
**Courtesy appointment in the Department of Psychology and the Department of Teaching and Learning.

WASHINGTON UNIVERSITY IN ST. LOUIS

Department of Computer Science

Programs of Study	Both M.S. and D.Sc. degrees are offered. The doctorate involves course work to obtain breadth of knowledge and original research that culminates in a high-quality dissertation. Students intending to obtain the doctorate may enter without a master's degree. Time to complete the doctorate is typically four to six years.
	The master's program can be completed by course work alone or can include a thesis or project. Approximately ten courses are required (about three semesters). Many master's students are encouraged to remain for the doctoral program.
Research Facilities	Faculty and graduate student offices are housed contiguously in two modern research buildings. Most graduate students are in 2-person offices with individual workstations and have full access to the Internet. The department has extensive ATM network connections that reach across the campus and has dedicated laboratory facilities in networking and communications, visualization, and parallel and distributed computing. Strong systems staff support is provided for the computing and network infrastructure.
Financial Aid	The tuition award for the first year is $35,210, which includes full-time tuition (9 units per semester) and an annual stipend of $17,460 ($1455 per month). The financial aid package is continued until graduation (up to five years), as long as the student continues to make satisfactory progress toward the degree with respect to both course work and research.
Cost of Study	For students in the Department of Computer Science at Washington University in St. Louis, tuition is $975 per unit in 1999–2000, and most courses are 3 units.
Living and Housing Costs	Compared to most large cities, St. Louis is very affordable. Students prefer to live off campus because of the variety of housing in attractive neighborhoods within walking or short driving distance. Two-bedroom apartments range from $450 to $700 a month.
Student Group	Of approximately 100 graduate students, about two thirds are full-time, and half of those are pursuing doctorates. GRE General Test scores for incoming doctoral students are usually in the 90th percentile range.
Student Outcomes	Among recent doctoral graduates, 5 chose to start their research careers with Bell Laboratories Research. Doctoral graduates of the department may be found on the faculty of the Georgia Institute of Technology, Columbia University, and the Naval Postgraduate School, among others. Graduates at all levels are highly sought after by major computer firms, such as Microsoft, Oracle, and Intel, and by companies that make extensive use of computing capabilities in their business and products.
Location	No city is more American than St. Louis, which mixes Eastern, Southern, and Western culture and is often highly ranked among the best places to live. St. Louis is easily accessible from anywhere in the U.S. and is the national hub for Trans World Airlines. Winning national acclaim are the art museum and sculpture park, science center, botanical gardens, and various zoological and municipal parks as well as the symphony, the opera, and dance and music companies; many buildings have been praised for their architectural design. There are a dozen Fortune 500 companies and three professional sports franchises.
The University and The Department	Washington University in St. Louis seeks excellence and prestige in everything that it does. It is an independent institution founded in 1853 and now has 12,000 students and nearly 2,500 faculty members. Parts of the 1904 World's Fair and Olympics were held on the 169-acre campus, which is bordered on the east by St. Louis's famed Forest Park and on the north, west, and south by well-established suburbs. Twenty Nobel Prize winners have been associated with the University.
	The department is an innovator in research relationships with industry. Research support is approximately $6 million per year (an average of $340,000 per full-time faculty person), and all faculty members are active in research. The adviser-student ratio for doctoral supervision averages 1:2; no adviser currently supervises more than five doctoral candidates. The department's small size and collegial atmosphere provide a supportive and friendly environment. The department has a history of strong ties to biomedical computing, and faculty members collaborate extensively with those from other departments. About 30 external speakers arrive each year, and visiting researchers come from all over the world.
Applying	Applications should include GRE General Test scores as well as scores on the appropriate Subject Test. A strong computing background is recommended regardless of undergraduate major. Students who desire to enter the computing field from other areas of science and engineering may take advantage of several conversion classes that are designed to prepare them for entry into the graduate programs.
	Applications for graduate assistantships must be received by January 15. Applicants whose native language is not English must submit a minimum score of 575 on the TOEFL and 4.5 on the TWE.
Correspondence and Information	Admissions Department of Computer Science Campus Box 1045 Washington University in St. Louis St. Louis, Missouri 63130-4899 Telephone: 314-935-6160 Fax: 314-935-7302 E-mail: admissions@cs.wustl.edu World Wide Web: http://www.cs.wustl.edu

Washington University in St. Louis

THE FACULTY AND THEIR RESEARCH

Jerome R. Cox Jr., Senior Professor; Sc.D., MIT, 1954. Computer visualization, digital communication, biomedical computing.

Ron K. Cytron, Associate Professor; Ph.D., Illinois, 1984. Programming language optimization, parallel computations.

Richard A. Dammkoehler, Senior Professor; M.S.I.E., Washington (St. Louis), 1958. Computer programming theory and systems, information retrieval.

Mark A. Franklin, Hugo F. and Ina Champ Urbauer Professor; Ph.D., Carnegie Mellon, 1970. Computer architecture and systems analysis, parallel processing, VLSI design.

Will D. Gillett, Associate Professor; Ph.D., Illinois, 1977. Compiler theory and implementation, algorithm analysis and DNA mapping.

Kenneth J. Goldman, Associate Professor; Ph.D., MIT, 1990. Distributed programming environments, distributed applications, distributed algorithms.

Sally A. Goldman, Associate Professor and Assistant Chair; Ph.D., MIT, 1990. Computational learning theory, algorithms and computational geometry.

Takayuki Dan Kimura, Professor; Ph.D., Pennsylvania, 1971. Communication and computation, visual programming languages, multimedia user interfaces.

Ronald P. Loui, Associate Professor; Ph.D., Rochester, 1988. Artificial intelligence, legal reasoning, philosophy of computing, Web agents.

Gurudatta M. Parulkar, Professor; Ph.D., Delaware, 1987. Computer networking and Internet, multimedia systems.

Gruia-Catalin Roman, Professor and Chairman; Ph.D., Pennsylvania, 1976. Distributed and concurrent systems, software engineering, mobile computing, visualization.

Seymour V. Pollack, Professor Emeritus; M.S., Polytechnic of Brooklyn, 1960. Information systems, intellectual property.

Tuomas W. Sandholm, Assistant Professor; Ph.D., Massachusetts Amherst, 1996. Artificial intelligence, multiagent systems, machine learning, resource-bounded reasoning, electronic commerce.

Douglas C. Schmidt, Associate Professor; Ph.D., California, Irvine, 1994. Communication middleware, real-time ORBs, object-oriented design patterns and frameworks.

Subhash Suri, Associate Professor; Ph.D., Johns Hopkins, 1987. Computational geometry, algorithms, network design and protocols, computer graphics.

Jonathan S. Turner, Henry Edwin Sever Professor; Ph.D., Northwestern, 1982. Networking and communications, algorithms and complexity, VLSI applications.

George Varghese, Associate Professor; Ph.D., MIT, 1993. Distributed algorithms, network protocols, fault tolerance.

Affiliate Faculty

Keith Bennett (system/software engineering, virtual reality and visualization, space applications), Robert J. Benson (technology and information management), G. James Blaine (biomedical computing, digital communications, Picture Archive and Communications Systems (PACS)), A. Maynard Engebretson (speech and hearing applications), Mark Frisse (information retrieval), Barry L. Kalman (neural networks), I. Norman Katz (numerical analysis), Stan C. Kwasny (natural language), Harold Mack (architecture, languages), L. Andrew Oldroyd (image understanding and remote sensing, robotics, programming languages), Robert A. Rouse (expert systems), Kenneth F. Wong (computer communications, networking, software architecture).

RESEARCH AREAS

Networking and Communications. The networking and communications group at the University engages in fundamental research on building faster and more reliable networking infrastructures and developing applications that can take advantage of the infrastructures it builds. The group performs research in all areas and layers of networking, including optical and ATM switching, gigabit Internet routers, efficient host-network interfaces, operating system support for networking, middleware, and multimedia and management applications. In 1997, an informal survey conducted by *Business Week* ranked Washington University among the top universities in the world for networking.

Computer Engineering. The Computer and Communications Research Center (CCRC), led by Professor Mark A. Franklin, is focused on promoting research and development in computer engineering, an overlap area between computer science and electrical engineering. The interdepartmental center has research spanning the areas of telecommunications systems and parallel computing and architectures. The goal is to consider the interacting issues associated with hardware, software, and algorithms together when designing complex, high-performance computer/communications systems. Current center research focuses on areas of new and promising computer and communications architecture.

Multimedia Communication. The Computer Visualization Laboratory was established with the mission to carry out advanced research in the areas of software visualization, multimedia communication, and computer graphics and to provide technical and intellectual leadership in the adoption and use of multimedia technology. The laboratory is equipped with a unique blend of computational, communication, multimedia, and video capabilities. Recent projects involving multimedia technology cover computational geometry, time-critical rendering of 3-D scenes, software visualization, visual languages, pen-based interfaces, and Web-based presentations. Multimedia research in other departmental centers and groups complements the laboratory's activities by focusing on the underlying technology needed to deliver multimedia presentations.

Artificial Intelligence. Artificial intelligence (AI) is a vibrant research focus at the University, consisting of a number of related projects with both basic and applied research objectives. The research methodology covers a broad spectrum of theoretical work, prototype system building, and experimentation. The faculty is at the forefront of the modern, more formal discipline of AI. At the same time, its research agendas are shaped by the current application pull, such as the Internet and electronic commerce.

Algorithms and Languages. Algorithms and programming language research at the University is broadly focused on advanced research in the foundations of computing. Current interests lie in the areas of computational geometry, computational learning theory, compilers, programming languages, high-speed networking, and computational biology. The group's mission is to address important algorithmic and theoretical problems raised by cutting-edge technologies, to develop conceptual frameworks for studying these problems, and to invent methodologies that transcend specific problems.

Computational Science. The mission of the Laboratory for Computational Science (LCS), led by Associate Professor Will Gillett, is to find unique software and hardware solutions to difficult, real-world problems, in particular those combinatorially explosive problems that tend to resist standard algorithmic and hardware solutions. Such problems traditionally stem from the applied sciences rather than from the computer and information sciences. The three important areas currently investigated in LCS all have biological roots, although the scope of LCS is intended to encompass other sciences, such as physics and chemistry.

WASHINGTON UNIVERSITY IN ST. LOUIS

Department of Systems Science and Mathematics

Programs of Study	The department offers courses of study that lead to the Master of Control Engineering (M.C.E.) and the Master of Science (M.S.) and the Doctor of Science (D.Sc.) in systems science and mathematics. The M.C.E. and M.S. degrees require 30 semester credit hours, which may include an optional 6 semester credit hours for a thesis, project, or independent study. The D.Sc. requires 48 semester credit hours of course work plus 24 additional semester credit hours of doctoral research. Opportunities for advanced study and research are currently available in such areas as linear and nonlinear dynamics and control, scheduling and transportation systems, robotics and automation, discrete event dynamical systems, identification and estimation, multisensor fusion and navigation, machine vision and control, computational mathematics, finite elements, optimal control, mathematics of large-scale power systems, and intelligent systems. The department also participates in a joint doctoral program with the Department of Economics.
Research Facilities	There are three research Centers in the Department: the Center for Robotics and Automation, the Center for Optimization and Semantic Control (COSC), and the Center for Bio-Cybernetics and Intelligent Systems.
	The Center for Robotics and Automation carries out sponsored research in intelligent robotic systems. The activity of the Center emphasizes robotic workstation modeling, controller design, intelligent planning and distributed-control architecture for multirobot control, and underwater robotic systems. The ultimate goal is to achieve a design of sensor/event-referenced action planning and real-time control of robotic systems. The problems studied involve both theoretical development in robot control and implementation problems in sensory measurement, computer control, and software development. The research performed in the Center has been recognized in many international forums.
	The research efforts of the Center for Optimization and Semantic Control are directed towards the solution of large-scale, complex, and usually time-dependent problems for which the existing and classical methodologies of solution are either not available or not sufficiently powerful. Since the mathematical models of such systems are typically vague (or sometimes do not even exist), Center researchers use a judicious combination of classical mathematical methodologies, such as control theory, differential equations, and stochastics, together with artificial intelligence paradigms, such as neural networks, rule-based systems, and logic programming. Applications include transportation scheduling, aircraft control in emergencies, and autonomous vehicle control. COSC has had in the past and currently has research contracts with Systems and Electronics, Inc., a subsidiary of ESCO Electronics (ESCO is an offshoot of MEMC).
	The main emphasis of the Center for Bio-Cybernetics and Intelligent Systems is to train a new generation of students from biology, engineering, and medicine in modern techniques in computation and control, nonlinear control, systems identification, and estimation. Such an emphasis is clearly lacking in present day curricula, both nationally and internationally. Research areas in the Center include computational neuroscience, modeling computation and simulation in biosystems, and multisensor fusion and movement control. Research is currently funded by the National Science Foundation (NSF) and supports collaborative efforts with Texas Tech University and the University of Chicago.
	The 1.8-million-volume John M. Olin Library houses the engineering collection. There are also thirteen separate school and department libraries. Faculty members, students, and staff members have access to a wide range of computing resources. The department supports a Laboratory for Computation and Control with extensive computational power. This facility also provides e-mail and network services to the faculty and staff members and students of the department. There are also computer facilities in each of the centers.
Financial Aid	Aid is available for qualified students in the form of research assistantships. These are financed by University funds and sponsored research projects. Typically, the level of support for the nine-month academic year is $1000 per month plus tuition for 9 semester credit hours per semester. Another $1000 for one summer month is available for students who have no other summer employment. Research assistants are usually required to assist faculty members in teaching activities. The available support is offered during the spring on the basis of merit. Applications are accepted at all times, but it is desirable to apply by March 1 for support starting in September. Applicants are also strongly encouraged to seek fellowships sponsored by such agencies as the NSF, and international students are urged to apply for financial assistance that may be available through their home countries.
Cost of Study	Tuition for full-time graduate study is $975 per semester credit hour in 1999–2000. Part-time students pay $830 per semester credit hour (master's candidates only).
Living and Housing Costs	The University has no separate housing for married graduate students. Most graduate students, single or married, live off campus. Rooms and apartments are available in the area surrounding the University. The Campus Housing Referral Service (Campus Box 1059) provides assistance in locating suitable accommodations and maintains a list of off-campus housing. One- or two-bedroom apartments can be rented near the University starting at $300 per month. Rooms in private homes are available, sometimes in exchange for work, and usually include kitchen privileges.
Student Group	The department's graduate enrollment is about 48; 11 of these are part-time students who are employed full-time in the St. Louis area. About 31 of the current group are D.Sc. candidates.
Location	St. Louis offers social and cultural attractions and entertainment possibilities such as the St. Louis Symphony, the Loretto-Hilton Repertory Theatre, the Missouri Botanical Garden, and professional hockey, football, and baseball. Immediately adjacent to the campus, in Forest Park, are the St. Louis Art Museum, the St. Louis Science Center, the St. Louis Zoo, and the Municipal Opera. Near the western side of the campus is the business center of the city of Clayton (the St. Louis County seat), and within walking distance are movie theaters, stores, banks, snack shops, restaurants, and the shopping area of University City.
The University	Washington University in St. Louis is an independent, privately endowed and supported institution located on 169 acres within the greater St. Louis area. It was founded in 1853. The University has about 9,400 full-time students, approximately half of whom are undergraduates. There are about 1,280 part-time students. The University has 2,083 full-time and 193 part-time faculty members. It is composed of several major divisions: Arts and Sciences; the Graduate Schools of Arts and Sciences, Architecture, Business and Public Administration, Engineering and Applied Science, Fine Arts, Law, Medicine, and Social Work; the School of Continuing Education; and the Summer School. The University ranks among the top ten in the nation in the number of Nobel laureates associated with it.
Applying	There are no specific deadlines for application for admission, although it is wise to apply before March 1 for the following fall semester. Applicants must have received a bachelor's degree in engineering, mathematics, or science or show by reason of their record that they will receive it before matriculation. GRE scores are required for all full-time applications.
Correspondence and Information	Director of Graduate Programs Campus Box 1040 Washington University in St. Louis One Brookings Drive St. Louis, Missouri 63130 Telephone: 314-935-6001 Fax: 314-935-6121 E-mail: sandra@zach.wustl.edu World Wide Web: http://zach.wustl.edu/

Washington University in St. Louis

THE FACULTY AND THEIR RESEARCH

The faculty members in the department participate in both the undergraduate and graduate programs. Many of the faculty members have significant industrial experience and contacts.

Christopher I. Byrnes, Professor and Dean, School of Engineering and Applied Science; Ph.D., Massachusetts, 1975. Linear and nonlinear systems, adaptive control, dynamical systems.

Liyi Dai, Assistant Professor; Ph.D., Harvard, 1993. Discrete event dynamic systems, control theory, manufacturing systems.

James W. Dille, Affiliate Professor (Eclipse Capital Management, Inc.); Ph.D., Harvard, 1988. Queueing theory, discrete event simulation.

Bijoy K. Ghosh, Professor; Ph.D., Harvard, 1983. Computer vision, linear and nonlinear system theory, robotics, signal processing.

David S. Gilliam, Affiliate Professor (Texas Tech University); Ph.D., Utah, 1977. Control of distributed parameter systems, partial differential equations.

Xian-Zhong Guo, Affiliate Professor (Engineering Software Research and Development, Inc); Ph.D., Maryland, 1992. Numerical analysis.

Alberto Isidori, Professor; Libera Docenza, Rome, 1969. Linear and nonlinear systems, stability, dynamical systems.

Babu Joseph, Affiliate Professor (Washington University, Chemical Engineering); Ph.D., Case Western Reserve, 1975. Process network analysis, simulation and process control.

I. Norman Katz, Professor and Chair; Ph.D., MIT, 1959. Computational mathematics, differential equations, finite element methods, algorithms for parallel computations, solutions of Hamilton-Jacobi equations.

Kiyong Kim, Affiliate Professor (Basler Electric); D.Sc., Washington (St. Louis), 1995. Power system analysis and simulation, large-scale computational methods.

James D. Lang, Affiliate Professor (Boeing Corporation); Ph.D., Cranfield Institute of Technology, 1975. Aircraft performance, stability, and control.

Anders Lindquist, Affiliate Professor (Royal Institute of Technology); Ph.D., Royal Institute of Technology (Stockholm), 1972. Optimization and system theory, stochastic realization and control.

Edward P. Loucks, Affiliate Professor (Mallinckrodt Chemical, Inc.); D.Sc., Washington (St. Louis), 1994. Parameter identification, machine vision.

Zhi-Zhong Mou, Affiliate Professor (Marquette Co.); B.S.M.E., Northeast University of Technology (China), 1967. Quality control, reliability and maintainability engineering.

Hiroaki Mukai, Professor and Director of the Undergraduate Program; Ph.D., Berkeley, 1974. Theory and computational methods for optimization, optimal control, electric power system operation, probabilistic simulation, and optimization.

William J. Murphy, Affiliate Professor (Washington University, Electrical Engineering); D.Sc., Washington (St. Louis), 1967. Control systems.

Ervin Y. Rodin, Professor; Ph.D., Texas at Austin, 1964. Optimization, differential games, artificial intelligence, mathematical modelling.

Heinz M. Schättler, Associate Professor; Ph.D., Rutgers, 1986. Optimal control, nonlinear systems, stochastic calculus.

Jian Song, Honorary Distinguished Visiting Professor; Ph.D., Moscow, 1960. Aerospace systems, population systems, control theory.

Vaidyanathan Sundarapandian, Assistant Professor; D.Sc., Washington (St. Louis), 1996. Linear and nonlinear systems, stability, dynamical systems.

Tzyh-Jong Tarn, Professor; D.Sc., Washington (St. Louis), 1968. Quantum mechanical systems, bilinear and nonlinear systems, robotics and automation.

Alan C. Wheeler, Affiliate Professor; Ph.D., Stanford, 1968. Statistics and operations research.

Kevin A. Wise, Affiliate Professor (Boeing Company); Ph.D., Illinois, 1987. Control systems.

John Zaborszky, Professor; D.Sc., Royal Hungarian Technical University, 1943. Control theory, electric power systems.

Recent Research Projects

Optimal Control of Navier-Stokes Equations
Feedback Designs for Nonlinear Systems
Nonlinear Control Systems
An Approach of Ordinal Comparison to the Design and Optimization of Discrete Event Dynamic Systems
Visually Guided Sensor Integration, Control and Planning in Robotics and Automation
Knowledge Based Action Planning and Control Problems in Engineering and Biology
Visionics—An Integral Approach to Analysis and Design
Robust Control of Non-Minimum Phase Non-Linear Systems
Iterative Algorithms for the p-Version of the Finite Element Method
Artificial Intelligence Methodologies in Air Transportation Network Routing and Scheduling
Artificial Intelligence Methodologies in Air Transportation (general)
Variational Methods in the Control of Nonlinear Systems
Theoretical Study of a Reconfigurable Intelligent Manufacturing Workcell
Task Synchronization via Integration of Scheduling, Planning and Control for Underwater Robotic Vehicles
Structure in Power System Analysis and Control
Brake and Tire Wear Control Study Analysis
Layered Defense Project
Neural Network Parameter Identification
An Approach of Ordinal Comparison to the Design and Optimization of Discrete Event Dynamic Systems
Modeling and Diagnosis Methods for Large-Scale Complex Networks

WAYNE STATE UNIVERSITY

Department of Computer Science

Programs of Study

The department offers programs leading to the Master of Arts, Master of Science, and Doctor of Philosophy degrees. Courses offered in the department cover a broad range of interests, with particular emphasis on the areas of artificial intelligence, computational modeling of biological systems, computer graphics, data mining, database systems, image and video computing, multimedia information systems, natural language processing, neural networks, numerical analysis, parallel and distributed systems, pattern recognition, and software engineering.

Doctoral students are required to complete 90 credits beyond the bachelor's degree. This includes 30 credits of a Ph.D. dissertation that describes original, independent research. In addition, all doctoral students are required to pass a comprehensive proficiency examination in computer science.

For the M.A. degree, students must complete 31 credits of course work and a comprehensive exit exam. Those seeking the M.S. degree are required to complete 33 credits of course work, including a master's thesis of 8 credits.

Research Facilities

The Department of Computer Science operates seven research and two instructional laboratories. The research laboratories include the Artificial Intelligence Laboratory, the Biocomputing Laboratory, the Computer Graphics and Animation Laboratory, the Multimedia Information Systems Laboratory, the Parallel and Distributed Computing Laboratory, the Software Engineering Laboratory, and the Vision and Neural Networks Laboratory.

The instructional laboratories are equipped with Pentium-based computers that run both Windows NT Workstation 4.0 and Linux. One instructional laboratory is dedicated to instructor-led classes and has thirty-two student workstations. The other instructional laboratory is open to all computer science students for class work and general purpose use and has forty student workstations.

Financial Aid

Twenty-five graduate teaching assistantships are offered by the department on a nine-month basis. Teaching assistants are assigned part-time duties in the instructional program. The department also employs a number of graduate assistants during the summer for instructional and laboratory development purposes. Furthermore, students support themselves through research assistantships with faculty members who are pursuing various projects funded by government and private agencies.

Cost of Study

For 1999–2000, Michigan state resident graduate tuition is $247 for the first credit hour and $163 for subsequent hours per term. Nonresident graduate tuition is $439 for the first credit hour and $370 for subsequent hours per term. These fees include a $69 registration fee and a $15 Omnibus Credit Hour Fee to a maximum of 12 credit hours per term; both fees are required for residents and nonresidents. (Tuition and fees are subject to change without notice by action of the Board of Governors.)

Living and Housing Costs

A single, nonresident student needs approximately $19,650 for living expenses for the academic year, including tuition, housing, and other costs, and approximately $23,000 for the calendar year (if courses are taken throughout the year).

Student Group

The graduate student enrollment of the department consists of approximately 320 full- and part-time students. The majority of the part-time students are working professionals whose special skills enrich the classroom experience.

Location

Wayne State University is located in the heart of Detroit's cultural center. The Detroit Institute of the Arts, Detroit Historical Museum, Detroit Science Center, the Museum of African American History, and the main branch of the Detroit Public Library are all within easy walking distance.

The University

The main campus with its 203 acres of landscaped pedestrian malls is the meeting ground for more than 30,000 students enrolled in its thirteen schools and colleges. More than 13,000 students are enrolled in Wayne's graduate and professional programs. There are 136 master's and sixty-one doctoral majors and thirty certificate, specialist, and professional programs. More than 1,200 master's and 150 doctoral degrees are awarded each year. The University enjoys an excellent international reputation and enrolls more than 1,100 students from other countries. Wayne State is committed to a leadership role in research and creative activity. It offers students the opportunity to work closely with a variety of research centers and institutes. Examples in the physical sciences are the Center for Automotive Research, the Bioengineering Center, the Center for Molecular Biology, the Institute of Chemical Toxicology, and the Institute for Manufacturing Research.

Applying

Application forms for admission and financial aid may be obtained by writing to the address given below. Completed forms should be received in the department by February 15 for the fall semester and by October 15 for the winter semester. Admission and financial aid decisions are made after careful review of the applicant's research goals, undergraduate background in computer science, scores from a recent GRE General Test (verbal, quantitative, and analytical), and letters of recommendation. International students whose native language is not English must submit scores from the Test of English as a Foreign Language (TOEFL) or the Michigan English Language Assessment Battery (MELAB).

Correspondence and Information

Director of Graduate Studies
Department of Computer Science
Wayne State University
5143 Cass Avenue, Room 431 State Hall
Detroit, Michigan 48202
Telephone: 313-577-2477
Fax: 313-577-6868
World Wide Web: http://www.cs.wayne.edu

Wayne State University

THE FACULTY AND THEIR RESEARCH

Michael Conrad, Professor; Ph.D., Stanford, 1970. Biological information processing, computational modeling, molecular computing, brain models and intelligence, adaptability theory, evolutionary programming.

Sorin Draghici, Assistant Professor; Ph.D., St. Andrews (UK), 1995. Intelligent systems, neural networks, computer vision, pattern recognition, genetic algorithms, biological modeling, nonconventional computing.

Farshad Fotouhi, Associate Professor; Ph.D., Michigan State, 1988. Multimedia/hypermedia information systems, object-oriented databases, data warehouses, database interfaces.

Narendra Goel, Professor; Ph.D., Maryland, 1965. Computer simulation; computer graphics and animation; modeling of physical, engineering, and biological systems; remote sensing; numerical solutions of differential equations.

William I. Grosky, Professor and Chairman; Ph.D., Yale, 1971. Multimedia information systems, databases, Web technology.

Lucja Iwanska, Assistant Professor; Ph.D., Illinois at Urbana-Champaign, 1992. Artificial intelligence: natural language processing (computational models of semantics, pragmatics and context, logical and nonlogical reasoning, discourse processing), knowledge representation systems based on natural language.

Vaclav Rajlich, Professor; Ph.D., Case Western Reserve, 1971. Software engineering: evolution and maintenance.

Robert G. Reynolds, Associate Professor; Ph.D., Michigan, 1979. Artificial intelligence, intelligent agents, machine learning, genetic algorithms, cultural algorithms, evolution-based software development environments, software reuse, software metrics, expert systems, evolutionary computation, evolutionary programming, genetic programming.

Ishwar Sethi, Professor; Ph.D., Indian Institute of Technology (Kharagpur), 1978. Computer vision, data mining, pattern recognition, neural nets, multimedia information retrieval.

Frank Stomp, Assistant Professor; Ph.D., Eindhoven University of Technology (the Netherlands), 1989. Formal verification, distributed algorithms, security.

Nai-Kuan Tsao, Associate Professor; Ph.D., Hawaii, 1970. Analysis and implementation of numerical algorithms in analysis and linear algebra.

Richard Weinand, Lecturer; M.S.C.S., Wayne State, 1985. Information processing and problem solving in the immune system, modeling and simulation of complex systems.

Seymour Wolfson, Associate Professor; Ph.D., Wayne State, 1965. Internet, numerical methods, computer systems.

WORCESTER POLYTECHNIC INSTITUTE

Department of Computer Science

Programs of Study	The Department of Computer Science offers programs leading to the Master of Science (M.S.) and Doctor of Philosophy (Ph.D.) degrees. Graduate courses are scheduled in the late afternoon or evening to accommodate both full- and part-time students.

In the M.S. program, students take courses in the core areas of computer science to provide a strong basis for further study. Additional courses are required to expose students to the major areas of computer science research. The master's program consists of eleven computer science courses or of eight courses and a thesis. Some students undertake thesis research in cooperation with local industry or are supported by research grants. Full-time students may complete their degree requirements in about two years, while part-time students typically take longer. A practice-oriented specialization in computer and communications networks (CCN), consisting of nine courses plus an industrial internship, is also available.

The Ph.D. program consists of an additional nine courses in computer science, qualifying and comprehensive examinations, and completion of research leading to a dissertation.

Many applied and theoretical areas of computer science are available for course work and research. Research interests of the faculty include analysis of algorithms, artificial intelligence, computer graphics, computer vision, database systems, distributed systems, electronic publishing, expert systems, graph theory, networks, performance evaluation, programming languages and compilers, software engineering, user interfaces, and visualization.

Research Facilities

The department is housed in Fuller Laboratories, WPI's newest academic building, which was designed specifically for multimedia, high-technology education. A wide variety of computing equipment is available to graduate students in computer science. Offices are equipped with X-terminals or high-end PCs, with access to appropriate fileservers and the Internet. Research labs provide additional machines and more specialized software. In addition, the College Computer Center, which shares Fuller Labs with the Department of Computer Science, has many general-access labs. These resources are interconnected by a campuswide voice/data communications network providing easy access from every classroom, lab, and office.

Financial Aid

Financial aid is available to qualified students in the form of fellowships and assistantships. Robert H. Goddard Fellowships provide a stipend in the amount of $12,330 per years plus remission of tuition. Assistantships are also available and require half-time teaching or research duties. Stipends for teaching assistantships start at $1370 per month for the academic year plus remission of tuition; stipends for research assistantships vary. Additional assistance may be available for the summer. Many part-time students receive tuition support from their employers.

Cost of Study

Graduate tuition for the 1999–2000 academic year is $661 per credit hour. There are nominal extra charges for thesis, health insurance, and other fees.

Living and Housing Costs

Although graduate students do not generally live in dormitory rooms, they may use the Institute's dining facilities. Apartments and rooms in private homes are available near the campus at varying costs. Graduate students should plan on living expenses of around $9600 for the academic year, not including tuition.

Student Group

The computer science department has about 60 full-time and 50 part-time graduate students. Approximately half of the full-time students receive financial aid from the Institute or from faculty research grants. Recent graduates are employed in a variety of jobs in industry and in academia.

Location

The Institute is located in an attractive residential section of Worcester, which is a community of about 170,000 in central Massachusetts. Worcester is located near Massachusetts's high-technology corridor, creating many opportunities for industrial-academic cooperation.

The city, the second-largest in New England, has many colleges and a wide variety of cultural opportunities. The nationally famous Worcester Art Museum is located three blocks from campus. A civic center, the Worcester Centrum, is a regular venue of touring music groups and sports events. The cultural and academic activities of Boston are nearby, and there are recreational activities in Cape Cod to the east, the Berkshires to the west, and the ski areas to the north.

The Institute

Worcester Polytechnic Institute, founded in 1865, is the third-oldest college of engineering and science in the United States. Graduate study has been part of the Institute's activity for more than ninety years. The computer science department was established in 1969 and began its graduate program in the same year. Classes are small and provide for close student-faculty relationships. In keeping with the Institute's dedication to project-based education, many graduate courses involve a significant component of individual or group projects. The Institute's modern facilities are located on a pleasant 80-acre campus within walking distance of the center of Worcester. Complete athletic and recreational facilities and a program of concerts, movies, and special events are available to graduate students.

Applying

Applications for admission and financial assistance should be submitted by February 15 but are considered at any time. An applicant must submit an application form, official transcripts from all colleges attended, and three letters of recommendation. The Graduate Record Examinations General Test is required of all applicants. International students who have not completed an undergraduate degree from an English-speaking country are required to take the Test of English as a Foreign Language (TOEFL). Applicants for the Ph.D. program should have a B.S. or an M.S. degree in computer science.

Correspondence and Information

For program information and application forms, interested students should contact:

Graduate Coordinator
Department of Computer Science
Worcester Polytechnic Institute
Worcester, Massachusetts 01609
Telephone: 508-831-5357
Fax: 508-831-5776
E-mail: graduate@cs.wpi.edu
World Wide Web: http://www.cs.wpi.edu/

Worcester Polytechnic Institute

THE FACULTY AND THEIR RESEARCH

Lee A. Becker, Associate Professor; Ph.D., Illinois, 1978. Artificial intelligence, object-oriented analysis and design, visual communication. Professor Becker is interested in machine learning and cognitive modeling. He is currently investigating cooperative induction and a visual interlingua.

David C. Brown, Professor; Ph.D., Ohio State, 1984. Knowledge-based design systems, artificial intelligence. Professor Brown's research is concerned with modeling the way humans design things, such as mechanical components, electronics, or buildings. He is also investigating the use of AI techniques for systems used in manufacturing.

Mark L. Claypool, Assistant Professor; Ph.D., Minnesota, 1997. Distributed systems, networking, multimedia and collaborative filtering. Professor Claypool is interested in the performance of multimedia applications, particularly when such applications are distributed across networked machines. He concentrates on building software systems and evaluating their performance through experiments.

Isabel F. Cruz, Assistant Professor; Ph.D., Toronto, 1994. Database systems, multimedia information systems, digital libraries, user interfaces, information visualization, visual languages, graph drawing, constraints. Professor Cruz's current research interests include the design and implementation of systems to query and visualize multimedia information on distributed repositories, including the Web.

David Finkel, Professor; Ph.D., Chicago, 1971. Computer system performance evaluation, distributed computing systems. Professor Finkel's current research is concerned with the performance of distributed and multiprocessor operating systems. He is also investigating computer networking systems, specifically networking systems for multimedia applications.

Michael A. Gennert, Associate Professor; Sc.D., MIT, 1987. Computer vision, spatio-temporal databases, theoretical computer science. Professor Gennert conducts research in computational vision, including stereo, motion, and fractal image modeling, with applications in biomedicine and automated inspection; image and multimedia databases, including temporal data modeling, with applications in global change analysis; and category theory and abstraction theory, with applications in programming language design.

Nabil I. Hachem, Associate Professor; Ph.D., Syracuse, 1988. Management of very large data and knowledge bases, computer architecture and database machines. Professor Hachem is investigating work flow and document management systems as well as very large distributed data management systems.

George T. Heineman, Assistant Professor; Ph.D., Columbia, 1996. Software engineering, object-oriented design, component-based software engineering. Professor Heineman is interested in developing models for adaptable software components. He also investigates extended transaction models.

Micha Hofri, Professor; Ph.D., Technion (Israel), 1972. Analysis of algorithms, performance evaluation, applied probability and the use of statistics in algorithms, asymptotics.

Robert E. Kinicki, Associate Professor; Ph.D., Duke, 1978. Network management, computer system performance evaluation, computer networks. Professor Kinicki's current research involves congestion control algorithms for TCP/IP, network management, and electronic commerce.

Karen A. Lemone, Associate Professor; Ph.D., Northeastern, 1979. Electronic documents, language translation. Professor Lemone is interested in electronic documents and learning environments for the World Wide Web. Research includes systems that allow a document to be described separately from its content for the efficient production of high-quality documents.

Carolina Ruiz, Assistant Professor; Ph.D., Maryland, 1996. Deductive databases, logic programming, data mining, machine learning. Professor Ruiz's most recent work is in logic programming for knowledge discovery in databases. She is also working on applications of automated theorem-proving techniques to inductive learning.

Elke A. Rundensteiner, Associate Professor; Ph.D., California, Irvine, 1992. Database and information systems. Professor Rundensteiner is currently leading research projects in schema evolution and transformation technology, object-oriented databases, data warehousing over distributed sources, Web-based database tools, and visual data exploration.

Gabor N. Sarkozy, Assistant Professor; Ph.D., Rutgers, 1994. Graph theory, combinatorics, algorithms. Professor Sarkozy's research is concerned with extremal graph theory, especially its relationship with other fields such as theoretical computer science and number theory. His current research includes the development of a new method for finding certain special subgraphs in dense graphs. He is also investigating algorithmic implementations of this method.

Stanley Selkow, Professor; Ph.D., Pennsylvania, 1970. Combinatorial algorithms, graph theory, analysis of algorithms. Professor Selkow is studying polynomial-time algorithms to solve problems in graph theory and other discrete combinatorial problems. He is also investigating the analysis of algorithms and data structures.

Matthew O. Ward, Associate Professor; Ph.D., Connecticut, 1981. Data and information visualization, computer graphics, and spatial data analysis and management. Professor Ward is investigating methods for the visual exploration of large multivariate data sets. He is also interested in the integration of management, analysis, and display of scientific data.

Craig E. Wills, Associate Professor; Ph.D., Purdue, 1988. Distributed systems, networking, user interfaces. Professor Wills is interested in issues concerning a user's access to a computer system, particularly when the system is a distributed network of machines. His current work primarily involves Web performance and caching.

SELECTED RECENT PUBLICATIONS

Grecu, D. L., and **L. A. Becker.** Coactive learning for distributed data mining. In *Fourth International Conference on Knowledge Discovery in Databases, KDD-98,* pp. 209–13. New York: AAAI Press, 1998.

Brown, D. C., and M. E. Balazs. A preliminary investigation of design simplification by analogy. *Proc. Conf. Artif. Intell. Design '98,* Lisbon, Portugal, July 1998.

Claypool, M. L., and T. J. Riedl. End end-to-end quality in multimedia applications. In *Handbook on Multimedia Computing,* chap. 40. Boca Raton, Fla.: CRC Press, 1998.

Cruz, I. F., and W. T. Lucas. Virtual document generation from multimedia repositories. *Theory Pract. Object Syst.* 4(4):245–60, 1998.

Finkel, D., C. E. Wills, B. Brennan, and C. Brennan. Distriblets: Java-based distributed computing on the web. *Internet Res. Electronic Networking Applications Policy* 9(1), 1998.

Lisin, D. A., and **M. A. Gennert.** Optimal function approximation using fuzzy rules. *Proc. Int. Conf. North Am. Fuzzy Information Processing Soc.,* 1999.

Taylor, S. E., **N. I. Hachem,** and **S. M. Selkow.** The average height of a node in the BANG directory tree. *Inf. Proc. Lett.* 61:55–61, 1997.

Heineman, G. T. A model for designing adaptable software components. *22nd Annual International Computer Science and Application Conference, COMPSAC-98,* pp. 121–7, 1998.

Hofri, M. *Analysis of Algorithms: Computational Methods and Mathematical Tools.* New York: Oxford University Press, 1995.

Raghavendra, A., and **R. E. Kinicki.** A simulation peformance study of TCP vegas and random early detection. *Proc. 18th Int. Performance Computing Commun. Conf.,* Phoenix, Arizona, pp. 169–76, 1999.

Lemone, K. A. ReCourse: A system for retargetable course generation. *ITS '96,* Montreal, 1996.

Ruiz, C., and J. Minker. Logic knowledge bases with two default rules. *Ann. Math. Artif. Intell.* 22(3,4):333–61, 1998.

Kuno, H. A., and **E. A. Rundensteiner.** Incremental maintenance of materialized object-oriented views in MultiView: Strategies and performance evaluation. *IEEE Trans. Data Knowledge Eng.* 10(5):768–92, 1998.

Atkins, A. C., **G. N. Sarkozy,** and **S. M. Selkow.** Counting irregular multigraphs. *Discrete Math.* 195:235–7, 1999.

Resnick, R., **M. O. Ward,** and **E. A. Rundensteiner.** FED—A framework for iterative data selection in exploratory visualization. *Proc. Tenth Int. Conf. Sci. Stat. Database Manage.,* (SSDBM 98), 1998.

Wills, C. E., D. C. Brown, B. Dunskus, and J. Kemble. Evaluating network serviceability. *Comp. Networks ISDN Syst.* 30(24):2283–91, 1998.

WRIGHT STATE UNIVERSITY

Computer Science Program

WRIGHT STATE
UNIVERSITY™

Programs of Study	The Department of Computer Science and Engineering offers graduate programs leading to the Master of Science in Computer Science (M.S.C.S.) and the Doctor of Philosophy in computer science and engineering. A Master of Science in Computer Engineering program is also available. Interested students should also see the Computer Engineering Program in this volume.
	The M.S.C.S. program requires 48 graduate quarter credit hours in a department-approved program of study. Typically, students must complete ten courses and orally defend a thesis or complete twelve courses in the nonthesis option. The Ph.D. program requires 91 graduate quarter credit hours after completion of the M.S.C.S. or 136 hours after completion of a bachelor's degree. Students without a master's degree must complete the requirements for the M.S.C.S. as the first phase of their Ph.D. program.
	Ph.D. students must complete 76 hours of formal course work (courses completed in the master's program may partially satisfy this requirement), pass the Ph.D. qualifying examinations and candidacy examination, and successfully complete and defend a dissertation.
Research Facilities	Modern laboratory facilities provide ample equipment for research in a number of areas. The College of Engineering and Computer Science provides access to DEC Alpha servers and workstations, a Silicon Graphics (SGI) Reality Engine 2, and SGI, DEC, and Sun workstations as well as numerous networked PCs and X-Windows terminals. In addition to these machines, dedicated research labs containing an SGI Onyx 2 with eight parallel processors as well as other SGI, DEC, and Sun equipment are available. Wright State is a member of OARnet, which provides access to the resources of the Ohio Supercomputer Center. Wright State's Information Technology Research Institute, a collaboration of academia, industry, and government, provides opportunities for research in all areas of modern information technology.
	Laboratories specifically dedicated to student and faculty research support studies in artificial intelligence, networking, human-computer interaction, parallel and concurrent computing, operating systems, control and robotics, digital communications, graphics, computer vision and imaging, computer-aided design, digital design, VLSI, software engineering, and optical computing. Wright Patterson Air Force Base and several Dayton industries participate in sponsored research, and their facilities are frequently available to graduate students and faculty. In addition, Wright Patterson's Major Shared Resource Center provides an opportunity for collaborative research with a focus on supercomputing.
Financial Aid	Financial aid available to graduate students includes research assistantships, teaching assistantships, Federal Perkins Loans, Federal Stafford Student Loans, short-term loans, and academic tuition fellowships. Scholarships are also available through the Dayton Area Graduate Studies Institute. Competitive stipends for research and teaching assistantships are available.
Cost of Study	Tuition in 1999 for residents of Ohio is $161 per quarter hour for part-time study (1 to 10½ hours) and $1703 for full-time study (11 to 18 hours). Tuition for out-of-state students is $282 per quarter hour for part-time study and $3013 for full-time study.
Living and Housing Costs	Many students commute to classes; however, campus housing is available. Room and board cost about $1600 per quarter. There are a variety of apartments and houses for rent in the area. The cost of living in Dayton is low compared with that in most other metropolitan areas.
Student Group	Approximately 3,200 graduate students are enrolled at Wright State University in forty graduate degree programs that lead to master's, Ed.S., Ph.D., M.D., and Psy.D. degrees.
Student Outcomes	Graduates are employed in a variety of professional positions both in and out of the state. Typical positions include software engineer, computer engineer, systems analyst, and systems programmer.
Location	Although Wright State has a suburban setting, it is closely tied to Dayton, the fourth-largest metropolitan area in Ohio. Dayton is a manufacturing center and is also the home of Wright-Patterson Air Force Base, the center of Air Force research and procurement. This combination of industrial and military development has produced an unusual concentration of scientific, technical, and research activity. Dayton has a considerable and varied cultural life. In the surrounding area there are at least fifteen other colleges and universities.
The University	Wright State University is an exciting and expanding university that continues in the scientific and engineering innovative spirit of aviation pioneers Orville and Wilbur Wright. The University has modern buildings that facilitate access for all, and the 557-acre main campus has accommodations for academics, support programs, sports, housing, and arts activities. Wright State's phenomenal growth as an institution—from one building, 92 employees, and 3,200 students in 1964 to forty-two modern buildings, nearly 2,100 employees, and more than 15,000 students in 1998—has necessarily stimulated a comparable growth in the areas that surround the University.
Applying	The basic requirement for admission to the master's degree program in computer science is a bachelor's degree in computer science or a related area with an overall undergraduate grade point average of at least 3.0 (on a 4.0 scale) or 2.7 with an average of 3.0 or better in the major field. A minimum grade point average of 3.3 is expected for admission to the Ph.D. program. Scores on the GRE General Test are required. Specific prerequisites for admission include calculus, linear algebra, discrete math for computing, and physics. The student's background should include knowledge of a higher-level language, data structures, concurrent programming, computer organization, operating systems, and digital hardware design. Minor background deficiencies may be made up after admission to the program by taking appropriate courses.
	There is no deadline for admission applications, since admission decisions are made on a rolling basis, and programs may be started in any quarter. The deadline to apply for teaching assistantships, however, is February 1.
Correspondence and Information	Oscar N. Garcia, Chair Department of Computer Science and Engineering Wright State University 3640 Colonel Glenn Highway Dayton, Ohio 45435-0001 Telephone: 937-775-5131 Fax: 937-775-5133 E-mail: cse-dept@.cs.wright.edu World Wide Web: http://www.cs.wright.edu/cse/general.html

Wright State University

THE FACULTY AND THEIR RESEARCH

A. A. S. Awwal, Associate Professor; Ph.D., Dayton, 1989. Digital optical computing, novel learning and architectures for neural networks, computer arithmetic, digital systems, computer-aided optical systems design, optical data storage/retrieval algorithms, fingerprint/character recognition, automated target recognition.

P. Bruce Berra, Professor; Ph.D., Purdue, 1968. Optical and electronic computer architectures, very large multimedia data/knowledge bases.

C.-I. Henry Chen, Associate Professor; Ph.D., Minnesota, 1989. CAD, simulation and testing of VLSI circuits, aerospace and electronics systems (testability synthesis and design verification of digital systems, signal processing, and microwave electronic circuits).

C. L. Philip Chen, Associate Professor; Ph.D., Purdue, 1988. Neural networks and applications, CAD/CAM and robotics, intelligent systems and interfaces, knowledge-based systems.

Jer-Sen Chen, Research Assistant Professor; Ph.D., USC, 1989. Computer graphics, image and video compression, human-computer interactions.

Soon M. Chung, Associate Professor; Ph.D., Syracuse, 1990. Database multimedia, parallel processing, computer architecture.

Michael T. Cox, Assistant Professor; Ph.D., Georgia Tech, 1996. Intelligent interfaces, case-based reasoning, automated planning, machine learning, natural-language processing.

Guozhu Dong, Associate Professor; Ph.D., USC, 1988. Database systems, database, data mining and knowledge discovery, data warehousing and integration, workflow.

Travis E. Doom, Assistant Professor; Ph.D., Michigan State, 1998. Computer architecture, computer systems, design automation, computational mathematics and theory.

Oscar N. Garcia, Professor; Ph.D., Maryland, 1969. Speech recognition and articulatory synthesis, knowledge-based systems, computer architecture, human-computer interaction, intelligent interfaces, machine intelligence.

A. Ardeshir Goshtasby, Associate Professor; Ph.D., Michigan State, 1983. Intelligent interfaces, machine vision, computer graphics and visualization, geometric modeling, medical image analysis.

Ricardo Gutierrez-Osuna, Assistant Professor; Ph.D., North Carolina State, 1998. Electronic olfaction, mobile robotics, pattern recognition, artificial intelligence.

Jack S. Jean, Associate Professor; Ph.D., USC, 1988. High-performance computer architectures, machine intelligence.

Prabhaker Mateti, Associate Professor; Ph.D., Illinois at Urbana-Champaign, 1976. Distributed computing, Internet security, formal methods in software design.

Kuldip S. Rattan, Professor; Ph.D., Kentucky, 1975. Fuzzy control, robotics, digital control systems, prosthetic/orthotics and microprocessor applications.

Mateen M. Rizki, Associate Professor; Ph.D., Wayne State, 1985. Modeling, simulation, biological information processing, machine intelligence, pattern recognition, image processing.

Robert C. Shock, Associate Professor; Ph.D., North Carolina at Chapel Hill, 1969. Software engineering, database systems.

Raymond E. Siferd, Professor; Ph.D., Air Force Tech, 1977. VLSI design and parallel computing.

Thomas A. Sudkamp, Professor; Ph.D., Notre Dame, 1978. Approximate reasoning, machine intelligence.

Krishnaprasad Thirunarayan (T. K. Prasad), Associate Professor; Ph.D., SUNY at Stony Brook, 1989. Knowledge representation and reasoning, object-oriented programming, specification of programming languages.

Karen A. Tomko, Assistant Professor; Ph.D., Michigan, 1995. Parallel computing, application optimization, graph partitioning, reconfigurable computing.

YALE UNIVERSITY

Department of Computer Science

Programs of Study

The department features four areas of concentration: artificial intelligence, scientific computation (numerical analysis), systems and programming languages, and theory of computation, but research often crosses area boundaries.

The graduate degree program in computer science stresses original research by the student both as an individual and as a member of the community of scholars. To this end, course requirements are minimal, and students normally begin research by the fall term of their second year of graduate study.

Research Facilities

The department operates a high-bandwidth, local area computer network based mainly on distributed workstations and servers, with connections to worldwide networks. Workstations include Sun SPARCstations, workstation PCs (NT and/or Linux), and NeXTstations. A vision laboratory contains specialized equipment for vision and robotics research. Various printers, including color printers, as well as image scanners are also available. The primary educational facility consists of thirty-seven PC workstations supported by a large Intel PC server. This facility is used for courses and unsponsored research by computer science majors and first-year graduate students. Access to computing, through both the workstations and remote login facilities, is available to everyone in the department.

Financial Aid

Most Ph.D. students receive financial aid in the form of a fellowship or research assistantship. For 1999–2000, fellowship and assistantship stipends are $13,000 for nine months. Financial aid includes tuition in addition to the stipend. Students may supplement their income with teaching assistantships.

Cost of Study

Tuition for the two-semester academic year is $23,330 for 1999–2000.

Living and Housing Costs

During the 1999–2000 academic year, the cost of living is approximately $12,875 for a single student and $19,980 for a married student. A wide range of privately owned accommodations is available within easy commuting distance.

Student Group

There are 31 students studying for the Ph.D. and 10 students studying for the master's degree.

Location

A small Yankee town of 136,000 lies outside the Yale campus. New Haven dates back to 1638, and, in the midst of a busy urban center, several areas of the city retain the atmosphere of earlier days. The city has a rich cultural life, independent of that provided by the University. Furthermore, there is hourly train service to New York City, which is only 75 miles away.

The University and The Department

Yale was established in 1701 and today is one of the leading universities in the world. It draws students from every part of the United States and from many other countries.

Founded in 1969 as a small graduate program, the Yale computer science department is now a rapidly expanding academic community dedicated to education and research in computer science. The department numbers 14 faculty members, 10 research associates, 41 graduate students, and 72 undergraduate majors.

Applying

An applicant should have a strong preparation in mathematics, engineering, or science. He or she should be competent in programming but does not need to know computer science beyond that basic level. Application for admission in fall 2000 should begin in fall 1999.

Correspondence and Information

Graduate Admissions
Department of Computer Science
Yale University
P.O. Box 208285
New Haven, Connecticut 06520-8285
Telephone: 203-432-1283

Yale University

THE FACULTY AND THEIR RESEARCH

James Aspnes, Associate Professor; Ph.D., Carnegie Mellon, 1992. Randomized, distributed, online algorithms.

Stanley C. Eisenstat, Professor; Ph.D., Stanford, 1972. Numerical linear and nonlinear algebra, direct and iterative methods for solving sparse linear systems, eigenvalue problems, parallel computing.

Michael J. Fischer, Professor; Ph.D., Harvard, 1968. Cryptography and computer security, distributed systems and protocols, communication networks, analysis of algorithms and data structures, complexity theory.

David Gelernter, Professor; Ph.D., SUNY at Stony Brook, 1982. Operating system interfaces, software architecture, parallel programming and AI.

Paul R. Hudak, Professor; Ph.D., Utah, 1982. Functional programming, formal methods, compilers and interpreters, computer music.

Ravi Kannan, Professor; Ph.D., Cornell, 1980. Mathematical algorithms, theoretical computer science.

Arvind Krishnamurthy, Assistant Professor; Ph.D., Berkeley, 1999. Computer systems, compilers, parallel and cluster computing, computer architecture.

Bradley C. Kuszmaul, Assistant Professor; Ph.D., MIT, 1994. High-performance computer architecture, algorithmic parallel programming, computer systems.

László Lovász, Professor; Ph.D., Eötvös Loránd (Budapest), 1971. Combinatorial optimization, discrete mathematics, randomized algorithms, complexity theory.

Drew McDermott, Professor; Ph.D., MIT, 1976. Reasoning about space and time, planning, vision, robotics.

Vladimir Rokhlin, Professor; Ph.D., Rice, 1983. Numerical scattering theory, elliptic partial differential equations, numerical solution of integral equations.

Martin H. Schultz, Arthur K. Watson Professor of Computer Science; Ph.D., Harvard, 1965. Numerical analysis, scientific computing, parallel computation.

Zhong Shao, Assistant Professor; Ph.D., Princeton, 1994. Compilers, programming languages, and run-time environments and their interaction with modern architectures and operating systems.

Steven Zucker, Professor of Computer Science and Electrical Engineering; Ph.D., Drexel, 1975. Computational vision, computational neuroscience, robotics, psychophysics.

Section 9
Electrical and Computer Engineering

This section contains a directory of institutions offering graduate work in electrical and computer engineering, followed by in-depth entries submitted by institutions that chose to prepare detailed program descriptions. Additional information about programs listed in the directory but not augmented by an in-depth entry may be obtained by writing directly to the dean of a graduate school or chair of a department at the address given in the directory.

For programs offering related work, see also in this book Computer Science and Information Technology, Energy and Power Engineering, Engineering and Applied Sciences, Industrial Engineering, and Mechanical Engineering and Mechanics. In Book 4, see Mathematical Sciences and Physics.

CONTENTS

CONTENTS

Computer Engineering

Air Force Institute of Technology, School of Engineering, Department of Electrical and Computer Engineering, Program in Computer Engineering, Wright-Patterson AFB, OH 45433-7765. Offers MS, PhD. *Accreditation:* ABET (one or more programs are accredited). Part-time programs available. *Faculty:* 9 full-time (0 women). *Students:* 8 full-time. In 1998, 2 doctorates awarded (100% found work related to degree). *Degree requirements:* For master's and doctorate, computer language, thesis/dissertation required, foreign language not required. *Entrance requirements:* For master's, GRE General Test (minimum score of 500 on verbal section, 600 on quantitative required), minimum GPA of 3.0, must be military officer or U.S. citizen; for doctorate, GRE General Test (minimum score of 550 on verbal section, 650 on quantitative required), minimum GPA of 3.0, must be military officer or U.S. citizen. Application fee: $0. *Faculty research:* Computer networks. *Unit head:* Maj. Rick Raines, Chief, 937-255-3636 Ext. 4527, Fax: 937-656-4055, E-mail: rraines@afit.af.mil.

Arizona State University East, College of Technology and Applied Sciences, Department of Electronics and Computer Engineering Technology, Mesa, AZ 85212. Offers MS. Part-time and evening/weekend programs available. *Faculty:* 7 full-time (1 woman). *Students:* 25 full-time (9 women), 27 part-time (5 women); includes 8 minority (2 African Americans, 4 Asian Americans or Pacific Islanders, 2 Hispanic Americans), 22 international. Average age 35. 24 applicants, 75% accepted. In 1998, 13 degrees awarded. *Degree requirements:* For master's, thesis or applied project and oral defense required. *Entrance requirements:* For master's, minimum GPA of 3.0. *Average time to degree:* Master's–2.9 years full-time, 4 years part-time. *Application deadline:* Applications are processed on a rolling basis. Application fee: $45. *Financial aid:* In 1998–99, 9 research assistantships with partial tuition reimbursements (averaging $10,506 per year) were awarded.; teaching assistantships, career-related internships or fieldwork, Federal Work-Study, grants, scholarships, and tuition waivers (full and partial) also available. Aid available to part-time students. Financial aid application deadline: 3/1; financial aid applicants required to submit FAFSA. *Faculty research:* Digital signal processing, image communication, reconfiguring computer architectures, insulator contamination/spark over, pulmonary function measurement. Total annual research expenditures: $361,209. *Unit head:* Dr. Robert W. Nowlin, Chair, 602-727-1137, Fax: 602-727-1723, E-mail: robert.nowlin@asu.edu. *Application contact:* Dr. Robert W. Nowlin, Chair, 602-727-1137, Fax: 602-727-1723, E-mail: robert.nowlin@asu.edu.

Auburn University, Graduate School, College of Engineering, Department of Computer Science and Engineering, Auburn, Auburn University, AL 36849-0002. Offers MCSE, MS, PhD. Part-time programs available. *Faculty:* 12 full-time (2 women). *Students:* 38 full-time (11 women), 29 part-time (11 women); includes 12 minority (8 African Americans, 3 Asian Americans or Pacific Islanders, 1 Hispanic American), 24 international. 61 applicants, 39% accepted. In 1998, 25 master's, 5 doctorates awarded. *Degree requirements:* For master's, thesis (MS) required; for doctorate, dissertation required, foreign language not required. *Entrance requirements:* For master's, GRE General Test, GRE Subject Test; for doctorate, GRE General Test (minimum score of 400 on each section required), GRE Subject Test. *Application deadline:* For fall admission, 9/1; for spring admission, 3/1. Applications are processed on a rolling basis. Application fee: $25 ($50 for international students). Tuition, state resident: full-time $2,760; part-time $76 per credit hour. Tuition, nonresident: full-time $8,280; part-time $228 per credit hour. *Financial aid:* Research assistantships, teaching assistantships, Federal Work-Study available. Aid available to part-time students. Financial aid application deadline: 3/15. *Faculty research:* Parallelizable, scalable software translations; graphical representations of algorithms, structures, and processes; graph drawing. Total annual research expenditures: $400,000. *Unit head:* Dr. James Cross, Head, 334-844-4330. *Application contact:* Dr. John F. Pritchett, Dean of the Graduate School, 334-844-4700.

Boston University, College of Engineering, Department of Electrical and Computer Engineering, Boston, MA 02215. Offers computer engineering (PhD); computer systems engineering (MS); electrical engineering (MS, PhD); systems engineering (PhD). Part-time programs available. *Faculty:* 40 full-time (4 women), 4 part-time (0 women). *Students:* 117 full-time (20 women), 26 part-time (6 women); includes 9 minority (3 African Americans, 6 Asian Americans or Pacific Islanders), 83 international. Average age 27. 600 applicants, 18% accepted. In 1998, 32 master's, 15 doctorates awarded. Terminal master's awarded for partial completion of doctoral program. *Degree requirements:* For master's, thesis or alternative required, foreign language not required; for doctorate, dissertation required, foreign language not required. *Entrance requirements:* For master's, GRE General Test, TOEFL (minimum score of 500 required; 213 for computer-based); for doctorate, GRE General Test, TOEFL. *Application deadline:* For fall admission, 4/1; for spring admission, 10/1. Applications are processed on a rolling basis. Application fee: $50. Tuition: Full-time $23,770; part-time $743 per credit. Required fees: $220. Tuition and fees vary according to class time, course level, campus/location and program. *Financial aid:* In 1998–99, 113 students received aid, including 12 fellowships with full tuition reimbursements available (averaging $13,000 per year), 57 research assistantships with full tuition reimbursements available (averaging $11,500 per year), 24 teaching assistantships with full tuition reimbursements available (averaging $11,500 per year); career-related internships or fieldwork, Federal Work-Study, institutionally-sponsored loans, and scholarships also available. Financial aid application deadline: 12/15; financial aid applicants required to submit FAFSA. *Faculty research:* Signal processing, multimedia communications, quantum optics, VLSI, high-performance computing, magnetic and optical devices. Total annual research expenditures: $4 million. *Unit head:* Dr. Bahaa Saleh, Chairman, 617-353-7176, Fax: 617-353-6440. *Application contact:* Cheryl Kelley, Graduate Programs Director, 617-353-9760, Fax: 617-353-0259, E-mail: enggrad@bu.edu.

See in-depth description on page 977.

California State University, Long Beach, Graduate Studies, College of Engineering, Department of Computer Engineering and Computer Science, Long Beach, CA 90840. Offers computer engineering (MS); computer science (MS). Part-time programs available. *Faculty:* 17 full-time (3 women), 8 part-time (0 women). *Students:* 127 full-time (35 women), 189 part-time (51 women); includes 137 minority (5 African Americans, 124 Asian Americans or Pacific Islanders, 8 Hispanic Americans), 101 international. Average age 31. 386 applicants, 55% accepted. In 1998, 48 degrees awarded. *Degree requirements:* For master's, computer language, thesis or alternative required, foreign language not required. *Entrance requirements:* For master's, TOEFL (minimum score of 550 required). *Application deadline:* For fall admission, 8/1; for spring admission, 12/1. Application fee: $55. Electronic applications accepted. Tuition, nonresident: part-time $246 per unit. Required fees: $569 per semester. Tuition and fees vary according to course load. *Financial aid:* Teaching assistantships, Federal Work-Study, grants, institutionally-sponsored loans, and unspecified assistantships available. Financial aid application deadline: 3/2. *Faculty research:* Artificial intelligence, software engineering, computer simulation and modeling, user-interface design, networking. *Unit head:* Dr. Sandra Cynar, Chair, 562-985-4285, Fax: 562-985-7561, E-mail: cynar@csulb.edu. *Application contact:* Dr. Dar-Biau Liu, Graduate Adviser, 562-985-1594, Fax: 562-985-7561, E-mail: liu@csulb.edu.

California State University, Northridge, Graduate Studies, College of Engineering and Computer Science, Department of Electrical and Computer Engineering, Northridge, CA 91330. Offers biomedical engineering (MS); communications/radar engineering (MS); control engineering (MS); digital/computer engineering (MS); electronics engineering (MS); microwave/antenna engineering (MS). Part-time and evening/weekend programs available. *Faculty:* 17 full-time, 3 part-time. *Students:* 20 full-time (2 women), 77 part-time (8 women); includes 33 minority (3 African Americans, 24 Asian Americans or Pacific Islanders, 6 Hispanic Americans), 9 international. Average age 34. 58 applicants, 71% accepted. In 1998, 30 degrees awarded. *Degree requirements:* For master's, thesis or alternative required, foreign language not required. *Entrance requirements:* For master's, GRE General Test, TOEFL, minimum GPA of 2.5. *Application deadline:* For fall admission, 11/30. Application fee: $55. Tuition, nonresident: part-time $246 per unit. International tuition: $7,874 full-time. Required fees: $1,970. Tuition and fees vary according to course load. *Financial aid:* Application deadline: 3/1. *Faculty research:* Reflector antenna study, radome study. *Unit head:* Dr. Nagwa Bekir, Chair, 818-677-2190. *Application contact:* Nagi El Naga, Graduate Coordinator, 818-677-2180.

Carnegie Mellon University, Carnegie Institute of Technology, Department of Electrical and Computer Engineering, Pittsburgh, PA 15213-3891. Offers biomedical engineering (MS, PhD); electrical and computer engineering (MS, PhD). Part-time programs available. *Faculty:* 47 full-time (2 women), 2 part-time (0 women). *Students:* 248 full-time (33 women), 25 part-time (1 woman); includes 23 minority (4 African Americans, 19 Asian Americans or Pacific Islanders), 148 international. 691 applicants, 17% accepted. In 1998, 70 master's awarded (3% entered university research/teaching, 77% found other work related to degree, 20% continued full-time study); 24 doctorates awarded (21% entered university research/teaching, 79% found other work related to degree). *Degree requirements:* For master's, thesis required, foreign language not required; for doctorate, computer language, dissertation, qualifying exam, teaching experience required, foreign language not required. *Entrance requirements:* For master's and doctorate, GRE General Test, TOEFL. *Average time to degree:* Master's–1.5 years full-time, 4 years part-time; doctorate–4.5 years full-time, 6 years part-time. *Application deadline:* For fall admission, 1/15; for spring admission, 10/15. Application fee: $50. *Financial aid:* In 1998–99, fellowships with full tuition reimbursements (averaging $17,100 per year), 191 research assistantships with full tuition reimbursements (averaging $17,100 per year), 16 teaching assistantships with full tuition reimbursements (averaging $12,800 per year) were awarded.; institutionally-sponsored loans also available. Financial aid application deadline: 1/15. *Faculty research:* Computer-aided design, solid-state devices, VLSI, processing, robotics and controls, signal processing, data systems storage. Total annual research expenditures: $12.2 million. *Unit head:* Robert M. White, Head, 412-268-7400, Fax: 412-268-5787, E-mail: white@ece.cmu.edu. *Application contact:* Lynn E. Philibin, Assistant Head for Graduate Studies, 412-268-3291, Fax: 412-268-2860, E-mail: lynn@ece.cmu.edu.

See in-depth description on page 979.

Case Western Reserve University, School of Graduate Studies, The Case School of Engineering, Department of Electrical, Systems, Computer Engineering and Science, Cleveland, OH 44106. Offers computer engineering and science (MS, PhD), including computer engineering, computing and information science; electrical engineering (MS, PhD); systems and control engineering (MS, PhD). Part-time and evening/weekend programs available. Postbaccalaureate distance learning degree programs offered (minimal on-campus study). *Faculty:* 28 full-time (2 women). *Students:* 73 full-time (13 women), 101 part-time (17 women). Average age 25. 484 applicants, 38% accepted. In 1998, 68 master's, 20 doctorates awarded. Terminal master's awarded for partial completion of doctoral program. *Degree requirements:* For master's and doctorate, thesis/dissertation required, foreign language not required. *Entrance requirements:* For master's and doctorate, GRE General Test, TOEFL (minimum score of 550 required). *Average time to degree:* Master's–2 years full-time, 3 years part-time; doctorate–4 years full-time, 5 years part-time. *Application deadline:* For fall admission, 3/1; for spring admission, 11/1. Applications are processed on a rolling basis. Application fee: $25. Electronic applications accepted. *Financial aid:* In 1998–99, 96 students received aid, including 70 research assistantships with full and partial tuition reimbursements available (averaging $15,600 per year), 33 teaching assistantships with full and partial tuition reimbursements available (averaging $10,170 per year); fellowships, career-related internships or fieldwork, Federal Work-Study, and institutionally-sponsored loans also available. Aid available to part-time students. Financial aid application deadline: 3/1. *Faculty research:* Microelectromechanical systems, control, artificial intelligence, mixed signals. Total annual research expenditures: $5.8 million. *Unit head:* Dr. Robert V. Edwards, Acting Chairman, 216-368-2800, Fax: 216-368-6888, E-mail: rve2@po.cwru.edu. *Application contact:* Elizabethanne M. Fuller, Department Assistant, 216-368-4080, Fax: 216-368-2668, E-mail: emf4@po.cwru.edu.

Clarkson University, Graduate School, School of Engineering, Department of Electrical and Computer Engineering, Potsdam, NY 13699. Offers computer engineering (ME, MS); electrical and computer engineering (PhD); electrical engineering (ME, MS). Part-time programs available. *Faculty:* 13 full-time (1 woman). *Students:* 28 full-time (6 women); includes 1 minority (Native American), 24 international. Average age 27. 168 applicants, 61% accepted. In 1998, 25 master's, 1 doctorate awarded. *Degree requirements:* For master's, thesis required, foreign language not required; for doctorate, dissertation, departmental qualifying exam required, foreign language not required. *Entrance requirements:* For master's, GRE, TOEFL. *Application deadline:* For fall admission, 5/15 (priority date); for spring admission, 10/15 (priority date). Applications are processed on a rolling basis. Application fee: $25 ($35 for international students). Tuition: Part-time $661 per credit hour. Required fees: $215 per semester. *Financial aid:* In 1998–99, 1 fellowship, 8 research assistantships, 6 teaching assistantships were awarded. *Faculty research:* Robotics, electrical machines, electrical circuits, dielectric liquids, controls, power systems. Total annual research expenditures: $285,589. *Unit head:* Dr. Susan E. Conry, Chair, 315-268-6511, Fax: 315-268-7600, E-mail: conry@clarkson.edu. *Application contact:* Dr. Philip K. Hopke, Dean of the Graduate School, 315-268-6447, Fax: 315-268-7994, E-mail: hopkepk@clarkson.edu.

Clemson University, Graduate School, College of Engineering and Science, Department of Electrical and Computer Engineering, Program in Computer Engineering, Clemson, SC 29634. Offers MS, PhD. *Students:* 27 full-time (8 women), 5 part-time (1 woman), 17 international. 74 applicants, 36% accepted. In 1998, 5 master's awarded. *Degree requirements:* For master's, thesis or alternative required, foreign language not required; for doctorate, dissertation, departmental qualifying exam required, foreign language not required. *Entrance requirements:* For master's and doctorate, GRE General Test, TOEFL. *Application deadline:* For fall admission, 6/1. Application fee: $35. *Financial aid:* Fellowships, research assistantships, teaching assistantships, career-related internships or fieldwork available. Financial aid applicants required to submit FAFSA. *Faculty research:* Interface applications, software development, multisystem communications, artificial intelligence, robotics. *Unit head:* Dr. Ernest Baxa, Coordinator, 864-656-5900, Fax: 864-656-5910, E-mail: ece-grad-program@clemson.edu. *Application contact:* Dr. Ernest Baxa, Coordinator, 864-656-5900, Fax: 864-656-5910, E-mail: ece-grad-program@clemson.edu.

Cleveland State University, College of Graduate Studies, Fenn College of Engineering, Department of Electrical and Computer Engineering, Cleveland, OH 44115-2440. Offers MS, D Eng. Part-time programs available. *Faculty:* 11 full-time (0 women). *Students:* 6 full-time (0 women), 47 part-time (5 women); includes 5 minority (1 African American, 2 Asian Americans or Pacific Islanders, 2 Hispanic Americans), 20 international. Average age 30. 99 applicants, 66% accepted. In 1998, 20 master's, 4 doctorates awarded. *Degree requirements:* For master's, exam or thesis required; for doctorate, dissertation, candidacy and qualifying exams required, foreign language not required. *Entrance requirements:* For master's, GRE General Test, GRE Subject Test, TOEFL (minimum score of 550 required), minimum GPA of 3.0; for doctorate, GRE General Test, GRE Subject Test, TOEFL, minimum GPA of 3.5. *Application deadline:* For fall admission, 7/15 (priority date). Applications are processed on a rolling basis. Application fee: $25. *Financial aid:* In 1998–99, 1 research assistantship, 3 teaching assistantships were awarded.; career-related internships or fieldwork also available. *Faculty research:* Computer networks, knowledge-based control systems, artificial intelligence, electromagnetic interference, semiconducting materials for solar cell digital communications. *Unit head:* Dr. George L. Kramerich, Chairperson, 216-687-2586, Fax: 216-687-9280.

Colorado State University, Graduate School, College of Engineering, Department of Electrical and Computer Engineering, Fort Collins, CO 80523-0015. Offers MS, PhD. *Faculty:* 17 full-time (1 woman). *Students:* 37 full-time (2 women), 28 part-time (4 women); includes 4 minority (1 African American, 1 Asian American or Pacific Islander, 2 Hispanic Americans), 40 international. Average age 29. 310 applicants, 63% accepted. In 1998, 13 master's, 15

Computer Engineering

Colorado State University (continued)

doctorates awarded. *Degree requirements:* For master's, thesis (for some programs), final exam required, foreign language not required; for doctorate, dissertation, qualifying, preliminary, and final exams required, foreign language not required. *Entrance requirements:* For master's, GRE General Test (minimum combined score of 1100 required), TOEFL, minimum GPA of 3.0; for doctorate, GRE General Test (minimum combined score of 1100 required), TOEFL, minimum GPA of 3.5. *Application deadline:* For fall admission, 2/1 (priority date). Applications are processed on a rolling basis. Application fee: $30. Electronic applications accepted. *Financial aid:* In 1998–99, 18 research assistantships, 9 teaching assistantships were awarded.; fellowships, career-related internships or fieldwork, Federal Work-Study, institutionally-sponsored loans, and traineeships also available. *Faculty research:* Optoelectronics and microelectronics, radar remote sensing and nuclear device testing, controls. Total annual research expenditures: $3.2 million. *Unit head:* Derek L. Lile, Head, 970-491-6600, Fax: 970-491-2249, E-mail: lile@engr.colostate.edu. *Application contact:* Elisabeth L. Wadman, Graduate Coordinator, 970-491-6600, Fax: 970-491-6706, E-mail: ewadman@engr.colostate.edu.

Colorado Technical University, Graduate Studies, Program in Computer Engineering, Colorado Springs, CO 80907-3896. Offers MSCE. Part-time and evening/weekend programs available. *Degree requirements:* For master's, computer language, thesis or alternative required, foreign language not required. *Entrance requirements:* For master's, minimum undergraduate GPA of 3.0.

Colorado Technical University Denver Campus, Program in Computer Engineering, Greenwood Village, CO 80111. Offers MSCE.

Concordia University, School of Graduate Studies, Faculty of Engineering and Computer Science, Department of Electrical and Computer Engineering, Montréal, PQ H3G 1M8, Canada. Offers M Eng, MA Sc, PhD. *Students:* 157 full-time (29 women), 18 part-time (2 women). *Degree requirements:* For master's, computer language required, thesis optional, foreign language not required; for doctorate, computer language, dissertation, comprehensive exam required, foreign language not required. *Application deadline:* For fall admission, 6/1; for spring admission, 10/1. Application fee: $50. *Faculty research:* Computer communications and protocols, circuits and systems, graph theory, VLSI systems, microelectronics. *Unit head:* Dr. J. C. Giguere, Chair, 514-848-3104, Fax: 514-848-2802. *Application contact:* Dr. A. J. Al-Khalili, Director, 514-848-3103, Fax: 514-848-2802.

Cornell University, Graduate School, Graduate Fields of Engineering, Field of Electrical Engineering, Ithaca, NY 14853-0001. Offers computer engineering (M Eng, PhD); electrical engineering (M Eng, PhD); electrical systems (M Eng, PhD); electrophysics (M Eng, PhD). *Faculty:* 46 full-time. *Students:* 196 full-time (27 women); includes 38 minority (6 African Americans, 24 Asian Americans or Pacific Islanders, 8 Hispanic Americans), 94 international. *Degree requirements:* For doctorate, dissertation required, foreign language not required, foreign language not required. *Entrance requirements:* For master's and doctorate, GRE General Test, TOEFL (minimum score of 600 required). *Application deadline:* For fall admission, 1/15. Application fee: $65. Electronic applications accepted. *Unit head:* Director of Graduate Studies, 607-255-4304. *Application contact:* Graduate Field Assistant, 607-255-4304, E-mail: ee_msphd@cornell.edu.

See in-depth description on page 983.

Dalhousie University, Faculty of Graduate Studies, DalTech, Faculty of Engineering, Department of Electrical and Computer Engineering, Halifax, NS B3H 3J5, Canada. Offers M Eng, MA Sc, PhD. *Faculty:* 9 full-time (0 women), 4 part-time (0 women). *Students:* 45 full-time (7 women), 6 part-time. Average age 29. 92 applicants, 48% accepted. In 1998, 6 master's, 1 doctorate awarded (100% entered university research/teaching). *Degree requirements:* For master's and doctorate, thesis/dissertation required, foreign language not required. *Entrance requirements:* For master's and doctorate, TOEFL (minimum score of 580 required). *Application deadline:* For fall admission, 6/1; for winter admission, 10/1; for spring admission, 2/1. Applications are processed on a rolling basis. Application fee: $55. *Financial aid:* Fellowships, research assistantships, teaching assistantships, scholarships and unspecified assistantships available. *Faculty research:* Communications, computer engineering, power engineering, electronics, systems engineering. *Unit head:* Dr. Sherwin Nugent, Head, 902-494-3106, Fax: 902-422-7535, E-mail: edce@dal.ca. *Application contact:* Shelley Parker, Admissions Coordinator, Graduate Studies and Research, 902-494-1288, Fax: 902-494-3149, E-mail: shelley.parker@dal.ca.

Dartmouth College, Thayer School of Engineering, Program in Computer Engineering, Hanover, NH 03755. Offers MS, PhD. *Degree requirements:* For master's, thesis required; for doctorate, dissertation, candidacy oral exam required. *Entrance requirements:* For master's and doctorate, GRE General Test. *Application deadline:* For fall admission, 1/15 (priority date). Application fee: $20 ($40 for international students). *Financial aid:* Fellowships, research assistantships, teaching assistantships, career-related internships or fieldwork, Federal Work-Study, institutionally-sponsored loans, and tuition waivers (full and partial) available. Financial aid application deadline: 1/15. *Faculty research:* Networking and communications, computer architecture, distributed agents, parallel and distributed computing simulation and performance analysis, VLSI design and testing. Total annual research expenditures: $300,000. *Unit head:* Shelley Parker, Admissions Coordinator, Graduate Studies and Research, 902-494-1288, Fax: 902-494-3149, E-mail: shelley.parker@dal.ca. *Application contact:* Candace S. Potter, Admissions Coordinator, 603-646-3844, Fax: 603-646-3856, E-mail: candace.potter@dartmouth.edu.

Drexel University, Graduate School, College of Engineering, Department of Electrical and Computer Engineering, Philadelphia, PA 19104-2875. Offers electrical and computer engineering (PhD); electrical engineering (MSEE); telecommunications engineering (MSEE). Part-time and evening/weekend programs available. *Faculty:* 29 full-time, 6 part-time. *Students:* 85 full-time (19 women), 101 part-time (12 women); includes 17 minority (4 African Americans, 10 Asian Americans or Pacific Islanders, 3 Hispanic Americans), 83 international. Average age 32. 570 applicants, 63% accepted. In 1998, 48 master's, 9 doctorates awarded. Terminal master's awarded for partial completion of doctoral program. *Degree requirements:* For master's, thesis required (for some programs), foreign language not required; for doctorate, dissertation required, foreign language not required. *Entrance requirements:* For master's, TOEFL (minimum score of 570 required), minimum GPA of 3.0, BS in electrical engineering or physics; for doctorate, TOEFL (minimum score of 570 required), minimum GPA of 3.5, MS in electrical engineering. *Application deadline:* For fall admission, 8/21. Applications are processed on a rolling basis. Application fee: $35. Tuition: Full-time $15,795; part-time $585 per credit. Required fees: $375; $67 per term. Tuition and fees vary according to program. *Financial aid:* In 1998–99, 21 research assistantships, 26 teaching assistantships were awarded.; career-related internships or fieldwork and unspecified assistantships also available. Financial aid application deadline: 2/1. *Faculty research:* Power systems planning and control, semiconductors, photovoltaics, electronic systems, image processing. *Unit head:* Dr. Nihat M. Bilgutay, Head, 215-895-6806, Fax: 215-895-1695, E-mail: bilgutay@ece.drexel.edu. *Application contact:* Dr. Mohana Shankar, Graduate Adviser, 215-895-6632.

See in-depth description on page 985.

Duke University, Graduate School, School of Engineering, Department of Electrical and Computer Engineering, Durham, NC 27708-0586. Offers MS, PhD. Part-time programs available. *Faculty:* 26 full-time, 13 part-time. *Students:* 80 full-time, 1 part-time; includes 3 minority (1 Asian American or Pacific Islander, 2 Hispanic Americans), 52 international. 216 applicants, 20% accepted. In 1998, 17 master's, 6 doctorates awarded. Terminal master's awarded for partial completion of doctoral program. *Degree requirements:* For doctorate, dissertation required, foreign language not required, foreign language not required. *Entrance requirements:* For master's and doctorate, GRE General Test. *Application deadline:* For fall admission, 12/31; for spring admission, 11/1. Application fee: $75. *Financial aid:* Fellowships, research assistantships, Federal Work-Study available. Financial aid application deadline: 12/31. *Unit head:* Hisham Massoud, Director of Graduate Studies, 919-660-5245, Fax: 919-660-5293, E-mail: steph@ee.duke.edu.

See in-depth description on page 987.

École Polytechnique de Montréal, Graduate Programs, Department of Electrical and Computer Engineering, Montréal, PQ H3C 3A7, Canada. Offers automation (M Eng, M Sc A, PhD); computer science (M Eng, M Sc A, PhD); electrotechnology (M Eng, M Sc A, PhD); microelectronics (M Eng, M Sc A, PhD); microwave technology (M Eng, M Sc A, PhD). Part-time and evening/weekend programs available. *Degree requirements:* For master's and doctorate, one foreign language, computer language, thesis/dissertation required. *Entrance requirements:* For master's, minimum GPA of 2.75; for doctorate, minimum GPA of 3.0. *Faculty research:* Microwaves, telecommunications, software engineering.

Fairleigh Dickinson University, Teaneck–Hackensack Campus, University College: Arts, Sciences, and Professional Studies, Department of Systems Science, Teaneck, NJ 07666-1914. Offers computer engineering (MS); environmental studies (MS). *Degree requirements:* For master's, foreign language and thesis not required. *Entrance requirements:* For master's, GRE General Test.

Florida Atlantic University, College of Engineering, Department of Computer Science and Engineering, Program in Computer Engineering, Boca Raton, FL 33431-0991. Offers MS, PhD. Part-time and evening/weekend programs available. *Faculty:* 11 full-time (1 woman). *Students:* 28 full-time (8 women), 31 part-time (6 women); includes 10 minority (3 African Americans, 7 Asian Americans or Pacific Islanders), 37 international. Average age 31. In 1998, 28 master's, 1 doctorate awarded. Terminal master's awarded for partial completion of doctoral program. *Degree requirements:* For master's, thesis optional, foreign language not required; for doctorate, dissertation, qualifying exam required, foreign language not required. *Entrance requirements:* For master's, GRE General Test (minimum combined score of 1000 required), TOEFL (minimum score of 550 required), minimum GPA of 3.0; for doctorate, TOEFL (minimum score of 550 required), GRE General Test (minimum combined score of 1100 required) and minimum GPA of 3.0 or GRE General Test (minimum combined score of 1000 required) and minimum GPA of 3.5. *Application deadline:* For fall admission, 4/10 (priority date); for spring admission, 10/1. Applications are processed on a rolling basis. Application fee: $20. Tuition, state resident: part-time $148 per credit hour. Tuition, nonresident: part-time $509 per credit hour. *Financial aid:* Fellowships, research assistantships, teaching assistantships, career-related internships or fieldwork, Federal Work-Study, and unspecified assistantships available. Aid available to part-time students. Financial aid application deadline: 4/1; financial aid applicants required to submit FAFSA. *Faculty research:* VLSI and neural networks, data communications, fault tolerance, data security, parallel systems. *Unit head:* Marlene Mansfield, Student Coordinator, 212-854-3104, Fax: 212-932-9421, E-mail: mansfld@ee.columbia.edu. *Application contact:* Patricia Capozziello, Graduate Admissions Coordinator, 561-297-2694, Fax: 561-297-2659, E-mail: capozzie@fau.edu.

Florida Institute of Technology, Graduate School, College of Engineering, Division of Electrical and Computer Science and Engineering, Program in Computer Engineering, Melbourne, FL 32901-6975. Offers MS, PhD. Part-time and evening/weekend programs available. *Faculty:* 6 full-time (0 women), 1 part-time (0 women). *Students:* 14 full-time (1 woman), 34 part-time (4 women); includes 5 minority (1 Asian American or Pacific Islander, 4 Hispanic Americans), 25 international. Average age 29. 49 applicants, 65% accepted. In 1998, 8 master's, 2 doctorates awarded. *Degree requirements:* For master's, thesis optional, foreign language not required; for doctorate, dissertation, comprehensive exam required, foreign language not required. *Entrance requirements:* For master's, minimum GPA of 3.0; for doctorate, minimum GPA of 3.2. *Application deadline:* Applications are processed on a rolling basis. Application fee: $50. Electronic applications accepted. Tuition: Part-time $575 per credit hour. Required fees: $100. Tuition and fees vary according to campus/location and program. *Financial aid:* In 1998–99, 12 students received aid, including 2 research assistantships with full and partial tuition reimbursements available (averaging $3,600 per year), 9 teaching assistantships with full and partial tuition reimbursements available (averaging $4,894 per year); tuition remissions also available. Financial aid application deadline: 3/1; financial aid applicants required to submit FAFSA. *Faculty research:* Neural networks, parallel processing, reliability testing, image processing. Total annual research expenditures: $105,240. *Unit head:* Dr. John Hadjilogiou, Chair, 407-674-7217, Fax: 407-674-8192, E-mail: jh@ee.fit.edu. *Application contact:* Carolyn P. Farrior, Associate Dean of Graduate Admissions, 407-674-7118, Fax: 407-723-9468, E-mail: cfarrior@fit.edu.

Florida International University, College of Engineering, Department of Electrical Engineering, Program in Computer Engineering, Miami, FL 33199. Offers MS. Part-time and evening/weekend programs available. *Students:* 8 full-time (2 women), 14 part-time (1 woman); includes 12 minority (1 African American, 1 Asian American or Pacific Islander, 10 Hispanic Americans), 8 international. Average age 30. 27 applicants, 44% accepted. In 1998, 5 degrees awarded. *Degree requirements:* For master's, thesis optional. *Entrance requirements:* For master's, GRE General Test (minimum combined score of 1000 required), TOEFL (minimum score of 550 required). *Application deadline:* For fall admission, 4/1 (priority date); for spring admission, 10/1. Applications are processed on a rolling basis. Application fee: $20. Tuition, state resident: part-time $145 per credit hour. Tuition, nonresident: part-time $506 per credit hour. Required fees: $158; $158 per year. *Unit head:* Dr. Malek Adjouadi, Acting Chairperson, Department of Electrical Engineering, 305-348-2807, Fax: 305-348-3707, E-mail: makel@vision.fiu.edu.

The George Washington University, School of Engineering and Applied Science, Department of Electrical and Computer Engineering, Washington, DC 20052. Offers MS, D Sc, App Sc, Engr, MEM/MS. Part-time and evening/weekend programs available. *Degree requirements:* For master's, thesis optional, foreign language not required; for doctorate and other advanced degree, dissertation defense, qualifying exam required; for other advanced degree, foreign language and thesis not required. *Entrance requirements:* For master's, TOEFL (minimum score of 550 required; average 580) or George Washington University English as a Foreign Language Test, appropriate bachelor's degree, minimum GPA of 3.0; for doctorate, TOEFL (minimum score of 550 required; average 580) or George Washington University English as a Foreign Language Test, appropriate bachelor's or master's degree, minimum GPA of 3.3, GRE required if highest earned degree is BS; for other advanced degree, TOEFL (minimum score of 550 required; average 580) or George Washington University English as a Foreign Language Test, appropriate master's degree, minimum GPA of 3.0. *Application deadline:* For fall admission, 3/1 (priority date); for spring admission, 10/1. Applications are processed on a rolling basis. Application fee: $55. Tuition: Full-time $17,328; part-time $722 per credit hour. Required fees: $828; $35 per credit hour. Tuition and fees vary according to campus/location and program. *Financial aid:* Fellowships, research assistantships, teaching assistantships, career-related internships or fieldwork and institutionally-sponsored loans available. Financial aid application deadline: 3/1; financial aid applicants required to submit FAFSA. *Faculty research:* Computer graphics, multimedia systems. *Unit head:* Dr. Murray Loew, Chair, 202-994-6083, Fax: 202-994-0227, E-mail: eecs@seas.gwu.edu. *Application contact:* Howard M. Davis, Manager, Office of Admissions and Student Records, 202-994-6158, Fax: 202-994-0909, E-mail: data:adms@seas.gwu.edu.

See in-depth description on page 991.

Georgia Institute of Technology, Graduate Studies and Research, College of Engineering, School of Electrical and Computer Engineering, Atlanta, GA 30332-0001. Offers MS, MSEE, PhD. Part-time programs available. Postbaccalaureate distance learning degree programs offered. Terminal master's awarded for partial completion of doctoral program. *Degree requirements:* For master's, thesis optional, foreign language not required; for doctorate, dissertation required, foreign language not required. *Entrance requirements:* For master's, GRE General Test (minimum score of 500 on verbal section, 700 on quantitative and analytical required), TOEFL (minimum score of 550 required), minimum GPA of 3.0; for doctorate, GRE General Test (minimum score of 500 on verbal section, 700 on quantitative and analytical required), TOEFL (minimum score of 550 required), minimum GPA of 3.5. *Faculty research:* Telecommunications, computer systems, microelectronics, optical engineering, digital signal processing.

Illinois Institute of Technology, Graduate College, Armour College of Engineering and Sciences, Department of Electrical and Computer Engineering, Chicago, IL 60616-3793. Offers computer systems engineering (MS); electrical and computer engineering (MECE); electrical engineering (MS, PhD); manufacturing engineering (MME, MS). Part-time and evening/weekend programs available. *Faculty:* 21 full-time (1 woman), 11 part-time (0 women). *Students:* 76 full-time (10 women), 258 part-time (38 women); includes 83 minority (17 African Americans, 50 Asian Americans or Pacific Islanders, 15 Hispanic Americans, 1 Native American), 118 international. 759 applicants, 46% accepted. In 1998, 86 master's, 7 doctorates awarded. Terminal master's awarded for partial completion of doctoral program. *Degree requirements:* For master's, thesis (for some programs), comprehensive exam required, foreign language not required; for doctorate, dissertation, comprehensive exam required, foreign language not required. *Entrance requirements:* For master's, GRE (minimim score of 1200 required), TOEFL (minimum score of 550 required), undergraduate GPA of 3.0 required; for doctorate, GRE (minimum score of 1200 required), TOEFL (minimum score of 550 required), undergraduate GPA of 3.0 required. *Application deadline:* For fall admission, 7/1; for spring admission, 12/1. Applications are processed on a rolling basis. Application fee: $30. Electronic applications accepted. *Financial aid:* In 1998–99, 1 fellowship, 15 research assistantships, 19 teaching assistantships were awarded.; Federal Work-Study, institutionally-sponsored loans, scholarships, and graduate assistantships also available. Financial aid application deadline:3/1. *Faculty research:* Computer system design, photonics, biomedical engineering, power systems analysis, communications theory. Total annual research expenditures: $766,272. *Unit head:* Dr. D. Ucci, Interim Chairman, 312-567-3402, Fax: 312-567-8976, E-mail: du@ece.iit.edu. *Application contact:* Dr. S. Mohammad Shahidehpour, Dean of Graduate College, 312-567-3024, Fax: 312-567-7517, E-mail: grad@minna.cns.iit.edu.

See in-depth description on page 995.

Indiana State University, School of Graduate Studies, School of Technology, Department of Electronics and Computer Technology, Terre Haute, IN 47809-1401. Offers MA, MS. *Faculty:* 5 full-time (0 women). *Students:* 18 full-time (3 women), 5 part-time; includes 6 minority (5 African Americans, 1 Asian American or Pacific Islander), 13 international. Average age 31. 33 applicants, 61% accepted. In 1998, 9 degrees awarded. *Degree requirements:* For master's, thesis required, foreign language not required. *Entrance requirements:* For master's, TOEFL (minimum score of 550 required), bachelor's degree in industrial technology or related field, minimum undergraduate GPA of 2.5. *Average time to degree:* Master's–2 years full-time, 5 years part-time. *Application deadline:* For fall admission, 7/1 (priority date); for spring admission, 11/1 (priority date). Applications are processed on a rolling basis. Application fee: $20. Electronic applications accepted. *Financial aid:* In 1998–99, 5 research assistantships with partial tuition reimbursements, 4 teaching assistantships with partial tuition reimbursements were awarded.; fellowships, Federal Work-Study and institutionally-sponsored loans also available. Financial aid application deadline: 3/1; financial aid applicants required to submit FAFSA. *Unit head:* Dr. Robert English, Chairperson, 812-237-3456.

Instituto Tecnológico y de Estudios Superiores de Monterrey, Campus Chihuahua, Graduate Programs, Chihuahua, 31110, Mexico. Offers computer systems engineering (Ingeniero); electrical engineering (Ingeniero); electromechanical engineering (Ingeniero); electronic engineering (Ingeniero); engineering administration (MEA); industrial engineering (MIE, Ingeniero); international trade (MIT); mechanical engineering (Ingeniero).

Iowa State University of Science and Technology, Graduate College, College of Engineering, Department of Electrical and Computer Engineering, Ames, IA 50011. Offers computer engineering (MS, PhD); electrical engineering (MS, PhD). *Faculty:* 43 full-time, 3 part-time. *Students:* 161 full-time (37 women), 63 part-time (7 women); includes 7 minority (4 African Americans, 2 Asian Americans or Pacific Islanders, 1 Hispanic American), 165 international. 776 applicants, 13% accepted. In 1998, 65 master's, 19 doctorates awarded. *Degree requirements:* For master's, thesis or alternative required, foreign language not required; for doctorate, dissertation required, foreign language not required. *Entrance requirements:* For master's and doctorate, GRE General Test, TOEFL (minimum score of 570 required). *Application deadline:* For fall admission, 1/15 (priority date); for spring admission, 9/15. Application fee: $20 ($50 for international students). Electronic applications accepted. Tuition, state resident: full-time $3,308. Tuition, nonresident: full-time $9,744. Part-time tuition and fees vary according to course load, campus/location and program. *Financial aid:* In 1998–99, 127 research assistantships with partial tuition reimbursements (averaging $11,868 per year), 46 teaching assistantships with partial tuition reimbursements (averaging $12,018 per year) were awarded.; fellowships, scholarships also available. *Unit head:* Dr. S. S. Venkata, Chair, 515-294-2667, E-mail: ecegrad@ee.iastate.edu. *Application contact:* Dr. Satish Udpa, Director of Graduate Education, 515-294-2667, E-mail: ecegrad@ee.iastate.edu.

Johns Hopkins University, G. W. C. Whiting School of Engineering, Department of Electrical and Computer Engineering, Baltimore, MD 21218-2699. Offers MSE, PhD. Part-time programs available. *Faculty:* 16 full-time (0 women), 8 part-time (0 women). *Students:* 70 full-time (11 women), 16 part-time (3 women); includes 4 minority (1 African American, 3 Asian Americans or Pacific Islanders), 39 international. Average age 25. 274 applicants, 34% accepted. In 1998, 8 master's awarded (100% found work related to degree); 7 doctorates awarded. Terminal master's awarded for partial completion of doctoral program. *Degree requirements:* For master's, foreign language and thesis not required; for doctorate, dissertation required, foreign language not required. *Entrance requirements:* For master's, GRE General Test, TOEFL (minimum score of 560 required); for doctorate, GRE General Test, TOEFL. *Application deadline:* For fall admission, 1/1. Applications are processed on a rolling basis. Application fee: $50. Tuition: Full-time $23,660. Tuition and fees vary according to program. *Financial aid:* In 1998–99, 5 fellowships, 27 research assistantships, 20 teaching assistantships were awarded.; Federal Work-Study, grants, and institutionally-sponsored loans also available. Financial aid application deadline: 1/1. *Faculty research:* Communications systems, quantum electronics and optics, signal and image processing, VLSI, microwaves. Total annual research expenditures: $3.2 million. *Unit head:* Dr. Frederic M. Davidson, Chair, 410-516-7007, Fax: 410-516-5566, E-mail: fmd@light.ece.jhu.edu. *Application contact:* Gail M. O'Connor, Academic Coordinator II, 410-516-4808, Fax: 410-516-5566, E-mail: gradadm@ece.jhu.edu.

See in-depth description on page 997.

Kansas State University, Graduate School, College of Engineering, Department of Electrical and Computer Engineering, Manhattan, KS 66506. Offers bioengineering (MS, PhD); communications (MS, PhD); computer engineering (MS, PhD); control systems (MS, PhD); electric energy systems (MS, PhD); instrumentation (MS, PhD); signal processing (MS, PhD). Postbaccalaureate distance learning degree programs offered (no on-campus study). *Faculty:* 21 full-time (3 women). *Students:* 26 full-time (2 women), 24 part-time (2 women), 18 international. 163 applicants, 23% accepted. In 1998, 11 master's, 1 doctorate awarded (100% found work related to degree). *Degree requirements:* For master's, thesis optional; for doctorate, dissertation required. *Entrance requirements:* For master's, GRE General Test (minimum score of 400 on verbal section, 600 on quantitative, 600 on analytical required); for doctorate, GRE General Test (minimum score of 400 on verbal section, 600 on quantitative required). *Average time to degree:* Master's–1.5 years full-time; doctorate–4 years full-time. *Application deadline:* For fall admission, 3/1; for spring admission, 9/1. Applications are processed on a rolling basis. Application fee: $0 ($25 for international students). Electronic applications accepted. *Financial aid:* In 1998–99, 17 research assistantships with partial tuition reimbursements (averaging $9,900 per year), 7 teaching assistantships with full tuition reimbursements (averaging $9,900 per year) were awarded.; career-related internships or fieldwork also available. *Faculty research:* Digital signal processing, communications systems. Total annual research expenditures: $910,000. *Unit head:* Dr. David Soldan, Head, 785-532-5600, E-mail: grad@eece.ksu.edu.

Lehigh University, College of Engineering and Applied Science, Department of Electrical Engineering, Computer Science and Computer Engineering, Program in Computer Engineering, Bethlehem, PA 18015-3094. Offers MS. Part-time programs available. *Faculty:* 3 full-time (0 women), 1 part-time (0 women). *Students:* 7 full-time (2 women), 4 part-time; includes 1 minority (Asian American or Pacific Islander), 3 international. Average age 24. 8 applicants, 63% accepted. In 1998, 5 degrees awarded. *Degree requirements:* For master's, oral presentation of thesis required. *Entrance requirements:* For master's, GRE General Test (minimum combined score of 1600 on three sections required), TOEFL (minimum score of 550 required), minimum GPA of 3.0. *Application deadline:* For fall admission, 7/15; for spring admission, 12/1. Applications are processed on a rolling basis. Application fee: $40. Electronic applications accepted. *Financial aid:* Fellowships, research assistantships, teaching assistantships available. Financial aid application deadline: 1/15. *Unit head:* Dr. S. Mohammad Shahidehpour, Dean of Graduate College, 312-567-3024, Fax: 312-567-7517, E-mail: grad@minna.cns.iit.edu. *Application contact:* Anne Nierer, Graduate Coordinator, 610-758-4072, Fax: 610-758-6279, E-mail: aln3@lehigh.edu.

Louisiana State University and Agricultural and Mechanical College, Graduate School, College of Engineering, Department of Electrical and Computer Engineering, Baton Rouge, LA 70803. Offers MSEE, PhD. *Faculty:* 25 full-time (2 women). *Students:* 79 full-time (15 women), 13 part-time (2 women); includes 6 minority (2 African Americans, 4 Asian Americans or Pacific Islanders), 75 international. Average age 27. 260 applicants, 35% accepted. In 1998, 23 master's, 4 doctorates awarded. Terminal master's awarded for partial completion of doctoral program. *Degree requirements:* For master's, thesis optional, foreign language not required; for doctorate, dissertation required, foreign language not required. *Entrance requirements:* For master's, GRE General Test, TOEFL (minimum score of 550 required), minimum GPA of 3.0; for doctorate, GRE General Test, TOEFL (minimum score of 550 required), minimum GPA of 3.5. *Application deadline:* For fall admission, 1/25 (priority date). Applications are processed on a rolling basis. Application fee: $25. *Financial aid:* In 1998–99, 3 fellowships, 31 research assistantships with partial tuition reimbursements, 20 teaching assistantships with partial tuition reimbursements were awarded.; institutionally-sponsored loans and unspecified assistantships also available. Financial aid application deadline: 2/28. *Faculty research:* Electronics, power engineering, systems and signal processing, communications. *Unit head:* Dr. Alan H. Marshak, Chair, 225-388-5241, Fax: 225-388-5200. *Application contact:* Dr. Jorge L. Aravena, Graduate Adviser, 225-388-5478, E-mail: aravena@ee.lsu.edu.

See in-depth description on page 999.

Manhattan College, Graduate Division, School of Engineering, Program in Computer Engineering, Riverdale, NY 10471. Offers MS. Part-time and evening/weekend programs available. *Degree requirements:* For master's, computer language, thesis or alternative required, foreign language not required. *Entrance requirements:* For master's, GRE, TOEFL, minimum GPA of 3.0.

Marquette University, Graduate School, College of Engineering, Department of Electrical and Computer Engineering, Milwaukee, WI 53201-1881. Offers MS, PhD. Part-time and evening/weekend programs available. *Faculty:* 13 full-time (2 women), 2 part-time (0 women). *Students:* 28 full-time (4 women), 67 part-time (8 women); includes 4 minority (1 African American, 1 Asian American or Pacific Islander, 2 Hispanic Americans), 33 international. Average age 28. 57 applicants, 35% accepted. In 1998, 15 master's awarded (100% found work related to degree); 4 doctorates awarded (100% found work related to degree). Terminal master's awarded for partial completion of doctoral program. *Degree requirements:* For master's, thesis optional, foreign language not required; for doctorate, computer language, dissertation defense, qualifying exam required. *Entrance requirements:* For master's, TOEFL (minimum score of 575 required), GRE General Test or minimum GPA of 3.0; for doctorate, GRE General Test (minimum score of 700 on quantitative section required), TOEFL (minimum score of 575 required). *Application deadline:* For fall admission, 7/15 (priority date); for spring admission, 11/15. Applications are processed on a rolling basis. Application fee: $40. Tuition: Part-time $510 per credit hour. Tuition and fees vary according to program. *Financial aid:* In 1998–99, 40 students received aid, including 5 fellowships, 5 research assistantships, 20 teaching assistantships; Federal Work-Study, institutionally-sponsored loans, and scholarships also available. Financial aid application deadline: 2/15. *Faculty research:* Electric machines, drives, and controls; applied solid-state electronics; computers and signal processing; microwaves and antennas; electronic and ultrasonic sensors. *Unit head:* Dr. Jeffrey L. Hock, Chairman, 414-288-6820, Fax: 414-288-5579, E-mail: hock@marquette.edu. *Application contact:* Dr. James E. Richie, Director of Graduate Studies, 414-288-5326, Fax: 414-288-5579, E-mail: richiej@marquette.edu.

Michigan State University, Graduate School, College of Engineering, Department of Electrical and Computer Engineering, East Lansing, MI 48824-1020. Offers MS, PhD. *Faculty:* 25. *Students:* 134 (16 women); includes 19 minority (5 African Americans, 9 Asian Americans or Pacific Islanders, 5 Hispanic Americans) 87 international. In 1998, 27 master's, 10 doctorates awarded. *Degree requirements:* For master's, exit exam required, thesis optional, foreign language not required; for doctorate, dissertation, comprehensive and qualifying exams required, foreign language not required. *Entrance requirements:* For master's and doctorate, GRE General Test (minimum combined score of 1800 on three sections required), TOEFL (minimum score of 580 required). *Application deadline:* Applications are processed on a rolling basis. Application fee: $30 ($40 for international students). *Financial aid:* In 1998–99, 35 research assistantships with tuition reimbursements (averaging $12,461 per year), 17 teaching assistantships with tuition reimbursements (averaging $12,185 per year) were awarded.; fellowships Financial aid applicants required to submit FAFSA. *Unit head:* Dr. Jes Asmussen, Chairperson, 517-355-5066, Fax: 517-353-1980.

See in-depth description on page 1003.

Mississippi State University, College of Engineering, Department of Electrical and Computer Engineering, Mississippi State, MS 39762. Offers computer engineering (MS, PhD); electrical engineering (MS, PhD). Part-time programs available. Postbaccalaureate distance learning degree programs offered (minimal on-campus study). *Students:* 85 full-time (10 women), 20 part-time (3 women); includes 44 minority (4 African Americans, 40 Asian Americans or Pacific Islanders), 28 international. Average age 28. 161 applicants, 31% accepted. In 1998, 28 master's awarded. *Degree requirements:* For master's, computer language required (for some programs), comprehensive oral exam required, thesis optional, foreign language not required; for doctorate, computer language (for some programs), dissertation, comprehensive oral and written exam required, foreign language not required. *Entrance requirements:* For master's, GRE General Test, TOEFL (minimum score of 550 required), minimum GPA of 2.75; for doctorate, GRE, TOEFL (minimum score of 550 required). *Application deadline:* For fall admission, 7/1; for spring admission, 11/1. Applications are processed on a rolling basis. Application fee: $25 for international students. *Financial aid:* Federal Work-Study, institutionally-sponsored loans, and unspecified assistantships available. Financial aid applicants required to submit FAFSA. *Faculty research:* Digital computing, power, controls, communication systems, microelectronics. Total annual research expenditures: $1.5 million. *Unit head:* Dr. G. Marshall Molen, Head, 662-325-3912, Fax: 662-325-2298, E-mail: molen@ece.msstate.edu. *Application contact:* Jerry B. Inmon, Director of Admissions, 662-325-2224, Fax: 662-325-7360, E-mail: admit@admissions.msstate.edu.

Announcement: The Graduate Program in Computer Science (MS and PhD) is directed toward the design and development of computer systems. Areas of research include information and signal processing; graphics and visualization; high performance, distributed, and parallel computing; microelectronics; and CAD. See the in-depth description in the computer science section or visit the Web site at http://www.cs.msstate.edu/.

National Technological University, Programs in Engineering, Fort Collins, CO 80526-1842. Offers chemical engineering (MS); computer engineering (MS); computer science (MS); electrical engineering (MS); engineering management (MS); hazardous waste management (MS); health physics (MS); management of technology (MS); manufacturing systems engineering (MS); materials science and engineering (MS); software engineering (MS); special majors

Computer Engineering

National Technological University (continued)
(MS); transportation engineering (MS); transportation systems engineering (MS). Part-time programs available. *Faculty:* 600 part-time (20 women). *Entrance requirements:* For master's, BS in engineering or related field. *Application deadline:* Applications are processed on a rolling basis. Application fee: $50. *Unit head:* Lionel V. Baldwin, President, 970-495-6400, Fax: 970-484-0668, E-mail: baldwin@mail.ntu.edu.

Naval Postgraduate School, Graduate Programs, Department of Electrical and Computer Engineering, Monterey, CA 93943. Offers MS, PhD, Eng. Program only open to commissioned officers of the United States and friendly nations and selected United States federal civilian employees. *Accreditation:* ABET (one or more programs are accredited). Part-time programs available. Postbaccalaureate distance learning degree programs offered (minimal on-campus study). *Students:* 105 full-time, 34 international. In 1998, 55 master's, 1 doctorate, 2 other advanced degrees awarded. *Degree requirements:* For master's and Eng, computer language, thesis required, foreign language not required; for doctorate, one foreign language, computer language, dissertation required. *Unit head:* Dr. Jeff Knorr, Chairman, 831-656-2081. *Application contact:* Theodore H. Calhoon, Director of Admissions, 831-656-3093, Fax: 831-656-2891, E-mail: tcalhoon@nps.navy.mil.

New Jersey Institute of Technology, Office of Graduate Studies, Department of Electrical and Computer Engineering, Newark, NJ 07102-1982. Offers computer engineering (MS); electrical engineering (MS, PhD, Engineer), including biomedical systems (MS, PhD), communication and signal processing (MS, PhD), computer systems (MS, PhD), control systems (MS, PhD), microwave and lightwave engineering (MS, PhD), solid-state materials and devices (MS, PhD); power engineering (MS); telecommunications (MS). Part-time and evening/weekend programs available. Terminal master's awarded for partial completion of doctoral program. *Degree requirements:* For master's, thesis required (for some programs), foreign language not required; for doctorate, dissertation, residency required, foreign language not required. *Entrance requirements:* For master's, GRE General Test (minimum score of 450 on verbal section, 600 on quantitative, 550 on analytical required); for doctorate, GRE General Test (minimum score of 450 on verbal section, 600 on quantitative, 550 on analytical required), minimum graduate GPA of 3.5. Electronic applications accepted. *Faculty research:* Communications systems design, digital signal processing.

New Mexico State University, Graduate School, College of Engineering, Department of Electrical and Computer Engineering, Las Cruces, NM 88003-8001. Offers MSEE, PhD. Part-time programs available. *Faculty:* 21 full-time (1 woman), 2 part-time (0 women). *Students:* 65 full-time (8 women), 45 part-time (4 women); includes 16 minority (2 African Americans, 2 Asian Americans or Pacific Islanders, 12 Hispanic Americans), 44 international. Average age 33. 94 applicants, 78% accepted. In 1998, 34 master's, 5 doctorates awarded. *Degree requirements:* For master's, thesis required (for some programs); for doctorate, 2 foreign languages (computer language can substitute for one), dissertation required. *Entrance requirements:* For master's, minimum GPA of 3.0; for doctorate, departmental qualifying exam, minimum GPA of 3.0. *Application deadline:* For fall admission, 7/1 (priority date); for spring admission, 11/1. Applications are processed on a rolling basis. Application fee: $15 ($35 for international students). Electronic applications accepted. Tuition, state resident: full-time $2,682; part-time $112 per credit. Tuition, nonresident: full-time $8,376; part-time $349 per credit. Tuition and fees vary according to course load. *Financial aid:* Fellowships, research assistantships, teaching assistantships, career-related internships or fieldwork and Federal Work-Study available. Aid available to part-time students. Financial aid application deadline: 3/1. *Faculty research:* Software engineering, high performance computing, telemetering and space telecommunications, computational electromagnetics. *Unit head:* Dr. Steven P. Castillo, Interim Head, 505-646-3115, Fax: 505-646-1435, E-mail: scastill@nmsu.edu. *Application contact:* Dr. Javin M. Taylor, Associate Dean, 505-646-3115, Fax: 505-646-1435, E-mail: jtaylor@nmsu.edu.

North Carolina State University, Graduate School, College of Engineering, Department of Electrical and Computer Engineering, Raleigh, NC 27695. Offers MS, PhD. Part-time programs available. *Faculty:* 101 full-time (6 women), 78 part-time (8 women). *Students:* 266 full-time (45 women), 125 part-time (10 women); includes 67 minority (29 African Americans, 30 Asian Americans or Pacific Islanders, 8 Hispanic Americans), 157 international. Average age 29. 380 applicants, 28% accepted. In 1998, 104 master's, 43 doctorates awarded. Terminal master's awarded for partial completion of doctoral program. *Degree requirements:* For master's, thesis optional, foreign language not required; for doctorate, dissertation required, foreign language not required. *Entrance requirements:* For master's, GRE, TOEFL (minimum score of 575 required), minimum GPA of 3.2 in electrical engineering course work; for doctorate, GRE, TOEFL (minimum score of 625 required), minimum GPA of 3.5 in electrical engineering course work. *Application deadline:* For fall admission, 6/25; for spring admission, 11/25. Applications are processed on a rolling basis. Application fee: $45. *Financial aid:* In 1998–99, 51 fellowships (averaging $2,702 per year), 183 research assistantships (averaging $4,863 per year), 162 teaching assistantships (averaging $5,067 per year) were awarded.; career-related internships or fieldwork also available. Financial aid application deadline: 3/1. *Faculty research:* Microwave devices, communications, signal processing, solid-state, power and control systems, VLSI design, wireless communications. Total annual research expenditures: $13.8 million. *Unit head:* Dr. Robert Kolbas, Head, 919-515-7350, Fax: 919-515-5523, E-mail: kolbas@eos.ncsu.edu. *Application contact:* Dr. Richard Kuehn, Graduate Coordinator, 919-515-5090, Fax: 919-515-5601, E-mail: rkuehn@eos.ncsu.edu.

See in-depth description on page 1005.

Northeastern University, College of Engineering, Computer Systems Engineering Program, Boston, MA 02115-5096. Offers MS. Part-time programs available. *Students:* 27 full-time (9 women), 34 part-time (4 women); includes 3 minority (1 African American, 2 Asian Americans or Pacific Islanders), 28 international. Average age 25. 66 applicants, 76% accepted. In 1998, 24 degrees awarded. *Degree requirements:* For master's, computer language required, thesis optional, foreign language not required. *Entrance requirements:* For master's, GRE General Test. *Average time to degree:* Master's–2.28 years full-time, 5 years part-time. *Application deadline:* For fall admission, 4/15. Applications are processed on a rolling basis. Application fee: $50. *Financial aid:* In 1998–99, 14 students received aid, including 3 research assistantships with full tuition reimbursements available (averaging $12,450 per year), 2 teaching assistantships with full tuition reimbursements available (averaging $12,450 per year); fellowships, career-related internships or fieldwork, Federal Work-Study, tuition waivers (full), and unspecified assistantships also available. Aid available to part-time students. Financial aid application deadline: 2/15; financial aid applicants required to submit FAFSA. *Faculty research:* Engineering software design, CAD/CAM, robotics. *Unit head:* Dr. Chung Yu, Graduate Coordinator, 336-334-7760 Ext. 213, Fax: 336-334-7716, E-mail: yu@genesis.ncat.edu. *Application contact:* Stephen L. Gibson, Associate Director, 617-373-2711, Fax: 617-373-2501, E-mail: grad-eng.@coe.neu.edu.

Northeastern University, College of Engineering, Department of Electrical and Computer Engineering, Boston, MA 02115-5096. Offers MS, PhD. Part-time programs available. *Faculty:* 38 full-time (4 women), 8 part-time (0 women). *Students:* 148 full-time (27 women), 219 part-time (31 women); includes 25 minority (5 African Americans, 9 Hispanic Americans, 11 Native Americans), 131 international. Average age 25. 683 applicants, 32% accepted. In 1998, 59 master's, 10 doctorates awarded. Terminal master's awarded for partial completion of doctoral program. *Degree requirements:* For master's, thesis optional, foreign language not required; for doctorate, dissertation, departmental qualifying exam required, foreign language not required. *Entrance requirements:* For master's and doctorate, GRE General Test. *Average time to degree:* Master's–2.07 years full-time, 4.68 years part-time; doctorate–4.89 years full-time. *Application deadline:* For fall admission, 4/15. Applications are processed on a rolling basis. Application fee: $50. *Financial aid:* In 1998–99, 110 students received aid, including 63 research assistantships with full tuition reimbursements available (averaging $12,450 per year), 34 teaching assistantships with full tuition reimbursements available (averaging $12,450 per year); fellowships, career-related internships or fieldwork, Federal Work-Study, tuition waivers (full), and unspecified assistantships also available. Aid available to part-time students. Financial aid application deadline: 2/15; financial aid applicants required to submit FAFSA. *Faculty research:* Digital communications, signal processing and sensor data fusion, electromagnetics, control systems, plasma science. *Unit head:* Dr. Arvin Grabel, Interim Chairman, 617-373-4159, Fax: 617-373-8970. *Application contact:* Stephen L. Gibson, Associate Director, 617-373-2711, Fax: 617-373-2501, E-mail: grad-eng@coe.neu.edu.

See in-depth description on page 1007.

Northwestern Polytechnic University, School of Engineering, Fremont, CA 94539-7482. Offers computer science (MS); computer systems engineering (MS); electrical engineering (MS). Part-time and evening/weekend programs available. *Faculty:* 8 full-time, 52 part-time. *Students:* 262. *Degree requirements:* For master's, computer language, thesis required, foreign language not required. *Entrance requirements:* For master's, TOEFL. *Application deadline:* For fall admission, 8/15; for winter admission, 12/15; for spring admission, 7/15. Applications are processed on a rolling basis. Application fee: $50 ($75 for international students). Tuition: Full-time $6,750; part-time $375 per unit. Required fees: $135 per term. Tuition and fees vary according to course load and program. *Unit head:* Dr. Pochang Hsu, Dean, 510-657-5911, Fax: 510-657-8975, E-mail: npuadm@npu0.npu.edu. *Application contact:* Dr. Fred Kuttner, Dean of Academic Affairs and Admissions, 510-657-5911, Fax: 510-657-8975, E-mail: npuadm@npu0.npu.edu.

Northwestern University, The Graduate School, Robert R. McCormick School of Engineering and Applied Science, Department of Electrical and Computer Engineering, Evanston, IL 60208. Offers electrical and computer engineering (MS, PhD); information technology (MIT). MS and PhD admissions and degrees offered through The Graduate School. Part-time programs available. *Faculty:* 31 full-time (2 women). *Students:* 154 full-time (31 women), 9 part-time (1 woman); includes 17 minority (4 African Americans, 12 Asian Americans or Pacific Islanders, 1 Hispanic American), 98 international. 656 applicants, 12% accepted. In 1998, 28 master's, 15 doctorates awarded. Terminal master's awarded for partial completion of doctoral program. *Degree requirements:* For master's, thesis or project required; for doctorate, dissertation required, foreign language not required. *Entrance requirements:* For master's and doctorate, GRE General Test, TOEFL (minimum score of 560 required). *Application deadline:* For fall admission, 6/1. Applications are processed on a rolling basis. Application fee: $50 ($55 for international students). *Financial aid:* In 1998–99, 105 students received aid, including 14 fellowships with full tuition reimbursements available (averaging $11,673 per year), 62 research assistantships with partial tuition reimbursements available (averaging $16,285 per year), 27 teaching assistantships with full tuition reimbursements available (averaging $12,042 per year); career-related internships or fieldwork, Federal Work-Study, institutionally-sponsored loans, and scholarships also available. Financial aid application deadline: 1/15; financial aid applicants required to submit FAFSA. *Faculty research:* Solid-state engineering networks and communications, optical systems, parallel and distributed computing, VLSI design and computer-aided design. Total annual research expenditures: $9.2 million. *Unit head:* Dr. Prithvaraj Banerjee, Chair, 847-491-3641, Fax: 847-491-4455, E-mail: banerjee@ece.nwu.edu. *Application contact:* Lawrence Henschen, Admission Officer, 847-491-3338, Fax: 847-491-4455, E-mail: henschen@ece.nwu.edu.

See in-depth description on page 1009.

Oakland University, Graduate Studies, School of Engineering and Computer Science, Program in Computer Science and Engineering, Rochester, MI 48309-4401. Offers computer science (MS); software engineering (MS). Part-time and evening/weekend programs available. *Faculty:* 12 full-time, 3 part-time. *Students:* 48 full-time (22 women), 118 part-time (30 women); includes 21 minority (3 African Americans, 18 Asian Americans or Pacific Islanders), 58 international. Average age 31. 113 applicants, 71% accepted. In 1998, 43 degrees awarded. *Degree requirements:* For master's, foreign language and thesis not required. *Entrance requirements:* For master's, minimum GPA of 3.0 for unconditional admission. *Application deadline:* For fall admission, 7/15; for spring admission, 3/15. Application fee: $30. Tuition, state resident: part-time $221 per credit hour. Tuition, nonresident: part-time $488 per credit hour. Required fees: $214 per semester. Part-time tuition and fees vary according to program. *Financial aid:* Federal Work-Study, institutionally-sponsored loans, and tuition waivers (full) available. Financial aid application deadline: 3/1; financial aid applicants required to submit FAFSA. *Unit head:* Dr. Christian Wagner, Chair, 248-370-2200.

Oakland University, Graduate Studies, School of Engineering and Computer Science, Program in Electrical and Computer Engineering, Rochester, MI 48309-4401. Offers MS. Part-time and evening/weekend programs available. *Faculty:* 7 full-time. *Students:* 16 full-time (6 women), 38 part-time (3 women); includes 6 minority (1 African American, 5 Asian Americans or Pacific Islanders), 5 international. Average age 30. 36 applicants, 75% accepted. In 1998, 17 degrees awarded. *Degree requirements:* For master's, foreign language and thesis not required. *Entrance requirements:* For master's, minimum GPA of 3.0 for unconditional admission. *Application deadline:* For fall admission, 7/15; for spring admission, 3/15. Application fee: $30. Tuition, state resident: part-time $221 per credit hour. Tuition, nonresident: part-time $488 per credit hour. Required fees: $214 per semester. Part-time tuition and fees vary according to program. *Financial aid:* Federal Work-Study, institutionally-sponsored loans, and tuition waivers (full) available. Financial aid application deadline: 3/1; financial aid applicants required to submit FAFSA. *Unit head:* Dr. Naim A. Kheir, Chair, 248-370-2177.

Oklahoma State University, Graduate College, College of Engineering, Architecture and Technology, School of Electrical and Computer Engineering, Stillwater, OK 74078. Offers M En, MS, PhD. *Faculty:* 16 full-time (0 women), 1 part-time (0 women). *Students:* 60 full-time (4 women), 47 part-time (5 women); includes 10 minority (1 African American, 8 Asian Americans or Pacific Islanders, 1 Hispanic American), 64 international. Average age 28. In 1998, 51 master's, 4 doctorates awarded. *Degree requirements:* For master's, thesis or alternative required, foreign language not required; for doctorate, dissertation required, foreign language not required. *Entrance requirements:* For master's and doctorate, TOEFL (minimum score of 575 required). *Application deadline:* For fall admission, 7/1 (priority date). Application fee: $25. *Financial aid:* In 1998–99, 43 students received aid, including 25 research assistantships (averaging $10,825 per year), 18 teaching assistantships (averaging $7,820 per year); career-related internships or fieldwork, Federal Work-Study, and tuition waivers (partial) also available. Aid available to part-time students. Financial aid application deadline: 3/1. *Unit head:* Dr. Michael A. Soderstrand, Interim Head, 405-744-5151.

See in-depth description on page 1015.

Old Dominion University, College of Engineering and Technology, Department of Electrical and Computer Engineering, Norfolk, VA 23529. Offers ME, MS, PhD. Part-time programs available. Postbaccalaureate distance learning degree programs offered (minimal on-campus study). *Faculty:* 10 full-time (1 woman), 1 part-time (0 women). *Students:* 112 full-time (16 women), 18 part-time (4 women); includes 2 minority (both African Americans), 78 international. Average age 24. In 1998, 13 master's, 4 doctorates awarded. *Degree requirements:* For master's, comprehensive exam, thesis (MS) required; for doctorate, computer language, dissertation, candidacy exam, diagnostic exam required, foreign language not required. *Entrance requirements:* For master's, GRE, TOEFL (minimum score of 550 required), minimum GPA of 3.0; for doctorate, GRE, TOEFL (minimum score of 550 required), minimum GPA of 3.25. *Application deadline:* For fall admission, 7/1; for spring admission, 10/1. Applications are processed on a rolling basis. Application fee: $30. *Financial aid:* In 1998–99, 38 research assistantships (averaging $3,941 per year), 30 teaching assistantships (averaging $2,633 per year) were awarded.; fellowships, career-related internships or fieldwork, grants, and tuition waivers (partial) also available. Aid available to part-time students. Financial aid application deadline: 2/15; financial aid applicants required to submit FAFSA. *Faculty research:* Digital signal processing, control engineering, gaseous electronics, ultrafast (femtosecond) laser applications, interaction of fields with living organisms, modeling and simulation. Total annual research expenditures: $1.3 million. *Unit head:* Dr. Stephen Zahorian, Chair, 757-683-3741, Fax: 757-

683-3220, E-mail: zahorian@ece.odu.edu. *Application contact:* Dr. Amin N. Dharamsi, Graduate Program Director, 757-683-3741, Fax: 757-683-3220, E-mail: dharamsi@ece.odu.edu.

Oregon Graduate Institute of Science and Technology, Graduate Studies, Department of Computer Science and Engineering, Portland, OR 97291-1000. Offers computational finance (MS, Certificate); computer science and engineering (MS, PhD). Part-time and evening/weekend programs available. *Faculty:* 16 full-time (2 women), 31 part-time (6 women). *Students:* 70 full-time (16 women), 61 part-time (20 women); includes 5 Asian Americans or Pacific Islanders, 1 Native American, 54 international. Average age 31. 257 applicants, 43% accepted. In 1998, 29 master's, 2 doctorates awarded. Terminal master's awarded for partial completion of doctoral program. *Degree requirements:* For master's, computer language required, thesis optional, foreign language not required; for doctorate, computer language, comprehensive exam, oral defense of dissertation required. *Entrance requirements:* For master's and doctorate, GRE General Test, TOEFL (minimum score of 600 required). *Average time to degree:* Master's–1.8 years full-time, 3.9 years part-time; doctorate–5.6 years full-time. *Application deadline:* For fall admission, 3/1 (priority date). Applications are processed on a rolling basis. Application fee: $50. Electronic applications accepted. *Financial aid:* In 1998–99, 41 students received aid, including 38 research assistantships, 3 teaching assistantships; fellowships, grants and scholarships also available. Financial aid application deadline: 3/1. *Faculty research:* Computer systems architecture, intelligent and interactive systems, programming models and systems, theory of computation. *Unit head:* Dr. Andrew P. Black, Head, 503-690-1250, E-mail: black@cse.ogi.edu. *Application contact:* Shirley Kapsch, Enrollment Manager, 503-690-1255, Fax: 503-690-1285, E-mail: kapsch@cse.ogi.edu.

Oregon Graduate Institute of Science and Technology, Graduate Studies, Department of Electrical and Computer Engineering, Portland, OR 97291-1000. Offers computational finance (Certificate); computer engineering (MS, PhD); electrical engineering (MS, PhD). Part-time programs available. *Faculty:* 13 full-time (0 women), 34 part-time (1 woman). *Students:* 48 full-time (15 women), 67 part-time (14 women); includes 16 minority (all Asian Americans or Pacific Islanders) Average age 29. 124 applicants, 31% accepted. In 1998, 22 master's, 1 doctorate awarded. Terminal master's awarded for partial completion of doctoral program. *Degree requirements:* For master's, thesis optional, foreign language not required; for doctorate, comprehensive exam, oral defense of dissertation required. *Entrance requirements:* For master's, TOEFL (minimum score of 550 required); for doctorate, GRE General Test, GRE Subject Test, TOEFL (minimum score of 550 required). *Average time to degree:* Master's–1.5 years full-time, 1.8 years part-time; doctorate–4.8 years full-time. *Application deadline:* For fall admission, 3/1 (priority date). Applications are processed on a rolling basis. Application fee: $50. Electronic applications accepted. *Financial aid:* In 1998–99, 20 students received aid, including 20 research assistantships; fellowships, Federal Work-Study also available. Financial aid application deadline: 3/1. *Faculty research:* Semiconductor materials, microwave circuits, atmospheric optics, surface physics, electron and ion optics. *Unit head:* Dr. Dan Hammerstrom, Head, 503-690-4037, Fax: 503-690-1406. *Application contact:* Don Johansen, Enrollment Manager, 503-690-1315, E-mail: johansen@ece.ogi.edu.

See in-depth description on page 1017.

Oregon Institute of Technology, Department of Computer Engineering Technology, Klamath Falls, OR 97601-8801. Offers MS. Part-time programs available. Postbaccalaureate distance learning degree programs offered (minimal on-campus study). *Faculty:* 2 full-time (0 women). *Students:* 3 full-time (1 woman), 1 part-time. In 1998, 1 degree awarded. *Degree requirements:* For master's, one foreign language, computer language, project required, thesis not required. *Entrance requirements:* For master's, GRE General Test (minimum combined score of 1500 on three sections required), TOEFL (minimum score of 600 required). *Application deadline:* For fall admission, 5/1 (priority date); for winter admission, 10/1 (priority date); for spring admission, 2/1 (priority date). Applications are processed on a rolling basis. Application fee: $50. Electronic applications accepted. *Financial aid:* Federal Work-Study and institutionally-sponsored loans available. Aid available to part-time students. Financial aid application deadline: 3/1; financial aid applicants required to submit FAFSA. *Unit head:* Don Metzler, Chair, 541-885-1604, E-mail: metzler@oit.edu. *Application contact:* Saichi Oba, Director of Admissions, 541-885-1150, Fax: 541-885-1115, E-mail: oit@oit.edu.

Oregon State University, Graduate School, College of Engineering, Department of Electrical and Computer Engineering, Corvallis, OR 97331. Offers MAIS, MS, PhD. Part-time programs available. *Faculty:* 26 full-time (1 woman), 2 part-time (0 women). *Students:* 67 full-time (11 women), 20 part-time; includes 9 minority (1 African American, 8 Asian Americans or Pacific Islanders), 56 international. Average age 27. In 1998, 30 master's, 6 doctorates awarded. *Degree requirements:* For doctorate, dissertation, departmental qualifying exam required, foreign language not required, foreign language not required. *Entrance requirements:* For master's and doctorate, TOEFL (minimum score of 575 required), minimum GPA of 3.0 in last 90 hours. *Application deadline:* For fall admission, 3/1. Applications are processed on a rolling basis. Application fee: $50. *Financial aid:* Research assistantships, teaching assistantships, institutionally-sponsored loans available. Aid available to part-time students. Financial aid application deadline: 2/1. *Faculty research:* Analog and mixed mode IC's; materials, devices, and electroluminescence; microwave and optoelectrics; control systems and signal processing; novel electrical machines. *Unit head:* Alan K. Wallace, Head, 541-737-3617, Fax: 541-737-1300, E-mail: wallace@orst.edu. *Application contact:* Alan K. Wallace, Head, 541-737-3617, Fax: 541-737-1300, E-mail: wallace@orst.edu.

Pennsylvania State University University Park Campus, Graduate School, College of Engineering, Department of Computer Science and Engineering, State College, University Park, PA 16802-1503. Offers M Eng, MS, PhD. *Students:* 91 full-time (20 women), 32 part-time (9 women). In 1998, 32 master's, 6 doctorates awarded. *Degree requirements:* For doctorate, dissertation required, foreign language not required. *Entrance requirements:* For master's and doctorate, GRE General Test. Application fee: $50. *Unit head:* Dr. Dale A. Miller, Head, 814-865-9505.

Portland State University, Graduate Studies, School of Engineering and Applied Science, Department of Electrical and Computer Engineering, Portland, OR 97207-0751. Offers MS, PhD. Part-time and evening/weekend programs available. *Faculty:* 11 full-time (1 woman), 6 part-time (0 women). *Students:* 53 full-time (7 women), 71 part-time (12 women); includes 9 minority (8 Asian Americans or Pacific Islanders, 1 Hispanic American), 71 international. Average age 30. 108 applicants, 61% accepted. In 1998, 19 master's, 2 doctorates awarded. *Degree requirements:* For master's, variable foreign language requirement, computer language, thesis or alternative, oral exam required; for doctorate, one foreign language, computer language, dissertation, oral and written exams required. *Entrance requirements:* For master's, TOEFL (minimum score of 550 required), minimum GPA of 3.0 in upper-division course work or 2.75 overall; for doctorate, GRE General Test, GRE Subject Test, minimum GPA of 3.0 in upper-division course work. *Application deadline:* For fall admission, 3/1 (priority date); for spring admission, 11/1. Applications are processed on a rolling basis. Application fee: $50. *Financial aid:* In 1998–99, 4 research assistantships, 28 teaching assistantships were awarded.; career-related internships or fieldwork, Federal Work-Study, and institutionally-sponsored loans also available. Aid available to part-time students. Financial aid application deadline: 3/1; financial aid applicants required to submit FAFSA. *Faculty research:* Optics and laser systems, design automation, VLSI design, computer systems, power electronics. Total annual research expenditures: $214,568. *Unit head:* Dr. Rolf Schaumann, Chair, 503-725-3806, Fax: 503-725-3807, E-mail: schaumann@ee.pdx.edu. *Application contact:* Dr. Y. C. Jenq, Coordinator, 503-725-3806, Fax: 503-725-3807, E-mail: jenq@ee.pdx.edu.

Princeton University, Graduate School, School of Engineering and Applied Science, Department of Electrical Engineering, Princeton, NJ 08544-1019. Offers computer engineering (MSE, PhD); electrical engineering (M Eng); electronic materials and devices (MSE, PhD); information sciences and systems (MSE, PhD); optoelectronics (MSE, PhD). Part-time programs available. *Faculty:* 27 full-time (3 women). *Students:* 135 full-time (18 women), 10 part-time (1 woman). *Degree requirements:* For master's, thesis optional; for doctorate, dissertation required. *Entrance requirements:* For master's and doctorate, GRE General Test, TOEFL. *Application deadline:* For fall admission, 1/3. Electronic applications accepted. *Unit head:* Prof. Wayne Wolf, Director of Graduate Studies, 609-258-3335, Fax: 609-258-3745, E-mail: dgs@ee.princeton.edu. *Application contact:* Prof. Wayne Wolf, Director of Graduate Studies, 609-258-3335, Fax: 609-258-3745, E-mail: dgs@ee.princeton.edu.

See in-depth description on page 1021.

Purdue University, Graduate School, Schools of Engineering, School of Electrical and Computer Engineering, West Lafayette, IN 47907. Offers biomedical engineering (MS Bm E, PhD); computer engineering (MS, PhD); electrical engineering (MS, PhD). Part-time programs available. Postbaccalaureate distance learning degree programs offered (no on-campus study). *Faculty:* 60 full-time (4 women), 9 part-time (0 women). *Students:* 330 full-time (68 women), 4 part-time (3 women); includes 6 African Americans, 11 Hispanic Americans Average age 25. 1558 applicants, 36% accepted. In 1998, 76 master's, 41 doctorates awarded. *Degree requirements:* For master's, thesis optional, foreign language not required; for doctorate, dissertation required, foreign language not required. *Entrance requirements:* For master's and doctorate, GRE General Test (combined average 2070 on three sections), TOEFL (minimum score of 575 required; average 635). *Average time to degree:* Master's–1.75 years full-time, 4 years part-time; doctorate–5.5 years full-time. *Application deadline:* For fall admission, 1/15 (priority date); for spring admission, 9/1. Applications are processed on a rolling basis. Application fee: $30. Electronic applications accepted. *Financial aid:* In 1998–99, 298 students received aid, including 18 fellowships, 186 research assistantships, 96 teaching assistantships Financial aid applicants required to submit FAFSA. *Faculty research:* Biomedical communications and signal processing, solid-state materials and devices fields and optics, automatic controls, energy sources and systems, VLSI and circuit design. Total annual research expenditures: $12.5 million. *Unit head:* Dr. W. K. Fuchs, Head, 765-494-3539, Fax: 765-494-3544, E-mail: fuchs@purdue.edu. *Application contact:* Dr. A. M. Weiner, Director of Admissions, 765-494-3392, Fax: 765-494-3393, E-mail: ecegrad@ecn.purdue.edu.

Announcement: The School of Electrical and Computer Engineering at Purdue University in West Lafayette includes a large program in computer engineering, with 27 faculty members, 81 master's students, and approximately 75 PhD students. In 1999, 76 MS and 41 PhD degrees were awarded. Research areas include computer architecture; parallel and distributed processing, including architecture, algorithms, software, and applications; fault-tolerant computing; VLSI systems; artificial intelligence; and multimedia systems.

See in-depth description on page 1023.

Rensselaer Polytechnic Institute, Graduate School, School of Engineering, Department of Electrical, Computer, and Systems Engineering, Program in Computer and Systems Engineering, Troy, NY 12180-3590. Offers M Eng, MS, D Eng, PhD, MBA/M Eng. Part-time programs available. *Faculty:* 36 full-time (1 woman), 4 part-time (0 women). *Students:* 49 full-time (7 women), 12 part-time (1 woman); includes 8 minority (6 Asian Americans or Pacific Islanders, 2 Hispanic Americans), 20 international. 134 applicants, 35% accepted. In 1998, 16 master's, 7 doctorates awarded. Terminal master's awarded for partial completion of doctoral program. *Degree requirements:* For master's, thesis required (for some programs), foreign language not required; for doctorate, dissertation required, foreign language not required. *Entrance requirements:* For master's and doctorate, GRE, TOEFL (minimum score of 600 required). *Average time to degree:* Master's–1.5 years full-time; doctorate–3 years full-time. *Application deadline:* For fall admission, 2/1; for spring admission, 10/1. Applications are processed on a rolling basis. Application fee: $35. *Financial aid:* Fellowships, research assistantships, teaching assistantships, career-related internships or fieldwork and institutionally-sponsored loans available. Financial aid application deadline: 2/1. *Faculty research:* Multimedia via ATM, mobile robotics, thermophotovoltaic devices, microelectronic interconnections, agile manufacturing. Total annual research expenditures: $2.4 million. *Unit head:* Ann Bruno, Manager of Graduate Admissions and Financial Aid, 518-276-2554, Fax: 518-276-2433, E-mail: bruno@ecse.rpi.edu. *Application contact:* Ann Bruno, Manager of Graduate Admissions and Financial Aid, 518-276-2554, Fax: 518-276-2433, E-mail: bruno@ecse.rpi.edu.

Rice University, Graduate Programs, George R. Brown School of Engineering, Department of Electrical and Computer Engineering, Houston, TX 77251-1892. Offers bioengineering (MS, PhD); circuits, controls, and communication systems (MS, PhD); computer science and engineering (MS, PhD); electrical engineering (MEE); lasers, microwaves, and solid-state electronics (MS, PhD). Part-time programs available. *Degree requirements:* For master's, thesis required (for some programs), foreign language not required; for doctorate, dissertation required, foreign language not required. *Entrance requirements:* For master's and doctorate, GRE General Test, GRE Subject Test, TOEFL (minimum score of 550 required), minimum GPA of 3.0. *Faculty research:* Physical electronics.

Rochester Institute of Technology, Part-time and Graduate Admissions, College of Engineering, Department of Computer Engineering, Rochester, NY 14623-5604. Offers MS. *Students:* 3 full-time (1 woman), 14 part-time (1 woman); includes 2 minority (1 African American, 1 Asian American or Pacific Islander), 1 international. 33 applicants, 45% accepted. In 1998, 7 degrees awarded. *Degree requirements:* Foreign language not required. *Entrance requirements:* For master's, TOEFL, minimum GPA of 3.0. *Application deadline:* For fall admission, 3/1 (priority date). Applications are processed on a rolling basis. Application fee: $40. *Unit head:* Dr. Roy Czernikowski, Head, 716-475-2987. *Application contact:* Dr. Richard Reeve, Associate Dean, 716-475-7048, E-mail: nrreie@rit.edu.

Rutgers, The State University of New Jersey, New Brunswick, Graduate School, Program in Electrical and Computer Engineering, New Brunswick, NJ 08903. Offers communications and solid-state electronics (MS, PhD); computer engineering (MS, PhD); control systems (MS, PhD); digital signal processing (MS, PhD). Part-time programs available. *Faculty:* 28 full-time (2 women), 2 part-time (0 women). *Students:* 60 full-time (4 women), 128 part-time (18 women); includes 85 minority (3 African Americans, 81 Asian Americans or Pacific Islanders, 1 Hispanic American), 9 international. 533 applicants, 24% accepted. In 1998, 42 master's awarded (80% found work related to degree, 20% continued full-time study); 14 doctorates awarded. Terminal master's awarded for partial completion of doctoral program. *Degree requirements:* For master's, thesis optional, foreign language not required; for doctorate, dissertation required, foreign language not required. *Entrance requirements:* For master's and doctorate, GRE General Test (minimum score of 600 on verbal section , 730 on quantitative, 660 on analytical required). *Average time to degree:* Master's–2 years full-time, 3 years part-time; doctorate–3 years full-time, 5 years part-time. *Application deadline:* For fall admission, 2/1 (priority date); for spring admission, 11/1. Applications are processed on a rolling basis. Application fee: $50. Electronic applications accepted. *Financial aid:* In 1998–99, 84 students received aid, including 2 fellowships with partial tuition reimbursements available (averaging $12,000 per year), 52 research assistantships with full tuition reimbursements available (averaging $12,000 per year), 30 teaching assistantships with full tuition reimbursements available (averaging $12,000 per year); Federal Work-Study and tuition waivers (full) also available. Financial aid application deadline: 2/1; financial aid applicants required to submit FAFSA. *Faculty research:* Communication and information processing, wireless information networks, micro-vacuum devices, machine vision, VLSI design. Total annual research expenditures: $8.7 million. *Unit head:* Dr. David G. Daut, Director, 732-445-2578, Fax: 732-445-2820.

See in-depth description on page 1027.

St. Mary's University of San Antonio, Graduate School, Department of Engineering, Program in Electrical/Computer Engineering, San Antonio, TX 78228-8507. Offers MS. *Students:* 23 (4 women). Average age 25. In 1998, 4 degrees awarded. *Degree requirements:* For master's, computer language, thesis required, foreign language not required. *Entrance requirements:* For master's, GRE General Test, BS in science or engineering. *Application deadline:* For fall admission, 8/1. Application fee: $15. *Financial aid:* Teaching assistantships, Federal Work-

Computer Engineering

St. Mary's University of San Antonio (continued)
Study available. *Faculty research:* Robotics, artificial intelligence, manufacturing engineering. *Unit head:* Dr. Abe Yazdani, Adviser, Department of Engineering, 210-436-3305.

San Jose State University, Graduate Studies, College of Engineering, Department of Computer, Information and Systems Engineering, Program in Computer Engineering, San Jose, CA 95192-0001. Offers computer engineering (MS); computer software (MS); computerized robots and computer applications (MS); microprocessors and microcomputers (MS). *Faculty:* 5 full-time (0 women), 12 part-time (1 woman). *Students:* 57 full-time (28 women), 90 part-time (19 women); includes 106 minority (1 African American, 103 Asian Americans or Pacific Islanders, 2 Hispanic Americans), 27 international. Average age 29. 198 applicants, 53% accepted. In 1998, 31 degrees awarded. *Degree requirements:* For master's, computer language, thesis, comprehensive exam required, foreign language not required. *Entrance requirements:* For master's, GRE General Test (minimum combined score of 1500 on three sections required), BS in computer science or 24 credits in related area. *Application deadline:* For fall admission, 6/1. Applications are processed on a rolling basis. Application fee: $59. Tuition, nonresident: part-time $246 per unit. Required fees: $1,939; $1,309 per year. *Financial aid:* Teaching assistantships, career-related internships or fieldwork, Federal Work-Study, and institutionally-sponsored loans available. Aid available to part-time students. Financial aid application deadline: 5/1. *Faculty research:* Robotics, database management systems, computer networks. *Unit head:* Dr. Haluk Ozemek, Coordinator, 408-924-4100. *Application contact:* Dr. Haluk Ozemek, Coordinator, 408-924-4100.

Santa Clara University, School of Engineering, Department of Computer Science and Engineering, Santa Clara, CA 95053-0001. Offers computer science and engineering (MSCSE, PhD); high performance computing (Certificate); software engineering (Certificate). Part-time and evening/weekend programs available. *Students:* 141 full-time (76 women), 229 part-time (86 women); includes 91 minority (88 Asian Americans or Pacific Islanders, 3 Hispanic Americans), 222 international. Average age 34. 264 applicants, 58% accepted. In 1998, 173 master's, 1 doctorate awarded. *Degree requirements:* For master's, computer language, thesis or alternative required, foreign language not required; for doctorate and Certificate, computer language, dissertation required, foreign language not required. *Entrance requirements:* For master's, GRE General Test, TOEFL (minimum score of 550 required), minimum GPA of 2.75; for doctorate, GRE General Test, GRE Subject Test, TOEFL (minimum score of 550 required), master's degree or equivalent; for Certificate, master's degree, published paper. *Application deadline:* For fall admission, 6/1; for spring admission, 1/1. Applications are processed on a rolling basis. Application fee: $40. *Financial aid:* Fellowships, research assistantships, teaching assistantships, Federal Work-Study available. Aid available to part-time students. Financial aid application deadline: 2/1; financial aid applicants required to submit CSS PROFILE or FAFSA. *Unit head:* Dr. Daniel W. Lewis, Chair, 408-554-5281. *Application contact:* Tina Samms, Assistant Director of Graduate Admissions, 408-554-4313, Fax: 408-554-5474, E-mail: engrgrad@scu.edu.

Southern Methodist University, School of Engineering and Applied Science, Department of Computer Science and Engineering, Dallas, TX 75275. Offers computer engineering (MS Cp E, PhD); computer science (MS, PhD); engineering management (MSEM, DE); operations research (MS, PhD); software engineering (MS). Part-time programs available. Postbaccalaureate distance learning degree programs offered (no on-campus study). *Faculty:* 13 full-time (2 women), 12 part-time (1 woman). *Students:* 57 full-time (22 women), 294 part-time (60 women); includes 85 minority (24 African Americans, 44 Asian Americans or Pacific Islanders, 16 Hispanic Americans, 1 Native American), 69 international. Average age 32. 236 applicants, 44% accepted. In 1998, 87 master's, 7 doctorates awarded. *Degree requirements:* For master's, thesis optional, foreign language not required; for doctorate, dissertation, oral and written qualifying exams, oral final exam (PhD) required. *Entrance requirements:* For master's, GRE General Test (minimum score of 650 on quantitative section required), TOEFL (minimum score of 550 required), minimum GPA of 3.0 in last 2 years; bachelor's degree in engineering, mathematics, or sciences; for doctorate, preliminary counseling exam (PhD), minimum GPA of 3.0, bachelor's degree in related field, MA (DE). *Application deadline:* For fall admission, 8/1 (priority date); for spring admission, 12/15. Applications are processed on a rolling basis. Application fee: $25. Tuition: Full-time $9,216; part-time $512 per credit hour. Required fees: $88 per credit hour. Part-time tuition and fees vary according to course load and campus/location. *Financial aid:* Fellowships, research assistantships, teaching assistantships available. Financial aid applicants required to submit FAFSA. *Faculty research:* Computer arithmetic, distributed and fault-tolerant computing, main memory databse systems, natural language processing. *Unit head:* Dr. Richard V. Helgason, Interim Chair, 214-768-3278, E-mail: helgason@seas.smu.edu. *Application contact:* Dr. Zeynep Celik-Butler, Assistant Dean for Graduate Studies and Research, 214-768-3979, Fax: 214-768-3845, E-mail: zcb@seas.smu.edu.

Southern Polytechnic State University, College of Technology, Department of Electrical and Computer Engineering Technology, Marietta, GA 30060-2896. Offers engineering technology (MS). Part-time and evening/weekend programs available. *Faculty:* 2 full-time (0 women). *Students:* 9 full-time (0 women), 23 part-time (2 women); includes 8 minority (3 African Americans, 3 Asian Americans or Pacific Islanders, 2 Hispanic Americans), 6 international. Average age 34. 3 applicants, 67% accepted. In 1998, 8 degrees awarded (100% found work related to degree). *Degree requirements:* For master's, foreign language and thesis not required. *Entrance requirements:* For master's, GRE General Test. *Application deadline:* For fall admission, 7/15 (priority date); for spring admission, 12/1. Applications are processed on a rolling basis. Application fee: $20. Tuition, state resident: full-time $2,146; part-time $119 per credit hour. Tuition, nonresident: full-time $7,586; part-time $421 per credit hour. *Financial aid:* In 1998–99, 15 students received aid; teaching assistantships, career-related internships or fieldwork and Federal Work-Study available. Aid available to part-time students. Financial aid application deadline: 5/1; financial aid applicants required to submit FAFSA. *Unit head:* Kim Davis, Acting Head, 770-528-7246, Fax: 770-528-7285, E-mail: kdavis0@spsu.edu.

State University of New York at Buffalo, Graduate School, College of Arts and Sciences, Department of Computer Science and Engineering, Buffalo, NY 14260. Offers MS, PhD. *Faculty:* 25 full-time (6 women), 8 part-time (2 women). *Students:* 80 full-time (14 women), 22 part-time (3 women); includes 3 minority (all Asian Americans or Pacific Islanders), 74 international. Average age 25. 373 applicants, 48% accepted. In 1998, 29 master's, 4 doctorates awarded. Terminal master's awarded for partial completion of doctoral program. *Degree requirements:* For master's, computer language, thesis or alternative required, foreign language not required; for doctorate, computer language, dissertation, comprehensive qualifying exam required, foreign language not required. *Entrance requirements:* For master's and doctorate, GRE General Test, GRE Subject Test (computer science), TOEFL (minimum score of 550 required). *Application deadline:* For fall admission, 12/31. Application fee: $35. Electronic applications accepted. Tuition, state resident: full-time $5,100; part-time $213 per credit hour. Tuition, nonresident: full-time $8,416; part-time $351 per credit hour. Required fees: $870; $75 per semester. Tuition and fees vary according to course load and program. *Financial aid:* In 1998–99, 100 students received aid, including 35 research assistantships with tuition reimbursements available (averaging $10,350 per year), 65 teaching assistantships with tuition reimbursements available (averaging $10,350 per year); fellowships with tuition reimbursements available, Federal Work-Study, institutionally-sponsored loans, and unspecified assistantships also available. Financial aid application deadline: 12/31; financial aid applicants required to submit FAFSA. *Faculty research:* Artificial intelligence, computer vision, theoretical computer science, parallel architecture, operating systems. Total annual research expenditures: $1.4 million. *Unit head:* Dr. Stuart C. Shapiro, Chairman, 716-645-3180 Ext. 125, Fax: 716-645-3464, E-mail: cse-chair@cse.buffalo.edu. *Application contact:* Dr. Raj Acharya, Director of Graduate Studies, 716-645-3180 Ext. 141, Fax: 716-645-3464, E-mail: cse-dgs@cse.buffalo.edu.

State University of New York at Stony Brook, Graduate School, College of Engineering and Applied Sciences, Department of Electrical and Computer Engineering, Stony Brook, NY 11794. Offers MS, PhD. Evening/weekend programs available. *Faculty:* 21 full-time (2 women), 4 part-time (0 women). *Students:* 56 full-time (12 women), 48 part-time (4 women); includes 10 minority (1 African American, 9 Asian Americans or Pacific Islanders), 61 international. 389 applicants, 58% accepted. In 1998, 31 master's, 8 doctorates awarded. *Degree requirements:* For master's, thesis or alternative required, foreign language not required; for doctorate, dissertation, comprehensive exams required, foreign language not required. *Entrance requirements:* For master's and doctorate, GRE General Test, TOEFL. *Application deadline:* For fall admission, 1/15. Application fee: $50. *Financial aid:* In 1998–99, 10 research assistantships, 25 teaching assistantships were awarded.; fellowships *Faculty research:* System science, solid-state electronics, computer engineering. Total annual research expenditures: $1.4 million. *Unit head:* Dr. Serge Luryi, Chairman, 516-632-8420. *Application contact:* Dr. Chi-Tsong Chen, Director, 516-632-8400, Fax: 516-632-8494, E-mail: ctchen@sbee.sunysb.edu.

See in-depth description on page 1035.

Stevens Institute of Technology, Graduate School, Charles V. Schaefer Jr. School of Engineering, Department of Electrical and Computer Engineering, Program in Computer Engineering, Hoboken, NJ 07030. Offers computer and communications security (Certificate); computer and information engineering (M Eng, PhD, Engr); computer architecture and digital system design (M Eng, PhD, Engr); digital systems and VLSI design (Certificate); image and signal processing (M Eng, PhD, Engr); information networks (Certificate); robotics and automation (M Eng, PhD, Engr); software engineering (M Eng, PhD, Engr). Part-time and evening/weekend programs available. Terminal master's awarded for partial completion of doctoral program. *Degree requirements:* For master's and other advanced degree, computer language required, foreign language not required; for doctorate, computer language, dissertation required. *Entrance requirements:* For master's and doctorate, GRE, TOEFL; for other advanced degree, GRE. Electronic applications accepted.

Syracuse University, Graduate School, L. C. Smith College of Engineering and Computer Science, Department of Electrical Engineering and Computer Science, Program in Computer Engineering, Syracuse, NY 13244-0003. Offers MS, PhD, CE. *Faculty:* 36 full-time, 8 part-time. *Students:* 58 full-time (9 women), 68 part-time (14 women); includes 12 minority (1 African American, 10 Asian Americans or Pacific Islanders, 1 Hispanic American), 66 international. Average age 29. 179 applicants, 72% accepted. In 1998, 80 master's, 2 doctorates awarded. *Degree requirements:* For master's, foreign language and thesis not required; for doctorate, computer language, dissertation required, foreign language not required. *Entrance requirements:* For master's and doctorate, GRE General Test, GRE Subject Test. *Application deadline:* Applications are processed on a rolling basis. Application fee: $40. Tuition: Full-time $13,992; part-time $583 per credit hour. *Financial aid:* Fellowships, research assistantships, teaching assistantships, Federal Work-Study and tuition waivers (partial) available. Financial aid application deadline: 3/1. *Faculty research:* Hardware, software, computer applications. *Unit head:* Garth Foster, Graduate Director, 315-443-4370.

Temple University, Graduate School, College of Science and Technology, College of Engineering, Program in Electrical and Computer Engineering, Philadelphia, PA 19122-6096. Offers MSE. Part-time programs available. *Faculty:* 12 full-time (0 women). *Students:* 28; includes 13 minority (1 African American, 11 Asian Americans or Pacific Islanders, 1 Hispanic American), 6 international. 80 applicants, 30% accepted. *Degree requirements:* For master's, thesis required, foreign language not required. *Entrance requirements:* For master's, GRE General Test (minimum combined score of 1500 required), TOEFL (minimum score of 575 required). *Application deadline:* For fall admission, 7/1; for spring admission, 11/1. Applications are processed on a rolling basis. Application fee: $40. *Financial aid:* Fellowships, research assistantships, teaching assistantships, Federal Work-Study and institutionally-sponsored loans available. Financial aid application deadline: 2/15. *Faculty research:* Neural networks, adaptive control, robotics, multiprocessor systems, vacuum microelectronics. Total annual research expenditures:$425,000. *Unit head:* Dr. John J. Helferty, Director, 215-204-4523, Fax: 215-204-5960.

Texas A&M University, College of Engineering, Department of Computer Science, College Station, TX 77843. Offers computer engineering (MCE, MS, PhD); computer science (MCS, MS, PhD). Part-time programs available. *Faculty:* 26 full-time (3 women), 1 part-time (0 women). *Students:* 162 full-time (24 women), 68 part-time (14 women); includes 8 minority (5 African Americans, 3 Hispanic Americans), 155 international. *Degree requirements:* For master's, computer language, thesis (MS) required; for doctorate, computer language, dissertation required, foreign language not required. *Entrance requirements:* For master's and doctorate, GRE General Test, TOEFL. *Application deadline:* For fall admission, 5/1 (priority date). Application fee: $50 ($75 for international students). *Unit head:* Dr. Wei Zhao, Head, 409-845-5534, Fax: 409-847-8578, E-mail: csdept@cs.tamu.edu. *Application contact:* S. Bart Childs, Graduate Adviser, 409-845-8981, E-mail: csdept@cs.tamu.edu.

Université de Sherbrooke, Faculty of Applied Sciences, Department of Electrical Engineering and Computer Engineering, Sherbrooke, PQ J1K 2R1, Canada. Offers M Sc A, PhD. *Degree requirements:* For master's and doctorate, thesis/dissertation required. *Faculty research:* Minielectronics, biomedical engineering, digital signal prolonging and telecommunications, software engineering and artificial intelligence.

The University of Alabama at Birmingham, Graduate School, School of Engineering, Department of Electrical and Computer Engineering, Birmingham, AL 35294. Offers MSEE, PhD. Evening/weekend programs available. *Students:* 8 full-time (2 women), 13 part-time (3 women); includes 2 minority (1 African American, 1 Asian American or Pacific Islander), 6 international. 27 applicants, 93% accepted. In 1998, 9 master's awarded. *Degree requirements:* For master's, thesis or alternative required, foreign language not required; for doctorate, dissertation required, foreign language not required. *Entrance requirements:* For master's, GRE General Test (minimum score of 500 on each section required), BSEE. *Application deadline:* Applications are processed on a rolling basis. Application fee: $30 ($60 for international students). Electronic applications accepted. *Financial aid:* In 1998–99, 1 student received aid, including 1 research assistantship with full tuition reimbursement available (averaging $9,500 per year); tuition waivers (full and partial) also available. *Unit head:* Dr. Gregg L. Vaughn, Interim Chair, 205-934-8440, Fax: 205-975-3337, E-mail: gvaughn@uab.edu.

The University of Alabama in Huntsville, School of Graduate Studies, College of Engineering, Department of Electrical and Computer Engineering, Huntsville, AL 35899. Offers computer engineering (PhD); electrical and computer engineering (MSE); electrical engineering (PhD); optical science and engineering (PhD). Part-time and evening/weekend programs available. *Faculty:* 22 full-time (1 woman), 2 part-time (0 women). *Students:* 64 full-time (17 women), 64 part-time (7 women); includes 13 minority (5 African Americans, 6 Asian Americans or Pacific Islanders, 2 Hispanic Americans), 38 international. Average age 31. 94 applicants, 80% accepted. In 1998, 29 master's, 9 doctorates awarded. *Degree requirements:* For master's, oral and written exams required, thesis optional, foreign language not required; for doctorate, dissertation, oral and written exams required, foreign language not required. *Entrance requirements:* For master's, GRE General Test (minimum combined score of 1500 on three sections required), appropriate bachelor's degree, minimum GPA of 3.0; for doctorate, GRE General Test (minimum combined score of 1500 on three sections required), minimum GPA of 3.0. *Application deadline:* For fall admission, 7/24 (priority date); for spring admission, 11/15 (priority date). Applications are processed on a rolling basis. Application fee: $20. Tuition and fees vary according to course load. *Financial aid:* In 1998–99, 39 students received aid, including 10 research assistantships with full and partial tuition reimbursements available (averaging $8,379 per year), 29 teaching assistantships with full and partial tuition reimbursements available (averaging $7,894 per year); fellowships with full and partial tuition reimbursements available, career-related internships or fieldwork, Federal Work-Study, grants, institutionally-sponsored loans, scholarships, and tuition waivers (full and partial) also available. Aid available to part-time students. Financial aid application deadline: 4/1; financial aid applicants required to submit FAFSA. *Faculty research:* Optical signal processing, electromagnetics, photonics, nonlinear waves, computer architecture. Total annual research expenditures: $1.4 million. *Unit head:* Dr. Reza Adhami, Chair, 256-890-6316, Fax: 256-890-6803, E-mail: adhami@eb.uah.edu.

See in-depth description on page 1041.

University of Alberta, Faculty of Graduate Studies and Research, Department of Electrical and Computer Engineering, Edmonton, AB T6G 2E1, Canada. Offers computational optics (PhD); computer engineering (M Eng, M Sc, PhD); control systems (M Eng, M Sc, PhD); engineering management (M Eng); laser physics (M Sc, PhD); oil sands (M Eng, M Sc, PhD); plasma physics (M Sc, PhD); power engineering (M Eng, M Sc, PhD); telecommunications (M Eng, M Sc, PhD). Terminal master's awarded for partial completion of doctoral program. *Degree requirements:* For master's and doctorate, thesis/dissertation required, foreign language not required. *Entrance requirements:* For master's and doctorate, TOEFL (minimum score of 580 required; average 610). Electronic applications accepted. *Faculty research:* Controls, communications, microelectronics, electromagnetics.

The University of Arizona, Graduate College, College of Engineering and Mines, Department of Electrical and Computer Engineering, Tucson, AZ 85721. Offers MS, PhD. Part-time programs available. *Faculty:* 64. *Students:* 135 full-time (18 women), 73 part-time (8 women). Average age 29. 314 applicants, 46% accepted. In 1998, 32 master's, 12 doctorates awarded. *Degree requirements:* For master's, thesis required (for some programs), foreign language not required; for doctorate, dissertation required, foreign language not required. *Entrance requirements:* For master's, GRE General Test, TOEFL (minimum score of 575 required); for doctorate, GRE General Test, TOEFL (minimum score of 600 required). *Application deadline:* For fall admission, 6/15. Applications are processed on a rolling basis. Application fee: $35. *Financial aid:* Fellowships, research assistantships, teaching assistantships, institutionally-sponsored loans and scholarships available. Financial aid application deadline: 3/15. *Faculty research:* Communication systems, control systems, signal processing, computer-aided logic. *Unit head:* Kenneth F. Galloway, Head, 520-621-6193. *Application contact:* Barbie Horton, Graduate Academic Adviser, 520-621-6195, Fax: 520-621-8076.

See in-depth description on page 1043.

University of Arkansas, Graduate School, College of Engineering, Department of Computer Systems Engineering, Fayetteville, AR 72701-1201. Offers MSCSE, MSE, PhD. *Faculty:* 7 full-time (0 women). *Students:* 26 full-time (7 women), 7 part-time (1 woman); includes 6 minority (1 African American, 5 Asian Americans or Pacific Islanders), 7 international. 21 applicants, 76% accepted. In 1998, 12 master's awarded. *Degree requirements:* For master's, computer language required, thesis optional, foreign language not required; for doctorate, one foreign language, dissertation required. Application fee: $40 ($50 for international students). Tuition, state resident: full-time $3,186. Tuition, nonresident: full-time $7,560. Required fees: $378. *Financial aid:* In 1998–99, 4 research assistantships, 7 teaching assistantships were awarded.; career-related internships or fieldwork and Federal Work-Study also available. Aid available to part-time students. Financial aid application deadline: 4/1; financial aid applicants required to submit FAFSA. *Unit head:* Dr. David Andrews, Chair, 501-575-6036. *Application contact:* Dr. Robert Crisp, Graduate Coordinator, E-mail: grad-info@engr.uark.edu.

University of Bridgeport, College of Graduate and Undergraduate Studies, School of Science, Engineering, and Technology, Department of Computer Science and Engineering, Bridgeport, CT 06601. Offers computer engineering (MS); computer science (MS). *Faculty:* 8 full-time (0 women), 5 part-time (0 women). *Students:* 87 full-time (15 women), 148 part-time (57 women); includes 20 minority (19 Asian Americans or Pacific Islanders, 1 Hispanic American), 212 international. Average age 31. 236 applicants, 64% accepted. In 1998, 23 degrees awarded. *Degree requirements:* For master's, thesis optional, foreign language not required. *Entrance requirements:* For master's, TOEFL. *Application deadline:* Applications are processed on a rolling basis. Application fee: $35 ($50 for international students). *Financial aid:* In 1998–99, 58 students received aid; research assistantships, teaching assistantships, career-related internships or fieldwork, Federal Work-Study, institutionally-sponsored loans, and tuition waivers (partial) available. Aid available to part-time students. Financial aid application deadline: 6/1; financial aid applicants required to submit FAFSA. *Unit head:* Dr. Stephen F. Grodzinsky, Chairman, 203-576-4145.

University of Calgary, Faculty of Graduate Studies, Faculty of Engineering, Department of Electrical and Computer Engineering, Calgary, AB T2N 1N4, Canada. Offers M Eng, M Sc, PhD. Part-time programs available. *Faculty:* 24 full-time (0 women), 10 part-time (0 women). *Students:* 70 full-time (2 women), 31 part-time (2 women). 200 applicants, 26% accepted. In 1998, 8 master's, 8 doctorates awarded. *Degree requirements:* For master's, thesis, thesis (M Sc) required, foreign language not required; for doctorate, dissertation, candidacy exam required, foreign language not required. *Entrance requirements:* For master's and doctorate, TOEFL (minimum score of 550 required; 213 for computer-based), minimum GPA of 3.0. *Average time to degree:* Master's–2 years full-time; doctorate–3.2 years full-time. *Application deadline:* For fall admission, 4/30; for winter admission, 9/30. Applications are processed on a rolling basis. Application fee: $60. *Financial aid:* In 1998–99, 59 students received aid; fellowships, research assistantships, teaching assistantships available. Financial aid application deadline: 5/30. *Faculty research:* Control, electronics, power systems, signal and image processing, telecommunications, software engineering. Total annual research expenditures: $1.8 million. *Unit head:* Dr. R. H. Johnston, Head, 403-220-5003, Fax: 403-282-6855, E-mail: grad-studies@enel.ucalgary.ca. *Application contact:* Dr. A. Sesay, Associate Head, 403-220-6163, Fax: 403-282-6855, E-mail: grad_studies@enel.ucalgary.ca.

University of California, Davis, Graduate Studies, College of Engineering, Program in Electrical and Computer Engineering, Davis, CA 95616. Offers MS, PhD. *Faculty:* 33 full-time (2 women), 2 part-time (0 women). *Students:* 116 full-time (13 women), 4 part-time; includes 26 minority (2 African Americans, 21 Asian Americans or Pacific Islanders, 3 Hispanic Americans), 37 international. 505 applicants, 25% accepted. In 1998, 11 master's, 12 doctorates awarded. Terminal master's awarded for partial completion of doctoral program. *Degree requirements:* For master's, thesis optional, foreign language not required; for doctorate, dissertation, preliminary and qualifying exams, thesis defense required, foreign language not required. *Entrance requirements:* For master's, GRE General Test, minimum GPA of 3.2; for doctorate, GRE, minimum graduate GPA of 3.5. *Application deadline:* For fall admission, 2/15. Application fee: $40. Electronic applications accepted. *Financial aid:* In 1998–99, 26 fellowships with full and partial tuition reimbursements, 35 research assistantships with tuition reimbursements, 29 teaching assistantships with full and partial tuition reimbursements were awarded.; Federal Work-Study also available. Financial aid application deadline: 1/15; financial aid applicants required to submit FAFSA. *Unit head:* Bernard Levy, Chairperson, 530-752-8251. *Application contact:* Anita Morales, Graduate Staff Assistant, 530-752-8251, E-mail: gradinfo@ucdavis.edu.

University of California, Irvine, Office of Research and Graduate Studies, School of Engineering, Department of Electrical and Computer Engineering, Irvine, CA 92697. Offers computer networks and distributed computing (MS, PhD); computer systems and software (MS, PhD); electrical engineering (MS, PhD). Part-time programs available. *Faculty:* 24 full-time (1 woman), 1 part-time (0 women). *Students:* 92 full-time (17 women), 15 part-time (1 woman); includes 24 minority (1 African American, 23 Asian Americans or Pacific Islanders), 63 international. 308 applicants, 37% accepted. In 1998, 30 master's, 9 doctorates awarded. Terminal master's awarded for partial completion of doctoral program. *Degree requirements:* For doctorate, dissertation required, foreign language not required, foreign language not required. *Entrance requirements:* For master's, GRE General Test, minimum GPA of 3.0; for doctorate, GRE General Test. *Application deadline:* For fall admission, 1/15 (priority date). Applications are processed on a rolling basis. Application fee: $40. Electronic applications accepted. *Financial aid:* Fellowships, research assistantships, teaching assistantships, institutionally-sponsored loans and tuition waivers (full and partial) available. Financial aid application deadline: 3/2; financial aid applicants required to submit FAFSA. *Faculty research:* Optical and solid-state devices, systems and signal processing. *Unit head:* Dr. Nader Bagherzadeh, Chair, 949-824-8720, Fax: 949-824-3779. *Application contact:* Ronnie A. Gran, Graduate Admissions Coordinator, 949-824-5489, Fax: 949-824-1853, E-mail: ragran@uci.edu.

See in-depth description on page 1045.

University of California, San Diego, Graduate Studies and Research, Department of Computer Science and Engineering, La Jolla, CA 92093-5003. Offers computer engineering (MS, PhD); computer science (MS, PhD). *Faculty:* 18. *Students:* 144 (30 women). 428 applicants, 30% accepted. In 1998, 19 master's, 9 doctorates awarded. *Degree requirements:* For master's, foreign language and thesis not required; for doctorate, dissertation required. *Entrance requirements:* For master's and doctorate, GRE General Test. *Application deadline:* For fall admission, 1/15. Application fee: $40. *Faculty research:* Analysis of algorithms, combinatorial algorithms, discrete optimization. *Unit head:* Jeanne Ferrante, Chair. *Application contact:* Graduate Coordinator, 619-534-6005.

University of California, San Diego, Graduate Studies and Research, Department of Electrical and Computer Engineering, La Jolla, CA 92093-5003. Offers applied ocean science (MS, PhD); applied physics (MS, PhD); communication theory and systems (MS, PhD); computer engineering (MS, PhD); electrical engineering (M Eng, MS, PhD); electronic circuits and systems (MS, PhD); intelligent systems, robotics and control (MS, PhD); photonics (MS, PhD); signal and image processing (MS, PhD). *Faculty:* 35. *Students:* 251 (24 women). 590 applicants, 29% accepted. In 1998, 24 master's, 27 doctorates awarded. *Entrance requirements:* For master's and doctorate, GRE General Test. Application fee: $40. *Unit head:* William Coles, Chair. *Application contact:* Graduate Coordinator, 619-534-6606.

See in-depth description on page 1051.

University of California, San Diego, Graduate Studies and Research, Interdisciplinary Program in Cognitive Science, La Jolla, CA 92093-5003. Offers cognitive science/anthropology (PhD); cognitive science/communication (PhD); cognitive science/computer science and engineering (PhD); cognitive science/linguistics (PhD); cognitive science/neuroscience (PhD); cognitive science/philosophy (PhD); cognitive science/psychology (PhD); cognitive science/sociology (PhD). Admissions through affiliated departments. *Faculty:* 51 full-time (6 women). *Students:* 10 full-time (3 women). *Degree requirements:* For doctorate, dissertation required. *Entrance requirements:* For doctorate, GRE General Test. *Application deadline:* Applications are processed on a rolling basis. Application fee: $40. *Unit head:* Walter J. Savitch, Director, 619-534-7141, Fax: 619-534-1128, E-mail: wsavitch@ucsd.edu. *Application contact:* Gris Arellano-Ramirez, Graduate Coordinator, 619-534-7141, Fax: 619-534-1128, E-mail: gradinfo@cogsci.ucsd.edu.

University of California, Santa Barbara, Graduate Division, College of Engineering, Department of Electrical and Computer Engineering, Santa Barbara, CA 93106. Offers MS, PhD. *Students:* 196 full-time (32 women). 489 applicants, 35% accepted. In 1998, 50 master's, 29 doctorates awarded. *Degree requirements:* For master's, thesis or alternative required, foreign language not required; for doctorate, dissertation required, foreign language not required. *Entrance requirements:* For master's and doctorate, GRE General Test, TOEFL (minimum score of 560 required). *Application deadline:* For fall admission, 12/15. Application fee: $40. Electronic applications accepted. *Financial aid:* Fellowships, research assistantships, teaching assistantships, career-related internships or fieldwork, Federal Work-Study, institutionally-sponsored loans, and tuition waivers (full and partial) available. Financial aid application deadline: 12/15; financial aid applicants required to submit FAFSA. *Faculty research:* Solid-state device theory, physics of solid-state materials, computer architecture. *Unit head:* Steve Butner, Chair, 805-893-3821, E-mail: butner@ece.ucsb.edu. *Application contact:* Dawn Keiling, Graduate Program Assistant, 805-893-3114, E-mail: dawn@ece.ucsb.edu.

See in-depth description on page 1053.

University of California, Santa Cruz, Graduate Division, Division of Natural Sciences, Program in Computer Engineering, Santa Cruz, CA 95064. Offers MS, PhD. *Faculty:* 12 full-time. *Students:* 79 full-time (15 women); includes 12 minority (1 African American, 10 Asian Americans or Pacific Islanders, 1 Hispanic American), 31 international. 87 applicants, 52% accepted. In 1998, 23 master's, 4 doctorates awarded. *Degree requirements:* For doctorate, dissertation, comprehensive and oral exams required. *Entrance requirements:* For master's and doctorate, GRE General Test, GRE Subject Test. *Application deadline:* For fall admission, 2/1. Application fee: $40. *Financial aid:* Fellowships, research assistantships, teaching assistantships, career-related internships or fieldwork, Federal Work-Study, and institutionally-sponsored loans available. Financial aid application deadline: 2/1. *Faculty research:* Computer-aided design of digital systems. *Unit head:* Dr. Joel Ferguson, Chairperson, 831-459-4172. *Application contact:* Graduate Admissions, 831-459-2301.

See in-depth description on page 1055.

University of Central Florida, College of Engineering, Department of Electrical and Computer Engineering, Orlando, FL 32816. Offers computer engineering (MS Cp E, PhD, Certificate); electrical engineering (MSEE, PhD, Certificate); optical science and engineering (MS, PhD). Part-time and evening/weekend programs available. *Faculty:* 28 full-time, 14 part-time. *Students:* 144 full-time (24 women), 110 part-time (19 women); includes 51 minority (9 African Americans, 24 Asian Americans or Pacific Islanders, 18 Hispanic Americans), 106 international. Average age 31. 160 applicants, 39% accepted. In 1998, 50 master's, 9 doctorates awarded. *Degree requirements:* For master's, computer language, thesis or alternative required, foreign language not required; for doctorate, computer language, dissertation, departmental qualifying exam, candidacy exam required, foreign language not required. *Entrance requirements:* For master's, GRE General Test (minimum combined score of 1000 required), TOEFL (minimum score of 550 required; 213 computer-based), minimum GPA of 3.0 in last 60 hours; for doctorate, GRE General Test (minimum combined score of 1100 required), TOEFL (minimum score of 550 required; 213 computer-based), minimum GPA of 3.5 in last 60 hours. *Application deadline:* For fall admission, 7/15; for spring admission, 12/15. Application fee: $20. Tuition, state resident: full-time $2,054; part-time $137 per credit. Tuition, nonresident: full-time $7,207; part-time $480 per credit. Required fees: $47 per term. *Financial aid:* In 1998–99, 164 students received aid, including 31 fellowships with partial tuition reimbursements available (averaging $2,337 per year), 102 teaching assistantships with partial tuition reimbursements available (averaging $1,940 per year); research assistantships with partial tuition reimbursements available, career-related internships or fieldwork, Federal Work-Study, institutionally-sponsored loans, tuition waivers (partial), and unspecified assistantships also available. Financial aid application deadline: 3/1; financial aid applicants required to submit FAFSA. *Faculty research:* Communication theory, solid-state devices, electromagnetics, electrooptics, digital signal processing. *Unit head:* Dr. W. B. Mikhael, Chair, 407-823-3210, E-mail: mikhael@ucf1vm.cc.ucf.edu. *Application contact:* Dr. Juin J. Liou, Coordinator, 407-823-2610.

See in-depth description on page 1057.

University of Cincinnati, Division of Research and Advanced Studies, College of Engineering, Department of Electrical and Computer Engineering and Computer Science, Program in Computer Engineering, Cincinnati, OH 45221-0091. Offers MS. *Students:* 86 full-time (14 women), 12 part-time (2 women); includes 7 minority (4 African Americans, 2 Asian Americans or Pacific Islanders, 1 Hispanic American), 84 international. In 1998, 23 degrees awarded. *Degree requirements:* For master's, thesis required, foreign language not required. *Entrance requirements:* For master's, GRE General Test, TOEFL (minimum score of 550 required), BS in electrical engineering or related field. *Average time to degree:* Master's–2.4 years full-time. *Application deadline:* For fall admission, 2/1 (priority date). Application fee: $40. *Financial aid:* Fellowships, tuition waivers (full) and unspecified assistantships available. Financial aid application deadline: 2/1. *Faculty research:* Digital signal processing, large-scale systems, picture processing. *Unit head:* Barbie Horton, Graduate Academic Adviser, 520-621-6195, Fax: 520-621-8076. *Application contact:* Dieter Schmidt, Graduate Program Director, 513-556-1816, Fax: 513-556-7326, E-mail: dieter.schmidt@uc.edu.

Computer Engineering

University of Cincinnati, Division of Research and Advanced Studies, College of Engineering, Department of Electrical and Computer Engineering and Computer Science, Program in Computer Science and Engineering, Cincinnati, OH 45221-0091. Offers PhD. *Students:* 26 full-time (4 women), 3 part-time; includes 1 African American, 21 international. In 1998, 4 degrees awarded. *Degree requirements:* For doctorate, dissertation required, foreign language not required. *Entrance requirements:* For doctorate, GRE General Test, TOEFL. *Average time to degree:* Doctorate–4.9 years full-time. *Application deadline:* For fall admission, 2/1 (priority date). Application fee: $40. *Financial aid:* Fellowships, tuition waivers (full) and unspecified assistantships available. Financial aid application deadline: 2/1. *Unit head:* Graduate Student Affairs, 909-787-5639, Fax: 909-787-4643, E-mail: gradadmissions@cs.ucr.edu. *Application contact:* Dieter Schmidt, Graduate Program Director, 513-556-1816, Fax: 513-556-7326, E-mail: dieter.schmidt@uc.edu.

University of Colorado at Boulder, Graduate School, College of Engineering and Applied Science, Department of Electrical and Computer Engineering, Boulder, CO 80309. Offers electrical engineering (ME, MS, PhD), including computer engineering. *Degree requirements:* For master's, thesis or alternative, comprehensive exam required; for doctorate, dissertation, departmental qualifying exam required. *Entrance requirements:* For master's, GRE General Test, minimum undergraduate GPA of 3.0; for doctorate, GRE General Test, minimum undergraduate GPA of 3.5.

University of Colorado at Colorado Springs, Graduate School, College of Engineering and Applied Science, Department of Electrical and Computer Engineering, Colorado Springs, CO 80933-7150. Offers MS, PhD. Part-time and evening/weekend programs available. *Faculty:* 9 full-time (0 women), 12 part-time (0 women). *Students:* 38 full-time (3 women), 31 part-time (3 women); includes 6 Asian Americans or Pacific Islanders, 1 Hispanic American, 1 Native American, 12 international. Average age 29. 29 applicants, 83% accepted. In 1998, 12 master's awarded (100% found work related to degree); 3 doctorates awarded (100% found work related to degree). *Degree requirements:* For master's, thesis required (for some programs), foreign language not required; for doctorate, dissertation, comprehensive exams required, foreign language not required. *Entrance requirements:* For master's, GRE General Test (minimum combined score of 1200 required), TOEFL (minimum score of 550 required), minimum GPA of 3.0; for doctorate, GRE General Test (minimum combined score of 1200 required), TOEFL (minimum score of 550 required), minimum GPA of 3.3. Application fee: $40 ($50 for international students). Tuition, state resident: full-time $2,768; part-time $118 per credit. Tuition, nonresident: full-time $10,392; part-time $425 per credit. Required fees: $265; $7.5 per credit. One-time fee: $28. Tuition and fees vary according to program and student level. *Financial aid:* In 1998–99, 12 students received aid; fellowships, research assistantships, teaching assistantships, career-related internships or fieldwork and Federal Work-Study available. Financial aid application deadline: 5/1. *Faculty research:* Signal processing, neural networks, integrated ferroelectric devices, applied electromagnetics, circuit design. Total annual research expenditures: $497,000. *Unit head:* Dr. Ramaswami Dandapani, Chairman, 719-262-3044, Fax: 719-262-3589, E-mail: rdan@vlsid.uccs.edu. *Application contact:* Sue Bidlingmaier, Academic Adviser, 719-262-3351, Fax: 719-262-3589, E-mail: suebid@ecemail.uccs.edu.

University of Dayton, Graduate School, School of Engineering, Department of Electrical and Computer Engineering, Dayton, OH 45469-1300. Offers MSEE, DE, PhD. Part-time and evening/weekend programs available. *Faculty:* 16 full-time (0 women), 3 part-time (0 women). *Students:* 25 full-time (3 women), 34 part-time (4 women); includes 10 minority (1 African American, 6 Asian Americans or Pacific Islanders, 3 Hispanic Americans), 11 international. Average age 24. In 1998, 21 master's, 3 doctorates awarded. *Degree requirements:* For master's, thesis optional, foreign language not required; for doctorate, variable foreign language requirement (computer language can substitute for one), dissertation, departmental qualifying exam required. *Entrance requirements:* For master's, TOEFL. *Application deadline:* For fall admission, 8/1. Applications are processed on a rolling basis. Application fee: $30. *Financial aid:* In 1998–99, 15 students received aid, including 1 fellowship with full tuition reimbursement available (averaging $15,000 per year), 7 research assistantships with full tuition reimbursements available (averaging $12,000 per year), 3 teaching assistantships with full tuition reimbursements available (averaging $9,000 per year); institutionally-sponsored loans also available. Financial aid application deadline: 5/1. *Faculty research:* Analog and signal processing, electromagnetics, electro-optics, digital computer architectures. *Unit head:* Dr. Donald L. Moon, Associate Dean, 937-229-2241, Fax: 937-229-2471, E-mail: dmoon@engr.udayton.edu. *Application contact:* Dr. Donald L. Moon, Associate Dean, 937-229-2241, Fax: 937-229-2471, E-mail: dmoon@engr.udayton.edu.

University of Denver, Graduate Studies, Faculty of Natural Sciences, Mathematics and Engineering, Department of Engineering, Denver, CO 80208. Offers computer science and engineering (MS); electrical engineering (MS); management and general engineering (MSMGEN); materials science (PhD); mechanical engineering (MS). Part-time and evening/weekend programs available. *Faculty:* 15. *Students:* 23 (9 women) 8 international. Terminal master's awarded for partial completion of doctoral program. *Degree requirements:* For master's, thesis required (for some programs), foreign language not required; for doctorate, dissertation required, foreign language not required. *Entrance requirements:* For master's and doctorate, GRE General Test, TOEFL (minimum score of 570 required), TSE (minimum score of 230 required). *Application deadline:* Applications are processed on a rolling basis. Application fee: $40 ($45 for international students). *Unit head:* Dr. Albert J. Rosa, Chair, 303-871-2102. *Application contact:* Louise Carlson, Assistant to Chair, 303-871-2107.

University of Florida, Graduate School, College of Engineering, Department of Electrical and Computer Engineering, Gainesville, FL 32611. Offers ME, MS, PhD, Engr. Part-time programs available. *Faculty:* 64. *Students:* 265 full-time (30 women), 64 part-time (5 women); includes 52 minority (6 African Americans, 16 Asian Americans or Pacific Islanders, 29 Hispanic Americans, 1 Native American), 170 international. 457 applicants, 63% accepted. In 1998, 65 master's, 20 doctorates awarded. Terminal master's awarded for partial completion of doctoral program. *Degree requirements:* For master's, thesis optional, foreign language not required; for doctorate and Engr, dissertation required, foreign language not required. *Entrance requirements:* For master's, GRE General Test (minimum score of 350 on verbal section, 1000 combined required), TOEFL (minimum score of 550 required), minimum GPA of 3.0; for doctorate, GRE General Test (minimum score of 350 on verbal section, 1200 combined on three sections required), TOEFL (minimum score of 550 required), minimum GPA of 3.5; for Engr, GRE General Test. *Application deadline:* For fall admission, 6/1 (priority date). Applications are processed on a rolling basis. Application fee: $20. Electronic applications accepted. *Financial aid:* In 1998–99, 136 students received aid, including 41 fellowships, 113 research assistantships, 33 teaching assistantships; unspecified assistantships also available. Financial aid application deadline: 4/15. *Faculty research:* Communications, electronics, digital signal processing, photonics. *Unit head:* Dr. Martin A. Uman, Chair, 352-392-0913, Fax: 352-392-8671, E-mail: muman@admin.ee.ufl.edu. *Application contact:* Dr. Jacob Hammer, Graduate Coordinator, 352-392-6607, Fax: 352-392-8381, E-mail: hammer@leaf.ce.ufl.edu.

See in-depth description on page 1065.

University of Florida, Graduate School, Graduate Engineering and Research Center (GERC), Gainesville, FL 32611. Offers aerospace engineering (ME, MS, PhD, Engr); electrical and computer engineering (ME, MS, PhD, Engr); engineering mechanics (ME, MS, PhD, Engr); industrial and systems engineering (ME, MS, PhD, Engr). Part-time programs available. Postbaccalaureate distance learning degree programs offered. *Faculty:* 6 full-time (0 women), 16 part-time (1 woman). *Students:* 13 full-time (6 women), 159 part-time (29 women); includes 21 minority (8 African Americans, 8 Asian Americans or Pacific Islanders, 4 Hispanic Americans, 1 Native American) Terminal master's awarded for partial completion of doctoral program. *Degree requirements:* For master's, computer language required (for some programs), thesis optional, foreign language not required; for doctorate, computer language (for some programs), dissertation required; for Engr, computer language (for some programs), thesis required, foreign language not required. *Entrance requirements:* For master's, GRE General Test (minimum combined score of 1000 required), TOEFL, minimum GPA of 3.0; for doctorate, GRE General Test (minimum combined score of 1200 required), written and oral qualifying exams, TOEFL, minimum GPA of 3.0, master's degree in engineering; for Engr, GRE General Test (minimum combined score of 1000 required), TOEFL, minimum GPA of 3.0, master's degree in engineering. *Application deadline:* For fall admission, 6/1; for spring admission, 10/1. Applications are processed on a rolling basis. Application fee: $20. Electronic applications accepted. *Unit head:* Dr. Pasquale M. Sforza, Director, 850-833-9355, Fax: 850-833-9366, E-mail: sforza@gerc.eng.ufl.edu. *Application contact:* Judi Shivers, Program Assistant, 850-833-9350, Fax: 850-833-9366, E-mail: reginfo@gerc.eng.ufl.edu.

University of Houston, Cullen College of Engineering, Program in Computer and Systems Engineering, Houston, TX 77004. Offers MSCSE, PhD. Part-time and evening/weekend programs available. *Students:* 11 full-time (5 women), 9 part-time (3 women); includes 3 minority (1 African American, 2 Asian Americans or Pacific Islanders), 10 international. Average age 29. 81 applicants, 16% accepted. In 1998, 6 master's awarded. Terminal master's awarded for partial completion of doctoral program. *Degree requirements:* For master's, thesis required (for some programs), foreign language not required; for doctorate, dissertation, departmental qualifying exams required, foreign language not required. *Entrance requirements:* For master's and doctorate, GRE General Test, TOEFL. *Application deadline:* For fall admission, 7/3 (priority date); for spring admission, 12/4. Applications are processed on a rolling basis. Application fee: $25 ($75 for international students). *Financial aid:* Fellowships, research assistantships, teaching assistantships, Federal Work-Study and tuition waivers (partial) available. Financial aid application deadline: 7/1. *Faculty research:* Parallel processing, parallel algorithms and architectures, neural networks. *Unit head:* Dr. Pauline Markenscoff, Director, 713-743-4403, Fax: 713-743-4444, E-mail: markenscoff@uh.edu. *Application contact:* Mylyssa McDonald, Graduate Analyst, 713-743-4403, Fax: 713-743-4444, E-mail: mmm05866@jetson.uh.edu.

University of Houston–Clear Lake, School of Natural and Applied Sciences, Program in Computer Engineering, Houston, TX 77058-1098. Offers MS. *Faculty:* 8 full-time (1 woman), 1 part-time (0 women). *Students:* 16 full-time (1 woman), 44 part-time (10 women); includes 17 minority (4 African Americans, 12 Asian Americans or Pacific Islanders, 1 Hispanic American), 18 international. Average age 32. *Degree requirements:* Foreign language not required. *Entrance requirements:* For master's, GRE General Test. *Application deadline:* Applications are processed on a rolling basis. Application fee: $30 ($70 for international students). *Financial aid:* Research assistantships, teaching assistantships available. Financial aid application deadline: 5/1. *Unit head:* Dr. Tom Harman, Chair, 281-283-3850, Fax: 281-283-3703, E-mail: harman@uhcl4.cl.uh.edu. *Application contact:* Dr. Robert Ferebee, Associate Dean, 281-283-3700, Fax: 281-283-3707, E-mail: ferebee@uhcl4.cl.uh.edu.

University of Idaho, College of Graduate Studies, College of Engineering, Department of Electrical Engineering, Program in Computer Engineering, Moscow, ID 83844-4140. Offers M Engr, MS. *Students:* 3 full-time (0 women), 9 part-time; includes 1 minority (African American) *Degree requirements:* For master's, computer language, thesis required, foreign language not required. *Entrance requirements:* For master's, minimum GPA of 2.8. *Application deadline:* For fall admission, 8/1; for spring admission, 12/15. Application fee: $35 ($45 for international students). *Financial aid:* Federal Work-Study available. Financial aid application deadline: 2/15. *Unit head:* Dr. David Egolf, Chairman, Department of Electrical Engineering, 208-885-6554.

University of Illinois at Chicago, Graduate College, College of Engineering, Department of Electrical Engineering and Computer Science, Program in Computer Science and Engineering, Chicago, IL 60607-7128. Offers MS, PhD. Evening/weekend programs available. *Degree requirements:* For master's, computer language, thesis or alternative required, foreign language not required; for doctorate, computer language, dissertation, departmental qualifying exam required, foreign language not required. *Entrance requirements:* For master's, TOEFL (minimum score of 550 required), minimum GPA of 3.75 on a 5.0 scale, BS in related field; for doctorate, GRE General Test, TOEFL (minimum score of 550 required), minimum GPA of 3.75 on a 5.0 scale, MS in related field. *Application deadline:* For fall admission, 6/7; for spring admission, 11/1. Application fee: $40 ($50 for international students). *Financial aid:* Fellowships, research assistantships, teaching assistantships available. *Unit head:* Gyan Agarwal, Director of Graduate Studies, 312-996-8679.

University of Illinois at Urbana–Champaign, Graduate College, College of Engineering, Department of Electrical and Computer Engineering, Urbana, IL 61801. Offers computer engineering (MS, PhD); electrical engineering (MS, PhD). *Faculty:* 89 full-time (3 women). *Students:* 447 full-time (41 women); includes 63 minority (4 African Americans, 50 Asian Americans or Pacific Islanders, 9 Hispanic Americans), 212 international. 930 applicants, 11% accepted. In 1998, 99 master's, 41 doctorates awarded. *Degree requirements:* For master's and doctorate, thesis/dissertation required, foreign language not required. *Entrance requirements:* For master's, minimum GPA of 4.0 on a 5.0 scale. *Application deadline:* For fall admission, 1/15. Applications are processed on a rolling basis. Application fee: $40 ($50 for international students). Tuition, state resident: full-time $4,616. Tuition, nonresident: full-time $11,768. Full-time tuition and fees vary according to course load. *Financial aid:* In 1998–99, 29 fellowships, 274 research assistantships, 101 teaching assistantships were awarded.; tuition waivers (full and partial) also available. Financial aid application deadline: 1/15. *Unit head:* Sung Mo Kang, Head, 217-333-2301. *Application contact:* William Perkins, Director of Graduate Studies, 217-333-0207, Fax: 217-244-7075, E-mail: perkins@decision.csl.uiuc.edu.

See in-depth description on page 1071.

The University of Iowa, Graduate College, College of Engineering, Department of Electrical and Computer Engineering, Iowa City, IA 52242-1316. Offers MS, PhD. *Faculty:* 16 full-time, 3 part-time. *Students:* 39 full-time (8 women), 22 part-time (4 women); includes 5 minority (all Asian Americans or Pacific Islanders), 31 international. 267 applicants, 48% accepted. In 1998, 12 master's, 4 doctorates awarded. *Degree requirements:* For master's, thesis optional; for doctorate, dissertation, comprehensive exam required. *Entrance requirements:* For master's and doctorate, TOEFL. *Application deadline:* Applications are processed on a rolling basis. Application fee: $30 ($50 for international students). *Financial aid:* In 1998–99, 1 fellowship, 21 research assistantships, 21 teaching assistantships were awarded. Financial aid applicants required to submit FAFSA. *Faculty research:* Computer systems and applications, biomedical image processing, robotics, acousto-optics. *Unit head:* Dr. Sudhakar M. Reddy, Chair, 319-335-5196.

University of Louisville, Graduate School, Speed Scientific School, Department of Engineering Mathematics and Computer Science, Program in Computer Science and Engineering, Louisville, KY 40292-0001. Offers PhD. *Students:* 105 full-time (18 women), 39 part-time (11 women); includes 14 minority (1 African American, 13 Asian Americans or Pacific Islanders), 105 international. Average age 29. In 1998, 16 degrees awarded. *Degree requirements:* For doctorate, dissertation required, foreign language not required. *Entrance requirements:* For doctorate, GRE General Test (minimum combined score of 1200 required). *Application deadline:* Applications are processed on a rolling basis. Application fee: $25. *Unit head:* Dr. Peter Aronhime, Director.

University of Maine, Graduate School, College of Engineering, Department of Electrical and Computer Engineering, Orono, ME 04469. Offers computer engineering (MS); electrical engineering (MS, PhD). Part-time programs available. *Faculty:* 14 full-time. *Students:* 13 full-time (3 women), 7 part-time; includes 2 minority (1 African American, 1 Asian American or Pacific Islander), 4 international. Average age 25. 29 applicants, 41% accepted. In 1998, 6 master's awarded. *Degree requirements:* For master's, thesis required (for some programs), foreign language not required; for doctorate, dissertation required, foreign language not required. *Entrance requirements:* For master's and doctorate, GRE General Test, TOEFL (minimum score of 550 required). *Application deadline:* For fall admission, 2/1 (priority date); for spring

Part-time $768 per unit. Required fees: $350 per semester. *Financial aid:* In 1998–99, 1 fellowship, 7 research assistantships, 4 teaching assistantships were awarded.; Federal Work-Study and institutionally-sponsored loans also available. Aid available to part-time students. Financial aid application deadline: 2/15; financial aid applicants required to submit FAFSA. *Unit head:* Dr. Robert Scholtz, Co-Chairman, Department of Electrical Engineering, 213-740-7327.

University of South Florida, Graduate School, College of Engineering, Department of Computer Science and Engineering, Tampa, FL 33620-9951. Offers computer engineering (M Cp E, MS Cp E); computer science (MCS, MSCS); computer science and engineering (PhD). Part-time programs available. *Faculty:* 17 full-time (2 women). *Students:* 95 full-time (20 women), 72 part-time (14 women); includes 27 minority (5 African Americans, 17 Asian Americans or Pacific Islanders, 5 Hispanic Americans), 85 international. Average age 30. 198 applicants, 78% accepted. In 1998, 40 master's awarded (80% found work related to degree, 20% continued full-time study); 6 doctorates awarded (50% entered university research/teaching, 50% found other work related to degree). Terminal master's awarded for partial completion of doctoral program. *Degree requirements:* For master's, computer language required, foreign language not required; for doctorate, dissertation, 2 tools of research as specified by dissertation committee required, foreign language not required. *Entrance requirements:* For master's, GRE General Test (minimum combined score of 1200 required), minimum GPA of 3.0 during previous 2 years; for doctorate, GRE General Test (minimum combined score of 1200 required, 500 on verbal section). *Average time to degree:* Master's–2.5 years full-time, 5 years part-time; doctorate–4.5 years full-time, 7 years part-time. *Application deadline:* For fall admission, 6/1; for spring admission, 10/15. Application fee: $20. Electronic applications accepted. Tuition, state resident: part-time $148 per credit hour. Tuition, nonresident: part-time $509 per credit hour. *Financial aid:* In 1998–99, 2 fellowships with full tuition reimbursements (averaging $9,500 per year), 60 research assistantships with full tuition reimbursements (averaging $11,993 per year), 22 teaching assistantships with full tuition reimbursements (averaging $11,746 per year) were awarded.; career-related internships or fieldwork, Federal Work-Study, institutionally-sponsored loans, and tuition waivers (partial) also available. Aid available to part-time students. Financial aid applicants required to submit FAFSA. *Faculty research:* Computer vision, databases, VLSI design and test, networks, artificial intelligence. Total annual research expenditures: $1.1 million. *Unit head:* Dr. Abe Kandel, Chairperson, 813-974-3652, Fax: 813-974-5456, E-mail: kandel@csee.usf.edu. *Application contact:* Dr. Dmitry B. Goldgof, Graduate Director, 813-974-3033, Fax: 813-974-5456, E-mail: msphd@csee.usf.edu.

University of Southwestern Louisiana, Graduate School, College of Engineering, Center for Advanced Computer Studies, Lafayette, LA 70504. Offers computer engineering (MS, PhD); computer science (MS, PhD). Part-time programs available. *Faculty:* 28 full-time (5 women). *Students:* 106 full-time (26 women), 18 part-time (3 women); includes 4 minority (1 African American, 2 Asian Americans or Pacific Islanders, 1 Hispanic American), 97 international. 401 applicants, 64% accepted. In 1998, 53 master's, 9 doctorates awarded. Terminal master's awarded for partial completion of doctoral program. *Degree requirements:* For master's, computer language, thesis or alternative required, foreign language not required; for doctorate, computer language, dissertation, final oral exam required, foreign language not required. *Entrance requirements:* For master's, GRE General Test, TOEFL, minimum GPA of 2.75; for doctorate, GRE General Test, TOEFL, minimum GPA of 3.0. *Application deadline:* For fall admission, 5/15. Application fee: $5 ($15 for international students). *Financial aid:* In 1998–99, 10 fellowships (averaging $15,700 per year), 74 research assistantships with full tuition reimbursements (averaging $6,824 per year) were awarded.; teaching assistantships, Federal Work-Study and tuition waivers (full) also available. Financial aid application deadline: 3/1. *Unit head:* Dr. Magdy A. Bayoumi, Chair, 318-482-6147. *Application contact:* Dr. William Edwards, Graduate Coordinator, 318-482-6284.

University of Southwestern Louisiana, Graduate School, College of Engineering, Department of Electrical and Computer Engineering, Lafayette, LA 70504. Offers computer engineering (MS, PhD); telecommunications (MSTC). *Faculty:* 6 full-time (0 women). *Students:* 47 full-time (12 women), 17 part-time (1 woman); includes 5 minority (3 African Americans, 1 Asian American or Pacific Islander, 1 Hispanic American), 41 international. *Degree requirements:* Foreign language not required. *Entrance requirements:* For master's, GRE General Test, minimum GPA of 2.75. *Application deadline:* For fall admission, 5/15. Application fee: $5 ($15 for international students). *Unit head:* Dr. Robert Henry, Acting Head, 318-482-6568.

The University of Texas at Arlington, Graduate School, College of Engineering, Department of Computer Science and Engineering, Arlington, TX 76019. Offers M Engr, M Sw En, MCS, MS, PhD. *Faculty:* 16 full-time (2 women). *Students:* 218 full-time (44 women), 138 part-time (18 women); includes 35 minority (1 African American, 30 Asian Americans or Pacific Islanders, 4 Hispanic Americans), 252 international. 434 applicants, 36% accepted. In 1998, 121 master's, 3 doctorates awarded. *Degree requirements:* For master's, computer language, thesis required (for some programs), foreign language not required; for doctorate, computer language, dissertation required, foreign language not required. *Entrance requirements:* For master's, GRE General Test (minimum combined score of 1100 required), TOEFL (minimum score of 560 required); for doctorate, GRE General Test (minimum combined score of 1250 required), TOEFL (minimum score of 560 required). *Application deadline:* Applications are processed on a rolling basis. Application fee: $25 ($50 for international students). Tuition, state resident: full-time $1,368; part-time $76 per semester hour. Tuition, nonresident: full-time $5,454; part-time $303 per semester hour. Required fees: $66 per semester hour. $86 per term. Tuition and fees vary according to course load. *Financial aid:* Research assistantships, teaching assistantships, career-related internships or fieldwork and tuition waivers (partial) available. *Unit head:* Dr. Bill D. Carroll, Chairman, 817-272-3785, Fax: 817-272-3784, E-mail: carroll@cse.uta.edu. *Application contact:* Dr. Bob P. Weems, Graduate Adviser, 817-272-3785, Fax: 817-272-3784, E-mail: weems@cse.uta.edu.

The University of Texas at Austin, Graduate School, College of Engineering, Department of Electrical and Computer Engineering, Austin, TX 78712-1111. Offers MSE, PhD. Part-time programs available. *Faculty:* 65 full-time (5 women), 12 part-time (1 woman). *Students:* 396 full-time (59 women), 169 part-time (16 women); includes 63 minority (5 African Americans, 42 Asian Americans or Pacific Islanders, 15 Hispanic Americans, 1 Native American), 336 international. 1098 applicants, 33% accepted. In 1998, 124 master's, 37 doctorates awarded. *Entrance requirements:* For master's, GRE General Test (combined average 2000), minimum GPA of 3.3 in upper-division course work; for doctorate, GRE General Test (combined average 2000). *Application deadline:* For fall admission, 1/2. Applications are processed on a rolling basis. Application fee: $50 ($75 for international students). Electronic applications accepted. *Financial aid:* Fellowships, research assistantships, teaching assistantships available. Financial aid application deadline: 1/2. *Unit head:* Dr. Francis X. Bostick, Chairman, 512-471-6179, Fax: 512-471-5532, E-mail: bostick@ece.utexas.edu. *Application contact:* Dr. Tony P. Ambler, Graduate Adviser, 512-475-6153, Fax: 512-471-5532, E-mail: ambler@ece.utexas.edu.

See in-depth description on page 1105.

The University of Texas at El Paso, Graduate School, College of Engineering, Department of Electrical and Computer Engineering, El Paso, TX 79968-0001. Offers computer engineering (MS, PhD); electrical engineering (MS, PhD). Part-time and evening/weekend programs available. *Faculty:* 17 full-time (1 woman), 8 part-time (0 women). *Students:* 50 full-time (8 women), 29 part-time (3 women); includes 27 minority (2 Asian Americans or Pacific Islanders, 25 Hispanic Americans), 38 international. Average age 29. 58 applicants, 48% accepted. In 1998, 16 master's awarded. *Degree requirements:* For master's, thesis optional; for doctorate, dissertation required. *Entrance requirements:* For master's, GRE General Test, TOEFL (minimum score of 550 required), minimum GPA of 3.0; for doctorate, GRE General Test, TOEFL, qualifying exam, minimum graduate GPA of 3.0. *Application deadline:* Applications are processed on a rolling basis. Application fee: $15 ($65 for international students). Electronic applica-

tions accepted. Tuition, state resident: full-time $2,790. Tuition, nonresident: full-time $7,710. *Financial aid:* In 1998–99, 60 students received aid; fellowships, research assistantships, teaching assistantships, Federal Work-Study, institutionally-sponsored loans, and tuition waivers (partial) available. Financial aid applicants required to submit FAFSA. *Faculty research:* Signal and image processing, computer architecture, fiber optics, computational electromagnetics, electronic displays and thin films. Total annual research expenditures: $436,501. *Unit head:* Dr. Michael Austin, Chairperson, 915-747-5470, Fax: 915-747-5616. *Application contact:* Susan Jordan, Director, Graduate Student Services, 915-747-5491, Fax: 915-747-5788, E-mail: sjordan@utep.edu.

University of Toronto, School of Graduate Studies, Physical Sciences Division, Faculty of Applied Science and Engineering, Department of Electrical and Computer Engineering, Toronto, ON M5S 1A1, Canada. Offers M Eng, MA Sc, PhD. Part-time programs available. *Degree requirements:* For master's, thesis required (for some programs), foreign language not required; for doctorate, dissertation required, foreign language not required.

Villanova University, College of Engineering, Department of Electrical and Computer Engineering, Program in Computer Engineering, Villanova, PA 19085-1699. Offers MSCE. Part-time and evening/weekend programs available. *Students:* 7 full-time (1 woman), 24 part-time (2 women); includes 1 minority (Hispanic American), 6 international. Average age 27. 19 applicants, 58% accepted. In 1998, 10 degrees awarded (100% found work related to degree). *Degree requirements:* For master's, thesis optional, foreign language not required. *Entrance requirements:* For master's, GRE General Test (for applicants with degrees from foreign universities), TOEFL (minimum score of 550 required), BEE, minimum GPA of 3.0. *Application deadline:* For fall admission, 8/1 (priority date); for spring admission, 12/1. Applications are processed on a rolling basis. Application fee: $40. *Financial aid:* In 1998–99, 5 students received aid, including 2 teaching assistantships; research assistantships, scholarships also available. Financial aid application deadline: 3/15. *Faculty research:* Expert systems, computer vision, neuralnetworks, image processing, computer architectures. *Unit head:* Dr. S. S. Rao, Chairman, Department of Electrical and Computer Engineering, 610-519-4971, Fax: 610-519-4436, E-mail: rao@ee.vill.edu.

Virginia Polytechnic Institute and State University, Graduate School, College of Engineering, Department of Electrical and Computer Engineering, Blacksburg, VA 24061. Offers MS, PhD. *Students:* 298 full-time (41 women), 158 part-time (17 women); includes 56 minority (13 African Americans, 37 Asian Americans or Pacific Islanders, 6 Hispanic Americans), 223 international. 727 applicants, 35% accepted. In 1998, 121 master's, 22 doctorates awarded. *Degree requirements:* For master's, foreign language and thesis not required; for doctorate, dissertation required. *Entrance requirements:* For master's and doctorate, GRE General Test, GRE Subject Test, TOEFL (minimum score of 600 required). *Application deadline:* For fall admission, 12/1 (priority date). Applications are processed on a rolling basis. Application fee: $25. *Financial aid:* In 1998–99, 56 research assistantships, 47 teaching assistantships were awarded.; fellowships, career-related internships or fieldwork and unspecified assistantships also available. Financial aid application deadline: 4/1. *Unit head:* Dr. Leonard A. Ferrari, Interim Head, 540-231-6646, E-mail: ferrari@vt.edu.

See in-depth description on page 1113.

Wayne State University, Graduate School, College of Engineering, Department of Electrical and Computer Engineering, Program in Computer Engineering, Detroit, MI 48202. Offers MS, PhD. *Degree requirements:* For master's, thesis optional, foreign language not required; for doctorate, dissertation required, foreign language not required. *Faculty research:* Neural networks, parallel processing, pattern recognition, VLSI, computer architecture.

Wayne State University, Graduate School, College of Engineering, Interdisciplinary Program in Electronics and Computer Control Systems, Detroit, MI 48202. Offers MS. *Degree requirements:* For master's, thesis optional, foreign language not required.

Western Michigan University, Graduate College, College of Engineering and Applied Sciences, Department of Electrical Engineering and Computer Engineering, Kalamazoo, MI 49008. Offers computer engineering (MSE); electrical engineering (MSE). Part-time programs available. *Students:* 9 full-time (0 women), 46 part-time (2 women); includes 2 minority (1 Asian American or Pacific Islander, 1 Native American), 47 international. 138 applicants, 68% accepted. In 1998, 7 degrees awarded. *Degree requirements:* For master's, thesis optional. *Entrance requirements:* For master's, minimum GPA of 3.0. *Application deadline:* For fall admission, 2/15 (priority date). Applications are processed on a rolling basis. Application fee: $25. *Financial aid:* Fellowships, research assistantships, teaching assistantships, career-related internships or fieldwork and Federal Work-Study available. Financial aid application deadline: 2/15; financial aid applicants required to submit FAFSA. *Faculty research:* Fiber optics, computer architecture, bioelectromagnetics, acoustics. *Unit head:* Dr. S. Hossein Mousavinezhad, Chair, 616-387-4057. *Application contact:* Paula J. Boodt, Coordinator, Graduate Admissions and Recruitment, 616-387-2000, Fax: 616-387-2355, E-mail: paula.boodt@wmich.edu.

West Virginia University, College of Engineering and Mineral Resources, Department of Computer Science and Electrical Engineering, Program in Computer Engineering, Morgantown, WV 26506. Offers engineering (PhD). *Degree requirements:* For doctorate, computer language, dissertation, comprehensive exam required, foreign language not required. *Entrance requirements:* For doctorate, GRE General Test (score in 80th percentile or higher required), TOEFL (minimum score of 550 required), minimum GPA of 3.0. *Faculty research:* Software engineering, microprocessor applications, microelectronic systems, fault tolerance, advanced computer architectures and networks, neural networks.

Widener University, School of Engineering, Program in Computer and Software Engineering, Chester, PA 19013-5792. Offers ME, ME/MBA. Part-time and evening/weekend programs available. *Faculty:* 5 part-time (1 woman). *Students:* 8 full-time (2 women), 29 part-time (7 women); includes 6 minority (2 African Americans, 4 Asian Americans or Pacific Islanders), 15 international. 28 applicants, 93% accepted. In 1998, 16 degrees awarded. *Degree requirements:* For master's, thesis optional, foreign language not required. *Entrance requirements:* For master's, GMAT (ME/MBA). *Average time to degree:* Master's–2 years full-time, 4 years part-time. *Application deadline:* For fall admission, 8/1 (priority date); for spring admission, 12/1. Applications are processed on a rolling basis. Application fee: $25 ($300 for international students). *Financial aid:* In 1998–99, 1 research assistantship with full tuition reimbursement (averaging $7,500 per year) was awarded.; unspecified assistantships also available. Financial aid application deadline: 3/15. *Faculty research:* Computer and software engineering, computer network fault-tolerant computing, and optical computing. *Unit head:* Dr. Alfred T. Johnson, Chairman, Department of Electrical/Telecommunication Engineering, 610-499-4053, Fax: 610-499-4059, E-mail: alfred.t.johnson@widener.edu. *Application contact:* Dr. David H. T. Chen, Assistant Dean for Graduate Programs and Research, 610-499-4049, Fax: 610-499-4059, E-mail: david.h.chen@widener.edu.

Worcester Polytechnic Institute, Graduate Studies, Department of Electrical and Computer Engineering, Worcester, MA 01609-2280. Offers electrical and computer engineering (MS, PhD, Advanced Certificate, Certificate); power systems engineering (MS, PhD). Part-time and evening/weekend programs available. *Faculty:* 22 full-time (1 woman), 3 part-time (0 women). *Students:* 48 full-time (7 women), 47 part-time (6 women); includes 7 minority (1 African American, 3 Asian Americans or Pacific Islanders, 3 Hispanic Americans), 32 international. 186 applicants, 46% accepted. In 1998, 31 master's, 6 doctorates awarded. Terminal master's awarded for partial completion of doctoral program. *Degree requirements:* For master's, thesis optional, foreign language not required; for doctorate, dissertation required, foreign language not required. *Entrance requirements:* For master's and doctorate, GRE (required for non-native speakers of English; combined average of 2024 on three sections), TOEFL (minimum score of 550 required; average 618). *Application deadline:* For fall admission, 2/15 (priority date); for spring admission, 10/15 (priority date). Applications are processed on a rolling basis. Application fee: $50. Electronic applications accepted. *Financial aid:* In 1998–99, 46 students

Computer Engineering–Electrical Engineering

Worcester Polytechnic Institute (continued)
received aid, including 1 fellowship with full tuition reimbursement available (averaging $12,500 per year), 27 research assistantships with full tuition reimbursements available (averaging $15,000 per year), 18 teaching assistantships with full tuition reimbursements available (averaging $11,970 per year); career-related internships or fieldwork, grants, institutionally-sponsored loans, and scholarships also available. Financial aid application deadline: 2/15; financial aid applicants required to submit FAFSA. *Faculty research:* Communications and signal processing, ultrasonics and electromagnetic engineering, electronics and solid state, systems and controls. Total annual research expenditures: $1.1 million. *Unit head:* Dr. John A. Orr, Head, 508-831-5231, Fax: 508-831-5491, E-mail: orr@wpi.edu. *Application contact:* Fred J. Looft, Graduate Coordinator, 508-831-5231, Fax: 508-831-5491, E-mail: fjlooft@wpi.edu.

Wright State University, School of Graduate Studies, College of Engineering and Computer Science, Department of Computer Science and Engineering, Computer Engineering Program, Dayton, OH 45435. Offers MSCE. *Students:* 16 full-time (0 women), 6 part-time (2 women); includes 1 minority (Asian American or Pacific Islander), 12 international. 33 applicants, 42% accepted. In 1998, 13 degrees awarded. *Degree requirements:* For master's, thesis optional, foreign language not required. *Entrance requirements:* For master's, GRE General Test, TOEFL (minimum score of 550 required), minimum GPA of 3.0 in major, 2.7 overall. Application fee: $25. *Financial aid:* Fellowships, research assistantships, teaching assistantships available. Aid available to part-time students. Financial aid application deadline: 3/31; financial aid applicants required to submit FAFSA. *Faculty research:* Networking and digital communications, parallel and concurrent computing, robotics and control, computer vision, optical computing. *Unit head:* Fred J. Looft, Graduate Coordinator, 508-831-5231, Fax: 508-831-5491, E-mail: fjlooft@wpi.edu. *Application contact:* Dr. Jay E. Dejongh, Graduate Adviser, 937-775-5136, Fax: 937-775-5133.

See in-depth description on page 1119.

Wright State University, School of Graduate Studies, College of Engineering and Computer Science, Department of Computer Science and Engineering, Program in Computer Science and Engineering, Dayton, OH 45435. Offers PhD. *Students:* 20 full-time (2 women), 5 part-time; includes 1 minority (Asian American or Pacific Islander), 13 international. 67 applicants, 60% accepted. In 1998, 1 degree awarded. *Degree requirements:* For doctorate, dissertation, candidacy and general exams required, foreign language not required. *Entrance requirements:* For doctorate, GRE General Test, TOEFL (minimum score of 550 required), minimum GPA of 3.3. Application fee: $25. *Financial aid:* Application deadline: 3/31. *Unit head:* Dr. P. Bruce Berra, Director, 937-775-5138, Fax: 937-775-5133.

Electrical Engineering

Air Force Institute of Technology, School of Engineering, Department of Electrical and Computer Engineering, Program in Electrical Engineering, Wright-Patterson AFB, OH 45433-7765. Offers electrical engineering (MS, PhD); electro-optics (MS, PhD). *Accreditation:* ABET (one or more programs are accredited). Part-time programs available. *Faculty:* 10 full-time (0 women), 1 part-time (0 women). *Students:* 57 full-time, 8 part-time. In 1998, 6 master's awarded (100% found work related to degree); 9 doctorates awarded (100% found work related to degree). *Degree requirements:* For master's, computer language, thesis required, foreign language not required; for doctorate, dissertation required, foreign language not required. *Entrance requirements:* For master's, GRE General Test (minimum score of 500 on verbal section, 600 on quantitative required), minimum GPA of 3.0, must be military officer or U.S. citizen; for doctorate, GRE General Test (minimum score of 550 on verbal section, 650 on quantitative required), minimum GPA of 3.0, must be military officer or U.S. citizen. Application fee: $0. *Faculty research:* Remote sensing, microelectronics, pattern recognition, communications, guidance and control. *Unit head:* Maj. James A. Lott, Chief, 937-255-3636 Ext. 4527, Fax: 937-656-4055, E-mail: jlott@afit.af.mil.

Alfred University, Graduate School, Program in Electrical Engineering, Alfred, NY 14802-1205. Offers MS. Part-time programs available. *Faculty:* 4 full-time (0 women), 1 part-time (0 women). 26 applicants, 81% accepted. In 1998, 1 degree awarded. *Degree requirements:* For master's, thesis required, foreign language not required. *Entrance requirements:* For master's, TOEFL (minimum score of 590 required). *Application deadline:* Applications are processed on a rolling basis. Application fee: $50. *Financial aid:* Research assistantships, career-related internships or fieldwork, Federal Work-Study, and tuition waivers (full and partial) available. Aid available to part-time students. Financial aid applicants required to submit FAFSA. *Unit head:* Dr. James Lancaster, Director, 607-871-2130, E-mail: flancaster@bigvax.alfred.edu. *Application contact:* Cathleen R. Johnson, Coordinator of Graduate Admissions, 607-871-2141, Fax: 607-871-2198, E-mail: johnsonc@king.alfred.edu.

Arizona State University, Graduate College, College of Engineering and Applied Sciences, Department of Electrical Engineering, Tempe, AZ 85287. Offers MS, MSE, PhD. *Faculty:* 49 full-time (6 women), 7 part-time (1 woman). *Students:* 268 full-time (35 women), 186 part-time (35 women); includes 54 minority (8 African Americans, 28 Asian Americans or Pacific Islanders, 17 Hispanic Americans, 1 Native American), 250 international. Average age 28. 846 applicants, 57% accepted. In 1998, 117 master's, 21 doctorates awarded. *Degree requirements:* For master's, thesis or alternative required; for doctorate, dissertation required. *Entrance requirements:* For master's and doctorate, GRE General Test (recommended), TOEFL (minimum score of 550 required). Application fee: $45. *Financial aid:* Fellowships, research assistantships, teaching assistantships available. *Faculty research:* Solid-state electronics, computer applications, power systems, communications. *Unit head:* Dr. Stephen Goodnick, Chair, 480-965-3424. *Application contact:* Dr. Joseph C. Palais, Director of Graduate Studies, 480-965-3590, Fax: 480-965-3837, E-mail: eeinfo@asu.edu.

See in-depth description on page 975.

Arizona State University East, College of Technology and Applied Sciences, Department of Electronics and Computer Engineering Technology, Mesa, AZ 85212. Offers MS. Part-time and evening/weekend programs available. *Faculty:* 7 full-time (1 woman). *Students:* 25 full-time (9 women), 27 part-time (5 women); includes 8 minority (2 African Americans, 4 Asian Americans or Pacific Islanders, 2 Hispanic Americans), 22 international. Average age 35. 24 applicants, 75% accepted. In 1998, 13 degrees awarded. *Degree requirements:* For master's, thesis or applied project and oral defense required. *Entrance requirements:* For master's, minimum GPA of 3.0. *Average time to degree:* Master's–2.9 years full-time, 4 years part-time. *Application deadline:* Applications are processed on a rolling basis. Application fee: $45. *Financial aid:* In 1998–99, 9 research assistantships with partial tuition reimbursements (averaging $10,506 per year) were awarded.; teaching assistantships, career-related internships or fieldwork, Federal Work-Study, grants, scholarships, and tuition waivers (full and partial) also available. Aid available to part-time students. Financial aid application deadline: 3/1; financial aid applicants required to submit FAFSA. *Faculty research:* Digital signal processing, image communication, reconfiguring computer architectures, insulator contamination/spark over, pulmonary function measurement. Total annual research expenditures: $361,209. *Unit head:* Dr. Robert W. Nowlin, Chair, 602-727-1137, Fax: 602-727-1723, E-mail: robert.nowlin@asu.edu. *Application contact:* Dr. Robert W. Nowlin, Chair, 602-727-1137, Fax: 602-727-1723, E-mail: robert.nowlin@asu.edu.

Auburn University, Graduate School, College of Engineering, Department of Electrical Engineering, Auburn, Auburn University, AL 36849-0002. Offers MEE, MS, PhD. Part-time programs available. *Faculty:* 27 full-time (1 woman). *Students:* 33 full-time (3 women), 43 part-time (4 women); includes 4 minority (2 African Americans, 1 Asian American or Pacific Islander, 1 Hispanic American), 33 international. 75 applicants, 40% accepted. In 1998, 22 master's, 7 doctorates awarded. *Degree requirements:* For master's, thesis (MS), comprehensive exam required; for doctorate, dissertation required. *Entrance requirements:* For master's, GRE General Test, GRE Subject Test; for doctorate, GRE General Test (minimum score of 400 on each section required), GRE Subject Test. *Application deadline:* For fall admission, 9/1; for spring admission, 3/1. Applications are processed on a rolling basis. Application fee: $25 ($50 for international students). Tuition, state resident: full-time $2,760; part-time $76 per credit hour. Tuition, nonresident: full-time $8,280; part-time $228 per credit hour. *Financial aid:* Fellowships, research assistantships, teaching assistantships, Federal Work-Study available. Aid available to part-time students. Financial aid application deadline: 3/15. *Faculty research:* Power systems, energy conversion, electronics, electromagnetics, digital systems. *Unit head:* Dr. J. David Irwin, Head, 334-844-1800. *Application contact:* Dr. John F. Pritchett, Dean of the Graduate School, 334-844-4700.

Boston University, College of Engineering, Department of Electrical and Computer Engineering, Boston, MA 02215. Offers computer engineering (PhD); computer systems engineering (MS); electrical engineering (MS, PhD); systems engineering (PhD). Part-time programs available. *Faculty:* 40 full-time (4 women), 4 part-time (0 women). *Students:* 117 full-time (20 women), 26 part-time (6 women); includes 9 minority (3 African Americans, 6 Asian Americans or Pacific Islanders), 83 international. Average age 27. 600 applicants, 18% accepted. In 1998, 32 master's, 15 doctorates awarded. Terminal master's awarded for partial completion of doctoral program. *Degree requirements:* For master's, thesis or alternative required, foreign language not required; for doctorate, dissertation required, foreign language not required. *Entrance requirements:* For master's, GRE General Test, TOEFL (minimum score of 500 required; 213 for computer-based); for doctorate, GRE General Test, TOEFL. *Application deadline:* For fall admission, 4/1; for spring admission, 10/1. Applications are processed on a rolling basis. Application fee: $50. Tuition: Full-time $23,770; part-time $743 per credit. Required fees: $220. Tuition and fees vary according to class time, course level, campus/location and program. *Financial aid:* In 1998–99, 113 students received aid, including 12 fellowships with full tuition reimbursements available (averaging $13,000 per year), 57 research assistantships with full tuition reimbursements available (averaging $11,500 per year), 24 teaching assistantships with full tuition reimbursements available (averaging $11,500 per year); career-related internships or fieldwork, Federal Work-Study, institutionally-sponsored loans, and scholarships also available. Financial aid application deadline: 12/15; financial aid applicants required to submit FAFSA. *Faculty research:* Signal processing, multimedia communications, quantum optics, VLSI, high-performance computing, magnetic and optical devices. Total annual research expenditures: $4 million. *Unit head:* Dr. Bahaa Saleh, Chairman, 617-353-7176, Fax: 617-353-6440. *Application contact:* Cheryl Kelley, Graduate Programs Director, 617-353-9760, Fax: 617-353-0259, E-mail: enggrad@bu.edu.

See in-depth description on page 977.

Bradley University, Graduate School, College of Engineering and Technology, Department of Electrical Engineering, Peoria, IL 61625-0002. Offers MSEE. Part-time and evening/weekend programs available. *Degree requirements:* For master's, comprehensive exam required, foreign language and thesis not required. *Entrance requirements:* For master's, GRE, TOEFL (minimum score of 525 required), minimum GPA of 3.0.

Brigham Young University, Graduate Studies, College of Engineering and Technology, Department of Electrical and Computer Engineering, Provo, UT 84602-1001. Offers electrical engineering (MS); engineering (PhD). Part-time programs available. *Faculty:* 20 full-time (0 women). *Students:* 64 full-time (2 women); includes 1 minority (African American), 8 international. Average age 23. 94 applicants, 29% accepted. In 1998, 22 master's, 4 doctorates awarded (25% entered university research/teaching, 75% found other work related to degree). *Degree requirements:* For master's, thesis optional, foreign language not required; for doctorate, dissertation required, foreign language not required. *Entrance requirements:* For master's, GRE General Test, TOEFL (minimum score of 550 required), minimum GPA of 3.0 in last 60 hours; for doctorate, minimum GPA of 3.4 in last 60 hours. *Average time to degree:* Master's–2 years full-time; doctorate–3.4 years full-time. *Application deadline:* For fall admission, 2/15; for winter admission, 9/15. Applications are processed on a rolling basis. Application fee: $30. Electronic applications accepted. Tuition: Full-time $3,330; part-time $185 per credit hour. Tuition and fees vary according to program and student's religious affiliation. *Financial aid:* In 1998–99, 27 students received aid, including 10 fellowships, 5 research assistantships, 19 teaching assistantships; scholarships also available. Financial aid application deadline: 3/15. *Faculty research:* Microwave remote sensing, reconfigurable computing, VLSI design tools, microelectronics circuit design and fabrication, digital signal processing, intelligent multi-agent control. Total annual research expenditures: $1.8 million. *Unit head:* Dr. Richard L. Frost, Chair, 801-378-3930, Fax: 801-378-6586, E-mail: rickf@ee.byu.edu. *Application contact:* Ann Tanner, Graduate Secretary, 801-378-4013, Fax: 801-378-6586, E-mail: grad@ee.byu.edu.

Brown University, Graduate School, Division of Engineering, Program in Electrical Sciences, Providence, RI 02912. Offers Sc M, PhD. *Degree requirements:* For doctorate, dissertation, preliminary exam required, foreign language not required, foreign language not required.

Bucknell University, Graduate Studies, College of Engineering, Department of Electrical Engineering, Lewisburg, PA 17837. Offers MS, MSEE. *Faculty:* 7 full-time. *Students:* 5 (1 woman). *Degree requirements:* For master's, thesis required, foreign language not required. *Entrance requirements:* For master's, GRE General Test (minimum combined score of 1000 required), GRE Subject Test, TOEFL (minimum score of 550 required), minimum GPA of 2.8. *Application deadline:* For fall admission, 6/1 (priority date); for spring admission, 12/1 (priority date). Applications are processed on a rolling basis. Application fee: $25. Tuition: Part-time $2,600 per course. Tuition and fees vary according to course load. *Financial aid:* Unspecified assistantships available. Financial aid application deadline: 3/1. *Unit head:* Dr. Edward Mastascusa, Head, 570-577-1234.

California Institute of Technology, Division of Engineering and Applied Science, Option in Electrical Engineering, Pasadena, CA 91125-0001. Offers MS, PhD, Engr. *Faculty:* 16 full-time (2 women). *Students:* 102 full-time (18 women), 68 international. 776 applicants, 4% accepted. In 1998, 16 master's, 10 doctorates awarded. *Degree requirements:* For master's, foreign language and thesis not required; for doctorate, dissertation required, foreign language not required. *Application deadline:* For fall admission, 1/15. Application fee: $0. *Faculty research:* Solid-state electronics, power electronics, communications, controls, submillimeter-wave integrated circuits. *Unit head:* Dr. Robert J. McEliece, Executive Officer, 626-395-3891. *Application contact:* Dr. P. P. Vaidyanathan, Representative, 626-395-4681.

California Polytechnic State University, San Luis Obispo, College of Engineering, Department of Electrical Engineering, San Luis Obispo, CA 93407. Offers MS. Part-time programs available. *Faculty:* 19 full-time (1 woman), 13 part-time (0 women). *Students:* 8 full-time (0 women), 10 part-time. 23 applicants, 43% accepted. In 1998, 6 degrees awarded. *Degree requirements:*

For master's, thesis or alternative required, foreign language not required. *Entrance requirements:* For master's, GRE General Test, GRE Subject Test, minimum GPA of 3.0 in last 90 quarter units. *Average time to degree:* Master's–2 years full-time, 4 years part-time. *Application deadline:* For fall admission, 5/31 (priority date); for spring admission, 12/31. Applications are processed on a rolling basis. Application fee: $55. Tuition, nonresident: part-time $164 per unit. Required fees: $531 per quarter. *Financial aid:* In 1998–99, 6 fellowships, 7 research assistantships, 8 teaching assistantships were awarded.; career-related internships or fieldwork also available. Financial aid application deadline: 3/2; financial aid applicants required to submit FAFSA. *Faculty research:* Communications, systems analysis, control systems, electronic devices, microprocessors. *Unit head:* Dr. Martin Kaliski, Chair, 805-756-2781, Fax: 805-756-1458, E-mail: mkaliski@ohm.calpoly.edu. *Application contact:* Dr. Donley Winger, Coordinator, 805-756-1462, Fax: 805-756-1458, E-mail: dwinger@ohm.calpoly.edu.

California State Polytechnic University, Pomona, Graduate Studies, College of Engineering, Pomona, CA 91768-2557. Offers electrical engineering (MSEE); engineering (MSE). Part-time programs available. *Faculty:* 60. *Students:* 23 full-time (6 women), 153 part-time (26 women); includes 83 minority (1 African American, 75 Asian Americans or Pacific Islanders, 6 Hispanic Americans, 1 Native American), 27 international. *Degree requirements:* For master's, computer language, thesis or comprehensive exam required. *Entrance requirements:* For master's, TOEFL, GRE General Test or minimum GPA of 3.0 in upper-level course work. *Application deadline:* Applications are processed on a rolling basis. Application fee: $55. Tuition, nonresident: part-time $164 per unit. *Unit head:* Dr. Carl E. Rathmann, Interim Dean, 909-869-2600, Fax: 909-869-4370, E-mail: cerathmann@csupomona.edu. *Application contact:* Dr. Elhami T. Ibrahim, Graduate Director, 909-869-2476, Fax: 909-869-4370, E-mail: etibrahim@csupomona.edu.

California State University, Chico, Graduate School, College of Engineering, Computer Science, and Technology, Department of Electrical Engineering, Chico, CA 95929-0722. Offers MS. *Faculty:* 10 full-time (0 women), 1 part-time (0 women). *Students:* 2 full-time (0 women), 4 part-time; includes 1 minority (Asian American or Pacific Islander), 2 international. Average age 31. *Degree requirements:* For master's, thesis or alternative, oral exam required, foreign language not required. *Entrance requirements:* For master's, GRE General Test or MAT. *Application deadline:* For fall admission, 4/1. Applications are processed on a rolling basis. Application fee: $55. *Financial aid:* Fellowships, teaching assistantships available. *Unit head:* Dr. Louis Harrold, Chair, 530-898-5343. *Application contact:* Dr. Richard Bednar, Graduate Coordinator, 530-898-5098.

California State University, Fresno, Division of Graduate Studies, School of Engineering, Program in Electrical Engineering, Fresno, CA 93740-0057. Offers MS. Offered at Edwards Air Force Base. Part-time and evening/weekend programs available. *Faculty:* 1 part-time (0 women). In 1998, 1 degree awarded. *Degree requirements:* For master's, computer language, thesis or alternative required, foreign language not required. *Entrance requirements:* For master's, GRE General Test, TOEFL (minimum score of 550 required). *Application deadline:* For fall admission, 8/1 (priority date); for spring admission, 12/1. Applications are processed on a rolling basis. Application fee: $55. Electronic applications accepted. Tuition, nonresident: part-time $246 per unit. Required fees: $1,906; $620 per semester. *Financial aid:* Application deadline: 3/1; *Unit head:* Dr. Peter J. Kasvinsky, Dean of Graduate Studies, 330-742-3091, Fax: 330-742-1580, E-mail: amgrad03@ysub.ysu.edu. *Application contact:* James W. Smolka, Coordinator, 805-258-5936.

California State University, Fullerton, Graduate Studies, School of Engineering and Computer Science, Department of Electrical Engineering, Fullerton, CA 92834-9480. Offers MS. Part-time programs available. *Faculty:* 13 full-time (2 women). *Students:* 4 full-time (0 women), 66 part-time (10 women); includes 33 minority (2 African Americans, 29 Asian Americans or Pacific Islanders, 2 Hispanic Americans), 11 international. Average age 31. 46 applicants, 67% accepted. In 1998, 30 degrees awarded. *Degree requirements:* For master's, computer language, comprehensive exam, project or thesis required. *Entrance requirements:* For master's, GRE General Test, GRE Subject Test, minimum GPA of 2.5 (undergraduate), 3.0 (graduate). Application fee: $55. Tuition, nonresident: part-time $264 per unit. Required fees: $1,947; $1,281 per year. *Financial aid:* Career-related internships or fieldwork, Federal Work-Study, grants, and institutionally-sponsored loans available. Aid available to part-time students. Financial aid application deadline: 3/1. *Unit head:* Dr. Karim Hamidian, Chair, 714-278-3013.

California State University, Long Beach, Graduate Studies, College of Engineering, Department of Electrical Engineering, Long Beach, CA 90840. Offers MSE, MSEE. Part-time programs available. *Faculty:* 24 full-time (0 women), 6 part-time (0 women). *Students:* 16 full-time (0 women), 71 part-time (8 women); includes 48 minority (4 African Americans, 37 Asian Americans or Pacific Islanders, 7 Hispanic Americans), 23 international. Average age 31. 97 applicants, 52% accepted. In 1998, 26 degrees awarded. *Degree requirements:* For master's, comprehensive exam or thesis required. *Entrance requirements:* For master's, TOEFL (minimum score of 550 required). *Application deadline:* For fall admission, 8/1; for spring admission, 12/1. Application fee: $55. Electronic applications accepted. Tuition, nonresident: part-time $246 per unit. Required fees: $569 per semester. Tuition and fees vary according to course load. *Financial aid:* Teaching assistantships, career-related internships or fieldwork, Federal Work-Study, grants, institutionally-sponsored loans, and unspecified assistantships available. Financial aid application deadline: 3/2. *Faculty research:* Health care systems, IC, VLSI, communications, CAD/CAM. *Unit head:* Dr. Fumio Hamano, Chair, 562-985-5102, Fax: 562-985-5327. *Application contact:* Dr. Michael Singh-Chelian, Graduate Adviser, 562-985-1516, Fax: 562-985-5327, E-mail: mchelian@engr.csulb.edu.

California State University, Los Angeles, Graduate Studies, School of Engineering and Technology, Department of Electrical and Computer Engineering, Los Angeles, CA 90032-8530. Offers electrical engineering (MS). Part-time and evening/weekend programs available. *Faculty:* 8 full-time, 10 part-time. *Students:* 20 full-time (8 women), 55 part-time (11 women); includes 49 minority (4 African Americans, 35 Asian Americans or Pacific Islanders, 10 Hispanic Americans), 15 international. In 1998, 21 degrees awarded. *Degree requirements:* For master's, computer language, comprehensive exam or thesis required. *Entrance requirements:* For master's, GRE General Test, GRE Subject Test, TOEFL (minimum score of 550 required). *Application deadline:* For fall admission, 6/30; for spring admission, 2/1. Applications are processed on a rolling basis. Application fee: $55. *Financial aid:* In 1998–99, 21 students received aid. Federal Work-Study available. Aid available to part-time students. Financial aid application deadline: 3/1. *Unit head:* Dr. Helen Boussalis, Chair, 323-343-4470.

California State University, Northridge, Graduate Studies, College of Engineering and Computer Science, Department of Electrical and Computer Engineering, Northridge, CA 91330. Offers biomedical engineering (MS); communications/radar engineering (MS); control engineering (MS); digital/computer engineering (MS); electronics engineering (MS); microwave/antenna engineering (MS). Part-time and evening/weekend programs available. *Faculty:* 17 full-time, 3 part-time. *Students:* 20 full-time (2 women), 77 part-time (8 women); includes 33 minority (3 African Americans, 24 Asian Americans or Pacific Islanders, 6 Hispanic Americans), 9 international. Average age 34. 58 applicants, 71% accepted. In 1998, 30 degrees awarded. *Degree requirements:* For master's, thesis or alternative required, foreign language not required. *Entrance requirements:* For master's, GRE General Test, TOEFL, minimum GPA of 2.5. *Application deadline:* For fall admission, 11/30. Application fee: $55. Tuition, nonresident: part-time $246 per unit. International tuition: $7,874 full-time. Required fees: $1,970. Tuition and fees vary according to course load. *Financial aid:* Application deadline: 3/1. *Faculty research:* Reflector antenna study, radome study. *Unit head:* Dr. Nagwa Bekir, Chair, 818-677-2190. *Application contact:* Nagi El Naga, Graduate Coordinator, 818-677-2180.

California State University, Sacramento, Graduate Studies, School of Engineering and Computer Science, Department of Electrical and Electronic Engineering, Sacramento, CA 95819-6048. Offers electrical engineering (MS). Part-time and evening/weekend programs available. *Degree requirements:* For master's, writing proficiency exam required. *Entrance requirements:* For master's, TOEFL (minimum score of 550 required). *Application deadline:* For fall admission, 4/15; for spring admission, 11/1. Application fee: $55. *Financial aid:* Research assistantships, teaching assistantships, career-related internships or fieldwork and Federal Work-Study available. Aid available to part-time students. Financial aid application deadline: 3/1. *Unit head:* Dr. S. K. Ramesh, Chair, 916-278-6873. *Application contact:* Dr. C. Desmond, Graduate Coordinator, 916-278-6873.

Carleton University, Faculty of Graduate Studies, Faculty of Engineering and Design, Ottawa-Carleton Institute for Electrical and Computer Engineering, Department of Electronics, Ottawa, ON K1S 5B6, Canada. Offers M Eng, PhD. *Faculty:* 16 full-time (0 women). *Students:* 58 full-time (6 women), 33 part-time (3 women). In 1998, 15 master's, 3 doctorates awarded. *Degree requirements:* For master's, thesis optional; for doctorate, dissertation, comprehensive exam required. *Entrance requirements:* For master's, TOEFL (minimum score of 550 required), honors degree; for doctorate, TOEFL (minimum score of 550 required), MA Sc or M Eng. *Average time to degree:* Master's–2.4 years full-time, 4 years part-time; doctorate–2.7 years full-time. *Application deadline:* For fall admission, 3/1 (priority date). Applications are processed on a rolling basis. Application fee: $35. *Financial aid:* Application deadline: 3/1. Total annual research expenditures: $1.8 million. *Unit head:* J. S. Wight, Chair, 613-520-5754, Fax: 613-520-5708, E-mail: jsw@doe.carleton.ca. *Application contact:* David J. Walkey, Supervisor of Graduate Studies, 613-520-5754, Fax: 613-520-5708, E-mail: walkey@doe.carleton.ca.

Carleton University, Faculty of Graduate Studies, Faculty of Engineering and Design, Ottawa-Carleton Institute for Electrical and Computer Engineering, Department of Systems and Computer Engineering, Program in Electrical Engineering, Ottawa, ON K1S 5B6, Canada. Offers M Eng, PhD. *Degree requirements:* For master's, thesis optional; for doctorate, dissertation, comprehensive exam required. *Entrance requirements:* For master's, TOEFL (minimum score of 550 required), honors degree; for doctorate, TOEFL (minimum score of 550 required), MA Sc or M Eng. *Application deadline:* For fall admission, 3/1. Applications are processed on a rolling basis. Application fee: $35. *Financial aid:* Application deadline: 3/1. *Unit head:* Ann Tanner, Graduate Secretary, 801-378-4013, Fax: 801-378-6586, E-mail: grad@ee.byu.edu. *Application contact:* Howard M. Schwartz, Supervisor of Graduate Studies, 613-520-5740, Fax: 613-520-5727, E-mail: schwartz@sce.carleton.ca.

Carnegie Mellon University, Carnegie Institute of Technology, Department of Electrical and Computer Engineering, Pittsburgh, PA 15213-3891. Offers biomedical engineering (MS, PhD); electrical and computer engineering (MS, PhD). Part-time programs available. *Faculty:* 47 full-time (2 women), 2 part-time (0 women). *Students:* 248 full-time (33 women), 25 part-time (1 woman); includes 23 minority (4 African Americans, 19 Asian Americans or Pacific Islanders), 148 international. 691 applicants, 17% accepted. In 1998, 70 master's awarded (3% entered university research/teaching, 77% found other work related to degree, 20% continued full-time study); 24 doctorates awarded (21% entered university research/teaching, 79% found other work related to degree). *Degree requirements:* For master's, thesis required, foreign language not required; for doctorate, computer language, dissertation, qualifying exam, teaching experience required, foreign language not required. *Entrance requirements:* For master's and doctorate, GRE General Test, TOEFL. *Average time to degree:* Master's–1.5 years full-time, 4 years part-time; doctorate–4.5 years full-time, 6 years part-time. *Application deadline:* For fall admission, 1/15; for spring admission, 10/15. Application fee: $50. *Financial aid:* In 1998–99, fellowships with full tuition reimbursements (averaging $17,100 per year), 191 research assistantships with full tuition reimbursements (averaging $17,100 per year), 16 teaching assistantships with full tuition reimbursements (averaging $12,800 per year) were awarded.; institutionally-sponsored loans also available. Financial aid application deadline: 1/15. *Faculty research:* Computer-aided design, solid-state devices, VLSI, processing, robotics and controls, signal processing, data systems storage. Total annual research expenditures: $12.2 million. *Unit head:* Robert M. White, Head, 412-268-7400, Fax: 412-268-5787, E-mail: white@ece.cmu.edu. *Application contact:* Lynn E. Philibin, Assistant Head for Graduate Studies, 412-268-3291, Fax: 412-268-2860, E-mail: lynn@ece.cmu.edu.

See in-depth description on page 979.

Case Western Reserve University, School of Graduate Studies, The Case School of Engineering, Department of Electrical, Systems, Computer Engineering and Science, Cleveland, OH 44106. Offers computer engineering and science (MS, PhD), including computer engineering, computing and information science; electrical engineering (MS, PhD); systems and control engineering (MS, PhD). Part-time and evening/weekend programs available. Postbaccalaureate distance learning degree programs offered (minimal on-campus study). *Faculty:* 28 full-time (2 women). *Students:* 73 full-time (13 women), 101 part-time (17 women). Average age 25. 484 applicants, 38% accepted. In 1998, 68 master's, 20 doctorates awarded. Terminal master's awarded for partial completion of doctoral program. *Degree requirements:* For master's and doctorate, thesis/dissertation required, foreign language not required. *Entrance requirements:* For master's and doctorate, GRE General Test, TOEFL (minimum score of 550 required). *Average time to degree:* Master's–2 years full-time, 3 years part-time; doctorate–4 years full-time, 5 years part-time. *Application deadline:* For fall admission, 3/1; for spring admission, 11/1. Applications are processed on a rolling basis. Application fee: $25. Electronic applications accepted. *Financial aid:* In 1998–99, 96 students received aid, including 70 research assistantships with full and partial tuition reimbursements available (averaging $15,600 per year), 33 teaching assistantships with full and partial tuition reimbursements available (averaging $10,170 per year); fellowships, career-related internships or fieldwork, Federal Work-Study, and institutionally-sponsored loans also available. Aid available to part-time students. Financial aid application deadline: 3/1. *Faculty research:* Microelectromechanical systems, control, artificial intelligence, mixed signals. Total annual research expenditures: $5.8 million. *Unit head:* Dr. Robert V. Edwards, Acting Chairman, 216-368-2800, Fax: 216-368-6888, E-mail: rve2@po.cwru.edu. *Application contact:* Elizabethanne M. Fuller, Department Assistant, 216-368-4080, Fax: 216-368-2668, E-mail: emf4@po.cwru.edu.

The Catholic University of America, School of Engineering, Department of Electrical Engineering and Computer Science, Washington, DC 20064. Offers MEE, MS Engr, D Engr, PhD. Part-time and evening/weekend programs available. *Faculty:* 7 full-time (0 women), 4 part-time (0 women). *Students:* 11 full-time (1 woman), 20 part-time (4 women); includes 10 minority (4 African Americans, 4 Asian Americans or Pacific Islanders, 2 Hispanic Americans), 10 international. Average age 30. 40 applicants, 50% accepted. In 1998, 11 master's, 1 doctorate awarded. *Degree requirements:* For master's, thesis optional, foreign language not required; for doctorate, dissertation, comprehensive and oral exams required, foreign language not required. *Entrance requirements:* For master's, TOEFL (minimum score of 550 required), minimum GPA of 3.0; for doctorate, TOEFL (minimum score of 550 required), minimum GPA of 3.4. *Application deadline:* For fall admission, 8/1 (priority date); for spring admission, 12/1. Applications are processed on a rolling basis. Application fee: $50. *Financial aid:* Research assistantships, career-related internships or fieldwork, Federal Work-Study, institutionally-sponsored loans, tuition waivers (full and partial), and unspecified assistantships available. Aid available to part-time students. Financial aid application deadline: 2/1. *Faculty research:* Signal and image processing, computer communications, robotics, intelligent controls, bioelectromagnetics, properties of materials. Total annual research expenditures: $320,000. *Unit head:* Dr. Charles Nguyen, Chair, 202-319-5193.

City College of the City University of New York, Graduate School, School of Engineering, Department of Electrical Engineering, New York, NY 10031-9198. Offers ME, MS, PhD. Part-time programs available. *Students:* 16 full-time (1 woman), 47 part-time (6 women). In 1998, 26 degrees awarded. *Degree requirements:* For master's, computer language required, thesis optional, foreign language not required; for doctorate, one foreign language (computer language can substitute), dissertation, comprehensive exams required. *Entrance requirements:* For master's, TOEFL (minimum score of 500 required); for doctorate, GRE General Test, TOEFL. *Application deadline:* Applications are processed on a rolling basis. Application fee: $40. *Financial aid:* Fellowships, research assistantships, Federal Work-Study and tuition waivers (full and partial) available. Aid available to part-time students. Financial aid application deadline:

Electrical Engineering

City College of the City University of New York (continued)
5/1. *Faculty research:* Optical electronics, microwaves, communication, signal processing, control systems. *Unit head:* Fred Thau, Chairman, 212-690-7250. *Application contact:* Graduate Admissions Office, 212-650-6977.

Clarkson University, Graduate School, School of Engineering, Department of Electrical and Computer Engineering, Potsdam, NY 13699. Offers computer engineering (ME, MS); electrical and computer engineering (PhD); electrical engineering (ME, MS). Part-time programs available. *Faculty:* 13 full-time (1 woman). *Students:* 28 full-time (6 women); includes 1 minority (Native American), 24 international. Average age 27. 168 applicants, 61% accepted. In 1998, 25 master's, 1 doctorate awarded. *Degree requirements:* For master's, thesis required, foreign language not required; for doctorate, dissertation, departmental qualifying exam required, foreign language not required. *Entrance requirements:* For master's, GRE, TOEFL. *Application deadline:* For fall admission, 5/15 (priority date); for spring admission, 10/15 (priority date). Applications are processed on a rolling basis. Application fee: $25 ($35 for international students). Tuition: Part-time $661 per credit hour. Required fees: $215 per semester. *Financial aid:* In 1998–99, 1 fellowship, 8 research assistantships, 6 teaching assistantships were awarded. *Faculty research:* Robotics, electrical machines, electrical circuits, dielectric liquids, controls, power systems. Total annual research expenditures: $285,589. *Unit head:* Dr. Susan E. Conry, Chair, 315-268-6511, Fax: 315-268-7600, E-mail: conry@clarkson.edu. *Application contact:* Dr. Philip K. Hopke, Dean of the Graduate School, 315-268-6447, Fax: 315-268-7994, E-mail: hopkepk@clarkson.edu.

Clemson University, Graduate School, College of Engineering and Science, Department of Electrical and Computer Engineering, Program in Electrical Engineering, Clemson, SC 29634. Offers M Engr, MS, PhD. *Students:* 91 full-time (10 women), 10 part-time (4 women); includes 4 minority (2 African Americans, 1 Asian American or Pacific Islander, 1 Hispanic American), 62 international. 255 applicants, 57% accepted. In 1998, 33 master's, 7 doctorates awarded. *Degree requirements:* For master's, thesis or alternative required, foreign language not required; for doctorate, dissertation, departmental qualifying exam required, foreign language not required. *Entrance requirements:* For master's, GRE General Test (MS), TOEFL; for doctorate, GRE General Test, TOEFL. *Application deadline:* For fall admission, 6/1. Application fee: $35. *Financial aid:* Fellowships, research assistantships, teaching assistantships, career-related internships or fieldwork available. Financial aid applicants required to submit FAFSA. *Faculty research:* Microelectronics, robotics, signal processing/communications, power systems, control. *Unit head:* Dr. Ernest Baxa, Coordinator, 864-656-5900, Fax: 864-656-5910, E-mail: ece-grad-program@clemson.edu. *Application contact:* Dr. Ernest Baxa, Coordinator, 864-656-5900, Fax: 864-656-5910, E-mail: ece-grad-program@clemson.edu.

Cleveland State University, College of Graduate Studies, Fenn College of Engineering, Department of Electrical and Computer Engineering, Cleveland, OH 44115-2440. Offers MS, D Eng. Part-time programs available. *Faculty:* 11 full-time (0 women). *Students:* 6 full-time (0 women), 47 part-time (5 women); includes 5 minority (1 African American, 2 Asian Americans or Pacific Islanders, 2 Hispanic Americans), 20 international. Average age 30. 99 applicants, 66% accepted. In 1998, 20 master's, 4 doctorates awarded. *Degree requirements:* For master's, exam or thesis required; for doctorate, dissertation, candidacy and qualifying exams required, foreign language not required. *Entrance requirements:* For master's, GRE General Test, GRE Subject Test, TOEFL (minimum score of 550 required), minimum GPA of 3.0; for doctorate, GRE General Test, GRE Subject Test, TOEFL, minimum GPA of 3.5. *Application deadline:* For fall admission, 7/15 (priority date). Applications are processed on a rolling basis. Application fee: $25. *Financial aid:* In 1998–99, 1 research assistantship, 3 teaching assistantships were awarded.; career-related internships or fieldwork also available. *Faculty research:* Computer networks, knowledge-based control systems, artificial intelligence, electromagnetic interference, semiconducting materials for solar cell digital communications. *Unit head:* Dr. George L. Kramerich, Chairperson, 216-687-2586, Fax: 216-687-9280.

Colorado State University, Graduate School, College of Engineering, Department of Electrical and Computer Engineering, Fort Collins, CO 80523-0015. Offers MS, PhD. *Faculty:* 17 full-time (1 woman). *Students:* 37 full-time (2 women), 28 part-time (4 women); includes 4 minority (1 African American, 1 Asian American or Pacific Islander, 2 Hispanic Americans), 40 international. Average age 29. 310 applicants, 63% accepted. In 1998, 13 master's, 15 doctorates awarded. *Degree requirements:* For master's, thesis (for some programs), final exam required, foreign language not required; for doctorate, dissertation, qualifying, preliminary, and final exams required, foreign language not required. *Entrance requirements:* For master's, GRE General Test (minimum combined score of 1100 required), TOEFL, minimum GPA of 3.0; for doctorate, GRE General Test (minimum combined score of 1100 required), TOEFL, minimum GPA of 3.5. *Application deadline:* For fall admission, 2/1 (priority date). Applications are processed on a rolling basis. Application fee: $30. Electronic applications accepted. *Financial aid:* In 1998–99, 18 research assistantships, 9 teaching assistantships were awarded.; fellowships, career-related internships or fieldwork, Federal Work-Study, institutionally-sponsored loans, and traineeships also available. *Faculty research:* Optoelectronics and microelectronics, radar remote sensing and nuclear device testing, controls. Total annual research expenditures: $3.2 million. *Unit head:* Derek L. Lile, Head, 970-491-6600, Fax: 970-491-2249, E-mail: lile@engr.colostate.edu. *Application contact:* Elisabeth L. Wadman, Graduate Coordinator, 970-491-6600, Fax: 970-491-6706, E-mail: ewadman@engr.colostate.edu.

Colorado Technical University, Graduate Studies, Program in Electrical Engineering, Colorado Springs, CO 80907-3896. Offers MSEE. Part-time and evening/weekend programs available. *Degree requirements:* For master's, thesis or alternative required, foreign language not required. *Entrance requirements:* For master's, minimum undergraduate GPA of 3.0. *Faculty research:* Electronic systems design, communication systems design.

Colorado Technical University Denver Campus, Program in Electrical Engineering, Greenwood Village, CO 80111. Offers MSEE.

Columbia University, Fu Foundation School of Engineering and Applied Science, Department of Electrical Engineering, New York, NY 10027. Offers electrical engineering (MS, Eng Sc D, PhD, EE); solid state science and engineering (MS, Eng Sc D, PhD); telecommunications (MS). PhD offered through the Graduate School of Arts and Sciences. Part-time programs available. Postbaccalaureate distance learning degree programs offered (no on-campus study). *Faculty:* 17 full-time (0 women), 13 part-time (0 women). *Students:* 141 full-time (17 women), 73 part-time (4 women). 680 applicants, 35% accepted. In 1998, 68 master's, 12 doctorates awarded. Terminal master's awarded for partial completion of doctoral program. *Degree requirements:* For master's, foreign language and thesis not required; for doctorate, dissertation, qualifying exam required, foreign language not required. *Entrance requirements:* For master's and doctorate, GRE General Test, TOEFL. *Application deadline:* For fall admission, 1/5; for spring admission, 10/1. Application fee: $55. *Financial aid:* In 1998–99, 47 research assistantships with full tuition reimbursements (averaging $17,000 per year), 11 teaching assistantships with full tuition reimbursements (averaging $15,000 per year) were awarded.; Federal Work-Study and readerships, preceptorships also available. Financial aid application deadline: 1/5; financial aid applicants required to submit FAFSA. *Faculty research:* Communications/networking, signal/image processing, microelectronic circuits/VLSI design, microelectronic devices/optoelectronics/electromagnetics/plasma physics. Total annual research expenditures: $7.5 million. *Unit head:* Chairman, 212-854-5019, Fax: 212-932-9421. *Application contact:* Marlene Mansfield, Student Coordinator, 212-854-3104, Fax: 212-932-9421, E-mail: mansfld@ee.columbia.edu.

See in-depth description on page 981.

Concordia University, School of Graduate Studies, Faculty of Engineering and Computer Science, Department of Electrical and Computer Engineering, Montréal, PQ H3G 1M8, Canada. Offers M Eng, MA Sc, PhD. *Students:* 157 full-time (29 women), 18 part-time (2 women). *Degree requirements:* For master's, computer language required, thesis optional, foreign language not required; for doctorate, computer language, dissertation, comprehensive exam required, foreign language not required. *Application deadline:* For fall admission, 6/1; for spring admission, 10/1. Application fee: $50. *Faculty research:* Computer communications and protocols, circuits and systems, graph theory, VLSI systems, microelectronics. *Unit head:* Dr. J. C. Giguere, Chair, 514-848-3104, Fax: 514-848-2802. *Application contact:* Dr. A. J. Al-Khalili, Director, 514-848-3103, Fax: 514-848-2802.

Cornell University, Graduate School, Graduate Fields of Engineering, Field of Electrical Engineering, Ithaca, NY 14853-0001. Offers computer engineering (M Eng, PhD); electrical engineering (M Eng, PhD); electrical systems (M Eng, PhD); electrophysics (M Eng, PhD). *Faculty:* 46 full-time. *Students:* 196 full-time (27 women); includes 38 minority (6 African Americans, 24 Asian Americans or Pacific Islanders, 8 Hispanic Americans), 94 international. 585 applicants, 41% accepted. In 1998, 81 master's, 20 doctorates awarded. *Degree requirements:* For doctorate, dissertation required, foreign language not required, foreign language not required. *Entrance requirements:* For master's and doctorate, GRE General Test, TOEFL (minimum score of 600 required). *Application deadline:* For fall admission, 1/15. Application fee: $65. Electronic applications accepted. *Financial aid:* In 1998–99, 123 students received aid, including 22 fellowships with full tuition reimbursements available, 64 research assistantships with full tuition reimbursements available, 37 teaching assistantships with full tuition reimbursements available; institutionally-sponsored loans, scholarships, tuition waivers (full and partial), and unspecified assistantships also available. Financial aid applicants required to submit FAFSA. *Faculty research:* Communications, information theory, signal processing, and power and control; plasma science and technology; space and upper-atmospheric science; remote sensing; solid state electronics and optoelectronics. *Unit head:* Director of Graduate Studies, 607-255-4304. *Application contact:* Graduate Field Assistant, 607-255-4304, E-mail: ee_msphd@cornell.edu.

See in-depth description on page 983.

Dalhousie University, Faculty of Graduate Studies, DalTech, Faculty of Engineering, Department of Electrical and Computer Engineering, Halifax, NS B3H 3J5, Canada. Offers M Eng, MA Sc, PhD. *Faculty:* 9 full-time (0 women), 4 part-time (0 women). *Students:* 45 full-time (7 women), 6 part-time. Average age 29. 92 applicants, 48% accepted. In 1998, 6 master's, 1 doctorate awarded (100% entered university research/teaching). *Degree requirements:* For master's and doctorate, thesis/dissertation required, foreign language not required. *Entrance requirements:* For master's and doctorate, TOEFL (minimum score of 580 required). *Application deadline:* For fall admission, 6/1; for winter admission, 10/1; for spring admission, 2/1. Applications are processed on a rolling basis. Application fee: $55. *Financial aid:* Fellowships, research assistantships, teaching assistantships, scholarships and unspecified assistantships available. *Faculty research:* Communications, computer engineering, power engineering, electronics, systems engineering. *Unit head:* Dr. Sherwin Nugent, Head, 902-494-3106, Fax: 902-422-7535, E-mail: edce@dal.ca. *Application contact:* Shelley Parker, Admissions Coordinator, Graduate Studies and Research, 902-494-1288, Fax: 902-494-3149, E-mail: shelley.parker@dal.ca.

Dartmouth College, Thayer School of Engineering, Program in Electrical Engineering, Hanover, NH 03755. Offers MS, PhD. *Degree requirements:* For master's, thesis required; for doctorate, dissertation, candidacy oral exam required. *Entrance requirements:* For master's and doctorate, GRE General Test. *Application deadline:* For fall admission, 1/15 (priority date). Application fee: $20 ($40 for international students). *Financial aid:* Career-related internships or fieldwork, Federal Work-Study, institutionally-sponsored loans, and tuition waivers (full and partial) available. Financial aid application deadline: 1/15. *Faculty research:* Power electronics and microengineering, image and signal processing, optics, lasers and optoelectronics, electromagnetic fields, waves and antennas, space and ionospheric physics. Total annual research expenditures: $1.1 million. *Unit head:* Dr. Ernest Baxa, Coordinator, Program in Computer Engineering, 864-656-5900, Fax: 864-656-5910, E-mail: ece-grad-program@clemson.edu. *Application contact:* Candace S. Potter, Admissions Coordinator, 603-646-3844, Fax: 603-646-3856, E-mail: candace.potter@dartmouth.edu.

Drexel University, Graduate School, College of Engineering, Department of Electrical and Computer Engineering, Philadelphia, PA 19104-2875. Offers electrical and computer engineering (PhD); electrical engineering (MSEE); telecommunications engineering (MSEE). Part-time and evening/weekend programs available. *Faculty:* 29 full-time, 6 part-time. *Students:* 85 full-time (19 women), 101 part-time (12 women); includes 17 minority (4 African Americans, 10 Asian Americans or Pacific Islanders, 3 Hispanic Americans), 83 international. Average age 32. 570 applicants, 63% accepted. In 1998, 48 master's, 9 doctorates awarded. Terminal master's awarded for partial completion of doctoral program. *Degree requirements:* For master's, thesis required (for some programs), foreign language not required; for doctorate, dissertation required, foreign language not required. *Entrance requirements:* For master's, TOEFL (minimum score of 570 required), minimum GPA of 3.0, BS in electrical engineering or physics; for doctorate, TOEFL (minimum score of 570 required), minimum GPA of 3.5, MS in electrical engineering. *Application deadline:* For fall admission, 8/21. Applications are processed on a rolling basis. Application fee: $35. Tuition: Full-time $15,795; part-time $585 per credit. Required fees: $375; $67 per term. Tuition and fees vary according to program. *Financial aid:* In 1998–99, 21 research assistantships, 26 teaching assistantships were awarded.; career-related internships or fieldwork and unspecified assistantships also available. Financial aid application deadline: 2/1. *Faculty research:* Power systems planning and control, semiconductors, photovoltaics, electronic systems, image processing. *Unit head:* Dr. Nihat M. Bilgutay, Head, 215-895-6806, Fax: 215-895-1695, E-mail: bilgutay@ece.drexel.edu. *Application contact:* Dr. Mohana Shankar, Graduate Adviser, 215-895-6632.

See in-depth description on page 985.

Duke University, Graduate School, School of Engineering, Department of Electrical and Computer Engineering, Durham, NC 27708-0586. Offers MS, PhD. Part-time programs available. *Faculty:* 26 full-time, 13 part-time. *Students:* 80 full-time, 1 part-time; includes 3 minority (1 Asian American or Pacific Islander, 2 Hispanic Americans), 52 international. 216 applicants, 20% accepted. In 1998, 17 master's, 6 doctorates awarded. Terminal master's awarded for partial completion of doctoral program. *Degree requirements:* For doctorate, dissertation required, foreign language not required, foreign language not required. *Entrance requirements:* For master's and doctorate, GRE General Test. *Application deadline:* For fall admission, 12/31; for spring admission, 11/1. Application fee: $75. *Financial aid:* Fellowships, research assistantships, Federal Work-Study available. Financial aid application deadline: 12/31. *Unit head:* Hisham Massoud, Director of Graduate Studies, 919-660-5245, Fax: 919-660-5293, E-mail: steph@ee.duke.edu.

See in-depth description on page 987.

École Polytechnique de Montréal, Graduate Programs, Department of Electrical and Computer Engineering, Montréal, PQ H3C 3A7, Canada. Offers automation (M Eng, M Sc A, PhD); computer science (M Eng, M Sc A, PhD); electrotechnology (M Eng, M Sc A, PhD); microelectronics (M Eng, M Sc A, PhD); microwave technology (M Eng, M Sc A, PhD). Part-time and evening/weekend programs available. *Degree requirements:* For master's and doctorate, one foreign language, computer language, thesis/dissertation required. *Entrance requirements:* For master's, minimum GPA of 2.75; for doctorate, minimum GPA of 3.0. *Faculty research:* Microwaves, telecommunications, software engineering.

Fairleigh Dickinson University, Teaneck–Hackensack Campus, University College: Arts, Sciences, and Professional Studies, School of Engineering and Engineering Technology, Department of Electrical Engineering, Teaneck, NJ 07666-1914. Offers MSEE. *Degree requirements:* For master's, thesis optional, foreign language not required. *Entrance requirements:* For master's, GRE General Test. *Faculty research:* Adaptive signal processing, telecommunications, computer engineering, digital image processing, photonics, neural networks, fuzzy logic.

Florida Agricultural and Mechanical University, Division of Graduate Studies, Research, and Continuing Education, FAMU-FSU College of Engineering, Department of Electrical Engineering, Tallahassee, FL 32307-3200. Offers MS, PhD. *Students:* 11 (1 woman); includes 6 minority (all African Americans) 5 international. In 1998, 1 master's awarded. *Entrance requirements:* For master's, GRE General Test (minimum combined score of 1000 required), minimum GPA of 3.0. *Application deadline:* For fall admission, 7/1. Application fee: $20. *Unit head:* Dr. C. J. Chen, Dean, FAMU-FSU College of Engineering, 850-487-6100, Fax: 850-487-6486.

Florida Atlantic University, College of Engineering, Department of Electrical Engineering, Boca Raton, FL 33431-0991. Offers MS, PhD. Part-time and evening/weekend programs available. Postbaccalaureate distance learning degree programs offered (minimal on-campus study). *Faculty:* 18 full-time (3 women), 2 part-time (0 women). *Students:* 27 full-time (8 women), 35 part-time (6 women); includes 13 minority (3 African Americans, 2 Asian Americans or Pacific Islanders, 8 Hispanic Americans), 27 international. Average age 33. In 1998, 16 master's, 8 doctorates awarded. Terminal master's awarded for partial completion of doctoral program. *Degree requirements:* For master's, thesis optional, foreign language not required; for doctorate, dissertation, qualifying exam required, foreign language not required. *Entrance requirements:* For master's and doctorate, GRE General Test (minimum combined score of 1000 required), TOEFL (minimum score of 550 required), minimum GPA of 3.0. *Application deadline:* For fall admission, 4/10 (priority date); for spring admission, 10/1. Applications are processed on a rolling basis. Application fee: $20. Tuition, state resident: part-time $148 per credit hour. Tuition, nonresident: part-time $509 per credit hour. *Financial aid:* Fellowships, research assistantships, teaching assistantships, career-related internships or fieldwork, Federal Work-Study, and unspecified assistantships available. Aid available to part-time students. Financial aid application deadline: 4/1; financial aid applicants required to submit FAFSA. *Faculty research:* Telecommunications, signal processing, imaging systems, electromagnetics, controls. Total annual research expenditures: $1.7 million. *Unit head:* Dr. Salvatore D. Morgera, Chairman, 561-297-3412, Fax: 561-297-2336, E-mail: smorgera@fau.edu. *Application contact:* Patricia Capozziello, Graduate Admissions Coordinator, 561-297-2694, Fax: 561-297-2659, E-mail: capozzie@fau.edu.

Florida Institute of Technology, Graduate School, College of Engineering, Division of Electrical and Computer Science and Engineering, Program in Electrical Engineering, Melbourne, FL 32901-6975. Offers MS, PhD. Part-time and evening/weekend programs available. *Faculty:* 6 full-time (0 women), 3 part-time (0 women). *Students:* 25 full-time (3 women), 72 part-time (8 women); includes 15 minority (2 African Americans, 6 Asian Americans or Pacific Islanders, 7 Hispanic Americans), 37 international. Average age 32. 157 applicants, 60% accepted. In 1998, 15 master's, 6 doctorates awarded. *Degree requirements:* For master's, thesis optional, foreign language not required; for doctorate, dissertation, comprehensive exam required. *Entrance requirements:* For master's, minimum GPA of 3.0; for doctorate, minimum GPA of 3.2. *Application deadline:* Applications are processed on a rolling basis. Application fee: $50. Electronic applications accepted. Tuition: Part-time $575 per credit hour. Required fees: $100. Tuition and fees vary according to campus/location and program. *Financial aid:* In 1998–99, 22 students received aid, including 8 research assistantships (averaging $5,302 per year), 12 teaching assistantships (averaging $4,059 per year); tuition remissions also available. Financial aid application deadline: 3/1; financial aid applicants required to submit FAFSA. *Faculty research:* Electrooptics, electromagnetics, microelectronics, signals and controls, neural network applications. Total annual research expenditures: $541,840. *Unit head:* Dr. Barry G. Grossman, Chair, 407-674-7429, Fax: 407-674-8192, E-mail: grossman@ee.fit.edu. *Application contact:* Carolyn P. Farrior, Associate Dean of Graduate Admissions, 407-674-7118, Fax: 407-723-9468, E-mail: cfarrior@fit.edu.

See in-depth description on page 989.

Florida Institute of Technology, Graduate School, School of Extended Graduate Studies, Program in Electrical Engineering, Melbourne, FL 32901-6975. Offers MS. Part-time and evening/weekend programs available. Average age 31. 3 applicants, 33% accepted. In 1998, 8 degrees awarded (100% found work related to degree). *Degree requirements:* For master's, thesis optional, foreign language not required. *Entrance requirements:* For master's, minimum GPA of 3.0. *Average time to degree:* Master's–1 year full-time, 3 years part-time. *Application deadline:* Applications are processed on a rolling basis. Application fee: $50. Electronic applications accepted. Tuition: Part-time $270 per credit hour. Part-time tuition and fees vary according to campus/location. *Financial aid:* Application deadline: 3/1; *Unit head:* James W. Smolka, Coordinator, 805-258-5936. *Application contact:* Carolyn P. Farrior, Associate Dean of Graduate Admissions, 407-674-7118, Fax: 407-723-9468, E-mail: cfarrior@fit.edu.

Florida International University, College of Engineering, Department of Electrical Engineering, Program in Electrical Engineering, Miami, FL 33199. Offers MS, PhD. Part-time and evening/weekend programs available. *Students:* 17 full-time (4 women), 40 part-time (3 women); includes 32 minority (1 African American, 2 Asian Americans or Pacific Islanders, 29 Hispanic Americans), 15 international. Average age 32. 78 applicants, 37% accepted. In 1998, 13 master's, 2 doctorates awarded. *Degree requirements:* For master's, thesis optional; for doctorate, dissertation required. *Entrance requirements:* For master's, GRE General Test (minimum combined score of 1000 required), TOEFL (minimum score of 550 required); for doctorate, GRE General Test (minimum combined score of 1000 required), TOEFL (minimum score of 550 required), minimum graduate GPA of 3.3. *Application deadline:* For fall admission, 4/1 (priority date); for spring admission, 10/1. Applications are processed on a rolling basis. Application fee: $20. Tuition, state resident: part-time $145 per credit hour. Tuition, nonresident: part-time $506 per credit hour. Required fees: $158; $158 per year. *Unit head:* Dr. Malek Adjouadi, Acting Chairperson, Department of Electrical Engineering, 305-348-2807, Fax: 305-348-3707, E-mail: makel@vision.fiu.edu.

Florida State University, Graduate Studies, FAMU/FSU College of Engineering, Department of Electrical Engineering, Tallahassee, FL 32306. Offers MS, PhD. Part-time programs available. *Faculty:* 17 full-time (2 women), 2 part-time (0 women). *Students:* 44 full-time (11 women), 10 part-time (2 women); includes 17 minority (8 African Americans, 6 Asian Americans or Pacific Islanders, 3 Hispanic Americans), 9 international. Average age 26. 500 applicants, 14% accepted. In 1998, 20 master's, 1 doctorate awarded. *Degree requirements:* For master's, thesis optional, foreign language not required; for doctorate, dissertation, preliminary exam, qualifying exam required, foreign language not required. *Entrance requirements:* For master's, GRE General Test (minimum combined score of 1000 required), TOEFL (minimum score of 550 required), minimum GPA of 3.0, BS in electrical engineering; for doctorate, GRE General Test (minimum combined score of 1100 required), TOEFL (minimum score of 550 required), minimum GPA of 3.5 (graduate), MS in electrical engineering. *Average time to degree:* Master's–2 years full-time, 3.5 years part-time; doctorate–3.5 years full-time. *Application deadline:* For fall admission, 7/15; for spring admission, 11/23. Applications are processed on a rolling basis. Application fee: $20. Tuition, state resident: part-time $139 per credit hour. Tuition, nonresident: part-time $482 per credit hour. Tuition and fees vary according to program. *Financial aid:* In 1998–99, 1 fellowship, 8 research assistantships, 11 teaching assistantships were awarded.; career-related internships or fieldwork and institutionally-sponsored loans also available. Financial aid application deadline: 6/15. *Faculty research:* Electromagnetics, digital signal processing, computer systems, image processing, laser optics. Total annual research expenditures:$692,800. *Unit head:* Dr. Reginald J. Perry, Chair, 850-410-6465, Fax: 850-410-6479, E-mail: perry@eng.fsu.edu. *Application contact:* Adrianne F. Humes, Graduate Studies Assistant, 850-487-6454, Fax: 850-487-6479, E-mail: humes@eng.fsu.edu.

Gannon University, School of Graduate Studies, College of Sciences, Engineering, and Health Sciences, School of Sciences and Engineering, Program in Engineering, Erie, PA 16541-0001. Offers electrical engineering (MS); embedded software engineering (MS); mechanical engineering (MS). Part-time and evening/weekend programs available. *Students:* 23 full-time (4 women), 29 part-time (5 women); includes 1 minority (Asian American or Pacific Islander), 13 international. *Degree requirements:* For master's, thesis or alternative, comprehensive exam required. *Entrance requirements:* For master's, GRE Subject Test, bachelor's degree in engineering, minimum QPA of 2.5. *Application deadline:* Applications are processed on a rolling basis. Application fee: $25. *Unit head:* Dr. Mehmet Cultu, Co-Director, 814-871-7624. *Application contact:* Beth Nemenz, Director of Admissions, 814-871-7240, Fax: 814-871-5803, E-mail: admissions@gannon.edu.

George Mason University, School of Information Technology and Engineering, Department of Electrical and Computer Engineering, Fairfax, VA 22030-4444. Offers electrical engineering (MS). Part-time and evening/weekend programs available. *Faculty:* 21 full-time (1 woman), 9 part-time (0 women). *Students:* 18 full-time (4 women), 252 part-time (45 women); includes 72 minority (19 African Americans, 45 Asian Americans or Pacific Islanders, 8 Hispanic Americans), 34 international. Average age 31. 177 applicants, 84% accepted. In 1998, 45 degrees awarded. *Degree requirements:* For master's, computer language required, thesis optional, foreign language not required. *Entrance requirements:* For master's, GMAT or GRE General Test, TOEFL (minimum score of 575 required), bachelor's degree in electrical engineering or related field, minimum GPA of 3.0 in last 60 hours. *Application deadline:* For fall admission, 5/1; for spring admission, 11/1. Application fee: $30. Electronic applications accepted. Tuition, state resident: full-time $4,416; part-time $184 per credit hour. Tuition, nonresident: full-time $12,516; part-time $522 per credit hour. Tuition and fees vary according to program. *Financial aid:* Fellowships, research assistantships, teaching assistantships, career-related internships or fieldwork and Federal Work-Study available. Aid available to part-time students. Financial aid application deadline: 3/1; financial aid applicants required to submit FAFSA. *Faculty research:* Communication networks, signal processing, system failure diagnosis, multiprocessors, material processing using microwave energy. *Unit head:* Andre Manitius, Chairperson, 703-993-1570, Fax: 703-993-1601, E-mail: ece@bass.gmu.edu.

The George Washington University, School of Engineering and Applied Science, Department of Electrical and Computer Engineering, Washington, DC 20052. Offers MS, D Sc, App Sc, Engr, MEM/MS. Part-time and evening/weekend programs available. *Degree requirements:* For master's, thesis optional, foreign language not required; for doctorate and other advanced degree, dissertation defense, qualifying exam required; for other advanced degree, foreign language and thesis not required. *Entrance requirements:* For master's, TOEFL (minimum score of 550 required; average 580) or George Washington University English as a Foreign Language Test, appropriate bachelor's degree, minimum GPA of 3.0; for doctorate, TOEFL (minimum score of 550 required; average 580) or George Washington University English as a Foreign Language Test, appropriate bachelor's or master's degree, minimum GPA of 3.3, GRE required if highest earned degree is BS; for other advanced degree, TOEFL (minimum score of 550 required; average 580) or George Washington University English as a Foreign Language Test, appropriate master's degree, minimum GPA of 3.0. *Application deadline:* For fall admission, 3/1 (priority date); for spring admission, 10/1. Applications are processed on a rolling basis. Application fee: $55. Tuition: Full-time $17,328; part-time $722 per credit hour. Required fees: $828; $35 per credit hour. Tuition and fees vary according to campus/location and program. *Financial aid:* Fellowships, research assistantships, teaching assistantships, career-related internships or fieldwork and institutionally-sponsored loans available. Financial aid application deadline: 3/1; financial aid applicants required to submit FAFSA. *Faculty research:* Computer graphics, multimedia systems. *Unit head:* Dr. Murray Loew, Chair, 202-994-6083, Fax: 202-994-0227, E-mail: eecs@seas.gwu.edu. *Application contact:* Howard M. Davis, Manager, Office of Admissions and Student Records, 202-994-6158, Fax: 202-994-0909, E-mail: data:adms@seas.gwu.edu.

See in-depth description on page 991.

Georgia Institute of Technology, Graduate Studies and Research, College of Engineering, School of Electrical and Computer Engineering, Atlanta, GA 30332-0001. Offers MS, MSEE, PhD. Part-time programs available. Postbaccalaureate distance learning degree programs offered. Terminal master's awarded for partial completion of doctoral program. *Degree requirements:* For master's, thesis optional, foreign language not required; for doctorate, dissertation required, foreign language not required. *Entrance requirements:* For master's, GRE General Test (minimum score of 500 on verbal section, 700 on quantitative and analytical required), TOEFL (minimum score of 550 required), minimum GPA of 3.0; for doctorate, GRE General Test (minimum score of 500 on verbal section, 700 on quantitative and analytical required), TOEFL (minimum score of 550 required), minimum GPA of 3.5. *Faculty research:* Telecommunications, computer systems, microelectronics, optical engineering, digital signal processing.

Gonzaga University, Graduate School, School of Engineering, Program in Electrical Engineering, Spokane, WA 99258. Offers MSEE. *Degree requirements:* For master's, project required, foreign language and thesis not required. *Entrance requirements:* For master's, GRE General Test, TOEFL (minimum score of 550 required), minimum GPA of 3.0 during previous 2 years.

Graduate School and University Center of the City University of New York, Graduate Studies, Program in Engineering, New York, NY 10036-8099. Offers chemical engineering (PhD); civil engineering (PhD); electrical engineering (PhD); mechanical engineering (PhD). *Faculty:* 68 full-time (1 woman). *Students:* 105 full-time (16 women), 11 part-time (2 women); includes 12 African Americans, 5 Asian Americans or Pacific Islanders, 4 Hispanic Americans *Degree requirements:* For doctorate, dissertation required, dissertation required. *Entrance requirements:* For doctorate, GRE General Test. *Application deadline:* For fall admission, 4/15. Application fee: $40. *Unit head:* Dr. Mumtaz Kassir, Acting Executive Officer, 212-650-8030.

Howard University, College of Engineering, Architecture, and Computer Sciences, School of Engineering and Computer Science, Department of Electrical Engineering, Washington, DC 20059-0002. Offers M Eng, PhD. Offered through the Graduate School of Arts and Sciences. Part-time programs available. *Faculty:* 15 full-time (0 women), 1 part-time (0 women). *Students:* 31 full-time (2 women), 1 part-time; all minorities (all African Americans) Average age 25. 6 applicants, 67% accepted. In 1998, 10 master's, 2 doctorates awarded. *Degree requirements:* For master's, computer language, qualifying exam required, thesis optional, foreign language not required; for doctorate, one foreign language, computer language, dissertation, preliminary exam required. *Entrance requirements:* For master's, GRE General Test, TOEFL, bachelor's degree in electrical engineering, minimum GPA of 3.0; for doctorate, GRE General Test, TOEFL, minimum GPA of 3.0. *Application deadline:* For fall admission, 4/1; for spring admission, 11/1. Application fee: $45. Electronic applications accepted. *Financial aid:* In 1998–99, 15 research assistantships with full tuition reimbursements (averaging $8,000 per year), 2 teaching assistantships with full tuition reimbursements (averaging $8,000 per year) were awarded.; fellowships with full tuition reimbursements, career-related internships or fieldwork, grants, and institutionally-sponsored loans also available. Financial aid application deadline: 4/1. *Faculty research:* Solid-state electronics, antennas and microwaves, communications and signal processing, controls and powersystems, applied electromagnetics. Total annual research expenditures: $3.4 million. *Unit head:* Dr. James A. Momoh, Chair, 202-806-6585, Fax: 202-806-5288, E-mail: jm@scs.howard.edu. *Application contact:* Dr. Mohamed F. Chouikha, Graduate Studies Chairman, 202-806-4816, Fax: 202-805-5258, E-mail: cm@scs.howard.edu.

See in-depth description on page 993.

Illinois Institute of Technology, Graduate College, Armour College of Engineering and Sciences, Department of Electrical and Computer Engineering, Chicago, IL 60616-3793. Offers computer systems engineering (MS); electrical and computer engineering (MECE); electrical engineering (MS, PhD); manufacturing engineering (MME, MS). Part-time and evening/weekend programs available. *Faculty:* 21 full-time (1 woman), 11 part-time (0 women). *Students:* 76 full-time (10 women), 258 part-time (38 women); includes 83 minority (17 African Americans, 50 Asian Americans or Pacific Islanders, 15 Hispanic Americans, 1 Native American), 118 international. 759 applicants, 46% accepted. In 1998, 86 master's, 7 doctorates awarded. Terminal master's awarded for partial completion of doctoral program. *Degree requirements:* For master's, thesis (for some programs), comprehensive exam required, foreign language not required; for doctorate, dissertation, comprehensive exam required, foreign language not required. *Entrance requirements:* For master's, GRE (minimim score of 1200 required), TOEFL (minimum score of 550 required), undergraduate GPA of 3.0 required; for doctorate,

Electrical Engineering

Illinois Institute of Technology (continued)

GRE (minimum score of 1200 required), TOEFL (minimum score of 550 required), undergraduate GPA of 3.0 required. *Application deadline:* For fall admission, 7/1; for spring admission, 12/1. Applications are processed on a rolling basis. Application fee: $30. Electronic applications accepted. *Financial aid:* In 1998–99, 1 fellowship, 15 research assistantships, 19 teaching assistantships were awarded.; Federal Work-Study, institutionally-sponsored loans, scholarships, and graduate assistantships also available. Financial aid application deadline:3/1. *Faculty research:* Computer system design, photonics, biomedical engineering, power systems analysis, communications theory. Total annual research expenditures: $766,272. *Unit head:* Dr. D. Ucci, Interim Chairman, 312-567-3402, Fax: 312-567-8976, E-mail: du@ece.iit.edu. *Application contact:* Dr. S. Mohammad Shahidehpour, Dean of Graduate College, 312-567-3024, Fax: 312-567-7517, E-mail: grad@minna.cns.iit.edu.

See in-depth description on page 995.

Indiana University–Purdue University Indianapolis, School of Engineering and Technology, Department of Electrical Engineering, Indianapolis, IN 46202-2896. Offers biomedical engineering (MS Bm E, PhD); electrical engineering (MSEE). *Students:* 13 full-time (7 women), 16 part-time (3 women); includes 1 minority (Asian American or Pacific Islander), 9 international. Average age 27. 34 applicants, 29% accepted. In 1998, 3 degrees awarded. *Degree requirements:* For master's, thesis optional, foreign language not required. *Entrance requirements:* For master's, GRE, TOEFL (minimum score of 550 required), minimum B average. *Application deadline:* For fall admission, 7/1. Application fee: $25 ($50 for international students). Tuition, state resident: part-time $171 per credit hour. Tuition, nonresident: part-time $490 per credit hour. Required fees: $121 per year. *Financial aid:* Fellowships, research assistantships, Federal Work-Study and tuition waivers (full and partial) available. Financial aid application deadline: 3/1. *Faculty research:* Control and automation, signal processing, robotics, medical imaging, neural networks. Total annual research expenditures: $172,654. *Unit head:* Dr. Edward Berbari, Chair, 317-274-9721. *Application contact:* Toni Giffin, Department Secretary, 317-274-9726, Fax: 317-274-4493, E-mail: ee@engr.iupui.edu.

Instituto Tecnológico y de Estudios Superiores de Monterrey, Campus Chihuahua, Graduate Programs, Chihuahua, 31110, Mexico. Offers computer systems engineering (Ingeniero); electrical engineering (Ingeniero); electromechanical engineering (Ingeniero); electronic engineering (Ingeniero); engineering administration (MEA); industrial engineering (MIE, Ingeniero); international trade (MIT); mechanical engineering (Ingeniero).

Instituto Tecnológico y de Estudios Superiores de Monterrey, Campus Monterrey, Graduate and Research Division, Programs in Engineering, Monterrey, 64849, Mexico. Offers applied statistics (M Eng); artificial intelligence (PhD); automation engineering (M Eng); chemical engineering (M Eng); civil engineering (M Eng); electrical engineering (M Eng); electronic engineering (M Eng); environmental engineering (M Eng); industrial engineering (M Eng, PhD); manufacturing engineering (M Eng); mechanical engineering (M Eng); systems and quality engineering (M Eng). M Eng offered jointly with the University of Waterloo; PhD (industrial engineering) offered jointly with Texas A&M University. Part-time and evening/weekend programs available. Terminal master's awarded for partial completion of doctoral program. *Degree requirements:* For master's and doctorate, one foreign language, computer language, thesis/dissertation required. *Entrance requirements:* For master's, PAEG, TOEFL; for doctorate, GRE, TOEFL, master's in related field. *Faculty research:* Flexible manufacturing cells, materials, statistical methods, environmental prevention, control and evaluation.

Iowa State University of Science and Technology, Graduate College, College of Engineering, Department of Electrical and Computer Engineering, Ames, IA 50011. Offers computer engineering (MS, PhD); electrical engineering (MS, PhD). *Faculty:* 43 full-time, 3 part-time. *Students:* 161 full-time (37 women), 63 part-time (7 women); includes 7 minority (4 African Americans, 2 Asian Americans or Pacific Islanders, 1 Hispanic American), 165 international. 776 applicants, 13% accepted. In 1998, 65 master's, 19 doctorates awarded. *Degree requirements:* For master's, thesis or alternative required, foreign language not required; for doctorate, dissertation required, foreign language not required. *Entrance requirements:* For master's and doctorate, GRE General Test, TOEFL (minimum score of 570 required). *Application deadline:* For fall admission, 1/15 (priority date); for spring admission, 9/15. Application fee: $20 ($50 for international students). Electronic applications accepted. Tuition, state resident: full-time $3,308. Tuition, nonresident: full-time $9,744. Part-time tuition and fees vary according to course load, campus/location and program. *Financial aid:* In 1998–99, 127 research assistantships with partial tuition reimbursements (averaging $11,868 per year), 46 teaching assistantships with partial tuition reimbursements (averaging $12,018 per year) were awarded.; fellowships, scholarships also available. *Unit head:* Dr. S. S. Venkata, Chair, 515-294-2667, E-mail: ecegrad@ee.iastate.edu. *Application contact:* Dr. Satish Udpa, Director of Graduate Education, 515-294-2667, E-mail: ecegrad@ee.iastate.edu.

Johns Hopkins University, G. W. C. Whiting School of Engineering, Department of Electrical and Computer Engineering, Baltimore, MD 21218-2699. Offers MSE, PhD. Part-time programs available. *Faculty:* 16 full-time (0 women), 8 part-time (0 women). *Students:* 70 full-time (11 women), 16 part-time (3 women); includes 4 minority (1 African American, 3 Asian Americans or Pacific Islanders), 39 international. Average age 25. 274 applicants, 34% accepted. In 1998, 8 master's awarded (100% found work related to degree); 7 doctorates awarded. Terminal master's awarded for partial completion of doctoral program. *Degree requirements:* For master's, foreign language and thesis not required; for doctorate, dissertation required, foreign language not required. *Entrance requirements:* For master's, GRE General Test, TOEFL (minimum score of 560 required); for doctorate, GRE General Test, TOEFL. *Application deadline:* For fall admission, 1/1. Applications are processed on a rolling basis. Application fee: $50. Tuition: Full-time $23,660. Tuition and fees vary according to program. *Financial aid:* In 1998–99, 5 fellowships, 27 research assistantships, 20 teaching assistantships were awarded.; Federal Work-Study, grants, and institutionally-sponsored loans also available. Financial aid application deadline: 1/1. *Faculty research:* Communications systems, quantum electronics and optics, signal and image processing, VLSI, microwaves. Total annual research expenditures: $3.2 million. *Unit head:* Dr. Frederic M. Davidson, Chair, 410-516-7007, Fax: 410-516-5566, E-mail: fmd@light.ece.jhu.edu. *Application contact:* Gail M. O'Connor, Academic Coordinator II, 410-516-4808, Fax: 410-516-5566, E-mail: gradadm@ece.jhu.edu.

See in-depth description on page 997.

Kansas State University, Graduate School, College of Engineering, Department of Electrical and Computer Engineering, Manhattan, KS 66506. Offers bioengineering (MS, PhD); communications (MS, PhD); computer engineering (MS, PhD); control systems (MS, PhD); electric energy systems (MS, PhD); instrumentation (MS, PhD); signal processing (MS, PhD). Postbaccalaureate distance learning degree programs offered (no on-campus study). *Faculty:* 21 full-time (3 women). *Students:* 26 full-time (2 women), 24 part-time (2 women), 18 international. 163 applicants, 23% accepted. In 1998, 11 master's, 1 doctorate awarded (100% found work related to degree). *Degree requirements:* For master's, thesis optional; for doctorate, dissertation required. *Entrance requirements:* For master's, GRE General Test (minimum score of 400 on verbal section, 600 on quantitative, 600 on analytical required); for doctorate, GRE General Test (minimum score of 400 on verbal section, 600 on quantitative required). *Average time to degree:* Master's–1.5 years full-time; doctorate–4 years full-time. *Application deadline:* For fall admission, 3/1; for spring admission, 9/1. Applications are processed on a rolling basis. Application fee: $0 ($25 for international students). Electronic applications accepted. *Financial aid:* In 1998–99, 17 research assistantships with partial tuition reimbursements (averaging $9,900 per year), 7 teaching assistantships with full tuition reimbursements (averaging $9,900 per year) were awarded.; career-related internships or fieldwork also available. *Faculty research:* Digital signal processing, communications systems. Total annual research expenditures: $910,000. *Unit head:* Dr. David Soldan, Head, 785-532-5600, E-mail: grad@eece.ksu.edu.

Lamar University, College of Graduate Studies, College of Engineering, Department of Electrical Engineering, Beaumont, TX 77710. Offers ME, MES, DE. Part-time programs available. *Faculty:* 6 full-time (1 woman). *Students:* 5 full-time (1 woman), 2 part-time, (all international). 250 applicants, 40% accepted. In 1998, 5 degrees awarded (20% entered university research/teaching, 80% found other work related to degree). *Degree requirements:* For master's, thesis required (for some programs), foreign language not required; for doctorate, computer language, dissertation required, foreign language not required. *Entrance requirements:* For master's, GRE General Test (minimum combined score of 1100 required), TOEFL (minimum score of 575 required); for doctorate, GRE General Test (minimum combined score of 1300 on three sections required), TOEFL (minimum score of 600 required). *Average time to degree:* Master's–2 years full-time, 4 years part-time; doctorate–3.5 years full-time, 6 years part-time. *Application deadline:* For fall admission, 5/15 (priority date); for spring admission, 10/1 (priority date). Applications are processed on a rolling basis. Application fee: $0. *Financial aid:* In 1998–99, 2 fellowships with partial tuition reimbursements, 1 research assistantship with partial tuition reimbursement, 3 teaching assistantships with partial tuition reimbursements were awarded. Financial aid application deadline: 4/1. *Unit head:* Dr. Bernard Maxum, Chair, 409-880-8746, Fax: 409-880-8121, E-mail: maxumbj@hal.lamar.edu.

Lehigh University, College of Engineering and Applied Science, Department of Electrical Engineering, Computer Science and Computer Engineering, Program in Electrical Engineering, Bethlehem, PA 18015-3094. Offers M Eng, MS, PhD. Part-time programs available. *Faculty:* 14 full-time (1 woman). *Students:* 73 full-time (13 women), 37 part-time (5 women); includes 13 minority (2 African Americans, 11 Asian Americans or Pacific Islanders), 61 international. Average age 24. 208 applicants, 8% accepted. In 1998, 9 master's, 4 doctorates awarded. *Degree requirements:* For master's, oral presentation of thesis required; for doctorate, dissertation, qualifying, general, and oral exams required. *Entrance requirements:* For master's, GRE General Test (minimum combined score of 1600 on three sections required), TOEFL (minimum score of 550 required), minimum GPA of 3.0; for doctorate, GRE General Test (minimum combined score of 1600 on three sections required), TOEFL (minimum score of 550 required), MS, minimum GPA of 3.25. *Application deadline:* For fall admission, 7/15; for spring admission, 12/1. Applications are processed on a rolling basis. Application fee: $40. Electronic applications accepted. *Financial aid:* Fellowships, research assistantships, teaching assistantships available. Financial aid application deadline: 1/15. *Unit head:* Anne Nierer, Graduate Coordinator, 610-758-4072, Fax: 610-758-6279, E-mail: aln3@lehigh.edu. *Application contact:* Anne Nierer, Graduate Coordinator, 610-758-4072, Fax: 610-758-6279, E-mail: aln3@lehigh.edu.

Louisiana State University and Agricultural and Mechanical College, Graduate School, College of Engineering, Department of Electrical and Computer Engineering, Baton Rouge, LA 70803. Offers MSEE, PhD. *Faculty:* 25 full-time (2 women). *Students:* 79 full-time (15 women), 13 part-time (2 women); includes 6 minority (2 African Americans, 4 Asian Americans or Pacific Islanders), 75 international. Average age 27. 260 applicants, 35% accepted. In 1998, 23 master's, 4 doctorates awarded. Terminal master's awarded for partial completion of doctoral program. *Degree requirements:* For master's, thesis optional, foreign language not required; for doctorate, dissertation required, foreign language not required. *Entrance requirements:* For master's, GRE General Test, TOEFL (minimum score of 550 required), minimum GPA of 3.0; for doctorate, GRE General Test, TOEFL (minimum score of 550 required), minimum GPA of 3.5. *Application deadline:* For fall admission, 1/25 (priority date). Applications are processed on a rolling basis. Application fee: $25. *Financial aid:* In 1998–99, 3 fellowships, 31 research assistantships with partial tuition reimbursements, 20 teaching assistantships with partial tuition reimbursements were awarded.; institutionally-sponsored loans and unspecified assistantships also available. Financial aid application deadline: 2/28. *Faculty research:* Electronics, power engineering, systems and signal processing, communications. *Unit head:* Dr. Alan H. Marshak, Chair, 225-388-5241, Fax: 225-388-5200. *Application contact:* Dr. Jorge L. Aravena, Graduate Adviser, 225-388-5478, E-mail: aravena@ee.lsu.edu.

See in-depth description on page 999.

Louisiana Tech University, Graduate School, College of Engineering and Science, Department of Electrical Engineering, Ruston, LA 71272. Offers MS, D Eng. Part-time programs available. Terminal master's awarded for partial completion of doctoral program. *Degree requirements:* For master's and doctorate, thesis/dissertation required, foreign language not required. *Entrance requirements:* For master's, GRE General Test (minimum combined score of 1070 required), TOEFL (minimum score of 550 required), minimum GPA of 3.0 in last 60 hours; for doctorate, TOEFL (minimum score of 550 required), minimum graduate GPA of 3.25 (with MS) or GRE General Test (minimum combined score of 1270 required without MS). *Faculty research:* Communications, computers and microprocessors, electrical and power systems, pattern recognition, robotics.

Loyola Marymount University, Graduate Division, College of Science and Engineering, Department of Electrical Engineering and Computer Science, Program in Electrical Engineering, Los Angeles, CA 90045-8350. Offers MSE. Part-time and evening/weekend programs available. *Faculty:* 8 full-time (1 woman), 2 part-time (0 women). *Students:* 9 full-time (0 women), 5 part-time (1 woman); includes 9 minority (2 African Americans, 4 Asian Americans or Pacific Islanders, 3 Hispanic Americans), 1 international. 18 applicants, 33% accepted. In 1998, 1 degree awarded. *Degree requirements:* For master's, computer language, research seminar required, foreign language and thesis not required. *Entrance requirements:* For master's, TOEFL (minimum score of 550 required). Application fee: $35. Electronic applications accepted. Tuition: Part-time $525 per unit. Required fees: $143; $14 per semester. Tuition and fees vary according to program. *Financial aid:* In 1998–99, 2 students received aid. Scholarships available. Aid available to part-time students. Financial aid application deadline: 3/2; financial aid applicants required to submit FAFSA. *Unit head:* Dr. Paul A. Rude, Graduate Director, 310-338-5101. *Application contact:* Dr. Paul A. Rude, Graduate Director, 310-338-5101.

Manhattan College, Graduate Division, School of Engineering, Program in Electrical Engineering, Riverdale, NY 10471. Offers MS. Part-time and evening/weekend programs available. *Degree requirements:* For master's, computer language, thesis or alternative required, foreign language not required. *Entrance requirements:* For master's, GRE, TOEFL, minimum GPA of 3.0. *Faculty research:* Multimedia tools, neural networks, robotic control systems, magnetic resonance imaging, telemedicine, computer-based instruction.

Marquette University, Graduate School, College of Engineering, Department of Electrical and Computer Engineering, Milwaukee, WI 53201-1881. Offers MS, PhD. Part-time and evening/weekend programs available. *Faculty:* 13 full-time (2 women), 2 part-time (0 women). *Students:* 28 full-time (4 women), 67 part-time (8 women); includes 4 minority (1 African American, 1 Asian American or Pacific Islander, 2 Hispanic Americans), 33 international. Average age 28. 57 applicants, 35% accepted. In 1998, 15 master's awarded (100% found work related to degree); 4 doctorates awarded (100% found work related to degree). Terminal master's awarded for partial completion of doctoral program. *Degree requirements:* For master's, thesis optional, foreign language not required; for doctorate, computer language, dissertation defense, qualifying exam required. *Entrance requirements:* For master's, TOEFL (minimum score of 575 required), GRE General Test or minimum GPA of 3.0; for doctorate, GRE General Test (minimum score of 700 on quantitative section required), TOEFL (minimum score of 575 required). *Application deadline:* For fall admission, 7/15 (priority date); for spring admission, 11/15. Applications are processed on a rolling basis. Application fee: $40. Tuition: Part-time $510 per credit hour. Tuition and fees vary according to program. *Financial aid:* In 1998–99, 40 students received aid, including 5 fellowships, 5 research assistantships, 20 teaching assistantships; Federal Work-Study, institutionally-sponsored loans, and scholarships also available. Financial aid application deadline: 2/15. *Faculty research:* Electric machines, drives, and controls; applied solid-state electronics; computers and signal processing; microwaves and antennas; electronic and ultrasonic sensors. *Unit head:* Dr. Jeffrey L. Hock, Chairman, 414-288-6820, Fax: 414-288-5579, E-mail: hock@marquette.edu. *Application contact:* Dr. James E. Richie, Director of Graduate Studies, 414-288-5326, Fax: 414-288-5579, E-mail: richiej@marquette.edu.

Massachusetts Institute of Technology, School of Engineering, Department of Electrical Engineering and Computer Science, Cambridge, MA 02139-4307. Offers computer science (EE); electrical engineering (EE); electrical engineering and computer science (M Eng, SM, PhD, Sc D). *Faculty:* 101 full-time (6 women), 2 part-time (0 women). *Students:* 771 full-time (153 women), 25 part-time (4 women); includes 189 minority (18 African Americans, 149 Asian Americans or Pacific Islanders, 21 Hispanic Americans, 1 Native American), 179 international. Average age 26. 2185 applicants, 23% accepted. In 1998, 306 master's, 68 doctorates awarded. Terminal master's awarded for partial completion of doctoral program. *Degree requirements:* For master's and EE, thesis required; for doctorate and EE, dissertation, comprehensive exams required. *Application deadline:* For fall admission, 1/15; for spring admission, 11/1. Application fee: $55. *Financial aid:* In 1998–99, 724 students received aid, including 148 fellowships, 511 research assistantships, 117 teaching assistantships; career-related internships or fieldwork, Federal Work-Study, and institutionally-sponsored loans also available. Financial aid applicants required to submit FAFSA. *Faculty research:* Modem control and system theory, radio astronomy, knowledge-based application systems, artificial intelligence, electrohydrodynamics. Total annual research expenditures: $49.3 million. *Unit head:* Dr. John V. Guttag, Head, 617-253-6022, E-mail: guttag@eecs.mit.edu. *Application contact:* Peggy Carney, Administrator, 617-253-4603, E-mail: peggy@eecs.mit.edu.

McGill University, Faculty of Graduate Studies and Research, Faculty of Engineering, Department of Electrical Engineering, Montréal, PQ H3A 2T5, Canada. Offers M Eng, PhD. *Faculty:* 29 full-time (1 woman). *Students:* 158 full-time (25 women), 37 part-time (7 women), 43 international. 178 applicants, 49% accepted. In 1998, 50 master's, 22 doctorates awarded. *Degree requirements:* For master's, thesis optional; for doctorate, dissertation, qualifying exam required. *Entrance requirements:* For master's, TOEFL (minimum score of 600 required), minimum GPA of 3.0; for doctorate, TOEFL (minimum score of 600 required). *Application deadline:* For fall admission, 2/1. Applications are processed on a rolling basis. Application fee: $60. *Financial aid:* Teaching assistantships available. *Unit head:* J. Webb, Associate Chair and Graduate Program Director, 514-398-7126, Fax: 514-398-4470.

See in-depth description on page 1001.

McMaster University, School of Graduate Studies, Faculty of Engineering, Department of Electrical and Computer Engineering, Hamilton, ON L8S 4M2, Canada. Offers electrical engineering (M Eng, PhD). *Faculty:* 27 full-time. *Students:* 71 full-time, 36 part-time. *Degree requirements:* For master's, thesis required, foreign language not required; for doctorate, dissertation, comprehensive exam required, foreign language not required. *Application deadline:* For fall admission, 3/1 (priority date). Applications are processed on a rolling basis. Application fee: $50. *Financial aid:* In 1998–99, teaching assistantships (averaging $7,722 per year); fellowships, research assistantships *Unit head:* Dr. D. R. Conn, Chair, 905-525-9140 Ext. 24826.

McNeese State University, Graduate School, College of Engineering and Technology, Lake Charles, LA 70609-2495. Offers chemical engineering (M Eng); civil engineering (M Eng); electrical engineering (M Eng); mechanical engineering (M Eng). Part-time and evening/weekend programs available. *Faculty:* 13 full-time (1 woman). *Students:* 5 full-time (0 women), 3 part-time. *Degree requirements:* For master's, computer language, thesis or alternative required, foreign language not required. *Entrance requirements:* For master's, GRE General Test, TOEFL, minimum undergraduate GPA of 3.0. *Application deadline:* For fall admission, 7/15 (priority date). Applications are processed on a rolling basis. Application fee: $10 ($25 for international students). *Unit head:* Dr. O. C. Karkalits, Dean, 318-475-5875.

Memorial University of Newfoundland, School of Graduate Studies, Faculty of Engineering and Applied Science, St. John's, NF A1C 5S7, Canada. Offers civil engineering (M Eng, PhD); electrical engineering (M Eng, PhD); mechanical engineering (M Eng, PhD); ocean engineering (M Eng, PhD). Part-time programs available. *Students:* 75 full-time (11 women), 28 part-time (2 women), 31 international. *Degree requirements:* For master's, thesis optional; for doctorate, dissertation, comprehensive exam required. *Application deadline:* For fall admission, 3/1. Application fee: $40. *Unit head:* Dr. Rangaswany Seshadri, Dean, 709-737-8810, Fax: 709-737-8975, E-mail: sesh@engr.mun.ca. *Application contact:* Dr. J. J. Sharp, Associate Dean, 709-737-8901, Fax: 709-737-3480, E-mail: jsharp@engr.mun.ca.

Mercer University, School of Engineering, Macon, GA 31207-0003. Offers biomedical engineering (MSE); electrical engineering (MSE); engineering management (MSE); mechanical engineering (MSE); software engineering (MSE); software systems (MS); technical management (MS). Part-time and evening/weekend programs available. *Faculty:* 23 full-time (1 woman), 6 part-time (0 women). *Degree requirements:* For master's, computer language, thesis or alternative required, foreign language not required. *Entrance requirements:* For master's, GRE, minimum undergraduate GPA of 3.0. *Application deadline:* For fall admission, 7/1; for spring admission, 11/15. Applications are processed on a rolling basis. Application fee: $35 ($50 for international students). *Unit head:* Dr. Benjamin S. Kelley, Dean, 912-752-2459, Fax: 912-752-5593, E-mail: kelley_bs@mercer.edu. *Application contact:* Kathy Olivier, Coordinator, Special Programs, 912-752-2196, E-mail: oliver_kh@mercer.edu.

Mercer University, Cecil B. Day Campus, School of Engineering, Atlanta, GA 30341-4155. Offers electrical engineering (MSE); engineering management (MSE); software engineering (MSE); software systems (MS); technical communication management (MS). Part-time and evening/weekend programs available. Postbaccalaureate distance learning degree programs offered (no on-campus study). *Faculty:* 5 full-time (1 woman), 1 part-time (0 women). *Degree requirements:* For master's, computer language, thesis or alternative required, foreign language not required. *Entrance requirements:* For master's, GRE, minimum GPA of 3.0 in major. *Application deadline:* For fall admission, 7/1; for spring admission, 11/15. Applications are processed on a rolling basis. Application fee: $35 ($50 for international students). *Unit head:* Dr. Benjamin S. Kelley, Acting Dean, 912-752-2459, E-mail: kelley_bs@mercer.edu. *Application contact:* Dr. David Leonard, Director of Admissions, 770-986-3203.

Michigan State University, Graduate School, College of Engineering, Department of Electrical and Computer Engineering, East Lansing, MI 48824-1020. Offers MS, PhD. *Faculty:* 25. *Students:* 134 (16 women); includes 19 minority (5 African Americans, 9 Asian Americans or Pacific Islanders, 5 Hispanic Americans) 87 international. In 1998, 27 master's, 10 doctorates awarded. *Degree requirements:* For master's, exit exam required, thesis optional, foreign language not required; for doctorate, dissertation, comprehensive and qualifying exams required, foreign language not required. *Entrance requirements:* For master's and doctorate, GRE General Test (minimum combined score of 1800 on three sections required), TOEFL (minimum score of 580 required). *Application deadline:* Applications are processed on a rolling basis. Application fee: $30 ($40 for international students). *Financial aid:* In 1998–99, 35 research assistantships with tuition reimbursements (averaging $12,461 per year), 17 teaching assistantships with tuition reimbursements (averaging $12,185 per year) were awarded.; fellowships Financial aid applicants required to submit FAFSA. *Unit head:* Dr. Jes Asmussen, Chairperson, 517-355-5066, Fax: 517-353-1980.

See in-depth description on page 1003.

Michigan Technological University, Graduate School, College of Engineering, Department of Electrical Engineering, Houghton, MI 49931-1295. Offers electrical engineering (MS, PhD); sensing and signal processing (PhD). Part-time programs available. *Faculty:* 20 full-time (2 women). *Students:* 12 full-time (all women), 36 part-time; includes 1 minority (African American), 34 international. Average age 26. 111 applicants, 60% accepted. In 1998, 17 master's, 1 doctorate awarded. *Degree requirements:* For master's, thesis or alternative required, foreign language not required; for doctorate, dissertation required, foreign language not required. *Entrance requirements:* For master's, GRE General Test (combined average 1960 on three sections), TOEFL (minimum score of 600 required; average 605), BSEE or equivalent; for doctorate, GRE General Test (combined average 2020 on three sections), TOEFL (minimum score of 600 required; average 634). *Average time to degree:* Master's–2.9 years full-time. *Application deadline:* For fall admission, 3/15 (priority date). Applications are processed on a rolling basis. Application fee: $30 ($35 for international students). Tuition, state resident: full-time $4,377. Tuition, nonresident: full-time $9,108. Required fees: $126. Tuition and fees vary according to course load. *Financial aid:* In 1998–99, 6 fellowships (averaging $3,168 per year), 12 research assistantships (averaging $7,209 per year), 20 teaching assistantships (averaging $5,071 per year) were awarded.; career-related internships or fieldwork, Federal Work-Study, institutionally-sponsored loans, and unspecified assistantships also available. Aid available to part-time students. Financial aid application deadline: 3/1; financial aid applicants required to submit FAFSA. *Faculty research:* Signal and image processing, power systems, energetic materials, controls, microelectromechanical systems. Total annual research expenditures: $1.3 million. *Unit head:* Dr. John A. Soper, Chair, 906-487-2054, Fax: 906-487-2949, E-mail: jasoper@mtu.edu. *Application contact:* Gina Stevens, Graduate Coordinator, 906-487-2550, Fax: 906-487-2949, E-mail: gmsteven@mtu.edu.

Minnesota State University, Mankato, College of Graduate Studies, College of Science, Engineering and Technology, Program in Electrical Engineering and Electronic Engineering Technology, Mankato, MN 56002-8400. Offers MSE. *Faculty:* 9 full-time (0 women). *Students:* 8 full-time (1 woman), 2 part-time; includes 1 minority (Asian American or Pacific Islander) Average age 31. *Degree requirements:* For master's, thesis, comprehensive exam required. *Entrance requirements:* For master's, GRE General Test, minimum GPA of 3.0 during previous 2 years. *Application deadline:* For fall admission, 7/9 (priority date); for spring admission, 11/27. Applications are processed on a rolling basis. Application fee: $20. *Financial aid:* Research assistantships with partial tuition reimbursements, teaching assistantships with partial tuition reimbursements available. Financial aid application deadline: 3/15. *Unit head:* Dr. Carl Gruber, Chairman, 507-389-6536. *Application contact:* Joni Roberts, Admissions Coordinator, 507-389-2321, Fax: 507-389-5974, E-mail: grad@mankato.msus.edu.

Mississippi State University, College of Engineering, Department of Electrical and Computer Engineering, Mississippi State, MS 39762. Offers computer engineering (MS, PhD); electrical engineering (MS, PhD). Part-time programs available. Postbaccalaureate distance learning degree programs offered (minimal on-campus study). *Students:* 85 full-time (10 women), 20 part-time (3 women); includes 44 minority (4 African Americans, 40 Asian Americans or Pacific Islanders), 28 international. Average age 28. 161 applicants, 31% accepted. In 1998, 28 master's awarded. *Degree requirements:* For master's, computer language required (for some programs), comprehensive oral exam required, thesis optional, foreign language not required; for doctorate, computer language (for some programs), dissertation, comprehensive oral and written exam required, foreign language not required. *Entrance requirements:* For master's, GRE General Test, TOEFL (minimum score of 550 required), minimum GPA of 2.75; for doctorate, GRE, TOEFL (minimum score of 550 required). *Application deadline:* For fall admission, 7/1; for spring admission, 11/1. Applications are processed on a rolling basis. Application fee: $25 for international students. *Financial aid:* Federal Work-Study, institutionally-sponsored loans, and unspecified assistantships available. Financial aid applicants required to submit FAFSA. *Faculty research:* Digital computing, power, controls, communication systems, microelectronics. Total annual research expenditures: $1.5 million. *Unit head:* Dr. G. Marshall Molen, Head, 662-325-3912, Fax: 662-325-2298, E-mail: molen@ece.msstate.edu. *Application contact:* Jerry B. Inmon, Director of Admissions, 662-325-2224, Fax: 662-325-7360, E-mail: admit@admissions.msstate.edu.

Monmouth University, Graduate School, Department of Electronic Engineering, West Long Branch, NJ 07764-1898. Offers MS. Part-time and evening/weekend programs available. *Faculty:* 2 full-time (0 women), 1 part-time (0 women). *Students:* 7 full-time (2 women), 30 part-time (6 women); includes 5 minority (all Asian Americans or Pacific Islanders), 8 international. Average age 33. In 1998, 8 degrees awarded. *Degree requirements:* For master's, computer language, thesis required, foreign language not required. *Entrance requirements:* For master's, minimum GPA of 3.0 in major, 2.5 overall. *Application deadline:* For fall admission, 8/15 (priority date); for spring admission, 12/15 (priority date). Applications are processed on a rolling basis. Application fee: $35 ($40 for international students). Electronic applications accepted. *Financial aid:* Unspecified assistantships available. Financial aid application deadline: 3/1. *Faculty research:* Image processing, neural networks, computer networks, electromagnetics. *Unit head:* Dr. Harris Drucker, Director, 732-571-3698. *Application contact:* 732-571-3452, Fax: 732-571-5123, E-mail: gradadm@monmouth.edu.

Montana State University–Bozeman, College of Graduate Studies, College of Engineering, Department of Electrical and Computer Engineering, Bozeman, MT 59717. Offers electrical engineering (MS); engineering (PhD). Part-time programs available. *Students:* 11 full-time (2 women), 6 part-time; includes 1 minority (Hispanic American) Average age 31. 14 applicants, 86% accepted. In 1998, 8 master's, 1 doctorate awarded. *Degree requirements:* For master's, thesis or alternative required, foreign language not required; for doctorate, dissertation required, foreign language not required. *Entrance requirements:* For master's and doctorate, GRE General Test (minimum combined score of 1700 on three sections required), TOEFL (minimum score of 600 required). *Application deadline:* For fall admission, 6/1 (priority date); for spring admission, 11/1. Applications are processed on a rolling basis. Application fee: $50. *Financial aid:* In 1998–99, 14 research assistantships with full tuition reimbursements (averaging $12,000 per year), 1 teaching assistantship with full tuition reimbursement (averaging $9,000 per year) were awarded.; Federal Work-Study and scholarships also available. Financial aid application deadline: 3/1; financial aid applicants required to submit FAFSA. *Faculty research:* Optics, electromagnetic fields interacting with biological systems, communications, power/power systems, feedback and control on large scale systems, biomedical engineering. Total annual research expenditures: $180,482. *Unit head:* Dr. John Hantor, Head, 406-994-2505, Fax: 406-994-5958, E-mail: johnh@ee.montana.edu.

National Technological University, Programs in Engineering, Fort Collins, CO 80526-1842. Offers chemical engineering (MS); computer engineering (MS); computer science (MS); electrical engineering (MS); engineering management (MS); hazardous waste management (MS); health physics (MS); management of technology (MS); manufacturing systems engineering (MS); materials science and engineering (MS); software engineering (MS); special majors (MS); transportation engineering (MS); transportation systems engineering (MS). Part-time programs available. *Faculty:* 600 part-time (20 women). *Entrance requirements:* For master's, BS in engineering or related field. *Application deadline:* Applications are processed on a rolling basis. Application fee: $50. *Unit head:* Lionel V. Baldwin, President, 970-495-6400, Fax: 970-484-0668, E-mail: baldwin@mail.ntu.edu.

National University, Graduate Studies, School of Business and Technology, Department of Technology, La Jolla, CA 92037-1011. Offers e-commerce (MBA, MS); electronic engineering (MS); engineering management (MS); environmental management (MBA, MS); industrial engineering management (MS); software engineering (MS); technology management (MBA, MS); telecommunication systems management (MS). Part-time and evening/weekend programs available. Postbaccalaureate distance learning degree programs offered (minimal on-campus study). *Faculty:* 12 full-time, 125 part-time. *Students:* 305 (79 women); includes 122 minority (34 African Americans, 69 Asian Americans or Pacific Islanders, 17 Hispanic Americans, 2 Native Americans) 53 international. *Degree requirements:* For master's, foreign language and thesis not required. *Entrance requirements:* For master's, interview, minimum GPA of 2.5. *Application deadline:* Applications are processed on a rolling basis. Application fee: $60 ($100 for international students). Tuition: Full-time $7,830; part-time $870 per course. One-time fee: $60. Tuition and fees vary according to campus/location. *Unit head:* Dr. Leonid Preiser, Chair, 858-642-8425, Fax: 858-642-8716, E-mail: lpreiser@nu.edu. *Application contact:* Nancy Rohland, Director of Enrollment Management, 858-642-8180, Fax: 858-642-8709, E-mail: nrohland@nu.edu.

Naval Postgraduate School, Graduate Programs, Department of Electrical and Computer Engineering, Monterey, CA 93943. Offers MS, PhD, Eng. Program only open to commissioned officers of the United States and friendly nations and selected United States federal civilian employees. *Accreditation:* ABET (one or more programs are accredited). Part-time programs available. Postbaccalaureate distance learning degree programs offered (minimal

Electrical Engineering

Naval Postgraduate School (continued)
on-campus study). *Students:* 105 full-time, 34 international. In 1998, 55 master's, 1 doctorate, 2 other advanced degrees awarded. *Degree requirements:* For master's and Eng, computer language, thesis required, foreign language not required; for doctorate, one foreign language, computer language, dissertation required. *Unit head:* Dr. Jeff Knorr, Chairman, 831-656-2081. *Application contact:* Theodore H. Calhoon, Director of Admissions, 831-656-3093, Fax: 831-656-2891, E-mail: tcalhoon@nps.navy.mil.

New Jersey Institute of Technology, Office of Graduate Studies, Department of Electrical and Computer Engineering, Newark, NJ 07102-1982. Offers computer engineering (MS); electrical engineering (MS, PhD, Engineer), including biomedical systems (MS, PhD), communication and signal processing (MS, PhD), computer systems (MS, PhD), control systems (MS, PhD), microwave and lightwave engineering (MS, PhD), solid-state materials and devices (MS, PhD); power engineering (MS); telecommunications (MS). Part-time and evening/weekend programs available. Terminal master's awarded for partial completion of doctoral program. *Degree requirements:* For master's, thesis required (for some programs), foreign language not required; for doctorate, dissertation, residency required, foreign language not required. *Entrance requirements:* For master's, GRE General Test (minimum score of 450 on verbal section, 600 on quantitative, 550 on analytical required); for doctorate, GRE General Test (minimum score of 450 on verbal section, 600 on quantitative, 550 on analytical required), minimum graduate GPA of 3.5. Electronic applications accepted. *Faculty research:* Communications systems design, digital signal processing.

New Mexico State University, Graduate School, College of Engineering, Department of Electrical and Computer Engineering, Las Cruces, NM 88003-8001. Offers MSEE, PhD. Part-time programs available. *Faculty:* 21 full-time (1 woman), 2 part-time (0 women). *Students:* 65 full-time (8 women), 45 part-time (4 women); includes 16 minority (2 African Americans, 2 Asian Americans or Pacific Islanders, 12 Hispanic Americans), 44 international. Average age 33. 94 applicants, 70% accepted. In 1998, 34 master's, 5 doctorates awarded. *Degree requirements:* For master's, thesis required (for some programs); for doctorate, 2 foreign languages (computer language can substitute for one), dissertation required. *Entrance requirements:* For master's, minimum GPA of 3.0; for doctorate, departmental qualifying exam, minimum GPA of 3.0. *Application deadline:* For fall admission, 7/1 (priority date); for spring admission, 11/1. Applications are processed on a rolling basis. Application fee: $15 ($35 for international students). Electronic applications accepted. Tuition, state resident: full-time $2,682; part-time $112 per credit. Tuition, nonresident: full-time $8,376; part-time $349 per credit. Tuition and fees vary according to course load. *Financial aid:* Fellowships, research assistantships, teaching assistantships, career-related internships or fieldwork and Federal Work-Study available. Aid available to part-time students. Financial aid application deadline: 3/1. *Faculty research:* Software engineering, high performance computing, telemetering and space telecommunications, computational electromagnetics. *Unit head:* Dr. Steven P. Castillo, Interim Head, 505-646-3115, Fax: 505-646-1435, E-mail: scastill@nmsu.edu. *Application contact:* Dr. Javin M. Taylor, Associate Dean, 505-646-3115, Fax: 505-646-1435, E-mail: jtaylor@nmsu.edu.

New York Institute of Technology, Graduate Division, School of Engineering and Technology, Program in Electrical Engineering, Old Westbury, NY 11568-8000. Offers MS. Part-time and evening/weekend programs available. *Students:* 13 full-time (0 women), 34 part-time (3 women); includes 20 minority (9 African Americans, 6 Asian Americans or Pacific Islanders, 5 Hispanic Americans), 8 international. Average age 33. 56 applicants, 52% accepted. In 1998, 10 degrees awarded. *Degree requirements:* For master's, project required, foreign language and thesis not required. *Entrance requirements:* For master's, GRE General Test, TOEFL, BS in electrical engineering or related field, minimum QPA of 2.85. *Average time to degree:* Master's–2 years part-time. *Application deadline:* For fall admission, 8/1. Applications are processed on a rolling basis. Application fee: $50. Electronic applications accepted. *Financial aid:* Fellowships, research assistantships, institutionally-sponsored loans, tuition waivers (full and partial), and unspecified assistantships available. Aid available to part-time students. *Faculty research:* Computer networks, control theory, lightwaves and optics, robotics, signal processing. *Unit head:* Dr. Ayat Jafari, Chair, 516-686-7523. *Application contact:* Glenn Berman, Executive Director of Admissions, 516-686-7519, Fax: 516-626-0419, E-mail: gberman@iris.nyit.edu.

North Carolina Agricultural and Technical State University, Graduate School, College of Engineering, Department of Electrical Engineering, Greensboro, NC 27411. Offers MSEE, PhD. Part-time programs available. *Faculty:* 21 full-time (2 women), 1 part-time (0 women). *Students:* 55 full-time (8 women), 7 part-time (1 woman); includes 28 minority (27 African Americans, 1 Hispanic American), 25 international. Average age 24. 64 applicants, 63% accepted. In 1998, 25 master's, 4 doctorates awarded (20% entered university research/teaching, 80% found other work related to degree). *Degree requirements:* For master's, project, thesis defense required; for doctorate, dissertation required, foreign language not required. *Entrance requirements:* For master's, GRE General Test, (minimum combined score of 1000 required), GRE Subject Test, minimum GPA of 2.8; for doctorate, GRE General Test (minimum combined score of 1100 required), minimum GPA of 3.0. *Average time to degree:* Doctorate–4 years full-time. *Application deadline:* For fall admission, 7/1; for spring admission, 1/9. Applications are processed on a rolling basis. Application fee: $35. *Financial aid:* In 1998–99, 59 students received aid, including 10 fellowships with tuition reimbursements available (averaging $16,000 per year), 20 research assistantships with partial tuition reimbursements available (averaging $14,000 per year), 15 teaching assistantships with partial tuition reimbursements available (averaging $12,000 per year); institutionally-sponsored loans, scholarships, and tuition waivers (partial) also available. Financial aid application deadline: 3/30. *Faculty research:* Semiconductor compounds, VLSI design, image processing, optical systems and devices, fault-tolerant computing. Total annual research expenditures: $1.1 million. *Unit head:* Dr. Ward J. Collis, Interim Chairperson, 336-334-7760, E-mail: collis@genesis.ncat.edu. *Application contact:* Dr. Chung Yu, Graduate Coordinator, 336-334-7760 Ext. 213, Fax: 336-334-7716, E-mail: yu@genesis.ncat.edu.

North Carolina State University, Graduate School, College of Engineering, Department of Electrical and Computer Engineering, Raleigh, NC 27695. Offers MS, PhD. Part-time programs available. *Faculty:* 101 full-time (6 women), 78 part-time (8 women). *Students:* 266 full-time (45 women), 125 part-time (10 women); includes 67 minority (29 African Americans, 30 Asian Americans or Pacific Islanders, 8 Hispanic Americans), 157 international. Average age 29. 380 applicants, 28% accepted. In 1998, 104 master's, 43 doctorates awarded. Terminal master's awarded for partial completion of doctoral program. *Degree requirements:* For master's, thesis optional, foreign language not required; for doctorate, dissertation required, foreign language not required. *Entrance requirements:* For master's, GRE, TOEFL (minimum score of 575 required), minimum GPA of 3.2 in electrical engineering course work; for doctorate, GRE, TOEFL (minimum score of 625 required), minimum GPA of 3.5 in electrical engineering course work. *Application deadline:* For fall admission, 6/25; for spring admission, 11/25. Applications are processed on a rolling basis. Application fee: $45. *Financial aid:* In 1998–99, 51 fellowships (averaging $2,702 per year), 183 research assistantships (averaging $4,863 per year), 162 teaching assistantships (averaging $5,067 per year) were awarded.; career-related internships or fieldwork also available. Financial aid application deadline: 3/1. *Faculty research:* Microwave devices, communications, signal processing, solid-state, power and control systems, VLSI design, wireless communications. Total annual research expenditures: $13.8 million. *Unit head:* Dr. Robert Kolbas, Head, 919-515-7350, Fax: 919-515-5523, E-mail: kolbas@eos.ncsu.edu. *Application contact:* Dr. Richard Kuehn, Graduate Coordinator, 919-515-5090, Fax: 919-515-5601, E-mail: rkuehn@eos.ncsu.edu.

See in-depth description on page 1005.

North Dakota State University, Graduate Studies and Research, College of Engineering and Architecture, Department of Electrical Engineering, Fargo, ND 58105. Offers MS. *Faculty:* 16 full-time (1 woman), 2 part-time (0 women). *Students:* 8 full-time (2 women), 13 part-time, 10 international. Average age 28. 65 applicants, 66% accepted. In 1998, 3 degrees awarded (100% found work related to degree). *Degree requirements:* For master's, computer language, thesis or alternative required, foreign language not required. *Entrance requirements:* For master's, TOEFL (minimum score of 525 required). *Application deadline:* For fall admission, 3/1 (priority date). Applications are processed on a rolling basis. Application fee: $25. *Financial aid:* In 1998–99, research assistantships (averaging $8,100 per year), teaching assistantships (averaging $8,100 per year) were awarded.; career-related internships or fieldwork, Federal Work-Study, institutionally-sponsored loans, and tuition waivers (full) also available. Financial aid application deadline: 4/15. *Faculty research:* Computers, power and control systems, microwaves, communications and signal processing, bioengineering. *Unit head:* Dr. Orlando Baiocchi, Chair, 701-231-7608, Fax: 701-231-8677.

Northeastern University, College of Engineering, Department of Electrical and Computer Engineering, Boston, MA 02115-5096. Offers MS, PhD. Part-time programs available. *Faculty:* 38 full-time (4 women), 8 part-time (0 women). *Students:* 148 full-time (27 women), 219 part-time (31 women); includes 25 minority (5 African Americans, 9 Hispanic Americans, 11 Native Americans), 131 international. Average age 25. 683 applicants, 32% accepted. In 1998, 59 master's, 10 doctorates awarded. Terminal master's awarded for partial completion of doctoral program. *Degree requirements:* For master's, thesis optional, foreign language not required; for doctorate, dissertation, departmental qualifying exam required, foreign language not required. *Entrance requirements:* For master's and doctorate, GRE General Test. *Average time to degree:* Master's–2.07 years full-time, 4.68 years part-time; doctorate–4.89 years full-time. *Application deadline:* For fall admission, 4/15. Applications are processed on a rolling basis. Application fee: $50. *Financial aid:* In 1998–99, 110 students received aid, including 63 research assistantships with full tuition reimbursements available (averaging $12,450 per year), 34 teaching assistantships with full tuition reimbursements available (averaging $12,450 per year); fellowships, career-related internships or fieldwork, Federal Work-Study, tuition waivers (full), and unspecified assistantships also available. Aid available to part-time students. Financial aid application deadline: 2/15; financial aid applicants required to submit FAFSA. *Faculty research:* Digital communications, signal processing and sensor data fusion, electromagnetics, control systems, plasma science. *Unit head:* Dr. Arvin Grabel, Interim Chairman, 617-373-4159, Fax: 617-373-8970. *Application contact:* Stephen L. Gibson, Associate Director, 617-373-2711, Fax: 617-373-2501, E-mail: grad-eng@coe.neu.edu.

Announcement: Center for Electromagnetics Research and Center for Communications and Digital Signal Processing enable students to study and conduct research in electromagnetics, microwave materials, electrooptics, communications, and digital signal processing. Other areas of research include electronic circuits and devices, microelectromechanical systems (MEMS), computer architecture and software engineering, motion control systems, and power electronics.

See in-depth description on page 1007.

Northern Illinois University, Graduate School, College of Engineering and Engineering Technology, Department of Electrical Engineering, De Kalb, IL 60115-2854. Offers MS. Part-time programs available. *Faculty:* 12 full-time (1 woman). *Students:* 21 full-time (4 women), 31 part-time (5 women); includes 4 minority (1 African American, 3 Asian Americans or Pacific Islanders), 42 international. Average age 27. 98 applicants, 60% accepted. In 1998, 16 degrees awarded. *Degree requirements:* For master's, thesis or alternative, comprehensive exam required, foreign language not required. *Entrance requirements:* For master's, GRE General Test, TOEFL (minimum score of 550 required; 213 for computer-based), minimum GPA of 2.75. *Application deadline:* For fall admission, 6/1; for spring admission, 11/1. Applications are processed on a rolling basis. Application fee: $30. *Financial aid:* In 1998–99, 5 research assistantships, 1 teaching assistantship were awarded.; fellowships, career-related internships or fieldwork, Federal Work-Study, tuition waivers (full), and unspecified assistantships also available. Aid available to part-time students. *Unit head:* Dr. Vincent McGinn, Chair, 815-753-9962.

Northwestern Polytechnic University, School of Engineering, Fremont, CA 94539-7482. Offers computer science (MS); computer systems engineering (MS); electrical engineering (MS). Part-time and evening/weekend programs available. *Faculty:* 8 full-time, 52 part-time. *Students:* 262. *Degree requirements:* For master's, computer language, thesis required, foreign language not required. *Entrance requirements:* For master's, TOEFL. *Application deadline:* For fall admission, 8/15; for winter admission, 12/15; for spring admission, 7/15. Applications are processed on a rolling basis. Application fee: $50 ($75 for international students). Tuition: Full-time $6,750; part-time $375 per unit. Required fees: $135 per term. Tuition and fees vary according to course load and program. *Unit head:* Dr. Pochang Hsu, Dean, 510-657-5911, Fax: 510-657-8975, E-mail: npuadm@npu0.npu.edu. *Application contact:* Dr. Fred Kuttner, Dean of Academic Affairs and Admissions, 510-657-5911, Fax: 510-657-8975, E-mail: npuadm@npu0.npu.edu.

Northwestern University, The Graduate School, Robert R. McCormick School of Engineering and Applied Science, Department of Electrical and Computer Engineering, Evanston, IL 60208. Offers electrical and computer engineering (MS, PhD); information technology (MIT). MS and PhD admissions and degrees offered through The Graduate School. Part-time programs available. *Faculty:* 31 full-time (2 women). *Students:* 154 full-time (31 women), 9 part-time (1 woman); includes 17 minority (4 African Americans, 12 Asian Americans or Pacific Islanders, 1 Hispanic American), 98 international. 656 applicants, 12% accepted. In 1998, 28 master's, 15 doctorates awarded. Terminal master's awarded for partial completion of doctoral program. *Degree requirements:* For master's, thesis or project required; for doctorate, dissertation required, foreign language not required. *Entrance requirements:* For master's and doctorate, GRE General Test, TOEFL (minimum score of 560 required). *Application deadline:* For fall admission, 6/1. Applications are processed on a rolling basis. Application fee: $50 ($55 for international students). *Financial aid:* In 1998–99, 105 students received aid, including 14 fellowships with full tuition reimbursements available (averaging $11,673 per year), 62 research assistantships with partial tuition reimbursements available (averaging $16,285 per year), 27 teaching assistantships with full tuition reimbursements available (averaging $12,042 per year); career-related internships or fieldwork, Federal Work-Study, institutionally-sponsored loans, and scholarships also available. Financial aid application deadline: 1/15; financial aid applicants required to submit FAFSA. *Faculty research:* Solid-state engineering networks and communications, optical systems, parallel and distributed computing, VLSI design and computer-aided design. Total annual research expenditures: $9.2 million. *Unit head:* Dr. Prithvaraj Banerjee, Chair, 847-491-3641, Fax: 847-491-4455, E-mail: banerjee@ece.nwu.edu. *Application contact:* Lawrence Henschen, Admission Officer, 847-491-3338, Fax: 847-491-4455, E-mail: henschen@ece.nwu.edu.

See in-depth description on page 1009.

Oakland University, Graduate Studies, School of Engineering and Computer Science, Program in Electrical and Computer Engineering, Rochester, MI 48309-4401. Offers MS. Part-time and evening/weekend programs available. *Faculty:* 7 full-time. *Students:* 16 full-time (6 women), 38 part-time (3 women); includes 6 minority (1 African American, 5 Asian Americans or Pacific Islanders), 5 international. Average age 30. 36 applicants, 75% accepted. In 1998, 17 degrees awarded. *Degree requirements:* For master's, foreign language and thesis not required. *Entrance requirements:* For master's, minimum GPA of 3.0 for unconditional admission. *Application deadline:* For fall admission, 7/15; for spring admission, 3/15. Application fee: $30. Tuition, state resident: part-time $221 per credit hour. Tuition, nonresident: part-time $488 per credit hour. Required fees: $214 per semester. Part-time tuition and fees vary according to program. *Financial aid:* Federal Work-Study, institutionally-sponsored loans, and tuition waivers (full) available. Financial aid application deadline: 3/1; financial aid applicants required to submit FAFSA. *Unit head:* Dr. Naim A. Kheir, Chair, 248-370-2177.

The Ohio State University, Graduate School, College of Engineering, Department of Electrical Engineering, Columbus, OH 43210. Offers MS, PhD. *Faculty:* 45 full-time, 19 part-time. *Students:* 228 full-time (40 women), 25 part-time (5 women); includes 11 minority (9 Asian Americans or Pacific Islanders, 2 Hispanic Americans), 173 international. 935 applicants,

21% accepted. In 1998, 61 master's, 18 doctorates awarded. *Degree requirements:* For master's, computer language required, thesis optional, foreign language not required; for doctorate, computer language, dissertation required, foreign language not required. *Entrance requirements:* For master's and doctorate, GRE General Test, GRE Subject Test, or minimum GPA of 3.0. *Application deadline:* For fall admission, 8/15. Applications are processed on a rolling basis. Application fee: $30 ($40 for international students). *Financial aid:* Fellowships, research assistantships, teaching assistantships, career-related internships or fieldwork, Federal Work-Study, and institutionally-sponsored loans available. Aid available to part-time students. *Unit head:* Yuan F. Zheng, Chair, 614-292-2572, Fax: 614-292-7596, E-mail: zheng.5@osu.edu.

See in-depth description on page 1011.

Ohio University, Graduate Studies, College of Engineering and Technology, School of Electrical Engineering and Computer Science, Athens, OH 45701-2979. Offers electrical engineering (MS, PhD). *Faculty:* 22 full-time (1 woman), 11 part-time (3 women). *Students:* 85 full-time (22 women), 29 part-time (5 women); includes 3 minority (all Asian Americans or Pacific Islanders), 89 international. 397 applicants, 61% accepted. In 1998, 13 master's, 5 doctorates awarded. *Degree requirements:* For master's, thesis required, foreign language not required; for doctorate, dissertation, comprehensive and qualifying exams required. *Entrance requirements:* For master's, GRE, BSEE, minimum GPA of 3.0; for doctorate, GRE, MSEE, minimum GPA of 3.0. Application fee: $30. Tuition, state resident: full-time $5,754; part-time $238 per credit hour. Tuition, nonresident: full-time $11,055; part-time $457 per credit hour. Tuition and fees vary according to course load, campus/location and program. *Financial aid:* In 1998–99, 10 fellowships, 6 research assistantships, 4 teaching assistantships were awarded.; Federal Work-Study and institutionally-sponsored loans also available. *Faculty research:* Communication, control, circuits, industrial, electromagnetics. *Unit head:* Dr. Dennis Irwin, Chairman, 740-593-1566. *Application contact:* Dr. Douglas A. Lawrence, Graduate Chair, 740-593-1922.

Announcement: The Ohio University School of Electrical Engineering and Computer Science offers MS and PhD degrees in electrical engineering. A concentration in computer science is offered in both degree programs. Research opportunities exist in many areas of electrical engineering and computer science. Stocker Research Associateships and teaching and graduate assistantships are available on a competitive basis and carry highly competitive stipends together with a tuition waiver. For further information, contact Douglas A. Lawrence, Graduate Chair, School of Electrical Engineering and Computer Science, Ohio University, Athens, Ohio 45701; e-mail: gradinfo@homer.ece.ohiou.edu

See in-depth description on page 1013.

Oklahoma State University, Graduate College, College of Engineering, Architecture and Technology, School of Electrical and Computer Engineering, Stillwater, OK 74078. Offers M En, MS, PhD. *Faculty:* 16 full-time (0 women), 1 part-time (0 women). *Students:* 60 full-time (4 women), 47 part-time (5 women); includes 10 minority (1 African American, 8 Asian Americans or Pacific Islanders, 1 Hispanic American), 64 international. Average age 28. In 1998, 51 master's, 4 doctorates awarded. *Degree requirements:* For master's, thesis or alternative required, foreign language not required; for doctorate, dissertation required, foreign language not required. *Entrance requirements:* For master's and doctorate, TOEFL (minimum score of 575 required). *Application deadline:* For fall admission, 7/1 (priority date). Application fee: $25. *Financial aid:* In 1998–99, 43 students received aid, including 25 research assistantships (averaging $10,825 per year), 18 teaching assistantships (averaging $7,820 per year); career-related internships or fieldwork, Federal Work-Study, and tuition waivers (partial) also available. Aid available to part-time students. Financial aid application deadline: 3/1. *Unit head:* Dr. Michael A. Soderstrand, Interim Head, 405-744-5151.

See in-depth description on page 1015.

Old Dominion University, College of Engineering and Technology, Department of Electrical and Computer Engineering, Norfolk, VA 23529. Offers ME, MS, PhD. Part-time programs available. Postbaccalaureate distance learning degree programs offered (minimal on-campus study). *Faculty:* 10 full-time (1 woman), 1 part-time (0 women). *Students:* 112 full-time (16 women), 18 part-time (4 women); includes 2 minority (both African Americans), 78 international. Average age 24. In 1998, 13 master's, 4 doctorates awarded. *Degree requirements:* For master's, comprehensive exam, thesis (MS) required; for doctorate, computer language, dissertation, candidacy exam, diagnostic exam required, foreign language not required. *Entrance requirements:* For master's, GRE, TOEFL (minimum score of 550 required), minimum GPA of 3.0; for doctorate, GRE, TOEFL (minimum score of 550 required), minimum GPA of 3.25. *Application deadline:* For fall admission, 7/1; for spring admission, 10/1. Applications are processed on a rolling basis. Application fee: $30. *Financial aid:* In 1998–99, 38 research assistantships (averaging $3,941 per year), 30 teaching assistantships (averaging $2,633 per year) were awarded.; fellowships, career-related internships or fieldwork, grants, and tuition waivers (partial) also available. Aid available to part-time students. Financial aid application deadline: 2/15; financial aid applicants required to submit FAFSA. *Faculty research:* Digital signal processing, control engineering, gaseous electronics, ultrafast (femtosecom) laser applications, interaction of fields with living organisms, modeling and simulation. Total annual research expenditures: $1.3 million. *Unit head:* Dr. Stephen Zahorian, Chair, 757-683-3741, Fax: 757-683-3220, E-mail: zahorian@ece.odu.edu. *Application contact:* Dr. Amin N. Dharamsi, Graduate Program Drector, 757-683-3741, Fax: 757-683-3220, E-mail: dharamsi@ece.odu.edu.

Oregon Graduate Institute of Science and Technology, Graduate Studies, Department of Electrical and Computer Engineering, Portland, OR 97291-1000. Offers computational finance (Certificate); computer engineering (MS, PhD); electrical engineering (MS, PhD). Part-time programs available. *Faculty:* 13 full-time (0 women), 34 part-time (1 woman). *Students:* 48 full-time (15 women), 67 part-time (14 women); includes 16 minority (all Asian Americans or Pacific Islanders) Average age 29. 124 applicants, 31% accepted. In 1998, 22 master's, 1 doctorate awarded. Terminal master's awarded for partial completion of doctoral program. *Degree requirements:* For master's, thesis optional, foreign language not required; for doctorate, comprehensive exam, oral defense of dissertation required. *Entrance requirements:* For master's, TOEFL (minimum score of 550 required); for doctorate, GRE General Test, GRE Subject Test, TOEFL (minimum score of 550 required). *Average time to degree:* Master's–1.5 years full-time, 1.8 years part-time; doctorate–4.8 years full-time. *Application deadline:* For fall admission, 3/1 (priority date). Applications are processed on a rolling basis. Application fee: $50. Electronic applications accepted. *Financial aid:* In 1998–99, 20 students received aid, including 20 research assistantships; fellowships, Federal Work-Study also available. Financial aid application deadline: 3/1. *Faculty research:* Semiconductor materials, microwave circuits, atmospheric optics, surface physics, electron and ion optics. *Unit head:* Dr. Dan Hammerstrom, Head, 503-690-4037, Fax: 503-690-1406. *Application contact:* Don Johansen, Enrollment Manager, 503-690-1315, E-mail: johansen@ece.ogi.edu.

See in-depth description on page 1017.

Oregon State University, Graduate School, College of Engineering, Department of Electrical and Computer Engineering, Corvallis, OR 97331. Offers MAIS, MS, PhD. Part-time programs available. *Faculty:* 26 full-time (1 woman), 2 part-time (0 women). *Students:* 67 full-time (11 women), 20 part-time; includes 9 minority (1 African American, 8 Asian Americans or Pacific Islanders), 56 international. Average age 27. In 1998, 30 master's, 6 doctorates awarded. *Degree requirements:* For doctorate, dissertation, departmental qualifying exam required, foreign language not required, foreign language not required. *Entrance requirements:* For master's and doctorate, TOEFL (minimum score of 575 required), minimum GPA of 3.0 in last 90 hours. *Application deadline:* For fall admission, 3/1. Applications are processed on a rolling basis. Application fee: $50. *Financial aid:* Research assistantships, teaching assistantships, institutionally-sponsored loans available. Aid available to part-time students. Financial aid application deadline: 2/1. *Faculty research:* Analog and mixed mode IC's; materials, devices, and electroluminescence; microwave and optoelectrics; control systems and signal processing; novel electrical machines. *Unit head:* Alan K. Wallace, Head, 541-737-3617, Fax: 541-737-1300, E-mail: wallace@orst.edu. *Application contact:* Alan K. Wallace, Head, 541-737-3617, Fax: 541-737-1300, E-mail: wallace@orst.edu.

Pennsylvania State University Great Valley School of Graduate Professional Studies, Graduate Studies and Continuing Education, College of Engineering, Program in Electrical Engineering, Malvern, PA 19355-1488. Offers M Eng. Average age 34. In 1998, 6 degrees awarded. *Entrance requirements:* For master's, GRE. Application fee: $50. *Unit head:* Dr. David Russell, Adviser, 610-648-3243. *Application contact:* 610-648-3242, Fax: 610-889-1334.

Pennsylvania State University Harrisburg Campus of the Capital College, Graduate Center, School of Science, Engineering and Technology, Program in Electrical Engineering, Middletown, PA 17057-4898. Offers M Eng. *Degree requirements:* For master's, thesis required, foreign language not required. *Application deadline:* For fall admission, 7/26. Application fee: $50. *Unit head:* Dr. Jerry Shaup, Chair, 717-948-6114.

Pennsylvania State University University Park Campus, Graduate School, College of Engineering, Department of Electrical Engineering, State College, University Park, PA 16802-1503. Offers MS, PhD. *Students:* 153 full-time (24 women), 67 part-time (5 women). In 1998, 46 master's, 20 doctorates awarded. *Degree requirements:* For master's, foreign language and thesis not required; for doctorate, dissertation required, foreign language not required. *Entrance requirements:* For master's and doctorate, GRE General Test. Application fee: $50. *Unit head:* Dr. Larry C. Burton, Head, 814-863-2788. *Application contact:* Dr. James W. Robinson, Chair, 814-863-7295.

Polytechnic University, Brooklyn Campus, Department of Electrical Engineering, Major in Electrical Engineering, Brooklyn, NY 11201-2990. Offers MS, PhD. Part-time and evening/weekend programs available. *Students:* 32 full-time (5 women), 128 part-time (7 women); includes 28 minority (9 African Americans, 15 Asian Americans or Pacific Islanders, 4 Hispanic Americans), 71 international. Average age 33. 248 applicants, 33% accepted. In 1998, 67 master's, 20 doctorates awarded. *Degree requirements:* For master's, thesis optional; for doctorate, dissertation required. *Entrance requirements:* For master's, BS in electrical engineering; for doctorate, qualifying exam, MS in electrical engineering. *Application deadline:* Applications are processed on a rolling basis. Application fee: $45. Electronic applications accepted. *Financial aid:* Fellowships, research assistantships, teaching assistantships, institutionally-sponsored loans available. Aid available to part-time students. Financial aid applicants required to submit FAFSA. *Unit head:* 732-571-3452, Fax: 732-571-5123, E-mail: gradadm@monmouth.edu. *Application contact:* John S. Kerge, Dean of Admissions, 718-260-3200, Fax: 718-260-3446, E-mail: admitme@poly.edu.

See in-depth description on page 1019.

Polytechnic University, Farmingdale Campus, Graduate Programs, Department of Electrical Engineering, Major in Electrical Engineering, Farmingdale, NY 11735-3995. Offers MS, PhD. *Students:* 5 full-time (1 woman), 48 part-time (3 women); includes 4 minority (3 Asian Americans or Pacific Islanders, 1 Hispanic American), 9 international. Average age 33. 7 applicants, 43% accepted. In 1998, 24 master's awarded. *Degree requirements:* For master's, computer language, thesis required (for some programs); for doctorate, dissertation required. *Application deadline:* Applications are processed on a rolling basis. Application fee: $45. Electronic applications accepted. *Financial aid:* Institutionally-sponsored loans available. Aid available to part-time students. Financial aid applicants required to submit FAFSA. *Unit head:* Dr. Javin M. Taylor, Associate Dean, 505-646-3115, Fax: 505-646-1435, E-mail: jtaylor@nmsu.edu. *Application contact:* John S. Kerge, Dean of Admissions, 718-260-3200, Fax: 718-260-3446, E-mail: admitme@poly.edu.

Polytechnic University, Westchester Graduate Center, Graduate Programs, Department of Electrical Engineering, Major in Electrical Engineering, Hawthorne, NY 10532-1507. Offers MS, PhD. *Students:* 1 full-time (0 women), 22 part-time (3 women); includes 5 minority (1 African American, 2 Asian Americans or Pacific Islanders, 2 Hispanic Americans), 3 international. Average age 33. 4 applicants, 100% accepted. In 1998, 4 master's awarded. *Degree requirements:* For master's, computer language, thesis required (for some programs); for doctorate, dissertation required. *Application deadline:* Applications are processed on a rolling basis. Application fee: $45. Electronic applications accepted. *Unit head:* Jerry B. Inmon, Director of Admissions, 662-325-2224, Fax: 662-325-7360, E-mail: admit@admissions.msstate.edu. *Application contact:* John S. Kerge, Dean of Admissions, 718-260-3200, Fax: 718-260-3446, E-mail: admitme@poly.edu.

Portland State University, Graduate Studies, School of Engineering and Applied Science, Department of Electrical and Computer Engineering, Portland, OR 97207-0751. Offers MS, PhD. Part-time and evening/weekend programs available. *Faculty:* 11 full-time (1 woman), 6 part-time (0 women). *Students:* 53 full-time (7 women), 71 part-time (12 women); includes 9 minority (8 Asian Americans or Pacific Islanders, 1 Hispanic American), 71 international. Average age 30. 108 applicants, 61% accepted. In 1998, 19 master's, 2 doctorates awarded. *Degree requirements:* For master's, variable foreign language requirement, computer language, thesis or alternative, oral exam required; for doctorate, one foreign language, computer language, dissertation, oral and written exams required. *Entrance requirements:* For master's, TOEFL (minimum score of 550 required), minimum GPA of 3.0 in upper-division course work or 2.75 overall; for doctorate, GRE General Test, GRE Subject Test, minimum GPA of 3.0 in upper-division course work. *Application deadline:* For fall admission, 3/1 (priority date); for spring admission, 11/1. Applications are processed on a rolling basis. Application fee: $50. *Financial aid:* In 1998–99, 4 research assistantships, 28 teaching assistantships were awarded.; career-related internships or fieldwork, Federal Work-Study, and institutionally-sponsored loans also available. Aid available to part-time students. Financial aid application deadline: 3/1; financial aid applicants required to submit FAFSA. *Faculty research:* Optics and laser systems, design automation, VLSI design, computer systems, power electronics. Total annual research expenditures: $214,568. *Unit head:* Dr. Rolf Schaumann, Chair, 503-725-3806, Fax: 503-725-3807, E-mail: schaumann@ee.pdx.edu. *Application contact:* Dr. Y. C. Jenq, Coordinator, 503-725-3806, Fax: 503-725-3807, E-mail: jenq@ee.pdx.edu.

Princeton University, Graduate School, School of Engineering and Applied Science, Department of Electrical Engineering, Princeton, NJ 08544-1019. Offers computer engineering (MSE, PhD); electrical engineering (M Eng); electronic materials and devices (MSE, PhD); information sciences and systems (MSE, PhD); optoelectronics (MSE, PhD). Part-time programs available. *Faculty:* 27 full-time (3 women). *Students:* 135 full-time (18 women), 10 part-time (1 woman). Average age 21. 425 applicants, 19% accepted. In 1998, 11 master's, 14 doctorates awarded. *Degree requirements:* For master's, thesis optional; for doctorate, dissertation required. *Entrance requirements:* For master's and doctorate, GRE General Test, TOEFL. *Average time to degree:* Master's–1 year full-time, 2 years part-time; doctorate–5 years full-time. *Application deadline:* For fall admission, 1/3. Electronic applications accepted. *Financial aid:* In 1998–99, 26 fellowships, 71 research assistantships (averaging $14,000 per year), 29 teaching assistantships (averaging $14,500 per year) were awarded.; Federal Work-Study and institutionally-sponsored loans also available. Financial aid application deadline: 1/3. *Faculty research:* Nanostructures, computer architecture, multimedia. *Unit head:* Prof. Wayne Wolf, Director of Graduate Studies, 609-258-3335, Fax: 609-258-3745, E-mail: dgs@ee.princeton.edu. *Application contact:* Prof. Wayne Wolf, Director of Graduate Studies, 609-258-3335, Fax: 609-258-3745, E-mail: dgs@ee.princeton.edu.

See in-depth description on page 1021.

Purdue University, Graduate School, Schools of Engineering, School of Electrical and Computer Engineering, West Lafayette, IN 47907. Offers biomedical engineering (MS Bm E, PhD); computer engineering (MS, PhD); electrical engineering (MS, PhD). Part-time programs available. Postbaccalaureate distance learning degree programs offered (no on-campus study). *Faculty:* 60 full-time (4 women), 9 part-time (0 women). *Students:* 330 full-time (68 women), 4

Electrical Engineering

Purdue University (continued)

part-time (3 women); includes 6 African Americans, 11 Hispanic Americans Average age 25. 1558 applicants, 36% accepted. In 1998, 76 master's, 41 doctorates awarded. *Degree requirements:* For master's, thesis optional, foreign language not required; for doctorate, dissertation required, foreign language not required. *Entrance requirements:* For master's and doctorate, GRE General Test (combined average 2070 on three sections), TOEFL (minimum score of 575 required; average 635). *Average time to degree:* Master's–1.75 years full-time, 4 years part-time; doctorate–5.5 years full-time. *Application deadline:* For fall admission, 1/15 (priority date); for spring admission, 9/1. Applications are processed on a rolling basis. Application fee: $30. Electronic applications accepted. *Financial aid:* In 1998–99, 298 students received aid, including 18 fellowships, 186 research assistantships, 96 teaching assistantships Financial aid applicants required to submit FAFSA. *Faculty research:* Biomedical communications and signal processing, solid-state materials and devices fields and optics, automatic controls, energy sources and systems, VLSI and circuit design. Total annual research expenditures: $12.5 million. *Unit head:* Dr. W. K. Fuchs, Head, 765-494-3539, Fax: 765-494-3544, E-mail: fuchs@purdue.edu. *Application contact:* Dr. A. M. Weiner, Director of Admissions, 765-494-3392, Fax: 765-494-3393, E-mail: ecegrad@ecn.purdue.edu.

See in-depth description on page 1023.

Queen's University at Kingston, School of Graduate Studies and Research, Faculty of Applied Science, Department of Electrical Engineering, Kingston, ON K7L 3N6, Canada. Offers M Sc, M Sc Eng, PhD. Part-time programs available. *Students:* 74 full-time (12 women), 27 part-time (3 women). In 1998, 25 master's, 5 doctorates awarded. *Degree requirements:* For master's, thesis optional, foreign language not required; for doctorate, dissertation, comprehensive exam required, foreign language not required. *Entrance requirements:* For master's and doctorate, TOEFL (minimum score of 550 required). *Application deadline:* For fall admission, 2/28 (priority date). Application fee: $60. Electronic applications accepted. *Financial aid:* Fellowships, research assistantships, teaching assistantships, institutionally-sponsored loans available. Financial aid application deadline: 3/1. *Faculty research:* Communications and signal processing systems. *Unit head:* Dr. J. C. Cartledge, Head, 613-533-2947. *Application contact:* Dr. P. C. Sen, Graduate Coordinator, 613-533-2942.

Rensselaer at Hartford, School of Engineering, Program in Electrical Engineering, Hartford, CT 06120-2991. Offers MS. Part-time and evening/weekend programs available. *Degree requirements:* For master's, seminar required, thesis optional, foreign language not required. *Entrance requirements:* For master's, TOEFL (minimum score of 570 required).

Rensselaer Polytechnic Institute, Graduate School, School of Engineering, Department of Electrical, Computer, and Systems Engineering, Program in Electrical Engineering, Troy, NY 12180-3590. Offers M Eng, MS, D Eng, PhD, MBA/M Eng. Part-time programs available. Postbaccalaureate distance learning degree programs offered (no on-campus study). *Faculty:* 36 full-time (1 woman), 4 part-time (0 women). *Students:* 102 full-time (9 women), 37 part-time (2 women); includes 11 minority (3 African Americans, 7 Asian Americans or Pacific Islanders, 1 Hispanic American), 66 international. 569 applicants, 26% accepted. In 1998, 37 master's, 17 doctorates awarded. Terminal master's awarded for partial completion of doctoral program. *Degree requirements:* For master's, thesis required (for some programs), foreign language not required; for doctorate, dissertation required, foreign language not required. *Entrance requirements:* For master's and doctorate, GRE, TOEFL (minimum score of 600 required). *Average time to degree:* Master's–1.5 years full-time; doctorate–3 years full-time. *Application deadline:* For fall admission, 2/1; for spring admission, 10/1. Applications are processed on a rolling basis. Application fee: $35. *Financial aid:* In 1998–99, 4 fellowships, 64 research assistantships, 20 teaching assistantships were awarded.; career-related internships or fieldwork and institutionally-sponsored loans also available. Financial aid application deadline: 2/1. *Faculty research:* Networking and multimedia via ATM, thermophotovoltaic devices, microelectronic interconnections, agile manufacturing, mobile robotics. Total annual research expenditures: $2.4 million. *Unit head:* Dr. Y. C. Jenq, Coordinator, 503-725-3806, Fax: 503-725-3807, E-mail: jenq@ee.pdx.edu. *Application contact:* Ann Bruno, Manager of Graduate Admissions and Financial Aid, 518-276-2554, Fax: 518-276-2433, E-mail: bruno@ecse.rpi.edu.

See in-depth description on page 1025.

Rice University, Graduate Programs, George R. Brown School of Engineering, Department of Electrical and Computer Engineering, Houston, TX 77251-1892. Offers bioengineering (MS, PhD); circuits, controls, and communication systems (MS, PhD); computer science and engineering (MS, PhD); electrical engineering (MEE); lasers, microwaves, and solid-state electronics (MS, PhD). Part-time programs available. *Degree requirements:* For master's, thesis required (for some programs), foreign language not required; for doctorate, dissertation required, foreign language not required. *Entrance requirements:* For master's and doctorate, GRE General Test, GRE Subject Test, TOEFL (minimum score of 550 required), minimum GPA of 3.0. *Faculty research:* Physical electronics.

Rochester Institute of Technology, Part-time and Graduate Admissions, College of Engineering, Department of Electrical Engineering, Rochester, NY 14623-5604. Offers MSEE. *Students:* 13 full-time (1 woman), 34 part-time (2 women); includes 6 minority (2 African Americans, 4 Hispanic Americans), 11 international. 129 applicants, 67% accepted. In 1998, 18 degrees awarded. *Degree requirements:* For master's, thesis optional, foreign language not required. *Entrance requirements:* For master's, TOEFL, minimum GPA of 3.0. *Application deadline:* For fall admission, 3/1 (priority date). Applications are processed on a rolling basis. Application fee: $40. *Financial aid:* Research assistantships available. *Faculty research:* Integrated optics, control systems, digital signal processing, robotic vision. *Unit head:* Dr. Raman Unnikrishnan, Head, 716-475-2165. *Application contact:* Dr. Richard Reeve, Associate Dean, 716-475-7048, E-mail: nrreie@rit.edu.

Rochester Institute of Technology, Part-time and Graduate Admissions, College of Engineering, Department of Microelectronic Engineering, Rochester, NY 14623-5604. Offers ME, MS. *Students:* 10 full-time (3 women), 6 part-time; includes 4 minority (all Asian Americans or Pacific Islanders), 6 international. 22 applicants, 73% accepted. In 1998, 14 degrees awarded. *Degree requirements:* Foreign language not required. *Entrance requirements:* For master's, TOEFL, minimum GPA of 3.0. *Application deadline:* For fall admission, 3/1 (priority date). Applications are processed on a rolling basis. Application fee: $40. *Financial aid:* Fellowships, research assistantships, teaching assistantships, career-related internships or fieldwork, Federal Work-Study, and institutionally-sponsored loans available. Aid available to part-time students. *Faculty research:* Semiconductor device fabrication, lithography, materials, gallium arsenide. *Unit head:* Dr. Lynn Fuller, Director, 716-475-2035, E-mail: lffeee@rit.edu. *Application contact:* Dr. Richard Reeve, Associate Dean, 716-475-7048, E-mail: nrreie@rit.edu.

Rose-Hulman Institute of Technology, Faculty of Engineering and Applied Sciences, Department of Electrical and Computer Engineering, Terre Haute, IN 47803-3920. Offers electrical engineering (MS). Part-time programs available. Postbaccalaureate distance learning degree programs offered (minimal on-campus study). *Faculty:* 19 full-time (1 woman). *Students:* 28 full-time (12 women), 27 international. Average age 23. 30 applicants, 20% accepted. In 1998, 7 degrees awarded. *Degree requirements:* For master's, thesis required, foreign language not required. *Entrance requirements:* For master's, GRE, TOEFL (minimum score of 580 required), minimum GPA of 3.0. *Average time to degree:* Master's–2 years full-time, 5 years part-time. *Application deadline:* For fall admission, 2/1 (priority date). Applications are processed on a rolling basis. Application fee: $0. *Financial aid:* In 1998–99, 18 students received aid, including 3 fellowships with full and partial tuition reimbursements available (averaging $6,000 per year); research assistantships, teaching assistantships, grants, institutionally-sponsored loans, and tuition waivers (full and partial) also available. Financial aid application deadline:2/1. *Faculty research:* Network synthesis, digital electronics, control systems, image processing, electromagnetics. Total annual research expenditures: $777,147. *Unit head:* Dr. Barry J. Farbrother, Chairman, 812-877-8414, Fax: 812-877-8895, E-mail: barry.j.farbrother@rose-hulman.edu. *Application contact:* Dr. Buck F. Brown, Dean for Research and Graduate Studies, 812-877-8403, Fax: 812-877-8102, E-mail: buck.brown@rose-hulman.edu.

Rutgers, The State University of New Jersey, New Brunswick, Graduate School, Program in Electrical and Computer Engineering, New Brunswick, NJ 08903. Offers communications and solid-state electronics (MS, PhD); computer engineering (MS, PhD); control systems (MS, PhD); digital signal processing (MS, PhD). Part-time programs available. *Faculty:* 28 full-time (2 women), 2 part-time (0 women). *Students:* 60 full-time (4 women), 128 part-time (18 women); includes 85 minority (3 African Americans, 81 Asian Americans or Pacific Islanders, 1 Hispanic American), 9 international. 533 applicants, 24% accepted. In 1998, 42 master's awarded (80% found work related to degree, 20% continued full-time study); 14 doctorates awarded. Terminal master's awarded for partial completion of doctoral program. *Degree requirements:* For master's, thesis optional, foreign language not required; for doctorate, dissertation required, foreign language not required. *Entrance requirements:* For master's and doctorate, GRE General Test (minimum score of 600 on verbal section , 730 on quantitative, 660 on analytical required). *Average time to degree:* Master's–2 years full-time, 3 years part-time; doctorate–3 years full-time, 5 years part-time. *Application deadline:* For fall admission, 2/1 (priority date); for spring admission, 11/1. Applications are processed on a rolling basis. Application fee: $50. Electronic applications accepted. *Financial aid:* In 1998–99, 84 students received aid, including 2 fellowships with partial tuition reimbursements available (averaging $12,000 per year), 52 research assistantships with full tuition reimbursements available (averaging $12,000 per year), 30 teaching assistantships with full tuition reimbursements available (averaging $12,000 per year); Federal Work-Study and tuition waivers (full) also available. Financial aid application deadline: 2/1; financial aid applicants required to submit FAFSA. *Faculty research:* Communication and information processing, wireless information networks, micro-vacuum devices, machine vision, VLSI design. Total annual research expenditures: $8.7 million. *Unit head:* Dr. David G. Daut, Director, 732-445-2578, Fax: 732-445-2820.

See in-depth description on page 1027.

St. Mary's University of San Antonio, Graduate School, Department of Engineering, Program in Electrical/Computer Engineering, San Antonio, TX 78228-8507. Offers MS. *Students:* 23 (4 women). Average age 25. In 1998, 4 degrees awarded. *Degree requirements:* For master's, computer language, thesis required, foreign language not required. *Entrance requirements:* For master's, GRE General Test, BS in science or engineering. *Application deadline:* For fall admission, 8/1. Application fee: $15. *Financial aid:* Teaching assistantships, Federal Work-Study available. *Faculty research:* Robotics, artificial intelligence, manufacturing engineering. *Unit head:* Dr. Abe Yazdani, Adviser, Department of Engineering, 210-436-3305.

San Diego State University, Graduate and Research Affairs, College of Engineering, Department of Electrical and Computer Engineering, San Diego, CA 92182. Offers electrical engineering (MS). Evening/weekend programs available. *Students:* 21 full-time (2 women), 73 part-time (8 women); includes 45 minority (5 African Americans, 22 Asian Americans or Pacific Islanders, 18 Hispanic Americans), 11 international. Average age 29. 46 applicants, 61% accepted. In 1998, 13 degrees awarded. *Degree requirements:* Foreign language not required. *Entrance requirements:* For master's, GRE General Test (minimum combined score of 950 required), TOEFL (minimum score of 550 required). *Application deadline:* For fall admission, 7/1 (priority date); for spring admission, 12/1. Applications are processed on a rolling basis. Application fee: $55. *Financial aid:* Career-related internships or fieldwork available. *Faculty research:* Ultra-high speed integral circuits and systems, naval command control and ocean surveillance, signal processing and analysis. Total annual research expenditures: $525,000. *Unit head:* Andrew Szeto, Chair, 619-594-5718, Fax: 619-594-6005, E-mail: andrew.szeto@sdsu.edu. *Application contact:* M. Lin, Graduate Coordinator, 619-594-2493, Fax: 619-594-6005, E-mail: maolin@sdsu.edu.

San Jose State University, Graduate Studies, College of Engineering, Department of Electrical Engineering, San Jose, CA 95192-0001. Offers MS. *Faculty:* 15 full-time (2 women), 27 part-time (4 women). *Students:* 63 full-time (27 women), 174 part-time (23 women); includes 159 minority (4 African Americans, 153 Asian Americans or Pacific Islanders, 2 Hispanic Americans), 39 international. Average age 30. 218 applicants, 56% accepted. In 1998, 95 degrees awarded. *Degree requirements:* For master's, thesis required. *Entrance requirements:* For master's, GRE General Test (minimum combined score of 1500 on three sections required), minimum GPA of 3.0. *Application deadline:* For fall admission, 6/1. Applications are processed on a rolling basis. Application fee: $59. Tuition, nonresident: part-time $246 per unit. Required fees: $1,939; $1,309 per year. *Unit head:* Dr. Belle Wei, Chair, 408-924-3881, Fax: 408-924-3925. *Application contact:* Dr. Rangaiya Rao, Graduate Coordinator, 408-924-3914.

Santa Clara University, School of Engineering, Department of Electrical Engineering, Santa Clara, CA 95053-0001. Offers ASIC design and test (Certificate); data storage technologies (Certificate); electrical engineering (MSEE, PhD, Engineer). Part-time and evening/weekend programs available. *Students:* 34 full-time (7 women), 183 part-time (32 women); includes 89 minority (5 African Americans, 78 Asian Americans or Pacific Islanders, 6 Hispanic Americans), 53 international. Average age 35. 116 applicants, 78% accepted. In 1998, 59 master's, 1 doctorate awarded. *Degree requirements:* For master's, computer language, thesis or alternative required, foreign language not required; for doctorate and other advanced degree, computer language, dissertation required, foreign language not required. *Entrance requirements:* For master's, GRE General Test, TOEFL (minimum score of 550 required), minimum GPA of 2.75; for doctorate, GRE General Test, GRE Subject Test, TOEFL (minimum score of 550 required), master's degree or equivalent; for other advanced degree, master's degree, published paper. *Application deadline:* For fall admission, 6/1; for spring admission, 1/1. Applications are processed on a rolling basis. Application fee: $40. *Financial aid:* Fellowships, research assistantships, teaching assistantships, Federal Work-Study and institutionally-sponsored loans available. Aid available to part-time students. Financial aid application deadline: 2/1; financial aid applicants required to submit CSS PROFILE or FAFSA. *Unit head:* Dr. Sally Wood, Chair, 408-554-6867. *Application contact:* Tina Samms, Assistant Director of Graduate Admissions, 408-554-4313, Fax: 408-554-5474, E-mail: engr-grad@scu.edu.

South Dakota School of Mines and Technology, Graduate Division, Department of Electrical Engineering, Rapid City, SD 57701-3995. Offers MS. Part-time programs available. *Faculty:* 11 full-time (0 women), 1 part-time (0 women). *Students:* 25 full-time (1 woman), 21 international. Average age 27. In 1998, 5 degrees awarded. *Degree requirements:* For master's, thesis required, foreign language not required. *Entrance requirements:* For master's, GRE (for applicants from schools without ABET accreditation), TOEFL (minimum score of 520 required), TWE. *Application deadline:* For fall admission, 6/15 (priority date); for spring admission, 10/15. Applications are processed on a rolling basis. Application fee: $15. Electronic applications accepted. Tuition, state resident: part-time $89 per hour. Tuition, nonresident: part-time $261 per hour. Part-time tuition and fees vary according to program. *Financial aid:* In 1998–99, 15 students received aid, including 15 teaching assistantships; fellowships, research assistantships, Federal Work-Study and institutionally-sponsored loans also available. Aid available to part-time students. Financial aid application deadline: 5/15. *Faculty research:* Semiconductors, systems, digital systems, computers, superconductivity. Total annual research expenditures: $49,294. *Unit head:* Dr. Larry Simonson, Chair, 605-394-2451. *Application contact:* Brenda Brown, Secretary, 800-454-8162 Ext. 2493, Fax: 605-394-5360, E-mail: graduate_admissions@silver.sdmt.edu.

South Dakota School of Mines and Technology, Graduate Division, Division of Material Engineering and Science, Doctoral Program in Materials Engineering and Science, Rapid City, SD 57701-3995. Offers chemical engineering (PhD); chemistry (PhD); civil engineering (PhD); electrical engineering (PhD); mechanical engineering (PhD); metallurgical engineering (PhD); physics (PhD). Part-time programs available. *Students:* 14 full-time (2 women), 9 international. *Degree requirements:* For doctorate, dissertation required, foreign language not required. *Entrance requirements:* For doctorate, TOEFL (minimum score of 520 required), TWE, minimum graduate GPA of 3.0. *Application deadline:* For fall admission, 6/15 (priority date); for spring admission, 10/15. Applications are processed on a rolling basis. Application fee: $15. Electronic applications accepted. Tuition, state resident: part-time $89 per hour. Tuition, nonresident: part-time $261 per hour. Part-time tuition and fees vary according to program. *Unit head:* Dr. Chris Jenkins, Coordinator, 605-394-2406. *Application contact:* Brenda Brown, Secretary, 800-454-8162 Ext. 2493, Fax: 605-394-5360, E-mail: graduate_admissions@silver.sdmt.edu.

South Dakota State University, Graduate School, College of Engineering, Department of Electrical Engineering, Brookings, SD 57007. Offers MS. *Degree requirements:* For master's, thesis, oral exam required, foreign language not required. *Entrance requirements:* For master's, TOEFL (minimum score of 550 required). *Faculty research:* Image processing, electromagnetics communications, power systems, electrical materials and sensors.

Southern Illinois University Carbondale, Graduate School, College of Engineering, Department of Electrical Engineering, Carbondale, IL 62901-6806. Offers MS. *Faculty:* 15 full-time (1 woman). *Students:* 47 full-time (4 women), 11 part-time (2 women). 146 applicants, 51% accepted. In 1998, 13 degrees awarded. *Degree requirements:* For master's, thesis, comprehensive exam required, foreign language not required. *Entrance requirements:* For master's, TOEFL (minimum score of 550 required), minimum GPA of 2.7. *Application deadline:* Applications are processed on a rolling basis. Application fee: $20. *Financial aid:* In 1998–99, 21 students received aid, including 6 research assistantships with full tuition reimbursements available; fellowships with full tuition reimbursements available, teaching assistantships with full tuition reimbursements available, Federal Work-Study, institutionally-sponsored loans, and tuition waivers (full) also available. Aid available to part-time students. Financial aid application deadline: 1/15. *Faculty research:* Circuits and power systems, communications and signal processing, controls and systems, electromagnetics and optics, electronics instrumentation and bioengineering. Total annual research expenditures: $254,257. *Unit head:* Dr. Glafkos D. Galanos, Chair, 618-536-2364.

Southern Illinois University Carbondale, Graduate School, College of Engineering, Program in Engineering Sciences, Carbondale, IL 62901-6806. Offers electrical systems (PhD); fossil energy (PhD); mechanics (PhD). *Faculty:* 57 full-time (1 woman). *Students:* 16 full-time (2 women), 10 part-time (2 women); includes 3 minority (1 African American, 2 Asian Americans or Pacific Islanders), 17 international. *Degree requirements:* For doctorate, dissertation required. *Entrance requirements:* For doctorate, GRE General Test, TOEFL (minimum score of 600 required), minimum GPA of 3.5. Application fee: $20. *Unit head:* Dr. Hasan Sevim, Associate Dean, 618-453-4321, Fax: 618-453-4235.

Southern Illinois University Edwardsville, Graduate Studies and Research, School of Engineering, Program in Electrical Engineering, Edwardsville, IL 62026-0001. Offers MS. Part-time programs available. *Students:* 51 full-time (11 women), 43 part-time (9 women); includes 8 minority (4 African Americans, 4 Asian Americans or Pacific Islanders), 57 international. 166 applicants, 43% accepted. In 1998, 27 degrees awarded. *Degree requirements:* For master's, thesis or research paper, final exam required. *Entrance requirements:* For master's, TOEFL (minimum score of 550 required). *Application deadline:* For fall admission, 7/24. Application fee: $25. *Financial aid:* In 1998–99, 8 research assistantships with full tuition reimbursements, 10 teaching assistantships with full tuition reimbursements were awarded.; fellowships with full tuition reimbursements, Federal Work-Study, institutionally-sponsored loans, and unspecified assistantships also available. Aid available to part-time students. *Unit head:* Dr. Raghu Bollini, Chair, 618-650-2524, E-mail: rbollin@siue.edu. *Application contact:* Dr. Arjun Godhwani, Director, 618-650-2524, E-mail: agodhwa@siue.edu.

Southern Methodist University, School of Engineering and Applied Science, Department of Electrical Engineering, Dallas, TX 75275. Offers electrical engineering (MSEE, PhD); telecommunications (MS). Postbaccalaureate distance learning degree programs offered (no on-campus study). *Faculty:* 20 full-time (1 woman), 6 part-time (0 women). *Students:* 41 full-time (9 women), 355 part-time (55 women); includes 114 minority (19 African Americans, 73 Asian Americans or Pacific Islanders, 21 Hispanic Americans, 1 Native American), 63 international. Average age 33. 265 applicants, 48% accepted. In 1998, 112 master's, 7 doctorates awarded. *Degree requirements:* For master's, thesis optional, foreign language not required; for doctorate, dissertation, oral and written qualifying exams, oral final exam required. *Entrance requirements:* For master's, GRE General Test (minimum score of 650 on quantitative section required), TOEFL (minimum score of 550 required), minimum GPA of 3.0 in last 2 years; bachelor's degree in engineering, mathematics, or sciences; for doctorate, preliminary counseling exam, minimum GPA of 3.0, bachelor's degree in related field. *Application deadline:* For fall admission, 8/1 (priority date); for spring admission, 12/15. Applications are processed on a rolling basis. Application fee: $25. Tuition: Full-time $9,216; part-time $512 per credit hour. Required fees: $88 per credit hour. Part-time tuition and fees vary according to course load and campus/location. *Financial aid:* Fellowships, research assistantships, teaching assistantships, institutionally-sponsored loans and tuition waivers (full) available. Financial aid applicants required to submit FAFSA. *Faculty research:* Mobile communications, optical and millimeter wave communications, solid state devices and materials, digital signal processing. *Unit head:* Dr. Jerry D. Gibson, Chair, 214-768-3113, Fax: 214-768-3573, E-mail: gibson@seas.smu.edu. *Application contact:* Dr. Zeynep Celik-Butler, Assistant Dean for Graduate Studies and Research, 214-768-3979, Fax: 214-768-3845, E-mail: zcb@seas.smu.edu.

See in-depth description on page 1029.

Southern Polytechnic State University, College of Technology, Department of Electrical and Computer Engineering Technology, Marietta, GA 30060-2896. Offers engineering technology (MS). Part-time and evening/weekend programs available. *Faculty:* 2 full-time (0 women). *Students:* 9 full-time (0 women), 23 part-time (2 women); includes 8 minority (3 African Americans, 3 Asian Americans or Pacific Islanders, 2 Hispanic Americans), 6 international. Average age 34. 3 applicants, 67% accepted. In 1998, 8 degrees awarded (100% found work related to degree). *Degree requirements:* For master's, foreign language and thesis not required. *Entrance requirements:* For master's, GRE General Test. *Application deadline:* For fall admission, 7/15 (priority date); for spring admission, 12/1. Applications are processed on a rolling basis. Application fee: $20. Tuition, state resident: full-time $2,146; part-time $119 per credit hour. Tuition, nonresident: full-time $7,586; part-time $421 per credit hour. *Financial aid:* In 1998–99, 15 students received aid; teaching assistantships, career-related internships or fieldwork and Federal Work-Study available. Aid available to part-time students. Financial aid application deadline: 5/1; financial aid applicants required to submit FAFSA. *Unit head:* Kim Davis, Acting Head, 770-528-7246, Fax: 770-528-7285, E-mail: kdavis0@spsu.edu.

Stanford University, School of Engineering, Department of Electrical Engineering, Stanford, CA 94305-9991. Offers MS, PhD, Eng. *Faculty:* 49 full-time (1 woman). *Students:* 639 full-time (98 women), 208 part-time (34 women); includes 180 minority (14 African Americans, 144 Asian Americans or Pacific Islanders, 20 Hispanic Americans, 2 Native Americans), 412 international. Average age 26. 1194 applicants, 40% accepted. In 1998, 225 master's, 63 doctorates awarded. Terminal master's awarded for partial completion of doctoral program. *Degree requirements:* For master's, foreign language and thesis not required; for doctorate, dissertation required, foreign language not required; for Eng, thesis required. *Entrance requirements:* For master's, doctorate, and Eng, GRE General Test, TOEFL. *Application deadline:* For fall admission, 1/5. Application fee: $65 ($80 for international students). Electronic applications accepted. Tuition: Full-time $24,588. Required fees: $152. Part-time tuition and fees vary according to course load. *Financial aid:* Fellowships, research assistantships, teaching assistantships, Federal Work-Study and institutionally-sponsored loans available. Financial aid application deadline: 1/1. *Unit head:* James Plummer, Chair, 650-723-5782, Fax: 650-723-4659, E-mail: plummer@ee.stanford.edu. *Application contact:* Director of Graduate Admissions, 650-723-4115.

State University of New York at Binghamton, Graduate School, Thomas J. Watson School of Engineering and Applied Science, Department of Electrical Engineering, Binghamton, NY 13902-6000. Offers M Eng, MS, PhD. Part-time and evening/weekend programs available. *Students:* 33 full-time (6 women), 45 part-time (5 women); includes 16 minority (4 African Americans, 11 Asian Americans or Pacific Islanders, 1 Hispanic American), 22 international. Average age 30. 142 applicants, 51% accepted. In 1998, 31 master's, 3 doctorates awarded. *Degree requirements:* For master's, thesis or alternative required, foreign language not required; for doctorate, dissertation required, foreign language not required. *Entrance requirements:* For master's and doctorate, GRE General Test, GRE Subject Test, TOEFL (minimum score of 550 required). *Application deadline:* For fall admission, 4/15 (priority date); for spring admission, 11/1. Applications are processed on a rolling basis. Application fee: $50. Electronic applications accepted. Tuition, state resident: full-time $5,100; part-time $213 per credit. Tuition, nonresident: full-time $8,416; part-time $351 per credit. Required fees: $77 per credit. Part-time tuition and fees vary according to course load. *Financial aid:* In 1998–99, 27 students received aid, including 12 research assistantships with full tuition reimbursements available (averaging $8,514 per year), 12 teaching assistantships with full tuition reimbursements available (averaging $8,274 per year); fellowships, career-related internships or fieldwork, Federal Work-Study, institutionally-sponsored loans, and unspecified assistantships also available. Aid available to part-time students. Financial aid application deadline: 2/15. *Unit head:* Dr. Richard Plumb, Chairperson, 609-777-4856.

See in-depth description on page 1031.

State University of New York at Buffalo, Graduate School, School of Engineering and Applied Sciences, Department of Electrical Engineering, Buffalo, NY 14260. Offers M Eng, MS, PhD. Part-time programs available. *Faculty:* 19 full-time (1 woman), 2 part-time (0 women). *Students:* 70 full-time (14 women), 73 part-time (5 women); includes 12 minority (2 African Americans, 10 Asian Americans or Pacific Islanders), 81 international. Average age 24. 527 applicants, 63% accepted. In 1998, 50 master's, 13 doctorates awarded. Terminal master's awarded for partial completion of doctoral program. *Degree requirements:* For master's, exam, project required, thesis optional, foreign language not required; for doctorate, dissertation required, foreign language not required. *Entrance requirements:* For master's and doctorate, GRE General Test, TOEFL (minimum score of 550 required). *Application deadline:* For fall admission, 2/1 (priority date); for spring admission, 9/28. Applications are processed on a rolling basis. Application fee: $35. Tuition, state resident: full-time $5,100; part-time $213 per credit hour. Tuition, nonresident: full-time $8,416; part-time $351 per credit hour. Required fees: $870; $75 per semester. Tuition and fees vary according to course load and program. *Financial aid:* In 1998–99, 57 students received aid, including 5 fellowships with full tuition reimbursements available (averaging $8,200 per year), 25 research assistantships with full tuition reimbursements available (averaging $10,000 per year), 18 teaching assistantships with full tuition reimbursements available (averaging $10,360 per year); career-related internships or fieldwork, Federal Work-Study, institutionally-sponsored loans, tuition waivers (full and partial), and unspecified assistantships also available. Financial aid application deadline: 2/1; financial aid applicants required to submit FAFSA. *Faculty research:* High power electronics and plasmas, electronic materials signal and image processing, photonics and communications, optics. Total annual research expenditures: $916,709. *Unit head:* Dr. Pao-Lo Liu, Chairman, 716-645-2422 Ext. 2124, Fax: 716-645-3656, E-mail: paololiu@eng.buffalo.edu. *Application contact:* Dr. Donald Givone, Director of Graduate Admissions, 716-645-2422 Ext. 2129, Fax: 716-645-3656, E-mail: kosik@acsu.buffalo.edu.

See in-depth description on page 1033.

State University of New York at Stony Brook, Graduate School, College of Engineering and Applied Sciences, Department of Electrical and Computer Engineering, Stony Brook, NY 11794. Offers MS, PhD. Evening/weekend programs available. *Faculty:* 21 full-time (2 women), 4 part-time (0 women). *Students:* 56 full-time (12 women), 48 part-time (4 women); includes 10 minority (1 African American, 9 Asian Americans or Pacific Islanders), 61 international. 389 applicants, 58% accepted. In 1998, 31 master's, 8 doctorates awarded. *Degree requirements:* For master's, thesis or alternative required, foreign language not required; for doctorate, dissertation, comprehensive exams required, foreign language not required. *Entrance requirements:* For master's and doctorate, GRE General Test, TOEFL. *Application deadline:* For fall admission, 1/15. Application fee: $50. *Financial aid:* In 1998–99, 10 research assistantships, 25 teaching assistantships were awarded.; fellowships *Faculty research:* System science, solid-state electronics, computer engineering. Total annual research expenditures: $1.4 million. *Unit head:* Dr. Serge Luryi, Chairman, 516-632-8420. *Application contact:* Dr. Chi-Tsong Chen, Director, 516-632-8400, Fax: 516-632-8494, E-mail: ctchen@sbee.sunysb.edu.

See in-depth description on page 1035.

Stevens Institute of Technology, Graduate School, Charles V. Schaefer Jr. School of Engineering, Department of Electrical and Computer Engineering, Program in Electrical Engineering, Hoboken, NJ 07030. Offers computer architecture and digital system design (M Eng, PhD, Engr); robotics/control/instrumentation (M Eng, PhD, Engr); signal and image processing (M Eng, PhD, Engr); telecommunications engineering (M Eng, PhD, Engr); telecommunications management (MS, PhD, Certificate). MS, PhD, and Certificate offered in cooperation with the Program in Telecommunications Management. *Degree requirements:* For master's, computer language required, thesis optional, foreign language not required; for doctorate, variable foreign language requirement, computer language, dissertation required; for other advanced degree, computer language required, foreign language not required. *Entrance requirements:* For master's and doctorate, GRE, TOEFL; for other advanced degree, GRE. Electronic applications accepted.

Syracuse University, Graduate School, L. C. Smith College of Engineering and Computer Science, Department of Electrical Engineering and Computer Science, Program in Electrical Engineering, Syracuse, NY 13244-0003. Offers MS, PhD, EE. *Students:* 51 full-time (8 women), 50 part-time (3 women); includes 8 minority (2 African Americans, 6 Asian Americans or Pacific Islanders), 56 international. Average age 28. 373 applicants, 82% accepted. In 1998, 29 master's, 3 doctorates awarded. *Degree requirements:* For master's, foreign language and thesis not required; for doctorate, computer language, dissertation required, foreign language not required. *Entrance requirements:* For master's and doctorate, GRE General Test, GRE Subject Test. *Application deadline:* Applications are processed on a rolling basis. Application fee: $40. Tuition: Full-time $13,992; part-time $583 per credit hour. *Financial aid:* Fellowships, research assistantships, teaching assistantships, Federal Work-Study available. Financial aid application deadline: 3/1. *Faculty research:* Electromagnetics, electronic devices, systems. *Unit head:* Ercunent Arvas, Graduate Director.

See in-depth description on page 1037.

Temple University, Graduate School, College of Science and Technology, College of Engineering, Program in Electrical and Computer Engineering, Philadelphia, PA 19122-6096. Offers MSE. Part-time programs available. *Faculty:* 12 full-time (0 women). *Students:* 28; includes 13 minority (1 African American, 11 Asian Americans or Pacific Islanders, 1 Hispanic American), 6 international. 80 applicants, 30% accepted. *Degree requirements:* For master's, thesis required, foreign language not required. *Entrance requirements:* For master's, GRE General Test (minimum combined score of 1500 required), TOEFL (minimum score of 575 required). *Application deadline:* For fall admission, 7/1; for spring admission, 11/1. Applications are processed on a rolling basis. Application fee: $40. *Financial aid:* Fellowships, research assistantships, teaching assistantships, Federal Work-Study and institutionally-sponsored loans available. Financial aid application deadline: 2/15. *Faculty research:* Neural networks, adaptive control, robotics, multiprocessor systems, vacuum microelectronics. Total annual research expenditures:$425,000. *Unit head:* Dr. John J. Helferty, Director, 215-204-4523, Fax: 215-204-5960.

Tennessee Technological University, Graduate School, College of Engineering, Department of Electrical Engineering, Cookeville, TN 38505. Offers MS, PhD. Part-time programs available. *Faculty:* 19 full-time (0 women). *Students:* 30 full-time (4 women), 3 part-time; includes 27 minority (1 African American, 26 Asian Americans or Pacific Islanders) Average age 27. 239 applicants, 51% accepted. In 1998, 16 master's awarded. *Degree requirements:* For master's, thesis required, foreign language not required; for doctorate, one foreign language (computer

Electrical Engineering

Tennessee Technological University (continued)
language can substitute), dissertation required. *Entrance requirements:* For master's, GRE General Test, TOEFL (minimum score of 525 required); for doctorate, GRE Subject Test, TOEFL (minimum score of 525 required), minimum GPA of 3.5. *Application deadline:* For fall admission, 3/1 (priority date); for spring admission, 8/1. Application fee: $25 ($30 for international students). Tuition, state resident: part-time $137 per hour. Tuition, nonresident: part-time $361 per hour. Required fees: $17 per hour. Tuition and fees vary according to course load. *Financial aid:* In 1998–99, 1 fellowship, 11 research assistantships (averaging $6,300 per year), 14 teaching assistantships (averaging $6,300 per year) were awarded.; career-related internships or fieldwork also available. Financial aid application deadline: 4/1. *Faculty research:* Control, digital, and power systems. *Unit head:* Dr. P. K. Rajan, Chairperson, 931-372-3397, Fax: 931-372-3436, E-mail: pkrajan@tntech.edu. *Application contact:* Dr. Rebecca F. Quattlebaum, Dean of the Graduate School, 931-372-3233, Fax: 931-372-3497, E-mail: rquattlebaum@tntech.edu.

Texas A&M University, College of Engineering, Department of Electrical Engineering, College Station, TX 77843. Offers M Eng, MS, D Eng, PhD. *Faculty:* 49 full-time (3 women), 8 part-time (3 women). *Students:* 324 full-time (52 women), 68 part-time (10 women); includes 56 minority (10 African Americans, 34 Asian Americans or Pacific Islanders, 12 Hispanic Americans), 181 international. Average age 28. 391 applicants, 76% accepted. In 1998, 49 master's, 29 doctorates awarded. *Degree requirements:* For master's, thesis (MS) required; for doctorate, dissertation (PhD) required. *Entrance requirements:* For master's and doctorate, GRE General Test, TOEFL. Application fee: $50 ($75 for international students). *Financial aid:* Fellowships, research assistantships, teaching assistantships, career-related internships or fieldwork available. Financial aid application deadline: 4/1; financial aid applicants required to submit FAFSA. *Faculty research:* Solid-state, electric power systems, and communications engineering. *Unit head:* Dr. Chanan Singh, Head, 409-845-7441. *Application contact:* Norman Griswold, Graduate Adviser, 409-845-7441.

See in-depth description on page 1039.

Texas A&M University–Kingsville, College of Graduate Studies, College of Engineering, Department of Electrical Engineering and Computer Science, Program in Electrical Engineering, Kingsville, TX 78363. Offers ME, MS. *Students:* 8 full-time (1 woman), 19 part-time (2 women). *Degree requirements:* For master's, computer language, thesis or alternative, comprehensive exam required, foreign language not required. *Entrance requirements:* For master's, GRE General Test (minimum combined score of 1000 required), TOEFL (minimum score of 525 required), minimum GPA of 3.0. *Application deadline:* For fall admission, 6/1; for spring admission, 11/15. Applications are processed on a rolling basis. Application fee: $15 ($25 for international students). Tuition, state resident: full-time $2,062. Tuition, nonresident: full-time $7,246. *Financial aid:* Application deadline: 5/15. *Unit head:* Dr. Rajab Challoo, Coordinator, 361-593-2001. *Application contact:* H. D. Gorakhpurwalla, Graduate Coordinator, 361-593-2004.

Texas Tech University, Graduate School, College of Engineering, Department of Electrical Engineering, Lubbock, TX 79409. Offers MSEE, PhD. Part-time programs available. *Faculty:* 19 full-time (2 women), 1 part-time (0 women). *Students:* 77 full-time (12 women), 13 part-time; includes 3 minority (all Hispanic Americans), 74 international. Average age 26. 284 applicants, 35% accepted. In 1998, 9 master's, 6 doctorates awarded. *Degree requirements:* For master's and doctorate, computer language, thesis/dissertation required, foreign language not required. *Entrance requirements:* For master's, GRE General Test (minimum combined score of 1000 required; average 1235), minimum GPA of 3.0; for doctorate, GRE General Test (minimum combined score of 1000 required), minimum GPA of 3.0. *Application deadline:* For fall admission, 4/15 (priority date); for spring admission, 11/1 (priority date). Applications are processed on a rolling basis. Application fee: $25 ($50 for international students). Electronic applications accepted. *Financial aid:* In 1998–99, 67 research assistantships (averaging $8,598 per year) were awarded.; fellowships, teaching assistantships, Federal Work-Study and institutionally-sponsored loans also available. Aid available to part-time students. Financial aid application deadline: 5/15; financial aid applicants required to submit FAFSA. *Faculty research:* High-voltage space power, accuracy enhancement in optical computing, computer vision in image processing. Total annual research expenditures: $5.5 million. *Unit head:* Dr. Jon G. Bredeson, Chair, 806-742-3533.

Tufts University, Division of Graduate and Continuing Studies and Research, Graduate School of Arts and Sciences, College of Engineering, Department of Electrical Engineering and Computer Science, Medford, MA 02155. Offers computer science (MS, PhD); electrical engineering (MS, PhD). Part-time programs available. *Faculty:* 17 full-time, 5 part-time. *Students:* 99 (26 women); includes 9 minority (2 African Americans, 7 Asian Americans or Pacific Islanders) 32 international. 147 applicants, 41% accepted. In 1998, 19 master's, 6 doctorates awarded. Terminal master's awarded for partial completion of doctoral program. *Degree requirements:* For master's, thesis or alternative required, foreign language not required; for doctorate, dissertation required, foreign language not required. *Entrance requirements:* For master's and doctorate, GRE General Test, TOEFL (minimum score of 550 required). *Application deadline:* For fall admission, 3/1; for spring admission, 10/15. Applications are processed on a rolling basis. Application fee: $50. *Financial aid:* Research assistantships with full and partial tuition reimbursements, teaching assistantships with full and partial tuition reimbursements, Federal Work-Study, scholarships, and tuition waivers (partial) available. Financial aid application deadline: 2/15; financial aid applicants required to submit FAFSA. *Unit head:* Robert Gonsalves, Chair, 617-627-3217, Fax: 617-627-3220. *Application contact:* Anselm Blumer, 617-623-3217, Fax: 617-627-3220, E-mail: webmaster@eecs.tufts.edu.

Tufts University, Division of Graduate and Continuing Studies and Research, Professional and Continuing Studies, Microwave and Wireless Engineering Program, Medford, MA 02155. Offers Certificate. Part-time and evening/weekend programs available. Average age 24. 1 applicants, 100% accepted. In 1998, 2 degrees awarded. *Average time to degree:* 1 year part-time. *Application deadline:* For fall admission, 8/15 (priority date); for spring admission, 12/12. Applications are processed on a rolling basis. Application fee: $40. *Financial aid:* Available to part-time students. Application deadline: 5/1; *Unit head:* Norman Griswold, Graduate Adviser, 409-845-7441. *Application contact:* Norman Griswold, Graduate Adviser, 409-845-7441.

Tulane University, School of Engineering, Department of Electrical Engineering, New Orleans, LA 70118-5669. Offers MS, MSE, PhD, Sc D. MS and PhD offered through the Graduate School. Part-time programs available. *Students:* 19 full-time (7 women), 3 part-time (1 woman); includes 4 minority (2 African Americans, 1 Asian American or Pacific Islander, 1 Hispanic American), 11 international. 95 applicants, 9% accepted. In 1998, 6 master's, 2 doctorates awarded. Terminal master's awarded for partial completion of doctoral program. *Degree requirements:* For master's and doctorate, thesis/dissertation required. *Entrance requirements:* For master's, GRE General Test (combined average 1361), TOEFL, minimum B average in undergraduate course work; for doctorate, GRE General Test, TOEFL, minimum B average in undergraduate course work. *Application deadline:* For fall admission, 7/1. Application fee: $35. *Financial aid:* Research assistantships, teaching assistantships, Federal Work-Study, institutionally-sponsored loans, and tuition waivers (full and partial) available. Financial aid application deadline: 2/1. *Faculty research:* Control systems, digital signal processing, image processing, power systems. *Unit head:* Dr. Enrique Barbieri, Chairman, 504-865-5785. *Application contact:* Dr. E. Michaelides, Associate Dean, 504-865-5764.

Tuskegee University, Graduate Programs, College of Engineering, Architecture and Physical Sciences, Department of Electrical Engineering, Tuskegee, AL 36088. Offers MSEE. *Faculty:* 8 full-time (0 women). *Students:* 13 full-time (4 women), 6 part-time; includes 12 minority (all African Americans), 6 international. Average age 24. In 1998, 8 degrees awarded. *Degree requirements:* For master's, computer language, thesis or alternative required, foreign language not required. *Entrance requirements:* For master's, GRE General Test, GRE Subject Test. *Application deadline:* For fall admission, 7/15. Applications are processed on a rolling basis. Application fee: $25 ($35 for international students). *Financial aid:* Fellowships, research assistantships, teaching assistantships, career-related internships or fieldwork, Federal Work-Study, and institutionally-sponsored loans available. Aid available to part-time students. Financial aid application deadline: 4/15. *Faculty research:* Photovoltaic insulation, automatic guidance and control, wind energy. *Unit head:* Dr. Numam Dogen, Acting Head, 334-727-8298.

Union College, Graduate and Continuing Studies, Division of Engineering and Computer Science, Department of Electrical Engineering and Computer Science, Program in Electrical Engineering, Schenectady, NY 12308-2311. Offers MS. *Students:* 6 full-time (2 women), 8 part-time (1 woman); includes 2 minority (both Asian Americans or Pacific Islanders), 3 international. 8 applicants, 100% accepted. In 1998, 1 degree awarded. *Degree requirements:* For master's, computer language, comprehensive and departmental qualifying exams required, foreign language and thesis not required. *Entrance requirements:* For master's, minimum GPA of 3.0. *Application deadline:* Applications are processed on a rolling basis. Application fee: $50. Tuition: Part-time $1,786 per course. *Unit head:* Dr. Michael Rudko, Chair, 518-388-6316.

Universidad de las Américas–Puebla, Division of Graduate Studies, School of Engineering, Program in Electronic Engineering, Cholula, 72820, Mexico. Offers MS. Part-time and evening/weekend programs available. *Faculty:* 8 full-time (0 women), 2 part-time (0 women). *Students:* 27 full-time (2 women); all minorities (all Hispanic Americans) Average age 25. 16 applicants, 88% accepted. In 1998, 4 degrees awarded. *Degree requirements:* For master's, one foreign language, computer language, thesis required. *Average time to degree:* Master's–2.5 years full-time, 3.5 years part-time. *Application deadline:* For fall admission, 7/16. Applications are processed on a rolling basis. Application fee: $0. *Financial aid:* In 1998–99, 17 students received aid, including 5 research assistantships, 2 teaching assistantships Aid available to part-time students. Financial aid application deadline: 5/15. *Faculty research:* Telecommunications, data processing, digital systems. Total annual research expenditures: $40,000. *Unit head:* Dr. Rubén Alejos, Coordinator, 22-29-20-42, Fax: 22-29-20-32, E-mail: ralejos@mail.udlap.mx. *Application contact:* Mauricio Villegas, Chair of Admissions Office, 22-29-20-17, Fax: 22-29-20-18, E-mail: admision@mail.udlap.mx.

Université de Moncton, School of Engineering, Program in Electrical Engineering, Moncton, NB E1A 3E9, Canada. Offers M Sc A. *Faculty:* 2 full-time (0 women). *Students:* 3 full-time (0 women). 4 applicants, 75% accepted. *Degree requirements:* For master's, thesis, proficiency in French required. *Application deadline:* For fall admission, 6/1 (priority date); for winter admission, 11/15 (priority date). Application fee: $30. *Financial aid:* In 1998–99, 3 students received aid, including fellowships (averaging $17,200 per year), research assistantships (averaging $5,325 per year), teaching assistantships (averaging $1,000 per year) Financial aid application deadline: 5/31. *Faculty research:* Telecommunications, electronics and instrumentation, analog and digital electronics, electronic control of machines, energy systems, electronic design. Total annual research expenditures: $22,150. *Unit head:* Dr. Duc T. Phi, Chairman, 506-858-4346, Fax: 506-858-4082, E-mail: phid@umoncton.ca.

Université de Sherbrooke, Faculty of Applied Sciences, Department of Electrical Engineering and Computer Engineering, Sherbrooke, PQ J1K 2R1, Canada. Offers M Sc A, PhD. *Degree requirements:* For master's and doctorate, thesis/dissertation required. *Faculty research:* Minielectronics, biomedical engineering, digital signal prolonging and telecommunications, software engineering and artificial intelligence.

Université du Québec à Trois-Rivières, Graduate Programs, Program in Electrical Engineering, Trois-Rivières, PQ G9A 5H7, Canada. Offers M Sc A, PhD. Part-time programs available. *Students:* 19 full-time (3 women). 23 applicants, 70% accepted. In 1998, 1 degree awarded. *Degree requirements:* For master's and doctorate, thesis/dissertation required. *Entrance requirements:* For master's, appropriate bachelor's degree, proficiency in French; for doctorate, appropriate master's degree, proficiency in French. *Application deadline:* For fall admission, 2/1. Application fee: $30. *Financial aid:* Fellowships, research assistantships, teaching assistantships available. *Faculty research:* Industrial electronics. *Unit head:* Pierre Sicard, Director, 819-376-5071 Ext. 3931, Fax: 819-376-5012. *Application contact:* Suzanne Camirand, Admissions Officer, 819-376-5045 Ext. 2591, Fax: 819-376-5210, E-mail: suzanne_camirand@uqtr.uquebec.ca.

Université Laval, Faculty of Graduate Studies, Faculty of Sciences and Engineering, Department of Electrical and Computer Engineering, Sainte-Foy, PQ G1K 7P4, Canada. Offers M Sc, PhD. *Students:* 86 full-time (11 women), 23 part-time (3 women). 74 applicants, 42% accepted. In 1998, 20 master's, 7 doctorates awarded. *Application deadline:* For fall admission, 3/1. Application fee: $30. *Faculty research:* Telecommunication, frequency standards, biomedical engineering and electronic instrumentation. *Unit head:* Paul Fortier, Director, 418-656-2131 Ext. 3555, Fax: 418-656-3159, E-mail: paul.fortier@gel.ulaval.ca.

The University of Akron, Graduate School, College of Engineering, Department of Electrical Engineering, Akron, OH 44325-0001. Offers MSEE, PhD. Evening/weekend programs available. *Faculty:* 12 full-time, 2 part-time. *Students:* 13 full-time (5 women), 7 part-time (1 woman); includes 1 minority (Asian American or Pacific Islander), 5 international. Average age 28. In 1998, 10 master's awarded. *Degree requirements:* For master's, thesis or alternative required, foreign language not required; for doctorate, variable foreign language requirement (computer language can substitute for one), dissertation, candidacy exam, qualifying exam required. *Entrance requirements:* For master's, TOEFL (minimum score of 550 required), minimum GPA of 2.75; for doctorate, GRE, TOEFL (minimum score of 550 required). *Average time to degree:* Master's–2 years full-time, 4 years part-time. *Application deadline:* For fall admission, 3/1. Applications are processed on a rolling basis. Application fee: $25 ($50 for international students). Tuition, state resident: part-time $189 per credit. Tuition, nonresident: part-time $353 per credit. Required fees: $7.3 per credit. *Financial aid:* In 1998–99, 1 research assistantship with full tuition reimbursement, 18 teaching assistantships with full tuition reimbursements were awarded.; fellowships with full tuition reimbursements, career-related internships or fieldwork and tuition waivers (full) also available. Financial aid application deadline: 3/1. *Faculty research:* Signal processing, digital control systems, computer interfacing, nondestructive testing. *Unit head:* Dr. Nathan Ida, Chair, 330-972-7679, E-mail: nida@uakron.edu. *Application contact:* Dr. Nathan Ida, Chair, 330-972-7679, E-mail: nida@uakron.edu.

The University of Alabama, Graduate School, College of Engineering, Department of Electrical Engineering, Tuscaloosa, AL 35487. Offers MSEE, PhD. *Faculty:* 14 full-time (1 woman). *Students:* 27 full-time (2 women), 13 part-time (3 women); includes 2 minority (both African Americans), 28 international. Average age 28. 38 applicants, 68% accepted. In 1998, 7 master's awarded. *Degree requirements:* For master's, thesis or alternative required, foreign language not required; for doctorate, one foreign language (computer language can substitute), dissertation required. *Entrance requirements:* For master's and doctorate, GRE General Test (minimum combined score of 1500 on three sections required), minimum GPA of 3.0 in last 60 hours. *Application deadline:* For fall admission, 7/6. Applications are processed on a rolling basis. Application fee: $25. *Financial aid:* In 1998–99, 25 students received aid, including 3 research assistantships, 16 teaching assistantships; fellowships, Federal Work-Study also available. *Faculty research:* Computer engineering, microelectronics, power systems, electromagnetics, control systems. *Unit head:* Dr. Raghuendra K. Pandey, Head, 205-348-6812, Fax: 205-348-6959, E-mail: rpandey@coe.eng.ua.edu.

The University of Alabama at Birmingham, Graduate School, School of Engineering, Department of Electrical and Computer Engineering, Birmingham, AL 35294. Offers MSEE, PhD. Evening/weekend programs available. *Students:* 8 full-time (2 women), 13 part-time (3 women); includes 2 minority (1 African American, 1 Asian American or Pacific Islander), 6 international. 27 applicants, 93% accepted. In 1998, 9 master's awarded. *Degree requirements:* For master's, thesis or alternative required, foreign language not required; for doctorate, dissertation required, foreign language not required. *Entrance requirements:* For master's, GRE General Test (minimum score of 500 on each section required), BSEE. *Application deadline:* Applications are processed on a rolling basis. Application fee: $30 ($60 for international students). Electronic applications accepted. *Financial aid:* In 1998–99, 1 student received aid, including 1 research assistantship with full tuition reimbursement available (averaging $9,500 per year); tuition waivers (full and partial) also available. *Unit head:* Dr. Gregg L. Vaughn, Interim Chair, 205-934-8440, Fax: 205-975-3337, E-mail: gvaughn@uab.edu.

The University of Alabama in Huntsville, School of Graduate Studies, College of Engineering, Department of Electrical and Computer Engineering, Huntsville, AL 35899. Offers computer engineering (PhD); electrical and computer engineering (MSE); electrical engineering (PhD); optical science and engineering (PhD). Part-time and evening/weekend programs available. *Faculty:* 22 full-time (1 woman), 2 part-time (0 women). *Students:* 64 full-time (17 women), 64 part-time (7 women); includes 13 minority (5 African Americans, 6 Asian Americans or Pacific Islanders, 2 Hispanic Americans), 38 international. Average age 31. 94 applicants, 80% accepted. In 1998, 29 master's, 9 doctorates awarded. *Degree requirements:* For master's, oral and written exams required, thesis optional, foreign language not required; for doctorate, dissertation, oral and written exams required, foreign language not required. *Entrance requirements:* For master's, GRE General Test (minimum combined score of 1500 on three sections required), appropriate bachelor's degree, minimum GPA of 3.0; for doctorate, GRE General Test (minimum combined score of 1500 on three sections required), minimum GPA of 3.0. *Application deadline:* For fall admission, 7/24 (priority date); for spring admission, 11/15 (priority date). Applications are processed on a rolling basis. Application fee: $20. Tuition and fees vary according to course load. *Financial aid:* In 1998–99, 39 students received aid, including 10 research assistantships with full and partial tuition reimbursements available (averaging $8,379 per year), 29 teaching assistantships with full and partial tuition reimbursements available (averaging $7,894 per year); fellowships with full and partial tuition reimbursements available, career-related internships or fieldwork, Federal Work-Study, grants, institutionally-sponsored loans, scholarships, and tuition waivers (full and partial) also available. Aid available to part-time students. Financial aid application deadline: 4/1; financial aid applicants required to submit FAFSA. *Faculty research:* Optical signal processing, electromagnetics, photonics, nonlinear waves, computer architecture. Total annual research expenditures: $1.4 million. *Unit head:* Dr. Reza Adhami, Chair, 256-890-6316, Fax: 256-890-6803, E-mail: adhami@eb.uah.edu.

See in-depth description on page 1041.

University of Alaska Fairbanks, Graduate School, College of Science, Engineering and Mathematics, Department of Electrical Engineering, Fairbanks, AK 99775-7480. Offers MEE, MS. *Faculty:* 7 full-time (0 women), 2 part-time (1 woman). *Students:* 17 full-time (3 women), 2 part-time; includes 2 minority (both Asian Americans or Pacific Islanders), 4 international. Average age 30. 17 applicants, 65% accepted. In 1998, 4 degrees awarded. *Degree requirements:* For master's, thesis or alternative, comprehensive exam required, foreign language not required. *Entrance requirements:* For master's, GRE General Test, TOEFL (minimum score of 550 required). *Application deadline:* For fall admission, 8/1. Applications are processed on a rolling basis. Application fee: $35. *Financial aid:* Research assistantships, teaching assistantships, career-related internships or fieldwork available. *Faculty research:* Geomagnetically-induced currents in power lines, telecommunications. *Unit head:* Dr. Joe Hawkins, Head, 907-474-7137.

University of Alberta, Faculty of Graduate Studies and Research, Department of Electrical and Computer Engineering, Edmonton, AB T6G 2E1, Canada. Offers computational optics (PhD); computer engineering (M Eng, M Sc, PhD); control systems (M Eng, M Sc, PhD); engineering management (M Eng); laser physics (M Sc, PhD); oil sands (M Eng, M Sc, PhD); plasma physics (M Sc, PhD); power engineering (M Eng, M Sc, PhD); telecommunications (M Eng, M Sc, PhD). Terminal master's awarded for partial completion of doctoral program. *Degree requirements:* For master's and doctorate, thesis/dissertation required, foreign language not required. *Entrance requirements:* For master's and doctorate, TOEFL (minimum score of 580 required; average 610). Electronic applications accepted. *Faculty research:* Controls, communications, microelectronics, electromagnetics.

The University of Arizona, Graduate College, College of Engineering and Mines, Department of Electrical and Computer Engineering, Tucson, AZ 85721. Offers MS, PhD. Part-time programs available. *Faculty:* 64. *Students:* 135 full-time (18 women), 73 part-time (8 women). Average age 29. 314 applicants, 46% accepted. In 1998, 32 master's, 12 doctorates awarded. *Degree requirements:* For master's, thesis required (for some programs), foreign language not required; for doctorate, dissertation required, foreign language not required. *Entrance requirements:* For master's, GRE General Test, TOEFL (minimum score of 575 required); for doctorate, GRE General Test, TOEFL (minimum score of 600 required). *Application deadline:* For fall admission, 6/15. Applications are processed on a rolling basis. Application fee: $35. *Financial aid:* Fellowships, research assistantships, teaching assistantships, institutionally-sponsored loans and scholarships available. Financial aid application deadline: 3/15. *Faculty research:* Communication systems, control systems, signal processing, computer-aided logic. *Unit head:* Kenneth F. Galloway, Head, 520-621-6193. *Application contact:* Barbie Horton, Graduate Academic Adviser, 520-621-6195, Fax: 520-621-8076.

See in-depth description on page 1043.

University of Arkansas, Graduate School, College of Engineering, Department of Electrical Engineering, Fayetteville, AR 72701-1201. Offers MSE, MSEE, PhD. *Faculty:* 17 full-time (0 women). *Students:* 47 full-time (8 women), 22 part-time (2 women); includes 8 minority (1 African American, 5 Asian Americans or Pacific Islanders, 2 Hispanic Americans), 41 international. 110 applicants, 54% accepted. In 1998, 25 master's, 4 doctorates awarded. *Degree requirements:* For master's, thesis optional, foreign language not required; for doctorate, one foreign language, dissertation required. Application fee: $40 ($50 for international students). Tuition, state resident: full-time $3,186. Tuition, nonresident: full-time $7,560. Required fees: $378. *Financial aid:* In 1998–99, 11 research assistantships, 13 teaching assistantships were awarded.; fellowships, career-related internships or fieldwork and Federal Work-Study also available. Aid available to part-time students. Financial aid application deadline: 4/1; financial aid applicants required to submit FAFSA. *Unit head:* Neil Schmitt, Acting Chair, 501-575-3005. *Application contact:* Kraig Olejniczak, Graduate Coordinator, 501-575-3005, E-mail: ldr@engr.uark.edu.

University of Bridgeport, College of Graduate and Undergraduate Studies, School of Science, Engineering, and Technology, Department of Electrical Engineering, Bridgeport, CT 06601. Offers MS. *Faculty:* 2 full-time (0 women), 1 part-time (0 women). *Students:* 16 full-time (3 women), 6 part-time, 18 international. Average age 29. 132 applicants, 92% accepted. In 1998, 5 degrees awarded. *Degree requirements:* For master's, thesis optional, foreign language not required. *Entrance requirements:* For master's, TOEFL. *Application deadline:* Applications are processed on a rolling basis. Application fee: $35 ($50 for international students). *Financial aid:* In 1998–99, 12 students received aid; research assistantships, teaching assistantships, career-related internships or fieldwork, Federal Work-Study, institutionally-sponsored loans, and tuition waivers (partial) available. Aid available to part-time students. Financial aid application deadline: 6/1; financial aid applicants required to submit FAFSA. *Unit head:* Dr. Wenelin Janeff, Chairman, 203-576-4296.

University of British Columbia, Faculty of Graduate Studies, Faculty of Applied Science, Department of Electrical Engineering, Vancouver, BC V6T 1Z2, Canada. Offers M Eng, MA Sc, PhD. *Degree requirements:* For master's, thesis required (for some programs), foreign language not required; for doctorate, dissertation required, foreign language not required. *Entrance requirements:* For master's and doctorate, TOEFL (minimum score of 600 required). *Faculty research:* Applied electromagnetics, biomedical engineering, communications and signal processing, computer and software engineering, power engineering, robotics, solid-state, systems and control.

University of Calgary, Faculty of Graduate Studies, Faculty of Engineering, Department of Electrical and Computer Engineering, Calgary, AB T2N 1N4, Canada. Offers M Eng, M Sc, PhD. Part-time programs available. *Faculty:* 24 full-time (0 women), 10 part-time (0 women). *Students:* 70 full-time (2 women), 31 part-time (2 women). 200 applicants, 26% accepted. In 1998, 8 master's, 8 doctorates awarded. *Degree requirements:* For master's, thesis, thesis (M Sc) required, foreign language not required; for doctorate, dissertation, candidacy exam required, foreign language not required. *Entrance requirements:* For master's and doctorate, TOEFL (minimum score of 550 required; 213 for computer-based), minimum GPA of 3.0. *Average time to degree:* Master's–2 years full-time; doctorate–3.2 years full-time. *Application deadline:* For fall admission, 4/30; for winter admission, 9/30. Applications are processed on a rolling basis. Application fee: $60. *Financial aid:* In 1998–99, 59 students received aid; fellowships, research assistantships, teaching assistantships available. Financial aid application deadline: 5/30. *Faculty research:* Control, electronics, power systems, signal and image processing, telecommunications, software engineering. Total annual research expenditures: $1.8 million. *Unit head:* Dr. R. H. Johnston, Head, 403-220-5003, Fax: 403-282-6855, E-mail: grad-studies@enel.ucalgary.ca. *Application contact:* Dr. A. Sesay, Associate Head, 403-220-6163, Fax: 403-282-6855, E-mail: grad_studies@enel.ucalgary.ca.

University of California, Berkeley, Graduate Division, College of Engineering, Department of Electrical Engineering and Computer Sciences, Berkeley, CA 94720-1500. Offers computer science (MS, PhD); electrical engineering (M Eng, MS, D Eng, PhD). *Students:* 471 full-time (67 women); includes 138 minority (17 African Americans, 110 Asian Americans or Pacific Islanders, 10 Hispanic Americans, 1 Native American), 156 international. 1976 applicants, 12% accepted. In 1998, 51 master's, 34 doctorates awarded. *Degree requirements:* For master's, comprehensive exam or thesis (MS) required; for doctorate, dissertation, qualifying exam required, foreign language not required. *Entrance requirements:* For master's and doctorate, GRE General Test, TOEFL (minimum score of 570 required), minimum GPA of 3.0. *Application deadline:* For fall admission, 12/15. Application fee: $40. *Financial aid:* Fellowships, research assistantships, teaching assistantships, scholarships available. Financial aid application deadline: 12/15. *Unit head:* Dr. Richard Newton, Chair, 510-642-0253. *Application contact:* Admission Assistant, 510-642-3068, Fax: 510-642-2845, E-mail: gradadm@eecs.berkeley.edu.

University of California, Davis, Graduate Studies, College of Engineering, Program in Electrical and Computer Engineering, Davis, CA 95616. Offers MS, PhD. *Faculty:* 33 full-time (2 women), 2 part-time (0 women). *Students:* 116 full-time (13 women), 4 part-time; includes 26 minority (2 African Americans, 21 Asian Americans or Pacific Islanders, 3 Hispanic Americans), 37 international. 505 applicants, 25% accepted. In 1998, 11 master's, 12 doctorates awarded. Terminal master's awarded for partial completion of doctoral program. *Degree requirements:* For master's, thesis optional, foreign language not required; for doctorate, dissertation, preliminary and qualifying exams, thesis defense required, foreign language not required. *Entrance requirements:* For master's, GRE General Test, minimum GPA of 3.2; for doctorate, GRE, minimum graduate GPA of 3.5. *Application deadline:* For fall admission, 2/15. Application fee: $40. Electronic applications accepted. *Financial aid:* In 1998–99, 26 fellowships with full and partial tuition reimbursements, 35 research assistantships with tuition reimbursements, 29 teaching assistantships with full and partial tuition reimbursements were awarded.; Federal Work-Study also available. Financial aid application deadline: 1/15; financial aid applicants required to submit FAFSA. *Unit head:* Bernard Levy, Chairperson, 530-752-8251. *Application contact:* Anita Morales, Graduate Staff Assistant, 530-752-8251, E-mail: gradinfo@ucdavis.edu.

University of California, Irvine, Office of Research and Graduate Studies, School of Engineering, Department of Electrical and Computer Engineering, Irvine, CA 92697. Offers computer networks and distributed computing (MS, PhD); computer systems and software (MS, PhD); electrical engineering (MS, PhD). Part-time programs available. *Faculty:* 24 full-time (1 woman), 1 part-time (0 women). *Students:* 92 full-time (17 women), 15 part-time (1 woman); includes 24 minority (1 African American, 23 Asian Americans or Pacific Islanders), 63 international. 308 applicants, 37% accepted. In 1998, 30 master's, 9 doctorates awarded. Terminal master's awarded for partial completion of doctoral program. *Degree requirements:* For doctorate, dissertation required, foreign language not required, foreign language not required. *Entrance requirements:* For master's, GRE General Test, minimum GPA of 3.0; for doctorate, GRE General Test. *Application deadline:* For fall admission, 1/15 (priority date). Applications are processed on a rolling basis. Application fee: $40. Electronic applications accepted. *Financial aid:* Fellowships, research assistantships, teaching assistantships, institutionally-sponsored loans and tuition waivers (full and partial) available. Financial aid application deadline: 3/2; financial aid applicants required to submit FAFSA. *Faculty research:* Optical and solid-state devices, systems and signal processing. *Unit head:* Dr. Nader Bagherzadeh, Chair, 949-824-8720, Fax: 949-824-3779. *Application contact:* Ronnie A. Gran, Graduate Admissions Coordinator, 949-824-5489, Fax: 949-824-1853, E-mail: ragran@uci.edu.

See in-depth description on page 1045.

University of California, Los Angeles, Graduate Division, School of Engineering and Applied Science, Department of Electrical Engineering, Los Angeles, CA 90095. Offers electrical engineering (MS, PhD); operations research (MS, PhD). *Faculty:* 41 full-time, 25 part-time. *Students:* 338 full-time (45 women); includes 106 minority (6 African Americans, 86 Asian Americans or Pacific Islanders, 13 Hispanic Americans, 1 Native American), 132 international. 931 applicants, 25% accepted. In 1998, 74 master's, 26 doctorates awarded. *Degree requirements:* For master's, comprehensive exam or thesis required; for doctorate, dissertation, qualifying exams required, foreign language not required. *Entrance requirements:* For master's, GRE General Test, minimum GPA of 3.0; for doctorate, GRE General Test, minimum GPA of 3.25. *Application deadline:* For fall admission, 1/15. Application fee: $40. Electronic applications accepted. *Financial aid:* In 1998–99, 30 fellowships, 223 research assistantships, 40 teaching assistantships were awarded.; career-related internships or fieldwork, Federal Work-Study, institutionally-sponsored loans, and tuition waivers (full and partial) also available. Financial aid application deadline: 1/15; financial aid applicants required to submit FAFSA. *Unit head:* Dr. William J. Kaiser, Chair, 310-825-6465. *Application contact:* Maida Bassili, Student Affairs Officer, 310-825-9383, E-mail: mbassili@ea.ucla.edu.

See in-depth description on page 1047.

University of California, Riverside, Graduate Division, College of Engineering, Department of Electrical Engineering, Riverside, CA 92521-0102. Offers MS, PhD. *Students:* 9 (2 women) 8 international. *Degree requirements:* For doctorate, dissertation, qualifying exams required, foreign language not required, foreign language not required. *Entrance requirements:* For master's and doctorate, GRE General Test (minimum combined score of 1100 required), TOEFL (minimum score of 550 required). *Application deadline:* For fall admission, 5/1; for spring admission, 12/1. Applications are processed on a rolling basis. Application fee: $40. *Financial aid:* Application deadline: 2/1. *Unit head:* Dr. Satish K. Tripathi, Dean, College of Engineering, 909-787-2942.

See in-depth description on page 1049.

University of California, San Diego, Graduate Studies and Research, Department of Electrical and Computer Engineering, La Jolla, CA 92093-5003. Offers applied ocean science (MS, PhD); applied physics (MS, PhD); communication theory and systems (MS, PhD); computer engineering (MS, PhD); electrical engineering (M Eng, MS, PhD); electronic circuits and systems (MS, PhD); intelligent systems, robotics and control (MS, PhD); photonics (MS, PhD); signal and image processing (MS, PhD). *Faculty:* 35. *Students:* 251 (24 women). 590 applicants, 29% accepted. In 1998, 24 master's, 27 doctorates awarded. *Entrance requirements:* For master's and doctorate, GRE General Test. Application fee: $40. *Unit head:* William Coles, Chair. *Application contact:* Graduate Coordinator, 619-534-6606.

See in-depth description on page 1051.

University of California, Santa Barbara, Graduate Division, College of Engineering, Department of Electrical and Computer Engineering, Santa Barbara, CA 93106. Offers MS, PhD. *Students:* 196 full-time (32 women). 489 applicants, 35% accepted. In 1998, 50 master's, 29 doctorates awarded. *Degree requirements:* For master's, thesis or alternative required, foreign

Electrical Engineering

University of California, Santa Barbara (continued)

language not required; for doctorate, dissertation required, foreign language not required. *Entrance requirements:* For master's and doctorate, GRE General Test, TOEFL (minimum score of 560 required). *Application deadline:* For fall admission, 12/15. Application fee: $40. Electronic applications accepted. *Financial aid:* Fellowships, research assistantships, teaching assistantships, career-related internships or fieldwork, Federal Work-Study, institutionally-sponsored loans, and tuition waivers (full and partial) available. Financial aid application deadline: 12/15; financial aid applicants required to submit FAFSA. *Faculty research:* Solid-state device theory, physics of solid-state materials, computer architecture. *Unit head:* Steve Butner, Chair, 805-893-3821, E-mail: butner@ece.ucsb.edu. *Application contact:* Dawn Keiling, Graduate Program Assistant, 805-893-3114, E-mail: dawn@ece.ucsb.edu.

See in-depth description on page 1053.

University of Central Florida, College of Engineering, Department of Electrical and Computer Engineering, Orlando, FL 32816. Offers computer engineering (MS Cp E, PhD, Certificate); electrical engineering (MSEE, PhD, Certificate); optical science and engineering (MS, PhD). Part-time and evening/weekend programs available. *Faculty:* 28 full-time, 14 part-time. *Students:* 144 full-time (24 women), 110 part-time (19 women); includes 51 minority (9 African Americans, 24 Asian Americans or Pacific Islanders, 18 Hispanic Americans), 106 international. Average age 31. 160 applicants, 39% accepted. In 1998, 50 master's, 9 doctorates awarded. *Degree requirements:* For master's, computer language, thesis or alternative required, foreign language not required; for doctorate, computer language, dissertation, departmental qualifying exam, candidacy exam required, foreign language not required. *Entrance requirements:* For master's, GRE General Test (minimum combined score of 1000 required), TOEFL (minimum score of 550 required; 213 computer-based), minimum GPA of 3.0 in last 60 hours; for doctorate, GRE General Test (minimum combined score of 1100 required), TOEFL (minimum score of 550 required; 213 computer-based), minimum GPA of 3.5 in last 60 hours. *Application deadline:* For fall admission, 7/15; for spring admission, 12/15. Application fee: $20. Tuition, state resident: full-time $2,054; part-time $137 per credit. Tuition, nonresident: full-time $7,207; part-time $480 per credit. Required fees: $47 per term. *Financial aid:* In 1998–99, 164 students received aid, including 31 fellowships with partial tuition reimbursements available (averaging $2,337 per year), 102 teaching assistantships with partial tuition reimbursements available (averaging $1,940 per year); research assistantships with partial tuition reimbursements available, career-related internships or fieldwork, Federal Work-Study, institutionally-sponsored loans, tuition waivers (partial), and unspecified assistantships also available. Financial aid application deadline: 3/1; financial aid applicants required to submit FAFSA. *Faculty research:* Communication theory, solid-state devices, electromagnetics, electrooptics, digital signal processing. *Unit head:* Dr. W. B. Mikhael, Chair, 407-823-3210, E-mail: mikhael@ucf1vm.cc.ucf.edu. *Application contact:* Dr. Juin J. Liou, Coordinator, 407-823-2610.

See in-depth description on page 1057.

University of Cincinnati, Division of Research and Advanced Studies, College of Engineering, Department of Electrical and Computer Engineering and Computer Science, Program in Electrical Engineering, Cincinnati, OH 45221-0091. Offers MS, PhD. *Students:* 110 full-time (9 women), 22 part-time (3 women); includes 8 minority (2 African Americans, 3 Asian Americans or Pacific Islanders, 3 Hispanic Americans), 99 international. 93 applicants, 74% accepted. In 1998, 32 master's, 8 doctorates awarded. *Degree requirements:* For master's and doctorate, thesis/dissertation required, foreign language not required. *Entrance requirements:* For master's, GRE General Test, TOEFL (minimum score of 525 required), BS in electrical engineering or related field; for doctorate, GRE General Test, TOEFL. *Average time to degree:* Master's–2.6 years full-time; doctorate–5.1 years full-time. *Application deadline:* For fall admission, 2/1 (priority date). Application fee: $40. *Financial aid:* Fellowships, tuition waivers (full) and unspecified assistantships available. Aid available to part-time students. Financial aid application deadline: 2/1. *Faculty research:* Integrated circuits and optical devices, charge-coupled devices, photosensitive devices. *Unit head:* Maida Bassili, Student Affairs Officer, 310-825-9383, E-mail: mbassili@ea.ucla.edu. *Application contact:* Dieter Schmidt, Graduate Program Director, 513-556-1816, Fax: 513-556-7326, E-mail: dieter.schmidt@uc.edu.

See in-depth description on page 1059.

University of Colorado at Boulder, Graduate School, College of Engineering and Applied Science, Department of Electrical and Computer Engineering, Boulder, CO 80309. Offers electrical engineering (ME, MS, PhD), including computer engineering. *Degree requirements:* For master's, thesis or alternative, comprehensive exam required; for doctorate, dissertation, departmental qualifying exam required. *Entrance requirements:* For master's, GRE General Test, minimum undergraduate GPA of 3.0; for doctorate, GRE General Test, minimum undergraduate GPA of 3.5.

University of Colorado at Colorado Springs, Graduate School, College of Engineering and Applied Science, Department of Electrical and Computer Engineering, Colorado Springs, CO 80933-7150. Offers MS, PhD. Part-time and evening/weekend programs available. *Faculty:* 9 full-time (0 women), 12 part-time (0 women). *Students:* 38 full-time (3 women), 31 part-time (3 women); includes 6 Asian Americans or Pacific Islanders, 1 Hispanic American, 1 Native American, 12 international. Average age 29. 29 applicants, 83% accepted. In 1998, 12 master's awarded (100% found work related to degree); 3 doctorates awarded (100% found work related to degree). *Degree requirements:* For master's, thesis required (for some programs), foreign language not required; for doctorate, dissertation, comprehensive exams required, foreign language not required. *Entrance requirements:* For master's, GRE General Test (minimum combined score of 1200 required), TOEFL (minimum score of 550 required), minimum GPA of 3.0; for doctorate, GRE General Test (minimum combined score of 1200 required), TOEFL (minimum score of 550 required), minimum GPA of 3.3. Application fee: $40 ($50 for international students). Tuition, state resident: full-time $2,768; part-time $118 per credit. Tuition, nonresident: full-time $10,392; part-time $425 per credit. Required fees: $265; $7.5 per credit. One-time fee: $28. Tuition and fees vary according to program and student level. *Financial aid:* In 1998–99, 12 students received aid; fellowships, research assistantships, teaching assistantships, career-related internships or fieldwork and Federal Work-Study available. Financial aid application deadline: 5/1. *Faculty research:* Signal processing, neural networks, integrated ferroelectric devices, applied electromagnetics, circuit design. Total annual research expenditures: $497,000. *Unit head:* Dr. Ramaswami Dandapani, Chairman, 719-262-3044, Fax: 719-262-3589, E-mail: rdan@vlsid.uccs.edu. *Application contact:* Sue Bidlingmaier, Academic Adviser, 719-262-3351, Fax: 719-262-3589, E-mail: suebid@ecemail.uccs.edu.

University of Colorado at Denver, Graduate School, College of Engineering and Applied Science, Department of Electrical Engineering, Denver, CO 80217-3364. Offers MS. Part-time and evening/weekend programs available. *Faculty:* 9. *Students:* 6 full-time (1 woman), 38 part-time (3 women); includes 3 minority (1 African American, 1 Asian American or Pacific Islander, 1 Hispanic American), 5 international. Average age 32. 32 applicants, 81% accepted. In 1998, 18 degrees awarded. *Degree requirements:* For master's, thesis or alternative required. *Entrance requirements:* For master's, GRE. *Application deadline:* For fall admission, 3/15; for spring admission, 9/15. Applications are processed on a rolling basis. Application fee: $50 ($60 for international students). Electronic applications accepted. Tuition, state resident: part-time $217 per credit hour. Tuition, nonresident: part-time $783 per credit hour. Required fees: $3 per credit hour. $130 per year. One-time fee: $25 part-time. *Financial aid:* Research assistantships, teaching assistantships, career-related internships or fieldwork and Federal Work-Study available. Financial aid application deadline: 3/1; financial aid applicants required to submit FAFSA. Total annual research expenditures: $40,225. *Unit head:* Joe Hibey, Chair, 303-556-2872, Fax: 303-556-2383. *Application contact:* Christy Mourning, Program Assistant, 303-556-2872, Fax: 303-556-2383, E-mail: cmournin@carbon.cudenver.edu.

University of Connecticut, Graduate School, School of Engineering, Field of Electrical and Systems Engineering, Storrs, CT 06269. Offers biological engineering (MS); control and communication systems (MS, PhD); electromagnetics and physical electronics (MS, PhD). Terminal master's awarded for partial completion of doctoral program. *Degree requirements:* For master's, thesis or alternative required; for doctorate, dissertation required. *Entrance requirements:* For master's and doctorate, GRE General Test, TOEFL.

See in-depth description on page 1061.

University of Dayton, Graduate School, School of Engineering, Department of Electrical and Computer Engineering, Dayton, OH 45469-1300. Offers MSEE, DE, PhD. Part-time and evening/weekend programs available. *Faculty:* 16 full-time (0 women), 3 part-time (0 women). *Students:* 25 full-time (3 women), 34 part-time (4 women); includes 10 minority (1 African American, 6 Asian Americans or Pacific Islanders, 3 Hispanic Americans), 11 international. Average age 24. In 1998, 21 master's, 3 doctorates awarded. *Degree requirements:* For master's, thesis optional, foreign language not required; for doctorate, variable foreign language requirement (computer language can substitute for one), dissertation, departmental qualifying exam required. *Entrance requirements:* For master's, TOEFL. *Application deadline:* For fall admission, 8/1. Applications are processed on a rolling basis. Application fee: $30. *Financial aid:* In 1998–99, 15 students received aid, including 1 fellowship with full tuition reimbursement available (averaging $15,000 per year), 7 research assistantships with full tuition reimbursements available (averaging $12,000 per year), 3 teaching assistantships with full tuition reimbursements available (averaging $9,000 per year); institutionally-sponsored loans also available. Financial aid application deadline: 5/1. *Faculty research:* Analog and signal processing, electromagnetics, electro-optics, digital computer architectures. *Unit head:* Dr. Donald L. Moon, Associate Dean, 937-229-2241, Fax: 937-229-2471, E-mail: dmoon@engr.udayton.edu. *Application contact:* Dr. Donald L. Moon, Associate Dean, 937-229-2241, Fax: 937-229-2471, E-mail: dmoon@engr.udayton.edu.

University of Delaware, College of Engineering, Department of Electrical and Computer Engineering, Newark, DE 19716. Offers MEE, PhD. Part-time programs available. Postbaccalaureate distance learning degree programs offered (no on-campus study). *Faculty:* 17 full-time (0 women), 2 part-time (0 women). *Students:* 52 full-time (5 women), 28 part-time (7 women); includes 2 minority (1 African American, 1 Asian American or Pacific Islander), 52 international. Average age 25. 101 applicants, 19% accepted. In 1998, 13 master's, 6 doctorates awarded (17% entered university research/teaching, 83% found other work related to degree). Terminal master's awarded for partial completion of doctoral program. *Degree requirements:* For master's, thesis optional, foreign language not required; for doctorate, dissertation required, foreign language not required. *Entrance requirements:* For master's and doctorate, GRE General Test. *Average time to degree:* Master's–2 years full-time, 3 years part-time; doctorate–4 years full-time, 6 years part-time. *Application deadline:* For fall admission, 7/1; for spring admission, 12/1. Applications are processed on a rolling basis. Application fee: $45. Electronic applications accepted. *Financial aid:* In 1998–99, 51 students received aid, including 36 research assistantships with full tuition reimbursements available (averaging $11,500 per year), 15 teaching assistantships with full tuition reimbursements available (averaging $11,500 per year); fellowships, Federal Work-Study, institutionally-sponsored loans, and unspecified assistantships also available. Financial aid application deadline: 3/1. *Faculty research:* Signal and image processing, communications, devices and materials, opto-electronics, computer engineering. Total annual research expenditures: $3 million. *Unit head:* Dr. Neal Gallagher, Chairman, 302-831-3142, Fax: 302-831-4316, E-mail: gallaghe@ee.udel.edu. *Application contact:* Dr. Robert G. Hunsperger, Chair, Graduate Committee, 302-831-8031, Fax: 302-831-4316, E-mail: hunsperg@ee.udel.edu.

University of Denver, Graduate Studies, Faculty of Natural Sciences, Mathematics and Engineering, Department of Engineering, Denver, CO 80208. Offers computer science and engineering (MS); electrical engineering (MS); management and general engineering (MSMGEN); materials science (PhD); mechanical engineering (MS). Part-time and evening/weekend programs available. *Faculty:* 15. *Students:* 23 (9 women) 8 international. Terminal master's awarded for partial completion of doctoral program. *Degree requirements:* For master's, thesis required (for some programs), foreign language not required; for doctorate, dissertation required, foreign language not required. *Entrance requirements:* For master's and doctorate, GRE General Test, TOEFL (minimum score of 570 required), TSE (minimum score of 230 required). *Application deadline:* Applications are processed on a rolling basis. Application fee: $40 ($45 for international students). *Unit head:* Dr. Albert J. Rosa, Chair, 303-871-2102. *Application contact:* Louise Carlson, Assistant to Chair, 303-871-2107.

University of Detroit Mercy, College of Engineering and Science, Department of Electrical Engineering, Detroit, MI 48219-0900. Offers ME, DE. Evening/weekend programs available. *Degree requirements:* For master's, computer language required, foreign language not required; for doctorate, dissertation required. *Faculty research:* Electromagnetics, computer architecture, systems.

See in-depth description on page 1063.

University of Florida, Graduate School, College of Engineering, Department of Electrical and Computer Engineering, Gainesville, FL 32611. Offers ME, MS, PhD, Engr. Part-time programs available. *Faculty:* 64. *Students:* 265 full-time (30 women), 64 part-time (5 women); includes 52 minority (6 African Americans, 16 Asian Americans or Pacific Islanders, 29 Hispanic Americans, 1 Native American), 170 international. 457 applicants, 63% accepted. In 1998, 65 master's, 20 doctorates awarded. Terminal master's awarded for partial completion of doctoral program. *Degree requirements:* For master's, thesis optional, foreign language not required; for doctorate and Engr, dissertation required, foreign language not required. *Entrance requirements:* For master's, GRE General Test (minimum score of 350 on verbal section, 1000 combined required), TOEFL (minimum score of 550 required), minimum GPA of 3.0; for doctorate, GRE General Test (minimum score of 350 on verbal section, 1200 combined on three sections required), TOEFL (minimum score of 550 required), minimum GPA of 3.5; for Engr, GRE General Test. *Application deadline:* For fall admission, 6/1 (priority date). Applications are processed on a rolling basis. Application fee: $20. Electronic applications accepted. *Financial aid:* In 1998–99, 136 students received aid, including 41 fellowships, 113 research assistantships, 33 teaching assistantships; unspecified assistantships also available. Financial aid application deadline: 4/15. *Faculty research:* Communications, electronics, digital signal processing, photonics. *Unit head:* Dr. Martin A. Uman, Chair, 352-392-0913, Fax: 352-392-8671, E-mail: muman@admin.ee.ufl.edu. *Application contact:* Dr. Jacob Hammer, Graduate Coordinator, 352-392-6607, Fax: 352-392-8381, E-mail: hammer@leaf.ce.ufl.edu.

See in-depth description on page 1065.

University of Florida, Graduate School, Graduate Engineering and Research Center (GERC), Gainesville, FL 32611. Offers aerospace engineering (ME, MS, PhD, Engr); electrical and computer engineering (ME, MS, PhD, Engr); engineering mechanics (ME, MS, PhD, Engr); industrial and systems engineering (ME, MS, PhD, Engr). Part-time programs available. Postbaccalaureate distance learning degree programs offered. *Faculty:* 6 full-time (0 women), 16 part-time (1 woman). *Students:* 13 full-time (6 women), 159 part-time (29 women); includes 21 minority (8 African Americans, 8 Asian Americans or Pacific Islanders, 4 Hispanic Americans, 1 Native American) Terminal master's awarded for partial completion of doctoral program. *Degree requirements:* For master's, computer language required (for some programs), thesis optional, foreign language not required; for doctorate, computer language (for some programs), dissertation required; for Engr, computer language (for some programs), thesis required, foreign language not required. *Entrance requirements:* For master's, GRE General Test (minimum combined score of 1000 required), TOEFL, minimum GPA of 3.0; for doctorate, GRE General Test (minimum combined score of 1200 required), written and oral qualifying exams, TOEFL, minimum GPA of 3.0, master's degree in engineering; for Engr, GRE General Test (minimum combined score of 1000 required), TOEFL, minimum GPA of 3.0, master's degree in engineering. *Application deadline:* For fall admission, 6/1; for spring admission, 10/1. Applications are processed on a rolling basis. Application fee: $20. Electronic applications accepted. *Unit head:* Dr. Pasquale M. Sforza, Director, 850-833-9355, Fax: 850-833-9366, E-mail: sforza@gerc.eng.

ufl.edu. *Application contact:* Judi Shivers, Program Assistant, 850-833-9350, Fax: 850-833-9366, E-mail: reginfo@gerc.eng.ufl.edu.

University of Hawaii at Manoa, Graduate Division, College of Engineering, Department of Electrical Engineering, Honolulu, HI 96822. Offers MS, PhD. *Faculty:* 29 full-time (1 woman). *Students:* 45 full-time (13 women), 32 part-time (4 women); includes 11 minority (all Asian Americans or Pacific Islanders), 60 international. 129 applicants, 59% accepted. In 1998, 21 master's, 1 doctorate awarded. *Degree requirements:* For master's, thesis required, foreign language not required; for doctorate, one foreign language, dissertation, exams required. *Entrance requirements:* For master's, TOEFL (minimum score of 540 required; average 607); for doctorate, GRE General Test, GRE Subject Test, TOEFL (minimum score of 540 required; average 607). *Application deadline:* For fall admission, 3/1 (priority date); for spring admission, 8/1. Applications are processed on a rolling basis. Application fee: $25 ($50 for international students). *Financial aid:* In 1998–99, 42 research assistantships (averaging $15,190 per year), 9 teaching assistantships (averaging $12,873 per year) were awarded.; tuition waivers (full and partial) also available. *Faculty research:* Computers and artificial intelligence, communication and networking, control theory, physical electronics, VLSI design, micromillimeter waves. Total annual research expenditures: $2 million. *Unit head:* Dr. Kazutoshi Najita, Chairperson, 808-956-7586, Fax: 808-956-3427, E-mail: najita@spectra.eng.hawaii.edu. *Application contact:* Dr. Anthony Kuh, Graduate Field Chairperson, 808-956-7443, Fax: 808-956-3427, E-mail: kuh@spectra.eng.hawaii.edu.

University of Houston, Cullen College of Engineering, Department of Electrical and Computer Engineering, Houston, TX 77004. Offers MEE, MSEE, PhD. Part-time and evening/weekend programs available. *Faculty:* 29 full-time (3 women), 4 part-time (0 women). *Students:* 124 full-time (29 women), 53 part-time (9 women); includes 17 minority (4 African Americans, 11 Asian Americans or Pacific Islanders, 2 Hispanic Americans), 111 international. Average age 29. 180 applicants, 18% accepted. In 1998, 57 master's, 11 doctorates awarded. Terminal master's awarded for partial completion of doctoral program. *Degree requirements:* For master's, thesis required (for some programs), foreign language not required; for doctorate, dissertation, departmental qualifying exam required, foreign language not required. *Entrance requirements:* For master's and doctorate, GRE General Test, TOEFL. *Application deadline:* For fall admission, 7/3 (priority date); for spring admission, 12/4. Applications are processed on a rolling basis. Application fee: $25 ($75 for international students). *Financial aid:* Research assistantships, teaching assistantships, career-related internships or fieldwork, Federal Work-Study, institutionally-sponsored loans, and tuition waivers (partial) available. Financial aid application deadline: 7/1. *Faculty research:* Applied electromagnetics and microelectronics, signal and image processing, biomedical engineering, geophysical applications, control engineering. Total annual research expenditures: $2 million. *Unit head:* Dr. Wallace L. Anderson, Chairman, 713-743-4400, Fax: 713-743-4444, E-mail: wanderson@uh.edu. *Application contact:* Mylyssa McDonald, Graduate Analyst, 713-743-4403, Fax: 713-743-4444, E-mail: mmm05866@jetson.uh.edu.

See in-depth description on page 1067.

University of Idaho, College of Graduate Studies, College of Engineering, Department of Electrical Engineering, Moscow, ID 83844-4140. Offers computer engineering (M Engr, MS); electrical engineering (M Engr, MS, PhD). *Faculty:* 16 full-time (0 women), 2 part-time (1 woman). *Students:* 18 full-time (0 women), 69 part-time (3 women). In 1998, 18 master's, 1 doctorate awarded. *Degree requirements:* For master's, computer language, thesis required, foreign language not required; for doctorate, computer language, dissertation required. *Entrance requirements:* For master's, minimum GPA of 2.8; for doctorate, minimum undergraduate GPA of 2.8, 3.0 graduate. *Application deadline:* For fall admission, 8/1; for spring admission, 12/15. Application fee: $35 ($45 for international students). *Financial aid:* In 1998–99, 4 teaching assistantships (averaging $14,251 per year) were awarded.; fellowships, research assistantships, career-related internships or fieldwork and Federal Work-Study also available. Financial aid application deadline: 2/15. *Faculty research:* Digital systems, energy systems. *Unit head:* Dr. David Egolf, Chairman, 208-885-6554.

University of Illinois at Chicago, Graduate College, College of Engineering, Department of Electrical Engineering and Computer Science, Program in Electrical Engineering, Chicago, IL 60607-7128. Offers MS, PhD. Evening/weekend programs available. *Degree requirements:* For master's, computer language, thesis or alternative required, foreign language not required; for doctorate, computer language, dissertation, departmental qualifying exam required, foreign language not required. *Entrance requirements:* For master's, TOEFL (minimum score of 550 required), minimum GPA of 3.75 on a 5.0 scale, BS in related field; for doctorate, GRE General Test, TOEFL (minimum score of 550 required), minimum GPA of 3.75 on a 5.0 scale, MS in related field. *Application deadline:* For fall admission, 6/7; for spring admission, 11/1. Application fee: $40 ($50 for international students). *Financial aid:* Fellowships, research assistantships, teaching assistantships available. *Unit head:* Gyan Agarwal, Director of Graduate Studies, 312-996-8679.

See in-depth description on page 1069.

University of Illinois at Urbana–Champaign, Graduate College, College of Engineering, Department of Electrical and Computer Engineering, Urbana, IL 61801. Offers computer engineering (MS, PhD); electrical engineering (MS, PhD). *Faculty:* 89 full-time (3 women). *Students:* 447 full-time (41 women); includes 63 minority (4 African Americans, 50 Asian Americans or Pacific Islanders, 9 Hispanic Americans), 212 international. 930 applicants, 11% accepted. In 1998, 99 master's, 41 doctorates awarded. *Degree requirements:* For master's and doctorate, thesis/dissertation required, foreign language not required. *Entrance requirements:* For master's, minimum GPA of 4.0 on a 5.0 scale. *Application deadline:* For fall admission, 1/15. Applications are processed on a rolling basis. Application fee: $40 ($50 for international students). Tuition, state resident: full-time $4,616. Tuition, nonresident: full-time $11,768. Full-time tuition and fees vary according to course load. *Financial aid:* In 1998–99, 29 fellowships, 274 research assistantships, 101 teaching assistantships were awarded.; tuition waivers (full and partial) also available. Financial aid application deadline: 1/15. *Unit head:* Sung Mo Kang, Head, 217-333-2301. *Application contact:* William Perkins, Director of Graduate Studies, 217-333-0207, Fax: 217-244-7075, E-mail: perkins@decision.csl.uiuc.edu.

See in-depth description on page 1071.

The University of Iowa, Graduate College, College of Engineering, Department of Electrical and Computer Engineering, Iowa City, IA 52242-1316. Offers MS, PhD. *Faculty:* 16 full-time, 3 part-time. *Students:* 39 full-time (8 women), 22 part-time (4 women); includes 5 minority (all Asian Americans or Pacific Islanders), 31 international. 267 applicants, 48% accepted. In 1998, 12 master's, 4 doctorates awarded. *Degree requirements:* For master's, thesis optional; for doctorate, dissertation, comprehensive exam required. *Entrance requirements:* For master's and doctorate, TOEFL. *Application deadline:* Applications are processed on a rolling basis. Application fee: $30 ($50 for international students). *Financial aid:* In 1998–99, 1 fellowship, 21 research assistantships, 21 teaching assistantships were awarded. Financial aid applicants required to submit FAFSA. *Faculty research:* Computer systems and applications, biomedical image processing, robotics, acousto-optics. *Unit head:* Dr. Sudhakar M. Reddy, Chair, 319-335-5196.

University of Kansas, Graduate School, School of Engineering, Department of Electrical Engineering and Computer Science, Program in Electrical Engineering, Lawrence, KS 66045. Offers MS, DE, PhD. Terminal master's awarded for partial completion of doctoral program. *Degree requirements:* For master's, exam required, thesis optional, foreign language not required; for doctorate, one foreign language (computer language can substitute), dissertation, comprehensive and qualifying exams required. *Entrance requirements:* For master's and doctorate, GRE, TOEFL (minimum score of 600 required), minimum GPA of 3.0. *Application deadline:* For fall admission, 4/15 (priority date); for spring admission, 10/15. Applications are processed on a rolling basis. Application fee: $30. *Financial aid:* Fellowships, research assistantships, teaching assistantships, career-related internships or fieldwork available. *Application contact:* Victor Wallace, Graduate Director, 785-864-4487, Fax: 785-864-3226, E-mail: grad.admissions@eecs.ukans.edu.

See in-depth description on page 1073.

University of Kentucky, Graduate School, Graduate School Programs from the College of Engineering, Program in Electrical Engineering, Lexington, KY 40506-0032. Offers MSEE, PhD. *Degree requirements:* For master's, comprehensive exam required, thesis optional, foreign language not required; for doctorate, dissertation, comprehensive exam required. *Entrance requirements:* For master's, GRE General Test (minimum combined score of 1100 on three sections required), minimum undergraduate GPA of 3.0; for doctorate, GRE General Test (minimum combined score of 1200 on three sections required), minimum graduate GPA of 3.0. *Faculty research:* Signal processing, systems, and control; electromagnetic field theory; power electronics and machines; computer engineering and VLSI; materials and devices.

University of Louisville, Graduate School, Speed Scientific School, Department of Electrical Engineering, Louisville, KY 40292-0001. Offers M Eng, MS. *Accreditation:* ABET (one or more programs are accredited). *Faculty:* 15 full-time (0 women), 1 part-time (0 women). *Students:* 22 full-time (2 women), 99 part-time (9 women); includes 15 minority (4 African Americans, 6 Asian Americans or Pacific Islanders, 4 Hispanic Americans, 1 Native American), 19 international. Average age 29. In 1998, 22 degrees awarded. *Degree requirements:* For master's, thesis required, foreign language not required. *Entrance requirements:* For master's, GRE General Test (minimum combined score of 1200 required), minimum GPA of 2.75 required. *Application deadline:* Applications are processed on a rolling basis. Application fee: $25. Electronic applications accepted. *Unit head:* Darrel L. Chenoweth, Chair, 502-852-6289, Fax: 502-852-8851, E-mail: dlchen01@gwise.louisville.edu. *Application contact:* Joyce M. Hack, Administrative Assistant, 502-852-7517, Fax: 502-852-8851, E-mail: jmhack01@gwise.louisville.edu.

University of Maine, Graduate School, College of Engineering, Department of Electrical and Computer Engineering, Orono, ME 04469. Offers computer engineering (MS); electrical engineering (MS, PhD). Part-time programs available. *Faculty:* 14 full-time. *Students:* 13 full-time (3 women), 7 part-time; includes 2 minority (1 African American, 1 Asian American or Pacific Islander), 4 international. Average age 25. 29 applicants, 41% accepted. In 1998, 6 master's awarded. *Degree requirements:* For master's, thesis required (for some programs), foreign language not required; for doctorate, dissertation required, foreign language not required. *Entrance requirements:* For master's and doctorate, GRE General Test, TOEFL (minimum score of 550 required). *Application deadline:* For fall admission, 2/1 (priority date); for spring admission, 10/15. Applications are processed on a rolling basis. Application fee: $50. *Financial aid:* In 1998–99, 13 research assistantships with tuition reimbursements (averaging $9,450 per year), 2 teaching assistantships with tuition reimbursements (averaging $7,236 per year) were awarded.; Federal Work-Study, institutionally-sponsored loans, and tuition waivers (full and partial) also available. Financial aid application deadline: 3/1. *Unit head:* Dr. John Field, Chair, 207-581-2223, Fax: 201-581-2220. *Application contact:* Scott G. Delcourt, Director of the Graduate School, 207-581-3218, Fax: 207-581-3232, E-mail: graduate@maine.edu.

University of Manitoba, Faculty of Graduate Studies, Faculty of Engineering, Department of Electrical and Computer Engineering, Winnipeg, MB R3T 2N2, Canada. Offers M Eng, M Sc, PhD. *Degree requirements:* For master's and doctorate, thesis/dissertation required. *Unit head:* S. Onyshko, Head.

University of Maryland, Baltimore County, Graduate School, College of Engineering, Department of Computer Science and Electrical Engineering, Program in Electrical Engineering, Baltimore, MD 21250-5398. Offers MS, PhD. *Faculty:* 11 full-time (0 women), 8 part-time (0 women). *Students:* 30 full-time (9 women), 26 part-time (3 women); includes 5 minority (1 African American, 4 Asian Americans or Pacific Islanders), 39 international. 114 applicants, 25% accepted. In 1998, 9 master's, 8 doctorates awarded. *Entrance requirements:* For master's and doctorate, GRE General Test. *Application deadline:* For fall admission, 7/1. Applications are processed on a rolling basis. Application fee: $45. *Faculty research:* Integrated optics and optoelectronics, light propagation, communications and signal processing, medical imaging, semiconductor lasers. *Unit head:* Mylyssa McDonald, Graduate Analyst, 713-743-4403, Fax: 713-743-4444, E-mail: mmm05866@jetson.uh.edu. *Application contact:* Dr. Ray Chen, Director, 440-455-3788.

University of Maryland, College Park, Graduate School, A. James Clark School of Engineering, Department of Electrical and Computer Engineering, College Park, MD 20742-5045. Offers electrical and computer engineering (M Eng, MS, PhD); telecommunications (MS). Part-time and evening/weekend programs available. Postbaccalaureate distance learning degree programs offered. *Faculty:* 62 full-time. *Students:* 281 full-time, 82 part-time. 1165 applicants, 18% accepted. In 1998, 54 master's, 36 doctorates awarded. *Degree requirements:* For master's, thesis or alternative required, foreign language not required; for doctorate, dissertation, oral exam, qualifying exam required. *Entrance requirements:* For master's, minimum GPA of 3.0. *Application deadline:* Applications are processed on a rolling basis. Application fee: $50 ($70 for international students). Tuition, state resident: part-time $272 per credit hour. Tuition, nonresident: part-time $475 per credit hour. Required fees: $632; $379 per year. *Financial aid:* In 1998–99, 51 fellowships with full tuition reimbursements (averaging $4,000 per year), 179 research assistantships with tuition reimbursements (averaging $9,809 per year), 42 teaching assistantships with tuition reimbursements (averaging $8,791 per year) were awarded.; career-related internships or fieldwork also available. Financial aid applicants required to submit FAFSA. *Faculty research:* Communications and control, electrophysics, micro-electronics. Total annual research expenditures: $9.3 million. *Unit head:* Dr. Nariman Farvardin, Chairman, 301-405-3683, Fax: 301-314-9281. *Application contact:* Trudy Lindsey, Director, Graduate Admission and Records, 301-405-4198, Fax: 301-314-9305, E-mail: grschool@deans.umd.edu.

See in-depth description on page 1075.

University of Maryland, College Park, Graduate School, A. James Clark School of Engineering, Professional Program in Engineering, College Park, MD 20742-5045. Offers aerospace engineering (M Eng); chemical engineering (M Eng); civil engineering (M Eng); electrical engineering (M Eng); fire protection engineering (M Eng); materials science and engineering (M Eng); mechanical engineering (M Eng); reliability engineering (M Eng); systems engineering (M Eng). Part-time and evening/weekend programs available. Postbaccalaureate distance learning degree programs offered. *Faculty:* 11 part-time (0 women). *Students:* 20 full-time (3 women), 205 part-time (42 women); includes 58 minority (27 African Americans, 25 Asian Americans or Pacific Islanders, 5 Hispanic Americans, 1 Native American), 20 international. *Degree requirements:* For master's, foreign language and thesis not required. *Application deadline:* Applications are processed on a rolling basis. Application fee: $50 ($70 for international students). Tuition, state resident: part-time $272 per credit hour. Tuition, nonresident: part-time $475 per credit hour. Required fees: $632; $379 per year. *Unit head:* Dr. Patrick Cunniff, Associate Dean, 301-405-5256, Fax: 301-314-9477. *Application contact:* Trudy Lindsey, Director, Graduate Admission and Records, 301-405-4198, Fax: 301-314-9305, E-mail: grschool@deans.umd.edu.

University of Massachusetts Amherst, Graduate School, College of Engineering, Department of Electrical and Computer Engineering, Amherst, MA 01003. Offers MS, PhD. Part-time and evening/weekend programs available. *Faculty:* 31 full-time (1 woman). *Students:* 92 full-time (26 women), 52 part-time (4 women); includes 5 minority (1 African American, 2 Asian Americans or Pacific Islanders, 2 Hispanic Americans), 109 international. Average age 27. 630 applicants, 30% accepted. In 1998, 38 master's, 14 doctorates awarded. Terminal master's awarded for partial completion of doctoral program. *Degree requirements:* For master's, foreign language and thesis not required; for doctorate, dissertation required, foreign language not required. *Entrance requirements:* For master's and doctorate, GRE General Test. *Application deadline:* For fall admission, 2/1 (priority date); for spring admission, 10/1. Applications are processed on a rolling basis. Application fee: $40. Tuition, state resident: full-time $2,640; part-time $165 per credit. Tuition, nonresident: full-time $9,756; part-time $407 per credit.

Electrical Engineering

University of Massachusetts Amherst (continued)

Required fees: $1,221 per term. One-time fee: $110. Full-time tuition and fees vary according to course load, campus/location and reciprocity agreements. *Financial aid:* In 1998–99, 9 fellowships with full tuition reimbursements (averaging $3,296 per year), 80 research assistantships with full tuition reimbursements (averaging $9,353 per year), 38 teaching assistantships with full tuition reimbursements (averaging $9,833 per year) were awarded.; career-related internships or fieldwork, Federal Work-Study, grants, scholarships, traineeships, and unspecified assistantships also available. Aid available to part-time students. Financial aid application deadline: 2/1. *Unit head:* Dr. Seshu B. Desu, Head, 413-545-0962, Fax: 413-545-4611, E-mail: sdesu@ecs.umass.edu. *Application contact:* Dr. Donald E. Scott, Chair, Admissions Committee, 413-545-0937, Fax: 413-545-4611, E-mail: scott@ecs.umass.edu.

University of Massachusetts Dartmouth, Graduate School, College of Engineering, Department of Electrical and Computer Engineering, North Dartmouth, MA 02747-2300. Offers electrical engineering (MS, PhD). Part-time programs available. *Faculty:* 22 full-time (2 women), 1 (woman) part-time. *Students:* 27 full-time (3 women), 48 part-time (7 women); includes 2 minority (1 African American, 1 Asian American or Pacific Islander), 35 international. Average age 30. 150 applicants, 92% accepted. In 1998, 12 master's, 1 doctorate awarded. *Degree requirements:* For master's, thesis or alternative, culminating project or thesis required, foreign language not required; for doctorate, dissertation, comprehensive exam required, foreign language not required. *Entrance requirements:* For master's, GRE General Test, TOEFL. *Application deadline:* For fall admission, 4/20 (priority date); for spring admission, 11/15 (priority date). Applications are processed on a rolling basis. Application fee: $40 for international students. Tuition, area resident: Full-time $3,107; part-time $129 per credit. Tuition, state resident: full-time $2,071; part-time $86 per credit. Tuition, nonresident: full-time $7,845; part-time $327 per credit. Required fees: $2,888. Full-time tuition and fees vary according to program and reciprocity agreements. Part-time tuition and fees vary according to course load and reciprocity agreements. *Financial aid:* In 1998–99, 20 research assistantships with full tuition reimbursements (averaging $9,550 per year), 12 teaching assistantships with full tuition reimbursements (averaging $8,508 per year) were awarded.; Federal Work-Study and unspecified assistantships also available. Aid available to part-time students. Financial aid application deadline: 2/1; financial aid applicants required to submit FAFSA. *Faculty research:* Signal processing, systems analysis, underwater acoustics. Total annual research expenditures: $291,000. *Unit head:* Karen Payton, Director, 508-999-8434. *Application contact:* Carol A. Novo, Graduate Admissions Office, 508-999-8026, Fax: 508-999-8183, E-mail: graduate@umassd.edu.

University of Massachusetts Lowell, Graduate School, James B. Francis College of Engineering, Department of Electrical Engineering, Program in Electrical Engineering, Lowell, MA 01854-2881. Offers MS Eng, D Eng. Part-time and evening/weekend programs available. Terminal master's awarded for partial completion of doctoral program. *Degree requirements:* For master's, thesis required, foreign language not required; for doctorate, 2 foreign languages, computer language, dissertation required. *Entrance requirements:* For master's and doctorate, GRE General Test. *Application deadline:* For fall admission, 4/1 (priority date); for spring admission, 10/1. Applications are processed on a rolling basis. Application fee: $20 ($35 for international students). *Financial aid:* Career-related internships or fieldwork, Federal Work-Study, and institutionally-sponsored loans available. Aid available to part-time students. Financial aid application deadline: 4/1. *Application contact:* Dr. Ross Holmstrom, Coordinator, 978-934-3307, E-mail: ross_holmstrom@woods.uml.edu.

The University of Memphis, Graduate School, Herff College of Engineering, Department of Electrical Engineering, Memphis, TN 38152. Offers automatic control systems (MS); biomedical systems (MS); communications and propagation systems (MS); electrical engineering (PhD); engineering computer systems (MS). *Faculty:* 11 full-time (0 women), 3 part-time (0 women). *Students:* 16 full-time (2 women), 5 part-time; includes 1 minority (Asian American or Pacific Islander), 14 international. Average age 26. 49 applicants, 43% accepted. In 1998, 11 degrees awarded. *Degree requirements:* For master's, thesis or alternative, written comprehensive exam required. *Entrance requirements:* For master's, GRE General Test (minimum combined score of 1000 required) or MAT, minimum undergraduate GPA of 2.5. *Application deadline:* For fall admission, 8/1; for spring admission, 12/1. Application fee: $25 ($50 for international students). Tuition, state resident: full-time $3,410; part-time $178 per credit hour. Tuition, nonresident: full-time $8,670; part-time $408 per credit hour. Tuition and fees vary according to program. *Financial aid:* In 1998–99, 20 students received aid, including 6 research assistantships, 12 teaching assistantships; career-related internships or fieldwork also available. *Faculty research:* Ventricular arrhythmias, cerebral palsy, automatic computer troubleshooting, noninvasive monitoring of gastric motor functions. *Unit head:* Dr. Babajide Familoni, Chair, 901-678-2175. *Application contact:* Dr. Steven T. Griffin, Coordinator of Graduate Studies, 901-678-3250.

University of Miami, Graduate School, College of Engineering, Department of Electrical and Computer Engineering, Coral Gables, FL 33124. Offers MSECE, PhD. Part-time programs available. *Faculty:* 15. *Students:* 23 full-time (3 women), 18 part-time (3 women); includes 20 minority (2 African Americans, 12 Asian Americans or Pacific Islanders, 6 Hispanic Americans) Average age 25. 100 applicants, 40% accepted. In 1998, 10 master's, 5 doctorates awarded. *Degree requirements:* For master's, thesis optional, foreign language not required; for doctorate, dissertation, comprehensive exam required, foreign language not required. *Entrance requirements:* For master's, GRE General Test (minimum combined score of 1000 required), TOEFL (minimum score of 550 required), minimum GPA of 3.0; for doctorate, GRE General Test (minimum combined score of 1100 required), TOEFL (minimum score of 550 required), minimum GPA of 3.3 (undergraduate), 3.5 (graduate). *Application deadline:* For fall admission, 4/1 (priority date); for spring admission, 11/1. Application fee: $35. Tuition: Full-time $15,336; part-time $852 per credit. Required fees: $174. Tuition and fees vary according to program. *Financial aid:* In 1998–99, 17 students received aid, including 2 fellowships with full tuition reimbursements available, 5 research assistantships with full tuition reimbursements available, 13 teaching assistantships with full tuition reimbursements available; career-related internships or fieldwork, Federal Work-Study, institutionally-sponsored loans, scholarships, and unspecified assistantships also available. Aid available to part-time students. *Faculty research:* Computer network, computer vision and image processing, database systems, digital signal processing, machine intelligence. Total annual research expenditures: $203,792. *Unit head:* Dr. Tzay Young, Chairman, 305-284-3291. *Application contact:* Dr. Claude Lindquist, Graduate Adviser, 305-284-3291.

University of Michigan, Horace H. Rackham School of Graduate Studies, College of Engineering, Department of Electrical Engineering and Computer Science, Division of Electrical Science and Engineering, Ann Arbor, MI 48109. Offers electrical engineering (MS, MSE, PhD, EE). *Degree requirements:* For doctorate, dissertation, oral defense of dissertation, preliminary exams required. *Entrance requirements:* For master's, GRE General Test (minimum combined score of 1900 on three sections required; average 1934); for doctorate, GRE General Test (minimum combined score of 1900 on three sections required; average 1934), master's degree; for EE, GRE.

See in-depth description on page 1077.

University of Michigan–Dearborn, College of Engineering and Computer Science, Department of Electrical and Computer Engineering, Dearborn, MI 48128-1491. Offers computer engineering (MSE); electrical engineering (MSE). Part-time programs available. *Faculty:* 11 full-time (1 woman), 10 part-time (1 woman). *Students:* 3 full-time (0 women), 84 part-time (12 women); includes 10 minority (1 African American, 8 Asian Americans or Pacific Islanders, 1 Hispanic American), 1 international. Average age 29. In 1998, 44 degrees awarded. *Degree requirements:* For master's, computer language required, thesis optional, foreign language not required. *Entrance requirements:* For master's, bachelor's degree in electrical and computer engineering or equivalent, minimum GPA of 3.0. *Application deadline:* For fall admission, 8/1; for winter admission, 12/1; for spring admission, 4/1. Applications are processed on a rolling basis. Application fee: $55. Tuition, state resident: part-time $259 per credit hour. Tuition, nonresident: part-time $748 per credit hour. Required fees: $80 per course. Tuition and fees vary according to course level, course load and program. *Financial aid:* In 1998–99, 9 research assistantships were awarded.; fellowships, teaching assistantships, Federal Work-Study also available. *Faculty research:* Process control, fuzzy systems design, machine vision image processing and pattern recognition. *Unit head:* Dr. M. Shridhar, Chair, 313-593-5420, Fax: 313-593-9967, E-mail: shridhar@umdsun2.umd.umich.edu. *Application contact:* Kathy Gentry, Secretary II, 313-593-5420, Fax: 313-593-9967, E-mail: kwebb@engin.umd.umich.edu.

University of Minnesota, Twin Cities Campus, Graduate School, Institute of Technology, Department of Electrical and Computer Engineering, Minneapolis, MN 55455-0213. Offers MEE, MSEE, PhD. Part-time programs available. *Degree requirements:* For master's, thesis or alternative required, foreign language not required; for doctorate, dissertation required, foreign language not required. *Entrance requirements:* For master's and doctorate, TOEFL (minimum score of 550 required). *Faculty research:* Signal processing, microelectronics, computers, controls, power electronics.

See in-depth description on page 1079.

University of Missouri–Columbia, Graduate School, College of Engineering, Department of Electrical Engineering, Columbia, MO 65211. Offers MS, PhD. *Faculty:* 24 full-time (0 women). *Students:* 19 full-time (4 women), 26 part-time; includes 2 minority (1 African American, 1 Asian American or Pacific Islander), 38 international. 45 applicants, 36% accepted. In 1998, 16 master's, 6 doctorates awarded. *Degree requirements:* For master's, thesis or alternative required, foreign language not required; for doctorate, dissertation required, foreign language not required. *Entrance requirements:* For master's, GRE General Test, TOEFL, minimum GPA of 3.0; for doctorate, GRE General Test, GRE Subject Test, TOEFL, minimum GPA of 3.0. *Application deadline:* For fall admission, 7/1 (priority date). Applications are processed on a rolling basis. Application fee: $30 ($50 for international students). *Financial aid:* Research assistantships, teaching assistantships, institutionally-sponsored loans available. *Unit head:* Dr. Kenneth Unklesbay, Chair, 573-882-2781, Fax: 573-882-0397, E-mail: unk@ece.missouri.edu.

See in-depth description on page 1085.

University of Missouri–Rolla, Graduate School, School of Engineering, Department of Electrical and Computer Engineering, Program in Electrical Engineering, Rolla, MO 65409-0910. Offers MS, DE, PhD. Part-time and evening/weekend programs available. *Students:* 94 full-time (18 women), 16 part-time (2 women); includes 4 minority (1 African American, 3 Asian Americans or Pacific Islanders), 70 international. Average age 27. 499 applicants, 65% accepted. In 1998, 32 master's, 7 doctorates awarded. Terminal master's awarded for partial completion of doctoral program. *Degree requirements:* For master's, thesis or alternative required, foreign language not required; for doctorate, dissertation, departmental qualifying exam required. *Entrance requirements:* For master's and doctorate, GRE General Test (score in 85th percentile or higher required), TOEFL (minimum score of 580 required; average 600). *Application deadline:* For fall admission, 6/1; for spring admission, 11/1. Applications are processed on a rolling basis. Application fee: $25. Electronic applications accepted. *Financial aid:* Application deadline: 3/1. *Unit head:* Paul D. Stigall, Assistant Chairman, 573-341-4533, Fax: 573-341-4532, E-mail: stigall@ee.umr.edu. *Application contact:* Paul D. Stigall, Assistant Chairman, 573-341-4533, Fax: 573-341-4532, E-mail: stigall@ee.umr.edu.

University of Nebraska–Lincoln, Graduate College, College of Engineering and Technology, Department of Electrical Engineering, Lincoln, NE 68588. Offers electrical engineering (MS); engineering (PhD). *Faculty:* 19 full-time (0 women), 1 part-time (0 women). *Students:* 14 full-time (3 women), 15 part-time; includes 1 minority (Asian American or Pacific Islander), 18 international. Average age 27. 139 applicants, 21% accepted. In 1998, 9 degrees awarded. *Degree requirements:* For master's, thesis optional, foreign language not required; for doctorate, dissertation, comprehensive exams required. *Entrance requirements:* For master's and doctorate, GRE General Test, TOEFL (minimum score of 550 required). *Application deadline:* For fall admission, 4/15; for spring admission, 10/15. Application fee: $35. Electronic applications accepted. *Financial aid:* In 1998–99, 6 fellowships, 7 teaching assistantships were awarded.; research assistantships, Federal Work-Study also available. Aid available to part-time students. Financial aid application deadline: 2/15. *Faculty research:* Communication and signal processing systems, electromagnetics, gaseous electronics and plasmas, power/control/digital systems, remote sensing. *Unit head:* Dr. Rodney Soukup, Chair, 402-472-3771, Fax: 402-472-4732.

University of Nevada, Las Vegas, Graduate College, Howard R. Hughes College of Engineering, Department of Electrical and Computer Engineering, Las Vegas, NV 89154-9900. Offers MSE, PhD. *Faculty:* 12 full-time (1 woman). *Students:* 18 full-time (4 women), 11 part-time (1 woman); includes 4 minority (all Asian Americans or Pacific Islanders), 18 international. 42 applicants, 33% accepted. In 1998, 10 master's awarded. *Degree requirements:* For master's, comprehensive exam required, thesis optional, foreign language not required; for doctorate, dissertation required. *Entrance requirements:* For master's, TOEFL (minimum score of 550 required), bachelor's degree in electrical engineering or related field, GRE General Test or minimum GPA of 3.0; for doctorate, minimum GPA of 3.5. *Application deadline:* For fall admission, 6/15 (priority date); for spring admission, 11/15. Applications are processed on a rolling basis. Application fee: $40 ($95 for international students). *Financial aid:* In 1998–99, 7 research assistantships with full tuition reimbursements (averaging $8,834 per year), 14 teaching assistantships with partial tuition reimbursements (averaging $8,361 per year) were awarded.; tuition waivers (full) also available. Financial aid application deadline:3/1. *Unit head:* Dr. Ashok Iyer, Chair, 702-895-4184. *Application contact:* Admissions and Information, 702-895-4183.

University of Nevada, Reno, Graduate School, College of Engineering, Department of Electrical Engineering, Reno, NV 89557. Offers MS, PhD. Terminal master's awarded for partial completion of doctoral program. *Degree requirements:* For master's, thesis optional, foreign language not required; for doctorate, dissertation required, foreign language not required. *Entrance requirements:* For master's, GRE General Test, TOEFL (minimum score of 500 required), minimum GPA of 2.75; for doctorate, GRE General Test, TOEFL (minimum score of 500 required), minimum GPA of 3.0.

University of New Brunswick, School of Graduate Studies, Faculty of Engineering, Department of Electrical and Computer Engineering, Fredericton, NB E3B 5A3, Canada. Offers M Eng, M Sc E, PhD. Part-time programs available. *Faculty:* 20 full-time (3 women). *Students:* 31 full-time (5 women), 2 part-time. 45 applicants, 44% accepted. In 1998, 5 master's, 1 doctorate awarded. *Degree requirements:* For master's, thesis required (for some programs), foreign language not required; for doctorate, dissertation, qualifying exam required, foreign language not required. *Entrance requirements:* For master's and doctorate, TOEFL (minimum score of 550 required), TWE, minimum GPA of 3.0. *Application deadline:* Applications are processed on a rolling basis. Application fee: $25. *Financial aid:* Research assistantships, teaching assistantships available. *Faculty research:* Renewable energy, fiber optics, digital communications, nonlinear control microelectronics. *Unit head:* Dr. B. G. Colpitts, Director of Graduate Studies, 504-453-, Fax: 504-453-3589, E-mail: gradce@unb.ca.

University of New Hampshire, Graduate School, College of Engineering and Physical Sciences, Programs in Engineering, Department of Electrical and Computer Engineering, Durham, NH 03824. Offers electrical engineering (MS). *Faculty:* 19 full-time. *Students:* 9 full-time (0 women), 39 part-time (4 women); includes 6 minority (4 Asian Americans or Pacific Islanders, 2 Hispanic Americans), 7 international. Average age 30. 31 applicants, 84% accepted. In 1998, 9 degrees awarded. *Degree requirements:* For master's, thesis or alternative required, foreign language not required. *Application deadline:* For fall admission, 4/1 (priority date). Applications are processed on a rolling basis. Application fee: $50. Tuition, area resident: Full-time $5,750; part-time $319 per credit. Tuition, state resident: full-time $8,625. Tuition, nonresident: full-time $14,640; part-time $598 per credit. Required fees: $224 per semester.

Tuition and fees vary according to course load, degree level and program. *Financial aid:* In 1998–99, 10 research assistantships, 7 teaching assistantships were awarded.; Federal Work-Study, scholarships, and tuition waivers (full and partial) also available. Aid available to part-time students. Financial aid application deadline: 2/15. *Faculty research:* Biomedical engineering, communications systems and information theory, digital systems, illumination engineering. *Unit head:* John LaCourse, Chairperson, 603-862-1324. *Application contact:* Tom Miller, Graduate Coordinator, 603-862-1326.

University of New Hampshire, Graduate School, College of Engineering and Physical Sciences, Programs in Engineering, Doctoral Program in Engineering, Durham, NH 03824. Offers chemical engineering (PhD); civil engineering (PhD); electrical engineering (PhD); mechanical engineering (PhD); systems design engineering (PhD). *Students:* 16 full-time (3 women), 8 part-time (1 woman), 12 international. *Degree requirements:* For doctorate, dissertation required. *Entrance requirements:* For doctorate, GRE (for civil and mechanical engineering options). *Application deadline:* For fall admission, 4/1 (priority date). Applications are processed on a rolling basis. Application fee: $50. Tuition, area resident: Full-time $5,750; part-time $319 per credit. Tuition, state resident: full-time $8,625. Tuition, nonresident: full-time $14,640; part-time $598 per credit. Required fees: $224 per semester. Tuition and fees vary according to course load, degree level and program.

University of New Haven, Graduate School, School of Engineering and Applied Science, Program in Electrical Engineering, West Haven, CT 06516-1916. Offers MSEE. Part-time and evening/weekend programs available. *Students:* 4 full-time (1 woman), 24 part-time (4 women); includes 1 minority (Asian American or Pacific Islander), 10 international. 19 applicants, 79% accepted. In 1998, 14 degrees awarded. *Degree requirements:* For master's, thesis or alternative required, foreign language not required. *Entrance requirements:* For master's, bachelor's degree in electrical engineering. *Application deadline:* Applications are processed on a rolling basis. Application fee: $50. *Financial aid:* Federal Work-Study available. Aid available to part-time students. Financial aid application deadline: 5/1; financial aid applicants required to submit FAFSA. *Unit head:* Dr. Bijan Karimi, Coordinator, 203-932-7164.

University of New Mexico, Graduate School, School of Engineering, Department of Electrical and Computer Engineering, Albuquerque, NM 87131-2039. Offers computer engineering (MS, PhD); manufacturing engineering (ME, MS); microelectronics (MS, PhD); network and control systems (MS, PhD); optoelectronics (MS, PhD); pulsed power and plasma science (MS, PhD); signal processing and communications (MS, PhD). ME offered through the Manufacturing Engineering Program. Part-time and evening/weekend programs available. *Faculty:* 41 full-time (3 women), 14 part-time (1 woman). *Students:* 76 full-time (18 women), 53 part-time (10 women); includes 23 minority (1 African American, 7 Asian Americans or Pacific Islanders, 14 Hispanic Americans, 1 Native American), 44 international. Average age 32. 68 applicants, 59% accepted. In 1998, 45 master's, 8 doctorates awarded. *Degree requirements:* For master's, thesis required (for some programs), foreign language not required; for doctorate, dissertation required, foreign language not required. *Entrance requirements:* For master's, GRE General Test (minimum score of 400 on verbal section, 650 on quantitative required), minimum GPA of 3.0; for doctorate, GRE General Test (minimum score of 450 on verbal section, 690 on quantitative requireed), minimum GPA of 3.0. *Application deadline:* For fall admission, 7/15; for spring admission, 11/14. Applications are processed on a rolling basis. Application fee: $25. *Financial aid:* In 1998–99, 80 students received aid, including 6 fellowships (averaging $1,824 per year), 69 research assistantships with tuition reimbursements available (averaging $3,211 per year), 13 teaching assistantships with tuition reimbursements available (averaging $10,404 per year); career-related internships or fieldwork and Federal Work-Study also available. Financial aid application deadline: 5/31. *Faculty research:* Applied electromagnetics, high performance computing. Total annual research expenditures: $3.2 million. *Unit head:* Dr. Christos Christodovlov, Chair, 505-277-6580, Fax: 505-277-1439, E-mail: cgc@eece.unm.edu. *Application contact:* Dr. Kenneth Jungling, Graduate Coordinator, 505-277-1433, Fax: 505-277-1439, E-mail: ken@eece.unm.edu.

See in-depth description on page 1087.

University of New Orleans, Graduate School, College of Engineering, Concentration in Electrical Engineering, New Orleans, LA 70148. Offers MS. *Faculty:* 10 full-time (0 women), 2 part-time (0 women). *Students:* 16 full-time (1 woman), 16 part-time (3 women); includes 3 minority (2 African Americans, 1 Hispanic American), 14 international. Average age 27. 133 applicants, 36% accepted. In 1998, 10 degrees awarded. *Degree requirements:* For master's, thesis optional, foreign language not required. *Entrance requirements:* For master's, GRE General Test (minimum combined score of 1200 required), minimum GPA of 3.0. *Application deadline:* For fall admission, 7/1 (priority date). Applications are processed on a rolling basis. Application fee: $20. Tuition, state resident: full-time $2,362. Tuition, nonresident: full-time $7,888. Part-time tuition and fees vary according to course load. *Financial aid:* Research assistantships, teaching assistantships available. *Faculty research:* Optics, ellipsometry, power systems, power-harmonics, optimal controls. *Unit head:* Dr. Russell Trahan, Chairman, 504-280-6176, Fax: 504-286-3950, E-mail: retee@uno.edu. *Application contact:* Dr. Xiao-Rong Li, Graduate Coordinator, 504-280-7416, Fax: 504-286-3950, E-mail: xrlee@uno.edu.

University of North Carolina at Charlotte, Graduate School, The William States Lee College of Engineering, Department of Electrical and Computer Engineering, Charlotte, NC 28223-0001. Offers MSEE, PhD. Part-time and evening/weekend programs available. *Faculty:* 19 full-time (1 woman), 1 part-time (0 women). *Students:* 39 full-time (5 women), 34 part-time (6 women); includes 12 minority (1 African American, 10 Asian Americans or Pacific Islanders, 1 Hispanic American), 29 international. Average age 29. 73 applicants, 92% accepted. In 1998, 5 master's, 2 doctorates awarded. *Degree requirements:* For master's, thesis or project required. *Entrance requirements:* For master's, GRE General Test, minimum GPA of 3.0 in undergraduate major, 2.75 overall. *Application deadline:* For fall admission, 7/15; for spring admission, 11/15. Applications are processed on a rolling basis. Application fee: $35. Electronic applications accepted. *Financial aid:* In 1998–99, 2 fellowships (averaging $8,000 per year), 23 research assistantships, 28 teaching assistantships were awarded.; Federal Work-Study also available. Financial aid application deadline: 4/1. *Faculty research:* Power systems, dynamics and control of flexibility, power load management, microelectronics. *Unit head:* Dr. Farid M. Tranjan, Chair, 704-547-2302, Fax: 704-547-2352, E-mail: tranjan@email.uncc.edu. *Application contact:* Kathy Barringer, Assistant Director of Graduate Admissions, 704-547-3366, Fax: 704-547-3279, E-mail: gradadm@email.uncc.edu.

University of North Dakota, Graduate School, School of Engineering and Mines, Department of Electrical Engineering, Grand Forks, ND 58202. Offers M Engr, MS. Part-time programs available. *Faculty:* 5 full-time (0 women). *Students:* 5 full-time (1 woman), 1 part-time. 11 applicants, 64% accepted. In 1998, 4 degrees awarded. *Degree requirements:* For master's, thesis or alternative required. *Entrance requirements:* For master's, GRE General Test, TOEFL (minimum score of 550 required), minimum GPA of 3.0 (MS), 2.5 (M Engr). *Application deadline:* For fall admission, 3/1 (priority date). Applications are processed on a rolling basis. Application fee: $20. *Financial aid:* In 1998–99, 5 students received aid, including 1 research assistantship, 4 teaching assistantships; fellowships, Federal Work-Study, institutionally-sponsored loans, and tuition waivers (full and partial) also available. Financial aid application deadline: 3/15. *Faculty research:* Controls and robotics, signal processing, energy conversion, microwaves, computer engineering. *Unit head:* Dr. Nagy Bengiamin, Chairperson, 701-777-4331, Fax: 701-777-4838, E-mail: bengiami@plains.nodak.edu.

University of Notre Dame, Graduate School, College of Engineering, Department of Electrical Engineering, Notre Dame, IN 46556. Offers MS, PhD. Part-time programs available. *Faculty:* 22 full-time (0 women), 5 part-time (0 women). *Students:* 66 full-time (10 women), 1 part-time; includes 6 minority (3 African Americans, 1 Asian American or Pacific Islander, 2 Hispanic Americans), 50 international. 147 applicants, 24% accepted. In 1998, 25 master's, 7 doctorates awarded (87% found work related to degree). Terminal master's awarded for partial completion of doctoral program. *Degree requirements:* For master's and doctorate, thesis/dissertation required, foreign language not required. *Entrance requirements:* For master's and doctorate, GRE General Test, TOEFL (minimum score of 600 required; 250 for computer-based). *Average time to degree:* Master's–2 years full-time; doctorate–5.9 years full-time. *Application deadline:* For fall admission, 2/1 (priority date); for spring admission, 11/1. Applications are processed on a rolling basis. Application fee: $40. *Financial aid:* In 1998–99, 63 students received aid, including 6 fellowships with full tuition reimbursements available (averaging $16,000 per year), 24 research assistantships with full tuition reimbursements available (averaging $11,500 per year), 20 teaching assistantships with full tuition reimbursements available (averaging $11,500 per year); scholarships and tuition waivers (full) also available. Financial aid application deadline: 2/1. *Faculty research:* Electronic properties of materials and devices, signal and imaging processing, communication theory, control theory and applications, optoelectronics. Total annual research expenditures: $3 million. *Unit head:* Dr. Gerald J. Iafrete, Director of Graduate Studies, 219-631-4647, Fax: 219-631-4393, E-mail: e.egrad@nd.edu. *Application contact:* Dr. Terrence J. Akai, Director of Graduate Admissions, 219-631-7706, Fax: 219-631-4183, E-mail: gradad@nd.edu.

Announcement: Areas of specialization include communication systems, control systems, signal and image processing, microelectronic fabrication, optoelectronics, and quantum electronic materials and devices. Financial aid includes several competitive fellowships available to outstanding applicants, research assistantships, and teaching assistantships providing full tuition waivers plus annual stipends of $15,000 to $18,000.

See in-depth description on page 1089.

University of Oklahoma, Graduate College, College of Engineering, School of Electrical and Computer Engineering, Norman, OK 73019-0390. Offers MS, PhD. *Faculty:* 15 full-time (1 woman), 4 part-time (0 women). *Students:* 25 full-time (3 women), 33 part-time (4 women); includes 3 minority (1 African American, 2 Asian Americans or Pacific Islanders), 38 international. Average age 28. 61 applicants, 69% accepted. In 1998, 18 master's, 8 doctorates awarded. *Degree requirements:* For master's, thesis, oral exam required, foreign language not required; for doctorate, dissertation, general exam, oral exam, qualifying exam required, foreign language not required. *Entrance requirements:* For master's and doctorate, GRE General Test (minimum combined score of 1150 required), TOEFL (minimum score of 550 required). *Application deadline:* For fall admission, 4/1 (priority date); for spring admission, 9/1 (priority date). Applications are processed on a rolling basis. Application fee: $25. Tuition, state resident: part-time $86 per credit hour. Tuition, nonresident: part-time $275 per credit hour. Tuition and fees vary according to course level, course load and program. *Financial aid:* In 1998–99, 23 research assistantships, 15 teaching assistantships were awarded.; fellowships, Federal Work-Study, institutionally-sponsored loans, and tuition waivers (partial) also available. Financial aid application deadline: 4/15. *Faculty research:* Communications, biomedical engineering, digital systems, solid-state, power systems control. *Unit head:* Gerald E. Crain, Director, 405-325-4721, Fax: 405-325-7066. *Application contact:* Dr. Samuel Lee, Graduate Adviser, 405-325-4721.

See in-depth description on page 1091.

University of Ottawa, School of Graduate Studies and Research, Faculty of Engineering, Ottawa-Carleton Institute for Electrical and Computer Engineering, Ottawa, ON K1N 6N5, Canada. Offers M Eng, MA Sc, PhD. *Faculty:* 44 full-time, 7 part-time. *Students:* 129 full-time (24 women), 116 part-time (24 women), 13 international. Average age 31. In 1998, 56 master's, 15 doctorates awarded. *Degree requirements:* For master's, thesis or alternative required, foreign language not required; for doctorate, dissertation required, foreign language not required. *Entrance requirements:* For master's, honors degree or equivalent, minimum B average; for doctorate, minimum A- average. *Application deadline:* For fall admission, 3/1 (priority date). Application fee: $35. *Financial aid:* Fellowships, research assistantships, teaching assistantships, Federal Work-Study available. *Unit head:* Jean-Yves Chouinard, Director, 613-562-5800 Ext. 6218, Fax: 613-562-5175. *Application contact:* Lucette Lepage, Academic Assistant, 613-562-5800 Ext. 6212, Fax: 613-562-5175, E-mail: lepage@site.uottawa.ca.

University of Pennsylvania, School of Engineering and Applied Science, Department of Electrical Engineering, Philadelphia, PA 19104. Offers MSE, PhD. Part-time programs available. Terminal master's awarded for partial completion of doctoral program. *Degree requirements:* For master's, thesis optional, foreign language not required; for doctorate, dissertation required, foreign language not required. *Entrance requirements:* For master's and doctorate, TOEFL (minimum score of 600 required). *Faculty research:* Electro-optics, microwave and millimeter-wave optics, solid-state and chemical electronics, electromagnetic propagation, inverse scattering.

See in-depth description on page 1093.

University of Pittsburgh, School of Engineering, Department of Electrical Engineering, Pittsburgh, PA 15260. Offers MSEE, PhD. Part-time and evening/weekend programs available. *Faculty:* 21 full-time (2 women). *Students:* 68 full-time (10 women), 62 part-time (7 women); includes 16 minority (6 African Americans, 9 Asian Americans or Pacific Islanders, 1 Hispanic American), 57 international. 331 applicants, 85% accepted. In 1998, 20 master's, 14 doctorates awarded. Terminal master's awarded for partial completion of doctoral program. *Degree requirements:* For master's, computer language required, thesis optional, foreign language not required; for doctorate, computer language, dissertation, comprehensive and final oral exams required, foreign language not required. *Entrance requirements:* For master's and doctorate, GRE General Test, TOEFL (minimum score of 550 required), minimum QPA of 3.0. *Average time to degree:* Master's–2 years full-time, 3 years part-time; doctorate–5 years full-time, 8 years part-time. *Application deadline:* For fall admission, 8/1 (priority date); for spring admission, 12/1 (priority date). Applications are processed on a rolling basis. Application fee: $30 ($40 for international students). *Financial aid:* In 1998–99, 50 students received aid, including 29 research assistantships (averaging $9,800 per year), 21 teaching assistantships (averaging $10,728 per year); fellowships, grants, scholarships, and tuition waivers (full and partial) also available. Financial aid application deadline: 2/15. *Faculty research:* Computer engineering, image processing, signal processing, electro-optic devices, controls/power. Total annual research expenditures: $2.6 million. *Unit head:* Dr. Joel Falk, Chairman, 412-624-8002, Fax: 412-624-8003, E-mail: falk@ee.pitt.edu. *Application contact:* Luis F. Chaparro, Graduate Coordinator, 412-624-9665, Fax: 412-624-8003, E-mail: chaparro@ee.pitt.edu.

See in-depth description on page 1095.

University of Portland, Graduate School, Multnomah School of Engineering, Department of Electrical Engineering, Portland, OR 97203-5798. Offers MSEE. Part-time and evening/weekend programs available. *Students:* 4 full-time (1 woman), 1 part-time. 10 applicants, 50% accepted. In 1998, 4 degrees awarded. *Degree requirements:* For master's, computer language required, foreign language and thesis not required. *Entrance requirements:* For master's, GRE General Test, TOEFL (minimum score of 550 required), minimum GPA of 3.0. *Application deadline:* For fall admission, 8/1 (priority date); for spring admission, 12/1. Applications are processed on a rolling basis. Application fee: $40. Tuition: Part-time $563 per semester hour. *Financial aid:* Application deadline: 3/15. *Unit head:* Dr. Khalid Khan, Director, 503-943-7276.

University of Puerto Rico, Mayagüez Campus, Graduate Studies, College of Engineering, Department of Electrical Engineering, Mayagüez, PR 00681-5000. Offers computer engineering (M Co E, MS); electrical engineering (MEE, MS). Part-time programs available. *Degree requirements:* For master's, thesis, comprehensive exam required, foreign language not required. *Entrance requirements:* For master's, minimum GPA of 2.5, proficiency in English and Spanish. *Faculty research:* Microcomputer interfacing, control systems, power systems, electronics.

University of Rhode Island, Graduate School, College of Engineering, Department of Electrical and Computer Engineering, Kingston, RI 02881. Offers MS, PhD. Part-time programs available. *Degree requirements:* For master's, thesis or alternative required; for doctorate, dissertation, comprehensive exam required. *Entrance requirements:* For master's and doctorate, GRE

Electrical Engineering

University of Rhode Island (continued)
General Test, TOEFL. *Faculty research:* Digital signal processing, computer engineering, VLSI, fiber optics and materials, communication and control systems, biomedical engineering.

Announcement: The Department of Electrical and Computer Engineering at the University of Rhode Island has 19 full-time faculty members, a host of adjunct and visiting professors, and a substantial amount of funded research. Faculty research interests include signal processing and system theory, computer engineering, optics and optical fibers, thin films and electronic materials, VLSI and analog and digital design, and biomedical engineering. Graduate work in the department leads to master's and doctoral degrees in electrical engineering. The MS degree can be pursued purely by course work (minimum of 30 credits) including 1 course involving a substantial written report and a written comprehensive examination or by a combination of course work and an MS thesis. The PhD degree is more research-oriented and requires a minimum of 72 graduate credits beyond the baccalaureate or 42 credits beyond the MS degree. Although part-time status is possible, each PhD candidate is required to have at least 1 year (2 consecutive semesters) of full-time residence. Typically, the department has 90 full-time and 40 part-time graduate students. A number of teaching and research assistantships and a few fellowships become available each year. For further information, contact Professor Shashanka S. Mitra, Director of Graduate Studies, Department of Electrical and Computer Engineering, University of Rhode Island, Kingston, Rhode Island, 02881 (telephone: 402-874-2506; e-mail: mitra@ele.uri.edu).

University of Rochester, The College, School of Engineering and Applied Sciences, Department of Electrical and Computer Engineering, Rochester, NY 14627-0250. Offers MS, PhD. Part-time programs available. *Faculty:* 14. *Students:* 52 full-time (3 women), 13 part-time (3 women); includes 7 minority (5 African Americans, 2 Asian Americans or Pacific Islanders), 36 international. 388 applicants, 12% accepted. In 1998, 20 master's, 8 doctorates awarded. Terminal master's awarded for partial completion of doctoral program. *Degree requirements:* For master's, foreign language and thesis not required; for doctorate, dissertation, preliminary and oral exams required, foreign language not required. *Entrance requirements:* For master's and doctorate, GRE, TOEFL. *Application deadline:* For fall admission, 2/1. Application fee: $25. *Financial aid:* Fellowships, research assistantships, teaching assistantships, tuition waivers (full and partial) available. Financial aid application deadline: 2/1. *Unit head:* Philippe Fauchet, Chair, 716-275-4054. *Application contact:* Ruth Williams, Graduate Program Secretary, 716-275-7417.

See in-depth description on page 1097.

University of Saskatchewan, College of Graduate Studies and Research, College of Engineering, Department of Electrical Engineering, Saskatoon, SK S7N 5A2, Canada. Offers M Eng, M Sc, PhD. *Degree requirements:* For master's and doctorate, thesis/dissertation required, foreign language not required. *Entrance requirements:* For master's and doctorate, GRE, TOEFL.

University of South Alabama, Graduate School, College of Engineering, Department of Computer and Electrical Engineering, Mobile, AL 36688-0002. Offers MSEE. Part-time programs available. *Faculty:* 9 full-time (0 women). *Students:* 21 full-time (3 women), 9 part-time; includes 2 minority (1 African American, 1 Asian American or Pacific Islander), 20 international. 88 applicants, 59% accepted. In 1998, 7 degrees awarded. *Degree requirements:* For master's, project or thesis required. *Entrance requirements:* For master's, GRE General Test (minimum combined score of 1000 required), BS in engineering, minimum GPA of 3.0. *Application deadline:* For fall admission, 9/1 (priority date). Applications are processed on a rolling basis. Application fee: $25. Tuition, state resident: part-time $116 per semester hour. Tuition, nonresident: part-time $230 per semester hour. Required fees: $121 per semester. Part-time tuition and fees vary according to course load and program. *Financial aid:* In 1998–99, 1 research assistantship was awarded.; career-related internships or fieldwork and institutionally-sponsored loans also available. Aid available to part-time students. Financial aid application deadline: 4/1. *Unit head:* Dr. Martin Parker, Chairperson, 334-460-6117.

University of South Carolina, Graduate School, College of Engineering and Information Technology, Department of Electrical and Computer Engineering, Program in Electrical Engineering, Columbia, SC 29208. Offers ME, MS, PhD. Part-time and evening/weekend programs available. Postbaccalaureate distance learning degree programs offered (minimal on-campus study). *Students:* 35 full-time (2 women), 21 part-time (1 woman); includes 5 minority (4 African Americans, 1 Asian American or Pacific Islander), 25 international. Average age 31. In 1998, 9 master's, 1 doctorate awarded. *Degree requirements:* For master's, thesis or alternative required, foreign language not required; for doctorate, dissertation required, foreign language not required. *Entrance requirements:* For master's and doctorate, GRE General Test (minimum combined score of 1100 required), TOEFL (minimum score of 550 required). *Application deadline:* For fall admission, 3/1 (priority date); for spring admission, 11/1. Applications are processed on a rolling basis. Application fee: $35. Electronic applications accepted. Tuition, state resident: full-time $4,014; part-time $202 per credit hour. Tuition, nonresident: full-time $8,528; part-time $428 per credit hour. Required fees: $100; $4 per credit hour. Tuition and fees vary according to program. *Financial aid:* In 1998–99, research assistantships with partial tuition reimbursements (averaging $12,000 per year), teaching assistantships with partial tuition reimbursements (averaging $12,000 per year) were awarded.; fellowships, career-related internships or fieldwork also available. *Faculty research:* Visualization, electromagnetics, robotics, microelectronics. *Application contact:* Leck Mason, Graduate Administration, 803-777-7522, Fax: 803-777-8045, E-mail: mason@ece.sc.edu.

University of Southern California, Graduate School, School of Engineering, Department of Electrical Engineering, Program in Electrical Engineering, Los Angeles, CA 90089. Offers MS, PhD, Engr. *Students:* 339 full-time (45 women), 424 part-time (55 women); includes 131 minority (14 African Americans, 102 Asian Americans or Pacific Islanders, 15 Hispanic Americans), 479 international. Average age 27. 1213 applicants, 79% accepted. In 1998, 247 master's, 53 doctorates awarded. *Degree requirements:* For master's, thesis optional; for doctorate, dissertation required. *Entrance requirements:* For master's, doctorate, and Engr, GRE General Test. *Application deadline:* For fall admission, 2/1 (priority date); for spring admission, 9/1. Application fee: $55. Tuition: Part-time $768 per unit. Required fees: $350 per semester. *Financial aid:* In 1998–99, 21 fellowships, 166 research assistantships, 52 teaching assistantships were awarded.; Federal Work-Study, institutionally-sponsored loans, and scholarships also available. Aid available to part-time students. Financial aid application deadline: 2/15; financial aid applicants required to submit FAFSA. *Unit head:* Dr. Robert Scholtz, Co-Chairman, Department of Electrical Engineering, 213-740-7327.

See in-depth description on page 1099.

University of South Florida, Graduate School, College of Engineering, Department of Electrical Engineering, Tampa, FL 33620-9951. Offers ME, MEE, MSE, MSEE, PhD, MS/MS. Part-time programs available. *Faculty:* 23 full-time (0 women), 1 part-time (0 women). *Students:* 61 full-time (9 women), 55 part-time (4 women); includes 22 minority (2 African Americans, 8 Asian Americans or Pacific Islanders, 11 Hispanic Americans, 1 Native American), 59 international. Average age 31. 139 applicants, 81% accepted. In 1998, 34 master's, 8 doctorates awarded (50% entered university research/teaching, 50% found other work related to degree). Terminal master's awarded for partial completion of doctoral program. *Degree requirements:* For master's, thesis or alternative required, foreign language not required; for doctorate, dissertation, 2 tools of research as specified by dissertation committee required, foreign language not required. *Entrance requirements:* For master's, GRE General Test (minimum combined score of 1000 required), minimum GPA of 3.0 during previous 2 years; for doctorate, GRE General Test (minimum combined score of 1000 required). *Average time to degree:* Master's–2.5 years full-time, 4.5 years part-time; doctorate–4 years full-time, 7 years part-time. *Application deadline:* For fall admission, 6/1; for spring admission, 10/15. Application fee: $20. Electronic applications accepted. Tuition, state resident: part-time $148 per credit hour. Tuition, nonresident: part-time $509 per credit hour. *Financial aid:* In 1998–99, 62 students received aid, including 1 fellowship with full tuition reimbursement available (averaging $7,000 per year), 43 research assistantships with full tuition reimbursements available (averaging $11,960 per year), 18 teaching assistantships with full tuition reimbursements available (averaging $11,960 per year); career-related internships or fieldwork, Federal Work-Study, institutionally-sponsored loans, and tuition waivers (partial) also available. Aid available to part-time students. Financial aid applicants required to submit FAFSA. *Faculty research:* Controls, including system parameter identification; computer-aided design and modeling; microwave and hybrid circuits. Total annual research expenditures: $1.3 million. *Unit head:* Dr. E. K. Stefanakos, Chairperson, 813-974-2369, Fax: 813-974-5250, E-mail: stefanak@eng.usf.edu. *Application contact:* Dr. Kenneth A. Buckle, Graduate Coordinator, 813-974-2369, Fax: 813-974-5250, E-mail: buckle@eng.usf.edu.

See in-depth description on page 1101.

University of Tennessee, Knoxville, Graduate School, College of Engineering, Department of Electrical Engineering, Knoxville, TN 37996. Offers MS, PhD. Part-time programs available. *Faculty:* 26 full-time (0 women), 4 part-time (0 women). *Students:* 40 full-time (4 women), 37 part-time (4 women); includes 8 minority (3 African Americans, 3 Asian Americans or Pacific Islanders, 2 Native Americans), 26 international. Average age 33. 119 applicants, 46% accepted. In 1998, 18 master's, 3 doctorates awarded. *Degree requirements:* For master's, thesis or alternative required, foreign language not required; for doctorate, dissertation required, foreign language not required. *Entrance requirements:* For master's, TOEFL (minimum score of 580 required), minimum GPA of 2.7; for doctorate, GRE General Test, TOEFL (minimum score of 580 required), minimum GPA of 2.7. *Application deadline:* For fall admission, 2/1 (priority date). Applications are processed on a rolling basis. Application fee: $35. Electronic applications accepted. *Financial aid:* In 1998–99, 34 research assistantships were awarded.; fellowships, teaching assistantships, career-related internships or fieldwork, Federal Work-Study, institutionally-sponsored loans, and unspecified assistantships also available. Financial aid application deadline: 2/1; financial aid applicants required to submit FAFSA. *Unit head:* Dr. Mohammad Karim, Head, 423-974-3461, Fax: 423-974-5483, E-mail: karim@utk.edu. *Application contact:* Dr. Mohammad Karim, Head, 423-974-3461, Fax: 423-974-5483, E-mail: karim@utk.edu.

University of Tennessee Space Institute, Graduate Programs, Program in Electrical Engineering, Tullahoma, TN 37388-9700. Offers MS, PhD. *Faculty:* 4 full-time (0 women), 1 part-time (0 women). *Students:* 5 full-time (0 women), 6 part-time; includes 2 minority (1 African American, 1 Asian American or Pacific Islander), 3 international. 31 applicants, 81% accepted. In 1998, 1 master's awarded. *Degree requirements:* For master's, thesis required (for some programs), foreign language not required; for doctorate, dissertation required. *Application deadline:* Applications are processed on a rolling basis. Application fee: $35. *Financial aid:* Fellowships, research assistantships, career-related internships or fieldwork, Federal Work-Study, and tuition waivers (full and partial) available. Financial aid applicants required to submit FAFSA. *Unit head:* Dr. Roy Joseph, Degree Program Chairman, 931-393-7457, Fax: 931-454-2271, E-mail: rjoseph@utsi.edu. *Application contact:* Dr. Edwin M. Gleason, Assistant Dean for Admissions and Student Affairs, 931-393-7432, Fax: 931-393-7346, E-mail: egleason@utsi.edu.

The University of Texas at Arlington, Graduate School, College of Engineering, Department of Electrical Engineering, Arlington, TX 76019. Offers M Engr, MS, PhD. *Faculty:* 22 full-time (1 woman). *Students:* 91 full-time (15 women), 136 part-time (19 women); includes 43 minority (4 African Americans, 31 Asian Americans or Pacific Islanders, 6 Hispanic Americans, 2 Native Americans), 112 international. 361 applicants, 41% accepted. In 1998, 59 master's, 9 doctorates awarded. *Degree requirements:* For master's, thesis required (for some programs), foreign language not required; for doctorate, dissertation required, foreign language not required. *Entrance requirements:* For master's and doctorate, GRE General Test, TOEFL. *Application deadline:* Applications are processed on a rolling basis. Application fee: $25 ($50 for international students). Tuition, state resident: full-time $1,368; part-time $76 per semester hour. Tuition, nonresident: full-time $5,454; part-time $303 per semester hour. Required fees: $66 per semester hour. $86 per term. Tuition and fees vary according to course load. *Financial aid:* Fellowships, research assistantships, teaching assistantships available. *Unit head:* Dr. Robert Magnusson, Chair, 817-272-2672, Fax: 817-272-2253, E-mail: magnusson@uta.edu. *Application contact:* Dr. Ronald L. Carter, Graduate Adviser, 817-272-2671, Fax: 817-272-2253, E-mail: ronc@uta.edu.

Announcement: The department offers outstanding opportunities to work with industries in the Dallas–Fort Worth area. Strong programs exist in telecommunications, energy systems, controls and robotics, manufacturing, remote sensing, microelectronics, RF and microwave electronics, optoelectronics, optics, digital signal and image processing, neural networks, VLSI design and applications, virtual environment and prototyping, and human performance.

See in-depth description on page 1103.

The University of Texas at Austin, Graduate School, College of Engineering, Department of Electrical and Computer Engineering, Austin, TX 78712-1111. Offers MSE, PhD. Part-time programs available. *Faculty:* 65 full-time (5 women), 12 part-time (1 woman). *Students:* 396 full-time (59 women), 169 part-time (16 women); includes 63 minority (5 African Americans, 42 Asian Americans or Pacific Islanders, 15 Hispanic Americans, 1 Native American), 336 international. 1098 applicants, 33% accepted. In 1998, 124 master's, 37 doctorates awarded. *Entrance requirements:* For master's, GRE General Test (combined average 2000), minimum GPA of 3.3 in upper-division course work; for doctorate, GRE General Test (combined average 2000). *Application deadline:* For fall admission, 1/2. Applications are processed on a rolling basis. Application fee: $50 ($75 for international students). Electronic applications accepted. *Financial aid:* Fellowships, research assistantships, teaching assistantships available. Financial aid application deadline: 1/2. *Unit head:* Dr. Francis X. Bostick, Chairman, 512-471-6179, Fax: 512-471-5532, E-mail: bostick@ece.utexas.edu. *Application contact:* Dr. Tony P. Ambler, Graduate Adviser, 512-475-6153, Fax: 512-471-5532, E-mail: ambler@ece.utexas.edu.

Announcement: The department has one of the country's leading graduate programs. Specialization areas include biomedical engineering; computer engineering (including software engineering); electromagnetics and acoustics; energy systems; manufacturing systems engineering; plasma, quantum electronics, and optics; solid-state electronics; telecommunications and information systems engineering; and, for professionals, engineering management and software engineering degrees.

See in-depth description on page 1105.

The University of Texas at Dallas, Erik Jonsson School of Engineering and Computer Science, Programs in Electrical Engineering, Richardson, TX 75083-0688. Offers electrical engineering (MSEE, PhD); microelectronics (MSEE); telecommunications (MSEE). Part-time and evening/weekend programs available. *Students:* 90 full-time (23 women), 139 part-time (21 women); includes 53 minority (4 African Americans, 40 Asian Americans or Pacific Islanders, 9 Hispanic Americans), 94 international. Average age 29. In 1998, 54 master's, 7 doctorates awarded. *Degree requirements:* For master's, thesis (for some programs), minimum GPA of 3.0, thesis or major design project required, foreign language not required; for doctorate, dissertation, minimum GPA of 3.5 required, foreign language not required. *Entrance requirements:* For master's, GRE General Test (minimum score of 500 on verbal section, 700 on quantitative, 600 on analytical required), TOEFL (minimum score of 550 required), minimum GPA of 3.0 in related bachelor's degree; for doctorate, GRE General Test (minimum score of 500 on verbal section, 700 on quantitative, 600 on analytical required), TOEFL (minimum score of 550 required), minimum GPA of 3.5. *Application deadline:* For fall admission, 7/15; for spring admission, 11/15. Applications are processed on a rolling basis. Application fee: $25 ($75 for international students). *Financial aid:* Fellowships, research assistantships, teaching assistantships, Federal Work-Study, grants, institutionally-sponsored loans, and scholarships available. Aid available to part-time students. Financial aid application deadline: 4/30; financial aid applicants required to submit FAFSA. *Faculty research:* Communications and signal processing, solid-state devices and circuits, digital systems, optical devices, materials and systems, lasers and photonics. *Unit head:* Dr. William Frensley, Head, 972-883-2412, Fax: 972-883-2710, E-mail: frensley@utdallas.edu. *Application contact:* Cynthia Stewart, Secretary, 972-883-2993, Fax: 972-883-2710, E-mail: ee-grad-info@utdallas.edu.

The University of Texas at El Paso, Graduate School, College of Engineering, Department of Electrical and Computer Engineering, El Paso, TX 79968-0001. Offers computer engineering (MS, PhD); electrical engineering (MS, PhD). Part-time and evening/weekend programs available. *Faculty:* 17 full-time (1 woman), 8 part-time (0 women). *Students:* 50 full-time (8 women), 29 part-time (3 women); includes 27 minority (2 Asian Americans or Pacific Islanders, 25 Hispanic Americans), 38 international. Average age 29. 58 applicants, 48% accepted. In 1998, 16 master's awarded. *Degree requirements:* For master's, thesis optional; for doctorate, dissertation required. *Entrance requirements:* For master's, GRE General Test, TOEFL (minimum score of 550 required), minimum GPA of 3.0; for doctorate, GRE General Test, TOEFL, qualifying exam, minimum graduate GPA of 3.0. *Application deadline:* Applications are processed on a rolling basis. Application fee: $15 ($65 for international students). Electronic applications accepted. Tuition, state resident: full-time $2,790. Tuition, nonresident: full-time $7,710. *Financial aid:* In 1998–99, 60 students received aid; fellowships, research assistantships, teaching assistantships, Federal Work-Study, institutionally-sponsored loans, and tuition waivers (partial) available. Financial aid applicants required to submit FAFSA. *Faculty research:* Signal and image processing, computer architecture, fiber optics, computational electromagnetics, electronic displays and thin films. Total annual research expenditures: $436,501. *Unit head:* Dr. Michael Austin, Chairperson, 915-747-5470, Fax: 915-747-5616. *Application contact:* Susan Jordan, Director, Graduate Student Services, 915-747-5491, Fax: 915-747-5788, E-mail: sjordan@utep.edu.

The University of Texas at San Antonio, College of Sciences and Engineering, Division of Engineering, San Antonio, TX 78249-0617. Offers civil engineering (MS); electrical engineering (MS); mechanical engineering (MS). Part-time and evening/weekend programs available. *Faculty:* 22 full-time (2 women), 15 part-time (2 women). *Students:* 19 full-time (1 woman), 90 part-time (15 women); includes 38 minority (5 African Americans, 11 Asian Americans or Pacific Islanders, 21 Hispanic Americans, 1 Native American), 21 international. *Degree requirements:* For master's, thesis optional, foreign language not required. *Entrance requirements:* For master's, GRE General Test. *Application deadline:* For fall admission, 7/1; for spring admission, 12/1. Applications are processed on a rolling basis. Application fee: $25. *Unit head:* Dr. Lex Akers, Director, 210-458-4490.

University of Toledo, Graduate School, College of Engineering, Department of Electrical Engineering and Computer Science, Toledo, OH 43606-3398. Offers computer science (MSES); electrical engineering (MSEE); engineering sciences (PhD). Part-time and evening/weekend programs available. *Faculty:* 21 full-time (3 women), 1 part-time (0 women). *Students:* 123 full-time (25 women), 34 part-time (4 women); includes 5 minority (1 African American, 3 Asian Americans or Pacific Islanders, 1 Native American), 128 international. Average age 25. 528 applicants, 52% accepted. In 1998, 78 master's, 1 doctorate awarded. *Degree requirements:* For master's, thesis or alternative required, foreign language not required; for doctorate, dissertation required, foreign language not required. *Entrance requirements:* For master's, GRE General Test (minimum combined score of 1700 on three sections required), TOEFL (minimum score of 550 required), minimum GPA of 2.7; for doctorate, GRE General Test (minimum combined score of 1700 on three sections required), TOEFL (minimum score of 550 required). *Average time to degree:* Master's–2 years full-time; doctorate–4 years full-time. *Application deadline:* For fall admission, 5/31 (priority date). Applications are processed on a rolling basis. Application fee: $30. Electronic applications accepted. *Financial aid:* In 1998–99, 146 students received aid, including 1 fellowship with full tuition reimbursement available, 7 research assistantships with full tuition reimbursements available, 12 teaching assistantships with full tuition reimbursements available; Federal Work-Study, scholarships, and tuition waivers (full) also available. Aid available to part-time students. Financial aid application deadline: 4/1. *Faculty research:* Power electronics, digital television, satellite communications, computer networks, fault-tolerant computing, weather and intelligent transportation. Total annual research expenditures: $539,300. *Unit head:* Dr. Adel Ghandakly, Chairman, 419-530-8146, E-mail: aghanda2@uoft02.utoledo.edu. *Application contact:* Sylvia Pinkerman, Academic Program Coordinator, 419-530-8144, Fax: 419-530-8146, E-mail: spinkerm@eng.utoledo.edu.

University of Toronto, School of Graduate Studies, Physical Sciences Division, Faculty of Applied Science and Engineering, Department of Electrical and Computer Engineering, Toronto, ON M5S 1A1, Canada. Offers M Eng, MA Sc, PhD. Part-time programs available. *Degree requirements:* For master's, thesis required (for some programs), foreign language not required; for doctorate, dissertation required, foreign language not required.

University of Tulsa, Graduate School, College of Business Administration, Department of Engineering and Technology Management, Tulsa, OK 74104-3189. Offers chemical engineering (METM); computer science (METM); electrical engineering (METM); geological science (METM); mathematics (METM); mechanical engineering (METM); petroleum engineering (METM). Part-time and evening/weekend programs available. *Students:* 3 full-time (1 woman), 1 part-time, 3 international. *Degree requirements:* For master's, foreign language and thesis not required. *Entrance requirements:* For master's, GRE General Test (minimum score of 430 on verbal section, 600 on quantitative required), TOEFL (minimum score of 575 required). *Application deadline:* Applications are processed on a rolling basis. Application fee: $30. Electronic applications accepted. Tuition: Full-time $8,640; part-time $480 per hour. Required fees: $3 per hour. One-time fee: $200 full-time. Tuition and fees vary according to program. *Unit head:* Dr. Richard C. Burgess, Assistant Dean/Director of Graduate Business Studies, 918-631-2242, Fax: 918-631-2142.

University of Tulsa, Graduate School, College of Engineering and Applied Sciences, Department of Electrical Engineering, Tulsa, OK 74104-3189. Offers ME, MSE. Part-time programs available. *Faculty:* 7 full-time (0 women). *Students:* 17 full-time (4 women), 1 part-time, 14 international. Average age 27. 24 applicants, 96% accepted. In 1998, 7 degrees awarded. *Degree requirements:* For master's, computer language, design report (ME), thesis (MS) required. *Entrance requirements:* For master's, GRE General Test (minimum score of 650 on quantitative section required), TOEFL (minimum score of 550 required). *Application deadline:* Applications are processed on a rolling basis. Application fee: $30. Electronic applications accepted. Tuition: Full-time $8,640; part-time $480 per hour. Required fees: $3 per hour. One-time fee: $200 full-time. Tuition and fees vary according to program. *Financial aid:* In 1998–99, 18 students received aid, including 9 research assistantships with full and partial tuition reimbursements available (averaging $3,289 per year), 9 teaching assistantships with full and partial tuition reimbursements available (averaging $4,000 per year); fellowships, career-related internships or fieldwork, Federal Work-Study, and tuition waivers (partial) also available. Aid available to part-time students. Financial aid application deadline: 2/1; financial aid applicants required to submit FAFSA. *Faculty research:* Simulation, linear and digital electronics, VLSI microprocessors, radar scattering, computer-aided design. *Unit head:* Dr. Gerald R. Kane, Chairperson, 918-631-3270. *Application contact:* Dr. Heng-Ming Tai, Adviser, 918-631-3271, Fax: 918-631-3344.

University of Utah, Graduate School, College of Engineering, Department of Electrical Engineering, Salt Lake City, UT 84112-1107. Offers M Phil, ME, MS, PhD, EE. Part-time programs available. *Faculty:* 15 full-time (1 woman), 12 part-time (1 woman). *Students:* 40 full-time (7 women), 31 part-time (4 women); includes 4 minority (1 Asian American or Pacific Islander, 2 Hispanic Americans, 1 Native American), 38 international. Average age 29. In 1998, 18 master's, 7 doctorates awarded. Terminal master's awarded for partial completion of doctoral program. *Degree requirements:* For master's, thesis (for some programs), comprehensive exam (MS) required, foreign language not required; for doctorate, dissertation, comprehensive exam required, foreign language not required. *Entrance requirements:* For master's and doctorate, GRE, TOEFL (minimum score of 560 required), minimum GPA of 3.0. *Application deadline:* For fall admission, 2/1. Application fee: $30 ($50 for international students). *Financial aid:* In 1998–99, 14 teaching assistantships were awarded.; fellowships, research assistantships, Federal Work-Study and institutionally-sponsored loans also available. Financial aid application deadline: 2/1. *Faculty research:* Semiconductors, VLSI design, control systems, electromagnetics and applied optics, communication theory and digital signal processing. *Unit head:* Om P. Gandhi, Chair, 801-581-7743, Fax: 801-581-5281, E-mail: gandhi@ee.utah.edu. *Application contact:* Bob Benner, Director, Graduate Studies, 801-581-6684.

University of Vermont, Graduate College, College of Engineering and Mathematics, Department of Computer Science and Electrical Engineering, Program in Electrical Engineering, Burlington, VT 05405-0160. Offers MS, PhD. *Degree requirements:* For master's, thesis or alternative required, foreign language not required; for doctorate, dissertation required. *Entrance requirements:* For master's and doctorate, GRE General Test, TOEFL (minimum score of 550 required).

University of Victoria, Faculty of Graduate Studies, Faculty of Engineering, Department of Electrical Engineering, Victoria, BC V8W 2Y2, Canada. Offers M Eng, MA Sc, PhD. *Faculty:* 19 full-time (1 woman), 8 part-time (0 women). *Students:* 67 full-time (8 women), 4 part-time, 38 international. Average age 27. 190 applicants, 12% accepted. In 1998, 9 master's, 8 doctorates awarded. *Degree requirements:* For master's and doctorate, thesis/dissertation required, foreign language not required. *Entrance requirements:* For master's, TOEFL (minimum score of 575 required), bachelor's degree in engineering; for doctorate, TOEFL (minimum score of 575 required), master's degree. *Average time to degree:* Master's–2.55 years full-time; doctorate–4.49 years full-time. *Application deadline:* For fall admission, 3/15 (priority date); for spring admission, 9/15. Applications are processed on a rolling basis. Application fee: $50. *Financial aid:* In 1998–99, 4 fellowships, 41 research assistantships, 24 teaching assistantships were awarded.; career-related internships or fieldwork, institutionally-sponsored loans, and awards also available. Financial aid application deadline: 2/15. *Faculty research:* Communications and computers; electromagnetics, microwaves, and optics; electronics; power systems, signal processing, and control. *Unit head:* Dr. P. Agathoklis, Chair, 250-721-8618, Fax: 250-721-6052, E-mail: pan@ece.uvic.ca. *Application contact:* Dr. Kin Li, Graduate Adviser, 250-721-8683, Fax: 250-721-6052, E-mail: kinli@ece.uvic.ca.

University of Virginia, School of Engineering and Applied Science, Department of Electrical Engineering, Charlottesville, VA 22903. Offers ME, MS, PhD. Postbaccalaureate distance learning degree programs offered (no on-campus study). *Faculty:* 24 full-time (3 women), 1 part-time (0 women). *Students:* 70 full-time (11 women), 5 part-time (1 woman); includes 6 minority (3 African Americans, 3 Asian Americans or Pacific Islanders), 43 international. Average age 27. 283 applicants, 22% accepted. In 1998, 25 master's, 7 doctorates awarded. *Degree requirements:* For master's, thesis (MS) required; for doctorate, dissertation, comprehensive exam required, foreign language not required. *Entrance requirements:* For master's and doctorate, GRE General Test. *Application deadline:* For fall admission, 8/1; for spring admission, 12/1. Application fee: $60. *Financial aid:* Fellowships available. Financial aid application deadline: 2/1. *Unit head:* James H. Aylor, Chairman, 804-924-3960. *Application contact:* J. Milton Adams, Assistant Dean, 804-924-3897, E-mail: twr2c@virginia.edu.

See in-depth description on page 1107.

University of Washington, Graduate School, College of Engineering, Department of Electrical Engineering, Seattle, WA 98195. Offers MSEE, PhD. *Degree requirements:* For master's, thesis optional, foreign language not required; for doctorate, dissertation required, foreign language not required. *Entrance requirements:* For master's, GRE General Test, TOEFL (minimum score of 580 required), minimum GPA of 3.0; for doctorate, GRE General Test, TOEFL (minimum score of 580 required), MS, minimum GPA of 3.0. Tuition, state resident: full-time $5,196; part-time $475 per credit. Tuition, nonresident: full-time $13,485; part-time $1,285 per credit. Required fees: $387; $38 per credit. Tuition and fees vary according to course load. *Faculty research:* Controls and robotics, communications and signal processing, electromagnetics, optics and acoustics, electronic devices and photonics.

See in-depth description on page 1109.

University of Waterloo, Graduate Studies, Faculty of Engineering, Department of Electrical and Computer Engineering, Waterloo, ON N2L 3G1, Canada. Offers electrical engineering (MA Sc, PhD); electrical engineering (software engineering) (MA Sc). Part-time programs available. *Faculty:* 45 full-time (3 women), 10 part-time (0 women). *Students:* 159 full-time (34 women), 14 part-time (1 woman). 200 applicants, 30% accepted. In 1998, 18 master's, 11 doctorates awarded. *Degree requirements:* For master's, thesis (for some programs), research paper or thesis required, foreign language not required; for doctorate, dissertation, comprehensive exam required, foreign language not required. *Entrance requirements:* For master's, TOEFL (minimum score of 550 required), honors degree, minimum B+ average; for doctorate, TOEFL (minimum score of 550 required), master's degree. *Average time to degree:* Master's–2 years full-time; doctorate–4 years full-time. *Application deadline:* For fall admission, 2/1; for winter admission, 6/1; for spring admission, 10/1. Application fee: $50. *Expenses:* Tuition and fees charges are reported in Canadian dollars. Tuition, state resident: full-time $3,168 Canadian dollars; part-time $792 Canadian dollars per term. Tuition, nonresident: full-time $8,000 Canadian dollars; part-time $2,000 Canadian dollars. Required fees: $45 Canadian dollars per term. Tuition and fees vary according to program. *Financial aid:* Fellowships, research assistantships, teaching assistantships available. *Faculty research:* Communications, computers, systems and control, silicon devices, power engineering, antennas, microwaves, wave-optics, high-voltage, VLSI. *Unit head:* Dr. A. Vanelli, Chair, 519-888-4016 Ext. 4016, Fax: 519-746-3077, E-mail: vanelli@cheetah.vlsi.uwaterloo.ca. *Application contact:* Dr. D. W. L. Wang, Graduate Officer, 519-888-4567 Ext. 3330, Fax: 519-746-3077, E-mail: dwang@kingcong.uwaterloo.ca.

University of Windsor, College of Graduate Studies and Research, Faculty of Engineering, Electrical Engineering, Windsor, ON N9B 3P4, Canada. Offers MA Sc, PhD. Part-time programs available. *Degree requirements:* For master's and doctorate, thesis/dissertation required, foreign language not required. *Entrance requirements:* For master's, TOEFL (minimum score of 550 required), minimum B average; for doctorate, TOEFL, master's degree. *Faculty research:* Systems, signals, power.

University of Wisconsin–Madison, Graduate School, College of Engineering, Department of Electrical and Computer Engineering, Madison, WI 53706-1380. Offers electrical engineering (MS, PhD). Part-time programs available. Postbaccalaureate distance learning degree programs offered (minimal on-campus study). *Faculty:* 41 full-time (3 women), 1 part-time (0 women). *Students:* 198 full-time (13 women), 38 part-time (3 women); includes 6 minority (2 African Americans, 2 Asian Americans or Pacific Islanders, 2 Hispanic Americans), 150 international. 400 applicants, 81% accepted. In 1998, 62 master's, 16 doctorates awarded. Terminal master's awarded for partial completion of doctoral program. *Degree requirements:* For master's, thesis or alternative, exam required, foreign language not required; for doctorate, dissertation required, foreign language not required. *Entrance requirements:* For master's and doctorate, GRE General Test, TOEFL (minimum score of 580 required). *Average time to degree:* Master's–2 years full-time; doctorate–6 years full-time. *Application deadline:* For fall admission, 6/1; for spring admission, 11/1. Applications are processed on a rolling basis. Application fee: $45. Electronic applications accepted. *Financial aid:* In 1998–99, 20 fellowships, 88 research assistantships with full tuition reimbursements (averaging $14,600 per year), 109 teaching assistantships with full tuition reimbursements were awarded.; career-related internships or fieldwork, Federal Work-Study, and institutionally-sponsored loans also available. Aid available to part-time students. Financial aid application deadline: 12/1. *Faculty research:* Microelectronics, computer architecture, power electronics and systems, communications, signal processing. Total annual research expenditures: $10.3 million. *Unit head:* Willis J. Tompkins, Chair, 608-262-2745, Fax: 608-262-1267. *Application contact:* Kathy Monroe, Graduate Secretary, 608-262-2745, Fax: 608-262-1267, E-mail: gradapp@ece.wisc.edu.

Electrical Engineering

University of Wyoming, Graduate School, College of Engineering, Department of Electrical Engineering, Laramie, WY 82071. Offers MS, PhD. Part-time programs available. *Faculty:* 15. *Students:* 26 full-time (1 woman), 8 part-time (2 women); includes 2 minority (both Asian Americans or Pacific Islanders), 16 international. 138 applicants, 25% accepted. In 1998, 8 master's awarded (100% found work related to degree); 1 doctorate awarded (100% entered university research/teaching). *Degree requirements:* For master's and doctorate, thesis/dissertation required, foreign language not required. *Entrance requirements:* For master's, GRE General Test (minimum combined score of 1375 required), TOEFL (minimum score of 550 required), minimum GPA of 3.0; for doctorate, GRE General Test (minimum combined score of 1475 required), TOEFL (minimum score of 550 required), minimum GPA of 3.0. *Application deadline:* For fall admission, 6/1 (priority date); for spring admission, 10/1. Applications are processed on a rolling basis. Application fee: $40. Electronic applications accepted. Tuition, state resident: full-time $2,520; part-time $140 per credit hour. Tuition, nonresident: full-time $7,790; part-time $433 per credit hour. Required fees: $400; $7 per credit hour. Full-time tuition and fees vary according to course load and program. *Financial aid:* In 1998–99, 7 research assistantships, 15 teaching assistantships were awarded. Financial aid application deadline: 3/1. *Faculty research:* Robotics and controls, signal and speech processing, power electronics, power systems, fuzzy and neural systems, instrumentation. *Unit head:* Dr. John W. Steadman, Associate Dean of Engineering, 307-766-2240. *Application contact:* Dr. John E. McInroy, Graduate Coordinator, 307-766-6137.

Utah State University, School of Graduate Studies, College of Engineering, Department of Electrical and Computer Engineering, Logan, UT 84322. Offers electrical engineering (ME, MS, PhD, EE). Part-time programs available. *Faculty:* 18 full-time (1 woman), 2 part-time (both women). *Students:* 63 full-time (7 women), 17 part-time (1 woman), 54 international. Average age 26. 166 applicants, 31% accepted. In 1998, 30 master's, 3 doctorates awarded. *Degree requirements:* For master's, thesis required (for some programs), foreign language not required; for doctorate, dissertation required, foreign language not required. *Entrance requirements:* For master's and doctorate, GRE General Test (score in 40th percentile or higher required), TOEFL (minimum score of 560 required), minimum GPA of 3.0. *Application deadline:* For fall admission, 1/1 (priority date); for spring admission, 10/15. Application fee: $40. Tuition, state resident: full-time $1,492. Tuition, nonresident: full-time $5,232. Required fees: $434. Tuition and fees vary according to course load. *Financial aid:* In 1998–99, 2 fellowships with partial tuition reimbursements (averaging $11,000 per year), 50 research assistantships with partial tuition reimbursements (averaging $7,000 per year), 4 teaching assistantships with partial tuition reimbursements (averaging $4,000 per year) were awarded.; Federal Work-Study and institutionally-sponsored loans also available. *Faculty research:* Parallel processing, networking, control systems, digital signal processing, communications, real time systems. Total annual research expenditures: $3 million. *Unit head:* Gardiner S. Stiles, Head, 435-797-2840, Fax: 435-797-3054. *Application contact:* Gardiner S. Stiles, Head, 435-797-2840, Fax: 435-797-3054.

Vanderbilt University, School of Engineering, Department of Electrical and Computer Engineering, Program in Electrical Engineering, Nashville, TN 37240-1001. Offers M Eng, MS, PhD. MS and PhD offered through the Graduate School. *Faculty:* 24 full-time (1 woman). *Students:* 113 full-time (14 women), 2 part-time; includes 81 minority (6 African Americans, 75 Asian Americans or Pacific Islanders), 2 international. Average age 26. 136 applicants, 23% accepted. In 1998, 20 master's awarded (60% entered university research/teaching); 3 doctorates awarded (100% found work related to degree). *Degree requirements:* For master's and doctorate, thesis/dissertation required, foreign language not required. *Entrance requirements:* For master's, GRE General Test (minimum combined score of 1200 required), GRE Subject Test, TOEFL (minimum score of 550 required); for doctorate, GRE General Test (minimum combined score of 1200 required), GRE Subject Test, TOEFL (minimum score of 575 required). *Average time to degree:* Master's–2 years full-time; doctorate–4 years full-time. *Application deadline:* For fall admission, 1/15 (priority date); for spring admission, 11/1. Application fee: $40. *Financial aid:* In 1998–99, 88 students received aid, including 6 fellowships with tuition reimbursements available, 32 research assistantships with tuition reimbursements available, 23 teaching assistantships with tuition reimbursements available; institutionally-sponsored loans also available. Financial aid application deadline: 1/15. *Faculty research:* Robotics microelectronics, signal and image processing. Total annual research expenditures: $4.7 million. *Unit head:* Dr. John E. McInroy, Graduate Coordinator, 307-766-6137. *Application contact:* Francis Wells, Director of Graduate Studies, 615-343-7549, Fax: 615-343-6702, E-mail: jean.tidwell@vanderbilt.edu.

Villanova University, College of Engineering, Department of Electrical and Computer Engineering, Program in Electrical Engineering, Villanova, PA 19085-1699. Offers MSEE. Part-time and evening/weekend programs available. *Students:* 16 full-time (3 women), 21 part-time (4 women); includes 2 minority (1 African American, 1 Hispanic American), 17 international. Average age 27. 62 applicants, 65% accepted. In 1998, 14 degrees awarded. *Degree requirements:* For master's, thesis optional, foreign language not required. *Entrance requirements:* For master's, GRE General Test (for applicants with degrees from foreign universities), TOEFL (minimum score of 550 required), BEE, minimum GPA of 3.0. *Application deadline:* For fall admission, 8/1 (priority date); for spring admission, 12/1. Applications are processed on a rolling basis. Application fee: $40. *Financial aid:* In 1998–99, 22 students received aid, including 3 research assistantships, 8 teaching assistantships; Federal Work-Study and scholarships also available. Financial aid application deadline: 3/15. *Faculty research:* Signal processing, communications, antennas, devices. *Unit head:* Dr. S. S. Rao, Chairman, Department of Electrical and Computer Engineering, 610-519-4971, Fax: 610-519-4436, E-mail: rao@ee.vill.edu.

See in-depth description on page 1111.

Virginia Polytechnic Institute and State University, Graduate School, College of Engineering, Department of Electrical and Computer Engineering, Blacksburg, VA 24061. Offers MS, PhD. *Students:* 298 full-time (41 women), 158 part-time (17 women); includes 56 minority (13 African Americans, 37 Asian Americans or Pacific Islanders, 6 Hispanic Americans), 223 international. 727 applicants, 35% accepted. In 1998, 121 master's, 22 doctorates awarded. *Degree requirements:* For master's, foreign language and thesis not required; for doctorate, dissertation required. *Entrance requirements:* For master's and doctorate, GRE General Test, GRE Subject Test, TOEFL (minimum score of 600 required). *Application deadline:* For fall admission, 12/1 (priority date). Applications are processed on a rolling basis. Application fee: $25. *Financial aid:* In 1998–99, 56 research assistantships, 47 teaching assistantships were awarded.; fellowships, career-related internships or fieldwork and unspecified assistantships also available. Financial aid application deadline: 4/1. *Unit head:* Dr. Leonard A. Ferrari, Interim Head, 540-231-6646, E-mail: ferrari@vt.edu.

See in-depth description on page 1113.

Washington State University, Graduate School, College of Engineering and Architecture, School of Electrical Engineering and Computer Science, Program in Electrical Engineering, Pullman, WA 99164. Offers MS, PhD. *Students:* 58 full-time (12 women), 7 part-time (1 woman), 47 international. In 1998, 18 master's, 7 doctorates awarded. *Degree requirements:* For master's, thesis, oral exam required, foreign language not required; for doctorate, dissertation, oral exam, qualifying exam required, foreign language not required. *Entrance requirements:* For master's and doctorate, GRE General Test, minimum GPA of 3.0. *Average time to degree:* Master's–2 years full-time; doctorate–4 years full-time. *Application deadline:* For fall admission, 3/1 (priority date). Applications are processed on a rolling basis. Application fee: $35. *Financial aid:* In 1998–99, 32 research assistantships, 27 teaching assistantships were awarded.; career-related internships or fieldwork, Federal Work-Study, institutionally-sponsored loans, and tuition waivers (partial) also available. Financial aid application deadline: 4/1; financial aid applicants required to submit FAFSA. *Unit head:* Francis Wells, Director of Graduate Studies, 615-343-7549, Fax: 615-343-6702, E-mail: jean.tidwell@vanderbilt.edu. *Application contact:* Dr. Ali Saberi, Graduate Coordinator, 509-335-5222, Fax: 509-335-3818, E-mail: saberi@eecs.wsu.edu.

Washington University in St. Louis, School of Engineering and Applied Science, Sever Institute of Technology, Department of Electrical Engineering, St. Louis, MO 63130-4899. Offers MSEE, D Sc. Part-time programs available. *Faculty:* 18 full-time (2 women), 17 part-time (0 women). *Students:* 66 full-time (14 women), 45 part-time (5 women); includes 8 minority (3 African Americans, 5 Asian Americans or Pacific Islanders), 51 international. Average age 25. 368 applicants, 33% accepted. In 1998, 25 master's, 6 doctorates awarded. Terminal master's awarded for partial completion of doctoral program. *Degree requirements:* For master's, thesis optional, foreign language not required; for doctorate, variable foreign language requirement, dissertation, departmental qualifying exam required. *Entrance requirements:* For master's, minimum GPA of 3.0 during previous 2 years. *Average time to degree:* Master's–2 years full-time, 3 years part-time; doctorate–4.8 years full-time. *Application deadline:* For fall admission, 8/18 (priority date); for spring admission, 1/10. Applications are processed on a rolling basis. Application fee: $20. *Financial aid:* In 1998–99, 41 students received aid, including 6 fellowships with full tuition reimbursements available (averaging $12,500 per year), 26 research assistantships with full tuition reimbursements available (averaging $18,000 per year); Federal Work-Study and institutionally-sponsored loans also available. Financial aid application deadline: 2/1. *Faculty research:* Applied physics, signal and image processing, computer engineering, telecommunications, biomedical engineering. Total annual research expenditures: $1.6 million. *Unit head:* Dr. Barry Spielman, Chairman, 314-935-5565, Fax: 314-935-7500, E-mail: bes@ee.wustl.edu. *Application contact:* Graduate Admissions, 314-935-6138, Fax: 314-935-7500, E-mail: eeadmin@ee.wustl.edu.

See in-depth description on page 1115.

Wayne State University, Graduate School, College of Engineering, Department of Electrical and Computer Engineering, Program in Electrical Engineering, Detroit, MI 48202. Offers MS, PhD. *Degree requirements:* For master's, thesis optional, foreign language not required; for doctorate, dissertation required, foreign language not required. *Faculty research:* Biomedical systems, control systems, solid state materials, optical materials, hybrid vehicle.

See in-depth description on page 1117.

Wayne State University, Graduate School, College of Engineering, Interdisciplinary Program in Electronics and Computer Control Systems, Detroit, MI 48202. Offers MS. *Degree requirements:* For master's, thesis optional, foreign language not required.

Western Michigan University, Graduate College, College of Engineering and Applied Sciences, Department of Electrical Engineering and Computer Engineering, Kalamazoo, MI 49008. Offers computer engineering (MSE); electrical engineering (MSE). Part-time programs available. *Students:* 9 full-time (0 women), 46 part-time (2 women); includes 2 minority (1 Asian American or Pacific Islander, 1 Native American), 47 international. 138 applicants, 68% accepted. In 1998, 7 degrees awarded. *Degree requirements:* For master's, thesis optional. *Entrance requirements:* For master's, minimum GPA of 3.0. *Application deadline:* For fall admission, 2/15 (priority date). Applications are processed on a rolling basis. Application fee: $25. *Financial aid:* Fellowships, research assistantships, teaching assistantships, career-related internships or fieldwork and Federal Work-Study available. Financial aid application deadline: 2/15; financial aid applicants required to submit FAFSA. *Faculty research:* Fiber optics, computer architecture, bioelectromagnetics, acoustics. *Unit head:* Dr. S. Hossein Mousavinezhad, Chair, 616-387-4057. *Application contact:* Paula J. Boodt, Coordinator, Graduate Admissions and Recruitment, 616-387-2000, Fax: 616-387-2355, E-mail: paula.boodt@wmich.edu.

Western New England College, School of Engineering, Department of Electrical Engineering, Springfield, MA 01119-2654. Offers MSEE. Part-time and evening/weekend programs available. *Faculty:* 7 full-time (0 women). Average age 29. In 1998, 4 degrees awarded. *Degree requirements:* For master's, computer language, comprehensive exam required, thesis optional, foreign language not required. *Entrance requirements:* For master's, bachelor's degree in engineering or related field. *Application deadline:* Applications are processed on a rolling basis. Application fee: $30. *Financial aid:* Teaching assistantships available. Aid available to part-time students. Financial aid application deadline: 4/1; financial aid applicants required to submit FAFSA. *Faculty research:* Superconductors, microwave cooking, computer voice output, digital filters, computer engineering. *Unit head:* Dr. Asif Hassan, Chair, 413-782-1491. *Application contact:* Harry F. Neunder, Coordinator, Continuing Education, 413-782-1750, Fax: 413-782-1779, E-mail: hneunder@wnec.edu.

West Virginia University, College of Engineering and Mineral Resources, Department of Computer Science and Electrical Engineering, Program in Electrical Engineering, Morgantown, WV 26506. Offers electrical engineering (MSEE); engineering (MSE, PhD). Terminal master's awarded for partial completion of doctoral program. *Degree requirements:* For master's, computer language, thesis or alternative required, foreign language not required; for doctorate, computer language, dissertation, comprehensive exam required, foreign language not required. *Entrance requirements:* For master's and doctorate, GRE General Test (score in 80th percentile or higher required), TOEFL (minimum score of 550 required), minimum GPA of 3.0. *Faculty research:* Power and control systems, communications and signal processing, electromechanical systems, microelectronics and photonics.

Wichita State University, Graduate School, College of Engineering, Department of Electrical Engineering, Wichita, KS 67260. Offers MS, PhD. Part-time and evening/weekend programs available. *Faculty:* 13 full-time (1 women), 5 part-time (1 woman). *Students:* 84 full-time (8 women), 39 part-time (5 women); includes 2 minority (both Asian Americans or Pacific Islanders), 98 international. Average age 30. 235 applicants, 77% accepted. In 1998, 42 master's, 6 doctorates awarded. *Degree requirements:* For master's, computer language, thesis or alternative, oral exam required, foreign language not required; for doctorate, one foreign language (computer language can substitute), dissertation, comprehensive exam required. *Entrance requirements:* For master's and doctorate, GRE, TOEFL (minimum score of 550 required). *Application deadline:* For fall admission, 7/1 (priority date); for spring admission, 1/1. Applications are processed on a rolling basis. Application fee: $25 ($40 for international students). Electronic applications accepted. *Financial aid:* In 1998–99, 17 research assistantships (averaging $4,000 per year), 10 teaching assistantships with full tuition reimbursements (averaging $2,400 per year) were awarded.; fellowships, Federal Work-Study, institutionally-sponsored loans, and unspecified assistantships also available. Financial aid application deadline: 4/1; financial aid applicants required to submit FAFSA. *Faculty research:* Rehabilitation engineering, control systems, power systems, digital systems, communications/signal processing. Total annual research expenditures: $273,265. *Unit head:* Dr. Everett Johnson, Chairperson, 316-978-3415, Fax: 316-978-3307, E-mail: ejohnson@ee.twsu.edu. *Application contact:* Dr. E. Sawan, Graduate Coordinator, 316-978-3415, Fax: 316-978-3307, E-mail: ed@ee.twsu.edu.

Widener University, School of Engineering, Program in Electrical/Telecommunication Engineering, Chester, PA 19013-5792. Offers ME, ME/MBA. Part-time and evening/weekend programs available. *Faculty:* 4 part-time (1 woman). *Students:* 2 full-time (0 women), 7 part-time; includes 3 minority (1 African American, 2 Asian Americans or Pacific Islanders), 2 international. 5 applicants, 80% accepted. In 1998, 1 degree awarded. *Degree requirements:* For master's, thesis optional, foreign language not required. *Entrance requirements:* For master's, GMAT (ME/MBA). *Average time to degree:* Master's–2 years full-time, 4 years part-time. *Application deadline:* For fall admission, 8/1 (priority date); for spring admission, 12/1. Applications are processed on a rolling basis. Application fee: $25 ($300 for international students). *Financial aid:* In 1998–99, 1 teaching assistantship with full tuition reimbursement (averaging $7,500 per year) was awarded.; unspecified assistantships also available. Financial aid application deadline: 3/15. *Faculty research:* Signal and image processing, electromagnetics, telecommunications and computer network. *Unit head:* Dr. Alfred T. Johnson, Chairman, Department of Electrical/Telecommunication Engineering, 610-499-4053, Fax: 610-499-4059, E-mail: alfred.t.johnson@widener.edu.

Wilkes University, School of Science and Engineering, Wilkes-Barre, PA 18766-0002. Offers electrical engineering (MSEE). *Degree requirements:* Foreign language not required. *Entrance requirements:* For master's, GRE General Test.

Worcester Polytechnic Institute, Graduate Studies, Department of Electrical and Computer Engineering, Worcester, MA 01609-2280. Offers electrical and computer engineering (MS, PhD, Advanced Certificate, Certificate); power systems engineering (MS, PhD). Part-time and evening/weekend programs available. *Faculty:* 22 full-time (1 woman), 3 part-time (0 women). *Students:* 48 full-time (7 women), 47 part-time (6 women); includes 7 minority (1 African American, 3 Asian Americans or Pacific Islanders, 3 Hispanic Americans), 32 international. 186 applicants, 46% accepted. In 1998, 31 master's, 6 doctorates awarded. Terminal master's awarded for partial completion of doctoral program. *Degree requirements:* For master's, thesis optional, foreign language not required; for doctorate, dissertation required, foreign language not required. *Entrance requirements:* For master's and doctorate, GRE (required for non-native speakers of English; combined average of 2024 on three sections), TOEFL (minimum score of 550 required; average 618). *Application deadline:* For fall admission, 2/15 (priority date); for spring admission, 10/15 (priority date). Applications are processed on a rolling basis. Application fee: $50. Electronic applications accepted. *Financial aid:* In 1998–99, 46 students received aid, including 1 fellowship with full tuition reimbursement available (averaging $12,500 per year), 27 research assistantships with full tuition reimbursements available (averaging $15,000 per year), 18 teaching assistantships with full tuition reimbursements available (averaging $11,970 per year); career-related internships or fieldwork, grants, institutionally-sponsored loans, and scholarships also available. Financial aid application deadline: 2/15; financial aid applicants required to submit FAFSA. *Faculty research:* Communications and signal processing, ultrasonics and electromagnetic engineering, electronics and solid state, systems and controls. Total annual research expenditures: $1.1 million. *Unit head:* Dr. John A. Orr, Head, 508-831-5231, Fax: 508-831-5491, E-mail: orr@wpi.edu. *Application contact:* Fred J. Looft, Graduate Coordinator, 508-831-5231, Fax: 508-831-5491, E-mail: fjlooft@wpi.edu.

Wright State University, School of Graduate Studies, College of Engineering and Computer Science, Programs in Engineering, Program in Electrical Engineering, Dayton, OH 45435. Offers MSE. Part-time and evening/weekend programs available. *Students:* 66 full-time (9 women), 49 part-time (2 women); includes 8 minority (3 African Americans, 4 Asian Americans or Pacific Islanders, 1 Native American), 52 international. Average age 27. 231 applicants, 70% accepted. In 1998, 66 degrees awarded. *Degree requirements:* For master's, thesis or course option alternative required. *Entrance requirements:* For master's, TOEFL (minimum score of 550 required). *Application deadline:* Applications are processed on a rolling basis. Application fee: $25. *Financial aid:* Fellowships, research assistantships, teaching assistantships, unspecified assistantships available. Aid available to part-time students. Financial aid application deadline: 3/1; financial aid applicants required to submit FAFSA. *Faculty research:*

Robotics, circuit design, power electronics, image processing, communication systems. *Unit head:* Dr. Raymond E. Siferd, Chair, 937-775-5037, Fax: 937-775-5009. *Application contact:* Dr. Larry L. Smith, Graduate Adviser, 937-775-5037, Fax: 937-775-5009.

Yale University, Graduate School of Arts and Sciences, Programs in Engineering and Applied Science, Department of Electrical Engineering, New Haven, CT 06520. Offers MS, PhD. *Faculty:* 18. *Students:* 36 full-time (9 women), 1 (woman) part-time, 32 international. 128 applicants, 18% accepted. In 1998, 1 master's, 4 doctorates awarded. Terminal master's awarded for partial completion of doctoral program. *Degree requirements:* For master's, foreign language and thesis not required; for doctorate, dissertation, exam required, foreign language not required. *Entrance requirements:* For master's and doctorate, GRE General Test, TOEFL. *Average time to degree:* Doctorate–4.8 years full-time. *Application deadline:* For fall admission, 1/4. Application fee: $65. *Financial aid:* Federal Work-Study and institutionally-sponsored loans available. Aid available to part-time students. *Unit head:* Chair, 203-432-4300. *Application contact:* Admissions Information, 203-432-2770.

Youngstown State University, Graduate School, William Rayen College of Engineering, Department of Electrical Engineering, Youngstown, OH 44555-0001. Offers MSE. Part-time and evening/weekend programs available. *Faculty:* 4 full-time (0 women). *Students:* 10 full-time (1 woman), 10 part-time (2 women), 9 international. 6 applicants, 67% accepted. In 1998, 4 degrees awarded. *Degree requirements:* For master's, computer language required, thesis optional, foreign language not required. *Entrance requirements:* For master's, TOEFL (minimum score of 550 required), minimum GPA of 2.75 in field. *Application deadline:* For fall admission, 8/15 (priority date); for winter admission, 11/15 (priority date); for spring admission, 2/15 (priority date). Applications are processed on a rolling basis. Application fee: $30 ($75 for international students). Tuition, state resident: part-time $97 per credit hour. Tuition, nonresident: part-time $219 per credit hour. Required fees: $21 per credit hour. $41 per quarter. *Financial aid:* In 1998–99, 5 students received aid, including 2 research assistantships with full tuition reimbursements available (averaging $7,500 per year), 1 teaching assistantship with full tuition reimbursement available (averaging $7,500 per year); Federal Work-Study, institutionally-sponsored loans, and scholarships also available. Aid available to part-time students. Financial aid application deadline: 3/1. *Faculty research:* Computer-aided design, power systems, electromagnetic energy conversion, sensors, control systems. *Unit head:* Dr. Salvatore R. Pansino, Chair, 330-742-3012. *Application contact:* Dr. Peter J. Kasvinsky, Dean of Graduate Studies, 330-742-3091, Fax: 330-742-1580, E-mail: amgrad03@ysub.ysu.edu.

Cross-Discipline Announcements

Carnegie Mellon University, Information Networking Institute, Pittsburgh, PA 15213-3891.

The MS in Information Networking is a cooperative endeavor of the Schools of Engineering, Computer Science, and Business, providing an alternative to the conventional one-year computer science or electrical engineering graduate program by integrating both and adding some features of an MBA program. Now in its eleventh year, the program carefully selects 35 people from the engineering and computer science disciplines to form each new class. The program provides technical electives aimed at several areas of specialization including (a) the telecommunications and computing industries, (b) wireless and mobile computing, (c) the financial services industry, and (d) systems integrating and consulting.

Carnegie Mellon University, School of Computer Science, Robotics Institute, Pittsburgh, PA 15213-3891.

Carnegie Mellon's MS and PhD programs in robotics are highly research oriented and interdisciplinary in nature, drawing from such fields as computer science, electrical and computer engineering, and mechanical engineering. Students may specialize in machine perception, artificial intelligence, manipulation, autonomous vehicles, manufacturing automation, or other areas. See in-depth description in Section 8: Computer Science and Information Technology.

Carnegie Mellon University, School of Computer Science, Software Engineering Program, Pittsburgh, PA 15213-3891.

Master of Software Engineering is a unique 1-year program at Carnegie Mellon University. The program takes a hands-on approach to developing expertise under the guidance of experienced software engineers. Emphasizing practical results balanced by scientific underpinnings, the program concentrates on the engineering of superior software systems through the application of principles from computer science and related fields. The software development studio is a major component of the program. Working closely with a faculty member, student teams analyze a problem, plan and implement a software development project, and evaluate the outcome. Students have full access to the resources of the School of Computer Science, a world leader in

computer science research, as well as the broad base of experience in the University's Software Engineering Institute, the only one of its kind.

Princeton University, Graduate School, Princeton Materials Institute, Princeton, NJ 08540-5211.

The Princeton Materials Institute welcomes students interested in cross-disciplinary research involving any aspect of materials science and engineering. Faculty members from 8 academic departments collaborate in a remarkably broad range of programs, supported by outstanding facilities. Some fellowships are available. See in-depth description in Section 21 of this guide.

Rensselaer Polytechnic Institute, Graduate School, School of Engineering, Department of Decision Sciences and Engineering Systems, Troy, NY 12180-3590.

This interdisciplinary department offers programs in manufacturing systems engineering, industrial and management engineering, operations research and statistics, and information systems. It prepares students to model complex systems and to use analytical and computational techniques in problem solving and the design of decision support systems. See in-depth description in Industrial Engineering section.

Vanderbilt University, School of Engineering, Department of Biomedical Engineering, Nashville, TN 37240-1001.

The Department of Biomedical Engineering is especially interested in attracting graduate students with backgrounds in electrical engineering who wish to study the applications of their field to biomedical problems. Research is available in biomedical instrumentation, medical computing, imaging, biomedical optics, and neuroscience. See in-depth description in the Bioengineering, Biomedical Engineering, and Biotechnology section.

ARIZONA STATE UNIVERSITY

Department of Electrical Engineering

Programs of Study

The department offers graduate programs leading to the Master of Science in Engineering, the Master of Science, the Master of Engineering, and the Doctor of Philosophy degrees. Graduate courses and programs are offered in solid-state electronics, semiconductor processing and manufacturing, power engineering, electromagnetics, control systems, communications, signal processing, and coherent optics.

The Master of Science in Engineering degree (M.S.E.) is a professional degree. General requirements include 30 semester hours of graduate-level course work and a final examination.

The Master of Science degree (M.S.) is a research degree, culminating in a thesis. Course work includes 12 semester hours in the major and two 6-hour minors. Typically, two years of study are required to complete the degree. Only those who are granted graduate assistantships or who are outstanding students showing research potential are admitted to the M.S. program.

The Doctor of Philosophy degree (Ph.D.) is awarded based upon evidence of excellence in research leading to a scholarly dissertation that contributes to scientific knowledge. A total of 84 semester hours of graduate study beyond the bachelor's degree is required. A minimum of 30 semester hours of course work must be completed in residence at Arizona State University (ASU). Twenty-four semester hours of research and dissertation complete the 84-hour program. A departmental qualifying examination is taken near the beginning of the program, and an individualized comprehensive examination is administered after all course work is completed.

Research Facilities

Centers of research excellence have been established in several areas. Those closely affiliated with electrical engineering are the Center for Solid State Electronics Research, the Center for System Science and Engineering, the Center for Low Power Electronics, the Telecommunications Research Center, the Center for Advanced Control of Energy and Power Systems, and the Center for Innovation in Engineering Education.

The department maintains an active program of research and development supported by funds from federal agencies, private foundations, private corporations, and the University. Opportunities for research are offered to students whose goals are research, development, design, manufacturing, systems, engineering management, teaching, or other professional activities in electrical engineering. Significant research activities exist in solid-state electronics (nanoelectronics, optoelectronics, materials processing and science, neural VLSI networks, low-power electronics, and semiconductor theory), power engineering (power systems, power quality, system control, transmission and distribution, power electronics, high-voltage engineering, and computer applications), electromagnetics (antennas, propagation, scattering, microwaves, radio frequency, and radar), coherent optics (lasers, integrated optics, and fiber optics), control systems (linear, nonlinear, real-time, adaptive, and robust control systems) and communications (digital communications, signal processing, wireless, and coding).

Financial Aid

Teaching and research assistantships are available on a competitive basis. Academic-year stipends range from $8405 to $12,713. Summer stipends further increase the financial aid package for many students. Teaching assistants usually supervise undergraduate teaching labs, while research assistants perform research applicable to their theses and dissertations. Out-of-state tuition is waived for all graduate assistants. Graduate Tuition Scholarships (GTS), which cover out-of-state tuition, and Graduate Academic Scholarships (GAS), which cover in-state registration fees, are also available on a competitive basis. The Corporate Leaders program provides practical work opportunities. Industrial internship programs are available for U.S. and international students.

Application forms for GTS, GAS, and assistantships should be submitted to the Department of Electrical Engineering. Loans and college work-study support are available through the Student Financial Assistance Office and require filing the FAFSA. The University Graduate Fellows (UGF) award provides research assistantships with a stipend of more than $18,000 and payment of all tuition and fees. Information about loans, work-study, and employment is available through Student Financial Assistance (telephone: 602-965-3355).

Cost of Study

The registration fee for full-time Arizona residents is $2044 per academic year in 1999–2000. Out-of-state tuition is $9040.

Living and Housing Costs

The cost of room and board in University residences ranged from $2420 to $3995 for the 1998–99 school year. Abundant housing is also available near the campus at prices that vary widely.

Student Group

There are 467 graduate students in the department. About 130 of these are Ph.D. students. The remainder are master's degree candidates. The department televises many of its graduate courses to major local companies. The students attending the televised classes are included as regularly enrolled graduate students. The department also participates in the National Technological University (NTU) system of nationally televised classes.

Location

The University's main campus is located in Tempe, Arizona, a city of 156,000 in the Phoenix metropolitan area. Strong academic programs and faculty are complemented by the attractions of year-round sunshine, cultural diversity on campus and in the community, and the resources of one of the nation's fastest-growing cities.

The University

Arizona State University is the nation's fifth-largest university. Of ASU's 44,255 students, 10,758 are pursuing graduate study. ASU's main campus comprises nearly 700 acres and offers outstanding physical facilities to support the University's educational and research programs. Included within the more than 125 buildings are twelve colleges and schools, a University-wide computer system, seven libraries (including an $8-million building dedicated solely to the Noble Science and Engineering Library), and more than two dozen specialized centers of research. ASU's commitment to permanently establish itself as a major research institution is demonstrated by the construction and acquisition of research facilities and resources as well as the addition of new research faculty and staff members.

Applying

All students must apply for admission through the Graduate College. For application forms, students should contact Graduate Admissions, Arizona State University, Tempe, Arizona 85287-1003 or call 602-965-6113. An applicant whose undergraduate degree is not from a program accredited by the Accreditation Board of Engineering and Technology (ABET) must submit scores from the Graduate Record Examinations (GRE) General Test. Students whose first language is not English must achieve a minimum score of 550 (paper-based) or 213 (computer-based) on the TOEFL.

Correspondence and Information

Director of Graduate Studies
Department of Electrical Engineering
Arizona State University
Tempe, Arizona 85287-5706

Telephone: 602-965-3590
Fax: 602-965-3837
E-mail: eeinfo@asu.edu
World Wide Web: http://www.eas.asu.edu/~eee

Arizona State University

THE FACULTY AND THEIR RESEARCH

Control Systems

Peter E. Crouch, Dean, College of Engineering and Applied Sciences; Ph.D., Harvard. Nonlinear control systems, applied mathematics, semiconductor manufacturing, power systems.

Walter T. Higgins Jr., Professor; Ph.D., Arizona. Digital control systems.

Frank Hoppensteadt, Professor of Electrical Engineering and Mathematics and Director, Center for System Science and Engineering Research; Ph.D., Wisconsin. Neuroscience and mathematical biology.

Armando A. Rodriguez, Associate Professor; Ph.D., MIT. Robust and nonlinear distributed control theory.

Jennie Si, Associate Professor; Ph.D., Notre Dame. Control theory, learning systems.

Konstantinos S. Tsakalis, Associate Professor; Ph.D., USC. Adaptive and nonlinear control systems, semiconductor manufacturing.

Electromagnetics, Antennas, Microwaves, Radar, Coherent Optics

James T. Aberle, Associate Professor; Ph.D., Massachusetts. Computational electromagnetics, conformal antennas, electromagnetic scattering, radar cross section, numerical techniques.

Constantine A. Balanis, Professor, Regents' Professor of Engineering, and Director, Telecommunications Research Center; Ph.D., Ohio State. Antennas, propagation, scattering, computational electromagnetics, transients in MMICs.

Samir M. El-Ghazaly, Professor; Ph.D., Texas at Austin. Microwave active and passive devices, electromagnetics, semiconductor device simulation, numerical techniques.

El-Badawy A. El-Sharawy, Associate Professor; Ph.D., Massachusetts. Printed microwave circuits, anisotropic devices, numerical techniques, antennas.

Joseph C. Palais, Professor and Director, Graduate Studies in Electrical Engineering; Ph.D., Michigan. Fiber optics, optical communications, fiber sensors, holography.

George W. Pan, Professor; Ph.D., Kansas. Packaging and interconnections, MCMS/MMICS, rough surface scattering, computational EM.

Power Engineering

Richard Farmer, Faculty Associate; M.S.E.E., Arizona State. Power system transients, power system analysis, transmission and distribution.

Ravi S. Gorur, Professor; Ph.D., Windsor. High-voltage engineering, insulating materials and systems for electric power, smart materials, pulse power.

Gerald T. Heydt, Professor and Director, Center for Advanced Control of Energy and Power Systems; Ph.D., Purdue. Electric power quality, power systems.

Keith E. Holbert, Associate Professor; Ph.D., Tennessee, Knoxville. Power plant modeling, dynamics and diagnostics, signal validation, noise analysis, instrument calibration reduction.

George G. Karady, Professor and SRP Professor of Engineering; Ph.D., Budapest University for Technical Sciences. Power electronics, high-voltage engineering, variable speed drives, thyristor control, power supplies, electric insulation, neural networks for power, pollution flashover.

Daniel J. Tylavsky, Associate Professor; Ph.D., Penn State. Power systems analysis, computational methods, large electrical systems.

Signal Processing and Communication Systems

Glen Abousleman, Faculty Associate; Ph.D., Arizona. Video compression, telephony, mobile satellite communications.

Tinku Acharya, Adjunct Professor; Ph.D., Central Florida. VLSI architectures and algorithms and digital image processing.

Jeffrey M. Capone, Assistant Professor; Ph.D., Northeastern. Controlling quality of service in Integrated Service Networks, resource management in wireless ATMs, multihop packet radio networks.

Chaitali Chakrabarti, Associate Professor; Ph.D., Maryland. VLSI architectures for digital signal processing.

Bruce Cochran, Faculty Associate; Ph.D., Arizona State. Communications, signal processing.

Douglas Cochran, Associate Professor; Ph.D., Harvard. Harmonic and statistical signal analysis, signal detection, wavelet analysis, sonar.

Tolga Duman, Assistant Professor; Ph.D., Northeastern. Digital communications, channel coding, turbo codes and turbo-coded modulation systems.

Lina J. Karam, Assistant Professor; Ph.D., Georgia Tech. Multidimensional digital signal processing, image and video processing and coding.

Darryl R. Morrell, Associate Professor; Ph.D., Brigham Young. Estimation and detection, stochastic filtering, statistical signal processing, epistemology, target tracking, data compression, image compression.

Sethuraman Panchanathan, Affiliated Professor; Ph.D., Ottawa. Multimedia computing and communications.

Andreas S. Spanias, Professor; Ph.D., West Virginia. Digital signal processing, adaptive filters, speech processing and coding, active noise control.

Solid-State Electronics

David R. Allee, Associate Professor; Ph.D., Stanford. Nanometer scale fabrication, nanolithography, low-power analog circuit design.

David J. Allstot, Professor; Ph.D., Berkeley. Analysis and design of analog, digital, and mixed-signal integrated systems; CAD for electrothermal and substrate interactions in CMOS integrated circuits; design and computer-aided optimization of CMOS RF integrated circuits; low-power systems.

Jonathan P. Bird, Associate Professor; Ph.D., Sussex (England). Studies of quantum transport phenomena in semiconductor nanostructures.

David K. Ferry, Regents' Professor of Engineering; Ph.D., Texas at Austin. Nanoelectronics, lithography, and quantum structured devices.

Stephen Goodnick, Department Chair and Professor; Ph.D., Colorado State. Semiconductor transport, quantum and nanostructure device technology, and high frequency devices.

Edwin W. Greeneich, Associate Professor; Ph.D., Berkeley. Semiconductor device modeling, analog integrated circuits, BiCMOS devices and circuits.

Robert O. Grondin, Associate Professor; Ph.D., Michigan. Small high-speed devices, physics of electrical engineering.

Michael N. Kozicki, Professor and Director, Center for Solid State Electronics Research; Ph.D., Edinburgh. Silicon integrated circuit processing, nanoelectronics, switching in ion-conducting glasses, integrated field emission devices, biophotonic integrated systems.

Bassam Matar, Faculty Associate; M.S.E.E., Oklahoma State. Semiconductor manufacturing technology.

Ronald J. Roedel, Professor; Ph.D., UCLA. Semiconductor materials and devices.

Dieter K. Schroder, Professor and Director, Center for Low Power Electronics; Ph.D., Illinois at Urbana-Champaign. Semiconductor devices, characterization, low-power electronics, defects.

Jun Shen, Associate Professor; Ph.D., Notre Dame. Solid-state device physics.

Brian J. Skromme, Associate Professor; Ph.D., Illinois at Urbana-Champaign. Compound semiconductor materials and devices.

Trevor J. Thornton, Professor; Ph.D., Cambridge. Mesoscopic physics and silicon device processing.

Drajica Vasileska-Kafedziska, Assistant Professor; Ph.D., Arizona State. Solid-state devices, quantum structured devices.

Navid Yazdi, Assistant Professor; Ph.D., Michigan. VLSI circuits and architectures, microelectromechanical systems (MEMS).

Yong-Hang Zhang, Associate Professor; Ph.D., Max-Planck Institute. Semiconductor materials and devices.

BOSTON UNIVERSITY

College of Engineering
Department of Electrical and Computer Engineering

Programs of Study	Ph.D. and M.S. programs prepare students for the application of state-of-the-art analysis and design methods to research and development problems in electrical and computer systems engineering. Primary research areas in the department include photonics, signal processing and recognition, electronic materials and devices, high-performance computing applications, space physics, computer hardware testing and fault-tolerant design, communication networks, multimedia, microprocessor development, and software engineering. Both post-bachelor's and post-master's Ph.D. tracks are available, leading to degrees in electrical, computer, and systems engineering. Ph.D. study requires 32 credits (eight courses) beyond the M.S. requirements. Ph.D. study also requires at least two consecutive semesters of residence, passing qualifying exams, and the dissertation. M.S. degrees are offered in electrical and computer systems engineering. The M.S. programs require a minimum of 36 credits (nine courses), including the completion of a thesis or project, and can be completed by a full-time student in one calendar year.
Research Facilities	Facilities include the following research laboratories: applied electromagnetics; complex systems dynamics; design and testing of computer and communications systems; embedded systems; functorial electromagnetics; high-performance computing; integrated circuit fabrication; knowledge-based signal processing; magnetic and optical devices; microprocessor research; liquid crystal display; molecular beam epitaxy; multidimensional signal processing; multimedia communications; near-field optical microscopy/spectroscopy; picosecond spectroscopy; quantum optics; radio communications and plasma research; semiconductor device research; sensors, actuators, and micromechanics; signal processing and interpretation; space sciences; speech research; VLSI process modeling and characterization; and VLSI research. These laboratories are complemented by a superb computational/networking environment with equipment ranging from personal computers to multimedia workstations to a 38-processor SGI Power Challenge Array scalable parallel supercomputer. Many ECE students also conduct their research in interdisciplinary centers such as those in photonics, computational science, space physics, and adaptive systems. In 1997, the department's facilities expanded into a section of a new building that contains one quarter of a million square feet.
Financial Aid	A full range of financial aid is available, including Presidential University Graduate Fellowships, Dean's Fellows, GAANN Fellowships in Photonics, graduate teaching fellowships, research assistantships, and various scholarships. In 1999–2000, teaching fellowships provide stipends of $12,500 per academic year and require approximately 20 hours a week of instructional duties. Recipients receive a tuition waiver for 8 to 10 credits per semster and up to 8 additional credits the following summer. Research assistantship stipends are comparable to or higher than those of teaching fellowships and are also supplemented by tuition waivers. University and Dean's Fellows scholarships range up to $37,830 (including stipend and tuition) per year; the department encourages GEM scholars to apply. Federal Direct Student Loan and Work-Study applicants must submit a Free Application for Federal Student Aid (FAFSA) to the Federal Student Aid Programs Office. Work-Study and FAFSA forms may be obtained from the Graduate Programs Office.
Cost of Study	In 1999–2000, tuition and fees for full-time study are $23,770. For part-time students, the cost is $743 per credit hour.
Living and Housing Costs	Privately owned apartments or rooms are readily available. Living expenses for a single student are estimated at $10,730 for the 1999–2000 nine-month academic year.
Student Group	The department has 73 students in the M.S. programs and 68 students pursuing the Ph.D. In 1998–99, 20 ECE students were supported as graduate teaching fellows; another 46 were supported as research assistants.
Student Outcomes	Most graduates of the M.S. programs enter industry. With a growing base of funded research, increasing numbers are pursuing Ph.D. research at Boston University.
Location	Boston offers a sophisticated environment with world-renowned academic and scientific resources. The area's seminar and colloquium programs allow students to participate at the cutting edge of research. Access to nearby concentrations of high-technology industry provides another vital element.
The University and The Department	Boston University, incorporated in 1869, is an independent, coeducational, nonsectarian university, fully open to women and to all minority groups. Its approximately 23,500 full-time students and 3,130 faculty members make it one of the largest independent universities in the world. The department is the largest in the College of Engineering and has experienced remarkable growth over the last decade. The faculty of the department has sought to build a stronger research program while maintaining its traditional commitment to the educational experience of its students. Interaction with local industry has created a focus on state-of-the-art educational and research issues. A graduate distance learning program was reinitiated during spring 1999. The other departments in the College are Biomedical Engineering, Manufacturing Engineering, and Aerospace and Mechanical Engineering.
Applying	Applicants to the M.S., post-bachelor's Ph.D., and post-master's Ph.D. programs should show a high degree of scholarship in an undergraduate program in engineering or science at an accredited college or university. Outstanding post-bachelor's candidates can be admitted directly to Ph.D. studies. Students desiring financial aid should apply by January 15 for fall admission and October 1 for spring. Applications for admission without financial aid can be submitted until April 1 and October 15, respectively. Required credentials include official transcripts, specific letters of recommendation, and GRE General Test scores. There are more requirements for international applicants, including the TOEFL.

Correspondence and Information

For program information:
ECE Department-Graduate Programs
Boston University
8 St. Mary's Street
Boston, Massachusetts 02215

Telephone: 617-353-1048
Fax: 617-353-7337
E-mail: ecegrad@enga.bu.edu
World Wide Web: http://www-eng.bu.edu/ECE

For application forms:
Graduate Programs
Boston University
48 Cummington Street
Boston, Massachusetts 02215

Telephone: 617-353-9760
Fax: 617-353-0259
E-mail: enggrad@bu.edu
World Wide Web: http://www.bu.edu/eng/grad

Boston University

THE FACULTY AND THEIR RESEARCH

Dimiter Avresky, Assistant Professor; Ph.D., Russian Academy of Sciences. Software implemented fault tolerance in parallel and distributed systems, verification, testing and validation of software protocols, performance analysis of networks, parallel programming.

John Brackett, Professor; Ph.D., Purdue. Software requirements definition and software architecture; software testing, especially of large object-oriented software systems.

Richard Brower, Professor; Ph.D., Berkeley. Quantum field theory, strings, molecular dynamics.

Jeffrey B. Carruthers, Assistant Professor; Ph.D., Berkeley. Communications—infrared and mobile radio, signal processing.

Cristos Cassandras, Associate Professor; Ph.D., Harvard. Analysis and control of discrete-event dynamic systems, stochastic control and optimization, dynamic control of computer and communication networks.

David Castañon, Associate Professor; Ph.D., MIT. Stochastic control, estimation, optimization, image processing, parallel and distributed computing.

Scott Dunham, Associate Professor; Ph.D., Stanford. Modeling and simulation of VLSI fabrication processes, point defect interaction in semiconductors, kinetics of extended defect evolution.

Charles R. Eddy Jr., Assistant Professor; Ph.D., Johns Hopkins. III-V nitrides, semiconductor properties and devices (visible and UV emitters, detectors, and high-temperature transistors).

Solomon Eisenberg, Associate Professor; Ph.D., MIT. Electrokinetic and other electromagnetic interactions in connective tissues and membranes.

Carol Espy-Wilson, Assistant Professor; Ph.D., MIT. Speech recognition, speech processing, acoustic phonetics, and digital signal processing.

Azza Fahim, Research Assistant Professor; Ph.D., Cairo. Electric machines, computations in electromagnetics.

Leopold Felsen, Professor; D.E.E., Polytechnic of Brooklyn. Wave propagation and diffraction in various disciplines, high-frequency asymptotics, wave-oriented data processing and imaging.

Theodore A. Fritz, Assistant Professor; Ph.D., Iowa. Development and flying of satellite-borne instruments that measure energetic particles in Earth's magnetosphere.

Roscoe Giles, Associate Professor; Ph.D., Stanford. Advanced computer architectures, distributed and parallel computing.

Bennett Goldberg, Associate Professor and Associate Professor of Physics; Ph.D., Brown. Optical processes in semiconductors and devices, near-field scanning optical microscopy and spectroscopy, low-temperature magneto-optics and transport of quantum confined electron systems.

Mark Horenstein, Associate Professor; Ph.D., MIT. Applied electromagnetics and electrostatics, instrumentation and measurement, microelectronics.

Allyn Hubbard, Associate Professor; Ph.D., Wisconsin. VLSI circuit design: digital, analog, subthreshold analog, biCMOS, and CMOS; special-purpose integrated circuits: neural-net, image processing, sonar signal processing DNA and large-molecule analysis, neural tissue interface; auditory research: theory and models of mammalian auditory signal processing, electronic ears.

Floyd Humphrey, Research Professor; Ph.D., Caltech. Magnetic materials, magnetic digital storage.

W. Clem Karl, Assistant Professor; Ph.D., MIT. Multidimensional and multiscale statistical signal and image processing, geometric estimation, medical signal and image processing.

Mark Karpovsky, Professor; Ph.D., Leningrad Electrotechnical Institute. Testing and diagnosis of computer hardware, fault-tolerant computing, error correcting codes.

Thomas Kincaid, Professor; Ph.D., MIT. Signal and image processing, neurodynamics, nondestructive testing.

Robert Kotiuga, Associate Professor; Ph.D., McGill. Mathematical and computational methods in electromagnetics.

Min-Chang Lee, Professor; Ph.D., California, San Diego. Radio communications, applied plasma physics and ionospheric radio physics.

Lev Levitin, Distinguished Professor; Ph.D., Moscow. Information theory, physics of communication and computation, reliable computing.

Thomas D. C. Little, Assistant Professor; Ph.D., Syracuse. Multimedia computing, computer networking, software engineering.

Michael Mendillo, Professor and Professor of Astronomy; Ph.D., Boston University. Instrumentation for low-light imaging of astronomical targets, digital image processing, optical tomography, Monte Carlo simulations of extended atmospheres of Jupiter and the Moon.

Theodore Moustakas, Professor; Ph.D., Columbia. III-V nitrides, semiconductor properties and devices (visible-UV emitters, detectors, and high-temperature transistors).

Syed Hamid Nawab, Associate Professor; Ph.D., MIT. Digital signal processing and signal processing for real-time, low-power, and communications applications.

Truong Q. Nguyen, Associate Professor; Ph.D., Caltech. Digital signal processing, image processing, signal compression and analysis, wavelets, filter banks, applications.

William Oliver, Associate Professor; Ph.D., Illinois. Upper-atmosphere/ionosphere physics, radar experimentation and data analysis, global atmosphere modeling and simulation.

Mari Ostendorf, Associate Professor; Ph.D., Stanford. Statistical modeling for signal interpretation, enhancement, and data compression, particularly speech processing.

David Perreault, Professor; Ph.D., Purdue. Nonlinear networks, computer-aided design, microprocessors, distributed-signal networks.

Tatyana Roziner, Associate Professor; Ph.D., Moscow Scientific Research Institute. Digital design, testing and diagnostics of computer hardware, fault-tolerant computing.

Michael Ruane, Associate Professor; Ph.D., MIT. Magneto-optical materials and devices, optical systems, communications.

Bahaa Saleh, Professor and Department Chairman; Ph.D., MIT. Nonlinear and quantum optics, optical communication, liquid crystal displays, image processing.

Fred Schubert, Professor; Doktor Ingenieur, Stuttgart (Germany). Technology and physics of lasers and light-emitting diodes, concepts for novel semiconductor devices.

Eric Schwartz, Professor; Ph.D., Columbia. Computational neural sciences, machine vision, neuroanatomy and neural modeling.

Alexander Sergienko, Assistant Professor; Ph.D., Moscow. Femtosecond quantum optics, fundamental interactions of quantum light with matter, presize optical measurement and quantum metrology, quantum communications and quantum cryptography.

Thomas Skinner, Associate Professor; Ph.D., Boston University. Microprocessors, computer networks, operating systems, distributed systems, object-oriented programming.

William Skocpol, Professor; Ph.D., Harvard. Nanofabrication, device processing and transport experiments in materials.

Johannes Smits, Associate Professor; Ph.D., Twente University of Technology (Netherlands). Micromechanics, microsensors and actuators, device fabrication.

Neeraj Suri, Associate Professor; Ph.D., Massachusetts Amherst. Design and analysis of algorithms and architectures for distributed, dependable, real-time systems; operating systems; verification and validation; network routing; SW dependability.

Malvin Teich, Professor; Ph.D., Cornell. Quantum optics, photonics, lightwave systems, fractal point processes in physical and biological systems, information transmission in biological sensory systems.

Tommaso Toffoli, Research Professor; Ph.D., Michigan. Programmable matter: design and use of fine-grained mesh computers; information mechanics; fundamental connections between physics and computation; quantum computation; information theory of fine-grained processes.

M. Selim Unlü, Associate Professor; Ph.D., Illinois at Urbana-Champaign. Design, processing, characterization, and simulation of optoelectronic devices, resonant cavity enhanced photonic devices, near-field optical microscopy and spectroscopy.

Richard Vidale, Professor; Ph.D., Wisconsin. Modeling and simulation, software engineering, real-time systems.

Moe Wasserman, Associate Professor Emeritus; Ph.D., Michigan. Semiconductor processing, electronic circuits.

CARNEGIE MELLON UNIVERSITY

Department of Electrical and Computer Engineering

Programs of Study

The department offers several graduate degree programs. Students who have earned a B.S. degree may apply to the M.S. degree program, which has two options: the course option and the project option. Students who have earned an M.S. degree may apply to the Ph.D. program. (The Direct Ph.D. Program is for highly qualified students who are interested in pursuing a Ph.D. immediately after completing a B.S. degree.)

Students in the project option are required to submit a project report. The completion time for an M.S. degree is usually one year to eighteen months. Fulfillment of the Ph.D. requirements takes three to four years beyond the M.S. degree and requires passing a qualifying examination, completing an internship in university teaching, writing a thesis that describes the results of independent research, and passing an oral defense of the research. All full-time Ph.D. students and project option master's students engage in research under faculty guidance. Currently, the department has 50 faculty members; all are actively engaged in research in addition to their regular teaching responsibilities. While several faculty members hold joint appointments with other departments, many of the faculty members also have close ties with various interdisciplinary research centers in the University, such as the Institute of Complex Engineered Systems and the Information Networking Institute. The department is home to several internationally recognized research centers, including an NSF Engineering Research Center in Data Storage Systems and the SRC-CMU Research Center for Computer-Aided Design. Major areas of research include agent-based systems; biomedical engineering; communications/information engineering; computer system and network architecture; computer-aided design of VLSI circuits, systems, and technology, including synthesis, verification, simulation, test, manufacturing, custom analog and digital IC design, and semiconductor fabrication; control systems; design, fabrication, and characterization of microelectromechanical systems; design optimization; distributed and real-time/multimedia systems; data storage technology and systems, including magnetic and optical recording; electronic growth, physics, materials, and processing; electronic, magnetic, and optical phenomena and devices, including design and microfabrication; embedded systems; fault tolerance and affordable dependable computing; high-performance I/O systems; manufacturing systems and automatic assembly; mobile and wireless computing; neural networks; operating systems; rapid prototyping; robotics; sensory-based and supervisory control; signal processing in optical, video, image, speech, and storage systems; processor design; parallel processing; and technology and public policy.

Research Facilities

The department has extensive computational facilities that include more than 570 advanced workstations supporting research and education. Research projects also make frequent use of supercomputers, available to the department through the Pittsburgh Supercomputing Center. The facilities also include numerous laser printers and color-printing and slide-making capabilities. These systems are all connected to the CMU campuswide computer network via an Ethernet local area network. A fully equipped 4,000-square-foot, class 100 clean room supports research in semiconductor and magnetic and optical device research and can be used to produce state-of-the-art solid-state and recording devices. In addition to the clean room, the department has a molecular-beam epitaxy system and extensive facilities for measuring the electrical and optical properties of semiconducting and superconducting materials, structures, and devices. Computer-controlled systems for performing photoluminescence spectroscopy, deep-level transient spectroscopy, device parameter analysis, capacitance voltage measurements, and other functions are all available. The Data Storage Systems laboratories contain several test stands for evaluating both rigid and flexible magnetic and optical recording media and heads housed in custom-built RF-shielded labs, a high-resolution magnetooptic photometer with picosecond temporal resolution, and several polarized-light microscopes equipped with drive electronics for testing magnetooptic recording media. Facilities also include a vibrating sample magnetometer, an inductive hysteresis loop tracer, a torque magnetometer, a SQUID magnetometer, an alternating gradient magnetometer, an atomic magnetic force microscope, a 35-GHz ferromagnetic resonance spectrometer, and a spin polarized electron microscope for magnetic domain observation. The optical and digital processing labs contain large vibration-isolated tables, lasers, associated optical components, and extensive optical and electrical detection and measurement facilities. Optical systems include image processing, pattern recognition, feature generation, and optical neural networks.

Financial Aid

Graduate research and teaching assistantships are available to U.S. and international students and include a typical stipend of about $1425–$1525 per month plus tuition. In the award of financial aid, consideration is given to the student's undergraduate and graduate academic records, GRE scores, and letters of reference indicating outstanding academic potential.

Cost of Study

Tuition for graduate students enrolled in the Department of Electrical and Computer Engineering is $22,100 for the academic year 1999–2000. Books and supplies cost about $1450 per year.

Living and Housing Costs

Graduate accommodations are not provided, although there are various board plans available in nearby rooms or apartments. Approximate living expenses, including room and board, insurance, transportation, and miscellaneous expenses, average $15,640 per academic year, exclusive of tuition.

Student Group

The campus enrollment averages 7,800 students; 2,800 are graduate students. The University has 1,000 full-time faculty members. Within the Department of Electrical and Computer Engineering, there are 241 full-time graduate students, with 107 in the master's program and 134 in the Ph.D. program.

Location

The greater Pittsburgh metropolitan area has more than 2 million residents. The Carnegie Mellon campus encompasses approximately 100 acres and adjoins a 500-acre city park and quiet residential communities with abundant student housing.

The University

The University is composed of the Carnegie Institute of Technology (the Engineering School), Mellon College of Science, the College of Fine Arts, the Graduate School of Industrial Administration, the Heinz School, the College of Humanities and Social Sciences, and the School of Computer Sciences.

Applying

All applicants are required to take the GRE (General Test) at least six weeks prior to the application deadline. All students whose native language is other than English are required to take the Test of English as a Foreign Language (TOEFL). Applications for the fall semester must be received by January 15. Official transcripts, three letters of recommendation, and GRE scores must be provided. Application materials may be obtained by writing to the department.

Correspondence and Information

Graduate Admissions
Department of Electrical and Computer Engineering
Carnegie Mellon University
Pittsburgh, Pennsylvania 15213

Telephone: 412-268-3200
Fax: 412-268-2860
E-mail: apps@ece.cmu.edu
World Wide Web: http://www.ece.cmu.edu

Carnegie Mellon University

THE FACULTY AND THEIR RESEARCH

J. A. Bain, Research Engineer; Ph.D., Stanford, 1993. Thin-film magnetic device design, fabrication, and testing; tape recording.

V. Bhagavatula, Professor; Ph.D., Carnegie Mellon, 1980. Optical pattern recognition, neural networks, signal processing for storage.

R. P. Bianchini Jr., Adjunct Professor; Ph.D., Carnegie Mellon, 1989. Distributed fault-tolerant computing and computer networks.

S. Blanton, Assistant Professor; Ph.D., Michigan, 1985. Design and test of VLSI circuits, fault-tolerant computing, computer architecture.

R. E. Bryant, President's Professor of Computer Science; Ph.D., MIT, 1981. Digital system simulation and verification.

L. R. Carley, Professor; Ph.D., MIT, 1984. CAD and design of analog signal processing circuits and MEMS systems.

D. P. Casasent, George Westinghouse Professor; Ph.D., Illinois at Urbana–Champaign, 1969. Pattern recognition; neural nets; image processing; product inspection; face, fingerprint, and SAR processing; optical data processing.

Z. J. Cendes, Adjunct Professor; Ph.D., McGill, 1973. Computer-aided design of high-speed digital interconnects, microwave and antenna field simulation, modeling of magnetic and semiconductor devices, numerical methods.

S. H. Charap, Emeritus Professor; Ph.D., Rutgers, 1959. Magnetic phenomena and devices, theory and modeling, magnetic recording.

T. Chen, Associate Professor; Ph.D., Caltech, 1993. Multimedia communications, video coding, multimedia standards, audio-visual interaction, multirate signal processing.

E. Clarke, FORE Systems Professor of Computer Science; Ph.D., Cornell, 1976. Hardware and software verification, automatic theorem proving, symbolic computation.

S. W. Director, Adjunct Professor of Electrical and Computer Engineering; Ph.D., Berkeley, 1968. Computer-aided VLSI design, CAD frameworks, statistical design.

C. N. Faloutsos, Associate Professor of Computer Science; Ph.D., Toronto, 1987. Databases, data mining, multimedia indexing, performance evaluation.

G. K. Fedder, Associate Professor of Electrical and Computer Engineering and Robotics; Ph.D., Berkeley, 1994. Microelectromechanical systems (MEMS), MEMS CAD, micro-robotics.

R. M. Feenstra, Professor of Physics; Ph.D., Caltech, 1982. Molecular beam epitaxy, scanning tunneling microscopy.

A. Fisher, Principal Systems Scientist and Associate Dean of Computer Science; Ph.D., Carnegie Mellon, 1984. Network architecture, parallel computing.

K. Gabriel, Professor of Electrical and Computer Engineering and Robotics; Sc.D., MIT, 1983. Microelectromechanical systems.

G. Ganger, Assistant Professor; Ph.D., Michigan, 1995. Operating systems, computer architecture, distributed systems, storage/file systems, networking, computer architecture.

G. A. Gibson, Associate Professor of Computer Science; Ph.D., Berkeley, 1991. Computer architecture, operating systems, file systems, disk arrays.

S. C. Goldstein, Assistant Professor of Computer Science; Ph.D., Berkeley, 1997. Parallel systems, reconfigurable computing.

D. W. Greve, Professor; Ph.D., Lehigh, 1979. Epitaxy of germanium silicon alloys, semiconductor process technology, wide-gap semiconductors.

J. B. Hampshire II, Research Engineer; Ph.D., Carnegie Mellon, 1993. Learning theory, pattern recognition, neural networks.

J. F. Hoburg, Professor; Ph.D., MIT, 1975. Electromagnetics, electromechanics, magnetic shielding, applied electrostatics.

D. B. Johnson, Assistant Professor of Computer Science; Ph.D., Rice, 1990. Wireless and mobile networking, operating systems, distributed systems, fault tolerance.

A. G. Jordan, Emeritus Keithley University Professor of Electrical and Computer Engineering and Robotics; Ph.D., Carnegie Tech, 1959. Advanced video systems, robotics, management of technology, studies of computer industries.

T. Kanade, Director of the Robotics Institute and U. A. and Helen Whitaker University Professor of Computer Science, Robotics, and Electrical and Computer Engineering; Ph.D., Kyoto, 1973. Computer vision, autonomous systems, medical robotics.

P. K. Khosla, Philip and Marsha Dowd Professor of Engineering and Robotics and Head; Ph.D., Carnegie Mellon, 1986. Mechatronics, agent-based systems, S/W engineering for real-time systems, distributed robotic systems, gesture-based programming, distributed information systems.

H. S. Kim, Associate Professor; Ph.D., Toronto, 1990. Computer networks, advanced switch architectures, high-speed networks.

P. J. Koopman, Assistant Professor; Ph.D., Carnegie Mellon, 1989. Robust distributed embedded systems, computer architecture, methodical system design.

B. H. Krogh, Professor; Ph.D., Illinois at Urbana–Champaign, 1982. Discrete and continuous control systems.

M. H. Kryder, University Professor; Ph.D., Caltech, 1970. Magnetic and optical recording technologies.

D. N. Lambeth, Professor of Electrical and Computer Engineering and of Materials Science and Engineering and Associate Director, Data Storage Systems Center; Ph.D., MIT, 1973. Magnetism, recording systems, media, precision instrumentation, transducers and sensors.

W. Maly, U. A. and Helen Whitaker Professor; Ph.D., Polish Academy of Sciences, 1975. Computer-aided design and manufacturing of VLSICs.

W. C. Messner, Associate Professor of Mechanical Engineering; Ph.D., Berkeley, 1992. Control theory, control for data storage systems, robot control, highly distributed control, Web-based education.

A. G. Milnes, Emeritus Professor; D.Sc., Bristol, 1956. Semiconductor phenomena and devices.

M. G. Morgan, Lord Professor of Electrical and Computer Engineering and of Engineering and Public Policy and Department Head of Engineering and Public Policy; Ph.D., California, San Diego, 1969. Technology and public policy including risk analysis.

J. M. F. Moura, Professor; D.Sc., MIT, 1975. Communications; signal, image, video, and multimedia processing; multiresolution and wavelet transforms.

T. C. Mowry, Associate Professor of Computer Science; Ph.D., Stanford, 1994. Computer architecture, compilers, operating systems, parallel processing.

T. Mukherjee, Research Engineer, Electrical and Computer Engineering and Assistant Director, Center for Electronic Design Automation; Ph.D., Carnegie Mellon, 1995. CAD for MEMS, analog circuit synthesis.

D. Nagle, Assistant Professor; Ph.D., Michigan, 1994. Computer architecture, operating systems, storage technologies.

C. P. Neuman, Professor and Undergraduate Advisor; Ph.D., Harvard, 1968. Control engineering and robotics.

D. O' Hallaron, Associate Professor of Electrical and Computer Engineering and Computer Science; Ph.D., Virginia, 1986. High-performance distributed computing.

J. M. Paul, Visiting Assistant Professor; Ph.D., Pittsburgh, 1994. Hardware/software codesign and cosimulation; parallel processing.

J. M. Peha, Associate Professor; Ph.D., Stanford, 1991. Telecommunications, computer networks, wireless networks, telecom policy.

L. T. Pileggi, Professor; Ph.D., Carnegie Mellon, 1989. Circuit-level analysis and design automation for electronic systems.

R. Rajkumar, Senior Research Engineer and Senior Systems Scientist; Ph.D., Carnegie Mellon, 1989. Multimedia and real-time systems.

R. A. Rohrer, Adjunct Professor; Ph.D., Berkeley, 1963. Electronic circuits, systems design automation.

R. A. Rutenbar, Professor; Ph.D., Michigan, 1984. VLSI CAD, algorithms, analog and digital circuits.

T. E. Schlesinger, Professor, Associate Department Head; Ph.D., Caltech, 1985. III–V semiconductors, optoelectronics, nuclear detectors.

H. H. Schmit, Assistant Professor; Ph.D., Carnegie Mellon, 1995. Reconfigurable computing, low-power design.

J. P. Shen, Professor; Ph.D., USC, 1981. Modern superscaler processor design, instruction-level parallelism and code scheduling.

D. P. Siewiorek, Buhl Professor of Electrical and Computer Engineering and Computer Science and Associate Director, Institute for Complex Engineered Systems; Ph.D., Stanford, 1972. Computer architecture, reliability, CAD, mobile computing.

M. A. Sirbu, Professor of Engineering and Public Policy and Industrial Administration; Sc.D., MIT, 1973. Telecommunications policy and economics.

D. D. Stancil, Professor and Associate Dean of CIT; Ph.D., MIT, 1981. Wireless communications, integrated optics, optical data storage.

P. A. Steenkiste, Senior Research Scientist of Computer Science; Ph.D., Stanford, 1987. Computer networks, distributed systems, distributed computing.

R. M. Stern Jr., Professor and Associate Director, Information Networking Institute; Ph.D., MIT, 1976. Speech recognition, auditory perception, signal processing.

A. J. Strojwas, Professor; Ph.D., Carnegie Mellon, 1982. Statistically based CAD/CAM of VLSI circuits.

J. K. Strosnider, Adjunct Professor; Ph.D., Carnegie Mellon, 1988. E-business architecture, integrating the Web, IT, and automated workflow.

T. Sullivan, Lecturer; Ph.D., Carnegie Mellon, 1996. Audio signal processing, music and sound recording applications.

S. N. Talukdar, Professor; Ph.D., Purdue, 1970. Agent-based systems, distributed problem solving, power systems, organization design.

A. A. Thiele, Distinguished Scholar; Ph.D., MIT, 1965. Micromagnetics, topology, and statistics of domain wall motion.

D. E. Thomas Jr., Professor and Director, Center for Electronics Design Automation; Ph.D., Carnegie Mellon, 1977. Computer-aided design of mixed hardware and software systems.

R. M. Unetich, Adjunct Professor; B.S.E.E., Carnegie Mellon, 1968. Wireless communication systems.

R. M. White, University Professor and Director, Data Storage Systems Center; Ph.D., Stanford, 1964. Magnetic device phenomena, technology policy.

H. Zhang, Assistant Professor of Computer Science; Ph.D., Berkeley, 1993. Computer networks, Internet, resource management, multimedia.

J. G. Zhu, Professor; Ph.D., California, San Diego, 1989. Micromagnetics, magnetoelectronic devices, magnetic recording.

COLUMBIA UNIVERSITY

Department of Electrical Engineering

Programs of Study

The Department of Electrical Engineering offers programs of study leading to the degrees of Master of Science (M.S.), Electrical Engineer (E.E.), Doctor of Engineering Science (Eng.Sc.D.), and Doctor of Philosophy (Ph.D.). Registration as a nondegree candidate (special student) is also permitted.

There are no prescribed course requirements for these degrees. Students, in consultation with their faculty advisers, design their own programs focusing on particular fields. Among them are semiconductor physics materials and devices; telecommunication systems and computer networks; high-speed analog, RF analog, and mixed analog/digital integrated circuits and systems; image and video processing; electromagnetic theory and applications; plasma physics; quantum electronics; sensory perception; and medical electronics.

Graduate studies are closely associated with research. Faculty members are engaged in theoretical and experimental research in various areas of their disciplines (see reverse of this page).

Access also exists to a number of interdisciplinary programs such as Computer Engineering, Solid-State Science and Engineering, and Bioengineering. In addition, substantial research interactions occur with the Departments of Applied Physics, Computer Science, and Industrial Engineering and Operations Research and with the College of Physicians and Surgeons.

The requirements for the Ph.D. and Eng.Sc.D. degrees are identical. Both require a dissertation based on the candidate's original research, conducted under the supervision of a faculty member. The work may be theoretical or experimental or both. The E.E. degree program does not require a thesis. It provides specialization beyond the M.S. degree in a field chosen by the student and is particularly suited to those who wish to advance their professional development after a period of industrial employment.

Research Facilities

Every phase of current research activities is fully supported and carried out in one of more than a dozen well-equipped research laboratories run by the department.

Specifically, laboratory research is conducted in the following laboratories: Multimedia Networking Laboratory; Ultrafast Opto-Electronics Laboratory; Photonics Laboratory; Microelectronics Device Fabrication Laboratory; Molecular Beam Epitaxy Laboratory; Laser Processing and Surface Analysis Laboratory; VLSI Design Laboratory; Image and Advanced Television Laboratory; Lightwave Communications Laboratory; Mixed Analog-Digital VLSI Laboratory; and Plasma Physics Laboratory (in conjunction with the Department of Applied Physics).

Financial Aid

Teaching assistantships and graduate research assistantships are available. Stipends range from $1244 to $1450 per month plus tuition exemption.

Cost of Study

The annual tuition for 1999–2000 is estimated at $24,000, plus fees.

Living and Housing Costs

The University provides limited housing for graduate men and women who are registered either for an approved program of full-time academic study or for doctoral dissertation research. University residence halls include traditional dormitory facilities as well as suites and apartments for single and married students; furnishings and utilities may be included. An estimated minimum of $17,000 should be allowed for board, room, and personal expenses for the academic year.

University Real Estate properties include apartments owned and managed by the University in the immediate vicinity of the Morningside Heights campus. These are leased yearly, as they become available, to single and married students at rates that reflect the size and location of each apartment as well as whether furnishings or utilities are included.

Requests for additional information and application forms should be directed to the Assignments Office, 111 Wallach Hall, Columbia University, New York, New York 10027.

Student Group

In 1998–99, enrollment in the Department of Electrical Engineering totaled 333 students and included 90 undergraduates (juniors and seniors), 148 master's degree candidates, 85 doctoral candidates with master's degrees, and 10 professional and part-time special students. The student population has a diverse and international character.

Location

The proximity of many local industries provides strong student-industry contact and excellent job opportunities. Cooperative research projects are available in neighboring industrial laboratories, which are engaged in research and development in computers, telecommunications, electronics, defense, and health care. Adjunct faculty from industry provide courses in areas of current professional interest. Frequent colloquia are given on current research by distinguished speakers from industry and neighboring universities.

The University

Since its founding in 1754, Columbia University has attracted students interested in the issues of their times. Opened as King's College under charter of King George II to "prevent the growth of republican principles which prevail already too much in the colonies," it instead educated founders of a new and powerful nation: Alexander Hamilton, John Jay, Robert Livingston, and Gouverneur Morris. Since then such notable figures as Michael Pupin, Edwin Armstrong, and Jacob Millman have served as professors of electrical engineering at Columbia.

Applying

Application forms should be requested from the Office of Engineering Admissions, 520 Seeley W. Mudd. Applications should be filed by January 5 for admission the following September. Notification of admission decisions are mailed beginning March 1.

Correspondence and Information

Student Coordinator
Department of Electrical Engineering
Columbia University
500 West 120th Street, Room 1312
New York, New York 10027-6699

Telephone: 212-854-3104
World Wide Web: http://www.ee.columbia.edu

Columbia University

THE FACULTY AND THEIR RESEARCH

Dimitris Anastassiou, Professor, joined the E.E. faculty of Columbia University in 1983. Prior to that, he was with the IBM Thomas J. Watson Research Center (Yorktown Heights, New York) as a Research Staff member. He is the Director of the Columbia New Media Technology Center (CNMTC). He is an IEEE Fellow, the recipient of an IBM Outstanding Innovation Award, and a National Science Foundation Presidential Young Investigator Award winner. He has been Associate Editor of the *IEEE Transactions on Circuits and Systems for Video Technology*.

Andrew T. Campbell is an Assistant Professor in the Department of Electrical Engineering and a member of the COMET Group at the Center for Telecommunications Research at Columbia University in New York. His areas of interest include programmable networks, mobile networking, distributed systems, and QOS research. He is a past cochair of the International Workshop on Quality of Service (IWQOS97) and is cochair of the 6th International Workshop on Mobile Multimedia Communications (MOMUC99). Professor Campbell received his Ph.D. in computer science in 1996 and an NSF CAREER Award in 1999.

Shih-Fu Chang, Associate Professor, joined the E.E. faculty in 1993. He is currently working on compressed-domain image technology, content-based visual query, scalable video server design, and digital video coding/networking. Professor Chang is coleading the development of Columbia's Video-on-Demand/Multimedia Testbed and is one of the principal investigators of the ADVENT (All Digital Video Encoding Networking Transmission) Project at Columbia. He is the recipient of the Best Paper Award at the ACM 1st Multimedia Conference in 1993, the Young Investigator Award at the SPIE's Symposium on Visual Communications and Signal Processing 1993, an NSF CAREER award in 1995, and an IBM University Partnership/Faculty Development Award in 1995.

Paul Diament, Professor, joined the E.E. faculty in 1963. His teaching and research areas are in all phases of electromagnetics and wave propagation, including microwaves, antennas, optics, radiation statistics, plasmas, wave interactions, relativistic electron beams, and transient electromagnetic phenomena. He is the author of *Wave Transmission and Fiber Optics*, published in 1990 by Macmillan.

Alexandros Eleftheriadis, Assistant Professor, joined the E.E. faculty in 1995. He is leading a research team that is working in the areas of media representation, with special emphasis on video signal processing and compression, video communication systems (including video-on-demand and Internet video), distributed multimedia systems, and the fundamentals of compression. Key research activities include object-based audiovisual information creation and playback tools, the Flavor language (an extension of C++/Java for media representation), and Complexity Distortion Theory (a new mathematical framework for algorithmic content representation). Professor Eleftheriadis is a member of ISO/IEC JTC1/SC29/WG11 (MPEG) and ANSI X313.1, which develops national and international standards for audiovisual information representation.

Tony Heinz, Professor, joined Columbia in 1995 after twelve years at the IBM T. J. Watson Research Center in Yorktown Heights, New York. Professor Heinz holds a joint appointment in the Departments of Electrical Engineering and Physics. His teaching and research interests lay in the areas of ultrafast optoelectronics and spectroscopy and of surface dynamics and surface process control.

Aurel A. Lazar, Professor, joined the E.E. faculty in 1980. His research interests span both theoretical and experimental studies of multimedia networks. His theoretical research during the 1980s involved the modeling, analysis, and control of broadband networks. He was the chief architect of two experimental networks that are generically called MAGNET. He is currently leading the COMET project of the Center of Telecommunications Research from the foundations of the real-time control and management architecture of multimedia networks. Professor Lazar is a member of the editorial board, *Journal of Multimedia Tools and Applications*, Kluwer; editor of *Multimedia Systems*, ACM/Springer Verlag; member of the editorial board, *Telecommunications Systems*, Baltzer; and editor, *Telecommunication Networks and Computer Systems* (Monograph Series), Springer-Verlag, New York.

Richard M. Osgood Jr., Higgins Professor of Electrical Engineering and Applied Physics, joined Columbia University in 1981 and became Higgins Professor of Electrical Engineering and Applied Physics in 1988. Professor Osgood was a cofounder of the Columbia Microelectronics Sciences Laboratories (MSL) and has served as Director or Co-Director of MSL and the Columbia Radiation Laboratory (CRL). He is a member of the ACS and MRS and a fellow of the IEEE and OSA. He was Co-Editor of *Applied Physics* (1983–1995) and Associate Editor of the *IEEE Journal of Quantum Electronics* (1981–1988). He is on the ARPA Defense Sciences Research Council (Materials Research Council) and the Los Alamos National Laboratory Visiting Advisory Board (Chemical Sciences and Technology Division). Professor Osgood has served as Councilor of the Materials Research Society and as a member of the DOE Basic Energy Sciences Advisory Committee. In 1991, Dr. Osgood received the R. W. Wood Award from the Optical Society of America.

Henning Schulzrinne received his undergraduate degree in economics and electrical engineering from the Technische Hochschule in Darmstadt, Germany, in 1984; his M.S.E.E. degree as a Fulbright scholar from the University of Cincinnati, Ohio, in 1987; and his Ph.D. degree from the University of Massachusetts Amherst in 1992. From 1992 to 1994, he was a member of the technical staff at AT&T Bell Laboratories, Murray Hill. From 1994 to 1996, he was Associate Department Head at GMD-Fokus (Berlin), before joining the Departments of Computer Science and Electrical Engineering at Columbia University. His research interests encompass real-time, multimedia network services in the Internet and modeling and performance evaluation. He is an editor of the *Journal of Communications and Networks* and IEEE Communications Society editor of the *IEEE Internet Computing Magazine*. He cochairs the IEEE Communications Society Internet Technical Committee and is Vice Chair of the IEEE Communications Society Technical Committee on Computer Communications. He has been Vice General Chair of IEEE Infocom and will be co–technical chair of that conference in 2000.

Amiya K. Sen, Professor, joined the E.E. faculty in 1963. He has been engaged in theoretical and experimental research on a variety of basic problems in plasma physics and some outstanding problems in controlled thermonuclear fusion. His current research interests include studies of important plasma instabilities. He is also working on unique feedback control techniques of distributed parameter systems to stabilize these instabilities. He is a Fellow of the American Physical Society and IEEE. He has been a consultant/adviser to the Lawrence Livermore National Laboratory, Princeton Plasma Physics Laboratory, U.S. Department of Energy, and the National Science Foundation. He received the Great Teacher Award of Columbia University in 1984.

Kenneth Shepard, Assistant Professor, joined the E.E. faculty in fall 1997. Before joining Columbia, Professor Shepard was a research staff member and manager of the VLSI Design Department at the IBM T. J. Watson Research Center, where he was responsible for the design methodology for IBM's G4 S/390 microprocessors. His honors include the Hertz Foundation doctoral thesis prize, IBM Research Division awards for contributions to the G4 microprocessor design, and an NSF Early Faculty Development CAREER award. He is also Chief Technology Officer and co-founder of CadMOS Design Technology, a design automation start-up located in California.

Malvin C. Teich, Professor Emeritus.

Yannis P. Tsividis is Charles Batchelor Professor of Electrical Engineering. He has been working on the merging of precision analog and digital circuits on a single chip ever since he joined Columbia in 1976. He and his students are responsible for several contributions toward that goal, ranging from precision device modeling and novel circuit building blocks to new techniques for analog signal processing, self-correcting chips, switched-capacitor network theory, and the creation of computer simulation programs. He is the recipient of the 1984 IEEE W. R. G. Baker Best Paper Award, the 1986 European Solid-State Circuits Conference Best Paper Award, and the 1998 Guillemin-Cauer Best Paper Award and corecipient of the 1987 IEEE Circuits and Systems Society Darlington Best Paper Award. He is a Fellow of the IEEE. He has received the Great Teacher Award from the Columbia University Alumni Association and the Distinguished Faculty Teaching Award from the Columbia Engineering School Alumni Association.

Wen I. Wang joined the E.E. department in 1987, where he is Thayer Lindsley Professor. Between 1981 and 1982, he worked at the Rockwell Science Center, and between 1982 and 1987, he worked at the IBM T. J. Watson Research Center. His current research interests include high-temperature transistors and high-speed optoelectronic devices. He has contributed some 200 journal articles in the areas of heterostructure device physics, high-speed transistors, semiconductor lasers, photodetectors, molecular-beam epitaxy, and surface science. He is a Fellow of the IEEE and the American Physical Society and a Distinguished Lecturer of the IEEE Electron Device Society.

Charles A. Zukowski, Associate Professor, joined the E.E. faculty of Columbia University in 1985. He is the author of the book *The Bounding Approach to VLSI Circuit Simulation*, published in 1986 by Kluwer Academic Publishers. He is currently working on designing high-speed and low-energy digital ICs for communications networks. Examples include switches, buffers, packet error checkers, and time-division multiplexers. He consults in the field of CMOS IC design. He is a recipient of a Presidential Young Investigator Award from the NSF in the field of CAD. He has also been active in the IEEE, both as a reviewer and in the organization of conferences.

CORNELL UNIVERSITY

College of Engineering
School of Electrical Engineering

Programs of Study	The Graduate Field of Electrical Engineering at Cornell University offers study leading to the M.S. and Ph.D. degrees and to the professional Master of Engineering (Electrical) degree. A wide range of interests are covered. The major areas of concentration are communications, information theory, signal processing, and power and control systems; computer engineering; plasma physics, space plasma physics, and electromagnetics; solid-state electronics; and optoelectronics. There are two avenues by which a student may proceed: the one-year Master of Engineering (Electrical) degree program, which places major emphasis on design capability at a high level of professional competence, and the Master of Science/Doctor of Philosophy (M.S./Ph.D.) degree program, which requires several years of study and is oriented toward research. Students who want to limit their study to the master's degree should consider the M.Eng. (Electrical) program. The M.Eng. (Electrical) degree requires 30 credits of advanced technical work, including a design project. Degree requirements for the M.S./Ph.D. degrees are kept at a minimum to give the student maximum flexibility in choosing a program of study. Independent thesis research is an important part of the M.S. and Ph.D. programs. Candidates for the Ph.D. degree must pass an EE qualifying examination before the beginning of their second term; a comprehensive admission-to-candidacy examination, required for formal admission to doctoral candidacy, may be taken after a student has earned 2 units of residence credit (a unit is given for each term of full-time graduate study successfully completed) but must be taken before the student begins the seventh unit of residence. Final examinations are given after completion of the M.S. and the Ph.D. dissertations. They cover subject matter related to the topics of the dissertations.
Research Facilities	The College of Engineering has a substantial campus with well-equipped buildings. The classrooms, offices, undergraduate laboratories, and many of the graduate research laboratories of the School of Electrical Engineering are housed in Phillips Hall and in the adjacent Rhodes Hall. Among the graduate research laboratories are those devoted to communications, computer engineering, control systems, digital signal processing, high-energy particle beams, integrated circuits, ionospheric physics and radio-wave propagation, lasers and optoelectronics, microwave and semiconductor devices, and semiconductor material preparation and characterization. Electrical Engineering faculty members and graduate students use the facilities of the following centers, laboratories, and programs: the Center for Applied Mathematics, the Electronic Packaging Program, the Joint Services Electronics Program, the Kettering Energy Systems Laboratory, the Laboratory of Plasma Studies, the Materials Science Center, the Microscience and Technology Program, the Optoelectronics Technology Center, and the Center for Theory and Simulation in Science and Engineering. In addition, the facilities of the National Nanofabrication Facility (adjacent to Phillips Hall), the National Astronomy and Ionosphere Center (Arecibo, Puerto Rico), and the Jicamarca Radio Observatory (Peru) are available. All research areas are served by a variety of computing resources. These include networked multi-MIP workstations, PCs, and the Cornell National Supercomputing Facility, which is located in Rhodes Hall.
Financial Aid	Teaching assistantships, graduate research assistantships, and fellowships are available. Nearly all M.S./Ph.D. students in electrical engineering are enrolled full-time and receive full financial support. Students are strongly encouraged to apply for National Science Foundation and Department of Defense Fellowships. Application information is available at the 230 Phillips Hall address listed below.
Cost of Study	Tuition for a full-time program of study is $23,700 for the 1999–2000 academic year.
Living and Housing Costs	Living costs vary according to the family situation of the student, the particular housing accommodations secured, and other factors. A rough estimate for a single student, exclusive of tuition but including room and board, books, and medical insurance, is $11,800 for twelve months.
Student Group	The University has 13,400 undergraduate and 5,200 graduate and professional students. The Graduate Field of Electrical Engineering currently has 210 full-time students. Approximately 30 new M.S./Ph.D. and 60 M.Eng. (Electrical) students enter the Graduate Field of Electrical Engineering each fall. The EE School also has 33 postdoctoral associates and visiting scientists.
Location	Ithaca, a city of 45,000 on Cayuga Lake, is the home of both Ithaca College and Cornell and is one of the country's great educational communities, offering cultural advantages that rival those of many large cities. State parks and recreational facilities, including those for camping, boating, skiing, and hiking, are located nearby.
The University	Cornell University was founded in 1865. Studies in electrical engineering began in 1883. Today, the faculty of the Graduate Field of Electrical Engineering consists of 40 distinguished members.
Applying	Applicants should have a baccalaureate degree from an accredited institution, or its equivalent, in an area of study that adequately prepares them for advanced study in electrical engineering. They must have maintained a minimum undergraduate grade point average (GPA) of 3.5 (A = 4.0). A minimum TOEFL score of 600 and a preapplication are required of international students who are not studying in the United States. The preapplication should include the applicant's GPA and class standing. Scores on the GRE General Test are required of all M.S./Ph.D. and M.Eng. (Electrical) applicants (scores on a Subject Test are optional). Applications are considered for the fall term only. Applications are interactive or may be downloaded via the World Wide Web for the M.S./Ph.D. program at http://www.gradschool.cornell.edu and for the M.Eng. (Electrical) program at http://www.ee.cornell.edu/. Printed applications are not mailed out until after August 1 for consideration for the fall term of the following year. Application deadlines are January 15 for fellowship consideration and February 15 for other applicants.

Correspondence and Information

For M.S./Ph.D. degree information:
Director of Graduate Studies
School of Electrical Engineering
230 Phillips Hall
Cornell University
Ithaca, New York 14853-5401

Telephone: 607-255-4304
Fax: 607-254-4565
E-mail: ee_msphd@cornell.edu
World Wide Web: http://www.ee.cornell.edu

For M.Eng. (Electrical) degree information:
Master of Electrical Engineering Program
School of Electrical Engineering
222 Phillips Hall
Cornell University
Ithaca, New York 14853-5401

Telephone: 607-255-8414
Fax: 607-254-4565
E-mail: meng@ee.cornell.edu
World Wide Web: http://www.ee.cornell.edu

Cornell University

THE FACULTY AND THEIR RESEARCH

J. M. Ballantyne, Professor; Ph.D., MIT. Optoelectronic materials and devices, integrated optoelectronics, device nanofabrication, optical interconnects.

T. Berger, Professor; Ph.D., Harvard. Information theory; multimedia, wireless, data, and video compression.

A. W. Bojanczyk, Associate Professor; Ph.D., Warsaw. Parallel algorithms and architectures for signal processing and scientific computing, numerical analysis.

H. D. Chiang, Professor; Ph.D., Berkeley. Nonlinear circuits and systems, power systems, artificial neural networks, control systems, optimization theory.

H. .G. Craighead, Professor; Ph.D., Cornell. Nanofabrication, biotechnology, optical properties of nanostructures.

R. D'Andrea, Assistant Professor; Ph.D., Caltech. Dynamics, robust and optimal control.

D. F. Delchamps, Associate Professor; Ph.D., Harvard. Dynamical systems, control theory, hybrid systems, cognitive science, evolutionary computation.

L. F. Eastman, Professor; Ph.D., Cornell. Microwave, millimeter-wave, optical, and high-speed solid-state devices; compound semiconductor growth by molecular-beam epitaxy; submicron fabrication technology.

D. T. Farley, Professor; Ph.D., Cornell. Ionospheric physics, space-plasma physics, radar techniques.

T. L. Fine, Professor; Ph.D., Harvard. Neural networks, wireless and Internet communications, unconventional probabilistic modeling.

Z. J. Haas, Associate Professor; Ph.D., Stanford. Wireless networks, wireless communication, ad-hoc networks, mobile systems, computer networks, data communication, high-speed protocols, optical networks.

D. A. Hammer, Professor; Ph.D., Cornell. Plasma physics, controlled fusion, intense ion beams, plasma radiation sources.

C. Heegard, Associate Professor; Ph.D., Stanford. Information theory, coding theory, digital communications, VLSI systems.

M. A. Heinrich, Assistant Professor; Ph.D., Stanford. Parallel computer architecture, data-intensive computing, DSM protocols, distributed systems.

S. S. Hemami, Assistant Professor; Ph.D., Stanford. Image and video coding and transmission, signal processing.

C. R. Johnson Jr., Professor; Ph.D., Stanford. Adaptive digital signal processing in communication systems, blind equalization, multiuser interference rejection.

E. C. Kan, Assistant Professor; Ph.D., Illinois at Urbana-Champaign. VLSI processes, devices and circuits, technology CAD, functional nanostructures, scientific computing.

M. C. Kelley, Professor; Ph.D., Berkeley. Space/upper, atmospheric science, radar, lidar, rocket and satellite instrumentation.

S. Keshav, Associate Professor; Ph.D., Berkeley. Computer networking, Internet infrastructure, protocol design and evaluation, quality of service, network performance management.

P. M. Kintner, Professor; Ph.D., Minnesota. Global Positioning System, space physics and engineering, spacecraft systems.

K. T. Kornegay, Assistant Professor; Ph.D., Berkeley. VLSI, wide-bandgap devices and circuits, fabrication technology, integrated systems.

J. P. Krusius, Professor; Ph.D., Helsinki University of Technology. Nanoelectronics, nanofabrication, electronic system packaging, flat panel displays.

R. L. Liboff, Professor; Ph.D., NYU. Transport in metals and semiconductors, superlattice analysis, mesoscopic metal elements, localization, percolation and hopping phenomena, nonlinearity, classical and quantum chaos.

Y. H. Lo, Associate Professor; Ph.D., Berkeley. Optoelectronic materials and devices, semiconductor lasers, microelectronics, MicroElectroMechanical Systems (MEMS).

M. Y. Louge, Professor; Ph.D., Stanford. Capacitance measurements in effective media (snow, gas-solid suspensions), microgravity flows of grains and gas-solid suspensions, computer vision.

N. C. MacDonald, Professor; Ph.D., Berkeley. Nanostructure science; MEMS: microinstruments, micro-optics, and scanned-probe instruments in silicon; MEMS-based silicon processes.

R. Manohar, Assistant Professor; Ph.D., Caltech. Asynchronous VLSI, architecture, low-power design, concurrent systems, formal methods.

P. R. McIsaac, Professor; Ph.D., Michigan. Electromagnetic theory, microwave circuits and devices.

B. A. Minch, Assistant Professor; Ph.D., Caltech. Analog IC design, low-power analog and digital circuits and systems, analog information processing systems, information theory.

J. A. Nation, Professor; Ph.D., Imperial College (London). High-power microwave generation, applied electrodynamics, high-energy electron beams, accelerator physics.

T. W. Parks, Professor; Ph.D., Cornell. Digital signal processing, wavelets, pattern classification and image processing.

K. K. Pingali, Associate Professor; Ph.D., MIT. Programming languages and semantics, restructuring compilers, parallel architectures.

C. R. Pollock, Professor; Ph.D., Rice. Solid-state and tunable lasers, fiber optics, ultrashort pulse generation.

M. L. Psiaki, Associate Professor; Ph.D., Princeton. Estimation, Global Positioning System, guidance and control.

A. P. Reeves, Associate Professor; Ph.D., Kent (United Kingdom). Computer vision, biomedical image analysis, image processing, parallel processing.

C. E. Seyler, Professor; Ph.D., Iowa. Theoretical and computational plasma physics, space plasmas, modeling of satellite, sounding rocket, and radar observations.

J. R. Shealy, Professor; Ph.D., Cornell. Vapor-phase epitaxial growth of III-V compounds, optoelectronic devices and integration.

B. C. Smith, Assistant Professor; Ph.D., Berkeley. Distributed multimedia systems, networking, video and image processing.

M. G. Spencer, Professor; Ph.D., Cornell. Microwave, optical, high-speed, high-temperature, and high-power switching solid-state devices; compound semiconductor growth by molecular-beam epitaxy (MBE) and metal organic vapor epitaxy (MOVPE); high-temperature growth and frabrication technology; deep-level studies.

R. N. Sudan, Professor; Ph.D., Imperial College (London). Plasma physics, thermonuclear fusion, space and solar physics, high-power electron- and ion-beam physics, intense laser-plasma interaction.

C. L. Tang, Professor; Ph.D., Harvard. Lasers, quantum electronics, semiconductor materials and devices, ultrafast optical processes, nonlinear optics and devices.

R. J. Thomas, Professor; Ph.D., Wayne State. Systems engineering with applications to electric power systems.

J. S. Thorp, Professor and Director; Ph.D., Cornell. Applications of optimization and control theory to power systems.

N. C. Tien, Assistant Professor; Ph.D., California, San Diego. Silicon MEMS, design and fabrication of microactuators, microsensors, micromechanical structures, and systems.

L. Tong, Associate Professor; Ph.D., Notre Dame. Statistical signal processing, communication systems and networks, adaptive receiver design, estimation theory.

V. V. Veeravalli, Assistant Professor; Ph.D. Illinois. Mobile and wireless communications, communications theory, detection and estimation theory.

S. B. Wicker, Associate Professor; Ph.D., USC. Wireless information networks, artificial intelligence and error control coding.

RESEARCH AREAS

Communications, Information Theory, Signal Processing, and Power and Control. Pattern classification, neural networks, and signal processing; energy conversion and power systems; networks, coding, data compression, and information theory; adaptive and nonlinear dynamical systems; image and video processing and compression.

Computer Engineering. Modern computer architecture, embedded systems, VLSI and asynchronous VLSI systems, reconfigurable computing, parallel and distributed computing, software technology and applications, computer vision, wireless and high-speed information networks.

Plasma Physics, Space Science and Engineering, and Electomagnetics. Upper atmospheric, ionospheric, and magnetosperic science, radar, satellites, and sounding rockets; fusion, particle beams, solar system, and fundumental plasmas; pulsed power, electron, and ion beams and plasma radiation; plasma fabrication; electromagnetics.

Solid-State Electronics and Optoelectronics. Electronic and optoelectronic materials; microfabrication and nanofabrication technology; electronic devices, circuits, and system integration; optoelectronic, optical, laser, and wide-bandgap semiconductor devices; millimeter-wave devices and systems; sensors, MEMS, micromechanics, and nanoelectromechanics.

DREXEL UNIVERSITY

College of Engineering
Electrical and Computer Engineering

Program of Study

Graduate study in the department leads to a Master of Science in Electrical and Computer Engineering (M.S.E.E.), a Master of Science in Telecommunications Engineering (M.S.E.E./Telecommunications Engineering), and a Doctor of Philosophy (Ph.D.) in electrical and computer engineering.

The degree offerings for electrical and computer engineering are focused on four general areas: electrophysics, systems, power systems, and computers and digital circuits. The field of electrophysics includes electronic devices, electromagnetics, acoustics, lightwaves, and microwaves. The study of systems includes controls, communications, telecommunications, signal processing, and robotics. The study of power systems can include resources planning and allocation and system security and stability. The study of computers and digital circuits can entail parallel processing, digital circuits, fault-tolerant design, networks, and imaging.

The curriculum for the Master of Science in Electrical and Computer Engineering encompasses 45 approved quarter credits (approximately 15 courses). The curriculum is structured on predesigned sequences of foundation courses that provide a background for advanced studies. The Ph.D. degree in electrical and computer engineering requires 90 quarter credits beyond the B.S. and a dissertation. At least one year of full-time study on campus is required. In addition to the basic University guidelines, a written preliminary examination is also required by the department. All Ph.D. candidates must participate in teaching, research, and the department's seminar program. Prospective Ph.D. students are welcome to contact the department to discuss their research interests.

The Master of Science in Electrical Engineering/Telecommunications Engineering degree program provides a specialization in telecommunications engineering. The program responds to the tremendous and growing demand for engineers with telecommunications expertise. This demand is being generated by the spread of such technologies as computer networks, e-mail systems, mobile phone systems, and interactive cable television. The degree requires 45 or 48 approved course credits. The requirements include a 27-credit core, 12 to 15 credits of electives, and 3 to 6 credits for thesis work or a full-time professional internship through Drexel's Career-Integrated Education (CIE) option.

Research Facilities

The following are the major facilities in use in the department: Computer-Based Interactive Systems Laboratory (CBIS); Electrophysics Laboratories: Microwave Laboratory, Microfabrication Clean-Room Lab, Optoelectronics Lab, Thick-Film Hybrid Circuit Lab, and Thin-Film Lab; VLSI Design Center; Image and Computer Vision Center (ICVC); Laboratory of Applied Machine Intelligence and Robotics (LAMIR); Lightwave Engineering Laboratory; Emory Long Computer Center; Power Systems Laboratories; Signal Processing Laboratory (SPL); Systems Laboratory; and laboratories related to biomedical engineering: Bioelectrode Research Laboratory (BERL), Cardiovascular Dynamics Laboratory, Telemetry, Sensor and Instrumentation Laboratory, and Ultrasound Laboratories. Telecommunications and Wireless Systems labs, the Communications and Signal Processing Laboratory (CSPL), and the Multimedia Signal Processing Laboratory (MSPL), are being developed.

Financial Aid

Teaching, research, and graduate assistantships are available and are awarded on a full-time or half-time basis. Full-time appointments carry an average monthly stipend of $1250 and 9 credits of tuition per term. The Dean's Fellowship for full-time study provides 40 percent of a student's tuition for the first term when another form of assistantship is not available. This fellowship is renewable after the first term depending on the student's academic performance.

Cost of Study

Tuition is $585 per credit hour in the 1999–2000 academic year for both the M.S. and Ph.D. The University fee is $125 per term for full-time students and $67 per term for part-time graduate students.

Living and Housing Costs

University-approved residences cost $1300 to $1600 per term for room and board. Housing of all types is available around the Drexel community for single and married students.

Student Group

Drexel has 9,590 students from forty-two states and 102 countries. Of these, 1,925 are part-time graduate students and 860 are full-time graduate students. There are 145 part-time and 60 full-time graduate students within the graduate curriculum of the department.

Location

Drexel is located in Philadelphia, a city of 2 million people and a center of science and industry. The campus is easily reached by bus, subway, railroad, and car. It is only a few minutes walk from the heart of Philadelphia, close to its many centers of education, entertainment, culture, and industry. It is a convenient train or car ride to New York City; Washington, D.C.; the Atlantic City boardwalk; and the Pocono Mountain resort and ski area. Philadelphia International Airport is only 15 minutes away by car or high-speed rail link.

The University

Drexel was founded in 1891. Its students are enrolled in five academic units: engineering, arts and sciences, business administration, the Nesbitt College of Design Arts, and information science and technology. Drexel has one of the largest private undergraduate engineering colleges in the country, and it is one of the top twenty private universities in the granting of master's degrees.

Applying

Graduate students may apply with the intention of enrolling in any of Drexel's four academic terms (fall, winter, spring, summer). Completed applications for admission and supporting transcripts and references should be on file six weeks prior to the start of classes each term. Requests for financial aid should accompany applications for admission. Applications for financial aid must be received by February 1 for consideration for the next fall term. Inquiries may be directed to any member of the graduate faculty listed on the back of this page doing research in the student's area of interest.

Correspondence and Information

Graduate Advisor
Department of Electrical and Computer Engineering
Drexel University
3141 Chestnut Street
Philadelphia, Pennsylvania 19104
World Wide Web: http://coe.drexel.edu/ECE/ece_home.html

Drexel University

THE FACULTY

The following lists current full-time faculty members and their areas of research interests. Prospective students are welcome to contact individual faculty members.

Izhak Bar-Kana, Research Professor; Ph.D., Rensselaer. Systems and control, guidance, adoptive control. (E-mail: barkana@ece.drexel.edu)

Richard B. Beard, Professor Emeritus; Ph.D., Pennsylvania. Electrophysics and biomedical: bioelectrochemistry and acoustic properties of materials, biocompatibility studies. (E-mail: beardrb@post.drexel.edu)

Nihat M. Bilgutay, Professor and Department Head; Ph.D., Purdue. Systems and biomedical: communication theory, ultrasonic imaging, nondestructive testing, and signal processing. (E-mail: bilgutay@ece.drexel.edu)

Maja Bystrom, Assistant Professor; Ph.D., Rensselaer. Telecommunications, information theory, image and video processing. (E-mail: bystrom@ece.drexel.edu)

Fernand S. Cohen, Professor; Ph.D., Brown. Computer vision, medical imaging, ultrasonics, pattern recognition, applied stochastic processes. (E-mail: fscohen@coe.drexel.edu)

Richard L. Coren, Professor; Ph.D., Polytechnic of Brooklyn. Electromagnetic fields, antennas, shielding, cybernetics of evolving systems. (E-mail: coren@ece.drexel.edu)

Afshin Daryoush, Professor; Ph.D., Drexel. Electrophysics: electromagnetic fields, antennas, telecommunications, microwave and millimeter-wave solid state devices and circuits, heterostructures, electrooptics, optoelectronics, fiber optics, integrated optics, electromagnetic sensors. (E-mail: daryoush@ece.drexel.edu)

Bruce A. Eisenstein, Arthur A. Rowland Professor; Ph.D., Pennsylvania. Systems and biomedical: digital signal processing, pattern recognition, communication theory. (E-mail: eisenstein@ece.drexel.edu)

Robert Fischl, John Jarem Professor Emeritus; Ph.D., Michigan. Power: systems, networks, controls, computer-aided design, power systems, solar energy. (E-mail: fischl@cbis.ece.drexel.edu)

William Freedman, Associate Professor; Ph.D., Drexel. Biomedical systems, rehabilitation, neural systems, computer engineering. (E-mail: freedman@ece.drexel.edu)

Eli Fromm, Roy A Brothers University Professor and Vice President for Educational Research and Development; Ph.D., Thomas Jefferson. Bioengineering: biotelemetry, sensors, bioinstrumentation, communications, professional society activities. (E-mail: fromme@duvm.ocs.drexel.edu)

Edwin L. Gerber, Professor; Ph.D., Pennsylvania. Electrophysics: physical electronics, electronic devices, computerized instrumentation. (E-mail: gerber@ece.drexel.edu)

Allon Guez, Professor; Ph.D., Florida. Robotics, intelligent control, dynamic systems. (E-mail: guez@cbis.ece.drexel.edu)

Peter R. Herczfeld, Lester A. Kraus Professor; Ph.D., Minnesota. Microwaves and millimeter waves, lightwave engineering, fiber optics, solar energy, solid-state electronics. (E-mail: herczfeld@ece.drexel.edu)

Leonid Hrebien, Associate Professor and Associate Dean; Ph.D., Drexel. Systems and biomedical: cardiovascular system characterization; tissue excitability measurement; acceleration effects on cardiovascular and cerebrovascular functions. (E-mail: hrebien@ece.drexel.edu)

Dov Jaron, Calhoun Distinguished Professor of Engineering and Medicine; Ph.D., Pennsylvania. Development, physiologic evaluation, and clinical implementation of mechanical devices to assist the failing heart, control and optimization of circulatory devices, computer application to patient monitoring, properties of bioelectrodes, instrumentation. (E-mail: dov.jaron@coe.drexel.edu)

Paul R. Kalata, Associate Professor; Ph.D., IIT. Systems: estimation, identification and control theory, adaptive control and filtering, computer control systems.

Moshe Kam, Professor; Ph.D., Drexel. Systems: detection and estimation, decision theory, decision fusion, forensic pattern recognition, robot navigation. (E-mail: kam@lorelei.ece.drexel.edu)

Constantine Katsinis, Associate Professor; Ph.D., Rhode Island. Computer architecture, modeling and applications, parallel processing systems, fault-tolerant systems, operating systems image processing and pattern recognition. (E-mail: katsinis@ece.drexel.edu)

Stanislav B. Kesler, Associate Professor; Ph.D., McMaster. Systems: communication, satellite communications, signal processing, spectral analysis, array signal processing. (E-mail: keslerb@post.drexel.edu)

Peter A. Lewin, Professor and Director, Biomedical Ultrasound Research Center; Ph.D., Technical University of Denmark. Ultrasonic characterization of materials, propagation of ultrasonic waves in inhomogeneous media, electroacoustic transducers, biological effects of ultrasound, physical acoustics, underwater acoustics. (E-mail: lewin@ece.drexel.edu)

Alexander M. Meystel, Professor and Director of Laboratory of Applied Machine Intelligence and Robotics; Ph.D., ENIMS (Moscow). Intelligent control, machine intelligence, autonomous systems, robotics. (E-mail: meystel@ece.drexel.edu)

Karen Nan Miu, Assistant Professor; Ph.D., Cornell. System electric power distribution systems, distribution automation, network reconfiguration.

Bahram Nabet, Associate Professor and Assistant Department Head for Graduate Affairs; Ph.D., Washington (Seattle). Electrophysics: compound semiconductor devices and circuits; fabrication and modeling; neural networks; vision. (E-mail: nabet@ece.drexel.edu)

Prawat Nagvajara, Associate Professor; Ph.D., Boston University. Design and testing of computer hardware; fault-tolerant computing; error-correcting code. (E-mail: prawat_nagvajara@cbis.ece.drexel.edu)

Vernon L. Newhouse, Professor Emeritus; Ph.D., Leeds (England). Biomedical and electrophysics: ultrasonic flow measurement, imaging and texture analysis in medicine, ultrasonic nondestructive testing and robot sensing, clinical engineering. (E-mail: newhouse@ece.drexel.edu)

Dagmar Niebur, Assistant Professor; Ph.D., Swiss Federal Institute of Technology (Lausanne). Intelligent information processing techniques for power system monitoring and control. (E-mail: niebur@ece.drexel.edu)

Chikaodinaka O. D. Nwankpa, Associate Professor and Director, Electrical Power Research; Ph.D., IIT. Power: power systems planning and operation; systems: modeling and control of nonlinear systems; stochastic systems theory. (E-mail: chika_nwankpa@cbis.ece.drexel.edu)

Banu Onaral, H. H. Sun Professor and Director, School of Biomedical Engineering, Science, and Health Systems; Ph.D., Pennsylvania. Biomedical: bioelectrodes, biological signal processing, measurement of very low frequency phenomena, microcomputer applications, automated measurements, fractals, sealing. (E-mail: onaral@cbis.ece.drexel.edu)

Stewart D. Personick, E. Warren Colehower Chair Professor; Ph.D., MIT. Telecommunications, information networking, optical fiber communication systems.

Athina P. Petropulu, Associate Professor; Ph.D., Northeastern. Statistical signal processing, communications, higher-order statistics, fractional-order statistics, ultrasound imaging.

Kambiz Pourrezaei, Associate Professor; Ph.D., Rensselaer. Electrophysics: focused ion beams, lithography, plasma engineering. (E-mail: pourrezaei@ece.drexel.edu)

Robert G. Quinn, Francis C. Powell Professor; Ph.D., Catholic University. Optoelectronics and plasmas: optical fibers, devices and systems, cosmic and geophysical plasma phenomena. (E-mail: quinn@cbis.ece.drexel.edu)

John Reid, Professor Emeritus; Ph.D., Pennsylvania. Biomedical: ultrasonics and ultrasound, medical instrumentation, echocardiography. (E-mail: reid@coe.drexel.edu)

Kevin J. Scoles, Associate Professor and Assistant Department Head for Undergraduate Studies; Ph.D., Dartmouth. Electrophysics: device fabrication, photovoltaics, solid-state physics, digital circuit design, computer-aided design. (E-mail: scoles@ece.drexel.edu)

Harish Sethu, Assistant Professor; Ph.D., Lehigh. Computer networks, parallel computer architecture, switching networks and topologies, network computing, computer arithmetic, network traffic analysis.

P. Mohana Shankar, Professor; Ph.D., Indian Institute of Technology (Delhi). Telecommunications, mobile systems, fiber optics, speckle, biomedical ultrasonics. (E-mail: shankar@ece.drexel.edu)

Ernest L. Stagliano Jr., Lecturer; M.S.E.E., Drexel. Power, protective relaying, systems, circuits.

Hun H. Sun, Professor Emeritus; Ph.D., Cornell. Systems and biomedical: control systems, network analysis and synthesis.

Lazar Trachtenberg, Professor; Sc.D., Technion (Israel). Design and testing of hardware (multilevel gage arrays), fault tolerant computing, design of reliable suboptimal digital filters. (E-mail: lazar_trachtenberg@coe.drexel.edu)

Oleh J. Tretiak, Robert C. Disque Professor and Director of Image Processing Center; Sc.D., MIT. Computers and image processing: microcomputer image processing workstation, computer tomography, pattern recognition, computer systems. (E-mail: tretiak@ece.drexel.edu)

Dong Wei, Assistant Professor; Ph.D., Texas at Austin. Communications, signal and image processing, wavelets, time-frequency analysis.

DUKE UNIVERSITY

Department of Electrical and Computer Engineering

Programs of Study	The Department of Electrical and Computer Engineering offers graduate study leading to the degrees of Master of Science and Doctor of Philosophy. Programs of advanced course work and research are available in five major areas: computer engineering, control, electromagnetics, signal processing, and solid-state devices and circuits. Interdisciplinary programs are also available that combine other areas of engineering (biomedical engineering and materials science and engineering), the natural sciences (physics and chemistry), computational science (computer science and biology), and medicine (imaging). Under the reciprocal agreement with neighboring universities, a student may include some courses offered at the University of North Carolina at Chapel Hill and North Carolina State University in Raleigh.
Research Facilities	The department has its own advanced computing facilities, with more than seventy modern IBM, Sun, and Silicon Graphics workstations and a Parallel Processing Machine with advanced visualization capabilities. It also has access to supercomputers at the North Carolina Supercomputing Center (NCSC). Additional facilities include a Visualization Laboratory in the School of Engineering, which was established by a grant from Silicon Graphics, and computer workstation clusters throughout the School of Engineering. The department's faculty participates in the Duke NSF Engineering Research Center on Emerging Cardiovascular Technologies and the Engineering Research Center for Advanced Electronic Materials Processing at North Carolina State University. The Center for Advanced Computing and Communications (CACC) is an NSF Industry-University Research Center that was founded jointly with North Carolina State University. It supports research in various areas, including advanced networking (ATM (asynchronous transfer mode), packet switching, and wireless) and the performance and reliability of computer systems (hardware and software). The Design Automation Technology Center (DATC) supports research in the development of highly efficient CAD tools to support chip, board, and system-level design. A specific cluster of workstations supports VLSI chip design and other related research activities. The Department of Electrical and Computer Engineering at Duke is the lead for a five-university Multidisciplinary Research Initiative (MURI) in demining. Particular technologies being explored for mine detection and identification include electromagnetic, infrared, magnetic, acoustic, and olfactory sensors. The Electromagnetic Engineering Research Laboratory is equipped for simulation, fabrication, and testing of RF and microwave circuits and systems operating in the 0.1- to 12-GHz frequency range. The laboratory also possesses equipment for measuring the basic electromagnetic properties of materials (conductivity and permittivity). In addition, an automated 3-D positioning and electromagnetic measurement system allows the mapping of antenna patterns for communications and medical applications. An extensive inventory of microwave and optical components and test equipment provides the resources for numerous laboratory experiments. The Signal Processing Laboratory supports research in ocean acoustic signal processing, signal processing, signal detection and estimation theory, speech processing and audiology, adaptive filtering and spectral analysis, and digital signal processing. The laboratory has data acquisition capabilities and an extensive library of signal processing and simulation software. It works in close cooperation with the Research Triangle Institute, the Duke Medical Center, the Scripps Institution of Oceanography, and the Office of Naval Research. Graduate students interested in experimental materials and devices in silicon microelectronics have access to the clean-room facilities at North Carolina State University. The department has a leading effort in ULSI MOSFET device simulation in the deep-submicron regime. The Integrated Circuit Fabrication facility at Duke provides processes to fabricate test metal-gate n-channel MOS devices. The device characterization facilities include I(V), C(V), and G(V) computer-controlled device characterization on a semi-automated wafer prober, bias-temperature-stress measurements, spreading resistance measurements, and wire bonding. A Molecular-Beam Epitaxy (MBE) III-V compound semiconductor growth facility is available to conduct research in the development of novel III-V devices and heterostructures. The Terahertz Optoelectronics Laboratory supports research in ultrafast (sub-picosecond) terahertz bandwidth devices and phenomena in collaboration with the physics department. This laboratory is also equipped with femtosecond resolution autocorrelators and delay lines, terahertz sources and Fourier transform spectrometers, and complete instrumentation for computer control of experiments.
Financial Aid	Financial support is available to a select number of students. Fellowships valued at up to $33,000 for the calendar year and research and teaching assistantships of $30,000 for the academic year are awarded on an annual basis and can be maintained until a student receives a degree. Summer research assistantship appointments are also available. Fellowships and assistantships typically include all tuition and fees regularly required of graduate students in addition to monthly stipends.
Cost of Study	For 1999–2000, tuition is $8760 per semester (12 units at $730 per unit). In addition to tuition, a registration fee of $1250 and a health fee of $240 are required each semester.
Living and Housing Costs	Duke has two residential apartment facilities available through an application process. These apartments are available for continuous occupancy throughout the calendar year. All of the apartments are completely furnished. The fee for town-house apartments, including utilities, is $2700 per occupant for the fall and spring semesters, on the basis of 2 students to a two-bedroom apartment. Rates in central campus apartments range from $2300 per occupant for 3 students in a three-bedroom apartment to $3463 for a single student in an efficiency apartment. Several large apartment complexes are within walking distance of the campus.
Student Group	There are 236 graduate students enrolled in the School of Engineering, 80 of whom are in the Department of Electrical and Computer Engineering.
Location	Located in the rolling central Piedmont area of North Carolina, the Duke University campus is widely regarded as one of the most beautiful in the nation. The four-season climate is mild, but good winter skiing is available in the North Carolina mountains a few hours' drive to the west, and ocean recreation is a similar distance away to the east. Duke is readily accessible by Interstates 85 and 40 and from Raleigh-Durham International Airport, which is located about a 20-minute drive from the campus via Interstate 40 and the Durham expressway.
The University	Trinity College, founded in 1859, was selected by James B. Duke as the major recipient of a 1924 endowment that enabled a university to be organized around the college and to be named for Washington B. Duke, the family patriarch. A Department of Engineering was established at Trinity College in 1910 and, following the establishment of Duke University in 1924, the Department of Electrical Engineering was formed in 1920. Its name changed to the Department of Electrical and Computer Engineering in 1996. Duke University remains a privately supported university, with more than 11,000 students in degree programs.
Applying	Admission to the department is based on a review of previous education and experience, the applicant's statement of intent, letters of evaluation, standardized test scores (GRE and TOEFL), and grade point average. Applications for admission may be submitted at any time. Applications for financial awards, however, should generally be submitted along with all supporting documents before December 31 for admission the following fall semester. While offers of financial support may be made at any time, most are awarded in the winter and early spring for the following academic year.
Correspondence and Information	Director of Graduate Studies Department of Electrical and Computer Engineering School of Engineering Box 90291 Duke University Durham, North Carolina 27708-0291 Telephone: 919-660-5245 E-mail: massoud@ee.duke.edu

Duke University

THE FACULTY AND THEIR RESEARCH

The following are the faculty's current areas of active graduate research and interests.

Computer Engineering: High-performance scientific computing and simulation, novel computer architectures and parallel processing, operating systems, computer and system architecture, distributed systems, memory design for serial and parallel architectures, computer architecture, fault-tolerant computer design, software and hardware testing methodologies and reliability analysis, neural network–based compression, computer-communication networks, computer-aided design and testing of VLSI circuits and systems, VLSI design verification, dependable computing, digital and analog system design methodology and concurrent engineering, computer performance evaluation, systems modeling, fault-tolerant computing, reliability analysis.

Intelligent Controls: Nonlinear control systems, nonlinear dynamics, intelligent systems and control, machine intelligence, fuzzy logic and neural networks, pattern recognition.

Electromagnetics: Microwaves and antenna design, radar, materials characterization, electromagnetic circuits and antennas, electromagnetic properties of materials, field and wave interactions with materials, microwave and optical communication systems.

Signal Processing and Communications: Signal detection and estimation, ocean acoustics, digital signal processing, statistical communications, speech understanding, signal processing for and models of cochlear implants, hearing aids, auditory systems modeling and analysis, statistical signal processing, sensory prosthetics, speech processing, signal processing using computational physical models with applications in sonar and communications, sensor array design and analysis, adaptive filtering and spectrum analysis, radar signal processing, digital communications, cardiovascular modeling.

Solid-State Devices and Materials: Semiconductor devices and materials, optoelectronics, MOS device physics and modeling, carrier tunneling in MOS devices, MOS device simulation, electronic transport in semiconductor quantum well structures and experimental studies of quantum chaos, device and process modeling, integrated circuit design and fabrication, application of micromachined systems in medicine, microelectromechanical systems, microelectrofluidic systems, molecular-beam epitaxy, compound semiconductor materials and devices.

RECENT PUBLICATIONS

Schlick, T., et al. **(J. A. Board).** Algorithmic challenges in computational molecular biophysics. *J. Comp. Phys.* 151:9–48, 1999.

Carin, L., et al. Ultra-wide-band synthetic-aperture radar for mine-field detection. *IEEE Antennas Propagation Magazine* 41:18–33, 1999.

Zavada, J. M., **H. C. Casey Jr.** et al. Correlation of substitutional hydrogen to refractive index profiles in annealed proton-exchanged Z- and X-cut LiNbO$_3$. *J. Appl. Phys.* 77(6):2697–708, 1995.

Chakrabarty, K., B. T. Murray, and J. P. Hayes. Optimal zero-aliasing space compaction of test responses. *IEEE Trans. Comput.* 47(11):1171–87, 1998.

Throckmorton, C. S., and **L. M. Collins.** Investigation of the effects of temporal interactions under electrical stimulation on speech recognition skills in cochlear implant subjects. *J. Acoust. Soc. Am.* 105:861–73, 1999.

Li, H. X., **T. M. Daniels-Race,** and Z. G. Wang. Growth mode and strain relaxation of InAs on InP (111)A grown by molecular-beam epitaxy. *Appl. Phys. Lett.* 74(10):1388–90, 1999.

Dewey, A., V. Srinivasan, and E. Icoz. Towards a visual modeling approach to designing microelectromechanical system (MEMS) transducers. *J. Micromech. Microeng.,* in press.

Fair, R. B. History of some early developments in ion-implantation technology leading to silicon transistor manufacturing. *Proc. IEEE* 86(1):111–37, 1998.

Gelenbe, E., K. Harmanc, and **J. Krolik.** Learning neural networks for detection and classification of synchronous recurrent transient signals. *Signal Processing* 64(3):233–47, 1998.

Drozd, J. M., and **W. T. Joines.** Using parallel resonators to create improved maximally-flat quarter-wavelength transformer impedance matching networks. *IEEE Trans. Microwave Theory Tech.* 47(2):1–10, 1999.

Papazoglou, M., and **J. Krolik.** Matched-field estimation of aircraft altitude from multiple over-the-horizon radar revisits. *IEEE Trans. Signal Processing* 47(4):966–76, 1999.

Vasudevan, N., **H. Z. Massoud,** and **R. B. Fair.** A thermal model for the initiation of programming in metal-to-metal amorphous-silicon antifuses. *J. Electrochem. Soc.* 146(4):1536–9, 1999.

Tantum, S. L., and **L. W. Nolte.** Tracking and localizing a moving source in an uncertain shallow water environment. *J. Acoust. Soc. Am.* 103(1):362–73, 1998.

Luo, K. J., **S. W. Teitsworth** et al. Controllable bistabilities and bifurcations in a photoexcited GaAs/AlAs superlattice. *Appl. Phys. Lett.,* in press.

Garg, S., A. Puliafito, M. Telek, and **K. Trivedi.** Analysis of preventive maintenance in transactions based software systems. *IEEE Trans. Comput.* 47(1):96–107, 1998.

Wang, H. O., and E. H. Abed. Bifurcation control of a chaotic system. *Automatica* 31(9):1213–26, 1995.

Tyan, C.-Y., **P. P. Wang,** D. R. Bahler, and S. Rangaswamy. A new methodology of fuzzy constraint-based controller design via constraint-network processing. *IEEE Trans. Fuzzy Syst.* 4(2):166–78, 1996.

Wu, S. M., **G. A. Ybarra,** and W. E Alexander. A complex optimal signal processing algorithm for frequency-stepped CW data. *IEEE Trans. Circuits Syst. II* 45(6):754–7, 1998.

FLORIDA INSTITUTE OF TECHNOLOGY

College of Engineering
Programs in Electrical and Computer Engineering

Programs of Study

Florida Institute of Technology offers programs of study leading to the Master of Science and Doctor of Philosophy degrees in electrical engineering and computer engineering. These programs are designed to provide opportunities for students' development of professional engineering competence and scholarly achievement.

Research Facilities

There are more than 12,000 square feet of well-equipped laboratory facilities available for use by students, faculty members, and researchers. Networked computing facilities include more than twenty-five Sun SPARCstations and sixty Macintosh and PC-compatible microcomputers, as well as a VAX and a Harris superminicomputer. Computing resources are networked via Ethernet and allow supercomputer and worldwide computer access via Internet. Research facilities support basic and applied research in electronics, communications, microwave systems, microelectronics, photonics, controls and robotics, and neural networks, as well as VLSI and computer system design. The Centers for Electronics Manufacturability and Enhancement of Quality in Engineering Education provide challenging opportunities for students. Agencies of the federal and state governments, as well as major corporations support research efforts in these and other technical areas.

Financial Aid

Graduate teaching and research assistantships are available to qualified students. For 1998–99, typical stipends ranged upward from $9600 for twelve months for approximately half-time duties. Some assistantships include tuition.

Cost of Study

For 1999–2000, tuition is $575 per semester credit hour for all students. As noted above, tuition is included with some graduate assistantships.

Living and Housing Costs

Room and board on campus cost approximately $2200 per semester. On-campus housing (dormitories and apartments) is available for full-time single and married graduate students, but priority for dormitory rooms is given to undergraduate students. Many apartment complexes and rental houses are available near the campus.

Student Outcomes

Graduates of the programs in electrical and computer engineering are employed by such companies as IBM, Texas Instruments, NASA, Harris Corporation, AT&T, General Electric, Keane Inc., Matrox, Northrup Grumman, Lockheed Martin, McDonnell Douglas, DBA Systems, Rockwell International, Advanced Micro Devices, USF&G, United Technologies, Honeywell, Computer Sciences Raytheon, ITT Aerospace, U.S. Patent Office, CIA, KPMG Audit, Los Alamos National Lab, Hewlett-Packard, Intel, Naval Air Systems Command, Naval Undersea Warfare Center, Macintosh Software Development, and Computer Task Group.

Location

The campus is located in Melbourne, on Florida's east coast. It is an area 3 miles from Atlantic Ocean beaches, with a year-round subtropical climate. The area's economy is supported by a well-balanced mix of industries in electronics, aviation, light manufacturing, optics, communications, agriculture, and tourism. Many companies support activities at the Kennedy Space Center.

The Institute

Florida Institute of Technology, founded in 1958, has developed into a distinctive independent university that provides undergraduate and graduate education in engineering and sciences for students from throughout the United States and many other countries. Florida Tech is supported by local industry and is the recipient of many research grants and contracts.

Applying

Applicants for graduate study in electrical engineering should have an undergraduate degree in electrical engineering, while those applying for graduate study in computer engineering should have an undergraduate degree in computer engineering. An applicant whose degree is in another field of engineering or in the applied sciences will be reviewed; however, undergraduate course work in the field of study is generally required prior to starting the Master of Science program.

Forms and instructions for applying for admission and assistantships are sent upon request. Doctoral applicants are asked to submit three letters of recommendation from academic references and a statement of purpose giving their reason for graduate study. Although the GRE is not required, it is considered for students with marginal undergraduate academic performance. International students applying for assistantships must have a TOEFL score greater than 600 and a TSE score of 45. International students without English proficiency and with a TOEFL score of less than 550 may need to enroll in language courses before beginning their graduate studies. Separate application for an assistantship should be made on forms available from the Graduate School.

Correspondence and Information

Graduate Admissions Office
Florida Institute of Technology
150 West University Boulevard
Melbourne, Florida 32901-6988
Telephone: 407-674-8027
　　　　　800-944-4348 (toll-free)
Fax: 407-723-9468
World Wide Web: http://www.fit.edu

Dr. R. G. Deshmukh
Electrical and Computer Engineering Programs
Florida Institute of Technology
150 West University Boulevard
Melbourne, Florida 32901-6988
Telephone: 407-674-8060
Fax: 407-984-8461
E-mail: rgd@ee.fit.edu

Florida Institute of Technology

THE FACULTY AND THEIR RESEARCH

Electrical Engineering

F. M. Caini, Associate Professor. Acoustical signal processing, fiber-optic sensors, laser remote sensing, communications.

B. G. Grossman, Professor and Electrical Engineering Program Chair. Optic fiber-optic sensors; optical computing and signal processing; optical neural networks; optical communication systems; electrooptic, acoustooptic, and nonlinear optical devices/systems; optical artificial neural networks; smart structures.

F. Ham, Harris Professor. Linear and nonlinear control systems, robust control, optimal tracking systems, biosensor research, neural networks, robotics.

W. M. Nunn Jr., Professor. Electromagnetic theory, electromagnetic scattering, antennas, radar systems, microwave, electron physics.

T. J. Sanders, Harris Professor and Division Director, Electrical and Computer Science and Engineering. Microelectronics design for manufacturability, advanced IC process development, optoelectronics, radiation hardening, microelectronics packaging.

R. L. Sullivan, Professor and Vice President for Research and Graduate Programs, University of Florida. Power system planning and operations, electrotechnologies.

M. Thursby, Associate Professor. Artificial neural networks, smart antenna systems, nondestructive testing, automated RF measurements.

Computer Engineering

H. K. Brown, Associate Professor. Microelectronic devices and processes, ASIC design for supercomputers, neural networks, VLSI design and testing, computer architectures and networks.

R. H. Cofer, Associate Professor. VLSI design, image processing, high-performance computational structures, perception and biosensor research, neural networks and parallel processing.

R. G. Deshmukh, Associate Professor. High-performance computer architectures, neural networks and parallel processing, microprocessor applications, digital systems.

J. Hadjilogiou, Professor, Computer Engineering Program Chair, and Director of the Center for Enhancement of Quality in Engineering Education. Automata theory, computer organization and architecture, fault diagnosis and reliable system design, statistical process control, design for manufacturability, integration of quality issues into the curriculum.

S. Kozaitis, Associate Professor. Pattern recognition, optical signal processing, fiber-optic sensing.

M. M. Shahsavari, Assistant Professor. Computer-aided design, VLSI circuits, high-performance computer architectures, computer networks and communications, microprocessor applications.

THE GEORGE WASHINGTON UNIVERSITY

School of Engineering and Applied Science
Department of Electrical and Computer Engineering

Programs of Study	The department offers the M.S., Engineer (Engr.), and Applied Scientist (App.Sc.) professional degrees and the D.Sc. in three fields of study: computer engineering, electrical engineering, and telecommunication and computers (M.S. only). Computer engineering concentrations include computer engineering and architecture, parallel and distributed computing, and microelectronics and VLSI systems. Electrical engineering concentrations include biomedical engineering; communications; controls, systems, and signal processing; electrical power and engineering management; electromagnetic engineering; energy conversion, power, and transmission; and microelectronics and VLSI systems. The interdisciplinary field of telecommunication and computers combines study in the areas of computer science, electrical engineering, and engineering management. M.S. programs require a minimum of 24 semester hours of graduate-level courses and 6 semester hours of thesis or a minimum of 30 semester hours for the nonthesis option. Engr. and App.Sc. professional degree programs emphasize applied subject material. At least two years of professional experience beyond a master's degree are required for admission to these programs, which require at least 30 semester hours of approved post-master's courses. The D.Sc. program normally consists of one major and two minor areas of concentration, totaling a minimum of 30 course semester hours beyond the master's level or a minimum of 54 credit hours of approved graduate work for students whose highest earned degree is a baccalaureate, and 24 dissertation research semester hours. Each doctoral student is assigned an adviser whose research specialty is in the student's major area of study; together they determine the major and minor areas of study.
Research Facilities	The School provides UNIX, PC, and Macintosh resources, featuring a Sun Microsystems S1000 multiprocessor server and Ultra 3D workstations, Dell Pentium Pro II computers, and Apple Power Macintosh G3 computers. Systems are equipped with licensed software from leading manufacturers, such as CADENCE, MathWorks, Synopsys, AdaCore Technologies, Sun Microsystems, Autodesk, Asymetrix, Adobe, and Microsoft. A sample of capabilities includes a Cascade Microtech Parametric probe station to test VLSI wafers to 40 GHz, 3-D ultrasound, National Instruments LabView software, and S parameter remote sensing up to 40 GHz. The facility's labs support engineering design, artificial intelligence, multimedia, software development, Web technology, graphics, analysis, and project management and include high-quality laser and color printing and scanning equipment. In addition, the University provides extensive computing resources, with a Sun Microsystems SPARCcenter 2000E for academic information system services and a Sun Microsystems Enterprise 4000 with four processors dedicated to research computing support. A high-performance ATM network integrates these resources, and Internet connectivity is provided for research support. University libraries contain 1.7 million volumes. Their computer system provides online access to the shared catalog of all seven member libraries of the Washington Research Library Consortium, as well as to periodical and newspaper index databases. Gelman Library provides additional databases on CD-ROM, including several specific to engineering, and offers research consultation services. Students have ready access to the Library of Congress.
Financial Aid	Teaching assistantships provided remission of tuition for 9 hours per semester and a salary ranging from $1200 to $3500 for each course taught per semester in 1998–99. Research assistants received a salary of $8000 to $16,000 for the calendar year. School Graduate Fellow, Dean's Fellow, and Department Fellow awards range from $7500 to $15,000 for eligible full-time students. Full-time students who are U.S. citizens or permanent residents may be eligible for Graduate Engineering Honors Fellowships.
Cost of Study	For the 1998–99 academic year, tuition was charged at the rate of $680 per semester hour, payable on a course-by-course basis.
Living and Housing Costs	Apartments for students enrolled in the School of Engineering and Applied Science are available in the surrounding area at a wide range of costs. These costs start at about $600 a month.
Student Group	Students in the School of Engineering and Applied Science include graduates of most colleges and universities in the United States and a number of other countries. Approximately 1,000 students are working on master's degrees, 20 on professional degrees, and 440 on the D.Sc.
Location	The Washington area has the second-largest concentration of research and development activity in the nation. Library facilities are outstanding in scope. The campus is in the Foggy Bottom historic district of Washington, D.C.
The School	The School, organized in 1884, offers limited course work in engineering, engineering administration, operations research, physical science, mathematics, economics, and statistics during summer sessions.
Applying	Admission to the M.S. program requires an appropriate bachelor's degree from a recognized institution and evidence of a capacity for productive work in the field. Admission to professional degree programs requires an appropriate master's degree from a recognized institution and evidence of a capacity for productive work in the field. Applicants for doctoral study must have adequate preparation for advanced study, including a bachelor's or a master's degree or the equivalent and a capacity for creative scholarship. March 1 and October 1 are the respective priority deadlines for applications for the fall and spring; applications are accepted on a space-available basis thereafter.
Correspondence and Information	Professor Nicholas Kyriakopoulos, Interim Chair Department of Electrical and Computer Engineering School of Engineering and Applied Science The George Washington University Washington, D.C. 20052 Telephone: 202-994-6083 Fax: 202-994-0227 E-mail: eecs@seas.gwu.edu World Wide Web: http://www.seas.gwu.edu	Interim Dean Thomas A. Mazzuchi School of Engineering and Applied Science The George Washington University Washington, D.C. 20052 Telephone: 202-994-3096 800-537-7327 (toll-free) Fax: 202-994-4522

The George Washington University

THE FACULTY AND THEIR RESEARCH

Nikitas A. Alexandridis, Professor; Ph.D., UCLA, 1971. Microprocessors, parallel computer architectures, multiprocessor systems, adaptable computer architectures, multilevel vision systems.

Giorgio V. Borgiotti, Professor Emeritus; D.E.E., Rome, 1957. Electrophysics, acoustic radiation and scattering, structural acoustics, signal processing, modal analysis of vibration, radar systems.

Robert L. Carroll Jr., Professor; Ph.D., Connecticut, 1973. Control systems, adaptive and learning systems, multidimensional systems, robotic systems, multitarget tracking.

Milos Doroslovacki, Assistant Professor; Ph.D., Cincinnati, 1994. Communications theory, mobile communications, coding theory.

Edward Della Torre, Professor; D.Eng.Sc., Columbia, 1964. Numerical device modeling, magnetic recording and bubble memory, magnetic phenomena.

Sajjad H. Durrani, Research Professor; Sc.D., New Mexico, 1962. Space communications: new services, systems planning, policy issues, research and development, and research planning and management.

Burton I. Edelson, Research Professor; Ph.D., Yale, 1960. Space science and technology, satellite communications systems.

Marvin F. Eisenberg, Professor Emeritus; Ph.D., Florida, 1961. Medical engineering, thermography, evoked response, ultrasonics.

Kie-Bum Eom, Associate Professor; Ph.D., Purdue, 1986. Computer vision, pattern recognition, image modeling.

Zhenyu Guo, Associate Professor; Ph.D., McGill, 1993. Medical ultrasound, biomedical signal processing, three-dimensional Doppler imaging, medical virtual instrumentation, cardiac physiology.

Robert J. Harrington, Professor of Engineering and Applied Science and Chairman of the Department; Ph.D., Liverpool, 1965. Simulation and control of electrical machinery and power systems, transient stability, transmission line switching transients, power system planning and operation.

Hermann J. Helgert, Professor; Ph.D., SUNY at Buffalo, 1966. Communications and information theory, coding theory, data communications, computer networks.

Robert B. Heller, Professor Emeritus; Ph.D., Saint Louis, 1951. Information processing, computer-to-computer communication networks, satellite communications, electromagnetic radiation.

Walter K. Kahn, Professor; D.E.E., Polytechnic of Brooklyn, 1960. Antennas, microwave components, fiber optics, electrophysics.

Can E. Korman, Associate Professor; Ph.D., Maryland, 1990. Numerical modeling of semiconductor devices, VLSI, magnetics, signal processing.

Nicholas Kyriakopoulos, Professor; D.Sc., George Washington, 1968. Monitoring systems, digital signal processing, controls and systems theory, applications of technology to arms control, data communications and protection.

Roger H. Lang, Professor; Ph.D., Polytechnic of Brooklyn, 1968. Wave propagation in random media, remote sensing, and adaptive arrays.

Ting N. Lee, Professor; Ph.D., Wisconsin, 1972. Networks, linear systems.

Murray H. Loew, Professor; Ph.D., Purdue, 1972. Pattern recognition, medical engineering, image processing.

Martha Pardavi-Horvath, Professor; Ph.D., Hungarian Academy of Sciences, 1985. Magnetic phenomena, magnetic recording processes and materials, magnetooptic devices and materials.

Raymond L. Pickholtz, Professor; Ph.D., Polytechnic of Brooklyn, 1966. Data communications, computer communication networks, communications theory, secure communications.

Debabrata Saha, Associate Professor; Ph.D., Michigan, 1986. Communication theory, modulation and coding techniques.

Suresh Subramaniam, Assistant Professor; Ph.D., Washington (Seattle), 1997. Design and analysis of communication networks, optical and wireless networks.

Branimir R. Vojcic, Associate Professor; Ph.D., Belgrade, 1989. Communications theory, spread spectrum, mobile and fading communications.

Wasyl Wasylkiwskyi, Professor; Ph.D., Polytechnic of Brooklyn, 1968. Electromagnetic waves, propagation, signal processing, remote sensing.

Mona E. Zaghloul, Professor; Ph.D., Waterloo, 1975. Computer-aided analysis and design of integrated circuits; analog and digital VLSI modeling, design, and testing.

---

*Joint appointment with Department of Computer Science and Engineering

HOWARD UNIVERSITY

Department of Electrical Engineering

Programs of Study

The Department of Electrical Engineering offers programs leading to the Master of Engineering and Doctor of Philosophy degrees in electrical engineering. At the master's level, major areas of study and research are solid-state electronics, control engineering, power systems, antennas, microwaves, communications, and signal processing. The Master of Engineering program offers the thesis or nonthesis option. The thesis option requires 24 credit hours of courses and a 6-credit-hour thesis. The nonthesis option requires 33 credit hours of course work and a comprehensive exam. Some specialization areas also require an engineering project. In the Ph.D. program, major areas of study and research are solid-state electronics, control engineering, power systems, communications, signal processing, and applied electromagnetics. The Ph.D. program requires 72 credit hours beyond the bachelor's degree, four semesters of residence and full-time study (two of the four semesters must be consecutive), passing grades on preliminary and qualifying exams, and evidence of submittal of a manuscript based on the student's dissertation research to a refereed journal or professional conference.

Research Facilities

The Materials Science Research Center of Excellence (MSRCE) has an interdisciplinary research team dedicated to the resolution of problems associated with the growth, characterization, and fabrication of novel electronic and electrooptic materials and devices for high-power, high-frequency, and high-temperature applications. State-of-the-art laboratories for growth and epitaxy, fabrication, characterization, and testing and an ultrafast laser facility are housed in the MSRCE.

The Center for Energy Systems and Controls (CESC) is dedicated to research and development of efficient tools for analysis and design of power system operations and planning. System theories and emerging technologies such as artificial intelligence (expert system), artificial neural network, and fuzzy logic are investigated to improve analysis of decision making for power system design and studies. Investigators of the center apply modern control theories to solve large-scale engineering problems.

In addition to the laboratories housed within the centers, students have access to other departmental laboratories, including the Communications and Signal Processing Laboratory, which is equipped for simulation of communications and signal processing systems and functions, detection and estimation algorithms, adaptive filters and arrays, emulation and applications studies of digital signal processing chips, and time-domain and frequency-domain measurements and analysis of real-time signals. The Microwave Laboratory provides fixed and swept-frequency systems for a wide range of millimeter-wave, microwave, waveguide, and coaxial component studies and designs.

Students also have access to excellent computing resources at Howard, including the School of Engineering's centralized facility, the Computer Learning and Design Center. This center provides a full spectrum of computer resources for research, such as computer-aided designs, programming and engineering problem solving, and document publishing. The center's major systems include HP and DEC VAX microcomputers, various peripherals, and remote-access capability. Ten Sun workstations and an Alliant minisupercomputer are available for advanced research.

Financial Aid

Financial aid is available to qualified students on a competitive basis through graduate, teaching, and research assistantships. In 1998–99, graduate and teaching assistantships provided stipends ranging from $7000 (per semester) to $14,000 (per year), plus tuition waivers. Some research assistantships offer support of up to $20,000. Recipients are required to work up to 20 hours per week.

Cost of Study

Tuition and fees for full-time graduate study are $10,712.50 per year in 1999–2000.

Living and Housing Costs

Most graduate students live off campus. Howard Plaza Towers, a high-rise complex, has some accommodations for students. Those who desire assistance in locating housing in the Washington metropolitan area should write to the Supervisor of Off-Campus Housing, Howard University.

Student Group

Approximately 60 students, including students from various regions of the United States and several other countries, are enrolled in the electrical engineering graduate programs.

Location

Howard University is located in Washington, D.C., a city that offers opportunities for study and research.

The University

Founded in 1867, Howard University is a privately governed institution. It has eighteen schools and colleges, and its main campus occupies almost 90 acres. Howard is a member of the National Consortium for Graduate Degrees for Minorities in Engineering and the Consortium of Universities of the Washington Metropolitan Area, which includes Georgetown, George Washington, American, and Gallaudet universities; the University of the District of Columbia; The Catholic University of America; and Mount Vernon and Trinity colleges.

Applying

Regular admission to the master's program is considered for individuals who hold a bachelor's degree in electrical engineering from an accredited institution and whose undergraduate grade point average is at least 3.0 (out of 4.0). In special circumstances, individuals having a lower grade point average or those with degrees in other disciplines or branches of engineering may be provisionally admitted. Admission to the doctoral program may be sought by persons holding a degree in electrical engineering from a nationally or regionally accredited institution. These persons must have maintained a minimum grade point average of 3.0. Candidates with equivalent qualifications earned at foreign institutions will be considered.

Additional information and application forms are obtainable on request. Applicants should provide credentials before April 1 for August admission and before November 1 for January admission. All applicants to the Graduate School of Arts and Sciences are required to submit scores on the General Test of the Graduate Record Examinations as part of their application. Applicants requiring financial assistance should provide all supporting materials as soon as possible before the above deadlines. International students whose native language is not English must report scores from the TOEFL.

Correspondence and Information

Chairman
Graduate Studies Committee
Department of Electrical Engineering
Howard University
Washington, D.C. 20059

Howard University

THE GRADUATE FACULTY AND THEIR RESEARCH

Professors
Ajit K. Choudhury, Ph.D., UCLA. Controls and communications.

Tepper L. Gill, Ph.D., Wayne State. Mathematical physics, quantum field theory, magnetism.

Gary L. Harris, Ph.D., Cornell. Electrophysics, device fabrication, characterization in materials and devices.

S. Noor Mohammad, Ph.D., Calcutta. Semiconductor materials and devices.

James A. Momoh, Chairman, Electrical Engineering Department; Ph.D., Howard. Systems engineering, power systems and controls, expert systems, neural networks.

Steven L. Richardson, Ph.D., Ohio State. Condensed matter, theoretical physics, electronic structure theory of semiconductor crystals, surfaces and computational physics.

Michael G. Spencer, Ph.D., Cornell. Electrophysics, device fabrication, microwave devices and materials characterization.

Yen-chu Wang, Ph.D., NYU. Antennas, microwaves, applied superconductivity.

Associate Professors
Mohammed Chouikha, Ph.D., Colorado. Estimation theory and detection, image and signal processing.

Ahmed Rubaai, Ph.D., Cleveland State. Power systems and controls.

Raj C. Yalamanchili, Chairman, Graduate Studies Committee; Ph.D., Georgetown. Microwave amplifiers and antennas.

Assistant Professors
Peter Bofah, Ph.D., Howard. Power devices and systems and controls.

Jamshid Goshtasbi, Ph.D., SUNY at Binghamton. Computer architecture, digital design.

Chang Jong Kim, Ph.D., Texas A&M. Electric power systems operation, control and protection.

Adjunct Professors
Leonard Rockett, Ph.D., Columbia. Solid-state electronics, device and materials characterization, integrated circuit design and fabrication.

Woodford Zachary, Ph.D., Maryland. Theoretical physics.

ILLINOIS INSTITUTE OF TECHNOLOGY

Department of Electrical and Computer Engineering

Programs of Study

The Department of Electrical and Computer Engineering offers a full spectrum of graduate degree programs at the master's and doctoral levels, both full-time and part-time. The Master of Electrical and Computer Engineering (M.E.C.E.) is a one-year course of study that prepares students for professional practice in electrical and computer engineering. The Master of Science in Electrical Engineering (M.S.E.E.) is a two- to three-semester program combining breadth across several areas of study within electrical engineering and specialization within one area, with an option to pursue thesis research under the guidance of a faculty adviser. Areas of study include communication and signal processing; networks, electronics, and electromagnetics; power and control systems; and computer engineering. The Master of Science in Computer Systems Engineering (M.S.C.S.E.) is similar to the M.S.E.E., but the M.S.C.S.E. emphasizes computer engineering with areas of study that include computer hardware design, computer networking and communications, and system and application software. The Doctor of Philosophy (Ph.D.) degree is awarded to recognize a high level of mastery in electrical engineering and requires a research dissertation representing an original contribution to knowledge in the field. The Ph.D. program is appropriate for those students with M.S. degrees who are interested in pursuing an academic or industrial research career. To be admitted to candidacy, the student must pass an oral qualifying examination and an oral comprehensive examination, both conducted by an appointed committee. Research, conducted in consultation with a faculty adviser, culminates in an oral defense of the dissertation. The program length is typically three to four years beyond the M.S. degree.

Research Facilities

The department operates research laboratories for work in CAD for VLSI, communications, computer networking, image processing and medical imaging, microwave electronics, optical fiber components, power systems, signal processing, and ultrasonic diagnostics. The department also collaborates with and utilizes the research resources of the IIT Research Institute and the Pritzker Institute of Medical Engineering.

The department operates a Sun Workstation network and several minicomputers and personal computer networks for student use. University computer resources include an SGI Challenge L system, an SGI workstation cluster, DEC VAX's, and more than 350 personal computers.

Financial Aid

Financial support in the forms of fellowships, teaching assistantships, research assistantships, and scholarships are awarded on a competitive basis. Support varies from tuition only to tuition plus stipend for nine to twelve months. Approximately 80 percent of full-time students receive some form of financial aid. Primary consideration for graduate financial aid is given to applications received before March 1.

Cost of Study

Graduate tuition for 1999–2000 is $590 per credit hour. Full-time study is a minimum of 9 hours per semester. International students must register for a minimum of 12 credits or the equivalent per semester.

Living and Housing Costs

Housing is available for graduate students in IIT residence halls. In 1999–2000, the cost of room and board ranges from $5155 to $6880. Unfurnished IIT apartments are available at costs ranging from $458 to $927 per month, including utilities. Early application for apartments is recommended. Several off-campus apartment complexes are located within a mile of the Institute.

Student Group

IIT's total enrollment in 1998–99 was approximately 5,900, more than half of whom were graduate students. Graduate enrollment in electrical and computer engineering numbered 324. Five continents were represented in the student group, making for a diverse and stimulating atmosphere.

Student Outcomes

Graduates of the master's programs are typically employed by the engineering industry, at corporations both large and small, or at one of a wide range of technology companies located in the greater Chicago area. Graduates of the doctoral program most frequently enter research positions in industry at locations throughout the United States. Several recent doctoral graduates are pursuing academic careers at major universities.

Location

IIT's Main Campus is located near the heart of Chicago, just 3 miles south of the Loop, and is central to the greater Chicago area's thriving technological community of business, industry, and research institutions. Internationally known for its architecture, museums, symphony, and theater as well as its beautiful lakefront on the western shore of Lake Michigan and the unusually rich variety of its ethnic communities, Chicago offers a vast array of recreational and cultural opportunities. The main campus, designed by Ludwig Mies van der Rohe and regarded internationally as a landmark of twentieth-century architecture, occupies fifty buildings on a 120-acre site and includes research institutes, libraries, laboratories, residence halls, a sports center, and other facilities. Among its immediate neighbors are Comiskey Park, home of the Chicago White Sox; two major medical centers; and the McCormick Place Exposition Center. The downtown campus is in the Loop near the city's financial trading, banking, and legal centers. The Rice campus is in suburban Wheaton, convenient to the Interstate 88 research and technology corridor west of the city. The Moffett campus is in southwest suburban Summit-Argo.

The Institute

Illinois Institute of Technology was formed in 1940 by the merger of Armour Institute of Technology (founded in 1890) and Lewis Institute (founded 1896). IIT offers programs of study in engineering and the sciences, architecture, design, psychology, business, and law. IIT is a member of the prestigious Association of Independent Technological Universities (AITU).

Applying

Applications and supporting documents, including transcripts, test scores, and three letters of recommendation, should be received by the Office of Graduate Admissions (3300 South Federal Street, Room 301A, Chicago, Illinois 60616) by June 1 for fall, November 1 for spring, or May 1 for summer matriculation. Applications that include a request for consideration for financial aid should be received by March 1. Application forms and additional information are available on line at http://www.grad.iit.edu.

Applications for financial aid, which is usually awarded for the academic year beginning in the fall term, should be received by March 1.

Correspondence and Information

Department of Electrical and Computer Engineering
Illinois Institute of Technology
Chicago, Illinois 60616-3793

Telephone: 312-567-3400
Fax: 312-567-8976
E-mail: gradinfo@ece.iit.edu
World Wide Web: http://www.ece.iit.edu/

Illinois Institute of Technology

THE FACULTY AND THEIR RESEARCH

Robert C. Arzbaecher, Professor and Director of the Pritzker Institute of Medical Engineering; Ph.D., Illinois. Biomedical engineering, signal processing and control.

Guillermo E. Atkin, Associate Professor; Ph.D., Waterloo. Modulation and coding for bandwidth efficient communication, digital mobile and wireless communication, spread-spectrum and optical communication systems.

Wai-Yip Geoffrey Chan, Motorola Assistant Professor; Ph.D., California, Santa Barbara. Signal compression, audiovisual communication, vector quantization.

Alexander J. Flueck, Assistant Professor; Ph.D., Cornell. Power systems, computational methods, control systems.

Nikolas Galatsanos, Associate Professor; Ph.D., Wisconsin. Image processing, image restoration, image coding, multidimensional signal processing.

Joseph L. LoCicero, Professor; Ph.D., CUNY. Communication and digital signal processing, speech and image processing, discrete multi-tone communications, automatic speech recognition, ultra-wideband communications.

Jeffrey P. Mills, Visiting Assistant Professor; Ph.D., IIT. Electromagnetic compatibility.

John A. Nestor, Associate Professor and Associate Chairman, Computer Engineering Program; Ph.D., Carnegie Mellon. Computer-aided design, VLSI design.

V. C. Ramesh, Assistant Professor; Ph.D., Carnegie Mellon. Power systems, intelligent systems, Internet computing, optimization, economics.

Gerald F. Saletta, Associate Professor and Associate Dean of Undergraduate Studies; Ph.D., IIT. Electronics, digital systems.

Jafar Saniie, Professor; Ph.D., Purdue. Digital signal and image processing, ultrasonic imaging, pattern recognition, detection and estimation, diffraction tomography, nondestructive testing.

Marco Saraniti, Assistant Professor; Ph.D., Munich. Semiconductor device modeling, computational electronics, numerical methods.

Mohammad Shahidehpour, Professor and Dean of the Graduate College; Ph.D., Missouri. Large-scale power systems, nonlinear stochastic systems, optimization theory.

Henry Stark, Bodine Distinguished Professor; Ph.D., Columbia. Image reconstruction, medical imaging, pattern recognition, signal processing and sampling theory, optics.

Philip Troyk, Associate Professor; Ph.D., Illinois. Polymers for electronics, neural implants, solid-state power systems.

Donald R. Ucci, Associate Professor and Interim Chairman; Ph.D., CUNY Graduate Center. Adaptive systems, signal processing, communications, computer systems, stochastic processes.

Miles Wernick, Assistant Professor; Ph.D., Rochester. Medical imaging, image processing, pattern recognition.

Geoffrey A. Williamson, Associate Professor and Associate Dean of the Graduate College; Ph.D., Cornell. Adaptive filtering and control; signal processing, parameter estimation and system identification, control systems, robust control theory.

Albert Z. H. Wang, Assistant Professor; Ph.D., SUNY at Buffalo. Integrated circuit design and reliability, mixed-signal circuits, semiconductor devices.

Thomas T. Y. Wong, Professor; Ph.D., Northwestern. Microwave communications systems, nonlinear device measurement, semiconductor device theory, microwave electronics and instrumentation.

Huapeng Wu, Visiting Assistant Professor; Ph.D., Waterloo. Cryptography, computer network security, computer architecture and arithmetic, error-control codes.

Yongyi Yang, Assistant Professor; Ph.D., IIT. Signal and image processing, data compression, applied mathematical and statistical methods.

Research Areas

Active research programs are conducted in the general areas of communication systems and signal processing; power and control systems; electromagnetics, networks, and electronics; digital and computer systems; and bioengineering. Specific topics include adaptive control and signal processing; automated diagnosis; automatic speech recognition; biomedical engineering; bionics (including prosthesis and drug infusion); CAD for VLSI; coding theory; design automation; diffraction tomography; digital and data communication; digital filters; electrical machine analysis and optimization; electron devices; fiber and integrated optics; high-definition video; image processing; intelligent systems; microwave electronics; millimeter-wave communications systems; neural network; nondestructive testing; nonlinear device characterization; optimum communication system design; pattern recognition; photonic switching; power electronics; power system stability, operation, and control; sampling and signal recovery; signal analysis techniques; speech processing; spread spectrum techniques; and transient electromagnetics.

JOHNS HOPKINS UNIVERSITY

Department of Electrical and Computer Engineering

Programs of Study

Graduate study is oriented toward the Ph.D., with emphasis on applicable research and scholarship rather than on engineering practice. Research interests of the faculty are concentrated in communications, computer engineering, solid-state electronics, information processing, nonlinear and quantum optics and electronics, systems and control theory, and language and speech processing.

The program encourages students to become involved in research-oriented studies as early as possible. A departmental examination must be passed before the beginning of the fourth semester of graduate study. Then a member of the faculty is selected as the research sponsor, and the student and sponsor plan together the remainder of the graduate program. The program requirements include passing a Graduate Board oral examination given by a panel of 5 faculty members, preparing a brief research proposal, and presenting a departmental seminar. Finally, there is a dissertation defense open to the public.

The department also offers both full-time and part-time M.S.E. programs. Such programs require at least eight 1-semester courses at the graduate level, plus satisfactory completion of either a master's essay, a special project, or two additional 1-semester courses.

Research Facilities

The department occupies Barton Hall and portions of Latrobe Hall, both of which contain extensive laboratory and computing facilities. Major research laboratories include the Image Analysis and Communications, Parallel Processing, Solid State Electronics, Quantum Electronics, Optical Communications, Sensory Aids, and Nonlinear Optics Laboratories and the Center for Language and Speech Processing.

The department operates numerous computer servers, workstations, and personal computers that run UNIX, Windows NT, and Windows 98 operating systems. It also operates a network of Sun Microsystems computers, which are used for CAD software for VSLI. The Center for Language and Speech Processing operates a network of twenty-five display workstations and a network of Solaris and Linux computers that perform parallel processing of speech recognition algorithms. In addition, Homewood Academic Computing provides state-of-the-art PC and UNIX terminals for graduate and undergraduate student use. More information is available on the World Wide Web (http://www.jhu.edu/~hac).

Financial Aid

Financial assistance is available through tuition scholarships, teaching assistantships, and research assistantships. The nominal assistantship salary in 1999–2000 is $13,635 for nine months. Limited financial aid is available for terminal M.S.E. students. The department accepts candidates with a bachelor's degree directly into the Ph.D. programs and offers full financial aid support to outstanding applicants.

Cost of Study

Annual tuition for 1999–2000 is $23,660.

Living and Housing Costs

University-owned apartments are available for single and married students. Private off-campus housing in Baltimore is readily available in a wide range of prices.

Student Group

The department has about 80 graduate students, of whom about 65 are working toward the Ph.D. degree, and about 150 undergraduates. In Arts and Sciences and Engineering, there are 3,725 undergraduates, 1,342 full-time graduate students, and 356 faculty members.

Location

The University is located in Homewood, a residential area located only a short distance north of downtown Baltimore. The 140-acre campus is wooded and quiet, providing an appropriate atmosphere for study and research. The Johns Hopkins medical institutions are separately located in east Baltimore. The Johns Hopkins Applied Physics Laboratory is located 20 miles south of Baltimore.

The University

Founded in 1876, Johns Hopkins was the first university to offer graduate education as it is known today in America. From the start, the University was dedicated to the idea of "creative scholarship." Its faculty members were scholars as well as teachers. Students were to engage not merely in the absorption of knowledge dispensed by their elders but also in the creation of knowledge through their own research. Thus, although small in size, Johns Hopkins has had a profound influence on American higher education.

Applying

Qualified individuals are encouraged to visit the campus to discuss their plans for graduate study with the faculty. The deadline for applying for admission is January 15. Notification of awards is made by March 15. Submission of scores on the Graduate Record Examinations is required. Midyear entrance into the graduate programs is difficult and not recommended. Applications, forms, and instructions are available on the World Wide Web to download, complete, and submit. The address is http://www.ece.jhu.edu/, sub-heading "GRADUATE ADMISSIONS."

Correspondence and Information

Graduate Admissions Committee
Department of Electrical and Computer Engineering
Johns Hopkins University
Baltimore, Maryland 21218
Telephone: 410-516-4808
Fax: 410-516-5566
E-mail: gradadm@ece.jhu.edu

Johns Hopkins University

THE FACULTY AND THEIR RESEARCH

A. G. Andreou, Professor; Ph.D., Johns Hopkins, 1986. Electron devices, analog VLSI, sensors and micropower electronics, physics of computation.

G. Cauwenberghs, Assistant Professor; Ph.D., Caltech, 1994. Analog and digital VLSI systems, distributed parallel processing, analog neural computation.

F. M. Davidson, Professor and Chair; Ph.D., Rochester, 1969. Quantum optics, optical coherence, optical communications.

R. R. Etienne-Cummings, Assistant Professor; Ph.D., Pennsylvania, 1995. Mixed-signal VLSI systems, computational sensors, motor-sensory systems, robotics and neuromorphic engineering.

J. I. Goutsias, Professor; Ph.D., USC, 1986. Signal processing, image processing, and analysis.

P. A. Iglesias, Associate Professor; Ph.D., Cambridge, 1991. Linear control, H-infinity control, adaptive control.

F. Jelinek, Professor and Director, Center for Language and Speech Processing; Ph.D., MIT, 1962. Speech recognition, statistical methods of natural language processing, information theory.

R. I. Joseph, Professor; Ph.D., Harvard, 1962. Electromagnetic theory, nonlinear wave propagation, solitons.

J. U. Kang, Assistant Professor; Ph.D., Central Florida, 1996. Optoelectronics, nonlinear optics, fiber optics and lasers.

A. E. Kaplan, Professor; Ph.D., USSR Academy of Sciences, 1967. Quantum electronics, nonlinear optics, optical bistability.

J. B. Khurgin, Professor; Ph.D., Polytechnic, 1986. Quantum electronics, nonlinear optics.

G. G. L. Meyer, Professor; Ph.D., Berkeley, 1970. Parallel computing, computational methods, fault-tolerant computing.

J. L. Prince, Professor; Ph.D., MIT, 1988. Multidimensional signal processing, medical imaging, computational geometry.

W. J. Rugh, Professor; Ph.D., Northwestern, 1969. Linear and nonlinear systems theory, control theory.

T. D. Tran, Assistant Professor; Ph.D., Wisconsin, 1998. Filter banks, wavelets, multirate systems and applications.

H. L. Weinert, Professor; Ph.D., Stanford, 1972. Statistical signal and image processing.

C. H. Westgate, William B. Kouwenhoven Professor; Ph.D., Princeton, 1966. Solid-state electronics, design and circuit modeling, microwaves, high-speed circuits.

Associated Faculty: Joint Appointments and Part-Time Members

William R. Brody, Professor and President of the University. Medical imaging, magnetic resonance imaging.

William J. Byrne, Lecturer; Ph.D., Maryland, 1993. Speech recognition, information theory and statistics.

R. E. Glaser, Lecturer (part-time); Ph.D., Johns Hopkins, 1981. Advanced digital logic systems.

W. C. Gore, Professor (part-time); Dr.Engr., Johns Hopkins, 1952. Information theory, error-correcting codes, communication systems.

Douglas M. Green, Research Professor and Associate Dean for Research, Whiting School of Engineering; Ph.D., Texas at Austin, 1977. Microcirculation trauma, highly parallel architectures, computer design for microelectronics, fault-tolerant wafer-scale processors.

Robert E. Jenkins, Senior Lecturer; M.S., Maryland, 1967. Computer engineering.

Sanjeev P. Khudanpur, Lecturer; Ph.D., Maryland, 1997. Information theory, statistical modeling, language and pronunciation modeling for speech recognition.

Elliot McVeigh, Associate Professor, Biomedical Engineering; Ph.D., Toronto, 1988. Magnetic resonance imaging.

Michael I. Miller, Professor and Director, Center for Imaging Science, Biomedical Engineering/Electrical and Computer Engineering (Joint Appointment); Ph.D., Johns Hopkins, 1983. Image understanding, pattern theory, computer vision, medical imaging/computational anatomy, computational neuroscience.

T. O. Poehler, Research Professor and Vice Provost for Research; Ph.D., Johns Hopkins, 1961. Quantum electronics, solid-state physics.

Larry Wolff, Associate Professor, Computer Science; Ph.D., Columbia, 1991. Computer vision, computational sensors for vision/robotics, computer graphics.

LOUISIANA STATE UNIVERSITY

Department of Electrical and Computer Engineering

Programs of Study

The department is the focus for research and graduate education in electrical and computer engineering in the state of Louisiana. It offers programs of study leading to the M.S. and Ph.D. degrees. Areas of study are computers (including computer architectures, parallel and distributed processing, and fault-tolerant computing), electronics (including device-oriented electronics and solid-state materials), communication systems (including digital communication, data compression, and digital signal processing), control systems (including robust, adaptive, and time-varying control), and electric power (including power systems and power electronics). Students seeking the M.S. degree may select either a thesis (24 semester hours of course work plus 6 semester hours of thesis credit) or a nonthesis (36 semester hours of course work) option. The Ph.D. degree requires a dissertation and a minimum of 18 semester hours of course work beyond the requirements for the M.S. Students are encouraged to enter the Ph.D. program directly after receiving the B.S., in which case requirements are a dissertation and 48 semester hours of approved course work.

Research Facilities

The department possesses modern research facilities, including laboratories for materials science, VLSI, computer engineering, control systems, and power systems and electrical machines. The departmental computing facilities include a network of workstations and personal computers supporting a variety of operating systems and applications. A University-wide System Network Computing Center offers supercomputer and mainframe power. The department's Solid State Laboratory is the only one of its kind in Louisiana to carry out interdisciplinary research in the areas of semiconductor material growth, characterization, device fabrication, and measurements. The laboratory has a 200-square-foot class 100 clean room for photolithography. The department's faculty is involved in the Center for Advanced Microstructures and Devices (CAMD), established by the University for carrying out research in X-ray lithography for submicron devices. The VLSI systems design laboratory houses graphics terminals, state-of-the-art Sun Workstations, and Berkeley CAD VLSI tools. The laboratory is used for instructional and research purposes, both for the designing of smart silicon VLSI chips and for VLSI device modeling. Designs can be sent via electronic mail to Silicon Foundry (MOSIS) for chip fabrication. The Systems Laboratory supports research in automatic control, communications, and signal processing. The laboratory houses several high-powered Sun Workstations and personal computers, a television camera and signal acquisition hardware and software, process simulators, and advanced system simulation software. The Power Electronics Laboratory offers the student hands-on experience with several power electronic devices, many of them rated at more than 6 kilowatts. Also available are facilities for experimenting with new power electronic circuit designs capable of handling large amounts of power. The department has access to the Remote Sensing and Image Processing Laboratory, the Nuclear Science Center, and other University facilities. The Division of Engineering Research provides additional support for research activities. The University library has holdings of about 2.1 million volumes and receives more than 23,800 periodical titles.

Financial Aid

The department attempts to provide financial support to all qualified Ph.D. students and to outstanding M.S. students. Graduate assistantships in 1998–99 carried stipends ranging upward from $10,000 for approximately half-time duties performed during the academic year. Assistantships for the fiscal year carried proportionately higher stipends. Graduate assistants are considered Louisiana residents for the purpose of fee assessments. Research performed under an assistantship normally provides the basis for the thesis. Fellowships, with stipends of up to $20,000, are also available; they require no duties of the recipient and included the H. D. Perkins and Board of Regents' Fellowships. Summer employment can be found in private industry.

Cost of Study

Tuition and fees in 1998–99 for full-time graduate students who are residents of Louisiana totaled $1368 per semester; nonresidents paid $3318 per semester. Diploma and thesis-binding fees were $40 and $20, respectively, for the master's degree and $60 and $45 for the doctorate.

Living and Housing Costs

The cost of dormitory rooms for single students in 1998–99 ranged from $965 to $1150 per semester. Unfurnished married student apartments were available for $325 to $425 per month. Off-campus apartments are plentiful; they rented for $250 to $550 per month.

Student Group

Enrollment on the Baton Rouge campus is more than 27,000 (50 percent are women); graduate and professional enrollments total 4,472. Enrollment in the department for the fall of 1998 was 82 full-time graduate students, including 28 Ph.D. candidates. Students come to the University from every state in the Union and more than sixty other countries.

Location

Baton Rouge, the state capital, is a growing industrial metropolis of 350,000. It is centered on the oil and petrochemical industries and is a deepwater port on the Mississippi River. The semitropical climate makes outdoor activities, such as golf and tennis, popular throughout the year, and the area is widely known for its fishing, boating, and hunting. Social life in Baton Rouge, relaxed and informal, is enhanced by a symphony orchestra, a little theater, and a Civic Center Complex. New Orleans, renowned for its southern hospitality, historic sites, antebellum charm, and recreational facilities, is located just 70 miles to the southeast.

The University

Louisiana State, founded in 1860, is on the southern edge of Baton Rouge, ½ mile east of the Mississippi River. The principal buildings, grouped on a landscaped 300-acre tract, exhibit a blend of contemporary design and the older Italian style, with its tile roofs and colonnaded passageways. Growth and improvements in buildings, facilities, and staff have been steady, and the University is now a modern multicampus facility well equipped to meet the educational, social, and cultural needs of its students. More than 26,000 graduate degrees have been awarded. The University is the largest and most comprehensive in Louisiana.

Applying

Applications for admission are considered throughout the year but should be submitted as early as possible. If financial assistance is desired, applications should be completed by January 31 for the fall and November 1 for the spring semesters. An application fee of $25 is required. Applicants are required to submit scores on the GRE General Test, and international students must pass the TOEFL with a minimum score of 550.

Correspondence and Information

For formal applications:

Office of the Registrar
Louisiana State University
Baton Rouge, Louisiana 70803-2804
Telephone: 504-388-1686

For further information and financial assistance:

Graduate Coordinator
Department of Electrical and Computer Engineering
Louisiana State University
Baton Rouge, Louisiana 70803-5901
Telephone: 504-388-5241
E-mail: ecegrad@ee.lsu.edu
World Wide Web: http://www.ee.lsu.edu/

Louisiana State University

THE FACULTY AND THEIR RESEARCH

Pratul K. Ajmera, Professor; Ph.D., North Carolina State, 1975. Semiconductor materials and devices.

Jorge L. Aravena, Professor; Ph.D., Michigan, 1980. System theory, computer-based control systems, signal processing.

Jack C. Cho, Associate Professor; Ph.D., Iowa State, 1967. Magnetic devices, magnetic bubble memory.

Leszeck Czarnecki, Associate Professor; Ph.D., 1969, D.Sc., 1984, Silesia Technical (Poland). Power electronics, nonsinusoidal systems, network analysis and synthesis.

Fred I. Denny, Associate Professor; Ph.D., Mississippi State, 1973. Electrical power systems, energy efficiency, control centers, distributed resources.

Ahmed A. El-Amawy, Professor; Ph.D., Iowa State, 1983. Microprocessors, computer architecture, VLSI processor arrays.

Martin Feldman, Professor; Ph.D., Cornell, 1962. Applied optics, integrated circuit lithography, metrology.

Guoxiang Gu, Associate Professor; Ph.D., Minnesota, 1988. Digital signal processing, controls.

Charles H. Harlow, Professor; Ph.D., Texas at Austin, 1967. Digital systems, image analysis, software engineering.

Subhash C. Kak, Professor; Ph.D., Indian Institute of Technology (Delhi), 1970. Information structures, parallel processing, data security, artificial intelligence.

Ralph A. Kinney, Professor; Ph.D., Florida, 1967. Field theory, fiber optics, computer applications.

David M. Koppelman, Associate Professor; Ph.D., RPI, 1988. Advanced computer architectures, parallel processing, neural networks.

Gil S. Lee, Associate Professor; Ph.D., North Carolina State, 1987. Solid-state materials and devices, molecular-beam epitaxy quantum well structures, optoelectronics.

J. B. Lee, Assistant Professor; Ph.D., Georgia Tech, 1997. Design, fabrication, and characterization of MEMS, microsensors, and microactuators.

Alan H. Marshak, Professor and Chairman; Ph.D., Arizona, 1969. Semiconductor device physics, device analysis, transport theory, heterostructures.

Morteza Naraghi-Pour, Associate Professor; Ph.D., Michigan, 1986. Communication theory, image processing, signal processing.

Suresh Rai, Associate Professor; Ph.D., Kurukshetra (India), 1980. Computer networking, parallel processing, microprocessors, reliability.

Jagannathan Ramanujam, Associate Professor; Ph.D., Ohio State, 1990. Operating systems and programs for parallel machines, computer architecture.

Mohammad Saquib, Assistant Professor; Ph.D., Rutgers, 1998. Queueing networks, communication theory, spread spectrum, multiuser detection, synchronization and advanced signal processing techniques for wireless communication systems.

Alexander Skavantzos, Associate Professor; Ph.D., Florida, 1987. Computer architecture, digital signal processing, parallel processing.

Ashok Srivastava, Associate Professor; Ph.D., Indian Institute of Technology (Delhi), 1975. Semiconductor device modeling, IC design and fabrication.

Jerry Trahan, Associate Professor; Ph.D., Illinois at Urbana-Champaign, 1988. Theory of computation, parallel processing, computational complexity, algorithm design and analysis.

Ramachandran Vaidyanathan, Associate Professor; Ph.D., Syracuse, 1990. Parallel computation, problem decomposition, parallel algorithms, interconnection networks, algorithm complexity.

Charles H. Voss, Professor; Ph.D., North Carolina State, 1963. Electronics and electrical systems.

Kemin Zhou, Associate Professor; Ph.D., Minnesota, 1988. Control and system theory, computer-aided system design, signal processing, industrial applications of advanced control theory.

Jianchao Zhu, Associate Professor; Ph.D., Alabama in Huntsville, 1989. Control and system theory, communications, artificial intelligence, digital signal processing.

Publications and Research Grants

The graduate faculty of the department publishes an average of thirty-five refereed journal papers and forty-one conference presentations per year; graduate student collaborators often appear as coauthors. A complete list of publications of the last three years is available from the departmental graduate office.

Several faculty members hold associate editorships in prestigious scientific journals, and all faculty members are frequent reviewers for *IEEE Transactions* and other journals. Members of the faculty have also organized special sessions or given invited presentations at various events.

Faculty research is supported by grants from federal, state, and industrial sources, including LEQSF; NSF; U.S. Air Force; U.S. Army; U.S. Navy; Fingerhut Corporation; EPRI; LSU Utilities Power Research Consortium; Optek Technology, Inc.; Southern University and A&M; Meretel; Formosa Plastic Corporation; and Georgia Institute of Technology. During the last three years, sponsored research contracts exceeded $4.1 million. Recent research grants include "Parallelizing Programs for Distributed Memory Multiprocessors," "Fault-Tolerant Neural Networks: Design Theories and Applications," "Dynamic Clustering," "High Efficiency Monolithic Tandem Solar Cells," "Parallel Computing with Reconfigurable Buses," "High Speed Heterojunction Bipolar Transistors," "Modeling and Control of Uncertain Systems with Applications," "Theory and Design of Branch-and-Combine Clock Networks," "Loop Transformations and Scheduling Techniques for Distributed Memory Processors," "Robust Stability and Stabilization of Systems with Real Parameter Uncertainty," "Neural Network Approach Towards Logic Testing and Design-for-Testability," "Alignment System for Anorad X-Ray Exposure Tool," "Development of High Speed Complementary Heterojunction Bipolar Transistors," "Modeling and Control of Uncertain Systems with Applications to Air Force Problems," "Employing Instruction History in the Management of Shared Memory Coherent Caches," "Languages, Compilers, and Runtime Systems for Parallel Architectures," "Robust Adaptive Control & Its Application to Missile Autopilots," "Algorithmic Scalability in Reconfigurable Bus-Based Models," "Evaluation and Development of Methods and Equipment for Power Quality Improvement," "Multi-Channel Switching in Optical Fiber Networks," "Control of HSPS and SVC for Enhancing Power System Stability," "Fabrication of INSB Magnetoresistor," "High Performance Control Systems Design," "Learning and Generalization Using Neural Network with Applications to On-Line F-16 Reference Model," "A Coherent-Cache Computer for Research and Education," "Investigation of Capacity, Microwave Coexistence and Security in Wireless Network," "Power System Stability Study," and "Design and Analysis of Multicast ATM Switching Networks with Internal Flow Control."

McGILL UNIVERSITY

Department of Electrical Engineering

Programs of Study

The Department of Electrical Engineering offers programs of graduate study leading to the degrees of Master of Engineering and Doctor of Philosophy. The areas of concentration include biomedical engineering, communication systems, computational analysis and design for electromagnetics, computer engineering, computer vision, control systems, engineering software, microwaves, optical waveguides, photonic systems, power electronics, power systems, robotics, solid-state devices, and VLSI systems.

The M.Eng. program requires the equivalent of one calendar year of full-time study but is normally completed in 1½ years. The degree may be obtained by the completion of six graduate courses and an externally examined thesis or nine graduate courses and an internally examined project.

The Ph.D. program requires the equivalent of two years of full-time study beyond the master's degree. While there are no formal course requirements for the degree, which is awarded upon completion of a satisfactory thesis, students usually select courses applicable to their research interests.

Research Facilities

The Department of Electrical Engineering has extensive facilities for all its main research areas. Laboratories for research into robotics, control, and vision are available in the associated Center for Intelligent Machines. The telecommunications laboratories have been recently reequipped and are centered on work in signal compression and wireless communication. These laboratories form part of the Canadian Institute for Telecommunications Research (CITR)—a federally funded network of Centres of Excellence. Digital and analogue design test laboratories form part of the Microelectronics and Computer Systems Laboratory. There are extensive research laboratories for antenna and microwave work as well as a fully equipped facility for optical fiber and integrated optics research. There is a fully equipped laboratory for photonic systems research, which includes continuous wave and femtosecond Ti:Sapphire lasers, diode lasers, extensive optics and optomechanics, and sophisticated electronic and imaging equipment. Solid-state facilities include measurement equipment for optical and electronic properties of materials, vacuum deposition and RF sputtering systems. The Computational Analysis and Design Laboratory provides tools for numerical analysis, visualization, interface design, and knowledge-based system development, as does the power systems facility. There is a well-equipped laboratory for power electronics and machines research. The department has extensive computer facilities available. Almost all the research machines are networked, providing access to a vast array of different hardware. In addition, there are links to the Centre de Recherche Informatique de Montreal (CRIM) and the University Computing Centre.

Financial Aid

The department awards a number of graduate assistantships that carry an annual stipend of approximately Can$15,000 per year to qualified full-time graduate students. These are normally tenable for a maximum period of eighteen months for students registered for the M.Eng. degree and for thirty-six months for Ph.D. candidates. Graduate assistants and holders of other scholarships or fellowships may, with the approval of their supervisors, also undertake a teaching assistantship for an additional remuneration of between Can$400 and Can$3000 per year. Ample job opportunities exist for spouses because of McGill's proximity to Montreal's business section.

Cost of Study

In 1999–2000, tuition for international students is Can$8866.20 per year for the M.Eng. program and Can$8056.20 per year for the Ph.D. program; an additional session costs Can$2195.38 per year. For Canadian citizens and landed immigrants who are Quebec residents, the tuition fee is Can$2504.20 per year; an additional session costs Can$2083.38 per year. For Canadian citizens and permanent residents who are not Quebec residents, tuition fees are Can$4004.20 per year for the M.Eng. program and Can$2504.20 per year for the Ph.D. program; an additional session costs Can$2083.38 per year. Required health insurance is Can$578 per year for a single student (Can$2537 per year for family coverage). Books and supplies cost approximately Can$550 per year. These fees are set by the Quebec government and are subject to change.

Living and Housing Costs

The average living expenses are Can$800 per month for a single student and Can$1100 per month for a married student. A large variety of apartments in subdivided older houses and apartment blocks are available within walking distance of the University. Monthly rents range upward from Can$300 for a one-room apartment and Can$400 for a two-room apartment.

Student Group

The department has approximately 235 graduate students, almost all full-time, who come from all provinces of Canada and countries around the world to form a diverse and participative academic and social group. Most eventually take up careers in consulting engineering, university teaching, government professions, and industry, especially in high-technology sectors.

Location

The University is situated on a beautiful campus in the heart of Montreal. The city is attractive and cosmopolitan, with a population of 2.5 million, the majority of whom speak French and the remainder English. Cultural activities abound: the main museums and concert halls are all within a few blocks of the University. An excellent and inexpensive public transport system of buses and subway trains allows for easy travelling in and around the city. Outdoor activities—skiing, sailing, cycling, and walking—are easily accessible.

The University

McGill University was founded in 1821. It is a nondenominational institution and operates in the English language. Other than engineering, its faculties are agriculture, arts, dentistry, education, graduate studies and research, law, management, medicine, religious studies, and science. Of 15,000 full-time students, about 3,000 are studying for advanced degrees. The Faculty of Engineering is one of Canada's oldest. It also has Departments of Chemical Engineering, Civil Engineering and Applied Mechanics, and Mining and Metallurgical Engineering and Schools of Architecture, Computer Science, and Urban Planning.

Applying

Applications can be obtained from the department. Completed applications, including official transcripts, reference letters, and Test of English as a Foreign Language (TOEFL) scores, Graduate Record Examinations (GRE) scores, and Advanced Engineering Test scores, must be received in the departmental office by February 1 for September admission.

Scores on the GRE General Test and Advanced Engineering Test are required of all applicants from colleges and universities outside Canada. A minimum total GRE score of 1800 is required. International applicants must give evidence of adequate proficiency in the English language by achieving a score of at least 600 on the TOEFL. Admission and financial aid decisions are based on the student's transcripts, general academic background, letters of recommendation, and specific research goals. Applications are considered for admission for the fall session only.

Correspondence and Information

For further information:
Graduate Program Admissions
Department of Electrical Engineering
Room 603, McConnell Engineering Building
3480 University Street
McGill University
Montreal, Quebec H3A 2A7
Canada
Telephone: 514-398-7344
Fax: 514-398-4470
E-mail: grad@ee.lan.mcgill.ca

McGill University

THE FACULTY AND THEIR RESEARCH

P. R. Bélanger, Ph.D., MIT. Control systems, industrial process control, parameter identification, robotics.

M. L. Blostein, Ph.D., Illinois. Analysis and design of digital signal processing for telecommunication applications.

B. Boulet, Ph.D., Toronto. Robust H-infinity control, industrial process control, model validation, fuzzy logic control, manufacturing execution systems.

P. E. Caines, Ph.D., London; PE. Systems and control theory: stochastic systems, adaptive control, logic, hybrid and hierarchical systems.

J. J. Clark, Ph.D., British Columbia. Computer vision, smart sensors, analogue and digital VLSI sensors, sensors signal processing, robotics.

J. R. Cooperstock, Ph.D., Toronto. Human-computer interaction, intelligent systems, ubiquitous computing, telepresence, multimodal interfaces.

C. H. Champness, Ph.D., McGill. Electronic and optical properties of semiconductors, physical electronics of solid-state devices.

M. El-Gamal, Ph.D., McGill. Integrated circuits (ICs) for communication, high-frequency circuit design, low-voltage analog and mixed-signal ICs, device modeling.

F. Ferrie, Ph.D., McGill. Computer vision, robotics, image processing.

K. Fraser, M.Eng., McGill. Electronics.

F. D. Galiana, Ph.D., MIT. Power systems planning and control, expert systems, deregulation.

V. Hayward, Ph.D., Paris XI. Robotics, programming languages and trajectory calculation, design and applied control.

P. Kabal, Ph.D., Toronto. Digital signal/speech processing, adaptive filters/systems, data communications.

M. A. Kaplan, Ph.D., Cornell. Traffic and network management for wireline and wireless telecommunication.

A. G. Kirk, Ph.D., London. Photonics, optical interconnects for electronic systems, microoptics, diffractive optics.

K. Khordoc, Ph.D., McGill. Formal specification and verification, timing verification, real-time systems.

H. Leib, Ph.D., Toronto. Digital communication, telecommunication systems, signal processing, statistical communication and information theory.

M. D. Levine, Ph.D., London. Computer vision, robotics, image processing.

D. A. Lowther, Ph.D., CNAA (U.K.). Numerical methods in electromagnetics, interactive graphics and visualization of electromagnetic fields, knowledge-based techniques applied to electromagnetic design and analysis.

S. McFee, Ph.D., McGill. Computer modelling; simulation and visualization of electromagnetic fields in microwave, optical, and power-frequency devices.

H. Michalska, Ph.D., London. Control theory, nonlinear systems, stabilization, optimal control.

R. Negulescu, Ph.D., Waterloo. Automated verification, concurrency theory, CAD tools, asynchronous circuits.

B. T. Ooi, Ph.D., McGill. High-power electronics, electromecanical energy conversion.

D. V. Plant, Ph.D., Brown. Photonic/optoelectronic devices and systems, optomechanics, smart pixels, optical interconnects.

G. W. Roberts, Ph.D., Toronto. Analogue IC design and test, filter theory, synthesis and design.

N. C. Rumin, Ph.D., McGill. Circuit simulation and timing analysis, device modelling of VLSI systems.

I. S. Shih, Ph.D., McGill. Semiconductor materials and devices, applied high Tc superconductivity and field emission devices.

J. Webb, Ph.D., Cambridge. Computer modelling, simulation and visualization of electromagnetic fields in microwave, optical, and power-frequency devices.

G. L. Yip, Ph.D., Toronto. Electromagnetic wave theory, fiber and integrated optics, guided-wave photonic devices, optical communication systems.

Z. Zilic, Ph.D., Toronto. Digital system design, logic synthesis, multiprocessors.

Associate Faculty

J. H. T. Bates, Ph.D., Canterbury (New Zealand). Theoretical models of mammalian respiratory mechanics describing relationships between pressures and flows; biological signal processing, especially numerical techniques for deconvolving noisy signals.

E. Cerny, Ph.D., McGill. Computer-aided design and verification (formal and by simulation) of microelectronic systems.

E. Dubois, Ph.D., Toronto. Image coding and processing, multidimensional signal processing.

G. L. Dudek, Ph.D., Toronto. Computational vision, artificial intelligence, robotic navigation, operating systems.

A. Evans, Ph.D., Leeds (England). 3-D imaging of brain function (PET, SPECT) and anatomy (MRI, CT), kinetic analysis of tracers using PET, imaging physics of PET scanners, functional neuroanatomy of normal cognitive processing.

M. J. Ferguson, Ph.D., Stanford. Telecommunication software, design, specification, formal methods.

W. R. J. Funnell, Ph.D., McGill. Modeling of auditory mechanics using finite-element method, creation of 3-D finite-element models from serial sections, CT and MRI, use of computers and multimedia in medical education.

H. L. Galiana, Ph.D., McGill. Modeling of binocular eye-movement control, nonlinear structurally modulated network, coordination of eyes with head and body in orientation and tracking, detection of parameter jumps in physiological signals, clinical applications of control models in automated diagnosis, porting of biological controls to robots.

J. Gotman, Ph.D., McGill. Analysis of electrophysical signals, particularly the EEG; automation and quantification of EEG analysis in epilepsy, sleep, and neurological disorders; analysis of interactions between brain regions.

C. K. Jen, Ph.D., McGill. Ultrasonic sensors and systems for online process monitoring, ultrasonic nondestructive testing of materials and composites, ultrasonic imaging.

G. Joos, Ph.D., McGill. Power electronics, flexible AC transmission systems (FACTS).

R. E. Kearney, Ph.D., McGill. Human motor control: joint mechanics, reflex function and muscle dynamics; motor system pathologies; systems identification: nonlinear, time-varying, two-dimensional, and closed-loop.

J. Konrad, Ph.D., McGill. Image and video coding/processing, stereoscopic video, multidimensional signal processing.

M. A. Marin, Ph.D., UCLA. Computer systems, logic design, expert systems for maintenance and design, parallel computing.

D. McGillis, B.Eng., McGill. Power system planning and expert systems.

D. O'Shaughnessy, Ph.D., MIT. Digital speech coding, speech synthesis by rule, automatic speech recognition.

M. P. Slawnych, Ph.D., British Columbia. Experimental and theoretical muscle mechanics, from the single cross-bridge to the whole muscle; electromyographic estimation of numbers of motor units; analysis of delayed-onset muscle soreness.

M. R. Solyemani, Ph.D., Concordia. Information theory, digital signal processing, satellite communications.

L. A. Wegrowicz, Ph.D., Leningrad; D.Sc., IFTR (Warsaw). Applied electrodynamics, antenna and propagation engineering.

MICHIGAN STATE UNIVERSITY

College of Engineering
Department of Electrical and Computer Engineering

Programs of Study

The department offers graduate programs leading to the Master of Science and Doctor of Philosophy degrees. The student's graduate program is designed in consultation with a faculty adviser of his or her choice. There are excellent opportunities for advanced theoretical and experimental study in the areas of communication sciences, computer engineering, controls, electronic devices and circuits, electrophysics, power, and robotics.

The master's degree requires 30 semester credits and is normally completed in twelve to eighteen months. The student may choose either the nonthesis or the thesis option (6 to 8 thesis credits). The Ph.D. candidate must pass a qualifying examination and a comprehensive examination. Normally, 42 credits of course work beyond the bachelor's degree and a minimum of 24 research credits are taken in a typical Ph.D. program, which requires approximately three to four years to complete. The Ph.D. program prepares the student for a career in advanced research and/or teaching.

Research Facilities

The electrical and computer engineering department occupies approximately 9,000 square feet of research space in the Engineering Research Complex in addition to numerous research laboratories in the Engineering Building and other campus sites. There are excellent laboratories and facilities for experimental research in backscattering, plasma science, microfabrication, robotics, manufacturing science, computer design, embedded system design, thin films, signal and speech processing, neural networks, and electric machines. The College of Engineering computer facilities, administered by the Division of Engineering Computer Services, include numerous student workrooms that contain a large number of personal computers and workstations. Access to outside computing facilities, including supercomputers, is provided through connections to all major academic and research networks. All graduate students have a permanent computer account and an e-mail address. Students can keep in contact with colleagues, research groups, databases, and agencies throughout the world.

The College's engineering library contains 58,000 volumes and 630 journal subscriptions. The University library system has more than 3 million volumes and more than 21,000 journal subscriptions.

Financial Aid

Research and teaching assistantships and fellowships are available to qualified students. In 1998–99, annual half-time assistantship stipends were $15,191 for first-year M.S. students and $18,184 for senior M.S. students and Ph.D. students. Up to 6 credits of tuition per semester are waived for students with assistantships, and nonresident graduate assistants pay the resident tuition rate on all additional credits. Fellowships range from $100 per semester to $16,152 per calendar year.

Cost of Study

In 1998–99, graduate tuition was $222.50 per semester credit for Michigan residents and $450 per semester credit for nonresidents. In addition, per semester fees included a $288 registration fee for students enrolling for more than 4 credits ($238 for students enrolling for 4 or fewer credits), a $237 engineering program fee for students enrolling for more than 4 credits ($131 for students enrolling for 4 or fewer credits), a $3 FM radio tax, a $9 student information technology fee, a $4.25 student newspaper tax (for students enrolling for 10 or more credits), and a $4.50 Council of Graduate Students tax.

Living and Housing Costs

Owen Graduate Residence Hall offers comfortable living in an atmosphere conducive to advanced study and the exchange of ideas. The 1998–99 rates were $1928 per single-room occupancy and $1649 per double-room occupancy per semester, including residence hall tax. Furnished University apartments are available for married students. The 1998–99 family monthly rates were $390 for a one-bedroom apartment and $432 for a two-bedroom apartment.

Student Group

For the 1998 fall term, 43,189 students were enrolled on Michigan State University's East Lansing campus; 33,419 were undergraduates, 6,472 were graduate students, and 1,932 were professional students. Students come from every state in the Union and about 107 other countries. Enrollment in the College of Engineering in the 1998 fall semester was 4,480, of whom 3,888 were undergraduate students and 592 were graduate students.

Location

Michigan State University is located in the south-central part of the state in East Lansing, which is approximately 80 miles northwest of Detroit and about 210 miles northeast of Chicago. East Lansing is a city of about 30,000 people in a metropolitan area of about 270,000. Lansing, the capital of Michigan, is located just 4 miles to the west. Many summer and winter recreational activities are located within driving distance of East Lansing.

The University

Michigan State University, founded in 1855, was the prototype institution for the nation's land-grant colleges. The enrollment of 42,000 makes MSU one of the largest single-campus universities in the nation. Such a large campus provides the engineering student with wide exposure to and interdisciplinary opportunities with the natural sciences, medicine, and business. Engineering is one of the most popular and demanding majors on campus. The number of graduate students and faculty members and the amount of research expenditures have grown considerably in the past ten years.

Applying

Application forms and instructions for submitting transcripts, letters of recommendation, and related material may be obtained from the department. Study may begin in fall, spring, or summer semesters; applications are accepted at any time but should be received at least two months prior to the expected starting date. Most fellowship and assistantship opportunities arise in the fall semester; applications for the fall semester should be submitted by February 1 if financial aid is desired. The GRE General Test is required for international students and recommended for domestic students, and the TOEFL is required for international students. A minimum TOEFL score of 580 is necessary for consideration.

Correspondence and Information

Graduate Program Coordinator
Department of Electrical and Computer Engineering
2120 Engineering Building
Michigan State University
East Lansing, Michigan 48824-1226
Telephone: 517-353-6773
E-mail: eegradoff@egr.msu.edu

Michigan State University

THE FACULTY AND THEIR RESEARCH

Dean M. Aslam, Associate Professor; Ph.D., Rhenish-Westphalian Technical (Aachen), 1983. Diamond film microsensors, Si cold cathodes, MOS technology. (Telephone: 353-6329; e-mail: aslam@egr.msu.edu)

Jes Asmussen Jr., Professor and Chairperson; Ph.D., Wisconsin, 1967. Plasmas, microwave processing of materials, ion and electrothermal thrusters. (Telephone: 355-4620; e-mail: asmussen@egr.msu.edu)

Virginia Ayres, Associate Professor; Ph.D., Purdue, 1985. Electronic and thermal properties of diamond material, electron beam sources and devices, plasmas. (Telephone: 355-5066; e-mail: ayres@egr.msu.edu)

John R. Deller Jr., Professor; Ph.D., Michigan, 1979. Speech processing, system identification, neural network models in speech, biomedical signal processing. (Telephone: 353-8840; e-mail: deller@egr.msu.edu)

P. David Fisher, Professor; Ph.D., Johns Hopkins, 1967. Digital circuit design, instrumentation. (Telephone: 355-5241, e-mail: fisher@egr.msu.edu)

Erik D. Goodman, Professor; Ph.D., Michigan, 1972. Computer-aided design and manufacturing (CAD/CAM), interactive computer graphics, genetic algorithms. (Telephone: 353-9695; e-mail: goodman@egr.msu.edu)

Timothy Grotjohn, Associate Professor; Ph.D., Purdue, 1986. Semiconductor physics and devices, integrated-circuits modeling, plasmas modeling and diagnostics. (Telephone: 353-8906; e-mail: grotjohn@egr.msu.edu)

Timothy Hogan, Assistant Professor; Ph.D., Northwestern, 1996. Charged transport measurements, pulse laser deposition of new electronic materials. (Telephone: 432-3176; e-mail: hogant@egr.msu.edu)

Leo Kempel, Assistant Professor; Ph.D., Michigan, 1994. Electromagnetic theory, computational electromagnetics, finite element methods, large-scale scientific computing, antenna analysis and design, scattering. (Telephone: 353-9944; e-mail: kempel@egr.msu.edu)

Hassan Khalil, Professor; Ph.D., Illinois, 1978. Nonlinear control, singular perturbation methods, robust control. (Telephone: 355-6689; e-mail: khalil@egr.msu.edu)

Bruce Kim, Assistant Professor; Ph.D., Georgia Tech, 1996. VLSI, multichip module, and MEMS design, testing, and verification; mixed-signal IC design and testing; VLSI algorithms and CAD; deep submicron design and analysis; computer architecture; high-speed digital hardware; computer networks; fault-tolerant systems; BIST design for embedded systems. (Telephone: 432-2630; e-mail: kimb@egr.msu.edu)

Robert D. Nowak, Assistant Professor; Ph.D., Wisconsin, 1995. Signal and image processing, nonlinear and statistical signal processing, applications in medicine and communications. (Telephone: 355-5235; e-mail: nowak@egr.msu.edu)

Dennis P. Nyquist, Professor; Ph.D., Michigan State, 1966. Electromagnetic theory, radiation, scattering, layered media, integrated circuits, guided-wave optics. (Telephone: 355-1771; e-mail: nyquist@egr.msu.edu)

Percy A. Pierre, Professor; Ph.D., Johns Hopkins, 1967. Communications theory, stochastic processes, signal detection and estimation. (Telephone: 432-5148; e-mail: pierre@egr.msu.edu)

Donnie K. Reinhard, Professor; Ph.D., MIT, 1973. Electronic materials and devices, plasma-assisted etching and deposition. (Telephone: 355-5214; e-mail: reinhard@egr.msu.edu)

James A. Resh, Associate Professor; Ph.D., Illinois, 1963. Nonlinear systems, computer-aided circuit analysis and design, microcomputers. (Telephone: 355-7649; e-mail: resh@egr.msu.edu)

Edward J. Rothwell, Associate Professor; Ph.D., Michigan State, 1985. Transient electromagnetic scattering, antennas, radar target identification, electromagnetic theory. (Telephone: 355-5231; e-mail: rothwell@egr.msu.edu)

Diane Thiede Rover, Associate Professor; Ph.D., Iowa State, 1989. Computer engineering, advanced architecture computers, performance measurement and visualization. (Telephone: 353-7735; e-mail: rover@egr.msu.edu)

Fathi M. A. Salam, Professor; Ph.D., Berkeley, 1983. Nonlinear systems and circuits, analog VLSI and neural networks, adaptive and robust control. (Telephone: 355-7695; e-mail: salam@egr.msu.edu)

Robert A. Schlueter, Professor; Ph.D., Polytechnic of Brooklyn, 1972. Stability; planning, operation, and control; and nonlinear and adaptive control applied to power systems. (Telephone: 355-5244; e-mail: schluete@egr.msu.edu)

Michael A. Shanblatt, Professor; Ph.D., Pittsburgh, 1980. Computer engineering, VLSI architectures for enhanced control, neural networks, VLSI design methodologies. (Telephone: 353-7249; e-mail: mas@egr.msu.edu)

Marvin Siegel, Professor; Ph.D., Harvard, 1970. Electromagnetic theory, communication theory, biomedical signal processing. (Telephone: 355-7688; e-mail: siegel@egr.msu.edu)

Elias G. Strangas, Associate Professor; Ph.D., Pittsburgh, 1980. Electrical machinery, finite-element methods for electromagnetic fields, electrical drives, power electronics. (Telephone: 353-3517; e-mail: strangas@egr.msu.edu)

R. Lal Tummala, Professor; Ph.D., Michigan State, 1970. Robotics, digital control, and manufacturing systems. (Telephone: 355-7453; e-mail: tummala@egr.msu.edu)

Chin-Long Wey, Professor; Ph.D., Texas Tech, 1983. Fault-tolerant design of microelectronics structures, VLSI design and testing, logic synthesis for testability, and analog/digital circuits faults diagnosis. (Telephone: 353-0665; e-mail: wey@egr.msu.edu)

Gregory M. Wierzba, Associate Professor; Ph.D., Wisconsin, 1978. Analog electronics, macromodeling, computer-aided design, active filters. (Telephone: 355-5225; e-mail: wierzba@egr.msu.edu)

Ning Xi, Assistant Professor; D.Sc., Washington (St. Louis), 1993. Robotics, intelligent control and applications, manufacturing automation. (Telephone: 432-1925; e-mail: xin@egr.msu.edu)

H. Roland Zapp, Associate Professor and Associate Chairperson; Ph.D., Stanford, 1969. Signal processing, data acquisition, communication, ultrasonic imaging, alternative energy systems. (Telephone: 355-5230; e-mail: zapp@egr.msu.edu)

NORTH CAROLINA STATE UNIVERSITY

Department of Electrical and Computer Engineering

Programs of Study

The Department of Electrical and Computer Engineering offers programs leading to the Master of Science (M.S.), Master of Engineering (M.E.), and Doctor of Philosophy (Ph.D.) degrees in either computer engineering or electrical engineering. The M.S. degree may be earned under either the thesis or nonthesis option. Either option of the M.S. degree may be designed to provide the background for professional practice or prepare for Ph.D. study. The M.S. program requires 30 hours of course work, 15 of which must be from a designated area of specialization. If the thesis option is selected, up to 6 credit hours of research may be selected. Students seldom need more than two years to meet these requirements. Normally, a student pursuing a Ph.D. takes approximately 72 semester hours of course credits beyond the B.S. degree. Students must pass a qualifying examination (within three semesters of admission into the Ph.D. program), an in-depth examination set by the student's advisory committee, and preliminary and final oral examinations. The Ph.D. degree typically requires three to five years. The department offers the opportunity for study in a large variety of specialties and cross-disciplines. These specialties include digital, analog and microwave circuits, and microwave communications and signal processing, biomedical engineering, computer communications, digital systems, electromagnetics, power and control systems, power electronics, solid state, system software, and VLSI design. The department has highly qualified faculty members in all of the above areas. The department cooperates closely with computer science, materials engineering, and operations research. M.S. students following the nonthesis option may choose to either sharply focus their studies on a particular area or study in several of the above areas. M.S. students following the thesis option, as well as Ph.D. students, generally tailor their studies to match their research interests. The department also offers five-year programs leading to a B.S. in electrical and computer engineering and a master's degree in business or a master's degree in electrical or computer engineering.

Research Facilities

The department has numerous laboratories that support both education and research. These laboratories provide access to special-purpose hardware and software that are not available through other departmental and college computing resources. The department is organized into several research units whose disciplinary areas coincide with faculty research interests. The Solid State Electronics Laboratory emphasizes the study of devices, material processes, and manufacturing equipment for very advanced devices and circuits in both silicon and compound semiconductor technologies. The National Science Foundation's Center for Advanced Electronic Materials Processing (AEMP) has established collaborative research programs with other NCSU departments, North Carolina universities, and institutions to develop new generations of semiconductor manufacturing processes that will enable efficient manufacture of VLSI chips with submicron features. The Power Semiconductor Research Center (PSRC) is supported by an industrial consortium and focuses on advanced smart and integrated devices and circuits for power systems. The Center for Advanced Computing and Communication (CACC) addresses a range of challenges arising in the design of complex, very high performance computer telecommunications and signal processing systems. The Electronics Research Laboratory focuses on modeling, simulation, and validation of antennas, signal propagation, and micro- and millimeter-wave devices and systems. Research in the Center for Robotics and Intelligent Machines explores a wide range of technological and theoretical issues arising in the design and prototyping of machines that operate with minimal intervention. The Electric Power Research Center (EPRC), operated jointly with the Department of Nuclear Engineering, conducts studies in a range of areas that include distribution systems, load modeling, reliability, intelligent fault detection and diagnosis systems, and power quality.

Financial Aid

The department offers a number of graduate teaching, research, and industrial assistantships each year to qualified applicants. The current stipend ranges from $12,000 to $15,000 for nine months. A typical academic year assistantship requires 20 hours of work per week and carries a stipend of $8000–$14,000. The College of Engineering offers fellowships of $8000 and $4000. Full-time students holding teaching or research assistantships are eligible to receive tuition remission and health insurance and must register for 9 credit hours per semester.

Cost of Study

The 1999–2000 tuition and fees for full-time students are $1185 per semester for state residents and $5768 for nonresidents.

Living and Housing Costs

On-campus dormitory facilities are provided for unmarried graduate students. In 1999–2000, the rent in these dormitories starts at $995 per semester. There are a limited number of University-owned apartments for married students; rent for those start at $345 per month.

Student Group

Enrollment in the graduate program is approximately 400 students, with approximately two thirds of the students in the M.S. programs and the other one third studying for the Ph.D. degree.

Student Outcomes

All graduates of the department have found opportunities across the spectrum of industry, government, and education. The industrial employment ranges from self-employed consultants to positions at organizations such as AT&T, Motorola, and Texas Instruments. In government, employment has been gained at institutions such as the Army Research Office, and in education, graduates have obtained positions at various universities.

Location

Raleigh is a relatively small and uncrowded city with a population of about 250,000. The city defines one point of the Research Triangle. The area offers a wide range of cultural activities. The climate is moderate, with four distinct seasons.

The University

North Carolina State University is a major research university in the land-grant tradition. The University was founded in 1887, to provide education and encourage economic development in agriculture and engineering. Fall 1998 enrollment exceeded 28,000 students, including 19,500 full-time undergraduate, 5,000 graduate, and 3,400 lifelong education students.

Applying

Applicants must have a bachelor's degree from an accredited institution. Ideally this degree should be in electrical engineering or computer engineering. All applicants must have taken the GRE General Test within the past four years. The TOEFL is required for international students whose native language is not English. All application materials for the fall semester must be received by June 25 for U.S. students and by May 1 for international students. For the spring semester, application materials must be received by November 25 and September 15, respectively.

Correspondence and Information

Dr. A. Reisman, Professor
Director of Graduate Programs
Department of Electrical and Computer Engineering, Box 7911
North Carolina State University
Raleigh, North Carolina 27695-7911

Telephone: 919-515-5090
Fax: 919-515-5601
E-mail: reisman@eos.ncsu.edu
World Wide Web: http://www.ece.ncsu.edu/

North Carolina State University

THE FACULTY AND THEIR RESEARCH

D. P. Agrawal, D.Sc., Swiss Federal Institute of Technology. Computer engineering.
S. T. Alexander, Ph.D., North Carolina State. Signal processing.
W. E. Alexander, Ph.D., New Mexico. Signal processing.
W. D. Allen, Ph.D., North Carolina State. Computer engineering.
B. J. Baliga, Ph.D., Rensselaer. Semiconductor devices.
M. E. Baran, Ph.D., Berkeley. Power systems.
S. M. Bedair, Ph.D., Berkeley. Semiconductor materials and devices.
G. Bilbro, Ph.D., Illinois. Communications and signal processing.
D. Bitzer, Ph.D., Illinois. Computer science.
S. Blanchard, Ph.D., Duke. Biomedical engineering.
J. J. Brickley, Ph.D., Virginia. Biomedical devices and systems.
G. Byrd, Ph.D., Stanford. Computer engineering.
M.-Y. Chow, Ph.D., Cornell. Power systems.
T. M. Conte, Ph.D., Illinois. Computer engineering.
E. Davis, Ph.D., Illinois. Computer architecture.
A. Duel-Hallen, Ph.D., Cornell. Communications.
A. E. Eichenberger, Ph.D., Michigan. Computer engineering.
P. D. Franzon, Ph.D., Adelaide (Australia). Microelectronic systems.
E. D. Gehringer, Ph.D., Purdue. Computer systems, architecture.
T. H. Glisson, Ph.D., SMU. Solid-state device modeling.
C. S. Gloster, Ph.D., North Carolina State. Digital systems.
J. J. Grainger, Ph.D., Wisconsin. Power systems.
E. Grant, Ph.D., Sheffield (Scotland). Robotics.
J. R. Hauser, Ph.D., Duke. Semiconductor materials and devices.
B. Hughes, Ph.D., Maryland. Data and digital communication.
W. Jasper, Ph.D., Stanford. Real-time control.
J. F. Kauffman, Ph.D., North Carolina State. Antennas and electromagnetics.
A. W. Kelley, Ph.D., Duke. Power systems.
K. W. Kim, Ph.D., Illinois. Semiconductor devices.
R. M. Kolbas, Ph.D., Illinois. Semiconductor devices.
H. Krim, Ph.D., Northeastern. Data and digital communication.
R. T. Kuehn, Ph.D., North Carolina State. Semiconductor devices.
M. A. Littlejohn, Ph.D., North Carolina State. Semiconductor devices.
W. Liu, Ph.D., Michigan. Microelectronic systems.
G. Lucovsky, Ph.D., Temple. Solid-state physics.
N. A. Masnari, Ph.D., Michigan. Semiconductor materials and devices.
N. F. J. Matthews (Emeritus), Ph.D., Princeton. Electromagnetics.
D. McAllister, Ph.D., North Carolina at Chapel Hill. Computer graphics.
J. W. Mink, Ph.D., Wisconsin. Microwave devices and systems.
V. Misra, Ph.D., North Carolina State. Solid state.
T. Mitchell, Ph.D., North Carolina State. Communications.
A. Mortazawi, Ph.D., Texas at Austin. RF and microwave systems.
H. T. Nagle, Ph.D., Auburn; M.D., Miami (Florida). Medical electronics.
J. Narayan, Ph.D., Berkeley. Solid-state science.
A. A. Nilsson, Ph.D., Lunds Universitet (Sweden). Communications.
J. B. O'Neal, Ph.D., Florida. Communications.
C. M. Osburn, Ph.D., Purdue. Semiconductor devices.
H. O. Ozturk, Ph.D., North Carolina State. Communications/signal processing.
M. C. Ozturk, Ph.D., North Carolina State. Semiconductor devices.
H. Perros, Ph.D., Trinity (Dublin). Queuing theory.
S. A. Rajala, Ph.D., Rice. Communications and signal processing.
D. S. Reeves, Ph.D., Penn State. Computer systems.
A. Reisman, Ph.D., Polytechnic of Brooklyn. Semiconductor materials and devices.
W. Robbins, Ph.D., Syracuse. Computer graphics.
W. E. Snyder, Ph.D., Illinois. Signal and image processing.
M. Stallmann, Ph.D., Colorado. Network algorithms.
M. B. Steer, Ph.D., Queensland (Australia). Analog and microwave systems.
K. Tai, Ph.D., Cornell. Software systems.
C. Townsend, M.S., Kansas. Circuits.
J. K. Townsend, Ph.D., Kansas. Communications.
H. J. Trussell, Ph.D., New Mexico. Image and signal processing.
I. Viniotis, Ph.D., Maryland. Communications.
M. Vouk, Ph.D., King's College (London). Software testing.
M. K. White, Ph.D., Berkeley. Signal processing.
J. J. Wortman, Ph.D., Duke. Solid-state materials and devices.

NORTHEASTERN UNIVERSITY

Department of Electrical and Computer Engineering

Programs of Study

The department offers graduate programs on either a full-time or part-time basis leading to the Master of Science and Doctor of Philosophy degrees in six areas of concentration: communications and signal processing; computer engineering; control systems and signal processing; electromagnetics, plasma, and optics; electronic circuits and semiconductor devices; and power systems. Techniques from these disciplines are applied to a variety of research problems, including speech and image processing, integrated circuits and VLSI, computer architecture and software development, digital communications, adaptive filtering, electromagnetic scattering, antennas, radar systems, telemetry, and robotics.

The M.S. degree requires 44 quarter hours (equivalent to eleven 4-quarter-hour courses, each of which is one quarter or twelve weeks in duration), which may include at the student's option an 8-quarter-hour master's thesis. The Ph.D. degree requires 70 quarter hours of work beyond the B.S. degree, a dissertation representing an original contribution to knowledge in the field, and a technical writing requirement. Students must pass a Ph.D. qualifying examination and a comprehensive examination in the major field, and they must present a successful oral defense of the dissertation. The Ph.D. residence requirement is one academic year.

Research Facilities

The department and its two research centers maintain modern laboratory and computer facilities for graduate research. The Center for Electromagnetics Research enables students interested in electromagnetics, optics, microwaves, and plasmas to pursue graduate theses in industry- and government-sponsored research at the University. The Center for Communications and Digital Signal Processing offers students a variety of opportunities for research in digital signal processing, communications, control systems, and robotics. The department has a VLSI/CAD computer facility that includes more than seventy DEC and Sun workstations. The Center for Communications and Digital Signal Processing maintains a computing research laboratory for student research. There are also research groups and laboratories in electron devices and computer engineering.

Financial Aid

Northeastern University awards need-based financial aid to graduate students through the Federal Perkins Loan, Federal Work-Study, and Federal Stafford Student Loan programs and offers a limited number of minority fellowships and Martin Luther King Jr. Scholarships. The graduate schools offer financial assistance through teaching, research, and administrative assistantship awards that include tuition remission and a stipend of approximately $12,450 (departmentally specific). These assistantships require a maximum of 20 hours of work per week. There are also a limited number of tuition assistantships that provide partial or full tuition remission and require a maximum of 10 hours of work per week.

Cost of Study

The tuition rate for 1998–99 was $465 per quarter hour of credit. There are special tuition charges for theses and dissertations, where applicable. The Student Center fee and health and accident insurance fee required for all full-time students is approximately $1100 per academic year.

Living and Housing Costs

Living expenses both on campus and off are estimated to be between $1200 and $1500 per month, with on-campus housing available on a limited basis to newly accepted students. A public transportation system serves the Greater Boston area, and there are subway and bus services nearby.

Student Group

In fall 1998, the department had 143 full-time graduate students and 146 part-time graduate students.

Student Outcomes

The majority of graduates find employment in various high-technology industries across the United States. Several Ph.D. graduates are employed by academic institutions in teaching and research.

Location

Northeastern University is located in the heart of Boston, a city that has played a pioneering role in American education. Within a 25-mile radius of the campus there are more than fifty degree-granting institutions. Within walking distance of the campus there are numerous renowned cultural centers, such as the Museum of Fine Arts, Isabella Stewart Gardner Museum, Symphony Hall, Horticultural Hall, and Boston Public Library. Theater in Boston includes everything from pre-Broadway to experimental and college productions. The Boston area is also the site of all home games of the Red Sox, Celtics, Bruins, and New England Patriots.

The University and The Department

Northeastern University is among the nation's largest private universities, with an international reputation as a leader in cooperative education. The cooperative plan of education, initiated by the College of Engineering in 1909 and subsequently adopted by the other colleges of the University, enables students to alternate periods of work and study. The cooperative education plan is available to selected graduate students. Today, Northeastern has eight undergraduate colleges, eight graduate and professional schools, several suburban campuses, and an extensive research division. The Department of Electrical and Computer Engineering offers its full-time day graduate programs at the University's Boston campus and its part-time evening programs at both the Boston campus and the suburban Burlington campus.

Applying

All applicants to the M.S. degree program must have a B.S. degree in electrical engineering with an acceptable quality of work from an ABET-accredited undergraduate program. Applicants with a B.S. degree in other engineering or related scientific fields and an appropriate background and preparation may also pursue this program. Applicants to the Ph.D. programs should have either a B.S.E.E. or M.S.E.E. with a high quality of work. International students holding undergraduate degrees from recognized engineering institutions outside the United States must submit GRE and TOEFL scores. All applicants must submit a completed application form, including official transcripts and a nonrefundable $50 processing fee. The application deadline for fall admission to the full-time program is April 15. To be considered for an assistantship, the application must be submitted by February 15.

Correspondence and Information

Cynthia Bates, Coordinator
Office of Graduate Student Affairs
Department of Electrical and Computer Engineering
Northeastern University
360 Huntington Avenue
Boston, Massachusetts 02115

Northeastern University

THE FACULTY AND THEIR RESEARCH

D. Brady, Associate Professor; Ph.D., Princeton. Digital communications, multiuser communications, acoustic communications.

D. Brooks, Associate Professor; Ph.D., Northeastern. Digital signal processing, biomedical signal processing.

S. Buus, Professor; Ph.D., Northeastern. Speech processing, psychoacoustics.

C. Chan, Professor; Ph.D., Iowa. Plasmas, electromagnetics.

J. Crisman, Associate Professor; Ph.D., Carnegie Mellon. Robotics, robot vision.

A. Devaney, Professor; Ph.D., Rochester. Tomography, electromagnetic wave propagation, inverse scattering, signal processing.

A. Grabel, Professor and Associate Chair; Sc.D., NYU. Electronic circuits and active networks.

J. Hanania, Professor; Ph.D., Leeds. Power systems.

J. Hopwood, Associate Professor; Ph.D., Michigan State. Plasma processing, IC fabrication.

V. Ingle, Associate Professor; Ph.D., RPI. Signal processing, image processing.

D. Kaeli, Assistant Professor; Ph.D., Rutgers. Computer architecture, software engineering.

M. Kokar, Associate Professor; Ph.D., Technical University of Wroclaw. Artificial intelligence, operating systems.

M. Leeser, Associate Professor; Ph.D., Cambridge. CAD, VLSI design, rapid system prototyping.

B. Lehman, Assistant Professor; Ph.D., Georgia Tech. Control systems, power systems.

H. Lev-Ari, Associate Professor; Ph.D., Stanford. Digital signal processing, adaptive filtering.

F. Lombardi, Professor and Chair; Ph.D., University College (London). Digital systems, fault-tolerant computing, CAD, manufacturing of Ics, configurable computing.

E. Manolakos, Associate Professor; Ph.D., USC. Computer architecture, VLSI design, pattern recognition.

N. McGruer, Associate Professor; Ph.D., Michigan State. Solid-state devices, microfabrication, microelectromechanical systems (MEMS).

L. McIlrath, Assistant Professor; Ph.D., MIT. Analog electronics, VLSI.

S. McKnight, Associate Professor; Ph.D., Maryland. Solid-state devices, electromagnetics.

D. McLaughlin, Associate Professor; Ph.D., Massachusetts. Radar systems, remote sensing.

W. Meleis, Assistant Professor; Ph.D., Michigan. Computer engineering, computer architecture, performance optimization.

F. Meyer, Assistant Professor; Ph.D., Massachusetts Amherst. Computer engineering.

E. Miller, Assistant Professor; Ph.D., MIT. Signal and image processing, inverse problems, wavelet and multiscale methods.

S. Mulukutla, Professor; Ph.D., Colorado. Power systems, electromagnetic theory and its applications to electrical machines.

S. Prasad, Professor; Ph.D., Harvard. Microwave solid-state devices and circuits, high-frequency device modeling.

J. Proakis, Professor; Ph.D., Harvard. Digital communications, adaptive filtering, estimation, digital signal processing.

C. Rappaport, Associate Professor; Sc.D., MIT. Electromagnetics, microwaves.

M. Salehi, Associate Professor; Ph.D., Stanford. Information theory, coding.

S. Sandler, Professor; Ph.D., Harvard. Electromagnetics, antennas, pattern recognition, robotics.

M. Schetzen, Professor; Sc.D., MIT. Systems theory, control systems, theory of nonlinear systems.

P. Serafim, Professor; Sc.D., MIT. Electromagnetics, remote sensing.

B. Shafai, Associate Professor; Ph.D., George Washington. Control systems, digital signal processing.

M. Silevitch, Professor; Ph.D., Northeastern. Plasma theory, applications of plasma theory to auroral phenomena.

A. Stankovic, Associate Professor; Sc.D., MIT. Power electronics, control systems.

I. Stavrakakis, Associate Professor; Ph.D., Virginia. Communications, optical networks.

G. Tadmor, Associate Professor; Ph.D., Weizmann (Israel). Control systems.

M. Vai, Associate Professor; Ph.D., Michigan State. VLSI design, computer engineering.

C. Vittoria, Professor; Ph.D., Yale. Electromagnetics, magnetic materials, microwave circuits.

P. Zavracky, Associate Professor; Ph.D., Tufts. Microsensor devices, device fabrication, MEMS.

Research Areas

Nonlinear systems theory.
Control systems.
Large-scale systems.
Distributed parameter systems.
Machine vision and robotics.
Autonomous intelligent machines.
Speech processing.
Psychoacoustics.
Biomedical signal processing.
Signal and systems analysis.
Plasma theory.
Electromagnetics.
Antennas.
Software engineering.
High-vacuum phenomena.
Semiconductor devices.
Integrated circuits.
VLSI fabrication, design, and testing.
Active networks.
FPGAs.

Optical electronics and communications.
Motor control.
Digital image processing.
Upper-atmosphere phenomena.
MEMS.
Radio frequency phenomena and systems.
Computer systems.
Radar systems.
Computer architecture.
Power systems.
Electrical machines.
Digital communications.
Communication networks.
Random-access communications.
Coding theory.
Digital signal processing.
Numerical methods.
Microwave devices and circuits.
Stochastic systems.
Information theory.

NORTHWESTERN UNIVERSITY

Robert R. McCormick School of Engineering and Applied Science of the Technological Institute
Department of Electrical and Computer Engineering

Programs of Study	The Department of Electrical and Computer Engineering offers a broad range of programs in electrical and computer engineering, or a combination of both, leading to the degrees of Master of Science and Doctor of Philosophy. The M.S. requires a minimum of one academic year of full-time study, and the Ph.D. requires a minimum of three years beyond the B.S. degree. Research focuses on the following areas: solid-state engineering (III-V, II-VI, and IV-IV semiconductors; high-temperature superconductors, oxides, and polymers; MBE, MOCVD, and VPE growth; device characterization and modeling; quantum devices; lasers; infrared and near-infrared devices); optical systems and technology (microphotonics and nanostructure devices, nonlinear and quantum optics, optical communications and networks, computational electromagnetics, imaging through turbulence); networks, communication, and control (wireless and mobile communications; estimation and detection; resource allocation, modeling, and analysis; protocols; nonlinear and robust control; stochastic hybrid systems); signal processing (image and video processing, recovery and compression, multimedia signal processing, filter design, rank-order operators, medical signal processing and imaging, medical instrumentation); parallel and distributed computing (architectures and systems, parallel compilers, parallel IO and disk organization, measurement and performance, distributed databases, data mining, numerical computing, robotics, computer vision); and VLSI design and CAD (ASIC and digital design, microprocessors and embedded systems, FPGA systems and adaptive computing, CAD algorithms and theory, low power design).
Research Facilities	The department has well-equipped laboratories for electronic circuits, digital circuits, solid-state electronics, thin-film device development, biomedical electronics, microwave techniques, real-time control systems, guided-wave and nonlinear optics, fiber optics, biological control systems, digital systems design, digital signal processing, image and speech processing, MOCVD, MOMBE reactors for optoelectronic materials and devices fabrication, numerical analysis, computer architecture, distributed computing systems, robotics, VLSI/CAD, communication networks, and microprocessor systems design. The department also uses the facilities of the Center for Quantum Devices, the Materials Research Center, the Manufacturing Engineering Center, the Advanced Cement Based Materials Center, the Optimization Center, and the Center for Parallel and Distributed Computing.
	The department's computing facilities include a 16-processor IBM SP-2 distributed-memory message-passing multicomputer, an 8-processor IBM J-10 shared-memory multiprocessor, and 8-processor SGI Origin 2000 distributed shared-memory multiprocessor, a network of fifty HP workstations, and a number of interconnected Sun SPARC teleservers and workstations. The laboratory is part of the NSFnet/Internet. All department computers are linked to the University's 622-Mbps OC-12 backbone. A public ISDN network and a large modem pool give faculty and staff members and students access to the backbone from their homes, which makes the full range of computing facilities and network services available from offices, labs, and homes.
	The Robert R. McCormick School of Engineering and Applied Science (MEAS) is located in the Technological Institute. Connected to the Technological Institute is the Science and Engineering Library. The University's main library is also close to the Technological Institute.
Financial Aid	Many types of financial support, awarded on a competitive basis, are available for graduate students at Northwestern. Full institutional stipends range from $12,000 to $16,000 plus tuition for nine months. Veterans' benefits may be received simultaneously with other aid. University awards include the Walter P. Murphy fellowships, teaching assistantships, and research assistantships. Other fellowships provided by various government agencies and industry pay for tuition and offer stipends to cover living expenses. Stipends usually increase after the first year of study.
Cost of Study	Tuition for 1999–2000 is $7266 per quarter, except for students who have been admitted to candidacy for the Ph.D., for whom tuition is $2422 per quarter.
Living and Housing Costs	Northwestern operates two apartment buildings in Evanston for single and married graduate students. Additional accommodations at reasonable costs are available in Evanston and nearby communities for students who choose to rent rooms, apartments, or houses from private owners.
Student Group	The schools on the Evanston campus annually enroll approximately 7,600 full-time undergraduate and more than 4,000 full-time graduate students. The University's total annual enrollment of about 17,880 students includes more than 2,200 full-time students on the Chicago campus and all men and women registered in the summer session, evening divisions, and part-time programs on both campuses. The McCormick School of Engineering and Applied Science has a current graduate enrollment of more than 900 students, over 160 of whom are in the Department of Electrical and Computer Engineering.
Location	In addition to enjoying the extensive cultural activities and sports offered at the University, graduate students make good use of the cultural advantages of the Chicago area. The location provides a good combination of urban amenities and the pleasant environment of a suburban residential community.
The University and The School	Northwestern University, founded in 1851, is a coeducational institution and the only privately supported university in the Big Ten. Northwestern has two campuses on the shore of Lake Michigan, one in suburban Evanston and the other near the downtown center of Chicago. There are more than 2,000 full-time faculty members at Northwestern; approximately 90 percent of them hold doctorates or the equivalent.
	The McCormick School of Engineering and Applied Science is located on the Evanston campus. Within MEAS are the Departments of Biomedical Engineering, Chemical Engineering, Civil Engineering, Electrical and Computer Engineering, Computer Science, Engineering Sciences and Applied Mathematics, Industrial Engineering and Management Sciences, Materials Science and Engineering, and Mechanical Engineering. Currently, the Department of Electrical and Computer Engineering budget for research is more than $9 million per year.
	Interdisciplinary research opportunities are available at various centers, including the Center for Quantum Devices, the Center for Information and Telecommunication Technology, the Manufacturing Engineering Center, the Materials Research Center, and the Institute for the Learning Sciences.
Applying	Graduate students are admitted primarily for the fall quarter. Completed applications for admission and for financial aid should be received by January 15.
Correspondence and Information	Graduate Director Department of Electrical and Computer Engineering Technological Institute Northwestern University Evanston, Illinois 60208-3118 Telephone: 847-491-7092 E-mail: grad@ece.nwu.edu World Wide Web: http://www.ece.nwu.edu

Northwestern University

THE FACULTY AND THEIR RESEARCH

Prithviraj Banerjee, Walter P. Murphy Professor, Chairman of the Department, and Director of the Center for Parallel and Distributed Computing; Ph.D., Illinois at Urbana-Champaign. Parallel compilers and software, parallel architectures, parallel algorithms for VLSI CAD.

Morris E. Brodwin, Emeritus Professor; Ph.D., Johns Hopkins. Electromagnetic characterization and thermal processing of materials.

Arthur R. Butz, Associate Professor; Ph.D., Minnesota. Digital signal processing.

Robert P. H. Chang, Professor; Ph.D., Princeton. Thin films for electronic and optoelectronic device applications.

Alok N. Choudhary, Associate Professor; Ph.D., Illinois at Urbana-Champaign. High-performance computing and communication, input-output, compiler and runtime systems for HPCC, multimedia systems and databases.

Randy A. Freeman, Assistant Professor; Ph.D., California, Santa Barbara. Robust nonlinear control, optimal control.

Abraham H. Haddad, Henry and Isabelle Dever Professor; Ph.D., Princeton. Stochastic systems, modeling, estimation, detection, nonlinear filtering, singular perturbation, applications to communications and control.

Scott A. Hauck, Assistant Professor; Ph.D., Washington, Seattle. Multi-FPGA systems, FPGA architectures and CAD tools, rapid-prototyping, asynchronous circuit and VLSI design, parallel processing and parallel programming languages.

Lawrence J. Henschen, Professor; Ph.D., Illinois at Urbana-Champaign. AI, theorem proving, deductive databases.

Seng-Tiong Ho, Associate Professor; Ph.D., MIT. Quantum optics, nonlinear optics, ultrafast optical devices, microcavity lasers.

Michael L. Honig, Professor; Ph.D., Berkeley. Digital communications, wireless communications, networks, signal processing.

Christopher L. Jelen, Assistant Professor; Ph.D., Northwestern. Compound semiconductor device growth, processing, and characterization; intersubband photodetectors and lasers; quantum devices.

Scott A. Jordan, Associate Professor; Ph.D., Berkeley. Resource allocation in computer/telecommunication networks.

Carl R. Kannewurf, Professor; Ph.D., Northwestern. Electronic materials: electrical and optical phenomena in semiconductors and metals, high-T_c superconductors, low dimensional materials and devices.

Aggelos K. Katsaggelos, Ameritech Professor; Ph.D., Georgia Tech. Multidimensional digital signal processing, processing of moving images, video coding, computational vision, parallel implementation of signal processing algorithms.

Andrew E. Kertesz, Professor; Ph.D., Northwestern. Binocular information processing and oculomotor control by the human visual system, medical instrumentation.

Gilbert K. Krulee, Emeritus Professor; Ph.D., MIT. Natural language systems, two-level grammars, intelligent support systems.

Prem Kumar, Professor; Ph.D., SUNY at Buffalo. Nonlinear and quantum optics, picosecond and subpicosecond phenomena, laser and atomic physics, optical communications and networks.

Chung-Chieh Lee, Professor; Ph.D., Princeton. Digital communications, communication network performance modeling and analysis, distributed multisensor detection and estimation.

Der-Tsai Lee, Professor; Ph.D., Illinois at Urbana-Champaign. Design and analysis of algorithms, data structures, VLSI systems, computational geometry, computational complexity, algorithm visualization.

Wei-Chung Lin, Associate Professor; Ph.D., Purdue. Computer vision, pattern recognition, neural networks, computer graphics.

Andreas Moshovos, Assistant Professor; Ph.D., Wisconsin–Madison. Computer architecture and compilers, hardware-software techniques to explore/exploit parallelism, intelligent memory systems.

Gordon J. Murphy, Emeritus Professor; Ph.D., Minnesota. Integrated computer-control systems, automated manufacturing systems, digital signal processing, microprocessor-based systems.

Nathan Newman, Associate Professor; Ph.D., Stanford. Fabrication of novel solid-state materials and devices, characterization and modeling of surface, interface and bulk phenomena in solids.

Jorge Nocedal, Professor and Deputy Director of the Optimization Center; Ph.D., Rice. Numerical analysis, nonlinear optimization, applied linear algebra, numerical software.

Martin A. Plonus, Professor; Ph.D., Michigan. Electromagnetic theory, particularly propagation and scattering of electromagnetic waves and optical communication through the turbulent atmosphere, consumer electronics.

Mort Rahimi, Professor and Vice President of Information Technology; Ph.D., Iowa. Artificial languages, computer networks, IT management.

Manijeh Razeghi, Walter P. Murphy Professor and Director of the Center for Quantum Devices; Ph.D., Paris. Solid-state science and technology; semiconductors: materials growth, physics, optical, electrical and structural characterization; opto-electronic device modeling and fabrication.

Alan V. Sahakian, Associate Professor; Ph.D., Wisconsin-Madison. Instrumentation, signal processing for medical applications.

Majid Sarrafzadeh, Professor; Ph.D., Illinois at Urbana-Champaign. VLSI design, computer-aided design, high-performance architectural design, design and analysis of algorithms, computational complexity, low power systems.

Peter I. Scheuermann, Professor; Ph.D., SUNY at Stony Brook. Physical database design, pictorial databases, parallel I/O systems, parallel algorithms for data-intensive applications, distributed database systems.

Allen Taflove, Professor; Ph.D., Northwestern. Applied electromagnetic field theory and applications, computational electromagnetics, scattering and diffraction, supercomputing, Maxwell's equations–based computational nonlinear optics, electromagnetic waves in nonlinear dispersive media, femtosecond optical switches.

Valerie E. Taylor, Associate Professor; Ph.D., Berkeley. Computer architecture, parallel processing, hardware development and analysis for scientific computations, and special-purpose processors.

James E. Van Ness, Emeritus Professor; Ph.D., Northwestern. Use of the digital computer to study large dynamic systems, numerical analysis, control systems, power systems.

Bruce W. Wessels, Walter P. Murphy Professor; Ph.D., MIT. Compound semiconductors, semiconductor processing, integrated optical devices and circuits.

Chi-Haur Wu, Associate Professor; Ph.D., Purdue. Robotics, CAD/CAM, industrial control, neural network, surgical robots.

Horace P. Yuen, Professor; D.Sc., MIT. Optical communication, theoretical quantum optics, measurement theory.

THE OHIO STATE UNIVERSITY

Department of Electrical Engineering

Programs of Study

The Department of Electrical Engineering (EE), a 1998 Selective Investment Award winner at the Ohio State University (OSU), offers graduate programs leading to the Master of Science (M.S.) and the Doctor of Philosophy (Ph.D.) degrees. The department conducts vigorous and extensive research in all aspects of electrical and computer engineering. Graduate students have ample opportunities to participate in cutting-edge research, using state-of-the-art equipment and computing facilities. All faculty members participate in research and development, either individually or in organized laboratories. Research areas include controls, electromagnetics, electromechanical systems, electronic materials and devices, mixed signal electronics, high-performance computing, photonics and optics, power systems, robotics, signal analysis and machine perception, signal processing, and wireless communications systems.

Research Facilities

EE has extensive experimental facilities and computer resources. The department supports more than twelve independent, state-of-the-art laboratories that participate in various research activities, such as industrial, government, interdisciplinary, and cross-university projects.

In addition, the department has two separate computing facilities dedicated for student use. The first facility is a department-wide computing facility used solely for undergraduate and graduate student computing. Ninety Hewlett Packard workstations running HPUX 10.20 are distributed across two north campus buildings in four public labs. Two of these labs have 24-hour keycard access, and the others are always available for remote logins after normal business hours. Primary software packages used on the HPs include ESPS Waves, Framemaker, HP ADS (formerly HP EEsof), Labview, Maple, Mentor Graphics, Matlab, PVWave, Saber, and Visual Thought. The department also has two labs that run Windows NT on twenty Dell OptiPlex GX1 (PII/400), which are available for students in specific classes. The second set of computing facilities is available to students depending on their area of concentration and include the Controls, High Performance Computing, Information Processing, MISES, and Microlan Laboratories, with HP, Sun, and/or Linux systems; signal analysis; and machine perception as well as one Beowulf cluster.

In addition to the primary computing facilities of the department, the Electro-Science Laboratory (ESL) has a major research facility on the west campus that is completely instrumented with equipment that supports advanced electromagnetics research. The department also houses a high-voltage laboratory, a clean room facility, a wireless and electron microscope, and three electronic materials laboratories. Finally, the department has access to the Ohio Supercomputer Center's high-performance computers, which include a CRAY T94, CRAY T3E, SGI Origin 2000, Mass Storage System, and Beowulf cluster. Further information on the department's facilities can be found on the World Wide Web at http://eewww.eng.ohio-state.edu/.

Financial Aid

Graduate research and teaching associates earn at least $1350 per month. Litton and Lucent Fellows also earn a minimum of $1350 per month. University fellowships are supplemented to equal the departmental fellowships and associateships.

Cost of Study

Tuition at the Ohio State University totals $1944 per quarter for Ohio residents and $4844 per quarter for nonresidents.

Living and Housing Costs

The University maintains graduate residence halls and apartments that are designed to provide a mature atmosphere for advanced study. Students should contact the Office of Residence Life at 640 Lincoln Tower, 1800 Cannon Drive, Columbus, Ohio 43210 (telephone: 614-292-8266) for more information about on-campus facilities. Information about off-campus housing can be acquired by contacting 211 Ohio Union, 1739 North High Street, Columbus, Ohio 43210 (telephone: 614-292-0100). Columbus housing costs are very reasonable compared to many major metropolitan areas.

Student Group

Approximately 270 graduate students are enrolled in EE. The diverse student population results in a vibrant and stimulating academic environment. Student groups include the Electrical Engineering Student Council, Eta Kappa Nu, IEEE, Theta Tau, OSU Sunrayce, Tau Beta Pi, OSU Formula Lightening, the Association for Computing Machinery, and the American Society of Engineering Education.

Location

The Ohio State University is located about 3 miles north of downtown Columbus, the capital of Ohio. The city has a metropolitan population of approximately 1.4 million and is one of the fastest-growing urban areas in the United States. Columbus is a global center for high technology, especially in information services; is home to an active arts community; and serves as the corporate and divisional headquarters for a number of major corporations.

The University and The Department

OSU was founded in 1870 as a land-grant university and remains one of the largest universities in the nation, serving 50,000 students annually. EE has a local, national, and international history of excellence. The department provides unique opportunities for students, including fellowships, research and teaching assistantships, internships and co-ops, industry partnerships, and a combined B.S./M.S. degree. External research funding for the department exceeded $7.8 million in 1998.

Applying

Students who are interested in applying to the Department of Electrical Engineering may request an information packet and an application from the address below. Upon requesting information, students should indicate whether they are domestic or international students.

Correspondence and Information

Graduate Studies Chairperson
Department of Electrical Engineering
The Ohio State University
2015 Neil Avenue
Columbus, Ohio 43210-1272

Telephone: 614-292-1752
Fax: 614-292-7596
E-mail: eegrad@ee.eng.ohio-state.edu

The Ohio State University

THE FACULTY AND THEIR RESEARCH

SYSTEMS AND CONTROL: Professors Jose B. Cruz, Robert E. Fenton, Hooshang Hemami, Randolph L. Moses, Hitay Ozbay, Umit Ozguner, Kevin M. Passino, Vadim Utkin, Stephen Yurkovich, and Yuan F. Zheng.

This area focuses on the development, implementation, and theory of controllers that automate tasks for high-technology systems. Research is concentrated in the development of control theory and modifications of the theory for design and analysis in diverse applications, such as transportation systems, automotive systems, industrial process control, military command and control, aspects of communication networks control, robotics, and automation in manufacturing.

ELECTROMAGNETICS, WIRELESS, AND OPTICS: Professors Betty Lise Anderson, Walter D. Burnside, Bradley D. Clymer, Robert J. Garbacz, Joel T. Johnson, Robert Lee, Curt A. Levis, Benedikt A. Munk, Edward H. Newman, Prabhakar H. Pathak, Leon Peters Jr., Roberto G. Rojas, and Roger C. Rudduck.

Electromagnetics. This area deals with the understanding, analysis, measurement, and control of electromagnetic fields. Applications include antennas and arrays, control of radar signatures, remote sensing and imaging, microwave circuits, underground radar, and wireless systems.

Wireless. This area focuses on research, development, and testing, including basestations and mobile antenna design and evaluation, smart antennae, concealment, geolocation, propagation modeling, software radio, system evaluation, and ultrawide ban transceiver systems and applications.

Photonics. Research encompasses both optics and electronics and applies to any use of optics and light to solve engineering problems or to understand the properties of light and its interactions within materials and devices. Specific areas of research include fiber optics, optical sensing, optical computing, optical interconnections, theoretical optics, optoelectronic integrated circuits, quantum devices, semiconductor lasers, detectors and modulators, and optical properties of materials.

ELECTRONIC MATERIAL AND DEVICE RESEARCH/MIXED SIGNAL ELECTRONICS: Professors Betty Lise Anderson, Steven B. Bibyk, Leonard J. Brillson, Mohammed Ismail, Furrukh Khan, Steven A. Ringel, Patrick Roblin, and George J. Valco.

Electronic and Materials Device Research. The dramatic progress in information processing and communications that is being made relies on developments in the area of electronic materials and devices. Specific areas of materials and device research in the department include computer modeling of devices, defects in materials and heterostructures, device fabrication and measurements, high-temperature superconductors, lasers, detectors and other photonics devices, molecular beam epitaxy of compound semiconductors and other electronic materials, photovoltaics and thermophotovoltaics, pulsed laser deposition of electronic materials and quantum device structures, quantum transport and inelastic tunneling, sensors, structure, properties and control of electronic materials surfaces and interfaces, and wide bandgap semiconductors.

Mixed Signal Electronic Systems. The Mixed Signals Electronic Systems Laboratory pursues collaborative research in mixed signals electronics, including analog and digital integrated circuits design, RFICs design, System-On-A-Chip (SOC) design, microwave/RF electronics, and power electronics for applications such as wireless systems and smart power.

COMPUTER ENGINEERING: Professors Stanley C. Ahalt, Kim L. Boyer, Kenneth J. Breeding, Joanne E. DeGroat, Patrick J. Flynn, Chao-Ju Jennifer Hou, Jogikal M. Jagadeesh, Charles A. Klein, David E. Orin, Fusun Ozguner, and Yuan F. Zheng.

High-Performance Computing. The High Performance Laboratory is a collaborative research environment. The areas of research include high-performance architectures, QoS control in the next-generation Internet, resource management and message scheduling in high-speed networks, real-time task management, distributed real-time computing, fault tolerance, parallel algorithms, communication and routing in distributed memory multiprocessors and networks of workstations, heterogeneous distributed computing, video compression, real-time computation of the wavelet transform, efficient coding, and network and modem transmission.

Design Automation. This area of research includes VHDL modeling of architectural building blocks, reengineering of integrated circuits, single-chip multiprocessor architectures, mixed-signal processors, and DSP processors. Other research areas include computer vision, image processing and biomedical image processing, dynamic simulation, graphic simulation, and computer graphics.

ELECTRIC POWER SYSTEMS: Professors Donald G. Kasten, Ali Keyhani, Stephen A. Sebo, and Longya Xu.

This area encompasses a wide range of teaching and research topics. Current topics include those of traditional power system analysis and design, mechatronics, power electronics, high-voltage engineering, electric machine modeling and control, and electromagnetic compatibility. Areas of research include electric power systems, electric machines, high-voltage engineering, and power system economics.

ROBOTICS RESEARCH: Professors Kim L. Boyer, Hooshang Hemami, Charles A. Klein, David E. Orin, Umit Ozguner, Stephen Yurkovich, and Yuan F. Zheng.

This is an interdisciplinary program where courses are taught in the computer and control areas. While mobile robotic systems have received special attention, work is also ongoing in redundant manipulator, special architectures for robotic computations and the related areas of computer graphics, computer simulation, fault tolerance, large-scale system control theory, and neural networks. Areas of research include computer vision, computer graphics, dexterous hands, fuzzy control, human-robot coordination, kinematics redundancy, manufacturing, neural networks, and walking machines.

COMMUNICATIONS AND SIGNAL PROCESSING: Professors Stanley C. Ahalt, Kim L. Boyer, Steven B. Bibyk, Bradley D. Clymer, Michael P. Fitz, Patrick J. Flynn, Ashok Krishnamurthy, Curt A. Levis, Urbashi Mitra, Randolph L. Moses, and Lee C. Potter.

Signal Analysis and Machine Perception Research. Areas of research include aerial and satellite image understanding, large structural modelbase organization, motion analysis for collision avoidance, optimal methods in feature extraction, perpetual organization and Bayesian networks, robust estimators, surfaces in range data, stereo autonomous camera calibration, and video compression.

Information Processing Systems. The Information Processing Systems Laboratory is a collaborative research and education environment in which both basic research and applied projects are pursued in the areas of wireless communications, speech processing, data compression and classification, and statistical signal processing. Areas of research include auditory models, automatic target recognition, image compression, neural network algorithms and hardware, receiver design for wireless communications, robust parameter estimation, signal recovery and inverse problems, speech analysis, and synthesis and recognition.

Wireless Communications Systems. Wireless communications are used in a wide variety of applications, such as standard broadcast, cellular telephony, and indoor wireless networks. The projects ongoing in the wireless systems areas are developing the fundamental theory, studying advanced systems, and performing experimental work in support of the theory. Ongoing projects include fading/multipath environments, multiuser receivers for multirate communications, robust multiuser receivers, general adaptive multireceivers, blind equalizers for multiuser signals, narrowband wireless communications theory, equalization in rapid frequency selective fading, carrier synchronization at low SNR, wireless antennae systems, modem development for intelligent transportation applications, channel identification and equalization in video communications, and soft output demodulation algorithms.

OHIO UNIVERSITY

Fritz J. & Dolores H. Russ College of Engineering and Technology
School of Electrical Engineering and Computer Science

Programs of Study

Programs leading to the M.S. and Ph.D. degrees in electrical engineering are available. A concentration in computer science is offered in both degree programs. Major areas of study include avionics, computers, applied and theoretical computer science, communications, controls, information theory, solid-state electronics, energy conversion, power electronics, power systems, electromagnetics, signal processing, manufacturing, VLSI design, computer vision, electronic circuits, and optoelectronics. Successful applicants for the Ph.D. degree program are expected to hold an M.S. degree in electrical engineering, computer science, or a related field of engineering or the physical sciences. Typically, Ph.D. students complete two academic years of formal course work in a major area, a minor area, and either mathematics or physics, followed by a written comprehensive examination and an oral examination that includes the presentation of a dissertation research proposal. The remainder of the Ph.D. degree program consists of dissertation research, preparation of the dissertation, and the dissertation defense. The average duration of the program is four years. Ohio University regulations require that candidates for the Ph.D. degree be in residence for a minimum of three academic quarters. Recipients of the Ph.D. degree are prepared for research careers in the private, public, and academic sectors. Successful applicants for the M.S. degree are expected to hold a B.S degree in electrical engineering, computer science, or a related field of engineering or the physical sciences. The typical M.S. degree program consists of one year of formal course work followed by thesis research, preparation of the thesis, and a combined oral examination and thesis defense. The average duration of the M.S. program is two years. Recipients of the M.S. degree are prepared to enter the engineering profession at an advanced level or to pursue more advanced graduate work.

Research Facilities

The School of Electrical Engineering and Computer Science occupies the entire third floor of the five-story, 159,000-square-foot C. Paul and Beth K. Stocker Engineering and Technology Center and the fourth floor of Morton Hall. The Avionics Engineering Center, an administrative unit of the School, occupies a large portion of the second floor of Stocker Center. State-of-the-art laboratories are maintained to support research activities in avionics, computer networking, communications, computer vision, optoelectronics, controls, VLSI design, manufacturing, and large-scale software integration. Computing facilities include more than thirty-five UNIX workstations, access to the Ohio University mainframes and the Ohio Supercomputer, and numerous late-model personal computers. Licenses for recent releases of all major software tools in areas of current research activity are maintained. Historically, the School has been highly successful in maintaining its computer hardware and software capabilities at a state-of-the-art level. Roughly half of the $15-million Stocker Endowment is dedicated to the School and currently generates $500,000 annually in support of its research activities.

Financial Aid

All financial aid is awarded competitively based on standardized test scores and academic performance. In some cases, supplemental aid is available for highly qualified U.S. citizens. Financial aid consists of tuition scholarships, graduate assistantships, teaching assistantships, research assistantships, and Stocker Research Associateships. Stipends range from $6000 per academic year for graduate assistantships to as much as $15,000 per academic year for Stocker Research Associateships. All financial aid includes, at a minimum, a tuition scholarship. Financial aid for international students is contingent on placement by the Ohio Program of Intensive English (OPIE) in full-time academic study; the cost for remedying English-language deficiencies are borne by the student. International students are strongly encouraged to sit for the Test of Written English (TWE) before applying for admission. For more information regarding financial aid, including current stipends and the number of awards made annually, interested students should visit the School's World Wide Web site (http://www.ent.ohiou.edu/eecs/).

Cost of Study

In 1998–99, tuition and fees (9–18 credit hours) were $1810 per quarter for in-state students and $3477 per quarter for out-of-state students. Tuition and fees are subject to change without notice. The most current information on tuition and fees may be obtained at the Web site (http://www.cats.ohiou.edu/ohiou/about/factsandfigures.html).

Living and Housing Costs

The 1998–99 quarterly dormitory rates were $1023 for a standard single room and $832 for a standard double room. Board costs per quarter ranged from $584 for the seven-meals-a-week plan to $860 for the twenty-meals-a-week plan. University-owned apartments are also available. A one-bedroom unfurnished unit was $515 per month (furnished, $585); a two-bedroom unfurnished unit was $607 per month (furnished, $679). All apartments have a stove and refrigerator. Utilities, excluding telephone and cable television, are included in the costs. Students interested in these apartments should apply by January. Many private apartments are also available close by. For more information and the latest rates, students should call 740-593-4090 or visit the Web site (http://www.cats.ohiou.edu/~auxserv/houscost.htm#Costs).

Student Group

Total enrollment at Ohio University is 27,605. Enrollment at the Athens campus is 19,189 and includes students from more than ninety countries. The Russ College of Engineering and Technology enrolls more than 1,700 students, including approximately 260 graduate students. The graduate enrollment in the School of Electrical Engineering and Computer Science is 112, including about 35 Ph.D. students. Approximately 40 percent of the graduate students enrolled in the School are U.S. citizens. The School currently supports roughly 50 students on various forms of financial aid.

Location

Ohio University is located in the small city of Athens in scenic southeast Ohio. Athens is 75 miles southeast of Columbus, the state capital and a major metropolitan area. Cincinnati, Cleveland, and Pittsburgh are all roughly a 4-hour car ride away. The Athens area is rural, with beautiful rolling hills and woodlands and numerous state parks offering camping, hiking, and fishing. The Ohio River is only 30 miles from Athens and offers boating and waterskiing opportunities. Several professional bicycle races are held in the area on an annual basis, as is a major woodcraft festival.

The University and The School

Ohio University was chartered by the state of Ohio in 1804 and was the first university in the Northwest Territory. Today, the main campus in Athens consists of 1,700 acres and 197 buildings. Ohio University was recently designated a Research University II by the Carnegie Foundation for the Advancement of Teaching. Only 125 schools—3.4 percent—of the 3,600 schools assessed by the Carnegie Foundation are classified as research universities. Others in the Research II classification include Auburn University, Clemson University, Kansas State University, the University of Notre Dame, the University of Oklahoma, and Washington State University. According to the Carnegie Foundation definition, a Research University II "offers a full range of baccalaureate programs, is committed to graduate education through the doctorate, and gives high priority to research."

Applying

Applications are reviewed for admissions continuously. However, in order to be fully considered for financial aid, application files must be completed by the end of March. GRE scores are required for all international applicants and for applicants graduating from programs in the U.S. that are not ABET accredited. Although there is no minimum GRE score required for admission, most successful applicants score in the top 25th percentile on the quantitative and analytical portions. The Test of English as a Foreign Language (TOEFL) is required for nonnative speakers of English. Most successful applicants score 550 or above on the TOEFL. Nonnative speakers of English who are interested in financial aid are also strongly encouraged to take the TWE. Applications must be submitted to the Office of Graduate Student Services, Ohio University, Athens, Ohio 45701. For more information, applicants may visit the Web site (http://www.ohiou.edu/about/admit/index.html).

Correspondence and Information

Graduate Chair
School of Electrical Engineering and Computer Science
Ohio University
Athens, Ohio 45701-2979

Telephone: 740-593-1568
Fax: 740-593-0007
E-mail: gradinfo@homer.ece.ohiou.edu
World Wide Web: http://www.ent.ohiou.edu/eecs/

Ohio University

THE FACULTY AND THEIR RESEARCH

Chris Bartone, Assistant Professor; Ph.D., Ohio. Electronic navigation, GPS augmentation, electromagnetic wave propagation and antennas, communications and radar.

Michael S. Braasch, Assistant Professor; Ph.D., Ohio. Electronic navigation systems, navigation system simulation, GPS receiver design, GPS multipath analysis and mitigation.

Liming Cai, Assistant Professor; Ph.D., Texas A&M. Algorithms design and analysis, compiler systems, programming languages, theory of computation.

Mehmet Celenk, Associate Professor; Ph.D., Stevens. Digital image processing, computer and robot vision, multiprocessor systems and distributed computing, multimedia communications systems, parallel processing, computer architecture and digital design, pattern recognition.

David Chelberg, Associate Professor; Ph.D., Stanford. Computer vision, object recognition, medical image processing, artificial intelligence, scientific visualization, computer graphics.

Hollis C. Chen, Professor; Ph.D., Syracuse. Electromagnetic wave propagation and radiation in isotropic and anisotropic environment in moving media, plasmas, and ferrites; fiber optics; computer applications to electromagnetic problems; applied mathematics.

Robert A. Curtis, Associate Professor; Ph.D., NYU. Digital and analog electronic systems, semiconductor physics, microprocessors, charge-coupled devices.

Jeffrey C. Dill, Associate Professor; Ph.D., USC. Spread spectrum communications, error correcting/detecting codes, personal communications/multiple access, wavelet applications to communications.

Joseph E. Essman, Professor Emeritus (part-time); Ph.D., Purdue. Communication system—digital and analog, modulation and detection, adaptive systems, digital signal processing, image processing and data compression, adaptive arrays.

Voula Georgopoulos, Assistant Professor; Ph.D., Tufts. Time-frequency analysis, optical communications and optical signal processing, application of neural networks to signal processing, perception systems modeling.

Jeffrey J. Giesey, Associate Professor; Ph.D., Michigan. Ultrasonic imaging, image processing, biomedical applications.

Herman W. Hill, Professor; Ph.D., West Virginia. Power electronics, electromechanical energy conversion.

R. Dennis Irwin, Professor; Ph.D., Mississippi State. Reliable numerical algorithms for computer-aided analysis and design, sampled-data and digital robust control design, model identification, control system design for flexible structures.

Robert Judd, Cooper Industries Professor; Ph.D., Oakland. Control of manufacturing systems, discrete event systems, simulation, controls.

David W. Juedes, Assistant Professor; Ph.D., Iowa State. Complexity theory, automatic differentiation, information theory.

Harold Klock, Professor Emeritus (part-time); Ph.D., Northwestern. Digital systems design, computer-aided design of digital systems, digital computer architecture and design, microcomputer design and programming, particularly 16-bit and 32-bit systems.

Douglas A. Lawrence, Associate Professor; Ph.D., Johns Hopkins. Linear and nonlinear system theory, analytical aspects of gain scheduling, flight control systems.

Henryk J. Lozykowski, Professor; Ph.D., Copernicus (Poland). Fundamental optical properties of semiconductors, luminescence and optical absorption, exciton and impurity recombination, rare earth of heterostructures, semiconductors, lasers, integrated optics, fundamental properties of layered and alloyed semiconductors relevant to lasers, detectors, and other optoelectronic devices.

Brian Manhire, Professor; Ph.D., Ohio State. Electric power engineering, power system planning.

Cynthia Marling, Assistant Professor; Ph.D., Case Western Reserve. Artificial intelligence, software engineering.

David Matolak, Assistant Professor; Ph.D., Virginia. Wireless and mobile communications.

Richard H. McFarland, Russ Professor Emeritus (part-time); Ph.D., Ohio State. Avionics, aircraft navigation systems, radar antennas.

Jerrel R. Mitchell, Russ Professor; Ph.D., Mississippi State. Control system computer-aided analysis and design, frequency response system identification, analysis and design of sampled-data control systems, multivariable frequency response methods for control systems, parameter estimation techniques, optimal control theory, analysis and design of control systems for large space structures.

M. Ebrahim Mokari, Professor; Ph.D., Illinois. VLSI circuit simulations, computer-aided circuit design, microwave integrated circuits, analog and digital filters, device modeling and sensitivity analysis.

Joseph H. Nurre, Assistant Professor; Ph.D., Cincinnati. 3-D data processing, computer graphics, manufacturing and automation, computer-aided design for mechanical systems.

Shawn Ostermann, Assistant Professor; Ph.D., Purdue. Computer Internetworking and network protocols, data communications, operating systems.

Roger D. Radcliff, Professor; Ph.D., West Virginia. Antenna theory and design, computer solution of electromagnetic problems, electromagnetics and wave propagation.

Janusz A. Starzyk, Professor; Ph.D., Warsaw Technical. VLSI and VHDL design, design and applications of neural networks, analog and digital testing, CAD methods for large analog systems.

John A. Tague, Associate Professor; Ph.D., Penn State. Underwater signal processing, spectrum estimation, characterization of stochastic processes.

Brett C. Tjaden, Assistant Professor; Ph.D., Virginia. Cryptographic protocols, computer security, network security, electronic commerce.

Frank van Graas, Associate Professor; Ph.D., Ohio. Electronic navigation systems, satellite positioning, differential GPS, inertial navigation.

Constantinos Vassiliadis, Associate Professor; Ph.D., Mississippi State. Artificial intelligence, expert systems, inference engines and knowledge bases, knowledge acquisition, representation and programming, knowledge engineering, neural intelligence, learning algorithms, Internet intelligent agents.

Lonnie Welch, Professor; Ph.D., Ohio State. Real-time systems, distributed computing, software systems engineering, operating systems, network programs, computer architecture, compilers, QoS specification, benchmarking and dependability.

OKLAHOMA STATE UNIVERSITY

College of Engineering, Architecture, and Technology
School of Electrical and Computer Engineering

Programs of Study

The School of Electrical and Computer Engineering at Oklahoma State University (OSU) offers both M.S. and Ph.D. degrees. With the M.S. degree, there are two options: the Professional Path option requires 33 hours of course work, while the Traditional Path option involves 21 hours of course work and 9 hours of thesis research. Both degrees can be completed in approximately three semesters. For the Ph.D. degree, 24 course work hours and 36 thesis hours are required beyond the M.S. degree. Students without M.S. degrees can be admitted into the Ph.D. program, but the requirements are increased to 45 course work hours and 45 thesis hours in these cases. For more information, students can visit the Web site, listed below.

Several active research groups exist within the School. The areas of strength include controls, power and energy systems, photonics/optics/lasers, communications, and signal/speech/image/video processing and multimedia. The department is also expanding resources in computer engineering in conjunction with the new campus in Tulsa, Oklahoma.

In addition to the traditional programs in electrical and computer engineering, the School offers degree programs in control, photonics, engineering management, and telecommunications management. The interdisciplinary Master of Science in telecommunications management involves the participation of faculty members from the Departments of Management and Information Systems, Electrical and Computer Engineering, and Computer Science. For more information, students can visit the Web site (http://www.mstm.okstate.edu/webpage/).

Research Facilities

The School maintains a multimillion-dollar research budget each year. A wide variety of the latest computer equipment is available in the School. State-of-the-art laboratories in optics, lasers, signal processing, multimedia, controls, telecommunications, and energy systems are available to graduate students for research. The University has a large central research library covering more than 6 acres of floor space. A new $34-million research facility, the Advanced Technology Research Center, opened this year, providing more than 165,000 square feet of research space.

Financial Aid

Financial aid for graduate students includes fellowships, scholarships, teaching assistantships, grading positions, and research assistantships. The School offers a limited number of ECEN Graduate Scholar packages worth more than $18,000 per year. Approximately 70 percent of OSU students receive financial aid.

Cost of Study

Tuition for graduate-level courses is $80 per hour for residents and $254 per hour for nonresidents. Teaching assistants and research assistants pay tuition at the resident rate of $80 per hour.

Living and Housing Costs

The cost of a dormitory room is approximately $1870 per semester, including telephone service. Married student housing is available through OSU, and a number of nearby off-campus apartments are available. A nonresident single student can expect to spend $3200 per semester in housing, food, books, and miscellaneous expenses.

Student Group

The OSU campus enrollment is approximately 20,000 full-time students; approximately 4,700 of them are graduate students. The School of Electrical and Computer Engineering typically has between 100 and 200 graduate students in the department each year.

Student Outcomes

Students receiving graduate degrees in electrical and computer engineering at OSU are in demand within the international marketplace. Local companies that hire OSU students include Dallas-based companies TI/Raytheon, E-Systems, and Sprint; Tulsa-based companies Williams and MCI/Worldcom; and Oklahoma City–based companies Lucent Technologies, General Motors, and Seagate. Other students go on to prestigious positions in government and academia.

Location

Stillwater, located 65 miles from both Oklahoma City and Tulsa, has a population of 40,000 and is essentially a university town. It has been rated as one of the safest university locations in the Big 12 and nationally. The climate is mild and pleasant, and surrounding parks offer boating, fishing, hiking, and mountain biking.

The University

OSU was founded in 1890. Today, it is a comprehensive research university with an international reputation in telecommunications, laser research, and agriculture. The Stillwater campus has more than 100 buildings that are situated on 415 acres. The University participates in all major intercollegiate sports and ranks third nationally in the number of NCAA national championships. OSU is a member of the Big 12 conference.

Applying

Application forms are available from the Graduate College (http://www.osu-ours.okstate.edu/gradcoll/) and must be submitted in duplicate with official transcripts of all academic work completed. The informal deadline for applications and requests for financial aid is January 15 for admission for the fall semester. The TOEFL examination is required of international students. The GRE is not required, but it is encouraged for Ph.D. applicants.

Correspondence and Information

Professor Michael Soderstrand, Head
School of Electrical and Computer Engineering
Oklahoma State University
202 Engineering South
Stillwater, Oklahoma 74078
E-mail: sodersm@okway.okstate.edu
World Wide Web: http://elec-engr.okstate.edu/

Oklahoma State University

THE FACULTY AND THEIR RESEARCH

Michael Soderstrand, Professor and Head; Ph.D., California, Davis. Digital signal processing and communications, active-passive filters, VLSI.

Scott T. Acton, Associate Professor; Ph.D., Texas at Austin. Image processing, multimedia, video processing, computer vision.

Mark Allen, Adjunct Instructor; Ph.D., Oklahoma State. Telecommunications, signal processing.

H. Jack Allison, Professor; Ph.D., Oklahoma State. Unconventional energy systems, computer software.

R. Alan Cheville, Assistant Professor; Ph.D., Rice. Optics, electrooptics, lasers.

Seong-Jhin Choi, Visiting Assistant Professor; Ph.D., Kwangwoon. Multimedia, video storage and retrieval, digital signal processing.

Kenneth Church, Adjunct Instructor; Ph.D., Oklahoma State. Optoelectronics, lasers, photonics.

Guilhem Gallot, Visiting Assistant Professor; Ph.D., Ecole Polytechnic Palaiseau. Quantum optics, ultrafast optics, generation of ultrafast pulses.

Michael Gard, Adjunct Instructor; Ph.D., SMU. Geophysics, infrared imaging, sensors.

Thomas Gedra, Associate Professor; Ph.D., Berkeley. Power economics, stochastic control, nonlinear systems.

Daniel Grischkowsky, Bellmon Professor; Ph.D., Columbia. Electrooptics, ultrafast electrical pulses.

Martin Hagan, Professor; Ph.D., Kansas. Neural networks, continuous systems, time-series analysis, signal processing, parallel processing.

Chriswell Hutchens, Associate Professor; Ph.D., Missouri–Columbia. Analog CMOS, VLSI, electronic neural networks, bioengineering.

Louis Johnson, Associate Professor; Ph.D., MIT. Robot vision, computer architecture, digital electronics.

Jerzy S. Krasinski, Professor; Ph.D., Warsaw. Optics, spectroscopy, ultrafast lasers.

Carl Latino, Associate Professor; Ph.D., Penn State. Microprocessors, digital systems.

Roger McGowan, Visiting Assistant Professor; Ph.D., Colorado State. Optoelectronics, lasers, fiber optics.

Dipti Prasad Mukherjee, Visiting Assistant Professor; Ph.D., Indian Statistical Institute. Computer vision, image processing, computer graphics.

R. G. Ramakumar, Naeter Professor; Ph.D., Cornell; PSO. Renewable energy, power, reliability.

George Scheets Jr., Associate Professor; Ph.D., Kansas State. Communications, signal processing, computer simulations.

Scott R. Shepard, Associate Professor; Ph.D., MIT. Communications systems and devices, quantum estimation and measurement theory, quantum computation and encryption.

Keith A. Teague, Associate Professor; Ph.D., Oklahoma State. Signal, speech, and image processing; parallel processing.

James C. West, Associate Professor; Ph.D., Kansas. Radar remote sensing, electromagnetics, microwave systems.

R. K. Yarlagadda, Professor; Ph.D., Michigan State. Speech and seismic signal processing, communication theory.

Gary G. Yen, Assistant Professor; Ph.D., Notre Dame. Intelligent control, neural networks.

OREGON GRADUATE INSTITUTE OF SCIENCE & TECHNOLOGY

Department of Electrical and Computer Engineering

Programs of Study

The Department of Electrical and Computer Engineering offers programs of study leading to the M.S. and Ph.D. degrees in electrical engineering. Both thesis and nonthesis M.S. options are available, the former requiring 36 credit hours of classes and 12 credit hours of research and the latter requiring 44 course credits and 4 credits of research. An intensive M.S. program (minimum 48 credits for nonthesis) is also available. The Ph.D. requires successful completion of an oral qualifying examination covering core course topics before admission to Ph.D. candidacy. M.S. (thesis) and Ph.D. candidates must submit satisfactory theses and defend them in oral examinations. There is a two-year minimum residency requirement for the Ph.D. Full-time students are expected to complete their degree requirements within six years for the Ph.D. and four years for the M.S.

Research specialty areas include intelligent signal processing, with emphasis on human-like speech and image processing; adaptive neural net systems; advanced display technology; field emission; display technology; semiconductor materials, devices, and processing; biomedical engineering; and VLSI design and processing. The emphasis is on scientific and engineering investigations that have well-defined goals and real utility, pursued in an atmosphere resembling that of a working research and development laboratory. The relatively small size of the department, with a student-faculty ratio of about 4:1, and the overall educational philosophy guarantee that each student receives close individual supervision from his or her research adviser.

Research Facilities

For education and research support, the department has two Sun Workstation/Xterminal labs and an NT lab. Dedicated workstations (UNIX and NT) are used in the research labs. Another training lab is equipped for digital signal processing, chip testing, and embedded computing projects. The computer labs make extensive use of large computer-aided design (CAD) and mathematical tool packages. In addition, the department maintains three student computer labs with terminals stationed throughout the department. Semiconductor device, circuit, and applications facilities include MOCVD growth characterization laboratories (III-V, II-VI, and β-SiC), device and IC processing laboratories, and materials and device characterization laboratories. The department also houses laser characterization equipment. Laser application laboratories are available for the study of atmospheric optics and laser interactions with matter. High-resolution (submicrometer) focused ion beam facilities include a combination SEM-focused ion beam system used for optoelectronics device modification and a system for investigating ion beam–induced chemistry. Equipment supporting display technology research includes an atomic layer epitaxy reactor for growth of multilayered electroluminescent flat panel displays and a sputtering system with 3-inch RF and DC guns.

Financial Aid

Financial aid for entering students is awarded on a competitive basis to students with outstanding promise. Full-time Ph.D. students may be supported on research fellowships, which include full tuition scholarships. Full-time M.S. students may receive partial tuition scholarships. The Institute provides guidance for students who wish to secure external fellowships or student loans.

Cost of Study

Tuition for 1999–2000 is $495 per credit or $4465 for one quarter of full-time study.

Living and Housing Costs

OGI does not offer on-campus housing. Monthly rent for a two-bedroom apartment in the area typically ranges from $550 to $800. Housing information is available from OGI's Office of Academic and Student Services.

Student Group

As of January 1998, the department had 51 full-time students. Of these, 27 percent were Ph.D. students. Since 1994, the department has graduated 120 students—78 percent M.S. and 22 percent Ph.D.

Location

OGI is located 12 miles west of downtown Portland, a city that provides diverse cultural activities, including art, music, entertainment, and sports. The Oregon coast is about 60 miles to the west, and the Cascade Mountains are 50 miles to the east. Oregon has a large number of state parks, as well as national forest land and wilderness areas. These natural areas provide outstanding opportunities for skiing, hiking, fishing, and other outdoor activities.

The Institute and The Department

The goal of the department is to provide education of the highest quality in electrical and computer engineering through first-class research in areas of scientific importance. This philosophy ensures that graduates are actively recruited by internationally recognized industries and universities.

Applying

Applications are accepted at any time, but those requesting financial assistance should be submitted by March 1. Required admissions materials for both M.S. and Ph.D. applicants include three letters of recommendation, transcripts, and GRE scores. The OGI institutional code for the GRE is 4592. A minimum TOEFL score of 550 is also required of those whose native language is not English. The TOEFL requirement may be waived if the student received a prior degree in the United States.

Correspondence and Information

Admissions
Office of Academic and Student Services
Oregon Graduate Institute
P.O. Box 91000
Portland, Oregon 97291-1000

Telephone: 503-748-1027
      800-685-2423 (toll-free)
E-mail: admissions@admin.ogi.edu
World Wide Web: http://www.ogi.edu/

Oregon Graduate Institute of Science & Technology

THE FACULTY AND THEIR RESEARCH

Anthony E. Bell, Associate Professor; Ph.D. (physical chemistry), London, 1962. Development of liquid-metal field ion sources; field ionization, surface physics, and chemistry; field emission microscopy; energy distribution measurements; selected area processing for microcircuit fabrication, using focused electron beams.

C. Neil Berglund, Professor; Ph.D. (electrical engineering), Stanford, 1964. Optical and electron beam lithography, mask and reticle technology, metrology in semiconductor processing, management of technology, display technology.

* Ronald A. Cole, Professor and Director, Center for Spoken Language Understanding; Ph.D. (psychology), California, Riverside, 1971. Spoken language systems, integrating expert knowledge of human perception and communication into systems that recognize spoken language; speaker- and vocabulary-independent recognition of telephone speech in different languages, multilanguage speech data collection and transcription, automatic language identification.

V. S. Rao Gudimetla, Associate Professor; Ph.D. (applied physics), Oregon Graduate Center, 1982. Microwave circuits, device simulation, laser speckle, applied mathematics. (Leave of absence)

Dan Hammerstrom, Professor and Department Head; Ph.D. (electrical engineering), Illinois, 1977.

* Hynek Hermansky, Professor; Dr.Eng. (electrical engineering), Tokyo, 1983. Communication between human and machine; human perception and its computer simulation; speech production and perception; automatic recognition of speech, speech coding, synthesis, and enhancement; identification and extraction of linguistic information in realistic communication environments.

J. Fred Holmes, Professor; Ph.D. (electrical engineering), Washington (Seattle), 1968. Speckle propagation through turbulence, remote sensing of wind and turbulence, electrooptic systems for industrial process control applications, instrumentation signal processing.

Steven Jacques, Associate Professor; Ph.D. (biophysics and medical physics), Berkeley, 1984. Laser effects on biological systems using light-activated drugs, thermal processes, or photoacoustic shock waves; modeling of photon migration using Monte Carlo or Feynman path integrals.

* Todd K. Leen, Associate Professor; Ph.D. (physics), Wisconsin, 1982. Neural learning algorithms, architecture and theory, dynamics, noise, model complexity and pruning, applications to speech processing.

Michael Macon, Assistant Professor; Ph.D. (electrical engineering), Georgia Tech, 1996. Speech synthesis, speech and audio coding, speech enhancement, music synthesis, human auditory perception and modeling, digital signal processing.

* John E. Moody, Associate Professor; Ph.D. (physics), Princeton, 1984. Design and analysis of learning algorithms; statistical learning theory, including generalization and model selection; optimization methods (both deterministic and stochastic) and applications to signal processing, time series, macroeconomics, and finance.

James D. Parsons, Associate Professor; Ph.D. (engineering), UCLA, 1981. Relationships of semiconductor properties, solid-state device performance, and MOCVD growth; new semiconductors and device processing technologies; new solid-state device and monolithic IC design concepts in β-SiC, III-V, and II-VI semiconductors.

* Misha Pavel, Professor; Ph.D. (experimental psychology), NYU, 1980. Representation of uncertainty, decision making, and choice behavior; pattern recognition and categorization in humans and machines; information retrieval and decision support systems; image processing and sensor fusion. (Leave of absence)

Scott A. Prahl, Assistant Professor; Ph.D. (biomedical engineering), Texas at Austin, 1988. Interaction of light with tissue, pulsed photothermal radiometry, laser angioplasty, optical properties of biological materials, noninvasive medical diagnostics.

Shankar Rananavare, Associate Professor; Ph.D. (physical chemistry), Missouri, 1983. Structure and dynamics of microemulsions and nanoparticles; liquid crystal displays, antiferroelectric and amphitropic liquid crystals.

Raj Solanki, Associate Professor; Ph.D. (physics), Colorado State, 1982. Gas lasers, laser spectroscopy, gas plasmas, photon- and electron-beam-induced materials processing.

Pieter Vermeulen, Associate Professor; Ph.D. (electrical and computer engineering), Carnegie Mellon, 1989. Speech processing/recognition, pattern recognition, neural networks and computer architecture, realtime and embedded systems.

* Eric A. Wan, Assistant Professor; Ph.D. (electrical engineering), Stanford, 1993. Neural networks and adaptive signal processing, time series prediction, adaptive control, active noise cancellation, telecommunications.

The OGI campus and Science Park share a beautiful natural setting.

The Cooley Science Center.

POLYTECHNIC UNIVERSITY

Department of Electrical Engineering

Programs of Study	The Department of Electrical Engineering offers M.S. and Ph.D. degrees in electrical engineering and M.S. degrees in systems engineering, telecommunication networks, and electrophysics. Large enrollments allow courses to be offered in a wide variety of areas, including telecommunication networks, wireless networks, communication theory, multimedia, systems, control and robotics, image and signal processing, fiber optics, electromagnetics and microwaves, plasmas, electronics, VLSI design, and power systems. In addition to electrical engineering courses, students may also use courses in computer science, mathematics, physics, and other engineering disciplines toward the M.S. and Ph.D. degrees. A recent Gourman Report of M.S. programs in the United States ranks Polytechnic's electrical engineering program twenty-fourth in the nation. The electrophysics program is intended for students with a bachelor's degree in electrical engineering, physics, or related disciplines who are interested in studying the physical properties of devices and materials. The systems engineering program allows students with bachelor's degrees in mathematics, computer science, or other engineering disciplines to study the system and networking aspects of electrical sciences. The telecommunications program admits students with bachelor's degrees in computer science, computer engineering, or electrical engineering.
	The M.S. degree requires 36 units of course work, which is equivalent to twelve standard semester-long courses meeting 2¼ hours per week. Students may elect to do a 9-unit master's thesis, and up to 9 units may be transferred from other universities. For the telecommunications M.S. degree, students must complete a 3-unit project. In order to receive the degree, a student must maintain an average grade of at least a B. Students seeking the Ph.D. must take a minimum of 30 course units and 24 dissertation units past the M.S. In addition, they must pass a written and oral qualifying examination prior to taking dissertation credits and make a successful oral defense of the dissertation.
Research Facilities	The department has ongoing research programs that support the Ph.D. and M.S. programs. Research activities related to telecommunication networks, distributed systems, wireless networks, SONET and ATM networks, image processing and compression, video and multimedia transmission, and communication theory are organized through the New York State–funded Center for Advanced Technology in Telecommunications (CATT), in which many of the faculty members participate. Other research activities are under individual or groups of faculty members. The activities include signal processing, microwave-integrated circuits, control and robotics, power systems, underwater propagation, and plasmas and high-power sources. These research activities are supported by $3 million in grants and contracts from industry and from federal and state agencies. Research facilities are divided between the primary campus in Brooklyn and the Long Island campus in Farmingdale. The Brooklyn campus has research laboratories devoted to image processing, video networking, multimedia, signal processing, VLSI design, high-speed switching, LANs, control and robotics, wireless propagation, and power systems. On the Long Island campus, there are laboratories for studying microwave-integrated circuits, wireless communication systems, and plasmas and high-power sources.
Financial Aid	Financial aid includes half-tuition remission, research and teaching fellowships, graduate assistantships, and graduate traineeships. Stipends ranged from $1285 to $1470 per month for 1999.
Cost of Study	In 1998–99, tuition was $675 per unit.
Living and Housing Costs	In Brooklyn, there are dormitory facilities for single students. In addition, private rooming accommodations are available for all students. For students enrolled at the 25-acre Long Island campus in Farmingdale, accommodations are available in the residence hall. Private rooms for men, women, and families can be rented nearby. Basic living expenses for a single student are approximately $2000 per month.
Student Group	The student body consists of men and women who hold baccalaureate and graduate degrees from more than 350 institutions worldwide. The graduate students represent about fifty other countries. In 1998, 115 M.S. degrees and twenty Ph.D. degrees were awarded by the department.
Location	The main campus of Polytechnic is in the recently developed MetroTech Center in downtown Brooklyn, which is one of the five boroughs making up New York City. New York is a center for science, technology, finance, medicine, the arts, and theater. With its mix of people and cultures from all over the world, it is perhaps the most exciting city in the world. The Long Island campus is located 40 miles east in Farmingdale, a major center of the electronics industry. Some graduate courses are also offered at the Westchester campus in Hawthorne. The main campus is accessible by public transportation, and all campuses are easily reached by car.
The University	The University was formed in 1973 by the merger of the Polytechnic Institute of Brooklyn and the NYU School of Engineering and Science. Graduate programs are offered at the main campus in Brooklyn, the Long Island campus in Farmingdale, and the Westchester Center in Hawthorne. The evening sessions allow unusual latitude in adapting programs to the requirements of employment. The faculty has 120 full-time professors and a large adjunct staff. The University conducted more than $15.2 million of funded research in 1998. Since 1983, Polytechnic has been New York State's Center for Advanced Technology in Telecommunications.
Applying	For admission to the M.S. programs, the applicant must have a bachelor's degree from an accredited university with a grade point average indicative of success in graduate study. For the electrical engineering program, a B.S. in electrical engineering is required. Admission to the electrophysics program requires a B.S. in electrical engineering, physics, or a related discipline. Admission to the system engineering program requires a bachelor's degree in engineering, mathematics, or computer science. Admission to the telecommunications program requires a bachelor's degree in computer science, computer engineering, or electrical engineering. Students lacking some background may qualify for admission by taking specified deficiency courses. Students may be admitted directly into the Ph.D. programs with a bachelor's degree or with a master's degree. Those entering with a bachelor's will ordinarily satisfy the requirements for the M.S. on the way to the Ph.D. Students receiving a master's in systems engineering or electrophysics may go on for a Ph.D. in electrical engineering.
	Applicants should submit credentials as early as possible. Deadlines are April 1 for September registration, November 1 for January registration, and May 1 for the summer session, although late admission is possible. The financial aid deadline is April 1.
Correspondence and Information	Graduate Committee Department of Electrical Engineering Polytechnic University 6 Metrotech Center Brooklyn, New York 11201 Telephone: 718-260-3056 World Wide Web: http://www.poly.edu

Polytechnic University

THE FACULTY AND THEIR RESEARCH

S. Bergstein, Industry Professor; Ph.D. (electrical engineering), Polytechnic of New York. Communications.

H. Bertoni, Professor; Ph.D. (electrical engineering), Polytechnic of Brooklyn. Acoustics, electromagnetics, wireless communications.

R. Boorstyn, Professor; Ph.D. (electrical engineering), Polytechnic of Brooklyn. Computer communication networks, telecommunications.

M. Boukli, Assistant Industry Professor; Ph.D. (electrical engineering), Polytechnic of Brooklyn. Communication systems, fiber optics.

F. Cassara, Professor; Ph.D. (electrical engineering), Polytechnic of Brooklyn. Communication electronics, wireless communications.

J. Chao, Professor; Ph.D. (electrical engineering), Ohio. High-speed networks, ATM and photonic switch design, VLSI.

D. Czarkowski, Assistant Professor; Ph.D., Florida. Power electronics, power quality.

N. Das, Associate Professor; Ph.D. (electrical engineering), Massachusetts. Electromagnetics, antennas.

O. Guleryuz, Assistant Professor; Ph.D. (electrical engineering), Illinois at Urbana-Champaign. Image and video coding and processing, statistical signal models.

Z.-P. Jiang, Assistant Professor; Ph.D. (mathematics), École des Mines (Paris). Dynamic and control systems, robotics.

R. Karri, Associate Professor; Ph.D. (electrical engineering), California, San Diego. CAD, fault-tolerant computing, high-level synthesis, cryptography.

F. Khorrami, Associate Professor; Ph.D. (electrical engineering), Ohio State. Control systems and robotics.

M. Kouar, Associate Industry Professor; Ph.D. (electrical engineering), Polytechnic of Brooklyn. Electronic circuits, telecommunication systems.

S.-P. Kuo, Professor; Ph.D. (electrophysics), Polytechnic of New York. Magnetohydrodynamics, plasmas.

I.-T. Lu, Associate Professor; Ph.D. (electrical engineering), Polytechnic of New York. Acoustics, wireless communications.

S. Panwar, Associate Professor; Ph.D. (electrical engineering), Massachusetts. Communication networks.

S. Pillai, Professor and Department Head; Ph.D. (systems engineering), Pennsylvania. Signal processing and communications.

I. Selesnick, Assistant Professor; Ph.D. (electrical engineering), Rice. Signal processing.

L. Shaw, Professor; Ph.D. (electrical engineering), Stanford. Signal processing, reliability.

J. Snyder, Senior Industry Professor; M.S.E.E., Polytechnic of New York. Microprocessors, data acquisition, signal processing.

T. Tamir, University Professor; Ph.D. (electrophysics), Polytechnic of Brooklyn. Electromagnetics, electrooptics.

M. Veeraraghavan, Associate Professor; Ph.D. (electrical engineering), Duke. Networking protocols, wireless and optical communications.

P. Voltz, Associate Professor; Ph.D. (electrical engineering), Polytechnic. Communications, signal processing.

Y. Wang, Associate Professor; Ph.D. (electrical engineering), California, Santa Barbara. Medical imaging, computer vision, image and video signal processing.

D. Youla, University Professor; M.S.E.E., NYU. Networks, control systems.

Z. Zabar, Professor; Sc.D. (electrical engineering), Technion (Israel). Power electronics, electrical drives, power systems.

PRINCETON UNIVERSITY

School of Engineering and Applied Science
Department of Electrical Engineering

Programs of Study	The Department of Electrical Engineering offers programs of study leading to the Ph.D., M.S.E., and M.Eng. degrees. The requirements for the M.S.E. can normally be completed in one academic year of full-time graduate study. Selected students also may proceed on a half-time basis for two years in the industrial M.S.E. program. Requirements for the Ph.D. degree are completing at least one academic year of full-time residence as a degree candidate, presenting a research seminar and passing the Ph.D. general examination, submitting a doctoral dissertation and having it accepted by the faculty, and satisfactorily sustaining a final public oral examination, which includes a presentation and a defense of the dissertation. The Master of Engineering (M.Eng.) degree program meets the need for rigorous and advanced training in the applied aspects of modern technology beyond the typical four-year engineering degree program. For students with adequate preparation, the program can be completed in one year of full-time study. A thesis is not required; program requirements are met by successfully completing eight courses, six of which must be at the graduate level. Design projects, which count toward course requirements, are available. Part-time status is available for qualified students. Students can obtain further details and information or apply to the program via e-mail (MENG@princeton.edu) or telephone (609-258-2890).
Research Facilities	The Department of Electrical Engineering, located in the Engineering Quadrangle, has first-class facilities for experimental research in communication and information science, computer engineering, optical and optoelectronic engineering, and electronic materials and devices. The facilities include multiuser communication systems, VLSI array processors, computer-aided design and testing of digital systems, and digital signal processing. The image processing lab is equipped with a network of workstations, an image sequence storage and simulation system, and a laser videodisc recording system. Pulse-compressed mode-locked lasers, custom-built fiber-optic linear filters, optical bistable devices, and integrated-optic photonic switches are used for light-wave communications research. Instrumentation for electronic materials and devices includes molecular-beam epitaxy, CVD of crystalline and amorphous Si-Ge, electron diffraction and spectroscopy equipment, scanning tunneling microscopy, superconducting magnets, dilution refrigerators, and CW, picosecond, and femtosecond lasers. A clean-room laboratory is available for the processing, fabrication, and measurement of semiconductor structures and devices. Facilities include diffusion furnaces, electron-beam and thermal evaporators, and plasma etching and deposition, optical lithography, direct-write electron-beam patterning, and automated electrical measurement equipment.
	A NanoStructure lab has state-of-the-art facilities in nanofabrication and nanodevice characterization. The department's own computing facilities consist of a network of Sun and Sony workstations; HP 9000 series 200 and 550 workstations; an NSF-funded center for experimental computing research, which operates an SGI Challenge multiprocessor; a system of SGI Indigo workstations; and a Viewgraphs HDTV frame buffer. Free access to these computers is provided to graduate students for their research. Undergraduate microcomputer laboratories for systems and signal processing are also available for graduate student use. The research groups also have access to the Center for Information Technology (CIT) microcomputer clusters in the EQuad located near the campus.
	The Engineering Library is fully equipped with texts and reference material in all of the major areas of engineering, including both domestic and international periodicals and files of project research reports. The Fine Hall Library contains a large amount of reference material, including texts, periodicals, and reports in mathematics, physics, and statistics. An extensive collection of online journals is accessible through network computers.
Financial Aid	A substantial number of prestigious fellowships, including the Gordon Wu and Francis Upton Fellowships in Engineering, are available on a competitive basis. Financial assistance, equivalent to tuition plus approximately $14,500 to $15,000 for the 1999–2000 academic year, is also normally available in the form of assistantships in instruction and/or research. The assistantships are arranged as part of the student's training and permit a full program of study. Additional compensation is available in the summer.
Cost of Study	The tuition, approximately $25,100 for the 1999–2000 academic year, includes use of University facilities and comprehensive health and accident insurance coverage.
Living and Housing Costs	The Graduate College provides rooms for about 500 single students at $2500 to $4000 per academic year. Some rooms are available in the Princeton area for about $505 a month, and students may take meals at the Graduate College for the board rate. University housing is available for about 300 married students at approximately $520 to $830 a month plus utilities. Jobs for graduate students' spouses are available at the University or nearby.
Student Group	Of the 6,438 students at Princeton University, 1,913 are graduate students from many universities and countries. The Department of Electrical Engineering has 150 graduate students and 103 undergraduates.
Location	The University is located in the historic town of Princeton, 50 miles from both New York City and Philadelphia. The area offers sports, cultural, and entertainment activities, including an outstanding variety of concerts and plays at McCarter Theatre. The University has many facilities for such sports as golf, tennis, sailing, swimming, and squash, plus two large gymnasiums. The Pocono Mountains, 85 miles away, offer skiing and camping. The Jersey shore beaches are an hour's drive away.
The University and The School	Princeton's first engineering program was offered in 1873 in civil engineering. In 1922, the School of Engineering was formed by the Departments of Electrical Engineering, Civil Engineering, and Mechanical Engineering. The widening scope of today's engineering and the University's responsibilities for developing new areas of applied science, particularly those spreading across or falling outside traditional fields, led Princeton in 1962 to rename its engineering school the School of Engineering and Applied Science. There are currently five departments in the School: Chemical Engineering, Civil Engineering and Operations Research, Computer Science, Electrical Engineering, and Mechanical and Aerospace Engineering.
Applying	Completed applications for admission received before January 3 receive first consideration; fellowship and assistantship appointments are made from these applications. Applicants are notified of the results around March 15 and should include results of the Graduate Record Examinations (GRE) General Test as part of their application for admission. Students may request an application electronically (http://webware.Princeton.EDU/GSO/appform.htm) or access the online application.
Correspondence and Information	Director of Graduate Studies Department of Electrical Engineering Princeton University Princeton, New Jersey 08544 Telephone: 609-258-3335 World Wide Web: http://www.princeton.edu

Princeton University

THE FACULTY AND THEIR RESEARCH

Keren Bergman, Ph.D., MIT. Fiber optic systems and devices, nonlinear quantum noise, high bit-rate lasers.
Ravindra Bhatt, Ph.D., Illinois. Condensed-matter theory.
Stephen Y. Chou, Ph.D., MIT. Nanotechnology and nanoscale electronic, optoelectronic, and magnetic devices.
Bradley W. Dickinson, Ph.D., Stanford. Systems theory, signal processing.
Stephen R. Forrest, Ph.D., Michigan. Optoelectronics and organic semiconductors.
Niraj Jha, Ph.D., Illinois. Digital system testing, fault-tolerant computing, computer-aided design of integrated circuits, distributed computing.
Antoine Kahn, Ph.D., Princeton. Physics and chemistry of semiconductor surfaces and interfaces.
Hisashi Kobayashi, Ph.D., Princeton. Communication networks, digital communication, system performance analysis, queueing theory.
Sanjeev R. Kulkarni, Ph.D., MIT. Pattern recognition, machine learning, signal/image processing.
Sun-Yuan Kung, Ph.D., Stanford. VLSI signal processing, array processors, digital signal processing, modern spectrum analysis, neural computing.
Ruby B. Lee, Ph.D., Stanford. Computer architecture, multimedia architecture, security architecture.
Bede Liu, Ph.D., Polytechnic of Brooklyn. Image/video/signal processing.
Stephen A. Lyon, Ph.D., Caltech. Device physics and laser spectroscopy.
Sharad Malik, Ph.D., Berkeley. Design methodology and design tools for electronic systems.
Margaret R. Martonosi, Ph.D., Stanford. Performance analysis, architecture and compiler issues in high-performance and configurable computing.
Michael Orchard, Ph.D., Princeton. Image processing and video coding.
H. Vincent Poor, Ph.D., Princeton. Statistical signal processing, digital communications, multiuser communication systems.
Paul Prucnal, Ph.D., Columbia. Photonic switching, optical networks, VLSI optical interconnects and optical signal processing.
Peter Ramadge, Ph.D., Toronto. System theory, control theory.
Stuart C. Schwartz, Ph.D., Michigan. Signal and image processing, communication theory.
Mordechai Segev, Ph.D., Technion (Israel). Nonlinear optics.
Mansour Shayegan, Ph.D., MIT. Physics and technology of low-dimensional semiconductor structures.
James C. Sturm, Ph.D., Stanford. Physics and technology of semiconductors, fabrication processes.
Daniel C. Tsui, Ph.D., Chicago. Physics of thin films and interfaces.
Sergio Verdu, Ph.D., Illinois. Communication and information theory.
Sigurd Wagner, Ph.D., Vienna. Device materials and thin-film electronics, including solar cells.
Wayne H. Wolf, Ph.D., Stanford. VLSI-CAD, embedded computing, multimedia computing, video libraries.

RESEARCH AREAS

Computer Engineering

This program aims to prepare students for teaching and research in computer architecture, fault-tolerant computing, digital system testing, multimedia architecture, secure information processing, parallel processing, neural computing, high-level and system synthesis, embedded system design, reconfigurable computing, and related areas. Courses offered by the Department of Electrical Engineering include VLSI design, switching and sequential systems, digital system testing, computer architecture, multimedia architecture, neural computing, VLSI array processors, computer-aided design of integrated circuits, and reconfigurable computing. In addition, students elect appropriate courses offered by the Departments of Computer Science and Mathematics. Ongoing research focuses on multimedia architecture, digital system testing, synthesis for testability and low power, fault-tolerant computing, systolic and wavefront architectures, parallel processing, computer architecture, neural networks, Internet security, computer-aided design of digital systems, and adaptive systems.

Research in computer engineering has a strong theoretical component and often involves experimental work in design, testing, and simulation. Departmental facilities are available to support the research, and students gain substantial experience with hardware and software as part of an academic program in this area.

Electronic Materials and Devices

This program is directed toward preparing the student for research, advanced development, or teaching in the areas of solid-state electronics, physical electronics, electronic materials, or the physical aspects of electrotechnology in general. The student takes courses in the Department of Physics on appropriate topics, such as electromagnetic theory and quantum mechanics. Courses in solid-state and semiconductor physics, transport theory, semiconductor surface phenomena, solid-state devices, optical properties of solids, nanotechnology, electronic materials, and heterojunction structures are given within the Department of Electrical Engineering.

Research ranges from solid-state theory to experiments on quantum phenomena in semiconductors to the invention of new semiconductor processes and devices. Current research includes work on surfaces and interfaces of semiconductors; molecular-beam epitaxy growth of compound semiconductors and organic solids; materials for large-area electronics; optoelectronic devices for VLSI interconnects; optical processes in semiconductors, including subpicosecond phenomena; and transport along interfaces, especially two-dimensional magnetotransport.

Information Sciences and Systems

This program is broadly formulated to prepare the student for research and teaching in the general area of systems, with emphasis on stochastic systems, communication theory, systems theory, and signal and image processing. Current research activities include work in estimation and detection, stochastic processes, optimization of stochastic systems, adaptive and learning systems, artificial neural networks, control theory, nonlinear filtering, computer communication, data communication, digital signal processing, image processing, discrete event systems, wireless communications, information theory, multiuser communication systems, optical channels, and fiber-optic networks and photonic switching.

The research program is oriented mainly toward theoretical work and toward extensive use of computers for signal processing and for system simulation and optimization. Facilities are available to conduct a wide range of experimental research, particularly in the areas of digital and optical signal processing, random processes, and light-wave communication.

Optical and Optoelectronic Engineering

This program prepares students for research, advanced development, and teaching in the areas of optical communications, optoelectronic devices, and optical system design, with applications to fiber-optic networks, telecommunications systems, and multiprocessor interconnections. In addition to the core curriculum, the program of study can include selected courses in the Electronic Materials and Devices Group and the Information Sciences and Systems Group as well as courses in the Physics, Chemistry, and Mechanical and Aerospace Engineering departments.

The research program is both experimental and theoretical, and the projects can range from areas such as photonic switching, broadband optical networks, and optical computing to smart pixels. The laboratory facilities for optical communications systems are equipped with Argon, YAG, YLF, Titanium Sapphire, and Color-Center short-pulse laser systems, as well as a variety of semiconductor lasers, optical hardware, and systems test equipment. The facilities for sample growth are equipped with inorganic and organic MBE growth systems, a vapor deposition system, and an LPE growth system. The sample preparation and characterization facilities include reactive ion etching, an electron beam etching chamber, RHEED, STM, and a picosecond dye laser. For device processing, packaging, and microelectronics, the facilities include a mask aligner, a wire bonder, and a probe station.

PURDUE UNIVERSITY

School of Electrical and Computer Engineering

Programs of Study	The School of Electrical and Computer Engineering offers Master of Science (M.S.) and Ph.D. degrees with specialization in automatic control, biomedical engineering, communications and signal processing, computer engineering, energy sources and systems, fields and optics, solid-state devices and materials, and VLSI and circuit design. Interdisciplinary programs are also offered.

The master's program has both nonthesis and thesis options. The nonthesis option requires 30 hours of course work; the thesis option requires at least 18 hours of course work plus a master's thesis. Master's students typically take three to four semesters to complete their degrees.

The Ph.D. program requires 21 hours of course work beyond the master's degree or 42 hours beyond the bachelor's degree. Students must pass a qualifying exam covering graduate course work, a preliminary exam, and a dissertation defense. There is no foreign language requirement. The typical length of the Ph.D. program is five to six years beyond the date of entry into the graduate program.

Research Facilities
Laboratory facilities in the School of Electrical and Computer Engineering include the Advanced Digital Systems/Embedded Microcontroller Design Laboratory, the Applied Ultrasonics Laboratory, the Articulated Motion Laboratory, the Basil S. Turner Laboratory for Electroceramics, the Biomedical Acoustics Laboratory, the Communications Research Laboratory, the Dependable Computing Laboratory, the Digital Systems Laboratory, the Distributed Multimedia Systems Laboratory, the Electronic Imaging Systems Laboratory (EISL), the Energy Systems Analysis Consortium (ESAC), the Engineering Research Center for Collaborative Manufacturing, the High-Performance Computing Laboratory (HPCLab), the High Resolution Transmission Electron Microscopy Laboratory, the High-Speed Semiconductor Optical Characterization Laboratory, the III-V Molecular Beam Epitaxy Laboratory, the Low-Power VLSI Laboratory, the Magnetics Laboratory, the Materials Research Science and Engineering Center for Technology-Enabling Heterostructure Materials, the Microprocessor Systems and Interfacing Laboratory, the Microwave Laboratory, the Modern Optics Research Laboratory, the Multimedia Learning Laboratory, the Multispectral Image Processing (MIP) Laboratory, the Optical Information Processing Laboratory, the Parallel Distributed Processing Laboratory, the Purdue Electric Power Center (PEPC), the Purdue Machine Learning Lab, the Purdue Multimedia Testbed, the Purdue University Network Computing Hub (PUNCH), the Purdue–Notre Dame NSF Materials Research Group, the Robot Vision Laboratory, the Semiconductor Simulation Hub: A Network-Based Simulation Lab, the Sensor-Based Robot Control Laboratory, the Silicon Epitaxial Laboratory, the Solid-State Device and Materials Laboratory, the Spectral Imaging Systems Laboratory, the Spread Spectrum and Satellite Communications Research Laboratory, the Speech and Signal Processing Research Laboratory, the Scalable Parallel Research Applications Laboratory (SPiRAL), the Ultrafast Optics and Fiber Communications Laboratory, the VIADuct ATM Testbed, the Video and Image Processing Laboratory (VIPER), the VJ Multimedia Learning Lab, the VLSI Design Laboratory, the Wide Bandgap Photonics Molecular Beam Epitaxy Facility, and the Wide Bandgap Semiconductor Device Research at Purdue.

Extensive computing facilities for research are provided by the Engineering Computer Network (ECN).

Financial Aid
Various fellowships are awarded by Purdue University, by state and federal agencies, and by industrial sponsors. Departmental teaching and research assistantships are awarded on a competitive merit basis. Stipends range from approximately $1000 to $1500 per month in 1999–2000. The stipends also carry exemptions from University fees and tuition, except for approximately $315 per semester.

Cost of Study
In 1999–2000, Indiana residents pay approximately $1870 per semester for tuition and fees, and nonresidents pay approximately $6220 per semester. Books and supplies average $350 per semester.

Living and Housing Costs
In 1999–2000, University-supervised graduate residences cost an average of $3780 over a ten-month period (these figures are subject to change). The University operates 1,244 married-student apartments renting for an average of $4300 over a ten-month period. A variety of privately owned housing facilities are available for rent in the surrounding community. A helpful publication entitled *Housing and Financial Guide for Off-Campus Students* is available from the Office of the Dean of Students. Purdue estimates that the total cost of attending graduate school is approximately $26,815 per year for a non-Indiana resident.

Student Group
There are approximately 220 master's students and 250 Ph.D. students in the School of Electrical and Computer Engineering. Students represent all parts of the United States and many other countries. The electrical and computer engineering undergraduate program has approximately 1,050 students.

Location
The West Lafayette campus of Purdue University is located on 1,565 acres in north-central Indiana. The cities of Lafayette and West Lafayette, which are separated by the Wabash River, have a combined population of approximately 72,000. West Lafayette and the surrounding areas offer a variety of cultural activities as well as historic landmarks and recreational attractions. The Purdue airport was the first university-owned airport in the United States. The campus is located 60 miles from Indianapolis (population approximately 1.25 million in the metropolitan area) and 130 miles from Chicago.

The University
Purdue University was established as a land-grant institution in 1869. The main campus at West Lafayette has 35,715 students, of whom 6,013 are enrolled in graduate programs. Engineering students number 7,752, of whom 1,823 are graduate students. Purdue has graduated more female engineers than any other engineering school in the United States and has the founding chapter of the National Society of Black Engineers.

Applying
Applicants may apply for fall or spring admission (August or January, respectively). Applications and supporting material should be submitted at least four months prior to the beginning of the semester for which admission is sought. For consideration for fellowships or assistantships, applications should be submitted by January 15 (for fall admission) or by September 15 (for spring admission). The General Test of the Graduate Record Examinations (GRE) is required for all applicants. The Test of English as a Foreign Language (TOEFL) is required for all applicants whose native language is not English, unless the applicant has earned a complete degree in the United States. For admission, a minimum TOEFL score of 575 is required.

Correspondence and Information
Electrical and Computer Engineering Graduate Office
Purdue University
1285 Electrical Engineering Building
West Lafayette, Indiana 47907-1285

Telephone: 765-494-3392
Fax: 765-494-3393
E-mail: ecegrad@ecn.purdue.edu
World Wide Web: http://ECE.www.ecn.purdue.edu/ECE/Graduate/

Purdue University

THE FACULTY AND THEIR RESEARCH

Jan P. Allebach, Professor. Electronic imaging systems, image capture and rendering, color image processing, image quality, multispectral imaging.

Philip F. Bagwell, Associate Professor. Quantum mechanical electron transport, electron transport in small devices, physics of semiconductors and metals, superconductivity and superconducting devices.

V. Ragu Balakrishnan, Assistant Professor. Numerical methods and optimization for systems and control, control theory, control system analysis and design, signal processing.

Rashid Bashir, Assistant Professor. Advanced semiconductor fabrication techniques, microelectromechanical systems (MEMS), MEMS-based bioelectric interface devices.

Mark R. Bell, Associate Professor. Information theory, communication theory and systems, radar systems and signal processing, signal theory.

Arden Bement, Basil S. Turner Distinguished Professor of Engineering. Electroceramics, high-temperature superconducting thin films and hybrid structures, smart materials and functional gradient interfaces.

Charles A. Bouman, Associate Professor. Image processing, statistical modeling, pattern recognition, image database search, inverse problems.

Carla E. Brodley, Assistant Professor. Artificial intelligence, machine learning, computer vision, pattern recognition.

Chin-Lin Chen, Professor. Integrated optics, fiber optics.

Edwin K. P. Chong, Associate Professor. Discrete event systems, communication/computer networks, wireless systems, optimization methods.

James A. Cooper Jr., Professor. Semiconductor device physics, wide band gap semiconductor devices (SiC and AlGaN), electron transport in semiconductors, MOS interface characterization.

Edward J. Coyle, Professor. Computer networks: performance analysis of ATM networks, architecture and performance of all-optical networks, queuing theory and stochastic processes, characterization of video traffic on networks; nonlinear signal and image processing.

Supriyo Datta, Professor. Electronic transport in small devices, nanoelectronics, superconductivity.

Ray A. DeCarlo, Associate Professor. Large-scale systems, geometric multivariable control, decentralized pole placement and eigenvalue placement, variable structure control, fault analysis and diagnosis, system parameter identification, biomedical control.

Edward J. Delp, Professor. Image and video compression, multimedia systems, image processing, parallel processing, computer vision, medical imaging, communication and information theory.

Henry G. Dietz, Associate Professor. Optimizing and parallelizing compilers, computer architecture, linguistics, digital imaging, real time systems.

Peter C. Doerschuk, Associate Professor. Statistical signal processing, multidimensional signal processing, inverse problems, X-ray crystallography.

Rudolf Eigenmann, Associate Professor. Parallel computing, compilers, computational engineering, performance evaluation, parallel architectures.

Daniel S. Elliott, Professor. Nonlinear optics, multiphoton processes, photoionization, coherent control, laser coherence effects.

Okan K. Ersoy, Associate Professor. Digital signal and image processing, neural networks, information processing based on fields and optics, probability and statistics, Fourier-related transforms and convolution techniques, parallel processing, applied mathematics.

Babak Falsafi, Assistant Professor. Computer architecture, memory system design, shared-memory multiprocessors, single-chip VLSI multiprocessors, power-aware architecture, integrated network interfaces.

José A. B. Fortes, Professor and Assistant Head for Education. Parallel and distributed computing, computer architecture, network computing.

Fritz J. Friedlaender, Professor. Magnetics.

W. Kent Fuchs, Professor and Head. Dependable high-performance computing, VLSI CAD, diagnosis, test and failure analysis of integrated circuits.

Keinosuke Fukunaga, Professor. Information processing systems, pattern recognition, pattern processing, learning computer control systems.

Eric S. Furgason, Professor. Applied ultrasonics, acoustic emissions, nondestructive evaluation of materials, acoustic propagation and scattering.

Leslie A. Geddes, Showalter Distinguished Professor Emeritus of Bioengineering. Biomedical engineering (experimental physiology and cardiology).

Saul B. Gelfand, Associate Professor. Digital communications, statistical signal processing, optimization and pattern recognition.

Arif Ghafoor, Professor. Multimedia systems, databases, distributed computing systems, broadband multimedia networking.

Robert L. Givan, Assistant Professor. Decision theoretic planning, Markhov decision processes, stochastic modeling, Bayesian inference, knowledge representation, automated reasoning, functional programming, type inference, programming languages.

Jeffery L. Gray, Associate Professor. Computer modeling of semiconductor devices, semiconductor physics, solar cells.

Mary P. Harper, Associate Professor. Artificial intelligence, natural language processing, speech understanding, algorithms.

Leah H. Jamieson, Professor. Speech analysis and recognition, parallel processing, parallel algorithms, parallel speech processing, parallel image processing, algorithm complexity theory.

Avinash C. Kak, Professor. Computer and robotic vision, image databases, human-machine interaction, object-oriented languages and design.

Rangasami L. Kashyap, Professor. Pattern recognition, image processing, system identification, time series, database management systems.

Antti J. Koivo, Professor. Robotics, computer control, biomedical systems.

Cheng-Kok Koh, Assistant Professor. VLSI CAD for physical design, VLSI interconnect modeling/optimization, algorithm design/analysis.

Paul C. Krause, Professor. Electromechanical energy conversion, electric drive systems, analysis and control of electric energy systems.

James V. Krogmeier, Assistant Professor. Multidimensional signal processing, digital communications, equalization, adaptive filtering.

David A. Landgrebe, Professor. Signal representation, pattern recognition applications, image data processing, remote sensing systems.

C. S. George Lee, Professor. Neural fuzzy systems, robotics, assembly systems.

James S. Lehnert, Professor. Communication theory, information theory, spread spectrum signaling, packet radio systems, fading communication channels, channel equalization techniques, signal design and coding.

Mark S. Lundstrom, Professor. Semiconductor device physics, transport in ultrasmall devices, modeling and simulation.

Anthony A. Maciejewski, Associate Professor. Failure-tolerant and redundant robotic systems, image synthesis and computer graphic simulation.

Michael R. Melloch, Professor. Semiconductor physics, molecular beam epitaxy, heterostructures, superlattices, ultrasmall devices.

David G. Meyer, Associate Professor. Information technology, distance education, effectiveness of student learning in nontraditional environments.

Frederic J. Mowle, Professor. Data structures, software engineering, software testing, software metrics, testing tools, software management.

Gerold W. Neudeck, Professor. Semiconductor devices, silicon fabrication processes, advanced MOSFET, epitaxial lateral overgrowth, SOI devices, Si-Ge growth and advanced devices.

John A. Nyenhuis, Professor. Magnetic materials, magnetic sensors, magnetic bubbles, magnetic stimulation, electromagnetic calculations.

Lawrence L. Ogborn, Associate Professor. Electronic circuits, device applications and limitations, instrumentation, power electronics, circuit theory.

Chee-Mun Ong, Professor. Operation and control of electric power systems and electric machines.

Robert F. Pierret, Professor and Assistant Head for Facilities and Planning. Measurement of parameters characterizing solid-state materials and devices, metal-oxide/insulator-semiconductor (MOS and MIS) devices, semiconductor device physics.

Kaushik Roy, Assistant Professor. Low-power VLSI for portable computing and wireless communications, accurate power estimation techniques, low voltage design, VLSI signal processing, reconfigurable computing, field programmable gate arrays, VLSI testing and verification.

Richard J. Schwartz, Professor and Dean of the Schools of Engineering. Semiconductor devices, direct energy conversion, solar cells.

Ness B. Shroff, Assistant Professor. High-speed communication networks (B-ISDN and ATM), mobile communications, multimedia applications, network management, traffic management, performance modeling, congestion control, queuing theory, load balancing and routing, optimization.

H. J. Siegel, Professor. Parallel processing, interconnection networks, distributed processing, heterogenous computing, heterogeneous networks.

LeRoy F. Silva, Ball Brothers Professor of Engineering and Director of the Business and Industrial Development Center.

Scott D. Sudhoff, Associate Professor. Power electronics, energy conversion, finite inertia power systems.

Philip H. Swain, Professor, Director of the Center for Lifelong Learning, and Assistant Executive Vice President for Academic Affairs. Pattern recognition, image processing, distance learning, adult education.

Thomas M. Talavage, Assistant Professor. Functional neuroimaging, biomedical image and signal processing, audition, speech, neural prostheses.

Hong Tan, Assistant Professor. Contact-based human-machine interfaces, wearable HCI, haptic navigation displays, psychophysics.

T. N. Vijaykumar, Assistant Professor. Computer architecture, VLSI microarchitectures, instruction-level parallelism, processor and memory hierarchy hardware, compiler optimization.

Oleg Wasynczuk, Professor. Power systems, power system modeling and control, solid-state power conversion.

Kevin J. Webb, Associate Professor. Numerical electromagnetics, quantum electronic devices, microwave and optical measurements.

Andrew M. Weiner, Professor. Femtosecond optics, ultrafast photonics, ultrafast nonlinear optics, femtosecond pulse shaping, high-speed fiber communications, time-resolved spectroscopy in semiconductors, ultrafast laser control of quantum processes, imaging in scattering media.

George R. Wodicka, Associate Professor. Biomedical acoustics, biomedical signal processing, medical instrumentation, active noise reduction.

Stanislaw H. Zak, Associate Professor. Nonlinear systems, chaos, neural networks, fuzzy logic, optimization, dynamical system control.

Michael D. Zoltowski, Associate Professor. Space-time adaptive processing, mobile and wireless communications, adaptive antennas for GPS, spread spectrum (CDMA) communications, narrowband (TDMA) digital communications.

RENSSELAER POLYTECHNIC INSTITUTE

Department of Electrical, Computer, and Systems Engineering

Programs of Study
The department offers the M.Eng., M.S., D.E., and Ph.D. degrees in two curricula, electrical engineering and computer and systems engineering, which cover a wide range of disciplines and are sufficiently flexible to accommodate individual interests. The M.S. degree is for those interested in research careers and who may also want to pursue a Ph.D. The M.S. requires 30 credits, including a 6-credit thesis, and generally takes three academic semesters to complete. The M.Eng. degree is designed for students interested in becoming practicing professional engineers and also requires 30 credits, but no thesis, and can be completed in two academic semesters. The department jointly offers a 72-credit dual M.B.A./M.Eng. degree program with the Lally School of Management and Technology; this program can be completed in two calendar years.

Research Facilities
Much of the research in the department is conducted in the extensive facilities associated with Rensselaer's multidisciplinary research centers, including the Center for Integrated Electronics and Electronics Manufacturing; the New York State Center for Advanced Technology in Automation, Robotics and Manufacturing; the Center for Image Processing Research; the Scientific Computation Research Center; and the International Center for Multimedia Education. In addition, department facilities include the Speech Processing Laboratory, the Document Analysis and Geometric Modeling Laboratory, the Networking Laboratory, the Neural Net Computing Laboratory, the Advanced Imaging Systems Laboratory, the Solid State Microwave Laboratory, the Electronic Materials Laboratory, the Device Fabrication and Testing Laboratory, the Device Characterization Laboratory, and the Plasma Dynamics Laboratory. Research is also supported by Computing and Information Services, which operates and supports a sophisticated campuswide computing, information, and networking environment that includes campus site licenses for software, laptop hookups, desktop devices, advanced workstations, a visualization laboratory for scientific computation, a numerically intensive computing cluster, a 36-node SP2 parallel computer, and the Rensselaer Libraries, where modern library systems allow online access to collections, databases, and Internet resources.

Financial Aid
Financial aid is available in the form of fellowships, research or teaching assistantships, and scholarships. The stipend for graduate assistantships ranges from approximately $10,700 to $11,400 for the 1999–2000 academic year. In addition, full tuition is granted. Additional compensation for study in the summer months is usually available. Outstanding students may qualify for either industrial, Graduate School, or Rensselaer Scholar fellowships. These awards provide stipends of up to $15,000 and a full tuition and fees scholarship for the nine-month academic year. Low-interest, deferred-repayment graduate loans are also available to U.S. citizens with demonstrated need.

Cost of Study
Tuition for 1999–2000 is $665 per credit hour. Other fees amount to approximately $535 per semester. Books and supplies cost about $1700 per year.

Living and Housing Costs
The cost of rooms for single students in residence halls or apartments ranges from $3356 to $5298 for the 1999–2000 academic year. Family student housing, with a monthly rent of $592 to $720, is available.

Student Group
There are about 4,300 undergraduates and 1,750 graduate students representing all fifty states and more than eighty countries at Rensselaer.

Student Outcomes
Eighty-eight percent of the electrical, computer, and systems engineering department's 1998 graduate students were hired after graduation, with starting salaries that averaged $56,259 for master's degree recipients and $57,000–$75,000 for doctoral degree recipients.

Location
Rensselaer is situated on a scenic 260-acre hillside campus in Troy, New York, across the Hudson River from the state capital of Albany. Troy's central Northeast location provides students with a supportive, active, medium-sized community in which to live and an easy commute to Boston, New York, and Montreal and some of the country's finest outdoor recreation, including Lake George, Lake Placid, and the Adirondack, Catskill, Berkshire, and Green Mountains. The Capital Region has one of the largest concentrations of academic institutions in the United States. Sixty thousand students attend fourteen area colleges and benefit from shared activities and courses.

The University
Founded in 1824 and the first American college to award degrees in engineering and science, Rensselaer Polytechnic Institute today is accredited by the Middle States Association of Colleges and Schools and is a private, nonsectarian, coeducational university. Rensselaer has five schools—Architecture, Engineering, Management, Science, and Humanities and Social Sciences. The School of Engineering is ranked among the top twenty engineering schools nationally by the *U.S. News & World Report* survey and is ranked in the top ten by practicing engineers.

Applying
Admissions applications and all supporting credentials should be submitted well in advance of the preferred semester of entry to allow sufficient time for departmental review and processing. Scores on the General Test of the Graduate Record Examinations are required of all applicants. The application fee is $35. Due to the high level of competition for awards, the department requires that both U.S. and international applicants requesting financial aid submit all credentials by February 1 for fall and by October 1 for spring.

Correspondence and Information

For written information about graduate study:
Manager of Graduate Admissions
  and Financial Aid
Department of Electrical, Computer,
  and Systems Engineering
Rensselaer Polytechnic Institute
110 8th Street
Troy, New York 12180-3590
Telephone: 518-276-6225
E-mail: grad-info@ecse.rpi.edu
World Wide Web: http://www.ecse.rpi.edu

For applications and admissions information:
Director of Graduate Academic and Enrollment
  Services, Graduate Center
Rensselaer Polytechnic Institute
110 8th Street
Troy, New York 12180-3590
Telephone: 518-276-6789
E-mail: grad-services@rpi.edu
World Wide Web: http://www.rpi.edu

Rensselaer Polytechnic Institute

THE FACULTY AND THEIR RESEARCH

John B. Anderson, Professor; Ph.D., Cornell. Communications and coding theory.

Ishwara B. Bhat, Associate Professor; Ph.D., Rensselaer. Solid state, electronic materials.

A. Bruce Carlson, Professor and Curriculum Chair; Ph.D., Stanford. Communications systems, educational methods, social context of engineering.

Joseph H. Chow, Professor; Ph.D., Illinois. Large-scale modeling, multivariable control systems.

T. Paul Chow, Associate Professor; Ph.D., Rensselaer. Semiconductor device physics and processing technology, integrated circuits.

Kenneth A. Connor, Professor; Ph.D., Polytechnic of New York. Electromagnetic theory, wave propagation, plasmas for fusion research.

Thomas P. Crowley, Associate Professor; Ph.D., Princeton. Fusion and space plasmas, wave propagation, microwaves.

Alan A. Desrochers, Professor; Ph.D., Purdue. Nonlinear systems, robotics, control of automated manufacturing systems.

W. Randolph Franklin, Associate Professor; Ph.D., Harvard. Computational geometry, graphics and CAD, cartography, parallel algorithms, large databases, expert system verification.

Lester A. Gerhardt, Professor and Associate Dean of Engineering; Ph.D., SUNY at Buffalo. Communication systems, digital voice and image processing, adaptive systems and pattern recognition, computer-integrated manufacturing.

Ronald J. Gutmann, Professor; Ph.D., Rensselaer. Semiconductor devices, microwave monolithic integrated circuits, microwave nondestructive testing techniques, interconnect technology.

Timothy J. Holmes, Associate Professor of Biomedical Engineering and Affiliated Faculty of Electrical, Computer, and Systems Engineering; Ph.D., Washington (St. Louis). Internal imaging, signal processing and systems.

William C. Jennings, Professor; Ph.D., Rensselaer. Plasmas, gas lasers, microwaves.

Chuanyi Ji, Assistant Professor; Ph.D., Caltech. Neural networks, communications and information processing, pattern recognition.

Shivkumar Kalyanaraman, Assistant Professor; Ph.D., Ohio State. Computer networking, with an emphasis on traffic management, pricing, network management, internetworking, conformance, and interoperability.

Robert B. Kelley, Professor; Ph.D., UCLA. Robotic systems, machine intelligence, machine vision, expert systems.

Yannick L. LeCoz, Associate Professor; Ph.D., MIT. Solid-state devices.

Edward W. Maby, Associate Professor; Sc.D., MIT. Solid-state electronics, semiconductor-device simulation, circuit implementation of parallel algorithms.

John F. McDonald, Professor; Ph.D., Yale. Communication theory, coding and switching theory, computer architecture, digital signal processing.

James W. Modestino, Institute Professor and Director of the Center for Imaging Processing Research; Ph.D., Princeton. Information theory and coding, communications and signal processing, image processing and computer vision.

George Nagy, Professor; Ph.D., Cornell. Pattern recognition, computer vision, computational geometry, solid modeling, knowledge-based systems, digitizing camera performance.

William A. Pearlman, Professor; Ph.D., Stanford. Information theory and source coding, rate-distortion theory, image and optical data restoration, image coding, communication theory.

Kenneth Rose, Professor; Ph.D., Illinois. Semiconductor and superconductor materials and processing, VLSI design and testing.

Badrinath Roysam, Associate Professor; Ph.D., Washington (St. Louis). Massively parallel computation, joint symbolic/stochastic inference, hierarchical imaging and image processing.

Arthur C. Sanderson, Professor; Ph.D., Carnegie Mellon. Robotics, knowledge-based systems, computer vision.

Gary J. Saulnier, Associate Professor; Ph.D., Rensselaer. Circuits and electronics, communication systems, digital signal processing.

Michael Savic, Professor; Eng.Sc.D., Belgrade. Digital processing of speech and other signals, hardware and software implementation of fast signal processors, analog and digital electronics.

Paul M. Schoch, Associate Professor; Ph.D., Rensselaer. Plasma diagnostics, power electronics.

Michael S. Shur, Professor and Patricia W. and C. Sheldon Roberts Professor of Solid State Electronics; Ph.D., Ioffe (Russia). Semiconductor materials and devices, integrated circuit simulation, characterization and design.

Michael M. Skolnick, Associate Professor of Computer Science and Affiliated Faculty of Electrical, Computer, and Systems Engineering; Ph.D., Michigan. Image processing, computer vision, learning algorithms.

Harry E. Stephanou, Associate Professor and Director, Center for Advanced Technology in Automation and Robotics; Ph.D., Purdue. Robotics and automation, pattern recognition, neural networks.

James M. Tien, Professor and Chair of Decision Sciences and Engineering Systems and Affiliated Faculty of Electrical, Computer, and Systems Engineering; Ph.D., MIT. Information and decision systems methodology and applications.

David A. Torrey, Associate Professor of Electric Power Engineering and Affiliated Faculty of Electrical, Computer, and Systems Engineering; Ph.D., MIT. Power electronics, electric machine drives, and motion control.

Kenneth S. Vastola, Associate Professor; Ph.D., Illinois. Computer-communication networks, statistical signal processing and communications.

John T. Wen, Associate Professor; Ph.D., Rensselaer. Robot control, control of flexible structures, adaptive control.

John W. Woods, Professor; Ph.D., MIT. Digital signal processing; image processing, estimation theory.

Michael J. Wozny, Professor; Ph.D., Arizona. Computer graphics, computer-aided design, digital simulation, rapid prototyping systems.

Milos Zefran, Assistant Professor; Ph.D., Pennsylvania. Robotics, control of hybrid systems, control of mechanical systems.

RUTGERS, THE STATE UNIVERSITY OF NEW JERSEY, NEW BRUNSWICK

Department of Electrical and Computer Engineering

Program of Study

The graduate program in electrical and computer engineering, which leads to the M.S. and Ph.D. degrees, has facilities for education and research in computer engineering, integrated systems encompassing control systems, digital signal processing, and communications and solid-state electronics. Computer engineering involves the architecture and design of computing machines, information processing, and software engineering. Control systems is concerned with the design, analysis, simulation, and mathematical modeling of systems to ensure that an automatic process, such as that of a robot or spacecraft, meets and maintains certain criteria. Digital signal processing deals with discrete-time information processing, digital filter design, spectral analysis, and special-purpose signal processors. Electrical communications systems analysis and design is concentrated in the areas of source and channel encoding, analog and digital modulation methods, information theory, and telecommunication networks. Solid-state electronics encompasses the areas of microwave switching devices, semiconductor lasers, electrooptical modulation, solar cells, integrated circuits, and characterization of semiconductor materials and devices.

Master of Science degree candidates may elect either a thesis or nonthesis option. The thesis option consists of 24 credits of course work, 6 credits of research in a specialized area, and a final thesis presentation. In the nonthesis option, a candidate must complete 30 credits of course work, pass a written comprehensive examination, and submit a satisfactory tutorial paper in a course. Requirements for the M.S. degree may be satisfied for all options in a part-time evening program designed specifically for students employed in industry and other students whose obligations preclude full-time study.

Admission into the Ph.D. program requires an M.S. in electrical and/or computer engineering. Applicants having an M.S. in a closely related discipline may be admitted into the program provided their preparation has no significant deficiencies. A student is considered to be a Ph.D. candidate after satisfactory completion of the qualifying exam and presentation of the dissertation topic. A Ph.D. candidate, in conjunction with an adviser, is required to select a dissertation committee, submit a plan of study, and orally present a dissertation proposal. Minimum requirements for the Ph.D. degree include 48 credits beyond the baccalaureate in courses approved by the dissertation adviser and 24 credits of dissertation research beyond the M.S. degree. A public defense serves as the final Ph.D. dissertation exam. There is no foreign language requirement.

Research Facilities

There are three centers within the department: the Center for Digital Signal Processing, the Wireless Information Networks Laboratory (WINLAB), and the Microelectronics Research Laboratory (MERL), a semiconductor device processing clean room. Additional research is conducted with various Rutgers research centers established by the New Jersey Commission of Science and Technology, particularly the Center for Computer Aids for Industrial Productivity (CAIP) and the Fiber Optic Materials Research Program (FOMRP). In addition to the extensive facilities available at these centers, the department maintains the Digital Communications and Image Transmission Laboratory, Digital Signal Processing Laboratory, Digital Signal Processing Systems Laboratory, Digital Signal Processing Research and Graduate Instruction Laboratory, Local Area Network Laboratory, Machine Vision Laboratory, Robotics and Sensorics Laboratory, VLSI CAD Laboratory, Laboratory for Engineering Information Systems, and solid-state experimental facilities for crystal growth and preparation, electrical characterization, design and fabrication, optical measurements, structure and composition analysis, and thin-film deposition measurements. For detailed information on these and other facilities available to graduate students, contact the Graduate Director at the address given below.

Financial Aid

Assistantships and fellowships are available, each with a typical stipend of about $12,100 plus tuition for the academic year 1999–2000. In the award of financial aid, consideration is given to the student's undergraduate academic record, performance on the GRE, and letters of reference indicating outstanding ability.

Cost of Study

Tuition is $5900 for state residents and $8600 for out-of-state residents for the academic year 1999–2000. Fees, books, and supplies are about $1600 per year.

Living and Housing Costs

Graduate housing is available. Graduate students in the Department of Electrical and Computer Engineering may also reside in nearby rooms or apartments. Various board plans are available.

Student Group

The department has about 210 graduate students.

Location

New Brunswick (population 42,000), is in central New Jersey off Exit 9 of the New Jersey Turnpike and along the New York–Philadelphia railroad line. It is about 33 miles from New York City; frequent express bus service is available from a station near the College Avenue campus to terminals in central Manhattan. Princeton is 16 miles south, Philadelphia about 60 miles south, and Washington, D.C., within 200 miles.

The University

Rutgers, The State University of New Jersey, with more than 47,000 students on three campuses in Camden, Newark, and New Brunswick, is one of the major state university systems in the nation. The University comprises twenty-five degree-granting divisions: thirteen undergraduate colleges, eleven graduate schools, and one school offering both undergraduate and graduate degrees. Four are located in Camden, seven in Newark, and fourteen in New Brunswick. Rutgers has a unique history as a Colonial college, a land-grant institution, and a state university. Chartered in 1766 as Queen's College, the eighth institution of higher learning to be founded in the colonies before the Revolution, the school opened its doors in New Brunswick in 1771 with one instructor, one sophomore, and a handful of freshmen. In 1825, the name of the college was changed to Rutgers to honor a former trustee and revolutionary war veteran, Colonel Henry Rutgers.

Applying

Admission materials are available from the Office of Graduate and Professional Admissions, Van Nest Hall, Rutgers, The State University of New Jersey, New Brunswick, NJ 08903 (732-932-7711). A complete application consists of the application form, letters of recommendation, the application fee, official transcripts of previous academic work, a personal statement or essay, and scores on the GRE General Test. Detailed procedures and instructions accompany the application forms.

Correspondence and Information

Graduate Director
Department of Electrical and Computer Engineering
Rutgers University GPECE
94 Brett Road
Piscataway, New Jersey 08854
Telephone: 732-445-2578

Rutgers, The State University of New Jersey, New Brunswick

THE FACULTY AND THEIR RESEARCH

Grigore Burdea, Associate Professor of Electrical and Computer Engineering; Ph.D., NYU. Robotic systems, computer engineering.

Michael L. Bushnell, Associate Professor of Electrical and Computer Engineering; Ph.D., Carnegie Mellon. Computer engineering: computer-aided design of VLSI integrated circuits, silicon compilers, artificial intelligence techniques.

Michael F. Caggiano, Assistant Professor of Electrical and Computer Engineering; Ph.D., UCLA. High-performance and microwave IC device packaging.

David G. Daut, Professor of Electrical and Computer Engineering; Ph.D., RPI. Communications and information processing: digital communication system design and analysis, image coding and transmission.

James Flanagan, Board of Governors' Professor of Electrical and Computer Engineering and Director of the Center for Computer Aids for Industrial Productivity; Eng.Sc.D., MIT. Digital communications; speech processing; coding, recognition, and synthesis; acoustic systems; robotics; artificial intelligence; human/computer interfaces.

Herbert Freeman, Professor of Electrical and Computer Engineering; Eng.Sc.D., Columbia. Computer engineering: digital computer systems, computer architecture, image processing and graphics.

Zoran R. Gajic, Associate Professor of Electrical and Computer Engineering; Ph.D., Michigan State. Systems and controls: singular perturbation methods in control system analysis, linear stochastic estimation.

David J. Goodman, Professor of Electrical and Computer Engineering and Director of the Wireless Information Networks Laboratory; Ph.D., Imperial College, London. Communication systems, wireless access radio and systems.

Michael S. Hsiao, Assistant Professor of Electrical and Computer Engineering; Ph.D., Illinois. Computer engineering: VLSI circuits, automatic test pattern generation for combinational and sequential circuits, fault simulation, computer architecture.

Joseph Hui, Professor of Electrical and Computer Engineering; Ph.D., MIT. Communications networks: integrated broadband networks, switching and traffic theory, information and coding theory, parallel processing and adaptive computing.

Bogoljub Lalevic, Professor of Electrical and Computer Engineering; Ph.D., Temple. Solid-state electronics: gaseous and chemical semiconducting device sensors, high-power and microwave switching devices.

Yicheng Lu, Associate Professor of Electrical and Computer Engineering; Ph.D., Colorado. Semiconductor materials (GaAs and Si), metal-semiconductor contacts, device physics and fabrication.

Richard Mammone, Professor of Electrical and Computer Engineering; Ph.D., CUNY Graduate Center. Digital signal processing: image restoration, speech recognition, medical imaging.

Narayan B. Mandayam, Assistant Professor of Electrical and Computer Engineering; Ph.D., Rice. Communication theory, spread spectrum, wireless systems, multi-access protocols.

Thomas G. Marshall, Professor of Electrical and Computer Engineering and Director of the Center for Digital Signal Processing; Ph.D., Chalmers (Sweden). Digital signal processing: algorithms and specialized signal processing computers.

Ivan Marsic, Assistant Professor of Electrical and Computer Engineering; Ph.D., Rutgers. Computer engineering: distributed systems for collaborative/information processing and learning, image reconstruction, machine vision.

Sigrid R. McAfee, Associate Professor of Electrical and Computer Engineering; Ph.D., Polytechnic of New York. Solid-state electronics: deep levels in semiconductors, molecular beam epitaxy and MO-CVD gallium arsenide, AlGaAs and GaAs on silicon.

Peter Meer, Associate Professor of Electrical and Computer Engineering; D.Sc., Technion (Israel). Computer vision, pattern recognition, applied robust estimation, probabilistic algorithms for machine-vision problems.

Andrew Ogielski, Research Professor, DIMACS Center; Ph.D., Wroclaw (Poland). Modeling and distributed simulations of very large telecommunications systems.

Sophocles J. Orfanidis, Associate Professor of Electrical and Computer Engineering; Ph.D., Yale. Digital signal processing: signal estimation and modeling methods, adaptive signal processing and spectrum estimation.

Paul Panayotatos, Professor of Electrical and Computer Engineering; Eng.Sc.D., Columbia. Organic semiconductor solar cells, optical interconnects, microelectromechanical devices.

Manish Parashar, Assistant Professor of Electrical and Computer Engineering; Ph.D., Syracuse. Computer engineering: parallel and distributed computing, software engineering.

Narindra N. Puri, Professor of Electrical and Computer Engineering; Ph.D., Pennsylvania. Systems and controls: optimal adaptive control systems.

Christopher Rose, Associate Professor of Electrical and Computer Engineering; Ph.D., MIT. Complex systems (neural networks, communication networks), neurophysiology, communication network topologies, novel applications of superconducting materials.

Peddapullaiah Sannuti, Professor of Electrical and Computer Engineering; Ph.D., Illinois. Communication and control systems: singular perturbation analysis of Kalman filter with weak measurement noise.

Deborah Silver, Associate Professor of Electrical and Computer Engineering; Ph.D., Princeton. Computer graphics, computational geometry, numerical analysis.

Joseph Wilder, Research Professor of Electrical and Computer Engineering; Ph.D., Pennsylvania. Image processing, pattern recognition, machine vision.

Roy Yates, Associate Professor of Electrical and Computer Engineering; Ph.D., MIT. Routing and flow control for integrated broadband networks, reversible queuing systems, wireless cellular communication systems.

Jian Zhao, Associate Professor of Electrical and Computer Engineering; Ph.D., Carnegie Mellon. Heterojunctions and their optoelectronic device applications, computer modeling of III-V devices, deep traps in semiconductors, electromigration.

SOUTHERN METHODIST UNIVERSITY

Department of Electrical Engineering

Programs of Study

The department offers M.S. and Ph.D. degrees. The SMU electrical engineering department emphasizes the following major areas of interest: biomedical engineering (biomedical devices and instrumentation and biomedical signal capture, processing, and modeling); communications and information theory (detection and estimation theory, digital communications, spread spectrum, cellular communications, coding, encryption, radar/sonar, optical communications, and information theory); control systems and robotics (linear and nonlinear systems, robotics, and computer and robot vision); digital signal processing (digital filter design, system identification, spectral estimation, adaptive filters, and neural networks); image processing and computer vision (digital image processing, computer vision, and pattern recognition); lasers, optoelectronics, electromagnetic theory, and microwave electronics (classical optics, fiber optics, laser recording, integrated optics, dielectric waveguides, antennas, transmission lines, laser diodes, signal processors, and superconductive microwave and optoelectronic devices); solid-state circuits, computer-aided circuit design, and VLSI design (electronic circuits, computer-aided design, VLSI design, neural network implementation, parallel array architectures, and memory interfaces); electronic materials and solid-state devices (fabrication and characterization of devices and materials, device physics, microelectromechanical systems, noise in solid-state devices, infrared detectors, AlGaAs and GaAs devices and materials, thin films, superconductivity, superconductive devices and electronics, hybrid superconductor-semiconductor devices, ultrafast electronics, and applications of the scanning tunneling microscope); and telecommunications (telecommunication components and systems, data communications, digital telephony, and digital switching).

Research Facilities

The Biomedical Engineering Laboratory is equipped for the study of problems in biomedical engineering. The Cryoelectronics Laboratory includes Dewars, refrigerators, temperature controllers and sensors, computer workstations, and fiber-optic instrumentation to support low-temperature device characterization and superconductivity research. The Digital Signal Processing Laboratory is equipped with PC-based DSP workstations. The Image Processing and Analysis Laboratory incorporates two Recognition Concepts TRAPIX 5500 series real-time color image processing systems. The Microwave Electronics Laboratory has high-frequency oscilloscopes, synthesizers, power meters, generators, analyzers, and computer stations for design and analysis of microwave circuits. The Optical and Millimeter Wave Electronics Laboratory consists of lasers, optical equipment, millimeter-wave active and passive components, and system controllers. Major apparatus in the Robotics Laboratory are used for design and development of robot arm, robot hand, and vision systems. The Solid-State Device Characterization Laboratory is used for the computerized I-V, C-V, and noise characterization of devices. The lab contains a shielded room and various programmable and nonprogrammable instruments. The Solid-State Fabrication Laboratory consists of a class-10,000 clean room with class-100 laminar-flow work areas. It has photolithography, deposition, etching, and optical inspection equipment for the fabrication of solid-state devices.

Financial Aid

The Graduate Admissions Committee awards a limited number of merit-based research and teaching assistantships to incoming students, paying up to $1600 per month and covering tuition. Separate tuition assistantships are also available.

Cost of Study

Tuition and fees for graduate study in 1999–2000 are $600 per semester hour.

Living and Housing Costs

Dormitory housing charges each semester are approximately $1950 per person for double occupancy. Board on campus, including tax, costs $1510 per semester. Furnished efficiency and one- and two-bedroom apartments are available on campus, with utilities paid in some, at costs ranging from $2500 to $3000 per semester. The Dallas area offers inexpensive housing options within driving distance of the campus, with apartments starting from $500 per month.

Student Group

The department has about 50 doctoral students and 70 master's students. There are also postdoctoral research fellows. Approximately ten doctoral and twenty to thirty master's degrees are awarded each year. In addition, the department has about 300 students in the telecommunications program.

Student Outcomes

Recent graduates have obtained appointments at prestigious universities and industrial research positions at firms such as Texas Instruments.

Location

Dallas is the center of an attractive metropolitan area of 2 million people. It has fine parks, lakes, museums, theaters, orchestras, libraries, and places of worship. Clean and progressive, the city continues to grow as a center of business and light industry. The manufacturing, microelectronics, and telecommunications industries provide many opportunities for employment.

The University and The School

Southern Methodist University is a private, nonprofit coeducational institution located in suburban University Park, an incorporated residential district surrounded by Dallas, Texas. The School of Engineering and Applied Science (SEAS) traces its roots to 1925, when the Technical Club of Dallas, a professional organization of practicing engineers, petitioned SMU to fulfill the need for an engineering school in the Southwest.

Applying

Students may apply for admission at any time. However, initial review for admission in a given semester is dependent upon receipt by the Graduate Division of all requisite application materials no later than July 1 for fall admission, November 15 for spring admission, and April 15 for summer admission. International students should use the following dates: May 15 for fall admission, September 1 for spring admission, and February 1 for summer admission. GRE General Test scores are required.

Correspondence and Information

Graduate Admissions
School of Engineering and Applied Science
Southern Methodist University
Dallas, Texas 75275-0335

Telephone: 214-768-3900
World Wide Web: http://www.seas.smu.edu/

Southern Methodist University

THE FACULTY AND THEIR RESEARCH

The following is a partial list of faculty members affiliated with the Department of Electrical Engineering:

Kenneth L. Ashley, Professor; Ph.D., Carnegie Mellon; PE. Semiconductor optoelectronic devices, gallium arsenide circuits.

Donald P. Butler, Associate Professor; Ph.D., Rochester. Infrared detectors, microelectromechanics, optoelectronic and microwave devices, superconductive devices.

Jerome K. Butler, University Distinguished Professor; Ph.D., Kansas; PE. Integrated optical communications, solid-state injection lasers, surface-emitting lasers.

Zeynep Çelik-Butler, Associate Professor; Ph.D., Rochester. Infrared detectors, microelectromechanics, noise modeling of solid-state devices, high-T_c superconductors.

Thomas M. Chen, Associate Professor; Ph.D., Berkeley. ATM networks, active and programmable networks, Internet monitoring and management, traffic modeling and control.

Carlos E. Davila, Associate Professor; Ph.D., Texas at Austin. Adaptive signal processing, spectral estimation, system identification.

Scott C. Douglas, Associate Professor; Ph.D., Stanford. Adaptive filters, active noise control, digital signal processing systems, blind deconvolution and source separation.

James G. Dunham, Associate Professor; Ph.D., Stanford; PE. Data compression, cryptography, information and telecommunications theory.

Gary Evans, Professor; Ph.D., Caltech; PE. Design, fabrication, and analysis of surface-emitting lasers.

Robert R. Fossum, Professor; Ph.D., Oregon State. Communications.

Jerry D. Gibson, Professor and Chairman; Ph.D., SMU. Data, speech, image, and video compression; multimedia over networks; wireless communications; information theory.

W. Milton Gosney, Cecil and Ida Green Professor; Ph.D., Berkeley; PE. VLSI circuits and biomedical applications, scanning tunneling microscopy.

Someshwar C. Gupta, Cecil H. Green Professor; Ph.D., Berkeley; PE. Cellular, navigational, and personal communications.

Alireza Khotanzad, Associate Professor; Ph.D., Purdue; PE. Computer vision and pattern recognition, applications of neural networks.

Choon S. Lee, Associate Professor; Ph.D., Illinois at Urbana-Champaign. Electromagnetic scattering, reflector and microstrip antennas, millimeter-wave applications.

Geoffrey C. Orsak, Associate Professor; Ph.D., Rice. Communication theory, detection and estimation theory, information theory, statistical signal processing.

Behrouz Peikari, Professor; Ph.D., Berkeley; PE. Nonlinear circuits, filter design, image processing, robotics.

Mandyam D. Srinath, Professor; Ph.D., Illinois at Urbana-Champaign; PE. Adaptive filters, shape classification, neural networks.

André G. Vacroux, Professor; Ph.D., Purdue; PE. Microcomputer networks, switched networks, ISDN, biomedical systems.

Emeritus Professors

Kenneth W. Heizer, Ph.D., Illinois at Urbana-Champaign; PE.

Lorn L. Howard, Ph.D., Michigan State; PE. Neutral electronics and circuits.

John A. Savage, M.S.E.E., Texas at Austin; PE.

STATE UNIVERSITY OF NEW YORK AT BINGHAMTON

Thomas J. Watson School of Engineering and Applied Science
Department of Electrical Engineering

Programs of Study	The Department of Electrical Engineering in the Thomas J. Watson School of Engineering and Applied Science offers the B.S., M.S., M.Eng., and Ph.D. degrees. Course work for graduate degrees is available through EngiNet, the Watson School system for distance learning.
	The M.S.E.E. requires the student to complete either of two optional courses of study. The thesis option requires seven lecture courses plus a thesis and a seminar. The project option requires nine lecture courses plus a project and a seminar. The lecture courses selected must meet breadth and depth requirements. M.S.E.E. students specialize in one of the research areas listed below. The normal period for completion of a master's degree is 1½ years of full-time study.
	The M.Eng., with specialization in electrical engineering, requires four graduate courses in electrical engineering, four approved technical electives, and a two-course sequence in engineering design and development. One year of full-time study is required for completion of the degree program.
	Requirements for the Ph.D. degree are based on individual learning contracts and normally take three years beyond the master's degree. There is a broad preliminary examination and a comprehensive examination followed by the satisfactory defense of a dissertation. Residence requirement is 24 credit hours.
Research Facilities	The Watson School has excellent access to advanced computing facilities. An IBM 9121 linked to a network of Sun workstations and advanced PCs, as well as to the University computer networks and mainframes.
	Since the Watson School was first established in 1983, it has developed an international reputation in the multidisciplinary research speciality of electronics packaging. This research is housed in the Watson School's Integrated Electronics Engineering Center (IEEC). The IEEC is also a designated National Science Foundation state/industry/university cooperative research center and in 1993, it became a New York State Center for Advanced Technology (CAT). The IEEC also administers equipment that supports both local industry and University research.
	The Science and Engineering Library is part of the University library system, which has a total collection of more than 2 million items. Online access to other SUNY collections exists through the campus computer network. Resources are supplemented by membership in academic library consortia, notably the Research Libraries Group, Inc.
Financial Aid	Many students hold fellowships, traineeships, or assistantships (graduate, research, or teaching). Most awards include a full waiver of tuition and medical benefits. Other sources of financial aid include the New York State Tuition Assistance Program, the Federal Stafford Student Loan Program, the graduate and professional school College Work-Study Program, and campus jobs.
Cost of Study	For full-time matriculated graduate students, tuition in 1998–99 was $2550 per semester for state residents and $4208 per semester for nonresidents.
Living and Housing Costs	The campus graduate community housing has 3- and 4-person apartments, with living room, dining area, kitchen, and bath. Based on a 1998–99 academic-year lease, the semester rate for a single bedroom was $2050, and for a double bedroom, $1775 per person and $3085 per family apartment. The cost of meal plans per semester was as follows: basic, $787; standard, $1022; and ultra, $1097. Assistance in locating cheaper off-campus housing is provided by the listing services of the Off-Campus College.
Student Group	Of the 12,259 students enrolled at Binghamton University, 2,700 are graduate students. In the Watson School, there are 876 undergraduates and 368 graduate students.
	The Department of Electrical Engineering enrolled its first freshmen in 1995, having offered only upper-division undergraduate programs previously. Current approximate enrollments of 40 juniors, 40 seniors, 30 full-time and 35 part-time master's students, and 26 Ph.D. students permit the individual faculty attention for which the Department is known.
Location	The University's 606-acre campus is in a suburban setting just west of Binghamton, which is itself conveniently situated among the major cities of the Northeast: 200 miles from New York, Philadelphia, and Buffalo and 300 miles from Toronto, Boston, Montreal, Pittsburgh, and Washington, D.C. More than 300,000 people live within commuting distance of the campus. Cultural offerings in the community include the museum and programs of the Roberson Center for the Arts and Sciences, as well as performances by the Binghamton Symphony Tri-Cities Opera, Cider Mill Theater, and other groups.
The University and The School	The State University of New York at Binghamton is one of the four university centers in the 64-campus State University of New York system. The faculty numbers about 700. Graduate programs were initiated in 1961 with the establishment of Master of Arts programs in English and mathematics.
	The University's Art Gallery has a permanent collection representing all periods and also displays works from special loan exhibitions. The Anderson Center's annual concert series brings a wide variety of performing artists to campus. The Department of Theater stages more than twenty-five productions each year.
	The Watson School was created in 1983 by combining the established graduate programs in computer science and systems science from the School of Advanced Technology with new graduate programs in electrical, industrial, and mechanical engineering. The first electrical engineering juniors were admitted in 1984.
Applying	Holders of a bachelor's degree in electrical engineering or related discipline from any recognized college or university are eligible to apply. Requests for application forms can be sent via e-mail to the Office of Graduate Admissions at gradad@binghamton.edu. Applicants should submit GRE General Test scores. (Graduates of ABET–accredited engineering programs may apply for waiver of this requirement.) International applicants must submit TOEFL scores and provide proof of their ability to meet academic expenses. All credentials should be on file at least one month prior to anticipated enrollment. To ensure consideration for assistantship and fellowship awards, admission credentials should be received by February 15. Online application is available at http://www.gradschool.binghamton.edu.
Correspondence and Information	Director of Graduate Studies Department of Electrical Engineering Thomas J. Watson School of Engineering and Applied Science State University of New York at Binghamton Binghamton, New York 13902-6000 Telephone: 607-777-4856 Fax: 607-777-4464

State University of New York at Binghamton

THE FACULTY

Craig Bergman, Lecturer; M.S., Illinois at Urbana-Champaign, 1975. Digital design, microprocessors, human factors. (cbergman@binghamton.edu)

Nikolaos Bourbakis, Professor; Ph.D, Patras (Greece), 1983. Applied AI robotics, knowledge-based VLSI design, computer vision, text and image processing, multiprocessor system architectures, neural nets, automated software environment. (bourbaki@binghamton.edu)

Monish Chatterjee, Associate Professor; Ph.D, Iowa, 1985. Nonlinear wave phenomena, nonlinear modeling, quantum electronics, acoustooptics, fiber-optics and optical communications. (mrchat@binghamton.edu)

James Constable, Professor; Ph.D, Ohio State, 1969; PE. Instrumentation, cryogenics, electrical noise, contact resistance, electronics packaging. (constab@binghamton.edu)

Jose Delgado-Frias, Associate Professor; Ph.D, Texas A&M, 1986. Computer engineering, VLSI/WSI design, parallel computer architectures, reconfigurable computing, interconnection networks, novel computing paradigms. (delgado@binghamton.edu)

Mark Fowler, Assistant Professor; Ph.D., Penn State, 1991. Digital signal processing, video compression.

Harry Kroger, Professor; Ph.D., Cornell, 1962. Electronics packaging, physics and fabrication of superconductor and semiconductor devices, superconductor-semiconductor hybrid circuits. (hkroger@binghamton.edu)

James Morris, Professor; Ph.D, Saskatchewan, 1971; PE. Thick and thin films, semiconductor devices, electronics, engine sensors and control, electronics packaging, materials, engineering education. (jmorris@binghamton.edu)

Dhananjay Phatak, Assistant Professor; Ph.D, Massachusetts Amherst, 1994. Computer architectures, computer arithmetic, neural networks and applications. (phatak@ee.binghamton.edu)

Richard Plumb, Professor and Department Chair; Ph.D., Syracuse, 1988. Electromagnetics, ground-penetrating radar, scattering theory. (rplumb@binghamton.edu)

George Sackman, Professor; Ph.D, Stanford, 1964; PE. Signal processing, acoustic space-time array processing, digital audio, microwave electronics, engineering education. (gsackman@binghamton.edu)

Richard Schwartz, Professor Emeritus; Ph.D, Pennsylvania, 1959; PE. Microwave theory, antennas and propagation, acoustics, signal processing, engineering education. (schwartz@bingtjw.cc.binghamton.edu)

Victor Skormin, Professor; Ph.D, Moscow, 1975. Control engineering, operations research, computer simulation. (vskormin@binghamton.edu)

Douglas Summerville, Assistant Professor; Ph.D., SUNY at Binghamton, 1997. Computer engineering, parallel computer architectures, interconnection networks. (doug@parallel.ee.binghamton.edu)

Charles Taylor, Associate Professor; M.S., SUNY at Binghamton, 1970. Automatic controls, microprocessor applications, robotics. (ctaylor@binghamton.edu)

Peter Wagner, Professor; Ph.D, Berkeley, 1956. Semiconductor circuit elements, microwave resonance, surface electricity, applied optics. (pwagner@binghamton.edu)

N. Eva Wu, Associate Professor; Ph.D, Minnesota, 1987. Approximation, optimization, and stabilization of distributed parameter systems; robust control synthesis theory; control of robotic manipulators; signal processing. (evawu@binghamton.edu)

ADJUNCT FACULTY

Stephen Czarnecki, Loral Corporation; Ph.D, Princeton, 1983. Communications.

Mohammed Islam, IBM Corporation (Retired); Ph.D., Northeastern, 1964. Electromagnetics, microwaves.

John Pivnichny, IBM Corporation; Ph.D, Michigan, 1971. Control systems.

Theresa Sadeghi, M.S., Rensselaer, 1978. Control systems.

Charles Standish, IBM Corporation (Retired); Ph.D, Cornell, 1954. Systems theory.

RESEARCH AREAS OF SPECIALIZATION

Computer Engineering (Bergman, Bourbakis, Delgado-Frias, Phatak, Summerville, Taylor).

The unifying theme to the department's research in computer engineering is the study of computer architectures and digital systems. Novel computer concepts are first simulated to demonstrate potential capabilities and constructed for demonstration and verification of the special features of interest. Such machines developed in the department include a loosely coupled machine, RISC processors, multiprocessor vision system architectures, a synaptic connection machine, and a fast neural network architecture. There is also widespread activity in the applications of artificial intelligence concepts, including neural network applications, genetic algorithms, and the development of computer vision systems. Industrial applications of digital and microprocessor systems and microcontrollers are also supported by a general purpose development system.

Electrophysics (Chatterjee, Constable, Morris, Plumb, Wagner).

The group has a strong focus on research related to electronics packaging. This interest includes interconnect development and evaluation, including research on electrically conductive adhesives, metal-polymer systems, predictive testing for interconnect reliability, contact resistance of engineered surfaces, and noise analysis. Thin-film research is supported by a range of vacuum deposition techniques and focuses on the properties and applications of discontinuous (island) metal films and on noise and electromigration in thin-film VLSI interconnect lines. There are also noise studies of MESFETS and photovoltaics. Recent work in device modeling includes pulse thyristors and RTDs. Power electronics includes research on high-voltage and high-temperature modules, very high density multilayer power structures for telecommunications, power supplies, and motor drives.

The department is nationally known for its work in computational electromagnetics and in accurate and efficient modeling of VLSI and printed circuit board interconnections. Other recent work in electromagnetics includes cellular phone field strength modeling and near field radio propagation. Recent projects in optoelectronics have included fiber-optics, soliton theory, holographic optical interconnects, spatial multiplexing of 3-D scenes using pixelated computer-generated holography, optical logic gates using acoustooptic feedback, and electromagnetic propagation in complex media.

Systems (Bourbakis, Chatterjee, Fowler, Plumb, Sackman, Skormin, Taylor, Wagner, Wu).

Most recent research in signal processing has been concentrated in acoustics and digital audio. There is also an ongoing project in image processing (see computer vision). Within the communications category, work is concentrated in fiber optics and radio propagation.

The controls group enjoys strong ties to local industry, including aerospace, flight simulation, power generation and reticulation, electronics assembly, electrical vehicle, and materials processing the industries, among others. The control systems research group members have special interests in the applications of simulation, robotics, and embedded control in support of these areas and in the development of new theoretical techniques.

STATE UNIVERSITY OF NEW YORK AT BUFFALO

Department of Electrical Engineering

Programs of Study

The Department of Electrical Engineering offers M.S., M.Eng., and Ph.D. programs. The M.S. program requires 1 to 1½ years of full-time study. Students taking the M.S thesis option must complete eight graduate courses and thesis research. Students taking the M.S. project option must complete nine graduate courses and a project. Students in the M.Eng. program must complete eight graduate courses and a project. The Ph.D. program generally requires a minimum of two years of full-time study beyond the master's degree. Students must pass a qualifying examination and a comprehensive examination and complete dissertation research.

Research focus areas are represented by four groups: communications and signals; microelectronics, photonics and materials; digital electronics; and energy systems. These areas include telecommunications; mobile and personal communication systems; high-speed and optical networks; digital signal and image processing; neural networks; adaptive signal processing, detection, and estimation; adaptive antenna and radar arrays; inverse scattering, wave-propagation, and diffraction theory; electronic instrumentation and sensors; medical electronics and biomedical imaging; microelectronics and quantum electronics; semiconductors; heterostructures of III-V compounds, and photovoltaics; semiconductor photonic devices; computational photonics; laser spectroscopy; microscopy, microtomography, and lithography; holographic and laser-ablation techniques; superconducting and nanophase materials; energy systems and high-power electronics, pulsed power, dielectrics and insulation, and energy conservation; plasma physics, processing, and diagnostics; metal vapor plasmas, arc plasmas, and switching; and electromagnetic compatibility, RF, and microwaves.

Research Facilities

The Department of Electrical Engineering has an excellent theoretical, experimental, and applied research environment supported by an extensive computing infrastructure, including Sun and SGI workstations and servers and the following state-of-the-art research laboratories: Advanced Microscopy and Imaging Laboratory (AMIL); Communications Systems Laboratory (CSL); Computational Photonics Laboratory and Photonics Device Lab; Device Simulation and Characterization Laboratory; High Power Electronics Institute Laboratories (HPEI); High Resolution X-ray Diffraction Laboratory; Laboratory for Advanced Spectroscopic Evaluation (LASE); Microelectronic Fabrication Laboratory (MEFL); Plasma Processing and Spray Laboratory; RF & Microwave Laboratory; Superconductor Research and Environmental Research Laboratory; and Telecommunication Networks Research Laboratory. In addition, faculty members and graduate students collaborate in various research activities with University-wide centers, such as the Center for Advanced Photonic & Electronic Materials (CAPEM), the Center of Excellence for Document Analysis and Recognition (CEDAR), and the Center of Computational Research (CCR).

Financial Aid

A number of teaching assistantships are available, with a stipend of $10,700 plus a tuition scholarship. Research assistantships are also available through research grants. There are also a number of Presidential Fellowships available from the State University of New York with a stipend of $14,700.

Cost of Study

For 1998–99, full-time tuition and fees were $2985 per semester for New York State residents and $4643 per semester for nonresidents. Students receiving financial aid from the University or the department are normally granted a tuition scholarship.

Living and Housing Costs

Graduate students generally live off campus, and the Off-Campus Housing Office maintains a file of housing accommodations in various price ranges.

Student Group

The University student population numbers about 25,000 and includes about 6,800 graduate students. There are approximately 150 graduate students in the department.

Student Outcomes

Graduate students attract offers from academia, research organizations, government laboratories, and industry, particularly from high-technology companies in telecommunications and digital signal processing (both in hardware and software development) and in semiconductor devices and nanofabrication technology companies.

Location

Buffalo, in western New York along the Niagara River and the shores of Lake Erie, has a metropolitan-area population of 1 million. Cultural and recreational interests include museums, art galleries, a concert hall, a botanical garden, a zoological park, a sports auditorium, major-league football, a symphony orchestra, and several academic institutions. Western New York offers opportunities for many outdoor activities, and nearby Canada has numerous resources for cultural events and sports.

The University and The Department

Formerly named the University of Buffalo, the institution was founded in 1846 as a medical school. About 1947, the state of New York formed a state university system and in 1962, it selected the University of Buffalo as one of its units. The University has two principal campuses: the Main Street Campus (178 acres, located in the northeast corner of Buffalo) and the Amherst Campus (more than 1,000 acres, 3 miles away). The Department of Electrical and Computer Engineering, established in 1945, is located on the Amherst Campus.

Applying

Applications for admission with financial aid must be completed by February 1 for September admission in order to ensure consideration for aid and assistantships. If no financial assistance is desired, two months should be allowed for the processing of an application before the start of any semester; international students should allow more time. All new applicants (except those for the M.Eng. program) are required to take the GRE General Test. International applicants must also supply TOEFL scores.

Correspondence and Information

Graduate Admissions
Department of Electrical Engineering
201 Bell Hall
State University of New York at Buffalo
Buffalo, New York 14260

Telephone: 716-645-2422
Fax: 716-645-3656
E-mail: ee-grad-apps@eng.buffalo.edu
World Wide Web: http://www-ee.eng.buffalo.edu

State University of New York at Buffalo

THE FACULTY AND THEIR RESEARCH

Wayne A. Anderson, Professor; Ph.D., SUNY at Buffalo. Semiconductors, thin-film techniques, photovoltaics, microelectronics, defects in semiconductors.

Stella N. Batalama, Assistant Professor; Ph.D., Virginia. Wireless communications, adaptive signal processing, detection and estimation theory and applications.

David M. Benenson, Professor; Ph.D., Caltech. Arc plasmas, plasma physics, circuit breakers, switching, plasma diagnostics, plasma processing, power, gas dynamics, acoustics.

Alexander N. Cartwright, Assistant Professor; Ph.D., Iowa. Semiconductor photonic devices, optical nondestructive testing, laser spectroscopy.

Ping-chin Cheng, Professor; Ph.D., Illinois. Microscopy and microtomography, biomedical imaging and lithography.

Kasra Etemadi, Associate Professor; Ph.D., Minnesota. Thermal plasma processing, plasma diagnostics.

Adly T. Fam, Professor; Ph.D., California, Irvine. Digital signal processing.

Donald D. Givone, Professor; Ph.D., Cornell. Logic design, automata theory, digital systems.

Raj K. Kaul, Professor; Ph.D., Columbia. Mathematical analysis, wave-propagation and diffraction theory.

Pao-Lo Liu, Professor and Chairman; Ph.D., Harvard. Photonic devices and computational photonics.

Dennis P. Malone, Distinguished Professor; Ph.D., Yale. Metal vapor plasmas, plasma diagnostics, quantum electronics, atomic and molecular physics.

Hinrich R. Martens, Professor; Ph.D., Michigan State. Digital control systems, modeling of dynamic systems, computer interfacing.

Dimitris A. Pados, Assistant Professor; Ph.D., Virginia. Communications, adaptive antenna and radar arrays, neural networks.

Mohammed Safiuddin, Advanced Technology Professor; Ph.D., SUNY at Buffalo. Industrial automation and control, energy conservation, technology management.

Walter J. Sarjeant, Professor, Ph.D., Western Ontario. Energy systems and high-power electronics.

David T. Shaw, Professor; Ph.D., Purdue. Ultrathin films, superconducting materials processing, particle formation and measurement, nanophase materials, holographic techniques, laser-ablation techniques.

Mehrdad Soumekh, Associate Professor; Ph.D., Minnesota. Signal and image processing, medical and radar imaging.

Ozan K. Tonguz, Professor; Ph.D., Rutgers. Telecommunications, mobile and personal communication systems, high-speed networking, and optical networks.

James J. Whalen, Professor; Ph.D., Johns Hopkins. Electromagnetic compatibility, RF and microwaves, microelectronics, semiconductors.

Chu Ryang Wie, Professor; Ph.D., Caltech. Semiconductor materials, surfaces and interfaces, heterostructures of III-V compounds.

Darold C. Wobschall, Associate Professor; Ph.D., SUNY at Buffalo. Electronic instrumentation, sensors, and medical electronics.

Adjunct Faculty

Raj S. Acharya, Professor; Ph.D., Minnesota. Multimedia computing, visualization, image processing, fractals.

Victor Demjanenko, Assistant Professor; Ph.D., SUNY at Buffalo. Computer networks, operating systems, computer architecture.

James Llinas, Advanced Technology Professor; Ph.D., SUNY at Buffalo. Multisource information fusion, statistics, knowledge-based systems.

Nihar R. Mahapatra, Assistant Professor; Ph.D., Minnesota. Parallel and distributed processing, low-power and advanced VLSI design and computer architecture, mobile computing, networking, and fault tolerance.

Russ Miller, Professor; Ph.D., SUNY at Binghamton, Parallel algorithms, computational science, computational crystallography.

Chunming Qiao, Associate Professor; Ph.D., Pittsburgh. Parallel and distributed systems, advanced networks.

Peter D. Scott, Associate Professor; Ph.D., Cornell. Machine vision, systems analysis.

Ramalingam Sridhar, Associate Professor; Ph.D., Washington State. Digital systems, computer architecture, VLSI design.

Rohini Srihari, Associate Professor; Ph.D., SUNY at Buffalo. Artificial intelligence, knowledge-based vision, vision-language interfaces.

Sargur N. Srihari, Distinguished Professor; Ph.D., Ohio State. Artificial intelligence, computer vision, pattern recognition.

Shambhu J. Upadhyaya, Associate Professor; Ph.D., Newcastle (Australia). Fault-tolerant computing, reliability, VLSI testing, diagnostic reasoning.

Aidong Zhang, Associate Professor; Ph.D., Purdue. Databases, digital libraries, multimedia systems.

Graduate students at the Communications Systems Laboratory (CSL).

Graduate student at the Laboratory for Advanced Spectroscopic Evaluation (LASE).

STONY BROOK
STATE UNIVERSITY OF NEW YORK

STATE UNIVERSITY OF NEW YORK AT STONY BROOK

Department of Electrical and Computer Engineering

Programs of Study

The Department of Electrical and Computer Engineering offers programs of study leading to the degrees of Master of Science (M.S.) and Doctor of Philosophy (Ph.D.). Registration as a nondegree special student is also permitted. Active research areas include computer engineering, telecommunications, computer networks, digital image processing and machine vision, integrated circuit design, digital data compression and coding, signal processing, optical signal processing, fiber optics, physical electronics, solid-state electronic devices and circuits, systems and controls, and VLSI.

The M.S. program can be completed in one year, although many students elect to complete the program in three semesters (1½ years).

Graduate study is closely associated with research. Laboratories and associated faculty members are listed on the reverse of this page.

Research Facilities

Students have access to extensive computing facilities that include the department's own Sun SPARCserver 330s with workstations, a network of Sun and HP workstations, and a VAX (VMS) cluster and IBM mainframe in the computer center. The computing system and laboratories are interconnected through Ethernet. The department has the following laboratories: Digital Signal Processing, Lasers, Computer Vision, Microelectronics/VLSI, Microprocessors System Design, Computer-aided Design, and Telecommunications.

Financial Aid

The department makes an effort to support as many graduate students as possible. There are teaching and research assistantships and a limited number of fellowships. Assistantships normally pay up to $9908 per academic year; some also include tuition remission. Support levels range up to $14,300 for the calendar year, plus tuition remission, in 1999–2000. There are also opportunities for summer research support that pay up to $3700 for three summer months.

Cost of Study

Tuition for full-time graduate study in 1998–99 was $2550 per semester for New York State residents and $4208 for nonresidents. Part-time students paid tuition of $213 per semester hour for residents and $351 for nonresidents.

Living and Housing Costs

A limited number of three- and four-bedroom units in University residence halls are available for unmarried graduate students at a cost of $223 to $331 per month. Housing for married students is available on the same basis as housing for single students. A limited amount of off-campus housing is available in the area. The University Housing Office maintains a file to assist both single and married students in finding suitable accommodations. Meals may be purchased at a reasonable cost in campus cafeterias.

Student Group

There were 82 full-time and 40 part-time graduate students in the department in 1998–99. A total of 17,000 students are enrolled at the State University of New York at Stony Brook.

Student Outcomes

The student body is international in character. Some graduates return to their native countries to pursue teaching or industrial jobs at such places as National Chaio-Tung University, Jordan Science and Technological University, and Microelectronics Tech. Taiwan. Graduates are now employed by AT&T Bell Labs, NYNEX, Jet Propulsion Lab., Panasonic Tech., and many other technological companies. Some are employed by nontechnological companies such as Lehman Brothers, Citicorp, and Mt. Sinai Hospital. Most graduates find employment within four months of graduation.

Location

Stony Brook is located on the wooded North Shore of Long Island, about 60 miles east of Manhattan and at the geographical center of Long Island. There is a very substantial electronics industry nearby, and the department maintains excellent working relations with these firms. The location makes New York City's cultural life and Suffolk County's tranquil countryside and seashores conveniently accessible. Brookhaven National Laboratory and the Cold Spring Harbor (biological) Laboratory are close by.

The University

The State University of New York at Stony Brook was established as a comprehensive university center in 1960. Since that time, an internationally renowned faculty has been formed to offer courses in forty-five major areas and interdisciplinary programs. Externally funded research has steadily increased to reach the current level of more than $70 million. The campus now comprises more than 100 buildings located on approximately 1,100 acres.

Applying

Applications are due February 1 for the fall semester. The GRE General Test is required for all students. For international students the TOEFL, with a minimum score of 550 (600 for an assistantship), is also required.

Correspondence and Information

Professor Gregory Belenky, Graduate Program Director
Department of Electrical and Computer Engineering
State University of New York at Stony Brook
Stony Brook, New York 11794-2350

Telephone: 516-632-8397 or 8400
E-mail: grad@ece.sunysb.edu
World Wide Web: http://www.ee.sunysb.edu:8080/

State University of New York at Stony Brook

THE FACULTY AND THEIR RESEARCH

Gregory L. Belenky, Associate Professor; Ph.D., Institute of Semiconductors (Kiev); D.Sc., Baku. Semiconductor devices, physics and technology, lasers for telecommunication.

Sheldon S. L. Chang, Emeritus Professor; Ph.D., Purdue. Optimal control, computer architecture, robotics, artificial intelligence, signal processing, economic theory.

Chi-Tsong Chen, Professor; Ph.D., Berkeley. Systems and control theory, digital signal processing.

Harbans Singh Dhadwal, Associate Professor; Ph.D., London. Lasers and instrumentation.

Petar Djuric, Associate Professor; Ph.D., Rhode Island. Signal and systems analysis.

Mikhail Dorojevets, Assistant Professor; Ph.D., Computing Center, Novosibirsk (Russia). Computer architectures, systems design.

Gene Gindi, Associate Professor; Ph.D., Arizona. Medical image processing, computer vision. (Joint appointment with the Department of Radiology)

Vera Gorfinkel, Associate Professor; Ph.D., A.F. Ioffe Physical-Technical Institute, St. Petersburg (Russia). Semiconductor devices, including microwave and optoelectronics.

Cem Hocaoglu, Assistant Professor; Ph.D., Rensselaer. Evolutionary computation, parallel and distributed computing, robotics, manufacturing.

Ridha Kamoua, Assistant Professor; Ph.D., Michigan. Solid-state devices and circuits; microwave devices and integrated circuits.

Adrian Leuciuc, Assistant Professor; Ph.D., Technical University of Iasi (Romania). Analog and mixed-signal circuit design.

Serge Luryi, Professor and Chair; Ph.D., Toronto. High-speed electronic and photonic devices, semiconductor physics and technology.

John H. Marburger III, Professor; Ph.D., Stanford. Theoretical laser physics. (Joint appointment with the Department of Physics)

Velio A. Marsocci, Distinguished Service Professor; Eng.Sc.D., NYU. Solid-state electronics, integrated electronics, biomedical engineering. (Also Clinical Professor of Health Sciences)

John Murray, Associate Professor; Ph.D., Notre Dame. Systems, controls, signal processing, image processing and instrumentation.

Jayant Parekh, Professor; Ph.D., Polytechnic of Brooklyn. Microwave acoustics and magnetics, microwave electronics.

Theo Pavlidis, Professor; Ph.D., Berkeley. Machine vision, pattern recognition, computer graphics, robotics. (Joint appointment with the Department of Computer Science)

Nam Phamdo, Assistant Professor; Ph.D., Maryland. Digital communications, data compression and coding, speech processing.

Stephen S. Rappaport, Professor; Ph.D., NYU. Communications systems, telecommunications.

Thomas G. Robertazzi, Associate Professor; Ph.D., Princeton. Computer networks, parallel processing, performance evaluation.

Yacov Shamash, Professor and Dean; Ph.D., Imperial College (London). Control systems and robotics.

Kenneth L. Short, Professor; Ph.D., SUNY at Stony Brook. Digital system design, microprocessors, instrumentation.

David R. Smith, Professor; Ph.D., Wisconsin–Madison. Logic design, computer architecture. (Joint appointment with the Department of Computer Science)

Murali Subbarao, Associate Professor; Ph.D., Maryland. Machine vision, image processing, information processing.

Stephen Sussman-Fort, Associate Professor; Ph.D., UCLA. Electronic circuits, CAD, solid-state electronics, electromagnetics.

K. Wendy Tang, Associate Professor; Ph.D., Rochester. Parallel and distributed processing, massively parallel systems, computer architecture.

Hang-Sheng Tuan, Professor; Ph.D., Harvard. Electromagnetic theory, integrated and fiber optics, microwave acoustics.

Armen H. Zemanian, Distinguished Professor; Eng.Sc.D., NYU. Network theory, VLSI circuits.

Department Laboratories

Each of the following laboratories contains special research and teaching equipment.

Advanced IC Design and Simulation Laboratory: Professor Leuciuc.
Communications, Signal Processing, Speech and Vision Laboratory: Professors Rappaport, Djuric.
Computer Vision Laboratory: Professor Subbarao.
Digital Signal Processing: Professor Murray.
DNA Lab: Professors Luryi and Gorfinkel.
Embedded Systems Design Laboratory: Professor Short.
Microwave Electronics Laboratory: Professors Parekh, Sussman-Fort, Tuan.
Optical Signal Processing and Fiber Optics Sensors Laboratory: Professor Dhadwal.
Semiconductor Electronics and Optoelectronics Laboratory: Professor Luryi.

SYRACUSE UNIVERSITY

Department of Electrical Engineering and Computer Science

Programs of Study

The Department of Electrical Engineering and Computer Science (EECS) offers Master of Science (M.S.) degrees in computer engineering, computer science, electrical engineering, computational science, and systems and information science. Doctoral (Ph.D.) degrees are offered in computer engineering, computer and information science, and electrical engineering.

Master of Science degrees in computer engineering and electrical engineering require a minimum of 30 credit hours, and the others require 33 credit hours. All M.S. degrees have a thesis option and a nonthesis option. Students enrolled in the nonthesis option in computer engineering, computer science, or electrical engineering may finish their M.S. degree in one year. In order to accomplish this, students must take courses in the fall, spring, and summer terms. Students may also complete the degree in a less intensive fashion over three or four regular semesters.

For professionals who have baccalaureate degrees in fields other than computer engineering, computer science, or electrical engineering and who are interested in a career change, the Department of EECS provides an opportunity to obtain an M.S. degree in one of these fields by combining suitable remedial undergraduate course work with the regular program of graduate work. Students enrolled in these programs can complete the M.S. degree with approximately 60 credit hours beyond the baccalaureate degree.

The Ph.D. degrees require 78 credit hours beyond the baccalaureate in computer engineering and electrical engineering and 90 credit hours in computer and information science. These credit hours include graduate course work and dissertation credit. The dissertation must represent an original contribution to knowledge based upon the candidate's research and must be defended before an examination committee.

Research Facilities

Major research laboratories include high-performance distributed computing, electronic design automation, scalable concurrent processing, solid-state, fiber optics, microwave RF-wireless signal processing, photonics, and robotics. Syracuse University has excellent computing facilities for the use of graduate students. The Department of EECS provides both UNIX and Windows-NT systems for general use by graduate students. The UNIX system is comprised of a network of Sun SPARC and Ultra SPARCservers and workstations, while the NT system is built on a network of Pentium II servers and workstations. All are connected to the campus network and thus to the Internet. Various research laboratories in the department provide substantial additional resources for parallel and high-performance computing and high-performance networking.

In addition, students in the department can readily access facilities operated by the University's central Computing and Media Services organization, the Center for Advanced Technology in Computer Applications and Software Engineering (CASE), and the Northeast Parallel Architectures Center (NPAC). Detailed information concerning computer facilities available at the University is available via the World Wide Web (http://netsys.syr.edu/sunix/), the CASE Center (http://www.cat.syr.edu), and NPAC (http://www.npac.syr.edu).

Financial Aid

EECS graduate assistantships are awarded on a competitive basis from applications received by January 1; assistantships are usually not available at any other time of the year. Graduate assistants may be required to assist in instructional or research activities. They receive a stipend for the academic year plus 24 credit hours of remitted tuition. Summer support is available for selected students working on various research projects and for some instructional work. Fellowships are available for superior students. These are awarded on a competitive basis and currently carry a stipend of $11,275.41 and 30 credit hours of remitted tuition. Applications for fellowships must be received by December 1.

Cost of Study

Tuition in 1999–2000 is $583 per credit hour.

Living and Housing Costs

University housing for single students ranges from $2011 to $3675 per semester. Married students may rent a one-bedroom University apartment at a monthly cost of $588 (unfurnished) to $640 (furnished) or a two-bedroom apartment at a monthly cost of $645 (unfurnished) to $715 (furnished).

Student Group

Undergraduate enrollment averages 10,000, and graduate enrollment is more than 4,000. Students are drawn from every state in the union and more than 100 countries. About 1,600 international students are registered each year in the undergraduate and graduate programs. Current graduate enrollment in the Department of EECS is more than 400 students.

Location

Syracuse, New York, is a city of approximately 150,000 people. The population of the greater metropolitan Syracuse area is more than 450,000. Located in the center of the state, the city serves as a focal point for many business, cultural, and entertainment activities for the central New York area. To the west is the famed Finger Lakes region; to the north and northeast are the Thousand Islands and the Adirondack Mountains.

The University

Syracuse University is a medium-sized, private, coeducational university. Founded in 1870, it is a comprehensive research university that has thirteen degree-granting schools and colleges and several interdisciplinary and continuing-education programs. The University's 640-acre, beautifully landscaped campus is situated among the hills of central New York State. The University has a growing stature in the sciences and engineering and maintains outstanding traditions in music, art, drama, communications, and public affairs.

Applying

Most students begin their studies in August; however, students may also start in January or May. The verbal, quantitative, and analytical tests of the Graduate Record Examinations are required for admission. International applicants are required to take the TOEFL. Application forms may be obtained by writing to the address given below. Prospective students may also apply on line (http://uplink.syr.edu/EECS).

Correspondence and Information

Ms. Barbara Hazard
Department of Electrical Engineering and Computer Science
123 Link Hall
L.C. Smith College of Engineering and Computer Science
Syracuse University
Syracuse, New York 13244-1240
Telephone: 315-443-2655
Fax: 315-443-2583
E-mail: eecsgrad@ecs.syr.edu

Syracuse University

THE FACULTY AND THEIR RESEARCH

Ercument Arvas, Professor; Ph.D., Syracuse, 1983. Electromagnetic theory, transmission lines, microwave engineering, engineering math, digital signal processing.

Howard A. Blair, Associate Professor; Ph.D., Syracuse, 1980. Logic in computer science, foundations of logic programming, discrete dynamical systems.

Per Brinch Hansen, Distinguished Professor; Ph.D./Dr.Techn., Denmark Technical, 1978. Programming languages for concurrent systems.

C. Y. Roger Chen, Professor; Ph.D., Illinois, 1987. Multimedia, object-oriented databases; multimedia transport protocols; network performance modeling; VLSI computer-aided design.

Shiu-Kai Chin, Associate Professor; Ph.D., Syracuse, 1986. Computer-aided design, VLSI design, formal methods, computer security, software engineering.

Ehat Ercanli, Temporary-Term Assistant Professor; Ph.D., Case Western Reserve, 1997. VLSI computer-aided design, automation for digital systems, computer architecture.

James W. Fawcett, Part-Time Associate Professor; Ph.D., Syracuse, 1981. Software, control systems, computers, communications.

Garth Foster, Professor and Director of Computer Engineering Programs; Ph.D., Syracuse, 1966. APL, prototyping, computer applications.

Geoffrey Fox, Professor and Director, Northeast Parallel Architectures Center (NPAC); Ph.D., Cambridge, 1967. Supercomputers, parallel architectures, concurrent algorithms and their applications in industry and academia.

Prasanta K. Ghosh, Associate Professor; Ph.D., Penn State, 1986. Microelectronics, solid-state devices, optoelectronics, thin-film processes, power engineering.

Amrit L. Goel, Professor; Ph.D., Wisconsin, 1968. Software engineering: metrics, testing, reliability, object-oriented metrics and testing, fault-tolerant software, applications of neural networks in software engineering.

Salim Hariri, Associate Professor; Ph.D., USC, 1986. High-performance distributed computing, high-speed communication protocols and networks, software tools for parallel and distributed computing.

Carlos R. P. Hartmann, Professor and Chairperson; Ph.D., Illinois, 1970. Coding theory, fault detection in digital circuits, fault-tolerant computing.

Can Isik, Associate Professor and Director of Electrical Engineering Programs; Ph.D., Florida, 1985. Robotics, applications of neural nets and fuzzy logic, control theory, computational intelligence.

Kamal Jabbour, Associate Professor; Ph.D., Salford, 1982. Computer networks, computer applications to power systems.

Yaoguo Jia, Part-Time Instructor; Ph.D., Syracuse, 1985. Computer architecture, software and systems.

Douglas V. Keller Jr., Research Professor; Ph.D., Syracuse, 1958. Surface and interfacial science of materials, in particular adhesion and wetting phenomena, coal science, separation science, friction, lubrication, and wear.

Philipp Kornreich, Professor; Ph.D., Pennsylvania, 1967. Fiber light amplifiers, lasers, image propagation through fibers.

Jay Kyoon Lee, Associate Professor; Ph.D., MIT, 1985. Electromagnetic waves, microwave remote sensing, antennas.

Harold F. Mattson Jr., Research Professor; Ph.D., MIT, 1955. Coding theory, combinatorial mathematics, applied algebra.

Kishan Mehrotra, Professor and Director of Computer and Information Science Programs; Ph.D., Wisconsin–Madison, 1971. Neural networks, algorithms, applied statistics, reliability analysis.

Chilukuri K. Mohan, Associate Professor; Ph.D., SUNY at Stony Brook, 1988. Artificial intelligence, automated reasoning, neural networks, equational specifications.

Susan Older, Assistant Professor; Ph.D., Carnegie Mellon, 1996. Semantics of programming languages, logics of programs and formal methods, concurrency.

Daniel J. Pease, Associate Professor; Ph.D., Syracuse, 1981. Design and development of shared and distributed parallel systems, software and tools, and performance estimation of user's C, C++, Ada, and Fortran 90; applications on different parallel architectures.

Frederick Phelps, Professor, Director, Soling Program, and Director, Iconic Communication Laboratory; Ph.D., Johns Hopkins, 1967. Iconic communications, systems, teaching/learning methods.

James S. Royer, Professor; Ph.D., SUNY at Buffalo, 1984. Theory of computation, computational complexity, theory of machine learning, theory of programming languages, connections between logic and computation.

Robert G. Sargent, Research Professor; Ph.D., Michigan, 1966. Modeling methodologies; model validation; methodology areas of discrete event simulation, including model specification and efficient and parallel/distributed computation; strategical and tactical experimental design; data analysis.

Tapan K. Sarkar, Professor; Ph.D., Syracuse, 1975. Radiation, design and fabrication of printed circuits, antenna design (radio and TV communication), adaptive and real-time digital signal processing.

Ernest Sibert, Professor; Ph.D., Rice, 1967. Computational logic, logic programming, parallel computation.

Q. Wang Song, Associate Professor; Ph.D., Penn State, 1989. Electrooptics, optical interconnections, real-time holography, photonic switching.

Stephen Taylor, Associate Professor; Ph.D., Weizmann (Israel), 1989. Parallel architectures and concurrent programming, software engineering, computer graphics, concurrent simulation techniques.

Pramod K. Varshney, Professor; Ph.D., Illinois, 1976. Multisensor data fusion, detection theory, knowledge-based signal processing, digital and wireless communications, image processing, communication networks.

Hong Wang, Professor; Ph.D., Minnesota, 1985. Signal processing, communication engineering, radar/sonar systems.

Donald Weiner, Research Professor; Ph.D., Purdue, 1964. Communications systems, nonlinear networks, signal processing, radar systems, weak signal detection in non-Gaussian environments, analysis of random data.

TEXAS A&M UNIVERSITY

College of Engineering
Department of Electrical Engineering

Programs of Study	The Department of Electrical Engineering offers programs of graduate study leading to the degrees of Master of Engineering, Master of Science, Doctor of Engineering, and Doctor of Philosophy.
	The Master of Engineering is a professional, nonthesis degree that requires 36 hours of course work and includes an engineering report documenting an engineering/design project. The objective of the program is to give the student both advanced courses in electrical engineering concepts and experience in applying these concepts to actual engineering problems. The program can be completed in approximately one year.
	The Doctor of Engineering degree requires 96 hours of graduate-level course work in engineering, management, and other areas related to the professional practice of engineering. A one-year internship in industry is also required. The program can be completed in three years.
	The M.S. and Ph.D. are research degrees. The M.S. degree requires 32 hours of course work, including a thesis, and can be completed in one year. The Ph.D. requires approximately 96 hours of course work and research, including the dissertation, and requires at least three years to complete. There is no foreign language requirement.
	Opportunities for graduate study and research exist in the areas of communications, computer-aided design of electronic circuits (including expert systems), computers and digital systems, control systems, digital signal processing, electric power systems and power electronics, electromagnetics, electronic materials, electrooptics, image processing, microelectronic circuits, microwaves, solid-state electronics, and VLSI. Interdisciplinary programs are also available.
Research Facilities	The Department of Electrical Engineering is housed in a modern building with extensive, well-equipped laboratory facilities. The department operates two SPARCserver 1000 computers as well as three high-performance laser printers for general graduate student use. In addition, there are more than forty-five SPARCstations in research laboratories and forty SPARCstations in student laboratories. There are more than twenty-five networked 486 class PCs available to students on a 24-hour basis. A power electronics laboratory including CAD, CAM, and MC development facilities, a power automation laboratory equipped with one Concurrent 5450 computer, and several PCs are available for research in power. Other University computing facilities available to electrical engineering students include a SPARCcenter 2000, a large VAXcluster, an IBM 3090, and an SGI Power Challenge XL shared memory parallel computer. The million-volume University library provides seating for more than 2,000 people and maintains substantial literature in all areas of electrical engineering. The department has access to IEEE publications on disc (IPO), which allows searching and full-text printing of almost all IEEE and IEE publications. Extensive solid-state and electrooptics facilities are available for device fabrication and testing.
Financial Aid	Financial aid is available in the form of research or teaching assistantships, scholarships, and fellowships. Stipends for assistantships in 1998–99 varied between $800 and $1000 per month; they qualify out-of-state students for resident tuition. Assistantships require approximately 20 hours of work per week, and recipients must register for 9 to 12 hours of course work each semester. Research assistants work on one of the department's research projects; teaching assistants participate in undergraduate instruction. Various types of graduate fellowships are awarded to outstanding students; recipients are expected to register for 12 hours each semester.
Cost of Study	Tuition in 1998–99 for Texas residents was $68 per semester credit hour, with a minimum charge of $100 per semester. Other fees totaled approximately $200 per semester. Tuition for nonresident students was $285 per semester credit hour. (Costs are subject to change.)
Living and Housing Costs	For single students, fees for room and board on campus in 1998–99 ranged from $400 to $1000 per month. For married students, a limited number of University-owned apartments, both furnished and unfurnished, were available at $185 to $290 per month, plus electricity. A large number of privately owned apartments are available in the community.
Student Group	The University is coeducational, and enrollment is more than 42,000, including about 7,000 graduate students. The enrollment in the College of Engineering is 7,000 undergraduates and 1,900 graduate students. There are 342 graduate students enrolled in electrical engineering—122 Master of Science, 49 Master of Engineering, and 170 doctoral degree candidates and 1 nondegree student.
Location	Texas A&M University is located about 100 miles northwest of Houston and 170 miles south of Dallas. The Bryan–College Station area is growing rapidly and has a population of about 113,000. Numerous athletic, cultural, and recreational activities take place in the area. There are two excellent public school systems and more than fifty churches of various denominations.
The University and The College	Texas A&M was founded in 1876 as a land-grant college and is the state's oldest public institution of higher learning. Through its College of Engineering, Engineering Experiment Station, and Engineering Extension Service, it provides a wide range of high-quality programs of education, research, and public service. The campus and adjacent University facilities cover more than 5,200 acres of land and include a physical plant valued at $450 million. Texas A&M ranks higher than any other institution of higher learning in the South or Southwest in the total value of its sponsored research. Total expenditures for research during 1998–99 exceeded $100 million. Approximately one third of the University's research programs are in engineering.
Applying	To be admitted to the Department of Electrical Engineering, an applicant must hold a B.S. degree in engineering or physical science. The minimum grade point average for admission to the master's degree program is 3.0 (on a 4.0 scale); admission to the Ph.D. program requires a minimum grade point average of 3.6 (on a 4.0 scale). For both the M.S. and the Ph.D. programs, the applicant must have a minimum score of 525 on the verbal section and 700 on the quantitative section of the General Test of the Graduate Record Examinations. Prospective students whose native language is not English must submit a minimum score of 600 on the Test of English as a Foreign Language. Applications for U.S. citizens and permanent residents should be received no later than six weeks before the beginning of the semester in which admission is desired.
Correspondence and Information	Graduate Coordinator Department of Electrical Engineering College of Engineering Texas A&M University College Station, Texas 77843

Texas A&M University

THE FACULTY AND THEIR RESEARCH

Analog and Mixed Signals. Integrated circuit design, very large scale integrated circuits, analog VLSI design, operational amplifier design and applications, switched-capacitor filters, low-noise front-end electronics, instrumentation, measurements, active and passive filter design, neural networks, data converters, expert systems for electronics applications, design of integrated circuits, including design methodology, analog and digital VLSI design, computer-aided design, simulation, biomedical applications, radiation-hardened IC circuit design, telecommunication applications, statistical circuit design, verification test and evaluation, analog design and characterization, RF 1C and systems. Multiproject chips are fabricated by cooperating industries; department facilities include the VLSI CAD and AI Lab, with Apollo and Sun CAD/CAE color workstations, forty PCs connected to the Sun server, alphanumeric and color graphics terminals, digitizer and plotters, and the VLSI Diagnostic Lab, with probing, dicing, and bonding facilities and workstation-based semiautomated test systems for low-noise and high-frequency VLSI systems.

Faculty: S. Embabi, Ph.D., Waterloo, 1991; J. Pineda, Ph.D., University of Technology (Netherlands), 1991; E. Sanchez-Sinencio, Ph.D., Illinois at Urbana-Champaign, 1973.

Biomedical Imaging. Theory, design, and implementation of instrumentation and techniques for biomedical imaging acquisition; optical tomographic imaging techniques and software development for optical imaging of biological tissues; magnetic resonance imaging and spectroscopy; theory and implementation of radio frequency sensors and magnetic resonance hardware; methods for dynamic imaging, thermal imaging, and magnetic resonance microscopy; theory and application of sensor arrays in medical imaging, including microwave, magnetic resonance, and ultrasound; biomedical image processing and analysis; image analysis techniques and algorithms for extracting clinically useful information from medical images, including X-ray, MR, nuclear medicine, and ultrasound; study of various types of filtering compression, feature extraction, and classification techniques specially designed for medical image data, including DCT, wavelets, morphological analysis, and random image models.

Faculty: A. K. Chan, Ph.D., Washington, 1971; E. Dougherty, Ph.D., Rutgers, 1974; N. C. Griswold, D.Eng., Kansas, 1975; N. Kehtarnavaz, Ph.D., Rice, 1987; R. D. Nevels, Ph.D., Mississippi, 1978; P. Morgan, Ph.D., Stanford, 1996; C. B. Su, Ph.D., Brandeis, 1979; S. M. Wright, Ph.D., Illinois, 1984.

Computer Engineering. Computer engineering, digital system design and test, digital VLSI/WSI and ASIC design, minicomputers, microprocessors, digital signal processing, digital control, computer communications, fault-tolerant architectures, AI systems, real-time systems, parallel and distributed systems. Special laboratory facilities include CAD/CAE workstations, minicomputers, microprocessor system development facilities, and a well-equipped digital system design laboratory.

Faculty: P. E. Cantrell, Ph.D., Georgia Tech, 1981; G. S. Choi, Ph.D., Illinois, 1994; M. Lu, Ph.D., Rice, 1987; M. R. Mercer, Ph.D., Texas at Austin, 1980; J. H. Painter, Ph.D., SMU, 1972; A. L. N. Reddy, Ph.D., Illinois, 1990; L. Wang, Ph.D., Texas, 1998; K. Watson, Ph.D., Texas Tech, 1982.

Control Systems. Linear multivariable control systems, model following, distributed parameter systems, finite-element models, homomorphic digital filtering, nonlinear control systems, robust control and adaptive control.

Faculty: S. P. Bhattacharyya, Ph.D., Rice, 1971; A. Datta, Ph.D., USC, 1991; J. W. Howze, Ph.D., Rice, 1970; G. Huang, D.Sc., Washington (St. Louis), 1980; J. H. Painter, Ph.D., SMU, 1972.

Electric Power and Power Electronics. System planning, dynamic analysis, reliability evaluation, control, and protection applied to conventional as well as specialized terrestrial, airborne, and spaceborne power and energy systems; analysis and control of electrical machines and variable-speed drive systems; power electronics; energy storage and pulsed power systems; microcomputer control and monitoring of energy systems; applications of expert systems to power system problems. Research is supported by utility, aerospace, and other industrial companies as well as state, federal, and private research funding agencies.

Faculty: A. Abur, Ph.D., Ohio State, 1985; K. Butler, Ph.D., Howard, 1994; M. Ehsani, Ph.D., Wisconsin–Madison, 1981; P. Enjeti, Ph.D., Concordia (Montreal), 1987; G. Huang, D.Sc., Washington (St. Louis), 1980; M. Kezunovic, Ph.D., Kansas, 1980; A. D. Patton, Ph.D., Texas A&M, 1972; B. D. Russell, Ph.D., Oklahoma, 1975; C. Singh, Ph.D., Saskatchewan, 1972; H. Toliyat, Ph.D., Wisconsin–Madison, 1991.

Electromagnetics and Microwaves. Antennas, electromagnetic wave propagation, electromagnetic theory, microwave systems, microwave solid-state circuits and devices, guided-wave structures, millimeter-wave circuits, microstrip and waveguide discontinuities, coupled-mode theory, numerical methods, microstrip antennas, antennas in stratified media. Special laboratories and facilities include a microwave anechoic chamber, a rooftop antenna range, an automatic microwave analyzer (HP8510), and Touchstone, Supercompact, IE3D, Sonnet, and MW-Spice programs for microwave circuit design.

Faculty: A. K. Chan, Ph.D., Washington (Seattle), 1971; K. Chang, Ph.D., Michigan, 1976; K. A. Michalski, Ph.D., Kentucky, 1981; R. B. Nevels, Ph.D., Mississippi, 1979; C. Nguyen, Ph.D., Central Florida, 1990; S. M. Wright, Ph.D., Illinois at Urbana-Champaign, 1984.

Solid State/Electrooptics. Optical waveguides, integrated optics, fiber optical devices, diode laser properties, optical materials; application of these technologies in communications, signal processing, sensing, and microwave systems; solid-state materials; process-induced defects in silicon; X-ray topography; diffusion; ion implantation; solar cells; semiconductor memory devices; microwave devices; epitaxial growth of III-V compound semiconductor layers. Extensive facilities are available for electrooptics research as well as optical waveguide device fabrication. Laboratories of the Institute for Solid State Electronics are fully equipped for research in solid-state materials, semiconductor devices, integrated circuits, single-crystal growth and characterization, and bulk and thin-film ferroelectric, magnetic, and semiconductor materials. The Molecular Beam Epitaxy Facility is equipped with two MBE machines (VG Instruments) for epitaxial growth of III-V heterojunction materials and silicon/silicon alloy superlattices. The silicon MBE has a special feature of two ion-implanters attached to the growth chamber. The Electron-Beam Lithography Facility is equipped for patterning semiconducting nanostructures.

Faculty: J. Blake, Ph.D., Stanford, 1988; A. K. Chan, Ph.D., Washington (Seattle), 1971; O. Eknoyan, Ph.D., Columbia, 1975; D. L. Parker, Ph.D., Texas A&M, 1968; C. B. Su, Ph.D., Brandeis, 1979; H. F. Taylor, Ph.D., Rice, 1967; L. C. Wang, Ph.D., California, San Diego, 1991; M. H. Weichold, Ph.D., Texas A&M, 1983.

Telecommunications and Signal Processing. Digital communications systems, information theory, coding, data compression, estimation and detection theory, digital signal processing, image analysis, image processing, knowledge-based signal processing, architectures for signal processing, computer communication networks. Special facilities include the Wireless Communications Laboratory (WCL) and the Digital Signal Processing Laboratory. The WCL includes a number of high-end workstations and PCs equipped with data acquisition hardware and DSP boards that facilitate analysis and real-time implementation of telecommunications systems. In addition, the laboratory has a number of telecommunication-specific hardware, including RF modulators, a real-time RF channel simulator, and other equipment that allows students and faculty members to implement and test prototype systems. The Digital Signal Processing Laboratory contains a number of hardware and high-end PCs equipped with a variety of DSP boards. The laboratory focuses on teaching and research in efficient real-time DSP implementations.

Faculty: P. E. Cantrell, Ph.D., Georgia Tech, 1981; E. Dougherty, Ph.D., Rutgers, 1974; C. Georghiades, D.Sc., Washington (St. Louis), 1985; N. C. Griswold, D.Engr., Kansas, 1975; D. R. Halverson, Ph.D., Texas at Austin, 1979; N. Kehtarnavaz, Ph.D., Rice, 1986; S. Miller, Ph.D., California, San Diego, 1988; K. Narayanan, Ph.D., Georgia Tech, 1998; J. H. Painter, Ph.D., SMU, 1972; X. Wang, Ph.D., Princeton, 1998.

UAH *THE UNIVERSITY OF ALABAMA IN HUNTSVILLE*

Department of Electrical and Computer Engineering

Programs of Study	The Department of Electrical and Computer Engineering (ECE) offers the Master of Science in Engineering (M.S.E.) in electrical engineering, computer engineering, and optics and photonics technology. Software engineering is offered as a concentration under the M.S.E. in computer engineering. The Ph.D. is awarded in electrical engineering and in computer engineering.
	An interdisciplinary Ph.D. in optical science and engineering, offered jointly by the College of Science and the College of Engineering, enrolled its first students in 1993 and awarded its first degree in 1995. This flexible program prepares students with varied scientific backgrounds for research careers in such fields as classical optics, spectroscopy, optical processing and computing, optical inspection, optical materials, liquid crystal displays, and optoelectronics. For more information on this program, call 256-890-6030, ext. 474.
	The ECE department encourages collaborative research and offers opportunities for students to work on broad programs of research extending across traditional disciplinary boundaries. Among such projects are 3-D displays, optical interconnect systems, and satellite communications. Faculty members from within and outside the department are actively involved with graduate student mentoring, ensuring that students develop the technical and communications skills necessary for successful careers.
Research Facilities	Students and faculty members have access to the Alabama Supercomputer Network's CRAY C94A and nCUBE2 as well as a variety of minicomputers and workstations through Ethernet connections in every office and laboratory. Personal computers are widely available throughout the Engineering Building. The department supports PC, Macintosh, Sun systems, and a Harris Nighthawk multiprocessor. Research laboratories include those for VLSI design, optical computing, nonlinear optics, optical communications, optical information processing, satellite communications, advanced display research, thin-film transistor development, nanofabrication, polymeric and liquid crystal photonics, real-time systems, and signal processing.
Financial Aid	Teaching and research assistantships are available on a competitive basis. The assistantships carry a stipend and cover tuition. The optical science and engineering program offers NSF graduate traineeships for qualified U.S. permanent residents.
Cost of Study	Tuition for the 1999–2000 academic year for a full-time Alabama resident enrolled for 9 semester hours averages $1430 per semester. Rates for out-of-state students are double those for Alabama residents. The estimated average cost of books per semester for full-time students is $250.
Living and Housing Costs	UAH offers air-conditioned apartments. Each has its own entrance and is carpeted and equipped with a stove and a refrigerator. Interested students should apply at least two academic terms before anticipated enrollment. Three-bedroom furnished dormitory suites (accommodate 2 students per bedroom) and three-bedroom furnished apartments for student families are available. A new, private dormitory houses 4 students per suite. Numerous apartments are within walking distance of the campus.
Student Group	The department is the largest on campus, enrolling 425 undergraduates and 174 graduate students, almost half of whom work in local industrial or government facilities. Many undergraduates take advantage of the Cooperative Program to alternate academic study with practical experience. Because of the large high-technology sector in Huntsville, almost all cooperative students live and work there, allowing them great flexibility in integrating their work and study.
Student Outcomes	Most students remain in the Huntsville area, where many employment opportunities exist within the federal and private sectors in fields such as optoelectronics, communications, control systems, computer engineering, and signal processing.
Location	Huntsville is the center of a diverse, dynamic metropolitan area of 250,000 that has developed around the high-technology focus of Redstone Arsenal (Army Missile Command, NASA Marshall Space Flight Center, Space and Strategic Defense Command) and the Cummings Research Park (ADTRAN, Nichols Research, Dynetics, BDM, SAIC, Rockwell, Lockheed-Martin, Teledyne Brown, Intergraph, and SCI, among others). The city is located on verdant rolling hills in the Tennessee Valley on the north bank of the Tennessee River in north-central Alabama.
The University and The Department	The department faculty members are research-active, with degrees from premier U.S. universities. With modern facilities for instruction and research, the department is the largest on campus, occupying 45,000 square feet of high-quality space.
Applying	Students may be admitted for any semester. Applicants attending of student visas must apply at least three months in advance of anticipated enrollment; all others should apply at least one month in advance. If assistantships are desired, applicants should apply early. All applicants must submit a completed application with a $20 application fee, official transcripts form each university previously attended, and the appropriate test scores.
Correspondence and Information	Director of Graduate Studies Department of Electrical and Computer Engineering The University of Alabama in Huntsville Huntsville, Alabama 35899 Telephone: 256-890-6316 Fax: 256-890-6803 E-mail: ece@ebs330.eb.uah.edu World Wide Web: http://www.eb.uah.edu/ece

The University of Alabama in Huntsville

THE FACULTY AND THEIR RESEARCH

M. A. G. Abushagur. Optical signal processing, computing and metrology.

R. R. Adhami, Chair. Digital signal processing, digital systems design.

N. F. Audeh. Electromagnetics.

P. P. Banerjee. Nonlinear wave phenomena, optical processing.

T. B. Boykin. Modeling compound and quantum semiconductor devices.

W. E. Cohen. Computer engineering, compilers.

R. L. Fork. Photonics, optical communications.

R. K. Gaede. Computer architecture, design for test, VHDL.

F. D. Ho. Microelectronic devices and integrated circuits, photovoltaic devices, electronic materials.

J. M. Jarem. Electromagnetics, antenna theory, microwave theory, optics.

C. D. Johnson. Control and dynamic systems.

L. Joiner. Wireless communications systems.

K. Kavi. Computer architecture.

J. H. Kulick. Computer design, computer-generated holography, medical image processing.

G. P. Nordin. Volume holographic optical memories, diffractive 3-D displays.

D. B. Pollock. Infrared optical systems.

W. A. Porter. Array architectures, pattern recognition, system theory and applications.

A. D. Poularikas. Statistical optics, signal processing.

D. Shen. Electronic materials and devices, thin-film deposition.

Y. Shtessel. Automatic control theory, sliding modes, multicriteria control.

N. Singh, Director of Graduate Studies. Electromagnetics, plasma, and space research.

J. Stensby. Communication systems and signal processing.

B. E. Wells. Computer architecture, parallel processing, digital design.

THE UNIVERSITY OF ARIZONA

Department of Electrical and Computer Engineering

Programs of Study

The department offers graduate programs leading to the M.S. and Ph.D. degrees with a major in electrical and computer engineering. A nonmajors program is available for qualified U.S. citizens who do not hold degrees in electrical or computer engineering.

The Master of Science degree requires a minimum of 30 units. There are thesis and nonthesis options. Candidates must pass a final oral examination.

The Ph.D. program must contain a minimum of 54 units of course work (including the Master of Science degree) and 18 units of dissertation study. To satisfy the residence requirement, the student must spend a minimum of two regular semesters of full-time study on campus. Students must pass a qualifying examination, which is usually taken during the first semester of residence beyond the master's degree, and are admitted to candidacy after passing an oral comprehensive examination near the end of the study program. The final examination is a defense of the dissertation. There is no foreign language requirement.

Research Facilities

The department is active in research in the general areas of communications, controls, and signal processing; computer engineering; electromagnetics; and microelectronics. Specialized laboratories are available to support research efforts. Research programs are an integral part of the department's educational activities. All facilities are housed in a modern building that includes more than 50,000 square feet of laboratory space. Much of the research is supported by grants from federal agencies or by industrial contracts.

Financial Aid

Fellowships, research and teaching assistantships, and tuition and academic scholarships are available for qualified students. In 1999–2000, assistantships provide a stipend of $15,714 to $18,855 per academic year (ten months) plus a waiver of out-of-state tuition. Recipients devote 20 hours per week to research or teaching duties during the academic year and may significantly supplement their stipend by full-time research in the department during the summer. Supplemental scholarships are available for especially well qualified students.

Cost of Study

The registration fee for full-time Arizona residents was $2162 per academic year in 1998–99. Students who have not yet established Arizona residence paid an additional $6952 in out-of-state tuition; this is normally waived for students supported by an assistantship or fellowship.

Living and Housing Costs

The average cost of a room in the residence halls was $2850 per academic year in 1998–99. Comparably priced off-campus housing is available within easy walking distance.

Student Group

There are about 207 full-time graduate students in the department. Research assistantships, teaching assistantships, scholarships, and fellowships are available to well-qualified applicants. The department has an active Graduate Student Association.

Location

The University is located in Tucson, whose excellent climate, clean air, and mountain vistas have made it a magnet for visitors and new residents. The metropolitan area has a population of approximately 680,000. Tucson enjoys mild winters. In the Santa Catalina mountain range on the north edge of the city, ski slopes, ponderosa pines, canyons, and grassy meadows attract skiers, climbers, and hikers. Tucson has a symphony orchestra, an opera company, many art galleries, and the Arizona Sonora Desert Museum. The city is often visited by theater companies and musicians.

The University

Founded in 1885, the University of Arizona is ranked by the National Science Foundation as one of the top twenty research universities in the nation. The University has fourteen colleges and an enrollment of 35,000 students; of these, 2,884 are enrolled in the fourteen departments of the College of Engineering and Mines. In the past decade, the University of Arizona Wildcat athletic teams, which compete in the Pacific-10 Conference, played in five postseason football bowl games, won six men's basketball conference titles, went to the men's NCAA Final Four in basketball three times, won four Collegiate World Series softball titles, and won national championships in both men's and women's golf. Most recently, the Cats won their first-ever NCAA championship in men's basketball in 1997.

Applying

Domestic applicants should submit applications for admission and all supporting material to the Graduate College as early as possible and no later than May 1 for the summer terms, June 1 for the fall term, and October 1 for the spring term. International applicants should submit applications for admission and all supporting material by February 1 for the summer and fall terms and August 1 for the spring term. Applicants for financial assistance must be accepted for admission and should submit supporting material by January 15 for the fall term and October 15 for the spring term. Applicants are required to submit GRE General Test scores, a statement of purpose, and three letters of recommendation directly to the department. All students whose native language is other than English must submit TOEFL scores directly to the Graduate College.

Correspondence and Information

Graduate Studies Office
Department of Electrical and Computer Engineering
ECE Building, Room 230
The University of Arizona
Tucson, Arizona 85721
Telephone: 520-621-6195

The University of Arizona

THE FACULTY AND THEIR RESEARCH

Jennifer K. Barton, Assistant Professor (also Assistant Professor in Biolmedical Engineering); Ph.D., Texas at Austin, 1998. Optical imaging (optical coherence tomography), laser-tissue interaction, bioinstrumentation.

John R. Brews, Professor; Ph.D., McGill, 1965. Semiconductor device physics, MOSFET design, electromagnetics of interconnections.

Jo Dale Carothers, Associate Professor; Ph.D., Texas at Austin, 1989. VLSI design, physical design (partitioning, placement, and routing) for single and multichip modules, low-power design, engineering applications of graph theory.

Francois E. Cellier, Professor; Ph.D., Swiss Federal Institute of Technology, 1979. Modeling, simulation, control, software engineering.

Thomas C. Cetas, Professor (also Professor in Radiation/Oncology); Ph.D., Iowa State, 1970. Thermal ionizing and non-ionizing radiation research for cancer and other medical therapies.

Donald G. Dudley, Professor Emeritus; Ph.D., UCLA, 1968. Target identification, transient scattering, coupling and penetration, microwave pulsed power, geophysical modeling.

Steven L. Dvorak, Associate Professor; Ph.D., Colorado at Boulder, 1989. Electromagnetic transients, wave propagation, analytical and computational electromagnetics, optics, applied mathematics, microwave measurements.

Jack D. Gaskill, Professor (also Professor in Optical Sciences); Ph.D., Stanford, 1968. Fourier optics, diffraction theory.

Glen C. Gerhard, Professor; Ph.D., Ohio State, 1963. Medical instrumentation and systems, technology and innovation in education.

Salim Hariri, Associate Professor; Ph.D., USC, 1986. Distributive computing systems.

Charles M. Higgins, Assistant Professor; Ph.D., Caltech, 1993. Neuromorphic engineering, focusing on analog VLSI for vision and robotic systems; real-time VLSI systems for processing visual motion, disparity, and egomotion; integrated sensors and control systems for highly capable, inexpensive autonomous robotics.

Fredrick J. Hill, Professor Emeritus; Ph.D., Utah, 1963. Digital systems, design languages, design automation, test generation.

W. Timothy Holman, Assistant Professor; Ph.D., Georgia Tech, 1994. Analog microelectronics.

Lawrence P. Huelsman, Professor Emeritus; Ph.D., Berkeley, 1960. Active and passive filters, computer-aided design.

Bobby R. Hunt, Professor; Ph.D., Arizona, 1967. Signal processing, digital image processing.

Raymond K. Kostuk, Professor; Ph.D., Stanford, 1986. Optical interconnects, diffractive optical elements, fiber optics, optical data storage.

Marwan M. Krunz, Assistant Professor; Ph.D., Michigan State, 1995. Modeling and performance evaluation of high-speed computer networks, traffic characterization, video-on-demand architectures.

Ahmed Louri, Associate Professor; Ph.D., USC, 1988. Computer architecture, computer networks, parallel processing, parallel algorithms, optical computing.

Michael W. Marcellin, Professor; Ph.D., Texas A&M, 1987. Data compression, digital communication and storage systems, digital signal processing.

Michael M. Marefat, Associate Professor; Ph.D., Purdue, 1991. Machine intelligence, software engineering, robotics, computer vision, computer graphics, intelligent control, CAD/CAM.

Ralph Martinez, Associate Professor; Ph.D., Arizona, 1976. Computer systems application and design, distributed computing environments, computer networks, medical imaging and communications, telemedicine systems.

Mark A. Neifeld, Associate Professor; Ph.D., Caltech, 1990. Optical memories and information processing, parallel coding and signal processing, pattern recognition, neural networks.

John F. O'Hanlon, Professor; Ph.D., Simon Fraser, 1967. Vacuum physics, microcontamination, semiconductor processing, physical electronics.

Olgierd A. Palusinski, Professor; Ph.D., Silesian Technical (Poland), 1966. Computer-aided design of integrated circuits, electronic packaging, circuit simulation.

John Papapolymerou, Assistant Professor; Ph.D., Michigan, 1999. RF/microwave circuits and systems, wireless communications, MEMS, integrative passives.

Harold G. Parks, Associate Professor; Ph.D., RPI, 1980. Integrated circuit processing, defect studies, yield enhancement and modeling, semiconductor devices.

John L. Prince, Professor; Ph.D., North Carolina State, 1969. Microelectronics, electronic packaging.

John A. Reagan, Professor and Department Head; Ph.D., Wisconsin–Madison, 1967. Electromagnetic remote sensing, atmospheric radiation and optics, optoelectronic instrumentation.

Jeffrey J. Rodriguez, Associate Professor; Ph.D., Texas at Austin, 1990. Digital image processing, computer vision, digital signal processing, biomedical image processing.

Jerzy W. Rozenblit, Professor; Ph.D., Wayne State, 1985. Modeling, simulation, artificial intelligence, knowledge-based design, intelligent systems, engineering of computer-based systems, codesign.

William E. Ryan, Associate Professor, Ph.D., Virginia, 1988. Error control coding, turbo codes, trellis codes, algebraic codes, linear and nonlinear equalization, applications of communication theory and coding to data transmission and storage.

Larry C. Schooley, Professor; Ph.D., Kansas, 1968. Communications systems, digital communication networks, telemetry, telescience.

Robert A. Schowengerdt, Professor; Ph.D., Arizona, 1975. Remote sensing systems and image processing, multispectral sensor and scene modeling, digital mapping techniques.

Robin N. Strickland, Professor and Associate Department Head; Ph.D., Sheffield, 1979. Digital image processing, computer vision, signal processing.

Malur K. Sundareshan, Professor; Ph.D., Indian Institute of Science, 1973. Control systems, communication networks, statistical signal processing, neural network theory, applications to control system design.

Miklos N. Szilagyi, Professor; Ph.D., 1965, Leningrad Electrotechnical; D.Sc., Hungarian Academy of Sciences, 1979. Particle beams and optics, computer-aided synthesis of electron and ion optical systems, physical electronics, artificial intelligence, neural networks, nonlinear stochastic dynamic system simulation.

Hal S. Tharp, Associate Professor; Ph.D., Illinois at Urbana-Champaign, 1986. Multivariable, multirate, robust, adaptive, and neural network control theory.

Spyros Tragoudas, Associate Professor; Ph.D., Texas at Dallas, 1991. VLSI CAD, VLSI testing, data transmission in networks.

Kathleen L. Virga, Assistant Professor; Ph.D., UCLA, 1996. Antennas for wireless communications and radar systems, measurement-base characterization of high-density microwave and high-speed VLSI circuits.

Sarma Vrudhula, Professor; Ph.D., USC, 1985. Design automation and testing of digital systems.

Arthur F. Witulski, Associate Professor; Ph.D., Colorado at Boulder, 1988. Power electronics, resonant converters, distributed electronic power systems, low-voltage power converters.

James C. Wyant, Professor (also Professor in Optical Sciences); Ph.D., Rochester, 1968. Optical testing, interferometry, holography, computerized optical metrology.

Bernard P. Zeigler, Professor; Ph.D., Michigan, 1968. Modeling and simulation environments, high-performance discrete event simulation, distributed modeling and simulation.

Richard W. Ziolkowski, Professor; Ph.D., Illinois at Urbana-Champaign, 1980. Computational electromagnetics; transient linear and nonlinear electromagnetics, optics, and acoustic phenomena.

UNIVERSITY OF CALIFORNIA, IRVINE

Department of Electrical and Computer Engineering

Programs of Study	The Department of Electrical and Computer Engineering offers courses leading to the degrees of Master of Science and Doctor of Philosophy, with concentrations in electrical engineering, computer networks and distributed computing, and computer systems and software. The electrical engineering concentration includes communication systems, control systems, digital systems, optoelectronic devices, semiconductor devices, electronics, electrooptics, machine vision, and signal processing. The computer engineering concentrations cover VLSI design, computer architecture, parallel and distributed computer systems, fault-tolerant computing, real-time systems, computer networks, and system software. The programs are designed for those who plan to enter the professional practice of engineering as it relates to design, research, teaching, and development—in industry, private practice, education, or public service. The fundamentals of engineering are emphasized so that graduates can continue professional development throughout their careers. Individual programs permit wide latitude in course work and research.
	The M.S. degree may be attained by the successful completion of 36 approved units or by a combination of course work and a thesis.
	The Ph.D. degree requires passing a preliminary examination, preparing research, advancing to candidacy, completing significant research investigation, and submitting and getting approval of a dissertation.
Research Facilities	Optical and solid-state research is performed in six laboratories, including facilities for integrated optics and microfabrication, MEMS fabrication, quantum electronics, photonics, advanced semiconductor devices, optoelectronic devices, and optoelectronic materials research. These laboratories house cleanroom facilities; photolithography, thin-film deposition, and molecular beam epitaxy systems; solid-state, gas, dye, and semiconductor lasers; optical benches and microscopes; scanning acoustic and electron microscopes; computer-based data acquisition systems; cryogenic and high-current testing chambers; and a host of electronic testing equipment. Advanced research in analog/RF circuit design and simulation is performed in the Analog Techniques Fueling Advanced Research in Microelectronics (ANTFARM) Laboratory.
	Computer engineering research laboratories include the VLSI design automation laboratory; the fault-tolerant multicomputer laboratory; the advanced computer architecture laboratory; the Distributed Real-time Ever Available Microcomputing (DREAM) Laboratory; the numerical processors and multiprocessors laboratory; the real-time systems laboratory; and the database systems laboratory. They are equipped with heterogeneous networks of state-of-the-art Sun, DEC, and Hewlett-Packard workstations; microprocessor development systems; an Intel hypercube; several high-performance file and compute servers; and a variety of advanced software packages such as VLSI CAD tools for layout, simulation, and synthesis and tools for performance evaluation and parallel programming.
	Research in machine vision and image processing is carried out at several laboratories, including the image acquisition laboratory, the pattern recognition and image modeling laboratory, the visualization laboratory, and the imaging architectures laboratory. These laboratories are equipped with image acquisition equipment, digital video processing equipment, Sun Workstations, graphics displays, and graphics and image processing software packages. Within the department, there is access to high-performance parallel machines and tools for the development of custom chips.
	The Power Electronics Laboratory is equipped with state-of-the-art instrumentation for design, simulation, layout, prototyping, and testing of switching/analog circuits.
Financial Aid	Fellowships and teaching and research assistantships are available on a competitive basis. Except for students on visas, there are opportunities for part-time work in the engineering community of Orange County. With the same exception, financial aid may be obtained from UCI's Financial Aid Office.
Cost of Study	In 1999–2000, student fees are $1726 per quarter for California residents and an additional $3441 per quarter for nonresidents. These fees are subject to change.
Living and Housing Costs	On-campus housing is available. In 1999–2000, monthly apartment rents are from $277 to $725 for single students and from $554 to $1202 for married students and families. Early application is advised for on-campus housing. Privately owned apartments are available close to the campus, and many types of housing can be found in the surrounding communities of Santa Ana, Newport Beach, Costa Mesa, Irvine, Tustin, and Laguna Beach.
Student Group	Current campus enrollment is 18,209, including 1,263 undergraduate and 309 graduate students in the School of Engineering.
Student Outcomes	M.S. graduates typically fill leading engineering positions in high-technology industries such as electronics, communications, computers, software, and aerospace. Ph.D. graduates most often obtain research and development positions with large industrial or governmental research laboratories or take academic positions that involve both teaching and research in their areas of specialization.
Location	The 1,510-acre UCI campus is in Orange County, 40 miles south of Los Angeles. Irvine is one of the nation's fastest-growing residential, industrial, and business areas, yet within view of the campus is a wildlife sanctuary; Pacific Ocean beaches are nearby. Residential areas range from the beach communities of Newport Beach and Laguna Beach to the socially and economically diverse urban centers of Santa Ana, Tustin, and Costa Mesa.
The University	One of the nine campuses in the University of California system, UCI now enrolls 3,638 graduate and professional students. The University offers graduate degrees through the Schools of Biological Sciences, Engineering, Fine Arts, Humanities, Physical Sciences, Social Ecology, and Social Sciences; the Graduate School of Management; the College of Medicine; and the Department of Information and Computer Science.
Applying	Application forms may be obtained by writing to the department. The deadlines for applications are May 1 for the fall quarter, October 15 for the winter quarter, and January 15 for the spring quarter. Applicants who wish to be considered for fellowships or for teaching or research assistantships should apply by February 1. Applicants must submit official records covering all postsecondary academic work, three letters of recommendation, and official scores on the General Test of the Graduate Record Examinations. International students whose native language is not English must submit the results of the Test of English as a Foreign Language (TOEFL).
Correspondence and Information	For applications and information about the department:

Graduate Admissions
Department of Electrical and Computer Engineering
355 Engineering Tower
University of California
Irvine, California 92697-2625

Telephone: 949-824-5489
E-mail: ece_grad@ece.uci.edu
World Wide Web: http://www.eng.uci.edu/ece/ece_grad

University of California, Irvine

THE FACULTY AND THEIR RESEARCH

Nicolaos G. Alexopoulos, Professor; Ph.D., Michigan. Integrated microwave and millimeter-wave circuits and antennas, substrate materials and thin films, electromagnetic theory.

Nader Bagherzadeh, Professor of Electrical and Computer Engineering and Information and Computer Science and Chair of the Department; Ph.D., Texas at Austin. Parallel processing, computer architecture, VLSI design.

Harut Barsamian, Adjunct Professor; M.S., USSR Academy of Sciences. Computer architectures, software engineering.

Neil J. Bershad, Professor Emeritus; Ph.D., RPI. Communication and information theory, signal processing.

Lubomir Bic, Professor of Information and Computer Science and Electrical and Computer Engineering; Ph.D., California, Irvine. Parallel processing, distributed systems, database machines.

Douglas M. Blough, Associate Professor of Electrical and Computer Engineering and Information and Computer Science; Ph.D., Johns Hopkins. Parallel processing, fault-tolerant computing, computer architecture.

Lynn Choi, Assistant Professor of Electrical and Computer Engineering; Ph.D., Illinois at Urbana–Champaign. Computer architecture, microprocessor design, optimizing compilers.

Rui J. P. de Figueiredo, Professor; Ph.D., Harvard. Intelligent sensing and control, applied mathematics.

Franco De Flaviis, Assistant Professor of Electrical and Computer Engineering; Ph.D., UCLA. Wireless communications devices and systems.

Nikil Dutt, Associate Professor of Information and Computer Science and Electrical and Computer Engineering; Ph.D., Illinois at Urbana–Champaign. Design modeling, languages and synthesis, CAD tools, computer architecture.

Daniel D. Gajski, Professor of Information and Computer Science and Electrical and Computer Engineering; Ph.D., Pennsylvania. Parallel algorithms and architectures, design methodology, design science, CAD algorithms and tools, software/hardware codesign.

Hideya Gamo, Professor Emeritus; D.Sc., Tokyo. Quantum electronics, electromagnetics, optics.

Michael Green, Associate Professor; Ph.D., UCLA. Analog integrated-circuit design, circuit simulation, nonlinear circuits.

Glenn E. Healey, Associate Professor; Ph.D., Stanford. Machine vision, computer engineering, image processing, computer graphics, intelligent machines.

Daniel Hirschberg, Professor of Information and Computer Science and Electrical and Computer Engineering; Ph.D., Princeton. Analysis of algorithms, data structures, models of computation.

K. H. (Kane) Kim, Professor of Electrical and Computer Engineering and Information and Computer Science; Ph.D., Berkeley. Ultrareliable distributed and parallel computing, real-time object-based system engineering.

Fadi J. Kurdahi, Associate Professor of Electrical and Computer Engineering and Information and Computer Science; Ph.D., USC. VLSI system design, design automation of digital systems.

Tomas Lang, Professor of Electrical and Computer Engineering and Information and Computer Science; Ph.D., Stanford. Numerical processors and multiprocessors, parallel computer systems.

Chin C. Lee, Professor; Ph.D., Carnegie Mellon. Electronic packaging, thermal management, integrated optics, photonics.

Henry P. Lee, Associate Professor; Ph.D., Berkeley. Optoelectronics, semiconductor materials and devices.

Guann-Pyng Li, Professor; Ph.D., UCLA. High-speed semiconductor technology, optoelectronic devices, integrated circuit fabrication and testing.

Kwei-Jay Lin, Professor of Electrical and Computer Engineering and Information and Computer Science; Ph.D., Maryland College Park. Real-time systems, distributed systems.

Orhan Nalcioglu, Professor of Radiological Sciences and Electrical and Computer Engineering; Ph.D., Oregon. Nuclear magnetic resonance imaging and spectroscopy, digital radiography.

Richard D. Nelson, Adjunct Professor; Ph.D., Michigan State. Sensors, microelectronics, photonics, medical imaging.

Alexandru Nicolau, Professor of Information and Computer Science and Electrical and Computer Engineering; Ph.D., Yale. Architecture, parallel computation, programming languages and compilers.

Robert M. Saunders, Professor Emeritus; Dr.Eng., Tokyo Institute of Technology; PE. Electromechanics, power systems.

Issac D. Scherson, Professor of Information and Computer Science and Electrical and Computer Engineering; Ph.D., Weizmann (Israel). Parallel computing architectures, massively parallel systems, parallel algorithms, interconnection networks, performance evaluation.

Roland Schinzinger, Professor Emeritus; Ph.D., Berkeley; PE. Electromagnetics, power systems, operations research.

Phillip C.-Y. Sheu, Professor of Electrical and Computer Engineering and Information and Computer Science; Ph.D., Berkeley. Robotics, database systems.

Jack Sklansky, Professor Emeritus of Electrical and Computer Engineering and Radiological Sciences; D.Sc., Columbia; PE. Pattern recognition, machine vision, medical imaging, neural learning, computer engineering.

Keyue M. Smedley, Assistant Professor; Ph.D., Caltech. Power electronics.

Allen R. Stubberud, Professor; Ph.D., UCLA; PE. Control systems, digital signal processing, estimation and optimization.

Tatsuya Suda, Professor of Information and Computer Science and Electrical and Computer Engineering; Ph.D., Kyoto. Computer networks, distributed systems, performance evaluations.

Harry H. Tan, Associate Professor; Ph.D., UCLA. Communication systems, information theory, coding theory, stochastic processes.

Chen S. Tsai, Professor; Ph.D., Stanford. Integrated and fiber-optic devices and materials, acoustooptics, magnetooptics, acoustic microscopy.

Wei Kang (Kevin) Tsai, Associate Professor; Ph.D., MIT. Data communication networks, neural networks, parallel algorithms and architectures, CAD for VLSI systems engineering.

UNIVERSITY OF CALIFORNIA, LOS ANGELES

Electrical Engineering Department

Programs of Study	The department offers programs of study leading to the degrees of Master of Science and Doctor of Philosophy in electrical engineering in the areas listed under faculty research areas.
	The M.S. program offers specializations in nine major fields; it requires a total of nine courses and either a thesis or a comprehensive examination. In the thesis plan, two of the nine courses relate to the research needed for writing the thesis. The program lasts from one to two years; some major fields offer the thesis plan only.
	The program for the Ph.D. degree requires a course of study in one major field and two distinct but supporting minor fields, followed by research on a topic in the major field. Competence in the major field is determined by an 8-hour examination and in each of the two minor fields by passing the three prescribed courses with an adequate grade point average. The research topic is chosen following discussion with the student's adviser, who then guides the research as it progresses. The research is monitored by a doctoral committee, which administers two oral examinations and approves the dissertation describing the research. There is no foreign language requirement. The Ph.D. program is usually completed in four or five years after the award of the M.S. degree.
Research Facilities	Laboratories are available in the department for research in modern integrated semiconductor device processing, complete molecular beam epitaxy systems, analog and digital electronics, hybrid integrated circuits, microwaves and millimeter waves, fiber optics, speech and image processing, microelectromechanical systems, antennas, communications and networking, lasers and quantum electronics, and plasma electronics. The department is also associated with research centers for high-frequency and high-speed electronics and for plasma physics and fusion engineering, all located at UCLA.
	Computer facilities for research and instruction range from the supercomputer to desktop workstations. The Electrical Engineering Department maintains a large network of UNIX platforms consisting of Sun, HP, and IBM workstations for the exclusive use of its graduate students and faculty members. A campuswide computing service offers access through remote terminals. All departments in the School of Engineering and Applied Science are linked by a common Ethernet (SEASNET), which provides access to the IBM 3090, the UNIX workstations, IBM PS/2, and Apple Macintosh classrooms.
	The UCLA library, which ranks in the top three nationally, has more than 6 million volumes. One of its specialized branches, the Science & Engineering Library (SEL), contains more than 460,000 volumes and receives more than 7,000 serials and more than 1.9 million technical reports. SEL provides major access to library materials through Orion, the UCLA online information system, and Melvyl, the UC (nine campuses) online system.
Financial Aid	Fellowships are available from a variety of sources for full-time students and are awarded on merit, mainly for the first year of study. For 1999–2000, stipends range from $10,000 to $13,500. Many teaching and research assistantships are also offered, usually after the first year.
Cost of Study	For 1999–2000, California residents pay $4555.50 for registration and incidental fees per academic year. Nonresidents pay an additional fee of $9384. The academic year consists of three quarters.
Living and Housing Costs	A typical budget for a California resident living in an off-campus apartment is $18,387.50 a year and includes required books and supplies, board and room for the period classes are in session during the three quarters, and a minimum allowance for variable items. For students with families, the University has 1,112 apartments, which rent for $667 to $1502 per month in 1999–2000. In addition, approximately 250 apartments within walking distance of campus are available for single students.
Student Group	The department has approximately 350 graduate students, who come from all parts of the world. The great majority of them are in their mid-twenties and enter the department in the M.S. program. About 25 percent continue for the Ph.D. This is the preferred mode of entry to the Ph.D. program, but a few students are admitted directly after they receive an M.S. degree elsewhere. International students constitute about 40 percent of the total and women about 13 percent.
Location	UCLA is located 5 miles from the Pacific Ocean on the north side of Los Angeles in the Westwood area, immediately adjacent to the Santa Monica Mountains. Los Angeles offers all the cultural and recreational amenities one expects in a major metropolitan area, such as theaters, cinemas, concert halls, museums, sports arenas, amusement parks, and facilities for swimming, sailing, skiing, and hiking.
The University and The Department	UCLA, one of the nine campuses that form the University of California, ranks among the leading universities in the United States. Student enrollment is about 35,000. Many undergraduates and a few graduate students commute from their homes; others live in apartments nearby or in the University dormitories on campus. Extensive programs for cultural events are presented on campus, and there are recreational facilities of all kinds available through the University.
	The Electrical Engineering Department has been in existence as a separate entity since 1968, when the College of Engineering was divided into departments and became the School of Engineering and Applied Science. Since then it has risen rapidly in stature and now ranks among the top ten in the nation. It has a faculty of 44 full-time professors and a number of visiting and part-time professorial appointees.
Applying	Graduate students are admitted only in the fall quarter, and the application deadline is January 15. Applicants for financial aid are encouraged to apply earlier. Applications must include transcripts of all previous academic work, scores on the GRE General Test, three letters of recommendation, and a statement of study plans. The GRE should be taken no later than the previous October. Offers of fellowships are made by about March 15 and offers of admission by April 15. The UCLA Application for Graduate Admission is now available on the World Wide Web.
Correspondence and Information	Vice-Chairman for Graduate Affairs Electrical Engineering Department 56-125B Engineering IV Box 951594 University of California Los Angeles, California 90095-1594 Telephone: 310-825-9383 World Wide Web: http://www.ee.ucla.edu

University of California, Los Angeles

FACULTY RESEARCH AREAS

Communications and Telecommunications

Research is concerned with communications, telecommunications, networking, and information processing principles and their engineering applications. Communications research includes satellite, spread-spectrum, and digital communications systems. Fast estimation, detection, and optimization algorithms and processing techniques for communications, radar, and VLSI design are studied. Research is conducted in stochastic modeling of telecommunications engineering systems, switching, architectures, queuing systems, computer communications networks, local-area/metropolitan-area/long-haul communications networks, optical communications networks, packet-radio and cellular radio networks, and personal communications systems. Research in networking also includes studies of processor communications and synchronization for parallel and distributed processing in computer systems. Several aspects of communications networks and processing systems are thoroughly investigated, including system architectures, protocols, performance modeling and analysis, simulation studies, and analytical optimization. Investigations in information theory involve basic concepts and practices of channel and source coding. Significant multidisciplinary programs, including sensing and radio communication networks, exist.

Integrated Circuits and Systems

IC&S students and faculty members are engaged in research on communications and RF IC design; analog and digital signal processing microsystems; integrated microsensors, microelectromechanical systems, and the associated low-power microelectronics; reconfigurable computing systems; and multimedia and communications processors. Current projects include wireless transceiver ICs, including RF and baseband circuits; high-speed data communication ICs; A/D and D/A converters; networking electronics; distributed sensors with wireless networking; and digital processor design. M.S. and Ph.D. degrees require a thesis based on an ongoing IC&S project and full-time presence on campus. More information may be obtained via the World Wide Web (http://www.icsl.ucla.edu/general/profs.html).

Signal Processing

Signal processing encompasses the techniques, hardware, algorithms, and systems used to process one-dimensional and multidimensional sequences of data. Research being conducted in the Signal Processing Group reflects the broad, interdisciplinary nature of the field today. Areas of current interest include analysis, synthesis, and coding of speech signals; video signal processing; digital filter analysis and design; image compression; communications signal processing; synthetic aperture radar remote sensing; signal processing for hearing aids; auditory system modeling; automatic speech recognition; wireless communication; digital signal processor architectures; adaptive filtering; multirate signal processing; and the characterization and analysis of three-dimensional time-varying medical image data. M.S. and Ph.D. programs include a thesis project, and a full-time presence on campus is required.

Control Systems

Faculty members and students conduct research in control, estimation, filtering, and identification of dynamic systems, including deterministic and stochastic, linear and nonlinear, and finite- and infinite-dimensional systems. Topics of particular interest include adaptive, distributed, nonlinear, optimal, and robust control, with applications to autonomous systems, smart structures, flight systems, microrobotics, and microelectromechanical systems.

Solid-State Electronics

Research involves studies of new and advanced devices with picosecond switching times and high-frequency capabilities up to submillimeter-wave ranges. Topics being investigated are hot-electron transistors, quantum devices, heterojunction bipolar transistors, HEMTs, and MESFETs, as well as more conventional scaled-down MOSFETs, SOI devices, bipolar devices, and photovoltaic devices. The studies of basic materials, submicron structures, and device principles range from Si, Si-Ge, Si-silicides, and III-V molecular beam epitaxy to the modeling of electron transport in high fields and short temporal and spatial scales. The research in progress also includes fabrication, testing, and reliability of new types of VLSI devices and circuits.

Photonics and Optoelectronics

The area of photonics and optoelectronics includes the development and applications of new types of solid-state, gas, and semiconductor lasers. This research also extends into areas of nonlinear optics, ultrashort pulse generation and applications, very high speed detection, infrared detectors, optical logic, fiber optics, integrated optics, optoelectronics, and optical communications. The application of quantum electronics technology to the testing and control of solid-state devices is a fast-growing and particularly interesting project. Equipment used for research includes argon-ion lasers, mode-locked solid-state lasers, compressed femtosecond laser pulses, continuous-wave dye lasers, single-frequency tunable dye lasers, pulsed dye lasers, excimer lasers, and infrared TEA lasers, as well as the latest gigabit-rate electronics, fiber optics, and optical systems.

Electromagnetics

Research is being pursued on satellite and personal communication antennas, biological interactions, integrated microwave and millimeter-wave circuits and printed-circuit antennas, substrate material effects, photonic bandgap structures, and novel guiding structures; integrated optics and optical signal processing; antenna theory and design, mutual coupling effects, integrated antennas and antenna radar cross-section studies; scattering by complex bodies, radar cross-section reduction techniques, reflector antenna design and analysis, modern antenna measurement and diagnostic techniques, satellite/spacecraft antenna studies, atmospheric pollutant scattering, and multiple scattering; geometrical theory of diffraction and asymptotic techniques; advanced numerical techniques in electromagnetics; electromechanics; and nonlinear electrodynamics.

Operations Research

Research is being conducted in optimization theory (nonconvex programming, linear and nonlinear programming, and applications to networks with particular emphasis upon communications network problems and engineering design problems) and in stochastic processes (renewal and point processes, Markov processes, queuing theory, stochastic dynamic programming, and applications to communications and telecommunications engineering). The department offers a combined set of courses for those students interested in telecommunications engineering and operations research, and students of operations research are expected to have a strong interest in telecommunications networks and engineering. Students with backgrounds in engineering, physical science, or mathematics are encouraged to apply.

Plasma Electronics

Research is concerned with the electrodynamics of charged particles in electrified fluids. Originally developed for understanding the behavior of magnetically and inertially confined plasmas for controlled fusion energy, this field has moved to such active areas as the generation of high-power radiation sources, such as free-electron lasers, plasma-wave particle accelerators, far-infrared and submillimeter-wave plasma diagnostics, plasma processing, and alternative fusion concepts. Extensive laboratory facilities exist, including high-power laser systems, microwave and millimeter-wave sources and detectors, high- and low-density plasma sources for fundamental studies of nonlinear wave effects, and a variety of diagnostic instruments. The NEPTUNE laser laboratory, a high-intensity subpicosecond laser facility, a plasma processing laboratory, a FEL laboratory, and a laser-plasma interaction laboratory are available for student training and research. In addition, experiments are conducted at several national laboratories and major fusion devices such as DIII-D Tokamak.

UNIVERSITY OF CALIFORNIA, RIVERSIDE

College of Engineering
Department of Electrical Engineering

Program of Study

The Bourns College of Engineering offers programs leading to M.S. and Ph.D. degrees in electrical engineering.

Instruction and opportunity for directed study exist in a variety of areas, including coding, communications, computer vision, control, detection and estimation, distributed systems, system identification, image processing, information theory, intelligent systems, machine learning, modeling and simulation, multimedia, navigation, neural networks, pattern recognition, robotics, and signal processing.

The master's degree program is normally completed in two years but can be achieved in one year by well-prepared students. The student may either pass a qualifying examination or complete a research thesis.

The normal time to complete the Ph.D. degree is three years for students holding an M.S. degree in electrical engineering and five years for those who entered the program without an M.S. degree in electrical engineering. A typical program includes completing course work, passing a Preliminary Examination, taking six months for independent research, presenting a dissertation proposal and passing the associated Qualifying Exam, and completing and presenting the dissertation research.

Research Facilities

The newly completed science library and main campus library maintain extensive holdings of engineering, mathematics, and computer science books and journals, including back issues. The Department of Electrical Engineering maintains extensive computer facilities, including more than ten servers supporting more than 200 computers. All students receive computer accounts with Internet and e-mail access. Electrical engineering teaching laboratories include: the Circuit and Electronics Lab Dynamic Systems and Control Lab, the Electrical Engineering Electronics Shop, the Image Processing and Computer Vision Lab, the Logic Design and Digital Systems Lab, the Pentium and Sun Lab, the Robotics and Design Projects Lab, and the Signals and Communication Lab. Electrical engineering research laboratories include: the Distributed Robotics and Multimedia Lab, the Communication Research Lab, the Lab for Identification and Control, the Neural Networks and Pattern Recognition Lab, the Robotics Research Lab, and the Visualization and Intelligent Systems Lab. The Visualization and Intelligent Systems Laboratory (VISLab) is involved in basic and applied research developing intelligent systems for autonomous navigation, automatic object recognition, remote sensing, manufacturing, and various industrial and medical applications. It undertakes research in image processing, computer vision, computer graphics, CAD/CAM, pattern recognition, artificial intelligence, robotics, geometric modeling, perception, man–machine interfaces, machine learning, neural networks, transportation, the environment, and molecular modeling. The Center for Research in Intelligent Systems (CRIS) involves an interdisciplinary team of 18 UCR faculty members from seven departments to promote research and development of autonomous/semiautonomous systems with sensing capabilities that are able to communicate and interact with other intelligent (biological and artificial) systems. These intelligent systems perform tasks that require understanding of the environment through knowledge, learning, reasoning, and planning. The College of Engineering–Center for Environmental Research and Technology (CE-CERT) brings together academia, the regulatory community, and industry for cooperative research on the environment. CE-CERT's primary focus is on air pollution: the atmospheric processes that contribute to it and the technologies that can help to control it. The center helps to develop and analyze strategies and technologies for a cleaner, more energy-efficient future.

Financial Aid

Fellowships are awarded by the Graduate Division on a competitive basis, with stipends up to $15,000 for the nine-month academic year. These awards include payment of all assessed registration fees. Electrical engineering teaching assistantships are also available. A half-time appointment as a teaching assistant carried a stipend of $13,329 for the 1998–99 academic year. Research assistantships on supported programs are also available to well-prepared students.

Cost of Study

For 1998–99, California residents paid $4861 a year in tuition fees. Nonresidents were charged an additional tuition fee of $3128 per quarter. These amounts are subject to change.

Living and Housing Costs

Riverside offers graduate students one of the lowest costs of living of any city with a University of California campus. The cost of room and board in the residence halls averaged $6444 in 1998–99. The University owns 268 houses that are available to married students and single students with children. These houses rent for $430 to $470 per month. Rents for approximately 150 apartments and ninety-two suites that are available for single students range from $300 to $775 per month. Off-campus housing is available within walking distance of campus.

Student Group

The electrical engineering graduate program began in fall 1998. Out of 40 applicants, 9 students were admitted. Of those, 2 were women and 8 were international. The electrical engineering graduate admission committee seeks to expand on this group of talented and hardworking individuals and expects to admit 50 to 100 graduate students in the next five years. Faculty members look for high GRE scores and high GPAs and prefer a strong research background in students applying for this program.

Student Outcomes

University of California, Riverside, and the College of Engineering are working with industry in the Inland Empire to establish a strong technical workforce in this part of southern California. The Electrical Engineering Department also seeks to place graduates in top industrial and academic positions worldwide.

Location

Riverside is located between the golden coastline of Los Angeles and Orange Counties (40 miles west), the desert communities of Palm Springs (50 miles east), the San Bernardino Mountains (20 miles north), and San Diego (90 miles south). The location offers excellent opportunities for skiing, hiking, and biking. In addition, the climate offers the opportunity to take part in outdoor activities throughout the year. Riverside is a community of 250,000 people with excellent recreational facilities, a symphony orchestra, an opera association, a community theater, and several free community events.

The College

Founded in 1989, The Marlan and Rosemary Bourns College of Engineering stresses interdisciplinary collaboration. Faculty members and students work together inside Bourns Hall, a three-story, two-building architectural showpiece that opened in 1994. Instruction in the electrical engineering graduate programs began in fall 1998.

Applying

Application deadlines for international applicants are February 1 for the fall quarter, July 1 for the winter quarter, and October 1 for the spring quarter. Deadlines for domestic applicants are May 1 for the fall quarter, September 1 for the winter quarter, and December 1 for the spring quarter. To receive full consideration for financial support for fall, all application materials must be received by January 5. Applications postmarked after the published deadline are deferred to the following quarter. The University does not grant waivers of the graduate application fee. Specific requirements for admission can be found on the department's Web site, listed below.

The Graduate School Application, application fee ($40), official transcripts, three letters of recommendation, and official GRE and TOEFL scores (if applicable) should be submitted directly to the department.

Correspondence and Information

Electrical Engineering Graduate Student Affairs
Bourns College of Engineering
University of California, Riverside
Riverside, California 92521
Telephone: 909-787-2423
Fax: 909-787-2425
E-mail: grad-adm@ee.ucr.edu
World Wide Web: http://www.ee.ucr.edu

University of California, Riverside

THE FACULTY AND THEIR RESEARCH

Matthew J. Barth, Adjunct Associate Professor; Ph.D., California, Santa Barbara, 1990. Transportation Systems: transportation and emissions modeling, intelligent transportation systems (ITS), vehicle activity analysis, intelligent electric vehicles, intelligent sensing and control, mobile robot navigation. Robotics: active computer vision, panoramic sensing techniques, mobile robot navigation.

Gerardo Beni, Professor; Ph.D., UCLA, 1976. Robotics, swarm intelligence, distributed systems, multimedia.

Bir Bhanu, Professor; Ph.D., USC, 1981. Computer vision, image processing, pattern recognition, machine learning, artificial intelligence, robotics, multimedia databases, computer graphics and visualization, digital systems.

Jie Chen, Associate Professor; Ph.D., Michigan, 1990. Systems and control, system identification, robust control, linear multivariable systems theory, acoustic control.

Ilya Dumer, Professor; Ph.D., USSR Academy of Sciences, 1981. Data transmission and storage, coding theory, nonbinary and concatenated codes, decoding algorithms, code design.

Hossny El-Sherief, Adjunct Professor; Ph.D., McMaster, 1979. Digital signal processing, control systems, software engineering.

Jay Farrell, Associate Professor and Electrical Engineering Chair; Ph.D., Notre Dame, 1989. Adaptive and learning control, intelligent control, navigation, stability theory, autonomous systems.

Susan Hackwood, Professor; Ph.D., DeMontford, 1979. Robotics, distributed sensing systems, color vision and integrated manufacturing.

Ping Liang, Associate Professor; Ph.D., Pittsburgh, 1987. Image processing and analysis, medical image processing, pattern recognition, artificial neural networks, signal processing, pattern formation in distributed systems, decision support systems.

Cooperating Faculty

G. John Anderson, Professor, Psychology; Ph.D., California, Irvine, 1985. Perceptual, attentional, and cognitive limitations in the performance of complex perceptual tasks.

Subir Ghosh, Professor, Statistics; Ph.D, Colorado State, 1976. Statistical design and analysis of experiments, linear models, industrial statistics, sample surveys.

Qing Jiang, Professor, Mechanical Engineering; Ph.D., Caltech, 1990. Development of smart materials (electrically active materials specifically) for sensing and actuation applications; development of smart micro electromechanical systems (MEMS), including variable delays, tunable fibers and smart switches for radio frequency communications; and microrotation rate sensors for special applications that require sensors that withstand high levels of shock or impact.

Michael L. Lapidus, Professor, Mathematics; Ph.D., Université Pierre et Marie Curie (France). Mathematical physics.

Keh-Shin Lii, Professor, Statistics; Ph.D., California, San Diego, 1995. Spectral analysis of spatial/temporal weather variations.

Mart L. Molle, Professor, Computer Science; Ph.D., UCLA, 1981. Computer networking, performance evaluation, distributed algorithms.

J. Keith Oddson, Associate Professor, Mathematics; Ph.D., Maryland, 1965. Differential equations.

Thomas H. Payne, Associate Professor, Computer Science; Ph.D., Notre Dame, 1967. Efficient implementation of various programming language features related to issues in operating systems: concurrency, protection, dynamic binding.

S. James Press, Professor, Statistics; Ph.D., Stanford, 1964. Bayesian statistics, multivariable analysis, image classification and reconstruction, cognitive modeling in sample surveys.

Teodor C. Przymunsinski, Professor, Computer Science; Ph.D., Polish Academy of Sciences, 1974. Artificial intelligence.

Harry W.K. Tom, Professor, Physics; Ph.D., Berkeley, 1984. Nonlinear optics and femtosecond time-resolved laser techniques, surface dynamics, laser-induced surface chemical reactions, laser-induced phase transitions in bulk materials, nonlinear optics of the water/solid interface, terahertz spectroscopy.

Frank Vahid, Assistant Professor, Computer Science; Ph.D., California, Irvine, 1994. Embedded systems design.

Anders Wistrom, Assistant Professor, Environmental Engineering; Ph.D., California, Davis, 1993. Analysis of particulate systems for water pollution control, flocculation, sedimentation, and filtration processes; colloid stability and facilitated transport waste migration in the natural environment; physical and chemical unit processes for industrial, municipal water, and wastewater treatment.

UNIVERSITY OF CALIFORNIA, SAN DIEGO

Department of Electrical and Computer Engineering

Programs of Study

The Department of Electrical and Computer Engineering (ECE) offers graduate programs leading to the M.Eng., M.S., and Ph.D. degrees in electrical engineering and, jointly with the Department of Computer Science and Engineering, in computer engineering. The M.Eng. program is aimed at design engineers, for whom it would be a terminal professional degree. This program can be completed in one year of full-time study or in two years of half-time study. It is course work based, the course requirements are flexible, and it may include three courses in business, management, and finance. The M.S. program is intended primarily as preparation for advanced research. It may include a thesis (Plan I) or a project and a comprehensive examination (Plan II). Advanced research leading to the Ph.D. degree may be done in the following areas: communication theory and systems; computer engineering; electronic circuits and systems; electronic devices and materials; intelligent systems, robotics, and control; magnetic recording; photonics; radio and space science; and signal and image processing. In addition, there are interdepartmental curricula in advanced manufacturing, applied ocean sciences, and materials science. Course requirements for the Ph.D. and the M.S. programs are identical, and the Ph.D. preliminary examination serves as an M.S. comprehensive examination.

Research Facilities

The department has state-of-the-art research facilities in a wide range of areas. Facilities for materials and device research include several molecular beam epitaxy and organometallic vapor phase epitaxy reactors, electron beam lithography, a complete microfabrication facility, and laboratories for microelectronic and photonic device research. In the area of optical systems and photonics, a wide variety of lasers, optical tables, light valves, modulators, characterization equipment, computing platforms, and CAD tools are in use. The circuits and systems laboratories include computational platforms, software tools, and equipment for evaluation of microwave devices and circuits. The radio and space science group operates its own workstation network and makes extensive use of the San Diego Supercomputer Center. The Computer Vision and Resources Laboratories include optical systems for metric computer vision, a network of Sun and Silicon Graphics workstations, two Puma Arms, and a mobile golf cart under computer control. Communications and networking research activities are supported by laboratories providing modern software tools for analysis and simulation using a variety of computational platforms. The department operates or participates in a variety of research centers, including the NSF Industrial/University Cooperative Research Center for Ultra-High Speed Integrated Circuits and Systems, the ARPA-sponsored Optoelectronics Technology Center, the Center for Magnetic Recording Research, the Center for Astronomy and Space Science, the California Space Institute, and the Institute for Nonlinear Science. The San Diego Supercomputer Center, one of two NSF national centers for supercomputing research, is located on the University of California, San Diego (UCSD) campus and is heavily used for electrical and computer engineering research. The Center for Wireless Communications supports graduate-level research in communications theory, communications networks, multimedia applications, circuit design, antenna design, and propagation measurements/modeling.

Financial Aid

Financial aid is available in the form of fellowships, teaching assistantships, and research assistantships. The department normally supports all full-time graduate students, especially at the Ph.D. level. Most students are supported by a half-time research assistantship that provides approximately $16,770 during the calendar year plus tuition and fees. Entering students are often offered a combination of a fellowship and a teaching assistantship, which supports them until they find an adviser and obtain a research assistantship.

Cost of Study

In 1999–2000, full-time students who are California residents pay approximately $1630 per quarter in registration and incidental fees. Non-California residents pay $5100 per quarter for registration, tuition, and incidental fees. There is a reduced-fee structure for students enrolled on a half-time basis. Costs are subject to change.

Living and Housing Costs

University of California, San Diego (UCSD) provides 802 residential apartments for graduate students. Current monthly rates range from $300 for a single student to $650 for a family. For off-campus housing, prevailing rates range from $270 per month for a room in a private home to $900 or more for a two-bedroom apartment. Further information may be obtained from the UCSD Residential Housing Office (telephone: 858-534-2952).

Student Group

Current campus enrollment is about 19,370; of this number, 15,837 are undergraduates and 3,533 are graduate students. ECE has an undergraduate enrollment of about 700 and a graduate enrollment of about 250.

Location

The 2,040-acre campus spreads from the coastline, where the Scripps Institution of Oceanography is located, across a large wooded portion to the Torrey Pines Mesa overlooking the Pacific Ocean. To the east and north lie mountains, with Mexico to the south. The climate in San Diego is generally mild and pleasant year-round.

The University

One of nine campuses in the University of California System, UCSD comprises the General Campus, the School of Medicine, and the Scripps Institution of Oceanography. Established in La Jolla in 1960, it is one of the newer campuses, but in this short time it has become one of the major research universities in the country. The UCSD campus and the School of Engineering are ranked in the top ten nationwide by the National Academy of Sciences.

Applying

Applicants are considered for admission for the fall quarter only. All applicants are required to take the GRE General Test. International applicants whose native language is not English are required to take the TOEFL and obtain a minimum score of 550 on the paper-based test or 213 on the computer-based test (subject to change). A minimum GPA of 3.0 (on a 4.0 scale) is required for admission. The deadline for filing M.Eng., M.S., and Ph.D. applications is January 21, 2000.

Correspondence and Information

Department of Electrical and Computer Engineering
University of California, San Diego
9500 Gilman Drive
La Jolla, California 92093-0408
Telephone: 858-534-6606 or 4286
Fax: 858-534-2486
World Wide Web: http://www.ece.ucsd.edu

University of California, San Diego

THE FACULTY AND THEIR RESEARCH

Charles W. Tu, Professor and Department Chair; Ph.D., Yale. Molecular beam epitaxy, semiconductor materials and devices.

Anthony S. Acampora, Professor; Ph.D., Polytechnic of Brooklyn. Wireless communications, telecommunications networks.

Victor C. Anderson, Professor Emeritus; Ph.D., UCLA. Acoustics.

Peter M. Asbeck, Professor; Ph.D., MIT. Semiconductor device physics, MEMS, RF devices for wireless communications.

H. Neal Bertram, Professor; Ph.D., Harvard. Magnetic recording.

William S. C. Chang, Research Professor; Ph.D., Brown. Integrated optics, solid-state electronics, photonics.

Paul Chau, Associate Professor; Ph.D., Cornell. VLSI systems, digital signal processing, computer engineering, CAD.

C. K. Cheng, Adjunct Professor; Ph.D., Berkeley. Computer-aided design.

William A. Coles, Professor; Ph.D., California, San Diego. Radio astronomy, space physics, antennas and propagation.

Pamela C. Cosman, Assistant Professor; Ph.D., Stanford. Data compression, image processing.

Rene L. Cruz, Professor; Ph.D., Illinois. Communication networks.

Sujit Dey, Associate Professor; Ph.D., Duke. Hardware-software embedded systems, VLSI CAD, testing.

Sadik C. Esener, Professor; Ph.D., California, San Diego. Optoelectronic devices.

Shaya Fainman, Professor; Ph.D., Technion (Israel). Photonics, diffractive optics, information and image processing.

Jules A. Fejer, Professor Emeritus; D.Sc., Witwatersrand (South Africa). Space physics.

Ian Galton, Associate Professor; Ph.D., Caltech. Signal processing, mixed-mode integrated circuits for communication systems.

Clark C. Guest, Associate Professor; Ph.D., Georgia Tech. Computer vision, fiber optics, displays.

Robert Hecht-Nielsen, Adjunct Professor; Ph.D., Arizona State. Neural networks, neural computing.

Carl W. Helstrom, Professor Emeritus; Ph.D., Caltech. Communication theory, signal detection theory, optics.

John A. Hildebrand, Adjunct Professor; Ph.D., Stanford. Acoustics.

William B. Hodgkiss, Adjunct Professor; Ph.D., Duke. Digital signal processing, underwater acoustics.

Ramesh Jain, Professor; Ph.D., Indian Institute of Technology. Computer engineering, robotics.

Karen Kavanagh, Professor; Ph.D., Cornell. Materials science.

Kenneth Kreutz-Delgado, Associate Professor; Ph.D., California, San Diego. Systems science, machine intelligence, robotics.

Walter Ku, Professor; Ph.D., Polytechnic of Brooklyn. VLSI, IC design for signal processing and communications.

Lawrence E. Larson, Professor; Ph.D., UCLA. High-speed integrated circuit design in InP, GaAs, and Si technology, low-power circuit design, and RF design techniques for wireless communications; millimeterwave technology.

S. S. Lau, Professor; Ph.D., Berkeley. Electronic materials science.

Sing H. Lee, Professor; Ph.D., Berkeley. Photonics, microoptics, optoelectronic CAD and packaging.

James U. Lemke, Adjunct Professor; Ph.D., California, Santa Barbara. Magnetic recording.

Bill Lin, Assistant Professor; Ph.D., Berkeley. Computer engineering.

Yu Hwa Lo, Professor; Ph.D., Berkeley. Optoelectronics, MEMS, biosensors.

Robert Lugannani, Professor; Ph.D., Princeton. Stochastic processes, communication theory.

Huey-Lin Luo, Professor; Ph.D., Caltech. Solid-state physics, materials science, superconductivity.

Elias Masry, Professor; Ph.D., Princeton. Time-series analysis, communication theory.

D. Asoka Mendis, Professor Emeritus; Ph.D., D.Sc., Manchester (England). Solar system physics, cometary physics.

Laurence B. Milstein, Professor; Ph.D., Polytechnic of Brooklyn. Digital communication systems, communication theory.

Farrokh Najmabadi, Professor; Ph.D., Berkeley. Fusion.

Alon Orlitsky, Professor; Ph.D., Stanford. Information theory, learning theory, speech processing, signal processing.

Kevin B. Quest, Professor; Ph.D., UCLA. Solar system physics.

Bhaskar Rao, Professor; Ph.D., USC. Signal processing, estimation theory.

Ramesh Rao, Professor and Department Vice-Chair of Student Affairs; Ph.D., Maryland. Communication theory.

Barnaby J. Rickett, Professor; Ph.D., Manchester (England). Wave propagation in random media, radio, astronomy, solar wind.

Manuel Rotenberg, Research Professor; Ph.D., MIT. Numerical methods, population dynamics.

M. Lea Rudee, Research Professor; Ph.D., Stanford. Materials science.

Anthony Sebald, Associate Professor; Ph.D., Illinois. Adaptive control systems, neural networks, fuzzy control.

Vitaly Shapiro, Professor; Ph.D., Joint Institute for Nuclear Research (Russia). Space physics.

Paul H. Siegel, Professor; Ph.D., MIT. Communication theory, information theory, coding techniques for data storage and transmission.

Bang-Sup Song, Professor; Ph.D., Berkeley. Optoelectronic materials and devices.

David Sworder, Professor and Associate Dean, OGSR; Ph.D., UCLA. Systems control.

Mohan Trivedi, Professor; Ph.D., Utah State. Intelligent systems, machine vision, robotics, multimodal interfaces.

Alexander Vardy, Associate Professor; Ph.D., Tel-Aviv. Coding theory, decoding algorithms, modulation codes for storage systems.

Harry H. Wieder, Professor Emeritus; Ph.D., Colorado State. Optical electronics, solid-state physics.

Jack K. Wolf, Professor; Ph.D., Princeton. Communication theory, magnetic recording.

Edward T. Yu, Professor; Ph.D., Caltech. Semiconductor materials and devices.

Paul Yu, Professor; Ph.D., Caltech. Optoelectronic devices and materials.

Kenneth Y. Yun, Assistant Professor; Ph.D., Stanford. Asynchronous circuits, VLSI automation high-speed networks.

Kenneth A. Zeger, Professor; Ph.D., California, Santa Barbara. Communications, data compression, information theory.

James Zeidler, Adjunct Professor; Ph.D., Nebraska. Solid-state devices, adaptive signal processing.

UNIVERSITY OF CALIFORNIA, SANTA BARBARA

Department of Electrical and Computer Engineering

Programs of Study

Graduate studies leading to the M.S. and Ph.D. degrees in electrical and computer engineering are offered in the following areas of specialization: computer engineering; communications, control, and signal processing; and electronics and photonics.

Three quarters of residence are required in the M.S. program, and it is possible to complete the program in that time. Part-time students and those on assistantships usually require additional quarters. Both thesis and nonthesis options are available; a comprehensive examination is required in the nonthesis option.

The Ph.D. degree requires an approved program, including course work in a well-defined major area, demonstrated competence in two minor areas, and a dissertation. In addition, a student must pass a screening examination early in the program and a qualifying examination approximately two to three years after admission. Six quarters of residence are required; typically, the Ph.D. program is completed in about three years after completion of the M.S. program. There is no foreign language requirement.

Research Facilities

The Department of Electrical and Computer Engineering maintains a wide range of facilities for research and is closely associated with interdisciplinary campus research units, including the Compound Semiconductor Research Center, the Center for Quantized Electronic Structures, the Center for Computational Sciences and Engineering, the Center for Control Engineering and Computation, the Optoelectronic Technology Center, and the Multidisciplinary Optical Switching Technology Center. Facilities are available for all processes of device and integrated-circuit technology, focused ion beam, scanning electron microscopy, and metalorganic and molecular beam epitaxy; an optics laboratory for compound semiconductor and materials research; a microwave and millimeter-wave laboratory; a high-speed optical communication laboratory; an acoustic imaging facility; facilities for image digitization and processing; and individual laboratories for research and graduate instruction in communications, control and scientific computation, signal processing, computer architecture, software engineering, artificial intelligence, systolic computation, and VLSI CAD and VLSI testing, each with a state-of-the-art computing environment.

Financial Aid

Teaching and research assistantships and various fellowships are awarded on a competitive basis. The application deadline for awards is December 15. About 195 students received some form of financial aid in 1998–99.

Cost of Study

In 1999–2000, all graduate students pay registration fees of approximately $4935 per year ($1645 per quarter); nonresidents of California pay additional out-of-state tuition of $10,325 ($3128 per quarter).

Living and Housing Costs

One- and two-bedroom University-owned apartments are available for single and married students for rents of $424 to $932 per month. Privately owned apartments and houses are available in nearby areas. Most students share apartments or houses.

Student Group

The total campus enrollment is about 18,500 (51 percent women), with approximately 2,250 graduate students. The department has about 225 graduate students who come from all parts of the United States and many other nations. Approximately 44 percent of the graduate students in the department are international students. Approximately 55 students are M.S. candidates and 170 are doctoral candidates; sixty M.S. and twenty-five Ph.D. degrees are awarded annually.

Location

The University occupies a spacious site (815 acres) on a promontory bordered on two sides by the Pacific Ocean and on another by the Goleta Valley and Santa Ynez Mountains. The campus has a 7-mile system of bike paths that connect with those of surrounding communities. The Santa Barbara Airport and Goleta Beach State Park are immediately adjacent to the campus, which is 10 miles west of downtown Santa Barbara and 100 miles northwest of Los Angeles.

The University and The Department

UCSB is a research I institution offering undergraduate and graduate education in the arts, humanities, sciences and engineering, and social sciences. It is a member of the nine-campus University of California System, which is widely regarded to be the most distinguished system of public higher education in the United States. UCSB is now a member of the prestigious Association of American Universities (AAU), joining sixty leading institutions of higher learning in the U.S. and Canada that offer strong research and graduate education programs. The College of Engineering at UCSB ranks among the top four in the nation in terms of research funding per faculty member.

The Department of Electrical and Computer Engineering is a medium-size EE department with 34 faculty members, 280 undergraduate students, and 225 graduate students. There is a strong tradition of interdisciplinary education and research, with many multi-investigator research centers and programs, often involving faculty members and students from different areas in the department, other departments, and other universities. These projects also frequently include strong cooperation with industrial researchers.

Applying

The application deadline for those who desire financial award consideration is December 15. May 1 is the deadline for admission only. A bachelor's degree in electrical engineering, computer science, or some area of engineering or in mathematics, physics, or some related field of science is required. The GRE General Test is required of all applicants (the Subject Test is optional). All applicants who are non native speakers of English are required to submit TOEFL scores with their applications. Non-native speakers of English who have been awarded a bachelor's or master's degree by a U.S. institution are not required to submit TOEFL scores.

Correspondence and Information

Graduate Assistant
Department of Electrical and Computer Engineering
University of California
Santa Barbara, California 93106-9560
Telephone: 805-893-3114
Fax: 805-893-3262
E-mail: admit@ece.ucsb.edu
World Wide Web: http://www.ece.ucsb.edu

University of California, Santa Barbara

THE FACULTY AND THEIR RESEARCH

An asterisk (*) identifies faculty members who hold a joint appointment with the Department of Materials; a dagger (†) identifies those who hold a joint appointment with the Department of Computer Science; and a double dagger (††) identifies those who hold a joint appointment with the Department of Mathematics.

Daniel J. Blumenthal, Associate Professor; Ph.D., Colorado at Boulder. Fiber-optic networks, wavelength and subcarrier division multiplexing, photonic packet switching, signal processing in semiconductor optical devices, wavelength conversion, microwave photonics.

John E. Bowers, Professor; Ph.D., Stanford. High-speed photonic and electronic devices and integrated circuits, fiber-optic communication, semiconductors, laser physics and modelocking phenomena, compound semiconductor materials and processing.

Forrest D. Brewer, Associate Professor; Ph.D., Illinois. VLSI and computer system design, design automation, theory of design and design representations, symbolic techniques in high-level synthesis.

Steven E. Butner, Professor; Ph.D., Stanford. Computer architecture; VLSI design of CMOS and gallium arsenate ICs, with emphasis on distributed organizations and fault-tolerant structures.

Shivkumar Chandrasekaran, Assistant Professor; Ph.D., Yale. Numerical analysis, numerical linear algebra, scientific computation.

Kwang-Ting (Tim) Cheng, Professor; Ph.D., Berkeley. Design automation, VLSI and MCM testing, design synthesis, design verification, algorithms.

*Larry A. Coldren, Professor; Ph.D., Stanford. Semiconductor integrated optoelectronics, widely tunable lasers, vertical-cavity lasers, optical fiber communication, growth and planar processing techniques.

Nadir Dagli, Associate Professor; Ph.D., MIT. Design, fabrication, and modeling of photonic integrated circuits, ultrafast electrooptic modulators, solid-state microwave and millimeter-wave devices, experimental study of ballistic transport in quantum continued structures.

*Steven P. DenBaars, Associate Professor; Ph.D., USC. Metalorganic vapor phase epitaxy, optoelectronic materials, compound semiconductors, indium phosphide and gallium nitride, photonic devices.

^Arthur Gossard, Professor; Ph.D., Berkeley. Epitaxial crystal growth, artificially structured materials, semiconductor structures for optical and electronic devices, quantum confinement structures.

Evelyn Hu, Professor; Ph.D., Columbia. High-resolution fabrication techniques for semiconductor device structures, process-related materials damage, contact/interface studies, superconductivity.

Ronald Iltis, Professor; Ph.D., California, San Diego. Digital spread spectrum communications, spectral estimation and adaptive filtering.

Atac Imamoglu, Associate Professor; Ph.D., Stanford. Quantum optics, lasers without population inversion, quantum coherence in semiconductors, stochastic wave-function methods.

Petar V. Kokotovic, Professor and Director, Center for Control Engineering and Computation; Ph.D., USSR Academy of Sciences. Singular perturbations, nonlinear systems, adaptive control, automotive and jet engine control, sensitivity analysis, large-scale systems.

*Herbert Kroemer, Professor; Dr.rer.nat., Göttingen (Germany). General solid-state and device physics, heterostructures, molecular-beam epitaxy, compound semiconductor materials and devices, superconductivity.

Hua Lee, Professor; Director, Center for High-Speed Image Processing; Ph.D., California, Santa Barbara. High-performance image-formation algorithms, synthetic-aperture radar and sonar systems, acoustic microscopy, microwave nondestructive evaluation and dynamic vision systems, image system optimization.

Stephen I. Long, Professor; Ph.D., Cornell. Semiconductor devices and integrated circuits for high-speed digital and RF analog applications.

B. S. Manjunath, Associate Professor; Ph.D., USC. Image processing, computer vision, pattern recognition, neural networks, content-based retrieval in multimedia databases and learning algorithms.

Malgorzata Marek-Sadowska, Professor; Ph.D., Warsaw Technical. Design automation, computer-aided design, integrated circuit layout, logic synthesis.

†P. Michael Melliar-Smith, Professor; Ph.D., Cambridge. Distributed systems, fault tolerance, formal specification and verification, communication networks and protocols, asynchronous systems.

Umesh K. Mishra, Professor; Ph.D., Cornell. High-speed transistors, semiconductor device physics, quantum electronics, design and fabrication of millimeter-wave devices, in situ processing and integration techniques, wide band gap materials and devices.

Sanjit K. Mitra, Professor; Ph.D., Berkeley. Digital signal processing, image processing, computer-aided design and optimization.

Louise E. Moser, Associate Professor; Ph.D., Wisconsin. Distributed systems, computer networks, software engineering, fault tolerance, formal specification, and verification; performance evaluation.

Behrooz Parhami, Professor; Ph.D., UCLA. Computer design, computer arithmetic, dependable and fault-tolerant computing, parallel architectures and algorithms.

*Pierre M. Petroff, Professor; Director, Compound Semiconductor Research Center; Ph.D., Berkeley. Semiconductor device reliability, self-assembling nanostructures in semiconductors and ferromagnetic materials, spectroscopy of nanostructures, nanostructure devices.

Ian B. Rhodes, Professor; Ph.D., Stanford. Mathematical system theory and its applications, with emphasis on stochastic control, communication, and optimization problems, especially those involving decentralized information structures or parallel computational structures.

Mark J. W. Rodwell, Professor; Director, Compound Semiconductor Research Laboratories; Ph.D., Stanford. Heterojunction bipolar transistors, high-frequency integrated circuit design, electronics beyond 100 GHz.

Kenneth Rose, Associate Professor; Ph.D., Caltech. Information theory, source and channel coding, image coding, communications, pattern recognition.

John J. Shynk, Professor; Ph.D., Stanford. Adaptive filtering, blind equalization, wireless communications, neural networks, array processing.

Roy S. Smith, Associate Professor; Ph.D., Caltech. Robust control with an emphasis on the modeling, identification, and control of uncertain systems, applications and experimental work, including process control, flexible structures, automotive systems, semiconductor manufacturing, levitated magnetic bearings, and dynamic aeromaneuvering of interplanetary spacecraft.

Andrew Teel, Associate Professor; Ph.D., Berkeley. Control design and analysis for nonlinear dynamical systems, input-output methods, activator nonlinearities, applications to aerospace problems.

Emmanouel Varvarigos, Assistant Professor; Ph.D., MIT. Data networks, routing and communication aspects of parallel computations, communication systems, parallel processing architectures.

Pochi Yeh, Professor; Ph.D., Caltech. Optical computing, image processing, nonlinear optics, phase conjugation, dynamic holography, optical interconnection, neural networks.

Robert A. York, Associate Professor; Ph.D., Cornell. Electromagnetic theory, antennas, nonlinear circuits and dynamics, high-power/high-frequency devices and circuits, quasi-optics, microwave photonics.

UNIVERSITY OF CALIFORNIA, SANTA CRUZ

Graduate Program in Computer Engineering

Programs of Study

The Department of Computer Engineering (CE) at the University of California, Santa Cruz (UCSC) offers M.S. and Ph.D. degree programs and conducts research in computer-aided design of digital systems, including placement and routing, timing analysis, logic synthesis, specification languages, fault modeling, test generation, and multichip module design; computer systems design and applications, including VLSI, special-purpose processors, high-speed arithmetic circuits, and real-time systems; data compression, image and video coding, signal processing, and image processing, retrieval, and transmission; computer architecture and parallel processing, including massively parallel architecture, parallel programming and visualization, and memory and IO systems; and performance evaluation, communication, and networks, including queuing theory, high-speed networks and switching, network measurement, and simulation. CE enjoys a close relationship with the Department of Computer Science (CS), whose curriculum also covers the areas of machine learning, computer graphics and scientific visualization, operating systems, computational biology, distributed computing and debugging, theoretical computer science, and programming languages and environments. Faculty carry out joint research projects, supervise students, and teach courses for both departments. The department also has ties to nearby industry, employing computer professionals as visiting faculty and arranging for students to gain practical research experience through work in industrial labs. Students start the program with core courses in computer architecture and algorithms and then proceed to study thoroughly their area of specialization. The M.S. degree can be completed in one to two years. M.S. students may elect to complete a master's thesis. A Ph.D. degree is usually completed in five to six years. After completing the course requirements, students must pass an oral qualifying exam and write a dissertation. Part-time study is possible for students working in industry while going to school.

Research Facilities

The Jack Baskin School of Engineering operates several computer laboratories and facilities in support of research and graduate instruction in the Departments of Computer Science, Computer Engineering, and Electrical Engineering. The School of Engineering supports a network of more than 250 UNIX and NT systems, network printers, X-terminals, and various graphics devices. The School's network environment includes shared and switched 10Mb/sec and 100Mb/sec Ethernet. In support of general purpose computing, the School of Engineering operates three Sun servers that provide file service to all School of Engineering systems, several Sun servers that provide for general and specific computing needs (e.g., compute server, Web server, and print server), a general access DEC (Compaq) server, a general access SGI server, an NT domain server, and several graduate student computer labs that consist of a combination of Sun, SGI, DEC, and NT workstations, as well as X-terminals and printers. In addition to these facilities, the School of Engineering operates the following research labs with the indicated equipment: Machine Learning and Computational Biology Labs; Scientific Visualization, Computer Graphics, and Image Processing Labs; a Computer-Aided VLSI Design Lab; a Concurrent Systems Lab; and Computer Communications and High-Speed Networking Labs. In addition to the facilities provided by the Jack Baskin School of Engineering, students have access to the campus computing facilities of the UCSC Computer Center that provides computing to more than 2,000 users campuswide via a network of distributed UNIX workstations, an IBM ES9000, and a Microm data switch with 100 dial-in modems. The distributed UNIX computing facilities consist of a laboratory of twenty Sun workstation that are supported by several SunSPARC Station compute servers and file servers. The UNIX facilities are based on MIT's Project Athena distributed computing environment. The computer center maintains the University's network connection to the global Internet computer network.

Financial Aid

A limited number of fellowships provide a stipend of $9999 plus payment of all University fees except nonresident tuition to first-year students. A number of nonresident tuition waivers are awarded to students who are not residents of California. Half-time teaching assistantships provide a salary of $5233 per quarter, half-time research assistantships provide a salary of $5188, and both allow time for taking two lecture courses per quarter.

Cost of Study

Fees for the 1999–2000 academic year are approximately $5205. Students who are not California residents must pay the additional nonresident tuition fee of $10,320 per year.

Living and Housing Costs

Housing for single students living on the University campus costs $5406. For married students and single parents, housing costs $5086. For students living off campus, the amount is about $6000, not including board.

Student Group

Total enrollment at UC Santa Cruz is approximately 10,000 students; about 960 are graduate students. The number of graduate students currently in computer engineering is 106.

Location

Santa Cruz is one of America's most beautiful campuses. Overlooking Monterey Bay, it occupies 2,000 acres in protected redwood forests and meadows above the city of Santa Cruz. Nearby Santa Clara Valley, one of the most important centers for the computer industry, is a significant resource.

The University

Founded 1965, Santa Cruz is a small collegiate university devoted to excellence in undergraduate education and enriched by a select group of graduate programs and the presence of major research units.

Applying

Students must begin the program in the fall quarter. Completed applications must be received by February 1, 2000. Files of applicants for 2000–2001 are reviewed in late February 2000. Requests for applications should be directed to the Graduate Division, 150 Social Sciences II. The application fee is $40 and subject to increase without notification. Each applicant must take the GRE General Test and the Subject Test in either engineering or computer science and have the scores sent to the UCSC Graduate Division as part of the application file. The graduate brochure is available on line.

Correspondence and Information

Graduate Representative
Department of Computer Engineering
University of California, Santa Cruz
Santa Cruz, California 95064

Telephone: 831-459-2576
E-mail: soegradadm@cse.ucsc.edu
World Wide Web: http://www.cse.ucsc.edu

University of California, Santa Cruz

THE FACULTY AND THEIR RESEARCH

Computer Engineering

Alexandre Brandwajn, Professor; Ph.D. (computer science), Paris, 1975. Computer architecture, performance modeling, queuing network models of computer systems and operating systems.

Pak K. Chan, Associate Professor; Ph.D. (computer science), UCLA, 1987. Placement and routing, accelerators, partitioning, system prototyping using FPGAs.

Wayne Wei-ming Dai, Associate Professor; Ph.D. (electrical engineering), Berkeley, 1988. Computer-aided design of VLSI circuits, multichip modules, graph theory, computational geometry.

F. Joel Ferguson, Associate Professor; Ph.D. (computer engineering), Carnegie Mellon, 1987. Fault modeling, test generation and design-for-test of digital circuits and systems, fault-tolerant computing, VLSI design, computer-aided manufacturing for VLSI.

J. J. Garcia-Luna-Aceves, Professor; Ph.D. (electrical engineering), Hawaii at Manoa, 1983. Computer networks and multimedia information systems.

Richard Hughey, Associate Professor; Ph.D. (computer science), Brown, 1991. Computer architecture, parallel processing, parallel programming languages and environments, computer applications in biology, programmable systolic arrays.

Kevin Karplus, Associate Professor; Ph.D. (computer science), Stanford, 1983. Bioinformatic computational biology: applying information theory and stochastic modeling to biological sequence analysis, hidden Markov models, regularizers, mutual information, protein threading, multipurpose architecture for sequence analysis.

Harwood G. Kolsky, Adjunct Professor; Ph.D. (physics), Harvard, 1950. Scientific computing, compilers, image processing.

Glen G. Langdon Jr., Professor; Ph.D. (electrical engineering), Syracuse, 1968. Data compression, image and video coding, image segmentation, adaptive estimation of probability distributions.

Tracy Larrabee, Assistant Professor; Ph.D. (computer science), Stanford, 1990. Test pattern generation, design verification, logic synthesis.

Patrick E. Mantey, Professor and Dean, School of Engineering; Ph.D. (electrical engineering), Stanford, 1965. Image storage and retrieval; electronic libraries and multimedia; educational applications of computer technology; image and signal processing; graphics and workstation hardware; system architecture, design, and performance; simulation and modeling of complex systems; real-time data acquisition and control systems; graphics and database applications, including geographic information systems; and user-machine interaction.

Martine D. F. Schlag, Professor; Ph.D. (computer science), UCLA, 1986. VLSI design tools and algorithms; VLSI theory; formal specifications of VLSI circuits; Field-Programmable Gate Arrays.

Anujan Varma, Associate Professor; Ph.D. (computer engineering), USC, 1986. Computer networking, computer systems architecture, parallel processing.

Computer Science

Neil Balmforth, Associate Professor; Ph.D. (applied math and theoretical physics), Cambridge, 1990. Astrophysics, dynamical systems, fluid dynamics, mathematical biology, non-Newtonian fluids, plasma physics.

Scott Brandt, Assistant Professor; Ph.D. (computer science), Colorado, 1999. Operating systems, parallel and distributed systems, experimental systems, software systems, soft real-time processing, real-time and embedded systems, networking and Internet technology, asynchronous computer architectures, computer vision and robotics, image processing, virtual reality.

David Haussler, Professor; Ph.D. (computer science), Colorado, 1982. Machine learning, computational biology, neural networks, statistical decision theory, algorithms and complexity.

David Helmbold, Associate Professor; Ph.D. (computer science), Stanford, 1987. Machine learning, theoretical computer science, analysis of algorithms, intelligent systems.

David A. Huffman, Professor Emeritus; Sc.D. (electrical engineering), MIT, 1953. Information theory and coding, scene analysis, graph theory, signal design and processing, discrete systems, sequential circuits.

Phokion Kolaitis, Professor; Ph.D. (mathematics), UCLA, 1978. Logic in computer science, database theory, logic programming, complexity theory.

Robert A. Levinson, Associate Professor; Ph.D. (computer science), Texas at Austin, 1985. Artificial intelligence, machine learning, heuristic search, hierarchical reinforcement learning, associate pattern retrieval, computer chess.

Suresh K. Lodha, Associate Professor; Ph.D. (computer science), Rice, 1992. Computer graphics, scientific visualization, computer-aided geometric design, computer animation, image processing, scientific sonification.

Darrell Long, Associate Professor; Ph.D. (computer science), California, San Diego, 1988. Distributed computing systems, operating systems, performance evaluation, data management.

Tara Madhyastha, Assistant Professor; Ph.D. (computer science), Illinois, 1997. Parallel input/output, adaptive and predictive resource management policies, performance visualization and prediction, input/output characterization.

Charles E. McDowell, Professor; Ph.D. (computer science), California, San Diego, 1983. Computer architecture, parallel computing, microprogramming, compilers, operating systems.

Alex Pang, Associate Professor; Ph.D. (computer science), UCLA, 1990. Visualization (scientific, environmental, and uncertainty), computer graphics, virtual reality interfaces, and collaborative software.

Ira Pohl, Professor; Ph.D. (computer science), Stanford, 1969. Artificial intelligence, programming languages, heuristic methods, educational and social issues, combinatorial algorithms.

R. Michael Tanner, Professor; Ph.D. (electrical engineering), Stanford, 1971. Information theory, error-correcting codes, complexity, VLSI systems, fault tolerance.

Allen Van Gelder, Associate Professor; Ph.D. (computer science), Stanford, 1986. Logic programming algorithms, parallel algorithms, complexity, programming languages, automated theorem proving, scientific visualization.

Hongyun Wang, Assistant Professor; Ph.D. (applied math and statistics), Berkeley, 1996. Molecular modeling, fluid mechanics, numerical analysis, instability in physical systems, molecular structure analysis and computer visualization.

Manfred K. Warmuth, Professor; Ph.D. (computer science), Colorado, 1981. Machine learning, neural networks, parallel and distributed algorithms, online learning algorithms, complexity theory.

Jane P. Wilhelms, Associate Professor; Ph.D. (computer science), Berkeley, 1985. Computer graphics, computer animation, scientific visualization, modeling articulated bodies, physical simulation, behavioral animation.

Electrical Engineering

Jiayuan Fang, Associate Professor; Ph.D. (electrical engineering), Berkeley, 1989. Computational electromagnetics and microwaves, numerical techniques in computational electromagnetics, electromagnetic modeling of VLSI interconnections, simulation of electrical performance of electronic packaging, signal integrity and power supply noise analysis in electronic packaging, experimental characterization of integrated-circuit packages and printed circuit boards.

Claire Gu, Associate Professor; Ph.D. (electrical engineering), Caltech, 1989. Optical storage, optical fiber communications, optical information processing.

Peyman Milanfar, Assistant Professor; Ph.D. (electrical engineering), MIT, 1993. Statistical signal and image processing, inverse problems, detection and estimation, scientific computing, applied mathematics.

Ali Shakouri, Assistant Professor; Ph.D. (electrical engineering), Caltech, 1995. Integrated cooling of electronic components, electron and photon transport in semiconductor superlattices, photonic switching, fiber optics communication systems.

John Vesecky, Professor; Ph.D. (electrical engineering), Stanford, 1967. Remote sensing of ocean surface; ocean-current-measuring radar for coastal ecology and oceanography; radar and radar systems, especially synthetic aperture radar (SAR); wave scattering; remote sensing and public health; global change.

UNIVERSITY OF CENTRAL FLORIDA

Department of Electrical and Computer Engineering

Programs of Study

The Department of Electrical and Computer Engineering (ECE) offers programs leading to the Master of Science and Doctor of Philosophy degrees in electrical engineering, computer engineering, and optical sciences and engineering. Research areas include communication theory and systems, control and robotics, digital signal and image processing, digital systems and architecture, electrooptics, microelectronics and solid-state devices, microwaves and antennas, and software engineering and expert systems.

The master's degree with a thesis option requires a minimum of 30 semester hours, including 6 hours of thesis registration. All students completing a thesis on their research must undergo a final oral examination. A master's degree program without a thesis, requiring a minimum of 36 credit hours, is also available.

A maximum of 36 graduate semester hours taken in the master's degree can be accepted as credit toward the required minimum of 81 semester hours for the Ph.D. program. A minimum of 24 semester hours in basic sciences and engineering sciences is required with at least 9 hours taken outside the College of Engineering. A minimum of 36 hours within the field of specialization and a minimum of 24 dissertation hours must be earned to fulfill the requirements of 81 hours beyond the bachelor's degree.

Research Facilities

The department facilities include a Class 100 clean room, an RF/microwaves laboratory, a digital signal processing laboratory, and an image processing laboratory. Close interaction and research opportunities exist with the University of Central Florida (UCF) Center for Research and Education in Optics and Lasers (CREOL) and the Institute for Simulation and Training (IST). The laboratories in CREOL and IST have extensive state-of-the-art equipment.

Computer support is provided by a college-wide computer network with a Sun Ultra file server at its center. Distributed among the network are various IBM PC stations, Sun and IBM workstations, and an NCUBE (64001-E) 32-node supercomputer with a high-speed imaging display. The University library is well equipped with periodicals and books in electrical and computer engineering.

Financial Aid

Financial aid is available in the form of a limited number of teaching and research assistantships. These require from one-quarter- to one-half-time work loads, with compensation in the range of $3400 to $6700 for nine months. Also, the nonresident and resident tuition fees can be waived in many cases. Fellowships with an annual stipend of $10,000, including tuition remission, as well as Graduate Enhancement Awards are available to outstanding entering graduate students through the UCF Division of Sponsored Research. Graduate assistantships are also available through CREOL for students in the electrooptics program.

Cost of Study

Tuition in 1998–99 was $136.89 per semester hour; out-of-state students paid $480.45 per hour. General fees paid by all students amounted to $143 per term.

Living and Housing Costs

Double-occupancy rooms on campus rented for $1430 or $1495 per semester in 1998–99. Single-occupancy rooms rented for $1095 per semester. Single rooms in apartments are available at a cost of $1600 per semester. Meal plans in 1998–99 cost $875 per semester. There are many apartments near UCF, some within walking distance.

Student Group

The fall 1998 enrollment of the University was 29,838, the student body being almost equally divided between men and women. The enrollment of the department was 1,172, which included 243 graduate students. Most of the graduate courses are recorded on videotape and made available to five remote locations in the geographical area.

Student Outcomes

M.S. and Ph.D. graduates in electrical engineering, computer engineering, and optical sciences and engineering are employed with companies, both in Florida and nationally, such as Lockheed Martin, Loral, Lucent Technologies, Harris, Texas Instruments, Motorola, Vela Research, AT&T, Sawtek, Siemens, and Utilities.

Location

UCF is located 15 miles from downtown Orlando. Central Florida has recently shown dramatic industrial growth, particularly in such high-technology industries as aerospace, communications, and electronics. The Kennedy Space Center, with its launch site for satellites and the space shuttle, is nearby. Central Florida has become a major tourist area since the 1971 opening of Walt Disney World, just southwest of Orlando. The Atlantic Ocean, the Gulf of Mexico, and numerous rivers and spring-fed lakes provide many opportunities for outdoor recreation.

The University and The Department

Established as a state university in 1963, UCF admitted its first students in 1968. Today, the modern campus covers 1,227 wooded acres. The University's central location makes it accessible from all parts of the state. In addition, campuses are located in Cocoa, Daytona Beach, and South Orlando.

The ECE department occupies the fourth floor of the engineering building as well as portions of other floors. It is the largest department in the College of Engineering and one of the largest in the University.

Applying

Prospective students should apply to the Admissions Office at least five weeks before the start of classes for the term in which they plan to enroll. A $20 application fee, official transcripts from an accredited college, and GRE General Test scores are required. The minimum admissions requirements are based on an average of B or better of a baccalaureate program and a minimum combined score of 1000 on the verbal and quantitative portions of the GRE General Test. The deadline for financial aid applications is March 1.

Correspondence and Information

Requests for additional information should be directed to:

Graduate Coordinator
ECE Department
University of Central Florida
P.O. Box 162450
Orlando, Florida 32816-2450
Telephone: 407-823-3027

University of Central Florida

THE FACULTY AND THEIR RESEARCH

W. B. Mikhael, Chair; Ph.D., Concordia. Digital signal processing, adapative signal processing, multidimensional signal compression, filtering. (Telephone: 407-823-3210; e-mail: mikhael@mail.ucf.edu)

Professors

C. S. Bauer, Ph.D., Florida. Real-time simulation, computer graphics, urban systems engineering, software systems engineering. (Telephone: 407-823-2236; e-mail: bauer@mail.ucf.edu)

M. A. Belkerdid, Ph.D., Central Florida. Communication theory, spread spectrum communications, RF communications, surface acoustic wave systems, simulation systems. (Telephone: 407-823-5793; e-mail: mad@ece.engr.ucf.edu)

A. J. Gonzalez, Ph.D., Pittsburgh. Artificial intelligence, knowledge-based systems, automated diagnostics, intelligent simulations, validation and verification of knowledge-based systems. (Telephone: 407-823-5027; e-mail: ajg@ece.engr.ucf.edu)

W. L. Jones, Ph.D., Virginia Tech. Remote sensing, satellite communications, systems engineering. (Telephone: 407-823-6603; e-mail: ljones@pegasus.cc.ucf.edu)

J. J. Liou, Ph.D., Florida. Semiconductor device modeling, device simulation and characterization, computer-aided integrated circuit design. (Telephone: 407-823-5339; e-mail: jli@ece.engr.ucf.edu)

D. C. Malocha, Ph.D., Illinois at Urbana-Champaign. Surface acoustic wave technology, semiconductor device technology, solid-state device fabrication, RF communications. (Telephone: 407-823-2414; e-mail: dcm@ece.engr.ucf.edu)

R. Phillips, Ph.D., Arizona State. Optical propagation through random media. (Telephone: 407-823-6908; e-mail: phillips@mail.creol.ucf.edu)

N. S. Tzannes, Ph.D., Johns Hopkins. Communications, signal/image processing. (Telephone: 407-823-5768; e-mail: tzannes@pegasus.cc.ucf.edu)

Associate Professors

I. E. Batarseh, Ph.D., Illinois at Chicago. Power electronics, dc-to-dc power supplies, dynamic and control of power converters, power factor correction. (Telephone: 407-823-0185; e-mail: batarseh@mail.ucf.edu)

M. Georgiopoulos, Ph.D., Connecticut. Neural networks, pattern recognition, applications of neural networks in communications/electromagnetics/signal processing, communications systems. (Telephone: 407-823-5338; e-mail: mng@ece.engr.ucf.edu)

T. Kasparis, Ph.D., CUNY, City College. Digital signal processing, image processing, audio processing, adaptive median filtering, electronics. (Telephone: 407-823-5913; e-mail: tnk@ece.engr.ucf.edu)

H. Klee, Ph.D., Polytechnic. Simulation, driving simulators, systems and control. (Telephone: 407-823-2270; e-mail: hik@pegasus.cc.ucf.edu)

D. G. Linton, Ph.D., Florida. Software engineering, numerical methods, simulation, stochastic processes, reliability. (Telephone: 407-823-2320; e-mail: dgl@ece.engr.ucf.edu)

R. N. Miller, Ph.D., SUNY at Buffalo. Instrumentation, electronics. (Telephone: 407-823-2455; e-mail: rmiller@ucflvm.cc.ucf.edu)

H. R. Myler, Ph.D., New Mexico State. Intelligent control, machine intelligence, machine vision, image processing, control theory. (Telephone: 407-823-5098; e-mail: hrm@ece.engr.ucf.edu)

B. E. Petrasko, E.Eng., Detroit. Parallel architectures, distributed simulation, hardware description languages, digital system modeling, discrete-event simulation. (Telephone: 407-823-2549; e-mail: bep@ece.engr.ucf.edu)

Z. Qu, Ph.D., Georgia Tech. Controls, manufacturing, robotics, power systems, vision. (Telephone: 407-823-5976; e-mail: quz@ece.engr.ucf.edu)

S. M. Richie, Ph.D., Central Florida. Surface acoustic wave (SAW) device modeling, SAW device computer-aided design, transversal filter design theory, voice recognition systems. (Telephone: 407-823-5765; e-mail: sam@ece.engr.ucf.edu)

W. Shu, Ph.D., Illinois at Urbana-Champaign. Distributed and parallel systems. (Telephone: 407-823-2133; e-mail: shu@ece.engr.ucf.edu)

K. B. Sundaram, Ph.D., Indian Institute of Technology. Microelectronics, optoelectronic materials, thin films, micromachining. (Telephone: 407-823-5326; e-mail: kbs@ece.engr.ucf.edu)

P. Wahid, Ph.D., Indian Institute of Technology. Microwave and millimeter wave antennas, electromagnetics. (Telephone: 407-823-2610; e-mail: wahid@mail.ucf.edu)

G. Walton, Ph.D., Tennessee. Software engineering, software specification, software testing, software quality measurement and models. (Telephone: 407-823-3276; e-mail: walton@mail.ucf.edu)

A. R. Weeks, Ph.D., Central Florida. Color image processing, adaptive image processing, nonlinear digital filters, optical character recognition, image recognition. (Telephone: 407-823-5762; e-mail: arw@ece.engr.ucf.edu)

M. -Y. Wu, Ph.D., Santa Clara. Parallel and distributed computing systems, multimedia systems. (Telephone: 407-823-2151; e-mail: wu@ece.engr.ucf.edu)

J. S. Yuan, Ph.D., Florida. Semiconductor device modeling, device and circuit simulation, analog/digital analysis and design. (Telephone: 407-823-5719; e-mail: yuanj@pegasus.cc.ucf.edu)

J. Zalewski, Ph.D., Warsaw (Poland). Software engineering, real-time systems and distributed systems. (Telephone: 407-823-6171; e-mail: jza@ece.engr.ucf.edu)

Assistant Professors

R. F. DeMara, Ph.D., USC. Computer architecture, performance modeling, parallel and distributed processing, artificial intelligence, real-time speech processing. (Telephone: 407-823-5916; e-mail: rfd@ece.engr.ucf.edu)

F. Gonzalez, Ph.D., Illinois at Urbana-Champaign. Simulation and control of discrete-event systems. (Telephone: 407-823-3987; e-mail: fgonazale@pegasus.cc.ucf.edu)

M. G. Haralambous, D.Sc., George Washington. Robust stabilization and control of certain unstable plants. (Telephone: 407-823-2548; e-mail: mgh@ece.engr.ucf.edu)

Joint Appointees

G. Boreman, Associate Professor; Ph.D., Arizona. Infrared and electrooptical systems. (Telephone: 407-823-6815; e-mail: boreman@mail.creol.ucf.edu)

P. J. Delfyett, Associate Professor; Ph.D., CUNY. Ultrafast phototonics, fiber-optic networks, ultrafast measurements and signal processing. (Telephone: 407-823-6812; e-mail: delfyett@mail.creol.ucf.edu)

D. J. Hagan, Associate Professor; Ph.D., Heriot-Watt (Scotland). Nonlinear optics, lasers, ultrafast optics. (Telephone: 407-823-6817; e-mail: hagan@mail.creol.ucf.edu)

J. E. Harvey, Associate Professor; Ph.D., Arizona. Optical systems design and analysis. (Telephone: 407-823-6818; e-mail: harvey@mail.creol.ucf.edu)

P. Li Kam Wa, Associate Professor; Ph.D., Sheffield (England). High-speed optoelectronic devices using multiple quantum well semiconductors, all-optical switching and integrated optical circuits, short-range all-optical interconnects, ultrashort pulse generation and amplification in compact solid-state lasers. (Telephone: 407-823-6816; e-mail: patrick@mail.creol.ucf.edu)

M.. G. Moharam, Professor; Ph.D., British Columbia. Electrooptics, photorefractive devices, holographic data storage. (Telephone: 407-823-6833; e-mail: moharam@mail.creol.ucf.edu)

M. Richardson, Professor; Ph.D., London. New X-ray technologies, ultrahigh intensity lasers, condensed matter, high-power solid-state laser systems, X-ray and electron optics. (Telephone: 407-823-6819; e-mail: mcr@mail.creol.ucf.edu)

N. Riza, Associate Professor; Ph.D., Caltech. Photonic information processing systems for aerospace, medical, industrial, and telecommunication applications. (Telephone: 407-823-6829; e-mail: riza@mail.creol.ucf.edu)

J. P. Rolland, Assistant Professor; Ph.D., Arizona. Virtual environments, medical and biomedical imaging and optical system design. (Telephone: 407-823-6870; e-mail: rolland@mail.creol.ucf.edu)

W. T. Silfvast, Professor; Ph.D., Utah. Soft-X-ray sources, soft-X-ray microscopy, EUV lithograph, X-ray lasers. (Telephone: 407-823-6855; e-mail: silfvast@mail.creol.ucf.edu)

M. J. Soileau, Professor; Ph.D., USC. Laser-induced damage, optical limiting, nonlinear optics, self-focusing. (Telephone: 407-823-6834; e-mail: mj@mail.creol.ucf.edu)

G. I. Stegeman, Professor; Ph.D., Toronto. Nonlinear optical materials and their application in waveguide devices. (Telephone: 407-823-6915; e-mail: george@mail.creol.ucf.edu)

C. M. Stickley, Professor; Ph.D., Northeastern. Laser radar, radar materials, IR materials, optically pumped solid-state lasers. (Telephone: 407-823-6986; e-mail: cms@mail.creol.ucf.edu)

E. W. Van Stryland, Professor; Ph.D., Arizona. Nonlinear optics, ultrafast processes in semiconductors, laser-induced damage, nonlinear spectroscopy, optical limiting. (Telephone: 407-823-6814; e-mail: ewvs@mail.creol.ucf.edu)

UNIVERSITY OF CINCINNATI

College of Engineering
Department of Electrical & Computer Engineering and Computer Science

Programs of Study

The Department of Electrical & Computer Engineering and Computer Science (ECECS) offers three graduate degree programs leading to the Master of Science (M.S.) in computer engineering, computer science, and electrical engineering. It offers the Doctor of Philosophy (Ph.D.) degree in electrical engineering and in computer science and engineering. The department's graduate studies and research program are organized along the lines of faculty member interests and expertise. The faculty members are grouped together in interrelated but distinct research areas. This structure allows the critical mass necessary to facilitate innovative research efforts. Graduate programs, providing a balance between formal classroom instruction and research, are tailored to the student's professional goals. At present, the four research areas within the department are computer engineering, computer science, electronic materials and devices, and systems engineering.

Currently, there are 46 faculty members in the department. The graduate programs involve more than 270 full-time and 70 part-time graduate students. The graduate research environment, with more than $5 million in new funding each year, provides excellent opportunities to work on exciting and challenging projects. Funding comes from national and state research agencies, such as the National Science Foundation, National Institutes of Health, National Aeronautics and Space Administration, Defense Advanced Research Projects Agency, Army Research Office, Whitaker Foundation, Wright Patterson Air Force Base, Office of Naval Research, and Ohio Department of Transportation. Collaborative research projects are also well funded by industries, such as Dow Corning, General Electric, Harris Corporation, Lucent Technologies, Procter and Gamble, Taitech, AT&T, TMC, TRW, and Von Durpin.

Research Facilities

The department has excellent research and teaching facilities, including research laboratories with state-of-the-art equipment in the areas of computer science, software systems, artificial intelligence, neural networks, image processing, electronic design automation, MEMS (microelectromechanical systems), microsensors, millimeter waves and photonics, nanoelectronics, and optoelectronics. The departmental computing resources include a large number of workstations. The local area network allows faculty members and students to obtain permanent accounts on University computers, as well as access to supercomputers via Internet II.

Financial Aid

Fellowships, teaching and research assistantships, and tuition scholarships are available on a competitive basis to qualified full-time graduate students. Applicants are automatically considered for these awards during consideration for admission. Teaching assistantships include an average monthly stipend of $1100 and are accompanied by a tuition scholarship. Research assistantships are available with comparable stipends to incoming students who show exceptional promise and related research experience. Upper-level students are typically supported by externally sponsored research assistantships.

Cost of Study

For the 1999–2000 academic year, full-time tuition is $5139 for residents of Ohio and $10,326 for nonresidents. All students also pay a general fee, which amounts to $561 per academic year.

Living and Housing Costs

Room and board for a single student living on campus are approximately $8000. In addition, there are many apartments readily available within walking distance of the campus. Estimated annual living expenses for an international student are $11,900.

Location

Cincinnati is the twenty-third–largest city in the United States, with a greater metropolitan area population of 1.7 million. The city offers many sites of architectural and historic interest as well as a full range of cultural attractions, such as theaters, the symphony, and opera. The renowned College Conservatory of Music on campus offers frequent performances, many of them for free. The business and industrial base of the city is diverse, allowing qualified students to easily find part-time work. The largest companies in the city are Proctor & Gamble, General Electric, Merrell Dow, and Milacron. The city is served by the Greater Cincinnati International Airport, which is a major hub of Delta.

The University

The University of Cincinnati is a comprehensive state institution, with an average total enrollment of 35,000 students. Endowment funds, in excess of $390 million and $99 million in sponsored programs, place it in the top 2 percent of universities for funding. This allows it to provide the proper environment for innovative scholarship and research. An Engineering Research Center (ERC) was dedicated in 1995. It has 110 laboratories, offices for 200 graduate students, and thirteen conference rooms, plus extensive computer facilities.

Applying

A Bachelor of Science degree with a minimum GPA of 3.0 (on a 4.0 scale) is the norm for admission into a Master of Science program or a direct route Ph.D. degree program. Students with an appropriate Master of Science degree may be admitted into a Ph.D. program. The Graduate Record Examinations (GRE) General Test is required for all students applying for admission. International students are also required to take the Test of English as a Foreign Language (TOEFL), and they must achieve a minimum score of 550. The deadline for applications to be considered for financial aid is February 1.

Correspondence and Information

Director of Graduate Studies
Department of Electrical & Computer Engineering
    and Computer Science
University of Cincinnati, ML 30
Cincinnati, Ohio 45221-0030
Telephone: 513-556-0635
E-mail: grad_dir@ececs.uc.edu
World Wide Web: http://www.ececs.uc.edu

University of Cincinnati

THE FACULTY AND THEIR RESEARCH

Dharma Agrawal, OBR Distinguished Professor of Computer Science and Engineering; D.Sc., Swiss Federal Institute of Technology, 1975. Mobile networks, ad hoc networks, distributed systems, automatic parallelism detection and scheduling, systems reliability.

Chong H. Ahn, Associate Professor; Ph.D., Georgia Tech, 1993. Microelectromechanical systems (MEMS) biochemical microsensors, microfluidic systems, optoelectronic multichip modules.

Fred S. Annexstein, Associate Professor; Ph.D., Massachusetts Amherst, 1991. Theory of parallel distributed processing, routing in networks, parallel algorithms, scalable network design.

Kenneth A. Berman, Professor; Ph.D., Waterloo, 1979. Design and analysis of algorithms, scalable network design, network routing, applied graph theory, combinatorics.

Fred R. Beyette Jr., Assistant Professor; Ph.D., Colorado State, 1995. Optoelectronic device fabrication and testing, smart pixel device design and testing, photonic information processing systems design and implementation.

Dinesh K. Bhatia, Associate Professor; Ph.D., Texas at Dallas, 1990. CAD and architecture of field-programmable gate arrays, VLSI system, CAD, reconfigurable and adaptive computing, hardware prototyping.

Raj Bhatnagar, Associate Professor; Ph.D., Maryland, 1989. Artificial intelligence models for reasoning and decision making, pattern recognition, reasoning under uncertainty, machine learning, distributed artificial intelligence.

Punit Boolchand, Professor; Ph.D., Case Western Reserve, 1969. Molecular structure of noncrystalline materials-semiconducting glasses, optoelectronic materials, amorphous thin films.

Joseph T. Boyd, Professor; Ph.D., Ohio State, 1969. Integrated optoelectronics, optical characterization of materials, visible infrared detectors, optical processing of materials.

James J. Caffery Jr., Assistant Professor; Ph.D., Georgia Tech, 1998. Wireless communications (cellular/mobile/personal), spread spectrum systems, wireless location systems, multiuser communications.

Marc M. Cahay, Associate Professor; Ph.D., Purdue, 1987. Carrier transport in semiconductors, quantum mechanical effects in superlattices and quantum wells, nanostructure devices, Josephson junction arrays, implementation of artificial neural networks.

Harold W. Carter, Professor; Ph.D., USC, 1980. Mixed-signal modeling and simulation, parallel numerical analysis, design optimization, VLSI architecture and design methodologies.

Yizong Cheng, Associate Professor; Ph.D., Purdue, 1986. Machine learning, uncertainty and similarity measures, cluster analysis, self-organization of memory and databases.

Samir Ranjan Das, Associate Professor; Ph.D., Georgia Tech, 1994. Wireless networking and mobile computing, parallel and distributed simulation.

Karen C. Davis, Associate Professor; Ph.D., Southwestern Louisiana, 1990. Object database systems, query languages and optimization, data warehousing, data modeling for engineering applications.

Howard Fan, Professor; Ph.D., Illinois, 1985. Digital signal processing, array processing, adaptive signal processing, signal processing for communication, system identification.

Altan M. Ferendeci, Associate Professor; Ph.D., Case Western Reserve, 1969. Microwave and millimeter-wave devices and circuits, high T superconductivity, electrooptics.

John V. Franco, Geier Professor of Computer Science and Associate Professor; Ph.D., Rutgers, 1981. Design and analysis of algorithms, probabilistic analysis of algorithms, graph theory, combinatorics, numerical analysis.

Patrick H. Garrett, Associate Professor; Ph.D., Ohio, 1970; PE. Advanced instrumentation and digital control design for manufacturing processes integrating axiomatic function-to-process-to-product domains.

Chia-Yung Han, Associate Professor; Ph.D., Cincinnati, 1985. Knowledge engineering, computer vision, pattern recognition, CAD/CAM, computer graphics, multimedia systems.

Arthur J. Helmicki, Associate Professor; Ph.D., Rensselaer, 1989. Control-oriented modeling and identification of dynamical systems, robust multivariable control design, intelligent control system.

H. Thurman Henderson, Kartalia Professor; Ph.D., SMU, 1968. Microelectromechanical systems; microsensors for space, medical, and industrial applications; semiconductor device physics; microfluidics; biochemical and medical applications of microelectronics and MEMS.

Yiming Hu, Assistant Professor; Ph.D., Rhode Island, 1998. Computer architectures, high-performance I/O systems, operating systems, parallel and distributed computing, multimedia and Internet applications.

Peter B. Kosel, Professor; Ph.D., New South Wales (Australia), 1976. GaAs and diamond-based devices and circuits for photonics and electronics; CCD memories, ultrasound imaging devices, photodetectors, and radiation sensors; device modeling and circuit simulation.

Ravi Kothari, Associate Professor; Ph.D., West Virginia, 1991. Artificial neural networks, pattern recognition and image analysis.

Stephen Kowel, Professor and Dean; Ph.D., Pennsylvania, 1968. Liquid crystal adoptive optics, auto-stereoscopic displays, electrooptical applications of organic and polymer materials.

William Wei Lin, Assistant Professor; Ph.D., Caltech, 1997. Communication, information theory, error correcting codes, image coding and digital signal processing.

Thomas D. Mantei, Professor and Head; Ph.D., Stanford, 1967. High-density plasma source development and characterization, plasma-assisted etching of semiconductors, plasma-enhanced CVD of dielectric and metal nitride films.

Lawrence J. Mazlack, Associate Professor; D.Sc., Washington (Seattle), 1973. Approximate reasoning, learning, natural language, self-organization, representation, knowledge-based systems, unsupervised data mining, discovery, interfaces and heterogeneous management.

Ali A. Minai, Assistant Professor; Ph.D., Virginia, 1991. Complex adaptive systems, neural models of cognition and memory, mathematical biology, nonlinear dynamics, chaos and fractals, neuromorphic engineering.

Joseph H. Nevin, Professor; Ph.D., Cincinnati, 1974; PE. Microelectromechanical systems, integrated sensors, analog systems, IC design, analog design automation.

Santosh Pande, Assistant Professor; Ph.D., North Carolina State, 1993. Compiler optimizations for embedded and mobile systems, parallel and distributed systems, reconfigurable systems (transformations), object-oriented languages.

Jerome L. Paul, Professor; Ph.D., Case Western Reserve, 1965. Design and analysis of parallel and distributed algorithms, combinatorics, graph theory.

Marios M. Polycarpou, Associate Professor; Ph.D., USC, 1992. Systems and control, with emphasis on intelligent control, fault diagnosis, neural network learning, adaptive control, and nonlinear systems.

Carla C. Purdy, Associate Professor; Ph.D., Illinois, 1975; Ph.D., Texas A&M, 1986. Computer systems design and modeling, mixed technology design and simulation, MEMS, computer arithmetic, experimental CAD, benchmarking for VLSI, algorithms, women in science and engineering.

George Purdy, Professor; Ph.D., Illinois, 1972. Cryptography and data security, algorithms for VLSI, discrete and computational geometry, computational number theory.

Anca L. Ralescu, Professor; Ph.D., Indiana, 1983. Intelligent systems, soft computing, fuzzy information engineering, visual and auditory information processing and interpretation systems.

Panapakkam A. Ramamoorthy, Professor; Ph.D., Calgary, 1977. Digital signal processing and applications, neural networks and fuzzy expert systems, parallel processing, optical computing.

Kenneth P. Roenker, Professor; Ph.D., Iowa State, 1973. Compound semiconductor devices and fabrication, optoelectronic circuit applications, device physics, device reliability.

John Schlipf, Professor; Ph.D., Wisconsin, 1975. Logic programming, nonmonotonic inference and deductive databases, computability and complexity theory, model theory.

Dieter S. Schmidt, Professor; Ph.D., Minnesota, 1970. Symbolic computation, applications to problems in celestial mechanics, dynamical systems.

Andrew J. Steckl, Gieringer Professor and Ohio Eminent Scholar; Ph.D., Rochester, 1973. Semiconductor materials, devices, and fabrication: focused ion beam implantation, thin-film growth (CVD, MBE), wide bandgap semiconductors (SiC, GaN).

Ranga R. Vemuri, Associate Professor; Ph.D., Case Western Reserve, 1989. VLSI design, computer-aided design, VLSI design environments, formal verification, high-level synthesis, adaptive systems design, mixed-signal synthesis, hardware/software cosynthesis.

William G. Wee, Professor; Ph.D., Purdue, 1967. Artificial intelligence; neural networks; computer vision, including 3-D modeling and applications; picture coding and compression.

Philip A. Wilsey, Assistant Professor; Ph.D., Southwestern Louisiana, 1987. Computer architecture, hardware description of languages, parallel simulation, formal methods for design and analysis.

Hongwei Xi, Assistant Professor; Ph.D., Carnegie Mellon, 1998. Dependently typed assembly language, array-bound check elimination.

UNIVERSITY OF CONNECTICUT

Department of Electrical and Systems Engineering

Programs of Study	The department offers programs of study leading to the M.S. and Ph.D. degrees in two major areas: control and communication systems and electromagnetics and physical electronics. The department also offers a general M.S. in electrical engineering and cooperates in an interdisciplinary program in biomedical engineering. The program in control and communication systems focuses on control theory, computational methods, digital signal processing, communications, estimation and detection theory, group decision making, manufacturing scheduling, power system scheduling, optical computing, and neural networks. The program in electromagnetics and physical electronics includes electromagnetic wave propagation, antenna theory, microwave and optical radars, lasers, optoelectronics, fiber optics, semiconductor devices and circuits, and dielectric materials.
	The M.S. program offers two options. The thesis option requires at least 15 credit hours of graduate-level course work and completion of a thesis. The nonthesis option requires at least 24 credit hours of graduate-level course work. In either option, students must maintain at least a B average and pass a final M.S. exam.
	Each Ph.D. program is unique. Students typically take 30 credit hours of graduate-level course work beyond the master's degree. A dissertation contributing to the body of knowledge in the chosen area of research must be presented. Students must maintain a minimum B average, pass a two-part general examination near the end of formal course work, and successfully defend the dissertation.
Research Facilities	Departmental facilities include many well-equipped laboratories: the Biomedical Instrumentation Laboratory; the Central Laboratory for Imaging Research; the Cyberlab, for decision and control research; the Electrical Insulation Research Center; the Estimation and Signal Processing Laboratory; the High-Speed Network and Device Laboratory; the Information Processing Systems/Optical Computing Laboratory; the Lasers and Electrooptics Laboratory; the Manufacturing Systems Laboratory; the Micro/Optoelectronics Laboratory; the Optical Fiber Communications Laboratory; the Sub-Micron Device Fabrication Laboratory; the Systems Optimization Laboratory; and the Photonics Research Center. Numerous dedicated computers are associated with these laboratories, including IBM RISC 6000, Apollo, and Sun Workstations (networked to mainframe) supporting a number of software packages for simulation of devices, circuits, and systems. The Booth Research Center, which houses laboratories and computer facilities to support interdisciplinary work in such areas as computer science, CAD/CAM, robotics, and manufacturing systems, contains Sun-4 server computers. The University Computer Center has a large IBM ES-9000/580 system that supports local, interactive and batch, and remote computing throughout all University components.
Financial Aid	Financial aid is available in the form of fellowships and teaching and research assistantships, awarded individually or in combination. Students apply for financial aid through the graduate school. Stipends for fellowships are variable. Research and teaching assistantships require approximately 20 hours per week of work for the University during the academic year; stipends range from $14,155 to $16,555 in 1999–2000, depending on the student's experience. A tuition waiver and health benefits are included with an assistantship. Financial support for summer research is also available.
Cost of Study	In 1999–2000, tuition for residents of Connecticut enrolled as full-time students is $293 per credit hour up to a maximum of $2636 per semester. New England NEBHE Regional Student Program students (other than Connecticut residents) are charged $439 per credit hour up to a maximum of $3954 per semester; nonresidents are charged $761 per credit hour up to a maximum of $6848 per semester. Other mandatory fees for both residents and nonresidents are the general University fee ($116 up to a maximum of $346 per semester), the graduate activity fee of $13, the shuttlebus transit fee of $5, the graduate matriculation fee of $42, and an infrastructure fee of $105. International students applying through and funded by governmental, quasi-governmental, public, and private organizations are charged $300 per semester. Fees are subject to change without notice.
Living and Housing Costs	On-campus housing costs are $1563 per semester for room and $1363 per semester for board, excluding summer sessions, tuition, and fees. The cost for room and board for the academic year per graduate student is about $5852. University housing is available for single and married graduate students but is not sufficient to meet the demand. Rooms and apartments may be secured in the surrounding area. The University maintains a housing office, which offers assistance in finding suitable housing on or off campus.
Student Group	Approximately 6,600 students are pursuing graduate studies at the University of Connecticut. In fall 1998, the Electrical and Systems Engineering Department had 117 graduate students. Of these students, 66 were full-time (30 master's, 36 Ph.D.) and 51 were part-time (25 master's, 26 Ph.D.). In 1997–98, twenty-three M.S. degrees and nine Ph.D. degrees were awarded.
Location	The University of Connecticut is situated in picturesque northeastern Connecticut. Centrally located between Boston (1½ hours away by car) and New York (2½ hours away), the area offers rural living with ready access to these and other nearby urban centers. Swimming, fishing, and boating are popular in neighboring lakes and state parks. Excellent skiing may be found within a 2–4-hours' drive. Connecticut and Rhode Island beaches are hardly more than an hour from Storrs. Located in historic New England, the University offers many cultural opportunities, including drama, music, art, dance, lectures, and tours.
The University	Founded in 1881, the University of Connecticut now includes the main campus at Storrs, five regional campuses, and many offices and centers throughout the state. More than 25,000 students are enrolled. Numerous centers, institutes, and laboratories support research and outreach activities.
Applying	Application materials may be obtained from the Graduate Admissions Office, Box U-6A, Storrs, Connecticut 06269-1006. Complete applications must be received prior to June 1 for admission in the fall semester or November 1 for admission in the spring semester. International applicants must apply by April 1 or October 1 for fall or spring admission, respectively. Applications for financial aid should be received prior to February 1 for the following academic year. To be admitted, applicants must hold a bachelor's degree or its equivalent. Applicants are required to submit all previous transcripts, three letters of recommendation, and a personal letter of application. The General Test of the GRE is strongly recommended. International students from non-English-speaking countries are required to submit acceptable TOEFL scores.
Correspondence and Information	Graduate Admissions Chairman, Electrical Engineering Program, U-157 Department of Electrical and Systems Engineering University of Connecticut 260 Glenbrook Road Storrs, Connecticut 06269-2157 Fax: 860-486-2447 E-mail: rajeev@eng2.uconn.edu World Wide Web: http://www.ee.uconn.edu/

University of Connecticut

THE FACULTY AND THEIR RESEARCH

The faculty consists of scientists and engineers who are actively engaged in both teaching and research and have received numerous honors. A number of members have gained extensive national and international reputations, have held editorial positions on the leading journals of their fields, and have chaired major national and international conferences.

D. Abraham, Visiting Assistant Professor; Ph.D., Connecticut, 1993. Statistical and array signal processing for applications in underwater acoustics and communications.

A. F. M. Anwar, Associate Professor; Ph.D., Clarkson, 1988. Fabrication and modeling of quantum size effect devices.

J. E. Ayers, Associate Professor; Ph.D., RPI, 1990. Growth and characterization of semiconductors; III-V and II-VI materials.

R. Bansal, Professor; Ph.D., Harvard, 1981. Electromagnetic antennas and waves.

Y. Bar-Shalom, Professor and IEEE Fellow; Ph.D., Princeton, 1970. Estimation, multitarget-multisensor tracking, stochastic control and optimization.

S. A. Boggs, Research Professor and IEEE Fellow; Ph.D., Toronto, 1972. High-voltage dielectrics, high field properties of dielectrics, transient nonlinear finite element analysis, high-temperature superconducting transmission cable, SF6-insulated systems, high-voltage cable.

J. Bronzino, Professor-in-Residence and IEEE Fellow; Ph.D., Worcester Polytechnic, 1968. Quantification of bioelectric events and development of measures of brain maturation that can be utilized to evaluate the impact of various insults such as potential protein malnutrition and neonatal stress.

H. Chen, Research Assistant Professor; Xian Jiaotong, 1990. Production planning, scheduling, and control; discrete event dynamic systems.

P. Cheo, Professor and IEEE Fellow; Ph.D., Ohio State, 1964. Electrooptics, optoelectronics, and fiber optics.

E. Donkor, Associate Professor; Ph.D., Connecticut, 1988. Photonics switching, optical interconnects, semiconductor optics, VHDL.

J. Enderle, Professor and IEEE Fellow; Ph.D., RPI, 1980. Modeling physiological systems, system identification, signal processing, control theory.

M. D. Fox, Professor; Ph.D., Duke, 1972; M.D., Miami (Florida), 1983. Biomedical imaging, Doppler ultrasound, ultrasound and X-ray imaging.

F. C. Jain, Professor; Ph.D., Connecticut, 1973. Fabrication and modeling of semiconductor devices for micro/optoelectronics, blue-green lasers.

B. Javidi, Professor and IEEE Fellow; Ph.D., Penn State, 1986. Optical signal processing, information and pattern recognition, neural networks.

D. Jordan, Professor and Head of Department; Ph.D., Cornell, 1970. Classical and modern control theory and techniques; computational methods.

T. Kirubarajan, Research Assistant Professor; Ph.D., Connecticut, 1998. Data association and sensor fusion for multitarget tracking.

D. L. Kleinman, Professor Emeritus and IEEE Fellow; Sc.D., MIT, 1967. Man-machine systems, control systems, optimization algorithms.

P. B. Luh, Professor and IEEE Fellow; Ph.D., Harvard, 1980. Planning, scheduling, and coordination of manufacturing and power systems.

R. B. Northrop, Professor Emeritus; Ph.D., Connecticut, 1964. Biomedical instrumentation, neurophysiology.

K. R. Pattipati, Professor and Fellow, IEEE; Ph.D., Connecticut, 1980. Optimization, prognostics and diagnostics, multiobject tracking, adaptive organizations.

G. Taylor, Professor and IEEE Fellow; Ph.D., Toronto, 1971. Optoelectronics devices and integrated circuits; advanced materials.

P. Willett, Associate Professor; Ph.D., Princeton, 1986. Detection, target tracking, signal processing.

Q. Zhu, Assistant Professor; Ph.D., Pennsylvania, 1992. NIR-diffusive light imaging and ultrasound imaging.

RESEARCH LABORATORIES AND CURRENT RESEARCH AREAS

Biomedical Instrumentation Laboratory, J. Enderle and R. B. Northrop in charge. This laboratory contains the instrumentation used in electronic circuit development and physiology and the computational facilities for complex physiological modeling. Current research is focused on optical glucose sensors, closed-loop control of the self-administration of analgesic drugs, system identification and physiological modeling of the immune and oculomotor systems, and physiological neural networks.

Central Laboratory for Imaging Research, M. D. Fox in charge. This laboratory focuses on the visualization of three-dimensional medical images and scientific and engineering data. The lab has various computational imaging setups and Doppler ultrasound instrumentation.

Cyberlab, K. R. Pattipati and D. L. Kleinman in charge. This lab is dedicated to man-machine and systems research. The heart of the extensive computational facilities within the laboratory is an Ethernet cluster of UNIX-based high-performance Sun SPARCstations. The primary application of the state-of-the-art computer complex lies in conducting real-time simulations for the empirical study of multihuman distributed decision making, distributed computation and networking, and team multitask sequencing and scheduling in addition to systems and control research.

Electrical Insulation Research Center, S. Boggs in charge. This laboratory operates in an interdisciplinary fashion with the Institute of Materials Science. It has high-voltage DC, AC, and impulse test equipment, a time domain dielectric spectrometer for the measurement of dielectric loss and complex permittivity, optical microscopes, infrared microspectrophotometers, apparatus for measuring the threshold field for high carrier motility in solid dielectrics, proprietary software for transient nonlinear finite element analysis with coupled (e.g., electric and thermal) fields, and access to the wide range of analytical equipment in the Institute of Materials Science.

Estimation and Signal Processing Laboratory, Y. Bar-Shalom in charge. The ESP laboratory deals with signal and information processing for remote sensing and data fusion for multiple moving targets (as in air traffic control) using heterogenous sensors (radar, sonar, and electrooptical). The same technology has been successfully applied to biomedicine for tracking moving cells to assess their health.

High-Speed Network and Device Laboratory, E. Donkor in charge. This laboratory focuses on the research of network architectures and their enabling components for high-speed digital and high-frequency analog signal transmission over optical fibers. The laboratory is equipped with a 20-Ghz Bit-Error Rate (BER) test system, a UV/VIS/IR spectrophotometer, visible and infrared lasers, and work stations for testing and measurements of electrooptic devices.

Information Processing Systems/Optical Computing Laboratory, B. Javidi in charge. This laboratory is dedicated to research on information processing, imaging, signal and image processing, neural computing, pattern recognition, encryption, data security systems, data storage, real-time optical computing, 3-D visualization, and optoelectronic systems for communication and radar. The facilities include state-of-the-art computing systems with extensive software packages, information processing hardware, devices and systems for data acquisition, spatial light modulators for real-time optical computing, lasers, optical tables with accessories, and networking facilities.

Manufacturing Systems Laboratory, P. B. Luh in charge. This laboratory is dedicated to manufacturing systems research. Current research focuses on the development and implementation of high-performance planning, scheduling, and coordination systems. The facilities include an Ethernet cluster of seven Sun SPARC Workstations and PCs and extend to many industrial plants, including Pratt & Whitney, Cannondale, Toshiba, Delta, Sikorsky, and Northeast Utilities.

Micro/Optoelectronics Research Laboratory, F. C. Jain in charge. Current research is focused on mismatched heteroepitaxy of semiconductors, MQW optical modulators and tunable lasers at 1.55 micron, blue-green lasers, flat panel EL displays, quantum wire/dot optoelectronics devices, and quantum interference transistors. This lab is equipped with VPE reactors for Ge and Si growth; MOVPE of III-V and II-VI semiconductors; a high-resolution x-ray diffractometer, 10K photoluminescence setup; a photolithographic clean room; a reactive ion etcher and device processing setups for lasers and transistors; optoelectronic measurement, including a semiconductor laser/modulator test facility; and dedicated workstations for computer-aided device and VLSI design and simulation.

Optoelectronic Integrated Circuit Research Laboratory, G. Taylor in charge. This laboratory has MBE growth capability for III-V semiconductor optical and electronic devices, IC fabrication and characterization facilities, and test beds for basic optical communication systems. Current research is dedicated to optoelectronic integrated devices for communications.

System Optimization Laboratory, K. R. Pattipati in charge. This laboratory is dedicated to research on systems theory and optimization techniques to solve industrial problems. Current research is focused on automated testing, quality control, computer system performance optimization and scheduling, and multitarget tracking. The laboratory is equipped with seven Sun SPARC Workstations and enjoys a close relationship with local industries.

Ultrasound and Optical Imaging Laboratory, Q. Zhu in charge. Research focuses on NIR-diffusive light imaging and ultrasound imaging.

UNIVERSITY OF DETROIT MERCY

College of Engineering and Science
Department of Electrical and Computer Engineering

Programs of Study

The Department of Electrical and Computer Engineering offers programs leading to the Master of Engineering and Doctor of Engineering. The areas of concentration include computer engineering (specifically computer architecture, parallel processing, and microprocessors) and signals and systems (specifically digital signal processing, electromagnetic compatibility, automotive electronics, and control systems). Students who are interested in management can pursue a Master of Engineering Management degree with an emphasis on electrical engineering. The Master of Engineering in electrical engineering may be completed by either a thesis or nonthesis plan. The thesis plan includes 24 credit hours of course work and 6 credit hours of thesis. The nonthesis plan consists of 30 credit hours of course work. The Doctor of Engineering in electrical engineering requires the following postbaccalaureate components: four core courses, 30 hours of course work in a specific discipline, 9 credit hours of approved technical electives, and 36 hours of dissertation. The doctoral candidate must pass qualifying examinations and a comprehensive examination. Upon entry into the graduate program, each student is assigned an academic adviser. This adviser works with the student to develop a course program that best corresponds to the student's career objectives.

Research Facilities

The program takes advantage of the rich automotive electronics environment of Detroit. State-of-the-art computing and faculty research facilities are also available for the graduate programs. In addition, the advanced computing laboratory, which consists of a network of workstations, provides simulation and analysis tools for research activities. All electrical engineering laboratories utilize computer-aided engineering technology via PC-based systems.

Financial Aid

A variety of teaching and research fellowships/assistantships are available in the College of Engineering and Science. This financial aid is competitively awarded on a yearly basis. In addition, the Scholarship and Financial Aid Office accepts applications for grants, loans, and work-study assistance. Aid includes the Michigan Tuition Grant (for Michigan residents only), Federal Work-Study, and a variety of loans. The University also accepts third-party payments from employers and government agencies as well as offering payment plans of its own. For information regarding financial aid programs, students should call 313-993-3350.

Cost of Study

Tuition in 1999–2000 is $545 per credit hour. Registration fees are $100 for full-time students.

Living and Housing Costs

Housing is available on campus. Double-occupancy rates range from $1410 to $3280. Single-occupancy rates range from $2420 to $2720. The University's meal plans cost about $1085. All rates are for a sixteen-week term. For more information, students should call the Residence Life Office at 313-993-1230.

Student Group

The graduate students in the electrical and computer engineering department represent many groups: men, women, and minority and international students. Many students are pursuing their degree part-time in the evening while working full-time in industry. Approximately 6,700 students attend classes on three UDM campuses located in northwest and downtown Detroit.

Location

Students enjoy a variety of activities offered on campus and throughout the metropolitan Detroit area, including sports, theater, concerts, and more.

The University

As Michigan's largest Catholic university, the University of Detroit Mercy has an outstanding tradition of academic excellence firmly rooted in a strong liberal arts curriculum. This tradition dates back to the formation of two Detroit institutions, the University of Detroit, founded in 1877 by the Society of Jesus (Jesuits), and Mercy College of Detroit, founded in 1944 by the Religious Sisters of Mercy of the Americas. In 1990, these schools consolidated to become the University of Detroit Mercy. Today, UDM offers more than 120 majors and programs in nine different schools and colleges and is widely recognized for its programs in engineering, law, dentistry, nursing, and architecture. Faculty members are known for their excellence; more than 87 percent have a Ph.D. or comparable terminal degree.

Applying

Applications for admission should be completed at least six weeks before the beginning of a term. Applications for financial aid should be submitted by April 1. International students are urged to complete their applications at least three months before classes begin. Admission requirements are a bachelor's degree from an accredited college; a B average in the total undergraduate program and in the proposed field of study; and, normally, an undergraduate major or the equivalent in the proposed field. Official transcripts are required from all colleges attended. Applicants with less than a B average who present other evidence of ability to perform graduate-level work may be admitted as probationary students upon the recommendation of the director of the program.

Correspondence and Information

Records Office
College of Engineering and Science
University of Detroit Mercy
4001 West McNichols
P.O. Box 19900
Detroit, Michigan 48219-0900
Telephone: 313-993-3335
Fax: 313-993-1187

Professor C. J. Lin, Chairman
Department of Electrical and Computer Engineering
University of Detroit Mercy
4001 West McNichols
P.O. Box 19900
Detroit, Michigan 48219-0900
Telephone: 313-993-3365
Fax: 313-993-1187
E-mail: lincj@udmercy.edu

University of Detroit Mercy

THE FACULTY AND THEIR RESEARCH

Nizar Al-Holou, Associate Professor; Ph.D., Dayton. Microprocessors, digital logic, parallel processing, computer architecture.

Armand Ashrafzadeh, Associate Professor; Ph.D., Oklahoma. Electrical engineering modeling and simulation (CAE), biomedical engineering, nonlinear filtering, sensor modeling.

Mohan Krishnan, Associate Professor; Ph.D., Windsor. Pattern recognition, speech processing, digital signal processing, communications.

Chun-Ju Lin, Professor; Ph.D., Michigan State. Electromagnetic compatibilities, circuits and systems, power distribution networks and machines.

Mark Paulik, Associate Professor; Ph.D., Oakland. Microprocessor design and implementation, microcontrollers, pattern recognition, digital signal processing.

Dipak Sengupta, Professor; Ph.D., Toronto. Microwave and radar, electromagnetic compatibility, communication.

Sandra Yost, Assistant Professor; Ph.D., Notre Dame. Robust control stability theory, digital control systems.

Research Areas

The Department of Electrical and Computer Engineering has highly qualified faculty members with excellent academic and professional backgrounds. Some of the faculty members are internationally acclaimed for their research. Currently, faculty research interests are in the following areas: image processing and analysis, one- and two-dimensional sensor modeling and simulation, speech processing, parallel processing and distribution, electromagnetic interference, automotive electronics and antennas, and modeling and simulation (CAE).

UNIVERSITY OF FLORIDA

Department of Electrical and Computer Engineering

Programs of Study	The Department of Electrical and Computer Engineering offers programs of study leading to degrees of Master of Engineering, Master of Science, Engineer, and Doctor of Philosophy. Three main divisions are computer-information engineering, electronics, and electromagnetics-energy systems. These are partitioned into ten areas of major concentration: communications, computer systems and networks, device and physical electronics, digital signal processing, electric energy systems, electromagnetics, electronic circuits, intelligent and information systems, photonics, and systems and control. Course work and research opportunities are offered in many specializations, including computer networks control systems; digital communications; digital hardware and signal processing; electric energy and power electronics; electronic noise; image processing; integrated and fiber optics; laser electronics and fabrication; lightning; microprocessors; multimedia; radar; reliability of semiconductor devices and circuits; robotics and machine intelligence; signal estimation; solid-state and semiconductor devices; spread spectrum systems; telecommunication systems; VLSI circuit design, fabrication, and characterization; and wireless communications.
	The master's degree candidate may choose a thesis or nonthesis program. Although both programs require 33 semester hours of credit, the student who pursues the nonthesis option must pass a written examination that covers two of the above areas of concentration. The student who chooses the thesis option is expected to write a thesis and defend it during a final oral examination. The Master of Engineering degree is awarded to students with a baccalaureate in engineering, while the Master of Science degree is for those who have earned a bachelor's degree in engineering, math, or the sciences.
	The Engineer degree requires a thesis and a minimum of 30 semester hours of course work beyond the master's degree. The Engineer degree is a terminal degree and should not be considered as a partial fulfillment of requirements for the Ph.D.
	The doctoral candidate is required to pass a qualifying examination that is both written and oral. The student is also expected to complete a dissertation that reflects independent research and pass a related final examination. Ninety semester hours, which may include some or all of the master's degree course work, are required for the Ph.D. degree.
	The Department of Electrical and Computer Engineering also offers an off-campus nonthesis master's degree program for employees of various Florida industries and government agencies through the statewide Florida Engineering Education Delivery System. Classes originating at the University of Florida are offered by videotape and through Web-based courses on the Internet.
Research Facilities	The Department of Electrical and Computer Engineering occupies 85,000 square feet of space, much of which is devoted to funded research. The research facilities available include laboratories for IC processing; VLSI circuit design and characterization; communications; computational neuroengineering; digital signal processing; electronic noise; lightning; optical measurement, photodetection, and optical waveguides; robotics; speech analysis and synthesis; and power system transient analysis and power electronics. All facilities contain up-to-date equipment either purchased via funded research or donated by supportive companies. The research computing facilities are excellent and include seventy workstations, hundreds of PCs, and extensive industrial software packages. The teaching facilities include an additional thirty UNIX workstations and forty Pentium-class PCs that run a wide variety of technical software. The computers are networked to provide access to other university computing resources and to the Internet.
Financial Aid	Graduate teaching and research assistantships are available from the department and are awarded on the basis of academic performance, GRE scores, college transcripts, and letters of recommendation. In addition, they usually include a tuition fee waiver. Fellowships are also available.
Cost of Study	In spring 1999, the registration fee for most graduate course work was $137.75 per credit hour for Florida residents and $481.31 per credit hour for out-of-state students. The tuition fee may be waived for students holding graduate assistantships and fellowships. Students register for 9 to 12 credit hours per semester.
Living and Housing Costs	Rents for apartments provided by the University for single graduate and professional students begin at $261 per person per month. The University operates five apartment villages for families. There are also many apartment complexes in the area, with rent for one-bedroom apartments starting at approximately $300 per month, not including utilities.
Student Group	The total enrollment at the University of Florida is approximately 43,000. In the Department of Electrical and Computer Engineering, there are about 350 graduate students enrolled on campus.
Location	The University of Florida is located in Gainesville, a city of approximately 95,000 inhabitants (190,500 in the metropolitan area) in north-central Florida. Gainesville was rated as the most livable city in the U.S. in 1996 by *Money* magazine. Gainesville is just over an hour's drive from both the Gulf of Mexico and the Atlantic Ocean, and there are facilities close by for sailing, canoeing, waterskiing, tennis, fishing, and golf. The city offers many cultural events, including professional theater, art exhibits, open-air festivals, and concerts.
The University	The University of Florida is one of the nation's ten largest universities and ranks among the top three in the number of academic programs offered. Undergraduate students can take classes in 140 departments, while graduate degrees are offered in more than 100 fields. There are sixteen upper-division colleges and schools and four professional colleges (Law, Dentistry, Medicine, and Veterinary Medicine).
Applying	Application forms may be obtained from the Graduate Studies Office of the Department of Electrical and Computer Engineering. Admission to the graduate program requires a baccalaureate degree from an accredited college and an upper-division grade point average of 3.0 or better on a 4.0 scale. Students must submit satisfactory scores on the General Test of the GRE and a score of 550 or better on the TOEFL (if applicable). Students may be admitted in any semester, but application forms, transcripts, and test scores should be submitted six months before the desired date of registration.
Correspondence and Information	Graduate Studies Office Department of Electrical and Computer Engineering P.O. Box 116200 225 Larsen Hall University of Florida Gainesville, Florida 32611 Telephone: 352-392-4945 E-mail: info@graduate.ece.ecu World Wide Web: http://www.ece.ufl.edu

University of Florida

THE FACULTY AND THEIR RESEARCH

Communications
Leon W. Couch II, Associate Chairman (Academic Affairs); Ph.D., Florida, 1968. Communication systems and applications.
Peyton Z. Peebles Jr., Ph.D., Pennsylvania, 1967. Communication systems, radar system theory.
Tan F. Wong, Ph.D., Purdue, 1997. Wireless communications, spread spectrum and multiuser communications, adaptive signal processing.

Computer Systems and Networks
Yen-kuang Chen, Ph.D., Princeton, 1998. Multimedia architecture and algorithms, processor design, memory design, video compression and processing.
Alan D. George, Ph.D., Florida State, 1991. Parallel and distributed computing, high-performance computer networks, fault-tolerant computing.
Michel A. Lynch, Ph.D., Florida, 1972. Microprocessor applications, digital system design, computer sound synthesis.

Device and Physical Electronics
Gijs Bosman, Associate Chairman (Student Affairs); Ph.D., Utrecht (Netherlands), 1981. Electronic noise, quantum devices.
Jerry G. Fossum, Ph.D., Arizona, 1971. Semiconductor device theory, modeling, simulation, SOI circuits, IC technology, CAD.
Sheng S. Li, Ph.D., Rice, 1968. Photodetectors, IR imaging arrays, SOI materials and devices, photonic and quantum effect devices.
Fredrik A. Lindholm, Ph.D., Arizona, 1963. Semiconductor device physics.
Arnost Neugroschel, Ph.D., Technion (Israel), 1973. Semiconductor device physics and characterization.
Toshikazu Nishida, Ph.D., Illinois, 1988. Semiconductor sensors and devices.
Chih-Tang Sah, Graduate Research Professor and Eminent Scholar; Ph.D., Stanford, 1956. Semiconductor device physics, semiconductor reliability.

Digital Signal Processing
John M. M. Anderson, Ph.D., Virginia, 1992. Medical imaging, statistical signal processing, image processing.
Donald G. Childers, Ph.D., USC, 1964. Algorithms, digital signal/speech/image processing.
William W. Edmonson, Ph.D., North Carolina State, 1990. Adaptive filtering, communications.
John G. Harris, Ph.D., Caltech, 1991. Analog and digital signal processing, VLSI, neural networks.
Jian Li, Ph.D., Ohio State, 1991. Array processing, radar, imaging and target feature extraction.
Jose C. Principe, Ph.D., Florida, 1979. Adaptive non-Gaussian signal processing, neural networks, biomedical signal analysis.
Fred J. Taylor, Ph.D., Colorado, 1969. Digital signal processing, digital computer architecture, and design.

Electric Energy Systems
Dennis P. Carroll, Ph.D., Wisconsin, 1969. Electric energy systems, high-voltage transmission, power electronics, power system simulation.
Alexander Domijan Jr., Ph.D., Texas at Arlington, 1986. Electric energy systems, power quality, modeling and simulation, instrumentation.
Khai Ngo, Ph.D., Caltech, 1984. Power electronics, low-profile magnetics, power integrated circuits.

Electromagnetics
Vladimir A. Rakov, Ph.D., Tomsk (Russia), 1983. Lightning, atmospheric electricity, lightning protection.
Ewen M. Thomson, Ph.D., Queensland (Australia), 1985. Lightning.
Martin A. Uman, Chairman; Ph.D., Princeton, 1961. Lightning, atmospheric electricity, electromagnetics.

Electronic Circuits
William R. Eisenstadt, Ph.D., Stanford, 1984. Microwave integrated circuits, VLSI design.
Robert M. Fox, Ph.D., Auburn, 1986. Analog and digital VLSI circuits.
Mark Law, Ph.D., Stanford, 1988. VLSI process and device simulation.
Kenneth K. O, Ph.D., MIT, 1989. Electronic circuit devices, IC wireless technology.
Jack R. Smith, Ph.D., USC, 1964. Bioengineering, electronics, computers and signal detection.

Intelligent and Information Systems
A. Antonio Arroyo, Ph.D., Florida, 1981. Artificial intelligence, microprocessing, digital design, speech processing.
Herman Lam, Ph.D., Florida, 1979. Computer engineering, database management, computer architecture.
Michael C. Nechyba, Ph.D., Carnegie Mellon, 1998. Human-centered robotics, robot control, neural networks, hidden Markov models.
Stanley Y. W. Su, Ph.D., Wisconsin, 1968. Database management, database machines, software systems, parallel architecture and object-oriented knowledge base management.

Photonics
Chris S. Anderson, Ph.D., North Carolina State, 1991. Optics, optical signal processing, holography/interferometry, radar and communications.
Ramu V. Ramaswamy, Ph.D., Northwestern, 1969. Passive, active, linear, and nonlinear guided wave optical and optoelectronic devices.
Ramakant Srivastava, Ph.D., Indiana, 1973. Guided wave sensors, integrated optics.
Henry Zmuda, Ph.D., Cornell, 1984. Microwave system design, photonics.
Peter S. Zory, Ph.D., Carnegie Tech, 1964. Lasers.

Systems and Control
Thomas E. Bullock, Ph.D., Stanford, 1966. Mathematical systems theory, filtering, digital control with microprocessors.
Jacob Hammer, Graduate Coordinator; D.Sc., Technion (Israel), 1980. Mathematical systems theory, control systems.
Haniph A. Latchman, D.Phil., Oxford, 1986. Control systems and communications, multimedia information systems.

UNIVERSITY OF HOUSTON

Department of Electrical and Computer Engineering

Programs of Study

The department offers the Master of Electrical Engineering (M.E.E.), Master of Science (M.S.), and Doctor of Philosophy (Ph.D.) degrees. Fields of specialization include antennas and applied electromagnetics, bioelectromagnetics, biomedical engineering, communications, computers, control systems, high-temperature superconductivity, microelectronics, nondestructive evaluation, optics, pattern recognition, power systems, seismic exploration, signal and image analysis, systems analysis, ultrasonics, and well logging. The department also administers interdisciplinary graduate programs in biomedical engineering and computer and systems engineering.

The M.S. degree requires 24 hours of course work and a research thesis. The M.E.E. degree is a nonthesis option and requires 36 course hours. Requirements for the Ph.D. degree include a dissertation and 24 course hours beyond the master's degree. Candidates must also meet specific requirements with respect to both breadth and depth in their professional knowledge and may be assigned remedial work if these are not satisfied.

In response to the rapid growth in the telecommunications industry, a nonthesis Master of Electrical Engineering in telecommunications is offered. The degree requirements are similar to those of the M.E.E. and additionally may include a 6-credit-hour internship at one of many participating telecommunications companies.

Research Facilities

The department has extensive research facilities for work in antenna engineering; biomedical engineering; digital systems; electron beam and ion optics; high-temperature superconductivity; microwaves; power systems; seismic data analysis; semiconductor fabrication, including rapid thermal processing, advanced lithography, and defect characterization; well logging; and other areas. The University computing center maintains an AS/9000 mainframe and a cluster of VAX computers. A central campus facility maintains 140 workstations available 24 hours a day, 7 days a week. The Engineering Computing Center in the College of Engineering has more than 125 personal computers and workstations available for student use. In addition, personal computers and workstations are located throughout ECE department offices and laboratories.

Financial Aid

Graduate students with teaching or research assistantships are provided a stipend of up to $1200 per month, depending on qualifications and level. At present, about 70 percent of full-time M.S. and Ph.D. students are supported. All outstanding students will be considered for financial aid. There are no additional forms to be filled out; however, application deadlines may be earlier for those desiring to be considered for aid.

Cost of Study

Tuition and fees vary, but typical payments for 12 semester credit hours are $1358 for Texas residents and $3638 for nonresidents. Teaching and research assistants as well as scholarship recipients qualify for the resident rate.

Living and Housing Costs

On-campus living expenses are about $600 per month. Off-campus apartments (room only) begin at about $250 per month.

Student Group

There are about 2,300 students enrolled in the Cullen College of Engineering and about 800 in the Department of Electrical and Computer Engineering.

Location

The University's 390-acre campus is only a short drive from downtown Houston, the fourth-largest city in the nation. Also within a short distance is the internationally known Texas Medical Center, a 200-acre area in which are located the Baylor College of Medicine, the University of Texas College of Medicine, and the University of Texas Health Science Center at Houston, to name a few of its components. These institutions, together with the Johnson Space Center and other organizations, offer numerous interdisciplinary research opportunities. Texas Instruments and Compaq are among major high-technology companies with headquarters in Houston. The world-renowned Houston Symphony and Alley Theater present two of many exciting cultural opportunities. Houston is also home to many professional sports teams, including the Astros baseball team and the two-time World Champion Houston Rockets basketball team.

The University and The Department

A large majority of the University's 30,000 students commute. Although daytime enrollment now predominates, a tradition of providing night courses for part-time students at both the undergraduate and graduate levels is maintained. The wide scope of academic and research opportunities provided by the 44 doctoral and 106 master's-degree programs attracts a highly heterogeneous group of students and faculty. An international flavor results from a large enrollment of students from other nations.

Established in 1947, the Department of Electrical and Computer Engineering has experienced its greatest growth since 1963, when the previously private university became state supported. The electrical and computer engineering faculty members, most of whom have international reputations in their specialties, reflect an emphasis on excellence in research as well as teaching. As a consequence, the department enjoys a high level of research funding from outside agencies.

Applying

For unconditional admission to either master's degree program, a grade point average of at least 3.0 in the previous 60 hours of study, a satisfactory GRE score (verbal, math, and analytical), a bachelor's degree in electrical engineering (or a related discipline) from an accredited department, and three recommendations are required. Admission to the Ph.D. program requires a master's degree or 30 hours of graduate credit in electrical engineering (or a related field). GRE scores and a GPA appreciably above those specified for the master's level are expected.

Applications for admission, GRE scores, three recommendations, and two copies of all transcripts must be submitted directly to the Department of Electrical and Computer Engineering. International students and those seeking financial assistance are urged to apply several months in advance of their anticipated enrollment date. Results of the TOEFL are required of international applicants from non-English-speaking countries.

Correspondence and Information

Graduate Admissions Analyst
Department of Electrical and Computer Engineering
University of Houston
4800 Calhoun
Houston, Texas 77204-4793
Telephone: 713-743-4403
Fax: 713-743-4444
E-mail: mmcdonald@uh.edu
World Wide Web: http://www.egr.uh.edu/ECE

University of Houston

THE FACULTY AND THEIR RESEARCH

Wallace L. Anderson, Professor; Sc.D., New Mexico; PE. Ultrasonics, nondestructive evaluation, signal analysis, statistical estimation, pattern recognition, Fourier optics.

Betty J. Barr, Assistant Professor; Ph.D., Houston. Applied mathematics.

Earl J. Charlson, Professor; Ph.D., Carnegie Mellon; PE. Solid-state devices, electronic materials, integrated circuits.

Elaine M. Charlson, Professor and Associate Vice President/Associate Vice Chancellor; Ph.D., Missouri–Columbia. Solid-state, integrated circuits, plasma polymers.

Guanrong Chen, Professor; Ph.D., Texas A&M. Nonlinear systems—dynamics and control, robotics, fuzzy systems control, chaotic systems control.

Frank Claydon, Professor and Chair; Ph.D., Duke. Cardiac mapping, computer modeling of atrial and ventricular defibrillation, applied electromagnetics.

Ovidiu Crisan, Professor; D.Eng., Polytechnic Institute of Timisoara (Romania). Power systems operation, control and optimization, electric machine modeling, superconductivity in power systems.

John R. Glover Jr., Professor; Ph.D., Stanford. Adaptive systems, digital signal processing, biomedical signal processing, expert systems applications, educational software.

Thomas J. Hebert, Associate Professor; Ph.D., USC. Image and signal processing.

Martin C. Herbordt, Assistant Professor; Ph.D., Massachusetts. Computer architecture, high-performance systems, parallel processing, computer system simulation.

David R. Jackson, Professor; Ph.D., UCLA. Electromagnetic theory, microstrip antennas, microwave and millimeter-wave antennas, bioelectromagnetics.

Ben H. Jansen, Professor; Ph.D., Free University (Amsterdam). Biomedical signal analysis, pattern recognition, artificial intelligence, biomedical engineering, complex dynamic systems modeling, nonlinear dynamics, self-organizing systems.

Nicolaos B. Karayiannis, Associate Professor; Ph.D., Toronto. Artificial neural networks, supervised and unsupervised learning, fuzzy pattern recognition and vector quantization, image processing and analysis.

Tony L. King, Assistant Professor; Ph.D., Illinois. Control systems, electronic circuit design, instrumentation.

Periklis Y. Ktonas, Professor; Ph.D., Florida. Biomedical engineering, bioelectrical signal analysis, random data analysis.

Han Quang Le, Professor; Ph.D., MIT. Compound semiconductor devices, semiconductor optoelectronics, lasers, optics, photonic instrumentation and systems applications.

Chih-Hsiang Thompson Lin, Research Associate Professor; Ph.D., Missouri–Columbia. MBE growth of III-V materials, materials characterization, device modeling and simulations, mid-IR lasers and LEDs, MIR photodetectors, universal compliant substrates, resonant tunneling diodes, HEMTs, VCSELs, chemical sensing.

Ce Liu, Associate Professor; Ph.D., Jiaotong (China). Well logging, ground-penetrating radar, EM tomography.

Stuart A. Long, Professor; Ph.D., Harvard; PE. Applied electromagnetics, printed-circuit and millimeter-wave antennas, high-temperature superconducting materials and devices.

Pauline Markenscoff, Associate Professor; Ph.D., Minnesota. Computer architecture, performance evaluation of computer systems, distributed processing.

Haluk Ogmen, Associate Professor; Ph.D., Laval. Neural networks, biological and computer vision, biological sensory-motor control and adaptive robotics.

David M. Pai, Associate Professor; Ph.D., British Columbia. Lightwave communications, photonic devices.

Gerhard F. Paskusz, Professor; Ph.D., UCLA; PE. Computer-aided circuit analysis and design.

Steven Pei, Professor; Ph.D., SUNY at Stony Brook. Heterostructure FETs and heterojunction bipolar transistors based on wide bandgap III-V compounds, optoelectronic ICs.

William P. Schneider, Professor; S.M., MIT; PE. Controls, electronic instrumentation, dynamic positioning systems, sonar drill-pipe control systems, marine systems, control of unstable systems.

David P. Shattuck, Associate Professor; Ph.D., Duke; PE. Acoustic imaging and well logging.

Liang C. Shen, Professor; Ph.D., Harvard; PE. Antennas and wave propagation, underground antennas, printed-circuit antennas, electromagnetic techniques in well logging, subsurface sensing.

Leang S. Shieh, Professor; Ph.D., Houston; PE. Control systems; model reduction, identification, and design; optimal control; adaptive control; digital control; multivariable control systems.

Leonard Trombetta, Associate Professor; Ph.D., Lehigh. Electronic materials, MOS oxide reliability, semiconductor defect studies, electron devices, wide bandgap semiconductors.

James Wasson, Research Assistant Professor; Ph.D., Houston. Electronic materials, thin-film deposition, film stress analysis, reactive ion etching, ion beam lithography.

Jeffery T. Williams, Associate Professor; Ph.D., Arizona. Applied electromagnetics, wave propagation, numerical techniques, antenna measurements and design, high-frequency superconductor characterization and applications.

Donald R. Wilton, Professor; Ph.D., Illinois. Electromagnetic theory, mathematical methods, numerical techniques.

John C. Wolfe, Professor; Ph.D., Rochester. Materials research, electron and ion-beam devices, microfabrication.

Jaroslaw (Jarek) Wosik, Research Associate Professor; Ph.D., Warsaw. Microwave characterization and applications of superconductors, MRI rf receiver coils, technology of superconducting thin films, ESR spectroscopy, cryogenic and related techniques.

Wanda Zagozdzon-Wosik, Associate Professor; Ph.D., Warsaw Technical. Semiconductor integrated circuit processing technology, electron devices.

UIC

UNIVERSITY OF ILLINOIS AT CHICAGO

Department of Electrical Engineering and Computer Science
Program in Electrical Engineering

Programs of Study	The Department of Electrical Engineering and Computer Science offers a broad range of programs in electrical engineering and computer science and engineering leading to the M.S. and Ph.D. degrees. The M.S. degree requires 36 semester hours of study, including an optional thesis. The Ph.D. degree requires an additional 76 semester hours of credit. Ph.D. students are required to pass a written qualifying examination, an oral thesis proposal examination, and a final examination defending the thesis at the conclusion of the research.
	While a wide range of courses comprising many areas in electrical engineering is offered, the department has special strengths in the areas of microelectronics and microfabrication, electromagnetics and optics, power electronics, communications, controls, networks, biomedical applications, and signal processing.
Research Facilities	The department maintains two large, modern instructional computing facilities, including ten UNIX file servers that manage a total of more than 95 GB of disk space, four UltraSPARC compute servers, and 130 student-accessible Sun workstations for software development, database programming, VLSI design, HTML development, and numerical computing. There are also four Silicon Graphics workstations equipped with 24-bit graphics and thirty-four Macintosh computers that are used for assembly language programming and HTML development. All computers are networked via a 10-megabit switched Ethernet.
	In addition to the modern instructional computing facility, the department contains several specialized research laboratories, most of which are housed in the $30-million Engineering Research Facility. The Microfabrication Applications Laboratory, the Communications Laboratory, the Electromagnetics and Optics Laboratory, the Power Electronics Laboratory, the Visual/Motor Laboratory, and the Biomedical Functional Imaging and Computational Laboratory contain a wide array of specialized equipment for advanced research in electrical engineering. The Electronic Visualization Laboratory (EVL), the Interactive Computing Environments Laboratory, the Concurrent Software Systems Laboratory, the Intelligent Vehicle Highway Systems and Artificial Intelligence Laboratory, the Software Engineering Laboratory, the Knowledge and Database Systems Laboratory, the Distributed Real-Time Intelligent Systems Laboratory, the Parallel Systems Laboratory, the Laboratory for Advanced Computing, the Vision Interface and Systems Laboratory, and the Signal and Image Research Laboratory contain well more than 100 additional workstations and servers as well as an extensive array of computer-based multimedia equipment for advanced research in computer science and engineering.
	The departmental computing facilities are networked to general University computing resources and national networks, permitting high-speed access to specialized computing facilities, such as Connection Machine, Power Challenge Array, and Convex supercomputers at the National Center for Supercomputing Applications (NCSA) at the University of Illinois at Urbana-Champaign and to the IBM SP-2 at Argonne National Laboratory.
Financial Aid	Financial aid is available in the form of teaching and research assistantships as well as various University fellowships, and opportunities are excellent for research-oriented students with solid undergraduate backgrounds in electrical engineering. Half-time assistantships for new graduate students require 19 hours per week and pay $11,300 for the academic year ($13,816 for the calendar year). In 1998–99, the electrical engineering and computer science department supported more than 140 graduate students, about a third of whom were teaching assistants. Approximately 75 graduate students were supported by external research grants totaling more than $5 million.
Cost of Study	For 1998–99, Illinois residents paid $2746 and nonresidents paid $6064 per semester if registered for 12 or more hours, with a sliding scale for registration of fewer than 12 hours. Tuition and fees are waived for recipients of teaching and research assistantships, tuition and fee awards, and some University fellowships. Fees are subject to change.
Living and Housing Costs	In addition to University of Illinois at Chicago (UIC) residence halls, off-campus apartments and rooms are available. Meals may be taken in campus dining facilities or a variety of nearby commercial establishments. Food, housing, transportation, medical care, personal items, clothing, and incidentals are estimated to cost $11,000 per calendar year.
Student Group	The department has a student body of more than 1,100 undergraduate students and more than 500 graduate students. The graduate student body includes about 250 full-time (average of 12 or more hours) students, of whom more than 100 are pursuing the Ph.D. degree. Most graduate students are affiliated with one of the research laboratories listed above.
Location	Chicago is a vibrant, friendly, beautiful city where there is never a dull moment. The city has countless restaurants, signature blues clubs, excellent museums, and 12 miles of lakefront beach. The University, a 5-minute train ride from the center-city Loop, is situated on the original site of Jane Addams' Hull House, the first social settlement house in the United States. The University is bordered by Greektown to the north and Little Italy to the west. Chinatown is a 10-minute drive away.
The University	The University of Illinois at Chicago, with approximately 25,000 students, is the largest institution of higher learning in the Chicago area. UIC is listed among the Research-I universities in the country in external research funding. The University offers master's degrees in eighty-seven fields and doctorates in fifty-four areas.
Applying	Application materials for admission and financial aid may be obtained by writing to the address below. Requests for applications for teaching assistantships and fellowships (including tuition and fee waivers) should also be directed to the address below. Students seeking research assistantships are advised to write directly to faculty members in their area of interest. All international applicants are required to take the Graduate Record Examinations for admission. The GRE is recommended for all applicants for financial aid. Applications for teaching and research assistantships should be received no later than March 1. To be considered for a fellowship, applications are due no later than January 1. For fall 2000, the application for admission deadlines are March 15 for international students and June 1 for all other students. Applicants from non-English-speaking countries must take the TOEFL and score at least 570 to be considered for admission.
Correspondence and Information	Director of Graduate Admissions (M/C 154) Department of Electrical Engineering and Computer Science University of Illinois at Chicago 851 South Morgan Street (1120 SEO) Chicago, Illinois 60607-7053 Telephone: 312-996-2290 312-413-2291 Fax: 312-413-0024 E-mail: grad-info@eecs.uic.edu World Wide Web: http://www.eecs.uic.edu

University of Illinois at Chicago

THE FACULTY AND THEIR RESEARCH

Artificial Intelligence and Computer Vision

Barbara Di Eugenio, Assistant Professor; Ph.D., Pennsylvania, 1993. Artificial intelligence and natural language processing; knowledge presentation, planning, and dialogue interfaces; NLP for human-computer interaction and educational technology.

Simon Kasif, Associate Professor; Ph.D., Maryland, 1985. High-performance intelligent systems machine learning, data mining, computational biology, computational neuroscience, parallel computation.

Peter C. Nelson, Associate Professor; Ph.D., Northwestern, 1988. AI research in heuristic search and applied AI research in the areas of transportation, optimizing manufacturing scheduling problems (genetic algorithms and rule-based systems), and computational biology.

Francis K. H. Quek, Associate Professor; Ph.D., Michigan, 1990. Human-computer interaction, computer vision, robot navigation.

Boaz J. Super, Assistant Professor; Ph.D., Texas at Austin, 1992. Computer and biological vision, image processing, pattern recognition.

Computer Architecture and VLSI Systems

Wai-Kai Chen, Professor; Ph.D., Illinois at Urbana-Champaign, 1964. Broadband matching, applied graph theory, networks, systems, filters, VLSI placement, routing, and layout.

Shantanu Dutt, Associate Professor; Ph.D., Michigan, 1990. Parallel and distributed computing, VLSI CAD, fault-tolerant computing and computer architecture.

Joseph Hummel, Assistant Professor; Ph.D., California, Irvine, 1997. Compilers, programming languages, parallelism.

Jon A. Solworth, Associate Professor; Ph.D., NYU, 1987. Computer architecture, programming language design and compilation techniques for parallel processors, I/O, object stores and file systems.

Database Systems

Jorge Lobo, Associate Professor; Ph.D., Maryland, 1990. Knowledge representation, logic and databases, logic programming, nonmonotonic reasoning.

Ouri Wolfson, Associate Professor; Ph.D., NYU, 1984. Database systems, distributed systems, rule processing, mobile computing.

Clement T. Yu, Professor; Ph.D., Cornell, 1973. Database management, information retrieval and knowledge-base management, multimedia retrieval.

Electromagnetics and Power Electronics

Wolfgang-M. Boerner, Professor; Ph.D., Pennsylvania, 1967. Electromagnetics, inverse scattering, modern optics, geoelectromagnetism, electromagnetic imaging, remote sensing, wideband radar, optical polarimetry.

Sharad R. Laxpati, Associate Professor; Ph.D., Illinois at Urbana-Champaign, 1965. Antennas, electromagnetic theory, computational electromagnetic scattering, microwaves, wave propagation and communication.

James C. Lin, Professor; Ph.D., Washington (Seattle), 1971. Electromagnetics in biology and medicine, biomedical instrumentation, telemedicine, wireless communications.

Piergiorgio L. E. Uslenghi, Professor; Ph.D., Michigan, 1967. Electromagnetics, scattering theory, modern optics, solid-state, applied mathematics.

Hung-Yu Yang, Associate Professor; Ph.D., UCLA, 1988. Applied electromagnetics in ferromagnetic integrated circuits and antennas, novel microstrip antennas and arrays, planar integrated transformers for power electronics, advanced computational techniques.

Graphics and Interactive Systems

Richard Beigel, Associate Professor; Ph.D., Stanford, 1988. Human-computer interfaces, distributed data mining, algorithm design, complexity theory, fault diagnosis, molecular computing.

Thomas A. DeFanti, Professor; Ph.D., Ohio State, 1973. Computer graphics and video animation, virtual reality, electronic art, educational technology, operating systems, scientific and volume visualization, networks, parallel computers, supercomputers.

Andrew Johnson, Assistant Professor; Ph.D., Wayne State, 1995. Collaborative virtual reality, human-computer interaction, educational applications.

Robert V. Kenyon, Associate Professor; Ph.D., Berkeley, 1978. Human visual and motor systems, human spatial orientation, computer graphics and display technology, virtual environments, flight simulation, spaceflight and human adaptation.

Thomas G. Moher, Associate Professor; Ph.D., Minnesota, 1983. Human-computer interaction, programming environments, cognitive models.

Signal and Image Processing

Gyan C. Agarwal, Professor; Ph.D., Purdue, 1965. Automatic control, systems analysis, bioengineering, neuroscience, neural networks.

Rashid Ansari, Associate Professor; Ph.D., Princeton, 1981. Image/video processing, coding, transmission, layered video coding schemes for ATM networks, packet video transmissions, theoretical work on general framework for multirate processing of two-dimensional signals.

Jezekiel Ben-Arie, Associate Professor; Ph.D., Technion (Israel), 1986. Object and target recognition, image understanding and processing, shape and signal representation, wavelets and nonorthogonal expansions, neural networks, biomedical engineering.

Daniel Graupe, Professor; Ph.D., Liverpool, 1963. Control systems, time-series analysis, signal processing, neural networks, wavelets and electrical stimulation.

Bin He, Assistant Professor; Ph.D., Tokyo Institute of Technology, 1988. Imaging systems, signal and image processing, bioelectric phenomena.

Arye Nehorai, Professor; Ph.D., Stanford, 1983. Signal and image processing, biomedicine and communications.

William D. O'Neill, Professor; Ph.D., Notre Dame, 1965. Signal and image processing, bioengineering, and information theory.

Roland Priemer, Associate Professor; Ph.D., IIT, 1969. Optimal and adaptive digital signal processing, digital filters, control systems, microprocessor-based system design.

Dan Schonfeld, Associate Professor; Ph.D., Johns Hopkins, 1990. Signal and image processing, pattern recognition, and computer vision.

Oliver Yu, Assistant Professor; Ph.D., British Columbia, 1997. Wireless packet data network, mobile ATM and IP, telecommunications network architectures and protocols, internetworking among fixed and mobile networks.

Software Engineering

Ugo A. Buy, Associate Professor; Ph.D., Massachusetts, 1990. Software engineering, concurrency and real-time analysis.

Carl K. Chang, Associate Professor; Ph.D., Northwestern, 1982. Software engineering: specification, verification, and validation; testing: real-time systems, distributed computer systems, telecommunication software, and object-oriented methods.

Tadao Murata, Professor; Ph.D., Illinois at Urbana-Champaign, 1966. Petri net modeling and analysis of concurrent computer systems.

Sol M. Shatz, Associate Professor; Ph.D., Northwestern, 1983. Distributed computing systems, software engineering, operating systems.

Jeffrey J. P. Tsai, Professor; Ph.D., Northwestern, 1986. Knowledge-based software systems, artificial intelligence, expert systems, requirements specification language, distributed real-time software testing, debugging, and software metrics.

Solid State and Microfabrication

Davorin Babic, Assistant Professor; Ph.D., Pennsylvania, 1988. Physical principles of nanoscale Si particles, their modeling and application to optical modulator, physics and chemistry of silicon interfaces and surfaces.

Alan D. Feinerman, Associate Professor; Ph.D., Northwestern, 1987. Miniaturization of the scanning electron microscope, linear accelerator/undulator and other analytical instruments, 3-D fabrication techniques.

Gary D. Friedman, Associate Professor; Ph.D., Maryland, 1989. Electromagnetics, magnetic and dielectric materials, microfabrication.

Peter J. Hesketh, Associate Professor; Ph.D., Pennsylvania, 1987. Solid-state sensors and actuators, biosensors and microfabrication.

David L. Naylor, Associate Professor; Ph.D., USC, 1988. Development and application of microfabricated photonic materials and devices, diffractive optics: design, fabrication, and systems applications.

Krishna Shenai, Associate Professor; Ph.D., Stanford, 1986. Solid-state and microelectronics, power electronics, semiconductor manufacturing.

Theory and Algorithms

Ajay Kshemkalyani, Assistant Professor; Ph.D., Ohio State, 1991. Computer networks, distributed algorithms, theory of distributed computing, distributed systems, operating systems.

John Lillis, Assistant Professor; Ph.D., California, San Diego, 1996. Computer-aided design for VLSI circuits and combinatorial optimization.

A. Prasad Sistla, Associate Professor; Ph.D., Harvard, 1989. Distributed systems, semantics and verification of concurrent systems and database management systems.

Robert H. Sloan, Associate Professor; Ph.D., MIT, 1989. Design and analysis of algorithms, computational learning theory, machine learning, cryptography, software engineering, analysis of real-time programs, automatic program verification.

UNIVERSITY OF ILLINOIS AT URBANA–CHAMPAIGN

Department of Electrical and Computer Engineering

Programs of Study	Graduate work in the Department of Electrical and Computer Engineering is offered in four general areas: circuits and systems, which includes control, power and communication systems, and networks; computers and information systems, which includes computer architecture, computational complexity, fault-tolerant computing, VLSI systems, digital signal and image processing, coding and information theory, and artificial intelligence; electromagnetic fields, including aeronomy, the upper atmosphere and ionosphere, antennas, holography, propagation, and radio and optical remote sensing; and physical and quantum electronics, which includes the areas of semiconductor devices, optical electronics, nanoelectronics, electrophysics, charged particles, and gaseous electronics and plasmas. In addition to conducting work in these areas, many faculty members and graduate students in electrical engineering participate in interdisciplinary programs in other departments and laboratories, including bioengineering, biophysics, computer systems, decision and control, nuclear engineering, radio astronomy, and electronic music. The M.S. and the Ph.D. are awarded; each requires an acceptable thesis.
	Classes are generally small, with approximately 30 students in a class. The graduate students are more or less equally divided among the four general areas, with a smaller number of students in the interdisciplinary areas.
Research Facilities	The Department of Electrical and Computer Engineering has extensive state-of-the-art laboratories and facilities for research in the areas of acoustics, aeronomy, the upper atmosphere and ionosphere, antennas and electromagnetic theory, bioengineering, bioacoustics, computer-aided design of VLSI systems, control systems, electrophysics, fusion technology, laser physics and molecular spectroscopy, gaseous electronics and plasmas, optical electronics, power and energy systems, radio astronomy, radio-wave propagation, and semiconductors and solid-state devices, including molecular beam epitaxy, metalorganic chemical vapor deposition, and scanning tunneling microscopy–based nanofabrication. The department is also the home of multiple national research centers, which include the NSF Center for Computational Electronics, the Air Force Center for Computational Electromagnetics, and the Army Federated Laboratory for Human-Computer Interface.
Financial Aid	Various forms of financial aid are available, including University and industrial fellowships. Fellowships range from $5000 to $15,000 per academic year. They are normally tax-free and include exemption from tuition and fees. Many provide dependency allowances.
	Part-time teaching and research assistantships are available; for 1999–2000, a half-time teaching assistantship pays a minimum of $12,000 for nine months, plus exemption from tuition and fees. Other assistantships with differing degrees of work responsibility are available.
Cost of Study	For students who have fellowships or staff appointments of 25 percent to 67 percent time, tuition and fees are waived and only the health and insurance fee of $589 per semester must be paid. Full-time students without appointments or tuition and fee waivers paid $2820 per semester in 1998–99 if they were Illinois residents, $6258 per semester if they were nonresidents. Summer session charges were $1778 for residents and $3927 for nonresidents. These amounts are subject to upward revision.
Living and Housing Costs	For single students, University graduate residence halls have double rooms for $2416 and $2872 per academic year. Board is an additional $3368 per academic year. Married student housing rents for $350 to $523 per month (gas and electricity are not included). These amounts are subject to upward revision. Privately owned rooms and apartments are available at similar and higher rents.
Student Group	There are approximately 450 graduate students in the Department of Electrical and Computer Engineering. The 88 budgeted faculty members and the graduate students come from all over the United States and the world. There are approximately 2,000 students enrolled in the undergraduate and graduate programs.
Location	The University is located 130 miles south of Chicago in the twin cities of Urbana and Champaign. Willard Airport, Amtrak, and three interstate highways provide rapid access to all points. Many cultural and recreational facilities are available that would normally be found only in a very large city. The cities have an excellent public school system, with thirty-four elementary and secondary schools.
The University and The Department	Each year the University Star Course series brings outstanding entertainers to the campus, many of whom perform in the $20-million Krannert Center for the Performing Arts, while others appear in the 16,000-seat Assembly Hall. Allerton Park, a 1,500-acre estate owned by the University, is one of a group of public parks and recreation areas. Many conferences and symposia are held in Allerton's lovely and graceful setting.
	Because of the extended nature of the research programs in electrical and computer engineering, the department's activities are housed in a number of buildings on the Urbana campus, including the Everitt Laboratory, the Computer and Systems Research Laboratory, the Gaseous Electronics Laboratory, and the Microelectronics Laboratory. In addition, the department has research sites within 30 miles of the campus for research programs that require isolation from electromagnetic interference. A large proportion of the graduate students are engaged in work on the research projects of the department. Other graduate students assist in the undergraduate teaching program. The department cooperates closely with the Departments of Physics, Mathematics, Computer Science, Music, and Physiology and Biophysics and with the Beckman Institute for Advanced Science and Technology, the Coordinated Science Laboratory, the Materials Research Laboratory, the Atmospheric Research Laboratory, and the National Center for Supercomputing Applications, in order to complement the interdisciplinary programs. The University is an affirmative action, equal opportunity employer.
Applying	Information and application forms are available upon request. Applicants are required to take the General Test of the Graduate Record Examinations. Applications must be completed by January 15 for August admission. Awards are announced April 1. Applications for January admission are due by October 1, and awards are announced November 15. Women and members of minority groups are encouraged to apply.
Correspondence and Information	N. Narayana Rao, Associate Head Department of Electrical and Computer Engineering University of Illinois at Urbana-Champaign 1406 West Green Street Urbana, Illinois 61801
	Telephone: 217-333-2302 or 0207 E-mail: application@ece.uiuc.edu World Wide Web: http://www.ece.uiuc.edu

University of Illinois at Urbana–Champaign

THE FACULTY AND THEIR RESEARCH

Professors

I. Adesida: semiconductor electronics, microfabrication technology. N. Ahuja: computer vision, robotics, artificial intelligence. M. T. Basar: control systems, dynamic games, stochastic control and estimation. S. Bishop: semiconductor physics, optical spectroscopy. M. Blahut: signal processing, digital communications systems, statistical information processing. Y. Bresler: signal and image processing. R. Campbell: software engineering, operating systems. A. Cangellaris: computational electromagnetics, high-speed interconnect and package modeling, RF/microwave circuit design, antenna design. K. Y. Cheng: molecular beam epitaxy, optoelectronic devices and integrated circuits, high-speed devices. W. C. Chew: electromagnetic scattering, geophysical probing, remote sensing, microwave integrated circuits. S. L. Chuang: electromagnetics, integrated optics, quantum electronics. J. J. Coleman: semiconductor materials and devices. G. DeJong: artificial intelligence, natural-language processing, machine learning. J. G. Eden: laser physics, quantum electronics, molecular spectroscopy, semiconductors. M. Feng: high-frequency and high-speed integrated circuits. S. J. Franke: wave propagation, remote sensing, microwaves. L. A. Frizzell: ultrasonic biophysics, bioengineering. C. S. Gardner: optical communications, laser radar, fiber optic systems. G. Gross: power and energy systems. B. E. Hajek: communication networks, stochastic process. I. N. Hajj: computer-aided design, VLSI circuits and systems. K. Hess: semiconductors. N. Holonyak Jr.: solid-state devices (semiconductors). T. S. Huang: computers and pattern recognition. W.-M. Hwu: computer architecture. R. K. Iyer: computers, reliability and fault tolerance, measurement and experimentation. W. K. Jenkins: circuits and signal analysis. D. L. Jones: signal and image processing. S. M. Kang: reliable VLSI circuits, computer-aided design and layout of VLSI, device and circuit modeling of high-speed IC. K. Kim: fusion plasma engineering, lasers, charged-particle dynamics. P. T. Krein: power electronics. W. J. Kubitz: computers. E. Kudeki: radar studies of the atmosphere and ionosphere, ionospheric plasmas. P. R. Kumar: systems and control theory, stochastic systems. M. Kushner: simulation in plasma physics. D. H. Lawrie: high-performance computer architecture and software. P. Lauterbur: medical information sciences. J. P. Leburton: theory of semiconductor devices. S. Levinson: artificial intelligence, automatic speech recognition, natural-language processing. J. W. S. Liu: computer networks, distributed systems, databases, software engineering. M. C. Loui: parallel and distributed computation, computational complexity theory. J. W. Lyding: charge transport properties, reduced-dimensional materials and devices. J. V. Medanic: systems analysis, multivariable control systems. G. H. Miley: fusion, high-temperature plasmas, energy conversion. D. C. Munson Jr.: digital signal and image processing. S. Muroga: logic design of computers, computer-aided design. B. Oakley II: bioengineering. W. D. O'Brien Jr.: biological effects, bioengineering, biophysics, measurements and dosimetry of ultrasound. M. A. Pai: power and energy systems. J. H. Patel: computers. W. R. Perkins: control systems, systems theory and applications. D. Pines: condensed-matter theory, theoretical astrophysics. C. Polychronopoulos: computer systems. P. L. Ransom: holography, optical processing. N. N. Rao: ionosphere, radio-wave propagation. U. Ravaioli: simulation and plasma physics. S. R. Ray: computers. W. H. Sanders: computers, performance/dependability evaluation, fault-tolerant computing, computer networks and protocols. D. V. Sarwate: communication theory, error-control coding. P. W. Sauer: power and energy systems. P. D. Schomer: architectural acoustics, special measurements, noise control, environmental noise. M. W. Spong: control theory, robotics. G. E. Stillman: semiconductors, physics and device physics. G. Swenson: atmospheric remote sensing, atmospheric dynamics and chemistry, space environment measurements and modeling. T. N. Trick: integrated circuits, computer-aided analysis and design. J. R. Tucker: superconductive devices. R. J. Turnbull: energy conversion technology. B. W. Wah: computer architecture, parallel processing, computer networks, distributed databases, artificial intelligence, operating systems, VLSI systems.

Associate Professors

J. Bentsman: automatic control systems. D. Brady: optics, artificial neural networks, volume holographic information processing using photorefractive crystals. D. J. Brown: VLSI design, analysis of algorithms. K. Hsieh: characterization and development of optoelectronic materials growth. S. Hutchinson: robotics, computer vision. J. Jin: computational electromagnetics, MRI instrumentation. L. Kale: design of parallel execution schemes and architectures for unpredictably structured computations. T. Kerkhoven: semiconductor simulation, mathematical analysis of algorithms for physical problems, nonlinear partial differential equations, scientific computation. Z.-P. Liang: MRI and spectroscopy techniques, image processing and neural networks. U. Madhow: communication systems, communication networks, wireless channels. S. Meyn: communications, stochastic control theory. E. Michielssen: electromagnetic simulation, high-speed circuits. P. Moulin: signal processing. F. Najm: circuits. T. Overbye: power and energy systems. G. Papen: optical information processing, quantum optics, application of nonlinear wave mixing. J. Ponce: computer vision, robotics. E. Rosenbaum: reliability physics, IC reliability. J. Schutt-Aine: electromagnetic measurements, computer-controlled measurements. A. Webb: MRI, temperature measurement. B. C. Wheeler: bioengineering, sensor arrays, neural recording.

Assistant Professors

D. Beebe: microelectromechanical systems for biological applications, including tactile interfaces and microfluidic instruments. J. Bernhard: active and passive antennas, wireless communication, electromagnetic measurements and packaging. N. Carter: computer architecture, shared-memory systems, VLSI circuits. R. M. Fish: bioengineering, bioacoustics. R. Koetter: communications. C. Liu: microfabrication, integrated sensors and actuators, microelectromechanical systems. S. Lumetta: high-performance networking and computing, hierarchical systems, parallel runtime software. M. Medard: communications. E. Rudnick: VLSI verification, test, and diagnosis. N. Shanbag: VLSI design, algorithm development, signal processing. A. Singer: statistical signal processing, communications, estimation and decision. J. Torellas: computer architecture, memory hierarchies, parallel processing. B. Vaduvur: computer engineering.

Lecturers

T. Basar: communication systems. M.-C. Brunet: computer engineering. R. B. Uribe: cybernetics and digital systems. P. E. Weston: computers. J. Zhang: electronic circuits.

UNIVERSITY OF KANSAS

Department of Electrical Engineering and Computer Science
Programs in Electrical Engineering and Computer Science

Programs of Study

The Department of Electrical Engineering and Computer Science at the University of Kansas offers graduate programs of study and research leading to the degrees of M.S., Ph.D., and D.E. in electrical engineering and the M.S. and Ph.D. in computer science.

Specializations include algorithms, artificial intelligence, computer-aided design, computer systems, digital signal processing, distributed and parallel computing, electromagnetics, expert systems, graphics, high-speed networking, image processing, information retrieval, information systems, language processing, lightwave systems, performance modeling and analysis, programming languages, radar remote sensing, radar systems, semiconductor processing, telecommunications, theory of computing, and wireless communications.

The M.S. programs require 30 semester hours of graduate work with thesis and nonthesis options. The Ph.D. programs require 24 hours of course work beyond the M.S. and 18 hours of dissertation. The D.E. program requires 30 hours of course work beyond the M.S. and 12 to 18 hours of industrial internship.

Research Facilities

Almost all of the graduate research is conducted using facilities of four major laboratories: the Telecommunications and Information Sciences Laboratory (TISL), the Radar Systems and Remote Sensing Laboratory (RSL), The Center for Excellence in Computer-Aided Systems Engineering (CECASE), and the Design Technologies Laboratory (DesignLab).

All research is supported by an extensive network of workstations. Special facilities include a multigigabit experimental telecommunications network called "MAGIC," radar systems ranging from VHF through 140 GHz and lightwaves, wide-screen wall-panel graphics, and a new scalable multiprocessor computer for research in computer-aided design and graphics computing.

Financial Aid

Fellowships, research assistantships, teaching assistantships, graduate assistantships, and scholarships are available to the best qualified applicants. Teaching assistants receive a tuition waiver, and research assistants receive resident tuition. The department offers a competitive package of financial support. More than 135 of the graduate students receive some support.

Cost of Study

During the 1998–99 academic year, tuition and fees were approximately $94 per credit hour for residents and $309 per credit hour for nonresidents. An equipment fee of $15 per credit hour is assessed on all School of Engineering students and is used solely for student-accessible laboratory equipment.

Living and Housing Costs

Housing is available both on and off campus. Lawrence is a relatively inexpensive place to live compared to the national average.

Student Group

There are approximately 160 graduate students in the department, including 40 in doctoral programs. The student body is composed of both traditional and nontraditional students, who may also be engaged in industry or the military. The student population is diverse, coming from all parts of the world and from all parts of the United States.

Location

Kansas is located in the heart of the United States. The city of Lawrence is in the eastern part of the state, and has a cosmopolitan population of 78,000 people.

The University and The Department

The University of Kansas is a major educational and research institution with nearly 29,000 students and 2,100 faculty members. The University includes the main campus in Lawrence and several satellite campuses throughout the state.

The Department of Electrical Engineering began in 1887. It has continued to grow and develop along with the state-of-the-art, incorporating programs in computer science, begun in 1967, and computer engineering, begun in 1987. The department has cutting-edge facilities and 31 faculty members who conduct research with funding exceeding $3 million per year.

Applying

To qualify for admission, the applicant must hold a four-year baccalaureate degree or its equivalent from an accredited institution and meet other criteria demonstrating potential for success. Application forms, transcripts, GRE general test scores, three letters of recommendation, a statement of objectives, and an application fee of $30 are required. In addition, for persons whose native language is not English, a minimum TOEFL score of 600 is required. Deadlines and other details are explained in the application materials.

Correspondence and Information

Department of Electrical Engineering and Computer Science
Graduate Admissions
415 Snow Hall
University of Kansas
Lawrence, Kansas 66045
Telephone: 913-864-4487
Fax: 913-864-3226
E-mail: grad_admissions@eecs.ukans.edu

University of Kansas

THE FACULTY AND THEIR RESEARCH

W. Perry Alexander, Associate Professor; Ph.D., Kansas, 1993. Formal methods, software engineering, formal synthesis, architectures, software reuse.

Christopher T. Allen, Associate Professor; Ph.D., Kansas, 1984. Radar systems, high-speed digital circuits.

Allen L. Ambler, Professor; Ph.D., Wisconsin, 1973. Programming paradigms and languages; program design principles, approaches, and tools; visual languages; end-user programming systems; scientific visualization; functionally distributed programming.

Frank M. Brown, Associate Professor; Ph.D., Edinburgh, 1978. Automatic deductions, artificial intelligence.

Swapan Chakrabarti, Associate Professor; Ph.D., Nebraska, 1986. Neural networks and fuzzy systems, pattern classification, signal processing.

Don G. Daugherty, Professor; Ph.D., Wisconsin, 1964. Electronic circuits for communications and control.

Raymond H. Dean, Professor Emeritus; Ph.D., Princeton, 1966. Digital control, systems modeling, simulation, alternative energy sources.

Kenneth R. Demarest, Professor; Ph.D., Ohio State, 1980. Electromagnetic theory and computational techniques, antennae, electromagnetic interference and compatibility.

Harvey H. Doemland, Associate Professor Emeritus; Ph.D., Illinois, 1963. Electronic circuits, pulse circuits, biomedical engineering.

Joseph B. Evans, Professor; Ph.D., Princeton, 1988. Networking, communications, signal processing, VLSI design.

Victor S. Frost, Dan F. Servey Distinguished Professor; Ph.D., Kansas, 1982. Telecommunications, communications network control, modeling and analysis.

John M. Gauch, Associate Professor and Acting Chair; Ph.D., North Carolina, 1989. Image processing, computer vision, computer graphics.

Susan E. Gauch, Associate Professor; Ph.D., North Carolina, 1990. Information retrieval, corpus linguistics, multimedia databases.

Siva Prasad Gogineni, Dean E. Ackers Distinguished Professor; Ph.D., Kansas, 1984. Radar systems, microwave engineering.

Jerzy W. Grzymala-Busse, Professor; Ph.D., Technical University of Poznan (Poland), 1969. Expert systems, reasoning under uncertainty, knowledge acquisition, machine learning, rough set theory.

Jeremiah W. James, Assistant Professor; Ph.D., California, Santa Barbara, 1999. Distributed systems, concurrent objects, data consistency, fault tolerance.

Nancy G. Kinnersley, Associate Professor; Ph.D., Washington State, 1989. Design and analysis of algorithms, graph algorithms, discrete mathematics, tree automata, computational complexity.

Man C. Kong, Associate Professor; Ph.D., Nebraska, 1986. Design and analysis of algorithms, graph and network algorithms, parallel algorithms, combinatorial optimization, computational complexity, operations research.

Stephen P. Lohmeier, Assistant Professor; Ph.D., Massachusetts, 1996. Remote sensing, radar, microwave engineering, electromagnetics, signal processing, communications.

James R. Miller, Associate Professor; Ph.D., Purdue, 1979. Geometric modeling and computer-aided design, computer graphics, scientific visualization, object-oriented programming.

Gary J. Minden, Professor; Ph.D., Kansas, 1982. Digital systems, microprocessors, artificial intelligence.

Richard K. Moore, Distinguished Professor Emeritus; Ph.D., Cornell, 1951. Radar remote sensing, radio wave propagation, communications systems, antennas, radar imaging and backscatter systems.

Douglas Niehaus, Associate Professor; Ph.D., Massachusetts, 1994. Operating systems, real-time and distributed systems, programming environments.

David W. Petr, Associate Professor; Ph.D., Kansas, 1990. Communications, networking, digital signal processing.

Glenn E. Prescott, Associate Professor; Ph.D., Georgia Tech, 1984. Communications theory, digital signal processing.

James A. Roberts Jr., Professor and Vice-Chancellor; Ph.D., Santa Clara, 1979. Telecommunications, information theory, wireless communications.

James R. Rowland, Professor; Ph.D., Purdue, 1966. Stochastic systems modeling and analysis, control systems, radar detection and estimation.

Dale I. Rummer, Professor Emeritus; Ph.D., Kansas, 1963. Design of digital systems, design of microprocessor-based systems, computer-aided design tools.

Earl J. Schweppe, Professor; Ph.D., Illinois, 1955. Computer architecture, computer organization, computer history, personal computers, computer communications, interactive systems, concurrent processes, computer science curriculum.

K. Sam Shanmugan, Southwestern Bell Distinguished Professor; Ph.D., Oklahoma State, 1970. Telecommunications, pattern recognition, general systems theory, statistical communications theory, image processing, digital signal processing techniques.

William P. Smith, Professor Emeritus; Ph.D., Texas, 1950. Electrical power systems and alternate energy sources.

James M. Stiles, Assistant Professor; Ph.D., Michigan, 1995. Radar remote-sensing, propagation and scattering in random media, electromagnetic theory.

Harry E. Talley, Professor Emeritus; Ph.D., Kansas, 1954. Semiconductor devices, solid-state physics and electronics.

Costas Tsatsoulis, Professor; Ph.D., Purdue, 1987. Artificial intelligence, expert systems.

Hillel Unz, Professor Emeritus; Ph.D., Berkeley, 1957. Electromagnetic theory, antenna arrays, plasma propagation, acoustic waves, applied mathematics.

Victor L. Wallace, Professor; Ph.D., Michigan, 1969. Operating systems, computer graphics, user-machine interaction, computer networks, distributed systems, queuing theory, numerical analysis, performance evaluation.

UNIVERSITY OF MARYLAND, COLLEGE PARK

Department of Electrical and Computer Engineering

Programs of Study

The Department of Electrical and Computer Engineering offers graduate study leading to the Master of Science and Doctor of Philosophy degrees. The department's research and educational activities can be broadly divided into two areas: information sciences and systems and electronic sciences and devices. Within information sciences and systems, concentration is possible in communications and signal processing (random processes, detection and estimation, coding and information theory, digital signal processing, image processing, signal compression, communication networks, wireless and cellular systems, satellite communications, and optical communications), computer engineering (digital system design, design automation, parallel algorithms and architectures, VLSI architectures, fault-tolerant computing, neural networks, computer networking, operating systems, software engineering, and computer security), and controls (adaptive control, intelligent control, stochastic control, robust control, control of bifurcations and chaos, geometric control theory and robotics, control of discrete event systems, smart structure control, numerical optimization and optimization-based design, and control applications, including biomedical). Within electronic sciences and devices, concentration is possible in electrophysics (electromagnetic theory, plasmas, intense charged-particle beams and applications to accelerators, relativistic electronics and high-power microwave generation, high-power microwave components, nonlinear dynamics and chaos, quantum electronics, millimeter waves, optical engineering, lasers, nonlinear optics, ultrafast optoelectronics, femtosecond phenomena, RF photonics, optical-microwave interaction, optoelectronic devices, integration, assembly and packaging, photonic networks for computing and communication, optical communication, optical control of phased array antenna, and chemical physics and biophysics) and microelectronics (circuits, classical and quantum devices, VLSI, semiconductor modeling and computer-aided design, neural networks, microwave and integrated circuits, semiconductor materials and technology, and ion beam lithography). The M.S. program consists of course work plus a scholarly paper or somewhat less course work plus a research thesis with oral defense. The Ph.D. program comprises course work, a core course–based qualifying requirement, a dissertation based on original research, and an oral examination on the dissertation. Joint programs are maintained with other departments within the College of Engineering and the mathematics, physics, and computer science departments as well as with the Institute for Plasma Research, the Institute for Systems Research, the Institute for Advanced Computer Studies, the Institute for Physical Science and Technology, the Engineering Research Center, the Center for Superconductivity Research, the laboratory for physical sciences, and the chemical physics and transportation programs. Opportunities also exist for programs of study in conjunction with many national and international laboratories and technical facilities.

Research Facilities

The department is equipped with an extensive computer facility consisting of state-of-the-art mainframes, workstations, and personal computers located in several open laboratories and in a large number of specialized research laboratories. Faculty members and students affiliated with the Institute for Advanced Computer Studies have access to a Connection Machine that is housed in that institute. In addition, there are more than thirty specialized research laboratories supporting activities in speech and image processing, communication networks, robotics, control systems, VLSI design and testing, semiconductor materials and devices, photonics, fiber optics, microwave sources, ion beam lithography, and plasma science, among others. A complete engineering library is housed nearby.

Financial Aid

A significant number of graduate fellowships and teaching and research assistantships are available for well-qualified applicants. In 1999–2000, typical stipends for first-year students without an M.S. degree are $11,250 for an academic-year assistantship, $17,161 for a twelve-month assistantship, and $21,763 for a twelve-month "superfellowship." Government traineeships are awarded through the University to exceptionally well qualified students. Part-time support is available through many national laboratories and technical facilities located nearby. The University also has resident assistantships, summer dissertation fellowships, and various types of loans.

Cost of Study

In 1999–2000, tuition and fees for full-time study (9 credit hours) are $2765 per semester for Maryland residents and $4051 for nonresidents.

Living and Housing Costs

Board and lodging are available in many private homes and apartments in College Park and the vicinity. Rooms in private homes range in cost from $250 to $350 a month, and one-bedroom apartments rent for an average of $600 per month. A list of accommodations, both University and private, is maintained by the University's housing bureau.

Student Group

In fall 1998, there were 363 graduate students in electrical engineering; 236 were Ph.D. candidates. There were 281 full-time students and 82 part-time students.

Student Outcomes

During the past few years, the department has placed its Ph.D. graduates on the faculties of such academic institutions as Harvard University, Princeton University, Penn State University, SUNY at Stony Brook, University of Virginia, University of Wisconsin, and Queen's University at Kingston (Canada) as well as at corporate and national research labs, such as AT&T Bell Laboratories, IBM Research Laboratories, General Electric, Texas Instruments, Hewlett-Packard, Philips Laboratory, Microsoft, Allied Signal, Advance Micro Devices, Micron Technologies, Bell Northern Research, Silicon Valley Research, Fore Systems, Comsearch, LCC, SAIC, Los Alamos National Laboratory, Oak Ridge National Laboratory, Laboratory for Physical Sciences, and Army Research Laboratory.

Location

The central campus of the University of Maryland is located in College Park, Maryland, a suburban area roughly between and within easy commuting distance of Washington and Baltimore. The campus is 12 miles from the White House. The museums, galleries, theaters, federal and special libraries, universities, concert halls, and abundant cultural activities of both cities offer students unlimited opportunities to participate in the culture and social life of this thriving area. The immediate presence of many great national laboratories and technical facilities offers a particularly good opportunity to the graduate student in electrical engineering.

The University

The University is one of the oldest and largest state universities in the country. The College of Engineering is located on the central campus, which has a total student population of approximately 33,000. The University offers many cultural and entertainment activities and operates its own golf course and athletic facilities.

Applying

Applicants seeking admission should hold a B.S. degree with a B+ average or better from an accredited institution. Applications should be filed early; for best consideration, those seeking financial aid should apply by December 1 for fall admission and September 1 for spring admission (June 1 for international applicants). The submission of three letters of recommendation and scores on the General Test of the Graduate Record Examinations is required.

Correspondence and Information

Office of Graduate Studies
Department of Electrical and Computer Engineering
University of Maryland, College Park
College Park, Maryland 20742
Telephone: 301-405-3681

University of Maryland, College Park

THE FACULTY AND THEIR RESEARCH

Communications and Signal Processing

J. S. Baras, Ph.D., Harvard: intelligent control, hybrid communication network management, hybrid Internet, telemedicine systems, speech recognition, image understanding, control and systems theory, nonlinear robust control, manufacturing process planning. R. Chellappa, Ph.D., Purdue: signal/image processing, computer vision, pattern recognition. L. Davisson (Emeritus), Ph.D., UCLA: communications theory, information theory, signal processing. A. Ephremides, Ph.D., Princeton: communication theory, multiuser communication systems and networks, wireless networks, performance evaluation, protocol, satellite communications, system optimization and design. N. Farvardin, Ph.D., RPI: information theory and coding, data compression and application to image and speech coding, speech recognition, digital communications, wireless communication systems. T. Fuja, Ph.D., Cornell: digital communications, coding and information theory. J. Gansman, Ph.D., Purdue: communications and signal processing, wireless technologies, array signal processing, PSAM frame synchronization and structure, wireless modem architecture. E. Geraniotis, Ph.D., Illinois at Urbana-Champaign: communication networks, spread-spectrum systems, coding, robust signal processing. R. Harger (Emeritus), Ph.D., Michigan: signal/image processing. K. J. Liu, Ph.D., UCLA: signal/image processing, VLSI, communications. A. Makowski, Ph.D., Kentucky: stochastic control, queuing systems, applied stochastic processes. P. Narayan, D.Sc., Washington (St. Louis): information theory and applications, digital communications, communication networks, system identification and modeling. B. Papadopoulos, Ph.D., MIT: nonlinear signal processing and wireless communications. A. Papamarcou, Ph.D., Cornell: statistical communications. S. A. Shamma, Ph.D., Stanford: biological aspects of speech analysis, neural signal processing. L. Tassiulas, Ph.D., Maryland. Wireless communication networks, personal communications, high-speed multimedia networks, resource allocation, stochastic systems. S. Tretter, Ph.D., Princeton: communication theory, coding, signal processing.

Computer Engineering

S. S. Bhattacharyya, Ph.D., Berkeley: embedded systems, software synthesis, VLSI signal processing, parallel computation. N. DeClaris, Sc.D., MIT: mathematical system theory, system integration, computer engineering, information technology, computational intelligence, medical science and practice and medical informatics. M. Franklin, Ph.D., Wisconsin: computer architecture, instruction-level parallel processing. V. Gligor, Ph.D., Berkeley: operating systems, computer security, distributed systems. B. Jacob, Ph.D., Michigan: computer architecture and microarchitecture, operating systems, embedded systems, electronic music. J. F. JáJá, Ph.D., Harvard: parallel algorithms, high-performance computing, CAD tools for VLSI, computational complexity. P. Ligomenides (Emeritus), Ph.D., Stanford: information processing systems, AI, cybernetic and cognitive systems. D. Marculescu, Ph.D., USC: computer-aided design of digital systems, probabilistic analysis of digital systems. K. Nakajima, Ph.D., Northwestern: VLSI design, design automation, high-performance computing, applied graph theory. A. Oruc, Ph.D., Syracuse: computer architecture, interconnection network theory, multiprocessing systems. J. Pugsley (Emeritus), Ph.D., Illinois at Urbana-Champaign: computer systems, multiple-valued logic. C. Silio, Ph.D., Notre Dame: computer engineering, multivalued digital systems, computer networks. D. Stewart, Ph.D., Carnegie Mellon: software engineering, real-time systems, robotics and automated systems. U. Vishkin, D.S., Technion (Israel): parallel computation, design and analysis of algorithms, pattern matching, theory of computing. D. Yeung, Ph.D., MIT: computer architecture, large-scale distributed shared memory systems, scalable implementations of superscalar processors, performance evaluation of computer systems, applications for parallel architectures, interactions between operating systems and architectures, the MIT Alewife Multiprocessor, VLSI design.

Controls

E. Abed, Ph.D., Berkeley: nonlinear systems, singular perturbations, power and aerospace systems. G. Blankenship, Ph.D., MIT: stochastic and nonlinear control, adaptive control, AI in engineering design. F. Emad (Emeritus), Ph.D., Northwestern: power systems and control. P. S. Krishnaprasad, Ph.D., Harvard: control and filtering theory, control of multibody and flexible spacecraft, robotic manipulator control, signal processing and machine vision. W. Levine, Ph.D., MIT: control theory and its applications in neurophysiology, aerospace, and networks. S. I. Marcus, Ph.D., MIT: control and systems theory, stochastic systems, discrete event systems, intelligent control, manufacturing systems, communication networks. M. A. Shayman, Ph.D., Harvard: discrete event dynamic systems, communication network management, control theory. A. Tits, Ph.D., Berkeley: optimization-based design, nonlinear programming, robust linear system stability.

Electrophysics

T. Antonsen, Ph.D., Cornell: high-power sources of coherent radiation, controlled fusion and plasma physics. M. Dagenais, Ph.D., Rochester: integrated optoelectronics, optical communication, photonic integrated circuits, photonic switching, optoelectronic packaging. C. C. Davis, Ph.D., Manchester (England): quantum electronics, biophysics, laser interferometry, fiber sensors, atmospheric optics, near-field microscopy. W. Destler, Ph.D., Cornell: microwave and millimeter-wave sources, accelerator technology. J. Goldhar, Ph.D., MIT: high-power lasers, nonlinear optics. R. Gomez, Ph.D., Maryland: information storage technology, experimental micromagnetics, physics of magnetism, scanned probe microscopy and experimental surface science. V. Granatstein, Ph.D., Columbia: free-electron lasers and gyrotrons, plasma physics. P.-T. Ho, Sc.D., MIT: lasers and optics. U. Hochuli (Emeritus), Ph.D., Catholic University: gas laser technology, cold cathodes. D. N. Langenberg, Ph.D., Berkeley: condensed matter physics. W. Lawson, Ph.D., Maryland: high-power microwave source development, accelerator technology. C. Lee, Ph.D., Harvard: quantum electronics, nonlinear optics, picosecond optical electronics, millimeter waves. I. Mayergoyz, Doktor Nauk, Ukrainian Academy of Sciences: power, electromagnetic theory, semiconductor device modeling. H. Milchberg, Ph.D., Princeton: subpicosecond lasers and plasma interactions, atomic physics. P. O'Shea, Ph.D., Maryland: electron and ion beam technology, free electron light sources. E. Ott, Ph.D., Polytechnic of Brooklyn: chaos in dynamic systems, applications of chaotic dynamics in engineering and physics. H. Rabin, Ph.D., Maryland: Nonlinear optics, space science. M. P. Reiser (Emeritus), Ph.D., Mainz (Germany): physics of intense charged particle beams, high-brightness electron and ion sources, particle accelerators. M. Rhee, Ph.D., Catholic University: particle dynamics, plasma accelerators. C. Striffler, Ph.D., Michigan: plasma physics. L. Taylor (Emeritus), Ph.D., New Mexico State: Biomedical engineering, electromagnetic theory, atmospheric optics. T. Venkatesan, Ph.D., CUNY, Brooklyn: superconducting electronics, epitaxial metal-oxide thin films and devices. C.-H. Yang, Ph.D., Princeton: semiconductor physics and devices, quantum transport, silicon laser, ultrasmall MOSFET scaling and simulation. K. Zaki, Ph.D., Berkeley: microwaves, millimeter waves and optical devices, computer-aided design.

Microelectronics

D. Barbe, Ph.D., Johns Hopkins: solid-state devices, integrated circuits, electronic images, radiation effects. J. Frey, Ph.D., Berkeley: physics and technology of semiconductor devices for displays, communications, and VLSI digital applications, including methods for mass production. N. Goldsman, Ph.D., Cornell: device physics, electron transport in high-electric fields, microelectronic device reliability, device modeling. A. A. Iliadis, Ph.D., Manchester (England): molecular beam epitaxy, elemental and compound semiconductor devices and circuits. H. Lin (Emeritus), D.E.E., Polytechnic of Brooklyn: integrated circuits, semiconductor devices. J. Melngailis, Ph.D., Carnegie Mellon: microfabrication, ion lithography, focused ion beams. R. Newcomb, Ph.D., Berkeley: microsystems, network theory, robotics, biomedical engineering. J. Orloff, Ph.D., Oregon Graduate Center: high-brightness ion and electron sources, charged particle optics, micromachining with ion beams. M. Peckerar, Ph.D., Maryland: microelectronics, integrated circuits, microstructures. T. C. Gordon Wagner (Emeritus), Ph.D., Maryland, College Park: advanced electronic systems.

UNIVERSITY OF MICHIGAN

Department of Electrical Engineering and Computer Science

Programs of Study	The Department of Electrical Engineering and Computer Science (EECS) offers graduate programs leading to the degrees of Master of Science (M.S.), Master of Science in Engineering (M.S.E.), and Doctor of Philosophy (Ph.D.). EECS comprises three divisions: Computer Science and Engineering (CSE), Electrical Science and Engineering (ESE), and System Science and Engineering (SSE). The graduate programs in the CSE division are organized into the following four broad areas: hardware systems, intelligent systems, software and programming languages, and the theory of computation. This program is intended for students wishing to major in topics such as programming languages, software engineering, operating systems, computer architecture, computer networks, databases, fault tolerance, reliable computing, computer-aided design, artificial intelligence, robotics, computer vision, graphics, distributed systems, VLSI, and all aspects of theoretical computer science. The ESE division is organized into five broad areas: electromagnetics, optics, VLSI, circuits and microsystems, and solid-state. This program is intended for students wishing to major in topics such as circuits, electronics, electrodynamics, electromagnetics, electrooptics, microwave systems, remote sensing, solid-state materials, technologies, devices, and integrated circuits. The VLSI option bridges the areas of electrical and computer engineering. The SSE division is organized into four broad areas: communications, control, signal processing, and bioelectrical sciences. SSE programs are intended for students wishing to major in topics such as bioelectrical sciences, communications, networks, control, manufacturing, signal processing, information theory, random processes, and systems theory. This distinctive academic structure allows students to pursue diverse academic programs, which enables them to study and hone their research skills in more than one area.
Research Facilities	EECS departmental academic units, faculty members, and most of the research laboratories are housed in the modern EECS Building and in several nearby research buildings. EECS is home to more than a dozen state-of-the-art research laboratories, and it supports other interdepartmental research laboratories. The optics laboratories include the country's foremost center for research in ultrafast optical science. The Solid-State Laboratory has one of the most advanced facilities for solid-state device research in the world. The Radiation Laboratory conducts research in circuit, antenna, and system technology, which covers a wide range of EM-related problems, using theoretical, numerical, and experimental methods. The area of systems and control is involved in theoretical and applied projects that run the gamut from aerospace vehicles and automotive systems, to manufacturing and highway systems, to computer and communications networks. Communications research spans a broad spectrum of interests, including digital modulation, channel coding, source coding, information theory, optical communications, detection and estimation, spread spectrum, and multiuser communications and techniques. The Biosystems Laboratory collaborates with faculty members in the Medical School of the University of Michigan to offer excellent research opportunities. Signal processing, which focuses on the representation, manipulation, and analysis of signals, particularly natural ones, overlaps with many research areas. Computer hardware systems research, including architecture and CAD, is centered in the Advanced Computer Architecture Laboratory. The Artificial Intelligence Laboratory houses a multidisciplinary group of researchers involved in theoretical, experimental, and applied aspects of intelligent systems. Areas of excellence in theoretical computer science research include logic and parallel algorithms. Software research draws on all areas of computer science from architecture, databases, networks, and distributed systems to fault-tolerant and real-time systems and is housed in the Software Systems and Real-Time Computing laboratories. The EECS research environment is strengthened by a University-wide computer network infrastructure. The College of Engineering's CAEN network, one of the largest campus networks, supports both instructional and research computing and has links to research facilities throughout Michigan, the nation, and the world.
Financial Aid	A variety of fellowships, teaching assistantships, and research assistantships are available to students. Departmental aid is awarded on a merit basis. For recipients of NSF Fellowships, the University provides an additional stipend and the balance of tuition.
Cost of Study	For 1998–99, precandidate tuition and fees were $5749 per term for state residents and $10,848 per term for nonresidents. Candidacy rates were $3936, regardless of residency status. (Costs are subject to change each year.)
Living and Housing Costs	University-owned residence facilities provide a variety of single rooms and suites in dormitories, co-ops, and apartments. Many privately owned apartments are within walking distance of the North and Central campuses, and many others are serviced by the Ann Arbor Transit Authority's extensive bus system. A no-cost University bus system is available to provide service between the campuses and student housing. Food and living costs are equivalent with the national level.
Student Group	Approximately 600 graduate students are enrolled in the EECS department. More than half of the enrolled students are pursuing a Ph.D. degree. The student population includes excellent scholars from throughout the United States and the world whose diverse backgrounds and interests add much to the educational environment.
Location	The University of Michigan is located in the heart of Ann Arbor, a cosmopolitan community that has retained its small-town atmosphere and friendliness. It has an environment rich in opportunities to enjoy the fine arts, popular entertainment, sports, and widely varied recreational activities.
The University	The University of Michigan, founded in 1817, was the nation's first public university and has a long-standing tradition of excellence in engineering.
Applying	Applications and additional information may be obtained electronically or by writing to the EECS department.
Correspondence and Information	Department of Electrical Engineering and Computer Science 3314 EECS Building University of Michigan Ann Arbor, Michigan 48109-2122 E-mail: admit@eecs.umich.edu World Wide Web: http://www.eecs.umich.edu

University of Michigan

THE FACULTY AND THEIR RESEARCH

COMPUTER SCIENCE AND ENGINEERING DIVISION

Artificial Intelligence. Professors William Birmingham, Edmund Durfee, Stephen Kaplan, David Kieras, Daniel Koditschek, John Laird, Sang Wook Lee, William Rounds, Elliot Soloway, and Michael Wellman.

Many projects are now under way combining research in machine vision, natural-language understanding, distributed problem solving, machine learning, cognitive modeling, AI-aided design, autonomous and teleautonomous robotic systems, automated knowledge acquisition, AI-supported software development, rational decision making, digital libraries, computers and the arts, computers and education, and computer games.

Systems and Hardware. Professors Daniel Atkins, William Birmingham, Peter Chen, Edward Davidson, John Hayes, Pinaki Mazumder, John Meyer, Trevor Mudge, Marios Papaefthymiou, Steven Reinhardt, Karem Sakallah, and Gary Tyson.

Research on computer systems broadly covers the analysis and design of computers, computer-based systems, and their major components, with a focus on computer architecture and logic design. The computer systems hardware field has strong links with software (operating systems, programming languages), solid-state circuits (VLSI design), and several computer application areas (robotics, artificial intelligence, instrumentation, numerical methods). Research into VLSI design ranges from CAD tools such as logic simulation programs to the design of components for advanced computer systems.

Theoretical Computer Science. Professors Kevin Compton, Yuri Gurevich, William Rounds, and Quentin Stout.

Research projects include abstract state machines (formal specification and verification of computer systems), adaptive designs, algorithms and data structures, average case complexity, default domain theory, feature logic, finite model theory, graph theory, combinatorics, coding, parallel computing, and security.

Software & Distributed Computing Systems. Professors Peter Chen, Larry Flanigan, Farnam Jahanian, Sugih Jamin, Brian Noble, Atul Prakash, Steven Reinhardt, Kang Shin, Nandit Soparkar, Toby Teorey, and Gary Tyson.

Research in software bridges the gap between sophisticated applications and the challenging raw power of machines. This is difficult because large software systems are among the most complex systems ever built. Their enormous complexity can be managed only through the development of new abstractions, techniques, structures, and languages. There is a major focus on distributed computing from the perspective of computer network design and analysis, distributed file and storage systems, database systems, and collaborative computing. Another focus area is the specification, design, analysis, and implementation of real-time computing systems and applications.

ELECTRICAL SCIENCE AND ENGINEERING DIVISION

Applied Electromagnetics. Professors Anthony England, Brian Gilchrist, Linda Katehi, Gabriel Rebeiz, Kamal Sarabandi, Fawwaz Ulaby, and John Volakis.

Research in circuit, antenna, and system technology covers a wide range of EM-related problems using theoretical, numerical, and experimental methods. Areas of focus include microwave and millimeter-wave remote sensing, wireless systems, antennas, scattering by composite material structures, automotive sensors, and plasma propagation.

Optical Science. Professors Mohammed Islam, Henry Kapteyn, Emmett Leith, Gerard Mourou, Margaret Murnane, Theodore Norris, Stephen Rand, Duncan Steel, Herbert Winful, and Kim Winick.

Research is conducted in information processing, electrooptics, quantum optoeoectronics, and optical physics. Specific activities include image processing, interferometry, tomography, and holography; laser sources, grating structures, and waveguides; soliton-based fiber-optic communications; quantum studies of semiconductor heterostructures and other doped crystals; the study of ultrafast phenomena in chemistry, physics, and biophysics; and ultraintense pulses applied to coherent X-ray generation, laser acceleration of electrons, and medicine.

Solid-State Electronics. Professors Pallab Bhattacharya, Richard Brown, George Haddad, Jerzy Kanicki, Ronald Lomax, Carlos Mastrangelo, Leo McAfee, Khalil Najafi, Clark Nguyen, Stella Pang, Dimitris Pavlidis, Jasprit Singh, Fred Terry, and Kensall Wise.

Research is conducted in silicon circuits and microelectromechanical systems (MEMS), III/V semiconductors, semiconductor processing, and amorphous silicon devices. Major silicon projects include solid-state physical and chemical sensors, microelectromechanical actuators, microfluidics, micromechanical resonators for RF circuits, and displays. The III/V area is focused on growth and characterization of narrow and wide bandgap semiconductors, novel devices, monolithic analog and digital IC design, and optoelectronic integrated circuits. Process development is a part of all of these activities, and research is done in automated semiconductor manufacturing.

VLSI. Professors Richard Brown, Stephen Director, John Hayes, Ronald Lomax, Carlos Mastrangelo, Pinaki Mazumder, Trevor Mudge, Khalil Najafi, Karem Sakallah, and Kensall Wise.

The VLSI program covers both Electrical Science and Engineering and Computer Science and Engineering Divisions. Research in this area deals with many aspects of VLSI circuit design: analog, digital, and mixed signal circuits, with emphasis on low-power and high-performance circuits; computer-aided design, including clocking and timing, logic synthesis, physical design, and design verification; testing and design for testability; advanced semiconductor processes, logic families, and packaging; integrated circuit microarchitectures; and system integration.

SYSTEMS SCIENCE AND ENGINEERING DIVISION

Biosystems. Professors David Anderson, Spencer BeMent, Jeffrey Fessler, Janice Jenkins, and Matthew O'Donnell.

Ongoing projects include computerized electrocardiography, medical images from several areas, therapeutic and diagnostic ultrasound, processing of complex sounds by the central auditory system, coding of images by the retina, pattern recognition and classification of visual evoked potentials, and evaluation of multichannel recordings from small neural circuits. This area includes the NIH Center for Neural Communications.

Communications and Signal Processing. Professors John Coffey, Alfred Hero III, David Neuhoff, Wayne Stark, Demosthenis Teneketzis, Gregory Wakefield, Kimberly Wasserman, William Williams, Kim Winick, and Andrew Yagle.

Communications research focuses on system design, optimization, and performance analysis as well as on the development of theory to characterize the fundamental limits of communication system performance, including its mathematical foundations. Techniques include digital modulation, channel coding, source coding, information theory, optical communications, detection and estimation, spread spectrum, and multiuser communications and networks. Signal processing overlaps with many other research activities, particularly those of communication and bioengineering. Signal processing focuses on the representation, manipulation, and analysis of signals, particularly natural signals.

Systems and Control. Professors James Freudenberg, Jessy Grizzle, Pramod Khargonekar, Daniel Koditschek, Stéphane Lafortune, Semyon Meerkov, and Demosthenis Teneketzis.

This area is concerned with the investigation of fundamental properties of dynamic systems and the development of methods for their modification. It emphasizes the use of system-theoretical approaches to problems in robotics and automation, manufacturing, semiconductor processing, automotive systems, computer systems, and communications networks.

UNIVERSITY OF MINNESOTA

Electrical Engineering Graduate Program

Programs of Study
Four master's degree plans and one doctoral degree plan are available. All master's degree plans include a total of 30 semester credits. Minimum course requirements for all plans include 14 credits from electrical engineering courses at the 5xxx level or higher and 6 credits from courses outside electrical engineering at the 4xxx level or higher in a related field or minor. The Master of Science in Electrical Engineering (M.S.E.E.) degree can be earned via Plan A or Plan B. M.S.E.E. Plan A includes a total of 20 course credits and 10 thesis credits. M.S.E.E. Plan B includes a total of 30 course credits, including a Plan B paper. The Master of Electrical Engineering (M.E.E.) degree can be earned via the design project track or course work only track. The M.E.E. design project track includes a total of 20 course credits and 10 design project credits. The M.E.E. course work only track includes a total of 30 course credits. The Ph.D. plan includes a total of 40 course credits and 24 thesis credits. Minimum course requirements include 6 credits from 7xxx-level courses, 14 credits in electrical engineering courses, and 12 credits in the supporting program or minor. A written preliminary examination based on undergraduate material must be passed within the first or second year of study. An oral preliminary examination is taken when the course work is substantially complete. There is an oral paper presentation requirement, which is satisfied through a seminar offered within the department. A student with an assistantship can complete the master's program in two years and the Ph.D. program in five or six years after the B.S. degree.

Research Facilities
The department's research utilizes both computational and experimental facilities. Experimental support includes access to a state-of-the-art microelectronic laboratory run by the Institute of Technology, which is vibration-isolated and has a class-10 clean room. Both silicon and III-V compound semiconductor devices and circuits, as well as optical devices, are routinely fabricated in the laboratory; micromechanical devices are also fabricated. The laboratory has epitaxial growth facilities and processing equipment for fabrication. The nanostructure lab houses state-of-the-art equipment for fabrication and characterization of nanodevices, including an ultrahigh-resolution e-beam lithography system, an atomic force microscope, ferritosecond pulse lasers, and a terahertz electrooptical sampling system. The signal processing lab includes sensor arrays and data and image processing systems. The coherent optics laboratory includes lasers, optical tables, and computer facilities. The gaseous electronics laboratory has several ultrahigh-vacuum systems that can be modified to perform studies of plasma collisions and plasma-surface interactions. The magnetics laboratory includes sputtering and evaporation systems for thin-film fabrication and various diagnostic and test equipment. The microwave laboratory includes network and spectrum analyzers and a compact antenna test range. Diverse computing facilities are available to students in the program. There are workstations, personal computers, and central computing facilities.

Financial Aid
Exceptional students may receive fellowship support. Most fellowships support either the first year of study or the final year of Ph.D. dissertation writing. Research assistantships are available from faculty members who have research grants; the number and sponsorship of these projects may vary. Research assistants may be able to use their research projects for their theses. During the 1998–99 academic year, there were approximately 100 research assistants and 50 teaching assistants in the program. Students with fellowships or 50 percent appointments as assistants are given full tuition waivers and personal health insurance.

Cost of Study
Tuition is based on the number of credits, with a plateau of 7 to 15 credits. For 1998–99, resident graduate tuition was $1770 per quarter for 7 to 14 credits. Nonresident tuition was $3260 per quarter. Assistants and fellows may obtain total or partial tuition waivers, but all students must pay certain fees each quarter. More detailed information may be obtained via the World Wide Web (http://www1.umn.edu/tc/students/finances/aid98/basics3.html). Rates for tuition and fees are subject to increases by the Regents.

Living and Housing Costs
Housing options include University residence halls, married student housing cooperatives, and rental properties throughout the Twin Cities. Detailed housing information can be obtained via the World Wide Web (http://www1.umn.edu/housing/) or from the Housing Bureau in Comstock Hall (telephone: 612-624-2994, fax: 612-624-6987, e-mail: housing@tc.umn.edu).

Student Group
The Twin Cities campus of the University is one of the largest campuses in the country; there are approximately 10,000 graduate students in diverse programs on the campus. The electrical engineering program has about 230 graduate students on campus and another 70 taking courses in industry via real-time interactive television. The department's graduate group includes people from many states and countries.

Student Outcomes
Recent graduates have been offered positions in industry and on university faculties. Some remain in the upper Midwest, but others have relocated throughout the U.S. and other countries. Most graduates find positions immediately upon completion of their degree programs.

Location
The Twin Cities area offers diverse cultural and recreational opportunities, many within easy travel from the campus. Two nationally known orchestras, the Minnesota Orchestra and the St. Paul Chamber Orchestra, have full seasons, and the Minnesota Orchestra holds a summer festival in downtown Minneapolis. There are many professional and community theaters in the cities, including the world-renowned Guthrie Theater. Minnesota has more than 10,000 lakes offering both winter and summer fishing and summer boating; there are more than 10 lakes within the Twin Cities area. The Twin Cities are clean and attractive and have a low crime rate among metropolitan areas. The campus is within the city of Minneapolis and straddles the Mississippi River.

The University
The University of Minnesota was established in 1869 and has four campuses. The majority of the students study on the Twin Cities campus, where most of the engineering departments are located.

Applying
Applicants should contact the Director of Graduate Studies by mail to request official application forms. Since many course sequences start in the fall, students are strongly encouraged to enter the program at that time. The application deadline for those desiring financial aid is December 15 for fall quarter admission. This date concerns the arrival of the completed forms; applicants are encouraged to send review material well in advance of December 15. Students who do not require aid may submit completed applications up to July 1. All students from non-English-speaking countries must submit a recent TOEFL score, which must exceed 550 (213 for the computer-based exam). All applicants desiring financial aid must submit scores on the GRE General Test. Applicants must have a bachelor's degree from an engineering or science program; people with engineering technology degrees are not accepted.

Correspondence and Information
Director of Graduate Studies
Department of Electrical and Computer Engineering
University of Minnesota, Twin Cities
200 Union Street
Minneapolis, Minnesota 55455

Telephone: 612-625-3564
E-mail: graduate_studies@ece.umn.edu
World Wide Web: http://www.ece.umn.edu

University of Minnesota

THE FACULTY AND THEIR RESEARCH

Slim Alouini, Assistant Professor; Ph.D., Caltech, 1998. Wireless communications.

Stephen A. Campbell, Associate Professor; Ph.D., Northwestern, 1981. Fabrication and characterization of high-speed devices, microelectronics.

Vladimir S. Cherkassky, Associate Professor; Ph.D., Texas at Austin, 1985. Pattern recognition, neural networks, computer networks.

Philip I. Cohen, Professor; Ph.D., Wisconsin, 1975. Molecular-beam epitaxy of artificially structured microelectronic materials.

Rhonda Drayton, Assistant Professor; Ph.D., Michigan, 1995. Microelectronics, microwaves, optics.

Emad Ebbini, Associate Professor; Ph.D., Illinois at Urbana-Champaign, 1990. Signal processing.

Douglas W. Ernie, Associate Professor; Ph.D., Minnesota, 1980. Plasma physics and plasma chemistry, DC and radio-frequency plasma discharges.

Tryphon T. Georgiou, Professor; Ph.D., Florida, 1983. System theory and control engineering.

Georgios Giannakis, Professor; Ph.D., South Carolina, 1986. Statistical signal processing and applications to wired and wireless communications.

Anand Gopinath, Professor; Ph.D., 1965, D.Eng., 1978, Sheffield (England). Optoelectronic devices and circuits, including lasers, switches, modulators, guided wave structures, microwave devices, and circuits.

Ramesh Harjani, Associate Professor; Ph.D., Carnegie Mellon, 1989. Analog and mixed-signal design, CAD for analog and mixed-signal design.

Ted Higman, Associate Professor; Ph.D., Illinois, 1989. Atomic force microscopy, nanolithography.

James E. Holte, Associate Professor; Ph.D., Minnesota, 1960. Bioelectrical science with applications in medicine, aids for handicapped persons.

Jack H. Judy, Professor; Ph.D., Minnesota, 1965. Magnetic thin films, magnetic recording media and heads, micromagnetics, magnetic measurements, magnetic sensors, magnetic memories.

Mostafa Kaveh, Professor; Ph.D., Purdue, 1974. Communication theory, signal processing, parameter estimation, image processing.

John C. Kieffer, Professor; Ph.D., Illinois, 1970. Information theory, communication theory, techniques for reliable data transmission.

Larry L. Kinney, Professor; Ph.D., Iowa, 1968. Fault-tolerant computer design, test vector generation, fault simulation, concurrent error detection.

K. S. P. Kumar, Professor; Ph.D., Purdue, 1964. Control and systems.

E. Bruce Lee, Professor; Ph.D., Minnesota, 1960. Control and systems, mathematical models, synthesis techniques for control systems.

Thomas S. Lee, Associate Professor; Ph.D., Minnesota, 1961. Applied electrostatics, electrodynamics, and fluid-mechanical phenomena.

James Leger, Professor; Ph.D., California, San Diego, 1980. Electrooptics, Fourier optics and holography, microoptical devices, diffractive optics, semiconductor and solid-state lasers.

David J. Lilja, Associate Professor; Ph.D., Illinois at Urbana-Champaign, 1991. High-performance computing, parallel and distributed computing, computer architecture, compilers.

Radu Marculescu, Assistant Professor; Ph.D., South Carolina, 1998. Computer-aided design.

Christine Maziar, Professor; Ph.D., Purdue, 1986. Semiconductor devices.

Ned Mohan, Professor; Ph.D., Wisconsin, 1973. Electric power systems, power electronics, motion control.

Jay Moon, Associate Professor; Ph.D., Carnegie Mellon, 1990. Communications, magnetic recording, signal processing.

Marshall I. Nathan, Professor; Ph.D., Harvard, 1958. Molecular-beam epitaxy and its use to fabricate layered semiconductors, semiconductor device physics, large gap semiconductors.

Matthew T. O'Keefe, Associate Professor; Ph.D., Purdue, 1990. Computer architecture, compilers, parallel processing, applications, mass storage.

Keshab K. Parhi, Professor; Ph.D., Berkeley, 1988. Signal processing systems on VLSI chips, computer-aided design, signal processing, computer arithmetic.

William T. Peria, Professor; Ph.D., British Columbia, 1957. Physical electronics, semiconductor epitaxy, IC fabrication.

Dennis L. Polla, Professor; Ph.D., Berkeley, 1985. Microelectromechanical systems, biomedical microinstruments, solid-state materials and devices, integrated circuits.

William P. Robbins, Professor; Ph.D., Washington (Seattle), 1971. Acoustic sensing, ultrasonics, surface-wave devices, microactuators, power electronics.

P. Paul Ruden, Professor; Ph.D., Stuttgart, 1982. Semiconductor materials and physics of novel electronic devices.

Sachin Sapatnekar, Associate Professor; Ph.D., Illinois at Urbana-Champaign, 1992. CAD of VLSI systems, VLSI design.

Guillermo Sapiro, Assistant Professor; Ph.D., Technion (Israel), 1993. Computer vision, systems, image processing.

Gerald E. Sobelman, Associate Professor; Ph.D., Harvard, 1979. Digital VLSI design with applications in digital signal processing and digital communications.

Joseph Talghader, Assistant Professor; Ph.D., Berkeley, 1995. Optoelectronics, microelectronics.

Allen R. Tannenbaum, Professor; Ph.D., Harvard, 1976. Robust feedback control systems, mathematical analysis of control systems, computer vision, image processing.

Ahmed H. Tewfik, Professor; Sc.D., MIT, 1987. Multimedia data coding and management, wavelets in signal processing, medical and radar imaging, solitons, stochastic resonance.

Randall Victora, Associate Professor; Ph.D., Berkeley, 1985. Magnetics.

Bapiraju Vinnakota, Associate Professor; Ph.D., Princeton, 1991. Fault-tolerant computing, digital system testing, CAD for testing.

Bruce Wollenberg, Professor; Ph.D., Pennsylvania, 1974. Electric power system economic studies and analytical methods.

UNIVERSITY OF MINNESOTA

Institute of Technology
Graduate Program in Computer Engineering

Program of Study

Computer engineering is an interdisciplinary graduate program offered jointly by the Department of Electrical and Computer Engineering and the Department of Computer Science and Engineering. Students in this program develop a broad understanding of both hardware and software design issues. Two different degree options are available. The Master of Science (M.S.) in computer engineering degree is a traditional research-oriented graduate degree that prepares students to work in industry or to continue with their graduate studies in either electrical engineering or computer science. The Master of Computer Engineering (M.Comp.E.) degree is a course work-only professional engineering degree tailored to practicing computer scientists and engineers. Faculty members in the program work closely with graduate students conducting research in a wide variety of computer engineering topics, including computer architecture and system design, computer graphics, distributed systems, fault-tolerant computing, optimizing and parallelizing compilers, computer-aided design, databases, networks, operating systems, parallel computing, software engineering, and VLSI design and testing. Students begin their studies with a core program of courses in system software, computer architecture and networking, VLSI and digital design, and data structures and algorithms and then proceed to in-depth study in their area of specialization. M.S. students may elect to complete a master's thesis or an independent project. The comprehensive final exam for the M.S. degree is oral; no final exam is required for the M.Comp.E. degree. These degrees typically require one to two years of full-time study. Part-time study is encouraged for students who work in industry while attending classes.

Research Facilities

The Departments of Electrical and Computer Engineering and Computer Science and Engineering provide access to numerous specialized and general-purpose computing facilities and laboratories. Current computing resources include several hundred workstations and personal computers from vendors such as Hewlett-Packard, Silicon Graphics, Sun, Apple, and IBM. Research programs have access to departmental computing resources, including a cluster of multiprocessor SGI Challenge Computers connected via HIPPI and Fibre Channel, a cluster of IBM RS6000/590 computers connected via ATM, and an IBM SP-2 parallel computing system. Workstation access is also provided by the Institute of Technology public labs, which have several hundred Sun and SGI computers. The Minnesota Supercomputer Institute supports research that is carried out using the supercomputers and other resources of the Minnesota Supercomputer Center, Inc. The systems available include a Cray C-90, a Cray T3E, and numerous graphics workstations. The Laboratory for Computational Science and Engineering provides access to extensive computing, storage, and graphics and visualization facilities. University library facilities are excellent and have online access to the catalog. Extensive networking and dial-in facilities are provided to access all of these systems.

Financial Aid

Both fellowships and half-time teaching assistantships are available through the Departments of Electrical and Computer Engineering and Computer Science and Engineering, and research assistantships are available through individual faculty members. Typically, a half-time teaching assistant is expected to teach laboratory sections for one of the introductory courses or to assist with grading and other duties. These appointments typically require up to 20 hours per week, including preparation, student consulting, and grading. Research assistants are normally expected to devote an equivalent amount of time to funded research work. Because of the concentration of computer and computer-based industries in the Twin Cities area, graduate students often can find a variety of interesting part-time and summer job opportunities. Students with fellowships or half-time assistantships are given tuition waivers and personal health insurance.

Cost of Study

In 1998–99, resident tuition for graduate students was $1710 per quarter (7–14 credits); nonresident tuition was $3358 per quarter (7–14 credits). While assistants and fellows obtain tuition waivers for up to 14 credits per quarter, all students must pay certain fees each quarter. Rates for tuition and fees are subject to increases by the Board of Regents. The University is moving to a semester system beginning in fall 1999. Costs are expected to change accordingly.

Living and Housing Costs

For information about housing, students should contact University Housing Services, Comstock Hall-East, 210 Delaware Street SE, Minneapolis, Minnesota 55455 (telephone: 612-624-2994; fax: 612-624-6987; E-mail: housing @cafe.tc.umn.edu). Students who intend to live in a residence hall on campus or in nearby married-student housing should contact this office as soon as possible because housing is very limited.

Student Group

The Twin Cities campus of the University is one the largest campuses in the country, with approximately 10,000 graduate students in diverse programs. In addition to students taking courses on campus, many students working in industry take courses via real-time interactive television. The program includes people from many states and countries.

Student Outcomes

Many M.S. students continue into Ph.D. programs in electrical engineering or computer science. Most students graduating in computer engineering find positions in industry immediately upon completion of their degree programs.

Location

The Twin Cities area offers diverse cultural and recreational opportunities, many within easy travel from campus. Two nationally known orchestras have full seasons, and there are many professional and community theaters in the cities, including the world-renowned Guthrie Theater. Minnesota has more than 10,000 lakes, including several within the metropolitan area, that offer both winter and summer fishing and summer boating. The Twin Cities are clean and attractive and have a low crime rate among metropolitan areas. The campus is within the city of Minneapolis and straddles the Mississippi River.

The University

The University of Minnesota was established in 1869 as Minnesota's land-grant university. It consists of four campuses, with the majority of students studying on the Twin Cities campus, where most of the engineering departments are located.

Applying

Graduate study in computer engineering is open to students with an undergraduate degree in computer engineering, electrical engineering, computer science, or a closely related field, such as mathematics or physics. In some instances, additional preparatory work may be required. Applicants should contact the Director of Graduate Studies to request official application forms. The application deadline for students requesting financial aid is December 15 for admission for the following fall. All applicants requesting financial aid must submit scores from the GRE General Test, and all students from non-English-speaking countries must submit a recent TOEFL score.

Correspondence and Information

Director of Graduate Studies
Graduate Program in Computer Engineering
4-174 Electrical Engineering/Computer Science Building
University of Minnesota
200 Union Street S.E.
Minneapolis, Minnesota 55455

Telephone: 612-625-3300
Fax: 612-625-4583
E-mail: gradinfo@compengr.umn.edu
World Wide Web: http://www.compengr.umn.edu/

University of Minnesota

THE FACULTY AND THEIR RESEARCH

Vladimir Cherkassky, Associate Professor; Ph.D., Texas at Austin, 1985. Pattern recognition, neural networks, computer networks.

David H.-C. Du, Professor; Ph.D., Washington (Seattle), 1981. High-speed networking, multimedia applications, high-performance computing over clusters of workstations, database design, CAD for VLSI.

Mats Heimdahl, Assistant Professor; Ph.D., California, Irvine, 1994. Software engineering, formal specification languages, automated analysis of specifications, software development for safety critical control systems.

Larry Kinney, Professor; Ph.D., Iowa, 1968. Testing of digital systems, built-in-self-test, computer design, microprocessor-based systems, error-correcting codes.

Vipin Kumar, Professor; Ph.D., Maryland College Park, 1982. Parallel computing for sparse linear systems, scalability analysis.

David J. Lilja, Associate Professor; Ph.D., Illinois at Urbana-Champaign, 1991. Computer architecture, parallel and distributed computing, hardware-software interactions, performance analysis.

Lori Lucke, Adjunct Assistant Professor; Ph.D., Minnesota, 1992. VLSI architectures and algorithms for signal processing and image processing, high-level synthesis, CAD algorithms.

Farnaz Mounes-Toussi, Adjunct Assistant Professor; Ph.D., Minnesota, 1995. Computer architecture, memory systems for multiprocessors, performance analysis, hardware-software interactions.

Matthew T. O'Keefe, Associate Professor; Ph.D., Purdue, 1990. Computer architecture, compilers, parallel processing, applications, mass storage.

Sachin Sapatnekar, Associate Professor; Ph.D., Illinois at Urbana-Champaign, 1992. Computer-aided design of VLSI circuits, optimization algorithms.

Shashi Shekhar, Associate Professor; Ph.D., Berkeley, 1989. Databases, geographic information systems, parallel computing, real-time computing.

Eugene B. Shragowitz, Professor; Ph.D., National Scientific Research Laboratory (Moscow), 1971. Computer-aided design of VLSI, fuzzy logic, nonlinear network theory, combinatorial optimization, learning algorithms.

Gerald E. Sobelman, Associate Professor; Ph.D., Harvard, 1979. VLSI design, digital signal processing, error-correcting codes.

Jaideep Srivastava, Associate Professor; Ph.D., Berkeley, 1988. Databases, distributed systems, multimedia computing.

Wei-tek Tsai, Professor; Ph.D., Berkeley, 1982. Software engineering.

Bapiraju Vinnakota, Associate Professor and Director of Graduate Studies; Ph.D., Princeton, 1991. VLSI design and test, CAD for testing, fault-tolerant computing.

Richard Voyles, Assistant Professor; Ph.D., Carnegie Mellon, 1998. Robotics, computer vision, tactile sensors/activators.

Pen-Chung Yew, Professor; Ph.D., Illinois at Urbana-Champaign, 1981. Computer architecture, parallel machine organization, compilers, performance evaluation, parallel processing.

UNIVERSITY OF MISSOURI–COLUMBIA

Department of Computer Engineering and Computer Science

Programs of Study

The Department of Computer Engineering and Computer Science at the University of Missouri–Columbia offers graduate programs of study leading to Master of Science degrees in computer science and computer engineering and the Doctor of Philosophy in computer engineering and computer science. Faculty research areas include fuzzy set theory and fuzzy logic, neural networks, computer vision, parallel and distributed computing, computer networking, advanced computing and high-speed networking systems and applications, digital libraries, artificial intelligence, computer graphics and scientific visualization, multimedia systems, information systems and design, biomedical engineering, and database theory and design.

An M.S. candidate must complete a minimum of 30 semester credit hours and pass a final examination to demonstrate mastery of the work included in a thesis or substantial independent project. To achieve Ph.D. candidacy, a student must pass a qualifying examination. The Ph.D. program requires a minimum of 72 semester hours beyond the B.S. degree. The candidate must pass both a written and an oral examination, complete a doctoral dissertation on a topic approved by the candidate's advisory committee, and defend the dissertation in an oral final examination.

Research Facilities

A wide range of computing and networking resources are available to the students in the department, the college, and the campus. These resources provide ready access to state-of-the-art computing systems and networking facilities ranging from small desktop systems to large computational systems that are interconnected by traditional and advanced networking facilities. These connections also provide links to the Internet for global access to information, software, machines, and colleagues. All of these facilities provide a wealth of opportunity for students to use and study state-of-the-art computing.

Financial Aid

Financial aid is available through departmental teaching and research assistantships as well as campus fellowship programs. All assistantships and fellowships carry a full tuition waiver. Stipends vary based on level of employment and level of achievement in the degree program.

Cost of Study

For the 1998–99 academic year, the graduate educational fee was $162.60 per credit hour for Missouri residents and $489.10 per credit hour for nonresidents. An activities fee of $114.41 per semester is charged to students enrolling in 12 or more credit hours. For part-time enrollment, the student activities fee is calculated per credit hour. An additional fee of $35.70 per credit hour is charged to students enrolled in courses offered by the College of Engineering.

Living and Housing Costs

Housing is available for graduate students both on and off campus. Columbia living conditions are pleasant and of moderate cost comparable to the regional (Big 12 institutions) average.

Student Group

There are currently 76 graduate students in the Department of Computer Engineering and Computer Science; 48 are enrolled in the M.S. in computer science program, 10 in the M.S. in computer engineering program, and 18 in the Ph.D. program. The average age is approximately 26, and approximately 18 percent of the students are women. Students come from all over the United States and from sixteen other countries.

Student Outcomes

M.S. graduates typically find employment with a wide range of business, scientific, and research organizations as engineers, software designers, and application developers. Ph.D. graduates typically take research positions in industry or positions as faculty members in major universities across the country. International graduates often find employment in industry or academia in their home countries.

Location

The University is located in Columbia, a city of approximately 75,000. It is located in the center of Missouri, near the Missouri River valley. It is approximately a 2-hour drive from either St. Louis or Kansas City on Interstate 70.

The University and The Department

The University of Missouri–Columbia is the flagship campus of the University of Missouri System and has more than 22,000 students and 1,800 faculty members. It is a member of the American Association of Universities (AAU) and a Carnegie Research I institution. In addition to engineering, the University offers degrees in agriculture, business, education, journalism, law, medicine, nursing, veterinary medicine, and the arts and sciences.

The department was created as an independent department in the College of Engineering in 1995. The department was formed by combining the strengths in computational science from elements of the electrical and computer engineering department and the computer science department. Both electrical and computer engineering and computer science have long-standing histories at MU, and the new department was formed to capitalize upon the strengths of these two programs in a single unit.

Applying

Applicants should have earned a B.S. degree in computer engineering or computer science; if a B.S. in another field was obtained, accepted applicants will most likely have to complete preparatory courses in addition to their degree program requirements. To apply, an application form, GRE General Test scores, three confidential letters of recommendation, a personal statement describing the applicant's background and academic objectives, transcripts of all previous college work, and an application fee ($25 for U.S. residents or $50 for international applicants) must be submitted. Students whose native language is not English must also submit TOEFL scores.

Correspondence and Information

Director of Graduate Studies
Department of Computer Engineering and Computer Science
201 Engineering Building West
University of Missouri–Columbia
Columbia, Missouri 65211-2060
Telephone: 573-882-3842
Fax: 573-882-8318
E-mail: gradsec@cecs.missouri.edu
World Wide Web: http://www.cecs.missouri.edu

University of Missouri–Columbia

THE FACULTY AND THEIR RESEARCH

Professors

Su-Shing Chen, Chair; Ph.D., Maryland, 1970. Digital libraries, intelligent agent architectures, digital information environments, content-based indexing for spatial (information) objects.

Harry Tyrer, Associate Chair; Ph.D., Duke, 1972. Computer architecture, object-oriented languages, software engineering, biomedical engineering, biophysics, instrumentation.

James Keller, Ph.D., Missouri–Columbia, 1978. Fuzzy set theory and fuzzy logic, computer vision, pattern recognition, neural networks.

Otho R. Plummer, Ph.D., Texas at Austin, 1966. Application systems, database systems, scientific programming, graphics applications and virtual reality.

Frederick N. Springsteel, Ph.D., Washington (Seattle), 1967. Database design, theory of parallel algorithms, foundations of computing.

Xinhua Zhuang, Ph.D., Peking, 1963. Computer vision, image processing, pattern recognition, neural net computing, speech recognition, artificial intelligence.

Associate Professors

Paul Gader, Ph.D., Florida, 1986. Character and handwriting recognition, image algebra and math, morphology, pattern recognition.

Youran Lan, Ph.D., Michigan State, 1988. Parallel and distributed systems, parallel algorithm design, computer networking.

Kannappan Palaniappan, Ph.D., Illinois, 1991. Computer graphics, scientific visualization, remote sensing, stereo and nonrigid motion analysis, parallel algorithms for image analysis, sequence analysis.

Youssef G. Saab, Ph.D., Illinois, 1990. Combinatorial optimization, design automation, graph and geometric algorithms, stochastic algorithms.

Gordon K. Springer, Director of Graduate Studies; Ph.D., Penn State, 1970. Computer networking, advanced computing and high-speed networking systems and applications, distributed computing, software system design.

Yunxin Zhao, Ph.D., Washington (Seattle), 1988. Spoken language processing, multimedia interface, statistical identification and estimation, speech and signal processing.

Assistant Professors

Michael Jurczyk, Ph.D., Stuttgart (Germany), 1996. Interconnection networks for communication and parallel processing systems, ATM-networks, networked multimedia, parallel and distributed systems.

Yi Shang, Ph.D., Illinois at Urbana-Champaign, 1997. Artificial intelligence, computational science and engineering, nonlinear optimization, parallel and distributed computing.

Hongchi Shi, Ph.D., Florida, 1994. Parallel and distributed computing, image processing, computer vision.

Marjorie Skubic, Ph.D., Texas A&M, 1997. Sensory perception, robotics, virtual reality, human-machine interaction.

UNIVERSITY OF MISSOURI–COLUMBIA

Department of Electrical Engineering

Programs of Study

The Department of Electrical Engineering at the University of Missouri–Columbia offers graduate programs of study and research leading to the degrees of Master of Science and Doctor of Philosophy in electrical engineering. Faculty research areas include signal processing, wireless communication, computer communication, antenna design and remote sensing, solid-state device and optoelectronics, physical electronics, control, power electronics, digital power measurements, and biomedical engineering.

An M.S. candidate must complete a minimum of 30 total semester credit hours and pass a final examination to demonstrate mastery of the work included in a thesis or a substantial independent project. To achieve Ph.D. candidacy, a student must pass a qualifying exam. The Ph.D. program requires a minimum of 72 semester hours beyond the B.S., with research on the doctoral dissertation generally taking about one full year. The candidate must pass both a written and an oral comprehensive examination, complete a doctoral dissertation on a topic approved by his or her advisory committee, and defend the dissertation in an oral final examination.

Research Facilities

Research activities in electrical engineering are conducted in laboratories designed to support work in microcircuit design and processing, power electronics and power quality, electron beam radiation, molecular beam epitaxy, optoelectronics, microcomputer development, and high-frequency measurements. Research is enhanced by an extensive network of computer facilities. The department, college, and University computing facilities are connected to one another and the Internet through both fiber-optic and wire cables.

Financial Aid

Financial aid is available through departmental teaching and research assistantships as well as campus fellowship programs. All assistantships and fellowships carry a full tuition waiver. Stipends vary based on level of employment and level of achievement in degree program.

Cost of Study

For the 1998–99 academic year, the graduate educational fee was $162.60 per credit hour for Missouri residents and $489.10 per credit hour for nonresidents. An activities fee of $114.56 per semester was charged to students enrolling in 12 or more credit hours. For part-time enrollment, the student activities fee is calculated per credit hour. An additional fee of $35.70 per credit hour was charged to all students enrolled in courses offered by the College of Engineering.

Living and Housing Costs

Housing is available for graduate students both on and off campus. Columbia living conditions are pleasant and of moderate cost, comparable to the regional (Big 12 institutions) average.

Student Group

There are currently 57 graduate students in the Department of Electrical Engineering: 29 are enrolled in the Ph.D. program, and the remaining students are seeking M.S. degrees. Approximately 12 percent of the students are women. Students come from all over the United States and from nineteen countries.

Student Outcomes

M.S. graduates typically find employment in companies such as McDonnell Douglas, Intel, Emerson Electric, Texas Instruments, and Motorola. Ph.D. graduates take research positions in organizations such as IBM and Bell Laboratories or as faculty members in major universities across the country. International graduates often find employment in industry or academia in their home countries.

Location

The main location of the department is in Columbia; however, several of the faculty members are located in Kansas City as part of the Coordinated Engineering Program that is administered by the Columbia branch. Columbia, a city of approximately 75,000, is located in the center of Missouri, near the Missouri River valley. It is approximately a 2-hour drive from either St. Louis or Kansas City and only 1 hour from the scenic Ozark recreational areas. Kansas City has a population of approximately 1 million.

The University and The Department

The University of Missouri–Columbia is the flagship campus of the University of Missouri System and has 22,500 students and 1,993 faculty members. It is a member of the American Association of Universities (AAU) and a Carnegie Research I institution. In addition to engineering, the University offers degrees in agriculture, business, education, journalism, law, medicine, nursing, veterinary medicine, and the arts and sciences.

One of the pioneering electrical engineering programs in the country, the department at UMC was founded in 1885 and was the first such program west of the Mississippi. The department consists of 22 faculty members, 4 of whom are located at the Kansas City campus. Several more faculty members are expected to be recruited in fall 1999.

Applying

Applicants should have earned a B.S. degree in electrical engineering; if a B.S. in a field other than electrical engineering was obtained, accepted applicants will most likely be required to complete preparatory courses in addition to their degree program requirements. To apply, an application form, GRE General Test scores, three confidential letters of recommendation, a personal statement describing the applicant's background and academic objectives, transcripts of all previous college work, and an application fee ($25 for U.S. resident applicants or $50 for international applicants) must be submitted. Students whose native language is not English must also submit TOEFL scores.

Correspondence and Information

Director of Graduate Studies
Department of Electrical Engineering
University of Missouri–Columbia
219 Engineering Building West
Columbia, Missouri 65211
Telephone: 573-882-3539
Fax: 573-882-0397
E-mail: eegrsec@risc1.ecn.missouri.edu (Graduate Secretary)

University of Missouri–Columbia

THE FACULTY AND THEIR RESEARCH

Columbia Campus

Andrew J. Blanchard, Professor; Ph.D., Texas A&M, 1977. Microwave engineering, radar systems, remote sensing, long wavelength imaging, antennas.

Chang Wen Chen, Assistant Professor; Ph.D., Illinois at Urbana-Champaign, 1992. Image and signal processing, wireless communications.

Randy Curry, Assistant Professor; Ph.D., St. Andrews (Scotland), 1992. Physical electronics, pulse power applications, diagnostics.

Curt Davis, Associate Professor; Ph.D., Kansas. 1992. Microwave engineering, radar systems, satellite remote sensing.

Michael Devaney, Associate Professor; Ph.D., Missouri–Columbia, 1971. Digital systems, digital power measurement, power electronics, computer simulation.

T. Greg Engel, Assistant Professor; Ph.D., Texas Tech, 1990. Pulse power, high-power switching concepts, physical electronics.

Kevin Gillis, Assistant Professor; Ph.D., Washington (St. Louis), 1993. Biomedical problems.

Huber Graham, Associate Professor; Ph.D., MIT, 1969. Digital system design, circuit design.

Dominic K. C. Ho, Assistant Professor; Ph.D., Chinese University of Hong Kong, 1991. Digital signal processing, wireless communication.

Robert Leavene, Associate Professor; Ph.D., Missouri–Columbia, 1972. Digital system design, microprocessor applications, digital signal processing.

Kai-Fong Lee, Professor and Chair; Ph.D., Cornell, 1966. Antenna theory and design, applied electromagnetics.

Chun-Shin Lin, Associate Professor; Ph.D., Purdue, 1980. Robotics, computer vision.

Jon Meese, Professor; Ph.D., Purdue, 1970. Irradiations, optics, semiconductor theory, solid-state physics.

William Nunnally, Professor; Ph.D., Texas Tech, 1975. Physical electronics, pulsed power, antennas, applied electromagnetics.

Robert M. O'Connell, Associate Professor; Ph.D., Illinois at Urbana-Champaign, 1975. Semiconductor device modeling, laser effects in optical networks, power electronics.

Wes B. Sherman, Professor; Ph.D., Missouri–Columbia, 1966. Electromagnetics, communications, instrumentation.

Charles Slivinsky, Professor; Ph.D., Arizona, 1969. Power systems, digital signal processing, multimedia.

Kenneth Unklesbay, Professor; Ph.D., Missouri–Columbia, 1972. Control system design, food processing engineering.

Coordinated Engineering Program In Kansas City

Ghulam M. Chaudhry, Associate Professor; Ph.D., Wayne State, 1989. Computer networking, parallel processing.

Mohsen Guizani, Associate Professor; Ph.D., Syracuse, 1990. Computer communication and networking.

Jerome Knopp, Associate Professor; Ph.D., Texas at Austin, 1976. Optical systems, information processing, holography, high-energy laser systems, scalar diffraction theory, dimensional analysis, modeling.

David G. Skitek, Assistant Professor; Ph.D., Arizona State, 1973. Active network synthesis, simulations, digital signal processing.

UNIVERSITY OF NEW MEXICO

Department of Electrical and Computer Engineering

Programs of Study

Graduate work leading to the M.S. and Ph.D. degrees is offered by the department in the areas of circuits, communications, computer engineering, control systems, electromagnetics, electrooptics, image processing, microelectronics fabrication and design, plasma science, and signal processing. The M.S. degree is also offered in manufacturing engineering. The master's degree program requires 30 semester credit hours for a thesis option and 33 semester credit hours for a non-thesis option. M.S. students funded by either a teaching or research assistantship must pursue the thesis option. The Ph.D. program requires that a minimum of 24 graduate credit hours beyond the master's degree be completed at the University of New Mexico (UNM). Additional course work and research leading to the dissertation are geared to the individual student's needs and interests. As a potential candidate for the Ph.D. program, each student must pass the Ph.D. qualifying examination to establish levels and areas of scholastic capabilities.

Research Facilities

The department maintains state-of-the-art laboratories for computer vision and image processing, high-performance computing and networking, lasers and electrooptics, microprocessors (including advanced DSP platforms and emerging architectures), microwaves and antennas, pulsed power and plasma science, real-time computing and embedded systems (including a number of mobile robots and advanced real-time development systems), solid-state fabrication, and virtual reality/advanced human-computer interfaces. In addition, the department has a close affiliation with world-class research laboratories and terascale supercomputing platforms at the UNM-operated Maui and Albuquerque High Performance Computing Centers as well as Sandia National Laboratories and Los Alamos National Laboratory. Sponsored research expenditures for the 1997–98 academic year were approximately $12 million.

Financial Aid

Support is available in the form of teaching assistantships and research assistantships. Graduate internship programs are conducted with local industries, such as Sandia National Laboratories and the Air Force Research Laboratories. Annual stipends for full-time teaching assistantships require no more than 20 hours of service per week for the academic year. A tuition waiver for 12 credit hours per semester is also made. Research assistants are paid on a scale of $1000 to $1500 per month (half-time).

Cost of Study

In 1998–99, tuition and other fees for students carrying 12 or more credit hours were $1221.20 for state residents and $4345.40 for nonresidents. All residents carrying 11 or fewer credit hours paid $103.10 per semester credit hour, and nonresidents paid $103.10 per semester credit hour for up to 6 credit hours. For more than 6 semester credit hours, nonresidents paid $363.45 per credit hour. Domestic students can meet the requirements for resident status by living continuously in New Mexico for at least one year prior to registration for the following semester and by providing satisfactory evidence of their intent to retain residence in New Mexico. International students must provide proof of financial competence prior to admission to UNM.

Living and Housing Costs

Living costs in Albuquerque are somewhat lower than those in other cities of comparable size. In addition to tuition and fees, a single student's expenses are estimated at $10,250 per year; expenses for a single international student are approximately $17,250 per year, including tuition and fees.

Student Group

Students are drawn from all parts of the United States as well as from many other countries. The graduate enrollment in the department, including part-time students, is about 150, of whom 60 are Ph.D. candidates. During the past two years, the department has awarded 67 master's degrees and 21 Ph.D. degrees.

Student Outcomes

The current demand for graduate engineers is excellent, and the employment rate for electrical engineering and computer engineering graduates has been almost 100 percent. Graduates of the department have been employed in various positions, such as senior engineer; vice president, manufacturing; electronics engineer; and systems engineer. Examples of companies that hire the department's graduates include Intel, Motorola, Array Communications, Honeywell, and Philips Semiconductors as well as small entrepreneurial companies.

Location

Albuquerque, with a metropolitan population of more than 670,000, is the largest city in New Mexico. With an unusual blend of three cultures—Native American, Spanish-American, and Western—it is able to offer a wide variety of cultural, artistic, and aesthetic events. Several of these take place on campus, while others are in the city and neighboring pueblos. The Indian Pueblo Cultural Center, the National Atomic Museum, and the UNM Maxwell Museum of Anthropology offer facilities of particular interest. The city lies between the lowland of the Rio Grande and the towering, 11,000-foot Sandia Mountains. In this "Land of Enchantment" environment, the sun shines every day, and warm days are followed by cool nights. Hunting, fishing, ballooning, mountain climbing, and skiing are only a few of the recreational activities available.

The University

The University of New Mexico, ranked twenty-ninth among "Rising Research Public Universities" (Johns Hopkins Press, 1997), is the largest university in the state, with more than 30,000 students, and is a Carnegie I Research University. It was established in 1889 and is situated on 600 acres in the center of metropolitan Albuquerque. The University of New Mexico's School of Engineering has an enrollment of 1,500 undergraduate students and 500 graduate students. The resources of the University and its proximity to Sandia National Laboratories, Kirtland Air Force Base, and Los Alamos National Laboratories provide an excellent environment for advanced studies and research.

Applying

Prospective applicants should contact the Office of Graduate Studies as well as the Department of Electrical and Computer Engineering. The GRE General Test is required for admission to both the M.S. and Ph.D. programs. Applications, fees, and transcripts should be on file with the Office of Graduate Studies by July 16 for the fall semester, by November 13 for the spring semester, and by April 30 for the summer session. International applicants' materials should arrive six months prior to the semester for which the applicant is applying. Financial aid applications are due by March 1.

Correspondence and Information

Coordinator of Graduate Studies
Department of Electrical and Computer Engineering
University of New Mexico
Albuquerque, New Mexico 87131-1356
Telephone: 505-277-2600
E-mail: gradinfo@eece.unm.edu
World Wide Web: http://www.eece.unm.edu/

University of New Mexico

THE FACULTY AND THEIR RESEARCH

Chaouki T. Abdallah, Associate Professor; Ph.D., Georgia Tech. Control systems, adaptive and nonlinear systems, robot control, wireless communications and theory of computation.

Nasir Ahmed, Professor, Associate Provost for Research, and Interim Dean of Graduate Studies; Ph.D., New Mexico. Digital signal processing.

David A. Bader, Assistant Professor; Ph.D., Maryland, College Park. High-performance computing, parallel algorithms and architecture, remote sensing, image processing.

Steven R. J. Brueck, Professor and Director of the Center for High Technology Materials; Ph.D., MIT. Laser-material interactions, electrooptic devices, laser spectroscopy.

Thomas P. Caudell, Associate Professor; Ph.D., Arizona. Neural networks, virtual reality, machine vision, robotics, genetic algorithms, high-performance computing.

Julian Cheng, Professor; Ph.D., Harvard. Optical communication systems, optical data networks, optoelectronic devices and integrated circuits.

Christos Christodoulou, Professor and Chair; Ph.D., North Carolina State. Modeling of electromagnetic systems, phased array antennas, antennas for wireless communications, microwave systems and applications of neural networks in electromagnetics.

Peter Dorato, Professor; D.E.E., Polytechnic of Brooklyn. Optimal control, robust design in feedback control systems.

Charles B. Fleddermann, Professor; Ph.D., Illinois at Urbana-Champaign. Plasma processing, laser diagnostics, physical electronics, photovoltaics.

Paul A. Fleury, Professor and Dean of School of Engineering; Ph.D., MIT. Spectroscopy and nonlinear optics of condensed-matter physics.

John Michel Gahl, Associate Professor; Ph.D., Texas Tech. Physical electronics, rf technology, plasma science.

Charles F. Hawkins, Professor; Ph.D., Michigan. VLSI design, testability, reliability, fault analysis.

Gregory L. Heileman, Associate Professor; Ph.D., Central Florida. Data structures and algorithmic analysis, machine learning, theory of computing.

Stephen D. Hersee, Professor; Ph.D., Brighton Polytechnic (England). Semiconductor materials and optoelectronics devices.

Stanley Humphries, Professor; Ph.D., Berkeley. Numerical simulations in electromagnetics and bioengineering.

Don Hush, Associate Professor; Ph.D., New Mexico. Neural networks, pattern recognition, computer vision.

Ravi Jain, Professor; Ph.D., Berkeley. Lasers and nonlinear optics, ultrafast lasers and optoelectronics, optical fiber devices for sensors and communications, fiber lasers and amplifiers.

Mohammad Jamshidi, Professor and Director of the Center for Autonomous Control Engineering; Ph.D., Illinois at Urbana-Champaign. Large-scale system theory and applications, soft computing.

Ramiro Jordan, Associate Professor; Ph.D., Kansas State. Data communications, software development.

Kenneth C. Jungling, Professor; Ph.D., Illinois at Urbana-Champaign. Lasers and thin films, microanalytical thin-film diagnostics, optical detection, high-power laser damage testing.

Don L. Kendall, Professor; Ph.D., Stanford. Semiconductor diffusion, micromachining.

Luke F. Lester, Assistant Professor; Ph.D., Cornell. Semiconductor lasers and device processing.

Neeraj Magotra, Associate Professor; Ph.D., New Mexico. Digital signal and image processing, algorithms—theory to implementation, applied areas—speech, radar (SAR), seismic, and biomedical.

Gary K. Maki, Professor and Director of the Microelectronics Research Center; Ph.D., Missouri–Rolla. Digital design, fault-tolerant digital design, error correction codes, VLSI design and architectures.

Kevin Malloy, Associate Professor; Ph.D., Purdue. Semiconductor physics, device physics.

John R. McNeil, Professor; Ph.D., Colorado State. Optics and physical electronics, scatterometry for microelectronics manufacturing process control, electrooptical instrumentation, thin-film deposition processes, ion beam applications.

Donald A. Neamen, Professor and Associate Chairman of Electrical Engineering; Ph.D., New Mexico. Semiconductor devices and electronics.

Marek Osinski, Professor; Ph.D., Warsaw. Semiconductor lasers, optoelectronic devices and materials, group-III nitrides, degradation mechanisms and reliability, computer simulation.

L. Howard Pollard, Assistant Professor; Ph.D., Illinois at Urbana-Champaign. Computer architecture, digital design, fault tolerance, microprocessors.

Edl Schamiloglu, Associate Professor; Ph.D., Cornell. Plasma physics, charged particle beam propagation, applied electromagnetics.

John Sobolewski, Associate Professor and Associate Vice President of Computer and Information Research and Technology; Ph.D., Washington State. Data communications, networking, computer architecture, system information and design, medical application of computers.

Joint Appointees

Edward Angel, Professor; Ph.D., USC. Image processing, computer graphics, computer vision, virtual reality, robotics, massively parallel computing.

Jean-Claude Diels, Professor; Ph.D., Brussels. Laser physics and nonlinear optics, ultrafast phenomena.

Ronald Lumia, Professor; Ph.D., Virginia. Open architecture control, sensory-interactive robot control, software for manufacturing, automation and robotics.

John McIver, Professor; Ph.D., Rochester. Laser physics and nonlinear optics, quantum optics, nonlinear science.

Timothy Ross, Professor; Ph.D., Stanford. Computational and experimental mechanics, scanning electron microscopy, hazard survivability, structural dynamics, stochastic processes, risk assessment, fuzzy logic.

Wolfgang Rudolph, Professor; Ph.D., Jena (Germany). High-resolution spectroscopy and imaging, laser physics and nonlinear optics, ultrashort light pulses, biophysics.

Professor Emeriti

Lewellyn Boatwright, Ph.D., Illinois.
Victor Bolie, Ph.D., Iowa State.
Martin Bradshaw, Ph.D., Carnegie Tech.
William Byatt, Ph.D., Alabama.
Ronald DeVries, Ph.D., Arizona.
Ahmed Erteza, Ph.D., Carnegie Tech.
Wayne Grannemann, Ph.D., Texas at Austin.
Shyam Gurbaxani, Ph.D., Rutgers.
Shlomo Karni, Ph.D., Illinois at Urbana-Champaign.
Ruben Kelly, Ph.D., Oklahoma State.
Daniel Petersen, D.Eng.Sc., Rensselaer.
Russell Seacat, Ph.D., Texas A&M.
Harold Southward, Ph.D., Texas at Austin.
Richard H. Williams, Sc.D., New Mexico.

Research Centers Associated with the Department

Center for High Technology Materials (CHTM). Creating a leading optoelectronics and laser research center is the primary goal of the CHTM, an interdisciplinary organization that sponsors and encourages research efforts in the Departments of Electrical and Computer Engineering, Physics and Astronomy, Chemistry, and Chemical and Nuclear Engineering. CHTM's multilateral mission involves both research and education. It is dedicated to encouraging and strengthening interactions and the flow of technology among the University, government laboratories, and private industry while promoting economic development in the state.

Microelectronics Research Center (MRC). The goal of the MRC is to advance special-purpose, very largescale integrated (VLSI) processors and VLSI electronics to benefit the electronics industry and the nation. Industrial needs are addressed through close interaction with major electronic companies, and national needs are addressed through involvement with national research laboratories in NASA, DOD, and DOE.

Center for Autonomous Control Engineering (ACE). The Center for Autonomous Control Engineering is an interdisciplinary and committed program for research and education in autonomous control engineering and relevant areas. An outgrowth of a strong support and financial commitment from NASA and the Jet Propulsion Laboratory, ACE is a vital resource for cost-effective education and research in control technology that is related to NASA's mission and to U.S. industrial needs.

UNIVERSITY OF NOTRE DAME

College of Engineering
Department of Electrical Engineering

Programs of Study	The department offers programs leading to the Ph.D. and the M.S. degrees in electrical engineering, with emphasis on the former. Research areas include electronic circuits and systems—communications systems, control systems, and signal and image processing and electronic materials and devices—solid-state nanoelectronic and optoelectronic materials and devices and ultrahigh-speed and microwave ICs. A research M.S. degree requires 24 hours of course credits beyond the bachelor's degree and 6 hours for a thesis. The nonresearch M.S. degree requires 30 course credits. All continuing students must pass the qualifying examination that is administered at the end of the second semester. Students who show potential for doctoral-level work may apply for admission into the Ph.D. program after their second semester. Doctoral students are required to accumulate a minimum of 36 semester hours of satisfactory course credit beyond the bachelor's degree, pass the qualifying and candidacy examinations, spend at least two years in resident study, and write and defend a Ph.D. dissertation.
Research Facilities	Several major research laboratories for the study of electronic and photonic materials and devices and for the analysis and design of circuits and systems serve the department. The Microelectronics Lab houses extensive facilities for IC and device fabrication, including 10-nm 50-kV electron-beam lithography, photomask generators, mask aligners, wafer stepper, sixteen furnace tubes, six evaporators, ion implanter, plasma etcher, PECVD, RIE, and RTA. Inspection includes JEOL SEM and Hitachi S-4500 FESEM, prism coupler, ellipsometer, surface profiler, and 4-point probe. Advanced measurements utilize 300-mK–11T cryostat, 10-mK–11T dilution refrigerator, HP 4145B SPA, DLTS, Hall effect, and Keithley I-V and C-V systems. The High Speed Circuits and Devices Laboratory features device and circuit testing to 50-GHz that uses an HP 8722 network analyzer and HP 8565 spectrum analyzer. Digital text capability to 14 Gb/s is provided by an Anritsu MP 1763 pulse pattern generator. The Device Simulation Laboratory has a cluster of high-end Sun Workstations and supercomputer access for large-scale computations and visualization. The Nanospectroscopy Lab includes a 15-W Ar$^+$ laser, femtosecond mode-locked Ti:sapphire laser, He-Cd laser, He cryostats with high spatial resolution and magnetic fields to 12T, and AFM and NSOM systems. The Optoelectronics Lab has a 10-W Ar$^+$ laser and CW Ti:sapphire laser, spectrometers, and related optical characterization instrumentation. The Laboratory for Image and Signal Analysis features a dozen high-end Sun Workstations, equipment for the processing and real-time display of HDTV sequences, cameras, frame grabbers, flat-bed scanner, and several high-definition 24-bit color monitors and specialized printers. The Controls Systems Research Laboratory consists of several Sun Workstations and additional facilities required for the design and prototyping of control systems. The Structural Dynamics and Control/Earthquake Engineering Laboratory, jointly operated with the Department of Civil Engineering and Geological Sciences, employs a 2-inch displacement, 35 in/s, +4g acceleration, 0-50 Hz slip table for 1,000-pound test loads. The communications system research laboratory includes several high-speed workstations dedicated to the simulation and analysis of novel communication systems, as well as a full complement of RF measurement equipment, signal generators and analyzers, wideband digitizers, and connections to rooftop antennas. The department has its own electronics shop run by a full-time technician. The Solid-State Laboratories are overseen by a full-time professional and a full-time technician. The College supports a cluster of ninety-nine Sun UltraSPARC and fifteen PowerMac Workstations for research and instruction. The University Computing Center supports an SGI Origin supercomputing, IBM SP-2, and several RS/6000 systems. A College research library receives 850 engineering-related journals and provides easy access to numerous databases. Its 55,000 volumes augment the University's Theodore M. Hesburgh Library collection of more than 2 million volumes.
Financial Aid	Several prestigious fellowships are available to highly qualified first-time applicants, women, and students from minority groups. Also available are about twenty teaching assistantships and several research assistantships that provide stipends of at least $1350 per month each. All appointments include full remission of academic-year tuition.
Cost of Study	Tuition for graduate students is $10,920 per semester for full-time study in 1999–2000 (waived for fellowship and assistantship recipients).
Living and Housing Costs	Two large modern apartment complexes are available on campus for single graduate students. Married student housing and apartments adjacent to the campus in South Bend are also available, renting for $250 to $800 per month. The cost of living is below the national average.
Student Group	The department has about 65 undergraduates and 78 graduate students. It awards about fifteen M.S. degrees and ten Ph.D. degrees per year.
Location	The University is the cultural center of the northern Indiana–southwestern Michigan area and offers extensive cultural, social, and sports events throughout the year. Its 2,150-acre campus is just north of South Bend, a city of about 130,000 people, and approximately 90 miles east of Chicago (a 2-hour trip by car or train). South Bend's Morris Civic Auditorium hosts performances of Broadway plays and is the home of a first-rate symphony orchestra.
The University and The College	The University was founded in 1842 by the Reverend Edward Frederick Sorin and 6 brothers of the Congregation of Holy Cross. It was chartered as a university in 1844, and engineering studies were begun in 1873. The campus's twin lakes and many wooded areas provide a setting of natural beauty for more than 102 University buildings. The engineering buildings, Cushing and Fitzpatrick Halls, were erected in 1931 and 1979, respectively.
Applying	GRE General Test scores, TOEFL scores for international students, two transcripts showing academic credits and degrees, and letters of recommendation from 3 or 4 college faculty members should be sent to the Graduate Admissions Office, University of Notre Dame, 502 Main Building, Notre Dame, Indiana 46556. The GRE should be taken no later than January preceding the academic year of enrollment, particularly if financial aid is desired. The application deadline is February 1 for fall admission and November 1 for spring. The application fee for fall admission is $25 for applications submitted by December 1 and $40 for applications submitted after this date.
Correspondence and Information	Graduate Admissions Department of Electrical Engineering University of Notre Dame Notre Dame, Indiana 46556-5637 Telephone: 219-631-5480 E-mail: eegrad@nd.edu World Wide Web: http://www.nd.edu/~ee/

University of Notre Dame

THE FACULTY AND THEIR RESEARCH

Panos J. Antsaklis, Professor; Ph.D., Brown, 1977. Systems and control theory, intelligent control, control of hybrid systems, discrete event systems, neural networks.

Peter H. Bauer, Professor; Ph.D., Miami (Florida), 1988. System theory, digital signal processing, stability theory, multidimensional systems.

Gary H. Bernstein, Professor and Associate Chair; Ph.D., Arizona State, 1987. Nanostructure fabrication, electron beam lithography, high-speed circuits.

William B. Berry, Professor; Ph.D., Purdue, 1964. Solid-state energy conversion, thermoelectrics, photovoltaics.

David L. Cohn, Professor; Ph.D., MIT, 1970. Information and coding theory, communications, speech processing, microprocessors.

Oliver O. Collins, Associate Professor; Ph.D., Caltech, 1989. Information theory, coding, communications.

Daniel J. Costello Jr., Professor; Ph.D., Notre Dame, 1969. Information theory, channel coding, and digital communications.

Patrick J. Fay, Assistant Professor; Ph.D., Illinois at Urbana-Champaign, 1996. High-speed optoelectronics, microwave-integrated circuits, devices for ultrahigh-speed digital circuits.

Thomas E. Fuja, Associate Professor and Director of Graduate Studies; Ph.D., Cornell, 1987. Digital communications, channel coding and wireless communications.

Douglas C. Hall, Assistant Professor; Ph.D., Illinois at Urbana-Champaign, 1991. Optoelectronics device characterization, fabrication, and materials studies.

Eugene W. Henry, Professor; Ph.D., Stanford, 1960. Computers, controls, simulation, computer-aided design.

Yih-Fang Huang, Professor and Chair; Ph.D., Princeton, 1982. Statistical signal processing and communications, image coding.

Gerald J. Iafrate, Professor; Ph.D., Polytechnic of Brooklyn, 1970. Solid-state physics, nano-dimensional quantum engineering.

Thomas H. Kosel, Associate Professor; Ph.D., Berkeley, 1975. Wear, erosion, electron microscopy.

Michael D. Lemmon, Associate Professor; Ph.D., Carnegie Mellon, 1990. Control systems, parameter estimation, pattern recognition, neural networks.

Craig S. Lent, Professor; Ph.D., Minnesota, 1983. Solid-state physics and devices.

Ruey-wen Liu, Freimann Professor; Ph.D., Illinois at Urbana-Champaign, 1960. Large-scale system theory, nonlinear circuits and systems, feedback control theory, stability theory, fault diagnosis.

James L. Merz, Freimann Professor, Vice President for Graduate Studies and Research, and Dean of the Graduate School; Ph.D., Harvard, 1967. Semiconductor physics, materials, and devices; optical properties of solids; defects; nanostructures.

Anthony N. Michel, Freimann Professor; Ph.D., Marquette, 1968; D.Sc., Graz (Austria), 1973. Circuit and system theory, large-scale systems.

Wolfgang Porod, Professor; Ph.D., Graz (Austria), 1981. Solid-state devices, computational electronics, nanoelectronics.

Michael K. Sain, Freimann Professor; Ph.D., Illinois at Urbana-Champaign, 1965. Multivariable control systems, engine control, applied algebraic system theory.

Ken D. Sauer, Associate Professor; Ph.D., Princeton, 1989. Tomographic imaging, multivariate detection and estimation, image compression.

Gregory L. Snider, Assistant Professor; Ph.D., California, Santa Barbara, 1991. Design and fabrication of microelectromechanical devices and mesoscopic devices.

Robert L. Stevenson, Associate Professor; Ph.D., Purdue, 1990. Statistical and multidimensional signal and image processing, computer vision.

John J. Uhran Jr., Professor and Senior Associate Dean; Ph.D., Purdue, 1967. Communication theory, digital processing, large-scale simulation, computer applications for path planning and the disabled.

RESEARCH AREAS

Electronic Circuits and Systems. Approximately half of the faculty members have research interests in this area, which includes control systems and control, signal and image processing, and communications. Projects are conducted in the following areas: turbo coding and iterative decoding, bandwidth efficient coding and modulation—design of efficient coding and modulation schemes for reliable transmission over band-limited channels; radio architecture and codes for deep space and satellite communications; multimedia communication—combined source and channel coding and restoration techniques for robust transmission of video/audio; statistical signal processing—array signal processing (radar, sonar) and adaptive interference mitigation in wireless communications; identification and estimation—blind identification, set membership estimation, adaptive equalization, and spectral analysis; digital filtering—analysis and design of multidimensional filters, floating point realizations, robust stability of discrete-time systems, and nonlinear discrete-time systems; digital image processing—data compression for image sequences, video data processing, tomographic image reconstruction, and image restoration/enhancement; control systems—investigations of stability, robust control, restructurable control, zero dynamics, modeling, and nonlinear servomechanism design; control of communication networks; autonomous control systems—theoretical developments for realization of control systems with enhanced operational capabilities; hybrid and discrete event systems; and large-scale dynamic systems—qualitative properties of large-scale dynamical systems addressing Lyapunov stability, input-output properties, and decomposition problems.

Electronic Materials and Devices. The other half of the faculty members have research interests in this area, which includes solid-state, nanoelectronic, and optoelectronic materials and devices. Current research projects include quantum device phenomena—optical properties, localization, universal conductance fluctuations, transport, interference, and resonant tunneling; nanoelectronic systems—novel circuits-and-systems architectures for the nanoelectronic regime; experimental nanoelectronics—nanofabrication of quantum dots, cryogenic characterization of single-electron effects, and ultra-small resonant tunneling diodes for ultrahigh-speed digital ICs; nanospectroscopy—high-spatial, spectral, and temporal resolution investigations of quantum dots via atomic force microscopy and near-field scanning optical microscopy; device degradation—studies of the electromigration behavior of ultrasmall metal interconnects and hot carrier effects in MOS oxide breakdown phenomena; optoelectronic materials—studies of the optical and material properties of compound semiconductor native oxides; optoelectronic devices—fabrication and characterization of waveguides and optical components for integrated photonic ICs, semiconductor lasers, and optical amplifiers; and micromachining—fabrication of microelectromechanical devices utilizing Si processing, particularly reactive ion etching.

Golden dome atop historic Main Building, Basilica of the Sacred Heart, and other campus buildings.

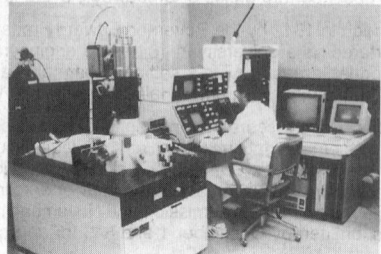

Electron-beam nanolithography system capable of writing 10-nm feature sizes.

Pergola and fountains outside of the Fitzpatrick Hall of Engineering.

UNIVERSITY OF OKLAHOMA

School of Electrical and Computer Engineering

Programs of Study	The degree programs offered by the School of Electrical and Computer Engineering (ECE) are the Master of Science in electrical engineering and the Doctor of Philosophy in electrical engineering.
	The M.S. degree in electrical engineering involves a minimum of one year of full-time study. The thesis program requires 30 semester hours of credit, including 6 semester hours devoted to the master's thesis. A nonthesis option, which requires 32 semester hours of course credit and a written comprehensive examination, is also offered. Course work for the degree may be selected from a variety of ECE and nondepartmental courses. Each student is able to design a program to meet his or her individual interests. Students may specialize in areas such as biomedical engineering, communications, computer architecture, power systems engineering, image and signal processing, solid-state devices and materials, and systems and control.
	The Ph.D. degree requires a minimum of two years of full-time study beyond the master's degree. Generally, a master's degree is required for admission into the Ph.D. program. Students must earn 90 semester hours beyond the B.S. degree to complete the Ph.D. The total number of semester hours includes credit for a doctoral dissertation and a minimum of 18 semester hours in a scientific area as a minor subject. All candidates must pass a qualifying examination and a general examination.
Research Facilities	The School maintains a number of instructional and research laboratories for graduate education. Facilities are available in digital systems, power systems, solid-state electronics, communications, signal processing, telecomputing, and artificial intelligence expert systems. The solid-state electronics laboratory has photolithography, vacuum deposition, MOCVD, MBE, and diffusion facilities for device fabrication. Materials analysis facilities for such functions as scanning electron microscopy and transmission electron microscopy are available at the University Electron Microscope and Electron Microprobe laboratories.
	The School operates and maintains modern computing facilities to support research. These include UNIX workstations, multimedia systems, personal computers, and specialized equipment for image processing. Additional up-to-date machines, including Macs, personal computers, workstations, and mainframe access, are provided by the Engineering Computing Network and University Computing Services.
Financial Aid	Financial aid is available in the form of fellowships, scholarships, graduate teaching assistantships, and research assistantships. Graduate research and teaching assistants serving 20 hours per week receive full out-of-state tuition waivers and partial in-state tuition waivers. In 1998–99, stipends for nine-month teaching assistantships ranged from $7227 to $8766 (50 percent full-time equivalent); stipends for twelve-month research assistantships ranged from $9636 to $14,400 (50 percent full-time equivalent). Assistantship awards are based on scholastic excellence, availability of funds, and satisfactory progress toward a degree. Fellowships are periodically available and typically provide stipends ranging from $12,000 to $18,000 per year, plus tuition and fees.
Cost of Study	In 1998–99, tuition for graduate students at the University was $80 per semester hour for in-state residents and $254.50 for out-of-state residents. Additional fees for full-time students are $350–$400 per semester (9 semester hours).
Living and Housing Costs	Single students can obtain housing in University dormitories, several of which are reserved for graduate students. In 1998–99, dormitory rates varied from $1798 to $2678 per semester (fifteen to nineteen meals included). Apartments are available for all graduates. The 1998–99 rates were $408–$539 per month for a two-bedroom apartment, depending on which apartment complex and whether the apartment is unfurnished or furnished. Privately owned housing is available at moderate rates near the University.
Student Group	There are about 23,000 students at the University of Oklahoma (OU), Norman Campus, including 4,000 graduate students. About 100 master's and 30 Ph.D. degrees are awarded each year in the College of Engineering. There are approximately 65 graduate students currently enrolled in electrical engineering programs.
Location	Norman has a population of about 80,000 and is located 18 miles south of Oklahoma City, which complements the excellent cultural, recreational, and sports entertainment opportunities in the area. The main attractions are the Oklahoma Art Center, the Oklahoma Historical Society Museum, and several theaters and professional sports teams. Numerous annual arts and special theme festivals in Norman and Oklahoma City are very popular. The Oklahoma Museum of Natural History, located on OU's campus, will open its new facility in 1999.
The University	The University of Oklahoma was chartered by the legislature of the territory of Oklahoma in 1890, seventeen years prior to statehood. The main campus occupies 1,000 acres of beautifully landscaped grounds. A great variety of music, art, and theater events and other recreational, cultural, and social activities are available on campus.
	The University's academic year consists of two 15-week semesters; there is also an 8-week summer term.
Applying	Domestic applications for graduate admission can be submitted at any time. However, students should allow several months for processing the application and transcripts (required from all previous colleges and universities attended). Admission into the M.S. or Ph.D. programs requires submission of GRE results, three letters of recommendation, and, for international students whose native language is not English, a minimum TOEFL score of 550. The Admissions Office application deadline for international students is April 1 for fall admission and September 1 for spring admission. Applications for financial aid for the fall semester should be received by March 15 and for the spring semester by October 1.

Correspondence and Information

For application forms:
Dean
Graduate College
University of Oklahoma
Norman, Oklahoma 73019
Telephone: 405-325-3811

For additional information
(e-mail is preferred):
Graduate Advisor
School of Electrical and Computer Engineering
University of Oklahoma
202 West Boyd, Room 219
Norman, Oklahoma 73019
Telephone: 405-325-4721
Electronic mail: eegrad@mailhost.ecn.ou.edu

University of Oklahoma

THE FACULTY AND THEIR RESEARCH

Arthur M. Breipohl, Oklahoma Gas & Electric Professor; Sc.D., New Mexico, 1964. Electric power systems analysis, probabilistic systems, decision and reliability analysis.

John Y. Cheung, Professor; Ph.D., Washington (Seattle), 1975. Digital signal processing, computer architecture, neural networks, biomedical engineering.

Jerry E. Crain, Professor and Director; Ph.D., Colorado, 1970. Antennas, rf/microwaves, electromagnetics compatibility.

Robert K. Crane, Professor; Ph.D., Worcester Polytechnic, 1970. Antennas, radar.

J. R. Cruz, Professor; Ph.D., Houston–University Park, 1980. Digital communications, signal processing.

Linda DeBrunner, Assistant Professor; Ph.D., Virginia Tech, 1991. Computer architecture, fault-tolerant computing.

Victor DeBrunner, Associate Professor; Ph.D., Virginia Tech, 1990. Digital signal processing, system identification, modeling.

M. Y. El-Ibiary, Professor; Ph.D., Imperial College (London), 1954. Microwaves, optical communications.

John E. Fagan, Professor; Ph.D., Texas at Arlington, 1977. Power systems, instrumentation, data acquisition and control, renewable energy, solar power.

Joseph Havlicek, Assistant Professor; Ph.D., Texas at Austin, 1996. Signal and image processing.

Fred N. Lee, Professor; Ph.D., Kansas, 1983. Power systems analysis, probabilistic systems, optimization theory, time-series analysis.

Samuel C. Lee, Professor; Ph.D., Illinois, 1965. Pattern recognition and computer vision, robotics and manufacturing, advanced computer systems, neural networks, expert systems applications.

Patrick J. McCann, Associate Professor; Ph.D., MIT, 1990. Solid-state electronic devices, tunable diode lasers, heteroepitaxial crystal growth.

Zhisheng Shi, Assistant Professor; Ph.D., Freiburg (Germany), 1995. Solid-state electronics, semiconductor lasers and detectors, solid-state lasers, molecular beam epitaxy and material characterization.

James J. Sluss Jr., Associate Professor; Ph.D., Virginia, 1989. Fiber-optic transmission systems, telecommunications, broad band networks.

Monte Tull, Associate Professor; Ph.D., Oklahoma, 1980. Digital design, computer architecture and alternate architectures, fuzzy logic, simulation, genetic algorithms and complexity, operations research.

UNIVERSITY OF PENNSYLVANIA

School of Engineering and Applied Science
Moore School of Electrical Engineering
Department of Electrical Engineering

Program of Study

The graduate program in electrical engineering encompasses the physical, device, telecommunications, and signal-processing aspects of electrical engineering. There are four areas of specialization in which the course work and research are coordinated: electromagnetic field phenomena, including diffraction scattering, propagation, guided waves, electromagnetic wave interaction with complex media (such as chiral, bianisotropic, fractal, and knotted media), imaging sciences, mathematical aspects of electromagnetic theory (such as fractional calculus and symmetry in electromagnetism), remote sensing of the environment, microwave and long-wavelength holographic imaging, and electrooptics; nonlinear dynamics, neurodynamics and communication theory, spectrum estimation and adaptive techniques, image processing, statistical techniques, digital signal processing, and neural networks; telecommunications, including packet and circuit switching, IP and ATM networks and protocols, network design, wireless communication, performance modeling in particular quality-of-service issues, communication architectures, and protocols and solid-state electronics, including integrated sensors, interface phenomena, mixed analog and digital integrated circuit and other devices, and electronic properties of materials and their applications. The department also offers a Professional Master of Science Program involving an internship in industry.

The minimum requirements for the M.S.E. are either 8 course units of formal course work and a 2-course-unit thesis describing the results of independent research or 10 course units of formal course work.

The Ph.D. requirements are 20 course units of studies, including the core areas of electrical engineering; a written qualifying examination; a departmental seminar; a dissertation, which includes a final oral presentation and defense of the work; and fulfillment of the equivalent of 2 course units as a teaching assistant for the department.

Research Facilities

The research program utilizes several modern laboratory facilities in the Moore School and other parts of the University. Major facilities include the Microfabrication Laboratory, the VLSI Laboratory, the Center for Telecommunications, the Electro-Optics/Microwave-Optics Holography/Photonic Neuroengineering Laboratory, the Optical Spectroscopy Laboratory, and the Valley Forge Research Center. Phased-array facilities are available at Valley Forge. There are excellent facilities for the study of synthetic metals, and the Laboratory for Research on the Structure of Matter is available for special measurements. Extensive optical measurement equipment covering the spectrum from 40 μm to less than 0.1 μm is available. Standard semiconductor processing facilities exist for support of the solid-state and electronics materials program. In collaboration with the physics department, submicron device processing and studies utilize laser holography and electron-beam GHz range that are available for antenna, radar, cross-section, imaging, microwave component, and other simulation aspects of signal-processing research, including facilities for image processing on dedicated minicomputers and microcomputers. The multimedia and computer networking research lab houses state-of-the-art networking equipment and multimedia computing facilities to support research efforts in integrated high-speed communications.

Financial Aid

Financial aid is available to qualified students in the Ph.D. program in the form of fellowships and research assistantships. In 1999–2000, the awards cover tuition and fees plus a stipend of $17,500 for twelve months.

Cost of Study

Tuition and fees for the academic year 1999–2000 for full-time study are approximately $32,200. For part-time study, the tuition and fees are approximately $3150 per course unit (one course).

Living and Housing Costs

On-campus housing is available for both single and married students and costs from approximately $420 to $850 per month, depending on the type of accommodation and the length of the lease. Detailed information is available from Housing and Conference Services at 3901 Locust. There are also numerous privately owned apartments in the immediate area.

Student Group

There are 33 students in the electrical engineering program.

Location

The University is located in West Philadelphia, just a few blocks from the heart of the city. Philadelphia is a twentieth-century city with seventeenth-century origins. Renowned museums, concert halls, theaters, and sports arenas provide cultural outlets for students. Fairmount Park extends through large sections of Philadelphia, occupying both banks of the Schuylkill River. The New Jersey shore is not far to the east, Pennsylvania Dutch country to the west, and the Poconos to the north. Equidistant from New York City and Washington, D.C., the city of Philadelphia is a patchwork of distinctive neighborhoods ranging from Colonial Society Hill to Chinatown.

The School

The School of Engineering and Applied Science has a distinguished reputation for the quality of its programs. Its alumni have achieved international distinction in research, management, industrial development, government service, and engineering education. Its faculty leads a research program that is at the forefront of modern technology and has made major contributions in a wide variety of fields. The School is in fact the birthplace of the modern computer, for it was at its Moore School of Electrical Engineering that ENIAC, the world's first electronic large-scale, general-purpose digital computer was created.

Applying

Candidates who have obtained a bachelor's degree may apply for admission by submitting an application in writing to the Office of Academic Programs, Graduate Admissions, School of Engineering and Applied Science, 111 Towne Building, University of Pennsylvania, 220 South 3rd Street, Philadelphia, Pennsylvania 19104-6391. Admission is based on the student's past record as well as on letters of recommendation. Scores on the Graduate Record Examinations are required. All international students whose native language is not English must arrange to take the Test of English as a Foreign Language (TOEFL) prior to applying; the minimum acceptable score is 600.

Correspondence and Information

Dr. Kenneth R. Laker, Graduate Group Chair
Department of Electrical Engineering
University of Pennsylvania
200 South 33rd Street
Philadelphia, Pennsylvania 19104-6390
Telephone: 215-898-5340

University of Pennsylvania

THE FACULTY AND THEIR RESEARCH

Kwabena Boahen, Ph.D., Caltech, 1997. Designing mixed analog-digital, multichip, and microelectronic systems that model the structure and function of early stages of the visual and auditory pathways. (Primary appointment in Bioengineering)

Joseph Bordogna, Ph.D., Pennsylvania, 1964. Electrooptics, optical recording materials, educational technology.

Takeshi Egami, Ph.D., Pennsylvania, 1971. Physics of solids, magnetic and superconducting oxides, x-ray and neutron diffraction, modeling and theory of local structure. (Primary appointment in Materials Science and Engineering)

Magda El Zarki, Ph.D., Columbia, 1987. Integrated services over the Internet, multimedia services and wireless networking.

Nader Engheta, Ph.D., Caltech, 1982. Applied and theoretical electromagnetics, wave interaction with unconventional complex media, polarization imaging and applications, EM materials, fractional paradigm in electromagnetic theory, optics, microwave, antennas, waveguides.

David Farber, M.S., Stevens, 1962. Telecommunications and information systems, computer communications and local area networks (LANs). (Primary appointment in Computer and Information Science)

Nabil H. Farhat, Ph.D., Pennsylvania, 1963. Microwave and acoustic holography and imaging, inverse scattering, electrooptics, optical computing.

Kenneth R. Foster, Ph.D., Indiana, 1971. Biomedical applications of nonionizing radiation from audio through microwave frequency ranges, including hyperthermia and clinical impedance techniques; dielectric properties of tissues, biological molecules, and suspensions; transport properties of complex suspensions; environmental risk assessment. (Primary appointment in Bioengineering)

William R. Graham, D.Phil., Oxford, 1965. Surfaces and interfaces: atomic structure and adsorption on metal surfaces, thin-film surface alloys, metal semiconductor interfaces. (Primary appointment in Materials Science and Engineering)

Roch Guerin, Ph.D., Caltech, 1986. High-speed networks, with emphasis on quality-of-service issues; IP and ATM protocol design and evaluation; modeling and performance analysis; scheduling algorithms; congestion control; QoS routing.

Dwight L. Jaggard, Ph.D., Caltech, 1976. Fractal electrodynamics, electromagnetic chirality, field symmetry and topology, optics and applied electromagnetic fields, imaging and inverse scattering.

Saleem A. Kassam, Ph.D., Princeton, 1975. Signal processing and communication theory: nonlinear filters, sensor array processing, adaptive schemes, spectrum estimation, detection and estimation.

Haralambos N. Kritikos, Ph.D., Pennsylvania, 1961. Applied electromagnetic fields, remote sensing, electromagnetic radiation hazards.

Kenneth R. Laker, Ph.D., NYU, 1973. Analog and digital signal processing, sampled data systems, VLSI design.

Sohrab Rabii, Chair; Ph.D., MIT, 1966. Theory of electronic properties of solids, defects in solids, alloy theory, relativistic effects in molecules and solids.

Jorge J. Santiago, Ph.D., Penn State, 1971. Materials for electronics, thin-film physics.

Camillo Jose Taylor, Ph.D., Yale, 1994. Computer vision and robotics. (Primary appointment in Computer Information Science)

Jan Van der Spiegel, Chair; Ph.D., Leuven (Belgium), 1979. Integrated sensors, signal conditioning, phoneme-based speech recognition, photosensitive devices such as CCD.

Santosh S. Venkatesh, Ph.D., Caltech, 1986. Pattern recognition, neural networks and cellular systems, distributed computing systems, digital signal processing, image processing, systems theory.

UNIVERSITY OF PITTSBURGH

Department of Electrical Engineering

Programs of Study	The department offers programs leading to the degrees of Master of Science (M.S.) and Doctor of Philosophy (Ph.D.) in electrical engineering. Graduate studies and research are concentrated in six major areas: bioengineering, computer engineering, control, electronics, image processing/computer vision, and signal processing/communications. The department has 27 faculty members.
	The M.S. degree has both research and professional tracks. The research track provides the student the opportunity to work on a thesis (applied or basic in nature) under the close supervision of a faculty adviser. The minimum requirements for the research track are 24 credits of graduate course work and preparation and defense of a thesis on a topic in the student's primary area of interest. For the professional option, the minimum requirement is 30 credits of graduate course work. The M.S. degree program can usually be completed in 1 to 1½ years on a full-time basis.
	Students who have an M.S. in electrical engineering and pass the Ph.D. Preliminary Exam are admitted to the Ph.D. program. After a student has been formally admitted, a faculty program committee is established for the purpose of advising and approving an appropriate plan of study for the student. A minimum of 72 credits, including 18 credits of research, beyond the B.S. degree is required. The Ph.D. student is expected to pass a comprehensive exam and complete a dissertation embodying an independent and original investigation of a problem of significance in his or her major area of specialization. The validity and contributions of the dissertation work are then defended in a final oral examination.
	Completion of the Ph.D. degree usually takes three years beyond the M.S. degree.
Research Facilities	The University of Pittsburgh has extensive computational facilities that include VMS and UNIX services. A very powerful computing capability is also available through CRAY and Alpha machines, which are located in one of five National Science Foundation Supercomputer Centers in the United States. Access to the University's computers is provided by a fiber-optic network with ports located at numerous points around the campus, including the Benedum Hall of Engineering. All students have access to the central computing facility.
	The Department of Electrical Engineering has numerous personal computers and workstations. The department has research and instructional laboratories in computer vision and pattern recognition, VLSI design and CAD, optical computing, lasers and nonlinear optics, microprocessor systems, neural networks, optoelectronics, and signal processing. The optoelectronics laboratory contains a metal-organic chemical vapor deposition (MOCVD) system for epitaxial growth of GaAs, AlAs, and InAs for fabrication of quantum wells, superlattices, and electronic devices.
	The University Library System maintains collections totaling 7.7 million volumes and microforms, as well as a digital library with a large number of databases and other electronic resources. The Bevier Engineering Library currently houses 63,000 volumes, 83,000 microforms, and 963 serials.
Financial Aid	Teaching assistantships are available from the department and are awarded on the basis of scholastic record, GRE General Test scores, and letters of reference. Research assistantships are typically awarded to graduate students who have been in the department for at least one term and who have distinguished themselves by superior performance in course work or project work. A limited number of fellowships are also available. In 1998–99, typical stipends for teaching and research assistantships ranged from \$1286 to \$1378 per month, plus tuition and medical benefits.
Cost of Study	For the 1998–99 academic year, tuition was \$9204 for state residents and \$18,952 for out-of-state students.
Living and Housing Costs	Students can rent furnished and unfurnished rooms and apartments near campus. The Department of Property Management (412-624-4317) provides University housing.
Student Group	The department has 140 graduate students, 43 of whom are Ph.D. students. Roughly one half of the graduate students are part-time, employed by industry in the Pittsburgh area.
Location	Pittsburgh is a cultural city and has been a major center for industrial activity and technological innovation for many decades. In recent years, the city has been undergoing a phase of corporate growth, emphasizing high technology. Pittsburgh is one of the nation's largest corporate headquarters cities. It offers many cultural and recreational resources, such as the Pittsburgh Symphony, the Pittsburgh Opera Company, the Pittsburgh Ballet Theatre, and major-league sports.
The University	The University of Pittsburgh is located in the Oakland section of the city, about 3 miles from downtown. The University began in 1787 and thus is one of the oldest in the country. Today, the University has approximately 32,000 students, of whom 9,430 are graduate students, enrolled in sixteen professional schools and the arts and sciences, including the School of Medicine, which is a world leader in organ transplantation.
Applying	An outstanding scholastic record and GRE General Test scores are required of all applicants. International students, except those who attended U.S. schools, must submit TOEFL scores. Women, African Americans, and Hispanics are encouraged to apply. If financial support is requested, completed applications must be received by February 1 for the fall term. Application materials can be obtained by writing to the Graduate Program Coordinator.
Correspondence and Information	Graduate Program Coordinator Department of Electrical Engineering 348 Benedum Hall University of Pittsburgh Pittsburgh, Pennsylvania 15261 Telephone: 412-624-8001 Fax: 412-624-8003 E-mail: eedept@ee.pitt.edu

University of Pittsburgh

THE FACULTY AND THEIR RESEARCH

Individual e-mail addresses are listed in parentheses after each faculty member's name.

J. Robert Boston, Associate Professor and Undergraduate Program Coordinator; Ph.D., Northwestern, 1971. Knowledge-based signal processing, control of artificial organs, modeling coordination of movement in people with chronic pain, representation of uncertainty using fuzzy logic and Dempster-Shafer theory. (boston@ee.pitt.edu)

David M. Brienza, Associate Professor of Health and Rehabilitation Sciences and of Electrical Engineering; Ph.D., Virginia, 1991. Control theory, soft tissue biomechanics, assistive technology, rehabilitation science. (dab3@pitt.edu)

J. Thomas Cain, Associate Professor; Ph.D., Pittsburgh, 1970. Algorithm development, digital implementation of real-time systems. (cain@ee.pitt.edu)

Shi-Kuo Chang, Professor of Computer Science, Electrical Engineering, and Information Science and Intelligent Systems; Ph.D., Berkeley, 1969. Pictorial information systems, visual languages, knowledge-based systems. (chang@cs.pitt.edu)

Luis F. Chaparro, Associate Professor and Graduate Program Coordinator; Ph.D., Berkeley, 1980. Statistical signal processing, time frequency, multidimensional system theory, image processing. (chaparro@ee.pitt.edu)

Panos K. Chrysanthis, Associate Professor of Computer Science and Electrical Engineering; Ph.D., Massachusetts, 1991. Database systems, distributed systems, operating systems, real-time systems. (panos@cs.pitt.edu)

R. Gerald Colclaser, Professor; D.Sc., Pittsburgh, 1968. Electrical transients in power systems, pulse power components and systems. (rgc@ee.pitt.edu)

Amro A. El-Jaroudi, Associate Professor; Ph.D., Northeastern, 1988. Digital processing of speech signals, spectral estimation, neural networks. (amro@ee.pitt.edu)

Mahmoud El-Nokali, Associate Professor; Ph.D., McGill, 1980. Microelectronics, semiconductor device modeling, computer-aided design, analog circuit design. (elnokali@ee.pitt.edu)

Joel Falk, Professor and Chair, Department of Electrical Engineering; Ph.D., Stanford, 1971. Linear and nonlinear optical devices, solid-state lasers, high speed electrooptic modulators, electrooptic field sensors, phase conjugation. (falk@ee.pitt.edu)

Ilan Gravé, Assistant Professor; Ph.D., Caltech, 1993. Optoelectronic integrated devices, low dimensional structures, resonant tunneling, quantum well infrared detectors, nonlinear optics, semiconductor lasers. (grave@ee.pitt.edu)

Richard W. Hall, Associate Professor; Ph.D., Northwestern, 1975. Computer vision, parallel algorithms and architectures for image processing, digital topology. (hall@ee.pitt.edu)

Raymond R. Hoare, Assistant Professor; Ph.D., Purdue, 1999. Parallel and distributed architectures, object-oriented intelligent networks, network-aware microdevices, reconfigurable computing, Internet application development. (hoare@ee.pitt.edu)

Ronald G. Hoelzeman, Associate Professor; Ph.D., Pittsburgh, 1970. Multiprocessor systems, parallel computer architectures, education innovation, computer-aided engineering. (hoelzema@ee.pitt.edu)

Steven P. Jacobs, Visiting Assistant Professor; D.Sc., Washington (St. Louis), 1997. Model-based estimation, automated systems for joint tracking and recognition, high-resolution radar. (spj1+@pitt.edu)

Hong Koo Kim, Associate Professor; Ph.D., Carnegie Mellon, 1989. Semiconductor materials and devices, optoelectronic devices, integrated optics. (kim@ee.pitt.edu)

George L. Kusic, Associate Professor; Ph.D., Carnegie Mellon, 1967. Real-time computer control of power systems. (kusic@ee.pitt.edu)

Dietrich W. Langer, Professor; Dr.Ing., Berlin Technical, 1961. Devices for optoelectronic applications. (dwl@ee.pitt.edu)

Steven P. Levitan, Wellington C. Carl Faculty Fellow and Professor; Ph.D., Massachusetts, 1984. Parallel computer architecture, optical computing, VLSI architectures, computer-aided design for VLSI. (steve@ee.pitt.edu)

Ching-Chung Li, Professor of Electrical Engineering and Computer Science; Ph.D., Northwestern, 1961. Computer vision, pattern recognition, biomedical image/signal processing, applications of wavelet transform. (ccl@ee.pitt.edu)

Patrick J. Loughlin, Fulton C. Noss Faculty Fellow and Associate Professor; Ph.D., Washington (Seattle), 1992. Nonstationary signal processing, time-frequency distributions, biomedical signal analysis, machine fault monitoring. (pat@ee.pitt.edu)

Rami G. Melhem, Professor of Computer Science and Electrical Engineering; Ph.D., Pittsburgh, 1983. Design and verification of parallel fault-tolerant and optical systems. (melhem@cs.pitt.edu)

Marlin H. Mickle, Professor of Electrical Engineering and Telecommunications; Ph.D., Pittsburgh, 1967. Microprocessor systems, parallel architectures, homogenous and heterogeneous architectures, parallel performance modeling and analysis, computer and communication networks.

Robert J. Sclabassi, Professor of Neurological Surgery, Electrical Engineering, Mechanical Engineering, and Behavioral Neuroscience; Ph.D., USC, 1971; M.D., Pittsburgh, 1981. Acquisition and analysis of electrical and magnetic data from the central nervous system. (bob@neuronet.pitt.edu)

Chris C. Shaw, Associate Professor of Radiology, Environmental and Occupational Health, and Electrical Engineering; Ph.D., Wisconsin–Madison, 1981. Physics and instrumentation of digital radiography, digital mammography, quantitative image processing and analysis, dual-energy subtraction imaging. (shaw@rad.arad.upmc.edu)

Marwan A. Simaan, Bell of PA/Bell Atlantic Professor; Ph.D., Illinois at Urbana-Champaign, 1972. Signal processing, array signal processing, geophysical applications, knowledge-based signal processing and control, statistical process control. (simaan@ee.pitt.edu)

Richard Thompson, Professor of Telecommunications and Electrical Engineering; Ph.D., Connecticut, 1971. Communication switching: system architecture, photonic switching, switching network architectures and control algorithms, intelligent networks; communication terminals: integrated services, human-computer interaction, and multimedia services. (rat@icarus.lis.pitt.edu)

UNIVERSITY OF ROCHESTER

Department of Electrical and Computer Engineering

Programs of Study

The department offers programs of study leading to the M.S. and Ph.D. degrees. Special features of graduate study in electrical and computer engineering at the University of Rochester are flexible degree programs, opportunities for interdisciplinary study, close faculty-student cooperation, major sponsored projects, excellent computing resources, and leading research facilities. Research emphases include biomedical ultrasound and medical imaging, solid-state devices, optoelectronics, superconductivity, VLSI circuits and systems, computer architecture, and signal and image processing. The M.S. degree requires 30 credit hours of graduate study and may be earned in one year of full-time study. Both thesis (Plan A) and nonthesis (Plan B) options exist. The Ph.D. degree requires 90 credit hours of graduate study, or 60 credit hours beyond the master's degree. Each student must pass a qualifying examination, submit a satisfactory written thesis proposal in his or her third year of full-time graduate study, and serve as a teaching assistant. Teaching experience involves a maximum of 15 hours of total time per week for two semesters and includes lecturing in problem sessions and laboratories.

Research Facilities

The University of Rochester is ranked in the Research I category and maintains outstanding research facilities. The faculty members of Electrical and Computer Engineering are directors or key researchers in a number of specialized research centers, including the Center for Superconducting Digital Electronics, the Center for Biomedical Ultrasound, the Center for Electronic Imaging Systems, and the Laboratory for Laser Energetics.

The Center for Superconducting Digital Electronics, sponsored by the Department of Defense and NSF, is creating digital integrated circuits for future ultrafast and quantum coherent computation. Novel electrooptic sampling techniques have been developed to measure the performance of these circuits.

The Center for Biomedical Ultrasound, funded by NIH and industry, brings together academic clinicians, scientists, and engineers to advance medical imaging and ultrasound instrumentation.

The Center for Electronic Imaging Systems, sponsored by NSF, NYS, and industry, has superb laboratories covering electronic imaging sensors, displays, image and video processing, and computer facilities for image encoding, transmission, and restoration techniques.

The Center for Future Health brings together engineers, scientists, and physicians to develop personal medical devices.

The Laboratory for Laser Energetics, a laser fusion lab, is supported by DOE and has developed some of the most advanced laser and optoelectronic devices in the world.

These outstanding, on-campus research centers provide students with access to leading experimental and computer facilities. In addition, other electrical and computer engineering department laboratory clusters are in place with extensive workstations and specialized equipment. These research laboratories include the high-performance VLSI/IC laboratory, the advanced computer architecture laboratory, the ultrasound laboratories, the solid-state device laboratories and clean room facility, and the 3-D image processing laboratory. The University library system contains more than 2 million volumes. The Carlson Library maintains complete collections in the research areas of engineering and applied sciences.

Financial Aid

Financial aid is available in the form of research or teaching assistantships and fellowships. Graduate assistants and fellows currently receive a stipend of up to $15,600 for twelve months and a tuition scholarship.

Cost of Study

Tuition for 1999–2000 is $697 per credit hour; the maximum charge is $11,152 for a 16-credit-hour semester. The annual health fee of approximately $950 includes medical insurance.

Living and Housing Costs

University housing (furnished and unfurnished) for single and married graduate students is available near the River Campus. Rents for apartments range from $350 to $600 per month for single students and from $550 to $900 per month for married students. The Housing Office maintains a listing of students seeking shared accommodations. A board plan is available, and several dining areas offer meals on a cash basis.

Student Group

The University's full-time enrollment is approximately 6,472, including 4,262 undergraduate and 2,210 graduate students. The part-time enrollment is approximately 1,355, including 242 undergraduate and 1,113 graduate students. The Department of Electrical and Computer Engineering enrolls an average of 52 full-time and 12 part-time graduate students.

Location

The city of Rochester, situated on the falls of the Genesee River about 10 miles south of Lake Ontario, is the heart of an urban-suburban community of more than 600,000. Noted for its high-tech industries, Rochester is the site of the Eastman Kodak Research Laboratories, the Xerox Webster Research Center, and many other R&D laboratories such as those of Bausch & Lomb. Many cultural and social activities are available in the community, with special emphasis on music in all forms.

The University

The University of Rochester is an independent, nonsectarian, coeducational institution of higher learning and research. Founded in 1850, it is one of the nation's most distinguished small universities. Academic and research programs are conducted by eight schools and colleges situated on three campuses. Programs ranging from the undergraduate to the postdoctoral level are offered in the humanities, the social sciences, the natural sciences, education, engineering, management, medicine, music, and nursing. The River Campus, which includes the School of Engineering and Applied Sciences, is situated on the Genesee River about 3 miles south of the city. The Medical Center is adjacent to the River Campus.

Applying

Applications are invited from students with a bachelor's degree in electrical and/or computer engineering or in a related field, such as physics, mathematics, computer science, or another engineering discipline. Full-time study is normally started at the beginning of the fall semester. Part-time students may begin study in either semester. Applications should be submitted by February 1 for fall admission with financial aid. Students should submit a completed application form, a transcript, three letters of recommendation, and a $25 fee. The GRE tests are required, and international students must submit scores on the TOEFL. To arrange a visit to see the department and meet with the faculty, students should call 716-275-4054.

Correspondence and Information

Graduate Admissions Committee
Department of Electrical and Computer Engineering
204 Hopeman Building
University of Rochester
P.O. Box 270126
Rochester, New York 14627-0126
E-mail: gradinfo@ece.rochester.edu
World Wide Web: http://www.ece.rochester.edu

University of Rochester

THE FACULTY AND THEIR RESEARCH

Professors
Alexander Albicki, Ph.D., Warsaw, 1973. Logic design, VLSI, power systems, data communications.

Mark F. Bocko, Ph.D., Rochester, 1984. Quantum electronics, superconducting electronics, quantum noise, quantum computing, electromechanical transducers.

Edwin L. Carstensen (Professor Emeritus), Ph.D., Pennsylvania, 1955. Biomedical ultrasound, bioelectric phenomena, studies of the interaction of acoustic and electric fields with biological materials.

Phillipe M. Fauchet, Ph.D., Stanford, 1984. Optoelectronics and photonic materials and devices, semiconductor physics, light-emitting porous and nanoscale silicon, biomedical sensors.

Eby G. Friedman, Ph.D., California, Irvine, 1989. VLSI circuits and systems, synchronization, clock distribution, pipelining, signal integrity, speed/power/area tradeoffs, CMOS circuits, WSI.

Thomas Y. Hsiang, Ph.D., Berkeley, 1977. Optoelectronics, ultrafast phenomena, superconductivity, electronic noise.

Thomas B. Jones, Ph.D., MIT, 1970. Electromechanics of particles, microelectromechanics, biological dielectrophoresis, industrial electrostatic hazards, nuisances.

Edwin Kinnen (Professor Emeritus), Ph.D., Purdue, 1958. VLSI systems, routing and placement, CAD tools.

Charles W. Merriam, Sc.D., MIT, 1958. Computer architecture, computer organization, programming languages.

Kevin J. Parker, Ph.D., MIT, 1981. Medical imaging, Doppler imaging techniques, digital halftoning.

A. Murat Tekalp, Ph.D., Rensselaer, 1984. Image processing, digital video processing, image restoration, image compression, pattern recognition.

Edward L. Titlebaum, Ph.D., Cornell, 1964. Multiple access communications, radar, sonar, signal design and coding psychoacoustics, echolocation, computer languages.

Robert C. Waag, Ph.D., Cornell, 1965. Biomedical ultrasound, ultrasonic signal processing, tissue characterization, nondestructive testing.

Associate Professors
Alan M. Kadin, Ph.D., Harvard, 1979. Superconducting thin films and devices, nonequilibrium effects, high-temperature superconductors, thin-film fabrication and processing, optoelectronic switching, magnetic thin films.

Jack G. Mottley, Ph.D., Washington (St. Louis), 1985. Quantitative ultrasonic tissue and materials characterization, biomedical ultrasound, anisotropy of ultrasonic parameters, contractile-state-dependence of ultrasonic parameters, ultrasonic contrast agents.

Assistant Professor
David Albonesi, Ph.D., Massachusetts, 1996. Computer architecture, microprocessor design, multiprocessor systems, computer systems performance analysis, experimental systems, high-speed reconfigurable architectures, digital systems design.

Research Professors
Diane Dalecki, Ph.D., Rochester, 1993. Biomedical ultrasound, nonlinear acoustics, lithotripsy, biological effects of ultrasound.

Marc J. Feldman, Ph.D., Berkeley, 1975. Superconducting digital electronics, ultra-low-noise electronics, radio astronomy instrumentation, quantum computation.

Roman Sobolewski, Ph.D., Warsaw, 1983. Optoelectronics, solid-state technology, nonequilibrium and ultrafast phenomena in condensed matter, superconductivity, microwaves and millimeter waves.

Leonid Tsybeskov, Ph.D., Odessa, 1986. Electronic materials and nanofabrication, nanoelectrics and optoelectronics, quantum devices.

Research Scientists
Xucai Chen, Ph.D., Yale, 1991. Acoustics, medical imaging, echocardiography, contrast agents.

A. Tanju Erdem, Ph.D., Rochester, 1990. Digital video processing, motion tracking, object-based video analysis and compositing, video data compression, image and video restoration and reconstruction.

Igor Vernik, Ph.D., Chernogolovka, 1997. Superconductivity, Josephson junctions and arrays, quantum electronic systems.

Research Associates
James Lacefield, Ph.D., Duke, 1999. Biomedical ultrasound, ultrasonic imaging, aberration correction, acoustic scattering.

Daniel Phillips, Ph.D., Rochester, 1998. Biomedical signal processing, medical ultrasound, computer simulation, electrophysiology.

Adjuncts
David Blackstock, Ph.D., Harvard, 1960. Acoustics, shock waves, nonlinear acoustics.

Chang Wen Chen, Ph.D., Illinois, 1992. Low bit rate image and video coding, video over wireless channel, ATM and cellular, HDTV and video phone, biomedical image understanding, image sequence analysis, image synthesis and visualization, motion, deformation and shape analysis.

Victor V. Derefinko, M.S., Virginia, 1967. Electronic design, including analog and digital processors and microprocessors.

Erich C. Everbach, Ph.D., Yale, 1989. Biomedical ultrasound, lithotripsy, echocardiography, transduction, nonlinear effects.

Michael Kriss, Ph.D., UCLA, 1969. Digital imaging systems, image quality, image processing, imaging.

Alex Pentland, Ph.D., MIT, 1982. Perceptional computing, wearable computing, machine vision, machine audition, interactive human-machine systems.

Lecturers
Eli Saber, Ph.D., Rochester, 1996. Digital image and video processing, pattern recognition, image quality, color characterization/printing, xerography.

Joint Appointments
Sandhya Dwarkadas, Ph.D., Rice, 1993. Parallel and distributed computing, compiler and run-time support for parallelism, computer architecture, networks, simulation methodology, performance evaluation, parallel applications research.

Nicholas George, Ph.D., Caltech, 1959. Optoelectronic systems, electronic imaging, automatic pattern recognition, speckle, holography.

Stephen F. Levinson, Ph.D., Purdue, 1981; M.D., Indiana, 1983. Ultrasonic characterization of muscle and other soft tissues, exercise physiology and biomechanics of human motion.

Ruola Ning, Ph.D., Utah, 1989. Three-dimensional tomographic angiography, volume CT, cone beam reconstruction algorithms, nondestructive testing, three-dimensional medical imaging, image-based explosive detection.

Denham S. Ward, Ph.D., UCLA, 1975; M.D., Miami (Florida), 1977. Control systems in respiration and bioengineering.

RESEARCH SPECIALTIES
Computer systems. (Albicki, Albonesi, Dwarkadas, Merriam)

Electromechanics and electrostatics. (Jones)

Microelectronics. (Friedman, Kinnen)

Optoelectronics and nanoelectronics. (Fauchet, Hsiang, Sobolewski, Tsybeskov)

Signal processing and biomedical imaging. (Carstensen, Dalecki, Mottley, Parker, Tekalp, Titlebaum, Waag)

Superconductivity and quantum electronics. (Bocko, Feldman, Kadin)

UNIVERSITY OF SOUTHERN CALIFORNIA

Department of Electrical Engineering

Programs of Study	The Department of Electrical Engineering consists of two separate areas, Systems and Electrophysics. The Systems area offers the M.S. degree in systems architecture engineering and the M.S. and Ph.D. degrees in computer engineering with concentrations in architecture, computer-aided design, computer networks, fault-tolerant computing, parallel processing, and VLSI systems. It also offers the M.S., Engineer, and Ph.D. degrees in electrical engineering in the following areas of concentration: biomedical engineering (M.S. only), communications, control theory, multimedia systems, signal and image processing, optics, and systems theory. The Electrophysics area offers the M.S., Engineer, and Ph.D. degrees in the following areas of concentration: electrical machines, electromagnetics and plasmas, integrated circuits, laser systems (M.S. and Engineer only), optics, photonics, power systems, quantum electronics, and solid-state electronics.

Almost all Systems and many Electrophysics graduate courses are broadcast over the Interactive Instructional Television Network, a one-way video, two-way audio broadcast system designed to enable part-time students working in companies in the area to take courses for credit without commuting frequently to campus.

Research Facilities With approximately $12 million in funded research annually, the University of Southern California (USC) Department of Electrical Engineering ranks as one of the most prominent university research centers in the nation. Research institutes connected with the department are the Center for Laser Studies, the Center for Photonics Technology, the Southern California Center for Advanced Transportation Technologies, the Communication Sciences Institute, the Institute for Robotics and Intelligent Systems, the Integrated Media Systems Center, and the Signal and Image Processing Institute. The department is also affiliated with the Biomedical Engineering Center and the Information Sciences Institute at Marina Del Rey.

The Systems area has an extensive and diverse computing environment. The primary component is a large number of UNIX-based systems from a variety of vendors, such as Sun, Hewlett-Packard, and DEC. Most of these are desktop workstations, supported by a collection of file servers and time-sharing systems and located in individual offices or common rooms. All of these systems are connected by Ethernet to the main campus network and to the Internet, which provides access to resources on a worldwide basis. In addition to the UNIX systems, the department has a large installed base of Apple Macintosh and IBM PC–type systems. These are used extensively for both word processing and research activities. Most are connected in some manner to the campus network. A large number of output devices, such as laser printers, color plotters, and film recorders, are available. A variety of input devices, such as drum scanners, flatbed scanners, rangefinders, and video cameras, are used for both research and word processing applications.

Financial Aid Teaching and research assistantships, fellowships, and work-study programs are available. Assistantship stipends range from $5700 to $12,800 per academic year and carry tuition remission. U.S. citizens may apply for fellowships sponsored by the American Electronics Association and the U.S. Army Research Office. Fellowships available to all entering graduate students include the USC Predoctoral Merit Fellowship. Many local industries offer full- or part-time employee fellowships for graduate study at USC. Department fellowships are offered to outstanding U.S. citizens to supplement teaching and research awards.

Cost of Study In 1998–99, tuition is $748 per semester unit and mandatory fees are $365 per semester. Graduates typically take 8–9 units per semester.

Living and Housing Costs The cost of on-campus room and board is approximately $7000 per year. Comparable housing off campus is available in the immediate vicinity.

Student Group The University enrolls 27,558 students, 46 percent of whom are women. In 1998–99, the department's enrollment was 1,600, including about 1,300 graduate students.

Student Outcomes Ph.D. graduates have taken university positions world-wide as well as positions in government and private research laboratories. M.S. and Ph.D. graduates compete successfully for jobs across the spectrum of electronic industries, including Silicon Valley computer firms, multimedia technology industries, telephonic network service providers, photonic and wireless communication system manufacturers, and Southern California aerospace companies.

Location USC is only 5 minutes by freeway from the center of Los Angeles and is secluded on an extensive, landscaped campus with a quiet academic atmosphere. Situated midway between the mountains and the sea—less than an hour from each—USC offers easy access to many outdoor diversions, such as surfing, boating, hiking, and skiing.

The University and The Department The University of Southern California, the oldest major university in the West, is private, nonsectarian, and coeducational. It is a member of the Association of American Universities, which requires members to have educational excellence in general and graduate and research excellence in particular. The variety and quality of the University's programs and resources attract faculty members and students from every corner of the globe.

The Department of Electrical Engineering has more than 60 full-time faculty members, many of whom have international reputations in their area of specialty. The faculty has been honored with numerous awards and distinctions: 8 are members of the National Academy of Engineering, and 28 have been elected Fellows of the IEEE.

Applying To be admitted to full graduate standing, the student must have a bachelor's degree and acceptable scores on the General Test of the Graduate Record Examinations. Application forms for financial aid awards may be obtained from the department upon request and should be returned by January 1.

Correspondence and Information

Dr. Martin Gundersen, Co-Chairman
Department of Electrical Engineering—Electrophysics
University of Southern California
Los Angeles, California 90089-0271
Telephone: 213-740-4700

Dr. Robert A. Scholtz, Co-Chairman
Department of Electrical Engineering—Systems
University of Southern California
Los Angeles, California 90089-2560
Telephone: 213-740-1788
Fax: 213-740-4449
E-mail: eesystem@ceng.usc.edu
World Wide Web: http://www.usc.edu/dept/ee/

University of Southern California

THE FACULTY AND THEIR RESEARCH

Electrophysics

Joe E. Baker, Instructor; Engineer, USC, 1970. Power electronics.
Milton Birnbaum, Research Professor; Ph.D., Maryland, 1953. Quantum electronics.
Tsen-Chung Cheng, Lloyd Hunt Professor; Sc.D., MIT, 1974. Power systems.
John Choma Jr., Professor; Ph.D., Pittsburgh, 1969. Integrated circuits.
Clarence Crowell, Professor; Ph.D., McGill, 1955. Solid state.
P. Daniel Dapkus, W. M. Keck Professor; Ph.D., Illinois at Urbana-Champaign, 1970. Quantum electronics.
Jack Feinberg, Professor; Ph.D., Berkeley, 1977. Nonlinear optics, lasers.
Srinivasiengar Govind, Adjunct Associate Professor; Ph.D., Mississippi, 1978. Electromagnetics.
Martin Gundersen, Professor and Co-Chairman; Ph.D., USC, 1972. Quantum electronics.
Robert Hellwarth, George Pfleger Professor; D.Phil., Oxford, 1955. Quantum electronics, lasers.
Kirby Holte, Adjunct Professor; Ph.D., Washington State, 1971. Electromagnetic fields.
Thomas C. Katsouleas, Professor; Ph.D., UCLA, 1984. Plasma physics, advanced acceleration, light sources.
Hans H. Kuehl, Professor; Ph.D., Caltech, 1959. Plasmas and electromagnetics.
Anthony F. J. Levi, Professor; Ph.D., Cambridge, 1983. Quantum electronics.
Virendra Mahajan, Adjunct Professor; Ph.D., Arizona, 1974. Optical sciences.
Lute Maleki, Adjunct Assistant Professor; Ph.D., LSU, 1975. Lasers.
Alan McCurdy, Associate Professor; Ph.D., Yale, 1987. Applied physics.
Ram C. Mukherji, Adjunct Associate Professor; M.S.E.E., USC, 1970. Economic operation of power systems.
Richard N. Nottenburg, Associate Professor; Ph.D., Swiss Federal Institute of Technology, 1984. Quantum electronics, photonics.
John O'Brian, Assistant Professor; Ph.D., Caltech, 1996. Photonics.
Aluizio Prata Jr., Associate Professor; Ph.D., USC, 1990. Applied electromagnetics.
Hanna Reisler, Research Assistant Professor; Ph.D., Weizmann (Israel), 1972. Lasers, quantum electronics.
Steven B. Sample, Professor and President; Ph.D., Illinois, 1965. Electromagnetics and antennas.
Bing J. Sheu, Professor; Ph.D., Berkeley, 1985. VLSI, signal and image processing.
Keith Soohoo, Adjunct Associate Professor; Ph.D., USC, 1964. Electromagnetics.
William H. Steier, Professor; Ph.D., Illinois, 1960. Lasers, optical devices.
Armand R. Tanguay, Associate Professor; Ph.D., Yale, 1976. Solid state, electronic devices.
William G. Wagner, Professor; Ph.D., Caltech, 1962. Quantum electronics.
Curt F. Wittig, Professor; Ph.D., Illinois at Urbana-Champaign, 1970. Quantum electronics.

Systems

Michael A. Arbib, Professor; Ph.D., MIT, 1963. Computer science.
Peter Beerel, Assistant Professor; Ph.D., Stanford, 1994. Computer engineering.
George A. Bekey, Professor; Ph.D., UCLA, 1962. Control systems, signal processing.
Melvin A. Breuer, Professor; Ph.D., Berkeley, 1965. Computer engineering.
Keith M. Chugg, Assistant Professor; Ph.D., USC, 1995. Digital communications.
Alvin M. Despain, Professor; Ph.D., Utah, 1966. Computer engineering.
Michel Dubois, Professor; Ph.D., Purdue, 1982. Computer engineering.
Robert Gagliardi, Professor; Ph.D., Yale, 1960. Communications.
Jean-Luc Gaudiot, Professor; Ph.D., UCLA, 1982. Computer engineering.
Seymour Ginsburg, Emeritus Professor; Ph.D., Michigan, 1952. Computer science.
Solomon W. Golomb, Professor; Ph.D., Harvard, 1957. Communications.
Sandeep K. Gupta, Associate Professor; Ph.D., Massachusetts at Amherst, 1991. Computer engineering.
Ellis Horowitz, Professor and Chairman of the Department of Computer Science; Ph.D., Wisconsin, 1970. Computer science.
Kai Hwang, Professor; Ph.D., Berkeley, 1972. Computer engineering.
Petros Ioannou, Professor; Ph.D., Illinois at Urbana-Champaign, 1982. Adaptive and feedback control systems.
Keith Jenkins, Associate Professor; Ph.D., USC, 1984. Signal and image processing.
Edmond Jonckheere, Professor; Ph.D., USC, 1978. Control theory.
Bart Kosko, Associate Professor; Ph.D., California, Irvine, 1987. Signal processing.
P. Vijay Kumar, Professor; Ph.D., USC, 1983. Communications.
Chung-Chieh (Jay) Kuo, Professor; Ph.D., MIT, 1987. Signal processing.
Chris Kyriakakis, Assistant Professor; Ph.D., USC, 1993. Immersive audio.
Richard Leahy, Professor; Ph.D., Newcastle, 1984. Signal and image processing.
Daniel C. Lee, Assistant Professor; Ph.D., MIT, 1992. Electrical engineering and computer science.
Victor O. K. Li, Professor; Ph.D., MIT, 1981. Communications.
William C. Lindsey, Professor; Ph.D., Purdue, 1962. Communications.
Toyone Mayeda, Adjunct Assistant Professor; M.S.E.E., USC, 1979. Computer engineering.
Gerard G. Medioni, Associate Professor; Ph.D., USC, 1983. Computer science.
Jerry M. Mendel, Professor; Ph.D., Polytechnic of Brooklyn, 1963. Signal processing.
Ramakant Nevatia, Professor; Ph.D., Stanford, 1975. Computer science.
Chrysostomos L. Nikias, Professor and Associate Dean; Ph.D., SUNY at Buffalo, 1982. Signal and image processing.
Antonio Ortega, Assistant Professor; Ph.D., Columbia, 1994. Signal and image processing.
George Papavassilopoulos, Professor; Ph.D., Illinois at Urbana-Champaign, 1979. Control systems.
Alice C. Parker, Professor; Ph.D., North Carolina State, 1975. Computer engineering.
Massoud Pedram, Associate Professor; Ph.D., Berkeley, 1991. Computer engineering.
Timothy Pinkston, Assistant Professor; Ph.D., Stanford, 1993. Computer engineering.
Andreas Polydoros, Professor; Ph.D., USC, 1982. Communication theory.
Viktor Prasanna, Professor; Ph.D., Penn State, 1983. Computer engineering.
Keith E. Price, Research Associate Professor; Ph.D., Carnegie Mellon, 1976. Computer science.
Gandhi Puvvada, Adjunct Assistant Professor; M.S.E.E., Houston, 1987. Computer engineering.
Eberhardt Rechtin, Emeritus Professor; Ph.D., Caltech, 1950. System architecting.
Irving S. Reed, Emeritus Professor; Ph.D., Caltech, 1949. Communications.
Aristides Requicha, Professor; Ph.D., Rochester, 1970. Computer science.
Michael G. Safonov, Professor; Ph.D., MIT, 1977. Control systems.
Alexander A. Sawchuk, Professor; Ph.D., Stanford, 1972. Signal processing.
Robert A. Scholtz, Professor and Co-Chairman; Ph.D., Stanford, 1964. Communications.
Leonard M. Silverman, Professor and Dean of the School of Engineering; Ph.D., Columbia, 1966. Control systems.
John A. Silvester, Professor and Vice-Provost for Scholarly Technology; Ph.D., UCLA, 1979. Computer engineering.
Monte Ung, Adjunct Professor; Ph.D., USC, 1970. Computer engineering.
Charles L. Weber, Professor; Ph.D., UCLA, 1964. Communications.
Lloyd R. Welch, Professor; Ph.D., Caltech, 1958. Communications.
Alan Willner, Associate Professor; Ph.D., Columbia, 1988. Communications.
Zhen Zhang, Professor; Ph.D., Cornell, 1984. Communications.

UNIVERSITY OF SOUTH FLORIDA

College of Engineering
Department of Electrical Engineering

Programs of Study

Ph.D. and master's (M.E., M.S.E., and M.S.E.E.) degrees in electrical engineering are granted by the department. Master's program options include circuit theory, control theory, communications and signal processing, electric power, microelectronics, and wireless systems. Digital architectures and design and Ph.D. studies emphasize microelectronic design and test (VLSI, VHSIC, MMIC, and ASIC design; microwave and high-frequency analog and digital circuit modeling and testability; interconnection systems; and reliability and failure mode studies); communications and signal processing (digital communications, networks, packet switching, digital video and HDTV, ISDN, optical-fiber, and comm-software); systems and controls; solid-state material and device processing and characterization; electrooptics; electromagnetics, microwave, and millimeter-wave engineering (antennas, devices, and systems); CAD and microprocessors; and biomedical engineering. Special interdisciplinary study is available through the engineering science program.

Master's degree programs require a year's full-time study beyond the bachelor's degree. Master's degrees may be pursued with or without a thesis (30 semester hours minimum with a thesis, 33 semester hours minimum without a thesis). The thesis accounts for 6 semester hours. The doctoral degree program normally requires three years of full-time study and research beyond the bachelor's degree or two years beyond the master's degree. The program culminates in the submission of a dissertation that demonstrates the student's capacity for considerable original thought, talent for significant research and/or design, and ability to organize and present his or her findings.

Students must achieve and maintain a minimum grade point average of 3.0 (on a 4.0 scale) in all courses taken for graduate credit.

Research Facilities

The Department of Electrical Engineering is housed in the College of Engineering building, which provides excellent graduate research facilities and computing capabilities. Students have access to Ardents, Suns, and PC laboratories as well as to the Ethernet and ISN networks. Laboratories include compound semiconductor materials processing and microfabrication facilities with MOCVD reactors and CVD and plasma processing equipment; thin-film and hybrid circuit facilities; extensive semiconductor characterization equipment (DC through optical, including a semiconductor parameter tester, a network analyzer, a scanning electron microscope, a deep-level transient spectroscope, and electrochemical profilers and beam characterization tools); a noise research laboratory with cryogenics for superconductor studies; a communications and signal processing laboratory focusing on integrated-services (data/voice/video) telecommunications, information transport networks, VLSI and microprocessor-based algorithm implementations, speech and image processing, and digital communications; a coherent fiber optics laboratory with extensive instrumentation; a Restructurable VLSI lab; and CAD and modeling laboratories for analog, digital, and microwave monolithic circuits and devices, as well as antennas and scattering characteristics. The VLSI CAD facility includes extensive software and an automated system.

Financial Aid

Teaching and research assistantships and fellowships are available to qualified students. In 1998–99, assistantships and fellowships carried estimated stipends of $9000 to $12,000 for twelve months for master's students and $10,000 to $16,000 for twelve months for Ph.D. students. (A thesis may be written on the subject covered by the research assistantship.) Duties normally require 20 hours per week. A major portion of the fees may be waived for employed students.

Cost of Study

For 1998–99, the in-state tuition fee was $142 per semester hour. Out-of-state students paid $485 per semester hour. Tuition and fee waivers may be available for graduate assistants.

Living and Housing Costs

The cost of living in the Tampa Bay area compares favorably with most other parts of the United States. There are limited facilities on campus for unmarried students. Off-campus housing is available immediately adjacent to the University. Meal plans are available in the University's dining halls. Excluding tuition, books, and transportation, minimum living costs for single students are about $1600 per semester.

Student Group

There are more than 37,000 students at the University of South Florida. About 5,000 are enrolled in graduate programs. Students come from virtually all states of the Union and many other countries. The Department of Electrical Engineering has an enrollment of about 500 undergraduate and 200 graduate students.

Student Outcomes

Students completing master's degrees have opted for employment in the growing microelectronics, wireless systems, and communications segments of industry. Recruitment by companies such as Texas Instruments, AT&T, Motorola, Honeywell, E-Systems, Intel, Harris, and Raytheon have resulted in challenging technical opportunities. Students completing Ph.D. degrees have also been in great demand, primarily by industrial organizations.

Location

The Tampa Bay area is rich in recreational, cultural, and athletic activities. The University is located about 10 miles from downtown Tampa and 35 miles from Clearwater and St. Petersburg. Symphonies, operas, theaters, chamber music orchestras, and sports events are regularly available.

The University and The College

The University opened its doors in 1960 and now has more than 36,000 students on its four campuses. The Tampa campus is the largest, with approximately 30,000 students. It comprises the Colleges of Engineering, Arts and Sciences, Business Administration, Education, Fine Arts, Medicine, Nursing, and Public Health. The campus is modern, airy, and attractive. The College of Engineering maintains close contact with local industry. Many engineering students find part-time or full-time employment with these companies.

Applying

Applicants for a master's program in engineering must have a bachelor's degree in an accredited engineering or related program. They must present a GPA of 3.0 or higher (for the last two years of undergraduate work) or a combined (verbal and quantitative) GRE General Test score of 1000 or better. All applicants must submit recent GRE General Test scores when applying. Applicants for the doctoral program should have an academic record that exceeds the foregoing minimum requirements. Applications for admission should be received by the Office of Graduate Admissions at least two months prior to expected enrollment. International applications need to be submitted at least four months prior to expected enrollment. For more information, students should see the World Wide Web site (http://ee.eng.usf.edu). Applicants can download the USF Graduate School Application Form (http://www.grad.usf.edu) and can communicate directly via e-mail.

Correspondence and Information

Dr. Kenneth A. Buckle, Graduate Program Coordinator
Department of Electrical Engineering
University of South Florida
Tampa, Florida 33620-5350

Telephone: 813-974-2369
Fax: 813-974-5250
E-mail: eegrad@eng.usf.edu

University of South Florida

THE FACULTY AND THEIR RESEARCH

Kenneth Buckle, Associate Professor; Ph.D., Wisconsin, 1984. Plasma processing, electromagnetics, microwave engineering in the deposition of thin films and their etching.

Yun-Leei Chiou, Professor; Ph.D., Purdue, 1969. Solid-state devices, hybrids, thin oxide films, breakdown mechanism, GaAs photonic devices, thick film, VLSI devices, silicon oxidation.

Lawrence P. Dunleavy, Associate Professor; Ph.D., Michigan, 1988. Microwave and millimeter wave circuits, MMIC implementation, microwave CAD, measurements and modeling of passive and active microwave components, quasi-optic millimeter-wave grid array analysis.

Christos Ferekides, Assistant Professor; Ph.D., South Florida, 1991. Thin-film electronic materials and devices for optoelectronic applications, thin-film depositions and properties, device fabrication and characterization.

Samuel Garrett, Professor; Sc.D., Pittsburgh, 1963. Control theory, power electronics, computer-aided design, circuit theory.

Horace Gordon, Lecturer; M.S.E., South Florida, 1970. Low-frequency and microwave circuit theory and design, communication theory and system design, digital signal processing theory and hardware design.

Worth Henley, Assistant Professor; Ph.D., South Florida, 1993. Electronic materials processing and characterization, advanced VLSI MOSFET device design and fabrication.

Rudolf E. Henning, Professor; D.Eng.Sc., Columbia, 1954; PE. Microwave and millimeter-wave microelectronic design; application of microwave theory and electromagnetics to instrumentation, sensing, and detection; properties and applications of microwave and millimeter-wave propagation; microwave/optical interactions and subsystems.

Andrew M. Hoff, Associate Professor; Ph.D., Penn State, 1988. Introduction and technology transfer of surface-potential mapping as a monitor of charging in plasma etch, ion implantation, plasma deposition, chemical processing for dielectric and wafer contamination.

Vijay K. Jain, Professor; Ph.D., Michigan State, 1964. Communication systems and networks (data/voice/video/multimedia); digital signal, image, and speech processing; VLSI and WSI implementations; parallel architectures; computer arithmetic and algorithms; pattern recognition by computers.

Lubek Jastrzebski, Professor; Ph.D., Polish Academy of Sciences, 1974. Silicon technology, material synthesis and characterization, material and devices modeling.

Firman Dean King, Assistant Professor; Ph.D., Florida, 1980. Control, estimation, and systems theory; random processes; robotics.

Michael Kovac, Dean and Professor; Ph.D., Northwestern, 1970. Solid-state devices, reliability, VLSI, optical image sensors, novel transducers.

Gerard Lachs, Professor; Ph.D., Syracuse, 1964. Fiber-optic communication theory, digital communication, quantum electronics, laser applications, signal detection and processing in human sensory perception.

P. K. Lala, Professor; Ph.D., City University (London), 1976. VLSI design and test, digital systems and architectures, fault-tolerant computing.

James Leffew, Assistant Professor; Ph.D., South Florida, 1985. Electromagnetics, microwave engineering, microwave CAD, network synthesis.

Don L. Morel, Professor; Ph.D., Tulane, 1971. Photovoltaic solar energy conversion, compound semiconductor and Group IV amorphous thin-film electronic materials and devices, large-area display and memory technology.

Wilfrido A. Moreno, Assistant Professor; Ph.D., South Florida, 1993. Laser restructuring for quick prototyping of electronic circuits, laser machining, experimental design, data acquisition and analysis, process control, circuit theory, computer-aided design.

Ravi Sankar, Professor; Ph.D., Penn State, 1985. Communication/computer networking (high-speed local and wide area networks), speech processing and recognition, signal processing and its applications, telecommunications, applications of neural networks, simulation and modeling.

Arthur David Snider, Professor; Ph.D., NYU, 1971. Spectral analysis, optimization, math modeling in electromagnetics, communications, electronics, heat transfer.

Elias (Lee) Stefanakos, Professor and Chairman; Ph.D., Washington State, 1969; PE. Electric vehicles, clean energy, photovoltaics.

Thomas Wade, Professor; Ph.D., Florida, 1974. Solid-state microelectronics, VLSI multilevel interconnection systems, test structure development, solid-state material characterization (e.g., polyimides and refractory silicides).

Thomas M. Weller, Assistant Professor; Ph.D., Michigan, 1995. Numerical electromagnetics, applications of micromachining to microwave/millimeter-wave circuit/system design.

Paris Wiley, Associate Professor; Ph.D., Virginia Tech, 1973. Analog and digital electronics, biomedical instrumentation, theoretical electromagnetics, microwave communications.

UNIVERSITY OF TEXAS AT ARLINGTON

Department of Electrical Engineering

Programs of Study

The Department of Electrical Engineering offers graduate programs leading to the M.S. and Ph.D. degrees. Course work and research are offered in the following areas: applied physical electronics; control and robotics; digital signal and image processing; digital and microcomputer systems; electromagnetic fields and remote sensing; energy systems; manufacturing systems; microelectronics and semiconductors; microwave, millimeter-wave, and optoelectronic devices; optics and electrooptics; telecommunications; and VLSI design and implementation.

The master's degree is normally completed in two years and requires 24 hours of course work and 6 hours of thesis work. Nonthesis (37 hours) and thesis substitute (33 hours) options are available.

The Ph.D. program requires additional course work, the passing of a diagnostic exam and a qualifying exam, and successful defense of the Ph.D. dissertation. The average time for completion is about four years beyond the Master of Science degree in electrical engineering.

Research Facilities

Equipment for graduate student research in electrical engineering is spread over 67,000 square feet of space in four buildings, plus another 48,000 square feet in the Automation and Robotics Research Institute. Specific equipment is listed on the reverse side of this page. Major research centers and faculty groups include the following: Center for Electronic, Materials, Devices, and Systems; Wave Scattering Research Center; Energy Systems Research Center; Human Performance Institute; Electro-Optics Research Center; Telecommunications and Signal and Image Processing Laboratories; Image Processing and Neural Networks Laboratory; and the Automation and Robotics Research Institute. In addition, several graduate projects involve work at local industries and at the University of Texas Southwestern Medical School at Dallas. Computing facilities are accessible over campuswide Ethernet and dial-up modems and include Sun, HP, and SGI workstations, high-end PC systems, and direct access to high-performance parallel mainframes that include SGI Origin 2000 and DEC. Department faculty members participate in the Texas Telecommunications Engineering Consortium and the Metroplex Research Consortium for Electronic Devices and Materials.

Financial Aid

Aid is available in the form of fellowships, research assistantships, and teaching assistantships. Stipends vary from $6500 to $24,000 for the calendar year, with recipients paying Texas resident tuition. Approximately ninety stipends are awarded each year.

Cost of Study

In 1999–2000, tuition and fees for 12 credit hours were $1631 for Texas residents and $4307 for nonresident students each semester. Half-time teaching or research assistants and holders of competitive fellowships are entitled to Texas resident rates.

Living and Housing Costs

University apartments are available at rates ranging from $270 to $532 per month. Dormitory rooms range from $702 to $812 per semester. University food service plans are available. Numerous private apartments are available at a wide range of prices. Rates are subject to change.

Student Group

There are approximately 180 M.S. and 75 Ph.D. students in electrical engineering and a total of 18,000 students at UT Arlington. Approximately half of the electrical engineering graduate students attend full-time, while many others are employed part-time in local industry.

Student Outcomes

Students completing advanced degrees are actively recruited by the area's sizeable electronic, telecommunications, and aerospace industries. Recent graduates have been employed by Advanced Micro Devices, Motorola, Alcatel, AT&T, Fujitsu, MCI, Nokia, Nortel, Lucent Technologies, SBC, Texas Instruments, Ericsson, Hughes, Raytheon E-Systems, National Semiconductor, and Lockheed Martin.

Location

The city of Arlington, 15 miles south of the Dallas–Fort Worth Airport, is located in the heart of the Dallas–Fort Worth metroplex, the nation's third-largest high-technology region. Dallas, 20 miles to the east, is a cosmopolitan city of tall buildings, distribution centers, fashion and technology-based industries, and the arts. Fort Worth, once a city of the Old West, 10 miles to the west of Arlington, now has art museums and many cultural activities as well as a variety of industries. Arlington is the home of the Texas Rangers and Six Flags over Texas. The area offers, in addition to campus activities, a full range of cultural and recreational opportunities, including museums, concerts, ballet, opera, theater, amusement parks, and professional sports.

The University and The Department

The University of Texas at Arlington is located on a modern 300-acre campus a few blocks from downtown Arlington. The University originated in 1895 as Arlington College, a private liberal arts institution; its founders located it "far from the temptations of city life." The school changed with the times and its surroundings. In 1967, the institution became the University of Texas at Arlington. The high-technology industries in telecommunications, electronics, and aerospace are helping to shape the College of Engineering into a premier high-tech research and education center. The Department of Electrical Engineering has 23 faculty members.

Applying

Students holding a B.S. degree in electrical engineering with a minimum upper-level grade point average of 3.0 (on a scale of 4.0) and a minimum combined GRE score of 1050 on the verbal and quantitative sections of the General Test are invited to apply for admission. A TOEFL score of 550 is required for students whose primary language is not English.

Applications for admission should be submitted by April 1 for the fall semester and November 1 for the spring semester. Admission materials can be obtained from the Graduate School office. Applicants for financial aid should submit a financial aid application directly to the department by March 1 to be considered for fall awards; applicants should also include copies of transcripts and GRE scores.

Correspondence and Information

For catalogs and application forms:
Graduate School
University of Texas at Arlington
Arlington, Texas 76019
Telephone: 817-272-2688
E-mail: graduate.school@uta.edu
World Wide Web: http://www.uta.edu

For department information:
Graduate Advisor
Department of Electrical Engineering
University of Texas at Arlington
Arlington, Texas 76019-0016
Telephone: 817-272-2671
Fax: 817-272-2253
E-mail: grad.adv@ee.uta.edu
World Wide Web: http://www-ee.uta.edu

University of Texas at Arlington

THE FACULTY AND THEIR RESEARCH

Kambiz Alavi, Professor; Ph.D., MIT, 1981. Molecular-beam epitaxy, heterojunction, quantum well, superlattice devices.

Jonathan W. Bredow, Associate Professor; Ph.D., Kansas, 1989. Radar systems design and remote sensing.

Ronald L. Carter, Professor; Ph.D., Michigan State, 1971. Semiconductor device physics, modeling and simulation, semiconductor electronics manufacturing.

Mo-Shing Chen, Professor; Ph.D., Texas, 1962. Power systems transmission, distribution, and generation.

Michal P. Chwialkowski, Associate Professor; Ph.D., Warsaw Technical, 1982. Digital instrumentation, magnetic resonance imaging.

W. Alan Davis, Associate Professor; Ph.D., Michigan, 1971. Microwave devices and circuits.

Venkat Devarajan, Professor; Ph.D., Texas at Arlington, 1980. Image processing applications, including flight simulation, digital photogrammetry, virtual reality, and virtual prototyping.

William E. Dillon, Associate Professor; Ph.D., Texas at Arlington, 1972. Energy conversion, space power.

Jack Fitzer, Professor and Associate Chairman; D.Sc., Washington (St. Louis), 1962. Controls and robotics, energy conversion.

Adrian K. Fung, Professor; Ph.D., Kansas, 1965. Wave scattering, radar image simulation and interpretation, remote sensing of air pollution.

George V. Kondraske, Professor; Ph.D., Texas at Arlington and Texas Health Science Center at Dallas, 1982. Microprocessor-based instrumentation, software systems, human sensory and motor functions.

Wei-Jen Lee, Professor; Ph.D., Texas at Arlington, 1985. Power system stability, power system protection, load flow analysis.

Frank L. Lewis, Professor; Ph.D., Georgia Tech, 1981. Systems and controls, robotics, adaptive systems, nonlinear systems, neural networks, manufacturing.

Robert Magnusson, Professor and Chairman; Ph.D., Georgia Tech, 1976. Diffractive optics, lasers, holography, electrooptics, integrated optics.

Theresa A. Maldonado, Associate Professor; Ph.D., Georgia Tech, 1990. Electrooptics, integrated optics, nonlinear optics.

Michael T. Manry, Professor; Ph.D., Texas at Austin, 1976. Digital signal and image processing, neural networks, pattern recognition.

John H. McElroy, Professor; Ph.D., Catholic University, 1978. Communication satellites, earth observations.

Vasant K. Prabhu, Professor; Sc.D., MIT, 1966. Telecommunications, digital portable and cellular radio networks.

K. R. Rao, Professor; Ph.D., New Mexico, 1966. Digital signal and image processing, discrete transforms.

Raymond R. Shoults, Professor; Ph.D., Texas at Arlington, 1974. Electric power systems, computer methods.

Charles V. Smith Jr., Professor; Ph.D., MIT, 1968. Continuum electromagnetics, energy conversion.

Saibun Tjuatja, Associate Professor; Ph.D., Texas at Arlington, 1992. Remote sensing of the environment, wave scattering and propagation, wireless communications, radar image processing and interpretation.

Kai-Shing Yeung, Professor; Dr.Ing., Karlsruhe (Germany), 1977. Nonlinear control, systems theory, robotics.

RESEARCH LABORATORIES AND FACILITIES

Controls, Robotics, and Manufacturing. UTA's Automation and Robotics Research Institute (ARRI) is a 48,000-square-foot off-campus extension housing robotics labs, intelligent material handling stations, and controls labs with industrial motion system test beds and real-time controllers capable of implementing advanced algorithms, including feedback linearization, adaptive control, neural net and fuzzy logic control. Supervisory controllers are capable of implementing discrete event rule-based control algorithms for dispatching shared resources and part routing with deadlock avoidance.

Electro-Optics Research Center. The EORC laboratories contain a large variety of lasers, including argon-ion, pulsed, and cw Nd:YAG, HeCd, HeNe, Cu-vapor, CO_2, tunable dye, Ti:sapphire, semiconductor, and high-power semiconductor array lasers. Numerous vibration-isolated tables with optical components, mounts, and computer-controlled test equipment are available. Experimental processes, including dielectric and metal thin-film deposition, holographic grating recording, reactive-ion etching, metallization, scanning-electron microscope inspection, and ellipsometric characterization, can be implemented.

High-Frequency Circuits and Devices. Facilities include a class 10/100 GaAs clean room with Karl Suss UV aligner, rapid thermal processor, reactive ion etch, scanning electron microscope, e-beam evaporator and sputter deposition; Varian GEN II modular molecular-beam epitaxy with e-beam evaporator and Auger analysis; automated microwave network analyzers; time domain reflectometry; cascade wafer probes; 250X step and repeat photomask facilities; DC measurement and analysis facilities; and CAD design and simulation facilities: HP, Super Compact, Touchstone, Harmonica, etc.

Image Processing and Neural Networks. Facilities include several PC workstations with connections to the Internet, a flat-bed image scanner, a TV camera with frame grabber, and extensive libraries of image processing and neural network software, which have been developed in the lab. The lab has access to a model 6400 nCUBE 2 parallel processing computer, a Meiko computing surface, and Sun Workstations.

Medical Imaging. Facilities are available through University of Texas Southwestern Medical Center in Dallas and include state-of-the-art magnetic resonance imaging instruments from GE, Philips, and Picker, along with high-performance image processing workstations and a high-speed image delivery network.

Power Systems. Facilities include a unique physical scale model power system simulation laboratory complete with an Energy Management System (EMS), protective relaying, 1,050 miles of AC transmission lines, 700 miles of DC transmission lines, power plant models (M-G sets), actual load models, and digital computer-based load control; a state-of-the-art digital computer-based Black-Start Training Simulator; and numerous sophisticated computer analysis programs, including power flow analysis, voltage stability analysis, transient stability analysis, and dynamic stability analysis.

Remote Sensing and Wave Scattering. There are a bistatic RCS measurement facility, 2–20 GHz, fully polarimetric imaging capability; a 75–110 GHz spectrometer; a bistatic laser scattering (663 nm) measurement system; an open-path Fourier transform infrared (FT-IR) spectrometer and a long-path ultraviolet spectrometer for environmental monitoring.

Robotics and Controls. There are digital control, vision, laser ranging, and force sensing systems; Puma and Adept robots; modular assembly components; abrasive water jet cutting; automated machining and reverse engineering; and CAD and simulation software: CADAM, CATIA, SILMA, custom human performance instrumentation, and DAP.

Semiconductor Device Modeling, Characterization, and Simulation. Facilities include HP device characterization laboratory instrumentation for dc to 26.5 Ghz, a circuit and device simulation laboratory with an HP 700 series workstation and an HP Advanced Design System, IC-CAP, and Silvaco simulation suites.

Signal Processing and Telecommunications. Facilities include extensive communication system simulation software on UNIX workstations, a TAAC-1 image and array processing module, high-resolution color displays, 560 Mb/s long-haul fiber-optic terminals, and an operational 150 Mb/s fiber-optic system including two duplex video channels with additional T1 Service over a 10-mile path.

Virtual Environment. An International Imaging Systems' M-75/S-600 suite and a Sun Workstation running a current version of Khoros and KBVISION provide the visual database tools. Two Hughes Micropoly-II image generation systems and Rend386 running on a 486DX-66 allow real-time rendering. A PowerGlove and stereo viewer are also available.

THE UNIVERSITY OF TEXAS AT AUSTIN

Department of Electrical and Computer Engineering

Programs of Study

The Department of Electrical and Computer Engineering (E.C.E.) offers programs leading to the degrees of Master of Science in Engineering and Doctor of Philosophy. Advanced studies may be pursued in the following technical areas: biomedical engineering, computer engineering (including software engineering), electromagnetics and acoustics, energy systems, manufacturing systems engineering, plasma/quantum electronics/optics, solid-state electronics, and telecommunications and information systems engineering.

The master's degree program is normally completed in 1 to 1½ years with the completion of either 30 semester hours of course work, including a 6-semester-hour thesis, or 33 semester hours of course work, including a 3-semester-hour report. A no-thesis/no-report option is also available upon completion of 36 hours of course work.

The Ph.D. program requires the passing of an oral and/or written qualifying examination, the completion of an individualized program of course work as set by the Qualifying Examination Committee, and the completion and final oral defense of the Ph.D. dissertation. The average time for completion is about three years beyond the Master of Science in Engineering degree.

Research Facilities

Equipment with capabilities approaching the state of the art is available for research in the computer and information systems, power, quantum electronics and optics, electromagnetics-acoustics, biomedical, and solid-state electronics areas. Major research centers where ECE graduate students perform thesis- and dissertation-related work include the Electronics Research Center, the Center for Energy Studies, the Center for Electromechanics, the Center for Fusion Engineering, the Electrical Engineering Research Laboratory, the Computer and Vision Research Center, the Laboratory for Intelligent Processes and Systems, the Microelectronics Research Center, the Computer Engineering Research Center, and the Applied Research Laboratories. The annual rate of state, federal, and industrial funding for research in the department is approximately $12 million.

Financial Aid

Financial support is available in the form of fellowships, research assistantships, and teaching assistantships. The large number of highly qualified applicants makes fellowship awards extremely competitive. These awards are mainly for the fall semester, and recipients are eligible for Texas-resident tuition. Research and teaching assistantships normally require the recipient to work 20 hours per week and take a course load of 9 semester hours. These stipends vary from approximately $8000 to $16,000 per calendar year. Recipients of 20-hours-per-week appointments are eligible for Texas-resident tuition.

Cost of Study

In 1999–2000, E.C.E. Texas residents pay $1946 per semester in tuition, required fees, and program and course fees for 9 semester credit hours, while E.C.E. nonresidents pay $3755 for 9 semester credit hours. Students awarded competitive fellowships or holding a half-time teaching assistantship or research assistantship automatically qualify for resident tuition. These amounts do not include lab fees or service-related fees; these may add several hundred dollars to the total.

Living and Housing Costs

Both University and privately owned housing accommodations are available. University dormitory rates range from $4523 to $6868 for the 1999–2000 academic year (nine months), depending on the type of room and board selected. University family apartments for married students vary in cost from $385 per month for an unfurnished one-bedroom unit to $558 for a modern three-bedroom apartment. Privately owned housing may be found in all price ranges. The temperate climate allows for casual, inexpensive dress. Prices for food and other necessities are reasonable.

Student Group

There are more than 500 graduate students in electrical and computer engineering at the University of Texas at Austin, representing nearly every part of the United States and the world. Although the majority of these are full-time students, a sizable number are employed in the growing electronics industries in and around Austin.

Location

Located on the edge of the Texas hill country—a region with abundant lakes, forests, wildlife, recreational opportunities, and a Sun Belt climate—Austin has been cited in a number of national publications as one of the nation's most desirable places in which to live. With a population around 614,000, Austin is large enough to provide widely varied leisure and entertainment opportunities without some of the congestion characteristic of larger cities. Culturally, the city is a fascinating mix of Western, Southern, and Spanish influences. In earlier years, the economic activity in Austin stemmed primarily from the state government and the University, but in recent years there has been a rapid expansion of industrial activities dominated by plants and laboratories of major corporations, primarily in the electronics area.

The University and The Department

The University, with an enrollment of about 49,000, has all the intellectual and social characteristics one would associate with the main campus of the state university system in a large, populous state. There is a wide diversity in student backgrounds, life-styles, and goals. Many of the University's graduate and professional programs, including those in the Department of Electrical and Computer Engineering, have a high national standing.

The department consists of 66 faculty members whose research interests and expertise encompass nearly every specialty area within electrical and computer engineering.

Applying

January 2 is the departmental deadline date for summer, fall, and spring (of the following year) admissions and for consideration for financial assistance. The admission application, transcripts, GRE General Test scores, and TOEFL scores (international students only) must be submitted directly to the Graduate and International Admissions Center by the above deadline date. Copies of these official documents, plus additional materials required by the department, should be sent directly to the department by the January 2 deadline.

Correspondence and Information

Graduate Advisor
Department of Electrical and Computer Engineering
The University of Texas at Austin, MC C0803
Austin, Texas 78712-1084
Telephone: 512-471-8511

The University of Texas at Austin

THE FACULTY AND THEIR RESEARCH

Biomedical Engineering
L. E. Baker, Ph.D. Biomedical engineering, physiology.
T. E. Milner, Ph.D. Biomedical engineering, noninvasive optical tomography, laser surgical procedures and diagnostics, tissue wound healing response.
J. A. Pearce, Ph.D. Biomedical engineering, tissue thermal damage, thermographic imaging, electrosurgery, electrophysiology.
R. Richards-Kortum, Ph.D. Biomedical engineering, laser spectroscopy, real-time tissue imaging with microscopic resolution.
H. G. Rylander III, M.D. Biomedical engineering, transducers, visual information processing, lasers in medicine.
J. W. Valvano, Ph.D. Biomedical instrumentation, heat transfer in biological systems, applications of microcomputers.
A. J. Welch, Ph.D. Biomedical engineering, lasers in medicine, optical-thermal laser-tissue interaction.

Computer Engineering
J. A. Abraham, Ph.D. Fault-tolerant computing, VLSI design and test, formal verification, software engineering.
J. K. Aggarwal, Ph.D. Computer vision, image processing, and pattern recognition for the automatic recognition of human motion and activities for security and surveillance applications.
A. P. Ambler, Ph.D. Computer engineering, CAD for test and design for test, economics of test, safety-critical systems.
A. Aziz, Ph.D. Design automation for VLSI systems: logic synthesis and formal verification.
K. S. Barber, Ph.D. Software engineering, agent-based systems, distributed artificial intelligence, knowledge-based planning.
C. M. Chase, Ph.D. Parallel computer architecture, distributed computing environments.
H. G. Cragon, B.S. Digital computer architecture.
B. L. Evans, Ph.D. Embedded systems; CAD tools; signal, image, and video processing.
V. K. Garg, Ph.D. Distributed systems, software engineering, control of discrete event systems.
J. Ghosh, Ph.D. Neural networks, data mining and knowledge discovery, smart engineering systems.
M. J. Gonzalez, Ph.D. Computer architecture, distributed systems.
M. F. Jacome, Ph.D. H/S codesign, ASIP design, design reuse.
L. K. John, Ph.D. Computer architecture, high-performance processors, memory systems, performance evaluation and benchmarking.
G. J. Lipovski, Ph.D. Logic-in-memory, digital systems architecture, microcontrollers.
S. M. Nettles, Ph.D. Systems and programming languages, storage management/garbage collection, active networks.
Y. N. Patt, Ph.D. Computer architecture, microarchitecture, high-performance implementation, experimental computer systems, digital logic, systems software, computer engineering education.
A. Ricciardi, Ph.D. Fault-tolerant distributed systems; high-performance, wide-area applications; formal methods.
C. H. Roth, Ph.D. Microcomputer systems and applications, theory and design of digital systems.
E. E. Swartzlander, Ph.D. Architecture for application-specific processors, computer arithmetic, computer architecture.
N. A. Touba, Ph.D. VLSI design and test, CAD, fault-tolerant computing.
J. W. Valvano, Ph.D. Microcomputer-based instrumentation, embedded microcomputer hardware/software simulators.
T. J. Wagner, Ph.D. Information theory, computers.

Electromagnetics and Acoustics
F. X. Bostick, Ph.D. Electronic systems, electromagnetic theory, electrical geoscience.
J. H. Davis, Ph.D. Radio astronomy instrumentation, microwave systems engineering.
M. D. Driga, Ph.D. Electromagnetic field theory, advanced computational electromagnetics.
E. L. Hixson, Ph.D. Electroacoustics, acoustics, noise control.
H. Ling, Ph.D. Computational electromagnetics, radar signal processing, synthetic aperture radar imaging, automatic target identification, wireless propagation channel modeling.
D. P. Neikirk, Ph.D. Electromagnetics in integrated circuits and solid-state devices.
H. W. Smith, Ph.D. Electrical geophysics, digital data processing.

Energy Systems
R. Baldick, Ph.D. Optimization of power system operations, transmission regulatory policy, optimization of engineering systems.
M. L. Baughman, Ph.D. Engineering economics, electrical power economics, transmission pricing and planning.
J. R. Cogdell, Ph.D. Electromechanical systems.
M. D. Driga, Ph.D. Electromechanical systems, pulsed power systems, electromagnetic macroparticle accelerators.
W. M. Grady, Ph.D. Power systems analysis, electromagnetics, power system harmonics.
W. F. Weldon, M.S. Design of electromechanical systems, pulsed power technology, electromechanical accelerators.

Manufacturing Systems Engineering
K. S. Barber, Ph.D. Software engineering, agent-based systems, distributed artificial intelligence, knowledge-based planning.
M. D. Driga, Ph.D. Intelligent robotics, application of electromagnetism to manufacturing processes.
R. H. Flake, Ph.D. Computer vision systems applied to manufacturing, nondestructive circuit test and characterization.

Plasma/Quantum Electronics/Optics
M. F. Becker, Ph.D. Optical materials, electrooptics.
J. C. Campbell, Ph.D. Optoelectronic devices, photonic integrated-circuit lightwave systems.
R. T. Chen, Ph.D. Optical interconnects, photonic integrated circuits, holographic optical elements, solid-state electronics and microelectronics.
G. A. Hallock, Ph.D. Thermonuclear fusion, plasma diagnostics, plasma turbulence and transport, plasma processing.
E. J. Powers, Ph.D. Applications of digital time-series analysis to nonlinear wave and turbulence phenomena.

Solid-State Electronics
S. K. Banerjee, Ph.D. Solid-state devices and materials.
A. B. Buckman, Ph.D. Optical sensors/transducers, fiber and guided-wave optics.
D. Deppe, Ph.D. Optoelectronic semiconductor devices.
R. D. Dupuis, Ph.D. Compound semiconductor materials and devices, metalorganic chemical vapor deposition, epitaxial growth.
J. B. Goodenough, Ph.D. Physics and chemistry of transition-metal compounds and their application, fast ionic transport, fuel cells.
A. L. Holmes Jr., Ph.D. Semiconductor materials growth and devices.
D.-L. Kwong, Ph.D. VLSI technology, semiconductor process and device modeling, solid-state materials and devices.
J. C. Lee, Ph.D. Semiconductor devices, materials, and technologies, including advanced submicron structures.
B. G. Streetman, Ph.D. Semiconductor materials and devices.
A. F. Tasch, Ph.D. Semiconductor processes and devices, VLSI/ULSI technology, device/process modeling.
R. M. Walser, Ph.D. Solid-state devices and materials research.

Telecommunications and Information Systems Engineering
A. Arapostathis, Ph.D. Systems, nonlinear dynamical systems, controlled Markov chains, stochastic hybrid systems.
A. C. Bovik, Ph.D. Image processing, digital video, computer vision.
G. de Veciana, Ph.D. Analysis and design of telecommunication systems, applied probability, information theory.
T. Konstantopoulos, Ph.D. Applied and theoretical probability, stochastic processes and systems, high-speed communication networks, operations research, queueing theory, point processes, dynamical systems and control.
S.-Q. Li, Ph.D. Network control and design, queueing theory, resource management and traffic engineering for next generation telecommunication networks.
E. J. Powers, Ph.D. Applications of higher-order statistical and wavelet-based signal processing.
I. W. Sandberg, Ph.D. Nonlinear systems, network and system theory.
G. L. Wise, Ph.D. Statistical communication theory, random processes, signal processing, probability theory, real analysis, counterexamples in many areas.
B. F. Womack, Ph.D. Computer engineering, biomedical engineering, adaptive control and cybernetics.
G. Xu, Ph.D. Wireless communications, digital signal processing, fast algorithms, microwave and RF engineering.

UNIVERSITY OF VIRGINIA

Department of Electrical Engineering

Programs of Study

The Department of Electrical Engineering offers programs leading to the degrees of Master of Science, Master of Engineering, and Doctor of Philosophy. The M.S. degree requires a minimum of 24 semester hours of graduate courses and submission of a thesis. The M.E. degree requires a minimum of 30 semester hours. Either master's degree can be completed in a calendar year of full-time study, but most students spend three to four semesters in the program. For the Ph.D. degree, students are expected to complete a minimum of 24 semester hours beyond the master's degree, pass an oral qualifying examination, and submit and defend a dissertation. The normal full-time graduate course load is three or four courses, depending upon the research load. The principal areas of instruction in the department are communication theory, computer engineering, control systems, microwaves, optoelectronics, pattern recognition and image processing, signal processing, and solid-state devices and materials. Research is in progress in these fields, and the active involvement of every graduate student is encouraged.

A program in computer engineering is available at the master's level. The program includes courses offered by the computer science department and computer engineering courses offered by the electrical engineering department.

The department also offers a special program that encourages students with a bachelor's degree in a related area, such as physics or mathematics, to earn a master's degree in electrical engineering. The program is designed to be completed in fifteen to twenty-four months.

Research Facilities

The Department of Electrical Engineering is well endowed with modern electronic research equipment. The Semiconductor Device Laboratory (SDL) has a 3,500-square-foot clean-room facility capable of complete fabrication of submicron semiconductor and superconductor devices. Major processing capabilities include deep UV lithography, reactive ion etching, ion milling, RF and DC sputtering, e-beam deposition, CVD oxide grown, molecular beam epitaxy (MBE) of Si\Ge and III-V compounds, and field emission scanning electron microscopy. The SDL also shares a Far-Infrared Receiver Laboratory with the Department of Physics. This facility houses laser local oscillator sources to 3 THz and a variety of submillimeter wavelength sources and test equipment. The Laboratory for Optics and Quantum Electronics (LOQE) is fully equipped for optical spectroscopy and the optical characterization of semiconductor devices. Major equipment includes an MBE system for III-V compound semiconductors, a photoluminescence system, and a tunable Ti:sapphire laser system. The Millimeter-wave Research Laboratory (MRL) is equipped with a variety of test and fabrication equipment, including a microwave probe station and an HP-8510 vector network analyzer to 100 GHz and Apollo workstations. The Center for Semicustom Integrated Systems, a research center within the department, operates Sun and Hewlett-Packard computers that contain a number of VLSI design tools and software packages, including the Full Mentor Graphics Tool Suite. The center also operates a Calcomp 68436 electrostatic plotter, an HP 82000-100 IC test system, high-speed logic analyzers, and high-speed digital oscilloscopes. In addition, the center supports digital designs incorporating programmable parts (e.g., PLD, FPGA, EPROM, and custom integrated circuits). The Communication Control and Signal Processing Laboratory utilizes computing resources available on Sun and RS6000 workstations, MATLAB signal processing software, and Gould and Pixar image processing hardware for still picture and image sequence research. The University's computing center operates numerous IBM RS6000, Silicon Graphics, and Sun computers.

Financial Aid

Financial aid is available in the form of University fellowships, national fellowships, and research and teaching assistantships. Fellowships in 1998–99 carried stipends of $9000 and up, plus tuition and fees, for the academic year. Research assistantships, supported by nonclassified sponsored research projects, provided stipends of $10,000 to $15,000 for the calendar year.

Cost of Study

In 1998–99, tuition and fees were $4796 per academic year for in-state residents and $15,040 per academic year for nonresidents.

Living and Housing Costs

Dormitory facilities were available for single students at $2100 to $2400 per academic year for 1998–99. University-operated accommodations for student families ranged from one-bedroom apartments for $416 per month to three-bedroom apartments for $520 per month (apartments are furnished and include utilities).

Student Group

There are 18,420 students at the University of Virginia, including 6,106 graduate students. The School of Engineering and Applied Science enrolls 1,800 students; 560 are graduate students. There are 109 full-time graduate students in the electrical engineering department.

Location

Charlottesville and its environs, situated in the foothills of the Blue Ridge Mountains, constitute a community of approximately 100,000 people. The climate is relatively mild, and opportunities for outdoor recreation extend throughout most of the year. Shenandoah National Park is 20 miles away, and Washington, D.C., is 110 miles away. The area contains many places of historic interest.

The University

The University of Virginia, founded in 1819 by Thomas Jefferson, is a state-aided institution that recognizes the importance of having a student body drawn from many parts of the country. It is widely known for its outstanding programs in a variety of areas, including engineering and applied science.

Approximately 155 full-time faculty members are active in teaching, research, and public service in the School of Engineering and Applied Science.

Applying

Students seeking financial aid should apply by February 1 of the year in which they plan to enroll. All applicants are required to submit GRE scores, and international students must have a minimum TOEFL score of 600.

Correspondence and Information

EE Graduate Office
Thornton Hall
University of Virginia
Charlottesville, Virginia 22903-2442

Telephone: 804-924-6077
E-mail: eegrad@virginia.edu
World Wide Web: http://www.ee.virginia.edu

University of Virginia

THE FACULTY AND THEIR RESEARCH

J. Aylor, Ph.D., Professor and Chair of the Department. Design automation, digital systems, VLSI systems, test technology.

J. Bean, Ph.D., John Marshall Money Professor. Molecular beam epitaxy, novel electronic materials.

R. Bradley, Ph.D., Research Assistant Professor. Microwave and millimeter-wave semiconductor devices and integrated circuitry, radio astronomy instrumentation.

M. Brandt-Pearce, Ph.D., Assistant Professor. Communication theory, optical communications, multiuser networks.

T. Crowe, Ph.D., Research Professor. High-frequency solid-state devices, novel solid-state devices, terahertz sources and receivers.

J. Dugan, Ph.D., Professor. Dependability analysis of fault-tolerant systems, hardware and software reliability engineering.

T. Globus, Ph.D., Associate Professor. Electrical and optical characterization of electron materials and devices.

B. Johnson, Ph.D., Professor. Fault-tolerant systems, VLSI testing, VLSI systems.

A. Lichtenberger, Ph.D., Research Assistant Professor. Superconducting materials and devices.

Z. Lin, Ph.D., Assistant Professor. Nonlinear control theory, control systems subject to actuator saturation, robust control theory, computer-aided control system design, and control applications.

P. Marshall, Ph.D., Associate Professor. Electric power and machinery, power electronics, energy policy, renewable energy.

M. Reed, Associate Professor. Medical and industrial application of microelectromechanical systems, microfabrication technology and piezoelectronically-tuned electrooptic devices.

N. Sidiropoulos, Ph.D., Assistant Professor. Statistical and nonlinear signal processing and mathematical imaging, optimal filtering, estimation and detection, regression coding, deconvolution and similarity testing.

M. Stan, Ph.D., Assistant Professor. Low power VLSI, FPGAs, mixed mode analog and digital VLSI, embedded systems and hardware/software codesign.

G. Tao, Ph.D., Associate Professor. Adaptive control, nonlinear systems, control applications, multivariable control systems, robust adaptive systems, robotics.

E. Towe, Ph.D., Associate Professor. Optics, quantum electronics, solid-state devices.

R. Weikle, Ph.D., Assistant Professor. Microwave and millimeter-wave circuits and radiating structures.

R. Williams, Ph.D., Associate Professor. Computer design, real-time systems, VLSI design/VLSI testing, speech synthesis.

S. Wilson, Ph.D., Professor. Communications and information theory.

ACTIVE RESEARCH PROJECTS

Communications and Control
Adaptive control.
Communication network protocols and analysis.
Digital speech encoding and image encoding at low bit rates.
Distributed signal processing.
Error control coding.
Indoor wireless infrared systems.
Microelectronic system design, fabrication, and testing.
Multiuser communications.
Optical communications.
Robust signal processing.
Signal design for band- and power-limited satellite channels.
Statistical modeling and system identification.

Computer Engineering
A high-performance memory controller.
Automated design of digital systems.
Computer-aided design tools for circuit layout.
Concurrent error detection techniques.
Fault simulation using VHDL.
Fault-tolerant electronics for powered wheelchairs.
Integrated performance and reliability modeling of fault-tolerant systems.
Modeling and simulation using VHDL.
Reliable architectures for magnetic bearing systems.
Safety critical systems for automatic train control.
Test-pattern generation in a parallel environment.
Test-pattern generation using genetic algorithms.
Uninterpreted/interpreted modeling and simulation.

Millimeter-Wave Research
Integrated circuit antennas.
Microwave and millimeter-wave power combining.
Millimeter-wave diode characterization and applications.
Millimeter- and submillimeter-wave receivers and sources.
Planar transmission lines and active circuits.
Quasi-optical arrays and circuits.

Optics and Quantum Electronics
Integrated optics.
Intersubband transition quantum well detectors.
Optical modulators.
Optical neural networks.
Optical polymer materials.
Vertical cavity surface-emitting lasers.
Visible second-harmonic semiconductor lasers.

Solid-State Devices
Design and analysis of novel three-terminal devices based on GaAs and related compounds.
Epitaxial growth of GaAs and related compounds by OMCVD and MBE.
High-frequency semiconductor device physics.
Mixer and varactor diodes for submillimeter-wave applications.
Noise theory and microwave noise measurement.
Solid-state device fabrication and processing technology.
Superconductive electronics.
Superconductor-insulator-superconductor (SIS) structures for millimeter- and submillimeter-wave mixer applications.
Terahertz receiver systems.
Thin-film transistors.
Transferred electron oscillators.
Wide band gap semiconductor technology.

UNIVERSITY OF WASHINGTON

Department of Electrical Engineering

Programs of Study	The Department of Electrical Engineering offers graduate programs leading to the degrees of Master of Science (M.S.E.E.) and Doctor of Philosophy (Ph.D.). Graduate courses and research programs are offered in modern VLSI, sensors, and semiconductor technology; mechatronics and intelligent control; advanced digital systems and communication; applied signal and image processing; advanced power technology; and applied electromagnetics, optics, and remote sensing. Opportunities also exist for participation in research on medical instrumentation in the Bioengineering Program and in marine acoustics and instrumentation systems at the Applied Physics Laboratory.
	For the M.S.E.E. degree, a minimum of 45 credits is required. Students writing a thesis must register for 9 to 12 credits. Students selecting the nonthesis option can either complete their degree by total course work or by a one-term project of 4 to 8 credits. Course work for any of the above-mentioned options must be selected with each student's supervisory committee approval to prepare the student in an area of specialization. If more flexibility is desired than the M.S.E.E. requirements allow, the interdisciplinary degree of Master of Science in Engineering is available.
	The M.S.E.E. degree is also offered to part-time students, employed in local industries, through the Televised Instruction in Engineering (TIE) program. Regular graduate courses are offered over cable television or via videotape to enable working engineers to participate in the program without traveling to campus.
	For the Ph.D. degree, the student must pass the departmental qualifying examination, pass an advanced general examination, pursue an original research problem, and report the results of the research in a dissertation that must be a contribution to knowledge. At least one year of course work beyond the M.S.E.E. degree is usually desirable. Exceptionally qualified students are encouraged to pursue the Ph.D. degree directly without first earning a master's degree.
Research Facilities	Facilities in the Electrical Engineering Building include laboratories for solid-state materials, microtechnology, microwaves and millimeter waves, computer technology, computer systems, machine vision, analog and digital electronics, energy systems, power electronics and electric drives, bioelectronics, control systems, statistical data analysis, neural networks, and signal processing and classification. Extensive computer facilities are available, and there is an integrated circuit and semiconductor sensor fabrication facility as well as an interactive facility for speech and sonar analyses.
Financial Aid	Approximately 155 teaching and research assistantships, scholarships, and fellowships (optional graduate appointee health insurance included) are available for qualified graduate students in all areas of electrical engineering. Teaching assistantships pay $9990 per academic year in 1998–99 for an entering graduate student. Higher amounts are paid for research assistants and doctoral students. All graduate assistants must pay a modest resident operating fee.
Cost of Study	Tuition and fees are $1811 per quarter for state residents and $4493 for nonresidents in 1999–2000; these rates are subject to change without notice.
Living and Housing Costs	The cost of board and room in University residences is $4779 plus a $60 deposit for the 1999–2000 academic year. Housing for married students is available for qualified applicants. Apartments and rooms are also available off campus at varying costs.
Student Group	The department's on-campus graduate enrollment is 260 students, about 60 percent of whom are in the M.S.E.E. program and 40 percent of whom are in the Ph.D. program. An additional 50 students are enrolled in the off-campus televised M.S.E.E. program.
Location	The University is located on the shores of Lake Washington in a residential area near downtown Seattle. The metropolitan region has 2.5 million people. Seattle is a cosmopolitan city with many cultural and scenic attractions. It enjoys a temperate marine climate and is located on Puget Sound between the Cascade and Olympic mountain ranges, which offer a rich assortment of recreational opportunities.
The University	The University's campus comprises 2.8 square kilometers of beautifully landscaped evergreens, flowering shrubs, and buildings. The University is the premier institution of higher learning in the Northwest and is a major graduate education and research center. It offers a full spectrum of academic disciplines and is a sea-grant institution.
Applying	Prospective students should request additional information and admission application forms from the graduate program coordinator. A separate application form is required for financial aid. The application deadline is February 1 for admission and financial aid awards. The General Test of the Graduate Record Examinations (GRE) is required of all students. Students are normally admitted in the fall quarter of the academic year. A $48.50 application fee must accompany the application.
Correspondence and Information	Graduate Program Coordinator Department of Electrical Engineering, Box 352500 University of Washington Seattle, Washington 98195 Telephone: 206-543-4924 E-mail: grad@ee.washington.edu World Wide Web: http://www.ee.washington.edu

University of Washington

THE FACULTY AND THEIR RESEARCH

Circuits and Sensors
M. A. Afromowitz. Microfabrication, integrated and fiber-optic sensors, biomedical instrumentation.
J. Andersen. Circuits, systems, CAD.
K. Bohringer. MEMS, microrobotics, and microfluidics.
R. B. Darling. Solid state, semiconductor devices, optoelectronics, microelectronics.
W. J. Helms. Integrated circuits, analog and digital circuit design.
T. P. Pearsall. Technology and physics of semiconductor devices, materials growth and characterization.
C. M. Sechen. Design and computer-aided design of analog and digital integrated circuits.
M. Soma. Integrated circuits: design, test, and reliability; bioelectronics.
D. M. Wilson. Distributed sensing systems.
S. S. Yee. Semiconductor devices, optical sensors and microsensors and integrated optics.

Communications
M. Azizoglu. Optical communication networks, high-speed network protocols, communication theory.
H. Liu. Wireless personal communications, signal processing and DSP applications.
S. Roy. Analysis and design of communications systems and networks, statistical signal processing, numerical and statistical computing.
J. A. Ritcey. Communications, statistical signal processing for radar, underwater acoustics, biomedicine.

Controls and Robotics
R. W. Albrecht. Mobile robotics, nuclear systems.
F. J. Alexandro. Control systems.
H. J. Chizeck. Adaptive control and systems identification, intelligent and biologically inspired control systems.
B. Hannaford. Telerobotics, bioengineering, human-machine systems, neurological control systems and models.
D. R. Meldrum. Automated laboratory systems, robotics and control, applications in biotechnology and transportation systems.
R. B. Pinter. Cybernetics, nonlinear and adaptive control systems, biophysics.
F. A. Spelman (Adjunct, Department of Bioengineering). Biological systems and models, cochlear implants, tissue resistivity.
J. Vagners (Adjunct, Department of Aeronautics and Astronautics). Dynamics, control systems analysis and synthesis, and optimal control and estimation.

Electromagnetics
C. H. Chan. Computational electromagnetics, microwave integrated circuits, scattering and antennas, bioengineering.
L. A. Crum (Research). Physical acoustics, medical ultrasonics, underwater acoustics.
D. R. Jackson (Research). Underwater acoustics.
Y. Kuga. Remote sensing, electromagnetics, optics.
C. Ramon (Research Adjunct, Department of Bioengineering). Biomagnetic imaging, image reconstruction, inverse problems, bioelectromagnetics.
E. I. Thorsos (Research). Rough surface scattering, underwater acoustics, electromagnetic scattering.
L. Tsang. Electromagnetics, remote sensing, optics of semiconductors, wave scattering.
D. P. Winebrenner (Research). Microwave remote sensing, wave scattering, electromagnetics, polarimetry, scattering statistics.

Energy
R. D. Christie. Power system operations, real-time expert systems, software engineering.
M. J. Damborg. Control systems theory, power system dynamics, computer applications, expert systems and database applications.
M. A. El-Sharkawi. Intelligent system applications to power, dynamic analysis and control of power systems, electric drives and power electronics.
C. C. Liu. Power systems, expert systems, power electronics.
H. P. Yee (Research). Power semiconductor devices, intelligent power devices and integrated circuits.

Signal and Image Processing
L. E. Atlas. Digital signal processing and time-frequency representations, applications in speech processing and manufacturing.
D. J. Dailey (Research). Digital signal processing, time series analysis, networks, distributed computing.
R. M. Haralick. Computer vision, artificial intelligence, pattern recognition, image processing.
J.-N. Hwang. Digital signal/image processing, pattern recognition, artificial neural networks.
P. L. Katz (Research). Underwater acoustics, image processing, digital signal processing, automatic target recognition, classification.
T. K. Lewellen (Adjunct, Department of Radiology). Medical imaging, PET, detectors, reconstruction algorithms.
R. J. Marks. Artificial neural networks, fuzzy systems, signal analysis, statistical communication theory, optical processing.
E. A. Riskin. Data compression, image processing.
J. D. Sahr. Radar, signal processing, ionospheric physics.
L. G. Shapiro. Computer vision, robotics, artificial intelligence, database systems.
M. T. Sun. Video coding and systems, multimedia networking, VLSI.

VLSI and Digital Systems
Y. Kim. Computer architecture, imaging systems, medical imaging, multimedia systems.
C. J. R. Shi. VLSI design automation, mixed-signal testing and reliability, hardware description languages, modeling and simulation.
M. Soma. Integrated circuits: design, test, and reliability; bioelectronics.
S. L. Tanimoto (Adjunct, Department of Computer Science and Engineering). Image analysis, artificial intelligence, computer graphics.
G. L. Zick. Image and multimedia databases, medical imaging.

E-mail for electrical engineering faculty members is lastname@ee.washington.edu

VILLANOVA UNIVERSITY

College of Engineering
Department of Electrical and Computer Engineering

Programs of Study

The Department of Electrical and Computer Engineering offers programs of study leading to the degrees of Master of Science in Electrical Engineering and Master of Science in Computer Engineering. In electrical engineering, graduate study and research are concentrated in the following areas: systems and signal processing, electromagnetics, solid state devices and optics, and electronic systems. In computer engineering, graduate courses and research are offered in the following subjects: computer organization and design, advanced computer architectures, VLSI design, artificial intelligence, computer vision, and neural networks. The master's program requires 33 semester credit hours of study, including at least 24 course credits and up to 9 credits of research. Students may elect to write a thesis or choose a nonthesis option. The main objective of the graduate programs is to provide a balance of theory and practical knowledge needed either to practice in the profession or to advance to a doctoral program.

Research Facilities

The department laboratories reflect state-of-the-art software, hardware, instrumentation, and equipment and are updated yearly. Hardware includes Sun UNIX workstations, file servers, a dial-in terminal server, and graphics terminals; microcomputer development systems; a microcontroller development system; and an HP microwave network analyzer. CAD software includes digital systems design, VLSI design, and image processing. A high-frequency anechoic chamber is available.

Financial Aid

Teaching assistantships are available that provide a stipend for the academic year, payable in ten installments, plus a waiver of tuition and fees. In the awarding of assistantships, consideration is given to performance on the Graduate Record Examinations and/or Test of English as a Foreign Language, the undergraduate academic record, and letters of reference indicating outstanding ability. Research and tuition scholarships may also be available. Eligibility for low-interest loans is determined by the University's Office of Financial Aid.

Cost of Study

Tuition for 1998–99 was $595 per credit. Books and supplies were about $200 per semester. Miscellaneous fees and parking amounted to approximately $60 per semester.

Living and Housing Costs

Normally, there is no on-campus housing available for graduate students unless they are counselors employed in undergraduate dormitories. Applications for counselorships should be directed to the Director of Resident Living in Tolentine Hall, not to the College of Engineering. Costs of off-campus housing vary greatly depending on the type of accommodation. Listings of rooms and apartments are maintained by the University's Off-Campus Housing Service.

Student Group

The department has 35 full-time and 76 part-time graduate students.

Location

The University's handsomely landscaped 240-acre campus is located on suburban Philadelphia's historic Main Line, 12 miles from the city. This beautiful residential area contains excellent restaurants and shopping centers and is near Valley Forge National Park. Philadelphia, with its many cultural, historic, and sports attractions, is directly accessible by rail and bus from the campus. The University is within driving distance of the New Jersey and Delaware beaches, New York City, and the Poconos' recreational area.

The University

Villanova University was founded in 1842 by the Fathers of the Order of Saint Augustine. The engineering college was established in 1905. The University is composed of the College of Arts and Sciences, the College of Nursing, the College of Engineering, the College of Commerce and Finance, and the School of Law.

Applying

A B.S. in electrical engineering or computer engineering is normally required for admission, but students with a B.S. in non–electrical engineering areas such as physics, mathematics, and applied sciences are considered for admission. Accepted students may be required to complete undergraduate prerequisites as determined by the department. The Graduate Record Examinations (GRE) are recommended for all applicants. For those applicants who hold a degree from a U.S. college or university, the school must be a regionally or nationally accredited institution of higher education. In the case of applicants who hold a degree from an institution outside the United States, the GRE is required. In addition, if the native language of the applicant is not English, a minimum score of 550 on the test of English as a Foreign Language (TOEFL) is required. Application materials may be obtained from the department.

Correspondence and Information

Dr. S. S. Rao
Professor and Chairperson
Department of Electrical and Computer Engineering
Villanova University
800 Lancaster Avenue
Villanova, Pennsylvania 19085-1681

Telephone: 610-519-4228 or 4970
Fax: 610-519-4436
E-mail: rao@ece.vill.edu or gradadm@ece.vill.edu
World Wide Web: http://www.ece.vill.edu

Villanova University

THE FACULTY AND THEIR RESEARCH

Moeness G. Amin, Professor; Ph.D., Colorado, 1984. Multidimensional signal processing, time-varying spectral analysis, adaptive filtering, signal detection and enhancement, noise canceling, system identification.

Kevin M. Buckley, Professor; Ph.D., USC, 1986. Optimum and adaptive filtering, parameter and spectrum estimation, communications, digital signal processing applications.

Julia V. Bukowski, Associate Professor; Ph.D., Pennsylvania, 1979. Large-scale systems; network theory; hardware, network, and software reliability.

Robert H. Caverly, Associate Professor; Ph.D., Johns Hopkins, 1983. RF, microwave, and microelectronic device and system design.

Frank N. DiMeo, Associate Professor; Ph.D., Pennsylvania, 1969. Real-time control and automation, robotics, man-machine interfacing, image processing, FORTH language, neural networks.

Ahmad Hoorfar, Associate Professor; Ph.D., Colorado, 1984. Electromagnetic field theory, microwave and millimeter-wave antennas and circuits, transient electromagnetics, mathematical physics, numerical techniques.

Mark A. Jupina, Assistant Professor; Ph.D., Penn State, 1990. Physical and electrical characterization of microelectronic materials and devices, modeling and analysis of microwave devices.

Stephen Konyk Jr., Assistant Professor; Ph.D., Drexel, 1985. Nonlinearly constrained adaptive filtering and estimation with applications to detective systems, e.g., radar, sonar, etc.; intelligent control and dynamics with applications to robotic design and automation.

Joseph L. Kozikowski, Associate Professor; Ph.D., Pennsylvania, 1969. Analog and digital integrated electronics; analysis and design of filters, oscillators, regulators, PLLs, interface circuits, and others; digital pulse transmission and crosstalk.

Edward Kresch, Associate Professor; Ph.D., Pennsylvania, 1968. Computers: programming, microprocessors and microcomputer design, combinational and sequential circuits, computer architecture; biomedical engineering: nerve conduction, gait analysis, cardiovascular dynamics.

William E. Mattis Jr., Associate Professor; Ph.D., Drexel, 1972. Digital systems, digital design, integrated circuit design, microprocessors, real-time processing and applications, computer communications, neural systems.

Frank J. Mercede, Assistant Professor; Ph.D., Drexel, 1989. Power electronics and systems.

Bijan Mobasseri, Associate Professor; Ph.D., Purdue, 1978. Computer vision and machine intelligence, intelligent robotics, hierarchical representations, parallel architectures for vision, pattern recognition, image processing.

Richard J. Perry, Associate Professor; Ph.D., Drexel, 1981. Computational algorithms and software, VLSI design, multivariable systems.

S. S. Rao, Professor and Chairperson; Ph.D., Kansas, 1966. Digital signal processing; estimation, detection, and identification algorithms; spectral analysis and estimation; nonlinear signal processing; digital filtering and implementation; statistical communication theory.

Pritpal Singh, Associate Professor; Ph.D., Delaware, 1984. Fabrication and characterization of electronic materials and devices, especially solar cells, infrared detectors, and bulk high T_c superconductors; modeling and analysis of semiconductor devices.

Anthony Zygmont, Professor; Ph.D., Pennsylvania, 1971. Applications of artificial intelligence, expert systems, numerical computation, and symbolic computation to the analysis and design of electrical engineering devices and systems.

Adjunct Faculty

Prabhakar Rao Chitrapu, Ph.D., Delft University of Technology, 1985. Wireless communications.

Edward L. Hepler, Ph.D., Drexel, 1979. Computer architecture and VLSI design.

Steven T. Kacenjar, Ph.D., Rochester, 1977. Electrooptics and fiber-optic communication systems.

James L. Marshall, M.S., MIT, 1949. Audio engineering.

Kistareddy Pallegadda, M.S.E.E., Villanova, 1984. Computer communications.

Louis Pitale, Ph.D., Drexel, 1980. Electromagnetics and nonlinear wave propagation.

Louis P. Rubinfield, Ph.D., Washington (St. Louis), 1980. Digital systems and computer engineering.

Louis J. Ruggeri, M.S.E.E., Penn State, 1973. Computers and electronics.

Bradley Keith Taylor, Ph.D., Carnegie Mellon, 1988. Optical information processing.

Richard Teti, Ph.D., Pennsylvania, 1991. Electromagnetics, signal theory, communications, microwave remote sensing.

Luke J. Turgeon, Ph.D., Massachusetts, 1977. Analog and integrated circuits.

Sydney S. Weinstein, M.S., Massachusetts, 1978. Software engineering.

VIRGINIA POLYTECHNIC INSTITUTE AND STATE UNIVERSITY

Bradley Department of Electrical and Computer Engineering

Programs of Study

The Bradley Department of Electrical and Computer Engineering offers programs of graduate study leading to the degrees of Master of Science and Doctor of Philosophy in electrical engineering and computer engineering.

The M.S. degree usually requires the equivalent of one year of full-time study, with a minimum of 30 semester credits of work acceptable for graduate credit. Although an M.S. degree may be earned with or without a thesis, research assistantships are awarded only to those who elect the thesis option. A final oral exam is required of all M.S. nonthesis candidates, while students electing the thesis option must take an oral examination on the thesis.

The Ph.D. program usually requires a minimum of three years of full-time work beyond the master's degree. Each student in the Ph.D. program must pass a written qualifying examination early in the program, a preliminary written and oral examination before dissertation work is begun, and a final oral examination on the dissertation and related topics.

Research Facilities

The University Computing Center has an IBM 9121 model 480E/VF, an IBM 9672 model R32, a MicroVAX 3100, and an IBM RS/6000. The University Computing Center also has an IBM SP2 available for numerically intensive applications. The University's scientific data visualization lab has several SGI workstations with Internet connections. Software available in the visualization lab includes PV-WAVE and AVS. Hard-copy devices include postscript laser printers and thermal transfer color printers. The College of Engineering has a Silicon Graphics Power Challenge XL. The department has a computing facility that includes networked Sun and HP workstations, personal computers, and software packages to support graduate work and research. Other computers, instrumentation, and specialized equipment are available to support specific research activities. Laboratories for research in acoustooptics; alternative energy; antenna design; computers; control systems; design automation; digital and image processing; digital signal processing; electric power; electronic materials and device fabrication; energy systems; fiber optics and electrooptics; high-frequency device modeling and testing; hybrid microelectronics; mobile and portable radio; motor drives; power electronics; radio frequency and microwave communication systems; robotics/AI lab, including a MERLIN 6540 industrial robot; satellite communications; VLSI design and computer-aided design; and wideband time domain and microwave characterization and modeling are also available.

Financial Aid

The department offers teaching and research assistantships. For 1999–2000, typical stipends are $1245 to $1975 per month, depending upon the student's academic record and previous experience. Research assistantships are available in acoustooptics, applied electromagnetics, computer engineering, control, electric power, fiber optics, image processing, microelectronics, mobile and portable radio, power electronics, robotics, satellite communications, and signal processing. In-state tuition scholarships are granted to students who are appointed to an assistantship.

Bradley Fellowships of $16,000 to $18,000 are available from the department. The department participates in a graduate co-op program with various companies and government agencies throughout the country.

Cost of Study

For 1999–2000, the full-time in-state instructional fee for graduate students is $2475 per semester ($3879 if out-of-state); in-state students who take fewer than 9 hours pay $229 per credit hour ($385 per credit hour if out-of-state). The above full-time fees include a $396 comprehensive fee and an $18 educational and general fee per semester, which are required of all students.

Living and Housing Costs

Only limited campus housing is available for graduate students, but there is an ample supply of suitable off-campus apartments available in Blacksburg and the surrounding area. Apartment rents range from approximately $350 to $500 per month, depending upon size and location. Meals are available in the student center dining areas at reasonable prices.

Student Group

There are about 500 graduate students in electrical and computer engineering, of whom about 350 are on the Blacksburg campus and the remainder are taking courses on a part-time basis at Falls Church, VCU, Dahlgren, and other locations within Virginia.

Location

Virginia Tech is located 40 miles west of Roanoke in the town of Blacksburg, on a plateau (elevation 2,100 feet) between the Blue Ridge and Allegheny mountains. Interstate 81 provides easy access to Roanoke. Recreational facilities in the area include many opportunities for picnicking, swimming, fishing, hunting, and camping in state parks and national forests.

The University

Virginia Tech was founded in 1872 as a public land-grant university and awarded it first graduate degree in 1892. The University has particularly extensive offerings in the fields of engineering and agriculture. The University sponsors a film series and performances by visiting lecturers, musicians, and drama groups in addition to offering productions by the University orchestra, chorus, and drama department. The world-renowned Audubon String Quartet is in residence at Virginia Tech. Nearby Radford University also sponsors a series of concerts and plays, to which students may subscribe. Various social activities are sponsored by many University and town organizations.

Applying

Each applicant must submit a $25 nonrefundable application fee, a completed application form, three letters of recommendation from former professors or employers, and official transcripts of undergraduate and graduate records to date. Scores from the GRE General Test are required for all applicants. The University requires all international students to submit TOEFL scores; the department requires a minimum score of 600. Applications can be submitted to the Graduate School at any time. At least two months should be allowed for processing an application and obtaining references. Completed application materials for admission and financial assistance for the 2000–01 academic year must be submitted to the Virginia Tech Graduate School no later than January 15, 2000.

Correspondence and Information

For application forms and admission information:
Holly Traweek, Graduate Advising Assistant
Bradley Department of Electrical and Computer
  Engineering
Mail Code 0111
Virginia Polytechnic Institute and State University
Blacksburg, Virginia 24061
Telephone: 540-231-7262
E-mail: tra@vt.edu

For specific program information:
Cynthia Hopkins, Graduate Counselor
Bradley Department of Electrical and Computer
  Engineering
Mail Code 0111
Virginia Polytechnic Institute and State University
Blacksburg, Virginia 24061
Telephone: 540-231-8393
E-mail: hopkins@vt.edu
World Wide Web: http://www.ee.vt.edu

Virginia Polytechnic Institute and State University

THE FACULTY AND THEIR RESEARCH

A. L. Abbott, Associate Professor; Ph.D., Illinois at Urbana-Champaign. Computer engineering.

J. Allnutt, Professor; Ph.D., London. Communications.

J. R. Armstrong, Professor; Ph.D., Marquette. Computer engineering.

P. M. Athanas, Associate Professor; Ph.D., Brown. Computers.

J. Baker, Assistant Professor; Ph.D., Georgia Tech. Computer engineering

W. T. Baumann, Associate Professor; Ph.D., Johns Hopkins. Systems and control.

J. S. Bay, Associate Professor; Ph.D., Ohio State. Robotics.

A. A. Beex, Professor; Ph.D., Colorado State. Signal processing.

A. E. Bell, Assistant Professor; Ph.D., Michigan. Signal processing.

I. M. Besieris, Professor; Ph.D., Case Tech. Theoretical and applied electromagnetics.

D. Borojevic, Professor; Ph.D., Virginia Tech. Power electronics.

C. W. Bostian, Clayton Ayre Professor; Ph.D., North Carolina State. Microwave systems, antennas and propagation.

R. P. Broadwater, Professor; Ph.D., Virginia Tech. Power systems.

G. S. Brown, Professor; Ph.D., Illinois at Urbana-Champaign. Electromagnetics.

D. D. Chen, Professor; Ph.D., Duke. Power electronics.

R. O. Claus, Willis G. Worcester Professor; Ph.D., Johns Hopkins. Fiber optics.

R. W. Conners, Associate Professor; Ph.D., Missouri–Columbia. Image processing.

W. R. Cyre, Associate Professor; Ph.D., Florida. Computer engineering.

L. Dasilva, Assistant Professor; Ph.D., Kansas. Computer engineering.

N. J. Davis, Associate Professor; Ph.D., Purdue. Computer communications and architecture.

W. A. Davis, Professor; Ph.D., Illinois at Urbana-Champaign. Electromagnetic fields and communication systems.

J. De La Ree López, Associate Professor; Ph.D., Pittsburgh. Power systems.

D. A. de Wolf, Professor; Ph.D., Eindhoven (Netherlands). Electromagnetics.

R. W. Ehrich, Professor; Ph.D., Northwestern. Computer science.

L. A. Ferrari, Professor and Department Head; Ph.D., California, Irvine. Image processing.

F. G. Gray, Professor; Ph.D., Michigan. Digital computers.

D. S. Ha, Associate Professor; Ph.D., Iowa. Fault-tolerant computing.

R. W. Hendricks, Professor; Ph.D., Cornell. Electronic materials.

Q. Huang, Assistant Professor; Ph.D., Cambridge (England). Power electronics.

I. Jacobs, Professor; Ph.D., Purdue. Fiber optics and communications.

M. T. Jones, Assistant Professor; Ph.D., Duke. Computer engineering.

P. Kachroo, Assistant Professor; Ph.D., Berkeley. Transportation and controls.

J. S. Lai, Associate Professor; Ph.D., Tennessee. Power electronics.

F. C. Lee, Lewis A. Hester Chair; Ph.D., Duke. Power electronics, nonlinear systems.

D. K. Lindner, Associate Professor; Ph.D., Illinois at Urbana-Champaign. Control theory.

Y. Liu, Associate Professor; Ph.D., Ohio State. Power systems, computer simulations and animations.

G. Q. Lu, Assistant Professor; Ph.D., Harvard. Electronic materials.

S. F. Midkiff, Associate Professor; Ph.D., Duke. Computer architecture.

L. M. Mili, Professor; Ph.D., Liège (Belgium). Power systems.

R. L. Moose, Associate Professor; Ph.D., Duke. Electronics and statistical communication theory.

K. A. Murphy, Associate Professor; Ph.D., Virginia Tech. Fiber optics.

C. E. Nunnally, Associate Professor and Assistant Department Head; Ph.D., Virginia. Digital computers and digital control.

A. G. Phadke, American Electric Power Professor; Ph.D., Wisconsin–Madison. Power systems.

T. C. Poon, Professor; Ph.D., Iowa. Electrooptics.

T. Pratt, Professor; Ph.D., Birmingham (England). Electromagnetics.

S. Rahman, Professor; Ph.D., Virginia Tech. Energy and environment.

S. Raman, Assistant Professor; Ph.D., Michigan. High-frequency/mixed-signal integrated circuits, microelectronics, applied electromagnetics.

K. Ramu, Professor; Ph.D., Concordia (Montreal). Motor drives.

T. S. Rappaport, James S. Tucker Professor; Ph.D., Purdue. Wave propagation and communications.

B. Ravindran, Assistant Professor; Ph.D., Texas at Arlington. Computer engineering.

J. H. Reed, Associate Professor; Ph.D., California, Davis. Communications.

S. M. Riad, Professor; Ph.D., Toledo. Time domain techniques.

J. W. Roach, Associate Professor; Ph.D., Texas. Computer science.

A. Safaai-Jazi, Associate Professor; Ph.D., McGill. Optical fibers.

W. A. Scales, Associate Professor; Ph.D., Cornell. Electromagnetics.

F. W. Stephenson, Professor and Dean, College of Engineering; Ph.D., Newcastle (England). Active networks, hybrid microelectronics.

R. Stolen, Professor; Ph.D., Berkeley. Fiber optics.

W. L. Stutzman, Thomas L. Phillips Professor; Ph.D., Ohio State. Antennas and propagation.

K. S. Tam, Associate Professor; Ph.D., Wisconsin–Madison. Power systems.

Y. G. Tirat-Gefen, Assistant Professor; Ph.D., USC. Computer engineering.

W. H. Tranter, Harry Lynde Bradley Professor; Ph.D., Alabama at Tuscaloosa. Wireless communications.

J. G. Tront, Professor; Ph.D., SUNY at Buffalo. Digital computers.

H. F. VanLandingham, Professor; Ph.D., Cornell. Digital control and signal processing.

A. Wang, Professor; Ph.D., Dalian (China). Fiber optics.

B. D. Woerner, Associate Professor; Ph.D., Michigan. Wireless communications.

WASHINGTON UNIVERSITY IN ST. LOUIS

Department of Electrical Engineering

Programs of Study

The Department of Electrical Engineering offers courses of study leading to M.S.E.E. and D.Sc. degrees. Graduate students may pursue studies in the departmental faculty's areas of research. These include solid-state engineering (semiconductor and superconductor theory and devices, plasma processing and nonlinear plasma theory, optoelectronics, microwave and magnetic information devices and systems), computer engineering (architecture, parallel processing; special-purpose systems design and implementation, VLSI systems), communication theory and systems, information theory, and signal and image processing. There is also an interdepartmental program in imaging science and engineering.

Candidates for the master's degree may elect a course option (30 hours of graduate course credit) or a thesis option (30 credit hours, including 6 hours of research). Candidates for the doctoral degree accumulate a minimum of 72 hours of graduate credit, with 48 credit hours in course work and 24 credit hours in research. Flexible course requirements permit the selection of a program of study designed for individual needs and career goals.

Washington University in St. Louis's academic year begins August 25 and ends May 19.

Research Facilities

The department operates several modern research laboratories. Research in the Computer and Communications Research Center is focused on high-speed wide-area communications systems, applications and performance analysis of parallel processing, and discrete-event analysis of digital systems. The Microelectronics Systems Laboratory performs research in the areas of electronics packaging, superconducting devices, plasma processing, and microfabrication technology. The Electronic Systems and Signals Research Laboratory has projects in medical imaging algorithms and technology, information theory applications, radar and sonar systems, ultrasonic imaging, astronomical imaging, and the design of digital hearing aids. The Center for Imaging Science, a federally funded research center in the department, is devoted to the development of mathematical and algorithmic foundations for the representation and understanding of complex multidimensional scenes. Other laboratories include the Optoelectronics Research Laboratory, the Nanoelectronics Laboratory, and the Magnetic Information Systems Center.

The School of Engineering and Applied Science runs the Center for Engineering Computing, which has computer workstations dedicated to supporting course work. The laboratories have extensive computational facilities.

Olin Library, a modern, fully equipped structure located in the center of the campus, contains more than 2 million volumes. These materials are augmented by the collections of several departmental libraries.

Financial Aid

The Department of Electrical Engineering offers financial aid to full-time students enrolled in programs leading to the M.S.E.E. (thesis option) and D.Sc. degrees. Offers of financial aid are limited to available funds. For exceptionally qualified applicants, the department annually awards a number of doctoral fellowships that include tuition remission and a stipend for ten months each year. Women are eligible to apply for the prestigious Olin four-year fellowships. Financial aid in the form of research assistantships is also available through research grants and contracts administered by individual faculty members.

Cost of Study

Graduate tuition is $975 per credit hour in 1999–2000.

Living and Housing Costs

The average cost of room and board for single graduate students in University residence facilities is $8507 per academic year. Nearby off-campus apartments also are available for students at reasonable costs.

Student Group

The department's 1998–99 graduate enrollment of 66 full-time, 29 part-time, and 16 special students reflects the average composition of the graduate student body in recent years. Approximately 44 percent of the full-time graduate students currently enrolled in the department are working toward the doctoral degree.

Student Outcomes

M.S.E.E. graduates have attended the top doctoral programs in the nation, including the program at Washington University, as well as other professional degree programs in medicine, law, and business. They are employed by industry leaders from all regions and by national laboratories. A number of D.Sc. graduates have been appointed to faculty positions in top research or teaching institutions and have been sought after by industrial, academic, and national laboratories.

Location

Because of Washington University's location in metropolitan St. Louis, students are able to enjoy a stimulating, cosmopolitan environment. Among the city's many social and cultural assets are an excellent art museum, a science center, a history museum, an outstanding symphony, a planetarium, a highly regarded botanical garden, and the well-known zoo. For sports entertainment, there are professional teams in baseball, football, hockey, and soccer.

The University

Washington University in St. Louis is an independent, privately endowed and supported institution located on 169 acres in the center of the greater St. Louis area. It was founded in 1853.

Washington University has 12,035 full-time students, about half of whom are undergraduates. The evening division has 1,414 part-time students. The University has 1,191 full-time and 524 part-time faculty members and is composed of several major divisions: Arts and Sciences; the Graduate Schools of Arts and Sciences, Architecture, Business and Public Administration, Engineering and Applied Science, Fine Arts, Law, Medicine, and Social Work; the School of Continuing Education; and the Summer School. The University ranks among the top ten in the nation in the number of Nobel laureates associated with it.

Applying

Application materials may be obtained by writing to the address below. There is no deadline for applying for admission. The deadline for filing applications for financial aid is February 1 preceding the academic year for which assistance is sought. All applicants for financial aid must submit test scores on the General Test of the Graduate Record Examinations (GRE).

Correspondence and Information

Chairman
Department of Electrical Engineering
Box 1127
Washington University in St. Louis
St. Louis, Missouri 63130
E-mail: admissions@ee.wustl.edu
World Wide Web: http://ee.wustl.edu/ee

Washington University in St. Louis

THE FACULTY AND THEIR RESEARCH

The faculty members of the department participate in both undergraduate and graduate education and in interdepartmental graduate programs, interschool programs, and industrial projects. Therefore, students receive a program of education that is designed to meet their interests and is guided by the experience of an active resident faculty involved in state-of-the-art research and development. Many faculty members have significant industrial experience, and many are registered professional engineers.

R. Martin Arthur, Professor; Ph.D., Pennsylvania, 1968. Ultrasonic imaging, electrocardiography.

Roger D. Chamberlain, Associate Professor; D.Sc., Washington (St. Louis), 1989. Computer engineering, parallel computation, computer architecture, multiprocessor systems.

Jerome R. Cox Jr., Harold B. and Adelaide G. Welge Professor; Sc.D., MIT, 1954. Computer visualization, digital communication, biomedical computing.

Mark A. Franklin, Hugo F. and Ina Champ Urbauer Professor; Ph.D., Carnegie Mellon, 1970. Digital computers, systems analysis and simulation.

Daniel R. Fuhrmann, Associate Professor; Ph.D., Princeton, 1984. Statistical signal processing, image compression, numerical linear algebra.

Manju V. Hegde, Associate Professor; Ph.D., Michigan, 1987. Telecommunication networks, communication and information theory, wireless networks, stochastic optimization.

Ronald S. Indeck, Professor; Ph.D., Minnesota, 1987. Magnetics, microelectronic devices, thin-film technology.

Jia G. Lu, Assistant Professor; Ph.D., Harvard, 1997. Nanostructured materials and devices, low-temperature physics.

Paul S. Min, Associate Professor; Ph.D., Michigan, 1987. Routing and control of telecommunications networks, fault-tolerance and reliability, software systems, network management.

Robert E. Morley Jr., Associate Professor; D.Sc., Washington (St. Louis), 1977. Computer and communication systems, VLSI design, digital signal processing.

Marcel W. Muller, Research Professor and Professor Emeritus; Ph.D., Stanford, 1957. Solid-state physics, microwave electronics, magnetics.

Joseph A. O'Sullivan, Associate Professor; Ph.D., Notre Dame, 1986. Statistical signal processing, radar systems, information theory, nonlinear control theory.

William F. Pickard, Professor; Ph.D., Harvard, 1962. Biological transport, electrobiology, energy engineering.

William D. Richard, Associate Professor; Ph.D., Missouri, 1988. Computer engineering, machine vision, medical instrumentation.

Daniel L. Rode, Professor; Director, Optoelectronics Research Laboratory; Ph.D., Case Western Reserve, 1968. Optoelectronics and fiber optics; semiconductor materials, processing, and devices.

Frederick U. Rosenberger, Ph.D., Washington (St. Louis), 1969. Computer systems.

John C. Schotland, Associate Professor; M.D., 1984, Ph.D., 1997, Pennsylvania. Optical physics, medical imaging, microscopy, remote sensing.

Barbara A. Shrauner, Professor; Ph.D., Harvard (Radcliffe), 1962. Plasma processing, semiconductor transport, symmetries of nonlinear differential equations.

Donald L. Snyder, Samuel C. Sachs Professor; Ph.D., MIT, 1966. Communication theory, random process theory, signal processing, biomedical engineering, image processing, radar.

Stefano Soatto, Assistant Professor; Ph.D., Caltech, 1996. Nonlinear estimation and control theory and applications, computer vision, mathematical modeling of human vision.

Barry E. Spielman, Professor and Chairman of the Department; Ph.D., Syracuse, 1971. High-frequency/high-speed devices, integrated circuits, superconducting electronics.

Professors Emeriti

Lloyd R. Brown, D.Sc., Washington (St. Louis), 1960. Automatic control, electronic instrumentation.

Marvin J. Fisher, Ph.D., Illinois, 1957. Energy conversion, power electronics.

Robert O. Gregory, D.Sc., Washington (St. Louis), 1964. Electronic instrumentation, microwave theory, circuit design.

Raymond M. Kline, Professor; Ph.D., Purdue, 1962. Computer engineering, computer-aided design, control systems.

Harold W. Shipton, F.I.E.R.E., Shrewsbury Technical, 1949. Biomedical engineering, electronic instrumentation.

Charles M. Wolfe, Ph.D., Illinois, 1965. Semiconductor materials and devices, statistical physics, optimization.

Affiliated Faculty

Frank R. Agovino, J.D., Cincinnati, 1973. Intellectual property of electronics.

G. James Blaine, D.Sc., Washington (St. Louis), 1974. Biomedical computing, digital communications and information systems.

John D. Corrigan, Ph.D., Missouri, 1973. Control systems theory.

Julius L. Goldstein, Ph.D., Rochester, 1965. Mathematical models of sensory communication, normal and impaired processing.

H. Richard Grodsky, Assistant Professor; D.Sc., Washington (St. Louis), 1971. Artificial intelligence, expert systems, intelligent tutoring systems, object-oriented modeling.

Tom R. Miller, Ph.D., Stanford, 1971; M.D., Missouri, 1976. Medical imaging.

Stanley Misler, Ph.D./M.D., NYU, 1977. Patch clamp characterization of ion channel in stimulus-secretion coupling stimulus.

William J. Murphy, D.Sc., Washington (St. Louis), 1967. Control systems.

Gurudatta M. Parulkar, Ph.D., Delaware, 1987. Computer communications, local area networks, distributed processing.

Jonathan S. Turner, Ph.D., Northwestern, 1982. Communications systems, algorithms and complexity, VLSI applications.

WAYNE STATE UNIVERSITY

Department of Electrical and Computer Engineering

Programs of Study

The Department of Electrical and Computer Engineering (ECE) offers graduate programs leading to the Master of Science (M.S.) and the Doctor of Philosophy (Ph.D.) degrees in electrical and computer engineering. Electrical engineering is organized into five areas: controls, communications and circuits, optical engineering, solid-state electronics, and power systems. Computer engineering is segmented into three areas: computer architecture and digital design, parallel and distributed systems, and artificial intelligence and applications. Thirty-two credits are required for the master's degree, and thesis and nonthesis options are available. A minimum of 90 credits beyond the bachelor's degree is required for the Ph.D., of which 30 credits are designated for the thesis requirement. Incoming students with an M.S. degree may transfer up to 30 graduate credits toward the 90-credit Ph.D. requirement. Most ECE course offerings are scheduled to begin after 3:30 p.m. in order to accommodate both full-time and part-time students. A course load of eight credits or more per semester is considered to be a full-time load. ECE graduate courses are four credits each and are offered twice per week for a period of two hours.

Research Facilities

The department's laboratories focus on the following fields: applied optics, computation and neural networks, enabling technologies for the disabled, device and circuit simulation, microelectronic and photonic materials, parallel and distributed computing, and VLSI design. The five specialty research groups within the department are Smart Sensors and Integrated Devices (G. Auner); Parallel and Distributed Computing (V. Chaudhary); Materials, Devices, and Circuit Simulation (V. Mitin); Photonic Devices and Systems (Yang Zhao); and Control Systems (F. Lin and L. Wang). The Department has a cluster of four SUN Enterprise 3000s (four CPUs) and Enterprise 4000s (six CPUs) connected by fast ethernet and ATM OC-3. In addition, an eight-node Linux/NT cluster of 350-MHz dual Pentium IIs is connected by a fast and gigabit ethernet, and four SUN E3500s, each with four CPUs, are currently being acquired. The department also has access to the College Computer Center and the University's Division of Computer and Information Technology. The College of Engineering maintains an Ethernet TCP/IP-based LAN, UNIX-based servers and workstations, and more than 100 Windows-based PCs for student use. The University Computing Center maintains a Cray J916 supercomputer with sixteen vector processors, and the acquisition of another massively parallel computer is being planned. The new Undergraduate Library, adjacent to the Engineering Building, has more than 700 PCs and PowerMacs for student use.

Financial Aid

Seventeen graduate teaching assistantships are offered by the department, which require students to support the instructional and teaching laboratory programs on a part-time basis. In addition, faculty members employ graduate research assistants who are performing thesis research on programs funded by government and private agencies. A few highly competitive fellowships and housing allowances are also available. Fellowships, traineeships, and assistantships normally provide stipends of $7000 to $17,000 per academic year, tuition, full medical and dental insurance, and, in some instances, a housing allowance in University-affiliated facilities.

Cost of Study

For 1998–99, Michigan state resident graduate tuition and fees were $247 for the first credit hour and $178 for each subsequent credit hour per semester. Nonresident graduate tuition and fees were $439 for the first credit hour and $370 for each subsequent credit hour per semester. Residents of northwestern Ohio and the Canadian province of Ontario pay tuition at the Michigan resident rate. (Tuition and fees are subject to change by action of the Wayne State University Board of Governors.)

Living and Housing Costs

The University offers apartment housing for single and married graduate students. The University Housing Office can also provide some assistance with off-campus accommodations. A typical one-bedroom apartment near the University costs from $300 to $650 per month; a two-bedroom apartment costs from $500 to $850 per month. A single student needs between $18,000 and $22,000 per calendar year for living expenses, including tuition, housing, food, personal, and other costs.

Student Group

Nearly 32,000 students are enrolled at Wayne State University (WSU), of whom half attend part-time. Approximately 10,000 of these students are enrolled in graduate and professional postbachelor programs. The College of Engineering enrolls nearly 1,400 undergraduates and more than 1,300 graduate students. The Department of Electrical and Computer Engineering has more than 450 graduate students, sixty percent of whom are employed part-time students. The majority of the part-time students are working professionals whose special skills enrich the classroom experience.

Location

Located in Detroit's Cultural Center, the University is surrounded by an area that offers many industrial, business, and cultural opportunities. Nearby are major research and production centers devoted to computer applications; telecommunications; fuels; metal cutting and stamping; polymers; automobiles; automotive equipment, parts, and components; and mass production tooling. The area is also a center for highway and construction industries and has two of the largest hospital complexes in the nation. Among the cultural attractions are professional and University theaters, opera, and symphony orchestras; art galleries and institutes; museums; classical, jazz, and soul music; and top-level international and Broadway entertainment. Detroit is the home of new baseball and football stadiums, and three full-scale casinos are about to open close to the campus. The Henry Ford Museum and Greenfield Village, where American history is reenacted, are 10 miles away. The Detroit River, Lake St. Clair, Lake Erie, and Lake Huron attract watersport and fishing enthusiasts. Michigan is a tourist destination and has many lakes, beaches, forests, and hiking and skiing areas. Windsor, Ontario, Canada, is only 3 miles south of the campus, and the cities of Chicago and Toronto are both less than 300 miles away.

The University and the College

WSU began as a post–high school program in Detroit's Central High School, whose beautiful building is now the beloved "Old Main." Successive educational needs led to junior college, city college, and city university status. In 1956, Wayne became a state university. It now comprises eleven colleges and schools as well as the Graduate School and the Center for Urban Studies. Its buildings have been designed by world-class architects and are surrounded by attractive landscaping. The North Central Association of Colleges and Schools accredits the University. The College of Engineering's research programs attract more than $13 million per year in research grants and contracts from government agencies and from public and private institutions. The Department of Electrical and Computer Engineering conducts more than $3 million in funded research per year.

Applying

Applications for graduate admission, accompanied by a $20 fee ($30 for international students) and transcripts should reach the Office of University Admissions by July 1 for the fall term, November 1 for the winter term, and March 15 for the spring/summer term. The deadline for fellowships, assistantships, and scholarships is February 1. GRE scores are not required for admission but are recommended for applicants seeking financial support. A minimum current TOEFL score of 550, or the equivalent, is required of international students whose native language is not English.

Correspondence and Information

Department of Electrical and Computer Engineering
College of Engineering
Wayne State University
Detroit, Michigan 48202-3902

Telephone: 313-577-3920
Fax: 313-577-1101
E-mail: psiy@ece.eng.wayne.edu
World Wide Web: http://www.wayne.edu (University)
http://www.eng.wayne.edu (College)
http://www.ece.eng.wayne.edu (Department)

Wayne State University

THE FACULTY AND THEIR RESEARCH

Gregory Auner, Professor. Wide bandgap semiconductors, graded pyroelectric materials, magnetic materials for sensors and device development, smart sensors. (telephone: 313-577-3904; e-mail: gauner@eng.wayne.edu)

Ivan Avrutsky, Assistant Professor. Optoelectornics, theory and technology of optical waveguides and gratings, fiber and integrated optics, optics of nanostructures, semiconductor lasers. (telephone: 313-577-4801; e-mail: avrutsky@ece.eng.wayne.edu)

Jatinder Bedi, Associate Professor. Real-time distributed processing of applications using microprocessors, design of controllers using Fuzzy-Neuro systems, digital communications systems. (telephone: 313-577-3850; e-mail: jbedi@eng.wayne.edu)

Vipin Chaudhary, Associate Professor. Parallelizing compilers and run-time support systems, parallel and distributed systems (algorithms, architectures, and applications), ATM, computer vision and image processing, graphics, biomedical engineering. (telephone: 313-577-0605; e-mail: vipin@eng.wayne.edu)

Guy Edjlali, Assistant Professor. Parallel and distributed computing, operating systems, compilers, security. (telephone: 313-577-3738; e-mail: edjlali@ece.eng.wayne.edu)

Robert Erlandson, Associate Professor and head of department's biomedical engineering effort. System methodologies suitable for analysis and evaluation of large complex systems, particularly physiological structures; development of decision-making methodologies utilizing multivalued logic and nonparametric techniques. (telephone: 313-577-3900; e-mail: rerlands@eng.wayne.edu)

Zinovi S. Gribnikov, Professor (Research). Theory of semiconductors and semiconductor devices; micro, nano, and optoelectronics. (telephone: 313-577-0764; e-mail: zinovi@ciao.eng.wayne.edu)

Mohamad Hassoun, Professor. Artificial neural systems; associative memories; machine learning; pattern recognition; application of artificial neural networks to physiologic signal processing, optimization, and control. (telephone: 313-577-3966; e-mail: hassoun@brain.eng.wayne.edu)

Feng Lin, Associate Professor. Systems and control, hierarchical structure of discrete event systems, decision analysis for complex processes, control and optimization of flexible manufacturing systems. (telephone: 313-577-3428; e-mail: flin@eng.wayne.edu)

Syed M. Mahmud, Associate Professor. Microprocessor-based system design, digital system design, special purpose computer architectures, cache-based multiprocessor system design and performance analysis. (telephone: 313-577-3855; e-mail: smahmud@eng.wayne.edu)

Jerome Meisel, Professor. Electromechanical energy conversion, bulk power system planning and operation, power conditioning, power electronic circuitry, applied control theory. (telephone: 313-577-3530; e-mail: jmeisel@eng.wayne.edu)

Vladimir Mitin, Professor. Numerical simulations of laser based on low-dimensional semiconductor structures, multiterminal lasers, vertical cavity lasers, high-frequency devices and thyristors, heat dissipation in low-dimensional structures and devices. (telephone: 313-577-8944; e-mail: mitin@eng.wayne.edu)

Sumit Roy, Assistant Professor (Research). Parallel and distributed computing, compilers operating systems, networking technologies. (telephone: 313-577-2207; e-mail: sroy@eng.wayne.edu)

Shishir Shah, Assistant Professor. Computer vision, image processing, biomedical image analysis, pattern recognition, robotics, human-computer interaction, sensor fusion, hybrid systems, neural networks, statistical estimation theory, computer graphics. (telephone:313-577-3530; e-mail: shishir@eng.wayne.edu)

Donald Silversmith, Professor. Microelectromechanical system design and fabrication technology, solid-state and microsystem device design, integrated circuit fabrication technology, VLSI design. (telephone: 313-577-0248; e-mail: silversm@ece.eng.wayne.edu)

Harpreet Singh, Professor. State-variables and system theoretic and Petri Net approach to computer hardware and software, vehicle guidance, software engineering, expert systems, VLSI design. (telephone: 313-577-3917; e-mail: hsingh@eng.wayne.edu)

Pepe Siy, Associate Professor. Pattern recognition, image processing, parallel discrete computational problems, analog and digital VLSI, smart sensor technology. (telephone: 313-577-3841; e-mail: psiy@eng.wayne.edu)

Loren Schwiebert, Assistant Professor. Parallel computer architecture, wireless networking, interconnection network routing, ATM networking. (telephone: 313-577-3990; e-mail: loren@eng.wayne.edu)

Le Yi Wang, Associate Professor. H-infinity optimization, stabilization and optimization of time-varying systems, frequency-domain systems identification, hybrid control systems, automotive control systems, nonlinear and adaptive control. (telephone: 313-577-4715; e-mail: lywang@eng.wayne.edu)

Franklin Westervelt, Professor and Chair. Computer systems, VLSI design, special purpose computer architectures for handling very large databases. (telephone: 313-577-3764; e-mail: fwesterv@eng.wayne.edu)

James Woodyard, Associate Professor. Ion beam analysis and modification of thin-film devices and device materials; hydrogenation, dehydrogenation, and radiation resistance of amorphous semiconductor materials; optical and electrical characterization of device materials and device fabrication. (telephone: 313-577-3758, e-mail: woodyard@eng.wayne.edu)

Chengzhong Xu, Assistant Professor. Parallel computing, particularly run-time and operating system support for irregularly structured applications; distributed shared memory systems; multiprocessor server technologies. (telephone: 313-577-3856; e-mail: czxu@eng.wayne.edu)

Yang Zhao, Professor. Nonlinear optical devices for communications, novel optical materials, optical sensing, lasers. (telephone: 313-577-3404; e-mail: yzhao@eng.wayne.edu)

WRIGHT STATE UNIVERSITY

Computer Engineering Program

WRIGHT STATE
UNIVERSITY™

Programs of Study	The Department of Computer Science and Engineering offers graduate programs leading to the Master of Science in Computer Engineering (M.S.C.E.) and the Doctor of Philosophy in computer science and engineering. A Master of Science in Computer Science program is also available. Interested students should also see the Computer Science Program in this volume.

The M.S.C.E. program requires 48 graduate quarter credit hours in a department-approved program of study. Typically, students must complete ten courses and orally defend a thesis or complete twelve courses in the nonthesis option. The Ph.D. program requires 91 graduate quarter credit hours after completion of the M.S.C.E. or 136 hours after completion of a bachelor's degree. Students without a master's degree must complete the requirements for the M.S.C.E. as the first phase of their Ph.D. program.

Ph.D. students must complete 76 hours of formal course work (courses completed in the master's program may partially satisfy this requirement), pass the Ph.D. qualifying examinations and candidacy examination, and successfully complete and defend a dissertation. |
| **Research Facilities** | Modern laboratory facilities provide ample equipment for research in a number of areas. The College of Engineering and Computer Science provides access to DEC Alpha servers and workstations, a Silicon Graphics (SGI) Reality Engine 2, and SGI, DEC, and Sun workstations as well as numerous networked PCs and X-Windows terminals. In addition to these machines, dedicated research labs containing an SGI Onyx 2 with eight parallel processors as well as other SGI, DEC, and Sun equipment are available. Wright State is a member of OARnet, which provides access to the resources of the Ohio Supercomputer Center. Wright State's Information Technology Research Institute, a collaboration of academia, industry, and government, provides opportunities for research in all areas of modern information technology.

Laboratories specifically dedicated to student and faculty research support studies in artificial intelligence, networking, human-computer interaction, parallel and concurrent computing, operating systems, control and robotics, digital communications, graphics, computer vision and imaging, computer-aided design, digital design, VLSI, software engineering, and optical computing. Wright Patterson Air Force Base and several Dayton industries participate in sponsored research, and their facilities are frequently available to graduate students and faculty. In addition, Wright Patterson's Major Shared Resource Center provides an opportunity for collaborative research with a focus on supercomputing. |
Financial Aid	Financial aid available to graduate students includes research assistantships, teaching assistantships, Federal Perkins Loans, Federal Stafford Student Loans, short-term loans, and academic tuition fellowships. Scholarships are also available through the Dayton Area Graduate Studies Institute. Competitive stipends for research and teaching assistantships are available.
Cost of Study	Tuition in 1999 for residents of Ohio is $161 per quarter hour for part-time study (1–10½ hours) and $1703 for full-time study (11–18 hours). Tuition for out-of-state students is $282 per quarter hour for part-time study and $3013 for full-time study.
Living and Housing Costs	Many students commute to classes; however, campus housing is available. Room and board cost about $1600 per quarter. There are a variety of apartments and houses for rent in the area. The cost of living in Dayton is low compared with that in most other metropolitan areas.
Student Group	Approximately 3,200 graduate students are enrolled at Wright State University in forty graduate degree programs that lead to master's, Ed.S., Ph.D., M.D., and Psy.D. degrees.
Student Outcomes	Graduates are employed in a variety of professional positions both in- and out-of-state. Typical positions include software engineer, computer engineer, systems analyst, and systems programmer.
Location	Although Wright State has a suburban setting, it is closely tied to Dayton, the fourth-largest metropolitan area in Ohio. Dayton is a manufacturing center and is also the home of Wright-Patterson Air Force Base, the center of Air Force research and procurement. This combination of industrial and military development has produced an unusual concentration of scientific, technical, and research activity. Dayton has a considerable and varied cultural life. In the surrounding area there are at least fifteen other colleges and universities.
The University	Wright State University is an exciting and expanding university that continues in the scientific and engineering innovative spirit of aviation pioneers Orville and Wilbur Wright. The University has modern buildings that facilitate access for all, and the 557-acre main campus has accommodations for academics, support programs, sports, housing, and arts activities. Wright State's phenomenal growth as an institution—from one building, 92 employees, and 3,200 students in 1964 to forty-two modern buildings, nearly 2,100 employees, and more than 15,000 students in 1998—has necessarily stimulated a comparable growth in the areas that surround the University.
Applying	The basic requirement for admission to the master's degree program in computer engineering is a bachelor's degree in computer engineering or a related area with an overall undergraduate grade point average of at least 3.0 (on a 4.0 scale) or 2.7 with an average of 3.0 or better in the major field. A minimum grade point average of 3.3 is expected for admission to the Ph.D. program. Scores on the GRE General Test are required. Specific prerequisites for admission include calculus, differential equations, linear algebra, physics, and circuit analysis. The student's background should include a knowledge of a higher-level language, data structures, concurrent programming, computer organization, operating systems, digital hardware design, and electronics. Minor background deficiencies may be made up after admission to the program by taking appropriate courses.

There is no deadline for admission applications since admission decisions are made on a rolling basis and programs may be started in any quarter; however, the deadline to apply for teaching assistantships is February 1. |
| **Correspondence and Information** | Oscar N. Garcia, Chair
Department of Computer Science and Engineering
Wright State University
3640 Colonel Glenn Highway
Dayton, Ohio 45435-0001
Telephone: 937-775-5131
Fax: 937-775-5133
E-mail: cse-dept@.cs.wright.edu
World Wide Web: http://www.cs.wright.edu/cse/general.html |

Wright State University

THE FACULTY AND THEIR RESEARCH

A. A. S. Awwal, Associate Professor; Ph.D., Dayton, 1989. Digital optical computing, novel learning and architectures for neural networks, computer arithmetic, digital systems, computer-aided optical systems design, optical data storage/retrieval algorithms, fingerprint/character recognition, automated target recognition.

P. Bruce Berra, Professor; Ph.D, Purdue, 1968. Optical and electronic computer architectures, very large multimedia data/knowledge bases.

C. I. Henry Chen, Associate Professor; Ph.D., Minnesota, 1989. CAD, simulation and testing of VLSI circuits, aerospace and electronics systems (testability synthesis and design verification of digital systems, signal processing, and microwave electronic circuits).

C. L. Philip Chen, Associate Professor; Ph.D., Purdue, 1988. Neural networks and applications, CAD/CAM and robotics, intelligent systems and interfaces, knowledge-based systems.

Jer-Sen Chen, Research Assistant Professor; Ph.D., USC, 1989. Computer graphics, image/video compression, human-computer interactions.

Soon M. Chung, Associate Professor; Ph.D., Syracuse, 1990. Database multimedia, parallel processing, computer architecture.

Michael T. Cox, Assistant Professor; Ph.D., Georgia Tech, 1996. Intelligent interfaces, case-based reasoning, automated planning, machine learning, natural language processing.

Guozhu Dong, Associate Professor; Ph.D., USC, 1988. Database systems; database, data mining, and knowledge discovery; data warehousing and integration; workflow.

Travis E. Doom, Assistant Professor; Ph.D., Michigan State, 1998. Computer architecture, computer systems, design automation, computational mathematics and theory.

Oscar N. Garcia, Professor; Ph.D., Maryland, 1969. Speech recognition and articulatory synthesis, knowledge-based systems, computer architecture, human-computer interaction, intelligent interfaces, machine intelligence.

A. Ardeshir Goshtasby, Associate Professor; Ph.D., Michigan State, 1983. Intelligent interfaces, machine vision, computer graphics and visualization, geometric modeling, medical image analysis.

Ricardo Gutierrez-Osuna, Assistant Professor; Ph.D., North Carolina State, 1998. Electronic olfaction, mobile robotics, pattern recognition, artificial intelligence.

Jack S. Jean, Associate Professor; Ph.D., USC, 1988. High-performance computer architectures, machine intelligence.

Prabhakar Mateti, Associate Professor; Ph.D., Illinois at Urbana-Champaign, 1976. Distributed computing, Internet security, formal methods in software design.

Kuldip Rattan, Professor; Ph.D., Kentucky, 1975. Fuzzy control, robotics, digital control systems, prosthetic/orthotics and microprocessor applications.

Mateen M. Rizki, Associate Professor; Ph.D., Wayne State, 1985. Modeling, simulation, biological information processing, machine intelligence, pattern recognition, image processing.

Robert C. Shock, Associate Professor; Ph.D., North Carolina at Chapel Hill, 1969. Software engineering, database systems.

Raymond E. Siferd, Professor; Ph.D., Air Force Tech, 1977. VLSI design and parallel computing.

Thomas A. Sudkamp, Professor; Ph.D., Notre Dame, 1978. Approximate reasoning, machine intelligence.

Krishnaprasad Thirunarayan (T. K. Prasad), Associate Professor; Ph.D., SUNY at Stony Brook, 1989. Knowledge representation and reasoning, object-oriented programming, specification of programming languages.

Karen A. Tomko, Assistant Professor; Ph.D., Michigan, 1995. Parallel computing, application optimization, graph partitioning, reconfigurable computing.

Section 10
Energy and Power Engineering

This section contains a directory of institutions offering graduate work in energy and power engineering, followed by in-depth entries submitted by institutions that chose to prepare detailed program descriptions. Additional information about programs listed in the directory but not augmented by an in-depth entry may be obtained by writing directly to the dean of a graduate school or chair of a department at the address given in the directory.

For programs offering related work, see also in this book Computer Science and Information Technology, Engineering and Applied Sciences, Industrial Engineering, and Mechanical Engineering and Mechanics. In Book 4, see Physics and Mathematical Sciences.

CONTENTS

Energy and Power Engineering

Colorado State University, Graduate School, College of Engineering, Department of Mechanical Engineering, Program in Energy and Environmental Engineering, Fort Collins, CO 80523-0015. Offers MS, PhD. Postbaccalaureate distance learning degree programs offered (minimal on-campus study). *Faculty:* 18 full-time (0 women). *Degree requirements:* For doctorate, dissertation required, foreign language not required, foreign language not required. *Entrance requirements:* For master's and doctorate, GRE General Test (minimum combined score of 1850 on three sections required; average 1872), TOEFL (minimum score of 550 required; average 596), minimum GPA of 3.0. *Application deadline:* For fall admission, 2/1 (priority date). Applications are processed on a rolling basis. Application fee: $30. Electronic applications accepted. *Faculty research:* Indoor air quality, solar energy industry energy conservation, building and industrial energy conservation, engine pollution abatement optimal control. *Unit head:* Christine Piper, Graduate Coordinator, 864-656-7581, E-mail: cpiper@clemson.edu. *Application contact:* Dr. Doug Hittle, Graduate Committee Chairman, 970-491-8617, Fax: 970-491-3827, E-mail: hittle@lamar.colostate.edu.

New Jersey Institute of Technology, Office of Graduate Studies, Department of Electrical and Computer Engineering, Newark, NJ 07102-1982. Offers computer engineering (MS); electrical engineering (MS, PhD, Engineer), including biomedical systems (MS, PhD), communication and signal processing (MS, PhD), computer systems (MS, PhD), control systems (MS, PhD), microwave and lightwave engineering (MS, PhD), solid-state materials and devices (MS, PhD); power engineering (MS); telecommunications (MS). Part-time and evening/weekend programs available. Terminal master's awarded for partial completion of doctoral program. *Degree requirements:* For master's, thesis required (for some programs), foreign language not required; for doctorate, dissertation, residency required, foreign language not required. *Entrance requirements:* For master's, GRE General Test (minimum score of 450 on verbal section, 600 on quantitative, 550 on analytical required); for doctorate, GRE General Test (minimum score of 450 on verbal section, 600 on quantitative, 550 on analytical required), minimum graduate GPA of 3.5. Electronic applications accepted. *Faculty research:* Communications systems design, digital signal processing.

New York Institute of Technology, Graduate Division, School of Engineering and Technology, Program in Energy Management, Old Westbury, NY 11568-8000. Offers energy management (MS); energy technology (Certificate); environmental management (Certificate). Part-time and evening/weekend programs available. Postbaccalaureate distance learning degree programs offered. *Students:* 14 full-time (2 women), 130 part-time (27 women); includes 57 minority (35 African Americans, 9 Asian Americans or Pacific Islanders, 13 Hispanic Americans), 6 international. *Degree requirements:* For master's, thesis or alternative, oral or written comprehensive exam required, foreign language not required; for degree, foreign language not required. *Entrance requirements:* For master's, minimum QPA of 2.85. *Application deadline:* For fall admission, 8/1. Applications are processed on a rolling basis. Application fee: $50. Electronic applications accepted. *Unit head:* Dr. Robert Amundsen, Chair, 516-686-7578. *Application contact:* Glenn Berman, Executive Director of Admissions, 516-686-7519, Fax: 516-626-0419, E-mail: gberman@iris.nyit.edu.

Rensselaer Polytechnic Institute, Graduate School, School of Engineering, Department of Electric Power Engineering, Troy, NY 12180-3590. Offers M Eng, MS, D Eng, PhD, MBA/M Eng. Part-time and evening/weekend programs available. *Faculty:* 5 full-time (0 women), 5 part-time (0 women). *Students:* 39 full-time (3 women), 13 part-time (1 woman); includes 7 minority (1 Asian American or Pacific Islander, 6 Hispanic Americans), 25 international. 72 applicants, 57% accepted. In 1998, 19 master's, 2 doctorates awarded. Terminal master's awarded for partial completion of doctoral program. *Degree requirements:* For master's, computer language, thesis required (for some programs), foreign language not required; for doctorate, computer language, dissertation required, foreign language not required. *Entrance requirements:* For master's and doctorate, GRE, TOEFL (minimum score of 550 required). *Application deadline:* For fall admission, 2/1 (priority date). Applications are processed on a rolling basis. Application fee: $35. *Financial aid:* In 1998–99, 22 students received aid, including 8 research assistantships (averaging $7,900 per year), 9 teaching assistantships (averaging $6,657 per year); fellowships, career-related internships or fieldwork and institutionally-sponsored loans also available. Financial aid application deadline: 2/1. *Faculty research:* Power switching technology, electric and magnetic field computation, electrical transients, electrical insulation, power system analysis, power electronic and motor control systems. Total annual research expenditures: $588,000. *Unit head:* Dr. J. Keith Nelson, Chair, 518-276-6328, Fax: 518-276-6226. *Application contact:* Rose Carignan, Admissions Assistant, 518-276-6329, Fax: 518-276-6226, E-mail: carigr@rpi.edu.

See in-depth description on page 1137.

Southern Illinois University Carbondale, Graduate School, College of Engineering, Program in Engineering Sciences, Carbondale, IL 62901-6806. Offers electrical systems (PhD); fossil energy (PhD); mechanics (PhD). *Faculty:* 57 full-time (1 woman). *Students:* 16 full-time (2 women), 10 part-time (2 women); includes 3 minority (1 African American, 2 Asian Americans or Pacific Islanders), 17 international. *Degree requirements:* For doctorate, dissertation required. *Entrance requirements:* For doctorate, GRE General Test, TOEFL (minimum score of 600 required), minimum GPA of 3.5. Application fee: $20. *Unit head:* Dr. Hasan Sevim, Associate Dean, 618-453-4321, Fax: 618-453-4235.

University of Alberta, Faculty of Graduate Studies and Research, Department of Electrical and Computer Engineering, Edmonton, AB T6G 2E1, Canada. Offers computational optics (PhD); computer engineering (M Eng, M Sc, PhD); control systems (M Eng, M Sc, PhD); engineering management (M Eng); laser physics (M Sc, PhD); oil sands (M Eng, M Sc, PhD); plasma physics (M Sc, PhD); power engineering (M Eng, M Sc, PhD); telecommunications (M Eng, M Sc, PhD). Terminal master's awarded for partial completion of doctoral program. *Degree requirements:* For master's and doctorate, thesis/dissertation required, foreign language not required. *Entrance requirements:* For master's and doctorate, TOEFL (minimum score of 580 required; average 610). Electronic applications accepted. *Faculty research:* Controls, communications, microelectronics, electromagnetics.

University of Massachusetts Lowell, Graduate School, College of Arts and Sciences, Department of Physics and Applied Physics, Program in Energy Engineering, Lowell, MA 01854-2881. Offers PhD. *Degree requirements:* For doctorate, 2 foreign languages (computer language can substitute for one), dissertation required. *Entrance requirements:* For doctorate, GRE General Test. *Application deadline:* For fall admission, 4/1 (priority date); for spring admission, 10/1. Applications are processed on a rolling basis. Application fee: $20 ($35 for international students). *Financial aid:* Fellowships, research assistantships, teaching assistantships, career-related internships or fieldwork, Federal Work-Study, and institutionally-sponsored loans available. Aid available to part-time students. Financial aid application deadline: 4/1. *Application contact:* Dr. Gus Couchell, Coordinator, 978-934-3772, E-mail: gus_couchell@uml.edu.

University of Massachusetts Lowell, Graduate School, James B. Francis College of Engineering, Department of Energy Engineering, Lowell, MA 01854-2881. Offers MS Eng. *Students:* 6 full-time (0 women), 15 part-time (2 women); includes 1 minority (Asian American or Pacific Islander), 4 international. In 1998, 10 degrees awarded. *Degree requirements:* For master's, thesis optional. *Entrance requirements:* For master's, GRE General Test. *Application deadline:* For fall admission, 4/1 (priority date); for spring admission, 10/1. Applications are processed on a rolling basis. Application fee: $20 ($35 for international students). *Financial aid:* In 1998–99, 2 research assistantships, 4 teaching assistantships were awarded. Financial aid application deadline: 4/1. *Unit head:* Dr. John Duffy, Graduate Coordinator, 978-934-2968, E-mail: john_duffy@woods.uml.edu. *Application contact:* Dr. John Duffy, Graduate Coordinator, 978-934-2968, E-mail: john_duffy@woods.uml.edu.

The University of Memphis, Graduate School, Herff College of Engineering, Department of Mechanical Engineering, Memphis, TN 38152. Offers design and mechanical engineering (MS); energy systems (MS); mechanical engineering (PhD); mechanical systems (MS); power systems (MS). Part-time programs available. *Faculty:* 11 full-time (0 women). *Students:* 9 full-time (2 women), 8 part-time (1 woman), 9 international. Terminal master's awarded for partial completion of doctoral program. *Degree requirements:* For master's and doctorate, thesis/dissertation, comprehensive exam required. *Entrance requirements:* For master's, GRE General Test (minimum combined score of 1100 required), BS in mechanical engineering, minimum undergraduate GPA of 3.0. *Application deadline:* For fall admission, 8/1; for spring admission, 12/1. Application fee: $25 ($50 for international students). Tuition, state resident: full-time $3,410; part-time $178 per credit hour. Tuition, nonresident: full-time $8,670; part-time $408 per credit hour. Tuition and fees vary according to program. *Unit head:* Dr. John I. Hochstein, Chair, 901-678-2173, Fax: 901-678-5459, E-mail: jhochste@memphis.edu. *Application contact:* Dr. Teong E. Tan, Coordinator of Graduate Studies, 901-678-3264, Fax: 901-678-5459, E-mail: ttan@memphis.edu.

University of North Dakota, Graduate School, School of Engineering and Mines, Program in Energy Engineering, Grand Forks, ND 58202. Offers PhD. *Faculty:* 8 full-time (0 women). *Students:* 2 full-time (0 women), 3 part-time (1 woman). In 1998, 1 degree awarded. *Degree requirements:* For doctorate, dissertation required. *Entrance requirements:* For doctorate, TOEFL (minimum score of 550 required), minimum GPA of 3.5. *Application deadline:* For fall admission, 3/1 (priority date). Applications are processed on a rolling basis. Application fee: $20. *Financial aid:* In 1998–99, 1 student received aid, including 1 research assistantship; fellowships Financial aid application deadline: 3/15. *Faculty research:* Combustion science, energy conversion, power transmission, environmental engineering. *Unit head:* Dr. Rashid Hasan, Director, 701-777-3411, Fax: 701-777-4838, E-mail: rashid_hasan@mail.und.nodak.edu.

Worcester Polytechnic Institute, Graduate Studies, Department of Electrical and Computer Engineering, Worcester, MA 01609-2280. Offers electrical and computer engineering (MS, PhD, Advanced Certificate, Certificate); power systems engineering (MS, PhD). Part-time and evening/weekend programs available. *Faculty:* 22 full-time (1 woman), 3 part-time (0 women). *Students:* 48 full-time (7 women), 47 part-time (6 women); includes 7 minority (1 African American, 3 Asian Americans or Pacific Islanders, 3 Hispanic Americans), 32 international. Terminal master's awarded for partial completion of doctoral program. *Degree requirements:* For master's, thesis optional, foreign language not required; for doctorate, dissertation required, foreign language not required. *Entrance requirements:* For master's and doctorate, GRE (required for non-native speakers of English; combined average of 2024 on three sections), TOEFL (minimum score of 550 required; average 618). *Application deadline:* For fall admission, 2/15 (priority date); for spring admission, 10/15 (priority date). Applications are processed on a rolling basis. Application fee: $50. Electronic applications accepted. *Unit head:* Dr. John A. Orr, Head, 508-831-5231, Fax: 508-831-5491, E-mail: orr@wpi.edu. *Application contact:* Fred J. Looft, Graduate Coordinator, 508-831-5231, Fax: 508-831-5491, E-mail: fjlooft@wpi.edu.

Nuclear Engineering

Air Force Institute of Technology, School of Engineering, Department of Engineering Physics, Program in Nuclear Engineering, Wright-Patterson AFB, OH 45433-7765. Offers MS, PhD. *Accreditation:* ABET (one or more programs are accredited). Part-time programs available. *Faculty:* 4 full-time (0 women). *Students:* 5 full-time (0 women). In 1998, 1 doctorate awarded (100% found work related to degree). *Degree requirements:* For master's and doctorate, thesis/dissertation required, foreign language not required. *Entrance requirements:* For master's, GRE General Test (minimum score of 500 on verbal section, 600 on quantitative required), minimum GPA of 3.0, must be military officer or U.S. citizen; for doctorate, GRE General Test (minimum score of 550 on verbal section, 650 on quantitative required), minimum GPA of 3.0, must be military officer or U.S. citizen. Application fee: $0. *Faculty research:* Nuclear weapon effects, nuclear nonproliferation, medical imaging. *Unit head:* Dr. Kirk A. Mathews, Curriculum Chair, 937-255-3636 Ext. 4508, E-mail: kmathews@afit.af.mil.

Cornell University, Graduate School, Graduate Fields of Engineering, Field of Nuclear Science and Engineering, Ithaca, NY 14853-0001. Offers nuclear engineering (M Eng); nuclear science and engineering (MS, PhD). *Faculty:* 10 full-time. *Students:* 4 full-time (0 women), 2 international. 7 applicants, 14% accepted. In 1998, 2 master's, 2 doctorates awarded. Terminal master's awarded for partial completion of doctoral program. *Degree requirements:* For master's, thesis (MS) required; for doctorate, dissertation required, foreign language not required. *Entrance requirements:* For master's, TOEFL (minimum score of 550 required); for doctorate, GRE General Test, TOEFL (minimum score of 550 required). *Application deadline:* For fall admission, 1/15. Application fee: $65. Electronic applications accepted. *Financial aid:* In 1998–99, 4 students received aid, including 1 fellowship with full tuition reimbursement available, 2 research assistantships with full tuition reimbursements available, 1 teaching assistantship with full tuition reimbursement available; institutionally-sponsored loans, scholarships, tuition waivers (full and partial), and unspecified assistantships also available. Financial aid applicants required to submit FAFSA. *Faculty research:* Reactor engineering, low-energy nuclear physics, fusion physics and technology. *Unit head:* Director of Graduate Studies, 607-255-3480. *Application contact:* Graduate Field Assistant, 607-255-3480, E-mail: ward_lab@cornell.edu.

See in-depth description on page 1127.

École Polytechnique de Montréal, Graduate Programs, Institute of Nuclear Engineering, Montréal, PQ H3C 3A7, Canada. Offers nuclear engineering (M Eng, PhD); nuclear engineering, socio-economics of energy (M Sc A). *Degree requirements:* For master's and doctorate, one foreign language, computer language, thesis/dissertation required. *Entrance requirements:* For master's, minimum GPA of 2.75; for doctorate, minimum GPA of 3.0. *Faculty research:* Nuclear technology, thermohydraulics.

Georgia Institute of Technology, Graduate Studies and Research, College of Engineering, George W. Woodruff School of Mechanical Engineering, Nuclear Engineering and Health Physics Programs, Atlanta, GA 30332-0001. Offers health physics (MSHP); nuclear engineering (MSNE, PhD). Part-time programs available. Terminal master's awarded for partial completion of doctoral program. *Degree requirements:* For master's, thesis optional, foreign language not required; for doctorate, dissertation, exams required, foreign language not required. *Entrance requirements:* For master's and doctorate, GRE General Test (recommended), TOEFL (minimum score of 580 required). *Faculty research:* Reactor engineering and systems, radiation technology, nuclear materials, environmental engineering, plasma physics.

See in-depth description on page 1129.

Idaho State University, Graduate School, College of Engineering, Pocatello, ID 83209. Offers engineering and applied science (PhD); environmental engineering (MS); hazardous waste management (MS); measurement and control engineering (MS); nuclear science and engineering (MS). MS (hazardous waste management), PhD offered jointly with the University of Idaho. Part-time programs available. *Degree requirements:* For master's, thesis required, foreign language not required; for doctorate, dissertation required. *Entrance requirements:* For master's and doctorate, GRE General Test, TOEFL. *Faculty research:* Isotope separation, control technology,two-phase flow, photosonolysis, criticality calculations.

Kansas State University, Graduate School, College of Engineering, Department of Mechanical and Engineering, Manhattan, KS 66506. Offers engineering (PhD); mechanical engineering (MS); nuclear engineering (MS). *Faculty:* 22 full-time (2 women). *Students:* 51 full-time (6 women), 10 part-time, 44 international. *Entrance requirements:* For master's and doctorate, GRE General Test (minimum combined score of 1350 on quantitative and analytical sections required), TOEFL (minimum score of 600 required). *Application deadline:* For fall admission, 9/1; for spring admission, 3/1. Applications are processed on a rolling basis. Application fee: $0 ($25 for international students). Electronic applications accepted. *Unit head:* J. Garth Thompson, Head, 785-532-5610, Fax: 785-532-7057. *Application contact:* Kirby Chapman, Graduate Coordinator, 785-532-5610, E-mail: dgs@ksume.me.ksu.edu.

Louisiana State University and Agricultural and Mechanical College, Graduate School, Center for Coastal, Energy and Environmental Resources, Nuclear Science Center, Baton Rouge, LA 70803. Offers nuclear science and engineering (MS). Part-time programs available. *Faculty:* 3 full-time (0 women), 2 part-time (0 women). *Students:* 5 full-time (2 women), 6 part-time (1 woman); includes 2 minority (1 Asian American or Pacific Islander, 1 Hispanic American), 1 international. Average age 33. 3 applicants, 100% accepted. In 1998, 5 degrees awarded. *Degree requirements:* For master's, computer language, thesis required, foreign language not required. *Entrance requirements:* For master's, GRE General Test (minimum combined score of 1000 required), TOEFL (minimum score of 550 required), minimum GPA of 3.0. *Average time to degree:* Master's–4 years full-time. *Application deadline:* For fall admission, 1/25 (priority date). Applications are processed on a rolling basis. Application fee: $25. *Financial aid:* In 1998–99, 5 research assistantships with partial tuition reimbursements were awarded.; teaching assistantships with partial tuition reimbursements, career-related internships or fieldwork and Federal Work-Study also available. *Faculty research:* Activation analysis, reactor analysis, heat and neutron transport, radiation therapy, biological application of radiation and isotopes. *Unit head:* Dr. Mark L. Williams, Director, 225-388-2745, Fax: 225-388-2894, E-mail: nsmark@nsmark.nucsci.lsu.edu.

Massachusetts Institute of Technology, School of Engineering, Department of Nuclear Engineering, Cambridge, MA 02139-4307. Offers nuclear engineering (SM, PhD, Sc D, NE); nuclear systems engineering (M Eng); radiological health and industrial radiation engineering (M Eng); radiological sciences (PhD, Sc D). *Faculty:* 17 full-time (1 woman). *Students:* 102 full-time (17 women); includes 6 minority (1 African American, 2 Asian Americans or Pacific Islanders, 3 Hispanic Americans), 44 international. Average age 27. 55 applicants, 64% accepted. In 1998, 32 master's, 14 doctorates awarded. *Degree requirements:* For master's and NE, thesis required; for doctorate and NE, dissertation, comprehensive exams required. *Entrance requirements:* For master's, doctorate, and NE, GRE General Test, TOEFL (minimum score of 577 required). *Application deadline:* For fall admission, 1/15 (priority date); for spring admission, 11/1. Applications are processed on a rolling basis. Application fee: $55. *Financial aid:* In 1998–99, 102 students received aid, including 8 fellowships, 76 research assistantships, 20 teaching assistantships; career-related internships or fieldwork, Federal Work-Study, and institutionally-sponsored loans also available. Financial aid application deadline: 2/15; financial aid applicants required to submit FAFSA. *Faculty research:* Reactor engineering, reactor safety analysis, nuclear materials engineering, applied plasma physics, applied radiation physics. Total annual research expenditures: $16.5 million. *Unit head:* Dr. Jeffrey P. Freidberg, Head, 617-253-8670, E-mail: jpfreid@mit.edu. *Application contact:* Clare Egan, Graduate Office Administrator, 617-253-3814, Fax: 617-258-7437, E-mail: cegan@mit.edu.

See in-depth description on page 1131.

McMaster University, School of Graduate Studies, Faculty of Engineering, Department of Engineering Physics, Hamilton, ON L8S 4M2, Canada. Offers engineering physics (M Eng, PhD); nuclear engineering (PhD). *Faculty:* 18 full-time, 1 part-time. *Students:* 29 full-time, 8 part-time. *Degree requirements:* For master's, thesis or alternative required, foreign language not required; for doctorate, dissertation, comprehensive exam required, foreign language not required. *Application deadline:* For fall admission, 3/1 (priority date). Applications are processed on a rolling basis. Application fee: $50. *Unit head:* Dr. Peter Mascher, Chair, 905-525-9140 Ext. 24548.

North Carolina State University, Graduate School, College of Engineering, Department of Nuclear Engineering, Raleigh, NC 27695. Offers MNE, MS, PhD. *Faculty:* 12 full-time (0 women), 11 part-time (0 women). *Students:* 32 full-time (5 women), 4 part-time; includes 6 minority (1 African American, 4 Asian Americans or Pacific Islanders, 1 Hispanic American), 17 international. Average age 28. 37 applicants, 27% accepted. In 1998, 2 master's, 3 doctorates awarded. *Degree requirements:* For master's, computer language, thesis required (for some programs), foreign language not required; for doctorate, computer language, dissertation required, foreign language not required. *Entrance requirements:* For master's, bachelor's degree in engineering or GRE; for doctorate, engineering degree or GRE. *Average time to degree:* Master's–2 years full-time; doctorate–5.5 years full-time. Application fee: $45. *Financial aid:* In 1998–99, 2 fellowships (averaging $2,002 per year), 5 research assistantships (averaging $6,486 per year), 2 teaching assistantships (averaging $4,054 per year) were awarded.; career-related internships or fieldwork, institutionally-sponsored loans, and traineeships also available. *Faculty research:* Reactor analysis methods, computational fluid models, nuclear materials, radiation measurements, plasma technology, spallation sources. Total annual research expenditures: $2.3 million. *Unit head:* Dr. Donald J. Dudziak, Head, 919-515-6289, Fax: 919-515-2301, E-mail: dudziak@eos.ncsu.edu. *Application contact:* Dr. Kuruvilla Verghese, Director of Graduate Programs, 919-515-3929, Fax: 919-515-5115, E-mail: verghese@eos.ncsu.edu.

The Ohio State University, Graduate School, College of Engineering, Department of Mechanical Engineering, Program in Nuclear Engineering, Columbus, OH 43210. Offers MS, PhD. *Faculty:* 5 full-time, 11 part-time. *Students:* 17 full-time (4 women), 3 part-time; includes 1 minority (African American), 6 international. 30 applicants, 20% accepted. In 1998, 7 master's, 4 doctorates awarded. *Degree requirements:* For master's, computer language required, thesis optional, foreign language not required; for doctorate, computer language, dissertation required, foreign language not required. *Entrance requirements:* For master's and doctorate, GRE General Test (international students) or minimum GPA of 3.0. *Application deadline:* For fall admission, 8/15. Applications are processed on a rolling basis. Application fee: $30 ($40 for international students). *Financial aid:* Fellowships, research assistantships, teaching assistantships, career-related internships or fieldwork, Federal Work-Study, and institutionally-sponsored loans available. Aid available to part-time students. *Unit head:* Richard N. Christensen, Graduate Studies Committee Chair, 614-292-0445, Fax: 614-292-3163, E-mail: christensen.3@osu.edu.

Oregon State University, Graduate School, College of Engineering, Department of Nuclear Engineering, Corvallis, OR 97331. Offers nuclear engineering (MAIS, MS, PhD); radiation health physics (MS). Part-time programs available. *Faculty:* 5 full-time (1 woman), 4 part-time (0 women). *Students:* 18 full-time (3 women), 4 part-time; includes 3 minority (2 Asian Americans or Pacific Islanders, 1 Native American), 5 international. Average age 29. In 1998, 4 master's, 1 doctorate awarded (100% found work related to degree). Terminal master's awarded for partial completion of doctoral program. *Degree requirements:* For master's and doctorate, thesis/dissertation required, foreign language not required. *Entrance requirements:* For master's and doctorate, GRE General Test, TOEFL (minimum score of 500 required), minimum GPA of 3.0 in last 90 hours. *Application deadline:* For fall admission, 6/15. Applications are processed on a rolling basis. Application fee: $50. *Financial aid:* Fellowships, research assistantships, teaching assistantships, institutionally-sponsored loans available. Aid available to part-time students. Financial aid application deadline: 2/1. *Faculty research:* Reactor thermal hydraulics and safety, applications of radiation and nuclear techniques, computational methods development, environmental transport of radioactive materials. *Unit head:* Dr. Andrew C. Klein, Acting Head, 541-737-2343, Fax: 541-737-0480, E-mail: kleina@ne.orst.edu. *Application contact:* Dr. Todd S. Palmer, Coordinator, 541-737-2343, Fax: 541-737-0480, E-mail: nuc_engr@ne.orst.edu.

Pennsylvania State University University Park Campus, Graduate School, College of Engineering, Department of Mechanical and Nuclear Engineering, State College, University Park, PA 16802-1503. Offers nuclear engineering (M Eng, MS, PhD). *Students:* 28 full-time (2 women), 10 part-time (1 woman). In 1998, 11 master's, 2 doctorates awarded. *Degree requirements:* For master's and doctorate, thesis/dissertation required, foreign language not required. *Entrance requirements:* For master's and doctorate, GRE General Test. Application fee: $50. *Financial aid:* In 1998–99, 3 fellowships, 24 research assistantships, 9 teaching assistantships were awarded. *Faculty research:* Reactor safety, radiation damage, advanced controls, radiation instrumentation, computational methods. *Unit head:* Dr. Anthony Baratta, Head, 814-865-4911. *Application contact:* Dr. Arthur T. Motta, Professor in Charge, 814-865-1341.

See in-depth description on page 1133.

Purdue University, Graduate School, Schools of Engineering, School of Nuclear Engineering, West Lafayette, IN 47907. Offers MS, MSNE, PhD. Part-time programs available. *Faculty:* 9 full-time (0 women), 2 part-time (0 women). *Students:* 37 full-time (4 women), 1 part-time; includes 1 minority (Asian American or Pacific Islander), 27 international. Average age 28. 53 applicants, 38% accepted. In 1998, 7 master's, 2 doctorates awarded (50% entered university research/teaching, 50% found other work related to degree). Terminal master's awarded for partial completion of doctoral program. *Degree requirements:* For master's, thesis or alternative required, foreign language not required; for doctorate, dissertation required, foreign language not required. *Entrance requirements:* For master's and doctorate, TOEFL (minimum score of 550 required), minimum GPA of 3.0. *Average time to degree:* Master's–1.8 years full-time; doctorate–5.5 years full-time. *Application deadline:* For fall admission, 12/31 (priority date); for spring admission, 8/15 (priority date). Applications are processed on a rolling basis. Application fee: $30. Electronic applications accepted. *Financial aid:* In 1998–99, 5 fellowships with full tuition reimbursements (averaging $1,100 per year), 30 research assistantships with full tuition reimbursements (averaging $1,200 per year), 4 teaching assistantships with full tuition reimbursements (averaging $1,100 per year) were awarded. Aid available to part-time students. Financial aid application deadline: 5/1; financial aid applicants required to submit FAFSA. *Faculty research:* Nuclear reactor safety, thermal hydraulics, fusion technology, reactor materials, reactor physics. *Unit head:* Dr. A. L. Bement, Head, 765-494-5742, Fax: 765-494-9570, E-mail: bement@ecn.purdue.edu. *Application contact:* Dr. C. K. Choi, Graduate Chairman, 765-494-6789, Fax: 765-494-9570, E-mail: choi@ecn.purdue.edu.

See in-depth description on page 1135.

Rensselaer Polytechnic Institute, Graduate School, School of Engineering, Department of Environmental and Energy Engineering, Program in Nuclear Engineering, Troy, NY 12180-3590. Offers nuclear engineering (M Eng, MS, D Eng); nuclear engineering and science (PhD). Part-time programs available. *Faculty:* 10 full-time (0 women), 14 part-time (0 women). *Students:* 13 full-time (1 woman), 4 part-time (1 woman); includes 3 minority (1 African American, 1 Asian American or Pacific Islander, 1 Hispanic American), 7 international. 17 applicants, 41% accepted. In 1998, 7 master's, 1 doctorate awarded. *Degree requirements:* For master's, thesis required (for some programs), foreign language not required; for doctorate, dissertation required, foreign language not required. *Entrance requirements:* For master's and doctorate, GRE, TOEFL (minimum score of 550 required). *Application deadline:* For fall admission, 2/1 (priority date). Applications are processed on a rolling basis. Application fee: $35. *Financial aid:* In 1998–99, 11 research assistantships, 9 teaching assistantships were awarded.; fellowships, career-related internships or fieldwork and institutionally-sponsored loans also available. Financial aid application deadline: 2/1. *Faculty research:* Multiphase phenomena, nuclear data measurements, radiation transport, biological effects of radiation, fusion reactor design. Total annual research expenditures: $896,000. *Unit head:* Pam Zepf, Senior Secretary, 518-276-6402, Fax: 518-276-3055, E-mail: zepf@rpi.edu. *Application contact:* Pam Zepf, Senior Secretary, 518-276-6402, Fax: 518-276-3055, E-mail: zepf@rpi.edu.

Announcement: This department offers graduate degrees in environmental engineering, nuclear engineering, and engineering physics. Environmental engineering research focuses on water quality, including bioremediation and physicochemical techniques. Nuclear engineering research focuses on fission and fusion power technology, nuclear data, and health physics. Engineering physics research focuses on multiphase phenomena and applied radiation. E-mail: denuej@rpi.edu, WWW: http://www.eng.rpi.edu/dept/neep/public_html/

Texas A&M University, College of Engineering, Department of Nuclear Engineering, College Station, TX 77843. Offers health physics/radiological health (MS), including health physics, industrial hygiene, safety engineering; nuclear engineering (M Eng, MS, PhD). *Faculty:* 20 full-time (0 women), 2 part-time (0 women). *Students:* 63 full-time, 16 part-time. Average age 29. 39 applicants, 79% accepted. In 1998, 28 master's, 5 doctorates awarded. *Degree requirements:* For master's, thesis or alternative required, foreign language not required; for doctorate, dissertation, departmental qualifying exams required, foreign language not required. *Entrance requirements:* For master's and doctorate, GRE General Test, TOEFL. Application fee: $50 ($75 for international students). *Financial aid:* Fellowships, research assistantships, teaching assistantships, career-related internships or fieldwork available. Financial aid application deadline: 4/1; financial aid applicants required to submit FAFSA. *Unit head:* Dr. Alan E. Waltar, Head, 409-845-4161. *Application contact:* Dr. Yassin A. Hassan, Graduate Coordinator, 409-845-7090.

Announcement: Facilities: 1-MW Triga reactor with pulsing capabilities and associated laboratories, AGN-201M reactor, subcritical facility nuclear measurement labs, radiochemistry labs, three charged-particle accelerators, 3-GeV cyclotron, plasma science/pulsed power laboratory, 2-phase flow research laboratories. Research areas: reactor safety, aerosols, thermal hydraulics, fusion reactor technology, advanced nuclear reactors, space reactors, zero-gravity 2-phase flow, laser flow visualization, numerical methods, radiation dosimetry, nuclear fuels/materials, neutron transmutation doping, ion-beam–solid interactions, waste management, and fissible materials disposition.

See in-depth description on page 1139.

The University of Arizona, Graduate College, College of Engineering and Mines, Department of Nuclear and Energy Engineering, Tucson, AZ 85721. Offers nuclear engineering (MS, PhD). Part-time programs available. *Faculty:* 9. *Students:* 2 full-time (0 women), 7 part-time (1 woman), 3 international. Average age 33. 1 applicants, 100% accepted. In 1998, 1 master's awarded. *Degree requirements:* For master's and doctorate, thesis/dissertation required, foreign language not required. *Entrance requirements:* For master's, TOEFL (minimum score of 550 required), minimum GPA of 3.0; for doctorate, TOEFL (minimum score of 550 required). *Application deadline:* For fall admission, 7/31. Applications are processed on a rolling basis. Application fee: $35. *Financial aid:* Fellowships, research assistantships, teaching assistant-

Nuclear Engineering

The University of Arizona (continued)
ships, scholarships available. *Faculty research:* Nuclear safety, waste management, energy resources, dynamics and control. *Unit head:* Dr. Morris Farr, Acting Head, 520-621-2311. *Application contact:* Dorothy Graves, Graduate Secretary, 520-621-2311, Fax: 520-621-8096.

University of California, Berkeley, Graduate Division, College of Engineering, Department of Nuclear Engineering, Berkeley, CA 94720-1500. Offers M Eng, MS, PhD. *Students:* 43 full-time (5 women); includes 5 minority (1 African American, 2 Asian Americans or Pacific Islanders, 2 Hispanic Americans), 20 international. 37 applicants, 35% accepted. In 1998, 4 master's, 11 doctorates awarded. *Degree requirements:* For master's, project or thesis required; for doctorate, dissertation, oral exam required. *Entrance requirements:* For master's and doctorate, GRE General Test, TOEFL, minimum GPA of 3.0. *Application deadline:* For fall admission, 2/10; for spring admission, 9/1. Application fee: $40. *Financial aid:* Fellowships, research assistantships, teaching assistantships available. Financial aid application deadline: 1/5. *Faculty research:* Applied nuclear reactions and instrumentation, fission reactor engineering, fusion reactor technology, nuclear waste and materials management, radiation protection and environmental effects. *Unit head:* Dr. William E. Kastenberg, Chair, 510-643-0574. *Application contact:* Sara Hill, Graduate Assistant for Admission, 510-642-5760, Fax: 510-643-9685, E-mail: gradinfo@nuc.berkeley.edu.

Announcement: Graduate program provides opportunities for specialization in nuclear energy (fission and fusion), nuclear materials and waste management, bionuclear and radiological science, medical imaging, reactor safety and risk, and numerical methods in nuclear engineering analysis and nuclear materials.

See in-depth description on page 1141.

University of Cincinnati, Division of Research and Advanced Studies, College of Engineering, Department of Mechanical, Industrial and Nuclear Engineering, Program in Nuclear Engineering, Cincinnati, OH 45221-0091. Offers MS, PhD. Part-time programs available. *Students:* 13 full-time (2 women), 4 part-time; includes 2 minority (1 African American, 1 Asian American or Pacific Islander), 11 international. In 1998, 9 master's, 2 doctorates awarded. Terminal master's awarded for partial completion of doctoral program. *Degree requirements:* For master's, project or thesis required; for doctorate, dissertation required, foreign language not required. *Entrance requirements:* For master's and doctorate, GRE General Test, TOEFL (minimum score of 550 required). *Average time to degree:* Master's–5.3 years full-time; doctorate–8.8 years full-time. *Application deadline:* For fall admission, 2/1 (priority date). Application fee: $40. *Financial aid:* Fellowships, career-related internships or fieldwork, tuition waivers (full), and unspecified assistantships available. Aid available to part-time students. Financial aid application deadline: 2/1. *Faculty research:* Nuclear fission reactor engineering, reduction and fusion effects, health and medical physics, radiological assessment. *Unit head:* Dorothy Graves, Graduate Secretary, 520-621-2311, Fax: 520-621-8096. *Application contact:* John Valentine, Graduate Program Director, 513-556-2482, Fax: 513-556-3390, E-mail: john.valentine@uc.edu.

University of Florida, Graduate School, College of Engineering, Department of Nuclear and Radiological Engineering, Gainesville, FL 32611. Offers engineering physics (ME, MS, PhD, Engr); health physics (MS, PhD); medical physics (MS, PhD); nuclear power engineering (ME, MS, PhD, Engr). *Faculty:* 34. *Students:* 47 full-time (5 women), 10 part-time (3 women); includes 6 minority (4 Asian Americans or Pacific Islanders, 1 Hispanic American, 1 Native American), 16 international. 95 applicants, 74% accepted. In 1998, 11 master's, 2 doctorates awarded. *Degree requirements:* For master's and doctorate, thesis/dissertation required; for Engr, thesis optional. *Entrance requirements:* For master's and doctorate, GRE General Test, TOEFL, minimum GPA of 3.0; for Engr, GRE General Test. *Application deadline:* For fall admission, 6/1 (priority date). Applications are processed on a rolling basis. Application fee: $20. Electronic applications accepted. *Financial aid:* In 1998–99, 32 students received aid, including 3 fellowships, 25 research assistantships, 2 teaching assistantships; institutionally-sponsored loans and unspecified assistantships also available. Financial aid application deadline: 2/1. *Faculty research:* Robotics, Florida radon mitigation, nuclear space power, radioactive waste management, internal dosimetry. *Unit head:* James S. Tulenko, Chair, 352-392-1401, Fax: 352-392-3380, E-mail: tulenko@sun-robot.nuceng.ufl.edu. *Application contact:* Dr. Wesley E. Bolch, Graduate Coordinator, 352-392-1361, E-mail: wbolch@apollo.nuceng.ufl.edu.

University of Idaho, College of Graduate Studies, College of Engineering, Department of Mechanical Engineering, Program in Nuclear Engineering, Moscow, ID 83844-4140. Offers M Engr, MS, PhD. *Degree requirements:* For master's, thesis or alternative required, foreign language not required; for doctorate, computer language, dissertation required. *Entrance requirements:* For master's, TOEFL, minimum GPA of 2.8; for doctorate, TOEFL, minimum undergraduate GPA of 2.8, 3.0 graduate. *Application deadline:* For fall admission, 8/1; for spring admission, 12/15. Application fee: $35 ($45 for international students). *Financial aid:* Application deadline: 2/15. *Unit head:* Rose Carignan, Admissions Assistant, 518-276-6329, Fax: 518-276-6226, E-mail: carigr@rpi.edu. *Application contact:* Dr. Alan G. Stephens, Director, 208-526-4907, Fax: 208-526-4970, E-mail: drislorr@is.is.isu.edu.

University of Illinois at Urbana–Champaign, Graduate College, College of Engineering, Department of Nuclear, Plasma, and Radiological Engineering, Urbana, IL 61801. Offers health physics (MS, PhD); nuclear engineering (MS, PhD). *Faculty:* 14 full-time (0 women). *Students:* 46 full-time (9 women); includes 6 minority (1 African American, 2 Asian Americans or Pacific Islanders, 3 Hispanic Americans), 20 international. 33 applicants, 27% accepted. In 1998, 12 master's, 10 doctorates awarded. *Degree requirements:* For master's and doctorate, thesis/dissertation required, foreign language not required. *Application deadline:* Applications are processed on a rolling basis. Application fee: $40 ($50 for international students). Tuition, state resident: full-time $4,616. Tuition, nonresident: full-time $11,768. Full-time tuition and fees vary according to course load. *Financial aid:* In 1998–99, 4 fellowships, 34 research assistantships, 7 teaching assistantships were awarded. Financial aid application deadline: 2/15. *Unit head:* Barclay G. Jones, Head, 217-333-3535, Fax: 217-333-2906, E-mail: bgjones@uiuc.edu.

See in-depth description on page 1143.

University of Maryland, College Park, Graduate School, A. James Clark School of Engineering, Department of Materials and Nuclear Engineering, Nuclear Engineering Program, College Park, MD 20742-5045. Offers MS, PhD. Part-time and evening/weekend programs available. Postbaccalaureate distance learning degree programs offered. *Students:* 16 full-time (3 women), 9 part-time; includes 7 minority (3 African Americans, 3 Asian Americans or Pacific Islanders, 1 Hispanic American), 7 international. 28 applicants, 29% accepted. In 1998, 1 master's, 3 doctorates awarded. *Degree requirements:* For master's, thesis optional, foreign language not required; for doctorate, dissertation, oral exam required. *Entrance requirements:* For master's and doctorate, GRE General Test, TOEFL, minimum GPA of 3.0. *Application deadline:* Applications are processed on a rolling basis. Application fee: $50 ($70 for international students). Tuition, state resident: part-time $272 per credit hour. Tuition, nonresident: part-time $475 per credit hour. Required fees: $632; $379 per year. *Financial aid:* In 1998–99, 4 fellowships were awarded.; research assistantships, teaching assistantships, tuition waivers (full) also available. Financial aid applicants required to submit FAFSA. *Faculty research:* Reliability and risk assessment, heat transfer and two-phase flow, reactor safety analysis, nuclear reactor, radiation/polymers. *Unit head:* Dr. Gary Pertmer, Associate Chair, 301-405-5209. *Application contact:* Trudy Lindsey, Director, Graduate Admission and Records, 301-405-4198, Fax: 301-314-9305, E-mail: grschool@deans.umd.edu.

University of Michigan, Horace H. Rackham School of Graduate Studies, College of Engineering, Department of Nuclear Engineering and Radiological Sciences, Ann Arbor, MI 48109. Offers nuclear engineering (MSE, PhD, Nuc E); nuclear science (MS, PhD); radiological health engineering (M Eng). Terminal master's awarded for partial completion of doctoral program. *Degree requirements:* For master's, thesis optional; for doctorate, dissertation, oral defense of dissertation, preliminary exams required. *Entrance requirements:* For master's and doctorate, GRE General Test. Electronic applications accepted. *Faculty research:* Reactor physics and engineering, instrumentation and measurement, nuclear materials, plasma physics and fusion, medical physics.

See in-depth description on page 1145.

University of Missouri–Columbia, Graduate School, College of Engineering, Nuclear Engineering Program, Columbia, MO 65211. Offers nuclear engineering (MS, PhD), including health physics (MS), medical physics (MS), nuclear engineering (MS). *Faculty:* 6 full-time (0 women). *Students:* 15 full-time (6 women), 12 part-time (3 women); includes 2 minority (1 African American, 1 Asian American or Pacific Islander), 11 international. 9 applicants, 78% accepted. In 1998, 4 master's, 6 doctorates awarded. *Degree requirements:* For master's, research project required; for doctorate, dissertation required. *Entrance requirements:* For master's and doctorate, GRE General Test, TOEFL. *Application deadline:* For fall admission, 3/15. Application fee: $30 ($50 for international students). *Financial aid:* In 1998–99, 6 fellowships were awarded.; research assistantships, teaching assistantships, institutionally-sponsored loans also available. *Unit head:* Dr. Tushar Ghosh, Director of Graduate Studies, 573-882-8201.

See in-depth description on page 1147.

University of Missouri–Rolla, Graduate School, School of Mines and Metallurgy, Department of Nuclear Engineering, Rolla, MO 65409-0910. Offers MS, DE, PhD. Part-time programs available. *Faculty:* 6 full-time (1 woman). *Students:* 11 full-time (0 women), 4 international. Average age 28. 7 applicants, 100% accepted. In 1998, 3 master's awarded (100% found work related to degree); 2 doctorates awarded. *Degree requirements:* For master's and doctorate, computer language, thesis/dissertation required, foreign language not required. *Entrance requirements:* For master's, GRE General Test (minimum combined score of 1100 required), TOEFL (minimum score of 550 required), minimum GPA of 3.0 in last 4 semesters; for doctorate, GRE General Test (minimum combined score of 1100 required), TOEFL (minimum score of 550 required). *Application deadline:* For fall admission, 7/1; for spring admission, 11/15. Applications are processed on a rolling basis. Application fee: $25. Electronic applications accepted. *Financial aid:* In 1998–99, 5 students received aid, including 1 fellowship with full tuition reimbursement available (averaging $10,000 per year), 3 research assistantships with partial tuition reimbursements available (averaging $12,986 per year), 1 teaching assistantship with partial tuition reimbursement available (averaging $12,986 per year); Federal Work-Study and institutionally-sponsored loans also available. Aid available to part-time students. Financial aid application deadline: 3/1; financial aid applicants required to submit FAFSA. *Faculty research:* Radiation transport, reactor safety, materials research, probabilistic risk assessment, reactor diagnostics. *Unit head:* Dr. Arvind Kumar, Chairman, 573-341-4720, Fax: 573-341-3609, E-mail: kumar@umr.edu.

University of New Mexico, Graduate School, School of Engineering, Department of Chemical and Nuclear Engineering, Program in Nuclear Engineering, Albuquerque, NM 87131-2039. Offers MS. *Faculty:* 8 full-time (0 women), 2 part-time (0 women). *Students:* 17 full-time (4 women), 31 part-time (2 women); includes 7 minority (2 Asian Americans or Pacific Islanders, 5 Hispanic Americans), 5 international. Average age 35. 5 applicants, 60% accepted. In 1998, 8 degrees awarded. *Degree requirements:* Foreign language not required. *Entrance requirements:* For master's, GRE General Test, minimum GPA of 3.0. *Application deadline:* For fall admission, 7/15; for spring admission, 11/14. Application fee: $25. *Financial aid:* In 1998–99, 24 students received aid, including 6 fellowships (averaging $933 per year), 4 research assistantships with tuition reimbursements available (averaging $2,981 per year) Financial aid application deadline: 3/15. Total annual research expenditures: $556,895. *Unit head:* Joseph Cecchi, Chair, Department of Chemical and Nuclear Engineering, 505-277-5431, Fax: 505-277-5433, E-mail: cecchi@unm.edu.

University of Tennessee, Knoxville, Graduate School, College of Engineering, Department of Nuclear Engineering, Knoxville, TN 37996. Offers nuclear engineering (MS, PhD), including radiological engineering (MS). Part-time programs available. *Faculty:* 11 full-time (1 woman), 2 part-time (0 women). *Students:* 25 full-time (5 women), 36 part-time (5 women); includes 3 minority (all Asian Americans or Pacific Islanders), 7 international. 20 applicants, 50% accepted. In 1998, 9 master's, 6 doctorates awarded. *Degree requirements:* For master's, thesis or alternative required, foreign language not required; for doctorate, dissertation required, foreign language not required. *Entrance requirements:* For master's and doctorate, GRE General Test, TOEFL (minimum score of 550 required), minimum GPA of 2.7. *Application deadline:* For fall admission, 2/1 (priority date). Applications are processed on a rolling basis. Application fee: $35. Electronic applications accepted. *Financial aid:* In 1998–99, 1 fellowship, 21 research assistantships, 1 teaching assistantship were awarded.; career-related internships or fieldwork, Federal Work-Study, institutionally-sponsored loans, and unspecified assistantships also available. Financial aid application deadline: 2/1; financial aid applicants required to submit FAFSA. *Unit head:* Dr. H. L. Dodds, Head, 423-974-2525, Fax: 423-974-0668, E-mail: hdj@utk.edu.

University of Utah, Graduate School, College of Engineering, Department of Mechanical Engineering, Program in Nuclear Engineering, Salt Lake City, UT 84112-1107. Offers ME, MS, PhD. *Students:* 3 full-time (0 women), 3 part-time (1 woman). Average age 30. Terminal master's awarded for partial completion of doctoral program. *Degree requirements:* For master's, computer language, special project (ME), thesis (MS) required; for doctorate, computer language, dissertation, comprehensive exam required, foreign language not required. *Entrance requirements:* For master's, GRE General Test, TOEFL (minimum score of 530 required), minimum GPA of 3.0; for doctorate, GRE, TOEFL (minimum score of 530 required), minimum GPA of 3.0. *Application deadline:* For fall admission, 7/1. Application fee: $30 ($50 for international students). *Financial aid:* Fellowships available. *Unit head:* Dr. G. M. Sandquist, Coordinator, 801-581-7372. *Application contact:* Dr. Patrick McMurtry, Director of Graduate Studies, 801-581-3889.

University of Virginia, School of Engineering and Applied Science, Program in Nuclear Engineering, Charlottesville, VA 22903. Offers ME, MS, PhD. Part-time programs available. *Faculty:* 4 full-time (0 women). *Students:* 10 full-time (1 woman), 3 international. Average age 30. 5 applicants, 40% accepted. In 1998, 2 master's, 3 doctorates awarded. Terminal master's awarded for partial completion of doctoral program. *Degree requirements:* For master's, thesis (MS) required; for doctorate, dissertation, comprehensive exam required, foreign language not required. *Entrance requirements:* For master's, GRE General Test, TOEFL; for doctorate, GRE General Test (minimum score of 700 on quantitative section required), TOEFL (minimum score of 600 required). *Application deadline:* For fall admission, 8/1 (priority date). Applications are processed on a rolling basis. Application fee: $60. *Financial aid:* Fellowships, research assistantships, teaching assistantships available. Financial aid application deadline: 2/1. *Faculty research:* Reactor design, nuclear reactor safety, radio isotope usage, radiation damage studies, neutron radiography. *Unit head:* Dr. Jeffrey Morton, Graduate Director, 804-924-6224, E-mail: jbm@virginia.edu. *Application contact:* J. Milton Adams, Assistant Dean, 804-924-3897, E-mail: twr2c@virginia.edu.

University of Wisconsin–Madison, Graduate School, College of Engineering, Department of Engineering Physics, Madison, WI 53706-1380. Offers engineering mechanics (MS, PhD); nuclear engineering and engineering physics (MS, PhD). Part-time programs available. Postbaccalaureate distance learning degree programs offered (minimal on-campus study). *Faculty:* 19 full-time (1 woman), 5 part-time (0 women). *Students:* 46 full-time (5 women), 5 part-time; includes 2 minority (both Asian Americans or Pacific Islanders), 19 international. Average age 26. 120 applicants, 51% accepted. In 1998, 23 master's awarded (4% entered university research/teaching, 39% found other work related to degree, 48% continued full-time study); 11 doctorates awarded (45% entered university research/teaching, 46% found other work related to degree). Terminal master's awarded for partial completion of doctoral program. *Degree requirements:* For master's, thesis optional, foreign language not required; for doctorate, dissertation required, foreign language not required. *Entrance requirements:* For master's and doctorate, GRE General Test, minimum GPA of 3.0 in last 60 hours, appropriate bachelor's degree. *Average time to degree:* Master's–2 years full-time, 3 years part-time; doctorate–6 years full-time. *Application deadline:* For fall admission, 1/15 (priority date). Applications are processed on a rolling basis. Application fee: $45. Electronic applications accepted. *Financial aid:* In 1998–99, 4 fellowships with full tuition reimbursements (averaging $15,000 per year), 24 research assistantships with full tuition reimbursements (averaging $14,600 per year), 10 teaching assistantships with full tuition reimbursements (averaging $15,500 per year) were awarded.; career-related internships or fieldwork, Federal Work-Study, and institutionally-sponsored loans also available. Aid available to part-time students. Financial aid application deadline: 1/15. *Faculty research:* Fission reactor engineering and safety, plasma physics and fusion technology, plasma processing and ion implantation, applied superconductivity and cryogenics, engineering mechanics and astronautics. Total annual research expenditures: $5.7 million. *Unit head:* Dr. Gilbert A. Emmert, Chair, 608-263-1646, Fax: 608-263-7451, E-mail: emmert@engr.wisc.edu.

See in-depth description on page 1149.

Cross-Discipline Announcement

University of Maryland, College Park, Graduate School, A. James Clark School of Engineering, Department of Materials and Nuclear Engineering, Materials Science and Engineering Program, College Park, MD 20742-5045.

The Materials Science and Engineering (MSE) Program offers educational opportunities leading to a Doctor of Philosophy degree in MSE and to a Master of Science degree in MSE. The program emphasizes thin films, semiconductors, polymers and composites, the processing of ceramics, and structural materials. See in-depth description in Materials Sciences and Engineering section.

CORNELL UNIVERSITY

Field of Nuclear Science and Engineering

Programs of Study	All programs are offered at the graduate level and allow specialization in basic nuclear and atomic science, in nuclear engineering, or in a combination of the two. Three degrees are offered: M.Eng., M.S., and Ph.D. Areas of particular interest are neutron activation analysis, neutron radiography, prompt gamma analysis, neutron depth profiling, and basic nuclear physics using neutrons; atomic collision and reaction studies with low-energy highly charged ions; plasma physics and fusion technology, including intense ion beams and their applications to magnetic and inertial confinement fusion; and nuclear power generation. Other areas of interest are radiation measurement, radiation effects in materials and in microelectronic devices, and engineering aspects of nuclear reactor power plants including severe accidents. In addition to the above traditional areas, a collaborative program of courses and research projects was inaugurated to develop and use nuclear analytical methods in non-nuclear research fields. The participating fields are archaeology, geology, and materials science and engineering. The M.Eng. (Nuclear) program is primarily a terminal professional degree program, but it is also a preparation for doctoral study. The two-term curriculum covers the basic principles of nuclear reactor systems, emphasizing reactor safety and radiation protection. A design project is required. The M.S. and Ph.D. programs are research oriented and require a thesis as well as course work. Students may choose a concentration in either nuclear science or nuclear engineering, and minors may be in any related engineering or scientific field.
Research Facilities	The Ward Center for Nuclear Sciences houses major facilities for reactor physics and engineering, basic and applied nuclear physics, radiation effects, and fundamental atomic and molecular processes. The largest facility is a TRIGA reactor with 500-kW steady-state power and a pulsing capability of up to 1,000 MW. It is most extensively used for neutron activation analysis. A special state-of-the-art low-background neutron beam for basic and applied uses will be available. It combines a cold-neutron source in a TRIGA beamport with a 13-meter-long curved neutron guide that acts as a filter to remove fast neutrons and gammas. This facility is the only one at a U.S. university that is designed for and devoted to low-background uses such as prompt gamma analysis and related basic nuclear physics research. A neutron depth profiling facility is being developed to measure depth profiles of light elements (^{10}B, ^{3}He, ^{6}Li, etc.) in any substrate material. A neutron radiography facility images objects up to 15 inches in diameter with an L/D ratio of 140. A 9-inch Thompson Tube is used for real-time radiography. A gamma irradiation cell with shielding for Co-60 sources up to 10-kilocurie strength is used for radiation chemistry, radiation damage, instrument testing, and radiation response of microelectronic devices. Alpha and Cf-252 sources are available for charged-particle-induced single-event-upset studies. Electron beam ion sources (EBIS) are valuable new tools in atomic physics studies with applications in plasma physics, astrophysics, and related areas. Cornell has one of the principal programs in the United States to develop EBIS sources and has constructed and used the first successful cryogenic EBIS in the country. Facilities for fusion physics and technology include several intense ion-beam generators, a Z-pinch device, and many high-speed diagnostic instruments. These are housed in the Laboratory of Plasma Studies and other locations outside Ward Center.
Financial Aid	Financial aid in the form of teaching and research assistantships, fellowships, loans, and part-time employment is available. Typically, students are enrolled full-time and M.S. and Ph.D. students receive some form of financial support. Most aid is awarded on the basis of merit.
Cost of Study	Tuition is $23,760 for the academic year, and books and supplies cost about $1000. Continuing students engaged in summer research are not charged summer tuition, but those taking courses pay $600 per credit.
Living and Housing Costs	Living expenses for a single student for the twelve-month academic year range from $10,800 to $13,700, including room, meals, medical insurance, and personal expenses. (Travel expenses are additional.) Married students should add about $7300.
Student Group	The number of graduate students registered in the Field of Nuclear Science and Engineering is small, usually less than 10. Each student has close contact with faculty members because of the small student group. Students have the opportunity to increase their range of association both through classes in other departments and through contact with the large number of students and researchers from other fields that use Ward Center facilities for research.
Student Outcomes	Recent graduates secured positions in a variety of industrial organizations, including electric utilities, as faculty members at universities, and at government laboratories.
Location	Ithaca, a city of about 40,000, is located on Cayuga Lake in the beautiful Finger Lakes region of upper New York State. Home for both Ithaca College and Cornell University, it is one of the country's premier educational communities, offering cultural advantages that rival those of many large cities. Nearby recreational facilities include ski areas and state parks.
The University and The Field	Cornell is a large and diverse university with an international reputation, and all of its resources are available to each graduate student. The total enrollment is about 18,000; about 5,500 are graduate students. Graduate study in nuclear science and engineering is conducted within that context; it both draws upon course offerings and facilities of other departments and provides training and facilities for other programs. The faculty is composed of 9 members whose area of research is in nuclear and plasma science and engineering.
Applying	Applicants should have a bachelor's degree in science, applied science, or engineering with an emphasis on mathematics and modern physics. Students seeking a fellowship should submit an application for fall admission by January 15 for the M.S. or Ph.D. program or by February 1 for the M.Eng. program. Students who submit applications later than these dates are considered for admission and for aid in the form of assistantships and part-time employment. GRE General Test scores are recommended (required for fellowship applicants); Subject Test scores are optional. TOEFL scores of 550 or better are required for international students whose native language is not English.
Correspondence and Information	Director of Graduate Studies Nuclear Science and Engineering Ward Center for Nuclear Sciences Cornell University Ithaca, New York 14853-7701 Telephone: 607-255-3480 Fax: 607-255-9417 E-mail: ward_lab_mailbox@cornell.edu

Cornell University

THE GRADUATE FACULTY AND THEIR RESEARCH

Faculty departmental affiliations are given in parentheses.

K. Bingham Cady (Theoretical and Applied Mechanics), Ph.D., MIT. Nuclear engineering, severe accident analysis, numerical methods and sensitivity analysis, reactor physics, marine transportation, economics of nuclear fuel recycling.

David A. Hammer (Electrical Engineering), Ph.D., Cornell. Plasma physics, nuclear fusion, high-power electron and ion-beam physics.

Bryan L. Isacks (Geological Sciences), Ph.D., Columbia. Seismology of nuclear reactor siting.

Robert W. Kay (Geological Sciences), Ph.D., Columbia. Nuclear methods in petrology and geochemistry.

Vaclav O. Kostroun (Applied and Engineering Physics), Ph.D., Oregon. Interaction of radiation in matter, atomic physics, EBIS sources.

Bruce R. Kusse (Applied and Engineering Physics), Ph.D., MIT. Plasma science, nuclear fusion, intense ion-beam physics.

Che-Yu Li (Materials Science and Engineering), Ph.D., Cornell. Nuclear materials, fast neutron damage.

Stephen C. McGuire (Nuclear Science and Engineering), Ph.D., Cornell. Nuclear physics, radiation effects on microelectronics, microelectronic materials characterization.

Michael O. Thompson (Materials Science and Engineering), Ph.D., Cornell. Fundamental mechanisms of point-defect diffusion in Si, energetic-beam processing of Si for ultra-shallow junctions, 100C processing of Si, dynamics of liquid- and solid-phase epitaxial processes.

Kenan Ünlü (Materials Science and Engineering), Ph.D., Michigan. Neutron activation analysis, neutron depth profiling, prompt gamma activation analysis, nuclear instrumentation.

SELECTED FACULTY PUBLICATIONS

The wide range of research carried on by Nuclear Science and Engineering faculty members is illustrated by the following list of publications.

Cady, K. B., et al. ORCA, a model for the economic analysis of the nuclear fuel cycle. COGEMA, Inc., 1999.

Cady, K. B. Marine transport of nuclear reactor fuel, plutonium, and vitrified residue—A review and analysis. *Spent Fuel Management Seminar XIV,* INMM, 1997.

Cady, K. B., et al. *Modular Accident Analysis Program,* 2 volumes (user's manual). Burr, Ridge, Ill.: Fauske and Associates, Inc., 1983.

Hammer, D. A., et al. Studies of plasma formation from exploding wires and multiwire arrays using X-ray backlightings. *Rev. Sci. Instrum.* 70:667, 1999.

Hammer, D. A., and **B. R. Kusse,** et al. Ion diode optics: Measurement of divergence and aiming of beams for transport to light-ion ICF targets. 11th International Conference on High Power Particle Beams, Prague, 1996.

Hammer, D. A., et al. Design and operation of a high pulse rate intense ion beam diode. *Rev. Sci. Instrum.* 66(5):3448–58, 1995.

Isacks, B. L., et al. Seasonal climatic forcing of alpine glaciers revealed with synthetic aperture radar. *J. Glaciology* 43(145):480–8, 1997.

Isacks, B. L., et al. Geophysical and geological databases and CTBT monitoring: A case study of the Middle East. In *Monitoring a Comprehensive Test Ban Treaty,* pp. 197–223. Kluwer Academic Press, 1996.

Isacks, B. L., et al. Spaceborne imaging radar (SIR-C/X-SAR) reveals near-surface properties of the South Patagonian icefield. *J. Geophys. Res. (Planets)* 101:23169–80, 1996.

Kay, R. W., et al. Magnesian andesite in the western Aleutian-Komandorsky region: Implications for slab melting and processes in the mantle wedge. *Geol. Soc. Am. Bull.* 107:505–19, 1995.

Kay, R. W., and S. M. Kay. Aleutian magmas in space and time. In *The Geology of America,* eds. G. Plafker and H. C. Berg., *Geol. Soc. America* G-1:687–722, 1994.

Kay, R. W., et al. Magmatic and tectonic development of the western Aleutians: An oceanic arc in a strike-slip setting. *J. Geophys. Res.* 98(11):807–34, 1993.

Kostroun, V. O., et al. Spectroscopic study of electrons emitted in Ar^{q+} ($8 \leq q \leq 16$) on Ar at 2.3 qkeV collision energy. *Phys. Rev. A* 53:2379, 1996.

Kostroun, V. O., et al. Scattering study of single charge transfer in Ar^{q+} ($11 \leq q \leq 14$) on Ar collisions at 72 qeV. *J. Phys. Chem.* 99:15669, 1995.

Kusse, B. R., and **D. A. Hammer,** et al. Studies of multiwire array plasma formation using X-ray backlighting. 12th International Conference on High Power Particle Beams, Haifa, 1998.

Kusse, B. R., et al. Effects of axial current in an extraction geometry applied-B ion diode. 12th International Conference on High Power Particle Beams, Haifa, 1998.

Kusse, B. R., and K. Kushelnick. Capillary discharges for channeling high intensity laser pulses. *Bull. Am. Phys. Soc.* 40(11):1662, 1995.

Li, C.-Y., et al. Effect of residual stress and adhesion on the hardness of copper films deposited on silicon. *J. Materials Res.* 5(4):776–83, 1990.

Li, C.-Y., et al. A damage integral approach to thermal fatigue of solder joints. *IEEE Trans. Components Hybrids Manufact. Technol.* 12(4):480–91, 1989.

McGuire, S. C., and J. D. Sulcer. Nickel aluminide thin film fabrication via ion beam sputtering of compound targets. *Surface Coatings Technol.,* 1998.

McGuire, S. C., et al. Impurity study of alumina and aluminum nitride ceramics: Microelectronics packaging applications. *Appl. Radiat. Isot.* 48(1):5–9, 1997.

McGuire, S. C., et al. Neutron activation for semiconductor materials characterization. In *Semiconductor Characterization: Present Status and Future Needs.* Woodbury, NY: AIP Press, 1996.

Thompson, M.O., et al. Thermal nitidation-enhanced diffusion of Sb and Si (100) doping superlattices. *Appl. Phys. Lett.* 69:1273–5, 1996.

Thompson, M.O., and K. M. Kramer. Pulsed-laser-induced epitaxial crystallization of carbon-silicon alloys. *J. Appl. Phys.* 79:4118–23, 1996.

Thompson, M.O., et al. Kinetics of crystal growth in SiGe alloys. *Phys. Rev. B* 53:8386–97, 1995.

Ünlü, K., et al. Helium-3 and boron-10 concentration and depth measurements in alloys and semiconductors using NDP. *Nucl. Instr. Meth. Phys. Res.* A:422, 1999.

Ünlü, K., et al. Prompt gamma activation analysis using a focused cold neutron beam. In *Proceedings of the 9th International Symposium on Capture Gamma-Ray Spectroscopy and Related Topics.* Budapest: Springer, 2:713, 1997.

Ünlü, K., et al. Effect of long-term alpha radiation on stainless steel used in plutonium encapsulation. *Trans. Am. Nucl. Soc.,* 1997.

GEORGIA INSTITUTE OF TECHNOLOGY
A Unit of the University System of Georgia

Nuclear and Radiological Engineering and Health Physics Programs

Programs of Study

Challenging programs at Georgia Tech await the graduate student of nuclear engineering or health physics. Master's and doctoral programs are designed to educate men and women who will contribute to the peaceful application of nuclear energy and to the protection of human beings and the environment from potentially harmful sources of radiation. An outstanding faculty is committed to guiding students into these areas of great national and international importance and to preparing them to be leaders within this vital field.

The programs lead to the Master of Science in Nuclear Engineering, Master of Science, Master of Science in Health Physics, and Doctor of Philosophy. The master's program in nuclear engineering includes options in both fission reactor engineering and fusion research. Opportunities exist for students to specialize by combining various nuclear engineering courses with courses from programs in other schools of the Institute. The program in health physics stresses applied health physics. The M.S.H.P. program is available to working professionals through video technologies and, beginning in fall 1999, via the Internet. A study-abroad program is also offered at Tech's European campus in Metz, France. The Georgia Tech Lorraine (GTL) program allows U.S. students to earn a degree from Georgia Tech and to study in France. A dual-degree program is available for study at GTL and ENSAM, a French school of mechanical engineering.

Research Facilities

In addition to the numerous resources shared with other programs at Georgia Tech, facilities specific to nuclear engineering and health physics that enhance graduate study and research include two subcritical assemblies, 3×10^5-curie cobalt 60 sources, and hot cells for handling radioactive materials. Extensive laboratories facilitate work in radiochemistry, materials preparation, neutron and spectroscopy, and radiobiology. Several related research centers, including the Nuclear Research Center, Fusion Research Center, and Environmental Resources Center, provide opportunities for students to work closely with faculty members in specialized programs. The Fusion Research Center collaboration with the national DIII-D plasma experiment provides an opportunity for student research at a world-class fusion physics facility. New facilities and resources enhance instruction and high-level research. The Institute's library has open stacks and is designated as one of twelve U.S. regional technical report centers. The collection is extensive and current in both technical books and periodicals. Only a few hours away from Georgia Tech are the Savannah River Site, known for its research in material behavior, and the Oak Ridge National Laboratory, one of the world's largest research establishments. It is possible for Ph.D. students to use the exceptional, state-of-the-art experimental facilities at Oak Ridge, Argonne, Savannah River Site, and other national laboratories to complement the facilities at Georgia Tech.

Financial Aid

Research and teaching assistantships, fellowships, and tuition waivers are available to graduate students. Graduate assistantships carry a twelve-month stipend ranging from $15,000 to $18,000 and include a waiver of out-of-state tuition. President's Fellowships and Woodruff Fellowships of up to $5500, which supplement graduate assistantships, are available to qualified students wishing to pursue the Ph.D. Federal, industrial, and private fellowships are also available. International students must guarantee their first-year support but are eligible to compete for awards on a per-term basis.

Cost of Study

The total fees in 1999–2000 for graduate students carrying a full academic load are $372 per term for GRAs and GTAs, $1448 per term for residents of Georgia, and $6139 per term for nonresidents. Part-time resident and nonresident graduate students are charged prorated fees. Students who are awarded scholarships or assistantships normally pay no tuition but must pay an athletic fee and other small fees. Fees are subject to change without notice.

Living and Housing Costs

Numerous contemporary suites for unmarried students and apartments for married students and their families are available at reasonable cost through the Institute. Three 350-unit Graduate Student Living Centers are open to graduate students. Rooms and apartments in privately owned dwellings within walking distance or a short driving distance from the campus are available at several price levels. Students should write to the Housing Office for details. Unmarried students should be able to meet minimum necessary expenses, exclusive of tuition and fees, of $10,000 for the calendar year.

Student Group

The Institute's total enrollment is approximately 14,500, including more than 3,500 graduate students. Students come from all fifty states and more than ninety other countries. Almost 20 percent of all graduate students are women. In fall 1998, the Nuclear Engineering and Health Physics Programs included 61 graduate students, 24 of whom were working at the doctoral level. Most graduate students in these programs receive financial aid.

Student Outcomes

Georgia Tech has one of the best career services programs in the country for engineering students. Furthermore, the Woodruff School has an in-house program for placement of Ph.D. students that includes a yearly, nationwide Graduate Student Symposium (Ph.D. Career Fair), *Ph.D. Résumé Book,* and numerous workshops. The endowed communications program offers assistance in preparing resumes and teaching and research portfolios. Since the summer of 1992, the Woodruff School has placed 40 of 41 of its nuclear and radiological engineering and health physics Ph.D. graduates. Roughly 70 percent of graduates go into industry, while 30 percent go into university positions. Recent graduates have been hired at Emory University, Medical College of Georgia, and the University of Nevada, Las Vegas as well as TVA, EG&G Applied Technologies, Siemens Power, and Rochester Gas and Electric.

Location

Atlanta is a city of more than 4 million people. Atlanta's 1,050-foot elevation results in freedom from climatic extremes, which allows for year-round recreational opportunities. The Atlanta area offers such sites of interest as the Martin Luther King Jr. National Historic Site, the state capitol, Stone Mountain Park, Fernbank Science Center, Carter Presidential Center, Centennial Olympic Park, Cyclorama, Zoo Atlanta, and Six Flags Over Georgia. There are also professional teams in football (the Falcons), baseball (the Braves), basketball (the Hawks), and hockey (the Thrashers). Atlanta hosted the 1996 Olympic Games, and the Georgia Tech campus was the site of the Olympic Village.

The Institute

Georgia Tech is a member of the University System of Georgia. The Institute is an accredited member of the Southern Association of Colleges and Schools. All four-year engineering curricula are accredited by the Accreditation Board for Engineering and Technology, the national engineering accrediting agency.

Applying

Forms for admission and financial aid may be obtained by writing to the address given below and should be returned, together with letters of recommendation, GRE scores, and official transcripts of previous academic work, at least six weeks before the beginning of the term of desired matriculation. Students wishing the fullest consideration for financial aid should have their applications completed by February 1 for admission in September.

Correspondence and Information

Graduate Studies
Nuclear Engineering and Health Physics Programs
The George W. Woodruff School of Mechanical
    Engineering
Georgia Institute of Technology
Atlanta, Georgia 30332-0405
Telephone: 404-894-3204
             800-543-2034 (toll-free)
Fax: 404-894-8336
E-mail: menehp.info@me.gatech.edu
World Wide Web: http://www.me.gatech.edu/ne_re_hp

Internet and Video-Based M.S. Programs
Center for Media-Based Instruction
Georgia Institute of Technology
Atlanta, Georgia 30332-0240
Telephone: 404-894-3378
E-mail: vbis@conted.gatech.edu (general)
      video.programs@me.gatech.edu
World Wide Web: http://www.conted.gatech.edu/vbis/
    vbis.html

Georgia Institute of Technology

THE FACULTY AND THEIR RESEARCH

Health Physics and Radiological Engineering

Kenneth W. Crase, Adjunct Professor; Ph.D., Tennessee, 1971. Radiation protection policy, personnel dosimetry, criticality dosimetry, environmental monitoring.

Nolan E. Hertel, Professor; Ph.D., Illinois, 1979. Radiation shielding, neutron dosimetry, radiological assessment, radioactive waste management, health risk assessment, accelerator sources and applications, high-energy particle transport, dry storage of spent fuel, skyshine.

Rodney D. Ice, Adjunct Professor; Ph.D., Purdue, 1967. Health physics, boron neutron capture therapy, radiopharmaceuticals, radionuclide methodology, hospital health physics.

Bernd Kahn, Professor Emeritus and Director, Environmental Radiation Laboratory; Ph.D., MIT, 1960. Health physics, analytical radiochemistry, environmental surveillance.

Jon H. Trueblood, Adjunct Professor; Ph.D., South Carolina, 1971. Digital imaging, mammography quality control, hospital health physics.

John D. Valentine, Associate Professor; Ph.D., Michigan, 1993. Radiation detection and measurements, medical imaging, environmental monitoring, nuclear waste monitoring, personnel monitoring, scintillator and semiconductor detector characterization and development.

C.-K. Chris Wang, Associate Professor; Ph.D., Ohio State, 1989. Radiation detection, radiation dosimetry, medical and industrial applications of ionizing radiations, spent nuclear fuel measurements.

F. Ward Whicker, Adjunct Professor; Ph.D., Colorado State, 1965. Radioecology, human and ecological risk assessment, dose reconstruction, radionuclide kinetics, radiation biology, radiotracer techniques.

Fusion

John Mandrekas, Research Scientist; Ph.D., Illinois, 1987. Nuclear fusion, plasma physics.

Weston M. Stacey, Callaway and Regents' Professor; Ph.D., MIT, 1966. Fusion engineering, plasma physics, reactor physics.

Reactor Physics

Ratib A. Karam, Professor Emeritus; Ph.D., Florida, 1963. Reactor physics, reactor design, reactor safety.

Farzad Rahnema, Associate Professor; Ph.D., UCLA, 1981. Reactor physics, perturbation and variational methods, reactor simulator and monitoring methods, criticality safety and benchmark methods.

Thermal Hydraulics

Said I. Abdel-Khalik, Southern Nuclear Distinguished Professor; Ph.D., Wisconsin, 1973. Reactor safety, thermal hydraulics, accident analysis.

S. Mostafa Ghiaasiaan, Associate Professor; Ph.D., UCLA, 1983. Heat transfer, multiphase flow, change-of-phase heat and mass transfer, aerosol and particle transport, thermal hydraulics of nuclear systems.

Sheldon M. Jeter, Associate Professor; Ph.D., Georgia Tech, 1979. Heat transfer, thermal hydraulics.

CURRENT RESEARCH

Health Physics and Radiological Engineering. Environmental radiological monitoring, control of radon in buildings, application of baseline data from environmental radiation monitoring, bioaccumulation in edible fish tissue, Radon-222 in water-supply wells, health impacts of waste facilities, computational neutron dosimetry, neutron instrument calibration, waste management, internal dosimetry, hospital health physics, x-ray, accelerator, teletherapy and radionuclide shielding, design and safety analysis, calibration and quality control, engineering of radioisotope laboratories and positron-emission-tomography (PET) facilities, emergency radiological internal dosimetry, emergency radiological response and acute radiation dose assessments, radiation protection, mammography dose reduction and quality control, planning and beam development for neutron capture therapy, spent fuel assay and storage, detector characterization, detector development, single photon emission, computer tomography.

Fusion. Plasma transport theory; neutral atom transport; plasma edge modeling; DIII-D experimental analysis; fusion reactor conceptual design; plasma engineering; participation in collaborative programs with the major fusion research laboratories, as well as with national and international conceptual design activities; fusion radioactive waste minimization.

Reactor Physics. Variational methods, diffusion theory, criticality, measurement methods of reactor parameters, neutron thermalization, heterogeneity effects, reactor kinetics, transport equation and methods to solve reactivity control, resonance self-shielding, perturbation and variational techniques in reactor physics, advanced treatment of neutron thermalization, Doppler broadening, Monte Carlo methods in solving the transport equation, fuel depletion, benchmarking calculations compared to measured reactor parameters, calculational uncertainties in reactor design.

Thermal Hydraulics. Surface-tension-driven flows, steam explosions, fluid-structure interaction, molten core–concrete interactions, transport phenomena in multiphase flow, direct contact condensation and condensing two-phase flow, change-of-phase heat transfer, transport of radioactive material during accidents, modeling of PWR steam generators during transients, accelerator targets design, microscale heat transfer, high-power density systems.

MASSACHUSETTS INSTITUTE OF TECHNOLOGY

Department of Nuclear Engineering

Programs of Study

The Department of Nuclear Engineering offers programs leading to the Master of Science, Master of Engineering, Nuclear Engineer, Doctor of Science, and Doctor of Philosophy degrees. The fields in which doctoral and master's degree candidates may elect to study are applied fusion technology, applied plasma physics, applied radiation physics, fission reactor engineering, fission reactor physics, health physics, nuclear and alternative energy systems and policy analysis, nuclear fuel management, nuclear material engineering, radiological sciences, and reactor safety analysis. The program in radiological sciences is offered jointly by the Department of Nuclear Engineering and the Harvard-MIT Division of Health Sciences and Technology.

Candidates for the S.M. degree are required to complete an acceptable thesis and at least 66 credit units in subjects more advanced than the required undergraduate preparation for nuclear engineering. Theses can be primarily theoretical or experimental, or they can combine both approaches. A master's thesis can be completed within twelve to eighteen months. The M.Eng. program prepares students for productive professional engineering careers by providing additional depth in nuclear-related subjects beyond the bachelor's degree together with the breadth of perspective necessary for engineering leadership in the field. In addition to a design/thesis project, the M.Eng. degree requirements include two courses in the fundamentals of the area, three planned electives, and one free elective. This is a fast-paced and intensive program that is normally completed in nine months. The object of the Nuclear Engineer degree program is to provide a broader knowledge of nuclear engineering than that required for the master's degree and to develop competence in engineering application or design, but with less emphasis on research than that characterizes a doctoral program. The program includes completion of both an extensive individually arranged academic course program and a special project of significant engineering value. Candidates are required to complete an acceptable thesis and at least 162 credit units in subjects more advanced than the required undergraduate preparation for nuclear engineering. A student normally needs two years to obtain the Nuclear Engineer degree. The doctoral degree program is designed to give the student a comprehensive knowledge of nuclear engineering and to develop competence in original research. The three principal parts of the doctoral program are the general examination, the core/major/minor program, and the doctoral thesis. Prior to starting doctoral research, the student is required to pass a comprehensive written and oral general exam, which tests knowledge of scientific and engineering principles, familiarity with nuclear engineering, and qualifications to undertake research in the chosen field. Candidates for the doctoral degree must also satisfactorily complete an approved program of advanced study, the core/major/minor program. The program requires that students take not less than 96 credit hours of subjects (excluding special problems), of which three subjects (36 units) are selected from specific department courses (the core). Three subjects (36 units) comprise a field of specialization (the major) that are closely related to the student's doctoral thesis topic (but beyond the subjects covered in the qualifying exam). Two subjects (24 units) must consist of coordinated subjects clearly outside the field of specialization (the minor). The completion and oral defense of a thesis on original research are required. Doctoral research may be undertaken in nuclear engineering or in a related field under the supervision of a faculty member of the Department of Nuclear Engineering or another department.

Research Facilities

The department has facilities to support an exceptional program of controlled fusion engineering studies and has its own well-equipped graduate laboratory for instruction on plasma laboratory techniques. The department has played a major role in the design and development of high magnetic-field fusion devices. Currently there are two major plasma experiments at MIT—the Alcator C-MOD tokamak and the Versatile Toroidal Facility (VTF)—both located in the Plasma Fusion Center. The tokamak, Alcator C-MOD, is a major plasma physics installation expected to function for many years as the plasma equivalent of the MIT Research Reactor and as an educational and research facility. MIT's 5-MW Research Reactor provides firsthand experience in the design, performance, and operation of nuclear reactors and serves as a source of radiations for use in laboratory instruction on radiation detection and measurement methods. Other research includes the development of accelerator-based facilities for neutron capture therapy of cancer and neutron tomography to identify aging in industrial structures and the development of positron emission tomography (PET) as a diagnostic technique for the brain. The department makes extensive use of the MIT Information Processing Center; its facilities include an IBM 370/168 for batch processing and an IBM 360/67 for time-sharing. Access to the time-sharing system is via consoles located around the Institute.

Financial Aid

Research and teaching assistantships are available for well-qualified students. Stipends for academic year 1999–2000 are $12,825 plus tuition for full-time teaching assistants and $12,825 (master's level) or $14,175 (doctoral level) plus tuition for full-time research assistants. Other financial support is available through government and private fellowships. Low-interest, deferred-repayment loans are also available.

Cost of Study

Tuition for the two-term 1999–2000 academic year is $25,000. The accident and hospital insurance fee is $636 (September through August).

Living and Housing Costs

On-campus, single-student housing is available at a cost of $376 to $443 per month in 1999–2000. For married students, rent per month for one-bedroom apartments ranges from $815 to $919; for two-bedroom apartments, $954 to $1060; for studio apartments, $702 to $731. A considerable range of private accommodations can be found within commuting distance.

Student Group

There were 103 graduate students in the department in fall 1998; all were full-time, 64 were candidates for master's or engineer's degrees, and 39 were candidates for doctoral degrees.

Location

MIT occupies a 125-acre campus on the north bank of the Charles River, facing the skyline of Boston. The cultural, scientific, and intellectual resources of Boston are easily accessible.

The Institute

MIT, founded in 1861 as a private, coeducational, endowed institution committed to the extension of knowledge through teaching and research, has grown to be one of the world's foremost institutes of technology. It is organized into five schools: Architecture and Planning, Engineering, Humanities and Social Science, Management, and Science. About 9,000 students are enrolled, half of them in the Graduate School.

Applying

For admission to the Graduate School, the student must complete an undergraduate program with sufficient background in mathematics, science, and engineering. Normally the GPA equivalent of B or better is required. The GRE General Test is required for all students. Students from non-English-speaking countries must earn a TOEFL score of at least 577. The application fee is $55. Applications should be submitted by January 15 for fall and by November 1 for spring admission. Financial aid deadlines are the same.

Correspondence and Information

Ms. Clare Egan
Graduate Program Administrator
Massachusetts Institute of Technology, Room 24-102
77 Massachusetts Avenue
Cambridge, Massachusetts 02139-4307
Telephone: 617-253-3814
E-mail: cegan@mit.edu

Massachusetts Institute of Technology

THE FACULTY AND THEIR RESEARCH

Professors

George Apostolakis, Ph.D., Caltech, 1973. Reliability and risk assessment, management of complex engineering systems, nuclear and toxic waste risk assessment and management, nuclear reactor safety.

Sow-Hsin Chen, Ph.D., McMaster, 1964. Applied neutron physics and spectroscopy, applications of laser light scattering to biological problems.

Jeffrey P. Freidberg, Head of Department; Ph.D., Polytechnic of Brooklyn, 1964. Theoretical plasma physics.

Michael W. Golay, Ph.D., Cornell, 1969. Reactor engineering, fluid mechanics, environmental and safety problems of nuclear power.

Kent F. Hansen, Sc.D., MIT, 1959. Nuclear energy policy and management, nuclear plant operations and simulation.

Otto K. Harling, Ph.D., Penn State, 1962. Neutron scattering, research reactor applications, experimental materials research, health physics.

Ian H. Hutchinson, Ph.D., Australian National, 1976. Experimental plasma physics and controlled fusion.

Mujid S. Kazimi, Ph.D., MIT, 1973. Reactor engineering, fusion and fast-reactor safety.

Ronald M. Latanision, Ph.D., Ohio State, 1968. Corrosion, materials processing, characterization and life-time prediction of materials in nuclear environments.

Richard K. Lester, Ph.D., MIT, 1979. Technology and policy assessment, nuclear-waste disposal.

Lawrence M. Lidsky, Ph.D., MIT, 1962. Fusion reactor design, advanced fusion systems.

John E. Meyer, Ph.D., Carnegie Tech, 1955. Structural mechanics, heat transfer and fluid flow.

Kenneth Russell, Ph.D., Carnegie Tech, 1964. Radiation effects on materials.

Neil E. Todreas, Sc.D., MIT, 1966. Reactor engineering, thermal analysis, heat transfer and fluid flow.

Sidney Yip, Ph.D., Michigan, 1962. Transport theory, neutron scattering, statistical mechanics.

Associate Professors

Ronald G. Ballinger, Sc.D., MIT, 1982. Corrosion and fatigue, stress corrosion cracking behavior in nuclear systems, fuel behavior modeling.

David G. Cory, Ph.D., Case Western Reserve, 1987. NMR imaging and spectroscopy, new methodology and applications of NMR techniques for the study of spatial properties of matter.

Kim Molvig, Ph.D., California, Irvine, 1975. Theoretical plasma physics.

Jacquelyn C. Yanch, Ph.D., London, 1988. Nuclear medical imaging, computational modeling in both therapy and image restoration, radiation health physics, neutron dosimetry.

Assistant Professors

Ken Czerwinski, Ph.D., Berkeley, 1992. Actinide spectroscopy, actinide thermodynamics, environmental chemistry of actinide elements, geochemical modeling.

Professors Emeriti

Manson Benedict, Ph.D.
Gordon L. Brownell, Ph.D.
Michael J. Driscoll, Sc.D.
Thomas H. Dupree, Ph.D.
Elias P. Gyftopoulos, Sc.D.
Allan F. Henry, Ph.D.
David D. Lanning, Ph.D.
Norman C. Rasmussen, Ph.D.

Research Staff

John Bernard, Ph.D., Principal Research Engineer.
Andrew Kadak, Ph.D., Visiting Senior Lecturer.
Richard Lanza, Ph.D., Senior Research Scientist.
Bruce Rosen, M.D., Ph.D., Senior Lecturer.

PENNSYLVANIA STATE UNIVERSITY

Department of Mechanical and Nuclear Engineering
Nuclear Engineering Program

Programs of Study

The Nuclear Engineering Program offers graduate programs leading to Master of Science (M.S.), Master of Engineering (M.Eng.), and Doctor of Philosophy (Ph.D.) degrees in nuclear engineering. Major areas of research in the program include fission reactor physics, transport theory methods development and application, nuclear fuel management, reactor safety and reliability, advanced reactor control, transient analysis modeling and accident analysis, thermal hydraulic code development, thermal hydraulic experiments, radiation monitoring and instrumentation, radiation effects on materials, materials development for radioactive waste confinement, and use of radioactive probes to study materials.

The M.S. program requires a minimum of 30 graduate credits, with at least 12 credits in the nuclear engineering department and at least 12 credits at the advanced graduate level. Two options are available for the M.S. degree. The thesis option includes 6 credits of thesis research. The nonthesis option requires 6 additional credits of advanced graduate-level courses plus a scholarly paper. The M.Eng. program is designed to provide depth in a nuclear engineering field that enhances the professional development of the student. It requires a 3-credit professional topics paper on an engineering subject.

The Ph.D. program emphasizes both breadth and depth in nuclear engineering to prepare the student to work in research and development. There are no specific credit requirements, but the typical program includes 45–55 academic credits beyond the bachelor's degree plus research. To become a doctoral candidate it is necessary, within eighteen months of starting graduate work to pass a candidacy examination covering basic knowledge in nuclear engineering areas. A comprehensive examination for dissertation approval is also required, generally a year before the defense of the thesis.

Research Facilities

Outstanding research facilities are housed in the department's Radiation Science and Engineering Center. Included are a 1-megawatt TRIGA reactor, with a modern digital control system; a gamma irradiation laboratory, with a gamma pool and a gamma irradiator; a neutron beam laboratory for neutron radiography; two hot cells; a materials research laboratory; an intelligent controls laboratory; a nuclear probe materials laboratory; and a thermal hydraulic test loop. Additional facilities include workstations in a transport analysis computations laboratory and a safety analysis computations laboratory as well as College of Engineering workstation laboratories. Other computationally intensive work is performed using the University's 62-node IBM SP2, or an IBM 3090-600S. Materials research equipment includes transmission electron microscopy (TEM) facilities and SEM facilities at the Materials Research Laboratory and the Materials Science Laboratories.

Financial Aid

Graduate assistantships for teaching and research are available for qualified students. For 1998–99, there was a stipend of approximately $1200 per month plus a tuition grant-in-aid. For outstanding students, a supplemental fellowship of up to $5000 was available from the Dean's Office to add to the graduate assistantship. University fellowships are awarded on a competitive basis. The department awards fellowships provided by the National Academy for Nuclear Training and administers graduate fellowships provided by the Department of Energy. Need-based financial aid, including loans or graduate work-study, is available through the Financial Aid Office.

Cost of Study

Tuition for 1998–99 was $3267 per semester for Pennsylvania residents and $6730 for nonresidents. For students who have completed their comprehensive examination, there was a charge of $780 per semester while completing the dissertation.

Living and Housing Costs

University and private housing are available to graduate students. The University's facilities range from dormitory rooms for single students to two-bedroom apartments for families. Dormitory rooms cost a minimum of $1050 per semester; room and full board, approximately $1950 to $2180; and apartments, a minimum of $1260 per semester. Most graduate students live in privately owned apartments that are available in the community.

Student Group

There are 38,219 students at the University Park campus, including 6,300 graduate students. Of these, 1,273 are in the College of Engineering. There are 40 graduate students in the Nuclear Engineering Program from various locations in the United States and abroad. Most students are financially assisted by fellowships or assistantships, although some are involved in part-time graduate study while employed. Many students are active in the student chapter of the American Nuclear Society, which won the outstanding student chapter award in 1997.

Location

Penn State's main campus, University Park, is located in the center of the state in the borough of State College. The town and its surrounding area, with a population of about 75,000, are located in low, rolling mountain country and offer a variety of recreational activities. The community and the University present a wide array of cultural and athletic events.

The University and The Program

Penn State is a land-grant university founded in 1855. Graduate work began in 1862. Today, the Graduate School has 2,200 faculty members and grants degrees in 149 majors. It awards about 1,749 master's degrees and 585 doctorates each year. Nuclear Engineering began its graduate program in 1959 and added an undergraduate program in 1968. There are 12 faculty members and 3 research faculty members. The Breazeale Nuclear Reactor, which began operation in 1955, is the first licensed operating reactor in the United States. It has had two power upgrades from the original 100 kW to its current power of 1-MW, and a digital control system was installed in 1991.

Applying

Admission to the program requires an undergraduate degree in engineering, chemistry, mathematics, or physics. Normally, a minimum GPA of B is required. The GRE General Test is required for all students. Students from non-English-speaking countries must earn a TOEFL score of at least 550. The application fee is $40. Applications for University Fellowships are due by January 31, and applications for the Dean's Fellowship Supplement I are due in early February.

Correspondence and Information

Dr. Arthur Motta, Graduate Admissions Officer
Nuclear Engineering Program
231 Sackett Building
Pennsylvania State University
University Park, Pennsylvania 16802

Telephone: 814-865-1341
E-mail: atm2@psu.edu
World Wide Web: http://www.nuce.psu.edu

Pennsylvania State University

THE FACULTY AND THEIR RESEARCH

Anthony J. Baratta, Professor and Chair, Nuclear Engineering Program; Ph.D., Brown, 1979. Reactor transient and accident analysis, severe core damage management, risk assessment, radiation effects on materials and devices, electron and neutron transport. (e-mail: ab2@psu.edu)

Jack S. Brenizer, Professor; Ph.D., Penn State, 1981. Radiation detection, neutron radiography, neutron activation analysis, nondestructive testing. (e-mail: jsbnuc@engr.psu.edu)

Gary L. Catchen, Professor; Ph.D., Columbia, 1979. Characterization of electronic, optical, and magnetic materials using hyperfine interaction techniques such as perturbed angular correlation spectroscopy; radiation detection and measurement; radiation dosimetry. (e-mail: g9c@psu.edu)

Robert M. Edwards, Associate Professor; Ph.D., Penn State, 1991. Power plant simulation and control; application of artificial intelligence, expert systems and robust control approaches to power plant operations. (e-mail: rmenuc@engr.psu.edu)

Madeline A. Feltus, Assistant Professor; Ph.D., Columbia, 1990. Nuclear safety analysis; neutronics and thermal hydraulics modeling of power plants; computational methods in reactor physics, kinetics, fuel management, and reactor transient simulations. (e-mail: maf7@psu.edu)

Alireza Haghighat, Professor; Ph.D., Washington (Seattle), 1986. Computational methods in reactor physics, development and implementation of vector and parallel algorithms for neutral particle transport methods, neutronics and thermal hydraulics modeling of power reactors. (e-mail: haghigha@gracie.nuce.psu.edu)

Lawrence E. Hochreiter, Professor; Ph.D., Purdue, 1971. Thermal hydraulic modeling of nuclear power plants, reactor safety analysis, experimental study of two-phase flow and heat transfer. (e-mail: lehnuc@nuce.psu.edu)

Kostadin Ivanov, Assistant Professor; Ph.D., Bulgarian Academy of Sciences, 1990. Nuclear fuel management, reactor physics, application of neutronic codes in safety analysis. (e-mail: kni1@psu.edu)

William A. Jester, Professor; Ph.D., Penn State, 1965. Utilizing radionuclear techniques, especially neutron activation analysis and radiation monitoring, in solving scientific and engineering problems; development of environmental radiation monitors and instrumentation. (e-mail: wajnuc@engr.psu.edu)

Edward S. Kenney, Professor Emeritus; Ph.D., Penn State, 1964. Nondestructive measurement techniques using radiation, radiation instrumentation design, radiation imaging. (e-mail: ek0@psu.edu)

Edward H. Klevans, Professor Emeritus; Ph.D., Michigan, 1962. Controlled fusion device modeling, plasma physics, radiation instrumentation for nondestructive evaluation. (e-mail: ehknuc@engr.psu.edu)

Samuel H. Levine, Professor Emeritus; Ph.D., Pittsburgh, 1954. Nuclear fuel management, reactor design, research reactor use for medical applications such as boron capture therapy. (e-mail: shl@psu.edu)

John H. Mahaffy, Associate Professor; Ph.D., Colorado, 1974. Computational fluid dynamics, two-phase flow and heat transfer, nuclear reactor safety analysis, and vector and parallel computational techniques. (e-mail: jhm@feynman.arl.psu.edu)

Arthur M. T. Motta, Associate Professor; Ph.D., Berkeley, 1988. Radiation damage to materials, characterization of microstructural evolution under irradiation using transmission electron microscopy and positron annihilation, kinetics and thermodynamics of phase transformation under irradiation, correlation of charged particle and neutron irradiation. (e-mail: atm2@psu.edu)

Gordon E. Robinson, Professor Emeritus; Ph.D., Penn State, 1970. Boiling heat transfer, reactor safety analysis, thermal hydraulic computer modeling of nuclear power plants. (e-mail: ger@psu.edu)

Barry E. Scheetz, Professor of Civil and Nuclear Engineering; Ph.D., Penn State, 1976. Waste form development, stability, and environmental interaction; cement and concrete for immobilization and isolation of wastes; cement chemistry; nuclear and hazardous chemical waste management; crystal chemistry and materials characterization. (e-mail: se6@psu.edu)

C. Fredrick Sears, Affiliate Associate Professor and Director, Breazeale Nuclear Reactor; Ph.D., Penn State, 1969. Reactor utilization in research. (e-mail: cfsnuc@engr.psu.edu)

Research Staff

Daniel E. Hughes, Senior Research Assistant; M.S., Penn State, 1986. Reactor operations and utilization, neutron radiography, reactor control. (e-mail: dehnuc@engr.psu.edu)

Bojan Petrovic, Research Associate; Ph.D., Penn State, 1995. Transport methods and applications to pressure vessel fluence, shielding, and reactor physics. (e-mail: petrovic@gracie.nuce.psu.edu)

RESEARCH AREAS

Reactor Safety Research

Severe accident analysis modeling; advanced thermal hydraulic computer code development, including rod bundle subchannel analysis, three-dimensional neutron kinetics; applications to BWR and PWR loss of coolant and transient analysis.

Reactor Core Analysis

Application of optimization techniques and expert systems to fuel management; design of low-leakage, long-life cores; use of statistical core analysis to increase thermal margin; coupled three-dimensional kinetics and thermal hydraulics analysis methods.

Reactor Operations Research

Development of expert systems for maintenance management and for preparation of high-quality safety evaluation reports; comparison of reliability-center-maintenance with preventative maintenance.

Neutronics and Transport Methods and Analysis

Development of parallel algorithms (domain decomposition methods differencing and iterative schemes) for neutron and gamma transport methods (discrete ordinates and Monte Carlo) in multiprocessor environments; improved models and computational methods (multigroup cross-section generation, uncertainty estimation, perturbation analysis, computational methods in reactor physics differencing scheme, acceleration of Monte Carlo) to determine neutron and gamma fluence to the reactor pressure vessel and other structural materials.

Advanced Control

Use of optima, adaptive, and robust control for reactor systems; development of fuzzy logic and neural network control and application of learning automata to select alternative control algorithms; real-time diagnostics based on AI techniques; human factors in control room design; experimental testing using the TRIGA reactor and the Intelligent Distributed Controls Laboratory.

Radiation Monitoring and Instrumentation

Development of new techniques to measure beta-emitting redionuclides in reactor effluent; wide-range instrument development for real-time beta dose measurement; gamma backscatter device to measure pipe wall thickness in insulated pipes.

Materials Research

Experimental and theoretical aspects of the effects of radiation on metals; stress-corrosion cracking of Ni-based alloys, deformation mechanisms of Zircaloy; pressure vessel embrittlement using positron annihilation; use of nuclear probes to characterize intermetallic compounds and electronic materials, including epitaxial growth of gallium arsenide.

PURDUE UNIVERSITY

School of Nuclear Engineering

Programs of Study

Graduate programs leading to the degrees of Master of Science in Nuclear Engineering (M.S.N.E.) and Doctor of Philosophy (Ph.D.) are offered for qualified students seeking advanced degrees. Information about the M.S.N.E. and Ph.D. programs can be found in the graduate school bulletin available from the School of Nuclear Engineering. Areas for graduate research and study include nuclear reactor theory and analysis, fast-reactor analysis, reactor simulation, reactor thermal hydraulics and safety, fusion plasma engineering and technology, design of advanced nuclear systems, radiation effects, nuclear materials, waste management, and applications of artificial intelligence in nuclear engineering.

A coordinated undergraduate program leading to an M.S.N.E. degree is available. Under this program, students can apply for admission to the Graduate School during the session in which the baccalaureate degree is being completed. Qualified students start planning their graduate program with their undergraduate counselors at the beginning of their junior year. The curriculum in nuclear engineering is ABET accredited.

Research Facilities

Research in nuclear engineering focuses on the design, development, manufacture, management, and control of engineering systems, subsystems, and their components. Research is conducted in reactor physics, fusion, thermal hydraulics and reactor safety, materials, and nuclear systems simulation. State-of-the-art computing systems are available for student use. The principal laboratories are the Thermal Hydraulics and Reactor Safety Laboratory, the Liquid-Metal Heat Transfer Laboratory, and the Energy Materials Laboratory.

Financial Aid

Each year numerous fellowships, scholarships, and traineeships are awarded by Purdue University. Many teaching and research assistantships are available each year, with stipends that ranged from $1100 to $1400 per month in 1998–99. The stipends also carry a remission of University fees and tuition except for $308 per semester. Further information may be obtained upon acceptance into the program.

Cost of Study

Fellowships, scholarships, and teaching and research assistantships provide tuition and fee waivers. Books and supplies average $355 per semester. For self-supported students in 1998–99, Indiana residents paid $1850 per semester for tuition and fees while nonresidents paid $5960 per semester for tuition and fees.

Living and Housing Costs

University-supervised graduate residences in 1998–99 cost from $270 to $375 per month. The University operates 1,384 married student apartments that rent at reasonable rates. Costs are subject to change. A variety of facilities exist in the surrounding communities. A helpful publication entitled *Housing and Financial Guide for Off-Campus Students* is available from the Office of the Dean of Students.

Student Group

Eighty-nine percent of the students are men and 11 percent are women, with 73 percent comprising international students. The faculty seeks students who are innovative and capable of extending their knowledge to become the finest in their field.

Student Outcomes

Graduates from the program are in high demand, and most receive multiple offers for employment at annual salaries that range from $55,000 to $70,000. Because of the diversified curriculum in nuclear engineering, graduates are employed in a wide variety of industries and government agencies, such as nuclear power utilities, nuclear fuel suppliers, nuclear system vendors, national laboratories, the U.S. Nuclear Regulatory Commission, the U.S. Department of Energy, and computer industries, as well as in academic faculty positions.

Location

The Purdue University School of Nuclear Engineering is located in West Lafayette, Indiana. The campus is located 60 miles from the Indianapolis state capital and 130 miles southeast of Chicago, Illinois.

The University and The School

Purdue University was established as a land-grant institution in 1869. The School of Nuclear Engineering was established in 1975, offering a Bachelor of Science degree. The School offers the B.S., N.E., M.S., M.S.N.E., and Ph.D. degrees.

Applying

All programs are open to qualified students with a bachelor's degree in engineering, but students with degrees in such areas as science and mathematics may also be admitted. Applications and supporting materials should be submitted by January 1 for fall admission and for all forms of financial assistance. When seeking information, students may contact the first address below for an appropriate response. The Graduate School office provides general information.

Correspondence and Information

Graduate Studies Office
1290 Nuclear Engineering Building
School of Nuclear Engineering
Purdue University
West Lafayette, Indiana 47907-1290
Telephone: 765-494-5749
Fax: 765-494-9570
World Wide Web: http://ne.www.ecn.purdue.edu/NE

Graduate School
Purdue University
West Lafayette, Indiana 47907
Telephone: 765-494-2600
World Wide Web: http://www.purdue.edu/GradSchool/gradhome.htm

Purdue University

THE FACULTY

Arden L. Bement Jr., Ph.D., Michigan, 1963 (Materials Engineering). Professor Bement heads the School of Nuclear Engineering. His research interests include nuclear materials and radiation effects. In addition, he is interested in national energy and technology policies and new applications of nuclear engineering.

Chan K. Choi, Ph.D., Southern Illinois, 1973 (Theoretical Physics). Professor Choi is active in the areas of fusion plasma engineering, computational and theoretical plasma physics, laser plasma interactions and target designs, nuclear medicine, and fusion space propulsion.

Thomas J. Downar, Ph.D., MIT, 1984 (Nuclear Engineering). Professor Downar's research interests include nuclear fuel management, LWR core physics analysis, and the application of advanced supercomputer methods to reactor physics problems.

Mamoru Ishii, Ph.D., Georgia Tech, 1971 (Mechanical Engineering). Professor Ishii's research interests include reactor thermal hydraulics, two-phase flow modeling, two-phase flow instrumentation, light water reactor safety, severe accident analysis, and heavy water isotope production reactor safety He is the author of *Thermo-Fluid Dynamic Theory of Two-Phase Flow.*

Martin A. Lopez de Bertodano, Ph.D., Rensselaer, 1991 (Nuclear Engineering). Professor Bertodano's research interests include two-phase experimental and computational fluid dynamics. He is particularly interested in multidimensional effects and two-phase turbulence.

Karl O. Ott, Ph.D., Gottingen (Germany), 1958 (Theoretical Physics). Professor Ott's research interests include reactor safety and risk assessment, reactor dynamics, fast-reactor theory, and fuel-cycle analysis. Professor Ott is a fellow of the American Nuclear Society. He is coauthor with W. A. Bezella of *Introductory Nuclear Reactor Statics* and co-author with R. J. Neuhold of *Introductory Nuclear Reactor Dynamics.*

Victor H. Ransom, Ph.D., Purdue, 1970 (Mechanical Engineering). Professor Ransom's research interests include modeling of multiphase flows, light water reactor transient simulation, and nuclear reactor safety.

Shripad T. Revankar, Ph.D., Karnatak (India), 1983 (Physics). Professor Revankar's research interests include experimental two-phase flow and heat transfer, two-phase flow diagnostics, integral system testing, nuclear reactor safety, and multiphase transport in porous media.

Alvin A. Solomon, Ph.D., Stanford, 1968 (Materials Science). Professor Solomon's research interests include high-temperature mechanical and physical properties of ceramic and metal matrix composites, behavior of nuclear fuels, processing and characterization of composites, nuclear waste materials, corrosion, and materials degradation in nuclear and energy-related materials.

Lefteri H. Tsoukalas, Ph.D., Illinois at Urbana-Champaign, 1989 (Nuclear Engineering). Professor Tsoukalas' research interests include fuzzy, neural, and other artificial intelligence approaches to existing and future-generation reactor systems, with emphases in modeling, diagnostics, and control. He is coauthor of *Fuzzy and Neural Approaches in Engineering* with R.E. Uhrig.

RENSSELAER POLYTECHNIC INSTITUTE

Department of Electric Power Engineering

Programs of Study

The Department of Electric Power Engineering is distinctive in that it is an independent, autonomous entity in the School of Engineering with its own faculty and facilities. It offers graduate programs leading to master's and doctoral degrees. The M.Eng. degree program has been developed over many years in close consultation with industry. It offers a wide range of graduate courses in such areas as power system analysis, surge phenomena, protective relaying, computer methods in power system computation, and power generation, operation, and control, which are complemented by courses from other engineering disciplines, management, and mathematics. The courses are taught by faculty members, many of whom have industrial experience in their areas of expertise. The master's program requires 30 credits, which may or may not include a thesis, and is frequently completed in one year. If possible, students should enter the program in the fall at the beginning of the academic year or else at the beginning of the preceding summer. Beginning in 1999, progressive introduction of distance versions of the key graduate courses is planned.

A prerequisite for the M.Eng. program is an engineering degree from an accredited institution. This strong, very practically oriented program attracts engineers from all over the United States and from every continent in the world. An average of twenty M.Eng. degrees in electric power engineering have been granted annually for the past ten years. Students whose undergraduate degree is not in engineering or not specifically in electrical or electric power engineering may pursue a program leading to an M.S. degree in electric power.

Both D.Eng. and Ph.D. degree programs are offered, depending upon the nature of the research undertaken. In either case a minimum of 90 credits, including the dissertation, is required beyond the bachelor's degree. A candidacy exam is also required.

Research Facilities

The department has experimental research facilities in the areas of power switching technology, gas physics, electrical machines, insulation systems and dielectrics, and electrical transient phenomena, as well as a power electronics laboratory. The department has its own Sun SPARC workstation cluster and a MicroVAX installation, which forms the basis for a power systems simulator for the development of protective relaying systems. Doctoral students can also take advantage of some of the finest industrial research facilities in the world, situated in the local area. Research is supported by such state-of-the-art facilities such as the computing facilities in the Center for Industrial Innovation, which provides graduate students with walk-in access to workstations and to programs from personal productivity aids to advanced computer-aided design and analysis packages; the Rensselaer Libraries, whose electronic information systems provide access to collections, databases, and library and Internet resources from campus and remote terminals; the Rensselaer Computing System, which permeates the campus with a coherent array of advanced workstations, a shared toolkit of applications for interactive learning and research, and high-speed Internet connectivity; a visualization laboratory for scientific computation; and a high-performance computing facility that includes a 36-node SP2 parallel computer.

Financial Aid

Traditionally, the department has been supported financially by utility and manufacturing industries. As a consequence, financial aid is available in a variety of forms to qualified students. These include scholarships, fellowships, and internships (wherein a student works for a company during the summer and continues with a project for the company during the year, receiving tuition and a stipend compensation). These awards are in addition to the usual teaching and research assistantships, which offer 24 credits of tuition and a stipend of about $11,000 for the 1999–2000 academic year. Outstanding students may qualify for University-supported Rensselaer Scholar Fellowships, which carry a stipend of $15,000 and a full waiver of tuition and fees. Low-interest, deferred-repayment graduate loans are also available to U.S. citizens with demonstrated need.

Cost of Study

Tuition for 1999–2000 is $665 per credit hour. Other fees amount to approximately $535 per semester. Books and supplies cost about $1700 per year.

Living and Housing Costs

The cost of rooms for single students in residence halls or apartments ranges from $3356 to $5298 for the 1999–2000 academic year. Family student housing is available, with monthly rents from $592 to $720.

Student Group

There are about 4,300 undergraduates and 1,750 graduate students, representing fifty states and more than eighty countries, at Rensselaer.

Student Outcomes

Eighty-eight percent of Rensselaer's 1998 graduating students were hired after graduation with starting salaries that averaged $56,259 for master's degree recipients and $57,000 to $75,000 for doctoral degree recipients.

Location

Rensselaer is situated on a scenic 260-acre hillside campus in Troy, New York, across the Hudson River from the state capital of Albany. Troy's central Northeast location provides students with a supportive, active, medium-sized community in which to live; an easy commute to Boston, New York, and Montreal; and some of the country's finest outdoor recreation, including Lake George, Lake Placid, and the Adirondack, Catskill, Berkshire, and Green Mountains. The Capital Region has one of the largest concentrations of academic institutions in the United States. Sixty thousand students attend fourteen area colleges and benefit from shared activities and courses.

The University

Founded in 1824 and the first American college to award degrees in engineering and science, Rensselaer Polytechnic Institute today is accredited by the Middle States Association of Colleges and Schools and is a private, nonsectarian, coeducational university. Rensselaer has five schools—Architecture, Engineering, Management, Science, and Humanities and Social Sciences. The School of Engineering is ranked among the top twenty engineering schools in the nation by the *U.S. News & World Report* survey and is ranked in the top ten by practicing engineers.

Applying

Applications for admission and two copies of all transcripts must be submitted to the Graduate Academic and Enrollment Services Office, which then forwards the material to the Department of Electric Power Engineering. Admissions applications and all supporting credentials should be submitted well in advance of the preferred semester of entry to allow sufficient time for departmental review and processing. The application fee is $35. Since the first departmental awards are usually made in March for the next full academic year, applicants requesting financial aid are encouraged to submit all required credentials by February 1 to ensure consideration. The TOEFL (minimum score 550) is required of international students, and the GRE General Test is recommended for all students. Decisions on admission are made promptly.

Correspondence and Information

For written information about graduate work:
Dr. J. Keith Nelson
Department of Electric Power Engineering
Rensselaer Polytechnic Institute
110 8th Street
Troy, New York 12180-3590

Telephone: 518-276-6329
Fax: 518-276-6226
E-mail: epe-info@rpi.edu
WWW: http://www.rpi.edu/dept/epe/WWW/index.html

For applications and admissions information:
Director of Graduate Academic and Enrollment
    Services, Graduate Center
Rensselaer Polytechnic Institute
110 8th Street
Troy, New York 12180-3590

Telephone: 518-276-6789
E-mail: grad-services@rpi.edu
World Wide Web: http://www.rpi.edu

Rensselaer Polytechnic Institute

THE FACULTY AND THEIR RESEARCH

George T. Berry, Adjunct Professor; M.E., Harvard; PE. Power system operation.
Ralph J. Caola, Adjunct Associate Professor; M.E., Rensselaer. Protective relaying.
Joe H. Chow, Professor; Ph.D., Illinois. Large-scale system modeling, multivariable control systems.
Robert C. Degeneff, Professor; Ph.D., Rensselaer; PE. Power system studies, HVDC, component modeling.
Allan Greenwood, Philip Sporn Professor of Engineering; Ph.D., Leeds. Power switching technology, electrical transients.
James M. Kokernak, Assistant Professor; Ph.D., Rensselaer. Power electronics, EMI/EMC, power quality.
J. Keith Nelson, Professor and Chair; Ph.D., London; PE. Insulation systems and dielectrics, gas physics.
Michael L. Reichard, Adjunct Associate Professor; M.E., Penn State; PE. Industrial power system design.
Sheppard J. Salon, Professor; Ph.D., Pittsburgh; PE. Machine design, system component modeling and simulation.
David A. Torrey, Niagara Mohawk Associate Professor of Power Electronics; Ph.D., MIT; PE. Power electronics and electromechanics.
Allen J. Wood, Adjunct Professor; Ph.D., Rensselaer; PE. Economic operation and control of power systems.

AREAS OF RESEARCH

Power Switching Technology

The continual growth of power systems through expansion and interconnection is placing increasingly stringent demands on the switching equipment that controls the flow of power and protects the systems under fault conditions. Work, therefore, is being addressed to the fundamental processes of circuit interruption, especially in vacuum and SF_6 circuit breakers. With the ever-increasing level of available short-circuit currents, means are being pursued to develop fault current limiters. Some of this technology is also being applied to the interactions between devices of the power system.

Electric and Magnetic Field Computation

The design of equipment to minimize losses, achieve compaction, or better utilize material frequently requires a sound knowledge of the electric and magnetic field configurations involved. Several projects in the recent past have adapted finite element methods to the solution of current problems in large machines and transformers.
Recent applications include high-efficiency machines, noise and vibration in electric motors, optimization and inverse problems, and the application of superconducting to power equipment.

Electrical Transients

Of current interest are those transients initiated by the switching of power plant auxiliaries and capacitor banks, especially by vacuum switching devices. The modeling of transients in transformer structures is also a focus as it provides insight into the problems of both design and operation. The techniques being developed are also finding application in new areas such as superconducting magnetic energy storage (SMES) devices.

Electrical Insulation

An electrical insulation system, be it solid, liquid, or gaseous or a combination of these, is an essential part of every piece of power equipment. Current research is directed toward understanding the fundamental behavior of insulation under a variety of operating conditions and toward the development of diagnostic instrumentation, particularly for generator insulation and static electrification events.

Power System Analysis

Optimization theory is used in the design of electric power systems to obtain highest efficiency at minimum cost. This has been extended to include the development of intelligent protective relaying utilizing the department's system simulator and EMTP studies.

Semiconductor Power Electronics and Motion Control

With improved power semiconductor devices, it is now possible to efficiently convert energy from one form to another electronically. At Rensselaer, a wide variety of disciplines are applied to electronic energy conversion, motion control, and power quality for the electric power and other automation industries. Current interests include propulsion systems for electric vehicles, generation systems suitable for wind turbines, the use of artificial intelligence (fuzzy logic, genetic algorithms, and neural networks) in the design and control of electronic power conversion and electric machines, the adaptive control of electric machines, and converter development for distributed operation systems.

The department specializes in many of the physical phenomena that limit power systems equipment.

TEXAS A&M UNIVERSITY

College of Engineering
Department of Nuclear Engineering

Programs of Study

Nuclear engineering is concerned with the release, control, and use of energy from nuclear sources; the production of thermal energy and radiation in these processes; and the interaction of radiation with matter. It is based on the principles of nuclear physics that govern radioactivity, fission, and fusion. The function of the nuclear engineer is to apply these principles to a wide range of challenging technological problems. The specialized study of health physics in the Department of Nuclear Engineering is soundly based on the fundamental aspects of radiation effects on matter.

The department offers the Master of Engineering, Master of Science, and Doctor of Philosophy degrees and also offers courses and faculty supervision for students pursuing the Doctor of Engineering degree. Students interested in doctoral-level studies in health physics can pursue these through the Ph.D. program in nuclear engineering. A professional education program in health physics is available at the master's level.

Admission to the nuclear engineering graduate program requires a bachelor's degree in engineering, chemistry, mathematics, or physics. Mathematics through differential equations is required and some nuclear physics background is highly desirable. Master's-level students normally spend eighteen months on campus, with the first eight months devoted primarily to course work and the latter ten months spent on thesis research. At the master's level in health physics, the student is required to spend the initial academic year taking formal course work in the Department of Nuclear Engineering and in other cooperating departments of the University. The summer is spent in special courses providing on-the-job training in health physics at the Texas A&M Cyclotron, the Nuclear Science Center Reactor, and the Radiological Safety Office. At least one additional semester is normally required to finish course work and complete a research project. Formal admission to the Ph.D. program requires successful completion of a qualifying examination.

The department has active research programs in the areas of ion-beam interactions with matter, transmutation doping, fusion technology, fuel-cycle analysis, radiation transport, internal and external dosimetry, whole-body phantoms, light water reactor core thermal hydraulic analysis, advanced nuclear reactor, thermal hydraulics and two-phase flow imaging techniques, heat-pipe modeling and experimental measurements, zero-gravity two-phase flow modeling and experimental measurements, space reactor systems modeling, MOX fuel, fuel performance, development of numerical analysis techniques, aerosol formation and interaction, and fissile materials disposition.

Research Facilities

Research facilities include a well-equipped radiation measurements laboratory, a radiochemistry laboratory, a thermal hydraulics laboratory, a subcritical reactor laboratory, an AGN-201 low-power nuclear reactor, a 1-megawatt thermal TRIGA research reactor, department microcomputers, a network of high-end workstations, a Cockcroft-Walton pulsed accelerator, and University mainframe supercomputers.

Financial Aid

Teaching and research assistantships are available. The stipend ranges from $1000 to $1400 per month in 1999–2000, depending on the student's qualifications. Fellowships are awarded by the University, and Texas A&M University strongly encourages fellowship applications from members of diversity groups. The department also administers Institute for Nuclear Power Operation Fellowships and is qualified for Department of Energy Fellowships in nuclear engineering, health physics, and waste management. Students may be employed by the Radiological Safety Office, the Cyclotron Institute, and the Nuclear Science Center.

Cost of Study

The 1999–2000 academic-year tuition for Texas resident students is $72 per semester credit hour. Tuition for nonresidents and international students is $285 per semester credit hour. Costs are subject to change.

Living and Housing Costs

The cost of living is low and the quality of life is high in the Bryan–College Station area. Graduate student housing is principally off campus and is reasonably priced, with a wide range as to the type of accommodation. Typically, a one-bedroom apartment can be found for under $500 per month, including utilities. The University runs its own bus transportation system to most of the off-campus housing complexes.

Student Group

The fall 1998 enrollment totaled 82 students, 60 full-time. Thirty-two were Ph.D. candidates. There are 11 women in the department, and one quarter of the student population is composed of international students. Graduates have found jobs with utilities, vendors, power plants, and universities throughout the country. Many are employed at national laboratories and a significant number have established their own companies. Health physics students also find employment opportunities at laboratories, universities, power plants, and medical facilities.

Location

Texas A&M is located on a 5,200-acre campus approximately at the center of a triangle formed by the cities of Houston, Austin, and Dallas–Fort Worth. Bryan–College Station is a rural community with a combined population of approximately 160,000 people.

The University and The Department

Texas A&M University was established in 1876 as a land-grant college. Currently, 41,461 students, including 6,774 in graduate studies, are enrolled in the University. There are eighty-four departments in the ten colleges of the University. The Texas A&M Department of Nuclear Engineering is one of the oldest nuclear engineering departments in the country, with the Nuclear Science Center having celebrated its thirtieth anniversary in 1991. The department has 14 tenured faculty members. Advanced nuclear power reactors, space applications of nuclear power, and fissile materials disposition have been areas of special interest to the department, with research on these topics taking up an increasing part of the department's expanding research program.

Applying

Admission to the department requires a bachelor's degree in engineering, chemistry, mathematics, or physics. Some nuclear physics background is highly desirable. Mathematics through differential equations is required. The GRE General Test is required, and students from non-English-speaking countries are required to demonstrate proficiency in English by submitting satisfactory TOEFL scores. An average of B or better is required; however, the final admission decision is based on a combination of grade point ratio, GRE scores, and recommendations.

Correspondence and Information

Dr. Yassin A. Hassan, Graduate Coordinator (99)
Department of Nuclear Engineering
Texas A&M University
College Station, Texas 77843-3133

Telephone: 409-845-7090
Fax: 409-845-6443
World Wide Web: http://trinity.tamu.edu

Texas A&M University

THE FACULTY AND THEIR RESEARCH

Professors
Jerome J. Congleton, Ph.D., Texas Tech, 1983. Ergonomics, safety engineering.
Ron R. Hart, Ph.D., Berkeley, 1967. Ion beam interactions, transmutation doping.
Yassin A. Hassan, Ph.D., Illinois, 1979. Experimental and computational thermal hydraulics.
William H. Marlow, Ph.D., Texas at Austin, 1973. Aerosol microphysics and applications.
J. Steven Moore, M.D., Texas Health Science Center at Dallas, 1978. Safety engineering.
Paul Nelson, Ph.D., New Mexico, 1969. Development of numerical methods.
Theodore A. Parish, Ph.D., Texas at Austin, 1973. Reactor physics, fuel cycle analysis, fusion technology.
Kenneth L. Peddicord, Ph.D., Illinois, 1972. Advanced fuels development.
John W. Poston Sr., Ph.D., Georgia Tech, 1971. Internal dosimetry, thermoluminescent dosimetry.
Alan E. Waltar, Ph.D., Berkeley, 1966. Fast-breeder reactor.

Associate Professors
Marvin L. Adams, Ph.D., Michigan, 1986. Computational methods.
Frederick R. Best, Ph.D., MIT, 1980. Heat-pipe and zero-gravity experimental thermal hydraulics.
Alexander Parlos, Ph.D., MIT, 1986. Automation and control.
Warren Daniel Reece, Ph.D., Georgia Tech, 1988. Dosimetry, radiation transport, breeder reactors, real-time location of breach of clad.

Adjunct Professors
Harry J. Ettinger, M.S., NYU, 1958.
Bryan L. Fearey, Ph.D., Iowa State, 1986.
Ronald E. Goans, M.D., George Washington, 1983.
William C. T. Inkret, Ph.D., Colorado State, 1986.
Milton E. McLain, Ph.D., Georgia Tech, 1972.
Jim E. Morel, Ph.D., New Mexico, 1979.
Koji Okamoto, Ph.D., Tokyo, 1992.
R. T. Perry, Ph.D., Texas A&M, 1974.
Woodrow W. Pitt, Ph.D., Tennessee, 1969.
Namir Saman, Ph.D., Arizona, 1989.
Gerald A. Schlapper, Ph.D., Missouri–Columbia, 1977.
Gregory D. Spriggs, Ph.D., Arizona, 1982.
Daniel J. Strom, Ph.D., North Carolina, 1984.

Other Faculty
Alfred A. Amendola, Ph.D., Texas A&M, 1989.
Dmitriy Y. Anistratov, Ph.D., Russian Academy of Sciences, 1993.
David R. Boyle, Ph.D., MIT, 1980.
Leslie A. Braby, Ph.D., Oregon State, 1972.
John R. Ford Jr., Ph.D., Tennessee, 1992.
Bruce L. Freeman, Ph.D., California, Davis, 1974.
Ian S. Hamilton, Ph.D., Texas A&M, 1995.
James C. Rock, Ph.D., Ohio State, 1972.
Michael J. Schuller, Ph.D., Texas A&M, 1985.
John P. Wagner, Ph.D., Johns Hopkins, 1966.
Rube B. Williams, Ph.D., Texas A&M, 1997.

UNIVERSITY OF CALIFORNIA, BERKELEY

College of Engineering
Department of Nuclear Engineering

Programs of Study	The graduate program in nuclear engineering at Berkeley offers instruction, research, and professional education in nuclear energy (fission and fusion), nuclear waste and materials management, and bionuclear and radiological science. Established in 1958, the department provides a graduate program consisting of the principal fields of reactor theory; reactor engineering, including thermal hydraulics and safety; nuclear materials; nuclear reactions and instrumentation; thermonuclear fusion; nuclear waste management; risk and systems analysis; medical imaging; and radiation physics and dosimetry.

The Master of Science degree is nominally a one-year program. Plan I requires a thesis and a minimum of 20 units of course work. Plan II requires 24 units of course work, a project culminating in a written report, and an oral examination on the project. The Master of Engineering program requires a minimum of 40 units: 20 units of courses in the professional major, oriented toward design and analysis; technical-breadth courses to acquaint the student with other technical fields; and a nonengineering breadth requirement. The doctoral program consists of formal course work and original research. A typical program includes approximately 34 units of course work in nuclear engineering and in two minor fields, one of which may be within the department and the other outside the department. Doctoral students must pass a written examination, normally taken during the first year and based on undergraduate subjects in applied thermodynamics, nuclear materials, heat transfer and fluid mechanics, nuclear physics, neutronics, fusion theory, and nuclear waste management. Upon passing an additional oral qualifying examination, the student is advanced to candidacy for the Ph.D. and devotes the remainder of his or her graduate career to conducting research that culminates in the doctoral dissertation. |
Research Facilities	The department's laboratories are equipped with instrumentation facilities for studying the effects and applications of nuclear radiation, including spectrometric analysis. Experimental loops are available for studying heat transfer and fluid-flow phenomena. The nuclear materials research program utilizes high-temperature equipment, mass spectrometry, and several high-vacuum stations for materials chemistry experiments. Research related to thermonuclear fusion is conducted in the department's laboratories using the rotating-target neutron source and the Berkeley Compact Toroid Experiment. The department has an advanced computer laboratory (ANECL) dedicated to both research and course work. The ANECL cluster consists of Sun UNIX workstations, Windows NT, and Apple PowerPCs. An extensive software library and access to University-operated mainframe computers and to the Internet are also available.
Financial Aid	University fellowships are awarded on a competitive basis. Fellowships include stipends, plus tuition and fees. Research and teaching assistants are eligible for up to half-time employment during the academic year; research assistantships can be full-time during the summer. Need-based financial aid, including grants, loans, and work-study, are available through the Office of Financial Aid.
Cost of Study	For 1999–2000, all graduate students pay fees of approximately $4400. Out-of-state and international students, in addition, pay tuition of approximately $9800 for the year.
Living and Housing Costs	To obtain information about housing options, students should refer to the Housing and Dining Services Web site (http://www.housing.berkeley.edu/) or write to Housing and Dining Services, University of California, Berkeley, 2401 Bowditch Street #2272, Berkeley, California 94720-2272 (telephone: 510-642-3642; e-mail: homeinfo@ berkeley.edu).Inquiries should include the following: indication of status as a prospective graduate student, the planned semester of entry into Berkeley, marital status, mailing address, and the type of housing the student is seeking. To obtain visas, newly admitted international students are required to show proof of having at their disposal a minimum of $28,000 for each year of study and residence.
Student Group	There are approximately 2,290 undergraduate and 1,425 graduate students enrolled in the College of Engineering. There are 45 graduate students in the Department of Nuclear Engineering. Approximately 33 percent are from other countries, and about 10 percent are women. Graduates find many opportunities for employment and professional careers in the United States and abroad. Recent graduates are employed in industry, academia, national laboratories, and state and federal agencies.
Location	The University is located at the base of the Berkeley hills, looking across the San Francisco Bay to San Francisco. The San Francisco Bay Area offers a variety of cultural and entertainment activities. Students have ready access to the Pacific Coast beaches and excellent skiing areas in the Sierra Nevada. The climate is cool in the summer, with no rainfall. Winters are mild, with intermittent rain and many sunny days.
The University	Berkeley is the parent campus of the University of California. It has an enrollment of 31,000, with 9,000 graduate students in 100 fields of study, and is noted for the academic distinction of its faculty, the quality and scope of its research activities, and the variety and vitality of student activities. It is ranked by its academic peers as one of the best graduate institutions in the United States.
Applying	Application forms should be obtained from the Web at http://www.grad.berkeley.edu/grad and submitted to the department. Applications should be received by February 10 for the fall semester and September 1 for the spring semester. The deadline for fellowships and scholarships is January 5. GRE scores are required of all applicants, and international students must submit TOEFL scores.
Correspondence and Information	Graduate Admissions Department of Nuclear Engineering University of California Berkeley, California 94720 E-mail: gradinfo@nuc.berkeley.edu World Wide Web: http://www.nuc.berkeley.edu

University of California, Berkeley

THE FACULTY AND THEIR RESEARCH

Joonhong Ahn, Assistant Professor; Ph.D., Berkeley, 1988; D.Eng., Tokyo, 1989. Performance assessment of geological disposal system for radioactive wastes, mass transport through heterogenous media. (E-mail: ahn@nuc.berkeley.edu)

Ehud Greenspan, Professor-in-Residence; Ph.D., Cornell, 1966. Neutronics of fission reactors, fusion reactors, radiation shielding, and medical applications: methods development and design optimization; advanced reactors and fuel cycle concepts; criticality safety. (E-mail: gehud@nuc.berkeley.edu)

William E. Kastenberg, Chancellor's Professor and Chairman; Ph.D., Berkeley, 1966. Nuclear reactor safety, risk assessment, and risk management for nuclear and nonnuclear technologies, high-level radioactive waste and nuclear materials management. (E-mail: kastenbe@nuc.berkeley.edu)

Edward C. Morse, Professor; Ph.D., Illinois, 1979. Fusion reactor design and applied plasma physics, experimental investigation of RF plasma heating, neutronics and neutron damage studies using the Rotating Target Neutron Source. (E-mail: morse@nuc.berkeley.edu)

Donald R. Olander, Professor; Sc.D., MIT, 1958. High-temperature kinetic and thermodynamic behavior of nuclear reactor fuels. (E-mail: fuelpr@cmsa.berkeley.edu)

Per Peterson, Associate Professor; Ph.D., Berkeley, 1988. Thermal-hydraulics in advanced reactors, inertial confinement fusion, and nuclear waste materials management. (E-mail: peterson@nuc.berkeley.edu)

Stanley G. Prussin, Professor; Ph.D., Michigan, 1964. Low-energy nuclear physics and applications, radiation and measurements, biomedical applications. (E-mail: prussin@nuc.berkeley.edu)

Keith Thomassen, Professor-in-Residence; Ph.D., Stanford, 1963. Plasma physics, confinement and stability of plasmas for thermonuclear fusion, fusion reactor design. (E-mail: thomassen@llnl.gov)

Jasmina L. Vujic, Assistant Professor; Ph.D., Michigan, 1990. Advanced numerical methods in radiation transport, neutronics of fission reactors, biomedical application of radiation, dosimetry and cancer therapy, optimization techniques for vector and parallel computers. (E-mail: vujic@nuc.berkeley.edu)

Emeritus Faculty

Emeritus faculty members may be contacted via e-mail at gradinfo@nuc.berkeley.edu.

Paul L. Chambré, Professor Emeritus; Ph.D., Berkeley, 1951. Analytical and numerical methods in radioactive waste management.

T. Kenneth Fowler, Professor Emeritus; Ph.D., Wisconsin–Madison, 1957. Plasma physics, confinement and stability of plasmas for thermonuclear fusion, fusion reactor design.

Lawrence M. Grossman, Professor Emeritus; Ph.D., Berkeley, 1948. Reactor physics, numerical approximation methods in neutron diffusion and transport theory, control and optimization theory in nuclear reactor engineering.

Selig N. Kaplan, Professor Emeritus; Ph.D., Berkeley, 1957. Nuclear instrumentation, electronic radiographic imaging.

Thomas H. Pigford, Professor Emeritus; Sc.D., MIT, 1952. Management of radioactive wastes, environmental effects of radionuclides.

Virgil E. Schrock, Professor Emeritus; M.S., Wisconsin, 1948; Mech.Eng., Berkeley, 1952. Thermal-hydraulic phenomena in the design and safety evaluation of nuclear power plants.

UNIVERSITY OF ILLINOIS AT URBANA–CHAMPAIGN

Department of Nuclear, Plasma, and Radiological Engineering

Programs of Study

The department offers curricula leading to the M.S. and Ph.D. degrees in nuclear, plasma, and radiological engineering. The M.S. program requires 8 units (1 unit equals 4 semester hours) of approved graduate credit, including a required thesis. A student must spend a minimum of two semesters in residence. The doctoral degree program can be divided into three stages, two of which must be completed in residence: a master's degree or an equivalent number of credits (8 units, or 32 semester hours), 8 units of course work and a preliminary examination, and 8 units of thesis research, a dissertation, and a final examination. Areas of research include both fission and fusion engineering and technology; plasma engineering and processing; shielding and radiation effects; thermal hydraulics and reactor safety; nuclear materials, corrosion, and irradiation damage; neutron scattering; neutron activation analysis; nuclear nonproliferation and public policy issues; waste management and site remediation; and biomedical imaging and health physics.

Research Facilities

A wide range of major research resources are available for nuclear, plasma, and radiological engineering research. A dense plasma focus fusion-related device for high-temperature plasma studies and an ultrahigh-vacuum laboratory for plasma-material interaction studies are available. Graduate students often perform interdisciplinary research work in the Materials Research Laboratory, Microelectronics Laboratory, Coordinated Science Laboratory, National Center for Supercomputing Applications, and Beckman Institute for Advanced Science and Technology. Faculty members in these laboratories hold affiliate appointments in the degree-granting departments. The mechanical behavior program provides a variety of facilities for studies of nuclear materials. Other radiological laboratories are also available for environmental studies and nuclear spectroscopy, health physics and radiation studies, nuclear-waste management, thermal hydraulics and reactor safety, reactor physics and reactor kinetics, controlled nuclear fusion, direct energy conversion, and lasers and plasma physics. The Department of Nuclear, Plasma, and Radiological Engineering also has a direct link to the National Magnetic Fusion Computer Center in Livermore, California, and is a participant in the Computational Science and Engineering Program on campus. In addition, a wide array of microcomputers and workstations are available for student use.

Financial Aid

Most graduate students receive some form of financial aid. For example, during 1998–99, there were 225 fellows and trainees and 1,187 part-time research assistants. Financial aid includes federally sponsored traineeships and fellowships and University and industrial fellowships. The University is approved for several fellowships, including those from the Department of Energy, the National Science Foundation, Hertz, and the Institute for Nuclear Power Operations. Also available are part- and full-time research and teaching assistantships. Research assistantships, which include exemption from tuition and partial fees, currently start at $12,330 for the academic year for half-time work (approximately 20 hours per week). The normal course load for a student working half-time is about three courses per semester.

Cost of Study

Tuition and fees vary according to the number of semester hours taken. For students who have fellowships or staff appointments, tuition and some fees are waived, but fees amounting to approximately $360 for health insurance and other miscellaneous fees must be paid by the student each semester. For a full program (3 or more units), tuition and fees per semester in 1998–99 were $2819 for Illinois residents and $6257 for nonresidents. Summer session charges for a full program (2.5 or more units) were $1778 for Illinois residents and $3927 for nonresidents.

Living and Housing Costs

University graduate residence halls have single rooms for $2696 and $3008 and double rooms for $2416 and $2872 per academic year. Board contracts are available for $1435 and $3368. University family housing rents for $350 to $510 per month. Privately owned rooms and apartments are available at similar and higher rents.

Student Group

Enrollment at the Urbana-Champaign campus is 36,303 students. The College of Engineering has 5,285 undergraduates and 2,007 graduate students. There are approximately 50 graduate students in nuclear, plasma, and radiological engineering. Of these, about 15 percent are supported by fellowships. Most others are supported as graduate research assistants.

Location

The University is located 150 miles south of Chicago in the twin cities of Urbana and Champaign (population 110,000). The area is primarily a university community, with excellent schools, parks, and modern shopping facilities. Willard Airport, major rail service, and three interstate highways provide rapid and accessible transportation. Many cultural and recreational facilities normally found only in a large city are available.

The University and The Department

The University of Illinois at Urbana-Champaign is in its second century of operation and is recognized as a national center of excellence in graduate education. The College of Engineering, founded in 1868, has grown and prospered with the University, establishing itself as a productive center of engineering research and education. Nuclear engineering was established in 1958 as an interdisciplinary program and was granted departmental status in 1986. Because of its original interdisciplinary nature, its research is carried out in several facilities. Fusion research is coordinated through the Fusion Studies Laboratory, with experimental facilities in the Nuclear Radiation Laboratory. Administrative offices, most classes, and the support shops are in the Nuclear Engineering Laboratory. Experimental facilities for thermal hydraulics, neutron activation analysis, plasma diagnostic and processing, imaging with ionizing radiation, and a variety of materials research are also available.

Applying

Students must have a strong background in mathematics, science, and/or engineering. An average of B or better is required. Scores on the GRE are not a formal requirement, although they are helpful for international applicants. Students from non-English-speaking countries must demonstrate proficiency in English, which is measured by TOEFL and SPEAK scores. Required minimum scores are 570 and 230, respectively, to enter nuclear, plasma, and radiological engineering. Additional English study may be necessary following a placement test. Those applying for a fellowship must submit applications prior to February 15. Other applications can be submitted up to three weeks prior to the start of the semester or until the application cycle is determined closed by the department's Graduate Admissions Committee, but early application is strongly advised.

Correspondence and Information

Dr. James F. Stubbins, Head
Department of Nuclear, Plasma, and Radiological Engineering
214 Nuclear Engineering Laboratory
University of Illinois at Urbana-Champaign
103 South Goodwin Avenue
Urbana, Illinois 61801-2984

Telephone: 217-333-3598
Fax: 217-333-2906
E-mail: nuclear@uiuc.edu
World Wide Web: http://www.ne.uiuc.edu

University of Illinois at Urbana–Champaign

THE FACULTY AND THEIR RESEARCH

Professors

R. A. Axford, Ph.D. Reactor physics, safety, and risk assessment; radiation hydrodynamics; heat transfer; optimal reactor control, synthesis, and nuclear fuel management; hydrodynamic stability; plasma physics; group invariant difference schemes.

B. G. Jones, Ph.D. Nuclear reactor safety, fluid mechanics, heat transfer, turbulence measurement, and modeling; flow-induced vibrations and hydroacoustics; multiphase flow; optimal function allocation for reactor control.

K. K. Kim, Ph.D. Fusion technology, fusion reactor fueling, high-energy lasers, electromagnetic railgun, laser fusion targets, ionized-cluster beam epitaxy, electrohydrodynamics.

G. H. Miley, Ph.D. Fusion, plasma engineering, fusion technology, computational methods for plasma analysis, nuclear-pumped lasers.

D. N. Ruzic, Ph.D. Experimental fusion research, modeling of edge-plasma atomic physics, atomic properties of potential first-wall materials, plasma-material interaction, plasma processing of semiconductors.

C. E. Singer, Ph.D. Analytical and numerical fusion plasma engineering and physics, as applied to tokamaks; advanced propulsion concepts.

J. F. Stubbins, Ph.D. Nuclear materials, irradiation damage and effects, mechanical properties, high-temperature corrosion, stress corrosion cracking, electron microscopy.

R. J. Turnbull, Ph.D. Controlled fusion, cluster beams, electrohydrodynamics.

Associate Professor

M. H. Ragheb, Ph.D. Computational methods, reactor theory, Monte Carlo methods, radiation protection and shielding, fusion systems, probabilistic risk assessment, probabilistic and possibilities decision making, applied artificial intelligence, supercomputing.

Assistant Professors

B. J. Heuser, Ph.D. Hydrogen in metals, hydrogen trapping at defects, metal hydrides, transmission electron microscopy studies of metal defects, neutron scattering, metal thin film multilayer structures.

R. Uddin, Ph.D. Advanced computational methods, theoretical and CFD, radiation transport and reactor physics, reactor engineering, multiphase flow, reliability and risk analysis.

E. C. Wiener, Ph.D. Cancer diagnosis and therapy, neutron capture therapy, magnetic resonance imaging, single-photon-emission computed tomography, radiotherapy, drug delivery.

Affiliate Faculty

In addition to the full-time faculty members, professors in a broad range of engineering disciplines are formally affiliated with the Department of Nuclear, Plasma, and Radiological Engineering. They provide a breadth of expertise that broadens opportunities for research activities.

R. J. Adrian, Department of Theoretical and Applied Mechanics.
R. S. Averback, Department of Materials Science and Engineering.
R. O. Buckius, Department of Mechanical and Industrial Engineering.
C. W. Bullard, Department of Mechanical and Industrial Engineering.
M. J. Kushner, Department of Electrical and Computer Engineering.
R. L. Magin, Department of Electrical and Computer Engineering.
R. F. Nelson, Radiation Oncology Department, Carle Clinic, Urbana, Illinois.
B. G. Thomas, Department of Mechanical and Industrial Engineering.
A. R. Twardock, Department of Veterinary Medicine.

Adjunct Faculty

Several adjunct faculty members are also interacting directly with the Department of Nuclear, Plasma, and Radiological Engineering, providing an even broader range of opportunities for research, including such areas as waste management, energy systems, reactor design, and health physics.

UNIVERSITY OF MICHIGAN

Department of Nuclear Engineering and Radiological Sciences

Programs of Study

The Department of Nuclear Engineering and Radiological Sciences at the University of Michigan offers graduate programs leading to the Master of Science and Doctor of Philosophy degrees in nuclear engineering and nuclear science and to the degree of Nuclear Engineer. Major areas of activity in the department are fission reactor physics, fusion and plasma physics, materials, radiation imaging and measurement, radiological health, radioactive materials management, and medical physics. A Master of Engineering in radiological health engineering is also offered.

The Master of Science program is designed to give students a working knowledge of a subdiscipline of nuclear engineering. The requirements for the master's degree are 30 hours of course work at the graduate level, including 20 hours in the Department of Nuclear Engineering and Radiological Sciences. All master's degree students must take a formal 400-level or higher laboratory course while enrolled as a graduate student. With the help of the student's adviser, additional courses are selected from nuclear engineering and radiological sciences, cognate fields of engineering, mathematics, physics, chemistry, and others. At least two courses outside nuclear engineering and radiological sciences are required.

The doctoral program in the Department of Nuclear Engineering and Radiological Sciences emphasizes a broad scientific background to prepare the engineer for work in research and development. This background is necessary not only because of the complexity of the problems facing the nuclear engineer but also because of a close liaison that must exist between the scientist and the engineer in all new and rapidly developing fields of technology. Dissertation topics may be theoretical or computational problems or experimental investigations making use of the extensive facilities of the department's laboratories and the Phoenix Memorial Laboratory.

Research Facilities

The department's facilities are among the finest in the nation and include the 2-megawatt Ford Nuclear Reactor, the Materials Dosimetry Reference Facility, the Michigan Ion Beam Laboratory, the Metastable Materials Laboratory, the Radiation Imaging Laboratory, the Glow Discharge Laboratory, the Intense Energy Beam Interaction Laboratory, the High Temperature Corrosion Laboratory, the Radioactive Materials Management Laboratory, the Materials Preparation Laboratory, and the Radiological Health Engineering Laboratory. The University library system includes the Graduate Library, the Undergraduate Library, and more than twenty departmental or divisional libraries, including the Engineering Library with more than 345,000 volumes. The department is part of the College of Engineering's Computer-Aided Engineering Network, which provides engineering students with the most advanced computing environment of all such engineering college facilities in the country.

Financial Aid

Students are eligible for a number of grants, fellowships, and traineeships provided by the federal government, industry, and the University of Michigan. The University administers Graduate Fellowships provided by the U.S. Department of Energy and awards fellowships provided by the National Academy for Nuclear Training. Historically, the department has ranked either first or second in the country in the total number of DOE fellows in residence. The department participates in programs sponsored by or in collaboration with several national laboratories, including Argonne, Brookhaven, Livermore, Los Alamos, Oak Ridge, and Sandia, some of which may provide financial assistance for thesis research carried out at these national laboratories. In addition, the department awards a number of research assistantships and teaching assistantships to qualified students.

Cost of Study

Tuition and fees for the 1998–99 school year for a full-time student were $6042 per academic term for Michigan residents, $10,943 per term for nonresidents, and $3936 per term for all doctoral candidates.

Living and Housing Costs

The University of Michigan maintains a Housing Information Office that serves as a clearinghouse for information and assists students in finding accommodations. On-campus 2-person residence hall housing, including room and board, costs $5488 per person for two terms. Family housing costs about $509 to $595 per month for a one-bedroom furnished apartment. The Housing Information Office administers University-operated facilities, which include residence halls, co-ops, apartments, and suites, and coordinates information about off-campus (privately owned) housing. The office can be contacted via e-mail (housing@umich.edu) or the World Wide Web (http://www.housing.umich.edu/).

Student Group

There are 79 students in the department's graduate programs, with 49 pursuing Ph.D.'s. The nation's first student branch of the American Nuclear Society was formed at the University in 1955. Graduate students play an active role in this branch, which serves as a focus for many department activities.

Location

The University of Michigan is located in the heart of Ann Arbor, just 40 miles west of Detroit. Ann Arbor—called Tree City, USA—has a population of 120,000, half of which consists of University people. The city has a rich tradition of hosting national and international cultural events and has become known as the cultural mecca of the Midwest. The rolling hills and changing seasons make outdoor recreation a popular activity. Ann Arbor is also the home of Michigan Stadium, the nation's largest football stadium.

The University

The University of Michigan, which was chartered in 1817, moved to its Ann Arbor campus in 1837. About 37,000 students are enrolled in the University's many colleges and schools in Ann Arbor. Students studying at two other campus locations—Dearborn and Flint—bring the total enrollment to more than 48,000. Of these, more than 36 percent are graduate and professional students. The Ann Arbor campus consists of the Central Campus, the Medical Campus, the Athletic Campus, and North Campus. North Campus, which is separated from Central Campus by the Huron River, is the home of the College of Engineering and site of the nuclear engineering and radiological sciences department's offices and the Ford Nuclear Reactor.

Applying

Students who apply for admission to graduate study at the master's level in the nuclear engineering program should have a bachelor's degree from a recognized engineering program. The nuclear science program is available to those with a bachelor's degree in physics, chemistry, or mathematics. Students with bachelor's degrees in a variety of other fields are welcome in the Radiological Health Engineering program.

Correspondence and Information

Graduate Program
Department of Nuclear Engineering and Radiological Sciences
Cooley Building
University of Michigan
Ann Arbor, Michigan 48109-2104
Telephone: 734-764-4260
Fax: 734-763-4540
World Wide Web: http://www.engin.umich.edu/~nuclear

University of Michigan

THE FACULTY

A. Ziya Akcasu, Professor Emeritus; Ph.D., Michigan, 1963. Dynamics of polymer solutions and blends, stochastic differential equations, reactor physics, kinetics. E-mail: ziya@umich.edu

Michael Atzmon, Associate Professor; Ph.D., Caltech, 1985. Kinetics and thermodynamics of materials, materials processing far from equilibrium, ion beam modification of materials. E-mail: atzmon@umich.edu

Alex F. Bielajew, Professor; Ph.D., Stanford, 1982. Theory of electron and photon transport, Monte Carlo theory and development, radiation dosimetry theory, radiotherapy treatment planning algorithms. E-mail: bielajew@umich.edu

Mary L. Brake, Associate Professor; Ph.D., Michigan State, 1983. Plasma physics, optical diagnostics, chemical kinetics, plasma-assisted materials processing. E-mail: brake@umich.edu

James J. Duderstadt, Professor and President, The University of Michigan; Ph.D., Caltech, 1967. Fission fusion reactors, statistical mechanics, plasma physics, computer simulation E-mail: jjd@j.imap.itd.umich.edu

Rodney C. Ewing, Professor; Ph.D., Stanford, 1974. Nuclear waste management, radiation effects in materials, ion beam modification of materials, mineralogy. E-mail: rodewing@umich.edu

Ronald F. Fleming, Professor and Director, Michigan Memorial Phoenix Project; Ph.D., Michigan, 1975. Neutron activation analysis, materials analysis using nuclear techniques, radiation measurements. E-mail: flemingr@umich.edu

Ronald M. Gilgenbach, Professor; Ph.D., Columbia, 1978. Plasmas, fusion, and laser and electron beam interactions with plasmas and materials. E-mail: rongilg@umich.edu

Zhong He, Assistant Research Scientist; Ph.D., Southampton (England), 1993. Room-temperature semiconductor gamma-ray detectors and gamma-ray imaging devices. E-mail: hezhong@engin.umich.edu

James P. Holloway, Associate Professor; Ph.D., Virginia, 1989. Computational physics, reactor physics, numerical methods for plasma kinetic theory, software engineering, radiation transport. E-mail: hagar@umich.edu

Terry Kammash, Stephen S. Attwood Professor Emeritus; Ph.D., Michigan, 1958. Fusion reactor physics and engineering, plasma physics, physics of intense charged particle beams, space nuclear power and propulsion. E-mail: tkammash@umich.edu

Kimberlee J. Kearfott, Professor; Sc.D., MIT, 1980. Health and medical physics, tomographic image reconstruction, digital image formation and processing, radiation detection and dosimetry, radiation protection. E-mail: kearfott@umich.edu

William Kerr, Professor Emeritus; Ph.D., Michigan, 1951. Reactor safety analysis, probabilistic risk analysis, radiation protection, reactor shielding, energy production. E-mail: wkerr@umich.edu

John S. King, Professor Emeritus; Ph.D., Michigan, 1953. Neutron spectroscopy, neutron physics. E-mail: jsking@umich.edu

Glenn F. Knoll, Professor; Ph.D., Michigan, 1963. Radiation measurements, neutron cross-sections, nuclear measurements, radiation imaging. E-mail: gknoll@umich.edu

Edward W. Larsen, Professor; Ph.D., RPI, 1971. Computational and analytical methods in neutron, electron, and photon transport theory. E-mail: edlarsen@engin.umich.edu

Y. Y. Lau, Professor; Ph.D., MIT, 1973. Plasma and beam physics, high power radiation sources, discharge and breakdown. E-mail: yylau@umich.edu

John C. Lee, Professor; Ph.D., Berkeley, 1969. Nuclear reactor physics, reactor safety, power plant simulation and control. E-mail: jcl@umich.edu

William R. Martin, Associate Dean, College of Engineering and Professor; Ph.D., Michigan 1976. Computational methods, reactor core analysis, reactor/thermal hydraulics, nonlinear radiation transport. E-mail: wrm@umich.edu

Donald P. Umstadter, Associate Professor; Ph.D., UCLA, 1989. Laser accelerators, ultrafast X rays, optics in relativistic plasmas, fusion.

Dietrich H. Vincent, Professor Emeritus; Dr.rer.nat., Gottingen (Germany), 1956. Gases in metals, ion beam analysis, radiation effects on materials.

Lu-Min Wang, Associate Research Scientist; Ph.D., Wisconsin–Madison, 1988. Ion-beam modification of materials, transmission electron microscopy, nanocrystalline materials, radiation effects in materials, and nuclear waste management. E-mail: lmwang@umich.edu

Gary S. Was, Chair and Professor; Sc.D., MIT, 1980. Radiation effects on materials, nuclear fuels, ion beam modification of materials, corrosion, hydrogen embrittlement, stress corrosion cracking. E-mail: gsw@umich.edu

David K. Wehe, Associate Professor; Ph.D., Michigan, 1984. Experimental neutron physics, radiation imaging and artificial intelligence applications. E-mail: dkw@umich.edu

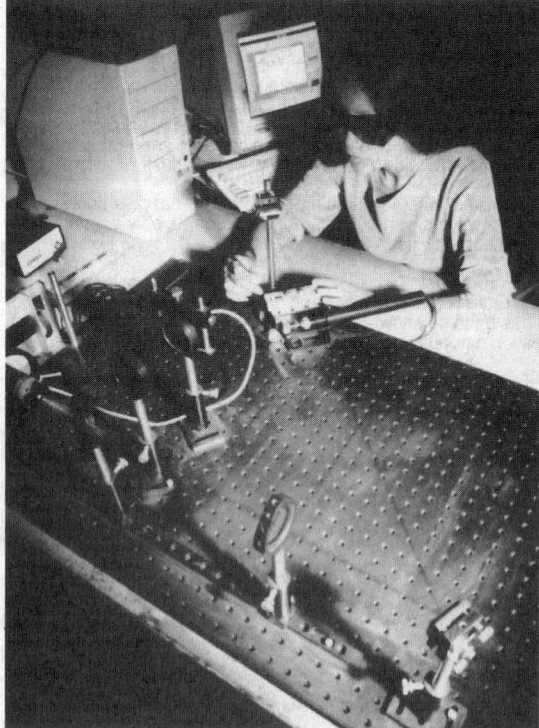

Student using a scanning laser reflection system in the Michigan Ion Beam Laboratory (MIBL) to measure thin film residual stresses.

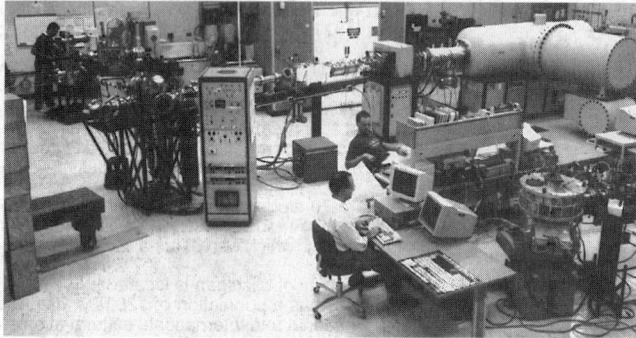

The Michigan Ion Beam Laboratory for Surface Modification and Analysis.

The Mortimer E. Cooley Building, home of the Department of Nuclear Engineering and Radiological Sciences.

UNIVERSITY OF MISSOURI–COLUMBIA

Nuclear Engineering Program

Program of Study

The Nuclear Engineering Program at the University of Missouri–Columbia (MU) offers degrees at the M.S. and Ph.D. levels. Special emphasis areas at the M.S. level include traditional nuclear engineering, health physics, and medical physics. Research areas at the Ph.D. level may include a diverse range of topics related to nuclear science. These areas are closely tied to the active research interests of the faculty, including power systems, health physics, medical physics, plasma physics, solid-state materials development, and particulate systems.

The M.S. degree program is designed for students with a B.S. degree in engineering or the sciences with an emphasis on math and modern physics. The curriculum includes a core of approximately five required courses followed by elective courses in the student's particular emphasis or research area for a minimum of 31 credits. A minimum of 3 hours of research and independent study is part of this 31-credit requirement.

The Ph.D. in nuclear engineering requires a minimum of 72 hours of course work beyond the baccalaureate degree. Students who have no postgraduate experience are advised to enroll initially as M.S. degree students and fulfill the minimum 30 credit hours required for this degree. The student then enrolls as a Ph.D. student for a minimum of 42 additional credits. A qualifying examination is administered during the first year of study to qualify a student as a Ph.D. candidate. A comprehensive examination for dissertation approval is also required at least one year prior to graduation.

Research Facilities

An important asset of the Nuclear Engineering Program is the Missouri University Research Reactor (MURR). At 10 MW, it is the highest-powered reactor on a university campus in the United States. The reactor operates 24 hours per day, seven days per week, with a full-time staff of more than 100 researchers and support personnel. Facilities include a flux tube with a maximum thermal flux of 6×10^{14} n/cm^2/sec, numerous irradiation positions, six beamtubes, a thermal column, and many research instruments, including a small angle neutron scattering (SANS) system, a neutron reflectometer facility, a residual stress facility, a neutron interferometry program, a filtered beam facility, a Mössbauer scattering facility (QUEGS), a gamma-ray diffractometer (MUGS), neutron activation analysis facilities, prompt gamma-ray neutron activation analysis, and depth profiling. Engineering studies are also in progress to increase the core power to approximately 25 MW. Thus, the MURR provides a focal point for many research efforts and a practical training ground for student involvement.

The interdisciplinary Particulate Systems Research Center (PSRC) is closely tied to the Nuclear Engineering Program. This center focuses on a range of topics, including aerosol mechanics in reactor safety analysis, environmental aspects of atmospheric effects of coal pollution and radon, and aerosols in manufacturing processes. New thrusts in nuclear engineering are also developing in semiconductor and diamond film research and in several laser research applications in the medical area. Health physics research is primarily on the effects and measurement of extremely low levels of radiation. This research is supported by the health physics staff at the MURR, the campus radiation safety office, and several outside agencies and utility companies.

Financial Aid

Students are eligible for a number of financial aid packages, including teaching and research assistantships administered by the Nuclear Engineering Program, graduate school fellowship programs, and national and industrial fellowships. These programs all include monthly stipends and a waiver of tuition. Submission of application materials makes the student eligible for consideration for financial aid. Typical financial packages (including stipends and tuition waivers) range from $8000 to $20,000 per year.

Cost of Study

In-state tuition for the 1999–2000 term is $167.80 per credit hour plus a $36.80 per hour supplemental fee for engineering courses. Out-of-state tuition is $504.80 per credit hour. Most students in the program receive tuition waivers as part of their financial aid package. Students also pay an activities fee of $9.53 per credit hour, a computer fee of $8.30 per credit hour, and a health fee of $57.50 per semester.

Living and Housing Costs

On-campus, single-student room and board cost $4655 per year, including twenty-one meals per week. Married student on-campus housing ranges from $286 to $380 per month. Private accommodations are available throughout Columbia at modest rates due to the low cost of living.

Student Group

The Nuclear Engineering Program averages approximately 40 graduate students representing a diverse and balanced group of ethnic and cultural backgrounds. There are about 25,000 students at the University of Missouri–Columbia.

Student Outcomes

Students graduating from the program in the past year have found positions in three primary areas. These include health physics positions in industry, hospitals, and government; medical physics positions in hospitals; and traditional nuclear engineering positions in a variety of industry and government settings.

Location

The University of Missouri–Columbia is located in Columbia, Missouri, on Interstate 70, halfway between Kansas City and St. Louis. The campus is near downtown Columbia and offers the advantages of a rural setting in a community of more than 75,000 people. Columbia is an educational and medical center and offers a wide range of cultural activities. It is routinely ranked in the top twenty best places to live by *Money* magazine.

The University

MU, established in 1839, is the oldest state university west of the Mississippi River. The Columbia campus is the largest of the four campuses of the University of Missouri System. Other campuses are in St. Louis, Kansas City, and Rolla. A member of the American Association of Universities (AAU) and a university classified Research I by the Carnegie Foundation for the Advancement of Teaching, MU is a premier provider of graduate and professional education.

Applying

Applicants should have a bachelor's degree in engineering, science, or applied science. All students are expected to have the equivalent of a full sequence of college calculus, including differential equations and at least two semesters of calculus-based physics. Applications are considered for both admission and financial support. Suggested application dates are March 1 for the fall term and October 1 for the winter term, although applications are considered (as funding permits) at later dates. GRE General Test scores are required by the Graduate School, and TOEFL scores of 500 or better are required of international students whose native language is not English.

Correspondence and Information

Director of Graduate Studies
Nuclear Engineering Program
E2434 Engineering Building East
University of Missouri–Columbia
Columbia, Missouri 65211

Telephone: 573-882-8201
Fax: 573-884-4801
E-mail: ne@risc1.ecn.missouri.edu

University of Missouri–Columbia

THE FACULTY AND THEIR RESEARCH

College of Engineering

Robert M. Brugger, Professor Emeritus and former director of the Missouri University Research Reactor; Ph.D., Rice, 1955; PE. Boron neutron capture therapy.

Robert L. Carter, Professor Emeritus; Ph.D., Duke, 1949; PE. Electrical and computer engineering, power systems.

Ardath H. Emmons, Professor Emeritus; Ph.D., Michigan, 1960; Certified Health Physicist. Hospital consultant concerning MRI, former Vice President of the University, original director of the Missouri University Research Reactor.

Tushar K. Ghosh, Associate Professor of Nuclear Engineering and Director of Graduate Studies; Ph.D., Oklahoma State, 1989. Aerosol kinetics; surface science, hazardous waste treatments, and mass transfer in adsorption and absorption processes (experimental and theoretical investigation); kinetics and reaction mechanisms of catalytic reactions; indoor air quality.

William R. Kimel, Professor Emeritus and Dean Emeritus of the College of Engineering; Ph.D., Berkeley, 1979; PE. Thermohydraulics, energy systems.

Sudarshan K. Loyalka, Professor; Ph.D., Stanford, 1967; PE. Kinetic theory of gases, aerosol mechanics (experimental and theoretical work relating to release and transmission of radioactive and nonradioactive aerosols), transmission of radiation in biological tissues, nuclear reactor physics and safety.

William H. Miller, Professor and Chair; Ph.D., Missouri–Columbia, 1976; PE. Radiation detection, neutron spectroscopy, neutron dosimetry, neutron radiation shielding, minicomputer and microcomputer applications and digital design, fast-neutron spectrometry by proton recoil techniques.

Mark A. Prelas, Professor; Ph.D., Illinois at Urbana-Champaign, 1979; PE. Laser physics, plasma physics, direct energy conversion.

Robert V. Tompson Jr., Associate Professor; Ph.D., Missouri–Columbia, 1988. Kinetic theory of gases, experimental and theoretical aerosol mechanics, indoor air quality, adsorption, neutron transport theory, nuclear reactor physics and safety.

Affiliated Faculty

Don M. Alger, Assistant Professor; Ph.D., Missouri–Columbia, 1974; PE. Private consultant, former Associate Director of the Missouri University Research Reactor.

Evan Boote, Assistant Professor of Radiology; Ph.D., Wisconsin–Madison, 1987. Medical physics.

Julie Dawson, Medical Physicist at St. Louis University Hospital; Ph.D., Missouri–Columbia, 1989; Certified American Board of Radiology. Radiation therapy.

Gary Ehrhardt, Senior Research Scientist at the Missouri University Research Reactor; Ph.D., Washington (St. Louis), 1976. Medical isotope development, production, and utilization.

Michael Glascock, Senior Research Scientist at the Missouri University Research Reactor; Ph.D., Iowa State, 1975. Neutron activation analysis.

Keith Hickey, Medical Physicist at Columbia Regional Radiation Therapy Center; Ph.D., Missouri–Columbia, 1989; Certified Health Physicist. Radiation therapy.

Gary Hughes, Supervisor of Research and Nuclear Safety at Union Electric's Callaway Nuclear Power Plant; Ph.D., Missouri–Columbia, 1981.

Jay F. Kunze, Professor Emeritus; Ph.D., Carnegie Tech, 1959; PE, Certified Health Physicist. Reactor physics calculations and experiments, health physics contamination control, electric power peaking/stored energy, shipping cask weeping analysis, reactor safety and thermalhydraulics, medical physics applications of radiation science.

Kiratadas Kutikkad, Research Scientist at the Missouri University Research Reactor; Ph.D., Florida, 1991. Reactor physics and safety.

Susan M. Langhorst, Assistant Professor and Manager of Health Physics at the Missouri University Research Reactor; Ph.D., Missouri–Columbia, 1982; Certified Health Physicist.

Philip K. Lee, Professor; Ph.D., Purdue, 1968; Certified Health Physicist.

Stephen Pickup, Assistant Research Professor, Department of Radiology; Ph.D., Drexel, 1987.

Wynn A. Volkert, Professor of Chemistry and Director of Radiation Sciences Division, Department of Radiology; Ph.D., Missouri–Columbia, 1968.

Student research at the Particulate Systems Research Center.

The 10 MW Missouri University Research Reactor, the most powerful reactor on a university campus in the United States.

UNIVERSITY OF WISCONSIN–MADISON

Department of Engineering Physics

Programs of Study

The Department of Engineering Physics offers graduate programs leading to the M.S. and Ph.D. degrees in nuclear engineering and engineering physics and to the M.S. and Ph.D. degrees in engineering mechanics.

The Nuclear Engineering and Engineering Physics Program incorporates course work and research in the areas of fission reactor engineering, reactor safety, multiphase phenomena, shock waves, plant life extension, probabilistic risk analysis, reactor operations, transmutation of nuclear waste, experimental and theoretical plasma physics for fusion and plasma processing applications, fusion technology, radiation effects on materials for fission and fusion applications, ion implantation, applied superconductivity and cryogenics, and large scale engineering computation. The Engineering Mechanics Program incorporates course work and research in the areas of structural and solid mechanics, continuum mechanics, fatigue, fracture mechanics, finite element methods, vibrations, dynamics and system identification, rheology, biomechanics, and computational mechanics. Similar programs at other universities are sometimes called theoretical and applied mechanics. Both programs are open to qualified students with bachelor's degrees in engineering, physics, and other physical sciences. The master's degrees require 24 credits of course work and are typically obtained in a calendar year; the M.S. thesis is optional. The doctoral degrees normally require a minimum of three years of course work and an original-research dissertation requiring one or more years; the course work is tailored to the individual student's interests; a minor is required in another field, such as physics, computer science, or mechanical engineering. Doctoral candidates must pass written and oral qualifying and preliminary examinations.

Research Facilities

Nuclear engineering facilities include a TRIGA reactor (1 MW steady-state and 1000 MW pulsed) with facilities for neutron radiography and neutron activation analysis, and reactor safety labs including a multiphase shock tube facility. Plasma physics facilities include a low aspect ratio tokamak, inertial electrostatic confinement devices, and basic plasma physics devices. The Plasma-Aided Manufacturing Center has several etching tools for graduate student research. The PSII laboratory includes a 100 kV implanter and access to extensive surface physics diagnostics. The applied superconductivity laboratory includes facilities for producing low- and high-temperature superconducting wire, and for research on cooling and refrigeration. A variety of superconducting magnets, power supplies up to 100 kA, and a large helium liquefier are available. Engineering mechanics facilities are consolidated in the Structures and Materials Testing Lab, which includes tension and compression machines, torsion machines, vibration shakers, and thermoelastic stress analysis (SPATE).

Financial Aid

Graduate students are eligible for fellowships granted by the University of Wisconsin, NSF, DOE (for nuclear engineering, magnetic fusion science, and magnetic fusion technology), National Academy of Nuclear Training, Nuclear Regulatory Commission, and other agencies and foundations. The department also awards many research and teaching assistantships; the research assistantship stipend in 1998–99 was $14,600 plus tuition for the calendar year. The teaching assistantship stipend in 1998–99 was $9633 plus tuition per academic year.

Cost of Study

Tuition for 1998–99 was $2464 per semester for state residents and $7595 per semester for nonresidents. Students with fellowships, research assistantships, and teaching assistantships normally do not pay tuition.

Living and Housing Costs

Living costs are relatively low in Madison. University-owned apartments for families cost $470 to $563 per month. On-campus housing for single graduate students is $363 per month and up. Numerous off-campus rooms and apartments are also available. Students should consult the World Wide Web for further information (http://www.wisc.edu/cac/housing/).

Student Group

The graduate student enrollment is typically 60–70 students in nuclear engineering and engineering physics and 10–20 students in engineering mechanics. An additional 10–15 students from other departments receive their research supervision from engineering physics faculty members. About 40 percent of the students come from abroad.

Student Outcomes

Students finishing with the M.S. degree normally obtain employment in industry. Most Ph.D. recipients go on to postdoctoral positions at other universities and research positions at national labs and companies. Some obtain faculty positions in universities.

Location

Madison is the capital of Wisconsin and has a population of 200,000. The city offers extensive cultural events and restaurants with varied cuisine. Madison's four lakes provide opportunities for swimming, fishing, and sailing. Madison has excellent air and bus transportation to Chicago, Milwaukee, Minneapolis, and other major metropolitan areas.

The University and The Department

The Madison campus has an enrollment of about 40,000 students, including 11,000 graduate students. The campus has 125 departments offering more than 4,500 courses in 10 major schools. The Madison campus is a major research university, with annual research expenditures of $370 million. The department is rated in the top five nuclear engineering departments in the U.S. and has externally funded research of about $6.6 million per year.

Applying

Application materials should be submitted to the Graduate School at least six weeks before the anticipated start of graduate study. International students whose native language is not English must include results of the Test of English as a Foreign Language (TOEFL). Applications for fellowships and research and teaching assistantships are normally due by January 15 for the following fall semester; however, assistantships are sometimes awarded during the year. Test scores for the GRE General Test are required.

Correspondence and Information

Department Chair
Department of Engineering Physics
1500 Engineering Drive
Madison, Wisconsin 53706

Telephone: 608-263-1646
Fax: 608-263-7451
E-mail: ep@engr.wisc.edu
World Wide Web: http://www.engr.wisc.edu/ep/

University of Wisconsin–Madison

THE FACULTY AND THEIR RESEARCH

Nuclear Engineering and Engineering Physics Faculty

Vicki M. Bier, Associate Professor (also Industrial Engineering); Ph.D. (operations research), MIT, 1983. Probabilistic risk analysis, reliability, decision analysis, treatment of uncertainty. (E-mail: bier@engr.wisc.edu)

James P. Blanchard, Associate Professor; Ph.D. (nuclear engineering), UCLA, 1988. Fission plant life extension, surface properties of materials, radiation damage, fusion reactor design. (E-mail: blanchard@engr.wisc.edu)

Riccardo Bonazza, Associate Professor (also Mechanical Engineering); Ph.D. (aeronautics), Caltech, 1992. Shock interface interactions, vapor explosion phenomena. (E-mail: bonazza@engr.wisc.edu)

James D. Callen, Kerst Professor (also Physics); Ph.D. (nuclear engineering), MIT, 1968. Development of plasma theory for confinement experiments and fusion reactor studies. (E-mail: callen@engr.wisc.edu)

Max W. Carbon, Professor Emeritus; Ph.D. (mechanical engineering), Purdue, 1949. Nuclear reactor safety, liquid metal breeder reactor development. (E-mail: carbon@engr.wisc.edu)

John R. Conrad, Wisconsin Distinguished Professor; Ph.D. (physics), Dartmouth, 1973. Plasma physics applications to industrial processes, particularly surface modification of materials. (E-mail: conrad@engr.wisc.edu)

Michael L. Corradini, Wisconsin Distinguished Professor (also Mechanical Engineering), Ph.D. (nuclear engineering), MIT, 1978. Thermal hydraulics and multiphase flow in reactor operation and safety, waste disposal, risk assessment. (E-mail: corradini@engr.wisc.edu)

Gilbert A. Emmert, Professor and Department Chair; Ph.D. (physics), Stevens, 1968. Theory of plasma-surface interactions, analysis of reactor-grade fusion plasmas, plasma processing. (E-mail: emmert@engr.wisc.edu)

Mohammed M. El-Wakil, Professor Emeritus (also Mechanical Engineering); Ph.D. (mechanical engineering), Wisconsin-Madison, 1954. Nuclear power plant thermal hydraulics. (E-mail: el-wakil@engr.wisc.edu)

Raymond J. Fonck, Professor; Ph.D. (physics), Wisconsin–Madison, 1978. Experimental research in plasma physics, atomic processes in high-temperature plasmas, applied optics. (E-mail: fonck@engr.wisc.edu)

Douglass L. Henderson, Associate Professor (also Biomedical Engineering); Ph.D. (nuclear engineering), Wisconsin–Madison, 1987. Radiation transport, transmutation of nuclear waste, fusion reactor neutronics and activation. (E-mail: henderson@engr.wisc.edu)

Noah Hershkowitz, Langmuir Professor; Ph.D. (physics), Johns Hopkins, 1966. Experimental studies of ICRF effects of fusion plasmas, basic plasma physics, plasma processing. (E-mail: hershkowitz@engr.wisc.edu)

Gerald L. Kulcinski, Grainger Professor; Ph.D. (nuclear engineering), Wisconsin–Madison, 1965. Nuclear materials, radiation damage, fusion reactor design studies. (E-mail: kulcinski@engr.wisc.edu)

Richard J. Matyi, Professor (also Materials Science and Engineering); Ph.D. (materials science and engineering), Northwestern, 1983. Ion implantation, epitaxial growth, materials analysis. (E-mail: matyi@engr.wisc.edu)

Gregory A. Moses, Professor; Ph.D. (nuclear engineering), Michigan, 1976. Modeling of dense plasmas for inertial confinement fusion, parallel computation of particle transport. (E-mail: moses@engr.wisc.edu)

John M. Pfotenhauer, Associate Professor (also Mechanical Engineering); Ph.D. (physics), Oregon, 1984. Applied superconductivity and cryogenics. (E-mail: pfot@engr.wisc.edu)

William F. Vogelsang, Professor Emeritus; Ph.D. (physics), Pittsburgh, 1956. Fusion reactor design, radioactivity in fusion systems. (E-mail: vogelsang@eng.wisc.edu)

Robert J. Witt, Associate Professor; Ph.D. (nuclear engineering), MIT 1987. Analytical/computational methods in fluid and solid mechanics, with applications in fission, fusion, and cryogenic systems. (E-mail: witt@engr.wisc.edu)

Engineering Mechanics Faculty

James P. Blanchard, Associate Professor; Ph.D. (nuclear engineering), UCLA, 1988. Mechanics of coatings, numerical methods. (E-mail: blanchard@engr.wisc.edu)

Riccardo Bonazza, Associate Professor (also Mechanical Engineering), Ph.D. (aeronautics), Caltech, 1992. Shock interface interactions, experimental fluid mechanics. (E-mail: bonazza@engr.wisc.edu)

Robert D. Cook, Professor Emeritus; Ph.D. (applied mechanics), Illinois, 1963. Stress analysis, structural mechanics, finite element methods.

Wendy C. Crone, Assistant Professor; Ph.D. (engineering mechanics), Minnesota, 1998. Experimental mechanics of materials; plasticity, fracture mechanics, and biomaterials. (E-mail: crone@engr.wisc.edu)

Walter J. Drugan, Professor; Ph.D. (solid mechanics), Brown, 1982. Nonlinear fracture mechanics, plasticity theory, advanced materials, shock waves in solids, continuum mechanics, applied mathematics. (E-mail: drugan@engr.wisc.edu)

Millard Johnson, Professor Emeritus, Ph.D. (mathematics), MIT, 1958. Continuum mechanics, rheology. (E-mail: millard@engr.wisc.edu)

Daniel C. Kammer, Associate Professor; Ph.D. (engineering mechanics), Wisconsin–Madison, 1983. Dynamics, stability and control of large structures, system identification, nonlinear dynamics. (E-mail: kammer@coefac.engr.wisc.edu)

Roderic S. Lakes, Wisconsin Distinguished Professor (also Biomedical Engineering); Ph.D. (physics), Rensselaer, 1975. Experimental mechanics, composite materials, biomechanics. (E-mail: lakes@engr.wisc.edu)

David S. Malkus, Professor; Ph.D. (mathematics), Boston University, 1976. Finite element methods for fluids, solids, and structures; theology; viscoelastic fluids. (E-mail: malkus@engr.wisc.edu)

Michael E. Plesha, Professor; Ph.D. (structural engineering and mechanics), Northwestern, 1983. Finite element and numerical methods, structural dynamics, contact-friction problems. (E-mail: plesha@engr.wisc.edu)

Bela I. Sandor, Professor Emeritus; Ph.D. (theoretical and applied mechanics), Illinois, 1968. Fatigue and fracture mechanics, experimental mechanics, differential infrared thermography. (E-mail: bsandor@macc.wisc.edu)

Alois L. Schlack Jr., Professor Emeritus; Ph.D. (engineering mechanics), Wisconsin–Madison, 1956. Dynamics, vibration, astrodynamics.

Ray Vanderby, Affiliate Associate Professor (also Orthopedic Surgery); Ph.D. (solid mechanics), Purdue, 1975. Biological connective tissue mechanics, orthopedic mechanics. (E-mail: vanderby@ortho.surgery.wisc.edu)

Fabian Waleffe, Associate Professor (also Mathematics); Ph.D. (applied mathematics), MIT, 1989. Applied mathematics, fluid dynamics, turbulence. (E-mail: waleffe@math.wisc.edu)

Robert J. Witt, Associate Professor; Ph.D. (nuclear engineering), MIT, 1987. Application of finite element methods and nodal methods in solid and fluid mechanics. (E-mail: witt@engr.wisc.edu)

Other Faculty

Richard J. Cashwell, Senior Lecturer and Reactor Director; B.S. (nuclear engineering), North Carolina State, 1958. Operator training, reactor operations. (E-mail: cashwell@engr.wisc.edu)

Lawrence Dresner, Adjunct Professor; Ph.D. (physics), Princeton, 1959. Theoretical aspects of superconductivity stability, cryogenic heat transfer.

Frederick E. Mills, Adjunct Professor; Ph.D. (physics), Illinois, 1955. Accelerator physics and technology, including electron cooling systems and free electron lasers.

Harrison H. Schmitt, Adjunct Professor; Ph.D. (geology), Harvard, 1964. Space resources and exploration.

Section 11
Engineering Design

This section contains a directory of institutions offering graduate work in engineering design, followed by in-depth entries submitted by institutions that chose to prepare detailed program descriptions. Additional information about programs listed in the directory but not augmented by an in-depth entry may be obtained by writing directly to the dean of a graduate school or chair of a department at the address given in the directory.

For programs offering related work, see also in this book Aerospace/Aeronautical Engineering; Bioengineering, Biomedical Engineering, and Biotechnology; Computer Science and Information Technology; Electrical and Computer Engineering; Energy and Power Engineering; Engineering and Applied Sciences; Industrial Engineering; Management of Engineering and Technology; and Mechanical Engineering and Mechanics. In Book 3, see Biological and Biomedical Sciences.

CONTENTS

Program Directory

In-Depth Description

Engineering Design

The Catholic University of America, School of Engineering, Department of Mechanical Engineering, Program in Design, Washington, DC 20064. Offers D Engr, PhD. *Degree requirements:* For doctorate, dissertation, comprehensive and oral exams required, foreign language not required. *Entrance requirements:* For doctorate, minimum GPA of 3.5. *Application deadline:* For fall admission, 8/1 (priority date); for spring admission, 12/1. Applications are processed on a rolling basis. Application fee: $50. *Financial aid:* Application deadline: 2/1. *Unit head:* Dr. John J. Gilheany, Chair, Department of Mechanical Engineering, 202-319-5170.

Kettering University, Graduate School, Mechanical Engineering Department, Flint, MI 48504-4898. Offers automotive engineering (MS Eng); mechanical design (MS Eng). *Faculty:* 26 full-time (3 women), 1 part-time (0 women). *Degree requirements:* For master's, foreign language and thesis not required. *Entrance requirements:* For master's, GRE General Test. *Application deadline:* For fall admission, 7/15. Applications are processed on a rolling basis. Application fee: $0. *Financial aid:* Fellowships with full tuition reimbursements, research assistantships with full tuition reimbursements, teaching assistantships with full tuition reimbursements, Federal Work-Study, institutionally-sponsored loans, and tuition waivers (partial) available. Aid available to part-time students. Financial aid application deadline: 7/15. *Unit head:* Dr. K. Joel Berry, Head, 810-762-7833, Fax: 810-762-7860, E-mail: jberry@kettering.edu. *Application contact:* Betty L. Bedore, Coordinator of Publicity, 810-762-7494, Fax: 810-762-9935, E-mail: bbedore@kettering.edu.

Stanford University, School of Engineering, Department of Mechanical Engineering, Program in Product Design, Stanford, CA 94305-9991. Offers MS. *Entrance requirements:* For master's, GRE General Test, TOEFL. *Application deadline:* For fall admission, 1/15. Application fee: $65 ($80 for international students). Electronic applications accepted. Tuition: Full-time $24,588. Required fees: $152. Part-time tuition and fees vary according to course load. *Financial aid:* Application deadline: 1/15. *Application contact:* Admissions Office, 650-723-3148.

Stevens Institute of Technology, Graduate School, Wesley J. Howe School of Technology Management, Program in Concurrent Design Management, Hoboken, NJ 07030. Offers M Eng. Offered in cooperation with the Department of Mechanical Engineering. *Degree requirements:* For master's, computer language required, thesis optional, foreign language not required. *Entrance requirements:* For master's, GMAT, GRE, TOEFL. Electronic applications accepted.

University of Illinois at Urbana–Champaign, Graduate College, College of Engineering, Department of General Engineering, Urbana, IL 61801. Offers systems engineering and engineering design (MS), including general engineering. *Faculty:* 18 full-time (1 woman). *Students:* 10 full-time (0 women); includes 3 minority (all Asian Americans or Pacific Islanders), 2 international. 18 applicants, 17% accepted. In 1998, 9 degrees awarded. *Degree requirements:* For master's, foreign language and thesis not required. *Entrance requirements:* For master's, minimum GPA of 3.0. *Application deadline:* For fall admission, 2/1. Applications are processed on a rolling basis. Application fee: $40 ($50 for international students). Tuition, state resident: full-time $4,616. Tuition, nonresident: full-time $11,768. Full-time tuition and fees vary according to course load. *Financial aid:* Fellowships, research assistantships, teaching assistantships, scholarships available. Financial aid application deadline: 2/15. *Unit head:* Dr. Harry E. Cook, Head, 217-333-2730. *Application contact:* Dr. Henrique Reis, Director of Graduate Studies, 217-333-1228, Fax: 217-244-5705, E-mail: h-reis@uiuc.edu.

University of New Haven, Graduate School, School of Engineering and Applied Science, Program in Environmental Engineering, West Haven, CT 06516-1916. Offers civil engineering design (Certificate); environmental engineering (MS). Part-time and evening/weekend programs available. *Students:* 7 full-time (0 women), 22 part-time (6 women); includes 3 minority (1 African American, 1 Asian American or Pacific Islander, 1 Native American), 8 international. *Degree requirements:* For master's, thesis or alternative required, foreign language not required. *Entrance requirements:* For master's, bachelor's degree in engineering. *Application deadline:* Applications are processed on a rolling basis. Application fee: $50. *Unit head:* Dr. Agamemnon D. Koutsospyros, Coordinator, 203-932-7398.

Section 12
Engineering Physics

This section contains a directory of institutions offering graduate work in engineering physics, followed by in-depth entries submitted by institutions that chose to prepare detailed program descriptions. Additional information about programs listed in the directory but not augmented by an in-depth entry may be obtained by writing directly to the dean of a graduate school or chair of a department at the address given in the directory.

For programs offering related work, see also in this book Electrical and Computer Engineering, Energy and Power Engineering (Nuclear Engineering), Engineering and Applied Sciences, and Materials Sciences and Engineering. In Book 3, see Biophysics; in Book 4, Physics; and in Book 6, Health Sciences (Medical Physics).

CONTENTS

Engineering Physics

Air Force Institute of Technology, School of Engineering, Department of Engineering Physics, Program in Engineering Physics, Wright-Patterson AFB, OH 45433-7765. Offers MS, PhD. *Accreditation:* ABET (one or more programs are accredited). Part-time programs available. *Faculty:* 13 full-time (0 women). *Students:* 26 full-time, 3 part-time. In 1998, 1 doctorate awarded (100% found work related to degree). *Degree requirements:* For master's and doctorate, thesis/dissertation required, foreign language not required. *Entrance requirements:* For master's, GRE General Test (minimum score of 500 on verbal section, 600 on quantitative required), minimum GPA of 3.0, must be military officer or U.S. citizen; for doctorate, GRE General Test (minimum score of 550 on verbal section, 650 on quantitative required), minimum GPA of 3.0, must be military officer or U.S. citizen. Application fee: $0. *Faculty research:* High-energy lasers, semiconductor physics, space environment, nonlinear optics. *Unit head:* Dr. Robert L. Hengehold, Head, Department of Engineering Physics, 937-255-3636 Ext. 4757, Fax: 937-255-2921, E-mail: rhengeho@afit.af.mil.

Cornell University, Graduate School, Graduate Fields of Engineering, Field of Applied Physics, Ithaca, NY 14853-0001. Offers applied physics (PhD); engineering physics (M Eng, PhD). *Faculty:* 42 full-time. *Students:* 62 full-time (13 women); includes 6 minority (2 African Americans, 2 Asian Americans or Pacific Islanders, 2 Hispanic Americans), 20 international. 101 applicants, 55% accepted. In 1998, 8 master's, 7 doctorates awarded. *Degree requirements:* For doctorate, dissertation, oral and written exams required, foreign language not required, foreign language not required. *Entrance requirements:* For master's, TOEFL (minimum score of 550 required); for doctorate, GRE General Test, GRE Subject Test, TOEFL (minimum score of 550 required). *Application deadline:* For fall admission, 1/15. Application fee: $65. *Financial aid:* In 1998–99, 55 students received aid, including 12 fellowships with full tuition reimbursements available, 35 research assistantships with full tuition reimbursements available, 8 teaching assistantships with full tuition reimbursements available; institutionally-sponsored loans, scholarships, tuition waivers (full and partial), and unspecified assistantships also available. *Faculty research:* Quantum and nonlinear optics; plasma physics; solid state physics, condensed matter physics, and nanotechnology; condensed matter physics; biophysics; electron and x-ray spectroscopy. *Unit head:* Graduate Faculty Representative, 607-255-0638. *Application contact:* Graduate Field Assistant, 607-255-0638, E-mail: aep_info@cornell.edu.

École Polytechnique de Montréal, Graduate Programs, Department of Engineering Physics, Montréal, PQ H3C 3A7, Canada. Offers optical engineering (M Eng, M Sc A, PhD); solid-state physics and engineering (M Eng, M Sc A, PhD). Part-time programs available. *Degree requirements:* For master's and doctorate, one foreign language, computer language, thesis/dissertation required. *Entrance requirements:* For master's, minimum GPA of 2.75; for doctorate, minimum GPA of 3.0. *Faculty research:* Optics, thin-film physics, laser spectroscopy, plasmas, photonic devices.

George Mason University, College of Arts and Sciences, Department of Physics, Fairfax, VA 22030-4444. Offers applied and engineering physics (MS). *Faculty:* 14 full-time (5 women), 4 part-time (1 woman). *Students:* 2 full-time (0 women), 11 part-time (4 women); includes 1 minority (Hispanic American) Average age 31. 8 applicants, 75% accepted. In 1998, 4 degrees awarded. *Degree requirements:* For master's, thesis optional. *Entrance requirements:* For master's, minimum GPA of 2.75 in last 60 hours. *Application deadline:* For fall admission, 5/1; for spring admission, 11/1. Application fee: $30. Electronic applications accepted. Tuition, state resident: full-time $4,416; part-time $184 per credit hour. Tuition, nonresident: full-time $12,516; part-time $522 per credit hour. Tuition and fees vary according to program. *Financial aid:* Research assistantships, teaching assistantships available. Aid available to part-time students. Financial aid application deadline: 3/1; financial aid applicants required to submit FAFSA. *Unit head:* Dr. Joseph Lieb, Chairman, 703-993-1280, Fax: 703-993-1269, E-mail: jlieb@osf1.gmu.edu.

McMaster University, School of Graduate Studies, Faculty of Engineering, Department of Engineering Physics, Hamilton, ON L8S 4M2, Canada. Offers engineering physics (M Eng, PhD); nuclear engineering (PhD). *Faculty:* 18 full-time, 1 part-time. *Students:* 29 full-time, 8 part-time. *Degree requirements:* For master's, thesis or alternative required, foreign language not required; for doctorate, dissertation, comprehensive exam required, foreign language not required. *Application deadline:* For fall admission, 3/1 (priority date). Applications are processed on a rolling basis. Application fee: $50. *Financial aid:* In 1998–99, teaching assistantships (averaging $7,722 per year); fellowships, research assistantships *Unit head:* Dr. Peter Mascher, Chair, 905-525-9140 Ext. 24548.

Polytechnic University, Farmingdale Campus, Graduate Programs, Department of Electrical Engineering, Major in Electrophysics, Farmingdale, NY 11735-3995. Offers MS. *Students:* 1 full-time (0 women). Average age 33. *Degree requirements:* For master's, computer language, thesis required (for some programs). *Application deadline:* Applications are processed on a rolling basis. Application fee: $45. Electronic applications accepted. *Financial aid:* Institutionally-sponsored loans available. Aid available to part-time students. Financial aid applicants required to submit FAFSA. *Application contact:* John S. Kerge, Dean of Admissions, 718-260-3200, Fax: 718-260-3446, E-mail: admitme@poly.edu.

Polytechnic University, Westchester Graduate Center, Graduate Programs, Department of Electrical Engineering, Major in Electrophysics, Hawthorne, NY 10532-1507. Offers MS. *Degree requirements:* For master's, computer language, thesis required (for some programs). *Application deadline:* Applications are processed on a rolling basis. Application fee: $45. Electronic applications accepted. *Unit head:* John S. Kerge, Dean of Admissions, 718-260-3200, Fax: 718-260-3446, E-mail: admitme@poly.edu. *Application contact:* John S. Kerge, Dean of Admissions, 718-260-3200, Fax: 718-260-3446, E-mail: admitme@poly.edu.

Rensselaer Polytechnic Institute, Graduate School, School of Engineering, Department of Environmental and Energy Engineering, Program in Engineering Physics, Troy, NY 12180-3590. Offers M Eng, MS, PhD, MBA/M Eng. Part-time programs available. *Faculty:* 10 full-time (0 women), 14 part-time (0 women). *Students:* 10 full-time (0 women); includes 3 minority (2 Asian Americans or Pacific Islanders, 1 Hispanic American), 6 international. 4 applicants, 75% accepted. In 1998, 1 master's, 2 doctorates awarded. Terminal master's awarded for partial completion of doctoral program. *Degree requirements:* For master's, computer language, thesis required (for some programs), foreign language not required; for doctorate, computer language, dissertation required, foreign language not required. *Entrance requirements:* For master's and doctorate, GRE, TOEFL (minimum score of 550 required). *Application deadline:* For fall admission, 2/1 (priority date). Applications are processed on a rolling basis. Application fee: $35. *Financial aid:* Fellowships, research assistantships, teaching assistantships, career-related internships or fieldwork, institutionally-sponsored loans, and tuition waivers (full and partial) available. Financial aid application deadline: 2/1. *Faculty research:* Multiphase science, applications of radiation, plasma physics. Total annual research expenditures: $1.2 million. *Unit head:* H. L. Cui, Chairman, Graduate Committee, 201-216-5637, Fax: 201-216-5638, E-mail: hcui@stevens-tech.edu. *Application contact:* Pam Zepf, Senior Secretary, 518-276-6402, Fax: 518-276-3055, E-mail: zepf@rpi.edu.

Stevens Institute of Technology, Graduate School, School of Applied Sciences and Liberal Arts, Department of Physics and Engineering Physics, Hoboken, NJ 07030. Offers applied optics (Certificate); engineering physics (M Eng), including engineering optics, engineering physics; physics (MS, PhD); surface physics (Certificate). Part-time and evening/weekend programs available. Terminal master's awarded for partial completion of doctoral program. *Degree requirements:* For master's, thesis optional, foreign language not required; for doctorate, dissertation required, foreign language not required; for Certificate, computer language required, foreign language not required. *Entrance requirements:* For master's and doctorate, GRE, TOEFL. Electronic applications accepted. *Faculty research:* Laser spectroscopy, physical kinetics, semiconductor-device physics, condensed-matter theory.

University of British Columbia, Faculty of Graduate Studies, Faculty of Science, Department of Physics and Astronomy, Program in Engineering Physics, Vancouver, BC V6T 1Z2, Canada. Offers MA Sc. *Degree requirements:* For master's, thesis required, foreign language not required. *Entrance requirements:* For master's, TOEFL (minimum score of 550 required), honors degree. *Faculty research:* Solid-state, nuclear, solar, and plasmaphysics; applied and nonlinear optics.

University of California, San Diego, Graduate Studies and Research, Department of Applied Mechanics and Engineering Sciences, Program in Engineering Physics, La Jolla, CA 92093-5003. Offers MS, PhD. Part-time programs available. *Students:* 7 full-time (1 woman); includes 2 minority (1 African American, 1 Hispanic American), 1 international. 5 applicants, 40% accepted. In 1998, 1 master's awarded. *Degree requirements:* For master's, comprehensive exam or thesis required; for doctorate, dissertation, qualifying exam required. *Entrance requirements:* For master's and doctorate, GRE General Test, TOEFL (minimum score of 550 required), minimum GPA of 3.0. *Application deadline:* For fall admission, 5/31. Application fee: $40. *Financial aid:* In 1998–99, fellowships with full tuition reimbursements (averaging $15,000 per year), research assistantships with full tuition reimbursements (averaging $15,000 per year), teaching assistantships with partial tuition reimbursements (averaging $13,000 per year) were awarded.; career-related internships or fieldwork also available. Financial aid application deadline: 1/31; financial aid applicants required to submit FAFSA. *Faculty research:* Combustion engineering, environmental mechanics, magnetic recording, materials processing, computational fluid dynamics. *Unit head:* Graduate Field Assistant, 607-255-0638, E-mail: aep_info@cornell.edu. *Application contact:* AMES Graduate Student Affairs, 619-534-4387, Fax: 619-534-1730, E-mail: bwalton@ames.ucsd.edu.

University of Florida, Graduate School, College of Engineering, Department of Nuclear and Radiological Engineering, Gainesville, FL 32611. Offers engineering physics (ME, MS, PhD, Engr); health physics (MS, PhD); medical physics (MS, PhD); nuclear power engineering (ME, MS, PhD, Engr). *Faculty:* 34. *Students:* 47 full-time (5 women), 10 part-time (3 women); includes 6 minority (4 Asian Americans or Pacific Islanders, 1 Hispanic American, 1 Native American), 16 international. *Degree requirements:* For master's and doctorate, thesis/dissertation required; for Engr, thesis optional. *Entrance requirements:* For master's and doctorate, GRE General Test, TOEFL, minimum GPA of 3.0; for Engr, GRE General Test. *Application deadline:* For fall admission, 6/1 (priority date). Applications are processed on a rolling basis. Application fee: $20. Electronic applications accepted. *Unit head:* James S. Tulenko, Chair, 352-392-1401, Fax: 352-392-3380, E-mail: tulenko@sun-robot.nuceng.ufl.edu. *Application contact:* Dr. Wesley E. Bolch, Graduate Coordinator, 352-392-1361, E-mail: wbolch@apollo.nuceng.ufl.edu.

University of Maine, Graduate School, College of Liberal Arts and Sciences, Department of Physics and Astronomy, Program in Engineering Physics, Orono, ME 04469. Offers M Eng. *Faculty:* 17 full-time, 1 part-time. *Degree requirements:* For master's, thesis or alternative required, foreign language not required. *Entrance requirements:* For master's, GRE General Test, GRE Subject Test, TOEFL (minimum score of 550 required). *Application deadline:* For fall admission, 2/1 (priority date); for spring admission, 10/15. Applications are processed on a rolling basis. Application fee: $50. *Financial aid:* Fellowships, teaching assistantships available. Financial aid application deadline: 3/1. *Unit head:* Pam Zepf, Senior Secretary, 518-276-6402, Fax: 518-276-3055, E-mail: zepf@rpi.edu. *Application contact:* Scott G. Delcourt, Director of the Graduate School, 207-581-3218, Fax: 207-581-3232, E-mail: graduate@maine.edu.

University of Oklahoma, Graduate College, College of Engineering, Program in Engineering Physics, Norman, OK 73019-0390. Offers M Nat Sci, MS, PhD. Part-time programs available. *Students:* 4 full-time (0 women), 3 part-time; includes 1 minority (Native American) Average age 27. 3 applicants, 0% accepted. In 1998, 1 master's awarded. Terminal master's awarded for partial completion of doctoral program. *Degree requirements:* For master's, thesis or alternative, departmental qualifying exam required, foreign language not required; for doctorate, dissertation, comprehensive, departmental qualifying, oral, and written exams required, foreign language not required. *Entrance requirements:* For master's and doctorate, GRE General Test, GRE Subject Test (physics), TOEFL (minimum score of 600 required), previous course work in physics. *Application deadline:* For fall admission, 3/1 (priority date); for spring admission, 10/1. Applications are processed on a rolling basis. Application fee: $25. Tuition, state resident: part-time $86 per credit hour. Tuition, nonresident: part-time $275 per credit hour. Tuition and fees vary according to course level, course load and program. *Financial aid:* Fellowships, research assistantships, teaching assistantships, Federal Work-Study, institutionally-sponsored loans, and tuition waivers (full and partial) available. Aid available to part-time students. Financial aid application deadline: 3/1. *Faculty research:* Applied physics, accelerator-based material characterization analysis, laboratory for artificially structured materials. *Unit head:* J. Milton Adams, Assistant Dean, 804-924-3897, E-mail: twr2c@virginia.edu. *Application contact:* Dr. John Furneaux, Chair, Graduate Selection Committee, 405-325-3961, Fax: 405-325-7557.

University of South Florida, Graduate School, College of Arts and Sciences, Department of Physics, Tampa, FL 33620-9951. Offers applied physics (MS); engineering science/physics (MS, PhD); physics (MA, MS). Part-time programs available. *Faculty:* 11 full-time (0 women), 4 part-time (0 women). *Students:* 16 full-time (3 women), 2 part-time (1 woman); includes 9 minority (2 African Americans, 6 Asian Americans or Pacific Islanders, 1 Hispanic American) *Degree requirements:* For master's, thesis optional, foreign language not required; for doctorate, 2 foreign languages (computer language can substitute for one), dissertation required. *Entrance requirements:* For master's, GRE General Test (minimum combined score of 1000 required), minimum GPA of 3.0 in last 60 hours; for doctorate, GRE General Test (minimum combined score of 1000 required), minimum graduate GPA of 3.2. *Application deadline:* For fall admission, 6/1 (priority date); for spring admission, 10/15. Applications are processed on a rolling basis. Application fee: $20. Electronic applications accepted. Tuition, state resident: part-time $148 per credit hour. Tuition, nonresident: part-time $509 per credit hour. *Unit head:* R. S. Chang, Chairperson, 813-974-2871, Fax: 813-974-5813, E-mail: chang@chuma.cas.usf.edu. *Application contact:* Pritish Mukherjee, Director, 813-974-5230, Fax: 813-974-5813, E-mail: pritish@chuma.cas.usf.edu.

University of Vermont, Graduate College, College of Arts and Sciences, Department of Physics, Program in Engineering Physics, Burlington, VT 05405-0160. Offers MS. *Degree requirements:* For master's, computer language required, foreign language not required. *Entrance requirements:* For master's, GRE General Test, TOEFL (minimum score of 550 required).

University of Virginia, School of Engineering and Applied Science, Department of Materials Science and Engineering, Program in Engineering Physics, Charlottesville, VA 22903. Offers MEP, MS, PhD. *Faculty:* 21 full-time (1 woman), 1 part-time (0 women). *Students:* 21 full-time (4 women), 2 part-time (1 woman); includes 2 minority (1 African American, 1 Asian American or Pacific Islander), 5 international. Average age 31. 27 applicants, 67% accepted. In 1998, 5 master's, 1 doctorate awarded. *Degree requirements:* For master's, comprehensive exam required; for doctorate, dissertation, comprehensive exams required, foreign language not required. *Entrance requirements:* For master's and doctorate, GRE General Test. *Application deadline:* For fall admission, 8/1; for spring admission, 12/1. Applications are processed on a rolling basis. Application fee: $60. *Financial aid:* Fellowships available. Financial aid application deadline: 2/1. *Faculty research:* Continuum and rarefied gas dynamics, ultracentrifuge isotope enrichment, solid-state physics, atmospheric physics, atomic collisions. *Unit head:* AMES Graduate Student Affairs, 619-534-4387, Fax: 619-534-1730, E-mail: bwalton@ames.ucsd.edu. *Application contact:* J. Milton Adams, Assistant Dean, 804-924-3897, E-mail: twr2c@virginia.edu.

University of Wisconsin–Madison, Graduate School, College of Engineering, Department of Engineering Physics, Madison, WI 53706-1380. Offers engineering mechanics (MS, PhD); nuclear engineering and engineering physics (MS, PhD). Part-time programs available. Postbaccalaureate distance learning degree programs offered (minimal on-campus study). *Faculty:* 19 full-time (1 woman), 5 part-time (0 women). *Students:* 46 full-time (5 women), 5 part-time; includes 2 minority (both Asian Americans or Pacific Islanders), 19 international. Average age 26. 120 applicants, 51% accepted. In 1998, 23 master's awarded (4% entered university research/teaching, 39% found other work related to degree, 48% continued full-time study); 11 doctorates awarded (45% entered university research/teaching, 46% found other work related to degree). Terminal master's awarded for partial completion of doctoral program. *Degree requirements:* For master's, thesis optional, foreign language not required; for doctorate, dissertation required, foreign language not required. *Entrance requirements:* For master's and doctorate, GRE General Test, minimum GPA of 3.0 in last 60 hours, appropriate bachelor's degree. *Average time to degree:* Master's–2 years full-time, 3 years part-time; doctorate–6 years full-time. *Application deadline:* For fall admission, 1/15 (priority date). Applications are processed on a rolling basis. Application fee: $45. Electronic applications accepted. *Financial aid:* In 1998–99, 4 fellowships with full tuition reimbursements (averaging $15,000 per year), 24 research assistantships with full tuition reimbursements (averaging $14,600 per year), 10 teaching assistantships with full tuition reimbursements (averaging $15,500 per year) were awarded.; career-related internships or fieldwork, Federal Work-Study, and institutionally-sponsored loans also available. Aid available to part-time students. Financial aid application deadline: 1/15. *Faculty research:* Fission reactor engineering and safety, plasma physics and fusion technology, plasma processing and ion implantation, applied superconductivity and cryogenics, engineering mechanics and astronautics. Total annual research expenditures: $5.7 million. *Unit head:* Dr. Gilbert A. Emmert, Chair, 608-263-1646, Fax: 608-263-7451, E-mail: emmert@engr.wisc.edu.

Yale University, Graduate School of Arts and Sciences, Programs in Engineering and Applied Science, Department of Applied Physics, New Haven, CT 06520. Offers MS, PhD. *Faculty:* 14 full-time (2 women). *Students:* 16 full-time (2 women); includes 1 minority (Asian American or Pacific Islander), 8 international. 36 applicants, 14% accepted. In 1998, 1 degree awarded. Terminal master's awarded for partial completion of doctoral program. *Degree requirements:* For master's, foreign language and thesis not required; for doctorate, dissertation, area exam required, foreign language not required. *Entrance requirements:* For master's and doctorate, GRE General Test, TOEFL. *Average time to degree:* Doctorate–6 years full-time. *Application deadline:* For fall admission, 1/4. Application fee: $65. *Financial aid:* Fellowships, Federal Work-Study and institutionally-sponsored loans available. Aid available to part-time students. *Unit head:* Chair, 203-432-4282. *Application contact:* Admissions Information, 203-432-2770.

Cross-Discipline Announcements

Columbia University, Fu Foundation School of Engineering and Applied Science, Department of Applied Physics and Applied Mathematics, New York, NY 10027.

BS, MS, Eng Sc D, and PhD degrees awarded. Areas of research: applied mathematics, atmospheric/space physics, computational fluid dynamics, condensed-matter physics, fluids, free-electron lasers, materials science, medical physics, oceanography, optical physics, plasma physics and fusion energy, surface physics. See in-depth description, Book 4, Physics section.

Columbia University, Fu Foundation School of Engineering and Applied Science, Department of Electrical Engineering, New York, NY 10027.

Department of Electrical Engineering offers BS, MS, PhD, and Eng Sc D in electrical engineering. Program emphases include photonics: optical materials, nonlinear optics, devices and system architectures, ultrafast optoelectronics, and semiconductor lasers, modulators, and integrated optics; microelectronics: molecular-beam epitaxy and quantum well device physics and GaAs, field-effect, and bipolar transistors thin-film science and electronics materials; and plasma physics.

Section 13
Geological, Mineral/Mining, and Petroleum Engineering

This section contains a directory of institutions offering graduate work in geological, mineral/mining, and petroleum engineering, followed by in-depth entries submitted by institutions that chose to prepare detailed program descriptions. Additional information about programs listed in the directory but not augmented by an in-depth entry may be obtained by writing directly to the dean of a graduate school or chair of a department at the address given in the directory.

For programs offering related work, see also in this book Chemical Engineering, Civil and Environmental Engineering, Electrical and Computer Engineering, Energy and Power Engineering, Engineering and Applied Sciences, Management of Engineering and Technology, and Materials Sciences and Engineering. In Book 4, see Geosciences and Marine Sciences and Oceanography.

CONTENTS

Geological Engineering

Arizona State University, Graduate College, College of Liberal Arts and Sciences, Department of Geology, Tempe, AZ 85287. Offers geological engineering (MS, PhD); natural science (MNS). *Faculty:* 21 full-time (4 women), 2 part-time (both women). *Students:* 43 full-time (17 women), 13 part-time (4 women); includes 4 minority (all Asian Americans or Pacific Islanders), 6 international. Average age 29. 70 applicants, 51% accepted. In 1998, 15 master's, 4 doctorates awarded. *Degree requirements:* For master's, thesis or alternative required; for doctorate, dissertation required. *Entrance requirements:* For master's and doctorate, GRE. Application fee: $45. *Faculty research:* Mechanics and deposits of volcanic eruptions, possible controls on global climate, electron microprobe studies of ore minerals. *Unit head:* Dr. James A. Tiburczy, Chair, 480-965-5081. *Application contact:* Graduate Secretary, 480-965-2213.

Colorado School of Mines, Graduate School, Department of Geology and Geological Engineering, Golden, CO 80401-1887. Offers engineering geology (Diploma); exploration geosciences (Diploma); geochemistry (MS, PhD); geological engineering (ME, MS, PhD, Diploma); geology (MS, PhD); hydrogeology (Diploma). Part-time programs available. *Faculty:* 27 full-time (5 women), 23 part-time (0 women). *Students:* 53 full-time (16 women), 68 part-time (19 women); includes 6 minority (3 Asian Americans or Pacific Islanders, 2 Hispanic Americans, 1 Native American), 26 international. 90 applicants, 56% accepted. In 1998, 15 master's awarded (100% found work related to degree); 6 doctorates awarded (100% found work related to degree). *Degree requirements:* For master's, thesis required, foreign language not required; for doctorate, dissertation, comprehensive exam required, foreign language not required; for Diploma, foreign language and thesis not required. *Entrance requirements:* For master's, doctorate, and Diploma, GRE General Test (combined average 1660 on three sections), GRE Subject Test, minimum GPA of 3.0. *Application deadline:* Applications are processed on a rolling basis. Application fee: $40. Electronic applications accepted. *Financial aid:* In 1998–99, 54 students received aid, including 6 fellowships, 18 research assistantships, 15 teaching assistantships; unspecified assistantships also available. Aid available to part-time students. Financial aid applicants required to submit FAFSA. *Faculty research:* Predictive sediment modeling, petrophysics, aquifer-contaminant flow modeling, water-rock interactions, geotechnical engineering. Total annual research expenditures: $1.2 million. *Unit head:* Dr. Roger Slatt, Head, 303-273-3800, E-mail: rslatt@mines.edu. *Application contact:* Marilyn Schwinger, Administrative Assistant, 303-273-3800, Fax: 303-273-3859, E-mail: mschwing@mines.edu.

Colorado School of Mines, Graduate School, Department of Geophysics, Golden, CO 80401-1887. Offers geophysical engineering (ME, MS, PhD); geophysics (MS, PhD, Diploma). Part-time programs available. *Faculty:* 11 full-time (0 women), 6 part-time (0 women). *Students:* 47 full-time (10 women), 8 part-time (3 women); includes 2 minority (1 Asian American or Pacific Islander, 1 Hispanic American), 27 international. *Degree requirements:* For master's, thesis required, foreign language not required; for doctorate, one foreign language, dissertation, comprehensive and oral exams required; for Diploma, foreign language and thesis not required. *Entrance requirements:* For master's and doctorate, GRE General Test (combined average 1790 on three sections), minimum GPA of 3.0; for Diploma, GRE General Test, minimum GPA of 3.0. *Application deadline:* Applications are processed on a rolling basis. Application fee: $40. Electronic applications accepted. *Unit head:* Dr. Thomas Davis, Acting Head, 303-273-3938. *Application contact:* Sara Summers, Program Assistant, 303-273-3935, Fax: 303-273-3478, E-mail: ssummers@mines.edu.

Columbia University, Fu Foundation School of Engineering and Applied Science, Department of Earth and Environmental Engineering, Program in Earth Resources Engineering, New York, NY 10027. Offers earth resources engineering (MS); earth systems engineering (PhD). Part-time programs available. *Faculty:* 7 full-time (0 women), 3 part-time (1 woman). *Students:* 2 full-time (0 women), 1 (woman) part-time, 2 international. In 1998, 8 master's awarded (50% found work related to degree, 50% continued full-time study); 6 doctorates awarded. *Degree requirements:* For doctorate, dissertation, qualifying exam required, foreign language not required. *Entrance requirements:* For master's and doctorate, GRE General Test, TOEFL. *Average time to degree:* Master's–6 years full-time. *Application deadline:* For fall admission, 2/15; for spring admission, 10/1. Application fee: $55. *Financial aid:* Federal Work-Study available. Financial aid application deadline: 1/5; financial aid applicants required to submit FAFSA. *Faculty research:* Industrial ecology, waste treatment and recycling, water resources, environmental remediation, hazardous waste disposal. *Unit head:* Dr. Barbara Algin, Departmental Administrator, 212-854-2905, Fax: 212-854-7081, E-mail: ba110@columbia.edu. *Application contact:* Dr. Barbara Algin, Departmental Administrator, 212-854-2905, Fax: 212-854-7081, E-mail: ba110@columbia.edu.

See in-depth description on page 1165.

Drexel University, Graduate School, College of Engineering, Department of Civil and Architectural Engineering, Program in Engineering Geology, Philadelphia, PA 19104-2875. Offers MS. *Faculty:* 1 full-time, 5 part-time. Average age 31. 16 applicants, 13% accepted. In 1998, 12 degrees awarded. *Degree requirements:* For master's, thesis optional, foreign language not required. *Entrance requirements:* For master's, TOEFL (minimum score of 570 required), BS in geology, civil engineering, or equivalent; minimum GPA of 3.0. *Application deadline:* For fall admission, 8/21. Applications are processed on a rolling basis. Application fee: $35. Tuition: Full-time $15,795; part-time $585 per credit. Required fees: $375; $67 per term. Tuition and fees vary according to program. *Financial aid:* Research assistantships, teaching assistantships, unspecified assistantships available. Financial aid application deadline: 2/1. *Unit head:* Dr. Edward Doheny, Director, 215-895-2344. *Application contact:* Kelli Kennedy, Director of Admissions, 215-895-6706, Fax: 215-895-5939, E-mail: crowlka@duvm.ocs.drexel.edu.

Announcement: The Engineering Geology Program specializes in engineering geology/geotechnics, hydrogeology/hydrology, and environmental geology. The program's goal is to improve the students' ability to understand and quantify the engineering and environmental problems and hazards created by geologic materials, processes, structure, and history. The curriculum emphasizes the importance of practical experience.

École Polytechnique de Montréal, Graduate Programs, Department of Mineral Engineering, Montréal, PQ H3C 3A7, Canada. Offers mining engineering and geological engineering (M Eng, M Sc A, PhD). Part-time programs available. *Degree requirements:* For master's and doctorate, one foreign language, computer language, thesis/dissertation required. *Entrance requirements:* For master's, minimum GPA of 2.75; for doctorate, minimum GPA of 3.0. *Faculty research:* Geostatistical evaluation, applied geophysics, hydrogeology, geomechanics, mining geology, soil decontamination.

Michigan Technological University, Graduate School, College of Engineering, Department of Geology, Geophysics and Geological Engineering, Program in Geological Engineering, Houghton, MI 49931-1295. Offers MS. Part-time programs available. *Faculty:* 3 full-time (0 women), 1 (woman) part-time. *Students:* 0 full-time (3 women). Average age 25. 7 applicants, 86% accepted. *Degree requirements:* For master's, thesis required, foreign language not required. *Entrance requirements:* For master's, GRE General Test (combined average 2000 on three sections), TOEFL (minimum score of 550 required). *Application deadline:* For fall admission, 3/15 (priority date). Applications are processed on a rolling basis. Application fee: $30 ($35 for international students). Tuition, state resident: full-time $4,377. Tuition, nonresident: full-time $9,108. Required fees: $126. Tuition and fees vary according to course load. *Financial aid:* In 1998–99, 3 research assistantships (averaging $7,198 per year), 4 teaching assistantships (averaging $6,891 per year) were awarded.; fellowships, career-related internships or fieldwork, Federal Work-Study, institutionally-sponsored loans, and unspecified assistantships also available. Aid available to part-time students. Financial aid applicants required to submit FAFSA. *Faculty research:* Enhanced recovery of petroleum, subsurface visualization, groundwater engineering, volcanic hazards, atmospheric remote sensing. *Application contact:* Dr. Jimmy Diehl, Graduate Coordinator, 906-487-2665, Fax: 906-487-3371, E-mail: jdiehl@mtu.edu.

Montana Tech of The University of Montana, Graduate School, Geoscience Program, Butte, MT 59701-8997. Offers geochemistry (MS); geological engineering (MS); geology (MS); geophysical engineering (MS); hydrogeological engineering (MS); hydrogeology (MS); mineral economics (MS). Part-time programs available. *Faculty:* 17 full-time (2 women). *Students:* 15 full-time (6 women), 7 part-time (1 woman); includes 1 minority (Native American) *Degree requirements:* For master's, thesis required (for some programs), foreign language not required. *Entrance requirements:* For master's, GRE General Test, TOEFL (minimum score of 525 required), minimum B average. *Application deadline:* For fall admission, 4/1 (priority date); for spring admission, 10/1 (priority date). Applications are processed on a rolling basis. Application fee: $30. Tuition, state resident: full-time $3,211; part-time $162 per credit hour. Tuition, nonresident: full-time $9,883; part-time $440 per credit hour. International tuition: $15,500 full-time. *Unit head:* Cindy Dunstan, Administrative Assistant, 406-496-4128, Fax: 406-496-4334, E-mail: cdunstan@mtech.edu. *Application contact:* Cindy Dunstan, Administrative Assistant, 406-496-4128, Fax: 406-496-4334, E-mail: cdunstan@mtech.edu.

South Dakota School of Mines and Technology, Graduate Division, Department of Geology and Geological Engineering, Rapid City, SD 57701-3995. Offers geology and geological engineering (MS, PhD); paleontology (MS). Part-time programs available. *Faculty:* 7 full-time (0 women), 3 part-time (0 women). *Students:* 28 full-time (10 women), 5 international. Average age 30. In 1998, 2 master's, 1 doctorate awarded. *Degree requirements:* For master's, thesis required, foreign language not required; for doctorate, dissertation required, dissertation required. *Entrance requirements:* For master's and doctorate, GRE General Test, GRE Subject Test, TOEFL (minimum score of 520 required), TWE. *Application deadline:* For fall admission, 6/15 (priority date); for spring admission, 10/15. Applications are processed on a rolling basis. Application fee: $15. Electronic applications accepted. Tuition, state resident: part-time $89 per hour. Tuition, nonresident: part-time $261 per hour. Part-time tuition and fees vary according to program. *Financial aid:* In 1998–99, 8 fellowships, 14 research assistantships, 14 teaching assistantships were awarded.; Federal Work-Study and institutionally-sponsored loans also available. Aid available to part-time students. Financial aid application deadline: 5/15. *Faculty research:* Contaminants in soil, nitrate leaching, environmental changes, fracture formations, greenhouse effect. Total annual research expenditures: $18,865. *Unit head:* Dr. James Fox, Dean, 605-394-2461. *Application contact:* Brenda Brown, Secretary, 800-454-8162 Ext. 2493, Fax: 605-394-5360, E-mail: graduate_admissions@silver.sdmt.edu.

The University of Akron, Graduate School, Buchtel College of Arts and Sciences, Department of Geology, Program in Engineering Geology, Akron, OH 44325-0001. Offers MS. *Students:* 1 full-time (0 women), 1 international. Average age 27. In 1998, 1 degree awarded. *Degree requirements:* For master's, thesis required, foreign language not required. *Entrance requirements:* For master's, minimum GPA of 2.75. *Average time to degree:* Master's–2 years full-time, 4 years part-time. *Application deadline:* For fall admission, 3/1. Applications are processed on a rolling basis. Application fee: $25 ($50 for international students). Tuition, state resident: part-time $189 per credit. Tuition, nonresident: part-time $353 per credit. Required fees: $7.3 per credit. *Unit head:* Dr. David McConnell, Director of Graduate Studies, 330-972-8047, E-mail: mcconnell@uakron.edu.

University of Alaska Fairbanks, Graduate School, College of Natural Resource Development and Management, School of Mineral Engineering, Department of Mining and Geological Engineering, Program in Geological Engineering, Fairbanks, AK 99775-7480. Offers MS, EM. 2 applicants, 100% accepted. *Degree requirements:* For master's, thesis, comprehensive exam required, foreign language not required. *Entrance requirements:* For master's, GRE General Test, TOEFL (minimum score of 550 required). *Application deadline:* For fall admission, 8/1. Applications are processed on a rolling basis. Application fee: $35. *Unit head:* Dr. Sukumar Bandopadhyay, Head, Department of Mining and Geological Engineering, 907-474-5120.

The University of Arizona, Graduate College, College of Engineering and Mines, Department of Mining and Geological Engineering, Program in Geological and Geophysical Engineering, Tucson, AZ 85721. Offers MS, PhD. Part-time programs available. *Students:* 9 full-time (3 women), 7 part-time (2 women); includes 1 minority (Hispanic American), 6 international. Average age 30. 14 applicants, 79% accepted. In 1998, 1 master's awarded. *Degree requirements:* For master's, thesis required, foreign language not required; for doctorate, computer language, dissertation required, foreign language not required. *Entrance requirements:* For master's and doctorate, GRE General Test, TOEFL (minimum score of 550 required). *Application deadline:* For fall admission, 3/1. Applications are processed on a rolling basis. Application fee: $35. *Financial aid:* Fellowships, research assistantships, teaching assistantships, institutionally-sponsored loans available. Financial aid application deadline: 3/1. *Faculty research:* High-resolution subsurface imaging, monitoring subsurface contamination, neural networks, remote sensing. *Unit head:* Elsie Nonaka, Graduate Adviser, 520-621-2147, Fax: 520-621-8330. *Application contact:* Elsie Nonaka, Graduate Adviser, 520-621-2147, Fax: 520-621-8330.

University of British Columbia, Faculty of Graduate Studies, Faculty of Science, Department of Earth and Ocean Sciences, Program in Geological Engineering, Vancouver, BC V6T 1Z2, Canada. Offers M Eng, MA Sc, PhD. Part-time programs available. *Degree requirements:* For master's, thesis required (for some programs); for doctorate, dissertation required. *Entrance requirements:* For master's and doctorate, TOEFL (minimum score of 600 required). Electronic applications accepted. *Faculty research:* Geohydrology, groundwater, mineral exploration.

University of California, Berkeley, Graduate Division, College of Engineering, Department of Materials Science and Mineral Engineering, Program in Engineering Geoscience, Berkeley, CA 94720-1500. Offers M Eng, MS, D Eng, PhD. *Degree requirements:* For master's, comprehensive exam or thesis (MS) required; for doctorate, dissertation, qualifying exam required. *Entrance requirements:* For master's and doctorate, GRE General Test, minimum GPA of 3.0. *Application deadline:* For fall admission, 2/10; for spring admission, 9/1. Application fee: $40. *Financial aid:* Application deadline: 1/5. *Application contact:* Carole James, Student Affairs Officer, 510-642-3801, Fax: 510-643-5792, E-mail: carolej@uclink4.berkeley.edu.

University of Idaho, College of Graduate Studies, College of Mines and Earth Resources, Department of Geology and Geological Engineering, Program in Geological Engineering, Moscow, ID 83844-4140. Offers MS. *Students:* 4 full-time (2 women), 12 part-time (1 woman); includes 2 minority (1 Asian American or Pacific Islander, 1 Native American) In 1998, 1 degree awarded. *Degree requirements:* For master's, thesis required. *Entrance requirements:* For master's, minimum GPA of 2.8. *Application deadline:* For fall admission, 8/1; for spring admission, 12/15. Application fee: $35 ($45 for international students). *Financial aid:* Application deadline: 2/15. *Unit head:* Dr. John Oldow, Head, Department of Geology and Geological Engineering, 208 885 7327.

University of Minnesota, Twin Cities Campus, Graduate School, Institute of Technology, Department of Civil Engineering, Minneapolis, MN 55455-0213. Offers civil engineering (MCE, MS, PhD); geological engineering (M Geo E, MS, PhD). Part-time programs available. *Faculty:* 34 full-time (4 women), 2 part-time (0 women). *Students:* 113 full-time (29 women), 23 part-time (5 women), 31 international. *Degree requirements:* For master's, thesis optional, foreign language not required; for doctorate, dissertation required, foreign language not required. *Entrance requirements:* For master's and doctorate, GRE General Test, TOEFL (minimum score of 550 required). *Application deadline:* For fall admission, 6/15 (priority date); for spring admission, 10/15. Applications are processed on a rolling basis. Application fee: $50 ($55 for international students). *Unit head:* John Gulliver, Head, 612-625-5522, Fax: 612-626-7750.

Application contact: Roxane McGlade, Student Personnel Worker for Graduate Studies, 612-625-9581, Fax: 612-626-7750, E-mail: gradsec@ce.umn.edu.

University of Missouri–Rolla, Graduate School, School of Mines and Metallurgy, Department of Geological and Petroleum Engineering, Program in Geological Engineering, Rolla, MO 65409-0910. Offers MS, DE, PhD. Part-time programs available. *Faculty:* 8 full-time (1 woman). *Students:* 12 full-time (6 women), 1 part-time, 2 international. Average age 32. 9 applicants, 78% accepted. In 1998, 3 master's, 2 doctorates awarded. *Degree requirements:* For master's, computer language required, foreign language and thesis not required; for doctorate, computer language, dissertation required, foreign language not required. *Entrance requirements:* For master's, GRE General Test (minimum combined score of 1100 required), TOEFL (minimum score of 550 required), minimum GPA of 3.0 in last 4 semesters; for doctorate, GRE General Test (minimum combined score of 1100 required), TOEFL (minimum score of 550 required). *Application deadline:* For fall admission, 7/1; for spring admission, 12/1. Applications are processed on a rolling basis. Application fee: $25. Electronic applications accepted. *Financial aid:* In 1998–99, 10 students received aid, including 2 fellowships with full tuition reimbursements available (averaging $11,250 per year), 3 research assistantships with partial tuition reimbursements available (averaging $6,493 per year), 5 teaching assistantships with partial tuition reimbursements available (averaging $6,493 per year); Federal Work-Study and institutionally-sponsored loans also available. Aid available to part-time students. Financial aid application deadline: 3/1; financial aid applicants required to submit FAFSA. *Faculty research:* Groundwater hydrology, hazardous waste management remote sensing and geographic information systems, geostatistics, geotechnical site selection. *Unit head:* Dr. John D. Rockaway, Chairman, Department of Geological and Petroleum Engineering, 573-341-4867, Fax: 573-341-6935, E-mail: rockaway@umr.edu.

University of Nevada, Reno, Graduate School, Mackay School of Mines, Department of Geological Sciences, Reno, NV 89557. Offers geochemistry (MS, PhD); geological engineering (MS, Geol E); geology (MS, PhD); geophysics (MS, PhD). *Degree requirements:* For master's, thesis optional, foreign language not required; for doctorate, one foreign language, dissertation required. *Entrance requirements:* For master's, GRE General Test, GRE Subject Test, TOEFL (minimum score of 500 required), minimum GPA of 2.75; for doctorate, GRE General Test, GRE Subject Test, TOEFL (minimum score of 500 required), minimum GPA of 3.0. *Faculty research:* Hydrothermal ore deposits, metamorphic and igneous petrogenesis, sedimentary rock record of earth history, field and petrographic investigation of magnetism, rock fracture mechanics.

University of Oklahoma, Graduate College, College of Engineering, School of Petroleum and Geological Engineering, Norman, OK 73019-0390. Offers MS, PhD. Part-time and evening/weekend programs available. *Faculty:* 9 full-time (0 women), 2 part-time (0 women). *Students:* 23 full-time (2 women), 27 part-time (5 women), 41 international. Average age 27. 33 applicants, 79% accepted. In 1998, 8 master's, 7 doctorates awarded. *Degree requirements:* For master's, computer language required, thesis optional, foreign language not required; for doctorate, one foreign language, computer language, dissertation, qualifying exam required. *Entrance requirements:* For master's and doctorate, GRE, TOEFL (minimum score of 550 required). *Application deadline:* For fall admission, 6/1 (priority date); for spring admission, 11/1. Applications are processed on a rolling basis. Application fee: $25. Tuition, state resident: part-time $86 per credit hour. Tuition, nonresident: part-time $275 per credit hour. Tuition and fees vary according to course level, course load and program. *Financial aid:* In 1998–99, 30 research assistantships, 9 teaching assistantships were awarded.; Federal Work-Study, institutionally-sponsored loans, and tuition waivers (full and partial) also available. Aid available to part-time students. Financial aid application deadline: 4/15. *Faculty research:* Production, enhanced oil recovery, rock mechanics, fracturing, natural gas. Total annual research expenditures: $3.6 million. *Unit head:* Dr. Keith K. Millheim, Director, 405-325-2921, Fax: 405-325-7477. *Application contact:* Dr. Anuj Gupta, Graduate Liaison, 405-325-2921, Fax: 405-325-7477.

University of Utah, Graduate School, College of Mines and Earth Sciences, Department of Geology and Geophysics, Program in Geological Engineering, Salt Lake City, UT 84112-1107. Offers ME, MS, PhD. *Students:* 3 full-time (1 woman), 1 part-time. In 1998, 1 master's awarded. Terminal master's awarded for partial completion of doctoral program. *Degree requirements:* For master's, computer language, thesis, qualifying exam required; for doctorate, one foreign language, computer language, dissertation required. *Entrance requirements:* For master's and doctorate, GRE General Test, TOEFL (minimum score of 500 required), minimum GPA of 3.25. *Application deadline:* For fall admission, 7/1. Application fee: $30 ($50 for international students). *Financial aid:* Fellowships, research assistantships, teaching assistantships, institutionally-sponsored loans available. Financial aid application deadline: 2/15. *Faculty research:* Electrical methods, mineral exploration, geothermal resources, groundwater. *Unit head:* Dr. Albert C. Reynolds, Adviser, 918-631-3043, Fax: 918-631-2059. *Application contact:* A. Anton Ekdale, Director of Graduate Studies, 801-581-7266, Fax: 801-581-7065.

University of Windsor, College of Graduate Studies and Research, Faculty of Science, Earth Sciences, Windsor, ON N9B 3P4, Canada. Offers geological engineering (MA Sc); geology (M Sc). Part-time programs available. *Degree requirements:* For master's, thesis required. *Entrance requirements:* For master's, TOEFL (minimum score of 550 required), minimum B average. *Faculty research:* Paleontology, geochemistry, sedimentology.

University of Wisconsin–Madison, Graduate School, College of Engineering, Geological Engineering Program, Madison, WI 53706-1380. Offers MS, PhD. *Faculty:* 11 full-time (3 women). *Students:* 10 full-time (0 women), 6 international. In 1998, 2 master's awarded (100% found work related to degree). *Degree requirements:* For doctorate, dissertation required. *Entrance requirements:* For master's and doctorate, GRE, TOEFL (minimum score of 580 required). *Application deadline:* For fall admission, 7/1 (priority date); for spring admission, 11/15 (priority date). Applications are processed on a rolling basis. Application fee: $45. Electronic applications accepted. *Financial aid:* Fellowships with tuition reimbursements, research assistantships with tuition reimbursements, teaching assistantships with tuition reimbursements available. Financial aid application deadline: 1/15. *Faculty research:* Constitute models for geomaterials, rock fracture, *in situ* stress determination, environmental geotechnics, site remediation. *Unit head:* Bezalel C. Haimson, Chair, 608-262-3732. *Application contact:* Diana J. Rhoads, Program Assistant 3, 608-263-1795, Fax: 608-262-8353, E-mail: matsciad@engr.wisc.edu.

Mineral/Mining Engineering

Colorado School of Mines, Graduate School, Department of Mining and Earth Systems Engineering, Golden, CO 80401-1887. Offers ME, MS, PhD. Part-time programs available. *Faculty:* 7 full-time (0 women), 3 part-time (0 women). *Students:* 13 full-time (1 woman), 21 part-time; includes 2 minority (1 Asian American or Pacific Islander, 1 Hispanic American), 22 international. 11 applicants, 73% accepted. In 1998, 3 master's awarded (100% found work related to degree); 1 doctorate awarded (100% found work related to degree). *Degree requirements:* For master's, thesis required (for some programs), foreign language not required; for doctorate, one foreign language, dissertation, comprehensive exam required. *Entrance requirements:* For master's and doctorate, GRE General Test, minimum GPA of 3.0. *Application deadline:* Applications are processed on a rolling basis. Application fee: $40. Electronic applications accepted. *Financial aid:* In 1998–99, 18 students received aid, including 4 fellowships, 8 teaching assistantships; research assistantships, unspecified assistantships also available. Aid available to part-time students. Financial aid applicants required to submit FAFSA. *Faculty research:* Mine evaluation and planning, geostatistics, mining robotics, water jet cutting, rock mechanics. Total annual research expenditures: $7,808. *Unit head:* Dr. Tibor Rozgonyi, Head, 303-273-3700, E-mail: trozgony@mines.edu. *Application contact:* Shannon Mann, Program Assistant, 303-273-3701, E-mail: smann@mines.edu.

Columbia University, Fu Foundation School of Engineering and Applied Science, Department of Earth and Environmental Engineering, Program in Minerals Engineering and Materials Science, New York, NY 10027. Offers metallurgical engineering (Engr); mineral engineering (Engr); minerals engineering and materials science (Eng Sc D, PhD); mines (Engr). Part-time programs available. *Faculty:* 12 full-time (2 women), 3 part-time (1 woman). *Students:* 3 full-time (0 women), 3 part-time; includes 1 minority (Asian American or Pacific Islander), 5 international. *Degree requirements:* For doctorate, dissertation, qualifying exam required. *Entrance requirements:* For doctorate and Engr, GRE General Test, TOEFL. Application fee: $55. *Financial aid:* Application deadline: 1/5; *Faculty research:* Comminution, mineral processing, flotation and flocculation, transport phenomena in heterogeneous reactions, extractive metallurgy, electrochemistry. Total annual research expenditures: $418,122. *Unit head:* Dr. Barbara Algin, Departmental Administrator, 212-854-2905, Fax: 212-854-7081, E-mail: ba110@columbia.edu. *Application contact:* Dr. Barbara Algin, Departmental Administrator, 212-854-2905, Fax: 212-854-7081, E-mail: ba110@columbia.edu.

Dalhousie University, Faculty of Graduate Studies, DalTech, Faculty of Engineering, Department of Mining and Metallurgical Engineering, Program in Mining, Halifax, NS B3H 3J5, Canada. Offers M Eng, MA Sc, PhD. In 1998, 1 master's, 3 doctorates awarded (33% entered university research/teaching, 67% found other work related to degree). *Degree requirements:* For master's and doctorate, thesis/dissertation required, foreign language not required. *Entrance requirements:* For master's and doctorate, TOEFL (minimum score of 580 required). *Application deadline:* For fall admission, 6/1; for winter admission, 10/1; for spring admission, 2/1. Application fee: $55. *Financial aid:* Fellowships, research assistantships, teaching assistantships, scholarships and unspecified assistantships available. *Unit head:* Lucille Boisselle-Roy, Admissions Officer, 514-987-3000 Ext. 3128, Fax: 514-987-7728. *Application contact:* Shelley Parker, Admissions Coordinator, Graduate Studies and Research, 902-494-1288, Fax: 902-494-3149, E-mail: shelley.parker@dal.ca.

École Polytechnique de Montréal, Graduate Programs, Department of Mineral Engineering, Montréal, PQ H3C 3A7, Canada. Offers mining engineering and geological engineering (M Eng, M Sc A, PhD). Part-time programs available. *Degree requirements:* For master's and doctorate, one foreign language, computer language, thesis/dissertation required. *Entrance requirements:* For master's, minimum GPA of 2.75; for doctorate, minimum GPA of 3.0. *Faculty research:* Geostatistical evaluation, applied geophysics, hydrogeology, geomechanics, mining geology, soil decontamination.

Laurentian University, School of Graduate Studies and Research, School of Engineering, Sudbury, ON P3E 2C6, Canada. Offers metallurgy (MA Sc); mining (M Eng). Part-time programs available. *Faculty:* 15 full-time (1 woman), 5 part-time (0 women). *Students:* 10 full-time (0 women), 7 part-time. *Degree requirements:* Foreign language not required. *Application deadline:* For fall admission, 9/1. Application fee: $50. *Unit head:* Dr. Paul Lindon, Director, 705-675-1151 Ext. 2244, Fax: 705-675-4862. *Application contact:* 705-675-1151 Ext. 3909, Fax: 705-675-4843.

McGill University, Faculty of Graduate Studies and Research, Faculty of Engineering, Department of Mining and Metallurgical Engineering, Program in Mining Engineering, Montréal, PQ H3A 2T5, Canada. Offers M Eng, M Sc, PhD, Diploma. *Students:* 25 full-time (1 woman). 10 applicants, 70% accepted. In 1998, 12 master's, 4 doctorates awarded. *Degree requirements:* For master's, thesis optional; for doctorate, dissertation required, foreign language not required. *Entrance requirements:* For master's and doctorate, TOEFL (minimum score of 550 required), minimum GPA of 3.0. *Application deadline:* For fall admission, 3/1; for winter admission, 8/1; for spring admission, 11/1. Applications are processed on a rolling basis. Application fee: $60. *Unit head:* H. Mitri, Director, 514-398-4755 Ext. 0540, Fax: 514-398-7099, E-mail: hanie@minmet.lan.mcgill.ca.

Michigan Technological University, Graduate School, College of Engineering, Department of Mining Engineering, Houghton, MI 49931-1295. Offers geotechnical engineering (PhD); mining engineering (MS, PhD). Part-time programs available. *Faculty:* 4 full-time (1 woman), 1 part-time (0 women). *Students:* 12 full-time (3 women); includes 2 minority (1 African American, 1 Asian American or Pacific Islander), 5 international. Average age 36. 9 applicants, 100% accepted. In 1998, 1 master's, 3 doctorates awarded. *Degree requirements:* For master's, thesis or alternative required, foreign language not required; for doctorate, dissertation required, foreign language not required. *Entrance requirements:* For master's and doctorate, TOEFL (minimum score of 550 required), BS in engineering or science. *Average time to degree:* Doctorate–5.8 years full-time. *Application deadline:* For fall admission, 3/15 (priority date). Applications are processed on a rolling basis. Application fee: $30 ($35 for international students). Tuition, state resident: full-time $4,377. Tuition, nonresident: full-time $9,108. Required fees: $126. Tuition and fees vary according to course load. *Financial aid:* In 1998–99, 6 fellowships (averaging $3,378 per year), 3 research assistantships (averaging $9,209 per year) were awarded.; teaching assistantships, career-related internships or fieldwork, Federal Work-Study, institutionally-sponsored loans, and unspecified assistantships also available. Aid available to part-time students. Financial aid application deadline: 4/1; financial aid applicants required to submit FAFSA. *Faculty research:* Rock fragmentation, rock mechanics, recycling and recovery of secondary materials, tunneling and underground construction. Total annual research expenditures: $82,326. *Unit head:* Dr. O. Francis Otuonye, Chair, 906-487-2610, Fax: 906-487-2495, E-mail: frotuony@mtu.edu. *Application contact:* Dr. Jiann-Yang Hwang, Associate Professor, 906-487-2600, Fax: 906-487-2495, E-mail: jhwang@mtu.edu.

Montana Tech of The University of Montana, Graduate School, Metallurgical/Mineral Processing Engineering Programs, Butte, MT 59701-8997. Offers MS. Part-time programs available. *Faculty:* 7 full-time (0 women). *Students:* 5 full-time (1 woman), 4 part-time. 8 applicants, 50% accepted. In 1998, 2 degrees awarded. *Degree requirements:* For master's, thesis optional, foreign language not required. *Entrance requirements:* For master's, GRE General Test, TOEFL (minimum score of 525 required), minimum B average. *Application deadline:* For fall admission, 4/1 (priority date); for spring admission, 10/1 (priority date). Applications are processed on a rolling basis. Application fee: $30. Tuition, state resident: full-time $3,211; part-time $162 per credit hour. Tuition, nonresident: full-time $9,883; part-time $440 per credit hour. International tuition: $15,500 full-time. *Financial aid:* In 1998–99, 2 research assistantships with partial tuition reimbursements (averaging $6,324 per year), 3 teaching assistantships with partial tuition reimbursements (averaging $3,400 per year) were awarded.; career-related internships or fieldwork, Federal Work-Study, institutionally-sponsored loans, and tuition waivers (full and partial) also available. Aid available to part-time students. Financial aid application deadline: 4/1; financial aid applicants required to submit FAFSA. *Faculty research:* Stabilizing hazardous waste, decontamination of metals by melt refining, ultraviolet enhancement of stabilization reactions. Total annual research expenditures: $400,000. *Unit head:* Dr. Courtney Young, Department Head, 406-496-4158, Fax: 406-496-4664, E-mail:

Mineral/Mining Engineering

Montana Tech of The University of Montana (continued)
cyoung@mtech.edu. *Application contact:* Cindy Dunstan, Administrative Assistant, 406-496-4128, Fax: 406-496-4334, E-mail: cdunstan@mtech.edu.

Montana Tech of The University of Montana, Graduate School, Mining Engineering Program, Butte, MT 59701-8997. Offers MS. Part-time programs available. *Faculty:* 4 full-time (1 woman). *Students:* 4 full-time (0 women), 1 part-time. 10 applicants, 50% accepted. In 1998, 6 degrees awarded. *Degree requirements:* For master's, thesis optional, foreign language not required. *Entrance requirements:* For master's, GRE General Test, TOEFL (minimum score of 525 required), minimum B average. *Application deadline:* For fall admission, 4/1 (priority date); for spring admission, 10/1 (priority date). Applications are processed on a rolling basis. Application fee: $30. Tuition, state resident: full-time $3,211; part-time $162 per credit hour. Tuition, nonresident: full-time $9,883; part-time $440 per credit hour. International tuition: $15,500 full-time. *Financial aid:* In 1998–99, 3 students received aid, including 1 teaching assistantship with partial tuition reimbursement available (averaging $3,400 per year); research assistantships, career-related internships or fieldwork, Federal Work-Study, institutionally-sponsored loans, and tuition waivers (full and partial) also available. Aid available to part-time students. Financial aid application deadline: 4/1; financial aid applicants required to submit FAFSA. *Faculty research:* Geostatistics, geomechanics, mine planning, economic models, equipment selection. Total annual research expenditures: $15,000. *Unit head:* Dr. H. Peter Knudsen, Dean, School of Mines, 406-496-4395, Fax: 406-496-4133, E-mail: hknudsen@mtech.edu. *Application contact:* Cindy Dunstan, Administrative Assistant, 406-496-4128, Fax: 406-496-4334, E-mail: cdunstan@mtech.edu.

New Mexico Institute of Mining and Technology, Graduate Studies, Department of Mineral and Environmental Engineering, Socorro, NM 87801. Offers environmental engineering (MS), including air quality engineering and science, hazardous waste engineering, water quality engineering and science; mineral engineering (MS). *Faculty:* 11 full-time (1 woman). *Students:* 12 full-time (3 women); includes 8 minority (all Hispanic Americans), 4 international. Average age 30. 13 applicants, 46% accepted. In 1998, 2 degrees awarded. *Degree requirements:* For master's, thesis required, foreign language not required. *Entrance requirements:* For master's, GRE General Test, TOEFL (minimum score of 540 required). *Average time to degree:* Master's–3 years full-time. *Application deadline:* For fall admission, 3/1 (priority date); for spring admission, 6/1. Applications are processed on a rolling basis. Application fee: $16. *Financial aid:* In 1998–99, 3 research assistantships (averaging $9,670 per year), 3 teaching assistantships (averaging $9,670 per year) were awarded.; fellowships, Federal Work-Study and institutionally-sponsored loans also available. Financial aid application deadline: 3/1; financial aid applicants required to submit CSS PROFILE or FAFSA. *Faculty research:* Rock mechanics, geological engineering, mining problems, blasting, shock waves. *Unit head:* Dr. Clinton P. Richardson, Chair, 505-835-5346, Fax: 505-835-5252. *Application contact:* Dr. David B. Johnson, Dean of Graduate Studies, 505-835-5513, Fax: 505-835-5476, E-mail: graduate@nmt.edu.

Pennsylvania State University University Park Campus, Graduate School, College of Earth and Mineral Sciences, Department of Energy and Geo-Environmental Engineering, Program in Mineral Processing, State College, University Park, PA 16802-1503. Offers MS, PhD. *Students:* 9 full-time (1 woman), 2 part-time. *Degree requirements:* For master's and doctorate, thesis/dissertation required. *Entrance requirements:* For master's and doctorate, GRE General Test, TOEFL. Application fee: $50. *Financial aid:* Application deadline: 12/31. *Unit head:* Dr. Richard Hogg, Chair, 814-865-3802.

Pennsylvania State University University Park Campus, Graduate School, College of Earth and Mineral Sciences, Department of Energy and Geo-Environmental Engineering, Program in Mining Engineering, State College, University Park, PA 16802-1503. Offers M Eng, MS, PhD. *Students:* 6 full-time (0 women). In 1998, 4 master's, 4 doctorates awarded. *Degree requirements:* For doctorate, dissertation required. *Entrance requirements:* For master's and doctorate, GRE General Test, TOEFL. Application fee: $50. *Financial aid:* Application deadline: 12/31. *Unit head:* Dr. Chris Bise, Chair, 814-863-1644.

Queen's University at Kingston, School of Graduate Studies and Research, Faculty of Applied Science, Department of Mining Engineering, Kingston, ON K7L 3N6, Canada. Offers M Sc, M Sc Eng, PhD. Part-time programs available. *Students:* 30 full-time (5 women), 3 part-time. In 1998, 15 master's, 3 doctorates awarded. *Degree requirements:* For master's, thesis optional, foreign language not required; for doctorate, dissertation, comprehensive exam required, foreign language not required. *Entrance requirements:* For master's and doctorate, TOEFL (minimum score of 550 required). *Application deadline:* For fall admission, 2/28 (priority date). Application fee: $60. Electronic applications accepted. *Financial aid:* Fellowships, research assistantships, teaching assistantships, institutionally-sponsored loans available. Financial aid application deadline: 3/1. *Faculty research:* Rock mechanics, drilling, blasting, explosives, ventilation/environmental control. *Unit head:* Dr. J. F. Archibald, Head, 613-533-2198. *Application contact:* Dr. W. T. Yen, Graduate Coordinator, 613-533-2206.

Southern Illinois University Carbondale, Graduate School, College of Engineering, Department of Mining Engineering, Carbondale, IL 62901-6806. Offers MS. *Faculty:* 4 full-time (0 women), 1 part-time (0 women). *Students:* 5 full-time (0 women), 4 part-time, 6 international. Average age 25. 2 applicants, 50% accepted. In 1998, 1 degree awarded. *Degree requirements:* For master's, thesis, comprehensive exam required, foreign language not required. *Entrance requirements:* For master's, TOEFL (minimum score of 550 required), minimum GPA of 2.7. *Application deadline:* Applications are processed on a rolling basis. Application fee: $20. *Financial aid:* In 1998–99, 8 students received aid; fellowships with full tuition reimbursements available, research assistantships with full tuition reimbursements available, teaching assistantships with full tuition reimbursements available, Federal Work-Study, institutionally-sponsored loans, and tuition waivers (full) available. Aid available to part-time students. Financial aid application deadline: 3/1. *Faculty research:* Rock mechanics and ground control, mine subsidence, mine systems analysis, fine coal cleaning, surface mine reclamation. Total annual research expenditures: $1.7 million. *Unit head:* Dr. Paul Chugh, Chairperson, 618-536-6637.

Université du Québec à Chicoutimi, Graduate Programs, Program in Mineral Resources, Chicoutimi, PQ G7H 2B1, Canada. Offers PhD. Offered jointly with the Université du Québec à Montréal. Part-time programs available. *Degree requirements:* For doctorate, dissertation required. *Entrance requirements:* For doctorate, appropriate master's degree, proficiency in French.

Université du Québec à Montréal, Graduate Programs, Program in Mineral Resources, Montréal, PQ H3C 3P8, Canada. Offers PhD. Offered jointly with the Université du Québec à Chicoutimi. Part-time programs available. *Degree requirements:* For doctorate, dissertation required. *Entrance requirements:* For doctorate, appropriate master's degree or equivalent and proficiency in French.

Université Laval, Faculty of Graduate Studies, Faculty of Sciences and Engineering, Department of Mining and Metallurgical Engineering, Sainte-Foy, PQ G1K 7P4, Canada. Offers metallurgical engineering (M Sc, PhD); mining (M Sc, PhD). *Students:* 31 full-time (6 women), 8 part-time (1 woman). 19 applicants, 58% accepted. In 1998, 6 master's awarded. *Application deadline:* For fall admission, 3/1. Application fee: $30. *Unit head:* Réal Tremblay, Director, 418-656-2131 Ext. 5047, Fax: 418-656-5343, E-mail: real.tremblay@gmn.ulaval.ca.

University of Alaska Fairbanks, Graduate School, College of Natural Resource Development and Management, School of Mineral Engineering, Department of Mining and Geological Engineering, Program in Mineral Preparation Engineering, Fairbanks, AK 99775-7480. Offers MS. 3 applicants, 100% accepted. In 1998, 2 degrees awarded. *Degree requirements:* For master's, thesis, comprehensive exam required, foreign language not required. *Entrance requirements:* For master's, GRE General Test, TOEFL (minimum score of 550 required). *Application deadline:* For fall admission, 8/1. Applications are processed on a rolling basis. Application fee: $35. *Financial aid:* Research assistantships, career-related internships or fieldwork available. *Faculty research:* Washability of coal, microbial mining, mineral leaching. *Unit head:* Dr. Sukumar Bandopadhyay, Head, Department of Mining and Geological Engineering, 907-474-5120.

University of Alaska Fairbanks, Graduate School, College of Natural Resource Development and Management, School of Mineral Engineering, Department of Mining and Geological Engineering, Program in Mining Engineering, Fairbanks, AK 99775-7480. Offers MS, EM. 2 applicants, 50% accepted. In 1998, 1 degree awarded. *Degree requirements:* For master's, thesis, comprehensive exam required, foreign language not required. *Entrance requirements:* For master's, GRE General Test, TOEFL (minimum score of 550 required). *Application deadline:* For fall admission, 8/1. Applications are processed on a rolling basis. Application fee: $35. *Unit head:* Dr. Sukumar Bandopadhyay, Head, Department of Mining and Geological Engineering, 907-474-5120.

University of Alberta, Faculty of Graduate Studies and Research, Department of Civil and Environmental Engineering, Edmonton, AB T6G 2E1, Canada. Offers construction engineering and management (M Eng, M Sc, PhD); environmental engineering (M Eng, M Sc, PhD); environmental science (M Sc, PhD); geoenvironmental engineering (M Eng, M Sc, PhD); geotechnical engineering (M Sc); geotechnical engineering (M Eng, PhD); mining engineering (M Eng, M Sc, PhD); petroleum engineering (M Eng, M Sc, PhD); structural engineering (M Eng, M Sc, PhD); water resources (M Eng, M Sc, PhD). Part-time programs available. *Degree requirements:* For master's, thesis required (for some programs), foreign language not required; for doctorate, dissertation required, foreign language not required. *Faculty research:* Mining.

The University of Arizona, Graduate College, College of Engineering and Mines, Department of Mining and Geological Engineering, Program in Mining Engineering, Tucson, AZ 85721. Offers MS, PhD. Part-time programs available. *Students:* 4 full-time (0 women), 4 part-time (1 woman), 7 international. Average age 33. 10 applicants, 50% accepted. In 1998, 5 master's awarded. *Degree requirements:* For master's, thesis required, foreign language not required; for doctorate, computer language, dissertation required, foreign language not required. *Entrance requirements:* For master's and doctorate, GRE General Test, TOEFL (minimum score of 550 required). *Application deadline:* For fall admission, 3/1. Applications are processed on a rolling basis. Application fee: $35. *Financial aid:* Fellowships, research assistantships, teaching assistantships, institutionally-sponsored loans available. Financial aid application deadline: 3/1. *Faculty research:* Mine system design, in-site leaching, fluid flow in rocks, geostatistics, rock mechanics. *Unit head:* Elsie Nonaka, Graduate Adviser, 520-621-2147, Fax: 520-621-8330. *Application contact:* Elsie Nonaka, Graduate Adviser, 520-621-2147, Fax: 520-621-8330.

See in-depth description on page 1171.

University of British Columbia, Faculty of Graduate Studies, Faculty of Applied Science, Department of Mining and Mineral Process Engineering, Vancouver, BC V6T 1Z2, Canada. Offers M Eng, MA Sc, PhD. *Degree requirements:* For master's and doctorate, thesis/dissertation required, foreign language not required. *Entrance requirements:* For master's and doctorate, TOEFL (minimum score of 570 required). *Faculty research:* Fiberglass rock bolting, expert systems development, environmental engineering and acid mine drainage, ventilation, sag mill modelling and simulation.

University of California, Berkeley, Graduate Division, College of Engineering, Department of Materials Science and Mineral Engineering, Berkeley, CA 94720-1500. Offers ceramic sciences and engineering (M Eng, MS, D Eng, PhD); engineering geoscience (M Eng, MS, D Eng, PhD); materials engineering (M Eng, MS, D Eng, PhD); mineral engineering (M Eng, MS, D Eng, PhD); petroleum engineering (M Eng, MS, D Eng, PhD); physical metallurgy (M Eng, MS, D Eng, PhD). *Students:* 110 full-time (23 women); includes 17 minority (11 Asian Americans or Pacific Islanders, 5 Hispanic Americans, 1 Native American), 42 international. 191 applicants, 22% accepted. In 1998, 21 master's, 11 doctorates awarded. *Degree requirements:* For master's, comprehensive exam or thesis (MS) required; for doctorate, dissertation, qualifying exam required. *Entrance requirements:* For master's and doctorate, GRE General Test, minimum GPA of 3.0. *Application deadline:* For fall admission, 2/10; for spring admission, 9/1. Application fee: $40. *Financial aid:* Fellowships, research assistantships, teaching assistantships available. Financial aid application deadline: 1/5. *Unit head:* Dr. Thomas M. Devine, Chair, 510-642-3801. *Application contact:* Carole James, Student Affairs Officer, 510-642-3801, Fax: 510-643-5792, E-mail: carolej@uclink4.berkeley.edu.

University of Idaho, College of Graduate Studies, College of Mines and Earth Resources, Department of Metallurgical and Mining Engineering, Programs in Mining Engineering, Moscow, ID 83844-4140. Offers metallurgical engineering (MS, PhD); mining engineering (MS, PhD). *Degree requirements:* For master's, computer language required, foreign language and thesis not required; for doctorate, one foreign language, computer language, dissertation required. *Entrance requirements:* For master's, minimum GPA of 2.8; for doctorate, minimum undergraduate GPA of 2.8, 3.0 graduate. *Application deadline:* For fall admission, 8/1; for spring admission, 12/15. Application fee: $35 ($45 for international students). *Financial aid:* Career-related internships or fieldwork available. Financial aid application deadline: 2/15. *Unit head:* Dr. Patrick Taylor, Head, Department of Metallurgical and Mining Engineering, 208-885-6769.

University of Kentucky, Graduate School, Graduate School Programs from the College of Engineering, Program in Mining Engineering, Lexington, KY 40506-0032. Offers MME, MS Min, PhD. *Degree requirements:* For master's, comprehensive exam required, thesis optional, foreign language not required; for doctorate, dissertation, comprehensive exam required. *Entrance requirements:* For master's, GRE General Test, minimum undergraduate GPA of 2.5; for doctorate, GRE General Test, minimum graduate GPA of 3.0. *Faculty research:* Benefication of fine and ultrafine particles, operation research in mining and mineral processing, land reclamation.

University of Missouri–Rolla, Graduate School, School of Mines and Metallurgy, Department of Mining Engineering, Rolla, MO 65409-0910. Offers MS, DE, PhD. Part-time programs available. *Faculty:* 8 full-time (0 women). *Students:* 8 full-time (1 woman), 2 international. Average age 33. 7 applicants, 57% accepted. In 1998, 3 master's awarded. Terminal master's awarded for partial completion of doctoral program. *Degree requirements:* For master's and doctorate, computer language, thesis/dissertation required, foreign language not required. *Entrance requirements:* For master's, GRE General Test (minimum combined score of 1200 required), TOEFL (minimum score of 550 required), minimum GPA of 3.0 in last 4 semesters; for doctorate, GRE General Test (minimum combined score of 1200 required), TOEFL (minimum score of 550 required). *Application deadline:* For fall admission, 7/1; for spring admission, 12/1. Applications are processed on a rolling basis. Application fee: $25. *Financial aid:* In 1998–99, 6 students received aid, including 3 fellowships with full tuition reimbursements available (averaging $11,250 per year), 3 research assistantships with partial tuition reimbursements available (averaging $12,986 per year); teaching assistantships with partial tuition reimbursements available, Federal Work-Study and institutionally-sponsored loans also available. Aid available to part-time students. Financial aid application deadline: 3/1; financial aid applicants required to submit FAFSA. *Faculty research:* Rock mechanics, coal preparation, explosives engineering, mine land reclamation, health and safety. *Unit head:* Dr. John W. Wilson, Chairman, 573-341-4753, Fax: 573-341-6934, E-mail: jwilson@umr.edu.

University of Nevada, Reno, Graduate School, Mackay School of Mines, Department of Mining Engineering, Reno, NV 89557. Offers MS, EM. *Degree requirements:* For master's, computer language, thesis required (for some programs), foreign language not required. *Entrance requirements:* For master's, GRE, TOEFL (minimum score of 500 required), minimum GPA of 2.75.

University of North Dakota, Graduate School, School of Engineering and Mines, Department of Civil Engineering, Grand Forks, ND 58202. Offers civil engineering (M Engr); sanitary engineering (M Engr), including soils and structures engineering, surface mining engineering. Part-time programs available. *Faculty:* 6 full-time (0 women). *Students:* 7 full-time (1 woman).

Degree requirements: For master's, thesis or alternative required, foreign language not required. *Entrance requirements:* For master's, GRE General Test, TOEFL (minimum score of 550 required), minimum GPA of 2.5. *Application deadline:* For fall admission, 3/1 (priority date). Applications are processed on a rolling basis. Application fee: $20. *Unit head:* Dr. Ronald Apanian, Chairperson, 701-777-3562, Fax: 701-777-4838, E-mail: ron_apanian@mail.und.nodak.edu.

University of Utah, Graduate School, College of Mines and Earth Sciences, Department of Mining Engineering, Salt Lake City, UT 84112-1107. Offers ME, MS, PhD, MBA/MS. *Faculty:* 5 full-time (0 women), 5 part-time (0 women). *Students:* 1 full-time (0 women), 8 part-time, 2 international. Average age 34. In 1998, 3 master's awarded. *Degree requirements:* For master's, computer language, comprehensive exam (ME), thesis (MS) required; for doctorate, one foreign language, computer language, dissertation required. *Entrance requirements:* For master's and doctorate, GRE General Test, TOEFL (minimum score of 500 required). *Application deadline:* For fall admission, 7/1. Application fee: $30 ($50 for international students). *Financial aid:* Fellowships, research assistantships, teaching assistantships, career-related internships or fieldwork and institutionally-sponsored loans available. Aid available to part-time students. Financial aid application deadline: 2/15. *Faculty research:* Mineral economics, rock mechanics, mine ventilation, computer economics, slope stability. *Unit head:* Dr. M. K. McCarter, Chair, 801-581-7198, Fax: 801-585-5410.

Virginia Polytechnic Institute and State University, Graduate School, College of Engineering, Department of Mining and Minerals Engineering, Blacksburg, VA 24061. Offers M Eng, MS, PhD. *Faculty:* 7 full-time (0 women). *Students:* 12 full-time (1 woman), 1 part-time, 7 international. 16 applicants, 69% accepted. In 1998, 5 master's, 3 doctorates awarded. *Degree requirements:* For master's, thesis required, foreign language not required. *Entrance requirements:* For master's and doctorate, TOEFL. *Application deadline:* For fall admission, 12/1 (priority date). Applications are processed on a rolling basis. Application fee: $25. *Financial aid:* In 1998–99, 3 research assistantships, 5 teaching assistantships were awarded.; fellowships, career-related internships or fieldwork and unspecified assistantships also available. Financial aid application deadline: 4/1. *Unit head:* Dr. S. Suboleski, Head, 540-231-6671, E-mail: stansub@vt.edu.

West Virginia University, College of Engineering and Mineral Resources, Department of Mining Engineering, Program in Engineering of Mines, Morgantown, WV 26506. Offers MSEM. Program being phased out; applicants no longer accepted. Part-time programs available. *Degree requirements:* For master's, thesis required, foreign language not required. *Faculty research:* Longwall mining, ground control, surface mining, respirable dust, expert systems.

West Virginia University, College of Engineering and Mineral Resources, Department of Mining Engineering, Program in Mineral Engineering, Morgantown, WV 26506. Offers PhD. Part-time programs available. *Degree requirements:* For doctorate, dissertation, comprehensive exam required, foreign language not required. *Entrance requirements:* For doctorate, GRE General Test (score in 75th percentile or higher on quantitative section required), TOEFL (minimum score of 550 required), MS in mineral engineering, minimum GPA of 3.5. *Faculty research:* Ground control, longwall mining, mine ventilation, respirable dust, expert systems.

Petroleum Engineering

Colorado School of Mines, Graduate School, Chemical Engineering and Petroleum Refining Department, Golden, CO 80401-1887. Offers ME, MS, PhD. Part-time programs available. *Faculty:* 24 full-time (3 women), 4 part-time (0 women). *Students:* 42 full-time (14 women), 14 part-time (4 women); includes 2 minority (1 Asian American or Pacific Islander, 1 Hispanic American), 29 international. 87 applicants, 32% accepted. In 1998, 12 master's awarded (92% found work related to degree); 1 doctorate awarded (100% found work related to degree). *Degree requirements:* For master's, thesis required (for some programs), foreign language not required; for doctorate, dissertation, comprehensive exam required, foreign language not required. *Entrance requirements:* For master's and doctorate, GRE General Test, minimum GPA of 3.0. *Application deadline:* Applications are processed on a rolling basis. Application fee: $40. Electronic applications accepted. *Financial aid:* In 1998–99, 37 students received aid, including 26 research assistantships, 8 teaching assistantships; fellowships, unspecified assistantships also available. Aid available to part-time students. Financial aid applicants required to submit FAFSA. *Faculty research:* Liquid fuels for the future, responsible management of hazardous substances, surface and interfacial engineering, advanced computational methods and process control, gas hydrates. Total annual research expenditures: $1 million. *Unit head:* Dr. Robert M. Baldwin, Head, 303-273-3720, E-mail: rbaldwin@mines.edu. *Application contact:* John Dorgan, Assistant Professor, 303-273-3539, Fax: 303-273-3730, E-mail: jdorgan@mines.edu.

Colorado School of Mines, Graduate School, Department of Petroleum Engineering, Golden, CO 80401-1887. Offers ME, MS, PhD. Part-time programs available. *Faculty:* 8 full-time (1 woman), 2 part-time (0 women). *Students:* 36 full-time (6 women), 8 part-time; includes 1 minority (Native American), 24 international. 41 applicants, 95% accepted. In 1998, 13 master's awarded (100% found work related to degree); 3 doctorates awarded (100% found work related to degree). *Degree requirements:* For master's, thesis required (for some programs), foreign language not required; for doctorate, one foreign language, dissertation, comprehensive exams required. *Entrance requirements:* For master's and doctorate, GRE General Test (combined average 1390 on three sections), minimum GPA of 3.0. *Application deadline:* Applications are processed on a rolling basis. Application fee: $40. Electronic applications accepted. *Financial aid:* In 1998–99, 21 students received aid, including 11 fellowships, 5 research assistantships, 5 teaching assistantships; career-related internships or fieldwork, institutionally-sponsored loans, and unspecified assistantships also available. Financial aid applicants required to submit FAFSA. *Faculty research:* Dynamic rock mechanics, deflagration theory, geostatistics, geochemistry, petrophysics. Total annual research expenditures: $40,061. *Unit head:* Dr. Craig W. Van Kirk, Head, 303-273-3740, E-mail: cvankirk@mines.edu. *Application contact:* Chris Trujillo, Administrative Assistant, 303-273-3188, Fax: 303-273-3189, E-mail: ctrujill@mines.edu.

Louisiana State University and Agricultural and Mechanical College, Graduate School, College of Engineering, Department of Petroleum Engineering, Baton Rouge, LA 70803. Offers MS Pet E, PhD. *Faculty:* 7 full-time (0 women). *Students:* 21 full-time (1 woman), 8 part-time; includes 3 minority (1 African American, 1 Hispanic American, 1 Native American), 18 international. Average age 33. 42 applicants, 60% accepted. In 1998, 2 master's awarded (100% found work related to degree); 2 doctorates awarded. *Degree requirements:* For master's, thesis or alternative required, foreign language not required; for doctorate, dissertation, exam required, foreign language not required. *Entrance requirements:* For master's and doctorate, GRE General Test, minimum GPA of 3.0. *Application deadline:* For fall admission, 1/25 (priority date). Applications are processed on a rolling basis. Application fee: $25. *Financial aid:* In 1998–99, 15 research assistantships with partial tuition reimbursements (averaging $9,400 per year) were awarded.; fellowships, teaching assistantships with partial tuition reimbursements, institutionally-sponsored loans also available. *Faculty research:* Enhanced oil and gas recovery, well control and blowout prevention, formation evaluation, environmental control. Total annual research expenditures: $1.3 million. *Unit head:* Dr. Z. Bassiouni, Chair, 225-388-5215, Fax: 225-388-6039, E-mail: pe.zab@unix1.sncc.lsu.edu. *Application contact:* Dr. Andrew Wojtanowicz, Graduate Adviser, 225-388-5215, E-mail: awojtan@unix1.sncc.lsu.edu.

Montana Tech of The University of Montana, Graduate School, Petroleum Engineering Program, Butte, MT 59701-8997. Offers MS. Part-time and evening/weekend programs available. *Faculty:* 5 full-time (1 woman). *Students:* 3 full-time (1 woman), 1 part-time. 4 applicants, 75% accepted. In 1998, 1 degree awarded. *Degree requirements:* For master's, thesis optional, foreign language not required. *Entrance requirements:* For master's, GRE General Test, TOEFL (minimum score of 525 required), minimum B average. *Application deadline:* For fall admission, 4/1 (priority date); for spring admission, 10/1 (priority date). Applications are processed on a rolling basis. Application fee: $30. Tuition, state resident: full-time $3,211; part-time $162 per credit hour. Tuition, nonresident: full-time $9,883; part-time $440 per credit hour. International tuition: $15,500 full-time. *Financial aid:* In 1998–99, 2 students received aid, including 2 teaching assistantships with partial tuition reimbursements available (averaging $3,400 per year); research assistantships, career-related internships or fieldwork, Federal Work-Study, institutionally-sponsored loans, and tuition waivers (full and partial) also available. Aid available to part-time students. Financial aid application deadline: 4/1; financial aid applicants required to submit FAFSA. *Faculty research:* Reservoir characterization, simulation, near well bore problems, PVT, environmental waste. Total annual research expenditures: $2.2 million. *Unit head:* Dr. Dan Bradley, Dean, College of Engineering, 406-496-4254, Fax: 406-496-4417, E-mail: dbradley@mtech.edu. *Application contact:* Cindy Dunstan, Administrative Assistant, 406-496-4128, Fax: 406-496-4334, E-mail: cdunstan@mtech.edu.

New Mexico Institute of Mining and Technology, Graduate Studies, Department of Petroleum Engineering, Socorro, NM 87801. Offers MS, PhD. *Faculty:* 9 full-time (0 women), 4 part-time (0 women). *Students:* 25 full-time (1 woman), 22 international. Average age 30. 28 applicants, 82% accepted. In 1998, 7 master's, 4 doctorates awarded. *Degree requirements:* For master's and doctorate, thesis/dissertation required, foreign language not required. *Entrance requirements:* For master's, GRE General Test, TOEFL (minimum score of 540 required); for doctorate, GRE General Test, GRE Subject Test, TOEFL (minimum score of 540 required). *Average time to degree:* Master's–3 years full-time; doctorate–6 years full-time. *Application deadline:* For fall admission, 3/1 (priority date); for spring admission, 6/1. Applications are processed on a rolling basis. Application fee: $16. *Financial aid:* In 1998–99, 20 research assistantships (averaging $9,670 per year), 4 teaching assistantships (averaging $9,670 per year) were awarded.; fellowships, Federal Work-Study and institutionally-sponsored loans also available. Financial aid application deadline: 3/1; financial aid applicants required to submit CSS PROFILE or FAFSA. *Faculty research:* Reservoir simulation, oil recovery processes, drilling and production. *Unit head:* Dr. Larry Teufel, Chairman, 505-835-5412, Fax: 505-835-5210, E-mail: petro@nmt.edu. *Application contact:* Dr. David B. Johnson, Dean of Graduate Studies, 505-835-5513, Fax: 505-835-5476, E-mail: graduate@nmt.edu.

Pennsylvania State University University Park Campus, Graduate School, College of Earth and Mineral Sciences, Department of Energy and Geo-Environmental Engineering, Program in Petroleum and Natural Gas Engineering, State College, University Park, PA 16802-1503. Offers MS, PhD. *Students:* 25 full-time (3 women), 2 part-time. In 1998, 6 master's, 1 doctorate awarded. *Degree requirements:* For master's, thesis required, foreign language not required; for doctorate, dissertation required. *Entrance requirements:* For master's and doctorate, GRE General Test, TOEFL. Application fee: $50. *Financial aid:* Application deadline: 12/31. *Unit head:* Dr. Turgay Ertekin, Chair, 814-865-6082.

Stanford University, School of Earth Sciences, Department of Petroleum Engineering, Stanford, CA 94305-9991. Offers MS, PhD, Eng. *Faculty:* 8 full-time (0 women). *Students:* 45 full-time (7 women), 10 part-time (1 woman); includes 2 minority (both Asian Americans or Pacific Islanders), 48 international. Average age 28. 56 applicants, 32% accepted. In 1998, 16 master's, 4 doctorates awarded. Terminal master's awarded for partial completion of doctoral program. *Degree requirements:* For masters, doctorate, and Eng, computer language, thesis/dissertation required, foreign language not required. *Entrance requirements:* For master's, doctorate, and Eng, GRE General Test, TOEFL. *Application deadline:* For fall admission, 1/15. Application fee: $65 ($80 for international students). Electronic applications accepted. Tuition: Full-time $23,058. Required fees: $152. Part-time tuition and fees vary according to course load. *Financial aid:* Fellowships, research assistantships, teaching assistantships, Federal Work-Study and institutionally-sponsored loans available. Aid available to part-time students. Financial aid application deadline: 2/15. *Unit head:* Roland Horne, Chair, 650-723-9595, Fax: 650-725-2099, E-mail: horne@pangea.stanford.edu. *Application contact:* 650-723-8314.

See in-depth description on page 1167.

Texas A&M University, College of Engineering, Harold Vance Department of Petroleum Engineering, College Station, TX 77843. Offers M Eng, MS, D Eng, PhD. Part-time programs available. Postbaccalaureate distance learning degree programs offered (no on-campus study). *Faculty:* 15 full-time (1 woman), 7 part-time (1 woman). *Students:* 72 full-time (7 women), 32 part-time (3 women); includes 3 minority (all Hispanic Americans), 76 international. Average age 25. In 1998, 29 master's, 10 doctorates awarded. *Degree requirements:* For master's, thesis (MS) required; for doctorate, dissertation (PhD) required. *Entrance requirements:* For master's, GRE General Test (minimum score 450 on verbal section; 650 on quantitative required), TOEFL (minimum score of 550 required); for doctorate, GRE General Test (minimum score of 450 on verbal section; 650 on quantitative required), TOEFL (minimum score of 550 required). *Average time to degree:* Master's–2 years full-time; doctorate–4 years full-time. Application fee: $50 ($75 for international students). *Financial aid:* In 1998–99, 6 fellowships (averaging $14,700 per year), 47 research assistantships (averaging $14,700 per year), 6 teaching assistantships (averaging $14,700 per year) were awarded.; career-related internships or fieldwork also available. Financial aid application deadline: 3/1; financial aid applicants required to submit FAFSA. *Faculty research:* Drilling and well stimulation, well completions and well performance, reservoir modeling and reservoir description, reservoir simulation, improved/enhanced recovery. Total annual research expenditures: $1.8 million. *Unit head:* Dr. Charles H. Bowman, Head, 409-845-2241, Fax: 409-845-1307, E-mail: bauman@spindletop.tamu.edu. *Application contact:* Dr. Thomas A. Blasingame, Graduate Adviser, 409-847-9095, Fax: 409-845-1307, E-mail: t-blasingame@spindletop.tamu.edu.

See in-depth description on page 1169.

Texas A&M University–Kingsville, College of Graduate Studies, College of Engineering, Department of Chemical Engineering and Natural Gas Engineering, Program in Natural Gas Engineering, Kingsville, TX 78363. Offers ME, MS. *Faculty:* 3 full-time, 2 part-time. *Students:* 6 full-time (1 woman), 1 part-time. *Degree requirements:* For master's, computer language, thesis or alternative, comprehensive exam required, foreign language not required. *Entrance requirements:* For master's, GRE General Test (minimum combined score of 1000 required), TOEFL (minimum score of 525 required), minimum GPA of 3.0. *Application deadline:* For fall admission, 6/1; for spring admission, 11/15. Applications are processed on a rolling basis. Application fee: $15 ($25 for international students). Tuition, state resident: full-time $2,062. Tuition, nonresident: full-time $7,246. *Financial aid:* Fellowships, research assistantships, Federal Work-Study, institutionally-sponsored loans, and tuition waivers (partial) available. Financial aid application deadline: 5/15. *Faculty research:* Gas processing, coal gasification and liquefaction, enhanced oil recovery, gas measurement, unconventional gas recovery. *Unit head:* Dr. Faleh Al-Sadoon, Coordinator, 361-593-2002.

Petroleum Engineering

Texas Tech University, Graduate School, College of Engineering, Department of Petroleum Engineering, Lubbock, TX 79409. Offers MS Pet E. Part-time programs available. *Faculty:* 4 full-time (0 women). *Students:* 8 full-time (1 woman), 1 part-time, 8 international. Average age 30. 18 applicants, 67% accepted. In 1998, 3 degrees awarded. *Degree requirements:* For master's, computer language, thesis or alternative required, foreign language not required. *Entrance requirements:* For master's, GRE General Test (minimum combined score of 1000 required; average 1147), minimum GPA of 3.0. *Application deadline:* For fall admission, 4/15 (priority date); for spring admission, 11/1 (priority date). Applications are processed on a rolling basis. Application fee: $25 ($50 for international students). Electronic applications accepted. *Financial aid:* In 1998–99, 8 research assistantships (averaging $8,438 per year) were awarded.; fellowships, teaching assistantships, career-related internships or fieldwork, Federal Work-Study, and institutionally-sponsored loans also available. Aid available to part-time students. Financial aid application deadline: 5/15; financial aid applicants required to submit FAFSA. *Faculty research:* Artificial intelligence for use in drilling/production, petroleum reservoir petrophysical studies, enhanced oil recovery process problems. Total annual research expenditures: $404,884. *Unit head:* Dr. Lloyd R. Heinze, Chairman, 806-742-3573, Fax: 806-742-3502, E-mail: theinze@coe.ttu.edu.

University of Alaska Fairbanks, Graduate School, College of Natural Resource Development and Management, School of Mineral Engineering, Department of Petroleum Engineering, Fairbanks, AK 99775-7480. Offers MS. *Faculty:* 5 full-time (0 women). *Students:* 5 full-time (0 women), 2 part-time (1 woman); includes 2 minority (both Asian Americans or Pacific Islanders), 5 international. Average age 33. 17 applicants, 94% accepted. In 1998, 2 degrees awarded. *Degree requirements:* For master's, thesis, comprehensive exam required, foreign language not required. *Entrance requirements:* For master's, GRE General Test, TOEFL (minimum score of 550 required). *Application deadline:* For fall admission, 8/1. Applications are processed on a rolling basis. Application fee: $35. *Financial aid:* Research assistantships, teaching assistantships, career-related internships or fieldwork available. *Faculty research:* Characterization of oil and gas. *Unit head:* Godwin Chukwu, Head, 907-474-7730.

University of Alberta, Faculty of Graduate Studies and Research, Department of Civil and Environmental Engineering, Edmonton, AB T6G 2E1, Canada. Offers construction engineering and management (M Eng, M Sc, PhD); environmental engineering (M Eng, M Sc, PhD); environmental science (M Sc, PhD); geoenvironmental engineering (M Eng, M Sc, PhD); geotechnical engineering (M Sc); geotechnical engineering (M Eng, PhD); mining engineering (M Eng, M Sc, PhD); petroleum engineering (M Eng, M Sc, PhD); structural engineering (M Eng, M Sc, PhD); water resources (M Eng, M Sc, PhD). Part-time programs available. *Degree requirements:* For master's, thesis required (for some programs), foreign language not required; for doctorate, dissertation required, foreign language not required. *Faculty research:* Mining.

University of Calgary, Faculty of Graduate Studies, Faculty of Engineering, Department of Chemical and Petroleum Engineering, Calgary, AB T2N 1N4, Canada. Offers M Eng, M Sc, PhD. Part-time programs available. *Faculty:* 19 full-time (2 women), 5 part-time (0 women). *Students:* 46 full-time (15 women), 22 part-time (7 women). Average age 24. 800 applicants, 2% accepted. In 1998, 13 master's awarded (60% found work related to degree, 40% continued full-time study); 9 doctorates awarded. *Degree requirements:* For master's, thesis required (for some programs), foreign language not required; for doctorate, dissertation, candidacy exam required, foreign language not required. *Entrance requirements:* For master's and doctorate, TOEFL (minimum score of 550 required). *Average time to degree:* Master's–2 years full-time, 4 years part-time; doctorate–4 years full-time. *Application deadline:* For fall admission, 5/31 (priority date); for winter admission, 9/30 (priority date); for spring admission, 1/31 (priority date). Applications are processed on a rolling basis. Application fee: $60. *Financial aid:* In 1998–99, 40 fellowships with partial tuition reimbursements, 25 research assistantships with partial tuition reimbursements, 30 teaching assistantships with partial tuition reimbursements were awarded.; grants also available. Financial aid application deadline: 5/31. *Faculty research:* Thermodynamics, transport phenomena, enhanced oil recovery, kinetics and fluidized beds, environmental engineering. Total annual research expenditures: $2 million. *Unit head:* R. G. Moore, Head, 403-220-5750, Fax: 403-284-4852, E-mail: moore@ench.ucalgary.ca. *Application contact:* A. K. Mehrotra, Associate Head, Graduate Studies, 403-220-7406, Fax: 403-284-4852, E-mail: mehrotra@acs.ucalgary.ca.

University of California, Berkeley, Graduate Division, College of Engineering, Department of Materials Science and Mineral Engineering, Program in Petroleum Engineering, Berkeley, CA 94720-1500. Offers M Eng, MS, D Eng, PhD. *Degree requirements:* For master's, comprehensive exam or thesis (MS) required; for doctorate, dissertation, qualifying exam required. *Entrance requirements:* For master's and doctorate, GRE General Test, minimum GPA of 3.0. *Application deadline:* For fall admission, 2/10; for spring admission, 9/1. Application fee: $40. *Financial aid:* Application deadline: 1/5. *Unit head:* Graduate Adviser, 604-822-2540, Fax: 604-822-5599. *Application contact:* Carole James, Student Affairs Officer, 510-642-3801, Fax: 510-643-5792, E-mail: carolej@uclink4.berkeley.edu.

University of Houston, Cullen College of Engineering, Program in Petroleum Engineering, Houston, TX 77004. Offers MSPE. Part-time and evening/weekend programs available. *Students:* 6 full-time (0 women), 19 part-time (2 women); includes 1 African American, 2 Hispanic Americans, 4 international. Average age 35. 7 applicants, 86% accepted. In 1998, 13 degrees awarded. *Degree requirements:* For master's, thesis required (for some programs), foreign language not required. *Entrance requirements:* For master's, GRE General Test, TOEFL. *Application deadline:* For fall admission, 7/3 (priority date); for spring admission, 12/4. Applications are processed on a rolling basis. Application fee: $25 ($75 for international students). *Financial aid:* Federal Work-Study available. Financial aid application deadline: 7/1. *Faculty research:* Supercomputer simulation of reservoirs, rock mechanics, reservoir property analysis. *Unit head:* Dr. Charles Arnold, Director, 713-743-4300, Fax: 713-743-4323.

University of Kansas, Graduate School, School of Engineering, Department of Chemical and Petroleum Engineering, Lawrence, KS 66045. Offers chemical engineering (MS); chemical/petroleum engineering (PhD); petroleum engineering (MS). Part-time programs available. *Faculty:* 11 full-time. *Students:* 15 full-time (4 women), 25 part-time (2 women); includes 1 minority (Asian American or Pacific Islander), 31 international. Average age 25. In 1998, 9 master's, 3 doctorates awarded. *Degree requirements:* For master's, computer language, thesis (for some programs), exam required, foreign language not required; for doctorate, computer language, dissertation, comprehensive and qualifying exams required, foreign language not required. *Entrance requirements:* For master's and doctorate, GRE General Test (minimum combined score of 2000 on three sections required for international students), TOEFL (minimum score of 600 required, TSE (minimum score of 40 required), minimum GPA of 3.0. *Application deadline:* For fall admission, 7/1 (priority date). Application fee: $30. *Financial aid:* In 1998–99, 1 fellowship, 6 research assistantships, 5 teaching assistantships were awarded.; Federal Work-Study also available. Financial aid application deadline: 1/31. *Faculty research:* Enhanced oil recovery, catalysis and kinetics, electrochemical engineering, biochemical engineering, semiconductor materials processing. *Unit head:* Don W. Green, Chairperson, 785-864-4965.

University of Missouri–Rolla, Graduate School, School of Mines and Metallurgy, Department of Geological and Petroleum Engineering, Program in Petroleum Engineering, Rolla, MO 65409-0910. Offers MS, DE, PhD. Part-time programs available. *Faculty:* 4 full-time (1 woman). *Students:* 5 full-time (0 women), 4 international. Average age 32. 8 applicants, 75% accepted. In 1998, 2 master's awarded. *Degree requirements:* For master's, computer language required, foreign language and thesis not required; for doctorate, computer language, dissertation required, foreign language not required. *Entrance requirements:* For master's, GRE General Test (minimum combined score of 1100 required), TOEFL (minimum score of 560 required), minimum GPA of 3.0 in last 4 semesters; for doctorate, GRE General Test (minimum combined score of 1250 required), TOEFL (minimum score of 560 required). *Application deadline:* For fall admission, 7/1; for spring admission, 12/1. Applications are processed on a rolling basis. Application fee: $25. Electronic applications accepted. *Financial aid:* In 1998–99, 4 students received aid, including 1 fellowship with full tuition reimbursement available (averaging $11,250 per year), 2 research assistantships with partial tuition reimbursements available (averaging $6,493 per year), 1 teaching assistantship with partial tuition reimbursement available (averaging $6,493 per year); Federal Work-Study and institutionally-sponsored loans also available. Aid available to part-time students. Financial aid application deadline: 3/1; financial aid applicants required to submit FAFSA. *Faculty research:* Oil recovery, drilling systems, petroleum reservoir, formation evaluation, core analysis/interpretation. *Unit head:* Dr. Daopu T. Numbere, Head, 573-341-4761, Fax: 573-341-6935, E-mail: numbere@umr.edu.

University of Oklahoma, Graduate College, College of Engineering, School of Petroleum and Geological Engineering, Norman, OK 73019-0390. Offers MS, PhD. Part-time and evening/weekend programs available. *Faculty:* 9 full-time (0 women), 2 part-time (0 women). *Students:* 23 full-time (2 women), 27 part-time (5 women), 41 international. Average age 27. 33 applicants, 79% accepted. In 1998, 8 master's, 7 doctorates awarded. *Degree requirements:* For master's, computer language required, thesis optional, foreign language not required; for doctorate, one foreign language, computer language, dissertation, qualifying exam required. *Entrance requirements:* For master's and doctorate, GRE, TOEFL (minimum score of 550 required). *Application deadline:* For fall admission, 6/1 (priority date); for spring admission, 11/1. Applications are processed on a rolling basis. Application fee: $25. Tuition, state resident: part-time $86 per credit hour. Tuition, nonresident: part-time $275 per credit hour. Tuition and fees vary according to course level, course load and program. *Financial aid:* In 1998–99, 30 research assistantships, 9 teaching assistantships were awarded.; Federal Work-Study, institutionally-sponsored loans, and tuition waivers (full and partial) also available. Aid available to part-time students. Financial aid application deadline: 4/15. *Faculty research:* Production, enhanced oil recovery, rock mechanics, fracturing, natural gas. Total annual research expenditures: $3.6 million. *Unit head:* Dr. Keith K. Millheim, Director, 405-325-2921, Fax: 405-325-7477. *Application contact:* Dr. Anuj Gupta, Graduate Liaison, 405-325-2921, Fax: 405-325-7477.

University of Pittsburgh, School of Engineering, Department of Chemical and Petroleum Engineering, Pittsburgh, PA 15260. Offers chemical engineering (MS Ch E, PhD); petroleum engineering (MSPE). Part-time and evening/weekend programs available. *Faculty:* 16 full-time (3 women), 3 part-time (0 women). *Students:* 60 full-time (19 women), 13 part-time (4 women); includes 4 minority (2 African Americans, 2 Asian Americans or Pacific Islanders), 36 international. 219 applicants, 16% accepted. In 1998, 15 master's, 8 doctorates awarded. *Degree requirements:* For master's, computer language, thesis required, foreign language not required; for doctorate, computer language, dissertation, comprehensive and final oral exams required, foreign language not required. *Entrance requirements:* For master's and doctorate, GRE General Test, TOEFL (minimum score of 550 required), minimum QPA of 3.2. *Average time to degree:* Master's–2 years full-time, 3 years part-time; doctorate–5 years full-time, 7 years part-time. *Application deadline:* For fall admission, 8/1 (priority date); for spring admission, 12/1 (priority date). Applications are processed on a rolling basis. Application fee: $30 ($40 for international students). *Financial aid:* In 1998–99, 1 fellowship (averaging $10,800 per year), 58 research assistantships (averaging $10,704 per year) were awarded.; teaching assistantships, grants, scholarships, traineeships, and tuition waivers (full and partial) also available. Financial aid application deadline: 2/15. *Faculty research:* Biotechnology, polymers, catalysis, energy and environment, computational modeling. Total annual research expenditures: $3.6 million. *Unit head:* Dr. Alan J. Russell, Chairman, 412-624-9630, Fax: 412-624-9639, E-mail: ajrche@vms.cis.pitt.edu. *Application contact:* James G. Goodwin, Graduate Coordinator, 412-624-9641, Fax: 412-624-9639, E-mail: goodwin@engrng.pitt.edu.

University of Southern California, Graduate School, School of Engineering, Department of Petroleum Engineering, Los Angeles, CA 90089. Offers MS, PhD, Engr. *Faculty:* 4 full-time, 1 part-time. *Students:* 10 full-time (1 woman), 9 part-time (3 women); includes 2 minority (1 Asian American or Pacific Islander, 1 Hispanic American), 12 international. Average age 29. 30 applicants, 63% accepted. In 1998, 1 master's, 2 doctorates awarded. *Degree requirements:* For master's, thesis optional; for doctorate, dissertation required. *Entrance requirements:* For master's, doctorate, and Engr, GRE General Test, GRE Subject Test. *Application deadline:* For fall admission, 7/1 (priority date); for spring admission, 12/1. Application fee: $55. Tuition: Part-time $768 per unit. Required fees: $350 per semester. *Financial aid:* In 1998–99, 6 fellowships, 7 research assistantships, 1 teaching assistantship were awarded.; Federal Work-Study and institutionally-sponsored loans also available. Aid available to part-time students. Financial aid application deadline: 2/15; financial aid applicants required to submit FAFSA. *Unit head:* Dr. Iraj Ershaghi, Chairperson, 213-740-0322.

University of Southwestern Louisiana, Graduate School, College of Engineering, Department of Petroleum Engineering, Lafayette, LA 70504. Offers MSE. Evening/weekend programs available. *Faculty:* 4 full-time (0 women). *Students:* 4 full-time (1 woman), 3 international. In 1998, 3 degrees awarded. *Degree requirements:* For master's, thesis or alternative, comprehensive exam required, foreign language not required. *Entrance requirements:* For master's, GRE General Test, minimum GPA of 2.85. *Application deadline:* For fall admission, 5/15. Application fee: $5 ($15 for international students). *Financial aid:* In 1998–99, 2 research assistantships with full tuition reimbursements (averaging $4,500 per year) were awarded.; Federal Work-Study and tuition waivers (full and partial) also available. Financial aid application deadline: 5/1. *Unit head:* Dr. Herman H. Rieke, Head, 318-482-6555.

The University of Texas at Austin, Graduate School, College of Engineering, Department of Petroleum and Geosystems Engineering, Austin, TX 78712-1111. Offers MSE, PhD. *Students:* 97 full-time (17 women), 10 part-time (1 woman); includes 32 minority (1 African American, 30 Asian Americans or Pacific Islanders, 1 Hispanic American) 82 applicants, 49% accepted. In 1998, 14 master's, 9 doctorates awarded. *Entrance requirements:* For master's and doctorate, GRE General Test (minimum combined score of 1000 required). Application fee: $50 ($75 for international students). *Financial aid:* Fellowships, research assistantships, teaching assistantships available. Financial aid application deadline: 2/1. *Unit head:* Dr. Ekwere J. Peters, Chairman, 512-471-3161. *Application contact:* Dr. Mukul Sharma, Graduate Adviser, 512-471-3257.

University of Tulsa, Graduate School, College of Business Administration, Department of Engineering and Technology Management, Tulsa, OK 74104-3189. Offers chemical engineering (METM); computer science (METM); electrical engineering (METM); geological science (METM); mathematics (METM); mechanical engineering (METM); petroleum engineering (METM). Part-time and evening/weekend programs available. *Students:* 3 full-time (1 woman), 1 part-time, 3 international. *Degree requirements:* For master's, foreign language and thesis not required. *Entrance requirements:* For master's, GRE General Test (minimum score of 430 on verbal section, 600 on quantitative required), TOEFL (minimum score of 575 required). *Application deadline:* Applications are processed on a rolling basis. Application fee: $30. Electronic applications accepted. Tuition: Full-time $8,640; part-time $480 per hour. Required fees: $3 per hour. One-time fee: $200 full-time. Tuition and fees vary according to program. *Unit head:* Dr. Richard C. Burgess, Assistant Dean/Director of Graduate Business Studies, 918-631-2242, Fax: 918-631-2142.

University of Tulsa, Graduate School, College of Engineering and Applied Sciences, Department of Petroleum Engineering, Tulsa, OK 74104-3189. Offers ME, MSE, PhD. Part-time programs available. *Faculty:* 9 full-time (0 women). *Students:* 48 full-time (7 women), 4 part-time; includes 1 minority (Asian American or Pacific Islander), 50 international. Average age 30. 50 applicants, 46% accepted. In 1998, 17 master's, 5 doctorates awarded. *Degree requirements:* For master's, thesis (MSE) required; for doctorate, computer language, dissertation required, foreign language not required. *Entrance requirements:* For master's, GRE General Test (minimum score of 600 on quantitative section required), TOEFL (minimum score of 550 required); for doctorate, GRE General Test (minimum score of 700 on quantitative section, 1100 combined required), TOEFL (minimum score of 550 required). *Application deadline:* Applications are processed on a rolling basis. Application fee: $30. Electronic applications accepted. Tuition: Full-time $8,640; part-time $480 per hour. Required fees:

$3 per hour. One-time fee: $200 full-time. Tuition and fees vary according to program. *Financial aid:* In 1998–99, 42 students received aid, including 2 fellowships (averaging $6,550 per year), 34 research assistantships (averaging $7,673 per year), 6 teaching assistantships (averaging $6,450 per year); career-related internships or fieldwork, Federal Work-Study, and tuition waivers (partial) also available. Aid available to part-time students. Financial aid application deadline: 2/1; financial aid applicants required to submit FAFSA. *Faculty research:* Drilling, artificial lift, reservoir simulation, reservoir characterization, well testing. *Unit head:* Dr. Stefan Z. Miska, Chairperson, 918-631-2533. *Application contact:* Dr. Albert C. Reynolds, Adviser, 918-631-3043, Fax: 918-631-2059.

University of Utah, Graduate School, College of Engineering, Department of Chemical and Fuels Engineering, Salt Lake City, UT 84112-1107. Offers chemical engineering (M Phil, ME, MS, PhD); fuels engineering (ME, MS, PhD). Part-time programs available. *Faculty:* 17 full-time (1 woman), 23 part-time (0 women). *Students:* 32 full-time (7 women), 12 part-time (1 woman), 27 international. Average age 30. In 1998, 11 master's, 7 doctorates awarded. *Degree requirements:* For master's and doctorate, foreign language and thesis required. *Entrance requirements:* For master's and doctorate, GRE, TOEFL (minimum score of 500 required), minimum GPA of 3.0. *Application deadline:* For fall admission, 7/1. Application fee: $30 ($50 for international students). *Financial aid:* In 1998–99, 4 teaching assistantships were awarded.; fellowships, research assistantships *Faculty research:* Computer-aided process synthesis and design, combustion of solid and liquid fossil fuels, oxygen mass transport in biochemical reactors. *Unit head:* Terry Ring, Chair, 801-585-5705, Fax: 801-581-8692. *Application contact:* Donald A. Dahlstrom, Director of Graduate Studies, 801-581-6934, Fax: 801-581-8692, E-mail: dadahlstrom@cc.utah.edu.

University of Wyoming, Graduate School, College of Engineering, Department of Chemical and Petroleum Engineering, Program in Petroleum Engineering, Laramie, WY 82071. Offers MS, PhD. *Faculty:* 5 full-time (0 women). *Students:* 2 full-time (0 women), 3 part-time, 4 international. 7 applicants, 14% accepted. In 1998, 5 master's awarded (100% found work related to degree); 1 doctorate awarded. Terminal master's awarded for partial completion of doctoral program. *Degree requirements:* For master's, thesis required, foreign language not required; for doctorate, computer language, dissertation required, foreign language not required. *Entrance requirements:* For master's, GRE General Test, TOEFL (minimum score of 550 required), minimum GPA of 3.0; for doctorate, GRE General Test, TOEFL, minimum GPA of 3.0. *Application deadline:* For fall admission, 4/15 (priority date). Applications are processed on a rolling basis. Application fee: $40. Electronic applications accepted. Tuition, state resident: full-time $2,520; part-time $140 per credit hour. Tuition, nonresident: full-time $7,790; part-time $433 per credit hour. Required fees: $400; $7 per credit hour. Full-time tuition and fees vary according to course load and program. *Financial aid:* Research assistantships, teaching assistantships, career-related internships or fieldwork, Federal Work-Study, and institutionally-sponsored loans available. Financial aid application deadline: 4/15. *Faculty research:* Oil recovery methods, oil production, coal bed methane. Total annual research expenditures: $300,000. *Application contact:* Dr. David O. Cooney, Graduate Student Coordinator, 307-766-6464, Fax: 307-766-6777, E-mail: cooney@uwyo.edu.

West Virginia University, College of Engineering and Mineral Resources, Department of Petroleum and Natural Gas Engineering, Morgantown, WV 26506. Offers MSPNGE. Part-time programs available. *Degree requirements:* For master's, thesis required, foreign language not required. *Entrance requirements:* For master's, TOEFL (minimum score of 550 required), minimum GPA of 3.0, BS or equivalent in petroleum or natural gas engineering. *Faculty research:* Gas reservoir engineering, well logging, environment artificial intelligence.

COLUMBIA UNIVERSITY

Fu Foundation School of Engineering and Applied Science
Earth and Environmental Engineering
Materials Science and Engineering

Programs of Study

Economic growth in the twentieth century has been based on the increased use of materials and the development of materials with extraordinary properties. It has also resulted in environmental problems for society and new challenges for engineers. The Department of Earth and Environmental Engineering of Columbia University is dedicated to the study of materials and the environment. It was formed around the historic Henry Krumb School of Mines (1864), the first mining and metallurgical school in the United States, and also involves civil engineering and earth sciences. The research and educational programs address two principal areas of engineering: environmentally sound technologies for extracting, processing, and recycling earth materials (minerals, energy, and water) and the design and fabrication of advanced materials, with full appreciation of their environmental impacts.

Technologies developed in the Department of Earth and Environmental Engineering (EAEE) for the identification, extraction, and physical and chemical processing of nonrenewable earth resources can also be applied to resource recovery from wastes, pollution prevention, and environmental remediation. The M.S. in earth resources engineering (ERE) is designed for engineers and scientists who evaluate, develop, or implement environmentally and economically sound technologies for identifying, extracting, processing, and recycling earth materials such as water, minerals, and fuels. Graduates are specifically qualified to work for the mineral, environmental, and energy industries and for government agencies responsible for energy and the environment. For students with a B.S. in engineering, at least 30 points, or ten courses, are required to graduate. For students with a B.S. or a B.A. degree, preferably with a science major, up to 48 points, or sixteen courses, are required to graduate. While students have a variety of options, the master's thesis is strongly encouraged and is required of students who are receiving any form of University funding. Part-time students are welcome, and some courses are available for distance learning through the Columbia Video Network.

Candidates for the new Ph.D. in earth systems engineering consult with a group of three advisers from the faculty of the department and from other disciplines that can contribute to the project to design their research program. This multidisciplinary faculty advisory group is also involved in the execution and defense of the research project.

Research subjects in the minerals engineering and materials science (MSME) program are selected from the entire spectrum of processing and design of inorganic materials. Candidates can also work on cross-disciplinary earth engineering projects, including policy and economic effects on resource extraction, processing, transport, and disposal and analytic modeling of the environmental impacts of industrial activities at different scales. The Eng.Sc.D. degree is administered by the School of Engineering (SEAS), and the Ph.D. is administered by SEAS and the Graduate School of Arts and Science. The qualifying examinations and all other performance requirements are the same for both degrees.

M.S. and doctoral programs in materials science and engineering (MSE) are offered in interdepartmental collaboration with the Departments of Applied Physics, Chemical/Electrical Engineering, and Physics and Chemistry. The professional degrees of Engineer of Mines, Metallurgical Engineer, and Mineral Engineer are offered to qualified graduate students.

Research Facilities

Information regarding general research, computer, and library facilities can be found in the Columbia School of Engineering page in Section 1 of this guide. The geomechanics laboratory is equipped with presses, diamond cutoff saws, and surface grinding machines. Laboratories are equipped with crushing, grinding, shredding, flotation, gravity, and other physical separation equipment and aqueous and high-temperature processing equipment. All the required modern facilities for characterization of physical and chemical properties, such as X-ray diffraction and fluorescence, SEM, TEM, ICD, and ESCA instruments, are available. The department shares a process design computer lab with chemical engineering and a hydrologic lab with civil engineering. A new laboratory was developed for subsurface imaging and contaminant transport studies. Mechanical testing equipment is also available at the Carleton Lab of Civil Engineering. Materials are prepared by a variety of techniques, including mechanical working, induction melting, electron-beam evaporation, and sputtering. Measurements can be carried out in controlled environments and temperature ranges from near absolute zero to ultrahigh temperatures. The facilities of the Lamont-Doherty Earth Observatory are used for water resource and near-surface geophysics research.

Many of the faculty members are associated with the Earth Engineering Center, the engineering arm of the Columbia Earth Institute. The institute includes among its member units the Lamont-Doherty Earth Observatory, the Center for Environmental Research and Conservation, Biosphere 2, and the International Research Institute for climate prediction. Students have full opportunity to participate in the diverse activities of the Earth Institute.

Financial Aid

Funds are available based on academic merit, financial need, and departmental requirements for research and teaching assistantships. Many students hold staff appointments, which form an integral part of students' training and allow rapid progress toward the degree. Loans and Federal Work-Study positions are available to needy students who are U.S. citizens, permanent resident aliens, or political refugees.

Cost of Study

Tuition for 1999–2000 is $24,150 (for 30 points), plus applicable fees.

Living and Housing Costs

University residence halls include traditional dormitory facilities, as well as suites and apartments for single and married students. On-campus living expenses for 1999–2000 are estimated at $10,800 (includes board). Limited graduate housing is available and is by application only. The cost averaged between $512 and $965 per month, depending on the type of accommodations desired. Other off-campus rooms and apartments are also available.

Student Group

Attending the schools and colleges that constitute Columbia are 20,438 students, 12,330 of whom are graduate students. The 1,000 graduate students attending the Fu Foundation School of Engineering and Applied Science represent almost 200 colleges and universities, about fifty of which are outside the United States.

Location

Columbia's Morningside Heights campus is located about 15 minutes from the heart of New York City. The student is offered a range of educational, cultural, and recreational opportunities.

The University

Originally designated King's College, Columbia opened its doors in 1754 under a grant issued by King George II. Over the years, professional and graduate schools were added. In 1896, it was redesignated Columbia University. The Fu Foundation School of Engineering and Applied Science is the outgrowth of one of these professional schools, the School of Mines, which was established in 1864 as the first school of its kind in the United States. The century-long tradition of the Fu Foundation School of Engineering and Applied Science has been the philosophy of combining a rich liberal education with the rigor of technical education. Many revolutionary advances of contemporary technology have been pioneered at Columbia.

Applying

The basic requirement for admission as a graduate student is a bachelor's degree in any field of engineering or a related field with a record that indicates the preparation and ability necessary for successful performance at Columbia. For the fall term, Ph.D., Eng.Sc.D., and financial aid applicants must apply by January 5. February 15 is the deadline for all other degrees. Applications that are received late are reviewed until August 1 if space is available. Those for the spring term should be submitted by October 1.

Correspondence and Information

Dr. Barbara Algin, Departmental Administrator
Department of Earth and Environmental Engineering
Henry Krumb School of Mines
918 Southwest Mudd, Mail Code 4711
Columbia University
500 West 120th Street
New York, New York 10027
E-mail: ba110@columbia.edu
World Wide Web: http://www.seas.columbia.edu/columbia/departments/krumb

Columbia University

THE FACULTY AND THEIR RESEARCH

Amvrossios (Ross) Bagtzoglou, Assistant Professor of Earth and Environmental Engineering and Civil Engineering; Ph.D., California, Irvine, 1990. Hydrogeology, contaminant transport, numerical analysis, stochastic processes, geostatistical simulation.

Daniel N. Beshers, Professor of Metallurgy; Ph.D., Illinois, 1956. Mechanical behavior of materials, magnetomechanical phenomena, composite materials.

Siu-Wai Chan, Associate Professor of Metallurgy and Materials Science; Ph.D., MIT, 1985. Electronic ceramics, thin films and interfaces, electron microscopy, grain boundaries, high-temperature superconductors.

Paul F. Duby, Professor of Mineral Engineering; Ph.D., Columbia, 1962. Aquatic chemistry, aqueous processing, electrochemistry.

Domenico Grasso, Professor of Earth and Environmental Engineering; Ph.D., Michigan, 1987. Aquatic chemistry, surface and interfacial phenomena, hazardous- and industrial-waste treatment, physicochemical processes.

James Im, Associate Professor of Metallurgy and Materials Science; Ph.D., MIT, 1989. Excimer laser crystallization of Si films for thin-film transistor applications, isothermal nucleation.

Gertrude Neumark, Professor of Materials Science; Ph.D., Columbia, 1951. Laser processing, luminescence, visible light–emitting materials and devices.

Peter Schlosser, Professor of Earth and Environmental Sciences and Engineering; Ph.D., Heidelberg, 1985. Water movement and variability in natural systems, ocean-atmosphere gas exchange, anthropogenic impacts on natural systems.

Ponisseril Somasundaran, La Von Duddleson Krumb Professor of Mineral Engineering; Ph.D., Berkeley, 1964. Colloid and interface science, particle technology, electrokinetics, flocculation, microbial interactions, ESR, NMR.

Richard Ian Stessel, Associate Professor of Earth and Environmental Engineering; Ph.D., Duke, 1983. Materials processing for recovery and recycling, solid- and hazardous-waste control engineering.

Nickolas J. Themelis, Stanley Thompson Professor of Chemical Metallurgy and Acting Chairman; Ph.D., McGill, 1961. Industrial ecology, high-temperature reactors, thermal plasma spraying, transport phenomena.

Nicholas J. Turro, Schweitzer Professor of Chemistry; Ph.D., Caltech, 1963. Surface chemistry, colloid chemistry, supramolecular photochemistry of organic compounds.

Roelof Versteeg, Associate Research Scientist and Adjunct Professor; Ph.D., Paris VII, 1991. Near-surface applied geophysics, subsurface imaging.

Tuncel M. Yegulalp, Professor of Mining Engineering; Eng.Sc.D., Columbia, 1968. Geostatistics, mine planning.

Adjunct Faculty

Vasilis Fthenakis, Adjunct Professor; Ph.D., NYU, 1991. Air pollution and prevention, chemical risk assessment, atmospheric dispersion modeling.

Berislav Markovic, Adjunct Professor; Ph.D., Zagreb, 1996. Applied surface chemistry, polymer behavior in aqueous/particle systems, instrumental chemical/physical analysis.

Ismail C. Noyan, Adjunct Professor; Ph.D., Northwestern, 1984. X-ray and neutron diffraction, deformation mechanics of mesoscopic domains and heterogeneous media, residual stresses.

Brian Pethica, Adjunct Senior Research Scientist; Ph.D., London, 1953; D.Sc., 1962, Sc.D., 1971, Cambridge. Fluid interfaces, monolayer and colloid thermodynamics, novel surfactants and surface active polymers.

Albert G. Tobin, Adjunct Associate Professor; Ph.D., Columbia, 1968. Ceramic and ceramic composite materials.

Iddo Wernick, Adjunct Professor; Ph.D., Columbia, 1992. Industrial ecology, material flows, separation technologies, land use.

Jung H. Yoon, Adjunct Assistant Professor; Eng.Sc.D., Columbia, 1998. Excimer laser crystallization, photosensitive materials.

Current Research Areas

EAEE and ERE: Use of geostatistical methodology for environmental assessment of mines and brownfields, modeling of mine drainage, constructed pilot wetland for storage and treatment of harbor sediments, GIS-based models that superimpose multiple sets of data on flows and sources/sinks of contaminants, and treatment of land and sediments contaminated by mining and processing activities. Research is conducted on a variety of geoenvironmental issues, with the intent to quantify, assess, and ultimately manage adverse human effects on the environment; contaminant transport in the subsurface; flow phenomena in saturated and unsaturated soils; probabilistic assessment of the effects of human activities on the environment; geostatistical simulation; and numerical modeling of estuarine flow and transport processes.

MSE and MSME: Research in materials science includes microscopic study of interfaces, grain boundaries, and thin films; lattice defects and electrical properties of ceramics; laser processing and solidification of silicon; and optical and electric properties of wide-band gap semiconductors. Research at the Langmuir Center for Colloids and Interfaces includes the enhancement of oil recovery by means of ficellar flooding of reservoirs, electroflotation of mineral particles, microbial interactions with minerals, selective flocculation of fine particles, and ultrafine grinding. There are many research projects in surface and colloid chemistry that involve both inorganic and organic materials, such as surfactants, polymers, and latexes.

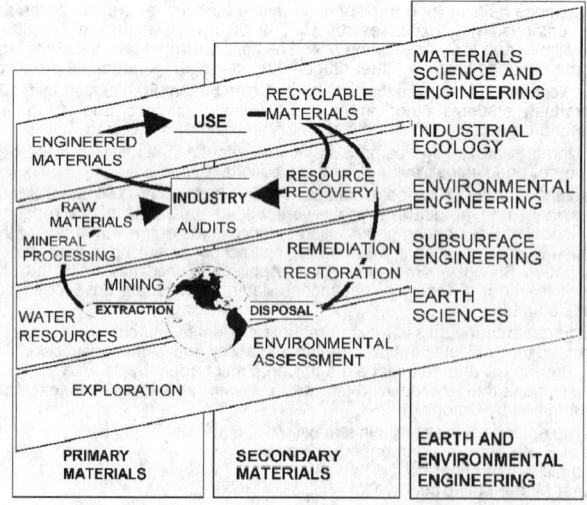

Earth and environmental engineering explores the interactions between the use of earth resources and the environment, from extraction to recycling or disposal.

STANFORD UNIVERSITY

School of Earth Sciences
Department of Petroleum Engineering

Programs of Study

The department offers programs of study leading to the degrees of Master of Science (M.S.), Engineer (Eng.), and Doctor of Philosophy (Ph.D.) in petroleum engineering. The M.S. degree is designed to provide advanced skills in engineering problem solving and to develop job skills at a higher level. Two degrees are available. The standard M.S. degree includes a research component and generally takes six quarters to complete. The M.S. degree for self-funded students requires no research report; students take more courses per quarter and can finish in three to four quarters. Engineer degrees include a broader and more in-depth range of skills and require about three years to complete. The Ph.D. degree offers significant depth in specific, original research areas and requires about 3½ to 5 years to complete, depending upon the background of the student.

Research Facilities

The department has major investments in equipment and research facilities housed in a modern building with excellent laboratories. Computer resources, both in the department and on campus, are also well developed. Research equipment includes an X-ray computerized tomographic scanner (CT scanner) and a gas chromatograph–mass spectrometer (GC/MS). The modern laboratories include combustion, steam injection, geothermal, adsorbtion, CT scan, miscible gas drive, and reservoir characterization facilities. Computerized data acquisition control is used extensively in all of the labs.

Financial Aid

Research and teaching assistantships are available to all qualified applicants. Such assistantships require the student to devote 20 hours per week to research prescribed by the funding project, which is usually closely related to the research requirements for the degree. Assistantship funds are supplemented with fellowship funds to provide equal funding to all applicants who are admitted with financial aid. Nine units of tuition are provided along with a stipend to cover normal living expenses. Students with aid can enroll in a maximum of 9 units of course work per quarter. Summer funding is usually available for students who choose to continue research projects during the summer. Students are also encouraged to work in industry during at least one summer to gain direct experience.

Cost of Study

In 1999–2000, full tuition for students without financial aid is $7686 per quarter. Normally, students must be present for three quarters of the academic year.

Living and Housing Costs

On-campus housing is available at a moderate cost for the geographical area. Newly entering students are guaranteed housing if they meet the first housing application deadline. Single-student apartment rentals on campus range from $1050 to $1500 per quarter. Married student housing ranges from $2600 to $3000 per quarter for couples with no children and from $3350 to $5500 per quarter for couples with children. Off-campus housing usually costs more. Beyond the first year, on-campus housing is only guaranteed for families.

Student Group

Graduate students comprise approximately one half of the Stanford University student body. The Department of Petroleum Engineering has about 55 graduate students.

Location

Stanford University is located 35 miles south of San Francisco and lies at the northernmost end of Silicon Valley. The progressive, highly educated communities that surround it on three sides offer world-class educational and community facilities. Stanford University is a beautiful and peaceful campus, with considerable open space surrounding the buildings. Opportunities to enjoy the exceptional climate and cultural, recreational, intellectual, and social activities available in the area abound.

The Department

The Department of Petroleum Engineering is consistently rated as one of the best programs in the country and the world. The department attracts a wide mix of students from many countries and academic backgrounds. Stanford students profit greatly from the international and cultural mix and enjoy friendly competition and cooperation with their colleagues. All graduates of the department have obtained jobs in recent years, and many find alumni of the department employed by the same companies they choose to work for.

Applying

The admission deadline to begin the following fall quarter is January 15. Applicants are considered by the Admissions Committee as a group, and decisions are made in March. Stanford University attracts gifted, highly motivated applicants from around the world, and admission is selective. GRE scores (and TOEFL scores for applicants from non-English-speaking countries) as well as letters of recommendation, transcripts, and a statement of purpose are all considered by the Admissions Committee. There are no minimum requirements, but the strength of a candidate's admission package influences his or her chances of being admitted.

Correspondence and Information

Professor Thomas Hewett
Graduate Admissions Chair
Department of Petroleum Engineering
Room 065
Green Earth Sciences Building
Stanford University
367 Panama Street
Stanford, California 94305-2220

Telephone: 650-723-8314
Fax: 650-725-2099
E-mail: peteng@pangea.stanford.edu
World Wide Web: http://ekofisk.stanford.edu/pe.html

Stanford University

THE FACULTY AND THEIR RESEARCH

Roland N. Horne, Professor and Chair; Ph.D., Auckland (New Zealand), 1975. Computer-aided well test analysis, optimization, geothermal reservoirs.

Khalid Aziz, Professor; Ph.D., Rice, 1966. Reservoir simulation and characterization, two-phase flow in pipes.

Martin Blunt, Associate Professor; Ph.D., Cambridge, 1988. Flow in porous media, mathematical physics, environmental applications.

William E. Brigham, Professor; Ph.D., Oklahoma, 1962. Thermal oil recovery, tracer analysis, reservoir engineering.

Louis Castanier, Senior Research Engineer; Ph.D., Politechnique Toulouse (France), 1978. Enhanced oil recovery, thermal recovery.

Louis J. Durlofsky, Research Associate Professor; Ph.D., MIT, 1988. Reservoir simulation and upscaling.

Thomas A. Hewett, Professor; Ph.D., MIT, 1970. Reservoir characterization and engineering.

Andre G. Journel, Professor (joint appointment with the Department of Geological and Environmental Sciences); D.Sc., Nancy (France), 1977. Geostatistics, reservoir characterization.

Anthony R. Kovscek, Assistant Professor; Ph.D., Berkeley, 1994. Enhanced oil recovery, heat and fluid transport, mechanisms in porous media.

Franklin M. Orr Jr., Professor and Dean, School of Earth Sciences; Ph.D., Minnesota, 1966. Miscible flooding, reservoir engineering, enhanced oil recovery.

Consulting Faculty

Warren K. Kourt, Consulting Professor. Oil and gas property evaluation.

Robert G. Lindblom, Consulting Professor. Production engineering, well logging, management.

Jane Woodward, Consulting Assistant Professor. Oil and gas property evaluation, energy resources and alternative fuels.

TEXAS A&M UNIVERSITY

Department of Petroleum Engineering

Programs of Study	The Department of Petroleum Engineering offers programs of study leading to the Master of Science (M.S.), Master of Engineering (M.E.), Doctor of Philosophy (Ph.D.), and Doctor of Engineering (D.Eng.) degrees. All students work with an advisory chair and committee. Advanced-level research and course study are offered in drilling, production, well stimulation, reservoir engineering, reservoir characterization, and reservoir management. Specific teaching and research areas include core analysis and flow imaging, drilling and drilling fluids, economic evaluation, field-scale reservoir studies, formation damage, formation evaluation, horizontal and extended-reach wells, improved oil and gas recovery, production and well test data analysis, reservoir characterization, reservoir management, reservoir simulation, rock mechanics, well logging, and well stimulation. New areas of research include multilateral wells, nonparametric optimization (including neural networks), streamtube flow simulation, and well performance modeling and interpretation.

The M.S. degree program requires a minimum of 32 semester hours beyond the baccalaureate degree. The student typically devotes two thirds of these hours to petroleum engineering graduate courses, including thesis-related research, and one third to graduate courses taught in other departments, such as mathematics, computer science, business, and geology. An acceptable thesis is required for the M.S. degree.

The M.E. degree program requires a minimum of 36 semester hours beyond the baccalaureate degree. Approximately two thirds of the hours are petroleum engineering courses, and one third are taken outside the department. A summary report is required for the M.E. degree.

The Ph.D. degree program requires a minimum of 96 semester hours beyond the baccalaureate degree. Students in the Ph.D. program devote at least one third of their time to petroleum engineering courses, approximately one third to research, and approximately one third to courses outside the department. An acceptable dissertation is required for the Ph.D. degree.

The D.Eng. degree program also requires a minimum of 96 semester hours beyond the baccalaureate degree, which include 80 hours of course work and 16 hours of a professional internship. An acceptable record of study is required for the D.Eng. degree.

Students entering the Graduate Program in Petroleum Engineering with degrees in areas other than petroleum engineering may be required to take prerequisite courses at the undergraduate level. These courses are required to ensure proficiency in petroleum engineering skills and academic success in graduate course work.

Research Facilities The Department of Petroleum Engineering is located in the Joe C. Richardson building, a ten-story structure that contains 50,000 square feet of space that is dedicated to petroleum engineering research and teaching. The department research facilities include laboratories for drilling, petrophysical analysis, formation evaluation, well stimulation, production engineering, reservoir engineering, reservoir modeling, enhanced oil recovery, and reservoir management. The department has a thirty-unit PC laboratory for undergraduate and graduate student use and a ten-unit PC laboratory devoted exclusively for use by graduate students. In addition, a number of workstation computers are available, and many graduate research groups have their own computing facilities.

Financial Aid Research and teaching assistantships may be available to qualified graduate students who have satisfied the requirements for a baccalaureate degree in petroleum engineering, other engineering disciplines, geology, or physical sciences. Requests for financial aid should be made on the University admissions application. Departmental aid is not guaranteed, but applicants are reviewed based on departmental resources and individual qualifications.

Cost of Study Graduate tuition for residents of Texas is $80 per semester credit hour with a $120 minimum; for nonresidents the tuition rate is $293 per semester credit hour. Additional fees of $500 to $650 per semester are required for student services, engineering equipment and computer access, and laboratory use. Nonresident tuition and fees are waived for eligible students receiving fellowships and/or assistantships.

Living and Housing Costs A limited number of University-owned apartments, both furnished and unfurnished, are available for graduate students. Rents range from $150 to $380 per month, plus electricity. A large number of private apartment complexes with one- to three-bedroom apartments are available, with rents ranging from $290 to $895 per month. The estimated living cost (not including school expenses) is approximately $350 to $500 per month.

Student Group The Department of Petroleum Engineering at Texas A&M University is the largest in the United States, with about 340 students (110 graduate students and 230 undergraduate students) currently enrolled. Virtually all graduate students hold fellowships, research or teaching assistantships, or corporate sponsorships.

Location Texas A&M is located in Bryan-College Station, two adjacent cities with a combined population of approximately 120,000. The University is about 90 miles northwest of Houston, 90 miles east of Austin, and 170 miles south of Dallas.

The University and The Department Texas A&M University has the largest college of engineering in the United States; about 9,800 engineering students are among the 43,000 students enrolled in the University. University and department graduates are well recognized in industry and are heavily recruited. Many graduates of the Department of Petroleum Engineering hold high-level positions in the petroleum industry as well as other leading industries.

Applying For graduate admission in petroleum engineering, an applicant must hold a baccalaureate degree from a college or university of recognized standing and be able to demonstrate fundamental skills in petroleum engineering. Acceptable scores on the Graduate Record Examinations (GRE) and the Test of English as a Foreign Language (TOEFL) are required. The combined verbal and quantitative GRE score must be more than 1100. TOEFL scores must be at least 550 on the paper test or at least 213 on the computer-based exam.

Applicants for admission to the petroleum engineering graduate program at Texas A&M should contact the Director of Admissions at Texas A&M University and return the completed application to that office (not the department). An admissions application can also be downloaded from the University Web site (http://www.tamu.edu/admissions/).

Correspondence and Information Dr. Tom Blasingame, Graduate Advisor
Ms. Eleanor Schuler, Administrative Assistant
Department of Petroleum Engineering
Texas A&M University
College Station, Texas 77843-3116

Telephone: 409-845-8402
Fax: 409-845-1307
E-mail: pete.graduate.program@tamu.edu
World Wide Web: http://pumpjack.tamu.edu/

Texas A&M University

THE FACULTY AND THEIR RESEARCH

Maria A. Barrufet, Associate Professor; Ph.D., Texas A&M, 1987. Phase equilibria of reservoir fluids, experimental and theoretical characterization of hydrocarbons, separation processes and optimization of production techniques, multiphase flow in volatile and condensate wells.

Thomas A. Blasingame, Associate Professor; Ph.D., Texas A&M, 1989. Production and well test data analysis, reservoir engineering, reservoir description, technical mathematics.

Charles H. Bowman, Professor and Department Head; Ph.D., Texas A&M, 1961. Petroleum business and finance, economic analysis, reservoir management, fluid properties.

David B. Burnett, Associate Research Scientist; M.S., Sam Houston State, 1974. Well completion technology, drill-in fluids for well completions, horizontal well stimulations, zone isolation techniques, enhanced oil recovery processes.

John C. Calhoun Jr., Professor Emeritus and Member of the National Academy of Engineering; Ph.D., Penn State, 1946. Petroleum engineering education, characterization of porous systems, management of petroleum reservoirs.

Paul B. Crawford, Professor Emeritus; Ph.D., Texas, 1949. Enhanced oil recovery research.

Akhil Datta-Gupta, Assistant Professor; Ph.D., Texas at Austin, 1992. High-resolution numerical schemes for reservoir simulation, geostatistics and stochastic reservoir characterization, modeling and scale-up of enhanced oil recovery processes, environmental remediation.

Stephen A. Holditch, Professor and Member of the National Academy of Engineering; Ph.D., Texas A&M, 1976. Formation evaluation in low-permeability gas reservoirs, hydraulic fracture treatment design, evaluation and optimization, reservoir simulation, well completions and workovers, coalbed methane development, horizontal well engineering.

Jerry L. Jensen, Associate Professor; Ph.D., Texas at Austin, 1986. Reservoir characterization, petrophysics, formation evaluation, geological statistics.

Hans C. Juvkam-Wold, Professor; Sc.D., MIT, 1969. Tubular mechanics in horizontal drilling, well control, Arctic and offshore drilling (including riserless drilling), rig equipment.

W. John Lee, Professor and Member of the National Academy of Engineering; Ph.D., Georgia Tech, 1962. Pressure transient testing, natural gas engineering, reservoir engineering.

Daulat D. Mamora, Assistant Professor; Ph.D., Stanford, 1993. Waterflood, thermal oil recovery, gas recycling, zone isolation in horizontal wells, reservoir simulation, reservoir management.

William D. McCain Jr., Visiting Professor; Ph.D., Georgia Tech, 1964. Reservoir engineering and reservoir management, reservoir rock and fluid properties, reservoir simulation (volatile oil and retrograde gas systems).

Larry D. Piper, Senior Lecturer; Ph.D., Texas A&M, 1984. Reservoir engineering, reservoir simulation, phase behavior, stochastic models.

James E. Russell, Professor; Ph.D., Northwestern, 1966. Rock mechanics, numerical methods, computer applications, waste disposal.

Stuart L. Scott, Associate Professor; Ph.D., Tulsa, 1987. Multiphase flow in pipes, oil and gas production systems, well performance and completions, production and reservoir engineering.

John P. Spivey, Visiting Assistant Professor; Ph.D., Texas A&M, 1984. Pressure transient analysis, reservoir simulation studies, petroleum engineering software development.

Richard A. Startzman, Professor; Ph.D., Texas A&M, 1969. Reservoir engineering, economic evaluation, artificial intelligence, operations research.

Peter P. Valkó, Visiting Associate Professor; Ph.D., Institute of Catalysis (Russia), 1981. Analysis of well performance, design and evaluation of hydraulic fracturing treatments.

Robert A. Wattenbarger, Professor; Ph.D., Stanford, 1967. Reservoir simulation, gas engineering, well test analysis, thermal recovery.

Robert L. Whiting, Professor Emeritus; M.S., Texas, 1943. Petroleum economics, enhanced recovery, reservoir engineering.

Ching H. Wu, Professor; Ph.D., Pittsburgh, 1968. Reservoir engineering, improved oil recovery, reservoir and process simulation.

THE UNIVERSITY OF ARIZONA

College of Engineering and Mines
Department of Mining and Geological Engineering

Programs of Study	The department offers programs leading to the M.S. and Ph.D. degrees with a major in mining, geological, and geophysical engineering. Advanced work in mining engineering is directed toward research and professional development in mine planning and design, geomechanics, mine ventilation, equipment automation, project feasibility, advanced mining systems, and excavation engineering. Advanced work in geological and geophysical engineering is directed toward ground stabilization, probabilistic geotechnics, earthquake engineering, applied geophysics, high-resolution subsurface imaging, remote sensing, mineral exploration, and environmental studies. Students with undergraduate majors in other engineering fields or in the physical sciences are encouraged to apply.
	The M.S. in mining, geological, and geophysical engineering requires at least 24 units of course work, a thesis (for which 6 units of credit are normally given), and a final examination covering both the thesis and the course work.
	The Master of Engineering program is intended to meet the educational needs of the practicing engineer. This program offers the practicing engineer the opportunity to design, in conjunction with an advisory committee, a program of study that can reflect his or her needs. This is a terminal master's degree program and requires the completion of 30 units of course work, with a selection of courses offered through a variety of distance delivery methods and flexible formats.
	The Ph.D. requires the completion of at least 66 units of course work, which is normally met through the 24 units from the master's degree, 12 additional units in the major area, 12 units in the minor area, and 18 units of dissertation.
Research Facilities	The department maintains well-equipped laboratories for research in a wide variety of mining, geological, and geophysical engineering fields. The Geomechanics Laboratory is equipped with state-of-the-art sample preparation and computer-controlled testing equipment and operates in conjunction with the Rock Mass Modeling and Computational Rock Mechanics Laboratories. The Advanced Mining Systems Lab (AMSL) provides a wide variety of facilities for research in mine automation; robotics and intelligent mining systems, which includes robotic mining lab devices; and an array of computers and computerized heavy mining equipment. Ancillary laboratories support research in mine ventilation, mine health and safety, mine operations, water-jet technology, and computer simulations and analysis. The Laboratory for Advanced Subsurface Imaging (LASI) maintains a world-class facility for research in high-resolution geophysics as applied to environmental engineering and mining surveys. Computer labs are networked with UNIX and NT workstations for imaging systems, which provide extensive capabilities in computational intelligence and visualization labs.
	The department also operates open-access computer laboratories that are equipped with the latest UNIX- and Windows-based workstations and related software packages. Field facilities at the San Xavier Mining Laboratory (SXML) provide underground and surface mine testing sites, while the Avra Valley facility is a first-class geophysical field test site.
Financial Aid	Financial assistance is available through fellowships, scholarships, and teaching or research assistantships. A half-time stipend for the academic year is $15,714. Additional summer support is available for most research assistantships. Tuition is waived for students who hold a teaching or research assistantship.
Cost of Study	In 1998–99, the yearly registration fee for 7 or more units was $2160 for both residents and nonresidents. Nonresident tuition for 7 or more units cost $2556 to $4356 per semester. Graduate assistants are exempt from paying tuition.
Living and Housing Costs	Most students choose to live in apartments in the University area or elsewhere in Tucson. Living costs are lower than in many other metropolitan areas. Students should contact the Center for Off-Campus Students at 520-621-7597 for off-campus housing information.
Student Group	Graduate enrollment in the department averages 21 M.S. and 15 Ph.D. candidates. The University's growing excellence attracts some 600 job recruiters annually from the fields of business, industry, government, and education; they represent the country's most prestigious companies and institutions.
Location	Tucson was founded in 1776 as a walled presidio outpost of the Spanish crown, and its identity has been shaped by the cultural groups that have dwelled there—Mexicans, Southwestern Native Americans, cowboys, and pioneers. The city is 60 miles north of Mexico, and it rests at an elevation of 2,400 feet in a valley that is surrounded by mountains. Tucson has mild winters and year-round warmth and sunshine. Opportunities for recreation and sports are abundant all year.
The University	Since its founding in 1885 as a land-grant institution, the University of Arizona has grown to become one of the outstanding research institutions in the world, achieving the status of a Research I university. The University of Arizona consistently ranks among the top fifteen public institutions in the U.S. in external research support. Only a few institutions in the country offer more fields of study at the graduate level. Among the many distinguished members of the faculty are 14 members of the National Academy of Science and a Nobel Prize winner.
Applying	Although applications are considered throughout the year, it is recommended that they be submitted by February 1 for fall entrance and by August 1 for spring entrance. GRE General Test scores are required. The Application for Graduate Admission is provided by either the Graduate College or the department and must be submitted prior to receipt of transcripts and GRE or TOEFL scores; otherwise, processing will be delayed. Two sets of transcripts from undergraduate and graduate work, if applicable, must be sent to the Graduate College in a sealed envelope by the institution at which the course work was completed.
Correspondence and Information	Dr. Paul J. A. Lever, Department Head Department of Mining and Geological Engineering The University of Arizona 1235 East North Campus Drive, Room 229 P.O. Box 210012 Tucson, Arizona 85721-0012 Telephone: 520-621-6063 Fax: 520-621-8330 E-mail: dept@mge.arizona.edu World Wide Web: http://w3.arizona.edu/~mge/

The University of Arizona

THE FACULTY AND THEIR RESEARCH

Richard D. Call, Adjunct Professor of Geological Engineering; Ph.D., Arizona. Mine planning, slope stability, geomechanics, engineering geology.

Charles E. Glass, Associate Professor of Geological Engineering; Ph.D., Berkeley. Remote sensing, spatial data analysis, earthquake and geologic hazards engineering. Current research, funded by NASA, uses NASA orbital platforms as a basis for developing a hazard and prediction system to mitigate the death and destruction caused by debris flows and lahars. Other interests include earthquake hazard estimation, geophysical studies in support of archaeological and historical projects, and scientific visualization.

Satya Harpalani, Associate Professor of Mining Engineering; Ph.D., Berkeley. Coalbed methane, in situ mining, gas and fluid flow in rocks and coal, carbon dioxide sequestration, mine ventilation and cooling.

J. Brent Hiskey, Professor of Material Science and Engineering and Mining and Geological Engineering and Director of the Center for Copper Recovery and Utilization; Ph.D., Utah. Hydrometallurgy, electrochemistry of mineral systems, in situ resource recovery systems, copper, gold, silver.

John M. Kemeny, Associate Professor of Mining Engineering; Ph.D., Berkeley. Rock mechanics; fracture mechanics; fault and earthquake mechanics; borehole breakout; microcrack imaging; damage modeling; measurement of fragmentation and fracture properties using digital image processing, which includes software to measure fragment size distributions in mining and mineral processing, characterizing rock masses using digital image processing, and 3-D stereo imaging.

Pinnaduwa H. S. W. Kulatilake, Professor of Geological Engineering; Ph.D., Ohio State. Rock mass fracture characterization and network modeling; rock joint geomechanics; jointed rock mass mechanical and hydraulic properties modeling; statistical, probabilistic, and numerical modeling in geoengineering; rock slope stability.

John C. Lacy, Adjunct Professor of Mining Engineering; J.D., Arizona. Mineral and public land law, mineral and land policy.

Paul J. A. Lever, Associate Professor of Mining Engineering and Head of the Department; Ph.D., Colorado School of Mines. Mine automation, mining robotics, mine plant design, computer systems, excavation system productivity, advanced mining systems.

Hugh B. Miller, Assistant Professor of Mining Engineering; Ph.D., Colorado School of Mines. Mine design and unit operations, hydroexcavation, water-jet technology, project feasibility, mine investment analysis.

William C. Peters, Professor Emeritus of Geological Engineering; Ph.D., Colorado. Mineral exploration, geologic mapping, regional resource evaluation.

Mary M. Poulton, Associate Professor of Geological Engineering; Ph.D., Arizona. Neural networks, pattern recognition, geophysics, environmental investigations, GIS, remote sensing, mineral and petroleum exploration.

Gopalan Ramadorai, Adjunct Professor of Metallurgical Engineering; Ph.D., Utah. Mineral processing, extractive metallurgy, process environmental studies.

Ben K. Sternberg, Professor of Geophysical Engineering and Director of the Laboratory for Advanced Subsurface Imaging (LASI); Ph.D., Wisconsin–Madison. Development of new measurement methods; electrical and electromagnetics geophysics; interpretation of geophysical data for environmental engineering, mining, oil, and gas applications.

Kenneth L. Zonge, Adjunct Professor of Geophysical Engineering; Ph.D., Arizona. Electrical methods of geophysics exploration.

AREAS OF RESEARCH

Advanced Mining Systems: Automation and robotics technologies in mining equipment and systems. Intelligent computer-based systems for mine operations and management (using artificial intelligence and knowledge engineering concepts). Autonomous mine excavation systems; mine/machine productivity analysis and control; space mining; and sensors, data acquisition, and real-time signal analysis in mine environments.

Advanced Subsurface Imaging: Research on electrical and electromagnetics geophysics. Interpretation of geophysical data for environmental engineering, mining, oil, and gas applications.

Computational Intelligence and Visualization for Earth Systems: Neural networks, pattern recognition, geophysics, environmental investigations, GIS, remote sensing, mineral and petroleum exploration. Geophysical studies in support of archaeological and historical projects and scientific visualization.

Mine Feasibility and Design: Research into the methodology and procedures for calculating resource and reserve estimates, assessing mine and project feasibility, and optimizing mine and plant design, which incorporates topics associated with project evaluation, investment analysis, and due diligence.

Rock Mass Fracture Characterization and Modeling: Includes rock joint geomechanical and hydraulic properties modeling; jointed rock mass mechanical and hydraulic properties modeling; statistical, probabilistic, and numerical modeling in geoengineering; and rock slope stability. The mechanics of fractures in rocks, which includes rock mechanics, rock fracture mechanics, fault and earthquake mechanics, microcrack growth modeling and testing, and micromechanical damage modeling using FEM and Mathematica.

Water-Jet Technology/Hydroexcavation: Research into the interaction between high-energy fluid streams and geomaterials, including jet dynamics and fluid impingement as well as target failure. Focuses on the integration of water-jet technology into mining and industrial applications as well as hardware design and instrumentation.

The Advanced Mining Systems Laboratory uses a state-of-the-art wheel loader for computer-based autonomous excavation research.

The Laboratory for Advanced Subsurface Imaging includes a mobile lab for acquisition and analysis of geophysical data at field sites.

Section 14
Industrial Engineering

This section contains a directory of institutions offering graduate work in industrial engineering, followed by in-depth entries submitted by institutions that chose to prepare detailed program descriptions. Additional information about programs listed in the directory but not augmented by an in-depth entry may be obtained by writing directly to the dean of a graduate school or chair of a department at the address given in the directory.

For programs offering related work, see also in this book Computer Science and Information Technology, Electrical and Computer Engineering, Energy and Power Engineering, Engineering and Applied Sciences, and Management of Engineering and Technology. In Book 4, see Mathematical Sciences and in Book 6, Business Administration and Management.

CONTENTS

Industrial/Management Engineering

Arizona State University, Graduate College, College of Engineering and Applied Sciences, Department of Industrial Engineering, Tempe, AZ 85287. Offers MS, MSE, PhD, MSE/MIMOT. *Faculty:* 22 full-time (3 women). *Students:* 118 full-time (21 women), 60 part-time (20 women); includes 8 minority (3 Asian Americans or Pacific Islanders, 5 Hispanic Americans), 101 international. Average age 29. 223 applicants, 67% accepted. In 1998, 57 master's, 9 doctorates awarded. *Degree requirements:* For master's, computer language, thesis or alternative required, foreign language not required; for doctorate, computer language, dissertation required, foreign language not required. *Entrance requirements:* For master's, GRE General Test (recommended), minimum GPA of 3.0; for doctorate, GRE General Test (recommended), minimum GPA of 3.5. Application fee: $45. *Financial aid:* Fellowships available. *Faculty research:* Computer-aided manufacturing, human factors, productivity improvement, computer-integrated manufacturing, operations research. *Unit head:* Dr. Gary L. Hogg, Chairman, 480-965-3185, E-mail: imse@www.eas.asu.edu. *Application contact:* Dr. Dwayne Rollier, Associate Chair, 480-965-6874.

See in-depth description on page 1195.

Auburn University, Graduate School, College of Engineering, Department of Industrial and Systems Engineering, Auburn, Auburn University, AL 36849-0002. Offers MIE, MS, PhD. Part-time programs available. *Faculty:* 11 full-time (2 women). *Students:* 27 full-time (5 women), 15 part-time (1 woman); includes 1 minority (Hispanic American), 16 international. 48 applicants, 40% accepted. In 1998, 17 master's, 7 doctorates awarded. *Degree requirements:* For master's, design project (MIE), thesis (MS) required; for doctorate, dissertation required, foreign language not required. *Entrance requirements:* For master's, GRE General Test; for doctorate, GRE General Test (minimum score of 400 on each section required). *Application deadline:* For fall admission, 9/1; for spring admission, 3/1. Applications are processed on a rolling basis. Application fee: $25 ($50 for international students). Tuition, state resident: full-time $2,760; part-time $76 per credit hour. Tuition, nonresident: full-time $8,280; part-time $228 per credit hour. *Financial aid:* Fellowships, research assistantships, teaching assistantships, Federal Work-Study available. Aid available to part-time students. Financial aid application deadline: 3/15. *Unit head:* Dr. V. E. Unger, Head, 334-844-1401. *Application contact:* Dr. John F. Pritchett, Dean of the Graduate School, 334-844-4700.

Bradley University, Graduate School, College of Engineering and Technology, Department of Industrial and Manufacturing Engineering and Technology, Peoria, IL 61625-0002. Offers MSIE, MSMFE. Part-time and evening/weekend programs available. *Degree requirements:* For master's, computer language, comprehensive exam, project required, foreign language and thesis not required. *Entrance requirements:* For master's, TOEFL (minimum score of 525 required), minimum GPA of 3.0.

California Polytechnic State University, San Luis Obispo, College of Engineering, Program in Engineering, San Luis Obispo, CA 93407. Offers biochemical engineering (MS); industrial engineering (MS); integrated technology management (MS); materials engineering (MS); mechanical engineering (MS); water engineering (MS). *Faculty:* 98 full-time (8 women), 82 part-time (14 women). *Students:* 25 full-time (3 women), 6 part-time. *Degree requirements:* Foreign language not required. *Entrance requirements:* For master's, GRE General Test, minimum GPA of 2.5 in last 90 quarter units. Application fee: $55. Tuition, nonresident: part-time $164 per unit. Required fees: $531 per quarter. *Unit head:* Dr. Paul E. Rainey, Associate Dean, 805-756-2131, Fax: 805-756-6503, E-mail: prainey@calpoly.edu. *Application contact:* Dr. Paul E. Rainey, Associate Dean, 805-756-2131, Fax: 805-756-6503, E-mail: prainey@calpoly.edu.

California State University, Fresno, Division of Graduate Studies, School of Agricultural Sciences and Technology, Department of Industrial Technology, Fresno, CA 93740-0057. Offers MS. Part-time and evening/weekend programs available. *Faculty:* 11 full-time (0 women). *Students:* 7 full-time (1 woman), 10 part-time (1 woman); includes 5 minority (2 Asian Americans or Pacific Islanders, 3 Hispanic Americans), 4 international. Average age 31. 8 applicants, 100% accepted. In 1998, 3 degrees awarded. *Degree requirements:* For master's, thesis or alternative required, foreign language not required. *Entrance requirements:* For master's, GRE General Test, TOEFL (minimum score of 550 required), minimum GPA of 2.5. *Average time to degree:* Master's–3.5 years full-time. *Application deadline:* For fall admission, 6/1 (priority date); for spring admission, 11/1. Applications are processed on a rolling basis. Application fee: $55. Electronic applications accepted. Tuition, nonresident: part-time $246 per unit. Required fees: $1,906; $620 per semester. *Financial aid:* In 1998–99, 6 students received aid, including 1 fellowship; career-related internships or fieldwork, Federal Work-Study, and scholarships also available. Financial aid application deadline: 3/1; financial aid applicants required to submit FAFSA. *Faculty research:* Fuels/pollution, energy, computer integrative manufacturing. *Unit head:* Dr. Tony Au, Chair, 559-278-2145, Fax: 559-278-5081, E-mail: tony_au@csufresno.edu. *Application contact:* Dr. Matthew Yen, Graduate Coordinator, 559-278-2145, Fax: 559-278-5081, E-mail: matthew_yen@csufresno.edu.

California State University, Northridge, Graduate Studies, College of Engineering and Computer Science, Department of Civil, Industrial and Applied Mechanics, Northridge, CA 91330. Offers applied mechanics (MSE); civil engineering (MS); engineering (MS); engineering management (MS); industrial engineering (MS); materials engineering (MS); mechanical engineering (MS), including aerospace engineering, applied engineering, machine design, mechanical engineering, structural engineering, thermofluids; mechanics (MS). Part-time and evening/weekend programs available. *Faculty:* 13 full-time, 2 part-time. *Students:* 10 full-time (2 women), 101 part-time (15 women); includes 38 minority (3 African Americans, 22 Asian Americans or Pacific Islanders, 11 Hispanic Americans, 2 Native Americans), 8 international. Average age 32. 58 applicants, 57% accepted. In 1998, 34 degrees awarded. *Degree requirements:* For master's, thesis required, foreign language not required. *Entrance requirements:* For master's, GRE General Test, TOEFL, minimum GPA of 2.5. *Application deadline:* For fall admission, 11/30. Application fee: $55. Tuition, nonresident: part-time $246 per unit. International tuition: $7,874 full-time. Required fees: $1,970. Tuition and fees vary according to course load. *Financial aid:* Teaching assistantships available. Financial aid application deadline: 3/1. *Faculty research:* Composite study. *Unit head:* Dr. Stephen Gadomski, Chair, 818-677-2166. *Application contact:* Dr. Ileana Costa, Graduate Coordinator, 818-677-3299.

Central Missouri State University, School of Graduate Studies, College of Applied Sciences and Technology, Department of Manufacturing and Construction, Warrensburg, MO 64093. Offers industrial management (MS); industrial technology (MS); technology (PhD), including technology management. Part-time programs available. *Faculty:* 8 full-time (0 women). *Students:* 16 full-time (3 women), 27 part-time (5 women); includes 3 minority (1 African American, 2 Asian Americans or Pacific Islanders), 14 international. In 1998, 13 degrees awarded. *Degree requirements:* For master's, comprehensive exam required, thesis not required. *Entrance requirements:* For master's, minimum GPA of 2.5; previous course work in mathematics, science, and technology. *Application deadline:* Applications are processed on a rolling basis. Application fee: $25 ($50 for international students). Tuition, state resident: full-time $3,576; part-time $149 per credit hour. Tuition, nonresident: full-time $7,152; part-time $298 per credit hour. *Financial aid:* In 1998–99, research assistantships with tuition reimbursements (averaging $3,750 per year), 5 teaching assistantships with tuition reimbursements (averaging $3,750 per year) were awarded.; fellowships with tuition reimbursements, Federal Work-Study, grants, scholarships, unspecified assistantships, and administrative and laboratory assistantships also available. Aid available to part-time students. Financial aid application deadline: 3/1; financial aid applicants required to submit FAFSA. *Unit head:* Dr. John Sutton, Chair, 660-543-4439, Fax: 660-543-4578, E-mail: sutton@cmsu1.cmsu.edu.

Clemson University, Graduate School, College of Engineering and Science, School of Mechanical and Industrial Engineering, Department of Industrial Engineering, Clemson, SC 29634. Offers MS, PhD. Part-time programs available. *Students:* 45 full-time (11 women), 4 part-time (1 woman); includes 2 minority (both African Americans), 34 international. Average age 25. 83 applicants, 76% accepted. In 1998, 18 master's, 1 doctorate awarded. Terminal master's awarded for partial completion of doctoral program. *Degree requirements:* For master's, computer language, thesis or alternative required, foreign language not required; for doctorate, computer language, dissertation required, foreign language not required. *Entrance requirements:* For master's and doctorate, GRE General Test, TOEFL (minimum score of 550 required). *Application deadline:* For fall admission, 6/1. Applications are processed on a rolling basis. Application fee: $35. *Financial aid:* Fellowships, research assistantships, teaching assistantships, career-related internships or fieldwork available. Financial aid applicants required to submit FAFSA. *Faculty research:* Computer-integrated manufacturing, human-computer interaction, ergonomics, quality engineering. Total annual research expenditures: $292,665. *Unit head:* Dr. Delbert L. Kimbler, Chair, 864-656-4716, Fax: 864-656-0795, E-mail: del.kimbler@ces.clemson.edu. *Application contact:* Dr. Anand Gramopadhye, Coordinator, 864-656-5540, Fax: 864-656-0795, E-mail: agramop@ces.clemson.edu.

Cleveland State University, College of Graduate Studies, Fenn College of Engineering, Department of Industrial Engineering, Cleveland, OH 44115-2440. Offers MS, D Eng. Part-time programs available. *Faculty:* 5 full-time (0 women). *Students:* 4 full-time (0 women), 48 part-time (7 women); includes 4 minority (1 African American, 2 Hispanic Americans, 1 Native American), 16 international. Average age 33. 64 applicants, 69% accepted. In 1998, 33 master's, 2 doctorates awarded. Terminal master's awarded for partial completion of doctoral program. *Degree requirements:* For master's, computer language, project or thesis required; for doctorate, computer language, dissertation, candidacy and qualifying exams required, foreign language not required. *Entrance requirements:* For master's, GRE General Test, GRE Subject Test, TOEFL, minimum GPA of 2.75; for doctorate, GRE General Test, GRE Subject Test, TOEFL, minimum GPA of 3.25. *Application deadline:* For fall admission, 7/15 (priority date). Applications are processed on a rolling basis. Application fee: $25. *Financial aid:* In 1998–99, 4 research assistantships, 1 teaching assistantship were awarded.; fellowships, career-related internships or fieldwork, institutionally-sponsored loans, and unspecified assistantships also available. Aid available to part-time students. *Faculty research:* Modeling of manufacturing systems, statistical process control, computerized production planning and facilities design, cellular manufacturing, artificial intelligence and sensors. *Unit head:* Dr. Andrew Ying-Hsin-Liou, Chairperson, 216-687-4667, Fax: 216-687-9330.

Colorado State University, Graduate School, College of Engineering, Department of Mechanical Engineering, Program in Industrial and Manufacturing Systems Engineering, Fort Collins, CO 80523-0015. Offers MS, PhD. *Faculty:* 18 full-time (0 women). *Degree requirements:* For doctorate, dissertation required, foreign language not required, foreign language not required. *Entrance requirements:* For master's and doctorate, GRE General Test (minimum combined score of 1850 on three sections required; average 1872), TOEFL (minimum score of 550 required; average 596), minimum GPA of 3.0. *Application deadline:* For fall admission, 2/1 (priority date). Applications are processed on a rolling basis. Application fee: $30. Electronic applications accepted. *Faculty research:* Robotic assembly, dynamics of robots, controls, manufacturability of mechanical devices and systems. *Unit head:* William Duff, Professor, 970-491-5859, Fax: 970-491-3827, E-mail: duff@engr.colostate.edu. *Application contact:* William Duff, Professor, 970-491-5859, Fax: 970-491-3827, E-mail: duff@engr.colostate.edu.

Columbia University, Fu Foundation School of Engineering and Applied Science, Department of Industrial Engineering and Operations Research, New York, NY 10027. Offers financial engineering (MS); industrial engineering (MS, Eng Sc D, PhD, Engr); operations research (MS, Eng Sc D, PhD). Part-time programs available. Postbaccalaureate distance learning degree programs offered (no on-campus study). *Faculty:* 11 full-time (1 woman), 10 part-time (0 women). *Students:* 68 full-time (19 women), 57 part-time (19 women). Average age 27. 332 applicants, 59% accepted. In 1998, 55 master's, 4 doctorates awarded. *Degree requirements:* For master's, foreign language and thesis not required; for doctorate, dissertation, oral and written qualifying exams required, foreign language not required. *Entrance requirements:* For master's, doctorate, and Engr, GRE General Test, TOEFL. *Average time to degree:* Master's–1 year full-time, 2 years part-time; doctorate–5 years full-time. *Application deadline:* For fall admission, 1/1; for spring admission, 10/1. Application fee: $55. *Financial aid:* In 1998–99, 20 students received aid, including 4 fellowships (averaging $5,000 per year), 7 research assistantships (averaging $15,600 per year), 9 teaching assistantships (averaging $15,600 per year); Federal Work-Study also available. Financial aid application deadline: 1/5; financial aid applicants required to submit FAFSA. *Faculty research:* Discrete event stochastic systems, optimization, productions planning and scheduling, inventory control, yield management, simulation, mathematical programming, combinatorial optimization, queuing. Total annual research expenditures: $1 million. *Unit head:* Dr. Donald Goldfarb, Chairman, 212-854-8011, Fax: 212-854-8103, E-mail: gold@ieor.columbia.edu. *Application contact:* Tamara Kachanov, Departmental Administrator, 212-854-1473, Fax: 212-854-8103, E-mail: tamara@ieor.columbia.edu.

See in-depth description on page 1199.

Cornell University, Graduate School, Graduate Fields of Engineering, Field of Operations Research and Industrial Engineering, Ithaca, NY 14853-0001. Offers applied probability and statistics (PhD); manufacturing systems engineering (PhD); mathematical programming (PhD); operations research and industrial engineering (M Eng). *Faculty:* 30 full-time. *Students:* 91 full-time (20 women); includes 22 minority (1 African American, 21 Asian Americans or Pacific Islanders), 36 international. 297 applicants, 52% accepted. In 1998, 71 master's, 7 doctorates awarded. Terminal master's awarded for partial completion of doctoral program. *Degree requirements:* For doctorate, dissertation required, foreign language not required, foreign language not required. *Entrance requirements:* For master's and doctorate, GRE General Test, TOEFL (minimum score of 550 required). *Application deadline:* For fall admission, 1/15. Application fee: $65. Electronic applications accepted. *Financial aid:* In 1998–99, 34 students received aid, including 9 fellowships with full tuition reimbursements available, 6 research assistantships with full tuition reimbursements available, 19 teaching assistantships with full tuition reimbursements available; institutionally-sponsored loans, scholarships, tuition waivers (full and partial), and unspecified assistantships also available. Financial aid applicants required to submit FAFSA. *Faculty research:* Combinatorial optimization, mathematical finance, stochastic processes, simulation. *Unit head:* Director of Graduate Studies, 607-255-9128. *Application contact:* Graduate Field Assistant, 607-255-9128, E-mail: orphd@cornell.edu.

Dalhousie University, Faculty of Graduate Studies, DalTech, Faculty of Engineering, Department of Industrial Engineering, Halifax, NS B3H 3J5, Canada. Offers M Eng, MA Sc, PhD. *Faculty:* 8 full-time (0 women), 1 part-time (0 women). *Students:* 13 full-time (5 women), 7 international. 20 applicants, 60% accepted. In 1998, 1 master's, 1 doctorate awarded (100% entered university research/teaching). *Degree requirements:* For master's and doctorate, thesis/dissertation required, foreign language not required. *Entrance requirements:* For master's and doctorate, TOEFL (minimum score of 580 required). *Application deadline:* For fall admission, 6/1; for winter admission, 10/1; for spring admission, 2/1. Applications are processed on a rolling basis. Application fee: $55. *Financial aid:* Fellowships, research assistantships, teaching assistantships, scholarships and unspecified assistantships available. *Faculty research:* Industrial ergonomics, operations research, production manufacturing systems, scheduling stochastic models. *Unit head:* Dr. E. Gunn, Head, 902-494-3281, Fax: 902-420-7858, E-mail: industrial.engineering@dal.ca. *Application contact:* Shelley Parker, Admissions Coordinator, Graduate Studies and Research, 902-494-1288, Fax: 902-494-3149, E-mail: shelley.parker@dal.ca.

Eastern Michigan University, Graduate School, College of Technology, Department of Industrial Technology, Program in Industrial Technology, Ypsilanti, MI 48197. Offers MS. *Faculty:* 18 full-time (1 woman). *Students:* 12 full-time, 134 part-time. 84 applicants, 68% accepted. In

1998, 25 degrees awarded. *Degree requirements:* For master's, thesis optional, foreign language not required. *Entrance requirements:* For master's, TOEFL (minimum score of 550 required). *Application deadline:* For fall admission, 5/15; for spring admission, 3/15. Applications are processed on a rolling basis. Application fee: $30. *Financial aid:* Available to part-time students. Application deadline: 3/15; *Unit head:* Dr. John Weeks, Coordinator, 734-487-2040.

École Polytechnique de Montréal, Graduate Programs, Department of Industrial Engineering, Montréal, PQ H3C 3A7, Canada. Offers ergonomy (M Eng, M Sc A, DESS); production (M Eng, M Sc A); technology management (M Eng, M Sc A). DESS offered jointly with Université de Montréal and École des Hautes Études Commerciales. Part-time programs available. *Degree requirements:* For master's, one foreign language, computer language, thesis required. *Entrance requirements:* For master's, minimum GPA of 2.75. *Faculty research:* Use of computers in organizations.

Florida Agricultural and Mechanical University, Division of Graduate Studies, Research, and Continuing Education, FAMU-FSU College of Engineering, Department of Industrial Engineering, Tallahassee, FL 32307-3200. Offers MS. *Students:* 12 (5 women); includes 10 minority (all African Americans) 2 international. In 1998, 3 degrees awarded. *Entrance requirements:* For master's, GRE General Test (minimum combined score of 1000 required), minimum GPA of 3.0. *Application deadline:* For fall admission, 7/1. Application fee: $20. *Unit head:* Dr. C. J. Chen, Dean, FAMU-FSU College of Engineering, 850-487-6100, Fax: 850-487-6486.

Florida International University, College of Engineering, Department of Industrial Engineering, Miami, FL 33199. Offers MS. Part-time and evening/weekend programs available. *Faculty:* 7 full-time (1 woman). *Students:* 5 full-time (4 women), 20 part-time (9 women); includes 12 minority (1 Asian American or Pacific Islander, 11 Hispanic Americans), 12 international. Average age 29. 50 applicants, 56% accepted. In 1998, 2 degrees awarded. *Degree requirements:* For master's, thesis not required. *Entrance requirements:* For master's, GRE General Test (minimum combined score of 1000 required), TOEFL (minimum score of 550 required). *Application deadline:* For fall admission, 4/1 (priority date); for spring admission, 10/1. Applications are processed on a rolling basis. Application fee: $20. Tuition, state resident: part-time $145 per credit hour. Tuition, nonresident: part-time $506 per credit hour. Required fees: $158; $158 per year. *Unit head:* Dr. Shih-Ming Lee, Chairperson, 305-348-2256, Fax: 305-348-3721, E-mail: leet@eng.fiu.edu.

Florida State University, Graduate Studies, FAMU/FSU College of Engineering, Department of Industrial Engineering, Tallahassee, FL 32306. Offers MS, PhD. *Faculty:* 10 full-time (0 women), 4 part-time (1 woman). *Students:* 32 full-time (10 women), 3 part-time; includes 16 minority (13 African Americans, 3 Hispanic Americans), 14 international. Average age 24. 18 applicants, 39% accepted. In 1998, 4 degrees awarded (50% found work related to degree, 50% continued full-time study). *Degree requirements:* For master's, computer language, thesis required, foreign language not required; for doctorate, dissertation, preliminary exam, qualifying exam required, foreign language not required. *Entrance requirements:* For master's, GRE General Test (minimum combined score of 1000 required; average 1150), minimum GPA of 3.0; for doctorate, GRE General Test (minimum combined score of 1100 required, average 1250), minimum GPA of 3.0 (without MS in industrial engineering), 3.4 (with MS in industrial engineering). *Average time to degree:* Master's–1.5 years full-time, 2.5 years part-time. *Application deadline:* For fall admission, 7/15; for spring admission, 11/23. Applications are processed on a rolling basis. Application fee: $20. Tuition, state resident: part-time $139 per credit hour. Tuition, nonresident: part-time $482 per credit hour. Tuition and fees vary according to program. *Financial aid:* In 1998–99, research assistantships with full tuition reimbursements (averaging $13,200 per year), teaching assistantships with full tuition reimbursements (averaging $13,200 per year) were awarded. Financial aid application deadline: 6/15. *Faculty research:* Precision manufacturing, composite manufacturing, green manufacturing, applied optimization, simulation. Total annual research expenditures: $3.2 million. *Unit head:* Dr. Ben Wang, Chair, 850-410-6339, Fax: 850-410-6342, E-mail: indwangl@eng.fsu.edu. *Application contact:* Stephanie Dickey, Office Assistant, 850-410-6345, Fax: 850-410-6342, E-mail: dickey@eng.fsu.edu.

Georgia Institute of Technology, Graduate Studies and Research, College of Engineering, School of Industrial and Systems Engineering, Program in Industrial and Systems Engineering, Atlanta, GA 30332-0001. Offers algorithms, combinatorics, and optimization (PhD); industrial and systems engineering (PhD); industrial engineering (MS, MSIE); statistics (MS Stat). Part-time programs available. Terminal master's awarded for partial completion of doctoral program. *Degree requirements:* For master's, computer language required, thesis optional, foreign language not required; for doctorate, computer language, dissertation required, foreign language not required. *Entrance requirements:* For master's and doctorate, GRE General Test, TOEFL (minimum score of 550 required), minimum GPA of 3.0. Electronic applications accepted. *Faculty research:* Computer-integrated manufacturing systems, materials handling systems, production and distribution.

See in-depth description on page 1201.

Illinois State University, Graduate School, College of Applied Science and Technology, Department of Industrial Technology, Normal, IL 61790-2200. Offers MS. *Faculty:* 10 full-time (2 women), 1 part-time (0 women). *Students:* 18 full-time (3 women), 22 part-time (5 women); includes 6 minority (5 African Americans, 1 Asian American or Pacific Islander), 5 international. 16 applicants, 94% accepted. In 1998, 15 degrees awarded. *Degree requirements:* For master's, thesis or alternative required. *Entrance requirements:* For master's, GRE General Test, minimum GPA of 2.8. *Application deadline:* Applications are processed on a rolling basis. Application fee: $0. Tuition, state resident: full-time $2,526; part-time $105 per credit hour. Tuition, nonresident: full-time $7,578; part-time $316 per credit hour. Required fees: $1,082; $38 per credit hour. Tuition and fees vary according to course load and program. *Financial aid:* In 1998–99, 10 teaching assistantships were awarded.; research assistantships, tuition waivers (full) and unspecified assistantships also available. Financial aid application deadline: 4/1. *Faculty research:* Technology preparation and curriculum integration; technical assistance for education-to-careers; synthesizing technological and business management skills for curriclum development and faculty enhancement; training design and development for Caterpillar, Inc.. Total annual research expenditures: $435,449. *Unit head:* Dr. Rodney Custer, Chairperson, 309-438-3661.

Indiana State University, School of Graduate Studies, School of Technology, Program in Industrial Technology, Terre Haute, IN 47809-1401. Offers MA, MS. *Faculty:* 12 full-time (0 women). *Students:* 6 full-time (1 woman), 7 part-time (1 woman); includes 1 minority (African American), 4 international. Average age 33. 14 applicants, 57% accepted. In 1998, 5 degrees awarded. *Degree requirements:* For master's, thesis optional, foreign language not required. *Entrance requirements:* For master's, TOEFL (minimum score of 550 required), bachelor's degree in industrial technology or related field, minimum undergraduate GPA of 2.5. *Average time to degree:* Master's–2 years full-time, 5 years part-time. *Application deadline:* For fall admission, 7/1 (priority date); for spring admission, 11/1 (priority date). Applications are processed on a rolling basis. Application fee: $20. Electronic applications accepted. *Financial aid:* In 1998–99, 3 research assistantships with partial tuition reimbursements were awarded.; fellowships, teaching assistantships Financial aid application deadline: 3/1; financial aid applicants required to submit FAFSA. *Unit head:* Dr. Joseph Freeze, Adviser, 812-237-3461.

Instituto Tecnológico y de Estudios Superiores de Monterrey, Campus Chihuahua, Graduate Programs, Chihuahua, 31110, Mexico. Offers computer systems engineering (Ingeniero); electrical engineering (Ingeniero); electromechanical engineering (Ingeniero); electronic engineering (Ingeniero); engineering administration (MEA); industrial engineering (MIE, Ingeniero); international trade (MIT); mechanical engineering (Ingeniero).

Instituto Tecnológico y de Estudios Superiores de Monterrey, Campus Estado de México, Graduate Division, Division of Engineering and Architecture, Atizapán de Zaragoza, 52500, Mexico. Offers computer science (MCS); environmental engineering (MEE); industrial engineering (MIE); manufacturing systems (MMS); materials engineering (PhD). *Degree requirements:* For master's; for doctorate, one foreign language, dissertation required. *Entrance requirements:* For master's, interview; for doctorate, research proposal. *Application deadline:* For fall admission, 1/13 (priority date); for spring admission, 4/4. Applications are processed on a rolling basis. Application fee: 750 Mexican pesos. *Unit head:* Juan López Díaz, Headmaster, 5-326-5530, Fax: 5-326-5531, E-mail: jlopez@campus.cem.itesm.mx. *Application contact:* Lourdes Turrubiates, Admissions Officer, 5-326-5776, Fax: 5-326-5788, E-mail: lturrubi@campus.cem.itesm.mx.

Instituto Tecnológico y de Estudios Superiores de Monterrey, Campus Laguna, Graduate School, Torreón, 27250, Mexico. Offers business administration (MBA); industrial engineering (MIE); management information systems (MS). Part-time programs available. *Degree requirements:* For master's, computer language required, foreign language and thesis not required. *Entrance requirements:* For master's, GMAT (minimum score of 450 required). *Faculty research:* Computer communications from home to the University.

Instituto Tecnológico y de Estudios Superiores de Monterrey, Campus Monterrey, Graduate and Research Division, Programs in Engineering, Monterrey, 64849, Mexico. Offers applied statistics (M Eng); artificial intelligence (PhD); automation engineering (M Eng); chemical engineering (M Eng); civil engineering (M Eng); electrical engineering (M Eng); electronic engineering (M Eng); environmental engineering (M Eng); industrial engineering (M Eng, PhD); manufacturing engineering (M Eng); mechanical engineering (M Eng); systems and quality engineering (M Eng). M Eng offered jointly with the University of Waterloo; PhD (industrial engineering) offered jointly with Texas A&M University. Part-time and evening/weekend programs available. Terminal master's awarded for partial completion of doctoral program. *Degree requirements:* For master's and doctorate, one foreign language, computer language, thesis/dissertation required. *Entrance requirements:* For master's, PAEG, TOEFL; for doctorate, GRE, TOEFL, master's in related field. *Faculty research:* Flexible manufacturing cells, materials, statistical methods, environmental prevention, control and evaluation.

Iowa State University of Science and Technology, Graduate College, College of Engineering, Department of Industrial and Manufacturing Systems Engineering, Ames, IA 50011. Offers industrial engineering (MS, PhD); operations research (MS). *Faculty:* 16 full-time, 2 part-time. *Students:* 44 full-time (4 women), 11 part-time (3 women); includes 1 minority (African American), 48 international. 118 applicants, 58% accepted. In 1998, 24 master's, 2 doctorates awarded. *Degree requirements:* For master's, computer language, thesis required (for some programs), foreign language not required; for doctorate, computer language, dissertation required, foreign language not required. *Entrance requirements:* For master's and doctorate, GRE General Test, TOEFL (minimum score of 550 required). *Application deadline:* For fall admission, 6/15 (priority date); for spring admission, 11/15 (priority date). Application fee: $20 ($50 for international students). Electronic applications accepted. Tuition, state resident: full-time $3,308. Tuition, nonresident: full-time $9,744. Part-time tuition and fees vary according to course load, campus/location and program. *Financial aid:* In 1998–99, 14 research assistantships with partial tuition reimbursements (averaging $8,775 per year), 13 teaching assistantships with partial tuition reimbursements (averaging $10,208 per year) were awarded.; fellowships, scholarships also available. *Faculty research:* Economic modeling, valuation techniques, robotics, digital controls, systems reliability. *Unit head:* Dr. Pius J. Egbelu, Chair, 515-294-1682, Fax: 515-294-3524. *Application contact:* Dr. John Jackman, Director of Graduate Studies, 515-294-0126.

Kansas State University, Graduate School, College of Engineering, Department of Industrial and Manufacturing Systems Engineering, Manhattan, KS 66506. Offers engineering management (MEM); industrial and manufacturing systems engineering (PhD); industrial engineering (MS); operations research (MS). Part-time programs available. Postbaccalaureate distance learning degree programs offered. *Faculty:* 10 full-time (1 woman). *Students:* 24 full-time (4 women), 37 part-time (6 women); includes 1 minority (Asian American or Pacific Islander), 29 international. 60 applicants, 57% accepted. In 1998, 19 master's awarded (5% entered university research/teaching, 95% found other work related to degree); 4 doctorates awarded (25% entered university research/teaching, 75% found other work related to degree). *Degree requirements:* For master's, computer language required, foreign language not required; for doctorate, computer language, dissertation required, foreign language not required. *Entrance requirements:* For master's and doctorate, GRE General Test (minimum score of 725 on quantitative section required). *Average time to degree:* Master's–1.75 years full-time, 2.19 years part-time; doctorate–3.6 years full-time. *Application deadline:* For fall admission, 3/1 (priority date); for spring admission, 9/1. Applications are processed on a rolling basis. Application fee: $0 ($25 for international students). Electronic applications accepted. *Financial aid:* In 1998–99, 15 research assistantships (averaging $5,850 per year), 1 teaching assistantship (averaging $8,460 per year) were awarded. *Faculty research:* Safety, quality operations research, tribology, economic analysis, expert systems. Total annual research expenditures: $346,000. *Unit head:* Dr. Bradley Kramer, Head, 785-532-5606.

Lamar University, College of Graduate Studies, College of Engineering, Department of Industrial Engineering, Program in Industrial Engineering, Beaumont, TX 77710. Offers ME, MES, DE. *Faculty:* 4 full-time (0 women). *Students:* 18 full-time (3 women), 2 part-time; includes 2 minority (1 African American, 1 Asian American or Pacific Islander), 17 international. 41 applicants, 95% accepted. In 1998, 5 degrees awarded. *Degree requirements:* For master's, computer language, thesis required (for some programs), foreign language not required; for doctorate, computer language, dissertation required, foreign language not required. *Entrance requirements:* For master's, GRE General Test (minimum combined score of 950 required), TOEFL (minimum score of 500 required); for doctorate, GRE General Test, TOEFL (minimum score of 530 required). *Average time to degree:* Master's–1.75 years full-time, 3 years part-time. *Application deadline:* For fall admission, 5/15 (priority date); for spring admission, 10/1 (priority date). Applications are processed on a rolling basis. Application fee: $0. *Financial aid:* In 1998–99, 2 fellowships, 1 research assistantship, 2 teaching assistantships were awarded. Financial aid application deadline: 4/1. *Faculty research:* Total quality management, CAD/CAM/CIM, scheduling. *Application contact:* Dr. Hsing-Wei Chu, Professor, 409-880-8804, Fax: 409-880-8121.

Lehigh University, College of Engineering and Applied Science, Department of Industrial and Manufacturing Systems Engineering, Program in Industrial Engineering, Bethlehem, PA 18015-3094. Offers M Eng, MS, PhD. Part-time programs available. Postbaccalaureate distance learning degree programs offered (no on-campus study). *Degree requirements:* For master's and doctorate, computer language, thesis/dissertation required, foreign language not required. *Entrance requirements:* For master's and doctorate, GRE General Test (score in 75th percentile or higher required), TOEFL (minimum score of 550 required), minimum GPA of 2.75.

Louisiana State University and Agricultural and Mechanical College, Graduate School, College of Engineering, Department of Industrial and Manufacturing Systems Engineering, Baton Rouge, LA 70803. Offers engineering science (PhD); industrial engineering (MSIE). *Faculty:* 10 full-time (0 women). *Students:* 27 full-time (5 women), 4 part-time (1 woman); includes 1 minority (African American), 27 international. Average age 26. 113 applicants, 58% accepted. In 1998, 9 master's awarded. Terminal master's awarded for partial completion of doctoral program. *Degree requirements:* For master's and doctorate, thesis/dissertation required, foreign language not required. *Entrance requirements:* For master's and doctorate, GRE General Test, minimum GPA of 3.0. *Application deadline:* For fall admission, 1/25 (priority date). Applications are processed on a rolling basis. Application fee: $25. *Financial aid:* In 1998–99, 20 research assistantships with partial tuition reimbursements, 1 teaching assistantship with partial tuition reimbursement were awarded.; fellowships, institutionally-sponsored loans and unspecified assistantships also available. Financial aid application deadline: 5/1. *Faculty research:* Ergonomics/human factors engineering, industrial hygiene, production systems, operations research, manufacturing engineering and maintenance. Total annual research expenditures: $400,000. *Unit head:* Dr. Dennis B. Webster, Chair, 225-388-5112, Fax: 225-

Industrial/Management Engineering

Louisiana State University and Agricultural and Mechanical College (continued)

388-5990, E-mail: dbw@imse.lsu.edu. *Application contact:* Dr. F. Aghazadeh, Graduate Adviser, 225-388-5112, E-mail: agha@imse.lsu.edu.

Louisiana Tech University, Graduate School, College of Engineering and Science, Department of Mechanical and Industrial Engineering, Ruston, LA 71272. Offers industrial engineering (MS, D Eng); manufacturing systems engineering (MS); mechanical engineering (MS, D Eng); operations research (MS). Part-time programs available. Terminal master's awarded for partial completion of doctoral program. *Degree requirements:* For master's and doctorate, thesis/dissertation required, foreign language not required. *Entrance requirements:* For master's, GRE General Test (minimum combined score of 1070 required), TOEFL (minimum score of 550 required), minimum GPA of 3.0 in last 60 hours; for doctorate, TOEFL (minimum score of 550 required), minimum graduate GPA of 3.25 (with MS) or GRE General Test (minimum combined score of 1270 required without MS). *Faculty research:* Engineering management, facilities planning, thermodynamics, automated manufacturing, micromanufacturing.

Loyola Marymount University, Graduate Division, College of Science and Engineering, Department of Mechanical Engineering, Program in Engineering and Production Management, Los Angeles, CA 90045-8350. Offers MS. Part-time and evening/weekend programs available. *Faculty:* 2 part-time (0 women). *Students:* 9 full-time (2 women), 28 part-time (1 woman); includes 15 minority (2 African Americans, 10 Asian Americans or Pacific Islanders, 2 Hispanic Americans, 1 Native American) 27 applicants, 70% accepted. In 1998, 1 degree awarded. *Degree requirements:* For master's, thesis or alternative, project required, foreign language not required. *Entrance requirements:* For master's, GMAT or GRE General Test, TOEFL (minimum score of 550 required). Application fee: $35. Electronic applications accepted. Tuition: Part-time $650 per unit. *Financial aid:* In 1998–99, 6 students received aid. Federal Work-Study, grants, and laboratory assistantships available. Aid available to part-time students. Financial aid application deadline: 3/2; financial aid applicants required to submit FAFSA. *Unit head:* Dr. Mel I. Mendelson, Director, 310-338-6020.

Mississippi State University, College of Engineering, Department of Industrial Engineering, Mississippi State, MS 39762. Offers MS. Part-time programs available. Postbaccalaureate distance learning degree programs offered (no on-campus study). *Students:* 16 full-time (8 women), 5 part-time (1 woman); includes 8 minority (4 African Americans, 4 Asian Americans or Pacific Islanders), 2 international. Average age 27. 27 applicants, 19% accepted. In 1998, 5 degrees awarded. *Degree requirements:* For master's, computer language, thesis (for some programs), comprehensive oral or written exam required, foreign language not required. *Entrance requirements:* For master's, GRE General Test, TOEFL (minimum score of 550 required), minimum GPA of 2.75. *Application deadline:* For fall admission, 7/1; for spring admission, 11/1. Applications are processed on a rolling basis. Application fee: $25 for international students. *Financial aid:* Fellowships with full tuition reimbursements, research assistantships with full tuition reimbursements, teaching assistantships with full tuition reimbursements, Federal Work-Study and institutionally-sponsored loans available. Financial aid applicants required to submit FAFSA. *Faculty research:* Operations research, ergonomics, production systems, management systems, transportation. Total annual research expenditures: $376,617. *Unit head:* Dr. Larry G. Brown, Head, 662-325-3865, Fax: 662-325-7618, E-mail: brown@engr.msstate.edu.

Montana State University–Bozeman, College of Graduate Studies, College of Engineering, Department of Chemical Engineering, Bozeman, MT 59717. Offers chemical engineering (MS); engineering (PhD); environmental engineering (MS, PhD); project engineering and management (MPEM). Part-time programs available. *Students:* 17 full-time (6 women), 12 part-time (8 women). *Degree requirements:* For master's, thesis or alternative required, foreign language not required; for doctorate, dissertation required, foreign language not required. *Entrance requirements:* For master's and doctorate, GRE General Test, TOEFL (minimum score of 550 required). *Application deadline:* For fall admission, 6/1 (priority date); for spring admission, 11/1. Applications are processed on a rolling basis. Application fee: $50. *Unit head:* Dr. John T. Sears, Head, 406-994-2221, Fax: 406-994-5308, E-mail: cheme@coe.montana.edu.

Montana State University–Bozeman, College of Graduate Studies, College of Engineering, Department of Mechanical and Industrial Engineering, Bozeman, MT 59717. Offers engineering (PhD); industrial and management engineering (MS); mechanical engineering (MS). Part-time programs available. *Students:* 28 full-time (6 women), 7 part-time. Average age 26. 37 applicants, 73% accepted. In 1998, 5 master's awarded. *Degree requirements:* For master's, thesis or alternative required, foreign language not required; for doctorate, dissertation required, foreign language not required. *Entrance requirements:* For master's and doctorate, GRE General Test, TOEFL. *Application deadline:* For fall admission, 6/1 (priority date); for spring admission, 11/1. Applications are processed on a rolling basis. Application fee: $50. *Financial aid:* Research assistantships, teaching assistantships available. Financial aid application deadline: 3/1; financial aid applicants required to submit FAFSA. *Faculty research:* Operations research, product processes and assessment, systems modeling/simulation, management planning, engineering economy. Total annual research expenditures: $922,633. *Unit head:* Dr. Vic Cundy, Head, 406-994-2203, Fax: 406-994-6292, E-mail: vcundy@me.montana.edu.

Montana Tech of The University of Montana, Graduate School, Project Engineering and Management Program, Butte, MT 59701-8997. Offers MPEM. Part-time and evening/weekend programs available. *Faculty:* 5 full-time (0 women). *Degree requirements:* Foreign language not required. *Entrance requirements:* For master's, GRE General Test, TOEFL (minimum score of 525 required), minimum B average. *Application deadline:* For fall admission, 4/1 (priority date); for spring admission, 10/1 (priority date). Applications are processed on a rolling basis. Application fee: $30. Tuition, state resident: full-time $3,211; part-time $162 per credit hour. Tuition, nonresident: full-time $9,883; part-time $440 per credit hour. International tuition: $15,500 full-time. *Financial aid:* Application deadline: 4/1. *Unit head:* Dr. Kumar Ganesan, Director, 406-496-4239, Fax: 406-496-4133, E-mail: kganesan@mtech.eu. *Application contact:* Cindy Dunstan, Administrative Assistant, 406-496-4128, Fax: 406-496-4334, E-mail: cdunstan@mtech.edu.

National University, Graduate Studies, School of Business and Technology, Department of Technology, La Jolla, CA 92037-1011. Offers e-commerce (MBA, MS); electronic engineering (MS); engineering management (MS); environmental management (MBA, MS); industrial engineering management (MS); software engineering (MS); technology management (MBA, MS); telecommunication systems management (MS). Part-time and evening/weekend programs available. Postbaccalaureate distance learning degree programs offered (minimal on-campus study). *Faculty:* 12 full-time, 125 part-time. *Students:* 305 (79 women); includes 122 minority (34 African Americans, 69 Asian Americans or Pacific Islanders, 17 Hispanic Americans, 2 Native Americans) 53 international. *Degree requirements:* For master's, foreign language and thesis not required. *Entrance requirements:* For master's, interview, minimum GPA of 2.5. *Application deadline:* Applications are processed on a rolling basis. Application fee: $60 ($100 for international students). Tuition: Full-time $7,830; part-time $870 per course. One-time fee: $60. Tuition and fees vary according to campus/location. *Unit head:* Dr. Leonid Preiser, Chair, 858-642-8425, Fax: 858-642-8716, E-mail: lpreiser@nu.edu. *Application contact:* Nancy Rohland, Director of Enrollment Management, 858-642-8180, Fax: 858-642-8709, E-mail: nrohland@nu.edu.

New Jersey Institute of Technology, Office of Graduate Studies, Department of Industrial and Manufacturing Engineering, Program in Industrial Engineering, Newark, NJ 07102-1982. Offers MS, PhD. Part-time and evening/weekend programs available. Terminal master's awarded for partial completion of doctoral program. *Degree requirements:* For master's, thesis or alternative required, foreign language not required; for doctorate, dissertation, residency required, foreign language not required. *Entrance requirements:* For master's, GRE General Test (minimum score of 450 on verbal section, 600 on quantitative, 550 on analytical required); for doctorate, GRE General Test (minimum score of 450 on verbal section, 600 on quantitative, 550 on analytical required), minimum graduate GPA of 3.5. Electronic applications accepted. *Faculty research:* Human factors engineering, manufacturing systems, materials, manufacturing automation and computer integration.

New Jersey Institute of Technology, Office of Graduate Studies, School of Management, Newark, NJ 07102-1982. Offers MS, PhD, M Arch/MS. PhD offered jointly with Rutgers, The State University of New Jersey, Newark. Part-time and evening/weekend programs available. Terminal master's awarded for partial completion of doctoral program. *Degree requirements:* For master's, thesis optional, foreign language not required; for doctorate, dissertation, residency required, foreign language not required. *Entrance requirements:* For master's, GRE General Test (minimum score of 450 on verbal section, 600 on quantitative, 550 on analytical required); for doctorate, GRE General Test (minimum score of 450 on verbal section, 600 on quantitative, 550 on analytical required), minimum graduate GPA of 3.5. Electronic applications accepted. *Faculty research:* Management of new technologies, information systems management, operations management systems, marketing management, human resource management.

New Mexico State University, Graduate School, College of Engineering, Department of Industrial Engineering, Las Cruces, NM 88003-8001. Offers MSIE, PhD. Part-time programs available. *Faculty:* 6 full-time (1 woman). *Students:* 16 full-time (4 women), 23 part-time (4 women); includes 7 minority (2 Asian Americans or Pacific Islanders, 5 Hispanic Americans), 11 international. Average age 33. 74 applicants, 58% accepted. In 1998, 36 master's, 3 doctorates awarded. *Degree requirements:* For master's, thesis optional; for doctorate, dissertation required. *Entrance requirements:* For doctorate, qualifying exam. *Application deadline:* For fall admission, 7/1 (priority date); for spring admission, 11/1. Applications are processed on a rolling basis. Application fee: $15 ($35 for international students). Electronic applications accepted. Tuition, state resident: full-time $2,682; part-time $112 per credit. Tuition, nonresident: full-time $8,376; part-time $349 per credit. Tuition and fees vary according to course load. *Financial aid:* Research assistantships, teaching assistantships, career-related internships or fieldwork and Federal Work-Study available. Aid available to part-time students. Financial aid application deadline: 3/1. *Faculty research:* Quality control, reliability, operations research, economic analysis, manufacturing systems. *Unit head:* Norris B. Green, Acting Head, 505-646-4923, Fax: 505-646-2976, E-mail: budgreen@nmsu.edu.

North Carolina Agricultural and Technical State University, Graduate School, College of Engineering, Department of Industrial Engineering, Greensboro, NC 27411. Offers MSIE, PhD. PhD offered jointly with North Carolina State University. Part-time programs available. *Faculty:* 9 full-time (0 women), 1 part-time (0 women). *Students:* 28 full-time (16 women), 18 part-time (9 women); includes 40 minority (34 African Americans, 6 Asian Americans or Pacific Islanders), 4 international. Average age 25. 27 applicants, 93% accepted. In 1998, 8 master's awarded (100% found work related to degree). *Degree requirements:* For master's, computer language, thesis, project required, foreign language not required. *Entrance requirements:* For master's, GRE General Test (recommended). *Average time to degree:* Master's–2 years full-time. *Application deadline:* For fall admission, 7/1; for spring admission, 1/9. Applications are processed on a rolling basis. Application fee: $35. *Financial aid:* Fellowships, research assistantships, teaching assistantships available. *Faculty research:* Human-machine systems engineering, management systems engineering, operations research and systems analysis, production systems engineering. Total annual research expenditures: $912,000. *Unit head:* Dr. Eui Park, Chairperson, 336-334-7780, Fax: 336-334-7729, E-mail: park@ncat.edu. *Application contact:* Dr. Bala Ram, Graduate Coordinator, 336-334-7780, Fax: 336-334-7729, E-mail: ram@ncat.edu.

North Carolina State University, Graduate School, College of Engineering, Department of Industrial Engineering, Raleigh, NC 27695. Offers MIE, MSIE, PhD. Part-time programs available. *Faculty:* 25 full-time (1 woman), 6 part-time (0 women). *Students:* 51 full-time (16 women), 20 part-time (9 women); includes 6 minority (4 African Americans, 2 Asian Americans or Pacific Islanders), 20 international. Average age 29. 119 applicants, 25% accepted. In 1998, 18 master's, 4 doctorates awarded. Terminal master's awarded for partial completion of doctoral program. *Degree requirements:* For master's, computer language, thesis required (for some programs), foreign language not required; for doctorate, computer language, dissertation required, foreign language not required. *Entrance requirements:* For master's, GRE General Test (minimum score of 450 on verbal section, 680 on quantitative, 550 on analytical required), TOEFL (minimum score of 575 required), minimum GPA of 3.0; for doctorate, GRE General Test (minimum score of 450 on verbal section, 680 on quantitative, 550 on analytical required), TOEFL (minimum score of 575 required). *Application deadline:* For fall admission, 6/25; for spring admission, 11/25. Applications are processed on a rolling basis. Application fee: $45. *Financial aid:* In 1998–99, 4 fellowships (averaging $3,003 per year), 34 research assistantships (averaging $3,753 per year), 30 teaching assistantships (averaging $4,504 per year) were awarded.; career-related internships or fieldwork and institutionally-sponsored loans also available. Financial aid application deadline: 3/1. *Faculty research:* Computer-integrated manufacturing production, scheduling, industrial ergonomics, simulation and modeling. Total annual research expenditures: $3.7 million. *Unit head:* Dr. Stephen D. Roberts, Head, 919-515-2362, Fax: 919-515-5281, E-mail: roberts@eos.ncsu.edu. *Application contact:* Dr. James R. Wilson, Director of Graduate Programs, 919-515-6415, Fax: 919-515-5281, E-mail: jwilson@eos.ncsu.edu.

See in-depth description on page 1205.

North Carolina State University, Graduate School, College of Management, Program in Management, Raleigh, NC 27695. Offers biotechnology (MS); computer science (MS); engineering (MS); forest resources management (MS); general business (MS); management information systems (MS); operations research (MS); statistics (MS); telecommunications systems engineering (MS); textile management (MS); total quality management (MS). Part-time programs available. *Faculty:* 40 full-time (9 women), 4 part-time (0 women). *Students:* 48 full-time (15 women), 156 part-time (43 women); includes 33 minority (16 African Americans, 15 Asian Americans or Pacific Islanders, 1 Hispanic American, 1 Native American), 4 international. *Degree requirements:* For master's, computer language required, foreign language and thesis not required. *Entrance requirements:* For master's, GRE or GMAT, TOEFL (minimum score of 550 required), minimum undergraduate GPA of 3.0. *Application deadline:* For fall admission, 6/25; for spring admission, 11/25. Applications are processed on a rolling basis. Application fee: $45. *Unit head:* Dr. Jack W. Wilson, Director of Graduate Programs, 919-515-4327, Fax: 919-515-6943, E-mail: jack_wilson@ncsu.edu. *Application contact:* Dr. Steven G. Allen, Director of Graduate Programs, 919-515-6941, Fax: 919-515-5073, E-mail: steve_allen@ncsu.edu.

North Dakota State University, Graduate Studies and Research, College of Engineering and Architecture, Department of Industrial and Manufacturing Engineering, Fargo, ND 58105. Offers MS. Part-time programs available. *Faculty:* 8 full-time (0 women). *Students:* 3 full-time (1 woman), 5 part-time (1 woman); includes 6 minority (all Asian Americans or Pacific Islanders) Average age 26. 38 applicants, 74% accepted. *Degree requirements:* For master's, computer language, thesis required, foreign language not required. *Entrance requirements:* For master's, GRE General Test (minimum combined score of 1650 on three sections required), TOEFL (minimum score of 550 required). *Average time to degree:* Master's–3 years full-time, 4 years part-time. *Application deadline:* For fall admission, 6/1; for spring admission, 10/1. Applications are processed on a rolling basis. Application fee: $25. *Financial aid:* In 1998–99, 1 research assistantship with tuition reimbursement, 5 teaching assistantships with tuition reimbursements were awarded.; Federal Work-Study and institutionally-sponsored loans also available. Financial aid application deadline: 4/15. *Faculty research:* Automation, engineering economy, artificial intelligence, human factors engineering, production and inventory control. *Unit head:* Dr. Fong-Yuen Ding, Interim Chair, 701-231-7223, Fax: 701-231-7195.

Northeastern University, College of Engineering, Department of Mechanical, Industrial, and Manufacturing Engineering, Boston, MA 02115-5096. Offers engineering management (MS); industrial engineering (MS, PhD); mechanical engineering (MS, PhD); operations research (MS).

Part-time programs available. *Faculty:* 30 full-time (1 woman), 13 part-time (1 woman). *Students:* 105 full-time (19 women), 118 part-time (25 women); includes 14 minority (5 African Americans, 7 Asian Americans or Pacific Islanders, 1 Hispanic American, 1 Native American), 84 international. Average age 25. 348 applicants, 59% accepted. In 1998, 48 master's, 8 doctorates awarded. *Degree requirements:* For master's, thesis required (for some programs), foreign language not required; for doctorate, one foreign language, dissertation, departmental qualifying exam required. *Entrance requirements:* For master's and doctorate, GRE General Test. *Average time to degree:* Master's–3.38 years full-time, 4.05 years part-time; doctorate–6.5 years full-time, 8 years part-time. *Application deadline:* For fall admission, 4/15. Applications are processed on a rolling basis. Application fee: $50. *Financial aid:* In 1998–99, 49 students received aid, including 14 research assistantships with full tuition reimbursements available (averaging $12,450 per year), 26 teaching assistantships with full tuition reimbursements available (averaging $12,450 per year); fellowships, career-related internships or fieldwork, Federal Work-Study, tuition waivers (full), and unspecified assistantships also available. Aid available to part-time students. Financial aid application deadline: 2/15; financial aid applicants required to submit FAFSA. *Faculty research:* Dry sliding instabilities, droplet deposition, helical hydraulic turbines, combustion, manufacturing systems. *Unit head:* Dr. John W. Cipolla, Chairman, 617-373-3810, Fax: 617-373-2921. *Application contact:* Stephen L. Gibson, Associate Director, 617-373-2711, Fax: 617-373-2501, E-mail: grad-eng@coe.neu.edu.

Northern Illinois University, Graduate School, College of Engineering and Engineering Technology, Department of Industrial Engineering, De Kalb, IL 60115-2854. Offers MS. Part-time programs available. *Faculty:* 5 full-time (0 women), 1 part-time (0 women). *Students:* 6 full-time (1 woman), 8 part-time (1 woman), 9 international. Average age 28. 37 applicants, 32% accepted. In 1998, 6 degrees awarded. *Degree requirements:* For master's, thesis or alternative, comprehensive exam required, foreign language not required. *Entrance requirements:* For master's, GRE General Test, TOEFL (minimum score of 550 required; 213 for computer-based), minimum GPA of 2.75. *Application deadline:* For fall admission, 6/1; for spring admission, 11/1. Applications are processed on a rolling basis. Application fee: $30. *Financial aid:* In 1998–99, 3 research assistantships were awarded.; fellowships, teaching assistantships, Federal Work-Study, tuition waivers (full), and unspecified assistantships also available. Aid available to part-time students. *Unit head:* Dr. Richard Marcellus, Chair, 815-753-9971.

Northwestern University, The Graduate School, Robert R. McCormick School of Engineering and Applied Science, Department of Industrial Engineering and Management Sciences, Evanston, IL 60208. Offers engineering management (MEM); industrial engineering and management science (MS, PhD); operations research (MS, PhD). MS and PhD admissions and degrees offered through The Graduate School. Part-time programs available. *Faculty:* 16 full-time (1 woman), 4 part-time (0 women). *Students:* 47 full-time (15 women), 130 part-time (30 women). 119 applicants, 48% accepted. In 1998, 15 master's, 5 doctorates awarded. Terminal master's awarded for partial completion of doctoral program. *Degree requirements:* For master's, computer language, comprehensive exam required, foreign language and thesis not required; for doctorate, computer language, dissertation, comprehensive exam required, foreign language not required. *Entrance requirements:* For master's, GRE General Test, TOEFL (minimum score of 560 required), 3 years of work experience (MEM); for doctorate, GRE General Test, TOEFL (minimum score of 560 required). *Application deadline:* For fall admission, 8/30. Application fee: $50 ($55 for international students). *Financial aid:* In 1998–99, 6 fellowships with full tuition reimbursements (averaging $11,673 per year), 14 research assistantships with partial tuition reimbursements (averaging $16,285 per year), 5 teaching assistantships with full tuition reimbursements (averaging $12,042 per year) were awarded.; career-related internships or fieldwork, Federal Work-Study, institutionally-sponsored loans, and scholarships also available. Financial aid application deadline: 1/15; financial aid applicants required to submit FAFSA. *Faculty research:* Production, logistics, optimization, simulation, statistics. Total annual research expenditures: $900,000. *Unit head:* Mark Daskin, Chair, 847-491-8796, Fax: 847-491-8005, E-mail: daskin@iems.nwu.edu. *Application contact:* June Wayne, Admission Contact, 847-491-4394, Fax: 847-491-8005, E-mail: wayne@iems.nwu.edu.

See in-depth description on page 1207.

The Ohio State University, Graduate School, College of Engineering, Department of Industrial, Welding, and Systems Engineering, Program in Industrial and Systems Engineering, Columbus, OH 43210. Offers MS, PhD. *Faculty:* 19 full-time, 14 part-time. *Students:* 115 full-time (21 women), 17 part-time (4 women); includes 9 minority (1 African American, 7 Asian Americans or Pacific Islanders, 1 Hispanic American), 86 international. 264 applicants, 27% accepted. In 1998, 29 master's, 8 doctorates awarded. *Degree requirements:* For master's, computer language required, thesis optional, foreign language not required; for doctorate, computer language, dissertation required, foreign language not required. *Entrance requirements:* For master's and doctorate, GRE General Test. *Application deadline:* For fall admission, 8/15. Applications are processed on a rolling basis. Application fee: $30 ($40 for international students). *Financial aid:* Fellowships, research assistantships, teaching assistantships, career-related internships or fieldwork, Federal Work-Study, institutionally-sponsored loans, and unspecified assistantships available. Aid available to part-time students. *Unit head:* Jerald Brevick, Graduate Studies Committee Chair, 614-292-0117, Fax: 614-292-7852, E-mail: brevick@ccl2.eng.ohio-state.edu.

Ohio University, Graduate Studies, College of Engineering and Technology, Department of Industrial and Manufacturing Systems Engineering, Athens, OH 45701-2979. Offers MS. *Faculty:* 8 full-time (0 women). *Students:* 20 full-time (2 women), 8 part-time (1 woman); includes 1 minority (Asian American or Pacific Islander), 23 international. Average age 24. 76 applicants, 68% accepted. In 1998, 6 degrees awarded. *Degree requirements:* For master's, computer language, non-thesis research project required, thesis optional, foreign language not required. *Entrance requirements:* For master's, GRE General Test. *Application deadline:* For fall admission, 3/1 (priority date). Applications are processed on a rolling basis. Application fee: $30. Tuition, state resident: full-time $5,754; part-time $238 per credit hour. Tuition, nonresident: full-time $11,055; part-time $457 per credit hour. Tuition and fees vary according to course load, campus/location and program. *Financial aid:* In 1998–99, 1 fellowship, 10 research assistantships, 8 teaching assistantships were awarded.; Federal Work-Study, institutionally-sponsored loans, and tuition waivers (full) also available. Financial aid application deadline: 3/1. *Faculty research:* Human factors engineering, MIS, quality systems. *Unit head:* Dr. Charles M. Parks, Chairman, 740-593-1540, Fax: 740-593-0778, E-mail: cparks@bobcat.ent.ohiou.edu. *Application contact:* Dr. Richard J. Gerth, Graduate Chairman, 740-593-1545, Fax: 740-593-0778, E-mail: gerth@bobcat.ent.ohiou.edu.

Oklahoma State University, Graduate College, College of Engineering, Architecture and Technology, School of Industrial Engineering and Management, Stillwater, OK 74078. Offers M En, MIE Mgmt, MS, PhD. *Faculty:* 10 full-time (1 woman), 1 part-time (0 women). *Students:* 60 full-time (5 women), 53 part-time (26 women); includes 5 minority (2 African Americans, 2 Asian Americans or Pacific Islanders, 1 Native American), 68 international. Average age 29. In 1998, 28 master's, 4 doctorates awarded. *Degree requirements:* For master's, creative component or thesis required; for doctorate, dissertation required, foreign language not required. *Entrance requirements:* For master's and doctorate, TOEFL (minimum score of 570 required). *Application deadline:* For fall admission, 7/1 (priority date). Application fee: $25. *Financial aid:* In 1998–99, 36 students received aid, including 12 research assistantships (averaging $10,950 per year), 24 teaching assistantships (averaging $8,341 per year); career-related internships or fieldwork, Federal Work-Study, and tuition waivers (partial) also available. Aid available to part-time students. Financial aid application deadline: 3/1. *Unit head:* Dr. C. Patrick Koelling, Head, 405-744-6055.

Oregon State University, Graduate School, College of Engineering, Department of Industrial and Manufacturing Engineering, Corvallis, OR 97331. Offers industrial engineering (MAIS, MS, PhD); manufacturing engineering (M Eng). Part-time programs available. Postbaccalaureate distance learning degree programs offered (minimal on-campus study). *Faculty:* 10 full-time (1 woman). *Students:* 33 full-time (9 women), 16 part-time (3 women); includes 1 minority (Hispanic American), 32 international. Average age 29. In 1998, 22 master's, 1 doctorate awarded. *Degree requirements:* For master's, thesis or alternative required, foreign language not required; for doctorate, dissertation required, foreign language not required. *Entrance requirements:* For master's, TOEFL (minimum score of 550 required), placement exam, minimum GPA of 3.0 in last 90 hours; for doctorate, GRE, TOEFL (minimum score of 550 required), placement exam, minimum GPA of 3.0 in last 90 hours. *Application deadline:* Applications are processed on a rolling basis. Application fee: $50. *Financial aid:* Fellowships, research assistantships, teaching assistantships, institutionally-sponsored loans and instructorships available. Aid available to part-time students. Financial aid application deadline: 2/1. *Faculty research:* Computer-integrated manufacturing, human factors, robotics, decision support systems, simulation modeling and analysis. *Unit head:* Dr. Sabah U. Randhawa, Head, 541-737-2365, Fax: 541-737-5241, E-mail: randhaws@ccmail.orst.edu. *Application contact:* Dr. Edward D. McDowell, Graduate Committee Chair, 541-737-2875, Fax: 541-737-5241, E-mail: mcdowele@ccmail.orst.edu.

Pennsylvania State University Great Valley School of Graduate Professional Studies, Graduate Studies and Continuing Education, College of Engineering, Program in Industrial Engineering, Malvern, PA 19355-1488. Offers M Eng. Average age 34. In 1998, 6 degrees awarded. *Entrance requirements:* For master's, GRE General Test. Application fee: $50. *Unit head:* Dr. John McCool, Coordinator, 610-648-3243. *Application contact:* 610-648-3242, Fax: 610-889-1334.

Pennsylvania State University University Park Campus, Graduate School, College of Engineering, Department of Industrial and Manufacturing Engineering, Program in Industrial Engineering, State College, University Park, PA 16802-1503. Offers MS, PhD. *Students:* 68 full-time (14 women), 30 part-time (3 women). In 1998, 26 master's, 10 doctorates awarded. *Degree requirements:* For master's, thesis required, foreign language not required; for doctorate, dissertation required. *Entrance requirements:* For master's and doctorate, GRE General Test. Application fee: $50. *Unit head:* Dr. A. Ravindran, Head, Department of Industrial and Manufacturing Engineering, 814-865-7601.

Polytechnic University, Brooklyn Campus, Department of Mechanical, Aerospace and Manufacturing Engineering, Major in Industrial Engineering, Brooklyn, NY 11201-2990. Offers MS. *Students:* 6 full-time (1 woman), 8 part-time; includes 2 minority (1 Asian American or Pacific Islander, 1 Hispanic American), 4 international. Average age 33. 24 applicants, 33% accepted. In 1998, 8 degrees awarded. *Entrance requirements:* For master's, BE or BS in engineering, physics, chemistry, mathematical sciences, or biological sciences or MBA. *Application deadline:* Applications are processed on a rolling basis. Application fee: $45. Electronic applications accepted. *Unit head:* Dr. Edward D. McDowell, Graduate Committee Chair, 541-737-2875, Fax: 541-737-5241, E-mail: mcdowele@ccmail.orst.edu. *Application contact:* John S. Kerge, Dean of Admissions, 718-260-3200, Fax: 718-260-3446, E-mail: admitme@poly.edu.

Polytechnic University, Farmingdale Campus, Graduate Programs, Department of Mechanical, Aerospace and Manufacturing Engineering, Major in Industrial Engineering, Farmingdale, NY 11735-3995. Offers MS. Average age 33. 1 applicants, 100% accepted. *Degree requirements:* For master's, computer language required. *Application deadline:* Applications are processed on a rolling basis. Application fee: $45. Electronic applications accepted. *Unit head:* John S. Kerge, Dean of Admissions, 718-260-3200, Fax: 718-260-3446, E-mail: admitme@poly.edu. *Application contact:* John S. Kerge, Dean of Admissions, 718-260-3200, Fax: 718-260-3446, E-mail: admitme@poly.edu.

Polytechnic University, Westchester Graduate Center, Graduate Programs, Department of Mechanical, Aerospace and Manufacturing Engineering, Major in Industrial Engineering, Hawthorne, NY 10532-1507. Offers MS. Average age 33. *Degree requirements:* For master's, computer language required. *Application deadline:* Applications are processed on a rolling basis. Application fee: $45. Electronic applications accepted. *Unit head:* Cindy Dunstan, Administrative Assistant, 406-496-4128, Fax: 406-496-4334, E-mail: cdunstan@mtech.edu. *Application contact:* John S. Kerge, Dean of Admissions, 718-260-3200, Fax: 718-260-3446, E-mail: admitme@poly.edu.

Purdue University, Graduate School, Schools of Engineering, School of Industrial Engineering, West Lafayette, IN 47907. Offers human factors in industrial engineering (MS, MSIE, PhD); manufacturing engineering (MS, MSIE, PhD); operations research (MS, MSIE, PhD); systems engineering (MS, MSIE, PhD). Part-time programs available. *Faculty:* 27 full-time (2 women), 3 part-time (0 women). *Students:* 124 full-time (26 women), 33 part-time (8 women); includes 13 minority (2 African Americans, 4 Asian Americans or Pacific Islanders, 6 Hispanic Americans, 1 Native American), 107 international. Average age 25. 299 applicants, 44% accepted. In 1998, 41 master's awarded (88% found work related to degree, 12% continued full-time study); 17 doctorates awarded. Terminal master's awarded for partial completion of doctoral program. *Degree requirements:* For master's, computer language required, thesis optional, foreign language not required; for doctorate, computer language, dissertation required, foreign language not required. *Entrance requirements:* For master's, GRE General Test (minimum score of 470 on verbal section, 700 on quantitative, 600 on analytical required; average 520 verbal, 710 quantitative, 660 analytical), TOEFL (minimum score of 570 required; average 600), minimum GPA of 3.0; for doctorate, GRE General Test (minimum score of 470 on verbal section, 700 on quantitative, 600 on analytical required; average 540 verbal, 750 quantitative, 675 analytical), TOEFL (minimum score of 570 required; average 610), MS thesis. *Average time to degree:* Master's–1.5 years full-time, 3 years part-time; doctorate–3 years full-time, 7 years part-time. *Application deadline:* For fall admission, 3/15; for spring admission, 9/1. Application fee: $30. Electronic applications accepted. *Financial aid:* In 1998–99, 89 students received aid, including 6 fellowships with full tuition reimbursements available (averaging $12,000 per year), 41 research assistantships with full tuition reimbursements available (averaging $11,880 per year), 42 teaching assistantships with full tuition reimbursements available (averaging $9,900 per year) Aid available to part-time students. Financial aid application deadline: 3/15; financial aid applicants required to submit FAFSA. *Faculty research:* Precision manufacturing process, computer-aided manufacturing, computer-aided process planning, knowledge-based systems, combinatorics. Total annual research expenditures: $1.7 million. *Unit head:* Dr. W. D. Compton, Interim Head, 765-494-5444, Fax: 765-494-1299, E-mail: dcompton@ecn.purdue.edu. *Application contact:* Dr. J. W. Barany, Associate Head, 765-494-5406, Fax: 765-494-1299, E-mail: jwb@ecn.purdue.edu.

Rensselaer Polytechnic Institute, Graduate School, School of Engineering, Department of Decision Sciences and Engineering Systems, Program in Industrial and Management Engineering, Troy, NY 12180-3590. Offers M Eng, MS, MBA/M Eng. Part-time and evening/weekend programs available. Postbaccalaureate distance learning degree programs offered (no on-campus study). *Faculty:* 12 full-time (0 women), 6 part-time (1 woman). *Students:* 44 full-time (11 women), 13 part-time (2 women); includes 8 minority (5 Asian Americans or Pacific Islanders, 3 Hispanic Americans), 29 international. 68 applicants, 66% accepted. In 1998, 33 degrees awarded. *Degree requirements:* For master's, thesis required (for some programs), foreign language not required. *Entrance requirements:* For master's, GRE General Test, TOEFL (minimum score of 550 required). *Application deadline:* For fall admission, 2/1 (priority date). Applications are processed on a rolling basis. Application fee: $35. *Financial aid:* In 1998–99, 8 research assistantships with full tuition reimbursements (averaging $10,600 per year), 7 teaching assistantships with full tuition reimbursements (averaging $10,600 per year) were awarded.; fellowships, career-related internships or fieldwork and institutionally-sponsored loans also available. Financial aid application deadline: 2/1. *Faculty research:* Manufacturing, MIS, statistical consulting, education/services, production/logistics/inventory. Total annual research expenditures: $2.2 million. *Unit head:* Lee Vilardi, Graduate Coordinator, 518-276-6681, Fax: 518-276-8227, E-mail: dsesgr@rpi.edu. *Application contact:* Lee Vilardi, Graduate Coordinator, 518-276-6681, Fax: 518-276-8227, E-mail: dsesgr@rpi.edu.

See in-depth description on page 1211.

Industrial/Management Engineering

Rochester Institute of Technology, Part-time and Graduate Admissions, College of Engineering, Department of Industrial and Manufacturing Engineering, Rochester, NY 14623-5604. Offers engineering management (ME); industrial engineering (ME); manufacturing engineering (ME); systems engineering (ME). *Students:* 6 full-time (3 women), 13 part-time (4 women); includes 3 minority (all African Americans), 4 international. 71 applicants, 56% accepted. In 1998, 11 degrees awarded. *Degree requirements:* For master's, internship required. *Entrance requirements:* For master's, TOEFL, minimum GPA of 3.0. *Application deadline:* For fall admission, 3/1 (priority date). Applications are processed on a rolling basis. Application fee: $40. *Financial aid:* Research assistantships, career-related internships or fieldwork and tuition waivers (partial) available. *Faculty research:* Safety, manufacturing (CAM), simulation. *Unit head:* Dr. Jasper Shealy, Head, 716-475-2134, E-mail: jeseie@rit.edu.

Rutgers, The State University of New Jersey, New Brunswick, Graduate School, Program in Industrial and Systems Engineering, New Brunswick, NJ 08903. Offers industrial and systems engineering (MS, PhD); manufacturing systems (MS); quality and reliability engineering (MS). Part-time and evening/weekend programs available. *Faculty:* 10 full-time (2 women). *Students:* 47 full-time (15 women), 14 part-time (9 women); includes 31 minority (2 African Americans, 29 Asian Americans or Pacific Islanders) Average age 26. 94 applicants, 40% accepted. In 1998, 16 master's awarded (100% found work related to degree); 3 doctorates awarded. Terminal master's awarded for partial completion of doctoral program. *Degree requirements:* For master's, computer language required, thesis optional, foreign language not required; for doctorate, computer language, dissertation required, foreign language not required. *Entrance requirements:* For master's and doctorate, GRE General Test, TOEFL. *Average time to degree:* Master's–1.5 years full-time, 3 years part-time; doctorate–4 years full-time. *Application deadline:* For fall admission, 1/1 (priority date); for spring admission, 1/1. Applications are processed on a rolling basis. Application fee: $50. *Financial aid:* In 1998–99, 26 students received aid, including 5 fellowships with full tuition reimbursements available (averaging $13,500 per year), 13 research assistantships with full tuition reimbursements available (averaging $13,500 per year), 5 teaching assistantships with full tuition reimbursements available (averaging $13,500 per year); Federal Work-Study and tuition waivers (partial) also available. Financial aid application deadline: 1/1; financial aid applicants required to submit FAFSA. *Faculty research:* Production and manufacturing systems, quality and reliability engineering, systems engineering and aviation safety. *Unit head:* Susan L. Albin, Director, 732-445-2238, Fax: 732-445-5467, E-mail: salbin@rci.rutgers.edu. *Application contact:* Cindy Ielmini, 732-445-3654, Fax: 732-445-5467, E-mail: ielmini@rci.rutgers.edu.

St. Mary's University of San Antonio, Graduate School, Department of Engineering, San Antonio, TX 78228-8507. Offers electrical engineering (MS); electrical/computer engineering (MS); engineering administration (MS); engineering computer application (MS); industrial engineering (MS); operations research (MS). Part-time and evening/weekend programs available. *Students:* 3 full-time (2 women), 49 part-time (6 women); includes 10 minority (2 African Americans, 8 Hispanic Americans), 8 international. *Degree requirements:* For master's, computer language, thesis required, foreign language not required. *Entrance requirements:* For master's, GRE General Test. *Application deadline:* For fall admission, 8/1. Application fee: $15. *Unit head:* Dr. Abe Yazdani, Adviser, 210-436-3305.

South Dakota State University, Graduate School, College of Engineering, Industrial Management Program, Brookings, SD 57007. Offers MS. *Degree requirements:* For master's, thesis, oral exam required, foreign language not required. *Entrance requirements:* For master's, TOEFL (minimum score of 580 required). *Faculty research:* Query, language development, statistical process control, foreign business plans.

Southern Polytechnic State University, College of Technology, Department of Industrial Engineering Technology, Marietta, GA 30060-2896. Offers quality assurance (MS). Part-time and evening/weekend programs available. Postbaccalaureate distance learning degree programs offered (minimal on-campus study). *Faculty:* 2 full-time (1 woman), 2 part-time (1 woman). *Students:* 16 full-time (7 women), 51 part-time (19 women); includes 14 minority (10 African Americans, 1 Asian American or Pacific Islander, 2 Hispanic Americans, 1 Native American) Average age 35. 22 applicants, 100% accepted. In 1998, 15 degrees awarded. *Degree requirements:* For master's, thesis optional, foreign language not required. *Entrance requirements:* For master's, GRE General Test. *Application deadline:* For fall admission, 7/15 (priority date); for spring admission, 12/1. Applications are processed on a rolling basis. Application fee: $20. Tuition, state resident: full-time $2,146; part-time $119 per credit hour. Tuition, nonresident: full-time $7,586; part-time $421 per credit hour. *Financial aid:* In 1998–99, 30 students received aid; teaching assistantships, career-related internships or fieldwork and Federal Work-Study available. Aid available to part-time students. Financial aid application deadline: 5/1; financial aid applicants required to submit FAFSA. *Unit head:* Thomas Carmichael, Head, 770-528-7339, Fax: 770-528-4991, E-mail: tcarmich@spsu.edu.

Southwest Texas State University, Graduate School, School of Applied Arts and Technology, Department of Technology, Program in Industrial Technology, San Marcos, TX 78666. Offers MST. Part-time and evening/weekend programs available. *Faculty:* 6 full-time (0 women), 3 part-time (0 women). *Students:* 3 full-time (0 women), 19 part-time (4 women); includes 6 minority (1 African American, 3 Asian Americans or Pacific Islanders, 2 Hispanic Americans), 3 international. Average age 32. In 1998, 4 degrees awarded. *Degree requirements:* For master's, comprehensive exam required, thesis optional, foreign language not required. *Entrance requirements:* For master's, GRE General Test (minimum combined score of 900 required), TOEFL (minimum score of 550 required), minimum GPA of 2.75 in last 60 hours. *Application deadline:* For fall admission, 6/15 (priority date); for spring admission, 10/15 (priority date). Applications are processed on a rolling basis. Application fee: $25 ($50 for international students). Tuition, state resident: full-time $684; part-time $38 per semester hour. Tuition, nonresident: full-time $4,572; part-time $254 per semester hour. *Financial aid:* Research assistantships, teaching assistantships, career-related internships or fieldwork, Federal Work-Study, and institutionally-sponsored loans available. Aid available to part-time students. Financial aid application deadline: 4/1; financial aid applicants required to submit FAFSA. *Unit head:* Dr. Andy Batey, Graduate Adviser, 512-245-2137, Fax: 512-245-3052, E-mail: ab08@swt.edu.

Stanford University, School of Engineering, Department of Industrial Engineering and Engineering Management, Stanford, CA 94305-9991. Offers engineering management (MS); industrial engineering (MS, PhD, Eng). *Faculty:* 12 full-time (3 women). *Students:* 96 full-time (31 women), 27 part-time (6 women); includes 35 minority (3 African Americans, 26 Asian Americans or Pacific Islanders, 6 Hispanic Americans), 52 international. Average age 27. 177 applicants, 24% accepted. In 1998, 69 master's, 4 doctorates awarded. *Degree requirements:* For master's, foreign language and thesis not required; for doctorate, dissertation required, foreign language not required; for Eng, thesis required. *Entrance requirements:* For master's, doctorate, and Eng, GRE General Test, TOEFL. *Application deadline:* For fall admission, 2/1. Application fee: $65 ($80 for international students). Electronic applications accepted. Tuition: Full-time $24,588. Required fees: $152. Part-time tuition and fees vary according to course load. *Financial aid:* Fellowships, research assistantships, teaching assistantships, Federal Work-Study and institutionally-sponsored loans available. Financial aid application deadline: 2/15. *Unit head:* M. Elisabeth Paté-Cornell, Chair, 650-723-3823 Ext. ', Fax: 650-725-8799, E-mail: mep@stanford.edu. *Application contact:* Graduate Admissions Coordinator, 650-725-1633.

See in-depth description on page 1215.

State University of New York at Binghamton, Graduate School, Thomas J. Watson School of Engineering and Applied Science, Department of Systems Science and Industrial Engineering, Binghamton, NY 13902-6000. Offers M Eng, MS, MSAT, PhD. Part-time and evening/weekend programs available. *Faculty:* 6 full-time, 3 part-time. *Students:* 57 full-time (5 women), 36 part-time (8 women); includes 8 minority (5 African Americans, 1 Asian American or Pacific Islander, 2 Hispanic Americans), 46 international. Average age 31. 82 applicants, 73% accepted. In 1998, 20 master's, 5 doctorates awarded. Terminal master's awarded for partial completion of doctoral program. *Degree requirements:* For master's, thesis or alternative required, foreign language not required; for doctorate, dissertation required, foreign language not required. *Entrance requirements:* For master's and doctorate, GRE General Test, GRE Subject Test, TOEFL (minimum score of 550 required). *Application deadline:* For fall admission, 4/15 (priority date); for spring admission, 11/1. Applications are processed on a rolling basis. Application fee: $50. Electronic applications accepted. Tuition, state resident: full-time $5,100; part-time $213 per credit. Tuition, nonresident: full-time $8,416; part-time $351 per credit. Required fees: $77 per credit. Part-time tuition and fees vary according to course load. *Financial aid:* In 1998–99, 49 students received aid, including 27 research assistantships with full tuition reimbursements available (averaging $8,833 per year), 8 teaching assistantships with full tuition reimbursements available (averaging $8,536 per year); fellowships, career-related internships or fieldwork, Federal Work-Study, institutionally-sponsored loans, and unspecified assistantships also available. Aid available to part-time students. Financial aid application deadline: 2/15. *Faculty research:* Problem restructuring, protein modeling. *Unit head:* Dr. Robert Emerson, Chair, 607-777-6509.

See in-depth description on page 1219.

State University of New York at Buffalo, Graduate School, School of Engineering and Applied Sciences, Department of Industrial Engineering, Buffalo, NY 14260. Offers M Eng, MS, PhD. Part-time programs available. Postbaccalaureate distance learning degree programs offered (minimal on-campus study). *Faculty:* 10 full-time (1 woman), 4 part-time (0 women). *Students:* 36 full-time (3 women), 60 part-time (10 women); includes 7 minority (1 African American, 6 Asian Americans or Pacific Islanders), 54 international. Average age 26. 240 applicants, 78% accepted. In 1998, 28 master's, 1 doctorate awarded. Terminal master's awarded for partial completion of doctoral program. *Degree requirements:* For master's, computer language, thesis or alternative required, foreign language not required; for doctorate, computer language, dissertation required, foreign language not required. *Entrance requirements:* For master's and doctorate, GRE General Test, TOEFL (minimum score of 550 required). *Application deadline:* For fall admission, 2/1 (priority date); for spring admission, 8/1. Applications are processed on a rolling basis. Application fee: $35. Tuition, state resident: full-time $5,100; part-time $213 per credit hour. Tuition, nonresident: full-time $8,416; part-time $351 per credit hour. Required fees: $870; $75 per semester. Tuition and fees vary according to course load and program. *Financial aid:* In 1998–99, 35 students received aid, including 5 fellowships, 14 research assistantships, 11 teaching assistantships; Federal Work-Study, institutionally-sponsored loans, tuition waivers (full and partial), and unspecified assistantships also available. Financial aid application deadline: 2/1; financial aid applicants required to submit FAFSA. *Faculty research:* Ergonomics, operations research, production systems. Total annual research expenditures: $516,026. *Unit head:* Dr. Rajan Batta, Chairman, 716-645-2357 Ext. 2110, Fax: 716-645-3302, E-mail: indgrad@ubvms.cc.buffalo.edu. *Application contact:* Dr. Colin G. Drury, Director of Graduate Studies, 716-645-2357 Ext. 2117, Fax: 716-645-3302, E-mail: maloney@acsu.buffalo.edu.

See in-depth description on page 1221.

State University of New York College at Buffalo, Graduate Studies and Research, Faculty of Applied Science and Education, Department of Technology, Program in Industrial Technology, Buffalo, NY 14222-1095. Offers MS. *Students:* 1 (woman) full-time, 32 part-time (4 women); includes 5 minority (4 African Americans, 1 Asian American or Pacific Islander) Average age 35. 10 applicants, 80% accepted. In 1998, 11 degrees awarded. *Degree requirements:* For master's, thesis or project required. *Entrance requirements:* For master's, minimum GPA of 2.5 in last 60 hours, New York teaching certificate. *Application deadline:* For fall admission, 5/1; for spring admission, 10/1. Application fee: $50. *Financial aid:* Fellowships available. Financial aid application deadline: 3/1. *Unit head:* Graduate Admissions Coordinator, 650-725-1633.

Tennessee Technological University, Graduate School, College of Engineering, Department of Industrial Engineering, Cookeville, TN 38505. Offers MS, PhD. Part-time programs available. *Faculty:* 7 full-time (0 women). *Students:* 11 full-time (1 woman), 3 part-time; all minorities (all Asian Americans or Pacific Islanders) Average age 26. 82 applicants, 44% accepted. In 1998, 6 master's awarded. *Degree requirements:* For master's, thesis required, foreign language not required; for doctorate, one foreign language (computer language can substitute), dissertation required. *Entrance requirements:* For master's, GRE General Test, TOEFL (minimum score of 525 required); for doctorate, GRE Subject Test, TOEFL (minimum score of 525 required), minimum GPA of 3.5. *Application deadline:* For fall admission, 3/1 (priority date); for spring admission, 8/1. Application fee: $25 ($30 for international students). Tuition, state resident: part-time $137 per hour. Tuition, nonresident: part-time $361 per hour. Required fees: $17 per hour. Tuition and fees vary according to course load. *Financial aid:* In 1998–99, 13 students received aid, including 8 research assistantships (averaging $7,000 per year); fellowships, teaching assistantships, career-related internships or fieldwork also available. Financial aid application deadline: 4/1. *Faculty research:* Economic analysis, manufacturing systems, quality control. *Unit head:* Dr. Jessica Matson, Chairperson, 931-372-3466, Fax: 931-372-6352. *Application contact:* Dr. Rebecca F. Quattlebaum, Dean of the Graduate School, 931-372-3233, Fax: 931-372-3497, E-mail: rquattlebaum@tntech.edu.

Texas A&M University, College of Engineering, Department of Industrial Engineering, College Station, TX 77843. Offers biomedical engineering (M Eng, MS, D Eng, PhD); industrial engineering (M Eng, MS, D Eng, PhD). Part-time programs available. Postbaccalaureate distance learning degree programs offered (no on-campus study). *Faculty:* 21 full-time (2 women), 2 part-time (0 women). *Students:* 86 full-time (13 women), 33 part-time (8 women); includes 4 minority (1 African American, 3 Hispanic Americans), 83 international. Average age 28. 230 applicants, 50% accepted. In 1998, 28 master's awarded (96% found work related to degree, 4% continued full-time study); 6 doctorates awarded (17% entered university research/teaching, 83% found other work related to degree). *Degree requirements:* For master's, computer language required, thesis optional, foreign language not required; for doctorate, computer language, dissertation (PhD), record of study (D Eng) required. *Entrance requirements:* For master's and doctorate, GRE General Test, TOEFL. *Application deadline:* Applications are processed on a rolling basis. Application fee: $50 ($75 for international students). Electronic applications accepted. *Financial aid:* In 1998–99, 3 fellowships with partial tuition reimbursements (averaging $12,000 per year), 24 research assistantships with partial tuition reimbursements (averaging $12,000 per year), 17 teaching assistantships with partial tuition reimbursements (averaging $12,000 per year) were awarded.; career-related internships or fieldwork also available. Financial aid application deadline: 4/1; financial aid applicants required to submit FAFSA. *Faculty research:* Human factors and ergonomics, manufacturing systems, computer integration, operations research. *Unit head:* Dr. Way Kuo, Head, 409-845-5535, Fax: 409-847-9005, E-mail: way@acs.tamu.edu. *Application contact:* Dr. Richard M. Feldman, Graduate Adviser, 409-845-5536, Fax: 409-847-9005, E-mail: richf@tamu.edu.

Texas A&M University–Kingsville, College of Graduate Studies, College of Engineering, Department of Mechanical and Industrial Engineering, Program in Industrial Engineering, Kingsville, TX 78363. Offers ME, MS. *Faculty:* 3. *Students:* 6 full-time (0 women), 19 part-time. *Degree requirements:* For master's, computer language, thesis or alternative, comprehensive exam required, foreign language not required. *Entrance requirements:* For master's, GRE General Test (minimum combined score of 1050 required), TOEFL (minimum score of 525 required), minimum GPA of 3.0. *Application deadline:* For fall admission, 6/1; for spring admission, 11/15. Applications are processed on a rolling basis. Application fee: $15 ($25 for international students). Tuition, state resident: full-time $2,062. Tuition, nonresident: full-time $7,246. *Financial aid:* Fellowships available. Financial aid application deadline: 5/15. *Faculty research:* Robotics and automation, neural networks and fuzzy logic, systems engineering/simulation modeling, integrated manufacturing and production systems. *Application contact:* Dr. Hayder Abdul Razzak, Coordinator, 361-593-2001.

Texas Tech University, Graduate School, College of Engineering, Department of Industrial Engineering, Lubbock, TX 79409. Offers MSIE, PhD. Part-time programs available. *Faculty:* 9 full-time (0 women). *Students:* 36 full-time (5 women), 3 part-time (1 woman); includes 1 minority (Asian American or Pacific Islander), 32 international. Average age 27. 109 applicants,

64% accepted. In 1998, 10 master's, 6 doctorates awarded. *Degree requirements:* For master's and doctorate, computer language, thesis/dissertation required, foreign language not required. *Entrance requirements:* For master's, GRE General Test (minimum combined score of 1000 required; average 1130), minimum GPA of 3.0; for doctorate, GRE General Test (minimum combined score of 1000 required), minimum GPA of 3.0. *Application deadline:* For fall admission, 4/15 (priority date); for spring admission, 11/1 (priority date). Applications are processed on a rolling basis. Application fee: $25 ($50 for international students). Electronic applications accepted. *Financial aid:* In 1998–99, 28 research assistantships (averaging $9,446 per year), 2 teaching assistantships (averaging $5,175 per year) were awarded.; fellowships, Federal Work-Study and institutionally-sponsored loans also available. Aid available to part-time students. Financial aid application deadline: 5/15; financial aid applicants required to submit FAFSA. *Faculty research:* Low-energy cotton gin conveying/drying techniques, ergonomic review of U.S. postal service carrier bags, development models to predict optimal lifting motion. Total annual research expenditures: $305,356. *Unit head:* Dr. James L. Smith, Chairman, 806-742-3543, Fax: 806-742-3411, E-mail: jlsmith@coe.ttu.edu.

Universidad de las Américas–Puebla, Division of Graduate Studies, School of Engineering, Program in Industrial Engineering, Cholula, 72820, Mexico. Offers industrial engineering (MS); production management (M Adm). Part-time and evening/weekend programs available. *Faculty:* 9 full-time (2 women), 2 part-time (1 woman). *Students:* 23 full-time (6 women), 2 part-time; all minorities (all Hispanic Americans) Average age 25. 12 applicants, 67% accepted. In 1998, 6 degrees awarded. *Degree requirements:* For master's, one foreign language, computer language, thesis required. *Average time to degree:* Master's–2.5 years full-time, 3.5 years part-time. *Application deadline:* For fall admission, 7/16. Applications are processed on a rolling basis. Application fee: $0. *Financial aid:* In 1998–99, 20 students received aid, including 2 research assistantships, 2 teaching assistantships Aid available to part-time students. Financial aid application deadline: 5/15. *Faculty research:* Textile industry, quality control. Total annual research expenditures: $33,000. *Unit head:* Dr. Dolores E. Luna, Coordinator, 22-29-20-30, Fax: 22-29-20-32, E-mail: dluna@mail.udlap.mx. *Application contact:* Mauricio Villegas, Chair of Admissions Office, 22-29-20-17, Fax: 22-29-20-18, E-mail: admision@mail.udlap.mx.

Université de Moncton, School of Engineering, Program in Industrial Engineering, Moncton, NB E1A 3E9, Canada. Offers M Sc A. *Faculty:* 1 (woman) full-time. *Students:* 1 (woman) full-time. Average age 25. *Degree requirements:* For master's, thesis, proficiency in French required. *Application deadline:* For fall admission, 6/1 (priority date); for winter admission, 11/15 (priority date). Application fee: $30. *Financial aid:* In 1998–99, 1 student received aid, including 1 research assistantship (averaging $4,800 per year); fellowships, teaching assistantships Financial aid application deadline: 5/31. *Faculty research:* Production systems, optimization, simulation and expert systems, modeling and warehousing systems, quality control. Total annual research expenditures: $26,150. *Unit head:* Dr. Gilles Cormier, Chairman, 506-858-4387, Fax: 506-858-4082, E-mail: cormieg@umoncton.ca.

Université du Québec à Trois-Rivières, Graduate Programs, Program in Industrial Engineering, Trois-Rivières, PQ G9A 5H7, Canada. Offers DESS. *Students:* 6 full-time (1 woman), 6 part-time (1 woman). 10 applicants, 90% accepted. *Degree requirements:* For DESS, thesis not required. *Entrance requirements:* For degree, appropriate bachelor's degree, proficiency in French. *Application deadline:* For fall admission, 2/1. Application fee: $30. *Financial aid:* Fellowships, research assistantships, teaching assistantships available. *Faculty research:* Production. *Unit head:* Georges Abdul-Nour, Director, 819-376-5070 Ext. 3936, Fax: 819-376-5012, E-mail: georges_abdulnour@uqtr.uquebec.ca. *Application contact:* Suzanne Camirand, Admissions Officer, 819-376-5045 Ext. 2591, Fax: 819-376-5210, E-mail: suzanne_camirand@uqtr.uquebec.ca.

Université Laval, Faculty of Graduate Studies, Faculty of Sciences and Engineering, Program in Industrial Engineering, Sainte-Foy, PQ G1K 7P4, Canada. Offers Diploma. *Students:* 4 full-time (1 woman), 28 part-time (5 women). 34 applicants, 82% accepted. In 1998, 11 degrees awarded. *Application deadline:* For fall admission, 3/1. Application fee: $30. *Unit head:* Daoud Aït-Kadi, Director, 418-656-2131 Ext. 2378, Fax: 418-656-7415, E-mail: gmc@gmc.ulaval.ca.

The University of Alabama, Graduate School, College of Engineering, Department of Industrial Engineering, Tuscaloosa, AL 35487. Offers MSE, MSIE. Postbaccalaureate distance learning degree programs offered. *Faculty:* 9 full-time (1 woman). *Students:* 21 full-time (5 women), 6 part-time (1 woman); includes 4 minority (3 African Americans, 1 Asian American or Pacific Islander), 16 international. Average age 24. 35 applicants, 60% accepted. In 1998, 8 degrees awarded. *Degree requirements:* For master's, thesis or alternative required, foreign language not required. *Entrance requirements:* For master's, GRE General Test (minimum combined score of 1500 on three sections required), minimum GPA of 3.0 in last 60 hours. *Average time to degree:* Master's–2 years full-time, 3 years part-time. *Application deadline:* For fall admission, 7/6. Applications are processed on a rolling basis. Application fee: $25. Electronic applications accepted. *Financial aid:* In 1998–99, 19 students received aid, including 2 fellowships (averaging $10,000 per year), 4 research assistantships (averaging $8,000 per year), 10 teaching assistantships (averaging $8,000 per year); Federal Work-Study and institutionally-sponsored loans also available. Financial aid application deadline: 2/15. *Faculty research:* Systems engineering and operations research, human factors engineering, manufacturing engineering, quality engineering. Total annual research expenditures: $300,000. *Unit head:* Dr. Robert G. Batson, Head, 205-348-7160, Fax: 205-348-7160, E-mail: rbatson@coe.eng.ua.edu. *Application contact:* Teresa A. McGhee, Secretary, 205-348-7160, Fax: 205-348-7162, E-mail: tmcghee@coe.eng.ua.edu.

The University of Alabama in Huntsville, School of Graduate Studies, College of Engineering, Department of Industrial and Systems Engineering/Engineering Management, Huntsville, AL 35899. Offers industrial engineering (MSE, PhD); operations research (MSOR). Part-time and evening/weekend programs available. Postbaccalaureate distance learning degree programs offered (no on-campus study). *Faculty:* 15 full-time (1 woman), 3 part-time (0 women). *Students:* 27 full-time (6 women), 166 part-time (31 women); includes 21 minority (14 African Americans, 2 Asian Americans or Pacific Islanders, 3 Hispanic Americans, 2 Native Americans), 3 international. *Degree requirements:* For master's, oral and written exams required, thesis optional, foreign language not required; for doctorate, dissertation, oral and written exams required, foreign language not required. *Entrance requirements:* For master's and doctorate, GRE General Test (minimum combined score of 1500 on three sections required), minimum GPA of 3.0. *Application deadline:* For fall admission, 7/24 (priority date); for spring admission, 11/15 (priority date). Applications are processed on a rolling basis. Application fee: $20. Tuition and fees vary according to course load. *Unit head:* Dr. Jerry Westbrook, Chair, 256-890-6256, Fax: 256-890-6608, E-mail: westbroo@eb.uah.edu.

The University of Arizona, Graduate College, College of Engineering and Mines, Department of Systems and Industrial Engineering, Program in Industrial Engineering, Tucson, AZ 85721. Offers MS. Part-time programs available. *Students:* 15 full-time (0 women), 4 part-time (1 woman), 17 international. Average age 27. 26 applicants, 54% accepted. In 1998, 8 degrees awarded. *Degree requirements:* For master's, computer language required, foreign language not required. *Entrance requirements:* For master's, GRE General Test (minimum combined score of 1150 required), TOEFL (minimum score of 550 required), minimum GPA of 3.0. *Application deadline:* For fall admission, 7/1. Applications are processed on a rolling basis. Application fee: $35. *Financial aid:* Fellowships, research assistantships, teaching assistantships, institutionally-sponsored loans and scholarships available. *Faculty research:* Operations research, manufacturing systems, quality and reliability, statistical/engineering design. *Unit head:* Dr. Rebecca F. Quattlebaum, Dean of the Graduate School, 931-372-3233, Fax: 931-372-3497, E-mail: rquattlebaum@tntech.edu. *Application contact:* Celia Stenzel, Graduate Secretary, 520-621-6551, Fax: 520-621-6555.

The University of Arizona, Graduate College, College of Engineering and Mines, Department of Systems and Industrial Engineering, Program in Systems and Industrial Engineering, Tucson, AZ 85721. Offers PhD. *Students:* 12 full-time (3 women), 7 part-time; includes 1 minority (Native American), 11 international. 17 applicants, 59% accepted. In 1998, 3 degrees awarded. *Degree requirements:* For doctorate, dissertation required. *Entrance requirements:* For doctorate, GRE General Test (minimum combined score of 1150 required), TOEFL (minimum score of 550 required), minimum GPA of 3.0. *Application deadline:* For fall admission, 7/1. Application fee: $35. *Unit head:* Celia Stenzel, Graduate Secretary, 520-621-6551, Fax: 520-621-6555. *Application contact:* Celia Stenzel, Graduate Secretary, 520-621-6551, Fax: 520-621-6555.

University of Arkansas, Graduate School, College of Engineering, Department of Industrial Engineering, Program in Industrial Engineering, Fayetteville, AR 72701-1201. Offers MSE, MSIE, PhD. *Students:* 13 full-time (4 women), 6 part-time; includes 2 minority (1 African American, 1 Asian American or Pacific Islander), 8 international. 50 applicants, 64% accepted. In 1998, 14 master's, 3 doctorates awarded. *Degree requirements:* For master's, thesis optional, foreign language not required; for doctorate, one foreign language, dissertation required. Application fee: $40 ($50 for international students). Tuition, state resident: full-time $3,186. Tuition, nonresident: full-time $7,560. Required fees: $378. *Financial aid:* Research assistantships, teaching assistantships, career-related internships or fieldwork and Federal Work-Study available. Aid available to part-time students. Financial aid application deadline: 4/1; financial aid applicants required to submit FAFSA. *Unit head:* Dr. Eric M. Malstrom, Chair, Department of Industrial Engineering, 501-575-3157.

University of Bridgeport, College of Graduate and Undergraduate Studies, School of Science, Engineering, and Technology, Department of Mechanical Engineering, Program in Management Engineering, Bridgeport, CT 06601. Offers MS. *Faculty:* 1 full-time (0 women), 5 part-time (0 women). Average age 34. 24 applicants, 71% accepted. In 1998, 51 degrees awarded. *Degree requirements:* For master's, thesis optional, foreign language not required. *Entrance requirements:* For master's, TOEFL. *Application deadline:* Applications are processed on a rolling basis. Application fee: $35 ($50 for international students). *Financial aid:* In 1998–99, 3 students received aid; fellowships, research assistantships, teaching assistantships, career-related internships or fieldwork, Federal Work-Study, institutionally-sponsored loans, and tuition waivers (partial) available. Aid available to part-time students. Financial aid application deadline: 6/1; financial aid applicants required to submit FAFSA. *Faculty research:* CAD/CAM. *Unit head:* Dr. Paul Bauer, Coordinator, 203-576-4379.

University of California, Berkeley, Graduate Division, College of Engineering, Department of Industrial Engineering and Operations Research, Berkeley, CA 94720-1500. Offers M Eng, MS, D Eng, PhD. *Students:* 69 full-time (14 women); includes 14 minority (2 African Americans, 10 Asian Americans or Pacific Islanders, 2 Hispanic Americans), 35 international. 230 applicants, 34% accepted. In 1998, 43 master's, 6 doctorates awarded. *Degree requirements:* For master's, comprehensive exam or thesis (MS) required; for doctorate, dissertation, qualifying exam required. *Entrance requirements:* For master's and doctorate, GRE General Test, minimum GPA of 3.0. *Average time to degree:* Master's–1.5 years full-time; doctorate–5.5 years full-time. *Application deadline:* For fall admission, 2/10; for spring admission, 9/1. Application fee: $40. *Financial aid:* Fellowships, research assistantships, teaching assistantships, career-related internships or fieldwork, Federal Work-Study, and tuition waivers (full and partial) available. Financial aid application deadline: 1/5. *Faculty research:* Mathematical programming, robotics and manufacturing, linear and nonlinear optimization, production planning and scheduling, queuing theory. *Unit head:* Dr. Candace A. Yano, Chair, 510-642-4992, Fax: 510-642-1403, E-mail: yano@ieor.berkeley.edu. *Application contact:* Marion Brackett, Student Affairs Officer, 510-642-5485, Fax: 510-642-1403, E-mail: std-svcs@ieor.berkeley.edu.

See in-depth description on page 1223.

University of Central Florida, College of Engineering, Department of Industrial Engineering and Management Systems, Orlando, FL 32816. Offers computer-integrated manufacturing (MS); engineering management (MS); industrial engineering (MSIE); industrial engineering and management systems (PhD); manufacturing engineering (MS Mfg E); operations research (MS); product assurance engineering (MS); simulation systems (MS). Part-time and evening/weekend programs available. *Faculty:* 20 full-time, 11 part-time. *Students:* 143 full-time (27 women), 112 part-time (20 women); includes 43 minority (6 African Americans, 12 Asian Americans or Pacific Islanders, 25 Hispanic Americans), 51 international. Average age 34. 112 applicants, 51% accepted. In 1998, 99 master's, 7 doctorates awarded. *Degree requirements:* For master's, computer language, thesis or alternative required, foreign language not required; for doctorate, computer language, dissertation, departmental qualifying exam, candidacy exam required, foreign language not required. *Entrance requirements:* For master's, GRE General Test (minimum combined score of 1000 required), TOEFL (minimum score of 550 required; 213 computer-based), minimum GPA of 3.0 in last 60 hours; for doctorate, TOEFL (minimum score of 550 required; 213 computer-based), minimum GPA of 3.5 in last 60 hours. *Application deadline:* For fall admission, 7/15; for spring admission, 12/15. Application fee: $20. Tuition, state resident: full-time $2,054; part-time $137 per credit. Tuition, nonresident: full-time $7,207; part-time $480 per credit. Required fees: $47 per term. *Financial aid:* In 1998–99, 23 fellowships with partial tuition reimbursements (averaging $1,901 per year), 165 research assistantships with partial tuition reimbursements (averaging $2,397 per year), 5 teaching assistantships with partial tuition reimbursements (averaging $1,567 per year) were awarded.; career-related internships or fieldwork, Federal Work-Study, institutionally-sponsored loans, tuition waivers (partial), and unspecified assistantships also available. Financial aid application deadline: 3/1; financial aid applicants required to submit FAFSA. *Unit head:* Dr. Charles Reily, Chair, 407-823-2204. *Application contact:* Dr. Linda Malone, Coordinator, 407-823-2204.

See in-depth description on page 1225.

University of Cincinnati, Division of Research and Advanced Studies, College of Engineering, Department of Mechanical, Industrial and Nuclear Engineering, Program in Industrial Engineering, Cincinnati, OH 45221-0091. Offers MS, PhD, MBA/MS. Part-time and evening/weekend programs available. *Students:* 70 full-time (7 women), 28 part-time (8 women); includes 6 minority (2 African Americans, 4 Asian Americans or Pacific Islanders), 70 international. In 1998, 13 master's, 3 doctorates awarded. *Degree requirements:* For master's, computer language, oral exam, thesis defense required; for doctorate, variable foreign language requirement, computer language, dissertation, oral exam required. *Entrance requirements:* For master's, GRE General Test, TOEFL, TSE; for doctorate, GRE General Test, TOEFL (minimum score of 575 required), TSE (minimum score of 250 required). *Average time to degree:* Master's–3.3 years full-time; doctorate–4 years full-time. *Application deadline:* For fall admission, 2/1 (priority date). Application fee: $40. *Financial aid:* Fellowships, career-related internships or fieldwork, tuition waivers (full), and unspecified assistantships available. Aid available to part-time students. Financial aid application deadline: 2/1. *Faculty research:* Operations research, engineering administration, safety. *Application contact:* Sam Anand, Graduate Program Director, 513-556-5596, Fax: 516-556-4999, E-mail: sam.anand@uc.edu.

University of Dayton, Graduate School, School of Engineering, Program in Management Science, Dayton, OH 45469-1300. Offers MSMS. Part-time and evening/weekend programs available. *Faculty:* 4 full-time (0 women), 4 part-time (1 woman). *Students:* 17 full-time (4 women), 28 part-time (7 women); includes 16 minority (6 African Americans, 2 Asian Americans or Pacific Islanders, 8 Hispanic Americans), 2 international. Average age 28. In 1998, 20 degrees awarded. *Degree requirements:* For master's, computer language required, foreign language and thesis not required. *Entrance requirements:* For master's, GRE General Test, TOEFL. *Application deadline:* For fall admission, 8/1. Application fee: $30. *Financial aid:* Research assistantships, teaching assistantships available. *Faculty research:* Artificial intelligence, program management, reliability, simulation modeling. *Unit head:* Dr. Patrick J. Sweeney, Chairman, 937-229-2238, Fax: 937-229-2756, E-mail: psweeney@engr.udayton.edu. *Application contact:* Dr. Donald L. Moon, Associate Dean, 937-229-2241, Fax: 937-229-2471, E-mail: dmoon@engr.udayton.edu.

University of Florida, Graduate School, College of Engineering, Department of Industrial and Systems Engineering, Gainesville, FL 32611. Offers engineering management (ME, MS);

Industrial/Management Engineering

University of Florida (continued)
facilities layout decision support systems energy (PhD); health systems (ME, MS); industrial engineering (PhD, Engr); manufacturing systems engineering (ME, MS, PhD, Certificate); operations research (ME, MS, PhD, Engr); production planning and control engineering management (PhD); quality and reliability assurance (ME, MS); systems engineering (PhD, Engr). *Faculty:* 22. *Students:* 56 full-time (14 women), 69 part-time (18 women); includes 17 minority (5 African Americans, 4 Asian Americans or Pacific Islanders, 8 Hispanic Americans), 48 international. 301 applicants, 65% accepted. In 1998, 64 master's, 3 doctorates awarded. *Degree requirements:* For master's, computer language, core exam required, thesis optional; for doctorate, dissertation, comprehensive exam required; for other advanced degree, computer language, thesis required. *Entrance requirements:* For master's and doctorate, GRE General Test, TOEFL, minimum GPA of 3.0; for other advanced degree, GRE General Test. *Application deadline:* For fall admission, 6/1 (priority date). Applications are processed on a rolling basis. Application fee: $20. Electronic applications accepted. *Financial aid:* In 1998–99, 27 students received aid, including 12 fellowships, 22 research assistantships, 2 teaching assistantships; career-related internships or fieldwork, Federal Work-Study, and unspecified assistantships also available. Financial aid application deadline: 1/10. *Unit head:* Dr. Donald Hearn, Chair, 352-392-1464, Fax: 352-392-3537, E-mail: hearn@ise.ufl.edu. *Application contact:* Dr. D. J. Elzinga, Graduate Coordinator, 352-392-1464, Fax: 352-392-3537, E-mail: elzinga@ise.ufl.edu.

See in-depth description on page 1227.

University of Florida, Graduate School, Graduate Engineering and Research Center (GERC), Gainesville, FL 32611. Offers aerospace engineering (ME, MS, PhD, Engr); electrical and computer engineering (ME, MS, PhD, Engr); engineering mechanics (ME, MS, PhD, Engr); industrial and systems engineering (ME, MS, PhD, Engr). Part-time programs available. Postbaccalaureate distance learning degree programs offered. *Faculty:* 6 full-time (0 women), 16 part-time (1 woman). *Students:* 13 full-time (6 women), 159 part-time (29 women); includes 21 minority (8 African Americans, 8 Asian Americans or Pacific Islanders, 4 Hispanic Americans, 1 Native American) Terminal master's awarded for partial completion of doctoral program. *Degree requirements:* For master's, computer language required (for some programs), thesis optional, foreign language not required; for doctorate, computer language (for some programs), dissertation required; for Engr, computer language (for some programs), thesis required, foreign language not required. *Entrance requirements:* For master's, GRE General Test (minimum combined score of 1000 required), TOEFL, minimum GPA of 3.0; for doctorate, GRE General Test (minimum combined score of 1200 required), written and oral qualifying exams, TOEFL, minimum GPA of 3.0, master's degree in engineering; for Engr, GRE General Test (minimum combined score of 1000 required), TOEFL, minimum GPA of 3.0, master's degree in engineering. *Application deadline:* For fall admission, 6/1; for spring admission, 10/1. Applications are processed on a rolling basis. Application fee: $20. Electronic applications accepted. *Unit head:* Dr. Pasquale M. Sforza, Director, 850-833-9355, Fax: 850-833-9366, E-mail: sforza@gerc.eng.ufl.edu. *Application contact:* Judi Shivers, Program Assistant, 850-833-9350, Fax: 850-833-9366, E-mail: reginfo@gerc.eng.ufl.edu.

University of Houston, Cullen College of Engineering, Department of Industrial Engineering, Houston, TX 77004. Offers MIE, MSIE, PhD, MBA/MIE. Part-time and evening/weekend programs available. *Faculty:* 9 full-time (1 woman), 3 part-time (0 women). *Students:* 41 full-time (10 women), 48 part-time (10 women); includes 14 minority (1 African American, 9 Asian Americans or Pacific Islanders, 4 Hispanic Americans), 32 international. Average age 32. 103 applicants, 45% accepted. In 1998, 16 master's, 4 doctorates awarded. Terminal master's awarded for partial completion of doctoral program. *Degree requirements:* For master's, thesis required (for some programs), foreign language not required; for doctorate, dissertation, departmental qualifying exam required, foreign language not required. *Entrance requirements:* For master's and doctorate, GRE General Test, TOEFL. *Application deadline:* For fall admission, 7/3 (priority date); for spring admission, 12/4. Applications are processed on a rolling basis. Application fee: $25 ($75 for international students). *Financial aid:* In 1998–99, 2 fellowships, 12 research assistantships, 14 teaching assistantships were awarded.; career-related internships or fieldwork, Federal Work-Study, and institutionally-sponsored loans also available. Financial aid application deadline: 7/1. *Faculty research:* CAD/CAM, artificial intelligence, applied operations research, engineering management, human factors and safety. Total annual research expenditures: $368,859. *Unit head:* Dr. Jen-Gwo Chen, Chairman, 713-743-4180, Fax: 713-743-4190, E-mail: jgchen@uh.edu. *Application contact:* Mary Patronella, Graduate Analyst, 713-743-4188, Fax: 713-743-4190, E-mail: hineamip@admin.uh.edu.

University of Illinois at Chicago, Graduate College, College of Engineering, Department of Mechanical Engineering, Program in Industrial Engineering, Chicago, IL 60607-7128. Offers industrial engineering (MS), including safety engineering. *Students:* 4 full-time (0 women), 8 part-time (1 woman); includes 4 minority (1 African American, 1 Asian American or Pacific Islander, 2 Hispanic Americans), 5 international. Average age 27. 37 applicants, 30% accepted. In 1998, 2 degrees awarded. *Degree requirements:* For master's, thesis required, foreign language not required. *Entrance requirements:* For master's, GRE General Test, TOEFL (minimum score of 550 required), minimum GPA of 3.75 on a 5.0 scale. *Application deadline:* For fall admission, 6/1; for spring admission, 10/15. Application fee: $40 ($50 for international students). *Financial aid:* In 1998–99, 4 students received aid; fellowships, research assistantships, teaching assistantships available. *Faculty research:* Systems modeling. *Unit head:* Krishna Gupta, Director of Graduate Studies. *Application contact:* Krishna Gupta, Director of Graduate Studies.

University of Illinois at Chicago, Graduate College, College of Engineering, Department of Mechanical Engineering, Program in Industrial Engineering and Operations Research, Chicago, IL 60607-7128. Offers PhD. *Students:* 6 full-time (0 women), 3 part-time, 8 international. Average age 27. 7 applicants, 0% accepted. In 1998, 2 degrees awarded. *Degree requirements:* For doctorate, dissertation required, foreign language not required. *Entrance requirements:* For doctorate, GRE General Test, TOEFL (minimum score of 550 required), minimum GPA of 3.75 on a 5.0 scale. *Application deadline:* For fall admission, 6/1; for spring admission, 10/15. Application fee: $40 ($50 for international students). *Financial aid:* Fellowships, research assistantships, teaching assistantships available. *Unit head:* Dr. Selcuk Guceri, Head, Department of Mechanical Engineering, 312-996-5096.

University of Illinois at Urbana–Champaign, Graduate College, College of Engineering, Department of Mechanical and Industrial Engineering, Urbana, IL 61801. Offers industrial engineering (MS, PhD); mechanical engineering (MS, PhD). *Faculty:* 47 full-time (4 women). *Students:* 247 full-time (24 women); includes 40 minority (6 African Americans, 25 Asian Americans or Pacific Islanders, 9 Hispanic Americans), 78 international. 503 applicants, 15% accepted. In 1998, 79 master's, 22 doctorates awarded. Terminal master's awarded for partial completion of doctoral program. *Degree requirements:* For master's, thesis, non-thesis available by petition required, foreign language not required; for doctorate, dissertation required, foreign language not required. *Entrance requirements:* For master's, GRE General Test, TOEFL (minimum score of 610 required; 257 for computer-based), minimum GPA of 4.25 on a 5.0 scale; for doctorate, GRE General Test, TOEFL (minimum score of 610 required; 257 for computer-based). *Application deadline:* For fall admission, 3/1; for spring admission, 10/1. Applications are processed on a rolling basis. Application fee: $40 ($50 for international students). Tuition, state resident: full-time $4,616. Tuition, nonresident: full-time $11,768. Full-time tuition and fees vary according to course load. *Financial aid:* In 1998–99, 247 students received aid, including 99 fellowships, 214 research assistantships, 38 teaching assistantships; Federal Work-Study, institutionally-sponsored loans, and tuition waivers (full and partial) also available. Financial aid application deadline: 3/1. *Faculty research:* Combustion and propulsion, design methodology, dynamic systems and controls, energy transfer, materials behavior and processing, manufacturing systems operations, management. Total annual research expenditures: $8 million. *Unit head:* Dr. Richard O. Buckius, Head, 217-333-1079. *Application contact:* Celia Snyder, Graduate Program Coordinator, 217-333-4390, Fax: 217-244-6534, E-mail: cgsnyder@uiuc.edu.

The University of Iowa, Graduate College, College of Engineering, Department of Industrial Engineering, Iowa City, IA 52242-1316. Offers engineering design and manufacturing (MS, PhD); ergonomics (MS, PhD); information and engineering management (MS, PhD); operations research (MS, PhD); quality engineering (MS, PhD). *Faculty:* 7 full-time, 1 part-time. *Students:* 18 full-time (4 women), 16 part-time (1 woman); includes 4 minority (1 African American, 2 Asian Americans or Pacific Islanders, 1 Hispanic American), 19 international. 214 applicants, 54% accepted. In 1998, 6 master's, 3 doctorates awarded. *Degree requirements:* For master's, thesis optional; for doctorate, dissertation, comprehensive exam required. *Entrance requirements:* For master's and doctorate, GRE General Test, GRE Subject Test, TOEFL. *Application deadline:* Applications are processed on a rolling basis. Application fee: $30 ($50 for international students). *Financial aid:* In 1998–99, 18 research assistantships, 10 teaching assistantships were awarded.; fellowships Financial aid applicants required to submit FAFSA. *Faculty research:* Artificial intelligence, planning and analysis, reliability and maintainability, human factors (ergonomics), manufacturing. *Unit head:* Peter J. O'Grady, Chair, 319-335-5939, Fax: 319-335-5424.

University of Louisville, Graduate School, Speed Scientific School, Department of Industrial Engineering, Louisville, KY 40292-0001. Offers M Eng, MS, PhD. *Accreditation:* ABET (one or more programs are accredited). *Faculty:* 10 full-time (2 women), 1 part-time (0 women). *Students:* 24 full-time (5 women), 65 part-time (13 women); includes 10 minority (7 African Americans, 2 Asian Americans or Pacific Islanders, 1 Hispanic American), 19 international. Average age 31. In 1998, 12 master's, 3 doctorates awarded. *Degree requirements:* For master's and doctorate, thesis/dissertation required, foreign language not required. *Entrance requirements:* For master's and doctorate, GRE General Test (minimum combined score of 1200 required). *Application deadline:* Applications are processed on a rolling basis. Application fee: $25. *Financial aid:* In 1998–99, 2 fellowships with full tuition reimbursements (averaging $12,000 per year), 2 research assistantships with full tuition reimbursements (averaging $12,000 per year), 6 teaching assistantships with full tuition reimbursements (averaging $12,000 per year) were awarded. *Faculty research:* Operations research, quality control, facilities planning, and material housing; industrial ergonomics, expert systems, and engineering economy. *Unit head:* Dr. Waldemar Karwowski, Chair, 502-852-6342, Fax: 502-852-5633, E-mail: w0karw01@gwise.louisville.edu. *Application contact:* Nancy D. White, Executive Secretary, 502-852-6342, Fax: 502-852-5633, E-mail: ndwhit01@gwise.louisville.edu.

University of Manitoba, Faculty of Graduate Studies, Faculty of Engineering, Department of Mechanical and Industrial Engineering, Winnipeg, MB R3T 2N2, Canada. Offers M Eng, M Sc, PhD. *Degree requirements:* For master's and doctorate, thesis/dissertation required. *Unit head:* D. W. Ruth, Head.

University of Massachusetts Amherst, Graduate School, College of Engineering, Department of Mechanical and Industrial Engineering, Program in Industrial Engineering and Operations Research, Amherst, MA 01003-2210. Offers MS, PhD. *Students:* 15 full-time (2 women), 17 part-time (6 women); includes 4 minority (1 African American, 3 Hispanic Americans), 21 international. Average age 29. 103 applicants, 44% accepted. In 1998, 10 master's, 8 doctorates awarded. *Degree requirements:* For master's, project required, foreign language and thesis not required; for doctorate, dissertation required, foreign language not required. *Entrance requirements:* For master's and doctorate, GRE General Test. *Application deadline:* For fall admission, 2/1 (priority date). Applications are processed on a rolling basis. Application fee: $35. Tuition, state resident: full-time $2,640; part-time $165 per credit. Tuition, nonresident: full-time $9,756; part-time $407 per credit. Required fees: $1,221 per term. One-time fee: $110. Full-time tuition and fees vary according to course load, campus/location and reciprocity agreements. *Financial aid:* Fellowships with full tuition reimbursements, research assistantships with full tuition reimbursements, teaching assistantships with full tuition reimbursements, career-related internships or fieldwork, Federal Work-Study, grants, scholarships, traineeships, and unspecified assistantships available. Aid available to part-time students. Financial aid application deadline: 2/1. *Unit head:* Dr. Donald Fisher, Director, 413-545-0955, Fax: 413-545-1027, E-mail: dfisher@ecs.umass.edu.

University of Massachusetts Lowell, Graduate School, James B. Francis College of Engineering, Department of Work Environment, Lowell, MA 01854-2881. Offers MS, Sc D. *Faculty:* 12 full-time (4 women). *Students:* 41 full-time (17 women), 85 part-time (28 women); includes 7 minority (5 African Americans, 2 Hispanic Americans), 5 international. 50 applicants, 88% accepted. In 1998, 11 master's, 1 doctorate awarded. Terminal master's awarded for partial completion of doctoral program. *Degree requirements:* For master's, thesis required, foreign language not required; for doctorate, computer language, dissertation required. *Entrance requirements:* For master's and doctorate, GRE General Test. *Application deadline:* For fall admission, 4/1 (priority date); for spring admission, 10/1. Applications are processed on a rolling basis. Application fee: $20 ($35 for international students). *Financial aid:* In 1998–99, 6 fellowships, 7 research assistantships, 3 teaching assistantships were awarded. Financial aid application deadline: 4/1. *Faculty research:* Ergonomics, industrial hygiene, epidemiology, work environment policy, toxic use reduction. Total annual research expenditures: $854,000. *Unit head:* Dr. Rafael Moure-Eraso, Chair, 978-934-3271. *Application contact:* Dr. Michael Ellenbecker, Coordinator, 978-934-3272.

The University of Memphis, Graduate School, Herff College of Engineering, Program in Industrial and Systems Engineering, Memphis, TN 38152. Offers MS. Part-time programs available. *Faculty:* 3 full-time (1 woman). *Students:* 3 full-time (0 women), 4 part-time; includes 1 minority (Asian American or Pacific Islander), 2 international. Average age 32. 38 applicants, 13% accepted. In 1998, 5 degrees awarded. *Degree requirements:* For master's, thesis or alternative, comprehensive exam required. *Entrance requirements:* For master's, GRE General Test (minimum combined score of 1000 required) or MAT, minimum undergraduate GPA of 2.5. *Application deadline:* For fall admission, 8/1; for spring admission, 12/1. Application fee: $25 ($50 for international students). Tuition, state resident: full-time $3,410; part-time $178 per credit hour. Tuition, nonresident: full-time $8,670; part-time $408 per credit hour. Tuition and fees vary according to program. *Financial aid:* In 1998–99, 5 students received aid, including 1 fellowship, 4 research assistantships; teaching assistantships, career-related internships or fieldwork also available. *Faculty research:* Integer programming, ergonomics, scheduling. Total annual research expenditures: $125,000. *Unit head:* Dr. Michael Racer, Coordinator, 901-678-3285, E-mail: mracer@memphis.edu.

University of Miami, Graduate School, College of Engineering, Department of Industrial Engineering, Coral Gables, FL 33124. Offers environmental health and safety (MS, MSEH), including environmental health and safety (MSEH), occupational ergonomics and safety (MS); ergonomics (PhD); industrial engineering (MSIE, PhD); management of technology (MS). *Faculty:* 10 full-time (1 woman), 3 part-time (0 women). *Students:* 50 full-time (6 women), 3 part-time (2 women); includes 14 minority (3 African Americans, 3 Asian Americans or Pacific Islanders, 8 Hispanic Americans), 23 international. Average age 28. 79 applicants, 75% accepted. In 1998, 5 master's, 1 doctorate awarded. *Degree requirements:* For master's, thesis required (for some programs), foreign language not required; for doctorate, dissertation required, foreign language not required. *Entrance requirements:* For master's, GRE General Test (minimum combined score of 1000 required), TOEFL (minimum score of 550 required), minimum GPA of 3.0; for doctorate, GRE General Test, TOEFL (minimum score of 550 required), minimum GPA of 3.0. *Application deadline:* For fall admission, 5/1 (priority date). Applications are processed on a rolling basis. Application fee: $35. Tuition: Full-time $15,336; part-time $852 per credit. Required fees: $174. Tuition and fees vary according to program. *Financial aid:* In 1998–99, 4 fellowships, 5 research assistantships, 4 teaching assistantships were awarded.; Federal Work-Study also available. Financial aid application deadline: 3/1; financial aid applicants required to submit FAFSA. *Faculty research:* Industrial applications of biomechanics and ergonomics, technology management, back pain, aging and the elderly, operations research, manufacturing, safety, human reliability. Total annual research expenditures: $306,383. *Unit head:* Dr. Norman Einspruch, Chairman, 305-284-2344, Fax: 305-284-4040. *Application contact:* Dr. Sara Czaja, Professor, 305-284-2344, Fax: 305-284-4040.

Industrial/Management Engineering

University of Michigan, Horace H. Rackham School of Graduate Studies, College of Engineering, Department of Industrial and Operations Engineering, Ann Arbor, MI 48109. Offers MS, MSE, PhD, IOE, MBA/MS, MBA/MSE, MHSA/MS. Part-time programs available. *Faculty:* 26 full-time (2 women). *Students:* 206 full-time (50 women). 255 applicants, 57% accepted. Terminal master's awarded for partial completion of doctoral program. *Degree requirements:* For master's, computer language required, foreign language and thesis not required; for doctorate, computer language, dissertation, oral defense of dissertation, preliminary exams required, foreign language not required. *Entrance requirements:* For master's, GRE General Test (minimum combined score of 1850 on three sections required), TOEFL, minimum GPA of 3.2; for doctorate, GRE General Test (minimum combined score of 1950 on three sections required), TOEFL, minimum GPA of 3.5. *Application deadline:* For fall admission, 5/1; for winter admission, 11/1. Applications are processed on a rolling basis. Application fee: $55. Electronic applications accepted. *Financial aid:* In 1998–99, 15 fellowships, 40 research assistantships, 25 teaching assistantships were awarded.; Federal Work-Study and institutionally-sponsored loans also available. Financial aid application deadline: 2/1. *Faculty research:* Optimization and stochastic processes, human performance and ergonomics, information systems, engineering management, quality engineering. *Unit head:* Dr. John R. Birge, Chair, 734-764-6480, Fax: 734-764-3451, E-mail: jrbirge@umich.edu. *Application contact:* Frances Bourdas, Graduate Program Assistant, 734-764-6480, Fax: 734-764-3451, E-mail: bourdas@engin.umich.edu.

University of Michigan–Dearborn, College of Engineering and Computer Science, Department of Industrial and Systems Engineering, Dearborn, MI 48128-1491. Offers engineering management (MS); industrial and systems engineering (MSE). Part-time and evening/weekend programs available. *Faculty:* 10 full-time (0 women), 7 part-time (1 woman). *Students:* 8 full-time (4 women), 294 part-time (80 women). Average age 29. In 1998, 68 degrees awarded. *Degree requirements:* For master's, computer language required, thesis optional, foreign language not required. *Entrance requirements:* For master's, bachelor's degree in applied mathematics, computer science, engineering, or physical science; minimum GPA of 3.0. *Application deadline:* For fall admission, 8/1 (priority date); for winter admission, 12/1 (priority date); for spring admission, 4/1. Applications are processed on a rolling basis. Application fee: $55. Electronic applications accepted. Tuition, state resident: part-time $259 per credit hour. Tuition, nonresident: part-time $748 per credit hour. Required fees: $80 per course. Tuition and fees vary according to course level, course load and program. *Financial aid:* Fellowships, research assistantships, teaching assistantships, Federal Work-Study available. *Faculty research:* Health care systems, databases, human factors, machine diagnostics, precision machining. *Unit head:* Dr. S. K. Kachhal, Chair, 313-593-5361, Fax: 313-593-3692. *Application contact:* Lisa M. Beach, Administrative Assistant, 313-593-5361, Fax: 313-593-3692, E-mail: lmbeach@umich.edu.

University of Minnesota, Twin Cities Campus, Graduate School, Institute of Technology, Department of Mechanical Engineering, Program in Industrial Engineering, Minneapolis, MN 55455-0213. Offers MIE, MSIE, PhD. Part-time programs available. *Degree requirements:* For master's, foreign language and thesis not required; for doctorate, dissertation required, foreign language not required. *Entrance requirements:* For master's, GRE General Test (score in 80th percentile or higher required; average 85th percentile), minimum GPA of 3.0; for doctorate, GRE General Test (score in 80th percentile or higher required; average 85th percentile).

See in-depth description on page 1233.

University of Missouri–Columbia, Graduate School, College of Engineering, Department of Industrial and Manufacturing Systems Engineering, Columbia, MO 65211. Offers MS, PhD. *Faculty:* 8 full-time (1 woman). *Students:* 17 full-time (1 woman), 11 part-time (1 woman); includes 2 minority (both African Americans), 24 international. 31 applicants, 48% accepted. In 1998, 8 master's, 6 doctorates awarded. *Degree requirements:* For master's, computer language, thesis or alternative required, foreign language not required; for doctorate, dissertation required. *Entrance requirements:* For master's and doctorate, GRE General Test, TOEFL, minimum GPA of 3.0. *Application deadline:* For fall admission, 4/1 (priority date). Applications are processed on a rolling basis. Application fee: $30 ($50 for international students). *Financial aid:* Research assistantships, teaching assistantships, institutionally-sponsored loans available. *Unit head:* Dr. Cerry Klein, Director of Graduate Studies, 573-882-9566.

See in-depth description on page 1235.

University of Nebraska–Lincoln, Graduate College, College of Engineering and Technology, Department of Industrial and Management Systems Engineering, Lincoln, NE 68588. Offers engineering (PhD); industrial and management systems engineering (MS). *Faculty:* 9 full-time (2 women), 1 part-time (0 women). *Students:* 9 full-time (2 women), 19 part-time (6 women); includes 1 minority (Asian American or Pacific Islander), 10 international. Average age 32. 42 applicants, 48% accepted. In 1998, 8 degrees awarded. *Degree requirements:* For master's, thesis optional, foreign language not required; for doctorate, dissertation, comprehensive exams required. *Entrance requirements:* For master's and doctorate, GRE General Test, TOEFL (minimum score of 525 required). *Application deadline:* For fall admission, 3/1 (priority date). Applications are processed on a rolling basis. Application fee: $35. Electronic applications accepted. *Financial aid:* In 1998–99, 21 research assistantships, 4 teaching assistantships were awarded.; fellowships, Federal Work-Study also available. Aid available to part-time students. Financial aid application deadline: 2/15. *Faculty research:* Operations research, ergonomics, manufacturing systems, nontraditional manufacturing and packaging. *Unit head:* Dr. Michael W. Riley, Chair, 402-472-3495.

University of New Haven, Graduate School, School of Engineering and Applied Science, Program in Industrial Engineering, West Haven, CT 06516-1916. Offers industrial engineering (MSIE); logistics (Certificate). Part-time and evening/weekend programs available. *Students:* 6 full-time (0 women), 9 part-time; includes 3 minority (1 African American, 1 Asian American or Pacific Islander, 1 Hispanic American), 7 international. 13 applicants, 69% accepted. In 1998, 22 degrees awarded. *Degree requirements:* For master's, computer language, thesis or alternative required, foreign language not required. *Entrance requirements:* For master's, bachelor's degree in engineering. *Application deadline:* Applications are processed on a rolling basis. Application fee: $50. *Financial aid:* Federal Work-Study available. Aid available to part-time students. Financial aid application deadline: 5/1; financial aid applicants required to submit FAFSA. *Unit head:* Dr. Ronald Wentworth, Coordinator, 203-932-7434.

University of Oklahoma, Graduate College, College of Engineering, School of Industrial Engineering, Norman, OK 73019-0390. Offers MS, PhD. Part-time and evening/weekend programs available. *Faculty:* 10 full-time (3 women), 3 part-time (0 women). *Students:* 43 full-time (7 women), 43 part-time (5 women); includes 6 minority (3 African Americans, 2 Asian Americans or Pacific Islanders, 1 Hispanic American), 61 international. Average age 30. 75 applicants, 87% accepted. In 1998, 27 master's, 4 doctorates awarded. Terminal master's awarded for partial completion of doctoral program. *Degree requirements:* For master's, computer language, comprehensive exam required, thesis optional, foreign language not required; for doctorate, computer language, dissertation, qualifying exam required, foreign language not required. *Entrance requirements:* For master's and doctorate, TOEFL (minimum score of 550 required). *Application deadline:* For fall admission, 6/1 (priority date). Applications are processed on a rolling basis. Application fee: $25. Tuition, state resident: part-time $86 per credit hour. Tuition, nonresident: part-time $275 per credit hour. Tuition and fees vary according to course level, course load and program. *Financial aid:* In 1998–99, 39 students received aid, including 12 research assistantships, 11 teaching assistantships; fellowships, career-related internships or fieldwork, Federal Work-Study, institutionally-sponsored loans, and tuition waivers (full and partial) also available. Financial aid application deadline: 5/1. *Faculty research:* Human factors engineering, operations research, production and manufacturing systems, quality and reliability engineering. *Unit head:* Thomas L. Landers, Director, 405-325-3721. *Application contact:* Dr. Shivakuman Raman, Graduate Liaison, 405-325-3721.

University of Pittsburgh, School of Engineering, Department of Industrial Engineering, Pittsburgh, PA 15260. Offers MSIE, PhD. Part-time and evening/weekend programs available. *Faculty:* 12 full-time (2 women), 3 part-time (1 woman). *Students:* 53 full-time (10 women), 16 part-time (3 women); includes 5 minority (3 African Americans, 1 Asian American or Pacific Islander, 1 Hispanic American), 40 international. 109 applicants, 73% accepted. In 1998, 20 master's, 3 doctorates awarded. Terminal master's awarded for partial completion of doctoral program. *Degree requirements:* For master's, computer language required, thesis optional, foreign language not required; for doctorate, computer language, dissertation, comprehensive and final oral exams required, foreign language not required. *Entrance requirements:* For master's and doctorate, GRE General Test, TOEFL (minimum score of 550 required), minimum QPA of 3.0. *Average time to degree:* Master's–2 years full-time, 3 years part-time; doctorate–4 years full-time, 6 years part-time. *Application deadline:* For fall admission, 8/1 (priority date); for spring admission, 12/1 (priority date). Applications are processed on a rolling basis. Application fee: $30 ($40 for international students). *Financial aid:* In 1998–99, 34 students received aid, including 34 research assistantships (averaging $10,536 per year); fellowships, teaching assistantships, grants, scholarships, and tuition waivers (full and partial) also available. Financial aid application deadline: 2/15. *Faculty research:* Operations research, engineering management, computational intelligence, manufacturing, information systems. Total annual research expenditures: $1.5 million. *Unit head:* Dr. Harvey Wolfe, Chairman, 412-624-9830, Fax: 412-624-9831, E-mail: hwolfe@engrng.pitt.edu. *Application contact:* Dr. Mainak Mazumdar, Graduate Coordinator, 412-624-9839, Fax: 412-624-9831, E-mail: mmaz.umdar@engrng.pitt.edu.

University of Puerto Rico, Mayagüez Campus, Graduate Studies, College of Engineering, Department of Industrial Engineering, Mayagüez, PR 00681-5000. Offers MMSE. Part-time programs available. *Degree requirements:* For master's, thesis, comprehensive exam, project required, foreign language not required. *Entrance requirements:* For master's, minimum GPA of 2.5, proficiency in English and Spanish. *Faculty research:* Quality control operations research.

University of Regina, Faculty of Graduate Studies and Research, Faculty of Engineering, Program in Industrial Systems Engineering, Regina, SK S4S 0A2, Canada. Offers M Eng, MA Sc, PhD. *Students:* 4 full-time (1 woman), 20 part-time (5 women). 19 applicants, 47% accepted. In 1998, 5 degrees awarded. *Degree requirements:* For master's, thesis required (for some programs), foreign language not required; for doctorate, dissertation required, foreign language not required. *Entrance requirements:* For master's, TOEFL (minimum score of 550 required); for doctorate, TOEFL (minimum score of 550 required), master's degree. *Application deadline:* Applications are processed on a rolling basis. Application fee: $0. *Expenses:* Tuition and fees charges are reported in Canadian dollars. Tuition, state resident: full-time $1,688 Canadian dollars; part-time $94 Canadian dollars per credit hour. International tuition: $3,375 Canadian dollars full-time. Required fees: $65 Canadian dollars per course. Tuition and fees vary according to course load and program. *Financial aid:* In 1998–99, 5 research assistantships, 4 teaching assistantships were awarded.; fellowships, career-related internships or fieldwork and scholarships also available. Financial aid application deadline: 6/15. *Faculty research:* Gas separation and purification, welding weldability. *Unit head:* Dr. P. Catania, Head, 306-585-4364, Fax: 306-585-4855, E-mail: peter.catania@uregina.ca. *Application contact:* Dr. S. Bhole, Coordinator, 306-585-4703, Fax: 306-585-4855, E-mail: sanjeev.bhole@uregina.ca.

University of Regina, Faculty of Graduate Studies and Research, Interdisciplinary Studies, Program in Advanced Manufacturing and Processing, Regina, SK S4S 0A2, Canada. Offers M Sc, MA Sc. *Students:* 1 full-time (0 women), 5 part-time. 1 applicants, 100% accepted. In 1998, 1 degree awarded. *Degree requirements:* For master's, computer language, thesis required, foreign language not required. *Entrance requirements:* For master's, TOEFL (minimum score of 580 required). *Application deadline:* Applications are processed on a rolling basis. Application fee: $0. *Expenses:* Tuition and fees charges are reported in Canadian dollars. Tuition, state resident: full-time $1,688 Canadian dollars; part-time $94 Canadian dollars per credit hour. International tuition: $3,375 Canadian dollars full-time. Required fees: $65 Canadian dollars per course. Tuition and fees vary according to course load and program. *Financial aid:* In 1998–99, 2 research assistantships, 1 teaching assistantship were awarded.; scholarships also available. Financial aid application deadline: 6/15. *Unit head:* Dr. M. Chen, Coordinator, 306-585-4736, Fax: 306-585-4556, E-mail: mingyuan.chen@uregina.ca.

University of Rhode Island, Graduate School, College of Engineering, Department of Industrial and Manufacturing Engineering, Program in Industrial Engineering, Kingston, RI 02881. Offers MS.

University of Southern California, Graduate School, School of Engineering, Department of Industrial and Systems Engineering, Program in Industrial and Systems Engineering, Los Angeles, CA 90089. Offers MS, PhD, Engr, MBA/MS. *Students:* 39 full-time (2 women), 29 part-time (3 women); includes 10 minority (all Asian Americans or Pacific Islanders), 38 international. Average age 31. 160 applicants, 73% accepted. In 1998, 15 master's, 4 doctorates awarded. *Degree requirements:* For master's, thesis optional; for doctorate, dissertation required. *Entrance requirements:* For master's, doctorate, and Engr, GRE General Test. *Application deadline:* For fall admission, 6/1 (priority date); for spring admission, 12/1. Application fee: $55. Tuition: Part-time $768 per unit. Required fees: $350 per semester. *Financial aid:* In 1998–99, 8 fellowships, 11 research assistantships, 7 teaching assistantships were awarded.; Federal Work-Study, institutionally-sponsored loans, and scholarships also available. Aid available to part-time students. Financial aid application deadline: 2/15; financial aid applicants required to submit FAFSA. *Unit head:* Dr. F. Stan Settles, Chairman, Department of Industrial and Systems Engineering, 213-740-4893.

University of Southern Colorado, College of Applied Science and Engineering Technology, Program in Industrial and Systems Engineering, Pueblo, CO 81001-4901. Offers MS. Part-time and evening/weekend programs available. *Faculty:* 5 full-time (1 woman), 2 part-time (0 women). *Students:* 21 full-time (1 woman), 16 part-time (1 woman); includes 3 minority (all Hispanic Americans), 24 international. Average age 29. 23 applicants, 74% accepted. In 1998, 16 degrees awarded. *Degree requirements:* For master's, computer language required, thesis optional, foreign language not required. *Entrance requirements:* For master's, GRE General Test, TOEFL. *Average time to degree:* Master's–1.5 years full-time, 3.5 years part-time. *Application deadline:* For fall admission, 7/19 (priority date); for spring admission, 11/30 (priority date). Applications are processed on a rolling basis. Application fee: $15 ($30 for international students). *Financial aid:* In 1998–99, 5 teaching assistantships were awarded.; career-related internships or fieldwork, Federal Work-Study, institutionally-sponsored loans, and scholarships also available. Financial aid application deadline: 3/1; financial aid applicants required to submit FAFSA. *Faculty research:* Computer-integrated manufacturing, reliability, economic development, design of experiments, scheduling, simulation. Total annual research expenditures: $178,000. *Unit head:* Dr. Jane M. Fraser, Chair, 719-549-2036, Fax: 719-549-2519, E-mail: jfraser@uscolo.edu. *Application contact:* Dr. Huseyin Sarper, Graduate Coordinator, 719-549-2889, Fax: 719-549-2519, E-mail: sarper@uscolo.edu.

University of South Florida, Graduate School, College of Engineering, Department of Industrial and Management Systems Engineering, Tampa, FL 33620-9951. Offers engineering management (ME, MSE, MSEM, MSIE); engineering science (PhD); industrial engineering (ME, MSE, MSEM, MSIE, PhD). Part-time programs available. Postbaccalaureate distance learning degree programs offered (minimal on-campus study). *Faculty:* 9 full-time (2 women), 1 (woman) part-time. *Students:* 26 full-time (5 women), 77 part-time (12 women); includes 18 minority (6 African Americans, 1 Asian American or Pacific Islander, 11 Hispanic Americans), 21 international. Average age 35. 91 applicants, 84% accepted. In 1998, 41 master's awarded (90% found work related to degree, 10% continued full-time study); 3 doctorates awarded (67% entered university research/teaching, 33% found other work related to degree). Terminal master's awarded for partial completion of doctoral program. *Degree requirements:* For master's, thesis required (for some programs), foreign language not required; for doctorate, computer language, dissertation, 2 tools of research as specified by dissertation committee required, foreign language not required. *Entrance requirements:* For master's, GRE General Test (minimum

Industrial/Management Engineering

University of South Florida (continued)
combined score of 1100 required), minimum GPA of 3.0 during previous 2 years; for doctorate, GRE General Test (minimum combined score of 1200 required). *Average time to degree:* Master's–2.5 years full-time, 4 years part-time; doctorate–4 years full-time, 7 years part-time. *Application deadline:* For fall admission, 6/1; for spring admission, 10/15. Application fee: $20. Electronic applications accepted. Tuition, state resident: part-time $148 per credit hour. Tuition, nonresident: part-time $509 per credit hour. *Financial aid:* In 1998–99, 1 fellowship with full tuition reimbursement (averaging $7,000 per year), 13 research assistantships with full tuition reimbursements (averaging $11,631 per year), 6 teaching assistantships with full tuition reimbursements (averaging $15,009 per year) were awarded.; career-related internships or fieldwork, Federal Work-Study, institutionally-sponsored loans, and tuition waivers (partial) also available. Aid available to part-time students. Financial aid application deadline: 3/31; financial aid applicants required to submit FAFSA. *Faculty research:* Quality control, optimization techniques, stochastic processes, computer automated manufacturing. Total annual research expenditures: $149,753. *Unit head:* Dr. Paul E. Givens, Chairperson, 813-974-2269, Fax: 813-974-5953, E-mail: givens@eng.usf.edu. *Application contact:* Dr. Michael X. Weng, Graduate Director, 813-974-5575, Fax: 813-974-5953, E-mail: weng@eng.usf.edu.

University of Tennessee, Knoxville, Graduate School, College of Engineering, Department of Industrial Engineering, Knoxville, TN 37996. Offers engineering management (MS); manufacturing systems engineering (MS); traditional industrial engineering (MS). Part-time programs available. Postbaccalaureate distance learning degree programs offered (no on-campus study). *Faculty:* 10 full-time (1 woman). *Students:* 25 full-time (7 women), 111 part-time (29 women); includes 14 minority (10 African Americans, 3 Asian Americans or Pacific Islanders, 1 Hispanic American), 11 international. 41 applicants, 29% accepted. In 1998, 28 degrees awarded. *Degree requirements:* For master's, thesis or alternative required, foreign language not required. *Entrance requirements:* For master's, TOEFL (minimum score of 550 required), minimum GPA of 2.7. *Application deadline:* For fall admission, 2/1 (priority date). Applications are processed on a rolling basis. Application fee: $35. Electronic applications accepted. *Financial aid:* In 1998–99, 2 research assistantships were awarded.; fellowships, teaching assistantships, career-related internships or fieldwork, Federal Work-Study, institutionally-sponsored loans, and unspecified assistantships also available. Financial aid application deadline: 2/1; financial aid applicants required to submit FAFSA. *Unit head:* Dr. Thomas Shannon, Head, 423-974-3333, Fax: 423-974-0588, E-mail: tshannon@utk.edu. *Application contact:* Dr. J. A. Bontadelli, Graduate Representative, E-mail: bontadelli@utk.edu.

University of Tennessee, Knoxville, Graduate School, College of Engineering, Department of Mechanical and Aerospace Engineering and Engineering Science, Program in Engineering Science, Knoxville, TN 37996. Offers applied artificial intelligence (MS); biomedical engineering (MS, PhD); composite materials (MS, PhD); computational mechanics (MS, PhD); engineering science (MS, PhD); fluid mechanics (MS, PhD); industrial engineering (MS, PhD); optical engineering (MS, PhD); solid mechanics (MS, PhD). Part-time programs available. *Students:* 33 full-time (9 women), 16 part-time (1 woman); includes 4 minority (1 African American, 2 Asian Americans or Pacific Islanders, 1 Native American), 10 international. *Degree requirements:* For master's, thesis or alternative required, foreign language not required; for doctorate, dissertation required, foreign language not required. *Entrance requirements:* For master's and doctorate, TOEFL (minimum score of 550 required), minimum GPA of 2.7. *Application deadline:* For fall admission, 2/1 (priority date). Applications are processed on a rolling basis. Application fee: $35. Electronic applications accepted. *Application contact:* Dr. Allen Yu, Graduate Representative, 923-974-4159, E-mail: nyu@utk.edu.

The University of Texas at Arlington, Graduate School, College of Engineering, Department of Industrial Engineering, Arlington, TX 76019. Offers M Engr, MS, PhD. *Faculty:* 9 full-time (1 woman). *Students:* 21 full-time (5 women), 16 part-time (5 women); includes 4 minority (1 African American, 3 Asian Americans or Pacific Islanders), 23 international. 85 applicants, 55% accepted. In 1998, 16 master's, 4 doctorates awarded. *Degree requirements:* For master's, thesis optional, foreign language not required; for doctorate, dissertation required, foreign language not required. *Entrance requirements:* For master's and doctorate, GRE General Test, TOEFL. *Application deadline:* Applications are processed on a rolling basis. Application fee: $25 ($50 for international students). Tuition, state resident: full-time $1,368; part-time $76 per semester hour. Tuition, nonresident: full-time $5,454; part-time $303 per semester hour. Required fees: $66 per semester hour. $86 per term. Tuition and fees vary according to course load. *Financial aid:* Research assistantships, teaching assistantships, institutionally-sponsored loans and tuition waivers (partial) available. *Unit head:* Dr. Donald H. Liles, Chair, 817-272-3092, Fax: 817-272-3406, E-mail: dliles@uta.edu.

The University of Texas at Austin, Graduate School, College of Engineering, Department of Mechanical Engineering, Program in Operations Research and Industrial Engineering, Austin, TX 78712-1111. Offers MSE, PhD. *Students:* 47 (13 women); includes 2 minority (1 African American, 1 Asian American or Pacific Islander) 18 international. 47 applicants, 66% accepted. In 1998, 16 degrees awarded. *Entrance requirements:* For master's, GRE General Test (minimum combined score of 1000 required), TOEFL (minimum score of 580 required); for doctorate, GRE General Test. Application fee: $50 ($75 for international students). *Financial aid:* Fellowships, research assistantships, teaching assistantships available. Financial aid application deadline: 2/1. *Unit head:* J. Wesley Barnes, Graduate Adviser, 512-471-3083, E-mail: wbarnes@mail.utexas.edu.

The University of Texas at El Paso, Graduate School, College of Engineering, Department of Mechanical and Industrial Engineering, Program in Industrial Engineering, El Paso, TX 79968-0001. Offers MS. Evening/weekend programs available. *Faculty:* 18 full-time (2 women), 3 part-time (1 woman). 18 applicants, 44% accepted. In 1998, 12 degrees awarded. *Degree requirements:* For master's, thesis optional, foreign language not required. *Entrance requirements:* For master's, GRE General Test, TOEFL (minimum score of 550 required), minimum GPA of 3.0 in major. *Application deadline:* Applications are processed on a rolling basis. Application fee: $15 ($65 for international students). Electronic applications accepted. Tuition, state resident: full-time $2,790. Tuition, nonresident: full-time $7,710. *Financial aid:* Fellowships, research assistantships, teaching assistantships, Federal Work-Study, institutionally-sponsored loans, and tuition waivers (partial) available. Financial aid applicants required to submit FAFSA. *Faculty research:* Computer vision, automated inspection, simulation and modeling. *Unit head:* Dr. Rolando Quintana, Graduate Adviser, 915-747-5450, Fax: 915-747-5016. *Application contact:* Susan Jordan, Director, Graduate Student Services, 915-747-5491, Fax: 915-747-5788, E-mail: sjordan@utep.edu.

University of Toledo, Graduate School, College of Engineering, Department of Mechanical, Industrial, and Manufacturing Engineering, Toledo, OH 43606-3398. Offers engineering sciences (PhD); industrial engineering (MSIE); mechanical engineering (MSME). Part-time programs available. Postbaccalaureate distance learning degree programs offered (minimal on-campus study). *Faculty:* 23 full-time (1 woman), 2 part-time (0 women). *Students:* 86 full-time (12 women), 54 part-time (11 women); includes 2 minority (both African Americans), 92 international. Average age 27. 517 applicants, 24% accepted. In 1998, 66 master's, 5 doctorates awarded. *Degree requirements:* For master's, thesis optional, foreign language not required; for doctorate, dissertation required, foreign language not required. *Entrance requirements:* For master's, GRE General Test (minimum score of 350 on verbal section, 700 on quantitative required), TOEFL (minimum score of 550 required), minimum GPA of 2.7; for doctorate, GRE General Test (minimum score of 350 on verbal section, 700 on quantitative required), TOEFL (minimum score of 550 required), minimum GPA of 3.3. *Average time to degree:* Master's–2 years full-time; doctorate–4 years full-time. *Application deadline:* For fall admission, 5/31 (priority date). Applications are processed on a rolling basis. Application fee: $30. Electronic applications accepted. *Financial aid:* In 1998–99, 97 students received aid, including 1 fellowship with full tuition reimbursement available, 32 research assistantships with full tuition reimbursements available, 15 teaching assistantships with full tuition reimbursements available; Federal Work-Study, scholarships, tuition waivers (full), and unspecified assistantships also available. Financial aid application deadline: 4/1. *Faculty research:* Computational and experimental thermal sciences, manufacturing process and systems, mechanics, materials, design, quality and management engineering systems. Total annual research expenditures: $1.9 million. *Unit head:* Dr. Nagi G. Naganathan, Chair, 419-530-8210, Fax: 419-530-8214, E-mail: nagi.naganathan@utoledo.edu. *Application contact:* Dr. Abdollah A. Afjeh, Director of Graduate Program, 419-530-8208, Fax: 419-530-8206, E-mail: aafjeh@uoft02.utoledo.edu.

University of Toronto, School of Graduate Studies, Physical Sciences Division, Faculty of Applied Science and Engineering, Department of Mechanical and Industrial Engineering, Toronto, ON M5S 1A1, Canada. Offers M Eng, MA Sc, PhD. Part-time programs available. *Degree requirements:* For master's, thesis required (for some programs), foreign language not required; for doctorate, dissertation required, foreign language not required.

University of Windsor, College of Graduate Studies and Research, Faculty of Engineering, Industrial Engineering and Manufacturing Systems, Windsor, ON N9B 3P4, Canada. Offers MA Sc. Part-time programs available. *Degree requirements:* For master's, thesis required, foreign language not required. *Entrance requirements:* For master's, TOEFL (minimum score of 550 required), minimum B average. *Faculty research:* Human factors, operations research.

University of Wisconsin–Madison, Graduate School, College of Engineering, Department of Industrial Engineering, Madison, WI 53706-1380. Offers MS, PhD. Part-time programs available. *Faculty:* 13 full-time (3 women), 3 part-time (2 women). *Students:* 94 full-time (33 women), 19 part-time (12 women). 158 applicants, 57% accepted. In 1998, 42 master's, 11 doctorates awarded. Terminal master's awarded for partial completion of doctoral program. *Degree requirements:* For master's, thesis optional, foreign language not required; for doctorate, dissertation required, foreign language not required. *Entrance requirements:* For master's, GRE General Test (combined average 1882 on three sections), minimum GPA of 3.0; for doctorate, GRE General Test (combined average 1882 on three sections), minimum GPA of 3.5. *Application deadline:* For fall admission, 4/1 (priority date); for spring admission, 10/1. Applications are processed on a rolling basis. Application fee: $45. Electronic applications accepted. *Financial aid:* In 1998–99, 56 students received aid, including 4 fellowships with partial tuition reimbursements available (averaging $18,639 per year), 24 research assistantships with full tuition reimbursements available (averaging $11,620 per year), 14 teaching assistantships with full tuition reimbursements available (averaging $9,549 per year); career-related internships or fieldwork, Federal Work-Study, and unspecified assistantships also available. *Faculty research:* Human factors, manufacturing and production systems, quality, health systems, decision sciences/operations research. Total annual research expenditures: $7.1 million. *Unit head:* Harold J. Steudel, Chair, 608-262-2686, Fax: 608-262-8454. *Application contact:* Suzanne M. Bader, Student Status Examiner II, 608-263-3955, Fax: 608-262-8454, E-mail: sbader@engr.wisc.edu.

Virginia Polytechnic Institute and State University, Graduate School, College of Engineering, Department of Industrial and Systems Engineering, Program in Industrial Engineering, Blacksburg, VA 24061. Offers M Eng, MS, PhD. *Degree requirements:* For master's and doctorate, computer language, thesis/dissertation required, foreign language not required. *Entrance requirements:* For master's, TOEFL (minimum score of 600 required); for doctorate, TOEFL (minimum score of 600 required), minimum GPA of 3.0. *Application deadline:* For fall admission, 12/1 (priority date). Applications are processed on a rolling basis. Application fee: $25. *Financial aid:* Application deadline: 4/1. *Unit head:* Dr. John Casali, Head, Department of Industrial and Systems Engineering, 540-231-6656, E-mail: jcasali@vt.edu.

Wayne State University, Graduate School, College of Engineering, Department of Industrial and Manufacturing Engineering, Program in Industrial Engineering, Detroit, MI 48202. Offers MS, PhD. *Degree requirements:* For master's, thesis optional, foreign language not required; for doctorate, dissertation required, foreign language not required. *Entrance requirements:* For master's, minimum undergraduate GPA of 2.8; for doctorate, minimum graduate GPA of 3.5. *Faculty research:* Reliability and quality, technology management, manufacturing systems, operations research, concurrent engineering.

Western Carolina University, Graduate School, College of Applied Science, Department of Industrial and Engineering Technology, Cullowhee, NC 28723. Offers MS. Part-time and evening/weekend programs available. *Faculty:* 6. *Students:* 1 (woman) full-time, 6 part-time; includes 1 minority (African American) 2 applicants, 100% accepted. In 1998, 1 degree awarded. *Degree requirements:* For master's, comprehensive exam required, foreign language and thesis not required. *Entrance requirements:* For master's, GRE General Test. *Application deadline:* For fall admission, 5/1 (priority date); for spring admission, 10/1 (priority date). Applications are processed on a rolling basis. Application fee: $35. Tuition, state resident: full-time $918. Tuition, nonresident: full-time $8,188. Required fees: $881. *Financial aid:* In 1998–99, 1 student received aid, including 1 research assistantship with full and partial tuition reimbursement available (averaging $5,000 per year); fellowships, teaching assistantships, Federal Work-Study, grants, and institutionally-sponsored loans also available. Financial aid application deadline: 3/15; financial aid applicants required to submit FAFSA. *Unit head:* George W. DeSain, Head, 828-227-7272. *Application contact:* Kathleen Owen, Assistant to the Dean, 828-227-7398, Fax: 828-227-7480, E-mail: kowen@wcu.edu.

Western Michigan University, Graduate College, College of Engineering and Applied Sciences, Department of Industrial and Manufacturing Engineering, Program in Industrial Engineering, Kalamazoo, MI 49008. Offers MSE. *Students:* 13 full-time (1 woman), 36 part-time (6 women); includes 3 minority (1 African American, 2 Asian Americans or Pacific Islanders), 35 international. 64 applicants, 70% accepted. In 1998, 12 degrees awarded. *Degree requirements:* Foreign language not required. *Entrance requirements:* For master's, minimum GPA of 3.0. *Application deadline:* For fall admission, 2/15 (priority date). Applications are processed on a rolling basis. Application fee: $25. *Financial aid:* Application deadline: 2/15. *Application contact:* Paula J. Boodt, Coordinator, Graduate Admissions and Recruitment, 616-387-2000, Fax: 616-387-2355, E-mail: paula.boodt@wmich.edu.

Western New England College, School of Engineering, Department of Industrial and Manufacturing Engineering, Springfield, MA 01119-2654. Offers MSEM. Part-time and evening/weekend programs available. *Faculty:* 4 full-time (0 women), 2 part-time (0 women). Average age 29. In 1998, 22 degrees awarded. *Degree requirements:* For master's, computer language, comprehensive exam required, thesis optional, foreign language not required. *Entrance requirements:* For master's, bachelor's degree in engineering or related field. *Application deadline:* Applications are processed on a rolling basis. Application fee: $30. *Financial aid:* Teaching assistantships available. Aid available to part-time students. Financial aid application deadline: 4/1; financial aid applicants required to submit FAFSA. *Faculty research:* Project scheduling, flexible manufacturing systems, facility layout, energy management. *Unit head:* Dr. J. Byron Nelson, Chair, 413-782-1289. *Application contact:* Harry F. Neunder, Coordinator, Continuing Education, 413-782-1750, Fax: 413-782-1779, E-mail: hneunder@wnec.edu.

West Virginia University, College of Engineering and Mineral Resources, Department of Industrial and Management Systems Engineering, Program in Industrial Engineering, Morgantown, WV 26506. Offers engineering (MSE, PhD); industrial engineering (MSIE). Part-time programs available. *Degree requirements:* For master's, computer language, thesis or alternative required, foreign language not required; for doctorate, computer language, dissertation, comprehensive exam required, foreign language not required. *Entrance requirements:* For master's, GRE General Test, TOEFL (minimum score of 550 required), minimum GPA of 3.0; for doctorate, GRE General Test, TOEFL (minimum score of 550 required), minimum GPA of 3.5. *Faculty research:* Production planning and control, quality control, robotics and CIMS, ergonomics, castings.

Wichita State University, Graduate School, College of Engineering, Department of Industrial and Manufacturing Engineering, Wichita, KS 67260. Offers MEM, MS, PhD. Part-time programs available. *Faculty:* 9 full-time (2 women), 1 part-time (0 women). *Students:* 52

full-time (4 women), 64 part-time (7 women); includes 5 minority (1 African American, 3 Asian Americans or Pacific Islanders, 1 Hispanic American), 75 international. Average age 33. 175 applicants, 83% accepted. In 1998, 22 master's, 2 doctorates awarded. *Degree requirements:* For master's, thesis optional, foreign language not required; for doctorate, dissertation, comprehensive exam required. *Entrance requirements:* For master's, GRE, TOEFL (minimum score of 550 required); for doctorate, GRE General Test, TOEFL (minimum score of 550 required). *Application deadline:* For fall admission, 7/1 (priority date); for spring admission, 1/1. Applications are processed on a rolling basis. Application fee: $25 ($40 for international students). Electronic applications accepted. *Financial aid:* In 1998–99, 28 research assistantships (averaging $4,000 per year), 10 teaching assistantships with full tuition reimbursements (averaging $4,000 per year) were awarded.; fellowships, Federal Work-Study, institutionally-sponsored loans, and unspecified assistantships also available. Aid available to part-time students. Financial aid application deadline: 4/1; financial aid applicants required to submit FAFSA. *Faculty research:* Ergonomics, rehabilitation, operations research, reverse engineering, assembly design/planning. Total annual research expenditures: $327,107. *Unit head:* Dr. Abu Masud, Chairperson, 316-978-3425, Fax: 316-978-3742, E-mail: masud@twsuvm.uc.twsu.edu. *Application contact:* Dr. Mark Kaiser, Graduate Coordinator, 316-978-5904, Fax: 316-978-3742, E-mail: kaiser@twsuvm.uc.twsu.edu.

Announcement: The IMfgE Department offers Master of Engineering Management and MS and PhD degrees in IE with 3 emphasis areas: engineering systems (optimization, expert systems, ANN, TQM, reliability, decision analysis, simulation, and management of production/service systems), ergonomics/human factors (industrial and cognitive ergonomics, human-machine systems, biomechanics, and rehabilitation), and manufacturing systems engineering (CIM/CAD/CAM, GD&T, enterprise engineering, supply chain management, process control/automation, and manufacturing processes and forming). The department maintains laboratories in ergonomics/human factors, graphics, CMM/metrology, CIM, process control, and manufacturing processes. For more information, contact the department by phone: 316-978-3425, fax: 316-978-3742, e-mail: gradco@ie.twsu.edu, or WWW: http://www.engr.twsu.edu/imfge/

Youngstown State University, Graduate School, William Rayen College of Engineering, Department of Mechanical and Industrial Engineering, Youngstown, OH 44555-0001. Offers MSE. Part-time and evening/weekend programs available. *Faculty:* 8 full-time (0 women), 1 (woman) part-time. *Students:* 16 full-time (5 women), 8 part-time (2 women); includes 2 minority (both African Americans), 6 international. 7 applicants, 100% accepted. In 1998, 9 degrees awarded. *Degree requirements:* For master's, computer language required, thesis optional, foreign language not required. *Entrance requirements:* For master's, TOEFL (minimum score of 550 required), minimum GPA of 2.75 in field. *Application deadline:* For fall admission, 8/15 (priority date); for winter admission, 11/15 (priority date); for spring admission, 2/15 (priority date). Applications are processed on a rolling basis. Application fee: $30 ($75 for international students). Tuition, state resident: part-time $97 per credit hour. Tuition, nonresident: part-time $219 per credit hour. Required fees: $21 per credit hour. $41 per quarter. *Financial aid:* In 1998–99, 5 students received aid, including 2 research assistantships with full tuition reimbursements available (averaging $7,500 per year); teaching assistantships, Federal Work-Study, institutionally-sponsored loans, and scholarships also available. Aid available to part-time students. Financial aid application deadline: 3/1. *Faculty research:* Kinematics and dynamics of machines, computational and experimental heat transfer, machine controls and mechanical design. *Unit head:* Dr. H. W. Shawn Kim, Chair, 330-742-3016. *Application contact:* Dr. Peter J. Kasvinsky, Dean of Graduate Studies, 330-742-3091, Fax: 330-742-1580, E-mail: amgrad03@ysub.ysu.edu.

Manufacturing Engineering

Arizona State University East, College of Technology and Applied Sciences, Department of Manufacturing and Aeronautical Engineering Technology, Mesa, AZ 85212. Offers MS. Part-time and evening/weekend programs available. *Faculty:* 2 full-time (0 women), 1 part-time (0 women). *Students:* 21 full-time (14 women), 64 part-time (25 women); includes 5 minority (3 Asian Americans or Pacific Islanders, 2 Hispanic Americans), 15 international. Average age 33. 42 applicants, 79% accepted. In 1998, 30 degrees awarded. *Degree requirements:* For master's, thesis or applied project and oral defense required. *Entrance requirements:* For master's, minimum GPA of 3.0. *Average time to degree:* Master's–2.8 years full-time, 5.4 years part-time. *Application deadline:* Applications are processed on a rolling basis. Application fee: $45. *Financial aid:* In 1998–99, 16 students received aid, including 5 research assistantships with partial tuition reimbursements available (averaging $8,556 per year); teaching assistantships, career-related internships or fieldwork, Federal Work-Study, grants, scholarships, and tuition waivers (full and partial) also available. Aid available to part-time students. Financial aid application deadline: 3/1; financial aid applicants required to submit FAFSA. *Faculty research:* Robotics, wind tunnel testing, propulsion systems, statistical process control. *Unit head:* Dr. Dale Palmgren, Chair, 602-727-1584, Fax: 602-727-1549, E-mail: palmgren@asu.edu. *Application contact:* Dr. Dale Palmgren, Chair, 602-727-1584, Fax: 602-727-1549, E-mail: palmgren@asu.edu.

Boston University, College of Engineering, Department of Manufacturing Engineering, Boston, MA 02215. Offers MS, PhD, MBA/MS. Part-time programs available. Postbaccalaureate distance learning degree programs offered (no on-campus study). *Faculty:* 18 full-time (1 woman), 2 part-time (0 women). *Students:* 44 full-time (12 women), 35 part-time (1 woman); includes 3 minority (all Asian Americans or Pacific Islanders), 30 international. Average age 30. 127 applicants, 28% accepted. In 1998, 42 master's, 3 doctorates awarded. Terminal master's awarded for partial completion of doctoral program. *Degree requirements:* For master's, foreign language and thesis not required; for doctorate, dissertation required, foreign language not required. *Entrance requirements:* For master's, GRE General Test, TOEFL (minimum score of 500 required; 213 for computer-based); for doctorate, GRE General Test, TOEFL. *Application deadline:* For fall admission, 4/1; for spring admission, 10/1. Applications are processed on a rolling basis. Application fee: $50. Tuition: Full-time $23,770; part-time $743 per credit. Required fees: $220. Tuition and fees vary according to class time, course level, campus/location and program. *Financial aid:* In 1998–99, 45 students received aid, including 12 fellowships with full tuition reimbursements available (averaging $13,000 per year), 14 research assistantships with full tuition reimbursements available (averaging $11,500 per year), 14 teaching assistantships with full tuition reimbursements available (averaging $11,500 per year); Federal Work-Study, institutionally-sponsored loans, and scholarships also available. Financial aid application deadline: 12/15; financial aid applicants required to submit FAFSA. *Faculty research:* Control and stochastic systems, materials processing, communication networks, electromagnetics. Total annual research expenditures: $2.3 million. *Unit head:* Dr. John Baillieul, Chairman, 617-353-9848, Fax: 617-353-5548. *Application contact:* Cheryl Kelley, Graduate Programs Director, 617-353-9760, Fax: 617-353-0259, E-mail: enggrad@bu.edu.

See in-depth description on page 1197.

Bowling Green State University, Graduate College, College of Technology, Program in Technology Systems, Bowling Green, OH 43403. Offers manufacturing technology (MIT). Part-time and evening/weekend programs available. *Faculty:* 11 full-time (1 woman). *Students:* 5 full-time (1 woman), 18 part-time (2 women); includes 3 minority (2 African Americans, 1 Hispanic American), 2 international. 24 applicants, 58% accepted. In 1998, 10 degrees awarded. *Degree requirements:* For master's, thesis or alternative required, foreign language not required. *Entrance requirements:* For master's, GRE General Test, TOEFL (minimum score of 550 required), minimum GPA of 3.0. *Application deadline:* For fall admission, 3/1. Application fee: $30. Electronic applications accepted. *Financial aid:* Research assistantships with full tuition reimbursements, teaching assistantships with full tuition reimbursements, career-related internships or fieldwork, Federal Work-Study, institutionally-sponsored loans, tuition waivers (full and partial), and unspecified assistantships available. Financial aid applicants required to submit FAFSA. *Unit head:* Dr. Sudershan Jetley, Chair, 419-372-2439. *Application contact:* Dr. Ernie Savage, Graduate Coordinator, 419-372-7613.

Bradley University, Graduate School, College of Engineering and Technology, Department of Industrial and Manufacturing Engineering and Technology, Peoria, IL 61625-0002. Offers MSIE, MSMFE. Part-time and evening/weekend programs available. *Degree requirements:* For master's, computer language, comprehensive exam, project required, foreign language and thesis not required. *Entrance requirements:* For master's, TOEFL (minimum score of 525 required), minimum GPA of 3.0.

Brigham Young University, Graduate Studies, College of Engineering and Technology, School of Technology, Provo, UT 84602-1001. Offers engineering technology (MS); technology teacher education (MS). *Faculty:* 21 full-time (0 women). *Students:* 15 full-time (0 women), 2 international. Average age 25. 10 applicants, 70% accepted. In 1998, 14 degrees awarded. *Degree requirements:* For master's, thesis required (for some programs), foreign language not required. *Entrance requirements:* For master's, GRE General Test, minimum GPA of 3.0 in last 60 hours. *Average time to degree:* Master's–1.5 years full-time, 3.5 years part-time. *Application deadline:* For fall admission, 2/28. Application fee: $30. Tuition: Full-time $3,330; part-time $185 per credit hour. Tuition and fees vary according to program and student's religious affiliation. *Financial aid:* Research assistantships, teaching assistantships, career-related internships or fieldwork available. Financial aid application deadline: 3/15. *Faculty research:* Composites, plasma treatment, geometric modeling, robotics, manufacturing control. *Unit head:* Thomas L. Erekson, Director, 801-378-6300, Fax: 801-378-7575, E-mail: erekson@byu.edu. *Application contact:* Graduate Coordinator, 801-378-6300, Fax: 801-378-7575, E-mail: ralowe@byu.edu.

Colorado State University, Graduate School, College of Engineering, Department of Mechanical Engineering, Program in Industrial and Manufacturing Systems Engineering, Fort Collins, CO 80523-0015. Offers MS, PhD. *Faculty:* 18 full-time (0 women). *Degree requirements:* For doctorate, dissertation required, foreign language not required, foreign language not required. *Entrance requirements:* For master's and doctorate, GRE General Test (minimum combined score of 1850 on three sections required; average 1872), TOEFL (minimum score of 550 required; average 596), minimum GPA of 3.0. *Application deadline:* For fall admission, 2/1 (priority date). Applications are processed on a rolling basis. Application fee: $30. Electronic applications accepted. *Faculty research:* Robotic assembly, dynamics of robots, controls, manufacturability of mechanical devices and systems. *Unit head:* William Duff, Professor, 970-491-5859, Fax: 970-491-3827, E-mail: duff@engr.colostate.edu. *Application contact:* William Duff, Professor, 970-491-5859, Fax: 970-491-3827, E-mail: duff@engr.colostate.edu.

Cornell University, Graduate School, Graduate Fields of Engineering, Field of Operations Research and Industrial Engineering, Ithaca, NY 14853-0001. Offers applied probability and statistics (PhD); manufacturing systems engineering (PhD); mathematical programming (PhD); operations research and industrial engineering (M Eng). *Faculty:* 30 full-time. *Students:* 91 full-time (20 women); includes 22 minority (1 African American, 21 Asian Americans or Pacific Islanders), 36 international. Terminal master's awarded for partial completion of doctoral program. *Degree requirements:* For doctorate, dissertation required, foreign language not required, foreign language not required. *Entrance requirements:* For master's and doctorate, GRE General Test, TOEFL (minimum score of 550 required). *Application deadline:* For fall admission, 1/15. Application fee: $65. Electronic applications accepted. *Unit head:* Director of Graduate Studies, 607-255-9128. *Application contact:* Graduate Field Assistant, 607-255-9128, E-mail: orphd@cornell.edu.

Drexel University, Graduate School, College of Engineering, Department of Mechanical Engineering and Mechanics, Program in Manufacturing Engineering, Philadelphia, PA 19104-2875. Offers MS, PhD. *Degree requirements:* For master's, thesis optional; for doctorate, dissertation required. *Entrance requirements:* For master's, TOEFL (minimum score of 570 required), minimum GPA of 3.0, BS in engineering or science; for doctorate, TOEFL (minimum score of 570 required), minimum GPA of 3.5, MS in engineering or science. *Application deadline:* For fall admission, 8/21. Applications are processed on a rolling basis. Application fee: $35. Tuition: Full-time $15,795; part-time $585 per credit. Required fees: $375; $67 per term. Tuition and fees vary according to program. *Financial aid:* Application deadline: 2/1. *Unit head:* Dr. Anand Gramopadhye, Coordinator, 864-656-5540, Fax: 864-656-0795, E-mail: agramop@ces.clemson.edu. *Application contact:* Dr. Anand Gramopadhye, Coordinator, 864-656-5540, Fax: 864-656-0795, E-mail: agramop@ces.clemson.edu.

Announcement: The Department of Mechanical Engineering and Mechanics at Drexel University offers a curriculum with a concentration in manufacturing engineering. The Manufacturing Engineering Program is designed to provide students with modern knowledge and skills needed to excel in solving interdisciplinary problems in manufacturing engineering. Students in the Manufacturing Engineering Program learn solid foundations of mechanical engineering and take a selection of manufacturing courses tailored to fit the specific interests of each student. Students graduate with an accredited degree in mechanical engineering, with a concentration in manufacturing engineering. Qualified students can enroll in programs leading to the BS, MS, or PhD degrees. Visit the Web site at http://www.mem.drexel.edu/.

Eastern Kentucky University, The Graduate School, College of Applied Arts and Technology, Department of Technology, Program in Manufacturing Technology, Richmond, KY 40475-3101. Offers MS. Part-time programs available. *Students:* 44. In 1998, 26 degrees awarded. *Degree requirements:* For master's, thesis not required. *Entrance requirements:* For master's, GRE General Test (minimum combined score of 1250 required), minimum GPA of 2.5. Application fee: $0. *Financial aid:* Research assistantships, teaching assistantships, Federal Work-Study available. Aid available to part-time students. *Unit head:* Clyde O. Craft, Chair, Department of Technology, 606-622-3232.

East Tennessee State University, School of Graduate Studies, College of Applied Science and Technology, Department of Technology, Johnson City, TN 37614-0734. Offers MS. Part-time programs available. *Degree requirements:* For master's, computer language, thesis or alternative, final oral exam required, foreign language not required. *Entrance requirements:* For master's, TOEFL (minimum score of 550 required), bachelor's degree in technical or related area, minimum GPA of 3.0. *Faculty research:* Computer-integrated manufacturing, technology education, CAD/CAM, organizational change.

Florida Atlantic University, College of Engineering, Department of Mechanical Engineering, Program in Manufacturing Systems Engineering, Boca Raton, FL 33431-0991. Offers MS. Part-time and evening/weekend programs available. *Faculty:* 3 full-time (0 women). *Students:*

Manufacturing Engineering

Florida Atlantic University (continued)
3 full-time (1 woman), 3 part-time (1 woman); includes 3 minority (1 African American, 1 Asian American or Pacific Islander, 1 Hispanic American), 1 international. Average age 29. In 1998, 4 degrees awarded. *Degree requirements:* For master's, thesis optional, foreign language not required. *Entrance requirements:* For master's, GRE General Test (minimum combined score of 1000 required), TOEFL (minimum score of 550 required), minimum GPA of 3.0. *Application deadline:* For fall admission, 4/10 (priority date); for spring admission, 10/1. Applications are processed on a rolling basis. Application fee: $20. Tuition, state resident: part-time $148 per credit hour. Tuition, nonresident: part-time $509 per credit hour. *Financial aid:* Research assistantships, teaching assistantships, career-related internships or fieldwork and Federal Work-Study available. Aid available to part-time students. Financial aid application deadline: 4/1; financial aid applicants required to submit FAFSA. *Faculty research:* Packaging, materials handling, design for manufacture, robotics, automation. *Unit head:* Patricia Capozziello, Graduate Admissions Coordinator, 561-297-2694, Fax: 561-297-2659, E-mail: capozzie@fau.edu. *Application contact:* Patricia Capozziello, Graduate Admissions Coordinator, 561-297-2694, Fax: 561-297-2659, E-mail: capozzie@fau.edu.

Illinois Institute of Technology, Graduate College, Armour College of Engineering and Sciences, Department of Electrical and Computer Engineering, Chicago, IL 60616-3793. Offers computer systems engineering (MS); electrical and computer engineering (MECE); electrical engineering (MS, PhD); manufacturing engineering (MME, MS). Part-time and evening/weekend programs available. *Faculty:* 21 full-time (1 woman), 11 part-time (0 women). *Students:* 76 full-time (10 women), 258 part-time (38 women); includes 83 minority (17 African Americans, 50 Asian Americans or Pacific Islanders, 15 Hispanic Americans, 1 Native American), 118 international. Terminal master's awarded for partial completion of doctoral program. *Degree requirements:* For master's, thesis (for some programs), comprehensive exam required, foreign language not required; for doctorate, dissertation, comprehensive exam required, foreign language not required. *Entrance requirements:* For master's, GRE (minimim score of 1200 required), TOEFL (minimum score of 550 required), undergraduate GPA of 3.0 required; for doctorate, GRE (minimum score of 1200 required), TOEFL (minimum score of 550 required), undergraduate GPA of 3.0 required. *Application deadline:* For fall admission, 7/1; for spring admission, 12/1. Applications are processed on a rolling basis. Application fee: $30. Electronic applications accepted. *Unit head:* Dr. D. Ucci, Interim Chairman, 312-567-3402, Fax: 312-567-8976, E-mail: du@ece.iit.edu. *Application contact:* Dr. S. Mohammad Shahidehpour, Dean of Graduate College, 312-567-3024, Fax: 312-567-7517, E-mail: grad@minna.cns.iit.edu.

Illinois Institute of Technology, Graduate College, Armour College of Engineering and Sciences, Department of Mechanical, Materials and Aerospace Engineering, Chicago, IL 60616-3793. Offers manufacturing engineering (MME, MS); mechanical and aerospace engineering (MMAE, MS, PhD); metallurgical and materials engineering (MMME, MS, PhD). Part-time and evening/weekend programs available. *Faculty:* 26 full-time (3 women), 12 part-time (0 women). *Students:* 52 full-time (4 women), 82 part-time (10 women); includes 20 minority (7 African Americans, 9 Asian Americans or Pacific Islanders, 4 Hispanic Americans), 51 international. Terminal master's awarded for partial completion of doctoral program. *Degree requirements:* For master's, thesis (for some programs), comprehensive exam required, foreign language not required; for doctorate, dissertation, comprehensive exam required, foreign language not required. *Entrance requirements:* For master's, GRE General Test (minimum combined score of 1200 required), TOEFL (minimum score of 550 required), undergraduate GPA of 3.0; for doctorate, GRE (minimum score of 1200 required), TOEFL (minimum score of 550 required), undergraduate GPA of 3.0 required. *Application deadline:* For fall admission, 7/1; for spring admission, 11/1. Applications are processed on a rolling basis. Application fee: $30. Electronic applications accepted. *Unit head:* Dr. Marek Dollar, Chair, 312-567-3175, Fax: 312-567-7230, E-mail: dept@mmae.iit.edu. *Application contact:* Dr. S. Mohammad Shahidehpour, Dean of Graduate College, 312-567-3024, Fax: 312-567-7517, E-mail: grad@minna.cns.iit.edu.

Instituto Tecnológico y de Estudios Superiores de Monterrey, Campus Estado de México, Graduate Division, Division of Engineering and Architecture, Atizapán de Zaragoza, 52500, Mexico. Offers computer science (MCS); environmental engineering (MEE); industrial engineering (MIE); manufacturing systems (MMS); materials engineering (PhD). *Degree requirements:* For master's; for doctorate, one foreign language, dissertation required. *Entrance requirements:* For master's, interview; for doctorate, research proposal. *Application deadline:* For fall admission, 1/13 (priority date); for spring admission, 4/4. Applications are processed on a rolling basis. Application fee: 750 Mexican pesos. *Unit head:* Juan López Díaz, Headmaster, 5-326-5530, Fax: 5-326-5531, E-mail: jlopez@campus.cem.itesm.mx. *Application contact:* Lourdes Turrubiates, Admissions Officer, 5-326-5776, Fax: 5-326-5788, E-mail: lturrubi@campus.cem.itesm.mx.

Instituto Tecnológico y de Estudios Superiores de Monterrey, Campus Monterrey, Graduate and Research Division, Programs in Engineering, Monterrey, 64849, Mexico. Offers applied statistics (M Eng); artificial intelligence (PhD); automation engineering (M Eng); chemical engineering (M Eng); civil engineering (M Eng); electrical engineering (M Eng); electronic engineering (M Eng); environmental engineering (M Eng); industrial engineering (M Eng, PhD); manufacturing engineering (M Eng); mechanical engineering (M Eng); systems and quality engineering (M Eng). M Eng offered jointly with the University of Waterloo; PhD (industrial engineering) offered jointly with Texas A&M University. Part-time and evening/weekend programs available. Terminal master's awarded for partial completion of doctoral program. *Degree requirements:* For master's and doctorate, one foreign language, computer language, thesis/dissertation required. *Entrance requirements:* For master's, PAEG, TOEFL; for doctorate, GRE, TOEFL, master's in related field. *Faculty research:* Flexible manufacturing cells, materials, statistical methods, environmental prevention, control and evaluation.

Kansas State University, Graduate School, College of Engineering, Department of Industrial and Manufacturing Systems Engineering, Manhattan, KS 66506. Offers engineering management (MEM); industrial and manufacturing systems engineering (PhD); industrial engineering (MS); operations research (MS). Part-time programs available. Postbaccalaureate distance learning degree programs offered. *Faculty:* 10 full-time (1 woman). *Students:* 24 full-time (4 women), 37 part-time (6 women); includes 1 minority (Asian American or Pacific Islander), 29 international. 60 applicants, 57% accepted. In 1998, 19 master's awarded (5% entered university research/teaching, 95% found other work related to degree); 4 doctorates awarded (25% entered university research/teaching, 75% found other work related to degree). *Degree requirements:* For master's, computer language required, foreign language not required; for doctorate, computer language, dissertation required, foreign language not required. *Entrance requirements:* For master's and doctorate, GRE General Test (minimum score of 725 on quantitative section required). *Average time to degree:* Master's–1.75 years full-time, 2.19 years part-time; doctorate–3.6 years full-time. *Application deadline:* For fall admission, 3/1 (priority date); for spring admission, 9/1. Applications are processed on a rolling basis. Application fee: $0 ($25 for international students). Electronic applications accepted. *Financial aid:* In 1998–99, 15 research assistantships (averaging $5,850 per year), 1 teaching assistantship (averaging $8,460 per year) were awarded. *Faculty research:* Safety, quality operations research, tribology, economic analysis, expert systems. Total annual research expenditures: $346,000. *Unit head:* Dr. Bradley Kramer, Head, 785-532-5606.

Kettering University, Graduate School, Industrial and Manufacturing Systems Engineering Department, Flint, MI 48504-4898. Offers manufacturing systems engineering (MS Eng). *Faculty:* 26 full-time (3 women), 1 part-time (0 women). *Degree requirements:* For master's, foreign language and thesis not required. *Entrance requirements:* For master's, GRE General Test. *Application deadline:* For fall admission, 7/15. Applications are processed on a rolling basis. Application fee: $0. *Financial aid:* Fellowships, research assistantships with full tuition reimbursements, teaching assistantships with full tuition reimbursements, Federal Work-Study, institutionally-sponsored loans, and tuition waivers (partial) available. Aid available to part-time students. Financial aid application deadline: 7/15. *Unit head:* Dr. David W. Poock, Head, 810-762-7940, Fax: 810-762-9924, E-mail: dpoock@kettering.edu. *Application contact:* Betty L. Bedore, Coordinator of Publicity, 810-762-7494, Fax: 810-762-9935, E-mail: bbedore@kettering.edu.

See in-depth description on page 1203.

Lawrence Technological University, College of Engineering, Southfield, MI 48075-1058. Offers automotive engineering (MAE); civil engineering (MCE); manufacturing systems (MEMS). Part-time and evening/weekend programs available. *Faculty:* 8 full-time (1 woman), 6 part-time (0 women). *Degree requirements:* For master's, foreign language and thesis not required. *Application deadline:* For fall admission, 8/1 (priority date); for spring admission, 1/1. Applications are processed on a rolling basis. Application fee: $50. Electronic applications accepted. Tuition: Full-time $5,128; part-time $419 per credit hour. Required fees: $100; $100 per year. $50 per semester. Tuition and fees vary according to course level. *Unit head:* Dr. George Kartsounes, Dean, 248-204-2500, Fax: 248-204-2509, E-mail: kartsounes@ltu.edu. *Application contact:* Lisa Kujawa, Director of Admissions, 248-204-3160, Fax: 248-204-3188, E-mail: admission@hu.edu.

Lehigh University, College of Engineering and Applied Science, Department of Industrial and Manufacturing Systems Engineering, Program in Manufacturing Systems Engineering, Bethlehem, PA 18015-3094. Offers MS. Part-time and evening/weekend programs available. *Students:* 3 full-time (0 women), 16 part-time (2 women); includes 1 Hispanic American, 1 international. 2 applicants, 100% accepted. In 1998, 19 degrees awarded. *Degree requirements:* For master's, computer language, project or thesis required. *Entrance requirements:* For master's, GRE General Test (score in 75th percentile or higher required), TOEFL (minimum score of 550 required), minimum GPA of 2.75. *Average time to degree:* Master's–1 year full-time, 2 years part-time. *Application deadline:* For fall admission, 7/15; for spring admission, 12/1. Applications are processed on a rolling basis. Application fee: $40. *Financial aid:* In 1998–99, 1 student received aid, including 1 research assistantship; fellowships, career-related internships or fieldwork and tuition waivers (full and partial) also available. Financial aid application deadline: 1/15. *Faculty research:* Manufacturing systems development and computer integration, CAD/CAM robotics and automation, management aspects of manufacturing systems. *Unit head:* Dr. Keith M. Gardiner, Director, 610-758-5070, Fax: 610-758-6527, E-mail: kg03@lehigh.edu. *Application contact:* Jeannette I. MacDonald, Graduate Coordinator, 610-758-4667, Fax: 610-758-6527, E-mail: jim1@lehigh.edu.

Louisiana Tech University, Graduate School, College of Engineering and Science, Department of Mechanical and Industrial Engineering, Ruston, LA 71272. Offers industrial engineering (MS, D Eng); manufacturing systems engineering (MS); mechanical engineering (MS, D Eng); operations research (MS). Part-time programs available. Terminal master's awarded for partial completion of doctoral program. *Degree requirements:* For master's and doctorate, thesis/dissertation required, foreign language not required. *Entrance requirements:* For master's, GRE General Test (minimum combined score of 1070 required), TOEFL (minimum score of 550 required), minimum GPA of 3.0 in last 60 hours; for doctorate, TOEFL (minimum score of 550 required), minimum graduate GPA of 3.25 (with MS) or GRE General Test (minimum combined score of 1270 required without MS). *Faculty research:* Engineering management, facilities planning, thermodynamics, automated manufacturing, micromanufacturing.

Marquette University, Graduate School, College of Engineering, Department of Mechanical and Industrial Engineering, Milwaukee, WI 53201-1881. Offers engineering management (MS); materials science and engineering (MS, PhD); mechanical engineering (MS, PhD), including manufacturing systems engineering. Part-time and evening/weekend programs available. *Faculty:* 18 full-time (0 women), 6 part-time (0 women). *Students:* 32 full-time (3 women), 52 part-time (4 women); includes 6 minority (1 African American, 2 Asian Americans or Pacific Islanders, 2 Hispanic Americans, 1 Native American), 30 international. Terminal master's awarded for partial completion of doctoral program. *Degree requirements:* For master's, thesis, comprehensive exam required, foreign language not required; for doctorate, dissertation, proficiency exam, qualifying exam required, foreign language not required. *Entrance requirements:* For master's and doctorate, GRE General Test, TOEFL (minimum score of 550 required), minimum GPA of 3.0. *Application deadline:* For fall admission, 8/1 (priority date); for spring admission, 1/1 (priority date). Applications are processed on a rolling basis. Application fee: $40. Tuition: Part-time $510 per credit hour. Tuition and fees vary according to program. *Unit head:* Dr. G. E. O. Widera, Chairman, 414-288-7259, Fax: 414-288-1647, E-mail: geo.widera@marquette.edu. *Application contact:* Dr. William E. Brower, Director of Graduate Studies, 414-288-1717, Fax: 414-288-7790, E-mail: 9322browerw@rms.csd.mu.edu.

Massachusetts Institute of Technology, School of Engineering, Leaders for Manufacturing Program, Cambridge, MA 02139-4307. Offers engineering (SM); management (MBA, SM). *Degree requirements:* For master's, thesis, off-site internship required. *Entrance requirements:* For master's, GMAT or GRE General Test, 2 years of work experience. *Application deadline:* For fall admission, 1/15. Application fee: $55. *Financial aid:* Fellowships, career-related internships or fieldwork, institutionally-sponsored loans, and scholarships available. Financial aid application deadline: 1/15. *Faculty research:* Scheduling logistics, product life cycle, variation reduction, product development, manufacturing operations. *Unit head:* Don Rosenfield, Director, 617-253-1064. *Application contact:* Sarah Shohet, Recruiting and Admissions Coordinator, 617-253-1055, E-mail: lfm@mit.edu.

Announcement: The Leaders for Manufacturing Program (LFM) is a partnership between the Massachusetts Institute of Technology (MIT) and leading US manufacturing firms to discover and translate into teaching and practice the principles that produce world-class manufacturing and manufacturing leaders. LFM graduates earn 2 master's degrees: one in engineering, awarded by one of seven departments in MIT's School of Engineering (Aeronautics and Astronautics, Chemical Engineering, Civil and Environmental Engineering, Electrical Engineering and Computer Science, Materials Science and Engineering, Mechanical Engineering, and Ocean Engineering), and one in management, awarded by the Sloan School of Management. LFM, launched in 1988, is one of the largest cooperative ventures ever undertaken by a major engineering school, a major management school, and industry. Web site: http://web.mit.edu/lfm/www.

Michigan State University, Graduate School, College of Agriculture and Natural Resources, School of Packaging, East Lansing, MI 48824-1020. Offers MS, PhD. Part-time programs available. *Faculty:* 12. *Students:* 28 full-time (12 women), 26 part-time (7 women); includes 6 minority (4 African Americans, 1 Asian American or Pacific Islander, 1 Hispanic American), 34 international. Average age 28. 17 applicants, 53% accepted. In 1998, 17 master's, 1 doctorate awarded. *Degree requirements:* For master's, thesis required, foreign language not required; for doctorate, dissertation required. *Entrance requirements:* For master's, GRE, minimum GPA of 3.0; for doctorate, GRE, MS, minimum GPA of 3.4. *Application deadline:* Applications are processed on a rolling basis. Application fee: $30 ($40 for international students). *Financial aid:* In 1998–99, 28 students received aid, including 18 research assistantships with tuition reimbursements available (averaging $10,260 per year), 13 teaching assistantships with tuition reimbursements available (averaging $10,050 per year); fellowships, Federal Work-Study and tuition waivers (partial) also available. Aid available to part-time students. Financial aid applicants required to submit FAFSA. *Faculty research:* Barrier packaging–food and health care, physical protection packaging, ergonomics, environmental management. Total annual research expenditures: $13,000. *Unit head:* Dr. Bruce Harte, Director, 517-355-9580. *Application contact:* Dr. Susan Selke, Professor, 517-353-4801, Fax: 517-353-8999, E-mail: sselke@pilot.msu.edu.

Minnesota State University, Mankato, College of Graduate Studies, College of Science, Engineering and Technology, Department of Manufacturing, Mankato, MN 56002-8400. Offers MS. *Faculty:* 4 full-time (0 women). *Students:* 1 full-time (0 women), 4 part-time (1 woman). Average age 30. In 1998, 3 degrees awarded. *Degree requirements:* For master's, thesis, comprehensive exam required, foreign language not required. *Entrance requirements:*

For master's, minimum GPA of 3.0 during previous 2 years. *Application deadline:* For fall admission, 7/9 (priority date); for spring admission, 11/27. Applications are processed on a rolling basis. Application fee: $20. *Financial aid:* Research assistantships with partial tuition reimbursements, teaching assistantships with partial tuition reimbursements available. Financial aid application deadline: 3/15; financial aid applicants required to submit FAFSA. *Unit head:* Kirk Ready, Chairperson, 507-389-6383. *Application contact:* Joni Roberts, Admissions Coordinator, 507-389-2321, Fax: 507-389-5974, E-mail: grad@mankato.msus.edu.

National Technological University, Programs in Engineering, Fort Collins, CO 80526-1842. Offers chemical engineering (MS); computer engineering (MS); computer science (MS); electrical engineering (MS); engineering management (MS); hazardous waste management (MS); health physics (MS); management of technology (MS); manufacturing systems engineering (MS); materials science and engineering (MS); software engineering (MS); special majors (MS); transportation engineering (MS); transportation systems engineering (MS). Part-time programs available. *Faculty:* 600 part-time (20 women). *Entrance requirements:* For master's, BS in engineering or related field. *Application deadline:* Applications are processed on a rolling basis. Application fee: $50. *Unit head:* Lionel V. Baldwin, President, 970-495-6400, Fax: 970-484-0668, E-mail: baldwin@mail.ntu.edu.

New Jersey Institute of Technology, Office of Graduate Studies, Department of Industrial and Manufacturing Engineering, Program in Manufacturing Systems Engineering, Newark, NJ 07102-1982. Offers MS. Part-time and evening/weekend programs available. *Degree requirements:* For master's, thesis or alternative required, foreign language not required. *Entrance requirements:* For master's, GRE General Test (minimum score of 450 on verbal section, 600 on quantitative, 550 on analytical required). Electronic applications accepted. *Faculty research:* Automated production systems, system and product design, manufacturing management systems, computer control of manufacturing systems.

North Carolina State University, Graduate School, College of Engineering, Integrated Manufacturing Systems Engineering Institute, Raleigh, NC 27695. Offers MIMS. Part-time programs available. *Faculty:* 39 full-time (1 woman), 3 part-time (0 women). *Students:* 22 full-time (6 women), 12 part-time (2 women); includes 6 minority (1 African American, 4 Asian Americans or Pacific Islanders, 1 Hispanic American), 13 international. Average age 29. 21 applicants, 71% accepted. In 1998, 11 degrees awarded. *Degree requirements:* For master's, thesis or alternative required, foreign language not required. *Entrance requirements:* For master's, TOEFL (minimum score of 550 required). *Average time to degree:* Master's–2 years full-time, 4 years part-time. *Application deadline:* For fall admission, 6/25; for spring admission, 11/25. Applications are processed on a rolling basis. Application fee: $45. *Financial aid:* In 1998–99, 19 research assistantships (averaging $4,740 per year) were awarded.; fellowships, teaching assistantships, career-related internships or fieldwork and scholarships also available. *Faculty research:* Programmable automation, concurrent engineering, rapid prototyping, mechatronics, manufacturing systems modeling. *Unit head:* Dr. Larry M. Silverberg, Director of Graduate Programs, 919-515-5282, Fax: 919-515-1675, E-mail: silver@eos.ncsu.edu.

Announcement: The IMSE Institute offers an interdisciplinary master's degree in manufacturing systems, mechatronics, and logistics. The 33-credit-hour program consists of 15 credit hours of common core, 12 credit hours of concentration electives, and 6 credit hours of project. Concentration areas include artificial intelligence; information systems; manufacturing automation; manufacturing materials systems; manufacturing operations management; mechanical product design; modeling and simulation of manufacturing systems; quality control; sensors, control, and robotics; and textile manufacturing systems. IMSE Institute offers competitive financial packages, including 2-year fellowships and 7½-month industrial internships. Visit the Institute's Web site: http://www.imse.ncsu.edu.

North Dakota State University, Graduate Studies and Research, College of Engineering and Architecture, Department of Industrial and Manufacturing Engineering, Fargo, ND 58105. Offers MS. Part-time programs available. *Faculty:* 8 full-time (0 women). *Students:* 3 full-time (1 woman), 5 part-time (1 woman); includes 6 minority (all Asian Americans or Pacific Islanders) Average age 26. 38 applicants, 74% accepted. *Degree requirements:* For master's, computer language, thesis required, foreign language not required. *Entrance requirements:* For master's, GRE General Test (minimum combined score of 1650 on three sections required), TOEFL (minimum score of 550 required). *Average time to degree:* Master's–3 years full-time, 4 years part-time. *Application deadline:* For fall admission, 6/1; for spring admission, 10/1. Applications are processed on a rolling basis. Application fee: $25. *Financial aid:* In 1998–99, 1 research assistantship with tuition reimbursement, 5 teaching assistantships with tuition reimbursements were awarded.; Federal Work-Study and institutionally-sponsored loans also available. Financial aid application deadline: 4/15. *Faculty research:* Automation, engineering economy, artificial intelligence, human factors engineering, production and inventory control. *Unit head:* Dr. Fong-Yuen Ding, Interim Chair, 701-231-7223, Fax: 701-231-7195.

Northeastern University, College of Engineering, Department of Mechanical, Industrial, and Manufacturing Engineering, Boston, MA 02115-5096. Offers engineering management (MS); industrial engineering (MS, PhD); mechanical engineering (MS, PhD); operations research (MS). Part-time programs available. *Faculty:* 30 full-time (1 woman), 13 part-time (1 woman). *Students:* 105 full-time (19 women), 118 part-time (25 women); includes 14 minority (5 African Americans, 7 Asian Americans or Pacific Islanders, 1 Hispanic American, 1 Native American), 84 international. Average age 25. 348 applicants, 59% accepted. In 1998, 48 master's, 8 doctorates awarded. *Degree requirements:* For master's, thesis required (for some programs), foreign language not required; for doctorate, one foreign language, dissertation, departmental qualifying exam required. *Entrance requirements:* For master's and doctorate, GRE General Test. *Average time to degree:* Master's–3.38 years full-time, 4.05 years part-time; doctorate–6.5 years full-time, 8 years part-time. *Application deadline:* For fall admission, 4/15. Applications are processed on a rolling basis. Application fee: $50. *Financial aid:* In 1998–99, 49 students received aid, including 14 research assistantships with full tuition reimbursements available (averaging $12,450 per year), 26 teaching assistantships with full tuition reimbursements available (averaging $12,450 per year); fellowships, career-related internships or fieldwork, Federal Work-Study, tuition waivers (full), and unspecified assistantships also available. Aid available to part-time students. Financial aid application deadline: 2/15; financial aid applicants required to submit FAFSA. *Faculty research:* Dry sliding instabilities, droplet deposition, helical hydraulic turbines, combustion, manufacturing systems. *Unit head:* Dr. John W. Cipolla, Chairman, 617-373-3810, Fax: 617-373-2921. *Application contact:* Stephen L. Gibson, Associate Director, 617-373-2711, Fax: 617-373-2501, E-mail: grad-eng@coe.neu.edu.

Northwestern University, The Graduate School, Robert R. McCormick School of Engineering and Applied Science, Department of Mechanical Engineering, Program in Manufacturing Engineering, Evanston, IL 60208. Offers MME. *Entrance requirements:* For master's, GRE General Test, TOEFL (minimum score of 560 required). *Application deadline:* For fall admission, 8/30. Applications are processed on a rolling basis. *Financial aid:* Institutionally-sponsored loans available. Financial aid application deadline: 1/15; financial aid applicants required to submit FAFSA. *Unit head:* Henry Stoll, Director. *Application contact:* Pat Dyess, Admission Contact, 847-491-7190, Fax: 847-491-3915, E-mail: j-dyess@nwu.edu.

Ohio University, Graduate Studies, College of Engineering and Technology, Department of Mechanical Engineering, Athens, OH 45701-2979. Offers manufacturing engineering (MS); mechanical engineering (MS, PhD), including CAD/CAM (MS), manufacturing (MS), mechanical systems (MS), technology management (MS), thermal systems (MS). *Faculty:* 11 full-time (0 women). *Students:* 50 full-time (5 women), 29 part-time (2 women); includes 1 minority (African American), 67 international. *Degree requirements:* For master's, thesis required, foreign language not required; for doctorate, dissertation required. *Entrance requirements:* For master's, BS in engineering or science, minimum GPA of 2.8; for doctorate, GRE. *Application deadline:* For fall admission, 3/15 (priority date). Applications are processed on a rolling basis. Application fee: $30. Tuition, state resident: full-time $5,754; part-time $238 per credit hour. Tuition, nonresident: full-time $11,055; part-time $457 per credit hour. Tuition and fees vary

according to course load, campus/location and program. *Unit head:* Dr. Jay S. Gunasekara, Chairman, 740-593-0563, Fax: 740-593-0476, E-mail: gsekera@bobcat.ent.ohiou.edu. *Application contact:* Dr. M. Khairul Alam, Graduate Chairman, 740-593-1598, Fax: 740-593-0476, E-mail: agrawal@bobcat.ent.ohiou.edu.

Oklahoma State University, Graduate College, College of Engineering, Architecture and Technology, School of Industrial Engineering and Management, Interdisciplinary Master of Manufacturing Systems Engineering Program, Stillwater, OK 74078. Offers M En. *Degree requirements:* For master's, creative component or thesis required. *Entrance requirements:* For master's, TOEFL (minimum score of 570 required). *Application deadline:* For fall admission, 7/1 (priority date). Application fee: $25. *Financial aid:* Research assistantships, teaching assistantships, career-related internships or fieldwork, Federal Work-Study, and tuition waivers (partial) available. Aid available to part-time students. Financial aid application deadline: 3/1. *Faculty research:* Integrated manufacturing systems, engineering practice in management, hardware aspects. *Unit head:* John W. Nazemetz, Director, 405-744-6055.

Old Dominion University, College of Engineering and Technology, Department of Mechanical Engineering, Norfolk, VA 23529. Offers design manufacturing (ME); engineering mechanics (ME, MS, PhD); mechanical engineering (ME, MS, PhD). Part-time and evening/weekend programs available. Postbaccalaureate distance learning degree programs offered (no on-campus study). *Faculty:* 14 full-time. *Students:* 61 full-time (3 women), 36 part-time (1 woman); includes 5 African Americans, 1 Hispanic American, 56 international. *Degree requirements:* For master's, computer language, comprehensive exam required, thesis optional, foreign language not required; for doctorate, computer language, dissertation, candidacy exam required, foreign language not required. *Entrance requirements:* For master's, GRE, TOEFL (minimum score of 550 required), minimum GPA of 3.0; for doctorate, GRE, TOEFL (minimum score of 550 required), minimum GPA of 3.25. *Application deadline:* For fall admission, 7/1; for spring admission, 10/1. Applications are processed on a rolling basis. Application fee: $30. Electronic applications accepted. *Unit head:* Dr. Jen Kuang Huang, Chair, 757-683-6363, Fax: 757-683-5344, E-mail: megpd@odu.edu. *Application contact:* Dr. Jen Kuang Huang, Chair, 757-683-6363, Fax: 757-683-5344, E-mail: megpd@odu.edu.

Oregon State University, Graduate School, College of Engineering, Department of Industrial and Manufacturing Engineering, Program in Manufacturing Engineering, Corvallis, OR 97331. Offers M Eng. Part-time programs available. Postbaccalaureate distance learning degree programs offered (minimal on-campus study). Average age 33. In 1998, 4 degrees awarded. *Degree requirements:* For master's, thesis or alternative required, foreign language not required. *Entrance requirements:* For master's, TOEFL (minimum score of 550 required), placement exam, minimum GPA of 3.0 in last 90 hours, BS from accredited engineering institution. *Application deadline:* For fall admission, 6/15 (priority date). Applications are processed on a rolling basis. Application fee: $50. *Financial aid:* Application deadline: 2/1. *Unit head:* Dr. Jen Kuang Huang, Chair, Department of Mechanical Engineering, 757-683-6363, Fax: 757-683-5344, E-mail: megpd@odu.edu. *Application contact:* Dr. Edward D. McDowell, Graduate Committee Chair, 541-737-2875, Fax: 541-737-5241, E-mail: mcdowele@ccmail.orst.edu.

Pennsylvania State University University Park Campus, Graduate School, College of Engineering, Department of Industrial and Manufacturing Engineering, Program in Manufacturing Engineering, State College, University Park, PA 16802-1503. Offers M Eng. *Degree requirements:* For master's, thesis required, foreign language not required. *Entrance requirements:* For master's, GRE General Test. Application fee: $50. *Unit head:* Dr. A. Ravindran, Head, Department of Industrial and Manufacturing Engineering, 814-865-7601.

Polytechnic University, Brooklyn Campus, Department of Mechanical, Aerospace and Manufacturing Engineering, Major in Manufacturing Engineering, Brooklyn, NY 11201-2990. Offers MS. *Students:* 3 full-time (0 women), 8 part-time; includes 1 minority (Hispanic American), 3 international. Average age 33. 12 applicants, 58% accepted. In 1998, 9 degrees awarded. *Entrance requirements:* For master's, BE or BS in engineering, physics, chemistry, mathematical sciences, or biological sciences or MBA. *Application deadline:* Applications are processed on a rolling basis. Application fee: $45. Electronic applications accepted. *Unit head:* Kathy Kelly, Director of Admissions, 973-596-3300, Fax: 973-596-3461, E-mail: admissions@njit.edu. *Application contact:* John S. Kerge, Dean of Admissions, 718-260-3200, Fax: 718-260-3446, E-mail: admitme@poly.edu.

See in-depth description on page 1209.

Polytechnic University, Farmingdale Campus, Graduate Programs, Department of Mechanical, Aerospace and Manufacturing Engineering, Major in Manufacturing Engineering, Farmingdale, NY 11735-3995. Offers MS. *Students:* 1 full-time (0 women). Average age 33. 5 applicants, 40% accepted. *Degree requirements:* For master's, computer language required. *Application deadline:* Applications are processed on a rolling basis. Application fee: $45. Electronic applications accepted. *Unit head:* John S. Kerge, Dean of Admissions, 718-260-3200, Fax: 718-260-3446, E-mail: admitme@poly.edu. *Application contact:* John S. Kerge, Dean of Admissions, 718-260-3200, Fax: 718-260-3446, E-mail: admitme@poly.edu.

Polytechnic University, Westchester Graduate Center, Graduate Programs, Department of Mechanical, Aerospace and Manufacturing Engineering, Major in Manufacturing Engineering, Hawthorne, NY 10532-1507. Offers MS. Average age 33. 3 applicants, 33% accepted. In 1998, 11 degrees awarded. *Degree requirements:* For master's, computer language required. *Application deadline:* Applications are processed on a rolling basis. Application fee: $45. Electronic applications accepted. *Unit head:* John S. Kerge, Dean of Admissions, 718-260-3200, Fax: 718-260-3446, E-mail: admitme@poly.edu. *Application contact:* John S. Kerge, Dean of Admissions, 718-260-3200, Fax: 718-260-3446, E-mail: admitme@poly.edu.

Portland State University, Graduate Studies, School of Engineering and Applied Science, Program in Manufacturing Engineering, Portland, OR 97207-0751. Offers ME. Part-time and evening/weekend programs available. Average age 30. 4 applicants, 50% accepted. *Entrance requirements:* For master's, TOEFL (minimum score of 550 required), minimum GPA of 3.0 in upper-division course work or 2.75 overall. *Application deadline:* For fall admission, 4/1 (priority date); for spring admission, 11/1. Applications are processed on a rolling basis. Application fee: $50. *Financial aid:* Research assistantships, teaching assistantships, career-related internships or fieldwork, Federal Work-Study, and institutionally-sponsored loans available. Aid available to part-time students. Financial aid application deadline: 3/1; financial aid applicants required to submit FAFSA. *Faculty research:* Quality assurance, concurrent engineering, production scheduling and control, manufacturing automation. *Unit head:* Dr. Graig Spolek, Head, 503-725-4631, Fax: 503-725-4298, E-mail: graig@eas.pdx.edu. *Application contact:* David Turcic, Coordinator, 503-725-4631, Fax: 503-725-4298, E-mail: davet@eas.pdx.edu.

Purdue University, Graduate School, Schools of Engineering, School of Industrial Engineering, West Lafayette, IN 47907. Offers human factors in industrial engineering (MS, MSIE, PhD); manufacturing engineering (MS, MSIE, PhD); operations research (MS, MSIE, PhD); systems engineering (MS, MSIE, PhD). Part-time programs available. *Faculty:* 27 full-time (2 women), 3 part-time (0 women). *Students:* 124 full-time (26 women), 33 part-time (8 women); includes 13 minority (2 African Americans, 4 Asian Americans or Pacific Islanders, 6 Hispanic Americans, 1 Native American), 107 international. Terminal master's awarded for partial completion of doctoral program. *Degree requirements:* For master's, computer language required, thesis optional, foreign language not required; for doctorate, computer language, dissertation required, foreign language not required. *Entrance requirements:* For master's, GRE General Test (minimum score of 470 on verbal section, 700 on quantitative, 600 on analytical required; average 520 verbal, 710 quantitative, 660 analytical), TOEFL (minimum score of 570 required; average 600), minimum GPA of 3.0; for doctorate, GRE General Test (minimum score of 470 on verbal section, 700 on quantitative, 600 on analytical required; average 540 verbal, 750 quantitative, 675 analytical), TOEFL (minimum score of 570 required; average 610), MS thesis. *Application deadline:* For fall admission, 3/15; for spring admission, 9/1. Application fee:

Manufacturing Engineering

Purdue University (continued)
$30. Electronic applications accepted. *Unit head:* Dr. W. D. Compton, Interim Head, 765-494-5444, Fax: 765-494-1299, E-mail: dcompton@ecn.purdue.edu. *Application contact:* Dr. J. W. Barany, Associate Head, 765-494-5406, Fax: 765-494-1299, E-mail: jwb@ecn.purdue.edu.

Rensselaer Polytechnic Institute, Graduate School, School of Engineering, Department of Decision Sciences and Engineering Systems, Program in Manufacturing Systems Engineering, Troy, NY 12180-3590. Offers M Eng, MS, MBA/M Eng. Part-time and evening/weekend programs available. *Faculty:* 12 full-time (0 women), 6 part-time (1 woman). *Students:* 9 full-time (1 woman), 2 part-time (1 woman), 8 international. 15 applicants, 53% accepted. In 1998, 11 degrees awarded. *Degree requirements:* For master's, thesis required (for some programs), foreign language not required. *Entrance requirements:* For master's, GRE General Test, TOEFL (minimum score of 550 required). *Application deadline:* For fall admission, 2/1 (priority date). Applications are processed on a rolling basis. Application fee: $35. *Financial aid:* In 1998–99, 1 fellowship, 2 teaching assistantships were awarded.; research assistantships, career-related internships or fieldwork and institutionally-sponsored loans also available. Financial aid application deadline: 2/1. *Faculty research:* Information systems, statistical consulting, education/services, production/logistics/inventory. Total annual research expenditures: $2.2 million. *Unit head:* Dr. M. Khairul Alam, Graduate Chairman, 740-593-1598, Fax: 740-593-0476, E-mail: agrawal@bobcat.ent.ohiou.edu. *Application contact:* Lee Vilardi, Graduate Coordinator, 518-276-6681, Fax: 518-276-8227, E-mail: dsesgr@rpi.edu.

Rochester Institute of Technology, Part-time and Graduate Admissions, College of Applied Science and Technology, Department of Engineering Technology, Program in Computer Integrated Manufacturing, Rochester, NY 14623-5604. Offers MS. *Students:* 8 full-time (1 woman), 12 part-time (2 women); includes 2 minority (1 African American, 1 Hispanic American), 4 international. 13 applicants, 85% accepted. In 1998, 10 degrees awarded. *Degree requirements:* For master's, computer language, thesis required. *Entrance requirements:* For master's, minimum GPA of 3.0. *Application deadline:* For fall admission, 3/1 (priority date). Applications are processed on a rolling basis. Application fee: $40. *Financial aid:* Unspecified assistantships available. *Unit head:* Clyde M. Creveling, Graduate Adviser, 716-475-5813, E-mail: cmcmet@rit.edu.

Rochester Institute of Technology, Part-time and Graduate Admissions, College of Applied Science and Technology, Department of Packaging Science, Rochester, NY 14623-5604. Offers MS. *Students:* 5 full-time (1 woman), 13 part-time (2 women); includes 1 minority (Hispanic American), 4 international. 2 applicants, 100% accepted. In 1998, 10 degrees awarded. *Degree requirements:* For master's, computer language, thesis required. *Entrance requirements:* For master's, minimum GPA of 3.0. *Application deadline:* For fall admission, 3/1 (priority date). Applications are processed on a rolling basis. Application fee: $40. *Financial aid:* Research assistantships available. *Unit head:* Dr. Daniel Goodwin, Director, 716-475-2278, E-mail: dlgipk@rit.edu.

Rochester Institute of Technology, Part-time and Graduate Admissions, College of Engineering, Department of Industrial and Manufacturing Engineering, Rochester, NY 14623-5604. Offers engineering management (ME); industrial engineering (ME); manufacturing engineering (ME); systems engineering (ME). *Students:* 6 full-time (3 women), 13 part-time (4 women); includes 3 minority (all African Americans), 4 international. 71 applicants, 56% accepted. In 1998, 11 degrees awarded. *Degree requirements:* For master's, internship required. *Entrance requirements:* For master's, TOEFL, minimum GPA of 3.0. *Application deadline:* For fall admission, 3/1 (priority date). Applications are processed on a rolling basis. Application fee: $40. *Financial aid:* Research assistantships, career-related internships or fieldwork and tuition waivers (partial) available. *Faculty research:* Safety, manufacturing (CAM), simulation. *Unit head:* Dr. Jasper Shealy, Head, 716-475-2134, E-mail: jeseie@rit.edu.

Southern Illinois University Carbondale, Graduate School, College of Engineering, Department of Technology, Carbondale, IL 62901-6806. Offers manufacturing systems (MS). *Faculty:* 9 full-time (1 woman). *Students:* 21 full-time (3 women), 7 part-time; includes 6 minority (5 African Americans, 1 Asian American or Pacific Islander), 8 international. Average age 25. 13 applicants, 69% accepted. In 1998, 11 degrees awarded. *Degree requirements:* For master's, thesis, comprehensive exam required, foreign language not required. *Entrance requirements:* For master's, TOEFL (minimum score of 550 required), minimum GPA of 2.7. *Application deadline:* Applications are processed on a rolling basis. Application fee: $20. *Financial aid:* In 1998–99, 15 students received aid, including 1 fellowship with full tuition reimbursement available, 3 research assistantships with full tuition reimbursements available, 9 teaching assistantships with full tuition reimbursements available; tuition waivers (full) also available. Financial aid application deadline: 7/1. *Faculty research:* Computer-aided manufacturing, robotics,quality assurance. Total annual research expenditures: $205,198. *Unit head:* Dr. Gary Butson, Chair, 618-536-3396. *Application contact:* Robert R. Ferketich, Director, 618-536-3396.

Southern Methodist University, School of Engineering and Applied Science, Department of Mechanical Engineering, Dallas, TX 75275. Offers manufacturing systems management (MS); mechanical engineering (MSME, PhD). Part-time programs available. Postbaccalaureate distance learning degree programs offered (no on-campus study). *Faculty:* 11 full-time (0 women), 2 part-time (1 woman). *Students:* 24 full-time (3 women), 45 part-time (1 woman); includes 9 minority (4 African Americans, 2 Asian Americans or Pacific Islanders, 3 Hispanic Americans), 25 international. Terminal master's awarded for partial completion of doctoral program. *Degree requirements:* For master's, thesis optional, foreign language not required; for doctorate, dissertation, oral and written qualifying exams, oral final exam required, foreign language not required. *Entrance requirements:* For master's, GRE General Test (minimum score of 650 on quantitative section required), TOEFL (minimum score of 550 required), minimum GPA of 3.0 in last 2 years; bachelor's degree in engineering, mathematics, or sciences; for doctorate, preliminary counseling exam, minimum graduate GPA of 3.0, bachelor's degree in related field. *Application deadline:* For fall admission, 8/1 (priority date); for spring admission, 12/15. Applications are processed on a rolling basis. Application fee: $25. Tuition: Full-time $9,216; part-time $512 per credit hour. Required fees: $88 per credit hour. Part-time tuition and fees vary according to course load and campus/location. *Unit head:* Dr. Osita Nwokah, Chair, 214-768-3200, Fax: 214-768-1473, E-mail: nwokah@seas.smu.edu. *Application contact:* Dr. Zeynep Celik-Butler, Assistant Dean for Graduate Studies and Research, 214-768-3979, Fax: 214-768-3845, E-mail: zcb@seas.smu.edu.

Stanford University, School of Engineering, Department of Mechanical Engineering, Program in Manufacturing Systems Engineering, Stanford, CA 94305-9991. Offers MS, MBA/MS. *Entrance requirements:* For master's, GRE General Test, TOEFL, GMAT. *Application deadline:* For fall admission, 1/14. Application fee: $65 ($80 for international students). Electronic applications accepted. Tuition: Full-time $24,588. Required fees: $152. Part-time tuition and fees vary according to course load. *Financial aid:* Application deadline: 1/15. *Application contact:* Admissions Office, 650-723-3148.

See in-depth description on page 1217.

Syracuse University, Graduate School, L. C. Smith College of Engineering and Computer Science, Program in Manufacturing Engineering, Syracuse, NY 13244-0003. Offers MS. *Students:* 9 full-time (1 woman), 7 part-time, 9 international. Average age 30. 33 applicants, 100% accepted. In 1998, 12 degrees awarded. *Degree requirements:* Foreign language not required. *Entrance requirements:* For master's, GRE General Test, GRE Subject Test. *Application deadline:* Applications are processed on a rolling basis. Application fee: $40. Tuition: Full-time $13,992; part-time $583 per credit hour. *Financial aid:* Fellowships, research assistantships, teaching assistantships, Federal Work-Study and tuition waivers (partial) available. Financial aid application deadline: 3/1. *Unit head:* Alan Levy, Graduate Director. *Application contact:* Eric Spina, Contact, 315-443-2341.

Tufts University, Division of Graduate and Continuing Studies and Research, Professional and Continuing Studies, Manufacturing Engineering Program, Medford, MA 02155. Offers Certificate. Part-time and evening/weekend programs available. Average age 40. 3 applicants, 100% accepted. In 1998, 12 degrees awarded. *Average time to degree:* 1 year part-time. *Application deadline:* For fall admission, 8/15 (priority date); for spring admission, 12/12. Applications are processed on a rolling basis. Application fee: $40. *Financial aid:* Available to part-time students. Application deadline: 5/1; *Unit head:* Dr. Yassin A. Hassan, Graduate Coordinator, 409-845-7090. *Application contact:* Dr. Yassin A. Hassan, Graduate Coordinator, 409-845-7090.

University of California, Los Angeles, Graduate Division, School of Engineering and Applied Science, Department of Mechanical and Aerospace Engineering, Program in Manufacturing Engineering, Los Angeles, CA 90095. Offers MS. *Students:* 8 full-time (1 woman); includes 5 minority (all Asian Americans or Pacific Islanders), 2 international. 16 applicants, 63% accepted. In 1998, 4 degrees awarded. *Degree requirements:* For master's, comprehensive exam or thesis required. *Entrance requirements:* For master's, GRE General Test, GRE Subject Test (required for foreign students), minimum GPA of 3.0. *Application deadline:* For fall admission, 1/5; for spring admission, 12/31. Application fee: $40. Electronic applications accepted. *Financial aid:* Fellowships, research assistantships, teaching assistantships, Federal Work-Study, institutionally-sponsored loans, and tuition waivers (full and partial) available. Financial aid application deadline: 1/5; financial aid applicants required to submit FAFSA. *Application contact:* Student Affairs Officer, E-mail: maeapp@ea.ucla.edu.

University of Central Florida, College of Engineering, Department of Industrial Engineering and Management Systems, Orlando, FL 32816. Offers computer-integrated manufacturing (MS); engineering management (MS); industrial engineering (MSIE); industrial engineering and management systems (PhD); manufacturing engineering (MS Mfg E); operations research (MS); product assurance engineering (MS); simulation systems (MS). Part-time and evening/weekend programs available. *Faculty:* 20 full-time, 11 part-time. *Students:* 143 full-time (27 women), 112 part-time (20 women); includes 43 minority (6 African Americans, 12 Asian Americans or Pacific Islanders, 25 Hispanic Americans), 51 international. *Degree requirements:* For master's, computer language, thesis or alternative required, foreign language not required; for doctorate, computer language, dissertation, departmental qualifying exam, candidacy exam required, foreign language not required. *Entrance requirements:* For master's, GRE General Test (minimum combined score of 1000 required), TOEFL (minimum score of 550 required; 213 computer-based), minimum GPA of 3.0 in last 60 hours; for doctorate, TOEFL (minimum score of 550 required; 213 computer-based), minimum GPA of 3.5 in last 60 hours. *Application deadline:* For fall admission, 7/15; for spring admission, 12/15. Application fee: $20. Tuition, state resident: full-time $2,054; part-time $137 per credit. Tuition, nonresident: full-time $7,207; part-time $480 per credit. Required fees: $47 per term. *Unit head:* Dr. Charles Reily, Chair, 407-823-2204. *Application contact:* Dr. Linda Malone, Coordinator, 407-823-2204.

See in-depth description on page 1225.

University of Detroit Mercy, College of Engineering and Science, Department of Mechanical Engineering, Detroit, MI 48219-0900. Offers automotive engineering (DE); engineering management (M Eng Mgt), including engineering management, mechanical engineering; manufacturing engineering (ME, DE); mechanical engineering (ME, DE). Evening/weekend programs available. *Degree requirements:* For master's, computer language required, foreign language not required; for doctorate, dissertation required. *Faculty research:* CAD/CAM.

University of Florida, Graduate School, College of Engineering, Department of Industrial and Systems Engineering, Program in Manufacturing Systems Engineering, Gainesville, FL 32611. Offers ME, MS, PhD, Certificate. Offered in cooperation with the Departments of Aerospace Engineering, Mechanics, and Engineering Science; Computer and Information Sciences; ElectricalEngineering; Industrial and Systems Engineering; Materials Science and Engineering; and Mechanical Engineering. *Degree requirements:* For master's, computer language, core exam, project or thesis required, thesis optional; for doctorate, dissertation, comprehensive exam required; for Certificate, computer language, thesis required. *Entrance requirements:* For master's and doctorate, GRE General Test, TOEFL, minimum GPA of 3.0; for Certificate, GRE General Test. *Application deadline:* For fall admission, 6/1 (priority date). Applications are processed on a rolling basis. Application fee: $20. Electronic applications accepted. *Financial aid:* Fellowships available. *Application contact:* Dr. D. J. Elzinga, Graduate Coordinator, 352-392-1464, Fax: 352-392-3537, E-mail: elzinga@ise.ufl.edu.

University of Houston, College of Technology, Houston, TX 77004. Offers construction management (MT); manufacturing systems (MT); microcomputer systems (MT); occupational technology (MSOT). Part-time and evening/weekend programs available. *Faculty:* 23 full-time (7 women), 3 part-time (0 women). *Students:* 17 full-time (11 women), 75 part-time (41 women); includes 27 minority (15 African Americans, 4 Asian Americans or Pacific Islanders, 8 Hispanic Americans), 6 international. *Degree requirements:* Foreign language not required. *Entrance requirements:* For master's, GMAT, GRE, or MAT (MSOT); GRE (MT), minimum GPA of 3.0 in last 60 hours. *Application deadline:* For fall admission, 7/1; for spring admission, 11/1. Application fee: $35 ($110 for international students). *Unit head:* Bernard McIntyre, Dean, 713-743-4028, Fax: 713-743-4032, E-mail: bmcintyre@uh.edu. *Application contact:* Holly Rosenthal, Graduate Academic Adviser, 713-743-4098, Fax: 713-743-4032, E-mail: hrosenthal@uh.edu.

The University of Iowa, Graduate College, College of Engineering, Department of Industrial Engineering, Iowa City, IA 52242-1316. Offers engineering design and manufacturing (MS, PhD); ergonomics (MS, PhD); information and engineering management (MS, PhD); operations research (MS, PhD); quality engineering (MS, PhD). *Faculty:* 7 full-time, 1 part-time. *Students:* 18 full-time (4 women), 16 part-time (1 woman); includes 4 minority (1 African American, 2 Asian Americans or Pacific Islanders, 1 Hispanic American), 19 international. *Degree requirements:* For master's, thesis optional; for doctorate, dissertation, comprehensive exam required. *Entrance requirements:* For master's and doctorate, GRE General Test, GRE Subject Test, TOEFL. *Application deadline:* Applications are processed on a rolling basis. Application fee: $30 ($50 for international students). *Unit head:* Peter J. O'Grady, Chair, 319-335-5939, Fax: 319-335-5424.

University of Kentucky, Graduate School, Graduate School Programs from the College of Engineering, Program in Manufacturing Systems Engineering, Lexington, KY 40506-0032. Offers MSMSE. *Degree requirements:* For master's, comprehensive exam required. *Entrance requirements:* For master's, GRE General Test. *Faculty research:* Manufacturing processes and equipment, manufacturing systems and control, computer-aided design and manufacturing, automation in manufacturing, electric manufacturing and packaging.

University of Maryland, College Park, Graduate School, A. James Clark School of Engineering, Department of Mechanical Engineering, College Park, MD 20742-5045. Offers electronic packaging and reliability (MS, PhD); manufacturing and design (MS, PhD); mechanical engineering (M Eng); mechanics and materials (MS, PhD); thermal and fluid sciences (MS, PhD). Part-time and evening/weekend programs available. Postbaccalaureate distance learning degree programs offered. *Faculty:* 61 full-time (5 women), 15 part-time (0 women). *Students:* 133 full-time (13 women), 60 part-time (7 women); includes 18 minority (11 African Americans, 6 Asian Americans or Pacific Islanders, 1 Hispanic American), 123 international. *Degree requirements:* For master's, thesis optional, foreign language not required; for doctorate, dissertation, qualifying exam required, foreign language not required. *Entrance requirements:* For master's, GRE, minimum GPA of 3.0. *Application deadline:* Applications are processed on a rolling basis. Application fee: $50 ($70 for international students). Tuition, state resident: part-time $272 per credit hour. Tuition, nonresident: part-time $475 per credit hour. Required fees: $632; $379 per year. *Unit head:* Dr. Davinder Anand, Chairman, 301-405-5294, Fax: 301-314-9477. *Application contact:* Dr. James M. Wallace, Graduate Director, 301-405-4216.

Manufacturing Engineering

University of Massachusetts Amherst, Graduate School, College of Engineering, Department of Mechanical and Industrial Engineering, Program in Manufacturing Engineering, Amherst, MA 01003. Offers MS. *Accreditation:* ABET. *Students:* 2 full-time (0 women), 3 part-time (1 woman), (all international). Average age 26. 41 applicants, 44% accepted. In 1998, 5 degrees awarded. *Degree requirements:* For master's, foreign language and thesis not required. *Entrance requirements:* For master's, GRE General Test. *Application deadline:* For fall admission, 2/1 (priority date); for spring admission, 10/1. Applications are processed on a rolling basis. Application fee: $40. Tuition, state resident: full-time $2,640; part-time $165 per credit. Tuition, nonresident: full-time $9,756; part-time $407 per credit. Required fees: $1,221 per term. One-time fee: $110. Full-time tuition and fees vary according to course load, campus/location and reciprocity agreements. *Financial aid:* Fellowships with full tuition reimbursements, research assistantships with full tuition reimbursements, teaching assistantships with full tuition reimbursements, career-related internships or fieldwork, Federal Work-Study, grants, scholarships, traineeships, and unspecified assistantships available. Aid available to part-time students. Financial aid application deadline: 2/1. *Unit head:* Dr. Donald Fisher, Director, 413-545-0955, Fax: 413-545-1027, E-mail: dfisher@ecs.umass.edu.

The University of Memphis, Graduate School, Herff College of Engineering, Department of Engineering Technology, Memphis, TN 38152. Offers architectural technology (MS); electronics engineering technology (MS); manufacturing engineering technology (MS). Part-time programs available. *Faculty:* 6 full-time (2 women). *Students:* 19 full-time (4 women), 8 part-time (2 women); includes 3 minority (2 African Americans, 1 Asian American or Pacific Islander), 21 international. *Degree requirements:* For master's, comprehensive exam required. *Entrance requirements:* For master's, GRE General Test (minimum combined score of 1000 required) or MAT, interview, minimum undergraduate GPA of 2.5. *Application deadline:* For fall admission, 8/1; for spring admission, 12/1. Applications are processed on a rolling basis. Application fee: $25 ($50 for international students). Electronic applications accepted. Tuition, state resident: full-time $3,410; part-time $178 per credit hour. Tuition, nonresident: full-time $8,670; part-time $408 per credit hour. Tuition and fees vary according to program. *Unit head:* Ronald L. Day, Chairman, 901-678-2238, Fax: 901-678-5145, E-mail: rday@memphis.edu. *Application contact:* Dr. Dean L. Smith, Coordinator of Graduate Studies, 901-678-3300, Fax: 901-678-5145, E-mail: dlsmith@cc.memphis.edu.

University of Michigan, Horace H. Rackham School of Graduate Studies, College of Engineering, Program in Manufacturing, Ann Arbor, MI 48109. Offers M Eng, D Eng, MBA/M Eng. Part-time programs available. Postbaccalaureate distance learning degree programs offered (no on-campus study). *Faculty:* 1 part-time (0 women). *Students:* 36 full-time (3 women), 27 part-time (5 women); includes 24 minority (8 African Americans, 13 Asian Americans or Pacific Islanders, 3 Hispanic Americans), 8 international. 83 applicants, 45% accepted. In 1998, 30 degrees awarded. Terminal master's awarded for partial completion of doctoral program. *Degree requirements:* For master's, team project required; for doctorate, dissertation, preliminary and qualifying exams required. *Entrance requirements:* For doctorate, GRE General Test (combined average 1798 on three sections). *Average time to degree:* Master's–1 year full-time, 3 years part-time. *Application deadline:* For fall admission, 7/15 (priority date). Applications are processed on a rolling basis. Application fee: $55. *Financial aid:* In 1998–99, 11 students received aid, including 11 fellowships with full and partial tuition reimbursements available (averaging $6,349 per year); research assistantships, career-related internships or fieldwork and institutionally-sponsored loans also available. Financial aid application deadline: 2/1; financial aid applicants required to submit FAFSA. *Unit head:* Debasish Dutta, Director, 734-764-3312, Fax: 734-647-0079, E-mail: pim@engin.umich.edu.

University of Michigan–Dearborn, College of Engineering and Computer Science, Interdisciplinary Programs, Program in Manufacturing Systems Engineering, Dearborn, MI 48128-1491. Offers MSE, D Eng. Part-time and evening/weekend programs available. *Faculty:* 1 full-time (0 women). *Students:* 4 full-time (2 women), 52 part-time (13 women); includes 9 minority (3 African Americans, 1 Asian American or Pacific Islander, 5 Hispanic Americans), 2 international. Average age 29. In 1998, 24 degrees awarded. *Degree requirements:* For master's, computer language required, thesis optional, foreign language not required. *Entrance requirements:* For master's, bachelor's degree in applied mathematics, computer science, engineering, or physical science; minimum GPA of 3.0. *Application deadline:* For fall admission, 8/1 (priority date); for winter admission, 12/1 (priority date); for spring admission, 4/1. Applications are processed on a rolling basis. Application fee: $55. Electronic applications accepted. Tuition, state resident: part-time $259 per credit hour. Tuition, nonresident: part-time $748 per credit hour. Required fees: $80 per course. Tuition and fees vary according to course level, course load and program. *Financial aid:* In 1998–99, 2 research assistantships were awarded. *Faculty research:* Toolwear metrology, paper handling, grinding wheel imbalance, machine mission. *Unit head:* Lisa M. Beach, Administrative Assistant, 313-593-5361, Fax: 313-593-3692, E-mail: lmbeach@umich.edu. *Application contact:* Deborah Parker, Administrative Assistant I, 313-593-5582, Fax: 313-593-5386, E-mail: debbie@umdsun2.umd.umich.edu.

University of Minnesota, Twin Cities Campus, Graduate School, Institute of Technology, Center for the Development of Technological Leadership, Program in Manufacturing Systems, Minneapolis, MN 55455-0213. Offers MS. Part-time and evening/weekend programs available. *Faculty:* 15 part-time (2 women). *Students:* 43 (6 women); includes 2 minority (both Hispanic Americans) 34 applicants, 88% accepted. *Degree requirements:* For master's, thesis, capstone project required, foreign language not required. *Entrance requirements:* For master's, 1 year of work experience in manufacturing, minimum undergraduate GPA of 3.0. *Application deadline:* For fall admission, 7/15 (priority date). Applications are processed on a rolling basis. Application fee: $50 ($55 for international students). Electronic applications accepted. *Financial aid:* Institutionally-sponsored loans available. Aid available to part-time students. Financial aid applicants required to submit FAFSA. *Faculty research:* Thermal systems, micro-electrical mechanical systems, plasma technology, composite manufacturing, manufacturing automation. *Unit head:* Student Services Associate, 734-764-9387. *Application contact:* Toni Limon, Admissions.

University of Missouri–Columbia, Graduate School, College of Engineering, Department of Industrial and Manufacturing Systems Engineering, Columbia, MO 65211. Offers MS, PhD. *Faculty:* 8 full-time (1 woman). *Students:* 17 full-time (1 woman), 11 part-time (1 woman); includes 2 minority (both African Americans), 24 international. 31 applicants, 48% accepted. In 1998, 8 master's, 6 doctorates awarded. *Degree requirements:* For master's, computer language, thesis or alternative required, foreign language not required; for doctorate, dissertation required. *Entrance requirements:* For master's and doctorate, GRE General Test, TOEFL, minimum GPA of 3.0. *Application deadline:* For fall admission, 4/1 (priority date). Applications are processed on a rolling basis. Application fee: $30 ($50 for international students). *Financial aid:* Research assistantships, teaching assistantships, institutionally-sponsored loans available. *Unit head:* Dr. Cerry Klein, Director of Graduate Studies, 573-882-9566.

See in-depth description on page 1235.

University of Nebraska–Lincoln, Graduate College, College of Engineering and Technology, Interdepartmental Area of Manufacturing Systems Engineering, Lincoln, NE 68588. Offers engineering (PhD); manufacturing systems engineering (MS). *Students:* 16 full-time (1 woman), 12 part-time; includes 1 minority (African American), 19 international. Average age 28. 29 applicants, 59% accepted. In 1998, 11 degrees awarded. *Degree requirements:* For master's, thesis optional; for doctorate, dissertation, comprehensive exams required. *Entrance requirements:* For master's and doctorate, GRE General Test, TOEFL (minimum score of 525 required). *Application deadline:* For fall admission, 3/1 (priority date). Applications are processed on a rolling basis. Application fee: $35. Electronic applications accepted. *Financial aid:* Fellowships, research assistantships, teaching assistantships, Federal Work-Study available. Aid available to part-time students. Financial aid application deadline: 2/15. *Unit head:* Dr. Kamlakar Rajurkar, Chair, 402-472-3495.

University of New Mexico, Graduate School, School of Engineering, Department of Electrical and Computer Engineering, Albuquerque, NM 87131-2039. Offers computer engineering (MS, PhD); manufacturing engineering (ME, MS); microelectronics (MS, PhD); network and control systems (MS, PhD); optoelectronics (MS, PhD); pulsed power and plasma science (MS, PhD); signal processing and communications (MS, PhD). ME offered through the Manufacturing Engineering Program. Part-time and evening/weekend programs available. *Faculty:* 41 full-time (3 women), 14 part-time (1 woman). *Students:* 76 full-time (18 women), 53 part-time (10 women); includes 23 minority (1 African American, 7 Asian Americans or Pacific Islanders, 14 Hispanic Americans, 1 Native American), 44 international. *Degree requirements:* For master's, thesis required (for some programs), foreign language not required; for doctorate, dissertation required, foreign language not required. *Entrance requirements:* For master's, GRE General Test (minimum score of 400 on verbal section, 650 on quantitative required), minimum GPA of 3.0; for doctorate, GRE General Test (minimum score of 450 on verbal section, 690 on quantitative requireed), minimum GPA of 3.0. *Application deadline:* For fall admission, 7/15; for spring admission, 11/14. Applications are processed on a rolling basis. Application fee: $25. *Unit head:* Dr. Christos Christodovlov, Chair, 505-277-6580, Fax: 505-277-1439, E-mail: cgc@eece.unm.edu. *Application contact:* Dr. Kenneth Jungling, Graduate Coordinator, 505-277-1433, Fax: 505-277-1439, E-mail: ken@eece.unm.edu.

University of New Mexico, Graduate School, School of Engineering, Department of Mechanical Engineering, Albuquerque, NM 87131-2039. Offers engineering (PhD); manufacturing engineering (ME, MS); mechanical engineering (MSME). ME offered through the Manufacturing Engineering Program. Part-time programs available. *Faculty:* 17 full-time (0 women), 18 part-time (2 women). *Students:* 20 full-time (2 women), 39 part-time (6 women); includes 18 minority (1 African American, 1 Asian American or Pacific Islander, 13 Hispanic Americans, 3 Native Americans), 8 international. *Degree requirements:* For master's, thesis required (for some programs), foreign language not required; for doctorate, dissertation required, foreign language not required. *Entrance requirements:* For master's and doctorate, GRE General Test, minimum GPA of 3.0. *Application deadline:* For fall admission, 7/15; for spring admission, 11/14. Applications are processed on a rolling basis. Application fee: $25. *Unit head:* Dr. David E. Thompson, Chair, 505-277-2761, Fax: 505-277-1571, E-mail: dthomp@me.unm.edu.

University of New Mexico, Graduate School, School of Engineering, Manufacturing Engineering Program, Albuquerque, NM 87131-2039. Offers ME. Program supports concentration in manufacturing engineering available in the MS programs in mechanical engineering and electrical engineering and computer engineering. Part-time programs available. *Faculty:* 1 full-time (0 women), 1 part-time (0 women). *Students:* 3 full-time (1 woman), 7 part-time (1 woman); includes 4 minority (all Hispanic Americans) Average age 33. 7 applicants, 86% accepted. *Degree requirements:* Foreign language not required. *Entrance requirements:* For master's, GRE General Test, minimum GPA of 3.0. *Application deadline:* For fall admission, 7/15; for spring admission, 11/14. Applications are processed on a rolling basis. Application fee: $25. *Financial aid:* In 1998–99, 2 fellowships (averaging $4,375 per year) were awarded.; research assistantships Total annual research expenditures: $4 million. *Unit head:* Dr. J. E. Wood, Director, 505-277-1420. *Application contact:* Emma Blythe, Administrative Assistant, 505-277-1420.

See in-depth description on page 1237.

University of Pittsburgh, School of Engineering, Program in Manufacturing Systems Engineering, Pittsburgh, PA 15260. Offers MSMfSE. Part-time and evening/weekend programs available. Postbaccalaureate distance learning degree programs offered (no on-campus study). 11 applicants, 100% accepted. In 1998, 13 degrees awarded. *Degree requirements:* For master's, computer language, thesis, internship or project required, foreign language not required. *Entrance requirements:* For master's, GRE General Test, TOEFL (minimum score of 550 required). *Average time to degree:* Master's–2 years full-time, 3 years part-time. *Application deadline:* For fall admission, 8/1 (priority date); for spring admission, 12/1 (priority date). Applications are processed on a rolling basis. Application fee: $30 ($40 for international students). *Financial aid:* Application deadline: 2/15. *Faculty research:* Manufacturing information systems, manufacturing process simulation, product design methods. *Unit head:* Dr. Harvey Wolfe, Chairman, 412-624-9830, Fax: 412-624-9831, E-mail: hwolfe@engrng.pitt.edu. *Application contact:* John H. Manley, Director, 412-624-9846, Fax: 412-624-9831, E-mail: jmanley@vms.cis.pitt.edu.

University of Rhode Island, Graduate School, College of Engineering, Department of Industrial and Manufacturing Engineering, Program in Manufacturing Engineering, Kingston, RI 02881. Offers MS. *Accreditation:* ABET.

University of St. Thomas, Graduate Studies, Graduate School of Applied Science and Engineering, Program in Manufacturing Systems Engineering, St. Paul, MN 55105-1096. Offers MMSE, MS, Certificate. *Accreditation:* ABET (one or more programs are accredited). Part-time and evening/weekend programs available. *Faculty:* 5 full-time (0 women), 13 part-time (0 women). *Students:* 6 full-time (1 woman), 204 part-time (41 women). Average age 34. 77 applicants, 95% accepted. In 1998, 22 master's awarded. *Degree requirements:* For master's, computer language, thesis required (for some programs). *Entrance requirements:* For master's, GMAT or GRE General Test. *Application deadline:* For fall admission, 8/1 (priority date); for spring admission, 1/1 (priority date). Applications are processed on a rolling basis. Application fee: $30. Electronic applications accepted. Tuition: Part-time $437 per credit. Tuition and fees vary according to degree level, program and student level. *Financial aid:* In 1998–99, 12 students received aid; fellowships, research assistantships, grants and institutionally-sponsored loans available. Aid available to part-time students. Financial aid application deadline: 4/1; financial aid applicants required to submit FAFSA. *Unit head:* Ron Bennett, Director, 651-962-5750, Fax: 651-962-6419, E-mail: rjbennett@stthomas.edu. *Application contact:* Marlene L. Houliston, Student Services Coordinator, 651-962-5750, Fax: 651-962-6419, E-mail: technology@stthomas.edu.

University of Southern California, Graduate School, School of Engineering, Department of Industrial and Systems Engineering, Program in Manufacturing Engineering, Los Angeles, CA 90089. Offers MS. *Students:* 6 full-time (1 woman), 3 part-time, 5 international. Average age 31. 15 applicants, 67% accepted. In 1998, 5 degrees awarded. *Degree requirements:* For master's, thesis optional. *Entrance requirements:* For master's, GRE General Test. *Application deadline:* For fall admission, 6/1 (priority date); for spring admission, 12/1. Application fee: $55. Tuition: Part-time $768 per unit. Required fees: $350 per semester. *Financial aid:* In 1998–99, 3 research assistantships were awarded.; fellowships, teaching assistantships, Federal Work-Study and institutionally-sponsored loans also available. Aid available to part-time students. Financial aid application deadline: 2/15; financial aid applicants required to submit FAFSA. *Unit head:* Dr. F. Stan Settles, Chairman, Department of Industrial and Systems Engineering, 213-740-4893.

University of Tennessee, Knoxville, Graduate School, College of Engineering, Department of Industrial Engineering, Knoxville, TN 37996. Offers engineering management (MS); manufacturing systems engineering (MS); traditional industrial engineering (MS). Part-time programs available. Postbaccalaureate distance learning degree programs offered (no on-campus study). *Faculty:* 10 full-time (1 woman). *Students:* 25 full-time (7 women), 111 part-time (29 women); includes 14 minority (10 African Americans, 3 Asian Americans or Pacific Islanders, 1 Hispanic American), 11 international. *Degree requirements:* For master's, thesis or alternative required, foreign language not required. *Entrance requirements:* For master's, TOEFL (minimum score of 550 required), minimum GPA of 2.7. *Application deadline:* For fall admission, 2/1 (priority date). Applications are processed on a rolling basis. Application fee: $35. Electronic applications accepted. *Unit head:* Dr. Thomas Shannon, Head, 423-974-3333, Fax: 423-974-0588, E-mail: tshannon@utk.edu. *Application contact:* Dr. J. A. Bontadelli, Graduate Representative, E-mail: bontadelli@utk.edu.

The University of Texas at Austin, Graduate School, College of Engineering, Program in Manufacturing Systems Engineering, Austin, TX 78712-1111. Offers MSE, MBA/MSE. *Faculty:* 69 full-time (7 women), 1 part-time (0 women). *Students:* 24 full-time (4 women), 5 part-time (1 woman); includes 4 minority (3 Asian Americans or Pacific Islanders, 1 Hispanic American), 19

Manufacturing Engineering–Reliability Engineering

The University of Texas at Austin (continued)
international. In 1998, 8 degrees awarded. *Degree requirements:* For master's, foreign language and thesis not required. *Entrance requirements:* For master's, GRE General Test (minimum combined score of 1000 required). Application fee: $50 ($75 for international students). *Financial aid:* In 1998–99, 15 students received aid, including 1 fellowship; research assistantships, teaching assistantships, career-related internships or fieldwork also available. Financial aid application deadline: 1/1. *Unit head:* Director, 512-471-1271. *Application contact:* Michael Bryant, Graduate Adviser, 512-471-3610.

The University of Texas at El Paso, Graduate School, College of Engineering, Department of Mechanical and Industrial Engineering, Program in Manufacturing Engineering, El Paso, TX 79968-0001. Offers MS. *Faculty:* 18 full-time (2 women), 3 part-time (1 woman). 15 applicants, 60% accepted. *Degree requirements:* For master's, thesis optional, foreign language not required. *Entrance requirements:* For master's, GRE General Test, TOEFL (minimum score of 550 required), minimum GPA of 3.0 in major. *Application deadline:* Applications are processed on a rolling basis. Application fee: $15 ($65 for international students). Electronic applications accepted. Tuition, state resident: full-time $2,790. Tuition, nonresident: full-time $7,710. *Financial aid:* Fellowships, research assistantships, teaching assistantships, Federal Work-Study, institutionally-sponsored loans, and tuition waivers (partial) available. *Unit head:* Dr. Rene Villalobos, Graduate Adviser, 915-747-5450, Fax: 915-747-5019, E-mail: rene@ulobos.me.ep.utexas.edu. *Application contact:* Susan Jordan, Director, Graduate Student Services, 915-747-5491, Fax: 915-747-5788, E-mail: sjordan@utep.edu.

University of Toronto, School of Graduate Studies, Physical Sciences Division, Faculty of Applied Science and Engineering, Collaborative Program in Integrated Manufacturing, Toronto, ON M5S 1A1, Canada. Offers M Eng. Part-time programs available. *Degree requirements:* For master's, thesis not required.

University of Wisconsin–Madison, Graduate School, College of Engineering, Manufacturing Systems Engineering Program, Madison, WI 53706-1380. Offers MS. Part-time programs available. *Faculty:* 44 part-time (3 women). *Students:* 38 full-time (7 women), 9 part-time (2 women); includes 6 minority (3 African Americans, 1 Asian American or Pacific Islander, 2 Hispanic Americans), 31 international. In 1998, 20 degrees awarded. *Degree requirements:* For master's, computer language, thesis required (for some programs), foreign language not required. *Entrance requirements:* For master's, GRE General Test. *Application deadline:* For fall admission, 6/15 (priority date); for spring admission, 10/31 (priority date). Applications are processed on a rolling basis. Application fee: $45. Electronic applications accepted. *Financial aid:* In 1998–99, 21 students received aid, including 7 fellowships, 11 research assistantships; career-related internships or fieldwork, Federal Work-Study, institutionally-sponsored loans, and unspecified assistantships also available. *Faculty research:* CAD/CAM, rapid prototyping, lead time reduction, quick response manufacturing. *Unit head:* Rajan Suri, Director, 608-262-0921, Fax: 608-265-4017, E-mail: suri@engr.wisc.edu. *Application contact:* Carol Enseki, Administrative Assistant, 608-262-0921, Fax: 608-265-4017.

See in-depth description on page 1243.

Villanova University, College of Engineering, Department of Mechanical Engineering, Villanova, PA 19085-1699. Offers manufacturing (Certificate); mechanical engineering (MME); virtual manufacturing (Certificate). Part-time and evening/weekend programs available. *Faculty:* 12 full-time (0 women), 2 part-time (0 women). *Students:* 7 full-time (0 women), 14 part-time (2 women); includes 1 minority (Asian American or Pacific Islander), 5 international. *Degree requirements:* For master's, computer language required, thesis optional, foreign language not required. *Entrance requirements:* For master's, GRE General Test (for applicants with degrees from foreign universities), TOEFL (minimum score of 575 required), BME, minimum GPA of 3.0. *Application deadline:* For fall admission, 8/1 (priority date); for spring admission, 12/1. Applications are processed on a rolling basis. Application fee: $40. *Unit head:* Dr. Alan M. Whitman, Chairperson, 610-519-4980, E-mail: awhitman@email.vill.edu.

Wayne State University, Graduate School, College of Engineering, Department of Industrial and Manufacturing Engineering, Program in Manufacturing Engineering, Detroit, MI 48202. Offers MS. *Degree requirements:* For master's, thesis optional, foreign language not required. *Entrance requirements:* For master's, minimum undergraduate GPA of 2.8. *Faculty research:* Design for manufacturing, machine tools, manufacturing processes, material selection for manufacturing, manufacturing systems.

Western Michigan University, Graduate College, College of Engineering and Applied Sciences, Department of Industrial and Manufacturing Engineering, Program in Manufacturing Science, Kalamazoo, MI 49008. Offers MS. *Students:* 1 full-time (0 women), 12 part-time (4 women). 10 applicants, 50% accepted. *Degree requirements:* Foreign language not required. *Entrance requirements:* For master's, GRE General Test, minimum GPA of 3.0. *Application deadline:* For fall admission, 2/15 (priority date). Applications are processed on a rolling basis. Application fee: $25. *Financial aid:* Application deadline: 2/15; *Unit head:* Carol Enseki, Administrative Assistant, 608-262-0921, Fax: 608-265-4017. *Application contact:* Paula J. Boodt, Coordinator, Graduate Admissions and Recruitment, 616-387-2000, Fax: 616-387-2355, E-mail: paula.boodt@wmich.edu.

Western New England College, School of Engineering, Department of Industrial and Manufacturing Engineering, Springfield, MA 01119-2654. Offers MSEM. Part-time and evening/weekend programs available. *Faculty:* 4 full-time (0 women), 2 part-time (0 women). Average age 29. In 1998, 22 degrees awarded. *Degree requirements:* For master's, computer language, comprehensive exam required, thesis optional, foreign language not required. *Entrance requirements:* For master's, bachelor's degree in engineering or related field. *Application deadline:* Applications are processed on a rolling basis. Application fee: $30. *Financial aid:* Teaching assistantships available. Aid available to part-time students. Financial aid application deadline: 4/1; financial aid applicants required to submit FAFSA. *Faculty research:* Project scheduling, flexible manufacturing systems, facility layout, energy management. *Unit head:* Dr. J. Byron Nelson, Chair, 413-782-1289. *Application contact:* Harry F. Neunder, Coordinator, Continuing Education, 413-782-1750, Fax: 413-782-1779, E-mail: hneunder@wnec.edu.

Wichita State University, Graduate School, College of Engineering, Department of Industrial and Manufacturing Engineering, Wichita, KS 67260. Offers MEM, MS, PhD. Part-time programs available. *Faculty:* 9 full-time (2 women), 1 part-time (0 women). *Students:* 52 full-time (4 women), 64 part-time (7 women); includes 5 minority (1 African American, 3 Asian Americans or Pacific Islanders, 1 Hispanic American), 75 international. Average age 33. 175 applicants, 00% accepted. In 1990, 22 master's, 2 doctorates awarded. *Degree requirements:* For master's, thesis optional, foreign language not required; for doctorate, dissertation, comprehensive exam required. *Entrance requirements:* For master's, GRE, TOEFL (minimum score of 550 required); for doctorate, GRE General Test, TOEFL (minimum score of 550 required). *Application deadline:* For fall admission, 7/1 (priority date); for spring admission, 1/1. Applications are processed on a rolling basis. Application fee: $25 ($40 for international students). Electronic applications accepted. *Financial aid:* In 1998–99, 28 research assistantships (averaging $4,000 per year), 10 teaching assistantships with full tuition reimbursements (averaging $4,000 per year) were awarded.; fellowships, Federal Work-Study, institutionally-sponsored loans, and unspecified assistantships also available. Aid available to part-time students. Financial aid application deadline: 4/1; financial aid applicants required to submit FAFSA. *Faculty research:* Ergonomics, rehabilitation, operations research, reverse engineering, assembly design/planning. Total annual research expenditures: $327,107. *Unit head:* Dr. Abu Masud, Chairperson, 316-978-3425, Fax: 316-978-3742, E-mail: masud@twsuvm.uc.twsu.edu. *Application contact:* Dr. Mark Kaiser, Graduate Coordinator, 316-978-5904, Fax: 316-978-3742, E-mail: kaiser@twsuvm.uc.twsu.edu.

Worcester Polytechnic Institute, Graduate Studies, Department of Manufacturing Engineering, Worcester, MA 01609-2280. Offers MS, PhD, Certificate. Part-time and evening/weekend programs available. *Faculty:* 3 full-time (0 women), 2 part-time (0 women). *Students:* 8 full-time (1 woman), 9 part-time; includes 2 minority (both Asian Americans or Pacific Islanders), 2 international. 34 applicants, 71% accepted. In 1998, 7 master's, 3 doctorates awarded. *Degree requirements:* For master's, thesis required, foreign language not required; for doctorate, dissertation, comprehensive exam, research proposal required, foreign language not required. *Entrance requirements:* For master's, TOEFL (minimum score of 550 required; average 600); for doctorate, GRE, TOEFL (minimum score of 550 required; average 600). *Average time to degree:* Master's–1.5 years full-time, 4 years part-time; doctorate–4 years full-time. *Application deadline:* For fall admission, 2/15 (priority date); for spring admission, 10/15 (priority date). Applications are processed on a rolling basis. Application fee: $50. Electronic applications accepted. *Financial aid:* In 1998–99, 1 student received aid, including 1 fellowship with full tuition reimbursement available (averaging $10,620 per year); research assistantships, teaching assistantships, career-related internships or fieldwork, grants, institutionally-sponsored loans, and scholarships also available. Financial aid application deadline: 2/15; financial aid applicants required to submit FAFSA. *Faculty research:* Nondestructive testing, intelligent design, robotics, surface engineering. *Unit head:* Dr. Shaukat Mirza, Head, 508-831-5355, Fax: 508-831-5178, E-mail: smirza@wpi.edu.

See in-depth description on page 1245.

Reliability Engineering

The University of Arizona, Graduate College, College of Engineering and Mines, Department of Systems and Industrial Engineering, Program in Reliability and Quality Engineering, Tucson, AZ 85721. Offers MS. *Students:* 5 full-time (1 woman), (all international). Average age 28. 6 applicants, 100% accepted. In 1998, 2 degrees awarded. *Degree requirements:* Foreign language not required. *Entrance requirements:* For master's, GRE General Test (minimum combined score of 1150 required), TOEFL (minimum score of 550 required), minimum GPA of 3.0. *Application deadline:* For fall admission, 7/1. Applications are processed on a rolling basis. Application fee: $35. *Financial aid:* Fellowships, research assistantships, teaching assistantships available. *Unit head:* Teresa A. McGhee, Secretary, 205-348-7160, Fax: 205-348-7162, E-mail: tmcghee@coe.eng.ua.edu. *Application contact:* Celia Stenzel, Graduate Secretary, 520-621-6551, Fax: 520-621-6555.

University of Maryland, College Park, Graduate School, A. James Clark School of Engineering, Department of Materials and Nuclear Engineering, Reliability Engineering Program, College Park, MD 20742-5045. Offers M Eng, MS, PhD. Part-time and evening/weekend programs available. Postbaccalaureate distance learning degree programs offered. *Students:* 17 full-time (1 woman), 32 part-time (8 women), 23 international. 25 applicants, 52% accepted. In 1998, 7 master's, 2 doctorates awarded. *Degree requirements:* For master's, thesis optional, foreign language not required; for doctorate, dissertation, oral exam required. *Entrance requirements:* For master's and doctorate, GRE General Test, TOEFL, minimum GPA of 3.0. *Application deadline:* For fall admission, 2/1. Applications are processed on a rolling basis. Application fee: $50 ($70 for international students). Tuition, state resident: part-time $272 per credit hour. Tuition, nonresident: part-time $475 per credit hour. Required fees: $632; $379 per year. *Financial aid:* Fellowships, research assistantships, teaching assistantships, career-related internships or fieldwork available. Financial aid applicants required to submit FAFSA. *Faculty research:* Electron linear acceleration, x-ray and imaging. *Unit head:* Dr. Marvin Roush, Associate Chair, 301-405-7299. *Application contact:* 301-405-7299.

See in-depth description on page 1231.

University of Maryland, College Park, Graduate School, A. James Clark School of Engineering, Professional Program in Engineering, College Park, MD 20742-5045. Offers aerospace engineering (M Eng); chemical engineering (M Eng); civil engineering (M Eng); electrical engineering (M Eng); fire protection engineering (M Eng); materials science and engineering (M Eng); mechanical engineering (M Eng); reliability engineering (M Eng); systems engineering (M Eng). Part-time and evening/weekend programs available. Postbaccalaureate distance learning degree programs offered. *Faculty:* 11 part-time (0 women). *Students:* 20 full-time (3 women), 205 part-time (42 women); includes 58 minority (27 African Americans, 25 Asian Americans or Pacific Islanders, 5 Hispanic Americans, 1 Native American), 20 international. *Degree requirements:* For master's, foreign language and thesis not required. *Application deadline:* Applications are processed on a rolling basis. Application fee: $50 ($70 for international students). Tuition, state resident: part-time $272 per credit hour. Tuition, nonresident: part-time $475 per credit hour. Required fees: $632; $379 per year. *Unit head:* Dr. Patrick Cunniff, Associate Dean, 301-405-5256, Fax: 301-314-9477. *Application contact:* Trudy Lindsey, Director, Graduate Admission and Records, 301-405-4198, Fax: 301-314-9305, E-mail: grschool@deans.umd.edu.

Safety Engineering

Murray State University, College of Industry and Technology, Department of Occupational Safety and Health, Murray, KY 42071-0009. Offers MS. *Accreditation:* ABET. Part-time programs available. *Students:* 32 full-time (8 women), 19 part-time (6 women); includes 8 minority (6 African Americans, 2 Hispanic Americans), 1 international. 13 applicants, 100% accepted. *Degree requirements:* Foreign language not required. *Entrance requirements:* For master's, GRE General Test, TOEFL (minimum score of 600 required). *Application deadline:* Applications are processed on a rolling basis. Application fee: $20. *Financial aid:* Research assistantships, teaching assistantships, Federal Work-Study available. Financial aid application deadline: 4/1. *Unit head:* Dr. Bassam Atieh, Graduate Coordinator, 502-762-6652, Fax: 502-762-3630, E-mail: bassam.atieh@murraystate.edu.

New Jersey Institute of Technology, Office of Graduate Studies, Department of Industrial and Manufacturing Engineering, Program in Occupational Safety and Health Engineering, Newark, NJ 07102-1982. Offers MS. Part-time and evening/weekend programs available. *Degree requirements:* For master's, thesis or alternative required, foreign language not required. *Entrance requirements:* For master's, GRE General Test (minimum score of 450 on verbal section, 600 on quantitative, 550 on analytical required). Electronic applications accepted. *Faculty research:* Realistic building codes, optimization of training programs, effect of physical and mental fatigue on training.

Texas A&M University, College of Engineering, Department of Nuclear Engineering, Program in Health Physics/Radiological Health, College Station, TX 77843. Offers health physics (MS); industrial hygiene (MS); safety engineering (MS). *Students:* 8 full-time (3 women), 11 part-time (2 women); includes 6 minority (3 Asian Americans or Pacific Islanders, 3 Hispanic Americans), 5 international. *Degree requirements:* For master's, thesis or alternative required, foreign language not required. *Entrance requirements:* For master's, GRE General Test, TOEFL. Application fee: $50 ($75 for international students). *Unit head:* Dr. Hayder Abdul Razzak, Coordinator, 361-593-2001. *Application contact:* Dr. Yassin A. Hassan, Graduate Coordinator, 409-845-7090.

University of Wisconsin–Stout, Graduate Studies, College of Technology, Engineering, and Management, Program in Risk Control, Menomonie, WI 54751. Offers MS. Part-time programs available. *Students:* 41 (18 women); includes 3 minority (2 African Americans, 1 Hispanic American) 2 international. In 1998, 23 degrees awarded. *Degree requirements:* For master's, thesis required, foreign language not required. *Application deadline:* Applications are processed on a rolling basis. Application fee: $45. *Financial aid:* Research assistantships, teaching assistantships, Federal Work-Study and tuition waivers (full and partial) available. Aid available to part-time students. Financial aid application deadline: 4/1; financial aid applicants required to submit FAFSA. *Unit head:* Dr. Elbert Sorrell, Interim Director, 715-232-2630.

West Virginia University, College of Engineering and Mineral Resources, Department of Industrial and Management Systems Engineering, Program in Occupational Hygiene and Occupational Safety, Morgantown, WV 26506. Offers MS. *Accreditation:* ABET. Part-time programs available. *Degree requirements:* For master's, computer language, thesis or alternative required, foreign language not required. *Entrance requirements:* For master's, TOEFL (minimum score of 550 required), minimum GPA of 2.75. *Faculty research:* Ergonomics, industrial hygiene, human factors engineering, back injuries.

West Virginia University, College of Engineering and Mineral Resources, Department of Safety and Environmental Management, Morgantown, WV 26506. Offers MS. *Accreditation:* ABET. Part-time programs available. *Degree requirements:* For master's, comprehensive written exam required, thesis optional, foreign language not required. *Entrance requirements:* For master's, GRE or MAT, TOEFL (minimum score of 550 required), minimum GPA of 2.75, 23 undergraduate hours in science and mathematics, 40 undergraduate hours in safety-related course work. *Faculty research:* Chemical hazard scoring, motorcycle safety, eyeguards in racquet sports, roofing shoes, ergonomics.

Systems Engineering

Air Force Institute of Technology, School of Engineering, Department of Aeronautics and Astronautics, Program in Systems Engineering, Wright-Patterson AFB, OH 45433-7765. Offers MS, PhD. *Accreditation:* ABET (one or more programs are accredited). Part-time programs available. *Faculty:* 4 full-time (0 women). *Students:* 5 full-time, 2 part-time. In 1998, 4 master's awarded (100% found work related to degree). *Degree requirements:* For master's and doctorate, thesis/dissertation required, foreign language not required. *Entrance requirements:* For master's, GRE General Test (minimum score of 500 on verbal section, 600 on quantitative required), minimum GPA of 3.0, must be military officer or U.S. citizen; for doctorate, GRE General Test (minimum score of 550 on verbal section, 650 on quantitative required), minimum GPA of 3.0, must be military officer or U.S. citizen. Application fee: $0. *Faculty research:* Reliability, maintainability, system optimization. *Unit head:* Lt. Col. Stu Kramer, Curriculum Chair, 937-255-3636 Ext. 4578, Fax: 937-656-7621, E-mail: skramer@afit.af.mil.

Auburn University, Graduate School, College of Engineering, Department of Industrial and Systems Engineering, Auburn, Auburn University, AL 36849-0002. Offers MIE, MS, PhD. Part-time programs available. *Faculty:* 11 full-time (2 women). *Students:* 27 full-time (5 women), 15 part-time (1 woman); includes 1 minority (Hispanic American), 16 international. 48 applicants, 40% accepted. In 1998, 17 master's, 7 doctorates awarded. *Degree requirements:* For master's, design project (MIE), thesis (MS) required; for doctorate, dissertation required, foreign language not required. *Entrance requirements:* For master's, GRE General Test; for doctorate, GRE General Test (minimum score of 400 on each section required). *Application deadline:* For fall admission, 9/1; for spring admission, 3/1. Applications are processed on a rolling basis. Application fee: $25 ($50 for international students). Tuition, state resident: full-time $2,760; part-time $76 per credit hour. Tuition, nonresident: full-time $8,280; part-time $228 per credit hour. *Financial aid:* Fellowships, research assistantships, teaching assistantships, Federal Work-Study available. Aid available to part-time students. Financial aid application deadline: 3/15. *Unit head:* Dr. V. E. Unger, Head, 334-844-1401. *Application contact:* Dr. John F. Pritchett, Dean of the Graduate School, 334-844-4700.

Boston University, College of Engineering, Department of Electrical and Computer Engineering, Boston, MA 02215. Offers computer engineering (PhD); computer systems engineering (MS); electrical engineering (MS, PhD); systems engineering (PhD). Part-time programs available. *Faculty:* 40 full-time (4 women), 4 part-time (0 women). *Students:* 117 full-time (20 women), 26 part-time (6 women); includes 9 minority (3 African Americans, 6 Asian Americans or Pacific Islanders), 83 international. Terminal master's awarded for partial completion of doctoral program. *Degree requirements:* For master's, thesis or alternative required, foreign language not required; for doctorate, dissertation required, foreign language not required. *Entrance requirements:* For master's, GRE General Test, TOEFL (minimum score of 500 required; 213 for computer-based); for doctorate, GRE General Test, TOEFL. *Application deadline:* For fall admission, 4/1; for spring admission, 10/1. Applications are processed on a rolling basis. Application fee: $50. Tuition: Full-time $23,770; part-time $743 per credit. Required fees: $220. Tuition and fees vary according to class time, course level, campus/location and program. *Unit head:* Dr. Bahaa Saleh, Chairman, 617-353-7176, Fax: 617-353-6440. *Application contact:* Cheryl Kelley, Graduate Programs Director, 617-353-9760, Fax: 617-353-0259, E-mail: enggrad@bu.edu.

California Institute of Technology, Division of Engineering and Applied Science, Option in Control and Dynamical Systems, Pasadena, CA 91125-0001. Offers PhD. *Faculty:* 1 full-time (0 women). *Students:* 28 full-time (5 women), 12 international. 92 applicants, 3% accepted. *Degree requirements:* For doctorate, dissertation required, foreign language not required. *Application deadline:* For fall admission, 1/15. Application fee: $0. *Faculty research:* Robustness, multivariable and nonlinear systems, optimal control, decentralized control, modeling and system identification for robust control. *Unit head:* Dr. Richard Murray, Representative, 626-395-6460.

California State University, Fullerton, Graduate Studies, School of Engineering and Computer Science, Program in Engineering, Fullerton, CA 92834-9480. Offers engineering science (MS); systems engineering (MS). Part-time programs available. *Degree requirements:* For master's, computer language, comprehensive exam, project or thesis required. *Entrance requirements:* For master's, minimum undergraduate GPA of 2.5. Application fee: $55. Tuition, nonresident: part-time $264 per unit. Required fees: $1,947; $1,281 per year. *Unit head:* Dr. David Falconer, Associate Dean, 714-278-3362.

Capitol College, Graduate School, Laurel, MD 20708-9759. Offers electronic commerce management (MS); information and telecommunications systems management (MS); systems management (MS). Part-time and evening/weekend programs available. *Faculty:* 1 full-time (0 women), 24 part-time (4 women). *Students:* 6 full-time (3 women), 145 part-time (39 women); includes 23 minority (9 African Americans, 7 Asian Americans or Pacific Islanders, 6 Hispanic Americans, 1 Native American), 2 international. Average age 31. 50 applicants, 86% accepted. In 1998, 43 degrees awarded (100% found work related to degree). *Degree requirements:* For master's, foreign language and thesis not required. *Entrance requirements:* For master's, GRE General Test (minimum score of 500 on each section required), minimum GPA of 3.0. *Average time to degree:* Master's–1 year full-time, 2.5 years part-time. *Application deadline:* For fall admission, 7/1 (priority date); for winter admission, 12/1 (priority date); for spring admission, 3/1 (priority date). Applications are processed on a rolling basis. Application fee: $25 ($100 for international students). Electronic applications accepted. Tuition: Full-time $5,328; part-time $888 per course. Tuition and fees vary according to campus/location and program. *Financial aid:* In 1998–99, 2 students received aid. Available to part-time students. Applicants required to submit FAFSA. *Unit head:* Dr. Joe Goldsmith, Dean of Graduate Studies, 301-369-2800, Fax: 301-953-3876. *Application contact:* Sandy Perriello, Coordinator of Graduate Administration, 703-998-5503, Fax: 703-379-8239, E-mail: gradschool@capitol-college.edu.

Carleton University, Faculty of Graduate Studies, Faculty of Engineering and Design, Ottawa-Carleton Institute for Electrical and Computer Engineering, Department of Systems and Computer Engineering, Ottawa, ON K1S 5B6, Canada. Offers electrical engineering (M Eng, PhD); information and systems science (M Sc); telecommunications technology management (M Eng). *Faculty:* 23 full-time (3 women). *Students:* 64 full-time (14 women), 51 part-time (9 women). *Degree requirements:* For master's, thesis optional; for doctorate, dissertation, comprehensive exam required. *Entrance requirements:* For master's, TOEFL (minimum score of 550 required), honors degree; for doctorate, TOEFL (minimum score of 550 required), MA Sc or M Eng. *Application deadline:* For fall admission, 3/1. Applications are processed on a rolling basis. Application fee: $35. *Unit head:* Rafik Goubran, Chair, 613-520-5740, Fax: 613-520-5742, E-mail: goubran@sce.carleton.ca. *Application contact:* D. G. Swartz, Supervisor of Graduate Studies, 613-520-5740, Fax: 613-520-5727, E-mail: swartz@sce.carleton.ca.

Case Western Reserve University, School of Graduate Studies, The Case School of Engineering, Department of Electrical, Systems, Computer Engineering and Science, Cleveland, OH 44106. Offers computer engineering and science (MS, PhD), including computer engineering, computing and information science; electrical engineering (MS, PhD); systems and control engineering (MS, PhD). Part-time and evening/weekend programs available. Postbaccalaureate distance learning degree programs offered (minimal on-campus study). *Faculty:* 28 full-time (2 women). *Students:* 73 full-time (13 women), 101 part-time (17 women). Average age 25. 484 applicants, 38% accepted. In 1998, 68 master's, 20 doctorates awarded. Terminal master's awarded for partial completion of doctoral program. *Degree requirements:* For master's and doctorate, thesis/dissertation required, foreign language not required. *Entrance requirements:* For master's and doctorate, GRE General Test, TOEFL (minimum score of 550 required). *Average time to degree:* Master's–2 years full-time, 3 years part-time; doctorate–4 years full-time, 5 years part-time. *Application deadline:* For fall admission, 3/1; for spring admission, 11/1. Applications are processed on a rolling basis. Application fee: $25. Electronic applications accepted. *Financial aid:* In 1998–99, 96 students received aid, including 70 research assistantships with full and partial tuition reimbursements available (averaging $15,600 per year), 33 teaching assistantships with full and partial tuition reimbursements available (averaging $10,170 per year); fellowships, career-related internships or fieldwork, Federal Work-Study, and institutionally-sponsored loans also available. Aid available to part-time students. Financial aid application deadline: 3/1. *Faculty research:* Microelectromechanical systems, control, artificial intelligence, mixed signals. Total annual research expenditures: $5.8 million. *Unit head:* Dr. Robert V. Edwards, Acting Chairman, 216-368-2800, Fax: 216-368-6888, E-mail: rve2@po.cwru.edu. *Application contact:* Elizabethanne M. Fuller, Department Assistant, 216-368-4080, Fax: 216-368-2668, E-mail: emf4@po.cwru.edu.

Colorado School of Mines, Graduate School, Division of Engineering, Golden, CO 80401-1887. Offers engineering systems (ME, MS, PhD). Part-time programs available. *Faculty:* 33 full-time (3 women), 19 part-time (4 women). *Students:* 35 full-time (5 women), 28 part-time (5 women); includes 8 minority (4 Asian Americans or Pacific Islanders, 3 Hispanic Americans, 1 Native American), 8 international. 44 applicants, 80% accepted. In 1998, 12 master's awarded (92% found work related to degree); 1 doctorate awarded (100% found work related to degree). *Degree requirements:* For master's, thesis required, foreign language not required; for doctorate, one foreign language, dissertation, comprehensive exams required. *Entrance requirements:* For master's and doctorate, GRE General Test (combined average 1880 on three sections), GRE Subject Test, minimum GPA of 3.0. *Application deadline:* Applications are processed on a rolling basis. Application fee: $40. Electronic applications accepted. *Financial aid:* In 1998–99, 49 students received aid, including 30 research assistantships, 11 teaching assistantships; fellowships, Federal Work-Study and unspecified assistantships also available. Financial aid applicants required to submit FAFSA. *Faculty research:* Geotechnical engineering, offshore mechanics, analytical design, process simulation, health monitoring. Total annual research expenditures: $936,455. *Unit head:* Dr. Joan Gosink, Head, 303-273-3650, Fax: 303-273-3602, E-mail: jgosink@mines.edu. *Application contact:* Barby Halliday, Student Services Specialist, 303-273-3211, Fax: 303-273-3278, E-mail: bhallida@mines.edu.

Embry-Riddle Aeronautical University, Daytona Beach Campus Graduate Program, Department of Human Factors and Systems, Daytona Beach, FL 32114-3900. Offers human factors

Systems Engineering

Embry-Riddle Aeronautical University (continued)
engineering (MSHFS); systems engineering (MSHFS). *Faculty:* 5 full-time (0 women), 1 part-time (0 women). *Students:* 16 full-time (8 women), 9 part-time (6 women); includes 6 minority (1 African American, 2 Asian Americans or Pacific Islanders, 3 Hispanic Americans), 4 international. Average age 31. 17 applicants, 94% accepted. *Degree requirements:* For master's, thesis, practicum, qualifying oral exam required, foreign language not required. *Entrance requirements:* For master's, TOEFL (minimum score of 550 required), minimum GPA of 2.5. *Application deadline:* Applications are processed on a rolling basis. Application fee: $30 ($50 for international students). Tuition: Full-time $8,190; part-time $455 per credit. Required fees: $105 per semester. Tuition and fees vary according to program. *Financial aid:* In 1998–99, 1 fellowship with tuition reimbursement (averaging $8,640 per year), 8 research assistantships with tuition reimbursements (averaging $8,640 per year), 1 teaching assistantship with tuition reimbursement (averaging $8,640 per year) were awarded.; career-related internships or fieldwork and unspecified assistantships also available. Financial aid application deadline: 4/15; financial aid applicants required to submit FAFSA. *Faculty research:* Consulting on aviation operations with FAA, human factors of airport security, effect of training order on flight performance of new pilots, human factors evaluation of a human computer interface of commercial database, evaluation of operator workload in future air traffic systems. Total annual research expenditures: $185,000. *Unit head:* Dr. Daniel Garland, Chair, 904-226-6790, Fax: 904-226-7050, E-mail: garlandd@db.erau.edu. *Application contact:* Ginny Tait, Graduate Admissions Specialist, 904-226-6115, Fax: 904-226-6299, E-mail: taitg@cts.db.erau.edu.

Florida Atlantic University, College of Engineering, Department of Mechanical Engineering, Program in Manufacturing Systems Engineering, Boca Raton, FL 33431-0991. Offers MS. Part-time and evening/weekend programs available. *Faculty:* 3 full-time (0 women). *Students:* 3 full-time (1 woman), 3 part-time (1 woman); includes 3 minority (1 African American, 1 Asian American or Pacific Islander, 1 Hispanic American), 1 international. Average age 29. In 1998, 4 degrees awarded. *Degree requirements:* For master's, thesis optional, foreign language not required. *Entrance requirements:* For master's, GRE General Test (minimum combined score of 1000 required), TOEFL (minimum score of 550 required), minimum GPA of 3.0. *Application deadline:* For fall admission, 4/10 (priority date); for spring admission, 10/1. Applications are processed on a rolling basis. Application fee: $20. Tuition, state resident: part-time $148 per credit hour. Tuition, nonresident: part-time $509 per credit hour. *Financial aid:* Research assistantships, teaching assistantships, career-related internships or fieldwork and Federal Work-Study available. Aid available to part-time students. Financial aid application deadline: 4/1; financial aid applicants required to submit FAFSA. *Faculty research:* Packaging, materials handling, design for manufacture, robotics, automation. *Unit head:* Patricia Capozziello, Graduate Admissions Coordinator, 561-297-2694, Fax: 561-297-2659, E-mail: capozzie@fau.edu. *Application contact:* Patricia Capozziello, Graduate Admissions Coordinator, 561-297-2694, Fax: 561-297-2659, E-mail: capozzie@fau.edu.

George Mason University, School of Information Technology and Engineering, Department of Systems Engineering and Operations Research, Fairfax, VA 22030-4444. Offers operations research and management science (MS); systems engineering (MS). Part-time and evening/weekend programs available. *Students:* 14 full-time (2 women), 183 part-time (50 women); includes 43 minority (20 African Americans, 17 Asian Americans or Pacific Islanders, 5 Hispanic Americans, 1 Native American), 17 international. Average age 34. 122 applicants, 84% accepted. In 1998, 46 degrees awarded. *Degree requirements:* For master's, computer language required, thesis optional, foreign language not required. *Entrance requirements:* For master's, GRE General Test, TOEFL (minimum score of 575 required), minimum GPA of 3.0 in last 60 hours. *Application deadline:* For fall admission, 5/1; for spring admission, 11/1. Application fee: $30. Electronic applications accepted. Tuition, state resident: full-time $4,416; part-time $184 per credit hour. Tuition, nonresident: full-time $12,516; part-time $522 per credit hour. Tuition and fees vary according to program. *Financial aid:* Fellowships, research assistantships, teaching assistantships, career-related internships or fieldwork and Federal Work-Study available. Aid available to part-time students. Financial aid application deadline: 3/1; financial aid applicants required to submit FAFSA. *Faculty research:* Requirements engineering, signal processing, systems architecture, data fusion. *Unit head:* Dr. Karla L. Hoffman, Chairman, 703-993-1670, Fax: 703-993-1521, E-mail: khoffman@gmu.edu.

The George Washington University, School of Engineering and Applied Science, Department of Engineering Management and Systems Engineering, Washington, DC 20052. Offers MEM, MS, D Sc, App Sc, Engr, MEM/MS. Part-time and evening/weekend programs available. *Faculty:* 12 full-time (0 women), 28 part-time (2 women). *Students:* 83 full-time (23 women), 338 part-time (90 women); includes 76 minority (39 African Americans, 24 Asian Americans or Pacific Islanders, 11 Hispanic Americans, 2 Native Americans), 132 international. Average age 35. 208 applicants, 88% accepted. In 1998, 163 master's, 12 doctorates awarded. *Degree requirements:* For master's, thesis optional, foreign language not required; for doctorate, one foreign language, dissertation, final and qualifying exams required; for other advanced degree, professional project required, foreign language and thesis not required. *Entrance requirements:* For master's, TOEFL (minimum score of 550 required; average 580) or George Washington University English as a Foreign Language Test, appropriate bachelor's degree; for doctorate, TOEFL (minimum score of 550 required; average 580) or George Washington University English as a Foreign Language Test, appropriate master's degree, minimum GPA of 3.5; for other advanced degree, TOEFL (minimum score of 550 required; average 580) or George Washington University English as a Foreign Language Test, appropriate master's degree, minimum GPA of 3.5, GRE required if highest earned degree is BS. *Application deadline:* For fall admission, 3/1; for spring admission, 10/1. Applications are processed on a rolling basis. Application fee: $50. Tuition: Full-time $17,328; part-time $722 per credit hour. Required fees: $828; $35 per credit hour. Tuition and fees vary according to campus/location and program. *Financial aid:* In 1998–99, 6 fellowships, 4 research assistantships, 18 teaching assistantships were awarded.; career-related internships or fieldwork and institutionally-sponsored loans also available. Financial aid application deadline: 3/1; financial aid applicants required to submit FAFSA. *Faculty research:* Artificial intelligence and expert systems, human factors engineering and systems analysis. Total annual research expenditures: $421,800. *Unit head:* Dr. Robert Waters, Chair, 202-994-7541. *Application contact:* Howard M. Davis, Manager, Office of Admissions and Student Records, 202-994-6158, Fax: 202-994-0909, E-mail: data: adms@seas.gwu.edu.

Georgia Institute of Technology, Graduate Studies and Research, College of Engineering, School of Industrial and Systems Engineering, Program in Industrial and Systems Engineering, Atlanta, GA 30332-0001. Offers algorithms, combinatorics, and optimization (PhD); industrial and systems engineering (PhD); industrial engineering (MS, MSIE); statistics (MS Stat). Part-time programs available. Terminal master's awarded for partial completion of doctoral program. *Degree requirements:* For master's, computer language required, thesis optional, foreign language not required; for doctorate, computer language, dissertation required, foreign language not required. *Entrance requirements:* For master's and doctorate, GRE General Test, TOEFL (minimum score of 550 required), minimum GPA of 3.0. Electronic applications accepted. *Faculty research:* Computer-integrated manufacturing systems, materials handling systems, production and distribution.

See in-depth description on page 1201.

Instituto Tecnológico y de Estudios Superiores de Monterrey, Campus Chihuahua, Graduate Programs, Chihuahua, 31110, Mexico. Offers computer systems engineering (Ingeniero); electrical engineering (Ingeniero); electromechanical engineering (Ingeniero); electronic engineering (Ingeniero); engineering administration (MEA); industrial engineering (MIE, Ingeniero); international trade (MIT); mechanical engineering (Ingeniero).

Instituto Tecnológico y de Estudios Superiores de Monterrey, Campus Monterrey, Graduate and Research Division, Programs in Engineering, Monterrey, 64849, Mexico. Offers applied statistics (M Eng); artificial intelligence (PhD); automation engineering (M Eng); chemical engineering (M Eng); civil engineering (M Eng); electrical engineering (M Eng); electronic engineering (M Eng); environmental engineering (M Eng); industrial engineering (M Eng, PhD); manufacturing engineering (M Eng); mechanical engineering (M Eng); systems and quality engineering (M Eng). M Eng offered jointly with the University of Waterloo; PhD (industrial engineering) offered jointly with Texas A&M University. Part-time and evening/weekend programs available. Terminal master's awarded for partial completion of doctoral program. *Degree requirements:* For master's and doctorate, one foreign language, computer language, thesis/dissertation required. *Entrance requirements:* For master's, PAEG, TOEFL; for doctorate, GRE, TOEFL, master's in related field. *Faculty research:* Flexible manufacturing cells, materials, statistical methods, environmental prevention, control and evaluation.

Iowa State University of Science and Technology, Graduate College, Interdisciplinary Programs, Program in Systems Engineering, Ames, IA 50011. Offers M Eng. *Students:* 3 full-time (1 woman), 19 part-time (2 women); includes 1 minority (Asian American or Pacific Islander), 1 international. 13 applicants, 54% accepted. In 1998, 7 degrees awarded. *Degree requirements:* For master's, thesis not required. *Entrance requirements:* For master's, TOEFL (minimum score of 550 required). *Application deadline:* For fall admission, 6/15 (priority date); for spring admission, 11/15. Applications are processed on a rolling basis. Application fee: $20 ($50 for international students). Electronic applications accepted. Tuition, state resident: full-time $3,308. Tuition, nonresident: full-time $9,744. Part-time tuition and fees vary according to course load, campus/location and program. *Financial aid:* In 1998–99, 2 research assistantships (averaging $10,215 per year) were awarded.; teaching assistantships *Unit head:* Dr. Douglas D. Gemmill, Supervisory Committee Chair, 515-294-8731, Fax: 515-294-3524, E-mail: syseng@iastate.edu.

Lehigh University, College of Engineering and Applied Science, Department of Industrial and Manufacturing Systems Engineering, Program in Manufacturing Systems Engineering, Bethlehem, PA 18015-3094. Offers MS. Part-time and evening/weekend programs available. *Students:* 3 full-time (0 women), 16 part-time (2 women); includes 1 Hispanic American, 1 international. 2 applicants, 100% accepted. In 1998, 19 degrees awarded. *Degree requirements:* For master's, computer language, project or thesis required. *Entrance requirements:* For master's, GRE General Test (score in 75th percentile or higher required), TOEFL (minimum score of 550 required), minimum GPA of 2.75. *Average time to degree:* Master's–1 year full-time, 2 years part-time. *Application deadline:* For fall admission, 7/15; for spring admission, 12/1. Applications are processed on a rolling basis. Application fee: $40. *Financial aid:* In 1998–99, 1 student received aid, including 1 research assistantship; fellowships, career-related internships or fieldwork and tuition waivers (full and partial) also available. Financial aid application deadline: 1/15. *Faculty research:* Manufacturing systems development and computer integration, CAD/CAM robotics and automation, management aspects of manufacturing systems. *Unit head:* Dr. Keith M. Gardiner, Director, 610-758-5070, Fax: 610-758-6527, E-mail: kg03@lehigh.edu. *Application contact:* Jeannette I. MacDonald, Graduate Coordinator, 610-758-4667, Fax: 610-758-6527, E-mail: jim1@lehigh.edu.

Louisiana State University in Shreveport, College of Sciences, Shreveport, LA 71115-2399. Offers systems technology (MST). Part-time and evening/weekend programs available. *Faculty:* 22 full-time (4 women). Average age 30. 12 applicants, 100% accepted. In 1998, 14 degrees awarded (100% found work related to degree). *Degree requirements:* For master's, computer language, comprehensive exam required, foreign language and thesis not required. *Entrance requirements:* For master's, GRE General Test (minimum combined score of 1400 on three sections required). *Average time to degree:* Master's–2.5 years part-time. *Application deadline:* For fall admission, 8/5 (priority date); for spring admission, 12/15. Applications are processed on a rolling basis. Application fee: $10. *Financial aid:* Teaching assistantships with partial tuition reimbursements available. *Faculty research:* Graphics, software quality, programming languages, tutoring systems. *Unit head:* Dr. Alfred McKinney, Dean, 318-797-5231, Fax: 318-797-5230, E-mail: amckinne@pilot.lsus.edu.

Massachusetts Institute of Technology, School of Engineering, System Design and Management Program, Cambridge, MA 02139-4307. Offers engineering and management (SM); system design and management (CAS). *Students:* 85 full-time (9 women), 6 international. Average age 35. 95 applicants, 68% accepted. In 1998, 6 degrees awarded. *Application deadline:* For winter admission, 6/15. Application fee: $55. *Financial aid:* In 1998–99, 5 research assistantships, 1 teaching assistantship were awarded. Financial aid applicants required to submit FAFSA. *Unit head:* John R. Williams, Co-Director, 617-253-7201, E-mail: jrw@mit.edu. *Application contact:* Marcia Chapman, Program Manager, 617-253-0254, E-mail: sdm@mit.edu.

National Technological University, Programs in Engineering, Fort Collins, CO 80526-1842. Offers chemical engineering (MS); computer engineering (MS); computer science (MS); electrical engineering (MS); engineering management (MS); hazardous waste management (MS); health physics (MS); management of technology (MS); manufacturing systems engineering (MS); materials science and engineering (MS); software engineering (MS); special majors (MS); transportation engineering (MS); transportation systems engineering (MS). Part-time programs available. *Faculty:* 600 part-time (20 women). *Entrance requirements:* For master's, BS in engineering or related field. *Application deadline:* Applications are processed on a rolling basis. Application fee: $50. *Unit head:* Lionel V. Baldwin, President, 970-495-6400, Fax: 970-484-0668, E-mail: baldwin@mail.ntu.edu.

Northeastern University, College of Engineering, Computer Systems Engineering Program, Boston, MA 02115-5096. Offers MS. Part-time programs available. *Students:* 27 full-time (9 women), 34 part-time (4 women); includes 3 minority (1 African American, 2 Asian Americans or Pacific Islanders), 28 international. Average age 25. 66 applicants, 76% accepted. In 1998, 24 degrees awarded. *Degree requirements:* For master's, computer language required, thesis optional, foreign language not required. *Entrance requirements:* For master's, GRE General Test. *Average time to degree:* Master's–2.28 years full-time, 5 years part-time. *Application deadline:* For fall admission, 4/15. Applications are processed on a rolling basis. Application fee: $50. *Financial aid:* In 1998–99, 14 students received aid, including 3 research assistantships with full tuition reimbursements available (averaging $12,450 per year), 2 teaching assistantships with full tuition reimbursements available (averaging $12,450 per year); fellowships, career-related internships or fieldwork, Federal Work-Study, tuition waivers (full), and unspecified assistantships also available. Aid available to part-time students. Financial aid application deadline: 2/15; financial aid applicants required to submit FAFSA. *Faculty research:* Engineering software design, CAD/CAM, robotics. *Unit head:* Dr. Chung Yu, Graduate Coordinator, 336-334-7760 Ext. 213, Fax: 336-334-7716, E-mail: yu@genesis.ncat.edu. *Application contact:* Stephen L. Gibson, Associate Director, 617-373-2711, Fax: 617-373-2501, E-mail: grad-eng.@coe.neu.edu.

Oakland University, Graduate Studies, School of Engineering and Computer Science, Program in Systems Engineering, Rochester, MI 48309-4401. Offers MS, PhD. *Faculty:* 7 full-time, 1 part-time. *Students:* 47 full-time (8 women), 49 part-time (8 women); includes 9 minority (4 African Americans, 4 Asian Americans or Pacific Islanders, 1 Hispanic American), 38 international. Average age 36. 39 applicants, 74% accepted. In 1998, 11 master's, 11 doctorates awarded. *Degree requirements:* For master's, foreign language and thesis not required; for doctorate, dissertation required, foreign language not required. *Entrance requirements:* For master's and doctorate, minimum GPA of 3.0 for unconditional admission. *Application deadline:* For fall admission, 7/15; for spring admission, 3/15. Application fee: $30. Tuition, state resident: part-time $221 per credit hour. Tuition, nonresident: part-time $488 per credit hour. Required fees: $214 per semester. Part-time tuition and fees vary according to program. *Financial aid:* Federal Work-Study, institutionally-sponsored loans, and tuition waivers (full) available. Financial aid application deadline: 3/1; financial aid applicants required to submit FAFSA. *Unit head:* Dr. Naim A. Kheir, Chair, 248-370-2177.

The Ohio State University, Graduate School, College of Engineering, Department of Industrial, Welding, and Systems Engineering, Program in Industrial and Systems Engineering, Columbus, OH 43210. Offers MS, PhD. *Faculty:* 19 full-time, 14 part-time. *Students:* 115 full-time (21 women), 17 part-time (4 women); includes 9 minority (1 African American, 7 Asian

Americans or Pacific Islanders, 1 Hispanic American), 86 international. 264 applicants, 27% accepted. In 1998, 29 master's, 8 doctorates awarded. *Degree requirements:* For master's, computer language required, thesis optional, foreign language not required; for doctorate, computer language, dissertation required, foreign language not required. *Entrance requirements:* For master's and doctorate, GRE General Test. *Application deadline:* For fall admission, 8/15. Applications are processed on a rolling basis. Application fee: $30 ($40 for international students). *Financial aid:* Fellowships, research assistantships, teaching assistantships, career-related internships or fieldwork, Federal Work-Study, institutionally-sponsored loans, and unspecified assistantships available. Aid available to part-time students. *Unit head:* Jerald Brevick, Graduate Studies Committee Chair, 614-292-0117, Fax: 614-292-7852, E-mail: brevick@ccl2.eng.ohio-state.edu.

Ohio University, Graduate Studies, College of Engineering and Technology, Department of Industrial and Manufacturing Systems Engineering, Athens, OH 45701-2979. Offers MS. *Faculty:* 8 full-time (0 women). *Students:* 20 full-time (2 women), 8 part-time (1 woman); includes 1 minority (Asian American or Pacific Islander), 23 international. Average age 24. 76 applicants, 68% accepted. In 1998, 6 degrees awarded. *Degree requirements:* For master's, computer language, non-thesis research project required, thesis optional, foreign language not required. *Entrance requirements:* For master's, GRE General Test. *Application deadline:* For fall admission, 3/1 (priority date). Applications are processed on a rolling basis. Application fee: $30. Tuition, state resident: full-time $5,754; part-time $238 per credit hour. Tuition, nonresident: full-time $11,055; part-time $457 per credit hour. Tuition and fees vary according to course load, campus/location and program. *Financial aid:* In 1998–99, 1 fellowship, 10 research assistantships, 8 teaching assistantships were awarded.; Federal Work-Study, institutionally-sponsored loans, and tuition waivers (full) also available. Financial aid application deadline: 3/1. *Faculty research:* Human factors engineering, MIS, quality systems. *Unit head:* Dr. Charles M. Parks, Chairman, 740-593-1540, Fax: 740-593-0778, E-mail: cparks@bobcat.ent.ohiou.edu. *Application contact:* Dr. Richard J. Gerth, Graduate Chairman, 740-593-1545, Fax: 740-593-0778, E-mail: gerth@bobcat.ent.ohiou.edu.

Oklahoma State University, Graduate College, College of Engineering, Architecture and Technology, School of Industrial Engineering and Management, Interdisciplinary Master of Manufacturing Systems Engineering Program, Stillwater, OK 74078. Offers M En. *Degree requirements:* For master's, creative component or thesis required. *Entrance requirements:* For master's, TOEFL (minimum score of 570 required). *Application deadline:* For fall admission, 7/1 (priority date). Application fee: $25. *Financial aid:* Research assistantships, teaching assistantships, career-related internships or fieldwork, Federal Work-Study, and tuition waivers (partial) available. Aid available to part-time students. Financial aid application deadline: 3/1. *Faculty research:* Integrated manufacturing systems, engineering practice in management, hardware aspects. *Unit head:* John W. Nazemetz, Director, 405-744-6055.

Pennsylvania State University Great Valley School of Graduate Professional Studies, Graduate Studies and Continuing Education, School of Graduate Professional Studies, Department of Engineering and Information Science, Program in Systems Engineering, Malvern, PA 19355-1488. Offers M Eng. *Students:* 1 full-time (0 women), 1 (woman) part-time. Application fee: $50. *Unit head:* John S. Kerge, Dean of Admissions, 718-260-3200, Fax: 718-260-3446, E-mail: admitme@poly.edu. *Application contact:* 610-648-3242, Fax: 610-889-1334.

Polytechnic University, Brooklyn Campus, Department of Electrical Engineering, Major in Systems Engineering, Brooklyn, NY 11201-2990. Offers MS. Part-time and evening/weekend programs available. *Students:* 3 full-time (0 women), 1 part-time; includes 1 minority (Asian American or Pacific Islander), 1 international. Average age 33. 2 applicants, 50% accepted. In 1998, 4 degrees awarded. *Degree requirements:* For master's, thesis optional. *Entrance requirements:* For master's, BS in electrical engineering. *Application deadline:* Applications are processed on a rolling basis. Application fee: $45. Electronic applications accepted. *Financial aid:* Fellowships, research assistantships, teaching assistantships, institutionally-sponsored loans available. Aid available to part-time students. Financial aid applicants required to submit FAFSA. *Unit head:* Kathy Kelly, Director of Admissions, 973-596-3300, Fax: 973-596-3461, E-mail: admissions@njit.edu. *Application contact:* John S. Kerge, Dean of Admissions, 718-260-3200, Fax: 718-260-3446, E-mail: admitme@poly.edu.

Polytechnic University, Farmingdale Campus, Graduate Programs, Department of Electrical Engineering, Major in Systems Engineering, Farmingdale, NY 11735-3995. Offers MS. Average age 33. *Degree requirements:* For master's, computer language required. *Application deadline:* Applications are processed on a rolling basis. Application fee: $45. Electronic applications accepted. *Unit head:* John S. Kerge, Dean of Admissions, 718-260-3200, Fax: 718-260-3446, E-mail: admitme@poly.edu. *Application contact:* John S. Kerge, Dean of Admissions, 718-260-3200, Fax: 718-260-3446, E-mail: admitme@poly.edu.

Polytechnic University, Westchester Graduate Center, Graduate Programs, Department of Electrical Engineering, Major in Systems Engineering, Hawthorne, NY 10532-1507. Offers MS. 1 applicants, 100% accepted. *Degree requirements:* For master's, computer language, thesis required (for some programs). *Application deadline:* Applications are processed on a rolling basis. Application fee: $45. Electronic applications accepted. *Unit head:* 610-648-3242, Fax: 610-889-1334. *Application contact:* John S. Kerge, Dean of Admissions, 718-260-3200, Fax: 718-260-3446, E-mail: admitme@poly.edu.

Purdue University, Graduate School, Schools of Engineering, School of Industrial Engineering, West Lafayette, IN 47907. Offers human factors in industrial engineering (MS, MSIE, PhD); manufacturing engineering (MS, MSIE, PhD); operations research (MS, MSIE, PhD); systems engineering (MS, MSIE, PhD). Part-time programs available. *Faculty:* 27 full-time (2 women), 3 part-time (0 women). *Students:* 124 full-time (26 women), 33 part-time (8 women); includes 13 minority (2 African Americans, 4 Asian Americans or Pacific Islanders, 6 Hispanic Americans, 1 Native American), 107 international. Terminal master's awarded for partial completion of doctoral program. *Degree requirements:* For master's, computer language required, thesis optional, foreign language not required; for doctorate, computer language, dissertation required, foreign language not required. *Entrance requirements:* For master's, GRE General Test (minimum score of 470 on verbal section, 700 on quantitative, 600 on analytical required; average 520 verbal, 710 quantitative, 660 analytical), TOEFL (minimum score of 570 required; average 600), minimum GPA of 3.0; for doctorate, GRE General Test (minimum score of 470 on verbal section, 700 on quantitative, 600 on analytical required; average 540 verbal, 750 quantitative, 675 analytical), TOEFL (minimum score of 570 required; average 610), MS thesis. *Application deadline:* For fall admission, 3/15; for spring admission, 9/1. Application fee: $30. Electronic applications accepted. *Unit head:* Dr. W. D. Compton, Interim Head, 765-494-5444, Fax: 765-494-1299, E-mail: dcompton@ecn.purdue.edu. *Application contact:* Dr. J. W. Barany, Associate Head, 765-494-5406, Fax: 765-494-1299, E-mail: jwb@ecn.purdue.edu.

Rensselaer Polytechnic Institute, Graduate School, School of Engineering, Department of Decision Sciences and Engineering Systems, Troy, NY 12180-3590. Offers decision sciences and engineering systems (PhD), including industrial and management engineering, manufacturing systems engineering, operations research and statistics; industrial and management engineering (M Eng, MS); manufacturing systems engineering (M Eng, MS); operations research and statistics (M Eng, MS). Part-time and evening/weekend programs available. Postbaccalaureate distance learning degree programs offered (no on-campus study). *Faculty:* 12 full-time (0 women), 6 part-time (1 woman). *Students:* 108 full-time (31 women), 22 part-time (5 women); includes 17 minority (2 African Americans, 9 Asian Americans or Pacific Islanders, 6 Hispanic Americans), 68 international. 210 applicants, 42% accepted. In 1998, 57 master's, 4 doctorates awarded. *Degree requirements:* For doctorate, dissertation required, foreign language not required, foreign language not required. *Entrance requirements:* For master's and doctorate, GRE General Test, TOEFL (minimum score of 550 required). *Application deadline:* For fall admission, 2/1 (priority date). Applications are processed on a rolling basis. Application fee: $35. *Financial aid:* In 1998–99, 6 fellowships with full tuition reimbursements (averaging $11,000 per year), 24 research assistantships with full tuition reimbursements (averaging $10,600 per year), 25 teaching assistantships with full tuition reimbursements (averaging $10,600 per year) were awarded.; career-related internships or fieldwork and institutionally-sponsored loans also available. Financial aid application deadline: 2/1. *Faculty research:* Information systems, statistical education/services, production/logistics/inventory. Total annual research expenditures: $2.2 million. *Unit head:* Dr. Charles J. Malmborg, Acting Chair, 518-276-2773, Fax: 518-276-8227, E-mail: dsesgr@rpi.edu. *Application contact:* Lee Vilardi, Graduate Coordinator, 518-276-6681, Fax: 518-276-8227, E-mail: dsesgr@rpi.edu.

Rensselaer Polytechnic Institute, Graduate School, School of Engineering, Department of Electrical, Computer, and Systems Engineering, Program in Computer and Systems Engineering, Troy, NY 12180-3590. Offers M Eng, MS, D Eng, PhD, MBA/M Eng. Part-time programs available. *Faculty:* 36 full-time (1 woman), 4 part-time (0 women). *Students:* 49 full-time (7 women), 12 part-time (1 woman); includes 8 minority (6 Asian Americans or Pacific Islanders, 2 Hispanic Americans), 20 international. 134 applicants, 35% accepted. In 1998, 16 master's, 7 doctorates awarded. Terminal master's awarded for partial completion of doctoral program. *Degree requirements:* For master's, thesis required (for some programs), foreign language not required; for doctorate, dissertation required, foreign language not required. *Entrance requirements:* For master's and doctorate, GRE, TOEFL (minimum score of 600 required). *Average time to degree:* Master's–1.5 years full-time; doctorate–3 years full-time. *Application deadline:* For fall admission, 2/1; for spring admission, 10/1. Applications are processed on a rolling basis. Application fee: $35. *Financial aid:* Fellowships, research assistantships, teaching assistantships, career-related internships or fieldwork and institutionally-sponsored loans available. Financial aid application deadline: 2/1. *Faculty research:* Multimedia via ATM, mobile robotics, thermophotovoltaic devices, microelectronic interconnections, agile manufacturing. Total annual research expenditures: $2.4 million. *Unit head:* Ann Bruno, Manager of Graduate Admissions and Financial Aid, 518-276-2554, Fax: 518-276-2433, E-mail: bruno@ecse.rpi.edu. *Application contact:* Ann Bruno, Manager of Graduate Admissions and Financial Aid, 518-276-2554, Fax: 518-276-2433, E-mail: bruno@ecse.rpi.edu.

Richmond, The American International University in London, Program in Systems Engineering and Management, Richmond, TW10 6JP, United Kingdom. Offers MS. Part-time and evening/weekend programs available. *Faculty:* 8 full-time (2 women), 8 part-time (1 woman). *Students:* 14 full-time (8 women), 7 part-time (4 women). Average age 27. In 1998, 2 degrees awarded (100% found work related to degree). *Entrance requirements:* For master's, TOEFL (minimum score of 550 required), minimum GPA of 3.0. *Average time to degree:* Master's–1.5 years full-time, 3 years part-time. *Application deadline:* For fall admission, 9/1 (priority date); for spring admission, 1/14. Applications are processed on a rolling basis. Application fee: $58. Electronic applications accepted. *Expenses:* Tuition and fees charges are reported in British pounds. Tuition: Full-time 14,095 British pounds; part-time 995 British pounds per course. Required fees: 105 British pounds per semester. *Financial aid:* Career-related internships or fieldwork, scholarships, and tuition waivers (partial) available. Aid available to part-time students. *Faculty research:* Critical systems safety, parallel processing systems, models and measures, software engineering, MIS. Total annual research expenditures: $20,000. *Unit head:* Dr. Margaret Myers, Director, 171-368-8416, Fax: 171-376-0836, E-mail: grad@richmond.ac.uk. *Application contact:* Nancy Barrett, Director of Graduate Admissions, 171-368-8489, Fax: 171-376-0836, E-mail: grad@richmond.ac.uk.

Rochester Institute of Technology, Part-time and Graduate Admissions, College of Engineering, Department of Industrial and Manufacturing Engineering, Rochester, NY 14623-5604. Offers engineering management (ME); industrial engineering (ME); manufacturing engineering (ME); systems engineering (ME). *Students:* 6 full-time (3 women), 13 part-time (4 women); includes 3 minority (all African Americans), 4 international. *Degree requirements:* For master's, internship required. *Entrance requirements:* For master's, TOEFL, minimum GPA of 3.0. *Application deadline:* For fall admission, 3/1 (priority date). Applications are processed on a rolling basis. Application fee: $40. *Unit head:* Dr. Jasper Shealy, Head, 716-475-2134, E-mail: jeseie@rit.edu.

Rutgers, The State University of New Jersey, New Brunswick, Graduate School, Program in Industrial and Systems Engineering, New Brunswick, NJ 08903. Offers industrial and systems engineering (MS, PhD); manufacturing systems (MS); quality and reliability engineering (MS). Part-time and evening/weekend programs available. *Faculty:* 10 full-time (2 women). *Students:* 47 full-time (15 women), 14 part-time (9 women); includes 31 minority (2 African Americans, 29 Asian Americans or Pacific Islanders) Average age 26. 94 applicants, 40% accepted. In 1998, 16 master's awarded (100% found work related to degree); 3 doctorates awarded. Terminal master's awarded for partial completion of doctoral program. *Degree requirements:* For master's, computer language required, thesis optional, foreign language not required; for doctorate, computer language, dissertation required, foreign language not required. *Entrance requirements:* For master's and doctorate, GRE General Test, TOEFL. *Average time to degree:* Master's–1.5 years full-time, 3 years part-time; doctorate–4 years full-time. *Application deadline:* For fall admission, 1/1 (priority date); for spring admission, 1/1. Applications are processed on a rolling basis. Application fee: $50. *Financial aid:* In 1998–99, 26 students received aid, including 5 fellowships with full tuition reimbursements available (averaging $13,500 per year), 13 research assistantships with full tuition reimbursements available (averaging $13,500 per year), 5 teaching assistantships with full tuition reimbursements available (averaging $13,500 per year); Federal Work-Study and tuition waivers (partial) also available. Financial aid application deadline: 1/1; financial aid applicants required to submit FAFSA. *Faculty research:* Production and manufacturing systems, quality and reliability engineering, systems engineering and aviation safety. *Unit head:* Susan L. Albin, Director, 732-445-2238, Fax: 732-445-5467, E-mail: salbin@rci.rutgers.edu. *Application contact:* Cindy Ielmini, 732-445-3654, Fax: 732-445-5467, E-mail: ielmini@rci.rutgers.edu.

San Jose State University, Graduate Studies, College of Engineering, Department of Computer, Information and Systems Engineering, Program in Information and Systems Engineering, San Jose, CA 95192-0001. Offers MS. Part-time programs available. *Faculty:* 5 full-time (0 women), 9 part-time (1 woman). *Students:* 14 full-time (5 women), 27 part-time (4 women); includes 19 minority (3 African Americans, 13 Asian Americans or Pacific Islanders, 3 Hispanic Americans), 7 international. Average age 32. 30 applicants, 60% accepted. In 1998, 9 degrees awarded. *Degree requirements:* For master's, comprehensive exam required. *Entrance requirements:* For master's, minimum GPA of 3.0. *Application deadline:* For fall admission, 6/1. Applications are processed on a rolling basis. Application fee: $59. Tuition, nonresident: part-time $246 per unit. Required fees: $1,939; $1,309 per year. *Financial aid:* Federal Work-Study available. *Unit head:* 800-677-9270, Fax: 303-964-5538, E-mail: admarg@regis.edu. *Application contact:* Dr. Louis Freund, Graduate Coordinator, 408-924-3890.

Southern Methodist University, School of Engineering and Applied Science, Center for Special Studies, Dallas, TX 75275. Offers applied science (MS, PhD); hazardous and waste materials management (MS); material science and engineering (MS); systems engineering (MS). *Faculty:* 12 part-time (0 women). *Students:* 10 full-time (2 women), 112 part-time (32 women); includes 22 minority (5 African Americans, 7 Asian Americans or Pacific Islanders, 10 Hispanic Americans), 7 international. *Degree requirements:* For master's, thesis optional, foreign language not required; for doctorate, dissertation, oral and written qualifying exams, oral final exam required. *Entrance requirements:* For master's, GRE General Test (minimum score of 650 on quantitative section required), TOEFL (minimum score of 550 required), minimum GPA of 3.0 in last 2 years, bachelor's degree in related field; for doctorate, preliminary counseling exam, minimum GPA of 3.0, bachelor's degree in related field. *Application deadline:* For fall admission, 8/1 (priority date); for spring admission, 12/15. Applications are processed on a rolling basis. Application fee: $25. Tuition: Full-time $9,216; part-time $512 per credit hour. Required fees: $88 per credit hour. Part-time tuition and fees vary according to course load and campus/location. *Unit head:* Dr. Buck F. Brown, Dean for Research and Graduate Studies, 812-877-8403, Fax: 812-877-8102, E-mail: buck.brown@rose-hulman.edu. *Application contact:* Dr. Zeynep Celik-Butler, Assistant Dean for Graduate Studies and Research, 214-768-3979, Fax: 214-768-3845, E-mail: zcb@seas.smu.edu.

Stanford University, School of Engineering, Department of Engineering-Economic Systems and Operations Research, Stanford, CA 94305-9991. Offers MS, PhD, Eng. *Faculty:* 20

Systems Engineering

Stanford University (continued)

full-time (0 women). *Students:* 260 full-time (60 women), 53 part-time (10 women); includes 35 minority (4 African Americans, 26 Asian Americans or Pacific Islanders, 5 Hispanic Americans), 194 international. Average age 26. 274 applicants, 76% accepted. In 1998, 135 master's, 17 doctorates awarded. Terminal master's awarded for partial completion of doctoral program. *Degree requirements:* For master's, thesis not required; for doctorate and Eng, dissertation required. *Entrance requirements:* For master's and Eng, GRE General Test, TOEFL; for doctorate, GRE General Test, GRE Subject Test, TOEFL. *Application deadline:* For fall admission, 2/1. Application fee: $65 ($80 for international students). Electronic applications accepted. Tuition: Full-time $24,588. Required fees: $152. Part-time tuition and fees vary according to course load. *Financial aid:* Research assistantships, Federal Work-Study and institutionally-sponsored loans available. Financial aid application deadline: 2/15. *Faculty research:* Mathematical systems analysis, decision analysis, systems economics, organizational economics, information policy. *Unit head:* James Sweeney, Chair, 650-723-2847, Fax: 650-723-1614, E-mail: sweeney@soe.stanford.edu. *Application contact:* Graduate Admissions Coordinator, 650-723-4168.

See in-depth description on page 1213.

Stanford University, School of Engineering, Department of Mechanical Engineering, Program in Manufacturing Systems Engineering, Stanford, CA 94305-9991. Offers MS, MBA/MS. *Entrance requirements:* For master's, GRE General Test, TOEFL, GMAT. *Application deadline:* For fall admission, 1/14. Application fee: $65 ($80 for international students). Electronic applications accepted. Tuition: Full-time $24,588. Required fees: $152. Part-time tuition and fees vary according to course load. *Financial aid:* Application deadline: 1/15. *Application contact:* Admissions Office, 650-723-3148.

See in-depth description on page 1217.

Université du Québec, École de technologie supérieure, Graduate Programs, Program in Systems Technology, Montréal, PQ H3C 1K3, Canada. Offers systems engineering (M Eng); systems technology (Diploma). *Degree requirements:* For master's and Diploma, thesis not required. *Entrance requirements:* For master's and Diploma, appropriate bachelor's degree, proficiency in French.

University of Alberta, Faculty of Graduate Studies and Research, Department of Electrical and Computer Engineering, Edmonton, AB T6G 2E1, Canada. Offers computational optics (PhD); computer engineering (M Eng, M Sc, PhD); control systems (M Eng, M Sc, PhD); engineering management (M Eng); laser physics (M Sc, PhD); oil sands (M Eng, M Sc, PhD); plasma physics (M Sc, PhD); power engineering (M Eng, M Sc, PhD); telecommunications (M Eng, M Sc, PhD). Terminal master's awarded for partial completion of doctoral program. *Degree requirements:* For master's and doctorate, thesis/dissertation required, foreign language not required. *Entrance requirements:* For master's and doctorate, TOEFL (minimum score of 580 required; average 610). Electronic applications accepted. *Faculty research:* Controls, communications, microelectronics, electromagnetics.

The University of Arizona, Graduate College, College of Engineering and Mines, Department of Systems and Industrial Engineering, Program in Systems and Industrial Engineering, Tucson, AZ 85721. Offers PhD. *Students:* 12 full-time (3 women), 7 part-time; includes 1 minority (Native American), 11 international. 17 applicants, 59% accepted. In 1998, 3 degrees awarded. *Degree requirements:* For doctorate, dissertation required. *Entrance requirements:* For doctorate, GRE General Test (minimum combined score of 1150 required), TOEFL (minimum score of 550 required), minimum GPA of 3.0. *Application deadline:* For fall admission, 7/1. Application fee: $35. *Unit head:* Celia Stenzel, Graduate Secretary, 520-621-6551, Fax: 520-621-6555. *Application contact:* Celia Stenzel, Graduate Secretary, 520-621-6551, Fax: 520-621-6555.

The University of Arizona, Graduate College, College of Engineering and Mines, Department of Systems and Industrial Engineering, Program in Systems Engineering, Tucson, AZ 85721. Offers MS. Part-time programs available. *Students:* 1 full-time (0 women), 23 part-time (4 women); includes 5 minority (1 African American, 4 Hispanic Americans) Average age 30. 17 applicants, 65% accepted. In 1998, 20 degrees awarded. *Degree requirements:* For master's, computer language required, foreign language not required. *Entrance requirements:* For master's, GRE General Test (minimum combined score of 1150 required), TOEFL (minimum score of 550 required), minimum GPA of 3.0. *Application deadline:* For fall admission, 7/1. Applications are processed on a rolling basis. Application fee: $35. *Financial aid:* Fellowships, research assistantships, teaching assistantships, institutionally-sponsored loans and scholarships available. *Faculty research:* Man/machine systems, optimal control, algorithmic probability. *Application contact:* Celia Stenzel, Graduate Secretary, 520-621-6551, Fax: 520-621-6555.

University of Connecticut, Graduate School, School of Engineering, Field of Electrical and Systems Engineering, Storrs, CT 06269. Offers biological engineering (MS); control and communication systems (MS, PhD); electromagnetics and physical electronics (MS, PhD). Terminal master's awarded for partial completion of doctoral program. *Degree requirements:* For master's, thesis or alternative required; for doctorate, dissertation required. *Entrance requirements:* For master's and doctorate, GRE General Test, TOEFL.

University of Florida, Graduate School, College of Engineering, Department of Industrial and Systems Engineering, Gainesville, FL 32611. Offers engineering management (ME, MS); facilities layout decision support systems energy (PhD); health systems (ME, MS); industrial engineering (PhD, Engr); manufacturing systems engineering (ME, MS, PhD, Certificate); operations research (ME, MS, PhD, Engr); production planning and control engineering management (PhD); quality and reliability assurance (ME, MS); systems engineering (PhD, Engr). *Faculty:* 22. *Students:* 56 full-time (14 women), 69 part-time (18 women); includes 17 minority (5 African Americans, 4 Asian Americans or Pacific Islanders, 8 Hispanic Americans), 48 international. 301 applicants, 65% accepted. In 1998, 64 master's, 3 doctorates awarded. *Degree requirements:* For master's, computer language, core exam required, thesis optional; for doctorate, dissertation, comprehensive exam required; for other advanced degree, computer language, thesis required. *Entrance requirements:* For master's and doctorate, GRE General Test, TOEFL, minimum GPA of 3.0; for other advanced degree, GRE General Test. *Application deadline:* For fall admission, 6/1 (priority date). Applications are processed on a rolling basis. Application fee: $20. Electronic applications accepted. *Financial aid:* In 1998–99, 27 students received aid, including 12 fellowships, 22 research assistantships, 2 teaching assistantships; career-related internships or fieldwork, Federal Work-Study, and unspecified assistantships also available. Financial aid application deadline: 1/10. *Unit head:* Dr. Donald Hearn, Chair, 352-392-1464, Fax: 352-392-3537, E-mail: hearn@ise.ufl.edu. *Application contact:* Dr. D. J. Elzinga, Graduate Coordinator, 352-392-1464, Fax: 352-392-3537, E-mail: elzinga@ise.ufl.edu.

See in-depth description on page 1227.

University of Florida, Graduate School, Graduate Engineering and Research Center (GERC), Gainesville, FL 32611. Offers aerospace engineering (ME, MS, PhD, Engr); electrical and computer engineering (ME, MS, PhD, Engr); engineering mechanics (ME, MS, PhD, Engr); industrial and systems engineering (ME, MS, PhD, Engr). Part-time programs available. Postbaccalaureate distance learning degree programs offered. *Faculty:* 6 full-time (0 women), 16 part-time (1 woman). *Students:* 13 full-time (6 women), 159 part-time (29 women); includes 21 minority (8 African Americans, 8 Asian Americans or Pacific Islanders, 4 Hispanic Americans, 1 Native American) Terminal master's awarded for partial completion of doctoral program. *Degree requirements:* For master's, computer language required (for some programs), thesis optional, foreign language not required; for doctorate, computer language (for some programs), dissertation required; for Engr, computer language (for some programs), thesis required, foreign language not required. *Entrance requirements:* For master's, GRE General Test (minimum combined score of 1000 required), TOEFL, minimum GPA of 3.0; for doctorate, GRE General Test (minimum combined score of 1200 required), written and oral qualifying exams, TOEFL,

minimum GPA of 3.0, master's degree in engineering; for Engr, GRE General Test (minimum combined score of 1000 required), TOEFL, minimum GPA of 3.0, master's degree in engineering. *Application deadline:* For fall admission, 6/1; for spring admission, 10/1. Applications are processed on a rolling basis. Application fee: $20. Electronic applications accepted. *Unit head:* Dr. Pasquale M. Sforza, Director, 850-833-9355, Fax: 850-833-9366, E-mail: sforza@gerc.eng.ufl.edu. *Application contact:* Judi Shivers, Program Assistant, 850-833-9350, Fax: 850-833-9366, E-mail: reginfo@gerc.eng.ufl.edu.

University of Houston, Cullen College of Engineering, Program in Computer and Systems Engineering, Houston, TX 77004. Offers MSCSE, PhD. Part-time and evening/weekend programs available. *Students:* 11 full-time (5 women), 9 part-time (3 women); includes 3 minority (1 African American, 2 Asian Americans or Pacific Islanders), 10 international. Average age 29. 81 applicants, 16% accepted. In 1998, 6 master's awarded. Terminal master's awarded for partial completion of doctoral program. *Degree requirements:* For master's, thesis required (for some programs), foreign language not required; for doctorate, dissertation, departmental qualifying exams required, foreign language not required. *Entrance requirements:* For master's and doctorate, GRE General Test, TOEFL. *Application deadline:* For fall admission, 7/3 (priority date); for spring admission, 12/4. Applications are processed on a rolling basis. Application fee: $25 ($75 for international students). *Financial aid:* Fellowships, research assistantships, teaching assistantships, Federal Work-Study and tuition waivers (partial) available. Financial aid application deadline: 7/1. *Faculty research:* Parallel processing, parallel algorithms and architectures, neural networks. *Unit head:* Dr. Pauline Markenscoff, Director, 713-743-4403, Fax: 713-743-4444, E-mail: markenscoff@uh.edu. *Application contact:* Mylyssa McDonald, Graduate Analyst, 713-743-4403, Fax: 713-743-4444, E-mail: mmm05866@jetson.uh.edu.

University of Illinois at Urbana–Champaign, Graduate College, College of Engineering, Department of General Engineering, Urbana, IL 61801. Offers systems engineering and engineering design (MS), including general engineering. *Faculty:* 18 full-time (1 woman). *Students:* 10 full-time (0 women); includes 3 minority (all Asian Americans or Pacific Islanders), 2 international. 18 applicants, 17% accepted. In 1998, 9 degrees awarded. *Degree requirements:* For master's, foreign language and thesis not required. *Entrance requirements:* For master's, minimum GPA of 3.0. *Application deadline:* For fall admission, 2/1. Applications are processed on a rolling basis. Application fee: $40 ($50 for international students). Tuition, state resident: full-time $4,616. Tuition, nonresident: full-time $11,768. Full-time tuition and fees vary according to course load. *Financial aid:* Fellowships, research assistantships, teaching assistantships, scholarships available. Financial aid application deadline: 2/15. *Unit head:* Dr. Harry E. Cook, Head, 217-333-2730. *Application contact:* Dr. Henrique Reis, Director of Graduate Studies, 217-333-1228, Fax: 217-244-5705, E-mail: h-reis@uiuc.edu.

See in-depth description on page 1229.

University of Maryland, College Park, Graduate School, A. James Clark School of Engineering, Professional Program in Engineering, College Park, MD 20742-5045. Offers aerospace engineering (M Eng); chemical engineering (M Eng); civil engineering (M Eng); electrical engineering (M Eng); fire protection engineering (M Eng); materials science and engineering (M Eng); mechanical engineering (M Eng); reliability engineering (M Eng); systems engineering (M Eng). Part-time and evening/weekend programs available. Postbaccalaureate distance learning degree programs offered. *Faculty:* 11 part-time (0 women). *Students:* 20 full-time (3 women), 205 part-time (42 women); includes 58 minority (27 African Americans, 25 Asian Americans or Pacific Islanders, 5 Hispanic Americans, 1 Native American), 20 international. *Degree requirements:* For master's, foreign language and thesis not required. *Application deadline:* Applications are processed on a rolling basis. Application fee: $50 ($70 for international students). Tuition, state resident: part-time $272 per credit hour. Tuition, nonresident: part-time $475 per credit hour. Required fees: $632; $379 per year. *Unit head:* Dr. Patrick Cunniff, Associate Dean, 301-405-5256, Fax: 301-314-9477. *Application contact:* Trudy Lindsey, Director, Graduate Admission and Records, 301-405-4198, Fax: 301-314-9305, E-mail: grschool@deans.umd.edu.

University of Maryland, College Park, Graduate School, A. James Clark School of Engineering, Systems Engineering Program, College Park, MD 20742-5045. Offers M Eng, MS. Part-time and evening/weekend programs available. *Students:* 14 full-time (1 woman), 9 part-time (5 women); includes 6 minority (4 African Americans, 2 Asian Americans or Pacific Islanders), 13 international. 53 applicants, 40% accepted. In 1998, 6 degrees awarded. *Degree requirements:* For master's, thesis optional, foreign language not required. *Entrance requirements:* For master's, GRE General Test, minimum GPA of 3.0. *Application deadline:* Applications are processed on a rolling basis. Application fee: $50 ($70 for international students). Tuition, state resident: part-time $272 per credit hour. Tuition, nonresident: part-time $475 per credit hour. Required fees: $632; $379 per year. *Financial aid:* Fellowships with tuition reimbursements, research assistantships with tuition reimbursements, teaching assistantships with tuition reimbursements, Federal Work-Study, grants, and scholarships available. Aid available to part-time students. Financial aid applicants required to submit FAFSA. *Faculty research:* Automation, computer, information, manufacturing, and process systems. Total annual research expenditures: $7.6 million. *Unit head:* Dr. Mark Shayman, Director, 301-405-6613. *Application contact:* Trudy Lindsey, Director, Graduate Admission and Records, 301-405-4198, Fax: 301-314-9305, E-mail: grschool@deans.umd.edu.

University of Massachusetts Lowell, Graduate School, James B. Francis College of Engineering, Department of Electrical Engineering, Program in Systems Engineering, Lowell, MA 01854-2881. Offers MS Eng, D Eng. Part-time programs available. *Degree requirements:* For master's, thesis optional, foreign language not required; for doctorate, 2 foreign languages, computer language, dissertation required. *Entrance requirements:* For master's and doctorate, GRE General Test. *Application deadline:* For fall admission, 4/1 (priority date); for spring admission, 10/1. Applications are processed on a rolling basis. Application fee: $20 ($35 for international students). *Financial aid:* Career-related internships or fieldwork available. Financial aid application deadline: 4/1. *Unit head:* Celia Snyder, Graduate Program Coordinator, 217-333-4390, Fax: 217-244-6534, E-mail: cgsnyder@uiuc.edu. *Application contact:* Dr. Ross Holmstrom, Coordinator, 978-934-3307, E-mail: ross_holmstrom@woods.uml.edu.

The University of Memphis, Graduate School, Herff College of Engineering, Program in Industrial and Systems Engineering, Memphis, TN 38152. Offers MS. Part-time programs available. *Faculty:* 3 full-time (1 woman). *Students:* 3 full-time (0 women), 4 part-time; includes 1 minority (Asian American or Pacific Islander), 2 international. Average age 32. 38 applicants, 13% accepted. In 1998, 5 degrees awarded. *Degree requirements:* For master's, thesis or alternative, comprehensive exam required. *Entrance requirements:* For master's, GRE General Test (minimum combined score of 1000 required) or MAT, minimum undergraduate GPA of 2.5. *Application deadline:* For fall admission, 8/1; for spring admission, 12/1. Application fee: $25 ($50 for international students). Tuition, state resident: full-time $3,410; part-time $178 per credit hour. Tuition, nonresident: full-time $8,670; part-time $408 per credit hour. Tuition and fees vary according to program. *Financial aid:* In 1998–99, 5 students received aid, including 1 fellowship, 4 research assistantships; teaching assistantships, career-related internships or fieldwork also available. *Faculty research:* Integer programming, ergonomics, scheduling. Total annual research expenditures: $125,000. *Unit head:* Dr. Michael Racer, Coordinator, 901-678-3285, E-mail: mracer@memphis.edu.

University of Michigan, Horace H. Rackham School of Graduate Studies, College of Engineering, Department of Electrical Engineering and Computer Science, Division of Systems Science and Engineering, Ann Arbor, MI 48109. Offers electrical engineering: systems (MS, MSE, PhD). Terminal master's awarded for partial completion of doctoral program. *Degree requirements:* For master's, thesis not required; for doctorate, dissertation, oral defense of dissertation, preliminary exams required. *Entrance requirements:* For master's, GRE General Test (minimum combined score of 1900 on three sections required; average 2002); for doctorate, GRE General Test (minimum combined score of 1900 on three sections required; average 2002), master's degree.

Systems Engineering

University of Michigan–Dearborn, College of Engineering and Computer Science, Department of Industrial and Systems Engineering, Dearborn, MI 48128-1491. Offers engineering management (MS); industrial and systems engineering (MSE). Part-time and evening/weekend programs available. *Faculty:* 10 full-time (0 women), 7 part-time (1 woman). *Students:* 8 full-time (4 women), 294 part-time (80 women). Average age 29. In 1998, 68 degrees awarded. *Degree requirements:* For master's, computer language required, thesis optional, foreign language not required. *Entrance requirements:* For master's, bachelor's degree in applied mathematics, computer science, engineering, or physical science; minimum GPA of 3.0. *Application deadline:* For fall admission, 8/1 (priority date); for winter admission, 12/1 (priority date); for spring admission, 4/1. Applications are processed on a rolling basis. Application fee: $55. Electronic applications accepted. Tuition, state resident: part-time $259 per credit hour. Tuition, nonresident: part-time $748 per credit hour. Required fees: $80 per course. Tuition and fees vary according to course level, course load and program. *Financial aid:* Fellowships, research assistantships, teaching assistantships, Federal Work-Study available. *Faculty research:* Health care systems, databases, human factors, machine diagnostics, precision machining. *Unit head:* Dr. S. K. Kachhal, Chair, 313-593-5361, Fax: 313-593-3692. *Application contact:* Lisa M. Beach, Administrative Assistant, 313-593-5361, Fax: 313-593-3692, E-mail: lmbeach@umich.edu.

University of New Hampshire, Graduate School, College of Engineering and Physical Sciences, Programs in Engineering, Doctoral Program in Engineering, Durham, NH 03824. Offers chemical engineering (PhD); civil engineering (PhD); electrical engineering (PhD); mechanical engineering (PhD); systems design engineering (PhD). *Students:* 16 full-time (3 women), 8 part-time (1 woman), 12 international. *Degree requirements:* For doctorate, dissertation required. *Entrance requirements:* For doctorate, GRE (for civil and mechanical engineering options). *Application deadline:* For fall admission, 4/1 (priority date). Applications are processed on a rolling basis. Application fee: $50. Tuition, area resident: Full-time $5,750; part-time $319 per credit. Tuition, state resident: full-time $8,625. Tuition, nonresident: full-time $14,640; part-time $598 per credit. Required fees: $224 per semester. Tuition and fees vary according to course load, degree level and program.

University of Pennsylvania, School of Engineering and Applied Science, Department of Systems Engineering, Philadelphia, PA 19104. Offers environmental resources engineering (MSE); environmental/resources engineering (PhD); systems engineering (MSE, PhD); technology and public policy (MSE, PhD); transportation (MSE, PhD). Part-time programs available. *Faculty:* 12 full-time (0 women), 9 part-time (2 women). *Students:* 26 full-time (10 women), 16 part-time (2 women); includes 4 minority (2 African Americans, 2 Hispanic Americans), 16 international. Terminal master's awarded for partial completion of doctoral program. *Degree requirements:* For master's, computer language required, foreign language not required; for doctorate, one foreign language, computer language, dissertation required. *Entrance requirements:* For master's and doctorate, TOEFL (minimum score of 600 required). *Application deadline:* For fall admission, 1/2 (priority date). Applications are processed on a rolling basis. Application fee: $65. Electronic applications accepted. *Financial aid:* Fellowships, research assistantships, teaching assistantships, scholarships available. *Faculty research:* Systems methodology, operations research, decision sciences, telecommunication systems, logistics, transportation systems, manufacturing systems, infrastructure systems. *Unit head:* Dr. G. Anandalingam, Chair, 215-898-8790, Fax: 215-898-5020, E-mail: anand@seas.upenn.edu. *Application contact:* Dr. Tony E. Smith, Graduate Group Chair, 215-898-9647, Fax: 215-898-5020, E-mail: tesmith@seas.upenn.edu.

See in-depth description on page 1239.

University of Pittsburgh, School of Engineering, Program in Manufacturing Systems Engineering, Pittsburgh, PA 15260. Offers MSMfSE. Part-time and evening/weekend programs available. Postbaccalaureate distance learning degree programs offered (no on-campus study). 11 applicants, 100% accepted. In 1998, 13 degrees awarded. *Degree requirements:* For master's, computer language, thesis, internship or project required, foreign language not required. *Entrance requirements:* For master's, GRE General Test, TOEFL (minimum score of 550 required). *Average time to degree:* Master's–2 years full-time, 3 years part-time. *Application deadline:* For fall admission, 8/1 (priority date); for spring admission, 12/1 (priority date). Applications are processed on a rolling basis. Application fee: $30 ($40 for international students). *Financial aid:* Application deadline: 2/15. *Faculty research:* Manufacturing information systems, manufacturing process simulation, product design methods. *Unit head:* Dr. Harvey Wolfe, Chairman, 412-624-9830, Fax: 412-624-9831, E-mail: hwolfe@engrng.pitt.edu. *Application contact:* John H. Manley, Director, 412-624-9846, Fax: 412-624-9831, E-mail: jmanley@vms.cis.pitt.edu.

University of Regina, Faculty of Graduate Studies and Research, Faculty of Engineering, Program in Industrial Systems Engineering, Regina, SK S4S 0A2, Canada. Offers M Eng, MA Sc, PhD. *Students:* 4 full-time (1 woman), 20 part-time (5 women). 19 applicants, 47% accepted. In 1998, 5 degrees awarded. *Degree requirements:* For master's, thesis required (for some programs), foreign language not required; for doctorate, dissertation required, foreign language not required. *Entrance requirements:* For master's, TOEFL (minimum score of 550 required); for doctorate, TOEFL (minimum score of 550 required), master's degree. *Application deadline:* Applications are processed on a rolling basis. Application fee: $0. *Expenses:* Tuition and fees charges are reported in Canadian dollars. Tuition, state resident: full-time $1,688 Canadian dollars; part-time $94 Canadian dollars per credit hour. International tuition: $3,375 Canadian dollars full-time. Required fees: $65 Canadian dollars per course. Tuition and fees vary according to course load and program. *Financial aid:* In 1998–99, 5 research assistantships, 4 teaching assistantships were awarded.; fellowships, career-related internships or fieldwork and scholarships also available. Financial aid application deadline: 6/15. *Faculty research:* Gas separation and purification, welding weldability. *Unit head:* Dr. P. Catania, Head, 306-585-4364, Fax: 306-585-4855, E-mail: peter.catania@uregina.ca. *Application contact:* Dr. S. Bhole, Coordinator, 306-585-4703, Fax: 306-585-4855, E-mail: sanjeev.bhole@uregina.ca.

University of Rhode Island, Graduate School, College of Engineering, Department of Mechanical Engineering and Applied Mechanics, Program in Design/Systems, Kingston, RI 02881. Offers MS, PhD. *Entrance requirements:* For master's and doctorate, GRE General Test (for foreign applicants only).

University of St. Thomas, Graduate Studies, Graduate School of Applied Science and Engineering, Program in Manufacturing Systems Engineering, St. Paul, MN 55105-1096. Offers MMSE, MS, Certificate. *Accreditation:* ABET (one or more programs are accredited). Part-time and evening/weekend programs available. *Faculty:* 5 full-time (0 women), 13 part-time (0 women). *Students:* 6 full-time (1 woman), 204 part-time (41 women). Average age 34. 77 applicants, 95% accepted. In 1998, 22 master's awarded. *Degree requirements:* For master's, computer language, thesis required (for some programs). *Entrance requirements:* For master's, GMAT or GRE General Test. *Application deadline:* For fall admission, 8/1 (priority date); for spring admission, 1/1 (priority date). Applications are processed on a rolling basis. Application fee: $30. Electronic applications accepted. Tuition: Part-time $437 per credit. Tuition and fees vary according to degree level, program and student level. *Financial aid:* In 1998–99, 12 students received aid; fellowships, research assistantships, grants and institutionally-sponsored loans available. Aid available to part-time students. Financial aid application deadline: 4/1; financial aid applicants required to submit FAFSA. *Unit head:* Ron Bennett, Director, 651-962-5750, Fax: 651-962-6419, E-mail: rjbennett@stthomas.edu. *Application contact:* Marlene L. Houliston, Student Services Coordinator, 651-962-5750, Fax: 651-962-6419, E-mail: technology@stthomas.edu.

University of Southern California, Graduate School, School of Engineering, Department of Electrical Engineering, Program in Systems Architecture and Engineering, Los Angeles, CA 90089. Offers MS. *Students:* 2 full-time (1 woman), 22 part-time (5 women); includes 13 minority (9 Asian Americans or Pacific Islanders, 4 Hispanic Americans), 1 international. Average age 33. 9 applicants, 89% accepted. In 1998, 12 degrees awarded. *Degree requirements:* For master's, thesis optional. *Entrance requirements:* For master's, GRE General Test. *Application deadline:* For fall admission, 6/1 (priority date); for spring admission, 9/1. Application fee: $55. Tuition: Part-time $768 per unit. Required fees: $350 per semester. *Financial aid:* Fellowships, research assistantships, teaching assistantships, Federal Work-Study, institutionally-sponsored loans, and scholarships available. Aid available to part-time students. Financial aid application deadline: 2/15; financial aid applicants required to submit FAFSA. *Unit head:* Dr. Robert Scholtz, Co-Chairman, Department of Electrical Engineering, 213-740-7327.

University of Southern California, Graduate School, School of Engineering, Department of Electrical Engineering, Program in VLSI Design, Los Angeles, CA 90089. Offers MS. *Students:* 25 full-time (2 women), 16 part-time (1 woman); includes 9 minority (8 Asian Americans or Pacific Islanders, 1 Hispanic American), 24 international. Average age 26. 87 applicants, 82% accepted. In 1998, 4 degrees awarded. *Degree requirements:* For master's, thesis optional. *Entrance requirements:* For master's, GRE General Test. *Application deadline:* For fall admission, 6/1 (priority date); for spring admission, 11/1. Application fee: $55. Tuition: Part-time $768 per unit. Required fees: $350 per semester. *Financial aid:* In 1998–99, 1 fellowship, 4 research assistantships were awarded.; teaching assistantships, Federal Work-Study and institutionally-sponsored loans also available. Aid available to part-time students. Financial aid application deadline: 2/15; financial aid applicants required to submit FAFSA. *Unit head:* Dr. Robert Scholtz, Co-Chairman, Department of Electrical Engineering, 213-740-7327.

University of Southern California, Graduate School, School of Engineering, Department of Industrial and Systems Engineering, Program in Industrial and Systems Engineering, Los Angeles, CA 90089. Offers MS, PhD, Engr, MBA/MS. *Students:* 39 full-time (2 women), 29 part-time (3 women); includes 10 minority (all Asian Americans or Pacific Islanders), 38 international. Average age 31. 160 applicants, 73% accepted. In 1998, 15 master's, 4 doctorates awarded. *Degree requirements:* For master's, thesis optional; for doctorate, dissertation required. *Entrance requirements:* For master's, doctorate, and Engr, GRE General Test. *Application deadline:* For fall admission, 6/1 (priority date); for spring admission, 12/1. Application fee: $55. Tuition: Part-time $768 per unit. Required fees: $350 per semester. *Financial aid:* In 1998–99, 8 fellowships, 11 research assistantships, 7 teaching assistantships were awarded.; Federal Work-Study, institutionally-sponsored loans, and scholarships also available. Aid available to part-time students. Financial aid application deadline: 2/15; financial aid applicants required to submit FAFSA. *Unit head:* Dr. F. Stan Settles, Chairman, Department of Industrial and Systems Engineering, 213-740-4893.

University of Southern Colorado, College of Applied Science and Engineering Technology, Program in Industrial and Systems Engineering, Pueblo, CO 81001-4901. Offers MS. Part-time and evening/weekend programs available. *Faculty:* 5 full-time (1 woman), 2 part-time (0 women). *Students:* 21 full-time (1 woman), 16 part-time (1 woman); includes 3 minority (all Hispanic Americans), 24 international. Average age 29. 23 applicants, 74% accepted. In 1998, 16 degrees awarded. *Degree requirements:* For master's, computer language required, thesis optional, foreign language not required. *Entrance requirements:* For master's, GRE General Test, TOEFL. *Average time to degree:* Master's–1.5 years full-time, 3.5 years part-time. *Application deadline:* For fall admission, 7/19 (priority date); for spring admission, 11/30 (priority date). Applications are processed on a rolling basis. Application fee: $15 ($30 for international students). *Financial aid:* In 1998–99, 5 teaching assistantships were awarded.; career-related internships or fieldwork, Federal Work-Study, institutionally-sponsored loans, and scholarships also available. Financial aid application deadline: 3/1; financial aid applicants required to submit FAFSA. *Faculty research:* Computer-integrated manufacturing, reliability, economic development, design of experiments, scheduling, simulation. Total annual research expenditures: $178,000. *Unit head:* Dr. Jane M. Fraser, Chair, 719-549-2036, Fax: 719-549-2519, E-mail: jfraser@uscolo.edu. *Application contact:* Dr. Huseyin Sarper, Graduate Coordinator, 719-549-2889, Fax: 719-549-2519, E-mail: sarper@uscolo.edu.

University of Virginia, School of Engineering and Applied Science, Department of Systems Engineering, Charlottesville, VA 22903. Offers ME, MS, PhD, ME/MBA. Postbaccalaureate distance learning degree programs offered (no on-campus study). *Faculty:* 13 full-time (1 woman), 3 part-time (1 woman). *Students:* 48 full-time (13 women), 3 part-time; includes 4 minority (3 African Americans, 1 Hispanic American), 11 international. Average age 30. 54 applicants, 37% accepted. In 1998, 14 master's, 4 doctorates awarded. *Degree requirements:* For doctorate, dissertation, comprehensive exam required, foreign language not required, foreign language not required. *Entrance requirements:* For master's and doctorate, GRE General Test. *Application deadline:* For fall admission, 8/1; for spring admission, 12/1. Applications are processed on a rolling basis. Application fee: $60. *Financial aid:* Fellowships available. Financial aid application deadline: 2/1. *Faculty research:* Operations research; systems methodology and design; systems management, decision, and control. *Unit head:* Donald E. Brown, Chairman, 804-924-5393. *Application contact:* J. Milton Adams, Assistant Dean, 804-924-3897, E-mail: twr2c@virginia.edu.

See in-depth description on page 1241.

University of Waterloo, Graduate Studies, Faculty of Engineering, Department of Systems Design Engineering, Waterloo, ON N2L 3G1, Canada. Offers MA Sc, PhD. Part-time programs available. *Faculty:* 21 full-time (2 women), 20 part-time (4 women). *Students:* 52 full-time (15 women), 6 part-time (1 woman). 30 applicants, 40% accepted. In 1998, 16 master's, 11 doctorates awarded. *Degree requirements:* For master's, research paper or thesis required; for doctorate, dissertation, comprehensive exam required. *Entrance requirements:* For master's, TOEFL (minimum score of 550 required), honors degree, minimum B average; for doctorate, TOEFL (minimum score of 550 required), master's degree, minimum A- average. *Application deadline:* Applications are processed on a rolling basis. Application fee: $50. *Expenses:* Tuition and fees charges are reported in Canadian dollars. Tuition, state resident: full-time $3,168 Canadian dollars; part-time $792 Canadian dollars per term. Tuition, nonresident: full-time $8,000 Canadian dollars; part-time $2,000 Canadian dollars. Required fees: $45 Canadian dollars per term. Tuition and fees vary according to program. *Financial aid:* Fellowships, research assistantships, teaching assistantships available. *Faculty research:* Socioeconomic systems, physical systems, human systems, pattern analysis and machine intelligence, computer systems and design. *Unit head:* Dr. M. E. Jernigan, Chair, 519-888-4567 Ext. 4644, E-mail: grad@sysoffice.watstar.uwaterloo.ca. *Application contact:* Carol Kendrick, Graduate Secretary, 519-888-4567 Ext. 3498, E-mail: grad@sysoffice.watstar.uwaterloo.ca.

University of West Florida, College of Science and Technology, Department of Computer Science, Pensacola, FL 32514-5750. Offers computer science (MS); systems and control engineering (MS). Part-time and evening/weekend programs available. *Students:* 37 full-time (14 women), 98 part-time (19 women); includes 21 minority (7 African Americans, 9 Asian Americans or Pacific Islanders, 5 Hispanic Americans), 10 international. *Degree requirements:* For master's, computer language required, thesis optional, foreign language not required. *Entrance requirements:* For master's, GRE General Test (minimum combined score of 1000 required). *Application deadline:* For fall admission, 7/1; for spring admission, 11/1. Applications are processed on a rolling basis. Application fee: $20. Tuition, state resident: full-time $3,582; part-time $149 per credit hour. Tuition, nonresident: full-time $12,240; part-time $510 per credit hour. *Unit head:* Dr. Ed Rodgers, Chairperson, 850-474-2542.

Virginia Polytechnic Institute and State University, Graduate School, College of Engineering, Program in Systems Engineering, Blacksburg, VA 24061. Offers M Eng, MS. Part-time and evening/weekend programs available. *Faculty:* 5 full-time (0 women). *Students:* 6 full-time (0 women), 95 part-time (27 women); includes 17 minority (5 African Americans, 10 Asian Americans or Pacific Islanders, 2 Hispanic Americans), 1 international. Average age 29. 19 applicants, 53% accepted. In 1998, 40 degrees awarded. *Degree requirements:* For master's, computer language, thesis or alternative required, foreign language not required. *Entrance requirements:* For master's, GRE General Test (minimum score of 600 on verbal section, 600 on quantitative, 450 on analytical required), TOEFL (minimum score of 600 required). *Application deadline:* For fall admission, 12/1 (priority date). Applications are processed on a roll-

Systems Engineering–Cross-Discipline Announcements

Virginia Polytechnic Institute and State University (continued)
ing basis. Application fee: $25. *Financial aid:* Application deadline: 4/1. *Unit head:* Dr. H. Kurstedt, Chairman, 540-231-5458, E-mail: hak@vt.edu.

Washington University in St. Louis, School of Engineering and Applied Science, Sever Institute of Technology, Department of Civil Engineering, Program in Transportation and Urban Systems Engineering, St. Louis, MO 63130-4899. Offers D Sc. *Degree requirements:* For doctorate, variable foreign language requirement, dissertation, departmental qualifying exam required.

West Virginia University Institute of Technology, College of Engineering, Program in Control Systems Engineering, Montgomery, WV 25136. Offers MS. Part-time programs available. *Faculty:* 13 full-time (1 woman). *Students:* 7 full-time (0 women), 5 part-time, 9 international. Average age 27. 52 applicants, 85% accepted. In 1998, 4 degrees awarded. *Degree*

requirements: For master's, thesis or alternative, fieldwork required, foreign language not required. *Entrance requirements:* For master's, GRE General Test (minimum combined score of 1100 required), TOEFL (minimum score of 550 required), minimum GPA of 3.0. *Average time to degree:* Master's–2 years full-time. *Application deadline:* For fall admission, 3/15 (priority date). Applications are processed on a rolling basis. Application fee: $10. Tuition, state resident: full-time $2,816; part-time $312 per credit. Tuition, nonresident: full-time $6,964; part-time $774 per credit. *Financial aid:* In 1998–99, 7 teaching assistantships were awarded.; career-related internships or fieldwork, Federal Work-Study, and institutionally-sponsored loans also available. Financial aid application deadline: 3/15. *Faculty research:* Process control. *Unit head:* S. E. Thornton, Associate Dean, 304-442-3162, Fax: 304-442-1006, E-mail: sethor@wvit.wvnet.edu. *Application contact:* Robert P. Scholl, Registrar, 304-442-3167, Fax: 304-442-3097, E-mail: rpscho@wvit.wvnet.edu.

Cross-Discipline Announcements

Carnegie Mellon University, Information Networking Institute, Pittsburgh, PA 15213-3891.

The MS in Information Networking is a cooperative endeavor of the Schools of Engineering, Computer Science, and Business, providing an alternative to the conventional one-year computer science or electrical engineering graduate program by integrating both and adding some features of an MBA program. Now in its eleventh year, the program carefully selects 35 people from the engineering and computer science disciplines to form each new class. The program provides technical electives aimed at several areas of specialization including (a) the telecommunications and computing industries, (b) wireless and mobile computing, (c) the financial services industry, and (d) systems integrating and consulting.

Carnegie Mellon University, School of Computer Science, Robotics Institute, Pittsburgh, PA 15213-3891.

Carnegie Mellon's MS and PhD programs in robotics are highly research oriented and interdisciplinary in nature, drawing from such fields as computer science, electrical and computer engineering, and mechanical engineering. Students may specialize in machine perception, artificial intelligence, manipulation, autonomous vehicles, manufacturing automation, or other areas. See in-depth description in Section 8: Computer Science and Information Technology.

Kettering University, Graduate School, Flint, MI 48504-4898.

Kettering offers teaching and research assistantships during all 4 quarters each year. Assistants are paid a stipend ($1000 per month) and tuition is waived. Departments select assistants according to their qualifications and the department's needs. Teaching assistants must have an excellent command of English.

ARIZONA STATE UNIVERSITY

Department of Industrial Engineering

Programs of Study

The Department of Industrial Engineering offers graduate programs leading to the Master of Science in Engineering degree, the Master of Science degree, and the Doctor of Philosophy degree in industrial engineering. The overall educational objective of graduate study in industrial engineering is to improve each student's ability to understand, analyze, and resolve problems within complex organizations. Industrial engineers must develop qualitative and quantitative abilities to assist management in such diverse organizations as banks, government, hospitals, military, and manufacturing operations. Areas of study within industrial engineering include manufacturing processes and control, enterprise information systems/management of technology, quality/reliability engineering, operations research and production systems, and special programs in semiconductor manufacturing and human factors. In addition, there is flexibility for students to tailor a program to meet their specific professional needs.

There are two options in the program for the Master of Sciences in Engineering degree. The first option requires the satisfactory completion of prerequisite courses in mathematics, science, computer programming, and industrial engineering, plus 30 semester hours of credit, including a 6-hour research thesis. The second option is similar but requires a final written examination and no thesis. Students in these two options may have a baccalaureate degree with a major other than engineering, although engineering, mathematics, science, or business are recommended. The student's qualifications are reviewed on an individual basis by the faculty.

The Doctor of Philosophy degree generally requires 30 semester hours of course work beyond the master's degree plus 24 hours for the dissertation.

Research Facilities

Research facilities within the department include established laboratories in virtual computer-integrated enterprise, quality and reliability engineering, networking, facility planning, rapid prototyping, simulation, integrated manufacturing engineering, industrial engineering design, and semiconductor modeling and analysis. Department laboratories are supplemented by laboratories in the Manufacturing Institute, the Center for Solid State Electronics Research, the System Science and Engineering Research Center, and the Telecommunications Research Center, all of which provide technical support for students and faculty members in the industrial engineering program.

Engineering Computer Services has a staff of 24 full-time computer specialists who support the wide array of computers and electronic equipment within the University. The University also maintains complete machine, carpentry, electrical, and paint shops to support graduate research.

The University libraries house more than 2 million volumes, 1.7 million microfilms, and 24,000 journal titles. The Noble Science and Engineering Library, a designated patent depository, houses the entire U.S. patent collection.

Financial Aid

A wide variety of financial support is available for graduate students, including teaching and research assistantships and University and industrial fellowships. Out-of-state tuition is waived for all graduate assistants and fellows. In addition, there are a number of Graduate Tuition Scholarships (GTS) covering out-of-state tuition and Graduate Academic Scholarships (GAS) covering in-state registration fees available on a competitive basis.

Cost of Study

In fall 1999, the registration and tuition fees for 12 hours or more are $1080 per semester for Arizona residents; nonresidents pay an additional $3476 (prorated for fewer than 12 hours).

Living and Housing Costs

Limited on-campus housing is available for unmarried students. In 1999–2000, room rates range from about $2900 to $3800 for the academic year. Twelve different meal plans are available (per week or semester). Moderately priced off-campus housing is available within walking distance of the campus.

Student Group

There are more than 46,000 students at Arizona State University, including more than 10,700 graduate students. More than 1,700 of the 5,547 students in the College of Engineering and Applied Sciences are enrolled in graduate programs. In fall 1998, there were 185 graduate students in the department.

Student Outcomes

Significant numbers of industrial engineering graduates are employed in manufacturing firms and service industries. The industrial engineering job market is expanding to include areas such as sales and marketing, finance, information systems, and personnel. Other industries employing industrial engineers include health care, transportation, banking, and social services. Placement of students has developed by maintaining close ties to the local professional chapter of the Institute of Industrial Engineers and by the contacts faculty members pursue with their colleagues in industry. The department averages 4–6 contacts per month from employers who are looking for full- and part-time employees.

Location

Tempe is a suburb bordering the city of Phoenix. The metropolitan area is the population, economic, and industrial center of the State of Arizona. Entertainment centers, art and anthropology museums, and sports arenas are among the wide variety of facilities available in the Valley of the Sun. The area's mild climate also provides for numerous year-round outdoor activities. The urban community surrounding ASU thrives on high-technology industry and is one of the fastest-growing communities in the country. As a result, many students find employment in the region during and after completion of their studies.

The University

Arizona State University traces its origin to 1885 and is the largest and oldest institution of higher learning in the state of Arizona. Its present enrollment of 46,000 places it as the fifth-largest central-campus university in the United States. ASU's campus comprises 700 acres and offers outstanding physical facilities to support the University's research and educational mission. Included within the more than 125 buildings are twelve colleges and schools, a University-wide computer system, seven libraries (including the Noble Science and Engineering Library), and more than two dozen specialized centers of research. ASU was recently awarded Research I status as a result of its past and continuing commitment to academic research.

Applying

All students must apply for admission through the Graduate College. For application forms, students should write to Graduate Admissions, Arizona State University, Tempe, Arizona 85287-1003 or call 480-965-6113. Forms may also be obtained via the World Wide Web (http://www.asu.edu/forms/adm.html). Submission of Graduate Record Examinations (GRE) scores, three letters of recommendation, and a statement of purpose are required for all applicants. International students whose native language is not English must submit a Test of English as a Foreign Language (TOEFL) score.

Correspondence and Information

Director of Graduate Studies
Industrial Engineering
P.O. Box 875906
Arizona State University
Tempe, Arizona 85287-5906

Telephone: 480-965-3185
E-mail: imse@www.eas.asu.edu
World Wide Web: http://www.eas.asu.edu/~imse/

Arizona State University

THE FACULTY AND THEIR RESEARCH

Mary R. Anderson, Associate Professor and Associate Dean of Student and Business Affairs of the College of Engineering and Applied Sciences; Ph.D., Iowa. Applied statistics and probability for quality control, career change programs for women and men in engineering.

James E. Bailey, Professor; Ph.D., Wayne State. Project management, management information systems, production planning and control, concurrent engineering.

W. Matthew Carlyle, Assistant Professor; Ph.D., Stanford. Optimization, mathematical programming, applied operations research.

Jeffery K. Cochran, Associate Professor; Ph.D., Purdue. Simulation, applied operations research, operations modeling with objects, production planning with quality constraints.

Kevin J. Dooley, Professor; Ph.D., Illinois. Management of technology, quality management, applications of complex systems theory.

Charles S. Elliott, Adjunct Associate Professor and Director of Center for Professional Development for the College of Engineering; Ph.D., Michigan State. Engineering management and manufacturing, management of technology.

John W. Fowler, Assistant Professor; Ph.D., Texas A&M. Semiconductor manufacturing systems analysis, discrete-event simulation, applied operations research.

Mark R. Henderson, Professor; Ph.D., Purdue. Feature-based product modeling, computer-aided design and manufacturing, computer graphics, solid modeling, geometric pattern recognition, geometric modeling, engineering design and manufacturing, rapid prototyping.

Gary L. Hogg, Professor and Chair; Ph.D., Texas at Austin. Applied optimization, simulation, manufacturing planning and control.

Norma F. Hubele, Professor; Ph.D., Rensselaer. Applied statistics, statistical process control, time-series analysis and forecasting, automated inspection.

J. Bert Keats, Professor; Ph.D., Oklahoma State. Reliability engineering, applied statistics, statistical process control, time series analysis forecasting and control, quality management.

Gerald T. Mackulak, Associate Professor; Ph.D., Purdue. Simulation, applied operations research to manufacturing and artificial intelligence for virtual manufacturing and production control.

Douglas C. Montgomery, Professor; Ph.D., Virginia Tech. Statistical design of experiments, optimization and response surface methodology, empirical stochastic modeling, applied operations research.

William C. Moor, Associate Professor; Ph.D., Northwestern. Organization control, engineering management, economics analysis, human factors and ergonomics.

Jong-I Mou, Assistant Professor; Ph.D., Purdue. Precision manufacturing and inspection, robotics and automation, modeling and controlling of computer-integrated manufacturing processes, control system design and synthesis.

Chell A. Roberts, Associate Professor; Ph.D., Virginia Tech. Computer-aided manufacturing, real-time control, discrete-event modeling, manufacturability assessment.

Dwayne A. Rollier, Associate Professor and Associate Chair; Ph.D., Florida State. Experimental design, manufacturing engineering, applied statistics.

George C. Runger, Associate Professor; Ph.D., Minnesota. Applied statistics, process control and optimization, application in nontraditional manufacturing such as semiconductor, electronics, chemical and process industries.

Dan L. Shunk, Associate Professor; Ph.D., Purdue. Agile, enterprise and CIM systems, group technology, planning systems, economics of computer-integrated manufacturing (CIM), flexible systems.

J. Rene Villalobos, Associate Professor; Ph.D., Texas A&M. Manufacturing systems, automated visual inspection, real-time quality control, intelligent manufacturing systems.

Philip M. Wolfe, Professor; Ph.D., Arizona State. Production control, manufacturing systems, automation, planning for computer-aided manufacturing, applications of client-server in enterprises, management of technology, design for the environment.

Nong Ye, Associate Professor; Ph.D., Purdue. Information fusion and intelligent systems, production planning and control, agile manufacturing, advanced interfacing technology, information and system engineering.

Emeritus Professors

David D. Bedworth, Professor Emeritus; Ph.D., Purdue. Production systems planning, analysis and control, computer-aided manufacturing, real-time control computer applications, time-series analysis and forecasting, industrial engineering, analysis.

Arthur G. Dean, Professor Emeritus; Ph.D., Texas A&M. Applied statistics, quality control, operations research, reliability engineering.

Richard L. Smith, Professor Emeritus; Ph.D., Arizona State. Information systems and analysis, database and knowledge-base systems, applications of client-server in enterprises, management of technology.

William R. Uttal, Professor Emeritus; Ph.D., Ohio State. Human factors, ergonomics, computer and human vision, computational models, neural and psychological foundation of human-computer interactions.

Hewitt H. Young, Professor Emeritus; Ph.D., Arizona State. Computer-aided design and manufacturing, industrial engineering, analysis and design, industrial mechanization and automatic control, production systems, discrete control (optimization), large-scale systems analysis, decision processes, human engineering and productivity.

BOSTON UNIVERSITY

College of Engineering
Department of Manufacturing Engineering

Programs of Study	The Department of Manufacturing Engineering offers graduate studies at the Ph.D. and M.S. levels. The application of modern engineering methods to problems of manufacturing is the major objective of the department's graduate program. In an internationally competitive manufacturing environment, flexibility and productivity are major challenges that can be met only by properly educated engineers and managers. The programs are aimed at meeting the manufacturing productivity challenge. The department's graduate program draws students from various disciplines, including mechanical, computer, materials, electrical, systems, and industrial engineering. M.S. candidates can select from six concentration areas: manufacturing systems and operations research; computer-integrated design, analysis, and manufacture; manufacturing operations management; automation and control in manufacturing; engineered materials and processes; and process design. In addition to the traditional on-campus course offerings available to full-time day students, alternative means of delivery, course schedules, and programs have been devised for either part-time students from industry or advanced students with work experience who are interested in applied research-intensive study. These include the dual M.S./M.B.A. degree program, delivered in collaboration with the Graduate School of Management; international exchange programs with European universities, including the international M.S. in global manufacturing program delivered in collaboration with the Fraunhofer Institute and the University of Aachen in Germany; special degree programs designed for industrial managers and practitioners who are interested in accelerated part-time study; and the distance learning M.S. program, which is offered during evening time slots to students attending class at remote sites through Boston University's interactive videoconferencing facilities and Internet-based two-way audiovisual information delivery. The M.S. program requires at least 36 credits (nine semester courses), of which at least 28 must be earned at Boston University. A thesis is an option but is not required. Both post-bachelor's and post-master's Ph.D. programs are available. Ph.D. study requires an additional 32 credits (eight semester courses) beyond the M.S. requirement. Ph.D. study also requires at least two consecutive semesters of residence, written and oral exams, a dissertation, and a dissertation defense.
Research Facilities	The department's research laboratories are complemented by the research facilities at the Metcalf Center for Science and Engineering. The Metcalf Center represents a $110-million commitment by the University to expand programs in applied science and engineering. The laboratories include facilities for computer-aided design and computer-aided manufacturing, integrated processing, robotics and vision, manufacturing systems and productivity, metal cutting, chemical vapor deposition, process control, automated manufacturing systems, rapid prototyping, and high-temperature chemical processing of materials. The department is also closely affiliated with the Fraunhofer Resource Center, which is part of the largest applied industrial research network in the world, focused on deploying state-of-the-art technology to industry.
Financial Aid	Financial aid opportunities are available, including Presidential University Graduate Fellowships, Dean's Fellowships, graduate teaching fellowships, research assistantships, and scholarships. In 1999–2000, teaching fellowships provide stipends of $12,500 per academic year and require approximately 20 hours a week of instructional and other duties. Recipients receive a tuition waiver for 8 to 10 credits per semester and up to 8 additional credits for the following summer. Assistantship stipend levels are comparable to those of teaching fellowships and are also supplemented by tuition waivers. University and Dean's Fellows scholarships range up to $37,830, including stipend and tuition, per academic year; the department encourages GEM scholars to apply. Federal Direct Student Loan and work-study applicants must send a Free Application for Federal Student Aid (FAFSA) to the Federal Student Aid Programs Office. Work-study and FAFSA forms may be obtained from the Graduate Programs Office.
Cost of Study	In 1999–2000, tuition and fees for full-time study are $23,770. For part-time study, the cost is $743 per credit hour.
Living and Housing Costs	Privately owned apartments or rooms are readily available. Living expenses for a single student are estimated at $10,730 for the nine-month academic year in 1999–2000.
Student Group	The department has 38 M.S. students and 9 students in the M.S./M.B.A. program; 32 students are pursuing the Ph.D. degree. In addition, the department has 43 students in its video-based manufacturing program for off-campus students working in major manufacturing companies. In 1998–99, there were 13 manufacturing engineering students supported as graduate teaching fellows, 20 as research assistants, 2 as Dean's Fellows, 1 as a Presidential University Fellow, and 5 as GAANN Fellows. Most M.S. graduates enter industry; however, many are pursuing research in Ph.D. programs.
Student Outcomes	Most graduates of the M.S. programs enter industry. However, with a growing base of funded research, increasing numbers are pursuing Ph.D. research at Boston University.
Location	Boston is a sophisticated metropolitan city with world-renowned academic and scientific resources. The facilities of other universities in the area are easily accessible, and excellent seminar and colloquium programs afford students rare opportunities to participate in cutting-edge research. Nearby high-technology industry provides another vital element in an exciting, academically oriented region.
The University and The Department	Boston University, incorporated in 1869, is an independent, coeducational, nonsectarian university, fully open to women and to all minorities. Its approximately 23,500 full-time students and 3,130 faculty members make it one of the largest independent universities in the world. The Department of Manufacturing Engineering was the first department in the country with an ABET-accredited B.S. program in manufacturing engineering. Interaction with local industry through research and part-time study corporate programs has created a focus on state-of-the-art educational and research issues.
Applying	M.S. or M.S./M.B.A. program applicants should have demonstrated a high degree of scholarship in an undergraduate program in engineering or science at an accredited school. Admission to the Ph.D. program normally requires an M.S., M.A., or the equivalent; unusually well-prepared students may be admitted directly after receiving a bachelor's degree. Students desiring financial aid should apply by January 15 for fall and October 1 for spring. Applications without financial aid requests can be submitted as late as April 1 and October 15, respectively. Required credentials include official transcripts, two specific letters of recommendation, GRE General Test scores for M.S. and Ph.D. applicants and GMAT or GRE General Test scores for M.S./M.B.A. applicants, and additional requirements, including TOEFL scores for international students.
Correspondence and Information	For program information: Graduate Programs Department of Manufacturing Engineering College of Engineering Boston University 15 St. Mary's Street Boston, Massachusetts 02215 Telephone: 617-353-2842 World Wide Web: http://www.bu.edu/eng/grad	For admission application forms: Graduate Programs College of Engineering Boston University 48 Cummington Street Boston, Massachusetts 02215 Telephone: 617-353-9760 Fax: 617-353-0259 E-mail: enggrad@bu.edu World Wide Web: http://www.bu.edu/eng/grad

Boston University

THE FACULTY AND THEIR RESEARCH

Dorothy C. Attaway, Assistant Professor; Ph.D., Boston University. Manufacturing systems, computer-aided design, boundary element methods.

John Baillieul, Professor and Chairman of Aerospace and Mechanical Engineering; Ph.D., Harvard. Robotics, control of mechanical systems, mathematical system theory.

Eytan Barouch, Professor; Ph.D., SUNY at Stony Brook. Simulation of industrial processes, numerical analysis, algorithm development.

Soumendra N. Basu, Associate Professor; Ph.D., MIT. Materials science with emphasis on structure/property relationships in thin films and coatings.

Michael Caramanis, Professor; Ph.D., Harvard. Mathematical programming, control and stochastic systems.

Christos C. Cassandras, Professor; Ph.D., Harvard. Discrete systems, control and optimization of manufacturing systems, simulation, communication networks.

Daniel Cole, Associate Professor; Ph.D., CUNY. Simulation of semiconductor device manufacturing and performance, numerical analysis, computational number theory.

Theo A. de Winter, Associate Professor; S.M., Mech.E., MIT; PE. Superconductivity, cryogenics, heat transfer, product design, magnetic systems applications.

Merrill L. Ebner, Professor; Sc.D., MIT; PE. Polymer solidification, integrated processing cells, capacity simulation, expert manufacturing systems.

Michael A. Gevelber, Associate Professor; Ph.D., MIT. Control system dynamics, modeling, design and control of materials processing application systems.

Yehonathan Hazony, Professor; Ph.D., Hebrew (Jerusalem). Computer methods for industrial automation, robotics and education.

Jian-Qiang Hu, Associate Professor; Ph.D., Harvard. Manufacturing applications of stochastic systems and queueing theories.

Uday B. Pal, Associate Professor; Ph.D., Penn State. High-temperature chemical and electrochemical processing of materials (metals and ceramics), with emphases on green manufacturing, sensors, fuel cells, and batteries.

Ioannis Ch. Paschalidis, Assistant Professor; Ph.D., MIT. Analysis and control of stochastic systems in manufacturing systems, communication networks and the service industry.

James R. Perkins, Assistant Professor; Ph.D., Illinois at Urbana-Champaign. Manufacturing systems, control and production scheduling.

Vinod K. Sarin, Professor; Sc.D., MIT. Material science, chemical vapor deposition of structural coatings, surface modifications.

Andre Sharon, Professor and Executive Director, Fraunhofer Resource Center; Ph.D., MIT. Electromechanical design, servo control, software development.

Pirooz Vakili, Associate Professor; Ph.D., Harvard. Discrete event dynamic systems, perturbation analysis, queueing networks.

Michael Yeung, Assistant Professor; Ph.D., Berkeley. Simulation of integrated circuits processing, parasitics modeling of integrated circuits, electromagnetics.

COLUMBIA UNIVERSITY

Department of Industrial Engineering and Operations Research

Programs of Study	The Department of Industrial Engineering and Operations Research offers programs of study leading to the degrees of Master of Science (M.S.), Doctor of Engineering Science (Eng.Sc.D.), and Doctor of Philosophy (Ph.D.) in either industrial engineering (IE) or operations research (OR). Combined M.S./M.B.A. programs are offered in conjunction with Columbia's Graduate Business School. The professional degree in industrial engineering is also offered.
	The Master of Science degree can be completed within one academic year of full-time classwork or longer if the student chooses to study part-time. Registration as a nondegree candidate (special student) is also possible. Required courses for the industrial engineering program include production management, operations research, deterministic models, and probability and statistics. Required courses for the operations research program include probability, statistical inference, deterministic models, simulation, and stochastic models. Individual programs are designed in consultation with a faculty adviser, and a number of concentrations are available. The department also offers a concentration in financial engineering for each of its Master of Science programs. The joint M.S./M.B.A. can be obtained after five terms of full-time study.
	The Ph.D. program is geared toward the exceptional student. The first year is used to prepare for the Ph.D. examinations, which requires advanced course work in probability theory and mathematical programming. In addition to advanced electives, each student must take mathematical analysis, statistical inference, and simulation, unless equivalent courses were taken prior to admittance. Thereafter, the focus is on research designed to prepare students for research and teaching careers.
Research Facilities	The department's PC computer laboratory contains Pentium and Pentium-Pro processor machines with supporting software and printers. Networked PCs and X terminals connected to departmental Sun Workstations as well as University computers are available in all teaching and graduate assistant offices. The department also has X terminals and Sun Workstations available in a convenient location for graduate students, as well as extensive PC and terminal facilities for all students. In addition, Columbia University's Academic Information Systems (AcIS) supports the academic computing and data communications needs of the University. The Columbia University library system is among the nation's ten largest academic libraries and maintains an excellent collection in industrial engineering and operations research and related disciplines.
Financial Aid	Financial support is awarded on a competitive basis in the form of assistantships that provide a stipend and a tuition allowance that covers a full 15-point program each semester. For 1999–2000, stipends for teaching assistants are $12,375 for nine months; for research assistants, stipends are $16,500 for twelve months.
Cost of Study	For 1999–2000, tuition for full-time students for the academic year is approximately $24,450; for part-time study, the cost is approximately $815 per credit. Annual fees are approximately $1225, and the cost of books is approximately $800.
Living and Housing Costs	The University provides limited housing for graduate men and women who are registered either for an approved program of full-time academic study or for doctoral dissertation research. University residence halls include traditional dormitory facilities as well as suites and apartments for single and married students; furnishings and utilities may be included. An estimated minimum of $13,000 should be allowed for board, room, and personal expenses for the academic year. Rooms are also available at International House; these cost from $2000 to $2500 per semester. University Real Estate properties include apartments owned and managed by the University in the immediate vicinity of the Morningside Heights campus. These are leased yearly, as they become available, to single and married students at rates that reflect the size and location of each apartment as well as whether furnishings or utilities are included. Requests for additional information and application forms should be directed to the Assignments Office, 111 Wallach Hall.
Student Group	In 1998–99, enrollment in the Department of Industrial Engineering and Operations Research totaled 315 students and included 179 undergraduates (juniors and seniors), 102 master's degree candidates, 25 doctoral candidates, and 9 part-time special students. The student population has a diverse and international character; 30 percent were women in 1998–99.
Location	New York City is the intellectual, artistic, cultural, gastronomic, corporate, financial, and media center of the United States and perhaps of the world. The city is renowned for its theaters, museums, libraries, restaurants, opera, and music. Inexpensive student tickets for cultural and sports events are frequently available, and the museums are open to students at very modest cost or are free. The ethnic variety of the city adds to its appeal.
	The city is bordered by uncongested areas of great beauty that provide varied types of recreation, such as hiking, camping, skiing, and ocean and lake swimming. There are superb beaches on Long Island and in New Jersey, while to the north lie the Catskill, Green, Berkshire, and Adirondack mountains. Close at hand is the beautiful Hudson River valley.
	The proximity of many local industries provides strong student-industry contact and excellent job opportunities. Adjunct faculty from industry provide courses in areas of current professional interest.
The University	Columbia University was established as King's College in 1754. Today it consists of sixteen schools and faculties and is one of the leading universities in the world. The University draws students from many countries. The high caliber of the students and faculty makes it an intellectually stimulating place to be. Columbia University is located on Morningside Heights, close to Lincoln Center for the Performing Arts, Greenwich Village, Central Park, and midtown Manhattan. Columbia athletic teams compete in the Ivy League.
Applying	For maximum consideration for admission, doctoral students should submit the following before January 1 for the autumn term and before October 1 for the spring term: an official application, transcripts, recommendations, and a $45 application fee. The deadline for application to the master's degree program is February 15 for the autumn term and October 1 for the spring. The General Test of the Graduate Record Examinations (GRE) is required for all graduate students; scores must be received prior to the deadline. Prospective students for the joint M.S./M.B.A. program must submit separate applications to the School of Engineering and Applied Science and the Graduate School of Business.
Correspondence and Information	For information about the programs: Admissions Committee Department of Industrial Engineering and Operations Research 331 Seeley W. Mudd Building Columbia University Mail Code #4704 New York, New York 10027 Telephone: 212-854-2942 E-mail: seasinfo.ie@columbia.edu	For applications: Office of Engineering Admissions 527 Seeley W. Mudd Building Columbia University New York, New York 10027 Telephone: 212-854-6438 E-mail: seasgradmit@columbia.edu

Columbia University

THE FACULTY AND THEIR RESEARCH

Department of Industrial Engineering and Operations Research

Daniel Bienstock, Ph.D. Combinatorial optimization and integer programming, parallel computing, applications to telecommunications.

Awi Federgruen, Ph.D. (Graduate School of Business). Modeling of stochastic systems, physical distribution management, dynamic programming.

Lisa Fleischer, Ph.D. Network flows, combinatorial optimization, approximation theory.

Guillermo Gallego, Ph.D. Inventory control, yield management, production planning, scheduling, semiconductor manufacturing.

Paul Glasserman, Ph.D. (Graduate School of Business). Stochastic systems, Monte Carlo simulation, mathematical and computational finance.

Donald Goldfarb, Ph.D. Algorithms for linear, quadratic, and nonlinear programming; network flows; large sparse systems.

Ioannis Karatzas, Ph.D. (Department of Mathematics). Stochastic differential equations, finance applications.

Morton Klein, Eng.Sc.D. Production planning, scheduling, statistical quality control.

Peter J. Kolesar, Ph.D. (Graduate School of Business). Service system management in the public and private sectors, quality management.

Steven Kou, Ph.D. Mathematical and computational finance, simulation, queuing theory, mathematical statistics.

Perwez Shahabuddin, Ph.D. Simulation, stochastic systems, fast simulation techniques, applications to communication systems.

Karl Sigman, Ph.D. Queuing theory, queuing networks, applied stochastic processes, point processes.

David Yao, Ph.D. Optimization and control of discrete event stochastic systems, queuing networks, manufacturing systems.

Associated Faculty in the Graduate School of Business

Julien Bramel, Ph.D. Vehicle routing, probabilistic analysis of algorithms, combinatorial optimization.

Sidney Browne, Ph.D. Stochastic systems, queuing.

Sebastian Ceria, Ph.D. Combinatorial optimization, integer programming, polyhedral combinatorics.

Linda Green, Ph.D. Stochastic systems, queuing, applications to service and production systems.

Garud Iyengar, Ph.D. Convex optimization, mathematical and computational finance, information theory, signal processing.

David W. Miller, Ph.D. Decision theory, utility theory, ruin models, explanatory statistical models.

Associated Faculty in the Department of Statistics

Victor de la Pena, Ph.D. Martingales, stopping times, sequential analysis, U-statistics.

Chris Heyde, Ph.D. Stochastic modeling, applied probability, asymptotic theory, inference for stochastic processes.

Adjunct Faculty

Kosrow Dehnad, Ph.D. Forecasting and finance.

A. Blanton Godfrey, Ph.D. Quality control and management.

Leon Gold, Ph.D. Human factors.

Peter Norden, Ph.D. Organization theory.

Gabor Pataki, Ph.D. Facilities layout and planning, applied integer programming.

Lucius Riccio, Ph.D. Operations research models in the public sector.

Paul Shapiro, Ph.D. Industrial information systems.

Dong Shaw, Ph.D. Deterministic models.

Henry Shum, Ph.D. Industrial economics.

Sheldon Weinig, Ph.D. Technology and policy issues in manufacturing enterprises.

David Williamson, Ph.D. Approximation algorithms.

Visiting Faculty

Xiuli Chao, Ph.D. Queuing theory, queuing networks and production management.

CURRENT AREAS OF RESEARCH

In industrial engineering, research is conducted in the design and control of manufacturing and service systems, including supply-chain management, inventory control, yield management, scheduling, and logistics. In operations research, new developments in mathematical programming, combinatorial optimization, queuing, reliability, simulation, mathematical and computational finance, and in both deterministic and stochastic network flows are being explored.

Projects are sponsored and supported by leading private firms and government agencies. In addition, students and faculty members are involved in the work of two NSF-funded research and educational centers: the Center for Applied Probability (CAP) and the Computational and Optimization Research Center (CORC). Both of these centers are principally supported by major grants from the National Science Foundation.

CAP is a cooperative center involving the School of Engineering and Applied Science, several departments in the Graduate School of Arts and Sciences, and the Graduate School of Business. Its main interests are in four applied areas: mathematical and computational finance, stochastic networks, logistics and distribution, and population dynamics. The center maintains a laboratory of workstations for exclusive use by graduate students. CORC is a center involving the IBM-Watson Research Center, Cornell University, the School of Engineering and Applied Science, the Graduate School of Arts and Sciences, and the Graduate School of Business. Its mission is the development of new theory and tools for the solution of computationally intensive optimization problems. It has its own parallel computer and workstations for graduate students and faculty members.

GEORGIA INSTITUTE OF TECHNOLOGY
A Unit of the University System of Georgia

School of Industrial and Systems Engineering

Programs of Study

The School offers five master's degrees and the Ph.D. The Master of Science in Industrial Engineering (M.S.I.E.) degree is available for students with undergraduate degrees in industrial engineering and for other engineers who have taken requisite courses. The Master of Science in Operations Research (M.S.O.R.) is a degree program intended for graduate students with strong interests in optimization, stochastic systems analysis, and operations research applications. The Master of Science in Statistics (M.S.S.) is administered by the School of Industrial and Systems Engineering, jointly with the School of Mathematics. The Master of Science in Health Systems (M.S.H.S.) degree program is intended for students who have a background in a scientific field such as engineering, mathematics, statistics, computer science, physical science, social science, or management science and who are interested in professional careers involving the analysis, design, and improvement of health-care management systems. The Master of Science (M.S.) is for students who desire to follow a program in human integrated systems. The primary objective of the Ph.D. program is to provide an educational experience for unusually gifted individuals that will help to prepare them for creative and productive careers. In addition to the Ph.D. in ISyE, qualified students may also seek a Ph.D. in algorithms, combinatorics, and optimization, which is administered jointly with the College of Computing and the School of Mathematics.

The focus of graduate educational and research activities is on structuring decision processes, systems analysis, and the design of complex systems that integrate technical, human, and economic resources within various constraints and environments. Graduate programs and research may emphasize specific areas of application, such as information and decision control systems, facilities planning and design, logistics, health systems, urban systems, economic decision analysis, production and distribution systems, human-machine systems, and materials handling.

Research Facilities

All students have on-campus and off-campus access to a variety of centrally maintained computers, including Sun Workstations and servers, IBM mainframes, IBM RISC workstations, supercomputers, and Macintosh computers. All systems are available on the campus network. The mainframes and workstations are also accessible via modem through Georgia Tech's dial-up network. In addition, the School maintains more than 250 UNIX workstations, Macintoshes, and PCs. ISyE's computer support is provided by 4 full-time systems support specialists and several part-time students who work for ISyE's Computer Coordinator's Office (CCO). CCO supports ISyE specific applications in the graduate and undergraduate computer laboratories. ISyE is committed to providing a state-of-the-art computing environment.

Six research centers are housed within the School: the Human-Machine Systems Research Center, the Health Systems Research Center, the Logistics Institute, the Statistics Center, the Technology Policy and Assessment Center, and the Center for Applied Probability (CAP).

Financial Aid

Financial assistance is available for highly qualified graduate students. Awards provide a basic stipend plus a stipend equal to the cost of current resident matriculation. Awards are given in one of the following forms: research assistantships, teaching assistantships, industrial fellowships and traineeships, and President's Fellowships.

Cost of Study

Tuition and fees are $5955 per academic semester in 1999–2000 for nonresidents of Georgia and $1749 per academic semester for residents. The rates for tuition and fees are subject to change at the end of any semester. Ordinarily, the nonresident tuition fee is waived for assistants, fellows, and trainees. Graduate students with assistantships pay reduced fees.

Living and Housing Costs

Dormitory rooms for unmarried students and apartments for married students are available at reasonable cost through the Institute. Rooms and apartments in privately owned dwellings within walking distance or a short driving distance are available for a range of prices. Assistance in locating housing is offered by the Institute's Housing Office.

Student Group

Approximately 12,000 students are enrolled at Georgia Tech. About 9,000 of these are undergraduate students and 3,000 are graduate students. The School of Industrial and Systems Engineering has an enrollment of 1,150 undergraduate students and 230 graduate students.

Student Outcomes

Graduates from the graduate program in ISyE take positions at many levels of industry, government, and academia. In the past few years, Ph.D. graduates are about evenly divided between academic positions and high-level (usually staff) positions in research organizations and groups as well as in consulting firms that have the need for people with specialized high-level skills. The master's degrees are viewed as professional (as opposed to research) degrees and are directed toward preparation for professional practice in areas such as manufacturing, logistics, production systems, applied statistics, operations research, industrial engineering, and health systems. Positions are with public- and private-sector organizations as well as with consulting firms.

Location

Atlanta is one of the most beautiful and exciting cities in the United States. Situated at an altitude of 1,050 feet above sea level, it is the second-highest major city in the country. Its topography is responsible for a favorable climate of moderate summers and mild winters. The annual mean temperature is 61.2°F. Atlanta's location has been an important factor in its development into the transportation, financial, and communications hub of the Southeast.

The School

The School of Industrial and Systems Engineering recognizes that study and research programs must not only prepare graduates to solve present-day problems in their chosen area but must also provide a sound basis for continued lifelong learning. The School has again been ranked as the number one IE graduate school by *U.S. News & World Report*. The School strives to educate students in such a way that they can keep pace with the rapidly developing technology of this age. Emphasis is placed upon fundamentals of science and mathematics, with necessary engineering design content bridging the theoretical courses and real-world applications. Within this framework, each student is encouraged to structure a program that meets his or her own goals.

Applying

Further information and application forms will be sent upon request.

Correspondence and Information

Dr. R. Gary Parker
Professor and Director of Academic Programs
School of Industrial and Systems Engineering
Georgia Institute of Technology
Atlanta, Georgia 30332-0205

Telephone: 404-894-4289
E-mail: gradstudies@isye.gatech.edu
World Wide Web: http://www.isye.gatech.edu

Georgia Institute of Technology

THE FACULTY AND THEIR RESEARCH

A. Russell Chandler III Endowed Chair
George L. Nemhauser, Institute Professor; Ph.D., Northwestern, 1961. Optimization.

Coca Cola Endowed Chair
Ellis L. Johnson, Professor; Ph.D., Berkeley, 1968. Integer programming, combinatorial optimization.

UPS and Regents' Professor
H. Donald Ratliff, Ph.D., Johns Hopkins, 1970; PE (Florida). Scheduling, network optimization.

Professors
Jerry Banks, Ph.D., Oklahoma State, 1966. Socioeconomic systems design and analysis.
Earl R. Barnes, Ph.D., Maryland, 1968. Linear programming and combinatorial optimization.
John J. Bartholdi III, Ph.D., Florida, 1977. Combinatorial optimization and scheduling.
Jiangang Dai, Ph.D., Stanford, 1990. Applied probability, processing networks.
Augustine O. Esogbue, Ph.D., USC, 1968. Operations research.
John J. Jarvis, Chair of the School; Ph.D., Johns Hopkins, 1968; PE (Georgia). Operations research.
Jack R. Lohmann, Associate Dean, College of Engineering; Ph.D., Stanford, 1979. Engineering economics, engineering education.
Leon F. McGinnis Jr., Ph.D., North Carolina State, 1974. Mathematical programming, engineering economic analysis.
Christine M. Mitchell, Ph.D., Ohio State, 1980. Human-machine systems.
R. Gary Parker, Director of Academic Programs; Ph.D., Kansas State, 1972. Discrete optimization.
Alan L. Porter, Ph.D., UCLA, 1972. Technology forecasting and assessment, social science methodology.
Richard F. Serfozo, Ph.D., Northwestern, 1969. Stochastic processes.
Alexander Shapiro, Ph.D., Ben Gurion (Israel), 1981. Statistics, stochastic systems and optimization.
Michael E. Thomas, Executive Vice President; Ph.D., Johns Hopkins, 1965. Operations research, mathematical programming.
Craig A. Tovey, Ph.D., Stanford, 1981. Discrete optimization, computational complexity. (Jointly appointed with the College of Computing).
John H. Vande Vate, Ph.D., MIT, 1984. Mathematical programming, computer science.

Associate Professors
Christos Alexopoulos, Ph.D., North Carolina, 1988. Stochastic systems and simulation.
Faiz A. Al-Khayyal, Ph.D., George Washington, 1977. Mathematical programming, global optimization.
Jane C. Ammons, Ph.D., Georgia, 1982. Industrial engineering, applied optimization.
Sigrun Andradottir, Ph.D., Stanford, 1990. Stochastic optimization, simulation, applied probability.
Willard R. Fey, M.S.E.E., MIT, 1961. System dynamics, economic decision analysis.
Robert D. Foley, Ph.D., Michigan, 1979. Stochastic systems.
Marc Goetschalckx, Ph.D., Georgia Tech, 1983. Logistics, supply chain, material handling, facilities design.
David Goldsman, Ph.D., Cornell, 1984. Operations research, applied statistics and simulation.
T. Govindaraj, Ph.D., Illinois at Urbana-Champaign, 1979. Man-machine systems, computerized automation.
Paul M. Griffin, Ph.D., Texas A&M, 1988. Automated visual inspection systems, image technologies, and manufacturing systems.
Steven T. Hackman, Ph.D., Berkeley, 1983. Production systems.
Anthony J. Hayter, Ph.D., Cornell, 1985. Statistics.
Russell G. Heikes, Ph.D., Texas Tech, 1972; PE (Georgia). Engineering statistics, simulation, optimization.
Alexander C. Kirlik, Ph.D., Ohio State, 1988. Human-machine systems.
Milena Mihail, Ph.D., Harvard, 1989. Theoretical computer science.
Renato Monteiro, Ph.D., Berkeley, 1988. Mathematical programming.
Justin A. Myrick, Ph.D., Missouri, 1974. Health-care and systems design.
Frank E. Roper Jr., Registrar; M.S.I.E., Georgia Tech, 1962. Industrial engineering.
Martin Savelsbergh, Ph.D., Rotterdam, 1988. Computational optimization.
Gunter P. Sharp, Ph.D., Georgia Tech, 1973; PE (Georgia). Operations research, feedback dynamics.
Mark Spearman, Ph.D., Texas A&M, 1986. Industrial engineering, manufacturing.
Kwok Tsui, Ph.D., Wisconsin, 1986. Statistics, quality control.
Chen Zhou, Ph.D., Penn State, 1988. Computer-integrated manufacturing.

Assistant Professors
Hayriye Ayhan, Ph.D., Texas A&M, 1995. Applied probability, stochastic processes.
Victoria Chung-Ping Chen, Ph.D., Cornell, 1993. Applied statistics, dynamic programming, operations research.
Lloyd W. Clarke, Ph.D., Pennsylvania, 1992. Modeling, logistics, computational optimization.
Douglas G. Down, Ph.D., Illinois at Urbana-Champaign, 1994. Stochastic process and systems.
Pinar Keskinocak, Ph.D., Carnegie Mellon, 1997. Operations research.
Anton Kleywegt, Ph.D., Purdue, 1996. Operations research.
Paul H. Kvam, Ph.D., California, Davis, 1990. Statistics.
Amy Pritchett, Ph.D., MIT, 1996. Human-machine systems.
Spyros Reveliotis, Ph.D., Illinois, 1996. Discrete event systems, manufacturing systems control, machine intelligence.

Adjunct Faculty
William B. Rouse, Adjunct Professor; Ph.D., MIT, 1972. Human-machine systems.

KETTERING UNIVERSITY

Master of Science in Engineering

Programs of Study	Focusing on practice rather than theory, the Master of Science in Engineering program emphasizes the application of knowledge and skills in modern industry and prepares students for rapidly changing technology. It is designed to be a terminal professional degree.

The three specialties of manufacturing systems, automotive systems, and mechanical design are supported by Kettering's long history of educating engineers for work in manufacturing and by the industrial experience of the faculty. The program requires 45 quarter credits and is available both full- and part-time. It can be completed in as little as one year of full-time study or in two years of part-time study. A thesis is not required. |
| **Research Facilities** | As a school that emphasizes the practical application of knowledge, Kettering takes pride in its laboratories. They include the Computer Integrated Manufacturing Laboratory, High Voltage Laboratory, Electrical Machines Laboratory, Laser Laboratories, Advanced Chemistry Laboratory, Design for Manufacturing Laboratory, Engineering Mechanics Laboratory, and the Automotive Laboratory.

There are three computer-aided design laboratories. One contains a group of Sun IPC Workstations using graphic software ranging from mechanical design to solid modeling. Another laboratory contains DEC GPX workstations, which are used for VLSI chip design and simulation. The third laboratory is equipped with 80386-based personal computers that handle graphics and CAD/CAM classes using AutoCAD software. |
Financial Aid	Assistantships and fellowships are available. Assistants are paid a stipend, and tuition is waived. Fellowships cover half of the student's tuition up to four graduate courses per term and normally include a room in Thompson Hall without charge. Also, a variety of loan programs are available to assist students.
Cost of Study	Costs are the same for in-state and out-of-state students—$410 per credit in 1999–2000, regardless of load. Textbooks are furnished without charge.
Living and Housing Costs	Graduate students can live in Thompson Hall or opt for off-campus housing. Thompson Hall ($1270 per quarter) is an attractive living facility located on campus and is a very short walk from classes. Plenty of off-campus housing is available through the Student Housing Network, most of it for singles. Furnished apartments are scarce. A cafeteria board contract ($710 per quarter) is available to all students.
Student Group	There are 87 students in the Master of Science in Engineering program—50 in manufacturing systems and 37 in mechanical design—and 754 students in the Master of Science in Operations Management program and the part-time Master of Science in Manufacturing Management program. The University has more than 2,200 undergraduates in totally cooperative programs in engineering and management—about 90 percent of them in engineering. These students are employed at more than 600 business locations throughout North America and several overseas countries.
Location	The University is located in Flint, Michigan, at the northern end of the bustling southeast Michigan industrial corridor. A good expressway system gives easy access to all of southeast Michigan, including Ann Arbor, metropolitan Detroit, and Lansing as well as Windsor, Ontario.
The University	Kettering (formerly GMI Engineering & Management Institute) is a private, independent college. (Until 1982, Kettering was General Motors Institute, a proprietary college.) Since 1919, Kettering has prepared students to become leaders in business and industry. The master's program is conducted by the academic departments with administrative support by various staff offices. There is no separate graduate school.
Applying	For regular admission to the degree program, applicants should hold a bachelor's degree in an appropriate engineering field from an ABET-accredited engineering program. Graduates of technology programs are not eligible for admission.

Applicants for full-time study should apply as early as possible, preferably before June. The absolute deadline is July 15. Kettering does not discriminate based on an individual's race, color, sex, creed, age, handicap, or national origin.

Prospective students should submit the following materials to the Admissions Office: an application form, official transcripts of all prior college work, two recommendations, and scores from the General Test of the GRE. Nonnative speakers of English must also submit TOEFL scores. |
| **Correspondence and Information** | Graduate Office
Kettering University
1700 West Third Avenue
Flint, Michigan 48504-4898
Telephone: 810-762-7494
 888-464-4723 (toll-free)
Fax: 810-762-9935
E-mail: bbedore@kettering.edu |

Kettering University

THE FACULTY AND THEIR RESEARCH

Kettering has a highly qualified faculty for the Master of Science in Engineering degree. Of the 30 professors who teach in the program, 26 hold a doctoral degree. Most have had industrial or extensive consulting experience. All have supervised theses jointly with representatives of business as part of the University's cooperative baccalaureate program and, through that activity, are familiar with industrial problems and needs.

Areas of research include lasers and optics, manufacturing systems, manufacturing processes, composite materials, production tools, vehicle suspension systems, automotive engine systems, and metallurgical imaging.

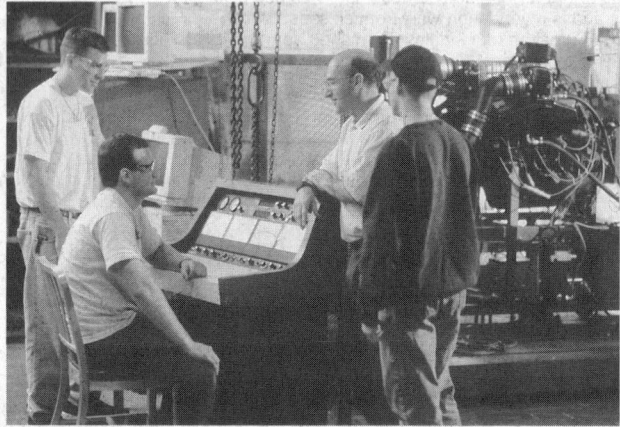

Kettering Automotive Engineering Laboratory.

Campus Center Building and Kettering Bell Tower.

Special events and activities occur in the Great Court of the Campus Center Building.

NORTH CAROLINA STATE UNIVERSITY

Department of Industrial Engineering

Programs of Study

The members of the graduate faculty in industrial engineering support a variety of academic and research interests leading to M.S. and Ph.D. degrees in the following areas: flexible manufacturing systems, production planning and control, simulation, optimization, real-time control, design for quality, inventory and scheduling theory, facilities design, furniture manufacturing and management, ergonomics, and occupational safety. Close working relationships exist with the Operations Research Program, the Integrated Manufacturing Systems Engineering Institute, and the Departments of Statistics, Computer Science, Economics, Textiles, Business Management, and Psychology.

The M.S. degree requires a total of 30 credit hours, including thesis. The M.I.E. degree requires 33 hours, with 6 hours of project work optional. The Ph.D. requires 30 hours of course work beyond the M.S., exclusive of hours for dissertation research. The above credit-hour requirements include a minor of 9 hours for the M.S. and 15 hours for the Ph.D. outside of but complementary to the area of the major. At the doctoral level, the minor can be distributed across related courses from business, psychology, public health, and other engineering disciplines. For superior students, a direct-track Ph.D. program is available; possession of the M.S. degree is not a prerequisite. Candidates for the M.S. and Ph.D. degrees are required to possess some breadth of exposure to core areas of industrial engineering. Apart from thesis and dissertation work, students are encouraged to pursue other independent study and research. Academic guidelines allow the student considerable latitude in the development of the plan of graduate work.

Research Facilities

The Management Systems Lab has a 4-station Hytrol conveyor system, a 4-station Lanco automatic assembly system, and a range of PLCs and MMIs. The Manufacturing Processes Lab provides CNC turning and vertical machining centers. The Rapid Prototyping/Metrology Lab includes a Sanders 3-D (inkjet technology) rapid prototyping machine, a JP systems laminated object machine, and a Brown & Sharpe coordinate measurement machine. The Logistics Lab consists of networked PCs running proprietary CAPS Logistics Supply Chain Designer and Route Pro and other logistics software. The Ergonomics Lab is fully equipped with state-of-the-art systems for conducting basic and field research in biomechanics, work physiology, and human performance. All research labs have access to a wide variety of software through a campus network of 1100 SUN and NT workstations running UNIX and Windows NT.

Financial Aid

The majority of U.S. applicants receive financial awards. Teaching assistantship stipends are $9000 per academic year. Research assistantship stipends are $12,000 per academic year. U.S. applicants with superior credentials are eligible for $3000 or $6000 fellowships that supplement assistantship stipends. Scholarship and fellowship support is available to qualified minority applicants.

Cost of Study

For the 1998–99 academic year, tuition and fees were $2200 for North Carolina residents and $11,800 for nonresidents. An out-of-state student appointed as a teaching or research assistant who is a U.S. citizen qualifies for in-state tuition rates. All research and teaching assistants with half-time appointments qualify for in-state tuition and health insurance. International applicants must certify the availability of at least $19,371 for the first year of study.

Living and Housing Costs

Dormitory facilities for unmarried graduate students and University housing for married students are available on campus. Numerous apartment complexes exist near campus with convenient bus service. The cost of living in Raleigh is moderate and below the national average.

Student Group

Campus enrollment currently exceeds 27,000. The College of Engineering enrolls approximately 5,600 undergraduate and 1,200 graduate students. Within the department, there were 69 active graduate students during the 1998–99 academic year (31 for the Ph.D., 23 for the M.S., and 15 for the M.I.E.); of these, 17 were women, and 17 were international students. All full-time, U.S. students desiring financial aid were supported. Students come from across the U.S. and from eight other countries. The department's placement record for its graduates, as well as for summer internships and co-op opportunities, is excellent. Most master's-level graduates are employed in industry and business; Ph.D. graduates are often employed in academic and research positions.

Location

Raleigh, the state capital, with a population of about 400,000, is located at one corner of the Research Triangle, defined by the location of Duke University in Durham, the University of North Carolina at Chapel Hill, and NCSU. Nearby is the Research Triangle Park, one of the fastest-growing research and high-technology industry centers in the world. The state of North Carolina ranks tenth in the nation in industrial activity. The Atlantic Ocean and the Blue Ridge Mountains are within a 3-hour drive.

The University and The Department

NCSU is the land-grant university of North Carolina and the largest of its state-supported institutions. Both faculty members and students have unique opportunities to pursue research with industrial sponsors in the Research Triangle area. Seminars and social events involving both students and faculty members are an integral part of the program. The close proximity of NCSU to ten area colleges and universities, as well as to a host of high-technology research and manufacturing facilities, offers student spouses an uncommon variety of opportunities for academic study or employment.

Applying

The deadline for fall semester entry is June 25 for U.S. citizens and May 1 for international applicants. Priority for financial awards goes to applicants as of March 1. GRE scores are required of all applicants, as is the TOEFL for all international applicants whose native language is not English. Applications from minorities are welcomed. The application fee cannot be waived for international applicants. International applicants must meet minimum standards of scores of 450, 680, and 550, respectively, on the verbal, quantitative, and analytical portions of the GRE General Test; 575 on the TOEFL; and 5.0 on the Test of Written English. For application materials, students should contact the graduate secretary at 919-515-6410.

Correspondence and Information

Associate Director of Graduate Programs
Department of Industrial Engineering
North Carolina State University
Raleigh, North Carolina 27695-7906
Telephone: 919-515-2364
Fax: 919-515-5281

For application packet, please contact:
Ms. Linda White
Department of Industrial Engineering
Box 7906
Raleigh, North Carolina 27695-7906
E-mail: white@page.ncsu.edu
World Wide Web: http://www.ie.ncsu.edu

North Carolina State University

THE FACULTY AND THEIR RESEARCH

Mahmoud A. Ayoub, Professor; Ph.D., Texas Tech, 1970. Biomechanics, work physiology, occupational safety, simulation, human performance.

Richard H. Bernhard, Professor; Ph.D., Cornell, 1961. Engineering economics, cost analysis and control, decision theory.

Denis R. Cormier, Assistant Professor; Ph.D., North Carolina State, 1995. Concurrent engineering, computer-integrated manufacturing, computer-aided process planning.

C. Thomas Culbreth, Henry A. Foscue Professor and Director, Furniture Manufacturing and Management Center; Ph.D., North Carolina State, 1984. Manufacturing processes, facilities design, computer-assisted process planning.

Salah E. Elmaghraby, University Professor; Ph.D., Cornell, 1958. Production systems, scheduling, activity networks, dynamic programming.

Shu-Cherng Fang, Walter Clark Professor; Ph.D., Northwestern, 1979. Computer-aided manufacturing, factory data networking, communications network design, operations research.

Yahya Fathi, Associate Professor; Ph.D., Michigan, 1979. Production planning, scheduling, inventory theory, mathematical programming.

Thom J. Hodgson, James T. Ryan Professor; Ph.D., Michigan, 1970. Scheduling theory, production control, inventory theory, automated production.

Michael G. Kay, Associate Professor, Ph.D., North Carolina State, 1992. Facilities layout and location, material handling, real-time control, intelligent machines and systems.

Russell E. King, Professor; Ph.D., Florida, 1986. Production control, automated production, stochastic processes, real-time control.

Yuan-Shin Lee, Associate Professor; Ph.D., Purdue, 1993. Computer-aided and computer-integrated manufacturing, process planning, multiaxis numerical controlled machining, computational geometry for design and manufacturing.

Wilbur L. Meier Jr., Professor; Ph.D., Texas, 1967. Systems engineering, logistics, planning, systems management, operations research.

Gary A. Mirka, Associate Professor; Ph.D., Ohio State, 1992. Biomechanical analysis, dynamic asymmetric lifting, spine biomechanics, trunk electromyography.

Henry L. W. Nuttle, Professor; Ph.D., Johns Hopkins, 1968. Planning, scheduling, and control; operations research.

Richard G. Pearson, Professor; Ph.D., Carnegie Mellon, 1961. Skilled performance, fatigue, safety, occupational and environmental stress.

Stephen D. Roberts, Professor; Ph.D., Purdue, 1968. Simulation language design, simulation modeling, software engineering.

Ezat T. Sanii, Associate Professor; Ph.D., Purdue, 1982. Manufacturing processes, flexible manufacturing systems, automation, computer-aided process planning, CAD/CAM integration, industrial robotics.

William A. Smith Jr., Professor; D.Eng.Sc., NYU, 1966. Decision and control systems, organizational planning, productivity/quality.

Carolyn M. Sommerich, Assistant Professor; Ph.D., Ohio State, 1994. Occupational ergonomics, biomechanics, biomechanical modeling, musculoskeletal disorders.

James B. Taylor, Assistant Professor; Ph.D., Purdue, 1995. Manufacturing processes, precision engineering, metrology, agile manufacturing systems, positioning systems, concurrent engineering.

James R. Wilson, Professor and Head; Ph.D., Purdue, 1979. Design and analysis of large-scale simulation experiments, analysis of production systems.

Robert E. Young, Professor; Ph.D., Purdue, 1977. Computer-integrated manufacturing (CIM) system design, manufacturing automation.

RESEARCH AREAS

Ergonomics. Areas of current research include human performance, biomechanical evaluation of workspaces, cumulative trauma disorders, ergonomic tool design, postural stress, prevention of hearing loss, and environmental stressors. A portion of the research in occupational biomechanics is conducted at the industrial site of corporate sponsors.

Ergonomics Resource Center. The Ergonomics Resource Center (ERC) seeks to improve productivity, safety, and well-being of the people of North Carolina in all sectors of business and industry, including government. The ERC supports business and labor in the pursuit of humanizing the workplace and in maintaining viable and competitive jobs. By emphasizing applied research and timely delivery of programs, the ERC identifies, analyzes, and corrects ergonomic deficiencies in the workplace. Its primary goal is to act as a bridge for technology transfer and information exchange between the University, state agencies, and industry. ERC is jointly sponsored by the North Carolina Department of Labor and North Carolina State University.

Furniture Manufacturing and Management Center. This is an industry/University cooperative organization involved in applied research and technology transfer efforts on behalf of the furniture industry. North Carolina is the leading producer of furniture in the United States, with annual shipments exceeding $4 billion. Financial support for the Center is provided by the Furniture Foundation in collaboration with the American Furniture Manufacturers Association. Research involves the development of flexible automation, computer-aided engineering and design systems, manufacturing scheduling/control systems, and applied ergonomic studies.

Manufacturing Systems. Faculty members in this area have funded projects in flexible manufacturing, CAD/CAM, robotics, artificial-intelligence-based scheduling for shop floor control, process control, CIM, CIM systems design, and concurrent engineering.

Production Systems. In this area, current research activity is focused on the topics of production planning, scheduling, control of automated manufacturing systems, facility location and design, vehicle routing and scheduling, project scheduling, and the design of production systems. Faculty members in this area interact on several projects with faculty members from operations research and the Integrated Manufacturing Systems Engineering Institute. Other active projects involve quality control, simulation, and logistics applications in manufacturing.

Systems Analysis and Optimization. Faculty members in this area investigate general problem-solving methodology with strong emphasis on the use of mathematics, statistics, and computer science. Recent developments include interior point methods for linear programming, heuristics for combinatorial optimization, stochastic models for intelligent vehicles, nonlinear programs to minimize product variability, object-oriented simulation, metamodeling, neural networks, fuzzy systems, and aggregate decision modeling.

THE RESEARCH TRIANGLE

North Carolina State University, Duke University, and the University of North Carolina at Chapel Hill are the vertices of the triangle after which the Research Triangle is named. Within the triangle is the 6,700-acre Research Triangle Park, which is the largest planned research park in the world. In the park, more than 30,000 people work in more than fifty highly acclaimed institutions and corporations. The recent addition of 2,700 acres provides space for forty more companies and 20,000 more employees. There is significant cooperation between NCSU and the Triangle companies and institutions. The Research Triangle Institute is a subsidiary of the three universities and is one of the largest and most successful contract-research organizations in the country. The Research Triangle has the highest concentration of Ph.D.'s of any comparable region in the United States. The region is consistently rated as one of the most desirable places to live in the United States.

NORTHWESTERN UNIVERSITY

Robert R. McCormick School of Engineering and Applied Science
Department of Industrial Engineering and Management Sciences

Programs of Study

The Department of Industrial Engineering and Management Sciences offers three graduate degrees: the Doctor of Philosophy (Ph.D.), the Master of Science (M.S.), and the Master of Engineering Management (M.E.M.).

The Ph.D. program prepares students for research- and technology-based careers in industry and academia. Graduates hold a variety of jobs in industrial and consulting firms, as well as faculty positions at numerous universities. Ph.D. students are accepted with a bachelor's or master's degree in a relevant discipline. During their first year, students take a set of core courses to provide fundamental knowledge; then, in consultation with their advisers, they design the balance of their course program in accordance with their background, objectives, and research interests, specializing in one of the following major areas of study: applied probability and simulation, economics and decision analysis, production and logistics, optimization, organization theory and systems analysis, and statistics and decision analysis. The program often includes graduate courses from other University departments in subjects such as mathematics, economics, engineering, management, and the social sciences. Students must also satisfactorily complete a departmental comprehensive examination at the end of the first year, a University qualifying examination for admission to candidacy, and a Ph.D. dissertation.

Those entering with a bachelor's degree may receive an M.S. after completing the first year of Ph.D. studies and exams. Students may also be admitted to pursue only the M.S. degree, in which case they take one year of course work—half devoted to core courses and half tailored to their particular interests—plus the first-year comprehensive exam. No M.S. thesis is required.

The M.E.M. program prepares people currently working in engineering, research and development, and related fields for increased responsibility in engineering management. Northwestern's approach blends the basics of management (accounting, economics, finance, and marketing), quantitative analysis (operations research, simulation, systems theory, and statistics), and behavioral science (organization theory and social systems) with engineering electives in the student's area of interest. The program is designed for part-time students, with courses scheduled mainly in the evening.

Research Facilities

The department maintains an extensive research computing network, as well as specialized facilities that include a Quality Engineering Laboratory and a Production and Logistics Laboratory. Faculty members and students from a variety of backgrounds are brought together at interdisciplinary centers established by the University and the School of Engineering and Applied Science, including the Transportation Center and the Optimization Technology Center. Argonne National Laboratory, located in suburban Chicago, also maintains ties with the department in such fields as optimization and decision analysis and has employed a number of Ph.D. students during or after their studies.

Financial Aid

Most full-time students in the Ph.D. program receive some form of financial aid. First-year Ph.D. students are eligible for Walter P. Murphy Fellowships and Royal E. Cabell Fellowships, which provide a tuition scholarship and a monthly stipend. Financial aid after the first year generally takes the form of a research or teaching assistantship. Neither fellowships, teaching assistantships, nor research assistantships are available to students in the Master of Science or the Master of Engineering Management programs. However, a loan program is available for all U.S. graduate students.

Cost of Study

In 1999–2000, tuition is $7266 per quarter; there are three quarters in the academic year. Books and materials cost an estimated $1300 per year.

Living and Housing Costs

The University operates Engelhart Hall, an apartment building for married and single graduate students. Students interested in other accommodations are assisted by the Off-Campus Housing Office. Living costs for a single student should be budgeted at about $9480 per year.

Student Group

The department's graduate enrollment is approximately 50 students, about half of whom are international students and about a quarter of whom are women. In addition, the department has about 100 part-time M.E.M. students and more than 200 undergraduates.

Location

Northwestern University has lakefront campuses in Chicago and in Evanston, a suburban residential community north of Chicago. The Technological Institute building, which houses the Department of Industrial Engineering and Management Sciences, is located on the Evanston campus.

The University

Northwestern University is a major private research university with a faculty of 2,100 and a student body of 13,000.

Applying

When applying to the Ph.D. or the M.S. program, students should use the Graduate School's Application for Admission and Financial Aid, which is available from the department at the address below. Under "Matriculation Data" on the application form, the department should be specified as Industrial Engineering and Management Sciences. The "Area of Specialization" is optional; it may be used to indicate an interest in a particular major area of study. Individuals may apply and be admitted at any time, although winter application for admission for the following fall quarter is preferred. Scores on the GRE General Test and two letters of recommendation are also required; applicants whose native language is not English must take the Test of English as a Foreign Language (TOEFL). As soon as all application materials are received, the application is processed, and the applicant is promptly informed of the decision.

Applicants planning on Ph.D. study may indicate on the Application for Admission and Financial Aid that they wish to be considered for financial aid. To ensure full consideration for financial aid, the application must be submitted by February 15. Murphy and Cabell Fellowship awards are announced in March or early April.

Applicants for the Master of Engineering Management program should request or download the M.E.M. Application Form from the address below.

Correspondence and Information

For M.S. and Ph.D. program information:

Graduate Program Coordinator
Department of Industrial Engineering
   and Management Sciences
Northwestern University
2145 Sheridan Road, Room C210
Evanston, Illinois 60208-3119

Telephone: 847-491-4394
Fax: 847-491-8005
E-mail: graduate@iems.nwu.edu
World Wide Web: http://www.iems.nwu.edu/programs/
   doctoral/

For M.E.M. program information:

Associate Director, M.E.M. Program
Department of Industrial Engineering
   and Management Sciences
Northwestern University
2145 Sheridan Road, Room C120
Evanston, Illinois 60208-3119

Telephone: 847-491-5584
Fax: 847-491-8005
E-mail: mem@iems.nwu.edu
World Wide Web: http://www.iems.nwu.edu/MEM/

Northwestern University

THE FACULTY AND THEIR RESEARCH

Bruce Ankenman, Ph.D. (industrial engineering), Wisconsin. Quality engineering and quality improvement, design of experiments, response surface methodology, applied statistical methods, engineering design, manufacturing.

Collette R. Coullard, Ph.D. (industrial engineering and management sciences), Northwestern. Operations research, combinatorial optimization, polyhedral combinatorics, graph theory, matroid theory.

Mark S. Daskin, Department Chair; Ph.D. (civil engineering), MIT. Logistics, facility location, vehicle routing, production planning.

Robert Fourer, Ph.D. (operations research), Stanford. Large-scale optimization, modeling languages and systems for optimization.

Donald N. Frey, Ph.D. (metallurgical engineering), Michigan. Information science and management of high-technology industry, organization theory, systems engineering, manufacturing.

Aaron J. Gellman, Director, Transportation Center; Ph.D. (economics), MIT. Management and utilization of technology, the process of innovation, transportation economics and policy.

Gordon Hazen, Ph.D. (industrial engineering), Purdue. Multiple-objective decision making, decision analysis, medical decision making.

Wallace J. Hopp, Ph.D. (industrial and operations engineering), Michigan. Stochastic optimization, applied operations research, production control.

Arthur P. Hurter Jr., Ph.D. (economics), Northwestern. Economics, production, facility investment, replacement, location.

Seyed M. R. Iravani, Ph.D. (industrial engineering), Toronto. Stochastic modeling and its applications in production control and analysis of manufacturing systems, queuing theory.

Gilbert K. Krulee, Ph.D. (psychology), MIT. Artificial intelligence, mathematical linguistics, psycholinguistics, computer processing of natural language.

Sanjay Mehrotra, Ph.D. (operations research), Columbia. Large-scale mathematical programming, parallel computation, production systems design and control.

Barry Nelson, Ph.D. (industrial engineering), Purdue. Computer simulation of stochastic systems, stochastic processes, statistics.

Jorge Nocedal, Ph.D. (mathematical sciences), Rice. Nonlinear optimization, numerical analysis, scientific computing.

David Simchi-Levi, Ph.D. (operations research), Tel-Aviv. Design and control of distribution and production systems, inventory theory, location theory, telecommunication networks, vehicle routing problems.

Ajit C. Tamhane, Ph.D. (statistics and operations research), Cornell. Mathematical and applied statistics.

Charles W. N. Thompson, Ph.D. (industrial engineering and management sciences), Northwestern. Organizational sciences, systems engineering.

Mark P. Van Oyen, Ph.D. (electrical engineering/systems), Michigan. Stochastic analysis and control of queuing networks, scheduling, information networks.

REPRESENTATIVE FACULTY PUBLICATIONS

Bisgaard, S., and **B. E. Ankenman.** Analytic parameter design. *Qual. Eng.* 8:75–91, 1995.

Owen, J. H., **C. R. Coullard,** and D. S. Dilworth. GIDEN: A graphical environment for network optimization. In *Sixth Industrial Engineering Research Conference Proceedings,* pp. 507–12, eds. G. L. Curry, B. Bidanda, and S. Jagdale. Institute of Industrial Engineers, 1997.

Hsu, V. N., **M. S. Daskin,** P. C. Jones, and T. J. Lowe. Tool selection for optimal part production: A Lagrangian relaxation approach. *IIIE Trans.* 27:417–26, 1995.

Fourer, R. Extending a general-purpose algebraic modeling language to combinatorial optimization: A logic programming approach. In *Advances in Computational and Stochastic Optimization, Logic Programming, and Heuristic Search: Interfaces in Computer Science and Operations Research,* pp. 31–74, ed. D. L. Woodruff. Kluwer Academic Publishers, 1998.

Frey, D. The new dynamism: Part 1, *Interfaces* 24(2):87–91, 1994. Part 2, *Interfaces* 24(3):105–8, 1994. Part 3, *Interfaces* 24(5):36–40, 1994.

Felli, J. C., and **G. B. Hazen.** Sensitivity analysis and the expected value of perfect information. *Med. Decis. Making* 18:95–109, 1998.

Hopp, W. J., M. L. Spearman, and R. Q. Zhang. Easily implementable (Q, r) inventory control policies. *Operations Res.* 45:327–40, 1997.

Hurter, A. P., and M. Van Buer. The newspaper distribution problem. *J. Business Logistics* 17:85–96, 1996.

Iravani, S. M. R., and M. J. M. Posner. An M/G/1 queue with cyclic service times. *Queueing Syst.* 22:145–69, 1996.

Mehrotra, S., and R. A. Stubbs. Predictor-corrector methods for a class of linear complementarity problems. *SLAM J. Optimization* 4:441–53, 1994.

Cario, M. C., and **B. L. Nelson.** Numerical methods for fitting and simulating autoregressive-to-anything processes. *INFORMS J. Comput.* 10:72–81, 1997.

Nocedal, J. Large scale unconstrained optimization. In *The State of the Art in Numerical Analysis,* pp. 311–38, eds. A. Watson and I. Duff. Oxford University Press, 1997.

Bertsimas, D., and **D. Simchi-Levi.** The new generation of vehicle routing research: Robust algorithms addressing uncertainty. *Operations Res.* 44:286–304, 1996.

Tamhane, A. C., Y. Hochberg, and C. W. Dunnett. Multiple test procedures for dose finding. *Biometrics* 52:21–37, 1996.

Thompson, C. W. N. Intermediate performance measures in engineering projects. In *Proceedings of the Portland International Conference on Management of Engineering and Technology,* 1997.

Van Oyen, M. P. Monotonicity of optimal performance measures for polling systems. *Probability Eng. Info. Sci.* 11:219–28, 1997.

POLYTECHNIC UNIVERSITY

Manufacturing Engineering Program

Program of Study

The M.S. programs in manufacturing and industrial engineering are offered on both the Brooklyn and the Westchester campuses of Polytechnic University. Some course work may have to be completed on the Brooklyn campus for the industrial engineering program. The programs may be taken either full-time or part-time. All classes are held in the evening. A full-time student can finish in one calendar year. Credit may be granted for up to three graduate courses taken elsewhere with a grade of B or better. The programs draw upon the talents of the manufacturing and industrial engineering faculty members, all of whom have extensive industry experience, and upon Polytechnic's long-established strengths in other fields of engineering and science and in management. The programs are open to those holding an accredited bachelor's degree in engineering (B.S. or B.E.); to graduates in physics, chemistry, materials sciences, the biological sciences, operations research, and economics; and to M.B.A. graduates.

The programs offer an internship option to full-time students. Students gain real-world experience and 6 of the 36 credits required for the master's degree by working in a cooperating firm in either the Brooklyn or the Westchester areas. Students who elect the internship option begin their program in the fall or spring semester with one full semester of academic work at Polytechnic. During that semester, they interview for an internship assignment.

Polytechnic's approach is based on the premise that a fully effective manufacturing or industrial engineer must understand all facets of the modern American manufacturing enterprise: manufacturing financials, design, quality engineering, managing and optimizing production, and overcoming cultural barriers to change. Each of these topics is covered in a core course and in follow-up courses that develop each subject in depth. In each subject area, the focus is on methods that are generally applicable and portable to most manufacturing circumstances.

The successful introduction of new methods requires understanding resistance to the introduction of new methods. This problem is addressed specifically in courses such as managing the human side of technological change, building high performance teams, and product realization process.

To an increasing extent, firms expect their engineers to take the initiative in improving productivity. Hence Polytechnic's programs emphasize methods for productivity improvement that students can identify and implement themselves. Indeed, a number of part-time students at Polytechnic have made significant contributions to their firms during the course of their studies. The program requirement consists of 36 credits (twelve 3-credit courses or the equivalent). Most students in the Manufacturing Engineering Program take the six core courses in manufacturing engineering (manufacturing systems engineering, computer integrated manufacturing, design for manufacturability, quality control and improvement, production science, and managing the human side of technological change, 15 credits) and complete a master's project (3 or 6 credits) based on work done in an internship program or a project at Polytechnic University. The balance of credits come from a concentration that the student designs with the help of a faculty adviser. Most students in the Industrial Engineering Program take the five core courses in industrial engineering (quality control and improvement, production planning and control, project planning and control, facility planning and design, and factory simulation) and complete the degree by taking electives from specific concentration areas or other departments at Polytechnic University. Students may choose electives that are pertinent to the student's needs or intended career path.

Research Facilities

Students have access to modern computational and simulation facilities as well as to laboratory equipment.

Financial Aid

For more information on how to apply for fellowships under this program, students should write to Professor Bartlett at the address given below.

Cost of Study

In 1998–99, graduate tuition was $675 per credit.

Living and Housing Costs

Dormitory rooms for single students are maintained in facilities near the Brooklyn campus. The cost in 1998–99 was approximately $600 to $700 per month. Private housing rentals are available in the Brooklyn area for single and married students in nearby Brooklyn Heights from $400 per person per month.

Student Group

Of the 65 graduate students in the program in 1998–99, 55 were part-time. Many are employed in the greater New York metropolitan area. Many states and countries are represented in the graduate student body.

Location

The Brooklyn campus is located in the new $1-billion MetroTech complex in downtown Brooklyn, just across the river from downtown Manhattan and within easy reach of one of the world's greatest concentrations of cultural, educational, entertainment, and sports activities.

The Westchester campus is located in Hawthorne, New York, in pleasant suburban surroundings. Several major highways provide easy access to New York City and various cultural and recreational areas in upstate New York. All of the courses in the Manufacturing Engineering Program are offered on both the Brooklyn and Westchester campuses.

The University

Polytechnic is a private, coeducational, technological university founded in 1854 as the Polytechnic Institute of Brooklyn. Polytechnic is one of the major technological universities in the country.

Applying

Applicants are encouraged to apply at any time and should include a resume with their applications.

Correspondence and Information

For more information, students can write to the address below or look up the department on Polytechnic's home page.

Manufacturing Engineering Program
Six MetroTech Center
Polytechnic University
Brooklyn, New York 11201
Telephone: 718-260-3106
Fax: 718-260-3216
World Wide Web: http://www.poly.edu

C. J. Bartlett, Director
Graduate Program in Manufacturing Engineering
Polytechnic University
Six Metrotech Center
Brooklyn, New York 11201
Telephone: 718-260-3106
Fax: 718-260-3216

Polytechnic University

THE FACULTY AND THEIR RESEARCH

All members of the faculty have extensive industrial experience and are currently active in their areas of expertise.

Charles B. Hoover Jr., Professor and Program Head; Ph.D., Yale, 1954; former Executive Director, AT&T Bell Labs. Manufacturing systems engineering, manufacturing education improvement, manufacturing economics, production flows, design.

Nathan Levine, Industry Professor; Ph.D., Illinois at Urbana-Champaign, 1957; former Director, AT&T Bell Labs. Quality control, quality improvement, process control, robust design.

Charles J. Bartlett, Industry Professor; Ph.D., MIT, 1966; former director, AT&T Bell Labs. Electronics systems design and manufacture, production science, manufacturing systems engineering.

Steven Bernstein, Adjunct Professor; M.S., Michigan, 1965; Member, Technical Staff, AT&T Bell Labs. Physical design.

David Fleck, Adjunct Professor; M.A.A.B.S, The Leadership Institute of Seattle, 1992. High-performance teams.

Michael Greenstein, Adjunct Professor; M.B.A., Louisville, 1977; Engineering Manager, AT&T Submarine Cable Systems. Design for manufacturability, engineering standards, safety systems.

S. Rajaram, Adjunct Professor; Ph.D., SUNY at Buffalo, 1980; Distinguished Member, Technical Staff, AT&T Bell Labs. Thermal design, thermal performance and reliability, environmental stress testing.

Reuven Shapira, Adjunct Professor; M.S., M.B.A., Tel Aviv, 1982. Quality process management, production planning and control.

David Soukup, Adjunct Professor; M.S., Tennessee, 1980. Factory simulation, project planning and control.

John Thomas, Adjunct Professor; M.B.A., Rochester, 1983; M.S., Polytechnic, 1992; Manager, Tool Engineering, Gould Division, Risdon, North America. Product realization process, manufacturing resources planning, project planning and control.

Blair Williams, Adjunct Professor; M.B.A., Loyola-Chicago, 1979; AT&T Bell Labs. Manufacturing and engineering.

Polytechnic University campus in Hawthorne, New York.

RENSSELAER POLYTECHNIC INSTITUTE

Department of Decision Sciences and Engineering Systems

Programs of Study

Rensselaer recognized the need for a new endeavor in decision sciences by forming a unique interdisciplinary department, Decision Sciences and Engineering Systems (DSES), within the Schools of Engineering, Management, and Science. The objectives of the department are to conduct research that leads to a better understanding of how information technology and quantitative analysis and modeling can support individuals, groups, and organizations in problem solving and decision making and to prepare engineers to design, develop, and implement complex decision support systems. In order to accomplish these objectives, knowledge from disciplines such as systems and industrial engineering, statistics, probability, operations research, information systems, artificial intelligence, computer science, and economics must be extended and integrated. The department combines degree programs in industrial and management engineering, manufacturing systems engineering, and operations research and statistics.

Concentrations available through DSES master's programs include data mining, financial engineering, information systems, management, quality engineering, service systems, simulation, systems engineering, systems modeling, and transportation systems. The Master of Science program requires 30–33 credit hours, including required and elective courses. Among the required courses is a 6-credit-hour master's project or thesis. The M.Eng. program is a nonthesis degree intended for professional practice. A student with an accredited B.S. or its equivalent can typically complete this 30–33 credit program in one year.

The Master of Science and Master of Engineering program in operations research and statistics focus on mathematical modeling and statistical techniques applicable to a wide range of practical problems connected with business, economic, social, and engineering systems. Students learn the theories, methodologies, and applications that underlie a range of analytical and optimization approaches. The program requires 30–33 credit hours, which may include 6 credit hours of a research project.

The Master of Science and Professional Master of Engineering programs in manufacturing systems engineering require 30–33 credit hours of study and focus on manufacturing issues, with concentrations in manufacturing quality systems, manufacturing systems modeling, manufacturing processes and technology, manufacturing information systems, and management systems. The program prepares students to deal with problems related to product design for manufacturability, operations analysis, the design of integrated manufacturing systems (planning, scheduling, and control), financial, organizational, and strategic management issues, and many elements of automation.

The Doctor of Philosophy program in decision sciences and engineering systems requires 90 credit hours of graduate studies (60 credits after a master's degree) with specialization either in industrial engineering, information systems, manufacturing systems engineering, operations research, or statistics. In addition to the course work, the program includes a qualifying examination, concentration area requirements, a candidacy examination, and a dissertation requirement.

Research Facilities

The DSES department has two strategic research thrusts: intelligent manufacturing and service systems, and data mining and decision support. The department is located in the George M. Low Center for Industrial Innovation, which also houses the Center for Advanced Technology in Automation, Robotics and Manufacturing; Center for Integrated Electronics and Electronics Manufacturing; and Rensselaer's Electronic Agile Manufacturing Research Institute. This proximity provides a symbiotic environment conducive to excellence in research. Other resources include an advanced computer graphics facility, a general industrial engineering laboratory, a teaching factory emulation laboratory, and an artificial intelligence facility.

Financial Aid

Financial aid is available in the form of teaching assistantships and scholarships. The stipend for assistantships ranges up to $10,980 for the 1999–2000 academic year. Tuition waivers may also be granted. Outstanding students may qualify for University-supported Rensselaer Scholar Fellowships, which carry a stipend of $15,000 and a full tuition scholarship. Low-interest, deferred-repayment graduate loans are also available for U.S. citizens with demonstrated need.

Cost of Study

Tuition for 1999–2000 is $665 per credit hour. Other fees amount to approximately $535 per semester. Books and supplies cost about $1700 per year.

Living and Housing Costs

The cost of rooms for single students in residence halls or apartments ranges from $3356 to $5298 for the 1999–2000 academic year. Family student housing apartments, with monthly rents of $592 to $720, are available.

Student Group

There are about 4,300 undergraduates and 1,750 graduate students representing all fifty states and more than eighty countries at Rensselaer.

Student Outcomes

Eighty-eight percent of Rensselaer's 1998 graduate students were hired after graduation with starting salaries that averaged $56,259 for master's degree recipients and $57,000–$75,000 for doctoral degree recipients.

Location

Rensselaer is situated on a scenic 260-acre hillside campus in Troy, New York, across the Hudson River from the state capital of Albany. Troy's central Northeast location provides students with a supportive, active, medium-sized community in which to live; an easy commute to Boston, New York, and Montreal; and some of the country's finest outdoor recreation sites, including Lake George, Lake Placid, and the Adirondack, Catskill, Berkshire, and Green mountains. The Capital Region has one of the largest concentrations of academic institutions in the United States. Sixty thousand students attend fourteen area colleges and benefit from shared activities and courses.

The University

Founded in 1824 and the first American college to award degrees in engineering and science, Rensselaer Polytechnic Institute today is accredited by the Middle States Association of Colleges and Schools and is a private, nonsectarian, coeducational university. Rensselaer has five schools—Architecture, Engineering, Management, Science, and Humanities and Social Sciences. The School of Engineering ranks among the top twenty engineering schools nationally by the *U.S. News & World Report* survey and is ranked in the top ten by practicing engineers.

Applying

Admissions applications and all supporting credentials should be submitted well in advance of the preferred semester of entry. GRE General Test scores are required. The application fee is $35. Since the first departmental awards are made in February and March for the next full academic year, applicants requesting financial aid are encouraged to submit all required credentials by February 1 to ensure consideration.

Correspondence and Information

For written information about graduate study:

Student Operations Coordinator
Department of Decision Sciences
    and Engineering Systems
Rensselaer Polytechnic Institute
110 8th Street
Troy, New York 12180-3590
Telephone: 518-276-6681
E-mail: dses@rpi.edu
World Wide Web: http://www.rpi.edu/dept/
    dses/www/homepage.html

For applications and admissions information:

Director of Graduate Academic and Enrollment
    Services, Graduate Center
Rensselaer Polytechnic Institute
110 8th Street
Troy, New York 12180-3590
Telephone: 518-276-6789
E-mail: grad-services@rpi.edu
World Wide Web: http://www.rpi.edu/dept/dses/www/

Rensselaer Polytechnic Institute

THE FACULTY AND THEIR RESEARCH

J. L. Adler, Assistant Professor (civil engineering); Ph.D., California at Irvine. Intelligent transportation systems, traveler information systems, driver behavior models, network algorithms, knowledge-based systems.

D. Berg, Institute Professor; Ph.D., Yale. Management of technological organizations, innovation, policy, manufacturing strategy, robotics, policy issues of research and development in the service sector.

R. J. Burke, Associate Professor; Ph.D., Massachusetts. Total quality management, industrial statistics, production management and process improvement, process reengineering.

J. G. Ecker, Professor; Ph.D., Michigan. Mathematical programming, multiobjective programming, geometric programming, mathematical programming applications, ellipsoid algorithms.

M J. Embrechts, Associate Professor; Ph.D., Virginia Tech. Neural networks, mathematical programming, artificial intelligence, systems engineering, computational cybernetics, data analysis.

R. J. Graves, Professor; Ph.D., SUNY at Buffalo. Manufacturing systems modeling and analysis, facilities planning and material handling system design, scheduling systems, concurrent engineering and design for manufacture, continuous flow manufacturing systems design.

J. Haddock, Professor; Ph.D., Purdue. Simulation, manufacturing/production and robotics and financial systems.

S. S. Heragu, Associate Professor and Director of Undergraduate Program; Ph.D., Manitoba. Artificial intelligence, cellular manufacturing, expert systems, facility design, flexible manufacturing systems, group technology, manufacturing systems, mathematical programming, operations research, plant layout, process planning, production and operations management.

C. Hsu, Professor and Director of Doctoral Program; Ph.D., Ohio State. Large-scale information systems, database and knowledge-based systems, computerized manufacturing, probabilistic programming and scheduling.

G. List, Professor; Ph.D., Pennsylvania. Real-time control of transportation network operations; multiobjective routing, scheduling, and fleet sizing; operations planning; hazardous materials logistics.

C. J. Malmborg, Professor and Acting Chair; Ph.D., Georgia Tech. Applied mathematical modeling, facility design, materials handling, material flow systems, manufacturing systems modeling and analysis, vehicle routing and scheduling.

M. Raghavachari, Professor; Ph.D., Berkeley. Statistical inference, quality control, multivariate methods, scheduling problems.

N. Shang, Assistant Professor; Ph.D., Berkeley. Computational statistics, statistical inference, design of experiments, data analysis.

G. R. Simons, Professor; Ph.D., Rensselaer. Managerial policy and organization.

P. Sullo, Associate Professor; Ph.D., Florida State. Reliability, life testing, and quality assurance; biostatistics; policy and risk analysis.

J. M. Tien, Professor and Acting Dean, School of Engineering; Ph.D., MIT. Systems modeling, queuing theory, public policy and decision analysis, computer performance evaluation, information and decision support systems, expert systems, computational cybernetics.

W. A. Wallace, Professor; Ph.D., Rensselaer. Public management and systems, decision support systems, expert systems.

J. W. Wilkinson, Professor Emeritus; Ph.D., North Carolina. Applied statistics, design of experiments, regression modeling, data analysis, statistical quality control.

T. R. Willemain, Associate Professor; Ph.D., MIT. Probabilistic modeling, data analysis, forecasting.

Research and Adjunct Faculty

W. J. Foley, Clinical Associate Professor; Ph.D., Rensselaer. Engineering design, computer simulation modeling, health applications of operations research, health-care policy analysis.

M. Grabowski, Research Associate Professor; Ph.D., Rensselaer. Management information systems, expert systems.

M. Kupferschmid, Adjunct Associate Professor; Ph.D., Rensselaer; PE. Mathematical programming, algorithm performance evaluation, engineering applications.

D. Sandhu, Adjunct Associate Professor; Ph.D., Toronto. Stochastic models in operations research, complex queuing networks and their applications to communication and manufacturing systems.

P. Witbeck, Adjunct Lecturer, DSES; M.S., Rensselaer. Industrial safety and hygiene.

Affiliated Faculty

A. A. Desrochers, Professor; Ph.D., Purdue. Nonlinear systems, robotics, control of automated manufacturing systems.

F. Dicesare, Professor Emeritus; Ph.D., Carnegie Mellon. Mathematical modeling, information systems, microprocessor applications.

W. R. Franklin, Associate Professor; Ph.D., Harvard. Computational geometry, graphics, CAD, cartography, parallel algorithms, large databases, expert systems.

D. H. Goldenberg, Associate Professor; Ph.D., Florida. Corporation finance and investment.

D. A. Grivas, Professor; Ph.D., Purdue. Geostochastics, applications of probability to the description of soil behavior, statistical analysis of soil parameters, numerical methods in geotechnical engineering, reliability analysis of soil structures and foundations.

J. E. Mitchell, Associate Professor; Ph.D., Cornell. Mathematical programming, combinatorics, nonlinear programming.

J. R. Norsworthy, Professor; Ph.D., Virginia. Economics of productivity, productivity measurements, industrial economics.

A. S. Paulson, Professor; Ph.D., Virginia Tech. Risk management, financial models, multivariate statistics, time series and forecasting, survival data analysis.

Major Research Activities

Material flow systems: design of automated guided vehicle systems, material flow system design, cellular manufacturing systems, vehicle routing and scheduling, automated storage and retrieval systems.

Flexible assembly systems: concurrent engineering, flow-line scheduling, workstation-based assembly system design.

Quality systems: multivariable control charting, nonparametric process control methods, experimental design methods for quality systems, reliability engineering, statistical computing.

Manufacturing and service facility design: modeling the relationship between product, facility, and logistical system design; development of imbedded design algorithms; forecasting models and decision support systems; storage systems modeling.

Modeling: modeling science, simulation, queuing, optimization, stochastic programming, neural networks, computational intelligence.

Information systems: knowledge/expert systems, decision support systems, performance evaluation, multicriteria decision making.

Management systems: management of technological organizations, management information/organizational systems, disaster management, health management/policy, project management, public systems and transportation management.

STANFORD UNIVERSITY

Department of Engineering-Economic Systems and Operations Research

Programs of Study

The Department of Engineering-Economic Systems and Operations Research (EES&OR) develops, applies, and disseminates principles and engineering methods for improving decision making in operations, strategy, and policy. EES&OR provides special strength in theory and application within the areas of optimization; probability and stochastic processes; systems and simulation; economics, finance, and investment; and decisions. EES&OR offers Master of Science (M.S.), Engineer, and Doctor of Philosophy (Ph.D.) degree programs and participates in a Bachelor of Science program in Mathematical and Computational Science. The M.S. degree requires 45 units of course work beyond the B.S. but no thesis and can be completed in three to five quarters. The Engineer degree requires 45 units beyond the M.S. degree plus a thesis, can be completed in two years, and can only be entered through the M.S. degree program. The Ph.D. degree takes a minimum of three academic years; the program is normally completed in four or five years. In addition to completing an academic program (including core courses) approved by the academic adviser and dissertation adviser, the Ph.D. student must pass the department qualifying procedure, pass a University oral examination based on the dissertation, and successfully complete a dissertation based on original work.

Research Facilities

EES&OR has many on-going research programs. Many students also become involved in industry-conducted research or project work in the Silicon Valley or elsewhere. The Terman Engineering Center houses a modern engineering library and computer cluster, and the University has several libraries available to graduate students with extensive collections. The Data Center provides computer facilities for both sponsored and unsponsored research. Through a campuswide network, students also have access to a Sun Workstation and IBM mainframes.

Financial Aid

A limited number of merit-based fellowships are available on a competitive basis for doctoral candidates only. This includes a number of fellowships for first-year students as well as some research and course assistantships. Supplemental financial aid can sometimes be obtained by grading papers and performing other tasks, and a loan program is available to U.S. residents through the Office of Graduate Financial Aid.

Cost of Study

The 1999–2000 tuition for graduate study in the School of Engineering is $8196 per quarter or $24,588 for the nine-month academic year (September–June). Book and supplies cost approximately $1250.

Living and Housing Costs

In 1999–2000, the cost for a single student living on campus for nine months is approximately $11,478. Miscellaneous items cost approximately $2000 for the academic year. Married couples should budget at least $3000 per month for living expenses and an additional $500 per child.

Student Group

There are approximately 300 students in the department. Many students have prior work experience; some students are employees of local industry who attend on a part-time basis through the Honors Co-op Program. Approximately one third of the students are married, and 40 percent are international students.

Student Outcomes

Individuals are prepared for a variety of professional careers, and graduates have pursued successful careers in university teaching and research, consulting, industrial research, strategic planning, financial analysis, project management, product development, government policy analysis, line management, and enterprise management. Some have founded companies specializing in management systems consulting, high-technology products, software, mining, or financial services. Other graduates have helped establish new analytical capabilities in existing firms or government agencies.

Location

The Stanford campus is located 30 miles south of San Francisco. The Pacific Ocean and its beaches are a 45-minute drive directly west of campus over the coastal mountain range. The yearly climate is generally mild.

The University

Stanford University is a private university, founded in 1885 by Senator and Mrs. Leland Stanford as a memorial to their only son. The University is organized into the Schools of Earth Sciences, Education, Engineering, Humanities and Sciences, Law, and Medicine and the Graduate School of Business. Stanford is a nonsectarian, coeducational university with an international reputation for academic excellence. It operates on a quarter system, with a shortened summer quarter. Enrollment totals about 14,000, including 7,467 graduate students.

Applying

Applications are accepted for admission only for the start of the academic year, which begins in the autumn quarter. The department does not grant deferrals. Complete applications for the Ph.D. program, including requests for merit-based fellowships, are due February 1; all other degree program applications without fellowship consideration are due March 31. The General Test of the Graduate Record Examinations (GRE) is required (scores must be from an examination taken within the last five years). All applicants whose native language is not English must submit Test of English as a Foreign Language (TOEFL) scores from an examination given in the last eighteen months. The TOEFL is waived for applicants who have recently completed two or more years of study in an English-speaking country. To begin the M.S. degree, a minimum score of 575 is required on the paper-based test; a total score in the range of 220 to 300 is required on the computer-based test. To begin the Ph.D. degree, a minimum score of 600 is required on the paper-based test; a total score in the range of 250 to 300 is required for the computer-based test. Applicants should take the TOEFL at least nine months before their intended date of enrollment at Stanford. Application forms can be obtained from the Graduate Admissions Support Section, Office of the Registrar, Stanford University, Stanford, California 94305-3052 or on the World Wide Web (http://www.stanford.edu/dept/registrar/admissions/applyinfo.html).

Complete program information and faculty, directory, and admissions information are available on the Web (http://www.stanford.edu/dept/eesor). A reprint of the *Stanford Bulletin* of the School of Engineering courses and degrees is sent with the paper application forms.

The self-addressed postcard included in the admission packet must be fully completed and include sufficient postage in order to be mailed back to the applicant. This postcard is sent to notify an applicant of missing admission documents or when all required admissions documents have been received and the admissions file is complete. Applicants using the online form are notified of missing documents or the completeness of their application via e-mail if that address is included on their application. Only applications that are complete by the above deadlines are reviewed. To protect the confidentiality of applicant records, information about applicants and admission documents is not given over the telephone. Admission notification is by U.S. mail only.

Correspondence and Information

Admissions Committee
Engineering-Economic Systems and Operations Research
Terman Engineering Center, Room 310
Stanford University
Stanford, California 94305-4023
E-mail: eesor-admissions@stanford.edu

Stanford University

THE FACULTY AND THEIR RESEARCH

Kenneth J. Arrow, Professor Emeritus; Ph.D., Columbia, 1951. Economic theory, information and organization.

Nicholas Bambos, Associate Professor; Ph.D., Berkeley, 1989. Queueing theory and stochastic networks with application to communications (especially wireless telecommunications).

Samuel S. Chiu, Associate Professor; Ph.D., MIT, 1981. Quantitative methods in urban service systems, location problems in the presence of queuing, optimization, probabilistic modeling.

Richard W. Cottle, Professor; Ph.D., Berkeley, 1964. Linear and nonlinear programming, complementarity theory.

George B. Dantzig, Professor Emeritus; Ph.D., Berkeley, 1946. Mathematical programming, applications of mathematics, energy modeling.

Meiring deVilliers, Assistant Professor; Ph.D., Stanford, 1997. Finance, economics, law, and systems.

Donald A. Dunn, Professor Emeritus; Ph.D., Stanford, 1956. Telecommunications policy, organizational economics.

B. Curtis Eaves, Professor; Ph.D., Stanford, 1969. Mathematical programming, piecewise linear modes, solutions of equations with homotopies and complementary pivot theory.

Peter W. Glynn, Professor; Ph.D., Stanford, 1982. Stochastic systems, queuing theory, computational probability.

Frederick Hillier, Professor Emeritus; Ph.D., Stanford, 1961. Optimization of queuing systems, integer programming, capital budgeting.

Ronald A. Howard, Professor and Director, Decisions and Ethics Center; Sc.D., MIT, 1958. Decision analysis, social analysis, probabilistic modeling.

Donald L. Iglehart, Professor Emeritus; Ph.D., Stanford, 1961. Applied probability, inventory theory, queuing theory, simulation methodology.

Blake Johnson, Assistant Professor; Ph.D., Stanford, 1994. Asset valuation, optimal investment and financing strategies in highly dynamic markets, large-project financing methods such as infrastructure investments.

Gerald J. Lieberman, Professor Emeritus; Ph.D., Stanford, 1953. Reliability, inventory theory, engineering statistics, probabilistic models.

David G. Luenberger, Professor; Ph.D., Stanford, 1963. Systems economics, optimization, control theory, business systems, mathematical programming.

Alan S. Manne, Professor Emeritus; Ph.D., Harvard, 1950. Mathematical programming; energy, environmental, and economic modeling.

Michael May, Professor (Research); Ph.D., Berkeley, 1951. Strategic studies and international security.

Walter Murray, Professor (Research); Ph.D., London, 1969. Numerical optimization.

William J. Perry, Professor; Ph.D., Penn State, 1957. Science and technology policy.

Michael A. Saunders, Professor (Research); Ph.D., Stanford, 1972. Numerical optimization.

Ross D. Shachter, Associate Professor; Ph.D., Berkeley, 1982. Optimization, probabilistic modeling, decision systems, medical decision analysis.

James L. Sweeney, Professor and Chair; Ph.D., Stanford, 1971. Energy modeling and analysis, economics of natural resources, applied economics, energy/economic policy.

Edison T. S. Tse, Associate Professor; Ph.D., MIT, 1970. Computer-integrated manufacturing enterprise management systems, decision systems, systems optimization, control.

Benjamin Van Roy, Assistant Professor; Ph.D., MIT, 1998. Control of complex systems, dynamic programming, computational learning, financial economics.

Arthur F. Veinott Jr., Professor; Engr. Sc.D., Columbia, 1960. Inventory theory, lattice programming, dynamic programming, mathematical programming.

John P. Weyant, Professor (Research) and Director, Energy Modeling Forum; Ph.D., Berkeley, 1976. Application of quantitative methods to strategic planning in government and industry, energy modeling and policy analysis.

Consulting Faculty

Adam Borison, Consulting Associate Professor; Principal, Applied Decision Analysis, Inc. (ADA), Menlo Park, California; Ph.D., Stanford, 1982. Application of analytical techniques to management problems, development and implementation of mathematical models for energy and environmental planning.

Hung-Po Chao, Consulting Professor; Senior Scientific Advisor, EPRI, Palo Alto, California; Ph.D., Stanford 1978. Environmental risk analysis.

Charles D. Feinstein, Consulting Associate Professor; Associate Professor, Santa Clara University, Santa Clara, California; Ph.D., Stanford, 1980. Dynamic systems analysis and control, forecasting techniques, design and analysis of information systems, mathematical modeling, mathematical programming theory and algorithm development.

Samuel Holtzman, Consulting Associate Professor; Associate, Strategic Decisions Group, Menlo Park, California; Ph.D., Stanford, 1985. Intelligent decision systems, medical decision analysis, ethics.

Gerd Infanger, Consulting Associate Professor; CEO, Infanger Investment Technology; Ph.D., Vienna Technical, 1986. Planning under uncertainty.

James E. Matheson, Consulting Professor; Director, Strategic Decisions Group, Menlo Park, California; Ph.D., Stanford, 1964. Decision analysis and intelligent decision systems.

Robert R. Maxfield, Consulting Professor; Director, Software Publishing Corporation, Saratoga, California; Ph.D., Stanford, 1969. Business management, system modeling.

Peter A. Morris, Consulting Professor; Principal, Applied Decision Analysis, Inc., Menlo Park, California; Ph.D., Stanford, 1971. Mathematical modeling, dynamic probabilistic methods.

D. Warner North, Consulting Professor; Principal, Decision Focus Inc., Los Altos, California; Ph.D., Stanford, 1970. Decision and environmental risk analysis.

Samual Savage, Consulting Professor; Ph.D., Yale 1973. Quantitative analysis in spreadsheets.

Richard D. Smallwood, Consulting Professor; President, Applied Decision Analysis, Inc., Menlo Park, California, Sc.D., MIT, 1981. Mathematical modeling, market analysis and forecasting, market research, decision analysis.

Mukund Thapa, Consulting Associate Professor; President, Stanford Business Software, Inc.; Ph.D., Stanford 1981. Mathematical, economic financial, consumer choice modeling, market analysis, optimization, scheduling, design and development of software.

STANFORD UNIVERSITY

School of Engineering
Department of Industrial Engineering and Engineering Management

Programs of Study

The M.S. requires 45 quarter units of course work and is normally a one-year program. The Ph.D. program requires four to five years, 90 quarter units, and a dissertation reflecting original research that contributes to knowledge in a specific field. Ph.D. qualifying examinations are taken by the end of the second year, and a final oral examination must be passed before the Ph.D. is awarded. Stanford's Manufacturing Programs full description in the Industrial Engineering section of this guide has more information on the Future Professors of Manufacturing Program and the M.S.E. in Manufacturing Systems Engineering.

Ph.D. areas of specialization in both course work and research include manufacturing, organizations, technology management, risk analysis, and production and operations management. In addition, many students are involved in interdisciplinary research. Departmental goals are to perform research at the frontiers of the areas of specialization and to train practitioners, researchers, and teachers in these areas. All Ph.D. research is closely supervised by a full-time faculty member, and there is frequent interaction between the student and his or her adviser. The research must focus on a problem of practical significance. Recent problem areas include multicriterion decision making, plant capacity expansion, capital allocation under uncertainty, and the formation of alliances in high-technology industries.

Research Facilities

The Terman Engineering Center houses a modern engineering research library with an extensive collection of important journals. Other libraries on campus are also open to graduate students. Mainframe, workstation, and personal computers are widely available for course work and research. Perhaps most important, a large number of high-technology firms have operations in the Stanford Industrial Park, and many graduate students become involved with these firms in either project or research work.

Financial Aid

Engineering Graduate Honors Fellowships, which provide stipends of $13,482 plus tuition of $15,264, totaling $28,746 for 1999–2000, are limited to first-year doctoral students. Students who are U.S. citizens or permanent residents admitted to a one-year master's degree program in the School of Engineering are eligible to apply for loans to assist them in financing their studies at Stanford. Many international students are supported by their governments or their employers. In addition, the School's honors cooperative program enables persons employed in industry to obtain a degree with support from their company.

Cost of Study

For the 1999–2000 academic year, students pay $24,588 in tuition. Portions of stipends may be withheld for income taxes but not for Social Security or disability taxes.

Living and Housing Costs

For the nine month 1998–99 academic year, living and housing costs (books, supplies, rent, food, and medical insurance) were estimated at $11,478 for single students living on-campus, $16,713 for single students living off-campus, $9390 additional for a spouse, and $2750 additional for each child.

Student Group

The student body represents a wide variety of nationalities and American geographical areas and includes an increasing number of women.

Student Outcomes

Recipients of Ph.D.'s typically obtain teaching positions. The M.S. degree leads to positions in middle management with industrial firms. In a typical year the department awards seventy M.S. degrees and four Ph.D. degrees.

Location

Stanford University is situated in the Santa Clara Valley, 30 miles south of San Francisco. Cultural events of all kinds are offered in the area. The temperate climate encourages athletic activities, and there are opportunities for virtually all recreational pursuits.

The University and The Department

Stanford is a private university, founded in 1885 by Senator and Mrs. Leland Stanford as a memorial to their only child. The Department of Industrial Engineering and Engineering Management is one of ten departments in the School of Engineering. The department emphasizes the design, development, and implementation of productive systems, involving the organization of people, information, money, and materials to produce and distribute products and services in all sectors: government, public and quasi-public organizations, and private industry. The faculty members spend a large amount of time interacting with the students, individually and in groups. Cooperation among the faculty members of the several departments and schools makes it possible for the student to pursue his or her individual interests with the assistance of some of the most prominent educators and researchers in the world.

Applying

Admission and financial aid application forms are available from the Office of Graduate Admissions. Students are only admitted for the fall quarter. The deadline for application is February 1. Applicants are required to have a B.S. in engineering or science and should have graduated in the top 10 percent of their class. The GRE General Test is required; the TOEFL is required for international students whose native language is not English. Applicants are notified by April regarding acceptance for admission in the following autumn.

Correspondence and Information

Department of Industrial Engineering
    and Engineering Management
Stanford University
Stanford, California 94305-4024

Telephone: 650-725-1633
Fax: 650-725-8799
E-mail: ieem@soe.stanford.edu
World Wide Web: http://ieem.stanford.edu

Stanford University

THE FACULTY AND THEIR RESEARCH

James L. Adams, Professor Emeritus; Ph.D. (mechanical engineering), Stanford, 1961. Creative problem solving, management of design in organizations, interaction between engineering and society, role of emotions in the engineering process.

Diane E. Bailey, Assistant Professor; Ph.D. (industrial engineering and operations research), Berkeley, 1994. Work organization in high-technology manufacturing environments, with a particular focus on work group performance.

Stephen R. Barley, Professor; Ph.D. (organizational studies), MIT, 1984. Influence of technology on organizations.

Margaret L. Brandeau, Professor; Ph.D. (engineering-economic systems), Stanford, 1985. Production and operations management (manufacturing systems planning, resource allocation), health care (policy analysis of AIDS interventions).

Thomas H. Byers, Associate Professor (teaching); Ph.D. (management science), Berkeley, 1982. Creation and growth of high technology ventures.

Robert C. Carlson, Professor; Ph.D. (mathematical sciences), Johns Hopkins, 1976. Production theory and analysis, transportation systems, manufacturing strategy.

Kathleen M. Eisenhardt, Professor; Ph.D. (organizational behavior), Stanford, 1982. Management of technology-based firms, entrepreneurship, inductive methods.

Warren H. Hausman, Professor; Ph.D. (industrial management), MIT, 1966. Operations management, supply chain management.

Pamela J. Hinds, Assistant Professor; Ph.D. (organization and management science), Carnegie Mellon, 1997. Interplay between information technologies, information sharing, and human judgment.

James V. Jucker, Professor; Ph.D. (industrial engineering), Stanford, 1968. Design of manufacturing systems.

Hau L. Lee, Professor; Ph.D. (production management), Pennsylvania (Wharton), 1982. Supply chain and operations management, quality management, inventory applications.

Robert E. McGinn, Professor (teaching); Ph.D. (philosophy and humanities), Stanford, 1970. Work, technology, and society, ethical issues in contemporary engineering workplaces.

M. Elisabeth Paté-Cornell, Professor and Chair; Ph.D. (engineering-economic systems), Stanford, 1978. Risk and decision analysis, engineering risk management, reliability of technological systems, engineering economy.

Robert I. Sutton, Professor; Ph.D. (organizational behavior), Michigan, 1984. Organizational behavior and design.

Ulrich W. Thonemann, Assistant Professor; Ph.D. (industrial engineering), Stanford, 1994. Production and operations management, supply chain management, information systems.

STANFORD UNIVERSITY

School of Engineering and Graduate School of Business
Manufacturing Programs

Programs of Study

The School of Engineering and the Graduate School of Business in partnership with Stanford Integrated Manufacturing Association (SIMA) companies are committed to educating outstanding managers, engineers, and professors who will contribute creatively and substantively to meeting manufacturing challenges faced by industry in the twenty-first century.

The manufacturing systems engineering (MSE) master's degree program is jointly offered by the Departments of Mechanical Engineering and of Industrial Engineering and Engineering Management. The three-quarter curriculum integrates engineering design and management focused on manufacturing. Areas of emphasis include quality assurance, organizational behavior, analysis of production systems, manufacturing strategy, manufacturing systems design, supply chain management, design for manufacturability, robotics, smart product design, and integrated design for marketability and manufacture. Vigorous industrial interaction, team-based work, and an emphasis on product development distinguish this program.

The M.B.A./MSE dual-degree program combines study in manufacturing systems engineering with business management and leads to two master's degrees. This seven-quarter dual-degree program is jointly offered by the Schools of Engineering (MSE) and Business (M.B.A.). Students are admitted to and complete degree requirements for both schools.

The Future Professors of Manufacturing Program is a special Ph.D. program designed for men and women with engineering and business backgrounds who are committed to conducting research and teaching on advanced manufacturing topics at U.S. universities. Students must first be admitted to an established Ph.D. degree program (see departmental descriptions) in engineering or business, then nominated for the FPM Program.

Research Facilities

The Terman Engineering Center houses a modern engineering research library with an extensive collection of professional journals. The J. Hugh Jackson Library in Business offers an extensive collection of up-to-date research journals and other material, including electronic resources. Other libraries on campus are also open to graduate students. Mainframe, workstation, and personal computers are widely available for course work and research. While numerous research facilities are involved in manufacturing-related research, five laboratories (Manufacturing Models, Product Realization, Rapid Prototyping, Smart Product Design, and Robotics) and two centers (Thornton Center for Engineering Management and the Center for Design Research) specifically focus on manufacturing. Graduate students interact with manufacturing firms through course projects and research work and through SIMA, an industry-Stanford partnership.

Financial Aid

Manufacturing students may apply for financial aid through the Schools of Engineering and Business (students should refer to school and department listings). SIMA fellowships are available at the master's level; at the Ph.D. level, financial aid is offered through the Future Professors of Manufacturing Program.

Cost of Study

The 1999–2000 nine-month academic year tuition for the School of Engineering is $24,588 and for the School of Business, $27,243, plus approximately $2553 for books, supplies, and fees.

Living and Housing Costs

The 1999–2000 estimated cost of living for nine months is $10,932 for a single student living on campus, $15,963 for a single student living off campus, $21,120 for a married student living on campus, and $23,916 for a married student living off campus. Families with children should anticipate higher costs.

Student Group

The student body in the Schools of Engineering and Business represents a wide variety of nationalities and geographical areas. The Stanford Engineering Club for Automation and Manufacturing (SECAM) and the Graduate School of Business Manufacturing Club form an exceptionally strong community of manufacturing students at Stanford.

Student Outcomes

Graduates of the MSE and dual-degree master's programs typically enter industry in engineering and management positions in manufacturing (e.g., supply chain management, product management, product and process design, operations, quality, systems, and materials). Future Professors of Manufacturing Ph.D. graduates have obtained faculty positions at major U.S. universities where they focus their teaching and research on manufacturing.

Location

Stanford University is located on an 8,200-acre campus, adjacent to the residential communities of Palo Alto and Menlo Park in the Santa Clara Valley and 35 miles south of San Francisco. The University is in the heart of Silicon Valley's high-technology manufacturing center. The climate is temperate, with rainfall normally occurring in the winter months. This climate encourages athletic activities, with opportunities for virtually all recreational pursuits.

The University and The Schools

Stanford University is a private university, founded in 1885. Enrollment is approximately 14,000. Forty-seven percent are undergraduates, and 53 percent are graduate students. The School of Engineering includes the areas of aeronautics and astronautics, chemical engineering, civil and environmental engineering, computer science, electrical engineering, engineering–economic systems and operations research, industrial engineering and engineering management, materials science and engineering, and mechanical engineering. The Graduate School of Business offers courses in the areas of accounting, economic analysis and policy, finance, marketing, operations, information and technology, organizational behavior, and political economics. Cooperation among the faculty members of the different departments and schools makes it possible for students to pursue individual interests with the assistance of some of the most prominent educators and researchers in the world.

Applying

MSE program applicants request the mechanical engineering packet from the Graduate Admissions Office. Dual-M.B.A./MSE program applicants apply to both programs through the M.B.A. Admissions Office. Ph.D. applicants request materials for engineering from the Graduate Admissions Office or for business from the Doctoral Program Office. Students are admitted to begin study in the autumn quarter only. Application deadlines are as follows: for the MSE program, January 15, 2000; for the M.B.A./MSE program, November 3, 1999, and January 12, 2000; and for Ph.D. programs in business, January 3, 2000. For Ph.D. programs in engineering, applicants must contact the department. The GRE is required for the MSE program and the GMAT for the M.B.A./MSE program. The MSE program also requires an undergraduate degree in engineering or science. Applicants should contact individual departments for Ph.D. requirements and SIMA for the Future Professors of Manufacturing Program.

Correspondence and Information

Stanford Integrated Manufacturing Association (SIMA)
Stanford University, Building 02-530
Stanford, California 94305-3036
Telephone: 650-723-3016 (SIMA)
      650-723-4291 (Graduate Admissions)
      650-723-2766 (M.B.A. Admissions)
      650-725-7462 (Doctoral Program Office—Business)
E-mail: sima-info@sima.stanford.edu
    ck.gaa@forsythe.stanford.edu (Graduate Admissions Office)
    mbainquiries@gsb.stanford.edu (M.B.A. Admissions Office)
    phd_program_inquiries@gsb.stanford.edu (Doctoral Program Office—Business)
World Wide Web: http://www-sima.stanford.edu/
    http://soe.stanford.edu/programs/graduate/graduate.html
    http://www-gsb.stanford.edu

Stanford University

THE FACULTY AND THEIR RESEARCH

James L. Adams, Professor of Industrial Engineering and Engineering Management; Ph.D., Stanford, 1961. Creative problem solving and management of design in organizations.

Diane E. Bailey, Assistant Professor of Industrial Engineering and Engineering Management; Ph.D., Berkeley, 1994. Work organization in high-technology manufacturing environments, work group performance.

Stephen R. Barley, Professor of Industrial Engineering and Engineering Management; Ph.D., MIT, 1994. Organizational theory, technical labor force, technological change, network analysis, ethnography, management of research and development operations.

David W. Beach, Professor of Mechanical Engineering; Ph.D., Stanford, 1961. Design and problem solving; linkage between design and manufacturing; computer-aided prototyping; integrated marketing, design, and manufacturing.

Margaret L. Brandeau, Professor of Industrial Engineering and Engineering Management; Ph.D., Stanford, 1985. Analytical models of management problems in areas of manufacturing systems.

Robert C. Carlson, Professor of Industrial Engineering and Engineering Management, Graduate School of Business; Ph.D., Johns Hopkins, 1976. Manufacturing strategy, models of production scheduling and control systems, multiobjective decision systems.

Mark R. Cutkosky, Professor of Mechanical Engineering and Associate Chair for Design and Manufacturing; Ph.D., Carnegie Mellon, 1985. Mechanical design, computer-aided manufacturing, robotics.

Kathleen M. Eisenhardt, Professor of Industrial Engineering and Engineering Management; Ph.D., Stanford, 1982. Strategy and organization of firms in high-velocity markets, innovation, strategic decision making.

Martin A. Fischer, Assistant Professor of Civil and Environmental Engineering; Ph.D., Stanford, 1991. Creation of 4-D symbolic product and process models to represent information required in the life cycle of engineered facilities.

George Foster, Paul L. and Phyllis Wattis Foundation Professor of Management and Co-Director of the Executive Program for Smaller Companies; Ph.D., Stanford, 1975. Cost accounting and cost management.

Alice P. Gast, Professor of Chemical Engineering; Ph.D., Princeton, 1984. Thermodynamic and transport properties of dispersions of small particles, polymeric micelles, proteins, and emulsions.

Peter W. Glynn, Professor of Engineering–Economic Systems and Operations Research; Ph.D., Stanford, 1982. Stochastic modeling and simulation; performance analysis for computer, telecommunications, and manufacturing systems; risk analysis in finance and engineering.

J. Michael Harrison, Professor of Operations Management, Graduate School of Business; Ph.D., Stanford, 1970. Production and operations management, with emphasis on time dimension of system performance.

Warren H. Hausman, Professor of Industrial Engineering and Engineering Management, Graduate School of Business; Ph.D., MIT, 1966. Operations management, planning, and control; inventory control; supply chain management.

Pamela J. Hinds, Assistant Professor of Industrial Engineering and Engineering Management; Ph.D., Carnegie Mellon, 1997. Effect of technology on individuals and small groups, management of technical work and expertise.

Charles A. Holloway, Kleiner, Perkins, Caufield and Byers Professor of Management (Graduate School of Business) and founding Co-Chair of the Stanford Integrated Manufacturing Association; Ph.D., UCLA, 1969. Management of technology, manufacturing strategy, quantitative analysis.

Kosuke Ishii, Associate Professor of Mechanical Engineering; Ph.D., Stanford, 1987. Design for manufacturability, life-cycle quality of mechanical and electromechanical systems.

James V. Jucker, Professor of Industrial Engineering and Engineering Management; Ph.D., Stanford, 1968. Design and organization of manufacturing systems, impact on workforce.

Sunil Kumar, Assistant Professor of Operations, Information, and Technology, Graduate School of Business; Ph.D., Illinois, 1996. Performance evaluation and dynamic control of manufacturing processes and communication systems using stochastic network models.

Rajiv Lal, Professor of Marketing and Management Science; Ph.D., Carnegie Mellon, 1983. Retailing supply chain management.

Jean-Claude Latombe, Professor of Computer Science and Director of the Robotics Laboratory; Ph.D., Grenoble (France), 1977. Robotics and artificial intelligence.

Hau L. Lee, Professor of Industrial Engineering and Engineering Management, Graduate School of Business; Ph.D., Pennsylvania, 1983. Production and operations management, with emphasis on supply chain management, inventory planning, integrated design, and quality management.

Larry Leifer, Professor of Mechanical Engineering and Director of the Center for Design Research and the Design Affiliates Program; Ph.D., Stanford, 1969. Design methodology, rehabilitation engineering and programmable electromechanical systems.

Raymond E. Levitt, Professor of Civil and Environmental Engineering; Ph.D., Stanford, 1975. Theory and tools supporting systematic engineering of organizations engaged in fast-paced project-oriented work processes.

Haim Mendelson, James Irvin Miller Professor of Information Systems; Ph.D., Tel-Aviv (Israel), 1979. Information systems.

James M. Patell, Hoover Professor of Public and Private Management, Graduate School of Business; Ph.D., Carnegie Mellon, 1974. Manufacturing systems, performance measurement.

Evan L. Porteus, Professor of Industrial Engineering and Engineering Management and Co-Director of Product Development and Manufacturing Strategy Executive Program; Ph.D., Case Tech, 1967. Operations research.

Friedrich B. Prinz, Rodney H. Adams Professor of Engineering, Departments of Mechanical Engineering and of Material Science and Engineering; Ph.D., Vienna, 1975. Rapid part prototyping and rapid tool generation, geometric modeling and material processing for manufacturing and design.

William C. Reynolds, Professor of Mechanical Engineering; Ph.D., Stanford, 1958. Turbulence, thermodynamics, computational fluid dynamics.

Krishna C. Saraswat, Professor of Electrical Engineering; Ph.D., Stanford, 1974. Fabrication processes, device structures, materials and equipment for VLSI and flat panel display manufacturing.

Sheri D. Sheppard, Associate Professor of Mechanical Engineering; Ph.D., Michigan, 1985. Finite-element analysis and fracture mechanics.

George S. Springer, Professor and Departmental Chair of Aeronautics and Astronautics, and Director of the Structures and Composites Laboratory; Ph.D., Yale, 1962. Manufacture, design, and environmental effects of fiber-reinforced composite materials.

V. "Seenu" Srinivasan, Ernest C. Arbuckle Professor of Marketing and Management Science; Ph.D., Carnegie Mellon, 1971. Market research for product development and pricing decisions.

Robert I. Sutton, Professor of Industrial Engineering and Engineering Management, Graduate School of Business; Ph.D., Michigan, 1984. Organizational decline and death, technology and work, innovation and the product design process, organizational performance.

C. B. Tatum, Professor of Civil and Environmental Engineering; Ph.D., Stanford, 1983. Technological advancement in construction and design–construction integration.

Ulrich W. Thonemann, Assistant Professor of Industrial Engineering and Engineering Management; Ph.D., Stanford, 1994. Application of analytical and quantitative techniques to manufacturing systems modeling, supply chain modeling, product design, management information systems.

Seungjin Whang, Associate Professor of Operations, Information, and Technology, Graduate School of Business; Ph.D., Rochester, 1988. Economic analysis of management information systems, design of database management systems, performance evaluation of computer systems.

Samuel C. Wood, Assistant Professor of Manufacturing and Technology, Graduate School of Business; Ph.D., Stanford, 1992. Product and technology development, manufacturing technology, modeling and simulation of manufacturing systems.

STATE UNIVERSITY OF NEW YORK AT BINGHAMTON

Thomas J. Watson School of Engineering and Applied Science
Department of Systems Science and Industrial Engineering

Programs of Study

The Department of Systems Science and Industrial Engineering has two major concentrations in its educational program. The first area is systems science. This area concerns the use of mathematical expressions to model and understand the behavior of complex systems. The second area is industrial engineering. This area is concerned with the modeling, design, analysis, and improvement of a variety of systems, primarily the manufacturing system. The department offers programs that lead to a Master of Science degree both in systems science and in industrial engineering. The Doctor of Philosophy degree is offered in systems science, which also has a specialization in manufacturing systems. Major research emphasis is currently being placed on the areas of uncertainty, fuzzy logic, optimization, intelligent systems, and world-class manufacturing with a primary application to the electronics manufacturing process, particularly in the areas of printed circuit board production and automated assembly.

The doctoral program is structured around the use of a learning contract. Each doctoral degree student must prepare, with his or her adviser, a learning contract that defines the overall program. This learning contract provides the student with flexibility and encourages the integration of a wide range of topics into a single degree program. The program sets certain boundaries for the learning contract, and these cover such things as a comprehensive exam, a teaching experience, a prospectus, and a final defense. The Ph.D. degree is awarded in recognition of the student's high achievement in research and based upon the understanding that the specialization and the focus for each student will be different.

The M.S. degrees require at least ten courses, or the equivalent, to be complete. The M.S. in systems science has a core of six courses with four additional courses as electives. It also includes a termination requirement, which may be either an examination or a project. The M.S. degree in industrial engineering consists of a core of four courses, with an additional four courses as electives, and requires that a thesis be undertaken for all full-time students. Part-time students are allowed to take additional course work and substitute a project for the thesis.

Research Facilities

The department has outstanding facilities. These consist of both University laboratories and a cooperative arrangement with industry. The University has helped the department obtain three specific laboratories. The first is the manufacturing software lab, which has been built over the last five years. The second is a lab for the development of manufacturing controls and contains both the computers and manufacturing hardware to develop control systems. The third lab is a newly developed, NSF-supported computer facility dedicated to further research in the areas of uncertainty and fuzzy logic.

The cooperative relationship with local industry has made available a multimillion-dollar laboratory for the development of research in electronics manufacturing. A local company (Universal Instruments) is a major provider of manufacturing equipment to the electronics manufacturing industry, and many students do their research and thesis work in the SMT laboratory of this company. It provides the student with a facility that cannot be duplicated and maintained within a university setting. Since 1990, the lab has been the research site for more than fifteen master's and doctoral degree students each semester. This relationship has also led to many major research projects with industry, consortiums, and agencies. It is important to note that this relationship has allowed the department to fund all of its graduate students for the last few years.

Financial Aid

Many students hold fellowships, traineeships, or graduate, research, or teaching assistantships. Most awards include a full waiver of tuition. Other sources of financial aid include the New York State Tuition Assistance Program, the Federal Stafford Student Loan Program, the graduate and professional school Federal Work-Study Program, and campus jobs.

Cost of Study

For full-time matriculated graduate students, tuition in 1998–99 was $2550 per semester for state residents and $4208 per semester for nonresidents.

Living and Housing Costs

A recently completed apartment complex, the Graduate Community, has 3- and 4-person apartments with living room, dining area, kitchen, and bath. Based on a 1998–99 academic-year lease, the semester rate for a single bedroom was $2050 and for a double bedroom $1775 per person and $3085 per couple. The cost of meal plans per semester was as follows: basic, $787; standard, $1022; and ultra, $1097. Assistance in locating off-campus housing is provided by the listing services of Off-Campus College.

Student Group

Of the 12,259 students enrolled at Binghamton University, 2,700 are graduate students. In the Watson School, there are 876 undergraduates and 368 graduate students. Many obtain jobs in local high-technology enterprises during their enrollment at the School and after graduation.

Location

The University's 606-acre campus is in a suburban setting just west of Binghamton. More than 300,000 people live within commuting distance of the campus. Cultural offerings in the community include the museum and programs of the Roberson Center for the Arts and Sciences, as well as performances by the Binghamton Symphony, Tri-Cities Opera, Civic Theater, and other groups. The University's Art Gallery has a permanent collection representing all periods and also displays works from special loan exhibitions. The annual concert series of the Anderson Center brings a wide variety of performing artists to campus. The Department of Theater stages more than twenty-five productions each year.

The University and The School

The State University of New York at Binghamton is one of the four university centers in the State University of New York System. The faculty numbers about 700. Graduate programs were initiated in 1961 with the establishment of Master of Arts programs in English and mathematics.

The Watson School was created in 1983 by combining the established graduate programs in computer science and systems science from the School of Advanced Technology with new programs in electrical, industrial, and mechanical engineering.

Applying

Holders of an appropriate bachelor's degree from any recognized college or university are eligible to apply. Requests for application forms can be sent via e-mail to the Office of Graduate Admissions at gradad@binghamton.edu. Applicants should submit GRE General Test scores. International applicants must submit TOEFL scores and provide proof of their ability to meet academic expenses. All credentials should be on file at least one month prior to anticipated enrollment. To ensure consideration for assistantship and fellowship awards, admission credentials should be received by February 15. Online application is available at http://www.gradschool.binghamton.edu.

Correspondence and Information

For Systems Science:
Dr. George Klir
Department of Systems Science
    and Industrial Engineering
Thomas J. Watson School of Engineering
    and Applied Science
Binghamton University
Binghamton, New York 13902-6000

For Industrial Engineering:
Dr. Robert Emerson
Department of Systems Science
    and Industrial Engineering
Thomas J. Watson School of Engineering
    and Applied Science
Binghamton University
Binghamton, New York 13902-6000

State University of New York at Binghamton

THE FACULTY AND THEIR RESEARCH

Robert Emerson, Professor and Chairman; Ph.D., Purdue. Integrated manufacturing, quality assurance, decision support systems.

David Enke, Assistant Professor; Ph.D. Missouri–Rolla. Neural networks, artificial vision, optimization, applied statistics, decision support systems, cognitive modeling.

Donald Gause, Bartle Professor; M.S., Michigan State. General design processes, user-oriented systems design, problem resolution processes, adaptive programming.

George Klir, Distinguished Professor and Director, Center for Intelligent Systems; Ph.D., Czechoslovak Academy of Sciences. General systems methodology, logic design and computer architecture, information theory, fuzzy systems.

Sarah Lam, Assistant Professor; Ph.D., Pittsburgh. Intelligent systems, statistical analysis and design of experiments, neural networks modeling.

Harold W. Lewis III, Associate Professor; Ph.D., SUNY at Binghamton. Fuzzy expert systems, approximate reasoning.

Howard Pattee, Professor Emeritus; Ph.D., Stanford. Theoretical biology, evolutionary models, linguistic control of dynamic systems.

Daryl Santos, Assistant Professor; Ph.D., Houston. Production scheduling and control, engineering economics, engineering management, simulation.

Krishnaswami Srihari, Professor; Ph.D., Virginia Tech. Manufacturing systems, computer-aided process planning, expert systems, computer-integrated manufacture.

Adjunct Faculty

Douglas Elias, Ph.D., SUNY at Binghamton. Systems methodology, computer software engineering.

STATE UNIVERSITY OF NEW YORK AT BUFFALO

Department of Industrial Engineering

Programs of Study

The Department of Industrial Engineering offers courses of study leading to the degrees of Master of Engineering (M.Eng.), Master of Science (M.S.), and Doctor of Philosophy (Ph.D.). The M.Eng. degree with an engineering management concentration in production management prepares students for management positions within an engineering organization. Students enrolled on a full-time basis can complete the program in one year; part-time students generally take three to four years. All students are required to complete an engineering project linked to industry and related to practical problems. Students may pursue the M.S. or Ph.D. in three areas of concentration: human factors and ergonomics, operations research, or production systems and manufacturing engineering. Human factors and ergonomics focuses on applications of engineering, psychology, computer science, and physiology to the modeling, analysis, and design of environments for industry, the service sector, transportation, and personal living. Operations research applies mathematics, statistics, computer science, and engineering principles to the formulation and solution of mathematical models for problems in long-range planning, energy and urban systems, health systems, and manufacturing. Production systems and manufacturing engineering focuses on production planning and scheduling, computer-integrated manufacturing, enterprise management, quality assurance, facilities location and design, robotic systems, concurrent engineering, control of flexible manufacturing systems, materials handling, and storage systems. For both the M.S. and Ph.D. programs, each of the specialty areas requires completion of specified sets of core courses. The M.S. program may be completed in one of two ways: with a formal research thesis representing 6 hours of credit (required in the human factors area), satisfactory performance in an oral defense of thesis examination, and completion of at least 24 hours of nonthesis course work or by passing an oral comprehensive examination based on a degree program consisting of at least 30 hours of course work. The Ph.D. program provides an advanced level of study and training for the development of research scientists to pursue positions in academia, industry, and government. The equivalent of at least three years of full-time study beyond the baccalaureate degree is required for completion of the Ph.D. program. At least one year must be spent in full-time residence. Milestones in the program include the satisfactory completion of a core examination that tests mastery of material covered in fundamental graduate courses in the student's area of specialization, satisfactory completion of an advanced examination that tests a student's preparedness for research, satisfactory defense of a dissertation proposal presented in written form and defended orally, and completion of the dissertation that must be defended in an oral examination. Faculty members work very closely with students in guiding the identification and study of research problems.

Research Facilities

The Department of Industrial Engineering houses the department's Intelligent Manufacturing Systems Laboratory (IMSL), the Human Factors Laboratories (HFLs), and several other research facilities that total 7,140 square feet. The IMSL consists of equipment that includes robots, numerical control machines, computer-controlled material handling systems, and various control devices; computer hardware and software that includes Silicon Graphics and Sun SPARC workstations and microcomputers with software for CAD/CAM, simulation, and production planning and control applications. The HFLs consist of a human-computer interaction laboratory for simulating and analyzing a variety of research issues in human interaction with intelligent and automated systems and instrumentation that can support a variety of ergonomics studies in work physiology, biomechanics, and visual sciences. Eight specially equipped small rooms and several larger laboratories provide space for student experimentation. Also available to students are numerous Sun workstations and a computer-aided design laboratory with the full capability of 3-D solid modeling and computer-aided engineering software systems.

Financial Aid

Financial aid opportunities include Woodburn Fellowships, Presidential Fellowships sponsored by the program and awarded through engineering-wide competition, teaching assistantships, and research assistantships. Partial tuition scholarships are also available for full-time M.Eng. and M.S. students who are New York State residents. A number of additional fellowships are available for underrepresented students. New York State students may qualify for special tuition assistance programs. Academic-year stipends range from $10,700 to $14,000 plus a full-tuition scholarship.

Cost of Study

In 1998–99, tuition and special fees that included transportation and health services for full-time in-state students totaled $5970. Out-of-state tuition and fees were $9286.

Living and Housing Costs

The cost of living is very affordable compared with other northeastern American cities. Graduate students have access to University dormitory facilities and a variety of off-campus areas are also available. Attractive one- and two-bedroom apartments can readily be found for costs ranging from $250 to $450 per month.

Student Group

In 1998–99, there were 94 graduate students in the program, including 25 part-time students. Fourteen percent of graduate students are women, and 61 percent are international students. Forty percent of full-time students received financial aid.

Student Outcomes

The University's Ph.D. graduates have taken academic positions in such institutions as the University of Illinois at Urbana–Champaign, Naval Postgraduate School, Rochester Institute of Technology, and Rutgers University. Ph.D. graduates have also taken high-level research and development positions in corporations such as AT&T, Kodak, The Mitre Corporation, United Airlines, USWest, and US Steel. Graduates with master's degrees have taken a variety of management, research, and engineering positions at mid-size and large corporations both locally and nationally and with government agencies.

Location

The city of Buffalo is the hub of a metropolitan area of more than 1 million individuals and is one-half hour by car from Niagara Falls and a 2-hour drive from Toronto. Buffalo has a renowned philharmonic orchestra and a diversified downtown theater district. Canada's Stratford Shakespearean Festival and Niagara-on-the-Lake's Shaw Festival are nearby, as are New York State's ArtPark and the Chautauqua Institution, noted for their spring and summer festivals of art, drama, and music. The proximity to Lakes Erie and Ontario moderate temperatures both in summer and winter, and provide the facilities for a wide variety of water sports activities. Many downhill and cross-country ski areas are located a short drive south of the city. The city provides major-league professional football (Buffalo Bills), ice hockey (Buffalo Sabres), and other professional sports leagues.

The University and The Department

The University at Buffalo, which celebrated its 150th anniversary in 1996, is the principal research university in the statewide SUNY system. Graduate education is a primary mission of the University. The University provides opportunities for interaction with graduate students through graduate clubs, sports, and multidisciplinary and professional interest groups.

Applying

The fall semester deadline for all students requesting financial aid is February 1, and notification of financial aid awards begins March 1. The deadline for all students not requesting financial aid and who are not permanent residents is April 15, and August 1 for U.S. citizens and permanent residents. All students requesting financial aid are required to take the GRE; all international students from countries in which English is not the primary language must take the TOEFL.

Correspondence and Information

Director of Graduate Studies, Office of Graduate Affairs
Department of Industrial Engineering
307 Lawrence D. Bell Hall, Box 602050
State University of New York at Buffalo
Buffalo, New York 14260-2050
Telephone: 716-645-2357 Ext. 3
Fax: 716-645-3302
E-mail: maloney@acsu.buffalo.edu

State University of New York at Buffalo

THE FACULTY AND THEIR RESEARCH

Full-Time Faculty

Ramakrishna Akella, Associate Professor; Ph.D., Indian Institute of Science (India), 1982. Computer integrated manufacturing, yield management and learning.
 In-line yield prediction methodologies using patterned wafer inspection information. *IEEE Trans. Semiconductor Manufact.* Invited paper, special issue, 1998. With Nurani et al.
 In-line defect sampling methodology. *Yield Management: An Integrated Framework,* in *IEEE Trans. Semiconductor Manufact.* 9(4), 1996. With Nurani and Strojwas.
 Diversification under supply uncertainty. *Manage. Sci.* 39(8):944–63, 1993. With Anupindi.

Rajan Batta, Professor; Ph.D., MIT, 1984. Operations research, production systems.
 On the analysis of two new models for transporting hazardous materials. *Oper. Res.* 44:710–23, 1996. With Jin and Karwan.
 Scheduling repairs at Texas Instruments. *Interfaces* 23(4):68–74, 1993. With Sivakumar and Tehrani.
 Optimal obnoxious paths on a network: Transportation of hazardous materials. *Oper. Res.* 36:84–92, 1988. With Chiu.

Wayne Bialas, Associate Professor; Ph.D., Cornell, 1975. Operations research, decentralized systems control theory, environmental systems, statistics.
 Cooperative N-person Stackelberg games. *Proc. 28th IEEE Conf. Decis. Control,* 1989.
 A hybrid algorithm for the three-level linear resource control problem. *Comput. Oper. Res.* 13(4):367–77, 1986. With Wen.
 Two-level linear programming. *Manage. Sci.* 30(8):1004–20, 1984. With Karwan.

Ann M. Bisantz, Assistant Professor; Ph.D., Georgia Tech, 1997. Human-machine systems, human decision making.
 Modeling and analysis of dynamic judgment tasks using a lens model approach. *Proceedings of IEEE Conference on Systems, Man, and Cybernetics,* 1997. With Gay et al.
 Making the abstraction hierarchy concrete. *Int. J. Hum.-Comput. Studies* 40:83–117, 1994. With Vicente.
 The effects of feedback on performance and retention of skill in a natural language interface. *Behav. Inf. Tech.* 12(1):32–47, 1993. With Sharit.

Ismael de Farias Jr., Assistant Professor; Ph.D., Georgia Tech, 1995. Operations research, discrete and nonlinear optimization, scheduling.
 A generalized assignment problem with special ordered sets: A polyhedral approach. *Math. Programming,* in press. With Johnson and Nemhauser.
 Facets of the complementarity knapsack problem. *Math. Oper. Res., in press.* With Johnson and Nemhauser.
 A branch-and-cut approach without binary variables to combinatorial optimization problems with continuous variables and combinatorial constraints. Eng. Knowledge Rev., in press. With Nemhauser.

Colin G. Drury, Professor; Ph.D., Birmingham, 1968. Human factors, industrial ergonomics, human errors, quality control.
 Do people choose an optimal response criterion in an inspection task? *IIE Trans.* 30:257–66, 1998. With Chi.
 Training for aircraft visual inspection. *Hum. Factors Ergonomics Manufact.* 3:171–96, 1997. With Gramopadhye and Prabhu.
 Binocular rivalry as aid in visual search. *Hum. Factors* 39(4):642–50, 1997. With Mazumder and Helander.

Mark H. Karwan, Professor; Ph.D., Georgia Tech, 1976. Mathematical programming, applied operations research, multiple-criteria decision making.
 Routing automated guided vehicles in the presence of interruptions. *Int. J. Production Res.* 37(3):653–81, 1999. With Narasimhan and Batta.
 When haste makes sense: Cracking down on street markets for illicit drugs. *Socio-Economic Plann. Sci.* 31(4):293–306, 1997. With Baveja, Caulkins, and Batta.
 Use of convex cones in interactive multiple objective decision making. *Manage. Sci.* 43(5):723–34, 1997. With Prasad and Zionts.

Li Lin, Associate Professor; Ph.D., Arizona State, 1989. Manufacturing systems, process planning and control, simulation.
 An object-oriented FMS real-time and feedback control model. *Int. J. Comput.-Integrated Manufact.,* in press. With Lo.
 Reducing total tardiness cost in manufacturing cell scheduling by a multi-factor priority rule. *Int. J. Production Res.,* in press. With Chen.
 An integrated graphic user interface (GUI) for concurrent engineering design of mechanical parts. *Comput.-Integrated Manufact.* 11(1–2):91–112, 1998. With K. Chen, S. Chen, and Changchien.

Rakesh Nagi, Associate Professor; Ph.D., Maryland, 1991. Agile manufacturing, JIT, production planning and control, facilities design.
 Integrated scheduling of material handling and manufacturing activities for just-in-time production of complex assemblies. *Int. J. Production. Res.* 36(3):653–81, 1998. With Anwar.
 An integrated formulation of manufacturing cell formation with capacity planning and multiple routings. *Ann. Oper. Res.* 77:79–95, 1998. With Ramabhatta.
 Design and implementation of a virtual information system for agile manufacturing. *IIE Trans. Design Manufact.,* special issue on agile manufacturing, 29(10):839–57, 1997. With Song.

Victor Paquet, Assistant Professor; Sc.D., Massachusetts Lowell, 1998. Industrial ergonomics, occupational safety and health, musculoskeletal epidemiology.
 An evaluation of manual materials handling in highway construction work. *Int. J. Ind. Ergonomics,* special issue on MMH, in press. With Punnett and Buchholz.
 A validity study of fixed-interval observations for the assessment of body postures during construction work. In Proceedings of the Human Factors and Ergonomics Society 42nd Annual Meeting, pp. 945–9. 1998. With Punnett and Buchholz.
 Reducing hazards during highway tunnel construction. In Ergonomics in Health Care and Rehabilitation, 187–203, ed. V. Rice. Butterworth-Heinemann, 1998. With Buchholz.

Christopher M. Rump, Assistant Professor; Ph.D., North Carolina at Chapel Hill, 1995. Operations research, queuing design and control, probabilistic modeling.
 Stability and chaos in input pricing for a service facility with adaptive customer response to congestion. *Manage. Sci.* 44(2):246–61, 1998. With Stidham.
 Multiproduct production planning in the presence of work-force learning. *Eur. J. Oper. Res.* 106(2–3):336–56, 1998. With Mazzola and Neebe.
 Asymptotic behavior of a relaxed sequential flow-control algorithm for multiclass networks. *Proc. 33rd Ann. Allerton Conf. Commun. Control Comput., University of Illinois at Urbana-Champaign,* pp. 756–65. With Stidham.

Adjunct Faculty

Robert E. Barnes, Adjunct Associate Professor; Ph.D., SUNY at Buffalo, 1984. Engineering management, health systems and industrial safety, human factors.

James Llinas, Adjunct Research Professor; Ph.D., SUNY at Buffalo, 1976. Multisource information fusion, statistics, knowledge-based systems.

John Zahorjan, Adjunct Professor; Ph.D., SUNY at Buffalo, 1974. Production engineering, materials management, productivity, quality control.

Emeritus Faculty

Warren H. Thomas, Distinguished Teaching Professor; Ph.D., Purdue, 1963. Simulation, production control.

Lecturers

Betsy B. Anderson, Lecturer; M.B.A., SUNY at Buffalo, 1994.
Laurel Dutton, Lecturer; M.S., SUNY at Buffalo, 1990.
Thomas D. Hill, Lecturer; Ph.D., SUNY at Buffalo, 1993.

UNIVERSITY OF CALIFORNIA, BERKELEY

Department of Industrial Engineering and Operations Research

Programs of Study

Graduate programs leading to the M.S., M.Eng., Ph.D., and D.Eng. are offered. All programs emphasize the development and use of quantitative models and techniques for the design, analysis, control, and operation of complex systems in the industrial, service, and public sectors.

The M.S. and M.Eng. programs are designed to meet the needs of students with engineering, mathematics, or science backgrounds who wish to enhance their technical competence or to prepare for managerial positions by gaining a broader perspective on modern engineering practice. The Ph.D. and D. Eng. programs prepare students for advanced research in the theory and application of industrial engineering and operations research methodologies. In the most recent National Research Council survey, the department ranked first in the quality of doctoral education in the discipline.

Undergraduates from scientific disciplines other than engineering may be accepted into these programs. A master's degree may be earned by course work in conjunction with a thesis or a comprehensive examination. Doctoral degrees require an oral examination in the major and two minor fields followed by submission of a thesis demonstrating ability to conduct independent advanced research.

Research Facilities

The Industrial Engineering and Operations Research (IEOR) department has several computing facilities, including three instructional microcomputer laboratories equipped with a network of desktop computers and advanced computer applications laboratories equipped with desktop computers and Sun workstations. All the workstations are linked on an Ethernet connected to the central campus computers. IEOR students also have access to several other research laboratories on campus.

Financial Aid

University fellowships, scholarships, and tuition waivers are available each year, as is aid from industrial donors. There is a Graduate Opportunity Program for qualified students. Teaching and research assistantships are also available, as are loan programs. While financial aid is scarce in the first year of study, most qualified doctoral students receive aid after their first year.

Cost of Study

For the 1998–99 academic year, California residents paid $4408 ($2204 per semester), while nonresidents and international students paid $13,796 ($6898 per semester). Students should anticipate a fee increase.

Living and Housing Costs

Students should anticipate a monthly cost of $1190 for housing, food, books, personal expenses, and transportation. Married student housing is available for $358 to $559 per month. Because of eight- to twelve-month waiting periods, students are encouraged to apply for housing when they apply for admission.

Student Group

The department has approximately 80 graduate students, including many from other countries. Student chapters of the Institute for Industrial Engineers (IIE) and the Institute for Operations Research and the Management Sciences (INFORMS) provide opportunities for social activities as well as contact with the profession.

Location

The University is located at the base of the Berkeley Hills, directly across the bay from San Francisco. The San Francisco Bay Area has a tremendous variety of cultural and entertainment activities to suit all tastes and interests. Students have ready access to the Pacific coast and beaches, and excellent skiing areas are located 3½ hours to the east in the Sierra Nevada. The climate is cool in the summer, with no rainfall. Winters are mild, with intermittent rainy and sunny weather.

The University

The Berkeley campus of the University of California has a enrollment of 30,000, with 8,700 graduate students. The campus is noted for the academic distinction of its faculty, the high quality and scope of its research activities, and the variety and vitality of student activities. It is ranked by the academic community as one of the best graduate institutions in the United States.

Applying

Application deadlines are January 5 for fellowships and scholarships and February 10 for other applications for the fall semester. September 1 is the deadline for the spring semester. For those entering in the spring semester, financial aid is extremely limited.

Applicants should have a bachelor's degree in engineering, physical science, mathematics, or a closely related field.

The department requires all graduate applicants to submit verbal, quantitative, and analytical scores on the General Test of the Graduate Record Examinations. Further information on graduate programs may be obtained from the address below and from the department's World Wide Web site (http://www.ieor.berkeley.edu).

Correspondence and Information

Student Affairs Officer
Department of Industrial Engineering and Operations Research
4135 Etcheverry Hall
University of California
Berkeley, California 94720-1777
Telephone: 510-642-5485
E-mail: std-svcs@ieor.berkeley.edu

University of California, Berkeley

THE FACULTY AND THEIR RESEARCH

The department has close collaborations with faculty members in the Departments of Mechanical Engineering, Electrical Engineering and Computer Sciences, and Economics and in the School of Business Administration in such cross-disciplinary areas as robotics, integrated manufacturing, combinatorics, management information systems, and energy economics.

Ilan Adler, Professor; Ph.D., Stanford, 1971. Mathematical programming, stochastic programming, networks and graphs.

Alper Atamturk, Assistant Professor; Ph.D., Georgia Tech, 1998. Optimization logistics.

Richard E. Barlow, Professor; Ph.D., Stanford, 1960. Reliability theory, statistical data analysis, Bayesian decision theory.

Edward R. F. W. Crossman, Professor Emeritus; Ph.D., Birmingham (England), 1965. Human performance and skill, work systems and organization, database systems and design.

Stuart E. Dreyfus, Professor Emeritus; Ph.D., Harvard, 1964. Neural networks, dynamic programming, limits of operations research modeling, cognitive ergonomics.

C. Roger Glassey, Professor Emeritus; Ph.D., Cornell, 1965. Simulation of manufacturing systems, production planning and scheduling, mathematical optimization.

Kenneth Y. Goldberg, Associate Professor; Ph.D., Carnegie Mellon, 1990. Robotics for automated manufacturing; motion planning algorithms for feeding, sorting, fixturing, and assembly; design of parts and devices for automation; robotics in multimedia applications.

Raymond C. Grassi, Professor Emeritus; M.S., Berkeley, 1944. Facilities planning, materials handling, production systems, CAD.

Dorit S. Hochbaum, Professor; Ph.D., Pennsylvania, 1979. Discrete optimization, graph problems, nonlinear optimization, location and distribution problems, scheduling problems.

William S. Jewell, Professor Emeritus; Sc.D., MIT, 1958. Risk theory/actuarial science, reliability theory, Bayesian modeling, computer modeling.

Philip M. Kaminsky, Assistant Professor; Ph.D., Northwestern, 1997. Modeling and analysis of production and logistics systems.

Robert C. Leachman, Professor; Ph.D., Berkeley, 1979. Production planning and scheduling, transportation planning.

Robert M. Oliver, Professor Emeritus; Sc.D., MIT, 1957. Prediction and forecasting, risk analysis, decision analysis.

Shmuel S. Oren, Professor; Ph.D., Stanford, 1972. Mathematical modeling and analysis of systems, optimization, systems economics, electric utility planning.

Sheldon M. Ross, Professor and Director of Admissions; Ph.D., Stanford, 1967. Stochastic models, applied statistics, simulation, sequential decision processes, quality control.

Lee W. Schruben, Professor; Ph.D., Yale, 1975. Computer simulation.

J. George Shanthikumar, Professor; Ph.D., Toronto, 1979. Stochastic processes, manufacturing, production systems modeling.

Ronald W. Wolff, Professor Emeritus; Ph.D., Case Tech, 1963. Queuing theory, approximations and bounds of stochastic processes, queuing network application to communication and data transmission systems.

Candace A. Yano, Professor and Chair; Ph.D., Stanford, 1981. Production and distribution system planning.

UNIVERSITY OF CENTRAL FLORIDA

College of Engineering
Department of Industrial Engineering and Management Systems

Programs of Study

In addition to the Doctor of Philosophy degree, the department offers a Master of Science in Industrial Engineering (M.S.I.E.) degree and a Master of Science (M.S.) degree with specializations in engineering management, human engineering/ergonomics, manufacturing systems, operations research, precision engineering and manufacturing (which includes a special focus in high-performance engine optimization), product assurance engineering, simulation modeling and analysis, and interactive simulation and training systems. The M.S.I.E. requires a bachelor's degree in industrial engineering. The M.S. degree requires a bachelor's degree in engineering or a closely related field as well as specific prerequisites, depending on the area of specialization. Students pursuing the M.S.I.E. must complete at least 24 semester credit hours of course work and a 6-credit-hour thesis, while those students pursuing the M.S. degree may complete a course-only option with 36 semester credit hours.

The department has a major externally funded research program with an annual budget of about $2 million. Major research thrusts are in the areas of intelligent simulation and training systems, space shuttle processing productivity, manufacturing and design innovations for energy-efficient industrialized housing, microelectronics manufacturing systems, human-computer interaction and cybersickness, high-performance internal-combustion engine optimization and manufacturing, and alternative fuels. Students have the opportunity to participate as graduate research assistants in these as well as in other ongoing research activities under the supervision of the faculty.

Research Facilities

The University is connected to the Florida Instructional Resources Network, which provides access to an impressive array of computer hardware and software available within the State University System of Florida, including supercomputers and parallel processing machines. The department has several laboratories to support research and instruction. The Intelligent Simulation and Training Systems Lab is equipped with various computers and workstations. The Schrader Manufacturing Engineering Lab is equipped with a variety of machine tools, robots, numerically controlled machines, precision measuring instruments, automation manufacturing cell training systems, and CAD workstations. The Human Engineering/Ergonomics and Facilities Lab is stocked with a variety of equipment for ergonomic research and instruction. The Design and Rapid Prototyping Lab includes a laser rapid prototyping machine used for both research and instruction. The Engine Laboratory supports the unique program in high-performance engine optimization and manufacturing. The department Computer Laboratory has a broad mix of computers, workstations, printers, and plotters. Faculty offices are mostly equipped with Pentium machines and laser printers.

Financial Aid

Financial aid is available in the form of a limited number of research assistantships that require one-quarter- to one-half-time workloads, with compensation in the range of $3800 to $10,000 for nine months. Both the nonresident and resident tuition can be waived in many cases. The University's Financial Aid Office administers long-term loans and institutional emergency short-term loans, and the University's Office of Graduate Studies administers a limited number of fellowships. The department awards a Robert D. Kersten Fellowship annually to the best entering student who applies for admission prior to March 1. The Kersten Fellowship includes a stipend of $3250, a graduate assistantship, and a tuition waiver. A limited number of industrial fellowships are available from time to time.

Cost of Study

Tuition in 1999–2000 is $146.22 per semester hour for Florida residents and $506.95 per semester hour for non-Florida residents. Graduate fees paid by all students amount to $47.30 per semester plus lab fees where applicable. Costs are subject to change. Registration for 6 semester hours qualifies one as a full-time student.

Living and Housing Costs

On-campus housing in a dorm room ranged from $1430 to $1600 per semester in 1998–99. Meal plans range from $451 to $1140 per semester. There are many privately owned apartments near the University of Central Florida (UCF), some within walking distance.

Student Group

The University of Central Florida has a total enrollment exceeding 28,000 students. Of these, about 3,500 are enrolled in the College of Engineering. The Department of Industrial Engineering and Management Systems has approximately 125 undergraduate majors and 400 graduate students, including 75 Ph.D. students. Approximately 30 percent of the graduate students are women. International students represent Brazil, China, Chile, Egypt, France, Germany, India, Jordan, Malaysia, Thailand, Saudi Arabia, and Sweden, among others.

Student Outcomes

Graduates of the various programs are recruited by leading industrial, service, educational, and governmental organizations. Graduates are in faculty positions at Oklahoma, East Carolina, Old Dominion, Arizona State, Embry-Riddle, and the U.S. Military Academy, among others. Industry positions have been found with companies that include Lockheed Martin, Walt Disney, United Space Alliance, United Parcel Service, Andersen Consulting, Decision Focus, Deloitte Touche, and Cirent Technologies. Government positions have been obtained with such agencies as NASA, U.S. Army, U.S. Air Force, U.S. Navy, and the city of Orlando.

Location

UCF is located 15 miles east of downtown Orlando, in an area with an estimated population in excess of 1.7 million. The area has shown dramatic industrial growth, particularly in high-technology areas. Service-industry growth has also been dramatic with the increased presence of Disney World, EPCOT, MGM, and Universal Studios. The Atlantic Ocean, the Gulf of Mexico, and numerous rivers and spring-fed lakes provide diverse opportunities for outdoor recreation. Professional sports and numerous music and art venues provide many cultural and entertainment options.

The University and The College

The University of Central Florida opened in the fall of 1968 as Florida Technological University as a primary support for the Cape Kennedy space complex. The name was changed in 1978 to reflect the broader educational mission of the University. The University's central location makes it accessible from all parts of the state. The department is one of five departments in the College of Engineering, which is one of five colleges in the University. UCF has branch campuses in Cocoa, Daytona Beach, and South Orlando.

Applying

Prospective students should apply to the Graduate Admissions Office, University of Central Florida, Orlando, Florida 32816-0112, at least six weeks (six months for international students) before the start of classes for the term in which they plan to enroll. A $20 application fee, official transcripts from an accredited college, and GRE General Test scores are required. Minimum admissions requirements include a B average in the last two years of undergraduate work or a combined score of at least 1000 on the verbal and quantitative portions of the GRE. Students whose native language is not English must score at least 550 on the Test of English as a Foreign Language (TOEFL).

Correspondence and Information

Director of Graduate Affairs
College of Engineering
University of Central Florida
Orlando, Florida 32816-0459
Telephone: 407-823-2455

Dr. Charles H. Reilly, Chair
Department of Industrial Engineering
    and Management Systems
University of Central Florida
Orlando, Florida 32816-2450
Telephone: 407-823-2204
E-mail: creilly@mail.ucf.edu
World Wide Web: http://ie.engr.ucf.edu

University of Central Florida

THE FACULTY AND THEIR RESEARCH

Robert L. Armacost, Associate Professor; D.Sc., George Washington, 1976. Operations research, mathematical programming, nonlinear programming, sensitivity analysis, decision theory, analytic hierarchy process, group decision making, multicriteria optimization, resource-constrained project scheduling, machine scheduling, project management, response surface methodology optimization.

John E. Biegel, Professor Emeritus; Ph.D., Syracuse, 1972; PE (Florida). Intelligent simulation and training systems, knowledge-based systems, intelligent simulation.

Ahmad K. Elshennawy, Associate Professor; Ph.D., Penn State, 1987. Manufacturing systems/engineering, quality assurance, metrology.

Robert L. Hoekstra, Associate Professor; Ph.D., Cincinnati, 1992. Design for assembly and manufacturability, advanced systems manufacturing, concurrent engineering, materials handling, robotics, logistics.

Yasser A. Hosni, Professor; Ph.D., Arkansas, 1976; PE (Florida). Productivity, operations research, production planning, space operations, rapid prototyping.

Timothy G. Kotnour, Assistant Professor; Ph.D., Virginia Tech, 1995. Management systems engineering, organizational learning, organizational change and transitions, strategic planning, project management.

Dennis J. Kulonda, Associate Professor; Ph.D., North Carolina State, 1976. Engineering management, multichannel manufacturing, continuous simulations, factory models, interactive training systems for industry and government.

Gene C. H. Lee, Associate Professor; Ph.D., Texas Tech, 1986; PE (Florida, Texas). Ergonomics, human factors, occupational safety and health.

Linda C. Malone, Associate Professor and Graduate Coordinator; Ph.D., Virginia Tech, 1975. Regression, design of experiments, response surface analysis, applied statistics.

Pamela R. McCauley-Bell, Associate Professor; Ph.D., Oklahoma, 1993. Expert systems, fuzzy logic, knowledge acquisition, simulation, failure analysis, ergonomics.

Mansooreh Mollaghasemi, Associate Professor; Ph.D., Louisville, 1991. System simulation, decision theory, operations research, multicriteria optimization.

Michael A. Mullens, Associate Professor; Ph.D., Georgia Tech, 1979. Manufacturing systems, warehousing, materials handling, industrialized housing, concurrent engineering.

Julia J. A. Pet-Armacost, Associate Professor; Ph.D., Case Western Reserve, 1986. Operations research, systems engineering, risk assessment and management, multicriteria decision making, group decision making, multiobjective mathematical programming, sequencing and scheduling.

Michael D. Proctor, Assistant Professor; Ph.D., North Carolina State, 1991. Internet VR training simulation, training simulation engineering, organizational training simulations, behavioral training simulations, after-action review systems.

James M. Ragusa, Associate Professor; D.B.A., Florida State, 1973. Multimedia applications, artificial intelligence, expert systems, intelligent agents.

Charles H. Reilly, Professor and Chair; Ph.D., Purdue, 1983. Mathematical programming, especially integer programming; development of exact and heuristic solution procedures; empirical evaluation of solution methods; applications of mathematical programming, particularly to problems in manufacturing and telecommunications; applied operations research; simulation modeling; variate generation.

George F. Schrader, Professor Emeritus; Ph.D., Illinois, 1960; PE (Florida). Manufacturing, engineering management.

Jose A. Sepulveda, Associate Professor; Ph.D., Pittsburgh, 1981; PE (Florida). Simulation systems, health operations research, real-time and object-oriented simulation, production and operations management.

Kay M. Stanney, Assistant Professor; Ph.D., Purdue, 1992. Human factors, human-computer interaction, human/virtual environment interaction.

William J. Thompson, Assistant Professor; Ph.D., Arizona State, 1970. Engineering management, facilities planning, production control.

Gary E. Whitehouse, Professor and Provost; Ph.D., Arizona State, 1966; PE (Florida). Computer modeling, operations research, engineering management.

Kent E. Williams, Associate Professor; Ph.D., Connecticut, 1976. Simulator-based training systems design and development, training effectiveness and efficiency, intelligent tutoring systems, applications of cognitive science to training system design.

AREAS OF CURRENT SPONSORED RESEARCH

Organizational Change and Transitions. Currently funded by NASA, the research aims to better understand large-scale organizational change through the development, application, and refinement of organization transitional models and processes. NASA internal and external studies are both being conducted.

Intelligent Simulation and Training Systems. The department supports the Department of Defense's Center of Excellence in Simulation and Training. Research efforts include the integration of expert systems in simulators and simulations so as to develop generic technology-based training systems and simulators. Research activities are supported via the department's M.S. program in simulation systems.

Energy Efficient Industrialized Housing. Because housing for the twenty-first century must be more affordable and energy efficient, research is focused on the investigation of how modern industrial engineering/manufacturing technologies can contribute toward making housing a future export industry for the United States. Key aspects of the research are concurrent engineering, design for manufacturability, CAD/CAM integration, and simulation.

Multimedia Applications Laboratory. The research conducted by the lab focuses on how knowledge-based systems interfaced with multimedia software and hardware can provide intelligent information search, retrieval, and display.

Alternative Fuels and Engine Performance. Funded by major automobile and automobile-parts companies, this research focuses on improving engine performance and reducing engine emissions through work on the entire engine system, including research on reducing NO_x in engines and research on alternative fuels.

Virtual Reality. Research focuses on human factors issues in virtual reality, including human performance efficiency as well as health and safety. Research also extends to investigating the social impact of the technology.

Internet Virtual Reality Simulation. The department supports Cirent employee training through development of Internet-based virtual reality training simulations. The simulation system represents the fabrication of microelectronics on wafers and is used for personnel training, quality, and productivity improvements.

UNIVERSITY OF FLORIDA

College of Engineering
Department of Industrial and Systems Engineering

Programs of Study

The Department of Industrial and Systems Engineering offers the Master of Engineering and the Master of Science degrees, each with a thesis or nonthesis option, with programs in engineering management (utilizing business administration courses), manufacturing and production systems, operations research, quality and reliability assurance, and special interest options such as health systems. In addition, the department offers the Engineer degree and the Doctor of Philosophy degree with specialization in decision support systems, digital simulation, energy efficiency, engineering management, facilities layout and location, intelligent manufacturing, operations research, and production planning and control.

An undergraduate degree in one of the engineering disciplines or in mathematics, statistics, physics, quantitative management, computer science, or similar fields is a prerequisite. An articulation program of foundation courses is required if the student's background is deficient.

Programs of study may be tailored to emphasize the individual interests of graduate students. Many master's degree students work closely with individual faculty members on theoretical or applied research problems. Areas of specialization include production control, facilities layout, quality assurance, decision support systems, health systems, and reliability. The faculty carries out research in such methodological areas as mathematical programming, location theory, stochastic processes, and combinatorial optimization and in such applied areas as manufacturing control, production scheduling, capacity planning, and emergency evacuation.

Research Facilities

The department has a 1,500-square-foot emulated flexible manufacturing laboratory emphasizing modern manufacturing technology. Graduate students have access to a new computer lab for project work with faculty members. Remote and batch processing are available on the University's IBM mainframe computer, with terminals available throughout the department and the College of Engineering. An extensive network of PCs and workstations is also available. The Center for Applied Optimization has recently expanded the UNIX network with a grant from NSF.

Financial Aid

Teaching and research assistantships are available from the department, and the Graduate School provides additional funding for fellowships. Assistantship stipends begin at $9000 for the 1999–2000 academic year. A limited number of research fellowships are available for applicants seeking the Ph.D. degree.

Cost of Study

The 1998–99 tuition was $137.75 per semester credit hour for Florida residents and $481.32 for nonresidents. Students on assistantships are normally eligible for full tuition fee waivers.

Living and Housing Costs

The University has modestly priced housing available for both married and unmarried students. The city and University provide bus services, although many students find biking the simplest method for getting about the campus and city.

Student Group

Approximately 40,000 students attend the University, of whom about 10 percent are enrolled in the College of Engineering. The department has 292 undergraduates and 115 graduate students, including a rich mixture of students from Florida and from other states and countries.

Location

Gainesville is an attractive college city of approximately 100,000 people, located in north-central Florida, midway between the Atlantic Ocean and the Gulf of Mexico. Enjoying a temperate climate that supports many outdoor activities, Gainesville is within 2 hours of Jacksonville, Orlando, and Tampa, and it is close to many tourist and recreational facilities. In 1995, *Money* magazine named Gainesville the best city in the United States to live in.

The Department

The Department of Industrial and Systems Engineering originated in 1930 as a combined engineering and business program, becoming formalized as an industrial engineering department in 1934. The first master's degree was awarded in 1942. To reflect a reorientation toward more modern analytic techniques, as well as a broader range of interest, the department's name was changed to Industrial and Systems Engineering in 1964. Since that time, it has been very active in research-oriented activities and programs and has performed studies for a variety of state and federal agencies as well as industrial organizations.

The department is dedicated to academic excellence in teaching and research. Modern methods of industrial and systems engineering and operations research and their application to important problems in the public and private sectors provide the major areas of emphasis.

Applying

Application forms and additional information and brochures are available from the department.

Correspondence and Information

Graduate Coordinator
Department of Industrial and Systems Engineering
P.O. Box 116595
University of Florida
Gainesville, Florida 32611-6595
Telephone: 352-392-6729
E-mail: jackson@ise.ufl.edu
World Wide Web: http://www.ise.ufl.edu

University of Florida

THE FACULTY AND THEIR RESEARCH

Professors

Ravindra K. Ahuja, Ph.D., IIT Kanpur, 1982. Network and combinatorial optimization, linear and integer programming, routing and scheduling, heuristic optimization. Paths and flows. In *Annotated Bibliographies in Combinatorial Optimization*, pp. 283–309, eds. S. Martello, F. Maffioli, and M. Dell'Amico. John Wiley & Sons, 1998. A new pivot selection rule for the network simplex algorithm. *Math. Programming* 78:149–58, 1997 (with P. T. Sokkalingam and P. Sharma). *Network Flows: Theory, Algorithms, and Applications,* Prentice Hall, 1993 (with T. L. Magnanti and J. B. Orlin).

Barney L. Capehart, Ph.D., Oklahoma, 1967. Systems analysis, digital simulation, energy management, and energy efficiency. *Guide to Energy Management,* 2nd Edition. Fairmont Press, 1997 (with W. Turner and W. Kennedy). Energy auditing. In *Energy Management Handbook,* 3rd Edition, chap. 3, ed. W. C. Turner. Fairmont Press, 1996 (with Mark Spiller). Power factor benefits of high efficiency motors. *Energy Eng.* 93(3):6–18, 1996 (with Kevin D. Slack).

D. Jack Elzinga, Ph.D., Northwestern, 1968. Quality management, operations research, applied optimization. Planning for curriculum renewal and accreditation under ABET engineering criteria 2000. *Proc. ASEE Annu. Conf.,* 1998 (with Leonard, Beasley, and Scales). *Business Process Engineering: The State of the Art,* eds. Elzinga, Gulledge, and Lee. Kluwer, 1998. Business process management: Survey and methodology. *IEEE Trans. Eng. Manage.* 42(2), 1995 (with Horak, Lee, and Bruner).

Richard L. Francis, Ph.D., Northwestern, 1967. Facilities planning and robotic layout, location theory. Row-column aggregation for rectilinear distance p-median problems. *Transp. Sci.* 30(2), 1996. *Facility layout and location: An analytical approach,* 2nd ed., coauthors W.L.F. McGinnis and J.A. White. New York: Prentice Hall, 1992. On worst-case aggregation analysis for network location problems. *Ann. Oper. Res.* 40:229–46, 1992.

Donald W. Hearn, Chairman; Ph.D., Johns Hopkins, 1971. Operations research, mathematical programming, applied optimization. Congestion toll pricing of traffic networks. *Network Optimization,* Springer-Verlag series, *Lecture Note in Economics and Mathematical Systems,* pp. 51–71, 1997 (with P. Bergendorff and M. Ramana). Network equilibrium models and algorithms. In *Handbooks in Operations Research Management Science 8: Network Routing,* eds. M. O. Ball, T. L. Magnanti, C. L. Monma, and G. L. Nemhauser. North-Holland, 1995 (with Florian).

Panos M. Pardalos, Ph.D., Minnesota, 1985. Global optimization, integer programming and graph algorithms, complexity, parallel computing and optimization software. Global optimization problems in computer vision. In *State of the Art in Global Optimization: Computational Methods and Applications,* pp. 457–74, eds. C. Floudas and P. Pardalos. Kluwer, 1996 (with Sussner and Ritter). Global minimization of nonconvex energy functions: Molecular conformation and protein folding. *American Mathematical Society DIMAS Series* vol. 23, 1996 (with Shalloway and Xue). *Introduction to global optimization.* Kluwer, 1995 (with Horst and Thoai).

Boghos D. Sivazlian, Ph.D., Case Tech, 1966. Operations research, theory of inventory and replacement, applied stochastic processes, military applications. Distribution and first moments of the busy and idle periods in controllable M/G/l queuing models with simple and dyadic policies. *Stochastic Anal. Appl.* 12(1), 1995 (with Gakis and Rhee). The moments of the extended compound point process and applications. *Stochastic Anal. Appl.* 12(2), 1994 (with Gakis). The correlation of the backward and forward recurrence times in a renewal process. *Stochastic Anal. Appl.* 12(2), 1994 (with Gakis).

Ralph W. Swain, Ph.D., Cornell, 1971. Health systems analysis, information systems, manufacturing. Creating an intelligent building. *Healthcare Executive,* December 1988. Group technology for electronic assembly. *Proc. Int. Electron. Assem. Conf.,* October 1985.

Associate Professors

Süleyman Tüfekci, Ph.D., Georgia Tech, 1977. Integrated product/process design, manufacturing systems, decision support systems, production planning, emergency management and engineering. Special issue on emergency management and engineering. *IEEE Trans. Eng. Manage.,* May 1998 (with A. Wallace). $O(m^2)$ algorithms for the two and three sublot lot streaming problem. *Prod. Operations Manage.* 6(1):74–96, 1997 (with E. Williams and M. Akansel). Emergency management and engineering. *Safety Sci.* 20, May 1995 (with Beroggi and Wybo). An integrated emergency management decision support system for hurricane emergencies. *Safety Sci.* 20:39–48, 1995. A genetic algorithm for the talent scheduling program. *Commun. Operations Res.* 21(8):927–40, 1994 (with Nordstrom).

Stanislav Uryasev, Ph.D., Institute of Cybernetics (Kiev), 1982. Operations research: optimization of large systems with nonsmooth and stochastic behavior, probabilistic risk analysis, applications in energy, transportation, and environment. Analytic perturbation analysis for DEDS with discontinuous sample-path functions. *Stochastic Models* 13(3), 1997. Failure-dependent test, repair, and shutdown strategies: Reducing the impact of common-cause failures. *Nucl. Technol.* 116(3), 1996 (with P. Samanta). The score functions. In *Encyclopedia of Operations Research and Management Science,* pp. 614–7, eds. S. I. Gass and C. M. Harris. Boston/Dordrecht/London: Kluwer Academic Publishers, 1996 (with R. Y. Rubinstein and S. Shapiro). Derivatives of probability functions and some applications. *Ann. Operations Res.* 56:287–311, 1995.

Assistant Professors

Sherman X. Bai, Ph.D., MIT, 1991. Manufacturing systems, real-time production scheduling, optimal control/optimization, dynamic game models, digital simulation, supply chain management, information systems technology. Scheduling manufacturing systems with work-in-process inventory control: Single-part type systems. *IIE Trans.* 27:599–617, 1995. A production control problem in competition. *Comp. Math. Applications* 29(5):65–80, 1995.

Joseph P. Geunes, Ph.D., Penn State, 1999. Production planning and control models, inventory and supply chain management, applied optimization. The newsvendor model: A dynamic policy for infinite-horizon inventory analysis. *Proc. Dec. Sci. Inst. Conference,* 1997.

Diane A. Schaub, Ph.D., Arizona State, 1994. Experimental design, statistical process control, quality management, manufacturing engineering. Experimental design for estimating both mean and variance functions. *J. of Quality Technol.,* 1996. Using experimental design to optimize the stereolithography process. *Quality Eng.,* 1996.

Associate Engineer

Kenneth E. Dominiak, Ph.D., Polytechnic of Brooklyn, 1971. System theory.

Assistant Engineer

Thomas M. Kisko, M.E., Florida, 1973. Simulation, microprocessor applications, automation, software engineering; traffic modeling. Regional evaluation modeling system (REMS): A decision support system for emergency area evacuations. *Proc. Fourth Comput. Ind. Eng.,* 1991. Design of a regional evaluation decision support system: Integrating simulation and optimization. *Proc. Soc. Comput. Simulation Conf. Emerg. Mgmnt. Eng. Conf.,* 1991.

Courtesy Professors

Heinz K. Fridrich, M.S., MIT, 1965. Manufacturing education, management of technology, product realization process, intelligent manufacturing control, environmentally conscious manufacturing.

Sencer Yeralan, Ph.D., Florida, 1984. Applied stochastic processes, manufacturing systems. A new standard for industrial control languages. *Proc. 12th Comput. Ind. Eng. Conf.,* 1990. A survey of the use of intelligent automation in the food processing industry. *Proc. Third Conf. Recent Adv. Robotics,* 1990. The minimax centre estimation problem for automated roundness inspection. *Eur. J. Operational Res.* 41:64–72, 1989.

Professors Emeriti

James F. Burns, Ph.D., MIT, 1967. Industrial dynamics, behavioral decision theory, conflict resolution.

Richard S. Leavenworth, Ph.D., Stanford, 1964. Engineering economics, decision analysis, statistical quality control.

John F. Mahoney, Ph.D., Lehigh, 1960. Numerical methods, applied mathematics, tensors.

Eginhard J. Muth, Ph.D., Polytechnic of Brooklyn, 1967. Theory of reliability, stochastic modeling, microprocessor applications, robotics, automation in manufacturing.

John A. Nattress, D.E., Embry-Riddle, 1969. Engineering economy, engineering administration.

Donald B. Wilcox, J.D., Emory, 1952. Legal aspects of engineering, accident prevention.

UNIVERSITY OF ILLINOIS AT URBANA–CHAMPAIGN

Department of General Engineering
Program in Engineering Design

Programs of Study	The Department of General Engineering offers a program of graduate instruction leading to the degree of Master of Science in general engineering. Students may also enroll in a joint M.S.G.E./M.B.A. program. Areas of study and research include computer-aided design, computer graphics, optimization, telecommunication network design, simulation, manufacturing systems, real-time decision making, design systems, nondestructive evaluation, reliability, robotics, control systems, and biomechanics/rehabilitation engineering. The program offers an approach to systems engineering and engineering design that crosses disciplinary lines.
	Each student must take a core of courses in general engineering and must also complete a thesis or project. In addition, the student undertakes study in a particular disciplinary area. The thesis or project may be based upon some aspect of an actual design problem, and the student may interact with cooperating companies throughout the course of an academic year in order to complete this requirement.
	Graduates of the program are prepared to enter professional engineering positions in private industry, government, or as entrepreneurs.
Research Facilities	The College of Engineering has an extraordinary computing environment, equipped with everything from supercomputers to a wide variety of workstations. NCSA Mosaic™, a distributed hypermedia browser that provides a unified interface to the Internet, was developed at the National Center for Supercomputing Applications on campus. The department has the following laboratories in support of research: algorithm visualization, decision systems, engineering graphics, microelectromechanical systems, manufacturing systems, design systems, nondestructive testing and evaluation, robotics, digital controls, genetic algorithms, John Deere Mechatronics, GTE Telecommunications, solid modeling, and computer-aided engineering. In addition, machine shops and facilities for the construction of prototypes are available within the College of Engineering.
	The recently opened Grainger Library is the largest engineering library in the United States and is one of the most technologically advanced information management centers in the world. UIUC is also home to well-known research organizations, including Beckman Institute for Advanced Science and Technology, Coordinated Science Lab, Materials Research Lab, Microelectronics Lab, and Talbot Mechanical Testing Lab.
Financial Aid	Financial aid is available in the form of research or teaching assistantships and scholarships or fellowships. The stipend for a half-time assistantship is $11,826 for the nine-month academic year 1999–2000. Tuition and service fee requirements are waived for students holding assistantships of one quarter time or more.
Cost of Study	Tuition in 1999–2000 is $2308 per semester for residents of Illinois and $5882 for nonresidents. All figures are subject to change.
Living and Housing Costs	In 1999–2000, graduate residence hall rates are $2696 to $3008 for a single room per person per year and range from $2416 to $2872 for a double room per person per year. The board contract ranges from $1435 to $3640 per year. Off-campus housing is available at similar or higher rents. All rates are subject to change.
Student Group	Graduate College enrollment includes students from every state and from more than ninety countries, and it exceeds 8,000 out of a campus enrollment of more than 34,000. More than 90 percent of the graduate students receive some form of financial aid. Most students enter from undergraduate study, while a few return from service or industry, some on a released-time basis. There are currently 58 graduate students doing their thesis research with general engineering faculty members.
Student Outcomes	A degree in the Department of General Engineering allows the student flexibility in tailoring a technical degree with a specific area of research and study, maximizing employment opportunities. Recent graduates are employed in a wide variety of fields, including manufacturing, business, medical research, consulting, and entrepreneurial enterprises. Employment rates after graduation average 95 percent, with starting salaries ranging from approximately $50,000 to $70,000 per year.
Location	The University is located in the twin cities of Urbana-Champaign (population 100,000) in central Illinois, 130 miles south of Chicago. The community has excellent public school and park systems and numerous shopping facilities. Transportation links include three interstate highways, Amtrak service, and Willard Airport, which maintains regular flights on American Eagle, Northwest Airlink, TransWorld, and USAir.
The University and The Department	Cultural activities on the campus are fostered by the Krannert Center for the Performing Arts, which contains five separate theaters for orchestra, opera, choral groups, theater, and dance. The 16,000-seat Assembly Hall serves as the stage for some of the nation's foremost entertainers. In addition to Big Ten Conference football and basketball, the University offers a broad program of intramural athletics. The University's 1,500-acre Allerton Park serves numerous recreation purposes and is the site of many conferences and symposia.
	The Department of General Engineering was established seventy-seven years ago and administers more than $750,000 in grants for research. The graduate degree program graduated its first student in 1979, and a joint M.S.G.E./M.B.A. program was added in 1987. Alumni work in such diverse fields as manufacturing, consulting, law, and medicine.
Applying	Information and application forms are obtainable on request. Applications should be submitted by February 1 for August admission and by October 1 for January admission. The program is open to all graduates of accredited programs in engineering who have maintained a minimum grade point average of 3.0 (on a 4.0 scale) during the last two years of undergraduate study, but admission is competitive. Undergraduate preparation should include courses in differential equations and control systems theory, but any deficiency can be made up after enrollment.
Correspondence and Information	Dr. Harry E. Cook, Head Department of General Engineering 117 Transportation Building University of Illinois at Urbana-Champaign 104 South Mathews Avenue Urbana, Illinois 61801 Telephone: 217-333-2730 World Wide Web: http://www.ge.uiuc.edu

University of Illinois at Urbana-Champaign

THE FACULTY AND THEIR RESEARCH

Professors
T. F. Conry, Ph.D., Wisconsin. Tribology.
H. E. Cook, Head; Ph.D., Northwestern. Product management, Six Sigma, value analysis.
W. J. Davis, Ph.D., Purdue. Large-scale systems, real-time decision making, manufacturing systems.
D. E. Goldberg, Ph.D., Michigan. Genetic algorithms.
E. N. Kuznetsov, D.Sc., Central Research Institute of Structures (Moscow). Tensile structures.
J. V. Medanic, D.Sc., Belgrade. Systems and control.
J. I. Palmore, Ph.D., Berkeley, Ph.D., Yale. Dynamical systems theory, celestial mechanics.
R. L. Price, Ph.D., Stanford. Organizational behavior, engineering management.
M. W. Spong, D.Sc., Washington (St. Louis). Nonlinear control, robotics.

Associate Professors
S. A. Burns, Ph.D., Illinois. Structural optimization, numerical methods, visualization.
W. B. Hall, Ph.D., Waterloo. Structural analysis, reliability.
M. H. Moeinzadeh, Ph.D., Ohio State. Biomechanics.
M. H. Pleck, Ph.D., Illinois. Computer graphics, computer-aided design.
H. L. M. dos Reis, Ph.D., MIT. Nondestructive testing and evaluation of materials and structures.
R. S. Sreenivas, Ph.D., Carnegie Mellon. Discrete event systems, control theory.
D. L. Thurston, Ph.D., MIT. Operations research, materials selection and design.
L. Wozniak, Ph.D., Illinois. Computer-simulated control systems.

Assistant Professors
N. R. Aluru, Ph.D., Stanford. CAD for microelectromechanical systems, applied and computational mechanics.
F. Bullo, Ph.D., Caltech. Nonlinear controls.
R. Srikant, Ph.D., Illinois. Network design and operations research.

Adjunct Professors
J. V. Carnahan, Ph.D., Purdue. Systems engineering, simulation.
R. L. Ruhl, Ph.D., Cornell. Dynamic systems and manufacturing systems.

Adjunct Assistant Professor
M. G. Strauss, Ph.D., Texas. Rehabilitation engineering.

Adjunct Lecturers
C. H. Mottier, J.D., Illinois. Law.
B. D. Woodard, J.D., South Carolina, M.B.A., Illinois. Law.

Emeritus Professors
J. S. Dobrovolny, M.S., Illinois. Technical education.
R. D. Hugelman, Ph.D., Oklahoma State. Fluidic control systems.
H. W. Knoebel, B.S., Illinois. Dynamical systems.
G. E. Martin, Ph.D., Illinois. Systems design.
L. D. Metz, Ph.D., Cornell. Vehicle mechanics, computer-aided design.
D. C. O'Bryant, Ed.D., Illinois. Educational systems.
M. Scheinman, M.S., CUNY, City College. Structural analysis and design.
H. Streeter, LL.B., Ph.D., Iowa. Ergonomics, man-machine interaction, legal problems in design.
T. R. Woodley, Ph.D., Illinois. Applied mechanics.

UNIVERSITY OF MARYLAND, COLLEGE PARK

College of Engineering
Reliability Engineering Program

Programs of Study

The Reliability Engineering Program at the University of Maryland is an interdisciplinary program leading to the Master of Science and Doctor of Philosophy degrees in reliability engineering. The program is designed to meet industry needs for engineers who can design and develop new products that meet high reliability, safety, and quality requirements. Some background in probability and statistics is desired. Course work emphasizes a basic physics-of-failure approach to the design of electronic and mechanical systems and the prediction of reliability and life based on the variability of materials properties and manufacturing tolerances. The program involves faculty members from every department in the College of Engineering as well as those from the Department of Computer Science and the College of Business and Management. Courses also deal with the following subjects, which are of primary concern in reliability engineering: failure and life data analysis and interpretation, stress versus life characteristics of electronic and mechanical components, derating of components to improve reliability, accelerated testing, statistical experimental design for use in developing robust designs and reducing manufacturing process variability, computer-aided design tools to permit concurrent engineering for reliability analysis, expert systems as an aid in system design and failure analysis, life cycle engineering to factor reliability into total systems costs, and Bayesian statistics. Qualified students in any field of engineering who have an interest in design for reliability and quality are eligible for graduate study in this interdisciplinary program.

Reliability-related research activities are under way in many departments at the University. In the materials and nuclear engineering department, reliability and risks associated with the nuclear power program are analyzed, as are radiation effects on the performance of electronic devices. In the mechanical engineering department, the Center for the Reliability of Electronic Packaging develops analytical tools and software for the reliability analysis and improvement of all levels of electronic packaging. A new effort focuses on reliability of gallium arsenide devices. In the electrical engineering department, design methods are studied to mitigate specific failure modes associated with high-density digital electronic devices. In the civil engineering department, research is conducted on the probabilistic design of engineering structures and risk-based inspection of such structures. The Department of Computer Science is a member of the Software Engineering Laboratory, a university-industry-government consortium actively involved in programs design to assess and improve software reliability and quality.

Research Facilities

Research laboratories throughout the University are heavily involved in reliability- and quality-related activities. There are two failure analysis laboratories equipped with scanning and transmission electron microscopes. Extensive optical and infrared microscopes are also available for materials characterization and failure analysis. The Mechanics of Materials Laboratory is equipped for measuring basic materials properties as well as thermal and creep properties. The Vibrations Laboratory is available for evaluating various vibration and shock loading conditions and assessing the effectiveness of environmental stress screening. Extensive software development and simulation laboratories are available and employed for activities including thermal and vibration analysis, expert systems development and evaluation, and development of concurrent engineering design software. Other laboratories are available for exploring the reliability and life characteristics of ceramic and composite materials.

Financial Aid

The University of Maryland makes every effort to offer financial assistance to qualified students through a variety of programs. Seventy percent of all full-time graduate students receive financial support. Graduate school fellowships are awarded on a competitive basis to full-time students nominated by the department. Application materials should be submitted early for consideration for this program. Teaching and research assistantships are also available for qualified students. The Reliability Engineering Program has also been able to arrange part-time work at local companies for graduate students with relevant backgrounds.

Cost of Study

Graduate tuition fees per credit hour for the academic year 1998–99 were $272 for residents and $400 for nonresidents. A $50 fee is required with graduate applications.

Living and Housing Costs

Shared on-campus housing in an apartment unit is available for $1634 per semester. This unit is shared with 4 or 5 other students. The off-campus housing office maintains an extensive and up-to-date computerized list of rooms, apartments, and houses that are for rent in the area. Average monthly rates for housing are $200–$400 for a room in a private or student home; $400–$650 for an efficiency, basement apartment, or one-bedroom apartment; and $250–$400 for a shared apartment.

Student Group

There are 160 students currently enrolled in the Reliability Engineering Program, with 43 pursuing the Ph.D. degree. These students are active in four departments at the University. Enrollment has been increasing steadily since the inception of the program in 1988.

Location

The University of Maryland is located on a 1,378-acre campus between Washington and Baltimore and is only 30 miles from historic Annapolis and the Chesapeake Bay. Six major government research laboratories are within a 10-mile radius. Cultural activities on campus are extensive, and both cities offer major theater attractions and museums.

The University

The College Park campus of the University of Maryland has a total enrollment of approximately 32,000. The College of Engineering undergraduate enrollment is 2,900, and graduate enrollment is 1,200. The College Park campus is the flagship campus of the state university system. New and expanding programs include manufacturing, robotics, biotechnology, systems engineering, and reliability engineering. Recreational facilities on the campus include thirty-eight outdoor tennis courts, two swimming pools, handball and racquetball courts, and an eighteen-hole golf course.

Applying

The application deadline is February 1 for the fall semester and September 1 for the spring semester. International student applications should be completed by November 1 for the following September. Applicants are automatically considered for assistantship awards.

Correspondence and Information

Reliability Engineering Program
Department of Materials and Nuclear Engineering
University of Maryland, College Park
College Park, Maryland 20742-7531
Telephone: 301-405-7299

University of Maryland, College Park

THE FACULTY AND THEIR RESEARCH

Bilal Ayyub, Professor of Civil Engineering; Ph.D., Georgia Tech. Structural reliability, risk assessment, and failure investigation.

Michael Ball, Professor of Business Management; Ph.D., Cornell. Network reliability and optimization.

Victor Basili, Professor of Computer Science; Ph.D., Texas at Arlington. Software engineering and quality.

Joseph Bernstein, Assistant Professor of Reliability Engineering; Ph.D., MIT. Microcircuit device reliability.

Aristos Christou, Professor of Materials, Nuclear and Reliability Engineering; Ph.D., Pennsylvania. Reliability of optoelectronic devices, failure modes in MOSFETs and MMICs.

Jeffrey Frey, Professor of Electrical Engineering; Ph.D., Berkeley. VLSI design and reliability enhancement, device physics of failure.

Thomas Fuja, Assistant Professor of Electrical Engineering; Ph.D., Cornell. Error correction codes and fault-tolerant design.

Isabel Lloyd, Assistant Professor of Engineering Materials; Ph.D., MIT. High-performance electronic ceramics, ferroelectric devices.

Patricia Mead, Assistant Professor of Mechanical Engineering; Ph.D., Maryland. Optoelectronic device reliability.

James Milke, Assistant Professor of Fire Protection Engineering; Ph.D., Maryland. Fire risk assessment.

Mohammad Modarres, Professor of Nuclear and Reliability Engineering; Ph.D., MIT. Expert system applications in reliability and safety, probabilistic risk assessment.

Ali Mosleh, Associate Professor of Nuclear and Reliability Engineering; Ph.D., Berkeley. Common cause failures, Bayesian reliability analysis and computer security.

Martin Peckerar, Professor of Electrical Engineering; Ph.D., Maryland. Microcircuit reliability.

Marvin Roush, Professor of Nuclear and Reliability Engineering; Ph.D., Maryland. Reliability physics, risk assessment, human reliability engineering.

Charles Schwartz, Associate Professor of Civil Engineering; Ph.D., MIT. Fracture mechanics, software design for reliability engineering applications.

Carol Smidts, Assistant Professor of Reliability Engineering; Ph.D., Brussels (Belgium). Software reliability, human reliability analysis.

Paul Smith, Professor of Statistics; Ph.D., Case Western Reserve. Statistical process control, Taguchi methods.

UNIVERSITY OF MINNESOTA

Program of Industrial Engineering

Programs of Study
Both M.S. and Ph.D. degrees are offered in industrial engineering. Two master's degree plans are available. A Master of Science degree (M.S.) Plan A requires a thesis and 20 semester course credits. A minimum of 14 course credits is taken in the major field and a minimum of 6 in one or more related fields. A Master of Science degree (M.S.) Plan B requires 30 semester course credits and one to three papers based on independent study. A minimum of 14 credits is taken in the major field and a minimum of 6 in one or more related fields, with the remainder taken from either major or nonmajor categories. The division of credits between the major and minor is quite flexible. Considerable flexibility exists in customizing the student's degree program by taking courses from related programs in statistics, mathematics, computer science, operations and management science, information and decision science, design/manufacturing, and management of technology. There is no foreign language requirement. The Ph.D. requires preliminary written and oral examinations, 44 semester course credits, and a final oral defense of the thesis. A student may substitute for the minor a suitable supporting program, which may embrace several disciplines. The University of Minnesota is regulated under a semester system. Students may choose to specialize in one of the four following areas: manufacturing systems and logistics, operations research, engineering management, and human factors. The program offers opportunities for interdisciplinary programs with the School of Management, and the Departments of Mathematics, Statistics, and Computer Science.

Research Facilities
The research facilities include a human factors engineering laboratory; the Robotics and Human Motion Laboratory; and the Computer Integrated Manufacturing Laboratory. This state-of-the-art facility features a NeXT station, several Sun Workstations, Pentium II-based personal computers and Macintosh computers, and software such as LINDO, MPX, SLAM, and SIMAN. In addition, students and faculty have access through high-speed networks to the most powerful computer systems for research and instruction in the world: the CMZ and CM5 Connection Machines, CRAY-2, CRAY X-MP, CRAY C90, CRAY T3D, and CRAY T3E supercomputers. CAD facilities include numerous workstations (e.g., Sun, Silicon Graphics, IBM, and Apple) plus a wide complement of commercial software.

Financial Aid
Excellent opportunities for financial support are available to qualified candidates through teaching and research assistantships, graduate fellowships and student internships. Qualified students may receive assistance in the form of National Science Foundation (NSF), Department of Defense, Graduate School, and corporate fellowships and teaching assistantships. In addition, many faculty members conduct research programs supported by the NSF, the National Aeronautics and Space Administration, the U.S. Air Force, the Department of Energy, the Department of Defense, the National Institutes of Health, the Environmental Protection Agency, industrial sponsors, and others. These programs employ students on research that may also be used for their theses. Fellowships and assistantships provide tuition scholarships. The University is located in a thriving metropolitan area and a vibrant technological center with a strong opportunities for interaction with industry.

Cost of Study
For 1998–99, tuition for Minnesota residents was $1710 per quarter (covering 7–14 credits). Nonresident tuition was $3358 per quarter (7–14 credits). Student service fees were about $160 per quarter. Modest increases in these rates may occur for 1999–2000, pending Regents' action. Graduate (teaching or research) assistants holding quarter-time appointments pay half the resident tuition rate, and those with half-time appointments pay no tuition. Detailed information on tuition and service fees can be obtained from the Office of the Dean of the Graduate School.

Living and Housing Costs
Information about housing may be obtained from the Housing Bureau, Comstock Hall. Accommodations are available for single students in modern residence halls and for married students in permanent one- and two-bedroom units. Early application for family units is advisable. There is also a good selection of privately owned housing near the University or on University express bus lines.

Student Group
There are approximately 10,000 graduate students on the Twin Cities campus, about 60 of whom are in industrial engineering. The student body includes students from all sections of the country and from every continent.

Student Outcomes
Graduates of the program have found engaging opportunities with a diverse array of employers. Research positions range from local industry, such as Honeywell, 3M, and a number of engineering and technical firms, to national companies like Hewlett-Packard and Applied Materials, to overseas firms. Some candidates opt to return to academia to teach or do research at local, national, and international universities.

Location
The Twin Cities offer outstanding opportunities for cultural and recreational activities on or very near the campus. The Twin Cities are renowned for varied attractions such as the Minnesota Orchestra and St. Paul Chamber Orchestras and the Science Museum and Omnitheater. The Guthrie Theater, the University Theater, and many regional theater companies make the cities a rich drama center. More than 11,000 Minnesota lakes, many within the St. Paul or Minneapolis city limits, and the contiguous vacation lands of northern Wisconsin and the St. Croix River Valley are home to many summer and winter sports.

The University and The Program
The industrial engineering program provides an unusually stimulating environment for graduate study. Many members have advanced degrees in the basic sciences as well as in engineering. A strong program of seminars and visiting lectureships has brought many well-known scientists and engineers to the department to offer courses in their specialties. The department is famous for its research and interest in a broad spectrum of technological endeavors. Certain faculty members have received honorary degrees from foreign universities, and several have been elected to the National Academy of Engineering. University facilities for swimming, tennis, golf, basketball, skating, broomball, handball, squash, and other sports are available to all students. There is also an active intramural program.

Applying
Application deadlines for admission are as follows: fall semester, December 31; spring semester, October 25. Admission is granted for both fall and spring semesters. All admission applications submitted no later than December 15 of the preceding year for which a September appointment is requested are automatically considered for financial assistance. All correspondence concerning admission should be directed to the Graduate School of the University, in Minneapolis.

Correspondence and Information
For further information:

Dr. T. H. Kuehn, Director of Graduate Studies
Department of Mechanical Engineering
University of Minnesota
111 Church Street, SE
Minneapolis, Minnesota 55455

University of Minnesota

THE FACULTY AND THEIR RESEARCH

The Program of Industrial Engineering is a quantitative program with a strong research focus on manufacturing systems, production logistics, transportation systems, applied optimization, decision analysis, information systems, engineering management, human-computer interface, and ergonomics. Activities are supported by leading edge laboratories, including the Laboratory for Manufacturing Logistics, Decision Support Laboratory, Computer Integrated Manufacturing Laboratory, and the Human Factors Laboratory. Faculty members in the program are affiliated with a number of University centers, including the Center for Development of Technological Leadership, the Center for Transportation Studies, and the Center for Distributed Robotics. Research in the program is funded by several federal agencies and industries, including NSF, DARPA, USDOT, Intel, Northwest Airlines, and Ford, among many others.

Sant Ram Arora. Optimization concepts applied to the design and operational management of production systems, resource allocation, capacity sizing and layout of production facilities, specification of quality dimensions, logistics support, design of experiments.

Avram Bar-Cohen. Executive Director and James Renier Chair in Technological Leadership, Center for the Development of Technological Leadership. Electronic packaging and manufacturing process, management of technology.

Saifallah Benjaafar, Graduate Program Director. Modeling, design, and analysis of manufacturing systems; production planning and scheduling; plant layout and logistics; flexible manufacturing and applications of stochastic processes; queuing networks; simulation; transportation and traffic flow modeling.

Caroline Hayes. Intelligent decision support tools for assisting in complex engineering tasks such as manufacturing planning and design.

Diwakar Gupta. Stochastic models of manufacturing systems, production economics and inventory management, queuing theory, and health operations.

Tarald Kvålseth. Human factors and ergonomics; application of ergonomics to workstation design; job analysis and evaluation; safety issues.

Yechiel Shulman, H.W. Sweatt Chair in Technological Leadership and Director, Center for the Development of Technological Leadership. Management of technology, with emphasis on management of emerging technologies; operations management; issues in technology transfer.

Patrick Starr. Modeling and simulation as applied to manufacturing systems and vehicle dynamics; sensitivity analysis of dynamic systems; articulating the process of design in various domains; faculty adviser for the University of Minnesota solar car, Baja Buggy, and Formula SAE car.

Michael Taaffe. Stochastic processes, methodological issues in simulation modeling and analysis

Selected Recent Publications

Arora, S.R., and S. Kumar. Efficient work force scheduling for a serial processing environment: A case study at *Minneapolis Star Tribune*. *OMEGA* 27(1):115-22, December, 1998.

Arora, S. R., and S. Kumar. Re-engineering: Focus towards enterprise integration. *Interfaces.*

Arora, S. R., and S. Kumar. A model for risk classification of banks. *Int. J. Res. Prog. Manage.* 16:145–65, March-April 1995.

Arora, S. R., and S. Kumar. Improved man power utilization in live-run shifts of newspaper printing. *IIE Trans.* 26(2), March 1994.

Arora, S. R., and S. Kumar. Development of internal audit and cycle counting procedures for reducing inventory miscounts. *Int. J. Operation Res. Manage.* 12:61–70, 1992.

Arora, S. R., and S. Kumar. Effects of inventory miscounts and non-inclusion of lead time variability on inventory system performance. *IIE Trans.* 24(2):96–103, 1992.

Benjaafar, S., and D. Gupta. Scope versus focus: Issues of flexibility, capacity, and number of production facilities. *IIE Transactions,* 30(5):413-25, 1998.

Benjaafar, S., and M. Sheikhzadeh. Design of flexible plant layouts. *IIE Trans.,* in press.

Benjaafar, S., and M. Sheikhzadeh. Scheduling policies, batch sizes, and manufacturing lead times. *IIE Trans.* 29(2):159–66, 1997.

Benjaafar, S. On production batches, transfer batches and lead times. *IIE Trans.* 28(4):357–62, 1996.

Benjaafar, S., and R. Ramakrishnan. Modeling, measurement, and evaluation of sequencing flexibility in manufacturing systems. *Int. J. Production Res.* 34(5):1195–220, 1996.

Benjaafar, S. Modeling and analysis of machine sharing in automated manufacturing systems. *Eur. J. Operational Res.* 91(1):56–73, 1996.

Schlabach, J. L., and **C. C. Hayes.** Fox-GA: A generic algorithm for generating and analyzing battlefield courses of action. *J. Evolution. Comput.,* in press.

Hayes, C. C., and M. I. Parzen. Quem: An achievement test for knowledge-based systems. *IEEE Trans. Data Knowledge Eng.* 9(6):838–47, 1997.

Fu. M. C., and **C. C. Hayes.** SEDAR: An expert critiquing system. *J. Comput. Civil Eng.* 2(1):60–9, 1997.

Hayes, C. C., and D. M. Ganes. Operator construction for re-usable planners in complex practical domains. *Int. J. Expert Syst. Res. Appl.* 9(3):393–408, 1996.

Kvålseth, T. O. Square-root formula for choice reaction time. *Percept. Motor Skills* 83:475–8, 1996.

Kvålseth, T. O. A generalized measure of homogeneity for nominal categorical data. *Percept. Motor Skills* 83:957–8, 1996.

Kvålseth, T. O. The relative useful information measure: Some comments. *Inf. Sci.* 56:35–38, 1991.

Kvålseth, T. O. On generalized information measures of human performance. *Percept. Motor Skills* 72:1059–63, 1991.

Shulman, Y., and A. Bar-Cohen. Pivotal technologies: An essential element in an MOT curriculum. *Proceedings of the Fourth International Conference on Management of Technology,* February 1994.

Shulman, Y., A. Shenhar, and D. Dvir. A two-dimensional taxonomy of products and innovation. *Proceedings of the Annual Conference of the Product Development and Management Association,* October 1993.

Starr, P. J. A personal computer-based CAD system for investigating lateral stability of vehicle-dolly-trailer combinations. *Int. J. Vehicle Design* 14(2/3):194–204, 1993.

Starr, P. J. Integration of simulation and analytical submodels for supporting manufacturing decisions. *Int. J. Production Res.* 29(9):1733–46, 1991.

Starr, P. J., E. Wong, and J. P. Pouplard. Eigenvalue sensitivity and linear system structure. *Large Scale Syst.* 10(3):175–92, 1986.

UNIVERSITY OF MISSOURI–COLUMBIA

College of Engineering
Department of Industrial and Manufacturing Systems Engineering

Programs of Study	The Department of Industrial and Manufacturing Systems Engineering offers both the Master of Science and the Doctor of Philosophy degrees. The master's programs include (1) a six-course core common to all master's candidates; (2) a three-course concentration around one of three areas of study—operations research, integrated manufacturing systems, and supply chain management; and (3) either supervised thesis research or one technical elective and a supervised design project. Two dual master's degree programs are offered that combine industrial engineering with public health or with business administration.
	The doctoral program emphasizes the conduct of theoretical and applied research to solve complex system design and evaluation problems. Areas included in the current department research focus are automated inspection systems, manufacturing analysis and design, integrated production systems, combinatorial optimization, supply chain management, and operations research.
	Students must successfully complete a Ph.D. qualifying examination before they are officially accepted into the doctoral program.
Research Facilities	The Ellis Library, the Engineering Library, and eight other libraries on the Columbia campus contain 2.3 million volumes and 2.1 million microforms. Bibliographic specialists and dedicated databases aid graduate students in their research. The College of Engineering has access to the University computer system and also maintains its own dedicated Engineering Computer Network composed of the most modern computing facilities. The department maintains modern facilities for research in all areas. Equipment and facilities of special note include the operations research/simulation lab consisting of an integrated computer network and state-of-the-art software and the Computer Aided Manufacturing Lab, which has an industrial CNC lathe and a CNC milling machine, computer vision systems, an integrated manufacturing cell, and an integrated CAD/CAM system, including a conveyor, robots, lathes, and milling machines.
Financial Aid	The department offers its students outstanding opportunities for support. Many full-time graduate students receive support from teaching and research assistantships; in addition, the University offers a variety of fellowships.
Cost of Study	The 1998–99 fees for the engineering graduate division were $162.60 per credit hour for Missouri residents and $489.10 per credit hour for nonresidents.
Living and Housing Costs	Housing is available for single and married students. Married graduate students have a choice of an apartment in the community or housing provided by the University. For single students, there are rooms or apartments available in the community or rooms in the University dormitories. The cost of a double-occupancy room in a residence hall and twenty meals per week (no Sunday evening meal), including a social fee, ranges from $3915 to $4660 for an academic year (two semesters).
Student Group	There are currently 40 students pursuing graduate degrees in industrial and manufacturing systems engineering. Approximately one fifth of these are pursuing the Ph.D. degree. A variety of different countries are represented. There are currently numerous Fulbright scholars and government-sponsored scholars who have diverse educational backgrounds, including math, physics, business, ME, CE, EE, and IE.
Location	The University of Missouri–Columbia, situated near downtown Columbia, offers the advantages of a rural setting in a community of more than 70,000 people. Columbia is an educational and medical center and offers a wide range of cultural activities.
	The Lake of the Ozarks and other surrounding areas provide many outdoor recreational opportunities such as fishing, boating, hiking, canoeing, hunting, and camping. The metropolitan areas of Kansas City and St. Louis, the former 120 miles west of Columbia and the latter 120 miles east, provide all the opportunities offered by such large cities.
The University and The College	The University of Missouri is one university with four campuses—Columbia, Kansas City, Rolla, and St. Louis. Established in 1839 at Columbia (the oldest and largest of the four campuses, with more than 23,000 students), the University is recognized as the first state university west of the Mississippi River and was designated a land-grant university in 1870. The College of Engineering is one of seventeen colleges and schools located on the Columbia campus. The College has about 2,100 students. Campus facilities for education, recreation, and general welfare include the Memorial Student Union, the Student Health Clinic, a theater, a swimming pool, a multipurpose building, a football stadium, a newspaper, a television station, and a radio station. The University of Missouri–Columbia is one of only twenty-three AAU Research I Universities in the country.
Applying	Admission to the master's program requires demonstrated ability in quantitative analysis. The applicant must have a minimum of 13 semester credit hours of calculus, 3 hours of differential equations, and 12 hours of basic sciences and have an undergraduate GPA of at least 3.0 on a 4.0 scale. Applicants must take the General Test of the GRE. International students must also submit TOEFL scores, which should include the writing test. Admission to the doctoral program requires excellence in course work, GRE scores, and strong indications of potential research ability. Only the most outstanding candidates are encouraged to apply. A master's thesis or an equivalent research document must be submitted along with doctoral program application materials.
	Prospective international graduate students should apply to the International Admissions Office, 230 Jesse Hall, at least sixty days prior to registration. Other applicants should submit their applications directly to the Department of Industrial and Manufacturing Systems Engineering at E3437 Engineering Building East. The application deadline for most fellowships and scholarships is February 1. International applications are transmitted from the International Admissions Office to the department for review. International students should refer to information furnished by the International Admissions Office for application deadlines.

Correspondence and Information

For general program information:

Director of Graduate Studies
Department of Industrial and Manufacturing Systems
   Engineering
E3437 Engineering Building East
University of Missouri–Columbia
Columbia, Missouri 65211

Telephone: 573-882-2692
Fax: 573-882-2693
E-mail: schwartzs@missouri.edu
World Wide Web: http://www.missouri.edu/~inengwww/

For an international student preliminary
   application package:

International Admissions Office
230 Jesse Hall
University of Missouri–Columbia
Columbia, Missouri 65211

Telephone: 573-882-3754

University of Missouri–Columbia

THE FACULTY AND THEIR RESEARCH

C. Alec Chang, Associate Professor; Ph.D., Mississippi State. Automated measurement and inspection with computer vision, product design, and quality engineering; multisensor fusion; management information systems.

Thomas J. Crowe, Assistant Professor; Ph.D., Arizona State. Strategic manufacturing planning and control; static and dynamic systems modeling; manufacturing data organization, storage, and communications; decision support systems; fuzzy number theory.

Larry G. David, Professor; Ph.D., Purdue. Quality control systems, statistical applications to manufacturing, analysis of capital expenditures, human factors engineering, product liability and safety.

Wooseung Jang, Assistant Professor; Ph.D., Berkeley. Stochastic processes and modeling, systems reliability, production and quality control of semiconductor manufacturing, queueing and efficiency control of communication networks.

Cerry M. Klein, Professor; Ph.D., Purdue. Mathematical programming, combinatorial optimization, applied mathematics, fuzzy logic, OR applications to manufacturing and production systems and decision making.

Owen W. Miller, Professor Emeritus; D.Sc., Washington (St. Louis). Statistical process control/statistical quality control, productivity enhancement for small business, IE/OR applications, Taguchi methods for small business.

James S. Noble, Associate Professor; Ph.D., Purdue. Material flow systems, facilities design, industrial ecology, production economics, performance measurement, production planning and control.

Luis G. Occeña, Associate Professor; Ph.D., Purdue. Systems integration, computer-aided process modeling, product design and manufacturing, industrial control and automation, artificial intelligence.

Elin M. Wicks, Assistant Professor; Ph.D., Virginia Tech. Manufacturing systems design, production planning and control, engineering economics, scheduling, multi-attribute decision making.

UNIVERSITY OF NEW MEXICO

Manufacturing Engineering Program

Programs of Study

The University of New Mexico (UNM) Manufacturing Engineering Program (MEP), a multidisciplinary graduate-level program within the School of Engineering (SOE) since 1987, offers the degree of Master of Engineering (M.Engr.) via all SOE departments and the degree of Master of Science (M.S., with a manufacturing concentration) via the Electrical Engineering and Computer Engineering Department and via the Mechanical Engineering Department. For the MEP, both the M.Engr. degree and the M.S. degree require 36 semester hours and a three-month industrial internship in a manufacturing setting. Although a thesis is an option, MEP students typically elect to do a manufacturing project with industry.

The M.Engr. degree supports tracks in computer-integrated manufacturing, mechanical and equipment manufacturing, and semiconductor and electronics manufacturing. The semiconductor and electronics manufacturing track has a special core that covers semiconductor (S/C) process design, microelectronics design and processing, and factory design and operations. Other areas of MEP course emphasis include plastics manufacturing and precision engineering and metrology. The MEP also cross-lists several M.B.A.-level courses provided by the UNM Anderson Schools of Management (ASM), including the manufacturing organization and management course that is used in the MEP core curriculum.

Many of the courses supported by the MEP are offered to regional companies and federal laboratory sites via Instructional Television (ITV). In addition, courses on advanced quality control and manufacturing economic analysis are available from New Mexico State University via ITV. Other courses, such as microelectronics failure analysis and microelectronics reliability, are offered nationwide via satellite links.

As of spring 1998, the University of New Mexico MEP and ASM offer a dual-degree program leading to an M.Engr. degree and an M.B.A. An industrial internship and project are an integral part of the program.

Research Facilities

Beginning in spring 1999, the Manufacturing Engineering Program is housed in the new Manufacturing Training and Technology Center (MTTC), located at the UNM Science and Technology Park. The 56,000-square-foot facility supports teaching and training, research and development, start-up companies, manufacturing prototyping, and extension service activities. The MTTC houses offices, labs, classrooms, a multimedia production studio, prototyping bays, CAD rooms, and a 6,200-square-foot clean room.

The MEP has Sun and SGI workstations, PCs and Macs, robot hardware and control software, CAD/CAM packages, factory simulators, dynamic systems modeling software, injection molding equipment, and extensive semiconductor processing equipment.

Financial Aid

The MEP has a limited number of industrial fellowships for qualified students. The MEP and associated faculty members also have research projects in a number of areas, for which research assistantships and tuition support are sometimes available. The MEP also assists students in seeking industrial support for internships, co-ops, and projects.

Cost of Study

In 1999–2000, tuition for students carrying 12 or more semester hours is about $1220 per semester for state residents and about $4350 per semester for nonresidents. There is a Graduate Student Association fee of approximately $20 per semester. There is also a School of Engineering fee of $10 per credit hour, which supports student labs.

Living and Housing Costs

Minimum full-time expenses, excluding tuition, are approximately $6000 on campus and $9000 off campus per semester. This includes health insurance, books, supplies, board, room, clothing, laundry, and miscellaneous expenses. University housing is available for both individuals and families.

Student Group

The University of New Mexico has a total undergraduate enrollment of approximately 24,400 and graduate enrollment of 5,800. The School of Engineering has about 1,200 undergraduate students and 515 graduate students, of whom 64 percent are master's candidates and 36 percent are doctoral candidates. Women constitute 15 percent of the graduate engineering enrollment. About half of the engineering graduate students are studying on a half-time basis.

Student Outcomes

Recent recipients of master's degrees from the Manufacturing Engineering Program have found employment in a wide variety of manufacturing firms, such as Ford, General Motors, Honeywell, IBM, Intel, Motorola, Philips Semiconductors, and Samsung.

Location

The main campus of the University is situated in Albuquerque, which has a metropolitan population of more than 500,000. With a unique blend of three cultures—Native American, Hispanic, and Anglo—it is able to offer a wide variety of cultural and artistic events. The city is a mile above sea level, overlooking the Rio Grande, and it abuts the Sandia Mountains, which reach to 10,678 feet. The "Land of Enchantment" affords opportunities for cycling, boating, golfing, hiking, hunting, fishing, ballooning, rafting, mountain climbing, hang gliding, and skiing.

The University

The University of New Mexico is the largest university in the state, with more than 30,000 students. It was established in 1889 and is situated on 600 acres in the center of metropolitan Albuquerque. UNM is a Carnegie Foundation Level I (highest level) Research University. The resources of the University and its proximity to Sandia National Laboratories, Phillips Air Force Laboratory, Los Alamos National Laboratory, and numerous high-tech start-up and manufacturing companies provide an excellent environment for advanced studies and research.

Applying

Prospective applicants should contact either the MEP or respective SOE departments directly for curriculum information and application forms and instructions. The GRE General Test is required for admission to M.Engr. and M.S. programs in manufacturing engineering.

Correspondence and Information

Director, Manufacturing Engineering Program
MTTC Building, Suite 235
University of New Mexico
800 Bradbury Drive, SE
Albuquerque, New Mexico 87106-4346

Telephone: 505-272-7150
Fax: 505-272-7152
E-mail: blythe@slider.me.unm.edu
World Wide Web: http://www-mep.unm.edu

University of New Mexico

THE FACULTY AND THEIR RESEARCH

Since the UNM Manufacturing Engineering Program is multidisciplinary, supportive faculty members are derived from several departments within the School of Engineering and the Anderson Schools of Management. A few of the many participating faculty members are listed below:

J. E. Wood, Professor of Mechanical Engineering and Director; Ph.D., MIT. Multiscale manufacturing training simulators, robotics control and interfaces, clean-room automation and training, laser-based manufacturing, microsensors and microactuators, biotechnology manufacturing.

R. Lumia, Professor of Mechanical Engineering; Ph.D., Virginia. Software for manufacturing operations, open-architecture robot control, sensor processing and modeling, automation systems programming.

J. H. Mullins, Professor Emeritus of Mechanical Engineering; Ph.D., Caltech. Auditory processing, manufacturing methods.

M. Shahinpoor, Professor of Mechanical Engineering; Ph.D., Delaware. Artificial muscles, kinematic structures, robotics, manufacturing.

H. Tran, Assistant Professor of Mechanical Engineering and Electrical Engineering; Ph.D., Stanford. Control systems, precision engineering, metrology and sensors, optoelectronics and optomechanical systems.

R. H. Williams, Professor Emeritus of Electrical Engineering and Computer Engineering; Sc.D., New Mexico. Instrumentation, signal processing, reliability, design for manufacturing.

UNIVERSITY OF PENNSYLVANIA

School of Engineering and Applied Science
Department of Systems Engineering

Programs of Study

The core issue in systems engineering is dealing with complexity. The programs of the Department of Systems Engineering focus on design and operation of large-scale systems. The systems degree programs give specific attention to the methodology of problem formulation, to understanding of system structure or system architecture, and to effective system integration and control.

The systems engineering discipline embraces the tools and methods necessary to deal quantitatively and qualitatively with large-scale systems. The requirements of the academic degree programs are designed to provide a facility with these tools and methods and experience in using them in a specific application area.

The research activities of the faculty members are directed toward development of the systems engineering tools and methodology and toward integrated application of these tools and methodology in several interdisciplinary fields. The systems tools and methodology areas include modeling and simulation of systems and networks, decision making and optimization, human factors and man/machine symbiosis, reliability of large-scale systems and software, sensor-driven control and robotics, and system design and system integration. The application areas include telecommunications and information systems, infrastructure systems, logistics and transportation systems, business and economic systems, environmental systems, and manufacturing systems.

The requirements for the M.S.E. degree consist of the satisfactory completion of ten courses, including core requirements. The requirements can be completed in one calendar year. The M.S.E./M.B.A. program permits simultaneous pursuit of two degrees. Candidates must apply for admission directly to the Wharton Graduate Division and also to the School of Engineering and Applied Science (SEAS). While the M.B.A. requires 18 course units and the M.S.E. requires 10 course units, in this dual-degree program the course requirements for both degrees can generally be satisfied with about 24 course units. The requirements for the Ph.D. degree include a minimum of twenty courses, a written qualifying examination taken upon acceptance into the Ph.D. program, a preliminary (specialty) examination, a dissertation proposal, and the defense of a scholarly dissertation.

Research Facilities

Numerous laboratories support the courses and research of the program, including the Systems Integration Laboratory, Environment/Resources Laboratory, Robotics Laboratory, Telecommunication Research Laboratory, Transportation Logistics Research Laboratory, and Computer Graphics Laboratory. Computational facilities include 125 Pentium II machines, twenty Macintosh II machines, a Sun SPARCserver 490, SPARC 220, a SPARCstation 1, and a 3/60 UNIX.

Financial Aid

Aid is available in several forms. Fellowships and scholarships, primarily for Ph.D. candidates, typically cover both tuition and a stipend.

Cost of Study

For the academic year 1999–2000, tuition for full-time study is $23,670 plus a general fee of $1546 and a technology fee of $458 per year. For part-time study, tuition is $2997 plus a general fee of $189 and a technology fee of $58 per course.

Living and Housing Costs

On-campus housing is available for both single and married students. In 1998–99, a residence for single students cost approximately $815 per month for a single apartment (including one bedroom, a living area, a kitchen, and a bathroom) or $455 per month for a single room (one bedroom and a shared bathroom). The cost for married student housing was approximately $845 per month (one bedroom, a living area, a kitchen, and a bathroom). There are also numerous apartments in the immediate area.

Student Group

There are 11,508 undergraduates and 10,361 graduate and professional students at the University of Pennsylvania. Of these, 18,064 are full-time and 3,805 are part-time. In the most recent surveys, the University ranked ninth nationally and third in the Ivy League in the number of international students.

Location

The University is located in Philadelphia, just a few blocks west of the heart of the city. Philadelphia is a twentieth-century city with seventeenth-century origins. Renowned museums, concert halls, theaters, and sports arenas provide cultural and recreational outlets for students. Fairmount Park extends through a large section of Philadelphia, occupying both banks of the Schuykill River. Not far away are the New Jersey shore to the east, Pennsylvania Dutch country to the west, and the Pocono Mountains to the north. Equidistant from New York City and Washington, D.C., the city of Philadelphia is a patchwork of distinctive neighborhoods ranging from Colonial Society Hill to Chinatown.

The University

The 250th anniversary of the founding of the University of Pennsylvania by Benjamin Franklin was marked in 1990. It is an Ivy League institution of fourteen schools and colleges occupying a 260-acre campus. The School of Engineering and Applied Science has a distinguished reputation for the high quality of its programs. Its alumni have achieved international distinction in research, management, industrial development, government service, and engineering education. Its faculty leads a research program that is at the forefront of modern technology and has made major contributions in a wide variety of fields. The School is the birthplace of the modern computer: it was at Penn that ENIAC, the world's first electronic, large-scale, general-purpose digital computer was created.

Applying

Admission into the graduate program in systems engineering requires a baccalaureate degree (usually in engineering, mathematics, physical science, or quantitative social science) or its equivalent, with good-to-excellent performance in mathematics courses. Admission is based on the applicant's predicted performance in the systems engineering programs. It is recommended that applicants take the GRE General Test.

Correspondence and Information

Shelley Brown, Graduate Group Assistant
Department of Systems Engineering
University of Pennsylvania
Philadelphia, Pennsylvania 19104-6315
Telephone: 215-898-9390
E-mail: shelley@seas.upenn.edu
World Wide Web: http://www.seas.upenn.edu/sys

University of Pennsylvania

FACULTY HEADS AND AREAS OF RESEARCH

Systems Engineering G. Anandalingam, Ph.D., Chair of the Department; Tony E. Smith, Ph.D., Chair of the Graduate Group. The department's programs focus on design and operation of large-scale systems. The core of the systems engineering discipline consists of the tools and methods necessary to deal quantitatively and qualitatively with large-scale systems. Research activities are directed toward development of the core systems tools and methodology and toward integrated application of these tools and methodology in several interdisciplinary fields. The core systems concepts, tools, and methodology areas include system design, system integration, modeling and simulation, network analysis, optimization, human factors, system evaluation, reliability, and control and automation. Application areas include computer systems, environmental and resources systems, logistics systems, manufacturing systems, structural and construction systems, telecommunication systems, and transportation systems. Most of the research activities are interdisciplinary in nature, and systems engineering faculty members and students typically interact or collaborate with professors and students from other departments of SEAS, the School of Arts and Sciences, and the Wharton School.

G. Anandalingam, Professor of Systems Engineering and Operations Research and Chair of Systems Engineering; Ph.D., Harvard, 1981. Telecommunications, network design, information systems, optimization, decision analyses.

Joseph Bordogna, Alfred Fitler Moore Professor of Engineering and Dean Emeritus; Ph.D., Pennsylvania, 1964 (secondary appointment in Systems Engineering). Electrooptics, signal processing, systems methodology, manufacturing, environmental technologies, science and technology policy.

Chun-Hung Chen, Assistant Professor of Systems Engineering; Ph.D., Harvard, 1994. Simulation, infinitesimal perturbation analysis, stochastic control, ordinal optimization.

Zhi-Long Chen, Assistant Professor of Systems Engineering; Ph.D., Princeton, 1997. Scheduling, routing, logistics network, technology choice and capacity planning, column generation, stochastic programming.

Morris Cohen, Matsushita Professor of Manufacturing and Logistics and Professor of Operations Management (secondary appointment in Systems Engineering); Ph.D., Northwestern, 1974. Supply chain management, logistics, marketing/manufacturing interface.

Karen Donohue, Donald B. Stott Term Assistant Professor of Operations and Information Management (secondary appointment in Systems Engineering); Ph.D., Northwestern, 1993. Design and analysis of manufacturing systems, economics of process capacity and quality, supply chain relationships, marketing/manufacturing interface.

C. Nelson Dorny, Professor of Systems Engineering; Ph.D., Stanford, 1965. System integration, telecommunication networks, multisensor robotic systems, microwave and ultrasonic imaging systems. Former White House Fellow (Special Assistant to the Secretary of Agriculture) dealing with problems of the economic development of rural areas.

Magda El Zarki, Associate Professor of Electrical Engineering (secondary appointment in Systems Engineering); Ph.D., Columbia, 1988. Audio/video synchronization, error concealment schemes and quality of service measures, wireless networks.

William F. Hamilton, Landau Professor of Management and Technology (secondary faculty in Systems Engineering); Ph.D., London School of Economics, 1967. Decision analysis, managerial economics, technological innovation, corporate planning and strategy. Former White House Fellow (Special Assistant to the Secretary of Transportation) involved in studies of rail system reorganization, regulatory reform, and economic policy.

Patrick T. Harker, UPS Professor and Chairman of Operations and Information Management, the Wharton School; Ph.D., Pennsylvania, 1983. Optimization and fixed point theory, service operations management, adaptive work systems for knowledge work, transportation systems.

John D. Keenan, Professor of Civil Systems and Director, Faculty Advising; Ph.D., Syracuse, 1972. Environmental engineering, water resources, hazardous-waste management and control.

John A. Lepore, Professor of Civil Systems; Ph.D., Pennsylvania, 1967. Structural construction systems, earthquake knowledge base systems, vulnerability studies for disaster mitigation, renewable resources, materials.

Walter A. Lyon, Adjunct Professor of Civil Engineering; M.S.E., Johns Hopkins, 1948. Environmental control and management, water quality information systems, environmental standards.

Vijay Kumar, Professor of Mechanical Engineering (secondary faculty in Systems Engineering); Ph.D., Ohio State, 1987. Coordination of multiple robotic systems, simulation and control of dynamic systems with changing topology, actively coordinated mobility systems.

Joseph P. Martin, Adjunct Associate Professor; Ph.D., Colorado State, 1983. Civil and architectural engineering.

Edward K. Morlok, UPS Foundation Professor of Transportation and Professor of Systems Engineering; Ph.D., Northwestern, 1967. Transportation systems and logistics, engineering economics, systems design. Editor, McGraw-Hill *Series In Transportation*.

Wen K. Shieh, Professor of Environmental Engineering; Ph.D. Massachusetts, 1978. Environmental systems simulation, biodegradation of toxic substances, process control and simulation biological waste treatment, biological fluidized bed technology.

Barry Silverman, Professor of Systems Engineering; Ph.D., Pennsylvania, 1977. Knowledge management (acquisition, modeling, inference, maintenance), intelligent software agents for human task performance support, medical informatics.

Jonathan Smith, Associate Professor of Computer and Information Science (secondary faculty in Systems Engineering); Ph.D., Columbia, 1989. Distributed systems, operating systems, active networks, memory management, computer security.

Tony E. Smith, Professor of Regional Science and Systems Engineering; Ph.D., Pennsylvania, 1969. Spatial interaction modeling, transportation network analysis, spatial statistics.

Lyle Ungar, Associate Professor of Computer and Information Science and Associate Professor of Chemical Engineering and Systems Engineering (secondary appointment in Systems Engineering); Ph.D., MIT, 1984. Knowledge-based systems, qualitative physics, machine learning, real and artificial neural networks.

Vukan R. Vuchic, UPS Foundation Professor of Transportation; Ph.D., Berkeley, 1966. Urban and regional transportation systems; planning, design, operation, and management of highway, transit, and pedestrian systems; relationship of transportation, environment and livability, and urban transportation policy.

Enver Yucesan, Associate Professor of Systems Engineering; Ph.D., Cornell, 1989. Simulation modeling, simulation optimization, queueing and performance modeling, production planning, logistics, supply chain management.

Iraj Zandi, Professor of Systems Engineering and National Center Professor of Resource Management and Technology; Ph.D., Georgia Tech, 1959. Purposeful systems, social systems science, problem formulation, environmental/resources systems, solid-waste management, resource recovery.

UNIVERSITY OF VIRGINIA

Department of Systems Engineering

Programs of Study	The Department of Systems Engineering offers programs leading to the degrees of Master of Science, Master of Engineering, and Doctor of Philosophy. Both the M.E. and M.S. degrees require a minimum of 32 semester hours. For the M.S. these 32 hours include 6 semester hours of thesis research and the successful defense of the thesis before a 3-member faculty committee. The M.E. can normally be completed in a calendar year of full-time study; the M.S. typically requires 1½ years. For the Ph.D. degree, students are expected to complete a minimum of 26 semester hours beyond the master's degree, pass a comprehensive examination, and submit and defend a dissertation. The normal full-time graduate course load is three or four courses, depending upon the research load undertaken.
	All degree programs center around the two primary components of systems engineering: systems analysis and systems design/integration. Systems analysis involves the formulation and use of modeling techniques for problem solving and decision making. Systems design and integration concerns the development, implementation, and coordination of operations, processes, and production functions. Current research topics include applications of methods of artificial intelligence, including expert systems, neural networks, genetic algorithms, intelligent computer interfaces, data fusion, and data analysis; risk management; environmental systems; business process modeling and financial engineering; analysis and design of large-scale communications systems; multiple objective systems modeling; and analysis and control of discrete-event systems, such as manufacturing, communications, computer, and transportation systems.
	Since its creation, the department has placed a strong emphasis on teaching excellence, and several faculty members have received teaching awards. Teaching assistants do not teach courses in this department. The department also has an excellent record of achievement in research. Most faculty members have externally funded research programs that enable students to participate in work at the leading edge of the discipline. Recent sponsors include the Jet Propulsion Laboratory, National Science Foundation, National Aeronautics and Space Administration, Department of Interior, National Weather Service, General Electric, National Geographic Society, and Martin Marietta.
Research Facilities	The University's computer center operates a network of IBM RS/6000s, an IBM SP2 parallel computer, and multiple workstation labs featuring Sun, Windows/Intel, and Macintosh computers. The Department of Systems Engineering operates graduate and undergraduate labs with networked IBM RS/6000, Sun, Windows/Intel, and Macintosh computers.
Financial Aid	Financial aid is available in the form of University fellowships, national fellowships, and research and teaching assistantships. Fellowships in 1998–99 carried stipends of $5000 and up plus tuition for the academic year. Research assistantships, supported by nonclassified sponsored research projects, provided stipends of $11,000–$15,000 for the calendar year.
Cost of Study	In 1998–99, tuition and fees were $4876 per academic year for state residents and $15,824 per academic year for nonresidents.
Living and Housing Costs	Dormitory facilities were available for single students at $2260 to $2460 for the 1998–99 academic year. University-operated accommodations for student families ranged from one-bedroom apartments for $457 per month to three-bedroom apartments for $551 per month (apartments are furnished and include utilities). Total expenses for a year of study at the University ranged from $14,900 to $24,800.
Student Group	There are 18,463 students at the University of Virginia, including 5,762 graduate students. The School of Engineering and Applied Science enrolls 2,323 students, of whom 512 are graduate students. There are 50 full-time graduate students in the Department of Systems Engineering.
Student Outcomes	According to the Bureau of Labor Statistics, employment opportunities for systems engineers and operations research professionals are expected to increase by more than 50 percent between the years 1990 and 2005. About 10 percent of department graduates have gone into banking/financial services, 42 percent into consulting/industry, and 37 percent into military/government applications. The median salary has been $48,000. Recent employers include AT&T Bell Labs, Price Waterhouse, Anderson Consulting, and Lockheed Martin.
Location	Charlottesville, including its environs, is a community of approximately 100,000 people, situated in the foothills of the Blue Ridge Mountains. Shenandoah National Park is 20 miles away, and Washington, D.C., is 110 miles away. The area contains many places of historic interest, including Monticello and Ash Lawn—the homes of Jefferson and Monroe.
The University	The University of Virginia, founded in 1819 by Thomas Jefferson, is a state-aided institution that recognizes the importance of having a student body drawn from many parts of the country. It is widely known for its rich heritage, graceful atmosphere, 1,500 acres of beautiful grounds, effective student-controlled honor system, and outstanding programs in a variety of areas, including engineering and applied science.
	There are 158 full-time faculty members who are active in teaching, research, and public service in the School of Engineering and Applied Science.
Applying	Applications are considered at any time, but it is highly recommended that students applying for financial aid submit their applications by February 1 of the year in which they plan to enroll.
Correspondence and Information	Graduate Program Director Department of Systems Engineering University of Virginia Charlottesville, Virginia 22903 Telephone: 804-924-5393

University of Virginia

THE FACULTY AND THEIR RESEARCH

P. A. Beling, Assistant Professor; Ph.D., Berkeley. Theory: optimization, computational complexity, parallel computation, strongly polynomial algorithms, approximation algorithms. Applications: tomography, medical imaging, production scheduling.

D. E. Brown, Associate Professor and Associate Director, Institute for Parallel Computation; Ph.D., Michigan. Theory: classification and induction algorithms, global optimization and search, uncertainty management in knowledge-based systems. Applications: knowledge-based systems, surveillance and monitoring, multisensor systems, data fusion, design aiding, software reliability and risk, parallel computing.

Y. Y. Haimes, Lawrence R. Quarles Professor and Director of Center for Risk Management of Engineering Systems; Ph.D., UCLA. Theory: risk and impact analysis, hierarchical-multiobjective decision making, large-scale systems. Applications: risk of extreme events, water resources planning and management, groundwater contamination, risk management of engineering systems, transportation systems.

T. E. Hutchinson, William Stantsfield Calcott Professor; Ph.D., Virginia. Research topics include human-machine interface, particularly as related to eye gaze input to computer systems and analysis of ocular characteristics as related to psychological state, microanalysis of biological tissue at the subcellular level, and cognition in the process of invention.

J. P. Ignizio, Professor and Chairman; Ph.D., Virginia Tech. Theory: artificial intelligence (including expert systems, neural networks, evolutionary heuristics, pattern classification, and hybrid systems), optimization (single and multiple objective), and numerical search. Applications: expert systems, scheduling, facility location, plant layout, agile manufacturing, and decision support systems in the health-care, military, communications, and banking sectors.

R. Krzysztofowicz, Professor; Ph.D., Arizona. Theory: Bayesian decision theory, utility and subjective probability, information value, multicriterion decision making, cognitive theories of judgment and decision making. Applications: intelligent decision systems; decision-aiding systems; forecast-decision systems; multisensor systems; risk, decision, and systems analysis; water resources; economics of weather forecasts.

J. H. Lambert, Research Assistant Professor, Center for Risk Management of Engineering Systems; Ph.D., Virginia. Theory: rare and extreme events, risk-based decision making, reliability engineering. Applications: highway safety and project selection, water supply, infrastructure surety, flood warning and protection, river transportation, aerospace hazards, communication networks, system acquisition, embedded digital systems, and food safety.

G. E. Louis, Assistant Professor; Ph.D., Carnegie Mellon. Theory: environmental systems analysis and management, economics and environmental policy, sustainable development adaptation strategies for global change, technology and society, risk management, urban systems. Applications: environmental policy setting, urban sanitation systems management, policy interventions for sustainable development, technology innovation for economic development.

C. M. Mastrangelo, Assistant Professor; Ph.D., Arizona. Theory: stochastic processes, statistical process control, multivariate quality control. Applications: discrete and continuous manufacturing systems.

S. D. Patek, Assistant Professor; Ph.D., MIT. Theory: control of large-scale uncertain systems, neurodynamic programming, optimal and robust control, optimization (linear, nonlinear, and dynamic programming), stochastic processes, game theory. Applications: decision aids for the design and regulation of complex dynamic systems, including military logistical/tactical systems, sensor management and information fusion systems, and telecommunication and manufacturing systems.

M. D. Rossetti, Assistant Professor; Ph.D., Ohio State. Theory: simulation, stochastic processes, queueing theory, decision support systems, automatic data collection systems. Applications: manufacturing systems design and analysis, simulation output analysis, communication systems, intelligent simulation, bar code and RF/ID systems.

W. T. Scherer, Associate Professor; Ph.D., Virginia. Theory: decision theory, Markov decision processes, intelligent decision systems, simulation/optimization. Applications: intelligent decision systems for health care and diagnostics, computational analysis of algorithms, Markov decision process applications, multiobjective decision aiding, program planning.

K. P. White Jr., Associate Professor; Ph.D., Duke. Theory: computer simulation and discrete-event systems, dynamic systems modeling and control, scheduling and sequencing theory, intelligent systems, decision support systems, human factors and behavioral issues. Applications: simulation output analysis and optimization, production scheduling, manufacturing systems engineering, passenger vehicle safety systems, intelligent simulation and decision systems, eye-gaze computing.

UNIVERSITY OF WISCONSIN–MADISON

College of Engineering
Manufacturing Systems Engineering Program

Program of Study

The program offers an M.S. in manufacturing systems engineering. At the present time, no Ph.D. degree is offered. The program is multidisciplinary, drawing courses and faculty members from the Departments of Mechanical, Industrial, Electrical and Computer, Civil and Environmental, Chemical, and Nuclear Engineering and Material Science and Engineering along with the School of Business and the Department of Computer Science.

Course work addresses solutions to problems in the design, development, implementation, operation, evaluation, and management of modern manufacturing systems.

Students may take one of three degree options: a 24-credit thesis option, a 30-credit nonthesis independent study project, or the 30-credit Engineering Management Specialization option. Courses are taken from four general categories: manufacturing processes and control; product design and process planning; design, operation, and improvement of manufacturing systems; and managerial and organizational aspects. The program can be tailored to suit individual goals and interests and may be completed in three to four semesters.

Research Facilities

The University of Wisconsin–Madison and its College of Engineering are highly research-oriented. Students have access to numerous laboratories, including a state-of-the-art Flexible Manufacturing Cell and the Computer-Aided Design Laboratory, Numerical Control Laboratory, Polymer Processing Research Laboratory, Laboratory for Applied Manufacturing and Controls, and Manufacturing Systems Analysis Laboratory.

Financial Aid

Financial aid in the form of research and teaching assistantships is available on a competitive basis from the program. Faculty members associated with the program may award aid on an individual basis, and the Graduate School offers aid in the form of fellowships and grants for which students are nominated by their departments.

Cost of Study

In 1998–99 tuition and fees for a nonresident student were $7594 per semester. Wisconsin residents paid $2463 per semester. For part-time study, the cost was $308 per credit for residents and $949 for nonresidents.

Living and Housing Costs

University housing is available for married and single graduate students. However, most students find off-campus apartments, which are readily available. Estimated living costs, including books and transportation, for a single student living off campus in the Madison area are about $8500 a year.

Student Group

Enrollment averages 40 students per semester, approximately one-fifth of whom are women. The student body is quite cosmopolitan, with more than half of the students coming from other countries. In addition, roughly half of these students are returning engineers from industry. The other half have come directly from their undergraduate education. Since the program's inception, more than 200 M.S./M.S.E. degrees have been granted.

Student Outcomes

Nearly all of the program's graduates are solving problems in industry, with many selecting from numerous job offers. Most students find jobs with major manufacturing corporations and consulting firms, nationally and internationally. Specific job titles include quality engineer, process engineer, director of occupational health and safety, manufacturing engineer, director of manufacturing, design margin manager of procurement engineering, and assistant professor of operations management. Hiring companies include AT&T, Andersen Consulting, Caterpillar, Cummins, General Motors, Hewlett-Packard, IBM, Intel, John Deere, and Procter and Gamble. In 1998–99, the median starting salary for an MSE graduate was $52,000.

Location

The University is located in the heart of the city of Madison. The campus runs along the southern shore of Lake Mendota, one of four lakes surrounding Madison. Both the campus and the city offer a wide variety of activities for all tastes, including the performing arts, local music, cross-country and downhill skiing, sailing, galleries, and museums. Various ethnic, community, and civic associations, as well as student organizations on campus, provide numerous opportunities for involvement in the community at large.

The University

The University of Wisconsin–Madison is home to more than 40,000 students, 25 percent of whom are at the graduate level. Founded in 1849, the University is one of the world's leading institutions of higher education, and it is ranked among the top ten in the United States in most scholarly surveys since 1910.

Both the College of Engineering and the School of Business at the University of Wisconsin–Madison have been recognized for the caliber of their work. Individual departments within the College of Engineering are often ranked among the top five in the United States.

Applying

Students may request information about the program and apply on line at the program's Web site as well as by contacting the program directly. The deadline for applications is June 15 for fall, October 31 for spring, and March 15 for summer.

The minimum admission requirements include an undergraduate degree in engineering from an ABET-accredited program (students with physical science backgrounds who have substantial industry experience may also apply); an undergraduate GPA of 3.0 (exceptions may be made for applicants with significant industry experience); and for students with a B.S. degree from a non-U.S. university, GRE and TOEFL scores.

Correspondence and Information

Manufacturing Systems Engineering Program
253 Mechanical Engineering
University of Wisconsin–Madison
1513 University Avenue
Madison, Wisconsin 53706-1572

Telephone: 608-262-0921
Fax: 608-265-4017
E-mail: mse@engr.wisc.edu
World Wide Web: http://www.engr.wisc.edu/msep

University of Wisconsin–Madison

THE FACULTY AND THEIR RESEARCH

Fred Bradley, Associate Professor (Materials Science and Engineering). Mathematical modeling of solidification and thermal stresses, computer-aided design of castings, control of molten metal processing.

Francesco Cerrina, Professor (Electrical Engineering).

Marvin DeVries, Professor (Mechanical Engineering). Metal cutting, computer-aided manufacturing, manufacturing systems engineering.

Neil Duffie, Professor (Mechanical Engineering). Computer controls, manufacturing systems, automated mold production.

Roxanne Engelstad, Professor (Mechanical Engineering). Structural dynamics, vibrations, mechanical modeling, mechanical design.

Donald Ermer, Professor (Mechanical Engineering). Total quality improvement process, SPC.

Nicola Ferrier, Assistant Professor (Mechanical Engineering). Robotics, robot sensing (computer vision and tactile sensing), sensor fusion.

Mark P. Finster, Associate Professor (School of Business). Customer-focused improvement management, quality and productivity improvement, new product and service development.

Frank Fronczak, Professor (Mechanical Engineering). Machine design, fluid power, rehabilitative equipment design.

Rajit Gadh, Professor (Mechanical Engineering). Virtual design, feature recognition in 3-D solid geometric modeling, Internet-based design.

A. Jeffrey Giacomin, Professor (Mechanical Engineering). Plastics processing, rheometer development, measurement of rheological properties.

Henry Guckel, Professor (Electrical and Computer Engineering). Microelectronics, integrated circuits.

Thomas F. Kelly, Professor (Materials Science and Engineering). Rapid solidification of materials, liquid-to-crystal nucleation and growth.

Rafael Lazimy, Professor (School of Business). Information systems (IS) and information technologies (IT), applications of IS/IT to manufacturing.

Miron Livny, Professor (Computer Science). High-throughput computing, visual data exploration, experiment management environments.

Robert Lorenz, Professor (Mechanical Engineering and Electrical Computer Engineering). Electromagnetic and electrostatic actuator theory and development, AC drive control theory.

Vladimir Lumelsky, Professor (Mechanical Engineering). Robotics and automation.

Richard Marleau, Associate Professor (Electrical and Computer Engineering). Real-time online application of computers in control and instrumentation.

Ella Mae Matsumura, Associate Professor (School of Business). Interface of cost management, activity-based costing and quality improvement, customer profitability analysis, game theoretic modeling.

Wayne Milestone, Professor (Mechanical Engineering). Machine design, failure analysis, fatigue design analysis, CAD.

John J. Moskwa, Associate Professor (Mechanical Engineering). Vehicle powertrain modeling and analysis, vehicle dynamics and multivariable system control.

David Nembhard, Assistant Professor (Industrial Engineering). Productivity measurement, knowledge engineering, admissible search heuristics, Bayesian decision making.

Harriet Nembhard, Assistant Professor (Industrial Engineering).

Michael Oliva, Associate Professor (Civil and Environmental Engineering). Manufactured prefabricated concrete building systems for earthquake resistance.

Tim Osswald, Associate Professor (Mechanical Engineering). Polymer processing, composites and rheology.

Robert G. Radwin, Professor (Biomedical Engineering and Industrial Engineering). Ergonomics, human factors, occupational biomechanics.

Bin Ran, Assistant Professor (Civil and Environmental Engineering). Intelligent transportation systems, dynamic travel demand forecasting.

James A. Rappold, Assistant Professor (School of Business).

Stephen Robinson, Professor (Industrial Engineering and Computer Science). Quantitative methods in managerial ergonomics, mathematical problems (especially stochastic optimization).

Bob Rowlands, Professor (Mechanical Engineering). Composite and advanced materials, experimental mechanics, stress analysis.

Jeffrey Russell, Professor (Civil and Environmental Engineering). Construction contractor evaluation, knowledge engineering, expert systems.

Vadim Shapiro, Assistant Professor (Mechanical Engineering). Computer modeling (geometric, solid, physical); computational geometry and applications to mechanical analysis, simulation, design, and manufacturing.

Leyuan Shi, Assistant Professor (Industrial Engineering). Manufacturing systems, operations research, simulation modeling, sensitivity analysis, optimization.

J. Leon Shohet, Professor (Electrical and Computer Engineering). Plasma processing and technology, plasma-aided manufacturing.

Harold Steudel, Professor (Industrial Engineering). Design and analysis of flexible manufacturing flowline workcells and other strategies for just-in-time manufacturing.

Donald S. Stone, Associate Professor (Materials Science and Engineering). Deformation, fracture, and fatigue of solids; property measurement; property-structure relationships.

Rajan Suri, Professor (Industrial Engineering); Director of the M.S.E. Program. Analysis of manufacturing systems, quick-response manufacturing, flexible manufacturing, lead-time reduction, time-based competitive strategies.

Arne Thesen, Professor (Industrial Engineering). Selective assembly systems, simulation modeling and analysis of manufacturing systems.

John Uicker, Professor (Mechanical Engineering). Kinematics and dynamics of mechanical systems, CAD/CAM/CAE, solid geometric modeling.

Raj Veeramani, Associate Professor (Industrial Engineering). Electronic commerce business models, decision technologies and strategies, Internet-aided sales force automation.

Urban Wemmerlöv, Professor (School of Business). Group technology and cellular manufacturing, just-in-time systems, managing technological change.

WORCESTER POLYTECHNIC INSTITUTE

Manufacturing Engineering Program

Programs of Study

Graduate programs in manufacturing engineering lead to the degrees of Master of Engineering (M.Eng.), Master of Science (M.S.), and Doctor of Philosophy (Ph.D.). The M.Eng. degree requires 30 credits of course work. The M.S. degree requires 24 credits of course work and 6 credits of thesis research. Both master's degrees can be pursued full- or part-time. The Doctor of Philosophy degree requires 60 credits of graduate work beyond the master's degree; 30 of the 60 credits must be in dissertation research credits. A specific program of study is developed by each candidate in association with his or her adviser. Admission to candidacy for the Ph.D. is granted when the student has passed a comprehensive examination dealing with subject areas that have been made part of the program.

The normal full-time registration for students without teaching or research assistantship duties is 9–12 credits per semester. Students who hold assistantships can register for up to 10 credits per semester. Full-time students can complete the M.Eng. degree in three semesters, the M.S. degree in two years, and the Ph.D. in three to four years.

Research areas include biomechanics/biofluids, design, solid mechanics and dynamics, computational methods, fluid and plasma dynamics, heat transfer, combustion, laser metrology, and materials science. In addition, opportunities are available for interdisciplinary research involving faculty members from Materials Engineering, Mechanical Engineering, and other departments.

Research Facilities

The department is housed in Higgins Laboratories, which recently underwent an $8-million expansion and renovation. Facilities include the Aerospace Laboratory, Biomechanics/Biofluids Laboratory, Laser and Holography Laboratory, Design Studio, Fire and Combustion Laboratory, Heat Transfer Laboratory, Vibration/Dynamics and Control Laboratory, Fluid Dynamics Laboratory, and Surface Metrology Laboratory. Additional facilities are housed in Washburn Shops and include the Aluminum Casting Research Laboratory, the Powder Metallurgy Research Center, the Robotics/CIM Laboratory, and several manufacturing engineering and materials science laboratories.

In addition to the computing facilities that serve the entire campus, the Department of Mechanical Engineering has a dedicated Graduate Computer Laboratory and numerous high-end workstations for graduate research.

Financial Aid

M.S. and Ph.D. students may obtain financial aid through teaching and research assistantships. Teaching assistants are expected to devote a maximum of 20 hours per week to assigned duties. Full assistantships provide tuition plus a stipend of approximately $12,330 for the nine-month academic year. Summer support is available through individual faculty research grants as well as through Summer Support Research Grants provided by the department. The Institute offers Robert H. Goddard Research Fellowships to outstanding M.S. and Ph.D. applicants. Each fellowship includes a stipend of $12,000 for twelve months and a full tuition waiver.

Cost of Study

Tuition for full-time students is $661 per credit hour in 1999–2000.

Living and Housing Costs

Most graduate students obtain housing in private apartments near the campus. Typically, the annual cost of rent plus utilities for a three-bedroom apartment shared by 3 students is approximately $3200 per year for each student. Worcester is a small city and costs are reasonable.

Student Group

About 680 undergraduate students and 100 full- and part-time graduate students are enrolled in the mechanical engineering department, which includes the manufacturing engineering program.

Location

WPI is located on an attractive, 72-acre, hilly campus, 1 mile from downtown Worcester. Flanked by city parks on two sides, the campus has a feeling of openness even though it is located in a city of 160,000. Worcester is well known for its many colleges and its Art Museum, Museum of Armor, Science Center, and Worcester Centrum. Music is well represented by several excellent choruses and a symphony orchestra. Concerts are given by distinguished visiting performers in the restored acoustical gem, Mechanics Hall. Outstanding dramatic productions are presented by several local theater companies. Boston and Cape Cod to the east and the Berkshires to the west can be easily reached from Worcester, and there is good skiing nearby.

The University

Worcester Polytechnic Institute was founded in 1865 as the Worcester County Free Institute of Industrial Science and is the third-oldest college of engineering, science, and management in the United States. A unique educational program was developed—a combination of scientific and technical studies with practical work in model industrial shops. Engineering students at WPI have the opportunity to become familiar with modern manufacturing processes as part of their program of study.

Graduate degrees have been awarded at WPI since 1898. New programs have been added regularly in response to the growing capabilities of the college and the changing needs of the professions. At the present time, the master's degree and the doctorate are both offered in thirteen disciplines. The current student body of 3,650 includes 1,025 full- and part-time graduate students.

Applying

Applicants to the M.Eng. or M.S. programs should have a B.S. in manufacturing engineering or a related field. Applicants to the Ph.D. program must hold a master's degree in manufacturing engineering or in a related area. For admission to the M.Eng., M.S., and Ph.D. programs, the GRE is strongly recommended for all students, and the TOEFL is required for international students. Students wishing to be considered for the Goddard Research Fellowship must submit GRE scores. Applications are accepted at any time. In making awards of financial aid, preference is given to students whose applications are received by March 1.

Correspondence and Information

Graduate Committee
Manufacturing Engineering Program
Worcester Polytechnic Institute
100 Institute Road
Worcester, Massachusetts 01609-2280
Telephone: 508-831-6088
Fax: 508-831-5680

Worcester Polytechnic Institute

THE FACULTY AND THEIR RESEARCH

Diran Apelian, Howet Professor; Sc.D., MIT, 1972. Materials engineering, solidification, aluminum foundry, powder metallurgy, metal processing.

Holly K. Ault, Associate Professor; Ph.D., Worcester Polytechnic, 1988. CAD, engineering design, geometric modeling.

Ronald R. Biederman, Fuller Professor of Mechanical Engineering; Ph.D., Connecticut, 1968. Nuclear engineering and energy.

Christopher A. Brown, Associate Professor; Ph.D., Vermont, 1983. Fractal analysis, machining, surface finish, biomechanics, surface metrology.

David C. Brown, Professor of Computer Science; Ph.D., Ohio State, 1984. Application of AI to design, systems to support concurrent engineering, expert systems.

Paul D. Cotnoir, Visiting Lecturer and Director, Central Massachusetts Manufacturing Partnership; M.S., Worcester Polytechnic.

Michael Demetriou, Assistant Professor; Ph.D., USC, 1993. Intelligent control, fault detection and diagnosis, vibration and acoustic control.

Chrystanthe Demetry, Associate Professor; Ph.D., MIT, 1993. Materials science, ceramics.

Mustapha S. Fofana, Assistant Professor; Ph.D., Waterloo, 1993. CAD/CAM, CIM/networked manufacturing systems, nonlinear chatter dynamics, delay dynamical systems.

Sharon A. Johnson, Associate Professor; Ph.D., Cornell, 1989. Operations research, dynamic programming, capacity planning, production system design.

R. Nathan Katz, Associate Research Professor; Ph.D., MIT, 1969. Materials science, ceramics, physical metallurgy.

Reinhold Ludwig, Professor of Electrical and Computer Engineering; Ph.D., Colorado State, 1986. Electromagnetic and acoustic nondestructive evaluation, electromagnetic/acoustic sensors, electromechanical device modeling, piezoelectric array transducers, numerical simulation, inverse and optimization methods for magnetic resonance imaging (MRI).

Makhlouf M. Makhlouf, Associate Professor; Ph.D., Worcester Polytechnic, 1990. Materials processing, manufacturing.

Shaukat Mirza, Professor and Director, Manufacturing Engineering Program; Ph.D., Wisconsin, 1962. Composites, finite-element analysis.

Denise W. Nicoletti, Associate Professor of Electrical and Computer Engineering; Ph.D., Drexel, 1991. Ultrasonic attenuation measurements, rough surfaces, grain-size distributions.

Francis Noonan, Associate Professor of Management; Ph.D., Massachusetts, 1973. Operations management, decision/risk analysis, environmental management.

James C. O'Shaughnessy, Professor of Civil and Environmental Engineering; Ph.D., Penn State, 1973. Pollution prevention and treatment of industrial wastes, hazardous-waste treatment and management, biological treatment of wastewaters.

Yiming (Kevin) Rong, Associate Professor; Ph.D., Kentucky, 1989. CAD/CAM, manufacturing processes, computer-aided fixture design.

Richard D. Sisson Jr., Professor and Head, Materials Science and Engineering Program; Ph.D., Purdue, 1975. Materials processing and manufacturing.

John M. Sullivan, Associate Professor; D.E., Dartmouth, 1986. Numerical methods, transport phenomena, CAD, robotics.

RECENT THESES AND DISSERTATIONS

FIDOE: A proof-of-concept Martian robotic support cart.

The application of the genetic algorithm to the kinematic design of turbine blade fixtures.

Automatic three-dimensional mesh generation for anatomically accurate human bodies and organs.

An integrated fixturing accuracy verification system.

Fast interference checking algorithm development for automated fixture design.

Constraint-based variation design with applications to fixture design.

Effects of atomic force microscopy scanning parameters in the characterization of surface roughness.

Advances in the design of pavement surfaces.

Design and analysis of a device to acquire detailed topographic information from pavement surfaces.

Machine replacement and cell assignment in product layouts.

Reduction of scrap and rework for investment casting at Komtek Inc. through process improvement.

Adaptive control for robotic grinding of marine propellers.

Stress indices in composite vessels with lugs.

The application of parametric software into the undergraduate computer-aided manufacturing environment.

Section 15
Management of Engineering and Technology

This section contains a directory of institutions offering graduate work in management of engineering and technology, followed by in-depth entries submitted by institutions that chose to prepare detailed program descriptions. Additional information about programs listed in the directory but not augmented by an in-depth entry may be obtained by writing directly to the dean of a graduate school or chair of a department at the address given in the directory.

For programs offering related work, see also in Book 2 Applied Arts and Design, Architecture, Economics, and Sociology, Anthropology, and Archaeology. In Book 3, see Ecology, Environmental Biology, and Evolutionary Biology and Biophysics (Radiation Biology); and in Book 6, Business Administration and Management, Health Services (Health Services Management and Hospital Administration), Law, and Public Health.

CONTENTS

Energy Management and Policy

Boston University, Graduate School of Arts and Sciences, Program in Energy and Environmental Studies, Boston, MA 02215. Offers energy and environmental analysis (MA); environmental remote sensing and geographic information systems (MA); international relations and resource and environmental management (MA); resource science (MA). Part-time programs available. *Faculty:* 9 full-time (1 woman), 1 part-time (0 women). *Students:* 9 full-time (7 women), 2 part-time (1 woman), 4 international. Average age 27. 106 applicants, 58% accepted. In 1998, 15 degrees awarded. *Degree requirements:* For master's, one foreign language, research paper required. *Entrance requirements:* For master's, GRE General Test, TOEFL (minimum score of 550 required). *Application deadline:* For fall admission, 7/1; for spring admission, 11/15. Applications are processed on a rolling basis. Application fee: $50. Tuition: Full-time $23,770; part-time $743 per credit. Required fees: $220. Tuition and fees vary according to class time, course level, campus/location and program. *Financial aid:* In 1998–99, 5 students received aid, including 1 fellowship, 3 research assistantships; career-related internships or fieldwork and Federal Work-Study also available. Aid available to part-time students. Financial aid application deadline: 1/15; financial aid applicants required to submit FAFSA. *Faculty research:* Modeling and systems analysis, policy analysis and evaluation. *Unit head:* Cutler J. Cleveland, Director, 617-353-7552, Fax: 617-353-5986, E-mail: cutler@bu.edu. *Application contact:* Alo Roy, Administrative Assistant, 617-353-3083, Fax: 617-353-5986, E-mail: alpana@bu.edu.

Colorado School of Mines, Graduate School, Division of Economics and Business, Program in Petroleum Economics and Management, Golden, CO 80401-1887. Offers MS, Diplôme d'ingénieur. Diplôme d'ingénieur offered jointly with École Nationale Supérrieure du Pétrole et des Monteurs. *Students:* 22 full-time (2 women), 2 part-time; includes 2 minority (1 African American, 1 Native American), 20 international. *Degree requirements:* For master's, foreign language and thesis not required. *Entrance requirements:* For master's, GRE General Test, TOEFL (minimum score of 550 required), 1 semester of course work in calculus, microeconomics, macroeconomics, statistics, and financial accounting; minimum GPA of 3.0; for Diplôme d'ingénieur, GRE General Test, minimum GPA of 3.0. *Application deadline:* For fall admission, 6/1 (priority date). Applications are processed on a rolling basis. Application fee: $40. Electronic applications accepted. *Financial aid:* Fellowships, research assistantships available. Financial aid application deadline: 3/1. *Faculty research:* International energy markets, strategic decision analysis in energy industries, capital budgeting in the oil and gas industry, oil economics, oil and the environment. Total annual research expenditures: $100,000. *Unit head:* Dr. M. W. Dixon, Graduate Coordinator, Program in Engineering Mechanics, 864-656-5624, Fax: 864-656-4435, E-mail: mwdxn@clemson.edu. *Application contact:* Carol Dahl, Professor of Economics, 303-273-3921, Fax: 303-273-3416, E-mail: cadahl@mines.edu.

École des Hautes Études Commerciales, School of Business Administration, Program in Energy Sector Management, Montréal, PQ H3T 2A7, Canada. Offers Diploma. All courses are given in French. Part-time programs available. *Faculty:* 166 full-time (36 women). *Degree requirements:* For Diploma, one foreign language, thesis not required. *Application deadline:* For fall admission, 5/15. Application fee: $40. *Financial aid:* Fellowships, scholarships available. *Unit head:* Dr. Jean-Pierre Le Goff, Director, 514-340-6441, Fax: 514-340-5640, E-mail: jean-pierre.legoff@hec.ca. *Application contact:* Martial Yergeau, Registrar, 514-340-6110, Fax: 514-340-5640, E-mail: martial.yergeau@hec.ca.

New York Institute of Technology, Graduate Division, School of Engineering and Technology, Program in Energy Management, Old Westbury, NY 11568-8000. Offers energy management (MS); energy technology (Certificate); environmental management (Certificate). Part-time and evening/weekend programs available. Postbaccalaureate distance learning degree programs offered. *Students:* 14 full-time (2 women), 130 part-time (27 women); includes 57 minority (35 African Americans, 9 Asian Americans or Pacific Islanders, 13 Hispanic Americans), 6 international. Average age 40. 36 applicants, 50% accepted. In 1998, 48 degrees awarded. *Degree requirements:* For master's, thesis or alternative, oral or written comprehensive exam required, foreign language not required; for degree, foreign language not required. *Entrance requirements:* For master's, minimum QPA of 2.85. *Application deadline:* For fall admission, 8/1. Applications are processed on a rolling basis. Application fee: $50. Electronic applications accepted. *Financial aid:* Fellowships, research assistantships, institutionally-sponsored loans, tuition waivers (full and partial), and unspecified assistantships available. Aid available to part-time students. *Unit head:* Dr. Robert Amundsen, Chair, 516-686-7578. *Application contact:* Glenn Berman, Executive Director of Admissions, 516-686-7519, Fax: 516-626-0419, E-mail: gberman@iris.nyit.edu.

Université du Québec à Trois-Rivières, Graduate Programs, Program in Energy Sciences, Trois-Rivières, PQ G9A 5H7, Canada. Offers M Sc, PhD. Part-time programs available. *Students:* 32 full-time (6 women). 24 applicants, 63% accepted. In 1998, 3 degrees awarded. *Degree requirements:* For doctorate, dissertation required. *Entrance requirements:* For doctorate, appropriate master's degree, proficiency in French. *Application deadline:* For fall admission, 2/1. Application fee: $30. *Financial aid:* Fellowships, research assistantships, teaching assistantships available. *Unit head:* Jean-Marie St. Arnaud, Director, 819-376-5108 Ext. 3589, Fax: 819-376-5012, E-mail: jean-marie_st.arnaud@uqtr.uquebec.ca. *Application contact:* Suzanne Camirand, Admissions Officer, 819-376-5045 Ext. 2591, Fax: 819-376-5210, E-mail: suzanne_camirand@uqtr.uquebec.ca.

Université du Québec, Institut national de la recherche scientifique, Graduate Programs, Research Center—Energy, Ste-Foy, PQ G1V 4C7, Canada. Offers energy and materials science (M Sc, PhD). PhD offered jointly with the Université du Québec à Trois-Rivières. Part-time programs available. *Degree requirements:* For master's and doctorate, thesis/dissertation required. *Entrance requirements:* For master's, appropriate bachelor's degree, proficiency in French; for doctorate, appropriate master's degree, proficiency in French. *Faculty research:* New energy sources, plasmas, fusion.

University of California, Berkeley, Graduate Division, Group in Energy and Resources, Berkeley, CA 94720-1500. Offers MA, MS, PhD. *Students:* 50 full-time (22 women); includes 3 minority (1 Asian American or Pacific Islander, 1 Hispanic American, 1 Native American), 9 international. 59 applicants, 32% accepted. In 1998, 14 master's, 4 doctorates awarded. *Degree requirements:* For master's, project or thesis required; for doctorate, dissertation, qualifying exam required. *Entrance requirements:* For master's and doctorate, GRE General Test, minimum GPA of 3.0. *Application deadline:* For fall admission, 1/5. Application fee: $40. *Financial aid:* Application deadline: 1/5. *Faculty research:* Technical, economic, environmental, and institutional aspects of energy conservation in residential and commercial buildings; international patterns of energy use; renewable energy sources and barriers to their potential contribution to energy supplies; assessment of conventional and nonconventional valuation of energy and environmental resources pricing. *Unit head:* Per Peterson, Chair, 510-642-1640. *Application contact:* Kate Blake, Student Affairs Officer, 510-642-1750, Fax: 510-642-1085, E-mail: kblake2@socrates.berkeley.edu.

University of Maryland, Baltimore County, Graduate School, College of Engineering, Department of Mechanical Engineering, Concentration in Energy, Baltimore, MD 21250-5398. Offers MS, PhD. *Entrance requirements:* For master's, GRE General Test, minimum GPA of 3.0; for doctorate, GRE General Test, GRE Subject Test, TOEFL, minimum GPA of 3.5. *Application deadline:* For fall admission, 7/1. Applications are processed on a rolling basis. Application fee: $45. *Application contact:* Dr. Appa Anjanappa, Director, 410-455-3330.

Engineering Management

California Polytechnic State University, San Luis Obispo, College of Business, San Luis Obispo, CA 93407. Offers agribusiness management (MBA); engineering management (MBA/MS); industrial technology (MA), including industrial and technical studies. *Faculty:* 45 full-time (9 women), 40 part-time (12 women). *Students:* 86 full-time (23 women), 11 part-time (4 women); includes 6 Asian Americans or Pacific Islanders *Degree requirements:* Foreign language not required. *Entrance requirements:* For master's, GRE General Test (combined average of 1650 on three sections required), GMAT (minimum score of 530 required; average 558). *Application deadline:* For fall admission, 7/1. Applications are processed on a rolling basis. Application fee: $55. Tuition, nonresident: part-time $164 per unit. Required fees: $531 per quarter. *Unit head:* Dr. William Boynton, Dean, 805-756-2705, Fax: 805-756-1473, E-mail: bboynton@calpoly.edu. *Application contact:* Dr. David Peach, Director, Graduate Management Program, 805-756-7187, Fax: 805-756-0110, E-mail: dpeach@calpoly.edu.

California Polytechnic State University, San Luis Obispo, College of Engineering, Engineering Management Program, San Luis Obispo, CA 93407. Offers MBA/MS. *Students:* 12 full-time (2 women). 18 applicants, 50% accepted. *Application deadline:* For fall admission, 7/1. Applications are processed on a rolling basis. Application fee: $55. Tuition, nonresident: part-time $164 per unit. Required fees: $531 per quarter. *Financial aid:* Fellowships, research assistantships, teaching assistantships, career-related internships or fieldwork, Federal Work-Study, and institutionally-sponsored loans available. Financial aid application deadline: 3/2; financial aid applicants required to submit FAFSA. *Unit head:* Don White, Professor, Department of Industrial Engineering, 805-756-2342, Fax: 805-756-5439, E-mail: dewhite@oboe.calpoly.edu. *Application contact:* Dr. David Peach, Director, Graduate Management Program, 805-756-7187, Fax: 805-756-0110, E-mail: dpeach@calpoly.edu.

California State University, Northridge, Graduate Studies, College of Engineering and Computer Science, Department of Civil, Industrial and Applied Mechanics, Northridge, CA 91330. Offers applied mechanics (MSE); civil engineering (MS); engineering (MS); engineering management (MS); industrial engineering (MS); materials engineering (MS); mechanical engineering (MS), including aerospace engineering, applied engineering, machine design, mechanical engineering, structural engineering, thermofluids; mechanics (MS). Part-time and evening/weekend programs available. *Faculty:* 13 full-time, 2 part-time. *Students:* 10 full-time (2 women), 101 part-time (15 women); includes 38 minority (3 African Americans, 22 Asian Americans or Pacific Islanders, 11 Hispanic Americans, 2 Native Americans), 8 international. *Degree requirements:* For master's, thesis required, foreign language not required. *Entrance requirements:* For master's, GRE General Test, TOEFL, minimum GPA of 2.5. *Application deadline:* For fall admission, 11/30. Application fee: $55. Tuition, nonresident: part-time $246 per unit. International tuition: $7,874 full-time. Required fees: $1,970. Tuition and fees vary according to course load. *Unit head:* Dr. Stephen Gadomski, Chair, 818-677-2166. *Application contact:* Dr. Ileana Costa, Graduate Coordinator, 818-677-3299.

The Catholic University of America, School of Engineering, Program in Engineering Management, Washington, DC 20064. Offers MS Engr. Part-time and evening/weekend programs available. *Faculty:* 11 part-time (0 women). *Students:* 8 full-time (3 women), 50 part-time (3 women); includes 10 minority (4 African Americans, 4 Asian Americans or Pacific Islanders, 2 Hispanic Americans), 8 international. Average age 34. 26 applicants, 96% accepted. In 1998, 32 degrees awarded (100% found work related to degree). *Degree requirements:* For master's, foreign language and thesis not required. *Average time to degree:* Master's–2 years part-time. *Application deadline:* For fall admission, 8/15 (priority date); for spring admission, 12/10. Applications are processed on a rolling basis. Application fee: $50. *Financial aid:* Federal Work-Study, institutionally-sponsored loans, and tuition waivers (full) available. Aid available to part-time students. Financial aid application deadline: 2/1. *Unit head:* S. A. Mohsberg, Director, 202-319-5191, Fax: 202-319-4499, E-mail: mohsberg@cua.edu.

Clarkson University, Graduate School, Interdisciplinary Studies, Potsdam, NY 13699. Offers engineering and manufacturing management (MS). Part-time and evening/weekend programs available. *Faculty:* 1 full-time (0 women). Average age 34. In 1998, 12 degrees awarded. *Entrance requirements:* For master's, TOEFL. *Application deadline:* For fall admission, 5/15 (priority date); for spring admission, 10/15 (priority date). Applications are processed on a rolling basis. Application fee: $25 ($35 for international students). Tuition: Part-time $661 per credit hour. Required fees: $215 per semester. *Financial aid:* Fellowships, research assistantships, teaching assistantships available. *Unit head:* Dr. Farzad Mahmoodi, Graduate Director, 315-268-6613, Fax: 315-268-3810, E-mail: mahmoodi@clarkson.edu.

Colorado State University, Graduate School, College of Engineering, Department of Mechanical Engineering, Program in Engineering Management, Fort Collins, CO 80523-0015. Offers MS. *Faculty:* 18 full-time (0 women). *Degree requirements:* Foreign language not required. *Entrance requirements:* For master's, GRE General Test (minimum combined score of 1850 on three sections required; average 1872), TOEFL (minimum score of 550 required; average 596), minimum GPA of 3.0. *Application deadline:* For fall admission, 2/1 (priority date). Applications are processed on a rolling basis. Application fee: $30. Electronic applications accepted. *Faculty research:* Reliability, quality control systems for manufacturing. *Unit head:* Dr. Philip K. Hopke, Dean of the Graduate School, 315-268-6447, Fax: 315-268-7994, E-mail: hopkepk@clarkson.edu. *Application contact:* Vickie Jensen, 970-491-5597.

Dartmouth College, Thayer School of Engineering, Program in Engineering Management, Hanover, NH 03755. Offers MEM, MBA/MEM. *Degree requirements:* For master's, design experience required, thesis not required. *Entrance requirements:* For master's, GRE General Test. *Application deadline:* For fall admission, 1/15 (priority date). Applications are processed on a rolling basis. Application fee: $20 ($40 for international students). *Financial aid:* Fellowships, teaching assistantships, career-related internships or fieldwork, Federal Work-Study, institutionally-sponsored loans, and tuition waivers (full and partial) available. Financial aid application deadline: 1/15; financial aid applicants required to submit CSS PROFILE. *Unit head:* Shelley Parker, Admissions Coordinator, Graduate Studies and Research, 902-494-1288, Fax: 902-494-3149, E-mail: shelley.parker@dal.ca. *Application contact:* Candace S. Potter, Admissions Coordinator, 603-646-3844, Fax: 603-646-3856, E-mail: candace.potter@dartmouth.edu.

Drexel University, Graduate School, College of Engineering, Program in Engineering Management, Philadelphia, PA 19104-2875. Offers MS, PhD. All classes held in evening. Part-time programs available. *Faculty:* 15 part-time. *Students:* 6 full-time (0 women), 160 part-time (19 women); includes 14 minority (6 African Americans, 5 Asian Americans or Pacific Islanders, 3 Hispanic Americans), 9 international. Average age 35. 67 applicants, 75% accepted. In 1998, 35 degrees awarded. *Degree requirements:* For master's, thesis optional; for doctorate, dissertation required. *Entrance requirements:* For master's, TOEFL (minimum score of 570 required), minimum GPA of 3.0; for doctorate, TOEFL (minimum score of 570 required). *Application deadline:* For fall admission, 8/21. Applications are processed on a rolling basis. Application fee: $35. Tuition: Full-time $15,795; part-time $585 per credit. Required fees: $375; $67 per term. Tuition and fees vary according to program. *Financial aid:* Application deadline: 2/1. *Faculty research:* Quality, operations research and management, ergonomics, applied statistics. *Unit head:* Dr. Steve Smith, Acting Director, 215-895-5809. *Application contact:* Kelli Kennedy, Director of Admissions, 215-895-6700, Fax: 215-895-5939, E-mail: crowlka@duvm.ocs.drexel.edu.

Announcement: The Engineering Management Program integrates the study of management disciplines within the context of engineering or technical operations. The program is designed to provide the background in management science necessary to advance from purely technical positions to those that include supervisory responsibilities. All faculty members have backgrounds in industry as well as formal academic preparation in their fields. Courses are scheduled in the evenings. Students can earn a degree through part-time or full-time study at on- and off-campus sites as well as through the Internet. Visit the Web site at http://online.drexel.edu/topclass/index.html.

Duke University, Graduate School, School of Engineering, Program in Engineering Management, Durham, NC 27708-0586. Offers MEM. *Entrance requirements:* For master's, GRE General Test. *Application deadline:* For fall admission, 12/31; for spring admission, 11/1. Application fee: $75. *Financial aid:* Application deadline: 12/31. *Unit head:* Dr. Earl H. Dowell, Dean, School of Engineering, 919-660-5389, Fax: 919-684-4860.

See in-depth description on page 1271.

Florida Institute of Technology, Graduate School, College of Engineering, Division of Electrical and Computer Science and Engineering, Program in Engineering Management, Melbourne, FL 32901-6975. Offers MS. Part-time and evening/weekend programs available. *Faculty:* 1 full-time (0 women). *Students:* 17 full-time (2 women), 21 part-time (4 women); includes 3 minority (1 African American, 2 Hispanic Americans), 19 international. Average age 29. 22 applicants, 77% accepted. In 1998, 19 degrees awarded. *Degree requirements:* For master's, thesis optional, foreign language not required. *Entrance requirements:* For master's, GRE General Test (minimum combined score of 1000 required), BS in engineering, minimum GPA of 3.0. *Application deadline:* Applications are processed on a rolling basis. Application fee: $50. Electronic applications accepted. Tuition: Part-time $575 per credit hour. Required fees: $100. Tuition and fees vary according to campus/location and program. *Financial aid:* Research assistantships, teaching assistantships, tuition remissions available. Financial aid application deadline: 3/1; financial aid applicants required to submit FAFSA. *Faculty research:* Software engineering, simulation, project management, multimedia tools, quality. *Unit head:* Dr. Wade H. Shaw, Chair, 407-674-7173, Fax: 407-984-7412, E-mail: wshaw@fit.edu. *Application contact:* Carolyn P. Farrior, Associate Dean of Graduate Admissions, 407-674-7118, Fax: 407-723-9468, E-mail: cfarrior@fit.edu.

Florida Institute of Technology, Graduate School, School of Extended Graduate Studies, Program in Engineering Management, Melbourne, FL 32901-6975. Offers MS. Part-time and evening/weekend programs available. *Students:* 1 full-time (0 women), 39 part-time (12 women); includes 8 minority (1 African American, 2 Asian Americans or Pacific Islanders, 4 Hispanic Americans, 1 Native American) Average age 33. 11 applicants, 36% accepted. In 1998, 13 degrees awarded (100% found work related to degree). *Degree requirements:* For master's, thesis optional, foreign language not required. *Entrance requirements:* For master's, GRE General Test (minimum combined score of 1000 required), BS in engineering, minimum GPA of 3.0. *Average time to degree:* Master's–1 year full-time, 3 years part-time. *Application deadline:* Applications are processed on a rolling basis. Application fee: $50. Electronic applications accepted. Tuition: Part-time $270 per credit hour. Part-time tuition and fees vary according to campus/location. *Financial aid:* Application deadline: 3/1; *Unit head:* Cecile Mitchell, Director for Enrollment Services, 907-786-1558. *Application contact:* Carolyn P. Farrior, Associate Dean of Graduate Admissions, 407-674-7118, Fax: 407-723-9468, E-mail: cfarrior@fit.edu.

The George Washington University, School of Engineering and Applied Science, Department of Engineering Management and Systems Engineering, Washington, DC 20052. Offers MEM, MS, D Sc, App Sc, Engr, MEM/MS. Part-time and evening/weekend programs available. *Faculty:* 12 full-time (0 women), 28 part-time (2 women). *Students:* 83 full-time (23 women), 338 part-time (90 women); includes 76 minority (39 African Americans, 24 Asian Americans or Pacific Islanders, 11 Hispanic Americans, 2 Native Americans), 132 international. Average age 35. 208 applicants, 88% accepted. In 1998, 163 master's, 12 doctorates awarded. *Degree requirements:* For master's, thesis optional, foreign language not required; for doctorate, one foreign language, dissertation, final and qualifying exams required; for other advanced degree, professional project required, foreign language and thesis not required. *Entrance requirements:* For master's, TOEFL (minimum score of 550 required; average 580) or George Washington University English as a Foreign Language Test, appropriate bachelor's degree; for doctorate, TOEFL (minimum score of 550 required; average 580) or George Washington University English as a Foreign Language Test, appropriate master's degree, minimum GPA of 3.5; for other advanced degree, TOEFL (minimum score of 550 required; average 580) or George Washington University English as a Foreign Language Test, appropriate master's degree, minimum GPA of 3.5, GRE required if highest earned degree is BS. *Application deadline:* For fall admission, 3/1; for spring admission, 10/1. Applications are processed on a rolling basis. Application fee: $50. Tuition: Full-time $17,328; part-time $722 per credit hour. Required fees: $828; $35 per credit hour. Tuition and fees vary according to campus/location and program. *Financial aid:* In 1998–99, 6 fellowships, 4 research assistantships, 18 teaching assistantships were awarded.; career-related internships or fieldwork and institutionally-sponsored loans also available. Financial aid application deadline: 3/1; financial aid applicants required to submit FAFSA. *Faculty research:* Artificial intelligence and expert systems, human factors engineering and systems analysis. Total annual research expenditures: $421,800. *Unit head:* Dr. Robert Waters, Chair, 202-994-7541. *Application contact:* Howard M. Davis, Manager, Office of Admissions and Student Records, 202-994-6158, Fax: 202-994-0909, E-mail: data: adms@seas.gwu.edu.

See in-depth description on page 1277.

Instituto Tecnológico y de Estudios Superiores de Monterrey, Campus Chihuahua, Graduate Programs, Chihuahua, 31110, Mexico. Offers computer systems engineering (Ingeniero); electrical engineering (Ingeniero); electromechanical engineering (Ingeniero); electronic engineering (Ingeniero); engineering administration (MEA); industrial engineering (MIE, Ingeniero); international trade (MIT); mechanical engineering (Ingeniero).

Kansas State University, Graduate School, College of Engineering, Department of Industrial and Manufacturing Systems Engineering, Manhattan, KS 66506. Offers engineering management (MEM); industrial and manufacturing systems engineering (PhD); industrial engineering (MS); operations research (MS). Part-time programs available. Postbaccalaureate distance learning degree programs offered. *Faculty:* 10 full-time (1 woman). *Students:* 24 full-time (4 women), 37 part-time (6 women); includes 1 minority (Asian American or Pacific Islander), 29 international. *Degree requirements:* For master's, computer language required, foreign language not required; for doctorate, computer language, dissertation required, foreign language not required. *Entrance requirements:* For master's and doctorate, GRE General Test (minimum score of 725 on quantitative section required). *Application deadline:* For fall admission, 3/1 (priority date); for spring admission, 9/1. Applications are processed on a rolling basis. Application fee: $0 ($25 for international students). Electronic applications accepted. *Unit head:* Dr. Bradley Kramer, Head, 785-532-5606.

Lamar University, College of Graduate Studies, College of Engineering, Department of Industrial Engineering, Program in Engineering Management, Beaumont, TX 77710. Offers MEM. *Faculty:* 4 full-time (0 women). *Students:* 3 full-time (1 woman), 3 part-time; includes 3 minority (1 African American, 1 Asian American or Pacific Islander, 1 Hispanic American), 1 international. Average age 35. 6 applicants, 67% accepted. In 1998, 3 degrees awarded (100% found work related to degree). *Degree requirements:* For master's, computer language, thesis or alternative required, foreign language not required. *Entrance requirements:* For master's, GRE General Test (minimum combined score of 1000 required), TOEFL (minimum score of 500 required), 5 years of work experience, minimum GPA of 2.5 or 3.0 on last 60 hours of undergraduate course work. *Average time to degree:* Master's–1.75 years full-time, 3 years part-time. *Application deadline:* For fall admission, 5/15 (priority date); for spring admission, 10/1 (priority date). Applications are processed on a rolling basis. Application fee: $0. *Financial aid:* Research assistantships, teaching assistantships available. Financial aid application deadline: 4/1. *Faculty research:* Total quality management, CAD/CAM, scheduling. *Unit head:* Dr. Robert C. Cammarata, Chair, Graduate Program Committee, 410-516-5462, Fax: 410-516-5293, E-mail: rcc@jhu.edu. *Application contact:* Dr. Hsing-Wei Chu, Professor, 409-880-8804, Fax: 409-880-8121.

Long Island University, C.W. Post Campus, College of Liberal Arts and Sciences, Department of Computer Sciences, Brookville, NY 11548-1300. Offers computer science education (MS); information systems (MS); management engineering (MS). Part-time and evening/weekend programs available. *Faculty:* 9 full-time (3 women), 9 part-time (0 women). *Students:* 26 full-time (9 women), 104 part-time (34 women); includes 45 minority (10 African Americans, 15 Asian Americans or Pacific Islanders, 20 Hispanic Americans), 25 international. *Degree requirements:* For master's, computer language, thesis or alternative, comprehensive exam required, foreign language not required. *Entrance requirements:* For master's, bachelor's degree in science, mathematics, or engineering. *Application deadline:* Applications are processed on a rolling basis. Application fee: $30. Electronic applications accepted. *Unit head:* Dr. Susan Dorchak, Chair, 516-299-2293, E-mail: dorchak@homet.liunet.edu. *Application contact:* John Keane, Graduate Adviser, 516-299-2293.

Loyola Marymount University, Graduate Division, College of Science and Engineering, Department of Mechanical Engineering, Program in Engineering and Production Management, Los Angeles, CA 90045-8350. Offers MS. Part-time and evening/weekend programs available. *Faculty:* 2 part-time (0 women). *Students:* 9 full-time (2 women), 28 part-time (1 woman); includes 15 minority (2 African Americans, 10 Asian Americans or Pacific Islanders, 2 Hispanic Americans, 1 Native American) 27 applicants, 70% accepted. In 1998, 1 degree awarded. *Degree requirements:* For master's, thesis or alternative, project required, foreign language not required. *Entrance requirements:* For master's, GMAT or GRE General Test, TOEFL (minimum score of 550 required). Application fee: $35. Electronic applications accepted. Tuition: Part-time $650 per unit. *Financial aid:* In 1998–99, 6 students received aid. Federal Work-Study, grants, and laboratory assistantships, available. Aid available to part-time students. Financial aid application deadline: 3/2; financial aid applicants required to submit FAFSA. *Unit head:* Dr. Mel I. Mendelson, Director, 310-338-6020.

Marquette University, Graduate School, College of Engineering, Department of Mechanical and Industrial Engineering, Milwaukee, WI 53201-1881. Offers engineering management (MS); materials science and engineering (MS, PhD); mechanical engineering (MS, PhD), including manufacturing systems engineering. Part-time and evening/weekend programs available. *Faculty:* 18 full-time (0 women), 6 part-time (0 women). *Students:* 32 full-time (3 women), 52 part-time (4 women); includes 6 minority (1 African American, 2 Asian Americans or Pacific Islanders, 2 Hispanic Americans, 1 Native American), 30 international. Terminal master's awarded for partial completion of doctoral program. *Degree requirements:* For master's, thesis, comprehensive exam required, foreign language not required; for doctorate, dissertation, proficiency exam, qualifying exam required, foreign language not required. *Entrance requirements:* For master's and doctorate, GRE General Test, TOEFL (minimum score of 550 required), minimum GPA of 3.0. *Application deadline:* For fall admission, 8/1 (priority date); for spring admission, 1/1 (priority date). Applications are processed on a rolling basis. Application fee: $40. Tuition: Part-time $510 per credit hour. Tuition and fees vary according to program. *Unit head:* Dr. G. E. O. Widera, Chairman, 414-288-7259, Fax: 414-288-1647, E-mail: geo.widera@marquette.edu. *Application contact:* Dr. William E. Brower, Director of Graduate Studies, 414-288-1717, Fax: 414-288-7790, E-mail: 9322browerw@rms.csd.mu.edu.

Massachusetts Institute of Technology, School of Engineering, System Design and Management Program, Cambridge, MA 02139-4307. Offers engineering and management (SM); system design and management (CAS). *Students:* 85 full-time (9 women), 6 international. Average age 35. 95 applicants, 68% accepted. In 1998, 6 degrees awarded. *Application deadline:* For winter admission, 6/15. Application fee: $55. *Financial aid:* In 1998–99, 5 research assistantships, 1 teaching assistantship were awarded. Financial aid applicants required to submit FAFSA. *Unit head:* John R. Williams, Co-Director, 617-253-7201, E-mail: jrw@mit.edu. *Application contact:* Marcia Chapman, Program Manager, 617-253-0254, E-mail: sdm@mit.edu.

Mercer University, School of Engineering, Macon, GA 31207-0003. Offers biomedical engineering (MSE); electrical engineering (MSE); engineering management (MSE); mechanical engineering (MSE); software engineering (MSE); software systems (MS); technical management (MS). Part-time and evening/weekend programs available. *Faculty:* 23 full-time (1 woman), 6 part-time (0 women). *Degree requirements:* For master's, computer language, thesis or alternative required, foreign language not required. *Entrance requirements:* For master's, GRE, minimum undergraduate GPA of 3.0. *Application deadline:* For fall admission, 7/1; for spring admission, 11/15. Applications are processed on a rolling basis. Application fee: $35 ($50 for international students). *Unit head:* Dr. Benjamin S. Kelley, Dean, 912-752-2459, Fax: 912-752-5593, E-mail: kelley_bs@mercer.edu. *Application contact:* Kathy Olivier, Coordinator, Special Programs, 912-752-2196, E-mail: oliver_kh@mercer.edu.

Mercer University, Cecil B. Day Campus, School of Engineering, Atlanta, GA 30341-4155. Offers electrical engineering (MSE); engineering management (MSE); software engineering (MSE); software systems (MS); technical communication management (MS). Part-time and evening/weekend programs available. Postbaccalaureate distance learning degree programs offered (no on-campus study). *Faculty:* 5 full-time (1 woman), 1 part-time (0 women). *Degree requirements:* For master's, computer language, thesis or alternative required, foreign language not required. *Entrance requirements:* For master's, GRE, minimum GPA of 3.0 in major. *Application deadline:* For fall admission, 7/1; for spring admission, 11/15. Applications are processed on a rolling basis. Application fee: $35 ($50 for international students). *Unit head:* Dr. Benjamin S. Kelley, Acting Dean, 912-752-2459, E-mail: kelley_bs@mercer.edu. *Application contact:* Dr. David Leonard, Director of Admissions, 770-986-3203.

Milwaukee School of Engineering, School of Business, Milwaukee, WI 53202-3109. Offers engineering management (MS). Part-time and evening/weekend programs available. *Faculty:* 1 full-time (0 women), 20 part-time (2 women). *Students:* 34 full-time (7 women), 232 part-time (30 women); includes 13 minority (4 African Americans, 5 Asian Americans or Pacific Islanders, 3 Hispanic Americans, 1 Native American), 7 international. *Degree requirements:* For master's, thesis defense or capstone project required. *Entrance requirements:* For master's, GMAT. *Average time to degree:* Master's–4.5 years part-time. *Application deadline:* For fall admission, 8/15 (priority date); for spring admission, 2/1. Applications are processed on a rolling basis. Application fee: $30. Electronic applications accepted. *Financial aid:* Career-related internships or fieldwork available. Aid available to part-time students. *Unit head:* Dr. George Lephardt, Chairman, 414-277-7352, Fax: 414-277-7479, E-mail: lephardt@msoe.edu. *Application contact:* Helen Boomsma, Director, Lifelong Learning Institute, 800-321-6763, Fax: 414-277-7475, E-mail: boomsma@msoe.edu.

National Technological University, Programs in Engineering, Fort Collins, CO 80526-1842. Offers chemical engineering (MS); computer engineering (MS); computer science (MS); electrical engineering (MS); engineering management (MS); hazardous waste management (MS); health physics (MS); management of technology (MS); manufacturing systems engineering

Engineering Management

National Technological University (continued)
(MS); materials science and engineering (MS); software engineering (MS); special majors (MS); transportation engineering (MS); transportation systems engineering (MS). Part-time programs available. *Faculty:* 600 part-time (20 women). *Entrance requirements:* For master's, BS in engineering or related field. *Application deadline:* Applications are processed on a rolling basis. Application fee: $50. *Unit head:* Lionel V. Baldwin, President, 970-495-6400, Fax: 970-484-0668, E-mail: baldwin@mail.ntu.edu.

National University, Graduate Studies, School of Business and Technology, Department of Technology, La Jolla, CA 92037-1011. Offers e-commerce (MBA, MS); electronic engineering (MS); engineering management (MS); environmental management (MBA, MS); industrial engineering management (MS); software engineering (MS); technology management (MBA, MS); telecommunication systems management (MS). Part-time and evening/weekend programs available. Postbaccalaureate distance learning degree programs offered (minimal on-campus study). *Faculty:* 12 full-time, 125 part-time. *Students:* 305 (79 women); includes 122 minority (34 African Americans, 69 Asian Americans or Pacific Islanders, 17 Hispanic Americans, 2 Native Americans) 53 international. *Degree requirements:* For master's, foreign language and thesis not required. *Entrance requirements:* For master's, interview, minimum GPA of 2.5. *Application deadline:* Applications are processed on a rolling basis. Application fee: $60 ($100 for international students). Tuition: Full-time $7,830; part-time $870 per course. One-time fee: $60. Tuition and fees vary according to campus/location. *Unit head:* Dr. Leonid Preiser, Chair, 858-642-8425, Fax: 858-642-8716, E-mail: lpreiser@nu.edu. *Application contact:* Nancy Rohland, Director of Enrollment Management, 858-642-8180, Fax: 858-642-8709, E-mail: nrohland@nu.edu.

New Jersey Institute of Technology, Office of Graduate Studies, Department of Industrial and Manufacturing Engineering, Program in Engineering Management, Newark, NJ 07102-1982. Offers MS. Part-time and evening/weekend programs available. *Degree requirements:* For master's, thesis or alternative required, foreign language not required. *Entrance requirements:* For master's, GRE General Test (minimum score of 450 on verbal section, 600 on quantitative, 550 on analytical required). Electronic applications accepted. *Faculty research:* Knowledge-based systems, CAD/CAM simulation and interface, expert systems.

Northeastern University, College of Engineering, Department of Mechanical, Industrial, and Manufacturing Engineering, Boston, MA 02115-5096. Offers engineering management (MS); industrial engineering (MS, PhD); mechanical engineering (MS, PhD); operations research (MS). Part-time programs available. *Faculty:* 30 full-time (1 woman), 13 part-time (1 woman). *Students:* 105 full-time (19 women), 118 part-time (25 women); includes 14 minority (5 African Americans, 7 Asian Americans or Pacific Islanders, 1 Hispanic American, 1 Native American), 84 international. *Degree requirements:* For master's, thesis required (for some programs), foreign language not required; for doctorate, one foreign language, dissertation, departmental qualifying exam required. *Entrance requirements:* For master's and doctorate, GRE General Test. *Application deadline:* For fall admission, 4/15. Applications are processed on a rolling basis. Application fee: $50. *Unit head:* Dr. John W. Cipolla, Chairman, 617-373-3810, Fax: 617-373-2921. *Application contact:* Stephen L. Gibson, Associate Director, 617-373-2711, Fax: 617-373-2501, E-mail: grad-eng@coe.neu.edu.

Northwestern University, The Graduate School, Robert R. McCormick School of Engineering and Applied Science, Department of Industrial Engineering and Management Sciences, Program in Engineering Management, Evanston, IL 60208. Offers MEM. *Degree requirements:* For master's, computer language, comprehensive exam required, foreign language and thesis not required. *Entrance requirements:* For master's, GRE General Test, TOEFL (minimum score of 560 required), 3 years of work experience. *Application deadline:* For fall admission, 8/30. Application fee: $50 ($55 for international students). *Financial aid:* Institutionally-sponsored loans available. Financial aid applicants required to submit FAFSA. *Unit head:* Donald Frey, Director. *Application contact:* Bridgette Calendo, Assistant Director, 847-491-5584, E-mail: bridge13@nwu.edu.

Oakland University, Graduate Studies, School of Engineering and Computer Science, Program in Engineering Management, Rochester, MI 48309-4401. Offers MS. *Students:* 26 full-time (7 women), 120 part-time (15 women); includes 8 minority (1 African American, 7 Asian Americans or Pacific Islanders), 11 international. Average age 29. 52 applicants, 81% accepted. In 1998, 46 degrees awarded. *Degree requirements:* For master's, foreign language and thesis not required. *Entrance requirements:* For master's, minimum GPA of 3.0 for unconditional admission. *Application deadline:* For fall admission, 7/15; for spring admission, 3/15. Applications are processed on a rolling basis. Application fee: $30. Tuition, state resident: part-time $221 per credit hour. Tuition, nonresident: part-time $488 per credit hour. Required fees: $214 per semester. Part-time tuition and fees vary according to program. *Financial aid:* Federal Work-Study, institutionally-sponsored loans, and tuition waivers (full) available. Financial aid application deadline: 3/1; financial aid applicants required to submit FAFSA. *Unit head:* Dr. Naim A. Kheir, Chair, 248-370-2177.

Ohio University, Graduate Studies, College of Engineering and Technology, Department of Mechanical Engineering, Athens, OH 45701-2979. Offers manufacturing engineering (MS); mechanical engineering (MS, PhD), including CAD/CAM (MS), manufacturing (MS), mechanical systems (MS), technology management (MS), thermal systems (MS). *Faculty:* 11 full-time (0 women). *Students:* 50 full-time (5 women), 29 part-time (2 women); includes 1 minority (African American), 67 international. *Degree requirements:* For master's, thesis required, foreign language not required; for doctorate, dissertation required. *Entrance requirements:* For master's, BS in engineering or science, minimum GPA of 2.8; for doctorate, GRE. *Application deadline:* For fall admission, 3/15 (priority date). Applications are processed on a rolling basis. Application fee: $30. Tuition, state resident: full-time $5,754; part-time $238 per credit hour. Tuition, nonresident: full-time $11,055; part-time $457 per credit hour. Tuition and fees vary according to course load, campus/location and program. *Unit head:* Dr. Jay S. Gunasekara, Chairman, 740-593-0563, Fax: 740-593-0476, E-mail: gsekera@bobcat.ent.ohiou.edu. *Application contact:* Dr. M. Khairul Alam, Graduate Chairman, 740-593-1598, Fax: 740-593-0476, E-mail: agrawal@bobcat.ent.ohiou.edu.

Old Dominion University, College of Engineering and Technology, Department of Engineering Management, Norfolk, VA 23529. Offers engineering management (MEM, MS, PhD); operations research/systems analysis (ME). Part-time and evening/weekend programs available. Postbaccalaureate distance learning degree programs offered (no on-campus study). *Faculty:* 8 full-time (2 women). *Students:* 15 full-time (1 woman), 79 part-time (15 women); includes 16 minority (8 African Americans, 5 Asian Americans or Pacific Islanders, 3 Hispanic Americans), 13 international. Average age 35. In 1998, 38 master's, 2 doctorates awarded. *Degree requirements:* For master's, computer language, comprehensive exam, project required, thesis optional, foreign language not required; for doctorate, computer language, dissertation, candidacy exam required, foreign language not required. *Entrance requirements:* For master's, GRE, TOEFL (minimum score of 550 required), minimum GPA of 2.75; for doctorate, GRE, TOEFL (minimum score of 550 required), minimum GPA of 3.25. *Application deadline:* For fall admission, 7/1; for spring admission, 10/1. Applications are processed on a rolling basis. Application fee: $30. Electronic applications accepted. *Financial aid:* In 1998–99, 27 students received aid, including 5 research assistantships (averaging $15,056 per year), 2 teaching assistantships; fellowships, career-related internships or fieldwork, grants, and tuition waivers (partial) also available. Aid available to part-time students. Financial aid application deadline: 2/15; financial aid applicants required to submit FAFSA. *Faculty research:* Design optimization, networks, commercial space, infrastructure, manufacturing processes. Total annual research expenditures: $278,742. *Unit head:* Dr. Resit Unal, Interim Chair, 737-683-5541, Fax: 737-683-5640.

See in-depth description on page 1283.

Pennsylvania State University University Park Campus, Graduate School, College of Earth and Mineral Sciences, Department of Energy and Geo-Environmental Engineering, Program in Mineral Engineering Management, State College, University Park, PA 16802-1503. Offers M Eng. *Degree requirements:* For master's, thesis required, foreign language not required. Application fee: $50. *Financial aid:* Application deadline: 12/31. *Unit head:* Dr. R. V. Ramani, Chair, 814-865-3437.

Portland State University, Graduate Studies, School of Engineering and Applied Science, Engineering Management Program, Portland, OR 97207-0751. Offers MS, PhD. Part-time and evening/weekend programs available. *Faculty:* 2 full-time (0 women), 1 part-time (0 women). *Students:* 12 full-time (3 women), 24 part-time (6 women); includes 3 minority (1 Asian American or Pacific Islander, 2 Hispanic Americans), 17 international. Average age 31. 21 applicants, 67% accepted. In 1998, 22 degrees awarded. *Degree requirements:* For master's, thesis optional; for doctorate, one foreign language, computer language, dissertation, oral and written exams required. *Entrance requirements:* For master's, TOEFL (minimum score of 550 required), minimum GPA of 3.0 in upper-division course work or 2.75 overall; for doctorate, GRE General Test, GRE Subject Test, TOEFL (minimum score of 575 required), minimum GPA of 3.0 in upper-division course work. *Application deadline:* For fall admission, 4/1 (priority date); for spring admission, 11/1. Applications are processed on a rolling basis. Application fee: $50. *Financial aid:* In 1998–99, 4 research assistantships were awarded.; teaching assistantships, career-related internships or fieldwork, Federal Work-Study, and institutionally-sponsored loans also available. Aid available to part-time students. Financial aid application deadline: 3/1; financial aid applicants required to submit FAFSA. *Faculty research:* Scheduling, hierarchical decision modeling, operations research, knowledge-based information systems. Total annual research expenditures: $12,336. *Unit head:* Dr. Dundar Kocaoglu, Director, 503-725-4660, Fax: 503-725-4667, E-mail: kocaoglu@emp.pdx.edu. *Application contact:* Mary E. Wiltse, Coordinator, 503-725-4660, Fax: 503-725-4667, E-mail: maryw@emp.pdx.edu.

Portland State University, Graduate Studies, Systems Science Program, Portland, OR 97207-0751. Offers systems science/anthropology (PhD); systems science/business administration (PhD); systems science/civil engineering (PhD); systems science/economics (PhD); systems science/engineering management (PhD); systems science/general (PhD); systems science/mathematical sciences (PhD); systems science/mechanical engineering (PhD); systems science/psychology (PhD); systems science/sociology (PhD). *Faculty:* 3 full-time (0 women), 1 part-time (0 women). *Students:* 45 full-time (17 women), 23 part-time (6 women); includes 5 minority (1 African American, 3 Asian Americans or Pacific Islanders, 1 Hispanic American), 12 international. *Degree requirements:* For doctorate, variable foreign language requirement, computer language, dissertation required. *Entrance requirements:* For doctorate, GMAT (score in 75th percentile or higher required), GRE General Test (score in 75th percentile or higher required), TOEFL (minimum score of 575 required), minimum undergraduate GPA of 3.0. *Application deadline:* For fall admission, 2/1; for spring admission, 11/1. Application fee: $50. *Unit head:* Dr. Nancy Perrin, Director, 503-725-4960, E-mail: perrinn@pdx.edu. *Application contact:* Dawn Kuenle, Coordinator, 503-725-4960, E-mail: dawn@sysc.pdx.edu.

Rensselaer Polytechnic Institute, Graduate School, Lally School of Management and Technology, Troy, NY 12180-3590. Offers accounting/finance (PhD); applied economics (PhD); business administration (MBA, PhD), including finance and accounting (MBA), information systems management (MBA), management (PhD), management of technology and entrepreneurships (MBA), manufacturing management (MBA), marketing management (MBA), operations research (MBA), organizational behavior and human resource management (MBA), statistical methods for management (MBA); business policy and strategy (PhD); environmental management and policy (MS, PhD); human resource (PhD); management and technology (MBA, MS, PhD); management information systems (PhD); managerial economics (PhD); manufacturing (PhD). Part-time and evening/weekend programs available. Postbaccalaureate distance learning degree programs offered (no on-campus study). *Faculty:* 36 full-time (5 women), 6 part-time (0 women). *Students:* 288 full-time (83 women), 128 part-time (39 women); includes 34 minority (6 African Americans, 19 Asian Americans or Pacific Islanders, 9 Hispanic Americans), 154 international. 358 applicants, 61% accepted. In 1998, 193 master's, 4 doctorates awarded. *Degree requirements:* For master's, computer language required, foreign language and thesis not required; for doctorate, computer language, dissertation required, foreign language not required. *Entrance requirements:* For master's, GMAT, TOEFL (minimum score of 570 required); for doctorate, GMAT or GRE General Test, TOEFL (minimum score of 570 required). *Application deadline:* For fall admission, 2/1 (priority date). Applications are processed on a rolling basis. Application fee: $35. Electronic applications accepted. *Financial aid:* In 1998–99, 104 students received aid; fellowships, research assistantships, teaching assistantships, career-related internships or fieldwork, institutionally-sponsored loans, scholarships, and tuition waivers (full and partial) available. Financial aid application deadline: 2/1. *Faculty research:* Entrepreneurship, operations management, product development, information systems, financial technology. Total annual research expenditures: $600,000. *Unit head:* Dr. Joseph G. Ecker, Dean, 518-276-6802, Fax: 518-276-8661, E-mail: management@rpi.edu. *Application contact:* Michele Martens, Manager of Enrollment Services, 518-276-6586, Fax: 518-276-2665, E-mail: management@rpi.edu.

See in-depth description on page 1287.

Rochester Institute of Technology, Part-time and Graduate Admissions, College of Engineering, Department of Industrial and Manufacturing Engineering, Rochester, NY 14623-5604. Offers engineering management (ME); industrial engineering (ME); manufacturing engineering (ME); systems engineering (ME). *Students:* 6 full-time (3 women), 13 part-time (4 women); includes 3 minority (all African Americans), 4 international. *Degree requirements:* For master's, internship required. *Entrance requirements:* For master's, TOEFL, minimum GPA of 3.0. *Application deadline:* For fall admission, 3/1 (priority date). Applications are processed on a rolling basis. Application fee: $40. *Unit head:* Dr. Jasper Shealy, Head, 716-475-2134, E-mail: jesele@rit.edu.

Rose-Hulman Institute of Technology, Faculty of Engineering and Applied Sciences, Program in Engineering Management, Terre Haute, IN 47803-3920. Offers MS. Part-time programs available. *Faculty:* 11 full-time (1 woman). *Students:* 1 full-time (0 women), 57 part-time (2 women), 5 international. Average age 32. 1 applicants, 100% accepted. In 1998, 5 degrees awarded. *Degree requirements:* For master's, integrated project required, foreign language and thesis not required. *Entrance requirements:* For master's, GRE, TOEFL (minimum score of 580 required), minimum GPA of 3.0. *Average time to degree:* Master's–5 years part-time. *Application deadline:* For fall admission, 2/1 (priority date). Applications are processed on a rolling basis. Application fee: $0. Tuition: Full-time $19,305; part-time $540 per credit hour. Required fees: $800. *Financial aid:* Application deadline: 2/1. *Unit head:* Dr. Thomas W. Mason, Director, 812-877-8155, Fax: 812-877-3198, E-mail: thomas.mason@rose-hulman.edu. *Application contact:* Dr. Buck F. Brown, Dean for Research and Graduate Studies, 812-877-8403, Fax: 812-877-8102, E-mail: buck.brown@rose-hulman.edu.

Saint Martin's College, Graduate Programs, Program in Engineering Management, Lacey, WA 98503-7500. Offers M Eng Mgt. Part-time programs available. *Faculty:* 3 full-time (0 women). *Students:* 7 full-time (0 women), 7 part-time (3 women); includes 1 minority (Hispanic American) Average age 32. 10 applicants, 70% accepted. In 1998, 10 degrees awarded. *Degree requirements:* For master's, thesis optional, foreign language not required. *Entrance requirements:* For master's, minimum GPA of 3.0. *Average time to degree:* Master's–1.5 years full-time, 3 years part-time. *Application deadline:* For fall admission, 7/1 (priority date); for spring admission, 12/1. Applications are processed on a rolling basis. Application fee: $35. Tuition: Full-time $8,424; part-time $468 per semester hour. *Financial aid:* Research assistantships, Federal Work-Study available. Aid available to part-time students. Financial aid application deadline: 3/1. *Faculty research:* Highway safety management. Total annual research expenditures: $7,500. *Unit head:* Dr. Tom McCormack, Director, 360-438-4322, Fax: 360-438-4548. *Application contact:* Ron Vandergriffe, Administrative Assistant, 360-438-4320.

St. Mary's University of San Antonio, Graduate School, Department of Engineering, San Antonio, TX 78228-8507. Offers electrical engineering (MS); electrical/computer engineering (MS); engineering administration (MS); engineering computer application (MS); industrial engineering (MS); operations research (MS). Part-time and evening/weekend programs available. *Students:* 3 full-time (2 women), 49 part-time (6 women); includes 10 minority (2 African Americans, 8 Hispanic Americans), 8 international. *Degree requirements:* For master's, computer language, thesis required, foreign language not required. *Entrance requirements:* For master's, GRE General Test. *Application deadline:* For fall admission, 8/1. Application fee: $15. *Unit head:* Dr. Abe Yazdani, Adviser, 210-436-3305.

Santa Clara University, School of Engineering, Program in Engineering Management, Santa Clara, CA 95053-0001. Offers MSE Mgt. Part-time and evening/weekend programs available. *Students:* 12 full-time (2 women), 155 part-time (35 women); includes 62 minority (2 African Americans, 51 Asian Americans or Pacific Islanders, 8 Hispanic Americans, 1 Native American), 30 international. Average age 35. 70 applicants, 87% accepted. In 1998, 29 degrees awarded. *Degree requirements:* For master's, thesis or alternative required, foreign language not required. *Entrance requirements:* For master's, GRE General Test, TOEFL (minimum score of 550 required), minimum GPA of 2.75. *Application deadline:* For fall admission, 6/1; for spring admission, 1/1. Applications are processed on a rolling basis. Application fee: $40. *Financial aid:* Fellowships, research assistantships, teaching assistantships, Federal Work-Study and institutionally-sponsored loans available. Aid available to part-time students. Financial aid application deadline: 2/1; financial aid applicants required to submit CSS PROFILE or FAFSA. *Unit head:* Dr. Robert J. Parden, Chair, 408-554-4061. *Application contact:* Tina Samms, Assistant Director of Graduate Admissions, 408-554-4313, Fax: 408-554-5474, E-mail: engr-grad@scu.edu.

Southern Methodist University, School of Engineering and Applied Science, Department of Computer Science and Engineering, Dallas, TX 75275. Offers computer engineering (MS Cp E, PhD); computer science (MS, PhD); engineering management (MSEM, DE); operations research (MS, PhD); software engineering (MS). Part-time programs available. Postbaccalaureate distance learning degree programs offered (no on-campus study). *Faculty:* 13 full-time (2 women), 12 part-time (1 woman). *Students:* 57 full-time (22 women), 294 part-time (60 women); includes 85 minority (24 African Americans, 44 Asian Americans or Pacific Islanders, 16 Hispanic Americans, 1 Native American), 69 international. *Degree requirements:* For master's, thesis optional, foreign language not required; for doctorate, dissertation, oral and written qualifying exams, oral final exam (PhD) required. *Entrance requirements:* For master's, GRE General Test (minimum score of 650 on quantitative section required), TOEFL (minimum score of 550 required), minimum GPA of 3.0 in last 2 years; bachelor's degree in engineering, mathematics, or sciences; for doctorate, preliminary counseling exam (PhD), minimum GPA of 3.0, bachelor's degree in related field, MA (DE). *Application deadline:* For fall admission, 8/1 (priority date); for spring admission, 12/15. Applications are processed on a rolling basis. Application fee: $25. Tuition: Full-time $9,216; part-time $512 per credit hour. Required fees: $88 per credit hour. Part-time tuition and fees vary according to course load and campus/location. *Unit head:* Dr. Richard V. Helgason, Interim Chair, 214-768-3278, E-mail: helgason@seas.smu.edu. *Application contact:* Dr. Zeynep Celik-Butler, Assistant Dean for Graduate Studies and Research, 214-768-3979, Fax: 214-768-3845, E-mail: zcb@seas.smu.edu.

Stanford University, School of Engineering, Department of Industrial Engineering and Engineering Management, Stanford, CA 94305-9991. Offers engineering management (MS); industrial engineering (MS, PhD, Eng). *Faculty:* 12 full-time (3 women). *Students:* 96 full-time (31 women), 27 part-time (6 women); includes 35 minority (3 African Americans, 26 Asian Americans or Pacific Islanders, 6 Hispanic Americans), 52 international. Average age 27. 177 applicants, 24% accepted. In 1998, 69 master's, 4 doctorates awarded. *Degree requirements:* For master's, foreign language and thesis not required; for doctorate, dissertation required, foreign language not required; for Eng, thesis required. *Entrance requirements:* For master's, doctorate, and Eng, GRE General Test, TOEFL. *Application deadline:* For fall admission, 2/1. Application fee: $65 ($80 for international students). Electronic applications accepted. Tuition: Full-time $24,588. Required fees: $152. Part-time tuition and fees vary according to course load. *Financial aid:* Fellowships, research assistantships, teaching assistantships, Federal Work-Study and institutionally-sponsored loans available. Financial aid application deadline: 2/15. *Unit head:* M. Elisabeth Paté-Cornell, Chair, 650-723-3823 Ext. ', Fax: 650-725-8799, E-mail: mep@stanford.edu. *Application contact:* Graduate Admissions Coordinator, 650-725-1633.

Syracuse University, Graduate School, L. C. Smith College of Engineering and Computer Science, Program in Engineering Management, Syracuse, NY 13244-0003. Offers MS. *Students:* 7 full-time (2 women), 57 part-time (10 women); includes 2 minority (1 African American, 1 Hispanic American), 15 international. Average age 32. 54 applicants, 93% accepted. In 1998, 24 degrees awarded. *Degree requirements:* Foreign language not required. *Entrance requirements:* For master's, GRE General Test, GRE Subject Test. *Application deadline:* Applications are processed on a rolling basis. Application fee: $40. Tuition: Full-time $13,992; part-time $583 per credit hour. *Financial aid:* Application deadline: 3/1. *Unit head:* Thomas Vedder, Graduate Director. *Application contact:* Contact, 315-443-1065.

Tufts University, Division of Graduate and Continuing Studies and Research, Graduate School of Arts and Sciences, College of Engineering, The Gordon Institute, Medford, MA 02155. Offers MSEM. Part-time programs available. *Faculty:* 10 full-time, 5 part-time. *Students:* 33 (5 women) 7 international. 20 applicants, 100% accepted. In 1998, 15 degrees awarded. *Degree requirements:* For master's, computer language required, foreign language and thesis not required. *Entrance requirements:* For master's, TOEFL (minimum score of 550 required). *Application deadline:* For fall admission, 5/1 (priority date). Applications are processed on a rolling basis. Application fee: $0. *Financial aid:* Application deadline: 2/15. *Unit head:* Edward N. Aqua, Director, 617-627-3111, Fax: 617-627-3180, E-mail: eaqua@emerald.tufts.edu.

The University of Akron, Graduate School, College of Engineering, Program in Engineering (Management Specialization), Akron, OH 44325-0001. Offers MSE. *Students:* 1 full-time (0 women), 12 part-time; includes 2 minority (1 African American, 1 Asian American or Pacific Islander), 1 international. Average age 28. In 1998, 4 degrees awarded. *Degree requirements:* Foreign language not required. *Entrance requirements:* For master's, TOEFL. *Average time to degree:* Master's–2 years full-time, 4 years part-time. *Application deadline:* Applications are processed on a rolling basis. Application fee: $25 ($50 for international students). Tuition, state resident: part-time $189 per credit. Tuition, nonresident: part-time $353 per credit. Required fees: $7.3 per credit. *Financial aid:* Application deadline: 3/1. *Unit head:* Dr. S. Graham Kelly, Interim Dean, College of Engineering, 330-972-6978, E-mail: sgraham@uakron.edu.

University of Alaska Anchorage, School of Engineering, Program in Engineering Management, Anchorage, AK 99508-8060. Offers MS. Part-time and evening/weekend programs available. 8 applicants, 50% accepted. In 1998, 9 degrees awarded. *Degree requirements:* For master's, computer language required, foreign language not required. *Entrance requirements:* For master's, GRE General Test, BS in engineering or science, work experience in engineering or science. *Application deadline:* For fall admission, 5/1 (priority date). Applications are processed on a rolling basis. Application fee: $45. *Financial aid:* Research assistantships, Federal Work-Study available. Aid available to part-time students. Financial aid application deadline: 4/1; financial aid applicants required to submit FAFSA. *Faculty research:* Engineering economy, long-range forecasting, multicriteria design making, project management process and training. *Unit head:* Dr. Jang Ra, Head, 907-786-1862, Fax: 907-786-1079. *Application contact:* Cecile Mitchell, Director for Enrollment Services, 907-786-1558.

University of Alaska Anchorage, School of Engineering, Program in Science Management, Anchorage, AK 99508-8060. Offers MS. Part-time and evening/weekend programs available. In 1998, 2 degrees awarded. *Degree requirements:* For master's, computer language required, foreign language not required. *Entrance requirements:* For master's, GRE General Test, BS in engineering or scientific field. *Application deadline:* For fall admission, 5/1 (priority date). Applications are processed on a rolling basis. Application fee: $45. *Financial aid:* Research assistantships available. Financial aid application deadline: 4/1; financial aid applicants required to submit FAFSA. *Faculty research:* Engineering economy, long-range forecasting, multicriteria decision maing, project management process and training. *Unit head:* Dr. Jang Ra, Head, 907-786-1862, Fax: 907-786-1079. *Application contact:* Cecile Mitchell, Director for Enrollment Services, 907-786-1558.

University of Alaska Fairbanks, Graduate School, College of Science, Engineering and Mathematics, Department of Engineering and Science Management, Fairbanks, AK 99775-7480. Offers engineering management (MS); science management (MS). *Faculty:* 1 full-time (0 women), 1 (woman) part-time. *Students:* 2 full-time (1 woman), 3 part-time (2 women); includes 1 minority (African American) Average age 39. 1 applicants, 0% accepted. In 1998, 5 degrees awarded. *Degree requirements:* For master's, thesis or alternative, comprehensive exam required, foreign language not required. *Entrance requirements:* For master's, GRE General Test, TOEFL (minimum score of 550 required). *Application deadline:* For fall admission, 8/1. Applications are processed on a rolling basis. Application fee: $35. *Financial aid:* Research assistantships, teaching assistantships available. *Unit head:* Dr. Lufti Raad, Head, 907-474-6121.

University of Alberta, Faculty of Graduate Studies and Research, Department of Electrical and Computer Engineering, Edmonton, AB T6G 2E1, Canada. Offers computational optics (PhD); computer engineering (M Eng, M Sc, PhD); control systems (M Eng, M Sc, PhD); engineering management (M Eng); laser physics (M Sc, PhD); oil sands (M Eng, M Sc, PhD); plasma physics (M Sc, PhD); power engineering (M Eng, M Sc, PhD); telecommunications (M Eng, M Sc, PhD). Terminal master's awarded for partial completion of doctoral program. *Degree requirements:* For master's and doctorate, thesis/dissertation required, foreign language not required. *Entrance requirements:* For master's and doctorate, TOEFL (minimum score of 580 required; average 610). Electronic applications accepted. *Faculty research:* Controls, communications, microelectronics, electromagnetics.

University of Alberta, Faculty of Graduate Studies and Research, Department of Mechanical Engineering, Edmonton, AB T6G 2E1, Canada. Offers engineering management (M Eng); mechanical engineering (M Eng, M Sc, PhD); oil sands (M Eng, M Sc). Part-time programs available. *Degree requirements:* For master's and doctorate, thesis/dissertation required, foreign language not required. *Entrance requirements:* For master's and doctorate, TOEFL (minimum score of 580 required), minimum GPA of 7.0 on a 9.0 scale. *Faculty research:* Combustion and environmental issues, advanced materials, computational fluid dynamics, biomedical, acoustics and vibrations.

University of Central Florida, College of Engineering, Department of Industrial Engineering and Management Systems, Orlando, FL 32816. Offers computer-integrated manufacturing (MS); engineering management (MS); industrial engineering (MSIE); industrial engineering and management systems (PhD); manufacturing engineering (MS Mfg E); operations research (MS); product assurance engineering (MS); simulation systems (MS). Part-time and evening/weekend programs available. *Faculty:* 20 full-time, 11 part-time. *Students:* 143 full-time (27 women), 112 part-time (20 women); includes 43 minority (6 African Americans, 12 Asian Americans or Pacific Islanders, 25 Hispanic Americans), 51 international. *Degree requirements:* For master's, computer language, thesis or alternative required, foreign language not required; for doctorate, computer language, dissertation, departmental qualifying exam, candidacy exam required, foreign language not required. *Entrance requirements:* For master's, GRE General Test (minimum combined score of 1000 required), TOEFL (minimum score of 550 required; 213 computer-based), minimum GPA of 3.0 in last 60 hours; for doctorate, TOEFL (minimum score of 550 required; 213 computer-based), minimum GPA of 3.5 in last 60 hours. *Application deadline:* For fall admission, 7/15; for spring admission, 12/15. Application fee: $20. Tuition, state resident: full-time $2,054; part-time $137 per credit. Tuition, nonresident: full-time $7,207; part-time $480 per credit. Required fees: $47 per term. *Unit head:* Dr. Charles Reily, Chair, 407-823-2204. *Application contact:* Dr. Linda Malone, Coordinator, 407-823-2204.

University of Colorado at Boulder, Graduate School, College of Engineering and Applied Science, Engineering Management Program, Boulder, CO 80309. Offers ME.

University of Dallas, Graduate School of Management, Program in Engineering Management, Irving, TX 75062-4736. Offers MBA, MM. Part-time programs available. *Students:* 58 (10 women). In 1998, 9 degrees awarded. *Degree requirements:* For master's, foreign language and thesis not required. *Entrance requirements:* For master's, GMAT (minimum score of 400 required), TOEFL (average 520), minimum GPA of 3.0. *Average time to degree:* Master's–1.3 years full-time, 2.5 years part-time. *Application deadline:* For fall admission, 8/6 (priority date); for spring admission, 12/8. Applications are processed on a rolling basis. Application fee: $25 ($50 for international students). *Financial aid:* Application deadline: 2/15. *Unit head:* Dr. David Gordon, Director, 972-721-5354, Fax: 972-721-5130. *Application contact:* Roxanne Del Rio, Director of Graduate Admissions, 972-721-5174, Fax: 972-721-4009, E-mail: admiss@gsm.udallas.edu.

University of Dayton, Graduate School, School of Engineering, Program in Engineering Management, Dayton, OH 45469-1300. Offers MSEM. Part-time and evening/weekend programs available. *Faculty:* 4 full-time (0 women), 4 part-time (1 woman). *Students:* 26 full-time (7 women), 35 part-time (5 women); includes 9 minority (5 African Americans, 3 Asian Americans or Pacific Islanders, 1 Hispanic American), 1 international. Average age 26. In 1998, 32 degrees awarded. *Degree requirements:* For master's, computer language required, foreign language and thesis not required. *Entrance requirements:* For master's, GRE General Test, TOEFL. *Application deadline:* For fall admission, 8/1. Application fee: $30. *Financial aid:* In 1998–99, 1 research assistantship with partial tuition reimbursement was awarded.; teaching assistantships *Faculty research:* Artificial intelligence, program management. *Unit head:* Dr. Patrick J. Sweeney, Chairman, 937-229-2238, Fax: 937-229-2756, E-mail: psweeney@engr.udayton.edu. *Application contact:* Dr. Donald L. Moon, Associate Dean, 937-229-2241, Fax: 937-229-2471, E-mail: dmoon@engr.udayton.edu.

University of Denver, Daniels College of Business, General Business Administration Program, Denver, CO 80208. Offers business administration (MBA); education management (MSM); health care management (MSM); management and communications (MSMC); management and general engineering (MSMGEN); management and telecommunications (MSMC); public health management (MSM); sports management (MSM). Part-time and evening/weekend programs available. *Faculty:* 76. *Students:* 306 (111 women); includes 16 minority (6 African Americans, 3 Asian Americans or Pacific Islanders, 6 Hispanic Americans, 1 Native American) 45 international. *Degree requirements:* For master's, foreign language and thesis not required. *Entrance requirements:* For master's, GMAT (average 545), LSAT (JD/MBA). *Application deadline:* For fall admission, 5/1 (priority date); for spring admission, 1/1. Applications are processed on a rolling basis. Application fee: $50. *Unit head:* Dr. Tom Howard, Director, 303-871-4402. *Application contact:* Jan Johnson, Executive Director, Student Services, 303-871-3416, Fax: 303-871-4466, E-mail: dcb@du.edu.

University of Detroit Mercy, College of Engineering and Science, Department of Mechanical Engineering, Program in Engineering Management, Detroit, MI 48219-0900. Offers engineering management (M Eng Mgt); mechanical engineering (M Eng Mgt). Evening/weekend programs available. *Degree requirements:* For master's, computer language, thesis or alternative required, foreign language not required.

University of Florida, Graduate School, College of Engineering, Department of Industrial and Systems Engineering, Gainesville, FL 32611. Offers engineering management (ME, MS); facilities layout decision support systems energy (PhD); health systems (ME, MS); industrial engineering (PhD, Engr); manufacturing systems engineering (ME, MS, PhD, Certificate); operations research (ME, MS, PhD, Engr); production planning and control engineering management (PhD); quality and reliability assurance (ME, MS); systems engineering (PhD, Engr). *Faculty:* 22. *Students:* 56 full-time (14 women), 69 part-time (18 women); includes 17 minority (5 African Americans, 4 Asian Americans or Pacific Islanders, 8 Hispanic Americans), 48 international. *Degree requirements:* For master's, computer language, core exam required, thesis optional; for doctorate, dissertation, comprehensive exam required; for other advanced degree, computer language, thesis required. *Entrance requirements:* For master's and doctor-

Engineering Management

University of Florida (continued)
ate, GRE General Test, TOEFL, minimum GPA of 3.0; for other advanced degree, GRE General Test. *Application deadline:* For fall admission, 6/1 (priority date). Applications are processed on a rolling basis. Application fee: $20. Electronic applications accepted. *Unit head:* Dr. Donald Hearn, Chair, 352-392-1464, Fax: 352-392-3537, E-mail: hearn@ise.ufl.edu. *Application contact:* Dr. D. J. Elzinga, Graduate Coordinator, 352-392-1464, Fax: 352-392-3537, E-mail: elzinga@ise.ufl.edu.

University of Kansas, Graduate School, School of Engineering, Engineering Management Program, Lawrence, KS 66045. Offers MS. Part-time and evening/weekend programs available. *Faculty:* 6 full-time. *Students:* 3 full-time (2 women), 158 part-time (25 women); includes 15 minority (3 African Americans, 8 Asian Americans or Pacific Islanders, 3 Hispanic Americans, 1 Native American), 9 international. In 1998, 38 degrees awarded. *Degree requirements:* For master's, exam required. *Entrance requirements:* For master's, Michigan English Language Assessment Battery, TOEFL, minimum GPA of 3.0, 2 years of industrial experience. *Application deadline:* For fall admission, 7/1. Application fee: $30 ($45 for international students). *Financial aid:* Fellowships, teaching assistantships available. *Unit head:* Herb Tuttle, Director, 785-897-8560, Fax: 785-897-8682.

University of Maryland, Baltimore County, Graduate School, College of Engineering, Engineering Management Program, Baltimore, MD 21250-5398. Offers MS. 7 applicants, 43% accepted. In 1998, 6 degrees awarded. *Entrance requirements:* For master's, GRE General Test, minimum GPA of 3.0. *Application deadline:* For fall admission, 7/1. Applications are processed on a rolling basis. Application fee: $45. *Unit head:* Dr. Antonio R. Moreira, Vice-Provost and Director, 410-455-6576. *Application contact:* Kathleen Stitely, 410-455-6841.

University of Maryland University College, Graduate School of Management and Technology, Program in Engineering Management, College Park, MD 20742-1600. Offers MS. Offered evenings and weekends only. Part-time and evening/weekend programs available. 10 applicants, 90% accepted. In 1998, 11 degrees awarded. *Degree requirements:* For master's, thesis or alternative, comprehensive exam required, foreign language not required. *Entrance requirements:* For master's, GRE or, GMAT, BS in engineering, physical science, or computer science; 3 years of related experience. *Application deadline:* Applications are processed on a rolling basis. Application fee: $50. Electronic applications accepted. Tuition, state resident: full-time $5,058; part-time $281 per credit. Tuition, nonresident: full-time $6,876; part-time $382 per credit. Tuition and fees vary according to program. *Financial aid:* Federal Work-Study, grants, and scholarships available. Aid available to part-time students. Financial aid application deadline: 6/1; financial aid applicants required to submit FAFSA. *Unit head:* John Aje, Director, 301-985-7200, Fax: 301-985-4611, E-mail: jaje@nova.umuc.edu. *Application contact:* Coordinator, Graduate Admissions, 301-985-7155, Fax: 301-985-7175, E-mail: gradinfo@nova.umuc.edu.

University of Massachusetts Amherst, Graduate School, College of Engineering, Department of Mechanical and Industrial Engineering, Program in Engineering Management, Amherst, MA 01003. Offers MS. *Students:* 1 full-time (0 women), 35 part-time (7 women); includes 2 minority (both Hispanic Americans), 2 international. Average age 37. 20 applicants, 95% accepted. In 1998, 10 degrees awarded. *Degree requirements:* For master's, foreign language and thesis not required. *Entrance requirements:* For master's, GRE General Test. *Application deadline:* For fall admission, 2/1 (priority date); for spring admission, 10/1. Applications are processed on a rolling basis. Application fee: $40. Tuition, state resident: full-time $2,640; part-time $165 per credit. Tuition, nonresident: full-time $9,756; part-time $407 per credit. Required fees: $1,221 per term. One-time fee: $110. Full-time tuition and fees vary according to course load, campus/location and reciprocity agreements. *Financial aid:* Fellowships with full tuition reimbursements, research assistantships with full tuition reimbursements, teaching assistantships with full tuition reimbursements, career-related internships or fieldwork, Federal Work-Study, grants, scholarships, traineeships, and unspecified assistantships available. Aid available to part-time students. Financial aid application deadline: 2/1. *Unit head:* Dr. Donald Fisher, Director, 413-545-0955, Fax: 413-545-1027, E-mail: dfisher@ecs.umass.edu.

University of Michigan–Dearborn, College of Engineering and Computer Science, Department of Industrial and Systems Engineering, Dearborn, MI 48128-1491. Offers engineering management (MS); industrial and systems engineering (MSE). Part-time and evening/weekend programs available. *Faculty:* 10 full-time (0 women), 7 part-time (1 woman). *Students:* 8 full-time (4 women), 294 part-time (80 women). *Degree requirements:* For master's, computer language required, thesis optional, foreign language not required. *Entrance requirements:* For master's, bachelor's degree in applied mathematics, computer science, engineering, or physical science; minimum GPA of 3.0. *Application deadline:* For fall admission, 8/1 (priority date); for winter admission, 12/1 (priority date); for spring admission, 4/1. Applications are processed on a rolling basis. Application fee: $55. Electronic applications accepted. Tuition, state resident: part-time $259 per credit hour. Tuition, nonresident: part-time $748 per credit hour. Required fees: $80 per course. Tuition and fees vary according to course level, course load and program. *Unit head:* Dr. S. K. Kachhal, Chair, 313-593-5361, Fax: 313-593-3692. *Application contact:* Lisa M. Beach, Administrative Assistant, 313-593-5361, Fax: 313-593-3692, E-mail: lmbeach@umich.edu.

University of Missouri–Rolla, Graduate School, School of Engineering, Department of Engineering Management, Rolla, MO 65409-0910. Offers MS, PhD. Part-time and evening/weekend programs available. *Faculty:* 18 full-time (1 woman). *Students:* 72 full-time (15 women), 121 part-time (20 women); includes 19 minority (6 African Americans, 9 Asian Americans or Pacific Islanders, 4 Hispanic Americans), 40 international. Average age 30. 111 applicants, 83% accepted. In 1998, 160 master's, 13 doctorates awarded. Terminal master's awarded for partial completion of doctoral program. *Degree requirements:* For master's, computer language, comprehensive exam or thesis required; for doctorate, computer language, dissertation required, foreign language not required. *Entrance requirements:* For master's, GRE General Test (minimum combined score of 1050 required), TOEFL (minimum score of 550 required); for doctorate, GRE General Test (minimum combined score of 1150 required), TOEFL (minimum score of 575 required). *Average time to degree:* Master's–1 year full-time, 4 years part-time; doctorate–3 years full-time. *Application deadline:* For fall admission, 7/1; for spring admission, 11/15. Applications are processed on a rolling basis. Application fee: $25. Electronic applications accepted. *Financial aid:* In 1998–99, 6 fellowships with partial tuition reimbursements, 23 research assistantships with partial tuition reimbursements, 5 teaching assistantships with partial tuition reimbursements were awarded.; career-related internships or fieldwork, Federal Work-Study, institutionally-sponsored loans, and tuition waivers (partial) also available. Financial aid application deadline: 7/31. *Faculty research:* Intelligent manufacturing, MIS, total quality management packaging, management of technology. *Unit head:* Dr. Henry A. Wiebe, Chairman, 573-341-4579, Fax: 573-341-6567, E-mail: wiebe@umr.edu. *Application contact:* Krista Chambers, Teleconference Program Specialist, 573-341-4990, Fax: 573-341-6567, E-mail: krista@shuttle.cc.umr.edu.

University of New Orleans, Graduate School, College of Engineering, Program in Engineering Management, New Orleans, LA 70148. Offers MS, Certificate. *Students:* 7 full-time (1 woman), 24 part-time (7 women); includes 3 minority (1 Asian American or Pacific Islander, 1 Hispanic American, 1 Native American), 9 international. Average age 30. 17 applicants, 65% accepted. In 1998, 9 degrees awarded. *Degree requirements:* For master's, thesis optional, foreign language not required. *Entrance requirements:* For master's, GRE General Test (minimum combined score of 1200 required), minimum GPA of 3.0. *Application deadline:* For fall admission, 7/1 (priority date). Applications are processed on a rolling basis. Application fee: $20. Tuition, state resident: full-time $2,362. Tuition, nonresident: full-time $7,888. Part-time tuition and fees vary according to course load. *Financial aid:* Fellowships, research assistantships, institutionally-sponsored loans available. *Unit head:* Will Lannes, Director, 504-280-7122, E-mail: wjlee@uno.edu.

University of Ottawa, School of Graduate Studies and Research, Faculty of Engineering, Engineering Management Program, Ottawa, ON K1N 6N5, Canada. Offers M Eng. *Faculty:* 10 full-time. *Students:* 13 full-time (4 women), 29 part-time (3 women), 7 international. Average age 31. In 1998, 22 degrees awarded. *Degree requirements:* For master's, thesis or alternative required, foreign language not required. *Entrance requirements:* For master's, honors degree or equivalent, minimum B average. *Application deadline:* For fall admission, 3/1 (priority date). Applications are processed on a rolling basis. Application fee: $35. *Financial aid:* Federal Work-Study available. *Unit head:* Dan Necsulescu, Director, 613-562-5800 Ext. 6270, Fax: 613-562-5174. *Application contact:* Marie Rainville, Secretary, 613-562-5784, Fax: 613-562-5177, E-mail: rainvill@genie.uottawa.ca.

University of Southern California, Graduate School, School of Engineering, Department of Industrial and Systems Engineering, Program in Engineering Management, Los Angeles, CA 90089. Offers MS. *Students:* 14 full-time (4 women), 12 part-time (4 women); includes 10 minority (1 African American, 6 Asian Americans or Pacific Islanders, 3 Hispanic Americans), 13 international. Average age 29. 23 applicants, 70% accepted. In 1998, 19 degrees awarded. *Degree requirements:* For master's, thesis optional. *Entrance requirements:* For master's, GRE General Test. *Application deadline:* For fall admission, 6/1 (priority date); for spring admission, 12/1. Application fee: $55. Tuition: Part-time $768 per unit. Required fees: $350 per semester. *Financial aid:* In 1998–99, 1 research assistantship, 1 teaching assistantship were awarded.; fellowships, Federal Work-Study, institutionally-sponsored loans, and scholarships also available. Aid available to part-time students. Financial aid application deadline: 2/15; financial aid applicants required to submit FAFSA. *Unit head:* Dr. F. Stan Settles, Chairman, Department of Industrial and Systems Engineering, 213-740-4893.

University of South Florida, Graduate School, College of Engineering, Department of Industrial and Management Systems Engineering, Tampa, FL 33620-9951. Offers engineering management (ME, MSE, MSEM, MSIE); engineering science (PhD); industrial engineering (ME, MSE, MSEM, MSIE, PhD). Part-time programs available. Postbaccalaureate distance learning degree programs offered (minimal on-campus study). *Faculty:* 9 full-time (2 women), 1 (woman) part-time. *Students:* 26 full-time (5 women), 77 part-time (12 women); includes 18 minority (6 African Americans, 1 Asian American or Pacific Islander, 11 Hispanic Americans), 21 international. Average age 35. 91 applicants, 84% accepted. In 1998, 41 master's awarded (90% found work related to degree, 10% continued full-time study); 3 doctorates awarded (67% entered university research/teaching, 33% found other work related to degree). Terminal master's awarded for partial completion of doctoral program. *Degree requirements:* For master's, thesis required (for some programs), foreign language not required; for doctorate, computer language, dissertation, 2 tools of research as specified by dissertation committee required, foreign language not required. *Entrance requirements:* For master's, GRE General Test (minimum combined score of 1100 required), minimum GPA of 3.0 during previous 2 years; for doctorate, GRE General Test (minimum combined score of 1200 required). *Average time to degree:* Master's–2.5 years full-time, 4 years part-time; doctorate–4 years full-time, 7 years part-time. *Application deadline:* For fall admission, 6/1; for spring admission, 10/15. Application fee: $20. Electronic applications accepted. Tuition, state resident: part-time $148 per credit hour. Tuition, nonresident: part-time $509 per credit hour. *Financial aid:* In 1998–99, 1 fellowship with full tuition reimbursement (averaging $7,000 per year), 13 research assistantships with full tuition reimbursements (averaging $11,631 per year), 6 teaching assistantships with full tuition reimbursements (averaging $15,009 per year) were awarded.; career-related internships or fieldwork, Federal Work-Study, institutionally-sponsored loans, and tuition waivers (partial) also available. Aid available to part-time students. Financial aid application deadline: 3/31; financial aid applicants required to submit FAFSA. *Faculty research:* Quality control, optimization techniques, stochastic processes, computer automated manufacturing. Total annual research expenditures: $149,753. *Unit head:* Dr. Paul E. Givens, Chairperson, 813-974-2269, Fax: 813-974-5953, E-mail: givens@eng.usf.edu. *Application contact:* Dr. Michael X. Weng, Graduate Director, 813-974-5575, Fax: 813-974-5953, E-mail: weng@eng.usf.edu.

University of Southwestern Louisiana, Graduate School, College of Engineering, Program in Engineering Management, Lafayette, LA 70504. Offers MSET. Part-time and evening/weekend programs available. *Faculty:* 3 full-time (0 women). *Students:* 2 full-time (0 women), 3 part-time, 2 international. 14 applicants, 71% accepted. In 1998, 2 degrees awarded. *Degree requirements:* For master's, computer language, thesis or alternative, comprehensive exam required, foreign language not required. *Entrance requirements:* For master's, GRE General Test, minimum GPA of 2.85. *Application deadline:* For fall admission, 5/15. Application fee: $5 ($15 for international students). *Financial aid:* In 1998–99, 1 research assistantship with full tuition reimbursement (averaging $4,500 per year) was awarded.; teaching assistantships, Federal Work-Study and tuition waivers (full and partial) also available. Financial aid application deadline: 5/1. *Faculty research:* Mathematical programming, production management forecasting. *Unit head:* Dr. Frank Trocki, Head, 318-482-6188.

University of Tennessee at Chattanooga, Graduate Division, School of Engineering, Program in Engineering Management, Chattanooga, TN 37403-2598. Offers MS. *Faculty:* 5 full-time (0 women). *Students:* 4 full-time (0 women), 34 part-time (7 women); includes 6 minority (all African Americans), 3 international. Average age 33. 7 applicants, 100% accepted. In 1998, 15 degrees awarded. *Degree requirements:* For master's, thesis required, foreign language not required. *Entrance requirements:* For master's, GRE General Test (combined average 1692 on three sections). *Application deadline:* Applications are processed on a rolling basis. Application fee: $25. *Financial aid:* Fellowships, research assistantships available. Financial aid application deadline: 4/1. *Unit head:* Dr. Phil Kazmersky, Director, 423-755-5320, Fax: 423-755-5229, E-mail: phil-kazmersky@utc.edu. *Application contact:* Dr. Deborah E. Arfken, Assistant Provost for Graduate Studies, 423-755-4667, Fax: 423-755-4478, E-mail: deborah-arfken@utc.edu.

University of Tennessee, Knoxville, Graduate School, College of Engineering, Department of Industrial Engineering, Knoxville, TN 37996. Offers engineering management (MS); manufacturing systems engineering (MS); traditional industrial engineering (MS). Part-time programs available. Postbaccalaureate distance learning degree programs offered (no on-campus study). *Faculty:* 10 full-time (1 woman). *Students:* 25 full-time (7 women), 111 part-time (29 women); includes 14 minority (10 African Americans, 3 Asian Americans or Pacific Islanders, 1 Hispanic American), 11 international. *Degree requirements:* For master's, thesis or alternative required, foreign language not required. *Entrance requirements:* For master's, TOEFL (minimum score of 550 required), minimum GPA of 2.7. *Application deadline:* For fall admission, 2/1 (priority date). Applications are processed on a rolling basis. Application fee: $35. Electronic applications accepted. *Unit head:* Dr. Thomas Shannon, Head, 423-974-3333, Fax: 423-974-0588, E-mail: tshannon@utk.edu. *Application contact:* Dr. J. A. Bontadelli, Graduate Representative, E-mail: bontadelli@utk.edu.

University of Tennessee Space Institute, Graduate Programs, Program in Industrial Engineering (Engineering Management), Tullahoma, TN 37388-9700. Offers engineering management (MS). *Faculty:* 3 full-time (0 women). *Students:* 6 full-time (3 women), 69 part-time (15 women); includes 6 minority (5 African Americans, 1 Hispanic American) 5 applicants, 80% accepted. In 1998, 2 degrees awarded. *Degree requirements:* For master's, thesis required (for some programs), foreign language not required. *Application deadline:* Applications are processed on a rolling basis. Application fee: $35. *Financial aid:* Fellowships, research assistantships, career-related internships or fieldwork, Federal Work-Study, and tuition waivers (full and partial) available. Financial aid applicants required to submit FAFSA. *Unit head:* Dr. Max Hailey, Degree Program Chairman, 931-393-7378, Fax: 931-393-7201, E-mail: mhailey@utsi.edu. *Application contact:* Dr. Edwin M. Gleason, Assistant Dean for Admissions and Student Affairs, 931-393-7432, Fax: 931-393-7346, E-mail: egleason@utsi.edu.

University of Tulsa, Graduate School, College of Business Administration, Department of Engineering and Technology Management, Tulsa, OK 74104-3189. Offers chemical engineering (METM); computer science (METM); electrical engineering (METM); geological science (METM); mathematics (METM); mechanical engineering (METM); petroleum engineering (METM). Part-time and evening/weekend programs available. *Students:* 3 full-time (1 woman), 1 part-time, 3 international. Average age 30. 2 applicants, 100% accepted. In 1998, 2 degrees awarded. *Degree requirements:* For master's, foreign language and thesis not required. *Entrance requirements:* For master's, GRE General Test (minimum score of 430 on verbal section, 600 on quantitative required), TOEFL (minimum score of 575 required). *Application deadline:* Applications are processed on a rolling basis. Application fee: $30. Electronic applications accepted. Tuition: Full-time $8,640; part-time $480 per hour. Required fees: $3 per hour. One-time fee: $200 full-time. Tuition and fees vary according to program. *Financial aid:* Fellowships, research assistantships, teaching assistantships, Federal Work-Study and tuition waivers (partial) available. Aid available to part-time students. Financial aid application deadline: 2/1; financial aid applicants required to submit FAFSA. *Unit head:* Dr. Richard C. Burgess, Assistant Dean/Director of Graduate Business Studies, 918-631-2242, Fax: 918-631-2142.

University of Utah, Graduate School, College of Engineering, Department of Mechanical Engineering, Program in Engineering Administration, Salt Lake City, UT 84112-1107. Offers MEA. *Degree requirements:* Foreign language not required. *Entrance requirements:* For master's, GRE General Test, TOEFL (minimum score of 530 required), minimum GPA of 3.0. *Application deadline:* For fall admission, 7/1. Application fee: $30 ($50 for international students). *Unit head:* Lydia C. Griffith, Graduate Coordinator, 512-471-1504, Fax: 512-475-8482, E-mail: lcgriffith@mail.utexas.edu. *Application contact:* Dr. Patrick McMurtry, Director of Graduate Studies, 801-581-3889.

University of Waterloo, Graduate Studies, Faculty of Engineering, Department of Management Sciences, Waterloo, ON N2L 3G1, Canada. Offers management of technology (MA Sc, PhD); management sciences (MA Sc, PhD). Part-time programs available. Postbaccalaureate distance learning degree programs offered (no on-campus study). *Faculty:* 13 full-time (2 women), 13 part-time (4 women). *Students:* 39 full-time (10 women), 45 part-time (9 women). In 1998, 12 master's, 5 doctorates awarded. *Degree requirements:* For master's, research paper or thesis required; for doctorate, dissertation, comprehensive exam required, foreign language not required. *Entrance requirements:* For master's, GMAT or GRE, TOEFL (minimum score of 550 required), honors degree, minimum B average; for doctorate, GMAT or GRE, TOEFL (minimum score of 550 required), master's degree. *Application deadline:* Applications are processed on a rolling basis. Application fee: $50. *Expenses:* Tuition and fees charges are reported in Canadian dollars. Tuition, state resident: full-time $3,168 Canadian dollars; part-time $792 Canadian dollars per term. Tuition, nonresident: full-time $8,000 Canadian dollars; part-time $2,000 Canadian dollars. Required fees: $45 Canadian dollars per term. Tuition and fees vary according to program. *Financial aid:* Fellowships, research assistantships, teaching assistantships, career-related internships or fieldwork and institutionally-sponsored loans available. Financial aid application deadline: 3/31. *Faculty research:* Operations research, manufacturing systems, scheduling, information systems. Total annual research expenditures: $894,138. *Unit head:* Dr. J. D. Fuller, Chair, 519-888-4567 Ext. 2683, Fax: 519-746-7252, E-mail: dfuller@engmail.uwaterloo.ca. *Application contact:* Dr. J. Webster, Graduate Officer, 519-888-4567 Ext. 5683, Fax: 519-746-7252, E-mail: jwebster@mansci2.uwaterloo.ca.

See in-depth description on page 1293.

Virginia Polytechnic Institute and State University, Graduate School, College of Engineering, Department of Industrial and Systems Engineering, Program in Engineering Administration, Blacksburg, VA 24061. Offers MEA. *Degree requirements:* For master's, computer language, thesis required, foreign language not required. *Entrance requirements:* For master's, TOEFL (minimum score of 600 required). *Application deadline:* For fall admission, 12/1 (priority date). Applications are processed on a rolling basis. Application fee: $25. *Financial aid:* Application deadline: 4/1. *Unit head:* R. Nash, Director of Graduate Studies, 615-322-2122, Fax: 615-322-7996, E-mail: nash@vuse.vanderbilt.edu.

Wayne State University, Graduate School, College of Engineering, Department of Industrial and Manufacturing Engineering, Program in Engineering Management, Detroit, MI 48202. Offers MS. *Degree requirements:* For master's, thesis optional, foreign language not required. *Entrance requirements:* For master's, minimum undergraduate GPA of 2.8. *Faculty research:* Technology and change management, quality/reliability, manufacturing systems/infrastructure.

Western Michigan University, Graduate College, College of Engineering and Applied Sciences, Department of Industrial and Manufacturing Engineering, Program in Engineering Management, Kalamazoo, MI 49008. Offers MS. *Students:* 5 full-time (1 woman), 126 part-time (14 women); includes 8 minority (3 African Americans, 4 Asian Americans or Pacific Islanders, 1 Hispanic American), 7 international. 21 applicants, 52% accepted. In 1998, 27 degrees awarded. *Degree requirements:* Foreign language not required. *Entrance requirements:* For master's, minimum GPA of 3.0. *Application deadline:* For fall admission, 2/15 (priority date). Applications are processed on a rolling basis. Application fee: $25. *Financial aid:* Application deadline: 2/15. *Application contact:* Paula J. Boodt, Coordinator, Graduate Admissions and Recruitment, 616-387-2000, Fax: 616-387-2355, E-mail: paula.boodt@wmich.edu.

Widener University, School of Engineering, Program in Engineering Management, Chester, PA 19013-5792. Offers ME. *Faculty:* 3 part-time (0 women). *Students:* 2 full-time (0 women), 14 part-time (2 women); includes 3 minority (1 African American, 2 Asian Americans or Pacific Islanders), 2 international. 18 applicants, 89% accepted. In 1998, 14 degrees awarded. *Degree requirements:* For master's, thesis optional, foreign language not required. *Average time to degree:* Master's–2 years full-time, 4 years part-time. *Application deadline:* For fall admission, 8/1 (priority date); for spring admission, 12/1. Applications are processed on a rolling basis. Application fee: $25 ($300 for international students). *Financial aid:* Application deadline: 3/15. *Unit head:* Dr. David H. T. Chen, Assistant Dean for Graduate Programs and Research, School of Engineering, 610-499-4049, Fax: 610-499-4059, E-mail: david.h.chen@widener.edu.

Ergonomics and Human Factors

Bentley College, Graduate School of Business, Program in Human Factors in Information Design, Waltham, MA 02452-4705. Offers MSHFID. *Faculty:* 4 full-time, 1 part-time. *Application deadline:* For fall admission 6/1 (priority date), for spring admission 11/1. Applications are processed on a rolling basis. Application fee: $50. Electronic applications accepted. *Unit head:* Dr. William Gribbons, Director, 781-891-2926. *Application contact:* Holly L. Chase, Associate Director, 781-891-2108, Fax: 781-891-2464. E-mail: gradadm@bentley.edu.

Announcement: Bentley College offers a Master of Science in Human Factors in Information Design (MSHFID) which blends theory of human factors with the practice of information design for topics such as usability testing, contextual analysis, repurposing information, expert systems, project management, and localization. The target audience includes technical writers, web developers, and motivational designers. MBA concentration and graduate certificate are also offered. Contact Holly Chase at gradadm@bentley.edu or 781-891-2108. Visit our Web site at http://www.bentley.edu/graduate

Cornell University, Graduate School, Graduate Fields of Human Ecology, Field of Design and Environmental Analysis, Ithaca, NY 14853-0001. Offers applied research in human-environment relations (MS); facilities planning and management (MS); housing and design (MS); human factors and ergonomics (MS); interior design (MA). *Faculty:* 11 full-time. *Students:* 23 full-time (13 women); includes 2 minority (1 Asian American or Pacific Islander, 1 Hispanic American), 13 international. *Degree requirements:* For master's, thesis required, foreign language not required. *Entrance requirements:* For master's, GRE General Test, TOEFL (minimum score of 550 required), portfolio (design programs), bachelor's degree in interior design or architecture. *Application deadline:* For fall admission, 2/1. Application fee: $65. Electronic applications accepted. *Unit head:* Director of Graduate Studies, 607-255-2168, Fax: 607-255-0305. *Application contact:* Graduate Field Assistant, 607-255-2168, Fax: 607-255-0305, E-mail: jl27@cornell.edu.

Embry-Riddle Aeronautical University, Daytona Beach Campus Graduate Program, Department of Human Factors and Systems, Daytona Beach, FL 32114-3900. Offers human factors engineering (MSHFS); systems engineering (MSHFS). *Faculty:* 5 full-time (0 women), 1 part-time (0 women). *Students:* 16 full-time (8 women), 9 part-time (6 women); includes 6 minority (1 African American, 2 Asian Americans or Pacific Islanders, 3 Hispanic Americans), 4 international. Average age 31. 17 applicants, 94% accepted. *Degree requirements:* For master's, thesis, practicum, qualifying oral exam required, foreign language not required. *Entrance requirements:* For master's, TOEFL (minimum score of 550 required), minimum GPA of 2.5. *Application deadline:* Applications are processed on a rolling basis. Application fee: $30 ($50 for international students). Tuition: Full-time $8,190; part-time $455 per credit. Required fees: $105 per semester. Tuition and fees vary according to program. *Financial aid:* In 1998–99, 1 fellowship with tuition reimbursement (averaging $8,640 per year), 8 research assistantships with tuition reimbursements (averaging $8,640 per year), 1 teaching assistantship with tuition reimbursement (averaging $8,640 per year) were awarded.; career-related internships or fieldwork and unspecified assistantships also available. Financial aid application deadline: 4/15; financial aid applicants required to submit FAFSA. *Faculty research:* Consulting on aviation operations with FAA, human factors of airport security, effect of training order on flight performance of new pilots, human factors evaluation of a human computer interface of commercial database, evaluation of operator workload in future air traffic systems. Total annual research expenditures: $185,000. *Unit head:* Dr. Daniel Garland, Chair, 904-226-6790, Fax: 904-226-7050, E-mail: garlandd@db.erau.edu. *Application contact:* Ginny Tait, Graduate Admissions Specialist, 904-226-6115, Fax: 904-226-6299, E-mail: taitg@cts.db.erau.edu.

See in-depth description on page 1275.

Florida Institute of Technology, Graduate School, School of Aeronautics, Melbourne, FL 32901-6975. Offers aviation (MSA); aviation human factors (MS). Part-time and evening/weekend programs available. *Faculty:* 11 full-time (1 woman), 4 part-time (0 women). *Students:* 8 full-time (2 women), 13 part-time (5 women); includes 2 minority (both Hispanic Americans), 4 international. *Degree requirements:* For master's, thesis optional, foreign language not required. *Entrance requirements:* For master's, GRE General Test, minimum undergraduate GPA of 3.0. *Application deadline:* For fall admission, 8/1; for spring admission, 12/1. Applications are processed on a rolling basis. Application fee: $50. Electronic applications accepted. Tuition: Part-time $575 per credit hour. Required fees: $100. Tuition and fees vary according to campus/location and program. *Unit head:* Dr. Nathaniel Villaire, Program Chairman of Graduate Studies, 407-674-8120, Fax: 407-725-6974, E-mail: villaire@fit.edu. *Application contact:* Carolyn P. Farrior, Associate Dean of Graduate Admissions, 407-674-7118, Fax: 407-723-9468, E-mail: cfarrior@fit.edu.

Purdue University, Graduate School, Schools of Engineering, School of Industrial Engineering, West Lafayette, IN 47907. Offers human factors in industrial engineering (MS, MSIE, PhD); manufacturing engineering (MS, MSIE, PhD); operations research (MS, MSIE, PhD); systems engineering (MS, MSIE, PhD). Part-time programs available. *Faculty:* 27 full-time (2 women), 3 part-time (0 women). *Students:* 124 full-time (26 women), 33 part-time (8 women); includes 13 minority (2 African Americans, 4 Asian Americans or Pacific Islanders, 6 Hispanic Americans, 1 Native American), 107 international. Terminal master's awarded for partial completion of doctoral program. *Degree requirements:* For master's, computer language required, thesis optional, foreign language not required; for doctorate, computer language, dissertation required, foreign language not required. *Entrance requirements:* For master's, GRE General Test (minimum score of 470 on verbal section, 700 on quantitative, 600 on analytical required; average 520 verbal, 710 quantitative, 660 analytical), TOEFL (minimum score of 570 required; average 600), minimum GPA of 3.0; for doctorate, GRE General Test (minimum score of 470 on verbal section, 700 on quantitative, 600 on analytical required; average 540 verbal, 750 quantitative, 675 analytical), TOEFL (minimum score of 570 required; average 610), MS thesis. *Application deadline:* For fall admission, 3/15; for spring admission, 9/1. Application fee: $30. Electronic applications accepted. *Unit head:* Dr. W. D. Compton, Interim Head, 765-494-5444, Fax: 765-494-1299, E-mail: dcompton@ecn.purdue.edu. *Application contact:* Dr. J. W. Barany, Associate Head, 765-494-5406, Fax: 765-494-1299, E-mail: jwb@ecn.purdue.edu.

Rensselaer Polytechnic Institute, Graduate School, School of Humanities and Social Sciences, Department of Philosophy, Psychology and Cognitive Science, Program in Psychology, Troy, NY 12180-3590. Offers human factors (MS); industrial-organizational psychology (MS); psychopharmacology (MS). Part-time programs available. *Faculty:* 13 full-time (1 woman), 3 part-time (0 women). *Students:* 27 full-time (19 women), 1 part-time; includes 2 minority (1 African American, 1 Asian American or Pacific Islander), 1 international. *Degree requirements:* For master's, thesis required, foreign language not required. *Entrance requirements:* For master's, GRE General Test, GRE Subject Test, TOEFL (minimum score of 550 required). *Application deadline:* For fall admission, 2/1 (priority date). Applications are processed on a rolling basis. Application fee: $35. *Unit head:* Dr. A. M. Birk, Graduate Coordinator, 613-533-2570. *Application contact:* Dr. Matthew Champagne, Admissions Coordinator, 518-276-6472, Fax: 518-276-8268, E-mail: bestlj@rpi.edu.

San Jose State University, Graduate Studies, Graduate Studies Program, Program in Human Factors/Ergonomics, San Jose, CA 95192-0001. Offers MS. Part-time programs available. *Faculty:* 7. *Students:* 9 full-time (5 women), 23 part-time (14 women); includes 9 minority (1 African American, 5 Asian Americans or Pacific Islanders, 3 Hispanic Americans) 23 applicants, 96% accepted. In 1998, 10 degrees awarded. *Entrance requirements:* For master's, minimum GPA of 3.0, previous course work in basic statistics. *Application deadline:* For fall admission, 6/1 (priority date). Applications are processed on a rolling basis. Application fee: $59. Tuition, nonresident: part-time $246 per unit. Required fees: $1,939; $1,309 per year. *Faculty research:* Ergonomic design and engineering, occupational therapy. *Unit head:* Dr. Louis Freund, Director, 408-924-3890.

Tufts University, Division of Graduate and Continuing Studies and Research, Graduate School of Arts and Sciences, College of Engineering, Department of Mechanical Engineering, Medford, MA 02155. Offers human factors (MS); mechanical engineering (ME, MS, PhD). Part-time programs available. *Faculty:* 13 full-time, 2 part-time. *Students:* 59 (16 women);

Ergonomics and Human Factors–Management of Technology

Tufts University (continued)
includes 5 minority (2 African Americans, 2 Asian Americans or Pacific Islanders, 1 Hispanic American) 18 international. Terminal master's awarded for partial completion of doctoral program. *Degree requirements:* For master's and doctorate, thesis/dissertation required, foreign language not required. *Entrance requirements:* For master's and doctorate, GRE General Test, TOEFL (minimum score of 550 required). *Application deadline:* For fall admission, 2/15; for spring admission, 10/15. Applications are processed on a rolling basis. Application fee: $50. *Unit head:* Vincent Manno, Chair, 617-627-3239, Fax: 617-627-3058, E-mail: meinfo@tufts.edu. *Application contact:* Robert Greif, 617-627-3239, Fax: 617-627-3058, E-mail: meinfo@tufts.edu.

Université de Montréal, Faculty of Graduate Studies, Programs in Ergonomics, Montréal, PQ H3C 3J7, Canada. Offers DESS. 2 applicants, 50% accepted. *Degree requirements:* For DESS, thesis not required. Application fee: $30. *Unit head:* Jean-Marc Robert, Director, 514-340-4566.

Université du Québec à Montréal, Graduate Programs, Program in Ergonomics in Occupational Health and Safety, Montréal, PQ H3C 3P8, Canada. Offers Diploma. Part-time programs available. *Degree requirements:* For Diploma, thesis not required. *Entrance requirements:* For degree, appropriate bachelor's degree or equivalent and proficiency in French.

The University of Iowa, Graduate College, College of Engineering, Department of Industrial Engineering, Iowa City, IA 52242-1316. Offers engineering design and manufacturing (MS, PhD); ergonomics (MS, PhD); information and engineering management (MS, PhD); operations research (MS, PhD); quality engineering (MS, PhD). *Faculty:* 7 full-time, 1 part-time. *Students:* 18 full-time (4 women), 16 part-time (1 woman); includes 4 minority (1 African American, 2 Asian Americans or Pacific Islanders, 1 Hispanic American), 19 international. *Degree requirements:* For master's, thesis optional; for doctorate, dissertation, comprehensive exam required. *Entrance requirements:* For master's and doctorate, GRE General Test, GRE Subject Test, TOEFL. *Application deadline:* Applications are processed on a rolling basis. Application fee: $30 ($50 for international students). *Unit head:* Peter J. O'Grady, Chair, 319-335-5939, Fax: 319-335-5424.

University of Miami, Graduate School, College of Engineering, Department of Industrial Engineering, Coral Gables, FL 33124. Offers environmental health and safety (MS, MSEH), including environmental health and safety (MSEH), occupational ergonomics and safety (MS); ergonomics (PhD); industrial engineering (MSIE, PhD); management of technology (MS). *Faculty:* 10 full-time (1 woman), 3 part-time (0 women). *Students:* 50 full-time (6 women), 3 part-time (2 women); includes 14 minority (3 African Americans, 3 Asian Americans or Pacific Islanders, 8 Hispanic Americans), 23 international. *Degree requirements:* For master's, thesis required (for some programs), foreign language not required; for doctorate, dissertation required, foreign language not required. *Entrance requirements:* For master's, GRE General Test (minimum combined score of 1000 required), TOEFL (minimum score of 550 required), minimum GPA of 3.0; for doctorate, GRE General Test, TOEFL (minimum score of 550 required), minimum GPA of 3.0. *Application deadline:* For fall admission, 5/1 (priority date). Applications are processed on a rolling basis. Application fee: $35. Tuition: Full-time $15,336; part-time $852 per credit. Required fees: $174. Tuition and fees vary according to program. *Unit head:* Dr. Norman Einspruch, Chairman, 305-284-2344, Fax: 305-284-4040. *Application contact:* Dr. Sara Czaja, Professor, 305-284-2344, Fax: 305-284-4040.

Wright State University, School of Graduate Studies, College of Engineering and Computer Science, Programs in Engineering, Program in Biomedical and Human Factors Engineering, Dayton, OH 45435. Offers biomedical engineering (MSE); human factors engineering (MSE). Part-time programs available. *Students:* 22 full-time (7 women), 27 part-time (10 women); includes 8 minority (2 African Americans, 6 Asian Americans or Pacific Islanders), 7 international. Average age 27. 75 applicants, 69% accepted. In 1998, 18 degrees awarded. *Degree requirements:* For master's, thesis or course option alternative required. *Entrance requirements:* For master's, TOEFL (minimum score of 550 required). *Application deadline:* For fall admission, 5/30. Applications are processed on a rolling basis. Application fee: $25. *Financial aid:* Fellowships, research assistantships, teaching assistantships, Federal Work-Study, institutionally-sponsored loans, and unspecified assistantships available. Aid available to part-time students. Financial aid application deadline: 3/15; financial aid applicants required to submit FAFSA. *Faculty research:* Medical imaging, functional electrical stimulation, implantable aids, man-machine interfaces, expert systems. *Unit head:* Dr. Richard J. Koubek, Chair, 937-775-5044, Fax: 937-775-7364.

Management of Technology

Athabasca University, Centre for Innovative Management, Athabasca, AB T9S 3A3, Canada. Offers business administration (MBA); information technology management (MBA, Advanced Diploma); management (Advanced Diploma). Part-time programs available. *Faculty:* 7 full-time (4 women), 53 part-time. *Degree requirements:* For master's, thesis or alternative required, foreign language not required. *Application deadline:* For fall admission, 6/15; for winter admission, 10/15; for spring admission, 2/15. Applications are processed on a rolling basis. Application fee: $150. Electronic applications accepted. *Expenses:* Tuition and fees charges are reported in Canadian dollars. Tuition, state resident: part-time $24,350 Canadian dollars per degree program. Tuition, nonresident: part-time $29,320 Canadian dollars. One-time fee: $650 Canadian dollars part-time. Part-time tuition and fees vary according to degree level, campus/location and program. *Unit head:* Lindsay Redpath, Executive Director, 780-459-1144, Fax: 780-459-2093, E-mail: lindsayr@cs.athabascau.ca. *Application contact:* Shelley Lynes, Manager, Registrations, Records, and Graduate Student Services, 800-561-4650, Fax: 800-561-4660.

Bentley College, Graduate School of Business, Self-paced MBA Program, Waltham, MA 02452-4705. Offers accountancy (MBA); advanced accountancy (MBA); business administration (Certificate); business communication (MBA); business data analysis (MBA); business economics (MBA); business ethics (MBA); entrepreneurial studies (MBA); finance (MBA); international business (MBA); management (MBA); management information systems (MBA); management of technology (MBA); marketing (MBA); operations management (MBA); taxation (MBA). Part-time and evening/weekend programs available. *Faculty:* 138 full-time (37 women), 52 part-time (11 women). *Students:* 190 full-time (94 women), 759 part-time (336 women); includes 41 minority (10 African Americans, 21 Asian Americans or Pacific Islanders, 9 Hispanic Americans, 1 Native American), 137 international. *Degree requirements:* For master's, computer language required, foreign language and thesis not required. *Entrance requirements:* For master's, GMAT (average 540), TOEFL (minimum score of 580 required). *Application deadline:* For fall admission, 6/1 (priority date); for spring admission, 11/1. Applications are processed on a rolling basis. Application fee: $50. Electronic applications accepted. *Unit head:* Dr. Judith B. Kamm, Director, 781-891-3433, Fax: 781-891-2464. *Application contact:* Holly L. Chase, Associate Director, 781-891-2108, Fax: 781-891-2464, E-mail: gradadm@bentley.edu.

Boston University, Metropolitan College, Program in Administrative Studies, Boston, MA 02215. Offers financial economics (MSAS); innovation and technology (MSAS); multinational commerce (MSAS); organizational policy (MSAS). Part-time and evening/weekend programs available. *Faculty:* 7 full-time (2 women), 25 part-time (5 women). *Students:* 107 full-time (42 women), 159 part-time (60 women). *Degree requirements:* For master's, thesis optional, foreign language not required. *Entrance requirements:* For master's, TOEFL (minimum score of 550 required), 1 year of work experience. *Application deadline:* Applications are processed on a rolling basis. Application fee: $50. Tuition: Part-time $508 per credit. Required fees: $40 per semester. Part-time tuition and fees vary according to class time. *Unit head:* Dr. Kip Becker, Chairman, E-mail: adminsc@bu.edu. *Application contact:* Department of Administrative Sciences, 617-353-3016, Fax: 617-353-6840, E-mail: adminsc@bu.edu.

Brigham Young University, Graduate Studies, College of Engineering and Technology, School of Technology, Provo, UT 84602-1001. Offers engineering technology (MS); technology teacher education (MS). *Faculty:* 21 full-time (0 women). *Students:* 15 full-time (0 women), 2 international. *Degree requirements:* For master's, thesis required (for some programs), foreign language not required. *Entrance requirements:* For master's, GRE General Test, minimum GPA of 3.0 in last 60 hours. *Application deadline:* For fall admission, 2/28. Application fee: $30. Tuition: Full-time $3,330; part-time $185 per credit hour. Tuition and fees vary according to program and student's religious affiliation. *Unit head:* Thomas L. Erekson, Director, 801-378-6300, Fax: 801-378-7575, E-mail: erekson@byu.edu. *Application contact:* Graduate Coordinator, 801-378-6300, Fax: 801-378-7575, E-mail: ralowe@byu.edu.

California Polytechnic State University, San Luis Obispo, College of Engineering, Program in Engineering, San Luis Obispo, CA 93407. Offers biochemical engineering (MS); industrial engineering (MS); integrated technology management (MS); materials engineering (MS); mechanical engineering (MS); water engineering (MS). *Faculty:* 98 full-time (8 women), 82 part-time (14 women). *Students:* 25 full-time (3 women), 6 part-time. *Degree requirements:* Foreign language not required. *Entrance requirements:* For master's, GRE General Test, minimum GPA of 2.5 in last 90 quarter units. Application fee: $55. Tuition, nonresident: part-time $164 per unit. Required fees: $531 per quarter. *Unit head:* Dr. Paul E. Rainey, Associate Dean, 805-756-2131, Fax: 805-756-6503, E-mail: prainey@calpoly.edu. *Application contact:* Dr. Paul E. Rainey, Associate Dean, 805-756-2131, Fax: 805-756-6503, E-mail: prainey@calpoly.edu.

Carlow College, Division of Business Management, Pittsburgh, PA 15213-3165. Offers management and technology (MS). Part-time and evening/weekend programs available. *Faculty:* 4 full-time (all women), 2 part-time (1 woman). *Degree requirements:* For master's, foreign language and thesis not required. *Entrance requirements:* For master's, interview, minimum GPA of 3.0. *Application deadline:* For fall admission, 4/1 (priority date); for spring admission, 11/1 (priority date). Applications are processed on a rolling basis. Application fee: $35. *Unit head:* Dr. Mary Rothenberger, Chair, 412-578-6181, Fax: 412-587-6367. *Application contact:* Bonnie Potthoff, Officer Manager, Graduate Studies, 412-578-8764, Fax: 412-578-8822, E-mail: bpotthoff@carlow.edu.

Central Missouri State University, School of Graduate Studies, College of Applied Sciences and Technology, Department of Manufacturing and Construction, Warrensburg, MO 64093. Offers industrial management (MS); industrial technology (MS); technology (PhD), including technology management. Part-time programs available. *Faculty:* 8 full-time (0 women). *Students:* 16 full-time (3 women), 27 part-time (5 women); includes 3 minority (1 African American, 2 Asian Americans or Pacific Islanders), 14 international. *Degree requirements:* For master's, comprehensive exam required, thesis not required. *Entrance requirements:* For master's, minimum GPA of 2.5; previous course work in mathematics, science, and technology. *Application deadline:* Applications are processed on a rolling basis. Application fee: $25 ($50 for international students). Tuition, state resident: full-time $3,576; part-time $149 per credit hour. Tuition, nonresident: full-time $7,152; part-time $298 per credit hour. *Unit head:* Dr. John Sutton, Chair, 660-543-4439, Fax: 660-543-4578, E-mail: sutton@cmsu1.cmsu.edu.

Central Missouri State University, School of Graduate Studies, College of Applied Sciences and Technology, Program in Technology, Warrensburg, MO 64093. Offers technology management (PhD). *Faculty:* 5 full-time (0 women). *Degree requirements:* For doctorate, dissertation, internship required. *Entrance requirements:* For doctorate, GRE General Test (minimum score of 500 on each section required), 2 years of work experience, minimum GPA of 3.0 undergraduate, 3.5 graduate. *Application deadline:* Applications are processed on a rolling basis. Tuition, state resident: full-time $3,576; part-time $149 per credit hour. Tuition, nonresident: full-time $7,152; part-time $298 per credit hour. Tuition and fees vary according to course load and campus/location. *Financial aid:* In 1998–99, 2 fellowships with tuition reimbursements (averaging $1,000 per year) were awarded.; unspecified assistantships also available. Financial aid application deadline: 3/1; financial aid applicants required to submit FAFSA. *Unit head:* Shi-Qing Wang, 216-368-6374, Fax: 216-368-4202, E-mail: sxw13@po.cwru.edu.

Colorado State University, Graduate School, College of Applied Human Sciences, Department of Manufacturing Technology and Construction Management, Fort Collins, CO 80523-0015. Offers automotive pollution control (MS); construction management (MS); historic preservation (PhD); industrial technology management (MS); technology education and training (MS); technology of industry (PhD). *Faculty:* 17 full-time (1 woman), 5 part-time (1 woman). *Students:* 14 full-time (1 woman), 8 part-time (3 women); includes 2 minority (both Hispanic Americans), 2 international. *Degree requirements:* For master's, computer language, thesis required (for some programs), foreign language not required; for doctorate, dissertation required. *Entrance requirements:* For master's, GRE General Test (minimum combined score of 1250 required), TOEFL (minimum score of 550 required; 213 for computer-based); for doctorate, GRE General Test, TOEFL. *Application deadline:* For fall admission, 4/1 (priority date). Applications are processed on a rolling basis. Application fee: $30. Electronic applications accepted. *Unit head:* Dr. Larry Grosse, Head, 970-491-7958, Fax: 970-491-2473, E-mail: drfire@aol.com. *Application contact:* Linda Burrous, Secretary, 970-491-7355, Fax: 970-491-2473, E-mail: burrous@cahs.colostate.edu.

École Polytechnique de Montréal, Graduate Programs, Department of Industrial Engineering, Montréal, PQ H3C 3A7, Canada. Offers ergonomy (M Eng, M Sc A, DESS); production (M Eng, M Sc A); technology management (M Eng, M Sc A). DESS offered jointly with Université de Montréal and École des Hautes Études Commerciales. Part-time programs available. *Degree requirements:* For master's, one foreign language, computer language, thesis required. *Entrance requirements:* For master's, minimum GPA of 2.75. *Faculty research:* Use of computers in organizations.

Embry-Riddle Aeronautical University, Extended Campus, Graduate Resident Centers, Department of Business Administration, Daytona Beach, FL 32114-3900. Offers aviation administration and management (MBAA); technical management (MSTM). Part-time and evening/weekend programs available. Postbaccalaureate distance learning degree programs offered (minimal on-campus study). *Students:* 16 full-time (3 women), 635 part-time (102 women); includes 116 minority (45 African Americans, 27 Asian Americans or Pacific Islanders, 32 Hispanic Americans, 12 Native Americans), 9 international. *Degree requirements:* For master's, thesis optional, foreign language not required. *Entrance requirements:* For master's, GMAT (minimum score of 500 required). *Application deadline:* Applications are processed on a rolling basis. Application fee: $30 ($50 for international students). Electronic applications accepted. Tuition: Full-time $6,897; part-time $314 per credit. Tuition and fees vary according to course load and campus/location. *Unit head:* Dr. Vance Mitchell, Chair, 360-375-1986, E-mail: mitchelv@cts.db.erau.edu. *Application contact:* Pam Thomas, Director of Admissions and Records, 904-226-6910, Fax: 904-226-6984, E-mail: ecinfo@ec.db.erau.edu.

Management of Technology

Fairfield University, School of Engineering, Fairfield, CT 06430-5195. Offers management of technology (MS); software engineering (MS). Part-time and evening/weekend programs available. *Faculty:* 44 part-time (4 women). *Students:* 1 full-time (0 women), 39 part-time (4 women); includes 2 minority (1 African American, 1 Hispanic American), 2 international. *Degree requirements:* For master's, thesis, final exam required. *Entrance requirements:* For master's, interview, minimum GPA of 2.8. *Application deadline:* For fall admission, 6/30 (priority date). Applications are processed on a rolling basis. Application fee: $40. *Unit head:* Dr. Evangelos Hadjimichael, Dean, 203-254-4000 Ext. 4147, Fax: 203-254-4013, E-mail: hadjm@fair1.fairfield.edu.

The George Washington University, School of Business and Public Management, Department of Management Science, Washington, DC 20052. Offers human resources management (MBA); information systems (MSIS); information systems management (MBA); logistics, operations, and materials management (MBA, MPA); management and organizations (PhD); management decision making (MBA, PhD); management of science, technology, and innovation (MBA); organizational behavior and development (MBA); project management (MS). Part-time and evening/weekend programs available. *Faculty:* 29 full-time (3 women), 23 part-time (5 women). *Students:* 227 full-time (75 women), 565 part-time (226 women); includes 168 minority (82 African Americans, 60 Asian Americans or Pacific Islanders, 24 Hispanic Americans, 2 Native Americans), 179 international. *Degree requirements:* For master's, computer language required, foreign language and thesis not required; for doctorate, computer language, dissertation required, foreign language not required. *Entrance requirements:* For master's, GMAT, TOEFL (minimum score of 550 required); for doctorate, GMAT or GRE, TOEFL (minimum score of 550 required). *Application deadline:* For fall admission, 4/1 (priority date); for spring admission, 10/1. Applications are processed on a rolling basis. Application fee: $55. Tuition: Full-time $17,328; part-time $722 per credit hour. Required fees: $828; $35 per credit hour. Tuition and fees vary according to campus/location and program. *Unit head:* Dr. Erik K. Winslow, Chair, 202-994-7375. *Application contact:* Lilly Hastings, Graduate Admissions, 202-994-6584, Fax: 202-994-6382.

Georgia Institute of Technology, Graduate Studies and Research, Dupree College of Management, Program in Management of Technology, Atlanta, GA 30332-0001. Offers MSMOT. Part-time and evening/weekend programs available. *Degree requirements:* For master's, study abroad required, foreign language and thesis not required. *Entrance requirements:* For master's, GMAT (average 600), TOEFL (minimum score of 550 required), 5 years of professional work experience. Electronic applications accepted. *Faculty research:* Innovation management, technology analysis, operations management.

Illinois Institute of Technology, Graduate College, Stuart School of Business, Chicago, IL 60661-3691. Offers business administration (MBA); environmental management (MS); financial markets and trading (MS); management science (PhD); marketing communication (MS); operations and technology management (MS). Part-time and evening/weekend programs available. *Faculty:* 20 full-time (1 woman), 35 part-time (6 women). *Students:* 150 full-time (40 women), 430 part-time (111 women); includes 135 minority (39 African Americans, 82 Asian Americans or Pacific Islanders, 14 Hispanic Americans), 104 international. Terminal master's awarded for partial completion of doctoral program. *Degree requirements:* For master's, comprehensive exam required, foreign language and thesis not required; for doctorate, dissertation, comprehensive exam required, foreign language not required. *Entrance requirements:* For master's and doctorate, GRE or GMAT, TOEFL (minimum score of 550 required). *Application deadline:* For fall admission, 8/1; for spring admission, 4/15. Applications are processed on a rolling basis. Application fee: $30. Electronic applications accepted. *Unit head:* Dr. M. Zia Hassan, Dean, 312-906-6500, Fax: 312-906-6549, E-mail: hassan@stuart.iit.edu. *Application contact:* Lynn Miller, Director, Admission, 312-906-6544, Fax: 312-906-6549, E-mail: degrees@stuart.iit.edu.

Instituto Tecnológico y de Estudios Superiores de Monterrey, Campus Morelos, Programs in Information Science, Cuernavaca, 62000, Mexico. Offers administration of information technology (MATI); computer science (MCC, DCC); information technology (MTI).

Lehigh University, College of Business and Economics, Department of Business, Bethlehem, PA 18015-3094. Offers business administration (MBA); business and economics (PhD); management of technology (MS). Part-time and evening/weekend programs available. Postbaccalaureate distance learning degree programs offered (minimal on-campus study). *Faculty:* 37 full-time (5 women), 9 part-time (2 women). *Students:* 57 full-time (16 women), 274 part-time (89 women); includes 10 minority (6 African Americans, 4 Hispanic Americans), 14 international. Terminal master's awarded for partial completion of doctoral program. *Degree requirements:* For master's, thesis (MS) required; for doctorate, dissertation required, foreign language not required. *Entrance requirements:* For master's, GMAT (average 590 for MBA), GRE, TOEFL (minimum score of 570 required; average 616); for doctorate, GMAT or GRE, TOEFL (minimum score of 570 required; average 604). *Application deadline:* For fall admission, 7/15; for spring admission, 12/1. Applications are processed on a rolling basis. Application fee: $40. *Unit head:* Kathleen A. Trexler, Associate Dean and Director, 610-758-3418, Fax: 610-758-5283, E-mail: kat3@lehigh.edu.

Marquette University, Graduate School, College of Engineering, Department of Biomedical Engineering, Milwaukee, WI 53201-1881. Offers bioinstrumentation/computers (MS, PhD); biomechanics/biomaterials (MS, PhD); functional imaging (PhD); healthcare technologies management (MS); systems physiology (MS, PhD). Part-time and evening/weekend programs available. *Faculty:* 12 full-time (2 women), 31 part-time (4 women). *Students:* 41 full-time (15 women), 31 part-time (5 women); includes 9 minority (6 Asian Americans or Pacific Islanders, 3 Hispanic Americans), 11 international. Terminal master's awarded for partial completion of doctoral program. *Degree requirements:* For master's, computer language, thesis, comprehensive exam required, foreign language not required; for doctorate, computer language, dissertation defense, qualifying exam required. *Entrance requirements:* For master's and doctorate, GRE General Test, TOEFL (minimum score of 575 required). *Application deadline:* For fall admission, 2/15 (priority date); for spring admission, 11/15 (priority date). Applications are processed on a rolling basis. Application fee: $40. Tuition: Part-time $510 per credit hour. Tuition and fees vary according to program. *Unit head:* Dr. Dean C. Jeutter, Acting Chairman, 414-288-3375, Fax: 414-288-7938, E-mail: jeutterd@ums.csd.mu.edu.

Marshall University, Graduate College, Graduate School of Information, Technology and Engineering, Program in Technology Management, Huntington, WV 25755-2020. Offers MS. Part-time and evening/weekend programs available. *Students:* 2 full-time (1 woman), 41 part-time (9 women), 2 international. Average age 40. *Degree requirements:* For master's, final project, oral exam required. *Entrance requirements:* For master's, GRE General Test, minimum undergraduate GPA of 2.5. *Financial aid:* Tuition waivers (full) available. Aid available to part-time students. Financial aid application deadline: 8/1; financial aid applicants required to submit FAFSA. *Unit head:* Dr. Bernard Gillispie, Director, 304-696-6007, E-mail: gillespb@marshall.edu. *Application contact:* Ken O'Neal, Assistant Vice President, Adult Student Services, 304-746-2500 Ext. 1907, Fax: 304-746-1902, E-mail: oneal@marshall.edu.

Mercer University, School of Engineering, Macon, GA 31207-0003. Offers biomedical engineering (MSE); electrical engineering (MSE); engineering management (MSE); mechanical engineering (MSE); software engineering (MSE); software systems (MS); technical management (MS). Part-time and evening/weekend programs available. *Faculty:* 23 full-time (1 woman), 6 part-time (0 women). *Degree requirements:* For master's, computer language, thesis or alternative required, foreign language not required. *Entrance requirements:* For master's, GRE, minimum undergraduate GPA of 3.0. *Application deadline:* For fall admission, 7/1; for spring admission, 11/15. Applications are processed on a rolling basis. Application fee: $35 ($50 for international students). *Unit head:* Dr. Benjamin S. Kelley, Dean, 912-752-2459, Fax: 912-752-5593, E-mail: kelley_bs@mercer.edu. *Application contact:* Kathy Olivier, Coordinator, Special Programs, 912-752-2196, E-mail: oliver_kh@mercer.edu.

Mercer University, Cecil B. Day Campus, Stetson School of Business and Economics, Atlanta, GA 30341-4155. Offers business administration (MBA, XMBA); health care management (MS); technology management (MS). Part-time and evening/weekend programs available. *Faculty:* 27 full-time (11 women), 11 part-time (3 women). *Students:* 288 full-time (136 women), 352 part-time (196 women); includes 187 minority (147 African Americans, 28 Asian Americans or Pacific Islanders, 12 Hispanic Americans), 94 international. *Degree requirements:* For master's, oral exam (health care management) required, foreign language and thesis not required. *Entrance requirements:* For master's, GMAT (average 450). *Application deadline:* For fall admission, 8/1 (priority date); for spring admission, 12/1 (priority date). Applications are processed on a rolling basis. Application fee: $35 ($50 for international students). *Unit head:* Dr. W. Carl Joiner, Dean, 912-752-2832. *Application contact:* Dr. Donald D. Wilson, Associate Dean, 770-986-3324, Fax: 770-986-3337, E-mail: wilson_dd@mercer.edu.

Murray State University, College of Industry and Technology, Department of Industrial and Engineering Technology, Program in Management of Technology, Murray, KY 42071-0009. Offers MS. Part-time programs available. *Students:* 30 full-time (6 women), 42 part-time (10 women); includes 5 minority (3 African Americans, 2 Hispanic Americans), 15 international. 19 applicants, 100% accepted. *Degree requirements:* Foreign language not required. *Entrance requirements:* For master's, GRE General Test, TOEFL (minimum score of 600 required). *Application deadline:* Applications are processed on a rolling basis. Application fee: $20. *Financial aid:* Research assistantships, teaching assistantships available. Financial aid application deadline: 4/1. *Unit head:* Kathe Prince, Senior Documentation Specialist, 906-487-2039, Fax: 906-487-2934, E-mail: kaprince@mtu.edu.

National Technological University, Programs in Engineering, Fort Collins, CO 80526-1842. Offers chemical engineering (MS); computer engineering (MS); computer science (MS); electrical engineering (MS); engineering management (MS); hazardous waste management (MS); health physics (MS); management of technology (MS); manufacturing systems engineering (MS); materials science and engineering (MS); software engineering (MS); special majors (MS); transportation engineering (MS); transportation systems engineering (MS). Part-time programs available. *Faculty:* 600 part-time (20 women). *Entrance requirements:* For master's, BS in engineering or related field. *Application deadline:* Applications are processed on a rolling basis. Application fee: $50. *Unit head:* Lionel V. Baldwin, President, 970-495-6400, Fax: 970-484-0668, E-mail: baldwin@mail.ntu.edu.

National University, Graduate Studies, School of Business and Technology, Department of Technology, La Jolla, CA 92037-1011. Offers e-commerce (MBA, MS); electronic engineering (MS); engineering management (MS); environmental management (MBA, MS); industrial engineering management (MS); software engineering (MS); technology management (MBA, MS); telecommunication systems management (MS). Part-time and evening/weekend programs available. Postbaccalaureate distance learning degree programs offered (minimal on-campus study). *Faculty:* 12 full-time, 125 part-time. *Students:* 305 (79 women); includes 122 minority (34 African Americans, 69 Asian Americans or Pacific Islanders, 17 Hispanic Americans, 2 Native Americans) 53 international. *Degree requirements:* For master's, foreign language and thesis not required. *Entrance requirements:* For master's, interview, minimum GPA of 2.5. *Application deadline:* Applications are processed on a rolling basis. Application fee: $60 ($100 for international students). Tuition: Full-time $7,830; part-time $870 per course. One-time fee: $60. Tuition and fees vary according to campus/location. *Unit head:* Dr. Leonid Preiser, Chair, 858-642-8425, Fax: 858-642-8716, E-mail: lpreiser@nu.edu. *Application contact:* Nancy Rohland, Director of Enrollment Management, 858-642-8180, Fax: 858-642-8709, E-mail: nrohland@nu.edu.

North Carolina Agricultural and Technical State University, Graduate School, School of Technology, Department of Manufacturing Systems, Greensboro, NC 27411. Offers industrial technology (MS, MSIT); safety and driver education (MS). Part-time and evening/weekend programs available. *Faculty:* 5 full-time (1 woman). *Students:* 35 full-time (14 women), 20 part-time (4 women); includes 48 minority (47 African Americans, 1 Hispanic American), 3 international. Average age 31. 22 applicants, 77% accepted. In 1998, 25 degrees awarded. *Degree requirements:* For master's, thesis or alternative, comprehensive exam, qualifying exam required, foreign language not required. *Entrance requirements:* For master's, GRE General Test, minimum GPA of 3.0. *Application deadline:* For fall admission, 6/1 (priority date); for spring admission, 12/1. Applications are processed on a rolling basis. Application fee: $35. *Financial aid:* Career-related internships or fieldwork available. Financial aid application deadline: 6/1. *Unit head:* Dr. Abhay Trivedi, Chairperson, 336-334-7158, Fax: 336-334-7704.

Northern Kentucky University, School of Graduate Programs, Program in Technology, Highland Heights, KY 41099. Offers MST. *Faculty:* 2 full-time (1 woman). *Students:* 3 full-time (0 women), 19 part-time (3 women); includes 3 minority (1 African American, 2 Asian Americans or Pacific Islanders), 1 international. Average age 36. *Degree requirements:* For master's, thesis optional, foreign language not required. *Entrance requirements:* For master's, GRE General Test. *Application deadline:* For fall admission, 8/15 (priority date); for spring admission, 10/15. Applications are processed on a rolling basis. Application fee: $25. *Financial aid:* In 1998–99, 1 research assistantship was awarded.; fellowships, career-related internships or fieldwork and institutionally-sponsored loans also available. *Unit head:* Dr. Charles Pinder, Chairperson, 606-572-5440. *Application contact:* Peg Griffin, Graduate Coordinator, 606-572-6364, E-mail: griffinp@nku.edu.

Oregon Graduate Institute of Science and Technology, Graduate Studies, Department of Management in Science and Technology, Portland, OR 97291-1000. Offers computational finance (Certificate); management in science and technology (MS). Part-time and evening/weekend programs available. *Faculty:* 3 full-time (0 women), 9 part-time (1 woman). *Students:* 4 full-time (1 woman), 68 part-time (16 women); includes 5 minority (all Asian Americans or Pacific Islanders), 4 international. Average age 37. 27 applicants, 63% accepted. In 1998, 12 master's awarded. *Degree requirements:* For master's, foreign language and thesis not required. *Entrance requirements:* For master's, TOEFL (minimum score of 650 required). *Application deadline:* Applications are processed on a rolling basis. Application fee: $50. Electronic applications accepted. *Unit head:* Dr. Fred Young Phillips, Head, 503-690-1353, Fax: 503-690-1268, E-mail: fphillips@admin.ogi.edu. *Application contact:* Victoria Tyler, Enrollment Manager, 503-690-1335, Fax: 503-690-1285, E-mail: vtyler@admin.ogi.edu.

Pacific Lutheran University, Division of Graduate Studies, School of Business Administration and Management, Tacoma, WA 98447. Offers business administration (MBA), including technology and innovation management. Part-time and evening/weekend programs available. *Faculty:* 11 full-time (1 woman). *Students:* 53 full-time (20 women), 40 part-time (14 women); includes 6 minority (1 African American, 2 Asian Americans or Pacific Islanders, 2 Hispanic Americans, 1 Native American), 14 international. *Degree requirements:* For master's, foreign language and thesis not required. *Entrance requirements:* For master's, GMAT (minimum score of 470 required), TOEFL (minimum score of 550 required). *Application deadline:* Applications are processed on a rolling basis. Application fee: $35. *Unit head:* Dr. Donald Bell, Dean, 253-535-7251. *Application contact:* Catherine Pratt, Director, 253-535-7250.

Pacific States University, College of Business, Los Angeles, CA 90006. Offers finance (MBA); international business (MBA); management of technology (MBA). Part-time and evening/weekend programs available. *Faculty:* 4 full-time (0 women), 14 part-time (0 women). *Students:* 63 full-time (12 women), 8 part-time (4 women). *Entrance requirements:* For master's, TOEFL (minimum score of 450 required), minimum undergraduate GPA of 2.5 during last 90 hours. *Application deadline:* For fall admission, 6/15; for spring admission, 12/15. Applications are processed on a rolling basis. Application fee: $50 ($100 for international students). Electronic applications accepted. *Unit head:* Dr. Kamol Somvichian, Director, 888-200-0383, Fax: 323-731-2383, E-mail: admission@psuca.edu. *Application contact:* Woo Yeol Lee, Admissions Officer, 888-200-0383, Fax: 323-731-7276, E-mail: admission@psuca.edu.

Pepperdine University, School of Business and Management, Culver City, CA 90230-7615. Offers business (MBA), including business administration; executive business administration

Management of Technology

Pepperdine University (continued)
(MBAA); organizational development (MSOD); technology management (MSTM). Part-time and evening/weekend programs available. *Faculty:* 74 full-time (11 women), 36 part-time (5 women). *Students:* 173 full-time (53 women), 1,756 part-time (658 women); includes 461 minority (86 African Americans, 237 Asian Americans or Pacific Islanders, 135 Hispanic Americans, 3 Native Americans), 48 international. *Entrance requirements:* For master's, GMAT or MAT. *Application deadline:* For fall admission, 5/1. Applications are processed on a rolling basis. Application fee: $45. *Unit head:* Dr. Otis Baskin, Dean, 310-568-5500. *Application contact:* Director of Career Development and Student Recruitment, 310-568-5790.

Polytechnic University, Brooklyn Campus, Department of Management, Major in Management of Technology, Brooklyn, NY 11201-2990. Offers MS. Average age 33. 8 applicants, 88% accepted. *Degree requirements:* For master's, thesis or alternative required. *Entrance requirements:* For master's, GMAT, minimum B average in undergraduate course work. *Application deadline:* Applications are processed on a rolling basis. Application fee: $45. Electronic applications accepted. *Unit head:* John S. Kerge, Dean of Admissions, 718-260-3200, Fax: 718-260-3446, E-mail: admitme@poly.edu. *Application contact:* John S. Kerge, Dean of Admissions, 718-260-3200, Fax: 718-260-3446, E-mail: admitme@poly.edu.

Polytechnic University, Westchester Graduate Center, Graduate Programs, Division of Management, Major in Management of Technology, Hawthorne, NY 10532-1507. Offers MS. 36 applicants, 94% accepted. In 1998, 34 degrees awarded. *Degree requirements:* For master's, computer language required. *Application deadline:* Applications are processed on a rolling basis. Application fee: $45. Electronic applications accepted. *Application contact:* John S. Kerge, Dean of Admissions, 718-260-3200, Fax: 718-260-3446, E-mail: admitme@poly.edu.

Regis University, School for Professional Studies, Program in Computer Information Systems, Denver, CO 80221-1099. Offers data base technology (MSCIS); management of technology (Certificate); multimedia technology (Certificate); networking technology (Certificate); object oriented technology (Certificate). Offered at Boulder Campus, Northwest Denver Campus, Southeast Denver Campus, Fort Collins Campus, and Colorado Springs Campus. Part-time and evening/weekend programs available. *Students:* 587. *Degree requirements:* For master's, computer language, final research project required, foreign language and thesis not required. *Entrance requirements:* For master's, TOEFL (minimum score of 550 required), 3 years of related experience, interview. *Application deadline:* Applications are processed on a rolling basis. Application fee: $75. Tuition: Part-time $290 per credit hour. *Unit head:* Don Archer, Chair, 303-458-4302. *Application contact:* 800-677-9270, Fax: 303-964-5538, E-mail: admarg@regis.edu.

Rensselaer Polytechnic Institute, Graduate School, Lally School of Management and Technology, Program in Management and Technology, Troy, NY 12180-3590. Offers MBA, MS, PhD. Part-time and evening/weekend programs available. *Faculty:* 36 full-time (5 women), 6 part-time (0 women). *Degree requirements:* For master's and doctorate, computer language required, foreign language and thesis not required. *Entrance requirements:* For master's, GMAT, TOEFL (minimum score of 570 required); for doctorate, GMAT or GRE General Test, TOEFL (minimum score of 570 required). *Application deadline:* For fall admission, 2/1 (priority date). Application fee: $35. Electronic applications accepted. *Financial aid:* Fellowships, research assistantships, teaching assistantships, career-related internships or fieldwork, institutionally-sponsored loans, and tuition waivers (full and partial) available. Financial aid application deadline: 2/1. *Application contact:* Michele Martens, Manager of Enrollment Services, 518-276-6586, Fax: 518-276-2665, E-mail: management@rpi.edu.

Rhode Island College, School of Graduate Studies, Center for Management and Technology, Department of Management and Technology, Providence, RI 02908-1924. Offers industrial technology (MS). In 1998, 7 degrees awarded. Tuition, state resident: part-time $162 per credit. Tuition, nonresident: part-time $328 per credit. Required fees: $18 per credit. One-time fee: $40. Tuition and fees vary according to program and reciprocity agreements. *Unit head:* Dr. David Blanchette, Chairperson, 401-456-8036.

Saginaw Valley State University, College of Science, Engineering, and Technology, Program in Technological Processes, University Center, MI 48710. Offers MS. Part-time and evening/weekend programs available. *Faculty:* 7 full-time (0 women), 5 part-time (1 woman). *Students:* 3 full-time (0 women), 29 part-time (5 women); includes 4 minority (3 African Americans, 1 Asian American or Pacific Islander), 1 international. 55 applicants, 87% accepted. *Entrance requirements:* For master's, TOEFL (minimum score of 525 required), minimum GPA of 3.0. *Application deadline:* Applications are processed on a rolling basis. Application fee: $25. *Financial aid:* In 1998–99, 1 fellowship with partial tuition reimbursement, 1 research assistantship with full tuition reimbursement (averaging $2,500 per year) were awarded.; Federal Work-Study also available. Aid available to part-time students. Financial aid application deadline: 4/1; financial aid applicants required to submit FAFSA. *Faculty research:* Materials, manufacturing processes, environmental remediation. *Application contact:* Wynn P. McDonald, Director, Graduate Admissions, 517-249-1696, Fax: 517-790-0180, E-mail: gradadm@svsu.edu.

St. Ambrose University, College of Business, Program in Business Administration, Davenport, IA 52803-2898. Offers management generalist (MBA); technical management (MBA). Part-time and evening/weekend programs available. *Degree requirements:* For master's, Capstone Seminar required, foreign language and thesis not required. *Entrance requirements:* For master's, GMAT. Electronic applications accepted.

South Dakota School of Mines and Technology, Graduate Division, Program in Technology Management, Rapid City, SD 57701-3995. Offers MS. Part-time programs available. *Faculty:* 3 part-time (0 women). *Students:* 25 full-time (5 women), 2 international. Average age 29. In 1998, 7 degrees awarded. *Degree requirements:* For master's, foreign language and thesis not required. *Entrance requirements:* For master's, GMAT, TOEFL (minimum score of 520 required), TWE. *Application deadline:* For fall admission, 6/15 (priority date); for spring admission, 10/15. Applications are processed on a rolling basis. Application fee: $15. Electronic applications accepted. Tuition, state resident: part-time $89 per hour. Tuition, nonresident: part-time $261 per hour. Part-time tuition and fees vary according to program. *Financial aid:* In 1998–99, 2 students received aid, including 2 teaching assistantships; fellowships, research assistantships, Federal Work-Study and institutionally-sponsored loans also available. Aid available to part-time students. Financial aid application deadline: 5/15. *Unit head:* Dr. Stuart D. Kellogg, Director, 605-394-1271. *Application contact:* Brenda Brown, Secretary, 800-454-8162 Ext. 2493, Fax: 605-394-5360, E-mail: graduate_admissions@silver.sdmt.edu.

Southern Polytechnic State University, School of Management, Marietta, GA 30060-2896. Offers MS. Part-time and evening/weekend programs available. *Faculty:* 6 full-time (2 women), 2 part-time (1 woman). *Students:* 38 full-time (21 women), 63 part-time (24 women); includes 35 minority (32 African Americans, 2 Asian Americans or Pacific Islanders, 1 Hispanic American), 22 international. Average age 34. 12 applicants, 100% accepted. In 1998, 42 degrees awarded (100% found work related to degree). *Degree requirements:* For master's, thesis optional, foreign language not required. *Entrance requirements:* For master's, GMAT. *Application deadline:* For fall admission, 7/15 (priority date); for spring admission, 12/1. Applications are processed on a rolling basis. Application fee: $20. Tuition, state resident: full-time $2,146; part-time $119 per credit hour. Tuition, nonresident: full-time $7,586; part-time $421 per credit hour. *Financial aid:* In 1998–99, 50 students received aid; teaching assistantships, career-related internships or fieldwork and Federal Work-Study available. Aid available to part-time students. Financial aid application deadline: 5/1; financial aid applicants required to submit FAFSA. *Unit head:* Dr. Robert J. Yancy, Dean, 770-528-7440, Fax: 770-528-4967, E-mail: ryancy@spsu.edu.

Southwest Texas State University, Graduate School, School of Applied Arts and Technology, Department of Technology, San Marcos, TX 78666. Offers industrial technology (MST). Part-time and evening/weekend programs available. *Faculty:* 6 full-time (0 women), 3 part-time (0 women). *Students:* 3 full-time (0 women), 19 part-time (4 women); includes 6 minority (1 African American, 3 Asian Americans or Pacific Islanders, 2 Hispanic Americans), 3 international. Average age 32. In 1998, 4 degrees awarded. *Degree requirements:* For master's, comprehensive exam required, thesis optional, foreign language not required. *Entrance requirements:* For master's, GRE General Test (minimum combined score of 900 required), TOEFL (minimum score of 550 required), minimum GPA of 2.75 in last 60 hours. *Application deadline:* For fall admission, 6/15 (priority date); for spring admission, 10/15 (priority date). Applications are processed on a rolling basis. Application fee: $25 ($50 for international students). Tuition, state resident: full-time $684; part-time $38 per semester hour. Tuition, nonresident: full-time $4,572; part-time $254 per semester hour. *Financial aid:* Research assistantships, teaching assistantships, career-related internships or fieldwork, Federal Work-Study, and institutionally-sponsored loans available. Aid available to part-time students. Financial aid application deadline: 4/1; financial aid applicants required to submit FAFSA. *Faculty research:* Rapid prototyping, casting technology, statistical process control, spatial abilities and gender. *Unit head:* Dr. Robert B. Habingreither, Chair, 512-245-2137, Fax: 512-245-3052, E-mail: rh03@swt.edu. *Application contact:* Dr. Andy Batey, Graduate Adviser, 512-245-2137, Fax: 512-245-3052, E-mail: ab08@swt.edu.

State University of New York at Stony Brook, Graduate School, College of Engineering and Applied Sciences, Program in Technology and Society, Stony Brook, NY 11794. Offers technological systems management (MS). Evening/weekend programs available. *Faculty:* 7 full-time (1 woman), 5 part-time (0 women). *Students:* 24 full-time (16 women), 41 part-time (14 women); includes 16 minority (9 African Americans, 5 Asian Americans or Pacific Islanders, 2 Hispanic Americans), 8 international. 16 applicants, 75% accepted. In 1998, 20 degrees awarded. *Degree requirements:* For master's, thesis required, foreign language not required. *Entrance requirements:* For master's, GRE General Test, TOEFL. *Application deadline:* For fall admission, 1/15. Application fee: $50. *Financial aid:* In 1998–99, 2 research assistantships, 5 teaching assistantships were awarded.; fellowships *Faculty research:* Computer applications for technology and society issues. Total annual research expenditures: $1.6 million. *Unit head:* Dr. Thomas Liao, Director, 516-632-8770. *Application contact:* Dr. Sheldon Reaven, Graduate Director, 516-632-8765, Fax: 516-632-7809, E-mail: sreaven@ccmail.sunysb.edu.

State University of New York at Stony Brook, Graduate School, College of Engineering and Applied Sciences, W. Averell Harriman School for Management and Policy, Stony Brook, NY 11794. Offers management and policy (MS); technology management (MS). Part-time and evening/weekend programs available. *Faculty:* 10 full-time (2 women), 18 part-time (7 women). *Students:* 61 full-time (29 women), 34 part-time (13 women); includes 18 minority (1 African American, 17 Asian Americans or Pacific Islanders), 29 international. *Degree requirements:* For master's, internship required. *Entrance requirements:* For master's, GMAT or GRE General Test, TOEFL. *Application deadline:* For fall admission, 1/15. Application fee: $50. *Unit head:* Dr. Thomas Sexton, Director, 516-632-7180. *Application contact:* Thomas Gjerde, Director of Graduate Studies, 516-632-7163.

Stevens Institute of Technology, Graduate School, Wesley J. Howe School of Technology Management, Program in Technology Management, Hoboken, NJ 07030. Offers MS, MTM, PhD, Certificate. Part-time and evening/weekend programs available. Postbaccalaureate distance learning degree programs offered (minimal on-campus study). Terminal master's awarded for partial completion of doctoral program. *Degree requirements:* For master's, foreign language and thesis not required; for doctorate, variable foreign language requirement, computer language, dissertation required; for Certificate, computer language required, foreign language not required. *Entrance requirements:* For master's and doctorate, GMAT, GRE, TOEFL. Electronic applications accepted.

Texas A&M University–Commerce, Graduate School, College of Business and Technology, Department of Industrial and Engineering Technology, Commerce, TX 75429-3011. Offers industry and technology (MS). Part-time programs available. *Faculty:* 3 full-time (0 women), 1 part-time (0 women). *Students:* 8 full-time (3 women), 13 part-time (3 women); includes 1 minority (African American), 11 international. Average age 36. In 1998, 12 degrees awarded. *Degree requirements:* For master's, thesis (for some programs), oral comprehensive exam required, foreign language not required. *Entrance requirements:* For master's, GMAT (minimum score of 376 required), GRE General Test (minimum combined score of 850 required). *Average time to degree:* Master's–2 years full-time, 2.75 years part-time. *Application deadline:* For fall admission, 6/1 (priority date); for spring admission, 11/1 (priority date). Applications are processed on a rolling basis. Application fee: $0 ($25 for international students). Electronic applications accepted. *Financial aid:* In 1998–99, research assistantships (averaging $7,750 per year), teaching assistantships (averaging $7,750 per year) were awarded.; Federal Work-Study, institutionally-sponsored loans, and scholarships also available. Financial aid application deadline: 5/1; financial aid applicants required to submit FAFSA. *Faculty research:* Environmental science, engineering microelectronics, natural sciences. Total annual research expenditures: $128,781. *Unit head:* Dr. Jerry D. Parish, Head, 903-886-5474. *Application contact:* Betty Hunt, Graduate Admissions Adviser, 903-886-5167, Fax: 903-886-5165, E-mail: betty_hunt@tamu_commerce.edu.

Thunderbird, The American Graduate School of International Management, Graduate Programs, Program in International Technology Management, Glendale, AZ 85306-3236. Offers MIMOT. *Degree requirements:* For master's, one foreign language required, thesis optional. *Entrance requirements:* For master's, GMAT (minimum score of 500 required; average 601), TOEFL (minimum score of 550 required; average 590). *Application deadline:* For fall admission, 1/31 (priority date); for spring admission, 7/31 (priority date). Application fee: $100. *Financial aid:* Application deadline: 4/1; *Unit head:* Dr. Dale Davison, Head, 602-978-7150. *Application contact:* Judy Johnson, Director of Admissions, 602-978-7210, Fax: 602-439-5432, E-mail: johnsonj@t-bird.edu.

United States International University, College of Business Administration, San Diego, CA 92131-1799. Offers business administration (MBA); information and technology management (DBA); international business (MIBA, DBA), including finance (DBA), marketing (DBA); strategic business (DBA). Part-time and evening/weekend programs available. Terminal master's awarded for partial completion of doctoral program. *Degree requirements:* For master's, computer language required, foreign language and thesis not required; for doctorate, computer language, dissertation required, foreign language not required. *Entrance requirements:* For master's, GMAT (minimum score of 350 required; average 490), minimum GPA of 3.0; for doctorate, GMAT (minimum score of 490 required; average 560), minimum GPA of 3.3. Electronic applications accepted. *Faculty research:* Cross-cultural management.

Université Laval, Faculty of Graduate Studies, Faculty of Sciences and Engineering, Program in Technological Entrepreneurship, Sainte-Foy, PQ G1K 7P4, Canada. Offers Diploma. *Students:* 2 full-time (1 woman), 7 part-time (3 women). 9 applicants, 67% accepted. *Application deadline:* For fall admission, 3/1. Application fee: $30. *Unit head:* Yron Gasse, Director, 418-656-2131 Ext. 7960, Fax: 418-656-3337, E-mail: yvon.gasse@mng.ulaval.ca.

University of British Columbia, Faculty of Graduate Studies, Faculty of Commerce and Business Administration, Program in Advanced Technology Management, Vancouver, BC V6T 1Z2, Canada. Offers M Eng.

University of California, San Diego, Graduate Studies and Research, Graduate School of International Relations and Pacific Studies, Program in International Technology Management, La Jolla, CA 92093-5003. Offers MITM. *Faculty:* 19. *Degree requirements:* For master's, one foreign language, thesis not required. *Entrance requirements:* For master's, GMAT or GRE General Test, TOEFL (minimum score of 550 required). *Average time to degree:* Master's–2 years full-time. *Application deadline:* For fall admission, 2/15 (priority date). Applications are processed on a rolling basis. Application fee: $40. *Financial aid:* Application deadline: 1/15. *Faculty research:* International relations, comparative politics, international political economy, international communications and information industries. *Unit head:* Paradee Chularee, Student Affairs Officer, 310-825-8913, Fax: 310-206-7353, E-mail: paradee@ea.ucla.edu. *Application* *contact:* Jori Cincotta, Director of Admissions, 619-534-5914, Fax: 619-534-1135, E-mail: irps-apply@ucsd.edu.

University of Colorado at Boulder, Graduate School of Business Administration, Boulder, CO 80309. Offers accounting (MS); business administration (MBA, PhD); business self designed (MBA); finance (MBA, PhD); marketing (MBA, PhD); organization management (MBA, PhD); taxation (MS); technology and innovation management (MBA). Evening/weekend programs available. *Degree requirements:* For master's, computer language required, foreign language not required; for doctorate, dissertation, research internship required. *Entrance requirements:* For master's and doctorate, GMAT.

University of Denver, University College, Denver, CO 80208. Offers applied communication (MSS); computer information systems (MCIS); environmental policy and management (MEPM); healthcare systems (MHS); liberal studies (MLS); library and information services (MLIS); public health (MPH); technology management (MoTM); telecommunications (MTEL). Part-time and evening/weekend programs available. Postbaccalaureate distance learning degree programs offered (no on-campus study). *Faculty:* 1 (woman) full-time, 553 part-time (181 women). *Students:* 1,529 (804 women); includes 189 minority (67 African Americans, 33 Asian Americans or Pacific Islanders, 75 Hispanic Americans, 14 Native Americans) 61 international. *Entrance requirements:* For master's, minimum undergraduate GPA of 3.0. *Application deadline:* For fall admission, 8/10; for spring admission, 2/22. Applications are processed on a rolling basis. Application fee: $25. *Unit head:* Peter Warren, Dean, 303-871-3268, Fax: 303-871-4047, E-mail: pwarren@du.edu. *Application contact:* Bryan Ehrlich, Admission Coordinator, 303-871-3969, Fax: 303-871-3303, E-mail: behrlich@du.edu.

University of Maryland University College, Graduate School of Management and Technology, Program in Technology Management, College Park, MD 20742-1600. Offers Exec MS, MS. Offered evenings and weekends only. Part-time and evening/weekend programs available. Postbaccalaureate distance learning degree programs offered (no on-campus study). *Students:* 62 full-time (23 women), 326 part-time (108 women); includes 112 minority (77 African Americans, 24 Asian Americans or Pacific Islanders, 11 Hispanic Americans), 5 international. 123 applicants, 99% accepted. In 1998, 112 degrees awarded. *Degree requirements:* For master's, thesis or alternative required, foreign language not required. *Application deadline:* Applications are processed on a rolling basis. Application fee: $50. Electronic applications accepted. Tuition, state resident: full-time $5,058; part-time $281 per credit. Tuition, nonresident: full-time $6,876; part-time $382 per credit. Tuition and fees vary according to program. *Financial aid:* Federal Work-Study, grants, and scholarships available. Aid available to part-time students. Financial aid application deadline: 6/1; financial aid applicants required to submit FAFSA. *Unit head:* John Aje, Director, 301-985-7200, Fax: 301-985-4611, E-mail: jaje@nova.umuc.edu. *Application contact:* Coordinator, Graduate Admissions, 301-985-7155, Fax: 301-985-7175, E-mail: gradinfo@nova.umuc.edu.

University of Miami, Graduate School, College of Engineering, Department of Industrial Engineering, Coral Gables, FL 33124. Offers environmental health and safety (MS, MSEH), including environmental health and safety (MSEH), occupational ergonomics and safety (MS); ergonomics (PhD); industrial engineering (MSIE, PhD); management of technology (MS). *Faculty:* 10 full-time (1 woman), 3 part-time (0 women). *Students:* 50 full-time (6 women), 3 part-time (2 women); includes 14 minority (3 African Americans, 3 Asian Americans or Pacific Islanders, 8 Hispanic Americans), 23 international. *Degree requirements:* For master's, thesis required (for some programs), foreign language not required; for doctorate, dissertation required, foreign language not required. *Entrance requirements:* For master's, GRE General Test (minimum combined score of 1000 required), TOEFL (minimum score of 550 required), minimum GPA of 3.0; for doctorate, GRE General Test, TOEFL (minimum score of 550 required), minimum GPA of 3.0. *Application deadline:* For fall admission, 5/1 (priority date). Applications are processed on a rolling basis. Application fee: $35. Tuition: Full-time $15,336; part-time $852 per credit. Required fees: $174. Tuition and fees vary according to program. *Unit head:* Dr. Norman Einspruch, Chairman, 305-284-2344, Fax: 305-284-4040. *Application contact:* Dr. Sara Czaja, Professor, 305-284-2344, Fax: 305-284-4040.

University of Minnesota, Twin Cities Campus, Graduate School, Institute of Technology, Center for the Development of Technological Leadership, Program in Management of Technology, Minneapolis, MN 55455-0213. Offers MSMOT. Part-time and evening/weekend programs available. *Faculty:* 22 part-time (1 woman). *Students:* 54 (10 women); includes 3 minority (1 African American, 1 Asian American or Pacific Islander, 1 Hispanic American) Average age 34. 44 applicants, 70% accepted. In 1998, 27 degrees awarded. *Degree requirements:* For master's, thesis, capstone project required, foreign language not required. *Entrance requirements:* For master's, 5 years of work experience in high-tech company, preferably in Twin Cities area; GMAT (minimum score of 500 required), GRE General Test, or minimum GPA of 3.0. *Average time to degree:* Master's–2 years full-time. *Application deadline:* For fall admission, 7/15 (priority date). Applications are processed on a rolling basis. Application fee: $50 ($55 for international students). Electronic applications accepted. *Financial aid:* Institutionally-sponsored loans available. Financial aid applicants required to submit FAFSA. *Faculty research:* Operations management, strategic management, technology foresight, marketing, business analysis. *Unit head:* Dr. Yechiel Shulman, Director of Graduate Studies, 612-624-9807, Fax: 612-624-7510. *Application contact:* Ann Bechtel, Admission Assistant, 612-624-5747, Fax: 612-624-7510, E-mail: bann@cdtl.umn.edu.

University of New Mexico, Graduate School, Robert O. Anderson Graduate School of Management, Program in Management of Technology, Albuquerque, NM 87131-2039. Offers EMBA, MBA. *Faculty:* 4 full-time (2 women), 2 part-time (0 women). *Degree requirements:* For master's, computer language required, foreign language and thesis not required. *Entrance requirements:* For master's, GMAT (minimum score of 500 required), TOEFL (minimum score of 550 required). *Application deadline:* For fall admission, 6/1 (priority date); for spring admission, 11/1. Applications are processed on a rolling basis. Application fee: $25. *Financial aid:* Fellowships, research assistantships, career-related internships or fieldwork, Federal Work-Study, scholarships, and unspecified assistantships available. *Faculty research:* Technology transfer, technology commercialization, management of research and development, chaotic systems, environmental technology. Total annual research expenditures: $200,000. *Unit head:* Dr. Suleiman Kassicieh, Co-Director, 505-277-9870, Fax: 505-277-7108, E-mail: kasicieh@unm.edu. *Application contact:* Loyola Chastain, MBA Program Manager, 505-277-3147, Fax: 505-277-9356, E-mail: lchast@unm.edu.

University of Pennsylvania, School of Engineering and Applied Science, Executive Engineering in the Management of Technology Program, Philadelphia, PA 19104. Offers management of technology (MSE). Offered jointly with the Wharton School. Part-time and evening/weekend programs available. *Degree requirements:* Foreign language not required.

See in-depth description on page 1291.

University of Phoenix, Graduate Programs, Business Administration and Management Programs, Program in Technology Management, Phoenix, AZ 85072-2069. Offers MBA. Programs offered at campuses in Colorado, Colorado Springs, Florida, Louisiana, Michigan, New Mexico, Northern California, Phoenix, San Diego, Southern California, Utah, online, and at the Center for Distance Education. Postbaccalaureate distance learning degree programs offered (no on-campus study). *Students:* 1,318 full-time (349 women); includes 337 minority (117 African Americans, 148 Asian Americans or Pacific Islanders, 69 Hispanic Americans, 3 Native Americans) Average age 35. *Degree requirements:* For master's, thesis or alternative required, foreign language not required. *Entrance requirements:* For master's, TOEFL (minimum score of 580 required), minimum undergraduate GPA of 2.5, 3 years of work experience, comprehensive cognitive assessment (COCA). *Application deadline:* Applications are processed on a rolling basis. Application fee: $50. *Financial aid:* Applicants required to submit FAFSA. *Application contact:* Campus Information Center, 602-966-9577.

The University of Texas at San Antonio, College of Business, San Antonio, TX 78249-0617. Offers accounting (MP Acct); business administration (MBA); management of technology (MSMOT); taxation (MT). Part-time and evening/weekend programs available. *Faculty:* 69 full-time (20 women), 44 part-time (7 women). *Students:* 161 full-time (64 women), 363 part-time (135 women); includes 147 minority (14 African Americans, 26 Asian Americans or Pacific Islanders, 105 Hispanic Americans, 2 Native Americans), 24 international. *Degree requirements:* For master's, foreign language and thesis not required. *Entrance requirements:* For master's, GMAT (minimum score of 500 required). *Application deadline:* For fall admission, 7/1. Applications are processed on a rolling basis. Application fee: $25. *Unit head:* James F. Gaertner, Dean, 210-458-4313. *Application contact:* Dr. John H. Brown, Director of Admissions and Registrar, 210-458-4530.

University of Tulsa, Graduate School, College of Business Administration, Department of Engineering and Technology Management, Tulsa, OK 74104-3189. Offers chemical engineering (METM); computer science (METM); electrical engineering (METM); geological science (METM); mathematics (METM); mechanical engineering (METM); petroleum engineering (METM). Part-time and evening/weekend programs available. *Students:* 3 full-time (1 woman), 1 part-time, 3 international. Average age 30. 2 applicants, 100% accepted. In 1998, 2 degrees awarded. *Degree requirements:* For master's, foreign language and thesis not required. *Entrance requirements:* For master's, GRE General Test (minimum score of 430 on verbal section, 600 on quantitative required), TOEFL (minimum score of 575 required). *Application deadline:* Applications are processed on a rolling basis. Application fee: $30. Electronic applications accepted. Tuition: Full-time $8,640; part-time $480 per hour. Required fees: $3 per hour. One-time fee: $200 full-time. Tuition and fees vary according to program. *Financial aid:* Fellowships, research assistantships, teaching assistantships, Federal Work-Study and tuition waivers (partial) available. Aid available to part-time students. Financial aid application deadline: 2/1; financial aid applicants required to submit FAFSA. *Unit head:* Dr. Richard C. Burgess, Assistant Dean/Director of Graduate Business Studies, 918-631-2242, Fax: 918-631-2142.

University of Waterloo, Graduate Studies, Faculty of Engineering, Department of Management Sciences, Waterloo, ON N2L 3G1, Canada. Offers management of technology (MA Sc, PhD); management sciences (MA Sc, PhD). Part-time programs available. Postbaccalaureate distance learning degree programs offered (no on-campus study). *Faculty:* 13 full-time (2 women), 13 part-time (4 women). *Students:* 39 full-time (10 women), 45 part-time (9 women). *Degree requirements:* For master's, research paper or thesis required; for doctorate, dissertation, comprehensive exam required, foreign language not required. *Entrance requirements:* For master's, GMAT or GRE, TOEFL (minimum score of 550 required), honors degree, minimum B average; for doctorate, GMAT or GRE, TOEFL (minimum score of 550 required), master's degree. *Application deadline:* Applications are processed on a rolling basis. Application fee: $50. *Expenses:* Tuition and fees charges are reported in Canadian dollars. Tuition, state resident: full-time $3,168 Canadian dollars; part-time $792 Canadian dollars per term. Tuition, nonresident: full-time $8,000 Canadian dollars; part-time $2,000 Canadian dollars. Required fees: $45 Canadian dollars per term. Tuition and fees vary according to program. *Unit head:* Dr. J. D. Fuller, Chair, 519-888-4567 Ext. 2683, Fax: 519-746-7252, E-mail: dfuller@engmail.uwaterloo.ca. *Application contact:* Dr. J. Webster, Graduate Officer, 519-888-4567 Ext. 5683, Fax: 519-746-7252, E-mail: jwebster@mansci2.uwaterloo.ca.

See in-depth description on page 1293.

University of Wisconsin–Stout, Graduate Studies, College of Technology, Engineering, and Management, Program in Management Technology, Menomonie, WI 54751. Offers MS. Part-time programs available. *Students:* 12 full-time (2 women), 12 part-time (1 woman), 13 international. 23 applicants, 57% accepted. In 1998, 19 degrees awarded. *Degree requirements:* For master's, thesis required, foreign language not required. *Application deadline:* Applications are processed on a rolling basis. Application fee: $45. *Financial aid:* In 1998–99, 1 research assistantship was awarded.; teaching assistantships, Federal Work-Study and tuition waivers (full and partial) also available. Aid available to part-time students. Financial aid application deadline: 4/1; financial aid applicants required to submit FAFSA. *Unit head:* Dr. Zenon Smolarek, Director, 715-232-1144.

Vanderbilt University, School of Engineering, Management of Technology Program, Nashville, TN 37240-1001. Offers M Eng, MS, PhD, MBA/M Eng. MS and PhD offered through the Graduate School. Part-time programs available. *Faculty:* 11 full-time (0 women), 7 part-time (0 women). *Students:* 25 full-time (5 women), 17 part-time (7 women); includes 8 minority (5 African Americans, 3 Asian Americans or Pacific Islanders), 22 international. Average age 32. 7 applicants, 57% accepted. In 1998, 17 master's, 1 doctorate awarded. *Degree requirements:* For master's and doctorate, thesis/dissertation required, foreign language not required. *Entrance requirements:* For master's, GMAT or GRE General Test; for doctorate, GRE General Test. *Application deadline:* For fall admission, 1/15. Application fee: $40. *Financial aid:* In 1998–99, 3 research assistantships with full tuition reimbursements were awarded.; fellowships, teaching assistantships, career-related internships or fieldwork and tuition waivers (full and partial) also available. Financial aid application deadline: 1/15. *Faculty research:* Management of innovation, assessment and forecasting, project management and new venture development. *Unit head:* Kazuhiko Kawamura, Interim Director, 615-322-3479, Fax: 615-322-7996, E-mail: kawamura@vuse.vanderbilt.edu. *Application contact:* R. Nash, Director of Graduate Studies, 615-322-2122, Fax: 615-322-7996, E-mail: nash@vuse.vanderbilt.edu.

Operations Research

Air Force Institute of Technology, School of Engineering, Department of Operational Sciences, Program in Operational Analysis, Wright-Patterson AFB, OH 45433-7765. Offers MS, PhD. Part-time programs available. *Faculty:* 11 full-time (0 women). *Students:* 21 full-time. In 1998, 11 master's awarded (100% found work related to degree). *Degree requirements:* For master's and doctorate, thesis/dissertation required, foreign language not required. *Entrance requirements:* For master's, GRE General Test (minimum score of 500 on verbal section, 600 on quantitative required), minimum GPA of 3.0, must be military officer or U.S. citizen; for doctorate, GRE General Test (minimum score of 550 on verbal section, 650 on quantitative required), minimum GPA of 3.0, must be military officer or U.S. citizen. Application fee: $0. *Faculty research:* Combat modeling, simulation modeling and analysis, applied probability/stochastics, deterministic modeling/optimization. *Unit head:* Lt. Col. J. O. Miller, Director, 937-255-6565 Ext. 4333, Fax: 937-656-4943, E-mail: jmiller@afit.af.mil.

Air Force Institute of Technology, School of Engineering, Department of Operational Sciences, Program in Operations Research, Wright-Patterson AFB, OH 45433-7765. Offers MS, PhD. Part-time programs available. *Faculty:* 11 full-time (0 women). *Students:* 46 full-time, 1 part-time. In 1998, 2 master's awarded (100% found work related to degree); 2 doctorates awarded (100% found work related to degree). *Degree requirements:* For master's and doctorate, thesis/dissertation required, foreign language not required. *Entrance requirements:* For master's, GRE General Test (minimum score of 500 on verbal section, 600 on quantitative required), minimum GPA of 3.0, must be military officer or U.S. citizen; for doctorate, GRE General Test (minimum score of 550 on verbal section, 650 on quantitative required), minimum GPA of 3.0, must be military officer or U.S. citizen. *Average time to degree:* Master's–1.5 years full-time; doctorate–3 years full-time. Application fee: $0. *Faculty research:* Deterministic modeling/optimization, simulation modeling/analysis, applied probability/stochastic analysis, combat modeling, decision analysis. *Unit head:* Maj. Paul Murdock, Director, 937-255-6565 Ext. 4339, Fax: 937-656-9943, E-mail: wmurdock@afit.af.mil.

Baruch College of the City University of New York, Zicklin School of Business, Department of Statistics and Computer Information Systems, Program in Operations Research, New York, NY 10010-5585. Offers MBA, MS. Part-time and evening/weekend programs available. *Faculty:* 3 full-time (1 woman). *Students:* 2 full-time, 4 part-time. In 1998, 3 degrees awarded. *Degree requirements:* For master's, computer language required, foreign language not required. *Entrance requirements:* For master's, GMAT, GRE General Test (MS), TOEFL (minimum score of 570 required), TWE (minimum score of 4.5 required). *Average time to degree:* Master's–2 years full-time, 4 years part-time. *Application deadline:* For fall admission, 4/1; for spring admission, 11/1. Application fee: $40. *Financial aid:* Research assistantships, career-related internships or fieldwork and Federal Work-Study available. Aid available to part-time students. Financial aid application deadline: 5/3; financial aid applicants required to submit FAFSA. *Application contact:* Michael S. Wynne, Office of Graduate Admissions, 212-802-2330, Fax: 212-802-2335, E-mail: graduate_admissions@baruch.cuny.edu.

California State University, Fullerton, Graduate Studies, School of Business Administration and Economics, Department of Management Science, Fullerton, CA 92834-9480. Offers management information systems (MS); management science (MBA, MS); operations research (MS); statistics (MS). Part-time and evening/weekend programs available. *Faculty:* 24 full-time (2 women), 16 part-time. *Students:* 7 full-time (6 women), 55 part-time (21 women); includes 19 minority (18 Asian Americans or Pacific Islanders, 1 Hispanic American), 26 international. *Degree requirements:* For master's, computer language, project or thesis required. *Entrance requirements:* For master's, GMAT (minimum score of 950 required), minimum AACSB index of 950. Application fee: $55. Tuition, nonresident: part-time $264 per unit. Required fees: $1,947; $1,281 per year. *Unit head:* Dr. Barry Pasternack, Chair, 714-278-2221.

California State University, Hayward, Graduate Programs, School of Business and Economics, Department of Management and Finance, Option in Operations Research, Hayward, CA 94542-3000. Offers MBA. *Degree requirements:* For master's, comprehensive exam or thesis required. *Entrance requirements:* For master's, GMAT, minimum GPA of 2.75. Application fee: $55. Tuition, nonresident: part-time $164 per unit. Required fees: $587 per quarter. *Financial aid:* Federal Work-Study and institutionally-sponsored loans available. Aid available to part-time students. Financial aid application deadline: 3/1. *Unit head:* Dr. Alan Goldberg, Coordinator, 510-885-3304. *Application contact:* Dr. Donna L. Wiley, Director of Graduate Programs, 510-885-3964.

Carnegie Mellon University, Graduate School of Industrial Administration, Program in Operations Research, Pittsburgh, PA 15213-3891. Offers PhD. *Faculty:* 6 full-time (0 women). *Degree requirements:* For doctorate, dissertation required, foreign language not required. *Entrance requirements:* For doctorate, GMAT or GRE General Test. *Application deadline:* For fall admission, 2/1. Application fee: $50. *Financial aid:* Fellowships available. Financial aid application deadline: 5/1. *Unit head:* Bonnie Potthoff, Officer Manager, Graduate Studies, 412-578-8764, Fax: 412-578-8822, E-mail: bpotthoff@carlow.edu. *Application contact:* Jackie Cavendish, Administrative Assistant, 412-268-2301.

Case Western Reserve University, Weatherhead School of Management, Department of Operations Research and Operations Management, Cleveland, OH 44106. Offers management science (MS), including finance, information systems, marketing, operations management, operations research, quality management; operations management (MBA); operations research (PhD). MS and PhD offered through the School of Graduate Studies. Part-time programs available. *Faculty:* 12 full-time (0 women). *Students:* 9 full-time (2 women), 5 part-time (1 woman), 3 international. Average age 28. In 1998, 5 master's, 7 doctorates awarded. *Degree requirements:* For master's, foreign language and thesis not required; for doctorate, dissertation required, foreign language not required. *Entrance requirements:* For master's, GMAT (MBA), GRE General Test (MS); for doctorate, GMAT, GRE General Test. *Average time to degree:* Doctorate–4 years full-time. *Application deadline:* For fall admission, 4/15 (priority date). Applications are processed on a rolling basis. Application fee: $50. *Financial aid:* Tuition waivers (full and partial) available. Financial aid application deadline: 5/1. *Faculty research:* Mathematical finance, mathematical programming, scheduling, stochastic optimization, environmental/energy models. *Unit head:* Matthew J. Sobel, Chairman, 216-368-6003. *Application contact:* Christine L. Gill, Director of Marketing and Admissions, 216-368-2144, Fax: 216-368-4776, E-mail: clg3@po.cwru.edu.

See in-depth description on page 1267.

Claremont Graduate University, Graduate Programs, Department of Mathematics, Claremont, CA 91711-6163. Offers engineering mathematics (PhD); financial engineering (MS); operations research and statistics (MA, MS); physical applied mathematics (MA, MS); pure mathematics (MA, MS, PhD); scientific computing (MA, MS); systems and control theory (MA, MS). Part-time programs available. *Faculty:* 3 full-time (0 women), 3 part-time (1 woman). *Students:* 12 full-time (8 women), 41 part-time (7 women); includes 17 minority (2 African Americans, 10 Asian Americans or Pacific Islanders, 5 Hispanic Americans), 16 international. Terminal master's awarded for partial completion of doctoral program. *Degree requirements:* For master's, foreign language and thesis not required; for doctorate, 2 foreign languages (computer language can substitute for one), dissertation required. *Entrance requirements:* For master's and doctorate, GRE General Test. *Application deadline:* For fall admission, 2/15 (priority date). Applications are processed on a rolling basis. Application fee: $40. Electronic applications accepted. Tuition: Full-time $20,950; part-time $913 per unit. Required fees: $65 per semester. Tuition and fees vary according to program. *Unit head:* Robert Williamson, Chair, 909-621-8080, Fax: 909-621-8390, E-mail: robert.williamson@cgu.edu. *Application contact:* Mary Solberg, Program Secretary, 909-621-8080, Fax: 909-621-8390, E-mail: math@cgu.edu.

Clemson University, Graduate School, College of Engineering and Science, Department of Mathematical Sciences, Clemson, SC 29634. Offers applied and pure mathematics (MS, PhD); computational mathematics (MS, PhD); management science (PhD); operations research (MS, PhD); statistics (MS, PhD). Part-time programs available. *Students:* 60 full-time (24 women), 1 part-time; includes 3 minority (2 African Americans, 1 Hispanic American), 11 international. *Degree requirements:* For master's, computer language, final project required, thesis optional, foreign language not required; for doctorate, computer language, dissertation, qualifying exams required, foreign language not required. *Entrance requirements:* For master's and doctorate, GRE General Test, TOEFL. *Application deadline:* For fall admission, 6/1. Application fee: $35. *Unit head:* Dr. Robert Fennell, Chair, 864-656-3436, Fax: 864-656-5230, E-mail: mathsci@clemson.edu. *Application contact:* Dr. Douglas Shier, Graduate Coordinator, 864-656-1100, Fax: 864-656-5230, E-mail: shierd@clemson.edu.

College of William and Mary, Faculty of Arts and Sciences, Department of Computer Science, Program in Computational Operations Research, Williamsburg, VA 23187-8795. Offers MS. Part-time programs available. *Degree requirements:* For master's, computer language, research project required, thesis optional. *Entrance requirements:* For master's, GRE General Test, minimum GPA of 2.5. *Application deadline:* For fall admission, 3/1 (priority date); for spring admission, 11/1. Applications are processed on a rolling basis. Application fee: $30. *Unit head:* Dr. Georges Fadel, Coordinator, Program in Mechanical Engineering, 864-656-5620, Fax: 864-656-4435. *Application contact:* Vanessa Godwin, Administrative Director, 757-221-3455, Fax: 757-221-1717, E-mail: gradinfo@cs.wm.edu.

Announcement: The College offers a program of study leading to the MS/PhD in computational operations research. Course offerings include linear programming, discrete optimization, networks, deterministic and stochastic models, reliability, decision theory, algorithms, and simulation. Typically fewer than 10 students per course. Students should e-mail cor@cs.wm.edu for brochure and application materials. World Wide Web: http://www.math.wm.edu/~leemis/or.html

Columbia University, Fu Foundation School of Engineering and Applied Science, Department of Industrial Engineering and Operations Research, New York, NY 10027. Offers financial engineering (MS); industrial engineering (MS, Eng Sc D, PhD, Engr); operations research (MS, Eng Sc D, PhD). Part-time programs available. Postbaccalaureate distance learning degree programs offered (no on-campus study). *Faculty:* 11 full-time (1 woman), 10 part-time (0 women). *Students:* 68 full-time (19 women), 57 part-time (19 women). Average age 27. 332 applicants, 59% accepted. In 1998, 55 master's, 4 doctorates awarded. *Degree requirements:* For master's, foreign language and thesis not required; for doctorate, dissertation, oral and written qualifying exams required, foreign language not required. *Entrance requirements:* For master's, doctorate, and Engr, GRE General Test, TOEFL. *Average time to degree:* Master's–1 year full-time, 2 years part-time; doctorate–5 years full-time. *Application deadline:* For fall admission, 1/1; for spring admission, 10/1. Application fee: $55. *Financial aid:* In 1998–99, 20 students received aid, including 4 fellowships (averaging $5,000 per year), 7 research assistantships (averaging $15,600 per year), 9 teaching assistantships (averaging $15,600 per year); Federal Work-Study also available. Financial aid application deadline: 1/5; financial aid applicants required to submit FAFSA. *Faculty research:* Discrete event stochastic systems, optimization, productions planning and scheduling, inventory control, yield management, simulation, mathematical programming, combinatorial optimization, queuing. Total annual research expenditures: $1 million. *Unit head:* Dr. Donald Goldfarb, Chairman, 212-854-8011, Fax: 212-854-8103, E-mail: gold@ieor.columbia.edu. *Application contact:* Tamara Kachanov, Departmental Administrator, 212-854-1473, Fax: 212-854-8103, E-mail: tamara@ieor.columbia.edu.

Cornell University, Graduate School, Graduate Fields of Engineering, Field of Operations Research and Industrial Engineering, Ithaca, NY 14853-0001. Offers applied probability and statistics (PhD); manufacturing systems engineering (PhD); mathematical programming (PhD); operations research and industrial engineering (M Eng). *Faculty:* 30 full-time. *Students:* 91 full-time (20 women); includes 22 minority (1 African American, 21 Asian Americans or Pacific Islanders), 36 international. 297 applicants, 52% accepted. In 1998, 71 master's, 7 doctorates awarded. Terminal master's awarded for partial completion of doctoral program. *Degree requirements:* For doctorate, dissertation required, foreign language not required, foreign language not required. *Entrance requirements:* For master's and doctorate, GRE General Test, TOEFL (minimum score of 550 required). *Application deadline:* For fall admission, 1/15. Application fee: $65. Electronic applications accepted. *Financial aid:* In 1998–99, 34 students received aid, including 9 fellowships with full tuition reimbursements available, 6 research assistantships with full tuition reimbursements available, 19 teaching assistantships with full tuition reimbursements available; institutionally-sponsored loans, scholarships, tuition waivers (full and partial), and unspecified assistantships also available. Financial aid applicants required to submit FAFSA. *Faculty research:* Combinatorial optimization, mathematical finance, stochastic processes, simulation. *Unit head:* Director of Graduate Studies, 607-255-9128. *Application contact:* Graduate Field Assistant, 607-255-9128, E-mail: orphd@cornell.edu.

See in-depth description on page 1269.

École Polytechnique de Montréal, Graduate Programs, Department of Industrial Engineering, Montréal, PQ H3C 3A7, Canada. Offers ergonomy (M Eng, M Sc A, DESS); production (M Eng, M Sc A); technology management (M Eng, M Sc A). DESS offered jointly with Université de Montréal and École des Hautes Études Commerciales. Part-time programs available. *Degree requirements:* For master's, one foreign language, computer language, thesis required. *Entrance requirements:* For master's, minimum GPA of 2.75. *Faculty research:* Use of computers in organizations.

École Polytechnique de Montréal, Graduate Programs, Department of Mathematics, Montréal, PQ H3C 3A7, Canada. Offers mathematical method in CA engineering (M Eng, M Sc A, PhD); operational research (M Eng, M Sc A, PhD). Part-time programs available. *Degree requirements:* For master's and doctorate, one foreign language, computer language, thesis/dissertation required. *Entrance requirements:* For master's, minimum GPA of 2.75; for doctorate, minimum GPA of 3.0. *Faculty research:* Statistics and probability, fractal analysis, optimization.

Embry-Riddle Aeronautical University, Daytona Beach Campus Graduate Program, Program in Industrial Optimization, Daytona Beach, FL 32114-3900. Offers MSIO. *Faculty:* 2 full-time (1 woman). *Degree requirements:* For master's, thesis optional, foreign language not required. *Entrance requirements:* For master's, TOEFL (minimum score of 550 required). Application fee: $30 ($50 for international students). Tuition: Full-time $8,190; part-time $455 per credit. Required fees: $105 per semester. Tuition and fees vary according to program. *Financial aid:* Application deadline: 4/15; *Faculty research:* Multiple response optimization, risk analysis, statistical process control and improvement, multiple criteria decision making. Total annual research expenditures: $800,000. *Unit head:* Dr. Deborah Osborne, Chair, 904-226-7688, Fax: 904-226-6269, E-mail: osborned@cts.db.erau.edu. *Application contact:* Ginny Tait, Graduate Admissions Specialist, 904-226-6115, Fax: 904-226-6299, E-mail: taitg@cts.db.erau.edu.

See in-depth description on page 1273.

Florida Institute of Technology, Graduate School, College of Science and Liberal Arts, Department of Mathematical Sciences, Program in Operations Research, Melbourne, FL 32901-6975. Offers MS, PhD. Part-time and evening/weekend programs available. *Faculty:* 1 full-time (0 women), 2 part-time (0 women). *Students:* 6 full-time (all women), 2 part-time (both women); includes 1 minority (African American), 4 international. Average age 38. 17 applicants, 76% accepted. In 1998, 1 master's, 3 doctorates awarded. Terminal master's awarded for partial completion of doctoral program. *Degree requirements:* For master's, computer language, comprehensive exam required, thesis optional, foreign language not required; for doctorate, computer language, dissertation, comprehensive exam required, foreign language not required.

Entrance requirements: For master's, minimum GPA of 3.0, proficiency in FORTRAN, Pascal, or C; for doctorate, GRE General Test, master's degree in operations research or equivalent, minimum GPA of 3.2. *Application deadline:* Applications are processed on a rolling basis. Application fee: $50. Electronic applications accepted. Tuition: Part-time $575 per credit hour. Required fees: $100. Tuition and fees vary according to campus/location and program. *Financial aid:* In 1998–99, 1 student received aid, including 1 research assistantship with tuition reimbursement available (averaging $2,470 per year); teaching assistantships with tuition reimbursements available, tuition remissions also available. Financial aid application deadline: 3/1; financial aid applicants required to submit FAFSA. *Faculty research:* Numerical computation, applied statistics, simulation, optimization, scheduling, decision analysis, queueing processes. *Unit head:* Dr. Jay Yellen, Chair, 407-674-7485, Fax: 407-674-7412, E-mail: yellen@fit.edu. *Application contact:* Carolyn P. Farrior, Associate Dean of Graduate Admissions, 407-674-7118, Fax: 407-723-9468, E-mail: cfarrior@fit.edu.

Florida Institute of Technology, Graduate School, School of Extended Graduate Studies, Program in Management, Melbourne, FL 32901-6975. Offers acquisition and contract management (MS, PMBA); global management (PMBA); health management (MS); human resources management (MS, PMBA); information systems (PMBA); logistics management (MS); management (MS), including information systems, transportation; materials acquisition management (MS); operations research (MS); research (PMBA); space systems (MS); space systems management (MS). *Students:* 104 full-time (27 women), 1,089 part-time (494 women); includes 234 minority (161 African Americans, 36 Asian Americans or Pacific Islanders, 30 Hispanic Americans, 7 Native Americans), 9 international. *Degree requirements:* Foreign language not required. *Entrance requirements:* For master's, GMAT (minimum score of 425 required), minimum GPA of 3.0. *Application deadline:* Applications are processed on a rolling basis. Application fee: $50. Electronic applications accepted. Tuition: Part-time $270 per credit hour. Part-time tuition and fees vary according to campus/location. *Application contact:* Carolyn P. Farrior, Associate Dean of Graduate Admissions, 407-674-7118, Fax: 407-723-9468, E-mail: cfarrior@fit.edu.

Florida Institute of Technology, Graduate School, School of Extended Graduate Studies, Program in Operations Research, Melbourne, FL 32901-6975. Offers MS. Part-time and evening/weekend programs available. *Students:* 1 full-time (0 women), 16 part-time (4 women); includes 1 minority (African American) Average age 33. 10 applicants, 80% accepted. In 1998, 2 degrees awarded (100% found work related to degree). *Degree requirements:* For master's, computer language, comprehensive exam required, thesis optional, foreign language not required. *Entrance requirements:* For master's, minimum GPA of 3.0, proficiency in FORTRAN, Pascal, or C. *Average time to degree:* Master's–1 year full-time, 3 years part-time. *Application deadline:* Applications are processed on a rolling basis. Application fee: $50. Electronic applications accepted. Tuition: Part-time $270 per credit hour. Part-time tuition and fees vary according to campus/location. *Financial aid:* Application deadline: 3/1; *Unit head:* Carolyn P. Farrior, Associate Dean of Graduate Admissions, 407-674-7118, Fax: 407-723-9468, E-mail: cfarrior@fit.edu. *Application contact:* Carolyn P. Farrior, Associate Dean of Graduate Admissions, 407-674-7118, Fax: 407-723-9468, E-mail: cfarrior@fit.edu.

George Mason University, School of Information Technology and Engineering, Department of Systems Engineering and Operations Research, Fairfax, VA 22030-4444. Offers operations research and management science (MS); systems engineering (MS). Part-time and evening/weekend programs available. *Students:* 14 full-time (2 women), 183 part-time (50 women); includes 43 minority (20 African Americans, 17 Asian Americans or Pacific Islanders, 5 Hispanic Americans, 1 Native American), 17 international. Average age 34. 122 applicants, 84% accepted. In 1998, 46 degrees awarded. *Degree requirements:* For master's, computer language required, thesis optional, foreign language not required. *Entrance requirements:* For master's, GRE General Test, TOEFL (minimum score of 575 required), minimum GPA of 3.0 in last 60 hours. *Application deadline:* For fall admission, 5/1; for spring admission, 11/1. Application fee: $30. Electronic applications accepted. Tuition, state resident: full-time $4,416; part-time $184 per credit hour. Tuition, nonresident: full-time $12,516; part-time $522 per credit hour. Tuition and fees vary according to program. *Financial aid:* Fellowships, research assistantships, teaching assistantships, career-related internships or fieldwork and Federal Work-Study available. Aid available to part-time students. Financial aid application deadline: 3/1; financial aid applicants required to submit FAFSA. *Faculty research:* Requirements engineering, signal processing, systems architecture, data fusion. *Unit head:* Dr. Karla L. Hoffman, Chairman, 703-993-1670, Fax: 703-993-1521, E-mail: khoffman@gmu.edu.

The George Washington University, School of Engineering and Applied Science, Department of Operations Research, Washington, DC 20052. Offers MS, D Sc, App Sc. Part-time and evening/weekend programs available. *Faculty:* 6 full-time (1 woman), 5 part-time (0 women). *Students:* 25 full-time (7 women), 29 part-time (9 women); includes 7 minority (3 African Americans, 4 Hispanic Americans), 29 international. Average age 30. 61 applicants, 95% accepted. In 1998, 15 master's, 4 doctorates, 1 other advanced degree awarded. *Degree requirements:* For master's, thesis optional, foreign language not required; for doctorate, computer language, dissertation, comprehensive, final, and qualifying exams required, foreign language not required; for App Sc, comprehensive exam required, foreign language and thesis not required. *Entrance requirements:* For master's, TOEFL (minimum score of 550 required; average 580) or George Washington University English as a Foreign Language Test, appropriate bachelor's degree; for doctorate, TOEFL (minimum score of 550 required; average 580) or George Washington University English as a Foreign Language Test, appropriate bachelor's or master's degree, GRE required if highest earned degree is BS; for App Sc, TOEFL (minimum score of 550 required; average 580) or George Washington University English as a Foreign Language Test, appropriate master's degree. *Application deadline:* For fall admission, 3/1; for spring admission, 10/1. Applications are processed on a rolling basis. Application fee: $55. Tuition: Full-time $17,328; part-time $722 per credit hour. Required fees: $828; $35 per credit hour. Tuition and fees vary according to campus/location and program. *Financial aid:* In 1998–99, 4 fellowships, 8 research assistantships, 3 teaching assistantships were awarded.; career-related internships or fieldwork and institutionally-sponsored loans also available. Financial aid application deadline: 3/1; financial aid applicants required to submit FAFSA. *Faculty research:* Reliability and quality control, risk analysis, multicriteria decision analysis, theory and applications of mathematical programming. Total annual research expenditures: $251,100. *Unit head:* Dr. Richard Soland, Chair, 202-994-6084. *Application contact:* Howard M. Davis, Manager, Office of Admissions and Student Records, 202-994-6158, Fax: 202-994-0909, E-mail: data:adms@seas.gwu.edu.

Georgia Institute of Technology, Graduate Studies and Research, College of Engineering, School of Industrial and Systems Engineering, Program in Operations Research, Atlanta, GA 30332-0001. Offers MSOR. Part-time programs available. *Degree requirements:* For master's, computer language required, foreign language and thesis not required. *Entrance requirements:* For master's, GRE General Test, TOEFL (minimum score of 550 required), minimum GPA of 3.0. Electronic applications accepted. *Faculty research:* Linear and nonlinear deterministic models in operations research, mathematical statistics, design of experiments.

Georgia State University, College of Business Administration, Department of Decision Sciences, Atlanta, GA 30303-3083. Offers MBA, MS, PhD. Part-time and evening/weekend programs available. *Faculty:* 19 full-time (1 woman), 1 part-time (0 women). *Students:* 20 full-time (9 women), 21 part-time (9 women); includes 4 minority (1 African American, 2 Asian Americans or Pacific Islanders, 1 Hispanic American), 9 international. Average age 33. 19 applicants, 58% accepted. In 1998, 11 master's, 4 doctorates awarded. Terminal master's awarded for partial completion of doctoral program. *Degree requirements:* For master's, foreign language and thesis not required; for doctorate, dissertation required, foreign language not required. *Entrance requirements:* For master's, GMAT (average 566), TOEFL; for doctorate, GMAT (average 670), TOEFL. *Application deadline:* For fall admission, 5/1; for spring admission, 10/1. Applications are processed on a rolling basis. Application fee: $25. Tuition, state resident: full-time $3,510; part-time $130 per credit hour. Tuition, nonresident: full-time $13,986; part-time $518 per credit hour. *Financial aid:* Fellowships, research assistantships,

teaching assistantships, career-related internships or fieldwork and tuition waivers (partial) available. Aid available to part-time students. Financial aid applicants required to submit FAFSA. *Unit head:* Dr. Merwyn Elliott, Chair, 404-651-4061, Fax: 404-651-2804. *Application contact:* Dr. Robert Elrod, Director of Graduate Studies, 404-651-4061, Fax: 404-651-2804, E-mail: relrod@gsu.edu.

Idaho State University, Graduate School, College of Engineering, Pocatello, ID 83209. Offers engineering and applied science (PhD); environmental engineering (MS); hazardous waste management (MS); measurement and control engineering (MS); nuclear science and engineering (MS). MS (hazardous waste management), PhD offered jointly with the University of Idaho. Part-time programs available. *Degree requirements:* For master's, thesis required, foreign language not required; for doctorate, dissertation required. *Entrance requirements:* For master's and doctorate, GRE General Test, TOEFL. *Faculty research:* Isotope separation, control technology, two-phase flow, photosonolysis, criticality calculations.

Indiana University–Purdue University Fort Wayne, School of Arts and Sciences, Department of Mathematical Sciences, Fort Wayne, IN 46805-1499. Offers applied mathematics (MS); mathematics (MS); operations research (MS). Part-time and evening/weekend programs available. *Faculty:* 18 full-time (4 women), 1 (woman) part-time. *Students:* 3 full-time (1 woman), 6 part-time (3 women). *Degree requirements:* For master's, foreign language and thesis not required. *Entrance requirements:* For master's, minimum GPA of 3.0, major or minor in mathematics. *Application deadline:* For fall admission, 5/1 (priority date); for spring admission, 12/1. Applications are processed on a rolling basis. Application fee: $30. *Unit head:* Raymond E. Pippert, Chair, 219-481-6224, Fax: 219-481-6880, E-mail: pippert@cvax.ipfw.indiana.edu. *Application contact:* W. Douglas Weakley, Director of Graduate Studies, 219-481-6238, Fax: 219-481-6880, E-mail: weakley@cvax.ipfw.indiana.edu.

Iowa State University of Science and Technology, Graduate College, College of Engineering, Department of Industrial and Manufacturing Systems Engineering, Ames, IA 50011. Offers industrial engineering (MS, PhD); operations research (MS). *Faculty:* 16 full-time, 2 part-time. *Students:* 44 full-time (4 women), 11 part-time (3 women); includes 1 minority (African American), 48 international. *Degree requirements:* For master's, computer language, thesis required (for some programs), foreign language not required; for doctorate, computer language, dissertation required, foreign language not required. *Entrance requirements:* For master's and doctorate, GRE General Test, TOEFL (minimum score of 550 required). *Application deadline:* For fall admission, 6/15 (priority date); for spring admission, 11/15 (priority date). Application fee: $20 ($50 for international students). Electronic applications accepted. Tuition, state resident: full-time $3,308. Tuition, nonresident: full-time $9,744. Part-time tuition and fees vary according to course load, campus/location and program. *Unit head:* Dr. Pius J. Egbelu, Chair, 515-294-1682, Fax: 515-294-3524. *Application contact:* Dr. John Jackman, Director of Graduate Studies, 515-294-0126.

Kansas State University, Graduate School, College of Engineering, Department of Industrial and Manufacturing Systems Engineering, Manhattan, KS 66506. Offers engineering management (MEM); industrial and manufacturing systems engineering (PhD); industrial engineering (MS); operations research (MS). Part-time programs available. Postbaccalaureate distance learning degree programs offered. *Faculty:* 10 full-time (1 woman). *Students:* 24 full-time (4 women), 37 part-time (6 women); includes 1 minority (Asian American or Pacific Islander), 29 international. *Degree requirements:* For master's, computer language required, foreign language not required; for doctorate, computer language, dissertation required, foreign language not required. *Entrance requirements:* For master's and doctorate, GRE General Test (minimum score of 725 on quantitative section required). *Application deadline:* For fall admission, 3/1 (priority date); for spring admission, 9/1. Applications are processed on a rolling basis. Application fee: $0 ($25 for international students). Electronic applications accepted. *Unit head:* Dr. Bradley Kramer, Head, 785-532-5606.

Louisiana Tech University, Graduate School, College of Engineering and Science, Department of Mechanical and Industrial Engineering, Ruston, LA 71272. Offers industrial engineering (MS, D Eng); manufacturing systems engineering (MS); mechanical engineering (MS, D Eng); operations research (MS). Part-time programs available. Terminal master's awarded for partial completion of doctoral program. *Degree requirements:* For master's and doctorate, thesis/dissertation required, foreign language not required. *Entrance requirements:* For master's, GRE General Test (minimum combined score of 1070 required), TOEFL (minimum score of 550 required), minimum GPA of 3.0 in last 60 hours; for doctorate, TOEFL (minimum score of 550 required), minimum graduate GPA of 3.25 (with MS) or GRE General Test (minimum combined score of 1270 required without MS). *Faculty research:* Engineering management, facilities planning, thermodynamics, automated manufacturing, micromanufacturing.

Massachusetts Institute of Technology, Operations Research Center, Cambridge, MA 02139-4307. Offers SM, PhD. *Faculty:* 35 full-time (2 women). *Students:* 46 full-time; includes 2 minority (both African Americans), 24 international. Average age 24. 83 applicants, 35% accepted. In 1998, 12 master's awarded (75% found work related to degree, 25% continued full-time study); 10 doctorates awarded (40% entered university research/teaching, 60% found other work related to degree). Terminal master's awarded for partial completion of doctoral program. *Degree requirements:* For master's, computer language, thesis required, foreign language not required; for doctorate, computer language, dissertation, written exam required, foreign language not required. *Entrance requirements:* For master's and doctorate, GRE General Test, GRE Subject Test, TOEFL (minimum score of 577 required). *Average time to degree:* Master's–2 years full-time; doctorate–4 years full-time. *Application deadline:* For fall admission, 1/15. Application fee: $55. *Financial aid:* In 1998–99, 11 fellowships (averaging $12,000 per year), 26 research assistantships (averaging $12,000 per year), 9 teaching assistantships (averaging $12,000 per year) were awarded.; Federal Work-Study, institutionally-sponsored loans, scholarships, and tuition waivers also available. Financial aid application deadline: 1/15. *Faculty research:* Linear and nonlinear programming, combinatorial and network optimization, operations management, transportation, applied probabilistic systems. *Unit head:* Dr. James B. Orlin, Co-Director, 617-253-6606, Fax: 617-258-9214, E-mail: jorlin@mit.edu. *Application contact:* Laura A. Rose, Admissions Coordinator, 617-253-9303, Fax: 617-258-9214, E-mail: lrose@mit.edu.

Announcement: ORC is proud of its scholastic achievements and supportive learning environment. Faculty members and students have distinguished themselves as winners of prizes from professional societies, recipients of MIT teaching awards, editors of major OR professional journals, and officers of leading professional societies. Strongly cohesive and supportive student group adds important dimension to quality of life and education at the ORC.

See in-depth description on page 1279.

Miami University, Graduate School, College of Arts and Sciences, Department of Mathematics and Statistics, Program in Mathematics, Oxford, OH 45056. Offers mathematics (MA, MAT, MS); mathematics/operations research (MS); statistics (MS). Part-time programs available. *Faculty:* 20. *Students:* 23 full-time (7 women); includes 1 minority (Asian American or Pacific Islander), 3 international. *Degree requirements:* For master's, final exam required, thesis not required. *Entrance requirements:* For master's, minimum undergraduate GPA of 3.0 during previous 2 years or 2.75 overall. *Application deadline:* For fall admission, 3/1 (priority date); for spring admission, 12/1. Applications are processed on a rolling basis. Application fee: $35. *Unit head:* Dr. Sheldon Davis, Director of Graduate Studies, 513-529-3527.

Michigan State University, Graduate School, College of Natural Science, Department of Statistics and Probability, East Lansing, MI 48824-1020. Offers applied statistics (MS); computational statistics (MS); operations research-statistics (MS); statistics (MA, MS, PhD). Part-time programs available. *Students:* 27 full-time (8 women), 13 part-time (3 women); includes 1 minority (Asian American or Pacific Islander), 33 international. Terminal master's awarded for partial completion of doctoral program. *Degree requirements:* For master's, foreign language and thesis not required; for doctorate, dissertation required. *Entrance*

Operations Research

Michigan State University (continued)
requirements: For master's and doctorate, GRE, TOEFL. *Application deadline:* For fall admission, 1/1 (priority date). Applications are processed on a rolling basis. Application fee: $30 ($40 for international students). *Unit head:* Dr. Habib Salehi, Chairman, 517-353-3391, Fax: 517-432-1405. *Application contact:* James Stapleton, Graduate Program Director, 517-355-9678, Fax: 517-432-1405, E-mail: stapleton@stt.msu.edu.

Naval Postgraduate School, Graduate Programs, Department of Operations Research, Monterey, CA 93943. Offers MS, PhD. Program only open to commissioned officers of the United States and friendly nations and selected United States federal civilian employees. Part-time programs available. *Students:* 142 full-time, 23 international. In 1998, 60 master's, 1 doctorate awarded. *Degree requirements:* For master's, computer language, thesis required, foreign language not required; for doctorate, one foreign language, computer language, dissertation required. *Unit head:* Dr. Richard Rosenthal, Chairman, 831-656-2381. *Application contact:* Theodore H. Calhoon, Director of Admissions, 831-656-3093, Fax: 831-656-2891, E-mail: tcalhoon@nps.navy.mil.

New Mexico Institute of Mining and Technology, Graduate Studies, Department of Mathematics, Socorro, NM 87801. Offers mathematics (MS); operations research (MS). *Faculty:* 10 full-time (1 woman). *Students:* 8 full-time (3 women); includes 1 minority (Hispanic American), 1 international. *Degree requirements:* For master's, thesis optional, foreign language not required. *Entrance requirements:* For master's, GRE General Test, TOEFL (minimum score of 540 required). *Application deadline:* For fall admission, 3/1 (priority date); for spring admission, 6/1. Applications are processed on a rolling basis. Application fee: $16. *Unit head:* Curtis Barefoot, Chairman, 505-835-5393, Fax: 505-835-5366, E-mail: barefoot@nmt.edu. *Application contact:* Dr. David B. Johnson, Dean of Graduate Studies, 505-835-5513, Fax: 505-835-5476, E-mail: graduate@nmt.edu.

New York University, Leonard N. Stern School of Business, Department of Statistics and Operations Research, New York, NY 10012-1019. Offers MBA, MS, PhD, APC. *Faculty:* 14 full-time (1 woman), 5 part-time (0 women). *Students:* 40 full-time, 64 part-time. In 1998, 56 master's, 3 doctorates awarded. *Degree requirements:* For master's, computer language required, foreign language and thesis not required; for doctorate, computer language, dissertation required, foreign language not required; for APC, foreign language and thesis not required. *Entrance requirements:* For master's, GMAT, TOEFL (minimum score of 600 required); for doctorate, GMAT. *Average time to degree:* Master's–3 years part-time. *Application deadline:* For fall admission, 3/15. Applications are processed on a rolling basis. Application fee: $75. Tuition: Full-time $27,500. *Financial aid:* Federal Work-Study available. Financial aid application deadline: 1/15; financial aid applicants required to submit FAFSA. *Faculty research:* Time-series modeling, stochastic process and financial modeling, statistical modeling. *Unit head:* Edward Melnick, Chair, 212-998-0440, E-mail: emelnick@stern.nyu.edu. *Application contact:* Mary Miller, Assistant Dean, MBA Admissions and Student Services, 212-998-0600, Fax: 212-995-4231, E-mail: sternmba@stern.nyu.edu.

North Carolina State University, Graduate School, College of Engineering, Program in Operations Research, Raleigh, NC 27695. Offers MOR, MS, PhD. Part-time programs available. *Faculty:* 40 full-time (1 woman), 2 part-time (0 women). *Students:* 18 full-time (7 women), 7 part-time (2 women); includes 8 minority (5 African Americans, 3 Asian Americans or Pacific Islanders), 11 international. Average age 30. 17 applicants, 24% accepted. In 1998, 1 master's awarded. *Degree requirements:* For master's, computer language, thesis (MS) required; for doctorate, computer language, dissertation, comprehensive oral and written exams required, foreign language not required. *Entrance requirements:* For master's, GRE General Test, TOEFL (minimum score of 550 required), minimum GPA of 2.7; for doctorate, GRE General Test, TOEFL (minimum score of 550 required), minimum GPA of 3.0. *Average time to degree:* Master's–2 years full-time, 3 years part-time; doctorate–3 years full-time, 5 years part-time. *Application deadline:* For fall admission, 6/25; for spring admission, 11/25. Applications are processed on a rolling basis. Application fee: $45. *Financial aid:* In 1998–99, 6 fellowships (averaging $2,559 per year), 1 research assistantship (averaging $3,153 per year), 9 teaching assistantships (averaging $4,504 per year) were awarded. *Faculty research:* Mathematical programming, activity networks, stochastic systems control, theory of algorithms, quality control. *Unit head:* Dr. William J. Stewart, Director of Graduate Programs, 919-515-7824, Fax: 919-515-5281, E-mail: billy@csc.ncsu.edu.

North Carolina State University, Graduate School, College of Management, Program in Management, Raleigh, NC 27695. Offers biotechnology (MS); computer science (MS); engineering (MS); forest resources management (MS); general business (MS); management information systems (MS); operations research (MS); statistics (MS); telecommunications systems engineering (MS); textile management (MS); total quality management (MS). Part-time programs available. *Faculty:* 40 full-time (9 women), 4 part-time (0 women). *Students:* 48 full-time (15 women), 156 part-time (43 women); includes 33 minority (16 African Americans, 15 Asian Americans or Pacific Islanders, 1 Hispanic American, 1 Native American), 4 international. *Degree requirements:* For master's, computer language required, foreign language and thesis not required. *Entrance requirements:* For master's, GRE or GMAT, TOEFL (minimum score of 550 required), minimum undergraduate GPA of 3.0. *Application deadline:* For fall admission, 6/25; for spring admission, 11/25. Applications are processed on a rolling basis. Application fee: $45. *Unit head:* Dr. Jack W. Wilson, Director of Graduate Programs, 919-515-4327, Fax: 919-515-6943, E-mail: jack_wilson@ncsu.edu. *Application contact:* Dr. Steven G. Allen, Director of Graduate Programs, 919-515-6941, Fax: 919-515-5073, E-mail: steve_allen@ncsu.edu.

North Dakota State University, Graduate Studies and Research, College of Science and Mathematics, Department of Computer Science, Fargo, ND 58105. Offers computer science (MS, PhD); operations research (MS). Part-time programs available. *Faculty:* 12 full-time (1 woman), 2 part-time (0 women). *Students:* 73 full-time (11 women), 19 part-time (7 women); includes 4 minority (2 African Americans, 2 Asian Americans or Pacific Islanders), 72 international. *Degree requirements:* For master's, computer language, comprehensive exam required, thesis optional, foreign language not required; for doctorate, computer language, dissertation, qualifying exam required, foreign language not required. *Entrance requirements:* For master's, TOEFL (minimum score of 600 required), minimum GPA of 3.0, BS in computer science or related field; for doctorate, TOEFL (minimum score of 600 required), minimum GPA of 3.25, MS in computer science or related field. *Application deadline:* For fall admission, 8/15 (priority date); for spring admission, 12/15 (priority date). Application fee: $25. *Unit head:* Dr. Kendall E. Nygard, Chair, 701-231-8562, Fax: 701-231-8255, E-mail: nygard@plains.nodak.edu.

Northeastern University, College of Engineering, Department of Mechanical, Industrial, and Manufacturing Engineering, Boston, MA 02115-5096. Offers engineering management (MS); industrial engineering (MS, PhD); mechanical engineering (MS, PhD); operations research (MS). Part-time programs available. *Faculty:* 30 full-time (1 woman), 13 part-time (1 woman). *Students:* 105 full-time (19 women), 118 part-time (25 women); includes 14 minority (5 African Americans, 7 Asian Americans or Pacific Islanders, 1 Hispanic American, 1 Native American), 84 international. *Degree requirements:* For master's, thesis required (for some programs), foreign language not required; for doctorate, one foreign language, dissertation, departmental qualifying exam required. *Entrance requirements:* For master's and doctorate, GRE General Test. *Application deadline:* For fall admission, 4/15. Applications are processed on a rolling basis. Application fee: $50. *Unit head:* Dr. John W. Cipolla, Chairman, 617-373-3810, Fax: 617-373-2921. *Application contact:* Stephen L. Gibson, Associate Director, 617-373-2711, Fax: 617-373-2501, E-mail: grad-eng@coe.neu.edu.

Northwestern University, The Graduate School, Robert R. McCormick School of Engineering and Applied Science, Department of Industrial Engineering and Management Sciences, Evanston, IL 60208. Offers engineering management (MEM); industrial engineering and management science (MS, PhD); operations research (MS, PhD). MS and PhD admissions and degrees offered through The Graduate School. Part-time programs available. *Faculty:* 16 full-time (1 woman), 4 part-time (0 women). *Students:* 47 full-time (15 women), 130 part-time (30 women). Terminal master's awarded for partial completion of doctoral program. *Degree requirements:* For master's, computer language, comprehensive exam required, foreign language and thesis not required; for doctorate, computer language, dissertation, comprehensive exam required, foreign language not required. *Entrance requirements:* For master's, GRE General Test, TOEFL (minimum score of 560 required), 3 years of work experience (MEM); for doctorate, GRE General Test, TOEFL (minimum score of 560 required). *Application deadline:* For fall admission, 8/30. Application fee: $50 ($55 for international students). *Unit head:* Mark Daskin, Chair, 847-491-8796, Fax: 847-491-8005, E-mail: daskin@iems.nwu.edu. *Application contact:* June Wayne, Admission Contact, 847-491-4394, Fax: 847-491-8005, E-mail: wayne@iems.nwu.edu.

Old Dominion University, College of Engineering and Technology, Department of Engineering Management, Norfolk, VA 23529. Offers engineering management (MEM, MS, PhD); operations research/systems analysis (ME). Part-time and evening/weekend programs available. Postbaccalaureate distance learning degree programs offered (no on-campus study). *Faculty:* 8 full-time (2 women). *Students:* 15 full-time (1 woman), 79 part-time (15 women); includes 16 minority (8 African Americans, 5 Asian Americans or Pacific Islanders, 3 Hispanic Americans), 13 international. *Degree requirements:* For master's, computer language, comprehensive exam, project required, thesis optional, foreign language not required; for doctorate, computer language, dissertation, candidacy exam required, foreign language not required. *Entrance requirements:* For master's, GRE, TOEFL (minimum score of 550 required), minimum GPA of 2.75; for doctorate, GRE, TOEFL (minimum score of 550 required), minimum GPA of 3.25. *Application deadline:* For fall admission, 7/1; for spring admission, 10/1. Applications are processed on a rolling basis. Application fee: $30. Electronic applications accepted. *Unit head:* Dr. Resit Unal, Interim Chair, 737-683-5541, Fax: 737-683-5640.

See in-depth description on page 1283.

Oregon State University, Graduate School, College of Science, Department of Statistics, Corvallis, OR 97331. Offers applied statistics (MA, MS, PhD); biometry (MA, MS, PhD); environmental statistics (MA, MS, PhD); mathematical statistics (MA, MS, PhD); operations research (MA, MAIS, MS); statistics (M Agr, MA, MS, PhD). Part-time programs available. *Faculty:* 14 full-time (4 women). *Degree requirements:* For master's, consulting experience required; for doctorate, dissertation, consulting experience required, foreign language not required. *Entrance requirements:* For master's and doctorate, TOEFL (minimum score of 550 required; 213 for computer-based), minimum GPA of 3.0 in last 90 hours. *Application deadline:* For fall admission, 2/15. Applications are processed on a rolling basis. Application fee: $50. *Unit head:* Dr. Robert Smythe, Chair, 541-737-3366. *Application contact:* Dr. Dawn Peters, Director of Graduate Studies, 541-737-1991, Fax: 541-737-3489, E-mail: statoff@stat.orst.edu.

Princeton University, Graduate School, School of Engineering and Applied Science, Department of Civil Engineering and Operations Research, Program in Statistics and Operations Research, Princeton, NJ 08544-1019. Offers MSE, PhD. MSE offered jointly with the Department of Electrical Engineering. *Degree requirements:* For master's, thesis required; for doctorate, dissertation, qualifying exam required. *Entrance requirements:* For master's, GRE General Test, GRE Subject Test, bachelor's degree in engineering or science; for doctorate, GRE General Test, GRE Subject Test.

Purdue University, Graduate School, Schools of Engineering, School of Industrial Engineering, West Lafayette, IN 47907. Offers human factors in industrial engineering (MS, MSIE, PhD); manufacturing engineering (MS, MSIE, PhD); operations research (MS, MSIE, PhD); systems engineering (MS, MSIE, PhD). Part-time programs available. *Faculty:* 27 full-time (2 women), 3 part-time (0 women). *Students:* 124 full-time (26 women), 33 part-time (8 women); includes 13 minority (2 African Americans, 4 Asian Americans or Pacific Islanders, 6 Hispanic Americans, 1 Native American), 107 international. Terminal master's awarded for partial completion of doctoral program. *Degree requirements:* For master's, computer language required, thesis optional, foreign language not required; for doctorate, computer language, dissertation required, foreign language not required. *Entrance requirements:* For master's, GRE General Test (minimum score of 470 on verbal section, 700 on quantitative, 600 on analytical required; average 520 verbal, 710 quantitative, 660 analytical), TOEFL (minimum score of 570 required; average 600), minimum GPA of 3.0; for doctorate, GRE General Test (minimum score of 470 on verbal section, 700 on quantitative, 600 on analytical required; average 540 verbal, 750 quantitative, 675 analytical), TOEFL (minimum score of 570 required; average 610), MS thesis. *Application deadline:* For fall admission, 3/15; for spring admission, 9/1. Application fee: $30. Electronic applications accepted. *Unit head:* Dr. W. D. Compton, Interim Head, 765-494-5444, Fax: 765-494-1299, E-mail: dcompton@ecn.purdue.edu. *Application contact:* Dr. J. W. Barany, Associate Head, 765-494-5406, Fax: 765-494-1299, E-mail: jwb@ecn.purdue.edu.

Rensselaer Polytechnic Institute, Graduate School, Lally School of Management and Technology, Troy, NY 12180-3590. Offers accounting/finance (PhD); applied economics (PhD); business administration (MBA, PhD), including finance and accounting (MBA), information systems management (MBA), management (PhD), management of technology and entrepreneurships (MBA), manufacturing management (MBA), marketing management (MBA), operations research (MBA), organizational behavior and human resource management (MBA), statistical methods for management (MBA); business policy and strategy (PhD); environmental management and policy (MS, PhD); human resource (PhD); management and technology (MBA, MS, PhD); management information systems (PhD); managerial economics (PhD); manufacturing (PhD). Part-time and evening/weekend programs available. Postbaccalaureate distance learning degree programs offered (no on-campus study). *Faculty:* 36 full-time (5 women), 6 part-time (0 women). *Students:* 288 full-time (83 women), 128 part-time (39 women); includes 34 minority (6 African Americans, 19 Asian Americans or Pacific Islanders, 9 Hispanic Americans), 154 international. *Degree requirements:* For master's, computer language required, foreign language and thesis not required; for doctorate, computer language, dissertation required, foreign language not required. *Entrance requirements:* For master's, GMAT, TOEFL (minimum score of 570 required); for doctorate, GMAT or GRE General Test, TOEFL (minimum score of 570 required). *Application deadline:* For fall admission, 2/1 (priority date). Applications are processed on a rolling basis. Application fee: $35. Electronic applications accepted. *Unit head:* Dr. Joseph G. Ecker, Dean, 518-276-6802, Fax: 518-276-8661, E-mail: management@rpi.edu. *Application contact:* Michele Martens, Manager of Enrollment Services, 518-276-6586, Fax: 518-276-2665, E-mail: management@rpi.edu.

See in-depth description on page 1287.

Rensselaer Polytechnic Institute, Graduate School, School of Engineering, Department of Decision Sciences and Engineering Systems, Program in Operations Research and Statistics, Troy, NY 12180-3590. Offers M Eng, MS, MBA/M Eng. Part-time programs available. *Faculty:* 12 full-time (0 women), 6 part-time (1 woman). *Students:* 21 full-time (8 women), 4 part-time (2 women); includes 5 minority (1 African American, 3 Asian Americans or Pacific Islanders, 1 Hispanic American), 10 international. 17 applicants, 76% accepted. In 1998, 13 degrees awarded. *Degree requirements:* For master's, thesis required (for some programs), foreign language not required. *Entrance requirements:* For master's, GRE General Test, TOEFL (minimum score of 550 required). *Application deadline:* For fall admission, 2/1 (priority date). Applications are processed on a rolling basis. Application fee: $35. *Financial aid:* In 1998–99, 1 fellowship with full tuition reimbursement (averaging $11,000 per year), 1 research assistantship with full tuition reimbursement (averaging $10,600 per year), 3 teaching assistantships with full tuition reimbursements (averaging $10,600 per year) were awarded.; career-related internships or fieldwork and institutionally-sponsored loans also available. Financial aid application deadline: 2/1. *Faculty research:* Manufacturing, MIS, statistical consulting, education services, production, logistics, inventory. Total annual research expenditures: $2 million. *Unit head:* John S. Kerge, Dean of Admissions, 718-260-3200, Fax: 718-260-3446, E-mail: admitme@poly.edu. *Application contact:* Lee Vilardi, Graduate Coordinator, 518-276-6681, Fax: 518-276-8227, E-mail: dsesgr@rpi.edu.

Rutgers, The State University of New Jersey, New Brunswick, Graduate School, Program in Operations Research, New Brunswick, NJ 08903. Offers PhD. Part-time programs available. *Faculty:* 33 full-time (2 women), 2 part-time (0 women). *Students:* 21 full-time (7 women), 7 part-time (1 woman); includes 1 minority (Hispanic American), 21 international. Average age 25. 35 applicants, 11% accepted. In 1998, 4 doctorates awarded (75% entered university research/teaching, 25% found other work related to degree). *Degree requirements:* For doctorate, computer language, dissertation, qualifying exam required, foreign language not required. *Entrance requirements:* For doctorate, GRE General Test, GRE Subject Test. *Average time to degree:* Doctorate–5 years full-time. *Application deadline:* For fall admission, 5/1. Application fee: $50. *Financial aid:* In 1998–99, 24 students received aid, including 5 fellowships, 4 research assistantships, 10 teaching assistantships; scholarships also available. Financial aid application deadline: 3/1; financial aid applicants required to submit FAFSA. *Faculty research:* Mathematical programming, combinatorial optimization, graph theory, stochastic modeling, queuing theory. Total annual research expenditures: $565,973. *Unit head:* Peter L. Hammer, Graduate Director, 732-445-3041, Fax: 732-445-5472, E-mail: hammer@rutcor.rutgers.edu.

St. Mary's University of San Antonio, Graduate School, Department of Engineering, San Antonio, TX 78228-8507. Offers electrical engineering (MS); electrical/computer engineering (MS); engineering administration (MS); engineering computer application (MS); industrial engineering (MS); operations research (MS). Part-time and evening/weekend programs available. *Students:* 3 full-time (2 women), 49 part-time (6 women); includes 10 minority (2 African Americans, 8 Hispanic Americans), 8 international. *Degree requirements:* For master's, computer language, thesis required, foreign language not required. *Entrance requirements:* For master's, GRE General Test. *Application deadline:* For fall admission, 8/1. Application fee: $15. *Unit head:* Dr. Abe Yazdani, Adviser, 210-436-3305.

Seton Hall University, College of Arts and Sciences, Department of Mathematics, South Orange, NJ 07079-2697. Offers operations research (MA, MS). Program being phased out; applicants no longer accepted. Part-time and evening/weekend programs available. *Degree requirements:* For master's, one foreign language required, (computer language can substitute), thesis not required.

Southern Methodist University, School of Engineering and Applied Science, Department of Computer Science and Engineering, Dallas, TX 75275. Offers computer engineering (MS Cp E, PhD); computer science (MS, PhD); engineering management (MSEM, DE); operations research (MS, PhD); software engineering (MS). Part-time programs available. Postbaccalaureate distance learning degree programs offered (no on-campus study). *Faculty:* 13 full-time (2 women), 12 part-time (1 woman). *Students:* 57 full-time (22 women), 294 part-time (60 women); includes 85 minority (24 African Americans, 44 Asian Americans or Pacific Islanders, 16 Hispanic Americans, 1 Native American), 69 international. *Degree requirements:* For master's, thesis optional, foreign language not required; for doctorate, dissertation, oral and written qualifying exams, oral final exam (PhD) required. *Entrance requirements:* For master's, GRE General Test (minimum score of 650 on quantitative section required), TOEFL (minimum score of 550 required), minimum GPA of 3.0 in last 2 years; bachelor's degree in engineering, mathematics, or sciences; for doctorate, preliminary counseling exam (PhD), minimum GPA of 3.0, bachelor's degree in related field, MA (DE). *Application deadline:* For fall admission, 8/1 (priority date); for spring admission, 12/15. Applications are processed on a rolling basis. Application fee: $25. Tuition: Full-time $9,216; part-time $512 per credit hour. Required fees: $88 per credit hour. Part-time tuition and fees vary according to course load and campus/location. *Unit head:* Dr. Richard V. Helgason, Interim Chair, 214-768-3278, E-mail: helgason@seas.smu.edu. *Application contact:* Dr. Zeynep Celik-Butler, Assistant Dean for Graduate Studies and Research, 214-768-3979, Fax: 214-768-3845, E-mail: zcb@seas.smu.edu.

Stanford University, School of Engineering, Department of Engineering-Economic Systems and Operations Research, Stanford, CA 94305-9991. Offers MS, PhD, Eng. *Faculty:* 20 full-time (0 women). *Students:* 260 full-time (60 women), 53 part-time (10 women); includes 35 minority (4 African Americans, 26 Asian Americans or Pacific Islanders, 5 Hispanic Americans), 194 international. Average age 26. 274 applicants, 76% accepted. In 1998, 135 master's, 17 doctorates awarded. Terminal master's awarded for partial completion of doctoral program. *Degree requirements:* For master's, thesis not required; for doctorate and Eng, dissertation required. *Entrance requirements:* For master's and Eng, GRE General Test, TOEFL; for doctorate, GRE General Test, GRE Subject Test, TOEFL. *Application deadline:* For fall admission, 2/1. Application fee: $65 ($80 for international students). Electronic applications accepted. Tuition: Full-time $24,588. Required fees: $152. Part-time tuition and fees vary according to course load. *Financial aid:* Research assistantships, Federal Work-Study and institutionally-sponsored loans available. Financial aid application deadline: 2/15. *Faculty research:* Mathematical systems analysis, decision analysis, systems economics, organizational economics, information policy. *Unit head:* James Sweeney, Chair, 650-723-2847, Fax: 650-723-1614, E-mail: sweeney@soe.stanford.edu. *Application contact:* Graduate Admissions Coordinator, 650-723-4168.

The University of Alabama in Huntsville, School of Graduate Studies, College of Engineering, Department of Industrial and Systems Engineering/Engineering Management, Program in Operations Research, Huntsville, AL 35899. Offers MSOR. Part-time and evening/weekend programs available. Average age 39. 4 applicants, 75% accepted. *Degree requirements:* For master's, oral and written exams required, thesis optional, foreign language not required. *Entrance requirements:* For master's, GRE General Test (minimum combined score of 1500 on three sections, 500 on quantitative section required), minimum GPA of 3.0, 6 hours of applied or mathematical statistics and calculus. *Application deadline:* For fall admission, 7/24 (priority date); for spring admission, 11/15 (priority date). Applications are processed on a rolling basis. Application fee: $20. Tuition and fees vary according to course load. *Financial aid:* Fellowships with full and partial tuition reimbursements, research assistantships with full and partial tuition reimbursements, teaching assistantships with full and partial tuition reimbursements, career-related internships or fieldwork, Federal Work-Study, grants, institutionally-sponsored loans, scholarships, and tuition waivers (full and partial) available. Aid available to part-time students. Financial aid application deadline: 4/1; financial aid applicants required to submit FAFSA. *Faculty research:* Simulation, manufacturing systems, system ergonomics, logistics. *Unit head:* Dr. Jerry Westbrook, Chair, Department of Industrial and Systems Engineering/Engineering Management, 256-890-6256, Fax: 256-890-6608, E-mail: westbroo@eb.uah.edu.

University of Arkansas, Graduate School, College of Engineering, Department of Industrial Engineering, Program in Operations Research, Fayetteville, AR 72701-1201. Offers MSE, MSOR. *Students:* 5 full-time (2 women). 2 applicants, 100% accepted. In 1998, 1 degree awarded. *Degree requirements:* For master's, thesis optional, foreign language not required. Application fee: $40 ($50 for international students). Tuition, state resident: full-time $3,186. Tuition, nonresident: full-time $7,560. Required fees: $378. *Financial aid:* Career-related internships or fieldwork and Federal Work-Study available. Aid available to part-time students. Financial aid application deadline: 4/1; financial aid applicants required to submit FAFSA. *Unit head:* Dr. Eric M. Malstrom, Chair, Department of Industrial Engineering, 501-575-3157.

University of California, Berkeley, Graduate Division, College of Engineering, Department of Industrial Engineering and Operations Research, Berkeley, CA 94720-1500. Offers M Eng, MS, D Eng, PhD. *Students:* 69 full-time (14 women); includes 14 minority (2 African Americans, 10 Asian Americans or Pacific Islanders, 2 Hispanic Americans), 35 international. 230 applicants, 34% accepted. In 1998, 43 master's, 6 doctorates awarded. *Degree requirements:* For master's, comprehensive exam or thesis (MS) required; for doctorate, dissertation, qualifying exam required. *Entrance requirements:* For master's and doctorate, GRE General Test, minimum GPA of 3.0. *Average time to degree:* Master's–1.5 years full-time; doctorate–5.5 years full-time. *Application deadline:* For fall admission, 2/10; for spring admission, 9/1. Application fee: $40. *Financial aid:* Fellowships, research assistantships, teaching assistantships, career-related internships or fieldwork, Federal Work-Study, and tuition waivers (full and partial) available. Financial aid application deadline: 1/5. *Faculty research:* Mathematical programming, robotics and manufacturing, linear and nonlinear optimization, production planning and scheduling, queuing theory. *Unit head:* Dr. Candace A. Yano, Chair, 510-642-4992, Fax: 510-642-1403, E-mail: yano@ieor.berkeley.edu. *Application contact:* Marion Brackett, Student Affairs Officer, 510-642-5485, Fax: 510-642-1403, E-mail: std-svcs@ieor.berkeley.edu.

University of California, Los Angeles, Graduate Division, School of Engineering and Applied Science, Department of Electrical Engineering, Program in Operations Research, Los Angeles, CA 90095. Offers MS, PhD. *Degree requirements:* For master's, comprehensive exam or thesis required; for doctorate, dissertation, qualifying exams required, foreign language not required. *Entrance requirements:* For master's, GRE General Test, minimum GPA of 3.0; for doctorate, GRE General Test, minimum GPA of 3.25. *Application deadline:* For fall admission, 1/15. Application fee: $40. Electronic applications accepted. *Financial aid:* Fellowships, research assistantships, teaching assistantships, career-related internships or fieldwork, Federal Work-Study, institutionally-sponsored loans, and tuition waivers (full and partial) available. Financial aid application deadline: 1/15; financial aid applicants required to submit FAFSA. *Application contact:* Maida Bassili, Student Affairs Officer, 310-825-9383, E-mail: mbassili@ea.ucla.edu.

University of Central Florida, College of Engineering, Department of Industrial Engineering and Management Systems, Orlando, FL 32816. Offers computer-integrated manufacturing (MS); engineering management (MS); industrial engineering (MSIE); industrial engineering and management systems (PhD); manufacturing engineering (MS Mfg E); operations research (MS); product assurance engineering (MS); simulation systems (MS). Part-time and evening/weekend programs available. *Faculty:* 20 full-time, 11 part-time. *Students:* 143 full-time (27 women), 112 part-time (20 women); includes 43 minority (6 African Americans, 12 Asian Americans or Pacific Islanders, 25 Hispanic Americans), 51 international. *Degree requirements:* For master's, computer language, thesis or alternative required, foreign language not required; for doctorate, computer language, dissertation, departmental qualifying exam, candidacy exam required, foreign language not required. *Entrance requirements:* For master's, GRE General Test (minimum combined score of 1000 required), TOEFL (minimum score of 550 required; 213 computer-based), minimum GPA of 3.0 in last 60 hours; for doctorate, TOEFL (minimum score of 550 required; 213 computer-based), minimum GPA of 3.5 in last 60 hours. *Application deadline:* For fall admission, 7/15; for spring admission, 12/15. Application fee: $20. Tuition, state resident: full-time $2,054; part-time $137 per credit. Tuition, nonresident: full-time $7,207; part-time $480 per credit. Required fees: $47 per term. *Unit head:* Dr. Charles Reily, Chair, 407-823-2204. *Application contact:* Dr. Linda Malone, Coordinator, 407-823-2204.

University of Delaware, College of Agriculture and Natural Resources, Operations Research Program, Newark, DE 19716. Offers MS, PhD. Part-time programs available. *Degree requirements:* For master's, thesis, oral exam required, foreign language not required; for doctorate, dissertation, qualifying exam required, foreign language not required. *Entrance requirements:* For master's and doctorate, GRE General Test (minimum combined score of 1150 required), TOEFL (minimum score of 600 required). *Faculty research:* Transportation engineering, agricultural production and resource economics, optimization, simulation and modeling, production scheduling.

Announcement: The interdisciplinary Operations Research Program at the University of Delaware includes 8 departments in the Colleges of Agricultural Science, Engineering, Economics and Business Administration, Marine Studies, Arts and Science, and Human Resources, Education and Public Policy. Programs tailored to student subject matter interests emphasizing problem solving require a thesis, internship, or dissertation and courses divided between operations research and supporting courses.

University of Florida, Graduate School, College of Engineering, Department of Industrial and Systems Engineering, Gainesville, FL 32611. Offers engineering management (ME, MS); facilities layout decision support systems energy (PhD); health systems (ME, MS); industrial engineering (PhD, Engr); manufacturing systems engineering (ME, MS, PhD, Certificate); operations research (ME, MS, PhD, Engr); production planning and control engineering management (PhD); quality and reliability assurance (ME, MS); systems engineering (PhD, Engr). *Faculty:* 22. *Students:* 56 full-time (14 women), 69 part-time (18 women); includes 17 minority (5 African Americans, 4 Asian Americans or Pacific Islanders, 8 Hispanic Americans), 48 international. *Degree requirements:* For master's, computer language, core exam required, thesis optional; for doctorate, dissertation, comprehensive exam required; for other advanced degree, computer language, thesis required. *Entrance requirements:* For master's and doctorate, GRE General Test, TOEFL, minimum GPA of 3.0; for other advanced degree, GRE General Test. *Application deadline:* For fall admission, 6/1 (priority date). Applications are processed on a rolling basis. Application fee: $20. Electronic applications accepted. *Unit head:* Dr. Donald Hearn, Chair, 352-392-1464, Fax: 352-392-3537, E-mail: hearn@ise.ufl.edu. *Application contact:* Dr. D. J. Elzinga, Graduate Coordinator, 352-392-1464, Fax: 352-392-3537, E-mail: elzinga@ise.ufl.edu.

University of Houston, College of Business Administration, Program in Statistics and Operations Research, Houston, TX 77004. Offers MBA, PhD. Part-time and evening/weekend programs available. *Faculty:* 6 full-time (0 women), 1 part-time (0 women). In 1998, 1 degree awarded. *Degree requirements:* For master's, computer language required, foreign language and thesis not required; for doctorate, computer language, dissertation, comprehensive exam required, foreign language not required. *Entrance requirements:* For master's, GMAT (average 590), TOEFL (minimum score of 620 required); for doctorate, GMAT or GRE. *Average time to degree:* Master's–2 years full-time, 3.5 years part-time; doctorate–4.5 years full-time. *Application deadline:* For fall admission, 5/1; for spring admission, 10/1. Applications are processed on a rolling basis. Application fee: $50 ($125 for international students). *Financial aid:* Research assistantships, teaching assistantships, career-related internships or fieldwork and Federal Work-Study available. Aid available to part-time students. Financial aid application deadline: 3/1; financial aid applicants required to submit FAFSA. *Unit head:* Dr. Dennis Adams, Chair, 713-743-4747. *Application contact:* 713-743-4900, Fax: 713-743-4942, E-mail: oss@cba.uh.edu.

University of Illinois at Chicago, Graduate College, College of Engineering, Department of Mechanical Engineering, Program in Industrial Engineering and Operations Research, Chicago, IL 60607-7128. Offers PhD. *Students:* 6 full-time (0 women), 3 part-time, 8 international. Average age 27. 7 applicants, 0% accepted. In 1998, 2 degrees awarded. *Degree requirements:* For doctorate, dissertation required, foreign language not required. *Entrance requirements:* For doctorate, GRE General Test, TOEFL (minimum score of 550 required), minimum GPA of 3.75 on a 5.0 scale. *Application deadline:* For fall admission, 6/1; for spring admission, 10/15. Application fee: $40 ($50 for international students). *Financial aid:* Fellowships, research assistantships, teaching assistantships available. *Unit head:* Dr. Selcuk Guceri, Head, Department of Mechanical Engineering, 312-996-5096.

The University of Iowa, Graduate College, College of Engineering, Department of Industrial Engineering, Iowa City, IA 52242-1316. Offers engineering design and manufacturing (MS, PhD); ergonomics (MS, PhD); information and engineering management (MS, PhD); operations research (MS, PhD); quality engineering (MS, PhD). *Faculty:* 7 full-time, 1 part-time. *Students:* 18 full-time (4 women), 16 part-time (1 woman); includes 4 minority (1 African American, 2 Asian Americans or Pacific Islanders, 1 Hispanic American), 19 international. *Degree requirements:* For master's, thesis optional; for doctorate, dissertation, comprehensive exam required. *Entrance requirements:* For master's and doctorate, GRE General Test, GRE Subject Test, TOEFL. *Application deadline:* Applications are processed on a rolling basis. Application fee: $30 ($50 for international students). *Unit head:* Peter J. O'Grady, Chair, 319-335-5939, Fax: 319-335-5424.

University of Massachusetts Amherst, Graduate School, College of Engineering, Department of Mechanical and Industrial Engineering, Program in Industrial Engineering and Opera-

Operations Research–Technology and Public Policy

University of Massachusetts Amherst (continued)
tions Research, Amherst, MA 01003-2210. Offers MS, PhD. *Students:* 15 full-time (2 women), 17 part-time (6 women); includes 4 minority (1 African American, 3 Hispanic Americans), 21 international. Average age 29. 103 applicants, 44% accepted. In 1998, 10 master's, 8 doctorates awarded. *Degree requirements:* For master's, project required, foreign language and thesis not required; for doctorate, dissertation required, foreign language not required. *Entrance requirements:* For master's and doctorate, GRE General Test. *Application deadline:* For fall admission, 2/1 (priority date). Applications are processed on a rolling basis. Application fee: $35. Tuition, state resident: full-time $2,640; part-time $165 per credit. Tuition, nonresident: full-time $9,756; part-time $407 per credit. Required fees: $1,221 per term. One-time fee: $110. Full-time tuition and fees vary according to course load, campus/location and reciprocity agreements. *Financial aid:* Fellowships with full tuition reimbursements, research assistantships with full tuition reimbursements, teaching assistantships with full tuition reimbursements, career-related internships or fieldwork, Federal Work-Study, grants, scholarships, traineeships, and unspecified assistantships available. Aid available to part-time students. Financial aid application deadline: 2/1. *Unit head:* Dr. Donald Fisher, Director, 413-545-0955, Fax: 413-545-1027, E-mail: dfisher@ecs.umass.edu.

University of Michigan, Horace H. Rackham School of Graduate Studies, College of Engineering, Department of Industrial and Operations Engineering, Ann Arbor, MI 48109. Offers MS, MSE, PhD, IOE, MBA/MS, MBA/MSE, MHSA/MS. Part-time programs available. *Faculty:* 26 full-time (2 women). *Students:* 206 full-time (50 women). 255 applicants, 57% accepted. Terminal master's awarded for partial completion of doctoral program. *Degree requirements:* For master's, computer language required, foreign language and thesis not required; for doctorate, computer language, dissertation, oral defense of dissertation, preliminary exams required, foreign language not required. *Entrance requirements:* For master's, GRE General Test (minimum combined score of 1850 on three sections required), TOEFL, minimum GPA of 3.2; for doctorate, GRE General Test (minimum combined score of 1950 on three sections required), TOEFL, minimum GPA of 3.5. *Application deadline:* For fall admission, 5/1; for winter admission, 11/1. Applications are processed on a rolling basis. Application fee: $55. Electronic applications accepted. *Financial aid:* In 1998–99, 15 fellowships, 40 research assistantships, 25 teaching assistantships were awarded.; Federal Work-Study and institutionally-sponsored loans also available. Financial aid application deadline: 2/1. *Faculty research:* Optimization and stochastic processes, human performance and ergonomics, information systems, engineering management, quality engineering. *Unit head:* Dr. John R. Birge, Chair, 734-764-6480, Fax: 734-764-3451, E-mail: jrbirge@umich.edu. *Application contact:* Frances Bourdas, Graduate Program Assistant, 734-764-6480, Fax: 734-764-3451, E-mail: bourdas@engin.umich.edu.

The University of Montana–Missoula, Graduate School, College of Arts and Sciences, Department of Mathematical Sciences, Missoula, MT 59812-0002. Offers algebra (MA, PhD); analysis (MA, PhD); applied mathematics (MA, PhD); mathematics (MAT); mathematics education (PhD); operations research (MA, PhD); statistics (MA, PhD). Part-time programs available. *Faculty:* 20 full-time (3 women). *Students:* 25 full-time (11 women), 2 part-time (1 woman); includes 6 minority (5 Asian Americans or Pacific Islanders, 1 Native American) Terminal master's awarded for partial completion of doctoral program. *Degree requirements:* For master's, foreign language and thesis not required; for doctorate, dissertation required. *Entrance requirements:* For master's and doctorate, GRE General Test. *Application deadline:* For fall admission, 3/1 (priority date). Application fee: $45. *Unit head:* Dr. Gloria Hewitt, Chair, 406-243-5311.

University of New Haven, Graduate School, School of Business, Program in Business Administration, West Haven, CT 06516-1916. Offers accounting (MBA); business policy and strategy (MBA); computer and information science (MBA); finance (MBA); health care management (MBA); health care marketing (MBA); hotel and restaurant management (MBA); human resources management (MBA); international business logistics (MBA); management and organization (MBA); management science (MBA); marketing (MBA); operations research (MBA); public relations (MBA); telecommunications (MBA); travel and tourism administration (MBA). Part-time and evening/weekend programs available. *Students:* 65 full-time (18 women), 275 part-time (107 women); includes 26 minority (11 African Americans, 8 Asian Americans or Pacific Islanders, 7 Hispanic Americans), 67 international. *Degree requirements:* For master's, thesis or alternative required, foreign language not required. *Application deadline:* Applications are processed on a rolling basis. Application fee: $50. *Unit head:* Dr. Omid Nodoushani, Coordinator, 203-932-7123.

University of New Haven, Graduate School, School of Engineering and Applied Science, Program in Operations Research, West Haven, CT 06516-1916. Offers MS. Part-time and evening/weekend programs available. *Students:* 1 full-time (0 women), 2 part-time, 1 international. 1 applicants, 100% accepted. In 1998, 2 degrees awarded. *Degree requirements:* For master's, computer language, thesis or alternative required, foreign language not required. *Application deadline:* Applications are processed on a rolling basis. Application fee: $50. *Financial aid:* Federal Work-Study available. Aid available to part-time students. Financial aid

application deadline: 5/1; financial aid applicants required to submit FAFSA. *Unit head:* Dr. Ronald Wentworth, Coordinator, 203-932-7434.

The University of North Carolina at Chapel Hill, Graduate School, College of Arts and Sciences, Department of Operations Research, Chapel Hill, NC 27599. Offers MS, PhD. *Degree requirements:* For master's, comprehensive exam required; for doctorate, dissertation, comprehensive exams required. *Entrance requirements:* For master's and doctorate, GRE General Test (minimum combined score of 1000 required), minimum GPA of 3.0.

University of Southern California, Graduate School, School of Engineering, Department of Industrial and Systems Engineering, Program in Operations Research, Los Angeles, CA 90089. Offers MS. *Students:* 4 full-time (2 women), 3 part-time (2 women); includes 3 minority (all Asian Americans or Pacific Islanders), 4 international. Average age 26. 11 applicants, 91% accepted. In 1998, 6 degrees awarded. *Degree requirements:* For master's, thesis optional. *Entrance requirements:* For master's, GRE General Test. *Application deadline:* For fall admission, 6/1 (priority date); for spring admission, 12/1. Application fee: $55. Tuition: Part-time $768 per unit. Required fees: $350 per semester. *Financial aid:* Fellowships, research assistantships, teaching assistantships, Federal Work-Study, institutionally-sponsored loans, and scholarships available. Aid available to part-time students. Financial aid application deadline: 2/15; financial aid applicants required to submit FAFSA. *Unit head:* Dr. F. Stan Settles, Chairman, Department of Industrial and Systems Engineering, 213-740-4893.

The University of Texas at Austin, Graduate School, College of Engineering, Department of Mechanical Engineering, Program in Operations Research and Industrial Engineering, Austin, TX 78712-1111. Offers MSE, PhD. *Students:* 47 (13 women); includes 2 minority (1 African American, 1 Asian American or Pacific Islander) 18 international. 47 applicants, 66% accepted. In 1998, 16 degrees awarded. *Entrance requirements:* For master's, GRE General Test (minimum combined score of 1000 required), TOEFL (minimum score of 580 required); for doctorate, GRE General Test. Application fee: $50 ($75 for international students). *Financial aid:* Fellowships, research assistantships, teaching assistantships available. Financial aid application deadline: 2/1. *Unit head:* J. Wesley Barnes, Graduate Adviser, 512-471-3083, E-mail: wbarnes@mail.utexas.edu.

Virginia Commonwealth University, School of Graduate Studies, College of Humanities and Sciences, Department of Mathematical Sciences, Program in Operations Research, Richmond, VA 23284-9005. Offers MS. *Degree requirements:* Foreign language not required. *Entrance requirements:* For master's, GRE General Test, GRE Subject Test, TOEFL. *Application deadline:* For fall admission, 7/1; for spring admission, 11/15. Applications are processed on a rolling basis. Application fee: $30. Tuition, state resident: full-time $4,031; part-time $224 per credit hour. Tuition, nonresident: full-time $11,946; part-time $664 per credit hour. Required fees: $1,081; $40 per credit hour. Tuition and fees vary according to campus/location and program. *Unit head:* Sarah M. Ross, Administrative Assistant, 615-343-2415, Fax: 615-343-8645, E-mail: sross@vuse.vanderbilt.edu. *Application contact:* Dr. James A. Wood, Director of Graduate Studies, 804-828-1301, Fax: 804-828-8785, E-mail: jawood@vcu.edu.

Virginia Polytechnic Institute and State University, Graduate School, College of Engineering, Department of Industrial and Systems Engineering, Program in Operations Research, Blacksburg, VA 24061. Offers M Eng, MS, PhD. *Degree requirements:* For master's and doctorate, computer language, thesis/dissertation required, foreign language not required. *Entrance requirements:* For master's, TOEFL (minimum score of 600 required); for doctorate, TOEFL (minimum score of 600 required), minimum GPA of 3.0. *Application deadline:* For fall admission, 12/1 (priority date). Applications are processed on a rolling basis. Application fee: $25. *Financial aid:* Application deadline: 4/1. *Unit head:* Dr. John Casali, Head, Department of Industrial and Systems Engineering, 540-231-6656, E-mail: jcasali@vt.edu.

Wayne State University, Graduate School, College of Engineering, Department of Industrial and Manufacturing Engineering, Program in Operations Research, Detroit, MI 48202. Offers MS. *Degree requirements:* For master's, thesis optional, foreign language not required. *Entrance requirements:* For master's, minimum undergraduate GPA of 2.8. *Faculty research:* Reliability and quality, technology management, manufacturing systems, concurrent engineering.

Western Michigan University, Graduate College, College of Engineering and Applied Sciences, Department of Industrial and Manufacturing Engineering, Program in Operations Research, Kalamazoo, MI 49008. Offers MS. *Students:* 3 full-time (1 woman), 4 part-time (2 women), 5 international. 10 applicants, 70% accepted. In 1998, 5 degrees awarded. *Degree requirements:* For master's, oral exams required, foreign language and thesis not required. *Entrance requirements:* For master's, minimum GPA of 3.0. *Application deadline:* For fall admission, 2/15 (priority date). Applications are processed on a rolling basis. Application fee: $25. *Financial aid:* Fellowships, research assistantships, teaching assistantships, Federal Work-Study available. Financial aid application deadline: 2/15; financial aid applicants required to submit FAFSA. *Unit head:* Barbara Baily, Director of Graduate Studies, 309-298-1806, Fax: 309-298-2245, E-mail: barb_baily@ccmail.wiu.edu. *Application contact:* Paula J. Boodt, Coordinator, Graduate Admissions and Recruitment, 616-387-2000, Fax: 616-387-2355, E-mail: paula.boodt@wmich.edu.

Technology and Public Policy

California State University, Los Angeles, Graduate Studies, School of Engineering and Technology, Department of Technology, Major in Industrial and Technical Studies, Los Angeles, CA 90032-8530. Offers MA. *Students:* 6 full-time (1 woman), 27 part-time (6 women); includes 15 minority (4 African Americans, 3 Asian Americans or Pacific Islanders, 8 Hispanic Americans), 3 international. In 1998, 4 degrees awarded. *Degree requirements:* For master's, computer language, project or thesis required. *Entrance requirements:* For master's, TOEFL (minimum score of 550 required), minimum GPA of 2.5. *Application deadline:* For fall admission, 6/30; for spring admission, 2/1. Applications are processed on a rolling basis. Application fee: $55. *Financial aid:* In 1998–99, 8 students received aid. Application deadline: 3/1. *Faculty research:* Instructional improvement, new applications. *Unit head:* Dr. Don Maurizio, Chair, Department of Technology, 323-343-4550.

Carnegie Mellon University, Carnegie Institute of Technology, Department of Civil and Environmental Engineering, Program in Civil Engineering/Engineering and Public Policy, Pittsburgh, PA 15213-3891. Offers MS, PhD. *Degree requirements:* For master's, thesis required, foreign language not required; for doctorate, dissertation, qualifying exam required, foreign language not required. *Entrance requirements:* For master's and doctorate, GRE General Test, TOEFL. *Application deadline:* For fall admission, 2/1 (priority date) for spring admission, 10/15. Application fee: $45. *Financial aid:* Application deadline: 2/1. *Unit head:* Valerie Bridge, Student Specialist, 412-268-3150, Fax: 412-268-1061, E-mail: vbog@andrew.cmu.edu. *Application contact:* Maxine A. Leffard, Graduate Program Administrator, 412-268-8712, Fax: 412-268-7813, E-mail: ce-addmissions+@andrew.cmu.edu.

Carnegie Mellon University, Carnegie Institute of Technology, Department of Engineering and Public Policy, Pittsburgh, PA 15213-3891. Offers MS, PhD. *Faculty:* 19 full-time (4 women). *Students:* 30 full-time (9 women), 4 part-time (1 woman); includes 1 minority (Asian American or Pacific Islander), 18 international. Average age 30. In 1998, 9 master's, 9 doctorates awarded. *Degree requirements:* For doctorate, computer language, dissertation, qualifying exam required, foreign language not required. *Entrance requirements:* For master's, GRE General Test, TOEFL; for doctorate, GRE General Test, TOEFL, BS in physical sciences or

engineering. *Application deadline:* For fall admission, 2/1. Application fee: $45. *Financial aid:* Fellowships, research assistantships, teaching assistantships, tuition waivers (full and partial) available. Financial aid application deadline: 4/1. *Faculty research:* Restructuring the electric power industry, economics of network-based information services, mathematical modeling of environmental systems, risk assessment and analysis. Total annual research expenditures: $2.7 million. *Unit head:* M. Granger Morgan, Head, 412-268-2672. *Application contact:* Mitchell Small, Associate Head for Graduate Education, 412-268-8782.

See in-depth description on page 1265.

Colorado State University, Graduate School, College of Applied Human Sciences, Department of Manufacturing Technology and Construction Management, Fort Collins, CO 80523-0015. Offers automotive pollution control (MS); construction management (MS); historic preservation (PhD); industrial technology management (MS); technology education and training (MS); technology of industry (PhD). *Faculty:* 17 full-time (1 woman), 5 part-time (1 woman). *Students:* 14 full-time (1 woman), 8 part-time (3 women); includes 2 minority (both Hispanic Americans), 2 international. *Degree requirements:* For master's, computer language, thesis required (for some programs), foreign language not required; for doctorate, dissertation required. *Entrance requirements:* For master's, GRE General Test (minimum combined score of 1250 required), TOEFL (minimum score of 550 required; 213 for computer-based); for doctorate, GRE General Test, TOEFL. *Application deadline:* For fall admission, 4/1 (priority date). Applications are processed on a rolling basis. Application fee: $30. Electronic applications accepted. *Unit head:* Dr. Larry Grosse, Head, 970-491-7958, Fax: 970-491-2473, E-mail: drfire@aol.com. *Application contact:* Linda Burrous, Secretary, 970-491-7355, Fax: 970-491-2473, E-mail: burrous@cahs.colostate.edu.

Eastern Michigan University, Graduate School, College of Technology, Department of Interdisciplinary Technology, Program in Liberal Studies in Technology, Ypsilanti, MI 48197. Offers MLS. In 1998, 55 degrees awarded. *Degree requirements:* For master's, thesis optional, foreign language not required. *Entrance requirements:* For master's, GRE General Test,

TOEFL (minimum score of 500 required), minimum GPA of 2.6. *Application deadline:* For fall admission, 5/15; for spring admission, 3/15. Applications are processed on a rolling basis. Application fee: $30. *Financial aid:* Fellowships, teaching assistantships available. Aid available to part-time students. Financial aid application deadline: 3/15; financial aid applicants required to submit FAFSA. *Unit head:* Dr. Wayne Hanewicz, Coordinator, 734-487-1161.

The George Washington University, Elliott School of International Affairs, Program in Science, Technology, and Public Policy, Washington, DC 20052. Offers MA, JD/MA, LL M/MA. Part-time and evening/weekend programs available. *Students:* 18 full-time (7 women), 9 part-time (4 women); includes 3 minority (2 Asian Americans or Pacific Islanders, 1 Hispanic American), 4 international. Average age 28. 33 applicants, 70% accepted. In 1998, 10 degrees awarded. *Degree requirements:* For master's, one foreign language (computer language can substitute), capstone project required. *Entrance requirements:* For master's, GRE General Test, TOEFL, (minimum score of 600 required; 250 for computer-based), minimum B average. *Application deadline:* For fall admission, 2/1; for spring admission, 11/1. Application fee: $55. Electronic applications accepted. Tuition: Full-time $17,328; part-time $722 per credit hour. Required fees: $828; $35 per credit hour. Tuition and fees vary according to campus/location and program. *Financial aid:* In 1998–99, 12 fellowships were awarded.; research assistantships, career-related internships or fieldwork, Federal Work-Study, institutionally-sponsored loans, and tuition waivers (full and partial) also available. Financial aid application deadline: 1/15; financial aid applicants required to submit FAFSA. *Faculty research:* Science policy, space policy, risk assessment, technology transfer, energy policy. *Unit head:* Dr. Nicholas Vonortas, Director, 202-994-7292. *Application contact:* Jeff V. Miles, Director of Graduate Admissions, 202-994-7050, Fax: 202-994-9537, E-mail: esiagrad@gwu.edu.

Announcement: Graduate study at both the MA and PhD levels is available through The George Washington University's Science, Technology, and Public Policy Program. In addition to focusing on general issues of science, technology, and innovation policy, students can concentrate in a sector such as space, the environment, or information telecommunications. The program also includes a focus on public–private sector relationships. Students come to the program from diverse backgrounds, ranging from humanities and social sciences to natural sciences and engineering. Part-time study is offered. Fellowship support is available, and students often work as research assistants or as interns in policy-oriented organizations in the Washington area.

Massachusetts Institute of Technology, School of Engineering, Technology and Policy Program, Cambridge, MA 02139-4307. Offers SM, PhD. *Students:* 103 full-time (30 women), 2 part-time; includes 9 minority (7 Asian Americans or Pacific Islanders, 2 Hispanic Americans), 49 international. Average age 27. 209 applicants, 51% accepted. In 1998, 48 master's awarded. Terminal master's awarded for partial completion of doctoral program. *Degree requirements:* For master's, computer language, thesis required, foreign language not required; for doctorate, computer language, dissertation, comprehensive exams required, foreign language not required. *Entrance requirements:* For master's, GRE General Test (minimum score of 400 on verbal section, 600 on quantitative required), TOEFL (minimum score of 580 required). *Average time to degree:* Master's–2 years full-time; doctorate–5 years full-time. *Application deadline:* For fall admission, 1/15; for spring admission, 11/15. Application fee: $55. *Financial aid:* In 1998–99, 72 students received aid, including 10 fellowships, 51 research assistantships, 2 teaching assistantships; career-related internships or fieldwork, Federal Work-Study, institutionally-sponsored loans, and tuition waivers (partial) also available. Financial aid application deadline: 3/15; financial aid applicants required to submit FAFSA. Total annual research expenditures: $40,149. *Unit head:* Dr. Richard de Neufville, Chairman, 617-253-7694, Fax: 617-253-7140, E-mail: ardent@mit.edu. *Application contact:* Gail Hickey, Program Administrator, 617-253-7693, Fax: 617-253-7140, E-mail: ghickey@mit.edu.

See in-depth description on page 1281.

Massachusetts Institute of Technology, School of Humanities and Social Science, Program in Science, Technology, and Society, Cambridge, MA 02139-4307. Offers history and social study of science and technology (PhD). *Faculty:* 27 full-time (10 women). *Students:* 28 full-time (11 women); includes 3 minority (1 African American, 1 Hispanic American, 1 Native American), 6 international. Average age 29. 78 applicants, 9% accepted. In 1998, 2 degrees awarded (50% entered university research/teaching, 50% found other work related to degree). *Degree requirements:* For doctorate, 2 foreign languages (computer language can substitute for one), dissertation required. *Entrance requirements:* For doctorate, GRE General Test, TOEFL. *Average time to degree:* Doctorate–6 years full-time. *Application deadline:* For fall admission, 1/15. Application fee: $50. *Financial aid:* In 1998–99, 23 students received aid, including 7 fellowships (averaging $13,000 per year), 2 research assistantships, 6 teaching assistantships (averaging $13,000 per year); Federal Work-Study and institutionally-sponsored loans also available. Financial aid application deadline: 1/15. *Faculty research:* Cultural studies of science and technology. *Unit head:* Michael M. J. Fischer, Director, 617-253-2564, Fax: 617-258-8118, E-mail: mfischer@mit.edu. *Application contact:* Ben Brophy, Coordinator, 617-253-3452, Fax: 617-258-8118, E-mail: stsprogram@mit.edu.

Northwestern University, The Graduate School, Program in Telecommunications Science, Management, and Policy, Evanston, IL 60208. Offers MA, MEM, MS, Certificate. MS awarded through the Department of Electrical Engineering and Computer Science; MA awarded through the Departments of Comunication Studies and Radio/Television/Film; MEM awarded through the Department of Industrial Engineering and Management Sciences. Part-time programs available. *Faculty:* 22 full-time (2 women), 1 (woman) part-time. *Students:* 19 full-time (8 women); includes 7 minority (all Asian Americans or Pacific Islanders), 1 international. Average age 26. In 1998, 7 degrees awarded (14% entered university research/teaching, 86% found other work related to degree). *Entrance requirements:* For master's, GRE General Test, TOEFL. *Average time to degree:* Master's–2 years full-time. *Application deadline:* For fall admission, 1/15 (priority date). Applications are processed on a rolling basis. Application fee: $50 ($55 for international students). *Financial aid:* Fellowships, research assistantships, teaching assistantships, career-related internships or fieldwork, Federal Work-Study, and institutionally-sponsored loans available. Financial aid application deadline: 1/15; financial aid applicants required to submit FAFSA. *Faculty research:* Electronic media management. *Unit head:* Dr. Steven S. Wildman, Director, 847-491-3539, Fax: 847-467-1171, E-mail: s-wildman@nwu.edu. *Application contact:* Theomary Karamanis, Program Assistant, 847-491-3539, Fax: 847-467-1171, E-mail: telecomprog@nwu.edu.

Rensselaer Polytechnic Institute, Graduate School, School of Humanities and Social Sciences, Department of Science and Technology Studies, Troy, NY 12180-3590. Offers MS, PhD. Part-time programs available. *Faculty:* 12 full-time (5 women), 4 part-time (0 women). *Students:* 15 full-time (5 women), 4 part-time (2 women); includes 2 minority (1 African American, 1 Asian American or Pacific Islander), 4 international. 26 applicants, 77% accepted. In 1998, 12 master's, 1 doctorate awarded. *Degree requirements:* For master's, thesis required (for some programs), foreign language not required; for doctorate, dissertation required, foreign language not required. *Entrance requirements:* For master's, GRE General Test, TOEFL (minimum score of 550 required); for doctorate, TOEFL (minimum score of 550 required). *Application deadline:* For fall admission, 2/1 (priority date). Applications are processed on a rolling basis. Application fee: $35. *Financial aid:* In 1998–99, 1 fellowship with full tuition reimbursement (averaging $11,000 per year), 3 research assistantships with full and partial tuition reimbursements (averaging $10,600 per year), 9 teaching assistantships with full and partial tuition reimbursements (averaging $10,600 per year) were awarded.; career-related internships or fieldwork, institutionally-sponsored loans, and tuition waivers (partial) also available. Financial aid application deadline: 2/1. *Faculty research:* Science/government relations; sociology of science, mathematics, and mind; nature of inquiry in the sciences; social and political issues generated by technology change; information technology; design. Total annual research expenditures: $289,000. *Unit head:* Dr. John Schumacher, Chair, 518-276-6574. *Application contact:* Dr. Linda Layne, Director of Graduate Studies, 518-276-6115, Fax: 518-276-2659, E-mail: laynel@rpi.edu.

See in-depth description on page 1285.

St. Cloud State University, School of Graduate Studies, College of Science and Engineering, Department of Environmental and Technological Studies, St. Cloud, MN 56301-4498. Offers MS. *Faculty:* 7 full-time (1 woman), 2 part-time (0 women). In 1998, 2 degrees awarded. *Degree requirements:* For master's, thesis or alternative required, foreign language not required. *Entrance requirements:* For master's, GRE General Test, minimum GPA of 2.75. Application fee: $20. *Financial aid:* Federal Work-Study and unspecified assistantships available. Financial aid application deadline: 3/1. *Unit head:* Dr. Anthony Schwaller, Chairperson, 320-255-3235, Fax: 320-654-5122, E-mail: ets@stclousstate.edu. *Application contact:* Ann Anderson, Graduate Studies Office, 320-255-2113, Fax: 320-654-5371, E-mail: aeanderson@stcloudstate.edu.

Stanford University, School of Engineering, Department of Engineering-Economic Systems and Operations Research, Stanford, CA 94305-9991. Offers MS, PhD, Eng. *Faculty:* 20 full-time (0 women). *Students:* 260 full-time (60 women), 53 part-time (10 women); includes 35 minority (4 African Americans, 26 Asian Americans or Pacific Islanders, 5 Hispanic Americans), 194 international. Average age 26. 274 applicants, 76% accepted. In 1998, 135 master's, 17 doctorates awarded. Terminal master's awarded for partial completion of doctoral program. *Degree requirements:* For master's, thesis not required; for doctorate and Eng, dissertation required. *Entrance requirements:* For master's and Eng, GRE General Test, TOEFL; for doctorate, GRE General Test, GRE Subject Test, TOEFL. *Application deadline:* For fall admission, 2/1. Application fee: $65 ($80 for international students). Electronic applications accepted. Tuition: Full-time $24,588. Required fees: $152. Part-time tuition and fees vary according to course load. *Financial aid:* Research assistantships, Federal Work-Study and institutionally-sponsored loans available. Financial aid application deadline: 2/15. *Faculty research:* Mathematical systems analysis, decision analysis, systems economics, organizational economics, information policy. *Unit head:* James Sweeney, Chair, 650-723-2847, Fax: 650-723-1614, E-mail: sweeney@soe.stanford.edu. *Application contact:* Graduate Admissions Coordinator, 650-723-4168.

University of Minnesota, Twin Cities Campus, Graduate School, Hubert H. Humphrey Institute of Public Affairs, Program in Science, Technology, and Environmental Policy, Minneapolis, MN 55455-0213. Offers MS. *Degree requirements:* For master's, thesis, internship or equivalent work experience required, foreign language not required. *Entrance requirements:* For master's, GRE General Test, TOEFL (minimum score of 600 required), undergraduate training in the biological or physical sciences or engineering. *Application deadline:* For fall admission, 1/15 (priority date). Applications are processed on a rolling basis. Application fee: $50 ($55 for international students). Electronic applications accepted. *Financial aid:* Fellowships, research assistantships, teaching assistantships, career-related internships or fieldwork, scholarships, and tuition waivers (full and partial) available. Financial aid application deadline: 1/15. *Faculty research:* Economics, history, philosophy, and politics of science and technology; organization and management of science and technology. *Unit head:* Dr. Constance L. Wood, Associate Dean, 606-257-4613, Fax: 606-323-1928. *Application contact:* Lynda Wilson, Director of Admissions, 612-626-7229, Fax: 612-625-6351, E-mail: admissions@hhh.umn.edu.

See in-depth description on page 1289.

University of Pennsylvania, School of Engineering and Applied Science, Department of Systems Engineering, Philadelphia, PA 19104. Offers environmental resources engineering (MSE); environmental/resources engineering (PhD); systems engineering (MSE, PhD); technology and public policy (MSE, PhD); transportation (MSE, PhD). Part-time programs available. *Faculty:* 12 full-time (0 women), 9 part-time (2 women). *Students:* 26 full-time (10 women), 16 part-time (2 women); includes 4 minority (2 African Americans, 2 Hispanic Americans), 16 international. Terminal master's awarded for partial completion of doctoral program. *Degree requirements:* For master's, computer language required, foreign language not required; for doctorate, one foreign language, computer language, dissertation required. *Entrance requirements:* For master's and doctorate, TOEFL (minimum score of 600 required). *Application deadline:* For fall admission, 1/2 (priority date). Applications are processed on a rolling basis. Application fee: $65. Electronic applications accepted. *Unit head:* Dr. G. Anandalingam, Chair, 215-898-8790, Fax: 215-898-5020, E-mail: anand@seas.upenn.edu. *Application contact:* Dr. Tony E. Smith, Graduate Group Chair, 215-898-9647, Fax: 215-898-5020, E-mail: tesmith@seas.upenn.edu.

The University of Texas at Austin, Graduate School, Program in Science and Technology Commercialization, Austin, TX 78712-1111. Offers MS. Twelve month program, beginning in January, with classes held every other Friday and Saturday. Evening/weekend programs available. *Students:* 44 full-time (16 women); includes 6 minority (2 Asian Americans or Pacific Islanders, 4 Hispanic Americans), 5 international. Average age 35. In 1998, 44 degrees awarded. *Application deadline:* For spring admission, 10/1. Applications are processed on a rolling basis. Application fee: $50 ($75 for international students). Electronic applications accepted. *Financial aid:* Institutionally-sponsored loans available. Financial aid application deadline: 2/1; financial aid applicants required to submit FAFSA. *Faculty research:* Technology transfer, entrepreneurship. *Unit head:* Dr. Barbara M. Fossum, Director, Academic Programs, 512-475-8957, Fax: 512-475-8901, E-mail: bfossum@icc.utexas.edu.

Washington University in St. Louis, School of Engineering and Applied Science, Sever Institute of Technology, Department of Engineering and Policy, St. Louis, MO 63130-4899. Offers MA, MS, D Sc. Part-time programs available. Terminal master's awarded for partial completion of doctoral program. *Degree requirements:* For master's, thesis optional, foreign language not required; for doctorate, variable foreign language requirement, dissertation required.

Western Illinois University, School of Graduate Studies, College of Business and Technology, Department of Engineering Technology, Macomb, IL 61455-1390. Offers MS. Part-time programs available. *Faculty:* 10 full-time (0 women). *Students:* 14 full-time (5 women), 13 part-time (2 women); includes 5 minority (2 African Americans, 1 Asian American or Pacific Islander, 2 Hispanic Americans), 7 international. Average age 30. 15 applicants, 87% accepted. In 1998, 6 degrees awarded. *Degree requirements:* For master's, thesis or alternative required, foreign language not required. *Application deadline:* Applications are processed on a rolling basis. Application fee: $0 ($25 for international students). *Financial aid:* In 1998–99, 8 students received aid, including 8 research assistantships with full tuition reimbursements available (averaging $4,880 per year) Financial aid applicants required to submit FAFSA. *Faculty research:* Blended fuels training, production of *Illinois Journal of Technology*, occupational safety. *Unit head:* Dr. Thomas Bridge, Chairperson, 309-298-1091. *Application contact:* Barbara Baily, Director of Graduate Studies, 309-298-1806, Fax: 309-298-2245, E-mail: barb_baily@ccmail.wiu.edu.

Cross-Discipline Announcements

Duquesne University, Bayer School of Natural and Environmental Sciences, Environmental Science and Management Program, Pittsburgh, PA 15282-0001.

The MS program is designed to meet the educational needs of the contemporary environmental professional in industry, government, academe, and the public policy arena. The curriculum combines a strong foundation in the environmental sciences with courses in business and behavioral sciences, public policy, and law. Joint, 5-year BS/MS-ESM programs are offered with chemistry, biology, and microbiology. Joint MBA/MS-ESM and JD/MS-ESM programs are also offered. See Book 4 for an in-depth description.

Georgia Institute of Technology, Graduate Studies and Research, Ivan Allen College of Policy and International Affairs, School of Public Policy, Atlanta, GA 30332-0001.

Students in the School of Public Policy at Georgia Tech pursue study of technology-intensive public policies. At the graduate level, students specialize in formation policy and in one of 4 areas: science and technology policy, environmental and energy policy, information and management policy, and economic development. The faculty is diverse; among its 21 members, 17 have their primary homes in public policy, while the remaining 4 (jointly appointed) have theirs in city planning, earth and atmospheric sciences, and industrial and systems engineering. Approximately 65 students are enrolled in the master's program, and about one third are part-time students. The School also has a BS and a PhD program and a joint PhD program with Georgia Sate University.

Monterey Institute of International Studies, Graduate School of International Policy Studies, Monterey, CA 93940-2691.

The MA in international environmental policy trains policymakers to respond to environmental problems in government, intergovernmental organizations, nonprofit issue-oriented groups, and businesses. Interdisciplinary options include using electives to acquire management expertise or expand area studies. Write to admissions office or call 831-647-4123. See in-depth description in Political Science and International Affairs section in Book 2 of this series.

Rensselaer Polytechnic Institute, Graduate School, School of Humanities and Social Sciences, Program in Ecological Economics, Values, and Policy, Troy, NY 12180-3590.

The professional MS program in ecological economics, values, and policy (EEVP) is a 30- or 45-credit joint offering of the Departments of Economics and Science and Technology Studies. The program builds on Rensselaer's internationally recognized expertise and course offerings in the economic, political, social, cultural, and ethical implications and interactions of science, technology, environment, and society. A student can choose to add 15 credit hours of science and/or engineering, earning the equivalent of a minor in environmental science or engineering, to receive a Certificate in Multidisciplinary Environmental Studies. EEVP is aimed at recent graduates and midcareer professionals in state and local government, secondary education, business, and the nonprofit sector who are looking to upgrade their skills, advance their careers, and solve pressing problems.

CARNEGIE MELLON UNIVERSITY

Department of Engineering and Public Policy

Program of Study

A broad range of critical problems require analysis and skills at the interface between technology and society. The graduate program in the Department of Engineering and Public Policy is designed to prepare students to adapt and extend the perspectives and tools of engineering and science to these problems. While students receive a considerable amount of training and experience in social science and in techniques of social analysis, the department's philosophical and methodological roots remain in engineering.

The primary degree offered is a research-oriented Ph.D. In preparing for the Ph.D., students may elect to earn a joint M.S. with one of the traditional engineering departments. Terminal M.S. programs are not recommended.

Graduate students take three types of courses: engineering and science courses, primarily offered by the engineering departments of Carnegie Institute of Technology (the College of Engineering); courses in social sciences and social analysis offered by the H. John Heinz III School of Public Policy and Management, the School of Humanities and Social Sciences, and the Graduate School of Industrial Administration; and courses offered by the Department of Engineering and Public Policy, including, at a minimum, a one-semester course in project management and a four-course sequence on applied policy analysis. Candidates for the Ph.D. degree should expect to spend at least three years or the equivalent in graduate study. Part-time Ph.D. degree candidates must devote at least one academic year to full-time graduate study.

The Department of Engineering and Public Policy subscribes firmly to the belief that the Ph.D. is a research degree. Research undertaken in fulfillment of the requirements for a Ph.D. must make a fundamental and generalizable contribution toward the definition, understanding, or solution of a class of problems in the area of technology and public policy. While faculty members assist students in defining and developing their research problems, responsibility in this area lies primarily with the student.

Faculty research and interest are currently focused on policy problems in energy and environmental policy; global change; information and telecommunication technologies; technology policy, including innovation manufacturing and international technology transfer and development; technical aspects of international peace and security; conventional and AI-based decision support systems; and risk assessment, management, and communication. Faculty members maintain many close ties in their areas of interest with industrial and public-sector organizations, and many members act as consultants to industry and government agencies. Students are encouraged to make full use of these faculty ties in defining and developing their research problems.

Specific details on degree requirements are included in the descriptive literature available from the department.

Research Facilities

The University has excellent computer and library facilities. In addition, research facilities are available through the Heinz School, the Graduate School of Industrial Administration, the Robotics Institute, and the Institute for Complex Engineered Systems.

Financial Aid

The department strives to provide financial aid for the majority of its qualified graduate students.

Cost of Study

Tuition is $22,100 for the 1999–2000 academic year.

Living and Housing Costs

The estimated cost for living expenses for 1999–2000, including housing, transportation, books and supplies, and incidentals, is $15,640. All graduate students live off campus in areas near the University.

Student Group

The graduate enrollment at Carnegie Mellon University totals more than 2,800 students, who come from colleges and universities throughout the United States and from many countries. In 1998, more than 1,000 graduate degrees were awarded, including 209 doctorates. The current enrollment in the department is 36 graduate students and 50 undergraduates.

Student Outcomes

Since 1978, the department has awarded ninety-eight Ph.D. degrees and sixty-four M.S. degrees. Graduates are currently employed in policy research in think tanks and consulting firms (29 percent); as university faculty members and postdoctoral research fellows (37 percent); in private-sector firms (21 percent); and in government and national labs (13 percent).

Location

Pittsburgh offers many valuable resources to students working on problems in the area of technology and society. It is a dynamic, working city that has been revitalized in the past few decades and has developed an outstanding cultural life. Excellent cultural opportunities are available, including museums, orchestras, opera, drama, and dance, and there are active groups in the fine arts and folk crafts. The city has a good botanical garden, a zoo, many fine parks, and excellent public radio and television. The countryside around Pittsburgh provides opportunities for outdoor activities in all seasons.

The University and The Department

Carnegie Mellon University has developed an international reputation as an outstanding environment in which to conduct interdisciplinary research on problems in the area of technology and society. The Department of Engineering and Public Policy maintains a vigorous intellectual interaction with the engineering departments and special programs of the Carnegie Institute of Technology, the Department of Computer Sciences, the H. John Heinz III School of Public Policy and Management, the College of Humanities and Social Sciences, and the Graduate School of Industrial Administration. In addition to these, the University includes the Mellon College of Science and the College of Fine Arts.

Applying

All entering students must hold the equivalent of an undergraduate degree in engineering, physical science, or mathematics. Students with education or experience beyond the bachelor's degree are particularly encouraged to apply. Applications are welcome at any time, but they are usually processed between December and March for fall admission. Students are urged to take the General Test of the Graduate Record Examinations (the test is required for international students). In addition to filing the formal application, each student is asked to submit a letter describing his or her background and probable area(s) of research interest. Correspondence about the details of the program is encouraged.

Correspondence and Information

Victoria Massimino, Coordinator of Graduate Recruiting
Department of Engineering and Public Policy
Carnegie Mellon University
Pittsburgh, Pennsylvania 15213

Telephone: 412-268-2670
E-mail: eppadmt+@andrew.cmu.edu
World Wide Web: http://www.epp.cmu.edu/

Carnegie Mellon University

THE FACULTY AND THEIR RESEARCH

V. S. Arunachalam, Distinguished Service Professor of Engineering and Public Policy, Materials Science and Engineering, and the Robotics Institute; Ph.D., Wales, Dr.Eng. (h.c.), Roorkee. Issues on technology transfer, electric power, and communications: technology and policy.

Alfred Blumstein, J. Erik Jonsson University Professor of Urban Systems and Operations Research, H. John Heinz III School of Public Policy and Management, and Professor of Engineering and Public Policy; Ph.D., Cornell. Methodologies for public systems analysis, criminology, criminal justice.

Kathleen M. Carley, Professor of Social and Decision Sciences, H. John Heinz III School of Public Policy and Management, and of Engineering and Public Policy and Director of the Center for Computational Analysis of Social and Organizational Systems; Ph.D., Harvard. Issues related to social and organizational networks, organizational adaptation, adaptive and evolutionary models, and information diffusion.

Wesley M. Cohen, Professor of Economics and Social Science and of Engineering and Public Policy; Ph.D., Yale. Economics of technological change, industrial organization economics, technology policy.

Jared L. Cohon, Professor of Civil and Environmental Engineering and of Engineering and Public Policy and President of the University; Ph.D., MIT. Environmental and energy systems analysis, multiple critical decision making.

Cliff I. Davidson, Professor of Civil Engineering and of Engineering and Public Policy; Ph.D., Caltech. Long-range atmospheric transport and policy implications of fine particulates, with emphasis on heavy metals and acidic species; mechanisms of gas and particle deposition; emissions and ultimate fate of indoor air pollutants.

Otto A. Davis, William W. Cooper Professor of Economics and Public Policy; Ph.D., Virginia. Welfare economics, imperfect markets, regulation of economic activity, policy analysis, public choice, evolution of government and economic institutions.

Michael L. DeKay, Assistant Professor of Engineering and Public Policy and of Decision Science; Ph.D., Colorado. Values, judgment, and decision making in environmental and medical domains, including risk analysis, habitat protection, defensive medicine, and organ allocation; experimental design and analysis.

Urmila M. Diwekar, Principal Research Engineer, Engineering and Public Policy; Ph.D., Indian Institute of Technology (Bombay). Process design, synthesis, optimization, optimal control, stochastic modeling and optimization, dynamic simulation, inverse problems, batch process design, risk minimization.

Hadi Dowlatabadi, Director, Center for Integrated Study of Human Dimensions of Global Change; Ph.D., Cambridge. Quantification of uncertainties, exploration of the dynamics of integrated models of human and natural systems and their interactions.

Scott Farrow, Principal Research Economist and Director, Center for the Study and Improvement of Regulation; Ph.D., Washington State. Design, behavior, and performance of environmental regulatory system; economic program evaluation; real options and benefit-cost analysis.

Paul S. Fischbeck, Associate Professor of Social and Decision Sciences and of Engineering and Public Policy; Ph.D., Stanford. Bayesian decision theory, geographic information systems, subjective probability assessment, system reliability, military decision making.

Baruch Fischhoff, University Professor of Social and Decision Sciences and of Engineering and Public Policy; Ph.D., Hebrew (Jerusalem). Judgment and decision making, risk management, risk perception and communication, environmental benefits assessment, historical and expert judgment, adolescent decision making, political and behavioral foundations of formal analysis.

Richard L. Florida, H. John Heinz III Professor of Regional Economic Development, Center for Economic Development at the H. John Heinz III School of Public Policy and Management; Ph.D., Columbia. Technological innovation, regional economic development, foreign direct investment, science and technology policy.

H. Keith Florig, Senior Research Engineer; Ph.D., Carnegie Mellon. Risk analysis; radiation risk management; environmental, energy, and technology policy in developing economies.

Carol B. Goldburg, Research Economist; Ph.D., Carnegie Mellon. Environmental, health, and safety regulatory analysis, including the value of information, risk analysis and management, and use and design of performance measures.

James E. Goodby, Distinguished Service Professor. U.S.-Russian relations; conflict resolution and peace keeping; arms control; European, Middle Eastern, and Asian security; decision making and international negotiations; proliferation of weapons of mass destruction.

Alex Hills, Distinguished Service Professor of Engineering and Public Policy; Ph.D., Carnegie Mellon. Telecommunications policy, wireless telecommunications technology, remote and rural telecommunications systems.

David A. Hounshell, Henry R. Luce Professor of Technology and Social Change; Ph.D., Delaware. Innovation; industrial research and development; industrialization of regions; history of science, technology, and business.

Milind Kandlikar, Research Engineer; Ph.D., Carnegie Mellon. Science, technology, and public policy in less industrialized countries; energy and environmental issues in India; climate and global environmental change; quantitative approaches for coping with uncertainty.

Lester B. Lave, University Professor, Higgins Professor of Economics, and Professor of Engineering and Public Policy; Ph.D., Harvard. University-wide green design initiative, product and process design for the environment, risk analysis and management, global climate change, life cycle analysis software, life cycle analysis of alternative car fuels.

Francis C. McMichael, Professor of Civil Engineering and of Engineering and Public Policy and Walter J. Blenko, Sr., Professor of Environmental Engineering; Ph.D., Caltech. Industrial and municipal solid-waste management, landfills, product design for the environment, engineering economics, risk analysis, source reduction and recycling.

Sue McNeil, Professor of Civil and Environmental Engineering and of Engineering and Public Policy; Ph.D., Carnegie Mellon. Infrastructure management, facility condition assessment, applications of advanced technology in transportation, brownfield development.

Benoit Morel, Senior Lecturer of Engineering and Public Policy and of Physics; Ph.D., Geneva. Nonlinear dynamical modeling, environmental studies, technical aspects of arms control, international security, space policy, stochastic processes, complexity theory.

M. Granger Morgan, Lord Chair Professor of Engineering and Head, Department of Engineering and Public Policy; Ph.D., California, San Diego. Technology and public policy, quantitative methods for uncertainty analysis, integrated assessment of global change, risk analysis, communication and ranking, social impacts of electrical technologies.

Indira Nair, Associate Professor and Vice Provost for Education; Ph.D., Northwestern. Health risk assessment; public, precollege, and engineering education; environmental policy, with respect to low-frequency radiation.

Spyros N. Pandis, Elias Associate Professor of Chemical Engineering and of Engineering and Public Policy; Ph.D., Caltech. Air pollution, atmospheric chemistry, aerosol science.

Jon M. Peha, Associate Professor of Electrical and Computer Engineering and of Engineering and Public Policy; Ph.D., Stanford. Technical and policy issues of computers and telecommunications networks, technology policy, information technology for developing countries.

Henry R. Piehler, Professor of Materials Sciences and Engineering, Engineering and Public Policy, and Biomedical Engineering; Sc.D., MIT. Deformation processing and the mechanical behavior of materials; powder processing; clad, coated, and composite materials; biomaterials and medical devices; product liability litigation and standardization processes; productivity and innovation.

James Risby, Research Engineer; Ph.D., MIT. Climatology; integrated assessment of climate, water resources, and agriculture; value of climate information studies; methods and quality assessment of environmental research.

Allen L. Robinson, Assistant Professor of Mechanical Engineering and of Engineering and Public Policy; Ph.D., Berkeley. Combustion, formation and control of air pollution, power generation, renewable energy.

Edward S. Rubin, The Alumni Professor of Environmental Engineering and Science, Professor of Mechanical Engineering and of Engineering and Public Policy, and Director, Center for Energy and Environmental Studies; Ph.D., Stanford. Integrated modeling of energy, environmental, economic, and policy problems.

Marvin A. Sirbu, Professor of Engineering and Public Policy and of Industrial Administration; Sc.D., MIT. Telecommunications technology, policy, and management; regulation and industrial structure of computer and communication technologies; communications networks and standards; economics of information and networks.

Mitchell J. Small, Professor of Civil and Environmental Engineering and of Engineering and Public Policy; Ph.D., Michigan. Mathematical modeling of water and air quality, statistical analysis of soil and groundwater monitoring data, uncertainty analysis, indoor air quality and human exposure, risk assessment, risk communication, drinking water regulations, environmental policy.

Joel A. Tarr, Richard S. Caliguiri Professor of Urban and Environmental History and Policy; Ph.D., Northwestern. Development of the urban infrastructure; urban technologies; environmental trends, problems, and regulatory policies; regional economic development.

Herbert L. Toor, University Professor Emeritus and Mobay Professor of Chemical Engineering and of Engineering and Public Policy; Ph.D., Northwestern. Transport phenomena, heat and mass transfer and diffusion-reaction kinetics.

Robert M. White, Professor of Engineering and Public Policy, University Professor of Electrical and Computer Engineering, and Director, Data Storage Systems Center; Ph.D., Stanford. Magnetic device phenomena, technology policy.

CASE WESTERN RESERVE UNIVERSITY

Department of Operations Research and Operations Management

Programs of Study

The Department of Operations Research and Operations Management, which established the first academic graduate program in the field, offers programs leading to the Master of Science in Management in operations research (M.S.M.), the Master of Science in Management in supply chain (M.S.M.), as well as the Ph.D. degree.

The M.S.M. programs focus on techniques and applications that enable the student to add value to their employers' organizations. They consist of 36 credit hours of course work, including a 6 credit hour overview of the principal functional areas of business that enable effective communication with others in an organization. The M.S.M. in operations research adds 18 credit hours of deterministic and probabilistic model building, applied statistics, and use of computers and 12 credit hours in a concentration, chosen from operations research, operations management, management information systems, and finance, to provide in-depth marketable skills in the chosen area. The M.S.M. in supply chain adds 24 credit hours in a core, which combines model building with logistics and supply chain management, and 6 credit hours of specialty electives. The M.S.M. degree is typically completed in eighteen full-time months, but a well prepared student can complete all requirements in twelve months. The M.S.M. programs are available full-time and part-time.

The Ph.D. degree requires approximately 60 to 66 credit hours of course work—normally completed in about four years. The required and elective courses in the area of specialization are arranged to form a coherent program of study. Students are admitted to Ph.D. candidacy after passing a comprehensive examination. Students demonstrate the ability to do independent research through writing a dissertation that makes a significant contribution to knowledge in the field.

Research Facilities

The Weatherhead School of Management maintains its own computer facilities featuring a computer-supplemented classroom, several computer-supported group problem-solving rooms, and a laboratory with more than fifty personal computers linked by a local area network to an extensive software library. All of the microcomputers are also linked to the campus network, which in turn is linked to the Internet and the world's information services. In addition, various members of the department's faculty have converted or developed a variety of packages and case studies addressed to such problems as warehouse location, vehicle scheduling, project management, assembly-line balancing, quality control, and investment and modern portfolio management.

The University libraries contain more than a million volumes. All journals in operations research and related areas are available. In addition, the department maintains a list of research memoranda of studies conducted by the department for industry, government, and public organizations.

Financial Aid

Numerous financial assistance packages are available, including tuition fellowships (full and partial), graduate assistantships, and corporate fellowships that also provide valuable industrial experience with area companies. The corporate internships reflect the department's commitment to assisting businesses in the Cleveland area and beyond.

Cost of Study

Full tuition in 1999-2000 is $19,200 for the doctoral program and $21,900 for the masters program, based on a 24 credit hour annual course load. Books and supplies cost approximately $800.

Living and Housing Costs

On-campus housing includes graduate residential dorms. A large selection of off-campus housing is available throughout the University Circle area. Information may be obtained from the off-campus housing office.

Student Group

The department's active INFORMS student chapter advances educational, career, and social interests. The department also hosts a founding chapter of the Omega Rho (Operations Research) Honorary Society.

Student Outcomes

Recent graduates are pursuing careers in corporations locally, nationally, and internationally (including banking, manufacturing, and consulting) and in government. Most of the doctoral recipients enter academia.

Location

Case Western Reserve University is located 4 miles east of downtown Cleveland in University Circle, which has perhaps the most extensive concentration of educational, scientific, artistic, social, and cultural institutions in the United States.

The University and The Department

Case Western Reserve University is engaged in instruction and research at the undergraduate and graduate levels in the physical, biological, mathematical, and social sciences; the humanities and arts; and the professions of dentistry, engineering, law, management, medicine, nursing, and social work. The Department of Operations Research and Operations Management and its 12 faculty members are committed to solving the operational problems of industry in the Cleveland area and elsewhere and to advancing the state of knowledge in operations research and operations management.

Applying

Students who wish to pursue a graduate degree should write to the addresses below. Applications and supporting materials must be received by April 15 for the summer term, June 15 for the fall term, and November 15 for the spring term. Applications requesting financial aid should be received two months prior to these deadlines. Applicants must take the General Test of the GRE or the GMAT. Applicants whose native language is not English and who are not currently enrolled in a university in the United States or Canada are required to take the TOEFL.

Correspondence and Information

Regarding Ph.D. program in OR:
Admissions Officer
Department of Operations Research and Operations
    Management
Weatherhead School of Management
Case Western Reserve University
10900 Euclid Avenue
Cleveland, OH 44106-7235
Telephone: 216-368-3845
E-mail: jjb2@po.cwru.edu

Regarding M.S.M. programs:
Office of Admissions
Weatherhead School of Management
Case Western Reserve University
10900 Euclid Avenue
Cleveland, OH 44106-7235
Telephone: 216-368-2030
    800-723-0203 (toll-free)
E-mail: msmorsc@pyrite.cwru.edu
World Wide Web: http://weatherhead.cwru.edu/orom

Case Western Reserve University

THE FACULTY AND THEIR RESEARCH

Ronald H. Ballou, Professor; Ph.D. (business logistics), Ohio State. Planning, analysis, and control of supply chains, with particular emphases on facility location, transportation, and inventory issues.

Apostolos N. Burnetas, Assistant Professor; Ph.D. (operations management), Rutgers. Stochastic optimization and learning, adaptive Markov decision models of supply chain management, real options and investment under uncertainty.

Hamilton Emmons, Professor; Ph.D. (operations research), Johns Hopkins. Queueing control, workforce and jobshop scheduling.

A. Dale Flowers, Associate Professor; D.B.A. (production management and industrial engineering), Indiana. Enterprise resource planning, operational forecasting, quality control and management, manufacturing planning and control systems.

Thomas E. Love, Assistant Professor; Ph.D. (statistics), Pennsylvania. Statistics education, data analysis, diagnostics, survey analysis, statistical process control.

Kamlesh Mathur, Associate Professor; Ph.D. (operations research), Case Western Reserve. Management science and statistical applications, mathematical programming.

Peter Ritchken, Professor and Kenneth Walter Haber Professor in Finance; Ph.D. (operations research), Case Western Reserve. Financial economics, fixed income and derivatives, applied stochastic processes and statistics.

Harvey M. Salkin, Professor; Ph.D. (operations research and statistics), RPI. Analytical portfolio management, finance and investments, integer programming, linear programming.

Matthew J. Sobel, Professor, Chair, and E. Mandell deWindt Professor of Leadership and Enterprise Development; Ph.D. (operations research), Stanford. Product design and technology change; coordination of operations, finance, and marketing; supply chain management; environmental and energy management; large-scale structured Markov decision processes.

Daniel Solow, Associate Professor; Ph.D. (operations research), Stanford. Linear and nonlinear programming, combinatorial optimization, mathematics and computer science education.

George Vairaktarakis, Assistant Professor; Ph.D. (industrial engineering), Florida. Production and workforce planning, just-in-time scheduling, management of flexible manufacturing systems, management of virtual and traditional projects, product/process design, quality management, robust optimization.

Yunzong Wang, Assistant Professor; Ph.D. (operations management), Pennsylvania. Service parts logistics, supply chain coordination, new product development, production and inventory models.

RECENT FACULTY PUBLICATIONS

Ballou, R. *Business Logistics Management*, 4th ed., Upper Saddle River, NJ: Prentice Hall, 1999.

Ballou, R. Evaluating inventory management performance using a turnover curve. *Proceedings of the 18th Annual Transportation and Logistics Educator's Conference*, San Diego, October 11, 1998.

Agbegha, G., **R. Ballou** and **K. Mathur.** Optimizing Auto Carrier Loading. *Transportation Sci.* 32(2):174–88, 1998.

Ballou, R. Business logistics—Importance and some research opportunities. *Gestao Producao* 4(2):117–29, 1997.

Ballou, R., and H. Meshkat. Warehouse location with uncertain stock availability. *J. Business Logistics* 17(2):197–216, 1996.

Burnetas, A. and Katehakis, M. Dynamic allocation policies for the finite horizon one-armed bandit problem. *Stochastic Analysis Applications* 16(5):845–59, 1998.

Bauer, P., **A. Burnetas**, V. Cvsa, and G. Reynolds. Optimal employment of economies of scale for the Federal Reserve cash processing infrastructure. *Federal Reserve Financial Services*, Working Paper Series, 1998.

Burnetas, A., and M. Katehakis. Optimal adaptive policies for Markov decision processes. *Math. Operations Res.* 22(1):222–55, 1997.

Burnetas, A., and **P. Ritchken.** On rational jump diffusion models: An approach using potentials. *Rev. Derivatives Res.* 1:325–49, 1997.

Burnetas, A., D. Solow, and R. Agarwal. An analysis and implementation of an efficient in-place bucket sort. *Acta Informatica* 34:687–700, 1997.

Burnetas, A., and M. Katehakis. Optimal adaptive policies for sequential allocation problems. *Advan. Appl. Math.* 17(2):122–42, 1996.

Emmons, H., and **S. Gilbert.** The role of returns policies in pricing and inventory decisions for catalogue goods. *Manage. Sci.* 44(2):276–83, 1998.

Rabinowitz, G., and **H. Emmons.** Optimal and heuristic inspection schedules for multistage production systems. *IIE Trans.* 29:1063–71, 1997.

Emmons, H., and D. -S. Fu. Sizing and scheduling a full-time and part-time workforce with off-day and off-weekend constraints. *Ann. Operations Res.* 70:473–92, 1997.

Liaee, M., and **H. Emmons.** Scheduling families of jobs with setup times. *Int. J. Production Econ.* 51:165–76, 1997.

Emmons, H., A. D. Flowers, C. Khot, and **K. Mathur.** *Mini Reference Manual for Personal STORM.* Prentice-Hall, 1996.

Bouzina, K., and **H. Emmons.** Interval scheduling on identical machines. *J. Global Optim.* 9:379–93, 1996.

Dondeti, V. R., and **H. Emmons.** Max-Min matching problems with multiple assignments. *J. Optim. Theory Appl.* 91(2):491–511, 1996.

Moore, S. M., et al. **(A. D. Flowers).** Using learning cycles to build an interdisciplinary curriculum in CI for health professions students in Cleveland. *Joint Comm. J. Quality Improve.* 22(3):165–71, 1996.

Aka, M., **S. Gilbert**, and **P. Ritchken.** Joint inventory/replacement policies for parallel machines. *IIE Trans.* 29(6):441–9, 1997.

Love, T. A project-driven second course. *J. Stat. Educ.* [on line], 6(1), 1998. (http://www.stat.ncsu.edu/info/jse/v6n1/love.html).

Love, T. Compromises and success stories in a project-driven second course. In *ASA Proceedings of the Statistical Education Section*, pp. 117–22. Alexandria, Va.: American Statistical Association, 1997.

Love, T. Distractor selection ratios. *Psychometrika* 62(1):51–62, 1997.

Love, T. Teaching, reading, and doing statistical proofs. In *ASA Proceedings of the Statistical Education Section.* Alexandria, VA: American Statistical Association, 1996.

Mathur, K. An integer programming based heuristic for the balanced loading problem. *Operations Res. Lett. 22(1):19–26, 1998.*

Viswanathan, S., and **K. Mathur.** Integrating routing and inventory decisions in one-warehouse multiretailer multiproduct distribution systems. *Manage. Sci.* 43(3):294–312, 1997.

Ritchken, P. and I. Popova. On bounding option prices in paretian stable markets. *J. Derivatives* 5(4):32–44, 1998.

Ritchken, P., I. Popova, and J. Thompson. The changing role of banks and the changing value of deposit guarantees. *Adv. Int. Banking Fin.* 3:1–22, 1998.

Ritchken, P., P. Boyle, and G. Pennacchi, eds. *Advances in Futures and Options Research,* vol. 9. Greenwich, Conn.: JAI Press, Inc., 1997.

Janicki, A., I. Popova, **P. Ritchken,** and W. Woyczynski. Option pricing bounds in an alpha-stable security market. *Commun. Stat. Stochastic Models* 13(4):817–39, 1997.

Bliss, R., and **P. Ritchken.** Empirical tests of two-state variable HJM models. *J. Money, Credit, Bank.*, 1996.

Pennacchi, G., **P. Ritchken,** and L. Sankarasubramanian. On pricing kernels and finite state variable Heath Jarrow Morton models. *Rev. Deriv. Res.* 1(1):87–99, 1996.

Solow, D. The Keys to Linear Algebra. Cleveland, OH: Books Unlimited, 1998.

Vairaktarakis, G. and P. Kouvelis. Incorporating dynamic aspects and uncertainty in 1-median location problems. *Nav. Res. Logist.* 45:1–22, 1998.

Kouvelis P. and **G. Vairaktarakis.** Flowshops with processing flexibility across production stages. *IIE Trans.* 30(8):735–46, 1998.

Lee C.-Y. and **G. Vairaktarakis.** Performance comparison of some classes of flexible flowshops and job shops. *Int. J. Flex. Manufact. Syst.* 10(4):379–405, 1998.

Vairaktarakis, G. and J. Winch. Worker cross-training in paced assembly systems. *Proceedings of the International Conference in Optimization: Theory and Applications (ICOTA 98)*, Curtin University, Australia, 1998.

Vairaktarakis, G. Analysis of scheduling algorithms for master-slave systems. *IIE Trans.* 29(11):939–49, 1997.

Lee, C. -Y., and **G. Vairaktarakis.** Workforce planning in mixed model transfer lines. *Operations Res.* 45(4):553–67, 1997.

Sahni, S., and **G. Vairaktarakis.** The master-slave paradigm in parallel computer and industrial settings. *J. Global. Optim.* 9:357–77, 1996.

Wang, Y. and Y. Gerchak. Periodic review production models with variable capacity, random yield, and uncertain demand. *Manage. Sci.* 42:130–7, 1996.

Wang, Y. and Y. Gerchak. Continuous review inventory control when capacity is variable. *Int. J. Prod. Econ.* 45:381–8, 1996.

Parlar, M., **Y. Wang**, and Y. Gerchak. A periodic review inventory model with Markovian supply availability. *Int. J. Prod. Econ.* 42:131–6, 1995.

Gershak, Y., **Y. Wang,** and C. Yano. Periodic review inventory models with inventory-level-dependent demand. *Nav. Res. Logist.* 41:99–116, 1994.

Gerchak, Y., **Y. Wang,** and C. Yano. Lot sizing in assembly systems with random component yields. *IIE Trans.* 26:19–24, 1994.

CORNELL UNIVERSITY

School of Operations Research and Industrial Engineering

Programs of Study	The School has one of the largest and best-known graduate programs in operations research and industrial engineering. It has 18 full-time faculty members and two graduate programs leading to the degrees of Master of Engineering and Doctor of Philosophy.
	The emphasis in the Ph.D. program is on operations research as a mathematical science, providing the graduate student with a strong analytical basis for advanced research in the theory and methodology of operations research, as well as in the development of new approaches to applications. Concentrations are offered in the areas of applied probability and statistics, manufacturing systems engineering, and mathematical programming. There are many opportunities to pursue joint programs in such fields as applied mathematics, biometrics, business, civil and environmental engineering, computer science, and economics. The Ph.D. program normally requires four or five years of study and research.
	The Master of Engineering program emphasizes the practice of operations research and prepares students for professional careers in industry, government, and finance. The course work provides breadth and depth of technical knowledge. A project provides an opportunity for students to synthesize the skills acquired in their courses. An interdisciplinary manufacturing option is available, as is a financial engineering option. The program can be completed in one academic year.
Research Facilities	The faculty members and Ph.D. students are housed in Frank H. T. Rhodes Hall (which also houses Cornell's supercomputing facilities), the Center for Applied Mathematics, and the Center for Statistics. The School has several graduate student computing laboratories equipped with workstations and microcomputers that are networked to the supercomputer. The Cornell University library is one of the largest in the country and maintains an excellent collection in operations research.
Financial Aid	Almost all students admitted to the Ph.D. program receive full financial aid in the form of a full tuition waiver and a stipend, either from a fellowship, research assistantship, or teaching assistantship. In 1999–2000, all full stipends provide at least $12,400 for nine months. Additional summer support is usually available. A small amount of financial aid is available to candidates for the M.Eng. degree.
Cost of Study	Tuition for the two-term academic year in 1998–99 was $22,780.
Living and Housing Costs	The cost of living in Ithaca is lower than in most major urban centers. For the 1998–99 academic year, expenses (books, room and board, medical insurance, personal expenses) for a single graduate student were estimated at between $9000 and $12,500, not including tuition, travel, and summer expenses.
Student Group	Approximately 30 students from a dozen different countries are enrolled in the Ph.D. program. Approximately half have undergraduate degrees in engineering, half in mathematics and applied mathematics. The enrollment in the M.Eng. program is expected to be 55 to 60. Most M.Eng. students have undergraduate engineering degrees, but degrees in mathematics are common among those choosing the financial engineering option. Usually one fourth to one third of the M.Eng. students are women.
Student Outcomes	In roughly equal proportions, M.Eng. graduates are employed in the financial sector (e.g., modeling and analysis of financial instruments), manufacturing logistics companies, and consulting (general business as well as manufacturing and operations). Ph.D. graduates are employed in universities, in government and industrial laboratories, in the financial sector, and in a wide array of enterprises involving logistics (e.g., airline scheduling).
Location	Cornell is located in Ithaca, a city of 29,000 in the Finger Lakes region of New York State. The countryside is one of rolling hills traversed by gorges and waterfalls, lakes, and streams. Opportunities for outdoor recreation include sailing, windsurfing, swimming, skiing, and hiking; three state parks lie within 10 miles of the city. Complementing these surroundings are the community activities of Ithaca, whose character combines the intimacy and accessibility of a small city with cultural offerings usually available only in metropolitan areas. Ithaca's cuisines, theaters, and exhibits typify the creative vitality of this pluralistic community.
The University	Cornell University was founded in 1865 and is a leading research and teaching institution. The current student population is approximately 18,000, including graduate and professional students and undergraduates. The graduate faculty numbers almost 1,600 and includes Nobel laureates, Pulitzer Prize recipients, and members of the National Academy of Sciences. Campus life is rich in opportunities for enjoyment of and participation in art, athletics, cinema, music, and theater; the concert series and art exhibits feature artists of international stature, and the state-of-the-art Center for the Performing Arts hosts theater and dance productions.
	Cornell University is an Equal Opportunity/Affirmative Action educator and employer.
Applying	Applications for admission should be received by January 15 if possible in order to be considered for Cornell fellowship and assistantship support. Applicants for the Ph.D. program are expected to have a solid background in mathematics, including at least calculus and linear algebra and preferably additional courses such as advanced calculus, real analysis, probability, and statistics; lack of a strong calculus-based probability course may delay access to the required Ph.D. courses in stochastic processes and statistics. Some background in computing is also desirable.
	Scores on the General Test of the Graduate Record Examinations are required of all Ph.D. applicants. Scores on a GRE Subject Test (in engineering or mathematics, for example) are recommended. All applicants whose native language is not English should submit scores on the Test of English as a Foreign Language (TOEFL). Applicants for the one-year M.Eng. program must have the equivalent of a standard four-semester engineering calculus sequence. They are expected to have completed a course in statistics and to be proficient at computer programming in either C or Pascal; otherwise, an additional semester will be required to complete the program.
Correspondence and Information	Graduate Admissions School of Operations Research and Industrial Engineering Frank H. T. Rhodes Hall Cornell University Ithaca, New York 14853 Telephone: 607-255-9128 Fax: 607-255-9129 E-mail: admissions@orie.cornell.edu

Cornell University

THE FACULTY AND THEIR RESEARCH

An asterisk (*) identifies those whose primary appointment is in another unit of the University.

Athanassios Avramidis, Assistant Professor. Simulation.
* Louis J. Billera, Professor. Combinatorics, discrete and convex geometry.
Robert G. Bland, Professor. Mathematical programming, combinatorial optimization.
* Thomas F. Coleman, Professor. Nonlinear optimization, parallel computation.
* Richard T. Durrett, Professor. Probability theory, mathematical biology.
* Eugene B. Dynkin, Professor. Probability theory, mathematical economics.
Peter L. Jackson, Associate Professor. Production and inventory management, manufacturing economics.
* Robert Jarrow, Professor. Mathematical finance.
* Harry Kesten, Professor. Probability.
* Jon Kleinberg, Assistant Professor. Combinatorial optimization, network flows, molecular biology.
William L. Maxwell, Professor Emeritus. Scheduling, materials handling, simulation.
John A. Muckstadt, Professor. Inventory control, logistics.
Narahari U. Prabhu, Professor Emeritus. Stochastic processes, queuing and storage theory.
James Renegar, Professor. Mathematical programming, interior point methods.
Sidney I. Resnick, Professor. Teletraffic modeling, extremes, large sample theory, time-series analysis, heavy tails.
Robin O. Roundy, Associate Professor. Production and inventory management.
David Ruppert, Professor. Nonparametric statistics, environmental statistics, measurement error.
Gennady Samorodnitsky, Associate Professor. Gaussian random fields, stable processes, long-range dependence, applied probability.
Lee W. Schruben, Professor. Manufacturing systems, simulation.
David B. Shmoys, Professor. Design and analysis of algorithms, combinatorial optimization, scheduling.
* Christine A. Shoemaker, Professor. Environmental management.
Elizabeth H. Slate, Assistant Professor. Statistics.
* Éva Tardos, Professor. Combinatorial optimization, network flows.
Michael J. Todd, Professor. Mathematical programming, interior point methods.
Leslie E. Trotter Jr., Professor. Combinatorial optimization, discrete mathematics.
Bruce W. Turnbull, Professor. Biomedical statistics, quality control, reliability theory.
* Stephen A. Vavasis, Associate Professor. Complexity of optimization, numerical analysis.
Lionel I. Weiss, Professor Emeritus. Statistical decision theory, nonparametric statistics.

RESEARCH AREAS

Applied Probability and Statistics. This area stresses the techniques and associated underlying theory of probability and statistics, particularly as they are applied to problems in science and engineering. The techniques emphasized are those associated with applied stochastic processes (for example, mathematical finance, queuing theory, traffic theory, and inventory theory) and statistics; the statistical aspects of the design, analysis, and interpretation of experiments; reliability theory; analysis of life data; environmental statistics; nonparametric inference; and time-series analysis.

Manufacturing Systems Engineering. The analysis and design of complex manufacturing and distribution systems are the central concerns in this area. The problems studied include the establishment of inventory-control policies in multistage production and distribution systems, the design of manufacturing plants with optimal amounts of equipment and optimal materials-handling systems, the planning and scheduling of production in large-scale multi-item, multilocation systems, and the economic analysis of engineering processes. Students use modern analytic and computer techniques in the design and analysis of such systems. Course work deals with inventory theory, scheduling theory, database design, simulation, computer graphics, mathematical programming, stochastic processes, and statistical analysis. Students are also expected to understand the manufacturing processes associated with some type of industry. Research may involve development of new mathematical methodology and is often conducted directly with a cooperating company in, for example, automotive or semiconductor manufacturing.

Mathematical Programming. This area is broadly concerned with optimization, including linear, nonlinear, integer, and combinatorial programming; network flows; problems of scheduling and sequencing; and discrete and computational geometry. Research ranges from the development of computational algorithms (exact and approximate) and their applications to the associated studies of duality theory, convex analysis, polyhedra, combinatorics, and graph theory.

SELECTED FACULTY PUBLICATIONS

A. Avramidis. Correlation-induction techniques for estimating quantiles in simulation experiments. *Oper. Res.* 46:574–91, 1997.

L. J. Billera and Bernd Sturmfels. Fiber polytopes. *Annals of Math* 135:527–49, 1992.

R. G. Bland. The allocation of resources by linear programming. *Sci. Am.* 244:126–44, 1981.

M. Hariga and **P. Jackson.** The warehouse scheduling prlem: formulation and algorithms. *IIE Trans.* 28:115–27, 1996.

P. L. Jackson, J. A. Muckstadt, and **W. L. Maxwell.** Determining optimal reorder intervals in capacitated production-distribution systems. *Mgmt. Sci.* 1988.

H. Kesten. The critical probability of bond percolation on the square lattice equals ½. *Comm. Math. Phys.* 74:41–59, 1980.

J. Renegar. Linear programming, complexity theory and elementary functional analysis. *Math. Prog.* 70:279–351, 1995.

S. Resnick. *Adventures in Stochastic Processes.* Boston: Birkhauser, 1992.

S. Resnick. *A Probability Path.* Boston: Birkhauser, 1998.

Y. T. Herer and **R. Roundy.** Heuristics for a one-warehouse multi-retailer distribution problem with performance bounds. *Oper. Res.* 45:102–15, 1997.

D. Ruppert. Local polynomial regression and its applications in environmental statistics. In *Statistics for the Environment,* vol. 3, eds. V. Barnett and F. Turkman. Chichester: John Wiley and Sons, 1997.

G. Samorodnitsky and M. Taqqu. *Stable Non-Gaussian Random Processes.* New York: Chapman and Hall, 1994.

L. Schruben. *Graphical Simulation Modeling and Analysis Using SIGMA,* 3rd ed. Massachusetts: Scientific Press, 1995.

L. A. Hall, A. S. Schulz, **D. B. Shmoys,** and J. Wein. Scheduling to minimize average completion time: Off-line and on-line approximation algorithms. *Math. Oper. Res.* 22:513–44, 1997.

J. Kleinberg and **É. Tardos.** Disjoint paths in densely embedded graphs. *Proc. of the 34th Annual IEEE Symposium on the Foundations of Computer Science.* 52–61, 1995.

S. Mizuno, **M. J. Todd,** and Y. Ye. An $O(\sqrt{n}L)$-iteration homogeneous and self-dual linear programming algorithm. *Math. Oper. Res.* 19:53–67, 1994.

E. C. Sewell and **L. E. Trotter Jr.** Stability critical graphs and even subdivisions of K_4. *J. Combinatorial Theory (B)* 59:74–84, 1993.

L. A. Waller, **B. W. Turnbull,** L. C. Clark, and P. Nasca. Examining spatial patterns of disease incidence data to detect clusters in a rare disease: A case study. In *Case Studies in Biometry,* pp. 3–23, eds. N. Lange, L. Ryan, L. Billard, D. Brillinger, L. Conquest, and J. Greenhouse. New York: Wiley, 1994.

S. Vavasis and Y. Ye. A primal-dual interior point method whose running time depends only on the constraint matrix. *Math. Prog.* 74:79–120, 1996.

DUKE UNIVERSITY

School of Engineering
Master of Engineering Management

Programs of Study

The School of Engineering, in cooperation with the School of Law and the Fuqua School of Business, offers a twelve-month program leading to a Master of Engineering Management degree. This program requires all students to take three core business courses and one core law course, in addition to four graduate engineering courses. The four engineering courses may be individually selected to conform to each student's special area of engineering interest. For students who have not had professional engineering experience, a summer industrial internship is also required. The program director works with each student in arranging these internships. For engineers who have had appropriate industrial experience, the internship requirement may be waived.

Technology management has become increasingly important in the modern age. In order to keep pace, individuals in management at all levels need a substantial engineering and science background. Similarly, engineers and many scientists increasingly need a knowledge of the implications of their work on corporate success. Engineers have an increasingly powerful and important role in the advancement of the economic position that the United States will occupy in the world of the twenty-first century, and it is equally certain that graduate engineering education will be a vital link in this development. The increasing importance of the management of technology in sustaining, as well as in advancing, the nation's economy was discussed extensively in the National Research Council Task Force report entitled "Management of Technology—The Hidden Competitive Advantage." This report, with participants from both industry and academia, concluded that the management of technology in all its forms was destined to play an increasingly critical role in the well-being of national economic life. To an ever-increasing extent, advanced technologies are a pervasive and crucial factor in the success of private corporations, the effectiveness of many government operations, and the well-being of national economies. A focus of this report was on the general lack of programs aimed at education in the management of technology.

The goal of this degree program is to provide a professional degree focused on educating future leaders in a field of technology and its industrial utilization. By analogy to what an M.B.A. does to prepare future leaders in business, this program is intended to prepare engineering leaders for the high technology enterprises that are destined to dominate economic life in the twenty-first century. Increasingly, positions throughout the economy require education and intellectual development beyond the bachelor's level. Also, the increasing complexity of the industrial environment makes it appropriate that engineering personnel have a knowledge of accounting, marketing, law, and interpersonal skills. Most business schools now require that their entering students have two or more years of experience in the business world before their applications are considered. Thus, although many engineering students are interested in continuing directly after graduation into a program that has a major management component, they are effectively foreclosed from doing so. Furthermore, traditional M.B.A. programs place little or no emphasis on some of the issues important to high technology industries. It is to meet these educational needs that the School of Engineering has developed its Master of Engineering Management program.

Research Facilities

The School of Engineering operates many laboratories that involve a wide variety of specialized engineering equipment and facilities for use in graduate engineering courses. Such specialized facilities include a wind tunnel, a microelectronics clean room, an animal operating theater, an X-ray diffraction laboratory, a scanning electron microscope, and an integrated system of UNIX workstations, which provide links to the North Carolina Supercomputing Center in the Research Triangle Park. However, research with a thesis is not a requirement for completion of the Master of Engineering Management degree. The Vesic Engineering Library receives more than 800 periodicals and houses 75,000 volumes.

Financial Aid

The Master of Engineering Management program is an approved Federal Stafford Student Loan program.

Cost of Study

A tuition of $2060 is charged for each course taken, and students who wish to complete this program in one year need to take four courses each term, for a total tuition charge of $16,480. Up to two courses, approved by the admissions committee as fully meeting program requirements, may be transferred from other graduate programs.

Living and Housing Costs

A limited number of University-owned apartments are available at varying rates. Meals are available through various University dining facilities.

Student Group

The program expects to enroll a cohesive group of approximately 25 entering students. Enrollment may increase slightly over the next few years but will be strictly limited to ensure program integrity.

Student Outcomes

Graduates of the program have taken management-oriented positions at major national and international organizations.

Location

Duke University is located in the Research Triangle Park area of North Carolina. The region is known for its lack of big-city congestion and for having the highest ratio of Ph.D. holders to the general population of any area in the country.

The University

Duke University, a privately supported institution, is a multifaceted university of national stature that occupies a 9,000-acre campus. It has a competitively selected student body and a low student-faculty ratio. The outstanding professional schools, especially the Fuqua School of Business and the School of Law, are important program resources.

Applying

Applications for admission for the fall semester should be completed by December 31. However, later applications are given full consideration, subject to the availability of space, up until the final date for completion of applications, July 15. Applications must include official transcripts and GRE scores before they can be considered.

Correspondence and Information

Program Director
Master of Engineering Management Program
Duke University
Box 90304
Durham, North Carolina 27708-0304
E-mail: memp@mems.egr.duke.edu
World Wide Web: http://www.mem.egr.duke.edu

Duke University

THE FACULTY AND THEIR RESEARCH

The core faculty members listed below of the Master of Engineering Management Program are augmented by the more than 72 teaching faculty members in the School of Engineering.

Franklin H. Cocks, Professor and Director; Sc.D., MIT, 1972. Material science: crystal growth, high-temperature superconductivity, X-ray reflection and diffraction, patent applications and processing.

James D. Cox, Professor of Law; LL.M., Harvard, 1979. Corporate and securities law.

Matthew Kuhn, Professor of Engineering Management; Ph.D., 1997. Management and marketing in technological enterprises.

David Lange, Professor of Law; LL.B, Illinois, 1971. Intellectual property, entertainment and communication.

Charles J. Skender, Assistant Professor; M.B.A., Duke, 1979. Managerial accounting.

EMBRY-RIDDLE AERONAUTICAL UNIVERSITY

Office of Graduate Programs and Research
Department of Computing and Mathematics
Program in Industrial Optimization

Programs of Study

Embry-Riddle Aeronautical University's Master of Science in industrial optimization (MSIO) degree program is designed to provide recent engineering and science graduates, as well as midcareer engineers and scientists, with an opportunity to develop skills in optimization, statistics, and quality that can be applied to product and process design and improvement. Engineers and scientists who complete this program can assume key positions in engineering and scientific research.

The MSIO degree program achieves its purpose by the extensive use of case studies that enable the students to gain practical skills in analyzing and solving current aviation/aerospace problems that require the application of optimization tools and/or statistics. Application software and teams are used, enabling students to solve problems in an environment that simulates process and product design and improvement organizations.

MSIO students have the opportunity to strengthen and expand discipline-specific skills by taking several courses in their field of expertise (engineering, computer science, business, or human factors) while developing the mathematical foundation necessary to solve complex application problems within their field.

The curriculum includes required courses in mathematical foundations, optimization, mathematical programming and decision making, and statistical quality analysis. The MSIO program offers students three program options: a thesis option that requires 30 credit hours, a research project option that requires 33 credit hours, and a course-only option that requires 36 credit hours. For the thesis option, 6 credit hours of specified electives and a 6-credit-hour thesis are required. For the research project option, 12 credit hours of specified electives and a 3-credit-hour research report are required. For the course-only option, 18 credit hours of specified electives are required, and a department-administered comprehensive exam must be satisfactorily completed prior to graduation.

Research Facilities

MSIO students have access to a state-of-the-art graphics and computing laboratory that consists of ten SGI 02s and four SGI Indys with many modern graphics features. Commercial software is available to perform many tasks in statistics, computation, and visualization. The department also has a variety of flight simulators as well as an on-campus Flight Safety Simulation Center. The Airway Science Simulation Laboratory contains many tools for simulating the National Airspace System, including weather, air traffic control, and many pilot–human factors interactions. A rapid prototyping laboratory is available, as is a laboratory devoted to real-time simulations and processes. Multiple PC laboratories, most at the Pentium level, are available at central locations across the campus.

Financial Aid

Embry-Riddle makes every effort, within the limitations of the financial resources available, to ensure that no qualified student is denied the opportunity to obtain an education because of inadequate funds. However, the primary responsibility for financing an education must be assumed by the student. A number of graduate assistantships that provide a stipend and a tuition waiver are available on a competitive basis each year. Other financial aid programs are Federal Stafford Student Loans, short-term loans, scholarship and fellowship programs, and the Embry-Riddle Student Employment Program (available on the Daytona Beach Campus). All graduate programs are approved for Veterans Administration education benefits.

Cost of Study

In 1999–2000, tuition costs are $455 per credit hour. Books and supplies cost approximately $300 per semester.

Living and Housing Costs

Some on-campus housing is available to graduate students. The cost of a standard double-occupancy room is $1400 per semester. Off-campus housing is reasonably priced. Single students who share rental and utility expenses can expect yearly off-campus room and board expenses of $4000. Married students should expect a higher average for yearly expenses.

Student Group

The graduate programs currently enroll 250 students on the Daytona Beach Campus. The College of Career Education enrolls more than 3,000 students in off-campus graduate degree programs, and the Center for Distance Learning enrolls an additional 900 students. On the Daytona Beach Campus, 40 percent are international students, 41 percent are women, and 42 percent are members of minority groups. More than 10 percent of the campus-based graduate students are employed full-time.

Location

The Daytona Beach Campus is adjacent to the Daytona Beach International Airport and is 10 minutes from the Daytona beaches. Within an hour's drive are Disney World and EPCOT Center, the Kennedy Space Center, Sea World, and St. Augustine.

The University

The University comprises the eastern campus at Daytona Beach; a western campus in Prescott, Arizona; and the Extended Campus for off-campus programs. Within the field of aviation, Embry-Riddle Aeronautical University has built a reputation for the high quality of instruction in its programs since its founding in 1926.

Applying

Applicants must possess an earned baccalaureate degree in engineering or science or the equivalent. The minimum undergraduate cumulative GPA is 2.5 on a 4.0 scale and a cumulative GPA of 3.0 in the senior year. Applications from U.S. citizens and permanent residents should be received at least thirty days prior to the first day of the term in which the student plans to enroll. International students should submit all of their documents at least ninety days prior to the first day of the term in which they plan to enroll.

Correspondence and Information

Graduate Admissions
Embry-Riddle Aeronautical University
600 South Clyde Morris Boulevard
Daytona Beach, Florida 32114-3900
Telephone: 904-226-6115
          800-388-3728 (toll-free)
Fax: 904-226-7050
E-mail: admit@db.erau.edu
World Wide Web: http://www.db.erau.edu

Embry-Riddle Aeronautical University

THE FACULTY AND THEIR RESEARCH

The following are faculty members on the Daytona Beach campus.

David G. Caraballo, Assistant Professor; Ph.D., Princeton.
John H. George, Professor; Ph.D., Alabama.
Deborah M. Osborne, Associate Professor; Ph.D., Central Florida.
David L. Ross, Associate Professor; M.A., Kentucky.
John R. Watret, Associate Professor; Ph.D., Texas A&M.

The research interests of the faculty include multivariate optimization, statistical process control, experimental design, operations research, decision theory, statistics, mathematical modeling, scientific computing and visualization, product design, aviation/aerospace applications of optimization techniques, and quality control and improvement. Faculty members receive recognition for the quality of their research by regularly obtaining competitive grants from agencies such as the Federal Aviation Administration, the Naval Research Laboratory, the Department of Energy, and the National Aeronautics and Space Administration.

EMBRY-RIDDLE AERONAUTICAL UNIVERSITY

Office of Graduate Programs and Research
Department of Human Factors and Systems

Program of Study

Embry-Riddle Aeronautical University's Master of Science in human factors and systems program develops graduates with the capacity to design, conduct, and apply human factors or systems research in support of the design of simple and complex systems. The program also develops students' ability to work as human factors or systems professionals in the aviation and aerospace environments based on their academic preparation and their active participation in human factors and systems projects.

All students are required to complete a total of 36 semester hours of graduate study, including a 6-hour integrated human factors and systems core, a 6-hour quantitative and methods core, a 6-hour human factors engineering or systems engineering core, 12 hours of electives within an area of specialization, and a final integration requirement that is made up of a qualifying oral exam and a thesis (6 credits). The program can be completed in four or five semesters of full-time study.

The program adheres to guidelines established by the Committee for Education and Training of the American Psychological Association's Division 21 and the accreditation requirements of the Education Committee of the Human Factors and Ergonomics Society.

Research Facilities

The Human Performance Laboratory houses a fully instrumented research flight simulator, an eye tracker, physiological data collection tools, and an exact copy of the FAA's air traffic management information system. Other facilities include the Airway Science Simulation Laboratory, which simulates all elements of the national airspace system. More than sixty aircraft; fourteen flight simulators, including a full motion B-737-300 and Beech 1900; and an integrated air traffic control simulation facility are also available.

A wide variety of traditional engineering laboratories are available for use in human factors and systems engineering research, as well as computer facilities that include MicroVAX II's, Sun and Silicon Graphics workstations, and Macintosh and IBM desktop computers.

Financial Aid

Embry-Riddle makes every effort, within the limitations of the financial resources available, to ensure that no qualified student is denied the opportunity to obtain an education because of inadequate funds. However, the primary responsibility for financing an education must be assumed by the student. A number of graduate assistantships providing a stipend and tuition waiver are available on a competitive basis each year. Other financial aid programs are Federal Stafford Student Loans, the Embry-Riddle Student Employment Program, short-term loans, and scholarship and fellowship programs. All graduate programs are approved for Veterans' Affairs education benefits.

Cost of Study

In 1999–2000, tuition costs are $455 per semester hour. Books and supplies cost approximately $300 per semester.

Living and Housing Costs

Some on-campus housing is available to graduate students. The cost for a standard double-occupancy room is $1400 per semester.

Student Group

The Graduate Programs currently enroll 250 graduate students on the Daytona Beach campus. The College of Career Education enrolls more than 3,000 students in graduate degree programs off campus. On the Daytona Beach campus, 40 percent are from other countries, 41 percent are women, and 42 percent are members of minority groups. More than 10 percent of the campus-based graduate students are employed full-time.

Location

The Daytona Beach campus is adjacent to the Daytona Beach International Airport and 10 minutes from the Daytona beaches. Within an hour's drive are Disney World and EPCOT, the Kennedy Space Center, Sea World, and St. Augustine.

The University

The University comprises the main campus at Daytona Beach; a western campus in Prescott, Arizona; and the College of Career Education. Within the field of aviation, Embry-Riddle Aeronautical University has built a reputation for the high quality of instruction in its programs since its founding in 1926.

Applying

Entry into the human factors and systems program requires knowledge in one of the following areas depending on the track chosen: engineering, psychology, education, or industrial design. The minimum desired undergraduate cumulative GPA is 2.5 out of a possible 4.0, with a 3.0 in the senior year. Applications from U.S. citizens and permanent residents should be received at least thirty days prior to the first day of the term in which the student plans to enroll. International students should submit all of their documents at least ninety days prior the first day of the term in which they plan to enroll.

Correspondence and Information

Graduate Admissions
Embry-Riddle Aeronautical University
600 S. Clyde Morris Boulevard
Daytona Beach, Florida 32114-3900
Telephone: 904-226-6115
            800-388-3728 (toll-free)
Fax: 904-226-7050
E-mail: admit@db.erau.edu
World Wide Web: http://www.db.erau.edu

Embry-Riddle Aeronautical University

THE FACULTY AND THEIR RESEARCH

The following are faculty members at the Daytona Beach campus.

James W. Blanchard, Associate Professor; Sc.D., George Washington. Systems management, systems engineering, performance assessment.

Daniel J. Garland, Associate Professor; Ph.D., Georgia. Human factors, cognitive psychology.

Gerald D. Gibb, Associate Professor; Ph.D., Brigham Young. Selection, training, workload.

V. David Hopkin, Visiting Distinguished Professor; M.S., Aberdeen (Scotland). Human factors, air traffic control.

John W. Williams, Professor; Ph.D., Mississippi State. Human factors, social psychology.

John A. Wise, Professor; Ph.D., Pittsburgh. Human factors, display, HCI.

THE GEORGE WASHINGTON UNIVERSITY

School of Engineering and Applied Science
Department of Engineering Management and Systems Engineering

Programs of Study

Graduate study leads to the Master of Engineering Management (M.E.M.) and M.S., Applied Scientist (App.Sc.) and Engineer (Engr.), and D.Sc. degrees. The department offers a joint M.S. degree program in electrical power and engineering management in cooperation with the Department of Electrical and Computer Engineering and a joint master's degree in industrial engineering and engineering statistics in cooperation with the Department of Statistics. The M.E.M. program prepares students whose undergraduate degree is in engineering, natural science, mathematics, or statistics for positions of managerial responsibility in technological and scientific organizations in industry, government, or the armed services. Students may choose from areas of concentration in construction and facilities management, electrical power and engineering management, engineering management, environmental and energy management, management of research and development, marketing of technology, transportation management, and public works management. The M.S. programs develop technical expertise in the areas of engineering economics, general operations research, information management, management science, manufacturing management, mathematical optimization, stochastic modeling, and systems engineering and management. In cooperation with other departments in the School of Engineering and Applied Science, a joint master's degree in industrial engineering is offered. Study in systems engineering and management provides a broad understanding of the management and decision-making process and of the systems approach to managerial problem solving and develops competence in applying mathematical and statistical techniques to the solution of managerial problems. Engineering economics prepares students to analyze the feasibility of projects and systems on an economic level. Information management prepares students to apply computer science to the management process. The master's programs require at least 27 credit hours of graduate-level courses and 6 of thesis or, for the nonthesis option, at least 36 credit hours. A comprehensive exam is required for the M.S. in industrial engineering and engineering statistics. The professional degree programs, the Applied Scientist and Engineer degrees, require at least 30 credit hours beyond the master's degree and are for students who want to pursue course work with an emphasis on applied material. The D.Sc. program provides advanced study and research to enable students to extend knowledge and to enhance career opportunities. The doctoral program normally consists of one major area of concentration and one or two minor areas, totaling a minimum of 30 credit hours of courses beyond the master's level or a minimum of 54 credit hours of approved graduate work for students whose highest earned degree is a baccalaureate. D.Sc. candidates must pass a qualifying exam prior to preparation of the dissertation, which requires a minimum of 24 credit hours.

Research Facilities

The School provides UNIX, PC, and Macintosh resources, featuring a Sun Microsystems S1000 multiprocessor server and Ultra 3D workstations, Dell Pentium Pro II computers, and Apple Power Macintosh G3 computers. Systems are equipped with licensed software from leading manufacturers. The facility's labs support engineering design, artificial intelligence, multimedia, software development, Web technology, graphics, and analysis and project management and include high-quality laser and color printing and scanning equipment. In addition, the University provides extensive computing resources, with a Sun Microsystems SPARCcenter 2000E for academic information system services and a Sun Microsystems Enterprise 4000, with four processors, dedicated to research computing support. A high-performance ATM network integrates these resources, and Internet connectivity is provided for research support. University libraries contain 1.7 million volumes. Students have access on line to the shared catalog of all member libraries of the Washington Research Library Consortium as well as to periodical and newspaper index databases. Gelman Library provides additional databases on CD-ROM, including several specific to engineering, and offers research consultation services. Students also have ready access to the Library of Congress. The Institute for Crisis and Disaster Management, Research, and Education is an interdisciplinary University institute for research, education, and training. The objective of the institute is to support the emergency and crisis management activities of government, not-for-profit, and corporate organizations that must prepare for and respond to natural, technological, and political crises. The Center for Structural Dynamics operates a large earthquake simulator for the analysis of stochastic response of physical systems to dynamic loads. The Declassification Productivity Research Center does research into the use of computers for the declassification of documents. The Institute for Reliability and Risk Analysis does research in risk modeling, risk analysis and assessment, and reliability engineering.

Financial Aid

Teaching assistantships provide tuition remission for up to 9 credit hours per semester and a salary of $2000 per semester. Full-time research assistants receive a salary of $8000 to $15,000 for the calendar year. School Graduate Fellow, Dean's Fellow, and Department Fellow awards range from $6000 to $18,000 for full-time students. Full-time students who are U.S. citizens or permanent residents may be eligible for Graduate Engineering Honors Fellowships.

Cost of Study

Tuition is charged at the rate of $726 per credit hour for the 1999–2000 academic year and is payable on a course-by-course basis.

Living and Housing Costs

Apartments for students are available in the surrounding area at a wide range of costs. These costs start at approximately $600 a month.

Student Group

The School has about 1,400 students working on master's degrees, 31 on professional degrees, and 406 on the D.Sc. degree. The department has more than 500 master's, 10 professional, and 200 D.Sc. degree students.

Location

The Washington area has the nation's second-largest concentration of research and development activity. Library facilities are outstanding. The campus is located in the Foggy Bottom area of Washington, D.C.

The University and The School

The School, organized in 1884, operates on a two-semester academic year. Limited course work in engineering, engineering management, operations research, physical science, mathematics, economics, and statistics is available during the University's summer sessions.

Applying

Admission to master's and professional degree programs requires an appropriate bachelor's or master's degree from a recognized institution and evidence of a capacity for productive work in the field selected. Applicants for doctoral study must have adequate preparation for advanced study, including a satisfactory master's degree or the equivalent, and a demonstrated capacity for industrious and creative scholarship. Applications for the master's degree programs are accepted on a space-available basis under rolling admissions. Applications for the D.Sc. and professional degree programs are reviewed biannually and must be received by March 1 and October 1 for admission to the fall and spring semesters, respectively.

Correspondence and Information

E. Lile Murphree Jr.
Department of Engineering Management
and   Systems Engineering
School of Engineering and Applied Science
The George Washington University
Washington, D.C. 20052
Telephone: 202-994-7541
Fax: 202-994-4606
E-mail: murphree@seas.gwu.edu
World Wide Web: http://www.seas.gwu.edu

Interim Dean Thomas A. Mazzuchi
School of Engineering and Applied Science
The George Washington University
Washington, D.C. 20052
Telephone: 202-994-8061
        800-537-7327 (toll-free)
Fax: 202-994-3394

The George Washington University

THE FACULTY AND THEIR RESEARCH

RESEARCH ACTIVITIES

The Department of Engineering Management and Systems Engineering faculty includes both research engineers and faculty members who are experts in the various fields of science and engineering, research and development, and manpower logistics. They have been responsible for funded research projects in a wide range of technical and management areas and may provide opportunities for graduate students to engage in hands-on research as members of project teams. Representative research reflecting the broad scope of activities includes ocean engineering management, personnel management, program appraisal for the Department of Energy, energy invention appraisal, R&D management, and engineering economy analysis.

Herman G. Abeledo, Assistant Professor of Engineering and Applied Science; Ph.D., Rutgers, 1993. Mathematical programming, combinatorial optimization.

George R. Brier, Professor of Engineering Management; D.Sc., George Washington, 1990. Operations management, organizational development, general management.

Jonathan P. Deason, Professor of Engineering Management; Ph.D., Virginia, 1984. Environmental systems, environmental management, risk assessment, pollution prevention.

Michael L. Donnell, Associate Professor of Engineering Management; Ph.D., Michigan, 1977. AI, mathematical modeling, analytical methods in management.

Michael R. Duffey, Associate Professor of Engineering Management; Ph.D., Massachusetts Amherst, 1992. Engineering design, manufacturing processes, design and manufacturing management, engineering economics.

Howard Eisner, Professor of Engineering Management and Distinguished Research Professor; D.Sc., George Washington, 1966. Computer-aided systems engineering, entrepreneurship, program management, systems analysis and engineering, software engineering.

James E. Falk, Professor of Operations Research; Ph.D., Michigan, 1965. Mathematical programming, game theory, mathematical modeling.

Gideon Frieder, A. James Clark Professor of Engineering and Applied Science and Professor of Statistics; D.Sc., Technion (Israel), 1967. Computational methods, complex system design, hardware/software interface, intellectual property law.

John R. Harrald, Professor of Engineering Management; Ph.D., Rensselaer, 1982. Management information systems, crisis management, emergency response systems, risk and vulnerability analysis, management and communication.

Theresa Jefferson, Visiting Assistant Professor of Engineering Management; D.Sc., George Washington, 1998. Operations research, engineering economics, management of technology.

Walter K. Kahn, Professor of Operations Research; D.E.E., Polytechnic of Brooklyn, 1960. Antennas, microwave components, fiber optics, electrophysics.

Can E. Korman, Associate Professor of Operations Research; Ph.D., Maryland, 1990. Numerical modeling of semiconductor devices, VLSI, magnetics, signal processing.

Nicholas Kyriakopoulos, Professor of Operations Research; D.Sc., George Washington, 1968. Monitoring systems, digital signal processing, controls and systems theory, applications of technology to arms control, data communications and protection.

Roger H. Lang, Professor of Operations Research; Ph.D., Polytechnic of Brooklyn, 1968. Wave propagation in random media, remote sensing, adaptive arrays.

Ting N. Lee, Professor of Operations Research; Ph.D., Wisconsin, 1972. Networks, linear systems.

Murray H. Loew, Professor of Operations Research; Ph.D., Purdue, 1972. Pattern recognition, medical engineering, image processing.

Thomas A. Mazzuchi, Professor of Operations Research and of Engineering Management and Interim Dean of the School; D.Sc., George Washington, 1982. Bayesian statistics, analytical methods, reliability analysis and risk analysis, quality control, stochastic models of operations research, time-series analysis, applied statistics.

E. Lile Murphree Jr., Professor of Engineering Management; Ph.D., Illinois at Urbana-Champaign, 1967. Organization theory, operations research, computer systems, construction and facilities management.

Martha Pardavi-Horvath, Professor of Operations Research; Ph.D., Hungarian Academy of Sciences, 1985. Magnetic phenomena, magnetic recording processes and materials, magnetooptic devices and materials.

Raymond L. Pickholtz, Professor of Operations Research; Ph.D., Polytechnic of Brooklyn, 1966. Data communications, computer communication networks, communications theory, secure communications.

Debabrata Saha, Associate Professor of Operations Research; Ph.D., Michigan, 1986. Communication theory, modulation and coding techniques.

Shahram Sarkani, Professor of Systems Engineering; Ph.D., Rice, 1987. Probabilistic analysis of structural fatigue and vibration, stochastic systems, systems engineering and analysis.

Nozer D. Singpurwalla, Professor of Operations Research and of Statistics and Distinguished Research Professor; Ph.D., NYU, 1968. Applied probability and Bayesian statistics, reliability theory and quality control, time-series analysis, fault-tree analysis, filtering theory, uncertainty in expert systems.

Richard M. Soland, Professor of Operations Research; Ph.D., MIT, 1964. Mathematical modeling, discrete optimization, decision analysis, multiple-criteria decision making.

Michael Stankosky, Associate Professor of Engineering Management; D.Sc., George Washington, 1998. Knowledge engineering, entrepreneurship, marketing of technology.

Suresh Subramaniam, Assistant Professor of Operations Research; Ph.D., Washington (Seattle), 1997. Design and analysis of communication networks, optical and wireless networks.

Rene van Dorp, Visiting Assistant Professor of Engineering Management; D.Sc., George Washington, 1998. Risk analysis, risk and reliability management, stochastic systems.

Branimir R. Vojcic, Associate Professor of Operations Research; Ph.D., Belgrade, 1989. Communications theory, spread spectrum, mobile and fading communications.

Wasyl Wasylkiwskyi, Professor of Operations Research; Ph.D., Polytechnic of Brooklyn, 1968. Electromagnetic waves, propagation, signal processing, remote sensing.

Robert C. Waters, Professor of Engineering Management; D.B.A., USC, 1968. R&D management, productivity, economic analysis, water resources, transportation management, technological change and innovation.

Mona E. Zaghloul, Professor of Operations Research; Ph.D., Waterloo, 1975. Computer-aided analysis and design of integrated circuits; analog and digital VLSI modeling, design, and testing.

Additional areas of current research interest include accelerated life testing; acceptance sampling; Bayesian statistical methods; combinatorial optimization; complex system design; computational methods; construction and facilities management; cost engineering; decision analysis; discrete optimization in stochastic systems; dynamic linear models; economic planning; energy management; environmental protection; executive development; expert opinion in reliability analysis; extreme value theory and applications; fault-tree analysis; forecasting; foundational issues in statistics; game theory; global optimization; human factors engineering; industrial competitiveness; influence diagrams and applications to decision analysis; interactive multiple-criteria decision making; interior point methods; Kalman filtering and time-series analysis; logistics; maintenance policies; marine technology; mathematical programming; missile defense problems; multiple-criteria decision making; neural networks; nonconvex programming techniques; numerical methods; optimization methods applied to economics, energy, logistics, location, and military problems; polyhedral combinatorics; quality control; reasoning by integration of spatial and temporal relations; reliability growth models; sequential games; sequential optimization problems; software reliability; solution and applications of structured integer programming problems; solution and bounding techniques for nonlinear programs with parameters; systems effectiveness; systems engineering; technology assessment; and technology transfer.

MASSACHUSETTS INSTITUTE OF TECHNOLOGY

Operations Research Center

Programs of Study	Since its founding in 1953 as an academic birthplace of operations research, the Operations Research Center (ORC) has offered the oldest continuously operating interdepartmental graduate degree program at MIT, granting both S.M. and Ph.D. degrees. The Center was founded with the explicit objective of becoming the focus of interdepartmental and interdisciplinary activity and of meeting MIT's objective of educating men and women in the application as well as the acquisition of knowledge.

The ORC brings together faculty and students from the Schools of Engineering, Management, and Science, three of the five schools at MIT. Due to its interdisciplinary character and the dual affiliation of its faculty and students, the ORC's educational programs are recognized widely as achieving a balance between OR theory and applications.

The educational program aims to train students in fundamental OR research methods and to engage them in research as early in the program as possible. Both S.M. and Ph.D. degree students take a set of core courses in OR methods and electives in theory and applications, including such topics as public systems, transportation, production management, finance, economic systems, and marketing. Master's degree candidates are required to write a thesis based on independent, usually applied, research. Doctoral degree candidates take qualifying examinations in the beginning of their second year and take oral general examinations in advanced subjects at the end of their second year. The Ph.D. program culminates with the preparation of a thesis whose purpose is to contribute to original research.

Research Facilities Research facilities are housed in the Muckley Building (E40), which contains the Philip M. Morse Reading Room, computer facilities, and office space for students, faculty, visitors, and administrative staff. The ORC has state-of-the-art PC/NT and Linux workstations that are directly connected to the MIT network. The reading room contains a collection of operations research professional journals and technical reports and working papers from MIT and other OR programs throughout the United States. MIT's information-processing services provide the latest developments in computer technology, including a broad spectrum of processing power from personal computer laboratories to large multiuser access systems, as well as a full range of support services. MIT's library system comprises five libraries (each of which is devoted to a major subject area) and a number of branch libraries and reading rooms. Cooperation with other Boston area libraries and national library services provides access to substantial research collections.

Financial Aid Financial aid is awarded on a competitive basis to entering Ph.D. and S.M. students. Support generally consists of research and teaching assistantships, which in 1999–2000 carry a stipend of approximately $13,000 per academic year and a full tuition award for full-time appointments. The stipend and tuition award are prorated for part-time appointments. Fellowships may also be awarded to qualified students. Most continuing graduate students requiring support hold assistantships throughout their graduate careers at the ORC.

Cost of Study The 1999–2000 tuition for all regular graduate students is $25,000 per academic year. The health insurance fee is $265 for the fall semester and $371 for the spring semester.

Living and Housing Costs For 1999–2000, monthly living expenses are estimated to be $2100 for a single graduate student attending MIT. Rooms and/or apartments are available for single and married students.

Student Group The ORC enrolls 50 students—36 Ph.D. students and 14 S.M. students. Each year, it awards about six Ph.D. degrees and ten S.M. degrees. The small number of students is a result of a selective admissions process and reflects the faculty's desire to provide a highly interactive and well-supervised environment for education and research. Approximately 55 percent of the students are from other countries, and about 23 percent are women; several hold scholarships from their respective countries, and some hold fellowships from the National Science Foundation, the Office of Naval Research, and the Charles Stark Draper Laboratory. ORC students engage in many social and recreational events as well. The MIT Institute for Operations Research and Management Sciences (INFORMS) Student Chapter organizes seminars with ORC faculty members and OR practitioners from business and industry. It also monitors the collections of the Philip M. Morse Reading Room and organizes social events and participation in intramural athletic activities. A notable feature of the ORC graduate programs is the strongly supportive relationships among its students.

Student Outcomes Graduates of the program are in high demand. Over the past two decades, 50 percent have accepted faculty positions at top universities, 30 percent have taken positions in research laboratories such as AT&T Bellcore and USWest, and 20 percent have gone to work in industry with companies like McKinsey and USAir. Their responsibilities are varied and include research, teaching, project work, and analyzing and designing systems in many different industries, including finance, manufacturing, transportation, and telecommunications.

Location MIT is located in Cambridge on the north bank of the Charles River facing the city of Boston. Many students and young professionals live in the area, which has more than fifty postsecondary schools. Within a 2-mile radius of MIT are a number of fine arts museums, an aquarium, libraries, Fenway Park, and the Fleet Center. Boston is an exciting city with a diverse ethnic population and many cultural resources. Massachusetts and the neighboring New England states are picturesque in all seasons and offer a variety of outdoor and historical attractions.

The Institute MIT is an independent, coeducational, privately endowed university that began operation in the mid–nineteenth century. Today 9,900 students are evenly divided between undergraduates and graduate students, and the teaching staff totals more than 1,500. There are five schools and twenty-one academic departments and more than fifty interdepartmental laboratories and centers. This breadth of educational disciplines offers the student a vast range of research possibilities as well as depth of focus.

Applying The application deadline is January 15 for fall-term admission. Applications must include a GRE Subject Test score (in any quantitative subject) and the GRE General Test score, three letters of recommendation, complete academic transcripts, and the student's statement of objectives. International students whose native language is not English must submit a TOEFL score of 577 or better.

Correspondence and Information
Professor James B. Orlin
Operations Research Center
MIT E40-149A
77 Massachusetts Avenue
Cambridge, Massachusetts 02139-4307
Telephone: 617-253-3601

Massachusetts Institute of Technology

THE FACULTY AND THEIR RESEARCH

George Apostolakis, Professor of Nuclear Engineering; Ph.D. (engineering science and applied mathematics), Caltech. Probabilistic risk and reliability assessment of complex technological systems, software dependability, human reliability.

Arnold I. Barnett, George Eastman Professor of Management Science; Ph.D. (applied mathematics), MIT. Probability modeling and statistics, transportation systems, criminal behavior, health, risk analysis and risk perception.

Cynthia Barnhart, Mitsui Career Development Professor of Civil and Environmental Engineering and Codirector, Operations Research Center; Ph.D. (transportation), MIT. Development and application of linear, integer, and network optimization models and methods to large-scale transportation systems.

Dimitris Bertsimas, Boeing Professor of Operations Research; Ph.D. (operations research and applied mathematics), MIT. Combinatorial optimization and integer programming, applied probability, dynamic and stochastic optimization, performance analysis and optimization of queueing systems, OR applications in air transportation, manufacturing and finance.

Gabriel R. Bitran, Nippon Telegraph and Telephone Professor of Management; Ph.D. (operations research), MIT. Mathematical programming, operations planning and control, production scheduling, inventory management, design of service delivery systems.

Ismail Chabini, Assistant Professor of Civil and Environmental Engineering; Ph.D. (computer science), Montreal. Computer science and transportation science, methodological operations research, optimization methods for static and temporal network flows, real-time network optimization.

Eric Feron, Assistant Professor of Aeronautics and Astronautics; Ph.D. (aeronautics and astronautics), Stanford. Applications of optimization theory control problems, automatic control, signal processing and air traffic control.

Joseph Ferreira Jr., Professor of Urban Planning and Operations Research; Ph.D. (operations research), MIT. Analytic methods and computer-based modeling for urban planning and policy analysis.

Charles H. Fine, Professor of Management Science; Ph.D. (business administration), Stanford. Operations management, manufacturing policy, manufacturing technology development and technology supply chain.

Ernst Frankel, Emeritus Professor of Ocean Systems and Professor of Management; Ph.D. (transportation economics), London. Routing, scheduling, and allocation of resources in transportation and manufacturing systems.

Robert M. Freund, Seley Professor of Operations Research; Ph.D. (operations research), Stanford. Mathematics of optimization modeling, the efficient solution of optimization programs, applications of optimization modeling in scheduling, transportation, and financial engineering.

Stanley B. Gershwin, Senior Research Scientist, Mechanical Engineering; Ph.D. (applied mathematics), Harvard. Manufacturing systems, hierarchical control, dynamic programming in hybrid systems, approximation techniques and decomposition methods for large-scale systems.

Stephen C. Graves, Abraham J. Siegel Professor of Management and Codirector, Leaders for Manufacturing Program; Ph.D. (operations research), Rochester. Manufacturing and logistics systems, production planning and scheduling, inventory management, design of manufacturing systems and evaluation of flexibility in manufacturing systems.

John Hauser, Kirin Professor of Marketing; Sc.D. (operations research), MIT. Metrics for new product development; evaluating and managing the tiers of R,D&E; customer satisfaction measurement and incentive systems; internal customer issues; quality function deployment; customer-driven engineering; market measurement.

Gordon M. Kaufman, Professor of Operations Research and Management Science; D.B.A., Harvard. Theoretical and applied statistical decision analysis with applications to oil and gas exploration, finite population inference, applied multivariate analysis.

Richard C. Larson, Professor of Electrical Engineering and Director of the Center for Advanced Educational Services; Ph.D. (electrical engineering), MIT. Applied OR, with emphasis on model building, analysis of transactional data, queues, spatially distributed service systems, inventory/routing systems, and personnel assignment systems.

Steven R. Lerman, Professor of Civil and Environmental Engineering and Director of Center for Educational Computing Initiatives; Ph.D. (transportation systems analysis), MIT. Educational uses of computer and communications technologies, uses of multimedia.

John D. C. Little, Institute Professor and Professor of Management Science; Ph.D. (physics), MIT. Model building and data analysis to support managerial decision making, mostly dealing with marketing issues.

Andrew Lo, Harris & Harris Group Professor, Sloan School of Management; Ph.D. (economics), Harvard. Empirical validation and implementation of financial asset pricing models, pricing of options and other derivative securities, financial engineering and risk management, trading technology and market microstructure, statistical methods and stochastic processes, computer algorithms and numerical methods, nonparametric estimation and inference, artificial intelligence, mathematical biology.

Thomas L. Magnanti, Institute Professor and Dean of Engineering; Ph.D. (operations research), Stanford. Combinatorial and network optimization, transportation planning, logistics, communication systems, large-scale optimization, manufacturing, nonlinear programming.

Sanjoy K. Mitter, Professor of Electrical Engineering; Director, Laboratory for Information and Decision Systems; and Director, Center for Intelligent Control Systems; Ph.D. (electrical engineering), Imperial College (London). Structure, function, and organization of complex systems; image analysis and vision; theory of stochastic dynamical systems; nonlinear filtering; stochastic and adaptive control.

Amedeo R. Odoni, T. Wilson Professor of Aeronautics and Astronautics and Professor of Civil and Environmental Engineering; Ph.D. (electrical engineering), MIT. Applied probability theory, queuing theory, transportation systems, risk benefit analysis, routing problems, network location theory, probabilistic combinatorial optimization problems, applications of operations research in air transportation, air traffic control and urban systems.

James B. Orlin, E. Pennell Brooks Professor of Operations Research; Head of Management Science Area, Sloan School of Management; and Codirector, Operations Research Center; Ph.D. (operations research), Stanford. Combinatorial optimization, network optimization, and logistics.

Georgia Perakis, Assistant Professor of Operations Research; Ph.D. (applied math), Brown. Optimization, variational inequalities, fixed point problems, traffic equilibrium and dynamic traffic assignment integer programming.

Donald Rosenfield, Senior Lecturer, Sloan School of Management, and Director, Leaders for Manufacturing Fellows Program; Ph.D. (operations research), Stanford. Logistics system design, logistics strategy, manufacturing strategy.

Alexander M. Samarov, Principal Research Associate, Center for Computational Research in Economics and Management Science; Ph.D. (statistics), USSR Academy of Sciences. Robust and nonparametric modeling in regression, multivariate analysis, and time series; estimation in long-memory time series models; computational methods in statistics and data analysis; recursive estimation; regression diagnostics.

Andreas Schulz, Assistant Professor of Operations Research and Management Science; Ph.D. (math), Berlin Technical. Combinatorial optimization, integer programming, combinatorics of polytopes, approximation algorithms, scheduling theory and algorithms.

Jeremy F. Shapiro, Professor of Operations Research and Management Science; Ph.D. (operations research), Stanford. Integer programming, large-scale programming on parallel computers, integration of modeling with knowledge-based systems, mathematical programming models for supply chain management and portfolio optimization.

Yossi Sheffi, Professor of Civil and Environmental Engineering and Director of the Center for Transportation Studies; Ph.D. (civil engineering), MIT. Operations planning methods for carriers and shippers, transportation system analysis, logistics and transportation network design.

John N. Tsitsiklis, Professor of Electrical Engineering; Ph.D. (electrical engineering), MIT. Stochastic control, applied probability, stochastic systems, optimization, algorithms and neural networks.

Jiang Wang, Professor of Finance; Ph.D. (finance), Pennsylvania. Models of security market, asset pricing theory, and international finance.

Yashan Wang, Assistant Professor of Management Science; Ph.D. (management science), Columbia. Manufacturing and logistic systems, inventory control, efficient simulation, computational finance.

Lawrence M. Wein, Leaders for Manufacturing Professor of Management Science; Ph.D. (operations research), Stanford. Design and scheduling of manufacturing systems, queuing theory, stochastic control, quality control problems in manufacturing, health-care management, environmental issues in manufacturing, and stochastic models of transportation systems.

Roy E. Welsch, Professor of Management Science and Statistics; Ph.D. (mathematics), Stanford. Nonlinear regression, process control, design of experiments, robustness of statistical estimators, statistical graphics, techniques for using data in the construction of management science and technical models.

Nigel H. M. Wilson, Professor of Civil and Environmental Engineering; Ph.D. (transportation systems and civil engineering), MIT. Urban transport, public transport operations, planning, and management, transport systems analysis.

MASSACHUSETTS INSTITUTE OF TECHNOLOGY

Schools of Engineering, Science, and Humanities and Social Sciences
Technology and Policy Program

Programs of Study	The MIT Technology and Policy Program (TPP) educates men and women for leadership in the important technological issues confronting society. Graduates are prepared to excel in developing and implementing effective strategies for dealing with the risks and opportunities of technology. This kind of education is vital to the future. The Technology and Policy Program is an interdisciplinary enterprise composed of about 200 students, faculty members, and staff members who have worked together closely for many years. Their strong understanding and mutual respect for each other's disciplines provides the basis for timely, effective work in this area. The Master of Science (S.M.) in technology and policy is the main degree offered. It is a professional degree, comparable to an M.B.A., that prepares students to enter directly into practice in government or industry. Building on a common core of policy analysis, practice, and leadership, students can concentrate either in a specific field of engineering or science or in political science and public policy. The program's Ph.D. curriculum blends technology and policy with disciplinary studies. Technology and policy graduates with doctorates are now on the faculty at major universities, including Brown, the University of California, Harvard, Lehigh, Princeton, NYU, MIT, the University of Southern California, Texas A&M, and the University of Texas. They have appointments in a variety of engineering departments, in economics, and in planning. Members of the Technology and Policy Program twice received MIT's Most Significant Improvement to MIT Education award. The Technology and Policy Program is a constituent of the Engineering Systems Division in the School of Engineering at MIT.
Research Facilities	The opportunities for research and thesis work focus around the MIT Center for Technology, Policy and Industrial Development. These are complemented by openings throughout the campus, particularly in the Communications Program, the Energy Laboratory, the Materials Systems Laboratory, the Laboratory for Information and Decision Systems, and the MIT-Harvard Joint Project on Negotiation.
Financial Aid	MIT makes financial support available to graduate students from a variety of sources in several different forms—fellowships, scholarships, teaching and research assistantships, the Federal Work-Study Program, and the Technology Loan Fund. Essentially all students receive financial support to complete the Technology and Policy Program. Such support is typically in the form of a full-time research assistantship providing both a stipend of around $1400 a month and full tuition. Entering students may receive only partial support in the beginning. Students with work experience will have better opportunities for research or teaching assistantships.
Cost of Study	Tuition for the two-term academic year for graduate students enrolled in the Technology and Policy Program at Massachusetts Institute of Technology is $25,000 in 1999–2000.
Living and Housing Costs	Dormitory accommodations for 466 single graduate students were available at a cost of approximately $12,000 for the 1998–99 academic year. Total living costs for a single student are about $28,000 per academic year ($38,000 per calendar year). Dormitory accommodations for married couples are available. Approximately 40 percent of new single graduate and married students applying for campus housing are assigned. Off-campus apartments are available within commuting distance of the campus; typical prices (excluding heat and utility costs) are $800 and up per month for a one-bedroom apartment and $900 and up per month for a two-bedroom apartment. The number of apartments is limited, so an early search for housing is advised.
Student Group	The program admits approximately 55 new students each year. There are about 150 on campus at any time, including about 25 doctoral students. The program builds each class of students from applicants with a variety of interests and backgrounds in order to provide each participant with the best opportunities for learning. Approximately 45 percent of the students are international students from more than forty countries. A quarter are women. Around three quarters have had one or more years of experience before they begin study in the program.
Location	MIT occupies a 125-acre campus on the north bank of the Charles River with a fine view of Boston's skyline. The vacation areas of New Hampshire and Vermont, only a few hours from the campus, provide mountaineering, skiing, and hiking opportunities. The Boston area is world renowned for its cultural facilities. Performances by the Boston Symphony Orchestra, the Boston Pops, and the Boston Opera take place within a 5-minute drive or 30-minute walk of the campus. The area has many famous museums, restaurants, and historic sites, and more than a dozen universities and specialized schools are within a few miles of the campus.
The Institute	MIT, founded in 1861 as a private, coeducational, endowed institution committed to the extension of knowledge through teaching and research, is one of the foremost institutes of technology in the world. The Institute comprises five schools: Architecture and Planning, Engineering, Humanities and Social Sciences, Management, and Science. Of the 9,885 students enrolled in 1998–99, 5,513 were graduate students and 4,372 were undergraduates. The campus provides excellent athletic facilities for indoor and outdoor sports, sailing, swimming, and crew. Extensive cultural programs on campus are available as well.
Applying	Applications for entrance to the Technology and Policy Program in September should be made by January 15. Applicants for admission in February should file by November 15. All prospective applicants should access the TPP Web page at the address below for a complete program description or write to the program to request copies of the *TPP Brochure* and application materials. All applicants should have a strong preparation, if not a degree, in engineering or science. Candidates should present Graduate Record Examinations (GRE) scores above 600 on the quantitative section and above 400 or 550 on the verbal section, depending on whether or not the student is a native speaker of English. Nonnative speakers of English must also score above 580 on the Test of English as a Foreign Language (TOEFL). Demonstrated evidence of work experience and leadership is welcome. The program provides equal opportunity to all.
Correspondence and Information	Dr. Richard de Neufville School of Engineering Technology and Policy Program Room E40-242a Massachusetts Institute of Technology 1 Amherst Street Cambridge, Massachusetts 02139 Telephone: 617-253-7693 Fax: 617-253-7140 E-mail: tpp@mit.edu World Wide Web: http://web.mit.edu/tpp/www

Massachusetts Institute of Technology

THE FACULTY AND THEIR RESEARCH

Richard de Neufville, Professor of Engineering and Program Chair; Ph.D., MIT. Strategic planning for risky situations, decision analysis process, applications to airport systems planning.

Richard Tabors, Senior Research Associate and Lecturer and First-Year Adviser; Ph.D., Syracuse. Development of electric utilities, efficient distribution of power through spot pricing, photovoltaic energy.

Daniel Roos, Japan Steel Industry Professor of Engineering and Director of the Center for Technology, Policy and Industrial Development; Ph.D., MIT. International competitiveness, industrial strategy and development in the automobile industry.

Nicholas Ashford, Professor of Technology and Policy; Ph.D., J.D., Chicago. Environmental law, occupational safety and health, regulation of innovation in chemical and pharmaceutical industries.

Louis Bucciarelli, Associate Professor of Engineering and of Science, Technology and Society; Ph.D., MIT. Engineering design and photovoltaics.

Charles Caldart, Research Associate in the Center for Technology Policy and Industrial Development; J.D., Washington (Seattle). Environmental law, pollution control, regulation of chemical toxins, radiation and biotechnology, law, technology, and public policy.

Joel Clark, POSCO Professor of Materials Systems and Director of the International Materials Program; Sc.D., S.M. (management), MIT. Materials selection in automotive and other major industries, economic modeling of production processes and of the demand for new materials.

John Ehrenfeld, Senior Research Associate and Lecturer and Coordinator of the MIT Working Group on Business and the Environment; Sc.D., MIT. Hazardous-waste technology and policy, management of chemicals in the environment.

Frank Field III, Principal Research Associate, Center for Technology, Policy, and Industrial Development; Ph.D., MIT. Director of the Materials Systems Laboratory and admissions officer of the Technology and Policy Program.

Richard Lester, Professor of Nuclear Engineering and Director of the Industrial Performance Center; Ph.D., MIT. Technological innovation and industrial performance, nuclear energy economics and policy, passive safety in nuclear power plant systems, nuclear-waste management.

Stuart Madnick, John Norris Maguire Professor of Information Technology in the MIT Sloan School of Management; Ph.D., MIT. Connectivity among distributed information systems, computer-integrated manufacturing, database technology, and software project management.

David Marks, Professor and Director of the Program in Environmental Engineering Education and Research (PEER); Ph.D., Johns Hopkins. Environmental systems, hazardous substances management, infrastructure management.

Theodore Postol, Professor of Science Technology and National Security Policy; Ph.D., MIT. Defense and arms control, science technology and society, technical dimensions of national security policy.

Harvey Sapolsky, Professor of Public Policy and Organization and Director of the Center for International Studies; Ph.D., M.P.A., Harvard. Analysis of institutional behavior, with emphasis on military, health, and science organizations; technology acquisition process in both the defense and nondefense sectors.

Lawrence Susskind, Professor of Urban Studies and Planning and Executive Director of the Harvard Law School Program on Negotiation; Ph.D., M.C.P., MIT. Development of principled negotiation for environmental dispute resolution, environmental management.

Leon Trilling, Professor of Aeronautics and Astronautics and of Science, Technology and Society; Ph.D., Caltech. Social studies in technology, fluid mechanics and gas dynamics.

Kosta Tsipis, Principal Research Scientist of the Science, Technology and Society Program and Director of the Program in Science and Technology for International Society; Ph.D., Columbia. Technical analysis of space and military systems, scientific and technical aspects of arms control and defense policy.

OLD DOMINION UNIVERSITY

College of Engineering and Technology
Department of Engineering Management

Programs of Study

The Department of Engineering Management at Old Dominion University provides its graduates with the necessary skills, knowledge, abilities, and attitudes required to design and manage the technology-based, project-driven enterprise. Building on concepts in systems science and systems engineering and founded on the solid principles of project and program management, the engineering management graduate programs emphasize technological leadership. Core course work in the engineering management programs concentrates on developing the knowledge and skills required by graduates to provide the leadership and management necessary to focus their organizations toward the development and application of technologies that will create new products, processes, and services and which, in turn, will create new markets or enable domination of existing ones. Through design projects and exercises, students are led through alternative ways of thinking and communicating about complex systems and technology. The engineering management programs provide students the opportunities in the classroom and through involvement with industrial partners to gain the confidence and experience to effectively integrate and apply technology in enterprise operations. Additionally, and perhaps more importantly, engineering management at Old Dominion University emphasizes the development of a professional perspective that anticipates opportunities for competitive advantages that technology can provide to an enterprise.

The Department of Engineering Management offers graduate programs leading to the degrees of Master of Engineering (M.E.), Master of Science (M.S.) in engineering management, Master of Engineering in operations research, and Doctor of Philosophy (Ph.D.) in engineering management. The M.E. programs address the needs of working professionals in engineering and applied science who have moved or will soon move into positions of managerial or leadership responsibility. The M.S. program is designed for full-time students who wish to include a research-oriented thesis in their program. The Ph.D. focuses on preparing individuals to perform rigorous research in the areas related to the design and management of complex human-technological systems and for careers in teaching and research at academic institutions as well as in other public and private organizations.

Research Facilities

The department supports its own computer laboratory of networked computers. Additionally, engineering management students have access to the extensive computer laboratories and specialized simulation software at the Virginia Analysis, Modeling, and Simulation Center.

The Old Dominion University library provides a full complement of state-of-the-art services. The University library contains 1.8 million items, which are accessible through an online Public Access Catalog that can be searched from terminals in the library and from remote locations on and off campus. Automated services, such as CD-ROM indexes to journal literature; government publications and statistics on CD-ROM; ProQuest databases, which offer indexing to general, business, and academic research publications; and online database searching, are offered by the library's Reference and Research Services Department.

Financial Aid

Financial aid is available in the form of fellowships, research assistantships, teaching assistantships, industrial fellowships, internships, and grants. The department attempts to support all full-time graduate students, especially at the Ph.D. level. Financial awards range from $3000 to $15,000 for the academic year, plus tuition, and are competitive. First-year support of international students is extremely rare.

Cost of Study

Tuition for the 1998–99 academic year was $180 per credit hour for Virginia residents and $474 per credit hour for non-Virginia residents. Health services fees and general service fees are $48 per semester. Personal expenses, including books, are estimated at $1200 per academic year.

Living and Housing Costs

Room and board are estimated to be approximately $6000 per academic year. A limited number of University-owned apartments and dormitories are available, ranging in cost from $1751 to $2433 per semester.

Student Group

The department has approximately 20 full-time students and 150 part-time students. The department has about 25 Ph.D. students in various stages of their programs. Twenty students receive financial aid in the form of graduate assistantships, fellowships, scholarships, and grants.

Student Outcomes

Graduates of the department have found employment in leading industrial, service, educational, and governmental organizations. Organizations employing recent graduates include NASA, Newport News Shipbuilding, U.S. Army, U.S. Navy, the State of Virginia, Mitsubushi Chemical, Thomas Jefferson Laboratories, Lucus Control Systems, Contrad Industries, the Boeing Company, TWA Airlines, and TranAir Airlines. Department graduates are now in the positions of assistant professor, manager of engineering, QA engineering, systems analyst, cost analyst, project engineer, team leader, and program manager.

Location

Old Dominion University is located in Norfolk, Virginia, and has a population of 1.3 million. Norfolk serves as the commercial and cultural center of a vibrant metropolitan community composed of a variety of individual municipalities, including historic Williamsburg, Jamestown, Yorktown, and scenic Virginia Beach. Collectively known as Hampton Roads, the area is ideal for fishing, boating, surfing, and other recreational activities. Located within the Hampton Roads area are the NASA Langley Research Center, the Continuous Electron Beam Acceleration Facility (CEBAF) at the Thomas Jefferson Laboratories, U.S. Navy and U.S. Army bases, and Newport News Shipbuilding. Hampton Roads encompasses one of the largest harbors in the world and is within 600 miles of two thirds of the population of the United States.

The University

Old Dominion University had its formal beginning in 1930 as a branch of the College of William and Mary and became autonomous in 1962. It is a state-assisted university and currently enrolls about 17,000 students in the Colleges of Arts and Letters, Business and Public Administration, Health Science, Sciences, and Engineering and Technology; the Darden School of Education; and the doctoral program in urban services.

Applying

Applicants for the master's degrees (M.S. or M.E.) in engineering management must have a bachelor's degree from an ABET–accredited program in engineering or engineering technology or from a fully accredited program in applied science, with a GPA of at least 3.0 for regular admission. For the master's degree in operations research, applicants must have an undergraduate degree from an accredited institution with a GPA of at least 3.0 for regular admission. Applicants for the Ph.D. degree must have a bachelor's or master's degree from an accredited institution in engineering, engineering technology, applied science, or applied mathematics and at least 30 hours of graduate study approved by the graduate program director. Applicants to all programs must have mathematics course work through calculus and differential equations. GRE general aptitude scores are required. A minimum TOEFL score of 550 is required for all international students when English is not their first language.

Correspondence and Information

Dr. Resit Unal
Graduate Program Director
Department of Engineering Management
Old Dominion University
Norfolk, Virginia 23529-0248

Telephone: 757-683-4554
Fax: 757-683-5640
E-mail: runal@odu.edu
World Wide Web: http://www.odu.edu/~enma/

Old Dominion University

THE FACULTY AND THEIR RESEARCH

Ralph Rogers, Professor and Chair, Department of Engineering Management; Virginia, 1987. Simulation methodology and applications, object-oriented and knowledge-based simulation, simulation education, systems engineering, command and control systems analysis and design, distribution systems.

William Swart, Professor and Dean, College of Engineering and Technology; Georgia Tech, 1970. Optimization, simulation, total quality management, management of change, project management.

Resit Unal, Professor and Graduate Program Director; Ph.D., Missouri–Rolla, 1986. Multidisciplinary design optimization, robust design, quality engineering, design of experiments, response surface methods, design for lifecycle cost, genetic algorithms.

Derya A. Jacobs, Associate Professor and Director of Engineering Foundations Division; Ph.D., Missouri–Rolla, 1989. Modeling, simulation, applied artificial intelligence, neural networks and experts systems, process optimization and quality management.

Abel A. Fernandez, Assistant Professor; Ph.D., Central Florida, 1995; PE. Risk analysis and management, stochastic optimization, systems modeling, project/systems engineering, project management.

Paul Kauffman, Assistant Professor; Ph.D., Penn State, 1997. Managerial decision making and planning in a global environment, strategic and appropriate technology integration in enterprise operations, technology impacts on operational performance, analytical/quantitative decision methods for competitive technology strategies, research portfolios, quality, product development.

Charles B. Keating, Assistant Professor; Ph.D., Old Dominion, 1993. Systems for organizational analysis and design, applied systems methodologies, organizational knowledge engineering and learning, sociotechnical systems, design processes, computer-based knowledge systems.

Rochelle Young, Assistant Professor; Ph.D., Old Dominion, 1996. Management, design, and analysis of technology policy and social transformation and knowledge management for technical organizations.

Research Activities

The Department of Engineering Management is extremely active in research and industrial partnerships. Sponsored research activities involve various agencies and companies, including the National Aeronautics and Space Administration (NASA); Jefferson Laboratory; the Army's Training and Doctrine Command (TRADOC); the U.S. Navy Naval Undersea Warfare Center; the American Red Cross and the U.S. Army Joint Training; Analysis and Simulation Center (JTASC); the Logistics Management Institute; Resource Consultants, Inc.; and Siemens Automotive.

Recent sponsored research programs include human vigilance monitoring models using EEG data; battle outcome prediction models; transportation and resource optimization models for health organizations; systems failure prediction models; Quality Function Deployment (QFD) for large complex systems; high-technology operations redesign; organizational systems engineering and change analysis; redesign of organizational knowledge systems; performance measurement systems development; problem definition, learning, and knowledge systems design; systems analysis of cruise missile recertification; and the NASA-sponsored effort to develop and apply multidisciplinary design optimization (MDO) techniques applicable to launch vehicle conceptual design.

RENSSELAER POLYTECHNIC INSTITUTE

Department of Science and Technology Studies

Programs of Study

The department is one of the few in the world that offers the full range of degrees (B.S., M.S., and Ph.D.) in science and technology studies (STS). The field of science and technology studies asks fundamental questions about the role of science and technology in social change. The graduate program emphasizes the cultural, historical, economic, political, and social dimensions of scientific and technological society, with a focus on ethical and values issues. The faculty includes some of the best-known writers and researchers in the field; topical foci include information technologies, biotechnologies, medicine, the environment, and law. Historical and cross-cultural comparison is used to deepen STS insight into evolving politics and institutions of science and technology, as well as into the role of science and technology in everyday life. The STS curriculum links theory to practice, analyzing options for policy and political action. A special departmental focus is design; students explore the way that assumptions about human nature and society, heterogeneous values and interests, and a wide variety of technical and social factors influence the design of technologies and technological systems.

The department's location in the midst of one of the nation's premier technological universities strengthens the programs and prepares students for a wide variety of fulfilling and exciting careers in academic, governmental and non-governmental organizations and the private sector.

M.S. students must complete 30 credit hours, including a 6-credit master's thesis or internship. The Ph.D. entails completion of 60 credit hours beyond the master's degree, including a 30-credit-hour dissertation. Field examinations are required in science studies, technology studies, or policy studies (two of the three areas) in addition to the student's own area of concentration.

Research Facilities

Research is supported by state-of-the-art facilities, such as the George M. Low Center for Industrial Innovation; the Rensselaer Libraries, with an electronic information system that provides access to collections; databases and Internet access from campus and remote terminals; the Rensselaer Computing System, which permeates the campus with a coherent array of advanced workstations; a shared to toolkit of applications for interactive learning and research and high-speed Internet connectivity; a visualization laboratory for scientific computation; and a high-performance computing facility that includes a 32-node SP2 parallel computer. In addition, the academic departments have extensive research capabilities and equipment. Rensselaer's extensive library resources are augmented by the large New York State Education Library and by the facilities of many other colleges located in the Troy-Albany-Schenectady area. The Institute's location in what was the Silicon Valley of the nineteenth century provides excellent primary sources for the history of technology, including the Hudson Mohawk Industrial Gateway, numerous local historical societies, and the well-documented role of Rensselaer.

Financial Aid

An NSF grant allows the department to fund three students each year, with research assistantships for design studies from 1999–2002. In addition, the department offers some teaching and research assistantships (a stipend up to $10,000 for the 1999–2000 academic year) and tuition for up to 19 credits per academic year for full-time students. Outstanding students may qualify for University-supported Rensselaer Scholar Fellowships. The STS department has an excellent record in terms of these awards. Low-interest, deferred-repayment graduate loans are also available to U.S. citizens with demonstrated need.

Cost of Study

Tuition for 1999–2000 is $665 per credit hour. Other fees amount to approximately $535 per semester. Books and supplies cost about $1700 per year.

Living and Housing Costs

The cost of rooms for single students in residence halls or apartments range from $3356 to $5298 for the 1999–2000 academic year. Family student housing, with a monthly rent of $592 to $720, is available.

Student Group

Rensselaer's student body is composed of 4,300 undergraduates and 1,750 graduate students who represent all fifty states and include approximately 1,000 international students who represent more than seventy-five countries.

Student Outcomes

Eighty-five percent of Rensselaer's 1998 graduate students were hired after graduation with starting salaries that averaged $56,259 for master's degree recipients and $57,000 to $75,000 for doctoral degree recipients.

Location

Rensselaer is situated on a scenic, 26-acre hillside campus in Troy, New York, across the Hudson River from the state capital of Albany. Troy's central northeast location provides students with a supportive, active, medium-sized community in which to live and an easy commute to Boston, New York, and Montreal and some of the country's finest outdoor recreation, including Lake George, Lake Placid, and the Adirondack, Catskill, Berkshire, and Green Mountains. The Capital Region has one of the largest concentrations of academic institutions in the United States. Sixty-thousand students attend fourteen area colleges and benefit from shared activities and courses.

The University

Founded in 1824 and the first U.S. college to award degrees in engineering and science, Rensselaer Polytechnic Institute today is accredited by the Middle States Association of Colleges and Schools and is a private, nonsectarian, coeducational university. Rensselaer has five schools—Architecture, Engineering, Management and Technology, Science, and Humanities and Social Sciences—and offers a total of ninety-four graduate degrees in forty-seven fields.

Applying

Admissions applications and all supporting credentials should be submitted well in advance of the preferred semester of entry to allow sufficient time for departmental review and processing. The application fee is $35. Since the first departmental awards are made in February and March for the next full academic year, applicants requesting financial aid are encouraged to submit all required credentials by February 1 to ensure consideration.

Correspondence and Information

For written information about graduate study:
Department of Science and Technology Studies
Rensselaer Polytechnic Institute
110 8th Street
Troy, New York 12180-3590
E-mail: vumbak@rpi.edu
World Wide Web: http://www.rpi.edu/dept/sts

For application and admissions information:
Graduate Academic and Enrollment Services
Rensselaer Polytechnic Institute
110 8th Street
Troy, New York 12180-3590
E-mail: grad-services@rpi.edu
World Wide Web: http://www.rpi.edu

Rensselaer Polytechnic Institute

THE FACULTY AND THEIR RESEARCH

Sharon Anderson-Gold, Ph.D. (philosophy), New School, 1980. Kantian ethics, history of modern philosophy, social and political philosophy, human rights, bioethics. Dr. Anderson-Gold is the author of *Crimes Against Humanity: A Kantian Perspective on International Law, Autonomy and Community, Kant's Ethical Anthropology and the Critical Foundations of Kant's Philosophy of History,* and *History of Philosophy Quarterly.*

Steve Breyman, Director, Ecological Economics, Values and Policy; Ph.D. (political science), California, Santa Barbara, 1992. Social movements, environmental politics and policy, Green parties, politics and theory, institutional greening, environment and development, environmental science. Dr. Breyman is the author of *Movement Genesis: The West German Peace Movement and Social Movement Theory* and *Why Movements Matter: The West German Movement, the SPD and INF Negotiations* (forthcoming).

Linnda R. Caporael, Ph.D. (psychology), California, Santa Barbara, 1979. Evolutionary theory, psychology and culture, social identity, design. Dr. Caporael is a contributor to *Science, Behavioral and Brain Sciences, Journal of Social Issues, Personality and Social Psychology Review,* and other journals and books.

Ron Eglash, Ph.D. (history of consciousness), California, Santa Cruz, 1992. African studies, anthropology, architecture, black history, complexity theory, cybernetics, math and science education, economic development, virtual communities. Dr. Eglash is the author of *African Fractals: Indigenous Design and Modern Computing.*

Kim Fortun, Ph.D. (anthropology), Rice, 1993. Cultural and political-economic analysis of globalization, environmentalism, and information technology; science, technology and law. Dr. Fortun is the author of *Advocating Bhopal: Environmentalism, Disaster, New World Orders.*

Michael Fortun, Ph.D. (history of science), Harvard, 1993. Historical and ethnographic studies of genomics, biotechnology and life sciences, ethical scientific literacy. Dr. Fortun is co-author of *Muddling Through: Pursuing Science and Truths in the 21st Century.*

David Hess, Director of Undergraduate Studies; Ph.D. (anthropology), Cornell, 1987. Medical anthropology, science studies, Brazil, alternative medicine, cancer therapy. Dr. Hess is the author of *Can Bacteria Cause Cancer?, Science and Technology in a Multicultural World, Science Studies in the New Age,* and *Samba in the Night: Spiritism in Brazil.*

Linda Layne, Hale Professor and Director of Graduate Studies; Ph.D. (anthropology), Princeton, 1986. Pregnancy loss, new reproductive technologies, consumer culture, popular representations of nature, universal design. Dr. Layne is the author of *Home and Homeland: Dialogies of Tribal and National Identities in Jordan* and *Motherhood Lost: The Cultural Construction of Pregnancy Loss in the United States* and is editor of *The Rhetoric of the Gift: Transformative Motherhood in a Consumer Culture* and *Anthropological Approaches in STS.*

Andrea Rusnock, Ph.D. (history), Princeton, 1990. History of medicine, history of quantification, gender, science and technology. Dr. Rusnock is the author of *The Correspondence of James Jurin (1684–1750), Physician and Secretary to the Royal Society.*

Sal Restivo, Ph.D. (sociology), Michigan State, 1971. Social studies of science, mathematics and mind. Dr. Restivo is the author of *The Social Relations of Physics, Mysticism and Mathematics; The Sociological Worldview; Mathematics in Society and History;* and *Science, Society, and Values* and is co-editor of *Comparative Studies in Science and Society, Math Worlds,* and *Degrees of Compromise.*

John Schumacher, Chair; D.Phil (philosophy), Oxford, 1974. Nature of inquiry, design studies, environmental philosophy, theories of knowledge and metaphysics, gender studies, philosophy of mental health and education. Dr. Schumacher is the author of *Human Posture: The Nature of Inquiry* and *Science, Local Knowledge, and Community* in *Knowledge and Society.*

Langdon Winner, Ph.D. (political science), Berkeley 1973. Political theory, technology and politics, social dimensions of design. Dr. Winner is the author of *Autonomous Technology* and *The Whale and the Reactor* and editor of *Democracy in a Technological Society.* He is currently writing a book about the politics of design in the contexts of engineering, architecture, and political theory.

Edward Woodhouse, Ph.D. (political science), Yale, 1983. Democratic steering of technologies, risky technologies, nuclear power. Dr. Woodhouse is co-author of *The Demise of Nuclear Energy, The Policy Making Process,* and *Averting Catastrophe: Strategies for Regulating Risky Technologies.*

ADJUNCT FACULTY

P. Thomas Carroll, Executive Director of the Hudson Mohawk Industrial Gateway; Ph.D., Pennsylvania, 1982, history and sociology of science. American cultural history, history of American science and technology; science, technology, community, and cultural transformations. Dr. Carroll is the author of *American Science Transformed, American Scientist, An Annotated Calendar of the Letters of Charles Darwin, Designing Modern America in the Silicon Valley of the Nineteenth Century,* and *Rensselaer* and is co-author of *Chemistry in America 1876–1976.*

Richard Jensen, Ph.D., Yale, 1966, American studies. American political history, especially presidential politics and popular voting behavior. Dr. Jensen is author of *The Winning of the Midwest: Social and Political Conflict 1888–1896, The Evolution of American Electoral Systems,* and *Grass Roots Politics: Parties, Issues, and Voters, 1954–1983.*

Thomas Lobe, Ph.D., Michigan, 1975, political science. U.S. foreign policy, the Cold War, and the intelligence community. Dr. Lobe is the author of *United States National Security Policy and Aid to the Thailand Police* and articles on Nicaragua and the Office of Public Safety.

Thomas Phelan, Institute Historian; M.A., College of Holy Cross, 1945, English. Nineteenth century American material culture, particularly as it relates to industrialization. He is the author of *Rensselaer: An Illustrated History of Rensselaer Polytechnic Institute* and *The Hudson Mohawk Gateway.*

Jesse Tatum, Ph.D., California, Berkeley, 1988, energy and resources. Technology and ways of being in the world; democratic shaping of technology. Dr. Tatum is the author of *Energy Possibilities: Rethinking Alternatives and the Choice Making Process,* and *Muted Voices: The Recovery of Democracy in the Shaping of Technology.*

POSTDOCTORAL FELLOW IN DESIGN STUDIES (1999–2000)

Atsushi Akera, Ph.D., Pennsylvania, 1998, history and sociology of science. Early history of scientific and technical computing in the United States, history of invention and innovation. Dr. Akera is the author of *Engineers or Managers? The Systems Analysis of Electronic Data Processing in the Federal Bureaucracy,* eds. Hughes and Hughes; *The Spread of the Systems Approach; Discovering the Scientific User: Cuthbert Hurd, An Applied Science Field Man;* and *Proceedings of the Conference on the History of Computing and Information Processing.*

RENSSELAER POLYTECHNIC INSTITUTE

Lally School of Management and Technology

Programs of Study	The Lally School of Management and Technology is focused on the intersection of management and technology and built upon the conviction that for all firms in all future markets, sustainable competitive advantage will be built upon a technological foundation. The School's mission is to educate a new breed of managers who are prepared to lead their companies in the effective and strategic use of technology. The Lally School offers M.B.A., M.S., and Ph.D. degree programs in management and has an Executive M.B.A. program and an international exchange program. The School's degree programs are accredited by AACSB–The International Association for Management Education.
	The Management and Technology M.B.A. (M&T M.B.A.) Program is the flagship program of Rensselaer's Lally School of Management and Technology. Course modules are offered in a stream format that provides integration across disciplines and between traditional management and technical areas. The 60-credit-hour M&T M.B.A. is designed to achieve a dual purpose: the development of technical managers who understand and are able to perform effectively in general management functions, and the development of general managers who understand and are able to interact effectively within the technological environment. Approximately 20 percent of full-time students pursue a dual degree, linking their management studies with graduate work in technical fields and completing both degrees in a reduced amount of time. The Master of Science is a 30-credit-hour program leading to a specified degree in a field and is designed for those who have professional experience.
	The M&T M.B.A. and M.S. programs are offered to both full-time and part-time students. All students are required to complete a core curriculum and may choose a concentration. Concentration areas include management information systems, manufacturing systems, manufacturing management, product development and management, technological entrepreneurship, research and development management, environmental management and policy, and financial technology.
	The three-year Ph.D. program is designed for students with superior abilities and a technological orientation who wish to pursue careers as educators, researchers, or professional specialists. Interdisciplinary graduate programs include those leading to the Ph.D. degree in decision sciences and engineering systems and to M.S. degrees in operations research and statistics, manufacturing systems engineering, industrial and management engineering, and environmental management and policy. An M.B.A./J.D. program is also offered in cooperation with Albany Law School of Union University. A new, state-of-the-art facility in Troy, New York, serves as the home of the Lally School. This fully networked building contains computer-interactive, videoconferencing, and distance education classrooms. A second campus is located in Hartford, Connecticut. M.B.A. and M.S. degrees are offered at this campus.
Research Facilities	The Lally School of Management and Technology is linked with the Design and Manufacturing Institute (searches out solutions to productivity problems), the Paul J. '69 and Kathleen M. Severino Center for Technological Entrepreneurship (provides focus for research in new ventures), the Center for Services Research and Education (applies information and decision technologies to improve productivity in the service sector), and the Center for the Study of Financial Technology (financial engineering, the impact of information technology on financial markets, and entrepreneurial finance). Research is supported by such state-of-the-art facilities as the Rensselaer Libraries, whose library systems allow access to collections, databases, and resources via the Internet. A computer network permeates the campus with a coherent array of advanced workstations, a shared toolkit of applications for interactive learning and research, and high-speed Internet connectivity. The Lally School's new technologically adaptive building serves to integrate technology into the curriculum.
Financial Aid	Most support is in the form of tuition scholarships and research or teaching assistantships. Outstanding students may qualify for Rensselaer Scholar Fellowships ($15,000 plus full waiver of tuition and fees) or other specially designated fellowships (stipends up to $10,000 plus tuition). Low-interest, deferred-repayment graduate loans are also available to U.S. citizens with demonstrated need.
Cost of Study	Tuition for 1999–2000 is $665 per credit hour. Other fees amount to approximately $535 per semester. Books and supplies cost about $1700 per year.
Living and Housing Costs	The cost of a room for a single student in a residence hall or apartment ranges from $3356 to $5298 for the 1999–2000 academic year. Family student housing, with a monthly rent of $592 to $720, is available.
Student Group	Enrollment is about 4,300 undergraduates and 1,750 graduate students. The Lally School has approximately 400 master's and doctoral students with a wide range of academic and work backgrounds.
Student Outcomes	Ninety percent of Rensselaer's full-time M.B.A. graduate students were hired within three months of graduation, with starting salaries that averaged $56,259 for master's degree recipients.
Location	Troy, Albany, and Schenectady form an upstate metropolitan area with a population of approximately 750,000. The area is a major center of government, industrial, research, and academic activity. Within easy driving distance are the headquarters or major research centers of some of the world's largest technology-based firms.
The University	Rensselaer is eminently qualified to merge the disciplines of technology and management. One of America's first technological universities, Rensselaer was established in 1824. Rensselaer has five schools—Architecture, Engineering, Management, Science, and Humanities and Social Sciences—that offer a total of ninety-eight graduate degrees in forty-seven fields.
Applying	Online and paper applications should be received approximately three to six months before the start of the desired term. The application fee is $35 (paper) or $50 (online). Students should contact the School for application deadlines for the Ph.D. and Executive M.B.A. programs. Admissions decisions are based on academic performance, GMAT scores, references, proven leadership qualities, and employment history. A good mathematics background and basic computer skills are helpful. Online applications are available on the Web through http://www.GradAdvantage.org or http://www.CollegeEdge.com.
Correspondence and Information	For written information about graduate study:

For written information about graduate study:

Manager of Enrollment Services
Lally School of Management and Technology
Rensselaer Polytechnic Institute
110 8th Street
Troy, New York 12180-3590

Telephone: 518-276-6586
Fax: 518-276-2665
E-mail: management@rpi.edu
World Wide Web: http://lallyschool.rpi.edu

For applications and admissions information:

Director of Graduate Academic and Enrollment
  Services, Graduate Center
Rensselaer Polytechnic Institute
110 8th Street
Troy, New York 12180-3590

Telephone: 518-276-6789
E-mail: grad-services@rpi.edu
World Wide Web: http://www.rpi.edu

Rensselaer Polytechnic Institute

THE FACULTY AND THEIR RESEARCH

Joseph G. Ecker, Dean; Ph.D., Michigan. Mathematical programming, operations research.
Richard P. LeMay, Associate Dean; Ph.D., Iowa. Management engineering, production/operations management.
Gene R. Simons, Associate Dean; Ph.D., Rensselaer. Industrial and management engineering, production and operations management, project planning and control, manufacturing systems.

Professors—Troy
Robert A. Baron, Ph.D., Iowa. Organizational behavior, entrepreneurship.
Daniel Berg, Institute Professor of Science and Technology; Ph.D., Yale. Management of technological organizations, policy issues of research and development in the service sector.
Joseph G. Ecker, Ph.D., Michigan. Mathematical program, operations research.
Jorgé Haddock, Ph.D., Purdue. Modeling of production and service systems, including simulation and optimization techniques.
G. Judd, Ph.D., Rensselaer. Management of higher education organizations, strategy and change.
J. R. Norsworthy, Ph.D., Virginia. Economics, business economics.
Albert S. Paulson, Frank and Lillian Gilbreth Professor in the Technologies of Management; Ph.D., Virginia Tech. Operations research and statistics, risk management and investment analysis.
Gene R. Simons, Ph.D., Rensselaer. Industrial and management engineering, production and operations management, project planning and control, manufacturing systems.

Clinical Professors—Troy
Pier A. Abetti, Ph.D., IIT; PE. Management of technology, international business development and strategic planning, entrepreneurship.
William Stitt, M.B.A., Harvard. Entrepreneurship.
Shubo Xu, Director, Sino-U.S. Management and Technology M.B.A. Program; Ph.D., Tianjin (China). Decision sciences, quantitative methods.

Associate Professors—Troy
Wolfgang Bessler, Ph.D., Hamburg (Germany). Financial management and institutions, international finance.
Jeffrey F. Durgee, Ph.D., Pittsburgh. Marketing research and advertising.
David H. Goldenberg, Ph.D., Florida. Corporation finance and investments.
Richard Leifer, Ph.D., Wisconsin. Organizational behavior and organizational design, management information systems.
Richard P. LeMay, Ph.D., Iowa. Management engineering, production/operations management.
Lois S. Peters, Ph.D., NYU. Science and technology policy.
Phillip Phan, Warren H. Bruggeman '46 and Pauline Urban Chair in Entrepreneurship; Ph.D., Washington (Seattle). Entrepreneurship, business policy.
Bruce Piasecki, Ph.D., Cornell. Environmental management and policy, hazardous-waste management, analysis of environmental executive decisions.
Susan S. Sanderson, Ph.D., Pittsburgh. International business, manufacturing policy, new product development.

Clinical Associate Professor—Troy
Ralph Miccio, J.D., Albany Law. Law, ethics.
Thomas T. Triscari, Ph.D., Rensselaer. Information systems.

Assistant Professors—Troy
Richard Burke, Director, Master's Programs; Ph.D., Massachusetts Amherst. Statistics, operations research, quality management.
David Hollingworth, Ph.D., Ohio State. Operations management.
Moren Levesque, Ph.D., British Columbia. Operations research, entrepreneurship.
Gideon Markman, Ph.D., Colorado. Entrepreneurship.
Christopher McDermott, Ph.D., North Carolina at Chapel Hill. Manufacturing strategy, operations management.
Satish Nambisan, Ph.D., Syracuse. Information systems.
Gina O'Connor, Ph.D., NYU. Marketing, product management.
Thiagarajan Ravichandran, Ph.D., Southern Illinois at Carbondale. Management information systems.
Mark P. Rice, Director, Severino Center for Technological Entrepreneurship; Ph.D., Rensselaer. Entrepreneurship, new ventures.
Shikhar Sarin, Robert and Irene Bozzone Professorship of Management and Technology; Ph.D., Texas at Austin. New product development, high-technology marketing.
Robert Veryzer, Ph.D., Florida. Marketing and consumer behavior.

Clinical Assistant Professors—Troy
Robert Boylan, Ph.D., Duke. Accounting and economics.
Irvin Morgan, M.B.A., Chicago. Finance.
William St. John, Ph.D., Rensselaer. Accounting, finance.
Robert Sands, M.B.A., SUNY at Albany. Organizational behavior and human resources management.
Kathy Silvester, Ph.D., Maryland. Accounting.

Adjunct Faculty—Troy
Judith A. Barnes, Ph.D., Rensselaer. Communication.
Zenas Block, B.S., CUNY, City College. Clinical professor of management at New York University, Stern School of Business; corporate entrepreneurship.
Glenn Doell, M.B.A., Rensselaer. Entrepreneurship.
Michael Hurley, Ph.D., Rensselaer. Human resource management, organizational behavior.
Jules Jacquin, M.B.A., Rensselaer. Accounting information systems, auditing.
Hugh Johnson, A.B., Dartmouth. Financial markets and analysis.
Jerry Mahone, B.A., Howard. Entrepreneurship.
K. Gary McClure, Ph.D., Central Florida. Finance.
David B. Mitchell, J.D., Vermont Law, M.B.A., Rensselaer. External evaluation of business, acounting, finance.
James Murtagh, M.B.A., Northern Colorado. Finance.
Samuel Rabino, Ph.D., NYU. Marketing, international business.
Charles Rancourt, M.S., Rensselaer. Statistics.
Peter Skinner, M.E., Rensselaer. Management of environmental technology.
Steven Walsh, Ph.D., Rensselaer. Strategy, management of technology.
Frank Wright, M.S.E.E., Naval Postgraduate School. General management, manufacturing operations, international business.

For a list of faculty members from Rensselaer at Hartford, students should write to the address on the front of this page.

UNIVERSITY OF MINNESOTA

Program in Science, Technology, and Environmental Policy

Programs of Study	The Humphrey Institute offers a Master of Science (M.S.) in science, technology, and environmental policy. Advances in science, as well as technological development, affect the choices made in a wide range of public policies. However, government investment in and regulation of science and technology affect its direction and health. The objective of the Master of Science degree in science, technology , and environmental policy is to provide students with an understanding of the role of science and technology in the economy, in food production and health, in energy and the environment, in security, and in education; the impact of science and technology on the political and economic relationships among nations; and the design of policies for appropriate promotion and regulation of science and technology regionally, nationally, and internationally.
	M.S. students are expected to have completed an undergraduate degree program with a major in one of the natural or engineering sciences or to have taken courses beyond the introductory level in at least one natural or engineering science. Required and elective courses in policy analysis provide students with the social science tools necessary to work effectively on public policy issues. Other requirements and electives directly address the policy implications of questions related to science, technology, and the environment.
	In addition to the M.S. program, the Humphrey Institute offers a Master of Public Policy (M.P.P.), a Master of Urban and Regional Planning (M.U.R.P.), and an Executive Master of Public Affairs (M.P.A.). These programs prepare students for positions in public, nonprofit, and private organizations as planners, managers, policy analysts, advocates, and organizers.
Research Facilities	With a collection of 4 million volumes, the University of Minnesota's library system is one of the largest American university libraries. Wilson Library, the main campus library, is located adjacent to the Humphrey Institute. The University maintains exceptional computer facilities, and the Institute has its own microcomputer lab for student use.
	Students may participate in the Humphrey Institute's policy research, analysis, and community service projects. Projects are varied and have included Population Analysis and Policy, the Environmental Training Project, and the Humphrey Institute Policy Forum.
Financial Aid	The Institute allocates financial aid primarily on the basis of academic merit. Sources of aid include Hubert H. Humphrey fellowships and scholarships, University of Minnesota Graduate School fellowships, and teaching and research assistantships. Amounts range from $5000 to $20,000 per year.
Cost of Study	Tuition is $484 per semester for Minnesota residents and $848 for nonresidents. For those students taking 6–15 credits, the cost is $2905 per semester for residents and $5091 for nonresidents. Tuition for the Executive M.P.A. program is $440 per credit for residents and $770 per credit for nonresidents. Residents of North Dakota, South Dakota, Wisconsin, and Manitoba may be eligible for resident tuition rates if they apply to their Higher Education Coordinating Board prior to enrollment each academic year.
Living and Housing Costs	Living costs in the Minneapolis–St. Paul region vary with such factors as marital status, number of dependents, lifestyle, and locational preferences. Students interested in local rental rates should contact the Housing Office, Comstock Hall East, 210 Delaware Street, SE (telephone: 612-624-2994; Web site: http://www.umn.edu/housing).
Student Group	The Humphrey Institute has 180 graduate students enrolled in its M.S., M.P.P., and M.U.R.P. degree programs. About half are women, 10 percent are international students, and 15 percent are from minority groups. Many students have had previous work experience. About 40 percent come from Minnesota; the others come from twenty-five states and five other countries.
Student Outcomes	Humphrey Institute graduates enter widely varied careers in public service. Graduates work in local, state, and federal government and in nonprofit and private firms.
Location	Minnesota and the Minneapolis–St. Paul area are known as places where government is responsive to the wishes of a responsible citizenry, where businesses foster a high level of entrepreneurship and innovation, and where corporations have made major investments in the social and cultural life of their communities. The Twin Cities are rich in cultural and recreational opportunities. National surveys consistently rank the Minneapolis–St. Paul area as one of the most desirable places in the country to live and work.
The Institute	The Humphrey Institute is housed in the Humphrey Center, built in 1985. The location within one of the nation's great universities gives students access to a wide variety of courses and programs and offers them the chance to work with faculty members who have international reputations as scholars, researchers, and professional practitioners. The Institute was founded in 1977 as a tribute to Vice President and Senator Hubert H. Humphrey. As the direct descendant of the University's pioneering Public Administration Center (1936–68) and distinguished School of Public Affairs (1968–77), the Humphrey Institute represents more than half a century of community service and academic achievement.
Applying	The Institute accepts students for admission in the fall semester only, and early application is encouraged. Students must complete their applications by January 15 to receive full consideration for fall admission and financial aid or by April 1 for admission only. Applications received after April 1 are reviewed on a space-available basis.
	Admission to the M.S. program is based on each applicant's prior scholastic achievement, statement of purpose, and letters of recommendation. The GRE General Test is required, and TOEFL scores are required of students whose first language is not English. Applications are especially encouraged from women and from members of minority groups.
Correspondence and Information	Admissions Office Hubert H. Humphrey Institute of Public Affairs 225 HHH Center University of Minnesota 301 19th Avenue, South Minneapolis, Minnesota 55455 Telephone: 612-626-8909 E-mail: admissions@hhh.umn.edu World Wide Web: http://www.hhh.umn.edu

University of Minnesota

THE FACULTY AND THEIR RESEARCH

John S. Adams, Ph.D., Professor. Housing policy, housing finance, American cities, regional economic development policy.

Sandra O. Archibald, Ph.D., Professor. Social costs of technology, design of effective environmental policy.

Sheila Ards, Ph.D., Assistant Professor. Child welfare, family policy.

Ragui Assaad, Ph.D., Associate Professor. International economic development, labor market analysis.

John E. Brandl, Ph.D., Dean and Professor. Social policy, state finance, education policy.

Geraldine Brookins, Ph.D., Professor. Parental socialization patterns, sex-role acquisition and stereotyping, dual-career families and their children, developmental psychopathology.

John M. Bryson, Ph.D., Professor. Strategic planning; project planning, coordination, implementation, and evaluation.

Nancy N. Eustis, Ph.D., Associate Dean for Resident Instruction and Professor. Aging and disability policy, long-term care, evaluation research.

Edward Goetz, Ph.D., Associate Professor. Housing policy, community development, and politics of urban and regional planning.

Maria Hanratty, Ph.D., Associate Professor. Health economics, the economics of poverty, comparative social welfare institutions.

Stephen A. Hoenack, Ph.D., Professor. Economic behavior within organizations, econometrics, higher education policy.

Ethan Kapstein, Ph.D., Stassen Professor in International Peace. Global political economy, defense conversion in Eastern Europe, security, world labor markets.

Kenneth H. Keller, Ph.D., Professor. Science policy, global technology development, technology in health care.

Sally J. Kenney, Ph.D., Associate Professor and Co-Director, Center on Women and Public Policy. Women and politics, women and law, feminist theory, British politics.

Morris M. Kleiner, Ph.D., Professor and AFL-CIO Chair Professor of Labor Policy. Role of labor unions, human resource policies, economic impact of demographic change.

Kenneth Kriz, Ph.D., Assistant Professor. Public and nonprofit management, public finance and policy analysis.

Robert T. Kudrle, Ph.D., Professor. Industrial organization, international economic policy, health-care policy.

Deborah Levison, Ph.D., Assistant Professor. Poverty, child labor, child care, and women's work in developing countries.

Ann Markusen, Ph.D., Professor. Regional economics, public finance, industrial organization.

Samuel L. Myers Jr., Ph.D., Roy Wilkins Chair Professor in Human Relations and Social Justice. Race relations and human rights, impacts of social policies on the poor.

G. Edward Schuh, Ph.D., Orville and Jane Freeman Professor for International Trade and Investment Policy. Agricultural and food policy, economic development, international trade.

Melissa Stone, Ph.D., Associate Professor. Nonprofit management, state-nonprofit relationships.

Affiliated Faculty from other University of Minnesota Departments

Rutherford Aris, Ph.D., Regents Professor, Chemical Engineering and Materials Science.

Philip Regal, Ph.D., Professor and Curator, Bell Museum of Natural History.

Vernon Ruttan, Ph.D., Regents Professor, Agricultural and Applied Economics.

Yechiel Shulman, Ph.D., Professor, Mechanical Engineering and Director, Center for Development of Technological Leadership.

Senior Fellows

Zbigniew Bochniarz, Ph.D., Director, Central and Eastern European Training Project and Center for Nations in Transition.

Harry Boyte, Ph.D., Director, Project Public Life and Co-director, Center for Comparative Democracy and Citizenship.

Barbara Crosby, Ph.D. Leadership and public policy, women in leadership, media and public policy, strategic planning, leadership in transnational contexts.

William A. Diaz, Ph.D., Director, Public Policy, Philanthropy, and the Nonprofit Sector Program.

Marsha Freeman, J.D., Ph.D., Director, International Women's Rights Action Watch. Domestic and international issues of women, law and public policy.

Lee Munnich, M.A., Director, State and Local Policy Program.

Joe Nathan, Ph.D., Director, Center for School Change. Education reform and economic development.

Tim Penny, Senior Fellow and Co-Director, Humphrey Institute Policy Forum. Former six-term U.S. congressional representative from southeastern Minnesota.

Melor Sturua, Ph.D., Carlson International Lecturer in Public Affairs. Former Soviet journalist and adviser to Soviet leaders Nikita Khrushchev and Leonid Brezhnev.

Sharon Tolbert-Glover, Ph.D., President of the Communities of Color Institute for Leadership and Organizational Development (CCI) and its Pathways Collaborative.

Vin Weber, Senior Fellow and Co-Director, Humphrey Institute Policy Forum. Former six-term congressman from Minnesota.

UNIVERSITY OF PENNSYLVANIA

School of Engineering and Applied Science
Executive Master's in Technology Management Program

Programs of Study	The Executive Master's in Technology Management (EMTM) is designed to maximize the career potential of technology managers by bringing them into the forefront of managerial leadership and entrepreneurial thinking. In today's business world, change is rapid; global competition is fierce, and successful integration of technology is key to success. This program's dual emphasis on emerging technologies and management principles prepares engineers and scientists to respond to these dynamic market forces.
	The first program of this type in the nation, EMTM continues to be the only program that incorporates a multiplicity of technologies.
	In 1998, the renowned Wharton School of Business joined the School of Engineering and Applied Science as a cosponsor of this program. At the same time, the program's name was changed from Executive Master of Science in Engineering (ExMSE) to Executive Master's in Technology Management. Graduates receive a Master of Science in Engineering (M.S.E.) in the management of technology (MOT).
	This program's strong, integrated curriculum has four key components: courses in state-of-the-art technology, business and economics courses with real-world applications, a thought-provoking seminar series, and extensive student interaction.Technical courses help participants acquire expertise in advanced and emerging technologies. Business and economics courses develop skills for managing technology-based enterprises and translating technological innovation into business success. The seminar series, which explores the current state of practice in the management of technology and the vision driving that practice, keeps the program fresh by continually giving rise to new courses.
	Extensive interaction with the program's diverse base of mature professionals helps students develop an ongoing business network, leadership skills, and the foundation for good teamwork.
	Because the program is geared toward full-time professionals, classes are held every second Friday and/or Saturday from September through May, allowing students to continue their careers without disruption. The duration is two years for students who attend Fridays and Saturdays (full-time) and four years for those who attend on Fridays or Saturdays only (half-time). Students can change from full-time to half-time and vice versa.
	The large variety of courses students take—twenty courses (10 credits) are required, and more may be taken—give students a broad exposure to business and technology. Required courses include accounting, engineering economics (managerial economics), finance, management information systems, management of technology, marketing strategies, operations management, organizational behavior and design, product design and development, statistics, and systems.
	Technology electives, focusing on the development and application of the latest technologies, include advanced materials, computer visualization, computational mathematics, expert systems and neural networks, microelectronics, modern biotechnology, photonics, robotics and automation, software development, and telecommunications (introduction to networking, advanced networking, and technology and competitive strategy).
	Other electives, concentrating on the business and economic factors that impact strategic decisions, include business policy, concurrent and simultaneous engineering, corporate ethics, environmental management, foundations of teamwork and leadership, intellectual property, logistics, negotiations, strategic management of innovation, technology and public policy, technology entrepreneurship, and technology entrepreneurship project.
Research Facilities	Penn Engineering has extensive computer, library, and research facilities. The Engineering Library maintains 122,000 volumes and more than 850 journal subscriptions and provides access to key engineering databases, such as Compendex. At the library, students with laptops can connect to PennNet, Penn's computer network, which houses Penn's nationally recognized Digital Library on the World Wide Web. Throughout the school, both Macintosh and IBM-compatible computers are available for student use.
Financial Aid	Though financial aid is not available, students can apply for a variety of both subsidized and nonsubsidized loans with the assistance of University Student Financial Services. Also, students are encouraged to check on their company's plan for tuition reimbursement. Many companies cover all or part of the cost of the program.
Cost of Study	For 1999–2000, the annual cost for the two-year program (Fridays and Saturdays) is $36,540 and for the four-year program (Fridays or Saturdays only) is $18,357. The tuition is subject to annual increases. This comprehensive fee includes tuition, hotel accommodations, assigned texts, and reference materials.
Living and Housing Costs	On Thursday and Friday nights of course weekends, students stay at an on-campus hotel so that they can completely immerse themselves in the program. In the evenings, students work together on team projects and, often, even consult one another on work-related problems. Staying on campus also provides convenient access to campus resources. The program fee includes the cost of accommodations.
Student Group	Students benefit from the opportunity to interact with the program's diverse base of mature professionals. Currently, the program has 194 students and 220 alumni from 180 companies and agencies in sixteen different industries. Students hold full-time positions in engineering, information systems, operations, research and development, and technical marketing. Typically, students have been out of school for five to fifteen years, are between 28 to 38 years old, and are ready to climb the technology management ladder.
Location	The University is located in West Philadelphia, which is close to the heart of the city. Philadelphia's renowned museums, concert halls, theaters, sports arenas, and parks provide students with rich cultural and recreational outlets.
The School	The School of Engineering, which since its founding in 1860 has fostered an interdisciplinary approach to both research and training, originated this unique program. The program's vision represents the ideals of the University's founder, Ben Franklin, who was an entrepreneur, technologist, politician, and leader. The cosponsor of EMTM, the Wharton School of Business, is recognized around the world for its innovative leadership and broad academic strengths across every major discipline and at every level of management education.
Applying	Candidates should be fast-track professionals with the intellectual and personal characteristics required to succeed as creative leaders in the management of technology. Along with strong academic and industrial performance records, candidates should have a minimum of three to five years of training and experience in an engineering, math, or science-related profession. Students with other professional experience and the appropriate technical background are also considered. The deadline for September admissions is June 1. GRE or GMAT scores are not required but may be submitted, if the applicant feels that they strengthen his/her application.
Correspondence and Information	Dr. Joel Adler EMTM Office School of Engineering and Applied Science 119 Towne Building University of Pennsylvania Philadelphia, Pennsylvania 19104-6391 Telephone: 215-898-2897 Fax: 215-573-9673 E-mail: emtm@seas.upenn.edu World Wide Web: http://www.seas.upenn.edu/emtm

University of Pennsylvania

THE FACULTY AND THEIR RESEARCH

Director and Associate Director

Joel Adler, Adjunct Professor of Systems Engineering and Associate Director, Executive Master's in Technology Management; Ph.D., Pennsylvania (Wharton), 1980. Forecasting, distribution optimization, applications generators, logistics operations and manufacturing systems.

Lyle Ungar, Associate Professor of Computer and Information Science (with secondary appointments in Chemical Engineering, Systems Engineering, and in the Telecommunications and Networking Program) and Director, Executive Master's in Technology Management; Ph.D., MIT, 1984. Knowledge-based systems, qualitative physics, machine learning, real and artificial neural networks.

School of Engineering and Applied Science (SEAS)

G. Anandalingam, Professor and Chairman of Systems Engineering; Ph.D., Cambridge, 1975. Network optimization, telecommunications systems design, technology and public policy, location analysis and multi-criteria decision-making issues.

Norman L. Badler, Professor and Computer and Information Sciences and Director of the Center for Human Modeling and Simulation; Ph.D., Toronto, 1975. Computer graphics: human figure modeling, manipulation and animation.

Haim Bau, Professor of Mechanical Engineering and Applied Mechanics and of Electrical Engineering; Ph.D., Cornell, 1980. Bifurcation and instability phenomena in flows and feedback control strategies that can alter them.

Magda El Zarki, Associate Professor of Electrical Engineering and of Computer and Information Science and Director of the Telecommunications and Networking Program; Ph.D., Columbia. Packet video (audio/video synchronization, error concealment schemes and quality of service measures), wireless networks.

Carl Gunter, Associate Professor, Computer and Information Science; Ph.D., Wisconsin–Madison, 1985. Programming languages, mathematical logic and computing, software specifications and standards, computer software law related to contracts and intellectual property.

Daniel A. Hammer, Professor of Chemical Engineering and of Bioengineering and Director, Master of Biotechnology Program; Ph.D., Pennsylvania, 1987. Cellular bioengineering—cell adhesion, and separation and molecular mechanisms of viral-cell infection.

Dwight Jaggard, Professor of Electrical Engineering and Associate Dean for Graduate Education and Research; Ph.D., Caltech, 1976. Electromagnetic chirality, fractal electrodynamics, scattering from knots, imaging and inverse scattering.

John Keenan, Professor of Civil Engineering and Director of Faculty Advising for Undergraduate Education, SEAS; Ph.D., Syracuse. Biological and health effects of pollution, wastewater treatment, water resources engineering, environmental impact assessment, hazardous materials in the environment.

Vijay Kumar, Professor of Mechanical Engineering and Applied Mechanics and of Computer and Information Science; Ph.D., Ohio State, 1987. Robotics, legged locomotion, rapid prototyping.

Solomon Pollack, Professor of Bioengineering, of Orthopedic Surgery Research, and of Nursing; Ph.D, Pennsylvania, 1961. Bioelectrical properties of bone and connective tissue, electrical stimulation of bone growth.

Jonathan M. Smith, Associate Professor of Computer and Information Science; Ph.D., Columbia, 1989. Implementation of 1-Gbps WANs.

Val Tannen, Associate Professor of Computer and Information Science; Ph.D., MIT, 1987. Programming languages, databases and logic in computer science.

Jan Van der Spiegel, Professor of Electrical Engineering and Director, Center for Sensor Technologies; Ph.D., Leuven (Belgium), 1974. Integrated circuit technology and materials, solid-state image sensors, low noise/low power analog integrated circuits for particle detectors, sensory neural processing systems.

Wayne Worrell, Professor of Materials Science and Engineering; Ph.D., MIT, 1963. Properties and novel applications of mixed-conducting oxides, prevention of environmental degradation of advanced materials at elevated temperatures, solid-state electrochemistry at elevated temperatures.

Iraj Zandi, Emeritus Professor of Systems and Chair of the National Center; Ph.D., Georgia Tech, 1959. Systems approach, systems management, applications of systems thinking to environmental/resources issues, intelligent systems and methodologies.

The Wharton School of Business

W. Bruce Allen, Professor of Public Policy and Management; Vice Dean, Graduate Division; and Director, Wharton Transportation Program; Ph.D., Northwestern, 1969. Freight demand theory, impact of transit investments, physical distribution management/business logistics, railroad and motor carrier economics, transportation regulation.

Stanley Baiman, Professor of Accounting and Chairperson, Accounting Department; Ph.D., Stanford, 1974. Role of accounting information for contracting, decision-making within firms.

David J. Ellison, Assistant Professor in Operations Management; Ph.D., Harvard, 1986. New product development, operations strategy, productivity and technology development.

Stewart Friedman, Adjunct Professor of Management and Director of the Wharton Leadership Program; Ph.D., Michigan, 1984. Work-life integration, leadership development, work teams and the dynamics of individual and organizational change.

Anjani Jain, Associate Director, Wharton Graduate Division; Ph.D., California, 1987. Probabilistic analysis of combinatorial optimization problems, analysis of manufacturing systems, product variety and manufacturing flexibility.

Karen A. Jehn, Associate Professor of Management; Ph.D., Northwestern, 1992. Intragroup and intergroup conflict; norms and values; group goal setting; cross-industry and cross-national comparisons of values, beliefs, goals, and conflict styles; lying and deceit in organizations.

Steven Kimbrough, Associate Professor of Operations and Information Management; Ph.D, Wisconsin, 1982. Decision support systems, electronic commerce, computational theories of rationality, evolutionary computation (genetic algorithms and genetic programming) and the application of formal logic.

Johannes Pennings, Professor of Management; Ph.D., Michigan, 1973. Organizational design, executive compensation and innovation.

John Percival, Adjunct Associate Professor of Finance; Ph.D., SUNY. Corporate finance, the bond market, regulated industries.

David Schmittlein, Professor of Marketing and Chairperson of the Marketing Department; Ph.D., Columbia, 1980. Methods for market segmentation, approaches for forecasting the impact of a firm's marketing actions, designing market research for use in legal cases.

Keith Weigelt, Class of 1965 Endowed Term Associate Professor of Management; Ph.D., Northwestern, 1986. Game theory, managerial compensation, experimental economics, economics of sports, decision making.

Adjunct Faculty

Joel DeLuca, Consultant in Strategic Planning, Leadership Development, and Executive Coaching; Ph.D., Yale, 1981.

Robin Derry, Consultant and Partner, The Rangeley Group (consulting on ethical management practices); Ph.D., Massachusetts, 1987. Ethical decision making of women and men, work and family balance, the influence of reward systems on ethical behavior at work.

Alvin Lehnerd, Consultant to industry with the Shadowstone Group. New product development, including the DustbusterTM, the first electronic and global iron, orthopedic surgical tools, and a complete redesign of Black & Decker double-insulated power tools.

Walter Lyon, Adjunct Professor of Systems Engineering and Engineering Consultant, Ph.D, Johns Hopkins. Served as Deputy Secretary for Planning of the Pennsylvania Department of Environmental Resources; drafted the State's Environmental Master Plan and the State's Recreation Plan; and worked on federal environmental and resources legislation and policy analysis.

Victor McCrary, Technical Manager, Information Technology Laboratory, National Institute of Standards and Technology; Ph.D., Howard, 1985. Manages research group whose projects include advanced display systems, high-performance storage systems, and optical information processing and transmission.

Keith Ross, Professor of Multimedia Communications, Institute Eurecom, France; Associate Professor of Operations and Information Management (secondary appointment), The Wharton School; and Adjunct Professor of Systems Engineering, SEAS, University of Pennsylvania; Ph.D, Columbia, 1979. Performance modeling and simulation of telecommunication networks, queuing theory, and optimization.

Stephen Sammut, Managing Director, Access Management Services, Inc.; Wharton School WEMBA VII Graduate. Technology transfer, corporate venture capital, the structuring of domestic and international joint ventures, management of research intensive startup companies.

S. David Wu, Associate Professor, Industrial and Manufacturing Systems Engineering, Lehigh University and Codirector, the Manufacturing Logistics Institute (MLI); Ph.D., Penn State. Manufacturing production and logistics systems; modeling and analysis of distributed decision procedures and combinatorial optimization.

UNIVERSITY OF WATERLOO

Department of Management Sciences

Programs of Study

The Department of Management Sciences offers graduate programs that lead to the degrees of Master of Applied Science (M.A.Sc.) and Doctor of Philosophy (Ph.D.). Special fields of interest include applied operations research, management information systems, and management of technology. There is also a distance education M.A.Sc. in the area of management of technology (MOT@Distance); it is taken on a part-time basis and is available to students in Canada and the U.S. only. Delivery of MOT@Distance is by an enhanced distance learning method, with the additional element of optional tutorials delivered by the Internet (Web site: http://innovate.uwaterloo.ca/mot-de/info.html).

The M.A.Sc. program is designed for students with an analytical background and prepares them for management positions in technology-dependent organizations or staff member or advisory positions in organizations that utilize the disciplines represented in the program.

Students in the regular program have the option of writing a thesis or completing a project. They may, if they wish, obtain their degree by means of the cooperative program of studies, including two work terms at an appropriate place of employment. The Ph.D. program is primarily for persons who seek university teaching and/or research positions. Its content, which is quite flexible, reflects the research interests of the various members of the faculty.

Research Facilities

The department is the home of two important research groups. The Waterloo Management of Integrated Manufacturing Systems Research Group (WATMIMS) works directly with industry in performing systems studies, systems modelling, and performance evaluation and in developing decision support systems for the participating firms. The Institute for Innovation Research (IIR) conducts research and industrial collaboration that focuses on how to significantly reduce the time interval during the formulation phase of the planning of a new product or service while increasing the quality of investment decisions. Other research focuses on how to estimate the business worthiness of new technology-based or knowledge-intensive ventures. Other institutes and centres associated with the department include the Institute for Risk Research (IRR), the Institute for Improvement in Quality and Productivity (IIQP), the Telelearning Network Centre of Excellence, the Centre for Sight Enhancement (CSE), and the Network for the Evaluation of Education and Training Technologies (EVNET).

Excellent computer facilities are available to every student in the department. Each student office is equipped with a networked computer, and there are other extensive computing facilities in the department as well as elsewhere in the University. There are three University libraries with more than 3 million cataloged items and 6,600 journal titles.

Financial Aid

All applicants are automatically considered for several scholarships, including the International Student Scholarship and the El Gabbani Scholarship. Research assistantships are offered by individual faculty members to qualified students to assist in their specific research programs. Teaching assistantships are allocated each term to suitably qualified students in their second or later terms of study.

Cost of Study

The typical total program cost (as of 1998) for the full-time, on-campus M.A.Sc. program was $8851 for Canadian students and $26,133 for international students. Tuition was $4426 per year for Canadian students and $13,066 for international students. Tuition fees for MOT@Distance were $2500 per term of study (with five 2-term courses, the total cost was $25,000). All figures are in Canadian dollars.

Living and Housing Costs

Excluding tuition and fees, all necessary living expenses for a single person to live at local standards total about Can$3000 per term. Accommodation in University residences costs approximately $330 to $760 per month. Family housing is $127 to $235 per month, while a private apartment is $300 to $1200 per month.

Student Group

About 40 percent of the entering students have undergraduate degrees in engineering, 20 percent in sciences, and the remaining 40 percent in business and other disciplines. About 25 percent of the on-campus students are part-time, and about 10 percent are women. About 50 percent of the on-campus students are Canadian, with the remaining 50 percent are from countries around the world. Distance education students are typically 5–10 years into a technical career.

Student Outcomes

Starting salaries earned by management sciences graduates are among the highest in Canada. Historically, approximately one third of full-time M.A.Sc. graduates begin as in-house analysts, taking such positions as operations research analyst, service planning analyst, or systems analyst. About one quarter work as analysts in consulting firms. The remainder have line-management positions, such as production manager, or are teachers in colleges and universities. Many move into line management as their careers progress. MOT@Distance graduates have enhanced their career advancement with their degrees. Most Ph.D. graduates have joined faculties in universities in Canada and abroad.

Location

The University campus is located on a beautiful 1,000-acre site in the city of Waterloo, about 100 kilometres west of Toronto. The Waterloo region, with a population of about 350,000, contains areas with a strong German-Mennonite background, which lends a distinct flavour to its culture. The metropolitan region is one of the fastest growing in Canada and has many of the advantages of a larger city.

The University and The Department

The University has been rated the best overall university in Canada for the last seven years and has almost 16,000 full-time students, of whom 1,400 are graduate students. There are about 800 faculty members with a wide variety of interests. The thirty teaching and service buildings on campus include a large Physical Activities Complex, extensive library facilities, two theatres, four residential church colleges, and a variety of modern residential accommodations.

In its brief history, the Department of Management Sciences has gained a worldwide reputation for the excellence of its faculty and programs. For example, professors from the department have won a number of honours from national and international organizations for outstanding research, exceptional teaching, distinguished achievement, and other professional accomplishments. Students from the department have won numerous prizes in competitions sponsored by such organizations as the Institute for Operations Research and the Management Sciences (INFORMS), the Decision Sciences Institute, and the Social Sciences and Humanities Research Council.

Applying

Applicants should have an honours degree and an excellent academic record from a recognized university. For the M.A.Sc., applicants must normally have at least a 75 percent (B) overall standing; Ph.D. applicants need at least 80 percent (A-) and evidence of research potential (e.g., a master's thesis). All applicants (other than applicants to MOT@Distance) must submit recent GRE or GMAT test results, and applicants whose native language is not English need to score at least 550 (paper-based) or 213 (computer-based) on the Test of English as a Foreign Language (TOEFL).

Complete applications for on-campus admission must be received by December 31 for the following fall admission for international applicants residing outside Canada and by April 30 for the following fall admission for any applicant residing in Canada. Deadlines for MOT@Distance applications are available by contacting the department.

Correspondence and Information

Ms. Marion Reid
Administrative Co-ordinator for Graduate Studies
Department of Management Sciences
University of Waterloo
200 University Avenue West
Waterloo, Ontario N2L 3G1
Canada
Telephone: 519-888-4567 Ext. 3670
Fax: 519-746-7252
E-mail: mreid@uwaterloo.ca
World Wide Web: http://www.mansci.uwaterloo.ca/

University of Waterloo

THE FACULTY AND THEIR RESEARCH

Thomas B. Astebro, Assistant Professor; Ph.D., Carnegie Mellon. Economics of technological changes. Basic statistics on the success rate and profits for independent investors. *Entrepreneurship Theor. Pract.,* in press.

Clifford G. Blake, Adjunct Assistant Professor; Ph.D., Waterloo. Management of technology, organizational behavior. A model of entrepreneurial venture performance. *J. Entrepreneurship Small Bus.* 9(4):19–26, 1992 (with Saleh).

James H. Bookbinder, Professor; Ph.D., California, San Diego. Applied operations. Markovian decision processes in shipment consolidation. *Transport. Sci.* 29(3):242–55, 1995 (with Higginson).

David M. Dilts, Professor; Ph.D., Oregon. Integrated enterprise; redesign of accounting information systems; service quality and delivery, with applications in optometry. The manufacturing strategy formulation process: Linking multifunctional viewpoints. *J. Operations Manage.* 15(4):223–41, 1997 (with Menda).

Niall M. Fraser, Professor; Ph.D., Waterloo. Applied game theory, multiple-criteria decision making, negotiation analysis. Ordinal preference representations. *Theory Decis.* 36(1):45–68, 1994.

J. David Fuller, Professor; Ph.D., British Columbia. Energy economics, operations research. An algorithm for the multi-period market equilibrium model with geometric distributed lag demand. *Operations Res.* 44(6):1002–12, 1996 (with Wu).

Yigal Gerchak, Professor; Ph.D., British Columbia. Operations management in production and service organizations; supply chain economics. An inventory model embedded in designing a supply contract. *Manage. Sci.* 43:184–5m, 1997 (with Henig, Ernst, and Pyke).

Paul D. Guild, Professor; D.Phil., Oxford. Management of technology, innovation and change. Equity investment decisions for technology based ventures. *Int. J. Manage. Technol.* 12(7/8):787–95, 1996 (with Bachher).

Elizabeth M. Jewkes, Associate Professor; Ph.D., Waterloo. Operations research. Flow time distributions in queues with setup times and customer batching. *Inf. Syst. Operations Res.* 35(1):76–91, 1997 (with He).

Ji-Ye Mao, Assistant Professor; Ph.D., British Columbia. Evaluation of the impact of emerging information technologies. Contextualized access to knowledge in knowledge-based systems: A process tracing case study. In *Proceedings of the 5th European Conference on Information Systems,* pp. 178–95, 1997 (with Benbasat).

Frank Safayeni, Professor; Ph.D., Victoria. Behavioral and organizational aspects of information systems, management of technology. The role of language and formal structure in the construction and maintenance of organizational images. *Int. Stud. Manage. Organizations,* in press.

Raymond G. Vickson, Professor; Ph.D., MIT. Operations research. On the probability distribution of total quality under sublot formation and process restoration. *Operations Res. Lett.* 20:243–7, 1997.

Jane Webster, Associate Professor; Ph.D., NYU. Information systems, organizational behaviour. Desktop videoconferencing: Experiences of complete users, wary users, and non-users. *MIS Q.* 22:257–86, 1998.

Section 16
Materials Sciences and Engineering

This section contains a directory of institutions offering graduate work in materials sciences and engineering, followed by in-depth entries submitted by institutions that chose to prepare detailed program descriptions. Additional information about programs listed in the directory but not augmented by an in-depth entry may be obtained by writing directly to the dean of a graduate school or chair of a department at the address given in the directory.

For programs offering related work, see also in this book Bioengineering, Biomedical Engineering, and Biotechnology; Engineering and Applied Sciences; and Geological, Mineral/Mining, and Petroleum Engineering. In Book 4, see Chemistry and Geosciences.

CONTENTS

Ceramic Sciences and Engineering

Alfred University, Graduate School, New York State College of Ceramics, School of Ceramic Engineering and Materials Science, Alfred, NY 14802-1205. Offers ceramic engineering (MS); ceramics (PhD); glass science (MS, PhD); materials science (MS). *Students:* 53 full-time (10 women), 7 part-time (3 women). Average age 24. 111 applicants, 23% accepted. In 1998, 11 master's, 10 doctorates awarded. *Degree requirements:* For master's and doctorate, thesis/dissertation required, foreign language not required. *Entrance requirements:* For master's and doctorate, TOEFL (minimum score of 590 required). *Application deadline:* Applications are processed on a rolling basis. Application fee: $50. Electronic applications accepted. *Financial aid:* Fellowships, research assistantships, teaching assistantships, Federal Work-Study and tuition waivers (full and partial) available. Aid available to part-time students. Financial aid applicants required to submit FAFSA. *Faculty research:* Fine-particle technology, x-ray diffraction, superconductivity, electronic materials. *Unit head:* Dr. James Reed, Dean, 607-871-2441, E-mail: freed@bigvax.alfred.edu. *Application contact:* Cathleen R. Johnson, Coordinator of Graduate Admissions, 607-871-2141, Fax: 607-871-2198, E-mail: johnsonc@king.alfred.edu.

See in-depth description on page 1319.

Clemson University, Graduate School, College of Engineering and Science, School of Chemical and Materials Engineering, Department of Ceramic Engineering, Clemson, SC 29634. Offers M Engr, MS, PhD. *Students:* 14 full-time (3 women), 5 part-time (2 women), 9 international. 15 applicants, 67% accepted. In 1998, 4 master's awarded. *Degree requirements:* For master's and doctorate, thesis/dissertation required, foreign language not required. *Entrance requirements:* For master's and doctorate, GRE General Test, TOEFL. *Application deadline:* For fall admission, 6/1. Application fee: $35. *Financial aid:* Fellowships, research assistantships, teaching assistantships, career-related internships or fieldwork available. Financial aid application deadline: 5/1; financial aid applicants required to submit FAFSA. *Faculty research:* Ceramic processing, hazardous material mediation, ceramic and carbon fibers, ceramic matrix composites, electronic ceramic materials. *Unit head:* Dr. Henry J. Rack, Chair, 864-656-5636, Fax: 864-656-1453, E-mail: rack@clemson.edu.

Georgia Institute of Technology, Graduate Studies and Research, College of Engineering, School of Materials Science and Engineering, Program in Ceramic Engineering, Atlanta, GA 30332-0001. Offers MSMSE, PhD. Terminal master's awarded for partial completion of doctoral program. *Degree requirements:* For master's and doctorate, computer language, thesis/dissertation required, foreign language not required. *Entrance requirements:* For master's and doctorate, GRE General Test, TOEFL (minimum score of 550 required). Electronic applications accepted. *Faculty research:* Crystal growth, cathode materials refractories, superconductivity, thermodynamics.

The Ohio State University, Graduate School, College of Engineering, Department of Materials Science and Engineering, Columbus, OH 43210. Offers ceramic engineering (MS, PhD); metallurgical engineering (MS, PhD). *Faculty:* 24 full-time, 1 part-time. *Students:* 82 full-time (14 women); includes 1 minority (Asian American or Pacific Islander), 67 international. *Degree requirements:* For master's and doctorate, computer language, thesis/dissertation required, foreign language not required. *Entrance requirements:* For master's and doctorate, GRE General Test (international students). *Application deadline:* For fall admission, 8/15. Applications are processed on a rolling basis. Application fee: $30 ($40 for international students). *Unit head:* Robert L. Snyder, Chairman, 614-292-2553, Fax: 614-292-1537, E-mail: snyder.355@osu.edu.

Pennsylvania State University University Park Campus, Graduate School, College of Earth and Mineral Sciences, Department of Materials Science and Engineering, State College, University Park, PA 16802-1503. Offers ceramic science (MS, PhD); fuel science (MS, PhD); metals science and engineering (MS, PhD); polymer science (MS, PhD). *Students:* 108 full-time (15 women), 11 part-time (1 woman). *Entrance requirements:* For master's and doctorate, GRE. *Application deadline:* For fall admission, 7/1. Application fee: $50. *Unit head:* Dr. R. E. Tressler, Head, 814-865-0497.

Rensselaer Polytechnic Institute, Graduate School, School of Engineering, Department of Materials Science and Engineering, Troy, NY 12180-3590. Offers ceramics and glass science (M Eng, MS, D Eng, PhD); composites (M Eng, MS, D Eng, PhD); electronic materials (M Eng, MS, D Eng, PhD); metallurgy (M Eng, MS, D Eng, PhD); polymers (M Eng, MS, D Eng, PhD). Part-time and evening/weekend programs available. *Faculty:* 17 full-time (1 woman), 5 part-time (0 women). *Students:* 45 full-time (8 women), 24 part-time (4 women); includes 5 minority (1 African American, 2 Asian Americans or Pacific Islanders, 2 Hispanic Americans), 26 international. Terminal master's awarded for partial completion of doctoral program. *Degree requirements:* For master's, thesis required (for some programs), foreign language not required; for doctorate, dissertation required, foreign language not required. *Entrance requirements:* For master's and doctorate, GRE, TOEFL (minimum score of 550 required). *Application deadline:* For fall admission, 2/1 (priority date). Applications are processed on a rolling basis. Application fee: $35. *Unit head:* Dr. Richard W. Siegel, Chair, 518-276-6373, Fax: 518-276-8554. *Application contact:* Dr. Roger Wright, Admissions Coordinator, 518-276-6372, Fax: 518-276-8554, E-mail: fowlen@rpi.edu.

See in-depth description on page 1337.

University of California, Berkeley, Graduate Division, College of Engineering, Department of Materials Science and Mineral Engineering, Program in Ceramic Sciences and Engineering, Berkeley, CA 94720-1500. Offers M Eng, MS, D Eng, PhD. *Degree requirements:* For master's, comprehensive exam or thesis (MS) required; for doctorate, dissertation, qualifying exam required. *Entrance requirements:* For master's and doctorate, GRE General Test, minimum GPA of 3.0. *Application deadline:* For fall admission, 2/10; for spring admission, 9/1. Application fee: $40. *Financial aid:* Application deadline: 1/5. *Unit head:* Gail Anderson, Graduate Secretary, 403-492-3598, Fax: 403-492-2200, E-mail: mecegrad@gpu.srv.ualberta.ca. *Application contact:* Carole James, Student Affairs Officer, 510-642-3801, Fax: 510-643-5792, E-mail: carolej@uclink4.berkeley.edu.

University of California, Los Angeles, Graduate Division, School of Engineering and Applied Science, Department of Materials Science and Engineering, Program in Ceramics Engineering, Los Angeles, CA 90095. Offers MS, PhD. *Degree requirements:* For master's, comprehensive exam or thesis required; for doctorate, dissertation, qualifying exams required, foreign language not required. *Entrance requirements:* For master's, GRE General Test, minimum GPA of 3.0; for doctorate, GRE General Test, minimum GPA of 3.25. *Application deadline:* For fall admission, 1/15; for spring admission, 12/31. Application fee: $40. Electronic applications accepted. *Financial aid:* Fellowships, research assistantships, teaching assistantships, Federal Work-Study, institutionally-sponsored loans, and tuition waivers (full and partial) available. Financial aid application deadline: 1/15; financial aid applicants required to submit FAFSA. *Application contact:* Paradee Chularee, Student Affairs Officer, 310-825-8913, Fax: 310-206-7353, E-mail: paradee@ea.ucla.edu.

University of Cincinnati, Division of Research and Advanced Studies, College of Engineering, Department of Materials Science and Metallurgical Engineering, Cincinnati, OH 45221-0091. Offers ceramic science and engineering (MS, PhD); materials science and engineering (MS, PhD); metallurgical engineering (MS, PhD); polymer science and engineering (MS, PhD). Evening/weekend programs available. *Faculty:* 9 full-time. *Students:* 79 full-time (15 women), 15 part-time (2 women); includes 8 minority (3 African Americans, 4 Asian Americans or Pacific Islanders, 1 Hispanic American), 61 international. *Degree requirements:* For master's, thesis optional, foreign language not required; for doctorate, one foreign language, dissertation, comprehensive exams, oral English proficiency exam required. *Entrance requirements:* For master's and doctorate, GRE General Test, TOEFL (minimum score of 550 required), BS in related field, minimum undergraduate GPA of 3.0. *Application deadline:* For fall admission, 2/1 (priority date). Applications are processed on a rolling basis. Application fee: $40. *Unit head:* Raj Singh, Head, 513-556-3114, Fax: 513-556-2569, E-mail: rsingh@uceng.uc.edu. *Application contact:* Dr. R. J. Roe, Director of Graduate Studies, 513-556-3117, Fax: 513-556-2569, E-mail: r.j.roe@uc.edu.

University of Florida, Graduate School, College of Engineering, Department of Materials Science and Engineering, Program in Ceramic Science and Engineering, Gainesville, FL 32611. Offers ME, MS, PhD, Engr. *Degree requirements:* For master's, thesis optional, foreign language not required; for doctorate, dissertation required, foreign language not required; for Engr, thesis optional. *Entrance requirements:* For master's and doctorate, GRE General Test (minimum combined score of 1100 required), TOEFL, minimum GPA of 3.0; for Engr, GRE General Test. *Application deadline:* For fall admission, 6/1 (priority date). Applications are processed on a rolling basis. Application fee: $20. *Application contact:* Jodi Vanderheyden, Graduate Secretary, 352-392-1451, Fax: 352-392-5630, E-mail: jheyd@mail.mse.ufl.edu.

University of Missouri–Rolla, Graduate School, School of Mines and Metallurgy, Department of Ceramic Engineering, Rolla, MO 65409-0910. Offers MS, PhD. Part-time programs available. *Faculty:* 8 full-time (0 women), 1 part-time (0 women). *Students:* 30 full-time (5 women), 15 international. Average age 29. 25 applicants, 60% accepted. In 1998, 4 master's, 1 doctorate awarded. Terminal master's awarded for partial completion of doctoral program. *Degree requirements:* For master's and doctorate, computer language, thesis/dissertation required, foreign language not required. *Entrance requirements:* For master's, GRE General Test (minimum combined score of 1100 required), TOEFL (minimum score of 600 required), minimum GPA of 3.0 in last 4 semesters; for doctorate, GRE General Test (minimum combined score of 1100 required), TOEFL (minimum score of 600 required). *Application deadline:* For fall admission, 7/1; for spring admission, 12/1. Applications are processed on a rolling basis. Application fee: $25. Electronic applications accepted. *Financial aid:* In 1998–99, 30 students received aid, including 1 fellowship with full tuition reimbursement available (averaging $10,800 per year), 29 research assistantships with partial tuition reimbursements available (averaging $12,986 per year); teaching assistantships with partial tuition reimbursements available, Federal Work-Study and institutionally-sponsored loans also available. Aid available to part-time students. Financial aid application deadline: 3/1; financial aid applicants required to submit FAFSA. *Faculty research:* Composite sintering, refractory materials, glass structure and properties, dielectrics and ferroelectrics, surfaces and coatings. *Unit head:* Dr. Wayne Huebner, Chairman, 573-341-6129, Fax: 573-341-6934, E-mail: huebner@umr.edu.

University of Washington, Graduate School, College of Engineering, Department of Materials Science and Engineering, Program in Ceramic Engineering, Seattle, WA 98195. Offers MS, MSE. *Degree requirements:* For master's, thesis required, foreign language not required. *Entrance requirements:* For master's, GRE General Test (minimum combined score of 1400 on quantitative and analytical sections required), TOEFL (minimum score of 600 required), minimum GPA of 3.0. *Application deadline:* For fall admission, 7/1; for winter admission, 11/1; for spring admission, 2/1. Applications are processed on a rolling basis. Application fee: $50. Tuition, state resident: full-time $5,196; part-time $475 per credit. Tuition, nonresident: full-time $13,485; part-time $1,285 per credit. Required fees: $387; $38 per credit. Tuition and fees vary according to course load. *Financial aid:* Fellowships, research assistantships, teaching assistantships, Federal Work-Study and stipend supplements available. Financial aid application deadline: 3/1. *Unit head:* Dr. D. Huston, Coordinator, 802-656-3320. *Application contact:* Dr. T. G. Stoebe, Graduate Coordinator, 206-543-7090, Fax: 206-543-3100, E-mail: mse@u.washington.edu.

Electronic Materials

Princeton University, Graduate School, School of Engineering and Applied Science, Department of Electrical Engineering, Princeton, NJ 08544-1019. Offers computer engineering (MSE, PhD); electrical engineering (M Eng); electronic materials and devices (MSE, PhD); information sciences and systems (MSE, PhD); optoelectronics (MSE, PhD). Part-time programs available. *Faculty:* 27 full-time (3 women). *Students:* 135 full-time (18 women), 10 part-time (1 woman). *Degree requirements:* For master's, thesis optional; for doctorate, dissertation required. *Entrance requirements:* For master's and doctorate, GRE General Test, TOEFL. *Application deadline:* For fall admission, 1/3. Electronic applications accepted. *Unit head:* Prof. Wayne Wolf, Director of Graduate Studies, 609-258-3335, Fax: 609-258-3745, E-mail: dgs@ee.princeton.edu. *Application contact:* Prof. Wayne Wolf, Director of Graduate Studies, 609-258-3335, Fax: 609-258-3745, E-mail: dgs@ee.princeton.edu.

Materials Engineering

Arizona State University, Graduate College, College of Engineering and Applied Sciences, Department of Chemical, Bio and Materials Engineering, Program in Materials Science and Engineering, Tempe, AZ 85287. Offers engineering science (MS, MSE, PhD), including materials science and engineering. *Degree requirements:* For doctorate, dissertation required. *Entrance requirements:* For master's and doctorate, GRE General Test. Application fee: $45. *Financial aid:* Fellowships, research assistantships, teaching assistantships available. *Faculty research:* Fluid dynamics, mechanics of solids and structures, structural dynamics, structural stability, nonlinear mechanics. *Unit head:* Dr. Stephen J. Krause, Coordinator, 480-965-2050.

See in-depth description on page 1321.

Arizona State University, Graduate College, Interdisciplinary Program in Science and Engineering of Materials, Tempe, AZ 85287. Offers PhD. *Degree requirements:* For doctorate, dissertation required. *Entrance requirements:* For doctorate, GRE. Application fee: $45. *Faculty research:* Silicon and gallium arsenide, fabrication of ultra small solid state electronic devices, structure of free surfaces of crystalline solids, ion implantation on solids, effects of high pressures on solids. *Unit head:* Dr. William T. Petuskey, Co-Director, 480-965-4549. *Application contact:* Dr. William T. Petuskey, Co-Director, 480-965-4549.

See in-depth description on page 1323.

Auburn University, Graduate School, College of Engineering, Department of Mechanical Engineering, Program in Materials Engineering, Auburn, Auburn University, AL 36849-0002. Offers M Mtl E, MS, PhD. *Faculty:* 8 full-time (1 woman). *Students:* 24 full-time (8 women), 34 part-time (4 women); includes 7 minority (6 African Americans, 1 Asian American or Pacific Islander), 30 international. 30 applicants, 40% accepted. In 1998, 9 master's, 6 doctorates awarded. *Degree requirements:* For master's, thesis (MS), oral exam required; for doctorate, dissertation required. *Entrance requirements:* For master's, GRE General Test; for doctorate, GRE General Test (minimum score of 400 on each section required). *Application deadline:* For fall admission, 9/1; for spring admission, 3/1. Applications are processed on a rolling basis. Application fee: $25 ($50 for international students). Tuition, state resident: full-time $2,760; part-time $76 per credit hour. Tuition, nonresident: full-time $8,280; part-time $228 per credit hour. *Financial aid:* Fellowships, research assistantships, teaching assistantships, Federal Work-Study available. Aid available to part-time students. Financial aid application deadline: 3/15. *Faculty research:* Smart materials. *Unit head:* Dr. Bryan Chin, Head, 334-844-3322. *Application contact:* Dr. John F. Pritchett, Dean of the Graduate School, 334-844-4700.

California State University, Northridge, Graduate Studies, College of Engineering and Computer Science, Department of Civil, Industrial and Applied Mechanics, Northridge, CA 91330. Offers applied mechanics (MSE); civil engineering (MS); engineering (MS); engineering management (MS); industrial engineering (MS); materials engineering (MS); mechanical engineering (MS), including aerospace engineering, applied engineering, machine design, mechanical engineering, structural engineering, thermofluids; mechanics (MS). Part-time and evening/weekend programs available. *Faculty:* 13 full-time, 2 part-time. *Students:* 10 full-time (2 women), 101 part-time (15 women); includes 38 minority (3 African Americans, 22 Asian Americans or Pacific Islanders, 11 Hispanic Americans, 2 Native Americans), 8 international. *Degree requirements:* For master's, thesis required, foreign language not required. *Entrance requirements:* For master's, GRE General Test, TOEFL, minimum GPA of 2.5. *Application deadline:* For fall admission, 11/30. Application fee: $55. Tuition, nonresident: part-time $246 per unit. International tuition: $7,874 full-time. Required fees: $1,970. Tuition and fees vary according to course load. *Unit head:* Dr. Stephen Gadomski, Chair, 818-677-2166. *Application contact:* Dr. Ileana Costa, Graduate Coordinator, 818-677-3299.

Carleton University, Faculty of Graduate Studies, Faculty of Engineering and Design, Department of Mechanical and Aerospace Engineering, Ottawa, ON K1S 5B6, Canada. Offers aerospace engineering (M Eng, PhD); materials engineering (M Eng); mechanical engineering (M Eng, PhD). *Faculty:* 23 full-time (1 woman). *Students:* 51 full-time (4 women), 11 part-time (2 women). *Degree requirements:* For master's, thesis optional; for doctorate, dissertation required. *Entrance requirements:* For master's, TOEFL (minimum score of 550 required), honors degree; for doctorate, TOEFL (minimum score of 550 required), MA Sc or M Eng. *Application deadline:* For fall admission, 3/1. Applications are processed on a rolling basis. Application fee: $35. *Unit head:* Paul Straznicky, Director, 613-520-2600 Ext. 5684, Fax: 613-520-5715, E-mail: pstrazni@mae.carleton.ca. *Application contact:* Ata M. Khan, Associate Dean of Engineering, 613-520-5659, Fax: 613-520-5682, E-mail: ata_khan@carleton.ca.

Carnegie Mellon University, Carnegie Institute of Technology, Department of Materials Science and Engineering, Pittsburgh, PA 15213-3891. Offers ME, MS, PhD. Part-time programs available. *Faculty:* 27 full-time (3 women), 1 part-time (0 women). *Students:* 62 full-time (12 women), 6 part-time (1 woman); includes 3 minority (1 African American, 1 Asian American or Pacific Islander, 1 Hispanic American), 42 international. Average age 28. In 1998, 18 master's, 12 doctorates awarded. Terminal master's awarded for partial completion of doctoral program. *Degree requirements:* For master's, exam required, foreign language and thesis not required; for doctorate, dissertation, qualifying exam required, foreign language not required. *Entrance requirements:* For master's and doctorate, GRE General Test, TOEFL. *Application deadline:* For fall admission, 2/1; for spring admission, 10/15. Application fee: $45. *Financial aid:* Fellowships, research assistantships, career-related internships or fieldwork available. Financial aid application deadline: 2/1. *Faculty research:* Materials characterization, process metallurgy, high strength alloys, growth kinetics, ceramics. Total annual research expenditures: $5.5 million. *Unit head:* Dr. Anthony Rollett, Head, 412-268-2700, Fax: 412-268-7596. *Application contact:* Director of Graduate Studies, 412-268-7574.

See in-depth description on page 1325.

Clemson University, Graduate School, College of Engineering and Science, School of Chemical and Materials Engineering, Program in Materials Science and Engineering, Clemson, SC 29634. Offers MS, PhD. Part-time programs available. *Students:* 25 full-time (3 women), 2 part-time (1 woman), 22 international. Average age 24. 70 applicants, 31% accepted. In 1998, 2 master's, 2 doctorates awarded. Terminal master's awarded for partial completion of doctoral program. *Degree requirements:* For master's and doctorate, thesis/dissertation required, foreign language not required. *Entrance requirements:* For master's and doctorate, GRE General Test, TOEFL. *Application deadline:* For fall admission, 2/15. Applications are processed on a rolling basis. Application fee: $35. *Financial aid:* Research assistantships, career-related internships or fieldwork available. Financial aid applicants required to submit FAFSA. *Faculty research:* Composites, fibers, ceramics, metallurgy, biomaterials, semiconductors. Total annual research expenditures: $1.2 million. *Unit head:* Dr. R. Larry Dooley, Coordinator, 864-656-5562, Fax: 864-656-5910, E-mail: dooley@eng.clemson.edu.

Announcement: The Materials Science and Engineering Program admits students with a baccalaureate degree in any branch of engineering, as well as chemistry, physics, and biology majors with a strong mathematical background. Emphasis is placed on applying the fundamental principles that govern the structure of the solid state to produce optimum mechanical, electrical, optical, and other physical properties. The curriculum provides for specialization in semiconductor manufacturing, metallurgy, glasses and ceramics, electronic materials, biomaterials, polymer and fiber science, and composite materials. The program is designed to produce engineers and scientists whose degrees represent specialization coupled with a broad foundation in all materials. Financial support to qualified students.

See in-depth description on page 1329.

Colorado School of Mines, Graduate School, Department of Metallurgical and Materials Engineering, Golden, CO 80401-1887. Offers ME, MS, PhD. Part-time programs available. *Faculty:* 25 full-time (0 women), 1 part-time (0 women). *Students:* 78 full-time (17 women), 25 part-time (4 women); includes 8 minority (5 Asian Americans or Pacific Islanders, 3 Hispanic Americans), 31 international. 92 applicants, 40% accepted. In 1998, 9 master's awarded (100% found work related to degree); 3 doctorates awarded (100% found work related to degree). *Degree requirements:* For master's, thesis required, foreign language not required; for doctorate, dissertation, comprehensive exams required, foreign language not required. *Entrance requirements:* For master's and doctorate, GRE General Test, minimum GPA of 3.0. *Application deadline:* Applications are processed on a rolling basis. Application fee: $40. Electronic applications accepted. *Financial aid:* In 1998–99, 55 students received aid, including 3 fellowships, 37 research assistantships, 10 teaching assistantships; unspecified assistantships also available. Aid available to part-time students. Financial aid applicants required to submit FAFSA. *Faculty research:* Phase transformations, nonferrous alloy systems, pyrometallurgy, reactive metals, engineered materials. Total annual research expenditures: $141,430. *Unit head:* Dr. John J. Moore, Head, 303-273-3770, Fax: 303-273-3795, E-mail: jjmoore@mines.edu. *Application contact:* Sharon Kirts, Program Assistant I, 303-273-3660, Fax: 303-384-2189, E-mail: skirts@mines.edu.

Colorado State University, Graduate School, College of Engineering, Department of Mechanical Engineering, Program in Mechanics and Materials, Fort Collins, CO 80523-0015. Offers MS, PhD. *Faculty:* 18 full-time (0 women). *Degree requirements:* For doctorate, dissertation required, foreign language not required, foreign language not required. *Entrance requirements:* For master's and doctorate, GRE General Test (minimum combined score of 1850 on three sections required; average 1872), TOEFL (minimum score of 550 required; average 596), minimum GPA of 3.0. *Application deadline:* For fall admission, 2/1 (priority date). Applications are processed on a rolling basis. Application fee: $30. Electronic applications accepted. *Faculty research:* Structured modeling of dynamic systems,composites, ultrathin casings, superconductivity, ion implantation. *Unit head:* Vickie Jensen, 970-491-5597. *Application contact:* Dr. Donald Radford, Associate Professor, 970-491-8677, Fax: 970-491-3827, E-mail: don@engr.colostate.edu.

Columbia University, Fu Foundation School of Engineering and Applied Science, Department of Earth and Environmental Engineering, Program in Materials Science and Engineering, New York, NY 10027. Offers materials engineering (MS, Eng Sc D); materials science (Eng Sc D, PhD, EM); materials science and engineering (MS); metallurgy (Met E). Part-time programs available. Postbaccalaureate distance learning degree programs offered (no on-campus study). *Faculty:* 7 full-time (2 women), 2 part-time (1 woman). *Students:* 25 full-time (6 women), 9 part-time; includes 4 minority (all Asian Americans or Pacific Islanders), 16 international. In 1998, 4 master's, 2 doctorates awarded. Terminal master's awarded for partial completion of doctoral program. *Degree requirements:* For master's, foreign language and thesis not required; for doctorate, dissertation, qualifying exam required, foreign language not required. *Entrance requirements:* For master's, doctorate, and other advanced degree, GRE General Test, TOEFL. Application fee: $55. *Financial aid:* In 1998–99, 24 students received aid; fellowships, research assistantships, teaching assistantships, Federal Work-Study available. Financial aid application deadline: 1/5; financial aid applicants required to submit FAFSA. *Faculty research:* Thin films, inelasticity, plastic deformation, interstitial impurities, materials for fuel cells and corrosion. Total annual research expenditures: $824,000. *Application contact:* Dr. Barbara Algin, Departmental Administrator, 212-854-2905, Fax: 212-854-7081, E-mail: ba110@columbia.edu.

Cornell University, Graduate School, Graduate Fields of Engineering, Field of Materials Science and Engineering, Ithaca, NY 14853-0001. Offers materials engineering (M Eng, PhD); materials science (M Eng, PhD). *Faculty:* 29 full-time. *Students:* 53 full-time (15 women); includes 6 minority (all Asian Americans or Pacific Islanders), 38 international. 225 applicants, 16% accepted. In 1998, 9 master's, 7 doctorates awarded. Terminal master's awarded for partial completion of doctoral program. *Degree requirements:* For doctorate, dissertation required, foreign language not required, foreign language not required. *Entrance requirements:* For master's and doctorate, GRE General Test, TOEFL (minimum score of 550 required). *Application deadline:* For fall admission, 1/15. Application fee: $65. Electronic applications accepted. *Financial aid:* In 1998–99, 45 students received aid, including 10 fellowships with full tuition reimbursements available, 27 research assistantships with full tuition reimbursements available, 8 teaching assistantships with full tuition reimbursements available; institutionally-sponsored loans, scholarships, tuition waivers (full and partial), and unspecified assistantships also available. Financial aid applicants required to submit FAFSA. *Faculty research:* Ceramics, complex fluids, metals, polymers, and semiconductors; electrical, magnetic, mechanical, optical, and structural properties; thin films, organic optoelectronics, nano-composites, inorganic organic hybrids, and composites. *Unit head:* Director of Graduate Studies, 607-255-9159. *Application contact:* Graduate Field Assistant, 607-255-9159, E-mail: matsci@cornell.edu.

Dartmouth College, Thayer School of Engineering, Program in Materials Sciences and Engineering, Hanover, NH 03755. Offers MS, PhD. *Degree requirements:* For master's, thesis required; for doctorate, dissertation, candidacy oral exam required. *Entrance requirements:* For master's and doctorate, GRE General Test. *Application deadline:* For fall admission, 1/15 (priority date). Application fee: $20 ($40 for international students). *Financial aid:* Fellowships, research assistantships, teaching assistantships, career-related internships or fieldwork, Federal Work-Study, institutionally-sponsored loans, and tuition waivers (full and partial) available. Financial aid application deadline: 1/15. *Faculty research:* Metals, intermetallics and ice physics, microelectronic and magnetic materials, optical thin films, biomaterials and nanostructures, laser-material interactions. Total annual research expenditures: $1.1 million. *Unit head:* Sharon Kirts, Program Assistant I, 303-273-3660, Fax: 303-384-2189, E-mail: skirts@mines.edu. *Application contact:* Candace S. Potter, Admissions Coordinator, 603-646-3844, Fax: 603-646-3856, E-mail: candace.potter@dartmouth.edu.

Drexel University, Graduate School, College of Engineering, Department of Materials Engineering, Philadelphia, PA 19104-2875. Offers MS, PhD. Part-time and evening/weekend programs available. *Faculty:* 10 full-time. *Students:* 13 full-time (2 women), 33 part-time (5 women); includes 1 minority (Asian American or Pacific Islander), 26 international. Average age 30. 137 applicants, 49% accepted. In 1998, 10 master's, 8 doctorates awarded. Terminal master's awarded for partial completion of doctoral program. *Degree requirements:* For master's, thesis or alternative required; for doctorate, dissertation required. *Entrance requirements:* For master's, TOEFL (minimum score of 570 required), minimum GPA of 3.0; for doctorate, TOEFL (minimum score of 570 required), minimum GPA of 3.0, MS. *Application deadline:* For fall admission, 8/21. Applications are processed on a rolling basis. Application fee: $35. Tuition: Full-time $15,795; part-time $585 per credit. Required fees: $375; $67 per term. Tuition and fees vary according to program. *Financial aid:* In 1998–99, 20 research assistantships, 4 teaching assistantships were awarded; career-related internships or fieldwork and unspecified assistantships also available. Financial aid application deadline: 2/1. *Faculty research:* Composite science; polymer and biomedical engineering; solidification; near net shape processing, including powder metallurgy. *Unit head:* Dr. Alan Lawley, Head, 215-895-2322. *Application contact:* Dr. Wei-Heng Shih, Graduate Adviser, 215-895-6636.

Announcement: The Department of Materials Engineering awards BS, MS, and PhD degrees. Both the MS and the PhD degrees can be pursued on a part-time basis. Students must fulfill a 1-academic-year (9-month) residency requirement for the PhD degree. Graduate-level courses primarily meet once a week in the evening. For more information, visit the Web site at http://www.materials.drexel.edu.

École Polytechnique de Montréal, Graduate Programs, Department of Metallurgical and Materials Engineering, Montréal, PQ H3C 3A7, Canada. Offers advanced materials (M Eng, M Sc A, PhD); chemical metallurgy (M Eng, M Sc A, PhD); physical metallurgy (M Eng, M Sc A, PhD). Part-time programs available. *Degree requirements:* For master's and doctorate, one foreign language, computer language, thesis/dissertation required. *Entrance*

Materials Engineering

École Polytechnique de Montréal (continued)
requirements: For master's, minimum GPA of 2.75; for doctorate, minimum GPA of 3.0. *Faculty research:* Refractory materials, fatigue, thermochemical analysis, electrochemistry, material characterization.

Georgia Institute of Technology, Graduate Studies and Research, College of Engineering, School of Materials Science and Engineering, Atlanta, GA 30332-0001. Offers biomedical engineering (MS Bio E); ceramic engineering (MSMSE, PhD); materials engineering (MS); metallurgy (MSMSE, PhD); polymers (MS Poly). Terminal master's awarded for partial completion of doctoral program. *Degree requirements:* For master's, computer language required, foreign language not required; for doctorate, computer language, dissertation required, foreign language not required. *Entrance requirements:* For master's and doctorate, GRE General Test, TOEFL (minimum score of 550 required). Electronic applications accepted. *Faculty research:* Corrosion, composites, surface modifications, *in situ* oxide-metal composites.

Howard University, College of Engineering, Architecture, and Computer Sciences, School of Engineering and Computer Science, Program in Materials Science and Engineering, Washington, DC 20059-0002. Offers MS, PhD. Part-time programs available. *Faculty:* 14 full-time (1 woman). *Students:* 5 full-time (1 woman); all minorities (all African Americans) Average age 25. 5 applicants, 100% accepted. Terminal master's awarded for partial completion of doctoral program. *Degree requirements:* For master's, thesis required, foreign language not required; for doctorate, dissertation, comprehensive exam required. *Entrance requirements:* For master's and doctorate, GRE General Test, TOEFL, minimum GPA of 3.0. *Application deadline:* For fall admission, 4/1; for spring admission, 11/1. Applications are processed on a rolling basis. Application fee: $45. *Financial aid:* In 1998–99, 4 fellowships with full tuition reimbursements (averaging $15,000 per year), 1 research assistantship with full tuition reimbursement (averaging $19,000 per year) were awarded.; career-related internships or fieldwork, grants, scholarships, tuition waivers (partial), and unspecified assistantships also available. Financial aid application deadline: 4/1. *Faculty research:* Polymers, biomaterials, composites, high temperature semiconductors, mechanical behavior. Total annual research expenditures: $600,000. *Unit head:* Clayton W. Bates, Director, 202-806-6147, Fax: 202-806-5258, E-mail: bates@negril.msrce.howard.edu.

Illinois Institute of Technology, Graduate College, Armour College of Engineering and Sciences, Department of Mechanical, Materials and Aerospace Engineering, Metallurgical and Materials Engineering Division, Chicago, IL 60616-3793. Offers MMME, MS, PhD. Part-time and evening/weekend programs available. *Faculty:* 8 full-time (1 woman), 4 part-time (0 women). *Students:* 13 full-time (4 women), 16 part-time (3 women); includes 3 minority (2 African Americans, 1 Hispanic American), 10 international. 46 applicants, 54% accepted. In 1998, 7 master's, 6 doctorates awarded. Terminal master's awarded for partial completion of doctoral program. *Degree requirements:* For master's, thesis (for some programs), comprehensive exam required, foreign language not required; for doctorate, dissertation, comprehensive exam required, foreign language not required. *Entrance requirements:* For master's, GRE General Test (minimum combined score of 1200 required), TOEFL (minimum score of 550 required), undergraduate GPA of 3.0; for doctorate, GRE (minimum score of 1200 required), TOEFL (minimum score of 550 required), undergraduate GPA of 3.0 required. *Application deadline:* For fall admission, 7/1; for spring admission, 11/1. Applications are processed on a rolling basis. Application fee: $30. Electronic applications accepted. *Financial aid:* In 1998–99, 1 fellowship, 3 research assistantships, 3 teaching assistantships were awarded.; Federal Work-Study, institutionally-sponsored loans, scholarships, and graduate assistantships also available. Financial aid application deadline: 3/1. *Faculty research:* High-temperature materials, laser processing, fracture mechanics, powder metallurgy, alloy design. *Unit head:* Judith Todd, Associate Chair, 312-567-3175, Fax: 312-567-8875, E-mail: jtodd@charlie.iit.edu. *Application contact:* Dr. S. Mohammad Shahidehpour, Dean of Graduate College, 312-567-3024, Fax: 312-567-7517, E-mail: grad@minna.cns.iit.edu.

Instituto Tecnológico y de Estudios Superiores de Monterrey, Campus Estado de México, Graduate Division, Division of Engineering and Architecture, Atizapán de Zaragoza, 52500, Mexico. Offers computer science (MCS); environmental engineering (MEE); industrial engineering (MIE); manufacturing systems (MMS); materials engineering (PhD). *Degree requirements:* For master's; for doctorate, one foreign language, dissertation required. *Entrance requirements:* For master's, interview; for doctorate, research proposal. *Application deadline:* For fall admission, 1/13 (priority date); for spring admission, 4/4. Applications are processed on a rolling basis. Application fee: 750 Mexican pesos. *Unit head:* Juan López Díaz, Headmaster, 5-326-5530, Fax: 5-326-5531, E-mail: jlopez@campus.cem.itesm.mx. *Application contact:* Lourdes Turrubiates, Admissions Officer, 5-326-5776, Fax: 5-326-5788, E-mail: lturrubi@campus.cem.itesm.mx.

Iowa State University of Science and Technology, Graduate College, College of Engineering, Department of Materials Science and Engineering, Ames, IA 50011. Offers MS, PhD. *Faculty:* 22 full-time, 2 part-time. *Students:* 34 full-time (4 women), 15 part-time (6 women); includes 2 minority (1 African American, 1 Hispanic American), 24 international. 67 applicants, 28% accepted. In 1998, 8 master's, 5 doctorates awarded. *Degree requirements:* For master's and doctorate, thesis/dissertation required. *Entrance requirements:* For master's and doctorate, GRE General Test (foreign students), TOEFL (minimum score of 550 required). *Application deadline:* For fall admission, 2/15 (priority date); for spring admission, 8/15 (priority date). Application fee: $20 ($50 for international students). Electronic applications accepted. Tuition, state resident: full-time $3,308. Tuition, nonresident: full-time $9,744. Part-time tuition and fees vary according to course load, campus/location and program. *Financial aid:* In 1998–99, 37 research assistantships with partial tuition reimbursements (averaging $12,269 per year), 1 teaching assistantship with partial tuition reimbursement (averaging $12,150 per year) were awarded.; fellowships, scholarships also available. *Unit head:* Dr. Mufit Akinc, Chair, 515-294-0738, Fax: 515-204-5444, E-mail: mse@iastate.edu. *Application contact:* Dr. R. William McCallum, 515-294-1214, E-mail: mse@iastate.edu.

Johns Hopkins University, G. W. C. Whiting School of Engineering, Department of Materials Science and Engineering, Baltimore, MD 21218-2699. Offers MSE, PhD. Part-time and evening/weekend programs available. *Faculty:* 10 full-time (0 women), 11 part-time (0 women). *Students:* 29 full-time (9 women), 7 part-time (2 women); includes 4 minority (3 Asian Americans or Pacific Islanders, 1 Hispanic American), 10 international. Average age 26. 69 applicants, 17% accepted. In 1998, 4 master's, 5 doctorates awarded. Terminal master's awarded for partial completion of doctoral program. *Degree requirements:* For master's and doctorate, thesis/dissertation, cumulative exam required, foreign language not required. *Entrance requirements:* For master's, GRE General Test (minimum combined score of 1950 on three sections required; average 2050), TOEFL (minimum score of 560 required; average 620); for doctorate, GRE General Test (minimum combined score of 1950 on three sections required; average 2050), TOEFL (minimum score of 600 required; average 620). *Average time to degree:* Master's–2.5 years full-time, 4 years part-time; doctorate–5 years full-time, 7.5 years part-time. *Application deadline:* Applications are processed on a rolling basis. Application fee: $50. Tuition: Full-time $23,660. Tuition and fees vary according to program. *Financial aid:* In 1998–99, 10 fellowships (averaging $11,700 per year), 13 research assistantships (averaging $11,250 per year), 10 teaching assistantships (averaging $10,800 per year) were awarded.; Federal Work-Study and institutionally-sponsored loans also available. Financial aid application deadline: 3/14. *Faculty research:* Metallurgy, nanomaterials, biomaterials, nondestructive characterization, electrochemistry. Total annual research expenditures: $2.8 million. *Unit head:* Dr. Peter C. Searson, Chair, 410-516-8774, Fax: 410-516-5293, E-mail: searson@jhu.edu. *Application contact:* Dr. Robert C. Cammarata, Chair, Graduate Program Committee, 410-516-5462, Fax: 410-516-5293, E-mail: rcc@jhu.edu.

See in-depth description on page 1331.

Lehigh University, College of Engineering and Applied Science, Department of Materials Science and Engineering, Bethlehem, PA 18015-3094. Offers M Eng, MS, PhD. Part-time programs available. *Faculty:* 15 full-time (1 woman), 2 part-time (0 women). *Students:* 34 full-time (4 women), 6 part-time (2 women); includes 4 minority (all Asian Americans or Pacific Islanders), 15 international. 390 applicants, 2% accepted. In 1998, 7 master's, 8 doctorates awarded. *Degree requirements:* For master's and doctorate, thesis/dissertation required, foreign language not required. *Entrance requirements:* For master's and doctorate, GRE General Test, TOEFL. *Application deadline:* For fall admission, 7/15; for spring admission, 12/1. Applications are processed on a rolling basis. Application fee: $40. *Financial aid:* In 1998–99, 4 fellowships with full tuition reimbursements (averaging $11,040 per year), 24 research assistantships with full and partial tuition reimbursements (averaging $10,350 per year), 2 teaching assistantships with full and partial tuition reimbursements (averaging $11,000 per year) were awarded. Financial aid application deadline: 1/15. *Faculty research:* Metals, ceramics, crystals, polymers, fatigue crack propagation, electron microscopy. Total annual research expenditures: $3.2 million. *Unit head:* Dr. David B. Williams, Chairperson, 610-758-6120, Fax: 610-758-4244, E-mail: dbw1@lehigh.edu. *Application contact:* Maxine C. Mattie, Graduate Administrative Coordinator, 610-758-4222, Fax: 610-758-4244, E-mail: mcm1@lehigh.edu.

Marquette University, Graduate School, College of Engineering, Department of Mechanical and Industrial Engineering, Milwaukee, WI 53201-1881. Offers engineering management (MS); materials science and engineering (MS, PhD); mechanical engineering (MS, PhD), including manufacturing systems engineering. Part-time and evening/weekend programs available. *Faculty:* 18 full-time (0 women), 6 part-time (0 women). *Students:* 32 full-time (3 women), 52 part-time (4 women); includes 6 minority (1 African American, 2 Asian Americans or Pacific Islanders, 2 Hispanic Americans, 1 Native American), 30 international. Terminal master's awarded for partial completion of doctoral program. *Degree requirements:* For master's, thesis, comprehensive exam required, foreign language not required; for doctorate, dissertation, proficiency exam, qualifying exam required, foreign language not required. *Entrance requirements:* For master's and doctorate, GRE General Test, TOEFL (minimum score of 550 required), minimum GPA of 3.0. *Application deadline:* For fall admission, 8/1 (priority date); for spring admission, 1/1 (priority date). Applications are processed on a rolling basis. Application fee: $40. Tuition: Part-time $510 per credit hour. Tuition and fees vary according to program. *Unit head:* Dr. G. E. O. Widera, Chairman, 414-288-7259, Fax: 414-288-1647, E-mail: geo.widera@marquette.edu. *Application contact:* Dr. William E. Brower, Director of Graduate Studies, 414-288-1717, Fax: 414-288-7790, E-mail: 9322browerw@rms.csd.mu.edu.

Massachusetts Institute of Technology, School of Engineering, Department of Materials Science and Engineering, Cambridge, MA 02139-4307. Offers materials engineering (Mat E); materials science and engineering (SM, PhD, Sc D); metallurgical engineering (Met E). *Faculty:* 31 full-time (6 women). *Students:* 167 full-time (41 women), 3 part-time; includes 23 minority (1 African American, 20 Asian Americans or Pacific Islanders, 2 Hispanic Americans), 61 international. Average age 26. 262 applicants, 29% accepted. In 1998, 30 master's, 26 doctorates awarded. Terminal master's awarded for partial completion of doctoral program. *Degree requirements:* For master's, thesis required, foreign language not required; for doctorate, dissertation, qualifying exams required, foreign language not required; for other advanced degree, thesis required. *Entrance requirements:* For master's, doctorate, and other advanced degree, GRE General Test. *Average time to degree:* Master's–2.5 years full-time; doctorate–6 years full-time. *Application deadline:* For fall admission, 1/15 (priority date); for spring admission, 11/1. Applications are processed on a rolling basis. Application fee: $55. *Financial aid:* In 1998–99, 164 students received aid, including 26 fellowships, 132 research assistantships, 12 teaching assistantships; career-related internships or fieldwork, grants, institutionally-sponsored loans, scholarships, and traineeships also available. Financial aid application deadline: 1/1; financial aid applicants required to submit FAFSA. *Faculty research:* Photonic band gap materials, advanced spectroscopic and diffraction techniques, rapid solidification process, high-strength polymer materials, computer modeling of structures and processing. Total annual research expenditures: $14.2 million. *Unit head:* Dr. Thomas W. Eagar, Head, 617-253-0948, Fax: 617-252-1773, E-mail: tweagar@mit.edu. *Application contact:* Kenneth C. Russell, Graduate Admissions Chairman, 617-253-3329, Fax: 617-258-8836, E-mail: kenruss@mit.edu.

McMaster University, School of Graduate Studies, Faculty of Engineering, Department of Materials Science and Engineering, Hamilton, ON L8S 4M2, Canada. Offers materials engineering (M Eng, M Sc, PhD); materials science (M Eng, M Sc, PhD). *Faculty:* 23 full-time. *Students:* 30 full-time, 3 part-time. *Degree requirements:* For master's, thesis required, foreign language not required; for doctorate, dissertation, comprehensive exam required, foreign language not required. *Application deadline:* For fall admission, 3/1 (priority date). Applications are processed on a rolling basis. Application fee: $50. *Financial aid:* In 1998–99, teaching assistantships (averaging $7,722 per year); fellowships, research assistantships *Unit head:* Dr. M. B. Ives, Chair, 905-525-9140 Ext. 24293.

Michigan Technological University, Graduate School, College of Engineering, Department of Metallurgical and Materials Engineering, Houghton, MI 49931-1295. Offers MS, PhD. Part-time programs available. *Faculty:* 17 full-time (1 woman), 2 part-time (0 women). *Students:* 34 full-time (8 women); includes 1 minority (Asian American or Pacific Islander), 21 international. Average age 27. 68 applicants, 41% accepted. In 1998, 7 master's, 4 doctorates awarded. *Degree requirements:* For master's, thesis required, foreign language not required; for doctorate, dissertation, qualifying exam required, foreign language not required. *Entrance requirements:* For master's, GRE General Test (combined average 1826 on three sections), TOEFL (minimum score of 550 required; average 602); for doctorate, GRE General Test (combined average 1952 on three sections), TOEFL (minimum score of 550 required; average 616). *Average time to degree:* Master's–3 years full-time; doctorate–5.2 years full-time. *Application deadline:* For fall admission, 3/15 (priority date). Applications are processed on a rolling basis. Application fee: $30 ($35 for international students). Tuition, state resident: full-time $4,377. Tuition, nonresident: full-time $9,108. Required fees: $126. Tuition and fees vary according to course load. *Financial aid:* In 1998–99, 5 fellowships (averaging $3,270 per year), 24 research assistantships (averaging $7,419 per year), 2 teaching assistantships (averaging $4,594 per year) were awarded.; career-related internships or fieldwork, Federal Work-Study, institutionally-sponsored loans, and unspecified assistantships also available. Aid available to part-time students. Financial aid application deadline: 3/1; financial aid applicants required to submit FAFSA. *Faculty research:* Structure/property/processing relationships, microstructural characterization, alloy design, materials and manufacturing processes, mineral processing. Total annual research expenditures: $1.6 million. *Unit head:* Dr. Calvin White, Chair, 906-487-2631, Fax: 906-487-2934, E-mail: cwhite@mtu.edu. *Application contact:* Kathe Prince, Senior Documentation Specialist, 906-487-2039, Fax: 906-487-2934, E-mail: kaprince@mtu.edu.

National Technological University, Programs in Engineering, Fort Collins, CO 80526-1842. Offers chemical engineering (MS); computer engineering (MS); computer science (MS); electrical engineering (MS); engineering management (MS); hazardous waste management (MS); health physics (MS); management of technology (MS); manufacturing systems engineering (MS); materials science and engineering (MS); software engineering (MS); special majors (MS); transportation engineering (MS); transportation systems engineering (MS). Part-time programs available. *Faculty:* 600 part-time (20 women). *Entrance requirements:* For master's, BS in engineering or related field. *Application deadline:* Applications are processed on a rolling basis. Application fee: $50. *Unit head:* Lionel V. Baldwin, President, 970-495-6400, Fax: 970-484-0668, E-mail: baldwin@mail.ntu.edu.

New Jersey Institute of Technology, Office of Graduate Studies, Department of Physics, Program in Materials Science and Engineering, Newark, NJ 07102-1982. Offers MS, PhD. *Degree requirements:* For master's, thesis required, foreign language not required; for doctorate, dissertation, residency required, foreign language not required. *Entrance requirements:* For master's, GRE General Test (minimum score of 450 on verbal section, 600 on quantitative, 550 on analytical required); for doctorate, GRE General Test (minimum score of 450 on verbal section, 600 on quantitative, 550 on analytical required), minimum graduate GPA of 3.5.

New Mexico Institute of Mining and Technology, Graduate Studies, Department of Material Engineering, Socorro, NM 87801. Offers MS, PhD. *Faculty:* 7 full-time (2 women). *Students:* 23 full-time (4 women), 2 part-time (both women); includes 1 minority (Hispanic American), 13 international. Average age 30. 64 applicants, 77% accepted. In 1998, 9 master's, 2 doctorates awarded. *Degree requirements:* For master's and doctorate, thesis/dissertation required, foreign language not required. *Entrance requirements:* For master's, GRE General Test, TOEFL (minimum score of 540 required); for doctorate, GRE General Test, GRE Subject Test, TOEFL (minimum score of 540 required). *Average time to degree:* Master's–3 years full-time; doctorate–6 years full-time. *Application deadline:* For fall admission, 3/1 (priority date); for spring admission, 6/1. Applications are processed on a rolling basis. Application fee: $16. *Financial aid:* In 1998–99, 17 research assistantships (averaging $9,670 per year), 2 teaching assistantships (averaging $9,670 per year) were awarded.; fellowships, Federal Work-Study and institutionally-sponsored loans also available. Financial aid application deadline: 3/1; financial aid applicants required to submit CSS PROFILE or FAFSA. *Faculty research:* Materials, metallurgy, ceramic working. *Unit head:* Dr. Osman Inal, Chairman, 505-835-5519, Fax: 505-835-5626, E-mail: inal@nmt.edu. *Application contact:* Dr. David B. Johnson, Dean of Graduate Studies, 505-835-5513, Fax: 505-835-5476, E-mail: graduate@nmt.edu.

North Carolina State University, Graduate School, College of Engineering, Department of Materials Science and Engineering, Raleigh, NC 27695. Offers MMSE, MS, PhD. *Faculty:* 21 full-time (1 woman), 17 part-time (0 women). *Students:* 103 full-time (14 women), 21 part-time (5 women); includes 14 minority (3 African Americans, 6 Asian Americans or Pacific Islanders, 3 Hispanic Americans, 2 Native Americans), 39 international. Average age 29. 99 applicants, 31% accepted. In 1998, 12 master's, 22 doctorates awarded. *Degree requirements:* For master's and doctorate, thesis/dissertation required, foreign language not required. *Application deadline:* For fall admission, 6/25. Application fee: $45. *Financial aid:* In 1998–99, 1 fellowship (averaging $2,002 per year), 61 research assistantships (averaging $5,079 per year), 10 teaching assistantships (averaging $4,992 per year) were awarded. *Faculty research:* Processing and properties of wide band gap semiconductors, ferroelectric thin-film materials, ductility of nanocrystalline materials, computational materials science, defects in silicon-based devices. Total annual research expenditures: $7.5 million. *Unit head:* Dr. J. Michael Rigsbee, Head, 919-515-3568, Fax: 919-515-7724, E-mail: mike_rigsbee@ncsu.edu. *Application contact:* Dr. Carl Koch, Director of Graduate Programs, 919-515-7340, Fax: 919-515-7724, E-mail: carl_koch@ncsu.edu.

Announcement: The department has an annual research budget exceeding $5 million and participates in several interdisciplinary research centers. Most students are supported by teaching/research assistantships and fellowships. Well-recognized faculty members of diverse backgrounds and interests and modern facilities, particularly for materials processing and characterization. Visit the department Web site at http://www.mse.ncsu.edu

Northwestern University, The Graduate School, Robert R. McCormick School of Engineering and Applied Science, Department of Materials Science and Engineering, Evanston, IL 60208. Offers materials science (PhD); materials science and engineering (MS). Admissions and degrees offered through The Graduate School. Part-time programs available. *Faculty:* 25 full-time (3 women), 1 part-time (0 women). *Students:* 101 full-time (27 women), 1 part-time; includes 23 minority (3 African Americans, 17 Asian Americans or Pacific Islanders, 3 Hispanic Americans), 13 international. 233 applicants, 31% accepted. In 1998, 3 master's, 20 doctorates awarded. Terminal master's awarded for partial completion of doctoral program. *Degree requirements:* For master's, thesis, oral thesis defense required, foreign language not required; for doctorate, dissertation, oral defense of dissertation, preliminary evaluation, qualifying exam required, foreign language not required. *Entrance requirements:* For master's and doctorate, GRE General Test, TOEFL (minimum score of 560 required). *Application deadline:* For fall admission, 8/30. Application fee: $50 ($55 for international students). *Financial aid:* In 1998–99, 15 fellowships with full tuition reimbursements (averaging $11,673 per year), 65 research assistantships with partial tuition reimbursements (averaging $16,285 per year), teaching assistantships with full tuition reimbursements (averaging $12,042 per year) were awarded.; career-related internships or fieldwork, Federal Work-Study, and institutionally-sponsored loans also available. Financial aid application deadline: 1/15; financial aid applicants required to submit FAFSA. *Faculty research:* Metallurgy, ceramics, polymers, electronic materials, biomaterials. Total annual research expenditures: $9.3 million. *Unit head:* Katherine Faber, Chair, 847-491-3537, Fax: 847-491-7820. *Application contact:* Sharon Jacknow, Admissions Contact, 847-491-3587, Fax: 847-491-7820, E-mail: sjacknow@nwu.edu.

Oregon Graduate Institute of Science and Technology, Graduate Studies, Department of Materials Science and Engineering, Portland, OR 97291-1000. Offers MS, PhD. Part-time programs available. Terminal master's awarded for partial completion of doctoral program. *Degree requirements:* For master's, thesis optional, foreign language not required; for doctorate, comprehensive exam, oral defense of dissertation required. *Entrance requirements:* For master's and doctorate, GRE General Test, TOEFL (minimum score of 550 required). *Application deadline:* Applications are processed on a rolling basis. Application fee: $50. Electronic applications accepted. *Financial aid:* Fellowships, research assistantships, Federal Work-Study available. Financial aid application deadline: 3/1. *Application contact:* Enrollment Manager, 800-685-2423, Fax: 503-690-1285, E-mail: admissions@admin.ogi.edu.

See in-depth description on page 1335.

Pennsylvania State University University Park Campus, Graduate School, College of Earth and Mineral Sciences, Department of Materials Science and Engineering, State College, University Park, PA 16802-1503. Offers ceramic science (MS, PhD); fuel science (MS, PhD); metals science and engineering (MS, PhD); polymer science (MS, PhD). *Students:* 108 full-time (15 women), 11 part-time (1 woman). In 1998, 9 master's, 10 doctorates awarded. *Entrance requirements:* For master's and doctorate, GRE. *Application deadline:* For fall admission, 7/1. Application fee: $50. *Financial aid:* Fellowships, unspecified assistantships available. Financial aid application deadline: 2/28. *Unit head:* Dr. R. E. Tressler, Head, 814-865-0497.

Purdue University, Graduate School, Schools of Engineering, School of Materials Engineering, West Lafayette, IN 47907. Offers materials engineering (MS, MSE, PhD); metallurgical engineering (MS Met E). Part-time programs available. *Faculty:* 12 full-time (1 woman). *Students:* 31 full-time (5 women), 3 part-time; includes 1 minority (Hispanic American), 11 international. Average age 25. 107 applicants, 11% accepted. In 1998, 5 master's awarded (% continued full-time study); 3 doctorates awarded (33% entered university research/teaching). *Degree requirements:* For master's, thesis required (for some programs), foreign language not required; for doctorate, dissertation required. *Entrance requirements:* For master's and doctorate, TOEFL (minimum score of 550 required). *Average time to degree:* Master's–1.5 years full-time, 3 years part-time; doctorate–5 years full-time, 6 years part-time. *Application deadline:* For fall admission, 2/1 (priority date). Application fee: $30. Electronic applications accepted. *Financial aid:* In 1998–99, 31 research assistantships with full tuition reimbursements (averaging $16,500 per year) were awarded.; fellowships, teaching assistantships Aid available to part-time students. Financial aid applicants required to submit FAFSA. *Faculty research:* Electronic behavior, mechanical behavior, thermodynamics, kinetics, phase transformations. Total annual research expenditures: $4.6 million. *Unit head:* Dr. G. L. Liedl, Head, 765-494-4094, Fax: 765-494-1204, E-mail: liedl@ecn.purdue.edu. *Application contact:* Dr. K. J. Bowman, Graduate Committee Chair, 765-494-6316, Fax: 765-494-1204, E-mail: kbowman@ecn.purdue.edu.

Queen's University at Kingston, School of Graduate Studies and Research, Faculty of Applied Science, Department of Materials and Metallurgical Engineering, Kingston, ON K7L 3N6, Canada. Offers M Sc, M Sc Eng, PhD. Part-time programs available. *Students:* 17 full-time (2 women), 3 part-time. In 1998, 5 master's, 1 doctorate awarded. *Degree requirements:* For master's, thesis optional, foreign language not required; for doctorate, dissertation, comprehensive exam required, foreign language not required. *Entrance requirements:* For master's and doctorate, TOEFL (minimum score of 550 required). *Application deadline:* For fall admission, 2/28 (priority date). Application fee: $60. Electronic applications accepted. *Financial aid:* Fellowships, research assistantships, teaching assistantships, institutionally-sponsored loans available. Financial aid application deadline: 3/1. *Faculty research:* Physical metallurgy, chemical metallurgy and mineral engineering. *Unit head:* Dr. J. Cameron, Head, 613-533-2758. *Application contact:* Dr. S. Saimoto, Graduate Coordinator, 613-533-2747.

Rensselaer Polytechnic Institute, Graduate School, School of Engineering, Department of Materials Science and Engineering, Troy, NY 12180-3590. Offers ceramics and glass science (M Eng, MS, D Eng, PhD); composites (M Eng, MS, D Eng, PhD); electronic materials (M Eng, MS, D Eng, PhD); metallurgy (M Eng, MS, D Eng, PhD); polymers (M Eng, MS, D Eng, PhD). Part-time and evening/weekend programs available. *Faculty:* 17 full-time (1 woman), 5 part-time (0 women). *Students:* 45 full-time (8 women), 24 part-time (4 women); includes 5 minority (1 African American, 2 Asian Americans or Pacific Islanders, 2 Hispanic Americans), 26 international. 198 applicants, 13% accepted. In 1998, 17 master's, 5 doctorates awarded. Terminal master's awarded for partial completion of doctoral program. *Degree requirements:* For master's, thesis required (for some programs), foreign language not required; for doctorate, dissertation required, foreign language not required. *Entrance requirements:* For master's and doctorate, GRE, TOEFL (minimum score of 550 required). *Application deadline:* For fall admission, 2/1 (priority date). Applications are processed on a rolling basis. Application fee: $35. *Financial aid:* In 1998–99, 38 students received aid, including 1 fellowship with full tuition reimbursement available (averaging $22,000 per year), 27 research assistantships with full tuition reimbursements available (averaging $15,000 per year), 10 teaching assistantships with full tuition reimbursements available (averaging $15,000 per year); career-related internships or fieldwork and institutionally-sponsored loans also available. Financial aid application deadline: 2/1. *Faculty research:* Materials processing, nanostructural materials, materials for microelectronics. Total annual research expenditures: $3.1 million. *Unit head:* Dr. Richard W. Siegel, Chair, 518-276-6373, Fax: 518-276-8554. *Application contact:* Dr. Roger Wright, Admissions Coordinator, 518-276-6372, Fax: 518-276-8554, E-mail: fowlen@rpi.edu.

See in-depth description on page 1337.

Rochester Institute of Technology, Part-time and Graduate Admissions, College of Science, Center for Materials Science and Engineering, Rochester, NY 14623-5603. Offers MS. Part-time programs available. *Students:* 2 full-time (0 women), 7 part-time (1 woman); includes 3 minority (1 Asian American or Pacific Islander, 1 Hispanic American, 1 Native American), 1 international. 50 applicants, 28% accepted. In 1998, 8 degrees awarded. *Degree requirements:* For master's, thesis optional, foreign language not required. *Entrance requirements:* For master's, TOEFL (minimum score of 550 required), minimum GPA of 3.0. *Application deadline:* For fall admission, 3/1 (priority date). Applications are processed on a rolling basis. Application fee: $40. *Financial aid:* Teaching assistantships, career-related internships or fieldwork, institutionally-sponsored loans, and tuition waivers (partial) available. Aid available to part-time students. Financial aid application deadline: 7/29. *Faculty research:* VUV modification of polymers, stress and morphology of sputtered copper films, MRI applications to materials problems. *Unit head:* Dr. Peter A. Cardegna, Program Director, 716-475-6652, E-mail: pacsps@rit.edu.

San Jose State University, Graduate Studies, College of Engineering, Department of Chemical Engineering and Materials Engineering, Program in Materials Engineering, San Jose, CA 95192-0001. Offers MS. Part-time programs available. *Faculty:* 6 full-time (1 woman), 3 part-time (0 women). *Students:* 14 full-time (6 women), 24 part-time (3 women); includes 13 minority (1 African American, 10 Asian Americans or Pacific Islanders, 2 Hispanic Americans), 11 international. Average age 31. 23 applicants, 74% accepted. In 1998, 10 degrees awarded. *Degree requirements:* For master's, computer language, thesis or alternative required. *Entrance requirements:* For master's, GRE, TOEFL (minimum score of 550 required). *Application deadline:* For fall admission, 6/1. Applications are processed on a rolling basis. Application fee: $59. Tuition, nonresident: part-time $246 per unit. Required fees: $1,939; $1,309 per year. *Financial aid:* In 1998–99, 3 teaching assistantships were awarded.; career-related internships or fieldwork, Federal Work-Study, and institutionally-sponsored loans also available. Aid available to part-time students. *Faculty research:* Electronic materials, thin films, electron microscopy, fiber composites, polymeric materials. *Unit head:* Dr. Buck F. Brown, Dean for Research and Graduate Studies, 812-877-8403, Fax: 812-877-8102, E-mail: buck.brown@rose-hulman.edu. *Application contact:* Dr. Guna Selvaduray, Graduate Coordinator, 408-924-3874.

South Dakota School of Mines and Technology, Graduate Division, Division of Material Engineering and Science, Doctoral Program in Materials Engineering and Science, Rapid City, SD 57701-3995. Offers chemical engineering (PhD); chemistry (PhD); civil engineering (PhD); electrical engineering (PhD); mechanical engineering (PhD); metallurgical engineering (PhD); physics (PhD). Part-time programs available. *Students:* 14 full-time (2 women), 9 international. Average age 35. In 1998, 1 degree awarded. *Degree requirements:* For doctorate, dissertation required, foreign language not required. *Entrance requirements:* For doctorate, TOEFL (minimum score of 520 required), TWE, minimum graduate GPA of 3.0. *Application deadline:* For fall admission, 6/15 (priority date); for spring admission, 10/15. Applications are processed on a rolling basis. Application fee: $15. Electronic applications accepted. Tuition, state resident: part-time $89 per hour. Tuition, nonresident: part-time $261 per hour. Part-time tuition and fees vary according to program. *Financial aid:* In 1998–99, 1 fellowship, 9 research assistantships, 6 teaching assistantships were awarded.; Federal Work-Study and institutionally-sponsored loans also available. Aid available to part-time students. Financial aid application deadline: 5/15. *Faculty research:* Thermophysical properties of solids, development of multiphase materials and composites, concrete technology, electronic polymer materials. *Unit head:* Dr. Chris Jenkins, Coordinator, 605-394-2406. *Application contact:* Brenda Brown, Secretary, 800-454-8162 Ext. 2493, Fax: 605-394-5360, E-mail: graduate_admissions@silver.sdmt.edu.

See in-depth description on page 1339.

South Dakota School of Mines and Technology, Graduate Division, Division of Material Engineering and Science, Master's Program in Materials Engineering and Science, Rapid City, SD 57701-3995. Offers chemistry (MS); metallurgical engineering (MS); physics (MS). *Students:* 17. In 1998, 2 degrees awarded. *Degree requirements:* Foreign language not required. *Entrance requirements:* For master's, TOEFL (minimum score of 520 required), TWE. *Application deadline:* For fall admission, 6/15 (priority date); for spring admission, 10/15. Applications are processed on a rolling basis. Application fee: $15. Electronic applications accepted. Tuition, state resident: part-time $89 per hour. Tuition, nonresident: part-time $261 per hour. Part-time tuition and fees vary according to program. *Financial aid:* Application deadline: 5/15. *Unit head:* James W. Smolka, Coordinator, 805-258-5936. *Application contact:* Brenda Brown, Secretary, 800-454-8162 Ext. 2493, Fax: 605-394-5360, E-mail: graduate_admissions@silver.sdmt.edu.

Southern Methodist University, School of Engineering and Applied Science, Center for Special Studies, Dallas, TX 75275. Offers applied science (MS, PhD); hazardous and waste materials management (MS); material science and engineering (MS); systems engineering (MS). *Faculty:* 12 part-time (0 women). *Students:* 10 full-time (2 women), 112 part-time (32 women); includes 22 minority (5 African Americans, 7 Asian Americans or Pacific Islanders, 10 Hispanic Americans), 7 international. *Degree requirements:* For master's, thesis optional, foreign language not required; for doctorate, dissertation, oral and written qualifying exams, oral final exam required. *Entrance requirements:* For master's, GRE General Test (minimum score of 650 on quantitative section required), TOEFL (minimum score of 550 required), minimum GPA of 3.0 in last 2 years, bachelor's degree in related field; for doctorate, preliminary counseling exam, minimum GPA of 3.0, bachelor's degree in related field. *Application deadline:* For fall admission, 8/1 (priority date); for spring admission, 12/15. Applications are processed on a rolling basis. Application fee: $25. Tuition: Full-time $9,216; part-time $512 per credit hour. Required fees: $88 per credit hour. Part-time tuition and fees vary according to course load and campus/location. *Unit head:* Dr. Buck F. Brown, Dean for Research and Graduate Studies, 812-877-8403, Fax: 812-877-8102, E-mail: buck.brown@rose-hulman.edu. *Application contact:*

Materials Engineering

Southern Methodist University (continued)
Dr. Zeynep Celik-Butler, Assistant Dean for Graduate Studies and Research, 214-768-3979, Fax: 214-768-3845, E-mail: zcb@seas.smu.edu.

Stanford University, School of Engineering, Department of Materials Science and Engineering, Stanford, CA 94305-9991. Offers MS, PhD, Eng. *Faculty:* 10 full-time (0 women). *Students:* 115 full-time (32 women), 29 part-time (8 women); includes 29 minority (1 African American, 26 Asian Americans or Pacific Islanders, 1 Hispanic American, 1 Native American), 59 international. Average age 26. 210 applicants, 53% accepted. In 1998, 23 master's, 19 doctorates awarded. *Degree requirements:* For master's, foreign language and thesis not required; for doctorate and Eng, dissertation required, foreign language not required. *Entrance requirements:* For master's, doctorate, and Eng, GRE General Test, TOEFL. *Application deadline:* For fall admission, 1/15. Application fee: $65 ($80 for international students). Electronic applications accepted. Tuition: Full-time $24,588. Required fees: $152. Part-time tuition and fees vary according to course load. *Financial aid:* Fellowships, research assistantships, teaching assistantships, Federal Work-Study and institutionally-sponsored loans available. Financial aid application deadline: 2/15. *Unit head:* John C. Bravman, Chair, 650-723-3698, Fax: 650-725-0538, E-mail: bravman@sierra.stanford.edu. *Application contact:* Graduate Administrator, 650-725-2648.

State University of New York at Stony Brook, Graduate School, College of Engineering and Applied Sciences, Department of Materials Science and Engineering, Stony Brook, NY 11794. Offers MS, PhD. *Faculty:* 11 full-time (1 woman), 2 part-time (0 women). *Students:* 45 full-time (5 women), 22 part-time (6 women); includes 10 minority (3 African Americans, 4 Asian Americans or Pacific Islanders, 3 Hispanic Americans), 37 international. 75 applicants, 65% accepted. In 1998, 8 master's, 9 doctorates awarded. *Degree requirements:* For master's, thesis or alternative required, foreign language not required; for doctorate, dissertation, comprehensive exams required, foreign language not required. *Entrance requirements:* For master's and doctorate, GRE General Test, TOEFL, minimum undergraduate GPA of 3.0. *Application deadline:* For fall admission, 1/15. Application fee: $50. *Financial aid:* In 1998–99, 1 fellowship, 24 research assistantships, 18 teaching assistantships were awarded. *Faculty research:* Electronic materials, biomaterials, synchrotron topography. Total annual research expenditures: $3.6 million. *Unit head:* Dr. Michael Dudley, Chairman, 516-632-8484. *Application contact:* Dr. Miriam Rafailovich, Director, 516-632-8484, Fax: 516-632-8052, E-mail: mrafailovich.nsf.@notes.cc.sunysb.edu.

Stevens Institute of Technology, Graduate School, Charles V. Schaefer Jr. School of Engineering, Department of Materials Science and Engineering, Hoboken, NJ 07030. Offers materials engineering (M Eng, PhD); materials science (MS, PhD); structural analysis of materials (Certificate); surface modification of materials (Certificate). Part-time and evening/weekend programs available. Terminal master's awarded for partial completion of doctoral program. *Degree requirements:* For master's, computer language required, thesis optional, foreign language not required; for doctorate, computer language, dissertation required; for Certificate, computer language required, foreign language not required. *Entrance requirements:* For master's and doctorate, TOEFL. Electronic applications accepted. *Faculty research:* Surface modification techniques, structure and surface analysis, corrosion, tribology, superconductor materials.

Texas A&M University, College of Engineering, Department of Civil Engineering, Program in Materials Engineering, College Station, TX 77843. Offers M Eng, MS, D Eng, PhD. D Eng offered through the College of Engineering. *Students:* 33. *Degree requirements:* For master's, thesis (MS) required; for doctorate, dissertation (PhD), internship (D Eng) required. *Entrance requirements:* For master's and doctorate, GRE General Test, TOEFL. Application fee: $50 ($75 for international students). *Financial aid:* Fellowships, research assistantships, teaching assistantships available. Financial aid application deadline: 4/1; financial aid applicants required to submit FAFSA. *Faculty research:* Innovative design methods, pavement distress characterization, materials property, characterization and modeling recyclable materials. *Unit head:* Dr. Roger E. Smith, Head, 409-845-9967, Fax: 409-862-2800, E-mail: ce-grad@tamu.edu. *Application contact:* Dr. Amy L. Epps, 409-845-2498, Fax: 409-845-2800, E-mail: ce-grad@tamu.edu.

Tuskegee University, Graduate Programs, College of Engineering, Architecture and Physical Sciences, Program in Material Science Engineering, Tuskegee, AL 36088. Offers PhD. *Students:* 9. *Application deadline:* For fall admission, 7/15. Applications are processed on a rolling basis. Application fee: $25 ($35 for international students). *Financial aid:* Application deadline: 4/15. *Unit head:* Dr. Shaik Jeelani, Head, 334-727-8375.

The University of Alabama, Graduate School, College of Engineering, Department of Metallurgical and Materials Engineering, Tuscaloosa, AL 35487. Offers MS Met E, PhD. *Faculty:* 9 full-time (1 woman), 4 part-time (0 women). *Students:* 29; includes 2 minority (1 African American, 1 Hispanic American), 18 international. Average age 29. 40 applicants, 13% accepted. In 1998, 2 master's, 3 doctorates awarded. *Degree requirements:* For master's, thesis or alternative required, foreign language not required; for doctorate, one foreign language (computer language can substitute), dissertation required. *Entrance requirements:* For master's, GRE General Test (minimum combined score of 1500 on three sections required), minimum GPA of 3.0 in last 60 hours. *Application deadline:* For fall admission, 7/1 (priority date). Applications are processed on a rolling basis. Application fee: $25. Electronic applications accepted. *Financial aid:* In 1998–99, 3 fellowships, 14 research assistantships, 6 teaching assistantships were awarded.; Federal Work-Study also available. *Faculty research:* Metals casting and solidification, corrosion and alloy development, metal matrix composites, electronic and magnetic properties. Total annual research expenditures: $2 million. *Unit head:* Dr. Richard C. Bradt, Head, 205-348-1740, Fax: 205-348-2164, E-mail: rcbradt@coe.eng.ua.edu. *Application contact:* Dr. Garry W. Warren, Professor, 205-348-1740, Fax: 205-348-2164, E-mail: gwarren@coe.eng.ua.edu.

The University of Alabama at Birmingham, Graduate School, School of Engineering, Department of Materials and Mechanical Engineering, Program in Materials Engineering, Birmingham, AL 35294. Offers materials engineering (MS Mt E); materials/metallurgical engineering (PhD). *Students:* 18 full-time (3 women), 4 part-time; includes 3 minority (1 African American, 2 Asian Americans or Pacific Islanders), 11 international. 40 applicants, 60% accepted. In 1998, 2 master's, 5 doctorates awarded. *Degree requirements:* For master's, thesis or alternative, comprehensive exam and project/thesis required, foreign language not required; for doctorate, dissertation, comprehensive exam, MSMTE required, foreign language not required. *Entrance requirements:* For master's and doctorate, GRE General Test (minimum score of 500 on each section required), minimum B average. *Application deadline:* Applications are processed on a rolling basis. Application fee: $30 ($60 for international students). Electronic applications accepted. *Financial aid:* In 1998–99, 10 fellowships with full and partial tuition reimbursements (averaging $12,500 per year), 8 research assistantships with full tuition reimbursements (averaging $12,500 per year) were awarded.; career-related internships or fieldwork, Federal Work-Study, and institutionally-sponsored loans also available. Aid available to part-time students. *Faculty research:* Casting metallurgy, microgravity solidification, thin film techniques, ceramics/glass processing, biomedical materials processing. *Unit head:* Dr. Krishan Chawla, Interim Chair, Department of Materials and Mechanical Engineering, 205-975-9725, Fax: 205-934-8485, E-mail: kchawla@uab.edu.

The University of Alabama in Huntsville, School of Graduate Studies, College of Engineering, Department of Chemical and Materials Engineering, Huntsville, AL 35899. Offers MSE. Part-time and evening/weekend programs available. *Faculty:* 5 full-time (0 women). *Students:* 8 full-time (2 women), 4 part-time (1 woman); includes 1 minority (African American), 7 international. Average age 29. 10 applicants, 90% accepted. In 1998, 1 degree awarded. *Degree requirements:* For master's, oral and written exams required, thesis optional, foreign language not required. *Entrance requirements:* For master's, GRE General Test (minimum combined score of 1500 on three sections required), appropriate bachelor's degree, minimum GPA of 3.0. *Application deadline:* For fall admission, 7/24 (priority date); for spring admission, 11/15 (priority date). Applications are processed on a rolling basis. Application fee: $20. Tuition and fees vary according to course load. *Financial aid:* In 1998–99, 9 research assistantships with full and partial tuition reimbursements (averaging $9,060 per year), 4 teaching assistantships with full and partial tuition reimbursements (averaging $8,775 per year) were awarded.; fellowships with full and partial tuition reimbursements, career-related internships or fieldwork, Federal Work-Study, grants, institutionally-sponsored loans, scholarships, and tuition waivers (full and partial) also available. Aid available to part-time students. Financial aid application deadline: 4/1; financial aid applicants required to submit FAFSA. *Faculty research:* Turbulence modeling, computational fluid dynamics, microgravity processing, multiphase transport, blood materials transport. Total annual research expenditures: $126,346. *Unit head:* Dr. Ramon Cerro, Chair, 256-890-6810, Fax: 256-890-6839, E-mail: rlc@eb.uah.edu.

University of Alberta, Faculty of Graduate Studies and Research, Department of Chemical and Materials Engineering, Edmonton, AB T6G 2E1, Canada. Offers chemical engineering (M Eng, M Sc, PhD); materials engineering (M Eng, M Sc, PhD); process control (M Eng, M Sc, PhD); welding (M Eng). Part-time programs available. Postbaccalaureate distance learning degree programs offered (minimal on-campus study). Terminal master's awarded for partial completion of doctoral program. *Degree requirements:* For master's and doctorate, thesis/dissertation required, foreign language not required. *Faculty research:* Advanced materials and polymers, catalytic and reaction engineering, mineral processing, physical metallurgy,fluid mechanics.

The University of Arizona, Graduate College, College of Engineering and Mines, Department of Materials Science and Engineering, Tucson, AZ 85721. Offers MS, PhD. Part-time programs available. *Faculty:* 21. *Students:* 35 full-time (11 women), 11 part-time (1 woman); includes 6 minority (1 African American, 3 Asian Americans or Pacific Islanders, 2 Hispanic Americans), 18 international. Average age 28. 24 applicants, 42% accepted. In 1998, 4 master's, 9 doctorates awarded. *Degree requirements:* For master's and doctorate, thesis/dissertation required, foreign language not required. *Entrance requirements:* For master's and doctorate, TOEFL (minimum score of 550 required). *Application deadline:* For fall admission, 5/15. Applications are processed on a rolling basis. Application fee: $35. *Financial aid:* Fellowships, research assistantships, teaching assistantships, institutionally-sponsored loans and scholarships available. Financial aid application deadline: 12/31. *Faculty research:* High-technology ceramics, optical materials, electronic materials, chemical metallurgy, science of materials. *Unit head:* Dr. Donald R. Uhlmann, Head, 520-621-6070. *Application contact:* Geri Hardy, Graduate Secretary, 520-621-2531, Fax: 520-621-8059.

University of British Columbia, Faculty of Graduate Studies, Faculty of Applied Science, Department of Metals and Materials Engineering, Vancouver, BC V6T 1Z2, Canada. Offers materials and metallurgy (M Sc, PhD); metals and materials engineering (M Eng, MA Sc, PhD). *Degree requirements:* For master's and doctorate, thesis/dissertation required. *Entrance requirements:* For master's and doctorate, TOEFL. *Faculty research:* Electroslag melting, mathematical modelling, solidification and hydrometallurgy.

University of California, Berkeley, Graduate Division, College of Engineering, Department of Materials Science and Mineral Engineering, Berkeley, CA 94720-1500. Offers ceramic sciences and engineering (M Eng, MS, D Eng, PhD); engineering geoscience (M Eng, MS, D Eng, PhD); materials engineering (M Eng, MS, D Eng, PhD); mineral engineering (M Eng, MS, D Eng, PhD); petroleum engineering (M Eng, MS, D Eng, PhD); physical metallurgy (M Eng, MS, D Eng, PhD). *Students:* 110 full-time (23 women); includes 17 minority (11 Asian Americans or Pacific Islanders, 5 Hispanic Americans, 1 Native American), 42 international. *Degree requirements:* For master's, comprehensive exam or thesis (MS) required; for doctorate, dissertation, qualifying exam required. *Entrance requirements:* For master's and doctorate, GRE General Test, minimum GPA of 3.0. *Application deadline:* For fall admission, 2/10; for spring admission, 9/1. Application fee: $40. *Unit head:* Dr. Thomas M. Devine, Chair, 510-642-3801. *Application contact:* Carole James, Student Affairs Officer, 510-642-3801, Fax: 510-643-5792, E-mail: carolej@uclink4.berkeley.edu.

University of California, Irvine, Office of Research and Graduate Studies, School of Engineering, Program in Materials Science and Engineering, Irvine, CA 92697. Offers engineering (MS, PhD). Part-time programs available. *Faculty:* 11 full-time (1 woman). *Students:* 26 full-time (6 women), 5 part-time (3 women); includes 11 minority (1 African American, 9 Asian Americans or Pacific Islanders, 1 Hispanic American), 10 international. 42 applicants, 81% accepted. In 1998, 2 master's, 5 doctorates awarded. Terminal master's awarded for partial completion of doctoral program. *Degree requirements:* For doctorate, dissertation required, foreign language not required. *Entrance requirements:* For master's, GRE General Test, minimum GPA of 3.0; for doctorate, GRE General Test. *Application deadline:* For fall admission, 1/15 (priority date). Applications are processed on a rolling basis. Application fee: $40. Electronic applications accepted. *Financial aid:* Fellowships, research assistantships, teaching assistantships, institutionally-sponsored loans and tuition waivers (full and partial) available. Financial aid application deadline: 3/2; financial aid applicants required to submit FAFSA. *Faculty research:* Composite materials, solidification processing, sol-gel processing, microstructural characterization, deformation and damage processes in advanced materials. *Unit head:* Dr. Farghalli A. Mohamed, Director, 949-824-5807, Fax: 949-824-3440, E-mail: famohame@uci.edu. *Application contact:* Admissions Assistant, 949-824-3562, Fax: 949-824-3440.

University of California, Los Angeles, Graduate Division, School of Engineering and Applied Science, Department of Materials Science and Engineering, Los Angeles, CA 90095. Offers ceramics engineering (MS, PhD); metallurgy (MS, PhD). *Faculty:* 7 full-time, 3 part-time. *Students:* 59 full-time (16 women); includes 14 minority (1 African American, 12 Asian Americans or Pacific Islanders, 1 Hispanic American), 26 international. 89 applicants, 55% accepted. In 1998, 15 master's, 4 doctorates awarded. *Degree requirements:* For master's, comprehensive exam or thesis required; for doctorate, dissertation, qualifying exams required, foreign language not required. *Entrance requirements:* For master's, GRE General Test, minimum GPA of 3.0; for doctorate, GRE General Test, minimum GPA of 3.25. *Application deadline:* For fall admission, 1/15; for spring admission, 12/31. Application fee: $40. Electronic applications accepted. *Financial aid:* In 1998–99, 23 fellowships, 49 research assistantships, 21 teaching assistantships were awarded.; Federal Work-Study, institutionally-sponsored loans, and tuition waivers (full and partial) also available. Financial aid application deadline: 1/15; financial aid applicants required to submit FAFSA. *Unit head:* Dr. King-Ning Tu, Chair, 310-206-4838. *Application contact:* Paradee Chularee, Student Affairs Officer, 310-825-8913, Fax: 310-206-7353, E-mail: paradee@ea.ucla.edu.

University of California, Santa Barbara, Graduate Division, College of Engineering, Department of Materials, Santa Barbara, CA 93106. Offers MS, PhD. *Faculty:* 30 full-time (2 women), 3 part-time (0 women). *Students:* 70 full-time (16 women); includes 10 minority (1 African American, 7 Asian Americans or Pacific Islanders, 2 Hispanic Americans), 14 international. Average age 24. 175 applicants, 21% accepted. In 1998, 1 master's, 15 doctorates awarded. *Degree requirements:* For master's and doctorate, thesis/dissertation required, foreign language not required. *Entrance requirements:* For master's and doctorate, GRE General Test, TOEFL (minimum score of 600 required). *Application deadline:* For fall admission, 1/15 (priority date); for spring admission, 11/1. Applications are processed on a rolling basis. Application fee: $40. Electronic applications accepted. *Financial aid:* In 1998–99, 70 students received aid, including 22 fellowships, 48 research assistantships, 20 teaching assistantships; career-related internships or fieldwork, Federal Work-Study, institutionally-sponsored loans, and tuition waivers (full) also available. Financial aid application deadline: 1/15; financial aid applicants required to submit FAFSA. *Faculty research:* Electronically conducting polymer systems, growth and processing of advanced semiconductors, structure and properties of artificially structured materials–electronic packages and composites, fine structures by nonequilibrium solidification or advanced deposition techniques, sensors for intelligent control systems in materials processing. *Unit head:* David R. Clarke, Chair, 805-893-4362. *Application contact:* Katie Bridgewater, Graduate Program Assistant, 805-893-4601, E-mail: matrls-grad-advisor@engineering.ucsb.edu.

Announcement: The UCSB Materials Department educates graduate students in advanced materials and introduces them to novel ways of doing research in a collaborative, multidisciplinary environment. Faculty members from around the world have come to UCSB to develop a unique academic research atmosphere. The research in the department is loosely organized into 4 groups specializing in electronic, inorganic, macromolecular/biomolecular, and structural materials with a strong collaboration in related activities with mechanical and environmental engineering, chemical engineering, electrical and computer engineering, physics, biology, and chemistry. For more information, contact the department through e-mail: matrls-grad-advisor@engineering.ucsb.edu or WWW: http://www.materials.ucsb.edu.

University of Central Florida, College of Engineering, Department of Mechanical, Materials, and Aerospace Engineering, Orlando, FL 32816. Offers aerospace systems (MSME, PhD); materials science and engineering (MSME, PhD); mechanical systems (MSME, PhD); mechanical, materials, and aerospace engineering (Certificate); thermofluids (MSME, PhD). Part-time and evening/weekend programs available. *Faculty:* 18 full-time, 6 part-time. *Students:* 61 full-time (14 women), 30 part-time (4 women); includes 12 minority (3 African Americans, 6 Asian Americans or Pacific Islanders, 3 Hispanic Americans), 42 international. Average age 31. 48 applicants, 58% accepted. In 1998, 23 master's, 2 doctorates awarded. *Degree requirements:* For master's, thesis or alternative required, foreign language not required; for doctorate, dissertation, departmental qualifying exam required, foreign language not required. *Entrance requirements:* For master's, GRE General Test (minimum combined score of 1000 required), TOEFL (minimum score of 550 required; 213 computer-based), minimum GPA of 3.0 in last 60 hours; for doctorate, TOEFL (minimum score of 550 required; 213 computer-based), minimum GPA of 3.5 in last 60 hours. *Application deadline:* For fall admission, 7/15; for spring admission, 12/15. Application fee: $20. Tuition, state resident: full-time $2,054; part-time $137 per credit. Tuition, nonresident: full-time $7,207; part-time $480 per credit. Required fees: $47 per term. *Financial aid:* In 1998–99, 71 students received aid, including 17 fellowships with partial tuition reimbursements available (averaging $2,824 per year), 36 teaching assistantships with partial tuition reimbursements available (averaging $3,509 per year); research assistantships with partial tuition reimbursements available, career-related internships or fieldwork, Federal Work-Study, institutionally-sponsored loans, tuition waivers (partial), and unspecified assistantships also available. Financial aid application deadline: 3/1; financial aid applicants required to submit FAFSA. *Faculty research:* Aerospace systems, computation of methods, dynamics and control, laser applications, materials science. *Unit head:* Dr. Louis Chow, Chair, 407-823-2333. *Application contact:* Dr. A. J. Kassab, Coordinator, 407-823-2416, Fax: 407-823-0208.

University of Cincinnati, Division of Research and Advanced Studies, College of Engineering, Department of Materials Science and Metallurgical Engineering, Program in Materials Science and Engineering, Cincinnati, OH 45221-0091. Offers MS, PhD. Evening/weekend programs available. *Faculty:* 8 full-time. *Students:* 78 full-time (15 women), 15 part-time (2 women); includes 8 minority (3 African Americans, 4 Asian Americans or Pacific Islanders, 1 Hispanic American), 61 international. *Degree requirements:* For master's, thesis optional, foreign language not required; for doctorate, one foreign language, dissertation, comprehensive exams, oral English proficiency exam required. *Entrance requirements:* For master's and doctorate, GRE General Test, TOEFL (minimum score of 550 required), BS in related field, minimum undergraduate GPA of 3.0. *Application deadline:* For fall admission, 2/1 (priority date). Application fee: $40. *Financial aid:* Fellowships, career-related internships or fieldwork, tuition waivers (full), and unspecified assistantships available. Aid available to part-time students. Financial aid application deadline: 2/1. *Faculty research:* Polymer characterization, surface analysis, and adhesion; mechanical behavior of high-temperature materials; composites; electrochemistry of materials. *Unit head:* Student Affairs Officer, E-mail: maeapp@ea.ucla.edu. *Application contact:* Dr. R. J. Roe, Graduate Program Director, 513-556-3117, Fax: 513-556-2569, E-mail: r.j.roe@uc.edu.

University of Dayton, Graduate School, School of Engineering, Department of Materials Engineering, Dayton, OH 45469-1300. Offers MS Mat E, DE, PhD. Part-time and evening/weekend programs available. *Faculty:* 2 full-time (0 women), 14 part-time (2 women). *Students:* 23 full-time (11 women), 22 part-time (4 women); includes 3 minority (1 African American, 2 Asian Americans or Pacific Islanders), 2 international. Average age 26. In 1998, 8 master's, 3 doctorates awarded. *Degree requirements:* For master's, thesis optional, foreign language not required; for doctorate, dissertation, departmental qualifying exam required. *Entrance requirements:* For master's, TOEFL. *Application deadline:* For fall admission, 8/1. Applications are processed on a rolling basis. Application fee: $30. *Financial aid:* In 1998–99, 15 students received aid, including 1 fellowship with full tuition reimbursement available (averaging $18,000 per year), 10 research assistantships with full tuition reimbursements available (averaging $13,500 per year); teaching assistantships, institutionally-sponsored loans also available. *Unit head:* Dr. James A. Snide, Director, 937-229-2241, E-mail: jsnide@engr.udayton.edu. *Application contact:* Dr. Donald L. Moon, Associate Dean, 937-229-2241, Fax: 937-229-2471, E-mail: dmoon@engr.udayton.edu.

University of Delaware, College of Engineering, Department of Materials Science and Engineering, Newark, DE 19716. Offers MMSE, PhD. *Faculty:* 12 full-time (2 women), 8 part-time (1 woman). *Students:* 33 full-time (7 women), 3 part-time (1 woman); includes 1 minority (African American), 11 international. Average age 23. 245 applicants, 7% accepted. In 1998, 6 master's, 3 doctorates awarded. Terminal master's awarded for partial completion of doctoral program. *Degree requirements:* For master's and doctorate, thesis/dissertation required, foreign language not required. *Entrance requirements:* For master's and doctorate, GRE General Test (minimum combined score of 1150 required; average 1350) TOEFL (minimum score of 600 required). *Average time to degree:* Master's–2 years full-time; doctorate–4 years full-time. *Application deadline:* For fall admission, 2/28 (priority date). Applications are processed on a rolling basis. Application fee: $45. Electronic applications accepted. *Financial aid:* In 1998–99, 22 students received aid, including 2 fellowships, 18 research assistantships, 3 teaching assistantships; career-related internships or fieldwork and Federal Work-Study also available. Aid available to part-time students. Financial aid application deadline: 2/28. *Faculty research:* Metals, polymers, composites, solid-state electronics, advanced oxides, organic thin films. *Unit head:* Dr. John Rabolt, Chairman, 302-831-2062, Fax: 302-831-4545, E-mail: matsci@me.udel.edu.

University of Florida, Graduate School, College of Engineering, Department of Materials Science and Engineering, Gainesville, FL 32611. Offers ceramic science and engineering (ME, MS, PhD, Engr); materials science and engineering (ME, MS, PhD, Certificate, Engr); metallurgical and materials engineering (ME, MS, PhD, Engr); metallurgical engineering (ME, MS, PhD, Engr); polymer science and engineering (ME, MS, PhD, Engr). *Faculty:* 36. *Students:* 148 full-time (40 women), 26 part-time (7 women); includes 22 minority (5 African Americans, 7 Asian Americans or Pacific Islanders, 9 Hispanic Americans, 1 Native American), 65 international. 414 applicants, 26% accepted. In 1998, 31 master's, 21 doctorates awarded. *Degree requirements:* For master's, thesis optional, foreign language not required; for doctorate, dissertation required, foreign language not required; for other advanced degree, thesis optional. *Entrance requirements:* For master's and doctorate, GRE General Test (minimum combined score of 1100 required), TOEFL, minimum GPA of 3.0; for other advanced degree, GRE General Test. *Application deadline:* For fall admission, 6/1 (priority date). Applications are processed on a rolling basis. Application fee: $20. Electronic applications accepted. *Financial aid:* In 1998–99, 118 students received aid, including 12 fellowships, 114 research assistantships; teaching assistantships. *Faculty research:* Polymeric materials, electronic materials, glass, biomaterials, composites. Total annual research expenditures: $6 million. *Unit head:* Dr. Reza Abbaschian, Chair, 352-392-1453, Fax: 352-392-6359, E-mail: rabba@mse.ufl.edu. *Application contact:* Jodi Vanderheyden, Graduate Secretary, 352-392-1451, Fax: 352-392-5630, E-mail: jheyd@mail.mse.ufl.edu.

See in-depth description on page 1345.

University of Houston, Cullen College of Engineering, Program in Materials Engineering, Houston, TX 77004. Offers MS Mat, PhD. Part-time and evening/weekend programs available.

Students: 13 full-time (0 women), 4 part-time (1 woman), 10 international. Average age 28. 44 applicants, 70% accepted. In 1998, 4 master's, 1 doctorate awarded. Terminal master's awarded for partial completion of doctoral program. *Degree requirements:* For master's, thesis required (for some programs), foreign language not required; for doctorate, dissertation, departmental qualifying exam required, foreign language not required. *Entrance requirements:* For master's and doctorate, GRE General Test, TOEFL. *Application deadline:* For fall admission, 7/3 (priority date); for spring admission, 12/4. Applications are processed on a rolling basis. Application fee: $25 ($75 for international students). *Financial aid:* Research assistantships, teaching assistantships, career-related internships or fieldwork, Federal Work-Study, and tuition waivers (partial) available. Financial aid application deadline: 4/1. *Faculty research:* Processing and mechanical properties of high temperature superconductors, nondestructive characterization of metal-matrix composites, mechanical behavior of polymeric and ceramic materials, fracture toughening of fibrous ceramic composites, fretting wear. *Unit head:* Dr. Kamel Salama, Director, 713-743-4500, Fax: 713-743-4503. *Application contact:* Sheena Paul, Graduate Admissions Analyst, 713-743-4505, Fax: 713-743-4503, E-mail: megrad@mail.me.uh.edu.

Announcement: Graduate degrees (MS, PhD) are offered to qualified students holding baccalaureate degrees in engineering, science, or mathematics. Broad areas of research include ceramic and metal microstructures; mechanical behavior of structural and electrical materials; processing of structural ceramics, HT superconductors, and coatings; nondestructive characterization; fracture mechanics; polymers; and composites. The program has consistently maintained support for about 20 graduate students over the past decade and continues to award TA and RA financial support to qualified candidates. For additional information contact Prof. K. Salama (telephone: 713-743-4514) or visit the Web site at http://www.me.uh.edu.

University of Illinois at Chicago, Graduate College, College of Engineering, Department of Civil and Materials Engineering, Chicago, IL 60607-7128. Offers MS, PhD. Evening/weekend programs available. *Faculty:* 13 full-time (0 women), 2 part-time (0 women). *Students:* 37 full-time (8 women), 31 part-time (8 women); includes 9 minority (4 African Americans, 4 Asian Americans or Pacific Islanders, 1 Hispanic American), 29 international. Average age 29. 114 applicants, 29% accepted. In 1998, 9 master's, 2 doctorates awarded. *Degree requirements:* For master's, thesis required (for some programs), foreign language not required; for doctorate, dissertation, preliminary and qualifying exams required, foreign language not required. *Entrance requirements:* For master's and doctorate, GRE General Test, TOEFL, minimum GPA of 4.0 on a 5.0 scale. *Application deadline:* For fall admission, 7/3; for spring admission, 11/8. Application fee: $40 ($50 for international students). *Financial aid:* In 1998–99, 14 students received aid; fellowships, research assistantships, teaching assistantships, tuition waivers (full) available. *Faculty research:* Transportation and geotechnical engineering, damage and anisotropic behavior, steel processing. *Unit head:* Dr. Mohsen Issa, Director of Graduate Studies, 312-996-3432.

University of Illinois at Urbana–Champaign, Graduate College, College of Engineering, Department of Materials Science and Engineering, Urbana, IL 61801. Offers MS, PhD. *Faculty:* 29 full-time (2 women), 4 part-time (0 women). *Students:* 112 full-time (31 women); includes 14 minority (8 Asian Americans or Pacific Islanders, 5 Hispanic Americans, 1 Native American), 55 international. 302 applicants, 7% accepted. In 1998, 14 master's, 25 doctorates awarded. *Degree requirements:* For master's, thesis required, foreign language not required; for doctorate, dissertation, departmental qualifying exam required, foreign language not required. *Application deadline:* Applications are processed on a rolling basis. Application fee: $40 ($50 for international students). Tuition, state resident: full-time $4,616. Tuition, nonresident: full-time $11,768. Full-time tuition and fees vary according to course load. *Financial aid:* Fellowships, research assistantships, teaching assistantships, tuition waivers (full and partial) available. Financial aid application deadline: 2/15. *Unit head:* James Economy, Head, 217-333-1440. *Application contact:* Robert Averback, Director of Graduate Studies, 217-333-4302, Fax: 217-333-2736, E-mail: averback@uiuc.edu.

University of Maryland, College Park, Graduate School, A. James Clark School of Engineering, Department of Materials and Nuclear Engineering, Materials Science and Engineering Program, College Park, MD 20742-5045. Offers M Eng, MS, PhD. Part-time and evening/weekend programs available. Postbaccalaureate distance learning degree programs offered. *Students:* 33 full-time (8 women), 29 part-time (6 women); includes 10 minority (4 African Americans, 6 Asian Americans or Pacific Islanders), 31 international. 131 applicants, 18% accepted. In 1998, 9 master's, 8 doctorates awarded. *Degree requirements:* For master's, thesis or alternative, research paper, written comprehensive exam required, foreign language not required; for doctorate, dissertation, oral exam required. *Entrance requirements:* For master's and doctorate, GRE General Test, TOEFL, minimum B+ average in undergraduate course work. *Application deadline:* Applications are processed on a rolling basis. Application fee: $50 ($70 for international students). Tuition, state resident: part-time $272 per credit hour. Tuition, nonresident: part-time $475 per credit hour. Required fees: $632; $379 per year. *Financial aid:* Fellowships, research assistantships, teaching assistantships available. Financial aid applicants required to submit FAFSA. *Unit head:* Dr. Henrique Reis, Director of Graduate Studies, 217-333-1228, Fax: 217-244-5705, E-mail: h-reis@uiuc.edu. *Application contact:* Barret Cole, Graduate Admissions Secretary, 301-405-5211, Fax: 301-314-2029.

Announcement: Interdisciplinary MS and PhD program that emphasizes thin films (ferroelectrics, semiconductors, ceramics), polymers (thermodynamics of complex systems, properties of fibers, environmental effects), ceramics (processing and manufacturing, substrates for electronic packaging semiconductors and ferroelectric ceramics, statistical process control, microwave processing), composites (multilayer, filled and fiber-reinforced polymeric and ceramic composite structures, metal matrix composites), microelectronic and electronic packaging materials (optical interconnects performance simulation, reliability of packaging materials and structures, heterostructures and compound semiconductors), and smart materials and structures (integrated sensors, composites actuators, shape memory materials and ferromagnetic composite systems).

See in-depth description on page 1347.

University of Maryland, College Park, Graduate School, A. James Clark School of Engineering, Department of Mechanical Engineering, College Park, MD 20742-5045. Offers electronic packaging and reliability (MS, PhD); manufacturing and design (MS, PhD); mechanical engineering (M Eng); mechanics and materials (MS, PhD); thermal and fluid sciences (MS, PhD). Part-time and evening/weekend programs available. Postbaccalaureate distance learning degree programs offered. *Faculty:* 61 full-time (5 women), 15 part-time (0 women). *Students:* 133 full-time (13 women), 60 part-time (7 women); includes 18 minority (11 African Americans, 6 Asian Americans or Pacific Islanders, 1 Hispanic American), 123 international. *Degree requirements:* For master's, thesis optional, foreign language not required; for doctorate, dissertation, qualifying exam required, foreign language not required. *Entrance requirements:* For master's, GRE, minimum GPA of 3.0. *Application deadline:* Applications are processed on a rolling basis. Application fee: $50 ($70 for international students). Tuition, state resident: part-time $272 per credit hour. Tuition, nonresident: part-time $475 per credit hour. Required fees: $632; $379 per year. *Unit head:* Dr. Davinder Anand, Chairman, 301-405-5294, Fax: 301-314-9477. *Application contact:* Dr. James M. Wallace, Graduate Director, 301-405-4216.

University of Maryland, College Park, Graduate School, A. James Clark School of Engineering, Professional Program in Engineering, College Park, MD 20742-5045. Offers aerospace engineering (M Eng); chemical engineering (M Eng); civil engineering (M Eng); electrical engineering (M Eng); fire protection engineering (M Eng); materials science and engineering (M Eng); mechanical engineering (M Eng); reliability engineering (M Eng); systems engineering (M Eng). Part-time and evening/weekend programs available. Postbaccalaureate distance learning degree programs offered. *Faculty:* 11 part-time (0 women). *Students:* 20 full-time (3 women), 205 part-time (42 women); includes 58 minority (27 African Americans, 25 Asian Americans or Pacific Islanders, 5 Hispanic Americans, 1 Native American), 20 international.

Materials Engineering

University of Maryland, College Park (continued)

Degree requirements: For master's, foreign language and thesis not required. *Application deadline:* Applications are processed on a rolling basis. Application fee: $50 ($70 for international students). Tuition, state resident: part-time $272 per credit hour. Tuition, nonresident: part-time $475 per credit hour. Required fees: $632; $379 per year. *Unit head:* Dr. Patrick Cunniff, Associate Dean, 301-405-5256, Fax: 301-314-9477. *Application contact:* Trudy Lindsey, Director, Graduate Admission and Records, 301-405-4198, Fax: 301-314-9305, E-mail: grschool@deans.umd.edu.

University of Michigan, Horace H. Rackham School of Graduate Studies, College of Engineering, Department of Materials Science and Engineering, Ann Arbor, MI 48109. Offers MS, PhD. Part-time programs available. *Faculty:* 21 full-time (2 women), 3 part-time (0 women). *Students:* 68 full-time (14 women), 21 part-time (6 women), 37 international. 285 applicants, 16% accepted. In 1998, 8 master's, 8 doctorates awarded. *Degree requirements:* For master's, thesis, oral defense of thesis required, foreign language not required; for doctorate, dissertation, oral defense of dissertation, written exam required, foreign language not required. *Entrance requirements:* For master's, GRE General Test, TOEFL (minimum score of 560 required), minimum GPA of 3.0 in related field; for doctorate, GRE General Test, TOEFL (minimum score of 560 required), minimum GPA of 3.0 in related field, master's degree. *Average time to degree:* Master's–2.6 years full-time, 3.5 years part-time; doctorate–5.2 years full-time. *Application deadline:* For fall admission, 2/15. Applications are processed on a rolling basis. Application fee: $55. *Financial aid:* In 1998–99, 2 fellowships with full tuition reimbursements, 60 research assistantships with full tuition reimbursements, 10 teaching assistantships with full tuition reimbursements were awarded. Financial aid application deadline: 2/15; financial aid applicants required to submit FAFSA. *Faculty research:* Composite materials, polymer alloys, structural ceramics, materials processing and manufacturing, micromechanical and macromechanical behavior, surface modification and energy-beam interactions of materials, device materials, theoretical modeling and computer simulation, materials characterization. Total annual research expenditures: $5.1 million. *Unit head:* Albert F. Yee, Chair, 734-763-2445, E-mail: afyee@engin.umich.edu. *Application contact:* Renee Hilgendorf, Graduate Program Assistant, 734-763-9790, Fax: 734-763-4788, E-mail: reneeh@umich.edu.

See in-depth description on page 1351.

University of Minnesota, Twin Cities Campus, Graduate School, Institute of Technology, Department of Chemical Engineering and Materials Science, Program in Materials Science and Engineering, Minneapolis, MN 55455-0213. Offers M Mat SE, MS Mat SE, PhD. Part-time programs available. Terminal master's awarded for partial completion of doctoral program. *Degree requirements:* For master's and doctorate, thesis/dissertation required, foreign language not required. *Entrance requirements:* For master's and doctorate, GRE General Test. *Faculty research:* Fracture micromechanics, hydrogen embrittlement, polymer physics, microelectric materials, corrosion science.

University of Nebraska–Lincoln, Graduate College, College of Engineering and Technology, Department of Mechanical Engineering, Lincoln, NE 68588. Offers engineering (PhD); mechanical engineering (MS), including materials science engineering. *Faculty:* 15 full-time (2 women). *Students:* 21 full-time (2 women), 9 part-time (1 woman), 13 international. *Degree requirements:* For master's, thesis optional, foreign language not required; for doctorate, dissertation, comprehensive exams required. *Entrance requirements:* For master's and doctorate, GRE General Test, TOEFL (minimum score of 550 required). *Application deadline:* For fall admission, 3/1 (priority date). Applications are processed on a rolling basis. Application fee: $35. Electronic applications accepted. *Unit head:* Dr. David Lou, Chair, 402-472-2375.

University of Pennsylvania, School of Engineering and Applied Science, Department of Materials Science and Engineering, Philadelphia, PA 19104. Offers MSE, PhD, MSE/MBA. Part-time programs available. Terminal master's awarded for partial completion of doctoral program. *Degree requirements:* For master's and doctorate, thesis/dissertation required, foreign language not required. *Entrance requirements:* For master's and doctorate, TOEFL (minimum score of 600 required). Electronic applications accepted. *Faculty research:* Advanced metallic, ceramic, and polymeric materials for device applications; micromechanics and structure of interfaces; thin film electronic materials; physics and chemistry of solids.

See in-depth description on page 1355.

University of Pittsburgh, School of Engineering, Department of Materials Science and Engineering, Pittsburgh, PA 15260. Offers materials science and engineering (MSMSE, PhD); metallurgical engineering (MS Met E, PhD). Part-time and evening/weekend programs available. *Faculty:* 13 full-time (0 women), 2 part-time (0 women). *Students:* 12 full-time (5 women), 19 part-time (2 women); includes 1 minority (Asian American or Pacific Islander), 8 international. 114 applicants, 9% accepted. In 1998, 6 master's, 9 doctorates awarded. Terminal master's awarded for partial completion of doctoral program. *Degree requirements:* For master's, thesis required (for some programs), foreign language not required; for doctorate, dissertation, comprehensive and final oral exams required, foreign language not required. *Entrance requirements:* For master's and doctorate, TOEFL (minimum score of 550 required), minimum QPA of 3.0. *Average time to degree:* Master's–3 years full-time, 5 years part-time; doctorate–6 years full-time. *Application deadline:* For fall admission, 8/1 (priority date); for spring admission, 12/1 (priority date). Applications are processed on a rolling basis. Application fee: $30 ($40 for international students). *Financial aid:* In 1998–99, 13 students received aid, including 9 research assistantships (averaging $10,032 per year), 4 teaching assistantships (averaging $10,600 per year); grants, scholarships, and tuition waivers (full and partial) also available. Financial aid application deadline: 2/15. *Faculty research:* Hot corrosion and oxidation, physical metallurgy and thermomechanical processing of steels, ceramic processing, phase transformation. Total annual research expenditures: $2 million. *Unit head:* Dr. William A. Soffa, Chairman, 412-624-9720, Fax: 412-624-8069, E-mail: wsoffa+@pitt.edu. *Application contact:* Pradeep P. Phule, Graduate Coordinator, 412-624-9736, Fax: 412-624-8069, E-mail: ppp@vms.cis.pitt.edu.

University of Southern California, Graduate School, School of Engineering, Department of Materials Science and Engineering, Program in Materials Engineering, Los Angeles, CA 90089. Offers MS. *Students:* 17 full-time (4 women), 4 part-time (1 woman), 20 international. Average age 27. 21 applicants, 81% accepted. In 1998, 10 degrees awarded. *Degree requirements:* For master's, thesis optional. *Entrance requirements:* For master's, GRE General Test. *Application deadline:* For fall admission, 7/1 (priority date); for spring admission, 12/1. Application fee: $55. Tuition: Part-time $768 per unit. Required fees: $350 per semester. *Financial aid:* Fellowships, research assistantships, teaching assistantships, Federal Work-Study and institutionally-sponsored loans available. Aid available to part-time students. Financial aid application deadline: 2/15; financial aid applicants required to submit FAFSA. *Unit head:* Dr. Florian Mansfeld, Chairman, Department of Materials Science and Engineering, 213-740-6011.

The University of Texas at Arlington, Graduate School, College of Engineering, Department of Mechanical and Aerospace Engineering, Materials Science and Engineering Program, Arlington, TX 76019. Offers MS, PhD. *Students:* 19 full-time (4 women), 13 part-time (6 women); includes 4 minority (3 Asian Americans or Pacific Islanders, 1 Hispanic American), 17 international. 70 applicants, 34% accepted. In 1998, 6 master's, 2 doctorates awarded. *Degree requirements:* For master's, thesis required (for some programs), foreign language not required; for doctorate, dissertation required, foreign language not required. *Entrance requirements:* For master's and doctorate, GRE General Test, TOEFL. *Application deadline:* Applications are processed on a rolling basis. Application fee: $25 ($50 for international students). Tuition, state resident: full-time $1,368; part-time $76 per semester hour. Tuition, nonresident: full-time $5,454; part-time $303 per semester hour. Required fees: $66 per semester hour. $86 per term. Tuition and fees vary according to course load. *Financial aid:* Research assistantships, teaching assistantships available. *Unit head:* Dr. Ronald L. Elsenbaumer, Director, 817-272-2398, Fax: 817-272-2538. *Application contact:* Pranesh B. Aswath, Graduate Adviser, 817-272-2008, Fax: 817-272-2538, E-mail: aswath@uta.edu.

See in-depth description on page 1357.

The University of Texas at Austin, Graduate School, College of Engineering, Program in Materials Science and Engineering, Austin, TX 78712-1111. Offers MSE, PhD. Part-time programs available. *Faculty:* 44 full-time (2 women). *Students:* 53 full-time (8 women), 7 part-time (3 women); includes 3 minority (2 Asian Americans or Pacific Islanders, 1 Hispanic American), 35 international. Average age 26. 72 applicants, 31% accepted. In 1998, 7 master's awarded (57% found work related to degree, 43% continued full-time study); 8 doctorates awarded (100% found work related to degree). *Degree requirements:* For master's and doctorate, thesis/dissertation required (for some programs), foreign language not required. *Entrance requirements:* For master's and doctorate, GRE General Test. *Average time to degree:* Master's–2.1 years full-time; doctorate–4.8 years full-time. *Application deadline:* For fall admission, 1/2 (priority date); for spring admission, 10/1. Applications are processed on a rolling basis. Application fee: $50 ($75 for international students). Electronic applications accepted. *Financial aid:* In 1998–99, 54 students received aid, including 2 fellowships with partial tuition reimbursements available (averaging $15,000 per year), 41 research assistantships with full tuition reimbursements available (averaging $13,000 per year), 11 teaching assistantships with partial tuition reimbursements available (averaging $13,000 per year); career-related internships or fieldwork and institutionally-sponsored loans also available. Financial aid application deadline: 1/1; financial aid applicants required to submit FAFSA. *Faculty research:* Ceramics and glasses, microelectronic materials and processing, metal systems, nanostructures, polymers . Total annual research expenditures: $3 million. *Unit head:* Donald R. Paul, Director, 512-471-1504, Fax: 512-475-8482, E-mail: drp@che.utexas.edu. *Application contact:* Lydia C. Griffith, Graduate Coordinator, 512-471-1504, Fax: 512-475-8482, E-mail: lcgriffith@mail.utexas.edu.

The University of Texas at El Paso, Graduate School, Interdisciplinary Program in Materials Science and Engineering, El Paso, TX 79968-0001. Offers PhD. *Students:* 19 full-time (5 women), 9 part-time (4 women); includes 13 minority (1 Asian American or Pacific Islander, 12 Hispanic Americans), 12 international. Average age 33. 30 applicants, 13% accepted. *Degree requirements:* For doctorate, dissertation, scholarly article required, foreign language not required. *Entrance requirements:* For doctorate, GRE General Test, TOEFL (minimum score of 550 required), minimum GPA of 3.2. *Application deadline:* Applications are processed on a rolling basis. Application fee: $0 ($65 for international students). Tuition, state resident: full-time $2,790. Tuition, nonresident: full-time $7,710. *Financial aid:* In 1998–99, 7 fellowships, 7 research assistantships, 2 teaching assistantships were awarded.; Federal Work-Study, institutionally-sponsored loans, and tuition waivers (partial) also available. Financial aid applicants required to submit FAFSA. *Faculty research:* Material chemistry, materials physics, mineral science. Total annual research expenditures: $3.5 million. *Unit head:* Dr. Lawrence Murr, Director, 915-747-6929, Fax: 915-747-5616. *Application contact:* Susan Jordan, Director, Graduate Student Services, 915-747-5491, Fax: 915-747-5788, E-mail: sjordan@utep.edu.

University of Utah, Graduate School, College of Engineering, Department of Materials Science and Engineering, Salt Lake City, UT 84112-1107. Offers ME, MS, PhD. Part-time programs available. *Faculty:* 10 full-time (0 women), 11 part-time (2 women). *Students:* 40 full-time (6 women), 16 part-time (5 women); includes 1 minority (Asian American or Pacific Islander), 37 international. Average age 30. In 1998, 4 master's, 5 doctorates awarded. Terminal master's awarded for partial completion of doctoral program. *Degree requirements:* For master's, thesis (MS) required; for doctorate, dissertation, exam required, foreign language not required. *Entrance requirements:* For master's and doctorate, TOEFL (minimum score of 500 required), minimum GPA of 3.0. *Application deadline:* For fall admission, 7/1. Application fee: $30 ($50 for international students). *Financial aid:* In 1998–99, 3 teaching assistantships were awarded.; research assistantships, career-related internships or fieldwork and Federal Work-Study also available. Financial aid application deadline: 3/1. *Faculty research:* Biomaterials, ceramic composition, electrochemistry, semiconductor materials and devices, crystal growth. *Unit head:* Anil Virkar, Chair, 801-581-6863, Fax: 801-581-4816, E-mail: anil.virkar@m.cc.utah.edu. *Application contact:* Judy Martinez, Director of Graduate Studies, 801-581-6863.

University of Washington, Graduate School, College of Engineering, Department of Materials Science and Engineering, Seattle, WA 98195-2120. Offers ceramic engineering (MS, MSE); materials science (MSE); materials science and engineering (MS, MSMSE, PhD); metallurgical engineering (MS, MSE). Part-time programs available. *Faculty:* 12 full-time (1 woman), 3 part-time (0 women). *Students:* 43 full-time (12 women), 18 part-time (5 women); includes 8 minority (2 African Americans, 3 Asian Americans or Pacific Islanders, 3 Hispanic Americans), 17 international. Average age 32. 62 applicants, 53% accepted. In 1998, 3 master's awarded (67% found work related to degree, 33% continued full-time study); 2 doctorates awarded (100% found work related to degree). *Degree requirements:* For master's and doctorate, thesis/dissertation required, foreign language not required. *Entrance requirements:* For master's and doctorate, GRE General Test (minimum combined score of 1400 on quantitative and analytical sections required), TOEFL (minimum score of 600 required), minimum GPA of 3.0. *Average time to degree:* Master's–3.4 years full-time, 7.5 years part-time; doctorate–6.25 years full-time. *Application deadline:* For fall admission, 7/1; for winter admission, 11/1; for spring admission, 2/1. Applications are processed on a rolling basis. Application fee: $50. Electronic applications accepted. Tuition, state resident: full-time $5,196; part-time $475 per credit. Tuition, nonresident: full-time $13,485; part-time $1,285 per credit. Required fees: $387; $38 per credit. Tuition and fees vary according to course load. *Financial aid:* In 1998–99, 37 students received aid, including 3 fellowships with full tuition reimbursements available (averaging $8,167 per year), 22 research assistantships with full tuition reimbursements available (averaging $14,088 per year), 12 teaching assistantships with full tuition reimbursements available (averaging $9,900 per year); career-related internships or fieldwork, Federal Work-Study, grants, institutionally-sponsored loans, scholarships, and unspecified assistantships also available. Financial aid application deadline: 3/1. *Faculty research:* Processing characterization and properties of structural and electronic materials. Total annual research expenditures: $1.1 million. *Unit head:* Dr. Rajendra K. Bordia, Chair, 206-543-2600, Fax: 206-543-3100, E-mail: bordia@u.washington.edu. *Application contact:* Dr. T. G. Stoebe, Graduate Coordinator, 206-543-7090, Fax: 206-543-3100, E-mail: mse@u.washington.edu.

University of Windsor, College of Graduate Studies and Research, Faculty of Engineering, Mechanical Engineering and Materials Engineering, Program in Engineering Materials, Windsor, ON N9B 3P4, Canada. Offers MA Sc, PhD. *Degree requirements:* For master's and doctorate, thesis/dissertation required, foreign language not required. *Entrance requirements:* For master's, TOEFL (minimum score of 550 required), minimum B average; for doctorate, TOEFL, master's degree. *Faculty research:* Physical metallurgy.

Vanderbilt University, School of Engineering, Program in Materials Science and Engineering, Nashville, TN 37240-1001. Offers MS, PhD. Part-time programs available. *Faculty:* 13 full-time (0 women), 1 part-time (0 women). *Students:* 18 full-time (4 women); includes 11 minority (8 Asian Americans or Pacific Islanders, 3 Native Americans) Average age 28. 53 applicants, 28% accepted. In 1998, 2 master's, 2 doctorates awarded. Terminal master's awarded for partial completion of doctoral program. *Degree requirements:* For master's and doctorate, thesis/dissertation required, foreign language not required. *Entrance requirements:* For master's and doctorate, GRE General Test. *Average time to degree:* Master's–2 years full-time; doctorate–2 years full-time. *Application deadline:* For fall admission, 1/15. Application fee: $40. *Financial aid:* In 1998–99, 4 fellowships, 10 research assistantships, 7 teaching assistantships were awarded.; institutionally-sponsored loans and tuition waivers (partial) also available. Aid available to part-time students. Financial aid application deadline: 1/15. *Faculty research:* Tribology, solidification, electron microscopy, ion implantation, diamond deposition. Total annual research expenditures: $1.4 million. *Unit head:* Robert J. Bayuzick, Director, 615-343-6868, Fax: 615-343-8645, E-mail: bayuzick@vuse.vanderbilt.edu. *Application contact:* Sarah M. Ross, Administrative Assistant, 615-343-2415, Fax: 615-343-8645, E-mail: sross@vuse.vanderbilt.edu.

Virginia Polytechnic Institute and State University, Graduate School, College of Engineering, Department of Materials Science and Engineering, Blacksburg, VA 24061. Offers M Eng, MS. *Students:* 9 full-time (2 women), 3 part-time (1 woman), 5 international. 32 applicants, 28% accepted. In 1998, 7 degrees awarded. *Degree requirements:* For master's, thesis required, foreign language not required. *Entrance requirements:* For master's, TOEFL. *Application deadline:* For fall admission, 12/1 (priority date). Applications are processed on a rolling basis. Application fee: $25. *Financial aid:* Fellowships, research assistantships, teaching assistantships, unspecified assistantships available. Financial aid application deadline: 4/1. *Unit head:* Dr. R. S. Gordon, Head, 540-231-6655, E-mail: gordonrs@vt.edu.

Virginia Polytechnic Institute and State University, Graduate School, College of Engineering, Program in Materials Engineering Science, Blacksburg, VA 24061. Offers PhD. *Students:* 8 full-time (3 women), 1 part-time, 6 international. 61 applicants, 18% accepted. In 1998, 1 degree awarded. *Degree requirements:* For doctorate, dissertation required. *Entrance requirements:* For doctorate, TOEFL. *Application deadline:* For fall admission, 12/1 (priority date). Applications are processed on a rolling basis. Application fee: $25. *Financial aid:* Fellowships, research assistantships, teaching assistantships, career-related internships or fieldwork and unspecified assistantships available. Financial aid application deadline: 4/1. *Faculty research:* Polymer syntheses, adhesive and sealant science, composite materials and structures, surface chemistry. *Unit head:* Dr. G. Wilkes, Chairman, 540-231-5316, E-mail: gwilkes@vt.edu.

Washington State University, Graduate School, College of Engineering and Architecture, School of Mechanical and Materials Engineering, Program in Materials Science and Engineering, Pullman, WA 99164. Offers materials science (PhD); materials science and engineering (MS). *Faculty:* 10 full-time (0 women). *Students:* 4 full-time (0 women), 1 part-time, 4 international. Average age 25. In 1998, 4 master's awarded (25% found work related to degree, 75% continued full-time study); 4 doctorates awarded. *Degree requirements:* For master's and doctorate, thesis/dissertation, oral exam required, foreign language not required. *Entrance requirements:* For master's and doctorate, GRE General Test, minimum GPA of 3.0. *Average time to degree:* Master's–2 years full-time; doctorate–4 years full-time. *Application deadline:* For fall admission, 3/1 (priority date). Applications are processed on a rolling basis. Application fee: $35. *Financial aid:* In 1998–99, 1 research assistantship, 1 teaching assistantship were awarded.; fellowships, career-related internships or fieldwork, Federal Work-Study, institutionally-sponsored loans, and teaching associateships also available. Financial aid application deadline: 4/1. *Faculty research:* Fatigue and fracture of structural materials, mechanical behavior and corrosion of thin films, superplastic forming, fundamental aspects of deformation processes including numerical modeling. *Unit head:* Dr. Grant Norton, Chair. *Application contact:* Dr. Grant Norton, Chair.

Washington University in St. Louis, School of Engineering and Applied Science, Sever Institute of Technology, Program in Materials Science and Engineering, St. Louis, MO 63130-4899. Offers MS, D Sc. Part-time programs available. *Faculty:* 14 full-time (0 women), 12 part-time (0 women). *Students:* 5 full-time (1 woman), 10 part-time (2 women), 8 international. 152 applicants, 7% accepted. In 1998, 4 master's, 2 doctorates awarded. Terminal master's awarded for partial completion of doctoral program. *Degree requirements:* For master's, thesis optional, foreign language not required; for doctorate, dissertation, departmental qualifying exam required, foreign language not required. *Entrance requirements:* For master's and doctorate, GRE. *Average time to degree:* Master's–1.5 years full-time, 5 years part-time; doctorate–5 years full-time, 8 years part-time. *Application deadline:* For fall admission, 2/15 (priority date). Applications are processed on a rolling basis. Application fee: $20. *Financial aid:* In 1998–99, 3 students received aid, including 1 research assistantship, 1 teaching assistantship; fellowships, career-related internships or fieldwork, Federal Work-Study, and institutionally-sponsored loans also available. Financial aid application deadline: 2/15. *Faculty research:* Material processing, magnetic materials, composites, name-phase materials, high-temperature materials. Total annual research expenditures: $500,000. *Unit head:* Dr. Kenneth Jerina, Chairman, 314-935-6012.

Wayne State University, Graduate School, College of Engineering, Department of Chemical Engineering and Materials Science, Program in Materials Science and Engineering, Detroit, MI 48202. Offers materials science and engineering (MS, PhD); polymer engineering (Certificate). Part-time programs available. Terminal master's awarded for partial completion of doctoral program. *Degree requirements:* For master's, thesis optional, foreign language not required; for doctorate, dissertation required, foreign language not required. *Faculty research:* Polymer science, rheology, fatigue in metals, metal matrix composites, ceramics.

Western Michigan University, Graduate College, College of Engineering and Applied Sciences, Department of Construction Engineering, Materials Engineering and Industrial Design, Program in Materials Science and Engineering, Kalamazoo, MI 49008. Offers MS. *Students:* 4 full-time (0 women), 12 part-time (2 women), 13 international. 31 applicants, 81% accepted. *Entrance requirements:* For master's, minimum GPA of 3.0. *Application deadline:* For fall admission, 2/15 (priority date). Applications are processed on a rolling basis. Application fee: $25. *Financial aid:* Application deadline: 2/15; *Application contact:* Paula J. Boodt, Coordinator, Graduate Admissions and Recruitment, 616-387-2000, Fax: 616-387-2355, E-mail: paula.boodt@wmich.edu.

Worcester Polytechnic Institute, Graduate Studies, Department of Materials Science and Engineering, Worcester, MA 01609-2280. Offers MS, PhD, Certificate. Part-time and evening/weekend programs available. *Faculty:* 8 full-time (2 women). *Students:* 17 full-time (6 women), 10 part-time (2 women); includes 1 minority (Asian American or Pacific Islander), 7 international. 65 applicants, 46% accepted. In 1998, 5 master's, 1 doctorate awarded. *Degree requirements:* For master's and doctorate, thesis/dissertation required, foreign language not required. *Entrance requirements:* For master's, TOEFL (minimum score of 550 required; average 599); for doctorate, GRE General Test, TOEFL (minimum score of 550 required; average 599). *Average time to degree:* Master's–1.5 years full-time, 4 years part-time; doctorate–4.5 years full-time. *Application deadline:* For fall admission, 2/15 (priority date); for spring admission, 10/15 (priority date). Applications are processed on a rolling basis. Application fee: $50. Electronic applications accepted. *Financial aid:* In 1998–99, 3 students received aid, including 1 fellowship with full tuition reimbursement available (averaging $11,970 per year), 2 teaching assistantships with full tuition reimbursements available (averaging $11,970 per year); research assistantships, career-related internships or fieldwork, grants, institutionally-sponsored loans, and scholarships also available. Financial aid application deadline: 2/15; financial aid applicants required to submit FAFSA. *Faculty research:* Tribology, corrosion, electron microscopy, powder metallurgy and ceramics. *Unit head:* Richard D. Sisson, Head, 508-831-5335, Fax: 508-831-5178, E-mail: sisson@wpi.edu.

Worcester Polytechnic Institute, Graduate Studies, Department of Mechanical Engineering, Worcester, MA 01609-2280. Offers materials science and engineering (MS, PhD); mechanical engineering (M Eng, MS, PhD, Advanced Certificate). Part-time and evening/weekend programs available. *Faculty:* 18 full-time (1 woman), 1 part-time (0 women). *Students:* 58 full-time (6 women), 19 part-time (3 women); includes 5 minority (4 Asian Americans or Pacific Islanders, 1 Hispanic American), 22 international. *Degree requirements:* For master's, thesis optional, foreign language not required; for doctorate, dissertation required, foreign language not required. *Entrance requirements:* For master's, TOEFL (minimum score of 550 required; average 608); for doctorate, GRE General Test (combined average 1947 on three sections), TOEFL (minimum score of 550 required; average 608). *Application deadline:* For fall admission, 2/15 (priority date); for spring admission, 10/15 (priority date). Applications are processed on a rolling basis. Application fee: $50. Electronic applications accepted. *Unit head:* Mohammad N. Noori, Head, 508-831-5759, Fax: 508-831-5680, E-mail: mnnoori@wpi.edu. *Application contact:* Mark Richman, Graduate Coordinator, 508-831-5556, Fax: 508-831-5680, E-mail: mrichman@wpi.edu.

Wright State University, School of Graduate Studies, College of Engineering and Computer Science, Programs in Engineering, Program in Mechanical and Materials Engineering, Dayton, OH 45435. Offers materials science and engineering (MSE); mechanical engineering (MSE). *Students:* 32 full-time (9 women), 30 part-time (3 women); includes 4 minority (1 African American, 2 Asian Americans or Pacific Islanders, 1 Hispanic American), 26 international. Average age 29. 130 applicants, 55% accepted. In 1998, 23 degrees awarded. *Degree requirements:* For master's, thesis or course option alternative required. *Entrance requirements:* For master's, TOEFL (minimum score of 550 required). Application fee: $25. *Financial aid:* Fellowships, research assistantships, teaching assistantships, unspecified assistantships available. Aid available to part-time students. Financial aid application deadline: 3/15; financial aid applicants required to submit FAFSA. *Unit head:* Dr. Richard J. Bethke, Chair, 937-775-5040, Fax: 937-775-5009.

Materials Sciences

Alabama Agricultural and Mechanical University, School of Graduate Studies, School of Arts and Sciences, Department of Natural and Physical Sciences, Area in Physics, Normal, AL 35762-1357. Offers applied physics (PhD); materials science (PhD); optics (PhD); physics (MS). Part-time programs available. *Faculty:* 11 full-time (1 woman). *Students:* 15 full-time (3 women), 17 part-time (8 women); includes 22 minority (all African Americans), 7 international. *Degree requirements:* For master's, thesis optional, foreign language not required; for doctorate, computer language, dissertation required, foreign language not required. *Entrance requirements:* For master's, GRE General Test, BS in electrical engineering or physics; for doctorate, GRE General Test (minimum combined score of 1000 required). *Application deadline:* For fall admission, 5/1 (priority date). Applications are processed on a rolling basis. Application fee: $15 ($20 for international students). Tuition, state resident: full-time $1,932. Tuition, nonresident: full-time $3,864. Tuition and fees vary according to course load. *Unit head:* J. C. Wang, Chairperson, 256-851-6662.

Alfred University, Graduate School, New York State College of Ceramics, School of Ceramic Engineering and Materials Science, Alfred, NY 14802-1205. Offers ceramic engineering (MS); ceramics (PhD); glass science (MS, PhD); materials science (MS). *Students:* 53 full-time (10 women), 7 part-time (3 women). Average age 24. 111 applicants, 23% accepted. In 1998, 11 master's, 10 doctorates awarded. *Degree requirements:* For master's and doctorate, thesis/dissertation required, foreign language not required. *Entrance requirements:* For master's and doctorate, TOEFL (minimum score of 590 required). *Application deadline:* Applications are processed on a rolling basis. Application fee: $50. Electronic applications accepted. *Financial aid:* Fellowships, research assistantships, teaching assistantships, Federal Work-Study and tuition waivers (full and partial) available. Aid available to part-time students. Financial aid applicants required to submit FAFSA. *Faculty research:* Fine-particle technology, x-ray diffraction, superconductivity, electronic materials. *Unit head:* Dr. James Reed, Dean, 607-871-2441, E-mail: freed@bigvax.alfred.edu. *Application contact:* Cathleen R. Johnson, Coordinator of Graduate Admissions, 607-871-2141, Fax: 607-871-2198, E-mail: johnsonc@king.alfred.edu.

See in-depth description on page 1319.

Arizona State University, Graduate College, College of Engineering and Applied Sciences, Department of Chemical, Bio and Materials Engineering, Program in Materials Science and Engineering, Tempe, AZ 85287. Offers engineering science (MS, MSE, PhD), including materials science and engineering. *Degree requirements:* For doctorate, dissertation required. *Entrance requirements:* For master's and doctorate, GRE General Test. Application fee: $45. *Financial aid:* Fellowships, research assistantships, teaching assistantships available. *Faculty research:* Fluid dynamics, mechanics of solids and structures, structural dynamics, structural stability, nonlinear mechanics. *Unit head:* Dr. Stephen J. Krause, Coordinator, 480-965-2050.

See in-depth description on page 1321.

Arizona State University, Graduate College, Interdisciplinary Program in Science and Engineering of Materials, Tempe, AZ 85287. Offers PhD. *Degree requirements:* For doctorate, dissertation required. *Entrance requirements:* For doctorate, GRE. Application fee: $45. *Faculty research:* Silicon and gallium arsenide, fabrication of ultra small solid state electronic devices, structure of free surfaces of crystalline solids, ion implantation on solids, effects of high pressures on solids. *Unit head:* Dr. William T. Petuskey, Co-Director, 480-965-4549. *Application contact:* Dr. William T. Petuskey, Co-Director, 480-965-4549.

See in-depth description on page 1323.

Brown University, Graduate School, Division of Engineering, Program in Materials Science, Providence, RI 02912. Offers Sc M, PhD. *Degree requirements:* For doctorate, dissertation, preliminary exam required, foreign language not required, foreign language not required.

California Institute of Technology, Division of Engineering and Applied Science, Option in Materials Science, Pasadena, CA 91125-0001. Offers MS, PhD. *Faculty:* 4 full-time (1 woman). *Students:* 26 full-time (6 women), 6 international. 78 applicants, 8% accepted. In 1998, 6 master's, 4 doctorates awarded. *Degree requirements:* For master's, foreign language and thesis not required; for doctorate, dissertation required, foreign language not required. *Application deadline:* For fall admission, 1/15. Application fee: $0. *Faculty research:* Mechanical properties, physical properties, kinetics of phase transformations, metastable phases, transmission electron microscopy. *Unit head:* Dr. Brent T. Fultz, Representative, 626-395-2170.

California Polytechnic State University, San Luis Obispo, College of Engineering, Program in Engineering, San Luis Obispo, CA 93407. Offers biochemical engineering (MS); industrial engineering (MS); integrated technology management (MS); materials engineering (MS); mechanical engineering (MS); water engineering (MS). *Faculty:* 98 full-time (8 women), 82 part-time (14 women). *Students:* 25 full-time (3 women), 6 part-time. *Degree requirements:* Foreign language not required. *Entrance requirements:* For master's, GRE General Test, minimum GPA of 2.5 in last 90 quarter units. Application fee: $55. Tuition, nonresident: part-time $164 per unit. Required fees: $531 per quarter. *Unit head:* Dr. Paul E. Rainey, Associate Dean, 805-756-2131, Fax: 805-756-6503, E-mail: prainey@calpoly.edu. *Application contact:* Dr. Paul E. Rainey, Associate Dean, 805-756-2131, Fax: 805-756-6503, E-mail: prainey@calpoly.edu.

Carnegie Mellon University, Carnegie Institute of Technology, Department of Materials Science and Engineering, Pittsburgh, PA 15213-3891. Offers ME, MS, PhD. Part-time programs available. *Faculty:* 27 full-time (3 women), 1 part-time (0 women). *Students:* 62 full-time (12 women), 6 part-time (1 woman); includes 3 minority (1 African American, 1 Asian American or Pacific Islander, 1 Hispanic American), 42 international. Average age 28. In 1998, 18 master's, 12 doctorates awarded. Terminal master's awarded for partial completion of doctoral program.

Materials Sciences

Carnegie Mellon University (continued)
Degree requirements: For master's, exam required, foreign language and thesis not required; for doctorate, dissertation, qualifying exam required, foreign language not required. *Entrance requirements:* For master's and doctorate, GRE General Test, TOEFL. *Application deadline:* For fall admission, 2/1; for spring admission, 10/15. Application fee: $45. *Financial aid:* Fellowships, research assistantships, career-related internships or fieldwork available. Financial aid application deadline: 2/1. *Faculty research:* Materials characterization, process metallurgy, high strength alloys, growth kinetics, ceramics. Total annual research expenditures: $5.5 million. *Unit head:* Dr. Anthony Rollettt, Head, 412-268-2700, Fax: 412-268-7596. *Application contact:* Director of Graduate Studies, 412-268-7574.

See in-depth description on page 1325.

Case Western Reserve University, School of Graduate Studies, The Case School of Engineering, Department of Materials Science and Engineering, Cleveland, OH 44106. Offers materials science (MS, PhD). Part-time programs available. Postbaccalaureate distance learning degree programs offered (no on-campus study). *Faculty:* 12 full-time (0 women). *Students:* 23 full-time (7 women), 43 part-time (6 women). Average age 24. 60 applicants, 68% accepted. In 1998, 8 master's, 6 doctorates awarded. Terminal master's awarded for partial completion of doctoral program. *Degree requirements:* For doctorate, dissertation required, foreign language not required, foreign language not required. *Entrance requirements:* For master's and doctorate, TOEFL (minimum score of 550 required). *Average time to degree:* Master's–2 years full-time, 5 years part-time; doctorate–5 years full-time. *Application deadline:* For fall admission, 2/15 (priority date); for spring admission, 9/15. Application fee: $25. *Financial aid:* In 1998–99, 3 fellowships with full tuition reimbursements (averaging $13,500 per year), 21 research assistantships with full and partial tuition reimbursements (averaging $13,500 per year), 12 teaching assistantships (averaging $10,125 per year) were awarded. Financial aid application deadline: 4/30. *Faculty research:* Deformation and fracture; materials processing; environmental effects of materials; surfaces and interfaces; electronic, magnetic, and optical materials. Total annual research expenditures: $4.5 million. *Unit head:* Gary M. Michal, Chairman, 216-368-5070, Fax: 216-368-4224, E-mail: emse@po.cwru.edu. *Application contact:* Sarah Davies, Secretary, 216-368-6495, Fax: 216-368-3209, E-mail: sadll@po.cwru.edu.

Clemson University, Graduate School, College of Engineering and Science, School of Chemical and Materials Engineering, Program in Materials Science and Engineering, Clemson, SC 29634. Offers MS, PhD. Part-time programs available. *Students:* 25 full-time (3 women), 2 part-time (1 woman), 22 international. Average age 24. 70 applicants, 31% accepted. In 1998, 2 master's, 2 doctorates awarded. Terminal master's awarded for partial completion of doctoral program. *Degree requirements:* For master's and doctorate, thesis/dissertation required, foreign language not required. *Entrance requirements:* For master's and doctorate, GRE General Test, TOEFL. *Application deadline:* For fall admission, 2/15. Applications are processed on a rolling basis. Application fee: $35. *Financial aid:* Research assistantships, career-related internships or fieldwork available. Financial aid applicants required to submit FAFSA. *Faculty research:* Composites, fibers, ceramics, metallurgy, biomaterials, semiconductors. Total annual research expenditures: $1.2 million. *Unit head:* Dr. R. Larry Dooley, Coordinator, 864-656-5562, Fax: 864-656-5910, E-mail: dooley@eng.clemson.edu.

See in-depth description on page 1329.

Colorado School of Mines, Graduate School, Program in Materials Science, Golden, CO 80401-1887. Offers MS, PhD. Part-time programs available. *Students:* 24 full-time (9 women), 13 part-time (2 women); includes 6 minority (4 Asian Americans or Pacific Islanders, 2 Hispanic Americans), 6 international. 44 applicants, 52% accepted. In 1998, 9 master's awarded (100% found work related to degree); 3 doctorates awarded (100% found work related to degree). *Degree requirements:* For master's, thesis required, foreign language not required; for doctorate, dissertation, oral and written comprehensive exams required, foreign language not required. *Entrance requirements:* For master's and doctorate, GRE General Test (combined average 1850 on three sections), minimum GPA of 3.0. *Application deadline:* Applications are processed on a rolling basis. Application fee: $40. Electronic applications accepted. *Financial aid:* In 1998–99, 24 students received aid, including 4 fellowships, 16 research assistantships, 4 teaching assistantships; unspecified assistantships also available. Financial aid applicants required to submit FAFSA. *Faculty research:* Ceramics processing, solar and electronic materials, optical properties of surfaces and interfaces, materials synthesis, metal and alloy processing. *Unit head:* Dr. John J. Moore, Director, 303-273-3770, Fax: 303-273-3795, E-mail: jjmoore@mines.edu. *Application contact:* Sharon Kirts, Program Assistant I, 303-273-3660, Fax: 303-384-2189, E-mail: skirts@mines.edu.

Columbia University, Fu Foundation School of Engineering and Applied Science, Department of Earth and Environmental Engineering, Program in Materials Science and Engineering, New York, NY 10027. Offers materials engineering (MS, Eng Sc D); materials science (Eng Sc D, PhD, EM); materials science and engineering (MS); metallurgy (Met E). Part-time programs available. Postbaccalaureate distance learning degree programs offered (no on-campus study). *Faculty:* 7 full-time (2 women), 2 part-time (1 woman). *Students:* 25 full-time (6 women), 9 part-time; includes 4 minority (all Asian Americans or Pacific Islanders), 16 international. In 1998, 4 master's, 2 doctorates awarded. Terminal master's awarded for partial completion of doctoral program. *Degree requirements:* For master's, foreign language and thesis not required; for doctorate, dissertation, qualifying exam required, foreign language not required. *Entrance requirements:* For master's, doctorate, and other advanced degree, GRE General Test, TOEFL. Application fee: $55. *Financial aid:* In 1998–99, 24 students received aid; fellowships, research assistantships, teaching assistantships, Federal Work-Study available. Financial aid application deadline: 1/5; financial aid applicants required to submit FAFSA. *Faculty research:* Thin films, inelasticity, plastic deformation, interstitial impurities, materials for fuel cells and corrosion. Total annual research expenditures: $824,000. *Application contact:* Dr. Barbara Algin, Departmental Administrator, 212-854-2905, Fax: 212-854-7081, E-mail: ba110@columbia.edu.

Columbia University, Fu Foundation School of Engineering and Applied Science, Department of Earth and Environmental Engineering, Program in Minerals Engineering and Materials Science, New York, NY 10027. Offers metallurgical engineering (Engr); mineral engineering (Engr); minerals engineering and materials science (Eng Sc D, PhD); mines (Engr). Part-time programs available. *Faculty:* 12 full-time (2 women), 3 part-time (1 woman). *Students:* 3 full-time (0 women), 3 part-time; includes 1 minority (Asian American or Pacific Islander), 5 international. *Degree requirements:* For doctorate, dissertation, qualifying exam required. *Entrance requirements:* For doctorate and Engr, GRE General Test, TOEFL. Application fee: $55. *Financial aid:* Application deadline: 1/5; *Faculty research:* Comminution, mineral processing, flotation and flocculation, transport phenomena in heterogeneous reactions, extractive metallurgy, electrochemistry. Total annual research expenditures: $418,122. *Unit head:* Dr. Barbara Algin, Departmental Administrator, 212-854-2905, Fax: 212-854-7081, E-mail: ba110@columbia.edu. *Application contact:* Dr. Barbara Algin, Departmental Administrator, 212-854-2905, Fax: 212-854-7081, E-mail: ba110@columbia.edu.

Cornell University, Graduate School, Graduate Fields of Engineering, Field of Materials Science and Engineering, Ithaca, NY 14853-0001. Offers materials engineering (M Eng, PhD); materials science (M Eng, PhD). *Faculty:* 29 full-time. *Students:* 53 full-time (15 women); includes 6 minority (all Asian Americans or Pacific Islanders), 38 international. 225 applicants, 16% accepted. In 1998, 9 master's, 7 doctorates awarded. Terminal master's awarded for partial completion of doctoral program. *Degree requirements:* For doctorate, dissertation required, foreign language not required, foreign language not required. *Entrance requirements:* For master's and doctorate, GRE General Test, TOEFL (minimum score of 550 required). *Application deadline:* For fall admission, 1/15. Application fee: $65. Electronic applications accepted. *Financial aid:* In 1998–99, 45 students received aid, including 10 fellowships with full tuition reimbursements available, 27 research assistantships with full tuition reimbursements available, 8 teaching assistantships with full tuition reimbursements available; institutionally-sponsored loans, scholarships, tuition waivers (full and partial), and unspecified assistantships also available. Financial aid applicants required to submit FAFSA. *Faculty research:* Ceramics, complex fluids, metals, polymers, and semiconductors; electrical, magnetic, mechanical, optical, and structural properties; thin films, organic optoelectronics, nano-composites, inorganic organic hybrids, and composites. *Unit head:* Director of Graduate Studies, 607-255-9159. *Application contact:* Graduate Field Assistant, 607-255-9159, E-mail: matsci@cornell.edu.

Dartmouth College, Thayer School of Engineering, Program in Materials Sciences and Engineering, Hanover, NH 03755. Offers MS, PhD. *Degree requirements:* For master's, thesis required; for doctorate, dissertation, candidacy oral exam required. *Entrance requirements:* For master's and doctorate, GRE General Test. *Application deadline:* For fall admission, 1/15 (priority date). Application fee: $20 ($40 for international students). *Financial aid:* Fellowships, research assistantships, teaching assistantships, career-related internships or fieldwork, Federal Work-Study, institutionally-sponsored loans, and tuition waivers (full and partial) available. Financial aid application deadline: 1/15. *Faculty research:* Metals, intermetallics and ice physics, microelectronic and magnetic materials, optical thin films, biomaterials and nanostructures, laser-material interactions. Total annual research expenditures: $1.1 million. *Unit head:* Sharon Kirts, Program Assistant I, 303-273-3660, Fax: 303-384-2189, E-mail: skirts@mines.edu. *Application contact:* Candace S. Potter, Admissions Coordinator, 603-646-3844, Fax: 603-646-3856, E-mail: candace.potter@dartmouth.edu.

Duke University, Graduate School, School of Engineering, Department of Mechanical Engineering and Materials Science, Durham, NC 27708-0586. Offers materials science (MS, PhD); mechanical engineering (MS, PhD). Part-time programs available. *Faculty:* 25 full-time, 4 part-time. *Students:* 61 full-time, 1 part-time; includes 3 minority (2 African Americans, 1 Asian American or Pacific Islander), 28 international. 108 applicants, 38% accepted. In 1998, 5 master's, 8 doctorates awarded. Terminal master's awarded for partial completion of doctoral program. *Degree requirements:* For master's, thesis optional, foreign language not required; for doctorate, dissertation required, foreign language not required. *Entrance requirements:* For master's and doctorate, GRE General Test. *Application deadline:* For fall admission, 12/31; for spring admission, 11/1. Application fee: $75. *Financial aid:* Fellowships, research assistantships, teaching assistantships, Federal Work-Study available. Financial aid application deadline: 12/31. *Unit head:* Dr. Charles M. Harman, Director of Graduate Studies, 919-660-5308, Fax: 919-660-8963, E-mail: kmrogers@duke.edu.

École Polytechnique de Montréal, Graduate Programs, Department of Metallurgical and Materials Engineering, Montréal, PQ H3C 3A7, Canada. Offers advanced materials (M Eng, M Sc A, PhD); chemical metallurgy (M Eng, M Sc A, PhD); physical metallurgy (M Eng, M Sc A, PhD). Part-time programs available. *Degree requirements:* For master's and doctorate, one foreign language, computer language, thesis/dissertation required. *Entrance requirements:* For master's, minimum GPA of 2.75; for doctorate, minimum GPA of 3.0. *Faculty research:* Refractory materials, fatigue, thermochemical analysis, electrochemistry, material characterization.

The George Washington University, Columbian School of Arts and Sciences, Department of Chemistry, Washington, DC 20052. Offers analytical chemistry (MS, PhD); inorganic chemistry (MS, PhD); materials science (MS, PhD); organic chemistry (MS, PhD); physical chemistry (MS, PhD). Part-time and evening/weekend programs available. *Faculty:* 5 full-time (0 women). *Students:* 19 full-time (14 women), 7 part-time (4 women); includes 6 minority (1 African American, 3 Asian Americans or Pacific Islanders, 2 Hispanic Americans), 6 international. Terminal master's awarded for partial completion of doctoral program. *Degree requirements:* For master's, computer language, thesis or alternative, comprehensive exam required; for doctorate, computer language, dissertation, general exam required. *Entrance requirements:* For master's and doctorate, GRE General Test, interview, minimum GPA of 3.0. Application fee: $55. Tuition: Full-time $17,328; part-time $722 per credit hour. Required fees: $828; $35 per credit hour. Tuition and fees vary according to campus/location and program. *Unit head:* Dr. Michael King, Chair, 202-994-6121.

Howard University, College of Engineering, Architecture, and Computer Sciences, School of Engineering and Computer Science, Program in Materials Science and Engineering, Washington, DC 20059-0002. Offers MS, PhD. Part-time programs available. *Faculty:* 14 full-time (1 woman). *Students:* 5 full-time (1 woman); all minorities (all African Americans) Average age 25. 5 applicants, 100% accepted. Terminal master's awarded for partial completion of doctoral program. *Degree requirements:* For master's, thesis required, foreign language not required; for doctorate, dissertation, comprehensive exam required. *Entrance requirements:* For master's and doctorate, GRE General Test, TOEFL, minimum GPA of 3.0. *Application deadline:* For fall admission, 4/1; for spring admission, 11/1. Applications are processed on a rolling basis. Application fee: $45. *Financial aid:* In 1998–99, 4 fellowships with full tuition reimbursements (averaging $15,000 per year), 1 research assistantship with full tuition reimbursement (averaging $19,000 per year) were awarded.; career-related internships or fieldwork, grants, scholarships, tuition waivers (partial), and unspecified assistantships also available. Financial aid application deadline: 4/1. *Faculty research:* Polymers, biomaterials, composites, high temperature semiconductors, mechanical behavior. Total annual research expenditures: $600,000. *Unit head:* Clayton W. Bates, Director, 202-806-6147, Fax: 202-806-5258, E-mail: bates@negril.msrce.howard.edu.

Iowa State University of Science and Technology, Graduate College, College of Engineering, Department of Materials Science and Engineering, Ames, IA 50011. Offers MS, PhD. *Faculty:* 22 full-time, 2 part-time. *Students:* 34 full-time (4 women), 15 part-time (6 women); includes 2 minority (1 African American, 1 Hispanic American), 24 international. 67 applicants, 28% accepted. In 1998, 8 master's, 5 doctorates awarded. *Degree requirements:* For master's and doctorate, thesis/dissertation required. *Entrance requirements:* For master's and doctorate, GRE General Test (foreign students), TOEFL (minimum score of 550 required). *Application deadline:* For fall admission, 2/15 (priority date); for spring admission, 8/15 (priority date). Application fee: $20 ($50 for international students). Electronic applications accepted. Tuition, state resident: full-time $3,308. Tuition, nonresident: full-time $9,744. Part-time tuition and fees vary according to course load, campus/location and program. *Financial aid:* In 1998–99, 37 research assistantships with partial tuition reimbursements (averaging $12,269 per year), 1 teaching assistantship with partial tuition reimbursement (averaging $12,150 per year) were awarded.; fellowships, scholarships also available. *Unit head:* Dr. Mufit Akinc, Chair, 515-294-0738, Fax: 515-204-5444, E-mail: mse@iastate.edu. *Application contact:* Dr. R. William McCallum, 515-294-1214, E-mail: mse@iastate.edu.

Jackson State University, Graduate School, School of Science and Technology, Department of Technology and Industrial Arts, Jackson, MS 39217. Offers hazardous materials management (MS); industrial arts education (MS Ed). Part-time and evening/weekend programs available. *Faculty:* 3 full-time (0 women). *Students:* 9 full-time (3 women), 12 part-time (4 women); includes 19 minority (all African Americans), 1 international. In 1998, 10 degrees awarded. *Degree requirements:* For master's, thesis or alternative, comprehensive exam required. *Entrance requirements:* For master's, GRE General Test (minimum combined score of 1000 required), TOEFL (minimum score of 550 required). *Application deadline:* For fall admission, 3/1 (priority date); for spring admission, 10/1. Applications are processed on a rolling basis. Application fee: $20. *Financial aid:* In 1998–99, 6 students received aid. Career-related internships or fieldwork, Federal Work-Study, scholarships, and unspecified assistantships available. Aid available to part-time students. Financial aid application deadline: 3/1; financial aid applicants required to submit FAFSA. *Unit head:* Dr. Sonny Bolls, Chair, 601-968-2466, Fax: 601-968-2344. *Application contact:* Curtis Gore, Admissions Coordinator, 601-974-5841, Fax: 601-974-6196, E-mail: cgore@ccaix.jsums.edu.

Johns Hopkins University, G. W. C. Whiting School of Engineering, Department of Materials Science and Engineering, Baltimore, MD 21218-2699. Offers MSE, PhD. Part-time and evening/weekend programs available. *Faculty:* 10 full-time (0 women), 11 part-time (0 women). *Students:* 29 full-time (9 women), 7 part-time (2 women); includes 4 minority (3 Asian Americans or Pacific Islanders, 1 Hispanic American), 10 international. Average age 26. 69 applicants,

17% accepted. In 1998, 4 master's, 5 doctorates awarded. Terminal master's awarded for partial completion of doctoral program. *Degree requirements:* For master's and doctorate, thesis/dissertation, cumulative exam required, foreign language not required. *Entrance requirements:* For master's, GRE General Test (minimum combined score of 1950 on three sections required; average 2050), TOEFL (minimum score of 560 required; average 620); for doctorate, GRE General Test (minimum combined score of 1950 on three sections required; average 2050), TOEFL (minimum score of 600 required; average 620). *Average time to degree:* Master's–2.5 years full-time, 4 years part-time; doctorate–5 years full-time, 7.5 years part-time. *Application deadline:* Applications are processed on a rolling basis. Application fee: $50. Tuition: Full-time $23,660. Tuition and fees vary according to program. *Financial aid:* In 1998–99, 10 fellowships (averaging $11,700 per year), 13 research assistantships (averaging $11,250 per year), 10 teaching assistantships (averaging $10,800 per year) were awarded.; Federal Work-Study and institutionally-sponsored loans also available. Financial aid application deadline: 3/14. *Faculty research:* Metallurgy, nanomaterials, biomaterials, nondestructive characterization, electrochemistry. Total annual research expenditures: $2.8 million. *Unit head:* Dr. Peter C. Searson, Chair, 410-516-8774, Fax: 410-516-5293, E-mail: searson@jhu.edu. *Application contact:* Dr. Robert C. Cammarata, Chair, Graduate Program Committee, 410-516-5462, Fax: 410-516-5293, E-mail: rcc@jhu.edu.

See in-depth description on page 1331.

Lehigh University, College of Engineering and Applied Science, Department of Materials Science and Engineering, Bethlehem, PA 18015-3094. Offers M Eng, MS, PhD. Part-time programs available. *Faculty:* 15 full-time (1 woman), 2 part-time (0 women). *Students:* 34 full-time (4 women), 6 part-time (2 women); includes 4 minority (all Asian Americans or Pacific Islanders), 15 international. 390 applicants, 2% accepted. In 1998, 7 master's, 8 doctorates awarded. *Degree requirements:* For master's and doctorate, thesis/dissertation required, foreign language not required. *Entrance requirements:* For master's and doctorate, GRE General Test, TOEFL. *Application deadline:* For fall admission, 7/15; for spring admission, 12/1. Applications are processed on a rolling basis. Application fee: $40. *Financial aid:* In 1998–99, 4 fellowships with full tuition reimbursements (averaging $11,040 per year), 24 research assistantships with full and partial tuition reimbursements (averaging $10,350 per year), 2 teaching assistantships with full and partial tuition reimbursements (averaging $11,000 per year) were awarded. Financial aid application deadline: 1/15. *Faculty research:* Metals, ceramics, crystals, polymers, fatigue crack propagation, electron microscopy. Total annual research expenditures: $3.2 million. *Unit head:* Dr. David B. Williams, Chairperson, 610-758-6120, Fax: 610-758-4244, E-mail: dbw1@lehigh.edu. *Application contact:* Maxine C. Mattie, Graduate Administrative Coordinator, 610-758-4222, Fax: 610-758-4244, E-mail: mcm1@lehigh.edu.

Marquette University, Graduate School, College of Engineering, Department of Mechanical and Industrial Engineering, Milwaukee, WI 53201-1881. Offers engineering management (MS); materials science and engineering (MS, PhD); mechanical engineering (MS, PhD), including manufacturing systems engineering. Part-time and evening/weekend programs available. *Faculty:* 18 full-time (0 women), 6 part-time (0 women). *Students:* 32 full-time (3 women), 52 part-time (4 women); includes 6 minority (1 African American, 2 Asian Americans or Pacific Islanders, 2 Hispanic Americans, 1 Native American), 30 international. Terminal master's awarded for partial completion of doctoral program. *Degree requirements:* For master's, thesis, comprehensive exam required, foreign language not required; for doctorate, dissertation, proficiency exam, qualifying exam required, foreign language not required. *Entrance requirements:* For master's and doctorate, GRE General Test, TOEFL (minimum score of 550 required), minimum GPA of 3.0. *Application deadline:* For fall admission, 8/1 (priority date); for spring admission, 1/1 (priority date). Applications are processed on a rolling basis. Application fee: $40. Tuition: Part-time $510 per credit hour. Tuition and fees vary according to program. *Unit head:* Dr. G. E. O. Widera, Chairman, 414-288-7259, Fax: 414-288-1647, E-mail: geo.widera@marquette.edu. *Application contact:* Dr. William E. Brower, Director of Graduate Studies, 414-288-1717, Fax: 414-288-7790, E-mail: 9322browerw@rms.csd.mu.edu.

Massachusetts Institute of Technology, School of Engineering, Department of Materials Science and Engineering, Cambridge, MA 02139-4307. Offers materials engineering (Mat E); materials science and engineering (SM, PhD, Sc D); metallurgical engineering (Met E). *Faculty:* 31 full-time (6 women). *Students:* 167 full-time (41 women), 3 part-time; includes 23 minority (1 African American, 20 Asian Americans or Pacific Islanders, 2 Hispanic Americans), 61 international. Average age 26. 262 applicants, 29% accepted. In 1998, 30 master's, 26 doctorates awarded. Terminal master's awarded for partial completion of doctoral program. *Degree requirements:* For master's, thesis required, foreign language not required; for doctorate, dissertation, qualifying exams required, foreign language not required; for other advanced degree, thesis required. *Entrance requirements:* For master's, doctorate, and other advanced degree, GRE General Test. *Average time to degree:* Master's–2.5 years full-time; doctorate–6 years full-time. *Application deadline:* For fall admission, 1/15 (priority date); for spring admission, 11/1. Applications are processed on a rolling basis. Application fee: $55. *Financial aid:* In 1998–99, 164 students received aid, including 26 fellowships, 132 research assistantships, 12 teaching assistantships; career-related internships or fieldwork, grants, institutionally-sponsored loans, scholarships, and traineeships also available. Financial aid application deadline: 1/1; financial aid applicants required to submit FAFSA. *Faculty research:* Photonic band gap materials, advanced spectroscopic and diffraction techniques, rapid solidification process, high-strength polymer materials, computer modeling of structures and processing. Total annual research expenditures: $14.2 million. *Unit head:* Dr. Thomas W. Eagar, Head, 617-253-0948, Fax: 617-252-1773, E-mail: tweagar@mit.edu. *Application contact:* Kenneth C. Russell, Graduate Admissions Chairman, 617-253-3329, Fax: 617-258-8836, E-mail: kenruss@mit.edu.

McMaster University, School of Graduate Studies, Faculty of Engineering, Department of Materials Science and Engineering, Hamilton, ON L8S 4M2, Canada. Offers materials engineering (M Eng, M Sc, PhD); materials science (M Eng, M Sc, PhD). *Faculty:* 23 full-time. *Students:* 30 full-time, 3 part-time. *Degree requirements:* For master's, thesis required, foreign language not required; for doctorate, dissertation, comprehensive exam required, foreign language not required. *Application deadline:* For fall admission, 3/1 (priority date). Applications are processed on a rolling basis. Application fee: $50. *Financial aid:* In 1998–99, teaching assistantships (averaging $7,722 per year); fellowships, research assistantships *Unit head:* Dr. M. B. Ives, Chair, 905-525-9140 Ext. 24293.

Michigan State University, Graduate School, College of Engineering, Department of Materials Science and Mechanics, East Lansing, MI 48824-1020. Offers materials science (MS, PhD); mechanics (MS, PhD); metallurgy (MS, PhD). Part-time programs available. *Faculty:* 21. *Students:* 39 full-time (11 women), 34 part-time (7 women); includes 5 minority (3 African Americans, 2 Asian Americans or Pacific Islanders), 40 international. Average age 30. 151 applicants, 23% accepted. In 1998, 16 master's, 4 doctorates awarded. *Degree requirements:* For master's, foreign language and thesis not required; for doctorate, dissertation required. *Entrance requirements:* For master's, GRE, TOEFL; for doctorate, GRE, TOEFL, sample of technical work. *Application deadline:* Applications are processed on a rolling basis. Application fee: $30 ($40 for international students). *Financial aid:* In 1998–99, 30 research assistantships with tuition reimbursements (averaging $12,539 per year), 20 teaching assistantships with tuition reimbursements (averaging $12,290 per year) were awarded.; fellowships Financial aid application deadline: 3/1; financial aid applicants required to submit FAFSA. *Faculty research:* Composite materials, thin films, materials modeling, computational mechanics, biomechanics. Total annual research expenditures: $910,000. *Unit head:* Dr. Nicholas Altiero, Chairperson, 517-355-5141, Fax: 517-353-9842.

See in-depth description on page 1333.

National Technological University, Programs in Engineering, Fort Collins, CO 80526-1842. Offers chemical engineering (MS); computer engineering (MS); computer science (MS); electrical engineering (MS); engineering management (MS); hazardous waste management (MS); health physics (MS); management of technology (MS); manufacturing systems engineering (MS); materials science and engineering (MS); software engineering (MS); special majors (MS); transportation engineering (MS); transportation systems engineering (MS). Part-time programs available. *Faculty:* 600 part-time (20 women). *Entrance requirements:* For master's, BS in engineering or related field. *Application deadline:* Applications are processed on a rolling basis. Application fee: $50. *Unit head:* Lionel V. Baldwin, President, 970-495-6400, Fax: 970-484-0668, E-mail: baldwin@mail.ntu.edu.

New Jersey Institute of Technology, Office of Graduate Studies, Department of Physics, Program in Materials Science and Engineering, Newark, NJ 07102-1982. Offers MS, PhD. *Degree requirements:* For master's, thesis required, foreign language not required; for doctorate, dissertation, residency required, foreign language not required. *Entrance requirements:* For master's, GRE General Test (minimum score of 450 on verbal section, 600 on quantitative, 550 on analytical required); for doctorate, GRE General Test (minimum score of 450 on verbal section, 600 on quantitative, 550 on analytical required), minimum graduate GPA of 3.5.

Norfolk State University, School of Graduate Studies, School of Science and Technology, Department of Chemistry and Physics, Norfolk, VA 23504-3907. Offers materials science (MS). *Degree requirements:* Foreign language not required.

North Carolina State University, Graduate School, College of Engineering, Department of Materials Science and Engineering, Raleigh, NC 27695. Offers MMSE, MS, PhD. *Faculty:* 21 full-time (1 woman), 17 part-time (0 women). *Students:* 103 full-time (14 women), 21 part-time (5 women); includes 14 minority (3 African Americans, 6 Asian Americans or Pacific Islanders, 3 Hispanic Americans, 2 Native Americans), 39 international. Average age 29. 99 applicants, 31% accepted. In 1998, 12 master's, 22 doctorates awarded. *Degree requirements:* For master's and doctorate, thesis/dissertation required, foreign language not required. *Application deadline:* For fall admission, 6/25. Application fee: $45. *Financial aid:* In 1998–99, 1 fellowship (averaging $2,002 per year), 61 research assistantships (averaging $5,079 per year), 10 teaching assistantships (averaging $4,992 per year) were awarded. *Faculty research:* Processing and properties of wide band gap semiconductors, ferroelectric thin-film materials, ductility of nanocrystalline materials, computational materials science, defects in silicon-based devices. Total annual research expenditures: $7.5 million. *Unit head:* Dr. J. Michael Rigsbee, Head, 919-515-3568, Fax: 919-515-7724, E-mail: mike_rigsbee@ncsu.edu. *Application contact:* Dr. Carl Koch, Director of Graduate Programs, 919-515-7340, Fax: 919-515-7724, E-mail: carl_koch@ncsu.edu.

Northwestern University, The Graduate School, Robert R. McCormick School of Engineering and Applied Science, Department of Materials Science and Engineering, Evanston, IL 60208. Offers materials science (PhD); materials science and engineering (MS). Admissions and degrees offered through The Graduate School. Part-time programs available. *Faculty:* 25 full-time (3 women), 1 part-time (0 women). *Students:* 101 full-time (27 women), 1 part-time; includes 23 minority (3 African Americans, 17 Asian Americans or Pacific Islanders, 3 Hispanic Americans), 13 international. 233 applicants, 31% accepted. In 1998, 3 master's, 20 doctorates awarded. Terminal master's awarded for partial completion of doctoral program. *Degree requirements:* For master's, thesis, oral thesis defense required, foreign language not required; for doctorate, dissertation, oral defense of dissertation, preliminary evaluation, qualifying exam required, foreign language not required. *Entrance requirements:* For master's and doctorate, GRE General Test, TOEFL (minimum score of 560 required). *Application deadline:* For fall admission, 8/30. Application fee: $50 ($55 for international students). *Financial aid:* In 1998–99, 15 fellowships with full tuition reimbursements (averaging $11,673 per year), 65 research assistantships with partial tuition reimbursements (averaging $16,285 per year), teaching assistantships with full tuition reimbursements (averaging $12,042 per year) were awarded.; career-related internships or fieldwork, Federal Work-Study, and institutionally-sponsored loans also available. Financial aid application deadline: 1/15; financial aid applicants required to submit FAFSA. *Faculty research:* Metallurgy, ceramics, polymers, electronic materials, biomaterials. Total annual research expenditures: $9.3 million. *Unit head:* Katherine Faber, Chair, 847-491-3537, Fax: 847-491-7820. *Application contact:* Sharon Jacknow, Admissions Contact, 847-491-3587, Fax: 847-491-7820, E-mail: sjacknow@nwu.edu.

Ohio University, Graduate Studies, College of Engineering and Technology, Department of Integrated Engineering, Athens, OH 45701-2979. Offers geotechnical and environmental engineering (PhD); intelligent systems (PhD); materials processing (PhD). *Faculty:* 39 full-time (1 woman). *Students:* 9 full-time (0 women), 6 part-time; includes 1 minority (Asian American or Pacific Islander), 12 international. *Degree requirements:* For doctorate, computer language, dissertation required, foreign language not required. *Entrance requirements:* For doctorate, GRE General Test, MS in engineering or related field. *Application deadline:* For fall admission, 3/15. Applications are processed on a rolling basis. Application fee: $30. Tuition, state resident: full-time $5,754; part-time $238 per credit hour. Tuition, nonresident: full-time $11,055; part-time $457 per credit hour. Tuition and fees vary according to course load, campus/location and program. *Unit head:* Dr. Jerrel R. Mitchell, Associate Dean for Research and Graduate Studies, 740-593-1482, E-mail: mitchell@bobcat.ent.ohiou.edu.

Oregon Graduate Institute of Science and Technology, Graduate Studies, Department of Materials Science and Engineering, Portland, OR 97291-1000. Offers MS, PhD. Part-time programs available. Terminal master's awarded for partial completion of doctoral program. *Degree requirements:* For master's, thesis optional, foreign language not required; for doctorate, comprehensive exam, oral defense of dissertation required. *Entrance requirements:* For master's and doctorate, GRE General Test, TOEFL (minimum score of 550 required). *Application deadline:* Applications are processed on a rolling basis. Application fee: $50. Electronic applications accepted. *Financial aid:* Fellowships, research assistantships, Federal Work-Study available. Financial aid application deadline: 3/1. *Application contact:* Enrollment Manager, 800-685-2423, Fax: 503-690-1285, E-mail: admissions@admin.ogi.edu.

See in-depth description on page 1335.

Oregon State University, Graduate School, College of Engineering, Department of Mechanical Engineering, Program in Materials Science, Corvallis, OR 97331. Offers MAIS, MS. *Faculty:* 2 full-time (0 women). *Students:* 7 full-time (1 woman), 1 part-time, 4 international. Average age 28. In 1998, 1 degree awarded. *Degree requirements:* For master's, computer language, thesis or alternative required, foreign language not required. *Entrance requirements:* For master's, GRE General Test, TOEFL (minimum score of 550 required), minimum GPA of 3.0 in last 90 hours. *Application deadline:* For fall admission, 3/1. Applications are processed on a rolling basis. Application fee: $50. *Financial aid:* Fellowships, research assistantships, teaching assistantships, Federal Work-Study and institutionally-sponsored loans available. Aid available to part-time students. Financial aid application deadline: 2/1. *Unit head:* Dr. Michael E. Kassner, Director, 541-737-7023, Fax: 541-737-2600, E-mail: kassner@engr.orst.edu. *Application contact:* Graduate Admissions Clerk, 541-737-3441, Fax: 541-737-2600.

Pennsylvania State University University Park Campus, Graduate School, College of Earth and Mineral Sciences, Department of Materials Science and Engineering, State College, University Park, PA 16802-1503. Offers ceramic science (MS, PhD); fuel science (MS, PhD); metals science and engineering (MS, PhD); polymer science (MS, PhD). *Students:* 108 full-time (15 women), 11 part-time (1 woman). In 1998, 9 master's, 10 doctorates awarded. *Entrance requirements:* For master's and doctorate, GRE. *Application deadline:* For fall admission, 7/1. Application fee: $50. *Financial aid:* Fellowships, unspecified assistantships available. Financial aid application deadline: 2/28. *Unit head:* Dr. R. E. Tressler, Head, 814-865-0497.

Pennsylvania State University University Park Campus, Graduate School, Intercollege Graduate Programs, Intercollege Graduate Program in Materials, State College, University Park, PA 16802-1503. Offers MS, PhD. *Students:* 53 full-time (16 women), 8 part-time (2 women). *Entrance requirements:* For master's and doctorate, GRE General Test. Application fee: $50. *Unit head:* Dr. Robert N. Pangborn, Chair, 814-865-1451.

Materials Sciences

Polytechnic University, Brooklyn Campus, Department of Mechanical, Aerospace and Manufacturing Engineering, Major in Materials Science, Brooklyn, NY 11201-2990. Offers MS. Part-time and evening/weekend programs available. *Students:* 6 full-time (0 women), 8 part-time (1 woman), 6 international. Average age 33. 36 applicants, 22% accepted. In 1998, 4 degrees awarded. *Degree requirements:* For master's, project or thesis required. *Application deadline:* Applications are processed on a rolling basis. Application fee: $45. Electronic applications accepted. *Financial aid:* Fellowships, research assistantships, teaching assistantships, institutionally-sponsored loans available. Aid available to part-time students. Financial aid applicants required to submit FAFSA. *Faculty research:* Studies of materials for aerospace, electronics, and energy-related applications; alloy hardening; deformation and fracture; phase transformations. Total annual research expenditures: $307,362. *Unit head:* John S. Kerge, Dean of Admissions, 718-260-3200, Fax: 718-260-3446, E-mail: admitme@poly.edu. *Application contact:* John S. Kerge, Dean of Admissions, 718-260-3200, Fax: 718-260-3446, E-mail: admitme@poly.edu.

Polytechnic University, Westchester Graduate Center, Graduate Programs, Department of Mechanical, Aerospace and Manufacturing Engineering, Major in Materials Science, Hawthorne, NY 10532-1507. Offers MS. Average age 33. In 1998, 1 degree awarded. *Degree requirements:* For master's, computer language required. *Application deadline:* Applications are processed on a rolling basis. Application fee: $45. Electronic applications accepted. *Unit head:* Carolyn P. Farrior, Associate Dean of Graduate Admissions, 407-674-7118, Fax: 407-723-9468, E-mail: cfarrior@fit.edu. *Application contact:* John S. Kerge, Dean of Admissions, 718-260-3200, Fax: 718-260-3446, E-mail: admitme@poly.edu.

Rensselaer Polytechnic Institute, Graduate School, School of Engineering, Department of Materials Science and Engineering, Troy, NY 12180-3590. Offers ceramics and glass science (M Eng, MS, D Eng, PhD); composites (M Eng, MS, D Eng, PhD); electronic materials (M Eng, MS, D Eng, PhD); metallurgy (M Eng, MS, D Eng, PhD); polymers (M Eng, MS, D Eng, PhD). Part-time and evening/weekend programs available. *Faculty:* 17 full-time (1 woman), 5 part-time (0 women). *Students:* 45 full-time (8 women), 24 part-time (4 women); includes 5 minority (1 African American, 2 Asian Americans or Pacific Islanders, 2 Hispanic Americans), 26 international. 198 applicants, 13% accepted. In 1998, 17 master's, 5 doctorates awarded. Terminal master's awarded for partial completion of doctoral program. *Degree requirements:* For master's, thesis required (for some programs), foreign language not required; for doctorate, dissertation required, foreign language not required. *Entrance requirements:* For master's and doctorate, GRE, TOEFL (minimum score of 550 required). *Application deadline:* For fall admission, 2/1 (priority date). Applications are processed on a rolling basis. Application fee: $35. *Financial aid:* In 1998–99, 38 students received aid, including 1 fellowship with full tuition reimbursement available (averaging $22,000 per year), 27 research assistantships with full tuition reimbursements available (averaging $15,000 per year), 10 teaching assistantships with full tuition reimbursements available (averaging $15,000 per year); career-related internships or fieldwork and institutionally-sponsored loans also available. Financial aid application deadline: 2/1. *Faculty research:* Materials processing, nanostructural materials, materials for microelectronics. Total annual research expenditures: $3.1 million. *Unit head:* Dr. Richard W. Siegel, Chair, 518-276-6373, Fax: 518-276-8554. *Application contact:* Dr. Roger Wright, Admissions Coordinator, 518-276-6372, Fax: 518-276-8554, E-mail: fowlen@rpi.edu.

See in-depth description on page 1337.

Rice University, Graduate Programs, George R. Brown School of Engineering, Department of Mechanical Engineering and Materials Science, Program in Materials Science, Houston, TX 77251-1892. Offers MME, MMS, MS, PhD. *Degree requirements:* For master's, thesis required (for some programs), foreign language not required; for doctorate, dissertation required, foreign language not required. *Entrance requirements:* For master's and doctorate, GRE General Test, GRE Subject Test, TOEFL (minimum score of 550 required), minimum GPA of 3.0. *Faculty research:* Solid mechanics, thermal sciences.

Rochester Institute of Technology, Part-time and Graduate Admissions, College of Science, Center for Materials Science and Engineering, Rochester, NY 14623-5603. Offers MS. Part-time programs available. *Students:* 2 full-time (0 women), 7 part-time (1 woman); includes 3 minority (1 Asian American or Pacific Islander, 1 Hispanic American, 1 Native American), 1 international. 50 applicants, 28% accepted. In 1998, 8 degrees awarded. *Degree requirements:* For master's, thesis optional, foreign language not required. *Entrance requirements:* For master's, TOEFL (minimum score of 550 required), minimum GPA of 3.0. *Application deadline:* For fall admission, 3/1 (priority date). Applications are processed on a rolling basis. Application fee: $40. *Financial aid:* Teaching assistantships, career-related internships or fieldwork, institutionally-sponsored loans, and tuition waivers (partial) available. Aid available to part-time students. Financial aid application deadline: 7/29. *Faculty research:* VUV modification of polymers, stress and morphology of sputtered copper films, MRI applications to materials problems. *Unit head:* Dr. Peter A. Cardegna, Program Director, 716-475-6652, E-mail: pacsps@rit.edu.

Rutgers, The State University of New Jersey, New Brunswick, Graduate School, Program in Ceramic and Materials Science and Engineering, New Brunswick, NJ 08903. Offers MS, PhD. Part-time programs available. *Faculty:* 23 full-time (1 woman). *Students:* 46 full-time (6 women), 18 part-time (4 women); includes 8 minority (7 Asian Americans or Pacific Islanders, 1 Hispanic American), 28 international. Average age 25. 117 applicants, 24% accepted. In 1998, 13 master's, 5 doctorates awarded. *Degree requirements:* For master's and doctorate, thesis/dissertation required, foreign language not required. *Entrance requirements:* For master's and doctorate, GRE General Test. *Application deadline:* For fall admission, 3/1; for spring admission, 11/1. Application fee: $50. *Financial aid:* In 1998–99, 30 students received aid, including 27 research assistantships with full tuition reimbursements available, 3 teaching assistantships with full tuition reimbursements available; fellowships Financial aid application deadline: 3/1; financial aid applicants required to submit FAFSA. *Faculty research:* Ceramic processing, nanostructured materials, electrical and structural ceramcis, fiber optics. Total annual research expenditures: $7 million. *Unit head:* Dr. W. Roger Cannon, Director, 732-445-4718. *Application contact:* Claudia Kuchinow, Assistant to the Director, 732-445-2159, Fax: 732-445-3258.

Rutgers, The State University of New Jersey, New Brunswick, Graduate School, Program in Materials Science and Engineering, New Brunswick, NJ 08903. Offers physical metallurgy (MS, PhD); polymer science (MS, PhD). Part-time programs available. *Faculty:* 16 full-time (0 women). *Students:* 49 full-time (6 women), 15 part-time (3 women); includes 3 minority (1 African American, 2 Asian Americans or Pacific Islanders), 39 international. Average age 23. 175 applicants, 23% accepted. In 1998, 6 master's, 9 doctorates awarded. Terminal master's awarded for partial completion of doctoral program. *Degree requirements:* For master's, thesis optional, foreign language not required; for doctorate, dissertation required, foreign language not required. *Entrance requirements:* For master's and doctorate, GRE General Test. *Application deadline:* For fall admission, 3/1 (priority date). Applications are processed on a rolling basis. Application fee: $40. *Financial aid:* In 1998–99, 21 students received aid, including 2 fellowships, 15 research assistantships, 4 teaching assistantships; tuition waivers (partial) also available. Financial aid application deadline: 3/1. *Faculty research:* Nanostructured materials, electroactive polymers, phase transformation, polymeric properties. Total annual research expenditures: $2 million. *Unit head:* Dr. Jerry I. Scheinbeim, Director, 732-932-3669. *Application contact:* Kathy Finnerty, Administrative Assistant, 732-932-2245, Fax: 732-932-5977.

South Dakota School of Mines and Technology, Graduate Division, Division of Material Engineering and Science, Doctoral Program in Materials Engineering and Science, Rapid City, SD 57701-3995. Offers chemical engineering (PhD); chemistry (PhD); civil engineering (PhD); electrical engineering (PhD); mechanical engineering (PhD); metallurgical engineering (PhD); physics (PhD). Part-time programs available. *Students:* 14 full-time (2 women), 9 international. Average age 35. In 1998, 1 degree awarded. *Degree requirements:* For doctorate, dissertation required, foreign language not required. *Entrance requirements:* For doctorate, TOEFL (minimum score of 520 required), TWE, minimum graduate GPA of 3.0. *Application deadline:* For fall admission, 6/15 (priority date); for spring admission, 10/15. Applications are processed on a rolling basis. Application fee: $15. Electronic applications accepted. Tuition, state resident: part-time $89 per hour. Tuition, nonresident: part-time $261 per hour. Part-time tuition and fees vary according to program. *Financial aid:* In 1998–99, 1 fellowship, 9 research assistantships, 6 teaching assistantships were awarded.; Federal Work-Study and institutionally-sponsored loans also available. Aid available to part-time students. Financial aid application deadline: 5/15. *Faculty research:* Thermophysical properties of solids, development of multiphase materials and composites, concrete technology, electronic polymer materials. *Unit head:* Dr. Chris Jenkins, Coordinator, 605-394-2406. *Application contact:* Brenda Brown, Secretary, 800-454-8162 Ext. 2493, Fax: 605-394-5360, E-mail: graduate_admissions@silver.sdmt.edu.

See in-depth description on page 1339.

South Dakota School of Mines and Technology, Graduate Division, Division of Material Engineering and Science, Master's Program in Materials Engineering and Science, Rapid City, SD 57701-3995. Offers chemistry (MS); metallurgical engineering (MS); physics (MS). *Students:* 17. In 1998, 2 degrees awarded. *Degree requirements:* Foreign language not required. *Entrance requirements:* For master's, TOEFL (minimum score of 520 required), TWE. *Application deadline:* For fall admission, 6/15 (priority date); for spring admission, 10/15. Applications are processed on a rolling basis. Application fee: $15. Electronic applications accepted. Tuition, state resident: part-time $89 per hour. Tuition, nonresident: part-time $261 per hour. Part-time tuition and fees vary according to program. *Financial aid:* Application deadline: 5/15. *Unit head:* James W. Smolka, Coordinator, 805-258-5936. *Application contact:* Brenda Brown, Secretary, 800-454-8162 Ext. 2493, Fax: 605-394-5360, E-mail: graduate_admissions@silver.sdmt.edu.

Southern Methodist University, School of Engineering and Applied Science, Center for Special Studies, Dallas, TX 75275. Offers applied science (MS, PhD); hazardous and waste materials management (MS); material science and engineering (MS); systems engineering (MS). *Faculty:* 12 part-time (0 women). *Students:* 10 full-time (2 women), 112 part-time (32 women); includes 22 minority (5 African Americans, 7 Asian Americans or Pacific Islanders, 10 Hispanic Americans), 7 international. *Degree requirements:* For master's, thesis optional, foreign language not required; for doctorate, dissertation, oral and written qualifying exams, oral final exam required. *Entrance requirements:* For master's, GRE General Test (minimum score of 650 on quantitative section required), TOEFL (minimum score of 550 required), minimum GPA of 3.0 in last 2 years, bachelor's degree in related field; for doctorate, preliminary counseling exam, minimum GPA of 3.0, bachelor's degree in related field. *Application deadline:* For fall admission, 8/1 (priority date); for spring admission, 12/15. Applications are processed on a rolling basis. Application fee: $25. Tuition: Full-time $9,216; part-time $512 per credit hour. Required fees: $88 per credit hour. Part-time tuition and fees vary according to course load and campus/location. *Unit head:* Dr. Buck F. Brown, Dean for Research and Graduate Studies, 812-877-8403, Fax: 812-877-8102, E-mail: buck.brown@rose-hulman.edu. *Application contact:* Dr. Zeynep Celik-Butler, Assistant Dean for Graduate Studies and Research, 214-768-3979, Fax: 214-768-3845, E-mail: zcb@seas.smu.edu.

Southwest Missouri State University, Graduate College, College of Natural and Applied Sciences, Department of Physics and Astronomy, Springfield, MO 65804-0094. Offers materials science (MS). Part-time programs available. *Faculty:* 10 full-time (1 woman). *Students:* 10 full-time (4 women), 1 part-time, 6 international. 8 applicants, 88% accepted. In 1998, 2 degrees awarded. *Degree requirements:* For master's, thesis, comprehensive exam required, foreign language not required. *Entrance requirements:* For master's, GRE General Test, minimum undergraduate GPA of 3.0. *Application deadline:* For fall admission, 8/7 (priority date); for spring admission, 12/7 (priority date). Applications are processed on a rolling basis. Application fee: $25. Electronic applications accepted. *Financial aid:* In 1998–99, research assistantships with tuition reimbursements (averaging $6,000 per year), teaching assistantships with tuition reimbursements (averaging $6,000 per year) were awarded.; Federal Work-Study, scholarships, and unspecified assistantships also available. *Unit head:* Dr. Ryan Giedd, Head, 417-836-5131, Fax: 417-836-6934, E-mail: materialsscience@mail.smsu.edu.

Stanford University, School of Engineering, Department of Materials Science and Engineering, Stanford, CA 94305-9991. Offers MS, PhD, Eng. *Faculty:* 10 full-time (0 women). *Students:* 115 full-time (32 women), 29 part-time (8 women); includes 29 minority (1 African American, 26 Asian Americans or Pacific Islanders, 1 Hispanic American, 1 Native American), 59 international. Average age 26. 210 applicants, 53% accepted. In 1998, 23 master's, 19 doctorates awarded. *Degree requirements:* For master's, foreign language and thesis not required; for doctorate and Eng, dissertation required, foreign language not required. *Entrance requirements:* For master's, doctorate, and Eng, GRE General Test, TOEFL. *Application deadline:* For fall admission, 1/15. Application fee: $65 ($80 for international students). Electronic applications accepted. Tuition: Full-time $24,588. Required fees: $152. Part-time tuition and fees vary according to course load. *Financial aid:* Fellowships, research assistantships, teaching assistantships, Federal Work-Study and institutionally-sponsored loans available. Financial aid application deadline: 2/15. *Unit head:* John C. Bravman, Chair, 650-723-3698, Fax: 650-725-0538, E-mail: bravman@sierra.stanford.edu. *Application contact:* Graduate Administrator, 650-725-2648.

State University of New York at Buffalo, Graduate School, School of Dental Medicine, Graduate Programs in Dental Medicine, Interdisciplinary Program in Biomaterials, Buffalo, NY 14260. Offers MS. Part-time programs available. *Students:* 1 full-time (0 women), 4 part-time (1 woman), 1 international. Average age 26. 10 applicants, 20% accepted. In 1998, 1 degree awarded (100% found work related to degree). *Degree requirements:* For master's, thesis required, foreign language not required. *Entrance requirements:* For master's, TOEFL (minimum score of 650 required). *Average time to degree:* Master's–2 years full-time. *Application deadline:* For fall admission, 3/1 (priority date); for spring admission, 9/1 (priority date). Applications are processed on a rolling basis. Application fee: $35. Tuition, state resident: full-time $5,100; part-time $213 per credit hour. Tuition, nonresident: full-time $8,416; part-time $351 per credit hour. Required fees: $870; $75 per semester. *Financial aid:* In 1998–99, 2 students received aid, including 2 research assistantships; Federal Work-Study and institutionally-sponsored loans also available. Financial aid application deadline: 2/1; financial aid applicants required to submit FAFSA. *Faculty research:* Bioengineering, surface science, bioadhesion, regulatory sterilization. *Unit head:* Dr. Robert E. Baier, Director, 716-829-3560, Fax: 716-835-4872, E-mail: baier@acsu.buffalo.edu. *Application contact:* Lois R. Karhinen, Assistant Register, 716-829-2839, Fax: 716-833-3517, E-mail: karhinen@acsu.buffalo.edu.

State University of New York at Stony Brook, Graduate School, College of Engineering and Applied Sciences, Department of Materials Science and Engineering, Stony Brook, NY 11794. Offers MS, PhD. *Faculty:* 11 full-time (1 woman), 2 part-time (0 women). *Students:* 45 full-time (5 women), 22 part-time (6 women); includes 10 minority (3 African Americans, 4 Asian Americans or Pacific Islanders, 3 Hispanic Americans), 37 international. 75 applicants, 65% accepted. In 1998, 8 master's, 9 doctorates awarded. *Degree requirements:* For master's, thesis or alternative required, foreign language not required; for doctorate, dissertation, comprehensive exams required, foreign language not required. *Entrance requirements:* For master's and doctorate, GRE General Test, TOEFL, minimum undergraduate GPA of 3.0. *Application deadline:* For fall admission, 1/15. Application fee: $50. *Financial aid:* In 1998–99, 1 fellowship, 24 research assistantships, 18 teaching assistantships were awarded. *Faculty research:* Electronic materials, biomaterials, synchrotron topography. Total annual research expenditures: $3.6 million. *Unit head:* Dr. Michael Dudley, Chairman, 516-632-8484. *Application contact:* Dr. Miriam Rafailovich, 516-632-8484, Fax: 516-632-8052, E-mail: mrafailovich.nsf.@notes.cc.sunysb.edu.

Stevens Institute of Technology, Graduate School, Charles V. Schaefer Jr. School of Engineering, Department of Materials Science and Engineering, Hoboken, NJ 07030. Offers materials engineering (M Eng, PhD); materials science (MS, PhD); structural analysis of materials (Certificate); surface modification of materials (Certificate). Part-time and evening/weekend programs available. Terminal master's awarded for partial completion of doctoral program. *Degree requirements:* For master's, computer language required, thesis optional, foreign language not required; for doctorate, computer language, dissertation required; for Certificate, computer language required, foreign language not required. *Entrance requirements:* For master's and doctorate, TOEFL. Electronic applications accepted. *Faculty research:* Surface modification techniques, structure and surface analysis, corrosion, tribology, superconductor materials.

Syracuse University, Graduate School, L. C. Smith College of Engineering and Computer Science, Department of Chemical Engineering and Materials Sciences, Program in Solid-State Science and Technology, Syracuse, NY 13244-0003. Offers MS, PhD. *Faculty:* 17. *Students:* 1 full-time (0 women), 1 international. Average age 37. 12 applicants, 67% accepted. *Degree requirements:* For master's, thesis required (for some programs), foreign language not required; for doctorate, computer language, dissertation required, foreign language not required. *Entrance requirements:* For master's and doctorate, GRE General Test, GRE Subject Test. *Application deadline:* Applications are processed on a rolling basis. Application fee: $40. Tuition: Full-time $13,992; part-time $583 per credit hour. *Financial aid:* Fellowships, research assistantships, teaching assistantships, Federal Work-Study and tuition waivers (partial) available. Financial aid application deadline: 3/1. *Faculty research:* Solid-gas interactions, mechanical properties, magnetic memories, interface science, epitaxial films. *Unit head:* Klaus Schröder, Chair, 315-443-5860.

Université du Québec, Institut national de la recherche scientifique, Graduate Programs, Research Center—Energy, Ste-Foy, PQ G1V 4C7, Canada. Offers energy and materials science (M Sc, PhD). PhD offered jointly with the Université du Québec à Trois-Rivières. Part-time programs available. *Degree requirements:* For master's and doctorate, thesis/dissertation required. *Entrance requirements:* For master's, appropriate bachelor's degree, proficiency in French; for doctorate, appropriate master's degree, proficiency in French. *Faculty research:* New energy sources, plasmas, fusion.

The University of Alabama, Graduate School, College of Engineering, Joint Materials Science PhD Program, Tuscaloosa, AL 35487. Offers PhD. *Students:* 5 full-time (0 women), 4 international. Average age 33. 3 applicants, 33% accepted. In 1998, 2 degrees awarded (50% entered university research/teaching, 50% found other work related to degree). *Degree requirements:* For doctorate, dissertation, comprehensive exam required, dissertation, comprehensive exam required. *Entrance requirements:* For doctorate, GRE General Test. *Application deadline:* For fall admission, 7/6. Applications are processed on a rolling basis. Application fee: $25. *Financial aid:* Research assistantships available. Financial aid application deadline: 4/1. *Faculty research:* Magnetic multilayers, metals casting, molecular electronics, conducting polymers, metals physics. *Unit head:* Dr. Chester Alexander, Campus Coordinator, 205-348-6367, Fax: 205-348-2346, E-mail: calexand@aalan.ua.edu.

The University of Alabama at Birmingham, Graduate School, School of Engineering, Department of Materials and Mechanical Engineering, Joint Materials Science PhD Program, Birmingham, AL 35294. Offers PhD. *Students:* 5 full-time (1 woman), 1 part-time, 3 international. 16 applicants, 50% accepted. In 1998, 1 degree awarded. *Degree requirements:* For doctorate, dissertation required, foreign language not required. *Entrance requirements:* For doctorate, GRE General Test (minimum score of 550 on each section required), minimum B average. *Application deadline:* For fall admission, 8/6. Applications are processed on a rolling basis. Application fee: $30 ($60 for international students). Electronic applications accepted. *Financial aid:* In 1998–99, 4 fellowships with full and partial tuition reimbursements (averaging $12,500 per year), 1 research assistantship with full and partial tuition reimbursement (averaging $12,500 per year) were awarded.; career-related internships or fieldwork, Federal Work-Study, and institutionally-sponsored loans also available. Aid available to part-time students. Financial aid application deadline: 4/1. *Faculty research:* Biocompatibility studies with biomaterials, microgravity solidification of proteins and metals, analysis of microelectronic materials, and thin film analysis using TEM. *Unit head:* Dr. Krishan Chawla, Interim Chair, Department of Materials and Mechanical Engineering, 205-975-9725, Fax: 205-934-8485, E-mail: kchawla@uab.edu.

The University of Alabama in Huntsville, School of Graduate Studies, Interdisciplinary Program in Materials Science, Huntsville, AL 35899. Offers MS, PhD. Part-time and evening/weekend programs available. *Faculty:* 1 full-time (0 women). *Students:* 13 full-time (5 women), 4 part-time; includes 1 minority (Native American), 8 international. Average age 31. 11 applicants, 82% accepted. In 1998, 3 master's, 2 doctorates awarded. *Degree requirements:* For doctorate, dissertation, oral and written exams required, foreign language not required. *Entrance requirements:* For doctorate, GRE General Test (minimum combined score of 1500 on three sections required), bachelor's degree in engineering or physical science, minimum GPA of 3.0. *Application deadline:* For fall admission, 7/24 (priority date); for spring admission, 11/15 (priority date). Applications are processed on a rolling basis. Application fee: $20. Tuition and fees vary according to course load. *Financial aid:* In 1998–99, 13 students received aid, including 6 research assistantships with full and partial tuition reimbursements available (averaging $9,762 per year), 7 teaching assistantships with full and partial tuition reimbursements available (averaging $8,454 per year); fellowships with full and partial tuition reimbursements available, career-related internships or fieldwork, Federal Work-Study, grants, institutionally-sponsored loans, scholarships, and tuition waivers (full and partial) also available. Aid available to part-time students. Financial aid application deadline: 4/1; financial aid applicants required to submit FAFSA. *Faculty research:* Materials structure and properties; materials processing; mechanical behavior; macromolecular materials; electronic, optical, and magnetic materials. *Unit head:* Dr. Clyde Riley, Director, 256-890-6153, Fax: 256-890-6349, E-mail: criley@matsci.uah.edu.

The University of Arizona, Graduate College, College of Engineering and Mines, Department of Materials Science and Engineering, Tucson, AZ 85721. Offers MS, PhD. Part-time programs available. *Faculty:* 21. *Students:* 35 full-time (11 women), 11 part-time (1 woman); includes 6 minority (1 African American, 3 Asian Americans or Pacific Islanders, 2 Hispanic Americans), 18 international. Average age 28. 24 applicants, 42% accepted. In 1998, 4 master's, 9 doctorates awarded. *Degree requirements:* For master's and doctorate, thesis/dissertation required, foreign language not required. *Entrance requirements:* For master's and doctorate, TOEFL (minimum score of 550 required). *Application deadline:* For fall admission, 5/15. Applications are processed on a rolling basis. Application fee: $35. *Financial aid:* Fellowships, research assistantships, teaching assistantships, institutionally-sponsored loans and scholarships available. Financial aid application deadline: 12/31. *Faculty research:* High-technology ceramics, optical materials, electronic materials, chemical metallurgy, science of materials. *Unit head:* Dr. Donald R. Uhlmann, Head, 520-621-6070. *Application contact:* Geri Hardy, Graduate Secretary, 520-621-2531, Fax: 520-621-8059.

University of British Columbia, Faculty of Graduate Studies, Faculty of Applied Science, Department of Metals and Materials Engineering, Vancouver, BC V6T 1Z2, Canada. Offers materials and metallurgy (M Sc, PhD); metals and materials engineering (M Eng, MA Sc, PhD). *Degree requirements:* For master's and doctorate, thesis/dissertation required. *Entrance requirements:* For master's and doctorate, TOEFL. *Faculty research:* Electroslag melting, mathematical modelling, solidification and hydrometallurgy.

University of California, Berkeley, Graduate Division, College of Engineering, Department of Materials Science and Mineral Engineering, Berkeley, CA 94720-1500. Offers ceramic sciences and engineering (M Eng, MS, D Eng, PhD); engineering geoscience (M Eng, MS, D Eng, PhD); materials engineering (M Eng, MS, D Eng, PhD); mineral engineering (M Eng, MS, D Eng, PhD); petroleum engineering (M Eng, MS, D Eng, PhD); physical metallurgy (M Eng, MS, D Eng, PhD). *Students:* 110 full-time (23 women); includes 17 minority (11 Asian Americans or Pacific Islanders, 5 Hispanic Americans, 1 Native American), 42 international. 191 applicants, 22% accepted. In 1998, 21 master's, 11 doctorates awarded. *Degree requirements:* For master's, comprehensive exam or thesis (MS) required; for doctorate, dissertation, qualifying exam required. *Entrance requirements:* For master's and doctorate, GRE General Test, minimum GPA of 3.0. *Application deadline:* For fall admission, 2/10; for spring admission, 9/1. Application fee: $40. *Financial aid:* Fellowships, research assistantships, teaching assistantships available. Financial aid application deadline: 1/5. *Unit head:* Dr. Thomas M. Devine, Chair, 510-642-3801. *Application contact:* Carole James, Student Affairs Officer, 510-642-3801, Fax: 510-643-5792, E-mail: carolej@uclink4.berkeley.edu.

University of California, Davis, Graduate Studies, College of Engineering, Program in Chemical Engineering and Materials Science, Davis, CA 95616. Offers chemical engineering (MS, PhD); materials science (MS, PhD, Certificate). *Faculty:* 25 full-time (5 women). *Students:* 55 full-time (17 women); includes 12 minority (1 African American, 8 Asian Americans or Pacific Islanders, 2 Hispanic Americans, 1 Native American), 15 international. Average age 26. 120 applicants, 25% accepted. In 1998, 11 master's, 12 doctorates awarded. Terminal master's awarded for partial completion of doctoral program. *Degree requirements:* For master's and doctorate, thesis/dissertation required, foreign language not required. *Entrance requirements:* For master's, GRE General Test (minimum score of 500 on verbal section, 700 on quantitative, 500 on analytical required), minimum GPA of 3.0; for doctorate, GRE General Test (minimum score of 500 on verbal section, 720 on quantitative, 500 on analytical required), TOEFL (minimum score of 600 required), minimum GPA of 3.0. *Application deadline:* For fall admission, 1/15 (priority date). Application fee: $40. Electronic applications accepted. *Financial aid:* In 1998–99, 18 fellowships, 37 research assistantships, 13 teaching assistantships were awarded. Financial aid application deadline: 1/15; financial aid applicants required to submit FAFSA. *Faculty research:* Transport phenomena, colloid science, catalysis, biotechnology, materials. *Unit head:* Subhash Risbud, Chairperson, 530-752-5132.

University of California, Irvine, Office of Research and Graduate Studies, School of Engineering, Program in Materials Science and Engineering, Irvine, CA 92697. Offers engineering (MS, PhD). Part-time programs available. *Faculty:* 11 full-time (1 woman). *Students:* 26 full-time (6 women), 5 part-time (3 women); includes 11 minority (1 African American, 9 Asian Americans or Pacific Islanders, 1 Hispanic American), 10 international. 42 applicants, 81% accepted. In 1998, 2 master's, 5 doctorates awarded. Terminal master's awarded for partial completion of doctoral program. *Degree requirements:* For doctorate, dissertation required, foreign language not required, foreign language not required. *Entrance requirements:* For master's, GRE General Test, minimum GPA of 3.0; for doctorate, GRE General Test. *Application deadline:* For fall admission, 1/15 (priority date). Applications are processed on a rolling basis. Application fee: $40. Electronic applications accepted. *Financial aid:* Fellowships, research assistantships, teaching assistantships, institutionally-sponsored loans and tuition waivers (full and partial) available. Financial aid application deadline: 3/2; financial aid applicants required to submit FAFSA. *Faculty research:* Composite materials, solidification processing, sol-gel processing, microstructural characterization, deformation and damage processes in advanced materials. *Unit head:* Dr. Farghalli A. Mohamed, Director, 949-824-5807, Fax: 949-824-3440, E-mail: famohame@uci.edu. *Application contact:* Admissions Assistant, 949-824-3562, Fax: 949-824-3440.

University of California, Los Angeles, Graduate Division, School of Engineering and Applied Science, Department of Materials Science and Engineering, Los Angeles, CA 90095. Offers ceramics engineering (MS, PhD); metallurgy (MS, PhD). *Faculty:* 7 full-time, 3 part-time. *Students:* 59 full-time (16 women); includes 14 minority (1 African American, 12 Asian Americans or Pacific Islanders, 1 Hispanic American), 26 international. 89 applicants, 55% accepted. In 1998, 15 master's, 4 doctorates awarded. *Degree requirements:* For master's, comprehensive exam or thesis required; for doctorate, dissertation, qualifying exams required, foreign language not required. *Entrance requirements:* For master's, GRE General Test, minimum GPA of 3.0; for doctorate, GRE General Test, minimum GPA of 3.25. *Application deadline:* For fall admission, 1/15; for spring admission, 12/31. Application fee: $40. Electronic applications accepted. *Financial aid:* In 1998–99, 23 fellowships, 49 research assistantships, 21 teaching assistantships were awarded.; Federal Work-Study, institutionally-sponsored loans, and tuition waivers (full and partial) also available. Financial aid application deadline: 1/15; financial aid applicants required to submit FAFSA. *Unit head:* Dr. King-Ning Tu, Chair, 310-206-4838. *Application contact:* Paradee Chularee, Student Affairs Officer, 310-825-8913, Fax: 310-206-7353, E-mail: paradee@ea.ucla.edu.

University of California, San Diego, Graduate Studies and Research, Materials Science Program, La Jolla, CA 92093-5003. Offers MS, PhD. *Faculty:* 10. *Students:* 26 (7 women). 48 applicants, 44% accepted. In 1998, 4 master's, 5 doctorates awarded. *Degree requirements:* For master's, comprehensive exam or thesis required; for doctorate, dissertation, oral exam required. *Entrance requirements:* For master's and doctorate, GRE General Test, TOEFL, minimum GPA of 3.0. Application fee: $40. *Financial aid:* Fellowships, research assistantships, traineeships available. Financial aid application deadline: 1/15. *Unit head:* M. Lea Rudee, Director. *Application contact:* Graduate Coordinator, 619-534-7715.

See in-depth description on page 1343.

University of California, Santa Barbara, Graduate Division, College of Engineering, Department of Materials, Santa Barbara, CA 93106. Offers MS, PhD. *Faculty:* 30 full-time (2 women), 3 part-time (0 women). *Students:* 70 full-time (16 women); includes 10 minority (1 African American, 7 Asian Americans or Pacific Islanders, 2 Hispanic Americans), 14 international. Average age 24. 175 applicants, 21% accepted. In 1998, 1 master's, 15 doctorates awarded. *Degree requirements:* For master's and doctorate, thesis/dissertation required, foreign language not required. *Entrance requirements:* For master's and doctorate, GRE General Test, TOEFL (minimum score of 600 required). *Application deadline:* For fall admission, 1/15 (priority date); for spring admission, 11/1. Applications are processed on a rolling basis. Application fee: $40. Electronic applications accepted. *Financial aid:* In 1998–99, 70 students received aid, including 22 fellowships, 48 research assistantships, 20 teaching assistantships; career-related internships or fieldwork, Federal Work-Study, institutionally-sponsored loans, and tuition waivers (full) also available. Financial aid application deadline: 1/15; financial aid applicants required to submit FAFSA. *Faculty research:* Electronically conducting polymer systems, growth and processing of advanced semiconductors, structure and properties of artificially structured materials–electronic packages and composites, fine structures by nonequilibrium solidification or advanced deposition techniques, sensors for intelligent control systems in materials processing. *Unit head:* David R. Clarke, Chair, 805-893-4362. *Application contact:* Katie Bridgewater, Graduate Program Assistant, 805-893-4601, E-mail: matrls-grad-advisor@engineering.ucsb.edu.

University of Central Florida, College of Engineering, Department of Mechanical, Materials, and Aerospace Engineering, Orlando, FL 32816. Offers aerospace systems (MSME, PhD); materials science and engineering (MSME, PhD); mechanical systems (MSME, PhD); mechanical, materials, and aerospace engineering (Certificate); thermofluids (MSME, PhD). Part-time and evening/weekend programs available. *Faculty:* 18 full-time, 6 part-time. *Students:* 61 full-time (14 women), 30 part-time (4 women); includes 12 minority (3 African Americans, 6 Asian Americans or Pacific Islanders, 3 Hispanic Americans), 42 international. Average age 31. 48 applicants, 58% accepted. In 1998, 23 master's, 2 doctorates awarded. *Degree requirements:* For master's, thesis or alternative required, foreign language not required; for doctorate, dissertation, departmental qualifying exam required, foreign language not required. *Entrance requirements:* For master's, GRE General Test (minimum combined score of 1000 required), TOEFL (minimum score of 550 required; 213 computer-based), minimum GPA of 3.0 in last 60 hours; for doctorate, TOEFL (minimum score of 550 required; 213 computer-based), minimum GPA of 3.5 in last 60 hours. *Application deadline:* For fall admission, 7/15; for spring admission, 12/15. Application fee: $20. Tuition, state resident: full-time $2,054; part-time $137 per credit. Tuition, nonresident: full-time $7,207; part-time $480 per credit. Required fees: $47 per term. *Financial aid:* In 1998–99, 71 students received aid, including 17 fellowships with partial tuition reimbursements available (averaging $2,824 per year), 36 teaching assistantships with partial tuition reimbursements available (averaging $3,509 per year); research assistantships with partial tuition reimbursements available, career-related internships or fieldwork, Federal Work-Study, institutionally-sponsored loans, tuition waivers (partial), and unspecified assistantships also available. Financial aid application deadline: 3/1; financial aid applicants required to submit FAFSA. *Faculty research:* Aerospace systems, computation of methods, dynamics and control, laser applications, materials science. *Unit head:* Dr. Louis

Materials Sciences

University of Central Florida (continued)

Chow, Chair, 407-823-2333. *Application contact:* Dr. A. J. Kassab, Coordinator, 407-823-2416, Fax: 407-823-0208.

University of Cincinnati, Division of Research and Advanced Studies, College of Engineering, Department of Materials Science and Metallurgical Engineering, Program in Materials Science and Engineering, Cincinnati, OH 45221-0091. Offers MS, PhD. Evening/weekend programs available. *Faculty:* 8 full-time. *Students:* 78 full-time (15 women), 15 part-time (2 women); includes 8 minority (3 African Americans, 4 Asian Americans or Pacific Islanders, 1 Hispanic American), 61 international. *Degree requirements:* For master's, thesis optional, foreign language not required; for doctorate, one foreign language, dissertation, comprehensive exams, oral English proficiency exam required. *Entrance requirements:* For master's and doctorate, GRE General Test, TOEFL (minimum score of 550 required), BS in related field, minimum undergraduate GPA of 3.0. *Application deadline:* For fall admission, 2/1 (priority date). Application fee: $40. *Financial aid:* Fellowships, career-related internships or fieldwork, tuition waivers (full), and unspecified assistantships available. Aid available to part-time students. Financial aid application deadline: 2/1. *Faculty research:* Polymer characterization, surface analysis, and adhesion; mechanical behavior of high-temperature materials; composites; electrochemistry of materials. *Unit head:* Student Affairs Officer, E-mail: maeapp@ea.ucla.edu. *Application contact:* Dr. R. J. Roe, Graduate Program Director, 513-556-3117, Fax: 513-556-2569, E-mail: r.j.roe@uc.edu.

University of Connecticut, Graduate School, School of Engineering, Field of Material Science, Storrs, CT 06269. Offers MS, PhD. *Degree requirements:* For doctorate, dissertation required. *Entrance requirements:* For master's and doctorate, GRE General Test, GRE Subject Test.

University of Delaware, College of Engineering, Department of Materials Science and Engineering, Newark, DE 19716. Offers MMSE, PhD. *Faculty:* 12 full-time (2 women), 8 part-time (1 woman). *Students:* 33 full-time (7 women), 3 part-time (1 woman); includes 1 minority (African American), 11 international. Average age 23. 245 applicants, 7% accepted. In 1998, 6 master's, 3 doctorates awarded. Terminal master's awarded for partial completion of doctoral program. *Degree requirements:* For master's and doctorate, thesis/dissertation required, foreign language not required. *Entrance requirements:* For master's and doctorate, GRE General Test (minimum combined score of 1150 required; average 1350) TOEFL (minimum score of 600 required). *Average time to degree:* Master's–2 years full-time; doctorate–4 years full-time. *Application deadline:* For fall admission, 2/28 (priority date). Applications are processed on a rolling basis. Application fee: $45. Electronic applications accepted. *Financial aid:* In 1998–99, 22 students received aid, including 2 fellowships, 18 research assistantships, 3 teaching assistantships; career-related internships or fieldwork and Federal Work-Study also available. Aid available to part-time students. Financial aid application deadline: 2/28. *Faculty research:* Metals, polymers, composites, solid-state electronics, advanced oxides, organic thin films. *Unit head:* Dr. John Rabolt, Chairman, 302-831-2062, Fax: 302-831-4545, E-mail: matsci@me.udel.edu.

University of Denver, Graduate Studies, Faculty of Natural Sciences, Mathematics and Engineering, Department of Engineering, Denver, CO 80208. Offers computer science and engineering (MS); electrical engineering (MS); management and general engineering (MSMGEN); materials science (PhD); mechanical engineering (MS). Part-time and evening/weekend programs available. *Faculty:* 15. *Students:* 23 (9 women) 8 international. Terminal master's awarded for partial completion of doctoral program. *Degree requirements:* For master's, thesis required (for some programs), foreign language not required; for doctorate, dissertation required, foreign language not required. *Entrance requirements:* For master's and doctorate, GRE General Test, TOEFL (minimum score of 570 required), TSE (minimum score of 230 required). *Application deadline:* Applications are processed on a rolling basis. Application fee: $40 ($45 for international students). *Unit head:* Dr. Albert J. Rosa, Chair, 303-871-2102. *Application contact:* Louise Carlson, Assistant to Chair, 303-871-2107.

University of Florida, Graduate School, College of Engineering, Department of Materials Science and Engineering, Gainesville, FL 32611. Offers ceramic science and engineering (ME, MS, PhD, Engr); materials science and engineering (ME, MS, PhD, Certificate, Engr); metallurgical and materials engineering (ME, MS, PhD, Engr); metallurgical engineering (ME, MS, PhD, Engr); polymer science and engineering (ME, MS, PhD, Engr). *Faculty:* 36. *Students:* 148 full-time (40 women), 26 part-time (7 women); includes 22 minority (5 African Americans, 7 Asian Americans or Pacific Islanders, 9 Hispanic Americans, 1 Native American), 65 international. 414 applicants, 26% accepted. In 1998, 31 master's, 21 doctorates awarded. *Degree requirements:* For master's, thesis optional, foreign language not required; for doctorate, dissertation required, foreign language not required; for other advanced degree, thesis optional. *Entrance requirements:* For master's and doctorate, GRE General Test (minimum combined score of 1100 required), TOEFL, minimum GPA of 3.0; for other advanced degree, GRE General Test. *Application deadline:* For fall admission, 6/1 (priority date). Applications are processed on a rolling basis. Application fee: $20. Electronic applications accepted. *Financial aid:* In 1998–99, 118 students received aid, including 12 fellowships, 114 research assistantships; teaching assistantships *Faculty research:* Polymeric materials, electronic materials, glass, biomaterials, composites. Total annual research expenditures: $6 million. *Unit head:* Dr. Reza Abbaschian, Chair, 352-392-1453, Fax: 352-392-6359, E-mail: rabba@mse.ufl.edu. *Application contact:* Jodi Vanderheyden, Graduate Secretary, 352-392-1451, Fax: 352-392-5630, E-mail: jheyd@mail.mse.ufl.edu.

See in-depth description on page 1345.

University of Illinois at Urbana–Champaign, Graduate College, College of Engineering, Department of Materials Science and Engineering, Urbana, IL 61801. Offers MS, PhD. *Faculty:* 29 full-time (2 women), 4 part-time (0 women). *Students:* 112 full-time (31 women); includes 14 minority (8 Asian Americans or Pacific Islanders, 5 Hispanic Americans, 1 Native American), 55 international. 302 applicants, 7% accepted. In 1998, 14 master's, 25 doctorates awarded. *Degree requirements:* For master's, thesis required, foreign language not required; for doctorate, dissertation, departmental qualifying exam required, foreign language not required. *Application deadline:* Applications are processed on a rolling basis. Application fee: $40 ($50 for international students). Tuition, state resident: full-time $4,616. Tuition, nonresident: full-time $11,768. Full-time tuition and fees vary according to course load. *Financial aid:* Fellowships, research assistantships, teaching assistantships, tuition waivers (full and partial) available. Financial aid application deadline: 2/15. *Unit head:* James Economy, Head, 217-333-1440. *Application contact:* Robert Averback, Director of Graduate Studies, 217-333-4302, Fax: 217-333-2736, E-mail: averback@uiuc.edu.

University of Kentucky, Graduate School, Graduate School Programs from the College of Engineering, Program in Materials Science, Lexington, KY 40506-0032. Offers MSMAE, PhD. *Degree requirements:* For master's, comprehensive exam required, thesis optional, foreign language not required; for doctorate, dissertation, comprehensive exam required, foreign language not required. *Entrance requirements:* For master's, GRE General Test, minimum undergraduate GPA of 2.5; for doctorate, GRE General Test, minimum graduate GPA of 3.0. *Faculty research:* Physical and mechanical metallurgy, computational material engineering, polymers and composites, high-temperature ceramics, powder metallurgy.

University of Maryland, College Park, Graduate School, A. James Clark School of Engineering, Department of Materials and Nuclear Engineering, Materials Science and Engineering Program, College Park, MD 20742-5045. Offers M Eng, MS, PhD. Part-time and evening/weekend programs available. Postbaccalaureate distance learning degree programs offered. *Students:* 33 full-time (8 women), 29 part-time (6 women); includes 10 minority (4 African Americans, 6 Asian Americans or Pacific Islanders), 31 international. 131 applicants, 18% accepted. In 1998, 9 master's, 8 doctorates awarded. *Degree requirements:* For master's, thesis or alternative, research paper, written comprehensive exam required, foreign language not required; for doctorate, dissertation, oral exam required. *Entrance requirements:* For master's and doctorate, GRE General Test, TOEFL, minimum B+ average in undergraduate

course work. *Application deadline:* Applications are processed on a rolling basis. Application fee: $50 ($70 for international students). Tuition, state resident: part-time $272 per credit hour. Tuition, nonresident: part-time $475 per credit hour. Required fees: $632; $379 per year. *Financial aid:* Fellowships, research assistantships, teaching assistantships available. Financial aid applicants required to submit FAFSA. *Unit head:* Dr. Henrique Reis, Director of Graduate Studies, 217-333-1228, Fax: 217-244-5705, E-mail: h-reis@uiuc.edu. *Application contact:* Barret Cole, Graduate Admissions Secretary, 301-405-5211, Fax: 301-314-2029.

See in-depth description on page 1347.

University of Maryland, College Park, Graduate School, A. James Clark School of Engineering, Professional Program in Engineering, College Park, MD 20742-5045. Offers aerospace engineering (M Eng); chemical engineering (M Eng); civil engineering (M Eng); electrical engineering (M Eng); fire protection engineering (M Eng); materials science and engineering (M Eng); mechanical engineering (M Eng); reliability engineering (M Eng); systems engineering (M Eng). Part-time and evening/weekend programs available. Postbaccalaureate distance learning degree programs offered. *Faculty:* 11 part-time (0 women). *Students:* 20 full-time (3 women), 205 part-time (42 women); includes 58 minority (27 African Americans, 25 Asian Americans or Pacific Islanders, 5 Hispanic Americans, 1 Native American), 20 international. *Degree requirements:* For master's, foreign language and thesis not required. *Application deadline:* Applications are processed on a rolling basis. Application fee: $50 ($70 for international students). Tuition, state resident: part-time $272 per credit hour. Tuition, nonresident: part-time $475 per credit hour. Required fees: $632; $379 per year. *Unit head:* Dr. Patrick Cunniff, Associate Dean, 301-405-5256, Fax: 301-314-9477. *Application contact:* Trudy Lindsey, Director, Graduate Admission and Records, 301-405-4198, Fax: 301-314-9305, E-mail: grschool@deans.umd.edu.

University of Michigan, Horace H. Rackham School of Graduate Studies, College of Engineering, Department of Materials Science and Engineering, Ann Arbor, MI 48109. Offers MS, PhD. Part-time programs available. *Faculty:* 21 full-time (2 women), 3 part-time (0 women). *Students:* 68 full-time (14 women), 21 part-time (6 women), 37 international. 285 applicants, 16% accepted. In 1998, 8 master's, 8 doctorates awarded. *Degree requirements:* For master's, thesis, oral defense of thesis required, foreign language not required; for doctorate, dissertation, oral defense of dissertation, written exam required, foreign language not required. *Entrance requirements:* For master's, GRE General Test, TOEFL (minimum score of 560 required), minimum GPA of 3.0 in related field; for doctorate, GRE General Test, TOEFL (minimum score of 560 required), minimum GPA of 3.0 in related field, master's degree. *Average time to degree:* Master's–2.6 years full-time, 3.5 years part-time; doctorate–5.2 years full-time. *Application deadline:* For fall admission, 2/15. Applications are processed on a rolling basis. Application fee: $55. *Financial aid:* In 1998–99, 2 fellowships with full tuition reimbursements, 60 research assistantships with full tuition reimbursements, 10 teaching assistantships with full tuition reimbursements were awarded. Financial aid application deadline: 2/15; financial aid applicants required to submit FAFSA. *Faculty research:* Composite materials, polymer alloys, structural ceramics, materials processing and manufacturing, micromechanical and macromechanical behavior, surface modification and energy-beam interactions of materials, device materials, theoretical modeling and computer simulation, materials characterization. Total annual research expenditures: $5.1 million. *Unit head:* Albert F. Yee, Chair, 734-763-2445, E-mail: afyee@engin.umich.edu. *Application contact:* Renee Hilgendorf, Graduate Program Assistant, 734-763-9790, Fax: 734-763-4788, E-mail: reneeh@umich.edu.

See in-depth description on page 1351.

University of Minnesota, Twin Cities Campus, Graduate School, Institute of Technology, Department of Chemical Engineering and Materials Science, Program in Materials Science and Engineering, Minneapolis, MN 55455-0213. Offers M Mat SE, MS Mat SE, PhD. Part-time programs available. Terminal master's awarded for partial completion of doctoral program. *Degree requirements:* For master's and doctorate, thesis/dissertation required, foreign language not required. *Entrance requirements:* For master's and doctorate, GRE General Test. *Faculty research:* Fracture micromechanics, hydrogen embrittlement, polymer physics, microelectric materials, corrosion science.

The University of North Carolina at Chapel Hill, Graduate School, Curriculum in Applied and Materials Science, Chapel Hill, NC 27599. Offers materials science (MS, PhD). *Faculty:* 18 full-time. *Students:* 10 full-time (4 women); includes 6 minority (all Asian Americans or Pacific Islanders) Average age 23. 25 applicants, 12% accepted. In 1998, 2 degrees awarded (100% entered university research/teaching). Terminal master's awarded for partial completion of doctoral program. *Degree requirements:* For doctorate, dissertation required, foreign language not required, foreign language not required. *Entrance requirements:* For master's, GRE General Test (minimum combined score of 1000 required), minimum GPA of 3.0; for doctorate, GRE General Test (minimum combined score of 1000 required). *Average time to degree:* Master's–2 years full-time. *Application deadline:* For fall admission, 1/1 (priority date); for winter admission, 10/15; for spring admission, 4/15 (priority date). Applications are processed on a rolling basis. Application fee: $55. Electronic applications accepted. *Financial aid:* In 1998–99, 7 students received aid, including 4 research assistantships with full tuition reimbursements available (averaging $13,000 per year); teaching assistantships, grants, tuition waivers (full), and unspecified assistantships also available. Financial aid application deadline: 3/1. *Faculty research:* Scanning tunneling microscopy, magnetic resonance, carbon nanotubes, thin films, biomaterials. *Unit head:* Prof. Sean Washburn, Chair, 919-962-6293, Fax: 919-962-3341, E-mail: materials_science@unc.edu. *Application contact:* Elizabeth F. Craig, Coordinator, Curriculum in Applied and Materials Sciences, 919-962-6293, Fax: 919-962-3341, E-mail: materials_science@unc.edu.

See in-depth description on page 1353.

University of North Texas, Robert B. Toulouse School of Graduate Studies, College of Arts and Sciences, Department of Materials Science, Denton, TX 76203. Offers MS, PhD. *Faculty:* 5 full-time (1 woman). *Students:* 4 full-time (1 woman), 10 part-time (1 woman); includes 1 minority (Hispanic American), 7 international. *Degree requirements:* For doctorate, dissertation required. *Entrance requirements:* For master's, GRE General Test (minimum score of 400 on each section, 1000 combined required); for doctorate, GRE General Test. *Application deadline:* For fall admission, 7/17. Application fee: $25 ($50 for international students). *Unit head:* Dr. Bruce Gnade, Chair, 940-565-3260, Fax: 940-565-4824, E-mail: gnade@unt.edu. *Application contact:* Dr. Witold Brostow, Graduate Adviser, 940-565-3260, Fax: 940-565-4824, E-mail: brostow@unt.edu.

University of Pennsylvania, School of Engineering and Applied Science, Department of Materials Science and Engineering, Philadelphia, PA 19104. Offers MSE, PhD, MSE/MBA. Part-time programs available. Terminal master's awarded for partial completion of doctoral program. *Degree requirements:* For master's and doctorate, thesis/dissertation required, foreign language not required. *Entrance requirements:* For master's and doctorate, TOEFL (minimum score of 600 required). Electronic applications accepted. *Faculty research:* Advanced metallic, ceramic, and polymeric materials for device applications; micromechanics and structure of interfaces; thin film electronic materials; physics and chemistry of solids.

See in-depth description on page 1355.

University of Pittsburgh, School of Engineering, Department of Materials Science and Engineering, Pittsburgh, PA 15260. Offers materials science and engineering (MSMSE, PhD); metallurgical engineering (MS Met E, PhD). Part-time and evening/weekend programs available. *Faculty:* 13 full-time (0 women), 2 part-time (0 women). *Students:* 12 full-time (5 women), 19 part-time (2 women); includes 1 minority (Asian American or Pacific Islander), 8 international. 114 applicants, 9% accepted. In 1998, 6 master's, 9 doctorates awarded. Terminal master's awarded for partial completion of doctoral program. *Degree requirements:* For master's, thesis required (for some programs), foreign language not required; for doctorate, dissertation, comprehensive and final oral exams required, foreign language not required. *Entrance* *requirements:* For master's and doctorate, TOEFL (minimum score of 550 required), minimum QPA of 3.0. *Average time to degree:* Master's–3 years full-time, 5 years part-time; doctorate–6 years full-time. *Application deadline:* For fall admission, 8/1 (priority date); for spring admission, 12/1 (priority date). Applications are processed on a rolling basis. Application fee: $30 ($40 for international students). *Financial aid:* In 1998–99, 13 students received aid, including 9 research assistantships (averaging $10,032 per year), 4 teaching assistantships (averaging $10,600 per year); grants, scholarships, and tuition waivers (full and partial) also available. Financial aid application deadline: 2/15. *Faculty research:* Hot corrosion and oxidation, physical metallurgy and thermomechanical processing of steels, ceramic processing, phase transformation. Total annual research expenditures: $2 million. *Unit head:* Dr. William A. Soffa, Chairman, 412-624-9720, Fax: 412-624-8069, E-mail: wsoffa+@pitt.edu. *Application contact:* Pradeep P. Phule, Graduate Coordinator, 412-624-9736, Fax: 412-624-8069, E-mail: ppp@vms.cis.pitt.edu.

University of Rochester, The College, School of Engineering and Applied Sciences, Department of Mechanical Engineering, Program in Materials Science, Rochester, NY 14627-0250. Offers MS, PhD. Part-time programs available. *Students:* 33 full-time (7 women), 10 part-time (3 women); includes 1 minority (Asian American or Pacific Islander), 25 international. 61 applicants, 21% accepted. In 1998, 5 master's, 1 doctorate awarded. Terminal master's awarded for partial completion of doctoral program. *Degree requirements:* For master's, comprehensive exam required, thesis optional, foreign language not required; for doctorate, dissertation, preliminary and qualifying exams required, foreign language not required. *Entrance requirements:* For master's and doctorate, GRE, TOEFL. *Application deadline:* For fall admission, 2/1. Application fee: $25. *Financial aid:* Fellowships, research assistantships, teaching assistantships, tuition waivers (full and partial) available. Financial aid application deadline:2/1.

University of Southern California, Graduate School, School of Engineering, Department of Materials Science and Engineering, Program in Materials Science, Los Angeles, CA 90089. Offers MS, PhD, Engr. *Students:* 25 full-time (5 women), 11 part-time (3 women); includes 4 minority (all Asian Americans or Pacific Islanders), 30 international. Average age 28. 169 applicants, 59% accepted. In 1998, 2 master's, 10 doctorates awarded. *Degree requirements:* For master's, thesis optional; for doctorate, dissertation required. *Entrance requirements:* For master's, doctorate, and Engr, GRE General Test. *Application deadline:* For fall admission, 7/1 (priority date); for spring admission, 12/1. Application fee: $55. Tuition: Part-time $768 per unit. Required fees: $350 per semester. *Financial aid:* In 1998–99, 2 fellowships, 14 research assistantships, 3 teaching assistantships were awarded.; Federal Work-Study and institutionally-sponsored loans also available. Aid available to part-time students. Financial aid application deadline: 2/15; financial aid applicants required to submit FAFSA. *Unit head:* Dr. Florian Mansfeld, Chairman, Department of Materials Science and Engineering, 213-740-6011.

University of Tennessee, Knoxville, Graduate School, College of Engineering, Department of Mechanical and Aerospace Engineering and Engineering Science, Program in Engineering Science, Knoxville, TN 37996. Offers applied artificial intelligence (MS); biomedical engineering (MS, PhD); composite materials (MS, PhD); computational mechanics (MS, PhD); engineering science (MS, PhD); fluid mechanics (MS, PhD); industrial engineering (MS, PhD); optical engineering (MS, PhD); solid mechanics (MS, PhD). Part-time programs available. *Students:* 33 full-time (9 women), 16 part-time (1 woman); includes 4 minority (1 African American, 2 Asian Americans or Pacific Islanders, 1 Native American), 10 international. *Degree requirements:* For master's, thesis or alternative required, foreign language not required; for doctorate, dissertation required, foreign language not required. *Entrance requirements:* For master's and doctorate, TOEFL (minimum score of 550 required), minimum GPA of 2.7. *Application deadline:* For fall admission, 2/1 (priority date). Applications are processed on a rolling basis. Application fee: $35. Electronic applications accepted. *Application contact:* Dr. Allen Yu, Graduate Representative, 923-974-4159, E-mail: nyu@utk.edu.

The University of Texas at Arlington, Graduate School, College of Engineering, Department of Mechanical and Aerospace Engineering, Materials Science and Engineering Program, Arlington, TX 76019. Offers MS, PhD. *Students:* 19 full-time (4 women), 13 part-time (6 women); includes 4 minority (3 Asian Americans or Pacific Islanders, 1 Hispanic American), 17 international. 70 applicants, 34% accepted. In 1998, 6 master's, 2 doctorates awarded. *Degree requirements:* For master's, thesis required (for some programs), foreign language not required; for doctorate, dissertation required, foreign language not required. *Entrance requirements:* For master's and doctorate, GRE General Test, TOEFL. *Application deadline:* Applications are processed on a rolling basis. Application fee: $25 ($50 for international students). Tuition, state resident: full-time $1,368; part-time $76 per semester hour. Tuition, nonresident: full-time $5,454; part-time $303 per semester hour. Required fees: $66 per semester hour. $86 per term. Tuition and fees vary according to course load. *Financial aid:* Research assistantships, teaching assistantships available. *Unit head:* Dr. Ronald L. Elsenbaumer, Director, 817-272-2398, Fax: 817-272-2538. *Application contact:* Pranesh B. Aswath, Graduate Adviser, 817-272-2008, Fax: 817-272-2538, E-mail: aswath@uta.edu.

See in-depth description on page 1357.

The University of Texas at Austin, Graduate School, College of Engineering, Program in Materials Science and Engineering, Austin, TX 78712-1111. Offers MSE, PhD. Part-time programs available. *Faculty:* 44 full-time (2 women). *Students:* 53 full-time (8 women), 7 part-time (3 women); includes 3 minority (2 Asian Americans or Pacific Islanders, 1 Hispanic American), 35 international. Average age 26. 72 applicants, 31% accepted. In 1998, 7 master's awarded (57% found work related to degree, 43% continued full-time study); 8 doctorates awarded (100% found work related to degree). *Degree requirements:* For master's and doctorate, thesis/dissertation required (for some programs), foreign language not required. *Entrance requirements:* For master's and doctorate, GRE General Test. *Average time to degree:* Master's–2.1 years full-time; doctorate–4.8 years full-time. *Application deadline:* For fall admission, 1/2 (priority date); for spring admission, 10/1. Applications are processed on a rolling basis. Application fee: $50 ($75 for international students). Electronic applications accepted. *Financial aid:* In 1998–99, 54 students received aid, including 2 fellowships with partial tuition reimbursements available (averaging $15,000 per year), 41 research assistantships with full tuition reimbursements available (averaging $13,000 per year), 11 teaching assistantships with partial tuition reimbursements available (averaging $13,000 per year); career-related internships or fieldwork and institutionally-sponsored loans also available. Financial aid application deadline: 1/1; financial aid applicants required to submit FAFSA. *Faculty research:* Ceramics and glasses, microelectronic materials and processing, metal systems, nanostructures, polymers . Total annual research expenditures: $3 million. *Unit head:* Donald R. Paul, Director, 512-471-1504, Fax: 512-475-8482, E-mail: drp@che.utexas.edu. *Application contact:* Lydia C. Griffith, Graduate Coordinator, 512-471-1504, Fax: 512-475-8482, E-mail: lcgriffith@mail.utexas.edu.

The University of Texas at El Paso, Graduate School, Interdisciplinary Program in Materials Science and Engineering, El Paso, TX 79968-0001. Offers PhD. *Students:* 19 full-time (5 women), 9 part-time (4 women); includes 13 minority (1 Asian American or Pacific Islander, 12 Hispanic Americans), 12 international. Average age 33. 30 applicants, 13% accepted. *Degree requirements:* For doctorate, dissertation, scholarly article required, foreign language not required. *Entrance requirements:* For doctorate, GRE General Test, TOEFL (minimum score of 550 required), minimum GPA of 3.2. *Application deadline:* Applications are processed on a rolling basis. Application fee: $0 ($65 for international students). Tuition, state resident: full-time $2,790. Tuition, nonresident: full-time $7,710. *Financial aid:* In 1998–99, 7 fellowships, 7 research assistantships, 2 teaching assistantships were awarded.; Federal Work-Study, institutionally-sponsored loans, and tuition waivers (partial) also available. Financial aid applicants required to submit FAFSA. *Faculty research:* Material chemistry, materials physics, mineral science. Total annual research expenditures: $3.5 million. *Unit head:* Dr. Lawrence Murr, Director, 915-747-6929, Fax: 915-747-5616. *Application contact:* Susan Jordan, Director, Graduate Student Services, 915-747-5491, Fax: 915-747-5788, E-mail: sjordan@utep.edu.

University of Toronto, School of Graduate Studies, Physical Sciences Division, Faculty of Applied Science and Engineering, Department of Metallurgy and Materials Science, Toronto, ON M5S 1A1, Canada. Offers M Eng, MA Sc, PhD. Part-time programs available. *Degree requirements:* For master's, thesis required (for some programs); for doctorate, dissertation required.

University of Utah, Graduate School, College of Engineering, Department of Materials Science and Engineering, Salt Lake City, UT 84112-1107. Offers ME, MS, PhD. Part-time programs available. *Faculty:* 10 full-time (0 women), 11 part-time (2 women). *Students:* 40 full-time (6 women), 16 part-time (5 women); includes 1 minority (Asian American or Pacific Islander), 37 international. Average age 30. In 1998, 4 master's, 5 doctorates awarded. Terminal master's awarded for partial completion of doctoral program. *Degree requirements:* For master's, thesis (MS) required; for doctorate, dissertation, exam required, foreign language not required. *Entrance requirements:* For master's and doctorate, TOEFL (minimum score of 500 required), minimum GPA of 3.0. *Application deadline:* For fall admission, 7/1. Application fee: $30 ($50 for international students). *Financial aid:* In 1998–99, 3 teaching assistantships were awarded.; research assistantships, career-related internships or fieldwork and Federal Work-Study also available. Financial aid application deadline: 3/1. *Faculty research:* Biomaterials, ceramic composition, electrochemistry, semiconductor materials and devices, crystal growth. *Unit head:* Anil Virkar, Chair, 801-581-6863, Fax: 801-581-4816, E-mail: anil.virkar@m.cc.utah.edu. *Application contact:* Judy Martinez, Director of Graduate Studies, 801-581-6863.

University of Vermont, Graduate College, College of Engineering and Mathematics, Program in Materials Science, Burlington, VT 05405-0160. Offers MS, PhD. *Degree requirements:* For master's, thesis or alternative required, foreign language not required; for doctorate, dissertation required, foreign language not required. *Entrance requirements:* For master's and doctorate, GRE General Test, TOEFL (minimum score of 550 required).

University of Virginia, School of Engineering and Applied Science, Department of Materials Science and Engineering, Program in Materials Science, Charlottesville, VA 22903. Offers MMSE, MS, PhD. Postbaccalaureate distance learning degree programs offered (no on-campus study). *Faculty:* 21 full-time (1 woman), 1 part-time (0 women). *Students:* 54 full-time (12 women), 1 part-time; includes 4 minority (1 African American, 2 Asian Americans or Pacific Islanders, 1 Hispanic American), 14 international. Average age 27. 78 applicants, 23% accepted. In 1998, 17 master's, 9 doctorates awarded. *Degree requirements:* For master's, comprehensive exam required; for doctorate, dissertation, comprehensive exam required, foreign language not required. *Entrance requirements:* For master's and doctorate, GRE General Test. *Application deadline:* For fall admission, 8/1; for spring admission, 12/1. Applications are processed on a rolling basis. Application fee: $60. *Financial aid:* Application deadline: 2/1. *Application contact:* J. Milton Adams, Assistant Dean, 804-924-3897, E-mail: twr2c@virginia.edu.

University of Washington, Graduate School, College of Engineering, Department of Materials Science and Engineering, Seattle, WA 98195-2120. Offers ceramic engineering (MS, MSE); materials science (MSE); materials science and engineering (MS, MSMSE, PhD); metallurgical engineering (MS, MSE). Part-time programs available. *Faculty:* 12 full-time (1 woman), 3 part-time (0 women). *Students:* 43 full-time (12 women), 18 part-time (5 women); includes 8 minority (2 African Americans, 3 Asian Americans or Pacific Islanders, 3 Hispanic Americans), 17 international. Average age 32. 62 applicants, 53% accepted. In 1998, 3 master's awarded (67% found work related to degree, 33% continued full-time study); 2 doctorates awarded (100% found work related to degree). *Degree requirements:* For master's and doctorate, thesis/dissertation required, foreign language not required. *Entrance requirements:* For master's and doctorate, GRE General Test (minimum combined score of 1400 on quantitative and analytical sections required), TOEFL (minimum score of 600 required), minimum GPA of 3.0. *Average time to degree:* Master's–3.4 years full-time, 7.5 years part-time; doctorate–6.25 years full-time. *Application deadline:* For fall admission, 7/1; for winter admission, 11/1; for spring admission, 2/1. Applications are processed on a rolling basis. Application fee: $50. Electronic applications accepted. Tuition, state resident: full-time $5,196; part-time $475 per credit. Tuition, nonresident: full-time $13,485; part-time $1,285 per credit. Required fees: $387; $38 per credit. Tuition and fees vary according to course load. *Financial aid:* In 1998–99, 37 students received aid, including 3 fellowships with full tuition reimbursements available (averaging $8,167 per year), 22 research assistantships with full tuition reimbursements available (averaging $14,088 per year), 12 teaching assistantships with full tuition reimbursements available (averaging $9,900 per year); career-related internships or fieldwork, Federal Work-Study, grants, institutionally-sponsored loans, scholarships, and unspecified assistantships also available. Financial aid application deadline: 3/1. *Faculty research:* Processing characterization and properties of structural and electronic materials. Total annual research expenditures: $1.1 million. *Unit head:* Dr. Rajendra K. Bordia, Chair, 206-543-2600, Fax: 206-543-3100, E-mail: bordia@u.washington.edu. *Application contact:* Dr. T. G. Stoebe, Graduate Coordinator, 206-543-7090, Fax: 206-543-3100, E-mail: mse@u.washington.edu.

Announcement: Research programs and facilities are available in a wide range of advanced materials areas, including characterization, processing and properties of ceramic and polymeric composites, surface characterization in electronic materials, high-technology structural ceramics, smart materials, fuel cell materials, optical device materials, advanced metallic alloys, biomaterials, and computational materials science. Visit the Web site at http://weber.u.washington.edu/~material/

University of Wisconsin–Madison, Graduate School, College of Engineering, Materials Science Program, Madison, WI 53706-1380. Offers MS, PhD. Part-time programs available. *Faculty:* 42 full-time (4 women). *Students:* 45 full-time (13 women), 2 part-time (1 woman); includes 3 minority (2 African Americans, 1 Hispanic American), 23 international. Average age 30. In 1998, 7 master's, 9 doctorates awarded (100% found work related to degree). *Degree requirements:* For doctorate, dissertation required. *Entrance requirements:* For master's and doctorate, GRE General Test. *Average time to degree:* Master's–2 years full-time; doctorate–5 years full-time. *Application deadline:* For fall admission, 7/1 (priority date); for spring admission, 11/15 (priority date). Applications are processed on a rolling basis. Application fee: $45. Electronic applications accepted. *Financial aid:* Fellowships with full tuition reimbursements, research assistantships with full tuition reimbursements, teaching assistantships with tuition reimbursements available. Financial aid application deadline: 1/15. *Faculty research:* Superconductors, electronic materials, cast metals, vapor deposition, composites, surfaces, interfaces, x-ray lithography. *Unit head:* Eric E. Hellstrom, Chair, 608-263-1795, Fax: 608-262-8353, E-mail: matsciad@engr.wisc.edu. *Application contact:* Diana J. Rhoads, Program Assistant 3, 608-263-1795, Fax: 608-262-8353, E-mail: matsciad@engr.wisc.edu.

Vanderbilt University, School of Engineering, Program in Materials Science and Engineering, Nashville, TN 37240-1001. Offers MS, PhD. Part-time programs available. *Faculty:* 13 full-time (0 women), 1 part-time (0 women). *Students:* 18 full-time (4 women); includes 11 minority (8 Asian Americans or Pacific Islanders, 3 Native Americans) Average age 28. 53 applicants, 28% accepted. In 1998, 2 master's, 2 doctorates awarded. Terminal master's awarded for partial completion of doctoral program. *Degree requirements:* For master's and doctorate, thesis/dissertation required, foreign language not required. *Entrance requirements:* For master's and doctorate, GRE General Test. *Average time to degree:* Master's–2 years full-time; doctorate–2 years full-time. *Application deadline:* For fall admission, 1/15. Application fee: $40. *Financial aid:* In 1998–99, 4 fellowships, 10 research assistantships, 7 teaching assistantships were awarded.; institutionally-sponsored loans and tuition waivers (partial) also available. Aid available to part-time students. Financial aid application deadline: 1/15. *Faculty research:* Tribology, solidification, electron microscopy, ion implantation, diamond deposition. Total annual research expenditures: $1.4 million. *Unit head:* Robert J. Bayuzick, Director, 615-343-6868, Fax: 615-343-8645, E-mail: bayuzick@vuse.vanderbilt.edu. *Application contact:* Sarah M. Ross, Administrative Assistant, 615-343-2415, Fax: 615-343-8645, E-mail: sross@vuse.vanderbilt.edu.

Materials Sciences–Metallurgical Engineering and Metallurgy

Virginia Polytechnic Institute and State University, Graduate School, College of Engineering, Department of Materials Science and Engineering, Blacksburg, VA 24061. Offers M Eng, MS. *Students:* 9 full-time (2 women), 3 part-time (1 woman), 5 international. 32 applicants, 28% accepted. In 1998, 7 degrees awarded. *Degree requirements:* For master's, thesis required, foreign language not required. *Entrance requirements:* For master's, TOEFL. *Application deadline:* For fall admission, 12/1 (priority date). Applications are processed on a rolling basis. Application fee: $25. *Financial aid:* Fellowships, research assistantships, teaching assistantships, unspecified assistantships available. Financial aid application deadline: 4/1. *Unit head:* Dr. R. S. Gordon, Head, 540-231-6655, E-mail: gordonrs@vt.edu.

Virginia Polytechnic Institute and State University, Graduate School, College of Engineering, Program in Materials Engineering Science, Blacksburg, VA 24061. Offers PhD. *Students:* 8 full-time (3 women), 1 part-time, 6 international. 61 applicants, 18% accepted. In 1998, 1 degree awarded. *Degree requirements:* For doctorate, dissertation required. *Entrance requirements:* For doctorate, TOEFL. *Application deadline:* For fall admission, 12/1 (priority date). Applications are processed on a rolling basis. Application fee: $25. *Financial aid:* Fellowships, research assistantships, teaching assistantships, career-related internships or fieldwork and unspecified assistantships available. Financial aid application deadline: 4/1. *Faculty research:* Polymer syntheses, adhesive and sealant science, composite materials and structures, surface chemistry. *Unit head:* Dr. G. Wilkes, Chairman, 540-231-5316, E-mail: gwilkes@vt.edu.

Washington State University, Graduate School, College of Engineering and Architecture, School of Mechanical and Materials Engineering, Program in Materials Science and Engineering, Pullman, WA 99164. Offers materials science (PhD); materials science and engineering (MS). *Faculty:* 10 full-time (0 women). *Students:* 4 full-time (0 women), 1 part-time, 4 international. Average age 25. In 1998, 4 master's awarded (25% found work related to degree, 75% continued full-time study); 4 doctorates awarded. *Degree requirements:* For master's and doctorate, thesis/dissertation, oral exam required, foreign language not required. *Entrance requirements:* For master's and doctorate, GRE General Test, minimum GPA of 3.0. *Average time to degree:* Master's–2 years full-time; doctorate–4 years full-time. *Application deadline:* For fall admission, 3/1 (priority date). Applications are processed on a rolling basis. Application fee: $35. *Financial aid:* In 1998–99, 1 research assistantship, 1 teaching assistantship were awarded.; fellowships, career-related internships or fieldwork, Federal Work-Study, institutionally-sponsored loans, and teaching associateships also available. Financial aid application deadline: 4/1. *Faculty research:* Fatigue and fracture of structural materials, mechanical behavior and corrosion of thin films, superplastic forming, fundamental aspects of deformation processes including numerical modeling. *Unit head:* Dr. Grant Norton, Chair. *Application contact:* Dr. Grant Norton, Chair.

Washington State University, Graduate School, College of Sciences, Department of Physics, Pullman, WA 99164. Offers chemical physics (PhD); material science (MS, PhD); physics (MS, PhD). *Faculty:* 20 full-time (1 woman). *Students:* 31 full-time (2 women), 5 part-time (1 woman); includes 3 minority (1 African American, 1 Asian American or Pacific Islander, 1 Hispanic American), 12 international. *Degree requirements:* For master's, thesis (for some programs), oral exam required; for doctorate, dissertation, oral exam required. *Entrance requirements:* For master's and doctorate, GRE General Test, GRE Subject Test, minimum GPA of 3.0. *Application deadline:* For fall admission, 3/1 (priority date). Applications are processed on a rolling basis. Application fee: $35. *Unit head:* Dr. Miles Dresser, Chair, 509-335-1698, Fax: 509-335-7816, E-mail: physics@wsu.edu. *Application contact:* Dr. Y. Gupta, 509-335-3140.

Washington University in St. Louis, School of Engineering and Applied Science, Sever Institute of Technology, Program in Materials Science and Engineering, St. Louis, MO 63130-4899. Offers MS, D Sc. Part-time programs available. *Faculty:* 14 full-time (0 women), 12 part-time (0 women). *Students:* 5 full-time (1 woman), 10 part-time (2 women), 8 international. 152 applicants, 7% accepted. In 1998, 4 master's, 2 doctorates awarded. Terminal master's awarded for partial completion of doctoral program. *Degree requirements:* For master's, thesis optional, foreign language not required; for doctorate, dissertation, departmental qualifying exam required, foreign language not required. *Entrance requirements:* For master's and doctorate, GRE. *Average time to degree:* Master's–1.5 years full-time, 5 years part-time; doctorate–5 years full-time, 8 years part-time. *Application deadline:* For fall admission, 2/15 (priority date). Applications are processed on a rolling basis. Application fee: $20. *Financial aid:* In 1998–99, 3 students received aid, including 1 research assistantship, 1 teaching assistantship; fellowships, career-related internships or fieldwork, Federal Work-Study, and institutionally-sponsored loans also available. Financial aid application deadline: 2/15. *Faculty research:* Material processing, magnetic materials, composites, name-phase materials, high-temperature materials. Total annual research expenditures: $500,000. *Unit head:* Dr. Kenneth Jerina, Chairman, 314-935-6012.

Wayne State University, Graduate School, College of Engineering, Department of Chemical Engineering and Materials Science, Program in Materials Science and Engineering, Detroit, MI 48202. Offers materials science and engineering (MS, PhD); polymer engineering (Certificate). Part-time programs available. Terminal master's awarded for partial completion of doctoral program. *Degree requirements:* For master's, thesis optional, foreign language not required; for doctorate, dissertation required, foreign language not required. *Faculty research:* Polymer science, rheology, fatigue in metals, metal matrix composites, ceramics.

Western Michigan University, Graduate College, College of Engineering and Applied Sciences, Department of Construction Engineering, Materials Engineering and Industrial Design, Program in Materials Science and Engineering, Kalamazoo, MI 49008. Offers MS. *Students:* 4 full-time (0 women), 12 part-time (2 women), 13 international. 31 applicants, 81% accepted. *Entrance requirements:* For master's, minimum GPA of 3.0. *Application deadline:* For fall admission, 2/15 (priority date). Applications are processed on a rolling basis. Application fee: $25. *Financial aid:* Application deadline: 2/15; *Application contact:* Paula J. Boodt, Coordinator, Graduate Admissions and Recruitment, 616-387-2000, Fax: 616-387-2355, E-mail: paula.boodt@wmich.edu.

Worcester Polytechnic Institute, Graduate Studies, Department of Materials Science and Engineering, Worcester, MA 01609-2280. Offers MS, PhD, Certificate. Part-time and evening/weekend programs available. *Faculty:* 8 full-time (2 women). *Students:* 17 full-time (6 women), 10 part-time (2 women); includes 1 minority (Asian American or Pacific Islander), 7 international. 65 applicants, 46% accepted. In 1998, 5 master's, 1 doctorate awarded. *Degree requirements:* For master's and doctorate, thesis/dissertation required, foreign language not required. *Entrance requirements:* For master's, TOEFL (minimum score of 550 required; average 599); for doctorate, GRE General Test, TOEFL (minimum score of 550 required; average 599). *Average time to degree:* Master's–1.5 years full-time, 4 years part-time; doctorate–4.5 years full-time. *Application deadline:* For fall admission, 2/15 (priority date); for spring admission, 10/15 (priority date). Applications are processed on a rolling basis. Application fee: $50. Electronic applications accepted. *Financial aid:* In 1998–99, 3 students received aid, including 1 fellowship with full tuition reimbursement available (averaging $11,970 per year), 2 teaching assistantships with full tuition reimbursements available (averaging $11,970 per year); research assistantships, career-related internships or fieldwork, grants, institutionally-sponsored loans, and scholarships also available. Financial aid application deadline: 2/15; financial aid applicants required to submit FAFSA. *Faculty research:* Tribology, corrosion, electron microscopy, powder metallurgy and ceramics. *Unit head:* Richard D. Sisson, Head, 508-831-5335, Fax: 508-831-5178, E-mail: sisson@wpi.edu.

Worcester Polytechnic Institute, Graduate Studies, Department of Mechanical Engineering, Worcester, MA 01609-2280. Offers materials science and engineering (MS, PhD); mechanical engineering (M Eng, MS, PhD, Advanced Certificate). Part-time and evening/weekend programs available. *Faculty:* 18 full-time (1 woman), 1 part-time (0 women). *Students:* 58 full-time (6 women), 19 part-time (3 women); includes 5 minority (4 Asian Americans or Pacific Islanders, 1 Hispanic American), 22 international. *Degree requirements:* For master's, thesis optional, foreign language not required; for doctorate, dissertation required, foreign language not required. *Entrance requirements:* For master's, TOEFL (minimum score of 550 required; average 608); for doctorate, GRE General Test (combined average 1947 on three sections), TOEFL (minimum score of 550 required; average 608). *Application deadline:* For fall admission, 2/15 (priority date); for spring admission, 10/15 (priority date). Applications are processed on a rolling basis. Application fee: $50. Electronic applications accepted. *Unit head:* Mohammad N. Noori, Head, 508-831-5759, Fax: 508-831-5680, E-mail: mnnoori@wpi.edu. *Application contact:* Mark Richman, Graduate Coordinator, 508-831-5556, Fax: 508-831-5680, E-mail: mrichman@wpi.edu.

Wright State University, School of Graduate Studies, College of Engineering and Computer Science, Programs in Engineering, Program in Mechanical and Materials Engineering, Dayton, OH 45435. Offers materials science and engineering (MSE); mechanical engineering (MSE). *Students:* 32 full-time (9 women), 30 part-time (3 women); includes 4 minority (1 African American, 2 Asian Americans or Pacific Islanders, 1 Hispanic American), 26 international. *Degree requirements:* For master's, thesis or course option alternative required. *Entrance requirements:* For master's, TOEFL (minimum score of 550 required). Application fee: $25. *Unit head:* Dr. Richard J. Bethke, Chair, 937-775-5040, Fax: 937-775-5009.

Metallurgical Engineering and Metallurgy

Colorado School of Mines, Graduate School, Department of Metallurgical and Materials Engineering, Golden, CO 80401-1887. Offers ME, MS, PhD. Part-time programs available. *Faculty:* 25 full-time (0 women), 1 part-time (0 women). *Students:* 78 full-time (17 women), 25 part-time (4 women); includes 8 minority (5 Asian Americans or Pacific Islanders, 3 Hispanic Americans), 31 international. 92 applicants, 40% accepted. In 1998, 9 master's awarded (100% found work related to degree); 3 doctorates awarded (100% found work related to degree). *Degree requirements:* For master's, thesis required, foreign language not required; for doctorate, dissertation, comprehensive exams required, foreign language not required. *Entrance requirements:* For master's and doctorate, GRE General Test, minimum GPA of 3.0. *Application deadline:* Applications are processed on a rolling basis. Application fee: $40. Electronic applications accepted. *Financial aid:* In 1998–99, 55 students received aid, including 3 fellowships, 37 research assistantships, 10 teaching assistantships; unspecified assistantships also available. Aid available to part-time students. Financial aid applicants required to submit FAFSA. *Faculty research:* Phase transformations, nonferrous alloy systems, pyrometallurgy, reactive metals, engineered materials. Total annual research expenditures: $141,430. *Unit head:* Dr. John J. Moore, Head, 303-273-3770, Fax: 303-273-3795, E-mail: jjmoore@mines.edu. *Application contact:* Sharon Kirts, Program Assistant I, 303-273-3660, Fax: 303-384-2189, E-mail: skirts@mines.edu.

Columbia University, Fu Foundation School of Engineering and Applied Science, Department of Earth and Environmental Engineering, Program in Materials Science and Engineering, New York, NY 10027. Offers materials engineering (MS, Eng Sc D); materials science (Eng Sc D, PhD, EM); materials science and engineering (MS); metallurgy (Met E). Part-time programs available. Postbaccalaureate distance learning degree programs offered (no on-campus study). *Faculty:* 7 full-time (2 women), 2 part-time (1 woman). *Students:* 25 full-time (0 women), 9 part-time; includes 4 minority (all Asian Americans or Pacific Islanders), 16 international. Terminal master's awarded for partial completion of doctoral program. *Degree requirements:* For master's, foreign language and thesis not required; for doctorate, dissertation, qualifying exam required, foreign language not required. *Entrance requirements:* For master's, doctorate, and other advanced degree, GRE General Test, TOEFL. Application fee: $55. *Application contact:* Dr. Barbara Algin, Departmental Administrator, 212-854-2905, Fax: 212-854-7081, E-mail: ba110@columbia.edu.

Columbia University, Fu Foundation School of Engineering and Applied Science, Department of Earth and Environmental Engineering, Program in Minerals Engineering and Materials Science, New York, NY 10027. Offers metallurgical engineering (Engr); mineral engineering (Engr); minerals engineering and materials science (Eng Sc D, PhD); mines (Engr). Part-time programs available. *Faculty:* 12 full-time (2 women), 3 part-time (1 woman). *Students:* 3 full-time (0 women), 3 part-time; includes 1 minority (Asian American or Pacific Islander), 5 international. *Degree requirements:* For doctorate, dissertation, qualifying exam required. *Entrance requirements:* For doctorate and Engr, GRE General Test, TOEFL. Application fee: $55. *Unit head:* Dr. Barbara Algin, Departmental Administrator, 212-854-2905, Fax: 212-854-7081, E-mail: ba110@columbia.edu. *Application contact:* Dr. Barbara Algin, Departmental Administrator, 212-854-2905, Fax: 212-854-7081, E-mail: ba110@columbia.edu.

Dalhousie University, Faculty of Graduate Studies, DalTech, Faculty of Engineering, Department of Mining and Metallurgical Engineering, Program in Metallurgical Engineering, Halifax, NS B3H 3J5, Canada. Offers M Eng, MA Sc, PhD. *Faculty:* 5 full-time (0 women), 2 part-time (0 women). In 1998, 1 master's, 2 doctorates awarded (100% found work related to degree). *Degree requirements:* For master's and doctorate, thesis/dissertation required, foreign language not required. *Entrance requirements:* For master's and doctorate, TOEFL (minimum score of 580 required). *Application deadline:* For fall admission, 6/1; for winter admission, 10/1; for spring admission, 2/1. Applications are processed on a rolling basis. Application fee: $55. *Financial aid:* Fellowships, research assistantships, teaching assistantships, scholarships and traineeships available. *Faculty research:* Ceramic and metal matrix composites, electron microscopy, electrolysis in molten salt, fracture mechanics, electronic materials. *Unit head:* Dr. G. Kipouros, Head, 902-494-3954, Fax: 902-425-1037, E-mail: minmet@dal.ca. *Application contact:* Shelley Parker, Admissions Coordinator, Graduate Studies and Research, 902-494-1288, Fax: 902-494-3149, E-mail: shelley.parker@dal.ca.

École Polytechnique de Montréal, Graduate Programs, Department of Metallurgical and Materials Engineering, Montréal, PQ H3C 3A7, Canada. Offers advanced materials (M Eng, M Sc A, PhD); chemical metallurgy (M Eng, M Sc A, PhD); physical metallurgy (M Eng, M Sc A, PhD). Part-time programs available. *Degree requirements:* For master's and doctorate, one foreign language, computer language, thesis/dissertation required. *Entrance requirements:* For master's, minimum GPA of 2.75; for doctorate, minimum GPA of 3.0. *Faculty research:* Refractory materials, fatigue, thermochemical analysis, electrochemistry, material characterization.

Georgia Institute of Technology, Graduate Studies and Research, College of Engineering, School of Materials Science and Engineering, Program in Metallurgy, Atlanta, GA 30332-0001. Offers MSMSE, PhD. Terminal master's awarded for partial completion of doctoral program.

Degree requirements: For master's, computer language required, thesis optional, foreign language not required; for doctorate, computer language, dissertation required, foreign language not required. *Entrance requirements:* For master's and doctorate, GRE General Test, TOEFL (minimum score of 550 required). Electronic applications accepted. *Faculty research:* Surface science, corrosion, fracture and fatigue, surface modification, ion implantation and plating.

Illinois Institute of Technology, Graduate College, Armour College of Engineering and Sciences, Department of Mechanical, Materials and Aerospace Engineering, Metallurgical and Materials Engineering Division, Chicago, IL 60616-3793. Offers MMME, MS, PhD. Part-time and evening/weekend programs available. *Faculty:* 8 full-time (1 woman), 4 part-time (0 women). *Students:* 13 full-time (4 women), 16 part-time (3 women); includes 3 minority (2 African Americans, 1 Hispanic American), 10 international. 46 applicants, 54% accepted. In 1998, 7 master's, 6 doctorates awarded. Terminal master's awarded for partial completion of doctoral program. *Degree requirements:* For master's, thesis (for some programs), comprehensive exam required, foreign language not required; for doctorate, dissertation, comprehensive exam required, foreign language not required. *Entrance requirements:* For master's, GRE General Test (minimum combined score of 1200 required), TOEFL (minimum score of 550 required), undergraduate GPA of 3.0; for doctorate, GRE (minimum score of 1200 required), TOEFL (minimum score of 550 required), undergraduate GPA of 3.0 required. *Application deadline:* For fall admission, 7/1; for spring admission, 11/1. Applications are processed on a rolling basis. Application fee: $30. Electronic applications accepted. *Financial aid:* In 1998–99, 1 fellowship, 3 research assistantships, 3 teaching assistantships were awarded.; Federal Work-Study, institutionally-sponsored loans, scholarships, and graduate assistantships also available. Financial aid application deadline: 3/1. *Faculty research:* High-temperature materials, laser processing, fracture mechanics, powder metallurgy, alloy design. *Unit head:* Judith Todd, Associate Chair, 312-567-3175, Fax: 312-567-8875, E-mail: jtodd@charlie.iit.edu. *Application contact:* Dr. S. Mohammad Shahidehpour, Dean of Graduate College, 312-567-3024, Fax: 312-567-7517, E-mail: grad@minna.cns.iit.edu.

Laurentian University, School of Graduate Studies and Research, School of Engineering, Sudbury, ON P3E 2C6, Canada. Offers metallurgy (MA Sc); mining (M Eng). Part-time programs available. *Faculty:* 15 full-time (1 woman), 5 part-time (0 women). *Students:* 10 full-time (0 women), 7 part-time. *Degree requirements:* Foreign language not required. *Application deadline:* For fall admission, 9/1. Application fee: $50. *Unit head:* Dr. Paul Lindon, Director, 705-675-1151 Ext. 2244, Fax: 705-675-4862. *Application contact:* 705-675-1151 Ext. 3909, Fax: 705-675-4843.

Massachusetts Institute of Technology, School of Engineering, Department of Materials Science and Engineering, Cambridge, MA 02139-4307. Offers materials engineering (Mat E); materials science and engineering (SM, PhD, Sc D); metallurgical engineering (Met E). *Faculty:* 31 full-time (6 women). *Students:* 167 full-time (41 women), 3 part-time; includes 23 minority (1 African American, 20 Asian Americans or Pacific Islanders, 2 Hispanic Americans), 61 international. Terminal master's awarded for partial completion of doctoral program. *Degree requirements:* For master's, thesis required, foreign language not required; for doctorate, dissertation, qualifying exams required, foreign language not required; for other advanced degree, thesis required. *Entrance requirements:* For master's, doctorate, and other advanced degree, GRE General Test. *Application deadline:* For fall admission, 1/15 (priority date); for spring admission, 11/1. Applications are processed on a rolling basis. Application fee: $55. *Unit head:* Dr. Thomas W. Eagar, Head, 617-253-0948, Fax: 617-252-1773, E-mail: tweagar@mit.edu. *Application contact:* Kenneth C. Russell, Graduate Admissions Chairman, 617-253-3329, Fax: 617-258-8836, E-mail: kenruss@mit.edu.

McGill University, Faculty of Graduate Studies and Research, Faculty of Engineering, Department of Mining and Metallurgical Engineering, Program in Metallurgical Engineering, Montréal, PQ H3A 2T5, Canada. Offers M Eng, PhD. *Students:* 105 full-time (21 women). 40 applicants, 53% accepted. In 1998, 6 master's, 19 doctorates awarded. *Degree requirements:* For master's, thesis optional, foreign language not required; for doctorate, dissertation required, foreign language not required. *Entrance requirements:* For master's and doctorate, TOEFL (minimum score of 550 required), minimum GPA of 3.0. *Application deadline:* For fall admission, 3/1; for winter admission, 8/1; for spring admission, 11/1. Applications are processed on a rolling basis. Application fee: $60. *Unit head:* Laura A. Rose, Admissions Coordinator, 617-253-9303, Fax: 617-258-9214, E-mail: lrose@mit.edu. *Application contact:* Dr. S. Yue, Director, 514-398-4755 Ext. 0486, Fax: 514-398-4492, E-mail: steve@minmet.lan.mcgill.ca.

Michigan Technological University, Graduate School, College of Engineering, Department of Metallurgical and Materials Engineering, Houghton, MI 49931-1295. Offers MS, PhD. Part-time programs available. *Faculty:* 17 full-time (1 woman), 2 part-time (0 women). *Students:* 34 full-time (8 women); includes 1 minority (Asian American or Pacific Islander), 21 international. Average age 27. 68 applicants, 41% accepted. In 1998, 7 master's, 4 doctorates awarded. *Degree requirements:* For master's, thesis required, foreign language not required; for doctorate, dissertation, qualifying exam required, foreign language not required. *Entrance requirements:* For master's, GRE General Test (combined average 1826 on three sections), TOEFL (minimum score of 550 required; average 602); for doctorate, GRE General Test (combined average 1952 on three sections), TOEFL (minimum score of 550 required; average 616). *Average time to degree:* Master's–3 years full-time; doctorate–5.2 years full-time. *Application deadline:* For fall admission, 3/15 (priority date). Applications are processed on a rolling basis. Application fee: $30 ($35 for international students). Tuition, state resident: full-time $4,377. Tuition, nonresident: full-time $9,108. Required fees: $126. Tuition and fees vary according to course load. *Financial aid:* In 1998–99, 5 fellowships (averaging $3,270 per year), 24 research assistantships (averaging $7,419 per year), 2 teaching assistantships (averaging $4,594 per year) were awarded.; career-related internships or fieldwork, Federal Work-Study, institutionally-sponsored loans, and unspecified assistantships also available. Aid available to part-time students. Financial aid application deadline: 3/1; financial aid applicants required to submit FAFSA. *Faculty research:* Structure/property/processing relationships, microstructural characterization, alloy design, materials and manufacturing processes, mineral processing. Total annual research expenditures: $1.6 million. *Unit head:* Dr. Calvin White, Chair, 906-487-2631, Fax: 906-487-2934, E-mail: cwhite@mtu.edu. *Application contact:* Kathe Prince, Senior Documentation Specialist, 906-487-2039, Fax: 906-487-2934, E-mail: kaprince@mtu.edu.

Montana Tech of The University of Montana, Graduate School, Metallurgical/Mineral Processing Engineering Programs, Butte, MT 59701-8997. Offers MS. Part-time programs available. *Faculty:* 7 full-time (0 women). *Students:* 5 full-time (1 woman), 4 part-time. 8 applicants, 50% accepted. In 1998, 2 degrees awarded. *Degree requirements:* For master's, thesis optional, foreign language not required. *Entrance requirements:* For master's, GRE General Test, TOEFL (minimum score of 525 required), minimum B average. *Application deadline:* For fall admission, 4/1 (priority date); for spring admission, 10/1 (priority date). Applications are processed on a rolling basis. Application fee: $30. Tuition, state resident: full-time $3,211; part-time $162 per credit hour. Tuition, nonresident: full-time $9,883; part-time $440 per credit hour. International tuition: $15,500 full-time. *Financial aid:* In 1998–99, 2 research assistantships with partial tuition reimbursements (averaging $6,324 per year), 3 teaching assistantships with partial tuition reimbursements (averaging $3,400 per year) were awarded.; career-related internships or fieldwork, Federal Work-Study, institutionally-sponsored loans, and tuition waivers (full and partial) also available. Aid available to part-time students. Financial aid application deadline: 4/1; financial aid applicants required to submit FAFSA. *Faculty research:* Stabilizing hazardous waste, decontamination of metals by melt refining, ultraviolet enhancement of stabilization reactions. Total annual research expenditures: $400,000. *Unit head:* Dr. Courtney Young, Department Head, 406-496-4158, Fax: 406-496-4664, E-mail: cyoung@mtech.edu. *Application contact:* Cindy Dunstan, Administrative Assistant, 406-496-4128, Fax: 406-496-4334, E-mail: cdunstan@mtech.edu.

The Ohio State University, Graduate School, College of Engineering, Department of Industrial, Welding, and Systems Engineering, Program in Welding Engineering, Columbus, OH 43210. Offers MS, PhD. *Faculty:* 12 full-time, 26 part-time. *Students:* 36 full-time (0 women), 7 part-time, 32 international. 31 applicants, 26% accepted. In 1998, 14 master's, 5 doctorates awarded. *Degree requirements:* For master's, computer language required, thesis optional, foreign language not required; for doctorate, computer language, dissertation required, foreign language not required. *Application deadline:* For fall admission, 8/15. Applications are processed on a rolling basis. Application fee: $30 ($40 for international students). *Financial aid:* Fellowships, research assistantships, teaching assistantships, Federal Work-Study and institutionally-sponsored loans available. Aid available to part-time students. *Unit head:* Stanislav Rokhlin, Graduate Studies Committee Chair, 614-292-6841, Fax: 614-292-6842, E-mail: rokhlin.2@osu.edu.

The Ohio State University, Graduate School, College of Engineering, Department of Materials Science and Engineering, Columbus, OH 43210. Offers ceramic engineering (MS, PhD); metallurgical engineering (MS, PhD). *Faculty:* 24 full-time, 1 part-time. *Students:* 82 full-time (14 women); includes 1 minority (Asian American or Pacific Islander), 67 international. *Degree requirements:* For master's and doctorate, computer language, thesis/dissertation required, foreign language not required. *Entrance requirements:* For master's and doctorate, GRE General Test (international students). *Application deadline:* For fall admission, 8/15. Applications are processed on a rolling basis. Application fee: $30 ($40 for international students). *Unit head:* Robert L. Snyder, Chairman, 614-292-2553, Fax: 614-292-1537, E-mail: snyder.355@osu.edu.

Pennsylvania State University University Park Campus, Graduate School, College of Earth and Mineral Sciences, Department of Materials Science and Engineering, State College, University Park, PA 16802-1503. Offers ceramic science (MS, PhD); fuel science (MS, PhD); metals science and engineering (MS, PhD); polymer science (MS, PhD). *Students:* 108 full-time (15 women), 11 part-time (1 woman). *Entrance requirements:* For master's and doctorate, GRE. *Application deadline:* For fall admission, 7/1. Application fee: $50. *Unit head:* Dr. R. E. Tressler, Head, 814-865-0497.

Purdue University, Graduate School, Schools of Engineering, School of Materials Engineering, West Lafayette, IN 47907. Offers materials engineering (MS, MSE, PhD); metallurgical engineering (MS Met E). Part-time programs available. *Faculty:* 12 full-time (1 woman). *Students:* 31 full-time (5 women), 3 part-time; includes 1 minority (Hispanic American), 11 international. *Degree requirements:* For master's, thesis required (for some programs), foreign language not required; for doctorate, dissertation required. *Entrance requirements:* For master's and doctorate, TOEFL (minimum score of 550 required). *Application deadline:* For fall admission, 2/1 (priority date). Application fee: $30. Electronic applications accepted. *Unit head:* Dr. G. L. Liedl, Head, 765-494-4094, Fax: 765-494-1204, E-mail: liedl@ecn.purdue.edu. *Application contact:* Dr. K. J. Bowman, Graduate Committee Chair, 765-494-6316, Fax: 765-494-1204, E-mail: kbowman@ecn.purdue.edu.

Queen's University at Kingston, School of Graduate Studies and Research, Faculty of Applied Science, Department of Materials and Metallurgical Engineering, Kingston, ON K7L 3N6, Canada. Offers M Sc, M Sc Eng, PhD. Part-time programs available. *Students:* 17 full-time (2 women), 3 part-time. In 1998, 5 master's, 1 doctorate awarded. *Degree requirements:* For master's, thesis optional, foreign language not required; for doctorate, dissertation, comprehensive exam required, foreign language not required. *Entrance requirements:* For master's and doctorate, TOEFL (minimum score of 550 required). *Application deadline:* For fall admission, 2/28 (priority date). Application fee: $60. Electronic applications accepted. *Financial aid:* Fellowships, research assistantships, teaching assistantships, institutionally-sponsored loans available. Financial aid application deadline: 3/1. *Faculty research:* Physical metallurgy, chemical metallurgy and mineral engineering. *Unit head:* Dr. J. Cameron, Head, 613-533-2758. *Application contact:* Dr. S. Saimoto, Graduate Coordinator, 613-533-2747.

Rensselaer Polytechnic Institute, Graduate School, School of Engineering, Department of Materials Science and Engineering, Troy, NY 12180-3590. Offers ceramics and glass science (M Eng, MS, D Eng, PhD); composites (M Eng, MS, D Eng, PhD); electronic materials (M Eng, MS, D Eng, PhD); metallurgy (M Eng, MS, D Eng, PhD); polymers (M Eng, MS, D Eng, PhD). Part-time and evening/weekend programs available. *Faculty:* 17 full-time (1 woman), 5 part-time (0 women). *Students:* 45 full-time (8 women), 24 part-time (4 women); includes 5 minority (1 African American, 2 Asian Americans or Pacific Islanders, 2 Hispanic Americans), 26 international. Terminal master's awarded for partial completion of doctoral program. *Degree requirements:* For master's, thesis required (for some programs), foreign language not required; for doctorate, dissertation required, foreign language not required. *Entrance requirements:* For master's and doctorate, GRE, TOEFL (minimum score of 550 required). *Application deadline:* For fall admission, 2/1 (priority date). Applications are processed on a rolling basis. Application fee: $35. *Unit head:* Dr. Richard W. Siegel, Chair, 518-276-6373, Fax: 518-276-8554. *Application contact:* Dr. Roger Wright, Admissions Coordinator, 518-276-6372, Fax: 518-276-8554, E-mail: fowlen@rpi.edu.

See in-depth description on page 1337.

Rutgers, The State University of New Jersey, New Brunswick, Graduate School, Program in Materials Science and Engineering, New Brunswick, NJ 08903. Offers physical metallurgy (MS, PhD); polymer science (MS, PhD). Part-time programs available. *Faculty:* 16 full-time (0 women). *Students:* 49 full-time (6 women), 15 part-time (3 women); includes 3 minority (1 African American, 2 Asian Americans or Pacific Islanders), 39 international. Terminal master's awarded for partial completion of doctoral program. *Degree requirements:* For master's, thesis optional, foreign language not required; for doctorate, dissertation required, foreign language not required. *Entrance requirements:* For master's and doctorate, GRE General Test. *Application deadline:* For fall admission, 3/1 (priority date). Applications are processed on a rolling basis. Application fee: $40. *Unit head:* Dr. Jerry I. Scheinbeim, Director, 732-932-3669. *Application contact:* Kathy Finnerty, Administrative Assistant, 732-932-2245, Fax: 732-932-5977.

South Dakota School of Mines and Technology, Graduate Division, Division of Material Engineering and Science, Doctoral Program in Materials Engineering and Science, Rapid City, SD 57701-3995. Offers chemical engineering (PhD); chemistry (PhD); civil engineering (PhD); electrical engineering (PhD); mechanical engineering (PhD); metallurgical engineering (PhD); physics (PhD). Part-time programs available. *Students:* 14 full-time (2 women), 9 international. *Degree requirements:* For doctorate, dissertation required, foreign language not required. *Entrance requirements:* For doctorate, TOEFL (minimum score of 520 required), TWE, minimum graduate GPA of 3.0. *Application deadline:* For fall admission, 6/15 (priority date); for spring admission, 10/15. Applications are processed on a rolling basis. Application fee: $15. Electronic applications accepted. Tuition, state resident: part-time $89 per hour. Tuition, nonresident: part-time $261 per hour. Part-time tuition and fees vary according to program. *Unit head:* Dr. Chris Jenkins, Coordinator, 605-394-2406. *Application contact:* Brenda Brown, Secretary, 800-454-8162 Ext. 2493, Fax: 605-394-5360, E-mail: graduate_admissions@silver.sdmt.edu.

See in-depth description on page 1339.

South Dakota School of Mines and Technology, Graduate Division, Division of Material Engineering and Science, Master's Program in Materials Engineering and Science, Rapid City, SD 57701-3995. Offers chemistry (MS); metallurgical engineering (MS); physics (MS). *Students:* 17. *Degree requirements:* Foreign language not required. *Entrance requirements:* For master's, TOEFL (minimum score of 520 required), TWE. *Application deadline:* For fall admission, 6/15 (priority date); for spring admission, 10/15. Applications are processed on a rolling basis. Application fee: $15. Electronic applications accepted. Tuition, state resident: part-time $89 per hour. Tuition, nonresident: part-time $261 per hour. Part-time tuition and fees vary according to program. *Unit head:* James W. Smolka, Coordinator, 805-258-5936. *Application contact:* Brenda Brown, Secretary, 800-454-8162 Ext. 2493, Fax: 605-394-5360, E-mail: graduate_admissions@silver.sdmt.edu.

Université Laval, Faculty of Graduate Studies, Faculty of Sciences and Engineering, Department of Mining and Metallurgical Engineering, Sainte-Foy, PQ G1K 7P4, Canada. Offers

Metallurgical Engineering and Metallurgy

Université Laval (continued)

metallurgical engineering (M Sc, PhD); mining (M Sc, PhD). *Students:* 31 full-time (6 women), 8 part-time (1 woman). 19 applicants, 58% accepted. In 1998, 6 master's awarded. *Application deadline:* For fall admission, 3/1. Application fee: $30. *Unit head:* Réal Tremblay, Director, 418-656-2131 Ext. 5047, Fax: 418-656-5343, E-mail: real.tremblay@gmn.ulaval.ca.

The University of Alabama, Graduate School, College of Engineering, Department of Metallurgical and Materials Engineering, Tuscaloosa, AL 35487. Offers MS Met E, PhD. *Faculty:* 9 full-time (1 woman), 4 part-time (0 women). *Students:* 29; includes 2 minority (1 African American, 1 Hispanic American), 18 international. Average age 29. 40 applicants, 13% accepted. In 1998, 2 master's, 3 doctorates awarded. *Degree requirements:* For master's, thesis or alternative required, foreign language not required; for doctorate, one foreign language (computer language can substitute), dissertation required. *Entrance requirements:* For master's, GRE General Test (minimum combined score of 1500 on three sections required), minimum GPA of 3.0 in last 60 hours. *Application deadline:* For fall admission, 7/1 (priority date). Applications are processed on a rolling basis. Application fee: $25. Electronic applications accepted. *Financial aid:* In 1998–99, 3 fellowships, 14 research assistantships, 6 teaching assistantships were awarded.; Federal Work-Study also available. *Faculty research:* Metals casting and solidification, corrosion and alloy development, metal matrix composites, electronic and magnetic properties. Total annual research expenditures: $2 million. *Unit head:* Dr. Richard C. Bradt, Head, 205-348-1740, Fax: 205-348-2164, E-mail: rcbradt@coe.eng.ua.edu. *Application contact:* Dr. Garry W. Warren, Professor, 205-348-1740, Fax: 205-348-2164, E-mail: gwarren@coe.eng.ua.edu.

The University of Alabama at Birmingham, Graduate School, School of Engineering, Department of Materials and Mechanical Engineering, Program in Materials Engineering, Birmingham, AL 35294. Offers materials engineering (MS Mt E); materials/metallurgical engineering (PhD). *Students:* 18 full-time (3 women), 4 part-time; includes 3 minority (1 African American, 2 Asian Americans or Pacific Islanders), 11 international. *Degree requirements:* For master's, thesis or alternative, comprehensive exam and project/thesis required, foreign language not required; for doctorate, dissertation, comprehensive exam, MSMTE required, foreign language not required. *Entrance requirements:* For master's and doctorate, GRE General Test (minimum score of 500 on each section required), minimum B average. *Application deadline:* Applications are processed on a rolling basis. Application fee: $30 ($60 for international students). Electronic applications accepted. *Unit head:* Dr. Krishan Chawla, Interim Chair, Department of Materials and Mechanical Engineering, 205-975-9725, Fax: 205-934-8485, E-mail: kchawla@uab.edu.

University of British Columbia, Faculty of Graduate Studies, Faculty of Applied Science, Department of Metals and Materials Engineering, Vancouver, BC V6T 1Z2, Canada. Offers materials and metallurgy (M Sc, PhD); metals and materials engineering (M Eng, MA Sc, PhD). *Degree requirements:* For master's and doctorate, thesis/dissertation required. *Entrance requirements:* For master's and doctorate, TOEFL. *Faculty research:* Electroslag melting, mathematical modelling, solidification and hydrometallurgy.

University of California, Berkeley, Graduate Division, College of Engineering, Department of Materials Science and Mineral Engineering, Program in Physical Metallurgy, Berkeley, CA 94720-1500. Offers M Eng, MS, D Eng, PhD. *Degree requirements:* For master's, comprehensive exam or thesis (MS) required; for doctorate, dissertation, qualifying exam required. *Entrance requirements:* For master's and doctorate, GRE General Test, minimum GPA of 3.0. *Application deadline:* For fall admission, 2/10; for spring admission, 9/1. Application fee: $40. *Financial aid:* Application deadline: 1/5. *Application contact:* Carole James, Student Affairs Officer, 510-642-3801, Fax: 510-643-5792, E-mail: carolej@uclink4.berkeley.edu.

University of California, Los Angeles, Graduate Division, School of Engineering and Applied Science, Department of Materials Science and Engineering, Program in Metallurgy, Los Angeles, CA 90095. Offers MS, PhD. *Degree requirements:* For master's, comprehensive exam or thesis required; for doctorate, dissertation, qualifying exams required, foreign language not required. *Entrance requirements:* For master's, GRE General Test, minimum GPA of 3.0; for doctorate, GRE General Test, minimum GPA of 3.25. *Application deadline:* For fall admission, 1/15; for spring admission, 12/31. Application fee: $40. Electronic applications accepted. *Financial aid:* Fellowships, research assistantships, teaching assistantships, Federal Work-Study, institutionally-sponsored loans, and tuition waivers (full and partial) available. Financial aid application deadline: 1/15; financial aid applicants required to submit FAFSA. *Application contact:* Paradee Chularee, Student Affairs Officer, 310-825-8913, Fax: 310-206-7353, E-mail: paradee@ea.ucla.edu.

University of Cincinnati, Division of Research and Advanced Studies, College of Engineering, Department of Materials Science and Metallurgical Engineering, Program in Metallurgical Engineering, Cincinnati, OH 45221-0091. Offers MS, PhD. Evening/weekend programs available. *Faculty:* 1 full-time. *Students:* 1 full-time (0 women). *Degree requirements:* For master's, thesis optional, foreign language not required; for doctorate, one foreign language, dissertation, comprehensive exams, oral English proficiency exam required. *Entrance requirements:* For master's and doctorate, GRE General Test, TOEFL (minimum score of 550 required), BS in related field, minimum undergraduate GPA of 3.0. *Application deadline:* For fall admission, 2/1 (priority date). Application fee: $40. *Financial aid:* Fellowships, career-related internships or fieldwork, tuition waivers (full), and unspecified assistantships available. Aid available to part-time students. Financial aid application deadline: 2/1. *Faculty research:* Polymer characterization, surface analysis, and adhesion; high-temperature coatings and physical chemistry of materials. *Unit head:* Admissions Assistant, 949-824-3562, Fax: 949-824-3440. *Application contact:* Dr. R. J. Roe, Graduate Program Director, 513-556-3117, Fax: 513-556-2569, E-mail: r.j.roe@uc.edu.

University of Connecticut, Graduate School, School of Engineering, Field of Metallurgy, Storrs, CT 06269. Offers MS, PhD. Terminal master's awarded for partial completion of doctoral program. *Degree requirements:* For master's, thesis or alternative required; for doctorate, dissertation required. *Entrance requirements:* For master's and doctorate, GRE General Test, GRE Subject Test. *Faculty research:* Microsegregation and coarsening, fatigue crack, electron-dislocation interaction.

University of Florida, Graduate School, College of Engineering, Department of Materials Science and Engineering, Program in Metallurgical Engineering, Gainesville, FL 32611. Offers ME, MS, PhD, Engr. *Degree requirements:* For master's, thesis optional, foreign language not required; for doctorate, dissertation required, foreign language not required; for Engr, thesis optional. *Entrance requirements:* For master's and doctorate, GRE General Test (minimum combined score of 1100 required), TOEFL, minimum GPA of 3.0; for Engr, GRE General Test. *Application deadline:* For fall admission, 6/1 (priority date). Applications are processed on a rolling basis. Application fee: $20. *Application contact:* Jodi Vanderheyden, Graduate Secretary, 352-392-1451, Fax: 352-392-5630, E-mail: jheyd@mail.mse.ufl.edu.

University of Idaho, College of Graduate Studies, College of Mines and Earth Resources, Department of Metallurgical and Mining Engineering, Program in Metallurgy Engineering, Moscow, ID 83844-4140. Offers MS. *Students:* 4 full-time (2 women), 6 part-time (3 women), 5 international. In 1998, 1 degree awarded. *Degree requirements:* For master's, computer language required, foreign language not required. *Entrance requirements:* For master's, minimum GPA of 2.8. *Application deadline:* For fall admission, 8/1; for spring admission, 12/15. Application fee: $35 ($45 for international students). *Financial aid:* Application deadline: 2/15. *Unit head:* Dr. Patrick Taylor, Head, Department of Metallurgical and Mining Engineering, 208-885-6769.

University of Idaho, College of Graduate Studies, College of Mines and Earth Resources, Department of Metallurgical and Mining Engineering, Program in Mining Engineering: Metallurgy, Moscow, ID 83844-4140. Offers PhD. *Students:* 9 full-time (1 woman), 4 part-time (1 woman), 9 international. In 1998, 5 degrees awarded. *Degree requirements:* For doctorate,

dissertation required. *Entrance requirements:* For doctorate, minimum undergraduate GPA of 2.8, 3.0 graduate. *Application deadline:* For fall admission, 8/1; for spring admission, 12/15. Application fee: $35 ($45 for international students). *Financial aid:* Application deadline: 2/15. *Unit head:* Dr. Patrick Taylor, Head, Department of Metallurgical and Mining Engineering, 208-885-6769.

University of Idaho, College of Graduate Studies, College of Mines and Earth Resources, Department of Metallurgical and Mining Engineering, Programs in Metallurgy, Moscow, ID 83844-4140. Offers MS. In 1998, 1 degree awarded. *Degree requirements:* For master's, computer language, thesis required, foreign language not required. *Entrance requirements:* For master's, minimum GPA of 2.8. *Application deadline:* For fall admission, 8/1; for spring admission, 12/15. Application fee: $35 ($45 for international students). *Financial aid:* Career-related internships or fieldwork available. Financial aid application deadline: 2/15. *Faculty research:* Critical metals, mineral preparation, hydrometallurgy, trace metals. *Unit head:* Dr. Patrick Taylor, Head, Department of Metallurgical and Mining Engineering, 208-885-6769.

University of Idaho, College of Graduate Studies, College of Mines and Earth Resources, Department of Metallurgical and Mining Engineering, Programs in Mining Engineering, Moscow, ID 83844-4140. Offers metallurgical engineering (MS, PhD); mining engineering (MS, PhD). *Degree requirements:* For master's, computer language required, foreign language and thesis not required; for doctorate, one foreign language, computer language, dissertation required. *Entrance requirements:* For master's, minimum GPA of 2.8; for doctorate, minimum undergraduate GPA of 2.8, 3.0 graduate. *Application deadline:* For fall admission, 8/1; for spring admission, 12/15. Application fee: $35 ($45 for international students). *Unit head:* Dr. Patrick Taylor, Head, Department of Metallurgical and Mining Engineering, 208-885-6769.

University of Missouri–Rolla, Graduate School, School of Mines and Metallurgy, Department of Metallurgical Engineering, Rolla, MO 65409-0910. Offers MS, PhD. Part-time programs available. *Faculty:* 10 full-time (0 women). *Students:* 21 full-time (1 woman); includes 1 minority (Hispanic American), 10 international. Average age 29. 51 applicants, 55% accepted. In 1998, 4 master's, 6 doctorates awarded (100% found work related to degree). *Degree requirements:* For master's and doctorate, computer language, thesis/dissertation required, foreign language not required. *Entrance requirements:* For master's, GRE General Test (minimum combined score of 1100 required), TOEFL (minimum score of 600 required), minimum GPA of 3.0 in last 4 semesters; for doctorate, GRE General Test (minimum combined score of 1100 required), TOEFL (minimum score of 600 required). *Application deadline:* For fall admission, 6/15; for spring admission, 11/15. Applications are processed on a rolling basis. Application fee: $25. Electronic applications accepted. *Financial aid:* In 1998–99, 20 students received aid, including 3 fellowships with full tuition reimbursements available (averaging $13,195 per year), 16 research assistantships with partial tuition reimbursements available (averaging $12,986 per year), 1 teaching assistantship with partial tuition reimbursement available (averaging $12,986 per year); Federal Work-Study and institutionally-sponsored loans also available. Aid available to part-time students. Financial aid application deadline: 3/1; financial aid applicants required to submit FAFSA. *Faculty research:* Pyrometallurgy, electrometallurgy, material processing, advanced materials development and defect characterization. *Unit head:* Dr. John L. Watson, Chairman, 573-341-4724, Fax: 573-341-6934, E-mail: jwatson@umr.edu.

University of Nevada, Reno, Graduate School, Mackay School of Mines, Department of Chemical and Metallurgical Engineering, Reno, NV 89557. Offers metallurgical engineering (MS, PhD, Met E). Terminal master's awarded for partial completion of doctoral program. *Degree requirements:* For master's, thesis required, foreign language not required; for doctorate, dissertation required. *Entrance requirements:* For master's, GRE, TOEFL (minimum score of 500 required), minimum GPA of 2.75; for doctorate, GRE, TOEFL (minimum score of 500 required), minimum GPA of 3.0. *Faculty research:* Hydrometallurgy, applied surface chemistry, mineral processing, mineral bioprocessing, ceramics.

University of Pittsburgh, School of Engineering, Department of Materials Science and Engineering, Pittsburgh, PA 15260. Offers materials science and engineering (MSMSE, PhD); metallurgical engineering (MS Met E, PhD). Part-time and evening/weekend programs available. *Faculty:* 13 full-time (0 women), 2 part-time (0 women). *Students:* 12 full-time (5 women), 19 part-time (2 women); includes 1 minority (Asian American or Pacific Islander), 8 international. Terminal master's awarded for partial completion of doctoral program. *Degree requirements:* For master's, thesis required (for some programs), foreign language not required; for doctorate, dissertation, comprehensive and final oral exams required, foreign language not required. *Entrance requirements:* For master's and doctorate, TOEFL (minimum score of 550 required), minimum QPA of 3.0. *Application deadline:* For fall admission, 8/1 (priority date); for spring admission, 12/1 (priority date). Applications are processed on a rolling basis. Application fee: $30 ($40 for international students). *Unit head:* Dr. William A. Soffa, Chairman, 412-624-9720, Fax: 412-624-8069, E-mail: wsoffa+@pitt.edu. *Application contact:* Pradeep P. Phule, Graduate Coordinator, 412-624-9736, Fax: 412-624-8069, E-mail: ppp@vms.cis.pitt.edu.

University of Tennessee, Knoxville, Graduate School, College of Engineering, Department of Materials Science and Engineering, Program in Metallurgical Engineering, Knoxville, TN 37996. Offers MS, PhD. *Students:* 17 full-time (1 woman), 8 part-time (2 women); includes 1 minority (African American), 8 international. 48 applicants, 25% accepted. In 1998, 5 master's, 5 doctorates awarded. *Degree requirements:* For master's, thesis or alternative required, foreign language not required; for doctorate, dissertation required, foreign language not required. *Entrance requirements:* For master's and doctorate, TOEFL (minimum score of 550 required), minimum GPA of 2.7. *Application deadline:* For fall admission, 2/1 (priority date). Applications are processed on a rolling basis. Application fee: $35. Electronic applications accepted. *Financial aid:* Application deadline: 2/1; *Unit head:* Dr. J. E. Spruiell, Head, Department of Materials Science and Engineering, 423-974-5336, Fax: 423-974-4115, E-mail: spruiell@utk.edu.

University of Tennessee Space Institute, Graduate Programs, Program in Metallurgical Engineering, Tullahoma, TN 37388-9700. Offers MS, PhD. *Students:* 3 full-time (0 women), 1 part-time; includes 1 minority (Asian American or Pacific Islander), 1 international. 2 applicants, 50% accepted. In 1998, 1 master's awarded. *Degree requirements:* For doctorate, dissertation required, foreign language not required. *Application deadline:* Applications are processed on a rolling basis. Application fee: $35. *Unit head:* Dr. Mary Helen McCay, Degree Program Chair, 931-393-7473, Fax: 931-454-2271, E-mail: mmccay@utsi.edu. *Application contact:* Dr. Edwin M. Gleason, Assistant Dean for Admissions and Student Affairs, 931-393-7432, Fax: 931-393-7346, E-mail: egleason@utsi.edu.

The University of Texas at El Paso, Graduate School, College of Engineering, Department of Metallurgical Engineering, El Paso, TX 79968-0001. Offers metallurgical engineering (MS). Part-time and evening/weekend programs available. *Faculty:* 6 full-time (0 women). *Students:* 14 full-time (6 women), 7 part-time (1 woman); includes 15 minority (14 Hispanic Americans, 1 Native American), 5 international. Average age 27. 21 applicants, 38% accepted. In 1998, 4 degrees awarded. *Degree requirements:* For master's, thesis, scholarly article required, foreign language not required. *Entrance requirements:* For master's, GRE General Test, TOEFL (minimum score of 550 required), minimum GPA of 3.2. *Application deadline:* Applications are processed on a rolling basis. Application fee: $15 ($65 for international students). Electronic applications accepted. Tuition, state resident: full-time $2,790. Tuition, nonresident: full-time $7,710. *Financial aid:* In 1998–99, 20 students received aid, including 8 fellowships, 15 research assistantships, 3 teaching assistantships; career-related internships or fieldwork, Federal Work-Study, institutionally-sponsored loans, scholarships, and tuition waivers (partial) also available. Financial aid applicants required to submit FAFSA. *Faculty research:* Structure-property relationship, corrosion material in space, extreme deformation welding, composites. Total annual research expenditures: $1.5 million. *Unit head:* Dr. Lawrence Murr, Chairperson, 915-747-6929. *Application contact:* Susan Jordan, Director, Graduate Student Services, 915-747-5491, Fax: 915-747-5788, E-mail: sjordan@utep.edu.

University of Toronto, School of Graduate Studies, Physical Sciences Division, Collaborative Program in Welding Engineering, Toronto, ON M5S 1A1, Canada. Offers M Eng, MA Sc. Part-time programs available. *Degree requirements:* For master's, thesis required (for some programs).

University of Toronto, School of Graduate Studies, Physical Sciences Division, Faculty of Applied Science and Engineering, Department of Metallurgy and Materials Science, Toronto, ON M5S 1A1, Canada. Offers M Eng, MA Sc, PhD. Part-time programs available. *Degree requirements:* For master's, thesis required (for some programs); for doctorate, dissertation required.

University of Utah, Graduate School, College of Mines and Earth Sciences, Department of Metallurgy and Metallurgical Engineering, Salt Lake City, UT 84112-1107. Offers ME, MS, PhD. *Faculty:* 10 full-time (1 woman), 11 part-time (1 woman). *Students:* 15 full-time (2 women), 11 part-time, 18 international. Average age 31. In 1998, 1 master's, 5 doctorates awarded. *Degree requirements:* For master's, comprehensive exam (ME), thesis (MS) required; for doctorate, dissertation required, foreign language not required. *Entrance requirements:* For master's and doctorate, GRE General Test, TOEFL (minimum score of 500 required), minimum GPA of 3.0. *Application deadline:* For fall admission, 7/1. Application fee: $30 ($50 for international students). *Financial aid:* Fellowships, research assistantships, teaching assistantships, institutionally-sponsored loans available. Financial aid application deadline: 2/15. *Faculty research:* Physical metallurgy, mathematical modeling, mineral processing, thermodynamics, solvent extraction. *Unit head:* Dr. J. Gerald Byrne, Chair, 801-581-6386, Fax: 801-581-5560, E-mail: jgbyrne@mines.utah.edu. *Application contact:* Dr. R. K. Rajamani, Director of Graduate Admissions, 801-581-3107, Fax: 801-581-4937, E-mail: rajamani@mines.utah.edu.

University of Washington, Graduate School, College of Engineering, Department of Materials Science and Engineering, Program in Metallurgical Engineering, Seattle, WA 98195. Offers MS, MSE. *Degree requirements:* For master's, thesis required, foreign language not required. *Entrance requirements:* For master's, GRE General Test (minimum combined score of 1400 on quantitative and analytical sections required), TOEFL (minimum score of 600 required), minimum GPA of 3.0. *Application deadline:* For fall admission, 7/1; for winter admission, 11/1; for spring admission, 2/1. Applications are processed on a rolling basis. Application fee: $50. Tuition, state resident: full-time $5,196; part-time $475 per credit. Tuition, nonresident: full-time $13,485; part-time $1,285 per credit. Required fees: $387; $38 per credit. Tuition and fees vary according to course load. *Financial aid:* Fellowships with full tuition reimbursements, research assistantships with full tuition reimbursements, teaching assistantships with full tuition reimbursements, career-related internships or fieldwork, grants, institutionally-sponsored loans, scholarships, unspecified assistantships, and stipend supplements available. Financial aid application deadline: 3/1. *Unit head:* Michelle Trudeau, Student Services Manager, 206-616-1533, Fax: 206-685-0790, E-mail: michtru@u.washington.edu. *Application contact:* Dr. T. G. Stoebe, Graduate Coordinator, 206-543-7090, Fax: 206-543-3100, E-mail: mse@u.washington.edu.

University of Wisconsin–Madison, Graduate School, College of Engineering, Department of Materials Science and Engineering, Madison, WI 53706-1380. Offers metallurgical engineering (MS, PhD). *Faculty:* 12 full-time (1 woman). *Students:* 17 full-time (0 women), 13 international. Average age 32. In 1998, 4 doctorates awarded (100% found work related to degree). *Degree requirements:* For master's and doctorate, thesis/dissertation required. *Entrance requirements:* For master's and doctorate, GRE General Test. *Average time to degree:* Master's–2 years full-time; doctorate–5 years full-time. *Application deadline:* For fall admission, 7/1 (priority date); for spring admission, 11/15 (priority date). Applications are processed on a rolling basis. Application fee: $45. Electronic applications accepted. *Financial aid:* Fellowships with tuition reimbursements, research assistantships with tuition reimbursements, teaching assistantships with tuition reimbursements available. Financial aid application deadline: 1/15. *Faculty research:* Superconductors, electronic materials, cast metals, ceramics, materials characterization. Total annual research expenditures: $8 million. *Unit head:* Eric E. Hellstrom, Chair, 608-262-1821, Fax: 608-262-8353, E-mail: matsciad@engr.wisc.edu. *Application contact:* Diana J. Rhoads, Program Assistant 3, 608-263-1795, Fax: 608-262-8353, E-mail: matsciad@engr.wisc.edu.

Polymer Science and Engineering

Carnegie Mellon University, Carnegie Institute of Technology, Department of Chemical Engineering, Program in Colloids, Polymers and Surfaces, Pittsburgh, PA 15213-3891. Offers MS. Part-time and evening/weekend programs available. *Faculty:* 11 full-time (2 women), 1 (woman) part-time. Average age 28. *Degree requirements:* For master's, foreign language and thesis not required. *Entrance requirements:* For master's, GRE General Test, GRE Subject Test, TOEFL. *Average time to degree:* Master's–4 years part-time. *Application deadline:* For fall admission, 2/1; for spring admission, 10/15. Applications are processed on a rolling basis. Application fee: $50. *Faculty research:* Surface phenomena, polymer rheology, solubilization phenomena, colloid transport phenomena, polymer synthesis. *Unit head:* A. M. Jacobson, Director, 412-268-2244, Fax: 412-268-7139.

Carnegie Mellon University, Mellon College of Science, Department of Chemistry, Pittsburgh, PA 15213-3891. Offers chemical instrumentation (MS); chemistry (MS, PhD); colloids, polymers and surfaces (MS); polymer science (MS). Part-time programs available. *Faculty:* 40 full-time (7 women), 1 part-time (0 women). *Students:* 58 full-time (22 women), 3 part-time; includes 1 minority (Asian American or Pacific Islander), 29 international. Terminal master's awarded for partial completion of doctoral program. *Degree requirements:* For doctorate, dissertation, departmental qualifying and oral exams, teaching experience required, foreign language not required, foreign language not required. *Entrance requirements:* For master's, GRE General Test; for doctorate, GRE General Test, GRE Subject Test, TOEFL. *Application deadline:* For fall admission, 1/15 (priority date). Applications are processed on a rolling basis. Application fee: $0. Electronic applications accepted. *Unit head:* Krzystof Matyjaszewski, Head, 412-268-3209. *Application contact:* Valerie Bridge, Student Specialist, 412-268-3150, Fax: 412-268-1061, E-mail: vbog@andrew.cmu.edu.

Case Western Reserve University, School of Graduate Studies, The Case School of Engineering, Department of Macromolecular Science, Cleveland, OH 44106. Offers MS, PhD, MD/PhD. Part-time programs available. *Faculty:* 13 full-time (2 women), 17 part-time (0 women). *Students:* 22 full-time (10 women), 52 part-time (16 women). Average age 24. 209 applicants, 5% accepted. In 1998, 6 master's, 14 doctorates awarded. Terminal master's awarded for partial completion of doctoral program. *Degree requirements:* For master's and doctorate, thesis/dissertation required, foreign language not required. *Entrance requirements:* For master's and doctorate, GRE General Test, TOEFL (minimum score of 550 required). *Application deadline:* For fall admission, 2/28 (priority date). Applications are processed on a rolling basis. Application fee: $0 ($25 for international students). *Financial aid:* In 1998–99, 5 fellowships with full tuition reimbursements (averaging $15,000 per year), 48 research assistantships with full and partial tuition reimbursements (averaging $15,000 per year), 2 teaching assistantships with full and partial tuition reimbursements (averaging $11,250 per year) were awarded. *Faculty research:* Polymer chemistry, physics and engineering, polymer materials, structure, property relationships. Total annual research expenditures: $4.9 million. *Unit head:* Dr. Alexander M. Jamieson, Chairman, 216-368-4172, Fax: 216-368-4202, E-mail: amj@po.cwru.edu. *Application contact:* Shi-Qing Wang, 216-368-6374, Fax: 216-368-4202, E-mail: sxw13@po.cwru.edu.

See in-depth description on page 1327.

Clemson University, Graduate School, College of Engineering and Science, School of Textiles, Fiber and Polymer Science, Clemson, SC 29634. Offers textile and polymer science (PhD); textile chemistry (MS); textile science (MS). Part-time programs available. *Students:* 36 full-time (13 women), 3 part-time (1 woman); includes 1 minority (African American), 30 international. 47 applicants, 60% accepted. In 1998, 10 master's, 1 doctorate awarded. *Degree requirements:* For doctorate, dissertation required. *Entrance requirements:* For master's and doctorate, GRE General Test, TOEFL. Application fee: $35. *Financial aid:* Fellowships, research assistantships, teaching assistantships, career-related internships or fieldwork, institutionally-sponsored loans, and unspecified assistantships available. Financial aid applicants required to submit FAFSA. *Unit head:* Dr. D. V. Rippy, Director, 864-656-3176, Fax: 864-656-5973, E-mail: rippyd@clemson.edu.

Cornell University, Graduate School, Graduate Fields of Engineering, Field of Chemical Engineering, Ithaca, NY 14853-0001. Offers advanced materials processing (M Eng, MS, PhD); applied mathematics and computational methods (M Eng, MS, PhD); biochemical engineering (M Eng, MS, PhD); chemical reaction engineering (M Eng, MS, PhD); classical and statistical thermodynamics (M Eng, MS, PhD); fluid dynamics, rheology and biorheology (M Eng, MS, PhD); heat and mass transfer (M Eng, MS, PhD); kinetics and catalysis (M Eng, MS, PhD); polymers (M Eng, MS, PhD); surface science (M Eng, MS, PhD). *Faculty:* 18 full-time. *Students:* 71 full-time (17 women); includes 9 minority (1 African American, 7 Asian Americans or Pacific Islanders, 1 Hispanic American), 29 international. *Degree requirements:* For master's, thesis (MS) required; for doctorate, dissertation required, foreign language not required. *Entrance requirements:* For master's and doctorate, GRE General Test, TOEFL (minimum score of 580 required). *Application deadline:* For fall admission, 1/15. Application fee: $65. Electronic applications accepted. *Unit head:* Director of Graduate Studies, 607-255-4550, Fax: 607-255-9166. *Application contact:* Graduate Field Assistant, 607-255-4550, Fax: 607-255-9166, E-mail: dgs@cheme.cornell.edu.

Cornell University, Graduate School, Graduate Fields of Human Ecology, Field of Textiles, Ithaca, NY 14853-0001. Offers apparel design (MA, MPS); fiber science (MS, PhD); polymer science (MS, PhD); textile science (MS, PhD). *Faculty:* 14 full-time. *Students:* 15 full-time (6 women); includes 1 minority (Asian American or Pacific Islander), 10 international. *Degree requirements:* For master's, thesis (MA, MS), project paper (MPS) required; for doctorate, dissertation required, foreign language not required. *Entrance requirements:* For master's, GRE General Test (minimum combined score of 1800 on three sections required), TOEFL (minimum score of 600 required), portfolio (functional apparel design); for doctorate, GRE General Test (minimum combined score of 1800 on three sections required), TOEFL (minimum score of 600 required). *Application deadline:* For fall admission, 3/1. Application fee: $65. Electronic applications accepted. *Unit head:* Director of Graduate Studies, 607-255-3151, Fax: 607-255-1093. *Application contact:* Graduate Field Assistant, 607-255-3151, E-mail: textiles_grad@cornell.edu.

Eastern Michigan University, Graduate School, College of Technology, Department of Interdisciplinary Technology, Program in Polymer Technology, Ypsilanti, MI 48197. Offers MS. In 1998, 5 degrees awarded. *Degree requirements:* For master's, thesis optional, foreign language not required. *Entrance requirements:* For master's, GRE General Test, TOEFL (minimum score of 500 required), BS in chemistry, minimum GPA of 2.6. *Application deadline:* For fall admission, 5/15; for spring admission, 3/15. Applications are processed on a rolling basis. Application fee: $30. *Financial aid:* Fellowships, teaching assistantships available. Aid available to part-time students. Financial aid application deadline: 3/15; financial aid applicants required to submit FAFSA. *Unit head:* Dr. Wayne Hanewicz, Coordinator, 734-487-1235.

Georgia Institute of Technology, Graduate Studies and Research, College of Engineering, Multidisciplinary Program in Polymers, Atlanta, GA 30332-0001. Offers MS Poly. Offered jointly with the Schools of Chemical Engineering, Materials Science and Engineering, and Textile Engineering. *Degree requirements:* For master's, computer language, thesis required, foreign language not required. *Entrance requirements:* For master's, TOEFL (minimum score of 550 required), minimum GPA of 2.7.

Lehigh University, College of Engineering and Applied Science, Center for Polymer Science and Engineering, Bethlehem, PA 18015-3094. Offers MS, PhD. Programs are interdisciplinary. Part-time programs available. Terminal master's awarded for partial completion of doctoral program. *Degree requirements:* For master's and doctorate, thesis/dissertation required. *Entrance requirements:* For master's and doctorate, GRE General Test, TOEFL (minimum score of 550 required). *Faculty research:* Polymer colloids, polymer coatings, blends and composites, polymer interfaces, emulsion polymer.

North Dakota State University, Graduate Studies and Research, College of Science and Mathematics, Department of Polymers and Coatings, Fargo, ND 58105. Offers MS, PhD. Part-time programs available. *Faculty:* 4 full-time (0 women). *Students:* 21 full-time (4 women), 12 international. Average age 28. 66 applicants, 12% accepted. In 1998, 2 master's awarded (100% found work related to degree); 4 doctorates awarded (100% found work related to degree). Terminal master's awarded for partial completion of doctoral program. *Degree requirements:* For master's, thesis, cumulative exams required, foreign language not required; for doctorate, dissertation, comprehensive and cumulative exams required, foreign language not required. *Entrance requirements:* For master's, GRE General Test, TOEFL (minimum score of 525 required), BS in chemistry or chemical engineering; for doctorate, GRE Subject Test, TOEFL (minimum score of 525 required), BS in chemistry or chemical engineering. *Average time to degree:* Master's–2 years full-time; doctorate–4 years full-time. *Application deadline:* Applications are processed on a rolling basis. Application fee: $25. *Financial aid:* In 1998–99, 19 research assistantships with full tuition reimbursements (averaging $13,200 per year), 1 teaching assistantship with full tuition reimbursement (averaging $10,600 per year) were awarded.; fellowships, Federal Work-Study, institutionally-sponsored loans, and tuition waivers (full) also available. Aid available to part-time students. Financial aid application deadline: 3/15. *Faculty research:* Spectroscopy of adhesion and interfaces, corrosion and electrochemistry, polymer synthesis, organic/inorganic coatings, carbohydrate polymers. Total annual research expenditures: $248,534. *Unit head:* Dr. Marek W. Urban, Chair, 701-231-7633, Fax: 701-231-8439, E-mail: urban@plains.nodak.edu.

Announcement: North Dakota State University (NDSU) was the first university in the world to offer coatings science courses (1904). Today, the Department of Polymers and Coatings is the only formal department of its kind in the United States. Faculty members and students conduct research with leading companies and universities, both in the United States and abroad. Students from all over the world have enrolled in NDSU programs relating to polymers and coatings science. Faculty members are widely recognized as leaders in their research field. The department is a site for the National Science Foundation Coatings Research Center, one of the most prestigious academic-government-industry consortia in the United States.

Pennsylvania State University University Park Campus, Graduate School, College of Earth and Mineral Sciences, Department of Materials Science and Engineering, State College,

Polymer Science and Engineering

Pennsylvania State University University Park Campus (continued)
University Park, PA 16802-1503. Offers ceramic science (MS, PhD); fuel science (MS, PhD); metals science and engineering (MS, PhD); polymer science (MS, PhD). *Students:* 108 full-time (15 women), 11 part-time (1 woman). *Entrance requirements:* For master's and doctorate, GRE. *Application deadline:* For fall admission, 7/1. Application fee: $50. *Unit head:* Dr. R. E. Tressler, Head, 814-865-0497.

Polytechnic University, Brooklyn Campus, Department of Chemical Engineering, Chemistry and Materials Science, Major in Polymer Science and Engineering, Brooklyn, NY 11201-2990. Offers MS. *Students:* 3 full-time (2 women), 17 part-time (2 women); includes 2 minority (1 African American, 1 Asian American or Pacific Islander), 11 international. Average age 33. 38 applicants, 16% accepted. In 1998, 3 degrees awarded. *Application deadline:* Applications are processed on a rolling basis. Application fee: $45. Electronic applications accepted. *Application contact:* John S. Kerge, Dean of Admissions, 718-260-3200, Fax: 718-260-3446, E-mail: admitme@poly.edu.

Princeton University, Graduate School, Department of Chemistry, Program in Polymer Sciences and Materials, Princeton, NJ 08544-1019. Offers MSE, PhD. *Degree requirements:* For master's, thesis required; for doctorate, dissertation, cumulative and general exams required. *Entrance requirements:* For master's, GRE General Test; for doctorate, GRE General Test, GRE Subject Test.

Princeton University, Graduate School, School of Engineering and Applied Science, Department of Chemical Engineering, Princeton, NJ 08544-1019. Offers applied and computational mathematics (PhD); chemical engineering (M Eng, MSE, PhD); plasma science and technology (MSE, PhD); polymer sciences and materials (MSE, PhD). *Degree requirements:* For master's, thesis required; for doctorate, dissertation, general exam required. *Entrance requirements:* For master's and doctorate, GRE General Test, GRE Subject Test, TOEFL.

Rensselaer Polytechnic Institute, Graduate School, School of Engineering, Department of Materials Science and Engineering, Troy, NY 12180-3590. Offers ceramics and glass science (M Eng, MS, D Eng, PhD); composites (M Eng, MS, D Eng, PhD); electronic materials (M Eng, MS, D Eng, PhD); metallurgy (M Eng, MS, D Eng, PhD); polymers (M Eng, MS, D Eng, PhD). Part-time and evening/weekend programs available. *Faculty:* 17 full-time (1 woman), 5 part-time (0 women). *Students:* 45 full-time (8 women), 24 part-time (4 women); includes 5 minority (1 African American, 2 Asian Americans or Pacific Islanders, 2 Hispanic Americans), 26 international. Terminal master's awarded for partial completion of doctoral program. *Degree requirements:* For master's, thesis required (for some programs), foreign language not required; for doctorate, dissertation required, foreign language not required. *Entrance requirements:* For master's and doctorate, GRE, TOEFL (minimum score of 550 required). *Application deadline:* For fall admission, 2/1 (priority date). Applications are processed on a rolling basis. Application fee: $35. *Unit head:* Dr. Richard W. Siegel, Chair, 518-276-6373, Fax: 518-276-8554. *Application contact:* Dr. Roger Wright, Admissions Coordinator, 518-276-6372, Fax: 518-276-8554, E-mail: fowlen@rpi.edu.

See in-depth description on page 1337.

Rutgers, The State University of New Jersey, New Brunswick, Graduate School, Program in Materials Science and Engineering, New Brunswick, NJ 08903. Offers physical metallurgy (MS, PhD); polymer science (MS, PhD). Part-time programs available. *Faculty:* 16 full-time (0 women). *Students:* 49 full-time (6 women), 15 part-time (3 women); includes 3 minority (1 African American, 2 Asian Americans or Pacific Islanders), 39 international. Terminal master's awarded for partial completion of doctoral program. *Degree requirements:* For master's, thesis optional, foreign language not required; for doctorate, dissertation required, foreign language not required. *Entrance requirements:* For master's and doctorate, GRE General Test. *Application deadline:* For fall admission, 3/1 (priority date). Applications are processed on a rolling basis. Application fee: $40. *Unit head:* Dr. Jerry I. Scheinbeim, Director, 732-932-3669. *Application contact:* Kathy Finnerty, Administrative Assistant, 732-932-2245, Fax: 732-932-5977.

San Jose State University, Graduate Studies, College of Science, Department of Chemistry, San Jose, CA 95192-0001. Offers analytical chemistry (MS); biochemistry (MS); chemistry (MA); inorganic chemistry (MS); organic chemistry (MS); physical chemistry (MS); polymer chemistry (MS); radiochemistry (MS). Part-time and evening/weekend programs available. *Faculty:* 24 full-time (3 women), 5 part-time (2 women). *Students:* 7 full-time (5 women), 30 part-time (17 women); includes 24 minority (2 African Americans, 21 Asian Americans or Pacific Islanders, 1 Hispanic American), 4 international. *Degree requirements:* For master's, thesis or alternative required, foreign language not required. *Entrance requirements:* For master's, GRE, minimum B average. *Application deadline:* For fall admission, 6/1. Applications are processed on a rolling basis. Application fee: $59. Tuition, nonresident: part-time $246 per unit. Required fees: $1,939; $1,309 per year. *Unit head:* Dr. Pamela Stacks, Chair, 408-924-5000, Fax: 408-924-4945. *Application contact:* Dr. Roger Biringer, Graduate Adviser, 408-924-4961.

The University of Akron, Graduate School, College of Polymer Science and Polymer Engineering, Department of Polymer Engineering, Akron, OH 44325-0001. Offers MS, MSE, PhD. Part-time and evening/weekend programs available. *Faculty:* 9 full-time, 3 part-time. *Students:* 120 full-time (18 women), 12 part-time (2 women); includes 4 minority (2 African Americans, 2 Asian Americans or Pacific Islanders), 106 international. Average age 29. In 1998, 10 master's, 9 doctorates awarded. *Degree requirements:* For master's, thesis required, foreign language not required; for doctorate, variable foreign language requirement (computer language can substitute for one), dissertation required. *Entrance requirements:* For master's, TOEFL (minimum score of 550 required), bachelor's degree in engineering or physical science, minimum GPA of 2.75; for doctorate, TOEFL (minimum score of 550 required). *Average time to degree:* Master's–2 years full-time, 4 years part-time; doctorate–4 years full-time. *Application deadline:* For fall admission, 8/15. Applications are processed on a rolling basis. Application fee: $25 ($50 for international students). Tuition, state resident: part-time $189 per credit. Tuition, nonresident: part-time $353 per credit. Required fees: $7.3 per credit. *Financial aid:* In 1998–99, 108 students received aid, including 66 research assistantships with full tuition reimbursements available; fellowships with full tuition reimbursements available, teaching assistantships with full tuition reimbursements available, tuition waivers (full) and unspecified assistantships also available. *Faculty research:* Experimental numerical simulation of polymer processing, high-performance engineering plastics rheology, polymer blends. *Unit head:* Dr. Rudolph Scavuzzo, Interim Chair, 330-972-5904, E-mail: rscavuzzo@uakron.edu.

The University of Akron, Graduate School, College of Polymer Science and Polymer Engineering, Department of Polymer Science, Akron, OH 44325-0001. Offers MS, PhD. Part-time and evening/weekend programs available. *Faculty:* 16 full-time, 3 part-time. *Students:* 112 full-time (37 women), 27 part-time (9 women); includes 11 minority (2 African Americans, 8 Asian Americans or Pacific Islanders, 1 Native American), 69 international. Average age 29. In 1998, 10 master's, 35 doctorates awarded. Terminal master's awarded for partial completion of doctoral program. *Degree requirements:* For master's, computer language, thesis required, foreign language not required; for doctorate, one foreign language (computer language can substitute), dissertation required. *Entrance requirements:* For master's, TOEFL (minimum score of 550 required), minimum GPA of 2.75; for doctorate, TOEFL (minimum score of 550 required). *Average time to degree:* Master's–2 years full-time, 4 years part-time; doctorate–4 years full-time. *Application deadline:* For fall admission, 4/1. Applications are processed on a rolling basis. Application fee: $25 ($50 for international students). Tuition, state resident: part-time $189 per credit. Tuition, nonresident: part-time $353 per credit. Required fees: $7.3 per credit. *Financial aid:* In 1998–99, 105 students received aid, including 1 fellowship, 95 research assistantships; tuition waivers (full) and research contracts, grants also available. Financial aid application deadline: 3/30. *Faculty research:* Synthesis of polymers, structure of polymers, physical properties of polymers, engineering and technological properties of polymers, elastomers. *Unit head:* Dr. William J. Brittain, Chair, 330-972-5147, E-mail: wbrittain@uakron.edu.

See in-depth description on page 1341.

University of Cincinnati, Division of Research and Advanced Studies, College of Engineering, Department of Materials Science and Metallurgical Engineering, Cincinnati, OH 45221-0091. Offers ceramic science and engineering (MS, PhD); materials science and engineering (MS, PhD); metallurgical engineering (MS, PhD); polymer science and engineering (MS, PhD). Evening/weekend programs available. *Faculty:* 9 full-time. *Students:* 79 full-time (15 women), 15 part-time (2 women); includes 8 minority (3 African Americans, 4 Asian Americans or Pacific Islanders, 1 Hispanic American), 61 international. *Degree requirements:* For master's, thesis optional, foreign language not required; for doctorate, one foreign language, dissertation, comprehensive exams, oral English proficiency exam required. *Entrance requirements:* For master's and doctorate, GRE General Test, TOEFL (minimum score of 550 required), BS in related field, minimum undergraduate GPA of 3.0. *Application deadline:* For fall admission, 2/1 (priority date). Applications are processed on a rolling basis. Application fee: $40. *Unit head:* Raj Singh, Head, 513-556-3114, Fax: 513-556-2569, E-mail: rsingh@uceng.uc.edu. *Application contact:* Dr. R. J. Roe, Director of Graduate Studies, 513-556-3117, Fax: 513-556-2569, E-mail: r.j.roe@uc.edu.

University of Connecticut, Graduate School, School of Engineering, Polymer Science Program, Storrs, CT 06269. Offers MS, PhD. Terminal master's awarded for partial completion of doctoral program. *Degree requirements:* For doctorate, dissertation required. *Entrance requirements:* For master's and doctorate, GRE General Test, GRE Subject Test.

Announcement: The internationally renowned Polymer Program at the University of Connecticut has been a leader for 30 years in the science and engineering of polymeric materials. The 13 program faculty members, drawn from chemistry, chemical engineering, and molecular and cell biology, attract more than $3 million in annual funding and advise more than 50 graduate students from around the world. The broad interdisciplinary curriculum includes 6 fundamental core courses and 15 advanced topic courses. With a strong background in polymeric materials and cutting-edge research projects, program graduates find gainful employment both in academia and in a wide range of industries.

University of Detroit Mercy, College of Engineering and Science, Department of Chemical Engineering, Program in Polymer Engineering, Detroit, MI 48219-0900. Offers ME. Evening/weekend programs available. *Degree requirements:* For master's, computer language, thesis or alternative required, foreign language not required.

University of Florida, Graduate School, College of Engineering, Department of Materials Science and Engineering, Program in Polymer Science and Engineering, Gainesville, FL 32611. Offers ME, MS, PhD, Engr. *Degree requirements:* For master's, thesis optional, foreign language not required; for doctorate, dissertation required, foreign language not required; for Engr, thesis optional. *Entrance requirements:* For master's and doctorate, GRE General Test (minimum combined score of 1100 required), TOEFL, minimum GPA of 3.0; for Engr, GRE General Test. *Application deadline:* For fall admission, 6/1 (priority date). Applications are processed on a rolling basis. Application fee: $20. *Application contact:* Jodi Vanderheyden, Graduate Secretary, 352-392-1451, Fax: 352-392-5630, E-mail: jheyd@mail.mse.ufl.edu.

University of Massachusetts Amherst, Graduate School, College of Natural Sciences and Mathematics, Department of Polymer Science and Engineering, Amherst, MA 01003. Offers MS, PhD. Part-time programs available. *Faculty:* 15 full-time (0 women). *Students:* 17 full-time (5 women), 70 part-time (21 women); includes 11 minority (1 African American, 8 Asian Americans or Pacific Islanders, 2 Hispanic Americans), 24 international. Average age 26. 206 applicants, 17% accepted. In 1998, 17 master's, 20 doctorates awarded. Terminal master's awarded for partial completion of doctoral program. *Degree requirements:* For master's, foreign language and thesis not required; for doctorate, one foreign language (computer language can substitute), dissertation required. *Entrance requirements:* For master's and doctorate, GRE General Test. *Application deadline:* For fall admission, 2/1 (priority date). Applications are processed on a rolling basis. Application fee: $40. Tuition, state resident: full-time $2,640; part-time $165 per credit. Tuition, nonresident: full-time $9,756; part-time $407 per credit. Required fees: $1,221 per term. One-time fee: $110. Full-time tuition and fees vary according to course load, campus/location and reciprocity agreements. *Financial aid:* In 1998–99, 2 fellowships with full tuition reimbursements (averaging $10,346 per year), research assistantships with full tuition reimbursements (averaging $13,168 per year), 7 teaching assistantships with full tuition reimbursements (averaging $2,897 per year) were awarded.; career-related internships or fieldwork, Federal Work-Study, grants, scholarships, traineeships, and unspecified assistantships also available. Aid available to part-time students. Financial aid application deadline: 2/1. *Unit head:* Dr. Richard Farris, Head, 413-577-1127, Fax: 413-545-0082, E-mail: rjfarris@polysci.umass.edu.

See in-depth description on page 1349.

University of Massachusetts Lowell, Graduate School, College of Arts and Sciences, Department of Chemistry, Program in Polymer Sciences, Lowell, MA 01854-2881. Offers MS, PhD. *Students:* 8 full-time (2 women), 2 part-time (1 woman), 8 international. 49 applicants, 41% accepted. In 1998, 10 master's, 3 doctorates awarded. Terminal master's awarded for partial completion of doctoral program. *Degree requirements:* For master's, thesis required, foreign language not required; for doctorate, 2 foreign languages, computer language, dissertation required. *Entrance requirements:* For master's and doctorate, GRE General Test. *Application deadline:* For fall admission, 4/1 (priority date); for spring admission, 10/1. Applications are processed on a rolling basis. Application fee: $20 ($35 for international students). *Financial aid:* Research assistantships, teaching assistantships, career-related internships or fieldwork available. Financial aid application deadline: 4/1. *Unit head:* Dr. James Sherwood, Coordinator, 978-934-3313, E-mail: james_sherwood@woods.uml.edu. *Application contact:* Dr. Melissa McDonald, Coordinator, 978-934-3683, E-mail: melissa_mcdonald@woods.uml.edu.

University of Massachusetts Lowell, Graduate School, James B. Francis College of Engineering, Department of Plastics Engineering, Lowell, MA 01854-2881. Offers chemistry (PhD), including polymer science/plastics engineering; plastics engineering (MS Eng, D Eng), including coatings and adhesives (MS Eng), composites (MS Eng), plastics processing (MS Eng), product design (MS Eng). Part-time programs available. *Faculty:* 14 full-time (0 women). *Students:* 77 full-time (10 women), 159 part-time (23 women); includes 26 minority (4 African Americans, 19 Asian Americans or Pacific Islanders, 3 Native Americans), 57 international. 121 applicants, 81% accepted. In 1998, 48 master's awarded. Terminal master's awarded for partial completion of doctoral program. *Degree requirements:* For master's, thesis required, foreign language not required; for doctorate, 2 foreign languages (computer language can substitute for one), dissertation required. *Entrance requirements:* For master's and doctorate, GRE General Test. *Application deadline:* For fall admission, 4/1 (priority date); for spring admission, 10/1. Applications are processed on a rolling basis. Application fee: $20 ($35 for international students). *Financial aid:* In 1998–99, 25 research assistantships, 20 teaching assistantships were awarded.; career-related internships or fieldwork also available. Financial aid application deadline: 4/1. *Unit head:* Dr. Robert E. Nunn, Chair, 978-934-3432. *Application contact:* Dr. Ross Stacer, Coordinator, 978-934-3339.

University of Missouri–Kansas City, College of Arts and Sciences, Department of Chemistry, Kansas City, MO 64110-2499. Offers analytical chemistry (MS, PhD); inorganic chemistry (MS, PhD); organic chemistry (MS, PhD); physical chemistry (MS, PhD); polymer chemistry (MS, PhD). PhD offered through the School of Graduate Studies. Part-time programs available. *Faculty:* 12 full-time (2 women), 2 part-time (0 women). *Students:* 25 full-time (7 women), 7 part-time (2 women), 18 international. *Degree requirements:* For master's, thesis required (for some programs), foreign language not required; for doctorate, dissertation required, foreign language not required. *Entrance requirements:* For master's, TOEFL (minimum score of 550 required); for doctorate, GRE General Test (minimum combined score of 1500 on three sections required), TOEFL (minimum score of 550 required), TWE (minimum score of 4 required). *Application deadline:* For fall and spring admission, 2/1 (priority date); for winter admission, 9/1 (priority date). Applications are processed on a rolling basis. Application fee: $25. *Unit head:* Dr. Jerry Jean, Interim Chairperson, 816-235-2280, Fax: 816-235-5502, E-mail: jean@cctr.umkc.edu.

University of Southern Mississippi, Graduate School, College of Science and Technology, Department of Polymer Science, Hattiesburg, MS 39406-5167. Offers MS, PhD. *Faculty:* 11 full-time (0 women). *Students:* 68 full-time (15 women), 1 (woman) part-time; includes 5 minority (1 African American, 4 Asian Americans or Pacific Islanders) Average age 26. 87 applicants, 17% accepted. In 1998, 3 master's, 9 doctorates awarded. *Degree requirements:* For master's, thesis required, foreign language not required; for doctorate, 2 foreign languages (computer language can substitute for one), dissertation, comprehensive exam, original proposal required. *Entrance requirements:* For master's, GRE General Test, TOEFL, minimum GPA of 2.75; for doctorate, GRE General Test, TOEFL, minimum GPA of 3.0. *Application deadline:* For fall admission, 8/6 (priority date). Applications are processed on a rolling basis. Application fee: $0 ($25 for international students). Tuition, state resident: full-time $2,250; part-time $137 per semester hour. Tuition, nonresident: full-time $3,102; part-time $172 per semester hour. Required fees: $602. *Financial aid:* Fellowships, research assistantships, teaching assistantships available. Financial aid application deadline: 3/15. *Faculty research:* Water-soluble polymers, polymer composites, coatings, solid-state, laser-initiated polymerization. *Unit head:* Dr. Robert Lochhead, Chair, 601-266-4868.

University of Tennessee, Knoxville, Graduate School, College of Engineering, Department of Materials Science and Engineering, Program in Polymer Engineering, Knoxville, TN 37996. Offers MS, PhD. *Students:* 13 full-time (5 women), 10 part-time (2 women); includes 3 minority (all African Americans), 13 international. 43 applicants, 33% accepted. In 1998, 3 master's, 2 doctorates awarded. *Degree requirements:* For master's, thesis or alternative required, foreign

language not required; for doctorate, dissertation required, foreign language not required. *Entrance requirements:* For master's and doctorate, TOEFL (minimum score of 550 required), minimum GPA of 2.7. *Application deadline:* For fall admission, 2/1 (priority date). Applications are processed on a rolling basis. Application fee: $35. Electronic applications accepted. *Financial aid:* Application deadline: 2/1; *Unit head:* Dr. J. E. Spruiell, Head, Department of Materials Science and Engineering, 423-974-5336, Fax: 423-974-4115, E-mail: spruiell@utk.edu.

University of Wisconsin–Madison, Graduate School, College of Engineering, Department of Mechanical Engineering, Madison, WI 53706-1380. Offers mechanical engineering (MS, PhD); polymers (ME). Part-time programs available. Postbaccalaureate distance learning degree programs offered (no on-campus study). *Faculty:* 36 full-time (4 women), 4 part-time (0 women). *Students:* 191 full-time (7 women), 20 part-time, 94 international. Terminal master's awarded for partial completion of doctoral program. *Degree requirements:* For master's, thesis optional, foreign language not required; for doctorate, dissertation, qualifying exam, preliminary exam required, foreign language not required. *Entrance requirements:* For master's and doctorate, GRE, TOEFL, BS in mechanical engineering or related field, minimun GPA of 3.0 in last 60 hours. *Application deadline:* Applications are processed on a rolling basis. Application fee: $45. Electronic applications accepted. *Unit head:* Kenneth W. Ragland, Chair, 608-265-3576, Fax: 608-265-2316, E-mail: ragland@engr.wisc.edu. *Application contact:* Linda S. Aaberg, Student Status Examiner, 608-262-0666, Fax: 608-265-2316, E-mail: aaberg@engr.wisc.edu.

Wayne State University, Graduate School, College of Engineering, Department of Chemical Engineering and Materials Science, Program in Materials Science and Engineering, Detroit, MI 48202. Offers materials science and engineering (MS, PhD); polymer engineering (Certificate). Part-time programs available. Terminal master's awarded for partial completion of doctoral program. *Degree requirements:* For master's, thesis optional, foreign language not required; for doctorate, dissertation required, foreign language not required. *Faculty research:* Polymer science, rheology, fatigue in metals, metal matrix composites, ceramics.

Cross-Discipline Announcements

Columbia University, Fu Foundation School of Engineering and Applied Science, Department of Electrical Engineering, New York, NY 10027.

Department of Electrical Engineering offers BS, MS, PhD, and Eng Sc D in electrical engineering. Program emphases include photonics: optical materials, nonlinear optics, devices and system architectures, ultrafast optoelectronics, and semiconductor lasers, modulators, and integrated optics; microelectronics: molecular-beam epitaxy and quantum well device physics and GaAs, field-effect, and bipolar transistors thin-film science and electronics materials; and plasma physics.

Dartmouth College, Thayer School of Engineering, Hanover, NH 03755.

Thayer School offers MS and PhD programs concentrating on material sciences. The interdisciplinary character of the institution, modern laboratories, and the proximity to the Cold Regions

Research Engineering Laboratory provide unique opportunities for study and research in composites, intermetallics, thin films, and ice physics. See in-depth description in Section 1 of this guide.

Princeton University, Graduate School, Princeton Materials Institute, Princeton, NJ 08540-5211.

The Princeton Materials Institute welcomes students interested in cross-disciplinary research involving any aspect of materials science and engineering. Faculty members from 8 academic departments collaborate in a remarkably broad range of programs, supported by outstanding facilities. Some fellowships are available. See in-depth description in Section 21 of this guide.

ALFRED UNIVERSITY

New York State College of Ceramics
School of Ceramic Engineering and Materials Science
College of Engineering and Professional Studies

Programs of Study

Programs that lead to master's and doctoral degrees are open to graduates in the fields of ceramic engineering, chemical engineering, chemistry, electrical engineering, geology, glass science, industrial engineering, materials science, mechanical engineering, metallurgy, and physics.

The Master of Science degree is awarded in ceramic engineering, glass science, and materials science and engineering. The research track requires 15 credit hours of course work and 15 credit hours of an experimental thesis. The manufacturing option requires 24 credit hours of courses, including at least 6 credit hours in business and management, and a 6-credit dissertation. Successful completion of the program involves an oral defense of the thesis. The Master of Science degrees in electrical, industrial, and mechanical engineering are awarded upon the completion of 24 credit hours of course work and 6 credit hours of thesis. An oral presentation of the thesis project is required.

The Doctor of Philosophy degree is offered in the field of ceramics and glass science to well-qualified students. The normal residence requirement at Alfred is three years, but in no instance is it less than two years. Ninety hours of course credit beyond the requirements for the baccalaureate degree must be earned. Of these, a minimum of 40 credit hours must be in regular course work; the remainder may be earned as thesis credits.

Research Facilities

In addition to a complete stock of routine scientific equipment, the following major items and facilities are available for graduate research: transmission and scanning electron microscopes, an electron microprobe, X-ray diffraction equipment, X-ray fluorescence and IR analysis equipment, a laser-Raman spectrograph, a secondary-ion mass spectrometer, an Instron testing machine for determining mechanical properties of materials, machine and electronics workshop facilities with technical personnel, chemical analysis, powder processing and characterization, and microscopic-sample preparation facilities. Scholes Library is internationally recognized for the strength of its holdings in science, engineering, and technology; its traditional book and journal collections are supplemented by state-of-the-art access to digital information resources. The library occupies a new four-story building with a wide variety of computer equipment and private study carrels for graduate students. Computer facilities include two VAXclusters centered on VAX 6000 series computers, which are networked to numerous other workstations housed in research laboratories. Most graduate students have their own computer. Nysernet provides access to Internet and off-campus computers.

Financial Aid

Teaching and research assistantships sponsored by the College and fellowships sponsored by private industry and government agencies are available. Assistantships include stipends of up to $14,679 and remission of tuition for the 1999–2000 academic year. A limited number of postdoctoral fellowships are available.

Cost of Study

Yearly tuition and fees are $12,576 in 1999–2000 for the public-sector programs (ceramics and glass engineering and science) and $21,960 for the private-sector programs (electrical and mechanical engineering). For public-sector programs, laboratory fees of $500 per semester may also be payable.

Living and Housing Costs

Although expenses vary widely from one student to another, the estimated charge for room and board in the Alfred area is $7095 per school year. For assistance in locating housing, students should write to the director of residence life.

Student Group

Alfred University has 1,923 undergraduate and 336 graduate students drawn from thirty-four states and twelve countries. There are 97 graduate students in the School of Ceramic Engineering and Materials Science and Divisions of Electrical Engineering, Industrial Engineering, and Mechanical Engineering, allowing for close interaction between students and faculty members.

Student Outcomes

Most recent figures from the School's placement office indicate that program graduates are in demand throughout the United States. All of the program's Ph.D. graduates found employment as research scientists at such companies as Corning, Inc.; 3M Company; Florida Tile; Galileo Electrooptics; and Saint Gobain Industrial Ceramics. M.S. graduates found employment as research engineers or research scientists with places like Corning, Inc.; AVX Corporation; Corhart Corp.; CertainTeed Corp.; Ferro Corp.; and Unifrax Corp.

Location

Alfred University is located in Alfred, New York, a college town 70 miles south of Rochester, 90 miles southeast of Buffalo, and 60 miles west of Corning. Nestled among the pine-sheltered foothills of the Allegheny Mountains, it is a popular hunting area close both to ski slopes and to the water sports and fishing of the Finger Lakes region. New York City is 6 hours away via the Southern Tier Expressway.

The University

Founded in 1836, Alfred University, the second-oldest coed school in the United States, represents a special blending of private and public financing. The College of Ceramics is one of the statutory units of the State University of New York and, as such, is funded by the state; administratively, however, the College is an integral part of Alfred University. The remainder of the University is privately financed and consists of the College of Liberal Arts and Science, School of Business and Administration, and College of Professional Studies.

Applying

To be eligible for admission to the Graduate School, the undergraduate record must clearly indicate the ability to perform creditably at the graduate level. A nonrefundable charge of $50 is made to students applying for admission to the Graduate School.

Correspondence and Information

Cathleen Johnson
Coordinator of Graduate Admissions
Saxon Drive
Alfred University
Alfred, New York 14802-1232
Telephone: 607-871-2141

Alfred University

THE FACULTY AND THEIR RESEARCH

NEW YORK STATE COLLEGE OF CERAMICS

School of Ceramic Engineering and Materials Science

L. David Pye, Professor of Glass Science and Dean, New York State College of Ceramics; Ph.D., Alfred, 1968. Optical and electrical properties of glass.

James S. Reed, Professor of Ceramic Engineering and Dean, School of Ceramic Engineering and Materials Science; Ph.D., Alfred, 1965. Processing and shape-forming technology.

Vasantha R. W. Amarakoon, Professor of Ceramic and Electrical Engineering and Interim Director, Center for Advanced Ceramic Technology and Sponsored Programs; Ph.D., Illinois, 1984. Electronic ceramics, interfacial phenomena processing.

William B. Carlson, Associate Professor of Systems Engineering and Product Design; Ph.D., Penn State, 1987. Electronic materials and composites, multilayer component design for electronic or sensing applications, Stirling and Rankine cycle analysis.

William M. Carty, Associate Professor of Ceramic Engineering; Ph.D., Washington (Seattle), 1992. Ceramic processing, rheology, microstructural development, advanced materials.

Alexis G. Clare, Associate Professor of Glass Science; Ph.D., Reading (England), 1986. Optoelectronic glasses: structure and optics.

Robert A. Condrate Sr., Professor of Spectroscopy; Ph.D., IIT, 1966. Structure studies, infrared and laser-Raman spectroscopy, trace-element analysis, luminescence spectra.

Alastair N. Cormack, VanDerck Frechette Professor of Ceramic Science; Ph.D., Wales, 1980. Computer simulation of solids.

David A. Earl, Assistant Professor of Ceramic Engineering; Ph.D., Florida, 1998. Ceramic powder processing, applied statistics, ceramic whitewares and glazes.

Doreen D. Edwards, Assistant Professor of Ceramic Engineering and Materials Science; Ph.D., Northwestern, 1997. Electronic ceramics, impedance spectroscopy, defect chemistry, electrical characterization of materials.

Herbert Giesche, Assistant Professor of Ceramic Engineering; Ph.D., Mainz (Germany), 1987. Ceramic powder processing, microemulsions, colloidal science.

Paul F. Johnson, Associate Professor of Ceramic Engineering; Ph.D., Florida, 1977. Characterization of ceramics by electron-beam analysis, clays and clay products, computer automation and modeling.

Linda E. Jones, Associate Professor of Ceramic Engineering; Ph.D., Penn State, 1987. High-temperature structural composites, high-strength fibers, surface model of fibers, creep of single crystal oxide fibers, structure-property relations of composites.

William C. LaCourse, Professor of Glass Science; Ph.D., RPI, 1970. Structure and properties of chalcogenide glasses, surface-related properties of glass, strength and strengthening mechanism, durability composition development.

Alan M. Meier, Assistant Professor of Metallurgy and Materials Engineering; Ph.D., Colorado School of Mines, 1994. Spreading kinetics of reactive metal alloys on ceramic substrates, low-temperature bonding mechanisms, metal brazing process development, metal matrix composites.

Scott T. Misture, Assistant Professor of Materials Science and Engineering; Ph.D., Alfred, 1994. Diffraction and scattering studies of inorganics.

Steven M. Pilgrim, Associate Professor of Materials Science and Engineering; Ph.D., Penn State, 1987. Smart materials, relaxor ferroelectrics, microstructure-chemical effects on electromechanical properties.

Walter A. Schulze Jr., Professor of Ceramic and Electrical Engineering and Graduate Program Director; Ph.D., Penn State, 1973. Ferroelectrics, piezoelectrics, dielectrics.

Thomas P. Seward III, Professor of Glass Science and Director, NSF Industry-University Center for Glass Research; Ph.D., Harvard, 1968. Radiation effects in fused silica, nano-particles, densification and surface tension in glass.

James E. Shelby Jr., Professor of Glass Science; Ph.D., Missouri–Rolla, 1968. Transport properties of glass.

Jenifer A. T. Taylor, Associate Professor of Ceramic Engineering and Materials Science; Ph.D., Alfred, 1986. Fabrication and processing of ceramics for electronic applications, including ferroelectrics, sensors, dielectrics, and relaxors.

Rebecca Twite, Assistant Professor of Polymer Science and Coatings; Ph.D., North Dakota State, 1998. Polymer coatings on ceramics, glass, and metal substrates; polymer-assisted powder processing; corrosion protection by organic coatings.

James R. Varner, Professor of Ceramic Engineering and Undergraduate Program Director; Ph.D., Alfred, 1971. Mechanical behavior of ceramics, including contact damage and microindentation of glass and ceramics, strength and processing relationships of materials.

Arun K. Varshneya, Professor of Glass Science and Engineering; Ph.D., Case Western Reserve, 1970. Stresses in glass and glass-to-metal seals.

COLLEGE OF ENGINEERING AND PROFESSIONAL STUDIES

Division of Electrical Engineering

James T. Lancaster, Professor of Electrical Engineering and Head, Division of Electrical Engineering; Ph.D., Virginia Tech, 1971. Power generation, transmission, and distribution; PC applications in power systems.

Wallace B. Leigh, Associate Professor of Electrical Engineering; Ph.D., Northwestern, 1983. Optoelectronics and magnetooptics, flaw states in wide-gap semiconductors, physics of solids.

Jianxin Tang, Associate Professor of Electrical Engineering; Ph.D., Connecticut, 1989. Digital and analog electronics, control systems, parallel and distributed computation.

Xingwu Wang, Professor of Electrical Engineering; Ph.D., SUNY at Buffalo, 1987. Superconductivity, thin films, superfluidity, image processing, robotics, mathematical physics.

Division of Mechanical Engineering

William F. Hahn, Associate Professor of Mechanical Engineering; Ph.D., Illinois, 1969. Design engineering, modeling and control issues in manufacturing systems, production engineering, computers in engineering, kinematics of mechanisms.

Bruce R. Hollworth, Professor of Mechanical Engineering and Acting Chair, Division of Mechanical Engineering; Ph.D., Connecticut, 1974. Forced convection heat transfer, electronics cooling, gas turbine cooling.

Joseph Rosiczkowski, Associate Professor of Mechanical Engineering; Ph.D., Clarkson, 1989. Forced convection heat transfer.

John C. Williams, Assistant Professor of Mechanical Engineering; Ph.D., Clarkson, 1998. Materials structure and failure analysis.

ARIZONA STATE UNIVERSITY

Department of Chemical, Bio and Materials Engineering
Materials Science and Engineering Program

Programs of Study

The Materials Science and Engineering Program at Arizona State University (ASU) is in the Department of Chemical, Bio and Materials Engineering and offers graduate programs that lead to the Master of Science and Doctor of Philosophy degrees in engineering science with a specialization in materials science and engineering. The faculty offers a wide range of course work and research topics, allowing the student an opportunity to develop a program of study to satisfy his or her specific needs. The two major research areas are microelectronic materials and structural materials. With their choice of electives and thesis research topics, students are encouraged to take advantage of the synergy that exists between the three degree-granting programs in the department and between the departments within the college.

The Master of Science degree program requires a minimum of 30 credit hours, including 6 hours of thesis credit. An option is also offered for the Master of Science in Engineering, which requires 30 hours of course work only. All students are required to enroll in a 1-credit seminar during each semester in residence.

The Doctor of Philosophy degree program requires a minimum of 84 hours beyond the bachelor's degree. To satisfy the residency requirements, students must spend a minimum of two semesters of full-time study on campus. All candidates must take written and oral qualifying exams during their first year in the program. These exams cover the broad basics of materials science and engineering that are covered at the undergraduate level. Students are admitted to candidacy after passing a comprehensive exam, which is given near the completion of course work. The exam consists of an oral and written dissertation prospectus.

Students with a B.S. degree in a field other than materials science and engineering are encouraged to apply. The Graduate Committee determines, on an individual basis, which undergraduate courses must be taken to ensure success in the graduate program.

Research Facilities

Outstanding facilities and research equipment are available in individual faculty laboratories for conducting graduate student research. In addition, the materials faculty and students make use of specialized equipment housed in interdisciplinary research centers. The Center for Solid State Electronics Research has capabilities in all aspects of semiconductor materials processing. The center offers state-of-the-art capabilities in fabrication and the analysis of devices. The facilities include a 4,000-square-foot class 100 clean room. A University-based Center for Solid State Science features an extensive materials preparations facility and a high-resolution electron microscopy facility with capabilities in high-resolution transmission and scanning transmission electron microscopy along with other analytical instruments. These facilities support both students and faculty members in the materials science and engineering program.

Engineering Computer Services has a staff of 35 full-time computer specialists who support the wide array of computers and electronic equipment within the college. The college also maintains complete machine, carpentry, electrical, and paint shops to support graduate research.

The University libraries house more than 2 million volumes, 1.7 million microfilms, and 24,000 journal titles. The Noble Science and Engineering Library, a designated patent depository, houses the entire U.S. patent collection.

Financial Aid

A wide variety of financial support is available for graduate students, including teaching and research assistantships and University and industrial fellowships. Out-of-state tuition is waived for all graduate assistants and fellows. In addition, there are a number of Graduate Tuition Scholarships (GTS) that cover out-of-state tuition and Graduate Academic Scholarships (GAS) that cover in-state registration fees available on a highly competitive basis.

Cost of Study

In 1999–2000, the registration and tuition fee for 12 hours or more is $1044 per semester for Arizona residents; nonresidents pay an additional $3476 (prorated for fewer than 12 hours).

Living and Housing Costs

Limited on-campus housing is available for unmarried students. In 1998–99, room rates ranged from $2180 to $3805 for the academic year. Thirteen different meal plans are available (per week or semester). Moderately priced apartments are available within walking distance of the campus.

Student Group

There are more than 47,000 students at Arizona State, including more than 11,000 graduate students. More than 1,800 of the 6,300 students in the College of Engineering and Applied Sciences are enrolled in graduate programs. In fall 1998, there were approximately 140 graduate students in the department, with 47 of them enrolled in the materials science and engineering program.

Student Outcomes

Graduates from the Materials Science and Engineering Program gain nationwide employment in major manufacturing industries, such as semiconductors and structural materials, as well as in research laboratories, materials processing companies, utilities, consulting firms, and government agencies.

Location

Tempe is a suburb bordering the city of Phoenix. The metropolitan area is the population, economic, and industrial center of the state of Arizona. Entertainment centers, art and anthropology museums, and sports arenas are among the wide variety of facilities available in the Valley of the Sun. The area's mild climate also provides for numerous year-round outdoor activities. The urban community surrounding ASU thrives on high-technology industries and is one of the fastest-growing communities in the country.

The University

Arizona State University traces its origin to 1885 and is the largest and oldest institution of higher learning in the state of Arizona. Its current enrollment of 47,000 places it as the fifth-largest central-campus university in the United States. ASU's campus comprises 700 acres and offers outstanding physical facilities to support the University's research and educational mission. Included within the more than 125 buildings are twelve colleges and schools, a University-wide computer system, seven libraries (including the Noble Science and Engineering Library), and more than two dozen specialized centers of research. ASU was recently awarded Research I status as a result of its past and continuing commitment to academic research.

Applying

All students must apply for admission through the Graduate College. For application forms, students should contact Graduate Admissions, Arizona State University, Tempe, Arizona 85287-1003 or call 602-965-6113. Submission of Graduate Record Examinations (GRE) scores and a statement of purpose is required. International students whose native language is not English must submit a Test of English as a Foreign Language (TOEFL) score of 580 or better. Applications are reviewed continuously, but, for consideration for financial consideration, should be received by February 1 for the following academic year.

Correspondence and Information

Chair of Graduate Committee
Department of Chemical, Bio and Materials Engineering
Arizona State University
Box 876006
Tempe, Arizona 85287-6006
Telephone: 602-965-3313
E-mail: cbmerec@asuvax.eas.asu.edu

Arizona State University

THE FACULTY AND THEIR RESEARCH

Materials Science Engineering

James B. Adams, Ph.D., Wisconsin–Madison. Atomic-level modeling of structure, properties, and processing of materials; semiconductor processing (oxidation, metallization); adhesion; catalytic converters.

Terry L. Alford, Ph.D., Cornell. Microelectronic metallization and reliability, silicide formation, ion-beam modification of materials.

Sandwip K. Dey, Ph.D., Alfred, College of Ceramics. Thin-film processing science of electroceramics; characterization of electrical, microstructural, and microchemical properties; high-permittivity dielectrics for ULSI DRAMs and microelectronic packages.

Lester E. Hendrickson (Emeritus), Ph.D., Illinois. Corrosion, fracture and failure analysis, physical and chemical metallurgy.

Stephen J. Krause, Ph.D., Michigan. Processing and defects in semiconductor materials, ordered polymers, composite materials, X-ray diffraction, scanning and transmission electron microscopy, electron and X-ray diffraction.

Subhash Mahajan, Ph.D., Berkeley. Origins of defects in semiconductors and their influence on device behavior, deformation behavior of solids and development of contacts for high-temperature electronics.

James W. Mayer, Ph.D., Purdue. Electronic materials and metallization of integrated circuits; development of new semiconductor materials, such as the ternary alloy SiGeC grown on silicon; development of new metal systems for interconnectors; interdiffusion and reactions in thin films; analysis of paint pigments, art media, and metallic artifacts; ion-beam analysis and Rutherford backscattering analysis.

James T. Stanley (Emeritus), Ph.D., Illinois. Phase transformations, radiation effects on materials, erosion-corrosion.

Chemical Engineering

Stephen P. Beaudoin, Ph.D., North Carolina State. Transport phenomena; surface science concerning pollution prevention, waste minimization, and air pollution remediation.

James R. Beckman, Ph.D., Arizona. Unit operations, applied mathematics, crystallization control and nucleation, process simulation, data analysis, energy storage, solar cooling.

Lynn Bellamy, Ph.D., Tulane. Cognitive science, learning theory, learning style preferences, constructivist learning, classroom assessment, quality management principles, distancing theory (and representational competence), team dynamics, active learning and other classroom management strategies, curriculum (or course) development, design, specification, and assessment.

Neil S. Berman, Ph.D., Texas at Austin. Regional air pollution and global warming physical and numerical models, turbulent mixing experiments and calculations in atmospheric boundary layers and in liquid flows with polymeric additives, experimental measurements using laser induced fluorescence, transformations and deposition of air toxic materials.

Veronica A. Burrows, Ph.D., Princeton. Surface science; environmental sensors; semiconductor processing; interfacial chemical and physical processes in sensor processing, lubrication, and composite materials.

Timothy S. Cale, Ph.D., Houston. Semiconductor materials processing, deposition and etch processes, plasma processing, heterogeneous reactor analysis, heat and mass transfer, heterogeneous catalysis.

Antonio A. Garcia, Ph.D., Berkeley. Protein purification, acid-base molecular interactions in separations, solid-liquid interfacial phenomena, scanning probe microscopy, biocolloid chemistry, chromatography, biosensor immobilization.

James L. Kuester, Ph.D., Texas A&M. Chemical reactor analysis, thermochemical conversion processes, complex reaction systems, catalytic processes, process instrumentation and control, optimization, applied statistics, applied mathematics.

Gregory B. Raupp, Ph.D., Wisconsin. Gas-solid surface reaction mechanisms and kinetics, interaction between surface reactions and simultaneous transport processes, semiconductor materials processing, thermal and plasma-enhanced chemical vapor deposition (CVD), environmental pollution remediation and control, photocatalytic oxidation.

Daniel E. Rivera, Ph.D., Caltech. Control systems engineering, dynamic modeling via system identification, robust control, computer-aided control system design.

Vernon E. Sater, Ph.D., IIT. Heavy metal removal from wastewater, online process instrumentation.

Robert S. Torrest, Ph.D., Minnesota. Multiphase flow, filtration, polymer solution flow in porous media, in situ processes for energy and mineral recovery, pollution control.

Imre Zwiebel, Ph.D., Yale. Mass transfer, adsorption, surface phenomena, adsorption of proteins and macromolecules, protein engineering, separations and bioseparations, molecular sieves, applied mathematics.

Bioengineering

William J. Dorson (Emeritus), Ph.D., Cincinnati. Biomedical engineering, artificial organ problems and applications, rheology, physicochemical phenomena, transport processes, physiological models and analogs.

Eric J. Guilbeau, Ph.D., Louisiana Tech. Biomedical engineering, biomaterials, skeletal muscle cardiac assist, biological transport phenomena, development of biosensors, physiological systems analysis and simulation, artificial internal organs.

Jiping He, Ph.D., Maryland. Biomechanics, robotics, computational neuroscience, optimal control, system dynamics and control.

Daryl R. Kipke, Ph.D., Michigan. Computational neuroscience, auditory neurophysiology, electrical stimulation, speech recognition. Signal processing, biomedical instrumentation, neural modeling using parallel computers and analog VLSI technology.

Vincent B. Pizziconi, Ph.D., Arizona State. Biomedical engineering, molecular and cellular engineering/biological engineering/tissue engineering, artificial organs, biomaterials, biosensors, biotechnology, bioseparations, scanning tunneling microscopy (STM) and atomic force microscopy (AFM) of natural and synthetic biomaterials, fractals/nonlinear dynamical systems.

James D. Sweeney, Ph.D., Case Western Reserve. Biomedical engineering, rehabilitation engineering, applied neural control, neurophysiology, motor physiology, mathematical modeling.

Bruce C. Towe, Ph.D., Penn State. Bioinstrumentation, implantable biochemical sensors, medical ultrasound, bioelectric phenomena, bioimpedance imaging.

Gary T. Yamaguchi, Ph.D., Stanford. Biomechanics and rehabilitation engineering design, including joint mechanics; computer modeling of muscle, tendon, and joints; optimal control and dynamic analysis/simulation of movement, coordination, and functional neuromuscular stimulation.

ARIZONA STATE UNIVERSITY

Science and Engineering of Materials Interdisciplinary Graduate Program

Program of Study	The Science and Engineering of Materials (SEM) doctoral program is administered by the Graduate College with the concurrence of the College of Engineering and Applied Sciences and the College of Liberal Arts. The program is operated by its faculty, drawn from the Departments of Chemical, Bio, and Materials Engineering; Chemistry and Biochemistry; Electrical Engineering; Mechanical and Aerospace Engineering; and Physics and Astronomy. Upon completion of program requirements, students are awarded the Doctor of Philosophy degree with a major in science and engineering of materials.
	The program provides excellent graduate education in materials through both instruction and research, with concentrations in solid-state device materials design and high-resolution nanostructure analysis. Program graduates are well-prepared for careers in research, development, and teaching. Between 1990 and 1996, 22 students completed work for their doctorates in the program. The response of major industry and academic employers to the program graduates has been excellent. All graduates have obtained challenging positions in industry or universities in the U.S. and other countries.
	The program emphasizes synthesis of new materials and quantitative structure-property relations education and research. Six graduate courses treating characterization and analysis of microstructure, physical materials science, chemical thermodynamics and reaction kinetics, quantum theory of solids, and solid-state chemistry and physics provide the academic core for the program. Elective courses for students enrolled in the program are determined by the student's faculty supervisory committee in consultation with the student. Concentrations defined by special electives in solid-state device materials design and high-resolution nanostructure analysis are often selected by students. Students enter the program from diverse scientific and engineering disciplines. Program planning takes into account the breadth of the student's materials knowledge and experience, the student's primary materials interests, and career goals.
Research Facilities	Students use the specialized research laboratories of the faculty members and the larger multiuser laboratories in centers for their research. Multiuser laboratories are located in the Goldwater Materials Facility Center for Solid-State Science (CSSS) and the Center for Solid-State Electronics Research (CSSER). High-resolution electron microscopy instrumentation provides microstructural and nanochemical analysis capability down to atomic resolution and computer facilities for image simulation and image processing. Other CSSS laboratories are devoted to thin-film and surface analysis (including Auger, XPS, SIMS, RBS, and SEM), different types of synthesis and processing equipment for electronic, ceramic, and metallic materials, scanning-probe microscopes (STM and AFM), and high-speed graphics workstations for modeling and other computations (SGI). A 4,000-square-foot Class-100 clean room supporting wafer and device processing is located in CSSER. A device concept can be taken from computer-aided design through to packaging in this laboratory. CSSER also contains equipment for epitaxial thin-film growth and optical- or electron-beam lithography. The multiuser laboratories are operated and maintained by professional support staff members who assist students using the laboratories for research.
Financial Aid	Teaching and research assistantships are available for support of qualified students. These include a stipend for living expenses, typically about $12,000 for the academic year, plus a waiver of nonresident tuition. Self-supporting students may apply for Graduate Tuition Scholarships for nonresident tuition and fees, and all students may apply for the Graduate Academic Scholarships for resident fees. Students can also apply for a number of externally funded academic-year scholarships and summer industrial internships. Students in the program have won a number of these awards.
Cost of Study	For the 1999–2000 academic year (nine months), resident tuition fees are $115 per credit for 1 to 6 credits and $1094 total for 7 or more credits. For students who are not residents of Arizona, tuition fees are $389 per credit for 1 to 11 credits and $4670 total for 12 or more credits.
Living and Housing Costs	Most graduate students at Arizona State University (ASU) live off campus in readily available apartments within easy walking or bicycling distance. Rental fees vary but typically range from $500 to $700 per month for a one- to two-bedroom unfurnished or furnished unit, not including utilities. Students should plan on $500 to $600 per month for food, clothing, and incidentals. Automobile expenses are additional.
Student Group	There are typically 32 students in the program each year, a size that ensures individual attention by faculty members and collegiality. Elected student representatives advise the Director and University administrators of student needs, in addition to individual student interactions with mentors.
Location	The SEM program is on the ASU main campus in Tempe, a small city (population 165,000) in the Phoenix metropolitan area, noted for its congenial lifestyle. The University is a strong influence on the culture of Tempe. The downtown area, a few minutes walk from campus, has many favorite gathering places for students. The surrounding area offers many attractive outdoor activities such as backpacking, camping, and many water sports.
The University	ASU is a large metropolitan Research I University with abundant resources for learning and research. It has a campus of more than 700 acres and enrolls 43,000 students. Twelve colleges and schools, seven libraries (including the Noble Science and Engineering Library), and numerous specialized research centers are housed in more than 100 buildings connected by the University-wide computer system.
Applying	All students applying for graduate study at ASU must apply for admission through the Graduate College. Students can apply directly via the Internet through the Graduate College Web site (http://www.asu.edu/graduate). Students who want a paper application should contact Graduate College Admissions, Arizona State University, P.O. Box 871003, Tempe, Arizona 85287-1003 (telephone: 480-965-6113; E-mail: gradadmiss@asu.edu). The Graduate College requires that students submit the following documentation: the application form, an application fee of $45, and transcripts and, for international students, TOEFL scores. In addition to these requirements, the SEM program requires GRE scores, a resume, a statement of purpose, and three letters of recommendation. These materials should be submitted directly to the program by the deadlines. The deadline for fall entrance is February 15, and the deadline for spring entrance is October 15. To request a brochure for the SEM program, students should contact the program office. Students are encouraged to view their program Web site, listed below.
Correspondence and Information	Science and Engineering of Materials Program Arizona State University Box 871704, Tempe, Arizona 85287-1704 Telephone: 480-965-2460 Fax: 480-965-3559 E-mail: sem@asu.edu World Wide Web: http://www.asu.edu/graduate/SEM

Arizona State University

THE INTERDISCIPLINARY FACULTY AND THEIR RESEARCH

More information on the faculty members is listed on the program's Web site.

Program Co-Directors

James Adams, Professor of Chemical Biology and Materials Engineering; Ph.D., Wisconsin–Madison (jim.adams@asu.edu). Computer simulation of thin-film growth, surface science, oxidation of silicon, ion implantation, solar cells, catalytic converters, ceramic coatings for tools, adhesion/adhesive wear, development of computer simulation methods.

William Petuskey, Professor of Chemistry; Sc.D., MIT (bpetuskey@asu.edu). Chemistry of ceramic materials, solid-state reactions, high-pressure synthesis of new materials, electronic ceramics, structural ceramics.

Solid-State Science

Ray Carpenter, Professor and Past Director of the SEM Program; Ph.D., Berkeley (carpenter@asu.edu). Experimental synthesis of planar interfaces in structural and electronic materials, high-resolution analysis of structure and chemistry of interfaces by transmission electron microscopy, phase transformations and microstructure of solids.

Peter Crozier, Senior Research Scientist and Director of the Industrial Associates Program; Ph.D., Glascow (crozier@asu.edu). Microstructure and properties of heterogeneous catalysts and electronic materials; in situ, high-resolution, and analytical electron microscopy.

Moon Kim, Research Scientist; Ph.D., Arizona State (moon.kim@asu.edu). Atomic structure and chemistry of materials; structure/properties relations and phase transformations in solids; UHV synthesis of thin-film interface diffusion bonding; HRTEM and chemical imaging; analytical microscopy; RBS, AES, XPS, and computer simulation.

Martha (Molly) McCartney, Research Scientist; Ph.D., Arizona State (molly.mccartney@asu.edu). TEM off-axis electron holography of electrostatic and magnetic fields.

Michael McKelvy, Associate Research Scientist and Director of the Goldwater Materials Laboratories; Ph.D., Arizona State (mckelvy@asu.edu). Carbon dioxide mineral sequestration, intercalation chemistry, atomic-level real-time imaging of materials reaction processes, new materials synthesis, thermal chemistry and analysis.

Renu Sharma, Research Scientist; Ph.D., Stockholm (renu.sharma@asu.edu). Structure and defects in inorganic solids and in situ studies of gas-solid interactions at elevated temperature by high-resolution transmission electron microscopy.

David Smith, Professor and Director for the Center for High Resolution Electron Microscopy; D.Sc., Melbourne (david.smith@asu.edu). High-resolution electron microscopy and instrumentation; characterization of magnetic materials and semiconductor heterostructures, ceramics and small particles.

Materials Engineering

Terry Alford, Associate Professor; Ph.D., Cornell (alford@asu.edu).

Sandwip Dey, Professor; Ph.D., Alfred (sandwip.dey@asu.edu). Processing of electroceramics and thin films, chemical vapor deposition/etching (RF and ECR) and sol-gel, microstructural evolution, interfacial effects, electrical properties and modeling, dielectrics for ULSI memories and wireless communication systems.

Stephen Krause, Professor; Ph.D., Michigan (skrause@asu.edu). Structure and defects of semiconductor materials, microstructure of silicon-on-insulator materials, microstructure of contacts and implantation in GaN and AIN materials.

Subhash Mahajan, Professor; Ph.D., Berkeley (smahajan@asu.edu). Origins of growth-induced and processing-induced defects in semiconductors and their influence on device behavior; thermally stable contacts for devices; high-temperature electronics; deformation behavior of solids.

Chemistry

William Glaunsinger, Professor; Ph.D., Cornell (wglaunsinger@asu.edu). Chemical sensors, environmental analytical instrumentation, intercalation chemistry.

John Kouvetakis, Assistant Professor; Ph.D., Berkeley (jkouvetakis@asu.edu).

Paul McMillan, Professor and Director of the Center for Solid-State Science; Ph.D., Arizona State (pmcmillan@asu.edu). Vibrational and NMR spectroscopy; structure and dynamics of high-temperature liquids, glassy materials, high-pressure synthesis of new materials, oxynitrides, chalcogenides.

Electrical Engineering

Jonathan Bird, Associate Professor; D. Phil., Sussex (England) (bird@asu.edu). Studies of fundamental quantum transport phenomena in semiconductor nanostructure devices; effects studies include single electron tunneling, electron wave interference and spin-manipulation in both Si MOSFET and III-V heterostructure systems.

David Ferry, Regents Professor; Ph.D., Texas at Austin (ferry@asu.edu). Semiconductor transport, quantum transport, nanostructure electronics.

Stephen Goodnick, Professor and Chair of Electrical Engineering; Ph.D., Colorado State (stephen.goodnick@asu.edu).

Michael Kozicki, Professor and Director of the Center for Solid-State Electronics Research; Ph.D., Edinburgh (michael.kozicki@asu.edu). Integrated circuit processing, nanotechnology, low-power nonvolatile memories, integrated field emission devices, biohybrid systems.

Dieter Schroder, Professor; Ph.D., Illinois, Urbana-Champaign (schroder@asu.edu). Defects in semiconductors, semiconductor material and device characterization, low-power electronics, photovoltaics, imaging devices, hot electrons in MOS devices.

Yong-Hang Zhang, Associate Professor; Ph.D., Max Planck Institute (Stuttgart) (yhzhang@asu.edu). Optoelectronic devices and their applications, molecular beam epitaxy of III/V compound semiconductors.

Geology and Geochemistry

Peter Buseck, Regents Professor; Ph.D., Columbia (pbuseck@asu.edu). Transmission electron microscopy of minerals and their analogs, mineralogy and geochemistry of meteorites and cosmic dust, environmental and atmospheric geochemistry, electron microprobe analysis.

Mechanical Engineering

Karl Sieradzki, Professor; Ph.D., Syracuse (karl@icarus.eas.asu.edu). Fracture of solids, thermodynamics of solid structures, corrosion, mechanics of nanophase materials.

Physics

Peter Bennett, Professor; Ph.D., Wisconsin–Madison (peter.bennett@asu.edu). Surface structure and reactions in metal/semiconductor systems using STM, synchrotron X-ray diffraction, and low-energy electron microscopy.

Robert Culbertson, Associate Professor; Ph.D., Penn State (robert.culbertson@asu.edu). Ion-beam analysis and modification of materials; surface, interface, thin-film characterization.

Nicole Herbots, Associate Professor; Ph.D., Louvain (Belgium) (nicole.herbots@asu.edu).

Robert Marzke, Associate Professor; Ph.D., Columbia (robert.marzke@asu.edu). NMR studies of novel thermoelectrics, liquids at high pressure, oxide melts at ultrahigh temperatures, battery materials and electrolytes, catalysts.

Fernando Ponce, Professor; PhD., Stanford (480-727-6260; ponce@asu.edu). Atomic arrangement in semiconductors, optical and electronic properties, materials for microelectronics, optoelectronic applications.

Peter Rez, Professor; D.Phil., Oxford (peter.rez@asu.edu). Electron scattering theory for TEM, SEM, and analytical microscopy; electron energy loss spectroscopy; computer control of electron-beam instruments.

Otto Sankey, Professor; Ph.D., Washington (St. Louis) (otto.sankey@asu.edu). Theoretical studies of the electronic structure of solids and defects, molecular dynamics simulations of dynamical processes in solids, semiconductors, fullerenes, high-pressure materials synthesis.

Ignatius Tsong, Professor; Ph.D., London; D.Sc., Leeds (ig.tsong@asu.edu). Atomic structure of solid surfaces by scanning tunneling microscopy, nucleation and growth of epitaxial layers by low-energy electron microscopy, thin-film growth by chemical vapor deposition by energy-dispersive X-ray reflectivity and multibeam optical stress sensors.

John Venables, Professor; Ph.D., Cambridge (john.venables@asu.edu). Surface structure and surface processes, physics of thin films, analytical electron microscopy and diffraction.

CARNEGIE MELLON UNIVERSITY

Department of Materials Science and Engineering

Programs of Study

Graduate education in materials science and engineering at Carnegie Mellon is offered through both departmental and interdisciplinary programs.

The graduate program of the department can be divided into four areas that span the classical and modern fields of materials: electronic and magnetic materials, materials processing, mechanical behavior of materials, and surface and interface phenomena. The growing interest in nonmetallic materials, especially those used in electronic and magnetic applications, has made this one of the fastest-growing components of the graduate program. Over the last fifty years, the department has granted advanced degrees for graduate study to more than 1,500 engineers and scientists, many of whom now hold prominent positions in industry, government, and academic institutions. A Master of Science Course Option degree is awarded after the satisfactory completion of 96 units of courses and a comprehensive oral examination; a Master of Science Research Option degree is awarded after the completion of 60 units of course work, 96 units of project work, master thesis or paper, and an examination on the thesis work. The M.S. Research Option program is recommended for students who desire a terminal master's degree involving considerable independent research, irrespective of their intent to seek future employment in industry, government, or academia. The Ph.D. program requires a minimum of six graduate courses, passing the qualifying examination, satisfactory completion of research, and preparation of the doctoral thesis. An M.S. is not a prerequisite for the Ph.D.

Research Facilities

The department occupies 20,000 square feet in Wean Hall, which also houses the Science and Engineering Library. Laboratories are organized on the basis of a "central facilities" concept, with major centers for research augmented by smaller, more specialized facilities. Central facilities are available for characterization sample preparation, optical metallography, electron optics (TEM, STEM, SEM, HRTEM/GIF), X-ray, electron energy loss microanalysis, magnetic characterization facilities, Auger and scanning Auger spectrometry, heat-treating, mechanical testing, sputtering and thin-film deposition, X-ray diffraction, melting, deformation processing, powder production and consolidation, and physical chemistry of extractive metallurgy and steelmaking.

Financial Aid

Qualified applicants are eligible for available fellowships and assistantships. These pay tuition and an appropriate stipend to cover living expenses; the stipend increases upon passage of the Ph.D. qualifying examination. Summer appointments are generally available. Advanced students may obtain appointments that permit them to work full-time on their thesis. Assistantships (research and teaching) carry certain clearly defined responsibilities that are adjusted to the needs of the student and the department; details may be obtained upon request. Low-interest loans are also available.

Cost of Study

The tuition fee for full-time graduate students in 1999–2000 is approximately $22,175 per academic year. Part-time students are charged at a unit rate. Fees are subject to change.

Living and Housing Costs

Graduate students are responsible for obtaining their own housing; numerous apartments and rooms are located nearby, however, and meals are available at student cafeterias.

Student Group

The department has more than 50 full-time graduate students and about 5 part-time students from local industry and research organizations. The diversity of the student body, combined with the participation of graduate students from other departments, makes classes and seminars particularly stimulating.

Location

Pittsburgh, the headquarters of many of the nation's largest corporations, is a metropolitan area of more than 1.5 million people. There is an unusually large concentration of research laboratories in the area. Carnegie Mellon is located in Oakland, the cultural center of the city, on a tract of 90 acres adjacent to Schenley Park, the largest city park. The campus is close to the city's many cultural and sports activities and is 3 miles from the downtown business district.

The University

Through a gift from Andrew Carnegie, Carnegie Mellon was established in 1900 as the Carnegie Technical School. In 1912, the name of the school was changed to Carnegie Institute of Technology and in 1967 to Carnegie Mellon University. The University has assets of more than $500 million and a total enrollment of more than 7,150 students. There are about 865 full-time and 200 part-time members on the teaching faculty. The faculty is a diverse and productive group of scholars, artists, and scientists engaged in research, performing, and teaching.

Applying

Applications and complete credentials for graduate study at Carnegie Mellon University should be submitted by February 1 for fall enrollment. Students of high academic standing who have a bachelor's or master's degree in materials science or related fields of science and engineering are considered. Applications should include a record of undergraduate work for all but the final year, together with three letters of recommendation and scores on the General Test of the Graduate Record Examinations. International students should arrange for TOEFL scores to be forwarded to the department.

Correspondence and Information

Department of Materials Science and Engineering
3325 Wean Hall
Carnegie Mellon University
Pittsburgh, Pennsylvania 15213-3890
Telephone: 412-268-7574
Fax: 412-268-7596
E-mail: matsci@neon.mems.cmu.edu
World Wide Web: http://neon.mems.cmu.edu

Carnegie Mellon University

THE FACULTY AND THEIR RESEARCH

In the list below, the parenthetical date at the end of each entry is the year that the member joined the faculty.

Brent L. Adams, Professor of Materials Science and Engineering; Ph.D., Ohio State, 1979. Measurement and representation of polycrystalline microstructure and properties, grain boundaries. (1994)

Katayun Barmak, Associate Professor of Materials Science and Engineering; Ph.D., MIT, 1989. Phase transformations and associated microstructures in polycrystalline thin films, composite and layered coatings, sputtering, electrodeposition, thermal analysis, electron microscopy, X-ray diffraction. (1999)

Alan W. Cramb, POSCO Professor of Materials Science and Engineering and Co-Director, Center for Iron and Steel Research; Ph.D., Pennsylvania, 1979. Thermodynamics and kinetics, surface tension, continuous casting and solidification. (1986)

Marc De Graef, Associate Professor of Materials Science and Engineering; Ph.D., Catholic University Leuven (Belgium), 1989. Phase transformations and microstructures in intermetallics and magnetic materials, high-resolution and energy-filtered transmission electron microscopy. (1993)

Richard J. Fruehan, U.S. Steel Professor of Materials Science and Engineering; Ph.D., Pennsylvania, 1966. Thermodynamics and kinetics of iron and steelmaking reactions, kinetics of interfacial reactions, thermodynamics of metallurgical systems. (1981)

Warren M. Garrison, Professor of Materials Science and Engineering; Ph.D., Berkeley, 1979. Microstructural and environmental effects on deformation and fracture. (1984)

Prashant N. Kumta, Professor of Materials Science and Engineering; Ph.D., Arizona, 1990. Research in chemical processing, structure and properties of electrochemical and electronic ceramics, glasses, glass and ceramic composites, and bioceramic composites for bone tissue engineering. (1990)

David E. Laughlin, Professor of Materials Science and Engineering, Data Storage System Center; Editor, *Metallurgical and Materials Transactions*; Ph.D., MIT, 1973. Solid-state phase transitions, X-ray diffraction, electron microscopy, crystallography, magnetic materials, precipitation in aluminum alloys. (1974)

Michael E. McHenry, Professor of Materials Science and Engineering; Ph.D., MIT, 1988. Magnetic properties of materials, magnetic nanocrystals, high-temperature magnets, magnetic and electrical properties of superconductors, electronic structure calculations. (1989)

Henry R. Piehler, Professor of Materials Science and Engineering and of Engineering and Public Policy; Sc.D., MIT, 1967. Deformation processes and the mechanical behavior of materials, clad and coated metals, orthopedic materials and devices, legal and technical aspects of product liability litigation and standardization processes. (1967)

Tresa M. Pollock, Associate Professor of Materials Science and Engineering; Ph.D., MIT, 1989. Mechanical properties and processing of high-temperature structural materials, including superalloys, intermetallics, and composites. (1991)

Lisa M. Porter, Assistant Professor of Materials Science and Engineering; Ph.D., North Carolina State, 1994. Physics of metal semiconductor interfaces, ohmic and rectifying contacts, stability of electronically functional interfaces, and high-temperature devices. (1997)

Gregory S. Rohrer, Associate Professor of Materials Science and Engineering; Ph.D., Pennsylvania, 1989. Structure-property relationships for solid surfaces. (1990)

Anthony D. Rollett, Professor and Head of Materials Science and Engineering; Ph.D., Drexel, 1987. Microstructure-property relationships in materials, computer simulation of microstructural evolution and mechanical properties. (1995)

Paul A. Salvador, Assistant Professor of Materials Science and Engineering; Ph.D., Northwestern, 1997. Design of new electronic materials, characterization of their structure-property relationships, pulsed laser deposition of novel inorganic thin films. (1999)

Marek Skowronski, Professor of Materials Science and Engineering; Ph.D., Warsaw (Poland), 1982. Growth and characterization of compound semiconductors, fundamental issues in epitaxy and device processing. (1988)

Paul Wynblatt, Professor of Materials Science and Engineering; Ph.D., Berkeley, 1966. Compositional effects at surfaces and interfaces, two-dimensional phase transitions, equilibrium forms of crystals, wetting phenomena. (1981)

Affiliated and Emeritus Faculty

Hubert I. Aaronson, R. F. Mehl Professor Emeritus of Materials Science and Engineering; Ph.D., Carnegie Tech, 1954. Kinetics and mechanisms of diffusional transformations. (1979)

V. S. Arunachalam, Distinguished Service Professor of Materials Science and Engineering, Engineering and Public Policy and Robotics Institute; Ph.D., Wales (England), 1965. Integrated materials design, powder metallurgy, high-strength and high-temperature materials, materials and technology policies. (1995)

Charles L. Bauer, Professor Emeritus of Materials Science and Engineering; Dr.Eng., Yale, 1961. Grain boundary structure and related kinetic phenomena, production and characterization of surfaces and interfaces, structure and properties of thin films, tribology of magnetic recording media. (1961)

Thaddeus B. Massalski, Professor Emeritus of Materials Science and Engineering; Editor, *Progress in Materials Science*; Associate Editor, *Metallurgical Transactions*; Ph.D., Birmingham (England), 1954. Stability of alloy phases, imperfections in crystals, phase transformations and amorphous structures. (1959)

David M. Moon, Lecturer and Coordinator, Industrial Internship Option, Department of Materials Science and Engineering; Ph.D., Carnegie Mellon, 1967. (1998)

William W. Mullins, University Professor Emeritus of Applied Science; Ph.D., Chicago, 1955. Physical metallurgy; thermodynamics of solids; transport phenomena; electromigration, surface structure. (1960)

Harold W. Paxton, U.S. Steel University Professor Emeritus of Materials Science and Engineering; Ph.D., Birmingham (England), 1952. Research for basic industries, materials problems in engineering design. (1953, 1986)

S. G. Sankar, Adjunct Professor of Materials Science and Engineering; Ph.D., Pune (India), 1967. Superconductivity, electronic ceramics, and rare-earth magnetic materials. (1987)

W. E. Wallace, Adjunct Professor of Applied Science and Engineering; Ph.D., Pittsburgh, 1941. Synthesis and characterization of rare-earth magnets, hydrogen storage materials. (1983)

J. C. Williams, Adjunct Professor of Materials Science and Engineering; Ph.D., Washington (Seattle), 1968. Mechanical properties of structural materials. (1975)

CASE WESTERN RESERVE UNIVERSITY

Case School of Engineering
Department of Macromolecular Science

Programs of Study

The Department of Macromolecular Science, one of the largest departments of polymer science and engineering in the nation, offers programs of study leading to the M.S. and Ph.D. degrees in all branches of macromolecular science and engineering: synthesis, structural spectroscopy, morphology, mechanical properties, rheology, processing, and polymer physics. A strong research program in naturally occurring polymers includes a combined M.D./Ph.D. degree program offered in cooperation with the Case Western Reserve University School of Medicine. The department's goal is to educate outstanding polymer research scientists and engineers to meet the expanding requirements of industry and academia in the plastics, biomedical, and food-related fields. All students begin research within a few weeks of entering the department. To be admitted to candidacy for the Ph.D., however, students must pass a written and oral qualifying exam. The degree is awarded upon satisfactory completion of 36 credit hours of graduate course work and approval of a thesis based on original research. A large number of students present the results of their research work at national and international conferences. Students entering the Ph.D. program with a master's degree or its equivalent may be exempt from equivalent course requirements. The M.S. degree is awarded upon satisfactory completion of 18 credit hours of course work and approval of a thesis based on original research worth 9 credit hours. There is no foreign language requirement for either the M.S. or Ph.D. degree. A full-time graduate student normally registers for 9 credit hours per semester.

Research Facilities

The Kent H. Smith Engineering and Science Building, constructed with a budget of $24.1 million, has been the home of the department since 1994. Particularly noteworthy among the instrumentation available to students are several JEM transmission and scanning electron microscopes; small- and wide-angle X-ray diffractometers; laser-Raman and Fourier-transform infrared spectrometers; Nicolet NT-150 C and Bruker 300 MSL multinuclear Fourier-transform NMR spectrometers with "magic angle" and NMR imaging accessories for solid-state studies; fluorescence and circular dichroism spectrometers; static and dynamic light-scattering equipment; mechanical and rheological testing instruments, such as Instron testers and Weissenberg and Ferranti-Shirley rheometers; and a thermal analysis laboratory containing a Tian-Calvet microcalorimeter and a Perkin-Elmer DSC-7 thermal analysis system, a Rheometrics RMS-800 dynamic mechanical spectrometer, several capillary rheometers, and facilities for polymer fractionation (including a Waters high-pressure liquid chromatography instrument). The EPIC Center for Molecular Modeling of Polymers houses a Silicon Graphics 4D/220GTX Power Center with 32 MB of memory and a 1- to 2-GB hard disk and a Silicon Graphics 4D25G 20-MHz IRIS PC with 16 MB of memory and a 380-MB disk. A MicroVAX Server is the center of the departmental network of MicroVAX computers and workstations and provides a link to the campus network and to the Ohio Supercomputer Center (CRAY Y-MP).

Financial Aid

Incoming graduate students are eligible for graduate research assistantships. For the academic year 1999–2000 (nine months), the total amount of the stipend is $27,378, from which $14,328 is deducted for tuition. These awards involve research work with no teaching obligations.

Cost of Study

In 1999–2000, tuition for study during the academic year is $14,328 per year.

Living and Housing Costs

Case Western Reserve University is located in University Circle at the extreme eastern edge of Cleveland, adjacent to suburbs with a variety of types of housing. On-campus housing is also available. The cost of living in Cleveland is moderate, less than in most cities but higher than in many rural areas.

Student Group

The department usually enrolls approximately 20 undergraduates, 100 graduate students, and 15 postdoctoral associates. A good portion of the graduate students currently enrolled are American men and women. Ten percent of the students are married, and 60 percent are supported by fellowships or graduate assistantships. Many students are from midwestern and eastern states. Other students come from various parts of the world, some on international fellowships.

Student Outcomes

Graduates of the program obtain positions in all levels of the polymer science field. Approximately 75 percent of the graduates acquire jobs in industry, 15 percent in academia, and 3 percent in governmental positions. The industrial positions include chemists, advanced engineers, technical specialists, vice presidents, and presidents. Alumni are currently employed at companies such as Goodyear Tire & Rubber, B. F. Goodrich, N.I.S.T., Dow Chemical, Quantum Chemical, Avery Dennison, Siemens, and DuPont.

Location

At University Circle in Cleveland, one of the largest collections of academic, philanthropic, and cultural organizations in the United States is situated in pleasant, parklike environs. Among the institutions at University Circle are the Cleveland Museum of Art, the Natural History Museum, the Cleveland Institute of Music, and Severance Hall, home of the Cleveland Orchestra.

The University and The Department

Case Western Reserve University, with about 9,970 students, is a major private university in the United States and the largest in Ohio. Approximately two thirds of the students at Case Western Reserve are in the graduate and professional schools. Case School of Engineering, the technical college of the University, has a distinguished history in science and engineering. Professor A. Michelson's winning of the first Nobel Prize in physics to be awarded to an American, in 1907, was an auspicious start to Case's entry into the twentieth century. Western Reserve, the oldest college of the University, dates back to 1826. Its School of Medicine is widely recognized as one of the finest in the country, and the location of the medical school close to graduate science and engineering facilities has led to extensive interdisciplinary research on polymer materials.

Faculty members associated with the ALCOM Science and Technology Center on Liquid Crystals carry out research in polymer liquid crystals, and activities of the Edison Polymer Innovation Corporation and the Center for Applied Research include collaborative research projects of industrial researchers and education in this important area at the forefront of modern technologies.

Applying

Applicants should have an undergraduate degree in science or engineering. Most students currently enrolled have degrees in chemistry, chemical engineering, physics, materials science, or biochemistry, but degrees in other areas may be acceptable. Applications are reviewed as they are received. Graduate Record Examinations results are required.

Correspondence and Information

Application forms, brochures, and additional information may be obtained from:

Professor Shi-Qing Wang
Department of Macromolecular Science
Case Western Reserve University
Cleveland, Ohio 44106
Telephone: 216-368-4166
        800-368-6591 (toll-free)
Fax: 216-368-4202
E-mail: dxs12@po.cwru.edu

Case Western Reserve University

THE FACULTY AND THEIR RESEARCH

Eric Baer, Herbert Henry Dow Professor of Science and Engineering; Ph.D., Johns Hopkins, 1957. Irreversible microdeformation mechanisms, pressure effects on morphology and mechanical properties, relationships between hierarchical structure and mechanical function, mechanical properties of soft connective tissue, polymer composites and blends, polymerization and crystallization on crystalline surfaces, viscoelastic properties of polymer melts, damage and fracture analysis of polymers and their composites, structure-property relationships in biological systems.

John Blackwell, F. Alex Nason Professor; Ph.D., Leeds (England), 1967. Determination of the solid-state structure and morphology of polymers; X-ray analysis of the structure of thermotropic copolyesters, copolyimides, polyurethanes, and polysaccharides; supramolecular assemblies, fluoropolymers; molecular modeling of semicrystalline and liquid crystalline polymers; rheological properties of polysaccharides and glycoproteins.

Anne Hiltner, Professor; Ph.D., Oregon State, 1967. Structure-property relationships; irreversible deformation, crack propagation, and fracture of polymers, blends, and composites; microlayer processing of polymers; structure-function relationships in collagenous tissues; biostability of biomaterials.

Steven D. Hudson, Assistant Professor; Ph.D., Massachusetts, 1990. Development of polymeric materials with novel structure and properties; electron microscopy; diffraction; coalescence, aggregation, phase inversion, nanocomposites, liquid crystals, and supramolecular assemblies.

Hatsuo Ishida, Professor; Ph.D., Case Western Reserve, 1976. Processing of polymers and composite materials, structural analysis of surfaces and interfaces, molecular spectroscopy of synthetic polymers.

Alexander Jamieson, Professor and Chairman; D.Phil., Oxford, 1969. Quasielastic laser light scattering, relaxation and transport of macromolecules in solution and bulk, structure-function relationships of biological macromolecules.

Jack Koenig, J. Donnell Institute Professor; Ph.D., Nebraska, 1959. Polymer structure-property relationships using infrared, Raman, NMR spectroscopy, and spectroscopic imaging techniques.

Jerome B. Lando, Professor; Ph.D., Polytechnic of Brooklyn, 1963. Solid-state polymerization, X-ray crystallography of polymers, ultrathin polymer films.

Morton Litt, Professor; Ph.D., Polytechnic of Brooklyn, 1956. Kinetics and mechanisms of free radical and ionic polymerization, mechanical properties of polymers, fluorocarbon chemistry, synthesis of novel monomers and polymers, polymer electrical properties, cross-linked liquid crystal polymers.

Ica Manas-Zloczower, Professor; D.Sc., Technion (Israel), 1983. Structure and micromechanics of fine-particle clusters, interfacial engineering strategies for advanced materials processing, dispersive mixing mechanism and modeling, design and mixing optimization studies for polymer processing equipment through flow simulations.

Sergei Nazarenko, Assistant Professor; Ph.D., USSR Academy of Sciences, 1988. Diffusion and transport properties of polymeric materials, barrier structures, macromolecular interdiffusion, nonequilibrium behavior of polymer glasses.

Virgil Percec, Leonard Case, Jr. Professor of Engineering; Ph.D., Polytechnic Institute of Jassy (Romania), 1976. Polymer synthesis and modification; new polymerization reactions and reaction mechanisms; molecular recognition processes; self-assembled supramolecular systems; molecular, macromolecular, and supramolecular liquid crystals.

Charles E. Rogers, Professor Emeritus; Ph.D., Syracuse, 1957. Transport and mechanical properties of polymers; synthesis and properties of multicomponent polymeric systems; environmental effects on polymers; adhesion, adhesives, and coatings.

Robert Simha, Professor Emeritus; Ph.D., Vienna, 1935. Theory and experiment: thermal and pressure properties of polymer melt, crystal and liquid crystal, microstructure and macrostructure; composites as molecular structures; the glassy state: quasi-equilibrium, dynamics of physical aging processes, and correlations between thermodynamic, mechanical, and probe spectroscopy results; molecular dynamics and the glass transition.

Shi-Qing Wang, Professor; Ph.D., Chicago, 1987. Rheology and dynamics of polymeric and complex fluids, polymer interfaces, polymer processing behavior, molecular aspects of polymer mechanics.

Associate Faculty

James M. Anderson, Professor of Pathology and Macromolecular Science; Ph.D., Oregon State, 1967; M.D., Case Western Reserve, 1976. Development of polymers for medical and dental applications, biocompatibility and biostability of polymers, synthesis of random copolypeptides, implant retrieval and evaluation, polymeric slow-release systems for pharmaceuticals.

J. Adin Mann Jr., Professor of Chemical Engineering and Macromolecular Science; Ph.D., Iowa State, 1962. Surface chemistry; theories of interfacial phenomena; stochastic processes of adsorption and molecular rearrangement at interfaces; structure, dynamics, and function of ultrathin films, including surface polymerization; dispersions; surface light-scattering spectroscopy; surface diffraction using X-ray synchrotron sources.

Roger E. Marchant, Professor of Biomedical Engineering and Macromolecular Science; Ph.D., Case Western Reserve, 1984. Modification of polymer surfaces and biomaterials, interactions and behavior of polymers at biological interfaces, including measurement of intermolecular forces and three-dimensional structure of globular proteins by atomic force microscopy.

Syed Qutubuddin, Professor of Chemical Engineering and Macromolecular Science; Ph.D., Carnegie Mellon, 1983. Microemulsions, colloids and interfacial phenomena, polymer blends, composites and nanoparticles, laser light scattering, separations.

Charles Rosenblatt, Professor of Physics and Macromolecular Science; Ph.D., Harvard, 1978. Intense magnetic and electric field effects on soft condensed systems, magnetooptics and electrooptics, polymer liquid crystals, low-symmetry liquid crystals (including ferroelectrics), interfacial effects in macromolecules, complex fluids, and self-assembling systems; critical phenomena; phenomena in confined geometries; flat panel display devices.

Philip L. Taylor, Perkins Professor of Physics and Macromolecular Science; Ph.D., Cambridge, 1962. Phase transitions and equations of state for crystalline and liquid crystalline polymers, piezoelectricity and pyroelectricity, electrical conductivity.

CLEMSON UNIVERSITY

Materials Science and Engineering Program

Programs of Study

The Materials Science and Engineering Program prepares graduate students to apply science and engineering principles to solve problems related to the scientific understanding, characterization, and development of new technology necessary for the processing and manufacturing of different materials and related products. Students apply technical knowledge to solve short-term and long-term research problems and develop the necessary skills for independent learning. Current research represents various leading areas of materials science and engineering. One such area of the research spectrum focuses on materials and processing for manufacturing. The other area focuses on the investigation (both theoretical and experimental) of nanomaterials and atomic-level surface phenomena. Specifically, the principal thrust of this program is in the areas of semiconductors, polymers, ceramics, metals, biomaterials, fast-cycle manufacturing, thermoelectrics, composite materials, atomistic simulations, mathematical modeling of materials processing, and ceramic manufacturing. The program is geared toward educating the students to meet both the present and future needs of industry, academia, and government.

Interdisciplinary M.S. and Ph.D. degrees in materials science and engineering can be earned under the direction of materials science and engineering faculty members, who are in the Departments of Bioengineering, Ceramic and Materials Engineering, Chemical Engineering, Chemistry, Civil Engineering, Computer Science, Electrical Engineering, Forest Products, Mathematical Sciences, Mechanical Engineering, Physics, and Textiles and Fiber and Polymer Science. Close working relationships exist between the participating faculty members. Graduate students are mentored by faculty members who have diverse academic backgrounds, and they participate in cutting-edge research projects funded by federal and/or state agencies and industry.

Research Facilities

The research facilities of participating departments are available to the Materials Science and Engineering Program students. For complete listings of research facilities in various departments, students should consult specific departments' in-depth descriptions in Peterson's *Graduate Programs in Engineering & Applied Sciences.* Clemson University is also served by a central electron microscope facility, containing two transmission electron microscopes, two scanning electron microscopes, and an Auger microprobe with a second ion mass spectrometer. Applications such as voltage contrast and electron beam–induced current measurements are also available. The College of Engineering and Science maintains computational facilities that include more than 250 UNIX workstations and 125 microcomputers as well as state-of-the-art design and analysis software. Other on-campus facilities include machine shops with support technicians. Current resources and facilities of the Clemson University Libraries make it one of the most important research institutions in the Southeast. Graduate students are granted an extended six-week loan period for borrowing materials and are allowed to check out a total of three journals for a three-day period.

Financial Aid

A limited number of research assistantships requiring up to 20 hours of work per week are available. These assistantships carry a stipend of up to $15,000 per year and a reduction of tuition fees. There are also opportunities for exceptional students to obtain additional University fellowships of up to $5000 for twelve months.

Cost of Study

The tuition and fees for full-time students in 1998–99 were $1577 per semester for South Carolina residents and $3226 for nonresidents. All graduate assistants receive a reduction in tuition and fees and pay $493 per semester and $165 per summer session.

Living and Housing Costs

Rooms are available in dormitories for $810 to $1185 per semester or in University apartments for $1110 to $1280 per semester. Housing for married students is available for $290 to $495 a month; graduate assistants and fellows are given priority. Applications must be received at the housing office before May 1 for August housing and before November 1 for January housing.

Student Group

There are about 16,685 students at Clemson University, of whom approximately 22 percent are graduate students. There are 912 graduate students in the College of Engineering and Science. The Materials Science and Engineering Program has 27 graduate students. No undergraduate degree is offered in the Materials Science and Engineering Program.

Student Outcomes

After completion of an M.S. degree, recent graduates of the program have obtained employment as research/project/process engineers in industry, federal research laboratories, or academic laboratories. Some students earning an M.S. degree go on to work toward a Ph.D. degree. After completion of the Ph.D. degree, students go on to managerial research positions in industry or to postdoctoral fellow or faculty positions in academia.

Location

Clemson, South Carolina, is a residential community located approximately midway between Charlotte, North Carolina, and Atlanta, Georgia. Nearby Interstate Highway 85, Amtrak, and the Greenville-Spartanburg Airport link Clemson with major cities in the region. Clemson is 240 meters above sea level and has an average temperature of 17°C; the annual rainfall is about 130 centimeters. The University is located in the scenic foothills of the Blue Ridge Mountains on the sprawling 1,600-kilometer shoreline of Lake Hartwell.

The University

The campus consists of 2.4 square kilometers and represents an investment of approximately $270 million in academic buildings, student housing, and service facilities. It is surrounded by 81 square kilometers of farms and research lands for forestry, agriculture, and engineering. Clemson University is the state land-grant institution of South Carolina.

Applying

Applicants should have a B.S. degree from an accredited college or university in any of the major engineering disciplines. Students from chemistry or physics are also considered for admission if they demonstrate proficiency in certain prescribed engineering courses.

Direct admission to the Ph.D. degree (without completing the M.S. degree) is possible for exceptional students. All applicants must take the General Test of the GRE. A satisfactory score on the TOEFL is also required of international students whose native language is not English.

Correspondence and Information

Director
Materials Science and Engineering
201 Rhodes Research Center
Clemson University
Clemson, South Carolina 29634-0905

Telephone: 864-656-5559
Fax: 864-656-4466
E-mail: raw@ces.clemson.edu

Clemson University

THE FACULTY AND THEIR RESEARCH

S. Ahzi, Ph.D. Micromechanical modeling of materials behavior. (e-mail: sahzi@ces.clemson.edu)

A. Amirkhanian, Ph.D. Asphalt cement and asphalt concrete, portland cement, portland cement concrete, use of waste materials in highway construction. (e-mail: kcdoc@ces.clemson.edu)

J. M. Ballato, Ph.D. Optical properties of materials; photonic communication systems. (e-mail: jballat@ces.clemson.edu)

H. Behery, Ph.D. Fiber physics, yarn mechanics, geosynthetics, melt spinning. (e-mail: nassahb@ces.clemson.edu)

D. Carrol, Ph.D. Optical and electronic properties of nanophase materials and nanocomposites. (e-mail: dcarrol@clemson.edu)

J. A. Clayhold, Ph.D. Correlated electron systems, transport and thermal properties, superconductivity, magnetism, re-entrant superconductivity, precision measurements, thermoelectric effects. (e-mail: jclayho@clemson.edu)

S. E. Creager, Ph.D. Electrochemistry, electron transfer in monolayer films on electrodes, bioanalytical sensors using modified electrodes, electrochemical power sources, advanced batteries and fuel cells. (e-mail: screage@clemson.edu)

D. R. Dinger, Ph.D. Predictive process control, ceramic processing and modeling. (e-mail: ddennis@ces.clemson.edu)

M. J. Drews, Ph.D. Instrumental methods for polymer characterization. (e-mail: dmichae@ces.clemson.edu)

D. J. Dumin, Ph.D. New electron device/material interactions, electron device reliability. (e-mail: djdumin@ces.clemson.edu)

M. S. Ellison, Ph.D. Structure-property relationships in fiber-forming polymers. (e-mail: ellisom@ces.clemson.edu)

B. Goswami, Ph.D. Dynamics of processing and the mechanics of fibers, yarns, and fabrics. (e-mail: gbhuven@ces.clemson.edu)

R. V. Gregory, Ph.D. Conductive polymers. (e-mail: richar6@ces.clemson.edu)

M. Grujicic, Ph.D. Multilength scale computer simulation, processing, characterization and high-temperature materials. (e-mail: gmica@ces.clemson.edu)

B. Han, Ph.D. Electronic packaging. (e-mail: bhan@ces.clemson.edu)

D. N. Hon, Ph.D. Wood chemistry. (e-mail: dhon@clemson.edu)

S. J. Hwu, Ph.D. Synthesis of magnetic materials, thermoelectrical materials, optical materials. (e-mail: shwu@ces.clemson.edu)

P. F. Joseph, Ph.D. Linear and nonlinear fracture, contact mechanics, interface fracture, surface cracks, functionally graded materials, thermal barrier coatings. (e-mail: jpaul@ces.clemson.edu)

J. W. Kolis, Ph.D. Synthesis and chemistry of novel compounds with unusual structures and properties. (e-mail: kjoseph@ces.clemson.edu)

M. LaBerge, Ph.D. Tribology of artificial joints, design of alternative bearing materials, design and characterization of vascular stents. (e-mail: laberge@ces.clemson.edu)

R. A. Latour, Ph.D. Composite materials, orthopedic implant materials, macromolecular absorption. (e-mail: latourr@ces.clemson.edu)

B. I. Lee, Ph.D. Material synthesis, colloidal and sol-gel processing. (e-mail: lburtra@ces.clemson.edu)

H. D. Leigh, Ph.D. Refractory processing and utilization, production control techniques, high-temperature corrosion. (e-mail: lherber@ces.clemson.edu)

G. C. Lickfield, Ph.D. Synthesis and characterization of polymeric materials, polymer surface chemistry and modifications, computational chemistry. (e-mail: lgary@ces.clemson.edu)

J. R. Manson, Ph.D. Theory of surface characterization and analysis, theory of gas-surface interactions. (e-mail: jmanson@ces.clemson.edu)

R. K. Marcus, Ph.D. Plasma-based spectroscopic analysis of materials. (e-mail: marcusr@ces.clemson.edu)

P. J. McNulty, Ph.D. Radiation effect on thin gate oxides. (e-mail: mpeter@ces.clemson.edu)

A. Ogale, Ph.D. Processing and characterization of polymeric fibers, carbon fibers, and their composites. (e-mail: ogale@ces.clemson.edu)

W. T. Pennington, Ph.D. Synthesis and design of extended solids, X-ray diffraction. (e-mail: xraylab@ces.clemson.edu)

D. Perahia, Ph.D. Study of structure dynamics and forces of polymers and liquid crystals at interfaces. (e-mail: dperahi@ces.clemson.edu)

K. F. Poole, Ph.D. Processing and reliability of microelectronic devices. (e-mail: poole@ces.clemson.edu)

H. J. Rack, Ph.D. Phase transformation in titanium, metal matrix composites, deformation processing, fracture and fatigue, wear. (e-mail: rackh@ces.clemson.edu)

J. R. Ray, Ph.D. Molecular dynamics and Monte Carlo computer simulations of condensed-matter systems. (e-mail: jray@ces.clemson.edu)

S. Saha, Ph.D. Orthopedic biomechanics, biomaterials, bone mechanics, bioinstrumentation. (e-mail: ssaha@ces.clemson.edu)

R. W. Schwartz, Ph.D. Chemical-based processing routes to ceramics; thin-film fabrication; dielectric, ferroelectric, and piezoelectric ceramics and films; carbon fiber proc. (e-mail: schwar2@clemson.edu)

R. Singh, Ph.D. Low thermal budget processing of microelectronics, optoelectronics, and solar cells. (e-mail: srajend@ces.clemson.edu)

E. C. Skaar, Ph.D. Transport phenomena, automation and digital control, kinetics. (e-mail: ecskr@ces.clemson.edu)

D. Smith, Ph.D. Synthesis and properties of novel polymers, including fluoropolymers, polynapthlenes, and liquid crystalline thermosets for high-performance and optoelectronic applications. (e-mail: dwsmith@ces.clemson.edu)

D. E. Stevenson, Ph.D. Computational science and engineering, the science of developing and implementing scientific models on the computer. (e-mail: steve@ces.clemson.edu)

S. Stuart, Ph.D. Computational chemistry research on condensed phases at interfaces, using molecular dynamics and other simulation techniques. (e-mail: ss@ces.clemson.edu)

Y. Sun, Ph.D. Preparation and characterization of nanomaterials and electrical and nonlinear optical materials. (e-mail: syaping@clemson.edu)

T. D. Taylor, Ph.D. Mechanical properties of glass and ceramic materials, biocompatible glasses, glass-melting equipment, melting reactions: thermodynamics and kinetics. (e-mail: tdtyl@ces.clemson.edu)

G. X. Tessema, Ph.D. Superconductivity, low-dimensional materials. (e-mail: tguebre@ces.clemson.edu)

T. M. Tritt, Ph.D. Electronic and thermal transport related to solid-state materials for thermoelectric refrigeration and power generation applications. (e-mail: ttritt@ces.clemson.edu)

A. P. Wheeler, Ph.D. Development of biodegradable polymers, in particular polypeptides; study of biomineralized structures. (e-mail: wheeler@clemson.edu)

JOHNS HOPKINS UNIVERSITY

G. W. C. Whiting School of Engineering
Department of Materials Science and Engineering

Programs of Study

The Department of Materials Science and Engineering offers a graduate program that provides a broad-based, interdisciplinary approach to materials science and engineering, with a strong emphasis on research. The department is internationally recognized for its research programs in materials characterization, thin films, nanostructured materials, amorphous and metastable materials, computational materials science, and electronic materials. The department has also established a joint program in materials degradation and conservation with the Conservation Analytical Laboratory of the Smithsonian Institution in Washington, D.C. Because materials research is inherently interdisciplinary, the education and research activities of the department are enhanced by extensive collaborations with other departments. Students are encouraged to take advantage of the facilities and expertise that are available in other departments on campus. Graduate degrees in materials science, including the Ph.D., may be completed on either a full- or part-time basis. Students pursuing the Ph.D. on a part-time basis are required to spend at least one year in full-time residency on campus. The Master of Science in Engineering (M.S.E.) requires a research thesis and can be either a terminal program or preparation for doctoral study. The Master of Materials Science and Engineering (M.M.S.E.) is a part-time terminal degree program intended for practicing engineers, with classes primarily in the evening. Complete details are available under part-time programs in the engineering and applied science catalog.

Research Facilities

The department has facilities for synthesis, processing, characterization, and property measurements. Synthesis and processing facilities are available for casting and rapid solidification of metals and alloys, powder processing of metals and ceramics, microwave sintering, crystal growth, and sputtering, evaporation, and electrochemical deposition of thin films. Characterization facilities include scanning electron microscopy, scanning probe microscopy, transmission electron microscopy, field ion microscopy and spectroscopy, X-ray diffraction and small-angle scattering, X-ray photoelectron spectroscopy, Auger electron spectroscopy, and calorimetry and thermogravimetric analysis. Additional characterization capabilities include ultrasonics and acoustic resonance systems, laser ultrasonics equipment, microfocus X-ray imaging, topography and X-ray tomography, infrared imaging, eddy current imaging, impedance spectroscopy, and microwave reflectivity. Facilities are also available for the measurement of electrical, optical, mechanical, magnetic, and chemical properties of materials. Many other facilities are available through collaboration with government and industrial facilities off campus.

Financial Aid

Full-time graduate students are eligible for tuition remission. Full-time students pursuing a Ph.D. degree can receive an 80 percent tuition remission, with the student's research adviser paying the additional 20 percent of tuition and a stipend from their research contracts and grants. For entering full-time Ph.D. students, the stipend level is $1200 per month during the academic year and $1400 per month through the summer, for a total of $15,000 per year. Once Ph.D. students have completed all course requirements and passed the qualifying examination, the stipend is increased to $1400 per month year-round.

Cost of Study

Tuition for the 1998–99 academic year was $22,680. Modest yearly increases are to be expected.

Living and Housing Costs

The cost of living is comparable to that of other large northeastern American cities. Most students live in adjacent off-campus areas, either in privately run apartment facilities or in University subsidized housing. Additional information about residential facilities can be obtained from the Wolman Hall Housing Office, 3339 North Charles Street, Baltimore, Maryland 21218 (telephone: 410-516-7960).

Student Group

In 1998–99, there were 46 graduate students in the program; 4–8 students have been admitted annually in recent years. The Graduate Program in the School of Engineering has nearly 2,500 students.

Location

Johns Hopkins University is located in northwest Baltimore. Baltimore is one of the most dynamic parts of the fastest-growing high-technology corridor in America. Washington, D.C., to the south and both Philadelphia and New York to the north are easily accessible by car or train. Amtrak serves Baltimore through Penn Station, just south of Hopkins Homewood Campus. The Baltimore-Washington International Airport is located just a few miles south of the city.

The University and The Department

Johns Hopkins University is a private institution founded in 1876 as the first true American university on the European model: a graduate institution with an associated preparatory college. It boasts nearly eighty years of engineering history. The Department of Materials Science and Engineering is one of nine departments that comprise the G. W. C. Whiting School of Engineering at Hopkins.

Applying

To be admitted to a graduate program in the Department of Materials Science and Engineering, a student must submit a completed application form. Most applicants to the graduate program have undergraduate degrees in materials science and engineering, mechanical engineering, chemical engineering, or physics. Students with other science and engineering backgrounds are encouraged to apply to the programs as well. The application form can be obtained from the M.S.E. Web site (http://www.jhu.edu/~admis/grads/gradpage.html).

Correspondence and Information

Dr. Robert C. Cammarata, Chair
Graduate Program Committee
Department of Materials Science and Engineering
102 Maryland Hall
Johns Hopkins University
3400 North Charles Street
Baltimore, Maryland 21218-2689
Telephone: 410-516-5462
Fax: 410-516-5293
E-mail: rcc@jhu.edu
World Wide Web: http://www.jhu.edu/~matsci

Johns Hopkins University

THE FACULTY AND THEIR RESEARCH

Robert C. Cammarata, Professor; Ph.D., Harvard, 1985. Materials science of thin films and nanostructure materials, mechanical properties, phase transformations.

Robert E. Green Jr., Theophilus Halley Smoot Professor of Materials Science and Engineering; Ph.D., Brown, 1959. Materials science, solid-state physics, ultrasonics, acoustic emission, X-ray diffraction, electrooptical systems, nondestructive evaluation.

John D. Hoffman, Research Professor; Ph.D., Princeton, 1949. Crystallization of polymers, materials science and engineering.

Todd C. Hufnagel, Assistant Professor; Ph.D., Stanford, 1995. Solid-state phase transformations, amorphous and nanocrystalline materials, X-ray scattering, thin films, multilayers, magnetic materials.

Mo Li, Assistant Professor; Ph.D., Caltech, 1994. Computational materials science, defects, deformation mechanism of plasticity, fracture in glasses, atomistic simulation of nanocrystalline materials, glass transition, interfaces, irradiation effects on alloys and glasses, phase transition, simulation methods, software development for multimedia teaching tools.

Evan Ma, Associate Professor; Ph.D., Caltech, 1989. Nonequilibrium processing of materials, thermodynamics and kinetics of phase transformations, functional/structural materials for microelectromechanical systems (MEMS), thin films and surface modification.

Allan J. Melmed, Research Professor; Ph.D., Penn State, 1958. Atomic structure and composition analysis, nanograin material, thin-film composites, microscopy.

Dennis C. Nagle, Assistant Professor; Ph.D., Penn State, 1972. Materials science, ceramics, carbon/carbon composites, biomaterials, ceramic coatings, refractories, processing and characterization of materials.

Theodore O. Poehler, Research Professor and Vice Provost for Research; Ph.D., Johns Hopkins, 1961. Quantum electronics, solid-state physics.

Robert B. Pond Sr., Professor Emeritus; B.S., Virginia Tech, 1942. Physical metallurgy, materials science, solidification, superplasticity, solid mechanics.

Moshe Rosen, Professor; Ph.D., Weizmann (Israel), 1967. Materials science, phase transformations, elasticity and anelasticity, magnetoelasticity, ultrasonics, nondestructive evaluation.

Peter C. Searson, Professor; Ph.D., Manchester (England), 1982. Nanostructured materials, thin films, semiconductor surfaces and interfaces, electrochemical synthesis and modification of materials.

James B. Spicer, Assistant Professor; Ph.D., Johns Hopkins, 1991. Solid-state physics, materials science, optical nondestructive evaluation, process control, smart materials.

Timothy P. Weihs, Assistant Professor; Ph.D., Stanford, 1990. Processing and characterization of thin films, reactive multilayers, and microlaminates for structural applications: nanomechanical/micromechanical testing; biomaterials.

John M. Winter Jr., Associate Research Professor; Ph.D., Johns Hopkins, 1991. Electronic materials, physical metallurgy, X-ray diffraction, nondestructive evaluation.

MICHIGAN STATE UNIVERSITY

Department of Materials Science and Mechanics

Programs of Study

Graduate programs leading to the Master of Science and Doctor of Philosophy degrees in both materials science engineering and engineering mechanics are offered in the department. A wide range of course offerings are available in various colleges on campus, and an individualized graduate program can be developed with the assistance of a faculty adviser. For all fields, special emphasis is placed on the mastery of basic principles and methods. Courses and research are available in materials science and engineering: biomaterials, ceramic materials, composite materials, electron microscopy, high-temperature superconductors, impact damage, intermetallic alloys, laser processing of materials, mechanical and physical metallurgy, nondestructive evaluation, phase transformations, processing of ceramics, polymers and their composites, shape-memory and superelastic alloys, surface modification, and structural thin films and in mechanics engineering: applied mathematics, biomechanics, buckling, computational mechanics, continuum mechanics, dynamics, experimental mechanics, fracture mechanics, linear and nonlinear elasticity, mechanics of fatigue, mechanics of composite materials, micromechanics, optical methods of measurement, plasticity, damage mechanics, friction, thermoelasticity, vibration, and wave propagation.

Research Facilities

The department occupies approximately 20,000 square feet in the engineering building. In addition to several teaching laboratories, there are numerous experimental facilities for research in materials science, experimental mechanics, and biomechanics. Analytical facilities include an FTIR spectrometer, an optical metallograph with analysis system, SEM, FEG-SEM with digital image processing, TEM, a precision ion polishing system, an X-ray diffractometer, and an X-ray texture goniometer. Two laboratories devoted to surface modification are equipped with complete facilities for ion-beam-assisted deposition and ion implantation. Materials testing facilities include several servohydraulic and screw-drive loadframes, as well as a gas gun, an instrumented impact machine, microhardness testers, and units for creep and environmental effects testing. In NDE, facilities for acoustic emission, eddy-current, ESPI, ultrasonic scanning, and X-ray radiography are available. The department also houses facilities for strain and motion measurements using laser-optical techniques. The high-energy laser materials processing laboratory and biomechanics gait analysis laboratory are nationally recognized. The processing laboratory includes ceramic processing capabilities and hot and cold isostatic pressing capabilities. Faculty members and graduate students have access to the Composite Materials and Structures Center, which houses a wide variety of equipment for surface and interfacial analysis, experimental mechanics, and composite fabrication and characterization. The Case Center for Computer-Aided Engineering and Manufacturing provides an advanced computing environment for both graduate students and faculty members in the College of Engineering by virtue of its extensive array of computer hardware, software, and support services.

Financial Aid

Numerous fellowship and assistantship opportunities are available through the department, the College of Engineering, the Graduate Office, the Affirmative Action Program, the Composite Materials and Structures Center, and the Case Center for Computer-Aided Engineering and Manufacturing. Consideration for financial aid for highly qualified applicants is given as soon as the candidate has been accepted for admission in the graduate program. Early application for admission is highly recommended for candidates seeking financial assistance. All available funds for graduate support for the following academic year are usually finalized by March 15. Research assistantships are awarded only by individual faculty members and are competitive. Prospective students interested in a research assistantship should contact faculty members directly to discuss their research interests and the availability of funds. In 1998–99, annual half-time assistantship stipends were $15,192 for first-year M.S. students and $17,184 for senior M.S. students and Ph.D. students. Six credits of tuition per semester are waived for students with assistantships, and nonresident graduate assistants pay resident tuition on any additional credits. Registration/matriculation/infrastructure fees and health insurance coverage costs are included for students with assistantships. Fellowships range from $100 per semester to $17,184 per calendar year.

Cost of Study

In 1998–99, graduate tuition was $222.50 per semester credit for Michigan residents and $450 per semester credit for nonresidents. In addition, matriculation fees of $288 for students enrolling for more than 4 credits ($238 for students enrolling for 4 or fewer credits) are assessed each semester. Matriculation fees include the student information technology fee and the infrastructure/technology support fee. A $237 engineering program fee for students enrolling for more than 4 credits ($131 for students enrolling for 4 or fewer credits), a $3 FM radio tax, a $9 student information technology fee, a $4.25 student newspaper tax (for students enrolling for 10 or more credits), and a $4.50 Council of Graduate Students tax are also assessed.

Living and Housing Costs

Owen Graduate Residence Hall offers comfortable living in an atmosphere conducive to advanced study and the exchange of ideas. The 1998–99 rates per semester were $1928 for single occupancy and $1649 for double occupancy, including residence hall tax. This included a $280 credit toward food purchases. Furnished University apartments are available for married students. The 1998–99 family monthly rates were $420 for a one-bedroom apartment and $465 for a two-bedroom apartment.

Student Group

For the 1998 fall semester, 43,189 students were enrolled at Michigan State University's East Lansing campus; 33,419 were undergraduates, 6,472 were graduate students, and 1,932 were professional students. Students come from all over the United States and about 110 other countries. During this time period, 73 students were enrolled in M.S. and Ph.D. programs in materials science and mechanics.

Location

Michigan State University is located in the south-central part of Michigan, in East Lansing, which is approximately 80 miles northwest of Detroit and about 210 miles northeast of Chicago. East Lansing is a city of about 30,000 people in a metropolitan area of more than 270,000. Lansing, the capital of Michigan, is located just 4 miles to the west. Many summer and winter recreational opportunities are located within driving distance.

The University

Michigan State University, a pioneer land-grant A.A.U. institution, was founded in 1855. The curriculum includes more than 200 programs of undergraduate and graduate study, taught by more than 4,000 faculty and academic staff members in fourteen degree-granting colleges. The property holdings at East Lansing number 5,198 acres. Of this total, about 2,100 acres are in existing or planned campus development; the remaining 3,098 acres are devoted to experimental farms, outlying research facilities, and natural areas. Major buildings number about 150. The contiguous campus is noted for its beauty in all seasons of the year.

Applying

Application forms and instructions for submitting transcripts, letters of recommendation, and related materials may be obtained from the department. Study may begin in any semester. Applications are accepted at any time but should be received at least nine months prior to the expected starting semester. The GRE is required. A minimum TOEFL score of 570 is required for international students.

Correspondence and Information

Graduate Coordinator
Department of Materials Science and Mechanics
Michigan State University
East Lansing, Michigan 48824-1226

Telephone: 517-355-5141
Fax: 517-353-9842
E-mail: msmgradv@egr.msu.edu
World Wide Web: http://www.egr.msu.edu/MSM

Michigan State University

THE FACULTY AND THEIR RESEARCH

Nicholas J. Altiero, Professor and Chairperson; Ph.D., Michigan, 1974. Fracture mechanics, finite-element and boundary-element methods, biomechanics.

Ronald C. Averill, Associate Professor; Ph.D., Virginia Tech, 1992. Computational mechanics, finite-element method, composite plates and shells, failure of composites, crash analysis, micromechanics, optimal design.

Virginia M. Ayres, Associate Professor; Ph.D., Purdue, 1985. Electronic materials, diamond and amorphous tetrahedral carbon films, cold cathodes, scanning probe microscopies (AFM, STM, ScThn), FTIR, Raman, X-ray diffraction.

Thomas R. Bieler, Associate Professor; Ph.D., California, Davis, 1989. High-temperature deformation (creep, superplasticity, hot working), texture, Al, Ti, TiAl, lead-free solder alloys, electron microscopy.

Eldon D. Case, Associate Professor; Ph.D., Iowa State, 1980. Microcracking in ceramics; thermal and mechanical fatigue; ceramic composites; microwave sintering, joining, and binder-burnout of ceramics.

Gary L. Cloud, Professor; Ph.D., Michigan State, 1966. Experimental mechanics, optical techniques in strain measurement, fracture, fasteners, mechanics of composite materials, NDI-NDE.

Martin A. Crimp, Associate Professor; Ph.D., Case Western Reserve, 1987. Transmission electron microscopy, diffraction studies in scanning electron microscopy, dislocation substructures, intermetallic alloys, structure of magnetic multilayers.

Melissa J. Crimp, Assistant Professor; Ph.D., Case Western Reserve, 1988. Ceramic bone substitutes, colloidal processing techniques, processing/property relationships of ceramic matrix composites, cartilage regeneration, mechanical properties of biomaterials.

Lawrence T. Drzal, University Distinguished Professor and Director of the Composite Materials Center; Ph.D., Case Western Reserve, 1974. Adhesion, polymer-matrix composites, adhesion bonding, surface modification of polymers, powder composite processing, UV treatment of surfaces, coatings.

Erik D. Goodman, Professor and Director of the Case Center for Computer-Aided Engineering and Manufacturing; Ph.D., Michigan, 1972. Genetic algorithms and genetic programming in engineering design, improving performance of global engineering teams.

David S. Grummon, Professor; Ph.D., Michigan, 1986. Thin film shape-memory materials, surface modification, ion implantation, fatigue and fatigue crack initiation.

Roger C. Haut, Professor; Ph.D., Michigan State, 1971. Experimental biomechanics, injury biomechanics.

Tim Hogan, Assistant Professor; Ph.D., Northwestern, 1996. Charge transport characterization of electronic materials, pulsed laser deposition of novel electronic materials and quantum well structures, photolithographic processing of thin films for device fabrication and testing.

Robert P. Hubbard, Professor; Ph.D., Illinois, 1970. Biomedical measurement and modeling, biomechanics of injury, product design based on biomechanics, especially for seating and injury reduction.

Patrick Y. Kwon, Assistant Professor; Ph.D., Berkeley, 1994. Material issues in design and manufacturing, tool wear, machining, design and manufacturing automation and integration.

Andre Y. Lee, Associate Professor; Ph.D., Illinois, 1987. Thermodynamics and viscoelastic responses of polymers and polymer composites, rheology of inorganic-organic hybrid polymers.

Dahsin Liu, Professor; Ph.D., Virginia Tech, 1984. Mechanics of composite materials, micromechanics, impact dynamics, experimental mechanics, nonlinear computational mechanics.

James P. Lucas, Associate Professor; Ph.D., Minnesota, 1981. Environmental effects on materials behavior, composite materials, nanocharacterization of materials, Pb-free solders.

John F. Martin, Associate Professor; Ph.D., Illinois, 1973. Fatigue and fracture, cyclic plasticity, mechanics of materials.

Darren E. Mason, Assistant Professor; Ph.D., Minnesota, 1996. Phase transitions, variational methods, modeling of microstructure, mechanics of polycrystals.

Kali Mukherjee, University Distinguished Professor; Ph.D., Illinois, 1963. Phase transformation; laser processing of advanced materials; laser machining of wood, composites, and high T_c superconductors.

Jun Nogami, Associate Professor; Ph.D., Stanford, 1986. Scanning tunneling microscopy, electronic materials, epitaxial growth, nanostructures.

Cordell M. Overby, Associate Adjunct Professor and Assistant Vice President for Research Services; Sc.D., George Washington, 1986. Manufacturing systems, systems analysis.

Thomas J. Pence, Professor; Ph.D., Caltech, 1983. Linear and nonlinear elasticity, buckling and stability, phase transformations, composites, wave propagation in solids, mechanics of thin films.

Robert Wm. Soutas-Little, Professor; Ph.D., Wisconsin, 1962. Biodynamics of human movement, tissue biomechanics, continuum mechanics, boundary value problems in elasticity.

K. N. Subramanian, Professor; Ph.D., Michigan State, 1966. Mechanical properties of materials, composites, solders.

Hung-yu Tsai, Assistant Professor; Ph.D., Cornell, 1994. Continuum mechanics, elasticity, friction, mathematical modeling of phase transitions, active composites.

R. Lal Tummala, Professor; Ph.D., Michigan State, 1970. Robotics and manufacturing.

AREAS OF CURRENT RESEARCH

Polymers and Polymer Composites. Adhesive bonding, buckling and stability, crashworthiness simulation, dynamics of polymer melts, environmental effects, fracture and fatigue, interface and micromechanics, interferometric NDT, mechanical fasteners, theories of laminated composites, powder processing, processing of composites, random media and composites, repair techniques, structural performance of polymer glasses.

Ceramics and Ceramic Composites. Fracture and microcracking, impact damage, mechanical and thermal properties, processing behavior, high-temperature superconductor processing.

Metals and Metal Matrix Composites. Fabrication, high-temperature deformation, high strain rate deformation, micromechanical fracture and fatigue, welding and joining, surface treatment, properties of intermetallic alloys, superplastic forming of high-temperature alloys, texture analysis.

Thin Films, Superconductivity, and Phase Transformations. Diamond thin films, superconducting thin films, ion-beam-assisted deposition, phase transformations, mathematical analysis of phase transitions, superelasticity and shape-memory effects.

High-Energy Laser Processing of Materials. Laser machining of wood and composites, welding and cutting of metal matrix composites, laser-assisted thin-film high-temperature superconductor processing.

Computational Mechanics. Boundary integral methods, finite-element methods, inverse techniques.

Experimental Mechanics. Moire interferometry, video holography, laser Doppler, laser shearography, speckle methods, software development, three-dimensional fracture, geomechanics measurements.

Theoretical Mechanics. Buckling and stability, continuum mechanics, mechanics of fatigue, fracture mechanics, friction, linear elasticity, mechanics of composite materials, micromechanics, nonlinear elasticity, phase transitions, plasticity, plates and shells, wave propagation.

Biomaterials and Biomechanics. Athletic biomechanics, biodynamics, biological tissue mechanics, electrodiagnosis, impact-trauma biomechanics, mechanics of injury, spinal motion.

Manufacturing. Laser applications, machining processes, powder metallurgy, process management.

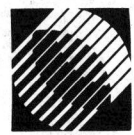

OREGON GRADUATE INSTITUTE OF SCIENCE AND TECHNOLOGY

Department of Materials Science and Engineering

Programs of Study

The Department of Materials Science and Engineering (MSE) offers M.S. and Ph.D. degree programs with flexible formats and broad research activity areas. Course requirements are designed to encourage broad as well as specialized expertise in materials science and engineering. The course of study is individually planned, and all students are advised by a department faculty-student program committee. Beginning in 1999, the department begins a three-year reorganization process, and prospective students are encouraged to explore joint or collaborative programs with MSE and Oregon Graduate Institute's (OGI) Department of Electrical and Computer Engineering (ECE). Students with undergraduate or advanced degrees in materials science, chemistry, physics, or related sciences, as well as in all branches of engineering, are invited to apply for admission.

Ph.D. students complete a minimum of 60 credit hours of course work during their stay at OGI. Students are admitted to candidacy upon successful completion of a comprehensive exam (consisting of both oral and written sections), which is administered within two years of being admitted to OGI. Upon successful completion of this exam, a student may commence work on a research dissertation topic. Ph.D. students are expected to complete their studies and dissertation work in three to four years.

M.S. students must complete 44 credit hours for their degree, usually in two years of full-time status. There are several options in the M.S. program. A thesis M.S. may be achieved through 28 credit hours of course work and 16 credit hours of thesis research, resulting in a written thesis. Nonthesis options are also available, with a degree being granted after 44 credit hours of course work or by taking 36 credit hours of course work and 8 credit hours of project research, which culminates in a written project report. M.S. students have the opportunity to focus their curriculum in such areas as electronic materials and packaging. The research programs emphasize the solidification and welding of engineering alloys, deposition and characterization of coatings and thin films, tribological characterization of materials, fracture characteristics, corrosion and stress corrosion phenomena, computational modeling of materials processing and materials transformations, defect structure and failure analysis of electronic materials, microchemistry and microstructural characterization of conventional and electronic materials, and production and characterization of nanoscale materials.

Research Facilities

The department has a number of unusual facilities that enhance its ability to conduct generic or company-specific research and development as well as technology transfer. Microstructural analysis facilities include Zeiss and Nikon research optical metallographs, a quantitative image analysis computer, a Zeiss 960 digital scanning electron microscope with a Tracor Northern 5500 Quantitative EDS, a Hitachi H-800 analytical scanning transmission electron microscope with a Tracor Northern 5500 Quantitative EDS and Gatan EELS, JEOL 2000 FX, ESI FIB micromachining work station, and facilities for sample preparation and darkroom printing. Alloy development facilities include equipment for small-scale (50–1,000 gm) induction/vacuum arc melting, pilot-scale (15–180 kg) vacuum arc remelting, and pilot-scale (15–550 kg) electroslag remelting. Mechanical testing equipment includes a 100-kip computerized Instron servohydraulic system, a 20-kip Instron servohydraulic system, 300-kip MTS servohydraulic system, a 120-kip SATEC test system, a 264-ft-lb instrumented impact test system, a 2,000-ft-lb instrumented impact drop tower, and 25,000-psi hydraulic test capability. Computational materials science research is conducted mainly with the aid of ANSYS and SYSWELD, commercial finite element analysis (FEA) programs, as well as an in-house custom FEA program. Computer platforms include an HP 9000-730 workstation, two IBM RS6000 workstations, an Apollo Domain 3000 workstation, X-terminals, Pentium II workstations, and other peripheral equipment, including a Tektronix Phaser III color printer. The department has access to other computing facilities on campus via Ethernet as well as to National Science Foundation supercomputer sites. Wear testing facilities include equipment to test surface profilometry and rolling contact, erosion, cavitation-erosion, and dry-sand rubber-wheel and pin-on-drom abrasive test systems. Simulation work is carried out on a Gleeble Thermal/Mechanical test system. Thermal spray research is conducted using a METCO flame spray gun and a Plazjet 200-KW high-energy hypersonic plasma facility (powder or wire, simultaneous feed). There is an assortment of furnaces for heat treatment work. The department also has a multichannel computer-controlled acoustic emission system. Sectioning ability on site utilizes abrasive cutoff saws, bandsaws, and oxyfuel cutting equipment. A Gleeble Thermal Simulator, DuPont Thermal Analysis System (1600 degrees C DTA, DSC), and Cahn TGA are used to perform thermal analysis research. The department is able to perform hydrostatic burst testing. Joining research includes complete facilities for electroslag welding, electroslag surfacing, submerged arc welding, GMAW, GTAW, and capacitive discharge processes; Varestraint, Lehigh, Tekken, and Implant testing; and consumable development. The department also has cold/hot pressing capabilities, sintering furnaces, isostatic presses, ball mills, and a dewaxing furnace.

Financial Aid

Fellowships and research assistantships are available. Full stipend awards for full-time Ph.D. students are most common, and some partial awards and tuition scholarships are also given.

Cost of Study

Tuition for 1999–2000 is $4465 per academic quarter, or $495 per credit hour.

Living and Housing Costs

OGI does not offer on-campus housing. Typical rentals in the area range between $550 and $800 for a two-bedroom apartment. More housing information is available from OGI's Office of Academic and Student Services.

Student Group

OGI enrolls about 350 students. Approximately 25 percent of the students are women, and approximately 45 percent of matriculated students are from other countries. Recent graduates have found employment as faculty members, researchers and regulators, and consultants.

Location

Situated in the beautiful Pacific Northwest, OGI is within a day's drive of the coast, mountains, and several national parks, making recreational opportunities varied and plentiful.

The Institute

OGI is a private, graduate-only institute dedicated to contemporary scientific and business research and education. Founded in 1963, OGI combines the vigorous research emphasis of a large university with the characteristics of personal interaction and collaboration found in a small research institute. OGI also offers Ph.D. and M.S. degrees in computer science and engineering, electrical engineering, environmental science and engineering, and biochemistry and molecular biology and M.S. degrees in management in science and technology and computational finance.

Applying

Applications are accepted at any time, but those requesting financial assistance should be submitted by March 1. Required admissions materials include three letters of recommendation, transcripts, and GRE test scores. The OGI institutional code for the GRE is 4592. A TOEFL score of at least 550 is required of applicants whose native language is not English. The TOEFL may be waived if the applicant received a prior degree in the United States. Due to the upcoming departmental reorganization, students are encouraged to discuss their particular areas of interest with the department or admissions office.

Correspondence and Information

Admissions
Office of Academic and Student Services
Oregon Graduate Institute
P.O. Box 91000
Portland, Oregon 97291-1000

Telephone: 503-748-1027
          800-685-2423 (toll-free)
E-mail: admissions@admin.ogi.edu
World Wide Web: http://www.ogi.edu

Oregon Graduate Institute of Science and Technology

THE FACULTY AND THEIR RESEARCH

David G. Atteridge, Professor of Materials Science and Engineering; Dr.Eng. (materials science), Berkeley, 1975. Measurement and modeling of thermomechanically induced material property changes, physical/mechanical metallurgy, processing/fabrication, microstructural evolution/prediction, production and characterization of nanoscale materials.

Martin Becker, Professor of Materials Science; Ph.D. (nuclear engineering), MIT, 1964. Radiation effects on semiconductors, linear and nonlinear diffusion and transport processes, nanoscale materials, mechanics and properties of coatings and thin films, heat transfer, residual stress and microstructure in welding, computational methods.

Paul Clayton, Professor of Materials Science and Provost; Ph.D. (metallurgy), Brunel (England), 1977. Wear, surface engineering, tribological aspects of wheel/rail contact, wood cutting, pulp refining.

Jack H. Devletian, Professor of Materials Science; Ph.D. (metallurgical engineering), Wisconsin, 1972. Solidification mechanics, physical metallurgy of weldments.

Jack M. McCarthy, Assistant Professor of Materials Science and Engineering; Ph.D. (materials science), Oregon Tech, 1996. Analytical electron microscopy, electronic materials, high-temperature alloys.

Lemmy L. Meekisho, Associate Professor of Materials Science and Engineering; Ph.D. (mechanical engineering), Carleton, 1988. Numerical analysis, modeling of materials processing.

Milton R. Scholl, Associate Professor of Materials Science and Engineering; Ph.D. (materials science and engineering), Oregon Graduate Center, 1987. Tribology and surface engineering, abrasive wear of materials, weld hardfacing alloy development, wear phenomena in pulp and paper processing, functional characterization of structure/performance relationships in thermally sprayed coatings, including nanoscale coatings.

William E. Wood, Professor of Materials Science and Engineering; Dr.Eng. (materials science), Berkeley, 1973. Structure-property relationships of materials, physical metallurgy, materials characterization, fracture, acoustic emission.

Margaret Ziomek-Moroz, Associate Professor of Materials Science and Engineering; Ph.D. (physical chemistry), Polish Academy of Sciences, 1986. Corrosion of metals, alloys, and advanced materials in aqueous solutions; deterioration of composite materials; synthesis of superconducting materials; electrochemical measurement techniques.

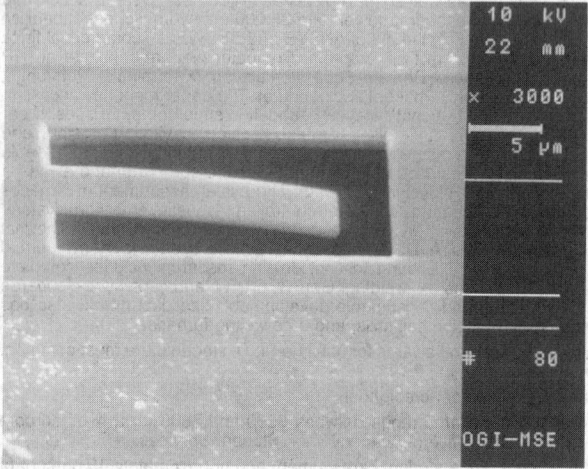

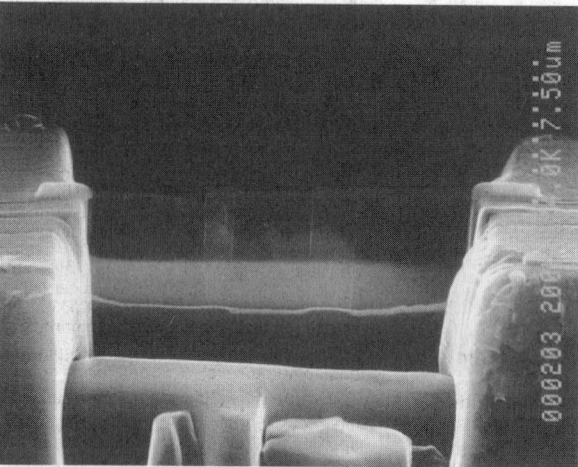

Sub-micrometer thickness cantilever beam machined from single crystal silicon on a 100 surface following aluminum film deposition, showing deflection of the cantilever due to residual stresses in the aluminum film.

Electron transparent 5 by 20 micrometer wall of single crystal silicon machined in a focused ion beam milling workstation for examination in the transmission electron microscope.

RENSSELAER POLYTECHNIC INSTITUTE

Department of Materials Science and Engineering

Programs of Study

The department offers programs leading to the M.S., M.Eng., D.Eng., and Ph.D. degrees in materials science and engineering. Active areas of coordinated faculty research include electronic materials, nanostructured materials, composite materials and structures, glass and ceramic science, and materials processing. Individual faculty program areas include melting and solidification, mechanical behavior, functional nanostructures, corrosion, welding and joining, and interface and surface science.

The M.S. degree is awarded after completion of 24 credits of course work and a 6-credit thesis. The M.Eng. degree requires 30 credits of course work and is a nonthesis degree intended for professional practice. A student with an accredited B.S. or its equivalent can typically complete this degree in one year. The doctoral programs require a total of 90 credits, of which at least 45 must be course credits, with the remainder assigned to a thesis based on original research. Admission to the doctoral programs is based on passing a comprehensive qualifying examination, generally taken after three or four semesters in residence.

Research Facilities

The bulk of the department's research facilities are housed in the 45,000-square-foot Materials Research Center. Central facilities include X-ray, electron optical, and surface analysis laboratories; a metallography laboratory; and mechanical testing facilities. This building also houses the Center for Glass Science and Technology, as well as specialized laboratories used in a number of individual faculty programs. Work in electronic materials is carried out in the Center for Integrated Electronics and Electronics Manufacturing, which includes a 10,000-square-foot, class-100 clean room. Research in composite materials is carried out in the Center for Composite Materials and Structures, which includes facilities for the fabrication and testing of a wide range of high-temperature composites.

Research is supported by such state-of-the-art facilities as the computing facilities in the Center for Industrial Innovation, which provides graduate students with walk-in access to workstations and to programs ranging from personal productivity aids to advanced computer-aided design and library analysis packages; the Rensselaer Libraries, whose electronic information systems provide access to collections, databases, and Internet resources from campus and remote terminals; the Rensselaer Computing System, which permeates the campus with a coherent array of advanced workstations, a shared toolkit of applications for interactive learning and research, and high-speed Internet connectivity; a visualization laboratory for scientific computation; and a high-performance computing facility that includes a 36-node SP2 parallel computer.

Financial Aid

Financial aid is available to full-time students in the form of fellowships, research or teaching assistantships, and scholarships. The stipend ranged between $9100 and $10,000 for the nine-month academic year in 1998–99. In addition, full tuition is usually granted. Additional compensation for study during the summer months may also be available. Outstanding students may qualify for University-supported Rensselaer Scholar Fellowships, which provide a $15,000 stipend plus tuition. Low-interest, deferred-repayment graduate loans are also available to U.S. citizens with demonstrated need.

Cost of Study

Tuition for 1999–2000 is $665 per credit hour. Other fees amount to approximately $535 per semester. Books and supplies cost about $1700 per year.

Living and Housing Costs

The cost of rooms for single students in residence halls or apartments ranges from $3356 to $5298 for the 1999–2000 academic year. Family student housing is available, with monthly rents ranging from $592 to $720.

Student Group

There are about 4,300 undergraduates and 1,750 graduate students at Rensselaer, representing all fifty states and more than eighty countries.

Student Outcomes

Eighty-eight percent of Rensselaer's 1998 graduating students were hired after graduation with starting salaries that averaged $56,259 for master's degree recipients and $57,000 to $75,000 for doctoral degree recipients.

Location

Rensselaer is situated on a scenic 260-acre hillside campus in Troy, New York, across the Hudson River from the state capital of Albany. Troy's central Northeast location provides students with a supportive, active, medium-sized community in which to live; an easy commute to Boston, New York City, and Montreal; and some of the country's finest outdoor recreation sites, including Lake George, Lake Placid, and the Adirondack, Catskill, Berkshire, and Green Mountains. The Capital Region has one of the largest concentrations of academic institutions in the United States. Sixty thousand students attend fourteen area colleges and benefit from shared activities and courses.

The University and The School

Founded in 1824 and the first U.S. college to award degrees in engineering and science, Rensselaer Polytechnic Institute today is accredited by the Middle States Association of Colleges and Schools and is a private, nonsectarian, coeducational university. All engineering graduate and undergraduate degree programs are also accredited by the Accreditation Board for Engineering and Technology (ABET). Rensselaer has five schools—Architecture, Engineering, Management, Science, and Humanities and Social Sciences. The School of Engineering is ranked among the top twenty engineering schools in the nation by the *U.S. News & World Report* survey and is ranked in the top ten by practicing engineers.

Applying

Admission applications and all supporting credentials should be submitted well in advance of the preferred semester of entry to allow sufficient time for departmental review and processing. The application fee is $35. Since the first departmental awards are made in February and March for the next full academic year, applicants requesting financial aid are encouraged to submit all required credentials by February 1 to ensure full consideration.

Correspondence and Information

For written information:
Prof. Roger Wright
Coordinator of Graduate Admissions
Materials Science and Engineering Department
Rensselaer Polytechnic Institute
110 8th Street
Troy, New York 12180-3590
Telephone: 518-276-6372
E-mail: wrighr@rpi.edu
World Wide Web: http://www.eng.rpi.edu:80/
  dept/materials/

For applications and admissions information:
Director of Graduate Academic and Enrollment
  Services, Graduate Center
Rensselaer Polytechnic Institute
110 8th Street
Troy, New York 12180-3590
Telephone: 518-276-6789
E-mail: grad-services@rpi.edu
World Wide Web: http://www.rpi.edu

Rensselaer Polytechnic Institute

THE FACULTY AND THEIR RESEARCH

Pulickel M. Ajayan, Assistant Professor of Materials Engineering; Ph.D., Northwestern, 1989. Carbon-based nanostructures, applications, electron microscopy.

Chan I. Chung, Professor of Polymer Engineering; Ph.D., Rutgers, 1969. Polymer processing, polymer melt rheology, block copolymers and blends, recycling binder-assisted powder processing.

Robert H. Doremus, New York State Professor of Glass and Science and Technology; Ph.D., Cambridge, 1953; Ph.D., Illinois, 1956. Glass science, crystallization, diffusion, crystal growth, ceramic processing, biomaterials.

David J. Duquette, Professor of Materials Engineering; Ph.D., MIT, 1968. Environmental and surface effects on the mechanical behavior of metals, corrosion, stress corrosion, corrosion fatigue, elevated temperature mechanical behavior of metallic materials.

Martin E. Glicksman, John Todd Horton Professor of Materials and Chemical Engineering; Ph.D., Rensselaer, 1970. Solidification processing of crystal growth, kinetics of phase transformations; purification of materials; measurement and modeling of microstructure evolution, microgravity research.

William B. Hillig, Research Professor; Ph.D., Michigan, 1953. Ceramic composites, behavior of structural ceramics and fibers at very high temperatures; strength-controlling processes in glasses and other brittle materials; kinetics of nucleation and crystal growth from liquids; coatings.

John B. Hudson, Professor of Materials Engineering; Ph.D., Rensselaer, 1960. Physics and chemistry of surfaces, gas-surface interactions and stability of interfaces, kinetics of thin-film growth by chemical vapor deposition, analysis of surfaces and films.

Pawel Keblinski, Assistant Professor of Materials Science; Ph.D., Penn State, 1995. Atomic-level modeling of materials, interfacial phenomena in metals, semiconductors and ceramics, structure-property relationship.

Matthew B. Koss, Research Assistant Professor; Ph.D., Tufts, 1990. Dendritic solidification, microgravity space flight experimentation and telescience.

Robert W. Messler Jr., Associate Professor of Materials Engineering and Director of the Warren F. Savage Materials Joining Laboratory; Ph.D., Rensselaer, 1970. Physical metallurgy of welding, brazing, and SHS joining; hybrid joining processes; innovative mechanical fastening; automation and intelligent control of welding; NDE of welds and weld defect prevention; environmentally sound joining.

Cornelius T. Moynihan, Professor of Materials Engineering; Ph.D., Princeton, 1965. Glass science, IR-transmitting fluoride glasses, electrical properties, structural relaxation, viscosity, crystallization, chemical durability.

Shyam P. Murarka, Professor of Materials Engineering and Director of the Center for Integrated Electronics; Ph.D., Agra, 1970; Ph.D., Minnesota, 1970. Metallization for the submicron SIC, silicides, metal-semiconductor interactions; low-temperature processing; rapid thermal annealing; dry etching processes, deposition and characterization of dielectric films, growth of thin insulating films, diffusion and defects in semiconductors and processes that reduce or eliminate defects.

Krishna Rajan, Professor of Materials Engineering; Sc.D., MIT, 1978. Electron microscopy and microanalysis, electronic/optoelectronic materials, thin films and superlattices, interfacial structure/property relationships in materials.

Ganapathiraman Ramanath, Assistant Professor of Materials Engineering; Ph.D., Illinois, 1996. Thin-film electronic materials: interconnects, diffusion barriers, low-k dielectrics; characterization of interfacial reactions, kinetics and mechanisms of microstructure and phase evolution during deposition and annealing; processing self-organized structures for microelectronics applications.

Eugene J. Rymaszewski, Research Professor of Materials Engineering and Associate Director of the Center for Integrated Electronics; Dipl. Ingenieur, Technical (Munich), 1950. Interdisciplinary aspects of microelectronics technology; VLSI chips and their packaging; relations of electrical, thermal, and topological characteristics of data processing equipment.

Linda Schadler, Assistant Professor of Materials Science; Ph.D., Pennsylvania, 1990. Polymer-based composites, interfaces, micro-Raman spectroscopy.

Richard W. Siegel, Robert W. Hunt Professor of Metallurgical Engineering and Department Head; Ph.D., Illinois, 1965. Nature and physical properties of defects in metals, atomic diffusion; synthesis and processing, characterization, and properties of nanophase materials, including ceramics and metals.

Christoph O. Steinbruchel, Associate Professor of Materials Engineering; Ph.D., Minnesota, 1974. Plasma and ion beam processing of microelectronic materials; thin films, surface science; sensor materials, micromachining and micromechanics.

Sanford S. Sternstein, William Weightman Walker Professor of Polymer Engineering and Director of the Center for Composite Materials and Structures; Ph.D., Rensselaer, 1961. Physical properties of polymers; rubber elasticity theory; fracture, yielding, and craze formation in glassy polymers and composites; viscoelastic properties; swelling in filled elastomers; high-performance composites.

Minoru Tomozawa, Professor of Materials Engineering; Ph.D., Penn State, 1968. Properties and structure of glasses; phase separation and crystallization; low-angle X-ray scattering and light scattering; electrical, chemical, and mechanical properties.

Roger N. Wright, Professor of Materials Engineering and Director of the High Temperature Technology Extension Program; Ph.D., MIT, 1969. Thermal and mechanical processing of metals, plastic flow and fracture of metals.

SOUTH DAKOTA SCHOOL OF MINES AND TECHNOLOGY

Materials Engineering and Science Program

Program of Study

The South Dakota School of Mines and Technology offers interdisciplinary graduate education and research in materials engineering and science (MES) leading to both M.S. and Ph.D. degrees. Individuals with baccalaureate degrees in chemistry, physics, metallurgy, and most types of engineering should be prepared for graduate study in the broad MES field. Students pursuing these MES degrees will have an opportunity to expand their knowledge and understanding of the science and technology of materials synthesis, behavior, and applications. Graduates of the MES programs should be capable of formulating innovative solutions to materials-based problems through the use of multidisciplinary knowledge gained from a combination of MES classroom and laboratory experiences in basic engineering and science. The M.S. degree program contains two options. In the thesis option, a minimum of 24 semester credit hours of course work and 6 credit hours of thesis research are required. In the nonthesis option, a minimum of 32 credit hours of course work and a supervised project are required. Every M.S. student is expected to complete at least 12 credit hours from core courses in atomic and molecular structure of materials, thermochemical processing fundamentals, structure-property relationships of materials, and advanced instrumental analysis of materials. These core courses are modularized so that students can select their 12 core credits utilizing those modules of most benefit to their specific programs of study. Within the Ph.D. program, candidates may select either a science or engineering emphasis. The Ph.D. program of study requires a minimum of 80 total semester credit hours, of which a minimum of 60 credits must be approved course work. However, a maximum of 24 credits of related course work from a prior master's degree may be applied to the Ph.D. credit requirement, although the master's degree is not required for admission. Doctoral students must pass a qualifying examination, a comprehensive examination, and an oral defense of their dissertation.

Research Facilities

Equipment available on campus includes an electron microprobe; scanning electron microscopes; an atomic adsorption spectrometer; X-ray diffraction equipment; a transmission electron microscope; AA/ICP; NMR, FT-IR, FT-Raman, and UV/visible spectrometers; a mass spectrometer; a microcalorimeter; gas chromatographs; a contact angle goniometer; an interfacial force microscope; a scanning laser-doppler velocimeter; and IBM RISC-6000 and Silicon Graphics workstations.

The Devereaux Library collection contains more than 180,000 books, more than 800 active periodicals, and about 209,000 microforms. Approximately 2,500 books are added annually. The library provides access to hundreds of online databases through utilities such as Dialog and WILSONLINE as well as numerous CD-ROM products. Access to a multitude of international and domestic library databases and journals is provided through the CARL UnCover System. Numerous government publications are received through the library's depository designation by the U.S. Government Printing Office.

Financial Aid

Qualified applicants are eligible for fellowships and graduate assistantships. Such awards are usually made on the basis of scholastic merit. Assistants and fellows must be registered for at least 9 credit hours to be eligible for reduced tuition. Graduate students may be eligible for other forms of financial aid such as Federal Stafford Student Loans, National Direct Student Loans, or Federal Work-Study programs. Typical compensation for a graduate student on a full-time assistantship ranges from about $8000 for nine months to about $16,000 for twelve months.

Cost of Study

Tuition for 1999–2000 is $85.25 per credit hour for state students and $251.45 for nonresident students. However, students receiving a minimum assistantship of $1820 and taking at least 9 credits per semester are eligible for a tuition rate of $28.42 per credit. Student fees range from $50 to $95 per credit hour.

Living and Housing Costs

At SDSM&T, assistance in finding off-campus rooms and apartments is available from the director of housing. Off-campus rooms range from $75 to $100 per week. Apartments rent for a minimum of $275 per month. On-campus board is payable by the meal or is available through various plans that range from $476 to $838 per semester for dormitory residents. When available to graduate students, dormitory rooms cost $658 per semester for double occupancy or $875 for a single room.

Student Group

Approximately 2,500 students are enrolled in engineering and science degree programs at the South Dakota School of Mines and Technology. About 250 of these are graduate students. Currently, 22 students are enrolled in the Materials Engineering and Science program.

Student Outcomes

Graduates with an M.S. in MES typically pursue applied research and management careers with industrial companies that focus on polymers, semiconductors, thin films, concrete technology, ceramics, and other multicomponent materials. Most graduates with a Ph.D. in MES proceed to careers in either academic institutions or national and international research laboratories.

Location

Rapid City (area population 70,000) is located on the east side of the Black Hills, on the banks of Rapid Creek, at an elevation of 3,200 feet. The Black Hills is a forested range measuring 100 miles by 50 miles and rising to 7,200 feet at Harney Peak. Mount Rushmore, a granite carving of 4 U.S. presidents, is 30 miles southwest of Rapid City in an area of state parks and national monuments and parks. The Black Hills is a popular tourist destination in the summer, offering many scenic and recreational attractions. In the winter, downhill skiing at two ski areas and cross-country skiing are popular. Summers are moderate, and winters are considerably milder than in other areas in the Midwest.

The School

The Dakota School of Mines was established in 1885 to provide instruction in mining engineering at a location where mining was the primary industry. The School of Mines opened for instruction in 1887. When North and South Dakota were admitted to statehood in 1889, the School was redesignated as the South Dakota School of Mines. In 1943, the name was changed to the South Dakota School of Mines and Technology, in recognition of the School's expanded role in new areas of science and technology. The School has a strong reputation for providing high-caliber graduates who are highly sought by industry.

Applying

International students should request an application form at least six months before the desired date of entry, while applications from U.S. residents should be submitted at least two months before the anticipated date of matriculation. The completed form, accompanied by a transcript of all undergraduate and graduate work and a nonrefundable fee of $15, should be submitted to the Admissions Office. Applicants must arrange for three letters of recommendation to be sent to the Admissions Office. A minimum TOEFL score of 520 is required of all applicants from non-English-speaking countries. Applicants for the Ph.D. program must submit scores from the Graduate Record Examinations.

Correspondence and Information

Program Director, MES Program
Graduate Education and Research Office
South Dakota School of Mines and Technology
501 East St. Joseph Street
Rapid City, South Dakota 57701-3995
Telephone: 605-394-2493
Fax: 605-394-5360
World Wide Web: http://www.sdsmt.edu

South Dakota School of Mines and Technology

THE FACULTY AND THEIR RESEARCH

Chemistry and Chemical Engineering

John T. Bendler, Ph.D., Yale. Statistical theory, molecular modeling, polymer properties, glass transition, surfaces of inorganic and polymeric materials.

David A. Boyles, Ph.D., Purdue. Synthetic organic chemistry; isolation, synthesis, and modification of natural products; medicinal chemistry; synthesis of plant growth regulators; synthesis of conjugated polymers; supramolecular assembly.

David J. Dixon, Ph.D., Texas at Austin. Supercritical fluid extraction, phase behavior and thermodynamics, polymer processing, surface science.

Sherry O. Farwell, Ph.D., Montana State. Surface analysis, solid sorption of gases, thin-film transducers, surface properties of precious metals, characterization of aerosols.

Steven M. McDowell, Ph.D., Iowa State. Synthesis and characterization of transition metal compounds, kinetics and mechanisms, small molecule interactions with metal compounds.

Jan A. Puszynski, Ph.D., Prague. Synthesis and processing of ceramics, combustion synthesis, reaction engineering, kinetics of heterogeneous reactions, mathematical modeling.

Robb M. Winter, Ph.D., Utah. Chemistry and micromechanics in polymer composites through application of ATR, FT-IR spectroscopy, and interfacial force microscopy.

Civil and Environmental Engineering

Sangchul Bang, Ph.D., California, Davis. Geotechnical engineering, numerical analysis, soil reinforcement.

M. R. Hansen, Ph.D., North Carolina State. High-performance concrete, especially the transition zone between aggregate and hydrated cement paste.

S. L. Iyer, Ph.D., South Dakota Mines and Tech. Composites, concrete reinforced structures.

Terje Preber, Ph.D., Wisconsin. Geotechnical engineering, fundamental soil properties, long-term behaviors of soils, experimental soil mechanics.

V. Ramakrishnan, Ph.D., London. Concrete materials, fiber concrete.

Engineering and Mining Experiment Station

Edward F. Duke, Ph.D., Dartmouth. Mineralogy, crystallography, electron microscopy, GIS, geochemistry.

Mechanical Engineering

Christopher H. M. Jenkins, Ph.D., Oregon State. Mechanics of composites, elastomers, and viscoelastic materials; dynamic material properties; constitutive modeling.

Sanjeev K. Khanna, Ph.D., Rhode Island. Experimental stress analysis, residual stress, welding, composite materials.

Lidvin Kjerengtroen, Ph.D., Arizona. Crack detection in beams and structures using vibration signatures, reliability-based design, fatigue, fasteners.

Metallurgical Engineering

Kenneth N. Han, Ph.D., Berkeley. Hydrometallurgy, metallurgical kinetics, fine-particle recovery, electrometallurgy, precious metallurgy, corrosion.

Stanley M. Howard, Ph.D., Colorado School of Mines. High-temperature thermodynamics, heat and mass transfer, computer modeling, chlorination metallurgy.

Jon J. Kellar, Ph.D., Utah. FT-IR spectroscopy, fiber-optic sensors, surface modification of reinforcements for polymer matrix composites.

Fernand D. S. Marquis, Ph.D., Imperial College (London). Strengthening mechanisms, fracture mechanics, materials design, metal and ceramic matrix composites.

Glen A. Stone, Ph.D., Berkeley. Transmission and scanning electron microscopy, X-ray diffraction, metallography, corrosion.

Mining Engineering

Eileen Ashworth, Ph.D., Arizona. Thermal properties of porous materials, effects of moisture and pore structure, heat flow in underground openings.

Zbigniew J. Hladysz, Ph.D., Central Mining Institute. Anisotropic properties of rocks.

Physics

T. Ashworth, Ph.D., Manchester. Experimental solid-state physics, low temperatures, thermal properties, porous materials, ice/water phase changes, highway de-icing, semiconductor materials.

Robert L. Corey, Ph.D., Washington (St. Louis). Experimental solid-state physics, nuclear magnetic resonance, electronic materials.

Michael Foygel, Ph.D., Odessa; D.Sc., St. Petersburg. Electronic properties of semiconductors and insulators.

Larry Meiners, Ph.D., Colorado State. MOCVD growth of III-V semiconductors, device processing, numerical modeling of semiconductor devices.

Andrey G. Petukhov, Ph.D., St. Petersburg. Electronic structure of materials, electronic properties of solid-state surfaces, interfaces and superlattices.

THE UNIVERSITY OF AKRON

College of Polymer Science and Polymer Engineering
Department of Polymer Science

Programs of Study

The Department of Polymer Science (DPS) offers an interdisciplinary graduate education leading to the Master of Science and Doctor of Philosophy degrees. Lecture and laboratory courses are available that cover most aspects of polymer science. Students specialize in organic polymer chemistry, polymer physical chemistry, polymer physics, molecular modeling, polymer engineering, or polymer technology, although they receive broad training in all these areas.

The minimal requirements for the Master of Science degree include 24 hours of course work, the completion of a research project and the writing of a thesis (6 credits), and demonstration of computer proficiency.

The Doctor of Philosophy degree requires a minimum of (but usually more than) 36 credits in graduate courses with satisfactory grades. At least 18 course credits and all dissertation credits (a minimum of 48) must be completed at the University of Akron. One year of full-time residence is required. In addition, the candidate must pass eight cumulative exams (given monthly during the academic year). Demonstrated computer proficiency is required of all doctoral candidates. Students may select any active faculty member in the department as a research adviser.

There are several national and international lectures and symposia sponsored by Polymer Science that feature cutting-edge research in academe and industry. Also, the H. A. Morton Visiting Professorship brings a world-renowned researcher in polymer science and engineering to Akron each semester for a semester of lectures and consultation with faculty, students, and local researchers.

Research Facilities

All of the research laboratories and classrooms in DPS are located in the Goodyear Polymer Center, which has 146,000 square feet of space. The labs are equipped with most of the modern tools necessary for the synthesis of polymers, the study of polymer structure and properties, and the fabrication of new composite materials. State-of-the-art research instrumentation is available for NMR, FTIR, and UV-visible spectroscopy; GPC; thermal analysis; rheo-optics; X-ray diffraction; and electron microscopy. An impressive array of research facilities are available through the department and the closely associated Maurice Morton Institute of Polymer Science. These include more than fifty laboratories for all major types of polymerizations and most conventional instruments for the measurement of polymer molecular structure and physical properties. Molecular modeling research is pursued using a range of powerful software and advanced computing hardware provided by the EPIC Macromolecular Modeling Center. A related contract research facility, the Edison Polymer Innovative Corporation (EPIC) Applied Research Lab, is also located in the Goodyear Polymer Center.

Financial Aid

The department supports more than 60 students. For 1999–2000, the standard stipends for departmental assistantships and industrial fellowships are $17,000 for twelve months. Tuition and fees are waived. More than 80 additional students are supported as research assistants by their research advisers from grants and contracts. Applications for financial support should be sent to the Department Chair with at least two letters of recommendation. Many other students are supported by their companies or governments.

Cost of Study

For the 1999–2000 academic year, tuition for Ohio residents is $178.10 per credit hour. Nonresidents pay an additional $155, for a total of $333.10 per credit hour. General fees are $81.65 per semester.

Living and Housing Costs

The cost of living is comparable to that found in other medium-sized northeastern American cities. Rooms in private houses are available in the University area and range from $200 to $300 per month. Apartments are more expensive. Off-campus residence housing for graduate students is available at the Judson House and the Ellis House (3 residents per complex). A meal ticket may be purchased for meals at the University Residence Hall (three meals daily, two meals on Sunday). For more information about off-campus housing, students should contact the Off-Campus Housing Office at 330-972-6936. Office hours are 8 a.m. to 4:30 p.m., Monday through Friday, Eastern time.

Student Group

The department has 130 full-time graduate students and about 20 part-time graduate students. Fifty percent of the full-time students are from the United States, and 50 percent are from other parts of the world.

Location

The city of Akron (population 240,000), which is situated in the northeastern part of Ohio, is known as the "Rubber Capital of the World." Although much of the rubber manufacturing by the big rubber companies has moved to the South, the headquarters and research centers for Bridgestone/Firestone, General, Goodrich, and Goodyear still remain in Akron, giving the city a "white collar" atmosphere. The region has more than 400 polymer-related companies with more than 30,000 employees. Numerous recreational and cultural facilities are available.

The University and The Department

The University of Akron was established in 1870 as Buchtel College, a church-related liberal arts college. In 1913 the college was turned over to the city of Akron and operated as a municipal university until 1967, when it became one of the state universities in Ohio. The University now has ten degree-granting colleges, with a total enrollment of about 27,000 students. The University operates on a two-semester system plus two summer sessions.

The Department of Polymer Science has been in operation since 1967. The first Ph.D. degrees in polymer chemistry, however, were awarded in the chemistry department in 1959. The Ph.D. degrees awarded in 1959 were also the first Ph.D. degrees granted at the University of Akron. The M.S. program in the chemistry of polymers has been in existence since the 1940s, and the teaching of rubber chemistry dates back another twenty years. The faculty of DPS is associated with the Maurice Morton Institute of Polymer Science, which administers government and industrial research contracts and the research facilities of the program. The Department of Polymer Science and the Maurice Morton Institute of Polymer Science, together with the Department of Polymer Engineering and the Institute of Polymer Engineering, make up the College of Polymer Science and Polymer Engineering, which was established in 1988.

Applying

Students with an undergraduate degree in chemistry, physics, or engineering are invited to apply. Students holding a degree in biology or natural sciences usually need additional courses on the undergraduate level in physical and analytical chemistry. For such students, a special nondegree admission may be given for one or two semesters, followed by a full admission upon a student's successful completion of the remedial undergraduate courses. All applications must be supported by at least one letter of recommendation stating that the candidate is able to engage in independent scientific research. GRE scores should be submitted with each application.

A student with an M.S. in the sciences from another university can be admitted to the Ph.D. program. In such cases, two letters of recommendation are required to be certain that the student is likely to be successful in doctoral research.

Correspondence and Information

Graduate Admissions
Department of Polymer Science
The University of Akron
Akron, Ohio 44325-3909

Telephone: 330-972-7542
Fax: 330-972-5290
E-mail: easter@frank.polymer.uakron.edu
World Wide Web: http://www.polymer.uakron.edu

The University of Akron

THE FACULTY AND THEIR RESEARCH

William J. Brittain, Professor and Chair, Department of Polymer Science; Ph.D., Caltech, 1982. Polymer synthesis, kinetics and mechanisms of polymerization, cyclic oligomers, self-assembled monolayers, nonlinear optical polymers, biomaterials, polyester and polymethacrylate synthesis.

Stephen Z. D. Cheng, Professor of Polymer Science; Ph.D., Rensselaer, 1985. Condensed state of polymeric materials; thermodynamics, kinetics of phase transitions; crystalline structure and morphology; liquid crystalline and rigid rod polymers; surface and interface of polymeric materials, fibers, films, and composites.

Ali N. Dhinojwala, Assistant Professor; Ph.D., Northwestern, 1994. Spectroscopic techniques to study structure and dynamics at polymeric and biologically active interfaces, dynamics in glassy and rubbery polymers, polymers and liquids at surfaces and in confined geometries, understanding diffusion of polymers and small molecules in heterogeneous systems and in restricted geometries.

Ronald K. Eby, R. C. Musson Professor of Polymer Science; Ph.D., Brown, 1958. Structure, morphology, and properties of polymers, advanced fibers elastomers, silks, laminates, and composites; computer modeling; physical acoustics; NDE; microscopy; X-ray scattering.

Mark D. Foster, Associate Professor of Polymer Science; Ph.D., Minnesota, 1987. Microstructure and dynamics of polymer systems, especially in thin films and near interfaces; novel X-ray and neutron-scattering techniques, especially reflectometry, for studying these phenomena.

John E. Frederick, Associate Professor of Polymer Science and Chemistry; Ph.D., Wisconsin, 1964. Light scattering from high polymer systems, viscoelastic properties of polymers.

Purushottam D. Gujrati, Professor of Physics and Polymer Science; Ph.D., Columbia, 1979. Statistical mechanics and field theory, phase transitions and critical phenomena, renormalization group theory, counting problems, enumerations of graphs, self-avoiding walks, glass transition.

Gary R. Hamed, Professor of Polymer Science; Ph.D., Akron, 1978. Rheology of adhesion, autohesion and tack, brass-rubber adhesion, fracture of polymers, processing of polymers, orientation of polymers, viscoelasticity.

Frank W. Harris, Distinguished Professor of Polymer Science and Director, Maurice Morton Institute of Polymer Science; Ph.D., Iowa, 1968. Condensation polymerization; specialty polymers, including polyimides and polyheterocyclic structure-property relationships in polymers; polymers for optical, microelectric, and fiber applications; controlled-release technology; polymers for medical applications.

H. James Harwood, Professor of Polymer Science and Chemistry; Ph.D., Yale, 1956. Free radical–initiated polymerization, highly alternating copolymerizations, polymer modification reactions, epimerization of stereoregular polymers, characterization of polymer and copolymer structure by NMR spectroscopy.

Frank N. Kelley, Dean of the College of Polymer Science and Engineering, Professor of Polymer Science; Ph.D., Akron, 1961. Structural characterization and fracture of thermosetting resins, time-dependent fracture of rubbery network polymers, properties of highly filled elastomers.

Joseph P. Kennedy, Distinguished Professor of Polymer Science and Chemistry; Ph.D., Vienna (Austria), 1954. Synthesis of well-defined macromolecules of potential industrial and biomaterial use by carbocationic polymerization, functionalization of polymers, grafts, blocks, networks, living polymerizations.

Wayne L. Mattice, Alex Schulman Professor of Polymer Science; Ph.D., Duke, 1968. Physical chemistry of macromolecules, dynamics, simulations, photophysics.

Coleen Pugh, Associate Professor of Polymer Science; Ph.D., Case Western Reserve, 1990. Synthesis and characterization of novel polymer materials, including side-chain liquid crystalline polymers and homopolyrotaxanes; living polymerizations; new polymerization methods.

Roderic P. Quirk, Distinguished and Kumho Professor of Polymer Science; Ph.D., Illinois, 1967. Organometallic polymerization catalysts; vinyl, diene, oxirane, and methacrylate monomers; functionalization reactions of polymeric carbanions; anionic synthesis of block copolymers, heteroarm star polymers, and macrocyclic polymers.

Darrell H. Reneker, Professor of Polymer Science; Ph.D., Chicago, 1959. Polymer physics, computer modeling of conformation defects in polymer crystals, morphology and scanning tunneling microscopy of polymers, electrospinning of polymer fibers, electrostriction of polymers.

Alexei P. Sokolov, Assistant Professor of Polymer Science; Ph.D., Physics, Russian Academy of Sciences, 1986. Nanostructure and dynamics of disordered systems, including polymers and liquids; glass transition; dynamics of biopolymers.

Marcia E. Weidknecht, Instructor of Polymer Science; B.S. (chemical engineering), New Hampshire, 1971. Solutions properties and physical properties of macromolecules and polymerization methods.

UNIVERSITY OF CALIFORNIA, SAN DIEGO

Materials Science Program

Programs of Study

The Materials Science Program is an interdisciplinary graduate program, with faculty members from several departments. Its governance is carried out by an Executive Committee that coordinates all affairs, including student admissions, degree requirements, graduate courses in materials science, maintenance of laboratory instructional facilities, seminars, special courses, part-time instructors, and related matters. Faculty members from the following departments participate in the program: Applied Mechanics and Engineering Sciences (AMES), Physics, Scripps Institution of Oceanography (SIO), Electrical and Computer Engineering (ECE), Bioengineering, and Chemistry. Students pursuing a graduate degree in materials science conduct research in one of the areas specified in the faculty and research list on the reverse of this page. A broad spectrum of areas, spanning the mechanical, chemical, magnetic, electronic, and optical behavior of materials, as well as their synthesis and processing, is encompassed by the research interests of the participating faculty members. Undergraduate preparation for the materials science M.S. and Ph.D. normally includes a degree in materials science and in engineering or physical sciences such as physics, chemistry, geology, and related disciplines. It is expected that interested students have an adequate background in mathematics, physics, chemistry, and related basic sciences. All M.S. students must complete 36 units of course work, of which 24 units comprise a core of required courses. In addition, they must either complete a thesis or pass a comprehensive examination. The Ph.D. program is designed to prepare students for a career in research and/or teaching in their area of specialization. After completing the M.S. degree or meeting equivalent requirements and meeting the minimum standard on the comprehensive examination for admission to the Ph.D. program, a student must successfully complete three advanced graduate courses approved by the dissertation adviser, pass an oral examination to be advanced to candidacy, and successfully complete and defend a dissertation that, in the opinion of the dissertation committee, contains original work that should lead to publication of at least one significant article in an appropriate refereed journal. In principle, it is possible to finish the M.S. degree in three quarters and a Ph.D. in an additional three years. More typically, a normally prepared B.S. student requires a total of about five years for the Ph.D.

Research Facilities

Facilities available for materials research include three centers for focused activities in addition to excellent general laboratories and computational and service centers. The Program for Dynamic Performance of Materials provides advanced facilities to study high-strain rate deformation. The equipment includes gas guns, Hopkinson bar high-power lasers, X-ray diffraction and optical equipment, and mechanical testing laboratories. This facility is part of the Center of Excellence for Advanced Materials, which focuses on coordinated macroscopic and microscopic testing and characterization of advanced engineering materials, including ceramics and ceramic composites, metallic and polymeric composites, high-strength alloys, and structural materials. Correlation of microstructures with properties is an integral part of research throughout the center. The Polymeric Composite Fabrication Laboratory includes a 10-foot-by-6-foot-diameter autoclave for thermoset/thermoplastic materials, a 70-ton hot press, filament winders, and complete characterization and testing capabilities. The Charles Lee Powell Structural Systems Laboratory houses one of the largest facilities for testing full-scale structures subjected to dynamic loading. The Center for Magnetic Recording Research is dedicated to the study of mechanical and materials aspects of magnetic storage devices. Complete laboratories for thin-film research are available, including molecular-beam epitaxy equipment, with a large group focusing on the electronic and optical properties of compound semiconductor devices. The biomaterials laboratories contain spectrophotometers for polymer characterization, gas and liquid chromatography equipment, and complete surgical facilities. Two electron microscopy facilities provide state-of-the-art scanning, transmission, and scanning-transmission capabilities; a Philips 300-kV transmission electron microscope with atomic resolution capability is part of the facility. Computational facilities include the San Diego Supercomputer Center.

Financial Aid

Financial aid is available in the form of fellowships, teaching assistantships, and research assistantships. The program attempts to support all full-time graduate students, especially at the Ph.D. level. Award of financial support is competitive, and stipends range from $5000 to a maximum of $17,000 for the calendar year, plus tuition and fees. While there are several types of support available, the most common is a half-time research assistantship, which pays approximately $17,000 plus tuition and fees during the calendar year.

Cost of Study

In 1999–2000, full-time students who are California residents pay approximately $1630 per quarter in registration and incidental fees. Non-California residents pay approximately $5070 per quarter for registration, tuition, and incidental fees. There is a reduced fee structure for students enrolled on a half-time basis. Costs are subject to change.

Living and Housing Costs

The University of California, San Diego (UCSD) provides 802 apartments for graduate students. Monthly rates range from $303 for a single student (studio) to $984 for a family. There is also a variety of off-campus housing available in the surrounding communities. Prevailing rates range from $270 per month for a room in a private home to $900 or more for a two-bedroom apartment. Information about housing may be obtained from the UCSD Housing Office (858-534-4723).

Student Group

Current campus enrollment is about 17,600: 14,600 undergraduates and 3,000 graduate students. The School of Engineering has an undergraduate enrollment of 2,980 and a graduate enrollment of about 600. The Materials Science Program has a current graduate enrollment of 30.

Location

The 2,000-acre campus spreads from the seashore, where the Scripps Institution of Oceanography is located, across a large wooded portion of the Torrey Pines Mesa overlooking the Pacific Ocean.

The University

One of nine campuses in the University of California System, UCSD comprises the General Campus, the School of Medicine, and the Scripps Institute of Oceanography. Albeit one of the newer campuses, it has become one of the top research universities in the country, with particular strengths in the physical and biological sciences.

Applying

Admission typically requires a B.S. and/or an M.S. degree in some branch of engineering, the physical sciences, or mathematics; a minimum GPA of 3.0; and strong letters of recommendation. GRE General Test scores are required. TOEFL scores are required from international applicants whose native language is not English. Applicants requesting financial aid should apply by January 15.

Correspondence and Information

Materials Science Program
Mail Code 0418
University of California, San Diego
La Jolla, California 92093-0418
Telephone: 858-534-7715
Fax: 858-534-2486
E-mail: matsci-info@soe.ucsd.edu
World Wide Web: http://www.soe.ucsd.edu/academic/matsci/matsci.html

University of California, San Diego

THE FACULTY AND THEIR RESEARCH

Gustaf Arrhenius, Professor of Oceanography; Ph.D., Stockholm. Structure, composition, and formation of oceanic minerals; solid-state physics and chemistry; interaction of organic molecules and crystals.

Robert J. Asaro, Professor of Materials Science; Ph.D., Stanford. Micromechanical modeling of materials.

Ami Berkowitz, Research Professor, Physics and Center for Magnetic Recording Research; Ph.D., Pennsylvania. Magnetic materials, correlation of microstructures with magnetic behavior, surface effects, relaxation phenomena.

John E. Crowell, Associate Professor of Chemistry; Ph.D., Berkeley. Surface chemistry, mechanistic and kinetic studies of dielectric and semiconductor thin-film growth by CVD, geometric and electronic structure of thin films and interfaces.

Robert C. Dynes, Professor of Physics and Chancellor; Ph.D., McMaster. Transport and thermodynamic properties of electronic materials at low temperatures, to include metals, semiconductors, and superconductors; effects of disorder and reduced dimensionality.

Sadik Esener, Professor of Electrical Engineering; Ph.D., California, San Diego. Electrooptical properties of materials, growth and characterization of ferroelectric thin films.

Yuan-Cheng Fung, Professor Emeritus of Bioengineering and Applied Mechanics; Ph.D., Caltech. Biomechanics, cardiopulmonary dynamics, relationship of tissue growth to stresses.

David A. Gough, Professor of Bioengineering; Ph.D., Utah. Mass transfer in biological systems, biotechnology, development of an implantable glucose sensor for diabetes, oxygen transport in the microcirculation, biotechnology reactor design, spatial and temporal development of neoplastic tissues.

Gilbert A. Hegemier, Professor of Applied Mechanics and Director, Charles Lee Powell Structural Systems Laboratory; Ph.D., Caltech. Theoretical and experimental solid and structural mechanics, nonlinear models for advanced composite materials, nonlinear constitutive theories for reinforced concrete, response of large structures to complex loading histories.

Frances Hellman, Associate Professor of Physics; Ph.D., Stanford. Experimental condensed-matter physics and materials science, magnetism, superconductivity, amorphous materials, thermodynamic and magnetic properties.

Richard K. Herz, Associate Professor of Chemical Engineering; Ph.D., Berkeley. Surface reaction kinetics, catalytic reactions.

Vistasp M. Karbhari, Associate Professor of Structural Engineering; Ph.D., Delaware. Mechanics and manufacturing science of composite materials, application of composites to infrastructure renewal, interphase science of dissimilar materials, durability of polymers and composites.

Karen L. Kavanagh, Professor of Electrical Engineering; Ph.D., Cornell. Electronic materials, characterization of semiconductor thin films and interfaces, structure-property correlations.

John B. Kosmatka, Associate Professor of Mechanical Engineering; Ph.D., UCLA. Structural and dynamic analysis of composite materials in aerospace vehicles, viscoelastic polymers, fiberous composites, analytical/experimental characterization.

S. S. Lau, Professor of Electrical Engineering; Ph.D., Berkeley. Microelectronics, particle-solid interactions, ion mixing, metal-semiconductor interactions, thin-film reactions.

Huey-Lin Luo, Professor of Electrical Engineering; Ph.D., Caltech. Applied solid-state physics, thin-film phenomena, physical metallurgy, thin-film materials for superconducting tunneling and Josephson junctions, interface studies.

M. Brian Maple, Professor of Physics; Ph.D., California, San Diego. Superconductivity, magnetism, properties of alloys, high-pressure physics, surface physics, catalysis.

Xanthippi Markenscoff, Professor of Mechanical Engineering; Ph.D., Princeton. Nonuniformly moving dislocations, analysis of singularities in elasticity, theory of singular asymptotics, interface mechanics, thin ligaments.

Joanna McKittrick, Associate Professor of Materials Science; Ph.D., MIT. Materials science and engineering, with an emphasis on glass and ceramics.

Marc A. Meyers, Professor of Materials Science and Associate Director, Institute for Mechanics and Materials; Ph.D., Denver. Dynamic synthesis and processing of materials, dynamic deformation and shock effects in materials.

David R. Miller, Professor of Chemical Engineering and Associate Vice-Chancellor of Academic Affairs; Ph.D., Princeton. Gas-surface interactions, molecular beams and gasdynamics, thin-film growth.

Hidenori Murakami, Professor of Applied Mechanics; Ph.D., California, San Diego. Development of advanced continuum models for composite materials, constitutive modeling, and FE analysis.

Sia Nemat-Nasser, Professor of Mechanics and Materials; Director, Center of Excellence for Advanced Materials; and Director, Institute for Mechanics and Materials; Ph.D., Berkeley. Micromechanics of materials; mechanisms of deformation, damage, and fracture in metals, ceramics, composites, geomaterials, and ionic polymer metal composites.

Vitali F. Nesterenko, Professor of Materials Science; Dr. of Physics and Mathematics, Russian Academy of Sciences. Powder materials processing, shear instability in inert and reactive materials, nonlinear micromechanics of heterogeneous materials.

Johann K. Oesterreicher, Professor of Chemistry; Ph.D., Vienna. Solid-state science, magnetic and superconducting systems, properties.

M. Lea Rudee, Professor of Materials Science and Coordinator, Graduate Program in Materials Science; Ph.D., Stanford. Application of electron microscopy and X-ray diffraction to the study of structure-property relations in magnetic and superconducting materials.

Michael Sailor, Professor of Chemistry; Ph.D., Northwestern. Chemistry of electronic materials, quantum-size particles, semiconductor thin films, electronically conductive polymers.

Geert W. Schmid-Schoenbein, Professor of Bioengineering; Ph.D., California, San Diego. Biomechanics and its application to cell mechanics, biorheology, and circulation.

Ivan K. Schuller, Professor of Physics; Ph.D., Northwestern. Experimental condensed-matter physics and materials science (thin films, heterostructures, and superconductivity).

Lu Jeu Sham, Professor of Physics; Ph.D., Cambridge. Solid-state theory: electron-phonon interaction; many-body theory of electrons in solids; density functional theory; electron properties in semiconductor heterojunctions, quantum wells, and superlattices.

Massoud Simnad, Adjunct Professor of Materials Science; Ph.D., Cambridge. Materials and nuclear engineering, high T_c superconductors.

Jan B. Talbot, Professor of Chemical Engineering; Ph.D., Minnesota. Corrosion, electroplating, display materials.

Frank E. Talke, CMRR Endowed Chair, Professor of Mechanical Engineering at Center for Magnetic Recording Research; Ph.D., Berkeley. Magnetic recording and printing technologies, tribology, surface roughness studies, friction and wear of materials, computer mechanics.

Yitzhak Tor, Assistant Professor of Chemistry; Ph.D., Weizmann (Israel). Organic, bioorganic, and materials chemistry; RNA recognition; organic chemistry of coordination compounds; metal-containing dendrimers and polymers.

Charles W. Tu, Professor of Electrical and Computer Engineering; Ph.D., Yale. Semiconductor materials, physics, and devices.

Kenneth Vecchio, Associate Professor of Materials Science; Ph.D., Lehigh. Electron microscopy, convergent-beam electron diffraction and X-ray microanalysis, fracture mechanics, fatigue and fractography.

Harry H. Wieder, Professor of Electrical Engineering; D.Sc., Colorado State. Semiconductor-device physics and applications, synthesis and characterization of compound semiconductor heterojunctions and quantum-confined structures.

Edward T. Yu, Professor of Electrical Engineering; Ph.D., Caltech. Semiconductor materials and devices.

Paul K. L. Yu, Professor of Electrical Engineering; Ph.D., Caltech. Semiconductor materials and device structures for electronic and optoelectronic applications, organometallic vapor-phase epitaxial and liquid-phase epitaxial techniques, microwave characterization of optoelectronic devices.

UNIVERSITY OF FLORIDA

College of Engineering
Department of Materials Science and Engineering

Programs of Study

The department offers M.S. and Ph.D. degrees in all areas of materials science and engineering, including biomaterials, ceramics, composites, electronic materials, glasses, metals, minerals, and polymers. A joint M.D./Ph.D. program in biomaterials, in collaboration with the medical school, is also available. The interdisciplinary research programs deal with the scientific and engineering aspects of the synthesis (thermodynamics, inorganic and organic reactions, phase transformations, and kinetics), structures (electronic, atomic, microstructure, and macrostructure), properties (electrical, optical, magnetic, chemical, and mechanical), processing (single crystals, polycrystals, amorphous, thin films, and composites), and applications (failure analysis, biomaterials, selection, and protection) of materials. Understanding the behavior of existing materials, developing advanced materials and novel processes, and selecting materials for the design of functional components are the general issues being addressed.

With current annual research expenditure of more than $10 million, the department ranks among the top ten materials, metallurgy, and ceramics departments in the nation. Supporting this research are 35 distinguished faculty members, approximately 200 graduate students, 32 scientists and research scholars, and 20 technical and supporting staff members.

Research Facilities

The department occupies approximately 80,000 square feet in Rhines Hall and the Materials Engineering Building, Materials Annex, and Advanced Materials Complex. A full range of modern facilities is available for research and instruction.

Analytical instruments valued at more than $6 million are housed within the Major Analytical Instrumentation Center, which provides research support and training not only to department and University researchers but also to industries throughout the southeastern United States. The major facilities include high-resolution TEM, TEM, STEM, SEM, AFM, STM (ambient and UHV), microprobe, scanning Auger, XPS, FIM, FT-IR, high-resolution X-ray and high-temperature X-ray diffractometer, DTA, TGA, DSC, LPE, MBE, and MOCVD. Other facilities are available for advanced composites, ceramic processing, electronic materials processing, mechanical testing, metals forming, optical microscopy, microwave processing, rapid solidification, single-crystal growth, hot isostatic press, vacuum hot press, sintering, sol-gel processing, fiber drawing, sputtering, and vapor deposition. Other specialized research facilities are housed within the Advanced Materials Research Center, Biomedical Engineering Center, Bioglass Research Center, and Mineral Resources Research Center.

Financial Aid

Financial aid is available in the form of teaching assistantships, graduate research assistantships, fellowships, and scholarships. For the 1999–2000 terms, full-time assistantships range from $12,500 to $16,000 for twelve months. Students with assistantships have been granted tuition waivers in past years.

Cost of Study

For 1998–99, the registration fee for most graduate course work was $138 per credit hour for Florida residents and $482 per credit hour for out-of-state students. When students hold assistantships, tuition waivers cover approximately 95 percent of their fees. This money has been provided by the state of Florida in the past, and it is fully expected that the state will continue to grant this waiver. Costs are subject to change for 1999–2000.

Living and Housing Costs

Living costs are moderate. For single students, a room in the coeducational residence halls is available for $1040 to $1744 per person per semester. (All costs are based on 1998–99 rates.) Private housing can be found within walking distance of the campus, and ample accommodations are available locally at a wide range of prices.

Student Group

The department's total enrollment is approximately 170 full-time graduate students and 100 full-time upper-division undergraduates who pursue programs of study in one of six specialties. The total University enrollment is approximately 42,000.

Student Outcomes

Since all commercial products rely on the properties and performance of materials of some type or another, materials scientists and engineers are in demand in a whole host of companies, both large and small. UF graduates, in particular, enjoy excellent employment prospects. For instance, recent graduates have found employment in a diverse selection of companies and industries such as aerospace, automotive, electronics, textiles, shipbuilding, petro-chemical, glass manufacturing, and biomedical.

Location

The University campus is located in Gainesville, a city of more than 90,000 in north-central Florida midway between the Atlantic Ocean and the Gulf of Mexico. There are many opportunities for swimming and boating at nearby lakes, springs, and rivers. Gainesville is served by several airlines, Amtrak, and bus lines. The city is located along I-75, only 1 hour south of I-10 and 2 hours north of Orlando.

The University, The College, and The Department

The University of Florida, a combined state university and land-grant college, was founded in 1906. The campus consists of 750 buildings of which 137 contain classrooms and laboratories. The College of Engineering consists of twelve degree-granting departments occupying fifteen buildings and has 288 full-time faculty members. The Department of Materials Science and Engineering is housed primarily in Rhines Hall and the Materials Engineering Building.

Applying

Admission to the graduate program requires a baccalaureate degree from an accredited college and an upper-division grade point average of 3.0 or better on a 4.0 scale. Students must submit satisfactory scores on the General Test of the Graduate Record Examinations. Students whose native language is other than English must also submit a score of 575 or better on the TOEFL. Students may be admitted at any time, but application forms, transcripts, and test scores should be submitted up to one year but no later than six months prior to the term of admission. All correspondence regarding admission and financial aid should be sent to the address given below.

Correspondence and Information

Academic Services Office
Department of Materials Science and Engineering
College of Engineering
University of Florida
108 Rhines Hall
P.O. Box 116400
Gainesville, Florida 32611-6400
Telephone: 352-846-3312
E-mail: academics@mse.ufl.edu
World Wide Web: http://www.mse.ufl.edu/

University of Florida

THE FACULTY AND THEIR RESEARCH

Dates of appointment are given in parentheses at end of entries.

Reza Abbaschian, Professor and Chairman; Ph.D., Berkeley, 1971. Solidification mechanisms, crystal growth, metals processing, composites. (1980)

Cammy R. Abernathy, Professor; Ph.D., Stanford, 1985. Metalorganic molecular beam epitaxy (MOMBE) of Si and III-V compound semiconductors for high-speed electronic and photonic devices, in situ processing, semiconductor characterization, precursor development, novel materials, crystal growth. (1993)

John R. Ambrose, Associate Professor; Ph.D., Maryland, 1972; PE. Corrosion, reaction kinetics, analysis of corrosion-related failures, electrochemical synthesis and modification of materials. (1978)

Stanley R. Bates, Associate Engineer and Director, Major Analytical Instrumentation Center; Ph.D., Florida, 1969; PE. Electron microscopy, characterization, surface characterization. (1969)

Christopher D. Batich, Professor; Ph.D., Rutgers, 1974. Surface and interface properties of polymers, polymer degradation, biomaterials (implants), XPS (ESCA). (1981)

Charles L. Beatty, Professor and Director, Polymer Processing and Properties Center; Ph.D., Massachusetts, 1972. Physical properties and processing of polymers, polymeric composites, room temperature curable ceramics and ceramic coatings. (1979)

Anthony B. Brennan, Associate Professor; Ph.D., Virginia Tech, 1990. Structure-property behavior of polymer/biopolymer interfaces, sol-gel processing of inorganic/organic hybrid composites, adhesives, stability of polymer composites. (1991)

David E. Clark, Professor; Ph.D., Florida, 1976; PE. Microwave processing and nuclear waste disposal, properties of glass and glass-ceramics, archaeological artifacts, coatings, sol-gel processing, composites. (1978)

Richard G. Connell Jr., Associate Professor and Assistant Chairman; Ph.D., Florida, 1973; PE. Metallography, joining, phase diagrams. (1973)

Robert T. DeHoff, Professor; Ph.D., Carnegie Mellon, 1959. Stereology, powder processing, microstructural evolution, diffusion, thermodynamics. (1959)

Elliot P. Douglas, Assistant Professor; Ph.D., Massachusetts Amherst, 1993. Structure-property-processing relationships of multiphase polymers, liquid crystalline polymers and thermosets, magnetic field processing of polymers, ionomers. (1996)

Fereshteh Ebrahimi, Associate Professor; Ph.D., Colorado School of Mines, 1982. Microstructure-mechanical behavior relationships, fracture mechanics, fracture paths, intermetallics, strengthening mechanisms. (1984)

Gerhard E. Fuchs, Assistant Professor; Ph.D., Rensselaer, 1986. Superalloys, intermetallics, composites, titanium and Ni-base alloys. (1998)

Eugene P. Goldberg, Professor and Director, Biomedical Engineering Center; Ph.D., Brown, 1953. Polymers, biopolymers, biosurfaces, biomedical polymers and devices. (1975)

Robert W. Gould, Professor Emeritus; Ph.D., Florida, 1964; PE. Failure analysis, fracture and fractography, product liability, X-ray diffraction. (1964)

Laurie B. Gower, Assistant Professor; Ph.D., Massachusetts Amherst, 1997. Engineered particulates, biomimetics, biomineralization, ceramic/polymer composites, crystal growth modifiers, nano- and meso-structures, hierarchical systems, ceramics thin films, biomedical materials. (1997)

Larry L. Hench, Graduate Research Professor and Director, Bioglass Research Center; Ph.D., Ohio State, 1964. Ceramics, electronic transport in inorganic systems, reaction kinetics in ceramic systems, sol-gel processing, prosthetic materials, optical materials, molecular orbital modeling of materials. (1964)

Paul H. Holloway, Professor; Ph.D., Rensselaer, 1972. Surface reactions and characterizations of surfaces, Auger spectroscopy, electronic materials, thin-film phenomena, optoelectronic materials and devices, electrical contacts, semiconductor passivation, tribological coatings. (1978)

Rolf E. Hummel, Professor and Graduate Co-Coordinator; Dr.rer.nat., Max Planck Institute (Germany), 1963. Electronic materials, optical properties of metals, alloys and semiconductor materials, reliability of semiconductor devices, electrotransport in thin films, thin-film phenomena/optoelectronics, light-emitting, spark-processed silicon. (1964)

Kevin S. Jones, Associate Professor; Ph.D., Berkeley, 1987. Electronic materials, semiconductor processing, ion implantation, defect characterization, transmission electron microscopy. (1987)

Michael J. Kaufman, Professor; Ph.D., Illinois at Urbana–Champaign, 1984. Structure-property-processing relationships in metals, intermetallics, and composites; physical metallurgy and phase transformations; conventional and rapid solidification; electron microscopy. (1989)

John J. Mecholsky Jr., Professor; Ph.D., Catholic University, 1973. Fracture of brittle materials; fractal fracture; fractography; mechanical behavior of metal/ceramic composites; microstructure-property relationships in ceramics, metal-ceramic, ceramic-ceramic, and glass-metal joining. (1990)

Brij M. Moudgil, Professor and Director, Mineral Resources Research Center; Dr.Eng.Sc., Columbia, 1981. Mineral processing, fine-particle technology, surface and colloid chemistry, crystal growth, suspension processing. (1981)

Stephen J. Pearton, Professor; Ph.D., Tasmania, 1983. Dry etching, implantation and rapid thermal processing of semiconductors, hydrogen in semiconductors, electronic properties of solids, electronic and photonic device processing. (1993)

Robert E. Reed-Hill, Professor Emeritus; D.Sc., Yale, 1956. Mechanical metallurgy, deformation mechanisms, fracture, creep. (1960)

Raymond A. Rummel, Associate Professor Emeritus; B.S., Pittsburgh, 1957. Shop and laboratory techniques. (1959)

Wolfgang M. Sigmund, Assistant Professor; Dr.Rer.Nat, Max Planck Institute (Germany), 1998. Colloid and interfacial chemistry, powder processing, surface chemical modification, thin films, polymers and ceramics, rapid prototyping. (1999)

Joseph H. Simmons, Professor; Ph.D., Catholic University, 1969. Physics of optical materials, nonlinear optics, spectroscopy of ultrafast phenomena, structure of amorphous materials, thin films for optical applications. (1984)

Rajiv K. Singh, Associate Professor; Ph.D., North Carolina State, 1989. Laser processing of materials, superconducting, diamond and semiconductor thin films, electronic materials characterization. (1990)

Ellis D. Verink Jr., Distinguished Service Professor Emeritus; Ph.D., Ohio State, 1965; PE. Corrosion of metals, physical metallurgy, surface chemistry. (1965)

Eric Wachsman, Assistant Professor; Ph.D., Stanford, 1990. Ionic and electronic conducting ceramics for sensor, fuel cell, battery, and gas separation applications; fundamental investigations of their transport properties and heterogeneous electrocatalytic activity. (1997)

E. Dow Whitney, Professor and Graduate Co-Coordinator; Ph.D., NYU, 1954; PE. Ceramics, high-temperature/high-pressure phase transformations, physical chemistry and tribology of refractory hard materials, ceramic cutting tools. (1970)

UNIVERSITY OF MARYLAND, COLLEGE PARK

Materials and Nuclear Engineering Department
Materials Science and Engineering Program

Programs of Study

The Materials Science and Engineering Program offers educational opportunities leading to a Doctor of Philosophy degree or Master of Science degree in materials science and engineering. The Master of Science degree in materials science and engineering is highly flexible, and students are encouraged combine the M.S. degree curriculum with specialization electives. A minimum of 30 credits is required for the M.S. degree and must include the core courses consisting of physics of materials, thermodynamics in materials science and kinetics of reactions in materials science, and defects in materials. The program is based on the student carrying out original research in the areas of materials specialization. The Materials Graduate Program offers students an interdisciplinary, science-based education and the research training necessary for the understanding of materials systems used in modern engineering design and manufacturing. The Ph.D. degree program requires a minimum of 48 credits for completion in addition to a research thesis based on original research. To qualify as a Ph.D. candidate, students must maintain a minimum 3.0 GPA (in materials science) and pass the qualifying exam. Completion of the equivalent of two full years of study beyond the B.S. degree is also highly recommended. This may be fulfilled by a program that includes at least 36 credit hours of course work. The courses taken for the M.S. degree are acceptable but should also include four technical electives in the student's selected area of research. The dissertation is the major portion of the Ph.D. program. The Ph.D. candidate must defend his or her dissertation prior to the final approval in an oral examination. This process includes a public presentation as a departmental seminar and a private oral examination. The areas of research specialization of the department are electronic materials, polymers and ceramics, and structural materials. In addition, for the working professional, opportunities leading to a Master of Engineering degree are available. Interested students are encouraged to contact the department for more details.

Research Facilities

The Thin Film Processing Laboratory emphasizes the growth of metal oxides, nitrides, and compound semiconductors by pulsed laser deposition (PLD) and by chemical vapor deposition (CVD). The department has a clean room facility for device fabrication and materials processing. This facility, named the Laboratory for Advanced Materials Processing (LAMP), includes lithographic capabilities. The Center for Microanalysis provides state-of-the-art analytical facilities that include an environmental scanning electron microscope (ESEM) equipped with energy dispersive spectroscopy (EDS) for quantitative as well as qualitative analyses; an electron microprobe with EDS and wavelength dispersive spectroscopy (WDS); and a variety of ancillary equipment for sample preparation and compositional analyses. The Mechanical Testing Central Facility can test materials over a wide range of temperature, strain rates, and sample configurations. The Polymer Characterization Laboratory has facilities for the advanced characterization of polymers and polymer-blend materials. Radiation and reactor facilities in the department include an open-pool TRIGA-type nuclear reactor that can be used to irradiate samples in either a pneumatic transfer system or large access ports close to the core; an electron linear accelerator (LINAC) with a unique exoatmospheric test chamber for the evaluation of space radiation effects on electronic and other materials; and a Cobalt-60 gamma ray source. The department facilities also include a 400 keV transmission electron microscope (TEM) and 100, 200 keV TEMs for materials structure research.

Financial Aid

A significant number of graduate fellowships and teaching and research assistantships are available for well-qualified applicants. In 1998–99, stipends began at $10,710 plus tuition for ten months. Government traineeships are awarded through the University to exceptionally well-qualified students. Part-time support is available through many national laboratories and technical facilities located nearby. The University also has resident assistantships, summer dissertation fellowships, and various types of loans.

Cost of Study

In 1998–99, tuition and fees for full-time study (10 credit hours) were $3002 for Maryland residents and $4282 for nonresidents.

Living and Housing Costs

Board and lodging are available in many private homes and apartments in College Park and the vicinity. Rooms in private homes range in cost from $250 to $350 a month, and one-bedroom apartments rent for an average of $600 per month. A list of accommodations, both University and private, is maintained by the University's housing bureau.

Student Group

In the department, there are a total of 312 students, of whom 57 are master's students.

Student Outcomes

During the past few years, the department has placed its Ph.D. graduates on the facilities of such academic institutions as University of Maryland, Howard University, Northwestern University, and Old Dominion University, as well as at corporate and national research labs such as Lockheed-Martin Corporation, COMSAT, Bendix Communications Corporation, General Electric Transportation Systems, Quantex, Computer Sciences Corporation, Hyundai Electronics Industries, Analog Devices, SAIC, PetroBras, National Semiconductor Corporation, Nuclear Regulatory Commission, NASA, National Security Agency, National Oceanic and Atmospheric Administration, National Institute of Standards and Technology, Department of Energy, U.S. Navy, Army Research Laboratory, Los Alamos National Laboratory, and Naval Research Laboratory.

Location

The central campus of the University of Maryland is located in College Park, Maryland, a suburban area roughly between and within easy commuting distance of Washington, D.C., and Baltimore. The campus is 12 miles from the White House. The museums, galleries, theaters, federal and special libraries, universities, concert halls, and abundant cultural activities of both cities offer students unlimited opportunities to participate in the cultural and social life of this thriving area. The immediate presence of many great national laboratories and technical facilities offers a particularly good opportunity for the graduate student in electrical engineering.

The University

The University is one of the oldest and largest state universities in the country. The College of Engineering is located on the central campus, which has a total student population of approximately 33,000. The University offers many cultural and entertainment activities and operates its own golf course and athletic facilities.

Applying

Applicants seeking admission should hold a B.S. degree with a B+ average or better from an accredited institution. Applications should be filed early; the deadline for assistantship applications is January 15. The submission of three recommendation letters and scores on the General Test of the Graduate Record Examinations is required.

Correspondence and Information

Barret Cole, Graduate Admissions Secretary
Department of Materials and Nuclear Engineering
Building 090, Room 2135
University of Maryland
College Park, Maryland 20742
Telephone: 301-405-5211
Fax: 301-314-2029
World Wide Web: http://www.mne.umd.edu

University of Maryland, College Park

THE FACULTY AND THEIR RESEARCH

Electronics Materials

Aris Christou, Chairman; Ph.D., Pennsylvania. Electronic packaging materials, thin-film semiconductors, reliability of electronic systems.

Luz Martinez-Miranda, Ph.D., MIT. Liquid crystals, X-ray diffraction techniques, materials characterization.

Martin Peckerar, Ph.D., Maryland. Semiconductors, process research and materials systems for lithography.

Ramamoorthy Ramesh, Ph.D., Berkeley. Ferroelectrics and metal oxides, thin-film materials, wide bandgap semiconductors.

Alexander L. Roytburd, D.Sc., Russian Academy of Science. Modulated structures in epitaxial layers and multilayer composites, formation and deformation of polydomain materials, effect of interfaces on magnetic properties of oxide conductors.

Gary Rubloff, Ph.D., Cornell. Silicon technology, manufacturing, CVD growth of nitrides, surface science.

Lourdes G. Salamanca-Riba, Ph.D., MIT. Structural studies of thin-film semiconductor heterostructures and superlattices, superconductors and metallic multilayers, optical properties of materials.

Manfred Wuttig, Ph.D., Technische Hochschule Dresden (Germany). Phase transformation in thin films, mechanics of thin films and membranes, smart materials.

Polymers and Ceramics

Mohamad Al-Sheikhly, Ph.D., Newcastle. Polymers, radiation engineering, electronic packaging materials, environmental effects.

Ira Block, Ph.D., Maryland. Polymer fibers, characterization.

Robert M. Briber, Ph.D., Massachusetts. Thermodynamics of complex polymer systems, structural solutions of novel molecules.

Walter Chappas, Ph.D., Maryland. Polymers radiation effects in electronics, composites, and elastomers.

Peter Kofinas, Ph.D., MIT. Polymer science and technology, polymer synthesis and processing.

Isabel K. Lloyd, Ph.D., MIT. Effects of processing on the behavior of advanced ceramics, ferroelectric ceramics.

Joseph Silverman (Emeritus), Ph.D., Columbia. Radiation engineering, polymer science and technology, elastomers, radiation manufacturing.

Betty F. Smith (Emeritus), Ph.D., Minnesota. Polymer and textile chemistry, formation of polymer fibers.

Otto Wilson, Ph.D., Rutgers. Ceramics biomimetic materials, advanced ceramic surfaces.

Kwan-Nan Yeh, Ph.D., Georgia. Modification and performance properties of fibrous materials.

Reliability and Radiation Science

Kazys Almenas, Ph.D., Warsaw. Thermal hydraulics, computer modeling.

Joseph Bernstein, Ph.D., MIT. Reliability of microelectronics, laser probing techniques, physics of failure.

Mirela Gavrilas, Ph.D., MIT. Thermal hydraulics, computer simulation, reactor design, heat transfer.

Mohammad Modarres, Ph.D., MIT. Expert systems in reliability and safety, probabilistic risk assessment.

Ali Mosleh, Ph.D., UCLA. Common cause failures, Bayesian reliability analysis, computer security.

Frank J. Munno, Ph.D., Florida. Nuclear reactor safety, reactor physics.

Gary A. Pertmer, Ph.D., Missouri. Reactor systems analysis, thermal hydraulics.

Marvin Roush, Ph.D., Maryland. Reliability physics, risk assessment.

Carol Smidts, Ph.D., Université Libre de Bruxelles (Belgium). Dynamic systems, human reliability, software reliability.

Lothar Wolf, Ph.D., Berlin. Thermal hydraulics, reliability and safety.

Structural Materials and Composites

Sreeramamurthy Ankem, Ph.D., Polytechnic of New York. Physical and mechanical metallurgy of titanium alloys, finite element method (FEM) of deformation and damping behavior of composite materials, microstructure evolution in multiphase materials.

Ronald Armstrong, Ph.D., Northwestern. Structural materials and dislocations science and engineering.

Richard J. Arsenault, Ph.D., Northwestern. Computer simulation of plastic deformation, composite strengthening, finite element method (FEM) analysis of microdeformation of composites.

Steven Hsu, Ph.D., Penn State. Materials tribology, ceramics and nanophase materials.

Brian Lawn, Ph.D., Western Australia. Fracture of brittle solids, ceramic composites, plasma-sprayed coatings.

UNIVERSITY OF MASSACHUSETTS AMHERST

Polymer Science and Engineering Department

Programs of Study

The Polymer Science and Engineering Department is an interdisciplinary unit that offers graduate studies leading to the Doctor of Philosophy degree. All aspects of polymer science and engineering are taught within the department. Research and course programs are available in polymer synthesis, characterization, morphology, rheology, physics, and engineering. All faculty members are dedicated solely to graduate teaching and research.

Research Facilities

The Polymer Science and Engineering Department occupies 172,000 square feet of modern office and laboratory space in the six-story Silvio O. Conte National Center for Polymer Research. This newly constructed building contains 150 fume hoods, machine and electronic shops, a microanalytical facility, and numerous centralized Shared Research Facilities organized through the NSF-sponsored Materials Research Science and Engineering Center (MRSEC). Near the Conte Center are the Physical Sciences Library (with a comprehensive collection of polymer literature), Scientific Glassblowing Laboratory, and University Computing Center. The department stresses the free access of students and staff members to research instruments; nearly all of the $20-million worth of research equipment located in the Conte Center is available free of user fees. The Spectroscopy Facility combines laboratories that are focused on NMR, vibrational spectroscopy, and X-ray photoelectron spectroscopy. Included in the NMR Laboratory are five high-field fluids and solid spectrometers, which allow virtually any NMR experiment to be conducted. The Vibrational Spectroscopy Laboratory houses a variety of Fourier transform and conventional IR and Raman instruments, ultraviolet and fluorescence spectrophotometers, and atomic-force microscopes. The Polymer Morphology Facility includes the W. M. Keck Electron Microscopy Laboratory, which hosts three transmission electron microscopes (100, 200, and 300 kV) and two scanning electron microscopes (one with a field emission gun), and the X-Ray and Scattering Laboratory, which includes four rooms with wide- and small-angle X-ray instruments (including a diffractometer, reflectometer, and two rotating anodes). The Characterization Facility combines the Mechanical Processing and Testing Laboratory and the Thermal Analysis Laboratory. Among the instruments are seven Instrons; a holographic interferometer for thin-film stress measurements; presses, calendars, and extruders; and numbers DSC, DMTA, and TGA models. The Rheology Facility contains commercial and custom instruments for measuring mechanical and optical properties of flowing liquids. Full dynamic and steady-shear characterizations are possible for melts and fluids. In addition, birefringence and light-scattering attachments are available for studies of the fluids in both shear and elongation. A unique high-pressure viscometer has been constructed for supercritical fluids investigations. The Mass Spectroscopy Facility is the newest of the central facilities, and several instrument acquisitions are underway. The facility currently includes a MALDI-TOF MS, electrospray MS, and GC-MS; both a quadrapole GC-MS and a second-sector MS are expected shortly. A Nanoprocessing and Characterization Facility is being constructed; this facility will focus on the preparation and characterization of thin-film devices, with features in the range of 10–100 nm. Instruments are expected to include ellipsometers, a surface plasmon device, scanning/tunneling microscopes, and etching/lithography equipment. A planned Molecular Weight Facility will collect and expand upon the department's collection of light-scattering instruments (dynamic, static, and chromatrography-coupled), liquid chromatographs, electrophoresis devices, osmometers, centrifuges, and densitometers.

Financial Aid

Fellowship stipends in 1999–2000 are $16,500 for twelve months, plus a waiver of tuition and certain fees. Basic health insurance is provided by the department. The fellowships have no teaching obligations; first-year students are concerned primarily with course work, choice of research adviser(s), and research orientation. First-year Ph.D. students are funded by the department, and subsequent support (at the same level) is provided by the student's adviser(s) from grant research funds.

Cost of Study

Tuition is waived for full-time Ph.D. students who receive departmental support. Nonwaivable fees for 1999–2000 are approximately $300 per semester for first-year graduate students and less for more senior students.

Living and Housing Costs

The cost of living in Amherst is comparable to other towns in the New England area. Cost-of-living estimates for an international student in residence are available through the University's Office for Foreign Students and Scholars. In 1998–99, the starting cost of residence hall room and board was $4000 per year. Off-campus housing is available.

Student Group

The department has 85 graduate students in residence and 25 postdoctoral associates. Students come from all parts of the United States and the world and majored in subjects such as chemistry, material science, and chemical engineering at the undergraduate level.

Location

Situated in one of the most picturesque sections of the state, the University of Massachusetts joins with its academic neighbors—Amherst, Smith, Mount Holyoke, and Hampshire Colleges—in maintaining the rich tradition of education and cultural activity associated with the Connecticut Valley region. The town of Amherst, founded in 1759, is in the Pioneer Valley. This valley has numerous dairy, poultry, and tobacco farms as well as many orchards. The principal industry in the area is education, with the University and Amherst College enrollments totaling more than 25,000. The resident population in the town of Amherst is about 15,000. Amherst, a town of tranquility, was the home of Emily Dickinson, Noah Webster, Helen Hunt Jackson, Henry Ward Beecher, Robert Frost, and others.

The University and The Department

The University of Massachusetts, the flagship university of the commonwealth, was founded under the Morrill Land Grant Act of 1863. Nine colleges and schools of the University are housed in more than 150 buildings on approximately 1,400 acres of land. There are about 23,000 students at the University, which operates on a semester system. The Polymer Science and Engineering Department is recognized by industry and academic institutions as one of the world's leading centers for polymer education and research. A program of fundamental polymer research has been in progress at the University for more than twenty-five years. The faculty is associated with the Materials Research Science and Engineering Center (MRSEC) and the Center for Massachusetts-Industry Research on Polymers (CUMIRP). MRSEC was established by the National Science Foundation and is partially responsible for the excellent central research facilities. CUMIRP is funded jointly by some of the leading U.S. manufacturers and users of polymers. Approximately 150 scientists and students are engaged in polymer research within the department.

Applying

Students who have majored in chemistry, physics, engineering, or materials science and have maintained at least a B average in their undergraduate major normally qualify for admission. Scores on the General Test of the GRE plus two letters of recommendation are required of all applicants. Application fees of $25 and $40 are charged to Massachusetts residents and nonresidents, respectively.

Correspondence and Information

Professor David A. Hoagland, Graduate Program Director
Polymer Science and Engineering Department
Silvio O. Conte National Center for Polymer Research
University of Massachusetts Amherst
Amherst, Massachusetts 01003
Telephone: 413-577-1120
E-mail: graduate@polysci.umass.edu
World Wide Web: http://www.pse.umass.edu

University of Massachusetts Amherst

THE FACULTY AND THEIR RESEARCH

Richard J. Farris, Distinguished University Professor of Polymer Science and Engineering and Head of the Department; Ph.D., Utah, 1970. Experimental mechanics, mechanics of coatings, high-modulus fibers, deformation, calorimetry and dilatometry, composite materials, microstructural modeling, viscoelasticity, constitutive theory, engineering mechanics.

Samuel P. Gido, Assistant Professor of Polymer Science and Engineering; Ph.D., MIT, 1993. Polymer morphology, electron microscopy, X-ray scattering, morphology of block copolymers and crystalline polymers.

Robert B. Hallock, Professor of Physics and Astronomy; Ph.D., Stanford, 1969. Adsorption of macromolecules, surface phase transitions, experimental low-temperature physics, liquid helium films, random media.

David A. Hoagland, Associate Professor of Polymer Science and Engineering and Graduate Program Director; Ph.D., Princeton, 1986. Molecular rheology, polyelectrolytes, electrophoresis and diffusion in complex media, polymer adsorption, light scattering, gels.

Shaw Ling Hsu, Professor of Polymer Science and Engineering; Ph.D., Michigan, 1975. Vibrational spectroscopic characterization of polymers, piezoelectricity and ferroelectricity of polymers, liquid crystalline polymers, phase separation of copolymers, spectroscopic characterization of polymer-metal interfaces.

Frank E. Karasz, Silvio O. Conte Distinguished Professor of Polymer Science and Engineering; Ph.D., Washington (Seattle), 1958. Electrooptically active polymers, polymer blend thermodynamics, order-disorder transitions, polymer-diluent systems.

Robert W. Lenz, Professor of Polymer Science and Engineering; Ph.D., SUNY College of Environmental Science and Forestry, 1956. Synthesis of new polymers, polymers produced by bacteria, biopolymers, biodegradation of polymers.

Alan J. Lesser, Assistant Professor of Polymer Science and Engineering; Ph.D., Case Western Reserve, 1989. Deformation and fracture of polymers and composites, fatigue endurance and durability of engineering polymers, micromechanics of polymer blends and composites, constitutive modeling of polymers in complex stress states.

C. Peter Lillya, Professor of Chemistry; Ph.D., Harvard, 1964. Self-organization of main chain liquid crystalline polymers, end-to-end associative polymers.

William J. MacKnight, Distinguished University Professor of Polymer Science and Engineering; Ph.D., Princeton, 1964. Property-structure relationships in microphase-separated polymers, polymer blends, segmented polyurethanes.

Thomas J. McCarthy, Professor of Polymer Science and Engineering and Coprincipal Investigator, CUMIRP; Ph.D., MIT, 1982. Polymer surface modification, adsorption of polymers at interfaces, polymer synthesis and modification, chemistry in supercritical fluid-swollen polymers.

Murugappan Muthukumar, Professor of Polymer Science and Engineering; Ph.D., Chicago, 1979. Statistical mechanics of polymers, polyelectrolytes, and liquid crystals; pattern recognition; self-assembly of hierarchical structures.

Jacques Penelle, Assistant Professor of Polymer Science and Engineering; Ph.D., Louvain (Belgium), 1989. Polymer synthesis, organic polymer chemistry, kinetics and mechanism determination.

Thomas P. Russell, Professor of Polymer Science and Engineering and Director of the Materials Research Science and Engineering Center; Ph.D., Massachusetts Amherst, 1979. Surface and interfacial properties of polymers, polymer morphology, kinetics of phase transitions, confinement effects on polymers, morphology control using supercritical fluids.

Richard S. Stein, Professor Emeritus of Chemistry; Ph.D., Princeton, 1949. Optical and mechanical properties of polymers, light-X-ray and neutron scattering from solid polymers, characterization of orientation, thermodynamics and kinetics of phase separation of polymer blends, neutron reflectivity studies of polymer surfaces and interfaces.

Helmut Strey, Assistant Professor of Polymer Science and Engineering; Ph.D., Munich Technical, 1993. Biophysics, biopolymers, liquid crystals, polyelectrolytes, polyelectrolyte-surfactant complexes, self-assembly, chiral interactions, structure and thermodynamics of biopolymer liquid crystals (small-angle X-ray scattering, polarizing microscopy, and osmotic stress).

H. Henning Winter, Distinguished University Professor of Chemical Engineering; Ph.D., Stuttgart (Germany), 1973. Rheometrical methods, rheology and material structure, phase transitions and rheology, modeling of processing flows.

UNIVERSITY OF MICHIGAN

College of Engineering
Department of Materials Science and Engineering

Programs of Study
The Department of Materials Science and Engineering offers Master of Science in Engineering (M.S.E.) and Ph.D. programs leading to degrees in materials science and engineering. Students may emphasize work in various materials categories or phenomena, although the department encourages a broad graduate educational experience. Course offerings include basic materials courses in structure of materials, thermodynamics, diffusion, phase transformations, mechanical behavior, and materials characterization. Courses also exist in many areas of special interest, such as corrosion, composites, deformation processing, and failure analysis. The M.S.E. degree, typically completed in one to two years, requires 30 credit hours of graduate study. A research project of up to 6 credit hours or a master's thesis of 9–11 credit hours is included within this total and often forms the basis for the student's Ph.D. qualifying examination. The Ph.D. degree, usually completed in four to five years beyond the B.S. degree, requires 18 credit hours of courses beyond the M.S.E. degree, passing grades on a research-based oral examination, a written examination based on advanced undergraduate and graduate-level course material, satisfactory completion of research, and defense of the doctoral dissertation. An academic year of residence and instructional experience are also required.

Faculty interests are diverse (see the reverse of this page) and fall into nine categories: composite materials, polymer alloys, structural ceramics, materials processing and manufacturing, micromechanical and macromechanical behavior, surface modification and energy-beam interactions of materials, device materials, theoretical modeling and computer simulation, and materials characterization. Many additional research activities exist in collaboration with other departments and graduate programs.

Research Facilities
The department occupies approximately 40,000 square feet, primarily in the H. H. Dow Building, but also in the adjacent G. G. Brown Building and the nearby Space Physics Research Laboratory. Research facilities include world-class laboratories for electron microscopy, ion-beam characterization and modification of materials, and solid-state device research. Modern instrumentation is added regularly. The department has installed its own computer network, to which computers in nearly all faculty and graduate student offices are connected. Some of the facilities include those for electron optics and surface science (five TEM/STEM units, three SEM units, ESCA/XPS, SAM, and electron microprobe analysis), X-ray diffraction and fluorescence, optical microscopy, metallography, specimen preparation facilities (single-crystal growers, heat treating, stereolithography apparatus (SLA), melt spinning, vacuum induction and plasma arc melting, and electron-beam zone melting), mechanical testing (ten servohydraulic and screw-driven machines), instrumented impact testers, deformation processing, a creep laboratory, powder compaction, hot and cold isostatic pressing, extruders, injection molders, several Fourier-transform infrared spectrometers, dynamic mechanical spectroscopy, arc-welding, and DTA, DSC, TGA, and dilatometry.

Financial Aid
Qualified applicants were eligible for fellowships and teaching or research assistantships that paid stipends of up to $17,735 per calendar year in 1998–99 plus tuition remission and some fringe benefits. Students funded by faculty advisers as research assistants work on research problems that are appropriate for their thesis topic. Teaching and research assistantships carry certain defined responsibilities that are adjusted to the needs of the student and the department.

Cost of Study
The 1998–99 tuition fee for full-time students was $6042 per term for Michigan residents and $10,943 per term for nonresidents.

Living and Housing Costs
A residence hall contract for room and board for the 1998–99 fall and winter terms ranged in cost from $5488 for a double to $6524 for a single. Family housing units cost from $552 per month for an unfurnished one-bedroom unit to $865 per month for a furnished three-bedroom unit. Prices include all utilities except telephone.

Many graduate students live in privately owned off-campus housing, which varies in expense depending on its proximity to the University. Food costs and local restaurant prices are typical of those in smaller cities in the Midwest.

Student Group
The department has approximately 70 full-time graduate students and about 15 part-time students from local industry and research laboratories. Approximately 60 percent of the students are from the United States, and 40 percent are from abroad. Most students receive financial aid from the department. The department also has about 100 undergraduate students. The College of Engineering enrolls approximately 6,600 students in nineteen degree programs. The total student enrollment on the Ann Arbor campus is about 36,000. The student-based Michigan Materials Society is very active.

Location
Ann Arbor is a cultural and cosmopolitan community of approximately 105,000 about 40 miles west of Detroit in southeastern Michigan. Ann Arbor offers world-class orchestras, dance companies, dramatic artists, and musical performers throughout the year. The internationally renowned May Festival of classical music and the Ann Arbor Folk Festival are held annually. Ann Arbor art fairs attract 500,000 patrons from across the nation every July. Recreational facilities are extensive, both on campus and throughout the community.

The University and The College
The University of Michigan, one of the nation's most distinguished state universities, is internationally recognized in all of its schools and colleges. The 2,400 faculty members and 36,000 students work in a modern environment that includes more than 250 research units. Michigan consistently ranks as a national leader in total research expenditures. The College of Engineering, of which the department is a part, awards about 1,000 B.S., 500 M.S., and 100 Ph.D. degrees annually. There are 310 faculty members, 630 supporting staff members, and more than 40,000 alumni. Many of the programs in the College are rated among the ten best in the nation, and the College itself is often ranked among the top five engineering schools and colleges.

Applying
Applications are accepted for either the fall (September) or winter (January) terms; however, most students are admitted in the fall term. Applications for fall admission should be received by February 15 if financial support is required. Additional information on admission may be obtained from the department or from the Horace H. Rackham School of Graduate Studies.

Correspondence and Information
Graduate Program Office
Department of Materials Science and Engineering
College of Engineering
University of Michigan
Ann Arbor, Michigan 48109-2136
Telephone: 734-763-9790
World Wide Web: http://msewww.engin.umich.edu

Horace H. Rackham School of Graduate Studies
Mail Office
University of Michigan
Ann Arbor, Michigan 48109

University of Michigan

THE FACULTY AND THEIR RESEARCH

Michael Atzmon, Associate Professor of Materials Science and Engineering and Nuclear Engineering; Ph.D., Caltech, 1985. Materials thermodynamics and kinetics, amorphous metal alloys, ion beam modification of materials.

John C. Bilello, Professor; Ph.D., Illinois, 1965. Synchrotron radiation and microstructure, thin films, fracture, mechanics of surfaces and interfaces.

Rodney C. Ewing, Professor of Materials Science and Engineering and Nuclear Engineering and Radiological Sciences; Ph.D., Stanford, 1974. Radiation effects in complex ceramics and minerals, crystal chemistry of actinides, nuclear materials.

Frank E. Filisko, Professor of Materials Science and Engineering and Macromolecular Science and Engineering; Ph.D., Case Western Reserve, 1969. Thermodynamics of polymers, biomaterials, polymer processing, composites, electrorheological fluids.

Amit K. Ghosh, Professor; Ph.D., MIT, 1972. Superplasticity, deformation processing, advanced metallic materials.

Ronald Gibala, Frances E. Van Vlack Professor of Materials Science and Engineering; Ph.D., Illinois, 1964. Mechanical behavior, defect solid state, hydrogen in metals, amorphous solids.

Rachel S. Goldman, Dow Corning Assistant Professor of Materials Science and Engineering; Ph.D., California, San Diego, 1995. Growth and characterization of structural, electronic, and optical properties of lattice-mismatched III-V compound semiconductors and metal-semiconductor interfaces, cross-sectional scanning tunneling microscopy of semiconductor heterostructures.

John W. Halloran, Professor; Ph.D., MIT, 1977. Ceramic processing, high-temperature superconductors, engineering ceramics.

William F. Hosford, Professor; Sc.D., MIT, 1959. Crystal plasticity, metal forming, anisotropic behavior.

J. Wayne Jones, Professor of Materials Science and Engineering and Associate Dean of Undergraduate Education, Engineering Administration; Ph.D., Vanderbilt, 1977. High-temperature materials, fracture, fatigue and creep properties.

Richard M. Laine, Associate Professor of Materials Science and Engineering, Macromolecular Science and Engineering, and Chemistry; Ph.D., USC, 1973. Inorganic and organometallic precursors, materials chemistry, catalysis.

John F. Mansfield, Associate Research Scientist; Ph.D., Bristol (England), 1983. Analytical electron microscopy of metals, semiconductors, and superconductors.

David C. Martin, Associate Professor of Materials Science and Engineering and Macromolecular Science and Engineering; Ph.D., Massachusetts, 1989. Structure-mechanical property relationships of polymers, polymer characterization, high-resolution electron microscopy.

Jyotirmoy Mazumder, Professor of Materials Science and Engineering and Robert H. Lurie Professor of Engineering in Mechanical Engineering and Applied Mechanics; Ph.D., Imperial College (London), 1978. Laser-aided manufacturing, atom to application for nonequilibrium synthesis, mathematical modeling, spectroscopic and optical diagnostics of laser materials interaction.

Joanna Mirecki Millunchick, Assistant Professor of Materials Science and Engineering; Ph.D., Northwestern, 1997. Correlation of structural and theoretical aspects of materials to optical and electrical properties via photoluminescence, hall mobility, and resistivity measurements; fabrication of novel microelectronic devices.

Xiaoqing Pan, Associate Professor of Materials Science and Engineering; Ph.D., Saarlandes (Germany), 1991. High-resolution electron microscopy, structural-property relationships of ceramics, thin-film processing and characterization.

Robert D. Pehlke, Professor of Materials Science and Engineering and Chemical Engineering; Sc.D., MIT, 1960. Chemical and process metallurgy, computer applications, heat transfer.

Richard E. Robertson, Professor of Materials Science and Engineering and Director of Macromolecular Science and Engineering; Ph.D., Caltech, 1960. Polymer structure, molecular dynamics and fracture, fiber composite properties, composite design and manufacturing.

John E. Sanchez Jr., Associate Professor of Materials Science and Engineering; Ph.D., Berkeley, 1990. Physical metallurgy and materials science of thin-film materials, crystallographic texture formation in layered materials, grain growth and second phase formation, microstructural evolution, and the mechanical properties of thin-film materials; metalligation of reliability of microelectronic devices.

David J. Srolovitz, Edward DeMille Campbell Professor of Materials Science and Engineering; Ph.D., Pennsylvania, 1981. Theoretical materials science and physics, computer simulation.

Michael D. Thouless, Associate Professor of Materials Science and Engineering and Mechanical Engineering and Applied Mechanics; Ph.D., Berkeley, 1984. Mechanical properties of materials, mechanics of thin films, coatings and interfaces, toughening of polymers, mechanical properties of adhesives.

Gary S. Was, Professor of Materials Science and Engineering and Nuclear Engineering; Sc.D., MIT, 1980. Ion-beam modification, radiation effects, stress corrosion cracking, hydrogen embrittlement.

Steven M. Yalisove, Associate Professor; Ph.D., Pennsylvania, 1986. Thin-film materials, surface analytical and ion-beam techniques, surface and interface structures.

Albert F. Yee, Professor and Chair of Materials Science and Engineering and Professor of Macromolecular Science and Engineering; Ph.D., Berkeley, 1971. Physics of polymers, mechanical properties of polymers and composites.

UNIVERSITY OF NORTH CAROLINA AT CHAPEL HILL

Curriculum in Applied and Material Sciences

Programs of Study

The Curriculum in Applied and Materials Sciences at the University of North Carolina at Chapel Hill is an interdisciplinary graduate program that brings together faculty members from physics, chemistry, and various departments in the health sciences, including dentistry, orthopedics, and biomedical engineering to engage in research and training in materials science. Faculty members from other departments, such as computer science, mathematics, environmental sciences and engineering, and biochemistry and biophysics also participate. Established in 1996 and administered through the Curriculum in Applied and Materials Sciences, the program is unique among materials science programs in that it is not located in an engineering school, but rather builds on the strengths in fundamental and applied science in the related disciplines at the University. The primary areas of emphasis in the program are electronic and optical materials, polymeric materials, and biomaterials. Students pursuing M.Sc. and Ph.D. degrees in materials science begin their studies with a core curriculum covering the fundamentals of materials and their structures, surfaces, fabrication, thermodynamics, and materials science laboratory techniques. They continue with elective courses as appropriate to their area of research concentration. Graduate students engage in research under the supervision of one of the participating faculty members in the Curriculum in Applied and Materials Sciences. The core curriculum consists of courses that are taken by all materials science students, unless they have received equivalent training elsewhere. All students must pass the following courses unless they have passed their equivalents elsewhere: APPL 141, 143 and MTSC 101, 102, 103, and 104. The courses cover fundamentals of materials science, structure of solids, materials fabrication, chemistry and physics of surfaces, thermodynamics, kinetics and diffusion, electronic and optical properties, and polymers. These courses are ordinarily completed in the first two years of study. Students with a background in a related field may find it necessary to take one or more preliminary courses in physics, chemistry, or mathematics as preparation for the core materials science courses. Elective courses are offered in a variety of materials science areas. Examples include Chemistry and Physics of Electronic Materials Processing, Physical Chemistry of Polymers, and Biocompatibility and Tissue Integration of Biomaterials. Decisions regarding course choices are guided by the student's research interests and the advice of the student's research supervisor and the Director of Graduate Studies. Each student, therefore, follows an individualized course of study.

Research Facilities

Students and faculty in the curriculum have access to the following central facilities located in various departments: NMR (2), computer modeling and computer graphics, confocal microscopy, electron microscopy, glass shop, machine shop (2), laser lab, mechanical testing, mass spectroscopy, and X-ray diffraction. In addition, a variety of equipment is located in individual research laboratories. This includes equipment for thermal analysis; polymer synthesis; FTIR, UV-Vis, Raman, and photluminescence spectroscopy; ellipsometry; CVD; MBE; thermal oxidation; AFM; RBS and ion channeling; electrical measurements; nonlinear optics; low temperatures; and high pressures. Facilities at other campuses (North Carolina State University) and MCNC are also available.

Financial Aid

Applicants to the program will automatically be considered for the limited number of assistantships and fellowships administered by the University, but acceptance to the program does not guarantee financial aid. Applicants are also urged to apply for the various national fellowships and scholarships in science and technology, such as those sponsored by the National Science Foundation, the National Institutes of Health, the Departments of Energy and Defense, and industrial sponsors.

Cost of Study

In 2000–2001, the tuition for a full-time student who is a North Carolina resident will be $1126 for a semester, and for a full-time out-of-state student, about $5750. Tuition does not include student fees, which are about $400. The College of the Arts and Sciences usually provides tuition remission for out-of-state students. With recent legislation, tuition waivers for qualifying graduate students have been adopted and recommended. As long as these recommendation for remission and waivers continue and the student remains in good standing, the student will be responsible only for their student fees. This policy may change in the future.

Living and Housing Costs

Costs vary widely depending on location and situation. The University has a limited amount of housing available for students with families, but most students live in privately rented apartments or houses.

Student Group

There are about a dozen graduate students with backgrounds in biology, chemistry, physics, and traditional materials science involved in a variety of projects, ranging from biomedical materials through CVD of fullerene materials. Five or more students are admitted each year.

Location

The University of North Carolina at Chapel Hill, the nation's first state university, is located near Research Triangle Park, home to more than 100 industries and foundations vital to the nation's ongoing research interests. The cities of Chapel Hill, Raleigh, and Durham, which form the points of the triangle, foster a unique metropolitan area offering abundant cultural and educational opportunities. *Money* magazine ranked the Triangle area first among the top 300 places to live in the United States in 1994. The town of Chapel Hill enjoys the best of small-town living in a unique cosmopolitan environment. The campus, which covers more than 700 acres, enrolls 24,000 undergraduate, graduate, and professional students and is one of sixteen constituent institutions of the multicampus state University. Study and research projects that span more than one campus are commonplace.

The University

The University of North Carolina at Chapel Hill was the first state university to admit students. It was chartered in 1789 and formally opened in 1795. The University's first building, Old East, is a national landmark. The University has a planetarium, an art museum, free concerts and movies, faculty and student art shows, University forums, and many other outstanding cultural activities. Excellent sports facilities are available to students. The University is committed to the principle of equal opportunity, and it does not discriminate on the basis of race, sex, color, national origin, religion, or handicap in its relationships with students, employees, or applicants for admission or employment.

Applying

To be admitted for study toward the M.Sc. or Ph.D. degree in materials science at UNC, an applicant must have completed an undergraduate degree in one or more of the chemical, physical, biological, or materials sciences or in engineering or mathematics. Exceptions to this policy are made at the discretion of the Graduate Admissions Committee. The verbal, quantitative, and analytical portions of the Graduate Record Examinations are required, and a GRE Subject Test in an appropriate subject (e.g. physics, chemistry, engineering) is recommended. If the applicant does not hold an undergraduate degree from an English-speaking university or college, acceptable scores on the Test of English as a Foreign Language (TOEFL) are required. Other materials to be submitted are detailed in the application forms.

Correspondence and Information

Professor Sean Washburn, Chairman, or
Elizabeth Craig, Coordinator
Curriculum in Applied and Material Sciences
CB#3287, 18-1A Venable Hall
University of North Carolina at Chapel Hill
Chapel Hill, North Carolina 27599-3287

Telephone: 919-962-6293
Fax: 919-962-3341
E-mail: materials_science@unc.edu
World Wide Web: http://www.unc.edu/depts/appl/ (Curriculum)
http://www.unc.edu (University)

University of North Carolina at Chapel Hill

THE FACULTY

Sean Washburn, Chair (physics and astronomy). Virtual reality interface for scanning probe microscopies for the manipulation of viruses and DNA, quantum transport, plasma reactor studies, virtual reality interfaces to instruments, electronic transport in conducting polymers doped either chemically or by ion implantation, plasma reactor studies aimed at providing clean, homogeneous large-area plasmas of moderate density and temperature for semiconductor applications, specifically for the growth of large-area diamond films.

Otto Zhou, Associate Chair for Graduate Studies (physics and astronomy). New methods of synthesis of carbon nanotubes and related materials, synthesis and properties of novel solid-state materials.

Stephen Quint, Associate Chair for Undergraduate Studies.

Steven Bayne (biomaterials). Dentin structure, bonding to dentin and polymer/collagen interactions, and fatigue analysis of biological structures using finite element analysis, mechanical property testing, and strain-gauge analysis of structural deformations.

John J. Boland (chemistry). Surface nucleation and growth.

Miles A. Crenshaw (pediatric dentistry). Control of nucleation and of post-nucleation growth by the matrices in biomineralization and the development of biomimetic composites.

Joseph M. DeSimone (chemistry). New methods of synthesis of engineering thermoplastics and fibers, in particular by using supercritical carbon dioxide; polymeric materials synthesis.

M. Gregory Forest (mathematics). Mathematics of crystallization in polymer flows, industrial fiber processes.

Eugene Irene (chemistry). Fabrication and luminescence behavior of silicon oxynitride films, with possible applications to flat-panel displays.

Robert P. Kusy (orthodontics and biomedical engineering). Fabrication and evaluation of micron-sized ultra-high-strength, ultra-high modulus composites for dental and medical applications; orthodontic materials—properties of materials.

Richard W. Linton (chemistry). Spectroscopic techniques for surface and microanalysis.

Jianping Lu (physics and astronomy). Monte Carlo simulations, molecular dynamics, and first-principles calculations to study self-assembly in a variety of systems relevant to living organisms, theoretical studies of materials, relationship between structure and properties in fullerene molecular solids such as KxC60 and carbon nanotubes.

Carol Lucas (biomedical engineering). Mathematical modeling of materials.

Laurie E. McNeil (physics and astronomy). Vibrational spectroscopy to study the relation between structural order and properties in self-organized block copolymer materials, structure-property relations, optical spectroscopy, fabrication and luminescence behavior of silicon oxynitride films, with possible applications to flat-panel displays.

Thomas J. Meyer (chemistry). Structured interfaces at electrodes.

Royce W. Murray (chemistry). Charge and mass transport in amorphous solids and semi-solids, electron transfer active polymers, metal clusters.

Nalin Parikh (physics and astronomy). Ion beam analysis to study doping of high-band-gap materials such as diamond and GaN for use in electronic devices, ion beam modifications and analysis.

Michael Rubinstein (chemistry). Theoretical analyses of polymer properties by building and solving corresponding molecular models, molecular models of polymers.

Edward T. Samulski (chemistry). Structure-property relations in liquid crystals and liquid crystal polymers using NMR and nonlinear optics, liquid crystals and liquid crystal polymers.

Richard Superfine (physics and astronomy). Virtual reality interface for scanning probe microscopies for the manipulation of viruses and DNA; interfacial ordering of molecules; scanning probe microscopies to investigate interfacial ordering of molecules, polypeptides, polymers, and colloidal particles.

Jeffrey Thompson (dentistry and biomedical engineering). Glass-ceramic materials as potential dental restoratives and their fracture using fractal analysis.

Frank Tsui (physics and astronomy). Synthesis of artificially structured materials.

Yue Wu (physics and astronomy). Using NMR to characterize quantum confinement in noninteracting GaA nanocrystals; quasicrystals, nanocrystals, and molecular motion in polymers.

UNIVERSITY OF PENNSYLVANIA

Department of Materials Science and Engineering

Programs of Study

The Department of Materials Science and Engineering offers programs leading to the degrees of Master of Science in Engineering (M.S.E.) and Doctor of Philosophy (Ph.D.). Details of the requirements for advanced degrees are found in *Procedures for Advanced Degrees,* which is given to all entering students or available from the department office upon request. The M.S.E. degree requires the completion of 10 course units, including a thesis based on original research or a research report based on a nonexperimental investigation of an approved topic. The M.S.E. degree is not a prerequisite for the Ph.D. degree. The requirements for the Ph.D. degree include completion of a minimum of twelve courses, passing of the Ph.D. qualifying examination and the oral research examination, an original dissertation, and passing of an oral dissertation defense. The goal of the program is to produce leaders in materials research, education, and application in industrial and academic worlds. Students may also take courses relevant to the research area from other departments, and interdisciplinary research is encouraged. Planning is underway to integrate the materials education in various departments. Details of the plan are available upon request.

Research Facilities

Modern research facilities exist at the Department of Materials Science and Engineering and the Laboratory for Research on the Structure of Matter (LRSM). They include an ion scattering facility (1.7 MV Pelletron accelerator for RBS, FRS, NRA, ion channeling, and TOF-MBE), a scanning probe imaging facility (STM and AFM), an electron microscopy facility (high-resolution 1.8Å TEM, analytical TEM, high-resolution SEM, regular TEM, and SEM), a thermal analysis facility (DSC, DMA, TGA, and DTA), a mechanical testing facility, a materials processing facility, a ceramics processing facility, an X-ray scattering facility, and a computing facility (Silicon Graphics, etc.). Direct access to national facilities through PRT or CAT include the Advanced Photon Source (Argonne National Laboratory), National Synchrotron Light Source (Brookhaven National Laboratory), and Neutron Scattering Facility (NIST). In addition, students carry out experiments at Intense Pulsed Neutron Source (Argonne National Laboratory) and other national facilities.

Financial Aid

A number of yearly fellowships and scholarships are available on a competitive basis. Provisions of these awards vary; the maximum benefits include payment of tuition and the general and technology fees plus a stipend. No financial aid is offered to part-time students.

Cost of Study

Tuition for full-time study for 1998–99 (including summer) was $28,908. For the summer, there was also a general fee of $1484 for full-time study. For part-time study, tuition was $2876 per course unit (one course); there were also a general fee of $170 per course and a technology fee of $55 per course. During the summer, the general fee was $137 per course and there is no technology fee.

Living and Housing Costs

On-campus housing is available for both single and married students. Residences for single students cost $445 and up per month; for a shared-living situation, $500 and up per month; and for a private apartment, $800 and up per month. Housing costs for married couples range from $650 to $1030 per month. There are numerous privately owned apartments in the immediate area.

Student Group

The graduate student body of about 50 includes students with diverse backgrounds and undergraduate degrees in materials science, physics, chemistry, mathematics, and other engineering fields.

Location

The University is located in West Philadelphia, just a few blocks from the heart of the city. Philadelphia is a twentieth-century city with seventeenth-century origins. Renowned museums, concert halls, theaters, and sports arenas provide cultural and recreational outlets for students. Fairmount Park extends through large sections of Philadelphia, occupying both banks of the Schuylkill River. Not far away are the Jersey shore to the east, Pennsylvania Dutch country to the west, and the Poconos to the north. Equidistant from New York City and Washington, D.C., the city of Philadelphia is a patchwork of distinctive neighborhoods ranging from colonial Society Hill to Chinatown.

The University and The Department

The University of Pennsylvania traces its origins more than 250 years ago to Benjamin Franklin. Today Penn is noted for the strength of its professional schools in medicine, law, and business (Wharton) and its science and engineering graduate programs. Materials science at Penn draws upon the excellence across campus by involving other departments, such as chemistry, physics, and chemical engineering. Many of the materials science faculty members are also members of the Laboratory for Research on the Structure of Matter (LRSM), which is an interdisciplinary research center at Penn. The LRSM encourages collaborative research projects, supports superior central equipment facilities, and provides connections to local industry.

Applying

Candidates who will obtain a bachelor's degree may apply for admission by submitting an application to the address below. Applications for fall admission must be received by January 15 to ensure consideration for financial aid. Admission is based on the student's past record and letters of recommendation. Scores on the Graduate Record Examinations are required. All students whose native language is not English must arrange to take the Test of English as a Foreign Language (TOEFL) prior to making application; the minimum accepted score is 600.

Correspondence and Information

Professor Peter K. Davies
Graduate Group Chair
Department of Materials Science and Engineering
University of Pennsylvania
3231 Walnut Street
Philadelphia, Pennsylvania 19104
Telephone: 215-898-8337
Fax: 215-573-2128
E-mail: msegrad@lrsm.upenn.edu
World Wide Web: http://www.seas.upenn.edu/mse/msehome.html

University of Pennsylvania

THE FACULTY AND THEIR RESEARCH

John A. Bassani, Professor and Chair, Mechanical Engineering and Applied Mechanics; Ph.D., Harvard. Continuum mechanics, plastic deformation of crystals, material instabilities, interface mechanics, atomic-level properties and photons, fracture mechanics, polymer deformation and fracture.

Dawn A. Bonnell, Associate Professor; Ph.D., Michigan. Nanometer-scale properties in ceramics, atomic structure of surfaces and interfaces, scanning probe microscopy, measurement and modeling of electron tunneling and space charge, magnetic and electric field gradients at interfaces, hierarchical functionality in ceramic systems.

Norman Brown, Professor; Ph.D., Berkeley. Fracture and crazing of structure polymers.

Robert Butera, Adjunct Professor; Ph.D., Case Western. Experimental physical chemistry and physics of polymeric materials and colloidal dispersions; use of rheological techniques to characterize the structure and dynamics of polymeric fluids and colloidal dispersions (flocculated and colloidally stable); special interest in polymerically stabilized colloids, latex stability and flocculation, colloidal gels, and hydrophobic association of amphiphilic polymers.

I-Wei Chen, Skirkanich Professor of Materials Innovation; Ph.D., MIT. Materials science of electronic and structural ceramics, including their thin films, heterostructures, and composites; materials design, synthesis, testing, and modeling.

Russell Composto, Associate Professor; Ph.D., Cornell. Polymer surfaces and interfaces, polymer blends, wetting, adhesion, thin films, diffusion, ion scattering.

Peter K. Davies, Professor; Ph.D., Arizona. Chemistry of solid-state crystal chemistry of inorganic materials, electronic ceramics, synthesis of new oxides, cation ordering reactions, nanostructural defects, high-resolution TEM.

John J. DeLuccia, Adjunct Professor; Ph.D., Pennsylvania. Electrochemical contributions to the micromechanics of dynamically loaded high-strength alloys; damaging interaction of environment and loading on alloys minimized through creation of new benign electrolytes; requisite passivity and electrochemical activity realization and electrochemical current associated with fatigue loading cycle used to assess fatigue damage in electrochemical fatigue sensor (EFS).

Takeshi Egami, Professor and Chairman of the Department; Ph.D., Pennsylvania. Structure and properties of transition metal oxides, superconductivity, neutron scattering, sychrontron radiation.

Gregory C. Farrington, Professor and Dean, School of Engineering and Applied Science; Ph.D., Harvard. Ceramics, advanced inorganic materials; preparation, structure, mechanical properties, solid-state electrochemical processes, conductive polymers.

John E. Fischer, Professor; Ph.D., Rensselaer. Solid-state synthesis, X-ray and neutron scattering, and other techniques to prepare and study novel carbon-based materials, including polymers, fullerenes, and anode materials for rechargeable batteries.

Roger French, Adjunct Professor; Ph.D., MIT. Electronic structure and optical properties of electronic and optical materials.

Louis Girifalco, Professor; Ph.D., Cincinnati. Statistical mechanics of solids, theory of cohesion, technological change.

Charles D. Graham Jr., Professor; Ph.D., Birmingham (England). Magnetic materials, including permanent magnet materials; soft magnetic materials and amorphous alloys; materials for magnetic recording.

William R. Graham, Professor; D.Phil., Oxford. Geometric, electronic, and vibrational structure and properties of surfaces and thin-film interface systems; medium-energy ion scattering, low-energy electron diffraction, Auger electron spectroscopy, molecular beam epitaxy, ultrahigh-vacuum thin-film growth, and characterization techniques.

Campbell Laird, Professor; Ph.D., Cambridge (England). Physical metallurgy, fatigue and fracture, heat treatment of alloys, corrosion, fatigue-sensing devices.

David E. Luzzi, Associate Professor; Ph.D., Northwestern. Material structure-property relationships at nanometer-length scales; carbon nanotubes; intermetallic compounds and composites; structure, diffusion, phase transformations, and mechanical properties of interfaces; electron microscopy.

Charles J. McMahon Jr., Professor; Ph.D., MIT. Physical metallurgy, micromechanisms of deformation and fracture, surface and interface phenomena.

David P. Pope, Professor; Ph.D., Caltech. Deformation of materials, high-temperature fracture, ordered alloys, brittle-to-ductile transition.

Vaclav Vitek, Professor; Ph.D., Czechoslovak Academy of Sciences. Theoretical and computer modeling of lattice defects and interfaces using molecular statics and dynamics; structure and properties of glasses and liquids; microscopic theory of plastic behavior and intergranular fracture, in particular in intermetallic compounds.

Karen I. Winey, Assistant Professor; Ph.D., Massachusetts. Structure-property relationships and thermodynamics of polymers; polymer morphology, polymer rheology, and phase behavior, especially in systems containing block copolymers and ionomers; polymer microscopy.

Wayne L. Worrell, Professor; Ph.D., MIT. Properties and novel applications (fuel cells, membranes, and sensors of mixed-conducting oxides); prevention of environmental degradation of advanced materials (composite and intermetallic compounds) at elevated temperatures; solid-state electochemistry at elevated temperatures.

COLLABORATING FACULTY MEMBERS

Nikolaos Aravas, Associate Professor of Mechanical Engineering and Applied Mechanics; Ph.D., Illinois. Various nonlinear phenomena in interfacial fracture, including plasticity, creep, and diffusion effects, and development of experimentally based models of decohesion in order to identify parameters that control crack growth resistance.

Raymond J. Gorte, Professor and Chair of Chemical Engineering; Ph.D., Minnesota. Heterogeneous catalysts, with particular emphasis on supported metals and zeolites; studies of the metal-oxide interface by using model systems and a wide variety of techniques, including various surface spectroscopies, TEM, and reaction measurements.

Paul A. Heiney, Professor of Physics; Ph.D., MIT. Techniques of high-resolution X-ray diffraction to study novel states of matter and their phase transitions.

Pedro Ponte Castaneda, Associate Professor of Mechanical Engineering and Applied Mechanics; Ph.D., Harvard. Interfacial crack growth under mixed mode conditions, the effective mechanical and physical properties of composite materials with nonlinear constitutive behavior, evolving microstructures, and the effect of porosity on localization and its implications for forming processes.

Jorge Santiago-Aviles, Associate Professor of Electrical Engineering; Ph.D., Penn State. Materials and devices for microelectronics, more specifically, metallization schemes for interconnections, rapid thermal processing for silicides, and the problem of homogeneous Schottky barriers in epitaxial silicides; electronic materials as biomembranes for sensors and amorphous diamond films.

Larry Sneddon, Professor of Chemistry; Ph.D., Indiana. Design and synthesis of new polymeric precursors to nonoxide ceramics that allow the controlled formation of the material in processed forms, such as fibers and films; use of new inorganic polymers first synthesized in the lab, including polyvinylborazine, polyborazylene, and polyborosilazanes, as precursors to a variety of boron- and/or silicon-based ceramics, such as boron carbide, boron nitride, metal borides, and SiNCB composites.

John Vohs, Associate Professor of Chemical Engineering and Associate Dean for Undergraduate Studies, SEAS; Ph.D., Delaware. Surface-sensitive spectroscopic techniques to study fundamental aspects of surface and interfacial phenomena occurring on ceramics and semiconductors.

THE UNIVERSITY OF TEXAS AT ARLINGTON

Materials Science and Engineering Program

Program of Study

The Materials Science and Engineering Program offers a multidisciplinary curriculum for master's and Ph.D. candidates. The program involves a set of core materials science and engineering courses that include physics of solids, chemical thermodynamics, mechanical behavior of materials, analysis of materials, and comprehensive survey courses of physical metallurgy, polymers, and ceramics. Subsequent to completion of a six-course core curriculum, Ph.D. candidates take supplemental and specialization courses in materials science and engineering or within a particular academic discipline. The course work of advanced study is tailored to the student in consultation with the student's research adviser. The training in this program is provided by the integrated efforts of faculty members in materials science and engineering, electrical engineering, civil engineering, mechanical engineering, aerospace engineering, biomedical engineering, biology, physics, mathematics, and chemistry. A minimum of 42 semester hours of graduate course work is required for students entering the Ph.D. program with a bachelor's degree. For those students entering with a master's degree, a minimum of 24 semester hours of graduate course work is required for the Ph.D. program. Students enrolled in the master's program take a minimum of 24 hours of graduate course work, predominantly in materials science and engineering courses.

Research Facilities

One of the major strengths of the Materials Science and Engineering Program is the research capabilities and facilities available. These include an electron microscope (SEM and TEM), class 10/100 clean rooms, autoclaves, a scanning tunneling microscope, and Cray supercomputers and facilities for polymer characterization, mechanical testing, thermal analysis, surface spectroscopy, molecular beam epitaxy, structural testing, electrooptics, optical spectroscopy, magnetic resonance, submicron lithography, and X-ray analysis. The laboratories are located in a number of modern, well-equipped buildings around campus. There are seven computer labs easily accessible to students.

There are a large number of research areas available to students in the Materials Science and Engineering Program, including areas for research in advanced composites, ceramics, semiconductors processing, metal matrix composites, positron annihilation, electrically conductive polymers, structural materials, electronic devices, biocompatible materials, surface physics, intermetallic systems, materials modeling, and optoelectronics. A number of research centers exist at UTA. These include the National Science Foundation Industry/University Cooperative Research Center for Advanced Electron Devices and Systems, the Center for Composite Materials, the Center for Advanced Polymer Research, the Center for Electron Microscopy, and the Center for Positron Studies.

Financial Aid

Teaching and research assistantships are available on a competitive basis. Also, work study, student loans, and grants are available on a need basis. For students on graduate assistantships, tuition is set at in-state rates.

Cost of Study

In 1999–2000, the tuition for a full-time student who is a Texas resident is about $1300 for a semester, and for a full-time out-of-state student, about $4000. Tuition does not include a general property deposit, a photo identification card fee, parking fees, and laboratory and supplemental fees. Graduation fees are paid in the semester in which the students graduates. Book costs vary depending on classes taken.

Living and Housing Costs

Rooms in residence halls range from $630 to $710 per semester (double) or from $945 to $1065 (single). The University also has efficiency one- and two-bedroom apartments that rent for $230 to $380 a month as well as two- and three-bedroom houses that rent from $400 to $454. Private rooms and apartments are available in off-campus residential areas, and listings are available in the housing office.

Student Group

The Materials Science and Engineering Program is composed of 15 master's and 18 doctoral students. About 10 students per semester receive research or teaching assistantships. Most students in the program pursue careers in research or industry; however, some pursue a teaching career. Nine students graduated in the 1998–99 academic year.

Location

Arlington is a city of more than 300,000 people and is located at the center of the Dallas–Fort Worth Metroplex. Electronics, aerospace, military, and other industries have provided national leadership in advanced materials development and application. UTA maintains close relationships with these industries, which provide exceptional opportunities for research. Due to the diversity of the population in the Dallas Metroplex area, there are many cultural activities for students to discover. Arlington is home to the Texas Rangers baseball team and is located near Irving, Texas, home of the 1993, 1994, and 1996 Super Bowl Champion Dallas Cowboys.

The University

UTA was founded in 1895 as Arlington College, and in 1959 it was elevated to senior college rank. In 1965 it was transferred from the Texas A&M University System to the University of Texas System. It is now the second-largest institution within the UT system. It has an enrollment of 19,000. UTA is located on a modern 365-acre campus in Arlington.

Applying

Application requirements for prospective students include a bachelor's degree, GRE scores, official undergraduate and graduate transcripts, three recommendation letters, an official application, and a nonrefundable $25 application fee. The deadlines for domestic applications are October 18 for the spring 2000 semester and June 19 for the fall 2000 semester.

International students and permanent residents interested in applying require a bachelor's degree, GRE and TOEFL scores, official undergraduate and graduate transcripts, three recommendation letters, an official application, an official graduate school financial statement form accompanied by an affidavit, and a $50 nonrefundable application fee. The deadlines for international applications are September 13 for the spring 2000 semester and April 12 for the fall 2000 semester.

Correspondence and Information

Dr. Ronald L. Elsenbaumer, Chairman or
Dr. Pranesh B. Aswath, Graduate Advisor
Materials Science and Engineering Program
The University of Texas at Arlington
500 West 1st Street
P.O. Box 19031
Arlington, Texas 76019-0031
Telephone: 817-272-2398
Fax: 817-272-2538
E-mail: aswath@uta.edu

The University of Texas at Arlington

THE FACULTY

Below is a listing of the faculty members associated with the Materials Science and Engineering Program and their current department affiliation. The date in parentheses is the year in which the faculty member joined the department.

Core Faculty

Pranesh B. Aswath, Associate Professor, Mechanical Engineering and Materials Science and Engineering (1990); Ph.D., Brown, 1990.
Wen S. Chan, Professor, Mechanical Engineering and Materials Science and Engineering (1988); Ph.D., Purdue, 1979.
Ronald L. Elsenbaumer, Professor, Chemistry, and Chairman of Materials Science and Engineering (1991); Ph.D., Stanford, 1978.
Roger D. Goolsby, Professor, Mechanical Engineering and Materials Science and Engineering (1980); Ph.D., Berkeley, 1971.
Robert M. Johnson, Professor, Mechanical Engineering and Materials Science and Engineering; Ph.D., Oklahoma, 1967.
Choong-un Kim, Assistant Professor, Mechanical Engineering and Materials Science and Engineering (1996); Ph.D., Berkeley, 1993.

Participating Faculty

Kambiz Alavi, Professor, Electrical Engineering (1988); Ph.D., MIT, 1981.
Truman D. Black, Professor, Physics (1965); Ph.D., Rice, 1964.
Ronald L. Carter, Professor, Electrical Engineering (1979); Ph.D., Iowa State, 1971.
W. Alan Davis, Associate Professor, Electrical Engineering (1983); Ph.D., Michigan, 1971.
Robert C. Eberhart, Professor and Chairman, Biomedical Engineering Program (1978); Ph.D., Berkeley, 1965.
John L. Fry, Professor, Physics (1971); Ph.D., California, Riverside, 1966.
Donald Greenspan, Professor, Mathematics (1978); Ph.D., Maryland, 1956.
A. Haji-Sheikh, Professor, Mechanical Engineering (1966); Ph.D., Minnesota, 1965.
Shiv P. Joshi, Professor, Aerospace Engineering (1988); Ph.D., Purdue, 1985.
Gary Kinsel, Assistant Professor, Chemistry (1994); Ph.D., Colorado, 1989.
Ali Riza Koymen, Associate Professor, Physics (1990); Ph.D., Michigan, 1984.
Frederick M. MacDonnell, Assistant Professor, Chemistry (1994); Ph.D., Northwestern, 1993.
Robert Magnusson, Associate Professor, Electrical Engineering (1984); Ph.D., Georgia Tech, 1976.
Dennis S. Marynick, Professor, Chemistry (1978); Ph.D., Harvard, 1973.
H. Keith McDowell, Professor, Chemistry (1991); Ph.D., Harvard, 1972, 1974.
Seiichi Nomura, Associate Professor, Mechanical Engineering (1982); Ph.D., Delaware, 1980; D.Eng., Tokyo, 1982.
Martin Pomerantz, Professor, Chemistry (1976); Ph.D., Yale, 1964.
Krishnan Rajeshwar, Professor, Chemistry (1983); Ph.D., Indian Institute of Science, 1974.
Asok K. Ray, Associate Professor, Physics (1984); Ph.D., Texas Tech, 1977.
Roy S. Rubins, Professor, Physics (1969); D.Phil., Oxford, 1961.
Zoltan A. Schelly, Professor, Chemistry (1977); D.Sc., Vienna Technical, 1967.
Suresh C. Sharma, Professor, Physics (1977); Ph.D., Brandeis, 1976.
Richard B. Timmons, Professor, Chemistry (1977); Ph.D., Catholic University, 1962.
Bo Ping Wang, Professor, Mechanical Engineering (1983); Ph.D., Virginia, 1974.
Alexander Weiss, Professor, Physics (1984); Ph.D., Brandeis, 1983.
Roy N. West, Professor, Physics (1987); Ph.D., London, 1965.
Robert L. Yuan, Professor, Civil Engineering and Engineering Mechanics (1968); Ph.D., Illinois, 1968.

Section 17
Mechanical Engineering and Mechanics

This section contains a directory of institutions offering graduate work in mechanical engineering and mechanics, followed by in-depth entries submitted by institutions that chose to prepare detailed program descriptions. Additional information about programs listed in the directory but not augmented by an in-depth entry may be obtained by writing directly to the dean of a graduate school or chair of a department at the address given in the directory.

For programs offering related work, see also in this book Engineering and Applied Sciences, Management of Engineering and Technology, and Materials Sciences and Engineering. In Book 4, see Geosciences and Physics.

CONTENTS

Mechanical Engineering

Alfred University, Graduate School, Graduate School, Program in Mechanical Engineering, Alfred, NY 14802-1205. Offers MS. Part-time programs available. *Faculty:* 4 full-time (0 women), 1 part-time (0 women). *Students:* 3 full-time (1 woman). 9 applicants, 78% accepted. *Degree requirements:* For master's, thesis required, foreign language not required. *Entrance requirements:* For master's, TOEFL. *Application deadline:* Applications are processed on a rolling basis. Application fee: $50. *Financial aid:* Research assistantships, career-related internships or fieldwork, Federal Work-Study, and tuition waivers (full and partial) available. Aid available to part-time students. Financial aid applicants required to submit FAFSA. *Unit head:* Dr. David Szczerbacki, Acting Dean, E-mail: fszczerbacki@king.alfred.edu. *Application contact:* Cathleen R. Johnson, Coordinator of Graduate Admissions, 607-871-2141, Fax: 607-871-2198, E-mail: johnsonc@king.alfred.edu.

Arizona State University, Graduate College, College of Engineering and Applied Sciences, Department of Mechanical and Aerospace Engineering, Tempe, AZ 85287. Offers aerospace engineering (MS, MSE, PhD); engineering science (MS, MSE, PhD); mechanical engineering (MS, MSE, PhD). *Faculty:* 35 full-time (3 women), 2 part-time (0 women). *Students:* 83 full-time (10 women), 39 part-time (6 women); includes 16 minority (1 African American, 11 Asian Americans or Pacific Islanders, 3 Hispanic Americans, 1 Native American), 64 international. Average age 28. 521 applicants, 49% accepted. In 1998, 37 master's, 8 doctorates awarded. *Degree requirements:* For master's, computer language, thesis or alternative required; for doctorate, computer language, dissertation required. *Entrance requirements:* For master's and doctorate, GRE General Test. Application fee: $45. *Financial aid:* Fellowships available. *Faculty research:* Aerodynamics, fluid mechanics, propulsion and space power, advanced structures and materials, robotics and automation. *Unit head:* Dr. Don L. Boyer, Chair, 480-965-3291, E-mail: mae@asu.edu. *Application contact:* Graduate Secretary, 480-965-4979.

Auburn University, Graduate School, College of Engineering, Department of Mechanical Engineering, Auburn, Auburn University, AL 36849-0002. Offers materials engineering (M Mtl E, MS, PhD); mechanical engineering (MME, MS, PhD). Part-time programs available. *Faculty:* 26 full-time (0 women). *Students:* 32 full-time (6 women), 28 part-time (4 women); includes 4 minority (3 African Americans, 1 Asian American or Pacific Islander), 30 international. 94 applicants, 44% accepted. In 1998, 26 master's, 15 doctorates awarded. *Degree requirements:* For master's, thesis (MS) required; for doctorate, dissertation required. *Entrance requirements:* For master's, GRE General Test; for doctorate, GRE General Test (minimum score of 400 on each section required). *Application deadline:* For fall admission, 9/1; for spring admission, 3/1. Applications are processed on a rolling basis. Application fee: $25 ($50 for international students). Tuition, state resident: full-time $2,760; part-time $76 per credit hour. Tuition, nonresident: full-time $8,280; part-time $228 per credit hour. *Financial aid:* Fellowships, research assistantships, teaching assistantships, Federal Work-Study available. Aid available to part-time students. Financial aid application deadline: 3/15. *Faculty research:* Engineering mechanics, experimental mechanics, engineering design, engineering acoustics, engineering optics. *Unit head:* Dr. David Dyer, Chair, 334-844-4820. *Application contact:* Dr. John F. Pritchett, Dean of the Graduate School, 334-844-4700.

Boston University, College of Engineering, Department of Aerospace and Mechanical Engineering, Boston, MA 02215. Offers aerospace engineering (MS, PhD); mechanical engineering (MS, PhD). Part-time programs available. *Faculty:* 26 full-time (3 women), 4 part-time (1 woman). *Students:* 41 full-time (8 women), 5 part-time (1 woman); includes 3 minority (1 African American, 2 Asian Americans or Pacific Islanders), 19 international. Average age 26. 178 applicants, 17% accepted. In 1998, 9 master's, 6 doctorates awarded. Terminal master's awarded for partial completion of doctoral program. *Degree requirements:* For master's, thesis or alternative required, foreign language not required; for doctorate, dissertation required, foreign language not required. *Entrance requirements:* For master's, GRE General Test, TOEFL (minimum score of 500 required; 213 for computer-based); for doctorate, GRE General Test, TOEFL. *Application deadline:* For fall admission, 4/1; for spring admission, 10/1. Applications are processed on a rolling basis. Application fee: $50. Tuition: Full-time $23,770; part-time $743 per credit. Required fees: $220. Tuition and fees vary according to class time, course level, campus/location and program. *Financial aid:* In 1998–99, 7 fellowships with full tuition reimbursements (averaging $13,000 per year), 19 research assistantships with full tuition reimbursements (averaging $11,500 per year), 20 teaching assistantships with full tuition reimbursements (averaging $11,500 per year) were awarded.; career-related internships or fieldwork, Federal Work-Study, institutionally-sponsored loans, and scholarships also available. Financial aid application deadline: 12/15; financial aid applicants required to submit FAFSA. *Faculty research:* Waves and acoustics, dynamics and controls, fluid mechanics, materials processing, precision engineering. Total annual research expenditures: $2.1 million. *Unit head:* Dr. Allan Pierce, Chairman, 617-353-2877, Fax: 617-353-5866. *Application contact:* Cheryl Kelley, Graduate Programs Director, 617-353-9760, Fax: 617-353-0259, E-mail: enggrad@bu.edu.

Bradley University, Graduate School, College of Engineering and Technology, Department of Mechanical Engineering, Peoria, IL 61625-0002. Offers MSME. Part-time and evening/weekend programs available. *Degree requirements:* For master's, comprehensive exam required, thesis optional, foreign language not required. *Entrance requirements:* For master's, TOEFL (minimum score of 525 required), minimum GPA of 3.0. *Faculty research:* Ground-coupled heat pumps, robotic end-effectors, power plant optimization.

Brigham Young University, Graduate Studies, College of Engineering and Technology, Department of Mechanical Engineering, Provo, UT 84602-1001. Offers MS, PhD, MBA/MS. *Faculty:* 23 full-time (0 women). *Students:* 49 full-time (3 women), 27 part-time (3 women); includes 4 minority (2 Asian Americans or Pacific Islanders, 2 Hispanic Americans), 6 international. Average age 24. 43 applicants, 72% accepted. In 1998, 23 master's, 4 doctorates awarded. *Degree requirements:* For master's, foreign language and thesis not required; for doctorate, 2 foreign languages (computer language can substitute for one), dissertation required. *Entrance requirements:* For master's, GRE General Test (minimum combined score of 1750 on three sections required), GRE Subject Test (minimum score of 575 required) or Utah State FE Exam, minimum GPA of 3.0 in last 60 hours; for doctorate, GRE General Test (minimum combined score of 1750 on three sections required), GRE Subject Test (minimum score of 575 required) or Utah State FE Exam. *Average time to degree:* Master's–2 years full-time; doctorate–4 years full-time. *Application deadline:* For fall and spring admission, 2/15; for winter admission, 9/15. Applications are processed on a rolling basis. Application fee: $30. Electronic applications accepted. Tuition: Full-time $3,330; part-time $185 per credit hour. Tuition and fees vary according to program and student's religious affiliation. *Financial aid:* In 1998–99, 4 fellowships with partial tuition reimbursements (averaging $13,000 per year), 20 research assistantships with partial tuition reimbursements (averaging $10,000 per year), 48 teaching assistantships with partial tuition reimbursements (averaging $8,000 per year) were awarded.; scholarships also available. Financial aid application deadline: 3/15; financial aid applicants required to submit FAFSA. *Faculty research:* Combustion, composite materials, advanced design methods and optimization, electronics heat transfer, acoustic noise controls and robotics, manufacturing. Total annual research expenditures: $1.5 million. *Unit head:* Dr. Alan R. Parkinson, Chair, 801-378-2625, Fax: 801-378-5037. *Application contact:* Dr. Craig C. Smith, Graduate Coordinator, 801-378-6545, Fax: 801-378-5037, E-mail: smith@byu.edu.

Brown University, Graduate School, Division of Engineering, Program in Mechanics of Solids and Structures, Providence, RI 02912. Offers Sc M, PhD. *Degree requirements:* For doctorate, dissertation, preliminary exam required, foreign language not required, foreign language not required.

Bucknell University, Graduate Studies, College of Engineering, Department of Mechanical Engineering, Lewisburg, PA 17837. Offers MS, MSME. *Faculty:* 6 full-time, 1 part-time. *Students:* 9 (2 women). *Degree requirements:* For master's, thesis required, foreign language not required. *Entrance requirements:* For master's, GRE General Test (minimum combined score of 1000 required), GRE Subject Test, TOEFL (minimum score of 550 required), minimum GPA of 2.8. *Application deadline:* For fall admission, 6/1 (priority date); for spring admission, 12/1 (priority date). Applications are processed on a rolling basis. Application fee: $25. Tuition: Part-time $2,600 per course. Tuition and fees vary according to course load. *Financial aid:* Unspecified assistantships available. Financial aid application deadline: 3/1. *Faculty research:* Heat pump performance, microprocessors in heat engine testing, computer-aided design. *Unit head:* Dr. David Cartwright, Head, 570-577-3193.

California Institute of Technology, Division of Engineering and Applied Science, Option in Mechanical Engineering, Pasadena, CA 91125-0001. Offers MS, PhD, Engr. *Faculty:* 7 full-time (1 woman). *Students:* 31 full-time (4 women), 17 international. 259 applicants, 2% accepted. In 1998, 4 master's, 5 doctorates, 2 other advanced degrees awarded. *Degree requirements:* For master's, foreign language and thesis not required; for doctorate, dissertation required, foreign language not required. *Application deadline:* For fall admission, 1/15. Application fee: $0. *Faculty research:* Design, mechanics, thermal and fluids engineering, jet propulsion. *Unit head:* Dr. Erik Antonsson, Executive Officer, 626-395-3790.

California Polytechnic State University, San Luis Obispo, College of Engineering, Program in Engineering, San Luis Obispo, CA 93407. Offers biochemical engineering (MS); industrial engineering (MS); integrated technology management (MS); materials engineering (MS); mechanical engineering (MS); water engineering (MS). *Faculty:* 98 full-time (8 women), 82 part-time (14 women). *Students:* 25 full-time (3 women), 6 part-time. *Degree requirements:* Foreign language not required. *Entrance requirements:* For master's, GRE General Test, minimum GPA of 2.5 in last 90 quarter units. Application fee: $55. Tuition, nonresident: part-time $164 per unit. Required fees: $531 per quarter. *Unit head:* Dr. Paul E. Rainey, Associate Dean, 805-756-2131, Fax: 805-756-6503, E-mail: prainey@calpoly.edu. *Application contact:* Dr. Paul E. Rainey, Associate Dean, 805-756-2131, Fax: 805-756-6503, E-mail: prainey@calpoly.edu.

California State University, Chico, Graduate School, Interdisciplinary Programs, Chico, CA 95929-0722. Offers applied mechanical engineering (MS); interdisciplinary studies (MA, MS); simulation science (MS). Part-time programs available. *Students:* 49 full-time (29 women), 45 part-time (32 women); includes 15 minority (3 African Americans, 11 Hispanic Americans, 1 Native American), 6 international. *Degree requirements:* For master's, thesis or alternative, oral exam required, foreign language not required. *Entrance requirements:* For master's, GRE General Test or MAT. *Application deadline:* For fall admission, 4/1. Applications are processed on a rolling basis. Application fee: $55. *Unit head:* Dr. Mark J. Morlock, Graduate Coordinator, 530-895-6171.

California State University, Fresno, Division of Graduate Studies, School of Engineering, Program in Mechanical Engineering, Fresno, CA 93740-0057. Offers MS. Offered at Edwards Air Force Base. *Faculty:* 3 part-time (0 women). 1 applicants, 100% accepted. In 1998, 4 degrees awarded. *Degree requirements:* For master's, computer language, thesis or alternative required, foreign language not required. *Entrance requirements:* For master's, GRE General Test, TOEFL (minimum score of 550 required). *Application deadline:* For fall admission, 8/1 (priority date); for spring admission, 12/1. Applications are processed on a rolling basis. Application fee: $55. Electronic applications accepted. Tuition, nonresident: part-time $246 per unit. Required fees: $1,906; $620 per semester. *Financial aid:* Application deadline: 3/1; *Unit head:* Wynn P. McDonald, Director, Graduate Admissions, 517-249-1696, Fax: 517-790-0180, E-mail: gradadm@svsu.edu. *Application contact:* James W. Smolka, Coordinator, 805-258-5936.

California State University, Fullerton, Graduate Studies, School of Engineering and Computer Science, Department of Mechanical Engineering, Fullerton, CA 92834-9480. Offers MS. Part-time programs available. *Faculty:* 8 full-time, 6 part-time. *Students:* 3 full-time (1 woman), 26 part-time (2 women); includes 11 minority (6 Asian Americans or Pacific Islanders, 5 Hispanic Americans), 6 international. Average age 30. 27 applicants, 48% accepted. In 1998, 9 degrees awarded. *Degree requirements:* For master's, computer language, comprehensive exam, project or thesis required. *Entrance requirements:* For master's, minimum undergraduate GPA of 2.5. Application fee: $55. Tuition, nonresident: part-time $264 per unit. Required fees: $1,947; $1,281 per year. *Financial aid:* Career-related internships or fieldwork, Federal Work-Study, grants, and institutionally-sponsored loans available. Aid available to part-time students. Financial aid application deadline: 3/1. *Unit head:* Dr. Timothy Lancey, Chair, 714-278-3014.

California State University, Long Beach, Graduate Studies, College of Engineering, Department of Mechanical Engineering, Long Beach, CA 90840. Offers MSE, MSME. Part-time programs available. *Faculty:* 18 full-time (1 woman), 14 part-time (2 women). *Students:* 10 full-time (1 woman), 75 part-time (6 women); includes 29 minority (1 African American, 22 Asian Americans or Pacific Islanders, 6 Hispanic Americans), 8 international. Average age 31. 63 applicants, 51% accepted. In 1998, 9 degrees awarded. *Degree requirements:* For master's, thesis required, foreign language not required. *Entrance requirements:* For master's, TOEFL (minimum score of 550 required). *Application deadline:* For fall admission, 8/1; for spring admission, 12/1. Application fee: $55. Electronic applications accepted. Tuition, nonresident: part-time $246 per unit. Required fees: $569 per semester. Tuition and fees vary according to course load. *Financial aid:* Career-related internships or fieldwork, Federal Work-Study, grants, institutionally-sponsored loans, and unspecified assistantships available. Financial aid application deadline: 3/2. *Faculty research:* Unsteady turbulent flows, solar energy, energy conversion, CAD/CAM, computer-assisted instruction. *Unit head:* Dr. Leonardo Perez y Perez, Chairman, 562-985-4407, Fax: 562-985-4408, E-mail: perez@engr.csulb.edu. *Application contact:* Dr. C. Barclay Gilpin, Graduate Coordinator, 562-985-1520, Fax: 562-985-4408, E-mail: gilpin@engr.csulb.edu.

California State University, Los Angeles, Graduate Studies, School of Engineering and Technology, Department of Mechanical Engineering, Los Angeles, CA 90032-8530. Offers MS. Part-time and evening/weekend programs available. *Faculty:* 7 full-time, 3 part-time. *Students:* 4 full-time (0 women), 20 part-time; includes 13 minority (6 Asian Americans or Pacific Islanders, 7 Hispanic Americans), 7 international. In 1998, 5 degrees awarded. *Degree requirements:* For master's, computer language, comprehensive exam or thesis required. *Entrance requirements:* For master's, TOEFL (minimum score of 550 required), minimum GPA of 2.75. *Application deadline:* For fall admission, 6/30; for spring admission, 2/1. Applications are processed on a rolling basis. Application fee: $55. *Financial aid:* In 1998–99, 9 students received aid. Federal Work-Study available. Aid available to part-time students. Financial aid application deadline: 3/1. *Faculty research:* Mechanical design, thermal systems, solar-powered vehicle. *Unit head:* Dr. Stephen Felszeghy, Chair, 323-343-4490.

California State University, Northridge, Graduate Studies, College of Engineering and Computer Science, Department of Civil, Industrial and Applied Mechanics, Department of Mechanical Engineering, Northridge, CA 91330. Offers aerospace engineering (MS); applied engineering (MS); machine design (MS); mechanical engineering (MS); structural engineering (MS); thermofluids (MS). Part-time and evening/weekend programs available. *Faculty:* 8 full-time, 4 part-time. *Students:* 3 full-time (0 women), 44 part-time (4 women); includes 17 minority (14 Asian Americans or Pacific Islanders, 2 Hispanic Americans, 1 Native American), 3 international. Average age 33. 16 applicants, 75% accepted. In 1998, 8 degrees awarded. *Degree requirements:* For master's, thesis or alternative required, foreign language not required. *Entrance requirements:* For master's, GRE General Test, TOEFL, minimum GPA of 2.5. *Application deadline:* For fall admission, 11/30. Application fee: $55. Tuition, nonresident: part-time $246 per unit. International tuition: $7,874 full-time. Required fees: $1,970. Tuition and fees vary according to course load. *Financial aid:* Application deadline: 3/1. *Unit head:* Dr. William J. Rivers, Chair, 818-677-2187. *Application contact:* Dr. Tom Mincer, Graduate Coordinator, 818-677-2007.

California State University, Sacramento, Graduate Studies, School of Engineering and Computer Science, Department of Mechanical Engineering, Sacramento, CA 95819-6048. Offers MS. Evening/weekend programs available. *Degree requirements:* For master's, thesis or alternative, writing proficiency exam required, foreign language not required. *Entrance requirements:* For master's, TOEFL (minimum score of 550 required). *Application deadline:* For fall admission, 4/15; for spring admission, 11/1. Application fee: $55. *Financial aid:* Research assistantships, teaching assistantships, career-related internships or fieldwork and Federal Work-Study available. Aid available to part-time students. Financial aid application deadline: 3/1. *Unit head:* Dr. Ngo Dinh Thinh, Chair, 916-278-6624. *Application contact:* Fred Reardon, Graduate Coordinator, 916-278-6727.

Carleton University, Faculty of Graduate Studies, Faculty of Engineering and Design, Department of Mechanical and Aerospace Engineering, Ottawa, ON K1S 5B6, Canada. Offers aerospace engineering (M Eng, PhD); materials engineering (M Eng); mechanical engineering (M Eng, PhD). *Faculty:* 23 full-time (1 woman). *Students:* 51 full-time (4 women), 11 part-time (2 women). Average age 29. In 1998, 18 master's, 1 doctorate awarded. *Degree requirements:* For master's, thesis optional; for doctorate, dissertation required. *Entrance requirements:* For master's, TOEFL (minimum score of 550 required), honors degree; for doctorate, TOEFL (minimum score of 550 required), MA Sc or M Eng. *Average time to degree:* Master's–2.2 years full-time; doctorate–6.3 years full-time. *Application deadline:* For fall admission, 3/1. Applications are processed on a rolling basis. Application fee: $35. *Financial aid:* In 1998–99, 1 student received aid. Application deadline: 3/1. *Faculty research:* Thermal fluids engineering, heat transfer, vehicle engineering. Total annual research expenditures: $1.9 million. *Unit head:* Paul Straznicky, Director, 613-520-2600 Ext. 5684, Fax: 613-520-5715, E-mail: pstrazni@mae.carleton.ca. *Application contact:* Ata M. Khan, Associate Dean of Engineering, 613-520-5659, Fax: 613-520-5682, E-mail: ata_khan@carleton.ca.

See in-depth description on page 1467.

Carnegie Mellon University, Carnegie Institute of Technology, Department of Mechanical Engineering, Pittsburgh, PA 15213-3891. Offers ME, MS, PhD. Part-time and evening/weekend programs available. *Faculty:* 23 full-time (2 women), 1 part-time (0 women). *Students:* 58 full-time (3 women), 6 part-time; includes 5 minority (1 Asian American or Pacific Islander, 3 Hispanic Americans, 1 Native American), 45 international. Average age 27. In 1998, 18 master's, 10 doctorates awarded. Terminal master's awarded for partial completion of doctoral program. *Degree requirements:* For master's, thesis required (for some programs), foreign language not required; for doctorate, dissertation (for some programs), qualifying exam required, foreign language not required. *Entrance requirements:* For master's and doctorate, GRE General Test, TOEFL. *Application deadline:* For fall admission, 2/1; for spring admission, 10/15. Application fee: $45. *Financial aid:* Fellowships, research assistantships, teaching assistantships, career-related internships or fieldwork available. Financial aid application deadline: 2/1. *Faculty research:* Combustion, design, fluid, and thermal sciences; computational fluid dynamics; energy and environment; solid mechanics; systems and controls; materials and manufacturing. Total annual research expenditures: $2.9 million. *Unit head:* Adnan Akay, Head, 412-268-2501. *Application contact:* Graduate Coordinator, 412-268-3175.

See in-depth description on page 1560.

Case Western Reserve University, School of Graduate Studies, The Case School of Engineering, Department of Mechanical and Aerospace Engineering, Cleveland, OH 44106. Offers fluid and thermal sciences (MS, PhD); mechanical engineering (MS, PhD). Part-time programs available. Postbaccalaureate distance learning degree programs offered (no on-campus study). *Faculty:* 17 full-time (1 woman), 2 part-time (1 woman). *Students:* 23 full-time (2 women), 72 part-time (10 women); includes 8 minority (2 African Americans, 5 Asian Americans or Pacific Islanders, 1 Hispanic American), 39 international. Average age 24. 88 applicants, 65% accepted. In 1998, 21 master's, 7 doctorates awarded. *Degree requirements:* For master's, thesis required (for some programs), foreign language not required; for doctorate, dissertation required, foreign language not required. *Entrance requirements:* For master's and doctorate, TOEFL (minimum score of 550 required). *Average time to degree:* Master's–2 years full-time; doctorate–4 years full-time. *Application deadline:* For fall admission, 7/1 (priority date). Applications are processed on a rolling basis. Application fee: $25. *Financial aid:* In 1998–99, 5 fellowships (averaging $15,000 per year), 23 research assistantships (averaging $14,400 per year), 14 teaching assistantships (averaging $10,800 per year) were awarded.; institutionally-sponsored loans and tuition waivers (full and partial) also available. Financial aid application deadline:3/1. *Faculty research:* Biomechanics, machinery diagnostics, robotics, microgravity fluid-mechanics, multiphase flows and combustion. Total annual research expenditures: $3.5 million. *Unit head:* Joseph M. Prahl, Chairman, 216-368-2940, Fax: 216-368-6445, E-mail: jmp@po.cwru.edu. *Application contact:* Angelika Szakacs, Secretary, 216-368-5403, Fax: 216-368-3007, E-mail: mechgac@po.cwru.edu.

The Catholic University of America, School of Engineering, Department of Mechanical Engineering, Program in Mechanical Design, Washington, DC 20064. Offers MME. *Degree requirements:* For master's, comprehensive exam required, thesis optional, foreign language not required. *Entrance requirements:* For master's, minimum GPA of 3.0. *Application deadline:* For fall admission, 8/1 (priority date); for spring admission, 12/1. Applications are processed on a rolling basis. Application fee: $50. *Financial aid:* Application deadline: 2/1. *Unit head:* Dr. John J. Gilheany, Chair, Department of Mechanical Engineering, 202-319-5170.

City College of the City University of New York, Graduate School, School of Engineering, Department of Mechanical Engineering, New York, NY 10031-9198. Offers ME, MS, PhD. Part-time programs available. *Students:* 4 full-time (0 women), 32 part-time (4 women). In 1998, 18 degrees awarded. *Degree requirements:* For master's, computer language required, thesis optional, foreign language not required; for doctorate, one foreign language (computer language can substitute), dissertation, comprehensive exams required. *Entrance requirements:* For master's, TOEFL (minimum score of 500 required); for doctorate, GRE General Test, TOEFL. *Application deadline:* Applications are processed on a rolling basis. Application fee: $40. *Financial aid:* Fellowships, research assistantships, teaching assistantships, Federal Work-Study and tuition waivers (full and partial) available. *Faculty research:* Bio-heat and mass transfer, bone mechanics, fracture mechanics, heat transfer in computer parts, mechanisms design. *Unit head:* Yanis Andreopoulos, Chairman, 212-650-5220. *Application contact:* Graduate Admissions Office, 212-650-6977.

Clarkson University, Graduate School, School of Engineering, Department of Mechanical and Aeronautical Engineering, Potsdam, NY 13699. Offers mechanical engineering (ME, MS, PhD). Part-time programs available. *Faculty:* 17 full-time (0 women). *Students:* 36 full-time (10 women), 1 part-time; includes 2 minority (1 African American, 1 Native American), 19 international. Average age 28. 121 applicants, 71% accepted. In 1998, 12 master's, 2 doctorates awarded (100% found work related to degree). *Degree requirements:* For master's, thesis required, foreign language not required; for doctorate, dissertation, departmental qualifying exam required, foreign language not required. *Entrance requirements:* For master's, GRE, TOEFL. *Application deadline:* For fall admission, 5/15 (priority date); for spring admission, 10/15 (priority date). Applications are processed on a rolling basis. Application fee: $25 ($35 for international students). Tuition: Part-time $661 per credit hour. Required fees: $215 per semester. *Financial aid:* In 1998–99, 5 fellowships, 17 research assistantships, 8 teaching assistantships were awarded. *Faculty research:* Turbulent flow, controls, fracture mechanics, materials science, fluid mechanics. Total annual research expenditures: $924,461. *Unit head:* Dr. M. Sathyamoorthy, Chairman, 315-268-6586, E-mail: sathy@clarkson.edu. *Application contact:* Dr. Philip K. Hopke, Dean of the Graduate School, 315-268-6447, Fax: 315-268-7994, E-mail: hopkepk@clarkson.edu.

Clemson University, Graduate School, College of Engineering and Science, School of Mechanical and Industrial Engineering, Department of Mechanical Engineering, Program in Mechanical Engineering, Clemson, SC 29634. Offers M Engr, MS, PhD. *Students:* 70 full-time (7 women), 22 part-time (1 woman); includes 1 minority (African American), 45 international. Average age 28. 130 applicants, 59% accepted. In 1998, 25 master's, 5 doctorates awarded. *Degree requirements:* For master's and doctorate, thesis/dissertation required, foreign language not required. *Entrance requirements:* For master's, GRE General Test (minimum combined score of 1500 on three sections required for MS), TOEFL (minimum score of 550 required); for doctorate, GRE General Test (minimum combined score of 1500 on three sections required), TOEFL (minimum score of 550 required). *Application deadline:* For fall admission, 6/1. Applications are processed on a rolling basis. Application fee: $35. *Financial aid:* Applicants required to submit FAFSA. *Unit head:* Dr. Georges Fadel, Coordinator, 864-656-5620, Fax: 864-656-4435. *Application contact:* Dr. Georges Fadel, Coordinator, 864-656-5620, Fax: 864-656-4435.

Cleveland State University, College of Graduate Studies, Fenn College of Engineering, Department of Mechanical Engineering, Cleveland, OH 44115-2440. Offers MS, D Eng. Part-time programs available. *Faculty:* 12 full-time (0 women). *Students:* 7 full-time (1 woman), 29 part-time (1 woman); includes 3 minority (1 Asian American or Pacific Islander, 1 Hispanic American, 1 Native American), 8 international. Average age 32. 118 applicants, 57% accepted. In 1998, 12 master's awarded (100% found work related to degree). *Degree requirements:* For master's, project or thesis required; for doctorate, dissertation, candidacy and qualifying exams required, foreign language not required. *Entrance requirements:* For master's, GRE General Test, GRE Subject Test, TOEFL, minimum GPA of 2.75; for doctorate, GRE General Test, GRE Subject Test, TOEFL, minimum GPA of 3.25. *Application deadline:* For fall admission, 7/15 (priority date). Applications are processed on a rolling basis. Application fee: $25. *Financial aid:* In 1998–99, 4 research assistantships, 2 teaching assistantships were awarded.; career-related internships or fieldwork, Federal Work-Study, institutionally-sponsored loans, and unspecified assistantships also available. Aid available to part-time students. *Faculty research:* Fluid piezoelectric sensors, laser-optical inspection simulation of forging and forming processes, multiphase flow and heat transfer, turbulent flows. *Unit head:* Dr. Mournir B. Ibrahim, Chair, 216-687-2595, Fax: 216-687-5375, E-mail: m.ibrahim@csuohio.edu.

Colorado State University, Graduate School, College of Engineering, Department of Mechanical Engineering, Fort Collins, CO 80523-0015. Offers bioengineering (MS, PhD); energy and environmental engineering (MS, PhD); energy conversion (MS, PhD); engineering management (MS); heat and mass transfer (MS, PhD); industrial and manufacturing systems engineering (MS, PhD); mechanical engineering (MS, PhD); mechanics and materials (MS, PhD). Part-time programs available. *Faculty:* 18 full-time (0 women). *Students:* 31 full-time (9 women), 28 part-time (4 women); includes 2 minority (1 Asian American or Pacific Islander, 1 Hispanic American), 12 international. Average age 29. 265 applicants, 49% accepted. In 1998, 30 master's, 3 doctorates awarded. Terminal master's awarded for partial completion of doctoral program. *Degree requirements:* For doctorate, dissertation required, foreign language not required, foreign language not required. *Entrance requirements:* For master's and doctorate, GRE General Test (minimum combined score of 1850 on three sections required; average 1872), TOEFL (minimum score of 550 required; average 596), minimum GPA of 3.0. *Application deadline:* For fall admission, 2/1 (priority date). Applications are processed on a rolling basis. Application fee: $30. Electronic applications accepted. *Financial aid:* In 1998–99, 1 fellowship, 23 research assistantships, 10 teaching assistantships were awarded.; traineeships also available. *Faculty research:* Space propulsion, computer-assisted engineering and design, controls and systems. *Unit head:* Dr. Tim W. Tong, Head, 970-491-6558, Fax: 970-491-3827, E-mail: tong@engr.colostate.edu. *Application contact:* Dr. Doug Hittle, Graduate Committee Chairman, 970-491-8617, Fax: 970-491-3827, E-mail: hittle@lamar.colostate.edu.

Columbia University, Fu Foundation School of Engineering and Applied Science, Department of Mechanical Engineering, New York, NY 10027. Offers ME, MS, Eng Sc D, PhD. PhD offered through the Graduate School of Arts and Sciences. Part-time programs available. *Faculty:* 10 full-time (0 women), 5 part-time (0 women). *Students:* 47 full-time (3 women), 27 part-time (4 women). Average age 25. 223 applicants, 27% accepted. In 1998, 14 master's awarded (92% found work related to degree); 4 doctorates awarded (100% found work related to degree). Terminal master's awarded for partial completion of doctoral program. *Degree requirements:* For master's, foreign language and thesis not required; for doctorate, dissertation, qualifying exam required, foreign language not required. *Entrance requirements:* For master's, GRE General Test, TOEFL, minimum GPA of 3.0; for doctorate, GRE General Test, TOEFL. *Average time to degree:* Master's–1 year full-time, 2.5 years part-time; doctorate–5 years full-time, 7 years part-time. *Application deadline:* For fall admission, 1/5; for spring admission, 10/1. Application fee: $55. *Financial aid:* In 1998–99, 22 students received aid, including 5 fellowships, 10 research assistantships, 7 teaching assistantships; career-related internships or fieldwork, Federal Work-Study, institutionally-sponsored loans, and research student support awards also available. Financial aid application deadline: 1/5; financial aid applicants required to submit FAFSA. *Faculty research:* Heat transfer and fluid mechanics, control theory and robotics, kinematics and dynamics of mechanisms, manufacturing engineering, orthopedic biomechanics. Total annual research expenditures: $776,700. *Unit head:* Dr. W. Michael Lai, Chairman, 212-854-4236, Fax: 212-854-3304, E-mail: lai@cvorma.orl.columbia.edu. *Application contact:* Xiamara Perez-Betances, Departmental Administrator, 212-854-2965, Fax: 212-854-3304, E-mail: xp1@columbia.edu.

Concordia University, School of Graduate Studies, Faculty of Engineering and Computer Science, Department of Mechanical Engineering, Montréal, PQ H3G 1M8, Canada. Offers composites (M Eng); mechanical engineering (M Eng, MA Sc, PhD, Certificate). *Students:* 99 full-time (15 women), 13 part-time. *Degree requirements:* For master's, variable foreign language requirement, computer language, thesis or alternative required; for doctorate, computer language, dissertation, comprehensive exam required, foreign language not required. *Application deadline:* For fall admission, 6/1; for spring admission, 10/1. Application fee: $50. *Faculty research:* Mechanical systems, fluid control systems, thermofluids engineering and robotics, industrial control systems. *Unit head:* Dr. V. S. Hoa, Chair, 514-848-3133, Fax: 514-848-3175. *Application contact:* Dr. R. Neemeh, Director, 514-848-3131, Fax: 514-848-3175.

Cornell University, Graduate School, Graduate Fields of Engineering, Field of Mechanical Engineering, Ithaca, NY 14853-0001. Offers biomechanical engineering (M Eng, MS, PhD); combustion (M Eng, MS, PhD); energy and power systems (M Eng, MS, PhD); fluid mechanics (M Eng, MS, PhD); heat transfer (M Eng, MS, PhD); materials and manufacturing engineering (M Eng, MS, PhD); mechanical systems and design (M Eng, MS, PhD); multiphase flows (M Eng, MS, PhD). *Faculty:* 35 full-time. *Students:* 71 full-time (8 women); includes 15 minority (11 Asian Americans or Pacific Islanders, 3 Hispanic Americans, 1 Native American), 32 international. 443 applicants, 29% accepted. In 1998, 44 master's, 7 doctorates awarded. Terminal master's awarded for partial completion of doctoral program. *Degree requirements:* For master's, project (M Eng), thesis (MS) required; for doctorate, dissertation, 2 semesters of teaching experience required. *Entrance requirements:* For master's and doctorate, GRE General Test, TOEFL (minimum score of 550 required). *Application deadline:* For fall admission, 1/15. Application fee: $65. Electronic applications accepted. *Financial aid:* In 1998–99, 48 students received aid, including 19 fellowships with full tuition reimbursements available, 18 research assistantships with full tuition reimbursements available, 11 teaching assistantships with full tuition reimbursements available; institutionally-sponsored loans, scholarships, tuition waivers (full and partial), and unspecified assistantships also available. Financial aid applicants required to submit FAFSA. *Faculty research:* System dynamics and control, biomechanics, manufacturing, deformation processes, robotics, CFD, turbulence. *Unit head:* Director of Graduate Studies, 607-255-5250, Fax: 607-255-1222. *Application contact:* Graduate Field Assistant, 607-255-5250, Fax: 607-255-1222, E-mail: maegrad@cornell.edu.

Dalhousie University, Faculty of Graduate Studies, DalTech, Faculty of Engineering, Department of Mechanical Engineering, Halifax, NS B3H 3J5, Canada. Offers M Eng, MA Sc, PhD. *Faculty:* 13 full-time (0 women), 2 part-time (0 women). *Students:* 48 full-time (5 women), 5 part-time. Average age 31. 46 applicants, 41% accepted. In 1998, 8 master's, 3 doctorates

Mechanical Engineering

Dalhousie University (continued)
awarded (66% entered university research/teaching, 33% found other work related to degree). *Degree requirements:* For master's and doctorate, thesis/dissertation required, foreign language not required. *Entrance requirements:* For master's and doctorate, TOEFL (minimum score of 580 required). *Application deadline:* For fall admission, 6/1; for winter admission, 10/1; for spring admission, 2/1. Applications are processed on a rolling basis. Application fee: $55. *Financial aid:* Fellowships, research assistantships, teaching assistantships, scholarships and unspecified assistantships available. *Faculty research:* Fluid dynamics and energy, system dynamics, smart materials, naval architecture, MEMS. *Unit head:* Dr. J. Militzer, Head, 902-494-3989, Fax: 902-423-6711, E-mail: mechanical.engineering@dal.ca. *Application contact:* Shelley Parker, Admissions Coordinator, Graduate Studies and Research, 902-494-1288, Fax: 902-494-3149, E-mail: shelley.parker@dal.ca.

Dartmouth College, Thayer School of Engineering, Program in Mechanical Engineering, Hanover, NH 03755. Offers MS, PhD. *Degree requirements:* For master's, thesis required; for doctorate, dissertation, candidacy oral exam required. *Entrance requirements:* For master's and doctorate, GRE General Test. *Application deadline:* For fall admission, 1/15 (priority date). Application fee: $20 ($40 for international students). *Financial aid:* Fellowships, research assistantships, teaching assistantships, career-related internships or fieldwork, Federal Work-Study, institutionally-sponsored loans, and tuition waivers (full and partial) available. Financial aid application deadline: 1/15. *Faculty research:* Solid mechanics and mechanical design, tribology, dynamics and control systems, thermal science and energy conversion, fluid mechanics and multi-phase flow. Total annual research expenditures: $450,000. *Unit head:* Shelley Parker, Admissions Coordinator, Graduate Studies and Research, 902-494-1288, Fax: 902-494-3149, E-mail: shelley.parker@dal.ca. *Application contact:* Candace S. Potter, Admissions Coordinator, 603-646-3844, Fax: 603-646-3856, E-mail: candace.potter@dartmouth.edu.

Drexel University, Graduate School, College of Engineering, Department of Mechanical Engineering and Mechanics, Philadelphia, PA 19104-2875. Offers manufacturing engineering (MS, PhD); mechanical engineering and mechanics (MS, PhD). Part-time and evening/weekend programs available. *Faculty:* 23 full-time, 9 part-time. *Students:* 31 full-time (2 women), 41 part-time (2 women); includes 12 minority (6 African Americans, 3 Asian Americans or Pacific Islanders, 3 Hispanic Americans), 30 international. Average age 32. 223 applicants, 69% accepted. In 1998, 27 master's, 9 doctorates awarded. Terminal master's awarded for partial completion of doctoral program. *Degree requirements:* For master's, thesis optional; for doctorate, dissertation required. *Entrance requirements:* For master's, TOEFL (minimum score of 570 required), minimum GPA of 3.0, BS in engineering or science; for doctorate, TOEFL (minimum score of 570 required), minimum GPA of 3.5, MS in engineering or science. *Application deadline:* For fall admission, 8/21. Applications are processed on a rolling basis. Application fee: $35. Tuition: Full-time $15,795; part-time $585 per credit. Required fees: $375; $67 per term. Tuition and fees vary according to program. *Financial aid:* In 1998–99, 17 research assistantships, 27 teaching assistantships were awarded.; unspecified assistantships also available. Financial aid application deadline: 2/1. *Faculty research:* Composites, dynamic systems and control, combustion and fuels, biomechanics, mechanics and thermal fluid sciences. *Unit head:* Dr. Nicholas Cernansky, Acting Head, 215-895-2352, Fax: 215-895-1478, E-mail: cernansk@coe.drexel.edu. *Application contact:* Dr. Tetn-Min Tan, Graduate Adviser, 215-895-2293.

See in-depth description on page 1385.

Duke University, Graduate School, School of Engineering, Department of Mechanical Engineering and Materials Science, Durham, NC 27708-0586. Offers materials science (MS, PhD); mechanical engineering (MS, PhD). Part-time programs available. *Faculty:* 25 full-time, 4 part-time. *Students:* 61 full-time, 1 part-time; includes 3 minority (2 African Americans, 1 Asian American or Pacific Islander), 28 international. 108 applicants, 38% accepted. In 1998, 5 master's, 8 doctorates awarded. Terminal master's awarded for partial completion of doctoral program. *Degree requirements:* For master's, thesis optional, foreign language not required; for doctorate, dissertation required, foreign language not required. *Entrance requirements:* For master's and doctorate, GRE General Test. *Application deadline:* For fall admission, 12/31; for spring admission, 11/1. Application fee: $75. *Financial aid:* Fellowships, research assistantships, teaching assistantships, Federal Work-Study available. Financial aid application deadline: 12/31. *Unit head:* Dr. Charles M. Harman, Director of Graduate Studies, 919-660-5308, Fax: 919-660-8963, E-mail: kmrogers@duke.edu.

See in-depth description on page 1387.

École Polytechnique de Montréal, Graduate Programs, Department of Mechanical Engineering, Montréal, PQ H3C 3A7, Canada. Offers aerothermics (M Eng, M Sc A, PhD); applied mechanics (M Eng, M Sc A, PhD); tool design (M Eng, M Sc A, PhD). Part-time and evening/weekend programs available. *Degree requirements:* For master's and doctorate, one foreign language, computer language, thesis/dissertation required. *Entrance requirements:* For master's, minimum GPA of 2.75; for doctorate, minimum GPA of 3.0. *Faculty research:* Noise control and vibration, fatigue and creep, aerodynamics, composite materials, biomechanics, robotics.

Florida Agricultural and Mechanical University, Division of Graduate Studies, Research, and Continuing Education, FAMU-FSU College of Engineering, Department of Mechanical Engineering, Tallahassee, FL 32307-3200. Offers MS, PhD. *Students:* 13 (6 women); includes 11 minority (10 African Americans, 1 Asian American or Pacific Islander) 2 international. In 1998, 2 master's awarded. *Entrance requirements:* For master's, GRE General Test (minimum combined score of 1000 required), minimum GPA of 3.0. *Application deadline:* For fall admission, 7/1. Application fee: $20. *Unit head:* Dr. C. J. Chen, Dean, FAMU-FSU College of Engineering, 850-487-6100, Fax: 850-487-6486.

See in-depth description on page 1389.

Florida Atlantic University, College of Engineering, Department of Mechanical Engineering, Program in Mechanical Engineering, Boca Raton, FL 33431-0991. Offers MS, PhD. Part-time and evening/weekend programs available. *Faculty:* 16 full-time (0 women). *Students:* 11 full-time (2 women), 17 part-time (2 women); includes 7 minority (1 Asian American or Pacific Islander, 6 Hispanic Americans), 12 international. Average age 32. In 1998, 6 master's, 2 doctorates awarded. Terminal master's awarded for partial completion of doctoral program. *Degree requirements:* For master's, thesis optional, foreign language not required; for doctorate, dissertation, qualifying exam required, foreign language not required. *Entrance requirements:* For master's and doctorate, GRE General Test (minimum combined score of 1000 required), TOEFL (minimum score of 550 required), minimum GPA of 3.0. *Application deadline:* For fall admission, 4/10 (priority date); for spring admission, 10/1. Applications are processed on a rolling basis. Application fee: $20. Tuition, state resident: part-time $148 per credit hour. Tuition, nonresident: part-time $509 per credit hour. *Financial aid:* Research assistantships, teaching assistantships, career-related internships or fieldwork available. Financial aid application deadline: 4/1. *Faculty research:* Fault detection and diagnostics, computational fluid mechanics, composite materials, two-phase flows, helicopter dynamics. *Unit head:* Candace S. Potter, Admissions Coordinator, 603-646-3844, Fax: 603-646-3856, E-mail: candace.potter@dartmouth.edu. *Application contact:* Patricia Capozziello, Graduate Admissions Coordinator, 561-297-2694, Fax: 561-297-2659, E-mail: capozzie@fau.edu.

Florida Institute of Technology, Graduate School, College of Engineering, Division of Engineering Science, Department of Mechanical Engineering, Melbourne, FL 32901-6975. Offers MS, PhD. Part-time programs available. *Faculty:* 6 full-time (0 women), 3 part-time (1 woman). *Students:* 6 full-time (0 women), 9 part-time; includes 1 minority (Asian American or Pacific Islander), 9 international. Average age 28. 70 applicants, 63% accepted. In 1998, 6 master's awarded. *Degree requirements:* For master's, thesis optional, foreign language not required; for doctorate, dissertation, comprehensive exam required, foreign language not required. *Entrance requirements:* For master's, GRE General Test, minimum GPA of 3.0; for doctorate, GRE General Test, GRE Subject Test, minimum GPA of 3.2. *Application deadline:* Applications are processed on a rolling basis. Application fee: $50. Electronic applications accepted. Tuition: Part-time $575 per credit hour. Required fees: $100. Tuition and fees vary according to campus/location and program. *Financial aid:* In 1998–99, 8 students received aid, including 4 research assistantships with full and partial tuition reimbursements available (averaging $3,560 per year), 3 teaching assistantships with full and partial tuition reimbursements available (averaging $3,040 per year); tuition remissions also available. Financial aid application deadline: 3/1; financial aid applicants required to submit FAFSA. *Faculty research:* Dynamic systems, robotics, and controls; structures, solid mechanics, and materials; thermal-fluid sciences. Total annual research expenditures: $125,600. *Unit head:* Dr. J. Engblom, Chair, 407-674-8092, Fax: 407-674-8813, E-mail: engblom@zach.fit.edu. *Application contact:* Carolyn P. Farrior, Associate Dean of Graduate Admissions, 407-674-7118, Fax: 407-723-9468, E-mail: cfarrior@fit.edu.

See in-depth description on page 1391.

Florida Institute of Technology, Graduate School, School of Extended Graduate Studies, Program in Mechanical Engineering, Melbourne, FL 32901-6975. Offers MS. Part-time programs available. Average age 30. 1 applicants, 0% accepted. In 1998, 1 degree awarded (100% found work related to degree). *Degree requirements:* For master's, thesis optional, foreign language not required. *Entrance requirements:* For master's, GRE General Test, minimum GPA of 3.0. *Average time to degree:* Master's–1 year full-time, 3 years part-time. *Application deadline:* Applications are processed on a rolling basis. Application fee: $50. Electronic applications accepted. Tuition: Part-time $270 per credit hour. Part-time tuition and fees vary according to campus/location. *Financial aid:* Application deadline: 3/1; *Unit head:* Carolyn P. Farrior, Associate Dean of Graduate Admissions, 407-674-7118, Fax: 407-723-9468, E-mail: cfarrior@fit.edu. *Application contact:* Carolyn P. Farrior, Associate Dean of Graduate Admissions, 407-674-7118, Fax: 407-723-9468, E-mail: cfarrior@fit.edu.

Florida International University, College of Engineering, Department of Mechanical Engineering, Miami, FL 33199. Offers MS, PhD. Part-time and evening/weekend programs available. *Faculty:* 14 full-time (0 women). *Students:* 20 full-time (0 women), 26 part-time (2 women); includes 19 minority (2 African Americans, 4 Asian Americans or Pacific Islanders, 13 Hispanic Americans), 25 international. Average age 31. 69 applicants, 38% accepted. In 1998, 28 master's, 1 doctorate awarded. *Degree requirements:* For master's, thesis optional; for doctorate, dissertation required. *Entrance requirements:* For master's and doctorate, GRE General Test (minimum combined score of 1000 required), TOEFL (minimum score of 500 required). *Application deadline:* For fall admission, 4/1 (priority date); for spring admission, 10/1. Applications are processed on a rolling basis. Application fee: $20. Tuition, state resident: part-time $145 per credit hour. Tuition, nonresident: part-time $506 per credit hour. Required fees: $158; $158 per year. *Unit head:* Dr. Richard K. Irey, Chairperson, 305-348-2569, Fax: 305-348-1932.

Florida State University, Graduate Studies, FAMU/FSU College of Engineering, Department of Mechanical Engineering, Tallahassee, FL 32306. Offers MS, PhD. Part-time programs available. Postbaccalaureate distance learning degree programs offered (minimal on-campus study). *Faculty:* 19 full-time (1 woman), 2 part-time (0 women). *Students:* 41 full-time (5 women), 8 part-time (1 woman); includes 11 minority (10 African Americans, 1 Asian American or Pacific Islander), 27 international. Average age 27. 102 applicants, 50% accepted. In 1998, 8 master's awarded (88% found work related to degree, 12% continued full-time study); 4 doctorates awarded (100% entered university research/teaching). *Degree requirements:* For master's, thesis or comprehensive exam required, thesis optional, foreign language not required; for doctorate, dissertation, preliminary exam, qualifying exam required, foreign language not required. *Entrance requirements:* For master's, GRE General Test (minimum combined score of 1150 required for U.S. applicants, 1350 for international applicants), TOEFL (minimum score of 550 required), minimum GPA of 3.0 (U.S. applicants), 3.5 (international applicants); BS in mechanical engineering; for doctorate, GRE General Test (minimum combined score of 1100 required for U.S. applicants, 1150 for international applicants), TOEFL (minimum score of 550 required), minimum GPA of 3.5 (graduate), MA (mechanical engineering), 3 units of directed research. *Average time to degree:* Master's–2 years full-time, 3.5 years part-time; doctorate–3.5 years full-time, 4.5 years part-time. *Application deadline:* For fall admission, 3/15; for spring admission, 11/1. Applications are processed on a rolling basis. Application fee: $20. Electronic applications accepted. Tuition, state resident: part-time $139 per credit hour. Tuition, nonresident: part-time $482 per credit hour. Tuition and fees vary according to program. *Financial aid:* In 1998–99, 30 students received aid, including 1 fellowship, 18 research assistantships, 13 teaching assistantships; career-related internships or fieldwork and institutionally-sponsored loans also available. Financial aid application deadline: 6/15. *Faculty research:* Fluid mechanics, super plastic metal forming, superconducting composite materials, dynamics and controls, magnet design. Total annual research expenditures: $3 million. *Unit head:* Dr. E. Collins, Associate Chair for Graduate Studies, 850-410-6373, Fax: 850-410-6337. *Application contact:* Lois Garner, Graduate Administrator, 850-410-6330, Fax: 850-410-6337, E-mail: garner@eng.fsu.edu.

See in-depth description on page 1389.

Gannon University, School of Graduate Studies, College of Sciences, Engineering, and Health Sciences, School of Sciences and Engineering, Program in Engineering, Erie, PA 16541-0001. Offers electrical engineering (MS); embedded software engineering (MS); mechanical engineering (MS). Part-time and evening/weekend programs available. *Students:* 23 full-time (4 women), 29 part-time (5 women); includes 1 minority (Asian American or Pacific Islander), 13 international. *Degree requirements:* For master's, thesis or alternative, comprehensive exam required. *Entrance requirements:* For master's, GRE Subject Test, bachelor's degree in engineering, minimum QPA of 2.5. *Application deadline:* Applications are processed on a rolling basis. Application fee: $25. *Unit head:* Dr. Mehmet Cultu, Co-Director, 814-871-7624. *Application contact:* Beth Nemenz, Director of Admissions, 814-871-7240, Fax: 814-871-5803, E-mail: admissions@gannon.edu.

The George Washington University, School of Engineering and Applied Science, Department of Mechanical and Aeronautical Engineering, Washington, DC 20052. Offers MS, D Sc, App Sc, Engr. Part-time and evening/weekend programs available. *Degree requirements:* For master's, thesis optional, foreign language not required; for doctorate, computer language, dissertation, final and qualifying exams required, foreign language not required; for other advanced degree, foreign language and thesis not required. *Entrance requirements:* For master's, TOEFL (minimum score of 550 required; average 580) or George Washington University English as a Foreign Language Test, appropriate bachelor's degree, minimum GPA of 3.0; for doctorate, TOEFL (minimum score of 550 required; average 580) or George Washington University English as a Foreign Language Test, appropriate bachelor's or master's degree, minimum GPA of 3.4, GRE required if highest earned degree is BS; for other advanced degree, TOEFL (minimum score of 550 required; average 580) or George Washington University English as a Foreign Language Test, appropriate master's degree, minimum GPA of 3.0. *Application deadline:* For fall admission, 3/1 (priority date); for spring admission, 10/1. Applications are processed on a rolling basis. Application fee: $55. Tuition: Full-time $17,328; part-time $722 per credit hour. Required fees: $828; $35 per credit hour. Tuition and fees vary according to campus/location and program. *Financial aid:* Fellowships, research assistantships, teaching assistantships, career-related internships or fieldwork and institutionally-sponsored loans available. Financial aid application deadline: 3/1; financial aid applicants required to submit FAFSA. *Unit head:* Dr. Michael K. Myers, Chair, 202-994-6749, Fax: 202-994-0238. *Application contact:* Howard M. Davis, Manager, Office of Admissions and Student Records, 202-994-6158, Fax: 202-994-0909, E-mail: data:adms@seas.gwu.edu.

See in-depth description on page 1393.

Georgia Institute of Technology, Graduate Studies and Research, College of Engineering, George W. Woodruff School of Mechanical Engineering, Program in Mechanical Engineering, Atlanta, GA 30332-0001. Offers biomedical engineering (MS Bio E); mechanical engineering (MS, MSME, PhD). Part-time programs available. Terminal master's awarded for partial completion of doctoral program. *Degree requirements:* For master's, thesis optional, foreign language not required; for doctorate, dissertation required, foreign language not required. *Entrance requirements:* For master's and doctorate, GRE General Test (recommended), TOEFL (minimum score of 580 required). *Faculty research:* Acoustics, automatic controls, CAD/CAM,fluid mechanics, lubrication.

See in-depth description on page 1395.

Graduate School and University Center of the City University of New York, Graduate Studies, Program in Engineering, New York, NY 10036-8099. Offers chemical engineering (PhD); civil engineering (PhD); electrical engineering (PhD); mechanical engineering (PhD). *Faculty:* 68 full-time (1 woman). *Students:* 105 full-time (16 women), 11 part-time (2 women); includes 12 African Americans, 5 Asian Americans or Pacific Islanders, 4 Hispanic Americans *Degree requirements:* For doctorate, dissertation required, dissertation required. *Entrance requirements:* For doctorate, GRE General Test. *Application deadline:* For fall admission, 4/15. Application fee: $40. *Unit head:* Dr. Mumtaz Kassir, Acting Executive Officer, 212-650-8030.

Howard University, College of Engineering, Architecture, and Computer Sciences, School of Engineering and Computer Science, Department of Mechanical Engineering, Washington, DC 20059-0002. Offers aerospace engineering/dynamics and controls (M Eng, PhD); applied mechanics (M Eng, PhD); CAD/CAM and robotics (M Eng, PhD); fluid and thermal sciences (M Eng, PhD). Part-time programs available. *Faculty:* 9 full-time (1 woman). *Students:* 17 full-time (7 women), 2 part-time; includes 7 African Americans, 1 Asian American or Pacific Islander, 7 international. Average age 25. 60 applicants, 20% accepted. In 1998, 2 master's awarded (100% found work related to degree); 1 doctorate awarded (100% entered university research/teaching). Terminal master's awarded for partial completion of doctoral program. *Degree requirements:* For master's, computer language, comprehensive exam required; for doctorate, one foreign language, computer language, dissertation, 2 terms of residency required. *Entrance requirements:* For master's and doctorate, GRE General Test, TOEFL, minimum GPA of 3.0. *Average time to degree:* Master's–2 years full-time; doctorate–4 years full-time. *Application deadline:* For fall admission, 4/1 (priority date); for spring admission, 11/1. Applications are processed on a rolling basis. Application fee: $45. Electronic applications accepted. *Financial aid:* In 1998–99, 9 students received aid, including 1 fellowship with full tuition reimbursement available (averaging $14,400 per year), 3 research assistantships (averaging $15,000 per year), 5 teaching assistantships with full tuition reimbursements available (averaging $10,696 per year); grants, institutionally-sponsored loans, and tuition waivers (partial) also available. Financial aid application deadline: 4/1; financial aid applicants required to submit FAFSA. *Faculty research:* The dynamics and control of large flexible space structures, optimization of space structures. Total annual research expenditures: $172,572. *Unit head:* Dr. Lewis Thigpen, Chair, 202-806-6600, Fax: 202-806-5258, E-mail: lthigpen@scs.howard.edu. *Application contact:* Dr. Sonya Smith, Graduate Director, 202-806-4837.

See in-depth description on page 1397.

Illinois Institute of Technology, Graduate College, Armour College of Engineering and Sciences, Department of Mechanical, Materials and Aerospace Engineering, Mechanical and Aerospace Engineering Division, Chicago, IL 60616-3793. Offers MMAE, MS, PhD. Part-time programs available. *Faculty:* 18 full-time (2 women), 8 part-time (0 women). *Students:* 39 full-time (0 women), 66 part-time (7 women); includes 17 minority (5 African Americans, 9 Asian Americans or Pacific Islanders, 3 Hispanic Americans), 41 international. 206 applicants, 44% accepted. In 1998, 18 master's, 3 doctorates awarded. Terminal master's awarded for partial completion of doctoral program. *Degree requirements:* For master's, thesis (for some programs), comprehensive exam required, foreign language not required; for doctorate, dissertation, comprehensive exam required, foreign language not required. *Entrance requirements:* For master's, GRE General Test (minimum combined score of 1200 required), TOEFL (minimum score of 550 required), undergraduate GPA of 3.0; for doctorate, GRE (minimum score of 1200 required), TOEFL (minimum score of 550 required), undergraduate GPA of 3.0 required. *Application deadline:* For fall admission, 7/1; for spring admission, 11/1. Applications are processed on a rolling basis. Application fee: $30. Electronic applications accepted. *Financial aid:* In 1998–99, 2 fellowships, 11 research assistantships, 14 teaching assistantships were awarded.; Federal Work-Study, institutionally-sponsored loans, scholarships, and graduate assistantships also available. Financial aid application deadline: 3/1. *Faculty research:* Solid and structural mechanics, fluid dynamics, thermal sciences, transportation engineering, design and manufacturing. *Unit head:* Dr. Kevin Meade, Associate Chair, Mechanical Engineering, 312-567-3175, Fax: 312-567-7230, E-mail: meade@mae.iit.edu. *Application contact:* Dr. S. Mohammad Shahidehpour, Dean of Graduate College, 312-567-3024, Fax: 312-567-7517, E-mail: grad@minna.cns.iit.edu.

See in-depth description on page 1399.

Indiana University–Purdue University Indianapolis, School of Engineering and Technology, Department of Mechanical Engineering, Indianapolis, IN 46202-2896. Offers biomedical engineering (MS Bm E); mechanical engineering (MSME). Part-time programs available. *Students:* 8 full-time (1 woman), 18 part-time (4 women); includes 4 minority (1 African American, 3 Asian Americans or Pacific Islanders), 11 international. Average age 26. 32 applicants, 31% accepted. In 1998, 10 degrees awarded. *Degree requirements:* For master's, thesis optional, foreign language not required. *Entrance requirements:* For master's, GRE, TOEFL (minimum score of 550 required), minimum B average. *Application deadline:* For fall admission, 7/1. Application fee: $25 ($50 for international students). Tuition, state resident: part-time $171 per credit hour. Tuition, nonresident: part-time $490 per credit hour. Required fees: $121 per year. *Financial aid:* In 1998–99, 9 research assistantships with full and partial tuition reimbursements were awarded.; fellowships, tuition waivers (full and partial) also available. Financial aid application deadline: 3/1. *Faculty research:* Computational fluid dynamics, heat transfer, finite-element methods, composites, biomechanics. *Unit head:* Dr. Jie Chen, Acting Chairman, 317-274-9717, Fax: 317-274-4567. *Application contact:* Vickie Lawrence, Graduate Program Secretary, 317-274-9740, Fax: 317-274-4567, E-mail: grad@engr.iupui.edu.

Institute of Paper Science and Technology, Graduate Programs, Program in Mechanical Engineering, Atlanta, GA 30318-5794. Offers MS, PhD. Terminal master's awarded for partial completion of doctoral program. *Degree requirements:* For master's, industrial experience, research project required, foreign language and thesis not required; for doctorate, dissertation required, foreign language not required. *Entrance requirements:* For master's and doctorate, GRE (score in 50th percentile or higher required), minimum GPA of 3.0. *Application deadline:* For fall admission, 3/1 (priority date). Application fee: $0. *Financial aid:* Career-related internships or fieldwork and institutionally-sponsored loans available. Financial aid applicants required to submit FAFSA. *Unit head:* Dr. David Orloff, Director, Engineering Division, 404-894-6649, Fax: 404-894-4778. *Application contact:* Dana Carter, Student Development Counselor, 404-894-5745, Fax: 404-894-4778, E-mail: dana.carter@ipst.edu.

Instituto Tecnológico y de Estudios Superiores de Monterrey, Campus Chihuahua, Graduate Programs, Chihuahua, 31110, Mexico. Offers computer systems engineering (Ingeniero); electrical engineering (Ingeniero); electromechanical engineering (Ingeniero); electronic engineering (Ingeniero); engineering administration (MEA); industrial engineering (MIE, Ingeniero); international trade (MIT); mechanical engineering (Ingeniero).

Instituto Tecnológico y de Estudios Superiores de Monterrey, Campus Monterrey, Graduate and Research Division, Programs in Engineering, Monterrey, 64849, Mexico. Offers applied statistics (M Eng); artificial intelligence (PhD); automation engineering (M Eng); chemical engineering (M Eng); civil engineering (M Eng); electrical engineering (M Eng); electronic engineering (M Eng); environmental engineering (M Eng); industrial engineering (M Eng, PhD); manufacturing engineering (M Eng); mechanical engineering (M Eng); systems and quality engineering (M Eng). M Eng offered jointly with the University of Waterloo; PhD (industrial engineering) offered jointly with Texas A&M University. Part-time and evening/weekend programs available. Terminal master's awarded for partial completion of doctoral program. *Degree requirements:* For master's and doctorate, one foreign language, computer language, thesis/dissertation required. *Entrance requirements:* For master's, PAEG, TOEFL; for doctorate, GRE, TOEFL, master's in related field. *Faculty research:* Flexible manufacturing cells, materials, statistical methods, environmental prevention, control and evaluation.

Iowa State University of Science and Technology, Graduate College, College of Engineering, Department of Mechanical Engineering, Ames, IA 50011. Offers MS, PhD. *Faculty:* 33 full-time, 1 part-time. *Students:* 49 full-time (5 women), 39 part-time (4 women); includes 6 minority (3 Asian Americans or Pacific Islanders, 3 Hispanic Americans), 46 international. 106 applicants, 24% accepted. In 1998, 32 master's, 8 doctorates awarded. *Degree requirements:* For master's, thesis or alternative required; for doctorate, dissertation required. *Entrance requirements:* For master's and doctorate, GRE General Test (foreign students), TOEFL (minimum score of 600 required). *Application deadline:* For fall admission, 4/1 (priority date); for spring admission, 10/1 (priority date). Application fee: $20 ($50 for international students). Electronic applications accepted. Tuition, state resident: full-time $3,308. Tuition, nonresident: full-time $9,744. Part-time tuition and fees vary according to course load, campus/location and program. *Financial aid:* In 1998–99, 51 research assistantships with partial tuition reimbursements (averaging $10,578 per year), 15 teaching assistantships with partial tuition reimbursements (averaging $12,099 per year) were awarded.; fellowships, scholarships also available. *Unit head:* Dr. Warren R. Devries, Chair, 515-294-5560, Fax: 515-294-3261, E-mail: megrad@iastate.edu. *Application contact:* Dr. Jon H. Van Gerpen, Director of Graduate Education, 515-294-5563, E-mail: megrad@iastate.edu.

Johns Hopkins University, G. W. C. Whiting School of Engineering, Department of Mechanical Engineering, Baltimore, MD 21218-2699. Offers mechanical engineering (MS, MSE, PhD), including mechanical engineering (MS), mechanics (MSE, PhD). Part-time programs available. *Faculty:* 13 full-time (1 woman), 6 part-time (0 women). *Students:* 63 full-time (8 women), 1 part-time; includes 5 minority (1 African American, 4 Asian Americans or Pacific Islanders), 32 international. Average age 25. 189 applicants, 16% accepted. In 1998, 7 master's, 9 doctorates awarded. Terminal master's awarded for partial completion of doctoral program. *Degree requirements:* For master's, thesis required (for some programs), foreign language not required; for doctorate, dissertation, oral exam required, foreign language not required. *Entrance requirements:* For master's and doctorate, GRE General Test (minimum score of 400 on verbal section, 780 on quantitative required), TOEFL (minimum score of 560 required). *Average time to degree:* Master's–2 years full-time; doctorate–5 years full-time. *Application deadline:* For fall admission, 2/15 (priority date). Application fee: $50. Tuition: Full-time $23,660. Tuition and fees vary according to program. *Financial aid:* In 1998–99, 6 fellowships (averaging $12,843 per year), 39 research assistantships (averaging $12,843 per year), 23 teaching assistantships (averaging $12,843 per year) were awarded.; Federal Work-Study, grants, and institutionally-sponsored loans also available. Aid available to part-time students. Financial aid application deadline: 2/15. *Faculty research:* Mechanics of composites, mechanical behavior of materials, complex flows and turbulence, robotics, heat and mass transfer. Total annual research expenditures: $3.5 million. *Unit head:* Dr. Andrew S. Douglas, Chair, 410-516-8789, Fax: 410-516-7254, E-mail: douglas@jhu.edu. *Application contact:* Dr. Charles Meneveau, Chair, Graduate Admissions Committee, 410-516-7132, Fax: 410-516-7254, E-mail: meneveau@titan.me.jhu.edu.

See in-depth description on page 1401.

Kansas State University, Graduate School, College of Engineering, Department of Mechanical and Engineering, Manhattan, KS 66506. Offers engineering (PhD); mechanical engineering (MS); nuclear engineering (MS). *Faculty:* 22 full-time (2 women). *Students:* 51 full-time (6 women), 10 part-time, 44 international. 56 applicants, 46% accepted. In 1998, 19 master's, 3 doctorates awarded. *Entrance requirements:* For master's and doctorate, GRE General Test (minimum combined score of 1350 on quantitative and analytical sections required), TOEFL (minimum score of 600 required). *Application deadline:* For fall admission, 9/1; for spring admission, 3/1. Applications are processed on a rolling basis. Application fee: $0 ($25 for international students). Electronic applications accepted. *Financial aid:* In 1998–99, 35 research assistantships (averaging $9,000 per year), 1 teaching assistantship (averaging $7,767 per year) were awarded.; career-related internships or fieldwork, Federal Work-Study, and institutionally-sponsored loans also available. *Faculty research:* Holographic particle velocimetry, optical measurements, thermal sciences, heat transfer, machine design. Total annual research expenditures: $2.2 million. *Unit head:* J. Garth Thompson, Head, 785-532-5610, Fax: 785-532-7057. *Application contact:* Kirby Chapman, Graduate Coordinator, 785-532-5610, E-mail: dgs@ksume.me.ksu.edu.

Kettering University, Graduate School, Mechanical Engineering Department, Flint, MI 48504-4898. Offers automotive engineering (MS Eng); mechanical design (MS Eng). *Faculty:* 26 full-time (3 women), 1 part-time (0 women). *Degree requirements:* For master's, foreign language and thesis not required. *Entrance requirements:* For master's, GRE General Test. *Application deadline:* For fall admission, 7/15. Applications are processed on a rolling basis. Application fee: $0. *Financial aid:* Fellowships with full tuition reimbursements, research assistantships with full tuition reimbursements, teaching assistantships with full tuition reimbursements, Federal Work-Study, institutionally-sponsored loans, and tuition waivers (partial) available. Aid available to part-time students. Financial aid application deadline: 7/15. *Unit head:* Dr. K. Joel Berry, Head, 810-762-7833, Fax: 810-762-7860, E-mail: jberry@kettering.edu. *Application contact:* Betty L. Bedore, Coordinator of Publicity, 810-762-7494, Fax: 810-762-9935, E-mail: bbedore@kettering.edu.

Lamar University, College of Graduate Studies, College of Engineering, Department of Mechanical Engineering, Beaumont, TX 77710. Offers ME, MES, DE. Part-time programs available. *Faculty:* 5 full-time (0 women). *Students:* 9 full-time (2 women), 1 part-time; includes 1 minority (Asian American or Pacific Islander), 8 international. Average age 30. In 1998, 9 degrees awarded. Terminal master's awarded for partial completion of doctoral program. *Degree requirements:* For master's, computer language, thesis required (for some programs), foreign language not required; for doctorate, computer language, dissertation required, foreign language not required. *Entrance requirements:* For master's, GRE General Test (minimum combined score of 1000 required), TOEFL (minimum score of 500 required); for doctorate, GRE General Test, TOEFL (minimum score of 530 required). *Average time to degree:* Master's–1.75 years full-time, 4 years part-time; doctorate–4 years full-time, 8 years part-time. *Application deadline:* For fall admission, 5/15 (priority date); for spring admission, 10/1 (priority date). Applications are processed on a rolling basis. Application fee: $0. *Financial aid:* In 1998–99, 1 fellowship, 1 teaching assistantship were awarded.; tuition waivers (partial) also available. Financial aid application deadline: 4/1. *Faculty research:* Aerospace systems, electrical power, robotics, solar energy, thermal control. Total annual research expenditures: $100,000. *Unit head:* Dr. Malur Srinivasan, Interim Chair, 409-880-8769, Fax: 409-880-8121. *Application contact:* Dr. Malur Srinivasan, Interim Chair, 409-880-8769, Fax: 409-880-8121.

Lehigh University, College of Engineering and Applied Science, Department of Mechanical Engineering and Mechanics, Bethlehem, PA 18015-3094. Offers applied mathematics (MS, PhD); mechanical engineering (M Eng, MS, PhD); mechanics (M Eng, MS, PhD). Part-time programs available. *Faculty:* 29 full-time (0 women). *Students:* 75 full-time (8 women), 11 part-time (1 woman); includes 7 minority (2 African Americans, 5 Hispanic Americans), 43 international. 51 applicants, 53% accepted. In 1998, 24 master's, 6 doctorates awarded. Terminal master's awarded for partial completion of doctoral program. *Degree requirements:* For master's and doctorate, thesis/dissertation required, foreign language not required. *Entrance requirements:* For master's and doctorate, TOEFL (minimum score of 550 required). *Average time to degree:* Master's–2 years full-time, 4 years part-time; doctorate–6 years full-time, 9 years part-time. *Application deadline:* For fall admission, 7/15; for spring admission, 12/1. Applications are processed on a rolling basis. Application fee: $40. *Financial aid:* In 1998–99, 17 fellowships with full and partial tuition reimbursements (averaging $11,000 per year), 15 research assistantships with full and partial tuition reimbursements (averaging $15,000 per year), 13 teaching assistantships with full and partial tuition reimbursements (averaging $11,000 per year) were awarded. Financial aid application deadline: 1/15. *Faculty research:* Thermofluids,

Mechanical Engineering

Lehigh University (continued)
dynamic systems, CAD/CAM. Total annual research expenditures: $2.5 million. *Unit head:* Dr. Charles Smith, Chairman, 610-758-4102, Fax: 610-758-6224, E-mail: crs1@lehigh.edu. *Application contact:* Donna Reiss, Graduate Coordinator, 610-758-4139, Fax: 610-758-6224, E-mail: dmr1@lehigh.edu.

Louisiana State University and Agricultural and Mechanical College, Graduate School, College of Engineering, Department of Mechanical Engineering, Baton Rouge, LA 70803. Offers MSME, PhD. Part-time programs available. *Faculty:* 16 full-time (0 women), 1 part-time (0 women). *Students:* 60 full-time (4 women), 6 part-time (1 woman); includes 3 minority (2 African Americans, 1 Asian American or Pacific Islander), 42 international. Average age 26. 210 applicants, 33% accepted. In 1998, 9 master's, 5 doctorates awarded. Terminal master's awarded for partial completion of doctoral program. *Degree requirements:* For master's and doctorate, thesis/dissertation required, foreign language not required. *Entrance requirements:* For master's and doctorate, GRE General Test, TOEFL, minimum GPA of 3.0. *Application deadline:* For fall admission, 1/25 (priority date). Applications are processed on a rolling basis. Application fee: $25. *Financial aid:* In 1998–99, 10 fellowships, 35 research assistantships with partial tuition reimbursements, 13 teaching assistantships with partial tuition reimbursements were awarded.; institutionally-sponsored loans, tuition waivers (full and partial), and unspecified assistantships also available. *Faculty research:* Computer-aided design, thermal and fluid sciences materials engineering, fluid mechanics, combustion and microsystems engineering. Total annual research expenditures: $1.2 million. *Unit head:* Dr. Mehdy Sabbaghiam, Chair, 225-388-5792, Fax: 225-388-5924, E-mail: sabbaghi@mail.eng.lsu.edu. *Application contact:* Dr. S. Acharya, Graduate Adviser, 225-388-5809, Fax: 225-388-5924, E-mail: acharya@me.lsu.edu.

Louisiana Tech University, Graduate School, College of Engineering and Science, Department of Mechanical and Industrial Engineering, Ruston, LA 71272. Offers industrial engineering (MS, D Eng); manufacturing systems engineering (MS); mechanical engineering (MS, D Eng); operations research (MS). Part-time programs available. Terminal master's awarded for partial completion of doctoral program. *Degree requirements:* For master's and doctorate, thesis/dissertation required, foreign language not required. *Entrance requirements:* For master's, GRE General Test (minimum combined score of 1070 required), TOEFL (minimum score of 550 required), minimum GPA of 3.0 in last 60 hours; for doctorate, TOEFL (minimum score of 550 required), minimum graduate GPA of 3.25 (with MS) or GRE General Test (minimum combined score of 1270 required without MS). *Faculty research:* Engineering management, facilities planning, thermodynamics, automated manufacturing, micromanufacturing.

Loyola Marymount University, Graduate Division, College of Science and Engineering, Department of Mechanical Engineering, Program in Mechanical Engineering, Los Angeles, CA 90045-8350. Offers MSE. *Faculty:* 6 full-time (0 women), 5 part-time (2 women). *Students:* 5 full-time (0 women), 7 part-time; includes 4 minority (3 Asian Americans or Pacific Islanders, 1 Hispanic American), 1 international. 10 applicants, 70% accepted. In 1998, 6 degrees awarded. *Degree requirements:* For master's, one foreign language required. *Entrance requirements:* For master's, TOEFL. Application fee: $35. Electronic applications accepted. Tuition: Part-time $525 per unit. Required fees: $143; $14 per semester. Tuition and fees vary according to program. *Financial aid:* In 1998–99, 2 students received aid. Federal Work-Study, grants, and laboratory assistantships available. Aid available to part-time students. Financial aid application deadline: 3/2; financial aid applicants required to submit FAFSA. *Unit head:* Dr. Bo Oppenheim, Director, 310-338-2825.

Manhattan College, Graduate Division, School of Engineering, Program in Mechanical Engineering, Riverdale, NY 10471. Offers MS. Part-time and evening/weekend programs available. *Degree requirements:* For master's, computer language, thesis or alternative required. *Entrance requirements:* For master's, GRE, TOEFL, minimum GPA of 3.0. *Faculty research:* Thermal analysis of rocket thrust chambers, stress analysis of torque transmission, quality of wood.

Marquette University, Graduate School, College of Engineering, Department of Mechanical and Industrial Engineering, Milwaukee, WI 53201-1881. Offers engineering management (MS); materials science and engineering (MS, PhD); mechanical engineering (MS, PhD), including manufacturing systems engineering. Part-time and evening/weekend programs available. *Faculty:* 18 full-time (0 women), 6 part-time (0 women). *Students:* 32 full-time (3 women), 52 part-time (4 women); includes 6 minority (1 African American, 2 Asian Americans or Pacific Islanders, 2 Hispanic Americans, 1 Native American), 30 international. Average age 27. 75 applicants, 73% accepted. In 1998, 9 master's, 5 doctorates awarded. Terminal master's awarded for partial completion of doctoral program. *Degree requirements:* For master's, thesis, comprehensive exam required, foreign language not required; for doctorate, dissertation, proficiency exam, qualifying exam required, foreign language not required. *Entrance requirements:* For master's and doctorate, GRE General Test, TOEFL (minimum score of 550 required), minimum GPA of 3.0. *Average time to degree:* Master's–2 years full-time, 5 years part-time; doctorate–5 years full-time. *Application deadline:* For fall admission, 8/1 (priority date); for spring admission, 1/1 (priority date). Applications are processed on a rolling basis. Application fee: $40. Tuition: Part-time $510 per credit hour. Tuition and fees vary according to program. *Financial aid:* In 1998–99, 19 students received aid, including 2 fellowships (averaging $9,540 per year), 7 research assistantships with tuition reimbursements available (averaging $9,540 per year), 10 teaching assistantships with tuition reimbursements available (averaging $9,540 per year); Federal Work-Study, grants, institutionally-sponsored loans, scholarships, and tuition waivers (full and partial) also available. Aid available to part-time students. Financial aid application deadline: 2/15. *Faculty research:* Computer-integrated manufacturing, energy conversion, simulation modeling and optimization, applied mechanics, metallurgy, ergonomics. Total annual research expenditures: $500,000. *Unit head:* Dr. G. E. O. Widera, Chairman, 414-288-7259, Fax: 414-288-1647, E-mail: geo.widera@marquette.edu. *Application contact:* Dr. William E. Brower, Director of Graduate Studies, 414-288-1717, Fax: 414-288-7790, E-mail: 9322browerw@rms.csd.mu.edu.

Massachusetts Institute of Technology, School of Engineering, Department of Mechanical Engineering, Cambridge, MA 02139-4307. Offers SM, PhD, Sc D, Mech E. *Faculty:* 55 full-time (4 women). *Students:* 387 full-time (54 women), 2 part-time (1 woman); includes 41 minority (6 African Americans, 29 Asian Americans or Pacific Islanders, 6 Hispanic Americans), 161 international. Average age 26. 632 applicants, 38% accepted. In 1998, 127 master's, 36 doctorates awarded. Terminal master's awarded for partial completion of doctoral program. *Degree requirements:* For master's, thesis required, foreign language not required; for doctorate, dissertation, comprehensive exams required, foreign language not required; for Mech E, thesis required. *Entrance requirements:* For master's, doctorate, and Mech E, GRE General Test. *Application deadline:* For fall admission, 12/15. Application fee: $55. *Financial aid:* In 1998–99, 360 students received aid, including 38 fellowships, 297 research assistantships, 40 teaching assistantships; career-related internships or fieldwork and institutionally-sponsored loans also available. Financial aid applicants required to submit FAFSA. *Faculty research:* Systems and control, fluid and thermal sciences, mechanics and materials, design and manufacturing. Total annual research expenditures: $20.7 million. *Unit head:* Dr. Nam P. Suh, Head, 617-253-2225, Fax: 617-258-6156, E-mail: npsuh@mit.edu. *Application contact:* Leslie M. Regan, Administrative Assistant, 617-253-2291, Fax: 617-258-5802, E-mail: megradoffice@mit.edu.

McGill University, Faculty of Graduate Studies and Research, Faculty of Engineering, Department of Mechanical Engineering, Montréal, PQ H3A 2T5, Canada. Offers manufacturing management (MMM); mechanical engineering (M Eng, PhD). *Faculty:* 19 full-time (0 women), 9 part-time (0 women). *Students:* 74 full-time (10 women), 24 part-time (7 women). Average age 25. 162 applicants, 49% accepted. In 1998, 23 master's, 4 doctorates awarded. *Degree requirements:* For master's and doctorate, thesis/dissertation required. *Entrance requirements:* For master's, TOEFL (minimum score of 550 required), B Eng or equivalent, minimum GPA of 3.0; for doctorate, TOEFL (minimum score of 550 required), M Eng or equivalent. *Application deadline:* For fall and spring admission, 2/1; for winter admission, 5/15. Applications are processed on a rolling basis. Application fee: $60. *Financial aid:* In 1998–99, 20 fellowships, 29 teaching assistantships were awarded.; research assistantships Financial aid application deadline: 2/1. *Faculty research:* Aerodynamics, automation and robotics, bioengineering, combustion and shock wave physics, flow-induced vibrations and dynamics. *Unit head:* Prof. N. Hori, Graduate Program Coordinator, 514-398-6282, Fax: 514-398-7365, E-mail: hori@mecheng.lan.mcgill.ca. *Application contact:* A. Cianci, Graduate Secretary, 514-398-6281, Fax: 514-398-7365, E-mail: annac@mecheng.mcgill.ca.

McMaster University, School of Graduate Studies, Faculty of Engineering, Department of Mechanical Engineering, Hamilton, ON L8S 4M2, Canada. Offers M Eng, PhD. *Faculty:* 15 full-time. *Students:* 39 full-time, 14 part-time. *Degree requirements:* For master's, thesis required, foreign language not required; for doctorate, dissertation, comprehensive exam required, foreign language not required. *Application deadline:* For fall admission, 3/1 (priority date). Applications are processed on a rolling basis. Application fee: $50. *Financial aid:* In 1998–99, teaching assistantships (averaging $7,722 per year); fellowships, research assistantships *Unit head:* Dr. M. Elbestawi, Chair, 905-525-9140 Ext. 24310.

McNeese State University, Graduate School, College of Engineering and Technology, Lake Charles, LA 70609-2495. Offers chemical engineering (M Eng); civil engineering (M Eng); electrical engineering (M Eng); mechanical engineering (M Eng). Part-time and evening/weekend programs available. *Faculty:* 13 full-time (1 woman). *Students:* 5 full-time (0 women), 3 part-time. *Degree requirements:* For master's, computer language, thesis or alternative required, foreign language not required. *Entrance requirements:* For master's, GRE General Test, TOEFL, minimum undergraduate GPA of 3.0. *Application deadline:* For fall admission, 7/15 (priority date). Applications are processed on a rolling basis. Application fee: $10 ($25 for international students). *Unit head:* Dr. O. C. Karkalits, Dean, 318-475-5875.

Memorial University of Newfoundland, School of Graduate Studies, Faculty of Engineering and Applied Science, St. John's, NF A1C 5S7, Canada. Offers civil engineering (M Eng, PhD); electrical engineering (M Eng, PhD); mechanical engineering (M Eng, PhD); ocean engineering (M Eng, PhD). Part-time programs available. *Students:* 75 full-time (11 women), 28 part-time (2 women), 31 international. *Degree requirements:* For master's, thesis optional; for doctorate, dissertation, comprehensive exam required. *Application deadline:* For fall admission, 3/1. Application fee: $40. *Unit head:* Dr. Rangaswany Seshadri, Dean, 709-737-8810, Fax: 709-737-8975, E-mail: sesh@engr.mun.ca. *Application contact:* Dr. J. J. Sharp, Associate Dean, 709-737-8901, Fax: 709-737-3480, E-mail: jsharp@engr.mun.ca.

Mercer University, School of Engineering, Macon, GA 31207-0003. Offers biomedical engineering (MSE); electrical engineering (MSE); engineering management (MSE); mechanical engineering (MSE); software engineering (MSE); software systems (MS); technical management (MS). Part-time and evening/weekend programs available. *Faculty:* 23 full-time (1 woman), 6 part-time (0 women). *Degree requirements:* For master's, computer language, thesis or alternative required, foreign language not required. *Entrance requirements:* For master's, GRE, minimum undergraduate GPA of 3.0. *Application deadline:* For fall admission, 7/1; for spring admission, 11/15. Applications are processed on a rolling basis. Application fee: $35 ($50 for international students). *Unit head:* Dr. Benjamin S. Kelley, Dean, 912-752-2459, Fax: 912-752-5593, E-mail: kelley_bs@mercer.edu. *Application contact:* Kathy Olivier, Coordinator, Special Programs, 912-752-2196, E-mail: oliver_kh@mercer.edu.

Michigan State University, Graduate School, College of Engineering, Department of Mechanical Engineering, East Lansing, MI 48824-1020. Offers MS, PhD. Part-time programs available. Postbaccalaureate distance learning degree programs offered (minimal on-campus study). *Faculty:* 24. *Students:* 39 full-time (3 women), 55 part-time (3 women); includes 3 minority (all Asian Americans or Pacific Islanders), 52 international. Average age 29. 324 applicants, 23% accepted. In 1998, 29 master's, 6 doctorates awarded. *Degree requirements:* For master's, thesis optional, foreign language not required; for doctorate, dissertation required, foreign language not required. *Entrance requirements:* For master's, GRE General Test (score in 90th percentile or higher on quantitative section required), TOEFL (minimum score of 570 required); for doctorate, GRE General Test (score in 80th percentile or higher on analytical section required), TOEFL (minimum score of 570 required). *Application deadline:* Applications are processed on a rolling basis. Application fee: $30 ($40 for international students). *Financial aid:* In 1998–99, 27 research assistantships with tuition reimbursements (averaging $12,777 per year), 33 teaching assistantships with tuition reimbursements (averaging $11,847 per year) were awarded.; fellowships Financial aid applicants required to submit FAFSA. *Faculty research:* Experimental fluid mechanics, nonlinear vibrations, control of ER fluids, flow diagnostics in IC engines, turbomachinery design. Total annual research expenditures: $1.5 million. *Unit head:* Dr. Ronald C. Rosenberg, Chairperson, 517-353-9861, Fax: 517-353-1750.

See in-depth description on page 1403.

Michigan Technological University, Graduate School, College of Engineering, Department of Mechanical Engineering-Engineering Mechanics, Program in Mechanical Engineering, Houghton, MI 49931-1295. Offers mechanical engineering (MS); mechanical engineering-engineering mechanics (PhD). Part-time programs available. *Faculty:* 42 full-time (2 women), 1 part-time (0 women). *Students:* 36 full-time (9 women), 80 part-time; includes 3 minority (all African Americans), 48 international. Average age 28. 185 applicants, 85% accepted. In 1998, 13 degrees awarded. *Degree requirements:* For master's and doctorate, thesis/dissertation required, foreign language not required. *Entrance requirements:* For master's, GRE General Test (combined average 1943 on three sections), TOEFL (minimum score of 550 required; average 602); for doctorate, GRE General Test (combined average 1957 on three sections), TOEFL (minimum score of 550 required; average 605). *Average time to degree:* Master's–2.6 years full-time; doctorate–4.4 years full-time. *Application deadline:* For fall admission, 3/15 (priority date). Applications are processed on a rolling basis. Application fee: $30 ($35 for international students). Tuition, state resident: full-time $4,377. Tuition, nonresident: full-time $9,108. Required fees: $126. Tuition and fees vary according to course load. *Financial aid:* In 1998–99, 16 fellowships (averaging $3,883 per year), 34 research assistantships (averaging $8,065 per year), 20 teaching assistantships (averaging $8,907 per year) were awarded.; Federal Work-Study, institutionally-sponsored loans, and unspecified assistantships also available. Aid available to part-time students. Financial aid applicants required to submit FAFSA. *Unit head:* Dr. William Predebon, Chair, Department of Mechanical Engineering-Engineering Mechanics, 906-487-2551, Fax: 906-487-2822, E-mail: wwpredebon@mtu.edu.

See in-depth description on page 1405.

Mississippi State University, College of Engineering, Department of Mechanical Engineering, Mississippi State, MS 39762. Offers MS. Part-time programs available. Postbaccalaureate distance learning degree programs offered (minimal on-campus study). *Students:* 14 full-time (3 women), 4 part-time; includes 7 minority (1 African American, 6 Asian Americans or Pacific Islanders), 1 international. Average age 27. 36 applicants, 25% accepted. In 1998, 7 degrees awarded. *Degree requirements:* For master's, oral exam required, thesis optional, foreign language not required. *Entrance requirements:* For master's, GRE General Test, TOEFL (minimum score of 550 required), minimum GPA of 2.75. *Application deadline:* For fall admission, 7/1; for spring admission, 11/1. Applications are processed on a rolling basis. Application fee: $25 for international students. *Financial aid:* Fellowships with full tuition reimbursements, research assistantships with full tuition reimbursements, teaching assistantships with full tuition reimbursements, Federal Work-Study and institutionally-sponsored loans available. Financial aid applicants required to submit FAFSA. *Faculty research:* Heat transfer, fluid dynamics, energy systems, manufacturing systems, materials. Total annual research expenditures: $931,266. *Unit head:* Dr. W. Glenn Steele, Head, 662-325-3260, Fax: 662-325-7223, E-mail: graduate@meng.msstate.edu. *Application contact:* Jerry B. Inmon, Director of Admissions, 662-325-2224, Fax: 662-325-7360, E-mail: admit@admissions.msstate.edu.

Montana State University–Bozeman, College of Graduate Studies, College of Engineering, Department of Mechanical and Industrial Engineering, Bozeman, MT 59717. Offers engineering (PhD); industrial and management engineering (MS); mechanical engineering (MS). Part-time programs available. *Students:* 28 full-time (6 women), 7 part-time. Average age 26. 37

applicants, 73% accepted. In 1998, 5 master's awarded. *Degree requirements:* For master's, thesis or alternative required, foreign language not required; for doctorate, dissertation required, foreign language not required. *Entrance requirements:* For master's and doctorate, GRE General Test, TOEFL. *Application deadline:* For fall admission, 6/1 (priority date); for spring admission, 11/1. Applications are processed on a rolling basis. Application fee: $50. *Financial aid:* Research assistantships, teaching assistantships available. Financial aid application deadline: 3/1; financial aid applicants required to submit FAFSA. *Faculty research:* Operations research, product processes and assessment, systems modeling/simulation, management planning, engineering economy. Total annual research expenditures: $922,633. *Unit head:* Dr. Vic Cundy, Head, 406-994-2203, Fax: 406-994-6292, E-mail: vcundy@me.montana.edu.

Naval Postgraduate School, Graduate Programs, Department of Mechanical Engineering, Monterey, CA 93943. Offers MS, D Eng, PhD, Eng. Program only open to commissioned officers of the United States and friendly nations and selected United States federal civilian employees. *Accreditation:* ABET (one or more programs are accredited). Part-time programs available. *Faculty:* 88 full-time. *Students:* 55 full-time, 6 international. In 1998, 27 master's, 2 doctorates, 4 other advanced degrees awarded. *Degree requirements:* For master's and Eng, computer language, thesis required, foreign language not required; for doctorate, one foreign language, computer language, dissertation required. *Unit head:* Dr. Terry R. McNelley, Chairman, 831-656-2589. *Application contact:* Theodore H. Calhoon, Director of Admissions, 831-656-3093, Fax: 831-656-2891, E-mail: tcalhoon@nps.navy.mil.

New Jersey Institute of Technology, Office of Graduate Studies, Department of Mechanical Engineering, Newark, NJ 07102-1982. Offers MS, PhD, Engineer. Part-time and evening/weekend programs available. Terminal master's awarded for partial completion of doctoral program. *Degree requirements:* For doctorate, dissertation, residency required, foreign language not required, foreign language not required. *Entrance requirements:* For master's, GRE General Test (minimum score of 450 on verbal section, 600 on quantitative, 550 on analytical required); for doctorate, GRE General Test (minimum score of 450 on verbal section, 600 on quantitative, 550 on analytical required), minimum graduate GPA of 3.5. Electronic applications accepted. *Faculty research:* Energy systems, structural mechanics, electromechanical systems.

New Mexico State University, Graduate School, College of Engineering, Department of Mechanical Engineering, Las Cruces, NM 88003-8001. Offers MSME, PhD. *Faculty:* 16 full-time (0 women). *Students:* 24 full-time (1 woman), 10 part-time (2 women); includes 5 minority (1 Asian American or Pacific Islander, 4 Hispanic Americans), 5 international. Average age 29. 62 applicants, 71% accepted. In 1998, 11 degrees awarded. *Degree requirements:* For master's, computer language, thesis required (for some programs); for doctorate, computer language, dissertation, 2 research tools required. *Entrance requirements:* For master's, minimum GPA of 3.0; for doctorate, qualifying exam, minimum GPA of 3.0. *Application deadline:* For fall admission, 7/1 (priority date); for spring admission, 11/1. Applications are processed on a rolling basis. Application fee: $15 ($35 for international students). Electronic applications accepted. Tuition, state resident: full-time $2,682; part-time $112 per credit. Tuition, nonresident: full-time $8,376; part-time $349 per credit. Tuition and fees vary according to course load. *Financial aid:* Fellowships, research assistantships, teaching assistantships, career-related internships or fieldwork and Federal Work-Study available. Aid available to part-time students. Financial aid application deadline: 3/1. *Faculty research:* Computational mechanics; robotics; CAD/CAM; control, dynamics, and solid mechanics; interconnection engineering. *Unit head:* Dr. Bahram Nassersharif, Head, 505-646-3501, Fax: 505-646-6111, E-mail: bn@nmsu.edu. *Application contact:* Dr. Ronald J. Pederson, Associate Head, 505-646-3501, Fax: 505-646-6111, E-mail: rpederso@nmsu.edu.

North Carolina Agricultural and Technical State University, Graduate School, College of Engineering, Department of Mechanical Engineering, Greensboro, NC 27411. Offers MSME, PhD. PhD offered jointly with North Carolina State University. Part-time programs available. *Faculty:* 21 full-time (1 woman), 8 part-time (1 woman). *Students:* 44 full-time (5 women), 6 part-time; includes 36 minority (35 African Americans, 1 Asian American or Pacific Islander), 12 international. Average age 26. 62 applicants, 71% accepted. In 1998, 16 master's awarded. *Degree requirements:* For master's, comprehensive exam, dual exam, qualifying exam, thesis defense required. *Entrance requirements:* For master's, GRE General Test, GRE Subject Test (recommended), TOEFL; for doctorate, GRE. *Average time to degree:* Master's–2 years full-time, 3 years part-time. *Application deadline:* For fall admission, 7/1 (priority date); for spring admission, 1/9. Applications are processed on a rolling basis. Application fee: $35. *Financial aid:* Fellowships with full tuition reimbursements, research assistantships with partial tuition reimbursements, teaching assistantships, tuition waivers (partial) and unspecified assistantships available. *Faculty research:* Composite materials, CAD, energy, boundary-layer theory, manufacturing and robotics. Total annual research expenditures: $3.9 million. *Unit head:* Dr. William J. Craft, Chairperson, 336-334-7620, Fax: 336-334-7417, E-mail: craft@ncat.edu. *Application contact:* Dr. Shih-Liang Wang, Graduate Coordinator, 336-334-7620, Fax: 336-334-7414, E-mail: slwang@ncat.edu.

North Carolina State University, Graduate School, College of Engineering, Department of Mechanical and Aerospace Engineering, Program in Mechanical Engineering, Raleigh, NC 27695. Offers MME, MS, PhD. Part-time programs available. *Faculty:* 42 full-time (1 woman), 47 part-time (0 women). *Students:* 78 full-time (5 women), 30 part-time (3 women); includes 14 minority (4 African Americans, 8 Asian Americans or Pacific Islanders, 1 Hispanic American, 1 Native American), 32 international. Average age 30. 124 applicants, 27% accepted. In 1998, 21 master's, 11 doctorates awarded. *Degree requirements:* For master's, thesis, oral exam required, foreign language not required; for doctorate, dissertation, oral and preliminary exams required, foreign language not required. *Entrance requirements:* For master's and doctorate, GRE General Test. *Average time to degree:* Master's–2 years full-time, 3 years part-time; doctorate–3 years full-time, 4 years part-time. *Application deadline:* For fall admission, 7/15; for spring admission, 12/15. Applications are processed on a rolling basis. Application fee: $45. *Financial aid:* Fellowships, research assistantships, teaching assistantships, career-related internships or fieldwork and institutionally-sponsored loans available. *Faculty research:* Vibration and control, fluid dynamics, thermal sciences, structure and materials, aerodynamics accoustics. *Unit head:* Dr. James C. Mulligan, Director of Graduate Programs, 919-515-2856, Fax: 919-515-7968, E-mail: mulligan@eos.ncsu.edu.

See in-depth description on page 1407.

North Dakota State University, Graduate Studies and Research, College of Engineering and Architecture, Department of Mechanical Engineering and Applied Mechanics, Fargo, ND 58105. Offers MS. Part-time and evening/weekend programs available. *Faculty:* 11 full-time (0 women). *Students:* 8 full-time (0 women), 5 part-time (1 woman); includes 1 minority (African American), 7 international. Average age 25. 38 applicants, 32% accepted. In 1998, 2 degrees awarded (100% found work related to degree). *Degree requirements:* For master's, thesis required, foreign language not required. *Entrance requirements:* For master's, TOEFL (minimum score of 550 required), minimum GPA of 3.0. *Application deadline:* For fall admission, 7/1 (priority date). Applications are processed on a rolling basis. Application fee: $25. *Financial aid:* In 1998–99, 6 students received aid, including 6 research assistantships with full tuition reimbursements available (averaging $8,000 per year); teaching assistantships, career-related internships or fieldwork, Federal Work-Study, and institutionally-sponsored loans also available. Financial aid application deadline: 2/15. *Faculty research:* Thermodynamics, finite element analysis, automotive systems, robotics. *Unit head:* Dr. Robert Pieri, Chair, 701-231-8671, Fax: 701-231-7195, E-mail: pieri@badlands.nodak.edu.

Northeastern University, College of Engineering, Department of Mechanical, Industrial, and Manufacturing Engineering, Boston, MA 02115-5096. Offers engineering management (MS); industrial engineering (MS, PhD); mechanical engineering (MS, PhD); operations research (MS). Part-time programs available. *Faculty:* 30 full-time (1 woman), 13 part-time (1 woman). *Students:* 105 full-time (19 women), 118 part-time (25 women); includes 14 minority (5 African Americans, 7 Asian Americans or Pacific Islanders, 1 Hispanic American, 1 Native American), 84 international. Average age 25. 348 applicants, 59% accepted. In 1998, 48 master's, 8 doctorates awarded. *Degree requirements:* For master's, thesis required (for some programs), foreign language not required; for doctorate, one foreign language, dissertation, departmental qualifying exam required. *Entrance requirements:* For master's and doctorate, GRE General Test. *Average time to degree:* Master's–3.38 years full-time, 4.05 years part-time; doctorate–6.5 years full-time, 8 years part-time. *Application deadline:* For fall admission, 4/15. Applications are processed on a rolling basis. Application fee: $50. *Financial aid:* In 1998–99, 49 students received aid, including 14 research assistantships with full tuition reimbursements available (averaging $12,450 per year), 26 teaching assistantships with full tuition reimbursements available (averaging $12,450 per year); fellowships, career-related internships or fieldwork, Federal Work-Study, tuition waivers (full), and unspecified assistantships also available. Aid available to part-time students. Financial aid application deadline: 2/15; financial aid applicants required to submit FAFSA. *Faculty research:* Dry sliding instabilities, droplet deposition, helical hydraulic turbines, combustion, manufacturing systems. *Unit head:* Dr. John W. Cipolla, Chairman, 617-373-3810, Fax: 617-373-2921. *Application contact:* Stephen L. Gibson, Associate Director, 617-373-2711, Fax: 617-373-2501, E-mail: grad-eng@coe.neu.edu.

See in-depth description on page 1409.

Northern Illinois University, Graduate School, College of Engineering and Engineering Technology, Department of Mechanical Engineering, De Kalb, IL 60115-2854. Offers MS. Part-time programs available. *Faculty:* 9 full-time (0 women), 1 part-time (0 women). *Students:* 9 full-time (0 women), 18 part-time (3 women), 19 international. Average age 28. 53 applicants, 26% accepted. In 1998, 6 degrees awarded. *Degree requirements:* For master's, comprehensive exam, thesis defense required. *Entrance requirements:* For master's, GRE General Test, TOEFL (minimum score of 550 required; 213 for computer-based), minimum GPA of 2.75. *Application deadline:* For fall admission, 6/1; for spring admission, 11/1. Applications are processed on a rolling basis. Application fee: $30. *Financial aid:* Fellowships, research assistantships, teaching assistantships, Federal Work-Study, tuition waivers (full), and unspecified assistantships available. Aid available to part-time students. *Unit head:* Dr. Parviz Payvar, Chair, 815-753-9970.

Northwestern University, The Graduate School, Robert R. McCormick School of Engineering and Applied Science, Department of Mechanical Engineering, Evanston, IL 60208. Offers manufacturing engineering (MME); mechanical engineering (MS, PhD). MS, PhD admissions and degrees offered through The Graduate School. Part-time programs available. *Faculty:* 24 full-time (2 women), 6 part-time (0 women). *Students:* 77 full-time (11 women), 2 part-time (1 woman); includes 8 minority (4 African Americans, 3 Asian Americans or Pacific Islanders, 1 Hispanic American), 47 international. 206 applicants, 52% accepted. In 1998, 9 master's, 16 doctorates awarded. Terminal master's awarded for partial completion of doctoral program. *Degree requirements:* For master's, thesis or alternative required, foreign language not required; for doctorate, dissertation required, foreign language not required. *Entrance requirements:* For master's and doctorate, GRE General Test, TOEFL (minimum score of 560 required). *Application deadline:* For fall admission, 8/30. Application fee: $50 ($55 for international students). *Financial aid:* In 1998–99, 8 fellowships with full tuition reimbursements (averaging $11,673 per year), 31 research assistantships with partial tuition reimbursements (averaging $16,285 per year), 10 teaching assistantships with full tuition reimbursements (averaging $12,042 per year) were awarded.; career-related internships or fieldwork, Federal Work-Study, institutionally-sponsored loans, and scholarships also available. Financial aid application deadline: 1/15; financial aid applicants required to submit FAFSA. *Faculty research:* Experimental, theoretical and finite element mechanics of materials, low gravity fluid mechanics and combustion, manufacturing processes, nondestructive testing, robotics and control. Total annual research expenditures: $5.4 million. *Unit head:* Ted Belytschko, Chair, 847-491-7470, Fax: 847-491-3915. *Application contact:* Pat Dyess, Admission Contact, 847-491-7190, Fax: 847-491-3915, E-mail: j-dyess@nwu.edu.

See in-depth description on page 1411.

Oakland University, Graduate Studies, School of Engineering and Computer Science, Program in Mechanical Engineering, Rochester, MI 48309-4401. Offers MS. Part-time and evening/weekend programs available. *Faculty:* 10 full-time, 5 part-time. *Students:* 63 full-time (16 women), 80 part-time (10 women); includes 14 minority (4 African Americans, 9 Asian Americans or Pacific Islanders, 1 Hispanic American), 12 international. Average age 28. 51 applicants, 76% accepted. In 1998, 61 degrees awarded. *Degree requirements:* For master's, foreign language and thesis not required. *Entrance requirements:* For master's, minimum GPA of 3.0 for unconditional admission. *Application deadline:* For fall admission, 7/15; for spring admission, 3/15. Application fee: $30. Tuition, state resident: part-time $221 per credit hour. Tuition, nonresident: part-time $488 per credit hour. Required fees: $214 per semester. Part-time tuition and fees vary according to program. *Financial aid:* Federal Work-Study, institutionally-sponsored loans, and tuition waivers (full) available. Financial aid application deadline: 3/1; financial aid applicants required to submit FAFSA. *Unit head:* Dr. Joseph Hovanesian, Chair, 248-370-2210.

The Ohio State University, Graduate School, College of Engineering, Department of Mechanical Engineering, Program in Mechanical Engineering, Columbus, OH 43210. Offers MS, PhD. *Faculty:* 31 full-time, 24 part-time. *Students:* 153 full-time (16 women), 46 part-time (4 women); includes 10 minority (1 African American, 8 Asian Americans or Pacific Islanders, 1 Hispanic American), 95 international. 406 applicants, 25% accepted. In 1998, 69 master's, 20 doctorates awarded. *Degree requirements:* For master's and doctorate, computer language, thesis/dissertation required, foreign language not required. *Application deadline:* For fall admission, 8/15. Applications are processed on a rolling basis. Application fee: $30 ($40 for international students). *Unit head:* Kenneth J. Waldron, Chair, Department of Mechanical Engineering, 614-292-2289, Fax: 614-292-3163, E-mail: waldron.3@osu.edu.

Ohio University, Graduate Studies, College of Engineering and Technology, Department of Mechanical Engineering, Athens, OH 45701-2979. Offers manufacturing engineering (MS); mechanical engineering (MS, PhD), including CAD/CAM (MS), manufacturing (MS), mechanical systems (MS), technology management (MS), thermal systems (MS). *Faculty:* 11 full-time (0 women). *Students:* 50 full-time (5 women), 29 part-time (2 women); includes 1 minority (African American), 67 international. 151 applicants, 81% accepted. In 1998, 16 degrees awarded. *Degree requirements:* For master's, thesis required, foreign language not required; for doctorate, dissertation required. *Entrance requirements:* For master's, BS in engineering or science, minimum GPA of 2.8; for doctorate, GRE. *Application deadline:* For fall admission, 3/15 (priority date). Applications are processed on a rolling basis. Application fee: $30. Tuition, state resident: full-time $5,754; part-time $238 per credit hour. Tuition, nonresident: full-time $11,055; part-time $457 per credit hour. Tuition and fees vary according to course load, campus/location and program. *Financial aid:* In 1998–99, 41 students received aid, including 9 research assistantships, 14 teaching assistantships; fellowships, career-related internships or fieldwork, Federal Work-Study, institutionally-sponsored loans, tuition waivers (full and partial), and unspecified assistantships also available. Financial aid application deadline: 3/15. *Faculty research:* Machine design, controls, aerodynamics, thermal manufacturing, robotics process modeling, air pollution, combustion. *Unit head:* Dr. Jay S. Gunasekara, Chairman, 740-593-0563, Fax: 740-593-0476, E-mail: gsekera@bobcat.ent.ohiou.edu. *Application contact:* Dr. M. Khairul Alam, Graduate Chairman, 740-593-1598, Fax: 740-593-0476, E-mail: agrawal@bobcat.ent.ohiou.edu.

Oklahoma State University, Graduate College, College of Engineering, Architecture and Technology, School of Mechanical and Aerospace Engineering, Stillwater, OK 74078. Offers mechanical engineering (M En, MS, PhD). *Faculty:* 17 full-time (0 women), 2 part-time (0 women). *Students:* 110 full-time (4 women), 47 part-time (3 women); includes 6 minority (1 African American, 2 Asian Americans or Pacific Islanders, 3 Hispanic Americans), 117 international. Average age 26. In 1998, 19 master's, 4 doctorates awarded. *Degree requirements:* For master's, thesis or alternative required, foreign language not required; for doctorate, dissertation required, foreign language not required. *Entrance requirements:* For master's, TOEFL; for doctorate, TOEFL (minimum score of 550 required). *Application deadline:* For fall

Mechanical Engineering

Oklahoma State University (continued)

admission, 7/1 (priority date). Application fee: $25. *Financial aid:* In 1998–99, 110 students received aid, including 68 research assistantships (averaging $7,767 per year), 42 teaching assistantships (averaging $6,481 per year); career-related internships or fieldwork, Federal Work-Study, and tuition waivers (partial) also available. Aid available to part-time students. Financial aid application deadline: 3/1. *Unit head:* Dr. Lawrence L. Hoberock, Head, 405-744-5900.

Old Dominion University, College of Engineering and Technology, Department of Mechanical Engineering, Norfolk, VA 23529. Offers design manufacturing (ME); engineering mechanics (ME, MS, PhD); mechanical engineering (ME, MS, PhD). Part-time and evening/weekend programs available. Postbaccalaureate distance learning degree programs offered (no on-campus study). *Faculty:* 14 full-time. *Students:* 61 full-time (3 women), 36 part-time (1 woman); includes 5 African Americans, 1 Hispanic American, 56 international. Average age 30. In 1998, 26 master's, 7 doctorates awarded. *Degree requirements:* For master's, computer language, comprehensive exam required, thesis optional, foreign language not required; for doctorate, computer language, dissertation, candidacy exam required, foreign language not required. *Entrance requirements:* For master's, GRE, TOEFL (minimum score of 550 required), minimum GPA of 3.0; for doctorate, GRE, TOEFL (minimum score of 550 required), minimum GPA of 3.25. *Application deadline:* For fall admission, 7/1; for spring admission, 10/1. Applications are processed on a rolling basis. Application fee: $30. Electronic applications accepted. *Financial aid:* In 1998–99, 61 students received aid, including 24 research assistantships (averaging $11,640 per year); fellowships, teaching assistantships, career-related internships or fieldwork, grants, institutionally-sponsored loans, and tuition waivers (partial) also available. Aid available to part-time students. Financial aid application deadline: 2/15; financial aid applicants required to submit FAFSA. *Faculty research:* Computational applied mechanics, manufacturing, experimental stress analysis, systems dynamics and control, mechanical design, computational fluid dynamics, optimization. Total annual research expenditures: $1.4 million. *Unit head:* Dr. Jen Kuang Huang, Chair, 757-683-6363, Fax: 757-683-5344, E-mail: megpd@odu.edu. *Application contact:* Dr. Jen Kuang Huang, Chair, 757-683-6363, Fax: 757-683-5344, E-mail: megpd@odu.edu.

Oregon State University, Graduate School, College of Engineering, Department of Mechanical Engineering, Corvallis, OR 97331. Offers materials science (MAIS, MS); mechanical engineering (MS, PhD). *Faculty:* 19 full-time (1 woman), 4 part-time (0 women). *Students:* 46 full-time (7 women), 10 part-time; includes 2 minority (both Asian Americans or Pacific Islanders), 28 international. Average age 28. In 1998, 13 master's, 2 doctorates awarded. *Degree requirements:* For master's, computer language, thesis or alternative required, foreign language not required; for doctorate, computer language, dissertation required, foreign language not required. *Entrance requirements:* For master's, GRE General Test, TOEFL (minimum score of 550 required), minimum GPA of 3.0 in last 90 hours; for doctorate, GRE General Test, GRE Subject Test, TOEFL (minimum score of 550 required), minimum GPA of 3.0 in last 90 hours. *Application deadline:* For fall admission, 3/1. Applications are processed on a rolling basis. Application fee: $50. *Financial aid:* Fellowships, research assistantships, teaching assistantships, Federal Work-Study and institutionally-sponsored loans available. Aid available to part-time students. Financial aid application deadline: 2/1. *Faculty research:* Design, thermal fluid sciences, energy conversion, mechanics. *Unit head:* Dr. Gordon M. Reistad, Head, 541-737-3441, Fax: 541-737-2600, E-mail: reistag@engr.orst.edu. *Application contact:* Graduate Admissions Clerk, 541-737-3441, Fax: 541-737-2600.

Pennsylvania State University University Park Campus, Graduate School, College of Engineering, Department of Mechanical Engineering, State College, University Park, PA 16802-1503. Offers M Eng, MS, PhD. *Students:* 154 full-time (17 women), 42 part-time (7 women). In 1998, 50 master's, 21 doctorates awarded. *Degree requirements:* For master's, foreign language and thesis not required; for doctorate, dissertation required, foreign language not required. *Entrance requirements:* For master's and doctorate, GRE General Test. Application fee: $50. *Unit head:* Dr. Richard C. Benson, Head, 814-865-2519. *Application contact:* Dr. Anil K. Kulkarni, Professor in Charge, 814-865-1345.

Polytechnic University, Brooklyn Campus, Department of Mechanical, Aerospace and Manufacturing Engineering, Major in Mechanical Engineering, Brooklyn, NY 11201-2990. Offers MS, PhD. *Students:* 12 full-time (0 women), 20 part-time (2 women); includes 5 minority (1 African American, 3 Asian Americans or Pacific Islanders, 1 Hispanic American), 9 international. Average age 33. 52 applicants, 40% accepted. In 1998, 9 master's, 1 doctorate awarded. *Entrance requirements:* For master's, BE or BS in engineering, physics, chemistry, mathematical sciences, or biological sciences or MBA. *Application deadline:* Applications are processed on a rolling basis. Application fee: $45. Electronic applications accepted. *Unit head:* William H. Firman, Director of Graduate Admissions, 215-951-2943, Fax: 215-951-2907, E-mail: gradadm@philau.edu. *Application contact:* John S. Kerge, Dean of Admissions, 718-260-3200, Fax: 718-260-3446, E-mail: admitme@poly.edu.

See in-depth description on page 1413.

Polytechnic University, Farmingdale Campus, Graduate Programs, Department of Mechanical, Aerospace and Manufacturing Engineering, Major in Mechanical Engineering, Farmingdale, NY 11735-3995. Offers MS, PhD. Average age 33. In 1998, 1 degree awarded. *Degree requirements:* For master's, computer language required. *Application deadline:* Applications are processed on a rolling basis. Application fee: $45. Electronic applications accepted. *Financial aid:* Institutionally-sponsored loans available. Aid available to part-time students. Financial aid applicants required to submit FAFSA. *Unit head:* John S. Kerge, Dean of Admissions, 718-260-3200, Fax: 718-260-3446, E-mail: admitme@poly.edu. *Application contact:* John S. Kerge, Dean of Admissions, 718-260-3200, Fax: 718-260-3446, E-mail: admitme@poly.edu.

Portland State University, Graduate Studies, School of Engineering and Applied Science, Department of Mechanical Engineering, Portland, OR 97207-0751. Offers MS, PhD. Part-time and evening/weekend programs available. *Faculty:* 8 full-time (0 women), 7 part-time (0 women). *Students:* 10 full-time (2 women), 12 part-time (1 woman); includes 3 minority (2 Asian Americans or Pacific Islanders, 1 Hispanic American), 8 international. Average age 28. 24 applicants, 63% accepted. In 1998, 4 degrees awarded. *Degree requirements:* For master's, thesis or alternative required, foreign language not required; for doctorate, one foreign language, computer language, dissertation, oral and written exams required. *Entrance requirements:* For master's, TOEFL (minimum score of 550 required), minimum GPA of 3.0 in upper-division course work or 2.75 overall; for doctorate, GRE General Test, GRE Subject Test, minimum GPA of 3.0 in upper-division course work. *Application deadline:* For fall admission, 4/1 (priority date); for spring admission, 11/1. Applications are processed on a rolling basis. Application fee: $50. *Financial aid:* In 1998–99, 1 research assistantship, 12 teaching assistantships were awarded.; Federal Work-Study and institutionally-sponsored loans also available. Aid available to part-time students. Financial aid application deadline: 3/1; financial aid applicants required to submit FAFSA. *Faculty research:* Mechanical system modeling, indoor air quality, manufacturing process, computational fluid dynamics, buildingscience. Total annual research expenditures: $50,193. *Unit head:* Dr. Graig Spolek, Head, 503-725-4290, Fax: 503-725-4298, E-mail: graig@eas.pdx.edu. *Application contact:* Gerald Recktenwald, Coordinator, 503-725-4290, Fax: 503-725-4298, E-mail: gerry@me.pdx.edu.

Portland State University, Graduate Studies, Systems Science Program, Portland, OR 97207-0751. Offers systems science/anthropology (PhD); systems science/business administration (PhD); systems science/civil engineering (PhD); systems science/economics (PhD); systems science/engineering management (PhD); systems science/general (PhD); systems science/mathematical sciences (PhD); systems science/mechanical engineering (PhD); systems science/psychology (PhD); systems science/sociology (PhD). *Faculty:* 3 full-time (0 women), 1 part-time (0 women). *Students:* 45 full-time (17 women), 23 part-time (6 women); includes 5 minority (1 African American, 3 Asian Americans or Pacific Islanders, 1 Hispanic American), 12 international. *Degree requirements:* For doctorate, variable foreign language requirement, computer language, dissertation required. *Entrance requirements:* For doctorate, GMAT (score in 75th percentile or higher required), GRE General Test (score in 75th percentile or higher required), TOEFL (minimum score of 575 required), minimum undergraduate GPA of 3.0. *Application deadline:* For fall admission, 2/1; for spring admission, 11/1. Application fee: $50. *Unit head:* Dr. Nancy Perrin, Director, 503-725-4960, E-mail: perrinn@pdx.edu. *Application contact:* Dawn Kuenle, Coordinator, 503-725-4960, E-mail: dawn@sysc.pdx.edu.

Princeton University, Graduate School, School of Engineering and Applied Science, Department of Mechanical and Aerospace Engineering, Princeton, NJ 08544-1019. Offers applied physics (M Eng, MSE, PhD); computational methods (M Eng, MSE); dynamics and control systems (M Eng, MSE, PhD); energy and environmental policy (M Eng, MSE, PhD); energy conversion, propulsion, and combustion (M Eng, MSE, PhD); flight science and technology (M Eng, MSE, PhD); fluid mechanics (M Eng, MSE, PhD). *Faculty:* 26 full-time (2 women). *Students:* 46 full-time (6 women); includes 5 minority (4 Asian Americans or Pacific Islanders, 1 Hispanic American), 17 international. Average age 23. 173 applicants, 22% accepted. In 1998, 6 master's awarded (83% found work related to degree, 17% continued full-time study); 14 doctorates awarded (29% entered university research/teaching, 71% found other work related to degree). *Degree requirements:* For master's and doctorate, thesis/dissertation required, foreign language not required. *Entrance requirements:* For master's and doctorate, GRE General Test. *Average time to degree:* Master's–2 years full-time; doctorate–5.53 years full-time. *Application deadline:* For fall admission, 1/3. Electronic applications accepted. *Financial aid:* Fellowships, research assistantships, teaching assistantships, Federal Work-Study and institutionally-sponsored loans available. Financial aid application deadline: 1/3. Total annual research expenditures: $5.3 million. *Unit head:* Prof. Richard B. Miles, Director of Graduate Studies, 609-258-4683, Fax: 609-258-6109, E-mail: maegrad@princeton.edu. *Application contact:* Etta Recke, Graduate Administrator, 609-258-4683, Fax: 609-258-6109, E-mail: etta@princeton.edu.

Purdue University, Graduate School, Schools of Engineering, School of Mechanical Engineering, West Lafayette, IN 47907. Offers biomedical engineering (MS Bm E, PhD); mechanical engineering (MS, MSE, MSME, PhD). *Faculty:* 45 full-time (2 women), 6 part-time (0 women). *Students:* 230 full-time (21 women), 22 part-time (1 woman). 809 applicants, 20% accepted. In 1998, 52 master's, 25 doctorates awarded. *Degree requirements:* For doctorate, dissertation required. *Entrance requirements:* For master's and doctorate, TOEFL (minimum score of 575 required). Application fee: $30. Electronic applications accepted. *Financial aid:* In 1998–99, 183 students received aid, including 31 fellowships with full tuition reimbursements available (averaging $15,000 per year), 111 research assistantships with full tuition reimbursements available (averaging $16,590 per year), 41 teaching assistantships with full tuition reimbursements available (averaging $13,375 per year); career-related internships or fieldwork also available. Aid available to part-time students. Financial aid applicants required to submit FAFSA. *Faculty research:* Design, manufacturing, thermal/fluid sciences, mechanics, electromechanical systems. *Unit head:* Dr. F. Dan Hirleman, Head, 765-494-5688.

See in-depth description on page 1415.

Queen's University at Kingston, School of Graduate Studies and Research, Faculty of Applied Science, Department of Mechanical Engineering, Kingston, ON K7L 3N6, Canada. Offers M Sc, M Sc Eng, PhD. Part-time programs available. *Students:* 44 full-time (5 women), 4 part-time (1 woman). In 1998, 15 master's, 11 doctorates awarded. *Degree requirements:* For master's, thesis optional, foreign language not required; for doctorate, dissertation, comprehensive exam required, foreign language not required. *Entrance requirements:* For master's and doctorate, TOEFL (minimum score of 550 required). *Application deadline:* For fall admission, 2/28 (priority date). Application fee: $60. Electronic applications accepted. *Financial aid:* Fellowships, research assistantships, teaching assistantships, institutionally-sponsored loans available. Financial aid application deadline: 3/1. *Faculty research:* Power generation and utilization, transportation, manufacturing and design engineering (including biomedical applications). *Unit head:* Dr. B. W. Surgenor, Head, 613-533-2568. *Application contact:* Dr. A. M. Birk, Graduate Coordinator, 613-533-2570.

Rensselaer at Hartford, School of Engineering, Program in Mechanical Engineering, Hartford, CT 06120-2991. Offers MS. Part-time and evening/weekend programs available. *Degree requirements:* For master's, seminar required, thesis optional, foreign language not required. *Entrance requirements:* For master's, TOEFL (minimum score of 570 required).

Rensselaer Polytechnic Institute, Graduate School, School of Engineering, Department of Mechanical Engineering, Aeronautical Engineering and Mechanics, Program in Mechanical Engineering, Troy, NY 12180-3590. Offers M Eng, MS, D Eng, PhD, MBA/M Eng. Part-time and evening/weekend programs available. Postbaccalaureate distance learning degree programs offered (no on-campus study). *Faculty:* 31 full-time (2 women), 5 part-time (0 women). *Students:* 116 full-time (9 women), 49 part-time (7 women); includes 24 minority (2 African Americans, 17 Asian Americans or Pacific Islanders, 5 Hispanic Americans), 72 international. 249 applicants, 33% accepted. In 1998, 53 master's, 17 doctorates awarded. *Degree requirements:* For master's, computer language, thesis required (for some programs), foreign language not required; for doctorate, computer language, dissertation required, foreign language not required. *Entrance requirements:* For master's and doctorate, GRE, TOEFL (minimum score of 550 required). *Application deadline:* For fall admission, 2/1 (priority date). Applications are processed on a rolling basis. Application fee: $35. *Financial aid:* In 1998–99, 100 students received aid; fellowships, research assistantships, teaching assistantships, career-related internships or fieldwork, institutionally-sponsored loans, and tuition waivers (partial) available. Financial aid application deadline: 2/1. *Faculty research:* Tribology, advanced composite materials, energy and combustion systems, computer-aided and optimal design, manufacturing and robotics. Total annual research expenditures: $2.5 million. *Unit head:* Dr. Roger Wright, Admissions Coordinator, 518-276-6372, Fax: 518-276-8554, E-mail: fowlen@rpi.edu. *Application contact:* Dr. Michael Jensen, Associate Head, 518-276-6432, Fax: 518-276-6025, E-mail: burged@rpi.edu.

Rice University, Graduate Programs, George R. Brown School of Engineering, Department of Mechanical Engineering and Materials Science, Program in Mechanical Engineering, Houston, TX 77251-1892. Offers MMS, MS, PhD. *Degree requirements:* For master's, thesis required (for some programs), foreign language not required; for doctorate, dissertation required, foreign language not required. *Entrance requirements:* For master's and doctorate, GRE General Test, GRE Subject Test, TOEFL (minimum score of 550 required), minimum GPA of 3.0. *Faculty research:* Solid mechanics, thermal sciences, materials science.

Rochester Institute of Technology, Part-time and Graduate Admissions, College of Engineering, Department of Mechanical Engineering, Rochester, NY 14623-5604. Offers MSME. *Students:* 11 full-time (2 women), 28 part-time (1 woman); includes 4 minority (1 African American, 2 Asian Americans or Pacific Islanders, 1 Hispanic American), 9 international. 51 applicants, 75% accepted. In 1998, 31 degrees awarded. *Degree requirements:* For master's, thesis optional, foreign language not required. *Entrance requirements:* For master's, TOEFL, minimum GPA of 3.0. *Application deadline:* For fall admission, 3/1 (priority date). Applications are processed on a rolling basis. Application fee: $40. *Financial aid:* Research assistantships, teaching assistantships available. *Unit head:* Dr. Charles Haines, Head, 716-475-2029, E-mail: cwheme@rit.edu. *Application contact:* Dr. Richard Reeve, Associate Dean, 716-475-7048, E-mail: nrreie@rit.edu.

Rose-Hulman Institute of Technology, Faculty of Engineering and Applied Sciences, Department of Mechanical Engineering, Terre Haute, IN 47803-3920. Offers MS. Part-time programs available. Postbaccalaureate distance learning degree programs offered (minimal on-campus study). *Faculty:* 22 full-time (2 women). *Students:* 14 full-time (3 women), 6 international. Average age 23. 18 applicants, 61% accepted. In 1998, 8 degrees awarded. *Degree requirements:* For master's, thesis required, foreign language not required. *Entrance requirements:* For master's, GRE, TOEFL (minimum score of 580 required), minimum GPA of

3.0. *Average time to degree:* Master's–2 years full-time. *Application deadline:* For fall admission, 2/1 (priority date). Applications are processed on a rolling basis. Application fee: $0. Tuition: Full-time $19,305; part-time $540 per credit hour. Required fees: $800. *Financial aid:* In 1998–99, 13 students received aid, including 7 fellowships (averaging $6,000 per year); research assistantships, teaching assistantships, grants, institutionally-sponsored loans, and tuition waivers (full and partial) also available. Financial aid application deadline: 2/1. *Faculty research:* Energy, robotics, noise and vibration, finite-element analysis. Total annual research expenditures: $85,455. *Unit head:* Dr. David J. Purdy, Chairman, 812-877-8320, Fax: 812-877-3198, E-mail: robert.steinhauser@rose-hulman.edu. *Application contact:* Dr. Buck F. Brown, Dean for Research and Graduate Studies, 812-877-8403, Fax: 812-877-8102, E-mail: buck.brown@rose-hulman.edu.

Rutgers, The State University of New Jersey, New Brunswick, Graduate School, Program in Mechanical and Aerospace Engineering, New Brunswick, NJ 08903. Offers computational fluid dynamics (MS, PhD); design and dynamics (MS, PhD); fluid mechanics (MS, PhD); heat transfer (MS, PhD); solid mechanics (MS, PhD). Part-time and evening/weekend programs available. *Faculty:* 29 full-time (0 women), 5 part-time (0 women). *Students:* 59 full-time (12 women), 21 part-time (4 women); includes 9 minority (1 African American, 6 Asian Americans or Pacific Islanders, 2 Hispanic Americans), 38 international. Average age 25. 214 applicants, 29% accepted. In 1998, 8 master's, 9 doctorates awarded. *Degree requirements:* For master's, thesis required (for some programs), foreign language not required; for doctorate, dissertation required, foreign language not required. *Entrance requirements:* For master's, GRE General Test, BS in mechanical/aerospace engineering or related field; for doctorate, GRE General Test, MS in mechanical/aerospace engineering or related field. *Application deadline:* For fall admission, 6/1. Application fee: $50. *Financial aid:* In 1998–99, 57 students received aid, including 8 fellowships, 25 research assistantships, 24 teaching assistantships; tuition waivers (full) also available. Financial aid application deadline: 3/1; financial aid applicants required to submit FAFSA. Total annual research expenditures: $8 million.

Saint Louis University, Graduate School, Department of Aerospace and Mechanical Engineering, St. Louis, MO 63103-2097. Offers MS, MS(R). *Faculty:* 13 full-time (2 women), 4 part-time (0 women). *Students:* 4 full-time (2 women), 11 part-time (2 women); includes 2 minority (1 Asian American or Pacific Islander, 1 Hispanic American), 9 international. Average age 27. 10 applicants, 70% accepted. In 1998, 10 degrees awarded. *Degree requirements:* For master's, comprehensive oral exam required, thesis optional, foreign language not required. *Entrance requirements:* For master's, GRE General Test. *Application deadline:* For fall admission, 7/1; for spring admission, 11/1. Applications are processed on a rolling basis. Application fee: $40. Tuition: Full-time $20,520; part-time $507 per credit hour. Required fees: $38 per term. Tuition and fees vary according to program. *Financial aid:* In 1998–99, 10 students received aid, including 1 research assistantship, 4 teaching assistantships Financial aid application deadline: 4/1; financial aid applicants required to submit FAFSA. *Faculty research:* Flight dynamics/control, structural dynamics, experimental aerodynamics, aircraft design/optimization. *Unit head:* Dr. Krishnaswamy Ravindra, Chairman, 314-977-8438, Fax: 314-977-8403, E-mail: ravindrak@slu.edu. *Application contact:* Dr. Marcia Buresch, Assistant Dean of the Graduate School, 314-977-2240, Fax: 314-977-3943, E-mail: bureschm@slu.edu.

San Diego State University, Graduate and Research Affairs, College of Engineering, Department of Mechanical Engineering, San Diego, CA 92182. Offers engineering sciences and applied mechanics (PhD); mechanical engineering (MS). Evening/weekend programs available. *Students:* 12 full-time (0 women), 21 part-time (1 woman); includes 10 minority (3 African Americans, 2 Asian Americans or Pacific Islanders, 5 Hispanic Americans), 4 international. Average age 29. 26 applicants, 46% accepted. In 1998, 10 degrees awarded. *Degree requirements:* For doctorate, dissertation required, foreign language not required. *Entrance requirements:* For master's, GRE General Test (minimum combined score of 950 required), TOEFL (minimum score of 550 required). *Application deadline:* Applications are processed on a rolling basis. Application fee: $55. *Financial aid:* Career-related internships or fieldwork available. *Faculty research:* Energy analysis and diagnosis, seawaterpump design, space-related research. Total annual research expenditures: $315,000. *Unit head:* Ronald Kline, Chair, 619-594-6067, Fax: 619-594-6005, E-mail: kline@kahuna.sdsu.edu.

San Jose State University, Graduate Studies, College of Engineering, Department of Mechanical and Aerospace Engineering, Program in Mechanical Engineering, San Jose, CA 95192-0001. Offers MS. Part-time programs available. *Faculty:* 11 full-time (0 women), 20 part-time (1 woman). *Students:* 16 full-time (3 women), 48 part-time (2 women); includes 40 minority (5 African Americans, 30 Asian Americans or Pacific Islanders, 5 Hispanic Americans), 7 international. Average age 30. 54 applicants, 72% accepted. In 1998, 22 degrees awarded. *Degree requirements:* For master's, computer language required, thesis optional, foreign language not required. *Entrance requirements:* For master's, GRE, TOEFL, BS in mechanical engineering or equivalent. *Application deadline:* For fall admission, 6/1. Applications are processed on a rolling basis. Application fee: $59. Tuition, nonresident: part-time $246 per unit. Required fees: $1,939; $1,309 per year. *Financial aid:* In 1998–99, 1 teaching assistantship was awarded. *Faculty research:* Gas dynamics, mechanics/vibrations, heat transfer, structural analysis, two-phase fluid flow. *Application contact:* Dr. William Seto, Graduate Adviser, 408-924-3947.

Santa Clara University, School of Engineering, Department of Mechanical Engineering, Santa Clara, CA 95053-0001. Offers MSME, PhD, Engineer. Part-time and evening/weekend programs available. *Students:* 3 full-time (0 women), 51 part-time (5 women); includes 15 minority (11 Asian Americans or Pacific Islanders, 4 Hispanic Americans), 2 international. Average age 35. 32 applicants, 72% accepted. In 1998, 5 master's, 1 doctorate, 1 other advanced degree awarded. *Degree requirements:* For master's, thesis or alternative required, foreign language not required; for doctorate and Engineer, dissertation required, foreign language not required. *Entrance requirements:* For master's, GRE General Test, GRE Subject Test, TOEFL (minimum score of 550 required), minimum GPA of 2.75; for doctorate, GRE General Test, GRE Subject Test, TOEFL (minimum score of 550 required), master's degree or equivalent; for Engineer, master's degree, published paper. *Application deadline:* For fall admission, 6/1; for spring admission, 1/1. Applications are processed on a rolling basis. Application fee: $40. *Financial aid:* Fellowships, research assistantships, teaching assistantships, Federal Work-Study and institutionally-sponsored loans available. Aid available to part-time students. Financial aid application deadline: 2/1; financial aid applicants required to submit CSS PROFILE or FAFSA. *Unit head:* Dr. Timothy Hight, Chair, 408-554-4937. *Application contact:* Tina Samms, Assistant Director of Graduate Admissions, 408-554-4313, Fax: 408-554-5474, E-mail: engr-grad@scu.edu.

South Dakota School of Mines and Technology, Graduate Division, Department of Mechanical Engineering, Rapid City, SD 57701-3995. Offers MS. Part-time programs available. *Faculty:* 9 full-time (0 women). *Students:* 17 full-time (2 women), 15 international. Average age 27. In 1998, 3 degrees awarded. *Degree requirements:* For master's, foreign language and thesis not required. *Entrance requirements:* For master's, TOEFL (minimum score of 520 required), TWE. *Application deadline:* For fall admission, 6/15 (priority date); for spring admission, 10/15. Applications are processed on a rolling basis. Application fee: $15. Electronic applications accepted. Tuition, state resident: part-time $89 per hour. Tuition, nonresident: part-time $261 per hour. Part-time tuition and fees vary according to program. *Financial aid:* In 1998–99, 2 fellowships, 5 research assistantships, 6 teaching assistantships were awarded.; Federal Work-Study and institutionally-sponsored loans also available. Aid available to part-time students. Financial aid application deadline: 5/15. *Faculty research:* Advanced composite materials, robotics, computer-integrated manufacturing, enhanced heat transfer, dynamic systems controls. Total annual research expenditures: $5,533. *Unit head:* Dr. Mike Langerman, Chair, 605-394-2401. *Application contact:* Brenda Brown, Secretary, 800-454-8162 Ext. 2493, Fax: 605-394-5360, E-mail: graduate_admissions@silver.sdmt.edu.

South Dakota School of Mines and Technology, Graduate Division, Division of Material Engineering and Science, Doctoral Program in Materials Engineering and Science, Rapid City, SD 57701-3995. Offers chemical engineering (PhD); chemistry (PhD); civil engineering (PhD); electrical engineering (PhD); mechanical engineering (PhD); metallurgical engineering (PhD); physics (PhD). Part-time programs available. *Students:* 14 full-time (2 women), 9 international. *Degree requirements:* For doctorate, dissertation required, foreign language not required. *Entrance requirements:* For doctorate, TOEFL (minimum score of 520 required), TWE, minimum graduate GPA of 3.0. *Application deadline:* For fall admission, 6/15 (priority date); for spring admission, 10/15. Applications are processed on a rolling basis. Application fee: $15. Electronic applications accepted. Tuition, state resident: part-time $89 per hour. Tuition, nonresident: part-time $261 per hour. Part-time tuition and fees vary according to program. *Unit head:* Dr. Chris Jenkins, Coordinator, 605-394-2406. *Application contact:* Brenda Brown, Secretary, 800-454-8162 Ext. 2493, Fax: 605-394-5360, E-mail: graduate_admissions@silver.sdmt.edu.

South Dakota State University, Graduate School, College of Engineering, Department of Mechanical Engineering, Brookings, SD 57007. Offers MS. *Degree requirements:* For master's, thesis, oral exam required, foreign language not required. *Entrance requirements:* For master's, TOEFL (minimum score of 540 required). *Faculty research:* Energy and system optimization, thermal transfer, mechanics of components, ground source heat pump, mechanical design.

Southern Illinois University Carbondale, Graduate School, College of Engineering, Department of Mechanical Engineering and Energy Processes, Carbondale, IL 62901-6806. Offers MS. *Faculty:* 14 full-time (0 women). *Students:* 35 full-time (5 women), 6 part-time (1 woman); includes 2 minority (1 African American, 1 Asian American or Pacific Islander), 27 international. Average age 23. 41 applicants, 63% accepted. In 1998, 9 degrees awarded. *Degree requirements:* For master's, thesis or alternative, comprehensive exam required, foreign language not required. *Entrance requirements:* For master's, GRE General Test, TOEFL (minimum score of 550 required), minimum GPA of 2.7. *Application deadline:* For fall admission, 1/31. Applications are processed on a rolling basis. Application fee: $20. *Financial aid:* In 1998–99, 34 students received aid; fellowships with full tuition reimbursements available, research assistantships with full tuition reimbursements available, teaching assistantships with full tuition reimbursements available, Federal Work-Study and institutionally-sponsored loans available. Aid available to part-time students. *Faculty research:* Coal conversion and processing, combustion, materials science and engineering, mechanical system dynamics. Total annual research expenditures: $1.4 million. *Unit head:* Acting Chair, 618-536-2396.

Southern Illinois University Edwardsville, Graduate Studies and Research, School of Engineering, Program in Mechanical Engineering, Edwardsville, IL 62026-0001. Offers MS. *Students:* 1 full-time (0 women), 3 part-time (1 woman), 1 international. 7 applicants, 100% accepted. *Degree requirements:* For master's, thesis or research paper, final exam required. *Entrance requirements:* For master's, TOEFL (minimum score of 550 required). *Application deadline:* For fall admission, 7/24. Application fee: $25. *Financial aid:* Fellowships with full tuition reimbursements, research assistantships with full tuition reimbursements, teaching assistantships with full tuition reimbursements, Federal Work-Study, institutionally-sponsored loans, and scholarships available. Aid available to part-time students. *Unit head:* Dr. Nader Saniei, Chair, 618-650-2584, E-mail: nsaniei@siue.edu.

Southern Methodist University, School of Engineering and Applied Science, Department of Mechanical Engineering, Dallas, TX 75275. Offers manufacturing systems management (MS); mechanical engineering (MSME, PhD). Part-time programs available. Postbaccalaureate distance learning degree programs offered (no on-campus study). *Faculty:* 11 full-time (0 women), 2 part-time (1 woman). *Students:* 24 full-time (3 women), 45 part-time (1 woman); includes 9 minority (4 African Americans, 2 Asian Americans or Pacific Islanders, 3 Hispanic Americans), 25 international. Average age 34. 58 applicants, 40% accepted. In 1998, 4 master's, 3 doctorates awarded. Terminal master's awarded for partial completion of doctoral program. *Degree requirements:* For master's, thesis optional, foreign language not required; for doctorate, dissertation, oral and written qualifying exams, oral final exam required, foreign language not required. *Entrance requirements:* For master's, GRE General Test (minimum score of 650 on quantitative section required), TOEFL (minimum score of 550 required), minimum GPA of 3.0 in last 2 years; bachelor's degree in engineering, mathematics, or sciences; for doctorate, preliminary counseling exam, minimum graduate GPA of 3.0, bachelor's degree in related field. *Application deadline:* For fall admission, 8/1 (priority date); for spring admission, 12/15. Applications are processed on a rolling basis. Application fee: $25. Tuition: Full-time $9,216; part-time $512 per credit hour. Required fees: $88 per credit hour. Part-time tuition and fees vary according to course load and campus/location. *Financial aid:* Research assistantships, teaching assistantships, Federal Work-Study, institutionally-sponsored loans, and tuition waivers (full and partial) available. Financial aid applicants required to submit FAFSA. *Faculty research:* Design, systems, and controls; thermal and fluid sciences. Total annual research expenditures: $1.3 million. *Unit head:* Dr. Osita Nwokah, Chair, 214-768-3200, Fax: 214-768-1473, E-mail: nwokah@seas.smu.edu. *Application contact:* Dr. Zeynep Celik-Butler, Assistant Dean for Graduate Studies and Research, 214-768-3979, Fax: 214-768-3845, E-mail: zcb@seas.smu.edu.

See in-depth description on page 1417.

Stanford University, School of Engineering, Department of Mechanical Engineering, Stanford, CA 94305-9991. Offers biomechanical engineering (MS); manufacturing systems engineering (MS); mechanical engineering (MS, PhD, Eng); product design (MS). *Faculty:* 33 full-time (1 woman). *Students:* 397 full-time (68 women), 123 part-time (23 women); includes 120 minority (20 African Americans, 70 Asian Americans or Pacific Islanders, 28 Hispanic Americans, 2 Native Americans), 109 international. Average age 27. 707 applicants, 50% accepted. In 1998, 180 master's, 35 doctorates awarded. *Degree requirements:* For doctorate and Eng, dissertation required. *Entrance requirements:* For master's, doctorate, and Eng, GRE General Test, TOEFL. *Application deadline:* For fall admission, 1/15. Application fee: $65 ($80 for international students). Electronic applications accepted. Tuition: Full-time $24,588. Required fees: $152. Part-time tuition and fees vary according to course load. *Financial aid:* Fellowships, research assistantships, teaching assistantships, Federal Work-Study and institutionally-sponsored loans available. Financial aid application deadline: 1/15. *Unit head:* Ronald Hanson, Chair, 650-723-1745, Fax: 650-725-4862, E-mail: hanson@cdr.stanford.edu. *Application contact:* Admissions Office, 650-723-3148.

See in-depth description on page 1419.

State University of New York at Binghamton, Graduate School, Thomas J. Watson School of Engineering and Applied Science, Department of Mechanical Engineering, Binghamton, NY 13902-6000. Offers M Eng, MS, PhD. Part-time and evening/weekend programs available. *Faculty:* 11 full-time, 2 part-time. *Students:* 40 full-time (2 women), 22 part-time (4 women); includes 6 minority (2 African Americans, 4 Asian Americans or Pacific Islanders), 27 international. Average age 29. 78 applicants, 51% accepted. In 1998, 4 master's, 1 doctorate awarded. *Degree requirements:* For master's, thesis or alternative required, foreign language not required; for doctorate, dissertation required, foreign language not required. *Entrance requirements:* For master's and doctorate, GRE General Test, GRE Subject Test, TOEFL (minimum score of 550 required). *Application deadline:* For fall admission, 4/15 (priority date); for spring admission, 11/1. Applications are processed on a rolling basis. Application fee: $50. Electronic applications accepted. Tuition, state resident: full-time $5,100; part-time $213 per credit. Tuition, nonresident: full-time $8,416; part-time $351 per credit. Required fees: $77 per credit. Part-time tuition and fees vary according to course load. *Financial aid:* In 1998–99, 35 students received aid, including 2 fellowships with full tuition reimbursements available (averaging $12,100 per year), 22 research assistantships with full tuition reimbursements available (averaging $8,800 per year), 9 teaching assistantships with full tuition reimbursements available (averaging $8,400 per year); career-related internships or fieldwork, Federal Work-Study, institutionally-sponsored loans, and unspecified assistantships also available. Aid available to part-time students. Financial aid application deadline: 2/15. *Unit head:* Dr. Ron Miles, Chairperson, 609-777-4747.

See in-depth description on page 1421.

Mechanical Engineering

State University of New York at Buffalo, Graduate School, School of Engineering and Applied Sciences, Department of Mechanical and Aerospace Engineering, Buffalo, NY 14260. Offers aerospace engineering (M Eng, MS, PhD); mechanical engineering (M Eng, MS, PhD). Part-time programs available. *Faculty:* 23 full-time (2 women), 8 part-time (0 women). *Students:* 74 full-time (2 women), 126 part-time (6 women); includes 12 minority (3 African Americans, 9 Asian Americans or Pacific Islanders), 91 international. Average age 24. 216 applicants, 76% accepted. In 1998, 40 master's, 8 doctorates awarded. Terminal master's awarded for partial completion of doctoral program. *Degree requirements:* For master's, comprehensive exam, project, or thesis required; for doctorate, dissertation required, foreign language not required. *Entrance requirements:* For master's and doctorate, GRE General Test, GRE Subject Test, TOEFL (minimum score of 550 required). *Average time to degree:* Master's–2 years full-time, 4 years part-time; doctorate–4 years full-time, 6 years part-time. *Application deadline:* For fall admission, 2/1; for spring admission, 10/1. Applications are processed on a rolling basis. Application fee: $35. Tuition, state resident: full-time $5,100; part-time $213 per credit hour. Tuition, nonresident: full-time $8,416; part-time $351 per credit hour. Required fees: $870; $75 per semester. Tuition and fees vary according to course load and program. *Financial aid:* In 1998–99, 63 students received aid, including 2 fellowships with tuition reimbursements available, 37 research assistantships with tuition reimbursements available (averaging $10,500 per year), 24 teaching assistantships with tuition reimbursements available (averaging $10,350 per year); Federal Work-Study, institutionally-sponsored loans, tuition waivers (full), and unspecified assistantships also available. Financial aid application deadline: 2/1; financial aid applicants required to submit FAFSA. *Faculty research:* Fluid and thermal sciences, systems anddesign, mechanics and materials. Total annual research expenditures: $924,273. *Unit head:* Dr. Christina L. Bloebaum, Chairman, 716-645-2593 Ext. 2231, Fax: 716-645-3875, E-mail: clb@eng.buffalo.edu. *Application contact:* Dr. Dale B. Taulbee, Director of Graduate Studies, 716-645-2593 Ext. 2307, Fax: 716-645-3875, E-mail: trldale@eng.buffalo.edu.

See in-depth description on page 1423.

State University of New York at Stony Brook, Graduate School, College of Engineering and Applied Sciences, Department of Mechanical Engineering, Stony Brook, NY 11794. Offers MS, PhD. Evening/weekend programs available. *Faculty:* 19 full-time (0 women), 4 part-time (0 women). *Students:* 46 full-time (10 women), 21 part-time; includes 6 minority (1 African American, 5 Asian Americans or Pacific Islanders), 46 international. 138 applicants, 37% accepted. In 1998, 11 master's, 10 doctorates awarded. *Degree requirements:* For master's, thesis or alternative required, foreign language not required; for doctorate, dissertation, comprehensive exams required, foreign language not required. *Entrance requirements:* For master's, GRE General Test, TOEFL, minimum GPA of 3.0; for doctorate, GRE General Test, TOEFL, minimum GPA of 3.5. *Application deadline:* For fall admission, 1/15. Application fee: $50. *Financial aid:* In 1998–99, 26 research assistantships, 16 teaching assistantships were awarded.; fellowships *Faculty research:* Atmospheric sciences, thermal fluid sciences, solid mechanics. Total annual research expenditures: $2.3 million. *Unit head:* Dr. Fu-Pen Chiang, Chairman, 516-632-8310. *Application contact:* Dr. John Kincaid, Director, 516-632-8305, Fax: 516-632-8720, E-mail: jkincaid@ccmail.sunysb.edu.

Stevens Institute of Technology, Graduate School, Charles V. Schaefer Jr. School of Engineering, Department of Mechanical Engineering, Hoboken, NJ 07030. Offers advanced manufacturing (Certificate); air pollution technology (Certificate); building energy systems (Certificate); computational methods in fluid mechanics and heat transfer (Certificate); concurrent design management (M Eng); controls in aerospace and robotics (Certificate); design and production management (MS, Certificate); finite-element analysis (Certificate); integrated production design (Certificate); mechanical engineering (M Eng, PhD, Engr); mechanism design (Certificate); power generation (Certificate); robotics and control (Certificate); stress analysis and design (Certificate); vibration and noise control (Certificate). MS and Certificate offered in cooperation with the Program in Design and Production Management; M Eng offered in cooperation with the Program in Concurrent Design Management. Part-time and evening/weekend programs available. Terminal master's awarded for partial completion of doctoral program. *Degree requirements:* For master's, computer language required, thesis optional, foreign language not required; for doctorate, variable foreign language requirement, computer language, dissertation required; for other advanced degree, computer language, project or thesis required. *Entrance requirements:* For master's, doctorate, and other advanced degree, TOEFL. Electronic applications accepted. *Faculty research:* Acoustics, incineration, CAD/CAM, computational fluid dynamics and heat transfer, robotics.

Syracuse University, Graduate School, L. C. Smith College of Engineering and Computer Science, Department of Mechanical and Aerospace Engineering, Program in Mechanical Engineering, Syracuse, NY 13244-0003. Offers MS, PhD. *Students:* 28 full-time (4 women), 12 part-time (2 women); includes 3 minority (1 African American, 1 Asian American or Pacific Islander, 1 Hispanic American), 21 international. Average age 29. 102 applicants, 94% accepted. In 1998, 7 master's, 6 doctorates awarded. *Degree requirements:* For master's, project or thesis required; for doctorate, computer language, dissertation required, foreign language not required. *Entrance requirements:* For master's and doctorate, GRE General Test, GRE Subject Test. *Application deadline:* Applications are processed on a rolling basis. Application fee: $40. Tuition: Full-time $13,992; part-time $583 per credit hour. *Financial aid:* Fellowships, research assistantships, teaching assistantships available. Financial aid application deadline: 3/1. *Unit head:* Alan Levy, Graduate Director.

Temple University, Graduate School, College of Science and Technology, College of Engineering, Program in Mechanical Engineering, Philadelphia, PA 19122-6096. Offers MSE. Part-time programs available. *Faculty:* 7 full-time (1 woman). *Students:* 13 (3 women); includes 6 minority (1 African American, 4 Asian Americans or Pacific Islanders, 1 Hispanic American) 1 international. 41 applicants, 34% accepted. In 1998, 18 degrees awarded. *Degree requirements:* For master's, thesis optional, foreign language not required. *Entrance requirements:* For master's, GRE General Test, TOEFL (minimum score of 575 required). *Application deadline:* For fall admission, 7/1; for spring admission, 11/1. Applications are processed on a rolling basis. Application fee: $40. *Financial aid:* Research assistantships, teaching assistantships, Federal Work-Study available. Financial aid application deadline: 2/15. *Faculty research:* Rapid solidification by melt spinning, microfracture analysis of dental materials, failure detection methods. Total annual research expenditures: $1 million. *Unit head:* Dr. Keya Sadeghipour, Director, 215-204-8624, Fax: 215-204-6936.

Tennessee Technological University, Graduate School, College of Engineering, Department of Mechanical Engineering, Cookeville, TN 38505. Offers MS, PhD. Part-time programs available. *Faculty:* 25 full-time (2 women). *Students:* 44 full-time (6 women), 9 part-time (1 woman); includes 34 minority (all Asian Americans or Pacific Islanders) Average age 28. 164 applicants, 89% accepted. In 1998, 9 master's awarded. *Degree requirements:* For master's, thesis required, foreign language not required; for doctorate, one foreign language (computer language can substitute), dissertation required. *Entrance requirements:* For master's, GRE General Test, TOEFL (minimum score of 525 required); for doctorate, GRE Subject Test, TOEFL (minimum score of 525 required), minimum GPA of 3.5. *Application deadline:* For fall admission, 3/1 (priority date); for spring admission, 8/1. Application fee: $25 ($30 for international students). Tuition, state resident: part-time $137 per hour. Tuition, nonresident: part-time $361 per hour. Required fees: $17 per hour. Tuition and fees vary according to course load. *Financial aid:* In 1998–99, 31 students received aid, including 22 research assistantships (averaging $7,000 per year), 15 teaching assistantships (averaging $7,000 per year); fellowships Financial aid application deadline: 4/1. *Faculty research:* Energy-related systems, design, acoustics and acoustical systems. *Unit head:* Dr. Edwin I. Griggs, Chairperson, 931-372-3254, Fax: 931-372-6340, E-mail: egriggs@tntech.edu. *Application contact:* Dr. Rebecca F. Quattlebaum, Dean of the Graduate School, 931-372-3233, Fax: 931-372-3497, E-mail: rquattlebaum@tntech.edu.

Texas A&M University, College of Engineering, Department of Mechanical Engineering, College Station, TX 77843. Offers M Eng, MS, D Eng, PhD. *Faculty:* 64 full-time (2 women), 5 part-time (0 women). *Students:* 175 full-time (10 women), 35 part-time (1 woman); includes 15 minority (6 African Americans, 2 Asian Americans or Pacific Islanders, 7 Hispanic Americans), 133 international. Average age 28. 140 applicants, 85% accepted. In 1998, 43 master's awarded (95% found work related to degree, 5% continued full-time study); 6 doctorates awarded. *Degree requirements:* For master's, thesis (MS) required; for doctorate, dissertation (PhD) required. *Entrance requirements:* For master's, GRE General Test (minimum combined score of 1200 required; average 1270), TOEFL (minimum score of 570 required; average 590), minimum undergraduate GPA of 3.0; for doctorate, GRE General Test (minimum combined score of 1200 required; average 1270), TOEFL (minimum score of 570 required; average 600), minimum graduate GPA of 3.5. *Average time to degree:* Master's–2.5 years full-time; doctorate–4 years full-time. *Application deadline:* For fall admission, 2/1 (priority date); for spring admission, 11/1. Applications are processed on a rolling basis. Application fee: $50 ($75 for international students). Electronic applications accepted. *Financial aid:* In 1998–99, 40 fellowships with partial tuition reimbursements (averaging $5,000 per year), 150 research assistantships with partial tuition reimbursements (averaging $14,000 per year), 75 teaching assistantships (averaging $14,000 per year) were awarded.; institutionally-sponsored loans also available. Financial aid application deadline: 3/1; financial aid applicants required to submit FAFSA. *Faculty research:* Thermal/fluid sciences, materials/manufacturing and controls systems. Total annual research expenditures: $7.5 million. *Unit head:* Dr. Suhada Jayasuriya, Head, 409-845-1251, Fax: 409-845-3081. *Application contact:* Kim Moses, Academic Adviser, 409-845-1270, Fax: 409-845-3081, E-mail: kmoses@menar.tamu.edu.

See in-depth description on page 1425.

Texas A&M University–Kingsville, College of Graduate Studies, College of Engineering, Department of Mechanical and Industrial Engineering, Program in Mechanical Engineering, Kingsville, TX 78363. Offers ME, MS. *Faculty:* 7. *Students:* 5 full-time (0 women), 9 part-time (3 women). *Degree requirements:* For master's, computer language, thesis or alternative, comprehensive exam required, foreign language not required. *Entrance requirements:* For master's, GRE General Test (minimum combined score of 1050 required), TOEFL (minimum score of 525 required), minimum GPA of 3.0. *Application deadline:* For fall admission, 6/1; for spring admission, 11/15. Applications are processed on a rolling basis. Application fee: $15 ($25 for international students). Tuition, state resident: full-time $2,062. Tuition, nonresident: full-time $7,246. *Financial aid:* Fellowships, research assistantships, teaching assistantships available. Financial aid application deadline: 5/15. *Faculty research:* Intelligent systems and controls; neural networks and fuzzy logic; robotics and automation; biomass, cogeneration, and enhanced heat transfer. *Unit head:* Dr. Hayder Abdul Razzak, Coordinator, 361-593-2001.

Texas Tech University, Graduate School, College of Engineering, Department of Mechanical Engineering, Lubbock, TX 79409. Offers MSME, PhD. Part-time programs available. *Faculty:* 15 full-time (0 women). *Students:* 62 full-time (6 women), 5 part-time; includes 6 minority (4 Asian Americans or Pacific Islanders, 2 Hispanic Americans), 44 international. Average age 27. 108 applicants, 62% accepted. In 1998, 12 master's, 3 doctorates awarded. *Degree requirements:* For master's and doctorate, computer language, thesis/dissertation required, foreign language not required. *Entrance requirements:* For master's, GRE General Test (minimum combined score of 1000 required; average 1149), minimum GPA of 3.0; for doctorate, GRE General Test (minimum combined score of 1000 required), minimum GPA of 3.0. *Application deadline:* For fall admission, 4/15 (priority date); for spring admission, 11/1 (priority date). Applications are processed on a rolling basis. Application fee: $25 ($50 for international students). Electronic applications accepted. *Financial aid:* In 1998–99, 33 students received aid, including 22 research assistantships (averaging $8,693 per year), 6 teaching assistantships (averaging $8,723 per year); fellowships, Federal Work-Study and institutionally-sponsored loans also available. Aid available to part-time students. Financial aid application deadline: 5/15; financial aid applicants required to submit FAFSA. *Faculty research:* Aerodynamics of automobiles and parachutes, natural gas, methanol and electric fueled vehicles, optomechanics, lubrication. Total annual research expenditures: $1 million. *Unit head:* Dr. Thomas D. Burton, Chairman, 806-742-3563, Fax: 806-742-3540, E-mail: tburton@coe.ttu.edu.

See in-depth description on page 1427.

Tufts University, Division of Graduate and Continuing Studies and Research, Graduate School of Arts and Sciences, College of Engineering, Department of Mechanical Engineering, Medford, MA 02155. Offers human factors (MS); mechanical engineering (ME, MS, PhD). Part-time programs available. *Faculty:* 13 full-time, 2 part-time. *Students:* 59 (16 women); includes 5 minority (2 African Americans, 2 Asian Americans or Pacific Islanders, 1 Hispanic American) 18 international. 28 applicants, 64% accepted. In 1998, 14 master's, 2 doctorates awarded. Terminal master's awarded for partial completion of doctoral program. *Degree requirements:* For master's and doctorate, thesis/dissertation required, foreign language not required. *Entrance requirements:* For master's and doctorate, GRE General Test, TOEFL (minimum score of 550 required). *Application deadline:* For fall admission, 2/15; for spring admission, 10/15. Applications are processed on a rolling basis. Application fee: $50. *Financial aid:* Research assistantships with full and partial tuition reimbursements, teaching assistantships with full and partial tuition reimbursements, Federal Work-Study, scholarships, and tuition waivers (partial) available. Financial aid application deadline: 2/15; financial aid applicants required to submit FAFSA. *Unit head:* Vincent Manno, Chair, 617-627-3239, Fax: 617-627-3058, E-mail: meinfo@tufts.edu. *Application contact:* Robert Greif, 617-627-3239, Fax: 617-627-3058, E-mail: meinfo@tufts.edu.

Tulane University, School of Engineering, Department of Mechanical Engineering, New Orleans, LA 70118-5669. Offers MS, MSE, PhD, Sc D. MS and PhD offered through the Graduate School. Part-time programs available. *Students:* 21 full-time (6 women), 3 part-time (1 woman), 17 international. 79 applicants, 32% accepted. In 1998, 2 master's, 1 doctorate awarded. Terminal master's awarded for partial completion of doctoral program. *Degree requirements:* For master's, thesis required, foreign language not required; for doctorate, 2 foreign languages, computer language, dissertation required. *Entrance requirements:* For master's and doctorate, GRE General Test, TOEFL, minimum B average in undergraduate course work. *Application deadline:* For fall admission, 7/15; for spring admission, 12/15. Application fee: $35. *Financial aid:* Research assistantships, teaching assistantships, Federal Work-Study and institutionally-sponsored loans available. Financial aid application deadline:2/1. *Unit head:* Dr. Paul Michael Lynch, Acting Dean, 504-865-5766. *Application contact:* Dr. E. Michaelides, Associate Dean, 504-865-5764.

See in-depth description on page 1429.

Tuskegee University, Graduate Programs, College of Engineering, Architecture and Physical Sciences, Department of Mechanical Engineering, Tuskegee, AL 36088. Offers MSME. *Faculty:* 11 full-time (0 women). *Students:* 16 full-time (2 women), 13 part-time (2 women); includes 16 minority (14 African Americans, 2 Asian Americans or Pacific Islanders), 9 international. Average age 24. In 1998, 6 degrees awarded. *Degree requirements:* For master's, computer language, thesis or alternative required, foreign language not required. *Entrance requirements:* For master's, GRE General Test, GRE Subject Test. *Application deadline:* For fall admission, 7/15. Applications are processed on a rolling basis. Application fee: $25 ($35 for international students). *Financial aid:* Fellowships, research assistantships, teaching assistantships, career-related internships or fieldwork, Federal Work-Study, and institutionally-sponsored loans available. Aid available to part-time students. Financial aid application deadline: 4/15. *Faculty research:* Superalloys, fatigue and surface machinery, energy management, solar energy. *Unit head:* Dr. Pradosh Ray, Head, 334-727-8989.

Union College, Graduate and Continuing Studies, Division of Engineering and Computer Science, Department of Mechanical Engineering, Schenectady, NY 12308-2311. Offers MS. *Students:* 1 full-time (0 women), 13 part-time (1 woman); includes 1 minority (Hispanic American) 10 applicants, 90% accepted. In 1998, 7 degrees awarded. *Degree requirements:* For master's, one foreign language, computer language, comprehensive exam required, thesis not required. *Entrance requirements:* For master's, minimum GPA of 3.0. *Application deadline:* Applications are processed on a rolling basis. Application fee: $50. Tuition: Part-time $1,786 per course. *Faculty research:* Metal fatigue testing, energy-related projects and equipment. *Unit head:* Dr. Richard D. Wilk, Chairman, 518-388-6268.

Université de Moncton, School of Engineering, Program in Mechanical Engineering, Moncton, NB E1A 3E9, Canada. Offers M Sc A. *Faculty:* 4 full-time (0 women). *Students:* 2 full-time (1 woman), 1 part-time, 1 international. Average age 25. 7 applicants, 86% accepted. *Degree requirements:* For master's, thesis, proficiency in French required. *Application deadline:* For fall admission, 6/1 (priority date); for winter admission, 11/15 (priority date). Application fee: $30. *Financial aid:* In 1998–99, 2 students received aid, including fellowships (averaging $17,200 per year), research assistantships (averaging $5,000 per year), teaching assistantships (averaging $1,700 per year) Financial aid application deadline: 5/31. *Faculty research:* Composite materials, thermal energy systems, control systems, fluid mechanics and heat transfer, CAD/CAM and robotics. Total annual research expenditures: $159,250. *Unit head:* Dr. Cong Tam Nguyen, Chairman, 506-858-4347, Fax: 506-858-4300, E-mail: nguyenc@umoncton.ca.

Université de Sherbrooke, Faculty of Applied Sciences, Department of Mechanical Engineering, Sherbrooke, PQ J1K 2R1, Canada. Offers M Sc A, PhD. *Degree requirements:* For master's and doctorate, thesis/dissertation required. *Faculty research:* Acoustics, aerodynamics, vehicle dynamics, composite materials, heat transfer.

Université Laval, Faculty of Graduate Studies, Faculty of Sciences and Engineering, Department of Mechanical Engineering, Program in Mechanical Engineering, Sainte-Foy, PQ G1K 7P4, Canada. Offers M Sc, PhD. *Students:* 61 full-time (8 women), 17 part-time (4 women). 42 applicants, 74% accepted. In 1998, 7 master's, 11 doctorates awarded. *Application deadline:* For fall admission, 3/1. Application fee: $30. *Unit head:* Alain de Champlain, Director, 418-656-2131 Ext. 2198, Fax: 418-656-7415, E-mail: alain.dechamplain@gmc.ulaval.ca.

The University of Akron, Graduate School, College of Engineering, Department of Mechanical Engineering, Akron, OH 44325-0001. Offers MS, PhD. Part-time and evening/weekend programs available. *Faculty:* 20 full-time, 7 part-time. *Students:* 39 full-time (7 women), 24 part-time (1 woman); includes 2 minority (1 Asian American or Pacific Islander, 1 Hispanic American), 39 international. Average age 29. In 1998, 18 master's awarded. Terminal master's awarded for partial completion of doctoral program. *Degree requirements:* For master's, thesis or alternative required, foreign language not required; for doctorate, variable foreign language requirement (computer language can substitute for one), dissertation, candidacy exam, qualifying exam required. *Entrance requirements:* For master's, TOEFL (minimum score of 550 required), minimum GPA of 2.75; for doctorate, GRE, TOEFL (minimum score of 550 required). *Average time to degree:* Master's–2 years full-time, 4 years part-time. *Application deadline:* For fall admission, 4/1. Applications are processed on a rolling basis. Application fee: $25 ($50 for international students). Tuition, state resident: part-time $189 per credit. Tuition, nonresident: part-time $353 per credit. Required fees: $7.3 per credit. *Financial aid:* In 1998–99, 62 students received aid, including 7 research assistantships with full tuition reimbursements available, 30 teaching assistantships with full tuition reimbursements available; fellowships with full tuition reimbursements available, tuition waivers (full) also available. Financial aid application deadline: 3/1. *Faculty research:* Computational fluid dynamics system control, energy systems, heat transfer, solid mechanics. *Unit head:* Dr. B. T. F. Chung, Chair, 330-972-7731, E-mail: bchung@uakron.edu. *Application contact:* Dr. B. T. F. Chung, Chair, 330-972-7731, E-mail: bchung@uakron.edu.

Announcement: Department has 18 full-time and 6 adjunct faculty members. Offers BS, MS, and PhD. Areas of specialty include mechanical design, CAD/CAM, materials, solid mechanics, dynamics, vibrations, heat transfer, fluid mechanics, system control, and robotics. Awards numerous graduate assistantships and tuition scholarships to qualified students. Research assistantships and fellowships are also available. International applicants for teaching assistantships need minimum scores of 550 TOEFL and 50 TSE. Within commuting distance of the campus there are numerous industrial firms in the fields of polymers, plastics, metals, power, aerospace, automotive, design, bearings, and chemicals. Job opportunities for graduates are excellent. Contact Dr. B. T. F. Chung, Chair, 330-972-7731 or fax: 330-972-6027.

The University of Alabama, Graduate School, College of Engineering, Department of Mechanical Engineering, Tuscaloosa, AL 35487. Offers MSME, PhD. Part-time and evening/weekend programs available. *Faculty:* 13 full-time (2 women), 2 part-time (0 women). *Students:* 32 full-time (2 women), 6 part-time; includes 2 minority (1 Asian American or Pacific Islander, 1 Native American), 15 international. Average age 27. 99 applicants, 81% accepted. In 1998, 20 master's, 5 doctorates awarded. Terminal master's awarded for partial completion of doctoral program. *Degree requirements:* For master's, thesis or alternative required, foreign language not required; for doctorate, dissertation required, foreign language not required. *Entrance requirements:* For master's and doctorate, GRE General Test (minimum combined score of 1500 on three sections required), minimum GPA of 3.0 in last 60 hours. *Application deadline:* For fall admission, 7/6 (priority date). Applications are processed on a rolling basis. Application fee: $25. *Financial aid:* In 1998–99, 2 fellowships, 15 research assistantships, 15 teaching assistantships were awarded.; career-related internships or fieldwork and Federal Work-Study also available. Financial aid application deadline: 1/15. *Faculty research:* Thermal/fluids, robotics, numerical modeling, energy conservation, manufacturing. *Unit head:* Dr. Stuart R. Bell, Head, 205-348-1644. *Application contact:* Dr. Will Schreiber, Graduate Supervisor, 205-348-1650, E-mail: wschreiber@coe.eng.ua.edu.

The University of Alabama at Birmingham, Graduate School, School of Engineering, Department of Materials and Mechanical Engineering, Program in Mechanical Engineering, Birmingham, AL 35294. Offers MSME, PhD. Evening/weekend programs available. *Students:* 13 full-time (3 women), 8 part-time (1 woman); includes 4 minority (all African Americans), 5 international. 23 applicants, 52% accepted. In 1998, 9 master's awarded. *Degree requirements:* For master's, thesis or alternative, comprehensive exam and project/thesis required, foreign language not required. *Entrance requirements:* For master's, GRE General Test (minimum score of 500 on each section required), minimum B average. *Application deadline:* Applications are processed on a rolling basis. Application fee: $30 ($60 for international students). Electronic applications accepted. *Financial aid:* In 1998–99, 1 fellowship with full tuition reimbursement (averaging $12,500 per year), 5 research assistantships with full tuition reimbursements (averaging $12,500 per year) were awarded.; career-related internships or fieldwork, Federal Work-Study, and institutionally-sponsored loans also available. Aid available to part-time students. *Faculty research:* Microfluid dynamics, rarefied gas dynamics, biofluid mechanics of the cardiovascular system, electro-optical diagnostic implementation. *Unit head:* Suzanne Camirand, Admissions Officer, 819-376-5045 Ext. 2591, Fax: 819-376-5210, E-mail: suzanne_camirand@uqtr.uquebec.ca.

The University of Alabama in Huntsville, School of Graduate Studies, College of Engineering, Department of Mechanical and Aerospace Engineering, Huntsville, AL 35899. Offers mechanical engineering (MSE, PhD). Part-time and evening/weekend programs available. *Faculty:* 14 full-time (0 women). *Students:* 40 full-time (6 women), 30 part-time (8 women); includes 7 minority (2 African Americans, 3 Asian Americans or Pacific Islanders, 2 Hispanic Americans), 15 international. Average age 31. 59 applicants, 78% accepted. In 1998, 14 master's, 12 doctorates awarded. *Degree requirements:* For master's, oral and written exams required, thesis optional, foreign language not required; for doctorate, dissertation, oral and written exams required, foreign language not required. *Entrance requirements:* For master's, GRE General Test (minimum combined score of 1500 on three sections required), BSE, minimum GPA of 3.0; for doctorate, GRE General Test (minimum combined score of 1500 on three sections required), minimum GPA of 3.0. *Application deadline:* For fall admission, 7/24 (priority date); for spring admission, 11/15 (priority date). Applications are processed on a rolling basis. Application fee: $20. Tuition and fees vary according to course load. *Financial aid:* In 1998–99, 26 students received aid, including 1 fellowship with full and partial tuition reimbursement available (averaging $14,400 per year), 14 research assistantships with full and partial tuition reimbursements available (averaging $11,180 per year), 11 teaching assistantships with full and partial tuition reimbursements available (averaging $9,115 per year); career-related internships or fieldwork, Federal Work-Study, grants, institutionally-sponsored loans, scholarships, and tuition waivers (full and partial) also available. Aid available to part-time students. Financial aid application deadline: 4/1; financial aid applicants required to submit FAFSA. *Faculty research:* Combustion, fluid dynamics, solar energy, propulsion, laser diagnostics. Total annual research expenditures: $75,055. *Unit head:* Dr. Gerald Karr, Chair, 256-890-6154, Fax: 256-890-6758, E-mail: karr@eb.uah.edu.

University of Alaska Fairbanks, Graduate School, College of Science, Engineering and Mathematics, Department of Mechanical Engineering, Fairbanks, AK 99775-7480. Offers MS. *Faculty:* 6 full-time (0 women), 1 part-time (0 women). *Students:* 8 full-time (2 women), 5 part-time (1 woman); includes 2 minority (1 African American, 1 Asian American or Pacific Islander), 9 international. Average age 32. 6 applicants, 100% accepted. In 1998, 4 degrees awarded. *Degree requirements:* For master's, thesis or alternative required, foreign language not required. *Entrance requirements:* For master's, GRE General Test, TOEFL (minimum score of 550 required). *Application deadline:* For fall admission, 8/1. Applications are processed on a rolling basis. Application fee: $35. *Financial aid:* Research assistantships, teaching assistantships available. *Unit head:* Dr. Ronald Johnson, Head, 907-474-7209.

University of Alberta, Faculty of Graduate Studies and Research, Department of Mechanical Engineering, Edmonton, AB T6G 2E1, Canada. Offers engineering management (M Eng); mechanical engineering (M Eng, M Sc, PhD); oil sands (M Eng, M Sc). Part-time programs available. *Degree requirements:* For master's and doctorate, thesis/dissertation required, foreign language not required. *Entrance requirements:* For master's and doctorate, TOEFL (minimum score of 580 required), minimum GPA of 7.0 on a 9.0 scale. *Faculty research:* Combustion and environmental issues, advanced materials, computational fluid dynamics, biomedical, acoustics and vibrations.

The University of Arizona, Graduate College, College of Engineering and Mines, Department of Aerospace and Mechanical Engineering, Program in Mechanical Engineering, Tucson, AZ 85721. Offers MS, PhD. Part-time programs available. *Students:* 60 full-time (7 women), 19 part-time (3 women); includes 4 minority (2 Asian Americans or Pacific Islanders, 2 Hispanic Americans), 42 international. Average age 28. 95 applicants, 71% accepted. In 1998, 17 master's, 5 doctorates awarded. *Degree requirements:* For master's, thesis or alternative required, foreign language not required; for doctorate, dissertation required. *Entrance requirements:* For master's and doctorate, GRE General Test, GRE Subject Test, TOEFL (minimum score of 550 required), minimum GPA of 3.0. *Application deadline:* For fall admission, 7/23. Applications are processed on a rolling basis. Application fee: $35. *Financial aid:* Fellowships, research assistantships, teaching assistantships available. *Faculty research:* Fluid mechanics, structures, computer-aided design, stability and control, probabilistic design. *Application contact:* Barbara Heefner, Graduate Secretary, 520-621-4692, Fax: 520-621-8191.

University of Arkansas, Graduate School, College of Engineering, Department of Mechanical Engineering, Fayetteville, AR 72701-1201. Offers MSE, MSME, PhD. *Faculty:* 12 full-time (0 women). *Students:* 17 full-time (3 women), 3 part-time (1 woman), 10 international. 25 applicants, 40% accepted. In 1998, 10 master's, 1 doctorate awarded. *Degree requirements:* For master's, thesis optional, foreign language not required; for doctorate, one foreign language, dissertation required. Application fee: $40 ($50 for international students). Tuition, state resident: full-time $3,186. Tuition, nonresident: full-time $7,560. Required fees: $378. *Financial aid:* In 1998–99, 10 research assistantships, 4 teaching assistantships were awarded.; career-related internships or fieldwork and Federal Work-Study also available. Aid available to part-time students. Financial aid application deadline: 4/1; financial aid applicants required to submit FAFSA. *Unit head:* Dr. W. F. Schmidt, Head, 501-575-3153, E-mail: ned@engr.uark.edu.

University of Bridgeport, College of Graduate and Undergraduate Studies, School of Science, Engineering, and Technology, Department of Mechanical Engineering, Program in Mechanical Engineering, Bridgeport, CT 06601. Offers MS. *Faculty:* 2 full-time (0 women), 4 part-time (0 women). *Students:* 4 full-time (0 women), 7 part-time (1 woman), 7 international. Average age 31. 67 applicants, 82% accepted. In 1998, 4 degrees awarded. *Degree requirements:* For master's, thesis optional, foreign language not required. *Entrance requirements:* For master's, TOEFL. *Application deadline:* Applications are processed on a rolling basis. Application fee: $35 ($50 for international students). *Financial aid:* In 1998–99, 3 students received aid; research assistantships, teaching assistantships, career-related internships or fieldwork, Federal Work-Study, and institutionally-sponsored loans available. Aid available to part-time students. Financial aid application deadline: 6/1; financial aid applicants required to submit FAFSA. *Faculty research:* Heat transfer. *Unit head:* Dr. Richard D. Schile, Chairman, Department of Mechanical Engineering, 203-576-4343.

University of British Columbia, Faculty of Graduate Studies, Faculty of Applied Science, Department of Mechanical Engineering, Vancouver, BC V6T 1Z2, Canada. Offers M Eng, MA Sc, PhD. Part-time programs available. *Degree requirements:* For master's, comprehensive exam, essay (M Eng), thesis (MA Sc) required; for doctorate, dissertation, qualifying exam required. *Entrance requirements:* For master's and doctorate, TOEFL (minimum score of 550 required; average 590). *Faculty research:* Applied mechanics, manufacturing, robotics and controls, thermodynamics and combustion, fluid/aerodynamics, acoustics.

University of Calgary, Faculty of Graduate Studies, Faculty of Engineering, Department of Mechanical and Manufacturing Engineering, Calgary, AB T2N 1N4, Canada. Offers mechanical engineering (M Eng, M Sc, PhD). *Faculty:* 23 full-time (3 women), 5 part-time (0 women). *Students:* 64 full-time (13 women), 10 part-time (1 woman). Average age 27. 110 applicants, 13% accepted. *Degree requirements:* For master's, thesis required (for some programs), foreign language not required; for doctorate, dissertation, candidacy exam required, foreign language not required. *Entrance requirements:* For master's, TOEFL (minimum score of 550 required), minimum GPA of 3.0; for doctorate, TOEFL (minimum score of 550 required), minimum GPA of 3.3. *Application deadline:* For fall admission, 5/31 (priority date). Applications are processed on a rolling basis. Application fee: $60. *Financial aid:* Research assistantships, teaching assistantships available. *Faculty research:* Thermofluids, solid mechanics, materials, biomechanics. Total annual research expenditures: $1.5 million. *Unit head:* G. T. Reader, Head, 403-220-5770, Fax: 403-282-8406, E-mail: greader@ucalgary.ca. *Application contact:* M. Epstein, Acting Associate Head, 403-220-5791, Fax: 403-282-8406, E-mail: epstein@enme.ucalgary.ca.

University of California, Berkeley, Graduate Division, College of Engineering, Department of Mechanical Engineering, Berkeley, CA 94720-1500. Offers M Eng, MS, D Eng, PhD. *Students:* 304 full-time (65 women); includes 62 minority (11 African Americans, 36 Asian Americans or Pacific Islanders, 13 Hispanic Americans, 2 Native Americans), 124 international. 521 applicants, 33% accepted. In 1998, 60 master's, 42 doctorates awarded. *Degree requirements:* For master's, comprehensive exam or thesis (MS) required; for doctorate, dissertation, preliminary and qualifying exams required. *Entrance requirements:* For master's and doctorate, GRE General Test, TOEFL (minimum score of 570 required), minimum GPA of 3.0. *Application deadline:* For fall admission, 12/23; for spring admission, 9/1. Application fee: $40. *Financial aid:* Fellowships available. Financial aid application deadline: 1/2. *Unit head:* Dr. David B. Bogy, Chair, 510-642-1339. *Application contact:* Rui Neves, Graduate Assistant for Admission, 510-642-5084, Fax: 510-642-6163, E-mail: mech@euler.berkeley.edu.

See in-depth description on page 1431.

University of California, Davis, Graduate Studies, College of Engineering, Program in Mechanical and Aeronautical Engineering, Davis, CA 95616. Offers aeronautical engineering (M Engr, MS, D Engr, PhD, Certificate); mechanical engineering (M Engr, MS, D Engr, PhD, Certificate). *Faculty:* 27 full-time (1 woman), 5 part-time (0 women). *Students:* 81 full-time (10 women), 4 part-time; includes 14 minority (11 Asian Americans or Pacific Islanders, 3 Hispanic Americans),

Mechanical Engineering

University of California, Davis (continued)
19 international. 125 applicants, 82% accepted. In 1998, 17 master's, 5 doctorates awarded. *Degree requirements:* For master's, thesis optional, foreign language not required; for doctorate, dissertation required, foreign language not required. *Entrance requirements:* For master's and doctorate, GRE General Test, minimum GPA of 3.0. *Application deadline:* For fall admission, 3/15. Application fee: $40. Electronic applications accepted. *Financial aid:* In 1998–99, 17 fellowships with full and partial tuition reimbursements, 36 research assistantships with full and partial tuition reimbursements, 18 teaching assistantships with full and partial tuition reimbursements were awarded. Financial aid application deadline: 1/15; financial aid applicants required to submit FAFSA. *Unit head:* Bahram Ravani, Chairperson, 530-752-0581, Fax: 530-752-4158. *Application contact:* Susan Fann, Academic Assistant, 530-752-0581, Fax: 530-752-4158, E-mail: sfann@ucdavis.edu.

University of California, Irvine, Office of Research and Graduate Studies, School of Engineering, Department of Mechanical and Aerospace Engineering, Irvine, CA 92697. Offers MS, PhD. Part-time programs available. *Faculty:* 18 full-time (1 woman). *Students:* 53 full-time (7 women), 8 part-time; includes 20 minority (1 African American, 15 Asian Americans or Pacific Islanders, 4 Hispanic Americans), 20 international. 134 applicants, 49% accepted. In 1998, 18 master's, 9 doctorates awarded. Terminal master's awarded for partial completion of doctoral program. *Degree requirements:* For doctorate, dissertation required, foreign language not required, foreign language not required. *Entrance requirements:* For master's, GRE General Test, minimum GPA of 3.0; for doctorate, GRE General Test. *Application deadline:* For fall admission, 1/15 (priority date). Applications are processed on a rolling basis. Application fee: $40. Electronic applications accepted. *Financial aid:* Fellowships, research assistantships, teaching assistantships, institutionally-sponsored loans and tuition waivers (full and partial) available. Financial aid application deadline: 3/2; financial aid applicants required to submit FAFSA. *Faculty research:* Thermal and fluid sciences, combustion and propulsion, control systems, robotics. *Unit head:* Dr. Said Elghobashi, Chair, 949-824-8451, Fax: 949-824-8585, E-mail: selghoba@uci.edu. *Application contact:* Dorothy Miles, Graduate Coordinator, 949-824-5469, Fax: 949-824-8585, E-mail: djmiles@uci.edu.

See in-depth description on page 1433.

University of California, Los Angeles, Graduate Division, School of Engineering and Applied Science, Department of Mechanical and Aerospace Engineering, Program in Mechanical Engineering, Los Angeles, CA 90095. Offers MS, PhD. *Students:* 142 full-time (18 women); includes 47 minority (3 African Americans, 38 Asian Americans or Pacific Islanders, 6 Hispanic Americans), 52 international. 156 applicants, 65% accepted. In 1998, 40 master's, 12 doctorates awarded. *Degree requirements:* For master's, comprehensive exam or thesis required; for doctorate, dissertation, qualifying exams required, foreign language not required. *Entrance requirements:* For master's, GRE General Test, GRE Subject Test (required for foreign students), minimum GPA of 3.0; for doctorate, GRE General Test, GRE Subject Test (required for foreign students), minimum GPA of 3.25. *Application deadline:* For fall admission, 1/5; for spring admission, 12/31. Application fee: $40. Electronic applications accepted. *Financial aid:* Fellowships, research assistantships, teaching assistantships, Federal Work-Study, institutionally-sponsored loans, and tuition waivers (full and partial) available. Financial aid application deadline: 1/5; financial aid applicants required to submit FAFSA. *Application contact:* Student Affairs Officer, E-mail: maeapp@ea.ucla.edu.

See in-depth description on page 1435.

University of California, San Diego, Graduate Studies and Research, Department of Applied Mechanics and Engineering Sciences, Program in Mechanical Engineering, La Jolla, CA 92093-5003. Offers MS, PhD. Part-time programs available. *Students:* 39 full-time (5 women), 5 part-time (2 women); includes 9 minority (5 Asian Americans or Pacific Islanders, 3 Hispanic Americans, 1 Native American), 11 international. 119 applicants, 28% accepted. In 1998, 1 master's, 6 doctorates awarded. *Degree requirements:* For master's, comprehensive exam or thesis required; for doctorate, dissertation, qualifying exam required. *Entrance requirements:* For master's and doctorate, GRE General Test, TOEFL (minimum score of 550 required), minimum GPA of 3.0. *Application deadline:* For fall admission, 5/31. Application fee: $40. *Financial aid:* In 1998–99, fellowships with full tuition reimbursements (averaging $15,000 per year), research assistantships with full tuition reimbursements (averaging $15,000 per year), teaching assistantships with partial tuition reimbursements (averaging $13,000 per year) were awarded.; career-related internships or fieldwork also available. Financial aid application deadline: 1/31; financial aid applicants required to submit FAFSA. *Faculty research:* Combustion engineering, environmental mechanics, magnetic recording, materials processing, computational fluid dynamics. *Unit head:* AMES Graduate Student Affairs, 619-534-4387, Fax: 619-534-1730, E-mail: bwalton@ames.ucsd.edu. *Application contact:* AMES Graduate Student Affairs, 619-534-4387, Fax: 619-534-1730, E-mail: bwalton@ames.ucsd.edu.

See in-depth description on page 1437.

University of California, Santa Barbara, Graduate Division, College of Engineering, Department of Mechanical and Environmental Engineering, Santa Barbara, CA 93106. Offers MS, PhD. *Students:* 64 full-time (8 women). 145 applicants, 47% accepted. In 1998, 23 master's, 10 doctorates awarded. *Degree requirements:* For master's, thesis or alternative required, foreign language not required; for doctorate, dissertation required, foreign language not required. *Entrance requirements:* For master's and doctorate, GRE General Test, TOEFL (minimum score of 550 required). *Application deadline:* For fall admission, 6/1. Application fee: $40. *Financial aid:* Fellowships, research assistantships, teaching assistantships, career-related internships or fieldwork, Federal Work-Study, institutionally-sponsored loans, and tuition waivers (full and partial) available. Financial aid application deadline: 1/15; financial aid applicants required to submit FAFSA. *Unit head:* G. Robert Odette, Chair, 805-893-3525. *Application contact:* Linda James, Graduate Program Assistant, 805-893-2239, E-mail: linda@engineering.ucsb.edu.

University of Central Florida, College of Engineering, Department of Mechanical, Materials, and Aerospace Engineering, Orlando, FL 32816. Offers aerospace systems (MSME, PhD); materials science and engineering (MSME, PhD); mechanical systems (MSME, PhD); mechanical, materials, and aerospace engineering (Certificate); thermofluids (MSME, PhD). Part-time and evening/weekend programs available. *Faculty:* 18 full-time, 6 part-time. *Students:* 61 full-time (14 women), 30 part-time (4 women); includes 12 minority (3 African Americans, 6 Asian Americans or Pacific Islanders, 3 Hispanic Americans), 42 international. Average age 31. 48 applicants, 58% accepted. In 1998, 23 master's, 2 doctorates awarded. *Degree requirements:* For master's, thesis or alternative required, foreign language not required; for doctorate, dissertation, departmental qualifying exam required, foreign language not required. *Entrance requirements:* For master's, GRE General Test (minimum combined score of 1000 required), TOEFL (minimum score of 550 required; 213 computer-based), minimum GPA of 3.0 in last 60 hours; for doctorate, TOEFL (minimum score of 550 required; 213 computer-based), minimum GPA of 3.5 in last 60 hours. *Application deadline:* For fall admission, 7/15; for spring admission, 12/15. Application fee: $20. Tuition, state resident: full-time $2,054; part-time $137 per credit. Tuition, nonresident: full-time $7,207; part-time $480 per credit. Required fees: $47 per term. *Financial aid:* In 1998–99, 71 students received aid, including 17 fellowships with partial tuition reimbursements available (averaging $2,824 per year), 36 teaching assistantships with partial tuition reimbursements available (averaging $3,509 per year); research assistantships with partial tuition reimbursements available, career-related internships or fieldwork, Federal Work-Study, institutionally-sponsored loans, tuition waivers (partial), and unspecified assistantships also available. Financial aid application deadline: 3/1; financial aid applicants required to submit FAFSA. *Faculty research:* Aerospace systems, computation of methods, dynamics and control, laser applications, materials science. *Unit head:* Dr. Louis Chow, Chair, 407-823-2333. *Application contact:* Dr. A. J. Kassab, Coordinator, 407-823-2416, Fax: 407-823-0208.

See in-depth description on page 1439.

University of Cincinnati, Division of Research and Advanced Studies, College of Engineering, Department of Mechanical, Industrial and Nuclear Engineering, Program in Mechanical Engineering, Cincinnati, OH 45221-0091. Offers MS, PhD. Evening/weekend programs available. *Students:* 127 full-time (13 women), 34 part-time (6 women); includes 6 minority (3 African Americans, 3 Asian Americans or Pacific Islanders), 101 international. In 1998, 46 master's, 5 doctorates awarded. Terminal master's awarded for partial completion of doctoral program. *Degree requirements:* For master's, computer language, oral exam or thesis defense required; for doctorate, variable foreign language requirement, computer language, dissertation required. *Entrance requirements:* For master's and doctorate, GRE General Test, TOEFL (minimum score of 575 required), TSE (minimum score of 250 required). *Average time to degree:* Master's–3.6 years full-time; doctorate–5.8 years full-time. *Application deadline:* For fall admission, 2/1 (priority date). Application fee: $40. *Financial aid:* Fellowships, career-related internships or fieldwork, tuition waivers (full), and unspecified assistantships available. Aid available to part-time students. Financial aid application deadline: 2/1. *Faculty research:* Signature analysis, structural analysis, energy, design, robotics. *Unit head:* Katie Bridgewater, Graduate Program Assistant, 805-893-4601, E-mail: matrls-grad-advisor@engineering.ucsb.edu. *Application contact:* Dr. Jay Kim, Graduate Program Director, 513-556-6300, Fax: 513-556-3390, E-mail: jay.kim@uc.edu.

University of Colorado at Boulder, Graduate School, College of Engineering and Applied Science, Department of Mechanical Engineering, Boulder, CO 80309. Offers ME, MS, PhD. Part-time programs available. Terminal master's awarded for partial completion of doctoral program. *Degree requirements:* For master's, comprehensive exam required, thesis optional, foreign language not required; for doctorate, dissertation, comprehensive, final, and preliminary exams required, foreign language not required. *Entrance requirements:* For master's and doctorate, TOEFL (minimum score of 550 required; average 600), minimum undergraduate GPA of 3.0. *Faculty research:* Thermal science, mechanics, materials research.

University of Colorado at Denver, Graduate School, College of Engineering and Applied Science, Department of Mechanical Engineering, Denver, CO 80217-3364. Offers MS. Part-time and evening/weekend programs available. *Faculty:* 11. *Students:* 7 full-time (0 women), 34 part-time (5 women); includes 6 minority (3 Asian Americans or Pacific Islanders, 3 Hispanic Americans), 6 international. Average age 30. 18 applicants, 78% accepted. In 1998, 5 degrees awarded. *Degree requirements:* For master's, thesis optional. *Entrance requirements:* For master's, GRE. *Application deadline:* For fall admission, 5/1; for spring admission, 11/1. Applications are processed on a rolling basis. Application fee: $50 ($60 for international students). Electronic applications accepted. Tuition, state resident: part-time $217 per credit hour. Tuition, nonresident: part-time $783 per credit hour. Required fees: $3 per credit hour. $130 per year. One-time fee: $25 part-time. *Financial aid:* Research assistantships, teaching assistantships, career-related internships or fieldwork and Federal Work-Study available. Financial aid application deadline: 3/1; financial aid applicants required to submit FAFSA. Total annual research expenditures: $101,944. *Unit head:* James Gerdeen, Chair, 303-556-2781, Fax: 303-556-6371, E-mail: jgerdeen@castle.cudenver.edu. *Application contact:* Loretta Duran, Program Assistant, 303-556-8516, Fax: 303-556-6371, E-mail: lduran@carbon.cudenver.edu.

University of Connecticut, Graduate School, School of Engineering, Department of Mechanical Engineering, Field of Mechanical Engineering, Storrs, CT 06269. Offers MS, PhD.

University of Dayton, Graduate School, School of Engineering, Department of Mechanical and Aerospace Engineering, Dayton, OH 45469-1300. Offers aerospace engineering (MSAE, DE, PhD); mechanical engineering (MSME, DE, PhD). Part-time programs available. *Faculty:* 20 full-time (0 women), 7 part-time (0 women). *Students:* 34 full-time (2 women), 27 part-time (3 women); includes 9 minority (3 African Americans, 3 Asian Americans or Pacific Islanders, 3 Hispanic Americans), 12 international. Average age 26. In 1998, 15 master's, 5 doctorates awarded. *Degree requirements:* For master's, foreign language and thesis not required; for doctorate, dissertation, departmental qualifying exam required. *Entrance requirements:* For master's, TOEFL. *Application deadline:* For fall admission, 8/1 (priority date). Applications are processed on a rolling basis. Application fee: $30. *Financial aid:* In 1998–99, 18 students received aid, including 1 fellowship with full tuition reimbursement available (averaging $18,000 per year), 15 research assistantships with full tuition reimbursements available (averaging $13,500 per year), 2 teaching assistantships with full tuition reimbursements available (averaging $9,000 per year); institutionally-sponsored loans and tuition waivers (full and partial) also available. *Faculty research:* Turbine blade convection, jet engine combustion, energy storage, heat pipes surface transfer, surface coating friction and wear. Total annual research expenditures: $400,000. *Unit head:* Dr. Glen E. Johnson, Chairperson, 937-229-2835, Fax: 937-229-2756, E-mail: gjohnson@engr.udayton.edu. *Application contact:* Dr. Donald L. Moon, Associate Dean, 937-229-2241, Fax: 937-229-2471, E-mail: dmoon@engr.udayton.edu.

University of Delaware, College of Engineering, Department of Mechanical Engineering, Newark, DE 19716. Offers MEM, MSME, PhD. Part-time programs available. *Faculty:* 19 full-time (1 woman), 18 part-time (0 women). *Students:* 77 full-time (8 women), 6 part-time, 61 international. Average age 28. 206 applicants, 42% accepted. In 1998, 9 master's, 9 doctorates awarded. Terminal master's awarded for partial completion of doctoral program. *Degree requirements:* For master's, thesis required (for some programs), foreign language not required; for doctorate, dissertation required, foreign language not required. *Entrance requirements:* For master's and doctorate, GRE General Test (minimum combined score of 1050 required), TOEFL (minimum score of 550 required). *Average time to degree:* Master's–2 years full-time, 3 years part-time; doctorate–4 years full-time, 5 years part-time. *Application deadline:* For fall admission, 7/1 (priority date); for spring admission, 1/16. Applications are processed on a rolling basis. Application fee: $45. *Financial aid:* In 1998–99, 4 fellowships with full tuition reimbursements, 47 research assistantships with full tuition reimbursements, 13 teaching assistantships with full tuition reimbursements were awarded. *Faculty research:* Biomechanics, composites, design and manufacturing processing, environmental and nonlinear dynamics, chaos. Total annual research expenditures: $3.6 million. *Unit head:* Dr. Andras Z. Szeri, Chair, 302-831-2421, Fax: 302-831-3619, E-mail: szeri@me.udel.edu. *Application contact:* Graduate Secretary, 302-831-2423, Fax: 302-831-3619, E-mail: grad_sec@me.udel.edu.

See in-depth description on page 1441.

University of Denver, Graduate Studies, Faculty of Natural Sciences, Mathematics and Engineering, Department of Engineering, Denver, CO 80208. Offers computer science and engineering (MS); electrical engineering (MS); management and general engineering (MSMGEN); materials science (PhD); mechanical engineering (MS). Part-time and evening/weekend programs available. *Faculty:* 15. *Students:* 23 (9 women) 8 international. Terminal master's awarded for partial completion of doctoral program. *Degree requirements:* For master's, thesis required (for some programs), foreign language not required; for doctorate, dissertation required, foreign language not required. *Entrance requirements:* For master's and doctorate, GRE General Test, TOEFL (minimum score of 570 required), TSE (minimum score of 230 required). *Application deadline:* Applications are processed on a rolling basis. Application fee: $40 ($45 for international students). *Unit head:* Dr. Albert J. Rosa, Chair, 303-871-2102. *Application contact:* Louise Carlson, Assistant to Chair, 303-871-2107.

University of Detroit Mercy, College of Engineering and Science, Department of Mechanical Engineering, Detroit, MI 48219-0900. Offers automotive engineering (DE); engineering management (M Eng Mgt), including engineering management, mechanical engineering; manufacturing engineering (ME, DE); mechanical engineering (ME, DE). Evening/weekend programs available. *Degree requirements:* For master's, computer language required, foreign language not required; for doctorate, dissertation required. *Faculty research:* CAD/CAM.

See in-depth description on page 1443.

University of Florida, Graduate School, College of Engineering, Department of Mechanical Engineering, Gainesville, FL 32611. Offers ME, MS, PhD, Certificate, Engr. Part-time programs available. *Faculty:* 39. *Students:* 94 full-time (5 women), 44 part-time (5 women); includes 14 minority (4 African Americans, 6 Asian Americans or Pacific Islanders, 4 Hispanic

Americans), 43 international. 283 applicants, 75% accepted. In 1998, 34 master's, 5 doctorates awarded. *Degree requirements:* For master's, thesis required (for some programs), foreign language not required; for doctorate and other advanced degree, dissertation required, foreign language not required. *Entrance requirements:* For master's and doctorate, GRE General Test, TOEFL, minimum GPA of 3.0; for other advanced degree, GRE General Test, TOEFL. *Application deadline:* For fall admission, 6/1 (priority date). Applications are processed on a rolling basis. Application fee: $20. *Financial aid:* In 1998–99, 79 students received aid, including 11 fellowships, 55 research assistantships, 12 teaching assistantships; institutionally-sponsored loans and unspecified assistantships also available. Aid available to part-time students. *Faculty research:* Robotics, energy conservation and conversions, thermal systems, mechanical systems, manufacturing. *Unit head:* Dr. William G. Tiederman, Chair, 352-392-0828, Fax: 352-392-1071, E-mail: kay@cimar.ufl.edu. *Application contact:* Dr. John C. Ziegert, Graduate Coordinator, 352-392-9930, Fax: 352-392-1071, E-mail: johnz@ufl.edu.

See in-depth description on page 1445.

University of Hawaii at Manoa, Graduate Division, College of Engineering, Department of Mechanical Engineering, Honolulu, HI 96822. Offers MS, PhD. *Faculty:* 18 full-time (1 woman). *Students:* 14 full-time (1 woman), 9 part-time (1 woman); includes 8 minority (1 African American, 7 Asian Americans or Pacific Islanders), 12 international. Average age 32. 28 applicants, 50% accepted. In 1998, 2 master's awarded (75% found work related to degree, 25% continued full-time study); 2 doctorates awarded. *Degree requirements:* For master's, thesis required, foreign language not required; for doctorate, dissertation, exams required. *Average time to degree:* Master's–2 years full-time; doctorate–3 years full-time. *Application deadline:* For fall admission, 1/15; for spring admission, 9/1. Applications are processed on a rolling basis. Application fee: $25 ($50 for international students). *Financial aid:* In 1998–99, 17 students received aid, including 9 research assistantships (averaging $15,530 per year), 4 teaching assistantships (averaging $13,169 per year); tuition waivers (full) also available. Financial aid application deadline: 8/31; financial aid applicants required to submit FAFSA. *Faculty research:* Materials and manufacturing; mechanics,systems and control; thermal and fluid sciences. Total annual research expenditures: $2 million. *Unit head:* Dr. Hi Chang Chai, Chairperson, 808-956-7167, Fax: 808-956-2373. *Application contact:* Junku Yuh, Graduate Chairperson, 808-956-6579, Fax: 808-956-2373, E-mail: yuh@wiliki.eng.hawaii.edu.

University of Houston, Cullen College of Engineering, Department of Mechanical Engineering, Houston, TX 77004. Offers MME, MSME, PhD. Part-time and evening/weekend programs available. *Faculty:* 26 full-time (0 women), 12 part-time (3 women). *Students:* 58 full-time (7 women), 42 part-time (5 women); includes 11 minority (1 African American, 7 Asian Americans or Pacific Islanders, 2 Hispanic Americans, 1 Native American), 56 international. Average age 28. 114 applicants, 64% accepted. In 1998, 17 master's, 8 doctorates awarded. Terminal master's awarded for partial completion of doctoral program. *Degree requirements:* For master's, thesis required (for some programs), foreign language not required; for doctorate, dissertation, departmental qualifying exam required, foreign language not required. *Entrance requirements:* For master's and doctorate, GRE General Test, TOEFL. *Application deadline:* For fall admission, 7/3 (priority date); for spring admission, 12/4. Applications are processed on a rolling basis. Application fee: $25 ($75 for international students). *Financial aid:* Fellowships, research assistantships, teaching assistantships, career-related internships or fieldwork and Federal Work-Study available. Financial aid application deadline: 2/15. *Faculty research:* Experimental and computational turbulence, composites, rheology, phase change/heat transfer, characterization of superconducting materials. Total annual research expenditures: $293,051. *Unit head:* Dr. Richard B. Bannerot, Chairman, 713-743-4500, Fax: 713-743-4503, E-mail: rbb@uh.edu. *Application contact:* Sheena Paul, Graduate Admissions Analyst, 713-743-4505, Fax: 713-743-4503, E-mail: megrad@mail.me.uh.edu.

See in-depth description on page 1447.

University of Idaho, College of Graduate Studies, College of Engineering, Department of Mechanical Engineering, Program in Mechanical Engineering, Moscow, ID 83844-4140. Offers M Engr, MS, PhD. *Students:* 18 full-time (2 women), 53 part-time (3 women); includes 6 minority (3 African Americans, 3 Asian Americans or Pacific Islanders), 1 international. In 1998, 13 master's awarded. *Degree requirements:* For master's, thesis or alternative required, foreign language not required; for doctorate, computer language, dissertation required. *Entrance requirements:* For master's, TOEFL, minimum GPA of 2.8; for doctorate, TOEFL, minimum undergraduate GPA of 2.8, 3.0 graduate. *Application deadline:* For fall admission, 8/1; for spring admission, 12/15. Application fee: $35 ($45 for international students). *Financial aid:* Application deadline: 2/15. *Unit head:* Dr. Steve Penoncello, Chair, Department of Mechanical Engineering, 208-885-6579.

University of Illinois at Chicago, Graduate College, College of Engineering, Department of Mechanical Engineering, Chicago, IL 60607-7128. Offers industrial engineering (MS), including industrial engineering; industrial engineering and operations research (PhD); mechanical engineering (MS, PhD), including fluids engineering, mechanical analysis and design, thermomechanical and power engineering. *Faculty:* 24 full-time (0 women). *Students:* 85 full-time (8 women), 54 part-time (6 women); includes 22 minority (4 African Americans, 16 Asian Americans or Pacific Islanders, 2 Hispanic Americans), 72 international. Average age 25. 197 applicants, 32% accepted. In 1998, 20 master's, 17 doctorates awarded. *Degree requirements:* For doctorate, dissertation required, foreign language not required, foreign language not required. *Entrance requirements:* For master's and doctorate, GRE General Test, TOEFL (minimum score of 550 required), minimum GPA of 3.75 on a 5.0 scale. *Application deadline:* For fall admission, 6/1; for spring admission, 10/15. Application fee: $40 ($50 for international students). *Financial aid:* In 1998–99, 36 students received aid; fellowships, research assistantships, teaching assistantships available. *Unit head:* Dr. Selcuk Guceri, Head, 312-996-5096.

See in-depth description on page 1449.

University of Illinois at Urbana–Champaign, Graduate College, College of Engineering, Department of Mechanical and Industrial Engineering, Urbana, IL 61801. Offers industrial engineering (MS, PhD); mechanical engineering (MS, PhD). *Faculty:* 47 full-time (4 women). *Students:* 247 full-time (24 women); includes 40 minority (6 African Americans, 25 Asian Americans or Pacific Islanders, 9 Hispanic Americans), 78 international. 503 applicants, 15% accepted. In 1998, 79 master's, 22 doctorates awarded. Terminal master's awarded for partial completion of doctoral program. *Degree requirements:* For master's, thesis, non-thesis available by petition required, foreign language not required; for doctorate, dissertation required, foreign language not required. *Entrance requirements:* For master's, GRE General Test, TOEFL (minimum score of 610 required; 257 for computer-based), minimum GPA of 4.25 on a 5.0 scale; for doctorate, GRE General Test, TOEFL (minimum score of 610 required; 257 for computer-based). *Application deadline:* For fall admission, 3/1; for spring admission, 10/1. Applications are processed on a rolling basis. Application fee: $40 ($50 for international students). Tuition, state resident: full-time $4,616. Tuition, nonresident: full-time $11,768. Full-time tuition and fees vary according to course load. *Financial aid:* In 1998–99, 247 students received aid, including 99 fellowships, 214 research assistantships, 38 teaching assistantships; Federal Work-Study, institutionally-sponsored loans, and tuition waivers (full and partial) also available. Financial aid application deadline: 3/1. *Faculty research:* Combustion and propulsion, design methodology, dynamic systems and controls, energy transfer, materials behavior and processing, manufacturing systems operations, management. Total annual research expenditures: $8 million. *Unit head:* Dr. Richard O. Buckius, Head, 217-333-1079. *Application contact:* Celia Snyder, Graduate Program Coordinator, 217-333-4390, Fax: 217-244-6534, E-mail: cgsnyder@uiuc.edu.

See in-depth description on page 1451.

The University of Iowa, Graduate College, College of Engineering, Department of Mechanical Engineering, Iowa City, IA 52242-1316. Offers MS, PhD. *Faculty:* 20 full-time. *Students:* 32 full-time (4 women), 34 part-time (4 women); includes 2 minority (1 Asian American or Pacific Islander, 1 Hispanic American), 47 international. 284 applicants, 27% accepted. In 1998, 14 master's, 11 doctorates awarded. *Degree requirements:* For master's, thesis optional; for doctorate, dissertation, comprehensive exam required. *Entrance requirements:* For master's and doctorate, GRE, TOEFL. *Application deadline:* For fall admission, 1/15; for spring admission, 7/15. Application fee: $30 ($50 for international students). *Financial aid:* In 1998–99, 3 fellowships, 41 research assistantships, 17 teaching assistantships were awarded. Financial aid applicants required to submit FAFSA. *Faculty research:* Machine dynamics, viscous and turbulent flows, ship hydrodynamics, numerical methods, convective-radiative heat transfer. *Unit head:* Dr. Lea-Der Chen, Chair, 319-335-5674, Fax: 319-335-5669.

Announcement: The department offers MS and PhD degrees with specializations in mechanical systems, fluid dynamics, and thermal systems. Departmental research is affiliated with the Iowa Institute of Hydraulic Research and Center for Computer-Aided Design. Excellent facilities exist in driving simulation, fluid dynamics, materials solidification, combustion, fatigue, and fracture mechanics of solids.

University of Kansas, Graduate School, School of Engineering, Department of Mechanical Engineering, Lawrence, KS 66045. Offers MS, DE, PhD. *Faculty:* 9 full-time. *Students:* 8 full-time (1 woman), 18 part-time (2 women); includes 1 minority (Asian American or Pacific Islander), 12 international. In 1998, 5 master's, 1 doctorate awarded. *Degree requirements:* For master's, thesis or alternative, exam required, foreign language not required; for doctorate, dissertation, comprehensive exam required. *Entrance requirements:* For master's and doctorate, Michigan English Language Assessment Battery, TOEFL, minimum GPA of 3.0. *Application deadline:* For fall admission, 7/1. Application fee: $30. *Financial aid:* Fellowships, research assistantships, teaching assistantships, career-related internships or fieldwork available. *Faculty research:* Heat transfer, energy analysis, computer-aided design, biomedical engineering. *Unit head:* Terry Faddis, Chair, 785-864-3181.

University of Kentucky, Graduate School, Graduate School Programs from the College of Engineering, Department of Mechanical Engineering, Lexington, KY 40506-0032. Offers MSME, PhD. *Degree requirements:* For master's, comprehensive exam required, thesis optional, foreign language not required; for doctorate, dissertation, comprehensive exam required, foreign language not required. *Entrance requirements:* For master's, GRE General Test, minimum undergraduate GPA of 2.8; for doctorate, GRE General Test, minimum graduate GPA of 3.0. *Faculty research:* Combustion, computational fluid dynamics, design and systems, manufacturing, thermal and fluid sciences.

University of Louisville, Graduate School, Speed Scientific School, Department of Mechanical Engineering, Louisville, KY 40292-0001. Offers M Eng, MS. *Accreditation:* ABET (one or more programs are accredited). *Faculty:* 11 full-time (1 woman), 1 part-time (0 women). *Students:* 18 full-time (0 women), 34 part-time (6 women); includes 5 minority (all Asian Americans or Pacific Islanders), 4 international. Average age 26. In 1998, 20 degrees awarded. *Degree requirements:* For master's, thesis required, foreign language not required. *Entrance requirements:* For master's, GRE General Test (minimum combined score of 1200 required). *Application deadline:* Applications are processed on a rolling basis. Application fee: $25. *Unit head:* Dr. Glen Prater, Chair, 502-852-6331, Fax: 502-852-6053, E-mail: gprater@louisville.edu.

University of Maine, Graduate School, College of Engineering, Department of Mechanical Engineering, Orono, ME 04469. Offers MS. *Faculty:* 9 full-time. *Students:* 9 full-time (1 woman), 1 part-time, 4 international. 4 applicants, 50% accepted. In 1998, 3 degrees awarded. *Degree requirements:* For master's, thesis required (for some programs), foreign language not required. *Entrance requirements:* For master's, GRE General Test, GRE Subject Test, TOEFL (minimum score of 550 required). *Application deadline:* For fall admission, 2/1 (priority date); for spring admission, 10/15. Applications are processed on a rolling basis. Application fee: $50. *Financial aid:* In 1998–99, 4 research assistantships with tuition reimbursements (averaging $13,750 per year), 4 teaching assistantships with tuition reimbursements (averaging $7,535 per year) were awarded.; Federal Work-Study and tuition waivers (full and partial) also available. Financial aid application deadline: 3/1. *Faculty research:* Higher order beam and plate theories, dynamic response of structural systems, heat transfer in window systems, forced convection heat transfer in gas turbine passages, effect of parallel space heating systems on utility load management. *Unit head:* Dr. Donald Grant, Chair, 207-581-2120, Fax: 207-581-2379. *Application contact:* Scott G. Delcourt, Director of the Graduate School, 207-581-3218, Fax: 207-581-3232, E-mail: graduate@maine.edu.

University of Manitoba, Faculty of Graduate Studies, Faculty of Engineering, Department of Mechanical and Industrial Engineering, Winnipeg, MB R3T 2N2, Canada. Offers M Eng, M Sc, PhD. *Degree requirements:* For master's and doctorate, thesis/dissertation required. *Unit head:* D. W. Ruth, Head.

University of Maryland, Baltimore County, Graduate School, College of Engineering, Department of Mechanical Engineering, Baltimore, MD 21250-5398. Offers computer-integrated manufacturing and design (MS, PhD); energy (MS, PhD); fluid mechanics (MS, PhD); solid mechanics (MS, PhD). *Faculty:* 16 full-time (1 woman), 3 part-time (1 woman). *Students:* 30 full-time (8 women), 19 part-time (4 women); includes 11 minority (2 African Americans, 8 Asian Americans or Pacific Islanders, 1 Hispanic American), 17 international. 77 applicants, 32% accepted. In 1998, 7 master's, 8 doctorates awarded. *Entrance requirements:* For master's, GRE General Test, minimum GPA of 3.0; for doctorate, GRE General Test, GRE Subject Test, TOEFL, minimum GPA of 3.5. *Application deadline:* For fall admission, 7/1. Applications are processed on a rolling basis. Application fee: $45. *Financial aid:* Fellowships, research assistantships, teaching assistantships available. *Faculty research:* Theoretical and applied mechanics, mechatronics (controls, sensors and actuators), dynamic plasticity. *Unit head:* Dr. Christian Von Kerczek, Interim Chair, 410-455-3300. *Application contact:* Dr. Appa Anjanappa, Director, 410-455-3330.

University of Maryland, College Park, Graduate School, A. James Clark School of Engineering, Department of Mechanical Engineering, College Park, MD 20742-5045. Offers electronic packaging and reliability (MS, PhD); manufacturing and design (MS, PhD); mechanical engineering (M Eng); mechanics and materials (MS, PhD); thermal and fluid sciences (MS, PhD). Part-time and evening/weekend programs available. Postbaccalaureate distance learning degree programs offered. *Faculty:* 61 full-time (5 women), 15 part-time (0 women). *Students:* 133 full-time (13 women), 60 part-time (7 women); includes 18 minority (11 African Americans, 6 Asian Americans or Pacific Islanders, 1 Hispanic American), 123 international. 341 applicants, 25% accepted. In 1998, 30 master's, 17 doctorates awarded. *Degree requirements:* For master's, thesis optional, foreign language not required; for doctorate, dissertation, qualifying exam required, foreign language not required. *Entrance requirements:* For master's, GRE, minimum GPA of 3.0. *Application deadline:* Applications are processed on a rolling basis. Application fee: $50 ($70 for international students). Tuition, state resident: part-time $272 per credit hour. Tuition, nonresident: part-time $475 per credit hour. Required fees: $632; $379 per year. *Financial aid:* In 1998–99, 22 fellowships with full tuition reimbursements (averaging $12,892 per year), 124 research assistantships with tuition reimbursements (averaging $11,853 per year), 25 teaching assistantships with tuition reimbursements (averaging $10,459 per year) were awarded.; Federal Work-Study, grants, and scholarships also available. Aid available to part-time students. Financial aid applicants required to submit FAFSA. *Faculty research:* Decomposition-based design, powder metallurgy, injection molding, kinematic synthesis, electronic packaging, dynamic deformation and fracture, turbulent flow. Total annual research expenditures: $5.7 million. *Unit head:* Dr. Davinder Anand, Chairman, 301-405-5294, Fax: 301-314-9477. *Application contact:* Dr. James M. Wallace, Graduate Director, 301-405-4216.

University of Maryland, College Park, Graduate School, A. James Clark School of Engineering, Professional Program in Engineering, College Park, MD 20742-5045. Offers aerospace engineering (M Eng); chemical engineering (M Eng); civil engineering (M Eng); electrical engineering (M Eng); fire protection engineering (M Eng); materials science and engineering

Mechanical Engineering

University of Maryland, College Park (continued)
(M Eng); mechanical engineering (M Eng); reliability engineering (M Eng); systems engineering (M Eng). Part-time and evening/weekend programs available. Postbaccalaureate distance learning degree programs offered. *Faculty:* 11 part-time (0 women). *Students:* 20 full-time (3 women), 205 part-time (42 women); includes 58 minority (27 African Americans, 25 Asian Americans or Pacific Islanders, 5 Hispanic Americans, 1 Native American), 20 international. *Degree requirements:* For master's, foreign language and thesis not required. *Application deadline:* Applications are processed on a rolling basis. Application fee: $50 ($70 for international students). Tuition, state resident: part-time $272 per credit hour. Tuition, nonresident: part-time $475 per credit hour. Required fees: $632; $379 per year. *Unit head:* Dr. Patrick Cunniff, Associate Dean, 301-405-5256, Fax: 301-314-9477. *Application contact:* Trudy Lindsey, Director, Graduate Admission and Records, 301-405-4198, Fax: 301-314-9305, E-mail: grschool@deans.umd.edu.

University of Massachusetts Amherst, Graduate School, College of Engineering, Department of Mechanical and Industrial Engineering, Program in Mechanical Engineering, Amherst, MA 01003. Offers MS, PhD. *Students:* 40 full-time (1 woman), 24 part-time; includes 1 minority (Hispanic American), 41 international. Average age 26. 229 applicants, 38% accepted. In 1998, 6 master's, 4 doctorates awarded. *Degree requirements:* For master's, thesis, project required, foreign language not required; for doctorate, dissertation required. *Entrance requirements:* For master's and doctorate, GRE General Test. *Application deadline:* For fall admission, 2/1 (priority date); for spring admission, 10/1. Applications are processed on a rolling basis. Application fee: $35. Tuition, state resident: full-time $2,640; part-time $165 per credit. Tuition, nonresident: full-time $9,756; part-time $407 per credit. Required fees: $1,221 per term. One-time fee: $110. Full-time tuition and fees vary according to course load, campus/location and reciprocity agreements. *Financial aid:* Fellowships with full tuition reimbursements, research assistantships with full tuition reimbursements, teaching assistantships with full tuition reimbursements, career-related internships or fieldwork, Federal Work-Study, grants, scholarships, traineeships, and unspecified assistantships available. Aid available to part-time students. Financial aid application deadline: 2/1. *Unit head:* Dr. James Rinderle, Director, 413-545-2505, Fax: 413-545-1027, E-mail: rinderle@ecs.umass.edu.

University of Massachusetts Dartmouth, Graduate School, College of Engineering, Program in Mechanical Engineering, North Dartmouth, MA 02747-2300. Offers MS. Part-time programs available. *Faculty:* 11 full-time (0 women). *Students:* 7 full-time (0 women), 2 part-time, 5 international. Average age 26. 20 applicants, 85% accepted. *Degree requirements:* For master's, thesis or alternative required, foreign language not required. *Entrance requirements:* For master's, GRE General Test, TOEFL. *Application deadline:* For fall admission, 4/20 (priority date); for spring admission, 11/15 (priority date). Application fee: $40 for international students. Tuition, area resident: Full-time $3,107; part-time $129 per credit. Tuition, state resident: full-time $2,071; part-time $86 per credit. Tuition, nonresident: full-time $7,845; part-time $327 per credit. Required fees: $2,888. Full-time tuition and fees vary according to program and reciprocity agreements. Part-time tuition and fees vary according to course load and reciprocity agreements. *Financial aid:* In 1998–99, 2 research assistantships with full tuition reimbursements (averaging $10,000 per year), 5 teaching assistantships with full tuition reimbursements (averaging $8,000 per year) were awarded.; Federal Work-Study and unspecified assistantships also available. Financial aid application deadline: 3/15. *Faculty research:* Treatment of hypertrophic burn scars with multi-beam laser, ground source heat pump research and development. Total annual research expenditures: $71,000. *Unit head:* John Rice, Director, 508-999-8498, Fax: 508-999-8881, E-mail: jrice@umassd.edu. *Application contact:* Carol A. Novo, Graduate Admissions Office, 508-999-8026, Fax: 508-999-8183, E-mail: graduate@umassd.edu.

University of Massachusetts Lowell, Graduate School, James B. Francis College of Engineering, Department of Mechanical Engineering, Lowell, MA 01854-2881. Offers MS Eng, D Eng. Part-time programs available. *Faculty:* 17 full-time (0 women). *Students:* 22 full-time (5 women), 45 part-time (2 women); includes 4 minority (1 African American, 2 Asian Americans or Pacific Islanders, 1 Native American), 17 international. 69 applicants, 55% accepted. In 1998, 7 master's, 2 doctorates awarded. Terminal master's awarded for partial completion of doctoral program. *Degree requirements:* For master's, thesis or alternative required, foreign language not required; for doctorate, 2 foreign languages (computer language can substitute for one), dissertation required. *Entrance requirements:* For master's and doctorate, GRE General Test. *Application deadline:* For fall admission, 4/1 (priority date); for spring admission, 10/1. Applications are processed on a rolling basis. Application fee: $20 ($35 for international students). *Financial aid:* In 1998–99, 3 research assistantships, 7 teaching assistantships were awarded.; career-related internships or fieldwork also available. Financial aid application deadline: 4/1. *Unit head:* Dr. Struan Robertson, Chair, 978-934-2950. *Application contact:* Dr. James Sherwood, Coordinator, 978-934-3313, E-mail: james_sherwood@woods.uml.edu.

The University of Memphis, Graduate School, Herff College of Engineering, Department of Mechanical Engineering, Memphis, TN 38152. Offers design and mechanical engineering (MS); energy systems (MS); mechanical engineering (PhD); mechanical systems (MS); power systems (MS). Part-time programs available. *Faculty:* 11 full-time (0 women). *Students:* 9 full-time (2 women), 8 part-time (1 woman), 9 international. Average age 31. 22 applicants, 32% accepted. In 1998, 4 degrees awarded. Terminal master's awarded for partial completion of doctoral program. *Degree requirements:* For master's and doctorate, thesis/dissertation, comprehensive exam required. *Entrance requirements:* For master's, GRE General Test (minimum combined score of 1100 required), BS in mechanical engineering, minimum undergraduate GPA of 3.0. *Application deadline:* For fall admission, 8/1; for spring admission, 12/1. Application fee: $25 ($50 for international students). Tuition, state resident: full-time $3,410; part-time $178 per credit hour. Tuition, nonresident: full-time $8,670; part-time $408 per credit hour. Tuition and fees vary according to program. *Financial aid:* In 1998–99, 17 students received aid, including 1 fellowship with full tuition reimbursement available, 8 research assistantships with full tuition reimbursements available, 9 teaching assistantships with full tuition reimbursements available; career-related internships or fieldwork and unspecified assistantships also available. *Faculty research:* Computational fluid dynamics, fracture mechanics, computational mechanics, composites, phase change. *Unit head:* Dr. John I. Hochstein, Chair, 901-678-2173, Fax: 901-678-5459, E-mail: jhochste@memphis.edu. *Application contact:* Dr. Teong E. Tan, Coordinator of Graduate Studies, 901-678-3264, Fax: 901-678-5459, E-mail: ttan@memphis.edu.

University of Miami, Graduate School, College of Engineering, Department of Mechanical Engineering, Coral Gables, FL 33124. Offers MS, MSME, DA, PhD. Part-time programs available. *Faculty:* 12 full-time (0 women). *Students:* 20 full-time (3 women), 1 part-time, 18 international. 75 applicants, 95% accepted. In 1998, 5 master's awarded (100% found work related to degree); 3 doctorates awarded (67% entered university research/teaching, 33% found other work related to degree). *Degree requirements:* For master's, thesis required (for some programs), foreign language not required; for doctorate, dissertation required, foreign language not required. *Entrance requirements:* For master's and doctorate, GRE General Test (minimum combined score of 1000 required), TOEFL (minimum score of 550 required), minimum GPA of 3.0. *Average time to degree:* Master's–2 years full-time; doctorate–4 years full-time. *Application deadline:* Applications are processed on a rolling basis. Application fee: $35. Tuition: Full-time $15,336; part-time $852 per credit. Required fees: $174. Tuition and fees vary according to program. *Financial aid:* In 1998–99, 12 students received aid, including 1 fellowship with tuition reimbursement available, 3 research assistantships with tuition reimbursements available, 8 teaching assistantships with tuition reimbursements available; career-related internships or fieldwork, Federal Work-Study, and tuition waivers (partial) also available. Financial aid application deadline: 2/1. *Faculty research:* Internal combustion engines, heat transfer, hydrogen energy, controls, mechanics. Total annual research expenditures: $1.9 million. *Unit head:* Dr. Singiresu S. Rao, Chairman, 305-284-2571, Fax: 305-284-2580, E-mail: ntangred@miami.edu. *Application contact:* Dr. Singiresu S. Rao, Chairman, 305-284-2571, Fax: 305-284-2580, E-mail: ntangred@miami.edu.

See in-depth description on page 1455.

University of Michigan, Horace H. Rackham School of Graduate Studies, College of Engineering, Department of Mechanical Engineering and Applied Mechanics, Program in Mechanical Engineering, Ann Arbor, MI 48109. Offers MSE, PhD. Part-time programs available. Terminal master's awarded for partial completion of doctoral program. *Degree requirements:* For master's, thesis optional, foreign language not required; for doctorate, dissertation, oral defense of dissertation, preliminary and qualifying exams required, foreign language not required. *Entrance requirements:* For master's, GRE General Test; for doctorate, GRE General Test, master's degree.

See in-depth description on page 1457.

University of Michigan–Dearborn, College of Engineering and Computer Science, Department of Mechanical Engineering, Dearborn, MI 48128-1491. Offers MSE. Part-time programs available. *Faculty:* 13 full-time (0 women), 9 part-time (1 woman). *Students:* 3 full-time (1 woman), 140 part-time (21 women); includes 17 minority (3 African Americans, 12 Asian Americans or Pacific Islanders, 1 Hispanic American, 1 Native American), 2 international. Average age 25. In 1998, 62 degrees awarded. *Degree requirements:* For master's, computer language required, thesis optional, foreign language not required. *Entrance requirements:* For master's, bachelor's degree in applied mathematics, computer science, engineering, or physical science; minimum GPA of 3.0. *Application deadline:* For fall admission, 8/1; for winter admission, 12/1 (priority date); for spring admission, 4/1 (priority date). Applications are processed on a rolling basis. Application fee: $55. Electronic applications accepted. Tuition, state resident: part-time $259 per credit hour. Tuition, nonresident: part-time $748 per credit hour. Required fees: $80 per course. Tuition and fees vary according to course level, course load and program. *Financial aid:* In 1998–99, 3 research assistantships, 2 teaching assistantships were awarded.; fellowships, Federal Work-Study also available. *Faculty research:* Combustion, fatigue, fracture/damage mechanics, noise and vibration, vehicle climate control. *Unit head:* Dr. C. L. Chow, Chair, 313-593-5465, Fax: 313-593-3851, E-mail: clchow@umich.edu. *Application contact:* Marguerite Bird, Academic Secretary, 313-593-5241, Fax: 313-593-3851, E-mail: mpbird@umich.edu.

University of Minnesota, Twin Cities Campus, Graduate School, Institute of Technology, Department of Mechanical Engineering, Minneapolis, MN 55455-0213. Offers industrial engineering (MIE, MSIE, PhD); mechanical engineering (MME, MSME, PhD). Part-time programs available. *Degree requirements:* For doctorate, dissertation required, foreign language not required. *Entrance requirements:* For master's, GRE General Test (score in 80th percentile or higher required; average 85th percentile), minimum GPA of 3.0; for doctorate, GRE General Test (score in 80th percentile or higher required; average 85th percentile).

See in-depth description on page 1459.

University of Missouri–Columbia, Graduate School, College of Engineering, Department of Mechanical and Aerospace Engineering, Columbia, MO 65211. Offers MS, PhD. *Faculty:* 21 full-time (1 woman). *Students:* 21 full-time (1 woman), 17 part-time (1 woman); includes 4 minority (2 African Americans, 1 Asian American or Pacific Islander, 1 Native American), 19 international. 23 applicants, 52% accepted. In 1998, 18 master's, 3 doctorates awarded. *Degree requirements:* For master's, thesis required, foreign language not required; for doctorate, dissertation required. *Entrance requirements:* For master's and doctorate, GRE General Test, TOEFL (minimum score of 550 required), minimum GPA of 3.0. *Application deadline:* Applications are processed on a rolling basis. Application fee: $30 ($50 for international students). *Financial aid:* Research assistantships, teaching assistantships, institutionally-sponsored loans available. *Unit head:* Dr. Uee Wan Cho, Director of Graduate Studies, 573-882-3778.

See in-depth description on page 1461.

University of Missouri–Rolla, Graduate School, School of Engineering, Department of Mechanical and Aerospace Engineering and Engineering Mechanics, Program in Mechanical Engineering, Rolla, MO 65409-0910. Offers MS, DE, PhD. Part-time programs available. *Faculty:* 22 full-time (1 woman). *Students:* 82 full-time (8 women), 16 part-time (2 women); includes 2 minority (both African Americans), 65 international. Average age 27. 224 applicants, 77% accepted. In 1998, 23 master's, 3 doctorates awarded. *Degree requirements:* For master's, thesis required (for some programs), foreign language not required; for doctorate, dissertation required, foreign language not required. *Entrance requirements:* For master's, GRE General Test (minimum combined score of 1150 required), TOEFL (minimum score of 570 required), minimum GPA of 3.0; for doctorate, GRE General Test (minimum combined score of 1150 required), TOEFL (minimum score of 570 required), minimum GPA of 3.5. *Average time to degree:* Master's–1.9 years full-time, 6 years part-time; doctorate–3.4 years full-time. *Application deadline:* For fall admission, 7/1; for spring admission, 12/1. Applications are processed on a rolling basis. Application fee: $25. Electronic applications accepted. *Financial aid:* In 1998–99, 16 fellowships with full and partial tuition reimbursements (averaging $9,739 per year), 20 research assistantships (averaging $10,829 per year), 30 teaching assistantships (averaging $11,380 per year) were awarded. Financial aid application deadline: 3/1. *Faculty research:* Fluid mechanics, thermal science, CAD/CAM and CAE, dynamics and vibrations, HVAC. *Unit head:* Coordinator, Graduate Admissions, 301-985-7155, Fax: 301-985-7175, E-mail: gradinfo@nova.umuc.edu. *Application contact:* Dr. James A. Drallmeier, Associate Professor, Associate Chair for Graduate Studies, 573-341-4710, Fax: 573-341-4607, E-mail: grad-students@gearbox.maem.umr.edu.

University of Nebraska–Lincoln, Graduate College, College of Engineering and Technology, Department of Mechanical Engineering, Lincoln, NE 68588. Offers engineering (PhD); mechanical engineering (MS), including materials science engineering. *Faculty:* 15 full-time (2 women). *Students:* 21 full-time (2 women), 9 part-time (1 woman), 13 international. Average age 28. 133 applicants, 26% accepted. In 1998, 5 degrees awarded. *Degree requirements:* For master's, thesis optional, foreign language not required; for doctorate, dissertation, comprehensive exams required. *Entrance requirements:* For master's and doctorate, GRE General Test, TOEFL (minimum score of 550 required). *Application deadline:* For fall admission, 3/1 (priority date). Applications are processed on a rolling basis. Application fee: $35. Electronic applications accepted. *Financial aid:* In 1998–99, 3 fellowships, 23 research assistantships, 13 teaching assistantships were awarded.; Federal Work-Study also available. Aid available to part-time students. Financial aid application deadline: 2/15. *Faculty research:* Computational thermal/fluid science, thin films, microcharacterization of materials, vehicle crashworthiness. Total annual research expenditures: $265,209. *Unit head:* Dr. David Lou, Chair, 402-472-2375.

See in-depth description on page 1463.

University of Nevada, Las Vegas, Graduate College, Howard R. Hughes College of Engineering, Department of Mechanical Engineering, Las Vegas, NV 89154-9900. Offers MSE, PhD. *Faculty:* 18 full-time (0 women). *Students:* 15 full-time (1 woman), 18 part-time (5 women); includes 5 minority (1 African American, 2 Asian Americans or Pacific Islanders, 2 Hispanic Americans), 9 international. 32 applicants, 59% accepted. In 1998, 13 master's awarded. *Degree requirements:* For master's, comprehensive exam required, thesis optional, foreign language not required; for doctorate, dissertation required. *Entrance requirements:* For master's, minimum GPA of 3.0; for doctorate, minimum GPA of 3.5. *Application deadline:* For fall admission, 6/15; for spring admission, 11/15. Application fee: $40 ($95 for international students). *Financial aid:* In 1998–99, 7 research assistantships with full tuition reimbursements (averaging $7,936 per year), 11 teaching assistantships with partial tuition reimbursements (averaging $8,973 per year) were awarded. Financial aid application deadline: 3/1. *Unit head:* Dr. Darrell Pepper, Chair, 702-895-1331.

University of Nevada, Reno, Graduate School, College of Engineering, Department of Mechanical Engineering, Reno, NV 89557. Offers MS, PhD. Terminal master's awarded for partial completion of doctoral program. *Degree requirements:* For master's, thesis optional, foreign language not required; for doctorate, dissertation required, foreign language not required. *Entrance requirements:* For master's, GRE General Test, TOEFL (minimum score of 500 required), minimum GPA of 2.75; for doctorate, GRE General Test, TOEFL (minimum score of 500 required), minimum GPA of 3.0. *Faculty research:* Composite, solid, fluid, thermal, and smart materials.

University of New Brunswick, School of Graduate Studies, Faculty of Engineering, Department of Mechanical Engineering, Fredericton, NB E3B 5A3, Canada. Offers applied mechanics (M Eng, M Sc E, PhD); mechanical engineering (M Eng, M Sc E, PhD). Part-time programs available. *Degree requirements:* For master's, thesis required, foreign language not required; for doctorate, dissertation, qualifying exam required, foreign language not required. *Entrance requirements:* For master's and doctorate, TOEFL, TWE, minimum GPA of 3.0.

University of New Hampshire, Graduate School, College of Engineering and Physical Sciences, Programs in Engineering, Department of Mechanical Engineering, Durham, NH 03824. Offers MS. *Faculty:* 14 full-time. *Students:* 5 full-time (2 women), 21 part-time (1 woman). Average age 27. 15 applicants, 87% accepted. In 1998, 5 degrees awarded. *Degree requirements:* For master's, thesis or alternative required, foreign language not required. *Entrance requirements:* For master's, GRE. *Application deadline:* For fall admission, 4/1 (priority date). Applications are processed on a rolling basis. Application fee: $50. Tuition, area resident: Full-time $5,750; part-time $319 per credit. Tuition, state resident: full-time $8,625. Tuition, nonresident: full-time $14,640; part-time $598 per credit. Required fees: $224 per semester. Tuition and fees vary according to course load, degree level and program. *Financial aid:* In 1998–99, 3 research assistantships, 7 teaching assistantships were awarded.; Federal Work-Study, scholarships, and tuition waivers (full and partial) also available. Aid available to part-time students. Financial aid application deadline: 2/15. *Faculty research:* Solid mechanics, dynamics, materials science, dynamic systems, automatic control. *Unit head:* Dave Limbert, Chairperson, 603-862-1748. *Application contact:* Dr. David W. Watt, Graduate Coordinator, 603-862-2555.

University of New Hampshire, Graduate School, College of Engineering and Physical Sciences, Programs in Engineering, Doctoral Program in Engineering, Durham, NH 03824. Offers chemical engineering (PhD); civil engineering (PhD); electrical engineering (PhD); mechanical engineering (PhD); systems design engineering (PhD). *Students:* 16 full-time (3 women), 8 part-time (1 woman), 12 international. *Degree requirements:* For doctorate, dissertation required. *Entrance requirements:* For doctorate, GRE (for civil and mechanical engineering options). *Application deadline:* For fall admission, 4/1 (priority date). Applications are processed on a rolling basis. Application fee: $50. Tuition, area resident: Full-time $5,750; part-time $319 per credit. Tuition, state resident: full-time $8,625. Tuition, nonresident: full-time $14,640; part-time $598 per credit. Required fees: $224 per semester. Tuition and fees vary according to course load, degree level and program.

University of New Haven, Graduate School, School of Engineering and Applied Science, Program in Mechanical Engineering, West Haven, CT 06516-1916. Offers MSME. 2 applicants, 50% accepted. *Degree requirements:* For master's, thesis or alternative required, foreign language not required. *Application deadline:* Applications are processed on a rolling basis. Application fee: $50. *Financial aid:* Federal Work-Study available. Aid available to part-time students. Financial aid application deadline: 5/1; financial aid applicants required to submit FAFSA. *Unit head:* Dr. Konstantine Lambrakis, Coordinator, 203-932-7408.

University of New Mexico, Graduate School, School of Engineering, Department of Mechanical Engineering, Albuquerque, NM 87131-2039. Offers engineering (PhD); manufacturing engineering (ME, MS); mechanical engineering (MSME). ME offered through the Manufacturing Engineering Program. Part-time programs available. *Faculty:* 17 full-time (0 women), 18 part-time (2 women). *Students:* 20 full-time (2 women), 39 part-time (6 women); includes 18 minority (1 African American, 1 Asian American or Pacific Islander, 13 Hispanic Americans, 3 Native Americans), 8 international. Average age 33. 14 applicants, 79% accepted. In 1998, 18 master's, 3 doctorates awarded. *Degree requirements:* For master's, thesis required (for some programs), foreign language not required; for doctorate, dissertation required, foreign language not required. *Entrance requirements:* For master's and doctorate, GRE General Test, minimum GPA of 3.0. *Application deadline:* For fall admission, 7/15; for spring admission, 11/14. Applications are processed on a rolling basis. Application fee: $25. *Financial aid:* In 1998–99, 33 students received aid, including 11 fellowships (averaging $2,241 per year), 18 research assistantships with tuition reimbursements available (averaging $3,192 per year), 2 teaching assistantships with tuition reimbursements available (averaging $14,040 per year); career-related internships or fieldwork and Federal Work-Study also available. Financial aid application deadline: 8/15. *Faculty research:* Modeling and simulation, materials, thermal fluids, artificial muscles. Total annual research expenditures: $628,000. *Unit head:* Dr. David E. Thompson, Chair, 505-277-2761, Fax: 505-277-1571, E-mail: dthomp@me.unm.edu.

See in-depth description on page 1465.

University of New Orleans, Graduate School, College of Engineering, Concentration in Mechanical Engineering, New Orleans, LA 70148. Offers MS. *Faculty:* 5 full-time (0 women). *Students:* 17 full-time (2 women), 13 part-time (2 women); includes 2 minority (1 African American, 1 Asian American or Pacific Islander), 20 international. Average age 26. 96 applicants, 51% accepted. In 1998, 8 degrees awarded. *Degree requirements:* For master's, thesis optional, foreign language not required. *Entrance requirements:* For master's, GRE General Test (minimum combined score of 1200 required), minimum GPA of 3.0. *Application deadline:* For fall admission, 7/1 (priority date). Applications are processed on a rolling basis. Application fee: $20. Tuition, state resident: full-time $2,362. Tuition, nonresident: full-time $7,888. Part-time tuition and fees vary according to course load. *Financial aid:* Research assistantships, teaching assistantships available. *Faculty research:* Two-phase flow instabilities, thermal-hydrodynamic modeling, solar energy, heat transfer from sprays, boundary integral techniques in mechanics. *Unit head:* Dr. Edwin Russo, Chairman, 504-280-6652, Fax: 504-280-5539, E-mail: eprme@uno.edu. *Application contact:* Dr. Kazim Akyuzlu, Graduate Coordinator, 504-280-6186, Fax: 504-280-7413, E-mail: kmame@uno.edu.

University of North Carolina at Charlotte, Graduate School, The William States Lee College of Engineering, Department of Mechanical Engineering and Engineering Science, Charlotte, NC 28223-0001. Offers MSME, PhD. Evening/weekend programs available. *Faculty:* 20 full-time (1 woman). *Students:* 21 full-time (3 women), 38 part-time (1 woman); includes 6 minority (3 African Americans, 3 Asian Americans or Pacific Islanders), 24 international. Average age 29. 39 applicants, 85% accepted. In 1998, 16 master's awarded. *Degree requirements:* For master's, thesis required. *Entrance requirements:* For master's, GRE General Test, minimum GPA of 3.0 in undergraduate major, 2.75 overall. *Application deadline:* For fall admission, 7/15; for spring admission, 11/15. Applications are processed on a rolling basis. Application fee: $35. Electronic applications accepted. *Financial aid:* In 1998–99, 1 fellowship (averaging $8,000 per year), 29 research assistantships, 20 teaching assistantships were awarded.; Federal Work-Study also available. Financial aid application deadline: 4/1. *Faculty research:* Microelectronics, friction/heat and wear, robotics, manufacturing, control systems. *Unit head:* Dr. Robert E. Johnson, Chair, 704-547-2303, Fax: 704-547-2352, E-mail: robejohn@email.uncc.edu. *Application contact:* Kathy Barringer, Assistant Director of Graduate Admissions, 704-547-3366, Fax: 704-547-3279, E-mail: gradadm@email.uncc.edu.

University of North Dakota, Graduate School, School of Engineering and Mines, Department of Mechanical Engineering, Grand Forks, ND 58202. Offers M Engr, MS. Part-time programs available. *Faculty:* 7 full-time (0 women). *Students:* 2 full-time (0 women). In 1998, 1 degree awarded. *Degree requirements:* For master's, thesis or alternative required, foreign language not required. *Entrance requirements:* For master's, GRE General Test, GRE Subject Test, TOEFL (minimum score of 550 required), minimum GPA of 3.0 (MS), 2.5 (M Engr). *Application deadline:* For fall admission, 3/1 (priority date). Applications are processed on a rolling basis. Application fee: $20. *Financial aid:* Fellowships, research assistantships, teaching assistantships, tuition waivers (partial) available. Financial aid application deadline: 3/15. *Unit head:* Dr. Donald Moen, Chairperson, 701-777-2571, Fax: 701-777-4838, E-mail: donald_moen@mail.und.nodak.edu.

University of Notre Dame, Graduate School, College of Engineering, Department of Aerospace and Mechanical Engineering, Program in Mechanical Engineering, Notre Dame, IN 46556. Offers MS, PhD. Part-time programs available. *Students:* 37 full-time (4 women), 1 (woman) part-time; includes 2 minority (1 Asian American or Pacific Islander, 1 Hispanic American), 27 international. 90 applicants, 33% accepted. In 1998, 8 master's, 5 doctorates awarded. Terminal master's awarded for partial completion of doctoral program. *Degree requirements:* For master's, thesis or alternative required, foreign language not required; for doctorate, dissertation required, foreign language not required. *Entrance requirements:* For master's and doctorate, GRE General Test, TOEFL (minimum score of 600 required; 250 for computer-based). *Average time to degree:* Master's–2 years full-time; doctorate–5 years full-time. *Application deadline:* For fall admission, 2/1 (priority date); for spring admission, 10/15. Applications are processed on a rolling basis. Application fee: $40. *Financial aid:* In 1998–99, fellowships with full tuition reimbursements (averaging $16,000 per year), research assistantships with full tuition reimbursements (averaging $11,500 per year), teaching assistantships with full tuition reimbursements (averaging $11,500 per year) were awarded.; tuition waivers (full) and unspecified assistantships also available. Financial aid application deadline: 2/1. *Faculty research:* Heat transfer and fluid mechanics, solid mechanics, mechanical systems and robotics. *Unit head:* Dr. David W. Watt, Graduate Coordinator, 603-862-2555. *Application contact:* Dr. Terrence J. Akai, Director of Graduate Admissions, 219-631-7706, Fax: 219-631-4183, E-mail: gradad@nd.edu.

University of Oklahoma, Graduate College, College of Engineering, School of Aerospace and Mechanical Engineering, Program in Mechanical Engineering, Norman, OK 73019-0390. Offers MS, PhD. *Students:* 37 full-time (6 women), 30 part-time; includes 4 minority (1 African American, 2 Asian Americans or Pacific Islanders, 1 Hispanic American), 40 international. Average age 28. 84 applicants, 83% accepted. In 1998, 13 master's, 1 doctorate awarded. *Degree requirements:* For master's, thesis or alternative, comprehensive exam required, foreign language not required; for doctorate, comprehensive exam, departmental qualifying exam required, thesis/dissertation optional, foreign language not required. *Entrance requirements:* For master's, GRE General Test, TOEFL (minimum score of 550 required), BS in engineering or physical sciences; for doctorate, GRE General Test, TOEFL (minimum score of 550 required), MS in mechanical engineering or equivalent. *Application deadline:* For fall admission, 6/1 (priority date). Applications are processed on a rolling basis. Application fee: $25. Tuition, state resident: part-time $86 per credit hour. Tuition, nonresident: part-time $275 per credit hour. Tuition and fees vary according to course level, course load and program. *Financial aid:* Fellowships, research assistantships, teaching assistantships, Federal Work-Study available. Financial aid application deadline: 3/1. *Faculty research:* Fluid mechanics, fracture mechanics, numerical methods in mechanics, combustion, nondestructive evaluation. *Unit head:* Elizabeth F. Craig, Coordinator, Curriculum in Applied and Materials Sciences, 919-962-6293, Fax: 919-962-3341, E-mail: materials_science@unc.edu. *Application contact:* Dr. Ronald Kline, Graduate Coordinator, 405-325-5011.

University of Ottawa, School of Graduate Studies and Research, Faculty of Engineering, Ottawa-Carleton Institute for Mechanical and Aerospace Engineering, Ottawa, ON K1N 6N5, Canada. Offers M Eng, MA Sc, PhD. *Faculty:* 57 full-time, 3 part-time. *Students:* 113 full-time (9 women), 20 part-time (2 women), 24 international. Average age 31. In 1998, 30 master's, 6 doctorates awarded. *Degree requirements:* For master's, thesis or alternative required, foreign language not required; for doctorate, dissertation required, foreign language not required. *Entrance requirements:* For master's, honors degree or equivalent, minimum B average; for doctorate, master's degree, minimum B+ average. *Application deadline:* For fall admission, 3/1. Application fee: $35. *Financial aid:* Fellowships, research assistantships, teaching assistantships, Federal Work-Study available. *Faculty research:* Fluid mechanics, heat transfer, solid mechanics, design, manufacturing. *Unit head:* David Redekop, Director, 613-562-5800 Ext. 6290, Fax: 613-562-5177. *Application contact:* Solange Lamontagne, Academic Assistant, 613-562-5834, Fax: 613-562-5177, E-mail: gradinfo@eng.uottawa.ca.

See in-depth description on page 1467.

University of Pennsylvania, School of Engineering and Applied Science, Department of Mechanical Engineering and Applied Mechanics, Philadelphia, PA 19104. Offers applied mechanics (MSE, PhD); mechanical engineering (MSE, PhD). Part-time programs available. *Degree requirements:* For master's, thesis optional, foreign language not required; for doctorate, dissertation required, foreign language not required. *Entrance requirements:* For master's and doctorate, TOEFL (minimum score of 600 required). Electronic applications accepted. *Faculty research:* Heat transfer, fluid mechanics, energy conversion, solid mechanics, dynamics of mechanisms and robots.

See in-depth description on page 1469.

University of Pittsburgh, School of Engineering, Department of Mechanical Engineering, Pittsburgh, PA 15260. Offers MSME, PhD. Part-time and evening/weekend programs available. *Faculty:* 13 full-time (1 woman), 2 part-time (1 woman). *Students:* 31 full-time (2 women), 26 part-time (2 women); includes 2 minority (1 African American, 1 Asian American or Pacific Islander), 26 international. 184 applicants, 26% accepted. In 1998, 33 master's, 3 doctorates awarded. Terminal master's awarded for partial completion of doctoral program. *Degree requirements:* For master's, computer language required, thesis optional, foreign language not required; for doctorate, computer language, dissertation, comprehensive and final oral exams required, foreign language not required. *Entrance requirements:* For master's and doctorate, TOEFL (minimum score of 550 required), minimum QPA of 3.0. *Average time to degree:* Master's–2 years full-time, 4 years part-time; doctorate–4 years full-time, 6 years part-time. *Application deadline:* For fall admission, 8/1 (priority date); for spring admission, 12/1 (priority date). Applications are processed on a rolling basis. Application fee: $30 ($40 for international students). *Financial aid:* In 1998–99, 29 students received aid, including 11 research assistantships (averaging $10,560 per year), 18 teaching assistantships (averaging $10,656 per year); fellowships, grants, scholarships, and tuition waivers (full and partial) also available. Financial aid application deadline: 2/15. *Faculty research:* Smart materials and structure solid mechanics, computational fluid dynamics, multiphase bio-fluid dynamics, mechanical vibration analysis. Total annual research expenditures: $330,858. *Unit head:* Dr. Fred S. Pettit, Interim Chairman, 412-624-9784, Fax: 412-624-4846, E-mail: pettit@engrng.pitt.edu. *Application contact:* Dr. William W. Clark, Graduate Coordinator, 412-624-9794, Fax: 412-624-4846, E-mail: wclark@engrng.pitt.edu.

University of Portland, Graduate School, Multnomah School of Engineering, Department of Mechanical Engineering, Portland, OR 97203-5798. Offers MSME. Part-time and evening/weekend programs available. *Students:* 2 full-time (0 women), 5 part-time. 3 applicants, 0% accepted. *Degree requirements:* For master's, computer language required, foreign language and thesis not required. *Entrance requirements:* For master's, GRE General Test, TOEFL (minimum score of 550 required), minimum GPA of 3.0. *Application deadline:* For fall admission, 8/1 (priority date); for spring admission, 12/1. Applications are processed on a rolling basis. Application fee: $40. Tuition: Part-time $563 per semester hour. *Financial aid:* Application deadline: 3/15. *Unit head:* Dr. Khalid Khan, Director, 503-943-7276.

University of Puerto Rico, Mayagüez Campus, Graduate Studies, College of Engineering, Department of Mechanical Engineering, Mayagüez, PR 00681-5000. Offers MME, MS. Part-time programs available. *Degree requirements:* For master's, thesis, comprehensive exam required, foreign language not required. *Entrance requirements:* For master's, minimum GPA of 2.5, proficiency in English and Spanish. *Faculty research:* Metallurgy, hybrid vehicles, manufacturing, thermal and fluid sciences, HVAC.

University of Rhode Island, Graduate School, College of Engineering, Department of Mechanical Engineering and Applied Mechanics, Kingston, RI 02881. Offers design/systems (MS, PhD); fluid mechanics (MS, PhD); solid mechanics (MS, PhD); thermal sciences (MS, PhD). *Entrance requirements:* For master's and doctorate, GRE General Test (for foreign applicants only).

Mechanical Engineering

University of Rochester, The College, School of Engineering and Applied Sciences, Department of Mechanical Engineering, Program in Mechanical Engineering, Rochester, NY 14627-0250. Offers MS, PhD. Part-time programs available. *Students:* 33 full-time (2 women), 18 part-time (1 woman); includes 1 minority (Asian American or Pacific Islander), 22 international. 83 applicants, 34% accepted. In 1998, 9 master's, 2 doctorates awarded. Terminal master's awarded for partial completion of doctoral program. *Degree requirements:* For master's, comprehensive exam required, thesis optional, foreign language not required; for doctorate, dissertation, preliminary and qualifying exams required, foreign language not required. *Entrance requirements:* For master's and doctorate, GRE, TOEFL. *Application deadline:* For fall admission, 2/1 (priority date). Application fee: $25. *Financial aid:* Fellowships, research assistantships, teaching assistantships, tuition waivers (full and partial) available. Financial aid application deadline: 2/1. *Application contact:* Donna Derks, Graduate Program Secretary, 716-275-2849.

University of Saskatchewan, College of Graduate Studies and Research, College of Engineering, Department of Mechanical Engineering, Saskatoon, SK S7N 5A2, Canada. Offers M Eng, M Sc, PhD. *Degree requirements:* For master's and doctorate, thesis/dissertation required, foreign language not required. *Entrance requirements:* For master's and doctorate, GRE, TOEFL.

University of South Alabama, Graduate School, College of Engineering, Department of Mechanical Engineering, Mobile, AL 36688-0002. Offers MSME. *Faculty:* 8 full-time (0 women). *Students:* 17 full-time (1 woman), 7 part-time, 14 international. 75 applicants, 63% accepted. In 1998, 5 degrees awarded. *Degree requirements:* For master's, project or thesis required. *Entrance requirements:* For master's, GRE General Test (minimum combined score of 1000 required), BS in engineering, minimum GPA of 3.0. *Application deadline:* For fall admission, 9/1 (priority date). Applications are processed on a rolling basis. Application fee: $25. Tuition, state resident: part-time $116 per semester hour. Tuition, nonresident: part-time $230 per semester hour. Required fees: $121 per semester. Part-time tuition and fees vary according to course load and program. *Financial aid:* In 1998–99, 2 research assistantships were awarded.; career-related internships or fieldwork and institutionally sponsored loans also available. Aid available to part-time students. Financial aid application deadline: 4/1. *Unit head:* Dr. Ali Engin, Chairperson, 334-460-6168. *Application contact:* Dr. Russell M. Hayes, Director of Graduate Studies, 334-460-6117.

University of South Carolina, Graduate School, College of Engineering and Information Technology, Department of Mechanical Engineering, Columbia, SC 29208. Offers ME, MS, PhD. Part-time and evening/weekend programs available. Postbaccalaureate distance learning degree programs offered. *Faculty:* 21 full-time (1 woman). *Students:* 38 full-time (6 women), 42 part-time (7 women); includes 11 minority (3 African Americans, 6 Asian Americans or Pacific Islanders, 2 Hispanic Americans), 29 international. Average age 31. In 1998, 28 master's, 1 doctorate awarded. *Degree requirements:* For master's, thesis required (for some programs), foreign language not required; for doctorate, dissertation required, foreign language not required. *Entrance requirements:* For master's and doctorate, GRE General Test (minimum combined score of 1100 required), TOEFL (minimum score of 525 required). *Application deadline:* For fall admission, 3/1 (priority date); for spring admission, 11/1. Applications are processed on a rolling basis. Application fee: $35. Electronic applications accepted. Tuition, state resident: full-time $4,014; part-time $202 per credit hour. Tuition, nonresident: full-time $8,528; part-time $428 per credit hour. Required fees: $100; $4 per credit hour. Tuition and fees vary according to program. *Financial aid:* In 1998–99, 14 research assistantships with partial tuition reimbursements, 21 teaching assistantships with partial tuition reimbursements were awarded.; fellowships, career-related internships or fieldwork also available. *Faculty research:* Heat exchangers, computer vision measurements in solid mechanics and biomechanics, robot dynamics and control. Total annual research expenditures: $1.2 million. *Unit head:* Dr. Abdel Moez Bayoumi, Chair, 803-777-4185, Fax: 803-777-0106, E-mail: ambayoum@engr.sc.edu. *Application contact:* Everose Alexander, Department Secretary, 803-777-4185, Fax: 803-777-0106, E-mail: alexande@eng.sc.edu.

University of Southern California, Graduate School, School of Engineering, Department of Mechanical Engineering, Los Angeles, CA 90089. Offers MS, PhD, Engr. *Faculty:* 12 full-time, 1 part-time. *Students:* 26 full-time (3 women), 32 part-time (3 women); includes 16 minority (2 African Americans, 10 Asian Americans or Pacific Islanders, 4 Hispanic Americans), 29 international. Average age 30. 225 applicants, 54% accepted. In 1998, 20 master's, 1 doctorate awarded. *Degree requirements:* For master's, thesis optional; for doctorate, dissertation required. *Entrance requirements:* For master's, doctorate, and Engr, GRE General Test. *Application deadline:* For fall admission, 2/1 (priority date); for spring admission, 2/1. Application fee: $55. Tuition: Part-time $768 per unit. Required fees: $350 per semester. *Financial aid:* In 1998–99, 4 fellowships, 16 research assistantships, 8 teaching assistantships were awarded.; Federal Work-Study and institutionally-sponsored loans also available. Aid available to part-time students. Financial aid application deadline: 2/15; financial aid applicants required to submit FAFSA. *Unit head:* Dr. Richard Kaplan, Acting Chairman, 213-743-0484.

See in-depth description on page 1471.

University of South Florida, Graduate School, College of Engineering, Department of Mechanical Engineering, Tampa, FL 33620-9951. Offers MME, MSE, MSME, PhD. Part-time programs available. *Faculty:* 9 full-time (0 women). *Students:* 14 full-time (1 woman), 19 part-time (2 women); includes 8 minority (3 Asian Americans or Pacific Islanders, 4 Hispanic Americans, 1 Native American), 10 international. Average age 27. 82 applicants, 93% accepted. In 1998, 12 master's awarded (75% found work related to degree, 25% continued full-time study); 3 doctorates awarded (67% entered university research/teaching, 33% found other work related to degree). Terminal master's awarded for partial completion of doctoral program. *Degree requirements:* For master's, thesis or alternative required, foreign language not required; for doctorate, dissertation, 2 tools of research as specified by dissertation committee required, foreign language not required. *Entrance requirements:* For master's, GRE General Test (minimum combined score of 1100 required), minimum GPA of 3.0 during previous 2 years; for doctorate, GRE General Test (minimum combined score of 1200 required). *Average time to degree:* Master's–2.5 years full-time, 4.5 years part-time; doctorate–4 years full-time, 7 years part-time. *Application deadline:* For fall admission, 6/1; for spring admission, 10/15. Application fee: $20. Electronic applications accepted. Tuition, state resident: part-time $148 per credit hour. Tuition, nonresident: part-time $509 per credit hour. *Financial aid:* In 1998–99, 18 students received aid, including 1 fellowship with full tuition reimbursement available (averaging $7,000 per year), 13 research assistantships with full tuition reimbursements available (averaging $10,320 per year), 4 teaching assistantships with full tuition reimbursements available (averaging $7,500 per year); career-related internships or fieldwork, Federal Work-Study, institutionally-sponsored loans, and tuition waivers (partial) also available. Financial aid applicants required to submit FAFSA. *Faculty research:* Aerodynamics, heat transfer, energy systems for space applications, building energy systems, robot sensors. Total annual research expenditures: $172,985. *Unit head:* Dr. Rajiu Dubey, Chairperson, 813-974-2280, Fax: 813-974-3539, E-mail: dubey@eng.usf.edu. *Application contact:* Dr. Muhammad M. Rahman, Graduate Coordinator, 813-974-5625, Fax: 813-974-3539, E-mail: rahman@eng.usf.edu.

University of Southwestern Louisiana, Graduate School, College of Engineering, Department of Mechanical Engineering, Lafayette, LA 70504. Offers MSE. Evening/weekend programs available. *Faculty:* 6 full-time (0 women). *Students:* 18 full-time (2 women), 2 part-time, 18 international. Average age 25. 97 applicants, 69% accepted. In 1998, 4 degrees awarded. *Degree requirements:* For master's, thesis or alternative, comprehensive exam required, foreign language not required. *Entrance requirements:* For master's, GRE General Test, BS in mechanical engineering, minimum GPA of 2.85. *Application deadline:* For fall admission, 5/15. Application fee: $5 ($15 for international students). *Financial aid:* In 1998–99, 4 research assistantships (averaging $4,500 per year) were awarded.; Federal Work-Study and tuition waivers (full and partial) also available. Financial aid application deadline: 5/1. *Faculty research:* CAD/CAM, machine design and vibration, thermal science. *Unit head:* Dr. William E. Simon, Head, 318-482-6517. *Application contact:* Dr. M. A. Elsayed, Graduate Coordinator, 318-482-5363.

University of Tennessee, Knoxville, Graduate School, College of Engineering, Department of Mechanical and Aerospace Engineering and Engineering Science, Program in Mechanical Engineering, Knoxville, TN 37996. Offers MS, PhD. Part-time programs available. *Students:* 46 full-time (6 women), 28 part-time (2 women); includes 4 minority (2 African Americans, 2 Asian Americans or Pacific Islanders), 15 international. 96 applicants, 41% accepted. In 1998, 13 master's, 5 doctorates awarded. *Degree requirements:* For master's, thesis or alternative required, foreign language not required; for doctorate, dissertation required, foreign language not required. *Entrance requirements:* For master's and doctorate, TOEFL (minimum score of 550 required), minimum GPA of 2.7. *Application deadline:* For fall admission, 2/1 (priority date). Applications are processed on a rolling basis. Application fee: $35. Electronic applications accepted. *Financial aid:* Application deadline: 2/1; *Application contact:* Dr. Allen Yu, Graduate Representative, 923-974-4159, E-mail: nyu@utk.edu.

University of Tennessee Space Institute, Graduate Programs, Program in Mechanical Engineering, Tullahoma, TN 37388-9700. Offers MS, PhD. Part-time programs available. *Faculty:* 3 full-time (0 women), 1 part-time (0 women). *Students:* 19 full-time (3 women), 17 part-time (2 women); includes 2 minority (both African Americans), 9 international. 39 applicants, 64% accepted. In 1998, 4 master's, 3 doctorates awarded. Terminal master's awarded for partial completion of doctoral program. *Degree requirements:* For master's, thesis required (for some programs), foreign language not required; for doctorate, dissertation required. *Entrance requirements:* For master's and doctorate, GRE General Test. *Application deadline:* Applications are processed on a rolling basis. Application fee: $35. *Financial aid:* Fellowships, research assistantships, Federal Work-Study available. Financial aid applicants required to submit FAFSA. *Unit head:* Dr. Ahmad Vakili, Degree Program Chairman, 931-393-7483, Fax: 931-393-7530, E-mail: avakili@utsi.edu. *Application contact:* Dr. Edwin M. Gleason, Assistant Dean for Admissions and Student Affairs, 931-393-7432, Fax: 931-393-7346, E-mail: egleason@utsi.edu.

The University of Texas at Arlington, Graduate School, College of Engineering, Department of Mechanical and Aerospace Engineering, Program in Mechanical Engineering, Arlington, TX 76019. Offers M Engr, MS, PhD. *Students:* 36 full-time (2 women), 45 part-time (4 women); includes 14 minority (2 African Americans, 7 Asian Americans or Pacific Islanders, 5 Hispanic Americans), 30 international. 143 applicants, 45% accepted. In 1998, 22 master's, 4 doctorates awarded. *Degree requirements:* For master's, thesis required (for some programs), foreign language not required; for doctorate, dissertation required, foreign language not required. *Entrance requirements:* For master's and doctorate, GRE General Test, TOEFL. *Application deadline:* Applications are processed on a rolling basis. Application fee: $25 ($50 for international students). Tuition, state resident: full-time $1,368; part-time $76 per semester hour. Tuition, nonresident: full-time $5,454; part-time $303 per semester hour. Required fees: $66 per semester hour. $86 per term. Tuition and fees vary according to course load. *Financial aid:* Research assistantships, teaching assistantships available. *Application contact:* Dr. Wen S. Chan, Graduate Adviser, 817-272-2561, Fax: 817-272-2538, E-mail: chan@mae.uta.edu.

The University of Texas at Austin, Graduate School, College of Engineering, Department of Mechanical Engineering, Austin, TX 78712-1111. Offers mechanical engineering (MSE, PhD); operations research and industrial engineering (MSE, PhD). *Students:* 185 full-time (18 women), 47 part-time (6 women); includes 32 minority (5 African Americans, 12 Asian Americans or Pacific Islanders, 14 Hispanic Americans, 1 Native American), 89 international. 344 applicants, 37% accepted. In 1998, 48 master's, 13 doctorates awarded. *Entrance requirements:* For master's, GRE General Test (minimum combined score of 1000 required), TOEFL (minimum score of 580 required); for doctorate, GRE General Test. Application fee: $50 ($75 for international students). *Financial aid:* Fellowships, research assistantships, teaching assistantships available. Financial aid application deadline: 2/1. *Unit head:* Dr. J. Parker Lamb, Acting Chairman, 512-471-1131. *Application contact:* Dr. Ronald E. Barr, Graduate Adviser, 512-471-7571.

The University of Texas at El Paso, Graduate School, College of Engineering, Department of Mechanical and Industrial Engineering, Program in Mechanical Engineering, El Paso, TX 79968-0001. Offers MS. Part-time and evening/weekend programs available. *Faculty:* 18 full-time (2 women), 3 part-time (1 woman). 30 applicants, 60% accepted. In 1998, 3 degrees awarded. *Degree requirements:* For master's, thesis required, foreign language not required. *Entrance requirements:* For master's, GRE General Test, TOEFL (minimum score of 550 required), minimum GPA of 3.0 in major. *Application deadline:* Applications are processed on a rolling basis. Application fee: $15 ($65 for international students). Tuition, state resident: full-time $2,790. Tuition, nonresident: full-time $7,710. *Financial aid:* Fellowships, research assistantships, teaching assistantships, Federal Work-Study, institutionally-sponsored loans, and tuition waivers (partial) available. Financial aid applicants required to submit FAFSA. *Faculty research:* Solar/wind energy, environment machine design. Total annual research expenditures: $381,314. *Unit head:* Dr. Rick Zadoks, Graduate Adviser, 915-747-5450, Fax: 915-747-5019. *Application contact:* Susan Jordan, Director, Graduate Student Services, 915-747-5491, Fax: 915-747-5788, E-mail: sjordan@utep.edu.

The University of Texas at San Antonio, College of Sciences and Engineering, Division of Engineering, San Antonio, TX 78249-0617. Offers civil engineering (MS); electrical engineering (MS); mechanical engineering (MS). Part-time and evening/weekend programs available. *Faculty:* 22 full-time (2 women), 15 part-time (2 women). *Students:* 19 full-time (1 woman), 90 part-time (15 women); includes 38 minority (5 African Americans, 11 Asian Americans or Pacific Islanders, 21 Hispanic Americans, 1 Native American), 21 international. *Degree requirements:* For master's, thesis optional, foreign language not required. *Entrance requirements:* For master's, GRE General Test. *Application deadline:* For fall admission, 7/1; for spring admission, 12/1. Applications are processed on a rolling basis. Application fee: $25. *Unit head:* Dr. Lex Akers, Director, 210-458-4490.

University of Toledo, Graduate School, College of Engineering, Department of Mechanical, Industrial, and Manufacturing Engineering, Toledo, OH 43606-3398. Offers engineering sciences (PhD); industrial engineering (MSIE); mechanical engineering (MSME). Part-time programs available. Postbaccalaureate distance learning degree programs offered (minimal on-campus study). *Faculty:* 23 full-time (1 woman), 2 part-time (0 women). *Students:* 86 full-time (12 women), 54 part-time (11 women); includes 2 minority (both African Americans), 92 international. Average age 27. 517 applicants, 24% accepted. In 1998, 66 master's, 5 doctorates awarded. *Degree requirements:* For master's, thesis optional, foreign language not required; for doctorate, dissertation required, foreign language not required. *Entrance requirements:* For master's, GRE General Test (minimum score of 350 on verbal section, 700 on quantitative required), TOEFL (minimum score of 550 required), minimum GPA of 2.7; for doctorate, GRE General Test (minimum score of 350 on verbal section, 700 on quantitative required), TOEFL (minimum score of 550 required), minimum GPA of 3.3. *Average time to degree:* Master's–2 years full-time; doctorate–4 years full-time. *Application deadline:* For fall admission, 5/31 (priority date). Applications are processed on a rolling basis. Application fee: $30. Electronic applications accepted. *Financial aid:* In 1998–99, 97 students received aid, including 1 fellowship with full tuition reimbursement available, 32 research assistantships with full tuition reimbursements available, 15 teaching assistantships with full tuition reimbursements available; Federal Work-Study, scholarships, tuition waivers (full), and unspecified assistantships also available. Financial aid application deadline: 4/1. *Faculty research:* Computational and experimental thermal sciences, manufacturing process and systems, mechanics, materials, design, quality and management engineering systems. Total annual research expenditures: $1.9 million. *Unit head:* Dr. Nagi G. Naganathan, Chair, 419-530-8210, Fax: 419-530-8214, E-mail: nagi.naganathan@utoledo.edu. *Application contact:* Dr. Abdollah A. Afjeh, Director of Graduate Program, 419-530-8208, Fax: 419-530-8206, E-mail: aafjeh@uoft02.utoledo.edu.

University of Toronto, School of Graduate Studies, Physical Sciences Division, Faculty of Applied Science and Engineering, Department of Mechanical and Industrial Engineering,

Toronto, ON M5S 1A1, Canada. Offers M Eng, MA Sc, PhD. Part-time programs available. *Degree requirements:* For master's, thesis required (for some programs), foreign language not required; for doctorate, dissertation required, foreign language not required.

University of Tulsa, Graduate School, College of Business Administration, Department of Engineering and Technology Management, Tulsa, OK 74104-3189. Offers chemical engineering (METM); computer science (METM); electrical engineering (METM); geological science (METM); mathematics (METM); mechanical engineering (METM); petroleum engineering (METM). Part-time and evening/weekend programs available. *Students:* 3 full-time (1 woman), 1 part-time, 3 international. *Degree requirements:* For master's, foreign language and thesis not required. *Entrance requirements:* For master's, GRE General Test (minimum score of 430 on verbal section, 600 on quantitative required), TOEFL (minimum score of 575 required). *Application deadline:* Applications are processed on a rolling basis. Application fee: $30. Electronic applications accepted. Tuition: Full-time $8,640; part-time $480 per hour. Required fees: $3 per hour. One-time fee: $200 full-time. Tuition and fees vary according to program. *Unit head:* Dr. Richard C. Burgess, Assistant Dean/Director of Graduate Business Studies, 918-631-2242, Fax: 918-631-2142.

University of Tulsa, Graduate School, College of Engineering and Applied Sciences, Department of Mechanical Engineering, Tulsa, OK 74104-3189. Offers ME, MSE, PhD. Part-time programs available. *Faculty:* 8 full-time (0 women). *Students:* 13 full-time (2 women), 6 part-time (1 woman); includes 3 minority (2 Asian Americans or Pacific Islanders, 1 Native American), 6 international. Average age 27. 24 applicants, 71% accepted. In 1998, 6 master's, 4 doctorates awarded. *Degree requirements:* For master's, computer language, thesis (MSE) required; for doctorate, computer language, dissertation required, foreign language not required. *Entrance requirements:* For master's, GRE General Test (minimum score of 600 on quantitative section required), TOEFL (minimum score of 550 required); for doctorate, GRE General Test (minimum score of 700 on quantitative section required, 1100 combined), TOEFL (minimum score of 550 required). *Application deadline:* Applications are processed on a rolling basis. Application fee: $30. Electronic applications accepted. Tuition: Full-time $8,640; part-time $480 per hour. Required fees: $3 per hour. One-time fee: $200 full-time. Tuition and fees vary according to program. *Financial aid:* In 1998–99, 18 students received aid, including 1 fellowship (averaging $9,333 per year), 8 research assistantships (averaging $4,171 per year), 9 teaching assistantships (averaging $5,353 per year); career-related internships or fieldwork, Federal Work-Study, and tuition waivers (partial) also available. Aid available to part-time students. Financial aid application deadline: 2/1; financial aid applicants required to submit FAFSA. *Faculty research:* Erosion and corrosion, solid mechanics, composite material, fatigue analysis, CAD/CAM, manufacturing. *Unit head:* Dr. Edmund F. Rybicki, Chairperson, 918-631-2521. *Application contact:* Dr. Siamack A. Shirazi, Adviser, 918-631-3001, Fax: 918-631-2397.

University of Utah, Graduate School, College of Engineering, Department of Mechanical Engineering, Program in Mechanical Engineering, Salt Lake City, UT 84112-1107. Offers M Phil, ME, MS, PhD. *Degree requirements:* For master's, special project (ME), thesis (MS) required; for doctorate, dissertation, comprehensive exam required, foreign language not required. *Entrance requirements:* For master's, GRE General Test, TOEFL (minimum score of 530 required), minimum GPA of 3.0; for doctorate, GRE, TOEFL (minimum score of 530 required), minimum GPA of 3.0. *Application deadline:* For fall admission, 7/1. Application fee: $30 ($50 for international students). *Application contact:* Dr. Patrick McMurtry, Director of Graduate Studies, 801-581-3889.

See in-depth description on page 1473.

University of Vermont, Graduate College, College of Engineering and Mathematics, Department of Mechanical Engineering, Burlington, VT 05405-0160. Offers MS, PhD. *Degree requirements:* For master's and doctorate, thesis/dissertation required, foreign language not required. *Entrance requirements:* For master's and doctorate, GRE General Test, TOEFL (minimum score of 550 required).

University of Victoria, Faculty of Graduate Studies, Faculty of Engineering, Department of Mechanical Engineering, Victoria, BC V8W 2Y2, Canada. Offers M Eng, MA Sc, PhD. Part-time programs available. *Faculty:* 19 full-time (2 women). *Students:* 42 full-time (2 women), 1 part-time; includes 9 minority (7 Asian Americans or Pacific Islanders, 2 Hispanic Americans), 3 international. Average age 27. 50 applicants, 14% accepted. In 1998, 9 master's, 1 doctorate awarded. *Degree requirements:* For master's, thesis required (for some programs), foreign language not required; for doctorate, dissertation, candidacy exam required, foreign language not required. *Entrance requirements:* For master's, minimum B average in undergraduate course work. *Average time to degree:* Master's–2.62 years full-time; doctorate–5.6 years full-time. *Application deadline:* For fall admission, 5/31 (priority date). Applications are processed on a rolling basis. Application fee: $50. *Financial aid:* In 1998–99, 7 fellowships, 36 research assistantships were awarded.; teaching assistantships, career-related internships or fieldwork and awards also available. Financial aid application deadline: 2/15. *Faculty research:* CAD/CAM, energy systems, cryofuels, fuel cell technology, computational mechanics. *Unit head:* Dr. S. Dost, Chair, 250-721-8900, Fax: 250-721-6051. *Application contact:* Dr. Inna Sharf, Graduate Adviser, 250-721-8291, Fax: 250-721-6051, E-mail: gradstudies@me.uvic.ca.

University of Virginia, School of Engineering and Applied Science, Department of Mechanical and Aerospace Engineering, Charlottesville, VA 22903. Offers ME, MS, PhD. *Faculty:* 30 full-time (2 women). *Students:* 65 full-time (4 women), 81 part-time (8 women); includes 8 minority (1 African American, 5 Asian Americans or Pacific Islanders, 2 Hispanic Americans), 17 international. Average age 27. 76 applicants, 18% accepted. In 1998, 25 master's, 4 doctorates awarded. *Degree requirements:* For master's, thesis (MS) required; for doctorate, dissertation, comprehensive exam required, foreign language not required. *Entrance requirements:* For master's, GRE General Test; for doctorate, GRE General Test (minimum score of 700 on quantitative section required), TOEFL (minimum score of 600 required). *Application deadline:* For fall admission, 8/1 (priority date). Applications are processed on a rolling basis. Application fee: $60. *Financial aid:* Fellowships, research assistantships, teaching assistantships available. Financial aid application deadline: 2/1. *Unit head:* Dr. Jeffrey Morton, Graduate Director, 804-924-6224, E-mail: jbm@virginia.edu. *Application contact:* J. Milton Adams, Assistant Dean, 804-924-3897, E-mail: twr2c@virginia.edu.

See in-depth description on page 1475.

University of Washington, Graduate School, College of Engineering, Department of Mechanical Engineering, Seattle, WA 98195. Offers MSE, MSME, PhD. Part-time programs available. *Faculty:* 28 full-time (1 woman). *Students:* 143 full-time (20 women), 34 part-time (1 woman); includes 30 minority (1 African American, 23 Asian Americans or Pacific Islanders, 4 Hispanic Americans, 2 Native Americans), 36 international. Average age 26. 190 applicants, 59% accepted. In 1998, 49 master's awarded (80% found work related to degree, 20% continued full-time study); 11 doctorates awarded (20% entered university research/teaching, 80% found other work related to degree). *Degree requirements:* For master's, thesis required (for some programs), foreign language not required; for doctorate, one foreign language, dissertation required. *Entrance requirements:* For master's, GRE General Test (minimum combined score of 1800 on three sections preferred), TOEFL (minimum score of 500 required), minimum GPA of 3.2 in last 2 years; for doctorate, GRE General Test (minimum combined score of 1800 on three sections preferred), TOEFL (minimum score of 500 required), minimum GPA of 3.5. *Average time to degree:* Master's–1.5 years full-time, 4 years part-time; doctorate–4.5 years full-time. *Application deadline:* For fall admission, 3/1 (priority date); for winter admission, 11/1; for spring admission, 2/1. Applications are processed on a rolling basis. Application fee: $50. Tuition, state resident: full-time $5,196; part-time $475 per credit. Tuition, nonresident: full-time $13,485; part-time $1,285 per credit. Required fees: $387; $38 per credit. Tuition and fees vary according to course load. *Financial aid:* In 1998–99, 6 fellowships, 48 research assistantships, 27 teaching assistantships were awarded. Financial aid application deadline: 2/1; financial aid applicants required to submit FAFSA. *Faculty research:* Manufacturing, automation, and controls; fluid and solid mechanics; heat transfer and combustion; fracture mechanics and composites. *Unit head:* Dr. William R. D. Wilson, Chair, 206-543-5090, Fax: 206-685-8047. *Application contact:* Graduate Secretary, 206-543-7963, Fax: 206-685-8047.

University of Waterloo, Graduate Studies, Faculty of Engineering, Department of Mechanical Engineering, Waterloo, ON N2L 3G1, Canada. Offers MA Sc, PhD. Part-time and evening/weekend programs available. *Faculty:* 32 full-time (2 women), 13 part-time (0 women). *Students:* 70 full-time (7 women), 21 part-time (1 woman). 78 applicants, 42% accepted. In 1998, 21 master's, 12 doctorates awarded. *Degree requirements:* For master's, computer language, research paper or thesis required; for doctorate, computer language, dissertation, comprehensive exam required, foreign language not required. *Entrance requirements:* For master's, TOEFL (minimum score of 550 required), TWE (minimum score of 4.0 required), honors degree, minimum B average; for doctorate, TOEFL (minimum score of 550 required), TWE (minimum score of 4.0 required), master's degree, minimum A- average. *Average time to degree:* Master's–2 years full-time, 4 years part-time; doctorate–4 years full-time, 5 years part-time. *Application deadline:* Applications are processed on a rolling basis. Application fee: $50. *Expenses:* Tuition and fees charges are reported in Canadian dollars. Tuition, state resident: full-time $3,168 Canadian dollars; part-time $792 Canadian dollars per term. Tuition, nonresident: full-time $8,000 Canadian dollars; part-time $2,000 Canadian dollars. Required fees: $45 Canadian dollars per term. Tuition and fees vary according to program. *Financial aid:* Research assistantships, teaching assistantships, institutionally-sponsored loans available. *Faculty research:* Fluid mechanics, thermal engineering, solid mechanics, automation and control, materials engineering. *Unit head:* Dr. R. Pick, Chair, 519-888-4567 Ext. 6740. *Application contact:* Dr. K. G. T. Hollands, Associate Chair, Graduate Studies, 519-888-4567 Ext. 3341.

University of Windsor, College of Graduate Studies and Research, Faculty of Engineering, Mechanical Engineering and Materials Engineering, Windsor, ON N9B 3P4, Canada. Offers engineering materials (MA Sc, PhD); mechanical engineering (MA Sc, PhD). Part-time programs available. *Degree requirements:* For master's and doctorate, thesis/dissertation required, foreign language not required. *Entrance requirements:* For master's, TOEFL (minimum score of 600 required), minimum B average; for doctorate, TOEFL, master's degree. *Faculty research:* Thermo-fluids, applied mechanics.

University of Wisconsin–Madison, Graduate School, College of Engineering, Department of Mechanical Engineering, Madison, WI 53706-1380. Offers mechanical engineering (MS, PhD); polymers (ME). Part-time programs available. Postbaccalaureate distance learning degree programs offered (no on-campus study). *Faculty:* 36 full-time (4 women), 4 part-time (0 women). *Students:* 191 full-time (7 women), 20 part-time, 94 international. Average age 23. 217 applicants, 51% accepted. In 1998, 48 master's, 24 doctorates awarded. Terminal master's awarded for partial completion of doctoral program. *Degree requirements:* For master's, thesis optional, foreign language not required; for doctorate, dissertation, qualifying exam, preliminary exam required, foreign language not required. *Entrance requirements:* For master's and doctorate, GRE, TOEFL, BS in mechanical engineering or related field, mininum GPA of 3.0 in last 60 hours. *Average time to degree:* Master's–1.5 years full-time, 3 years part-time; doctorate–3 years full-time, 6 years part-time. *Application deadline:* Applications are processed on a rolling basis. Application fee: $45. Electronic applications accepted. *Financial aid:* In 1998–99, 3 fellowships with tuition reimbursements, 99 research assistantships with tuition reimbursements, 24 teaching assistantships with tuition reimbursements were awarded.; institutionally-sponsored loans and scholarships also available. *Faculty research:* Dynamics, vibrations and acoustics, mechatronics, robotics and automation, design and manufacturing, materials processing, solar energy, combustion, fluids. Total annual research expenditures: $9.4 million. *Unit head:* Kenneth W. Ragland, Chair, 608-265-3576, Fax: 608-265-2316, E-mail: ragland@engr.wisc.edu. *Application contact:* Linda S. Aaberg, Student Status Examiner, 608-262-0666, Fax: 608-265-2316, E-mail: aaberg@engr.wisc.edu.

See in-depth description on page 1477.

University of Wyoming, Graduate School, College of Engineering, Department of Mechanical Engineering, Laramie, WY 82071. Offers MS, PhD. *Faculty:* 11 full-time (0 women), 1 (woman) part-time. *Students:* 16 full-time (2 women), 15 part-time (1 woman); includes 1 minority (Native American), 10 international. 27 applicants, 26% accepted. In 1998, 4 master's awarded. Terminal master's awarded for partial completion of doctoral program. *Degree requirements:* For master's, thesis required; for doctorate, dissertation required, foreign language not required. *Entrance requirements:* For master's and doctorate, GRE General Test, TOEFL, minimum GPA of 3.0. *Average time to degree:* Master's–1.8 years full-time, 3 years part-time; doctorate–4 years full-time. *Application deadline:* For fall admission, 6/1 (priority date). Applications are processed on a rolling basis. Application fee: $40. Tuition, state resident: full-time $2,520; part-time $140 per credit hour. Tuition, nonresident: full-time $7,790; part-time $433 per credit hour. Required fees: $400; $7 per credit hour. Full-time tuition and fees vary according to course load and program. *Financial aid:* In 1998–99, 10 research assistantships, 9 teaching assistantships were awarded.; career-related internships or fieldwork also available. Financial aid application deadline: 3/1. *Faculty research:* Composite materials, thermal fluid sciences, continuum mechanics. *Unit head:* Dr. William R. Lindberg, Head, 307-766-2122, Fax: 307-766-2695, E-mail: lindberg@uwyo.edu. *Application contact:* Brandy Gore, Assistant Coordinator, 307-766-2122, Fax: 307-766-2695, E-mail: thegore@uwyo.edu.

Utah State University, School of Graduate Studies, College of Engineering, Department of Mechanical and Aerospace Engineering, Logan, UT 84322. Offers aerospace engineering (MS, PhD); mechanical engineering (ME, MS, PhD). *Faculty:* 14 full-time (1 woman). *Students:* 31 full-time (2 women), 7 part-time; includes 1 minority (Asian American or Pacific Islander), 12 international. Average age 27. 74 applicants, 58% accepted. In 1998, 13 master's, 2 doctorates awarded. Terminal master's awarded for partial completion of doctoral program. *Degree requirements:* For master's, computer language, thesis required (for some programs), foreign language not required; for doctorate, computer language, dissertation required, foreign language not required. *Entrance requirements:* For master's, GRE General Test (score in 40th percentile or higher required), TOEFL (minimum score of 550 required), minimum GPA of 3.0; for doctorate, GRE General Test (score in 40th percentile or higher required), GRE Subject Test, TOEFL (minimum score of 550 required), minimum GPA of 3.0. *Application deadline:* For fall admission, 3/15 (priority date); for spring admission, 10/15. Applications are processed on a rolling basis. Application fee: $40. Tuition, state resident: full-time $1,492. Tuition, nonresident: full-time $5,232. Required fees: $434. Tuition and fees vary according to course load. *Financial aid:* In 1998–99, 20 students received aid, including 14 research assistantships with partial tuition reimbursements available (averaging $12,000 per year), 5 teaching assistantships with partial tuition reimbursements available (averaging $9,000 per year); fellowships with partial tuition reimbursements available, Federal Work-Study and institutionally-sponsored loans also available. Financial aid application deadline: 3/15. *Faculty research:* In-space instruments, cryogenic cooling, thermal science, space structures, composite materials. *Unit head:* J. Clair Batty, Head, 435-797-2868, Fax: 435-797-2417. *Application contact:* Joan P. Smith, Graduate Student Adviser, 435-797-0330, Fax: 435-797-2417, E-mail: jpsmith@mae.usu.edu.

Vanderbilt University, School of Engineering, Department of Mechanical Engineering, Nashville, TN 37240-1001. Offers M Eng, MS, PhD. MS and PhD offered through the Graduate School. Part-time programs available. *Faculty:* 16 full-time (1 woman), 7 part-time (0 women). *Students:* 50 full-time (6 women); includes 1 minority (African American), 18 international. Average age 24. 39 applicants, 28% accepted. In 1998, 14 master's awarded (100% found work related to degree); 8 doctorates awarded (100% found work related to degree). Terminal master's awarded for partial completion of doctoral program. *Degree requirements:* For master's and doctorate, thesis/dissertation required, foreign language not required. *Entrance requirements:* For master's and doctorate, GRE General Test. *Application deadline:* For fall admission, 1/15; for spring admission, 11/1. Application fee: $40. Electronic applications accepted. *Financial aid:* In 1998–99, 38 students received aid, including 11 fellowships with full tuition reimbursements available (averaging $16,000 per year), 15 research assistantships with full tuition reimbursements available (averaging $14,000 per year), 9 teaching assistantships with full

Mechanical Engineering

Vanderbilt University (continued)
tuition reimbursements available (averaging $13,200 per year); institutionally-sponsored loans and tuition waivers (full) also available. Aid available to part-time students. Financial aid application deadline: 1/15. *Faculty research:* Laser diagnostics, finite element analysis, rolling contact fracture, combustion, computer modeling, smart structures. Total annual research expenditures: $1.7 million. *Unit head:* Robert W. Pitz, Chair, 615-322-2413, Fax: 615-343-6687, E-mail: pitzrw@vuse.vanderbilt.edu. *Application contact:* Arthur M. Mellor, Director of Graduate Studies, 615-343-6214, Fax: 615-343-6687, E-mail: puccini@vuse.vanderbilt.edu.

Villanova University, College of Engineering, Department of Mechanical Engineering, Villanova, PA 19085-1699. Offers manufacturing (Certificate); mechanical engineering (MME); virtual manufacturing (Certificate). Part-time and evening/weekend programs available. *Faculty:* 12 full-time (0 women), 2 part-time (0 women). *Students:* 7 full-time (0 women), 14 part-time (2 women); includes 1 minority (Asian American or Pacific Islander), 5 international. Average age 27. 35 applicants, 77% accepted. In 1998, 13 degrees awarded (77% found work related to degree, 23% continued full-time study). *Degree requirements:* For master's, computer language required, thesis optional, foreign language not required. *Entrance requirements:* For master's, GRE General Test (for applicants with degrees from foreign universities), TOEFL (minimum score of 575 required), BME, minimum GPA of 3.0. *Application deadline:* For fall admission, 8/1 (priority date); for spring admission, 12/1. Applications are processed on a rolling basis. Application fee: $40. *Financial aid:* In 1998–99, 8 students received aid, including 1 research assistantship with full tuition reimbursement available (averaging $9,215 per year), 6 teaching assistantships with full tuition reimbursements available (averaging $9,215 per year); Federal Work-Study and tuition waivers (full) also available. Financial aid application deadline: 3/15. *Faculty research:* Composite materials, power plant systems, fluid mechanics, automated manufacturing, dynamic analysis. Total annual research expenditures: $65,000. *Unit head:* Dr. Alan M. Whitman, Chairperson, 610-519-4980, E-mail: awhitman@email.vill.edu.

See in-depth description on page 1479.

Virginia Polytechnic Institute and State University, Graduate School, College of Engineering, Department of Mechanical Engineering, Blacksburg, VA 24061. Offers M Eng, MS, PhD. *Faculty:* 41 full-time (0 women). *Students:* 134 full-time (17 women), 32 part-time (1 woman); includes 20 minority (7 African Americans, 3 Asian Americans or Pacific Islanders, 8 Hispanic Americans, 2 Native Americans), 62 international. 291 applicants, 42% accepted. In 1998, 47 master's, 13 doctorates awarded. *Degree requirements:* For master's, computer language, thesis (MS) required; for doctorate, computer language, dissertation required, foreign language not required. *Entrance requirements:* For master's, GRE General Test (minimum score of 600 on verbal section, 600 on quantitative, 480 on analytical required), TOEFL (minimum score of 600 required); for doctorate, GRE General Test (minimum score of 600 on verbal section, 600 on quantitative, 450 on analytical required), TOEFL (minimum score of 500 required). *Application deadline:* For fall admission, 12/1 (priority date). Applications are processed on a rolling basis. Application fee: $25. *Financial aid:* In 1998–99, 22 research assistantships, 11 teaching assistantships were awarded.; fellowships, career-related internships or fieldwork and unspecified assistantships also available. Financial aid application deadline: 4/1. *Faculty research:* Turbomachinery, CAD/CAM, thermofluid sciences, controls, mechanical system dynamics. *Unit head:* Dr. Walter O'Brien, Head, 540-231-6661, E-mail: walto@vt.edu.

Washington State University, Graduate School, College of Engineering and Architecture, School of Mechanical and Materials Engineering, Program in Mechanical Engineering, Pullman, WA 99164. Offers MS, PhD. *Faculty:* 24 full-time (2 women). *Students:* 32 full-time (3 women), 6 part-time (1 woman); includes 2 minority (both Asian Americans or Pacific Islanders), 22 international. Average age 25. In 1998, 15 master's, 1 doctorate awarded (100% found work related to degree). *Degree requirements:* For master's, oral exam required, thesis optional, foreign language not required; for doctorate, dissertation, oral exam required, foreign language not required. *Entrance requirements:* For master's and doctorate, GRE General Test, minimum GPA of 3.0. *Average time to degree:* Master's–2 years full-time; doctorate–4 years full-time. *Application deadline:* For fall admission, 3/1 (priority date). Applications are processed on a rolling basis. Application fee: $35. *Financial aid:* In 1998–99, 1 fellowship, 12 research assistantships, 19 teaching assistantships were awarded.; career-related internships or fieldwork, Federal Work-Study, institutionally-sponsored loans, and teaching associateships also available. Financial aid application deadline: 4/1; financial aid applicants required to submit FAFSA. *Faculty research:* Thermal fluids and heat transfer, combustion, solid mechanics virtual reality as applied to advanced manufacturing, advanced manufacturing and rapid prototyping. Total annual research expenditures: $1.3 million. *Application contact:* Lanni MacKenzie, Coordinator, 509-335-2727, Fax: 509-335-4662, E-mail: gradapp@mme.wsu.edu.

Washington University in St. Louis, School of Engineering and Applied Science, Sever Institute of Technology, Department of Mechanical Engineering, St. Louis, MO 63130-4899. Offers MS, D Sc. Part-time programs available. *Faculty:* 16 full-time (2 women), 12 part-time (0 women). *Students:* 16 full-time (4 women), 34 part-time (5 women); includes 1 minority (Asian American or Pacific Islander), 8 international. 120 applicants, 27% accepted. In 1998, 12 master's awarded. Terminal master's awarded for partial completion of doctoral program. *Degree requirements:* For master's, thesis optional, foreign language not required; for doctorate, dissertation required, foreign language not required. *Entrance requirements:* For doctorate, departmental qualifying exam. *Application deadline:* Applications are processed on a rolling basis. Application fee: $20. *Financial aid:* In 1998–99, 16 students received aid, including 12 research assistantships with full tuition reimbursements available, 4 teaching assistantships with full tuition reimbursements available *Faculty research:* Materials, combustion, flow control, biomechanics, nonlinear dynamics. *Unit head:* Dr. David A. Peters, Chairman, 314-935-4337, Fax: 314-935-4014, E-mail: dap@mecf.wustl.edu. *Application contact:* Mary C. Molloy, Graduate Admissions Secretary, 314-835-7096, Fax: 314-935-4014, E-mail: mcm@mecf.wustl.edu.

See in-depth description on page 1483.

Wayne State University, Graduate School, College of Engineering, Department of Mechanical Engineering, Detroit, MI 48202. Offers MS, PhD. *Degree requirements:* For master's, thesis optional, foreign language not required; for doctorate, dissertation required. *Entrance requirements:* For master's, minimum undergraduate GPA of 3.0; for doctorate, minimum graduate GPA of 3.5. *Faculty research:* Acoustic vibrations and noise control, trauma biomechanics, combustion, fluid mechanics.

Western Michigan University, Graduate College, College of Engineering and Applied Sciences, Department of Mechanical and Aeronautical Engineering, Kalamazoo, MI 49008. Offers mechanical engineering (MSE, PhD). Part-time programs available. *Students:* 11 full-time (0 women), 50 part-time (4 women); includes 4 minority (1 African American, 3 Asian Americans or Pacific Islanders), 28 international. 115 applicants, 57% accepted. In 1998, 9 degrees awarded. *Degree requirements:* For master's, thesis optional; for doctorate, dissertation, oral exam required. *Entrance requirements:* For master's, minimum GPA of 3.0; for doctorate, GRE General Test, minimum GPA of 3.0. *Application deadline:* For fall admission, 2/15 (priority date). Applications are processed on a rolling basis. Application fee: $25. *Financial aid:* Fellowships, research assistantships, teaching assistantships, career-related internships or fieldwork and Federal Work-Study available. Financial aid application deadline: 2/15; financial aid applicants required to submit FAFSA. *Faculty research:* Computational fluid dynamics, manufacturing process designs, composite materials, thermal fluid flow, experimental stress analysis. *Unit head:* Dr. Parviz Merati, Chair, 616-387-3366. *Application contact:* Paula J. Boodt, Coordinator, Graduate Admissions and Recruitment, 616-387-2000, Fax: 616-387-2355, E-mail: paula.boodt@wmich.edu.

Western New England College, School of Engineering, Department of Mechanical Engineering, Springfield, MA 01119-2654. Offers MSME. Part-time and evening/weekend programs available. *Faculty:* 6 full-time (0 women). Average age 29. In 1998, 4 degrees awarded. *Degree requirements:* For master's, computer language, comprehensive exam required, thesis optional, foreign language not required. *Entrance requirements:* For master's, bachelor's degree in engineering or related field. *Application deadline:* Applications are processed on a rolling basis. Application fee: $30. *Financial aid:* Teaching assistantships available. Aid available to part-time students. Financial aid application deadline: 4/1; financial aid applicants required to submit FAFSA. *Faculty research:* Low-loss fluid mixing, flow separation delay and alleviation, high-lift airfoils, ejector research, compact heat exchangers. *Unit head:* Dr. Robert C. Azar, Chair, 413-782-1334. *Application contact:* Harry F. Neunder, Coordinator, Continuing Education, 413-782-1750, Fax: 413-782-1779, E-mail: hneunder@wnec.edu.

West Virginia University, College of Engineering and Mineral Resources, Department of Mechanical and Aerospace Engineering, Program in Mechanical Engineering, Morgantown, WV 26506. Offers engineering (MSE, PhD); mechanical engineering (MSME). Part-time programs available. Terminal master's awarded for partial completion of doctoral program. *Degree requirements:* For master's, thesis required, foreign language not required; for doctorate, dissertation, comprehensive exam required, foreign language not required. *Entrance requirements:* For master's and doctorate, GRE Subject Test, TOEFL (minimum score of 550 required), minimum GPA of 3.0. *Faculty research:* Fluid mechanics, combustion, multiphaseflow, solid mechanics, structures, engines, alternate fuels.

Wichita State University, Graduate School, College of Engineering, Department of Mechanical Engineering, Wichita, KS 67260. Offers MS, PhD. Part-time programs available. *Faculty:* 12 full-time (2 women). *Students:* 32 full-time (3 women), 50 part-time (5 women); includes 4 minority (2 Asian Americans or Pacific Islanders, 2 Hispanic Americans), 52 international. Average age 31. 120 applicants, 58% accepted. In 1998, 15 master's, 4 doctorates awarded. *Degree requirements:* For master's, computer language, thesis or alternative, oral exam required, foreign language not required; for doctorate, one foreign language, computer language, dissertation, comprehensive, departmental qualifying, and oral exams required. *Entrance requirements:* For master's and doctorate, GRE, TOEFL (minimum score of 550 required), minimum GPA of 3.0 in last 2 years. *Application deadline:* For fall admission, 7/1 (priority date); for spring admission, 1/1. Applications are processed on a rolling basis. Application fee: $25 ($40 for international students). Electronic applications accepted. *Financial aid:* In 1998–99, 11 research assistantships (averaging $4,000 per year), 5 teaching assistantships with full tuition reimbursements (averaging $5,000 per year) were awarded.; fellowships, Federal Work-Study, institutionally-sponsored loans, and unspecified assistantships also available. Financial aid application deadline: 4/1; financial aid applicants required to submit FAFSA. *Faculty research:* Materials science, fluid mechanics, manufacturing systems, thermal science, combustion. Total annual research expenditures: $229,168. *Unit head:* Dr. Richard Johnson, Chairperson, 316-978-3402, Fax: 316-978-3236, E-mail: rjohnson@twsuvm.uc.twsu.edu. *Application contact:* Dr. Manesh Greywall, Graduate Coordinator, 316-978-3402, Fax: 316-978-3236, E-mail: greywall@twsuvm.uc.twsu.edu.

Announcement: MS and PhD work in thermal/fluid sciences, combustion, materials science, multibody dynamics, robotics, controls, and machine design. Significant experimental facilities in several areas are available through joint activities between the department and the National Institute for Aviation Research. Financial assistance, including resident tuition rates, is available for qualified students.

Widener University, School of Engineering, Program in Mechanical Engineering, Chester, PA 19013-5792. Offers ME, ME/MBA. *Faculty:* 4 part-time (1 woman). *Students:* 1 full-time (0 women), 9 part-time; includes 2 minority (1 Asian American or Pacific Islander, 1 Hispanic American), 2 international. 8 applicants, 88% accepted. In 1998, 2 degrees awarded. *Degree requirements:* For master's, thesis optional, foreign language not required. *Entrance requirements:* For master's, GMAT (ME/MBA). *Average time to degree:* Master's–2 years full-time, 4 years part-time. *Application deadline:* For fall admission, 8/1 (priority date); for spring admission, 12/1. Applications are processed on a rolling basis. Application fee: $25 ($300 for international students). *Financial aid:* In 1998–99, 1 teaching assistantship with full tuition reimbursement (averaging $7,500 per year) was awarded.; unspecified assistantships also available. Financial aid application deadline: 3/15. *Faculty research:* Computational fluid mechanics, thermal and solar engineering, energy conversion, composite materials, solid mechanics. *Unit head:* Dr. Anastas Lazaridis, Chairman, Department of Mechanical Engineering, 610-499-4192, Fax: 610-499-4059, E-mail: anastas.lazaridis@widener.edu. *Application contact:* Dr. David H. T. Chen, Assistant Dean for Graduate Programs and Research, 610-499-4049, Fax: 610-499-4059, E-mail: david.h.chen@widener.edu.

Worcester Polytechnic Institute, Graduate Studies, Department of Mechanical Engineering, Worcester, MA 01609-2280. Offers materials science and engineering (MS, PhD); mechanical engineering (M Eng, MS, PhD, Advanced Certificate). Part-time and evening/weekend programs available. *Faculty:* 18 full-time (1 woman), 1 part-time (0 women). *Students:* 58 full-time (6 women), 19 part-time (3 women); includes 5 minority (4 Asian Americans or Pacific Islanders, 1 Hispanic American), 22 international. Average age 25. 168 applicants, 70% accepted. In 1998, 14 master's, 3 doctorates awarded. *Degree requirements:* For master's, thesis optional, foreign language not required; for doctorate, dissertation required, foreign language not required. *Entrance requirements:* For master's, TOEFL (minimum score of 550 required; average 608); for doctorate, GRE General Test (combined average 1947 on three sections), TOEFL (minimum score of 550 required; average 608). *Application deadline:* For fall admission, 2/15 (priority date); for spring admission, 10/15 (priority date). Applications are processed on a rolling basis. Application fee: $50. Electronic applications accepted. *Financial aid:* In 1998–99, 62 students received aid, including 3 fellowships with full tuition reimbursements available (averaging $14,077 per year), 36 research assistantships with full tuition reimbursements available (averaging $15,000 per year), 23 teaching assistantships with full tuition reimbursements available (averaging $11,970 per year); career-related internships or fieldwork, grants, institutionally-sponsored loans, and scholarships also available. Financial aid application deadline: 2/15; financial aid applicants required to submit FAFSA. *Faculty research:* Fluid mechanics, rarified gas and plasma dynamics, solid mechanics, dynamics and vibrations, biomechanics. Total annual research expenditures: $2.4 million. *Unit head:* Mohammad N. Noori, Head, 508-831-5759, Fax: 508-831-5680, E-mail: mnnoori@wpi.edu. *Application contact:* Mark Richman, Graduate Coordinator, 508-831-5556, Fax: 508-831-5680, E-mail: mrichman@wpi.edu.

See in-depth description on page 1485.

Wright State University, School of Graduate Studies, College of Engineering and Computer Science, Programs in Engineering, Program in Mechanical and Materials Engineering, Dayton, OH 45435. Offers materials science and engineering (MSE); mechanical engineering (MSE). *Students:* 32 full-time (9 women), 30 part-time (3 women); includes 4 minority (1 African American, 2 Asian Americans or Pacific Islanders, 1 Hispanic American), 26 international. Average age 29. 130 applicants, 55% accepted. In 1998, 23 degrees awarded. *Degree requirements:* For master's, thesis or course option alternative required. *Entrance requirements:* For master's, TOEFL (minimum score of 550 required). Application fee: $25. *Financial aid:* Fellowships, research assistantships, teaching assistantships, unspecified assistantships available. Aid available to part-time students. Financial aid application deadline: 3/15; financial aid applicants required to submit FAFSA. *Unit head:* Dr. Richard J. Bethke, Chair, 937-775-5040, Fax: 937-775-5009.

Yale University, Graduate School of Arts and Sciences, Programs in Engineering and Applied Science, Department of Mechanical Engineering, New Haven, CT 06520. Offers applied mechanics and mechanical engineering (M Phil, MS, PhD). *Faculty:* 17. *Students:* 25 full-time (3 women), 15 international. 35 applicants, 23% accepted. In 1998, 1 master's, 3 doctorates awarded. Terminal master's awarded for partial completion of doctoral program. *Degree requirements:* For master's, foreign language and thesis not required; for doctorate, dissertation, exam required, foreign language not required. *Entrance requirements:* For master's and doctorate, GRE General Test, TOEFL. *Average time to degree:* Master's–5 years part-time; doctorate–5 years full-time. *Application deadline:* For fall admission, 1/4. Application fee: $65.

Financial aid: Federal Work-Study and institutionally-sponsored loans available. Aid available to part-time students. *Unit head:* Chair, 203-432-4344. *Application contact:* Admissions Information, 203-432-2770.

Youngstown State University, Graduate School, William Rayen College of Engineering, Department of Mechanical and Industrial Engineering, Youngstown, OH 44555-0001. Offers MSE. Part-time and evening/weekend programs available. *Faculty:* 8 full-time (0 women), 1 (woman) part-time. *Students:* 16 full-time (5 women), 8 part-time (2 women); includes 2 minority (both African Americans), 6 international. 7 applicants, 100% accepted. In 1998, 9 degrees awarded. *Degree requirements:* For master's, computer language required, thesis optional, foreign language not required. *Entrance requirements:* For master's, TOEFL (minimum score of 550 required), minimum GPA of 2.75 in field. *Application deadline:* For fall admission, 8/15 (priority date); for winter admission, 11/15 (priority date); for spring admission, 2/15

(priority date). Applications are processed on a rolling basis. Application fee: $30 ($75 for international students). Tuition, state resident: part-time $97 per credit hour. Tuition, nonresident: part-time $219 per credit hour. Required fees: $21 per credit hour. $41 per quarter. *Financial aid:* In 1998–99, 5 students received aid, including 2 research assistantships with full tuition reimbursements available (averaging $7,500 per year); teaching assistantships, Federal Work-Study, institutionally-sponsored loans, and scholarships also available. Aid available to part-time students. Financial aid application deadline: 3/1. *Faculty research:* Kinematics and dynamics of machines, computational and experimental heat transfer, machine controls and mechanical design. *Unit head:* Dr. H. W. Shawn Kim, Chair, 330-742-3016. *Application contact:* Dr. Peter J. Kasvinsky, Dean of Graduate Studies, 330-742-3091, Fax: 330-742-1580, E-mail: amgrad03@ysub.ysu.edu.

Mechanics

Brown University, Graduate School, Division of Engineering, Program in Mechanics of Solids and Structures, Providence, RI 02912. Offers Sc M, PhD. *Degree requirements:* For doctorate, dissertation, preliminary exam required, foreign language not required, foreign language not required.

California Institute of Technology, Division of Engineering and Applied Science, Option in Applied Mechanics, Pasadena, CA 91125-0001. Offers MS, PhD. *Faculty:* 3 full-time (0 women). *Students:* 9 full-time (0 women), 7 international. 23 applicants, 13% accepted. In 1998, 1 master's, 3 doctorates awarded. *Degree requirements:* For master's, foreign language and thesis not required; for doctorate, dissertation required, foreign language not required. *Application deadline:* For fall admission, 1/15. Application fee: $0. *Faculty research:* Elasticity, mechanics of quasi-static and dynamic fracture, dynamics and mechanical vibrations, stability and control. *Unit head:* Dr. John Hall, Executive Officer, 626-395-4160. *Application contact:* Dr. Erik Antonsson, Representative, 626-395-3790.

California State University, Fullerton, Graduate Studies, School of Engineering and Computer Science, Department of Civil Engineering and Engineering Mechanics, Fullerton, CA 92834-9480. Offers MS. *Faculty:* 7 full-time (0 women), 6 part-time. *Students:* 8 full-time (2 women), 25 part-time (4 women); includes 13 minority (1 African American, 11 Asian Americans or Pacific Islanders, 1 Hispanic American), 5 international. Average age 31. 18 applicants, 56% accepted. In 1998, 20 degrees awarded. *Degree requirements:* For master's, computer language, comprehensive exam, project or thesis required. *Entrance requirements:* For master's, minimum undergraduate GPA of 2.5. Application fee: $55. Tuition, nonresident: part-time $264 per unit. Required fees: $1,947; $1,281 per year. *Financial aid:* Career-related internships or fieldwork, Federal Work-Study, grants, and institutionally-sponsored loans available. Aid available to part-time students. Financial aid application deadline: 3/1. *Faculty research:* Soil-structure interaction, finite-element analysis, computer-aided analysis and design. *Unit head:* Dr. Chandra Putcha, Chair, 714-278-3012.

California State University, Northridge, Graduate Studies, College of Engineering and Computer Science, Department of Civil, Industrial and Applied Mechanics, Northridge, CA 91330. Offers applied mechanics (MSE); civil engineering (MS); engineering (MS); engineering management (MS); industrial engineering (MS); materials engineering (MS); mechanical engineering (MS), including aerospace engineering, applied engineering, machine design, mechanical engineering, structural engineering, thermofluids; mechanics (MS). Part-time and evening/weekend programs available. *Faculty:* 13 full-time, 2 part-time. *Students:* 10 full-time (2 women), 101 part-time (15 women); includes 38 minority (3 African Americans, 22 Asian Americans or Pacific Islanders, 11 Hispanic Americans, 2 Native Americans), 8 international. Average age 32. 58 applicants, 57% accepted. In 1998, 34 degrees awarded. *Degree requirements:* For master's, thesis required, foreign language not required. *Entrance requirements:* For master's, GRE General Test, TOEFL, minimum GPA of 2.5. *Application deadline:* For fall admission, 11/30. Application fee: $55. Tuition, nonresident: part-time $246 per unit. International tuition: $7,874 full-time. Required fees: $1,970. Tuition and fees vary according to course load. *Financial aid:* Teaching assistantships available. Financial aid application deadline: 3/1. *Faculty research:* Composite study. *Unit head:* Dr. Stephen Gadomski, Chair, 818-677-2166. *Application contact:* Dr. Ileana Costa, Graduate Coordinator, 818-677-3299.

The Catholic University of America, School of Engineering, Department of Civil Engineering, Program in Fluid and Solid Mechanics, Washington, DC 20064. Offers PhD. Average age 28. 1 applicants, 0% accepted. *Degree requirements:* For doctorate, dissertation, comprehensive and oral exams required, foreign language not required. *Entrance requirements:* For doctorate, minimum GPA of 3.5. *Application deadline:* For fall admission, 8/1 (priority date); for spring admission, 12/1. Applications are processed on a rolling basis. Application fee: $50. *Financial aid:* Research assistantships, teaching assistantships, career-related internships or fieldwork, Federal Work-Study, institutionally-sponsored loans, and tuition waivers (full and partial) available. Aid available to part-time students. Financial aid application deadline: 2/1. *Faculty research:* Wave propagation, geophysical fluid mechanics, composite material, fracture, biomechanics. *Unit head:* Dr. Timothy W. Kao, Chair, Department of Civil Engineering, 202-319-5163, Fax: 202-319-4499, E-mail: kao@cua.edu.

The Catholic University of America, School of Engineering, Department of Mechanical Engineering, Program in Fluid Mechanics and Thermal Science, Washington, DC 20064. Offers MME, D Engr, PhD. Average age 28. 2 applicants, 50% accepted. *Degree requirements:* For master's, comprehensive exam required, thesis optional, foreign language not required; for doctorate, dissertation, comprehensive and oral exams required, foreign language not required. *Entrance requirements:* For master's, minimum GPA of 3.0; for doctorate, minimum GPA of 3.5. *Application deadline:* For fall admission, 8/1 (priority date); for spring admission, 12/1. Applications are processed on a rolling basis. Application fee: $50. *Financial aid:* Research assistantships, teaching assistantships, career-related internships or fieldwork, Federal Work-Study, institutionally-sponsored loans, and tuition waivers (full and partial) available. Aid available to part-time students. Financial aid application deadline: 2/1. *Faculty research:* Turbulent flow cogeneration design for heating and cooling systems, solar wind slow shocks. Total annual research expenditures: $137,400. *Unit head:* Dr. Yun C. Whang, Director, 202-319-5170.

Clemson University, Graduate School, College of Engineering and Science, School of Mechanical and Industrial Engineering, Department of Mechanical Engineering, Program in Engineering Mechanics, Clemson, SC 29634. Offers MS, PhD. *Students:* 14 full-time (3 women), 1 (woman) part-time; includes 1 minority (African American), 12 international. Average age 28. 8 applicants, 88% accepted. In 1998, 1 master's, 1 doctorate awarded. *Degree requirements:* For master's and doctorate, thesis/dissertation required, foreign language not required. *Entrance requirements:* For master's and doctorate, GRE General Test, TOEFL (minimum score of 550 required). *Application deadline:* For fall admission, 6/1. Applications are processed on a rolling basis. Application fee: $35. *Financial aid:* Fellowships, research assistantships, teaching assistantships, career-related internships or fieldwork available. Financial aid applicants required to submit FAFSA. *Faculty research:* Composite materials, structural stability, vehicle dynamics, finite-element methods. *Unit head:* Dr. M. W. Dixon, Graduate Coordinator, 864-656-5624, Fax: 864-656-4435, E-mail: mwdxn@clemson.edu. *Application contact:* Dr. M. W. Dixon, Graduate Coordinator, 864-656-5624, Fax: 864-656-4435, E-mail: mwdxn@clemson.edu.

Colorado State University, Graduate School, College of Engineering, Department of Mechanical Engineering, Program in Mechanics and Materials, Fort Collins, CO 80523-0015. Offers MS, PhD. *Faculty:* 18 full-time (0 women). *Degree requirements:* For doctorate, dissertation required, foreign language not required, foreign language not required. *Entrance requirements:* For master's and doctorate, GRE General Test (minimum combined score of 1850 on three sections required; average 1872), TOEFL (minimum score of 550 required; average 596), minimum GPA of 3.0. *Application deadline:* For fall admission, 2/1 (priority date). Applications are processed on a rolling basis. Application fee: $30. Electronic applications accepted. *Faculty research:* Structured modeling of dynamic systems,composites, ultrathin casings, superconductivity, ion implantation. *Unit head:* Vickie Jensen, 970-491-5597. *Application contact:* Dr. Donald Radford, Associate Professor, 970-491-8677, Fax: 970-491-3827, E-mail: don@engr.colostate.edu.

Columbia University, Fu Foundation School of Engineering and Applied Science, Department of Civil Engineering and Engineering Mechanics, New York, NY 10027. Offers civil engineering (MS, Eng Sc D, PhD, Engr); mechanics (MS, Eng Sc D, PhD, Engr). Part-time programs available. *Faculty:* 11 full-time (0 women), 6 part-time (0 women). *Students:* 31 full-time (8 women), 29 part-time (5 women). 170 applicants, 34% accepted. In 1998, 24 master's, 5 doctorates, 2 other advanced degrees awarded. Terminal master's awarded for partial completion of doctoral program. *Degree requirements:* For master's, foreign language and thesis not required; for doctorate, dissertation, qualifying exam required, foreign language not required. *Entrance requirements:* For master's, doctorate, and Engr, GRE General Test, TOEFL. *Application deadline:* For fall admission, 1/5; for spring admission, 10/1. Application fee: $55. *Financial aid:* In 1998–99, 8 research assistantships, 6 teaching assistantships were awarded.; fellowships, Federal Work-Study and scholarships also available. Financial aid application deadline: 1/5; financial aid applicants required to submit FAFSA. *Faculty research:* Structural deterioration and control structural materials, damage mechanics, geoenvironmental engineering, construction engineering. *Unit head:* Dr. Rene B. Testa, Chairman, 212-854-6283, Fax: 212-854-6267, E-mail: testa@civil.columbia.edu. *Application contact:* Carolyn Waldo, Administrative Assistant, 212-854-3143, Fax: 212-854-6267, E-mail: clw1@columbia.edu.

Cornell University, Graduate School, Graduate Fields of Engineering, Field of Theoretical and Applied Mechanics, Ithaca, NY 14853-0001. Offers dynamics and spare mechanics (MS, PhD); engineering mechanics (M Eng); fluid mechanics (MS, PhD); mechanics of materials (MS, PhD); solid mechanics (MS, PhD). *Faculty:* 22 full-time. *Students:* 31 full-time (6 women); includes 4 minority (2 Asian Americans or Pacific Islanders, 2 Hispanic Americans), 19 international. 30 applicants, 43% accepted. In 1998, 1 master's, 5 doctorates awarded. *Degree requirements:* For master's, thesis, project report (M Eng) required, foreign language not required; for doctorate, dissertation required, foreign language not required. *Entrance requirements:* For master's and doctorate, GRE General Test (strongly recommended), TOEFL (minimum score of 600 required). *Application deadline:* For fall admission, 1/15. Application fee: $65. Electronic applications accepted. *Financial aid:* In 1998–99, 28 students received aid, including 7 fellowships with full tuition reimbursements available, 7 research assistantships with full tuition reimbursements available, 14 teaching assistantships with full tuition reimbursements available; institutionally-sponsored loans, scholarships, and unspecified assistantships also available. *Faculty research:* Continuum mechanics, composite materials, nonlinear dynamics, chaos and space mechanics, biomechanics, experimental and fracture mechanics. *Unit head:* Prof. Alan Zehnder, Director of Graduate Studies, 607-255-9181. *Application contact:* Graduate Field Assistant, 607-255-0988, E-mail: tam_grad@cornell.edu.

Drexel University, Graduate School, College of Engineering, Department of Mechanical Engineering and Mechanics, Philadelphia, PA 19104-2875. Offers manufacturing engineering (MS, PhD); mechanical engineering and mechanics (MS, PhD). Part-time and evening/weekend programs available. *Faculty:* 23 full-time, 9 part-time. *Students:* 31 full-time (2 women), 41 part-time (2 women); includes 12 minority (6 African Americans, 3 Asian Americans or Pacific Islanders, 3 Hispanic Americans), 30 international. Average age 32. 223 applicants, 69% accepted. In 1998, 27 master's, 9 doctorates awarded. Terminal master's awarded for partial completion of doctoral program. *Degree requirements:* For master's, thesis optional; for doctorate, dissertation required. *Entrance requirements:* For master's, TOEFL (minimum score of 570 required), minimum GPA of 3.0, BS in engineering or science; for doctorate, TOEFL (minimum score of 570 required), minimum GPA of 3.5, MS in engineering or science. *Application deadline:* For fall admission, 8/21. Applications are processed on a rolling basis. Application fee: $35. Tuition: Full-time $15,795; part-time $585 per credit. Required fees: $375; $67 per term. Tuition and fees vary according to program. *Financial aid:* In 1998–99, 17 research assistantships, 27 teaching assistantships were awarded.; unspecified assistantships also available. Financial aid application deadline: 2/1. *Faculty research:* Composites, dynamic systems and control, combustion and fuels, biomechanics, mechanics and thermal fluid sciences. *Unit head:* Dr. Nicholas Cernansky, Acting Head, 215-895-2352, Fax: 215-895-1478, E-mail: cernansk@coe.drexel.edu. *Application contact:* Dr. Tetn-Min Tan, Graduate Adviser, 215-895-2293.

See in-depth description on page 1385.

École Polytechnique de Montréal, Graduate Programs, Department of Mechanical Engineering, Montréal, PQ H3C 3A7, Canada. Offers aerothermics (M Eng, M Sc A, PhD); applied mechanics (M Eng, M Sc A, PhD); tool design (M Eng, M Sc A, PhD). Part-time and evening/weekend programs available. *Degree requirements:* For master's and doctorate, one foreign language, computer language, thesis/dissertation required. *Entrance requirements:* For master's, minimum GPA of 2.75; for doctorate, minimum GPA of 3.0. *Faculty research:* Noise control and vibration, fatigue and creep, aerodynamics, composite materials, biomechanics, robotics.

Georgia Institute of Technology, Graduate Studies and Research, College of Engineering, School of Civil and Environmental Engineering, Program in Engineering Science and Mechanics, Atlanta, GA 30332-0001. Offers MS, MSESM, PhD. Part-time programs available. Terminal master's awarded for partial completion of doctoral program. *Degree requirements:* For doctorate, computer language, dissertation required, foreign language not required, foreign language not required. *Entrance requirements:* For master's, GRE, TOEFL (minimum score of 550 required); for doctorate, GRE, TOEFL (minimum score of 550 required), minimum GPA of 3.2. *Faculty research:* Bioengineering, structural mechanics, solid mechanics, dynamics.

Mechanics

Howard University, College of Engineering, Architecture, and Computer Sciences, School of Engineering and Computer Science, Department of Mechanical Engineering, Washington, DC 20059-0002. Offers aerospace engineering/dynamics and controls (M Eng, PhD); applied mechanics (M Eng, PhD); CAD/CAM and robotics (M Eng, PhD); fluid and thermal sciences (M Eng, PhD). Part-time programs available. *Faculty:* 9 full-time (1 woman). *Students:* 17 full-time (7 women), 2 part-time; includes 7 African Americans, 1 Asian American or Pacific Islander, 7 international. Terminal master's awarded for partial completion of doctoral program. *Degree requirements:* For master's, computer language, comprehensive exam required; for doctorate, one foreign language, computer language, dissertation, 2 terms of residency required. *Entrance requirements:* For master's and doctorate, GRE General Test, TOEFL, minimum GPA of 3.0. *Application deadline:* For fall admission, 4/1 (priority date); for spring admission, 11/1. Applications are processed on a rolling basis. Application fee: $45. Electronic applications accepted. *Unit head:* Dr. Lewis Thigpen, Chair, 202-806-6600, Fax: 202-806-5258, E-mail: lthigpen@scs.howard.edu. *Application contact:* Dr. Sonya Smith, Graduate Director, 202-806-4837.

See in-depth description on page 1397.

Iowa State University of Science and Technology, Graduate College, College of Engineering, Department of Aerospace Engineering and Engineering Mechanics, Ames, IA 50011. Offers aerospace engineering (M Eng, MS, PhD); engineering mechanics (M Eng, MS, PhD). *Faculty:* 41 full-time. *Students:* 39 full-time (5 women), 14 part-time (3 women), 41 international. 79 applicants, 43% accepted. In 1998, 12 master's, 9 doctorates awarded. *Degree requirements:* For master's, thesis required (for some programs); for doctorate, dissertation required. *Entrance requirements:* For master's and doctorate, GRE General Test, TOEFL. *Application deadline:* For fall admission, 3/1 (priority date); for spring admission, 10/1 (priority date). Application fee: $20 ($50 for international students). Electronic applications accepted. Tuition, state resident: full-time $3,308. Tuition, nonresident: full-time $9,744. Part-time tuition and fees vary according to course load, campus/location and program. *Financial aid:* In 1998–99, 15 research assistantships with partial tuition reimbursements (averaging $11,076 per year), 25 teaching assistantships with partial tuition reimbursements (averaging $9,892 per year) were awarded.; fellowships, scholarships also available. *Unit head:* Dr. Thomas J. Rudolphi, Chair, 515-294-0095, E-mail: aeem_info@iastate.edu. *Application contact:* Dr. Ambar Mitra, Director of Graduate Education, 515-294-2694, E-mail: aeem_info@iastate.edu.

Johns Hopkins University, G. W. C. Whiting School of Engineering, Department of Mechanical Engineering, Baltimore, MD 21218-2699. Offers mechanical engineering (MS, MSE, PhD), including mechanical engineering (MS), mechanics (MSE, PhD). Part-time programs available. *Faculty:* 13 full-time (1 woman), 6 part-time (0 women). *Students:* 63 full-time (8 women), 1 part-time; includes 5 minority (1 African American, 4 Asian Americans or Pacific Islanders), 32 international. Terminal master's awarded for partial completion of doctoral program. *Degree requirements:* For master's, thesis required (for some programs), foreign language not required; for doctorate, dissertation, oral exam required, foreign language not required. *Entrance requirements:* For master's and doctorate, GRE General Test (minimum score of 400 on verbal section, 780 on quantitative required), TOEFL (minimum score of 560 required). *Application deadline:* For fall admission, 2/15 (priority date). Application fee: $50. Tuition: Full-time $23,660. Tuition and fees vary according to program. *Unit head:* Dr. Andrew S. Douglas, Chair, 410-516-8789, Fax: 410-516-7254, E-mail: douglas@jhu.edu. *Application contact:* Dr. Charles Meneveau, Chair, Graduate Admissions Committee, 410-516-7132, Fax: 410-516-7254, E-mail: meneveau@titan.me.jhu.edu.

See in-depth description on page 1401.

Lehigh University, College of Engineering and Applied Science, Department of Mechanical Engineering and Mechanics, Bethlehem, PA 18015-3094. Offers applied mathematics (MS, PhD); mechanical engineering (M Eng, MS, PhD); mechanics (M Eng, MS, PhD). Part-time programs available. *Faculty:* 29 full-time (0 women). *Students:* 75 full-time (8 women), 11 part-time (1 woman); includes 7 minority (2 African Americans, 5 Hispanic Americans), 43 international. 51 applicants, 53% accepted. In 1998, 24 master's, 6 doctorates awarded. Terminal master's awarded for partial completion of doctoral program. *Degree requirements:* For master's and doctorate, thesis/dissertation required, foreign language not required. *Entrance requirements:* For master's and doctorate, TOEFL (minimum score of 550 required). *Average time to degree:* Master's–2 years full-time, 4 years part-time; doctorate–6 years full-time, 9 years part-time. *Application deadline:* For fall admission, 7/15; for spring admission, 12/1. Applications are processed on a rolling basis. Application fee: $40. *Financial aid:* In 1998–99, 17 fellowships with full and partial tuition reimbursements (averaging $11,000 per year), 15 research assistantships with full and partial tuition reimbursements (averaging $15,000 per year), 13 teaching assistantships with full and partial tuition reimbursements (averaging $11,000 per year) were awarded. Financial aid application deadline: 1/15. *Faculty research:* Thermofluids, dynamic systems, CAD/CAM. Total annual research expenditures: $2.5 million. *Unit head:* Dr. Charles Smith, Chairman, 610-758-4102, Fax: 610-758-6224, E-mail: crs1@lehigh.edu. *Application contact:* Donna Reiss, Graduate Coordinator, 610-758-4139, Fax: 610-758-6224, E-mail: dmr1@lehigh.edu.

Louisiana State University and Agricultural and Mechanical College, Graduate School, College of Engineering, Department of Civil and Environmental Engineering, Baton Rouge, LA 70803. Offers environmental engineering (MSCE, PhD); geotechnical engineering (MSCE, PhD); structural engineering and mechanics (MSCE, PhD); transportation engineering (MSCE, PhD); water resources (MSCE, PhD). Part-time programs available. *Faculty:* 27 full-time (1 woman), 1 part-time (0 women). *Students:* 59 full-time (14 women), 24 part-time (2 women); includes 5 minority (1 African American, 3 Asian Americans or Pacific Islanders, 1 Hispanic American), 48 international. *Degree requirements:* For master's, thesis optional, foreign language not required; for doctorate, one foreign language (computer language can substitute), dissertation required. *Entrance requirements:* For master's and doctorate, GRE General Test, TOEFL, minimum GPA of 3.0. *Application deadline:* For fall admission, 1/25 (priority date). Applications are processed on a rolling basis. Application fee: $25. *Unit head:* Dr. Ronald F. Malone, Acting Chair, 225-388-8666, Fax: 225-388-8652, E-mail: rmalone@unix1.sncc.lsu.edu. *Application contact:* Dr. Vijaya K. A. Gopu, Graduate Coordinator, 225-388-8442, E-mail: cegopu@eng.lsu.edu.

McGill University, Faculty of Graduate Studies and Research, Faculty of Engineering, Department of Civil Engineering and Applied Mechanics, Program in Fluid Mechanics and Hydraulic Engineering, Montréal, PQ H3A 2T5, Canada. Offers M Eng, M Sc, PhD. Part-time and evening/weekend programs available. *Degree requirements:* For master's, computer language required, thesis optional, foreign language not required; for doctorate, computer language, dissertation required, foreign language not required. *Entrance requirements:* For master's, TOEFL (minimum score of 550 required), minimum GPA of 3.0; for doctorate, TOEFL (minimum score of 580 required). *Faculty research:* Transport processes in inland and coastal waters, experimental and computational methods.

Michigan State University, Graduate School, College of Engineering, Department of Materials Science and Mechanics, East Lansing, MI 48824-1020. Offers materials science (MS, PhD); mechanics (MS, PhD); metallurgy (MS, PhD). Part-time programs available. *Faculty:* 21. *Students:* 39 full-time (11 women), 34 part-time (7 women); includes 5 minority (3 African Americans, 2 Asian Americans or Pacific Islanders), 40 international. Average age 30. 151 applicants, 23% accepted. In 1998, 16 master's, 4 doctorates awarded. *Degree requirements:* For master's, foreign language and thesis not required; for doctorate, dissertation required. *Entrance requirements:* For master's, GRE, TOEFL; for doctorate, GRE, TOEFL, sample of technical work. *Application deadline:* Applications are processed on a rolling basis. Application fee: $30 ($40 for international students). *Financial aid:* In 1998–99, 30 research assistantships with tuition reimbursements (averaging $12,539 per year), 20 teaching assistantships with tuition reimbursements (averaging $12,290 per year) were awarded.; fellowships Financial aid application deadline: 3/1; financial aid applicants required to submit FAFSA. *Faculty research:* Composite materials, thin films, materials modeling, computational mechanics, biomechanics. Total annual research expenditures: $910,000. *Unit head:* Dr. Nicholas Altiero, Chairperson, 517-355-5141, Fax: 517-353-9842.

Michigan Technological University, Graduate School, College of Engineering, Department of Mechanical Engineering-Engineering Mechanics, Program in Engineering Mechanics, Houghton, MI 49931-1295. Offers MS. Part-time programs available. *Faculty:* 42 full-time (2 women), 1 part-time (0 women). *Students:* 3 full-time (0 women); includes 1 minority (African American) Average age 24. 2 applicants, 100% accepted. *Degree requirements:* For master's, thesis required, foreign language not required. *Entrance requirements:* For master's, GRE General Test, TOEFL (minimum score of 550 required). *Application deadline:* For fall admission, 3/15 (priority date). Applications are processed on a rolling basis. Application fee: $30 ($35 for international students). Tuition, state resident: full-time $4,377. Tuition, nonresident: full-time $9,108. Required fees: $126. Tuition and fees vary according to course load. *Financial aid:* In 1998–99, 3 teaching assistantships (averaging $9,189 per year) were awarded.; fellowships, research assistantships, Federal Work-Study and institutionally-sponsored loans also available. Aid available to part-time students. Financial aid application deadline: 4/1; financial aid applicants required to submit FAFSA. *Unit head:* Dr. William Predebon, Chair, Department of Mechanical Engineering-Engineering Mechanics, 906-487-2551, Fax: 906-487-2822, E-mail: wwpredebon@mtu.edu.

Mississippi State University, College of Engineering, Department of Aerospace Engineering, Mississippi State, MS 39762. Offers aerospace engineering (MS); engineering mechanics (MS). Part-time programs available. *Students:* 8 full-time (0 women), 2 part-time; includes 6 minority (all Asian Americans or Pacific Islanders), 2 international. *Degree requirements:* For master's, computer language required, foreign language not required. *Entrance requirements:* For master's, GRE General Test, TOEFL (minimum score of 550 required), minimum GPA of 2.75. *Application deadline:* For fall admission, 7/1; for spring admission, 11/1. Applications are processed on a rolling basis. Application fee: $25 for international students. *Unit head:* Dr. John C. McWhorter, Head, 662-325-3623, Fax: 662-325-7730, E-mail: mcwho@ae.msstate.edu. *Application contact:* Jerry B. Inmon, Director of Admissions, 662-325-2224, Fax: 662-325-7360, E-mail: admit@admissions.msstate.edu.

New Mexico Institute of Mining and Technology, Graduate Studies, Program in Engineering Science in Mechanics, Socorro, NM 87801. Offers MS. *Faculty:* 5 full-time (0 women). *Students:* 1 (woman) full-time. Average age 30. 12 applicants, 67% accepted. In 1998, 1 degree awarded. *Degree requirements:* For master's, thesis required, foreign language not required. *Entrance requirements:* For master's, GRE General Test, TOEFL (minimum score of 540 required). *Average time to degree:* Master's–3 years full-time. *Application deadline:* For fall admission, 3/1 (priority date); for spring admission, 6/1. Applications are processed on a rolling basis. Application fee: $16. *Financial aid:* In 1998–99, 1 research assistantship, 1 teaching assistantship were awarded.; fellowships, Federal Work-Study and institutionally-sponsored loans also available. Financial aid application deadline: 3/1; financial aid applicants required to submit CSS PROFILE or FAFSA. *Faculty research:* Vibrations, fluid-structure interactions. *Unit head:* Dr. Alan Miller, Chairman, 505-835-5318, E-mail: armiller@nmt.edu. *Application contact:* Dr. David B. Johnson, Dean of Graduate Studies, 505-835-5513, Fax: 505-835-5476, E-mail: graduate@nmt.edu.

North Dakota State University, Graduate Studies and Research, College of Engineering and Architecture, Department of Mechanical Engineering and Applied Mechanics, Fargo, ND 58105. Offers MS. Part-time and evening/weekend programs available. *Faculty:* 11 full-time (0 women). *Students:* 8 full-time (0 women), 5 part-time (1 woman); includes 1 minority (African American), 7 international. Average age 25. 38 applicants, 32% accepted. In 1998, 2 degrees awarded (100% found work related to degree). *Degree requirements:* For master's, thesis required, foreign language not required. *Entrance requirements:* For master's, TOEFL (minimum score of 550 required), minimum GPA of 3.0. *Application deadline:* For fall admission, 7/1 (priority date). Applications are processed on a rolling basis. Application fee: $25. *Financial aid:* In 1998–99, 6 students received aid, including 6 research assistantships with full tuition reimbursements available (averaging $8,000 per year); teaching assistantships, career-related internships or fieldwork, Federal Work-Study, and institutionally-sponsored loans also available. Financial aid application deadline: 2/15. *Faculty research:* Thermodynamics, finite element analysis, automotive systems, robotics. *Unit head:* Dr. Robert Pieri, Chair, 701-231-8671, Fax: 701-231-7195, E-mail: pieri@badlands.nodak.edu.

Northwestern University, The Graduate School, Robert R. McCormick School of Engineering and Applied Science, Department of Civil Engineering, Evanston, IL 60208. Offers biosolid mechanics (MS, PhD); environmental health engineering (MS, PhD); geotechnical engineering (MS, PhD); health physics/radiological health (MS, PhD); project management (MPM); structural engineering (MS, PhD); structural mechanics (MS, PhD); transportation systems engineering (MS, PhD). MS and PhD admissions and degrees offered through The Graduate School. Part-time programs available. *Faculty:* 26 full-time (2 women), 1 (woman) part-time. *Students:* 110 full-time (38 women), 7 part-time (2 women); includes 5 minority (2 African Americans, 3 Asian Americans or Pacific Islanders), 68 international. Terminal master's awarded for partial completion of doctoral program. *Degree requirements:* For master's, thesis required (for some programs); for doctorate, dissertation required. *Entrance requirements:* For master's and doctorate, GRE General Test, TOEFL (minimum score of 560 required). Application fee: $50 ($55 for international students). *Unit head:* Joseph L. Schofer, Chair, 847-491-3257, Fax: 847-491-4011, E-mail: j-schofer@nwu.edu. *Application contact:* Karm Kerwell, Secretary, 847-491-3176, Fax: 847-491-4011, E-mail: k-kerwell@nwu.edu.

Northwestern University, The Graduate School, Robert R. McCormick School of Engineering and Applied Science, Program in Theoretical and Applied Mechanics, Evanston, IL 60208. Offers fluid mechanics (MS, PhD); solid mechanics (MS, PhD). Admissions and degrees offered through The Graduate School. *Faculty:* 12 full-time (2 women). *Students:* 16 full-time (3 women); includes 4 minority (2 Asian Americans or Pacific Islanders, 2 Hispanic Americans), 6 international. 7 applicants, 43% accepted. In 1998, 2 master's, 2 doctorates awarded. *Degree requirements:* For master's, foreign language and thesis not required; for doctorate, dissertation required, foreign language not required. *Entrance requirements:* For master's and doctorate, GRE General Test, TOEFL (minimum score of 560 required). *Application deadline:* For fall admission, 8/31. Application fee: $50 ($55 for international students). *Financial aid:* In 1998–99, 1 fellowship with full tuition reimbursement (averaging $11,673 per year), 9 research assistantships with partial tuition reimbursements (averaging $16,285 per year), 1 teaching assistantship with full tuition reimbursement (averaging $12,042 per year) were awarded.; Federal Work-Study and institutionally-sponsored loans also available. Financial aid application deadline: 1/15; financial aid applicants required to submit FAFSA. *Faculty research:* Composite materials, computational mechanics, fracture and damage mechanics, geophysics, nondestructive evaluation. Total annual research expenditures: $4 million. *Unit head:* Issac Daniel, Director, 847-491-5649, Fax: 847-491-5227, E-mail: imdaniel@nwu.edu. *Application contact:* Karm Kerwell, Secretary, 847-491-3176, Fax: 847-491-4011, E-mail: k-kerwell@nwu.edu.

The Ohio State University, Graduate School, College of Engineering, Department of Aerospace Engineering, Applied Mechanics, and Aviation, Program in Engineering Mechanics, Columbus, OH 43210. Offers MS, PhD. *Faculty:* 12 full-time, 1 part-time. *Students:* 27 full-time (2 women), 1 part-time; includes 1 minority (Asian American or Pacific Islander), 24 international. 43 applicants, 35% accepted. In 1998, 4 master's, 1 doctorate awarded. *Degree requirements:* For master's, computer language required, thesis optional, foreign language not required; for doctorate, computer language, dissertation required, foreign language not required. *Entrance requirements:* For master's and doctorate, GRE General Test (international students). *Application deadline:* For fall admission, 8/15. Applications are processed on a rolling basis. Application fee: $30 ($40 for international students). *Financial aid:* Fellowships, research assistantships, teaching assistantships, career-related internships or fieldwork, Federal Work-Study, and institutionally-sponsored loans available. *Unit head:* Stephen E. Bechtel, Graduate Studies Committee Chair, 614-292-6570, Fax: 614-292-7369, E-mail: bechtel.3@osu.edu.

Old Dominion University, College of Engineering and Technology, Department of Aerospace Engineering, Norfolk, VA 23529. Offers aerospace engineering (ME, MS, PhD); engineering mechanics (ME, MS, PhD). Part-time and evening/weekend programs available. *Faculty:* 11 full-time. *Students:* 34 full-time (1 woman), 28 part-time (2 women); includes 4 minority (2 African Americans, 1 Asian American or Pacific Islander, 1 Hispanic American), 23 international. *Degree requirements:* For master's, computer language, thesis, comprehensive exam required, foreign language not required; for doctorate, computer language, dissertation, candidacy exam required, foreign language not required. *Entrance requirements:* For master's, TOEFL (minimum score of 550 required), minimum GPA of 3.0; for doctorate, TOEFL (minimum score of 550 required), minimum GPA of 3.25. *Application deadline:* For fall admission, 7/1; for spring admission, 10/1. Applications are processed on a rolling basis. Application fee: $30. Electronic applications accepted. *Unit head:* Dr. Thomas E. Alberts, Chair, 757-683-3736, Fax: 757-683-3200, E-mail: talberts@aero.odu.edu. *Application contact:* Dr. Thomas E. Alberts, Chair, 757-683-3736, Fax: 757-683-3200, E-mail: talberts@aero.odu.edu.

Old Dominion University, College of Engineering and Technology, Department of Mechanical Engineering, Norfolk, VA 23529. Offers design manufacturing (ME); engineering mechanics (ME, MS, PhD); mechanical engineering (ME, MS, PhD). Part-time and evening/weekend programs available. Postbaccalaureate distance learning degree programs offered (no on-campus study). *Faculty:* 14 full-time. *Students:* 61 full-time (3 women), 36 part-time (1 woman); includes 5 African Americans, 1 Hispanic American, 56 international. *Degree requirements:* For master's, computer language, comprehensive exam required, thesis optional, foreign language not required; for doctorate, computer language, dissertation, candidacy exam required, foreign language not required. *Entrance requirements:* For master's, GRE, TOEFL (minimum score of 550 required), minimum GPA of 3.0; for doctorate, GRE, TOEFL (minimum score of 550 required), minimum GPA of 3.25. *Application deadline:* For fall admission, 7/1; for spring admission, 10/1. Applications are processed on a rolling basis. Application fee: $30. Electronic applications accepted. *Unit head:* Dr. Jen Kuang Huang, Chair, 757-683-6363, Fax: 757-683-5344, E-mail: megpd@odu.edu. *Application contact:* Dr. Jen Kuang Huang, Chair, 757-683-6363, Fax: 757-683-5344, E-mail: megpd@odu.edu.

Pennsylvania State University University Park Campus, Graduate School, College of Engineering, Department of Engineering Science and Mechanics, Program in Engineering Mechanics, State College, University Park, PA 16802-1503. Offers M Eng, MS. *Students:* 15 full-time (1 woman), 3 part-time. In 1998, 11 degrees awarded. *Degree requirements:* Foreign language not required. *Entrance requirements:* For master's, GRE General Test. Application fee: $50. *Unit head:* Dr. S. I. Hayek, Chair, 814-863-7966.

Pennsylvania State University University Park Campus, Graduate School, College of Engineering, Department of Engineering Science and Mechanics, Program in Engineering Science and Mechanics, State College, University Park, PA 16802-1503. Offers PhD. *Students:* 51 full-time (4 women), 10 part-time (1 woman). *Degree requirements:* For doctorate, dissertation required. *Entrance requirements:* For doctorate, GRE General Test. Application fee: $50. *Unit head:* Dr. S. I. Hayek, Chair, 814-865-4523.

Rensselaer Polytechnic Institute, Graduate School, School of Engineering, Department of Mechanical Engineering, Aeronautical Engineering and Mechanics, Program in Mechanics, Troy, NY 12180-3590. Offers MS, PhD. Part-time and evening/weekend programs available. *Faculty:* 31 full-time (2 women), 5 part-time (0 women). *Students:* 2 full-time (0 women); includes 1 minority (Hispanic American), 1 international. 8 applicants, 25% accepted. In 1998, 1 doctorate awarded. *Degree requirements:* For master's, computer language required, foreign language not required; for doctorate, computer language, dissertation required, foreign language not required. *Entrance requirements:* For master's and doctorate, GRE, TOEFL (minimum score of 550 required). *Application deadline:* For fall admission, 2/1 (priority date). Applications are processed on a rolling basis. Application fee: $35. *Financial aid:* Fellowships, research assistantships, teaching assistantships, career-related internships or fieldwork, institutionally-sponsored loans, and tuition waivers (partial) available. Financial aid application deadline: 2/1. *Faculty research:* Biomechanics, applied and theoretical mechanics, mechanics of materials, fluid mechanics. Total annual research expenditures: $2.5 million. *Unit head:* John S. Kerge, Dean of Admissions, 718-260-3200, Fax: 718-260-3446, E-mail: admitme@poly.edu. *Application contact:* Dr. Michael Jensen, Associate Head, 518-276-6432, Fax: 518-276-6025, E-mail: burged@rpi.edu.

Rutgers, The State University of New Jersey, New Brunswick, Graduate School, Program in Mechanics, New Brunswick, NJ 08903. Offers MS, PhD. Part-time programs available. *Faculty:* 5 full-time (0 women). *Students:* 1 (woman) full-time, 1 part-time; includes 1 minority (Hispanic American), 1 international. Average age 25. 3 applicants, 0% accepted. In 1998, 1 doctorate awarded (100% entered university research/teaching). Terminal master's awarded for partial completion of doctoral program. *Degree requirements:* For master's, computer language, qualifying exam required, thesis optional, foreign language not required; for doctorate, computer language, dissertation, qualifying exam required, foreign language not required. *Entrance requirements:* For master's and doctorate, GRE General Test, TOEFL (mandatory for foreign students). *Application deadline:* For fall admission, 7/1; for spring admission, 12/1. Applications are processed on a rolling basis. Application fee: $50. *Financial aid:* Fellowships, research assistantships, teaching assistantships, tuition waivers (partial) available. Financial aid application deadline: 3/1; financial aid applicants required to submit FAFSA. *Faculty research:* Continuum mechanics, constitutive theory, thermodynamics, visolasticity, liquid crystal theory. Total annual research expenditures: $100,000. *Unit head:* Bernard D. Coleman, Graduate Director, 732-445-5558.

San Diego State University, Graduate and Research Affairs, College of Engineering, Department of Aerospace Engineering and Engineering Mechanics, San Diego, CA 92182. Offers aerospace engineering (MS); engineering mechanics (MS); engineering sciences and applied mechanics (PhD); flight dynamics (MS); fluid dynamics (MS). *Students:* 4 full-time (0 women), 2 part-time; includes 1 minority (Asian American or Pacific Islander), 3 international. 8 applicants, 50% accepted. In 1998, 3 degrees awarded. Terminal master's awarded for partial completion of doctoral program. *Degree requirements:* For doctorate, dissertation required, foreign language not required. *Entrance requirements:* For master's, GRE General Test (minimum combined score of 950 required), TOEFL (minimum score of 550 required). *Application deadline:* For fall admission, 7/1 (priority date); for spring admission, 12/1. Applications are processed on a rolling basis. Application fee: $55. *Faculty research:* Organized structures in post-stall flow over wings/three dimensional separated flow, airfoil growth effect, probabilities, structural mechanics. Total annual research expenditures: $35,000. *Unit head:* Joseph Katz, Chair, 619-594-6074, Fax: 619-594-6005, E-mail: jkatz@mail.sdsu.edu. *Application contact:* Allen Plotkin, Graduate Adviser, 619-594-7019, Fax: 619-594-6005, E-mail: allen.plotkin@sdsu.edu.

San Jose State University, Graduate Studies, College of Engineering, Department of Civil Engineering and Applied Mechanics, San Jose, CA 95192-0001. Offers MS. *Faculty:* 15 full-time (2 women), 9 part-time (0 women). *Students:* 29 full-time (6 women), 76 part-time (18 women); includes 39 minority (4 African Americans, 28 Asian Americans or Pacific Islanders, 7 Hispanic Americans), 9 international. Average age 30. 63 applicants, 73% accepted. In 1998, 41 degrees awarded. *Degree requirements:* For master's, thesis or alternative required. *Entrance requirements:* For master's, minimum GPA of 2.7. *Application deadline:* For fall admission, 6/1. Applications are processed on a rolling basis. Application fee: $59. Tuition, nonresident: part-time $246 per unit. Required fees: $1,939; $1,309 per year. *Unit head:* Dr. Thalia Anagnos, Chair, 408-924-3900, Fax: 408-924-4004. *Application contact:* Dr. Rhea Williamson, Graduate Adviser, 408-924-3849.

Southern Illinois University Carbondale, Graduate School, College of Engineering, Department of Civil Engineering and Mechanics, Carbondale, IL 62901-6806. Offers MS. *Faculty:* 12 full-time (1 woman), 1 part-time (0 women). *Students:* 21 full-time (1 woman), 4 part-time (3 women). Average age 26. 24 applicants, 63% accepted. In 1998, 9 degrees awarded. *Degree requirements:* For master's, thesis, comprehensive exam required, foreign language not required. *Entrance requirements:* For master's, TOEFL (minimum score of 550 required), minimum GPA of 2.7. *Application deadline:* Applications are processed on a rolling basis. Application fee: $20. *Financial aid:* In 1998–99, 21 students received aid, including 5 research assistantships with full tuition reimbursements available, 9 teaching assistantships with full tuition reimbursements available; fellowships with full tuition reimbursements available, Federal Work-Study, institutionally-sponsored loans, and tuition waivers (full) also available. Aid available to part-time students. Financial aid application deadline: 7/1. *Faculty research:* Composite materials, wastewater treatment, solid waste disposal, slurry transport, geotechnical engineering. Total annual research expenditures: $230,856. *Unit head:* Dr. Buck F. Brown, Dean for Research and Graduate Studies, 812-877-8403, Fax: 812-877-8102, E-mail: buck.brown@rose-hulman.edu.

Southern Illinois University Carbondale, Graduate School, College of Engineering, Program in Engineering Sciences, Carbondale, IL 62901-6806. Offers electrical systems (PhD); fossil energy (PhD); mechanics (PhD). *Faculty:* 57 full-time (1 woman). *Students:* 16 full-time (2 women), 10 part-time (2 women); includes 3 minority (1 African American, 2 Asian Americans or Pacific Islanders), 17 international. *Degree requirements:* For doctorate, dissertation required. *Entrance requirements:* For doctorate, GRE General Test, TOEFL (minimum score of 600 required), minimum GPA of 3.5. Application fee: $20. *Unit head:* Dr. Hasan Sevim, Associate Dean, 618-453-4321, Fax: 618-453-4235.

State University of New York at Buffalo, Graduate School, School of Engineering and Applied Sciences, Department of Civil, Structural, and Environmental Engineering, Buffalo, NY 14260. Offers computational engineering and mechanics (MS, PhD); construction (M Eng, MS, PhD); geoenvironmental and geotechnical engineering (M Eng, MS, PhD); structural and earthquake engineering (M Eng, MS, PhD); water resources and environmental engineering (M Eng, MS, PhD). Part-time programs available. Postbaccalaureate distance learning degree programs offered (minimal on-campus study). *Faculty:* 25 full-time (0 women), 7 part-time (1 woman). *Students:* 99 full-time (19 women), 71 part-time (13 women); includes 9 minority (2 African Americans, 5 Asian Americans or Pacific Islanders, 1 Hispanic American, 1 Native American), 101 international. Terminal master's awarded for partial completion of doctoral program. *Degree requirements:* For master's, computer language, project or thesis required; for doctorate, computer language, dissertation required, foreign language not required. *Entrance requirements:* For master's and doctorate, GRE General Test (minimum combined score of 1250 required), TOEFL (minimum score of 550 required). *Application deadline:* For fall admission, 1/15 (priority date); for spring admission, 10/1. Applications are processed on a rolling basis. Application fee: $35. Tuition, state resident: full-time $5,100; part-time $213 per credit hour. Tuition, nonresident: full-time $8,416; part-time $351 per credit hour. Required fees: $870; $75 per semester. Tuition and fees vary according to course load and program. *Unit head:* Dr. Andrei M. Reinhorn, Chairman, 716-645-2114 Ext. 2419, Fax: 716-645-3733, E-mail: reinhorn@civil.eng.buffalo.edu. *Application contact:* Dr. Joe Atkinson, Director of Graduate Admissions, 716-645-2114 Ext. 2326, Fax: 716-645-3667, E-mail: atkinson@acsu.buffalo.edu.

The University of Alabama, Graduate School, College of Engineering, Department of Aerospace Engineering and Mechanics, Tuscaloosa, AL 35487. Offers MSAE, MSESM, PhD. Part-time programs available. Postbaccalaureate distance learning degree programs offered (no on-campus study). *Faculty:* 14 full-time (0 women), 3 part-time (0 women). *Students:* 26 full-time (2 women), 42 part-time (2 women); includes 14 minority (1 African American, 13 Asian Americans or Pacific Islanders), 2 international. Average age 26. 40 applicants, 40% accepted. In 1998, 14 master's, 1 doctorate awarded. *Degree requirements:* For master's, thesis or alternative required, foreign language not required; for doctorate, dissertation required, foreign language not required. *Entrance requirements:* For master's, GRE General Test (minimum combined score of 1500 on three sections required). *Application deadline:* For fall admission, 7/6 (priority date). Applications are processed on a rolling basis. Application fee: $25. *Financial aid:* In 1998–99, 25 students received aid, including 5 fellowships, 12 research assistantships, 8 teaching assistantships; Federal Work-Study and institutionally-sponsored loans also available. Financial aid application deadline: 7/6. *Faculty research:* Flight simulation, advanced mechanical behavior in materials, fluid and solid computational mechanics, hypersonic aerodynamics, intelligent systems. Total annual research expenditures: $5 million. *Unit head:* Dr. Tom E. Novak, Head, 205-348-7300, Fax: 205-348-2094, E-mail: tnovak@coe.eng.ua.edu. *Application contact:* Dr. Amnon Katz, 205-348-7300.

The University of Arizona, Graduate College, College of Engineering and Mines, Department of Civil Engineering and Engineering Mechanics, Program in Engineering Mechanics, Tucson, AZ 85721. Offers MS, PhD. Part-time programs available. *Students:* 8 full-time (0 women), 4 part-time (1 woman); includes 3 minority (all Asian Americans or Pacific Islanders), 5 international. Average age 33. 12 applicants, 83% accepted. In 1998, 3 master's awarded. *Degree requirements:* For master's, computer language, thesis required, foreign language not required; for doctorate, computer language, dissertation, departmental qualifying exam required, foreign language not required. *Entrance requirements:* For master's, TOEFL (minimum score of 550 required), minimum GPA of 3.0; for doctorate, TOEFL (minimum score of 550 required), minimum GPA of 3.5. *Application deadline:* For fall admission, 8/1. Applications are processed on a rolling basis. Application fee: $35. *Financial aid:* Research assistantships, teaching assistantships available. Financial aid application deadline: 3/1. *Faculty research:* Constitutive modeling of solids and discontinuities, computational mechanics, fracture and composites stability, laboratory testing. *Application contact:* Mary Jankovsky, Graduate Secretary, 520-621-2266, Fax: 520-621-2550.

University of California, Berkeley, Graduate Division, College of Engineering, Department of Civil and Environmental Engineering, Berkeley, CA 94720-1500. Offers construction engineering and management (M Eng, MS, D Eng, PhD); environmental quality and environmental water resources engineering (M Eng, MS, D Eng, PhD); geotechnical engineering (M Eng, MS, D Eng, PhD); structural engineering, mechanics and materials (M Eng, MS, D Eng, PhD); transportation engineering (M Eng, MS, D Eng, PhD). *Students:* 326 full-time (97 women); includes 59 minority (3 African Americans, 42 Asian Americans or Pacific Islanders, 13 Hispanic Americans, 1 Native American), 113 international. *Degree requirements:* For master's, comprehensive exam or thesis (MS) required; for doctorate, dissertation, qualifying exam required. *Entrance requirements:* For master's, GRE General Test, minimum GPA of 3.0; for doctorate, GRE General Test, minimum GPA of 3.5. *Application deadline:* For fall admission, 2/10. Application fee: $40. *Unit head:* Dr. Adib Kanafani, Chair, 510-642-3261. *Application contact:* Mari Cook, Graduate Assistant for Admission, 510-643-8944, E-mail: mcook@ce.berkeley.edu.

University of California, San Diego, Graduate Studies and Research, Department of Applied Mechanics and Engineering Sciences, Program in Applied Mechanics, La Jolla, CA 92093-5003. Offers MS, PhD. Part-time programs available. *Students:* 10 full-time (0 women); includes 2 minority (both Asian Americans or Pacific Islanders), 5 international. 20 applicants, 25% accepted. In 1998, 3 doctorates awarded. *Degree requirements:* For master's, comprehensive exam or thesis required; for doctorate, dissertation, qualifying exam required. *Entrance requirements:* For master's and doctorate, GRE General Test, TOEFL (minimum score of 550 required), minimum GPA of 3.0. *Application deadline:* For fall admission, 5/31. Application fee: $40. *Financial aid:* In 1998–99, fellowships with full tuition reimbursements (averaging $15,000 per year), research assistantships with full tuition reimbursements (averaging $15,000 per year), teaching assistantships with partial tuition reimbursements (averaging $13,000 per year) were awarded.; career-related internships or fieldwork and scholarships also available. Financial aid application deadline: 1/31; financial aid applicants required to submit FAFSA. *Faculty research:* Combustion engineering, environmental mechanics, magnetic recording, materials processing, computational fluid dynamics. *Unit head:* Paradee Chularee, Student Affairs Officer, 310-825-8913, Fax: 310-206-7353, E-mail: paradee@ea.ucla.edu. *Application contact:* AMES Graduate Student Affairs, 619-534-4387, Fax: 619-534-1730, E-mail: bwalton@ames.ucsd.edu.

University of Cincinnati, Division of Research and Advanced Studies, College of Engineering, Department of Aerospace Engineering and Engineering Mechanics, Program in Engineer-

Mechanics

University of Cincinnati (continued)

ing Mechanics, Cincinnati, OH 45221-0091. Offers MS, PhD. *Students:* 11 full-time (1 woman), 7 part-time; includes 2 minority (1 African American, 1 Asian American or Pacific Islander), 5 international. In 1998, 2 master's, 1 doctorate awarded. *Degree requirements:* For master's, project or thesis required; for doctorate, one foreign language, dissertation required. *Entrance requirements:* For master's and doctorate, GRE General Test, TOEFL (minimum score of 525 required). *Average time to degree:* Master's–1.3 years full-time; doctorate–4.8 years full-time. *Application deadline:* For fall admission, 2/1 (priority date). Application fee: $40. *Financial aid:* Tuition waivers (full) available. Aid available to part-time students. Financial aid application deadline: 2/1. *Faculty research:* Constitutive modeling, finite elements, wave propagation, knee biomechanics, large space structures. *Unit head:* Graduate Coordinator, 619-534-7715. *Application contact:* Dr. Stan Rubin, Director of Graduate Studies, 513-556-3711, Fax: 513-556-5038, E-mail: srubin@uceng.uc.edu.

University of Connecticut, Graduate School, School of Engineering, Field of Applied Mechanics, Storrs, CT 06269. Offers PhD. *Degree requirements:* For doctorate, dissertation required. *Entrance requirements:* For doctorate, GRE General Test.

University of Connecticut, Graduate School, School of Engineering, Field of Fluid Dynamics, Storrs, CT 06269. Offers PhD. *Degree requirements:* For doctorate, dissertation required. *Entrance requirements:* For doctorate, GRE General Test.

University of Dayton, Graduate School, School of Engineering, Department of Civil Engineering, Program in Engineering Mechanics, Dayton, OH 45469-1300. Offers MSEM. *Degree requirements:* For master's, thesis or alternative required, foreign language not required. *Entrance requirements:* For master's, TOEFL. *Application deadline:* For fall admission, 8/1. Applications are processed on a rolling basis. Application fee: $30. *Application contact:* Dr. Donald L. Moon, Associate Dean, 937-229-2241, Fax: 937-229-2471, E-mail: dmoon@engr.udayton.edu.

University of Florida, Graduate School, College of Engineering, Department of Aerospace Engineering, Mechanics, and Engineering Science, Program in Engineering Science and Engineering Mechanics, Gainesville, FL 32611. Offers ME, MS, PhD, Engr. *Students:* 22 full-time (4 women), 10 part-time (1 woman); includes 2 minority (1 Asian American or Pacific Islander, 1 Hispanic American), 16 international. In 1998, 4 master's, 5 doctorates awarded. *Degree requirements:* For master's and Engr, thesis optional; for doctorate, dissertation required. *Entrance requirements:* For master's and doctorate, GRE General Test, TOEFL, minimum GPA of 3.0; for Engr, GRE General Test. *Application deadline:* For fall admission, 6/1 (priority date). Applications are processed on a rolling basis. Application fee: $20. Electronic applications accepted. *Financial aid:* Research assistantships available. *Application contact:* Dr. Chen-Chi Hsu, Graduate Coordinator, 352-392-9823, Fax: 352-392-7303, E-mail: cch@aero.ufl.edu.

University of Florida, Graduate School, Graduate Engineering and Research Center (GERC), Gainesville, FL 32611. Offers aerospace engineering (ME, MS, PhD, Engr); electrical and computer engineering (ME, MS, PhD, Engr); engineering mechanics (ME, MS, PhD, Engr); industrial and systems engineering (ME, MS, PhD, Engr). Part-time programs available. Postbaccalaureate distance learning degree programs offered. *Faculty:* 6 full-time (0 women), 16 part-time (1 woman). *Students:* 13 full-time (6 women), 159 part-time (29 women); includes 21 minority (8 African Americans, 8 Asian Americans or Pacific Islanders, 4 Hispanic Americans, 1 Native American) Terminal master's awarded for partial completion of doctoral program. *Degree requirements:* For master's, computer language required (for some programs), thesis optional, foreign language not required; for doctorate, computer language (for some programs), dissertation required; for Engr, computer language (for some programs), thesis required, foreign language not required. *Entrance requirements:* For master's, GRE General Test (minimum combined score of 1000 required), TOEFL, minimum GPA of 3.0; for doctorate, GRE General Test (minimum combined score of 1200 required), written and oral qualifying exams, TOEFL, minimum GPA of 3.0, master's degree in engineering; for Engr, GRE General Test (minimum combined score of 1000 required), TOEFL, minimum GPA of 3.0, master's degree in engineering. *Application deadline:* For fall admission, 6/1; for spring admission, 10/1. Applications are processed on a rolling basis. Application fee: $20. Electronic applications accepted. *Unit head:* Dr. Pasquale M. Sforza, Director, 850-833-9355, Fax: 850-833-9366, E-mail: sforza@gerc.eng.ufl.edu. *Application contact:* Judi Shivers, Program Assistant, 850-833-9350, Fax: 850-833-9366, E-mail: reginfo@gerc.eng.ufl.edu.

University of Illinois at Urbana–Champaign, Graduate College, College of Engineering, Department of Theoretical and Applied Mechanics, Urbana, IL 61801. Offers MS, PhD. *Faculty:* 17 full-time (1 woman). *Students:* 56 full-time (9 women); includes 4 minority (2 Asian Americans or Pacific Islanders, 2 Hispanic Americans), 35 international. 37 applicants, 35% accepted. In 1998, 7 master's, 14 doctorates awarded. *Degree requirements:* For master's, thesis or alternative required, foreign language not required; for doctorate, dissertation required, foreign language not required. *Entrance requirements:* For master's and doctorate, GRE General Test, TOEFL (minimum score of 607 required), minimum GPA of 4.0 on a 5.0 scale. *Application deadline:* Applications are processed on a rolling basis. Application fee: $40 ($50 for international students). Tuition, state resident: full-time $4,616. Tuition, nonresident: full-time $11,768. Full-time tuition and fees vary according to course load. *Financial aid:* In 1998–99, 3 fellowships, 23 research assistantships, 13 teaching assistantships were awarded. Financial aid application deadline: 2/15. *Faculty research:* Solid mechanics and materials, acoustics, computational mechanics. Total annual research expenditures: $1.8 million. *Unit head:* Dr. Hassan Aref, Head, 217-333-2329, Fax: 217-244-5707, E-mail: h-aref@uiuc.edu. *Application contact:* Dr. James W. Phillips, Graduate Coordinator, 217-333-4388, Fax: 217-244-5707, E-mail: jwp@uiuc.edu.

See in-depth description on page 1453.

University of Kentucky, Graduate School, Graduate School Programs from the College of Engineering, Program in Engineering Mechanics, Lexington, KY 40506-0032. Offers MSEM, PhD. *Degree requirements:* For master's, comprehensive exam required, thesis optional, foreign language not required; for doctorate, dissertation, comprehensive exam required. *Entrance requirements:* For master's, GRE General Test, minimum undergraduate GPA of 2.5; for doctorate, GRE General Test, minimum graduate GPA of 3.0. *Faculty research:* Computational methods in applied mechanics, stress analysis, wave propagation.

University of Maryland, Baltimore County, Graduate School, College of Engineering, Department of Mechanical Engineering, Concentration in Fluid Mechanics, Baltimore, MD 21250-5398. Offers MS, PhD. *Entrance requirements:* For master's, GRE General Test, minimum GPA of 3.0; for doctorate, GRE General Test, GRE Subject Test, TOEFL, minimum GPA of 3.5. *Application deadline:* For fall admission, 7/1. Applications are processed on a rolling basis. Application fee: $45. *Application contact:* Dr. Appa Anjanappa, Director, 410-455-3330.

University of Maryland, Baltimore County, Graduate School, College of Engineering, Department of Mechanical Engineering, Concentration in Solid Mechanics, Baltimore, MD 21250-5398. Offers MS, PhD. *Entrance requirements:* For master's, GRE General Test, minimum GPA of 3.0; for doctorate, GRE General Test, GRE Subject Test, TOEFL, minimum GPA of 3.5. *Application deadline:* For fall admission, 7/1. Applications are processed on a rolling basis. Application fee: $45. *Application contact:* Dr. Appa Anjanappa, Director, 410-455-3330.

University of Maryland, College Park, Graduate School, A. James Clark School of Engineering, Department of Mechanical Engineering, College Park, MD 20742-5045. Offers electronic packaging and reliability (MS, PhD); manufacturing and design (MS, PhD); mechanical engineering (M Eng); mechanics and materials (MS, PhD); thermal and fluid sciences (MS, PhD). Part-time and evening/weekend programs available. Postbaccalaureate distance learning degree programs offered. *Faculty:* 61 full-time (5 women), 15 part-time (0 women). *Students:* 133 full-time (13 women), 60 part-time (7 women); includes 18 minority (11 African Americans, 6 Asian Americans or Pacific Islanders, 1 Hispanic American), 123 international. *Degree requirements:* For master's, thesis optional, foreign language not required; for doctorate, dissertation, qualifying exam required, foreign language not required. *Entrance requirements:* For master's, GRE, minimum GPA of 3.0. *Application deadline:* Applications are processed on a rolling basis. Application fee: $50 ($70 for international students). Tuition, state resident: part-time $272 per credit hour. Tuition, nonresident: part-time $475 per credit hour. Required fees: $632; $379 per year. *Unit head:* Dr. Davinder Anand, Chairman, 301-405-5294, Fax: 301-314-9477. *Application contact:* Dr. James M. Wallace, Graduate Director, 301-405-4216.

University of Massachusetts Lowell, Graduate School, College of Arts and Sciences, Department of Physics and Applied Physics, Program in Applied Physics, Lowell, MA 01854-2881. Offers applied mechanics (PhD); applied physics (MS, PhD), including optical sciences (MS). Terminal master's awarded for partial completion of doctoral program. *Degree requirements:* For master's, thesis required; for doctorate, 2 foreign languages (computer language can substitute for one), dissertation required. *Entrance requirements:* For master's and doctorate, GRE General Test. *Application deadline:* For fall admission, 4/1 (priority date); for spring admission, 10/1. Applications are processed on a rolling basis. Application fee: $20 ($35 for international students). *Application contact:* Dr. Gus Couchell, Coordinator, 978-934-3772, E-mail: gus_couchell@uml.edu.

University of Michigan, Horace H. Rackham School of Graduate Studies, College of Engineering, Department of Mechanical Engineering and Applied Mechanics, Program in Applied Mechanics, Ann Arbor, MI 48109. Offers MSE, PhD. Part-time programs available. Terminal master's awarded for partial completion of doctoral program. *Degree requirements:* For master's, thesis optional, foreign language not required; for doctorate, dissertation, oral defense of dissertation, preliminary and qualifying exams required, foreign language not required. *Entrance requirements:* For master's, GRE General Test; for doctorate, GRE General Test, master's degree.

University of Minnesota, Twin Cities Campus, Graduate School, Institute of Technology, Department of Aerospace Engineering and Mechanics, Minneapolis, MN 55455-0213. Offers aerospace engineering (M Aero E, MS, PhD); mechanics (MS, PhD). Part-time programs available. *Faculty:* 18 full-time (2 women), 2 part-time (0 women). *Students:* 56 full-time (9 women); includes 15 minority (all Asian Americans or Pacific Islanders), 19 international. Average age 24. 109 applicants, 41% accepted. In 1998, 9 master's awarded (44% found work related to degree, 56% continued full-time study); 8 doctorates awarded (13% entered university research/teaching, 87% found other work related to degree). Terminal master's awarded for partial completion of doctoral program. *Degree requirements:* For master's, foreign language and thesis not required; for doctorate, dissertation required, foreign language not required. *Average time to degree:* Master's–2 years full-time, 5 years part-time; doctorate–5 years full-time. *Application deadline:* For fall admission, 7/15. Applications are processed on a rolling basis. Application fee: $40 ($50 for international students). *Financial aid:* In 1998–99, 27 research assistantships with full tuition reimbursements, 24 teaching assistantships with full tuition reimbursements were awarded. Financial aid application deadline: 2/1. *Faculty research:* Fluid mechanics, solid and continuum mechanics, dynamical systems and control therapy. Total annual research expenditures: $4.4 million. *Unit head:* William L. Garrard, Head, 612-625-8000, Fax: 612-626-1558, E-mail: dept@aem.umn.edu. *Application contact:* Ellen K. Longmire, Director of Graduate Studies, 612-625-8000, Fax: 612-626-1558, E-mail: dgs@aem.umn.edu.

University of Missouri–Rolla, Graduate School, School of Engineering, Department of Civil Engineering, Program in Fluid Mechanics, Rolla, MO 65409-0910. Offers MS, DE, PhD. *Degree requirements:* For master's, thesis or alternative required, foreign language not required; for doctorate, dissertation required, foreign language not required. *Entrance requirements:* For master's and doctorate, GRE General Test (minimum combined score of 1100 required), TOEFL (minimum score of 550 required), minimum GPA of 3.0.

University of Missouri–Rolla, Graduate School, School of Engineering, Department of Mechanical and Aerospace Engineering and Engineering Mechanics, Program in Engineering Mechanics, Rolla, MO 65409-0910. Offers MS, PhD. Part-time and evening/weekend programs available. *Faculty:* 6 full-time (0 women). *Students:* 7 full-time (2 women), 4 part-time; includes 2 minority (1 African American, 1 Asian American or Pacific Islander), 4 international. Average age 27. 11 applicants, 91% accepted. In 1998, 8 master's, 1 doctorate awarded. Terminal master's awarded for partial completion of doctoral program. *Degree requirements:* For master's, thesis required (for some programs), foreign language not required; for doctorate, dissertation required, foreign language not required. *Entrance requirements:* For master's, GRE General Test (minimum combined score of 1150 required), TOEFL (minimum score of 570 required), minimum GPA of 3.0; for doctorate, GRE General Test (minimum combined score of 1150 required), TOEFL (minimum score of 570 required), minimum GPA of 3.5. *Average time to degree:* Master's–1.9 years full-time, 6 years part-time; doctorate–3.4 years full-time. *Application deadline:* For fall admission, 7/1; for spring admission, 12/1. Applications are processed on a rolling basis. Application fee: $25. Electronic applications accepted. *Financial aid:* In 1998–99, 7 research assistantships with full and partial tuition reimbursements (averaging $9,739 per year) were awarded.; fellowships, teaching assistantships, Federal Work-Study and institutionally-sponsored loans also available. Aid available to part-time students. Financial aid application deadline: 3/1. *Faculty research:* Composite materials, finite element analysis, smart structures, solid mechanics. *Unit head:* Ann Bechtel, Admission Assistant, 612-624-5747, Fax: 612-624-7510, E-mail: bann@cdtl.umn.edu. *Application contact:* Dr. James A. Drallmeier, Associate Professor, Associate Chair for Graduate Studies, 573-341-4710, Fax: 573-341-4607, E-mail: grad-students@gearbox.maem.umr.edu.

University of Nebraska–Lincoln, Graduate College, College of Engineering and Technology, Department of Engineering Mechanics, Lincoln, NE 68588. Offers engineering (PhD); engineering mechanics (MS). *Faculty:* 12 full-time (0 women). *Students:* 6 full-time (0 women), 2 part-time, 6 international. Average age 31. 8 applicants, 63% accepted. In 1998, 3 degrees awarded. *Degree requirements:* For master's, thesis optional, foreign language not required; for doctorate, dissertation, comprehensive exams required. *Entrance requirements:* For master's and doctorate, GRE General Test, TOEFL (minimum score of 550 required). *Application deadline:* For fall admission, 3/1 (priority date). Applications are processed on a rolling basis. Application fee: $35. Electronic applications accepted. *Financial aid:* In 1998–99, 2 fellowships, 8 research assistantships, 6 teaching assistantships were awarded.; Federal Work-Study also available. Aid available to part-time students. Financial aid application deadline: 2/15. *Faculty research:* Analytical mechanics; experimental and computational solid mechanics; mechanics of materials including polymers, composites, and smart materials. *Unit head:* Dr. Millard Beatty, Chair, 402-472-2377, Fax: 402-472-8292.

University of New Brunswick, School of Graduate Studies, Faculty of Engineering, Department of Mechanical Engineering, Fredericton, NB E3B 5A3, Canada. Offers applied mechanics (M Eng, M Sc E, PhD); mechanical engineering (M Eng, M Sc E, PhD). Part-time programs available. *Degree requirements:* For master's, thesis required, foreign language not required; for doctorate, dissertation, qualifying exam required, foreign language not required. *Entrance requirements:* For master's and doctorate, TOEFL, TWE, minimum GPA of 3.0.

University of Pennsylvania, School of Engineering and Applied Science, Department of Mechanical Engineering and Applied Mechanics, Philadelphia, PA 19104. Offers applied mechanics (MSE, PhD); mechanical engineering (MSE, PhD). Part-time programs available. *Degree requirements:* For master's, thesis optional, foreign language not required; for doctorate, dissertation required, foreign language not required. *Entrance requirements:* For master's and doctorate, TOEFL (minimum score of 600 required). Electronic applications accepted. *Faculty research:* Heat transfer, fluid mechanics, energy conversion, solid mechanics, dynamics of mechanisms and robots.

See in-depth description on page 1469.

University of Rhode Island, Graduate School, College of Engineering, Department of Mechanical Engineering and Applied Mechanics, Kingston, RI 02881. Offers design/systems (MS,

PhD); fluid mechanics (MS, PhD); solid mechanics (MS, PhD); thermal sciences (MS, PhD). *Entrance requirements:* For master's and doctorate, GRE General Test (for foreign applicants only).

University of Southern California, Graduate School, School of Engineering, Program in Applied Mechanics, Los Angeles, CA 90089. Offers MS. *Degree requirements:* For master's, thesis optional. *Entrance requirements:* For master's, GRE General Test. Tuition: Part-time $768 per unit. Required fees: $350 per semester. *Financial aid:* Fellowships, research assistantships, teaching assistantships, Federal Work-Study and institutionally-sponsored loans available. Aid available to part-time students. Financial aid applicants required to submit FAFSA. *Unit head:* Dr. Leonard Silverman, Dean, School of Engineering, 213-740-0617.

University of Tennessee, Knoxville, Graduate School, College of Engineering, Department of Mechanical and Aerospace Engineering and Engineering Science, Program in Engineering Science, Knoxville, TN 37996. Offers applied artificial intelligence (MS); biomedical engineering (MS, PhD); composite materials (MS, PhD); computational mechanics (MS, PhD); engineering science (MS, PhD); fluid mechanics (MS, PhD); industrial engineering (MS, PhD); optical engineering (MS, PhD); solid mechanics (MS, PhD). Part-time programs available. *Students:* 33 full-time (9 women), 16 part-time (1 woman); includes 4 minority (1 African American, 2 Asian Americans or Pacific Islanders, 1 Native American), 10 international. *Degree requirements:* For master's, thesis or alternative required, foreign language not required; for doctorate, dissertation required, foreign language not required. *Entrance requirements:* For master's and doctorate, TOEFL (minimum score of 550 required), minimum GPA of 2.7. *Application deadline:* For fall admission, 2/1 (priority date). Applications are processed on a rolling basis. Application fee: $35. Electronic applications accepted. *Application contact:* Dr. Allen Yu, Graduate Representative, 923-974-4159, E-mail: nyu@utk.edu.

University of Tennessee Space Institute, Graduate Programs, Program in Engineering Sciences and Mechanics, Tullahoma, TN 37388-9700. Offers engineering sciences (MS, PhD); mechanics (MS, PhD). *Faculty:* 8 full-time (1 woman), 2 part-time (0 women). *Students:* 14 full-time (5 women), 5 part-time (1 woman); includes 1 minority (Native American), 1 international. 8 applicants, 50% accepted. In 1998, 1 master's, 1 doctorate awarded. *Degree requirements:* For master's, thesis required (for some programs), foreign language not required; for doctorate, dissertation required. *Application deadline:* Applications are processed on a rolling basis. Application fee: $35. *Financial aid:* Research assistantships, career-related internships or fieldwork, Federal Work-Study, and tuition waivers (full and partial) available. Financial aid applicants required to submit FAFSA. *Unit head:* Dr. Ahmad Vakili, Degree Program Chairman, 931-393-7483, Fax: 931-393-7530, E-mail: avakili@utsi.edu. *Application contact:* Dr. Edwin M. Gleason, Assistant Dean for Admissions and Student Affairs, 931-393-7432, Fax: 931-393-7346, E-mail: egleason@utsi.edu.

The University of Texas at Austin, Graduate School, College of Engineering, Department of Aerospace Engineering and Engineering Mechanics, Program in Engineering Mechanics, Austin, TX 78712-1111. Offers MSE, PhD. *Degree requirements:* For master's, foreign language and thesis not required; for doctorate, dissertation, qualifying exam required. *Entrance requirements:* For master's and doctorate, GRE General Test (minimum combined score of 1000 required). *Application deadline:* For fall admission, 2/1 (priority date); for spring admission, 10/1. Applications are processed on a rolling basis. Application fee: $50 ($75 for international students). *Financial aid:* Fellowships, research assistantships, teaching assistantships available. Financial aid application deadline: 2/1. *Unit head:* Dr. Edwin M. Gleason, Assistant Dean for Admissions and Student Affairs, 931-393-7432, Fax: 931-393-7346, E-mail: egleason@utsi.edu. *Application contact:* Dr. Eric Becker, Graduate Adviser, 512-471-7592.

University of Utah, Graduate School, College of Engineering, Department of Mechanical Engineering, Program in Applied Mechanics, Salt Lake City, UT 84112-1107. Offers MS. *Degree requirements:* For master's, thesis required, foreign language not required. *Entrance requirements:* For master's, GRE General Test, TOEFL (minimum score of 530 required), minimum GPA of 3.0. *Application deadline:* For fall admission, 7/1. Application fee: $30 ($50 for international students). *Unit head:* Dr. John H. Brown, Director of Admissions and Registrar, 210-458-4530. *Application contact:* Dr. Patrick McMurtry, Director of Graduate Studies, 801-581-3889.

University of Virginia, School of Engineering and Applied Science, Department of Civil Engineering, Program in Applied Mechanics, Charlottesville, VA 22903. Offers MAM, MS. Part-time programs available. *Faculty:* 5 full-time (0 women). *Students:* 1 full-time (0 women). Average age 24. 2 applicants, 50% accepted. *Degree requirements:* For master's, thesis required (for some programs), foreign language not required. *Entrance requirements:* For master's, GRE General Test. *Average time to degree:* Master's–1.7 years full-time. *Application deadline:* For fall admission, 2/1 (priority date). Applications are processed on a rolling basis. Application fee: $60. Electronic applications accepted. *Financial aid:* In 1998–99, 1 student received aid, including 1 fellowship with full tuition reimbursement available (averaging $12,000 per year), 1 research assistantship; teaching assistantships Financial aid application deadline: 2/1. *Faculty research:* Fluid mechanics, mechanics of composites, damage mechanics, finite element analysis, shell theory. Total annual research expenditures: $200,000. *Unit head:* Dr. Nicholas J. Garber, Chairman, Department of Civil Engineering, 804-924-7464, Fax: 804-982-2951, E-mail: njg@virginia.edu.

University of Wisconsin–Madison, Graduate School, College of Engineering, Department of Engineering Physics, Madison, WI 53706-1380. Offers engineering mechanics (MS, PhD); nuclear engineering and engineering physics (MS, PhD). Part-time programs available. Postbaccalaureate distance learning degree programs offered (minimal on-campus study). *Faculty:* 19 full-time (1 woman), 5 part-time (0 women). *Students:* 46 full-time (5 women), 5 part-time; includes 2 minority (both Asian Americans or Pacific Islanders), 19 international. Terminal master's awarded for partial completion of doctoral program. *Degree requirements:* For master's, thesis optional, foreign language not required; for doctorate, dissertation required, foreign language not required. *Entrance requirements:* For master's and doctorate, GRE General Test, minimum GPA of 3.0 in last 60 hours, appropriate bachelor's degree. *Application deadline:* For fall admission, 1/15 (priority date). Applications are processed on a rolling basis. Application fee: $45. Electronic applications accepted. *Unit head:* Dr. Gilbert A. Emmert, Chair, 608-263-1646, Fax: 608-263-7451, E-mail: emmert@engr.wisc.edu.

Virginia Polytechnic Institute and State University, Graduate School, College of Engineering, Department of Engineering Science and Mechanics, Blacksburg, VA 24061. Offers engineering mechanics (M Eng, MS, PhD). *Faculty:* 39 full-time (0 women), 2 part-time (1 woman). *Students:* 78 full-time (14 women), 8 part-time (2 women); includes 5 minority (1 African American, 4 Asian Americans or Pacific Islanders), 47 international. Average age 26. 50 applicants, 78% accepted. In 1998, 14 master's, 18 doctorates awarded. *Degree requirements:* For master's, thesis optional, foreign language not required; for doctorate, dissertation required, foreign language not required. *Entrance requirements:* For master's and doctorate, GRE General Test (minimum score of 400 on each section required), TOEFL (minimum score of 600 required). *Application deadline:* For fall admission, 12/1 (priority date). Applications are processed on a rolling basis. Application fee: $25. *Financial aid:* In 1998–99, 59 research assistantships, 17 teaching assistantships were awarded.; fellowships, career-related internships or fieldwork, Federal Work-Study, institutionally-sponsored loans, tuition waivers (full and partial), and unspecified assistantships also available. Aid available to part-time students. Financial aid application deadline: 4/1. *Faculty research:* Solid mechanics and materials, fluid mechanics, dynamics and vibrations, composite materials, computational mechanics and finite element methods. *Unit head:* Dr. Ed Henneke, Head, 540-231-6651, E-mail: henneke@vt.edu.

See in-depth description on page 1481.

Yale University, Graduate School of Arts and Sciences, Programs in Engineering and Applied Science, Department of Mechanical Engineering, New Haven, CT 06520. Offers applied mechanics and mechanical engineering (M Phil, MS, PhD). *Faculty:* 17. *Students:* 25 full-time (3 women), 15 international. Terminal master's awarded for partial completion of doctoral program. *Degree requirements:* For master's, foreign language and thesis not required; for doctorate, dissertation, exam required, foreign language not required. *Entrance requirements:* For master's and doctorate, GRE General Test, TOEFL. *Application deadline:* For fall admission, 1/4. Application fee: $65. *Unit head:* Chair, 203-432-4344. *Application contact:* Admissions Information, 203-432-2770.

Cross-Discipline Announcements

Carnegie Mellon University, School of Computer Science, Robotics Institute, Pittsburgh, PA 15213-3891.

Carnegie Mellon's MS and PhD programs in robotics are highly research oriented and interdisciplinary in nature, drawing from such fields as computer science, electrical and computer engineering, and mechanical engineering. Students may specialize in machine perception, artificial intelligence, manipulation, autonomous vehicles, manufacturing automation, or other areas. See in-depth description in Section 8: Computer Science and Information Technology.

Kettering University, Graduate School, Flint, MI 48504-4898.

Automotive Systems is a newly formed program under Kettering's MS in Engineering degree. The Mechanical Design major has recently been updated to better serve the needs of students interested in pursuing this degree. These majors are earned by attending classes on campus or by distance-learning videotape. Teaching and research assistantships are offered.

Pennsylvania State University University Park Campus, Graduate School, College of Engineering, Department of Architectural Engineering, State College, University Park, PA 16802-1503.

One of the major teaching and research groups in the department involves Building Mechanical and Energy Systems Engineering. Degrees offered include the M Eng, MS, and PhD. Areas of research include HVAC simulation and optimization, building energy analysis, building automation and control, solar and alternate energy sources, indoor air quality control, stratified chilled water, district heating and cooling, and smoke control. See in-depth description in the Architectural Engineering section.

Princeton University, Graduate School, Princeton Materials Institute, Princeton, NJ 08540-5211.

The Princeton Materials Institute welcomes students interested in cross-disciplinary research involving any aspect of materials science and engineering. Faculty members from 8 academic departments collaborate in a remarkably broad range of programs, supported by outstanding facilities. Some fellowships are available. See in-depth description in Section 21 of this guide.

Rensselaer Polytechnic Institute, Graduate School, School of Engineering, Department of Decision Sciences and Engineering Systems, Troy, NY 12180-3590.

This interdisciplinary department offers programs in manufacturing systems engineering, industrial and management engineering, operations research and statistics, and information systems. It prepares students to model complex systems and to use analytical and computational techniques in problem solving and the design of decision support systems. See in-depth description in Industrial Engineering section.

University of Wisconsin–Madison, Graduate School, College of Engineering, Department of Engineering Physics, Madison, WI 53706-1380.

The department offers graduate programs leading to the MS and PhD in engineering mechanics. Research in solid and structural mechanics, fracture mechanics, dynamics, fluid mechanics, rheology, biomechanics, and finite element methods. See in-depth description in Section 10, Energy and Power Engineering.

CARNEGIE MELLON UNIVERSITY

Carnegie Institute of Technology
Department of Mechanical Engineering

Programs of Study

Graduate research programs in the department span the range from projects in traditional mechanical engineering disciplines such as controls, dynamics, fluid and solid mechanics, heat transfer, and thermodynamics, to projects of a more applied nature, including combustion, composite materials, computational fluid dynamics, design, energy conversion, manufacturing, and robotics. The atmosphere of the department, together with its moderate size, encourages student-faculty interaction, sparking new research ideas and nurturing their development. Besides representing a range of subject areas, the programs involve varying combinations of experimental and theoretical activities, often featuring extensive use of computers. Students thus have a broad variety of choices to satisfy their professional objectives and interests. Some of the projects are carried out wholly within the department, while others involve considerable interaction with other departments or with centers and institutes established throughout the University—providing stimulation that contributes substantially to the vitality of the research. Each student's program of study is developed individually and is carefully tailored to his or her background, talents, and objectives. Typically, these programs provide a systematic development of the fundamentals of physical phenomena through course work based upon the various disciplines. At the same time, each student is brought into ongoing projects and research in which he or she participates with an increasing degree of responsibility. Completion of the student's degree program thus conjoins the principles basic to mechanical engineering and their application to significant, realistic problems. It is this duality of effort that distinguishes the department's graduate programs, and the extraordinary range of activities covered by mechanical engineering makes the field unusual among the traditional areas of engineering.

The Department of Mechanical Engineering offers three graduate degrees—the Master of Science/Course Work Option, the Master of Science/Project Option, and the Doctor of Philosophy—for work done either within the department or in conjunction with other programs in Carnegie Institute of Technology, the University's engineering college. In the latter case, requirements for the degree sought are developed jointly with the program.

Most students entering the Master of Science/Course Work Option degree program on a full-time basis hold a baccalaureate in engineering from an accredited institution. The Master of Science/Course Work Option degree is primarily a course work degree as the name suggests and typically takes about one year to complete.

The program for the Master of Science/Project Option degree includes a project and course work and requires a more extensive effort than for the Master of Science degree. Project activities in the Master of Science/Project Option program are of a substantive, professional nature and need approximately two years to complete. Most students choose to do their course work and project work concurrently.

Students admitted to the Doctor of Philosophy program should hold a master's degree in engineering or its equivalent, although some exceptional students are admitted directly into the Ph.D. program with a baccalaureate degree. The Ph.D. is a research-oriented degree, and admission to the program is limited to students who have demonstrated unusually high ability and who seek to make research a substantial part of their career. Ph.D. programs generally take between three and five years to complete, depending upon student interest, background, and project extent.

Research Facilities

A number of laboratories and other special facilities have been developed for the various research programs within the department. The Institute for Complex Engineered Systems (ICES) espouses four major thrusts: design and manufacturing, embedded and reliable information systems, MEMS and mechatronics, and tissue engineering and artificial organs, which work together to develop new methodologies, tools, and systems for effectively dealing with and reducing their ever-increasing complexity. Other laboratories available for research include the Fluid Mechanics Laboratory, the Thermal Sciences Laboratory, the Thermal Imaging Laboratory, a Combustion and Spray Laboratory, the Dynamic Systems Laboratory, the Acoustics/Vibrations Laboratory, a Manufacturing Sciences Laboratory, the Robotic Welding Laboratory, the Shape Deposition Laboratory, the Metrology Laboratory, the Controls Laboratory, and the Design Laboratory.

Financial Aid

Teaching assistantships and research assistantships are available to highly qualified full-time students pursuing either the Master of Science/Project Option degree or the Ph.D. degree. Teaching assistantships and research assistantships provide a tuition scholarship of $22,100 and a stipend of approximately $1400 per academic month, beginning fall 1999.

Cost of Study

In fall 1999, graduate student tuition is $22,100 per year, or $307 per unit for students enrolled for fewer than 36 units in any semester.

Living and Housing Costs

Living costs are moderate in the Pittsburgh area. Students generally find rentals in the Squirrel Hill, Shadyside, and Oakland areas satisfactory. In addition, the Housing Office maintains a current database of available local housing.

Student Group

Fifty-five full-time students and 15 part-time students are currently pursuing graduate work in the department and in the programs with which it is affiliated.

Location

The metropolitan area has more than 2 million people, many of them in local communities. Together, they support a broad range of cultural and sports activities, and many are staff members in the research laboratories and industries that are concentrated in large numbers in the Pittsburgh area.

The University

Carnegie Mellon is a small, private institution whose educational programs are professionally oriented. With a student-faculty ratio of about 10:1, there is a tradition of close interaction among the people in the various clusters of activity. A diverse, productive group of engineers, scientists, artists, and scholars is represented.

Applying

Application forms and instructions for submitting transcripts, letters of recommendation, and related materials should be obtained from the department. Applications for admission and financial aid should be made before February 1 for the fall semester (which begins in late August) and before October 15 for the spring semester (which begins in mid-January). All applicants are required to take the GRE General Test. International students who have not studied in the United States for at least two years must submit TOEFL scores. Further information is available upon inquiry.

Correspondence and Information

Graduate Program Coordinator
Department of Mechanical Engineering
Carnegie Mellon University
5000 Forbes Avenue
Pittsburgh, Pennsylvania 15213-3890

Telephone: 412-268-3175
Fax: 412-268-3348
E-mail: meche-admissions@andrew.cmu.edu

Carnegie Mellon University

THE FACULTY AND THEIR RESEARCH

Adnan Akay, Lord Professor and Head; Ph.D., North Carolina State, 1976. Acoustics, vibrations, noise control, and friction. Research on fundamental issues related to generation and transmission of sound and vibration, and engineering aspects of noise and vibration control. Emphasis on friction-induced vibrations with applications to automotive and aircraft braking systems and friction devices.

Cristina Amon, Professor; Sc.D., MIT, 1988. Computational fluid dynamics and heat transfer, stability and transition to turbulence, thermal modeling of electronic cooling, electronics packaging, concurrent thermal design, heat transfer enhancement techniques, thermal phenomena in manufacturing spray processes, transport in blood oxygenators and artificial lungs by experimental approaches and numerical techniques, including Spectral Element-Fourier Methods for DNS and Bayesian statistics.

Dwight M. B. Baumann, Professor; Sc.D., MIT, 1960. Design, entrepreneurship, transportation, microprocessor applications, command and control systems for the taxi, paratransit industry. Automation of the light aircraft flight. Projects include new biomedical devices and medical instrumentation and devices for laparoscopic surgery. Other topics include automation of net shape and lost foam casting.

Jack L. Beuth, Assistant Professor; Ph.D., Harvard, 1992. Solid mechanics, fracture mechanics, plasticity. Projects include the mechanics of layered manufacturing processes, fracture of thin-bonded films, interfacial fracture analysis of composite delamination, and the study of ductility limits in titanium aluminide aircraft engine components.

Jonathan Cagan, Professor; Ph.D., Berkeley, 1990. Design, optimization, artificial intelligence. Research focuses on design conceptualization and layout using advanced optimization/search techniques. Research pursues machine design, structural topology design, product layout, and integrated product development.

Norman A. Chigier, William J. Brown Professor and Director, Spray Systems Technology Center; Ph.D., 1961, Sc.D., 1977, Cambridge. Spray technology, atomization, combustion, particle size analysis, laser diagnostics; experiments in spray chambers, spray technology, molten metal sprays, automobile fuel injection emissions, planar mixing layers with density gradients; atomization and rheological properties of non-Newtonian liquids and slurries.

Howie Choset, Assistant Professor; Ph.D., Caltech, 1996. Robotics: sensor-based exploration, motion planning, highly articulated robots, mobile robotics, cooperating robots, inspection (of bridges), coverage algorithms, robotic de-mining, painting, sensors. Computational geometry: Voronoi diagrams, distance functions, nonsmooth analysis. Manufacturing, material transport.

Minking K. Chyu, Professor; Ph.D., Minnesota, 1986. Heat transfer, gas turbines, optical thermal sensing. Specific projects include cooling of gas turbine components, microscale flow and heat transfer, microelectromechanical systems.

Jerry H. Griffin, Professor; Ph.D., Caltech, 1973. Vibrations, damping, gas turbines, fatigue, dynamics, and friction. Projects include development of predictive models of damping due to microslip at friction interfaces, high-frequency tip mode vibration in low aspect ratio blades, and reduced-order models of blade mistuning.

William Messner, Associate Professor; Ph.D., Berkeley, 1992. Adaptive, nonlinear, and multi-input/multi-output controller design; data storage technology; and robotics. Research topics include control of dual stage actuator systems, distributed manipulation, position sensing, and optimal path planning.

Jayathi Y. Murthy, Associate Professor; Ph.D., Minnesota, 1984. Computational fluid dynamics and heat transfer. Research is a continuation of ten years of industrial CFD experience in developing fast, robust, and efficient solvers for problems of practical interest. Research includes development of finite

volume unstructured mesh methods; multigrid methods; numerical methods for radiation heat transfer; applications to industrial problems in automotive, chemical process industry, glass processing, and others; numerical modeling of multiphase flows and micro-manufacturing processes.

Allen L. Robinson, Assistant Professor; Ph.D., Berkeley, 1996. Combustion, heat transfer, and fluid dynamics. Experimental and analytical research focuses on energy and environmental problems. Current topics include carbon dioxide emissions, biomass fuels, fine particulate matter, and thermal properties of porous materials.

Wilfred T. Rouleau, Professor; Ph.D., Carnegie Tech, 1954. Fluid mechanics, thermodynamics, energy conversion, wave propagation, turbomachinery. Research includes aerodynamics of gas turbine erosion, wave propagation in Newtonian liquids and Bingham plastics with application to fluid transmission lines and bore-hole telemetry, efficiency improvement in Rankine cycle power plants, and automotive fuel cells.

Edward S. Rubin, Professor; Ph.D., Stanford, 1969. Design and analysis of energy conversion and environmental control systems, integrated assessment of energy-environmental issues, and interaction of technology and public policy. Current projects emphasize application to green design, electric power systems, and large-scale integrated assessments.

Kenji Shimada, Assistant Professor; Ph.D., MIT, 1993. Design, geometric modeling and simulation, CAD/CAM/CAE/CG. Topics include closer integration of design and analysis including automatic mesh generation, and application of computer networks, multimedia, and virtual reality to product design, analysis, and manufacturing.

Robert J. Simoneau, Distinguished Service Professor; Ph.D., North Carolina State, 1970. Heat transfer, fluid mechanics, and thermodynamics. Research interests in experiments in propulsion system phenomena such as turbulent heat transfer in highly disturbed flows, thermal management, and supercritical fluids.

Glenn B. Sinclair, Professor; Ph.D., Caltech, 1972. Solid mechanics, fatigue, fracture mechanics, tribology, and numerical methods. Research focuses on why various structural components break, fatigue, or wear out. Research concerns developing models that can predict when material will fail. New numerical techniques are often developed in the course of determining solutions.

Thomas F. Stahovich, Assistant Professor; Ph.D., MIT, 1995. Mechanical design and artificial intelligence. Research in computational support for design with emphasis on conceptual design and design rationale capture. Developing tools to read, understand, and use rough sketches engineers commonly use to record design information.

Paul S. Steif, Professor; Ph.D., Harvard, 1982. Applied mechanics, materials. Current projects include impact resistance of titanium aluminide in aircraft engines, damping due to rubber hysteresis and friction, machining due to particle-wall interactions in highly filled suspensions, stresses and tissue destruction during cyrosurgery.

Jonathan Wickert, Professor; Ph.D., Berkeley, 1989. Dynamics, dynamic systems, stability, mechanical vibration, active and passive vibration control. Experimental and analytical research focuses on vibration and stability of continuous or distributed-parameter mechanical systems, applications to computer disk and tape drives, rotating machinery, and manufacturing systems.

Shi-Chune Yao, Professor; Ph.D., Berkeley, 1974. Droplet and spray deposition, heat transfer and combustion, thermal modeling of material processes, and pollution reduction. Topics include spray modeling, paintings, impaction coolings in metallurgical processes, combustion and flame propagation in fuel sprays, and NO_x reduction using plasma-generated radicals.

DREXEL UNIVERSITY

Mechanical Engineering and Mechanics Department

Program of Study

Graduate study in the department leads to M.S. and Ph.D. degrees in mechanical engineering. The M.S. requires 45 quarter credits, which is the equivalent of one academic year of full-time study. The Ph.D. requires 90 quarter credits beyond the B.S. and a dissertation. At least one academic year of full-time study on campus is required.

The M.S. candidate develops a broad perspective and a mastery of fundamentals through a required sequence in mathematics and through sequences elected from fluid mechanics, heat transfer, dynamics, solid mechanics, thermodynamics, and systems analysis and control. Each student develops a plan of study in consultation with an adviser. The plan is flexible and allows electives in other curricula. A thesis (9 credits) is encouraged but is optional.

The Ph.D. is awarded on the attainment of a high level of accomplishment, as demonstrated by research achievements, and by passing written and oral examinations in a major study area and in applied mathematics. In general, it is expected that the student will take these qualifying examinations after successful completion of one year of graduate work at Drexel. The Ph.D. research is guided by an interdepartmental committee so that the student has wide latitude in selecting a research topic. Areas for research include biomechanics, fracture mechanics, heat transfer, solid mechanics, dynamics, energy, fluid mechanics, thermal science, combustion, fuels, active materials, composites, system dynamics, automatic controls, CAD/CAM, robotics, computational mechanics, and manufacturing. The dissertation research work must disclose a new solution to a significant scientific and technical problem. A public oral defense of the Ph.D. dissertation is required.

Research Facilities

Most of the department's experimental facilities are located at the Frederic O. Hess Engineering Research Laboratory at 34th Street and Lancaster Avenue. This modern well-equipped laboratory, which contains 35,000 square feet of usable space, provides excellent facilities for research activities and office space for full-time graduate students. Its facilities include machine and electronics shops and specialty laboratories in all of the major research areas, along with extensive support equipment such as computer data acquisition and control systems, oscilloscopes, high-speed cameras, an explosion chamber, a rapid prototyping machine, a CNC machining center, an injection molding machine, a coordinate measuring machine, a surface profilometer, a laser displacement meter, polariscopes, spectrum analyzers, testing machines (static and dynamic), gas chromatographs, GC/MSD, GC/FT-IR, laser anemometers and a phase Doppler particle analyzer, schlieren systems, holographic interferometry apparatus, a thermogravimetric analyzer, and several research internal combustion engines. A mainframe computer configured with vector processors in the University's Office of Computing Services and a large number of workstations in the departmental CAD Laboratory and Microcomputer Control Laboratory are accessible via networks to facilitate the department's teaching and research activities.

W. W. Hagerty library has an outstanding collection of more than 450,000 volumes, more than 6,800 serial titles, 17,000 microfilm reels, 600,000 microcard/fiche units, 100,000 documents, and numerous videotapes, audiotapes, records, and films. Several database searching services are available.

Financial Aid

Most full-time students receive financial aid. Half-time positions as research and teaching assistants pay tuition and fees, plus a monthly stipend, and allow a student to earn an M.S. in two academic years. Assistantships are ordinarily renewed for the second year of M.S. work and for further work toward a Ph.D.

Fellowships for full-time study toward a Ph.D. are initially granted for one year but are ordinarily renewed for one or two additional years of study. An award covers tuition and fees, plus a monthly stipend. These awards require no service to the University.

Under an industrial internship program, a student may earn an M.S. degree in twenty-one months by spending nine months at school, six months at work in a cooperating industry in an area consistent with the student's graduate program, and another six months at school. The internship program can carry a student toward the Ph.D.

Cost of Study

Tuition is $585 per credit for both M.S. and Ph.D. work; the University fee is $125 per term for full-time students and $67 per term for part-time students.

Living and Housing Costs

University-approved residences cost $1200 to $1400 per term for room and board. Housing of all types is available around the Drexel community for single and married students. Job opportunities are available for the spouses of married students.

Student Group

Drexel has more than 9,600 students from the U.S. and more than 100 countries. Of these, 1,900 are part-time graduate students and 900 are full-time graduate students. Within the graduate curriculum of the department, there are approximately 60 part-time and 40 full-time students.

Location

Drexel is located in Philadelphia, a city of 2 million and a center of science and industry. The campus is easily reached by bus, subway, railroad, and auto and is only a few minutes' walk from the heart of Philadelphia, close to its many centers of education, entertainment, culture, and industry. It is a convenient train or car ride to New York City, Washington, D.C., the Atlantic City boardwalk, and the Pocono Mountains resort and ski area. Philadelphia International Airport is only 15 minutes away by car or high-speed rail link.

The University

Drexel was founded in 1891. Its students are enrolled in six coeducational units: Engineering, Arts and Sciences, Business Administration, Nesbitt College of Design Arts, Evening College, and Information Science and Technology. Drexel has one of the largest private undergraduate engineering colleges in the country, and it is one of the top twenty private universities in the granting of master's degrees; it is rapidly expanding its doctoral degree offerings.

Applying

Classes start in mid-September and the beginning of January, April, and July. Completed applications for admission and supporting transcripts and references should be on file six weeks prior to the start of classes. Requests for financial aid should accompany applications for admission. Applications for financial aid must be received by February 15 for consideration for the next academic year. Inquiries may be directed to any member of the graduate faculty (listed on the back of this page) doing research in the student's area of interest.

Correspondence and Information

Office of Graduate Admissions, Box P
Drexel University
Philadelphia, Pennsylvania 19104
Telephone: 215-895-6700
E-mail: admissions@post.drexel.edu

Drexel University

THE FACULTY AND THEIR RESEARCH

Jonathan Awerbuch, Professor; D.Sc., Technion (Israel), 1972. Mechanics of composites, fracture and fatigue, impact and wave propagation, structural dynamics, nondestructive testing.

Leon Y. Bahar, Professor; Ph.D., Lehigh, 1963. Analytical methods in engineering, interaction between systems and control and analytical dynamics.

Nicholas P. Cernansky, Hess Chair Professor of Combustion and Interim Department Head; Ph.D., Berkeley, 1974. Combustion chemistry and kinetics, combustion generated pollution, utilization of alternative and synthetic fuels.

B. C. Chang, Professor; Ph.D., Rice, 1983. Computer-aided design of multivariable control systems, robust control systems, and optimal control systems.

Young I. Cho, Professor; Ph.D., Illinois at Chicago, 1980. Heat transfer, fluid mechanics, non-Newtonian flows, biofluid mechanics, rheology, phase-change heat transfer.

Kyung Choi, Research Professor; Ph.D., Wisconsin–Madison, 1983. Droplet two-phase flow and heat transfer, atomization and spray systems.

Pei C. Chou, Billings Professor Emeritus; Ph.D., NYU, 1951. Material response due to impulsive loading, dynamic properties of rubber and explosives, insensitive munitions.

Carlos F. M. Coimbra, Assistant Professor; Ph.D., California, Irvine, 1998. Applied math in fluid mechanics, heat and mass transfer, stokesian flows, turbulent particle dispersion, large-scale simulations, droplet/particle vaporization/devolatilization and combustion.

Jaydev P. Desai, Assistant Professor; Ph.D., Pennsylvania, 1998. Design and control of robotic systems, cybernetics, mechatronics, and biomechanics.

Baki Farouk, Professor; Ph.D., Delaware, 1981. Heat transfer, combustion, numerical methods, turbulence modeling, materials processing, two-phase flows.

Bernard B. Hamel, Research Professor; Ph.D., Princeton, 1963. Molecular and continuum gasdynamics, energy technology and utilization, waste disposal technology, artificial intelligence.

Harry G. Kwatny, Raynes Professor; Ph.D., Pennsylvania, 1967. Dynamic systems analysis, stochastic optimal control, control of electric power plants and systems.

Alan C. W. Lau, Associate Professor; Ph.D., MIT, 1982. Modern design and manufacturing, computational modeling and simulation, finite-element methods, fracture mechanics, damage-tolerant structural systems, composite materials.

David L. Miller, Associate Professor; Ph.D., LSU, 1984. Gas-phase reaction kinetics, thermodynamics, incineration of hazardous municipal and infectious wastes, engine combustion.

Gordon D. Moskowitz, Professor Emeritus; Ph.D., Princeton, 1964. Biomechanics, automated limb prosthesis, dynamics, design.

Gary A. Ruff, Associate Professor; Ph.D., Michigan, 1989. Two-phase flows, spray atomization, combustion, convective heat transfer, optical diagnostic techniques, image processing.

Rami Seliktar, Professor; Ph.D., Strathclyde (Scotland), 1971. Vehicle dynamics, orthopedic biomechanics, rehabilitation engineering, lower- and upper-limb prosthetics.

Sorin Siegler, Associate Professor and Graduate Advisor; Ph.D., Drexel, 1982. Orthopedic biomechanics, robotics, dynamics and control of human motion, applied mechanics.

Horacio Sosa, Associate Professor; Ph.D., Stanford, 1986. Theoretical and computational solid mechanics, electrodynamics of continuous media, active materials, fracture mechanics of piezoelectric ceramics.

Wei Sun, Assistant Professor and Director of Rapid Product Development Center; Ph.D., Drexel, 1992. Integrated CAD/CAE/CAM, concurrent engineering and rapid prototyping, FEM and design optimization, composites and smart materials.

Tein-min Tan, Associate Professor and Associate Department Head; Ph.D., Purdue, 1982. Mechanics of composites, computational mechanics and finite-elements methods, impact and wave propagation, structural dynamics.

Donald H. Thomas, Professor Emeritus; Ph.D., Case Tech, 1965. Control systems, biomechanics, product liability, vehicle dynamics, engineering design, computer-aided design.

Albert S. Wang, Soffa Professor; Ph.D., Delaware, 1967. Solid mechanics, mechanics of composite materials, fracture and fatigue of materials, engineering reliability.

Ajmal Yousuff, Associate Professor; Ph.D., Purdue, 1983. Optimal control, flexible structures, model and control simplifications.

Jack G. Zhou, Assistant Professor; Ph.D., NJIT, 1992. CAD/CAM, computer integrated manufacturing systems, manufacturing processes, rapid prototyping, system dynamics and automatic control, fractal geometry–engineering applications, tribology, precision instrumentation and measurement.

RESEARCH ACTIVITIES

The Mechanical Engineering and Mechanics Department is extremely active in research, the sponsors of which include, but are not limited to, the National Institutes of Health, the National Science Foundation, Electric Power Research Institute, U.S. Army, Navy, and Air Force, Philadelphia Electric Company, Sandia, Ford, Mobil Oil, General Motors Corporation, Department of Energy, Southwest Research, NASA, Westinghouse, Rockwell International Science Corporation, and Krusen Research Center. **Thermal/Fluid Sciences.** Energy conservation and utilization, building control systems, internal-combustion engine performance and emissions, pollutant formation and control, film cooling, heat transfer, buoyancy-driven flows, rotating flows, plasma depositions, hydrocarbon oxidation kinetics, turbulence, spray combustion, turbulent combustion, toxic-waste incineration, non-Newtonian flows, acoustic/fluid-flow interactions, high Kundsen number flows, multiphase (gas-droplet and liquid-bubble) flows. **Dynamic Systems and Controls.** Dynamic simulation of nuclear power plants; robust control; control theory modeling, simulation, and control of fossil power plants; optimal control of interconnected electric power systems; stability and bifurcations in nonlinear systems; aircraft flight mechanics and controls; trajectory optimization; control of flexible space structures; failure tolerance of fly-by-wire aircrafts. **Mechanics and Structures.** Structural dynamics and analysis; stress-wave mechanics; impact propagation in solid and composite materials; impact response of materials and structures; explosive and seismic loads; fracture, fatigue, and creep failure mechanisms in solids and composites; piezoelectrics ceramics, active materials; structural reliability; characterization of nonmetallic materials; numerical and computer-based methods of analysis (including finite-element codes); structural integrity of aging aircrafts. **Biomechanics.** Mechanical characteristics of bone and ligament; applications of robotics in orthopedic surgery; biomechanics of human joints; prosthetics and locomotion dynamics. **Computer-aided Design and Manufacturing (CAD/CAM).** Computer graphics, numerical simulation, manufacturing processes, computer numerically controlled (CNC) machinery, robotics, rapid prototyping, flexible manufacturing, rapid tooling, reverse engineering, virtual manufacturing.

DUKE UNIVERSITY

School of Engineering
Department of Mechanical Engineering and Materials Science

Programs of Study	The Department of Mechanical Engineering and Materials Science offers both the Master of Science and the Doctor of Philosophy degrees but places greater emphasis on the Ph.D. Flexible, highly individualized programs are offered at both levels. Each graduate student's program is developed individually and is tailored to meet his or her needs and objectives. Most master's degree aspirants write a research-based thesis, but a nonthesis option is available. There is no foreign language requirement for the Ph.D. degree.

Graduate study in the Department of Mechanical Engineering and Materials Science includes, for example, strong research projects in areas of aerodynamics and biomechanical engineering. Current research areas also include aeroelasticity and unsteady aerodynamics, especially as related to gas turbine engines; nonlinear dynamics and chaos; vortex dynamics; acoustics; structural acoustics; computational fluid dynamics; active magnetic suspension and vibration control; elastic, viscous, and adhesive properties of blood cells; structure and properties of biological membranes; drug delivery using microcarriers; acoustic and mechanical properties of kidney stones; thin-film processing; diffusion and point defects in semiconductors; light-emitting porous silicon; nondestructive testing through positron annihilation spectroscopy; the fine structure of metals and alloys; tribology; convective heat transfer; heat conduction and thermal radiation in microstructures; robotics; controls; active control; and adaptive structures. |
Research Facilities	The department's laboratories include a fully instrumented wind tunnel; materials laboratories with a scanning electron microscope; a high-vacuum sputtering system for thin-film synthesis; a scanning probe microscope; a pulsed-laser deposition system; a surface profiler; a low-temperature electrical transport measurement system with a superconducting magnet; a magnetometer; an AC susceptometer; a four-circle X-ray diffractometer; spectrometers; three positron annihilation systems; two cell biomechanics labs with inverted microscopes fitted with micromanipulators and video cameras; an electrohydraulic lithotripter; a vibrations laboratory; an acoustic laboratory with anechoic chamber; a microscale heat transfer laboratory with lasers, cryogenics, and analytic equipment; a low-temperature electrical transport measurement system; and a robotics laboratory. All computer workstations provide direct connections to the supercomputer facilities at the Microelectronics Center of North Carolina. The Engineering Library receives 800 periodicals and houses 75,000 volumes. The University has major research facilities in medicine, physics, and biology that interface with mechanical engineering and materials science research.
Financial Aid	Financial aid is available from the department to support highly qualified graduate students. Awards for beginning graduate students may provide for payment of tuition and fees and may also include a stipend typically of $1250 per month. James B. Duke Fellowships offer additional funding and are available on a competitive basis to qualified students. The teaching-assistance requirement is minimal, and the research assistance is related to research for the degree. Need is not a factor in graduate awards. Applicants from abroad must have outstanding credentials to be considered for initial support.
Cost of Study	Tuition for students in the Ph.D. program is $17,520 for the 1999–2000 year. Tuition for students in the master's program is $730 per unit in 1999–2000. The registration fee for both programs is $1250 per semester. Tuition for Ph.D. students is eliminated after the first five semesters for students entering Duke with a master's degree and after six semesters for students entering with a bachelor's degree.
Living and Housing Costs	The cost of University housing for a single student is expected to average about $4000 for the 1999–2000 academic year. Off-campus housing is plentiful. Meals are available through various University dining facilities.
Student Group	The department currently has about 60 full-time graduate students. Most are U.S. citizens, but a number of other countries are represented. The School of Engineering enrolls over 250 of the more than 3,500 graduate and professional students who attend the University.
Student Outcomes	Recent Ph.D. graduates have found employment opportunities ranging from joining faculties of major research universities to taking responsible positions in industrial research. Some have joined the professional staffs of federal research laboratories, particularly NASA.
Location	Duke University is located in the Research Triangle area of North Carolina. This region is known for its lack of big-city congestion and for having the highest ratio of Ph.D. holders to the general population of any area in the country.
The University	Duke is a private, multifaceted university of national stature that occupies a 9,000-acre campus. The graduate faculty numbers more than 1,400. It has a competitively selected student body and a low student-faculty ratio. There are many opportunities for interdisciplinary research within the University and with nearby laboratories and industry.
Applying	Applications for admission and financial aid awards for the fall semester should be completed by December 31. However, later applications receive full consideration within the limits of the existing funding then available. All applications for the fall semester must be completed prior to July 15; applications for the spring semester should be completed by November 1. Funding for spring semester awards is limited. Complete applications must include transcripts, GRE scores, and recommendations before they can be considered by the department. Applications may be obtained via the World Wide Web (http://www.gradschool.duke.edu/reqapp.htm).
Correspondence and Information	Charles M. Harman Director of Graduate Studies Department of Mechanical Engineering and Materials Science Duke University Durham, North Carolina 27708-0301 Telephone: 919-660-5310 E-mail: c.harman@duke.edu World Wide Web: http://www.egr.duke.edu/mems/index.html

Duke University

THE FACULTY AND THEIR RESEARCH

Listed below are the members of the faculty of the Department of Mechanical Engineering and Materials Science. The date in parentheses represents the year of appointment to the faculty.

Adrian Bejan, J. A. Jones Professor of Mechanical Engineering; Ph.D., MIT. Theoretical, numerical, and experimental investigation of heat transfer; thermodynamics and entropy generation minimization. (1984)

Donald B. Bliss, Associate Professor; Ph.D., MIT. Modeling the wake of a hovering rotor for numerical simulation, analysis of blade wake roll-up in inviscid and viscous flow, acoustic analysis with application to reducing aircraft interior noise. (1985)

Robert L. Clark, Assistant Professor; Ph.D., Virginia Tech. Active control of sound and vibrations, adaptive structures, acoustics and structural acoustics. (1992)

Franklin H. Cocks, Professor and Chairman; Sc.D., MIT. Materials science: mechanical properties of human calculi such as kidney stones and acoustical techniques for their comminution, high-temperature (90K) superconductors, crystal growth, X-ray reflection and diffraction. (1972)

Earl H. Dowell, Professor and Dean of Engineering; Sc.D., MIT. Dynamics of nonlinear fluid and structural systems; systems with many degrees of freedom and various damping mechanisms, with a focus on the dynamic instability (flutter) of such systems and their associated limit cycle and chaotic motions; applications to aerospace and other transportation vehicles. (1983)

Chang-Beom Eom, Associate Professor; Ph.D., Stanford. Thin-film processing and nanostructure fabrication of novel materials, such as high T_c superconductors, ferroelectrics, metallic oxides, and magnetic materials for device applications. (1993)

Linda P. Franzoni, Associate Professor; Ph.D., Duke. Innovative analysis methods for structural/acoustic systems, scattering and radiation from submarine hulls and vehicular interior noise. (1998)

Devendra P. Garg, Professor; Ph.D., NYU. Modeling, simulation, and control of linear and nonlinear automatic control systems; robotics and automated manufacturing systems. (1972)

Kenneth C. Hall, Associate Professor; Sc.D., MIT. Computational and analytic modeling of unsteady flows in turbomachinery and about aircraft; structural dynamics and aeroelasticity, with an emphasis on cascade flutter and forced response. (1990)

Charles M. Harman, Professor; Ph.D., Wisconsin. Energy analysis, thermodynamics. (1961)

Robert M. Hochmuth, Professor; Ph.D., Brown. Cellular biomechanics, viscoelasticity and adhesivity of cells. (1978)

Laurens E. Howle, Assistant Professor; Ph.D., Duke. Numerical simulation of nonlinear convection systems, nonlinear dynamics and control of chaotic systems. (1993)

Phillip L. Jones, Associate Professor; Ph.D., UCLA. Materials science: nanometer structural characterization of materials by positron annihilation lifetime and Doppler spectroscopy, effects of processing variables on the fine structure of a number of emerging materials systems examined by these means. (1977)

Josiah D. Knight, Associate Professor; Ph.D., Virginia. Lubrication, thermal and cavitation effects in bearings, vibration and rotor dynamic stability in turbomachinery, active control of vibration, magnetic bearings. (1985)

David Needham, Associate Professor; Ph.D., Nottingham. Biological-materials science, development of new encapsulation and coating technologies for micro and nano particles composed of gases, liquids, and solids. (1987)

George W. Pearsall, Professor; Sc.D., MIT. Materials science: failure analysis of metals, polymers, and glasses; structure-property relationships in engineering plastics; risk analysis and product safety in engineering design. (1964)

Edward J. Shaughnessy, Professor; Ph.D., Virginia. Flows in engine inlet protection devices; computational models of flow in the coronary arteries; heat, mass, and momentum transfer to fractal surfaces; conjugate convection in fractured rock. (1975)

Teh Y. Tan, Professor; Ph.D., Berkeley. Structural, analytical (electron microscopy and X-rays), thermodynamic, kinetic, and diffusional aspects of crystalline materials; point defects, line defects, and gettering in semiconductors. (1986)

H. Ping Ting-Beall, Research Assistant Professor; Ph.D., Tulane. Morphology and viscoelasticity of leukocytes and *in vitro* cytotoxicity testings and biomarkers. (1992)

Lawrence N. Virgin, Associate Professor; Ph.D., London. Experimental nonlinear dynamical systems, predicting instabilities, buckling, aeroelastic flutter, ship dynamics, chaos. (1989)

Donald Wright, Associate Professor; Ph.D., Purdue. Development of computer techniques for analysis of failed metallic components, using methods of artificial intelligence and expert systems. (1967)

Doncho V. Zhelev, Research Assistant Professor; Ph.D., Bulgarian Academy of Sciences. Cellular biomechanics and biomaterials: cellular polymers and cell motility, model membranes, membrane stability, and peptide-membrane interactions. (1994)

Pei Zhong, Research Associate Professor; Ph.D., Texas at Arlington. Shock-wave interaction with biological tissue and concrete such as kidney stones, acoustic cavitation. (1994)

FLORIDA AGRICULTURAL AND MECHANICAL UNIVERSITY / FLORIDA STATE UNIVERSITY

College of Engineering
Department of Mechanical Engineering

Programs of Study

Because mechanical engineering is a very broad discipline, encompassing basic sciences as well as different areas of engineering, the education, both undergraduate and graduate, and research in mechanical engineering at the FAMU/FSU College of Engineering are designed to provide an excellent technical background for a wide variety of careers in energy industries, product manufacturing industries, national and industrial laboratories, government agencies dealing with science and technology, academic and research institutions, and multidisciplinary fields. The department is organized broadly into the areas of thermal/fluid sciences, mechanics and materials, and controls and mechatronics. Graduate programs of study leading to a Master of Science and Ph.D. in mechanical engineering are offered. Current research interests include experimental and theoretical studies in the areas of fluid mechanics, aerodynamics, heat transfer, cryogenics, computational mechanics, automatic control, mechatronics, robotics, solid mechanics, materials, magnet technology, and computer-aided design. Cooperative research programs are carried out with the Supercomputer Computations Research Institute, Geophysical Fluid Dynamics Institute, Material Research and Technology Center, Department of Applied Mathematics, National High Magnetic Field Laboratory, and the Center for Nonlinear and Nonequilibrium Aeroscience.

The M.S. degree program consists of a thesis and a nonthesis option, each requiring the completion of 33 semester hours of course work. Of these, a maximum of 6 units taken at the graduate level can be transferred from other universities. In the nonthesis option, students substitute thesis work and thesis defense with equivalent course work and an oral comprehensive examination at the completion of their curriculum requirements. The Ph.D. degree program requires the completion of 45 semester hours beyond the M.S. degree. Details of course work and degree requirements are available from the address below.

Research Facilities

The Department of Mechanical Engineering conducts research in the areas of fluid mechanics, combustion, cryogenics, acoustics, solid mechanics, materials, controls, robotics, and sensor fusion. The facilities in the fluid mechanics research laboratory include a closed-circuit anechoic subsonic wind tunnel, a high-pressure blow down facility, a transonic wind tunnel, and aeroacoustic facilities. The laser diagnostics and optics laboratory has the latest equipment for advanced measurement techniques, such as PIV and LIF. The mechanics and materials laboratory contains an MTS 810 machine with computer control for data acquisition and analysis. The control and vibrations laboratory has vibration monitoring equipment, Pentium workstations, and several controls experiments. In the CAD/CAM laboratory, IBM RS/6000, SGI Indy and Pentium workstations, and software including AUTOCAD, ICM GMS, ICM Lynx2, are used.

Computer facilities include an MDI Adams, ETA FEMB, and CADSI DADS and a variety of workstations and Pentium PCs. The University Computing Center operates a 4-Processor CRAY Y-MP and other machines. The Supercomputer Computations Research Institute operates an eight-processor SGI PowerChallenge XL, an eight-processor IBM SP2, and a highly parallel CM2 64K-processor Connection Machine SIMD supercomputer.

Financial Aid

Teaching and research assistantships are available on a competitive basis. In the past, almost all graduate students have been provided with some form of financial assistantship. Assistantship stipends range from $9700 to $11,500 for full-time work for the nine-month academic year. Tuition waivers are also obtainable based on availability of funds and academic standing.

Cost of Study

Tuition in 1998–99 was $138.83 per credit hour for Florida residents and $482.39 per credit hour for out-of-state students.

Living and Housing Costs

On-campus housing is available at an estimated annual cost of $2220. Off-campus rental rates start at about $350 per month.

Student Group

Current enrollment in the College is approximately 2,000 students. In the fall 1998 semester, the Department of Mechanical Engineering had an enrollment of 19 Ph.D. and 24 M.S. students.

Location

Florida A&M University and Florida State University are located in Tallahassee, the state capital. Although Tallahassee is among the nation's fastest-growing cities, it has managed to preserve its natural beauty. Five large lakes surrounding Tallahassee, as well as the nearby Gulf of Mexico, offer numerous recreational opportunities including canoeing, fishing, waterskiing, boating, camping, and hunting.

The Universities

Florida A&M and Florida State University are public, coeducational institutions founded in 1887 and 1857, respectively, with current enrollments exceeding 38,000. The universities have great diversity in their cultural offerings and are rich in tradition. They have outstanding science departments and excellent schools and departments in such varied areas as law, music, theater, and religion. They are the homes of the renowned Rattlers and Seminoles.

Applying

Applications should be submitted as early as possible in the academic year prior to anticipated enrollment. The deadline for applications for the fall semester is March 1 for international students and July 1 for U.S. citizens. Students seeking assistantships are advised to apply by January 15. Master's candidates should have a bachelor's degree in mechanical engineering, while doctoral candidates should have both a bachelor's and master's degree in mechanical engineering.

Correspondence and Information

Completed FAMU application forms should be submitted to:
Graduate Admissions Office
400 Tucker Hall
Florida A&M University
Tallahassee, Florida 32307

FSU application forms should be sent to:
Graduate Admissions Office
2500 University Center
Florida State University
Tallahassee, Florida 32306-2400

For departmental information and financial aid applications:
Graduate Coordinator
Department of Mechanical Engineering
FAMU/FSU College of Engineering
2525 Pottsdamer Street, Room 229
Tallahassee, Florida 32310

Telephone: 850-410-6331
World Wide Web: http://www.eng.fsu.edu/

Florida Agricultural and Mechanical University/Florida State University

THE FACULTY AND THEIR RESEARCH

Farrukh Alvi, Visiting Assistant Professor of Mechanical Engineering; Ph.D., Penn State. Fluid dynamics, gasdynamics, optical fluid diagnostics.

George Buzyna, Professor of Mechanical Engineering; Ph.D., Yale. Physical gasdynamics, geophysical fluid dynamics: convection in rotating stratified fluids and transition to turbulence.

Namas Chandra, Professor of Mechanical Engineering; Ph.D., Texas A&M. Finite-element methods, solid mechanics, superplastic metal forming, composite materials.

Ching-Jen Chen, Professor of Mechanical Engineering and Dean; Ph.D., Case Tech. Heat transfer and computational fluid mechanics.

Emmanuel Collins, Associate Professor of Mechanical Engineering; Ph.D., Purdue. Controls and dynamics, numerical methods.

Frederick Foreman, Assistant Professor of Mechanical Engineering; Ph.D., Florida A&M. Computational structural mechanics, design.

Hamid Garmestani, Associate Professor of Mechanical Engineering; Ph.D., Cornell. Structural mechanics, material science, composite materials.

Peter Gielisse, Professor of Mechanical Engineering; Ph.D., Ohio State. Materials science, high-temperature composite materials, superconducting materials.

Patrick Hollis, Associate Professor of Mechanical Engineering; Ph.D., Cornell. Controls, nonlinear dynamical systems, robotics.

Louis Howard, Affiliate Professor of Mechanical Engineering and Professor of Mathematics; Ph.D., Princeton. Hydrodynamic stability, natural convection, bifurcation theory, geophysical fluid dynamics and fluid mechanics.

Joseph Johnson III, Affiliate Professor of Mechanical Engineering and Professor of Physics; Ph.D., Yale. Gasdynamics and reacting flows.

Peter Kalu, Assistant Professor of Mechanical Engineering; Ph.D., London. Mechanical and physical metallurgy, composites, deformation and recrystallization, superplasticity, texture and microtexture.

David Kopriva, Affiliate Professor of Mechanical Engineering and Associate Professor of Mathematics; Ph.D., Arizona. Computational fluid dynamics, numerical analysis, numerical solution of partial differential equations.

Anjaneyulu Krothapalli, Professor of Mechanical Engineering and Chairman; Ph.D., Stanford. Experimental fluid mechanics, aeroacoustics, aerodynamics, integrated design systems.

David Loper, Affiliate Professor of Mechanical Engineering and Professor of Mathematics; Ph.D., Case Tech. Heat transfer, non-Newtonian fluid mechanics, two-phase flows.

Luiz Lourenco, Professor of Mechanical Engineering; Ph.D., Brussels. Two-phase flows, laser diagnostics, instrumentation and heat transfer.

Arthur Mutambara, Assistant Professor of Mechanical Engineering; Ph.D., Oxford. Controls, robotics, sensor fusion.

Simone Peterson, Assistant Professor of Mechanical Engineering; Ph.D., MIT. Materials processing, ceramics, superconducting materials.

Justin Schwartz, Associate Professor of Mechanical Engineering; Ph.D., MIT. Magnetic engineering, superconductors, bulk processing, fluxpinning mechanisms and irradiation effects, nuclear engineering.

John Seely, Adjunct Professor of Mechanical Engineering; M.S., Stevens. Heat transfer and microelectronic equipment.

Chiang Shih, Associate Professor of Mechanical Engineering; Ph.D., USC. Unsteady aerodynamics, turbulent shear flows, laser diagnostics, instrumentation in fluid mechanics, high-lift rotor-blade aerodynamics, laminar and turbulent wall jets, thrust-induced effects on pitch-up delta wing.

Christopher Tam, Affiliate Professor of Mechanical Engineering and Professor of Mathematics; Ph.D., Caltech. Turbulence, aeroacoustics, perturbation methods, numerical methods.

Leon Van Dommelen, Professor of Mechanical Engineering; Ph.D., Cornell. Theoretical and computational fluid mechanics, computational mechanics and numerical methods.

Steven Van Sciver, Professor of Mechanical Engineering; Ph.D., Washington (Seattle). Cryogenics and heat transfer.

Research Fields

Fluid Dynamics and Thermal Sciences
Aeroacoustics
Biomedical fluid applications
Combustion
Computational fluid mechanics
Cryogenics
Experimental fluid mechanics
Gasdynamics
Geophysical fluid dynamics
Heat transfer
Hydrodynamic stability
Laser and optical diagnostics
Natural Convection

Mechanics and Materials
Bulk processing
Ceramics
Composite materials
Finite-element methods
Magnetic materials

Microscopy
Solid mechanics
Superconducting materials
Superplastic metal forming
Textures/microtexture

Controls and Mechatronics
Acoustic control
Decentralized estimation and control
Fixed-architecture control
Friction control
Fuzzy control
Mechatronic design
Navigation of autonomous vehicles
Neural network control
Numerical methods in control
Robotics
Robustness analysis and robust control
Sensor Fusion

FLORIDA INSTITUTE OF TECHNOLOGY

College of Engineering
Mechanical and Aerospace Engineering Programs

Programs of Study

The graduate program in mechanical engineering has three areas of specialization: dynamic systems, robotics, and controls; structures, solid mechanics, and materials; and thermal-fluid sciences. The graduate program in aerospace engineering also has three areas of specialization, which include aerodynamics and fluid dynamics, aerospace structures and materials, and combustion and propulsion. The master's degree requires 30 semester credit hours of course work, which may include 6 semester credit hours of thesis. The Ph.D. degree is offered in the same areas of specialization for each program for students who wish to carry out advanced research. The Ph.D. is conferred primarily in recognition of creative accomplishments and ability to independently investigate scientific or engineering problems. The work should consist of advanced studies and research leading to a significant contribution to the knowledge in a particular subject.

Research Facilities

Located in the Fluid Mechanics and Aerodynamics Laboratory are two low-speed wind tunnels, a smoke tunnel, a hydrostatic test bench, and a wave tank. Instrumentation includes hot-wire anemometry, laser Doppler velocimetry, pressure scanning systems, flow visualization, and computerized data acquisition. The Energy Laboratory includes experimentation of buoyancy effects and combustion, emission, and radiation effects in porous ceramic burners and internal combustion engines. The nearby Florida Solar Energy Center also provides the opportunity for collaborative energy research. The Mechanics and Materials Laboratory includes ovens, a platen press, grinding and polishing benches, and metallographic microscopes. Servo-hydraulic axial test, vibration, and instrumented low-energy impact systems are used to study the mechanical behavior of advanced composite and recycled plastic materials. The Dynamic Systems and Controls Laboratory provides a facility for research work on machinery diagnostics and mechanatronics. Also, the Robotics Laboratory provides advanced robot research capabilities.

Financial Aid

Graduate student assistantships, awarded each year to a limited number of highly qualified entering students, provide full tuition beginning in the fall semester, plus a stipend of up to $6460 for the 1999–2000 year. Assistants are assigned duties related to both undergraduate instruction and faculty research.

Cost of Study

Tuition is $575 per credit hour in 1999–2000. There is an entrance deposit of $300 for new students. Book costs are estimated at $550 per year.

Living and Housing Costs

Room and board on campus cost approximately $2400 per semester in 1999–2000. On-campus housing (dormitories and apartments) is available for full-time single and married graduate students, but priority for dormitory rooms is given to undergraduate students. Many apartment complexes and rental houses are available near the campus.

Student Group

Graduate students constitute more than one fourth of the approximately 4,000 students at Florida Tech's Melbourne campus. Only about one fourth of the students are from the state of Florida; the remainder are from all parts of the United States and from many other countries. There are 40–45 Mechanical and Aerospace Engineering graduate students.

Student Outcomes

Graduates of the program obtained positions in various companies, such as NASA, United Technologies, Lockheed Martin, Northrop Grumman, McDonnell Douglas, Harris Corp., Westinghouse, Rockwell International, The New Piper Aircraft, Honeywell Aircraft Systems, I-NET, Rosemount Inc., Johnson Controls World Services, Loral Aerospace, Boeing, Earth Tech, MCNC, North Carolina Supercomputing Center, U.S. Navy Civilian Personnel, U.S. Air Force Civilian Personnel, and EG&G.

Location

The greater Melbourne metropolitan area is located on Florida's "Space Coast," south of the Kennedy Space Center and Cape Canaveral Air Force Station. The climate is mild both winter and summer, with abundant sunshine and little variation in temperature. Opportunities for recreation include extensive ocean beaches; the Indian River, an attractive saltwater lagoon between the campus and the ocean; the St. Johns River and Lake Washington west of Melbourne; and central Florida's numerous commercial attractions. There are also outstanding entertainment, shopping, and housing facilities in the Melbourne area.

The Institute

In response to a need for specialized and advanced educational opportunities, Florida Institute of Technology was founded in 1958 by a group of scientists and engineers pioneering America's space program at Cape Canaveral. Florida Tech has rapidly developed into a residential institution that is the largest private technological university in the Southeast. Supported by community and industry, Florida Tech is currently the recipient of many research grants and contracts, a number of which provide financial support for graduate students. The campus is situated on 175 acres of partially wooded and beautifully landscaped grounds.

Applying

Forms for applying for admission and assistantships are sent on request. Admission in the fall semester is recommended, but full-time students may also enter in the spring semester, and part-time students may enter in any semester. Full-time students entering in the spring should plan a reduced course load in their first semester. International students should apply at least six months in advance. Assistantship applications and all supporting material must be received by March 1.

Correspondence and Information

Dr. J. Engblom, Program Chair, Mechanical
    Engineering
Dr. L. Krishnamurthy, Chair, Aerospace Engineering
Florida Institute of Technology
150 West University Boulevard
Melbourne, Florida 32901
Telephone: 407-674-8092
E-mail: engblom@zach.fit.edu (mechanical
    engineering)
    krishna@zach.fit.edu (aerospace engineering)

Graduate Admissions Office
Florida Institute of Technology
150 West University Boulevard
Melbourne, Florida 32901-6975
Telephone: 407-674-8027
World Wide Web: http://www.fit.edu

Florida Institute of Technology

THE FACULTY AND THEIR RESEARCH

Thomas E. Bowman, Professor; Ph.D., Northwestern. Fluid mechanics and thermal sciences; energy conversion, solar energy, appropriate solar technology; aerospace propulsion systems, low-gravity fluid mechanics, and heat transfer.

John J. Engblom, Professor and Program Chair, Mechanical Engineering; Ph.D., UCLA. Computational and experimental mechanics, finite element and boundary element methods development/modeling, mechanics of plastics/composites, computer-aided design, structural dynamics, stability.

David C. Fleming, Assistant Professor; Ph.D., Maryland. Structural mechanics, advanced composite materials, crashworthy aerospace vehicle design, finite-element analysis.

Hector Gutierrez, Assistant Professor; Ph.D., North Carolina State. Dynamic systems and control, design and analysis of real-time systems, design of system components: actuators/sensors, manufacturing systems, electromechanical systems.

Pei-feng Hsu, Associate Professor; Ph.D., Texas at Austin. Radiative and multimode heat transfer, numerical methods, premixed combustion modeling in porous ceramics, data acquisition and thermal conductivity, measurements, heat exchange and thermal systems design.

L. Krishnamurthy, Professor and Program Chair, Aerospace Engineering; Ph.D., California, San Diego. Fluid dynamics of chemically reacting flows, combustion theory, turbulence modeling, computational fluid dynamics, asymptotic analyses and perturbation techniques, fire research, aerospace propulsion.

Pierre Larochelle, Assistant Professor; Ph.D., California, Irvine. Theoretical kinematics, mechanism and machine design, robotics, dynamics and controls of mechanical systems, computer-aided design.

Kunal Mitra, Assistant Professor; Ph.D., Polytechnic. Thermal-fluid sciences, with emphasis on laser applications, thermal radiation, microscale heat transfer, materials processing, bioheat transfer modeling, heat conduction, and solar energy analysis.

John M. Russell, Associate Professor; Sc.D., MIT. Fluid dynamics, mathematical theory of shear flow instability, constitutive theory, dynamics of vortex tubes and filaments, applied aerodynamics, flight vehicle stability and control, applied mathematics.

Paavo Sepri, Associate Professor; Ph.D., California, San Diego. Fluid mechanics, turbulence, convective heat transfer, boundary layers, aerodynamics, wind-tunnel testing, droplet combustion, computational fluid dynamics.

Yahya I. Sharaf-Eldeen, Associate Professor; Ph.D., Ohio State; Ph.D., Oklahoma State. Modeling, simulation, and design of dynamic systems; advanced dynamics, vibration, and design of machinery; thermal-fluid sciences and energy/power systems.

Palmer C. Stiles, Assistant Professor; M.S., Illinois. Machine design, computer-aided manufacturing, robotics, all-terrain and human-powered vehicles, water conservation devices, small-aircraft design, earth-to-orbit vehicles.

Chelakara S. Subramanian, Associate Professor; Ph.D., Newcastle (Australia). Experimental fluid mechanics, turbulence measurements and modeling, data processing techniques, wind-tunnel experimentation, flow instabilities, structure of complex turbulent flows, boundary layer receptivity.

RESEARCH AREAS

Thermal Sciences

In the disciplines of energy, heat transfer, and combustion, research programs address aspects relevant to buoyancy-induced flows; density inversion phenomena; advanced computational methods for solving problems in radiative transport, mixed-mode heat transfer, and electronic cooling applications; combustion in porous ceramics; aerospace propulsion; automotive combustion; and issues related to environmental pollution and energy.

Fluid Mechanics and Aerodynamics

Turbulence research within the mechanical and aerospace engineering programs is being carried out experimentally in topics such as boundary layers with embedded vortices, spinning objects, passive control of flow separation, and the effect of unsteadiness on shear-layer instabilities. Theoretical and computational research is being pursued to characterize flow instabilities, leaks of cryogenic fluids, turbulent boundary-layer structure, aerodynamic interactions, moisture transport, and internal flow configurations. Research is also under way in wind engineering.

Structures, Solid Mechanics, and Materials

Efforts in mechanics and materials engineering focus on characterizing the mechanical behavior of composite materials (including recycled plastics) and on the design and manufacture of structures made of them. The relationship between microstructure and macroscopic behavior is being studied to better understand the effects of environmental conditions on the constituents of composite materials. Research is being conducted experimentally by the application of static and dynamic loads on recycled plastic and composite materials leading to an understanding of damage mechanisms. Computational models are being developed for predicting the mechanical performance of recycled plastic structures with reinforcement, for the propagation of delaminations, for the optimization of composite structure geometries, and for proving helicopter crashworthiness.

Systems Dynamics, Design, and Control

Research in the design and control of machine systems and other generalized systems is being pursued along both theoretical and practical avenues. Recent research contributions have been made in the synthesis and analysis of spatial and spherical mechanisms. Along the experimental avenue, research is being conducted in the monitoring and diagnosis of vibration and flutter in rotating machinery. Other research topics include robotics, optimal generation of prime-mover power, design of drive-trains, roadable aircraft, manufacturing systems, and high-speed transit concepts.

THE GEORGE WASHINGTON UNIVERSITY

School of Engineering and Applied Science
Department of Mechanical and Aerospace Engineering

Programs of Study

Programs are offered leading to the Master of Science degree (M.S.), the professional degrees of Applied Scientist (App.Sc.) and Engineer (Engr.), and the Doctor of Science degree (D.Sc.). The mechanical engineering field offers areas in aerospace engineering; design of mechanical engineering systems; fluid mechanics, thermal sciences, and energy; solid mechanics and materials science; and transportation safety engineering. Certain courses in aeroacoustics, aeronautics, and astronautics are usually offered only at NASA Langley Research Center in Hampton, Virginia, under the NASA-GW Joint Institute for the Advancement of Flight Sciences; however, full M.S. and D.Sc. degree programs in aerospace engineering are offered on the Washington, D.C., campus. Each student's program is individually planned with the adviser. The M.S. programs require satisfactory completion of 30 semester hours of graduate-level courses with a thesis (equivalent to 6 semester hours), or 33 semester hours without a thesis. Students who select the thesis option are required to defend their thesis. Professional degree programs consist of a minimum of 30 semester hours of approved graduate courses beyond a master's degree and a design project. The doctoral program prepares students for careers of creative scholarship. The program requires a minimum of 30 semester hours of formal study beyond the master's degree or a minimum of 54 credit hours of approved graduate work for students whose highest earned degree is a baccalaureate, prior to the qualifying examination. The research phase culminates in the presentation and oral defense of a dissertation. Work on the dissertation is equivalent to a minimum of 24 semester hours.

Research Facilities

The department has a wide variety of laboratory facilities that include three MTS random-cycle testing systems and other materials testing machines, a scanning electron microscope, and laboratories for mechanical engineering instrumentation, fluid mechanics and propulsion, combustion diagnostics, mechanical engineering design, computer-aided manufacturing, robotics, and general mechanical engineering and applied mechanics. The School provides UNIX, PC, and Macintosh resources, featuring a Sun Microsystems S1000 multiprocessor server and Ultra 3D workstations, Dell Pentium Pro II computers, and Apple Power Macintosh G3 computers. Systems are equipped with licensed software from leading manufacturers, such as CADENCE, MathWorks, Synopsys, AdaCore Technologies, Sun Microsystems, Autodesk, Asymetrix, Adobe, and Microsoft. In addition, the University provides extensive computing resources with a Sun Microsystems SPARC Center 2000E for academic information system services and a Sun Microsystems Enterprise 4000, with four processors, dedicated to research computing support. A high-performance ATM network integrates these resources, and Internet connectivity is provided for research support. The University libraries contain more than 1.7 million volumes, and students have access to the Library of Congress and the libraries of six other local universities.

Financial Aid

Teaching assistantships provided remission of tuition for 9 hours per semester and $1000–$3500 for each course taught in 1998–99. Research assistants received $8000–$15,000 for the calendar year. School Graduate Fellow, Dean's Fellow, and Department Fellow awards ranged from $7500 to $15,000 for full-time students. Full-time students who are U.S. citizens or permanent residents may be eligible for Graduate Engineering Honors Fellowships. The George Washington University–NASA Langley Research Center programs offer research scholar assistantships, including tuition, to students in the M.S. or D.Sc. programs in aerospace engineering; stipends were $14,500 a year for the M.S. and $15,500 a year for the D.Sc. The George Washington University–NASA Goddard Space Flight Center Cooperative Research and Education Program in Space Technology offers research scholar assistantships to students in the M.S. or D.Sc. programs in space systems and robotics; stipends ranged up to $15,000 a year plus tuition.

Cost of Study

In 1998–99, tuition was charged at the rate of $680 per semester hour, payable on a course-by-course basis.

Living and Housing Costs

Apartments for students attending the George Washington University are available in the area at a wide range of costs, starting at about $600 a month.

Student Group

Students in the School of Engineering and Applied Science include graduates of most colleges and universities in the United States and a number of other countries. Approximately 1,000 students are working on master's degrees, 20 on professional degrees, and 440 on the D.Sc.

Location

The Washington, D.C., area has the second-largest concentration of research and development activity in the United States. The department offers programs at both the main campus in the Foggy Bottom historic district and in the branch campuses in Virginia and Maryland.

The University and The School

The School was organized in 1884. It operates on a two-semester academic year. Limited course work in engineering, engineering administration, operations research, physical science, mathematics, economics, and statistics is available during the University's summer sessions.

Applying

Admission to a master's degree program generally requires an appropriate bachelor's degree from a recognized institution and evidence of a capacity for productive work in the field selected. Admission to professional degree programs requires an appropriate master's degree from a recognized institution and evidence of capacity for productive work in the field selected, as indicated by prior scholarship and/or professional experience. Applicants for doctoral study must have adequate preparation for advanced study, including demonstrated capacity for original scholarship and, if the highest degree earned is the baccalaureate, course work pertinent to the field, or else a master's degree or the equivalent in engineering or a related field. March 1 and October 1 are the respective fall and spring priority application deadlines.

Correspondence and Information

Professor Michael Myers
Department of Mechanical and Aerospace Engineering
School of Engineering and Applied Science
The George Washington University
Washington, D.C. 20052
Telephone: 202-994-6749
Fax: 202-994-0238
E-mail: cmee@seas.gwu.edu
World Wide Web: http://www.seas.gwu.edu

Interim Dean Thomas A. Mazzuchi
School of Engineering and Applied Science
The George Washington University
Washington, D.C. 20052
Telephone: 202-994-3096
       800-537-7327 (toll-free)
Fax: 202-994-4522

The George Washington University

THE FACULTY AND THEIR RESEARCH

Andrew D. Cutler, Assistant Professor of Engineering and Applied Science; Ph.D., Stanford, 1987. Low-speed boundary-layer flows, vortex flows, hot-wire anemometry.

Charles A. Garris, Professor of Engineering; Ph.D., SUNY at Stony Brook, 1971. Fluid mechanics, heat transfer, combustion, solar energy, propulsion.

Charles M. Gilmore, Professor of Engineering and Applied Science; Ph.D., Maryland, 1971. Metallurgy and materials science, materials engineering, solid mechanics, surface science.

Douglas L. Jones, Professor of Engineering; D.Sc., George Washington, 1970. Continuum mechanics, computer-integrated design and manufacturing, fracture mechanics, mechanical engineering design.

Roger E. Kaufman, Professor of Engineering; Ph.D., RPI, 1969. Kinemation synthesis, mechanisms, computer-aided design.

James Der-Yi Lee, Professor of Engineering and Applied Science; Ph.D., Princeton, 1971. Robotics, electromechanical control systems, solid mechanics, finite-element analysis.

Harold Liebowitz, L. Stanley Crane Professor of Engineering and Applied Science; D.Ae.E., Polytechnic of Brooklyn, 1948. Fracture mechanics, solid mechanics, structures.

Catherine Mavriplis, Associate Professor of Engineering and Applied Science; Ph.D., MIT, 1989. Theoretical and computational fluid dynamics, heat transfer, and aerodynamics; applied mathematics; numerical methods.

Michael K. Myers, Professor of Engineering and Applied Science and Chairman of the Department; Ph.D., Columbia, 1966. Acoustics, linear and nonlinear wave propagation, analytical methods in fluid mechanics.

Robert R. Sandusky, Professor of Engineering and Applied Science; M.S., Washington (Seattle), 1971. Aerospace vehicle design, synthesis optimization.

Yin-Lin Shen, Associate Professor of Mechanical Engineering; Ph.D., Wisconsin–Madison, 1991. Evaluation of coordinate measuring machine (CMM) based inspection results, error budgeting in machine design.

Robert H. Tolson, Professor of Engineering and Applied Science; Ph.D., Old Dominion, 1990. Astrodynamics, space missions, orbit determination, planetary gravity fields and atmospheres, optimal trajectories, large-space systems.

John L. Whitesides Jr., Professor of Engineering and Applied Science; Ph.D., Texas at Austin, 1968. Analytical methods in fluid mechanics, mathematical methods, acoustics.

The George Washington University is an equal opportunity/affirmative action institution.

GEORGIA INSTITUTE OF TECHNOLOGY
A Unit of the University System of Georgia

George W. Woodruff School of Mechanical Engineering

Programs of Study

The School has a vigorous graduate program that encompasses advanced study and research in several areas: acoustics, dynamics, and vibrations; automation, controls, and mechatronics; bioengineering; design, manufacturing, and computer-aided engineering; fluid mechanics; mechanics of materials, fracture mechanics, fatigue, and materials processing; thermal sciences, heat transfer, combustion, thermodynamics, power systems, HVAC, and refrigeration systems; and tribology, lubrication, and fluid sealing. Graduate programs based on these areas lead to the degrees of Master of Science in Mechanical Engineering, Master of Science, and Doctor of Philosophy for qualified graduates with backgrounds in engineering, mechanics, mathematics, physical sciences, and life sciences. Graduate courses are offered during every term, and students may study on a year-round basis. The MSME program is available to working professionals through video technologies and, beginning in fall 1999, via the Internet. A study-abroad program is also offered at Tech's European campus in Metz, France. The Georgia Tech Lorraine (GTL) program allows U.S. students to earn a degree from Georgia Tech and to study in France. A dual-degree program is available for study at GTL and ENSAM, a French school of mechanical engineering.

Research Facilities

The School has specialized instruments and equipment associated with the laboratories involving underwater acoustics, ultrasonics, bioacoustics, bioengineering, lubrication and rheology, deformation and fracture of materials and materials processing, combustion, emissions, fluid dynamics and hydrodynamic stability, turbulence, heat transfer and two-phase flow, vibration and thermal stress, structural analysis, systems design, automatic and digital control, and robotics. The Institute's library has open stacks and is designated as one of the twelve regional technical report centers in the United States. It is extensive and current in both technical books and periodicals.

Georgia Institute of Technology provides extensive computer facilities for student use, including access to supercomputers, mainframes, hundreds of workstations, and thousands of PCs. All computers are linked with each other and the Internet via an extensive fiber-optic campuswide network. The Woodruff School's research operation is serviced by its own computer specialists as well as machine and instrumentation shops with a full-time staff of technicians.

Financial Aid

Numerous research and teaching assistantships, fellowships, and tuition waivers are available to graduate students. Graduate assistantships carry a twelve-month stipend ranging from $15,000 to $18,000 and include a waiver of out-of-state tuition. Holders of assistantships pay fees in the amount of $372 each term. President's Fellowships and Woodruff Fellowships of up to $5500, which supplement graduate assistantships, are available to qualified students wishing to pursue the Ph.D. Many federal, industrial, and private fellowships are also available. International students must guarantee their first-year support but are eligible to compete for awards on a per-term basis.

Cost of Study

The total fees in 1999–2000 for graduate students carrying a full academic load are $372 per term for graduate research assistantships and teaching assistantships, $1448 per term for residents of Georgia, and $6139 for nonresidents. Part-time resident and nonresident graduate students are charged prorated fees. Fees are subject to change without notice.

Living and Housing Costs

Numerous contemporary suites for unmarried students and apartments for married students and their families are available at reasonable costs through the Institute. Three 350-unit Graduate Student Living Centers are open to graduate students. Rooms and apartments in privately owned dwellings within walking distance or a short driving distance are available in several price categories. Students should write to the Housing Office for details. Unmarried students should be able to meet minimum necessary expenses, exclusive of tuition and fees, of $10,000 for the calendar year.

Student Group

The Institute's total enrollment is approximately 13,000, including 3,500 graduate students. Students come from all states and more than ninety countries. Almost 20 percent of all graduate students are women. In fall 1998, the mechanical engineering program included 1,500 students; 479 were graduate students, of whom 192 were working at the doctoral level. Most graduate students in mechanical engineering receive financial aid.

Student Outcomes

Georgia Tech has one of the best career services programs in the country for engineering students. Furthermore, the Woodruff School has an in-house program for placement of Ph.D. students that includes a yearly, nationwide graduate student symposium (Ph.D. career fair), *Ph.D. Résumé Book,* and numerous workshops. The endowed communications program offers assistance in preparing resumes and teaching and research portfolios. Since summer 1992, the Woodruff School has placed 100 percent of its 210 Ph.D. graduates. Roughly 70 percent go into industry, while 30 percent go into university positions. Recent graduates have been hired at Cornell, Virginia, Boston, Kentucky, Virginia Tech, and Penn State as well as at Siemens, Milliken, Motorola, Raytheon, Los Alamos National Laboratory, Sandia, Hitachi, and Lucent Technologies. Six recent Ph.D. graduates have won NSF Career Awards.

Location

Atlanta is a city of nearly 4 million people. Its 1,050-foot elevation provides freedom from climatic extremes, allowing for year-round recreational opportunities. The Atlanta area offers such sites of interest as the Martin Luther King Jr. National Historic Site, the state capitol, Stone Mountain Park, Fernback Science Center, Carter Presidential Library, Cyclorama, Centennial Olympic Park, Zoo Atlanta, and Six Flags Over Georgia. There are also professional teams in football (the Falcons), baseball (the Braves), basketball (the Hawks), and hockey (the Thrashers). Atlanta hosted the 1996 Olympic Games, and the Georgia Tech campus was the site of the Olympic Village.

The Institute and The School

Georgia Tech is a member of the University System of Georgia. The Institute is an accredited member of the Southern Association of Colleges and Schools. All four-year engineering curricula are accredited by the Accreditation Board for Engineering and Technology, the national engineering accrediting agency.

Applying

Application forms for admission and financial aid may be obtained by writing to the address given below and should be returned, together with letters of recommendation, GRE scores, and official transcripts of previous academic work, at least six weeks before the beginning of the term in question. Students seeking financial aid for the fall term must submit their applications by February 1.

Correspondence and Information

Graduate Studies
George W. Woodruff School of Mechanical Engineering
Georgia Institute of Technology
Atlanta, Georgia 30332-0405

Telephone: 404-894-3204
      800-543-2034 (toll-free)
Fax: 404-894-8336
E-mail: menehp.info@me.gatech.edu
World Wide Web: http://www.me.gatech.edu

Internet and Video-Based MS Programs
Center for Media-Based Instruction
Georgia Institute of Technology
Atlanta, Georgia 30332-0240

Telephone: 404-894-3378
E-mail: vbis@conted.gatech.edu (general)
      video.programs@me.gatech.edu
World Wide Web: http://www.conted.gatech.edu/
      vbis/vbis.html

Georgia Institute of Technology

THE FACULTY AND THEIR RESEARCH

Acoustics and Dynamics
Yves H. Berthelot, Professor; Ph.D., Texas at Austin, 1985. Acoustics, laser instrumentation in acoustics, ultrasonics.
Kenneth A. Cunefare, Associate Professor; Ph.D., Penn State, 1990. Active/passive control, fluid-structure interaction, optimal acoustic design.
Aldo A. Ferri, Associate Professor; Ph.D., Princeton, 1985. Acoustics, structural dynamics, nonlinear dynamics and control.
Jerry H. Ginsberg, George W. Woodruff Chair in Mechanical Systems; E.Sc.D., Columbia, 1970. Vibrations, acoustics, dynamics, fluid-structure interaction.
Jacek Jarzynski, Professor; Ph.D., Imperial College (London), 1961. Acoustics, acousto-optics, transducers, ultrasonics.
Peter H. Rogers, Neely Distinguished Professor; Ph.D., Brown, 1970. Acoustics, bioacoustics.

Automation and Mechatronics
Wayne J. Book, Professor; Ph.D., MIT, 1974. Robotics, automation, modeling, and control of flexible, fluid power, and manufacturing systems.
Ye-Hwa Chen, Associate Professor; Ph.D., Berkeley, 1985. Controls, manufacturing systems, neural networks, fuzzy engineering.
Stephen L. Dickerson, Professor Emeritus; Sc.D., MIT, 1965. Manufacturing automation, machine vision sensors, robotics.
Imme Ebert-Uphoff, Assistant Professor; Ph.D., Johns Hopkins, 1997. Robotics, parallel platform manipulators, flight simulation, static balancing.
Kok-Meng Lee, Associate Professor; Ph.D., MIT, 1985. System dynamics, control, automation, optomechatronics.
Harvey Lipkin, Associate Professor; Ph.D., Florida, 1985. Design and analysis of mechanical systems, robotics, spatial mechanisms.
John G. Papastavridis, Associate Professor; Ph.D., Purdue, 1976. Analytical, structural, and nonlinear mechanics; vibrations; stability.
Nader Sadegh, Associate Professor; Ph.D., Berkeley, 1987. Controls, vibrations, design.
William E. Singhose, Assistant Professor; Ph.D., MIT, 1997. Vibration, flexible dynamics, command generation.

Bioengineering
Robert S. Cargill, Assistant Professor; Ph.D., Pennsylvania, 1994. Cell biomechanics, trauma and development, tissue engineering.
Andres Garcia, Assistant Professor; Ph.D., Pennsylvania, 1996. Cellular and tissue engineering, cell adhesion, biomaterials.
Robert E. Guldberg, Assistant Professor; Ph.D., Michigan, 1995. Biomechanics, image-based FEM, tissue engineering.
David N. Ku, Regents' Professor; Ph.D., Georgia Tech, 1983; M.D., Emory, 1984. Magnetic resonance, thrombosis, prostheses.
Marc E. Levenston, Assistant Professor; Ph.D., Stanford, 1995. Orthopedic biomechanics, soft tissue mechanics, tissue engineering.
Robert M. Nerem, Institute Professor and Parker H. Petit Distinguished Chair for Engineering in Medicine; Ph.D., Ohio State, 1964; NAE. Biomedical engineering, biomechanics, cellular engineering, tissue engineering.
Raymond P. Vito, Professor and Associate Chair for Undergraduate Studies; Ph.D., Cornell, 1971. Biomechanics, tissue mechanics, biomechanical design.
Timothy M. Wick, Associate Professor; Ph.D., Rice, 1988. Tissue and cellular engineering, bioreactor design, cell adhesion, blood rheology.
Ajit P. Yoganathan, Regents' Professor; Ph.D., Caltech, 1978. Cardiovascular fluid dynamics, rheology, Doppler ultrasound, MRI.
Cheng Zhu, Associate Professor; Ph.D., Columbia, 1988. Cell and molecular mechanics, applications to immunology and tumor biology.

Computer-Aided Engineering and Design
Bert Bras, Associate Professor; Ph.D., Houston, 1992. Environmentally conscious design, design for recycling, robust design.
Robert E. Fulton, Professor; Ph.D., Illinois, 1960. Finite-element methods, integrated CAD/CAM, information management, electronic commerce.
Farrokh Mistree, Professor; Ph.D., Berkeley, 1974. Design of open systems, decision-based design, product families, enterprise integration.
David W. Rosen, Associate Professor; Ph.D., Massachusetts, 1992. Virtual and rapid prototyping, intelligent CAD/CAM/CAE.
Suresh Sitaraman, Assistant Professor; Ph.D., Ohio State, 1989. CAD/CAE, electronic packaging, thermomechanics and reliability, FEM.

Fluid Mechanics
Cyrus Aidun, Adjunct Professor; Ph.D., Clarkson, 1985. Hydrodynamic stability, liquid coating, suspended particle hydrodynamics.
Prateen V. Desai, Professor; Ph.D., Tulane, 1967. Fluid mechanics, solidification, convection in materials processing.
Ari Glezer, Professor; Ph.D., Caltech, 1981. Fluid mechanics, turbulent shear flows, flow control, diagnostics.
Damir Juric, Assistant Professor; Ph.D., Michigan, 1996. Computational methods, multiphase flows, microscale materials processing.
G. Paul Neitzel, Professor; Ph.D., Johns Hopkins, 1979. Hydrodynamic stability, numerical methods, free-surface and rotating flows.
Marc K. Smith, Associate Professor; Ph.D., Northwestern, 1982. Flows, liquid films and droplets.
Minami Yoda, Assistant Professor; Ph.D., Stanford, 1992. Suspension flows, shear flows, flow-structure interactions, optimal diagnostics.

Heat Transfer, Combustion, and Energy Systems
Said I. Abdel-Khalik, Southern Nuclear Distinguished Professor; Ph.D., Wisconsin, 1973. Microscale, heat transfer, reactor safety, thermal hydraulics.
William Z. Black, Regents' and Georgia Power Distinguished Professor; Ph.D., Purdue, 1968. Heat transfer, thermodynamics, fluids.
S. Mostafa Ghiaasiaan, Associate Professor; Ph.D., UCLA, 1983. Multiphase flow, aerosol and particle transport, nuclear reactor engineering.
James G. Hartley, Professor; Ph.D., Georgia Tech, 1977. Heat transfer, thermodynamics, fluid mechanics.
Sheldon M. Jeter, Associate Professor; Ph.D., Georgia Tech, 1979. Thermal hydraulics, energy systems.
Prasanna V. Kadaba, Associate Professor; Ph.D., IIT, 1964. Heat transfer, I.C. engine design, energy and environmental systems, advanced cycles.
Alan V. Larson, Professor and Associate Chair for Administration; Ph.D., Illinois, 1961. Thermodynamics.
Sam V. Shelton, Associate Professor; Ph.D., Georgia Tech, 1969. Energy systems, HVAC systems, absorption, refrigeration.
Amyn S. Teja, Regents' Professor; Ph.D., Imperial College, 1972. Thermodynamics, fluid properties, supercritical fluid separations.
William J. Wepfer, Professor and Associate Chair for Graduate Studies; Ph.D., Wisconsin, 1979. Heat transfer, thermodynamics.
Ben T. Zinn, David S. Lewis Chair and Regents' Professor; Ph.D., Princeton, 1965; NAE. Combustion instability, pulse combustion, propulsion, acoustics.

Manufacturing
Daniel F. Baldwin, Assistant Professor; Ph.D., MIT, 1994. Manufacturing systems design, electronics manufacturing and packaging, polymer processing.
Jonathan S. Colton, Professor; Ph.D., MIT, 1986. Manufacturing, polymer and composites processing, design, rapid prototyping.
Steven Danyluk, Morris M. Bryan Jr. Chair for Advanced Manufacturing Systems; Ph.D., Cornell, 1974. Processing of materials, residual stresses, tribology, lubricant-surface interaction, chemomechanical polishing, sensors.
Thomas R. Kurfess, Associate Professor; Ph.D., MIT, 1989. System dynamics, control, metrology, CAD/CAM/CAE.
Steven Y. Liang, Associate Professor; Ph.D., Berkeley, 1987. Automated manufacturing, control systems, digital signal processing.
Shreyes N. Melkote, Assistant Professor; Ph.D., Michigan Tech, 1993. Machining process modeling, surfaces, CAM/CAPP, intelligent fixturing.
I. Charles Ume, Professor; Ph.D., South Carolina, 1985. Electronic packaging, mechatronics, laser moiré, laser ultrasonics.

Mechanics of Materials
Iwona Jasiuk, Associate Professor; Ph.D., Northwestern, 1986. Micromechanics, fracture, damage mechanics, composite materials, biomaterials.
W. Steven Johnson, Professor; Ph.D., Duke, 1979. Deformations, composite materials, joints.
W. Jack Lackey, Professor; Ph.D., North Carolina State, 1970. Ceramic and metallic coatings, composites, and rapid prototyping.
Christopher S. Lynch, Assistant Professor; Ph.D., California, Santa Barbara, 1992. Experimental mechanics, smart materials.
David L. McDowell, Regents' Professor and Carter Paden Chair in Metals Processing; Ph.D., Illinois, 1983. Fracture, fatigue, cyclic plasticity and viscoplasticity, finite strain effects, continuum damage, metals processing.
Richard W. Neu, Assistant Professor; Ph.D., Illinois, 1991. Fatigue, viscoplasticity, composite materials.
Jianmin Qu, Associate Professor; Ph.D., Northwestern, 1987. Fracture, materials, wave propagation, microelectronic packaging.
Min Zhou, Assistant Professor; Ph.D., Brown, 1993. Experimental mechanics, dynamic behavior, material failure and shear localization.

Tribology
Itzhak Green, Professor; Sc.D., Technion (Israel), 1984. Finite element method, rotordynamics, fluid sealing, design, integrated diagnostics.
Richard F. Salant, Professor; Sc.D., MIT, 1967. Fluid mechanics, fluid sealing.
Jeffrey L. Streator, Associate Professor; Ph.D., Berkeley, 1990. Computer-disk tribology, rheology, friction-induced vibration, capillarity.
Ward O. Winer, Eugene C. Gwaltney, Jr. Chair in Manufacturing and Chair, School of Mechanical Engineering; Ph.D., Michigan, 1961; Ph.D., Cambridge, 1964; NAE. High-pressure rheology, lubrication, tribology, thermomechanics, mechanical systems diagnostics.

HOWARD UNIVERSITY

Department of Mechanical Engineering

Programs of Study

The Department of Mechanical Engineering offers programs leading to the Master of Engineering degree and to the Ph.D. with concentrations in applied mechanics, aerospace engineering (dynamics and controls), CAD/CAM and robotics, and fluid-thermal sciences. With the approval of the department, special programs may be based on courses from other departments whenever such an arrangement contributes to an integrated program of graduate study in engineering. Both thesis and nonthesis options are available at the master's level.

Research Facilities

There are a variety of laboratories in the department, including a robotics laboratory, a mechanical measurement laboratory, a systems dynamics laboratory, an internal-combustion-engine laboratory, a thermomechanics and structures laboratory, an aerodynamics laboratory, and a computer laboratory for instruction and design in engineering. The department has numerous instruments for mechanical and electrical measurements and its own machine tool shop. The School of Engineering Computer and Learning and Design Center is equipped with an HP 3000 minicomputer, which is accessed by twenty-three alphanumeric terminals, five graphics terminals, and three plotters, all located in the center. An ECLIPSE MV/4000 with an additional ten terminals for graphics applications is being integrated into the system through a Gandalf Packnet matrix switching device. In addition, the School of Engineering houses a VAX-11/780 superminicomputer and a VAX-11/750.

Financial Aid

Several different types of financial aid are available for graduate students, including teaching and research assistantships, industrial fellowships, traineeships, and Federal Perkins Loans. A limited number of externally supported research assistantships offer support for the complete calendar year.

Cost of Study

The general fee for full-time study in 1999–2000 is $5482.50 per semester, including tuition, enrollment, and services fees.

Living and Housing Costs

Because of limited facilities, University residence halls are not able to accommodate all students who might wish to live on campus. Priority is given to those who are required to live on campus. Students enrolling in the Graduate School and professional schools who desire assistance in locating housing in the Washington area should write directly to the Supervisor of Off-Campus Housing. Arrangements will be made to help find University-approved off-campus accommodations.

Student Group

The current enrollment at Howard University is approximately 11,000 students; about 800 are in engineering. Some 1,250 are graduate students. Students come to Howard University from all parts of the United States as well as from eighty countries. With respect to the number of international students in the total student body, Howard University ranks first among all American colleges, with about 15 percent.

Location

Howard University's location in Washington, D.C., offers its students opportunities for study and research that few institutions in America can match. Howard students are frequent visitors to the Library of Congress, the Folger Shakespeare Library, the National Gallery of Art, the National Institutes of Health, the Botanical Gardens, the National Bureau of Standards, the Congress of the United States, and many other facilities of the federal government. Usually, several national meetings of different professional societies are held in Washington each year, and graduate students are encouraged to attend.

The University and The Department

Founded in 1867, Howard University has become one of America's most interesting institutions of higher learning, exceeding even the fondest dreams of its founders. From its humble beginnings, Howard has been transformed into a modern university with seventeen fully accredited schools and colleges and a well-equipped $50-million physical plant.

Student branches of the American Institute of Aeronautics and Astronautics and the American Society of Mechanical Engineers are active in the department. The student branch of the AIAA has been cited twice as outstanding. The student professional society branches visit various government and industrial facilities, such as the Goddard Space Flight Center, COMSAT Laboratories, Fairchild Space and Electronics Company, the Applied Physics Laboratory at Johns Hopkins University, and the National Air and Space Museum of the Smithsonian Institution.

Applying

Application for admission should be made directly to the Office of Admissions, while financial aid applications should be sent to the director of graduate studies. Both addresses are given below. Admission and financial aid decisions are based on the student's transcripts, letters of recommendation, and specific research goals. International students must submit their scores on the Test of English as a Foreign Language. The deadline for applications for financial aid is March 1.

Correspondence and Information

Office of Admissions
Howard University
Washington, D.C. 20059
Telephone: 202-806-6200

Director of Graduate Studies
College of Engineering, Architecture,
 and Computer Sciences
School of Engineering
Howard University
Washington, D.C. 20059
Telephone: 202-806-6600

Howard University

THE FACULTY AND THEIR RESEARCH

Professors

Peter M. Bainum, Ph.D., Catholic University; PE. Spacecraft attitude stability, astrodynamics, dynamics and control, large space structures.

Dah-Nien Fan, M.Aero.E., Ph.D., Cornell. LDV velocity bias, tensorial fluid mechanics, aerodynamics.

Emmanuel K. Glakpe, Ph.D., Arizona. Heat transfer, energy systems, nuclear engineering.

Robert Reiss, Sc.B., Brown; Ph.D., IIT. Optimal structural design, variational and extremum principles, biomechanics.

Lewis Thigpen, Ph.D., IIT; PE. Penetration mechanics, engineering education.

N. R. Vira, Ph.D., Howard. Robotics, CAD/CAM, manufacturing engineering, symbolic computation, optimal design.

M. Lucius Walker Jr., Ph.D., Carnegie Tech; PE. Engineering education, economic system design, engineering design.

Associate Professor

H. Algernon Whitworth, D.Sc., George Washington. Composite materials, fatigue and failure analysis.

Assistant Professors

Mohsen Mosleh, Ph.D., MIT. CAD/CAM, manufacturing processes, biomaterials tribology.

Sonya T. Smith, Ph.D., Virginia. Boundary layer stability, turbulent flow modeling and simulation, wave propagation.

DISSERTATIONS/THESES

Position assessment of the robot end effector using three linear cable transducers. August 1988.

Natural convection in a partially blocked vertical channel. August 1988.

Higher approximations to viscous flow in narrow curved gaps. August 1988.

Effects of torsion and curvature on viscous flow in helical tubes. August 1988.

Design sensitivity analysis and maximum frequency design of symmetric laminated angle-ply plates. May 1989.

Attitude and shape control of optimally designed large space structures. May 1989.

Dosimetry problems with double node microwave antennae in simultaneous intraoperative hyperthermia and intraoperative radiation therapy. August 1989.

Applied symbolic computation in robotics. December 1989.

Further development of a theory on the correction of statistical velocity bias in IRLV. December 1989.

A pyrolytic investigation of cotton stalks as an alternative source of energy. December 1989.

The estimation of attitude sensor misalignment due to spacecraft bending on the solar maximum mission. May 1990.

The optimal linear quadratic regulator digital control of a free-free orbiting platform. May 1990.

On the application of Green's Functional to distributed parameter sensitivity analysis and optimal structural design. July 1990.

Hybrid control system for space mast structures. December 1990.

Enhancement of IGES preprocessor for data exchange. December 1990.

Identification of robot error parameters. December 1990.

Effects of aspect ratio and Prandtl number of natural convection in irregular horizontal annuli. December 1990.

An alternate algebra for dynamic pose error modeling. December 1990.

Charts of aerodynamic parameters as a function of speed and altitude for various atmospheres. May 1991.

Fluid flow and heat transfer in an enclosure with a corrugated heated surface. May 1991.

Issues in control system design of large space structures. May 1991.

Rapid maneuvering of large, flexible orbiting systems. December, 1991.

Computer-aided formulation for position-dependent pose errors of robots. December 1991.

Development and analysis of flexible manufacturing system using fixed-increment time-interval mechanism. December 1991.

A finite-element approach to impact and penetration mechanics. December 1991.

Grid plate measurement. August 1992.

Combined natural convection and thermal radiation in square and corrugated enclosure. August 1992.

Three-dimensional laminar steady mixed convection in a horizontal concentric annulus with rotating inner cylinder. August 1992.

Effects of solar radiation pressure on the tethered antenna/reflector subsatellite system. December 1992.

Fatigue damage evaluation of a thermo-plastic matrix composite reinforced with continuous and long discontinuous (LDF) graphite fiber. December 1992.

Finite element modeling and analysis of gas metal arc welding process on plates using ANSYS. May 1993.

Eigensensitivity methods in composite structural analysis. August 1993.

Momentum exchange feedback control: A new approach to the control of flexible space tethered systems. August 1993.

On the dynamics and control of an orbiting shallow spherical shell in the presence of solar radiation. December 1993.

Three-dimensional natural convection in a corrugated enclosure. May 1994.

Dynamics and control of large flexible space-robotic systems. May 1994.

Identification and validation of robot error parameters. December 1994.

Control/structure interaction associated with flexible expandable launch vehicles (ELV). December 1994.

Dynamics and robust digital control for large orbiting space structures. July 1995.

Complex frequencies for viscously damped elastic structures determined from eigensensitivity analysis. December 1995.

A comparison of analytical and finite element heat transfer solution for arc welding simulations. May 1996.

Influence of a robot's kinematic error parameters on its end-effector dexterity. May 1996.

Application of speech recognition in the field of manufacturing. August 1996.

A benchmark method to evaluate the performance of supercomputers from the application point of view. May 1997.

Stress concentration and failure analysis of composite plates with circular opening. December 1997.

Numerical modeling of heat transfer in a piston/cylinder assembly. May 1998.

Thermal state-of-charge of solar heat receivers for space solar dynamic power. May 1998.

ILLINOIS INSTITUTE OF TECHNOLOGY

Department of Mechanical, Materials, and Aerospace Engineering

Programs of Study

The Department of Mechanical, Materials, and Aerospace Engineering offers several flexible programs in mechanical and aerospace engineering with four major areas of study: fluid dynamics, thermal sciences, solids and structures, dynamics and control, and design and manufacturing, leading to the Master of Science in mechanical and aerospace engineering, the Master of Mechanical and Aerospace Engineering, and the Doctor of Philosophy in mechanical and aerospace engineering. Programs in metallurgical and materials engineering lead to the Master of Science in metallurgical and materials engineering, the Master of Metallurgical and Materials Engineering, and the Doctor of Philosophy in metallurgical and materials engineering. In addition, the department offers interdisciplinary programs leading to the Master of Science in manufacturing engineering and the Master of Manufacturing Engineering.

The Master of Science program prepares students for careers in research and development and for admission to programs leading to the Doctor of Philosophy. Thirty-two credit hours of approved work, including 6 to 8 hours of thesis research, are required. The Master of Engineering is a professionally oriented program that prepares students for practice in industry. The degree requires 30 credit hours of course work, and courses are scheduled at a convenient time for working professionals and are offered at remote TV receiving sites. The Master of Engineering degree does not require a thesis or a comprehensive exam. Either master's program typically takes three full-time semesters to complete. The Doctor of Philosophy is awarded to students who demonstrate superior achievement in the field, and includes an original, independent research contribution. The program usually requires three to four semesters of course work beyond the master's level, culminating in a comprehensive examination conducted by an appointed committee, and one full year of research for the dissertation that is defended in a final public oral examination.

Research Facilities

The department operates several laboratories in support of the programs of study and research. Mechanical and aerospace engineering laboratories include the Fejer Unsteady Wind Tunnel; the Morkovin Low-Turbulence Tunnel; the National Diagnostic Facility; several other air and water facilities as well as supporting instrumentation; flow visualization systems; optical measuring equipment; the many computer-based data acquisition, processing, and display systems of the Fluid Dynamics Research Center; laboratories in the experimental mechanics and laboratories for research in tribology, robotics, computer-integrated manufacturing, footlink CAD lab, railroad engineering, materials and their processing, biomechanics and its instrumentation, combustion, internal combustion engines, and combined heat and mass transfer. The Department has numerous minicomputers and microcomputers and workstations available for computational research activities. Metallurgical and materials engineering laboratories include facilities for research on metallography, heat treatment, and mechanical testing; optical, scanning, and transmission electron microscopy; powder metallurgy; and laser machining facilities.

Financial Aid

Financial aid is available in the form of fellowships, teaching assistantships, and research assistantships that include a stipend and full or partial tuition. Applicants seeking financial aid from the department must pursue either the M.S. or Ph.D. degree, and are encouraged to submit all application materials by March 1. Department funds available to assist non-U.S. students are very limited and are awarded to students only after they have demonstrated superior capabilities in their first year of graduate study at IIT. The University has a financial aid program for graduate students that includes student loans and work-study. Applicants on F-1 student visas are not eligible for part-time employment during their first year at IIT. Interested students should contact the Office of Student Finance directly for further information about university financial aid.

Cost of Study

Tuition for the 1999–2000 academic year is $590 per credit hour.

Living and Housing Costs

Housing is available for graduate students in IIT's residence halls. The 1999–2000 cost of room and board ranges from $5155 to $6880 per semester. IIT also has apartments available at costs ranging from $458 to $927 per month, which includes some utilities. There are several off-campus apartment complexes within a mile of campus.

Student Group

Graduate enrollment in the Department of Mechanical, Materials and Aerospace Engineering in 1998 was 143 students. Five continents are represented in the student group, making for a diverse and stimulating atmosphere.

Student Outcomes

After completing the Master of Science program many graduates enter the Ph.D. program. Some students enter industry positions with corporations and agencies such as Boeing, McDonnell Douglas, Motorola, Amoco, General Dynamics, General Motors, AT&T, NASA, Inland Steel, Bethlehem Steel, and Argonne National Laboratory. Most Master of Engineering students attend IIT on a part-time basis and work in industry full-time while completing their degree.

Location

IIT's Main Campus is located near the heart of Chicago, just 3 miles south of the Loop and central to the greater Chicago area's thriving technological community of business, industry, and research institutions. Internationally known for its architecture, museums, symphony, and theater; its beautiful lakefront on the western shore of Lake Michigan; and the unusually rich variety of its ethnic communities, Chicago offers a vast array of recreational and cultural opportunities. The Main Campus, designed by Ludwig Mies van der Rohe and regarded internationally as a landmark of twentieth-century architecture, occupies fifty buildings on a 120-acre site and includes research institutes, libraries, laboratories, residence halls, a sports center, and other facilities. Among IIT's immediate neighbors are Comiskey Park, home of the Chicago White Sox; two major medical centers; and the McCormick Place Exposition Center. The Downtown Campus is in the Loop near the city's financial trading, banking, and legal centers. The Rice Campus is in suburban Wheaton, convenient to the Interstate 88 research and technology corridor west of the city. The Moffett Campus is in southwest suburban Summit-Argo.

The Institute

Illinois Institute of Technology was formed in 1940 by the merger of Armour Institute of Technology (founded in 1890) and Lewis Institute (founded in 1896). IIT offers programs of study in engineering and the sciences, architecture, design, public administration, technical communications and information design, psychology, business, and law. IIT is a member of the prestigious Association of Independent Technological Universities (AITU).

Applying

Applications and supporting documents, which include transcripts, test scores, and letters of recommendation, should be received by the Office of Graduate Admissions (3300 South Federal, Room 301A, Chicago, Illinois 60616) by June 1 for fall matriculation and by November 1 for spring matriculation. Applications that include requests for consideration for financial support should be received by March 1. Application forms and additional information are available on line at http://www.grad.iit.edu.

Correspondence and Information

Dr. Marek Dollar, Chair
Christine Eitel, Coordinator
Department of Mechanical, Materials, and Aerospace Engineering
Illinois Institute of Technology
Chicago, Illinois 60616-3793

Telephone: 312-567-3175
Fax: 312-567-7230
E-mail: dept@mmae.iit.edu
World Wide Web: http://mmae.iit.edu

Illinois Institute of Technology

THE FACULTY AND THEIR RESEARCH

V. Aronov, Associate Professor; Ph.D., Institute of Mechanization of Agriculture Kiev (Ukraine). Mechanical design, reliability and tribology.

R. L. Barnett, Professor; M.S., IIT. Structures and safety of machines.

L. J. Broutman, Research Professor; Ph.D., MIT. Composite materials and polymers.

K. W. Cassel, Assistant Professor; Ph.D., Lehigh. Fluid dynamics, analytical and computational techniques, techniques in the thermal-fluid sciences, unsteady boundary-layer flows.

T. C. Corke, Professor; Ph.D., IIT. Fluid mechanics, turbulence, transitional flows, wind engineering and atmospheric diffusion, instrumentation techniques, analog and digital data acquisition and signal processing.

R. C. Dix, Professor; Ph.D., Purdue. CAD/CAM, mechanical design and materials handling.

A. Dollar, Assistant Professor; Ph.D., Cracow (Poland). Solid mechanics, composite materials, computational mechanics and biomechanics.

M. Dollar, Professor, Chair; Ph.D., University of Mining and Metallurgy, Cracow (Poland). Materials science and engineering, high-temperature structural materials and electron microscopy.

R. Foley, Assistant Professor; Finkl Professor; Ph.D., Northwestern. Ferrous physical metallurgy, anelastic behavior of solids, formability, fracture and microstructure-property relationships.

M. R. Gosz, Assistant Professor; Ph.D., Northwestern. Solid Mechanics, finite element method, composite materials, 3-D fracture mechanics.

M. H. Hites, Research Assistant Professor; Ph.D., IIT. Fluid dynamics, turbulence, high-lift systems, Web-based digital data acquisition and processing.

M. J. Jennings, Lecturer; Ph.D., IIT. Fluid dynamics and thermal science.

J. S. Kallend, Professor; Ph.D., Cambridge. Computational methods of crystallographic texture analysis and properties of polycrystalline aggregates.

S. Kalpakjian, Professor; S.M., MIT. Manufacturing science and engineering, applied plasticity, mechanical metallurgy and tribology.

S. Kumar, Research Professor; Ph.D., Penn State. Railroad engineering, high-speed ground transportation, noise vibrations, friction and wear, and solid mechanics.

K. P. Meade, Associate Professor and Associate Chair, Mechanical Engineering; Ph.D., Northwestern. Solid mechanics, biomechanics, elasticity, fracture mechanics and computational mechanics.

S. Mostovoy, Associate Professor; Ph.D., IIT. Metallurgy, mechanical properties of materials, fatigue and fracture.

H. M. Nagib, John T. Rettaliata Professor, Dean of Armour College and Vice President of the Main Campus, and Director of the Fluid Dynamics Research Center; Ph.D., IIT. Fluid dynamics, heat transfer, applied turbulence, wind engineering and aeroacoustics.

S. E. Nair, Professor; Ph.D., California, San Diego. Solid mechanics, stress analysis of composite and inelastic material, dynamics of cable, fracture mechanics and wave propagation theory.

P. G. Nash, Professor; Ph.D., London. Physical metallurgy, intermetallics, powder metallurgy, composites, phase equilibria and transformations.

B. Pervan, Assistant Professor; Ph.D., Stanford. Dynamics, control, guidance, and navigation. Current research focuses on landing aircraft in zero visibility conditions using differential satellite-based navigation.

R. W. Porter, Professor and Director of the American Power Conference; Ph.D., Northwestern. Thermal engineering and power generation.

F. Ruiz, Associate Professor; Ph.D., Carnegie-Mellon. Combustion, atomization, pollution control of engines, fuel economy, alternative fuel, electronic cooling and special cooling.

C. A. Sciammarella, Research Professor; Ph.D., IIT. Experimental mechanics of solids with particular emphasis in optics applied to mechanics of materials and stress analysis and fracture mechanics.

M. N. Tarabishy, Visiting Assistant Professor; Ph.D., Wichita State. Dynamic systems and control, robotics, intelligent control, mechatronics.

J. A. Todd, Professor; Associate Chair, Metallurgical and Materials Engineering; and Iron and Steel Society Professor; Ph.D., Cambridge. Ferrous metallurgy and laser processing of ceramic materials.

T. C. Tszeng, Assistant Professor; Ph.D., Berkeley. Materials processing, phase transformation kinetics, heat treatment, powder metallurgy, composites, analytical and computational methods.

C. E. Wark, Associate Professor; Ph.D., IIT. Fluid dynamics, turbulence, digital data acquisition and processing.

J. L. Way, Professor and Associate Chair, Aerospace Engineering; Ph.D., Rensselaer. Fluid mechanics, aerodynamics, aerospace systems and control, and thermodynamics.

D. R. Williams, Professor; Ph.D., Princeton. Experimental fluid mechanics with emphasis on flow measurement and flow control techniques.

RESEARCH AREAS

Design and Manufacturing

The research activities in design and manufacturing are concentrated in the areas of computer-aided design: computer-based machine tool control and computer graphics in design; manufacturing processes: wear and fracture behavior of cutting tools; tribology: frictional wear characteristics of ceramics; and dynamic systems and mechanical vibrations: dynamic behavior of rail vehicles including self-excited oscillations, longitudinal dynamics and handling of trains, and interactive computer simulation of rail vehicles.

Dynamics and Control

Research in this area has been mainly in modeling, analysis and control of complex systems, intelligent control of flow around airfoils and aircraft body, laser machining control, and robotic devices.

Fluid Dynamics

Fluid dynamics is a particularly strong research activity in the department. The current interest is in the areas of turbulent flows, unsteady and separated flows, instabilities and transition, turbulence modeling, flow visualization techniques, computational fluid dynamics, turbulent combustion, and computer-based data acquisition, processing, and display. The newest addition to the department is the National Diagnostic Facility (NDF), a computer-controlled, high-speed, subsonic flow wind tunnel funded primarily by the Air Force Office of Research. In 1986, it was designated as a National Center of Excellence by the Department of Defense.

Metallurgical and Materials

Research concentrates on microstructural characterization, ferrous physical metallurgy, powder materials, laser processing and machining, high temperature structural materials, mechanical behavior, fatigue and fracture, environmental fatigue and fracture, computational X-ray diffraction analysis, texture, and recrystallization.

Solids and Structures

Research in this area is directed primarily along the lines of experimental mechanics: composites, fracture mechanics, design and testing of prosthetic devices, and holographic interferometry; computational mechanics: fracture mechanics, cable dynamics, and analysis of inelastic solids; and theoretical mechanics: wave propagation, fracture, elasticity, and models for scoliosis.

Thermal Sciences

Research in the field of thermal sciences currently follows two lines of investigation. The first area concerns the thermodynamic and economic aspects of industrial cogeneration of heat and electricity and waste heat generation. There is also interest in combustion, specifically high-speed fuel atomization, and alternative fuels.

JOHNS HOPKINS UNIVERSITY

Department of Mechanical Engineering

Programs of Study

The Department of Mechanical Engineering offers Master of Science and Doctor of Philosophy programs in the areas of the mechanics of materials, materials science, fluid mechanics, heat transfer, and robotics.

M.S. candidates must successfully complete a coordinated sequence of courses and independent study. Both thesis and nonthesis options are offered.

Ph.D. candidates must pass a departmental exam (taken after the third semester of study), pass the Graduate Board Oral exam (usually taken during the second or third year of full-time study), submit an original and high-quality doctoral dissertation, and pass the final thesis defense. The Ph.D. program normally requires approximately four years of full-time study beyond a bachelor's degree.

Graduate research and course work concentrate on the fundamentals of fluid mechanics, heat transfer, robotics, and mechanics and materials. Typically, graduate programs involve core courses in the Department of Mechanical Engineering and courses in related fields, such as mathematics, materials science, chemical engineering, civil engineering, environmental engineering, earth sciences, electrical engineering, computer science, physics, and chemistry.

Research Facilities

Experimental facilities for research in hydrodynamics include a jet cavitation facility and holographic and laser sheet flow visualization systems as well as image-processing equipment. A large wind tunnel with a test section of 10m x 1m x 1.3m is used for studies in turbulence. Close ties with the David Taylor Research Center provide access to towing tank facilities for quantitative visualization of 3-D fluid flow. New turbomachinery instrumentation is now in operation. Equipped with a wind tunnel for heat transfer measurements, the heat transfer laboratory is devoted to the study of heat transfer in complex geometries in single-phase and two-phase flows. In addition, a facility for holographic interferometry, used for density, temperature, and concentration measurements in complex systems, is part of the laboratory. PC-based data acquisition systems and a digital image processing system are used for data acquisition and quantitative analysis of visualization images obtained by holographic interferometry, thermochromatic liquid crystals, or infrared photography.

Robotics research facilities are concentrated in two laboratories: the Dynamical Systems and Control Laboratory and the Electro-mechanical Systems Laboratory. They are equipped to design, develop, and test novel robot prototypes for investigating advanced concepts in kinematics, dynamics, actuation, and sensing. Laboratory equipment includes a variety of high-end graphics workstations for CAD and simulations; PC-based data acquisition, vision, and control systems; motors, sensors, and drivers for prototype modular robots; extensive pneumatic actuation equipment; high-speed analog and digital test equipment; a robotics teaching station equipped with a dedicated teaching robot for student training; standard industrial robot arms; and a complete in-house machine shop.

Experimental facilities for research in mechanics and materials are concentrated within four laboratories. The Laboratory for Impact Dynamics and Rheology contains torsion, compression, and tension Kolsky bars, a 60mm gas gun, laser interferometer systems, and very high speed digital oscilloscopes for studying materials at ultra-high strain rates. The Laboratory for Active Materials and Biomimetics contains a biaxial testing system capable of measuring finite viscoelastic properties of active materials and a rheometrics solids analyzer. The Experimental Solid Mechanics Laboratory contains high-temperature creep frames, electrohydraulic test machines with high-temperature capabilities and laser extensometry, and unique microsample and bulge testers. Electron Microscopy facilities include an atomic resolution Philips 420 TEM, a high-resolution SEM, and various standard instruments. Image simulations use the Stadelmann EMS programs and a dedicated Silicon Graphics workstation. The Center for Non-Destructive Evaluation (CNDE) has outstanding facilities for ultrasonic, optical and X-ray characterization of materials.

Computer facilities in the Department of Mechanical Engineering include a Silicon Graphics Power Challenge L with six processors (rated at 1.8 Gflops peak) and 512 Mb RAM; a Power Indigo 2 visualization workstation; a number of SUN SPARCstations and IBM RISC 6000 research workstations; and Macintoshes and terminals in the department's computer room. Central computational facilities include several Silicon Graphics Power Challenges and access to national supercomputers.

Financial Aid

Nearly all entering graduate students are supported by departmental and research fellowships and by schoolwide Wolman fellowships. The nominal fellowship stipend for 1999–2000 is $18,533 for twelve months. Tuition support is provided through fellowships. Research assistantships cover the typical 4-year duration of the doctoral program. Research fellowships and assistantships are funded through programs with the National Science Foundation, the Office of Naval Research, the Army Research Office, the Department of Energy, the Air Force Office for Scientific Research, NASA, ARL, and industry.

Cost of Study

Tuition for 1999–2000 is $23,660 for the academic year.

Living and Housing Costs

University-owned apartments are available for single and married students at costs ranging from $325 to $650 per month. Early inquiry is advised. Privately owned apartments and houses are available in the Baltimore area in a wide range of prices. Most students find appropriate accommodations within walking distance of the University.

Student Group

There are approximately 65 graduate students in the program; almost all are pursuing their doctorate. The total undergraduate enrollment of the department is about 100, so those graduate students serving as teaching assistants interact with very small classes. There are 457 graduate students in the Whiting School of Engineering, which has a faculty of 112 and approximately 825 undergraduate students.

Location

The University is located in a residential area some thirty blocks north of downtown Baltimore and Harborplace. The 140-acre campus is wooded and quiet, providing an ideal atmosphere for study and research. The Johns Hopkins Medical Institutions and the world-renowned Peabody Conservatory are separately located and are connected by a free shuttle service. The Johns Hopkins Applied Physics Laboratory is 25 miles southwest of Baltimore. Washington is only an hour away by car or train.

The University

Founded in 1876, Johns Hopkins was the first university to offer graduate education as it is known today in the United States and thus has had a profound influence on American higher education. The University was dedicated from its inception to the concept of "creative scholarship." Students engage not merely in the absorption of existing knowledge but also in the creation of knowledge through their own research.

Applying

Interested individuals are encouraged to visit the campus to discuss their background and plans for graduate study. Applications are accepted at any time, although February 15 is the deadline for those seeking financial aid. Scores on the Graduate Record Examinations (both General Test and Subject Test in engineering) are strongly recommended for students with U.S. undergraduate degrees. Applicants with degrees from universities outside of the United States are required to submit TOEFL and GRE General Test scores. Application forms and further information are available from the address below.

Correspondence and Information

Graduate Committee Chairman
Department of Mechanical Engineering
Room 200, Latrobe Hall
Johns Hopkins University
Baltimore, Maryland 21218

Telephone: 410-516-7154
E-mail: megrad@titan.me.jhu.edu
World Wide Web: http://www.me.jhu.edu

Johns Hopkins University

THE FACULTY AND THEIR RESEARCH

Gang Bao, Associate Professor; Ph.D., Lehigh, 1987. Cellular biomechanics, protein and DNA deformation, brain damage during a car crash, biomedical MEMS, cardiac cell dysfunction, tissue engineering, mechanical testing of living cells, mechanics of composite materials and structures. Crack bridging models for fiber composites with slip dependence interfaces. *J. Mech. Phys. Solids* 41:1425, 1993 (with Song). Thermomechanical fatigue cracking in fiber reinforced metal matrix composites. *J. Mech. Phys. Solids* 43:1433, 1995 (with McMeeking). A heat transfer analysis for quartz microresonator IR sensors. *Int. J. Solids Struct.* 35:3635, 1998 (with Jiang).

Ilene J. Busch-Vishniac, Professor and Dean; Ph.D., MIT, 1981. Transduction, application of system dynamics and control techniques to sensors and actuators of multiple energy domains, electromechanical actuators, acoustic transducers, optical sensors.*Electromechanical Sensors and Actuators*, New York: Springer Verlag, 1999. Trends in electromechanical transduction. *Physics Today* 51:28, 1998. Noise reduction by a barrier with a random edge profile. *J. Acoust. Soc. Am.* 101:1, 1997 (with S. S. T. Ho and D. T. Blackstock).

Shiyi Chen, Professor; Ph.D., Beijing, 1987. Computational fluid dynamics, statistical theory and computation of fluid turbulence, mesoscopic physics and lattice Boltzman computational methods, multiphase flows and boiling. The scaling of pressure in isotropic turbulence. *Phys. Fluids* 10(9):2199, 1998. Lattice Boltzmann methods for fluid flows. *Annu. Rev. Fluid Mech.* 30:329, 1998 (with Doolen). Kinematic effects on local energy dissipation rate and local enstrophy in fluid turbulence. *Phys. Fluids* 10(1):312, 1998.

Gregory S. Chirikjian, Associate Professor; Ph.D., Caltech, 1992. Robotics; hyper-redundant robotic manipulators; design, kinematics, motion planning, dynamics, and control of mechanisms. A geometric approach to hyper-redundant manipulator obstacle avoidance. *ASME J. Mech. Design* 114:581–5, 1992 (with Burdick). A modal approach to hyper-redundant manipulator kinematics. *IEEE Trans. Robot. Automat.* 10(3):343–53, 1994 (with Burdick).

Andrew S. Douglas, Professor and Chair; Ph.D., Brown, 1982. Mechanics of compliant materials and soft tissues, including the kinematics of the heart and the tongue; active materials; dynamic ductile fracture mechanics. Description of the deformation of the left ventricle by a kinematic model. *J. Biomech.* 25:1119, 1992 (with Arts et al.). Loading rate induced changes in fracture fields. In *Recent Developments in Computational and Applied Mechanics*, ed. B. D. Reddy, p. 112, 1997 (with Premack).

Kevin J. Hemker, Associate Professor; Ph.D., Stanford, 1990. Correlating mechanical performance with transmission electron microscopy in a way that will identify the fundamental processes controlling deformation in advanced structural materials, thin films, MEMS and nanocrystalline materials, employing high resolution and analytical electron microscopy to characterize and study nanostructured materials and thin films. Correlating dislocation positions with intensity peaks in weak-beam TEM images. *Philos. Mag. A* 76:241–65, 1997. Tensile/compressive properties of single crystal gamma Ti-55.5% Al. *Metall. Mater. Trans.* 29A:65–71, 1998 (with Zupan).

Cila Herman, Associate Professor; Dr.Ing., Munich Technical, 1992. Experimental heat transfer and fluid mechanics, optical measurement techniques (holographic interferometry) applied to heat transfer measurements, heat transfer enhancement, boiling in microgravity and under the influence of electric fields, cooling of electronic equipment, thermoacoustic refrigeration, MEMS, heat transfer in oscillating flows, heat exchangers. Design optimization of thermoacoustic refrigerators. *Int. J. Refrig.* 20(1):3, 1997 (with Wetzel). Expanding the applications of holographic interferometry to the quantitative visualization of complex, oscillatory thermofluid processes. *Experiments Fluids* 24:431, 1998 (with Kang and Wetzel). Limitations of temperature measurements with holographic interferometry in the presence of pressure variations. *Exp. Therm. Fluid Sci.* 17:294, 1998 (with Wetzel).

Joseph Katz, Professor; Ph.D., Caltech, 1982. Experimental fluid mechanics, quantitative visualization of complex flows, turbomachines, breaking waves, bubbly and cavitating flows, stratified shear flows and oceanographic flows, development of advanced diagnostic techniques, PIV, holography for laboratory and field application. Lift and drag forces on microscopic bubbles entrained by a vortex. *Phys. Fluids* 7(2):389, 1995. On the structure of bow waves on a ship model. *J. Fluid Mech.* 346:77, 1997. Effect of modification to tongue and impeller geometry on unsteady flow, pressure fluctuations, and noise in a centrifugal pump. *J. Turbomachinery* 119:506, 1997. Turbulent flow measurement in a square duct with hybrid holographic PIV. *Exp. Fluids* 23:373, 1997.

Omar M. Knio, Associate Professor; Ph.D., MIT, 1990. Computational fluid mechanics, reacting shear flows, physical acoustics, atmospheric and oceanic flow, asymptotic techniques. Numerical simulation of a thermoacoustic refrigerator. Part II: Stratified flow around the stack. *J. Comput. Phys.* 144:299, 1998 (with Worlikar and Klein). Numerical simulation of large-amplitude internal solitary waves. *J. Fluid Mech.* 362:53, 1998 (with Terez). A dynamic LES scheme for the vorticity transport equation: Formulation and a priori tests. *J. Comput. Phys.* 145:693, 1998 (with Mansfield and Meneveau).

Charles Meneveau, Professor; Ph.D., Yale, 1989. Experimental, numerical, and theoretical studies in turbulence; subgrid scale modeling and large-eddy simulation; fractals; application of novel data-analysis techniques to shed new light on the long-standing turbulence problem; development of improved models for engineering applications. Statistics of turbulence subgrid scale stresses: Necessary conditions and experimental tests. *Phys. Fluids* 6:815, 1994. On the properties of similarity subgrid-scale models as deduced from measurements in a turbulent jet. *J. Fluid Mech.* 275:83–119, 1994 (with Liu and Katz). Subgrid-scale stresses and their modelling in a turbulent plane wake. *J. Fluid Mech.* 349:253–93, 1997 (with O'Neil).

Hasan N. Oguz, Associate Research Professor; Ph.D., USC, 1987. Fluid dynamics and underwater acoustics of bubbles, experimental and numerical study of bubble generation in liquids. Axisymmetric and three dimensional boundary integral simulations of bubble growth from an underwater orifice. *Eng. Anal. Bound. Elements J.* 19:319–30, 1997 (with Zeng). On the relaxation of laminar jets at high Reynolds numbers. *Phys. Fluids* 10:361–7, 1998. The natural frequency of oscillation of gas bubbles in tubes. *J. Acoust. Soc. Am.,* 103:3301–8, 1998 (with Prosperetti). The role of surface disturbances on the entrainment of bubbles. *J. Fluid Mech.,* 372:189-212, 1998.

Thomas R. Osborn, Professor (joint appointment with Department of Earth and Planetary Sciences); Ph.D., California, San Diego, 1969. Physical oceanography, turbulence, velocity and temperature microstructure, population dynamics, harmful algal blooms, coastal nutrient enrichment. Measurements in bubble plumes from a submarine. *Atmosphere-Ocean* 30(3):419–40, 1992 (with Farmer, Vagle, Thorpe, and Cure). Toward a theory of bio-physical control of harmful algal bloom dynamics and impacts. *Limnol. Oceanogr.* 42(5 part 2):1283–96, 1997. Finestructure, microstructure, and thin layers. *Oceanography* 11(1):36–43, 1998.

Marc B. Parlange, Professor (joint appointment with Department of Geography and Environmental Engineering); Ph.D., Cornell, 1990. Hydrology and fluid mechanics in the environment: land-atmosphere interaction, soil hydrology, evaporation, precipitation, atmospheric boundary layer, turbulence, lidar, transport in porous media. Hydrologic cycle explains the evaporation paradox. *Nature* 395:30, 1998 (with Brutsaert). On water vapor flow in field soils. *Water Resources Res.* 34(4):731–9, 1998 (with Cahill).

Andrea Prosperetti, Miller Professor; Ph.D., Caltech, 1974. Theoretical and computational fluid mechanics and acoustics: multiphase flow (phenomenology and direct numerical simulation), thermoacoustic processes (modeling and nonlinear effects), underwater sound (natural sources and scattering), bubbles in MEMS, air entrainment in liquid flows; limited experimental activity in the last two areas. Momentum and energy equations for disperse two-phase flows and their closure for dilute suspensions. *Int. J. Multiphase Flow* 23(3):425–53, 1997 (with Zhang). Air entrainment upon liquid impact. *Philos. Trans. R. Soc. London A* 355:491–506, 1997 (with Oguz). Linear thermoacoustic instability in the time domain. *J. Acoust. Soc. Am.* 103(6):3309, 1998 (with Karpov). Thermal processes in the oscillations of gas bubbles in tubes. *J. Acoust. Soc. Am.* 104(3):1389, 1998 (with Chen).

K. T. Ramesh, Professor; Ph.D., Brown, 1987. Tribology, rheology, micromechanisms of failure in composite materials, material behavior at very high strain rates, shear localization, biomimetics and active materials. Mechanical properties of compliant piezoelectric composites. *Smart Mater. Struct.* SPIE, 148, 1997 (with Marra and Douglas). The high-strain-rate response of alpha-titanium: Experiments deformation mechanisms and modeling. *Acta Materialia* 46(3):1, 1998 (with Chichili and Hemker). On the compressibility of a glass-forming lubricant: Experiments and molecular modeling. *J. Mech. Phys. Solids* 46:1699, 1998 (with Zhang). Influence of particle volume fraction, shape, and aspect ration on the behavior of particle-reinforced metal-matrix composites at high rates of strain. *Acta Materialia,* 46:5633, 1998 (with Li).

William N. Sharpe Jr., Decker Professor; Ph.D., Johns Hopkins, 1966. Experimental solid mechanics, especially laser-based interferometry over short gauge lengths; testing of small specimens; mechanical properties of MEMS materials. Mechanical properties of LIGA-deposited nickel for MEMS transducers. In *Proceedings Transducers '97*, Chicago, Illinois, pp. 607–10, 1997 (with LaVan and Edwards). Measurement of Young's Modulus, Poisson's Ratio, and tensile strength of polysilicon. In *Proceedings of the Tenth IEEE International Workshop on Microelectromechanical Systems*, Nagoya, Japan, pp. 424–9, January 1997 (with Yuan et al).

Louis L. Whitcomb, Assistant Professor; Ph.D., Yale, 1992. Dynamics, design, and model-based adaptive control of mechanical systems. Focus on dynamics and control of underwater robotics vehicles, dynamics and control of robot arms, and medical robotics. Adaptive model-based hybrid control of geometrically constrained robot arms. *IEEE Trans. on Robotics and Automation* 13(1):105, 1997 (with Arimoto et al). Towards precision robotic maneuvering, survey, and manipulation in unstructured undersea environments. *Robotics Research—The Eighth International Symposium*, In Y. Shirai and S. Hirose, eds., ch. 2, 45, London: Springer-Verlag, 1998 (with Yoerger et al). Comparative structural analysis of 2-dof semi-direct-drive linkages for robot arms.*IEEE/ASME Transactions on Mechatronics*, March 1999 (with Roy).

MICHIGAN STATE UNIVERSITY

Department of Mechanical Engineering

Programs of Study

The department offers graduate programs leading to the Master of Science and Doctor of Philosophy degrees. The student's graduate program is designed in consultation with a faculty adviser. There are excellent opportunities for analytical, computational, and experimental study in the areas of mechanical systems, thermal engineering, and fluid mechanics. Information about current areas of faculty research can be found on the reverse of this page.

The master's degree requires 30 semester credits and can be completed in one to two years of study. The student may choose a thesis option (6–8 credits) or a nonthesis option. Students have a unique opportunity to conduct master's degree research at the Rheinisch–Westfalische Technische Hochschule in Aachen, Germany. The Ph.D. candidate must pass a qualifying examination and a comprehensive examination. Typically, an additional 24 credits of course work beyond the M.S. degree and a minimum of 24 research credits are involved in a Ph.D. program. The Ph.D. program prepares a student for a career in advanced research and/or teaching.

Research Facilities

The mechanical engineering faculty occupies some 12,000 square feet of research space in the Engineering Research Facility in addition to laboratories in the Engineering Building and at two other sites. The facilities include Engine Research/Diagnostics, Heat Transfer, Nonlinear Vibrations, Control Systems, Computational Design, Composite Materials, Fluid Mechanics, Turbomachinery, and Bioengineering laboratories. The College's A. H. Case Center for Computer-Aided Engineering provides additional network, workstation, and software support for research activities.

Financial Aid

Fellowships, graduate teaching assistantships, and graduate research assistantships are available. Assistantship appointments may be for either the academic year or the calendar year and include health insurance coverage. Fellowship amounts and conditions vary according to their source; students may receive a waiver of all out-of-state tuition and fees and a stipend of up to $20,000 per calendar year. Most mechanical engineering graduate students receive financial support.

Cost of Study

In 1999–2000, graduate tuition is $223 per semester credit for Michigan residents and $450 per semester credit for nonresidents. In addition, per-semester fees include a $288 registration fee for students enrolling for more than 4 credits ($238 for students enrolling for 4 or fewer credits), a $237 engineering program fee for students enrolling for more than 4 credits ($128 for students enrolling for 4 or fewer credits), and $25.20 in other fees.

Living and Housing Costs

For residence halls, a standard double room is $2289 per semester per student. For University apartments, the range is $420 to $465 per month, depending upon the details.

Student Group

There are approximately 100 graduate students in the mechanical engineering program. About 50 percent are Ph.D. candidates. The College of Engineering has a total enrollment of about 600 graduate students.

Student Outcomes

Graduates of the M.S. and Ph.D. programs are actively pursued for employment by the automotive industry, specifically Ford and General Motors. Hewlett-Packard has also hired a number of M.S. graduates over the years. Several Ph.D. graduates have pursued academic careers at universities and colleges such as the University of Nebraska and Calvin College.

Location

The 5,100-acre East Lansing campus, one of the most beautiful in the nation, lies approximately 70 miles northwest of Detroit. The Red Cedar River traverses the northern 2,000 acres, which encompass the major developed area of the campus. The southern part of the campus is mainly used for agricultural activities. Much of the surrounding area is rural; the Greater Lansing and East Lansing areas, however, have a combined population of approximately 250,000. Lansing is the state capital and the home of Oldsmobile.

The University

Michigan State University, founded in 1855, was the prototype institution for the nation's land-grant colleges. Its enrollment of 42,000 makes MSU one of the largest single-campus universities in the nation. Such a large campus provides the graduate engineering student with wide exposure to additional academic areas such as natural science, medicine, and business. The number of graduate students and the amount of research expenditures have grown considerably in recent years. The University has excellent facilities and active programs for the performing arts in the Wharton Center.

Applying

Application forms and instructions for submitting required material may be obtained from the department. Study may begin in August, January, or May. Applications are accepted at any time but should be received at least two months prior to the expected starting date. Most fellowship and assistantship opportunities occur for the fall semester, and applications for the fall semester should be submitted by February if financial aid is desired. The GRE General Test, with a minimum score of 90 percent quantitative and 80 percent analytical, is required of all students. The TOEFL is required of international students. A minimum TOEFL score of 570 is necessary. All transcripts, a statement of purpose, and three letters of recommendation (in letter form addressed to the graduate advisor) are required.

Correspondence and Information

Graduate Advisor
Department of Mechanical Engineering
2555 Engineering Building
Michigan State University
East Lansing, Michigan 48824-1226
Telephone: 517-355-5220
Fax: 517-353-1750
E-mail: megradad@me.msu.edu

Michigan State University

THE FACULTY AND THEIR RESEARCH

Andre Benard, Assistant Professor; Ph.D., Delaware, 1995. Heat transfer, materials processing, solidification, polymer composites, numerical methods, rheology.

Giles J. Brereton, Associate Professor; Ph.D., Stanford, 1987. Fundamentals: turbulence, unsteady fluid mechanics, nonlinear acoustics, nonequilibrium thermodynamics; applications: internal combustion engines, emissions, cardiovascular processes.

Alejandro Diaz, Associate Professor; Ph.D., Michigan, 1982. Computational mechanics, structural design and optimization, finite element and meshless methods.

Abraham Engeda, Assistant Professor; Ph.D., Hannover (Germany), 1987. Applied thermofluids, internal and unsteady flow in turbomachines, design and performance prediction of turbomachines.

Brian Feeny, Assistant Professor; Ph.D., Cornell, 1990. Nonlinear dynamical systems and chaos, nonsmooth systems.

John F. Foss, Professor; Ph.D., Purdue, 1965. Fluid mechanics, experimental investigation of turbulence.

Mukesh V. Gandhi, Associate Professor; Ph.D., Michigan, 1984. Continuum mechanics, with emphasis on constitutive theories and large deformation problems; computer graphics and finite elements in CAD.

Erik D. Goodman, Professor; Ph.D., Michigan, 1972. Computer-aided design and manufacturing (CAD/CAM), genetic algorithms.

Alan G. Haddow, Associate Professor; Ph.D., Dundee (Scotland), 1983. Experimental and theoretical work on nonlinear dynamical systems, signal analysis.

Manoochehr Koochesfahani, Associate Professor; Ph.D., Caltech, 1984. Turbulent mixing, unsteady aerodynamics, turbulent shear flow control, optical diagnostics.

John R. Lloyd, University Distinguished Professor; Ph.D., Minnesota, 1971. Heat transfer, fire and combustion, turbomachinery, controllable fluids, heat exchangers.

John J. McGrath, Professor; Ph.D., MIT, 1977. Biophysics and bioengineering, composite materials, heat and mass transfer.

Matthew A. Medick, Professor; Ph.D., Columbia, 1958. Wave propagation, vibrations, applied mathematics.

Ranjan Mukherjee, Associate Professor; Ph.D., California, Santa Barbara, 1991. Dynamic systems and control.

Ahmed M. Naguib, Assistant Professor; Ph.D., IIT, 1992. Experimental fluid dynamics, turbulence, physics, flow diagnostics and control, MEMS application to fluid mechanics.

Farhang Pourboghrat, Assistant Professor; Ph.D., Minnesota, 1992. Computational solid mechanics and plasticity, shell theory, sheet and extrusion forming, material process design.

Clark J. Radcliffe, Professor; Ph.D., Berkeley, 1979. System dynamics, instrumentation design, control, acoustics.

Ronald C. Rosenberg, Professor and Chairperson; Ph.D., MIT, 1965. Dynamic systems modeling and simulation, multiport systems theory, computer-aided engineering design.

Harold Schock, Professor; Ph.D., Michigan Tech, 1979. Thermodynamics, combustion, optical diagnostics, turbulence, internal combustion engines.

Steven W. Shaw, Professor; Ph.D., Cornell, 1983. Dynamics and vibrations, applications of nonlinear dynamical systems theory, vibroimpact problems, vibration absorbers, ship stability, structural dynamics.

Craig W. Somerton, Associate Professor; Ph.D., UCLA, 1982. Heat and mass transfer in porous media, CAD of thermal systems.

Brian E. Thompson, Professor; Ph.D., Dundee (Scotland), 1976. High-speed machinery, composite materials, smart materials, design methodologies.

C. Y. Wang, Professor; Ph.D., MIT, 1966. Fluid mechanics, heat convection, biofluid mechanics.

Indrek Wichman, Associate Professor; Ph.D., Princeton, 1983. Combustion theory, fire research, ignition, extinction, and quenching, triple flames, sprays.

David Wiggert, Professor; Ph.D., Michigan, 1967. Fluid mechanics, fluid transients, groundwater contamination.

Mei Zhuang, Assistant Professor; Ph.D., Caltech, 1990. Computational fluid dynamics, mixing layer stability, transition simulation and modeling.

AREAS OF CURRENT RESEARCH

Acoustics and Vibration. Nonlinear dynamics of mechanical systems and structures, acoustic noise measurement, flow-induced vibrations, acoustic modeling, statistical energy analysis, design of vibration absorbers, structural dynamics, rotor imbalance and vibration.

Applied Computational Mechanics. Structural optimization, concurrent design of thin-walled structures, optimal design of microstructures, modeling of multiscale phenomena in phase change processes, computational aeroacoustics.

Automotive. Flow studies in internal combustion engines, active noise control, underhood thermal management, system modeling, sensor technology, numerical modeling of stamping operations, springback prediction, die design for stamping operations, numerical modeling of stretch forming and hydroforming of thin-walled hollow extrusions.

Bioengineering/Biophysics. Diagnostic and therapeutic use of stress waves in medicine, experimental and theoretical aspects of evaporation from the skin, in vivo measurement of local blood perfusion rates, irreversible thermodynamic analysis of coupled passive membrane transport properties, experimental characterization of coupled passive membrane transport properties, fundamental aspects of cryopreservation, medical robotics.

Combustion. Turbulent mixing, combustion and modeling in internal and external combustion engines, fundamental aspects of coal and wood combustion, soot formation in flames, basic studies of ignition, extinction, and quenching, triple flames, fire research modeling.

Computational Design. Structural and multidisciplinary optimization, meshless methods in topology optimization, advanced tools for engineering simulation and design, CAD of thermal systems.

Fluid Mechanics. Vorticity measurements in turbulent shear layers; turbulent mixing and flow control; shear layer and slit-jet flow instability; study of coherent motions in turbulent boundary layers; drag reduction; rough surface turbulence dynamics; unsteady separation; structure of unsteady airfoil and bluff body wakes; study of tabbed jets; dynamics of vortex interactions; optimization of induction system flows; in-cylinder fluid flows; mass airflow sensors; fan-induced flows; computational fluid dynamics, novel hot-wire anemometry techniques, optical diagnostics, molecular tagging techniques; fluid mechanics in material processing.

Heat/Mass Transfer and Thermodynamics. Three-dimensional radiation-convection interactions in complex enclosures, multicomponent mixture boiling and condensation, enhanced heat-transfer surfaces, porous media transport phenomena, heat and mass transfer in separated flows, heat exchanger performance, thermally smart composite materials, solidification phenomena, computational heat transfer, heat transfer in material processing.

Solid Mechanics and Design. Continuum mechanics, with emphasis on constitutive theories and large deformation problems; modeling phenomena associated with interacting media; optimal synthesis of composite materials for intelligent machinery; mechanisms fabricated with composite laminates; dynamics of flexible high-speed linkage machinery, macroscopically "smart" biomimetic materials, rigid and finite-strain elastoplasticity of anisotropic sheet metals, finite element modeling of anisotropic yield functions, strain-rate potentials and kinematic hardening.

Systems and Control Theory. Modeling of mixed-energy systems, bond graph theory, large-scale system simulation, robust control methods, active noise and vibration control, nonholonomic control systems, self-sensing and control of magnetic bearings.

Turbomachinery. Experimental work in centrifugal compressor aerodynamics and blade cooling heat-transfer; design, development, and testing of vaned diffusers.

MichiganTech.

MICHIGAN TECHNOLOGICAL UNIVERSITY

Department of Mechanical Engineering–Engineering Mechanics

Programs of Study

The M.S. is offered in both mechanical engineering and engineering mechanics, and the Ph.D. is offered in mechanical engineering–engineering mechanics. The purpose of the graduate programs is to prepare students to work in the design and development of advanced engineering technology and, in the case of the doctoral program, to teach or to carry out research. An option in manufacturing systems engineering is available in the M.S. program. The M.S. program requires the successful completion of at least 45 total quarter credits. There are three different plans for the degree. The thesis plan requires a thesis of at least 15 credits; the project plan involves a project report of 3–9 credits; and the course plan involves 45 credits of course work. A full-time student can expect to obtain an M.S. in five to seven quarters. The Ph.D. program requires the successful completion of course work and a research dissertation planned in consultation with the candidate's faculty adviser. The candidate must pass a qualifying examination in the early fall quarter or late spring quarter, pass a comprehensive examination, write and present a formal dissertation proposal, and present and defend a research dissertation. A year of course work beyond the M.S. courses is usually required in the Ph.D. program; the total time depends on the individual's background but is normally three years beyond the master's degree.

The department has four general areas of teaching and research: design and dynamic systems, energy and thermofluids, manufacturing/industrial, and solid mechanics. Within this framework, students select an adviser and develop a research topic and courses to suit their individual interests and needs. The faculty members from each technical area have established lists of courses from which students can build their program of study. Interdisciplinary endeavors and industrial interaction are also encouraged.

Research Facilities

The ME-EM department has well-equipped modern laboratories for experimental research in many areas, including heat transfer, fluid mechanics, engines, combustion, vibrations, experimental mechanics, material testing, fatigue, fracture, creep, plastics, composite materials, materials processing, manufacturing systems, machining, forming, robotics, acoustics, dynamics, and controls. Additional University-wide research facilities include a scanning electron microscope, a transmission electron microscope, extensive analytical chemistry instrumentation, and a graduate computing laboratory with extensive hardware and software capabilities. Computationally intensive machines are available. Research equipment and resources are also available at several University research institutes—among them, the Institute of Materials Processing, Institute of Wood Research, and Keweenaw Research Center. These institutes provide opportunities for cooperative and interdisciplinary research between the institute staff members, faculty members, and graduate students. The University library, a multimillion-dollar modern structure, incorporates the latest features in library planning and is particularly strong in the physical and natural sciences, mathematics, and engineering.

Financial Aid

Graduate teaching and research assistantships are available and provide $2680 per quarter plus tuition for M.S. students and $3100 per quarter plus tuition for Ph.D. students. Six department fellowships are available to outstanding students each year, along with a number of industrial fellowships.

Cost of Study

In 1998–99, full-time graduate students who were residents paid tuition and fees of $1469 per quarter plus $216 for each credit hour over 12; nonresidents paid $2856 per quarter plus $462 for every credit hour over 12. Tuition rates have not been determined for the 1999–2000 academic year.

Living and Housing Costs

The University maintains student housing and dormitory accommodations nearby. For 1999–2000, projected rates are $345 per month for a one-bedroom family apartment and $383 per month for a two-bedroom apartment. Private rooms and apartments are also available nearby. The cost of living is moderate.

Student Group

There are approximately 115 full-time graduate students in the department, with about 26 percent of them working toward the Ph.D. degree. Approximately 45 percent of the students are from overseas.

Location

The University is located in Houghton, next to Portage Lake, in Michigan's scenic Upper Peninsula near Lake Superior. Forests, parks, and clean lakes and streams make the area ideal for outdoor activities.

The University

Michigan Technological University was founded in 1885 as the Michigan School of Mines, in response to a growing need for mining and metallurgical engineers. In 1964, the name was changed to Michigan Technological University to reflect the broad spectrum of programs the institution offered in science, engineering, forestry, business, liberal arts, social sciences, science teacher education, and technology. Currently, twenty-seven programs grant a B.S. degree. Of these, twenty-two grant an M.S. and fifteen, a Ph.D. In addition, the University offers three B.A. degree programs, and seven programs culminate in a two-year Associate in Applied Science degree. MTU is a state-supported university. Of the approximately 6,250 students enrolled in 1998, about 62 percent were pursuing engineering studies, ranking this institution the seventeenth-largest engineering school in the nation and the third-largest in mechanical engineering. The College of Engineering is made up of eight departments covering the disciplines of mechanical engineering–engineering mechanics, metallurgy, chemistry, chemical engineering, civil and environmental engineering, electrical engineering, geology–geological engineering, general engineering, and mining engineering. Of the 149 members of the academic faculty in engineering, more than 93 percent have earned doctorates. University funding in both basic and applied research grew from $3 million in 1976 to more than $28 million in 1998.

Applying

Application forms and instructions for submitting transcripts, letters of recommendation, and related material should be obtained from the department. Study may begin in September, December, March, or June. Applications are accepted at any time but should be submitted six months before the expected starting quarter. Most fellowship and assistantship opportunities occur for the fall quarter. Applications for this quarter should be submitted by January 1 if financial aid is desired. International applicants must have a TOEFL score greater than 580 (at least 55 in each section), and they must take the GRE General Test. International students should be prepared to provide their own financial support for at least twelve months.

Correspondence and Information

Professor William W. Predebon, Chair
Department of Mechanical Engineering–
 Engineering Mechanics
Michigan Technological University
Houghton, Michigan 49931
Telephone: 906-487-2551
E-mail: megradap@mtu.edu

Professor John W. Sutherland, Associate
 Chair and Director of
 Graduate Studies
Department of Mechanical Engineering–
 Engineering Mechanics
Michigan Technological University
Houghton, Michigan 49931
Telephone: 906-487-2551

Michigan Technological University

THE FACULTY AND THEIR RESEARCH

D. L. Abata, Ph.D., Wisconsin. Thermodynamics, combustion, I-C engines.
E. Aifantis, Ph.D., Minnesota. Multiscale continuum mechanics, pattern-forming instabilities in plasticity and fracture.
C. L. Anderson, Ph.D., Wisconsin. Heat transfer, I-C engines.
O. Arici, Ph.D., Brown. Thermodynamics, heat transfer, solar energy, vehicle climate control.
J. E. Beard, Ph.D., Purdue. CAD, kinematics, bioengineering, manufacturing.
A. Chandra, Ph.D., Cornell. Materials processing, damage evolution, boundary element method, electronic packaging.
P. Cho, Ph.D., Northwestern. Combustion, fuels, air pollution.
H. A. Evensen, Ph.D., Syracuse. Noise, vibration, dynamic measurements.
L. W. Evers, Ph.D., Wisconsin. I-C engines, fuel injectors, sprays and atomization.
W. J. Frea, Ph.D., Washington (Seattle). Heat transfer, thermodynamics, fluid mechanics.
C. R. Friedrich, Ph.D., Oklahoma State. Micromanufacturing, micromechanical and precision machining, design.
T. R. Grimm, Ph.D., Michigan Tech. Engineering design, CAE, finite elements, biomechanics.
M. Gupta, Ph.D., Rutgers. Design with plastics and composites, polymer processing, electronic packaging, FEM.
G. Jayaraman, Ph.D., Iowa. Biomechanics.
J. H. Johnson, Ph.D., Wisconsin. Combustion, emissions, thermodynamics, engines, air pollution.
J. B. Ligon, Ph.D., Iowa State. Experimental mechanics, wave propagation, phytomechanics, polymer composite materials.
Z. K. Ling, Ph.D., Minnesota. Kinematics and dynamics, design and manufacturing automation, computer graphics, biomechanics.
E. Lumsdaine, Ph.D., New Mexico. Acoustics, vibration, heat transfer.
S. A. Majlessi, Ph.D., RPI. Manufacturing and materials processing, sheet metal forming, finite-element analysis, imaging analysis.
D. J. Michalek, Ph.D., Texas at Arlington. Computational fluid dynamics.
M. H. Miller, Ph.D., North Carolina State. Manufacturing, grinding, precision engineering.
I. Miskioglu, Ph.D., Iowa State. Experimental stress analysis, fracture mechanics, composite materials.
K. Moon, Ph.D., Illinois. Nanotechnology, manufacturing process modeling and optimization, surface texture characterization.
A. Narain, Ph.D., Minnesota. Heat transfer, fluid mechanics, film condensation flows, viscoelastic liquids.
D. A. Nelson, Ph.D., Duke. Electrohydrodynamics, bioheat transfer, biological effects of radio-frequency radiation.
S. M. Pandit, Ph.D., Wisconsin. Data-dependent systems modeling, forecasting, computer control, machine vision, laser interferometry.
G. G. Parker, Ph.D., SUNY at Buffalo. Dynamics, linear and nonlinear control, robotics.
C. E. Passerello, Ph.D., Cincinnati. Vibrations, dynamics, finite elements.
W. W. Predebon, Ph.D., Iowa State. Ceramic processing, behavior and characterization, wave propagation in solids, impact phenomena.
M. D. Rao, Ph.D., Auburn. Vibration, dynamic measurements, noise control and composites.
J. F. Schultze, Ph.D., Cincinnati. Control theory and application, structural dynamics, modal analysis.
W. R. Shapton, Ph.D., Cincinnati. Modal analysis, CAE, kinematics.
D. L. Sikarskie, Sc.D., Columbia. Structural mechanics, composite materials, sandwich structures.
G. Subhash, Ph.D., California, San Diego. Ceramics, micromechanics, wave propagation, experimental solid mechanics.
N. V. Suryanarayana, Ph.D., Michigan. Convective heat transfer, turbulent flows.
J. W. Sutherland, Ph.D., Illinois. Environmentally conscious manufacturing, metal cutting, quality engineering, machining dynamics.
M. Vable, Ph.D., Michigan. Computational structural mechanics.
C. D. Van Karsen, M.S.M.E., Cincinnati. Noise, vibration, experimental methods.
C. R. Vilmann, Ph.D., Northwestern. Fracture mechanics, finite elements.
R. O. Warrington, Ph.D., Montana. Micromanufacturing, microtransport processes, laser-based micromachining, heat transfer.
K. J. Weinmann, Ph.D., Illinois. Sheet-metal forming, tribology in metal forming, metal cutting.
S. L. Yang, Ph.D., Florida. Computational fluid dynamics, heat transfer.

Partial List of Active Research Projects

Finite-element analysis of sheet-metal forming.
Materials-processing modeling and optimization.
Nanotechnology: machining and assessment of precision surfaces.
Wavelet and DDS methods for monitoring and control of manufacturing processes.
Tool failure prediction.
Environmentally conscious design and manufacturing.
Mechanistic modeling, sensor integration, and intelligent control of machining processes.
Optimization sensor integration and feedback control in sheet-metal forming: the intelligent stamping die.
Modeling of high-speed machining: high strain rate perspective.
Microdrilling, micromilling models.
Micromechanical machining processes, mesoscopic systems.
Precision machining of brittle materials.
Adaptive control of structures, real-time parameter estimation, sensor and actuator dynamics, flexible body control.
Machine vision–based teleoperation of serial and parallel robots.
Nonlinear control and operator-in-the-loop simulation for crane systems.
Dynamic simulation and control of single- and multiple-stage stewart platforms.
Plastic encapsulation of microchips.
Fiber orientation in injection-molded plastic parts.
Vibrational energy flow through structures.
Processing experimental acoustic and vibration data.
Sound quality, experimental vibration analysis.
Boundary-element-method code for mechanical and adhesive fastening of composites.
Concentrated load transfer in composite plates and shells; failure and damage growth criteria for random fiber composites.
Hybrid methods (experimental and analytical) applied to fracture of bimaterial systems.
Strain-gaged plants as physiological transducers.
Gradient theory of plasticity, constitutive modeling; granular materials, nanophase materials.
Constitutive response, characterization of metals, ceramics, and composites; development of applications of high-strength ceramics.
Core failure in composite sandwich structures; fatigue of sandwich structures.
Fastening systems for thermoplastic polymer composites.
Efficacy of chest pads against baseball impacts in young athletes.
Efficacy of plate strengthening of osteotomized radius against bending and torsion.
Modeling of thermal characteristics of oil and coolant flows within a diesel cooling system.
High-speed photography of diesel combustion.
Control of diesel pollutants in underground mining; chemical characterization of diesel particulate emissions.
Analysis of parabolic solar collector based on real optical and thermal properties.
Condensation heat transfer.
Modeling and computation of buoyant and nonequilibrium turbulent flows.
Fuel sprays and fuel injectors.

NORTH CAROLINA STATE UNIVERSITY

College of Engineering
Department of Mechanical and Aerospace Engineering

Programs of Study

The Department of Mechanical and Aerospace Engineering (MAE) at North Carolina State University is the largest in the state and among the largest and most prominent in the nation. The strengths of the department lie in the thermal sciences, particularly thermal fluids, fluid mechanics, and combustion; mechanical sciences, including manufacturing mechanics, structural dynamics, and materials and controls; and the aerospace sciences, particularly hypersonics, propulsion, and computational fluid dynamics. The department offers the Master of Science (M.S.) degree in both mechanical engineering (ME) and aerospace engineering (AE) with a thesis or nonthesis option, the Master of Mechanical Engineering (M.M.E.) degree (nonthesis option), and the Doctor of Philosophy (Ph.D.) degree in both ME and AE. Approximately one-third of the graduate student population is international. Graduate courses are offered during the fall and spring semesters as well as during the two summer sessions, and students may study on a year-round basis.

Research Facilities

The department houses several facilities that support research activities. Included in these facilities are instruments and equipment associated with the seven centers and laboratories of the MAE department. The Center for Sound and Vibration encompasses a broad range of graduate research activities in the areas of acoustics and mechanical vibration. The Center has modern anechoic and reverberation facilities as well as state-of-the-art computational and data acquisition equipment. The Applied Energy Research Laboratory operates in conjunction with the NCSU Solar House and conducts research and development associated with ground-coupled heat pumps, air source systems, direct contact heat exchangers, and combustion.

The Precision Engineering Laboratory is a multidisciplinary research center that performs research in metrology (sensors and measurement systems), innovative precision fabrication processes, and real-time process control. As a facility established to promote the implementation of solar technologies in both residential and commercial components, the North Carolina Solar Center has a variety of resources that support this function. The Energy Analysis and Diagnostic Center conducts research and development in energy efficiency as well as alternative energy sources.

In response to national concerns over manufacturing in the U.S., the Integrated Manufacturing Systems Engineering Institute (IMSEI) was established to perform research in innovative manufacturing systems. IMSEI is an interdisciplinary institute involving faculty from the colleges of Management, Engineering, and Textiles. The Mars Mission Research Center has been established to broaden the nation's capabilities to meet the critical needs of the civilian space program. Areas of specialization include mission analysis and design, hypersonic aerodynamics and propulsion, spacecraft structures, navigation and control, and composite materials and fabrication.

Financial Aid

The Department of Mechanical and Aerospace Engineering offers a number of graduate research and teaching assistantships ranging from $9000 to $15,000 per year. A limited number of fellowships are also available.

Cost of Study

Tuition and fees for full-time study in 1999–2000 are $1185 per semester for North Carolina residents and $5768 per semester for nonresidents. Students taking fewer than 9 credits pay reduced amounts. An out-of-state student appointed as a teaching or research assistant will have all tuition and health insurance covered.

Living and Housing Costs

On-campus dormitory facilities are provided for unmarried graduate students. In 1999–2000, the rent for double rooms start at $1450 per semester. Accommodations in the newest residence hall for graduate students cost $1135 per semester. Apartments for married students in King Village rent for $305 per month for a studio, $320 for a one-bedroom apartment, and $345 for a two-bedroom apartment.

Student Group

The Department of Mechanical and Aerospace Engineering has an enrollment of 800 undergraduate students and 150 graduate students. Most graduate students find full- or part-time support through fellowships, assistantships, and special duties with research organizations in the area.

Location

Raleigh, the state capital, has a population of about 237,000. Nearby is the Research Triangle Park, one of the largest and fastest-growing research institutions of its type in the country. The University's concert series has more subscribers than any other in the United States. Excellent sports and recreational facilities are also available.

The University and The College

North Carolina State University is the principal technological institution of the University of North Carolina System. Its largest schools are the colleges of Engineering, Agriculture and Life Sciences, Physical and Mathematical Sciences, and Humanities and Social Sciences. In 1998–99, total enrollment in the Department of Mechanical and Aerospace Engineering was 150. A strong cooperative relationship exists with nearby Duke University and the University of North Carolina at Chapel Hill, as well as with the Research Triangle Park. The department has 40 faculty members with professorial rank. Some of their current research areas are listed on the back of this page.

Applying

Application submission deadlines for each semester may be obtained through the MAE Graduate Administrator's Office. An applicant desiring to visit the campus may request information concerning travel allowances by writing to the graduate administrator. Students may apply for fellowships or assistantships in their application for admission. For application forms and further information, students should write to the address given below.

Correspondence and Information

Dr. J. C. Mulligan, Director of Graduate Programs
Department of Mechanical and Aerospace Engineering
Box 7910
North Carolina State University
Raleigh, North Carolina 27695-7910
Telephone: 919-515-3026

North Carolina State University

THE FACULTY AND THEIR RESEARCH

E. M. Afify, Ph.D., Professor. Thermal sciences, internal combustion engines, combustion and exhaust emissions, air pollution control.

J. A. Bailey, Ph.D., Professor. Metal and wood machining, plastic deformation of materials, deformation processing, composite materials.

A. E. Bayoumi, Ph.D., Professor. Manufacturing systems.

M. A. Boles, Ph.D., Associate Professor. Thermodynamics, heat transfer, fluids.

N. Chokani, Ph.D., Associate Professor. Experimental and computational flow physics.

J. W. David, Ph.D., Associate Professor. Vibrations/dynamics, kinematics, automotive engineering, nonlinear mechanics.

F. R. DeJarnette, Ph.D., Professor, Department Head, and Program Director of Aerospace Engineering. Engineering and computational methods in aerothermodynamics.

T. A. Dow, Ph.D., Professor. Metrology systems, fabrication processes (machining, grinding, polishing) and real-time control.

H. M. Eckerlin, Ph.D., Professor. Industrial energy conservation/management, solar-active/passive/photovoltaic, steam generation and incineration.

J. R. Edwards, Jr., Ph.D., Assistant Professor. CFD, 2D and 3D compressible flows, thermochemical/nonequilibrium flow, turbulence modeling.

J. W. Eischen, Ph.D., Associate Professor. Computational solid mechanics, elasticity, fracture mechanics, structural dynamics.

M. M. Fikry, Ph.D., Visiting Professor. Heat transfer, energy and energy conservation (power plant engineering), refrigeration and air-conditioning.

R. D. Gould, Ph.D., Associate Professor. Experimental heat transfer, fluid mechanics, combustion, turbulence and nonintrusive optical diagnostics.

C. E. Hall, Jr., Ph.D., Associate Professor. Flight dynamics and control, nonlinear control theory, RPV system design and flight testing.

H. A. Hassan, Ph.D., Professor. Fluid mechanics, aerodynamics, combustion, transition, turbulence, Monte Carlo methods, CFD.

T. H. Hodgson, Ph.D., Professor. Theoretical acoustics, vibrations, fluid mechanics, turbulence, noise control, musical acoustics, digital signal processing.

R. R. Johnson, Ph.D., Associate Professor. Fluid mechanics, heat transfer, energy conversion, solar engineering and appropriate technology.

R. F. Keltie, Ph.D., Professor. Structural acoustics, vibration of rib-stiffened structures, acoustic radiation, mechanical vibrations.

E. C. Klang, Ph.D., Associate Professor. Analytical and experimental studies of composite materials, aerospace structural analysis.

C. Kleinstreuer, Ph.D., Professor. Computational biofluid mechanics, convection heat and mass transfer, system optimization.

A.V. Kuznetsov, Ph.D., Assistant Professor. Heat transfer and fluid flow in porous media, modeling of solidification processing, macrosegregation and process in the mushy zone during solidification of binary alloys.

J. W. Leach, Ph.D., Associate Professor. Energy conversion, heat transfer, fluid mechanics, thermophysical property evaluation.

G. K. F. Lee, Ph.D., Professor. Robotics, control theory.

K. L. Lyons, Ph.D., Assistant Professor. Experimental combustion, laser diagnostics, DFWM and PLIF, combustion engines, turbulent mixing.

C. J. Maday, Ph.D., Professor. Multivariable control systems, rotor dynamics, design optimization, bearing design, theory of constraints.

D. S. McRae, Ph.D., Professor. Computational fluid dynamics, unsteady flow, propulsion, environmental flows.

J. C. Mulligan, Ph.D., Professor. Heat transfer, fluid mechanics, thermodynamics, electronic cooling, incineration, combustion, and hazardous wastes.

R. T. Nagel, Ph.D., Professor. Experimental fluid mechanics, aeroacoustics, active noise control.

J. N. Perkins, Ph.D., Professor. Analysis, design and flight testing of unmanned aerial vehicles.

M. K. Ramasubramanian, Ph.D., Assistant Professor. Design, manufacturing, mechanics, behavior of paper and short fiber composites.

P. J. Ro, Ph.D., Associate Professor. Precision machining, robotics, manufacturing automation, control theory, system dynamics.

W. L. Roberts, Ph.D., Assistant Professor. Experimental combustion, laser diagnostics, DFWM and PLIF, propulsion, flame diffusion.

L. H. Royster, Ph.D., Professor. Effects of noise and vibration, noise and vibration control, vibrations of structures, machine design.

L. M. Silverberg, Ph.D., Professor. Structural electrodynamics and control, advanced antenna theory and design.

F. Y. Sorrell, Ph.D., Professor and Program Director of Mechanical Engineering. Fluid mechanics and heat transfer, heat transfer and chemical reactions in microelectronic manufacturing.

J. S. Strenkowski, Ph.D., Professor. Finite element analysis, nonlinear stresses, structural dynamics, computer-aided design and optimization.

G. D. Walberg, Ph.D., Professor. Spacecraft design, mission analysis, trajectory analysis and optimization, high-temperature gas dynamics and reentry vehicle thermal protection.

F. G. Yuan, Ph.D., Associate Professor. Failure, fracture and life prediction of advanced materials and structures, finite element methods.

M. A. Zikry, Ph.D., Associate Professor. Dynamics plasticity and fracture, constitutive relations for solids, computational solid mechanics.

NORTHEASTERN UNIVERSITY

Department of Mechanical, Industrial, and Manufacturing Engineering

Programs of Study

The Department of Mechanical, Industrial, and Manufacturing Engineering offers graduate programs on both a full-time and part-time basis, leading to the degrees of Master of Science and Doctor of Philosophy. The graduate programs have a strong core, supplemented by advanced courses in theoretical and applied engineering within an area of concentration in materials science and engineering, mechanics and design, thermofluids engineering, engineering management, and operations research. In addition, a full-time student may apply to participate in the cooperative plan.

Between 44 and 48 quarter hours of graduate study are required for the Master of Science degree, depending on the program.

The degree of Doctor of Philosophy is awarded to those candidates who demonstrate high attainment and research competence in the field of mechanical or industrial engineering. The student must pass the Ph.D. qualifying examination (consisting of written and oral portions), complete a program of course work, successfully complete a dissertation, and pass a final oral examination. A typical program includes at least 36 QH of course work beyond the M.S. degree, with at least 12 QH outside the area of concentration (or outside of the department). Although a reading knowledge of one foreign language of technological and/or scientific importance is required, the doctoral candidate may, with the approval of his or her dissertation adviser, petition to have this requirement waived.

Research Facilities

The department maintains broadbased research facilities. The Computing Resource Center (CRC) of the Division of Academic Computing provides access to computing resources. A high-speed data network links users and facilities on the central campus and on three satellite campuses. The campus network is also connected via the global Internet to resources around the world. Students have access to Digital VAX systems, Sun Workstations, labs of microcomputers, a computer-mail-and-conferencing system, and an array of specialized computing equipment. There is equipment for experimental research in combustion, heat transfer, fluid mechanics, biotechnology, composites, fatigue, tribology, high-temperature creep, thin-film fabrication, corrosion testing, and manufacturing.

The University library system has more than 650,000 volumes, 1.5 million microforms, 250,000 documents, 7,000 serial subscriptions, and 15,000 audio, video, and software titles. A large, central library contains technologically sophisticated library services, including an online catalog and circulation system, an information gateway, and a seventeen-station network of CD-ROM optical disk databases. Students also have access to other major research collections in the area through the Boston Library Consortium.

Financial Aid

Northeastern awards need-based financial aid to graduate students through the Federal Perkins Loan, Federal Work-Study, and Federal Stafford Student Loan programs. The University also offers a limited number of minority fellowships and Martin Luther King, Jr. Scholarships. The graduate schools offer financial assistance through teaching, research, and administrative assistantship awards that include tuition remission and a stipend of $12,450. These assistantships require a maximum of 20 hours of work per week. Also available are a limited number of tuition assistantships that provide partial or full tuition remission and require a maximum of 10 hours of work per week.

Cost of Study

The cost of tuition for the 1998–99 academic year in the Graduate School of Engineering was $465 per quarter hour of credit. A full-time program consists of at least 8 quarter hours. Where applicable, there are special tuition charges for theses and dissertations. Other charges include the Student Center fee and the health and accident insurance fee required of all full-time students.

Living and Housing Costs

The cost of room and board is estimated at $1200 per month, with some on-campus housing available to newly accepted students. Off-campus living expenses are estimated at $1500 per month. A public transportation system services the greater Boston area, and there are convenient subway and bus services.

Student Group

In fall 1998, 24,027 students were enrolled at Northeastern University, representing a wide variety of backgrounds. The department had 190 graduate students, 53 percent of whom attended on a full-time basis. There were 44 Ph.D. students in the department.

Student Outcomes

Over the last five years, roughly 50 percent of the Ph.D. graduates began faculty careers either directly or after postdoctoral research. The remainder found positions as research scientists in private industry and government laboratories. Master's students find work principally in local companies, although fully 40 percent go on to Ph.D. studies either at Northeastern or at other major research institutions.

Location

Boston, the capital city of Massachusetts, offers students a broad spectrum of academic, cultural, and recreational opportunities. In addition to the abundant resources available within Northeastern University, there are those provided by the other educational and cultural institutions of the Greater Boston area.

The University

Founded in 1898, Northeastern University is a privately endowed, nonsectarian institution of higher learning, and it is among the largest private universities in the country. The cooperative plan of education, initiated by the College of Engineering in 1909 and subsequently adopted by the other colleges of the University, enables students to alternate periods of work and study. The plan is available to selected graduate students. Today, Northeastern University has nine undergraduate schools and colleges, ten graduate and professional schools, two part-time undergraduate divisions, several suburban campuses, a number of research centers, and continuing and special education programs and institutes.

Applying

To be admitted, applicants must have a Bachelor of Science degree in mechanical engineering, industrial engineering, or a closely allied field. The degree, representing undergraduate work of acceptable quality, must be from a recognized institution. Applicants interested in full-time study should submit applications by February 15 to be considered for a graduate assistantship and by April 15 for fall admission without financial assistance. The GRE and the Test of English as a Foreign Language (TOEFL) are required of all applicants with undergraduate degrees from outside the United States.

Correspondence and Information

Director
Graduate School of Engineering
130 Snell Engineering Center
Northeastern University
Boston, Massachusetts 02115
Telephone: 617-373-2711

Northeastern University

THE FACULTY AND THEIR RESEARCH

Industrial Engineering

James C. Benneyan, Assistant Professor; Ph.D., Massachusetts. Quality engineering, statistical quality control, inspection error models, computer simulation, industrial experiments, application in manufacturing and health care, including semiconductor fabrication and cancer screening.

Thomas P. Cullinane, Professor; Ph.D., Virginia Tech. Manufacturing systems, facilities planning, project management.

Nasser Fard, Associate Professor; Ph.D., Arizona. Reliability analysis, quality engineering, stochastic modeling.

Surendra M. Gupta, Associate Professor; Ph.D., Purdue; PE. Simulation, operations research, production systems.

Sagar Kamarthi, Assistant Professor; Ph.D., Penn State. Neural networks and knowledge-based systems in design and manufacturing, processing monitoring and control.

Emanuel Melachrinoudis, Associate Professor; Ph.D., Massachusetts. Operations research, manufacturing systems.

Ronald R. Mourant, Professor; Ph.D., Ohio State. Simulation, human-computer interaction.

Ronald Perry, Associate Professor; Ph.D., Michigan. Simulation, management information systems.

Allen L. Soyster, Professor and Dean of the College; Ph.D., Carnegie Mellon. Production scheduling, mathematical modeling and optimization.

Gerard Voland, Associate Professor; Ph.D., Tufts. Engineering design, control theory, rehabilitation engineering.

Mechanical Engineering

George G. Adams, Professor; Ph.D., Berkeley. Response of elastic structures to moving loads, tribology, stress distributions at material interfaces, elasticity, stability.

Teiichi Ando, Associate Professor; Ph.D., Colorado School of Mines. Rapid solidification, materials processing and manufacturing processes.

Joseph T. Blucher, Associate Professor; Ph.D., MIT. Surface-treating processes: CVD, PVD, ion nitriding, and laser processing; metal-matrix composites; powder metallurgy; welding; cutting tools; manufacturing processes; failure analysis: fracture, fatigue, and wear.

John W. Cipolla Jr., Donald W. Smith Professor of Mechanical Engineering; Ph.D., Brown. Laser-aerosol interactions, including thermophoresis; heat and mass transfer; radiative transfer; kinetic theory.

Alexander M. Gorlov, Professor; Ph.D., Moscow Institute of Transport Engineers. Mechanical design of complex systems, mechanical apparatus for harnessing tidal and low-head hydro power, general applied mechanics problems.

Hamid N. Hashemi, Professor; Ph.D., MIT. Materials, composite materials, nondestructive evaluation, fatigue, wear, mechanics, finite elements, reliability-centered maintenance.

Olusegun J. Ilegbusi, Associate Professor; Ph.D., London. Turbulence modeling, with emphasis on transition, instability, and mixing; mathematical and physical modeling of multiphase phenomena and materials processing operations; processing and applications of metal-matrix composites.

Jacqueline A. Isaacs, Assistant Professor; Ph.D., MIT. Environmental and life-cycle issues in advanced emerging technologies, manufacturing economics of materials processing, automotive industry benchmarking studies.

Gregory J. Kowalski, Associate Professor; Ph.D., Wisconsin– Madison. Combined modes of heat transfer in participating media, solar energy, thermal electronic packaging, and combined heat and mass transfer.

Yiannis A. Levendis, Professor; Ph.D., Caltech. Combustion, incineration, air pollution, chemical kinetics, aerosol physics, internal combustion engines.

Achille Messac, Associate Professor; Ph.D., MIT. Structures, structural dynamics, multibody dynamics, control, multidisciplinary optimal design, optimal manufacturing, optimization, mechanics, analytical and computational CAD, finite-element analysis.

Mohammad Metghalchi, Associate Professor; Sc.D.; MIT. Laminar and turbulent flame propagation, stability in internal combustion engines, energy conversion, air pollution, chemical kinetics, advanced thermodynamics.

Richard J. Murphy, Professor; Ph.D., MIT. Production and consolidation of amorphous metal powder, high-temperature creep.

Uichiro Narusawa, Associate Professor; Ph.D., Michigan. Natural and double-diffusive convection in enclosures and saturated porous media, two-phase flows, thermocapillary flow.

Welville B. Nowak, Professor (Emeritus) and Senior Research Scientist; Ph.D., MIT. Materials science and engineering; thin films for resistance to corrosion, diffusion, and wear; photovoltaic solar cells; electronic materials.

John N. Rossettos, Professor; Ph.D., Harvard. Buckling and vibration of stiffened plates, mechanics of composite materials, applied mechanics.

Mohammad E. Taslim, Professor; Ph.D., Arizona. Computational and experimental fluid mechanics and heat transfer, double-diffusive convection.

Bruce H. Wilson, Assistant Professor; Ph.D., Michigan. Mechanical computer-aided engineering, automated modeling, dynamic systems and control, computer-aided design and control of drive train systems.

Yaman Yener, Professor; Ph.D., North Carolina State. Thermal radiation, heat and mass transfer, radiative transfer, aerosol thermophoresis with radiation.

Ibrahim Zeid, Professor; Ph.D. Akron. CAD/CAM, finite-element method, applied mechanics, manufacturing.

NORTHWESTERN UNIVERSITY

Robert R. McCormick School of Engineering and Applied Science
Department of Mechanical Engineering

Programs of Study	The department offers M.S. and Ph.D. degree programs that provide a strong background in mechanical engineering theory and practice. M.S. students must complete 12 units of credit (1 unit is equivalent to a one-quarter course), including 1–3 units of project. At least one of these courses must be in fluid mechanics and one in solid mechanics. In addition, at least six courses must be in mechanical engineering. M.S. degrees are generally completed in one to two years. The Ph.D. program emphasizes creative research and breadth of knowledge. Many of the department's graduates are represented on mechanical engineering faculties around the world, as well as in industry and government laboratories. Ph.D. students must complete eighteen courses beyond the B.S. Nine of these courses may be equivalent to those required for the M.S. degree, described above. The other nine courses must be graduate level and must include at least one fluid mechanics and one solid mechanics course. Each student must also pass a qualifying examination that includes an oral examination of his or her research proposal and a test of general academic preparation. The latter test, however, is exempted for any student who maintains at least a 3.5/4.0 grade point average in graduate course work. Ph.D. degrees generally require three to four years of study beyond the M.S. These curricula have been designed to be flexible so that each student may arrange a program to accommodate his or her individual needs, talents, and interests. Selection of research advisers (see reverse side for faculty research interests) normally occurs within the first two quarters of residence.
Research Facilities	Among the specialized equipment found in the department are line-focus and scanning acoustic microscopes; ultrasonic analyzers; a computer-aided acoustic emission system; a heterodyne dual-probe laser interferometer; a JEOL scanning electron microscope; an X-ray nondestructive evaluation facility; an optical interferometry laboratory equipped for Michelson interferometry, grating-shearing interferometry, holography, electronic speckle interferometry, and speckle shearography; a materials testing laboratory; a vertical wind tunnel; a laser-Doppler anemometer; a Cincinnati Milacron Sabre 750 2.5 axes CNC vertical machining center and a 2.5 axes CNC horizontal HPMC; a Cincinnati Milacron Vista 85-ton plastic injection molding machine; a Brown & Sharpe coordinate measuring machine; force-reflecting teleoperator interfaces; and multiple robotic surgery systems. Department-wide facilities include state-of-the-art teaching laboratories, workstation and PC laboratories, and a well-equipped, well-stocked model shop/prototyping laboratory. Other facilities found in the Technological Institute at Northwestern include electronics, glass, and machine shops; and the Seeley G. Mudd Science and Engineering Library. The library provides access to an extensive collection of journals and books in all fields of science and engineering, as well as the latest in computerized databases.
Financial Aid	The McCormick School of Engineering and Applied Science has approximately 80 fellowships available for the support of first-year students. Most students admitted to the department are supported by one of these fellowships or by a research assistantship. Continuing students are primarily supported by research assistantships and teaching assistantships. Stipends for 1999–2000 begin at $1385 per month, plus tuition.
Cost of Study	Tuition is $21,798 for the 1999–2000 academic year, with much lower rates during the summer quarter and after the student has completed three (for M.S.) or nine (for Ph.D.) quarters of full-time registration.
Living and Housing Costs	A wide variety of apartments, rooms, and houses are available within walking distance of the campus or within a short drive. Rents depend on the location and size of the apartment. University graduate student housing for 1999–2000 costs $538, $825, and $910 per month for twin studio, one-bedroom, and two-bedroom apartments, respectively. The Chicago area's extensive mass transit system allows students to avoid the expense of a car.
Student Group	Northwestern's 13,000 students come from across the United States and around the world. There are 5,500 graduate and professional students, including 925 in the McCormick School. About half of the approximately 90 graduate students in the Department of Mechanical Engineering are U.S. citizens or permanent residents.
Student Outcomes	The department's graduates have enjoyed great success in finding employment in industry, government research laboratories, and academia. Some have created their own employment, starting small businesses based upon thesis work. In recent years, doctoral students have gone on to research positions in industrial and government research laboratories (including Ford, Motorola, Alcoa, Argonne National Laboratory, and Sandia National Laboratory) and to faculty positions in major research universities (including Pennsylvania State University, University of Notre Dame, Johns Hopkins University, and Rensselaer Polytechnic Institute).
Location	The 240-acre Evanston campus is bounded on the east by Lake Michigan and on the west by the tree-lined streets of Evanston. Downtown Chicago is approximately 30 minutes away by car or public transportation, yet the campus has the feeling of spaciousness and serenity normally found on a geographically isolated campus. Most of Evanston's 3.5-mile lakefront is included in city parks, and the University campus is available for biking, swimming, sailing, and picnicking. Northwestern's outstanding music and theater programs offer a wide variety of performances throughout the year.
The University and The Department	Northwestern is one of the nation's largest private research universities, with eleven schools and colleges. The 30 members of the mechanical engineering faculty and 90 graduate students are actively involved in a wide variety of research sponsored by both federal agencies and private industry. The mechanical engineering faculty is highly recognized, holding six memberships in the National Academy of Engineering or National Academy of Science as well as editorships of a number of major journals.
Applying	Students with a strong background in mechanical engineering are encouraged to apply. Outstanding graduates in chemistry, mathematics, physics, or other fields of engineering are also accepted, but they may be required to take selected undergraduate courses without credit. All students must take the GRE General Test. Students whose native language is not English are also required to take the TOEFL. Applications are accepted throughout the year for admission in the fall, winter, or spring quarters. Applications received by February 15 will receive full consideration for financial aid beginning the following September.
Correspondence and Information	Director of Graduate Admissions Department of Mechanical Engineering Northwestern University Evanston, Illinois 60208-3111 Telephone: 847-491-7470 World Wide Web: http://www.mech.nwu.edu/

Northwestern University

THE FACULTY AND THEIR RESEARCH

Jan D. Achenbach, Ph.D., Stanford, 1962 (also civil engineering department). Mechanics of solids, quantitative nondestructive evaluation and fracture mechanics.

Ted Belytschko, Chairman, Ph.D., IIT, 1968 (also civil engineering department). Computational mechanics, finite elements and computer-aided engineering.

Catherine Brinson, Ph.D., Caltech, 1990. Macro- and micromechanics of solid materials; constitutive modeling, especially polymeric, composite, and smart materials.

Jian Cao, Ph.D., MIT, 1994. Modeling, design, and control of manufacturing processes; instability analysis; solid mechanics.

Herbert S. Cheng, Ph.D., Pennsylvania, 1961. Contact fatigue, thin-film lubrication, tribology of engine and transmission components.

J. Edward Colgate, Ph.D., MIT, 1988. Robotics, human-machine interaction, actuator design and control.

James Conley, Ph.D., Northwestern, 1987. Flexible manufacturing, miniature internal combustion engines.

Isaac M. Daniel, Ph.D., IIT, 1964 (also civil engineering department). Experimental mechanics (stress analysis, fracture mechanics, impact, wave propagation), mechanics of composite materials, nondestructive evaluation.

Stephen H. Davis, Ph.D., RPI, 1964 (also engineering sciences and applied mathematics department). Fluid dynamics, solidification, crystal growth.

Kornel F. Ehmann, Ph.D., Wisconsin, 1979. Machine-tool dynamics and control, metal-cutting processes, automation and robotics.

Leon M. Keer, Ph.D., Minnesota, 1962 (also civil engineering department). Fracture mechanics, contact theories, surface and subsurface contact fatigue life.

Alan L. Kistler, Ph.D., Johns Hopkins, 1955. Turbulent flow, rotating machinery, acoustics, energy conservation.

Arthur A. Kovitz, Ph.D., Princeton, 1956. Fluid dynamics, heat transfer, interfaces.

Sridhar Krishnaswamy, Ph.D., Caltech, 1989. Mechanics of solids and structures, experimental mechanics, fracture mechanics, nondestructive evaluation.

Elmer E. Lewis, Ph.D., Illinois, 1964. Radiation transport, reliability and risk analysis.

Seth Lichter, Ph.D., MIT, 1982. Fluid mechanics, contact-line physics, vorticity dynamics.

Wing K. Liu, Ph.D., Caltech, 1981. Finite-element and multiple-scale reproducing kernel methods, computational fluid mechanics, fluid-structure interaction, material forming processes, probabilistic finite elements.

Richard M. Lueptow, Ph.D., MIT, 1986. Turbulent boundary layers, fluid flow in biomedical devices.

Kevin M. Lynch, Ph.D., Carnegie Mellon, 1996. Robotics, manipulation and motion planning, nonholonomic dynamic systems, industrial and medical applications.

Moshe Matalon, Ph.D., Cornell, 1977 (also engineering sciences and applied mathematics department). Combustion theory, fluid dynamics.

Bernard J. Matkowsky, Ph.D., NYU (Courant), 1966 (also Chairman, Department of Engineering Sciences and Applied Mathematics). Combustion, combustion synthesis of advanced materials, nonlinear dynamics and pattern formation, analytical and computational methods.

Brian Moran, Ph.D., Brown, 1988 (also civil engineering department). Computational methods, continuum and fracture mechanics, material instability, micromechanics and composite materials.

Toshio Mura, Ph.D., Tokyo, 1954 (also civil engineering department). Fracture and fatigue, nondestructive measurement, composite materials, thin films.

Ferdinando A. Mussa-Ivaldi, Ph.D., Politecnico of Milan, 1987 (also physiology department). Planning, control, and learning of motor behaviors in biological and artificial systems.

Michael A. Peshkin, Ph.D., Carnegie Mellon, 1986. Robotics and intelligent mechanical systems, force-guided assembly, robot-assisted surgery.

John W. Rudnicki, Ph.D., Brown, 1977 (also civil engineering department). Fracture and inelastic behavior of solids, particularly geomaterials.

Siavash H. Sohrab, Ph.D., California, San Diego, 1981. Combustion, turbulent reactive flows, physicochemical thermodynamics.

Henry W. Stoll, Ph.D., Illinois, 1974. Design theory and methodology, design for manufacturability, mechanical system design and analysis.

Richard S. Tankin, Ph.D., Harvard, 1960. Combustion, heat transfer, fluid flow.

John A. Walker, Ph.D., Texas, 1964. Stability analysis and dynamic behavior.

Qian (Jane) Wang, Ph.D., Northwestern, 1993. Thermoelastic contacts, frictional heating and heat transfer in tribology, mixed-TEHD, thermal-tribological design of machine elements.

M. C. Yuen, Ph.D., Harvard, 1965. Heat transfer, fluid mechanics, multiphase flow.

POLYTECHNIC UNIVERSITY, BROOKLYN CAMPUS

Mechanical, Aerospace, Manufacturing, Materials, and Industrial Engineering Programs

Programs of Study

The department offers graduate programs in mechanical engineering, aerospace engineering, manufacturing engineering, materials engineering, and industrial engineering. The mechanical engineering program has three areas of specialty: mechanical analysis and design; thermal and fluid sciences; and systems, controls, and robotics. The degrees offered are Master of Science and Doctor of Philosophy. The Master of Science is offered in each discipline, and the Doctor of Philosophy is only offered in mechanical engineering.

The Master of Science degree requires 36 units of graduate study. A thesis or project is expected of all full-time students and can be counted for a maximum of 12 units (6 for the project). Each of the programs and specialties have certain required courses, and all programs permit students to satisfy specific needs through elective courses.

The doctoral degree program requires 36 units of graduate course work beyond an acceptable master's degree and a minimum of 24 dissertation units. In addition, a Ph.D. candidate must pass a written and oral qualifying examination, present a dissertation proposal, and complete and defend an acceptable doctoral dissertation.

Research Facilities

The department's research facilities are located in Brooklyn and on the Long Island Campus. Heat transfer laboratories include a thermal systems laboratory and a radiation/optics laboratory. The aerodynamics laboratory includes a supersonic wind tunnel facility with a Mach range of 1.75 to 4 and a shock-tube. The fluid mechanics laboratory is equipped with hot-wire anemometers and laser-Doppler velocimetry systems and has a non-Newtonian fluid mechanics facility with extensive visualization capabilities. The optical diagnostics laboratory houses several state-of-the-art pulsed lasers, interferometers, and CCD cameras for spectroscopic measurements in high speed flows. Controls and robotics research laboratories include the instrumentation laboratory, the digital feedback control and robotics laboratory, and the neural network and fuzzy control laboratory. Each research group has its own network of Sun and IBM SPARC workstations connected to personal computers, laser printers, and storage devices. The electron microscopy laboratory includes a 200-kv transmission electron microscope equipped with an energy dispersive X-ray microanalysis system and a scanning electron microscope equipped with an energy dispersive X-ray microanalysis system. The mechanical testing laboratory is equipped with Instron and MTS screw-driven and servo-hydraulic universal testing machines as well as creep, hardness, and impact testing units. Rapid solidification and composite processing units are also available, as are links to supercomputers.

Financial Aid

Financial aid in the form of teaching and research assistantships is available to highly qualified full-time students. Stipends vary but are in the range of $1145 to $1500 per month plus tuition. Other scholarships and internships are also available. Students should consult the department's Web site (listed below) for up-to-date listings.

Cost of Study

In 1998–99, tuition was $675 per credit.

Living and Housing Costs

Dormitory rooms for single students are maintained in facilities near the Brooklyn campus. The cost in 1998–99 was approximately $600–$900 per month. Private housing rentals are available for single and married students in nearby Brooklyn Heights and in other brownstone neighborhoods from $400 and up per person per month. Public transportation is plentiful, and cars are not required. On-campus dormitories are available on the Farmingdale campus.

Student Group

Of the 89 graduate students in the department in 1998–99, 29 were full-time. The rest studied part-time while employed in the New York region. Many states and countries are represented in the graduate student body.

Location

Located in the MetroTech complex in downtown Brooklyn across the river from downtown Manhattan, Polytechnic is within easy reach of one of the world's greatest concentrations of cultural, educational, entertainment, and sports activities. The New York metropolitan area provides an exciting complex of theaters, museums, historic sites, architectural wonders, and famous stores and shops. Nearby Long Island is a paradise for beach enthusiasts and for those who enjoy boating, fishing, and water sports. A system of parkways, bridges, and tunnels connects directly to the thruways and turnpikes of New Jersey, Pennsylvania, upstate New York, and New England.

The University

Polytechnic is a private, coeducational, technological university founded in 1854 as the Polytechnic Institute of Brooklyn. In 1973, Polytechnic Institute of Brooklyn and New York University's School of Engineering and Science merged to create the Polytechnic Institute of New York. In 1985, Polytechnic Institute of New York became Polytechnic University, thus reflecting Polytechnic's position as one of the major technological universities in the country. The Long Island Campus was established in 1961, the Westchester Campus in 1976. The graduate programs of the department are concentrated in the Brooklyn Campus. All graduate classes are held in the evenings, and part-time and full-time programs of study are available.

Applying

Applicants for the fall semester who are requesting assistantships or scholarships are encouraged to apply by March 1, since competition for these awards is keen. Candidates may also apply for the spring semester for admission only.

Correspondence and Information

Mechanical/Aerospace/Manufacturing/Materials/Industrial Engineering Programs
Polytechnic University
6 MetroTech Center
Brooklyn, New York 11201
Telephone: 718-260-3160
              718-260-3200 (general University telephone)
Fax: 718-260-3532
E-mail: mech@poly.edu
        maimeng@www.poly.edu
        admitme@poly.edu (to request applications)
World Wide Web: http://www.poly.edu
                http://mechanical.poly.edu/

Polytechnic University, Brooklyn Campus

THE FACULTY AND THEIR RESEARCH

The following is the list of faculty members at the beginning of the 1998–99 academic year.

Charles J. Bartlett, Industry Professor; Ph.D., MIT, 1966. Electronic systems design and manufacture, production science, and manufacturing systems engineering.

James Bentson, Industry Professor; Ph.D. Polytechnic of Brooklyn, 1968. Computational methods, energy systems, vehicle dynamics.

Charles W. Hoover, Industry Professor; Ph.D., Yale 1954. Manufacturing systems engineering, production processes, physical design.

Craig C. Jahnke, Visiting Assistant Professor; Ph.D., Caltech, 1990. Fluid mechanics, heat transfer, aircraft dynamics, nonlinear dynamics.

Iraj Kalkhoran, Associate Professor; Ph.D., Texas at Arlington, 1987. Gasdynamics, high speed aerodynamics, subsonic, transonic and supersonic wind tunnel testing, impulse facilities.

Vikram Kapila, Assistant Professor; Ph.D., Georgia Tech, 1996. Absolute stability theory, robust multivariable control, periodic and multirate control, fixed-architecture control, delay systems, actuator amplitude and rate saturation control.

Jerome M. Klosner, Professor; Ph.D., Polytechnic of Brooklyn, 1959. Structural dynamics, fluid structure interactions, thermal stress analysis.

Sunil Kumar, Associate Professor; Ph.D., Berkeley, 1987. Heat transfer, superconductor applications, microscale effects, radiation properties, bio-heat and mass transfer, applied mathematics.

Nathan Levine, Industry Professor; Ph.D., Illinois, 1957. Quality control, process control, quality engineering, robust design.

William R. McShane, Professor and Dean of Engineering; Ph.D., Polytechnic of Brooklyn, 1968. Quality control, controls and simulation, engineering economics.

Said Nourbakhsh, Professor; Ph.D., Leeds, 1978. Composite processing, mechanical behavior of materials, fracture, fatigue, intermetallic alloys, rapid solidification, electron microscopy, thin film ferroelectrics and piezoelectrics, smart materials, electronic materials, high-strain deformation, texture development, electron and X-ray diffraction, electron microscopy and microanalysis.

M. Volkan Otugen, Associate Professor; Ph.D., Drexel, 1986. Experimental and theoretical fluid mechanics, turbulence, optical diagnostics, high-speed aerodynamics, laser-based measurements.

Marcio S. de Queiroz, Visiting Assistant Professor; Ph.D., Clemson, 1997. Nonlinear and adaptive control, with emphasis on applications to mechatronic, electromechanical, and mechanical systems; control of distributed parameter systems; robotics.

Richard S. Thorsen, Vice President of Institutional Development; Ph.D., NYU, 1967. Heat transfer, nuclear reactor safety, alternate energy.

Anthony Tzes, Associate Professor; Ph.D., Ohio State, 1990. Robotics, adaptive control, neural networks and fuzzy logic, mechatronics.

William P. Vafakos, Professor, Ph.D., Polytechnic of Brooklyn, 1960. Solid mechanics, structures, vibrations.

George C. Vradis, Associate Professor; Ph.D. Polytechnic, 1987. Computational fluid dynamics and heat transfer, non–Newtonian fluid mechanics, combustion, energy systems.

Sung H. Whang, Professor; D. Eng. Sci., Columbia, 1979. Mechanical behavior of intermetallics at high temperatures; deformation mechanisms in titanium alumindes and silicides; strength, creep, and microstructure in nanostructured metals and alloys and rapidly solidified metals and superconducting ceramics; microstructural analysis of all these materials.

Recent Books Published by Faculty Members

McShane, W. R., and R. Roess, *Traffic Engineering*, Prentice Hall, 1990, 2nd Edition, 1998.

Hoover, C. W., and J. B. Jones, *Improving Engineering Design: Designing for Competitive Advantage*, National Academic Press, 1991.

Selected Faculty Research Results

Bartlett, C. J. Manufacturing across the curriculum at Polytechnic University. *ASEE/IEEE Frontiers in Educ. Conf.* Nov. 1998. With **Kumar.**

Bentson, J. Measurements of supersonic wing tip vortices. *AIAA J.*, 33, 1995. With Smart and **Kalkhoran.**

Bentson, J. Electrical breakdown of the space vacuum. *IEEE Trans. Plama Sci.*, 23, 1995. With Kunhardt and Popovic.

Hoover, C. W. Manufacturing engineering as an experimental science. *ASEE/IEEE Frontiers in Education* Nov. 1996. With **Bartlett.**

Jahnke, C. C., and D. T. Valentine. On the recirculation zones in a cylindrical container. *J. Fluid Eng.* 120:680–4, 1998.

Jahnke, C. C. On the roll-coupling instabilities of high performance aircraft. *Phil. Trans. R. Soc. London A* 356:1–17, 1998.

Kalkhoran, I. M. Planar laser-sheet visualization of oblique shock wave/vortex interaction. *Shock Waves* 8:243–55, 1998. With Smart and Popovic.

Kalkhoran, I. M. Supersonic vortex breakdown during vortex/cylinder interaction. *J. Fluid Mech.* 369:351–80, 1998. With Smart and Wang.

Kalkhoran, I. M. Flow model for predicting normal shock wave induced vortex breakdown. *AIAA J.* 35:1589–96, 1997. With Smart.

Kapila, V. Robust controller synthesis for systems with input-output nonlinearities: A trade-off between gain variation and parametric uncertainty. *Int. J. Robust Nonlinear Contr.* 8:567–83, 1998. With Haddad.

Kapila, V. Robust controller synthesis via shifted parameter-dependent quadratic cost bounds. *IEEE Trans. Autom. Contr.* 43:1003–7, 1998. With Haddad, Erwin, and Bernstein.

Kapila, V. Fixed-architecture controller synthesis for systems with input-output time-varying nonlinearities. *Int. J. Robust Nonlinear Contr.* 7:675–710, 1997. With Haddad.

Klosner, J. Effect of thermal stresses on the vibrations of composite cantilevered plates. *J. Compos. Tech. Res.* 15:123-35, 1993.

Kumar, S. Experimental modeling of circular hydraulic jump by the impingement of a water column on a horizontal disk. *ASME J. Fluids Eng.* 121:86–92, 1999. With Naraghi, Moallemi, and Naraghi.

Kumar, S. Development and comparison of models for light pulse transport through scattering-absorbing media. *Appl. Opt.* 38:188–96, 1999. With Mitra.

Kumar, S. Microscale aspects of thermal radiation transport and laser applications. *Adv. Heat Trans.* 33:187–294, 1998. With Mitra.

McShane, W. R. Recent developments in a mechatronics/design lab. *Controls Systems Mag.* 17:72–9, 1997. With **Tzes** and Guthy.

McShane, W.R. Applications of petri networks to transportation network modeling. *IEEE Trans. Vehicular Tech.* 45:391-400, 1996. With Kim and **Tzes.**

Nourbakhsh, S. Fabrication of Al_2O_3 reinforced Ni_3Al composites by a novel in-situ route. *Metall. Mater. Trans.* 28A:1069–77, 1997. With Sahin and Margolin.

Nourbakhsh, S. Microstructure of Al_2O_3 fiber reinforced superalloy (Inconel 718) composites. *Metall. Mater. Trans.* 27A:451–8, 1996. With Rhee, Sahin, and Margolin.

Nourbakhsh, S. Processing and characterization of fiber-reinforced intermetallic matrix composites. *J. Mater. Manuf. Processes* 11:283–305, 1996. With Margolin.

Otugen, M. V. A new wall pressure normalization for sudden expansion flows. *J. Fluids Eng.* 120:400–3, 1998. With Papadopoulos.

Otugen, M. V. Uncertainty estimates of turbulent temperature in Rayleigh scattering measurement. *Exp. Therm. Fluid Sci.* 15:25–32, 1997.

Otugen, M. V. An Nd:YAG laser-based dual-line detection Rayleigh scattering system. *AIAA J.* 35:776–81, 1997. With Kim and Popovic.

de Queiroz, M. S. Adaptive vibration control of an axially moving string. *ASME J. Vibration Acoust.* 121:41–9, 1999. With Dawson, Rahn, and Zhang.

de Queiroz, M. S. Adaptive position/force control of robot manipulators without velocity measurements: Theory and experimentation. *IEEE Trans. Syst. Man Cybern.* 27-B:796–809, 1997. With Hu, Dawson, Burg, and Donepudi.

Tzes, A. An Internet-based real-time control engineering laboratory. *IEEE Control Syst. Magazine*, 1999. With Overstreet.

Tzes, A. Modeling, plant uncertainties, and fuzzy logic sliding control of gaseous systems. In *IEEE Trans. Control Syst. Tech.*, vol. 7, no. 1, 1999. With Kiriakidis, Grivas, and Peng.

Tzes, A. Genetic-based fuzzy clustering for DC-motor friction identification and compensation. In *IEEE Trans. Control Syst. Tech.*, vol. 6, no. 4, 1998. With Peng and Guthy.

Vradis, G. C. Laminar flow of a non-linear viscoplastic fluid through an axisymmetric sudden expansion. *J. Fluids Eng.*, in press. With Hammad, **Otugen**, and Arik.

Vradis, G. C. The evolution of laminar jets of Herschel-Bulkley fluids. *Int. J. Heat Mass Trans.* 41:3575–88, 1998. With Jafri.

Vradis, G. C. A strongly-coupled bolck-implicit solution technique for non-Newtonian convective heat transfer problems. *Num. Heat Trans.- Fund. Part B* 33:79–97, 1998. With Hammad.

Whang, S. H. Cross-slip and glide behavior of ordinary dislocations in single crystal γ-Ti-65Al. *Intermetallics* 7:1–9, 1999. With Feng.

Whang, S. H. Dislocation core configurations of [101] superdislocations in single crystal Ti-56Al at 573K. *Intermetallics* 6:131–9, 1998. With Wang, Feng, and Allard.

Whang, S. H. Ordering, deformation and microstructure in $L1_0$ type FePt. *Acta Mater.* 46:6485–95, 1998. With Feng and Gao.

PURDUE UNIVERSITY

School of Mechanical Engineering

Programs of Study

The School of Mechanical Engineering offers graduate programs leading to the degrees of Master of Science and Doctor of Philosophy. M.S. and Ph.D. degrees in biomedical engineering are also offered through the School. Graduate courses cover the following topics: classical and statistical thermodynamics; combustion; air-breathing propulsion; fluid mechanics, gasdynamics, boundary-layer theory, turbulence, and computational fluid mechanics; turbomachinery; heat and mass transfer: conduction, diffusion, convection, radiation, computational heat transfer, and two-phase flow; material processing; applied optics; optical measurement technology; holography; vehicle dynamics, advanced dynamics, dynamic stability, vibrations of plates and shells; kinematics; dynamical problems in design; thermoelasticity; machine design; mechanics of machinery; lubrication; mechanical behavior of materials; computer graphics and computer-aided design; optimal design and system optimization techniques; reliability based design; design for manufacturability; nonlinear engineering systems; mechanical vibrations; engineering acoustics; discretized systems; theory and design of control systems; nonlinear control systems; microprocessors; signal processing; digital control; multivariable control system design; control of manufacturing processes; instrumentation for dynamical systems; robotics, automation, and manufacturing; numerical methods in mechanical engineering; and finite- and boundary-element analysis methods.

Opportunities for interdisciplinary study are available. Programs of study may include courses in other engineering disciplines as well as the sciences. Thesis and nonthesis options are available in the master's degree programs. Candidates must complete a minimum of 30 credit hours, 9 of which may be allocated for the thesis. A nonthesis master's degree program can be completed in approximately one year of full-time study; students with a half-time teaching or research assistantship can complete the thesis option in twenty-four months. A Ph.D. degree entails a total of 48 to 60 credit hours of course work, including courses taken on the master's level. A qualifying examination in the student's areas of study is required to qualify as a doctoral student and is taken during the first or second semester of residence. Preliminary and final examinations on the dissertation are required.

Research Facilities

The research activities of the School are conducted in fifteen laboratories, and annual expenditures are approximately $10 million. The facilities include Laboratories for Robotics, Automation and Manufacturing; Computer-Aided Design and Graphics; Composites Manufacturing; Structural Dynamics and Tribology; Automatic Control: Engine and Microprocessor Control; Fluid Mechanics; Aeromechanics and Propulsion; Heat Transfer; Boiling and Two-Phase Flow; Flame Diagnostics; Applied Optics; Materials; and Bioengineering. The Ray W. Herrick Laboratories provide facilities for basic and applied research in mechanical vibrations, noise control, acoustics, fluid mechanics and heat transfer for energy utilization, active and adaptive control, and electromechanical systems. Specialized facilities exist for the application of these basic engineering sciences to many areas, including heating and air conditioning, refrigeration, compressors, engines, vehicles, and other mechanical systems. The Maurice J. Zucrow Research Laboratories (formerly the Thermal Sciences and Propulsion Center) have facilities for research in large-scale turbomachinery, basic fluid dynamics, spray formation, basic combustion processes, and propulsion. Specialized facilities exist to support research in compressors, turbines, gas turbine combustors, spray formation, wind tunnels, pool flames, and various propulsion applications. Computational facilities within the School include three Sun Microsystems multiprocessor file/computer servers; many Sun, HP, IBM, and Silicon Graphics workstations running UNIX variants; and more than 160 Pentium PCs. These are networked with other computer systems throughout the schools of engineering as well as to the Internet. Approximately 100 terminals in offices and laboratories give users access to these departmental systems, as well as to various computer systems that the Purdue University Computing Center provides. In 1994, the Purdue University Computing Center (PUCC) added an Intel Paragon XP/S model 10 parallel supercomputer to its computing facilities. Existing PUCC resources include a cluster of seven IBM RISC Systems with a vector facility, two clusters of IBM RISC System/6000s that act as compute servers, with four additional 6000s and a pair of Sun 4/690 quad processors that serve students on 300 Sun Sparc 5's. Besides the general libraries, students have access to the Potter Engineering Library, which contains more than 350,000 volumes in engineering and related subjects. The library receives more than 2,000 periodicals, maintains an extensive collection of technical reports, and provides computer-based literature search services.

Financial Aid

Graduate teaching and research assistantships provide a minimum of $13,000 for a half-time appointment for the academic year 1999–2000. All tuition is waived, but students pay a fee of $310 per semester. A few University fellowships are available for qualified students, which pay a minimum of $12,000 for the calendar year, with tuition waived, and may be supplemented by a one-quarter-time assistantship.

Cost of Study

In 1998–99, full-time resident graduate students paid tuition and fees of $1750 per semester; nonresidents paid $5860 per semester. Summer-session rates are half of this amount.

Living and Housing Costs

University dormitories rent rooms for $280 to $480 per month in 1999–2000, without meals. Married student apartments are $409 to $518 per month in 1999–2000. Private rooms and apartments are also available near campus.

Student Group

The School currently has approximately 300 graduate students, 40 percent of whom are working toward the Ph.D. degree. Approximately 15 percent of the students are from other countries.

Location

Purdue University is located in West Lafayette, Indiana. The population of the Greater Lafayette area is about 130,000. The city is located 125 miles southeast of Chicago and 60 miles north of Indianapolis.

The University and The School

Purdue has nine schools of engineering and three divisions representing interdisciplinary research activities. The engineering schools have 5,881 undergraduate and 1,871 graduate students, and the total University enrollment is 36,878. The School of Mechanical Engineering has 815 undergraduates, 49 faculty members, and 46 professional, service, and clerical staff members.

Applying

Applicants should have a B.S. in mechanical engineering; others, if admitted, may be required to take certain undergraduate courses. Study may begin in January, June, or August. However, more fellowship and assistantship opportunities are available for the fall. The GRE General Test is recommended, especially for those seeking financial assistance. The TOEFL and GRE are required of international students.

Correspondence and Information

Professor E. Daniel Hirleman, Head
School of Mechanical Engineering
Purdue University
West Lafayette, Indiana 47907
Telephone: 765-494-5688

Professor Anil K. Bajaj, Graduate Chairman
School of Mechanical Engineering
Purdue University
West Lafayette, Indiana 47907
Telephone: 765-494-5730
Fax: 765-494-0539
E-mail: megrad@ecn.purdue.edu
WWW: http://www.ecn.purdue.edu/ME/Grad/

Purdue University

THE FACULTY AND THEIR RESEARCH

J. Abraham, Associate Professor; Ph.D., Princeton, 1986. Combustion, internal-combustion engines, computational fluid dynamics.

D. C. Anderson, Professor; Ph.D., Purdue, 1974. Design, computer-aided design, computer graphics.

A. K. Bajaj, Professor; Ph.D., Minnesota, 1981. Mechanics, dynamics, nonlinear oscillations, bifurcation phenomenon.

R. J. Bernhard, Professor; Ph.D., Iowa State, 1982. Acoustics and noise control, mechanics, numerical techniques.

J. S. Bolton, Professor; Ph.D., Southampton, 1984. Acoustics and noise control, mechanics, applied signal processing.

J. E. Braun, Associate Professor; Ph.D., Wisconsin, 1988. Thermal systems modeling, analysis, design and control.

G. Chiu, Assistant Professor; Ph.D., Berkeley, 1994. Integrated design and control of electromechanical systems, dynamic systems analysis and control, precision control, mechatronics.

R. J. Cipra, Associate Professor; Ph.D., Wisconsin, 1978. Mechanical systems, modeling, CAD, robotics and automation, kinematics.

P. Davies, Associate Professor; Ph.D., Southampton, 1985. Signal processing, system identification, sound quality, diagnostics.

D. P. DeWitt, Professor; Ph.D., Purdue, 1963. Heat transfer, thermophysical properties, radiation thermometry, bioengineering.

S. Fleeter, Professor; Ph.D., Case Western Reserve, 1970. Turbomachinery, fluid mechanics, flutter, aeromechanics.

M. A. Franchek, Associate Professor; Ph.D., Texas A&M, 1991. Multivariable and nonlinear control of transportation; power generating systems; electromechanical, hydraulic, and smart machines.

S. H. Frankel, Associate Professor; Ph.D., SUNY at Buffalo, 1993. Computational fluid dynamics, turbulence, combustion, aeroacoustics, multiphase flow.

V. W. Goldschmidt, Professor; Ph.D., Syracuse, 1965. Fluid mechanics, turbulence, flow and heat transfer in HVAC systems.

J. P. Gore, Professor; Ph.D., Penn State, 1986. Combustion, heat transfer, turbulent flames, pollution control, product development.

E. A. Groll, Assistant Professor; Ph.D., Hannover (Germany), 1994. Energy utilization, thermal sciences applied to HVAC&R equipment and systems.

K. H. Hawks, Associate Professor; Ph.D., Purdue, 1969. Power plant system analysis and simulation, energy management.

B. M. Hillberry, Professor; Ph.D., Iowa State, 1967. Fatigue and fracture mechanics, biomechanics.

E. D. Hirleman Jr., Professor; Ph.D., Purdue, 1977. Laser/optical measurement methods, engineering design process.

J. D. Hoffman, Professor; Ph.D., Purdue, 1963. Fluid mechanics, propulsion, numerical analysis, computational fluid dynamics.

J. D. Jones, Associate Professor; Ph.D., Virginia Tech, 1987. Acoustics, noise and vibration control, cooperative learning.

G. B. King, Associate Professor; Ph.D., Kansas State, 1983. Systems, measurement and control, materials processing.

K. Kokini, Professor; Ph.D., Syracuse, 1982. Thermal stresses, thermal fracture and fatigue of advanced materials, finite elements, biomaterials.

C. M. Krousgrill, Professor; Ph.D., Caltech, 1980. Mechanics, nonlinear oscillations, dynamics, friction-induced vibrations.

N. M. Laurendeau, Professor; Ph.D., Berkeley, 1972. Combustion diagnostics and kinetics, pollution abatement.

P. B. Lawless, Assistant Professor; Ph.D., Purdue, 1993. Turbomachinery aerodynamics, compressor stability, turbine cooling.

A. T. McDonald, Professor; Ph.D., Purdue, 1964. Fluid mechanics, design.

P. H. Meckl, Associate Professor; Ph.D., MIT, 1988. Dynamics and control, motion and vibration control, intelligent control and robotics.

L. Mongeau, Assistant Professor; Ph.D., Penn State, 1991. Acoustics, noise control, aeroacoustics, turbomachinery.

I. Mudawar, Professor; Ph.D., MIT, 1984. Multiphase heat transfer, electronic cooling, materials processing.

S. N. B. Murthy, Senior Researcher; D.I.C., Imperial College (London), 1953. Propulsion and aerothermodynamics.

J. T. Pearson, Associate Professor; Ph.D., SUNY at Stony Brook, 1967. Heat transfer, thermal systems modeling and design, product development, manufacturing.

G. R. Pennock, Associate Professor; Ph.D., California, Davis, 1983. Kinematic synthesis and analysis.

M. W. Plesniak, Associate Professor; Ph.D., Stanford, 1990. Fluid dynamics, turbulence, gas turbine cooling, mixing, particulate transport.

S. Ramadhyani, Professor; Ph.D., Minnesota, 1979. Numerical methods in heat transfer, combustion systems, heat transfer in sprays, gas turbines.

K. Ramani, Associate Professor; Ph.D., Stanford, 1991. Polymer processing, product design, information technology and CAD/CAM.

F. Sadeghi, Professor; Ph.D., North Carolina State, 1985. Tribology, stress analysis, lubrication, wear, sensor technology, CAD.

Y. C. Shin, Professor; Ph.D., Wisconsin, 1984. Manufacturing, intelligent and adaptive control, CIM, intelligent manufacturing systems.

T. Siegmund, Assistant Professor; Ph.D., Leoben (Austria), 1994. Micromechanics, computational fracture mechanics, design for structural integrity, finite element method.

J. G. Skifstad, Professor; Ph.D., Purdue, 1964. Physical fluid mechanics, atomization of liquids, gasdynamics, lasers.

W. Soedel, Professor; Ph.D., Purdue, 1967. Mechanics, vibrations, dynamics of shells, dynamics of machinery, stress analysis, acoustics.

P. E. Sojka, Associate Professor; Ph.D., Michigan State, 1983. Atomization, fluid instability, spray diagnostics.

J. M. Starkey, Associate Professor; Ph.D., Michigan State, 1982. Machine design, vibration isolation, vehicle dynamics.

W. H. Stevenson, Professor; Ph.D., Purdue, 1965. Applied optics, holography, laser velocimetry, optical measurement technology.

D. R. Tree, Professor; Ph.D., Purdue, 1966. Energy utilization, thermal system modeling, HVAC&R.

R. Viskanta, Goss Distinguished Professor; Ph.D., Purdue, 1960. Radiation heat transfer, heat transfer during melting and solidification, materials processing, combustion systems.

C. Wassgren, Assistant Professor; Ph.D., Caltech, 1997. Particulate systems, multicomponent fluid mechanics, computational models.

L. Xu, Associate Professor; Ph.D., Illinois, 1991. Bioheat transfer, hyperthermia, thermophysical properties of biological tissues.

X. Xu, Assistant Professor; Ph.D., Berkeley, 1994. Heat transfer, laser-assisted materials processing and diagnostics, thermophysical properties.

B. Yao, Assistant Professor; Ph.D., Berkeley, 1996. Nonlinear adaptive and robust control, control of electromechanical systems, robotics.

Purdue's program in mechanical engineering is more than 100 years old.

SOUTHERN METHODIST UNIVERSITY

Department of Mechanical Engineering

Programs of Study

The department offers M.S. and Ph.D. degrees. The programs are organized into three stress areas: design, systems, and control; mechanics and materials; and thermofluid sciences and environmental engineering. The design systems and control area also contains the rapidly growing area of manufacturing and material processing.

The design, systems, and control area covers mechanical and biomechanical systems, including impact dynamics, gait dynamics, and haptic interfaces, as well as robust multivariable control methodology and application to high-performance turbomachinery and manufacturing processes. Thermal system design and optimized mechanical design are also covered, as are manufacturing and related issues such as intelligent process monitoring, abrasive water jet machining, and welding research.

The mechanics and materials area covers material behavior under various applied stress, earthquake-resistant building, and equipment design. There is also work in fracture mechanics and fatigue. Studies related to nondestructive inspection testing and thermal nondestructive evaluations are performed, with applications in accident and disaster reconstruction and prevention.

The thermofluids and environmental engineering area covers research in fluid mechanics, including computational fluid dynamics with applications to free surface flow, thermohydraulic characterization of permeable media, and tsunami mitigation studies. Work in thermal sciences includes heat-absorption-based thermal systems analysis and electronic cooling technology. In biomedical engineering, work includes gene therapy and contact lens lubrication processes. In environmental engineering, work is done in adsorption of organic and inorganic pollutants, recycled sorbents, mathematical modeling of adsorption processes, industrial waste minimization, and biological denitrification of wastewater systems.

Research Facilities

There are excellent research facilities in each of the three areas. The design, systems, and control area has research laboratories equipped with high-speed camera systems for impact dynamics and a pneumatic haptic interface system for interaction with virtual objects. There is an injection molding machine for process control studies, as well as laboratories for welding, machining, and process monitoring. Ample computing facilities and design studios also exist.

The mechanics area has several heavy-duty testing machines and a nondestructive evaluation laboratory. In the thermal fluids area, there are virtual systems for flow visualization, instrumentation for heat transfer research, and a subsonic wind tunnel. In environmental engineering, there are several laboratories for isotherm studies, field water analysis, dynamic column studies, and low-level organics analysis.

Financial Aid

The Graduate Admissions Committee awards a limited number of merit-based research and teaching assistantships to incoming students, which pay up to $1600 per month and cover tuition. Separate tuition assistantships are also available.

Cost of Study

Tuition and fees for graduate study in 1999–2000 are $600 per semester hour.

Living and Housing Costs

Dormitory housing charges each semester are approximately $1950 per person for double occupancy. Board, including tax, costs $1510 per semester. Furnished efficiency and one- and two-bedroom apartments are available on campus, some with paid utilities, at costs ranging from $2500 to $3000 per semester. The Dallas area offers inexpensive housing options within driving distance of the campus, with apartments starting at $500 per month.

Student Group

The department has about 25 doctoral students and 40 master's degree students from many different states and countries. There are also several postdoctoral research fellows. Approximately two to three doctorates and ten master's degrees are awarded each year.

Student Outcomes

Recent graduates have obtained university appointments at Auburn, Johns Hopkins, and Tuskegee and industrial research positions at firms such as Texas Instruments.

Location

Dallas is the center of an attractive metropolitan area of 2 million people. It has fine parks, lakes, museums, theaters, orchestras, libraries, and places of worship. Clean and progressive, the city continues to grow as a center of business and light industry. The manufacturing, microelectronics, and telecommunications industries provide many opportunities for employment.

The University and The School

Southern Methodist University is a private, nonprofit, coeducational institution located in suburban University Park, an incorporated residential district surrounded by Dallas, Texas. The School of Engineering and Applied Science (SEAS) traces its roots to 1925, when the Technical Club of Dallas, a professional organization of practicing engineers, petitioned SMU to fulfill the need for an engineering school in the Southwest.

Applying

Students may apply for admission at any time. However, initial review for admission in a given semester depends upon receipt by the Graduate Division of all requisite application materials no later than July 1 for fall admission, November 15 for spring admission, or April 15 for summer admission. All international students must use the following dates: May 15 for fall admission, September 1 for spring admission, and February 1 for summer admission. GRE General Test scores are required.

Correspondence and Information

Graduate Admissions
School of Engineering and Applied Science
Southern Methodist University
Dallas, Texas 75275-0335
Telephone: 214-768-3900
World Wide Web: http://www.seas.smu.edu/

Southern Methodist University

THE FACULTY AND THEIR RESEARCH

The following is a partial list of faculty members affiliated with the Department of Mechanical Engineering.

Gemunu S. Happawana, Visiting Assistant Professor of Mechanical Engineering; Ph.D., Purdue, 1994. Modeling, stability analysis, and control of gas turbine engines, including vibration effects in mistuned linear and cyclic propulsion systems.

Jack P. Holman, Brown Foundation Professor of Mechanical Engineering; Ph.D., Oklahoma, 1958. Heat transfer, heating and air conditioning, energy management and conservation.

Yildirim Hurmuzlu, Associate Professor of Mechanical Engineering; Ph.D., Drexel, 1987. Controls, nonlinear dynamics, and biomechanics.

David B. Johnson, Associate Chair and Associate Professor of Mechanical Engineering; Ph.D., Stanford, 1968. Fluid dynamics, structural and vehicular dynamics, control system design and analysis, engineering design.

Radovan Kovacevic, Herman Brown Professor of Mechanical Engineering; Ph.D., Titograd (Yugoslavia), 1978. Modeling of abrasive waterjet machine, cooling/lubrication of high-pressure waterjets, sensing and control of welding processes, applied machine vision in quality control.

Jose L. Lage, J. Lindsay Embrey Professorship, Associate Professor of Mechanical Engineering; Ph.D., Duke, 1991. Heat transfer, fluid mechanics, thermodynamics, energy technology, experimental methods in thermal and fluid sciences.

Charles M. Lovas, Associate Professor of Mechanical Engineering; Ph.D., Notre Dame, 1968. Thermal system design, machine design, optimized mechanical design, computer-aided design, materials handling, thermal control of electronic equipment, integrated design and manufacturing.

Bijan Mohraz, Professor of Mechanical Engineering; Ph.D., Illinois at Urbana-Champaign, 1966. Structural analysis and design, structural dynamics, earthquake engineering.

Osita D. I. Nwokah, Professor and Chairman of Mechanical Engineering; Ph.D., London (Imperial College), 1975. Modeling, dynamic analysis, and feedback control of uncertain, complex mechanical and electromechanical processes.

Paul F. Packman, Professor of Mechanical Engineering; Ph.D., Syracuse, 1964. Materials engineering, fatigue, failure analysis, NDT, and fracture mechanics, particularly applied to mechanical and structural failures.

Peter E. Raad, Associate Professor of Mechanical Engineering; Ph.D., Tennessee, 1986. Free-surface fluid dynamics, modeling of nonlinear fluid-solid interactions, cooling of electronic equipment, flow in porous media, and contact lens dynamics during eye blinking, wave breaking, tsunami modeling and mitigation, adaptive thermal modeling of submicron integrated circuits.

Edward H. Smith, Associate Professor of Mechanical Engineering; Ph.D., Michigan, 1987. Mathematical modeling of the fate, transport, and treatment of pollutants in aqueous systems with a focus on adsorption processes; industrial waste minimization for metal finishing processes.

Hal Watson Jr., Associate Professor of Mechanical Engineering; Ph.D., Texas, 1967. Forensic engineering, machine design, acoustics and noise control, vehicle dynamics, robotics, automated machinery.

STANFORD UNIVERSITY

Department of Mechanical Engineering

Programs of Study	The Department of Mechanical Engineering is administratively organized into five divisions: Biomechanical Engineering, Design, Flow Physics and Computation, Mechanics and Computation, and Thermosciences. Academic programs range from the general program in mechanical engineering to programs with particular concentrations such as manufacturing systems engineering, product design, and biomechanical engineering.
	The Biomechanical Engineering (BME) Division has research programs that focus primarily on neuromuscular, musculoskeletal, and cardiovascular biomechanics. Research in other areas, including hearing, vision, ocean and plan biomechanics, biomaterials, biosensors, and imaging informatics, are conducted in collaboration with associated faculty members in medicine, biology, and engineering.
	The Design Division offers a comprehensive program in mechanical design, including emphases on the use of computers and microprocessors in mechanical devices, control systems, robotics, fatigue and applied fracture mechanics, experimental mechanics, applied finite-element analysis, mechanisms, product design, manufacturing, design fabrication and testing of microelectromechanical systems (MEMS), and applications of electrooptical technology and special programs that integrate two or more of these.
	The Manufacturing Systems Engineering Program is a joint offering of the Department of Mechanical Engineering and the Department of Industrial Engineering and Engineering Management leading to a Master of Science in Engineering with emphasis in manufacturing engineering. This program emphasizes mechanical design and engineering management as the key ingredients to manufacturing. There is also a dual M.S.E./M.B.A. program and a Ph.D. program in manufacturing. At the master's level, a joint program in design is offered with the Department of Art that leads to an M.S. in product design or an M.F.A. This program combines a thorough understanding of technology with a concern for human need and aesthetics. Experience and a portfolio are required for admission to this two-year program.
	The Flow Physics and Computation (FPC) Division is a joint laboratory of the Departments of Aeronautics and Astronautics and Mechanical Engineering. The FPC Division contributes new theories, models, and computational tools for accurate engineering design analysis and control of complex flows (including chemical reactions, acoustics, plasmas, and interactions with electromagnetic waves and other phenomena) of interest in aerodynamics, propulsion and power systems, materials processing, electronics cooling, environmental engineering, planetary entry, and other areas. Research in the FPC Division ranges from large eddy and direct simulation of complex flows to active flow control. FPC faculty members teach graduate and undergraduate courses in engineering computational mathematics, fluid mechanics, thermodynamics and propulsion, acoustics, aerodynamics, and computational fluid mechanics.
	The Mechanics and Computation Division covers the areas of computational mechanics, rigid and elastic body dynamics, finite-element analysis, continuum mechanics, dynamical systems, plasticity, fracture mechanics, and biomechanics.
	The Thermosciences Division covers the areas of combustion, air pollution, plasma sciences, heat transfer, fluid mechanics, high-temperature gasdynamics, fluid and thermal measurements, laser diagnostics, materials processing, flow and combustion control, and thermal engineering of microelectromechanical systems (MEMS).
Research Facilities	There are excellent facilities for research in fluid mechanics; turbulence physics; two-phase flows; heat transfer; combustion; coal conversion; spray dynamics; pollution control; laser diagnostics; plasma and combustion chemistry; plasma propulsion; nanoscale thermal characterization and processing; materials synthesis; smart products; robotics; experimental mechanics; fatigue and fracture mechanics; guidance, control, and precision instrumentation; manufacturing automation; design for manufacturability; rapid prototyping; microstructures; biomechanics; rehabilitation; structures and composite materials; graphics; and image processing. A number of these laboratories are unique in this country, and several are closely coupled with the Stanford Integrated Manufacturing Association, the Center for Design Research, the NASA-Ames Research Center, the Center for Integrated Systems, the Center for Turbulence Research, the Stanford Medical School, and the Veterans Administration Rehabilitation Research and Development Center.
Financial Aid	Each year the department awards several graduate fellowships, primarily to entering master's degree candidates. These awards are based on merit. Fellowships normally provide full tuition and a substantial living-expense stipend for the three-quarter period of study leading to the master's degree. Research assistantships are normally available to students at the post-master's degree level and occasionally for master's degree candidates. Loans based on financial need are available to U.S. citizens.
Cost of Study	In 1999–2000, tuition is $24,588 for the nine-month, three-quarter academic year ($8196 per quarter). Other fees, such as student health insurance, student association, and a one-time document fee, are also required.
Living and Housing Costs	Stanford provides more than 3,000 graduate students with on-campus housing. The 1999–2000 sample rates are based on an academic year single-occupancy co-op rate of $4310 to $4816 for a single dormitory room; a two-bedroom apartment is $5939, based on double occupancy. Off-campus housing is more expensive; the cost of living in the area is high.
Student Group	The department has approximately 200 doctoral students and 200 master's students from all parts of the nation and the world. Approximately 20–30 doctoral degrees and 160 master's degrees are awarded each year.
Location	The campus, extending from the wooded area surrounding Palo Alto to the foothills of the Coast Range, offers a great variety of recreational activities. The University is surrounded by the Stanford Industrial Park and San Francisco's busy suburbs. San Francisco, with its theaters, galleries, and restaurants, is 30 miles to the north. There is boating on nearby San Francisco Bay, and Pacific beaches are a 45-minute drive to the west. The Sierra Nevada snow country is a 4-hour drive east. The wine-producing areas of the state, the Gold Rush country, and Monterey, Carmel, and the Big Sur are within easy reach of the campus.
The University	Stanford University was founded in 1885 by Senator and Mrs. Leland Stanford and has an international reputation as an outstanding educational institution. Stanford has a long-standing tradition of academic excellence in the engineering and physical science fields and has produced many prominent engineers and scientists. The University's atmosphere is an unusual blend of a pleasant and uncrowded environment, a spirited and dynamic student body and faculty, and unswerving standards of academic excellence.
Applying	Completed applications, including transcripts, letters of recommendation, and GRE General Test scores, for admission with financial aid consideration must be received by January 15. Students should take the GRE no later than October to ensure that the scores are received by Stanford by this date. Applicants not competing for financial aid have until February 1 to submit the completed application. Decisions on financial awards are made on the basis of review, by the Admissions Committee, of the candidate's application and supplementary credentials. Applications for the Manufacturing Systems Engineering Program should be clearly marked IE/ME program.
Correspondence and Information	Judith Haccou, Manager of Student Services Mechanical Engineering Department Building 530, Room 125 Stanford University Stanford, California 94305-3030 Telephone: 650-725-7695 650-725-2075 Fax: 650-725-4862 E-mail: meinquiry@forsythe.stanford.edu WWW: http://cdr.stanford.edu/html/me/home.html

Stanford University

THE FACULTY AND THEIR AREAS OF RESEARCH

Biomechanical Engineering
Thomas Andriacchi, Professor of Mechanical Engineering and Functional Restoration; Ph.D., Illinois, 1974.
Gary S. Beaupre, Consulting Professor of Mechanical Engineering and Function Restoration; Ph.D., Stanford, 1983.
Dennis R. Carter, Professor of Mechanical Engineering; Ph.D., Stanford, 1992.
Scott L. Delp, Associate Professor of Mechanical Engineering; Ph.D., Stanford, 1990.
Jean H. Heegaard, Assistant Professor of Mechanical Engineering; Ph.D., Ecole Polytechnique Federale de Lausanne (Switzerland), 1993.
Charles Taylor, Assistant Professor (Research) of Surgery and courtesy Assistant Professor of Mechanical Engineering; Ph.D., Stanford, 1996.
Felix Zajac, Professor (Research) of Mechanical Engineering and Functional Restoration (School of Medicine); Ph.D., Stanford, 1968.

Research problems: musculoskeletal biomechanics, rehabilitation engineering, and other problems at the interface of mechanical engineering and medicine/biology.

Design
David W. Beach, Professor (Teaching) of Mechanical Engineering; M.S., Stanford, 1972.
J. Edward Carryer, Consulting Associate Professor of Mechanical Engineering; Ph.D., Stanford, 1992.
Mark R. Cutkosky, Professor of Mechanical Engineering and Associate Chair for Design and Manufacturing; Ph.D., Carnegie Mellon, 1985.
Rolf A. Faste, Associate Professor of Mechanical Engineering; M.S., Tufts, 1972; B.Arch., Syracuse, 1977.
J. Christian Gerdes, Assistant Professor of Mechanical Engineering; Ph.D., Berkeley, 1996.
Kosuke Ishii, Associate Professor of Mechanical Engineering; Ph.D., Stanford, 1987.
David M. Kelley, Associate Professor of Mechanical Engineering; M.S., Stanford, 1978.
Thomas W. Kenny, Assistant Professor of Mechanical Engineering; Ph.D., Berkeley, 1989.
Larry J. Leifer, Professor of Mechanical Engineering and of Neurology (School of Medicine); Ph.D., Stanford, 1969.
Drew V. Nelson, Professor of Mechanical Engineering; Ph.D., Stanford, 1978.
Friedrich B. Prinz, Professor of Mechanical Engineering and Materials Science; Ph.D., Vienna, 1975.
Bernard Roth, Professor of Mechanical Engineering; Ph.D., Columbia, 1962.
Sheri D. Sheppard, Associate Professor of Mechanical Engineering; Ph.D., Michigan, 1985.

Research problems: assistive device design-development, automated manufacturing, CAD, computational geometry, computer-aided prototyping, design theory and concurrent engineering methodology, experimental stress analysis, fatigue and fracture mechanics, human–machine interaction, kinematic synthesis, manufacturing engineering systems, mechanical system modeling and control, mechatronics, micromachining, product design, properties of microstructures, robotics, vehicle dynamics, visual thinking.

Flow Physics and Computation
Brian J. Cantwell, Professor of Aeronautics and Astronautics and of Mechanical Engineering; Ph.D., Caltech, 1976.
Paul A. Durbin, Professor (Research) of Mechanical Engineering; Ph.D., Cambridge, 1979.
Joel H. Ferziger, Professor of Mechanical Engineering and of Aeronautics and Astronautics; Ph.D., Michigan, 1962.
Sanjiva Lele, Associate Professor of Aeronautics and Astronautics and of Mechanical Engineering; Ph.D., Cornell, 1985.
Parviz Moin, Professor of Mechanical Engineering and of Aeronautics and Astronautics; Ph.D., Stanford, 1978.
William C. Reynolds, Professor of Mechanical Engineering and of Aeronautics and Astronautics; Ph.D., Stanford, 1958.

Research problems: acoustics, computational fluid dynamics, environmental fluid mechanics, flow control, internal flow, modeling complex turbulent flows, chemically reacting flows, high-speed flows, shock waves, turbomachinery, turbulence, control theory.

Mechanics and Computation
David M. Barnett, Professor of Mechanical Engineering and of Materials Science; Ph.D., Stanford, 1967.
Kychongjae Cho, Assistant Professor of Mechanical Engineering; Ph.D., MIT, 1994.
Huajian Gao, Associate Professor of Mechanical Engineering; Ph.D., Harvard, 1988.
Thomas J. R. Hughes, Professor of Mechanical Engineering; Ph.D., Berkeley, 1974.
Peter M. Pinsky, Professor of Mechanical Engineering; Ph.D., Berkeley, 1981.
George S. Springer, Professor of Aeronautics and Astronautics; Ph.D., Yale, 1962.
Charles R. Steele, Professor of Mechanical Engineering and of Aeronautics and Astronautics; Ph.D., Stanford, 1960.
Andrew M. Stuart, Associate Professor of Mechanical Engineering and Computer Science; D.Phil., Oxford, 1986.

Research problems: acoustics, aerospace structures, biomechanics, dynamic analysis, composite materials, computational mechanics, device simulation, finite element analysis, fracture and defect mechanics, hemodynamics, mathematical modeling, mechanics of deformable solids, nonlinear continuum mechanics, numerical analysis.

Thermosciences
Thomas Bowman, Professor of Mechanical Engineering; Ph.D., Princeton, 1966.
Mark A. Cappelli, Associate Professor of Mechanical Engineering; Ph.D., Toronto, 1987.
John K. Eaton, Professor and Associate Chairman of Mechanical Engineering; Ph.D., Stanford, 1980.
Christopher F. Edwards, Assistant Professor of Mechanical Engineering; Ph.D., Berkeley, 1985.
Kenneth E. Goodson, Assistant Professor of Mechanical Engineering; Ph.D., MIT, 1993.
Ronald K. Hanson, Professor and Chairman of Mechanical Engineering; Ph.D., Stanford, 1968.
Charles H. Kruger Jr., Professor of Mechanical Engineering and Dean of Research and Graduate Policy; D.I.C., Imperial College (London), 1957; Ph.D., MIT, 1960.
Reginald E. Mitchell, Associate Professor of Mechanical Engineering; Sc.D., MIT, 1975.
M. Godfrey Mungal, Associate Professor of Mechanical Engineering and Associate Chairman of Student Services; Ph.D., Caltech, 1983.
Juan Santiago, Assistant Professor of Mechanical Engineering; Ph.D., Illinois, 1995.

Research problems: physical gasdynamics, combustion, plasma processing of materials, convective heat transfer, optical diagnostics and sensors, microscale and nanoscale heat transfer.

STATE UNIVERSITY OF NEW YORK AT BINGHAMTON

Thomas J. Watson School of Engineering and Applied Science
Department of Mechanical Engineering

Programs of Study

The mechanical engineering department in the Thomas J. Watson School of Engineering and Applied Sciences offers programs of study leading to the degrees of Master of Science in Mechanical Engineering (M.S.M.E.), Master of Engineering (M.Eng.), and Doctor of Philosophy (Ph.D.) in mechanical engineering. Students matriculating in the mechanical engineering graduate programs will be those interested in following current research specializations in mechanics and design (with emphasis on solid mechanics/stress analysis, vibrations, tribology, and controls), thermofluids (with emphasis on heat transfer, lubrication, and physicochemical hydrodynamics), and manufacturing/materials (with emphasis on characterization and process analysis).

All M.S. degree programs in the Watson School require that the student complete at least eight courses plus a thesis. Ten courses are required for the M.Eng. degree, including a two-course engineering project. The normal period for completion of a master's degree is 1½ years of full-time study.

Graduation requirements for the Ph.D. degree include satisfactory completion of a comprehensive examination, based on an individual learning contract, and satisfactory defense of a dissertation. There is a 24-credit-hour residence requirement. The normal period for completion of the degree is two years beyond the master's degree. The learning contract permits the development of a highly individualized course of study in close cooperation with senior members of the academic faculty.

Doctoral programs in the Watson School focus on studies at the forefront of science and technology. Creative approaches to state-of-the-art problems in areas of faculty research interest are emphasized.

Research Facilities

The Watson School has developed an international reputation in the multidisciplinary research specialty of electronic packaging. This research is housed in the Watson School's Integrated Electronics Engineering Center (IEEC). The IEEC is also a designated National Science Foundation state/industry university cooperative research center, and in 1993 it became a New York State Center for Advanced Technology (CAT). State-of-the-art research and teaching laboratories support the efforts of faculty members to develop programs in a variety of other areas.

Departmental research facilities are concentrated in a number of special-purpose laboratories. In the Advanced Thermal/Fluid Mechanics Facility, flow modeling is aided by a DEC 3000 Mod 800 Workstation, while experiments incorporate viscometers and hot-stage microscopes with high-performance video microscope systems with frame grabber and image digitizing and processing capability. In the Heat Transfer Laboratory, measurements of convective and conductive heat flow are made with an IR thermography camera or holographic interferometry system, fluid flow studies are performed using techniques of flow visualization and laser Doppler velocimetry (LDV), and data acquisition and analysis is facilitated by PC-based systems. Recently added equipment in the Materials/Manufacturing Processes Laboratory includes an environmental scanning electron microscope incorporating cryo/heating/tensile stages, energy dispersive and digital image analysis, and residual gas analysis for studying materials behavior at high resolution in a variety of atmospheres; a digital microhardness tester; a multipurpose computerized data acquisition system; and a Starrett production computerized metrology system. Recent additions to the Opto-Mechanics Research Laboratory include a holographic interferometry and shadow moire facility for out-of-plane displacement measurements, an environmental chamber (180°C to 315°C), a 35mW laser, an image processing system, and several computers (Pentium and RISC workstations). The Tribology Laboratory is equipped with rotary and impact testers; a Modified Bowden-Leben ball/plane friction and wear tester with an Anorad x-y table, computerized; a WYKO 3-D Optical Topographer; and various computer data acquisition systems, while the Vibrations Laboratory offers extensive computational capabilities, including twelve networked PC workstations, and state-of-the-art instrumentation for inducing and measuring vibration, including accelerometers, photonic sensors, two laser vibrometers, and a variety of shakers.

Financial Aid

Many students hold fellowships; traineeships; or graduate, research, or teaching assistantships. Most awards include a full waiver of tuition. Other sources of financial aid include the New York State Tuition Assistance Program, the Federal Stafford Student Loan Program, the graduate and professional school College Work-Study Program, and campus jobs.

Cost of Study

For full-time matriculated graduate students, tuition in 1998–99 cost $2550 per semester for state residents and $4208 per semester for nonresidents.

Living and Housing Costs

A recently completed apartment complex, the Graduate Community, has 3- and 4-person apartments, with living room, dining area, kitchen, and bath. Based on a 1998–99 academic-year lease, the semester rate for a single bedroom was $2050 and for a double bedroom, $1775 per person and $3085 per couple. The cost of meal plans per semester is as follows: basic, $787; standard, $1022; and ultra, $1097. Assistance in locating off-campus housing is provided by the listing services of Off-Campus College.

Student Group

Of the 12,259 students enrolled at Binghamton University, 2,700 are graduate students. In the Watson School, there are 876 undergraduates and 368 graduate students. Many obtain jobs in local high-technology enterprises during their enrollment at the School and after graduation.

Location

The University's 606-acre campus is in a suburban setting just west of Binghamton. More than 300,000 people live within commuting distance of the campus. Cultural offerings in the community include the museum and programs of the Roberson Center for the Arts and Sciences, as well as performances by the Binghamton Symphony, Tri-Cities Opera, Civic Theater, and other groups. The University's Art Gallery has a permanent collection representing all periods and also displays works from special loan exhibitions. The annual concert series of the Anderson Center brings a wide variety of performing artists to campus. The Department of Theater stages more than twenty-five productions each year.

The University and The School

The State University of New York at Binghamton is one of the four university centers in the State University of New York System. The faculty numbers about 700. Graduate programs were initiated in 1961 with the establishment of Master of Arts programs in English and mathematics.

The Watson School was created in 1983 by combining the established graduate programs in computer science and systems science from the School of Advanced Technology with new programs in electrical, industrial, and mechanical engineering.

Applying

Holders of an appropriate bachelor's degree from any recognized college or university are eligible to apply. Application forms should be requested by e-mail from the Office of Graduate Admissions (gradad@binghamton.edu). Applicants should submit GRE General Test scores. International applicants must submit TOEFL scores and provide proof of their ability to meet academic expenses. All credentials should be on file at least one month prior to anticipated enrollment. To ensure consideration for assistantship and fellowship awards, admission credentials should be received by February 15. Online application is available at http://www.gradschool.binghamton.edu.

Correspondence and Information

Director of Graduate Studies
Department of Mechanical Engineering
Thomas J. Watson School of Engineering and Applied Science
State University of New York at Binghamton
P.O. Box 6000
Binghamton, New York 13902-6000
E-mail: atideren@binghamton.edu
World Wide Web: http://www.me.binghamton.edu

State University of New York at Binghamton

THE FACULTY AND THEIR RESEARCH

Frank Cardullo, Associate Professor; M.S., SUNY at Binghamton. Vehicle simulation, vehicle dynamics, man-machine systems.

Richard Culver, Professor; Ph.D., Cambridge. Dynamic instabilities in metal deformation, engineering education.

John Fillo, Professor and Associate Dean; Ph.D., Syracuse. Thermal fluid analysis, mathematical modeling, heat transfer in electronics, advanced technology.

Robert Frey, Lecturer; M.S., Syracuse. Experimental methods, instrumentation, vibration testing.

James Geer, Bartle Professor; Ph.D., NYU. Perturbation methods, nonlinear problems, slender body theory, symbolic computation.

Gary Lehmann, Associate Professor; Ph.D., Clarkson. Fluid dynamics, numerical and experimental heat transfer, cooling of electronics.

Ronald Miles, Professor and Chairman; Ph.D., Washington (Seattle). Vibrations, acoustics, fatigue, noise, biomechanics.

Bruce Murray, Assistant Professor; Ph.D., Arizona. Thermal and fluid sciences, computational fluid dynamics, materials processing.

James Pitarresi, Associate Professor; Ph.D., SUNY at Buffalo. Computational mechanics, vibration modeling and testing, electronic packaging.

Chittaranjan Sahay, Associate Professor; Ph.D., Indian Institute of Technology (Delhi). Solid mechanics, manufacturing and design.

Bahgat Sammakia, Professor and Director of the Integrated Electronics Engineering Center; Ph.D., SUNY at Buffalo. Thermal and fluid sciences, electronic packaging.

Timothy Singler, Associate Professor; Ph.D., Rochester. Experimental and analytical fluid mechanics, geophysical fluid mechanics, interfacial fluid mechanics, interfacial stability, applied mathematics.

D. C. Sun, Professor; Ph.D., Princeton. Mechanics, fluid and mechanical systems, tribology.

STATE UNIVERSITY OF NEW YORK AT BUFFALO

Department of Mechanical and Aerospace Engineering

Programs of Study

The department offers separate graduate degree programs in mechanical and aerospace engineering for both the Master of Science (M.S.) degree and the Doctor of Philosophy (Ph.D.) degree. The department also offers certificate programs and a Master of Engineering (M.Eng.) program that emphasizes applications. Although the basic graduate degree programs are in the fields of mechanical and aerospace engineering, specialization in particular areas is possible regardless of the program. There is a strong cluster of faculty members in the fluid and thermal sciences who cover areas in fluid mechanics, heat and mass transfer, computational fluid dynamics, combustion and propulsion, and thermodynamics and energy systems. Another strong area is systems and design, which includes concentrations in computer-aided design and optimization, dynamics and control systems, and vibration and stress analysis. Mechanics and materials is also a focus area in the department and offers concentrations in composite materials, metallurgy and material science, and solid mechanics and materials engineering. There is strong participation in biomedical engineering, which is an interdisciplinary program involving other engineering departments and the medical school. The Master of Science degree emphasizes course work and thesis research in one of the focus areas of the department. Although there is no core course requirement, students are normally expected to take certain designated courses in accordance with their chosen area of concentration. For a full-time student (taking four courses per semester), the completion of a master's program typically requires two or three semesters plus one summer. Course concentrations are designed to meet the needs of the practicing engineer with an M.S. degree as well as the student who intends to proceed to a Ph.D. program. The Ph.D. program requires at least three academic years of full-time graduate study beyond the baccalaureate degree. Two years are usually devoted to formal course work, while the third year is typically devoted to dissertation research. Ph.D. students are required to pass a qualifying examination during the second year of study.

Research Facilities

Modern, well-equipped laboratories support the research and graduate study activities of the department. These laboratories include the Assistive Device Design Laboratory; CAD facilities, including Sun Workstations with Proengineer and finite-element tools; the CFD laboratory, including Silicon Graphics and Sun Workstations with various Navier-Stokes solvers; a Combustion Research Laboratory, with an electrically heated furnace for ignition studies, a gas-heated furnace for liquid propellant combustion studies, a catalytic combustion reactor for low-NOX combustion studies, and an acoustic levitator for microgravity studies of drops and bubbles; the Heat Transfer Laboratory, with facilities for radiation heat transfer studies and heat exchanger testing; the Hemodynamics Laboratory, with a phantom replica casting facility of cerebrovascular and cardiovascular vessels, a mock circulation flow assessment unit, and a heart-valve testing facility; the Materials Laboratory, with equipment for materials processing and testing; the Multidisciplinary Optimization and Design Laboratory; and the Turbulence Research Laboratory, with a low-speed wind tunnel, a plume facility, and a turbine blade flow facility.

Financial Aid

A number of teaching assistantships are available, with a minimum stipend of $10,723. Research assistantships are available through research grants. Students receiving assistantships are normally granted full tuition scholarships. There are also a number of Presidential Fellowships in the amount of $4000, which are available from the State University of New York as supplements to the teaching and research assistantship stipends. Tuition scholarships are also available for full-time M.Eng. and M.S. students who are New York State residents.

Cost of Study

For 1998–99, full-time tuition was $2550 per semester for New York State residents and $4208 per semester for nonresidents.

Living and Housing Costs

The cost of living is very affordable compared to that of other northeastern American cities. Graduate students have access to University dormitory facilities, and a variety of off-campus areas are also available. Attractive one- and two-bedroom apartments can readily be found for costs ranging from $250 to $400 per month.

Student Group

The University's student population numbers about 25,000 and includes more than 6,800 graduate students. There are approximately 130 M.S. and M.Eng. and 50 Ph.D. students in mechanical and aerospace engineering.

Student Outcomes

Recent Ph.D. graduates have taken academic positions in such institutions as the University of Minnesota, Berkeley, Virginia Tech, Purdue University, Clarkson University, and Vanderbilt University. Ph.D. graduates have also taken high-level research and development positions in corporations such as Fluent Corporation, General Electric, General Motors, National Transportation Safety Board, NASA, Lockheed-Martin, Bell Labs, and JPL. Graduates with master's degrees have taken a variety of management, research, and engineering positions at midsize and large corporations both locally and nationally and with government agencies.

Location

The city of Buffalo is the hub of a metropolitan area of more than 1.2 million individuals and is ½ hour by car from Niagara Falls and a 2-hour drive from Toronto. Buffalo has a renowned philharmonic orchestra and a diversified downtown theater district. Canada's Stratford Shakespearean Festival and Niagara-on-the-Lake's Shaw Festival are nearby, as are New York State's ArtPark and the Chautauqua Institution, noted for their spring and summer festivals of art, drama, and music. The proximity to Lakes Erie and Ontario ensures moderate temperatures both in summer and winter and provides the facilities for a wide variety of water sports activities. Many downhill and cross-country ski areas are located a short drive south of the city. The city provides major-league professional football (Buffalo Bills) and ice hockey (Buffalo Sabres) and other professional sports leagues.

The University and The Department

Formerly named the University of Buffalo, the institution was founded in 1846 as a medical school. In 1962, the University joined the State University of New York System and has become the principal research university in the state system. The Department of Mechanical and Aerospace Engineering is part of the School of Engineering and Applied Sciences. The School and University provide opportunities for interaction with graduate students and faculty members through graduate clubs, sports, and multidisciplinary and professional interest groups.

Applying

Applications for admission with financial aid must be completed by February 1 for September admission in order to ensure consideration for aid and assistantships. If no financial assistance is desired, two months should be allowed for the processing of an application before the start of any semester; international students should allow more time. All new applicants are required to take the GRE General Test. International applicants must also supply TOEFL scores.

Correspondence and Information

Graduate Admissions
Department of Mechanical and Aerospace Engineering
318 Jarvis Hall, Box 604400
State University of New York at Buffalo
Buffalo, New York 14260-4400
Telephone: 716-645-2593 Ext. 2238
Fax: 716-645-3875
World Wide Web: http://www.mae.buffalo.edu/

State University of New York at Buffalo

THE FACULTY AND THEIR RESEARCH

Nasser Ashgriz, Ph.D. (mechanical engineering), Carnegie Mellon, 1984. Fluid mechanics, combustion and propulsion, pollution formation, computational fluid dynamics of free-surface flows, sprays and atomization, fluid interface instability. Current research: collective dynamics of bubble clouds, collision of liquid drops, combustion of fuel drops, instability of ferrofluid layers, fluid modeling of ink-jet printers.

Christina L. Bloebaum, Ph.D. (aerospace engineering, multidisciplinary design optimization), Florida, 1991. Large-scale engineering optimization and design, concurrent design methods, heuristic optimization methods. Current research: development of collaborative design methodologies for complex multidisciplinary systems, development of simulation-based convergence strategies for large-scale complex design, design visualization via graph morphing for multidisciplinary design optimization.

Harsh Deep Chopra, Ph.D. (materials science and engineering), Maryland, College Park, 1993. Functional materials; atomically engineered thin-film growth; magnetic multilayers; nanostructures, synthesis, characterization; surface and interfacial effects in thin-film magnetism. Current projects: giant magnetoresistive magnetic multilayers and spin valves; giant magnetostriction in bulk, single films, and multilayers; nondestructive evaluation: mechanical-magnetic relationship of soft magnetic materials.

Deborah D. L. Chung, Ph.D. (materials science and engineering), MIT, 1977. Composite materials, smart materials, electronic packaging materials, concrete, carbon. Current research: self-monitoring structural materials, concrete, structural composites, electronic packaging materials, activated carbon, battery electrodes.

James D. Felske, Ph.D. (mechanical engineering), Berkeley, 1974. Heat and mass transfer, fluid mechanics, thermodynamics, radiative transfer, spectroscopy, combustion. Current research: radiative properties of particles and droplets, radiative heat transfer, microscale thermophysics, modeling industrial thermal/fluid/chemical phenomena.

William K. George, Ph.D. (mechanical engineering), Johns Hopkins, 1971. Fluid mechanics, turbulence, experimental methods. Current research: asymptotic and similarity theories for wall-bounded turbulent flows, measurement of coherent structures in turbulent flow.

Peyman Givi, Ph.D. (mechanical engineering), Carnegie Mellon, 1984. Systems and control, vibration and shock, numerical methods, applied mathematics, statistical analysis, random data analysis, stochastic processes, thermal-fluid science and engineering, machine design. Current research: turbulence, combustion, computational methods and numerical algorithms, applied mathematics, magnetohydrodynamics, multiphase transport, theoretical statistics, spectral analysis, stochastic processes.

Thenkurussi Kesavadas, Ph.D. (industrial and manufacturing systems engineering), Penn State, 1995. CAD/CAM, robotics, virtual reality and manufacturing automation. Current research: virtual reality–based interactive automation of robotic manufacturing processes, automated assembly planning, knowledge-integrated virtual manufacturing laboratory development, Web-based distributed factory modeling.

Kemper E. Lewis, Ph.D. (design and optimization), Georgia Tech, 1996. Design methods, large-scale systems design, optimization, robust design, organizational design, game theory, open engineering systems, product and process quality control. Current research: modeling design processes for large-scale systems, modeling of product and process interactions, mixed discrete/continuous optimization of engineering systems.

B. Barry Lieber, Ph.D. (hemodynamics), Georgia Tech, 1985. Fluid mechanics, hemodynamics, biosignal processing, image processing. Current research: hemodynamics of cerebrovascular disease, blood flow in arteriovenous malformations, design and analysis of endovascular devices to combat strokes, angiographic image processing.

Ching Shi Liu, Ph.D. (mechanical engineering), Northwestern, 1961. Fluid mechanics, boundary layer stability, dynamic systems. Current research: bifurcation of stability in boundary layers, transition phenomena, shock-wave interactions.

Cyrus K. Madnia, Ph.D. (aerospace engineering), Michigan, 1989. Computational fluid dynamics, transport phenomena, turbulence, compressible flows. Current research: burning vortex rings, flame-vortex interaction in laminar and turbulent flows, numerical simulations of flames with inclusion of realistic chemistry schemes, turbulent mixing.

Roger W. Mayne, Ph.D. (mechanical engineering), Penn State, 1971. Systems, design optimization, mechanical design, computer graphics and CAD, modeling of dynamic systems, automatic control. Current research: applications of optimization methods, robot path planning, tools for computer-aided design, system dynamics.

John Medige, Ph.D. (mechanics), IIT, 1967. Biomechanics, properties of bone and soft tissue, joint mechanics. Current research: ultrasound as a predictor of bone properties, motor sensory mechanics, comparison forces in normal and artificial knees, screw fixation of fractured wrist bones.

Joseph C. Mollendorf, Ph.D. (mechanical engineering), Cornell, 1971. Heat transfer, fluid mechanics, design. Current research: multirecompression heating, assistive-device design.

D. Joseph Mook, Ph.D. (engineering mechanics), Virginia Tech, 1986. Controls, dynamics, system identification, estimation theory, modeling, nonlinear and chaotic dynamic systems. Current research: modeling, identification, and control of nonlinear vehicle dynamic tests; estimation and identification of satellite attitude dynamics; time-optimal feedback control; identification of gas-separation plants; large-order multi-input–multi-output bang-bang controllers.

Abani K. Patra, Ph.D. (computational and applied mathematics), Texas at Austin, 1995. Computational mechanics, solution adaptive finite element methods, parallel and high-performance computing. Current research: parallel adaptive FEM, applications to biomechanical implants, incompressible fluid flow.

William J. Rae, Ph.D. (aeronautical engineering), Cornell, 1960. Fluid mechanics, heat transfer, turbomachinery, dynamics of aircraft and road vehicles. Current research: turbomachinery, vehicle and missile dynamics.

Tarunraj Singh, Ph.D. (dynamics and control), Waterloo, 1991. Nonlinear control, optimal control, optimization, dynamics of flexible structures, vibration control. Current research: optimal control of flexing structures, system and identification of A/C systems, nonlinear control of electrohydraulic systems, neural-network modeling of dynamical systems.

James R. Sonnenmeier, Ph.D. (aerospace engineering), SUNY at Buffalo, 1994. Fluid mechanics, turbulence, aerodynamics, heat transfer computational fluid dynamics. Current research: applications of CFD to engineering problems in fluid/thermal science, design and analysis methods for axial and centrifugal fans.

Andres Soom, Ph.D. (mechanical engineering), Wisconsin–Madison, 1976. Tribology, noise and vibration, instrumentation, machinery diagnostics, mechanical design, dynamic systems. Current research: contact dynamics, mechanics of clutches, experimental dynamics.

Dale B. Taulbee, Ph.D. (theoretical and applied mechanics), Illinois, 1964. Fluid mechanics, turbulence, heat transfer, computational fluid dynamics, aerosol mechanics. Current research: modeling turbulent two-phase flows, turbulence modeling for near-wall flow and heat transfer, design and analysis methods for rotating machinery.

Robert C. Wetherhold, Ph.D. (mechanical engineering), Delaware, 1983. Composite materials, solid mechanics, materials engineering. Current research: fracture and reliability of materials, design of laminated composites, smart materials.

Chia-Ping Yu, Ph.D. (aeronautics, astronautics and engineering science), Purdue, 1964. Aerosol mechanics, electrodynamics of fluids, plasma physics. Current research: pulmonary gas and particle transport, dynamics of fibers, inhalation toxicology, therapeutic aerosols.

TEXAS A&M UNIVERSITY

Department of Mechanical Engineering

Programs of Study

The department confers four graduate degrees: Master of Engineering, Master of Science, Doctor of Engineering, and Doctor of Philosophy. The M.Eng. and D.Eng. degrees are oriented toward professional practice. The M.S. and Ph.D. degrees are oriented toward research and academic pursuits.

The M.Eng. degree requires a minimum of 36 credit hours, of which approximately one third must be taken in areas outside of mechanical engineering. Consequently, many of the courses taken by the student may be nontechnical and may emphasize communication and administration skills. This degree is offered in areas of mechanical engineering that have a prescribed plan of study on file in the department. The M.S. degree program is suited for students who have either a research or a design orientation; it has two options: thesis and nonthesis, respectively. If the nonthesis option is chosen, a technical report is required. The thesis option requires a minimum of 32 credit hours, while the nonthesis option requires 36. The D.Eng. program is especially designed for students who wish to practice the engineering profession at the highest level of competence. Recipients of the degree can become engineering specialists or eventually move into management. A typical D.Eng. program includes advanced courses in the analysis, synthesis, and design of engineering systems; in business administration; and in the humanities and social sciences. This professional graduate program consists of 96 credit hours beyond the bachelor's degree, of which 16 credit hours are awarded for a one-year internship in a practicing engineering capacity.

The Ph.D. program is structured for the student planning a career in independent research and teaching. Upon graduation, the student is expected to be aware of the frontiers of a particular area and to possess the analytical, numerical, and experimental skills necessary to create new knowledge in this area. The Ph.D. degree requires a minimum of 96 credit hours beyond the bachelor's degree or 64 beyond the master's degree. Of this number, a minimum of 48 credit hours of formal course work beyond the bachelor's degree is required. The remaining credits involve intensive research, and a dissertation is mandatory.

Research Facilities

The department is housed in an engineering building with modern office and laboratory space and has additional space in other A&M buildings. Active research facilities exist for fracture testing, composite materials, plastics, metallurgical studies, corrosion, experimental stress analysis, ultrasonics, nondestructive testing, vibration, shock, rotating machinery, turbomachinery, seals, fluid dynamics, gasdynamics, turbulence, heat transfer, power generation, tribology, combustion, in situ lignite gasification, solar energy, wind tunnel studies, aerosols, manufacturing processes, robotics and intelligent machinery, and computer-aided design and manufacturing. Computer facilities available to graduate students include the University's Cray and Silicon Graphics supercomputers, the College of Engineering's VAX cluster, and the department's PC lab with more than seventy-five high-performance PCs and twenty high-performance workstations. There are also several graphics devices, including an Evans & Sutherland PS 300 system. The University library contains more than 1 million books and journals.

Financial Aid

Graduate assistantships for teaching and research are available to qualified students and provide stipends of $1050 to $1300 per month. In addition, a growing number of fellowships and industry-supported assistantships are available to promising research students. Typically, more than 190 graduate students are supported as graduate research assistants and teaching assistants. Many job opportunities for students' spouses exist at the University and at firms in the fast-growing surrounding community.

Cost of Study

Full-time graduate students typically register for a minimum of 9 credit hours each regular semester. The estimated cost of tuition and required fees for graduate students for one semester in the College of Engineering is $1210 for resident students, $3118 for nonresident students, and $3142 for international students. Tuition and fees are subject to change without notice due to legislative action. Students supported by the department are required to register for 9 credit hours each regular semester and for 6 hours for the ten-week summer session. Students receiving assistantships or fellowships qualify for resident tuition rates.

Living and Housing Costs

The cost of living varies widely according to the type of accommodations sought, the student's marital status, and other conditions. Campus dormitory rooms are available to graduate students during the summer only. A limited number of University-owned apartments, both furnished and unfurnished, are available. A large number of privately owned apartments are available in the community. Further information regarding types of accessible housing and general living costs can be obtained through the Housing Office (409-845-2261).

Student Group

The projected graduate enrollment in mechanical engineering is about 240. Students in the department have a wide range of interests and backgrounds.

Location

Texas A&M University is located in College Station, Texas. College Station and the neighboring town of Bryan form a twin-city area that has a population of 125,000. Outdoor recreation abounds in nearby state forests and parks with several large lakes. Metropolitan recreational facilities can be enjoyed at one of several cities surrounding Bryan–College Station: Houston, Austin, Dallas, Fort Worth, and San Antonio. The climate is mild almost year-round. The two public school systems are excellent, and there are more than fifty churches of various denominations.

The University

The University was founded as a land-grant college in 1876 and is the state's oldest public institution of higher education. In recent years, it has become a sea-grant and space-grant university. Founded as a college of agriculture and mechanical studies, A&M has traditionally supported a strong mechanical engineering program. The campus is situated on 5,200 acres in the green rolling prairies of south-central Texas. University enrollment has averaged more than 42,000 in recent years, including approximately 7,500 graduate students. The University prides itself on offering excellent cultural programs during the school year and being the home of the George Bush Presidential Library.

Applying

Application forms and instructions for submitting grade transcripts, GRE scores, and related material can be obtained from the address below.

Correspondence and Information

N. K. Anand, Director
Graduate Program in Mechanical Engineering
Texas A&M University
College Station, Texas 77843-3123

Telephone: 409-845-1270
       800-874-4581 (toll-free)
E-mail: kmoses@mengr.tamu.edu
World Wide Web: http://wwwmengr.tamu.edu

Texas A&M University

THE FACULTY AND THEIR RESEARCH

R. M. Alexander, Professor and Head of Engineering Technology; Ph.D., Texas at Arlington. Design, vibrations.

N. K. Anand, Professor; Ph.D., Purdue. Computational heat transfer.

M. J. Andrews, Associate Professor; Ph.D., London. Heat transfer, combustion, computational fluid dynamics.

K. Annamalai, Professor; Ph.D., Georgia Tech. Burners, combustion, energy, fire.

A. Beskok, Assistant Professor; Ph.D., Princeton. Microscale fluid mechanics and heat transfer, numerical methods.

W. L. Bradley, Professor; Ph.D., Texas at Austin. Mechanical behavior of materials: fracture physics and fracture mechanics of metals, plastics, and composite materials.

D. E. Bray, Associate Professor; Ph.D., Oklahoma. Nondestructive testing, evaluation, ultrasonics.

C. P. Burger, Leland Jordan Professor; Ph.D., Cape Town (South Africa). Experimental mechanics, common nondestructive evaluation.

J. A. Caton, Professor; Ph.D., MIT. Internal combustion engines, combustion and fuels, cogeneration, boilers, energy systems.

D. W. Childs, Professor and Holder of Leland Jordan Chair; Ph.D., Texas at Austin. Rotor dynamics, rigid-body dynamics, vibrations.

R. Chona, Associate Professor; Ph.D., Maryland. Experimental and analytic fracture mechanics.

D. E. Claridge, Professor; Ph.D., Stanford. Energy management, building energy conservation.

A. Cohen, Zachry Professor; M.S., Stevens. Design and systems engineering.

L. J. Everett, Associate Professor; Ph.D., Texas A&M. Mechanics of multibody systems, mechatronics.

L. S. Fletcher, Regents and Dietz Professor; Ph.D., Arizona State. Heat transfer, fluid mechanics, aerothermodynamics.

R. B. Griffin, Associate Professor; Ph.D., Iowa State. Metallurgy, corrosion, tribology.

J. C. Han, HTRI Professor; D.Sc., MIT. Heat transfer in aircraft gas turbine engines, in space power and propulsion systems, and in electronic cooling and two-phase systems.

K. T. Hartwig, Professor; Ph.D., Wisconsin. Physical metallurgy, low-temperature materials, superconductivity.

W. M. Heffington, Associate Professor; Ph.D., California, San Diego. Energy-efficient equipment, combustion.

H. A. Hogan, Associate Professor; Ph.D., Texas A&M. Finite-element methods, biomechanics, solid mechanics.

C. L. Hough Jr., Associate Professor; Ph.D., Texas A&M. Manufacturing processes, machining.

S. Jayasuriya, Kotzebue Professor and Head; Ph.D., Wayne State. Robust control, active control of vibrations.

C. F. Kettleborough, Distinguished Professor; Ph.D., Sheffield (England). Tribology, numerical methods, solar energy.

K. D. Kihm, Associate Professor; Ph.D., Stanford. Liquid atomization and spray combustion, laser techniques, optical flow visualization.

T. J. Kozik, Professor; Ph.D., Ohio State. Analytical mechanics, theory of plates and shells.

T. R. Lalk, Associate Professor; Ph.D., Wisconsin. Internal combustion engines, energy systems.

R. Langari, Associate Professor; Ph.D., Berkeley. Intelligent control, fuzzy linguistic control, manufacturing.

S. C. Lau, Associate Professor; Ph.D., Minnesota. Heat and mass transfer, heat exchanger design.

R. P. Lucht, Professor; Ph.D., Purdue. Laser diagnostics, combustion, fluid mechanics.

H. Ma, Research Assistant Professor; Ph.D., Texas A&M. Microscale heat transfer.

J. E. Mayer Jr., Professor; D.Sc., MIT. Manufacturing processes, machining, metal cutting tools.

M. McDermott Jr., Associate Professor; Ph.D., Texas at Austin. Dynamics; automatic and manual control; real-time data acquisition, control, and simulation.

A. R. McFarland, Oscar Wyatt Professor; Ph.D., Minnesota. Aerosol mechanics, particle sampling, generation of mists and sprays, air cleaning.

G. L. Morrison, Nelson/Jackson Professor; Ph.D., Oklahoma State. Fluid dynamics, turbulence, 3-D laser anemometry.

S. T. Noah, Professor; Ph.D., West Virginia. Vibration, stability, tribology, nonlinear dynamics, chaos.

O. O. Ochoa, Professor; Ph.D., Texas A&M. Mechanics of composites, computational methods.

D. L. O'Neal, Professor; Ph.D., Purdue. Heat pump systems, HVAC system modeling, energy forecasting.

A. B. Palazzolo, Associate Professor; Ph.D., Virginia. Vibrations, finite and boundary elements, rotor dynamics, active vibration control.

G. P. Peterson, Professor and Executive Associate Dean; Ph.D., Texas A&M. Thermal control of electronic components, convection, heat pipes, two-phase heat transfer.

K. R. Rajagopal, Forsyth Professor; Ph.D., Minnesota. Mechanics and materials.

J. N. Reddy, Distinguished Professor and Holder of Oscar Wyatt Chair; Ph.D., Alabama in Huntsville. Continuum mechanics, numerical heat transfer, finite elements.

D. L. Rhode, Professor; Ph.D., Oklahoma State. Computational fluid dynamics and heat transfer, combustion aerodynamics.

H. H. Richardson, Distinguished Professor and Director of TTI; D.Sc., MIT. Transportation.

L. A. San Andres, Associate Professor; Ph.D., Texas A&M. Fluid film lubrication, rotordynamics.

T. Schobeiri, Professor; Ph.D., Technical University, Darmstadt (Germany). Dynamic behavior of turbomachinery systems, turbine performance.

J. Seyed-Yagoobi, Professor; Ph.D., Illinois. Electrohydrodynamic pumping, heat and mass transfer in porous media.

A. Srinivasa, Assistant Professor; Ph.D., Berkeley. Continuum mechanics, material processing, inelasticity.

H. J. Sue, Associate Professor; Ph.D., Michigan. Polymer science.

S. Suh, Assistant Professor; Ph.D., Texas A&M. Finite elements, design, wavelets, acoustics, electronic packaging.

D. V. Swaroop, Assistant Professor; Ph.D., Berkeley. Classical and advanced controls and stability, vehicle dynamic models, automated and intelligent highway systems.

W. D. Turner, Professor; Ph.D., Oklahoma. Energy management, energy systems, cogeneration.

J. M. Vance, Professor; Ph.D., Texas at Austin. Rotor dynamics of turbomachinery.

J. A. Weese, Regents Professor; Ph.D., Cornell. Elasticity, plasticity, structural dynamics, vibrations, experimental mechanics.

A. Wolfenden, Professor; D.Sc., Liverpool (England). Elastic modulus and damping measurements, composites, metallic glass.

TEXAS TECH UNIVERSITY

College of Engineering
Department of Mechanical Engineering

Programs of Study

The Department of Mechanical Engineering offers programs leading to the Master of Science in mechanical engineering and the Doctor of Philosophy degrees. Research is conducted within the traditional core areas of fluid dynamics, heat transfer, dynamics/controls, and solid mechanics/materials. In addition, a design-oriented specialization is available at the master's level, as is a special interdisciplinary program in advanced vehicle engineering.

The department offers two Master of Science options. The nonthesis option requires 36 hours of earned graduate credit. Under this option, the course work program of study is tailored by the student and research advising professor to prepare the student to conduct an independent research project requiring a report. Students should allow a minimum of fifteen months to complete this option.

The thesis option requires a minimum of 30 hours of graduate credit, of which 24 are earned by satisfactorily completing courses and a minimum of 6 are earned by thesis research. Under this option, the course work program of study is tailored by the student and research advising professor to prepare the student for conducting the thesis research. Students should allow a minimum of eighteen months to complete this option.

The Doctor of Philosophy degree is earned by completing a minimum of 90 hours of graduate credit, of which 60 consist of course work and a minimum of 30 consist of research credit. Graduate credits earned while completing a Master of Science program (either option) are applicable to the Ph.D. degree. Programs of study are developed by the thesis advising committee and student to meet the research objectives. Students should allow a minimum of thirty months beyond the Master of Science degree to complete this program. A departmental preliminary exam, thesis, and thesis defense are required.

Research Facilities

Laboratory and research facilities in the department are well equipped and modern. Major facilities include a tow tank of 80' x 15' x 10', which is among the largest university tow tanks in the country, for hydro/aerodynamics research. A new wind tunnel that has aerodynamic and boundary layer test sections began operation in fall 1998. In addition, research laboratories in optical methods in solid mechanics, dynamics and control of flexible robots, design and dynamics systems, thermal property characterization, IC engines, materials characterization, and a computer lab that supports the computational fluid dynamics program are actively used by graduate students. A fully equipped departmental shop is available. Computing facilities include PCs and workstations housed in the department, with access to an IBM parallel supercomputer.

Financial Aid

Teaching and research assistantships are available to qualified students. The monthly stipend for these assistantships is generally $950 to $1200 per month. In addition, a number of $1000 scholarships are available, and all admitted students are considered for such awards. Any of the preceding awards qualifies the student for in-state tuition.

Cost of Study

The 1999–2000 graduate tuition for Texas residents is $74 per semester credit hour; tuition for nonresidents is $285 per semester credit hour.

Living and Housing Costs

The cost of living in Lubbock is relatively low. University facilities are available for single students, and moderately priced apartments are available close to the University. Food and entertainment costs are also moderate.

Student Group

The department has approximately 400 undergraduate students and 60 graduate students. About 30 percent of the graduate students are pursuing the Ph.D. Approximately two thirds of the graduate students are receiving some type of financial support from the department.

Student Outcomes

The majority of M.S. and Ph.D. graduates find employment in private industry or in government laboratories. Recent graduates have been employed by Sandia, Chrysler, GM, John Deere, Texas Instruments, Shell Research, Amoco, Weber Aircraft, Apple Computer, Applied Materials, and Caterpillar.

Location

Lubbock is the center of a metropolitan area of approximately 200,000. Dry, crisp air and sunny days throughout most of the year provide a healthy and invigorating climate. An excellent recreation center, an Olympic-size swimming pool, an intramural sports field, and tennis courts are available on campus. Local cultural activities include a symphony orchestra, civic ballet, civic chorale, and an arts festival.

The University

Founded in 1923, Texas Tech University and Health Sciences Center is one of the four major state-supported comprehensive institutions of higher education in Texas. With 107 approved master's degree programs and sixty-four approved Ph.D. programs, Texas Tech is diverse in graduate degree opportunities. Enrollment in 1999–2000 was approximately 21,000 undergraduate and 4,000 graduate students. The library has more than 1.2 million volumes, 7,300 periodical subscriptions, and several unique archive collections. Texas Tech University and Health Sciences Center comprises seven colleges, a museum, a graduate school, and schools of law, medicine, nursing, and applied health, which combine to make Texas Tech a complete university.

Applying

Prospective students should have a cumulative GPA of at least 3.0 for their final 60 hours of undergraduate work and a GRE General Test score (verbal and quantitative) of at least 1000. Students not meeting these requirements will be considered on an individual basis. Students wishing to enter the program should also have an undergraduate degree in engineering, physics, or mathematics. Applications for admission and graduate assistantships for the fall term should be submitted no later than March 31, and for the spring term, no later than August 31. Applications from international students should be received at least one month earlier. For application materials, students should contact the Office of Graduate Admissions, Texas Tech University, P.O. Box 41030, Lubbock, Texas 79409-1030 (telephone: 806-742-2787, e-mail: gradschool@ttu.edu).

Correspondence and Information

Graduate Advisor
Department of Mechanical Engineering
Texas Tech University
Lubbock, Texas 79409-1021
Telephone: 806-742-3563

Texas Tech University

THE FACULTY AND THEIR RESEARCH

E. E. Anderson, Professor; Ph.D., Purdue, 1972. Radiative heat transfer, sprays, atomization, energy conversion.

A. A. Barhorst, Associate Professor; Ph.D., Texas A&M, 1991. Controls, dynamics, computer-assisted engineering.

J. M. Berg, Assistant Professor; Ph.D., Drexel, 1992. Control systems, modeling and control of metal forming and semiconductor manufacturing processes.

T. D. Burton, Professor and Chairman; Ph.D., Pennsylvania, 1976. Structural dynamics, nonlinear dynamics, chaos.

J. F. Cardenas-Garcia, Professor; Ph.D., Maryland, 1983. Holography, dynamic photoelasticity, projection moire, image processing, experimental stress analysis, fracture mechanics.

M. C. Chyu, Professor; Ph.D., Iowa State, 1984. Boiling heat transfer, two-phase phenomena, superconductors.

J. R. Dunn, Associate Professor; Ph.D., Georgia Tech, 1972. Fluid mechanics, aerodynamics, experimental methods, energy conversion.

S. Ekwaro-Osire, Assistant Professor; Ph.D., Texas Tech, 1993. Design, nonlinear and random vibrations.

A. Ertas, Professor; Ph.D., Texas A&M, 1984. Tribology, design, vibrations, dynamics.

J. Hashemi, Associate Professor; Ph.D., Drexel, 1988. Composite materials, metal forming, finite elements.

D. James, Associate Professor; Ph.D., Georgia Tech, 1992. Heat and mass transfer, experimental and computational fluid mechanics.

V. I. Levitas, Associate Professor; Ph.D., Institute for Superhard Materials (Kiev), 1981. Theoretical solid mechanics.

T. T. Maxwell, Associate Professor; Ph.D., London, 1977. Computational fluid mechanics, aerodynamics, energy conversion, vehicle systems.

J. W. Oler, Associate Professor; Ph.D., Purdue, 1980. Aerodynamics, potential flow, experimental methods.

S. Parameswaran, Associate Professor; Ph.D., London, 1986. Computational fluid dynamics, turbulence modeling.

J. Rasty, Associate Professor; Ph.D., Louisiana State, 1987. Residual stresses, forming processes, nondestructive testing.

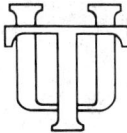

TULANE UNIVERSITY

School of Engineering
Department of Mechanical Engineering

Programs of Study

The Department of Mechanical Engineering offers programs leading to the Master of Science and Doctor of Philosophy degrees (administered by the Graduate School) and to the Master of Science in Engineering and Doctor of Science degrees (administered by the Graduate Division of the School of Engineering).

Master's programs require 24 semester hours of course work plus a thesis, or 30 semester hours of course work plus an exam. Only the first option is open to students who receive financial aid from the University. Students can usually complete a master's program in three semesters of full-time study. In the doctoral programs, which are tailored to suit individual needs and interests, students may specialize in a classical area or pursue studies in an interdisciplinary area. Doctoral programs require at least 48 semester hours beyond the B.S. as well as a dissertation. Doctoral degrees have been earned within three to five years of continuous study beyond the B.S.

Research Facilities

The School of Engineering maintains a variety of versatile precision instruments and several special facilities for research; all are available to faculty members and students in the Department of Mechanical Engineering. The department operates a UNIX network linking the laboratories, including data acquisition equipment that can be linked to research experiments requiring online data reduction, an Experimental Mechanics Laboratory, a Fluid Mechanics Laboratory, the Laboratory for Research in Intelligent Sensors, and the Robotics and Control Laboratory for the study of robotics and computer control. The departmental network is linked to the University computing network and then to various national computing networks. The University computing network includes a cluster of IBM RS/6000s consisting of a Model 540 PowerStation and two Model 340s. The robotics and control laboratory is equipped with a Unimation PUMA 560 industrial robot, an ADEPT industrial robot, a VICOM Digital Image Processor, a robotic MIG welder, and data acquisition equipment. The Laboratory for Research in Intelligent Sensing includes advanced ultrasonic ranging equipment used in applications such as mobile robots and robotic welding. The thermal science laboratory is equipped with an environmental chamber, which is currently being used to grow sea ice; several low-speed tunnels; mobile computer data acquisition units; and a water tunnel for sedimentation experiments.

Financial Aid

Financial support is awarded by the School of Engineering primarily on the basis of academic merit. Assistance is available in the form of tuition waivers and/or stipends, and such assistance usually demands about 15 hours per week of the student's time for participation in teaching or research, depending upon his or her academic load, level, and inclination and on the needs of the department. The University offers aid through various loan programs; it also has a plan for deferred tuition payment.

Cost of Study

Full-time tuition is $22,160 for 1999–2000, and fees are $1154. Annual increases are typically in the range of 3 to 5 percent. A full-time load implies a minimum of 9 semester hours. Tuition and fees are payable at registration. The majority of graduate students receive a full or partial tuition waiver. Other fees pertaining to the thesis or dissertation are described in appropriate University bulletins.

Living and Housing Costs

Apartments for single students (furnished) and for married students with families (unfurnished) are available through the Residential Life Office, but require early application. A one-bedroom unfurnished apartment rents for about $550 per month, including utilities. Privately owned housing may be found at lower prices.

Student Group

Full-time graduate students in mechanical engineering have been drawn from several states and countries. They are supported by teaching assistantships and research grant funds. In addition, there are part-time students who work in local industries. Most graduates of the doctoral programs have teaching positions at engineering schools; some lead corporate research groups.

Location

New Orleans offers a rich variety of cultural and recreational opportunities. The New Orleans Philharmonic–Symphony Orchestra, the New Orleans Opera Association, and the New Orleans Jazz and Heritage Festival schedule major events and performing artists throughout the concert season.

The University and The School

Tulane is a private, nonsectarian university offering a wide range of undergraduate, professional, and graduate courses for men and women. It dates from 1834, when it announced the formation of what is the present School of Medicine. The School of Engineering dates from the 1890s and is one of the oldest in the South. Tulane's policy is to recruit, retain, and promote outstanding students, faculty members, and staff members, regardless of sex, race, color, religion, or national origin. While located in one of the most fashionable neighborhoods and only 30 minutes by streetcar or bus from the central business district, Tulane has all the qualities of a live-away-from-home university, including many dorms and fraternities. Tulane enrolls about 9,000 full-time students, 2,000 part-time students, and 500 medical residents/fellows each year. The School of Engineering, which graduates about 160 students from all programs annually, offers four-year undergraduate programs leading to the B.S. in Engineering. These include curricula in biomedical, chemical, civil and environmental, electrical, and mechanical engineering and in computer science. Corresponding graduate degree programs are staffed by the departments. In 1990, the Regional Center for Global Environmental Change was established within the department and funded by the Department of Energy for the study of environmental problems related to climate changes.

Applying

Applicants with a B.S. from a recognized institution may be admitted to study for a graduate degree in mechanical engineering if their record and personal attributes indicate the ability to pursue advanced study successfully. Applicants may be required to make up undergraduate course deficiencies, for which graduate credit may not be awarded. Every applicant for admission to the Graduate School is expected to take the GRE General Test before acceptance; this requirement may be waived under very exceptional circumstances. Any applicant requesting financial aid must take the GRE. All international students whose native language is not English must show a satisfactory score on the TOEFL before formal acceptance to a degree program. Normally, no application for admission will be accepted after July 15 for the fall semester or after December 15 for the spring semester. Application forms may be obtained by writing to the address below. Students should indicate whether they wish to matriculate through the Graduate School or the Graduate Division of the School of Engineering.

Correspondence and Information

Graduate Advisor
Department of Mechanical Engineering
Tulane University
New Orleans, Louisiana 70118
Telephone: 504-865-5775
World Wide Web: http://www.Tulane.edu/~meche

Tulane University

THE FACULTY AND THEIR RESEARCH

Paul M. Lynch, Professor and Acting Dean; Ph.D., MIT. Control, robotics, intelligent systems, automation and optimal control, vibrations.

J. Fernando Figueroa, Associate Professor; Ph.D., Penn State. Ultrasonic sensing, robotics, control systems.

Michael C. Larson, Associate Professor; Ph.D., MIT. Machine design, three-dimensional analytical and computational fracture mechanics, composite materials, fracture in ceramic-fiber/ceramic-matrix composites.

Calvin Mackie, Assistant Professor; Ph.D., Georgia Tech. Heat transfer, fluid mechanics, hydrodynamic stability, solidification.

Morteza M. Mehrabadi, Professor and Acting Chair; Ph.D., Tulane. Continuum mechanics, constitutive modeling, mechanics of granular materials, micromechanics of sintering in the presence of a liquid phase.

Efstathios E. Michaelides, Professor and Associate Dean for Graduate Studies and Research; Ph.D., Brown. Two-phase flow, heat transfer, flow in membrane pores and porous media, turbulence.

Asher Rubinstein, Associate Professor; Ph.D., Brown. Solid mechanics, fracture mechanics, micromechanical aspects of material failure.

David J. Sailor, Associate Professor and Director, National Institute for Global Environmental Change, South Central region; Ph.D., Berkeley. Heat transfer, fluid mechanics, environmental modeling, energy-climate interactions.

Robert G. Watts, Professor; Ph.D., Purdue. Heat transfer, thermodynamics, fluid mechanics, analytic climatology, environmental analysis.

Research in Fluid and Thermal Sciences

Current research in thermal sciences covers a wide range of topics in heat transfer and fluid dynamics. Research is being conducted into measurement of particle suspension and deposition in boundary layers; transport of ions through membranes and porous media; solidification processes; heat transfer enhancement; turbulence modulation; double diffusion heat, mass, and momentum transfer; freezing and melting of sea ice; the effect of the history terms on the membrane and energy exchanges for particles; wall jets; plumes; flows around spinning objects (i.e., the Magnus effect); heat transfer from extended surfaces; specialized transducer development; paleoclimatology; ocean and lake dynamics; atmospheric modeling; regional climate change modeling; and the interactions between energy systems (supply and demand) and climate change/variability.

The projects described here often involve a high degree of interaction between experimentation, computational simulation, and theoretical analysis. Such projects find applications in a wide variety of areas, including manufacturing, materials processing, electronics cooling, pollutant dispersal, separation processes, and global warming and climate change.

The facilities available for thermal science research in the Department of Mechanical Engineering include several wind tunnels and water channels, an environmental chamber, a jet impingement heat transfer lab, and numerous computer workstations.

Research in Robotics and Control Systems

Robotics and control research is conducted in the areas of robotics, intelligent control, machine vision, ultrasonic sensing, and multisensor control.

Current research is centered on intelligent robotic control systems and intelligent sensing systems. A recent project that is an example of an intelligent control system is a vision-based robotic weld seam tracker, in which a video camera was attached to a robot carrying a welding torch. The camera is used to produce an image of the unwelded weld seam ahead of the torch, which is in turn used to compute the seam location and to guide the robot. This image is analyzed in order to accurately locate the position of the seam. The seam may have discontinuities or abrupt changes in direction. In practical industrial situations, the weld seam is not necessarily well prepared and so the image may be flawed. The intelligence in the system is used to resolve uncertainties in the image of the weld seam. In this way, the intelligence extends the capabilities of a conventional control system.

An example of an intelligent sensing system under development in the laboratory is an advanced ultrasonic navigational system. In this system, a transmitter is located on a moving device, such as a mobile robot. The transmitter emits a series of pulses received by an array of receivers. The time of flight of the pulse to each of the receivers is measured. (Advanced techniques are being tested to use pulse phase information to refine the time of flight measurement.) This information is combined with the known location of the receivers to determine an estimate of the position of the transmitter. Different methods are being investigated for performing this position estimate and assessing its accuracy. Intelligent methods are being used to cope with realities of practical workspaces, such as various objects blocking, or occluding, certain sensors and preventing their function, temperature variations affecting the speed of sound, and air currents and turbulence affecting the quality of the measurements. Here the intelligence acts to improve the quality of the sensor readings and to enhance the ability of the system to combine sensor readings to an improved state estimate.

Research in Solid Mechanics

Current research in the solid mechanics area includes the development of the constitutive theories for materials with microstructure, and fracture mechanics.

In the area of constitutive modeling, a fundamental problem of interest is the study of the influence of microstructure (fabric) and its evolution on the overall strength and failure of granular and other structured media. The results of such a model are applicable to complex phenomena such as flow through porous media, wave propagation through geological materials, powder metallurgy, etc. This study involves both analysis and computer simulation. Another problem in the area of constitutive modeling being studied is the development of an energy-based failure criterion for anisotropic composites subject to damage. The failure criterion being developed in this study is an extension of the (von Mises) distortional energy failure criterion for isotropic materials to anisotropic materials.

The research in the area of fracture mechanics addresses fundamental aspects of materials failure on micro and macro scales. Current projects deal with crack initiation, propagation, and crack path formation in high-tech aerospace materials with special properties. The effectiveness of the fiber reinforcement of ceramic matrix composites and the fracture process in these composites are subjects of recent projects as well. In addition, 3-D computational simulation of (fatigue) fracture of composite materials, especially high-temperature ceramic and metal matrix composites, is actively being pursued. The primary numerical tool for this purpose is the surface integral method, which has proven to be much more efficient than the traditional finite element or boundary element schemes for fracture problems. This computational scheme is also being implemented in a new nondestructive evaluation technique utilizing laser holographic interferometry.

UNIVERSITY OF CALIFORNIA, BERKELEY

Department of Mechanical Engineering

Programs of Study

The department offers programs leading to the M.S., M.Eng., Ph.D., and D.Eng. degrees. Graduate study and research in mechanical engineering are carried out in dynamics and controls, fluid and solid mechanics, materials and design, and thermosciences. Active programs are also under way in bioengineering, energy and resources, and manufacturing and robotics. A partial listing of specialty areas includes acoustics, aerodynamics, automatic control, bioengineering and biomechanics, CAD/CAM, combustion, composite materials, computational fluid dynamics, continuum mechanics, corrosion, cryosurgery, differential game theory, dynamic systems and stability, dynamics, elasticity, energy and resources, environmental restoration, experimental fluid mechanics, expert systems, fatigue, finite-element methods, fire science, fracture, geophysical fluid mechanics, heat transfer, kinematics, laser diagnostic methods, laser materials processing, machine tool control, mass transfer, material processing and cutting, microcomputers, micromechanical systems, nonlinear mechanics, optimal control and optimal design, pollution control, porous media, process control, propulsion, random processes, robotics, rock mechanics, thermal radiation, thermal systems, tool wear, tribology, transport phenomena, turbulence, vibrations, and wave propagation.

About a year and a half are required for the M.S. degree; a total of four to five years is required to complete the Ph.D. Each Ph.D. student has a major and two minor areas of study, one of which lies outside mechanical engineering. A satisfactory score on a preliminary examination and a strong academic record are required to enter the Ph.D. program, and an oral qualifying examination is required at the end of the course work and prior to full-time research. The qualifying examination is taken after two years of course work. Students are encouraged to start research early in their graduate careers.

Research Facilities

The department is equipped with truly exceptional laboratory and computational facilities for research and experimentation. The named laboratories include Automatic Control and Instrumentation, Automation and Robotics, Bioengineering, Combustion, Composite Materials, Computer-aided Engineering, Dynamic Stability, Enhanced Oil Recovery, Environmental Restoration, Experimental Mechanics, Expert Systems, Fire Safety Science, Fluid Mechanics, Heat Transfer and Thermodynamics, Impact and Wave Propagation, Laser Diagnostics, Magnetic Recording and Printing, Manufacturing, Mechanical Behavior of Materials, Subsonic and Supersonic Wind Tunnels, Vibration, and Welding and Welding Automation. In addition, facilities at the Lawrence Berkeley Laboratory; the Richmond Field Station; and the Medical Center of the University of California, San Francisco are available.

Financial Aid

Each year a number of graduate fellowships are awarded to students entering in the fall semester. Fellowships normally provide payment of tuition, fees, and an additional monthly stipend of $1201 in the form of research/teaching assistantships for the academic year. In addition, research/teaching assistantships are available on a competitive basis. Applicants can also contact members of the faculty directly to learn about research and aid possibilities. Graduate student instructors and researchers are paid $2613 per month for full-time employment.

Cost of Study

The fees for the 1999–2000 academic year are $4400. In addition, graduate students who are not California residents must pay a yearly tuition of $9384. Tuition fellowships are available for international doctoral students.

Living and Housing Costs

Off-campus housing costs vary. Graduate apartments rent for $624 to $676 per month. Family housing is available at costs ranging from $5980 to $6838 per year. Students are encouraged to contact the University housing office at 2401 Bowditch Street or call 510-642-3644.

Student Group

There are about 300 graduate students in the department; approximately half are doctoral students, 22 percent are international, and 15 percent are women.

Location

The University is located near the Berkeley Hills, across from Marin County, the Bay, and San Francisco. The Bay Area is known for its cultural diversity, recreational activities, and majestic natural beauty. Berkeley, with its mild weather, cosmopolitan environment, and political activism, is a dynamic and vibrant city with many fine restaurants, a rich cultural life, and a great variety of recreational activities. The Pacific Ocean and numerous beaches are a short drive away, and the Sierra Nevada, with its ski resorts, is a 3-hour drive to the east. The area is regarded by many as one of the most interesting and best places to live.

The University

The Berkeley campus of the University of California has a total enrollment of about 30,000 students; 9,000 are graduate students in more than 100 fields of study. The campus is noted for the academic distinction of its faculty, the quality and scope of its research, and the vitality and variety of its student activities. It has often been ranked as the most outstanding institution for graduate study in the United States.

Applying

All application materials for the 2000 fall semester must be submitted by Thursday, December 23, 1999. In addition, a few highly qualified applicants are admitted for the spring semester. The deadline for the 2001 spring semester is Friday, September 1, 2000. Scores on the GRE General Test (verbal, quantitative, and analytic) are required. GRE Subject Tests are not required. Applicants from countries in which the official language is not English are required to take the Test of English as a Foreign Language (TOEFL) unless the student has been enrolled full-time for one year in a university in the United States. A minimum TOEFL score of 570 is required.

Correspondence and Information

Graduate Matters
Mechanical Engineering
6143 Etcheverry Hall
University of California
Berkeley, California 94720-1742
Telephone: 510-642-5084
World Wide Web: http://www.me.berkeley.edu/

University of California, Berkeley

THE FACULTY AND THEIR RESEARCH

Alice M. Agogino, Professor; Ph.D., Stanford, 1984. Concurrent engineering, CAD/CAM, multiobjective and strategic optimal design, intelligent monitoring, probabilistic design, decision and expert systems, diagnostics and supervisory control.

David M. Auslander, Professor; Sc.D., MIT, 1966. Modeling and simulation of dynamic systems, automatic control system design, mechanical system control, microcomputer systems, bioengineering.

Stanley A. Berger, Professor; Ph.D., Brown, 1959. Theoretical fluid mechanics, physiological fluid mechanics.

David B. Bogy, Professor and Chairman; Ph.D., Brown, 1966. Analytical and numerical solutions of static and dynamic problems in solid and fluid mechanics, mechanics in computer technology.

Van P. Carey, Professor; Ph.D., SUNY at Buffalo, 1981. Thermophysics of multiphase systems, computational modeling of multiphase transport, prediction of transport using molecular simulation models and statistical mechanics.

James Casey, Professor; Ph.D., Berkeley, 1980. Continuum mechanics, plasticity, approximate nonlinear theories of elasticity, dynamics of nearly rigid bodies.

Jyh-Yuan Chen, Professor; Ph.D., Cornell, 1985. Computational modeling of reactive systems, turbulent flows, combustion chemical kinetics.

C. K. Hari Dharan, Professor; Ph.D., Berkeley, 1968. Mechanical behavior, mechanics, design and manufacturing, composite materials and structures, microelectronic packaging, spacecraft design, adaptive materials.

Robert W. Dibble, Professor; Ph.D., Wisconsin, 1975. Laser diagnostics in reactive systems, especially combustion.

David A. Dornfeld, Professor; Ph.D., Wisconsin, 1976. Flexible automation, analysis and control of manufacturing processes (machining and deburring, forming, welding), robotics, intelligent sensors, precision manufacturing.

A. Carlos Fernandez-Pello, Professor; Ph.D., California, San Diego, 1975. Heat and mass transfer effects on combustion, flame propagation, combustion of fuel droplets and sprays, ignition and extinction.

Michael Y. Frenklach, Professor; Ph.D., Hebrew (Jerusalem), 1976. Reaction mechanisms, kinetic and stochastic modeling, combustion chemistry, pollutant formation, chemical vapor deposition of diamond.

Ralph Greif, Professor; Ph.D., Harvard, 1962. Convection, thermal radiation, rotating flows, combustion, chemical vapor deposition, materials processing, phase change, particulate flows, welding, solar collection, reactor heat transfer.

Costas Grigoropoulos, Professor; Ph.D., Columbia, 1986. Heat transfer and fluid flow in laser melting, excimer laser materials processing and micromachining, microthermal property characterization In chip-scale integration.

J. Karl Hedrick, Professor; Ph.D., Stanford, 1971. Control systems, nonlinear systems, vehicle dynamics and control.

Roberto Horowitz, Professor; Ph.D., Berkeley, 1983. Model reference adaptive control, dynamic analysis and control of robotic manipulators, real-time applications of microcomputers, magnetic recording disk file servos.

George C. Johnson, Professor; Ph.D., Stanford, 1979. Ultrasonic nondestructive testing, finite deformation plasticity, effective property evaluation, characterization of thin solid films, micromechanics.

H. Kazerooni, Professor; Sc.D., MIT, 1985. Robotics, mechantronics, control systems, design, automated manufacturing, human-machine systems.

Tony M. Keaveny, Associate Professor; Ph.D., Cornell, 1991. Orthopedic biomechanics, mechanical behavior of bone, design and analysis of total joint prostheses, structural analysis of whole bones.

Kyriakos Komvopoulos, Professor; Ph.D., MIT, 1986. Tribology, contact mechanics, laser surface modification processes, mechanical behavior of materials, metal cutting and tool wear, finite-element methods.

Dennis Lieu, Professor; D.Eng., Berkeley, 1982. Dynamics of high-speed electromechanical devices, vibration and noise control of magnetically and electromagnetically excited structures, DC brushless motors, sports equipment design.

Fai Ma, Professor; Ph.D., Caltech, 1981. Vibration and control.

Arunava Majumdar, Associate Professor; Ph.D., Berkeley, 1989. Thermal and optical microscopy of nanostructures and nanodevices, micromechanical sensors and microelectromechanical systems (MEMS), thermal phenomena in electronic devices and packages, energetics in surface chemistry.

Philip Marcus, Professor; Ph.D., Princeton, 1978. Theoretical fluid mechanics, computational physics, astrophysical and geophysical flows, nonlinear dynamics and transition to chaos and turbulence.

S. Morris, Associate Professor; Ph.D., Johns Hopkins, 1980. Fluid mechanics and heat transfer in planetary interiors.

C. D. Mote Jr., Professor and Vice Chancellor, University Relations; Ph.D., Berkeley, 1963. Dynamic systems, vibration control, dynamic stability, finite element, instrumentation, design, sport mechanics, rotating disc stability, aerodynamic noise and vibration.

Oliver M. O'Reilly, Associate Professor; Ph.D., Cornell, 1990. Dynamics, especially dynamical systems theory, dynamics of constrained rigid bodies, and experimental dynamics of continuous media.

Andrew K. Packard, Professor; Ph.D., Berkeley, 1988. Control analysis and design, linear algebra and numerical algorithms in control problems, system theory and aerospace problems, flight control, fluid flow.

Patrick J. Pagni, Professor; Ph.D., MIT, 1970. Fire physics modeling: analyses of compartment fire development, ignition, smoldering, flame geometries, excess pyrolysate and soot formation.

Panayiotis Papadopoulos, Associate Professor; Ph.D., Berkeley, 1991. Solid mechanics, finite element methods, numerical mathematics.

Albert P. Pisano, Professor; Ph.D., Columbia, 1981. Microelectromechanical systems (MEMS), microsensors and microactuators, design of machinery, design optimization, automobile engine design.

Kameshwar Poolla, Professor; Ph.D., Florida, 1984. Robust multivariable control system synthesis, adaptive feedback systems, time-varying systems, control of flexible structures.

Lisa A. Pruitt, Associate Professor; Ph.D., Brown, 1993. Fatigue and fracture of advanced polymers, composites, and biomaterials; micromechanisms of cyclic damage.

Boris Rubinsky, Professor; Ph.D., MIT, 1980. Heat and mass transfer in biological media, bioprocesses and biotechnology, solidification processes, crystal growth, welding, finite-element methods, fluidized beds, transport phenomena.

Ömer Savas, Professor; Ph.D., Caltech, 1979. Fluid mechanics: rotating flows, turbulent flows, combustion.

Paul S. Sheng, Associate Professor; Ph.D., MIT, 1991. Laser material processing, nontraditional manufacturing processes, design-for-manufacturing, quality planning and control.

David J. Steigmann, Associate Professor; Ph.D., Brown, 1989. Continuum mechanics, shell theory, tensile structures, elastic stability, variational methods, capillary phenomena, mechanics of thin films.

Andrew J. Szeri, Associate Professor; Ph.D., Cornell, 1988. Convective/diffusive transport, nonlinear dynamics, perturbation methods, optimal control, mechanics; bubbles, sonoluminescence, surfactants, air-sea exchange, suspensions, polymer solutions.

Chang-Lin Tien, Professor; Ph.D., Princeton, 1959. Microscale heat transfer, thermal aspects in nonlinear optics, heat transfer in laser-materials interaction, fractal analysis in thermal phenomena.

Masayoshi Tomizuka, Professor; Ph.D., MIT, 1974. Optimal control, preview control, model reference adaptive systems, theory and applications to industrial processes, motion control, manufacturing problems and robotics.

Benson Tongue, Professor; Ph.D., Princeton, 1983. Nonlinear dynamics, vibrations, modal analysis, numerical modeling, acoustics.

Kent S. Udell, Professor; Ph.D., Utah, 1980. Transport in porous media: heat transfer with phase change, multiphase flow, enhanced oil recovery, removal of toxic contaminants from the subsurface, microscale heat transfer.

Paul K. Wright, Professor; Ph.D., Birmingham (England), 1971. Manufacturing automation, flexible manufacturing systems, artificial intelligence, rapid prototyping, mechanics of materials in relation to processing, engineering design.

Ronald W. Yeung, Professor; Ph.D., Berkeley, 1973. Hydrodynamics and hydromechanics, specifically wave-structure interaction, vorticity and free-surface interactions, numerical methods for separated flows and other hydrodynamic and hydromechanical problems, and numerous problems in ship and marine hydrodynamics.

UNIVERSITY OF CALIFORNIA, IRVINE

Department of Mechanical and Aerospace Engineering

Programs of Study

The Department of Mechanical and Aerospace Engineering offers courses leading to the degrees of Master of Science and Doctor of Philosophy in mechanical and aerospace engineering. The areas of emphasis offered at the graduate level include fluid and thermal sciences, combustion and propulsion, and systems and design. Study in fluid and thermal sciences emphasizes computational fluid dynamics and heat transfer, turbulence, atmospheric processes, supersonic shear flows, and non-Newtonian fluid mechanics. Combustion and propulsion includes the study of fuel-air mixing processes, turbulent transport, liquid sprays, and the formation of gaseous and solid pollutants in gas-, liquid-, and coal-fueled combustion systems. Study in systems and design encompasses nonlinear and robust control, parameter identification, computer-aided design, flight guidance and control, and the control and design of robotic systems. The programs are designed for those who plan to enter the professional practice of engineering as it relates to design, development, teaching, and research—in universities, industry, private practice, or public service. Fundamentals of engineering are emphasized so that graduates can continue professional development throughout their careers.

The M.S. degree may be attained by successful completion of 36 approved units and either a research thesis or a comprehensive examination (either including a project or courses only).

After a student has successfully completed an M.S. degree program (at UCI or at a comparable university), the Ph.D. degree requires a passing grade on a preliminary examination, research preparation, a qualifying examination and advancement to candidacy, completion of a significant research project, and the submission and defense of an acceptable dissertation.

Research Facilities

The combustion laboratory includes five high-bay test cells, a gas turbine cell, and support laboratories. One of the test cells is designed specifically for high-pressure experiments, and the second supports the operation of a supersonic tunnel. The test facilities include model and practical burners and gas turbine combustors, ramjet combustors, a coal combustion tunnel, a variety of spray test stands, and a variable geometry reactor for the study of turbulent transport in jet and recirculating flows. Diagnostics include conventional, standard laser, and advanced laser. The turbulence laboratory is equipped for constant temperature laser anemometry and the measurement of high-frequency temperature fluctuations. There are a low-turbulence wind tunnel, a small wind tunnel, and a vertical wind tunnel for the study of particle-turbulence interaction. A supersonic shear-flow facility employs advanced laser diagnostics for investigation of phenomena in high–Mach number turbulence. In the robotics laboratory, researchers use microprocessor-based control hardware to experiment with control components. Test beds, such as hydraulic and pneumatic actuators, DC motors, and mechanical hands, are available for experimentation. High-speed graphics workstations support research in robotics and computational geometry.

Three control laboratories provide computational resources and test beds that support the development of control design methods and applications to aerospace, automotive, and structural systems.

Additional computational resources include the campus Convex C240 supercomputer and access to the San Diego Supercomputing Center as well as a number of sophisticated graphics workstations and peripherals.

Financial Aid

Fellowships and teaching and research assistantships are available on a competitive basis. Except for students on visas, there are opportunities for part-time work in the engineering community of Orange County. With the same exception, financial aid may be obtained from UCI's Financial Aid Office.

Cost of Study

In 1999–2000, student fees are $1726 per quarter for California residents and an additional $3441 per quarter for nonresidents. Fees are subject to change.

Living and Housing Costs

On-campus housing is available. In 1999–2000, monthly apartment rents range from $277 to $725 for single students and from $554 to $1202 for married students and families. Early application is advised for on-campus housing. Privately owned apartments are available close to the campus, and many types of housing can be found in the surrounding communities of Santa Ana, Newport Beach, Costa Mesa, Irvine, Tustin, and Laguna Beach.

Student Group

Current campus enrollment is 18,209, including 1,263 undergraduate and 309 graduate students in the School of Engineering.

Student Outcomes

Graduates of the program move on to faculty positions at research universities, to postdoctoral research opportunities at universities or research laboratories, to teaching posts at colleges, and to positions in local, national, and international high-tech companies.

Location

The 1,510-acre UCI campus is in Orange County, 40 miles south of Los Angeles. Irvine is one of the nation's fastest-growing residential, industrial, and business areas, yet within view of the campus is a wildlife sanctuary; Pacific Ocean beaches are nearby. Residential areas range from the beach communities of Newport Beach and Laguna Beach to the socially and economically diverse urban centers of Santa Ana, Tustin, and Costa Mesa.

The University

One of the nine campuses in the University of California system, UCI now enrolls 3,638 graduate and professional students. The University offers graduate degrees through the Schools of Biological Sciences, Engineering, Fine Arts, Humanities, Physical Sciences, Social Ecology, and Social Sciences; the Graduate School of Management; the College of Medicine; and the Department of Information and Computer Science. The Department of Education offers courses and training leading to California teaching credentials.

Applying

Application forms and general information may be obtained by writing to the graduate coordinator at the Department of Mechanical and Aerospace Engineering. The deadlines for applications are June 1 for the fall quarter, October 15 for the winter quarter, and January 15 for the spring quarter. Applicants who wish to be considered for fellowships or for teaching or research assistantships should apply by February 1. Applicants must submit official records covering all postsecondary academic work, three letters of recommendation, and official scores on the General Test of the Graduate Record Examinations. International students whose native language is not English must submit the results of the Test of English as a Foreign Language (TOEFL).

Correspondence and Information

For information about the department:

Dr. Said Elghobashi, Chair
Department of Mechanical and Aerospace Engineering
University of California
Irvine, California 92697-3975
Telephone: 949-824-8451

For applications:

Dorothy Miles, Graduate Coordinator
Department of Mechanical and Aerospace Engineering
University of California
Irvine, California 92697-3975
Telephone: 949-824-5469
E-mail: djmiles@uci.edu
World Wide Web: http://www.eng.uci.edu/mae/

University of California, Irvine

THE FACULTY AND THEIR RESEARCH

James E. Bobrow, Professor; Ph.D., UCLA. Dynamic systems and control, robotics, fluid power.

Haris J. Catrakis, Assistant Professor; Ph.D., Caltech. Turbulence and mixing at high Reynolds numbers, flow control for aerospace and marine vehicles.

Donald Dabdub, Assistant Professor; Ph.D., Caltech. Mathematical modeling of air pollution dynamics by parallel computation.

Derek Dunn-Rankin, Professor; Ph.D., Berkeley. Combustion, aerosol sizing and transport, laser diagnostics and spectroscopy.

Donald K. Edwards, Professor Emeritus; Ph.D., Berkeley; PE. Heat and mass transfer.

Said E. Elghobashi, Professor and Chair of the Department; Ph.D., Imperial College (London). Direct simulation of turbulent, chemically reacting, and dispersed two-phase flows.

Carl A. Friehe, Professor; Ph.D., Stanford. Fluid mechanics, turbulence, micrometeorology, instrumentation.

Faryar Jabbari, Professor; Ph.D., UCLA. Robust and nonlinear control theory, adaptive parameter identification.

John C. LaRue, Professor and Associate Dean for Student Affairs; Ph.D., California, San Diego. Fluid mechanics; heat transfer; turbulence, instrumentation, and microelectromechanical systems.

Feng Liu, Associate Professor; Ph.D., Princeton. Computational fluid dynamics, aerodynamics and turbo machines.

J. Michael McCarthy, Professor; Ph.D., Stanford. Kinematic theory of spatial motion, design of mechanical systems, cooperating robots.

Kenneth D. Mease, Associate Professor; Ph.D., USC. Air and spacecraft guidance and control, geometric nonlinear control.

Melissa E. Orme, Associate Professor; Ph.D., USC. Droplet dynamics, fluid mechanics of materials synthesis, net-form manufacturing.

Dimitri Papamoschou, Associate Professor; Ph.D., Caltech. Compressible mixing and turbulence, supersonic jet noise reduction, diagnostics for compressible flow, acoustics in moving media, respiratory fluid mechanics.

Roger H. Rangel, Professor; Ph.D., Berkeley. Fluid mechanics and heat transfer of multiphase systems, including spray combustion, atomization, and metal spray solidification; applied mathematics.

David J. Reinkensmeyer, Assistant Professor; Ph.D., Berkeley. Biomedical engineering robotics.

G. Scott Samuelsen, Professor and Director, National Fuel Cell Research Center; Ph.D., Berkeley; PE. Energy, propulsion, combustion, and environmental conflict; turbulent transport in complex flows; spray physics; NO_x and soot formation; laser diagnostics and experimental methods; application of engineering science to practical propulsion and stationary systems; environmental ethics.

William E. Schmitendorf, Professor and Associate Dean for Academic Affairs; Ph.D., Purdue. Control theory and applications.

Athanasios Sideris, Associate Professor; Ph.D., USC. Control systems, neural networks.

William A. Sirignano, Professor; Ph.D., Princeton. Combustion theory and computational methods, multiphase flows, turbulent reacting flows, flame spread.

UNIVERSITY OF CALIFORNIA, LOS ANGELES

School of Engineering and Applied Science
Mechanical and Aerospace Engineering Department

Programs of Study	The Mechanical and Aerospace Engineering (MAE) Department encompasses a diversified program of graduate studies leading to the following degrees: M.S. and Ph.D. in mechanical engineering, M.S. and Ph.D. in aerospace engineering, Master of Engineering in Integrated Manufacturing, and M.S. in manufacturing engineering.
	The Master of Science degree program consists of nine courses of graduate and upper-division work in addition to a comprehensive examination or of seven courses and a thesis. The basic program of study for the Ph.D. is built around a major field and supporting minor fields as applicable. The Ph.D. requires a research dissertation.
	The MAE Department offers a broad spectrum of courses in a number of basic engineering disciplines that are common to all professional fields encompassed by the department. These are (1) aerodynamics, fluid mechanics, acoustics, and combustion; (2) dynamics and control, with applications to aerospace vehicles and mechanical engineering; (3) fusion reactor physics and engineering; (4) manufacturing and design engineering; (5) risk assessment and reliability; (6) structural and solid mechanics, with application to composites, structural optimization, aeroelasticity, wave propagation, nondestructive evaluation, and fracture mechanics; (7) thermal science and engineering; and (8) microelectromechanical systems (MEMS).
Research Facilities	Specialized laboratory facilities are available for research in the areas of combustion, composites, computational fluid dynamics, design and manufacturing, fluid mechanics, fusion, heat transfer, materials degradation, micromanufacturing and microsciences, plasma manufacturing, and thin films. The Nanoelectronics Research Facility has more than 8,000 square feet of clean room with a full line of equipment used for microelectromechanical systems (MEMS) study.
	The School of Engineering and Applied Science operates an advanced computer network of UNIX machines built around an IBM ES/9121 model 480 and a variety of RISC 6000 servers and workstations. This UNIX-based network provides cycle, file services, or general network services for more than 750 workstations and servers of all types. The School also maintains a computer classroom and five open computer labs with approximately 200 workstations (Apple Macintosh II, IBM PS/2, IBM PC/AT–compatible, and X-terminals) where computerized instruction occurs. The campus IBM ES/9000 model 900 supercomputer with vector facility provides AIX/ESA and MVS computing resources. An IBM SPX system is also available. Network access to the campus, regional, and national networks is also provided. Some sponsored research projects also use supercomputers such as CRAY C90 at NSF Supercomputer Centers and NASA. The UCLA library contains more than 6 million books and ranks fourth nationally in the overall quality of its resources. One of its specialized branches is the Engineering and Mathematical Sciences Library, which holds more than 225,000 volumes and receives more than 3,400 periodical and other serial titles. It has 1.9 million technical reports, including publications of the U.S. Department of Energy, NASA, and the National Technical Information Service. The library uses ORION, an online information system.
Financial Aid	Fellowships are available for students devoting full-time to study and research. Fellowships of $16,000 plus fees per year and teaching assistantships starting at $13,329 for half-time teaching assistance during the academic year are awarded. Graduate student research appointments are available at a base salary of approximately $23,000 per year. In addition, financial support is provided through full-time summer research appointments, as well as government and privately sponsored fellowships.
Cost of Study	For 1999–2000, California residents pay $1501.50 per quarter in registration and incidental fees. Nonresidents pay an additional fee of $3441 per quarter. The academic year consists of three quarters.
Living and Housing Costs	A typical budget for a California resident living in an off-campus apartment is approximately $20,000 a year, including required books and supplies, board and room for the period classes are in session during the three quarters, and a minimum allowance for other expenses.
Student Group	UCLA's total enrollment is approximately 34,000. The School of Engineering and Applied Science has approximately 1,000 graduate students and about 2,000 undergraduates. The MAE Department has 700 undergraduate and 200 graduate students.
Location	The greater Los Angeles area is a major site of industrial research, engineering, and business activity, primarily in aerospace, communications, electronics, energy, and other high-technology industries. The area within a 50-mile radius of UCLA comprises the economic equivalent of a country ranked eleventh internationally in terms of gross national product. It is a noted center for the performing arts, and it provides cultural, sport, and recreational facilities to meet varied tastes and budgets.
The University and The School	Ranking academically among the leading universities in the United States, UCLA has attracted distinguished scholars and researchers from all over the world. It shares with eight other campuses the history and prestige of the statewide University of California. The UCLA campus is located in the Westwood area, set against the Santa Monica Mountains, a few miles from the ocean. A large variety of cultural programs and sporting events are presented on the campus.
	The School of Engineering and Applied Science, established in 1968–69, is an outgrowth of the College of Engineering, founded in 1945. The curriculum of the School, covering the basic concepts of science and technology, provides a resource for all UCLA students and makes possible close interaction with the social sciences, humanities, fine arts, and other professional schools. Dr. A. R. Wazzan is dean of the School. The faculty has 140 full-time professors and 42 visiting and part-time professorial appointees. The School offers frequent seminars by noted scientists and engineers from the United States and abroad.
Applying	Applications for admission to graduate programs can be obtained by writing to the department or by calling the telephone number given below. For U.S. citizens and permanent residents, the application deadlines are January 5 (fall quarter), October 1 (winter quarter), and December 31 (spring quarter). Applicants on an F1 or J1 visa may apply for the fall quarter only.
Correspondence and Information	Student Services Office Mechanical and Aerospace Engineering Department 48-121 Engineering IV Box 951597 University of California Los Angeles, California 90095-1597 Telephone: 310-825-7793 E-mail: maeapp@ea.ucla.edu World Wide Web: http://www.mae.ucla.edu

University of California, Los Angeles

THE FACULTY AND THEIR RESEARCH

M. Abdou, Professor; Ph.D., Wisconsin. Fusion, nuclear, and mechanical engineering design, testing, and system analysis; thermomechanics; thermal hydraulics; neutronics; plasma-material interactions; blankets and high heat flux components; experiments, modeling and analysis.

O. Bendiksen, Professor; Ph.D., UCLA. Classical and computational aeroelasticity, structural dynamics, and unsteady aerodynamics.

G. Carman, Associate Professor; Ph.D., Virginia Tech. Electromagnetoelasticity models, fatigue characterization of piezoelectric ceramics, magnetostrictive composites, characterizing shape memory alloys, fiber-optic sensors, design of damage detection systems, micromechanical analysis of composite materials, experimentally evaluating damage in composites.

I. Catton, Professor; Ph.D., UCLA. Heat transfer and fluid mechanics, transfer phenomena in porous media, nucleonics heat transfer and thermal hydraulics, natural and forced convection, thermal/hydrodynamic stability, turbulence.

G. Chen, Associate Professor; Ph.D., Berkeley. Heat transfer in micro- and nanoscale-thinfilms and structures, electronic and photonic devices; thermoelectricity and thermoelectric devices; MEMS sensors, thermal phenomena in semiconductor manufacturing process.

V. Dhir, Professor; Ph.D., Kentucky. Two-phase heat transfer, boiling and condensation, thermal and hydrodynamic stability, thermal hydraulics of nuclear reactors, microgravity heat transfer, soil remediation.

J. Freund, Assistant Professor; Ph.D., Stanford. Aerodynamic sound, compressible turbulence, turbulent mixing, numerical methods.

P. Friedmann, Professor; Sc.D., MIT. Aeroelasticity of helicopters and fixed-wing aircraft, structural dynamics of rotating systems, rotor dynamics, unsteady aerodynamics, active control of structural dynamics, structural optimization with aeroelastic constraints.

N. Ghoniem, Professor; Ph.D., Wisconsin. Mechanical behavior of high-temperature materials, radiation interaction with material (e.g., laser, ions, plasma, electrons, and neutrons), material processing by plasma and beam sources, physics and mechanics of material defects, fusion energy.

J. Gibson, Professor; Ph.D., Texas. Control and identification of dynamical systems. Optimal and adaptive control of distributed systems, including flexible structures and fluid flows. Adaptive filtering, identification, and noise cancellation.

V. Gupta, Professor, Ph.D., MIT. Experimental mechanics, fracture of engineering solids, mechanics of thin films and interfaces, failure mechanisms and characterization of composite materials, ice mechanics.

H. Hahn, Professor; Ph.D., Penn State. Composites design and manufacturing, concurrent engineering, rapid prototyping, automation, mechanical behavior, nondestructive evaluation, smart structures.

C. Ho, Professor; Ph.D., Johns Hopkins. Large-scale integrated MEMS, control of turbulent flows, unsteady aerodynamics, experimental biofluid mechanics, jet and rotor-stator noise.

A. Karagozian, Professor; Ph.D., Caltech. Fluid mechanics of combustion systems with emphasis on numerical simulation and experimental interrogation of combustion control, acoustically driven reacting flows, and high-speed combustion systems.

R. Kelly, Professor; Sc.D., MIT. Thermal convection, thermocapillary convection, stability of shear flows, stratified and rotating flows, interfacial phenomena, microgravity fluid dynamics.

C.-J. Kim, Associate Professor; Ph.D., Berkeley. Microelectromechanical systems (MEMS), micromachining technologies, surface-tension-driven microactuation, microsensors and actuators, microdevices and systems, micromanufacturing, mechanics in microscale.

J. Kim, Professor; Ph.D., Stanford. Turbulence, numerical simulation of transitional and turbulent flows, turbulence and heat transfer control, numerical methods for direct and large-eddy simulations.

A. Lavine, Professor; Ph.D., Berkeley. Heat transfer: thermal aspects of manufacturing processes, plasma processing, shape memory alloys, natural and mixed convection, numerical heat transfer, microscale heat transfer.

A. Mal, Professor; Ph.D., Calcutta. Mechanics of solids, fractures and failure, wave propagation, nondestructive evaluation, composite materials.

R. M'Closkey, Assistant Professor; Ph.D., Caltech. Nonlinear control theory and design with application to mechanical and aerospace systems, real-time implementation.

W. Meecham, Professor; Ph.D., Michigan. Turbulence theory, aircraft noise, community noise.

A. Mills, Professor; Ph.D., Berkeley. Convective heat and mass transfer, condensation heat transfer, turbulent flows, ablation and transpiration cooling, perforated plate heat exchangers.

D. Mingori, Professor; Ph.D., Stanford. Dynamics and control, stability theory, nonlinear methods, applications to space and ground vehicles.

D. Okrent, Professor Emeritus; Ph.D., Harvard. Fast reactors, reactor physics, nuclear fuel element behavior, risk-benefit studies, nuclear safety, fusion reactor technology, societal risk.

J. Shamma, Professor; Ph.D., MIT. Feedback control theory and design with applications to mechanical, aerospace, and manufacturing systems.

Z. Shiller, Assistant Professor; Sc.D., MIT. Robotics, motion control of articulated machines, design of intelligent machines, dynamics and ground vehicles.

O. Smith, Professor; Ph.D., Berkeley. Combustion and combustion-generated air pollutants, hydrodynamics and chemical kinetics of combustion systems, semiconductor chemical vapor deposition.

J. Speyer, Professor; Ph.D., Harvard. Stochastic and deterministic optimal control and estimation with application to aerospace systems; guidance, flight control, and flight mechanics.

T.-C. Tsao, Professor; Berkeley. Modeling and control of dynamic systems with applications in mechanical systems, manufacturing processes, automotive systems and energy systems; digital control, repetitive and learning control, adaptive and optimal control; mechatronics.

D. Yang, Professor; Ph.D., Rutgers. Robotics and mechanisms; CAD/CAM systems, computer-controlled machines.

X. Zhong, Associate Professor; Ph.D., Stanford. Computational fluid dynamics, hypersonic flow, rarefied gas dynamics, numerical simulation of transient hypersonic flow with nonequilibrium real gas effects, and instability of hypersonic boundary layers.

UNIVERSITY OF CALIFORNIA, SAN DIEGO

Department of Mechanical and Aerospace Engineering

Programs of Study	The Department of Mechanical and Aerospace Engineering offers graduate instruction leading to the M.S. and Ph.D. degrees in engineering sciences with specializations in each of the following areas: aerospace engineering, applied mechanics, applied ocean sciences, chemical engineering, engineering physics, environmental engineering, dynamics and control, and mechanical engineering.
	The M.S. program extends and broadens an undergraduate background and equips practicing engineers with fundamental knowledge in their fields. The degree may be terminal or obtained on the way to earning the Ph.D. Two plans of study are offered, both requiring successful completion of 48 quarter units of credit: Plan I is a combination of course work and research, culminating in the preparation of a thesis; Plan II involves course work only and requires a comprehensive examination.
	The Ph.D. program prepares students for careers in research and/or teaching in their area of specialization. There are no formal course requirements; however, all students, in conjunction with their advisers, develop course programs that prepare them for the departmental Ph.D. qualifying examination, which tests students' capabilities in four areas of specialization and ascertains their potential for independent study, dissertation research, and completion and defense of that research.
	Research in fluid mechanics, combustion, and engineering physics encompasses a broad spectrum of problems in aerodynamics, ocean-related flows, turbulence, reacting flows, flow at low Reynolds numbers, mechanics of suspensions, dynamics of drops and cells, mixing, and chaos. The research is relevant to a variety of engineering disciplines including the design of airplanes and automobiles, aerospace guidance and control, prediction of the global climate, computation of the flow of suspensions, and studies of flow over magnetic tapes and disks. Research areas consist of a combination of experimental, theoretical, and computational programs addressing turbulent flows, mechanics of two-phase flow, laminar and turbulent combustion, theories in stirring and mixing, mechanics of drops and biological cells, flow in porous media, fluid-structure interaction, geophysical flows, propellent combustion, chaotic motion, microgravity flows, chemical kinetics of combustion systems, vortex dynamics, flow instabilities, and mechanics of flight both in the atmosphere and above, i.e., astrodynamics. Research in geophysical fluid mechanics, computation fluid dynamics, and mechanics of two-phase flows is conducted in the Fluid Mechanics Group. Research in the area of combustion is based at the Center for Energy and Combustion Research, directed by F. Williams.
	Research in solid mechanics and materials science is aimed at the development of advanced materials and the improvement of their processing. Basic knowledge of solid mechanics and materials science is applied to microstructurally tailor advanced materials for desired structural applications and to understand the failure of materials under static, repetitive, and dynamic loads as well as in hostile service environments. Research programs include interfacial properties and failure in composites, granular materials, computational methods for materials synthesis and compaction, and residual stresses in biological materials.
	The materials science program is offered as an interdepartmental program with specific admissions and academic requirements. MAE faculty members play an important role in the program, directed by Professor S. Nemat-Nasser. This program was established in 1988 and has a current enrollment of more than 30 graduate students.
Research Facilities	Among the research laboratories and equipment within the department are three wind tunnels and two water tunnels for boundary-layer and low-speed aerodynamics studies; a two-phase flow facility; a stratified flow channel facility; a counterflow spray combustion apparatus; spectroscopic equipment for high-temperature gasdynamics; potentiostats for electronic studies; a Rheometrics fluid rheometer; a molecular beam facility; ESCA/XPS and FT-IR surface spectroscopies; universal material testing machines; a dynamic shear soil testing apparatus; triaxial soil testing facilities; a facility to study photoelasticity and micromechanics of granular materials; a facility in the Center for Magnetic Recording Research to study mechanical and materials aspects of computer technology and magnetic disk and tape storage devices; several high-speed microcinematographic facilities; and two sets of major electron microscope facilities on the UCSD campus.
	The department maintains several CAD/CAM and computational fluid dynamics laboratories. Campus computing facilities include the CRAY C90 supercomputer at the San Diego Supercomputer Center.
Financial Aid	Financial aid is available in the form of fellowships, traineeships, teaching assistantships, and research assistantships. The department attempts to support all full-time graduate students, especially at the Ph.D. level. Award of financial support is competitive, and stipends range from $5000 to a maximum of $20,000 for the academic year, plus tuition and fees. While several types of support are available, the most common is a half-time research assistantship that provides about $16,140 during the twelve-month academic year plus tuition and fees. Funds for the support of international students are extremely limited, and selection is highly competitive.
Cost of Study	In 1999–2000, full-time students who are California residents pay an estimated $1630 per quarter in registration and incidental fees. Nonresidents pay an estimated total of $5100 per quarter for registration, tuition, and incidental fees. There is a reduced-fee structure for students enrolled on a half-time basis. Costs are subject to change.
Living and Housing Costs	UCSD provides 1,200 apartments for graduate students. Current monthly rates range from $320 for a single student in a shared apartment to $670 for a family. There is also a variety of off-campus housing in the surrounding communities. Prevailing rates range from $400 per month for a room in a private home to $950 or more for a two-bedroom apartment. Information in this regard may be obtained from the UCSD Housing Office.
Student Group	Current campus enrollment is about 18,825; of this number, 2,200 are graduate students. MAE has an undergraduate enrollment of about 500 and a graduate enrollment of about 110.
Location	The 1,200-acre campus spreads from the seashore, where the Scripps Institution of Oceanography is located, across a large wooded portion of the Torrey Pines Mesa overlooking the Pacific Ocean. To the east and north lie mountains, and to the south are Mexico and the almost uninhabited seacoast of Baja California.
The University	One of nine campuses in the University of California System, UCSD comprises the General Campus, the School of Medicine, and the Scripps Institution of Oceanography. Established in La Jolla in 1960, it is one of the newer campuses but has become a major university, with particular strengths in the physical and biological sciences.
Applying	A minimum GPA of 3.0 (on a 4.0 scale) is required for admission. All applicants are required to take the GRE General Test. International applicants whose native language is not English are required to take the TOEFL and obtain a minimum score of 550. In addition to test scores, applicants submit a completed Graduate Admission and Award Application, all official transcripts (English translation must accompany official transcripts written in other languages), a statement of purpose, and three letters of recommendation. The deadline for filing applications for international applicants and those requesting financial assistance is January 31. The deadline for domestic applicants not requesting financial assistance is May 31.
Correspondence and Information	Department of Mechanical and Aerospace Engineering, 0413 Irwin and Joan Jacobs School of Engineering University of California, San Diego La Jolla, California 92093-0413 Telephone: 858-534-4387 Fax: 858-534-1730 World Wide Web: http://www-mae.ucsd.edu/

University of California, San Diego

THE FACULTY AND THEIR RESEARCH

D. J. Benson, Associate Professor of Computational Mechanics; Ph.D., Michigan. Computational mechanics.

T. Bewley, Acting Assistant Professor of Mechanical Engineering; Ph.D., Stanford. Control and optimization of multiscale distributed systems such as fluid turbulence.

R. Bitmead, Professor of Electrical Engineering (Control Systems); Ph.D., Newcastle (Australia). Theory and applications of feedback control system design, modeling dynamical processes from data and the interaction between these two in adaptive systems for control and signal processing.

H. Bradner, Professor Emeritus of Engineering Physics and Geophysics; Ph.D., Caltech. Systems/controls, neutrino astronomy, genetic relationships in *Cypraea* mollusks.

R. J. Cattolica, Professor of Engineering Physics; Ph.D., Berkeley. Fluid mechanics, combustion, optical diagnostics, applied spectroscopy.

C. P. Caulfield, Assistant Professor of Applied Mathematics and Fluid Mechanics; Ph.D., Cambridge. Fluids flows of environmental, geophysical, or industrial relevance, typically where density variations play a significant dynamical role.

P. C. Chau, Professor of Chemical Engineering; Ph.D., Princeton. Catalysis, biochemical engineering.

R. W. Conn, Professor of Applied Mechanics and Engineering Sciences; Ph.D., Caltech. Plasma-material interactions, plasma physics, fusion technology and reactor design.

R. de Callafon, Assistant Professor of Mechanical Engineering; Ph.D., Delft University of Technology (Netherlands). Experimental dynamic modeling using system identification techniques, use of identified models in robust control design.

C. H. Gibson, Professor of Engineering Physics and Oceanography; Ph.D., Stanford. Turbulent flows, mixing, and diffusion in oceanography, astrophysics, combustion, and chemical engineering.

J. D. Goddard, Professor of Chemical Engineering; Ph.D., Berkeley. Fluid mechanics, transport processes, granular flows.

R. K. Herz, Associate Professor of Chemical Engineering; Ph.D., Berkeley. Heterogeneous catalysis, chemical reaction engineering.

A. Hoger, Associate Professor of Engineering Science; Ph.D., Illinois at Urbana-Champaign. Continuum mechanics, constitutive theory, residual stress, mechanics of biological tissues.

S. Krasheninnikov, Professor of Plasma Physics; Ph.D., Kurchatov Institute of Atomic Energy (Moscow). Physics of transport processes in low and high temperature plasmas.

M. Krstic, Associate Professor of Dynamics and Control; Ph.D., California, Santa Barbara. Nonlinear, adaptive, and robust theory and applications to aircraft engines, helicopters, high-angle-of-attack flight, satellites, and automotive systems.

J. C. Lasheras, Professor of Fluid Mechanics; Ph.D., Princeton. Turbulence, two-phase flows, combustion.

P. A. Libby, Professor Emeritus of Fluid Mechanics; Ph.D., Polytechnic of Brooklyn. Turbulence, combustion.

S.-C. Lin, Professor Emeritus of Engineering Physics; Ph.D., Cornell. High-temperature gasdynamics, radiation physics.

P. Linden, Professor of Environmental Engineering; Ph.D., Cambridge (England). Fluid mechanics.

S. G. Llewellyn-Smith, Assistant Professor of Applied Mathematics; Ph.D., Cambridge. Continuum mechanics, fluid dynamics, applied and industrial mathematics.

X. Markenscoff, Professor of Applied Mechanics; Ph.D., Princeton. Micromechanics, dislocation theory, composite materials, singular asymptotics.

J. M. McKittrick, Associate Professor of Materials Science; Ph.D., MIT. Experimental materials science, processing of ceramic materials.

M. A. Meyers, Professor of Materials Science; Ph.D., Denver. Dynamic behavior and processing of materials.

S. Middleman, Professor Emeritus of Chemical Engineering; D.Eng., Johns Hopkins. Applied fluid dynamics.

J. W. Miles, Professor Emeritus of Applied Mechanics and Geophysics; Ph.D., Caltech. Nonlinear waves and chaotic motions.

D. R. Miller, Professor of Chemical Engineering; Ph.D., Princeton. Gas-surface interactions, gasdynamics.

H. Murakami, Professor of Applied Mechanics; Ph.D., California, San Diego. Development of continuum models for heterogeneous composite materials, constitutive modeling of failure and damage, physically based virtual reality by using FEM.

W. Nachbar, Professor Emeritus of Applied Mechanics; Ph.D., Brown. Finite deformation and buckling of shell structures, strength and stability of structures under seismic excitation, elastic waves.

S. Nemat-Nasser, Professor of Solid and Structural Mechanics; Ph.D., Berkeley. Micromechanics; experimental/computational/analytical modeling of flow, fracture, failure modes, and constitutive relations of solids, including advanced composites, structural ceramics, metals, and geomaterials; earthquake analysis and ground failure.

V. F. Nesterenko, Associate Professor of Materials Science; Ph.D., Russian Academy of Sciences. Mechanics of dynamic processing of materials, nonlinear wave dynamics, shear instability in heterogeneous and reactive condensed materials.

K. K. Nomura, Assistant Professor of Fluid Mechanics; Ph.D., California, Irvine. Computational fluid mechanics, turbulence, combustion.

D. B. Olfe, Professor Emeritus of Engineering Physics; Ph.D., Caltech. Fluid mechanics, heat transfer.

S. S. Penner, Professor Emeritus of Engineering Physics; Ph.D., Wisconsin. Energy and combustion research, environmental issues.

C. Pozrikidis, Professor of Fluid Mechanics; Ph.D., Illinois at Urbana-Champaign. Fluid mechanics, applied mathematics.

S. Rand, Professor Emeritus of Engineering Physics; Ph.D., Berkeley. Dynamics of ionized gases, fundamentals of mechanics.

S. Sarkar, Associate Professor of Fluid Mechanics; Ph.D., Cornell. Computational fluid mechanics, simulation and modeling of turbulent flows.

A. M. Schneider, Professor Emeritus of Engineering Sciences; Sc.D., MIT. Navigation, guidance, and control of space- and earth-bound vehicles; modeling and control of physiological systems.

K. Seshadri, Professor of Chemical Engineering and Fluid Mechanics; Ph.D., California, San Diego. Combustion, fluid mechanics, applied mathematics.

R. E. Skelton, Professor of Engineering; Ph.D., UCLA. Dynamics and control, systems design, modeling, estimation and control of flexible structures.

H. W. Sorenson, Professor Emeritus of Engineering Sciences; Ph.D., UCLA. Nonlinear estimation, signal processing.

J. B. Talbot, Professor of Chemical Engineering; Ph.D., Minnesota. Corrosion, electrodeposition, electrophoretic deposition, electrochemical transport phenomena.

F. E. Talke, Professor in the Center for Magnetic Recording Research; Ph.D., Berkeley. Design of computer peripheral equipment, problems related to magnetic recording technology, tribology of head/disk and head/tape interface, ink-jet printing technology, mechanics and materials.

G. Tynan, Assistant Professor of Applied Plasma Physics; Ph.D., UCLA. Studies of ionized gases (plasmas) with applications to controlled fusion and to nanoscale manufacturing.

C. W. Van Atta, Professor Emeritus of Engineering Physics and Oceanography; Ph.D., Caltech. Geophysical fluid dynamics, turbulent flows.

K. S. Vecchio, Associate Professor of Materials Science; Ph.D., Lehigh. Electron microscopy of engineering materials, structure-property relations in materials at high rate, shock effects, fatigue and fracture.

F. A. Williams, Professor of Engineering Physics and Combustion; Ph.D., Caltech. Flame theory, combustion and turbulent flows, fire research, and other areas of combustion.

UNIVERSITY OF CENTRAL FLORIDA

Department of Mechanical, Materials, and Aerospace Engineering

Programs of Study

The Department of Mechanical, Materials, and Aerospace Engineering at the University of Central Florida (UCF) offers programs leading to master's and doctoral degrees, which provide graduate education in four options: Aerospace Systems, Materials Science and Engineering, Mechanical Systems, and Thermofluids. Areas covered by advanced courses include aerodynamics, astrodynamics, flight dynamics, materials characterization, X-ray diffraction and crystallography, corrosion, metallurgy, fractography, fracture mechanics, composite materials, tribology, experimental mechanics, CAD/CAM/FEM, rapid prototyping, vibrations, controls, finite elements, boundary elements, laser materials processing, advanced heat transfer, turbomachinery, gas kinetics and combustion, two-phase flow, and computational methods in fluid mechanics and heat transfer.

The master's degree programs are intended to integrate and extend the undergraduate training. The thesis option requires a 6-credit thesis and 24 credits of approved course work. The nonthesis option requires 36 credits of approved course work.

The doctoral program is intended primarily for students seeking a career in teaching or research. The Ph.D. degree recognizes excellent academic preparation and scholarly research. A doctoral student must pass a qualifying examination early in the program and later must pass a candidacy examination, including successful defense of a dissertation research proposal. The Ph.D. degree requires 81 credits, of which up to 30 credits may be applied from the master's program. It also requires that the candidate present and successfully defend a dissertation.

Research Facilities

Extensive laboratory facilities are available to support research. Two wind tunnels, laser-Doppler velocimeters, and other equipment support combustion, aerodynamics, and astrodynamics research. The Experimental Mechanics Laboratory contains laser vibrometer and modal testing equipment; holographic, laser speckle, polariscope and acoustic emission equipment for nondestructive evaluation; and various strain measurement equipment. The Dynamics and Controls Laboratory features several computer-driven actively controlled vibrational systems and several shakers. The Laser Machining and Materials Processing Laboratory features several powerful lasers, a five-axis computer-aided positioning system, and a computer vision system. The Materials Laboratory contains SEM, EDS, TEM, AES, mechanical test equipment, and electrochemical test systems. The materials characterization facility includes TEM, SEM, AES, SIMS, EPMA, and RBS. The Tribology Laboratory features a variety of wear- and contact-measurement equipment, including a UCF-developed chemomechanical wear test apparatus. The department maintains a CAD/CAM Laboratory containing a number of computer workstations and an SLA stereolithography machine for rapid prototyping. This laboratory is network-connected to other computers at UCF and elsewhere, including an Intel Paragon. The Electronic Cooling and Thermal Management Laboratory houses a phase-Doppler anemometer, advanced data acquisition systems, and cryogenic and liquid spray facilities.

Financial Aid

A number of teaching and research assistantships, fellowships, and scholarships are available on a competitive basis for students showing promise of good performance in the academic program and in assigned duties. In addition to a stipend, supported students receive the maximum available waiver of tuition and fees. Recipients are expected to devote up to 20 hours per week to assigned teaching or research duties. There are also opportunities for employment throughout UCF, in the research institutes affiliated with UCF, and in nearby industry.

Cost of Study

In 1998–99 tuition per credit was $136.89 for Florida residents and $480.45 for nonresidents.

Living and Housing Costs

UCF is located in Orlando, Florida, which enjoys relatively low living costs. Modestly priced apartments are available in the area. Residence hall accommodations cost from $925–$1480 per semester. Dining hall (board) plans are available at a cost of $800–$1050 per semester.

Student Group

In 1998, there were approximately 300 undergraduates in mechanical engineering and 100 in aerospace engineering. There were approximately 80 master's degree students and 24 Ph.D. degree students.

Student Outcomes

Graduates have obtained faculty positions at colleges and universities including Tuskegee University, the University of Southwestern Louisiana, and American University in Beirut. Other graduates have been employed as mechanical engineers by Siemens–Westinghouse, Lucent Technologies, Lockheed-Martin, and Walt Disney and as aerospace engineers at the NASA Kennedy Space Center.

Location

Orlando, Florida, one of the most rapidly growing urban areas in the United States, enjoys pleasant weather, low living costs, and a high quality of life. Central Florida, featuring the Kennedy Space Center, is one of the major high-technology areas in the United States.

The University

The University of Central Florida, which opened in 1968, is a comprehensive state-supported university with a strong emphasis on engineering. It has grown rapidly at both the undergraduate and the graduate levels and now has an enrollment of 26,000 students. It is near the Kennedy Space Center, the Central Florida Research Park, Westinghouse, Lockheed-Martin, and other major engineering centers. It is the home of several prominent research institutes, including the Center for Research and Education in Optics and Lasers, the Institute for Simulation and Training, the Florida Solar Energy Center, the Advanced Materials Processing and Characterization Center, and the Florida Space Institute. Its undergraduate degrees in aerospace engineering and mechanical engineering are accredited by the Accreditation Board for Engineering and Technology.

Applying

The application for admission, application fee, official transcripts, GRE scores, and TOEFL scores, if applicable, should be sent to the Graduate Program Coordinator of the department. Information regarding financial aid may be obtained from the Graduate Program Coordinator.

Correspondence and Information

Dr. Alain J. Kassab
Associate Professor and Graduate Program Coordinator
Department of Mechanical, Materials, and Aerospace Engineering
University of Central Florida
P.O. Box 162450
Orlando, Florida 32816-2450
Telephone: 407-823-5778
Fax: 407-823-0208
World Wide Web: http://www-mae.engr.ucf.edu

University of Central Florida

THE FACULTY AND THEIR RESEARCH

P. J. Bishop, Professor; Ph.D., Purdue, 1976. Heat transfer, laser-materials interactions.

R. H. Chen, Associate Professor; Ph.D., Michigan, 1988. Combustion, propulsion, fluid mechanics, experimental methods, aeroacoustics, electronic cooling.

L. Chew, Associate Professor; Ph.D., Washington (Seattle), 1990. Experimental aerodynamics, measurement.

L. C. Chow, Professor and Department Chair; Ph.D., Berkeley, 1978. Heat transfer, electronic packaging.

V. H. Desai, Professor; Ph.D., Johns Hopkins, 1984. High-temperature materials, failure analysis, corrosion, chemomechanical effects.

B. E. Eno, Professor; Ph.D., Cornell, 1971. Fluid mechanics, energy conversion, HVAC.

R. Evan-Iwanoski, Research Professor; Ph.D., Cornell, 1954. Dynamics, nonstationary vibrations, chaos.

L. A. Giannuzzi, Associate Professor; Ph.D., Penn State, 1992. Structure-property relationships in materials, electron microscopy, phase transformations, electronic materials.

A. H. Hagedoorn, Associate Professor; Ph.D., Cornell, 1973. Finite elements, structures, computer-aided design.

E. R. Hosler, Professor; Ph.D., Illinois, 1961. Two-phase flow, power cycles, flow visualization.

R. W. Johnson, Visiting Associate Professor; Ph.D., UCLA, 1966. Control systems, astrodynamics, applied mathematics.

J. S. Kapat, Assistant Professor; Sc.D., MIT, 1991. Fluid mechanics and heat transfer, turbulence and transition, gas turbines, transport of flow in material processing, MEMS for flow control.

A. Kassab, Associate Professor; Ph.D., Florida, 1989. Computational fluid mechanics and heat transfer, boundary element method, inverse problems.

K. Lin, Associate Professor; Ph.D., Michigan, 1990. Dynamics and controls, flight simulation, distributed interactive simulation.

J. D. McBrayer, Professor; D.Sc., Washington (St. Louis), 1968. Experimental aerodynamics, fluid mechanics, heat transfer.

A. Minardi, Associate Professor; Ph.D., Central Florida, 1991. Heat transfer, laser machining, energy conservation.

F. A. Moslehy, Professor; Ph.D., South Carolina, 1980. Applied and experimental mechanics, vibrations, NDE boundary element method.

J. Nayfeh, Associate Professor; Ph.D., Virginia Tech, 1990. Dynamics, vibrations, composite structures, CAD/CAM/FEM, rapid prototyping.

D. W. Nicholson, Professor; Ph.D., Yale, 1971. Finite elements, fracture mechanics, dynamics.

C. E. Nuckolls, Associate Professor; Ph.D., Oklahoma, 1970. Mechanisms, design, vehicle dynamics.

S. Seal, Assistant Professor; Ph.D., Wisconsin, 1996. Surface science, methods and instrumentation, high-temperature corrosion, coatings, nitrides, thin films, biomaterials, nanoparticles.

W. F. Smith, Professor; Sc.D., MIT, 1968. Materials science and engineering, physical metallurgy, electronic materials.

G. Ventre, Associate Professor and Director of Florida Solar Energy Center Photovoltaic Projects Office; Ph.D., Cincinnati, 1968. Space systems, orbital mechanics, photovoltaics.

D. Zhou, Assistant Professor; Ph.D., Arizona, 1995. Diamond films, carbon nanotubes, nanowhiskers, electron field emission, plasma processing, electrochemical processing, microstructural characterizations.

Joint Appointments

K. Belfield, Associate Professor; Ph.D., Syracuse, 1988. Polymer synthesis and characterization.

K. A. Cerqua-Richardson, Assistant Professor; Ph.D., Alfred, 1992. Ceramics, glass science.

B. Chai, Professor, Yale, 1975. Solid-state physics, single crystal growth.

M. B. Chopra, Associate Professor; Ph.D., SUNY at Buffalo, 1992. Thermoelasticity, boundary elements.

L. Debnath, Professor; Ph.D., London, 1967. Mathematical hydrodynamics, mathematical heat transfer.

N. G. Dhere, Research Professor; Ph.D., Poona (India), 1965. Thin films, solar cells, high-temperature superconductors.

A. Kar, Assistant Professor; Ph.D., Illinois at Urbana-Champaign, 1985. Laser-aided manufacturing, materials processing.

D. Malocha, Professor; Ph.D., Illinois, 1977. Solid-state devices, acoustoelectronics.

K. Vajravelu, Associate Professor; Ph.D., Indian Institute of Technology (Kharagpur), 1979. Mathematical hydrodynamics, mathematical heat transfer.

AREAS OF RESEARCH INTEREST

Aerospace Systems. Experimental aerodynamics, aeroacoustics, flight dynamics and simulation, propulsion, supersonic and subsonic combustion, optimal control of space vehicles, aerospace design, astrodynamics.

Computational Methods. Finite-element methods, boundary-element methods, finite-volume methods, finite-difference methods, automatic mesh generation, code development, coupling of finite difference and BEM for mixed fields.

Dynamics and Control. Modal analysis; adaptive, optimal, and launch control; unrestrained systems; flexible mechanisms; nonstationary oscillations; chaos; robotics.

Fracture. Stress corrosion cracking, experimental methods, surface analysis, fractography, finite elements in fracture, cracks in dynamic stress fields.

Laser Applications. Computer-aided laser machining, laser-materials interactions, laser-Doppler anemometry, laser-based measurements of strain and vibration, laser-aided manufacturing and materials processing, modeling of heat transfer interactions during laser material processing.

Materials Science and Engineering. Materials characterization, electronic materials, thin films, high-temperature materials, biomaterials, failure analysis, corrosion and electrochemistry, single crystals, photovoltaic materials, photonic crystals, glass and glass–ceramic materials for optical applications, phase transformations, diffusion, wear, laser materials processing and modeling.

Solid Mechanics. CAD/CAM/FEM, design, applied and experimental mechanics, nondestructive testing, vibration and modal analysis, constitutive modeling, design of machine components, inverse problems.

Thermofluids. Turbomachinery, miniaturization of thermal systems, two-phase flow, flow visualization, microelectronic packaging, photovoltaics, gas radiation heat transfer, conjugate heat transfer, inverse problems in heat transfer, combustion.

UNIVERSITY OF DELAWARE

Department of Mechanical Engineering

Programs of Study	The Department of Mechanical Engineering offers graduate programs leading to the degrees of Master of Science in Mechanical Engineering, Master of Engineering: Mechanical, and Doctor of Philosophy in mechanical engineering.
	The Master of Science in Mechanical Engineering degree requires a minimum of 24 credit hours of course work beyond the bachelor's degree and a thesis equivalent to 6 credit hours. The Master of Engineering: Mechanical degree requires the completion of 30 credit hours of course work beyond the bachelor's degree and does not require a thesis. Courses for both degrees include applied mathematics, engineering analysis, solid and fluid mechanics, dynamics and control, and materials science.
	The doctoral program in mechanical engineering allows considerable flexibility in setting up a plan of study that best suits the student's individual needs and interests. It is possible to pursue the Ph.D. degree directly after a bachelor's degree.
	The research of the mechanical engineering department is focused in six areas: composites, air pollution, biomedical engineering, manufacturing science, robotics and control, and advanced materials. The department faculty members are also involved in cross-disciplinary research programs. The Center for Composite Materials' work includes the mechanics and manufacture of advanced composite materials for the study of smart structures. Air pollution research involves high-performance computing techniques and advanced instrumentation to study particulate air pollutants and transport phenomena in combustion. The Center for Biomedical Engineering Research provides a framework for interdisciplinary research in the general area of bioengineering. Topics include the generation of force and motion in the human body, orthopedic and rehabilitation engineering, and the study of pulmonary and renal fluid mechanics. Manufacturing research is concerned with all aspects of flow in thin liquid films, including paints and other coatings; the behavior of fibers in concentrated suspensions; resin transfer mold-filling processes in composites manufacturing; rapid tooling; and lubrication and cooling during machining. Current research areas in robotics and control are design of novel robotic systems, coordination and control of multi-degree-of-freedom robot systems, smart materials and intelligent structures, and optimization of dynamic manufacturing processes. Advanced materials engineering is concerned with characterization and modeling of engineering materials, including polymer, metal, and ceramic matrix composites and high-strain-rate deformation, and split Hopkinson pressure bars for high-strain-rate testing.
Research Facilities	The department is housed in the Robert L. Spencer Laboratory, which contains modern facilities for a wide range of experimental programs. Among the facilities are a three-dimensional laser-Doppler velocimeter, scanning and transmission electron microscopes, mechanical- and ballistic-impact-testing systems, a rotating-mirror high-speed camera, a high-speed infrared thermographic camera, tension and compression split Hopkinson (Kolsky) bars, robots, a pilot-scale rotary coating machine, a 3-D printer, and extensive research-grade electronic instrumentation. A fully staffed and equipped machine shop with a CNC lathe and miller support the research programs. The Center for Composite Materials offers state-of-the-art experimental facilities and laboratories for research in polymer, metal, and ceramic matrix composites manufacturing and in nondestructive evaluation. The Applied Controls Laboratory is involved in projects related to precision forming and design of trim panels for aircraft interior noise suppression. The Mechanical Systems Laboratory houses active research in the areas of robotics, real-time planning and optimization of dynamic systems, and rehabilitation robotics. The Biomedical Engineering Laboratory is actively involved in musculoskeletal modeling and the study of rheumatoid arthritis. The Solid Mechanics Lab studies the impact response of solids using high-speed photography and infrared thermography. The Fluids Processing Laboratory is involved with manufacturing applications. The Air Pollution Laboratory measures and models atmospheric aerosols related to global warming and urban smog. The Turbulence and Multiphase Flow Lab conducts direct and large-eddy simulations of turbulence and particle-laden flows on in-house SGI/Sun UNIX computers and remote supercomputers. The Combustion Laboratory carries out a variety of experimental and computational work in the fields of combustion, combustion-generated pollutants, and high-temperature chemical kinetics. The Particle Image Velocimetry Laboratory develops nonintrusive imaging techniques, which are then applied to fundamental and applied problems in fluid mechanics and heat transfer.
Financial Aid	Stipends for graduate assistantships are approximately $1340 per month for the 1999–2000 academic year, plus tuition. Research assistantships, involving participation in a research project, can offer a larger annual stipend, depending on available funds. University Fellowships and special Minority Fellowships are available.
Cost of Study	Tuition and fees in 1998–99 were $4250 per year for Delaware residents and $12,250 per year for nonresidents.
Living and Housing Costs	Average annual room costs in the graduate apartment complex were $5000 in 1998–99. University-owned apartments and numerous privately owned apartments, both furnished and unfurnished, are available. Early inquiry is advised.
Student Group	More than 17,000 students are enrolled at the University of Delaware, 1,300 of whom are in the College of Engineering. The mechanical engineering department enrolls about 300 undergraduate and 85 graduate students.
Location	Newark, Delaware, is an hour's drive from both Philadelphia and Baltimore and a 2-hour drive from both Washington and New York. This proximity to large urban centers provides many cultural advantages, and the city of 30,000 people provides a pleasant environment for study and living.
The University	The University of Delaware began as a small liberal arts school in 1743 and became a land-grant college in 1867. It is situated in an area rich in technical talent and interests. Many firms employing mechanical engineers throughout the United States provide a beneficial interface between the University and industry.
Applying	There is no specific deadline for application for admission to graduate study. However, students requesting financial assistance should submit a completed application by February 1.
Correspondence and Information	Graduate Secretary Department of Mechanical Engineering University of Delaware Newark, Delaware 19716 Telephone: 302-831-2423 Fax: 302-831-3619 E-mail: grad_sec@me.udel.edu

University of Delaware

THE FACULTY AND THEIR RESEARCH

Professors

Suresh G. Advani, Professor and Acting Chair; Ph.D., Illinois at Urbana-Champaign. Computational fluid mechanics and heat transfer, composites and polymer processing, rheology.

Tsu-Wei Chou, Jerzy L. Nowinski Professor; Ph.D., Stanford. Polymer-, metal-, and ceramic-based composite materials; fracture of solids; applied mechanics; crystal defect theory.

Michael D. Greenberg, Ph.D., Cornell. Applied mathematics, fluid mechanics.

Leonard W. Schwartz, Ph.D., Stanford. Theoretical and computational fluid mechanics, applied mathematics.

Andras Z. Szeri, Robert L. Spencer Professor and Interim Dean; Ph.D., Leeds (England). Fluid mechanics, heat transfer, tribology.

Jack R. Vinson, H. Fletcher Brown Professor; Ph.D., Pennsylvania. Structural mechanics, aerospace structures, structures of anisotropic composite materials, plate and shell theory, structural optimization, sandwich structures.

Anthony S. Wexler, Ph.D., Caltech. Air pollution, aerosols, dosimetry/toxicology, functional electrical stimulation of muscles.

Dick J. Wilkins, Ph.D., Oklahoma. Learning, design integration, composites durability and damage control.

Associate Professors

Sunil K. Agrawal, Ph.D., Stanford. Robotics, optimization of dynamic systems, rehabilitation robotics, real-time planning.

Thomas S. Buchanan, Associate Professor and Director of the Center for Biomedical Engineering Research; Ph.D., Northwestern. Musculoskeletal biomechanics, neural control of movement, medical image processing, sports medicine.

Ian W. Hall, Ph.D., Leeds (England). Electron microscopy, metal matrix composites, metallurgy and materials science.

Michael Keefe, Ph.D., Minnesota. Computer-aided design, mechanical design, system modeling, rapid tooling.

Azar P. Majidi, Ph.D., Surrey (England). Fracture and failure of composites, ceramic matrix composites, textile composites, corrosion.

Ajay K. Prasad, Ph.D., Stanford. Particle imaging techniques for fluid mechanics and heat transfer, vortex dynamics, entrainment in heated jets, mixing studies.

R. Valéry Roy, Ph.D., Rice. Nonlinear stochastic and deterministic dynamics, chaotic dynamics, vibrations, numerical analysis, applied mathematics.

Michael H. Santare, Ph.D., Northwestern. Fracture mechanics, orthopedic biomechanics, composite structures.

Jian-Qiao Sun, Ph.D., Berkeley. Nonlinear dynamics, nonlinear control, smart materials, active vibration and noise control, manufacturing process control and automation, random and deterministic vibrations.

Assistant Professors

John Lambros, Ph.D., Caltech. Dynamic and quasi-static fracture, thermomechanical modeling and failure mechanics of composites, fracture of inhomogeneous and functionally graded solids.

Hai Wang, Ph.D., Penn State. Combustion, combustion-generated pollutants, high-temperature chemical kinetics.

Lian-Ping Wang, Ph.D., Washington State. Computational fluid dynamics, turbulence, multiphase flow, particulate processing.

Emeritus Faculty

James E. Danberg, Ph.D., Catholic University. Turbulent boundary layers, viscous flow, hypersonic aerodynamics.

Irwin G. Greenfield, Ph.D., Pennsylvania. Mechanical metallurgy, electron microscopy, effects of surfaces on mechanical properties, point defects in solids.

John D. Meakin, Institute of Energy Conversion; Ph.D., Leeds (England). Scanning and transmission electron microscopy, X-ray diffraction, photovoltaics.

Jerzy L. Nowinski, Ph.D., Warsaw. Mechanical behavior of bones, properties of waves in anisotropic media, statistical methods in computations of plates and shells, biomedical engineering, nonlocal and polyatomic continua.

John R. Zimmerman, Ph.D., Lehigh. Engineering design, automotive engineering, computer methods.

Secondary Appointments

James L. Glancey, University of Delaware Agriculture Department.

Pablo Huq, University of Delaware Marine Studies.

Research Professor

Tariq Rahman, A. I. duPont Institute, Applied Science & Engineering Laboratories, Wilmington, Delaware.

Adjunct Faculty

Richard A. Cairncross, Drexel University.

Shiyi Chen, Deputy Director, Center for Nonlinear Studies, Los Alamos National Laboratory, New Mexico.

Stephen Danforth, Professor, Department of Ceramic Engineering, Rutgers University.

Richard R. Eley, ICI Paints, Glidden Research Center, Ohio.

Bruce K. Fink, Materials Scientist, U.S. Army Research Lab, Aberdeen, Maryland.

Edward P. Garguilo, Dade Incorporated, Newark, Delaware.

Roger V. Gonzalez, Assistant Professor, Letourneau University.

Said Jahanmir, National Institute of Standards and Technology, Gaithersburg, Maryland.

Kostas Kontomaris, Senior Research Engineer, Dupont Research and Development, Experimental Station, Wilmington, Delaware.

Freeman Miller, A. I. duPont Institute, Wilmington, Delaware.

Peter Popper, E. I. duPont de Nemours & Co., Wilmington, Delaware.

Oliver E. Rodgers, Retired.

Karl V. Steiner, Executive Director, Center for Composite Materials, University of Delaware.

Recent Recognition

Faculty members of the Department of Mechanical Engineering serve as editors and/or editorial board members of *Acta Materiae Compositae, Composites Science and Technology, Journal of Applied Biomechanics, Journal of Biomechanics, Journal of Engineering Mathematics, Journal of Materials: Design and Application, Journal of Sandwich Structures and Materials, International Journal of Forming Processes,* and *Multibody System Dynamics.*

Faculty members of the Department of Mechanical Engineering have been named fellows of the American Society for Composites, American Society of Mechanical Engineers, American Institute of Aeronautics and Astronautics, and Alexander von Humboldt Foundation.

Faculty members of the Department of Mechanical Engineering have held national positions as President, American Society for Composites; Director, International Society for Coating Science and Technology; and Executive Board member, American Society of Biomechanics.

Faculty members of the Department of Mechanical Engineering have received prestigious awards, including the NSF Presidential Faculty Fellow Award, NSF Faculty Early Career Development (CAREER) Program Award, ONR-AIAA Research Award, Technomic National Award, Charles Russ Richards Award (ASME), and ASC Award for Outstanding Research.

UNIVERSITY OF DETROIT MERCY

College of Engineering and Science
Department of Mechanical Engineering

Programs of Study	Graduate programs leading to the degrees of Master of Engineering and Doctor of Engineering are offered by the department. Students with an interest in management can pursue a Master of Engineering Management degree with an emphasis on mechanical engineering. The programs of study are tailored to the student's individual interests in consultation with a faculty adviser. There are excellent opportunities for advanced theoretical and experimental study in the areas of mechanical systems, manufacturing, and thermo/fluid systems. Many research projects are simulated by the University's partnership with Chrysler, Ford, and General Motors as well as their supplier community. Substantial research support also comes from the Department of Defense and the National Science Foundation.
	The Master of Engineering in mechanical engineering may be completed through either a thesis or nonthesis plan. The thesis plan includes 24 credit hours of course work and 6 credit hours of thesis. The nonthesis plan consists of 33 credit hours of course work. All graduate courses are taught in the evening, making it possible to pursue a Master of Engineering degree while working full-time. The Doctor of Engineering in mechanical engineering requires the following postbaccalaureate components: four core courses, 30 hours of course work in a specific discipline, 9 credit hours of approved technical electives, and 36 hours of dissertation. The doctoral candidate must pass qualifying examinations and a comprehensive examination.
Research Facilities	There are excellent laboratories and facilities for experimental research in the department. The recently expanded Manufacturing Laboratory conducts research to advance the state of the art in four dominant areas: metalcutting, surface finish, computer-aided manufacturing, and quality control. The laboratory has facilities in CNC milling and turning, coordinate measuring machines, surface finish characterization, and machine tool dynamometry. The department has dedicated facilities to support research in airflow management in automotive systems, wind-driven material transport, and materials testing. The College of Engineering and Science's computer facilities include student laboratories with networked Windows-based and Sun workstations. The Ford Advanced Computing and Teaching Center supports both teaching and research in the College with a wide variety of software for geometric and solid modeling, design, finite element modeling, NC program generation, and flow visualization.
Financial Aid	A variety of teaching and research fellowships/assistantships are available in the College of Engineering and Science. This financial aid is competitively awarded on a yearly basis. A select number of full-time graduate students receive some financial support after their first semester of study. In addition, the Scholarship and Financial Aid Office accepts applications for grants, loans, and work-study assistance. Aid includes the Michigan Tuition Grant (for Michigan residents only), Federal Work-Study, and a variety of loans. The University also accepts third-party payments from employers and government agencies in addition to offering payment plans of its own. For information regarding financial aid programs, students should call 313-993-3350.
Cost of Study	Tuition in 1999–2000 is $545 per credit hour. Registration fees are $100 for full-time students.
Living and Housing Costs	Housing is available on campus. Double-occupancy rates range from $1410 to $3280. Single-occupancy rates range from $2420 to $2720. The University's meal plans cost about $1085. All rates are for a sixteen-week term. For more information, students should call Residence Life at 313-993-1230.
Student Group	There are 35 graduate students in the Department of Mechanical Engineering. Approximately 15 percent of the graduate student body is international. About one tenth of the graduate student body is pursuing a doctorate. The majority of Master of Engineering students are part-time. Approximately 6,700 students attend classes on three UDM campuses located in northwest and downtown Detroit.
Location	Students enjoy a variety of activities offered on campus and throughout the metropolitan Detroit area, including sports, theater, and concerts.
The University	As Michigan's largest Catholic university, the University of Detroit Mercy has an outstanding tradition of academic excellence firmly rooted in a strong liberal arts curriculum. This tradition dates back to the formation of two Detroit institutions: the Society of Jesus (Jesuits) and Mercy College of Detroit, founded by the Sisters of Mercy of the Americas. In 1990, these schools consolidated to become the University of Detroit Mercy. Today, UDM offers more than 120 majors and programs in nine different schools and colleges and is widely recognized for its programs in engineering, law, dentistry, nursing, and architecture. Faculty members are known for their excellence; more than 87 percent have a Ph.D. or comparable terminal degree.
Applying	Applications for admission normally should be completed at least six weeks before the beginning of a term. Applications for financial aid should be submitted by April 1. International students are urged to complete their applications at least three months before classes begin. Admission requirements are a bachelor's degree from an accredited college; a B average in the total undergraduate program and in the proposed field of study; and, normally, an undergraduate major or the equivalent in the proposed field. Official transcripts are required from all colleges attended. Applicants with less than a B average who present other evidence of ability to perform graduate-level work may be admitted as probationary students upon the recommendation of the director of the program.

| **Correspondence and Information** | Records Office
College of Engineering and Science
University of Detroit Mercy
P.O. Box 19900
Detroit, Michigan 48219-0900
Telephone: 313-993-3335
Fax: 313-993-1187 | Professor M. R. Schumack, Chairman
Department of Mechanical Engineering
University of Detroit Mercy
P.O. Box 19900
Detroit, Michigan 48219-0900
Telephone: 313-993-3370
Fax: 313-993-1187 |

University of Detroit Mercy

THE FACULTY AND THEIR RESEARCH

Yogendra S. Chadda, Professor; Ph.D., City University (London). Manufacturing, reliability.
Shuvra Das, Assistant Professor; Ph.D., Iowa State. Finite element modeling, modeling of manufacturing systems.
Paul J. Eagle, Associate Professor; D.Eng., Detroit. Manufacturing systems, nondestructive inspection, electronic information distribution.
Ricardo H. Espinosa, Visiting Professor (Monterrey Tech); D.Eng. candidate, Detroit Mercy. CAD, design of mechanical systems.
Arthur C. Haman, Professor and Associate Dean; M.Eng., Detroit. Internal combustion engines.
Karim H. Muci-Küchler, Associate Professor; Ph.D., Iowa State. Computational mechanics, design methodologies, biomechanics.
Carolyn J. Rimle, Instructor and Assistant Dean; M.Eng., Detroit. Minority engineering education.
Mark R. Schumack, Associate Professor and Chairman; Ph.D., Michigan. Fluid mechanics, thermal modeling.
Jonathan M. Weaver, Assistant Professor; Ph.D., RPI. Robotics, manufacturing systems.

Areas of Current Research
Acoustics and Vibrations
Forced vibrations of mechanical systems; noise, vibration, and harshness in automobiles; flow-induced vibrations.
Computer-Aided Design
Finite-element analysis, design of welded components, design of automotive components.
Fluid Mechanics
Aerodynamic effects of air flow on automotive systems, air management in vehicles, two-phase flow.
Heat/Mass Transfer
Transport of material by wind-driven waves, slug growth in water waves.
Manufacturing
Nondestructive inspection, laser welding, finite-element modeling of castings, design of cutting tools.
Solid Mechanics and Design
Design of composite parts, design of safety-critical systems, design of automotive components.
Systems and Control Theory
System simulation, control strategies for robotic manipulators, dynamic system behavior.

UNIVERSITY OF FLORIDA

Department of Mechanical Engineering

Program of Study	The Department of Mechanical Engineering offers programs of study leading to the degrees of Master of Science, Master of Engineering, Engineer, and Doctor of Philosophy. General areas within which students may specialize are biomechanical systems, energy conversion systems, mechanical systems, thermal systems, manufacturing, and robotics. Within these specializations are unique opportunities to complete theoretical and experimental research investigations in a wide variety of subspecialties, including automatic controls, biomechanics, combustion, cryogenics, energy conversion, environmental control, fluid dynamics, gas dynamics, heat transfer, kinematic synthesis, computer-aided design, finite-element methods, machine tools, machine dynamics, precision engineering, propulsion, solar energy, thermodynamics, and vibrations.

The master's degree candidate can choose a thesis or nonthesis program. The thesis degree requires 30 credit hours, and the student is expected to write a thesis and defend it during a final oral examination. The nonthesis degree requires 33 credit hours and a written comprehensive exam in the student's field of specialization. The Master of Engineering degree is available to students with a baccalaureate in engineering, while the Master of Science degree may be earned by those students with a bachelor's degree in engineering, math, or the sciences.

The Engineer degree requires a thesis and 30 credit hours of course work beyond the master's degree.

The Ph.D. degree requires 90 credit hours, which may include some or all of the master's degree course work. The doctoral student is required to pass both written and oral qualifying examinations. The student is expected to complete a dissertation, which reflects independent investigation, and defend it during an oral final examination.

The Department of Mechanical Engineering also offers off-campus master's degrees for employees of various Florida industries and government agencies through the statewide Florida Engineering Education Delivery System as well as at the Center for Advanced Studies in Engineering in Palm Beach Gardens, Florida.

Research Facilities	The mechanical engineering department occupies in excess of 43,000 square feet of modern, well-equipped space and houses a number of major laboratories, including the Center for Intelligent Machines and Robots, the Machine Tool Research Center, the Interdisciplinary Center for Aeronomy and Atmospheric Sciences, and laboratories for Computational Fluid Dynamics, Energy and Gas Dynamics, Energy Conversion, Heat Transfer, Internal Combustion Engines, Bouyant Flow, Automatic Controls, Refrigeration and Air Conditioning, Thermal Systems, Two-Phase Flow, Computer Graphics and CAD, Precision Metrology, and Systems Automation. An extensive network of modern computational facilities and software is available for student use. The department has complete machine shop and instrumentation facilities staffed by professional technicians.
Financial Aid	Graduate teaching and research assistantships are available from the department and are awarded on the basis of academic performance, GRE scores, college transcripts, and letters of recommendation. One-third-time and half-time assistantships are available. In 1998–99, stipends for half-time assistantships averaged $1083 per month. A limited number of fellowships are also available. Tuition and fees may be waived for students holding graduate assistantships and fellowships.
Cost of Study	In fall 1999, the registration fee for graduate course work is approximately $129 per credit hour for Florida residents and approximately $434 per credit hour for nonresidents.
Living and Housing Costs	Living costs are quite reasonable. In 1998–99, rents for apartment accommodations provided by the University for single graduate and professional students began at $967 per person per semester (double occupancy). For families, the University also operates five apartment villages, where rent is $251 per month and up, including some utilities. In addition to on-campus housing, there are many apartment complexes in the area, with rent for one-bedroom apartments starting at $450 per month, not including utilities.
Student Group	The total enrollment at the University of Florida is approximately 43,000. The Department of Mechanical Engineering currently has 122 graduate students enrolled on campus. These students comprise a stimulating mixture of men and women representing many regions of the United States plus a number of other countries.
The University and The Department	The University of Florida is a member of the Association of American Universities, whose members are preeminent in graduate and professional education and research. It is among the nation's largest universities and ranks in the top three in the number of academic programs offered. Graduate degrees are offered in more than 100 fields. The College of Engineering consists of twelve departments and enrolls approximately 2,500 undergraduate and 1,300 graduate students. The College of Engineering ranks among the top fifteen engineering colleges in the nation in research expenditures.

The Department of Mechanical Engineering is one of the largest departments in the College of Engineering, with approximately 540 undergraduate students and 122 graduate students. The department offers dedicated faculty members, excellent financial support for students, and a relaxed and congenial atmosphere, leading to outstanding opportunities for graduate education.

Applying	Application forms for admission and financial aid may be obtained from the department. For admission to the master's program, a baccalaureate degree in mechanical engineering from an accredited program with a B average, a combined score of 1100 for the verbal and quantitative portions of the GRE, and a score of 580 or better on the TOEFL (if applicable) are required. For admission to the doctoral program, a master's degree in mechanical engineering or a related engineering discipline, a GPA of 3.3 or higher, and a combined GRE score of 1200 for the verbal and quantitative portions are required. Students may be admitted for any semester, but applications, transcripts, and test scores should be submitted at least sixty days before the expected date of registration.
Correspondence and Information	Dr. John Ziegert, Graduate Coordinator or Ms. Becky Hoover, Graduate Secretary Graduate Studies Office Department of Mechanical Engineering 230 MEB University of Florida P.O. Box 116300 Gainesville, Florida 32611 Telephone: 352-392-0808 Fax: 352-392-1071 E-mail: megrad@cimar.me.ufl.edu

University of Florida

THE FACULTY AND THEIR RESEARCH

Nagaraj K. Arakere, Associate Professor; Ph.D., Arizona State, 1988. Mechanical systems design, tribology and surface engineering, fatigue and fracture mechanics.

Jacob Chung, Hines Eminent Scholar, Ph.D., Pennsylvania, 1979. Transport phenomena, microgravity boiling, microscale heat transfer, biomedical engineering.

Carl D. Crane, Associate Professor; Ph.D., Florida, 1987. Robotics, computer graphics, kinematics.

Joseph Duffy, Graduate Research Professor and Director, Center for Intelligent Machines and Robotics; Ph.D., Liverpool (England), 1963. Robotic manipulators, analytical geometry of machines, synthesis of mechanisms, machine design.

Roger A. Gater, Associate Professor; Ph.D., Purdue, 1969. Thermodynamics, fluid mechanics, heat transfer, combustion phenomena.

D. Yogi Goswami, Professor and Director, Solar Energy and Energy Conversion Laboratory; Ph.D., Auburn, 1975. Energy conversion, solar energy, heat transfer, fluid flow, HVAC.

Alex E. S. Green, Graduate Research Professor and Director, Interdisciplinary Center for Aeronomy and (Other) Atmospheric Sciences (ICAAS); Ph.D., Cincinnati, 1948. Radiological, atmospheric, nuclear, and atomic physics; clean combustion technologies.

David W. Hahn, Assistant Professor, Ph.D., Louisiana State, 1992. Laser-based diagnostics, combustion, heat transfer, particle analysis, biomaterials.

Herbert Ingley III, Associate Professor; Ph.D., Florida, 1971; PE. Heat transfer, acoustics, air conditioning, utilization of solar energy.

James F. Klausner, Associate Professor; Ph.D., Illinois, 1989. Boiling heat transfer, two-phase flow, slurry transport, bioengineering.

Ashok V. Kumar, Assistant Professor; Ph.D., MIT, 1993. Computer-aided design, design optimization, finite-element methods.

William E. Lear Jr., Assistant Professor; Ph.D., Stanford, 1984. Thermal sciences, high-energy gas dynamics.

Paul A. C. Mason, Assistant Professor; Ph.D., SUNY at Buffalo, 1995. Controls, dynamics, estimation theory, attitude determination.

Gary K. Matthew, Associate Professor; Ph.D., Florida, 1975. Kinematics, dynamics, computer-aided design.

Rajat Mittal, Assistant Professor; Ph.D., Illinois at Urbana-Champaign, 1995. Computational fluid mechanics and heat transfer, large scale simulations of turbulent and transitional flows.

Jill Peterson, Associate Professor; Ph.D., Rice, 1989. Gas dynamics, energy conversion.

Winfred M. Phillips, Associate Vice President EIES, Professor of Mechanical Engineering, Research Professor of Biomedical Engineering, and Dean of the College of Engineering; D.Sc., Virginia, 1968.

Kimberly D. Reisinger, Assistant Professor; Ph.D., Drexel, 1993. Biomechanics, analytical dynamics.

Vernon P. Roan Jr., Professor; Ph.D., Illinois, 1966. Fluid mechanics, gas dynamics, propulsion systems, transportation.

John K. Schueller, Associate Professor; Ph.D., Purdue, 1983. Manufacturing, controls, machine systems, off-highway vehicles.

Ali A. Seireg, Ebaugh Professor; Ph.D., Wisconsin–Madison, 1954. Machine design, biomechanics, optimization, robotics, computers in design engineering.

Sherif A. Sherif, Associate Professor; Ph.D., Iowa State, 1985. Thermal systems design, refrigeration, ventilation, air conditioning.

William G. Tiederman, Professor and Chairman; Ph.D., Stanford, 1965. Viscous fluid mechanics, turbulent wall flows.

Jiri Tlusty, Graduate Research Professor and Director, Machine Tool Research Center; Ph.D., Technical University (Czechoslovakia), 1958. Metal cutting, grinding and forming, technology and related failure analysis of metal cutting processes, reliability, productivity.

Gloria J. Wiens, Associate Professor; Ph.D., Michigan, 1986. System dynamics and controls, flexible multibody systems, robotics in manufacturing.

Otis R. Walton, Associate Professor; Ph.D., California, Davis, 1980. Computational physics, mechanics of granular materials, and equations of state.

Zhuomin Zhang, Assistant Professor; Ph.D., MIT, 1992. Microscale heat transfer, thin-film materials, thermal and optical systems.

John C. Ziegert, Associate Professor; Ph.D., Rhode Island, 1989. Robotics, manufacturing, design of mechanical systems, precision engineering.

UNIVERSITY OF HOUSTON

Cullen College of Engineering
Department of Mechanical Engineering

Program of Study

The Department of Mechanical Engineering has an active graduate program that encompasses advanced study and research in the major areas of control systems, design, heat transfer, fluid mechanics, mechanics, and materials science. Current research activities cover aerodynamics, turbulence and turbulent shear flows, and computational fluid dynamics; boiling heat transfer, highly turbulent forced convection, liquid crystal thermography, absorption heat pump designs, and applications of solar pond technology; linear and nonlinear elasticity, media with microstructure, rheology, continuum mechanics, fracture mechanics, biomechanics, and dynamics and vibration; metallurgy, tribology, ceramics, composites and superconductive materials, and nondestructive testing; control systems analysis and design; and design of engineering systems, computer-based decision making, instrumentation, and measurement systems. Research projects in these areas are funded by national and state agencies and industry.

Graduate programs, based on the areas listed above, lead to the degrees of Master of Mechanical Engineering, Master of Science, and Doctor of Philosophy for qualified applicants with backgrounds in engineering, mathematics, and physical sciences. Graduate courses are offered during all three semesters, and faculty members are available to supervise research throughout the year. Part-time graduate study and research leading to a degree are also possible, and sections of many courses are offered in the evening.

Research Facilities

The Department of Mechanical Engineering is in the Cullen College of Engineering, which is housed in two modern classroom-research buildings. The department's research facilities include laboratories to study aerodynamics, turbulence, shear flow and aeroacoustics; a laboratory for holographic velocimetry and combustion; a laboratory for computational fluid dynamics; laboratories to study the behavior of metals, polymers, ceramics, composite and superconducting materials and to develop new materials; laboratories for computational and experimental studies involving high-temperature mechanics, solid mechanics, and composite materials; a laboratory to study boiling and turbulent convective heat transfer; a laboratory to develop computer-based tools for designing systems and their components; a rheology laboratory to investigate flow properties of lubricants and polymer solutions; a dynamic fracture laboratory to investigate the mechanics of fracture; a laboratory for the micro-measurement of mechanical properties and the study of interfacial phenomena; a laboratory for modal analysis and controller design for fault-tolerant intelligent structural systems; and an anechoic chamber where research in sound-related activity is conducted. Well-equipped machine and electronics shops, manned by qualified technicians, are available in the department. Computational facilities available include workstation/Macintosh/PC clusters associated with various departmental laboratories, College of Engineering PC and DEC Alpha Workstation laboratories, as well as University-wide DEC clusters.

Financial Aid

Teaching fellowships, teaching assistantships, and research assistantships starting at $1200 per month are available. Full-time students who are receiving assistantships are eligible for a University fellowship covering the tuition of 9 semester credit hours.

Cost of Study

Tuition for Texas residents in 1999–2000 is $72 per semester credit hour (with a minimum payment of $120); for nonresidents, tuition is $274 per semester credit hour. In addition, each full-time student is charged fees ranging from $143 to $707 per semester. Lower rates apply for approved part-time students.

Living and Housing Costs

University-owned apartments are generally available to students. In addition, there are moderately priced apartments on the perimeter of the campus and within convenient distances of the campus. The cost of living in Houston is one of the lowest of all major cities in the nation.

Student Group

In fall 1998, there were 32,296 students enrolled at the University of Houston. Of the 2,172 engineering students, 588 were graduate students. The Department of Mechanical Engineering had 396 students, of whom 92 were graduate students.

Location

The University is located on a 390-acre campus in the southeastern section of Houston, one of the largest cities and ports in the United States. Houston is a modern, rapidly growing metropolis with a moderate climate that is pleasant the year round. The city is the home of many oil and petrochemical industries and of NASA's Johnson Space Center. Houston has excellent symphony orchestras, opera, ballet, and a number of theater companies. The area provides ample opportunity for fishing, hunting, golfing, and boating and offers college and professional sports.

The University

The University is a state-supported institution comprising the following colleges: Humanities and Fine Arts, Social Sciences, Natural Science and Mathematics, Business Administration, Architecture, Engineering, Law, Social Work, Education, Optometry, Pharmacy, Hotel and Restaurant Management, and Technology. The Department of Mechanical Engineering and the Departments of Chemical, Civil, Electrical, and Industrial Engineering constitute the Cullen College of Engineering.

Applying

Application forms for admission and financial aid may be obtained by contacting the address given below or by using the online application request feature of the department's web site. Applications should be returned together with letters of recommendation and official transcripts of previous academic work. GRE General Test scores are required, and international applicants must submit TOEFL scores.

Correspondence and Information

Graduate Admissions Analyst Assistant
Department of Mechanical Engineering
University of Houston
Houston, Texas 77204-4792

Telephone: 713-743-4505
Fax: 713-743-4503
E-mail: megrad@mail.me.uh.edu
World Wide Web: http://www.me.uh.edu

University of Houston

THE FACULTY AND THEIR RESEARCH

R. B. Bannerot, Professor and Chairman of the Department; Ph.D., Rice; PE. Thermal sciences: alternative energy and design of thermal systems.

Y. C. Chen, Associate Professor; Ph.D., Minnesota. Solid mechanics, composite materials.

C. Dalton, Professor; Ph.D., Texas at Austin; PE. Fluid mechanics: hydrodynamics and computational fluid dynamics.

R. Eichhorn, Professor; Ph.D., Minnesota; PE. Thermal sciences, natural convection, boiling heat transfer.

R. Glowinski, Professor; Ph.D., Paris. Finite elements in methods in continuum mechanics, applied mathematics.

K. M. Grigoriadis, Associate Professor; Ph.D., Purdue. Control systems analysis and design, robust and optimal control theory, modeling and identification of engineering systems.

D. K. Hollingsworth, Associate Professor; Ph.D., Stanford; PE. Convective heat transfer, turbulent flows, image processing.

F. Hussain, Cullen Professor; Ph.D., Stanford; PE. Fluid mechanics: turbulence, chaos, biofluids, aeroacoustics, and combustion.

S. J. Kleis, Associate Professor; Ph.D., Michigan State. Fluid mechanics: turbulence and biofluid mechanics.

I. Kunin, Professor; Ph.D., Leningrad Polytechnical. Solid mechanics, vortex flows, chaos, dynamics, gauge field theory.

J. H. Lienhard, M. D. Anderson Professor; Ph.D., Berkeley. Thermodynamics, heat transfer with phase change, history of technology.

R. W. Metcalfe, Professor; Ph.D., MIT. Computational fluid dynamics, spectral numerical methods.

A. Powell, Professor; Ph.D., Southampton (England). Aeroacoustics, flow-induced oscillations, acoustics, fluid mechanics.

J. R. Rao, Associate Professor; Ph.D., Michigan. Optimization, modeling of design and manufacturing systems, parametric nonlinear programming.

K. Ravi-Chandar, Professor; Ph.D., Caltech. Solid mechanics, fracture mechanics, behavior of superconducting materials.

K. Salama, Professor; Ph.D., Cairo. Materials science, metallurgy, superconductivity, mechanical properties, nondestructive characterization.

N. Shamsundar, Associate Professor; Ph.D., Minnesota. Heat transfer, thermodynamics, combustion.

W. E. VanArsdale, Associate Professor; Ph.D., Cornell. Continuum mechanics, polymer rheology, lubrication.

S. S. Wang, Distinguished University Professor; Ph.D., MIT. Thermomechanics and thermochemical behavior of composite materials, computational methods.

L. T. Wheeler, Professor; Ph.D., Caltech; PE. Solid mechanics, applied mathematics, elasticity.

K. White, Professor; Ph.D., Washington (Seattle). Materials science, ceramics, metallurgy, superconducting materials, fracture mechanics.

L. C. Witte, Professor; Ph.D., Oklahoma State; PE. Thermal sciences: boiling and condensation heat transfer and second-law analysis.

D. Zimmerman, Associate Professor; Ph.D., SUNY at Buffalo. Dynamic systems and control, structural health monitoring, vibration testing, experimental methods, structural computations, optimization.

THE UNIVERSITY OF ILLINOIS AT CHICAGO

Department of Mechanical Engineering

Programs of Study

The Department of Mechanical Engineering offers programs leading to the M.S. degree in mechanical engineering or industrial engineering and the Ph.D. degree in mechanical engineering or industrial engineering and operations research. The graduate program covers broad areas of mechanical and industrial engineering as well as related fields. In the primary areas of concentration, which include mechanical analysis and design, fluids engineering, thermal sciences, robotics, manufacturing, and industrial engineering, students may select basic and applied courses dealing with such topics as fluid mechanics, heat and mass transfer, thermodynamics, combustion, stress analysis, noise and vibrations, mechanical design, dynamics and control, computer-aided design and manufacturing, materials processing, production engineering, and human factors.

Two options are available for the M.S. degree. In the course option, 36 semester hours in course work are required. In the thesis option, 24 semester hours in course work and 12 semester hours in thesis work are required. Part-time graduate study through an off-campus program leading to the M.S. degree is possible.

For the Ph.D. degree, 32 additional semester hours in course work and 44 semester hours in thesis research must be completed. A written qualifying examination is required for doctoral students within the first year after admission.

M.S. and Ph.D. theses must be defended before an examination committee appointed by the Graduate College.

For further details, students should visit the department's Web site (http://www.me.uic.edu/).

Research Facilities

The department has laboratories for mechanical design, manufacturing, materials processing, materials characterization, vibrations, internal combustion engines, fluid mechanics, two-phase flow, CAD, precision engineering, control, optical measurement, heat transfer, rheology, combustion, solar energy, interferometry, robotics and automation, industrial simulation, and virtual reality applications in manufacturing.

Financial Aid

Financial aid is offered on a competitive basis and is available in the form of University fellowships, teaching or research assistantships, and tuition waivers. M.S. and Ph.D. program applicants with grade point averages of 4.5–5.0 and 4.75–5.0, respectively, are encouraged to apply for financial aid by January 1.

Cost of Study

Tuition and fees for full-time study in 1999–2000 are $5640 per academic year for Illinois residents and $12,478 for nonresidents.

Living and Housing Costs

On-campus housing for engineering graduate students is available but limited. Generally, students are able to find off-campus housing in nearby areas. The costs are therefore variable, but they are comparable to those in other large metropolitan areas.

Student Group

The University's total enrollment is about 25,000, and the College of Engineering has a graduate enrollment of 788. There are 153 graduate students in the Department of Mechanical Engineering. Of these, about 46 are full-time M.S. students, 46 are full-time Ph.D. students, and 61 are part-time M.S. or Ph.D. students.

Location

The campus is located just southwest of Chicago's downtown section. Many recreational and cultural facilities are available close by. The campus is easily reached by subway or bus and from five expressways covering the entire metropolitan area.

The University

The University of Illinois at Chicago was created in 1982 by the merger of the Chicago Circle campus (established in 1965) and the Medical Center campus (established in 1894) of the University of Illinois. The campus architecture is modern and compact. Two high-rise buildings house many administration and faculty offices. Many classrooms are clustered in the center of the campus. Laboratory facilities are housed in specially designed buildings.

Applying

An application for admission should be sent directly to the Office of Admission. An application for financial aid should be sent to the Department of Mechanical Engineering. Applicants from abroad are required to score at least 550 on the Test of English as a Foreign Language (TOEFL). Although not required for admission, submission of GRE scores is highly recommended for financial aid applicants.

Correspondence and Information

Director of Graduate Studies
Department of Mechanical Engineering
2041 ERF (M/C 251)
The University of Illinois at Chicago
842 West Taylor Street
Chicago, Illinois 60607-7022
Telephone: 312-996-6122
E-mail: megrad@uic.edu

The University of Illinois at Chicago

THE FACULTY AND THEIR RESEARCH

S. K. Aggarwal, Professor; Ph.D., Georgia Tech, 1979. Combustion, fluid mechanics, multiphase flows, mathematical modeling and computational methods, turbulent sprays, unsteady flames, emission control. Associate Fellow, AIAA; Technical Program Chairman, Propellants and Combustion, 1989 AIAA Aerospace Sciences Meeting; Air Force Fellow, 1987; Member, DOE Review Panel on Coal Combustion Research, 1990; Associate Editor, *AIAA Journal*.

F. M. L. Amirouche, Professor; Ph.D., Cincinnati, 1984. Multibody dynamics, CAD/CAE, robotics, biomechanics, human body vibrations, computational methods, stability, vibration and active control of vehicles, space structure dynamics, seat/cab suspension design, ER/MR fluid applications. Editor, ASEE Mechanics Division Newsletter, 1986–89; DOE Fellow, 1987; G7-NRC Fellow, 1993–94; SAE Teetor Award, 1994.

P. Banerjee, Associate Professor; Ph.D., Purdue, 1990. Facilities planning, virtual manufacturing software environment design, interactive system design, object-based manufacturing system, qualitative reasoning and optimization of manufacturing systems. Associate Editor, *IEEE Robotics and Automation*.

J. G. Boyd, Assistant Professor; Ph.D., Texas A&M, 1994. Micro-electro-mechanical devices (MEMs), microfluidic devices, thermomechanics of shape memory alloy composite materials, modeling of nonlinear microstructural evolution. NSF Career Award, 1999.

S. Cetinkunt, Professor; Ph.D., Georgia Tech, 1987. Automatic control, robotics, automation, CAD/CAM, microprocessor applications, expert systems, artificial intelligence.

S. Cha, Associate Professor; Ph.D., Michigan, 1980. Laser instrumentation; optical flow, heat, and combustion diagnostics; gross-field velocimetry; radiative heat transfer; laser machining and processing; photomechanics; holographic interferometry. DOE Research Fellow, 1985–87; NASA Research Fellow, 1992–94.

W. Chen, Assistant Professor; Ph.D., Georgia Tech, 1995. CAD, robust design, optimization. NSF Career Award, 1996; ASME Pi Tau Sigma Gold Medal, 1998.

M. Y. Choi, Associate Professor; Ph.D., Princeton, 1992. Combustion: radiative emission measurement and modeling of turbulent diffusion flames, optical and intrusive techniques for flame characterization and particulate measurements, droplet combustion characteristics.

P. M. Chung, Professor and Dean Emeritus; Ph.D., Minnesota, 1957. Chemically reacting flow, combustion, plasma-solid interactions, multiphase flows, hypersonic gas dynamics, statistical description of fluid turbulence.

B. D. Coller, Assistant Professor; Ph.D., Cornell, 1995. Nonlinear dynamics and control, bifurcation theory, chaos, fluids, aeroelasticity.

D. M. France, Professor and Associate Dean; Ph.D., Berkeley, 1969. Multiphase flow and heat transfer, including liquid-vapor boiling flows, solid-gas flow suspensions, and solid-liquid slurry flows; flow instabilities and heat transfer augmentation, including refrigerant system and large-scale steam generators and development of unique facilities for experimentation in internal and external boiling flows. DOE International Exchange Team, 1978; DOE Technical Review Team, 1988.

A. A. Fridman, Professor; Ph.D., Moscow Tech, 1979; D.Sc., Kurchatov Institute, 1982. Plasma processes, chemical kinetics, nonequilibrium processes, gas discharges, aerosols and catalysis. 1989 State Prize of U.S.S.R., Kurchatov Prizes 1980 and 1984, Soros Professor.

C. George, Assistant Professor; Ph.D., Minnesota, 1998. Numerical heat transfer, plasma processes.

Y. G. Gogotsi, Assistant Professor; Ph.D., Kiev Polytechnic, 1986; D.Sc., Ukrainian Academy of Science, 1994. Mechanical behavior and corrosion resistance of materials, ceramic matrix composites, diamond synthesis, use of Raman spectroscopy in materials research. A. von Humboldt Fellow, 1990–92; Frantsevich Prize, 1993; Director, NATO Advanced Studies Institute, 1998; NSF Career Award, 1999.

S. I. Guceri, Professor and Head; Ph.D., North Carolina State, 1976. Transport phenomena in materials processing; thermoplastic filament winding; composites manufacturing; injection molding; flow of polymers; resin transfer molding, structural reaction injection molding, micromechanics, thermal stresses; interphase characterization; ceramic rapid prototyping and ceramic-ceramic composites. Editor, *Journal of Materials Processing and Manufacturing Science*, 1992–present; editor, *Transport Phenomena in Processing*, Technomic Publishing Company, 1993.

K. C. Gupta, Professor and Director of Graduate Studies; Ph.D., Stanford, 1974. Kinematics, mechanism synthesis, robotics, cam dynamics, design optimization. Fellow, ASME; Henry Hess Award, ASME, 1979; South Pointing Chariot Award, AMR Conference, 1989; George N. Sandor Award, AM&R, 1997; Associate Editor, ASME *Journal of Mechanical Design*, 1981–82; Editorial Advisory Board, ASME *Applied Mechanics Reviews*, 1985–92; Editorial Advisory Board, *Journal of Applied Mechanisms and Robotics*; Associate Editor, *Mechanism and Machine Theory*; Conference Chairman, 1990 ASME Design Technical Conference and ASME Mechanisms Conference.

J. P. Hartnett, Professor and Director Emeritus, Energy Resources Center; Ph.D., Berkeley, 1954. Energy resources, heat transfer, non-Newtonian fluids. Guggenheim Fellowship, 1960–61; Fulbright Award, 1961; IIT Professional Achievement Award; A. V. Luikov Medal, 1981; Max Jacob Memorial Award, 1990; Member, International Higher Education Academy of Sciences, Moscow, 1993; Honorary Member, JSME, 1995; Foreign Fellow, Indian National Academy of Engineering, 1995; ASME Dedicated Service Award, 1995; Coeditor, *Heat Transfer—Japanese Research*; Coeditor, *Heat Transfer—Soviet Research*; Editor, *International Journal of Heat and Mass Transfer*; Coeditor, *International Communications in Heat and Mass Transfer*; Coeditor, *Handbook of Heat Transfer*; Coeditor, *Advances in Heat Transfer*.

J. O. M. Karlsson, Assistant Professor; Ph.D., MIT, 1994. Cryobiology, transport and phase transformations in biological systems, tissue engineering.

L. A. Kennedy, Professor and Dean, College of Engineering; Ph.D., Northwestern, 1964. Combustion, nonequilibrium processes, emission control, fluid mechanics, optical methods, heat transfer. Ralph Coats Roe Award, Fellow ASME, Associate Fellow AIAA, R. W. Kurtz Distinguished Professor at OSU, AT&T Foundation Award, Senior Fellow in Science—NATO, R. W. Teator Award, NSF Science Faculty Fellow; Associate Editor, *Journal of Propulsion and Power*; Editor-in-Chief, *International Journal of Experimental Methods in Thermal & Fluid Sciences*; Editor, *22nd International Symposium on Combustion*.

K. Kim, Associate Professor; Ph.D., Wisconsin–Madison, 1986. Computer-aided manufacturing, automation and robotics, sculptured surface machining, dynamics and control, precision engineering, mechanical vibrations. SME Outstanding Young Manufacturing Engineer Award, 1988.

F. L. Litvin, Professor Emeritus and Director, Gear Research Center; Doc.Tech.Sc., Leningrad Polytechnic Institute, 1954. Gear theory, generation, simulation, meshing, and contact; manipulators, spatial linkages. Silver Medal for Professional Work, USSR, 1971; Inventor of the USSR, 1978; Excellence and Distinction in Professional Work, USSR, 1978; Member, Honorary Editorial Advisory Board, *Mechanism and Machine Theory*; Associate Editor, *Computer Methods in Applied Mechanics and Engineering*; Best Paper Award, 1990 ASME Mechanism Conference.

F. Loth, Assistant Professor; Ph.D., Georgia Tech, 1993. Biofluids, hemodynamics, fluid, mechanics, heat transfer, transport phenomena.

C. M. Megaridis, Associate Professor; Ph.D., Brown, 1987. Soot formation and emission control, liquid-metal droplet dispensing in microelectronic packaging, droplet and spray combustion dynamics, multiphase heat and mass transfer, optical diagnostics in aerosols. Kenneth T. Whitby Award of the American Association of Aerosol Research (AAAR), 1997.

F. G. Miller, Associate Professor Emeritus; Ph.D., Illinois, 1961. Industrial engineering, facility analysis, maintenance and safety engineering.

W. J. Minkowycz, Professor; Ph.D., Minnesota, 1965. Heat transfer, porous media flows, two-phase flow, numerical heat transfer. Fellow, ASME; Editor, *International Journal of Heat and Mass Transfer*; Editor, *International Communications in Heat and Mass Transfer*; Editor-in-Chief, *Numerical Heat Transfer*; Editor, Wiley *Handbook of Numerical Heat Transfer*; Editor, Taylor and Francis Book Series in *Computational and Physical Processes in Mechanics and Thermal Sciences*; Editor, Taylor and Francis *Advances in Numerical Heat Transfer*; ASEE Ralph Coates Roe Award, 1988; ASME Heat Transfer Memorial Award, 1993.

I. K. Puri, Professor; Ph.D., California, San Diego, 1987. Combustion, flame ignition and extinction, fire spread and inhibition, pollutant formation, laser diagnostics. AAAS/EPA Environmental Fellow, 1993.

T. J. Royston, Assistant Professor; Ph.D., Ohio State, 1995. Acoustics and vibration; nonlinear dynamics of intelligent materials and structures, medical diagnostic techniques, nondestructive evaluation techniques, high-precision vibration isolation. NSF Career Award, 1998.

A. A. Shabana, Professor; Ph.D., Iowa, 1982. CAD, dynamic systems, vibrations, finite-element method. Fellow, ASME; Humboldt Prize, Germany, 1995; Fulbright Award, 1997; Contributing Editor, *Journal of Multibody System Dynamics*; Editorial Board, *Journal of Nonlinear Dynamics*.

S. M. Song, Associate Professor; Ph.D., Ohio State, 1984. Kinematics, dynamics, robotics, gripping, walking machines, gait study, CAD. Presidential Young Investigator Award, 1987.

W. M. Worek, Professor and Director, Energy Resources Center; Ph.D., IIT, 1980. Heat transfer and mass transfer processes, optical techniques, advanced energy systems, sorption processes in liquid and solid desiccant materials.

UNIVERSITY OF ILLINOIS AT URBANA–CHAMPAIGN

Department of Mechanical and Industrial Engineering

Programs of Study

Graduate study leading to the M.S. and Ph.D. degrees is offered. In mechanical engineering, studies are conducted in combustion, computer-aided design, control systems, mechatronics, microelectromechanical systems (MEMS), fluid mechanics, gasdynamics, heat and mass transfer, kinematics and dynamics of machinery, knowledge-based engineering expert systems, manufacturing systems, materials behavior, materials processing, multiphase flow, propulsion, system simulation and optimization, and tribology. Problems in energy systems include air pollution, combustion, energy logistics, internal combustion engines, propulsion, solar and renewable energy, and waste handling. Progress in the study of materials behavior and processing includes casting processes, ceramic-matrix composites, composite materials, creep, fatigue, fracture, high-temperature material behavior, laser processing, polymer processing, and thin films. Tribology studies include elastohydrodynamics and lubrication of oil and refrigerant mixtures. In industrial engineering, studies are conducted in human factors and engineering psychology, operations research, and production engineering. Study in the areas of cognitive engineering, computer-aided manufacturing, ergonomics, facilities planning, human-machine interaction, large-scale systems analysis, machine tool systems design, mathematical programming and optimization, production planning and control, and project management is aimed at improving the design and implementation of integrated systems of persons, materials, and equipment.

The department has several center-based research activities, including two NSF industry/university cooperative research centers, one in air-conditioning and refrigeration and one in machine tool systems. The University of Illinois is also the lead institution in the newly formed NSF/ARPA Agile Manufacturing Research Institute for Machine Tools. In addition, the department has the Fracture Control Program and the Institute for Competitive Manufacturing.

Eight units of study (32 semester hours) is a minimum requirement for the M.S. degree with thesis. A nonthesis option is available by petition. The Ph.D. degree requires 16 units beyond the master's degree. A direct Ph.D. program is also available, requiring 24 units beyond the bachelor's degree. For both Ph.D. programs, a written/oral qualifying exam and a thesis research proposal are required prior to the oral preliminary exam. An oral final exam based on the Ph.D. thesis is required.

Research Facilities

Facilities include laboratories for advanced automation, air conditioning and refrigeration, combustion, computer-aided design and simulation, computer-integrated manufacturing, control systems, design for manufacturing, flexible automation, gasdynamics, heat transfer, human factors and simulation of human-machine interaction, internal combustion engines, knowledge-based engineering systems, laser diagnostics for combustion, laser processing, machining and machine tool systems, mechanical behavior of materials, metrology, microelectromechanical systems, operations research, polymer and composite materials processing, precision engineering, propulsion, rapid prototyping, robotics, solar energy, thermal processing of materials, tribology, vehicle dynamics, and welding and heat treatment. Special facilities include a ½-acre solar pond, test facilities for refrigeration and air-conditioning systems and components, low- and high-speed wind tunnels, and laboratories for study of combustion, radiation, particulate and multiphase flow, complete specimen-scale mechanical testing equipment including an environmental testing chamber, thermomechanical and multiaxial loading capabilities, and laser processing facilities. The department has a construction shop with instrument makers and electronics technicians.

Financial Aid

About 90 percent of the graduate students receive some form of financial aid. Financial assistance includes fellowships, assistantships, and tuition and fee waivers. Fellowships and teaching and research assistantships provide a stipend plus tuition and service fee exemption. Stipends vary with entry level into the program; the minimum rate for half-time is $13,500 (for nine months) in 1999–2000.

Cost of Study

For 1999–2000, students with fellowships or assistantships pay only partial fees, totaling approximately $470 per semester. Full-time Illinois residents without appointments or waivers pay a base rate of $2308 per semester, and nonresidents pay a base rate of $5884 per semester. Summer session charges are approximately $1455 for residents and $3705 for nonresidents. All rates are subject to change.

Living and Housing Costs

For 1999–2000, single rooms in graduate dormitories range from $2696 to $3008 per academic year. For a double room, the cost varies from $2416 to $2872 for the academic year. Housing for married students ranges from $364 to $544 per month; utilities are included for most facilities. Off-campus housing is available at similar and higher rents.

Student Group

The department has about 250 graduate students, of whom about 40 percent are Ph.D. candidates. Most students enter graduate study directly after completing undergraduate work, although some come from work in industry. The majority of the M.S. candidates seek industrial positions involving research and development work; most Ph.D. recipients prefer research careers in industry or government or research and teaching at universities.

Location

The University is in the twin cities of Urbana and Champaign, 130 miles south of Chicago. Willard Airport, Amtrak, and two interstate highways provide convenient transportation to and from the area. In 1999, *Newsweek* rated Urbana-Champaign one of the top ten high-tech communities in the world.

The University

A land-grant university founded in 1867, the University enrolls about 34,000 students, including about 8,600 graduate students, and is well known for its role in higher education, scientific research, and public service.

Applying

Information and application forms can be obtained on request. Applications must be submitted by March 1 for August admission and by October 1 for January admission. The nonrefundable application fee is $40 for U.S. citizens and $50 for international applicants. Scores on the GRE General Test are required of all applicants. Students who have attended a school in a country where English is not the primary language must submit a minimum TOEFL score of 610 on the paper-based exam or 257 on the computer-based exam.

Correspondence and Information

Celia G. Snyder, Graduate Programs Coordinator
Department of Mechanical and Industrial Engineering
140 Mechanical Engineering Building
University of Illinois at Urbana-Champaign
1206 West Green Street
Urbana, Illinois 61801
Telephone: 217-333-4390
E-mail: cgsnyder@uiuc.edu
World Wide Web: http://www.mie.uiuc.edu

University of Illinois at Urbana-Champaign

THE FACULTY AND THEIR RESEARCH

A. L. Addy, Emeritus Professor; Ph.D., Illinois, 1963. Fluid and gas dynamics.

A. Alleyne, Assistant Professor; Ph.D., Berkeley, 1994. Control theory and application, vehicle dynamics, nonlinear and adaptive control, vibration isolation.

A. J. Beaudoin, Associate Professor; Ph.D., Cornell, 1993. Materials processing and behavior, finite element simulation, supercomputing.

J. Bentsman, Associate Professor; Ph.D., IIT, 1984. Automatic control systems.

M. Q. Brewster, Professor; Ph.D., Berkeley, 1981. Heat transfer, combustion, propulsion.

R. O. Buckius, Professor and Head of Department; Ph.D., Berkeley, 1975. Heat transfer, combustion.

C. W. Bullard, Professor; Ph.D., Illinois, 1971. Simulation and optimization of thermal systems, refrigeration and air-conditioning, technology policy analysis.

J. C. Chato, Emeritus Professor; Ph.D., MIT, 1960. Heat transfer, fluid mechanics, bioengineering, electrohydrodynamics.

A. M. Clausing, Emeritus Professor; Ph.D., Illinois, 1963. Heat transfer, solar thermal energy systems, computer modeling.

C. Cusano, Professor; Ph.D., Cornell, 1970. Friction, lubrication, and wear.

E. R. Damiano, Assistant Professor; Ph.D., Rensselaer, 1993. Biomechanics and biofluid dynamics, continuum mechanics, mixture theory, perturbation methods, microvascular and vestibular mechanics.

J. A. Dantzig, Professor; Ph.D., Johns Hopkins, 1977. Solidification, materials processing, computer modeling.

R. E. DeVor, Professor; Ph.D., Wisconsin–Madison, 1971. Machining and machine tool systems, quality control and industrial statistics.

G. E. Dullerud, Assistant Professor; Ph.D., Cambridge, 1994. Systems and control theory and applications.

W. E. Dunn, Associate Professor; Ph.D., Illinois, 1976. Heat and mass transfer, numerical modeling.

J. C. Dutton, Professor; Ph.D., Illinois, 1979. Gasdynamics, fluid mechanics, propulsion.

P. M. Ferreira, Professor and Associate Head of Department; Ph.D., Purdue, 1987. Factory automation, manufacturing processes, precision engineering and manufacturing applications of artificial intelligence.

J. G. Georgiadis, Professor; Ph.D., UCLA, 1987. Fluid mechanics and heat transfer, two-phase flow, quantitative visualization of complex interfaces, magnetic resonance imaging, bioengineering.

Y. Y. Huang, Associate Professor; Ph.D., Harvard, 1990. Mechanics of materials, fracture mechanics, microscale constitutive modeling.

A. M. Jacobi, Associate Professor; Ph.D., Purdue, 1989. Heat transfer, fluid mechanics, end-use energy applications, thermal systems.

P. M. Jones, Associate Professor; Ph.D., Georgia Tech, 1991. Human-machine systems engineering, knowledge-based support for supervisory control, intelligent tutoring systems, simulation.

S. G. Kapoor, Professor; Ph.D., Wisconsin–Madison, 1977. Manufacturing systems, CAD/CAM, robotics, engineering statistics.

H. Krier, Professor; Ph.D., Princeton, 1968. Combustion with solid and liquid rockets, multiphase flows with heat and mass transport, shock hydrodynamics, viscous gas dynamics and plasma dynamics.

C. F. Lee, Assistant Professor; Ph.D., Princeton, 1995. Modeling of two-phase turbulent reacting flows, internal combustion engines, liquid atomization and spray systems, laser diagnostics for concentration measurements.

C. Liu, Assistant Professor; Ph.D., Caltech, 1996. Microfabrication, micromachined sensors and actuators, microelectromechanical systems.

T. J. Mackin, Assistant Professor; Ph.D., Penn State, 1991. Mechanical properties of composites: experiment, theory, and computer simulation.

N. R. Miller, Associate Professor; Ph.D., Wisconsin–Madison, 1975. Advanced automation and instrumentation systems.

T. A. Newell, Associate Professor; Ph.D., Utah, 1980. Heat transfer, solar thermal-energy systems, solar ponds.

J. W. Nowak, Adjunct Professor; Ph.D., Illinois, 1988. Production operations management, quality of information.

U. S. Palekar, Associate Professor; Ph.D., SUNY at Buffalo, 1986. Integer programming, scheduling theory, facilities planning and location.

A. J. Pearlstein, Associate Professor; Ph.D., UCLA, 1983. Theoretical (i.e., computational and analytical) studies of incompressible flow, hydrodynamic stability, development of numerical methods, applications to materials processing.

J. E. Peters, Professor and Associate Head of Department; Ph.D., Purdue, 1981. Combustion, internal combustion engines, atomization.

M. L. Philpott, Associate Professor; Ph.D., Cranfield Institute of Technology (England), 1986. Design for manufacture, integration and automation of mold and die production, sculptured surface production, new and advanced manufacturing processes, nonhierarchical cell control, quality monitoring of robotic and automated arc welding, high-speed packaging machinery.

L. M. Phinney, Assistant Professor; Ph.D., Berkeley, 1997. Heat transfer and thermosciences, microscale thermophysics, thermal issues in microelectromechanical systems (MEMS).

M. T. A. Saif, Assistant Professor; Ph.D., Cornell, 1993. Design, analysis, and fabrication of microelectromechanical systems (MEMS); dynamical behavior of MEMS; mechanics of materials at nanometer scale.

H. Sehitoglu, Professor; Ph.D., Illinois, 1983. Cyclic deformation and fatigue, phase transformations, thermomechanical behavior of materials, constitutive equations.

M. A. Shannon, Assistant Professor; Ph.D., Berkeley, 1993. Fundamentals of laser energy interactions with materials with application to laser-material processing.

D. F. Socie, Professor; Ph.D., Illinois, 1977. Mechanical behavior of materials, fracture mechanics.

J. A. Stori, Assistant Professor; Ph.D., Berkeley, 1998. Manufacturing systems, operations research.

B. G. Thomas, Associate Professor; Ph.D., British Columbia, 1985. Metallurgical process engineering; computer modeling of fluid flow; heat transfer and stress analysis in solidification and casting processes, including continuous casting of steel and directional solidification of airfoils.

D. A. Tortorelli, Associate Professor; Ph.D., Illinois, 1989. Theoretical development and computer implementation of design and analysis methodologies in solid mechanics, heat transfer, and fluid mechanics.

T.-C. Tsao, Associate Professor; Ph.D., Berkeley, 1988. Modeling and control of mechanical systems and manufacturing processes, digital control, adaptive and learning control, precision engineering.

C. L. Tucker, Professor; Ph.D., MIT, 1978. Polymer processing, composite materials processing, computer-aided design.

A. F. Vakakis, Associate Professor; Ph.D., Caltech, 1990. Linear and nonlinear vibrations, chaotic dynamics, experimental and analytical modal analysis, large repetitive space structures.

S. P. Vanka, Professor; Ph.D., Imperial College (London), 1976. Computational fluid mechanics, heat transfer and reacting flows, large eddy and direct numerical simulations of turbulence.

J. S. Walker, Professor; Ph.D., Cornell, 1970. Magnetohydrodynamics, crystal growth, energy conversion, materials processing.

R. A. White, Emeritus Professor; Ph.D., Illinois, 1963. Gasdynamics, heat transfer, vehicle dynamics, internal combustion engines.

UNIVERSITY OF ILLINOIS
AT URBANA–CHAMPAIGN

Department of Theoretical and Applied Mechanics

Programs of Study

Graduate study leading to the M.S. and Ph.D. degrees is offered. There are six broadly defined areas of concentration: fluid mechanics (turbulence, flow visualization, coatings, foams, emulsions, combustion, chaos theory), solid mechanics (optimization, fracture mechanics, plasticity, residual stresses, materials processing), dynamics (acoustics, stochastic waves, nondestructive evaluation), materials engineering (metals, ceramics, composites, smart materials), computational science and engineering (computational fluid dynamics and solid mechanics), and applied mathematics (continuum thermomechanics, tensor analysis, asymptotic methods). Theoretical and computational research is done in all areas; in addition, experimental work is done in fluid mechanics, dynamics, and materials. An M.S. requires 8 units of credit with a thesis and 9 units without a thesis. A unit is equivalent to 4 semester hours. Most graduate courses carry ¾ of a unit or 1 unit of credit. A Ph.D. requires 8 units of course work and 8 units of thesis beyond the M.S. degree. An oral preliminary exam based on a proposed doctoral thesis topic and an oral final exam on the doctoral thesis are required for the Ph.D. degree.

Research Facilities

Facilities include experimental research laboratories, computer workstation laboratories, a research machine shop, and other support areas. The experimental research laboratories are in fluid mechanics (particle-image velocimetry, laser-Doppler velocimetry, hot-wire anemometry, turbulence, open-channel and conduit flow), composite materials testing (characterization, fatigue, fracture, temperature dependence), dynamics (acoustics, source characterization), and large-scale testing (structural mechanics). A separate Advanced Materials Testing and Evaluation Laboratory, housed in the same building as the department, provides additional multidisciplinary experimental research capabilities, including the preparation, examination, characterization, and testing of new composite materials. There are two departmental computer laboratories, one with Sun, NeXT, and HP workstations and another with Macs and PCs; all computers are networked to the campus mainframes (including those of UIUC Computing Services Office and the National Center for Supercomputing Applications). Additional in-house computing facilities include VAX and Silicon Graphics workstations associated with the fluid-mechanics research lab. The department has its own research machine shop and other professional and administrative staffs. Office space is provided for most graduate students.

Financial Aid

Financial assistance, available on a competitive basis, includes fellowships, assistantships, and waivers of tuition and fees. Fellowships carry an exemption from tuition and the service fee. Departmental appointments include teaching and research assistantships for quarter and half time and provide a stipend plus tuition and service fee exemption. Stipends vary with entry level into the program; the minimum rate for a half-time assistantship was $13,050 for the nine-month 1998–99 academic year.

Cost of Study

Students on fellowships or assistantships are exempt from tuition and the service fee. All students pay nonservice fees of $445 per semester. In 1998–99, full-time students not on fellowships or assistantships paid semester charges of $589 in total fees, plus tuition ($2230 for Illinois residents or $5668 for non-Illinois residents).

Living and Housing Costs

In the 1998–99 academic year, graduate residence halls had single rooms for $2592 and $2892 and double rooms for $2324 and $2762 per person. Optional meal contracts were available. University family apartments rented for $350 to $475 and houses rented for $511 to $687 per month. Off campus, two-bedroom apartments usually rent for $450 to $600 per month.

Student Group

The department has about 50 graduate students, of whom about two thirds are Ph.D. candidates. Students come from across the nation and from abroad. About 95 percent of the graduate students receive some form of financial aid, primarily half-time assistantships. The majority of the M.S. graduates seek industrial or government positions in research and development; most Ph.D. recipients prefer research careers in industry or government or to do research and teach at universities.

Student Outcomes

Master's students often continue on for the doctorate in mechanics, either at Illinois or at another university or they take advanced engineering positions in such companies as Ford Motor Co., Cummins Engine, Sundstrand, HKS (ABAQUS), and Stress Engineering Services.

Doctoral students usually become engineering specialists at such companies as IBM, Caterpillar, AC Technology, Structural Analysis, Ford Motor Co., Exxon, Schlumberger, and ALCOA or they take academic positions in mechanics in the U.S. or abroad—often after a postdoctoral appointment.

Location

The University is in the twin cities of Urbana and Champaign, 130 miles south of Chicago. Three interstate highways, several airlines, Amtrak, and an excellent mass-transit system serve the twin cities.

The University

A land-grant university founded in 1867, the University of Illinois enrolls approximately 58,000 students. About 34,000 of these, including about 8,500 graduate students, are at the Urbana-Champaign campus. Scientific, cultural, recreational, sports, and entertainment complexes abound on campus.

Applying

Application forms and information are obtainable on request. Applicants' files should be complete by February 15 if financial assistance is sought for August admission. GRE General Test scores are required; students whose native language is not English must also present recent TOEFL scores of 607 or higher. The application fee is $40 for domestic students and $50 for international students.

Correspondence and Information

Graduate Program Coordinator
216 Talbot Laboratory
University of Illinois
    at Urbana-Champaign
104 South Wright Street
Urbana, Illinois 61801
Telephone: 217-333-2322
Fax: 217-244-5707
World Wide Web: http://www.tam.uiuc.edu

University of Illinois at Urbana-Champaign

THE FACULTY AND THEIR RESEARCH

R. J. Adrian, Professor; Ph.D., Cambridge, 1972. Experimental fluid mechanics, instrumentation, optics, turbulence.

H. Aref, Professor and Head of Department; Ph.D., Cornell, 1980. Computational fluid dynamics, vortex structure, chaos, nonlinear dynamics.

S. Balachandar, Associate Professor; Ph.D., Brown, 1988. Fluid mechanics, computational fluid dynamics.

D. E. Carlson, Professor; Ph.D., Brown, 1965. Continuum thermomechanics, applied mathematics.

E. Fried, Assistant Professor; Ph.D., Caltech, 1991. Microstructural morphology, phase transitions, continuum mechanics, computational mechanics.

R. B. Haber, Professor; Ph.D., Cornell, 1980. Structural synthesis and shape optimization, fracture mechanics and dynamic crack propagation, computer simulation and visualization.

J. G. Harris, Professor; Ph.D., Northwestern, 1979. Elastic wave propagation, physical acoustics, asymptotic methods.

K. J. Hsia, Associate Professor; Ph.D., MIT, 1990. Micromechanics, plasticity and fracture of solids, composite materials.

R. D. Moser, Professor; Ph.D., Stanford, 1984. Turbulence physics, direct numerical simulation, spectral methods, chaos.

J. W. Phillips, Professor and Associate Head of Department; Ph.D., Brown, 1969. Experimental stress analysis, analysis of wire rope, mechanical testing.

D. N. Riahi, Professor; Ph.D., Florida State, 1974. Applied mathematics, fluid mechanics, convection and heat transfer, instability and turbulence, magnetohydrodynamics, hydrodynamics of crystal growth.

M. Short, Assistant Professor; Ph.D., Bristol, 1992. Dynamics of high-speed reacting flow, partial differential equations, asymptotic methods.

P. Sofronis, Associate Professor; Ph.D., Illinois, 1987. Materials modeling, fracture mechanics, toughening mechanisms, composite materials.

N. R. Sottos, Associate Professor; Ph.D., Delaware, 1990. Solid mechanics, thermal and residual stresses, laser interferometry, composite materials, materials interfaces.

D. S. Stewart, Professor; Ph.D., Cornell, 1981. Applied mathematics, fluid mechanics, combustion theory.

S. T. Thoroddsen, Assistant Professor; Ph.D., California, San Diego, 1991. Experimental fluid mechanics, turbulence.

R. L. Weaver, Professor; Ph.D., Cornell, 1977. Elastic wave propagation, acoustic emission, stochastic waves, ultrasonic nondestructive evaluation of materials.

CURRENT RESEARCH

Applied Mathematics: Nonlinear thermoelasticity, surface waves, chaotic systems, turbulence simulation, parabolic equations, cellular detonations, detonation instability.

Behavior of Engineering Materials: Damage in solids, brittle-to-ductile transition, adhesive bonding, dislocation initiation, fatigue in ceramics, energetic materials, creep resistance, composite materials, hydrogen embrittlement, powder consolidation, plastic deformation, interfacial properties, laser-generated stress waves, dimensional stability, damage tolerance, surface chemistry, microactuators, self-repair, health maintenance, magnetostrictive tagging, thick composites.

Computational Mechanics: Mantle convection, turbulent mixing, phase transitions, combustion-driven fracture, process simulation, extruded aluminum components, intergranular fracture, cellular detonation, combustion interfaces, level-set technology.

Dynamics, Vibrations, and Waves: Diffuse ultrasonics, materials characterization, noise generation, interfacial properties.

Mechanics of Fluids: Conditional eddies, wall turbulence, thermal convection, microscale fluidics, holographic PIV, plenum flow, coagulation, fragmentation, biomolecular MEMS, foam evolution, large-eddy simulation, corner flows, wake flows, aluminum droplet combustion, granular flow, solid-propellant ignition, solid-rocket applications, compressibility effects, boundary layers, plume convection, renormalization group theory, surface corrugation effects, microgravity thermocapillarity, alloy solidification, detonation initiation, reactive atmospheres, Taylor-Couette flow, two-phase flows.

Mechanics of Solids: Martensitic microstructure, nematic optical elastomers, linearized elastodynamics, thermoelasticity.

UNIVERSITY OF MIAMI

Department of Mechanical Engineering

Programs of Study

The Department of Mechanical Engineering offers programs leading to the Master of Science, Doctor of Philosophy, and Doctor of Arts degrees through the unified graduate program of the College of Engineering. The major areas of interest are fluid mechanics, computational fluid dynamics, heat transfer, energy conversion, internal combustion engines, alternative fuels, CAD, environmental engineering, robotics, controls, vibrations, materials science, solid mechanics, machine design, optimization, and reliability. In addition, specializations are available in the interdisciplinary fields of ocean engineering, biomedical engineering, systems design, and engineering management. Students select their own courses and research topics with the consent and advice of their advisers and within the regulations imposed by the University. Although most of their work is usually in subjects offered by the department, students are encouraged to take courses in mathematics, physics, chemistry, and other fields of engineering that fit into their overall program. There are both thesis and nonthesis master's options. Requirements for the Master of Science degree with a thesis are 24 semester credit hours of graduate course work and a 6-credit-hour equivalent of a thesis. In the nonthesis option, an approved integrated program of 36 semester credit hours of graduate course work is required. The student taking the thesis option must take an oral examination in defense of the thesis, and the student taking the nonthesis option must complete a 3-credit graduate project. The requirements for the Ph.D. degree normally consist of about 48 credit hours of graduate course work beyond the B.S. degree as well as a dissertation. There is no language requirement. The Doctor of Arts degree is designed primarily to prepare students for careers as college teachers. The program of study consists of 75 credit hours beyond the baccalaureate degree, including 6 credit hours of education courses and 12 credit hours of an internship and project. The master's degree can normally be completed in two semesters plus two summer sessions or in three semesters. The Ph.D. can be completed in five semesters after completion of the M.S. degree.

Research Facilities

The department has laboratories for research in the areas of fluid mechanics, computational fluid dynamics, heat transfer, solar energy, controls, HVAC, materials science (engines), design, and CAD. A variety of instruments and equipment is available in these laboratories. The College of Engineering also has a well-equipped machine shop. The University's central computer resources include three Alpha Server 4100s, an IBM 9672-R42, an IBM AS/400-F45, and two RS/6000s. The Engineering Computer Center has two DEC Alpha servers, a Pentium PC Novell server, and a DEC Station 5000/240. The three college-wide computer laboratories and various computer-supported classrooms include more than 200 Pentium-class personal computers with a range of scientific, engineering, and office software. The department has several workstations, graphics terminals, and stand-alone and mainframe interfacing devices. All computing facilities are connected to the University-wide network, which in turn is linked to the Internet. The department houses the Clean Energy Research Institute, which acts as a focal point for energy-related activities. The institute's goals are to conduct research to investigate clean-energy problems and to cooperate with other academic institutions, government, and private organizations in connection with these activities.

Financial Aid

A number of graduate fellowships and assistantships that provide a stipend and a tuition waiver are available to highly qualified students on a competitive basis each year. For the 1998–99 academic year, these ranged from $7600 plus tuition to $12,000 plus tuition.

Cost of Study

In 1998–99, the general fee for graduate students was $815 per semester credit hour. A student activity fee of $25 and a University fee of $62 were also required for full-time graduate students. All international students are also required to purchase medical insurance through the University of Miami Health Services. In 1998–99 the cost was $715.

Living and Housing Costs

Rooms for single students on campus were approximately $2500 to $2700 per semester in 1998–99. Proportionally lower fees were in effect for the summer sessions.

Location

Coral Gables, Miami, and south Florida offer many advantages to the graduate student who wants to concentrate on research. The Miami area is the largest pollution-free metropolitan area in the country and is well known as a recreational center that offers year-round outdoor activities, making it an attractive environment for study and living.

The University

The University of Miami was founded in 1925 and is nonprofit, nondenominational, and coeducational. It is free from religious and political control and derives its funds from tuition, alumni, and other gift contributions and teaching and research grants. Its students represent all fifty states and more than 125 other countries. Students of all religions, races, and nationalities work toward the objective of the University to produce graduates who can make significant contributions to society. The student body of the University consists of approximately 9,200 undergraduates and approximately 3,400 graduate and professional students. Of these, 138 undergraduates and 20 graduate students are in the Department of Mechanical Engineering.

Applying

Applications for admission and financial aid should be sent to the Office of Graduate Admissions. This office supplies the relevant forms and the graduate bulletin on request. Acceptance is determined by prospects of success as demonstrated by academic transcripts, references, and scores on the GRE General Test. Scores on the Test of English as a Foreign Language (TOEFL) or comparable evidence of competency in English is required of international applicants whose native language is not English. An application fee of $35 is required.

Correspondence and Information

Dr. Singiresu S. Rao
Chairman of Mechanical Engineering
Department of Mechanical Engineering
University of Miami
Coral Gables, Florida 33124-0624

Telephone: 305-284-2571
World Wide Web: http://www.eng.miami.edu

University of Miami

THE FACULTY AND THEIR RESEARCH

Jerome Catz, Professor; Sc.D., MIT, 1960. Measurement uncertainty, design.

Andrew Hsu, Associate Professor; Ph.D., Georgia Tech, 1986. Computational fluid dynamics, combustion, turbulence, aerodynamics. Current research projects: numerical simulation of turbulent combustion, numerical simulation of cardiovascular hemodynamics.

Sadik Kakac, Professor; Ph.D., Manchester, 1965. Heat transfer, nuclear engineering, thermal and fluid design. Current research projects: transient convective heat transfer in ducts, two-phase flow instabilities, electronic cooling.

Samuel S. Lee, Professor; Ph.D., Berkeley, 1965. Fluid mechanics and heat transfer.

Hongtan Liu, Assistant Professor; Ph.D., Miami (Florida), 1993. Two-phase flow instabilities, solar energy, fuel cells. Current research projects: models for simulation of fuel cell performance, transient convection in ducts with applications to electronics cooling, two-phase flow instabilities in boiling and condensing flow systems, solar energy for African countries, analysis and experimental study of hybrid solar collectors.

Ramarathnam Narasimhan, Research Assistant Professor; Ph.D., Miami (Florida), 1988. Heat transfer, computational fluid mechanics, flow visualization.

Singiresu S. Rao, Professor and Chairman; Ph.D., Case Western Reserve, 1972. Optimization and reliability in engineering design, structural control, design for manufacturability, fuzzy systems. Current research projects: fuzzy boundary element method for the analysis of imprecisely defined systems, internal analysis methods for design and manufacturing, game theory approaches for the design of mechanical and structural systems, condition monitoring of rotating machines using fuzzy logic, solution of large-scale engineering design optimization problems using interior point and parallel programming methods, game theory and fuzzy optimization approaches for air pollution control in thermal power plants.

Ali Shahin, Assistant Professor; Ph.D., Purdue, 1995. Robust control, smart actuators, adaptive structures. Current research projects: pro-engineer projects with Landis & Gyr/Siemens, robot design and manufacturing for pathology laboratory, control of vortex shedding, control of SMA actuators.

Narendra Simha, Assistant Professor; Ph.D., Minnesota, 1995. Biomechanics and continuum mechanics, martensitic phase transformations, biomaterials. Current research projects: fatigue properties of NiTi shape memory alloys, kinetics of phase boundaries, transformation toughening of zirconia ceramics.

Michael R. Swain, Associate Professor; Ph.D., Miami (Florida), 1979. Internal combustion engines, hydrogen safety analysis. Current research projects: safety analysis of gaseous fueled systems (comparisons of hydrogen with natural gas and lpg), hydrogen fueled engine development, V-8 high-performance engine development.

T. Nejat Veziroglu, Professor; Ph.D., London, 1959. Two-phase fluid flows, solar energy, hydrogen energy. Current research projects: hydrogen energy system, sustainable future, environmental impacts of energy systems, solar hydrogen production, hydrogen-hydride air conditioning, global warming: consequences and remedies.

Kau-Fui Wong, Professor; Ph.D., Case Western Reserve, 1977. Fluid/mass transfer in porous media, environment-energy systems, expert systems, ocean pollution, solid waste. Current research projects: spilled oil entrapment devices, national ambient air pollution quality standards-synthesis study, groundwater contamination and transport, exergetic analysis and optimization of complex energy systems, environmental chambers for indoor air quality testing.

REPRESENTATIVE SPONSORED PROJECTS

Development of Comprehensive Computer Models for Simulation of Fuel Cell Performance (Department of Energy): The objectives of the project are to identify all the important phenomena and parameters in fuel cell operation, to study their interrelationship, and to develop computer models that are capable of solving all the governing equations in unified domain simultaneously. The model's inputs are the same parameters as those in real-life fuel cell operations, thus eliminating errors due to arbitrary or simplified boundary conditions.

Experimental and Theoretical Analysis of Transient Convection in Ducts with Protruding Surfaces—Application to Electronic Cooling (U.S.–Brazil collaboration, NSF): The main objectives of this project are to study the fundamental mechanisms of transient-forced convection in channels with discrete heat sources and to develop semi-analytical expressions for correlating parameters of practical significance that can be used in future design applications.

Hydrogen-Fueled Engine Research (Ford Motor Company): The hydrogen-fueled engine research initially involved the conversion of a vehicle to operate on compressed gaseous hydrogen. Over the past several years, some twenty-five hydrogen-fueled engines have been constructed and tested. They utilized gas-mixing carburetors, port fuel injection, or direct fuel injection. In addition, throttled and unthrottled versions were constructed, employing water injection, air injection, or exhaust gas recirculation. Previous findings have shown that hydrogen-fueled engines can operate with emissions levels below detectability of standard emission equipment, and brake thermal efficiencies of 40 percent are possible.

Hydrogen Safety Research (Department of Energy): The gaseous fuel safety analysis research has been ongoing since 1988. The core of the work centers around comparisons of hydrogen, natural gas, LPG, and gasoline in various accident scenarios. The accident scenarios have varied from large pipeline leakage to residential stove leakage to vehicle leakage. Investigations are made both experimentally and through computer analysis. Full-scale models of garages, buses, and vehicles have been constructed to test the results of leakage rates up to 3,000 standard cubic feet per minute.

Control of Flexible Structures Due to Vortex Shedding: Sometimes, at various wind velocities, structures start to vibrate perpendicular to the direction of the wind. This highly nonlinear phenomena is know as vortex shedding. To control this phenomena, one must use nonlinear control techniques, such as describing fractions. The nature of the problem is such that linear robust control methodologies can be used to eliminate this undesired wind response.

Optimal H-infinity Control of Structural Vibrations Using Shape Memory Alloy Actuators: This work investigates robust control of structural vibrations using shape memory alloy (SMA) wires as actuators. The mathematical model for these SMA actuators is derived with emphasis on model uncertainty. The linearization of the relation between stress and temperature dynamics of SMA actuators is analyzed for active control. To handle the uncertainties caused by the linearization and the neglected high-frequency dynamics, optimal H-infinity control was employed to design a controller. An example is used to demonstrate the design procedures, and the control system is tested in a nonlinear environment.

UNIVERSITY OF MICHIGAN

Department of Mechanical Engineering and Applied Mechanics

Programs of Study

The Department of Mechanical Engineering and Applied Mechanics (MEAM) offers M.S.E. and Ph.D. degrees through the University of Michigan's Rackham Graduate School. MEAM also offers a College of Engineering Master of Engineering degree in automotive engineering. All master's degrees require the completion of 30 graduate credit hours. Students may choose from three options in order to satisfy the M.S.E. degree requirements: the course work option, the limited research option, and the master's thesis option. For students interested in multidisciplinary studies, MEAM offers an interdepartmental degree program entailing a single M.S.E. in two disciplines. Students may also complete a dual-degree program and earn two distinct M.S.E. degrees. The Ph.D. degree requires 18 graduate credit hours beyond the master's degree, the successful completion of qualifying examinations and a preliminary examination of the dissertation subject, and the completion and oral presentation of the dissertation.

Research Facilities

MEAM offers students a vast selection of outstanding research opportunities with its internationally renowned faculty, whose interests include automotive engineering; biomechanics; design; dynamics, vibrations, and acoustics; computational mechanics; fluid mechanics; heat transfer and combustion; manufacturing; materials; solid mechanics; and systems and controls. There are several state-of-the-art laboratories in each research area. MEAM's strong, far-reaching research funding base is testimony to the interest its research activities draw from around the world. Some of MEAM's many public funding sources include NSF, NIH, NASA, and the Department of Defense. Private sources of funding include Ford Motor Company, General Motors, Chrysler Corporation, General Dynamics, and Caterpillar, to name a few. The College of Engineering's new 225,000-square-foot Media Union houses the Engineering Library, the Computer Aided Engineering Network (CAEN), a virtual reality laboratory, and design and innovation studios. The Engineering Library collection, regarded as one of the finest in the world, includes more than 600,000 volumes and 1 million technical reports, historical records of early engineering research, U.S. and international industry standards, and resources on intellectual property, including U.S. patents and trademarks. The library also subscribes to more than 2,500 engineering journals. In addition, students have access to the University's network of twenty-five divisional libraries. The CAEN network is a premier computing environment for engineering-related research and education. CAEN maintains a fully integrated, multivendor network of more than 5,000 advanced-function workstations and specialized high-performance computers that serve students and faculty and staff members.

Financial Aid

A variety of financial support opportunities are available for MEAM graduate students. All recipients of financial aid from MEAM are required to be involved in departmental research activities with a faculty member. Three primary forms of financial aid are available: fellowships from the University of Michigan and industry endowments; research assistantships, in which graduate students work with a faculty member on a sponsored research project; and graduate student instructorships (a number of these appointments are available in undergraduate courses). Financial aid packages typically include full tuition and fees, a monthly stipend of at least $1450, and participation in the University's group health insurance programs.

Cost of Study

Tuition and fees for full-time study during 1998–99 were $5749 per term for Michigan residents and $10,848 per term for nonresidents, plus fees assessed for registration costs, student government, health services, intramural and recreational facilities, student facilities, and computer use. Costs are subject to change.

Living and Housing Costs

The estimated 1999–2000 living costs, including room and board, transportation, and miscellaneous personal needs, are $10,800 for the calendar year. University housing is available for both single and married students. Private off-campus housing within walking distance of the campus is also available.

Student Group

University of Michigan students come from all fifty states and more than 100 other countries. In the College of Engineering, graduate student enrollment is nearly 2,050 in more than twenty programs. Approximately 350 students are enrolled in MEAM programs, with approximately one half pursuing doctoral degrees.

Location

The Ann Arbor area, with a population of more than 250,000, is considered one of the most livable places in the country. Ann Arbor is a vital growing city, which provides the culture and opportunities of a major metropolitan area with the hometown family-oriented feeling of the Midwest. Acknowledged as the center of the state's booming high-technology industry and as a cultural mecca as well, Ann Arbor offers a landscape of parks, office buildings, boutiques, historic preservation areas, shopping malls, bike paths, and busy tree-lined streets. Ann Arbor is a delightful place in which to live, study, work, and play.

The University

The University of Michigan, recognized as the first model of a state university in America, was founded in 1817. The quality of its academic programs places Michigan in the top ten colleges and universities nationwide. In 1998, *U.S.News & World Report* ranked six of the University's graduate schools in the top twelve nationally, including the College of Engineering, which is ranked fourth. The Department of Mechanical Engineering and Applied Mechanics is particularly proud that its graduate programs have consistently ranked in the top five in the U.S. The University is the nation's leading public research university, with more than $458 million in research expenditures.

Application

The deadline for submitting an application is forty-five days before the start of the first semester of study. Students applying for financial aid for the fall semester must submit applications by January 15. MEAM requires the Graduate Record Examinations (GRE) General Test and three letters of recommendation. MEAM encourages applications from students with backgrounds in other engineering disciplines and in mathematics and physics. Outstanding students who have earned only an undergraduate degree are also encouraged to apply for direct admission to the doctoral program.

Correspondence and Information

MEAM Academic Services Office
Department of Mechanical Engineering and Applied Mechanics
2206 G. G. Brown Building
University of Michigan
Ann Arbor, Michigan 48109-2125
Telephone: 734-763-4277
E-mail: me.grad.application@umich.edu
World Wide Web: http://www.engin.umich.edu/dept/meam

University of Michigan

THE FACULTY AND THEIR RESEARCH

Faculty member e-mail addresses and further information may be obtained through the World Wide Web at http://www.engin.umich.edu/dept/meam/.

Rayhaneh Akhavan, Associate Professor; Ph.D., MIT, 1987. Fluid dynamics, turbulence, hydrostability, computational and experimental techniques.

Vedat S. Arpaci, Professor; Sc.D., MIT, 1958. Heat transfer, gas radiation and stability, applied mathematics.

Ellen M. Arruda, Assistant Professor; Ph.D., MIT, 1992. Mechanics of material deformation and failure in materials processing, deformation processes involved in strain-induced crystallization and in semicrystalline polymers.

Dennis M. Assanis, Arthur F. Thurman Professor; Ph.D., MIT, 1985. Engine design and processes, especially adiabatic engines.

Arvind Atreya, Professor; Ph.D., Harvard, 1983. Combustion, fire research, heat and mass transfer, flame radiation, pollutant formation.

James R. Barber, Professor; Ph.D., Cambridge, 1968. Solid mechanics, elasticity, thermoelasticity, design, crack and contact problems, stress analysis for design, friction and wear.

Claus Borgnakke, Associate Professor; Ph.D., Denmark Technical, 1977. Fluid mechanics, heat transfer, turbulence modeling flow calculations, turbulent combustion.

Diann Erbschloe Brei, Assistant Professor; Ph.D., Arizona State, 1993. Smart system design and analysis using materials such as piezoelectric and shape memory alloys, actuator and sensor design, microelectromechanical system design.

Michael M. Bridges, Assistant Professor; Ph.D., Clemson, 1994. Controls and robotics.

Steven L. Ceccio, Associate Professor; Ph.D., Caltech, 1990. Fluid mechanics, multiphase flows, cavitation, instrumentation.

Michael Chen, Professor; Ph.D., MIT, 1961. Heat transfer due to particle impact, measurement of solids in a fluidized bed.

David E. Cole, Joint Associate Professor; Ph.D., Michigan, 1966. Automotive engineering, future trends in automotive technology, strategic planning and policy analysis for the automotive industry, internal combustion engines.

Maria Comninou, Professor; Ph.D., Northwestern, 1973. Solid mechanics, fracture, wave propagation, contact problems.

David R. Dowling, Assistant Professor; Ph.D., Caltech, 1988. Turbulent mixing of mass, heat, and momentum in chemically reacting and inert flows; atmospheric and oceanic fluid mechanics and wave propagation.

Debasish Dutta, Associate Professor; Ph.D., Purdue, 1989. Computer-aided design, geometric modeling.

William J. Endres, Assistant Professor; Ph.D., Illinois at Urbana-Champaign, 1992. Machining processes; traditional and nontraditional machining systems modeling; applied dynamics and control theory; forming, design, and manufacturing methodology.

R. Brent Gillespie, Assistant Professor; Ph.D., Stanford, 1996. Dynamics systems and controls.

Steven A. Goldstein, Joint Professor; Ph.D., Michigan, 1981. Biomechanics of tissues, design and development of orthopedic implants.

Karl Grosh, Assistant Professor; Ph.D., Stanford, 1994. Structural acoustics, structural vibrations, smart structures and control, acoustic radiation and biomedical imaging.

Scott J. Hollister, Joint Professor; Ph.D., Michigan, 1991. Bone mechanics, design of total joint replacements, computational mechanics.

John W. Holmes, Associate Professor; Ph.D., MIT, 1986. Mechanical behavior of materials, fatigue of ceramic composites and intermetallics, oxidation and corrosion.

Shixin "Jack" Hu, Associate Professor; Ph.D., Michigan, 1990. Statistical process control, dynamic system identification, automotive design and manufacturing.

Gregory M. Hulbert, Associate Professor; Ph.D., Stanford, 1989. Computational mechanics, finite-element methods.

Stanley Jacobs, Joint Professor; Ph.D., Harvard, 1963. Geophysical fluid mechanics, applied mathematics.

Elijah Kannatey-Asibu Jr., Professor; Ph.D., Berkeley, 1980. Control of manufacturing processes, acoustic emission monitoring of manufacturing processes and structural integrity.

Bruce H. Karnopp, Associate Professor; Ph.D., Wisconsin–Madison, 1965. Dynamics, vibrations/vibration control, applied mathematics.

Massoud Kaviany, Professor; Ph.D., Berkeley, 1979. Fluid dynamics, heat transfer, convection behavior.

Noboru Kikuchi, Professor; Ph.D., Texas at Austin, 1977. Finite-element methods, metal forming, friction-contact problems, impact problems, numerical-grid generation.

Yoram Koren, Paul G. Gobel Endowed Professor; Ph.D., Technion (Israel), 1971. Robotics, control of manufacturing systems, optimization and adaptive control.

Sridhar Kota, Associate Professor; Ph.D., Minnesota, 1987. Machine design, mechanical computer-aided design, kinematics and mechanisms, application of artificial intelligence in mechanical design, conceptual design and experimental systems.

Arthur D. Kuo, Assistant Professor; Ph.D., Stanford, 1992. Biomechanics, dynamics, control systems, optical control.

Liwei Lin, Assistant Professor; Ph.D., Berkeley, 1993. Microelectromechanical systems (MEMS), microsensors and microactuators, micromachining processes, mechanics of microsystems.

Jyotirmoy Mazumder, Robert H. Lurie Professor of Engineering; Ph.D. Imperial College (London), 1978. Laser-aided manufacturing, atom to application for nonequilibrium synthesis, mathematical modeling, spectroscopic and optical diagnostics of laser-materials interaction.

David Mead, Associate Professor; Ph.D., Cambridge, 1988. Rheo-optical methods of studying material flow, polymers and polymer processing.

Herman Merte Jr., Professor; Ph.D., Michigan, 1960. Heat transfer, phase change dynamics, cryogenics.

Jun Ni, Professor; Ph.D, Wisconsin–Madison, 1987. Precision engineering, applications to manufacturing systems, manufacturing processes.

Jwo Pan, Associate Professor; Ph.D., Brown, 1981. Fracture mechanics, plasticity.

Panos Papalambros, Professor; Ph.D., Stanford, 1979. Design theory and methodology, nonlinear programming, design optimization, design education.

Huei Peng, Assistant Professor; Ph.D., Penn State, 1992. Optimal and adaptive control, learning control for nonlinear systems, IVHS.

Noel C. Perkins, Associate Professor; Ph.D., Berkeley, 1986. Vibration analysis, nonlinear cable dynamics, vibration and stability of axially moving materials, experimental dynamics.

Christophe Pierre, Professor; Ph.D., Duke, 1985. Structural dynamics, vibrations, nonlinear dynamics, aeroelasticity.

Kazuhiro Saitou, Assistant Professor; Ph.D., MIT, 1996. Design automation and optimization, discrete optimization, design for automated assembly, assembly in MEMS, evolutionary computation in design.

Ann Marie Sastry, Assistant Professor; Ph.D., Cornell, 1994. Mechanics of composite materials, transport in heterogenous media, structural analysis of composite components.

William W. Schultz, Associate Professor; Ph.D., Northwestern, 1982. Hydrodynamic stability, tribology, viscoelastic flows.

Richard A. Scott, Professor; Ph.D., Caltech, 1964. Macromechanics of composite media, nonlinear oscillations, vibrations, optimization.

Volker Sick, Associate Professor; Dr.rer.nat., Heidelberg, 1992. Laser techniques for combustion diagnostics, molecular spectroscopy, internal engine combustion, detailed modeling.

Steven Skerlos, Assistant Professor; Ph.D., Illinois at Urbana-Champaign, 1999. Environmental conscience manufacturing.

Jeffrey L. Stein, Professor; Ph.D., MIT, 1983. System dynamics, controls, flexible manufacturing systems, computer-aided manufacturing, servosystems, prosthetics.

John E. Taylor, Joint Professor; Ph.D., Michigan, 1964. Structural mechanics, optimal design.

Michael D. Thouless, Associate Professor; Ph.D., Berkeley, 1984. Materials, solid mechanics, ceramics, fiber composites, mixed-mode fracture of interfaces, adhesion and failure mechanics of thin films and coatings.

Dawn Tilbury, Assistant Professor; Ph.D., Berkeley, 1994. Control theory: nonholonomic systems, motion planning, exterior differential systems, nonlinear systems, intelligent control, hierarchical control systems, hybrid systems.

Gretar Tryggvason, Professor; Ph.D., Brown, 1985. Fluid mechanics, computational methods, multiphase flow.

A. Galip Ulsoy, William Clay Ford Professor of Manufacturing and Chairman; Ph.D., Berkeley, 1979. Modeling, analysis and control of mechanical systems, applications to manufacturing systems.

Alan S. Wineman, Professor; Ph.D., Brown, 1964. Continuum mechanics, nonlinear elasticity and viscoelasticity, large deformation effects in polymers.

Margaret Wooldridge, Assistant Professor; Ph.D., Stanford, 1995. Thermal and fluid sciences.

Wei-Hsuin Yang, Professor; Ph.D., Stanford, 1964. Mechanics of solids, plasticity, numerical methods.

Wen-Jei Yang, Professor; Ph.D., Michigan, 1960. Heat transfer, process dynamics, biothermal sciences, thermodynamics.

UNIVERSITY OF MINNESOTA

Department of Mechanical Engineering

Programs of Study

Both M.S. and Ph.D. degrees are offered in mechanical engineering. Two master's degree plans are available. A Master of Science degree (M.S.) Plan A requires a thesis and 20 semester course credits. A minimum of 14 course credits is taken in the major field and a minimum of 6 in one or more related fields. A Master of Science degree (M.S.) Plan B requires 30 semester course credits and one to three papers based on independent study. A minimum of 14 credits is taken in the major field and a minimum of 6 in one or more related fields, with the remainder taken from either major or nonmajor categories. The division of credits between the major and minor is quite flexible. There is no foreign language requirement. The Ph.D. requires preliminary written and oral examinations, 44 semester course credits, and a final oral defense of the thesis. A student may substitute for the minor a suitable supporting program that may embrace several disciplines. The University of Minnesota operates under the semester system as of fall 1999.

Research Facilities

The research facilities include computer-aided design and manufacturing and robotics facilities; environmental control and solar laboratories; power, propulsion, and combustion laboratories; a very high concentration ratio 4.2-meter solar furnace facility; a bioengineering laboratory; finite elements and computational mechanics, design, and instrumentation laboratories; a composites manufacturing laboratory; and a human factors engineering laboratory. The Particle Technology Laboratory and Heat-Transfer Laboratories are among the best equipped in the world. An Engineering Research Center on Plasma-Aided Manufacturing operates jointly with the University of Wisconsin–Madison. This center includes a sophisticated laser-scattering system for diagnostics of processing plasma. Students and faculty have access through high-speed networks to the most powerful computer systems for research and instruction in the world: the CMZ and CM5 Connection Machines, CRAY-2, CRAY X-MP, CRAY C90, CRAY T3D, and CRAY T3E supercomputers. CAD facilities include numerous workstations (e.g., Sun, Silicon Graphics, IBM, and Apple) plus a wide complement of commercial software.

Financial Aid

Qualified students may receive assistance in the form of National Science Foundation (NSF), Department of Defense, Graduate School, and corporate fellowships and teaching assistantships. In addition, many faculty members conduct research programs supported by the NSF, National Aeronautics and Space Administration, U.S. Air Force, Department of Energy, Department of Defense, National Institutes of Health, Environmental Protection Agency, industrial sponsors, and others. These programs employ students on research assistantships. The research topic may also be used for their theses. Fellowships and assistantships provide tuition scholarships and medical coverage.

Cost of Study

For 1998–99, tuition for Minnesota residents was $1710 per quarter (covering 7 to 14 credits). Nonresident tuition was $3358 per quarter (7 to 14 credits). Student service fees were about $160 per quarter. Modest increases in these rates may occur for 1999–2000 pending Regents' action. Graduate (teaching or research) assistants holding quarter-time appointments pay half the resident tuition rate, and those with half-time appointments pay no tuition. Detailed information on tuition and service fees can be obtained from the Office of the Dean of the Graduate School.

Living and Housing Costs

Information about housing may be obtained from the Housing Bureau, Comstock Hall. Accommodations are available for single students in modern residence halls and for married students in permanent one- and two-bedroom units. Early application for family units is advisable. There is also a good selection of privately owned housing near the University or on University express bus lines.

Student Group

There are approximately 10,000 graduate students on the Twin Cities campus, about 300 of whom are in mechanical engineering. The student body includes students from all sections of the country and from every continent.

Student Outcomes

Graduates of the program have found engaging opportunities with a diverse array of employers. Research positions may range from local industry, such as Honeywell, 3M, and a number of engineering and technical firms, to national companies such as Hewlett Packard and Applied Materials, to overseas firms. Some candidates opt to return to academia to teach or do research at local, national, and international universities.

Location

The Twin Cities offer outstanding opportunities for cultural and recreational activities on or very near the campus. The Twin Cities are renowned for varied attractions such as the Minnesota Orchestra and St. Paul Chamber Orchestras, and the Science Museum and Omnitheater. The Guthrie Theater, the University Theater, and many regional theater companies make the cities a rich drama center. More than 11,000 Minnesota lakes, many within the St. Paul or Minneapolis city limits, and the contiguous vacation lands of northern Wisconsin and the St. Croix River Valley are home to many summer and winter sports.

The University and The Department

The mechanical engineering department provides an unusually stimulating environment for graduate study. Many members have advanced degrees in the basic sciences as well as in engineering. A strong program of seminars and visiting lectureships has brought many well-known scientists and engineers to the department to offer courses in their specialties. The department is famous for its research and interest in a broad spectrum of technological endeavors. Certain faculty members have received honorary degrees from international universities, and several have been elected to the National Academy of Engineering. University facilities for swimming, tennis, golf, basketball, skating, broomball, handball, squash, and other sports are available to all students. There is also an active intramural program.

Applying

Application deadlines for admission are as follows: December 31 for the fall semester and October 25 for the spring semester. Admission is granted for both fall and spring semesters. All admission applications submitted no later than December 15 of the preceding year for which a September appointment is requested are automatically considered for financial assistance. All correspondence concerning admission should be directed to the Graduate School of the University, in Minneapolis.

Correspondence and Information

Dr. T. H. Kuehn, Director of Graduate Studies
Department of Mechanical Engineering
University of Minnesota
111 Church Street, SE
Minneapolis, Minnesota 55455

University of Minnesota

THE FACULTY AND THEIR RESEARCH AREAS

A. Bar-Cohen: boiling and two-phase flow, electronic packaging and manufacturing process, management of technology. J. C. Bischof: bioengineering, bioheat and mass transfer, cryobiology, hyperthermia. P. L. Blackshear Jr., Professor Emeritus: bioengineering, combustion, applied thermodynamics. T. R. Chase: computer-aided design, mechanical engineering database, kinematics, machine design. J. H. Davidson: solar energy, air filtration, environmental engineering. M. Donath: sensors and control systems as applied to vehicles and robotics. W. K. Durfee: product design, real-time control, biomechanics, rehabilitation engineering. E. R. G. Eckert, Regents' Professor Emeritus: heat and mass transfer, thermodynamics, gas turbines. M. Erdal: processing of high-performance materials, with emphasis on advanced ceramics and composites manufacturing; fluid mechanics; heat and mass transfer; advanced numerical analysis. A. G. Erdman: computer-aided design, kinematics, biomechanics, microelectromechanical systems. S. L. Girshick: plasma science and technology, aerosol science and technology. R. J. Goldstein: heat transfer, thermodynamics, fluid mechanics. J. V. R. Heberlein: plasma technology, electrode effects, plasma coating and waste-treatment processes, nanosize particle synthesis. W. E. Ibele: heat transfer, thermodynamics, power. D. B. Kittelson: energy conversion particle technology, combustion and propulsion. B. E. Klamecki: manufacturing process modeling and control, mechanical systems monitoring, tribology. U. R. Kortshagen: low-pressure processing plasmas, plasma contamination control, plasma modeling, electrical and optical diagnostics. T. H. Kuehn: HVAC and refrigeration, heat and mass transfer, filtration. F. A. Kulacki: convective transfer in porous and fractured media, geophysical heat transfer, modeling of nuclear/toxic waste disposal systems, technology-based educational methods. J. Lewis: biomechanics. P. Y. Li: Nonlinear and intelligent control, biomechanics, rehabilitation engineering, transportation systems, manufacturing. B. Y. H. Liu: particle technology, environmental control, solar energy; S. C. Mantell: manufacturing and design with composite materials; V. A. Marple: particle technology and aerosol science, environmental engineering. P. H. McMurry: aerosol science and engineering, environmental engineering. K. Ogata: control systems, optimization techniques. S. V. Patankar: heat and mass transfer, fluid mechanics, numerical methods. E. Pfender: arc technology, plasma heat transfer and plasma processing. D. Y. H. Pui: particle technology, environmental engineering. S. Ramalingam: manufacturing sciences, machining, metalworking, tribology, arc technology, coating technology. CAM. J. W. Ramsey: heat and mass transfer thermal environmental engineering. C. J. Scott: heat and mass transfer, fluid mechanics, thermodynamics. T. W. Simon: heat transfer, fluid mechanics, thermodynamics. E. M. Sparrow: applied heat transfer and thermal design, thermal issues in biomedical engineering. K. A. Stelson: manufacturing, system dynamics and controls. P. J. Strykowski: fluid mechanics, stability, mixing, shear flow control. K. K. Tamma: finite elements, computational mechanics, structural dynamics and contact-impact, fluid-thermal-structural interactions, computational microscale/macroscale heat transfer, process modeling and manufacturing simulations. M. R. Zachariah: optical diagnostics, aerosol science, nanostructured materials, combustion.

Applied Thermodynamics, Combustion, and Propulsion

Studies are conducted in thermodynamics, kinetics, and transport phenomena in chemically reacting systems. Graduate research includes fundamental combustion studies as well as studies of combustion in engines and combustion synthesis of materials. Problems of ignition, flame propagation, and the relationships between chemical kinetics and transport phenomena are studied. Study of both gaseous and particulate atmospheric emissions from combustion systems is being pursued in collaboration with industrial firms. Very high temperature solar thermochemical and solar electrochemical processes are also studied, using a 7-kW (thermal) solar furnace whose actual solar concentration ratio is 7,000 suns. The Center for Diesel Research specializes in developing and evaluating technology to reduce exhaust pollutants from heavy-duty diesel engines. The Center offers graduate and undergraduate students opportunities to work on interdisciplinary applied research projects, to interact closely with project sponsors, and to publish results.

Bioengineering

Current areas of focus include musculoskeletal engineering, heat and mass transfer in biological systems, human factors, and medical device design. Within musculoskeletal engineering, sport biomechanics studies include athletic performance in speed skating, in-line skating, soccer, and rowing, with the objective of improving performance and equipment design. Additional musculoskeletal engineering work focuses on orthopedic implants, material failure in arthritis, surgical reconstruction of ligaments, kinematics and dynamics of joint structures, musculoskeletal modeling, control systems for orthoses and prostheses, robotic manipulation and locomotion, and computer-aided diagnostic techniques. In heat and mass transfer, work is ongoing to measure, predict, and optimize biophysical and viability changes induced in cells and tissues during freezing and heating for clinical applications of cryosurgery, cryopreservation, and hyperthermia. In addition, the effects of heat and pressure on wound causation and healing, as well as the development and use of noninvasive heating or cooling to modulate human body thermoregulatory functions in a clinical setting, are being studied. In human factors, research in perceptual and human machine control issues related to driving vehicles and manipulating objects, as well as research in ergonomics, concerns health and safety at the human-machine interface in the workplace. These projects also involve cooperation with the biological sciences group. In medical device design, projects using MEMS (Micro-Electro Mechanical Systems) technology are being pursued in an interdisciplinary setting to design and build instrumentation for eye surgery and for many other medical applications.

Design and Control Systems

Current research activities include biomedical engineering; computer-based user interfaces; robotics; vehicle guidance and collision avoidance; fluid power systems; vibration; sensor technologies; control of nonlinear systems; real-time-embedded control systems; kinematics and computer-aided mechanism synthesis; micromechanisms and actuators; finite-element developments in modeling and analysis; development of computational methods for fluid-thermal-structural interactions; learning and intelligent systems; integration of intelligence, sensors, and actuators for materials processing and for transportation; engineering database; interface between design and manufacturing; human-machine interfaces; and virtual environments for product prototyping.

Environmental Control/Particle Technology

The Particle Technology Laboratory offers opportunities for studies on the properties and behavior of airborne particles (aerosols) and methods for their generation, measurement, sampling, and analysis. The laboratory is involved in research related to air quality modeling/measurement and emission control; contamination control in semiconductor manufacturing; air, gas, and liquid filtration; aerosol charging and collection by electrostatic precipitators; and airborne particles in the workplace and mining environments. Recent work on nanometer particles includes development of generation and detection methods, charging, and various industrial and biological applications. The Thermal Environmental Laboratory is involved in studies of ventilation, indoor air quality, energy conservation in building systems, and solar energy.

Heat Transfer and Thermodynamics

The Heat-Transfer Laboratory offers opportunities for study in a variety of subjects in thermodynamics, fluid mechanics, and heat transfer. Current research includes high-temperature heat transfer; plasma spraying; plasma synthesis, including diamond synthesis; arc technology; plasma waste treatment; thermomechanics; thermal convection; ebullient cooling; film cooling; enhanced heat transfer; heat transfer in turbulent flows; thermal radiation with participating media; optical measurements of fluid velocity; solidification and melting; mathematical modeling of turbulent flows; numerical techniques; equation of state and transport property determinations; and second-law analysis of thermodynamics devices and cycles. Many of these studies address critical thermal applications, such as turbine blade cooling; plasma spraying, thermal plasma CVD, and plasma sintering; mixing in high-speed combustion chambers and heat transfer in engines; cooling and thermostructural design of electronic and optical components, cryopreservation, and cryosurgery. Other studies are of a fundamental nature, leading to a basic understanding of the relevant thermofluid mechanisms.

Manufacturing

Research is being conducted in a number of areas with significant potential for improving productivity in manufacturing industries through technological innovations. Current research areas include robotics, intelligent machines, CAD/CAM, computer-adaptive control of manufacturing, process modeling, tribology, solid-state sensors, deformation processing, polymer processing, composite material processing, ceramic processing, and finite-element modeling. The interdependence of design and manufacturing processes is also recognized, and research toward the goal of completely integrated CAD/CAM systems is being pursued. Research topics are also available in the fields of flexible manufacturing systems, unmanned factories, artificial intelligence and knowledge-based systems, new processing technologies, and advanced material systems.

UNIVERSITY OF MISSOURI–COLUMBIA

Department of Mechanical and Aerospace Engineering

Programs of Study

The department offers graduate programs leading to M.S. and Ph.D. degrees. The master's degree program requires a minimum of 30 credit hours, including a maximum of 9 hours of thesis or 6 hours of project work. The candidate must submit a thesis or a project report to an examining committee, and a final oral examination is required. The doctoral degree program requires a minimum of 72 credit hours beyond the bachelor's degree. A written qualifying examination is required for students who do not meet the departmental exemption rule. The candidate must pass a comprehensive examination and submit and defend his or her dissertation at a final oral examination.

Major subject areas include AI/expert systems, automation, bioengineering, combustion, computational fluid dynamics, control, design optimization, finite and boundary element methods, fluid and aerosol mechanics, heat transfer, intelligent systems, manufacturing processes, materials science, mechanical synthesis, mechatronics, microprocessor applications, nonlinear structural mechanics, parallel computation, plasticity and fracture, residual stress, robotics, spectral and wavelet methods, thermal systems design, and ultrasonic nondestructive evaluation.

Research Facilities

In addition to a number of specialized laboratories in the major subject areas, the department has or has access to such specialty items as MTS and Instron material and structural test equipment, a wind tunnel, X-ray and electron microscope equipment, computer control systems, and the largest university nuclear research reactor in the United States.

The Engineering Computer Network (ECN) provides advanced engineering computation capability. CAD/CAM and graphics are the primary emphasis, although artificial intelligence, multiple high-level programming languages, and computational and simulation libraries are also available. The ECN operates three DEC super minicomputers and two remote terminal sites, including microcomputers and workstations.

Financial Aid

Graduate research and teaching assistantships provide stipends ranging from $4700 to $5200 per semester, plus a waiver of fees. Industrial fellowships provide $2000 per semester plus a waiver of fees. Most students receive either assistantships or fellowships. A few University fellowships, including one for minorities, are available for qualified students in addition to the assistantships.

Cost of Study

Fees for 1999–2000 are $167.80 per credit hour for residents or $504.80 for nonresidents. Supplemental fees include $8.30 per credit hour for computing and $36.80 for engineering courses. International students must take 9 credit hours (full-time) for the fall and winter semesters until they satisfy the credit hour requirement.

Living and Housing Costs

Residence halls (dormitories, including meals) are available at an average cost of $5500 per year. Married student housing rents for $265 to $350 per month. Privately owned rooms and apartments in the area are available at comparable and higher rents. Estimated living expenses (food, housing, books, supplies, medical insurance, and personal expenses) approximate $7000 per year.

Student Group

The department currently has approximately 40 graduate students and 280 undergraduate students. Approximately 40 percent of the graduate students are doctoral students, and 60 percent are international students.

Location

The University is located in Columbia, Missouri, midway between and 120 miles from St. Louis and Kansas City. The city population is approximately 75,000. Also located in Columbia are two small, private colleges: Stephens College and Columbia College. The city has consistently ranked high on *Money* magazine's list of most livable cities.

The University

The University of Missouri–Columbia (MU), established in 1839, is the oldest state university west of the Mississippi River. MU is the largest of the four campuses of the Missouri system. Other campuses are located in Kansas City, Rolla, and St. Louis. MU enrolls a total of 23,500 students, including 6,000 graduate students in ninety-seven graduate degree programs. Also located on the MU campus are the Schools of Journalism, Law, and Medicine as well as the College of Veterinary Medicine. With its diverse program, MU is one of the five most comprehensive universities in the nation.

Applying

Application forms and further information may be obtained from the department's Director of Graduate Studies. Applications for admission to graduate programs are accepted throughout the year. Students may begin study at the start of the fall, winter, or summer session. GRE General Test scores are required of all applicants. A minimum TOEFL score of 550 is recommended for international applicants.

Correspondence and Information

Director of Graduate Studies
Mechanical and Aerospace Engineering
E2412 Engineering Building East
University of Missouri–Columbia
Columbia, Missouri 65211
Telephone: 573-882-2085
Fax: 573-884-5090
E-mail: mae@showme.missouri.edu
World Wide Web: http://www.ecn.missouri.edu

University of Missouri–Columbia

THE FACULTY AND THEIR RESEARCH

William L. Carson, Professor; Ph.D, Iowa; PE. Mechanics, instrumentation, biomechanics, dynamics, controls, and design.

Uee Wan Cho, Associate Professor and Director of Graduate Studies; Ph.D., Brown. Solid mechanics, creep and plasticity, constitutive relations, continuum damage mechanics.

Roger C. Duffield, Professor; Ph.D., Kansas. Vibrations, solid mechanics, structural mechanics, structural and systems dynamics.

A. Sherif El-Gizawy, Associate Professor; Ph.D., Waterloo. Manufacturing design, process modeling, integrated computer-aided manufacturing, expert systems applications in manufacturing.

Roger Hill, Assistant Professor; Ph.D., Texas at Austin. Computational fluid dynamics and heat transfer, direct numerical simulation of transitional and turbulent flows.

Aaron D. Krawitz, Professor; Ph.D., Northwestern. Materials science, physical metallurgy, X-ray and neutron diffraction, composites.

Yuyi Lin, Associate Professor; Ph.D., Berkeley; PE. CAD/CAM, mechanical design optimization and automation, dynamics and control of mechanical systems.

Satish S. Nair, Associate Professor; Ph.D., Ohio State. Dynamic modeling and control of systems; robust, adaptive, neural network, and intelligent control; mechatronics; manufacturing and automation; design.

Steven P. Neal, Associate Professor; Ph.D., Iowa State. Ultrasonic nondestructive evaluation, ultrasonic tissue characterization, solid mechanics.

P. Frank Pai, Associate Professor; Ph.D., Virginia Tech; PE. Computational solid mechanics, highly flexible deployable structures, smart structures, modern nonlinear dynamics.

Robert Tzou, Professor and Chairman; Ph.D., Lehigh. High-rate, small-scale heat transfer and the associated failure in electronic materials, thermomechanical modeling for material damage, brittle and ductile fracture mechanics.

Oleg Vasilyev, Assistant Professor; Ph.D., Notre Dame. Computational fluid mechanics, wavelet methods for the solution of partial differential equations, large eddy simulation of turbulent flows.

Robert A. Winholtz, Associate Professor; Ph.D., Northwestern. Residual stresses, neutron and X-ray measurement of residual and applied stresses, composite and two-phase materials.

RESEARCH AREAS

Intelligent Control and Mechatronics

Emphasis areas are modeling, dynamics, and control of mechanical, electromechanical, thermal, robotic, and manufacturing systems; design of intelligent dynamic systems, including neural network, robust, adaptive, and hybrid techniques for the control of uncertain systems; automation system design; system integration issues using sensors, actuators, electronics, and control for complex systems; computer interfacing and real-time control algorithms and implementation issues.

Fast-transient Heat Transport in Small Scales and the Associated Material Failure

This area concentrates on the fundamental behavior of heat and mass transport in small scales. The focus is on the lagging behavior that becomes activated as the transient time is extremely short. Applications include the fast-transient responses in metals and disordered structures, including amorphous silicon, silicon dioxides, silica aerogels, and rapid laser machining. Ongoing research includes short-time interaction between thermal shock waves and gradient structures, laser-sensor interactions and the associated sensor failure, and laser-induced fracture and yielding on pressurized vessels.

Diffraction Stress Measurement

Neutron and X-ray facilities are available to measure residual and applied stresses with diffraction. The University of Missouri Research Reactor Center (MURR) is the largest reactor at a university in the country and houses an internationally known program for neutron residual stress measurement. Research programs include residual stresses in composite, metal, and ceramic materials as well as industrial components, residual stress effects in fatigue, and methodological studies.

Computational Fluid Dynamics and Heat Transfer

Research in this area involves development and application of spectral and wavelet computational techniques for a variety of flow and heat transfer problems as well as application of control volume techniques. Areas of interest include combustion processes, onset of flow instabilities and their control, transition to turbulence, identification of coherent structures, and massively parallel computing.

Manufacturing Systems Design

The work in this area is motivated by flexible manufacturing to extend and coordinate the capabilities of machine tools and material handling equipment. Current work involves means of integrating areas such as control, machine vision, and planning with a view to create a model-based integrated assembly system. Work also emphasizes the development of intelligent machines for manufacturing and assembly of precision components.

Design Process

The goal of research in this area is to gain a fundamental understanding of the design process and to link the product design with the process design (concurrent engineering approach) in order to optimize the development of new products. The focus areas include artificial intelligence and virtual reality application in the design process, design for manufacturing, design for performance and quality, robust design and Taguchi methods, design optimization, finite-element simulation, and parallel processing for large-scale problems.

Modeling and Optimization of Manufacturing Processes

Research in this area is focused on interactions of material properties with the environment imposed by process conditions. Physical, numerical, and heuristic modeling techniques are used to provide simulation, optimization, and control strategy for current and new processes involving metallic and composite materials.

Ultrasonics

Research areas include ultrasonic nondestructive evaluation (NDE) of engineering materials and ultrasonic tissue characterization. Both experimental techniques that rely on the ultrasonics lab in mechanical and aerospace engineering and analytical/computational techniques are employed. Areas of emphasis in NDE include wave propagation in random media, backscatter coefficient estimation in engineering materials, materials characterization, flaw signature estimation, statistical characterization of acoustic noise, and application of wavelets to problems in ultrasonics. Tissue characterization research focuses on prostate cancer detection with expanding interests in other tissue characterization areas.

Nonlinear Structural Mechanics

Research focuses on design and analysis of highly flexible, deployable, and inflatable isotropic and composite structures and on the development and experimental verification of an in-house nonlinear finite-element code based on new, geometrically exact structural theories using Jaumann strains and total-Lagrangian formulations.

Smart Structures

Research focuses on design and analysis of structures with integrated smart sensors and actuators for damage detection and alleviation, health monitoring, and dynamic characteristics regulation and on vibration suppression using nonlinear phenomena and active-passive dampings.

UNIVERSITY OF NEBRASKA–LINCOLN

College of Engineering and Technology
Department of Mechanical Engineering

Programs of Study	The Department of Mechanical Engineering offers programs leading to the M.S. and Ph.D. degrees. There are three primary areas of emphasis within the department: thermal/fluids engineering, systems and design engineering, and materials science engineering.
	The department offers a broad program of study leading to the M.S. in mechanical engineering. The thesis-based program requires 24 hours of course work and at least 6 hours of thesis credit. By following a prescribed program of study in the materials area, students may obtain an M.S. in mechanical engineering with an area of specialization in materials science engineering. It typically takes eighteen months to complete the master's program.
	Students in the doctoral program may obtain the Doctor of Philosophy in engineering with a designated field of either mechanical engineering or chemical and materials engineering. A Ph.D. supervisory committee, in consultation with the student, arranges an appropriate program of doctoral course work. After the course work is substantially complete, the graduate student must pass a written comprehensive exam administered by the supervisory committee. In addition to the course work, doctoral students must complete a written Ph.D. dissertation with an oral presentation and defense. It typically requires three years of study after the M.S. to complete the Ph.D.
Research Facilities	There are nine specialized research laboratories in the Department of Mechanical Engineering. The Computational Thermal-Fluid Sciences Laboratory is a state-of-the-art workstation facility for research in finite difference, finite element, and Green's functions methods applied to problems in fluid flow, heat transfer, and combustion. The Stratified Flow Laboratory supports experimental investigations of buoyant convection and mixing. Research in highway design and safety, especially roadside appurtenances, is conducted at the Midwest Roadside Safety Facility. The Robotics Laboratory has equipment for research on industrial robots and vision systems. The Central Facility for Electron Microscopy includes electron microscopes, sample preparation, data collection, and reduction instrumentation and is affiliated with the University's Center for Materials Research and Analysis. In the Powder Processing Laboratory, researchers investigate production of powder materials by mechanical alloying. Measurement and characterization of the mechanical and physical properties of materials is the purview of the Physical/Mechanical Materials Characterization Laboratory. In the Thin Film Laboratory research is conducted on thin film deposition and characterization, while work in the X-Ray Diffraction Laboratory focuses on powder and single crystal X-ray diffraction.
Financial Aid	Approximately 90 percent of the department's full-time graduate students are currently supported by research assistantships, teaching assistantships, or fellowships. Applicants with degrees from U.S. institutions and highly qualified international students are considered for such awards on a competitive basis. In addition to a monthly stipend, students holding research or teaching assistantships receive a tuition waiver and only pay program and facilities fees.
Cost of Study	Tuition in 1998–99 for Nebraska residents was $109.50 per credit hour. For nonresident students, it was $270.25 per credit hour. Lab fees ranged from $40 to $70 per semester. Program and facilities fees were $240 for full-time students.
Living and Housing Costs	On-campus room and board costs in the academic year 1998–99 ranged from $3725 to $4615. On-campus apartments for married students ranged in price from $335 per month to $505 per month, depending on size and location. Privately owned rental units are also readily available in Lincoln.
Student Group	The Department of Mechanical Engineering has approximately 40 full-time and 25 part-time graduate students. About 90 percent of the full-time students are supported by the department. The graduate population includes individuals from diverse backgrounds; approximately 50 percent of the full-time students are from countries other than the United States.
Student Outcomes	Although a number of M.S. graduates stay on to continue studies for the Ph.D. degree, some have moved to other programs to pursue Ph.D. degrees. Other students with graduate degrees from Nebraska have chosen to work in academic institutions, government agencies, or industry. Some graduates are employed by national or international companies such as Boeing, McDonnell-Douglas, General Motors, Ford, Black and Veatch, General Electric, Goodyear Tire and Rubber, Toyota, and Enron.
Location	Located midway between Denver and Chicago, Lincoln, Nebraska's capital, is a dynamic city of more than 200,000 people. The Lincoln and University communities offer many concerts, recitals, plays, art exhibits, and other cultural activities through museums, galleries, and performing arts facilities. Numerous recreational opportunities are provided by University facilities as well as the city's many parks and Salt Valley lakes.
The Department and The University	The University of Nebraska–Lincoln (UNL) is the largest component of the University of Nebraska system. UNL began as a land-grant university chartered in 1869 and granted its first engineering degree in 1882. Mechanical engineering is one of eight departments in the College and is the only mechanical engineering program in the state of Nebraska. There are currently more than 24,500 students on the Lincoln campus, with approximately 2,750 students in engineering and technology and 450 in mechanical engineering.
Applying	Applications should specify the Department of Mechanical Engineering and the area of interest (thermal/fluids, systems/design, materials). Students applying for the M.S. or Ph.D. programs in mechanical engineering should have a B.S. in mechanical engineering or equivalent. Ph.D. applicants must also have an M.S. in a closely related field. Students with other backgrounds may be required to take additional prerequisite course work.
	Students intending to study for the M.S. in mechanical engineering with an area of specialization in materials science engineering are expected to have undergraduate training in mechanical engineering, materials science, or related areas of chemistry or physics. Students applying for the Ph.D. program in chemical and materials engineering should have an M.S. in mechanical engineering, materials science, or related areas.
	There are no test requirements for students with degrees from U.S. institutions. International students without degrees from U.S. institutions are required to take the TOEFL and GRE General Test. Faculty members in the student's area of interest evaluate each case on an individual basis. Applications are evaluated as they arrive, and full processing of an application requires approximately two months.
Correspondence and Information	Dr. Lorraine G. Olson Professor and Graduate Chair Department of Mechanical Engineering University of Nebraska 255 WSEC, code P Lincoln, Nebraska 68588–0656 Telephone: 402-472-5082 Fax: 402-472-1465 E-mail: megrad@unl.edu World Wide Web: http://www.engr.unl.edu/me/

University of Nebraska–Lincoln

THE FACULTY AND THEIR RESEARCH

J. P. Barton, Associate Professor; Ph.D., Stanford, 1980. Laser beam/particle interactions, acoustics, electromagnetic wave theory, high-temperature gas dynamics, fluid mechanics, experimental methods, data acquisition and analysis, MHD power generation.

K. Coen-Brown, Lecturer; M.S., Nebraska, 1989. Engineering education in the field of engineering graphics and computer modeling: computer-aided drafting and design, solid modeling, rendering, animation, and other three-dimensional visualization techniques.

K. D. Cole, Associate Professor; Ph.D., Michigan State, 1986. Heat transfer and diffusion theory; Green's functions and symbolic computation; numerical modeling and thermal sensor technology; thermal conductivity measurements; experimental studies of bubble dissolution in plastic molding.

R. J. De Angelis, Professor; Ph.D., Northwestern, 1965. Advancement of the understanding of the relationships between the structure of a material and its engineering properties; role of thermomechanical processing on the stabilization or the modification of the microstructure; the improvement of analytical and experimental X-ray diffraction methods to perform detailed structural characterization of metallic nanoparticles made by various processing techniques; investigation of supported metal catalyst; Fe/Ag and Co/Ag composite thin films, electrodeposited thin films, and mechanically alloyed powders for processing to structural materials; role of the thermal stresses developed during solidification in generating dislocations; relationships between crystallographic texture, anisotropy, and mechanical deformation of metallic materials.

L. E. Ehlers, Associate Professor Emeritus; Ph.D., Oklahoma State, 1969. Fluid flow, wind energy, vibrations.

S. M. Farritor, Assistant Professor; Ph.D., MIT, 1998. Robotics for planetary exploration, design and control of field robotic systems, modular design, planning methods for planetary exploration and field robots.

G. Gogos, Associate Professor; Ph.D., Pennsylvania, 1986. Computational heat transfer and fluid flow; perturbation methods; fundamental processes associated with vaporizing/combusting sprays with applications in liquid-fueled rocket engines, gas turbines, diesel engines, and industrial furnaces (evaporation/combustion of moving droplets, subcritical and supercritical droplet evaporation, transition of envelop to wake flames in burning droplets, droplet interactions, interaction of sprays and buoyant diffusion flames); natural convection; heat transfer and material deposition in rotational molding.

K. O. Homan, Assistant Professor; Ph.D., Illinois, 1996. Fluid dynamics, heat transfer, and energy systems; computational and experimental studies of stratified flows; thermal energy storage systems.

D. L. Johnson, Professor Emeritus; Ph.D., Nebraska, 1968. Corrosion/degradation and hydrogen permeation and diffusion in metallic and nonmetallic systems, metallurgical thermochemistry.

Y. Liu, Research Assistant Professor; Ph.D., Tohoku (Japan), 1988. Crystal-structure and defect-structure study of metallic materials by various transmission electron microscopy (TEM) techniques, including high-resolution (0.19 nm), weak-beam, nanodiffraction, and convergent-beam electron diffraction; development of new TEM techniques, such as resolution extension by imaging processing; morphology imaging of nanostructured materials; magnetic domain imaging by differential phase contrast; development of new high-temperature structural materials; high-temperature nanostructured materials; high-temperature permanent magnet; nanostructure design of thin films by laser surface treatment.

D. Y. S. Lou, Professor; Sc.D., MIT, 1967. Rarefied gas dynamics, heat conduction in rarefied gases, thermal curing of composite materials, thermal manufacturing process analysis, thermal modeling of pulse combustors, heat transfer in phase change materials.

P. -C. Lu, Professor Emeritus; Ph.D., Case Tech, 1963. Computer-aided thermal systems analysis, design, and education; optimal design of thermal systems; numerical simulation of heat-exchanger dynamics; symbolic computation in thermal analysis; finite time/size thermodynamics.

R. C. Nelson, Professor Emeritus; D.Sc., Colorado School of Mines, 1951. Powder metallurgy; biomaterials; the mechanical behavior of materials, including failure analysis.

L. G. Olson, Professor; Ph.D., MIT, 1985. Application of finite element methodology to nontraditional areas, variational formulations for viscous-free surface flows, heat transfer and material deposition in rotational molding of thermoplastics, methods for gradually correcting infant skull growth abnormalities, inverse problem of electrocardiography (determining electrical patterns on the heart surface from body surface recordings). Previous research has involved fluid-structure interactions, ultrasonic cleaning, singular elements, surface tension, laser welding.

R. P. Ondracek, Research Assistant Professor; Ph.D., Johns Hopkins, 1994. Spray forming, intelligent control of manufacturing processes, materials education, development of materials laboratories for undergraduate and graduate students, sensor development for manufacturing processes.

A. R. Peters, Professor; Ph.D., Oklahoma State, 1967. Fluid mechanics; aerodynamics; heat transfer, combustion, and instrumentation; energy-related topics of two-phase flows; materials processing; heat-exchanger applications; product drying; engine performance.

J. D. Reid, Associate Professor; Ph.D., Michigan State, 1990. Vehicle crashworthiness and roadside safety design, analysis, and simulation; vehicle dynamics; nonlinear, large deformation, finite-element analysis; computer simulation; computer-aided engineering.

B. W. Robertson, Associate Professor; Ph.D., University of Glasgow, 1979. Materials science and engineering for information, microelectronic, and structural applications; fabrication of new materials and devices by plasma and electron-beam chemical vapor deposition methods and devices; nanoscale patterned magnets for nonvolatile magnetic memory and logic; dynamic study of electromigration in submicron interconnect wiring; development of electron-beam methods and instrumentation; novel boron-carbon semiconductor devices and metallic contacts for high-temperature electronics and neutron camera applications; transmission and scanning electron microscopy of materials.

S. L. Rohde, Associate Professor; Ph.D., Northwestern, 1991. Thin-film deposition and characterization, specifically in the areas of hard coatings, superlattice structures, and semiconductors; tribological (frictional), microstructural, and chemical analysis of thin films and modified surfaces; development of novel thin film sputtering equipment, deposition processes, ion implantation, and technology transfer.

G. R. Schade, Associate Professor; Ph.D., Iowa State, 1974. Stability and automation control, probabilistic design.

W. M. Szydlowski, Associate Professor; Ph.D., Technical University of Warsaw (Poland), 1975. Analysis and synthesis of mechanisms, computer simulation of mechanical systems, dynamics of machinery (mechanical impact and mechanisms of intermittent motion with clearances in particular), redundant constraints in large mechanical systems, some aspects of finite element analysis (residual stress problems).

C. W. S. To, Professor; Ph.D., Southampton, 1980. Sound and vibration studies (acoustic pulsation in pipelines, railway noise and vibration, signal analysis, structural dynamics, random vibration, nonlinear and chaotic vibration), solid mechanics (linear and nonlinear finite-element methods with application to laminated composite shell structures and modeling of aorta dissection), system dynamics (nonlinear and rigid-body dynamics), controls (deterministic and stochastic).

W. N. Weins, Associate Professor; Ph.D., Iowa State, 1980. Mechanical behavior of materials, internal friction damping in materials, friction and wear of materials, ferrous materials, failure analysis of materials.

UNIVERSITY OF NEW MEXICO

School of Engineering
Department of Mechanical Engineering

Programs of Study

The Department of Mechanical Engineering offers M.S. and Ph.D. degrees in the areas of biomechanics, computational mechanics, dynamic systems and control, energy/thermodynamics, fluid mechanics, heat transfer, manufacturing engineering, materials science, robotics, and solid mechanics.

The M.S. degree requires 31 semester hours with a thesis (6 hours) or 34 hours with a project (3–6 hours). Both the project and thesis are guided by a 3-member committee-on-studies. A student's course work is selected by the graduate advisor and the committee-on-studies. The M.S. program can be completed in one to two years of study.

The Ph.D. degree requires 54 semester hours beyond the B.S. degree. Up to 30 semester hours may be transferred from another institution. The Ph.D. degree is a research-oriented degree, and doctoral students are expected to have excelled in prior graduate studies and be capable of conducting and reporting original research suitable for publication. Ph.D. candidates must pass a departmental qualifying examination, present a dissertation proposal, and complete and defend a doctoral dissertation.

The M.S. or Ph.D. degrees in mechanical engineering may be obtained in conjunction with the Scientific and Engineering Computation Program, aimed at promoting advanced research in the computational sciences.

Research Facilities

The Mechanical Engineering building houses most research laboratories, including the Biomechanics, Controls, Fluid Mechanics, Heat Transfer, Material Science, Materials Test, Microprocessor, Robotics, and Vibrations Laboratories, and the Artificial Muscles Research Institute (AMRI). Other facilities supporting research are the High Performance Computing, Educational, and Research Center (HPCERC), which manages the Maui High Performance Computing Center, and the Manufacturing Technology and Training Center (MTTC).

The Department of Mechanical Engineering has close collaborations with the University of New Mexico (UNM) School of Medicine, nearby Los Alamos and Sandia National Laboratories, and the Air Force Research Laboratory/Phillips site. Research facilities at these laboratories are often used by graduate students.

The NASA-sponsored Center for Autonomous Control Engineering is a consortium of faculty members with shared interests in adaptive, fuzzy, and digital control systems and applications. The faculty members are drawn from electrical and computer engineering, mechanical engineering, chemical and nuclear engineering, and civil engineering. A few small fellowships are also awarded each year.

Financial Aid

Numerous research assistantships paying from $12,000 to $18,000 per year are available, along with teaching assistantships paying from $8600 to $10,000. Tuition is waived for assistantships. A few small fellowships are also awarded each year. Nearly all full-time graduates receive financial aid.

Cost of Study

In 1998–99, tuition was $1134 per semester for full-time New Mexico residents and $3998 per semester for nonresident students.

Living and Housing Costs

The cost of living in Albuquerque is moderate. The University has dormitories for single graduate students. Housing is also available for married graduate students. Local rental housing is generally plentiful.

Student Group

There are approximately 90 graduate students in the mechanical engineering graduate program; about 30 percent are Ph.D. students. The School of Engineering has a total enrollment of about 600 graduate students. Women constitute 20 percent of the graduate engineering enrollment.

Student Outcomes

Graduates of the M.S. and Ph.D. programs in mechanical engineering have found employment in a wide variety of organizations, including Sandia and Los Alamos National Laboratories, manufacturing firms such as Intel and General Motors, and at universities both in the U.S. and abroad.

Location

The University is located in Albuquerque, which has a metropolitan population of more than 500,000. The city is a mile above sea level, is divided by the Rio Grande, and abuts the Sandia Mountains, which reach to 10,678 feet and offer skiing and hiking opportunities.

Although Albuquerque undergoes seasonal changes, the dry, sunny climate rarely exhibits temperature extremes. Santa Fe, the first North American capital city, is nearby. The setting is rich with the traditions of Indian, Spanish, and Anglo cultures.

The University

The University of New Mexico was created by an act of the Territorial Legislature in 1889, twenty-three years before New Mexico became a state. Opened on June 15, 1892, it began full-term instruction on September 21 of that year.

UNM's mission includes offering comprehensive educational programs, conducting research and participating in other scholarly activities, and contributing to the quality of life in the state.

Applying

Admission decisions are made by the Graduate Advisor. Applicants must hold an accredited bachelor's degree and typically should have above a B average in their last two undergraduate years. All applicants must submit the results of the Graduate Record Examinations General Test prior to admission. International applicants must submit a recent TOEFL score. Application deadlines are June 30 (fall semester), November 15 (spring semester), and April 15 (summer semester).

Correspondence and Information

Graduate Advisor
Department of Mechanical Engineering
University of New Mexico
Albuquerque, New Mexico 87131

Telephone: 505-277-2761
Fax: 505-277-1571
E-mail: gradad@me.unm.edu
World Wide Web: http://www.me.unm.edu/

University of New Mexico

THE FACULTY AND THEIR RESEARCH

William E. Baker, Professor; Ph.D., Texas at Austin, 1966. Applied mechanics, design. Research interests: solid mechanics and design, with an emphasis in the area of experimental mechanics, dynamic properties of materials, and transducer development and evaluation.

Nader D. Ebrahimi, Assistant Professor; Ph.D., Wisconsin, 1983. Design and dynamic systems. Research interests: numerical optimization.

William A. Gross, Professor Emeritus; Ph.D., Berkeley, 1951. Energy. Research interests: tribology, especially as it relates to information storage and retrieval; renewable energy; technological entrepreneurship.

Marc S. Ingber, Professor; Ph.D., Michigan, 1984. Applied mechanics. Research interests: high-performance computational methods, computational fluid mechanics, stratified flows, hydrodynamic stability.

James R. Leith, Associate Professor; Ph.D., Texas at Arlington, 1980. Heat transfer, fluid mechanics, thermodynamics. Research interests: natural and mixed convection, hydrodynamic stability, bifurcation theory, dynamical phase transitions, stochastic processes, experimental methods.

Ronald Lumia, Professor; Ph.D., Virginia, 1979. Dynamic systems and control, software for manufacturing. Research interests: open systems technology, equipment interconnection over the Internet, factory automation, robotics, control systems.

Andrea A. Mammoli, Assistant Professor; Ph.D., Western Australia, 1995. Research interests: multiphase flow, composite materials, high-performance computing.

Joe H. Mullins, Professor Emeritus; Ph.D., Caltech, 1959. Manufacturing engineering. Research interests: systems, manufacturing processes, CAD/CAM, automation.

Arsalan Razani, Professor; Ph.D., Purdue, 1969. Thermal sciences, energy. Research interests: heat and mass transfer analysis of metal hydride energy systems, thermal system design and optimization, radiation and conduction heat transfer.

Charles G. Richards, Professor; Ph.D., Michigan, 1964. Fluid mechanics and thermal sciences. Research interests: computer simulation of incompressible flows and heat transfer processes.

Howard L. Schreyer, Professor; Ph.D., Michigan, 1965. Solid and computational mechanics. Research interests: constitutive equations based on plasticity and continuum damage mechanics, numerical algorithms for nonlinear problems in continuum mechanics.

Mohsen Shahinpoor, Professor; Ph.D., Delaware, 1970. Design, robotics, mechanics of materials. Research interests: smart/intelligent materials, structures and systems, ionic polymeric metal composites as sensors and actuators, artificial muscles, biomechanics, biomechatronics, heart and cardiac assist devices, mechanics of flexible structures, flow and characterization of granular materials.

Yu-Lin Shen, Assistant Professor; Ph.D., Brown, 1994. Materials science, solid mechanics. Research interests: thermomechanical integrity of modern integrated circuits and microelectronic packages, mechanical behavior of thin films and composite materials, micromechanical modeling of deformation and fracture.

Gregory P. Starr, Professor; Ph.D., Stanford, 1978. Dynamics systems and control, robotics, LEGO robots. Research interests: discrete-time control systems, multifingered robot hands, robot force and impedance control.

David E. Thompson, Professor; Ph.D., Purdue, 1970. Design and modeling. Research interests: biomechanics of the hand and spine, surface modeling of biological structures, nonlinear systems, modeling of adaptive systems.

Hy D. Tran, Assistant Professor; Ph.D., Stanford, 1993. Control systems, metrology, precision engineering. Research interests: electromechanical and optomechanical systems, MEMS, fiber optics, semiconductor manufacturing.

C. Randall Truman, Professor; Ph.D., Arizona State, 1983. Fluid mechanics, heat transfer. Research interests: aero-optical degradation by turbulent fluctuations in shear layers, turbulence and transition modeling, optical flow diagnostics.

John E. Wood, Professor; Ph.D., MIT, 1976. Manufacturing, design, bioengineering. Research interests: microsensors and microactuators, clean-room automation and robotics, biomechanics.

AREAS OF CURRENT RESEARCH

Artificial Muscles. Basic research and applications of electrically controllable artificial muscles made with ionic polymeric gel-noble metal composites. These materials are capable of both sensing and actuation and scale to MEMS levels.

Biomechanics/Biomechatronics. Spine biomechatronics, biomechanics of the hand, investigation of the forces and excursions of the tendons and muscles of the hand. Surface modeling of biological structures, specifically representations that apply to real structures, thereby extracting information about their anatomy and physiology. (Jointly with the UNM School of Medicine and UNM Department of Biology)

Computational and Continuum Mechanics. Constitutive equations for materials experiencing large deformation and failure. Numerical solutions to related problems based on the material point method.

Dynamic Systems and Control. Investigation of embedded piezoceramic actuators combined with adaptive sliding mode and neural net control to reduce vibration in flexible multibody dynamic systems. The use of adaptive control systems to control prostheses and other articulated systems. Precision engineering, metrology, microelectromechanical systems (MEMS), sensors, optomechanical systems, semiconductor lithography.

Energy/Thermodynamics/Heat Transfer. Heat and mass transfer analysis of metal hydride energy systems, thermal systems design and optimization, thermal ignition of energetic materials, second law and finite-time thermodynamic analyses of thermal systems, heat transfer in periodically varying phenomena, thermal analysis of microelectronic components and systems.

Fluid Mechanics. Optical propagation through turbulent shear flows, vortex structure of free shear layers produced by acoustic excitation, nonlinear control of adaptive optic correction of turbulence, optimal basis functions for tomographic reconstruction, bifurcation structure in thermal convection turbulence modeling, transport in disordered systems, numerical simulation of separation and transition, Rayleigh-Taylor mixing, study and simulation of complex and collective phenomena in multiphase flows and composite materials.

Materials Science. Mechanical properties of materials used in microelectronics and structural applications, micromechanical modeling of deformation and fracture of thin films and composite materials. Particular attention is devoted to gaining insights into the property-processing-microstructure relations by employing experimental and computational techniques.

Robotics. Open-architecture robot controllers applied to rapid development of intelligent systems, grasping with multifingered hands, design optimization of serial-parallel robot manipulators, force control by rigid and elastic robot manipulators.

Solid Mechanics. Thermomechanical responses of metal-ceramic composite materials and porous solids, mechanics issues in thin solid films, modeling of mechanical behavior in multiphase materials, experimental and numerical characteristics of deformation flow and fracture. Accelerometer evaluation for very high G applications.

Université d'Ottawa
University of Ottawa

Carleton University

UNIVERSITY OF OTTAWA / CARLETON UNIVERSITY

Ottawa–Carleton Institute for Mechanical and Aerospace Engineering

Programs of Study

The OCIMAE is a joint institute combining the graduate programs and research strengths of the Department of Mechanical and Aerospace Engineering at Carleton University and the Department of Mechanical Engineering at the University of Ottawa. The Institute offers one of the largest programs of graduate studies and research in mechanical and aerospace engineering in Canada. Admissions are administered through the Institute and each student registers in the University of the student's supervisor, from which the degree is to be granted, although courses may be taken at either University. Programs are offered leading to master's degrees in aerospace engineering, materials engineering, and mechanical engineering, as well as Ph.D. degrees in aerospace and mechanical engineering. Programs of combined analytical/numerical and experimental research are being carried out in such areas as thermo-fluid mechanics, solid and structural mechanics, systems and controls, combustion and power plants, materials and manufacturing, metallurgy and materials processing, industrial engineering, CAD/CAM/CIM, nuclear reactor engineering, transportation technology, building technology and wind engineering. More specifically, research in the field of turbomachinery includes two- and three-dimensional boundary layer behaviour; tip leakage and other loss effects; dynamics of gas turbine power plants; studies relating to the performance of highly loaded turbines; stress, deformation, and vibration of compressor and turbine blades; finite-element analysis; dynamics of high-speed rotors; and failure modes of materials in extreme environments. In computer-aided engineering, research includes computer-aided design and computer-integrated manufacturing, thermal (including heat and fluid flow) and mechanical (including stress and deformation) analysis of welding and casting processes, phase transformation and solidification processes, and hot working of metals and behaviour of composite materials. Transportation-related research includes computational solutions for steady and unsteady flows over complex configurations, effects of roughness on aerodynamic performance, boundary layer separation and control, and propeller and rotor aerodynamics and noise. It also includes design and optimization of off-road vehicles, vehicle safety, antilock braking systems, vehicle-terrain interactions, effect of vibration on vehicle performance, dynamics of air-cushion and magnetically levitated vehicles, composite and structural elements, and design and performance of lightweight structures. Building engineering research includes ventilation distribution, materials emission effects on air quality, and cladding designs as they affect performance of building envelopes. In the nuclear energy field, research focuses on heat transfer and two-phase flows (including the critical heat flux in CANDU and SLOWPOKE reactors using both experimental and numerical studies), reactor safety, and the diversified use of nuclear energy. Combustion studies involve computational and experimental examination of combustion phenomena, including droplet burning and systems such as internal combustion engines. Other projects involve the interaction of design parameters with fracture mechanics and fatigue, solid mechanics analyzed by boundary element methods, solid mechanics of advanced industrial materials, and experimental and computational fracture investigations. Control, robotics, and interfacing of these is a growing area of interest, including applications on earth and in space.

Research Facilities

Modern well-equipped laboratories include engine test cells, wind tunnels, two-phase flow loops, a computer-controlled filament winder, a composite braider, a computer-controlled tensile machine, a hydraulic fatigue testing machine and impact tester, high-speed data acquisition systems, photoelastic equipment, a shaker table, a vehicle performance testing system, a tire testing apparatus, computer-controlled machine tools, a low-temperature testing facility, and up-to-date computer workstations. In addition, extensive laboratory facilities of the National Research Council of Canada and other government research establishments are frequently used for research of mutual interest.

Financial Aid

Financial aid is available through scholarships, teaching assistantships, and research assistantships. The departments attempt to support all full-time graduate students admitted, unless they have well-documented sources of support apart from the Institute, which may be supplemented by teaching and/or research assistantships. Levels of support offered are intended to cover the cost of study.

Cost of Study

The tuition costs of permanent residents of Canada for 1998–99 were approximately Can$2000 per four-month term, including costs for various health insurance and other benefits. For nonpermanent residents, the basic tuition was approximately Can$4400 per term, plus about Can$500 per year for the additional benefits for single students' health insurance.

Living and Housing Costs

On-campus housing, including meals for single students, costs approximately Can$9000 for twelve months. Married students live off campus, where a variety of apartments and rooms are available at rates comparable to on-campus rates.

Student Group

The program has approximately 120 full-time students and 30 part-time students currently engaged in graduate studies in the two departments. Approximately 10 percent are women.

Student Outcomes

Doctoral graduates have found employment as university faculty members or as researchers in government or industrial research laboratories. Master's graduates have been employed in industry both in Canada and abroad as well as by government laboratories. Some have formed private businesses of their own with good success.

Location

Ottawa, Canada's capital, is part of a metropolitan area with a population of almost 1 million, situated on the Ontario-Quebec border with good cultural and recreational opportunities in each season. Its government offices and laboratories as well as the large and small high-technology industrial employers attract many professionals to live in the area.

The Universities

The overall enrollment in the two Universities is more than 35,000, of which the engineering departments represent more than 4,000 students in both undergraduate and graduate programs. The joint Institute was established in 1983, but both Universities have considerably longer histories. Graduate class sizes are, on the average, fewer than 10, and faculty members develop a close working relationship with their advisees.

Applying

Applications may be submitted at any time. To be considered for the first round of financial aid offers, which is generally based on a September entrance date, the application file must be completed by March 1. A minimum average of B or equivalent is required for admission to a master's program, and there are higher expectations for a Ph.D. program entry. Students from a non-English speaking background are required to achieve a score of at least 550 on the TOEFL, or equivalent, and although GRE scores are not mandatory, they can prove helpful in evaluating an application.

Correspondence and Information

Director
Ottawa-Carleton Institute for Mechanical and Aerospace Engineering
University of Ottawa
C406-161 Louis Pasteur
Ottawa, Ontario K1N 6N5
Canada
Telephone: 613-562-5800 Ext. 6189
Fax: 613-562-5174
E-mail: jallard@genie.uottawa.ca
World Wide Web: http://www.mae.carleton.ca/OCIMAE.html

University of Ottawa / Carleton University

THE FACULTY AND THEIR RESEARCH

F. F. Afagh, Professor (∗); Ph.D., Waterloo, 1986; P.Eng. Dynamics, vibrations, structural mechanics, smart structures.

M. Akben, Associate Professor (†); Ph.D., McGill, 1981; Eng. (O.I.Q.). Metallurgy, welding, hot working metals.

A. V. Artemev, Assistant Professor (∗); Ph.D., Moscow, 1987. Phase transformations, solidification processes.

J. C. Beddoes, Assistant Professor (∗); Ph.D., Carleton, 1993; P.Eng. Physical metallurgy and metal processing.

R. Bell, Professor (∗); Ph.D., Queen's (Belfast), 1975; P.Eng. Finite-element analysis, stress analysis, solid mechanics, fracture mechanics, biomechanics.

M. J. Bibby, Professor (∗); Ph.D., Alberta, 1966; P.Eng. Materials and manufacturing engineering, weld analysis.

S. C. Cheng, Professor (†); Ph.D., Auburn, 1968. Heat transfer, numerical methods.

B. S. Dhillon, Professor (†); Ph.D., Windsor, 1975; P.Eng. Reliability engineering.

A. E. F. Fahim, Professor (†); Ph.D., Concordia, 1983. CAD/CAM, controls.

J. A. Gaydos, Assistant Professor (∗); Ph.D., Toronto, 1992. Thermodynamics, continuum mechanics.

J. A. Goldak, Professor (∗); Ph.D., Alberta, 1964; P.Eng. Computer-integrated manufacturing, finite-element modelling of manufacturing processes.

D. J. Gorman, Professor Emeritus (†); Ph.D., Syracuse, 1965; P.Eng. Vibration.

Y. M. Haddad, Professor (†); Ph.D., McGill, 1975; P.Eng. Applied mechanics, solid mechanics, materials and design.

W. L. H. Hallett, Professor and Chairman (†); Ph.D., Karlsruhe (Germany), 1981. Fluid mechanics, combustion.

B. Jodoin, Professor (†); Ph.D., Sherbrooke, 1998; Ing. Thermofluids, thermal plasmas, numerical modelling.

R. J. Kind, Professor (∗); Ph.D., Cambridge, 1967; P.Eng. Aerodynamics of aircraft and turbomachinery, wind engineering.

Y. Lee, Professor Emeritus (†); D.Sc., Ph.D., Liverpool, 1964; P.Eng. Heat transfer, nuclear engineering.

M. Liang, Associate Professor (†); Ph.D., Windsor, 1991; P.Eng. Production and manufacturing systems.

L. Mallory, Assistant Professor (†); Ph.D., Ohio State, 1998. Welding.

M. J. McDill, Associate Professor (∗); Ph.D., Carleton, 1988; P.Eng. Adaptive methods for 3-D finite-element analysis.

R. E. Milane, Assistant Professor (†); Ph.D., Michigan, 1982. Combustion, fluid mechanics.

S. Mirza, Professor Emeritus (†); Ph.D., Wisconsin, 1962; P.Eng. Vibrations, stress analysis.

M. B. Munro, Professor (†); Ph.D., Waterloo, 1977; P.Eng. Composite materials.

D. S. Necsulescu, Professor (†); Ph.D., Polytechnic University (Bucharest), 1974; P.Eng. Control, dynamics, robotics, reliability.

A.K. Pilkey, Assistant Professor (∗); Ph.D., Carleton, 1997. Failure mechanisms, metaldographic image analysis, metal forming.

E. G. Plett, Professor (∗); Sc.D., MIT, 1966; P.Eng. Fluid mechanics, thermodynamics and heat/mass transfer, numeric modelling.

D. Redekop, Professor and Director of the Institute (†); Ph.D., Waterloo, 1977. Solid mechanics, finite-element analysis, robotics, vibrations.

W. G. Richarz, Associate Professor (∗); Ph.D., Toronto, 1978; P.Eng. Aeronautical engineering, acoustics, instrumentation.

J. T. Rogers, Professor Emeritus (∗); Ph.D., McGill, 1953; P.Eng. Heat transfer and nuclear safety.

D. L. Russell, Associate Professor (∗); Ph.D., MIT, 1990; P.Eng. (N.S.). Dynamics, control, biomechanics.

H. I. H. Saravanamuttoo, Professor Emeritus (∗); Ph.D., Bristol, 1968; P.Eng. Gas turbine performance, engine health monitoring.

J. Z. Sasiadek, Professor (∗); D.Sc., Wroclaw Technical (Poland), 1976; Ing. Control systems, robotics, microprocessor applications.

S. A. Sjolander, Professor (∗); Ph.D., Cambridge, 1981; P.Eng. Turbomachinery, aerodynamics, wind-tunnel engineering.

D. A. Staley, Professor (∗); Ph.D., Western Ontario, 1976. Spacecraft dynamics and control.

P. V. Straznicky, Professor and Chairman (∗); M.Eng., Toronto, 1972; P.Eng. Design, lightweight structures.

C. L. Tan, Professor (∗); Ph.D., London, 1979. Solid mechanics, boundary integral and finite-element methods.

S. Tavoularis, Professor (†); Ph.D., Johns Hopkins, 1978; P.Eng. Fluid mechanics, experimental techniques.

J. Y. Wong, Professor (∗); Ph.D., D.Sc., Newcastle upon Tyne (England), 1967; P.Eng., C.Eng. Transportation technology, vehicle engineering.

M. I. Yaras, Associate Professor (∗); Ph.D., Carleton, 1990; P.Eng. Turbomachinery, aerodynamics, computational fluid dynamics.

(∗) *Denotes faculty member at Carleton University.*
(†) *Denotes faculty member at the University of Ottawa.*

UNIVERSITY OF PENNSYLVANIA

School of Engineering and Applied Science
Department of Mechanical Engineering and Applied Mechanics

Programs of Study	Graduate study and research in the department lead to the Master of Science in Engineering and to the Doctor of Philosophy degrees. Areas of research include the mechanics of materials, fluid mechanics, mechanical systems, robotics, thermal sciences, energy conversion, and biomechanics. The research in the department is often interdisciplinary in nature and is done in close collaboration with faculty members from other departments, such as computer science, material science, and electrical engineering, and from other schools, such as the School of Medicine.
	The M.S.E. degree requires successful completion of 10 course units. Optional master's thesis research may count for up to 3 course units. Typically, M.S.E. students in good standing take up to five courses per semester.
	The Ph.D. degree requires a minimum of 20 course units beyond the B.S. degree and a dissertation. The M.S.E. degree is not a prerequisite for the Ph.D.
Research Facilities	The department offers excellent facilities for research in all areas of mechanical engineering.
	The fluid and thermal science laboratories occupy a total area of about 420 square meters and include modern equipment for flow visualization, interferometry, anemometry, and data acquisition.
	The Laboratory for the Research on the Structure of Matter (LRSM) provides ultrahigh-resolution transmission electron microscopy, an Advanced Mechanical Testing Facility, and a Materials Processing Facility.
	The Plasma Laboratory enables research on plasma arc discharge and heat transfer.
	The Microfabrication Facility is used for the microfabrication of mechanical, heat transfer, and flow elements on silicon chips and on other suitable materials.
	The General Robotics and Active Sensory Perception (GRASP) Lab houses a number of robot manipulators, mobile platforms, state-of-the-art vision and image processing systems, and CAD workstations.
	The newly renovated Manufacturing Technology Lab consists of a state-of-the-art CNC machining center, an injection-molding machine, a vacuum-forming machine, several desktop and numerically controlled mills and lathes, and several Silicon Graphics and Pentium-based workstations.
Financial Aid	Financial aid is available only to Ph.D. candidates. Awards are competitive and based on merit. Subject to continued good academic standing and the availability of funds, the department strives to provide financial aid to graduate students initially admitted with such aid throughout their tenure at the University.
Cost of Study	Tuition for the academic year 1998–99 was $22,716; there was a general fee of $1484 and an education technology fee of $440 for full-time study. For part-time study, per course unit (one course), tuition was $2876, the general fee was $170, and the education technology fee was $53.
Living and Housing Costs	On-campus housing is available for both single and married students. Residences for single students cost from $445 per month for a shared-living situation to $800 per month for a private apartment. Housing for married couples costs $665–$1050 per month. There are numerous privately owned apartments in the immediate area.
Student Group	Currently, there are 44 students enrolled in the graduate program; 38 are Ph.D. students, and 6 are M.S.E. students. Of the 44 graduate students, 41 are full-time students, and 3 are part-time students.
Student Outcomes	Many recent Ph.D. graduates have been employed in the research and development divisions of top firms such as IBM, AT&T Lucent Technologies, Lockheed Martin, and General Motors. Many have also become faculty members at top universities such as Northwestern University; University of California, Irvine; and Royal Institute of Technology, Sweden.
Location	The University of Pennsylvania, founded by Benjamin Franklin in 1740, is a member of the Ivy League and one of the major research universities in the United States. The University consists of fourteen colleges, schools, and institutions, all located in an attractive 260-acre campus in West Philadelphia, just a few blocks from the heart of Center City. Philadelphia is a twentieth-century city with seventeenth-century origins. The city consists of multicultural neighborhoods ranging from colonial Society Hill to Chinatown. Renowned museums, concert halls, theaters, restaurants, sports arenas, parks, and rivers provide a multitude of cultural and recreational opportunities. Not far from Philadelphia are a variety of popular sightseeing and recreational destinations: to the east, the Jersey shore; to the west, Pennsylvania Dutch country; and to the north, the Pocono Mountains with ski resorts and hiking trails. Philadelphia is located equidistant and within an easy commute from New York City and Washington, D.C. The city is served by an international airport and train service.
The Department	The Department of Mechanical Engineering and Applied Mechanics program was the second engineering curriculum established in the University (c. 1872), and its many alumni are noted for their distinguished careers and contributions in engineering, management, science, and education. The graduate program is designed to provide broad competence in basic mechanical engineering disciplines, in engineering mathematics and computation, and in major current interests in the field. The intellectual breadth of the field and the department's philosophy encourage interdisciplinary and interdepartmental collaboration in research.
Applying	A single application is required for both admission and financial aid. A complete application, including admission fee, must be submitted by July 1 for fall semester candidates and by November 1 for spring semester candidates; the deadline for financial aid consideration, however, is January 2. Those with financial need must abide by this earlier application deadline. International students whose native language is not English are required to submit an official TOEFL examination score of 600 or better, unless they have completed at least three years of academic work on a full-time basis at a university-level institution in an English-speaking country. Students with lower TOEFL scores are required to take English courses. International candidates for a teaching fellowship may have to provide further evidence of English fluency by taking the Educational Testing Service's Test of Spoken English (TSE). International students are encouraged to complete a preapplication form to facilitate initial screening. More detailed information and preapplication forms may be obtained via the World Wide Web (http://www.seas.upenn.edu/meam).
Correspondence and Information	Professor Haim Bau Graduate Group Chairman Department of Mechanical Engineering and Applied Mechanics University of Pennsylvania Philadelphia, Pennsylvania 19104-6315 Telephone: 215-898-4825 Fax: 215-573-6334 E-mail: meamgrad@seas.upenn.edu

University of Pennsylvania

THE FACULTY AND THEIR RESEARCH

G. K. Ananthasuresh, Assistant Professor; Ph.D., Michigan, 1994. Compliant mechanisms, microelectromechanical systems (MEMS), design optimization, kinematics of mechanisms.
Design and fabrication of micro-electro-mechanical systems. *ASME J. Mech. Design* 116(4):1081–8, 1994 (with Kota, Crary, and Wise).

Portonovo S. Ayyaswamy, Asa Whitney Professor of Dynamical Engineering; Ph.D., UCLA, 1971. Direct contact heat and mass transfer processes, bioheat transfer, arc-plasma heat transfer, hydrodynamic stability.
Oscillatory enhancement of the squeezing flow of yield stress fluids: A novel experimental result. *J. Fluid Mech.* 339:77–87, 1997 (with Zwick and Cohen).

John L. Bassani, Richard H. and S. L. Gabel Professor in Mechanical Engineering and Chairman; Ph.D., Harvard, 1978. Plastic deformation of crystals, atomic/continuum property relationships, interface mechanics, fracture mechanics, material stability at large strains, mechanics of living cells.
Plastic flow of crystals. *Adv. Appl. Mech.* 30:191–258, 1994.

Haim H. Bau, Professor and Graduate Group Chairman; Ph.D., Cornell, 1980. Bifurcation and instability phenomena in and feedback control of buoyancy-driven flows, transport phenomena in micron and submicron size structures, meso- and microelectromechanical systems, interaction between stress waves transmitted in a solid waveguide and adjacent medium.
Rendering subcritical Hopf bifurcation supercritical. *J. Fluid Mech.* 317:91–109, 1996 (with Yuen).

Pedro Ponte Castañeda, Professor; Ph.D., Harvard, 1986. Nonlinear composite materials, fracture mechanics, microstructure evolution and localization in manufacturing processes, nonlinear variational principles in mechanics.
Nonlinear composites. *Adv. Appl. Mech.* 34:171–302, 1998 (with Suquet).

Ira M. Cohen, Professor; Ph.D., Princeton, 1963. Basic problems in fluid mechanics and heat transfer as applied to microelectronic manufacturing.

Dennis E. Discher, Assistant Professor; Ph.D., Berkeley, 1993. Mechanics and structural assemblies of biomolecules, mechanochemistry of cells, mechanics and statistical mechanics of networks and complex fluids.

Howard H. Hu, Associate Professor; Ph.D., Minnesota, 1992. Modeling of complex flows with multiphase or polymeric fluids, computational fluid dynamics, hydrodynamic stability.
Direct simulation of flows of solid-liquid mixtures. *Int. J. Multiphase Flow* 22:335–52, 1996.

Vijay Kumar, Professor; Ph.D., Ohio State, 1987. Robotics, dynamics of systems with frictional contacts, actively coordinated mobility systems, mechanism design and control.
On the stability of grasped objects. *IEEE Trans. Robotics Automation* 12(6):904–17, 1996 (with Howard).

Noam Lior, Professor; Ph.D., Berkeley, 1973. Heat transfer and fluid mechanics, thermodynamics and Second-Law analysis, energy conversion, solar energy, combustion, flash evaporation and water desalination, destruction of hazardous wastes by photocatalysis and supercritical oxidation.
Sources of combustion irreversibility. *Combust. Sci. Technol.* 103:41–61, 1994 (with Dunbar).

James P. Ostrowski, Assistant Professor; Ph.D., Caltech, 1995. Dynamics and control of nonlinear mechanical systems, modular robotics, control of smart structures and underactuated systems, robotic locomotion.
The Geometric mechanics of undulatory robotic locomotion. *Int. J. Robotics Res.* 17(7):683–702, 1998 (with Burdick).

Sergio R. Turteltaub, Assistant Professor; Ph.D., Caltech, 1997. Dynamics of phase transitions in thermoelastic solids, optimization of structures and material properties, constitutive modeling of elastomeric foams.
Viscosity and strain gradient effects on the kinetics of propagating phase boundaries in solids. *J. Elasticity* 46(1):53–90, 1997.

AFFILIATED AND EMERITUS FACULTY

Norman Badler, Professor of Computer and Information Science; Ph.D., Toronto, 1975. Human figure modeling and animation. (World Wide Web page: http://www.cis.upenn.edu/~badler)

Ruzena Bajcsy, Professor of Computer and Information Science and Director of the General Robotics and Active Sensory Perception (GRASP) Laboratory; Ph.D., Slovak Technical, 1964; Ph.D., Stanford, 1972. Computer vision, robotics, and medical applications.

Daniel K. Bogen, Associate Professor of Bioengineering; Ph.D./M.D., Harvard, 1979. Engineering design, rehabilitation engineering, cognitive rehabilitation.

Stuart Churchill, Carl V. S. Patterson Professor Emeritus of Chemical Engineering; Ph.D., Michigan, 1952. Combustion, incineration, crystal growth, rate processes and correlation, computerized analysis.
New simplified models and formulations for turbulent flow and convection. *AIChE J.* 43:1125–40, 1997.

Benjamin Gebhart, Emeritus Samuel Landis Gabel Professor of Mechanical Engineering; Ph.D., Cornell, 1954. Heat rejection from compact integrated circuits, flow instability in buoyancy-driven flows, effect of motion pressure in buoyant flows.

Jerry P. Gollub, Professor of Physics at Haverford College and Adjunct Professor of Physics at University of Pennsylvania; Ph.D., Harvard, 1971. Hydrodynamic instabilities, pattern formation in fluids, mixing, spatiotemporal chaos, frictional dynamics, and turbulence.

Patrick Harker, Professor of Operations and Information Management and UPS Transportation Professor for the Private Sector; Ph.D., Pennsylvania, 1983. Continuous optimization, decision analysis, and complementarity theory and their use in the analysis of manufacturing, transportation, and service-sector operations.

Campbell Laird, Professor of Materials Science and Engineering; Ph.D., Cambridge, 1963. Physical metallurgy, electron microscopy, fatigue and fracture, heat treatment of alloys, corrosion, composites.

Burton Paul, Emeritus Asa Whitney Professor of Dynamical Engineering; Ph.D., Polytechnic, 1958. Contact stress with and without friction, dynamics of robots and linkage machinery, vibrations and stress analysis.

Richard Paul, Professor of Computer and Information Science; Ph.D., Stanford, 1972. Manipulation in hazardous environments, in outer space, underwater, and in microworlds as well as in environments in which it is necessary to work, such as the nuclear area.
Teleprogramming: Towards delay-invariant remote manipulation. *Presence: Teleoperators and Virtual Environments* 1(1):29–44.

Louis J. Soslowsky, Associate Professor of Orthopaedic Surgery and Bioengineering and Director of Orthopaedic Research; Ph.D., Columbia, 1991. Orthopedic biomechanics; structure-function studies of tendons and ligaments; models of tendon injury, repair, and healing; shoulder joint mechanics.

Karl T. Ulrich, Associate Professor of Operations and Information Management (Wharton School); Sc.D., MIT, 1988. Product design, product development, technology and operations management, computer-aided design, manufacturing.

Vaclav Vitek, Professor of Materials Science and Engineering; Ph.D., Czechoslovak Academy of Sciences, 1966. Computer modeling of the structure and properties of grain boundaries, metal-metal and metal-ceramic interfaces, dislocations and other lattice defects.

Jay Zemel, H. Nedwill Ramsey Professor Emeritus of Sensor Technologies; Ph.D., Syracuse, 1956. Gas and liquid transport in micromachined structures, development and use of pyroelectrics for gas flow measurements, thermal emission from microconfigured surfaces.

AREAS OF RESEARCH

Mechanics of Materials (Professors J. Bassani, D. Bogen, D. Discher, C. Laird, P. Ponte Castañeda, and V. Vitek): Crystal plasticity, effective properties of nonlinear composites, intermetallic compounds, localization studies, metal-forming processes, interfacial fracture, fatigue and high-temperature fracture, biomicrostructural and cell mechanics.

Fluid Mechanics (Professors P. S. Ayyaswamy, H. Bau, I. Cohen, D. Discher, B. Gebhart, J. Gollub, H. Hu, N. Lior, and J. Zemel): Buoyancy-induced flows and their control, microfabrication, meso- and microelectromechanical systems, moving water droplet interaction, combustion, multiphase flow, control of flow patterns, microelectronic manufacturing, complex fluids, biofluids, computational fluid dynamics.

Mechanical Systems (Professors G. Ananthasuresh, N. Badler, D. Bogen, P. Harker, V. Kumar, J. Ostrowski, B. Paul, R. Paul, and K. Ulrich): Robotics, computer-aided graphics and design, biomechanics, mechanisms and dynamics of machinery, microelectromechanical systems.

Thermal Sciences and Energy Conversion (Professors P. S. Ayyaswamy, H. Bau, S. Churchill, I. Cohen, B. Gebhart, H. Hu, N. Lior, and J. Zemel): Heat transfer, thermodynamics, convective flow, buoyancy-induced flow, bioheat transfer, flash evaporation, electronic packaging, use of solar energy for air-conditioning and dehumidification, passive solar heating, thermal storage, advance hybrid power cycles, salt gradient solar ponds, ocean–thermal energy conversion, combustion.

Biomechanics (Professors D. Discher and P. S. Ayyaswamy): Biomechanics research spans scales from the tissue through the molecular levels, with major efforts in cell mechanics, tendon and ligament properties, biomolecular simulation, and gravity effects on cells and tissues.

UNIVERSITY OF SOUTHERN CALIFORNIA

School of Engineering
Department of Mechanical Engineering

Programs of Study

The Department of Mechanical Engineering offers flexible programs in a variety of technical areas leading to the degrees of Master of Science in Mechanical Engineering, Engineer in Mechanical Engineering, and Doctor of Philosophy in mechanical engineering. Specialized programs for a Master of Science in Applied Mechanics, Master of Science in Materials Engineering, and Master of Science in Manufacturing Engineering are also available. An M.S. degree requires 27 semester units of approved course work either with or without a thesis. An additional 30 units of course work including a specialized research project are required for the Engineer in Mechanical Engineering degree. Nine of these 30 units must be in a minor area in engineering, and 6 units must be devoted to a research topic. No dissertation is required. However, a candidate for the E.M.E. degree must pass a qualifying examination at the conclusion of his or her course work. The E.M.E. requires approximately the same academic preparation as the Ph.D., and the E.M.E. course work may be applied toward a Ph.D. if the student is admitted to a doctoral program in the department. A Ph.D. student must pass a departmental screening examination at the beginning of the program and is admitted to candidacy after passing the qualifying examination at the conclusion of the formal course work. Finally, a candidate for the Ph.D. must complete an acceptable dissertation and defend it successfully. The Ph.D. in mechanical engineering is granted under the jurisdiction of the Graduate School in recognition of superior academic preparation and creative scholarly research.

Research Facilities

Specialized laboratory facilities in the traditional areas, such as heat transfer, combustion and propulsion, fluid mechanics, and materials form the central core of research work. In addition, interdisciplinary laboratories for biomechanics research, computer-aided design, concurrent engineering research, dynamics and controls, and robotics address relevant technological problems. A listing of current research areas appears on the reverse side of this page.

The University maintains a number of excellent libraries. Notable among them is the Seaver Science and Engineering Library, which contains major collections in engineering, physics, chemistry, and mathematics. The general collection is housed in the Doheny Memorial Library, which also includes the Undergraduate College Library. A recent addition to USC is the Leavey Library, which is an up-to-date and technologically advanced electronic resource center for education and research.

Financial Aid

A number of half-time teaching and research assistantships, fellowships, and scholarships are available to graduate students with a good academic record and an aptitude for either teaching or research. The stipends are adjusted frequently. Interested students should contact the department for information on current stipends and availability of scholarships.

Cost of Study

Estimated tuition and fees for 12 to 18 semester hours are $11,099 per semester in the 1999–2000 academic year; for those carrying fewer than 12 units per semester, the cost per unit is $748. The academic year consists of two semesters.

Living and Housing Costs

While the cost of living depends upon individual tastes and living habits, an average student attending the University of Southern California needs approximately $24,675 if living on campus or $25,600 if living off campus per calendar year. This figure includes the cost of books and supplies.

Student Group

USC's total enrollment, including part-time and full-time students, is 28,300. The School of Engineering has 1,700 undergraduate students and 2,200 graduate students enrolled.

Location

The greater Los Angeles area has an exceptional concentration of engineering research, development, and manufacturing. The University is situated at the heart of this activity on a 111-acre campus adjacent to Exposition Park. The availability of numerous cultural and recreational activities provides students with special curricular and extracurricular opportunities.

The University and The School

Founded in 1880, USC is the oldest nonsectarian private coeducational university in the western United States and has enjoyed a steady growth in research and enrollment. It offers excellence in education and is a major research university, with sixteen professional and specialized schools and the College of Letters, Arts, and Sciences. The faculty and the administration are dedicated to fulfilling a master plan for enterprise and excellence in education through continued basic research and high-quality teaching. The School of Engineering was founded in 1906. From a very modest beginning, the School has enjoyed continued growth; today, it consists of ten buildings providing classrooms, offices, and many modern laboratories and research institutes. Notable among these institutes are the Signal and Image Processing Institute, the National Center for Integrated Photonic Technology, Communications Sciences Institute, the Foundation for Cross Connection Control and Hydraulic Research, and the Information Sciences Institute located at Marina del Rey. Interdepartmental studies in biomechanics and bioengineering are possible through the full cooperation of the Orthopedic Hospital, the USC School of Medicine, the Los Angeles County Hospital, and Rancho Los Amigos Hospital. The Instructional Television Center enables many part-time students working in local companies to take televised courses for graduate credit without commuting frequently to campus. Some two dozen companies have television classrooms, and two regional television classrooms are operated by USC for those students who do not work for one of the subscribing companies.

Applying

The application for graduate work with complete credentials, the application fee, and scores on the General Test of the GRE should be filed with the Admissions Office. The department requires scores from both tests for evaluation. Applicants with questions regarding admissibility should contact the department prior to making a formal application for admission. Information concerning financial aid may be obtained from the department chairman.

Correspondence and Information

Chairman
Department of Mechanical Engineering
University of Southern California
Los Angeles, California 90089-1453
Telephone: 213-740-0484
Fax: 213-740-8071

University of Southern California

THE FACULTY AND THEIR AREAS OF CONCENTRATION

Charles Campbell, Professor; Ph.D., Caltech. Two-phase flow, flow of granular material, heat transfer, slurry flows, fluidized beds, comminutron, particle fracture.

Marijan Dravinski, Professor; Ph.D., IIT. Propagation and diffraction of elastic waves, geophysical wave propagation, earthquake engineering.

Fokion N. Egolfopoulos, Associate Professor; Ph.D., California, Davis. Combustion and propulsion.

Henryk Flashner, Professor (Aerospace); Ph.D., Berkeley. Dynamics and control of systems, control of structurally flexible systems, analysis of nonlinear systems.

Ramesh S. Guttalu, Research Associate Professor; Ph.D., Berkeley. Global behavior of nonlinear dynamical systems, point mapping and cell mapping methods, continuum mechanics.

Yan Jin, Assistant Professor; Ph.D., Tokyo. Collaborative engineering, design theory and methods, knowledge-based design and manufacturing systems, intelligent agents for engineering support.

Terence G. Langdon, Professor (Materials Science, Earth Sciences); D.Sc., Bristol (England). Mechanical properties of metals and ceramics, creep, superplasticity.

Stephen C.-Y. Lu, David Packard Professor in Manufacturing Engineering (Industrial and Systems Engineering, Computer Science); Ph.D., Carnegie Mellon. Manufacturing systems, concurrent engineering, knowledge-based expert systems, AI-based machine learning technologies.

Tony Maxworthy, Smith International Professor of Mechanical Engineering (Aerospace) and Member, National Academy of Engineering; Ph.D., Harvard. Geophysical fluid dynamics, unsteady aerodynamics, wave propagation, stability of fluid motion, bubble dynamics, turbulent flows, convective flows, interface dynamics.

Siavash Narimousa, Research Associate Professor; Ph.D., Johns Hopkins. Stratified flows, turbulent convective flows, rotating flows, air-sea interaction, mesoscale vortex dynamics, coastal currents and fronts, turbulent mixing.

Paul K. Newton, Professor (Aerospace); Ph.D., Brown. Nonlinear dynamical systems, waves, bifurcations, turbulence.

Steven R. Nutt, Professor (Materials Science); Ph.D., Virginia. Composite materials, deformation and damage mechanisms, interface structures and defects, electron microscopy.

Larry G. Redekopp, Professor (Aerospace Engineering); Ph.D., UCLA. Fluid dynamics, linear and nonlinear waves, stability of fluid motion.

Paul D. Ronney, Associate Professor; Sc.D., MIT. Combustion, fluid mechanics, internal combustion engines, low-gravity phenomena, radiative heat transfer.

Satwindar Sadhal, Professor; Ph.D., Caltech. Drops and bubbles in acoustic fields, thermocapillary flows with drops in low gravity, heat conduction in composite solids.

M. Oussama Safadi, Adjunct Professor; Ph.D., UCLA. Structural dynamics, finite element analysis, stress analysis, fracture mechanics.

Geoffrey R. Shiflett, Associate Professor; Ph.D., Berkeley. Kinematics and dynamics of mechanical systems, computer-aided design, optimal design techniques, microelectromechanical systems.

Firdaus E. Udwadia, Professor (Civil Engineering); Ph.D., Caltech. Structural dynamics, probabilistic mechanics, identification and control of large space structures, wave propagation.

Bingen Yang, Associate Professor; Ph.D., Berkeley. Distributed parameter systems, dynamics and control of flexible mechanical systems, smart structures, inflatable structures, numerical methods.

RESEARCH AREAS

Alloy Design. Current research involves the application of basic metallurgical principles to the design of alloys with useful properties and the design of dilute solid solution alloys suitable for creep applications.

Combustion. Current research topics include experimental and numerical studies of the effects of flow field nonuniformities and unsteadiness on the ignition, propagation, extinction, and kinetic structure of flames; experimental simulation of premixed turbulent combustion and extinction using aqueous autocatalytic chemical reactions; numerical simulation of turbulent combustion using front-tracking algorithms; microgravity combustion; flame spread over solid fuel beds; and absolute/convective instability of diffusion flames.

Computer-Aided Design. Analysis and synthesis of mechanisms, optimization techniques, computer graphics, and development of a microprocessor-based system to aid all phases of the design process and manufacturing.

Concurrent Engineering. Development of advanced information and computer technologies for manufacturing and design methodologies, knowledge-based expert systems for automation and systems integration.

Diffraction and Scattering of Elastic Waves. Two- and three-dimensional modeling of elastic-waves scattering by inclusions of arbitrary shape embedded in anisotropic media has been performed by using an indirect boundary integral equation method.

Droplet Combustion. Experimental and theoretical studies of the burning of two-phase slurry fuel droplets are being carried out.

Droplet Evaporation. An experimental and theoretical study is being made of liquid fuel evaporation under conditions of liquid phase pyrolysis. This work is applied to combustion-efficiency calculations and particulate formation.

Dynamics and Control. Interests are the control of robotic manipulators (including control of flexibility effects, attitude control, and momentum management for large spacecraft) and global analysis of nonlinear systems by cell- and point-mapping techniques.

Geophysical Fluid Mechanics. The study of fluid motion in which rotation and/or stratification are important dynamical influences. Among the aims is an interest in understanding dispersion and thermal structure of lakes, oceans, and atmospheres.

Heat Transfer in Composite Materials. Research activity includes analysis of average thermal properties, investigation of edge effects, and modeling of boundary conditions.

Identification and Control of Large Space Structures. Development of methodologies for optimally locating sensors and actuators for identification and control, trade-offs between identification and control, the development of near real-time identification algorithms, and the design of "active" structures.

Interface Dynamics. Experimental and numerical studies of the appropriate boundary conditions to be applied at miscible and immiscible interfaces between fluids of different physical and chemical properties.

Measurement of Long-Period Microtremors. Large-scale measurements of long-period microtremors have been performed in large metropolitan basins (Los Angeles, San Francisco, Anchorage, Northridge) in order to assess their applicability in assessing ground motion amplification during earthquakes.

Mechanical Properties. Areas of interest include high-temperature creep, superplasticity, superplastic forming, cyclic and static fatigue, and the fabrication of materials with ultrafine grain sizes. Experiments are performed on a wide range of metallic and nonmetallic materials, including ceramic- and metal-matrix composites.

Microgravity Combustion and Fluid Mechanics. Studies of weakly burning premixed and diffusion flames and reacting particle flows, flame spread over solid fuels, radiatively driven flows, investigations on the fluid mechanics and heat transfer associated with drops in low gravity, studies on thermocapillary flows in drops levitated by acoustic and/or electrostatic fields.

Nonlinear Systems. The global behavior of nonlinear dynamical systems and stability and bifurcation theory are being studied. Advanced computational techniques are used in the investigation of problems in nonlinear mechanics.

Robotic Research. Interests include the formulation of kinematic and dynamic equations for manipulators, control law development, and the application of differential games to control and coordinate a number of robotic manipulators.

Solidification. Stability of a freezing front in binary mixtures, the effects of an external flow on this stability, space applications.

Turbulence. Studies of turbulence in complex flows containing competing instability mechanisms. Approaches include DPIV imaging techniques and numerical simulation.

Two-Phase Flow. Investigation of the fracture flow and heat transfer properties of granular materials and dense fluid-particle systems, stability of sedimenting fronts.

UNIVERSITY OF UTAH

Department of Mechanical Engineering

THE UNIVERSITY OF UTAH

Programs of Study

Graduate student research includes a wide range of programs distributed among three departmental divisions: design, manufacturing, and controls; mechanics; and thermal, fluids, and energy systems. Active research programs include development of CAD and CAM tools; ergonomics and safety; prosthetic devices; automated manufacturing; rapid prototyping; free-form fabrication; mechatronics; inverse engineering problems; control of micro and macro systems; MEMS; fatigue, corrosion wear, and reliability; robotics; material response; failure of composites; degradation of polymers; energy resources and environmental problems; heat transfer (bio, gas turbine, and micro); computational mechanics; combustion; fluid mechanics; advanced flow measurement systems, turbulence modeling; and wind turbine design. Graduate degrees offered include the Master of Engineering, Master of Science, and Master of Philosophy in mechanical engineering. The M.E program provides a course of study to develop competencies in current engineering science and technology beyond the B.S. degree. There is no thesis requirement for the M.E. degree, but an engineering project is required. The M.S. degree requires a demonstrated research competency, generally in the form of a research thesis. A final oral comprehensive exam covering general aspects of the student's course work is required for both the M.E. and M.S. degrees. To be admitted to candidacy for the Ph.D. degree, each student must pass a written and oral qualifying exam. This exam is taken early in the student's program, generally within a year after acceptance into the Ph.D. program. A research comprehensive exam that includes a formal research proposal is generally scheduled one year after passing the Ph.D. qualifying exam. The Ph.D. requires a dissertation and an oral defense of the dissertation that documents innovative and independent research that contributes to the knowledge and advancement of the engineering profession.

Research Facilities

Research laboratories within the department include the Physical Fluid Dynamics Laboratory, Computational Fluid Dynamics Laboratory, Center for the Simulation of Accidental Fires and Explosions, Microscale Fluids and Heat Transfer Laboratory, Surface Layer Turbulence and Environmental Science Test Facility, Bioheat Transfer Laboratory, Convective Heat Transfer Laboratory, Fracture and Adhesive Laboratory, Materials Behavior Laboratory, Mechanics of Composites Laboratory, Structural Integrity Laboratory, Quality and Integrity Design Engineering Center, Center for Engineering Design, Robotics Laboratory, Controls and Design Laboratory, Center for Design Systems, Ergonomics and Safety Laboratory, and the Manufacturing Processes Laboratory. Numerous computing facilities are available, including the Center for High Performance Computing.

Financial Aid

Financial support for graduate students in the Department of Mechanical Engineering is provided primarily in the form of teaching and research assistantships. Each half-time teaching assistant and research assistant position also includes a full tuition waiver. A variety of federal, industrial, and private fellowships are available on a competitive basis.

Cost of Study

Total tuition and fees for graduate students in the 1999–2000 academic year are $1022.40 per semester (9 credit hours) for in-state residents and $3064.50 per semester (9 credit hours) for nonresidents. Students taking fewer than or more than 9 credit hours can subtract or add $76.70 for residents and $242.90 for nonresidents for each credit hour.

Living and Housing Costs

University housing costs range from $1792 per year for double rooms to $2405 per year for single rooms. Single students desiring on-campus housing should contact the Office of Residential Living, S114D Van Cott Hall, University of Utah, Salt Lake City, Utah 84112. Student family housing rates range from $330 to $565 per month. Information is available from the Student Family Housing Office, University of Utah, 1945 Sunnyside Avenue, Salt Lake City, Utah 84108.

Student Group

The University of Utah has a total student enrollment of approximately 26,000, including 5,000 graduate students. The graduate program in mechanical engineering has 119 students (82 M.S. and M.E. and 37 Ph.D.).

Location

The University of Utah's 1,500-acre campus stretches into the foothills of the Wasatch Range of the Rocky Mountains. Downtown Salt Lake City, offering first-rate art, sports, and entertainment, is 5 minutes from the University. Deep-powder skiing is readily accessible at seven major ski areas within 45 minutes of the campus.

The University

Founded in 1850, the University of Utah is composed of fifteen colleges and schools that offer ninety-two graduate programs. The Department of Mechanical Engineering is one of seven departments in the College of Engineering. The University of Utah is the major public higher education research institution in the region stretching from the Rockies to the Sierra Nevada Mountains. The University of Utah is classified as one of fifty comprehensive research universities among 3,400 U.S. institutions of higher education.

Applying

Students can be admitted in either the fall semester, with an application deadline of June 1, or spring semester, with an application deadline of November 1. Entrance requirements include a minimum 3.0 undergraduate GPA and a B.S. degree from an accredited engineering program in mechanical engineering or an allied field. International applicants are required to submit GRE General Test and TOEFL (530 minimum) scores.

Correspondence and Information

For departmental information:
Director of Graduate Studies
Department of Mechanical Engineering
University of Utah
50 South Campus Center Drive, Room 2202
Salt Lake City, Utah 84112-9208
E-mail: gradadms@stress.mech.utah.edu
World Wide Web: http://www.mech.utah.edu

Applications should be sent to:
Graduate Admissions
University of Utah
201 South 1460 E, Room 250S
Salt Lake City, Utah 84112-9057

University of Utah

THE FACULTY AND THEIR RESEARCH

Daniel O. Adams, Assistant Professor; Ph.D., Virginia Tech, 1991. Mechanics of composite materials, damage mechanics, experimental mechanics, moire-interferometry.

Tim A. Ameel, Assistant Professor; Ph.D., Arizona State, 1991. Microscale convective heat transfer, absorption refrigeration, heat exchangers.

Donald S. Bloswick, Associate Professor; Ph.D., Michigan, 1986. Ergonomics, occupational biomechanics, rehabilitation engineering.

Don R. Brown, Associate Professor; Ph.D., Stanford, 1988. Design systems.

V. Chandrasekaran, Research Assistant Professor; Ph.D., Utah, 1997. Fretting, fatigue, corrosion fatigue.

Kuan Chen, Associate Professor; Ph.D., Illinois, 1981. Thermodynamics, heat transfer, fluid mechanics, thermal diodes, plasma modeling.

Santosh Devasia, Assistant Professor; Ph.D. California, Santa Barbara, 1993. Chaotic and nonlinear dynamics, optimal and nonlinear controls, precision tracking, nanoscopy, active vibration control of smart structures.

K. L. DeVries, Distinguished Professor; Ph.D., Utah, 1962. Mechanical properties of materials; molecular, microscopic, and macroscopic phenomena associated with failure; adhesion; polymers; biomedical and dental materials.

Samuel Drake, Research Associate Professor; Sc.D., MIT, 1978. Integrated process planning and computer-aided manufacturing, design for manufacturing, design for assembly, industrial robotics.

Charles Elliot, Research Assistant Professor; Ph.D., Utah, 1993. Fatigue, fretting fatigue, corrosion fatigue.

E. S. Folias, Ph.D., Caltech, 1963. Application of solid and fracture mechanics to metal and composite material structures, ceramics, organic ceramics.

James E. Guilkey, Research Assistant Professor; Ph.D., Utah, 1997. Experimental fluid mechanics, computational structural mechanics, fluid-structure interaction.

A. Craig Hansen, Research Associate Professor; Ph.D., Colorado State, 1975. Aerodynamics, wind turbine engineering, wind engineering.

Todd B. Harman, Research Assistant Professor; Ph.D., Utah, 1998. Computational fluid dynamics, multimaterial flows, electrohydrodynamics in reacting flows.

David W. Hoeppner, Professor; Ph.D., Wisconsin, 1963. Engineering design of components and systems, with emphasis on holistic damage tolerance concepts, materials engineering, design methods, fatigue, fracture mechanisms, quality-based design, creep, reliability-based design, aircraft integrity issues.

Stephen C. Jacobsen, Distinguished Professor; Ph.D., MIT, 1973. Robotics, medical imaging systems, microelectromechanical systems (MEMS) sensors and actuators, micro drug delivery systems and medical electronic monitors.

Joseph C. Klewicki, Associate Professor; Ph.D., Michigan State, 1989. Fluid mechanics, vorticity dynamics, advanced experimental diagnostics, two-phase flows, atmospheric turbulence, stratified flows.

David J. Laino, Research Assistant Professor; Ph.D., Utah, 1997. Aerodynamics, wind turbine engineering, computer modeling of machine dynamics.

Phillip M. Ligrani, Professor; Ph.D., Stanford, 1980. Convective heat and mass transfer, laminar turbulent, and transitional flows; gas turbine blade cooling; experimental techniques; continuous SPLITT factionation.

Marshall Mattingly, Research Assistant Professor; Ph.D., Utah, 1998. Hyperthermia cancer treatment, thermal medical diagnostics, reduced-order modeling, learning control for robotic and biologic systems.

Patrick A. McMurtry, Associate Professor; Ph.D., Washington (Seattle), 1987. Fluid mechanics, turbulence modeling, computational fluid dynamics, multimaterial flows, reacting flows, turbulent mixing processes, stochastic modeling approaches.

Sanford G. Meek, Assistant Professor; Ph.D., Utah, 1982. Prosthetic arm design and control, biocontrol systems, robotics.

Christoph M. Pistor, Assistant Professor; Ph.D., Illinois at Chicago, 1988. Intelligent and rapid manufacturing, transport phenomena and property relations, functionally graded and composite materials, orthopedic implants.

Robert B. Roemer, Professor and Chair; Ph.D., Stanford, 1968. Application of thermal sciences to biomedical problems, hyperthermia cancer treatments and cancer detection, design in engineering and nature, micromechanical systems for medical robotics.

Gary M. Sandquist, Professor; Ph.D., Utah, 1964. Nuclear and environmental engineering, energy systems, waste management, risk assessment.

Elaine P. Scott, Associate Professor; Ph.D., Michigan State, 1989. Heat transfer, parameter estimation, inverse engineering problems.

Richard Shorthill, Research Professor; Ph.D., Utah, 1960. Lunar thermal and photometric properties, planetary surfaces, infrared radiometry.

David M. Slaughter, Research Assistant Professor; Ph.D., Utah, 1986. Nuclear and environmental engineering, energy and waste management, risk assessment, radioanalyses, process characterization, reactor dynamics and controls.

James K. Strozier, Research Professor; Ph.D., Michigan, 1966. Aerodynamics, flight mechanics, thermal systems design, gas dynamics.

Stephen R. Swanson, Professor; Ph.D. Utah, 1970. Solid mechanics, mechanics of composite materials, strength and fracture of fiber composites, fracture mechanics.

Charles L. Thomas, Assistant Professor; Ph.D., Drexel, 1993. Optical and ultrasonic sensor development for polymer manufacturing processes, control of manufacturing processes, rapid prototyping, manufacturing.

William K. Van Moorhem, Professor; Ph.D., Cornell, 1971. Unsteady fluid mechanics, acoustics, vibrations.

UNIVERSITY OF VIRGINIA

Department of Mechanical and Aerospace Engineering

Programs of Study

The department offers graduate study in advanced aspects of mechanical and aerospace engineering leading to the Master of Engineering (M.E.), Master of Science (M.S.), and Doctor of Philosophy (Ph.D.) degrees in mechanical and aerospace engineering. The master's programs are designed to strengthen and extend undergraduate training and competence and, for those working toward the M.S., offer an opportunity to pursue independent research. The M.S. degree requirements are 24 semester hours of approved graduate course work and submission of an acceptable thesis. The M.E. (nonthesis) degree requirement is 30 semester hours of approved course work. In addition, the program in nuclear engineering requires a final oral examination. The Ph.D. degree requirements include completion of approximately 24 semester hours of formal study beyond the master's degree, passage of a comprehensive written and oral examination, and submission of an acceptable dissertation. A typical student requires three semesters to complete the M.E. degree or two years for the M.S. degree. The typical doctoral student requires three to four years beyond the master's degree. Graduate students may concentrate their studies in fluid mechanics/gasdynamics, thermodynamics/heat transfer/combustion, dynamics/vibrations/rotating machinery, mechanical/computer-aided design and manufacturing, structural mechanics or dynamics, and control systems. A full line of course work is offered in these areas. Students are also expected to complete courses in advanced applied mathematics and in computational methods, and they may take advantage of a wide range of courses offered elsewhere in the School of Engineering and Applied Science and in the University. A variety of research opportunities exist in the department within the above subject areas, addressing theoretical, computational, and experimental problems. For research, a student works under the direct supervision of a department faculty member. Most graduate students in the department are associated with an ongoing funded research program. Programs in mechanical and aerospace engineering in which students currently participate include studies of internal flow patterns and vibrational dynamics in turbomachines and bearings; computer-aided design and manufacturing systems; gasdynamics and supersonic combustion using state-of-the-art laser diagnostic techniques; microgravity fluid mechanics; designs for improved wheelchairs and seating systems; automatic/digital vibration control systems for rotating machinery and for lightweight structures; magnetic journal bearing systems; automobile crash dynamics; thermal structures; and micro-scale heat transfer.

Research Facilities

A wide variety of computer systems, from single-user workstations and PCs to parallel machines, are available. UNIX systems available include IBM RS/6000 systems and Sun and Silicon Graphics Workstations. Access to supercomputers is available through connection to various Supercomputing Centers. Interactive-design graphics equipment is provided through the Center for Computer-Aided Engineering. Scientific and technical literature needs may be met through a full-service library system. Usual shop, instrumentation, supply, and other services are also provided. Department facilities include the Aerospace Research Laboratory, the Rotating Machinery and Controls Laboratory, the Automotive Crashworthiness Laboratory, and the Aerogel and Micro-scale Heat Transfer Laboratory.

Financial Aid

Aid is provided through fellowships, research assistantships, or teaching assistantships, or a combination of these. Twelve-month stipends for research assistants (88 hours per month) are $15,600 (M.S.) to $17,000 before the comprehensive examinations and $18,000 after the comprehensive examinations (Ph.D.). Nine-month stipends for teaching assistants (88 hours per month) are $10,714 (M.S.) and $12,056 (Ph.D.). Research and teaching assistantships also cover the cost of the student's tuition. Also available to top U.S. student applicants are Dean's Fellowships in the amount of $2000 (over nine months). This fellowship is renewable for up to three years. Students also finance their studies through national fellowships and traineeships, cooperative work, federally backed loans, and spouse employment.

Cost of Study

Tuition and fees for 1998–99 totaled $4896 for Virginia residents and $15,824 for nonresidents. Summer sessions cost about one fifth of these amounts. Books, materials, and services typically cost $800 per year.

Living and Housing Costs

In 1999–2000, $7000 per calendar year covers reasonable costs for housing, food, entertainment, and other living expenses for a single student. University dormitory rooms are $2100–$2300 per academic year; married student apartments, $400–$500 per month; and an all-meals cafeteria plan, $1744 per year.

Student Group

The department has about 65 full-time graduate students. Most have received their undergraduate degree in mechanical or aerospace engineering, and many have professional experience. The average age is 24. The percentages of women, married, and international students each range between 10 and 20 percent. Almost all of the department's graduate students accept positions in industrial firms, consulting firms, private or government labs, or engineering schools. The total enrollment of full-time graduate students in the school is 650 and in the University, 5,000.

Location

The University is located in Charlottesville. This city lies in the Piedmont region of Virginia, just to the east of the Blue Ridge Mountains and 110 miles southeast of Washington, D.C. The population of the city and its urban surroundings is about 100,000. The University and the Charlottesville area are widely known for their history and scenic beauty. The climate is generally mild, and numerous opportunities for outdoor recreation exist. The region offers a highly attractive variety of cultural activities and places of historic interest.

The University

The University of Virginia was founded in 1819 by Thomas Jefferson. It has national stature and is widely recognized for its high academic standards, its preparation of students for the higher professions, the integrity of its graduates (promoted through its student-run honor system), the high level of faculty-student contact, the flexibility and openness of its programs, and the attractive quality of its academic life. The total enrollment is 20,000. Almost half of these students come from outside the state. The men-women ratio is about 1:1.

Applying

Applications for admission may be submitted at any time. Although it is better to begin studies in the fall semester, studies may also be started with the spring or summer term. Applicants desiring first-round consideration for most types of financial aid should submit completed applications no later than February 1.

Correspondence and Information

Graduate Director
Department of Mechanical and Aerospace Engineering
Thornton Hall
University of Virginia
Charlottesville, Virginia 22901

Telephone: 804-924-7425
E-mail: mane-adm@virginia.edu
World Wide Web: http://www.mane.virginia.edu

University of Virginia

THE FACULTY AND THEIR RESEARCH

Ronald D. Flack, Professor and Chairman; Ph.D., Purdue, 1975. Turbomachinery, flow diagnostics, lubrication flows.

Paul E. Allaire, Professor; Ph.D., Northwestern, 1972. Magnetic bearings, rotor dynamics, fluid bearings, controls.

Lloyd E. Barrett, Professor; Ph.D., Virginia, 1978. Rotating machinery, lubrication, magnetic bearings.

J. Taylor Beard, Associate Professor; Ph.D., Oklahoma State, 1965. Thermal systems, heat transfer, solar energy.

Harsha K. Chelliah, Assistant Professor; Ph.D., Princeton, 1988. Combustion.

Jeff A. Crandall, Research Assistant Professor; Ph.D., Virginia, 1994. Auto safety and biomechanics.

John J. Dorning, Professor; Ph.D., Columbia, 1967. Bifurcation, nonlinear dynamics, and chaos.

Sam S. Fisher, Professor; Ph.D., UCLA, 1967. Turbulence, flow instabilities, turbulent flows, aerodynamics (experimental).

George T. Gillies, Research Professor; Ph.D., Virginia, 1980. Precision measurements, medical physics, gravitation.

Hossein Haj-Hariri, Associate Professor; Ph.D., MIT, 1987. Stability, fluid dynamics, mechanics, applied mathematics.

James L. Kelly, Professor; Ph.D., LSU, 1962. Nuclear chemical engineering.

Carl R. Knospe, Associate Professor; Ph.D., Virginia, 1989. Multivariable control, magnetic bearings.

Roland H. Krauss, Research Associate Professor; Ph.D., Virginia, 1972. Laser diagnostics, supersonic combustion.

Gabriel Laufer, Associate Professor; Ph.D., Princeton, 1979. Gasdynamics, laser diagnostics, supersonic combustion.

Eric H. Maslen, Associate Professor; Ph.D., Virginia, 1990. Magnetic bearings, controls.

James C. McDaniel Jr., Professor; Ph.D., Stanford, 1981. Gasdynamics, flow diagnostics, supersonic combustion.

Jeffrey B. Morton, Professor; Ph.D., Johns Hopkins, 1967. Fluid-flow stability, turbulence.

Robert U. Mulder, Associate Professor; Ph.D., Virginia, 1981. Neutron activation analysis, radioisotope applications, boron neutron cancer therapy.

Pamela M. Norris, Assistant Professor; Ph.D., Georgia Tech, 1992. Microscale heat transfer, sensor design, and aerogel applications.

Walter D. Pilkey, Professor; Ph.D., Penn State, 1962. Structural mechanics, vibrations, and computational mechanics, and biomechanics.

Robert J. Ribando, Associate Professor; Ph.D., Cornell, 1977. Computational fluid mechanics and heat transfer.

Larry G. Richards, Associate Professor; Ph.D., Illinois, 1971. Computer-aided design and manufacturing, statistics.

William W. Roberts Jr., Professor; Ph.D., MIT, 1969. Mathematical modeling, computer simulation, gas-fiber flows, industrial fiber-processing technologies.

Roger A. Rydin, Associate Professor; Sc.D., MIT, 1964. Computational methods, reactor physics, radiation shielding.

Timothy C. Scott, Associate Professor; Ph.D., Michigan, 1976. Automotive engineering, energy systems design.

Pradip N. Sheth, Associate Professor; Ph.D., Wisconsin, 1972. CAD/CAM, manufacturing systems, factory automation.

Susan E. Carlson Skalak, Assistant Professor; Ph.D., Georgia Tech, 1993. Concurrent engineering and design manufacturing.

John G. Thacker, Professor; Ph.D., Virginia, 1976. Machine design, manufacturing processes, biomechanics.

Earl A. Thornton, Professor; Ph.D., Virginia Tech, 1968. Finite elements, thermal structures.

Miles A. Townsend, Professor; Ph.D., Wisconsin, 1971. Design, dynamics, dynamical systems, optimization, biomechanics.

Houston G. Wood III, Associate Professor; Ph.D., Virginia, 1978. Analytical and computational fluid mechanics.

RESEARCH AREAS

Aerogels. Highly porous solids made out of materials such as silica, alumina, or zirconia.

Astrophysical Galactic Dynamics. Structure and dynamics of the gaseous component in galaxies.

Automatic Controls. Multivariable, nonlinear, and adaptive control for robotics, and magnetic suspension systems.

Biomechanics. Automotive safety, computational occupant simulation, structural dynamics, wheelchair integrity.

Boron Neutron-Capture Therapy. Treatment of melanomas with energy-filtered neutron from a research reactor.

Computer-Aided Engineering for Design and Manufacturing. Artificial intelligence and expert systems, modeling, and simulation; human factors and ergonomics.

Computational Methods for Nuclear Engineering. Development of new computational methods to solve nuclear engineering problems.

Control Systems. Research in automatic control of linear and nonlinear dynamical systems; applications in attitude control, mechatronics, robotics, and vibration isolation.

Experimental Fluid Mechanics. Flow instability in transition-to-turbulence processes in shear flows; the mechanics of diverging, narrow-channel flows.

Expert Systems. Research to produce software that can aid in diagnosing rotating machinery vibration problems.

Fluid Film Bearings. Characteristics of radial fluid film bearings and seals as applied to high-speed rotating equipment.

Industrial Fiber-Processing Technologies and Virtual Prototyping. Gas-fiber dynamics, mathematical-computational modeling of the motion, dynamics, transport, and deposition of fibers within specific classes of gas-fiber flow environments and machine designs.

Heat Transfer. Microscale heat transfer analysis of thin films and microstructures.

Magnetic Bearings. Magnetic bearings for rotating machinery suspensions and vibration isolation.

Medical Physics. New methods for the treatment of brain tumors, novel stereotactic surgical techniques, medical imaging.

Nuclear Reactor Stability. Analysis of nuclear reactors as nonlinear dynamical systems.

Polymer Aging. Effects of radiation and thermal aging on properties of polymers and electric cable insulation.

Precision Measurements. Thermal and electrical metrology, measurements of weak forces and small effects.

Rehabilitation Engineering. Research on wheeled mobility and specialized seating for the disabled population.

Rotor Dynamics. Computer simulation of the vibration characteristics of rotating machines.

Structural Dynamics. Research in structural mechanics, modification theory, and finite-element modeling.

Supersonic Mixing and Combustion. Study of the mixing of supersonic streams and the supersonic combustion of hydrogen in air using nonintrusive optical diagnostics, compressible fluid mechanics and finite-rate chemical kinetics of supersonic hydrogen and air.

Thermal Structures. Computational and experimental research for high-speed flight vehicles and orbiting structures.

Theoretical and Computational Fluid Dynamics. Hydrodynamic stability and heat-transfer interactions, supersonic combustion, finite-rate chemical kinetics, nuclear-reactor heat transfer, centrifuge flows, convective cooling of aerospace structures, two-phase flow dynamics, microgravity fluid mechanics, interfacial flows, phase transition.

Turbomachinery Fluid Dynamics. Internal flow in pumps, compressors, turbines, and torque converters; computational predictions and laser velocimetry and experimental techniques.

UNIVERSITY OF WISCONSIN–MADISON

Department of Mechanical Engineering

Programs of Study	Graduate study is offered leading to the M.S. and Ph.D. degrees. In general, the graduate program is tailored to meet the individual's needs and interests and is formulated in collaboration with the student's major professor. Students are encouraged to take courses inside and outside the department. Examples of areas in which a student may choose to specialize are automatic controls, bioengineering, biomechanics, combustion, composite materials, computer-aided design and manufacturing, energy conversion, fluid mechanics, fluid power, friction, gasdynamics, heat transfer, lubrication and wear, machine dynamics, materials joining, mechanism design, metal cutting, metal forming, noise control, ocean engineering, optimal design, plastics processing, quality control, robotics, solar energy, statistical experimental design, stochastic modeling, stress analysis, thermodynamics and vehicle powertrain systems, and vibrations.
	The M.S. degree requires 24 credits, including a thesis, or 30 credits without a thesis. The choice of a thesis or nonthesis program is made by the student in consultation with his or her major professor. The M.S. program can usually be completed in 1 to 1½ years of full-time study.
	The Ph.D. program typically consists of at least 48 formal course credits beyond the B.S. degree. Although there are no specific course requirements, students are expected to demonstrate strength in their chosen area. This strength is gained not only by an accumulation of course credits but also by a thorough investigation of a research problem. At least three years of study beyond the B.S. degree are normally required to complete a Ph.D. program.
Research Facilities	Facilities for research are housed in the Mechanical Engineering Building, the fourteen-story Engineering Research Building, and the Motor Vehicle Research Laboratory. Research laboratories in the department include the engine research center; solar energy lab; convective heat transfer lab; structural dynamics and vibrations research group; cryogenics and high-current labs; fluid power lab; laboratory for applied manufacturing; mechanical design lab; experimental mechanics and mechanical measurements lab; lab for integrated computer-aided research on virtual engineering design; molten metal lab; polymer processing lab; sensors, signal processing, and real-time controls integration lab; combustion processes lab; space automation and robotics lab; spatial automation lab; powertrain control lab; robotics lab; orthopedic research lab; and the biomechanics lab.
	In addition to the department laboratories, the facilities of the Computer-Aided Engineering Center and many well-equipped laboratories and shops within the University are available, as are numerous small computer systems for data acquisition and control of machines and experiments.
Financial Aid	Financial aid sources include scholarships, fellowships, research assistantships, teaching assistantships, and out-of-state tuition exemptions. In fall 1998, half-time research assistants received about $1215 per month each year. This stipend is supplemented by tuition exemption. Teaching assistants may receive the equivalent support. Half-time research assistants and teaching assistants are allowed to carry a full academic load. Research assistants are usually required to submit a thesis as part of the M.S. program. Stipends for scholarships and fellowships vary according to the nature of the award and source of funds but typically carry the same payments as those for research assistants and teaching assistants.
Cost of Study	Information about graduate fees and tuition costs per semester on the Madison campus for Wisconsin residents and for nonresident students may be obtained upon request. The tuition fees are waived for those nonresident research assistants, teaching assistants, and some fellowship holders who have a 33 percent or more appointment.
Living and Housing Costs	The University maintains unfurnished apartments for graduate students on campus in a wooded area adjacent to Lake Mendota. Monthly rents range from about $470 to $810 for one-, two-, and three-bedroom apartments, respectively. Heat and water are included, the kitchen is complete, and laundry facilities are available. Residence hall rooms are available from $2194 to $2860 per academic year. One- and two-bedroom furnished Harvey Street apartments are also available for single graduate students and range in price from $3163 to $4385 per year. The University maintains a Campus Assistance Center to help students locate private housing.
Student Group	Graduate students on the Madison campus come from fifty states and many other countries. The Department of Mechanical Engineering has approximately 200 full-time graduate students, of whom about one third are Ph.D. candidates.
Location	Madison, the state's capital, has four beautiful lakes and is known for its cultural and recreational opportunities. The city affords many educational advantages for engineering students.
The University and The Department	The Madison campus is the oldest of the fourteen 4-year campuses of the University of Wisconsin system. It is known as one of the world's great centers of learning. Graduate programs are offered in 150 departments; half of these programs are in the natural sciences. There are eight engineering departments.
	The Department of Mechanical Engineering has a faculty of 40 professors. Together, they represent a wide range of areas of expertise, experience, and research interests.
Applying	All applications for admission should be received at least 4–6 months before the semester in which the student plans to enroll. Applications for fellowships should be submitted no later than December 20 to be considered for fellowship competition during the following fall semester. Research assistantship support may be available at any time.
Correspondence and Information	Chairman, Department of Mechanical Engineering Room 240, Mechanical Engineering Building University of Wisconsin–Madison 1513 University Avenue Madison, Wisconsin 53706 Telephone: 608-262-3543 608-265-3576 (for applications) World Wide Web: http://engr.wisc.edu/ME/

University of Wisconsin–Madison

THE FACULTY AND THEIR RESEARCH

The Madison area code is 608. Faculty members' campus phone numbers and e-mail addresses are in parentheses.

William A. Beckman, Professor; Ph.D., Michigan, 1964. Solar energy, radiative heat transfer. (263-1590; beckman@engr.wisc.edu)

Riccardo Bonazza, Assistant Professor; Ph.D., Caltech, 1992. Vapor explosions; shock-interface interactions; unsteady, impulsive gas dynamics. (265-2337; bonazza@engr.wisc.edu) (also EP)

Michael L. Corradini, Professor; Ph.D., MIT, 1979. Two-phase flow, heat and mass transfer, risk and industrial safety. (262-3484; corradini@engr.wisc.edu) (also EP)

Marvin F. DeVries, Professor; Ph.D., Wisconsin, 1966. Metal cutting, CAM, manufacturing systems engineering. (262-1801; devries@engr.wisc.edu)

Neil A. Duffie, Professor; Ph.D., Wisconsin, 1980. Computer controls, manufacturing systems, sensors, closed-environment control systems, space-flight experiments. (262-9457; duffie@engr.wisc.edu)

Patrick D. Eagan, Assistant Professor; Ph.D., Wisconsin, 1995. Industrial environmental engineering and management. (263-7429; eagan@engr.wisc.edu) (also EPD)

Roxann L. Engelstad, Professor; Ph.D., Wisconsin, 1988. Structural dynamics, nonlinear vibrations, mechanical modeling and design. (262-5745; engelstad@engr.wisc.edu)

Donald S. Ermer, Professor; Ph.D., Wisconsin, 1966. Statistical quality/productivity improvements, engineering reliability, robust design. (262-2557; ermer@engr.wisc.edu) (also IE)

Patrick V. Farrell, Professor; Ph.D., Michigan, 1982. Fluid mechanics, heat and mass transfer, optical methods. (263-1686; farrell@engr.wisc.edu)

Nicola J. Ferrier, Assistant Professor; Ph.D., Harvard, 1992. Robotics, intelligent systems, sensing, computer vision, automation. (265-8793; ferrier@engr.wisc.edu)

David E. Foster, Professor; Ph.D., MIT, 1978. Combustion, chemical kinetics, thermodynamics. (263-1617; foster@engr.wisc.edu)

Frank J. Fronczak, Associate Professor; D.E., Kansas, 1978. Machine design, fluid power, vibrations. (262-1993; fronczak@engr.wisc.edu)

Rajit Gadh, Associate Professor; Ph.D., Carnegie Mellon, 1991. Virtual design, Internet-based CAD, collaborative design, 3-D multimedia, shape-features modeling, 3-D assembly/disassembly in design. (262-9058; gadh@engr.wisc.edu) (also MSE)

Jaal B. Ghandhi, Assistant Professor; Ph.D., Princeton, 1995. Combustion, optical methods, thermodynamics. (263-1684; ghandhi@engr.wisc.edu)

A. Jeffrey Giacomin, Associate Professor; Ph.D., McGill, 1987. Polymer processing, rheology, transport phenomena. (262-7473; giacomin@engr.wisc.edu)

Kreg G. Gruben, Assistant Professor; Ph.D., Johns Hopkins, 1993. Biomechanics. (262-2711; gruben@engr.wisc.edu) (also Kinesiology)

Sanford A. Klein, Professor; Ph.D., Wisconsin, 1976. Solar energy, thermodynamics, refrigeration, educational software development. (263-5626; klein@engr.wisc.edu)

Robert D. Lorenz, Professor; Ph.D., Wisconsin, 1984. High-performance system design, optics and sensor technology, computer control and manufacturing automation. (262-5343; lorenz@engr.wisc.edu) (also ECE)

Edward G. Lovell, Professor; Ph.D., Michigan, 1967. Structural mechanics, buckling, MEMS, thin-film stresses, shell vibrations. (262-0944; lovell@engr.wisc.edu)

Vladimir J. Lumelsky, Professor; Ph.D., Moscow, 1970. Robotics, control theory, industrial automation, kinematics. (263-1659; lumelsky@engr.wisc.edu)

Jay K. Martin, Professor; Ph.D., Michigan, 1984. Engines, combustion, fluid mechanics, diagnostic development. (263-9460; martin@engr.wisc.edu)

Wayne D. Milestone, Professor; Ph.D., Ohio State, 1966. Machine design, failure analysis, fatigue design analysis, CAD. (262-0023; milestone@engr.wisc.edu)

John W. Mitchell, Professor; Ph.D., Stanford, 1963. Heat transfer, building HVAC, energy use, solar energy. (262-5972; mitchell@engr.wisc.edu)

John J. Moskwa, Associate Professor; Ph.D., MIT, 1988. Powertrain system dynamic modeling, diagnostics and multivariable control. (263-2423; moskwa@engr.wisc.edu; http://www.erc.wisc.edu/powertrain/)

Glen E. Myers, Professor; Ph.D., Stanford, 1962. Heat transfer, conduction, finite-element analysis, thermodynamics. (262-0225; myers@engr.wisc.edu)

Tim A. Osswald, Associate Professor; Ph.D., Illinois, 1987. Polymer processing, computer modeling and simulation, CAD, composites. (263-9538; osswald@engr.wisc.edu)

John M. Pfotenhauer, Assistant Professor; Ph.D., Oregon. 1984. Cryogenics, applied superconductivity, heat transfer in liquid helium, refrigeration. (263-4082; pfotenhauer@engr.wisc.edu) (also EP)

Douglas T. Reindl, Assistant Professor; Ph.D., Wisconsin, 1992. Building energy, thermal energy storage, refrigeration, indoor air quality. (262-6381; reindl@engr.wisc.edu) (also EPD)

Rolf D. Reitz, Professor; Ph.D., Princeton, 1978. Computational fluid dynamics applied to engines and sprays, experimental engine and spray research. (262-0145; reitz@engr.wisc.edu)

Terry G. Richard, Professor; Ph.D., Wisconsin, 1973. Experimental mechanics, experimental dynamic structural analysis, and optical methods. (262-3439; richard@engr.wisc.edu)

Robert E. Rowlands, Professor; Ph.D., Illinois, 1967. Advanced materials, experimental mechanics, stress analysis, composites and fracture. (262-3205; rowlands@engr.wisc.edu)

Christopher J. Rutland, Associate Professor; Ph.D., Stanford, 1989. Turbulent reacting flows, engine simulation, turbulence modeling, computational fluid dynamics. (262-5853; rutland@engr.wisc.edu)

Vadim Shapiro, Assistant Professor; Ph.D., Cornell, 1991. Computer-aided design, geometric and physical modeling, computational geometry. (262-3591; vshapiro@engr.wisc.edu)

Kevin J. Shinners, Associate Professor; Ph.D., Wisconsin, 1985. Sensors to measure forage moisture and constituents; application of GPS and site-specific farming to forage equipment; systems to improve harvesting and handling of biomass and processing to enhance forage value. (263-0756; shinners@engr.wisc.edu) (also BSE)

Leslie M. Smith, Associate Professor; Ph.D., MIT, 1988. Turbulence and turbulence modeling, statistical physics, applied mathematics. (262-3852; smith@engr.wisc.edu) (also Math)

Richard J. Straub, Professor; Ph.D., Wisconsin, 1980. Machine design, agricultural equipment, and processing of biological materials. (262-0605; straub@engr.wisc.edu) (also BSE)

Karen A. Thole, Assistant Professor; Ph.D., Texas, 1992. Experimental and CFD studies of fluid mechanics and heat transfer, turbine blade flows, heat exchanger optimization. (262-0923; thole@engr.wisc.edu)

John J. Uicker Jr., Professor; Ph.D., Northwestern, 1965. Kinematics, dynamics, CAD/CAM, solid modeling. (262-3590; uicker@engr.wisc.edu)

Ray Vanderby, Associate Professor; Ph.D., Purdue, 1975. Biomechanics, tissue mechanics, micromechanics. (263-9593; vanderby@engr.wisc.edu) (also Orthopedic Surgery)

Dharmaraj Veeramani, Assistant Professor; Ph.D., Purdue, 1991. CAD/CAM, manufacturing systems control, Internet-aided collaborative design, manufacturing and supply-chain interaction. (262-0861; raj@ie.engr.wisc.edu) (also IE)

RESEARCH ACTIVITIES

Research Includes biomechanics, combustion, fluid dynamics, fluid power, heat transfer, integrated design and manufacturing, internal combustion engines, machine dynamics, vibrations and acoustics, manufacturing, mechatronics, mechanical systems modeling, polymer processing, powertrain design, robotics, and solar energy.

VILLANOVA UNIVERSITY

Department of Mechanical Engineering

Program of Study

The Department of Mechanical Engineering offers a graduate program of study leading to the degree of Master of Mechanical Engineering. The program is design-oriented and industry-related. It has been developed to meet the needs of both the practicing professional and the student planning further advanced study.

The program provides an opportunity for students with varied backgrounds and interests to develop a plan of study tailored to their needs. A minimum of 30 semester credit hours must be completed, including a course in advanced engineering analysis. The remaining credits may consist of approved course work, or the student can elect to submit a thesis based on 6 to 9 credit hours of research. All graduate assistants are required to write a thesis. The remaining courses may be selected from one or more of the following areas of mechanical engineering: thermal energy/fluids engineering, solid mechanics/materials, materials/manufacturing, dynamics and control, design, and CAD/CAM. A maximum of 6 credits from outside the department in related areas of engineering and mathematics may be accepted with the approval of the graduate committee chairman.

The department offers two certificate programs, one in manufacturing and one in virtual manufacturing, which provide flexibility for the working professionals. These programs offer a concentrated study of modern manufacturing principles. Both breadth and depth of coverage are emphasized. All courses that constitute these programs stress applications and manufacturing practice.

Research Facilities

The Mechanical Engineering Laboratories include both instructional and research facilities. Equipment is available for experimental work in optical heat transfer, fluid flow, vibrations, stress analysis, thermodynamics, combustion, power, and materials science. Instrumentation includes a laser-Doppler velocimeter, a scanning electron microscope, static and dynamic universal test machines, a magnetic bearing research rig, and an energy-dispersive spectrometer. The department participates in the College of Engineering's Computer-Aided Engineering Facility and has a Computer-Aided Manufacturing Laboratory with CNC machines and robots.

Financial Aid

A limited number of University assistantships paying tuition plus a stipend ($9215 per year in 1998–99) are available for graduate students in mechanical engineering. Tuition scholarships may also be available. Eligibility for low-interest student loans is determined by the Financial Aid Office.

Cost of Study

For 1998–99, tuition was $595 per credit. General University fees were $40 per semester. Tuition and fees are waived for graduate assistants and tuition scholars.

Living and Housing Costs

On-campus housing is not available for graduate students, but listings of furnished rooms and apartments are maintained by the University's Off-Campus Housing Service. Public transportation includes two rail lines and bus service. There is ample parking space, which costs $75 per year for students.

Student Group

The total enrollment at Villanova is approximately 11,000, which includes some 2,000 graduate students. About 30 graduate students are enrolled in the graduate program in mechanical engineering. Approximately one third of these are studying full-time; the remainder are professionals from industry and government who are pursuing part-time programs of study.

Location

Villanova University's handsomely landscaped 240-acre campus is located on suburban Philadelphia's historic Main Line, 12 miles from the city. This beautiful residential area has excellent restaurants and shopping districts and is near famed Valley Forge National Park. Direct rail and bus service is available from the campus to the center of Philadelphia, with its many cultural, historic, and sports attractions.

The University

Villanova University was founded in 1842 and is a privately supported, coeducational institution. It comprises the Colleges of Arts and Sciences, Commerce and Finance, Engineering, and Nursing; the Graduate School of Arts and Sciences; and the School of Law. The University and its programs are fully accredited by national and local accrediting agencies, including the Accreditation Board for Engineering and Technology (ABET). The high quality of Villanova's engineering programs is recognized by industry, government, and other institutions of higher education.

Applying

Application forms for admission and financial support may be obtained from the Department of Mechanical Engineering. Scores on the General Test of the Graduate Record Examinations are required for international applicants and graduates of institutions not accredited by ABET. International applicants must give evidence of adequate proficiency in the English language by achieving a score of at least 575 on the Test of English as a Foreign Language (TOEFL).

Correspondence and Information

Chairman, Graduate Committee
Department of Mechanical Engineering
Villanova University
Villanova, Pennsylvania 19085
Telephone: 610-519-4981

Villanova University

THE FACULTY AND THEIR RESEARCH

Listed below are members of the graduate faculty of the Department of Mechanical Engineering who are engaged in research or are available to guide the research of graduate students. All faculty members teach, and all hold full-time appointments.

Alan M. Whitman, Professor and Department Chairman; Ph.D., Pennsylvania, 1965. Plasma dynamics, asymptotic methods, dynamic analysis, wave propagation.

Hashem Ashrafiuon, Associate Professor; Ph.D., SUNY at Buffalo, 1988. Optimal design, multibody dynamics, robotics.

Young W. Chun, Associate Professor; Ph.D., Iowa, 1978. Solid mechanics, optimal design, numerical methods.

Kei-Peng Jen, Associate Professor; Ph.D., Tennessee Tech, 1983. Materials, fracture mechanics.

Gerard F. Jones, Associate Professor; Ph.D., Pennsylvania, 1981. Heat transfer, fluid mechanics, thermodynamics, solar energy utilization.

Kenneth A. Kroos, Associate Professor; Ph.D., Toledo, 1982; PE. Thermofluids, computer graphics.

John N. Majerus, Professor; Ph.D., Illinois, 1975. Computer-aided design, plastic materials.

Edward V. McAssey Jr., Professor; Ph.D., Pennsylvania 1968. Heat transfer, fluids, energy conversion.

Philip V. D. McLaughlin Jr., Professor; Ph.D., Pennsylvania 1969. Solid mechanics, composite materials, design.

Chandrasekhar Nataraj, Associate Professor; Ph.D., Arizona State, 1987. Controls, vibration, automation.

T. Radhakrishnan, Associate Professor; Ph.D., Wisconsin, 1980. Computer-aided manufacturing, robotics.

Sridhar Santhanam, Assistant Professor; Ph.D., Arizona State, 1989. Manufacturing processes, computer-aided manufacturing, tribology.

Computer-controlled milling machine in the CAM laboratory.

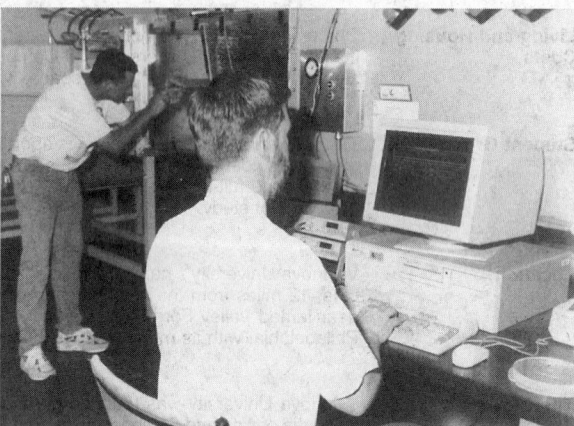

Subsonic wind tunnel with data acquisition system.

VIRGINIA POLYTECHNIC INSTITUTE AND STATE UNIVERSITY

Department of Engineering Science and Mechanics

Programs of Study

The Department of Engineering Science and Mechanics offers degree programs leading to the Master of Science (with thesis and nonthesis options), the professional-oriented Master of Engineering, and Doctor of Philosophy. All graduate students have the opportunity to study in the areas of mechanics of materials, composite materials, adhesion science, fluid mechanics, dynamics, vibrations, biomechanics, computational mechanics, and applied mathematics. Each program is designed to fit the student's interests, and minors are usually taken in mathematics, statistics, physics, and other engineering curricula. The graduate programs are modern and allow for independent study in specialized research areas. The M.S. and Master of Engineering degrees require a minimum of 30 semester credit hours, including course work in continuum mechanics and in each of the four areas of mathematics, solid mechanics, fluid mechanics, and mechanics of motion. The remaining courses necessary to complete the program may be free electives. Six hours of credit must be used for the thesis when that option is chosen. A 3-credit-hour project is required of all Master of Engineering candidates. The Ph.D. degree requires a minimum of 90 credit hours beyond the B.S. degree, including 30 to 46 hours of research and dissertation. The Ph.D. degree program must include a minimum of 15 hours of course work beyond the master's level, 6 hours of mathematics and 3 hours in each of the three areas of motion, solids, and fluids. No foreign language requirement currently exists.

Research Facilities

The department has a significant array of equipment for experimental investigations. Available equipment includes MTS, Instron, and Tinius-Olsen testing systems; a scanning acoustic microscope; two thermographic video cameras; a research photoelastic polariscope; a moiré plate apparatus; acoustic emission instrumentation; a photomechanics laboratory, including laser and optical systems; high-temperature vibration and shock equipment, including an MB 50 shaker; impact and creep-testing machines; thermal analysis equipment (DMA, DSC, TGA, and TMA); and electronic instrumentation, including PC-based data acquisition systems, a high-speed level recorder, oscilloscopes, strain gauge instrumentation, accelerometers, and transducers. In addition, housed in a fluid mechanics laboratory are a 100-foot towing basin with a wave maker and a towing carriage, a wind tunnel, two water tunnels, two shock tubes, biological fluid mechanics instrumentation, and various other assorted fluid mechanics equipment. Vibration, vibration control, and smart structures facilities also include a D-Base controller, a variety of sensor/actuator materials, and analyzers (Generad, Tektronix), including modal analysis software. For numerically intensive applications, the University Computing Center has an IBM SP2, and the College of Engineering has a Silicon Graphics Power Challenger XL. The University's scientific data visualization lab has several SGI workstations with Internet connections. Software in the visualization lab includes PV-WAVE and AVS. Hard-copy devices include postscript laser printers and thermal transfer color printers. For numerically intensive applications, the ESM department has its own Silicon Graphics Origin 200. The large number of department terminals, workstations, and PCs provide access to the University Computing Center's machines. The department's main computer lab features IBM, Silicon Graphics, and HP workstations with a large variety of application software, including PATRAN, ABAQUS, Mathematica, and MATLAB. Several PCs and Macintoshes with a variety of software are also available.

Financial Aid

The current rates for newly enrolled graduate students appointed on teaching and research assistantships and fellowships range between $1335 and $1750 per month. Students on assistantships are required to enroll for 12 credit hours per semester.

Cost of Study

Per semester, the in-state tuition in 1999–2000 is $2061, the student center fee is $105.50, the health fee is $84.50, the athletic fee is $116, the bus fee is $20, the technology fee is $18, and the recreational sports fee is $70. The department pays in-state tuition for students appointed on assistantships, and out-of-state tuition is waived.

Living and Housing Costs

Private housing, including both rooms and apartments, is available in Blacksburg and the surrounding area. Apartment rents range from approximately $350 (one bedroom) to $400 (two bedrooms) per month; normally these rates include all utilities except electricity and telephone.

Student Group

The fall 1998 graduate enrollment in the Department of Engineering Science and Mechanics consisted of 31 master's candidates and 55 Ph.D. candidates. During the past five years, 102 master's and 70 Ph.D. degrees were awarded.

Location

The Virginia Tech campus is located in Blacksburg, a town of about 40,000 situated in the Appalachian Mountains, west of the Blue Ridge Mountains, in southwestern Virginia. Forty miles to the east is the Roanoke-Salem metropolitan area, with a population of approximately 200,000. The high student population density in Blacksburg provides an excellent atmosphere for student life. Nearby Claytor Lake and Hungry Mother state parks provide boating, swimming, camping, and hiking opportunities. The Appalachian Trail is just minutes from Blacksburg and the Blue Ridge Parkway is only an hour away.

The University

Virginia Tech was founded in 1872 as a land-grant university and awarded its first graduate degree in 1892. Of the total of 25,608 on-campus students, 4,193 (including veterinary medicine students) are graduate students. About one fourth of the total enrollment is in engineering. Of the 2,095 students enrolled in the off-campus program, 2,093 are at the graduate level. For many years, Virginia Tech has ranked among the ten largest institutions in the United States in the number of engineering bachelor's degrees awarded each year. The College has technical interest groups in the areas of biomechanics, computer engineering, environmental science and engineering, materials, and systems engineering. The University campus provides facilities for outdoor and indoor sports, including an eighteen-hole golf course and an indoor swimming pool, and it offers a very active faculty–graduate student intramural program. Extensive cultural programs, including musical and dramatic events, lectures, and discussions, are held on campus.

Applying

Each applicant must submit a completed application form, three letters of recommendation from former professors or employers, and official transcript(s) of undergraduate and graduate records to date. GRE test scores are required for all applicants. Virginia Tech requires all international students to submit TOEFL scores; a minimum score of 600 is required. Application should be made as early as possible before the opening of the term for which admission is sought.

Correspondence and Information

Graduate Programs
Department of Engineering Science and Mechanics
Virginia Polytechnic Institute and State University,
   Mail Code 0219
Blacksburg, Virginia 24061

Telephone: 540-231-6743
Fax: 540-231-4574
E-mail: ltickle@vt.edu
World Wide Web: http://www.esm.vt.edu

Virginia Polytechnic Institute and State University

THE FACULTY AND THEIR RESEARCH

Listed here are the faculty members of the Department of Engineering Science and Mechanics who are engaged in research or who are available to guide the research of graduate students.

R. C. Batra, Clifton C. Garvin Professor; Ph.D., Johns Hopkins, 1972. Adiabatic shear bonding, penetration mechanics, finite element solution of nonlinear thermomechanical problems.

S. Case, Assistant Professor; Ph.D., Virginia Tech, 1996. Durability and life prediction of composite materials.

M. S. Cramer, Professor; Ph.D., Cornell, 1976. Fluid mechanics, analytical methods, nonlinear wave propagation.

D. A. Dillard, Professor; Ph.D., Virginia Tech, 1981. Adhesion science, experimental mechanics, viscoelasticity.

N. E. Dowling, Professor; Ph.D., Illinois at Urbana-Champaign, 1972. Fatigue of metals, fracture mechanics, engineering materials.

J. C. Duke Jr., Professor; Ph.D., Johns Hopkins, 1978. Nondestructive testing, materials science, measurement science, infrastructure assessment, health monitoring of structures.

J. R. Foy, Assistant Professor; Ph.D., Clemson, 1997. Tribology (natural and artificial joints and other locations in the body), biomaterials, surface engineering, cell-surface interactions.

J. W. Grant, Professor; D.Eng., Tulane, 1973. Biomechanics, vestibular system, aerospace medicine.

Z. Gurdal, Professor; Ph.D., Virginia Tech, 1985. Composite materials and structures, structural optimization.

M. R. Hajj, Associate Professor; Ph.D., Texas at Austin, 1990. Nonlinear dynamics in transitioning and turbulent fluid flows, hydrodynamic loadings and nonlinear response of offshore structures, higher-order digital time series analysis.

R. A. Heller, Francis J. Maher Professor (Emeritus); Ph.D., Columbia, 1958. Inelastic behavior of structures and materials, reliability and safety of structures, fatigue and failure, probabilistic mechanics, educational films.

S. L. Hendricks, Associate Professor; Ph.D., Virginia, 1979. Dynamics, hydrodynamics, stability, chaos.

E. G. Henneke II, Professor and Head; Ph.D., Johns Hopkins, 1968. Materials science, nondestructive evaluation of composite materials, wave propagation in anisotropic media.

M. W. Hyer, Professor; Ph.D., Michigan, 1974. Mechanics of composite materials and structures, solid mechanics.

R. M. Jones, Professor; Ph.D., Illinois, 1964. Solid mechanics, mechanics of composite materials and structures, plasticity, buckling.

L. G. Kraige, Professor; Ph.D., Virginia, 1975. Dynamics, vibrations.

R. D. Kriz, Associate Professor; Ph.D., Virginia Tech, 1979. Fracture mechanics and nondestructive evaluation of composite materials.

J. J. Lesko, Assistant Professor; Ph.D., Virginia Tech, 1994. Mechanics of composite materials and interphases, compression strength and durability, experimental mechanics.

L. Librescu, Professor; Ph.D., Polytechnic Institute (Romania), 1969. Theory of plates and shells, continuum mechanics, composite structures, buckling and postbuckling, viscoelasticity of structures, unsteady aerodynamics, aeroelasticity.

A. C. Loos, Professor; Ph.D., Michigan, 1982. Composite materials (environmental effects and manufacturing).

L. Meirovitch, University Distinguished Professor Emeritus; Ph.D., UCLA, 1960. Analytical dynamics, nonlinear analysis, vibration analysis, astrodynamics, space vehicle dynamics.

D. T. Mook, Waldo N. Harrison Professorship; Ph.D., Michigan, 1966. Hydrodynamic stability, non-Newtonian fluid flow.

D. H. Morris, Professor and Assistant Head; Ph.D., Iowa State, 1971. Experimental solid mechanics, composite materials.

A. H. Nayfeh, University Distinguished Professor; Ph.D., Stanford, 1964. Aerodynamics, perturbation methods, nonlinear dynamics and chaos, structural acoustics, dynamics of composites, seakeeping, transition.

S. A. Ragab, Professor; Ph.D., Virginia Tech, 1979. Computational fluid mechanics.

K. L. Reifsnider, Alexander F. Giacco Chair of Engineering; Ph.D., Johns Hopkins, 1968. Mechanics of material systems, durability, fatigue and fracture, performance simulation.

D. J. Schneck, Professor; Ph.D., Case Western Reserve, 1973. Biomechanics, blood flow, fluid mechanics, biological fluid mechanics.

M. P. Singh, Professor; Ph.D., Illinois at Urbana-Champaign, 1972. Probabilistic mechanics and reliability, structural dynamics and earthquake engineering.

C. W. Smith, Alumni Distinguished Professor Emeritus (part-time). Experimental fracture mechanics.

D. P. Telionis, Professor; Ph.D., Cornell, 1970. Unsteady aerodynamics, high-angle-of-attack aerodynamics, flow over fore-body configuration, delta wing flows, vortical flows and their interaction with solid surfaces, vortex shedding, free-surface effects.

S. Thangjitham, Associate Professor; Ph.D., Virginia Tech, 1984. Probabilistic methods and stochastic processes, vibrations and controls, composite structures, thermoelasticity, fracture mechanics, fatigue damage and life prediction.

L. A. Wojcik, Assistant Professor; Ph.D., Michigan, 1997. Biomechanics, musculoskeletal dynamics, neuromuscular control, dynamic systems modeling.

AREAS OF CURRENT RESEARCH AND INTEREST

Adhesion Science. Stress analysis of adhesive joints, adhesion test method development, fracture and fatigue of adhesive joints, interdisciplinary studies of adhesive durability, bioadhesion.

Applied Mathematics. Perturbation methods, nonlinear systems, systems stability, tensor analysis, approximate and numerical analyses, finite-element analyses, variational and energy methods, wave propagation.

Biomechanics. Mechanics of blood flow, fracture and strength of bones, muscles and tissues, vestibular system mechanics and motion sensing, physiologic modeling, hearing mechanics, aerospace medicine.

Composite Materials. Fracture mechanics, fatigue and failure, mathematical models for vibrations and wave propagation, delaminations, interlaminar shear, finite-element methods and numerical techniques, experimental mechanics, nondestructive testing, manufacturing, performance modeling, optimal design, design, nonlinear material modeling.

Continuum Mechanics. Rheology, thermodynamic stability, constitutive relations.

Dynamics and Vibrations. Nonlinear astrodynamics, space vehicle dynamics, random vibrations, acoustics, statistical analysis, wave propagation in solids, approximate methods, optimization methods, nonlinear dynamics and chaos, reentry dynamics, controls, earthquake engineering, structural acoustics, seakeeping, structural control, experimental methods.

Fluid Mechanics. Viscous flow; boundary-layer theory; gasdynamics, nonlinear dynamics of water waves, and wave propagation; wind engineering; hydrodynamic stability; non-Newtonian flow; perturbation methods; biological fluid mechanics; structure of boundary-layer flow; instrumentation; structure of the atmospheric boundary layer; laminar flow control; nonlinear acoustics; unsteady aerodynamics; high-angle-of-attack aerodynamics; flow over fore-body configuration; delta wing flows; vertical flows and their interaction with solid surfaces; flow-induced vibration, vortex shedding, unsteady separation, turbulent boundary-layer controls.

Mechanics of Materials. Ultrasonic testing of material properties, wave propagation in anisotropic media, theories and methods of nondestructive testing, time-resolved X-ray diffraction microscopy, continuum theory of material defects, embrittlement of materials, fatigue and failure, micromechanics, adhesion science, elasticity, viscoelasticity, fracture mechanics, experimental mechanics, continuum mechanics, plates and shells, plasticity, reliability and safety of structures, fatigue and failure, probabilistic mechanics, structural mechanics, composite materials and structures, stability, variational and finite-element methods, thermoelasticity, structural optimization, earthquake resistant structures, performance simulation, adiabatic shear bonding, penetration mechanics.

WASHINGTON UNIVERSITY IN ST. LOUIS

Department of Mechanical Engineering

Programs of Study

The department offers courses of study leading to the Master of Science and Doctor of Science degrees. Candidates for the M.S. may elect a course option (30 semester credits of graduate courses) or a thesis option (24 hours of course work plus 6 hours of research). For the D.Sc., a total of 42 credits beyond the master's is required, consisting of a minimum of 24 hours of course work and 18 hours of research. Opportunities for advanced study and research are currently available in such areas as aerosol science, air pollution, biomechanics and biomedical engineering, combustion science, composite materials, computational mechanics, finite elements, fluid mechanics, fracture mechanics, helicopter dynamics, machine vibrations, mechanical design, metallurgy, and turbomachinery. The department also participates in interdepartmental programs in biomedical engineering, environmental engineering, and materials science. Students interested in these fields and registered in the department may take courses offered by these programs and pursue a research project under the supervision of the associated faculty.

Research Facilities

Research facilities within the department and headed by department faculty members include the Center for Air Pollution Impact and Trend Analysis and the Center for Computational Mechanics. The Materials Science and Engineering Program at Washington University is an interdepartmental activity designed to encourage study and research in the areas of homogenous and multiphase materials for structural, electronic, magnetic, and biomedical applications. The degrees of M.S. and Ph.D. are awarded within the program. Graduate research in the Materials Research Laboratory currently covers the areas of structure and properties of metals; polymers, ceramics, and composites; advanced metal and polymer processing methods; polymer processing rheology; fracture mechanics of solids; and the fabrication, chemistry, physics, and mechanical behavior of composite materials. Laboratory facilities include equipment for electron and optical microscopy, mechanical property characterization, fabrication of nanophase materials, metals, composites, polymers, and measurements of a broad range of physical properties.

The 1.8-million-volume John M. Olin Library houses the engineering collection. There are also thirteen separate school and department libraries. Faculty, students, and staff have access to a wide range of computing resources. The Mechanical Engineering Computing Facility (MECF) provides computer, e-mail, and network services to the faculty, staff, and students of the department. The Center for Engineering Computing provides an educational computing environment.

Financial Aid

Aid is available for qualified students in the form of teaching and research assistantships. These are financed by University funds and sponsored research projects. Typically, the level of support for the nine-month academic year is $1250 per month, plus tuition for 9 units per semester. Normally, the duties assigned to research assistants coincide with their thesis research. There are also some tuition-only scholarships. The available support is offered during the spring on the basis of merit. Applications are accepted at all times, but it is desirable to apply by February 1 for support starting in September. Applicants are also strongly encouraged to seek fellowships sponsored by such agencies as the NSF, and international students are urged to apply for financial assistance that may be available through their home countries.

Cost of Study

Tuition for full-time graduate study is $975 per semester credit hour in 1999–2000. Part-time students pay $830 per unit (master's candidates only).

Living and Housing Costs

The University has no separate housing for married graduate students. Most graduate students, single or married, live off campus. Rooms and apartments are available in the area surrounding the University. The Campus Housing Referral Service (Campus Box 1059) provides assistance in locating suitable accommodations and maintains a list of off-campus housing. One- or two-bedroom apartments can be rented near the University starting at $300 per month. Rooms in private homes are available, sometimes in exchange for work, and usually include kitchen privileges.

Student Group

The department's graduate enrollment is about 60; more than half of these are part-time students who are employed full-time in the St. Louis area. Twenty-one of the current group are D.Sc. candidates.

Location

St. Louis offers social and cultural attractions and entertainment possibilities such as the St. Louis Symphony, the Loretto-Hilton Repertory Theatre, the Missouri Botanical Garden, and professional hockey, football, and baseball. Immediately adjacent to the campus, in Forest Park, are the St. Louis Art Museum, the St. Louis Science Center, the St. Louis Zoo, and the Municipal Opera. Near the western side of the campus is the business center of the city of Clayton (St. Louis County seat), and within walking distance are movie theaters, stores, banks, snack shops, restaurants, and the shopping area of University City.

The University

Washington University in St. Louis is an independent, privately endowed and supported institution located on 169 acres within the greater St. Louis area. It was founded in 1853. The University has 9,840 full-time students, more than half of whom are undergraduates. There are 2,193 part-time students. The University has 2,119 full-time and 170 part-time faculty members. It is composed of several major divisions: Arts and Sciences; the Graduate Schools of Arts and Sciences, Architecture, Business and Public Administration, Engineering and Applied Science, Fine Arts, Law, Medicine, and Social Work; the School of Continuing Education; and the Summer School. The University ranks among the top ten in the nation in the number of Nobel laureates associated with it.

Applying

There are no specific deadlines for application for admission, although it is wise to apply before March 1 for the following fall semester. Applicants must have received a bachelor's degree in engineering or science or show by reason of their record that they will receive it before matriculation. GRE scores are required for all full-time applicants.

Correspondence and Information

Director of Graduate Programs
Campus Box 1185
Washington University in St. Louis
One Brookings Drive
St. Louis, Missouri 63130

Washington University in St. Louis

THE FACULTY AND THEIR RESEARCH

The faculty members of the department participate in both undergraduate and graduate education and in the interdepartmental graduate programs, interschool programs, and industrial projects. Therefore, students receive a program of education designed to their interests and guided by the experiences of an active resident faculty involved in state-of-the-art research and development. Further, many of the faculty members have significant industrial experience, and many are registered professional engineers.

R. L. Axelbaum, Assistant Professor; Ph.D., California, Davis, 1988. Combustion theory, thermal sciences, fluid mechanics.

P. V. Bayly, Assistant Professor; Ph.D., Duke, 1993. Dynamics, vibrations.

H. J. Brandon, Affiliate Professor; D.Sc., Washington (St. Louis), 1969. Energetics, thermal systems.

W. B. Diboll, Professor Emeritus; M.S.M.E., RPI, 1951. Dynamics, vibrations, engineering design.

A. D. Dimarogonas, Palm Professor of Mechanical Design; Ph.D., RPI, 1970. Mechanical design, rotating machinery, manufacturing systems, biomechanical systems.

R. A. Gardner, Associate Professor and Assistant Chairman of the Department; Ph.D., Purdue, 1969. Magnetohydrodynamics, biofluid dynamics.

J. C. Georgian, Professor Emeritus; M.S., Cornell, 1941. Engineering design, dynamics of machinery, vibrations, turbomachinery.

M. P. Gomez, Affiliate Professor; Ph.D., Stanford, 1964. Materials science, fracture mechanics.

L. B. Gulbransen, Professor Emeritus; Ph.D., Utah, 1949. Physical metallurgy, materials science, solid state.

R. J. Hakkinen, Professor (part-time); Ph.D., Caltech, 1954. Experimental fluid mechanics.

K. H. Hohenemser, Professor Emeritus; Dr.Ing., Darmstadt, 1930. Continuum mechanics, vibrations, dynamic response, aerodynamics, wind-turbine design.

R. B. Husar, Professor and Director, Center for Air Pollution Impact and Trend Analysis; Ph.D., Minnesota, 1971. Air pollution, aerosol dynamics.

M. J. Jakiela, Lee Hunter Associate Professor of Mechanical Design; Ph.D., Michigan, 1988. Mechanical design, design for manufacturing robotics.

K. L. Jerina, Professor and Director, Materials Research Laboratory; D.Sc., Washington (St. Louis), 1974. Solid mechanics, composite materials, engineering design.

T. Korakianitis, Associate Professor; Sc.D., MIT, 1987. Analysis and design, piston engines, turbomachines, thermodynamics.

L. D. Kral, Associate Professor; Ph.D., Arizona, 1988. Fluid mechanics, flow control.

R. J. Okamoto, Assistant Professor; D.Sc., Washington (St. Louis), 1997. Biomechanics, design.

P. C. Paris, Professor; Ph.D., Lehigh, 1962. Solid mechanics, dynamics, fracture mechanics, stochastic processes.

M. T. Pauken, Assistant Professor; Ph.D., Georgia Tech, 1994. Thermal sciences, electrohydrodynamic heat transfer, environmental engineering.

D. A. Peters, Professor and Chairman and Director, Center for Computational Mechanics; Ph.D., Stanford, 1974. Aeroelasticity, vibrations, helicopter dynamics.

S. M. L. Sastry, Professor of Metallurgy and Materials Science; Ph.D., Toronto, 1974. Materials science, physical metallurgy.

B. A. Szabo, Greensfelder Professor of Mechanics; Ph.D., Buffalo, 1960. Numerical stress analysis, adaptive finite-element methods, fracture mechanics.

H. Tada, Affiliate Professor; Ph.D., Lehigh, 1972. Fracture mechanics, solid mechanics.

G. I. Zahalak, Professor; Eng.Sc.D., Columbia, 1972. Biomechanics, continuum mechanics.

Areas of Recent Research Activity

Biomechanics: Acoustical methods for bone diagnosis, microrheology of red blood cells, cellular micromechanics, muscle mechanics, cardiac mechanics, gait dynamics.

Combustion Science: Combustion synthesis of nanometer-sized reinforced composites, laser diagnostics in combustion systems, flame stability, soot formation in flames.

Computational Mechanics: Development of p and h-p finite element methods, error estimation and control, hierarchic models for plates and shells, adaptive methods, numerical techniques for compressible flows with time-dependent boundaries.

Computational Fluid Dynamics: Flow control, synthetic jets.

Design: Shape optimization, concurrent engineering.

Dynamic Systems: Rotor dynamics, smart structures, aeroelasticity, mechanics of helicopters, nonlinear and chaotic vibrations, space structures.

Environmental Science: Aerosol optics, analysis of atmospheric sulphur, synoptic scale air pollution.

Materials Science: Metal-matrix composites, intermetallic compounds, nanocrystalline materials, polymer composites, thermomechanical fatigue.

Mechanical Design: Design-decision algorithms, solid modeling and manufacturing interfaces, design for computer-aided manufacturing, formal models of the design process, electrorheological bearings.

Power and Propulsion Systems: Analysis and design of turbomachines, piston-assembly dynamics, valve-train dynamics, lubrication and wear, regenerative gas-turbine engines.

WORCESTER POLYTECHNIC INSTITUTE

Department of Mechanical Engineering

Programs of Study

Graduate programs in mechanical engineering lead to the degrees of Master of Engineering (M.Eng.), Master of Science (M.S.), and Doctor of Philosophy (Ph.D.). The M.Eng. degree requires 30 credits of course work. The M.S. degree requires 24 credits of course work and 6 credits of thesis research. Both master's degrees can be pursued full- or part-time. The Doctor of Philosophy degree requires 60 credits of graduate work beyond the master's degree; 30 of the 60 credits must be in dissertation research credits. A specific program of study is developed by each candidate in association with his or her adviser. Admission to candidacy for the Ph.D. is granted when the student has passed a comprehensive examination dealing with subject areas that have been made part of the program.

The normal full-time registration for students without teaching or research assistantship duties is 9–12 credits per semester. Students who hold assistantships can register for up to 10 credits per semester. Full-time students can complete the M.Eng. degree in three semesters, the M.S. degree two years, and the Ph.D. in three to four years.

Research areas include biomechanics/biofluids, design, solid mechanics and dynamics, computational methods, fluid and plasma dynamics, heat transfer, combustion, laser metrology, and materials science. In addition, opportunities are available for interdisciplinary research involving faculty from Materials Engineering, Manufacturing Engineering, and other departments.

Research Facilities

The department is housed in Higgins Laboratories, which recently underwent an $8-million expansion and renovation. Facilities include the Aerospace Laboratory, Biomechanics/Biofluids Laboratory, Laser and Holography Laboratory, Design Studio, Fire and Combustion Laboratory, Heat Transfer Laboratory, Hydrodynamics Laboratory, Surface Metrology Laboratory, Vibration/Dynamics and Control Laboratory, Fluid Dynamics Laboratory, and Surface Metrology Laboratory. Additional facilities are housed in Washburn Shops and include the Aluminum Casting Research Laboratory, the Powder Metallurgy Research Center, the Robotics/CIM Laboratory, several manufacturing engineering and materials science laboratories, and a 10-kw Nuclear Reactor Facility.

In addition to the computing facilities that serve the entire campus, the Department of Mechanical Engineering has a dedicated Graduate Computer Laboratory and numerous high-end workstations for graduate research.

Financial Aid

M.S. and Ph.D. students may obtain financial aid through teaching and research assistantships. Teaching assistants are expected to devote a maximum of 20 hours per week to assigned duties. Full assistantships provide tuition plus a stipend of approximately $9828 for the nine-month academic year. Summer support is available through individual faculty research grants as well as through Summer Support Research Grants provided by the department. The Institute offers Robert H. Goddard Research Fellowships to outstanding M.S. and Ph.D. applicants. Each fellowship includes a stipend of $12,000 for twelve months and a full tuition waiver. In addition, the department offers several industry-sponsored fellowships.

Cost of Study

Tuition for full-time students is $661 per credit hour in 1999–2000.

Living and Housing Costs

Most graduate students obtain housing in private apartments near the campus. Typically, the annual cost of rent plus utilities for a three-bedroom apartment shared by 3 students is approximately $3200 per year for each student. Worcester is a small city and costs are reasonable.

Student Group

About 580 undergraduate students and 100 full- and part-time graduate students are enrolled in the mechanical engineering department.

Location

WPI is located on an attractive, 72-acre, hilly campus, 1 mile from downtown Worcester. Flanked by city parks on two sides, the campus has a feeling of openness even though it is located in a city of 160,000. Worcester is well known for its many colleges and its Art Museum, Museum of Armor, Science Center, and Worcester Centrum. Music is well represented by several excellent choruses and a symphony orchestra. Concerts are given by distinguished visiting performers in the restored acoustical gem, Mechanics Hall. Outstanding dramatic productions are presented by several local theater companies. Boston and Cape Cod to the east and the Berkshires to the west can be easily reached from Worcester, and there is good skiing nearby.

The University

Worcester Polytechnic Institute was founded in 1865 as the Worcester County Free Institute of Industrial Science and is the third-oldest college of engineering, science, and management in the United States. A unique educational program was developed—a combination of scientific and technical studies with practical work in model industrial shops. Engineering students at WPI have the opportunity to become familiar with modern manufacturing processes as part of their program of study.

Graduate degrees have been awarded at WPI since 1898. New programs have been added regularly in response to the growing capabilities of the college and the changing needs of the professions. At the present time, the master's degree and the doctorate are both offered in thirteen disciplines. The current student body of 3,650 includes 1,025 full- and part-time graduate students.

Applying

Applicants to the M.Eng. or M.S. programs should have a B.S. in mechanical engineering or a related field. Applicants to the Ph.D. program must hold a master's degree in mechanical engineering or in a related area. For admission to the M.Eng., M.S., and Ph.D. programs, the GRE is strongly recommended for all students, and the TOEFL is required for international students. Students wishing to be considered for the Goddard Research Fellowship must submit GRE scores. Applications are accepted at any time. In making awards of financial aid, preference is given to students whose applications are received by March 1.

Correspondence and Information

Graduate Committee
Department of Mechanical Engineering
Worcester Polytechnic Institute
100 Institute Road
Worcester, Massachusetts 01609-2280
Telephone: 508-831-5026
Fax: 508-831-5680
World Wide Web: http://me.wpi.edu/

Worcester Polytechnic Institute

THE FACULTY AND THEIR RESEARCH

Andreas N. Alexandrou, Professor; Ph.D., Michigan, 1986. Computational fluid mechanics, materials processing.
Diran Apelian, Howmet Professor; Sc.D., MIT, 1972. Materials engineering, solidification, aluminum foundry, powder metallurgy, metal processing.
Holly K. Ault, Associate Professor; Ph.D., Worcester Polytechnic, 1988. Computer-aided design, engineering design, graphics.
Isa Bar-On, Associate Professor; Ph.D., Hebrew (Jerusalem), 1984. Materials engineering, fracture mechanics.
Ronald R. Biederman, Fuller Professor of Mechanical Engineering; Ph.D., Connecticut, 1968. Materials engineering.
E. Thomas Boulette, Professor and Director, Nuclear Engineering; Ph.D., Iowa State, 1968. Nuclear engineering and energy.
John M. Boyd, Professor Emeritus; Ph.D., Ohio State, 1962. Heat, power.
Christopher A. Brown, Associate Professor; Ph.D., Vermont, 1983. Fractal analysis, materials processing and manufacturing, biomechanics, surface metrology.
Eben Cobb, Visiting Assistant Professor; Ph.D., Connecticut, 1985. Applied mechanics, vibrations, engineering design.
Michael Demetriou, Assistant Professor; Ph.D., USC, 1993. Systems and control, structural/acoustic control, fault detection/diagnosis.
Chrysanthe Demetry, Associate Professor; Ph.D., MIT, 1993. Materials science, ceramics.
Mikhail F. Dimentberg, Professor; Ph.D., Moscow Power Institute, 1963. Applied mechanics, random vibrations, nonlinear dynamics, mechanical signature analysis.
William W. Durgin, Professor; Ph.D., Brown, 1970. Fluid mechanics, aerodynamics.
Vladimir Entov, Visiting Professor; Ph.D., Moscow Lomonosov, 1965. Fluid dynamics, engineering continuum mechanics, mathematical modeling, fracture mechanics, applied mechanics and mathematics.
Mustapha Fofana, Assistant Professor; Ph.D., Waterloo, 1993. Engineering mechanics, design.
Nikos A. Gatsonis, Assistant Professor; Ph.D., MIT, 1991. Computational fluid and plasma dynamics.
Hartley T. Grandin Jr., Professor Emeritus; Ph.D., Michigan State, 1972. Applied mechanics, design.
Raymond R. Hagglund, Professor; Ph.D., Illinois, 1962. Applied mechanics, dynamics.
James C. Hermanson, Associate Professor; Ph.D., Caltech, 1985. Experimental fluid mechanics, combustion, heat transfer.
Allen H. Hoffman, Professor; Ph.D., Colorado, 1970. Biomechanics, bioengineering, rehabilitation engineering.
Zhikun Hou, Associate Professor; Ph.D., Caltech, 1990. Applied mechanics, random vibrations and structural controls.
Hamid Johari, Associate Professor; Ph.D., Washington (Seattle), 1989. Experimental fluid mechanics, turbulence, aerodynamics.
R. Nathan Katz, Associate Research Professor; Ph.D., MIT, 1969. Materials science, ceramics, physical metallurgy.
Walter A. Kistler, Professor Emeritus; M.S., Worcester Polytechnic, 1968. Applied mechanics, engineering analysis.
Makhlouf M. Makhlouf, Associate Professor; Ph.D., Worcester Polytechnic, 1990. Materials processing, manufacturing.
John A. Mayer Jr., Professor Emeritus; M.S., Columbia, 1956. Nuclear engineering, energy.
Shaukat Mirza, Professor and Director, Manufacturing Engineering; Ph.D., Wisconsin, 1962. Computational solid mechanics, dynamics of composites.
Mohammad N. Noori, John Woodman Higgins Professor and Department Head; Ph.D., Virginia, 1984. Structural mechanics, random vibrations.
Robert L. Norton, Professor; M.S., Tufts, 1970. Engineering design.
David J. Olinger, Associate Professor; Ph.D., Yale, 1990. Fluid mechanics, nonlinear dynamics.
Ryszard J. Pryputniewicz, Professor; Ph.D., Connecticut, 1976. Laser holography.
Joseph J. Rencis, Associate Professor; Ph.D., Case Western Reserve, 1985. Applied mechanics, boundary- and finite-element methods.
Mark W. Richman, Associate Professor; Ph.D., Cornell, 1984. Applied mechanics, particulate flow.
Yiming (Kevin) Rong, Associate Professor; Ph.D., Kentucky, 1989. Computer-aided design, manufacturing dynamics and control; precision engineering.
Brian J. Savilonis, Professor; Ph.D., SUNY at Buffalo, 1976. Biofluid mechanics, fire modeling.
Kenneth E. Scott, Professor Emeritus; M.S., Worcester Polytechnic, 1954. Control engineering, heat transfer, CAD.
Satya S. Shivkumar, Associate Professor; Ph.D., Stevens, 1987. Materials science, biomaterials, plastics.
Richard D. Sisson Jr., Professor; Ph.D., Purdue, 1975. Materials processing and manufacturing.
John M. Sullivan, Associate Professor; Ph.D., Dartmouth, 1986. Numerical methods, heat and momentum transfer, CAD.
Donald N. Zwiep, Professor and Head Emeritus; M.S., Iowa State, 1951. Design.

Recent Theses and Dissertations

A decision model for risk management in new product development projects.
Aces study of the viscoelastic behavior of paper.
Advances in the design of pavement surfaces.
Analysis of mechanical hysteretic systems.
An automated system for design and analysis of total hip implants: A method of modeling the proximan endosteal canal using 3-D CT data.
An experimental investigation of torque balancing to reduce the torsional vibration of camshafts.
An extension of direct stochastic linearization technique for M.D.O.F. systems with general hysteresis.
Automatic three-dimensional mesh generation for anatomically accurate human body and organs.
Characterization of kinetic and mechanistic differences between free-surface and bulk grain growth in WC-Co materials.
Control of three-dimensional cylinder wake structures using coupled map lattices.
Design of a shape memory alloy activated gripper device.
Determination of the effective interfacial heat transfer coefficient between permanent molds and aluminum castings.
Development of a statistically optimized test method for the reduced pressure test.
Development of new composite materials for diamond grinding wheels.
Direct measurement of wing tip vortex circulation using ultrasound.
Discovery of design methodologies for the integration of multidisciplinary design problems.
Dynamic characterization of microcellular urethane jounce bumpers under impact conditions.
Effects of delay time on active controls.
Effects of superheat on liquid droplets in a supersonic freestream.
Electrical characterization of nanocrystalline TiO_2.
Exact solutions to boundary integrals for the two-dimensional Laplacian problem.
Feasibility study of using passive SMA absorbers to minimize secondary system structural response.
Grain size control and phase transformations in ZrO_2-$A1_2O_3$.
Investigation of permanent mold cooling methods to optimize part quality and cycle time~n phase II.
Kinetic theory of monosized particles and binary mixtures thermalized by vibrating seives.
Langmuir probe measurements in the plume of a pulsed plasma thruster.
Measurement of passive scalar concentration field in accelerating turbulent jets.
Mechatronics in tape casting: Implementation of precision machine design and supervisory control.
Penetration and mixing of fully modulated turbulent jets in crossflow.
Predict springback of cylindrical parts formed on fourslide equipment.
Process automation and insitu cutting of glass fibers.
Rheology of semisolid metal suspensions.
Strategic capacity planning for rapid prototyping service bureaus.
Subharmonic response of a vibroimpact system to narrow band random excitation.
The application of the genetic algorithm to the kinematic design of turbine blade fixtures.

Section 18
Ocean Engineering

This section contains a directory of institutions offering graduate work in ocean engineering, followed by in-depth entries submitted by institutions that chose to prepare detailed program descriptions. Additional information about programs listed in the directory but not augmented by an in-depth entry may be obtained by writing directly to the dean of a graduate school or chair of a department at the address given in the directory.

For programs offering related work, see also in this book Civil and Environmental Engineering and Engineering and Applied Sciences. In Book 3, see Marine Biology; and in Book 4, see Environmental Sciences and Management and Marine Sciences and Oceanography.

CONTENTS

Program Directory

In-Depth Descriptions

Ocean Engineering

Florida Atlantic University, College of Engineering, Department of Ocean Engineering, Program in Ocean Engineering, Boca Raton, FL 33431-0991. Offers MS, PhD. Part-time and evening/weekend programs available. *Faculty:* 14 full-time (0 women), 3 part-time (0 women). *Students:* 34 full-time (9 women), 10 part-time (1 woman); includes 2 minority (both Hispanic Americans), 29 international. Average age 26. In 1998, 16 master's, 4 doctorates awarded. Terminal master's awarded for partial completion of doctoral program. *Degree requirements:* For master's, thesis required (for some programs), foreign language not required; for doctorate, dissertation, qualifying exam required, foreign language not required. *Entrance requirements:* For master's and doctorate, GRE General Test (minimum combined score of 1000 required), TOEFL (minimum score of 550 required), minimum GPA of 3.0. *Application deadline:* For fall admission, 3/1 (priority date); for spring admission, 7/1. Applications are processed on a rolling basis. Application fee: $20. Tuition, state resident: part-time $148 per credit hour. Tuition, nonresident: part-time $509 per credit hour. *Financial aid:* In 1998–99, 44 research assistantships were awarded.; career-related internships or fieldwork, Federal Work-Study, and unspecified assistantships also available. Financial aid applicants required to submit FAFSA. *Faculty research:* Marine materials and corrosion, marine vehicles, autonomous underwater vehicles, acoustics and vibrations, hydrodynamics. *Application contact:* Patricia Capozziello, Graduate Admissions Coordinator, 561-297-2694, Fax: 561-297-2659, E-mail: capozzie@fau.edu.

Florida Institute of Technology, Graduate School, College of Engineering, Division of Marine and Environmental Systems, Program in Ocean Engineering, Melbourne, FL 32901-6975. Offers MS, PhD. Part-time programs available. *Faculty:* 4 full-time (0 women). *Students:* 7 full-time (3 women), 8 part-time; includes 1 minority (African American), 4 international. Average age 28. 24 applicants, 88% accepted. In 1998, 2 master's, 3 doctorates awarded. *Degree requirements:* For master's, thesis optional, foreign language not required; for doctorate, dissertation, comprehensive and departmental qualifying exams required, foreign language not required. *Entrance requirements:* For master's, minimum GPA of 3.0; for doctorate, minimum GPA of 3.2. *Application deadline:* Applications are processed on a rolling basis. Application fee: $50. Electronic applications accepted. Tuition: Part-time $575 per credit hour. Required fees: $100. Tuition and fees vary according to campus/location and program. *Financial aid:* In 1998–99, 9 students received aid, including 6 research assistantships (averaging $4,421 per year), 3 teaching assistantships (averaging $3,420 per year); tuition remissions also available. Financial aid application deadline: 3/1; financial aid applicants required to submit FAFSA. *Faculty research:* Underwater technology, materials and structures, coastal processes and engineering, marine vehicles and ocean systems, naval architecture. Total annual research expenditures: $485,090. *Unit head:* Dr. Andrew Zborowski, Chair, 407-674-7304, Fax: 407-674-7212, E-mail: zborowsk@fit.edu. *Application contact:* Carolyn P. Farrior, Associate Dean of Graduate Admissions, 407-674-7118, Fax: 407-723-9468, E-mail: cfarrior@fit.edu.

See in-depth description on page 1491.

Massachusetts Institute of Technology, Program in Oceanography/Applied Ocean Science and Engineering, Cambridge, MA 02139-4307. Offers biological oceanography (PhD, Sc D); chemical oceanography (PhD, Sc D); marine geochemistry (PhD, Sc D); marine geology (PhD, Sc D); oceanographic engineering (MS, PhD, Sc D, Eng); physical oceanography (PhD, Sc D). MS, PhD, and Sc D offered jointly with Woods Hole Oceanographic Institution. *Faculty:* 170 full-time. *Students:* 133 full-time, 43 international. Terminal master's awarded for partial completion of doctoral program. *Degree requirements:* For master's and Eng, thesis required (for some programs), foreign language not required; for doctorate, dissertation required, foreign language not required. *Entrance requirements:* For master's, GRE General Test; for doctorate, GRE General Test, GRE Subject Test. *Application deadline:* For fall admission, 1/15 (priority date). Application fee: $55. *Unit head:* Paola Rizzoli, Director, 617-253-2451. *Application contact:* Ronni Schwartz, Administrator, 617-253-7544, Fax: 617-253-9784.

Massachusetts Institute of Technology, School of Engineering, Department of Ocean Engineering, Cambridge, MA 02139-4307. Offers naval architecture and marine engineering (SM); naval engineering (Naval E); ocean engineering (M Eng, SM, PhD, Sc D, Ocean E); ocean systems management (SM). *Faculty:* 16 full-time (1 woman). *Students:* 103 full-time (10 women), 1 part-time; includes 1 minority (Asian American or Pacific Islander), 44 international. Average age 30. 68 applicants, 34% accepted. In 1998, 35 master's, 14 doctorates awarded. Terminal master's awarded for partial completion of doctoral program. *Degree requirements:* For master's and other advanced degree, computer language, thesis required; for doctorate, dissertation, comprehensive exams required. *Entrance requirements:* For master's, doctorate, and other advanced degree, GRE General Test, TOEFL (minimum score of 577 required). *Application deadline:* For fall admission, 1/15 (priority date); for spring admission, 11/1. Applications are processed on a rolling basis. Application fee: $55. *Financial aid:* In 1998–99, 50 students received aid, including 8 fellowships, 44 research assistantships, 7 teaching assistantships; institutionally-sponsored loans and tuition waivers also available. Financial aid application deadline: 1/15; financial aid applicants required to submit FAFSA. *Faculty research:* Computer-aided engineering, marine hydrodynamics, marine resource and environmental management, ocean instrumentation and measurement for environmental monitoring, ocean-related acoustics. Total annual research expenditures: $6.1 million. *Unit head:* Dr. C. Chryssostomidis, Head, 617-253-4330, Fax: 617-253-8125, E-mail: chrys@deslab.mit.edu. *Application contact:* Beth Tuths, Admissions Coordinator, 617-258-6157, Fax: 617-253-8125, E-mail: oe@mit.edu.

See in-depth description on page 1493.

Memorial University of Newfoundland, School of Graduate Studies, Faculty of Engineering and Applied Science, St. John's, NF A1C 5S7, Canada. Offers civil engineering (M Eng, PhD); electrical engineering (M Eng, PhD); mechanical engineering (M Eng, PhD); ocean engineering (M Eng, PhD). Part-time programs available. *Students:* 75 full-time (11 women), 28 part-time (2 women), 31 international. *Degree requirements:* For master's, thesis optional; for doctorate, dissertation, comprehensive exam required. *Application deadline:* For fall admission, 3/1. Application fee: $40. *Unit head:* Dr. Rangaswany Seshadri, Dean, 709-737-8810, Fax: 709-737-8975, E-mail: sesh@engr.mun.ca. *Application contact:* Dr. J. J. Sharp, Associate Dean, 709-737-8901, Fax: 709-737-3480, E-mail: jsharp@engr.mun.ca.

Oregon State University, Graduate School, College of Engineering, Department of Civil Engineering, Program in Ocean Engineering, Corvallis, OR 97331. Offers M Oc E. Part-time programs available. *Students:* 1 full-time (0 women), 1 part-time, 1 international. Average age 35. In 1998, 1 degree awarded. *Degree requirements:* For master's, thesis or alternative required, foreign language not required. *Entrance requirements:* For master's, GRE General Test (minimum combined score of 1000 required), TOEFL (minimum score of 550 required), minimum GPA of 3.0 in last 90 hours. *Application deadline:* For fall admission, 3/1 (priority date). Application fee: $50. *Financial aid:* Fellowships, research assistantships, teaching assistantships, career-related internships or fieldwork and institutionally-sponsored loans available. Aid available to part-time students. Financial aid application deadline: 2/1. *Faculty research:* Beach erosion and coastal protection, loads on sea-based structures, ocean wave mechanics, wave forces on structures, breakwater behavior. *Unit head:* Dr. Charles K. Sollitt, Director, 541-737-6887, Fax: 541-737-0485, E-mail: charles.sollitt@orst.edu.

Stevens Institute of Technology, Graduate School, Charles V. Schaefer Jr. School of Engineering, Department of Civil, Environmental, and Ocean Engineering, Program in Ocean Engineering, Hoboken, NJ 07030. Offers M Eng, PhD. *Degree requirements:* For master's, computer language required, thesis optional, foreign language not required; for doctorate, variable foreign language requirement, computer language, dissertation required. *Entrance requirements:* For master's, TOEFL (minimum score of 500 required); for doctorate, GRE, TOEFL (minimum score of 500 required). Electronic applications accepted. *Faculty research:* Estuarine oceanography, hydrodynamic and environmental processes, wave/ship interaction.

Texas A&M University, College of Engineering, Department of Civil Engineering, Program in Ocean Engineering, College Station, TX 77843. Offers M Eng, MS, D Eng, PhD. D Eng offered through the College of Engineering. *Students:* 41, 20 international. Average age 29. 49 applicants, 57% accepted. In 1998, 6 master's, 4 doctorates awarded. *Degree requirements:* For master's, thesis (MS) required; for doctorate, dissertation (PhD), internship (D Eng) required. *Entrance requirements:* For master's, GRE General Test, TOEFL; for doctorate, GRE General Test (minimum combined score of 1000 required), TOEFL. Application fee: $50 ($75 for international students). *Financial aid:* Fellowships, research assistantships, teaching assistantships available. Financial aid application deadline: 4/1; financial aid applicants required to submit FAFSA. *Faculty research:* Wave-structure interaction, marine mining and dredging, offshore pipelines. *Unit head:* Dr. Billy L. Edge, Head, 409-845-4515, Fax: 409-862-2800, E-mail: ce-grad@tamu.edu. *Application contact:* Dr. Hamm-Ching Chen, 409-845-2498, Fax: 409-862-2800, E-mail: ce-grad@tamu.edu.

See in-depth description on page 1495.

University of California, Berkeley, Graduate Division, Group in Ocean Engineering, Berkeley, CA 94720-1500. Offers M Eng, MS, D Eng, PhD. 20 applicants, 50% accepted. *Degree requirements:* For doctorate, dissertation required. *Entrance requirements:* For master's and doctorate, GRE General Test, minimum GPA of 3.0. *Application deadline:* For fall admission, 1/5; for spring admission, 9/1. Application fee: $40. *Financial aid:* Fellowships, research assistantships available. Financial aid application deadline: 1/5. *Unit head:* Prof. Alaa E. Mansour, Acting Chair, 510-642-5464, Fax: 510-642-6128, E-mail: alaa@uclink2.berkeley.edu. *Application contact:* Angela Corral, Graduate Assistant, 510-642-8790, Fax: 510-643-6103, E-mail: corral@coe.berkeley.edu.

See in-depth description on page 1497.

University of California, San Diego, Graduate Studies and Research, Department of Applied Mechanics and Engineering Sciences, Program in Applied Ocean Science, La Jolla, CA 92093-5003. Offers MS, PhD. Part-time programs available. *Students:* 1 full-time (0 women). 14 applicants, 14% accepted. *Degree requirements:* For master's, comprehensive exam or thesis required; for doctorate, dissertation, qualifying exam required. *Entrance requirements:* For master's and doctorate, GRE General Test, TOEFL (minimum score of 550 required), minimum GPA of 3.0. *Application deadline:* For fall admission, 5/31. Application fee: $40. *Financial aid:* In 1998–99, fellowships with full tuition reimbursements (averaging $15,000 per year), research assistantships with full tuition reimbursements (averaging $15,000 per year), teaching assistantships with partial tuition reimbursements (averaging $13,000 per year) were awarded. Financial aid application deadline: 1/31; financial aid applicants required to submit FAFSA. *Unit head:* Jori Cincotta, Director of Admissions, 619-534-5914, Fax: 619-534-1135, E-mail: irps-apply@ucsd.edu. *Application contact:* AMES Graduate Student Affairs, 619-534-4387, Fax: 619-534-1730, E-mail: bwalton@ames.ucsd.edu.

University of California, San Diego, Graduate Studies and Research, Department of Electrical and Computer Engineering, La Jolla, CA 92093-5003. Offers applied ocean science (MS, PhD); applied physics (MS, PhD); communication theory and systems (MS, PhD); computer engineering (MS, PhD); electrical engineering (M Eng, MS, PhD); electronic circuits and systems (MS, PhD); intelligent systems, robotics and control (MS, PhD); photonics (MS, PhD); signal and image processing (MS, PhD). *Faculty:* 35. *Students:* 251 (24 women). *Entrance requirements:* For master's and doctorate, GRE General Test. Application fee: $40. *Unit head:* William Coles, Chair. *Application contact:* Graduate Coordinator, 619-534-6606.

University of Connecticut, Graduate School, School of Engineering, Department of Mechanical Engineering, Field of Ocean Engineering, Storrs, CT 06269. Offers MS, PhD. Terminal master's awarded for partial completion of doctoral program. *Degree requirements:* For master's, thesis or alternative required, foreign language not required. *Entrance requirements:* For master's, GRE General Test, GRE Subject Test, TOEFL.

University of Delaware, College of Engineering, Department of Civil and Environmental Engineering, Program in Ocean Engineering, Newark, DE 19716. Offers MAS, MCE, PhD. Terminal master's awarded for partial completion of doctoral program. *Degree requirements:* For master's and doctorate, thesis/dissertation required, foreign language not required. *Entrance requirements:* For master's and doctorate, GRE General Test, TOEFL (minimum score of 550 required). *Faculty research:* Shoreline erosion, pollution of estuaries, offshore and near-shore wave motions, water and wave water mechanics, theoretical and experimental analysis.

University of Florida, Graduate School, College of Engineering, Department of Coastal and Oceanographic Engineering, Gainesville, FL 32611. Offers ME, MS, PhD, Engr. *Accreditation:* ABET (one or more programs are accredited). *Faculty:* 8. *Students:* 29 full-time (4 women), 8 part-time (2 women); includes 1 minority (Hispanic American), 14 international. 44 applicants, 70% accepted. In 1998, 6 master's, 1 doctorate awarded. *Degree requirements:* For master's and Engr, thesis optional; for doctorate, dissertation required. *Entrance requirements:* For master's and doctorate, GRE General Test, TOEFL, minimum GPA of 3.0; for Engr, GRE General Test. *Application deadline:* For fall admission, 6/1 (priority date). Applications are processed on a rolling basis. Application fee: $20. Electronic applications accepted. *Financial aid:* In 1998–99, 27 students received aid, including 25 research assistantships; fellowships, teaching assistantships *Faculty research:* Coastal processes, ocean process, coastal structures, ocean structure, wave forces. *Unit head:* Dr. Paul Thompson, Chair, 352-392-1436, Fax: 352-392-3466, E-mail: pthom@ce.ufl.edu. *Application contact:* Dr. Ashish J. Mehta, Graduate Coordinator, 352-392-4360, Fax: 352-392-3466, E-mail: mehta@coastal.ufl.edu.

University of Hawaii at Manoa, Graduate Division, School of Ocean and Earth Science and Technology, Department of Ocean and Resource Engineering, Honolulu, HI 96822. Offers MS, PhD. *Accreditation:* ABET (one or more programs are accredited). *Faculty:* 19 full-time (1 woman). *Students:* 19 full-time (3 women), 3 part-time; includes 2 minority (1 Asian American or Pacific Islander, 1 Hispanic American), 14 international. Average age 33. 26 applicants, 62% accepted. In 1998, 5 master's awarded (67% found work related to degree, 33% continued full-time study); 1 doctorate awarded (50% entered university research/teaching, 50% found other work related to degree). *Degree requirements:* For master's and doctorate, computer language, thesis/dissertation, exams required, foreign language not required. *Entrance requirements:* For master's and doctorate, GRE General Test, TOEFL (minimum score of 600 required). *Average time to degree:* Master's–3.6 years full-time; doctorate–5 years full-time, 8 years part-time. *Application deadline:* For fall admission, 3/1; for spring admission, 9/1. Application fee: $25 ($50 for international students). *Financial aid:* In 1998–99, 12 students received aid, including 6 research assistantships (averaging $17,312 per year), 2 teaching assistantships (averaging $14,691 per year); institutionally-sponsored loans and tuition waivers (full) also available. Financial aid application deadline: 3/1. *Faculty research:* Coastal and harbor engineering, nearshore environmental ocean engineering, marine structures/naval architecture. Total annual research expenditures: $436,100. *Unit head:* Ralph Moberly, Chairperson, 808-956-8765, Fax: 808-956-3498, E-mail: ralph@soest.hawaii.edu. *Application contact:* Alexander Malahoff, Graduate Chairperson, 808-956-7572, Fax: 808-956-3498, E-mail: malahoff@hawaii.edu.

University of Miami, Graduate School, Rosenstiel School of Marine and Atmospheric Science, Division of Applied Marine Physics, Coral Gables, FL 33124. Offers applied marine physics (MA, MS, PhD), including coastal ocean circulation dynamics (MS, PhD), dynamics and air-sea interaction physics (MA), ocean acoustics and geoacoustics (PhD), ocean acoustics and geoacoustics (MA), small-scale ocean surface (MA), small-scale ocean surface dynamics and air-sea interaction physics (PhD); ocean engineering (MS). Part-time programs available.

Ocean Engineering

Faculty: 11. *Students:* 10 full-time (0 women), 5 international. Terminal master's awarded for partial completion of doctoral program. *Degree requirements:* For master's and doctorate, thesis/dissertation required, foreign language not required. *Entrance requirements:* For master's and doctorate, GRE General Test, TOEFL (minimum score of 550 required). *Application deadline:* For fall admission, 2/1 (priority date). Applications are processed on a rolling basis. Application fee: $35. Electronic applications accepted. Tuition: Full-time $15,336; part-time $852 per credit. Required fees: $174. Tuition and fees vary according to program. *Unit head:* Dr. Harry A. De Ferrari, Chair, 305-361-4644, Fax: 305-261-4701. *Application contact:* Dr. Frank Millero, Associate Dean, 305-361-4155, Fax: 305-361-4771, E-mail: gso@rsmas.miami.edu.

University of Michigan, Horace H. Rackham School of Graduate Studies, College of Engineering, Department of Naval Architecture and Marine Engineering, Ann Arbor, MI 48109. Offers concurrent marine design (M Eng); naval architecture and marine engineering (MS, MSE, PhD, Mar Eng, Nav Arch). Part-time programs available. Postbaccalaureate distance learning degree programs offered (minimal on-campus study). *Faculty:* 12 full-time (2 women), 9 part-time (0 women). *Students:* 70 full-time (7 women), 2 part-time; includes 1 minority (Hispanic American), 29 international. Average age 26. 63 applicants, 95% accepted. In 1998, 33 master's, 6 doctorates awarded (100% found work related to degree). Terminal master's awarded for partial completion of doctoral program. *Degree requirements:* For master's, thesis required (for some programs), foreign language not required; for doctorate and other advanced degree, dissertation, oral defense of dissertation, preliminary exams required, foreign language not required; for other advanced degree, thesis, comprehensive written exam, oral defense of dissertation required, foreign language not required. *Entrance requirements:* For master's, GRE General Test (for financial aid), TOEFL (minimum score of 560 required) or Michigan English Language Assessment Battery (minimum score of 80 required); for doctorate, GRE General Test (minimum combined score of 1300 required; average 1352), master's degree; for other advanced degree, GRE General Test (minimum combined score of 1300 required). *Average time to degree:* Master's–1 year full-time; doctorate–4 years full-time. *Application deadline:* For fall admission, 2/1. Applications are processed on a rolling basis. Application fee: $55. Electronic applications accepted. *Financial aid:* In 1998–99, 24 students received aid, including 3 fellowships, 15 research assistantships, 2 teaching assistantships; career-related internships or fieldwork, Federal Work-Study, institutionally-sponsored loans, and scholarships also available. Financial aid application deadline: 2/1. *Faculty research:* Marine mechanics including hydrodynamics, structures, marine environmental engineering, marine design analysis, concurrent marine design, virtual reality. *Unit head:* Dr. Michael M. Bernitsas, Chair, 734-936-0566, Fax: 734-936-8820, E-mail: clv@engin.umich.edu. *Application contact:* Celia A. Eidex, Graduate Program Coordinator, 734-936-0566, Fax: 734-936-8820, E-mail: ceidex@engin.umich.edu.

University of New Hampshire, Graduate School, College of Engineering and Physical Sciences, Programs in Engineering, Programs in Ocean Engineering, Durham, NH 03824. Offers MS. *Faculty:* 13 full-time. *Students:* 5 full-time (1 woman), 6 part-time, 1 international. Average age 32. 6 applicants, 100% accepted. In 1998, 5 degrees awarded. *Degree requirements:* For master's, thesis required, foreign language not required. *Application deadline:* For fall admission, 4/1 (priority date). Applications are processed on a rolling basis. Application fee: $50. Tuition, area resident: Full-time $5,750; part-time $319 per credit. Tuition, state resident: full-time $8,625. Tuition, nonresident: full-time $14,640; part-time $598 per credit. Required fees: $224 per semester. Tuition and fees vary according to course load, degree level and program. *Financial aid:* In 1998–99, 1 research assistantship was awarded.; teaching assistantships, Federal Work-Study, scholarships, and tuition waivers (full and partial) also available. Aid available to part-time students. Financial aid application deadline: 2/15. *Unit head:* Dr. Kenneth Baldwin, Head, 603-862-1352.

University of Rhode Island, Graduate School, College of Engineering, Department of Ocean Engineering, Kingston, RI 02881. Offers MS, PhD.

University of Southern California, Graduate School, School of Engineering, Department of Civil Engineering, Program in Ocean Engineering, Los Angeles, CA 90089. Offers MS. *Students:* 1 full-time (0 women), 1 international. Average age 47. 3 applicants, 33% accepted. In 1998, 2 degrees awarded. *Degree requirements:* For master's, thesis optional. *Entrance requirements:* For master's, GRE General Test. *Application deadline:* For fall admission, 6/1 (priority date); for spring admission, 11/1. Application fee: $55. Tuition: Part-time $768 per unit. Required fees: $350 per semester. *Financial aid:* Fellowships, research assistantships, teaching assistantships, Federal Work-Study and institutionally-sponsored loans available. Aid available to part-time students. Financial aid application deadline: 2/15; financial aid applicants required to submit FAFSA. *Unit head:* Dr. L. Carter Wellford, Chair, Department of Civil Engineering, 213-740-0587.

Virginia Polytechnic Institute and State University, Graduate School, College of Engineering, Department of Aerospace and Ocean Engineering, Program in Ocean Engineering, Blacksburg, VA 24061. Offers MS. 4 applicants, 50% accepted. In 1998, 3 degrees awarded. *Degree requirements:* For master's, thesis required, foreign language not required. *Entrance requirements:* For master's, GRE (non-native speakers only), TOEFL. *Application deadline:* For fall admission, 12/1 (priority date). Applications are processed on a rolling basis. Application fee: $25. *Financial aid:* Application deadline: 4/1. *Unit head:* Dr. Bernard Grossman, Head, Department of Aerospace and Ocean Engineering, 540-231-6611, E-mail: grossman@aoe.vt.edu.

Webb Institute, Department of Ocean Technology and Commerce, Glen Cove, NY 11542-1398. Offers MS. *Degree requirements:* For master's, thesis required. *Entrance requirements:* For master's, GRE General Test (minimum combined score of 1200 required; average 1300).

FLORIDA INSTITUTE OF TECHNOLOGY

Division of Marine and Environmental Systems
Ocean Engineering Program

Programs of Study

The Florida Institute of Technology Division of Marine and Environmental Systems offers graduate courses and research opportunities leading to the Master of Science and Doctor of Philosophy in ocean engineering. The ocean engineering degree offers specialization in coastal zone processes and engineering, marine vehicles and ocean systems, materials and structures, and underwater technology. The division also offers graduate programs in oceanography, with options in chemical, biological, physical, and geological oceanography and in coastal zone management. The master's degree in oceanography is offered in thesis and nonthesis options.

Research Facilities

Florida Institute of Technology is conveniently located on the Atlantic coast in central Florida and has marine laboratories and field research sites both on the Indian River Lagoon and at an oceanfront marine research facility. The anchorage facility, just 5 minutes from campus, houses a fleet of small outboard-powered craft and medium-sized work boats. These boats are available to students and faculty members for teaching and research in the freshwater tributaries and the Indian River Lagoon. In addition, the 60-foot Research Vessel *Delphinus* is berthed at the anchorage facility. With her own captain and crew and the requisite marine and oceanographic cranes, winches, state-of-the-art sampling equipment, instrumentation, and laboratories, she is the focal point of marine and estuarine research in the region. The vessel accommodates a scientific team and crew for periods of seven to ten days. The *Delphinus* conducts short research and teaching cruises throughout the year.

The Institute's oceanfront marine research facility, the Vero Beach Marine Laboratory, located just 40 minutes from campus, provides facilities that include flowing seawater from the Atlantic Ocean and supports research in such areas as aquaculture, biofouling, and corrosion. There is also a permanent research platform, centrally located in the Indian River Lagoon, to support marine research projects such as materials and instrumentation testing.

On campus, divisional teaching and research facilities include a computer-aided design center and graduate research computer facilities for design, data analysis, numerical modeling, and other teaching and research activities. The ocean engineering test laboratory provides facilities for structural and pressure testing. The coastal processes laboratory provides facilities for core boring and sediment analysis, beach and hydrographic surveying, and oceanographic instrumentation for coastal research activities. Recent developments in the coastal engineering area include a circular wave basin and an ROV construction and test facility. Separate laboratories are available for biological, chemical, physical, geological, and instrumentation investigations. In addition, high-pressure, hydroacoustics, fluid dynamics, and optical facilities are available in the division. An electron microscope is also available for research work.

About 1 hour from campus is the Harbor Branch Oceanographic Institution; scientists and engineers there pursue their own research and development activities and interact with the Institute's students and faculty members on projects of mutual interest. Graduate students, especially those having an interest in submersibles, exploratory equipment, and instrumentation, frequently have the opportunity to conduct research with the HBOI staff and to utilize facilities at the institution. Similarly, students interested in hydroacoustics have the opportunity to work in association with scientists and engineers at the Naval Research Laboratory in nearby Orlando.

An exchange agreement with Herlot-Watt University in Scotland allows students to undertake research utilizing the University's extensive lasers, the wave/wind tank, and other facilities, including a 300-foot towing tank.

Financial Aid

Graduate teaching and research assistantships are available to qualified students. Typical stipends range from $13,000 upward for twelve months for approximately half-time duties; some assistantships include tuition.

Cost of Study

For 1999–2000, tuition is $575 per semester credit hour. As noted above, however, tuition is paid for some graduate students.

Living and Housing Costs

Room and board on campus cost approximately $2200 per semester in 1999–2000. On-campus housing (dormitories and apartments) is available for full-time single and married graduate students, but priority for dormitory rooms is given to undergraduate students. Many apartment complexes and rental houses are available near the campus.

Student Group

The College of Engineering has 450 graduate students.

Student Outcomes

Graduates of the Ocean Engineering Program are employed by such facilities as the Naval Sea Systems Command, the Naval Surface Warfare Center, the Naval Undersea Warfare Center, the Naval Research Lab, Newport News Shipbuilding, the U.S. Army Corps of Engineers, the American Bureau of Shipping, Digicon, LCT, NOAA, Autec, Tracor Marine, G. M. Selby & Associates, the U.S. Navy, the U.S. Coast Guard, Quantic Engineering & Logistics, Underwater Engineering Services, Biospherical Instruments Inc., Oceaneering International Inc., and EG&G.

Location

The campus is located in Melbourne, on Florida's east coast. The area, located 4 miles from Atlantic Ocean beaches, has a year-round subtropical climate. The area's economy is supported by a well-balanced mix of industries in electronics, aviation, light manufacturing, optics, communications, agriculture, and tourism. Many industries support activities at the Kennedy Space Center.

The Institute

Florida Institute of Technology is a distinctive, independent university founded in 1958 by a group of scientists and engineers to fulfill a need for specialized advanced educational opportunities on the Space Coast of Florida. Florida Tech is the only independent technological university in the Southeast. Supported by both industry and the community, Florida Tech is the recipient of many research grants and contracts, a number of which provide financial support for graduate students.

Applying

Forms and instructions for applying for admission and assistantships are sent on request. Admission is possible at the beginning of any semester, but admission in the fall semester is recommended. It is advantageous to apply early.

Correspondence and Information

Dr. Andrew Zborowski, Program Chair
Ocean Engineering Program
Florida Institute of Technology
Melbourne, Florida 32901-6975

Telephone: 407-674-8096
Fax: 407-674-7212
E-mail: zborowsk@marine.fit.edu

Graduate Admissions Office
Florida Institute of Technology
Melbourne, Florida 32901-6988

Telephone: 407-674-8027
 800-944-4348 (toll-free)
Fax: 407-723-9468
World Wide Web: http://www.fit.edu

Florida Institute of Technology

THE FACULTY AND THEIR RESEARCH

Lee Harris, Associate Professor; Ph.D., Florida; PE. Coastal engineering, coastal structures, beach erosion and control, physical oceanography, ports, harbors, marinas.

Geoffrey W. J. Swain, Associate Professor; Ph.D., Southampton. Materials corrosion, biofouling, offshore technology, ship operations.

Eric D. Thosteson, Assistant Professor; Ph.D., Florida. Coastal and nearshore engineering, coastal processes, wave mechanics, sediment transport.

Stephen L. Wood, Assistant Professor; Ph.D., Oregon State. Design of underwater vehicles, robotic systems development, navigation and control.

Andrew Zborowski, Professor and Program Chair; Ph.D., Gdansk (Poland). Ship hydrodynamics, marine structures, ship model tank studies, ship propulsion, dynamics of marine vehicles, high-speed craft design.

Deploying wave gauges offshore.

Florida Tech research vessel.

MASSACHUSETTS INSTITUTE OF TECHNOLOGY

Department of Ocean Engineering

Programs of Study

The Department of Ocean Engineering offers programs leading to the Master of Engineering, Master of Science, Ocean Engineer, Naval Engineer, Doctor of Philosophy, and Doctor of Science degrees. The fields in which master's degree candidates may elect to study are marine environmental systems, naval architecture and marine engineering, ocean engineering, and ocean systems management. Doctoral degrees are offered in ocean engineering, naval architecture and marine engineering, oceanographic engineering (MIT/Woods Hole Oceanographic Institution Joint Program), and ocean systems management. Major fields of study include hydrodynamics, structural mechanics, ocean-related acoustics, oceanographic engineering, underwater vehicle design, offshore engineering, ship/rig/platform design, marine resource development, ocean engineering and law, and marine transportation, to name a few. The department also participates in a number of interdepartmental degree programs, which include technology/policy and operations research. Students, working with their program advisers, can usually tailor their programs of graduate study to suit their individual interests and career objectives.

An acceptable course program plus an acceptable thesis leading to the Master of Engineering requires at least one year. Depending on background preparation, the Master of Engineering program in marine environmental systems can take from twelve to eighteen months to complete. The Master of Science degree usually requires from one to two academic years, depending upon undergraduate preparation. The Ocean Engineer and Naval Engineer degree programs require at least two years, including a substantial thesis. More than three years following the Bachelor of Science are usually required for the Doctor of Philosophy or Doctor of Science with a specification in an ocean-related field. Doctoral candidates must pass a qualifying examination, which is composed of two parts given at different stages of their academic program; complete a program of subjects; and successfully defend a doctoral thesis comprising original research.

Employment opportunities for graduates are available in industry, consulting firms, and government agencies.

Research Facilities

The department's research facilities include a Marine Hydrodynamics Water Tunnel, an Ocean Engineering Testing Tank, an Acoustics and Vibration Laboratory, a Computer-Aided Design Laboratory, a Marine Computation and Instrumentation Laboratory, a Fabrication Laboratory, a Marine Robotics Laboratory, a Vortical Flow Research Laboratory, a Laboratory for Ship and Platform Flows, an Impact and Crashworthiness Laboratory, an Undergraduate Teaching Laboratory, and an academic computing facility.

Financial Aid

A limited number of research and teaching assistantships are available for well-qualified students. Stipends for 1999–2000 are $12,825 for the nine-month academic year for full-time research assistants and $13,365 for the nine-month academic year for full-time teaching assistants. Not all appointments are available on a full-time basis. Tuition is prorated as to the percent-time appointment of the student; a full-time appointment includes full tuition remission.

Fellowships are available from a number of outside agencies. For specific information about these awards, students should write directly to the agency and/or contact the MIT Graduate School Office in Room 3-136.

Cost of Study

Tuition for the two-term 1999–2000 academic year is $25,000. No tuition is charged during the summer for registered students who are not taking classes.

Living and Housing Costs

On campus, single-student housing is available at a cost of $370 to $930 per month. For married students, rents per month range from $695 to $1050. A considerable range of private accommodations can be found off campus and within commuting distance.

Student Group

There are more than 100 graduate students in the department; all are full-time. Approximately one third are international students.

Location

MIT occupies a 125-acre campus on the north bank of the Charles River in Cambridge, facing the Boston skyline. The cultural, scientific, and intellectual resources of the Boston area are easily accessible. The area has many famous museums, restaurants, and historic sites as well as many college and university campuses. The ocean beaches and rural areas of New England are only a short drive from the campus.

The Institute and The Department

MIT was founded in 1861 as a private, coeducational, endowed institution committed to the extension of knowledge through teaching and research. It has grown to be one of the world's foremost institutes of technology. It is organized into five schools: Architecture and Planning, Engineering, Humanities and Social Science, Management, and Science. More than 9,000 students are enrolled, about half of them in the Graduate School.

In 1893, when it was founded, the department was called the Department of Naval Architecture and Marine Engineering, descriptive of the shipbuilding orientation of the curriculum. Nearly a century later, reflecting the growth of the marine industry and the comprehensive nature of the department, it was renamed the Department of Ocean Engineering.

Applying

To apply for admission to the graduate program, a student must complete an undergraduate program with sufficient background in mathematics, physics, and applied sciences and with an average of B or better. The GRE General Test is required of all applicants. Students from non-English-speaking countries must demonstrate proficiency in English by earning a TOEFL score of at least 577. Applications for the fall term are due on January 15. The application fee for admission is $50.

Correspondence and Information

Student Administration Office
Department of Ocean Engineering
Room 5-225
Massachusetts Institute of Technology
77 Massachusetts Avenue
Cambridge, Massachusetts 02139
Telephone: 617-253-1994

Massachusetts Institute of Technology

THE FACULTY AND THEIR RESEARCH

Professors

Arthur B. Baggeroer, Ford Professor and SECNAV Chair; Sc.D., MIT, 1968. Oceanographic and sonar systems, seismic exploration, acoustic communication systems.

Chryssostomos Chryssostomidis, 1993–97 Henry L. and Grace Doherty Professor of Ocean Science and Engineering, Head of the Department, and Director, MIT Sea Grant Program; Ph.D., MIT, 1970. Ship design, offshore structure design, computer-aided design, robotic submersible design.

Justin E. Kerwin, Professor of Naval Architecture; Ph.D., MIT, 1961. Marine hydrodynamics, propellers.

Dennis P. Mahoney, Professor of Naval Architecture; Ocean Engineer, MIT, 1977. Naval ship design and construction.

Henry S. Marcus, Professor of Marine Systems; D.B.A., Harvard, 1970. Ocean transportation management.

Koichi Masubuchi, Professor of Ocean Engineering and Materials Science; Ph.D., Tokyo, 1959. Materials and welding fabrication of marine and aerospace structures.

Jerome H. Milgram, William I. Koch Professor; Ph.D., MIT, 1965. Hydrodynamics, ocean waves and dynamics.

J. D. Nyhart, Professor of Ocean Engineering and Management; J.D., Harvard, 1958. Ocean and regulatory law.

Nicholas M. Patrikalakis, Kawasaki Professor of Engineering; Ph.D., MIT, 1983. Computer-aided ship and offshore structural design, marine riser dynamics.

Henrik Schmidt, Professor of Ocean Engineering and Associate Department Head; Associate Director, Sea Grant Program; Ph.D., Denmark Technical, 1978. Underwater acoustics, wave propagation, seismic exploration.

Paul D. Sclavounos, Professor of Naval Architecture; Ph.D., MIT, 1981. Hydrodynamics, analysis of wave effects on ocean vessels.

Michael S. Triantafyllou, Professor of Ocean Engineering; Sc.D., MIT, 1979. Design and control of ocean vessels, cable dynamics, remotely operated underwater vehicles.

J. Kim Vandiver, Professor of Ocean Engineering and Director, Edgerton Center; Ph.D., MIT, 1975. Dynamics of offshore structures, mechanical vibration, vortex-induced vibration.

Tomasz Wierzbicki, Professor of Applied Mechanics; Sc.D., Institute of Fundamental Technological Research (Warsaw), 1971. Structural mechanics, crashworthiness of structures.

Dick Kau-Ping Yue, Professor of Ocean Engineering; Sc.D./Ph.D., MIT, 1980. Numerical and theoretical wave hydrodynamics.

Associate Professor

Clifford Whitcomb, Associate Professor of Naval Construction and Engineering; Ph.D., Maryland, 1998. Naval ship design and construction.

Assistant Professor

John J. Leonard, Assistant Professor of Ocean Engineering; Ph.D., Oxford, 1991. Marine robotic systems, autonomous underwater vehicles.

Nicholas M. Makris, Assistant Professor of Ocean Engineering; Ph.D., MIT, 1990. Acoustical oceanography.

TEXAS A&M UNIVERSITY

Civil Engineering Department
Ocean Engineering Program

Program of Study

The Ocean Engineering Program is administered through the Civil Engineering Department and offers programs leading to the Master of Science, Master of Engineering, and Doctor of Philosophy degrees in ocean engineering. The College of Engineering offers the Doctor of Engineering degree with an emphasis in ocean engineering. Graduate courses are given in the areas of offshore and coastal structures, coastal engineering, dynamics of offshore structures, ocean wave mechanics, hydrodynamics, arctic offshore engineering, marine foundations, fluid-structure interactions, marine dredging, computational fluid dynamics, ocean vehicle dynamics, coastal sediment processes, and special topics in ocean engineering as new areas develop. Graduate students are involved in research in these areas.

The Master of Science degree program requires a minimum of two full semesters of approved courses and research (32 semester hours). This requirement is ordinarily met by completing at least 24 hours of course work and 8 hours of research. An acceptable thesis embodying original research is required. The student must pass a final examination covering his or her graduate program.

The Master of Engineering degree requires a minimum of 36 semester hours, of which one third must be taken in fields other than the major field. A thesis is not required for this degree, but work in the major field includes one or two written reports.

Students entering the Doctor of Philosophy program with a bachelor's degree must spend two academic years in resident study; those who enter with a master's degree must spend one academic year in resident study. A minimum of 96 credit hours beyond the bachelor's degree, or 64 credit hours beyond the master's degree, is normally required. A dissertation that demonstrates the student's ability to conduct independent, original research is required. The student must pass a qualifying exam during the first year of study, a preliminary written and oral exam after completion of course work, and a final defense of the dissertation.

The Doctor of Engineering degree normally requires a 96-credit-hour professional program beyond the bachelor's degree, or a minimum of 60 credit hours beyond the master's degree. A year of professional internship must be completed under the supervision of a practicing engineer in industry, business, or government, and candidates must pass a qualifying exam and a final oral exam, which includes the internship.

Research Facilities

The major laboratory facilities for research in ocean engineering are located in the Hydromechanics Laboratories and the Civil Engineering Laboratory Building. These facilities include two long two-dimensional glass-walled wave tanks, a three-dimensional wave basin, two variable-slope wave flumes, a towing tank, a dredge pump and pipe test loop, two data acquisition systems, and laser-Doppler and constant-temperature anemometry systems. The Offshore Technology Research Center, funded by the National Science Foundation as part of its Engineering Research Centers Program, offers excellent opportunities for deepwater technology research. A 3-D wave basin measuring 150 feet by 100 feet by 20 feet with a center pit 50 feet deep is available for deepwater technology research. The basin is equipped with a multisegment directional wave generator, capable of creating waves more than 3 feet high, and other state-of-the-art instrumentation. The research vessel R/V *Gyre* and a remotely controlled vehicle operated by the Oceanography Department are available for offshore and coastal research. A hyperbaric facility is available through the Biology Department. The University Computing Services Center houses an Amdahl 5860 and IBM 3090 and 4361 mainframe computers. The Engineering Computer Services facility operates a cluster of VAX computers and CRAY J-90 and SGI supercomputers. Access to other supercomputers is available through the Computing Services Center. Computer workstations and a microcomputer laboratory are available for use in the laboratory. The University library holds more than a million volumes and is well equipped for computer-based literature searches.

Financial Aid

Graduate teaching and research assistantships are available. Competitive fellowships are awarded by the University, the College of Engineering, and the Office of Graduate Studies. The Ocean Engineering Program also awards several competitive scholarships. Inquiries and application for financial aid should be made as early as possible.

Cost of Study

Tuition and fees in 1998–99 for Texas residents for a regular semester were $1350.60 plus laboratory fees. For nonresidents, the cost was $3300 plus laboratory fees. These fees are subject to change without notice.

Living and Housing Costs

Room and board on campus for single students cost approximately $400 to $800 per month. Numerous apartments for single students are also available surrounding the campus. For married students, a limited number of University-owned apartments, both furnished and unfurnished, are available. Rents range from $230 to $360 per month, excluding utilities.

Student Group

Texas A&M is a large university with an undergraduate enrollment of more than 35,000 and a graduate enrollment of 8,083. Students are drawn from every state in the Union and more than sixty other countries. There are more than 50 graduate students in Ocean Engineering and more than 350 graduate students in the Civil Engineering Department.

Location

The Bryan–College Station area is a progressive community with a population of about 100,000. Located centrally in relation to the metropolitan areas of Houston, Dallas, Fort Worth, and Austin, the area offers many recreational and cultural activities. The climate is relatively mild.

The University

The University is a land-grant and sea-grant college founded in 1876 as a military school. It is now a dynamic force in the advanced educational program of the state of Texas. The University has an excellent athletics program, and cultural activities are provided by the Town Hall series, the Aggie Players, the Singing Cadets, the Artists' Showcase, the Opera and Performing Arts Society, and many other groups and programs. The academic environment challenges and stimulates both faculty members and students to achieve their educational goals.

Applying

All communications concerning admission and registration should be addressed to the director of admissions. Applications for admission, including transcripts and scores on the Graduate Record Examinations, should be received at least four weeks before the beginning of the semester in which the student wishes to enroll.

Correspondence and Information

Dr. Billy L. Edge, Head
Ocean Engineering Program
Civil Engineering Department
Texas A&M University
College Station, Texas 77843

Telephone: 409-845-4515
E-mail: h-chen1@tamu.edu

Texas A&M University

THE FACULTY AND THEIR RESEARCH

*Aubrey L. Anderson, Professor of Oceanography; Ph.D., Texas at Austin, 1974. Underwater acoustics, instrumentation, field measurements.

Hamm-Ching Chen, Associate Professor of Ocean and Civil Engineering; Ph.D., Iowa, 1982. Computational fluid dynamics, viscous flows, turbulence modeling.

Daniel T. Cox, Assistant Professor of Civil and Ocean Engineering; Ph.D., Delaware, 1995. Surf zone dynamics, stratified flows, turbulence and transport processes, beach erosion.

Zachary T. Demirbilek, Visiting Assistant Professor of Ocean Engineering; Ph.D., Texas A&M, 1982. Nonlinear water waves, computational mechanics, ocean structures, probabilistic reliability theory.

*Wayne A. Dunlap, Professor of Civil Engineering; Ph.D., Texas A&M, 1966; PE. Marine foundations, offshore pipelines.

Billy L. Edge, Professor of Ocean and Civil Engineering and Head of the Ocean Engineering Program; Ph.D., Georgia Tech, 1969; PE. Applied hydrodynamics, coastal structures, dynamic coastal processes, mathematical modeling of natural systems, marine pollution control, physical modeling of hydraulic phenomena, sediment transport and estuarine analysis.

John E. Flipse, Distinguished Professor Emeritus of Ocean and Civil Engineering; M.M.E., NYU, 1948; PE. Deep-ocean mining, naval architecture, shipbuilding.

*Richard B. Griffin, Associate Professor of Mechanical Engineering; Ph.D., Iowa State, 1969. Marine corrosion, materials for ocean applications.

John B. Herbich, W. H. Bauer Professor Emeritus in Marine Dredging and Professor of Ocean and Civil Engineering; Ph.D., Penn State, 1963; PE. Coastal and ocean engineering, dredging, open channel flow, hydraulic engineering, wave forces on pipelines and structure.

Steve A. Hughes, Visiting Assistant Professor of Ocean Engineering; Ph.D., Florida, 1981. Wave dynamics, surf zone transformations, remote sensing and image analyses, sediment transport.

Chung H. Kim, Professor of Ocean and Civil Engineering; Ph.D., Hannover Tech, 1965; PE. Naval architecture, marine hydrodynamics, seakeeping theories, extreme wave loads, drift forces in directional seas.

Moo-Hyun Kim, Associate Professor of Ocean and Civil Engineering; Ph.D., MIT, 1988. Marine hydrodynamics, nonlinear stochastic analysis, computational fluid dynamics.

Jack Y. K. Lou, Professor Emeritus of Ocean and Civil Engineering; Ph.D., Polytechnic of Brooklyn, 1969; PE. Ocean vehicle dynamics, fluid-structure interactions, dynamic analysis of marine structures, towing and mooring analysis.

*Loren D. Lutes, Professor of Civil Engineering; Ph.D., Caltech, 1967. Dynamics and fatigue of offshore structures, probabilistic methods for structural analysis.

John M. Niedzwecki, Professor of Ocean and Civil Engineering; Ph.D., Catholic University, 1977; PE. Dynamics of offshore structures, arctic offshore engineering, ocean wave mechanics, stochastic and probabilistic analyses, finite-element and other numerical methods, expert systems.

Michael R. Palermo, Visiting Professor of Ocean Engineering; Ph.D., Vanderbilt, 1984. Environmental and water quality, marine dredging and dredged material disposal technology.

Robert E. Randall, Professor of Ocean and Civil Engineering; Ph.D., Rhode Island, 1972; PE. Fluid mechanics, underwater life support and diving technology, undersea acoustics, offshore and laboratory measurements, dilution and diffusion process, dredging technology.

*Robert O. Reid, Distinguished Professor of Oceanography; M.S., California, San Diego (Scripps), 1948. Physical oceanography, wave theory, numerical methods for oceanographic processes.

Norman W. Scheffner, Visiting Assistant Professor of Ocean Engineering; Ph.D., Florida, 1987. Coastal and ocean engineering, hydraulic/hydrology research.

Robert E. Schiller Jr., Professor Emeritus of Ocean and Civil Engineering, Ph.D., Colorado State, 1969. Hydraulics and hydrology, sediment transport and scour.

Richard J. Seymour, Professor Emeritus of Ocean and Civil Engineering; Ph.D., California, San Diego (Scripps), 1974. Ocean technology, wave mechanics, wave climatology, nearshore processes.

*Norris Stubbs, Associate Professor of Civil Engineering; Sc.D., Columbia, 1976. Structural reliability, structural dynamics, nondestructive evaluation, experimental mechanics.

Edward F. Thompson, Visiting Associate Professor of Ocean Engineering; Ph.D., George Washington, 1981. Wave climatology, shallow water waves, nearshore wave transformation.

Todd L. Walton Jr., Visiting Assistant Professor of Ocean Engineering; Ph.D., Florida, 1974. Sediment transport, coastal erosion, littoral drift, numerical modeling.

Jun Zhang, Associate Professor of Ocean and Civil Engineering; Ph.D., MIT, 1987. Hydrodynamics, wave dynamics, nonlinear surface-water wave interaction and wave breaking, digital signal processing, applied mathematics.

Faculty associates outside the Ocean Engineering Program.

Deepwater wave basin at the Offshore Technology Research Center.

UNIVERSITY OF CALIFORNIA, BERKELEY

Ocean Engineering Program

Program of Study	The objective of the Ocean Engineering Program is to educate engineers to design and analyze systems in the ocean and coastal environments and to conduct research for the efficient and environmentally sound utilization of ocean thermal and wave energy.

The current program includes marine structural mechanics, marine hydrodynamics, surface-wave mechanics, computational fluid mechanics, marine safety, floating-body dynamics, marine risk and reliability analysis, construction and maintenance, and ship and offshore operations. These areas are complemented by courses and faculty members in other departments in the areas of coastal and estuary dynamics, large-scale oceanic and atmospheric sciences, robotics and control technology, drilling technology, and business and management to form the Ocean Engineering Program.

Four graduate degrees are offered under the graduate program in ocean engineering. Students are admitted to all degree programs based on the degree goal stated on the application form. The Master of Science (M.S.) degree can be completed in two semesters of residence plus one summer; the Master of Engineering (M.Eng.) degree can be completed in three to four semesters. The Doctor of Philosophy (Ph.D.) and Doctor of Engineering (D.Eng.) degrees require a minimum residence of four semesters after the completion of a master's degree at Berkeley or another accredited institution. Each program's requirements consist of both core courses and electives. Core courses depend on student interest.

Electives can be chosen from a larger list of subjects, allowing the student to pursue areas of specific interest or to prepare for the doctoral degree. |
| **Research Facilities** | The Ocean Engineering Graduate Group laboratory facilities include equipment for conducting experiments on both hydrodynamic and structural problems of marine vehicles and ocean structures. The principle hydrodynamic facility is a towing tank that is 200 feet long, 8 feet wide, and 6 feet deep. It is equipped with a towing carriage capable of speeds up to approximately 1.8 meters per second (3.5 knots) and a servo-controlled wave generator that can produce either regular or random waves. Instruments are available for measuring fluid forces, pressures, motions, real-time digital images, and other properties of the model being tested.

Structural testing machines are available for conducting static bending experiments on intermediate-scale models. Strain gauges and deflection instrumentation installed on the model provide the means of recording the structural response to load systems that approximate the loads imposed by the sea on the full-scale structures. |
Financial Aid	Students are encouraged to apply for extramural (e.g., NSF) fellowships as well as for various others administered by the University. Research assistantships may be available in projects supported by extramural grants or contracts. An effort is made to align the research assistant's interests with those of a faculty member working with the extramural support.
Costs of Study	Fees, insurance, and tuition for 1999–2000 total $4409 for California residents and $14,731 for nonresidents.
Living and Housing Costs	Room and board in the San Francisco Bay area for the 1998–99 nine-month academic year averaged $8802. Books and supplies totaled approximately $578. Entertainment and miscellaneous expenses were approximately $2200. Costs are proportionally higher for the twelve-month academic year.
Student Group	There are about 24 students enrolled in the graduate group. Approximately 30,000 students, including 8,500 graduate students, are currently enrolled at Berkeley. The cosmopolitan student body includes about 1,900 international students from nearly ninety countries and about 3,500 students from states other than California.
Location	The campus is surrounded by wooded hills and by the business and residential districts of Berkeley (population 105,000). Despite its rapid growth, the campus retains much natural beauty, with wooded glens, spacious plazas, and picturesque Strawberry Creek running the length of the campus.
The University	In 1868, the legislature of California approved an act creating the University. Following a move from temporary facilities in Oakland, Berkeley became the site of the first permanent University of California campus in 1873. As enrollment increased, architecturally varied structures were added, and the present campus gradually took shape. Eight other campuses, each maintaining a separate administrative organization and style of academic life, were later established throughout the state.

Today, the University of California is one of the largest universities in the world. The University also maintains research and field stations, extension centers, and other facilities for research and instruction in more than eighty locations throughout California. With such extensive resources, the University enjoys a leading position among universities, offering advancement of knowledge in virtually every field of modern human endeavor and serving as an indispensable force in the development and growth of American society. |
| **Applying** | Complete applications, including official transcripts, GRE scores, three letters of recommendation, and a statement of purpose, are due January 5 for the following fall. To obtain application forms, students should contact the address below. |
| **Correspondence and Information** | Ocean Engineering Graduate Group
230 Bechtel Engineering Center, #1708
University of California
Berkeley, CA 94720-1708

Telephone: 510-642-8790
Fax: 510-643-6103
E-mail: oceaneng@coe.berkeley.edu
World Wide Web: http://www.oe.berkeley.edu |

University of California, Berkeley

THE FACULTY AND THEIR RESEARCH

Robert Bea, Civil and Environmental Engineering. Design, construction, maintenance, and requalification of marine systems; applications of reliability methods to coastal, harbor, offshore, and ship systems. (E-mail: bea@ce.berkeley.edu)

George A. Cooper, Materials Science and Mineral Engineering. Drilling for environmental monitoring and remediation, novel drilling methods and materials, petroleum engineering, rock destruction and drilling mechanics. (E-mail: gcooper@socrates.berkeley.edu)

Armen Der Kiureghian, Civil and Environmental Engineering. Structural reliability, risk analysis, stochastic structural dynamics, earthquake engineering. (E-mail: adk@ce.berkeley.edu)

C. K. Hari Dharan, Mechanical Engineering. Composite materials, mechanical behavior and manufacturing processes, electronic packaging, spacecraft design. (E-mail: dharan@euler.me.berkeley.edu)

Mostafa A. Foda, Civil and Environmental Engineering. Coastal processes, porous media dynamics, seafloor stability, submarine pipelines, tectonics and liquefaction, wave propagation. (E-mail: foda@euler.berkeley.edu)

J. Karl Hedrick, Mechanical Engineering. Automotive control systems, intelligent vehicle highway systems, vehicle dynamics, aircraft control systems, nonlinear adaptive control theory. (E-mail: khedrick@euler.me.berkeley.edu)

Alaa Mansour, Mechanical Engineering. Probabilistic dynamics of marine structures, structural reliability and safety, strength of ship and offshore structures, development of design criteria. (E-mail: alaa@uclink2.berkeley.edu)

Ömer Savas, Mechanical Engineering. Boundary layer, flame stability, flow visualization, helicity measurement, instrumentation, laminar and turbulent flames, turbulence, vorticity measurements. (E-mail: osavas@euler.me.berkeley.edu)

Rodney J. Sobey, Civil and Environmental Engineering. Coastal engineering, computational mechanics, estuarine circulation, fluid mechanics, harbors, hydrodynamics, ocean engineering, turbulence, wave propagation, wetlands. (E-mail: sobey@ce.berkeley.edu)

William C. Webster, Civil and Environmental Engineering. Nonlinear coupled motions of offshore structures, operations research, shallow water fluid mechanics, steep-water waves, wave energy. (E-mail: wwebster@socrates.berkeley.edu)

Ronald W. Yeung, Mechanical Engineering. Hydromechanics, mathematical modeling, numerical fluid mechanics, offshore mechanics, separated flows, wave-vorticity interaction, ship hydrodynamics. (E-mail: rwyeung@socrates.berkeley.edu)

Section 19
Paper and Textile Engineering

This section contains a directory of institutions offering graduate work in paper and textile engineering, followed by in-depth entries submitted by institutions that chose to prepare detailed program descriptions. Additional information about programs listed in the directory but not augmented by an in-depth entry may be obtained by writing directly to the dean of a graduate school or chair of a department at the address given in the directory.

For programs offering related work, see also in this book Engineering and Applied Sciences and Materials Sciences and Engineering. In Book 2, see Home Economics and Family Studies (Clothing and Textiles).

CONTENTS

Program Directories

Paper and Pulp Engineering

Georgia Institute of Technology, Graduate Studies and Research, College of Engineering, School of Chemical Engineering, Pulp and Paper Engineering Program, Atlanta, GA 30332-0001. Offers Certificate. *Degree requirements:* For degree, foreign language not required. *Faculty research:* Black liquor, oxygen delignification, high-yield pulping, synthetic fiber application, colloid-polymer flocculation.

Institute of Paper Science and Technology, Graduate Programs, Atlanta, GA 30318-5794. Offers MS, PhD. Part-time programs available. *Faculty:* 33 full-time (1 woman), 13 part-time (1 woman). *Students:* 72 full-time (16 women); includes 13 minority (2 African Americans, 7 Asian Americans or Pacific Islanders, 4 Hispanic Americans), 2 international. Average age 24. 66 applicants, 59% accepted. In 1998, 21 master's, 6 doctorates awarded (100% found work related to degree). Terminal master's awarded for partial completion of doctoral program. *Degree requirements:* For master's, industrial experience, research project required, foreign language and thesis not required; for doctorate, dissertation required, foreign language not required. *Entrance requirements:* For master's and doctorate, GRE (score in 50th percentile or higher required), minimum GPA of 3.0. *Average time to degree:* Master's–2 years full-time; doctorate–3.5 years full-time. *Application deadline:* For fall admission, 3/1 (priority date). Application fee: $0. Electronic applications accepted. *Financial aid:* In 1998–99, 70 fellowships with tuition reimbursements were awarded.; teaching assistantships, career-related internships or fieldwork and institutionally-sponsored loans also available. Financial aid applicants required to submit FAFSA. *Faculty research:* Bleaching; chemical pulping, impulse drying, biotechnology, physics of fluid flow; environmental improvement, recycling, closed mill technology; black liquor recovery. Total annual research expenditures: $16 million. *Unit head:* Dr. Barry W. Crouse, Dean, Academic Affairs, 404-894-5735, Fax: 404-894-4778, E-mail: barry.crouse@ipst.edu. *Application contact:* Dana Carter, Student Development Counselor, 404-894-5745, Fax: 404-894-4778, E-mail: dana.carter@ipst.edu.

Miami University, Graduate School, School of Applied Science, Department of Paper Science and Engineering, Oxford, OH 45056. Offers MS. *Faculty:* 7. *Students:* 11 full-time (1 woman), 2 part-time (1 woman), 7 international. 17 applicants, 65% accepted. In 1998, 6 degrees awarded. *Degree requirements:* For master's, thesis, comprehensive and final exams required, foreign language not required. *Entrance requirements:* For master's, GRE General Test (minimum combined score of 1100 required), MAT (minimum score of 54 required), minimum undergraduate GPA of 3.0 during previous 2 years or 2.75 overall. *Application deadline:* For fall admission, 3/1 (priority date); for spring admission, 12/1. Applications are processed on a rolling basis. Application fee: $35. *Financial aid:* Fellowships, research assistantships, teaching assistantships, Federal Work-Study and tuition waivers (full) available. Financial aid application deadline: 3/1. *Faculty research:* Papermaking and pulping, wet end chemistry, environmental control in the pulp and paper industry. *Unit head:* Dr. Allan Springer, Director of Graduate Studies, 513-529-2208.

North Carolina State University, Graduate School, College of Forest Resources, Department of Wood and Paper Science, Raleigh, NC 27695. Offers MS, MWPS, PhD. *Faculty:* 20 full-time (2 women), 12 part-time (0 women). *Students:* 22 full-time (4 women), 13 part-time (3 women); includes 3 minority (2 African Americans, 1 Asian American or Pacific Islander), 19 international. Average age 30. 11 applicants, 27% accepted. In 1998, 6 master's, 4 doctorates awarded. *Degree requirements:* For master's and doctorate, thesis/dissertation required, foreign language not required. *Entrance requirements:* For master's and doctorate, GRE General Test, TOEFL (minimum score of 550 required). *Application deadline:* For fall admission, 6/25; for spring admission, 11/25. Applications are processed on a rolling basis. Application fee: $45. *Financial aid:* Fellowships, research assistantships available. Financial aid application deadline: 2/15. *Faculty research:* Pulping, bleaching, recycling, papermaking, drying of wood, analysis/design of wood structures, wood machining and tooling. Total annual research expenditures: $2.4 million. *Unit head:* Dr. Michael J. Kocurek, Head, 919-515-5807, Fax: 919-515-6302, E-mail: mike_kocurek@ncsu.edu. *Application contact:* Dr. Richard D. Gilbert, Director of Graduate Programs, 919-515-5321, Fax: 919-515-6302, E-mail: gilbert@cfr.cfr.ncsu.edu.

Oregon State University, Graduate School, College of Forestry, Department of Forest Products, Corvallis, OR 97331. Offers forest products (MAIS, MF, MS, PhD); wood science and technology (MF, MS, PhD). *Accreditation:* SAF (one or more programs are accredited). Part-time programs available. *Faculty:* 6 full-time (1 woman), 1 part-time (0 women). *Students:* 18 full-time (2 women); includes 1 minority (Asian American or Pacific Islander), 9 international. *Degree requirements:* For master's, thesis required (for some programs), foreign language not required; for doctorate, dissertation required, foreign language not required. *Entrance requirements:* For master's and doctorate, GRE General Test, TOEFL (minimum score of 550 required), minimum GPA of 3.0 in last 90 hours. *Application deadline:* For fall admission, 3/1 (priority date). Applications are processed on a rolling basis. Application fee: $50. *Unit head:* Dr. Thomas E. McLain, Head, 541-737-4224, Fax: 541-737-3385, E-mail: forprod@fri.orst.edu. *Application contact:* Charles Brunner, Chair, 541-737-4205, Fax: 541-737-3385, E-mail: brunnerc@fri.orst.edu.

Université du Québec à Trois-Rivières, Graduate Programs, Program in Pulp and Paper Engineering, Trois-Rivières, PQ G9A 5H7, Canada. Offers M Sc A, PhD. *Students:* 36 full-time (16 women), 55 part-time (10 women). 24 applicants, 88% accepted. In 1998, 5 master's awarded. *Degree requirements:* For doctorate, dissertation required. *Entrance requirements:* For doctorate, appropriate master's degree, proficiency in French. *Application deadline:* For fall admission, 2/1 (priority date). Application fee: $30. *Financial aid:* Fellowships, research assistantships, teaching assistantships available. *Unit head:* Sylvain Robert, Director, 819-376-5075 Ext. 3260, Fax: 819-376-5148, E-mail: sylvain_robert@uqtr.quebec.ca. *Application contact:* Suzanne Camirand, Admissions Officer, 819-376-5045 Ext. 2591, Fax: 819-376-5210, E-mail: suzanne_camirand@uqtr.uquebec.ca.

University of British Columbia, Faculty of Graduate Studies, Faculty of Applied Science, Department of Chemical Engineering, Program in Pulp and Paper Engineering, Vancouver, BC V6T 1Z2, Canada. Offers M Eng. Part-time and evening/weekend programs available. *Degree requirements:* For master's, foreign language and thesis not required. *Entrance requirements:* For master's, TOEFL (minimum score of 550 required), minimum GPA of 3.0. *Faculty research:* Pulping, paper making, systems and control.

University of Washington, Graduate School, College of Forest Resources, Seattle, WA 98195. Offers forest economics (MS, PhD); forest ecosystem analysis (MS, PhD); forest engineering/forest hydrology (MS, PhD); forest products marketing (MS, PhD); forest soils (MS, PhD); pulp and paper science (MS, PhD); quantitative resource management (MS, PhD); silviculture (MFR); silviculture and forest protection (MS, PhD); social sciences (MS, PhD); urban horticulture (MFR, MS, PhD); wildlife science (MS, PhD). *Faculty:* 47 full-time (4 women), 16 part-time (2 women). *Students:* 161 full-time (65 women), 16 part-time (5 women); includes 15 minority (1 African American, 10 Asian Americans or Pacific Islanders, 1 Hispanic American, 3 Native Americans), 9 international. *Degree requirements:* For master's, thesis required (for some programs), foreign language not required; for doctorate, dissertation required, foreign language not required. *Entrance requirements:* For master's and doctorate, GRE, TOEFL (minimum score of 500 required), minimum GPA of 3.0. *Application deadline:* For fall admission, 2/1 (priority date); for winter admission, 11/1; for spring admission, 2/1. Applications are processed on a rolling basis. Application fee: $50. Electronic applications accepted. Tuition, state resident: full-time $5,196; part-time $475 per credit. Tuition, nonresident: full-time $13,485; part-time $1,285 per credit. Required fees: $387; $38 per credit. Tuition and fees vary according to course load. *Unit head:* Dr. David B. Thorud, Dean, 206-685-1928, Fax: 206-685-0790. *Application contact:* Michelle Trudeau, Student Services Manager, 206-616-1533, Fax: 206-685-0790, E-mail: michtru@u.washington.edu.

Western Michigan University, Graduate College, College of Engineering and Applied Sciences, Department of Paper, Imaging, and Chemical Engineering, Kalamazoo, MI 49008. Offers MS. *Students:* 8 full-time (1 woman), 13 part-time (2 women); includes 1 minority (African American), 14 international. 33 applicants, 70% accepted. In 1998, 6 degrees awarded. *Degree requirements:* For master's, thesis required, foreign language not required. *Entrance requirements:* For master's, minimum GPA of 3.0. *Application deadline:* For fall admission, 2/15 (priority date). Applications are processed on a rolling basis. Application fee: $25. *Financial aid:* Fellowships, research assistantships, teaching assistantships, Federal Work-Study available. Financial aid application deadline: 2/15; financial aid applicants required to submit FAFSA. *Faculty research:* Fiber recycling, paper machine wet end operations, paper coating. *Unit head:* Dr. Thomas Joyce, Chairperson, 616-387-2770. *Application contact:* Paula J. Boodt, Coordinator, Graduate Admissions and Recruitment, 616-387-2000, Fax: 616-387-2355, E-mail: paula.boodt@wmich.edu.

Textile Sciences and Engineering

Auburn University, Graduate School, Interdepartmental Programs, Interdepartmental Program in Textile Science, Auburn, Auburn University, AL 36849-0002. Offers MS. *Faculty:* 9 full-time. *Students:* 6 full-time (2 women), 9 part-time (4 women), 14 international. 16 applicants, 31% accepted. In 1998, 4 degrees awarded. *Degree requirements:* Foreign language not required. *Entrance requirements:* For master's, GRE General Test. *Application deadline:* For fall admission, 9/1; for spring admission, 3/1. Applications are processed on a rolling basis. Application fee: $25 ($50 for international students). Tuition, state resident: full-time $2,760; part-time $76 per credit hour. Tuition, nonresident: full-time $8,280; part-time $228 per credit hour. *Financial aid:* Research assistantships, Federal Work-Study available. Aid available to part-time students. Financial aid application deadline: 3/15. *Faculty research:* Design and utilization of textile products, engineering and technology of textile production, textile material science, textile chemistry, use of resources. *Unit head:* Dr. Royall M. Broughton, Graduate Program Officer, 334-844-4123, E-mail: royalb@eng.auburn.edu. *Application contact:* Dr. John F. Pritchett, Dean of the Graduate School, 334-844-4700, E-mail: hatchlb@mail.auburn.edu.

Clemson University, Graduate School, College of Engineering and Science, School of Textiles, Fiber and Polymer Science, Clemson, SC 29634. Offers textile and polymer science (PhD); textile chemistry (MS); textile science (MS). Part-time programs available. *Students:* 36 full-time (13 women), 3 part-time (1 woman); includes 1 minority (African American), 30 international. 47 applicants, 60% accepted. In 1998, 10 master's, 1 doctorate awarded. *Degree requirements:* For doctorate, dissertation required. *Entrance requirements:* For master's and doctorate, GRE General Test, TOEFL. Application fee: $35. *Financial aid:* Fellowships, research assistantships, teaching assistantships, career-related internships or fieldwork, institutionally-sponsored loans, and unspecified assistantships available. Financial aid applicants required to submit FAFSA. *Unit head:* Dr. D. V. Rippy, Director, 864-656-3176, Fax: 864-656-5973, E-mail: rippyd@clemson.edu.

Cornell University, Graduate School, Graduate Fields of Human Ecology, Field of Textiles, Ithaca, NY 14853-0001. Offers apparel design (MA, MPS); fiber science (MS, PhD); polymer science (MS, PhD); textile science (MS, PhD). *Faculty:* 14 full-time. *Students:* 15 full-time (6 women); includes 1 minority (Asian American or Pacific Islander), 10 international. 40 applicants, 35% accepted. In 1998, 2 master's, 4 doctorates awarded. *Degree requirements:* For master's, thesis (MA, MS), project paper (MPS) required; for doctorate, dissertation required, foreign language not required. *Entrance requirements:* For master's, GRE General Test (minimum combined score of 1800 on three sections required), TOEFL (minimum score of 600 required), portfolio (functional apparel design); for doctorate, GRE General Test (minimum combined score of 1800 on three sections required), TOEFL (minimum score of 600 required). *Application deadline:* For fall admission, 3/1. Application fee: $65. Electronic applications accepted. *Financial aid:* In 1998–99, 15 students received aid, including 4 fellowships with full tuition reimbursements available, 5 research assistantships with full tuition reimbursements available, 7 teaching assistantships with full tuition reimbursements available; institutionally-sponsored loans, scholarships, tuition waivers (full and partial), and unspecified assistantships also available. Financial aid applicants required to submit FAFSA. *Faculty research:* High performance fibers and composites, surface chemistry of fibers, geosynthetics, mechanics of fibrous assemblies, detergency, protective clothing, apparel sizing, mass customization, technology and management in apparel industry, international apparel production, historic costumes. *Unit head:* Director of Graduate Studies, 607-255-3151, Fax: 607-255-1093. *Application contact:* Graduate Field Assistant, 607-255-3151, E-mail: textiles_grad@cornell.edu.

Georgia Institute of Technology, Graduate Studies and Research, College of Engineering, School of Textile Engineering, Atlanta, GA 30332-0001. Offers biomedical engineering (MS Bio E); polymers (MS Poly); textile chemistry (MS, MST Ch); textile engineering (MS, MSTE, PhD); textiles (MS, MS Text). *Degree requirements:* For master's, thesis required (for some programs), foreign language not required; for doctorate, dissertation required, foreign language not required. *Entrance requirements:* For master's and doctorate, TOEFL (minimum score of 550 required), minimum GPA of 2.7. *Faculty research:* Energy conservation, environmental control, engineered fibrous structures, polymer synthesis and degradation, high performance organic-carbon-ceramic fibers.

Institute of Textile Technology, Program in Textile Technology, Charlottesville, VA 22903-4614. Offers MS. *Faculty:* 21 full-time (1 woman), 6 part-time (1 woman). *Students:* 27 full-time (11 women). Average age 25. 29 applicants, 52% accepted. In 1998, 12 degrees awarded (100% found work related to degree). *Degree requirements:* For master's, thesis, comprehensive exam required, foreign language not required. *Average time to degree:* Master's–2 years full-time. *Application deadline:* For fall admission, 3/15 (priority date). Application fee: $0. Electronic applications accepted. *Financial aid:* In 1998–99, fellowships (averaging $20,000 per year); career-related internships or fieldwork and tuition waivers (full) also available. Financial aid application deadline: 2/15; financial aid applicants required to submit FAFSA. *Faculty research:* Computer-aided manufacturing, raw materials, manufacturing process innovation. *Unit head:* Dr. William M. Fornadel, Vice President of Academic Affairs and Dean, 804-296-5511, Fax: 804-296-2957, E-mail: billf@itt.edu. *Application contact:* Tracey Templeton, Registrar, 804-296-5511 Ext. 275, Fax: 804-296-2957, E-mail: traceyt@itt.edu.

North Carolina State University, Graduate School, College of Textiles, Raleigh, NC 27695. Offers fiber and polymer science (PhD); textile chemistry (MS, MT); textile engineering (MS);

textile technology management (PhD); textiles (MS, MT, MTE). Part-time and evening/weekend programs available. Postbaccalaureate distance learning degree programs offered. *Faculty:* 37 full-time (3 women), 49 part-time (4 women). *Students:* 104 full-time (49 women), 28 part-time (11 women); includes 11 minority (5 African Americans, 6 Asian Americans or Pacific Islanders), 61 international. Average age 29. 76 applicants, 47% accepted. In 1998, 20 master's, 15 doctorates awarded. *Degree requirements:* For master's, thesis required; for doctorate, one foreign language, dissertation, cumulative exams required. *Entrance requirements:* For master's and doctorate, minimum B average. *Application deadline:* For fall admission, 6/25. Application fee: $45. *Financial aid:* In 1998–99, 1 fellowship, 82 research assistantships (averaging $5,537 per year) were awarded.; teaching assistantships, career-related internships or fieldwork, Federal Work-Study, and institutionally-sponsored loans also available. Aid available to part-time students. Financial aid application deadline: 6/1. *Faculty research:* Polymer modification, characterization, and extrusion; chemistry of textile wet processing; fiber science and composites. Total annual research expenditures: $9.8 million. *Unit head:* Dr. Robert A. Barnhardt, Dean, 919-515-6500, Fax: 919-515-6621, E-mail: robert_barnhardt@ncsu.edu. *Application contact:* Dawn Silsbee, Administrative Assistant, 919-515-6640, Fax: 919-515-6621.

Philadelphia University, School of Textiles and Materials Science, Program in Fibrous Materials Science, Philadelphia, PA 19144-5497. Offers MS. Part-time and evening/weekend programs available. *Degree requirements:* For master's, foreign language and thesis not required. *Entrance requirements:* For master's, GMAT, GRE, or MAT, minimum GPA of 2.75. *Application deadline:* Applications are processed on a rolling basis. Application fee: $35. *Financial aid:* Research assistantships, career-related internships or fieldwork, Federal Work-Study, and unspecified assistantships available. Financial aid applicants required to submit FAFSA. *Unit head:* Dr. Christopher Pastore, Director, 215-951-2683, Fax: 215-951-2651, E-mail: pastorec@philau.edu. *Application contact:* William H. Firman, Director of Graduate Admissions, 215-951-2943, Fax: 215-951-2907, E-mail: gradadm@philau.edu.

Philadelphia University, School of Textiles and Materials Science, Program in Textile Engineering, Philadelphia, PA 19144-5497. Offers MS. Part-time and evening/weekend programs available. *Degree requirements:* For master's, foreign language and thesis not required. *Entrance requirements:* For master's, GMAT, GRE, or MAT, minimum GPA of 2.75. *Application deadline:* Applications are processed on a rolling basis. Application fee: $35. *Financial aid:* Research assistantships, career-related internships or fieldwork, Federal Work-Study, and unspecified assistantships available. Financial aid applicants required to submit FAFSA. *Unit head:* Dr. Muthu Govindaraj, Director, 215-951-2684, Fax: 215-951-2651, E-mail: govindarajm@philau.edu. *Application contact:* William H. Firman, Director of Graduate Admissions, 215-951-2943, Fax: 215-951-2907, E-mail: gradadm@philau.edu.

University of Massachusetts Dartmouth, Graduate School, College of Engineering, Department of Textile Sciences, North Dartmouth, MA 02747-2300. Offers textile chemistry (MS); textile technology (MS). Part-time programs available. *Faculty:* 7 full-time (0 women). *Students:* 7 full-time (4 women), 2 part-time, (all international). Average age 28. 20 applicants, 85% accepted. In 1998, 4 degrees awarded. *Degree requirements:* For master's, thesis required, foreign language not required. *Entrance requirements:* For master's, GRE General Test, TOEFL. *Application deadline:* For fall admission, 4/20 (priority date); for spring admission, 11/15 (priority date). Applications are processed on a rolling basis. Application fee: $40 for international students. Tuition, area resident: Full-time $3,107; part-time $129 per credit. Tuition, state resident: full-time $2,071; part-time $86 per credit. Tuition, nonresident: full-time $7,845; part-time $327 per credit. Required fees: $2,888. Full-time tuition and fees vary according to program and reciprocity agreements. Part-time tuition and fees vary according to course load and reciprocity agreements. *Financial aid:* In 1998–99, 5 research assistantships with full tuition reimbursements (averaging $8,934 per year), 2 teaching assistantships with full tuition reimbursements (averaging $8,500 per year) were awarded.; Federal Work-Study and unspecified assistantships also available. Aid available to part-time students. Financial aid application deadline: 3/15; financial aid applicants required to submit FAFSA. *Faculty research:* Yarn manufacturing, microscopy, chemical treatment and light sensitivity of fabrics, colorimetry. *Unit head:* Dr. Kenneth D. Langley, Director, 508-999-8199, Fax: 508-999-9139, E-mail: klangley@umassd.edu. *Application contact:* Carol A. Novo, Graduate Admissions Office, 508-999-8026, Fax: 508-999-8183, E-mail: graduate@umassd.edu.

Section 20
Telecommunications

This section contains a directory of institutions offering graduate work in telecommunications, followed by in-depth entries submitted by institutions that chose to prepare detailed program descriptions. Additional information about programs listed in the directory but not augmented by an in-depth entry may be obtained by writing directly to the dean of a graduate school or chair of a department at the address given in the directory.

For programs offering related work, see also in this book Computer and Information Sciences and Engineering and Applied Sciences. In Book 2, see Communication and Media and in Book 6, Business Administration and Management.

CONTENTS

Program Directories

Announcement

Cross-Discipline Announcements

In-Depth Descriptions

Telecommunications

Azusa Pacific University, Graduate Studies, College of Liberal Arts and Sciences, Department of Computer Science, Azusa, CA 91702-7000. Offers applied computer science and technology (MS), including client/server technology, computer information systems, end-user support, inter-emphasis, software engineering, technical programming, telecommunications; client/server technology (Certificate); computer information systems (Certificate); computer science (Certificate); end-user training and support (Certificate); software engineering (MSE, Certificate); technical programming (Certificate); telecommunications (Certificate). Part-time and evening/weekend programs available. *Faculty:* 9 full-time (1 woman), 10 part-time (2 women). *Students:* 187. *Degree requirements:* For master's, computer language, thesis or alternative, project required, foreign language not required. *Entrance requirements:* For master's, minimum GPA of 3.0; proficiency in one programming language, college-level algebra, and applied calculus. *Application deadline:* For fall admission, 9/1 (priority date). Applications are processed on a rolling basis. Application fee: $45 ($65 for international students). *Unit head:* Dr. Samuel E. Sambasivam, Acting Chairman, 626-815-5476, Fax: 626-815-5323. *Application contact:* Dr. Samuel E. Sambasivam, Acting Chairman, 626-815-5476, Fax: 626-815-5323.

Boston University, Metropolitan College, Program in Computer Science, Boston, MA 02215. Offers computer information systems (MS); computer science (MS); telecommunications (MS). Part-time and evening/weekend programs available. *Faculty:* 8 full-time (2 women), 64 part-time. *Students:* 11 full-time (0 women), 406 part-time (84 women); includes 73 minority (11 African Americans, 59 Asian Americans or Pacific Islanders, 3 Hispanic Americans), 21 international. *Degree requirements:* For master's, computer language required, foreign language and thesis not required. *Application deadline:* Applications are processed on a rolling basis. Application fee: $50. Tuition: Part-time $508 per credit. Required fees: $40 per semester. Part-time tuition and fees vary according to class time. *Unit head:* Dr. Tanya Zlateva, Chairman, 617-353-2566. *Application contact:* Administrative Secretary, 617-353-2566, Fax: 617-353-2367, E-mail: csinfo@bu.edu.

Columbia University, Fu Foundation School of Engineering and Applied Science, Department of Electrical Engineering, New York, NY 10027. Offers electrical engineering (MS, Eng Sc D, PhD, EE); solid state science and engineering (MS, Eng Sc D, PhD); telecommunications (MS). PhD offered through the Graduate School of Arts and Sciences. Part-time programs available. Postbaccalaureate distance learning degree programs offered (no on-campus study). *Faculty:* 17 full-time (0 women), 13 part-time (0 women). *Students:* 141 full-time (17 women), 73 part-time (4 women). Terminal master's awarded for partial completion of doctoral program. *Degree requirements:* For master's, foreign language and thesis not required; for doctorate, dissertation, qualifying exam required, foreign language not required. *Entrance requirements:* For master's and doctorate, GRE General Test, TOEFL. *Application deadline:* For fall admission, 1/5; for spring admission, 10/1. Application fee: $55. *Unit head:* Chairman, 212-854-5019, Fax: 212-932-9421. *Application contact:* Marlene Mansfield, Student Coordinator, 212-854-3104, Fax: 212-932-9421, E-mail: mansfld@ee.columbia.edu.

DePaul University, School of Computer Science, Telecommunications, and Information Systems, Program in Telecommunications Systems, Chicago, IL 60604-2287. Offers MS. Part-time and evening/weekend programs available. *Students:* 127 full-time (31 women), 136 part-time (24 women); includes 126 minority (38 African Americans, 65 Asian Americans or Pacific Islanders, 23 Hispanic Americans), 13 international. Average age 32. 110 applicants, 76% accepted. In 1998, 47 degrees awarded. *Degree requirements:* For master's, computer language, thesis, comprehensive exam required, foreign language not required. *Entrance requirements:* For master's, passing grade on the department's Graduate Assessment Examination. *Application deadline:* For fall admission, 8/1 (priority date); for winter admission, 11/15 (priority date); for spring admission, 5/1 (priority date). Applications are processed on a rolling basis. Application fee: $25. *Financial aid:* Fellowships, Federal Work-Study and tuition waivers (partial) available. Financial aid application deadline: 4/1. *Unit head:* Dr. Greg Brewster, Director, 312-362-8381, Fax: 312-362-6116. *Application contact:* Anne B. Morley, Director of Student Services, 312-362-8714, Fax: 312-362-6116.

See in-depth description on page 1509.

Drexel University, Graduate School, College of Engineering, Department of Electrical and Computer Engineering, Program in Telecommunications Engineering, Philadelphia, PA 19104-2875. Offers MSEE. *Students:* 7 full-time (1 woman), 18 part-time (2 women); includes 2 minority (1 African American, 1 Hispanic American), 6 international. Average age 38. 42 applicants, 62% accepted. *Degree requirements:* Foreign language not required. *Entrance requirements:* For master's, TOEFL (minimum score of 570 required), minimum GPA of 3.0, BS in electrical engineering or physics. *Application deadline:* For fall admission, 8/21. Applications are processed on a rolling basis. Application fee: $35. Tuition: Full-time $15,795; part-time $585 per credit. Required fees: $375; $67 per term. Tuition and fees vary according to program. *Financial aid:* Research assistantships, teaching assistantships, unspecified assistantships available. Financial aid application deadline: 2/1. *Unit head:* Anne B. Morley, Director of Student Services, 312-362-8714, Fax: 312-362-6116. *Application contact:* Kelli Kennedy, Director of Admissions, 215-895-6706, Fax: 215-895-5939, E-mail: crowlka@duvm.ocs.drexel.edu.

Announcement: Fueled by the rapid spread of technologies such as e-mail, cellular and mobile phone systems, and the information superhighway, this program responds to the tremendous and growing demand for engineers with telecommunications expertise. Drexel's program in telecommunications engineering is interdisciplinary in its approach and combines a strong foundation in telecommunications engineering with training in other important issues such as global communications policy, business aspects of telecommunications, and information transfer and processing. For further information, see in-depth description in Electrical and Computer Engineering section.

George Mason University, College of Arts and Sciences, Program in Telecommunications, Fairfax, VA 22030-4444. Offers MA. Part-time and evening/weekend programs available. *Faculty:* 1 full-time (0 women). *Students:* 9 full-time (5 women), 72 part-time (30 women); includes 15 minority (8 African Americans, 2 Asian Americans or Pacific Islanders, 5 Hispanic Americans), 10 international. Average age 35. 68 applicants, 78% accepted. In 1998, 26 degrees awarded. *Degree requirements:* For master's, thesis optional. *Entrance requirements:* For master's, minimum GPA of 3.0 in last 60 hours. *Application deadline:* For fall admission, 5/1; for spring admission, 11/1. Application fee: $30. Electronic applications accepted. Tuition, state resident: full-time $4,416; part-time $184 per credit hour. Tuition, nonresident: full-time $12,516; part-time $522 per credit hour. Tuition and fees vary according to program. *Financial aid:* Career-related internships or fieldwork available. Financial aid application deadline: 3/1; financial aid applicants required to submit FAFSA. *Unit head:* Dr. Cynthia M. Lont, Director, 703-993-1100, Fax: 703-993-1096, E-mail: clont@gmu.edu. *Application contact:* Linda Atwell, Adviser, 703-993-1100, Fax: 703-993-1096, E-mail: latwell@gmu.edu.

The George Washington University, Columbian School of Arts and Sciences, Program in Telecommunication Studies, Washington, DC 20052. Offers MA. *Faculty:* 1 (woman) full-time, 1 part-time (0 women). *Students:* 7 full-time (2 women), 30 part-time (10 women); includes 9 minority (5 African Americans, 4 Asian Americans or Pacific Islanders), 6 international. Average age 36. 33 applicants, 88% accepted. In 1998, 22 degrees awarded. *Degree requirements:* For master's, thesis or alternative, comprehensive exam required, foreign language not required. *Entrance requirements:* For master's, GRE General Test, minimum GPA of 3.0. *Application deadline:* For fall admission, 5/1. Application fee: $55. Tuition: Full-time $17,328; part-time $722 per credit hour. Required fees: $828; $35 per credit hour. Tuition and fees vary according to campus/location and program. *Financial aid:* Federal Work-Study and institutionally-sponsored loans available. Financial aid application deadline: 2/1. *Unit head:* Dr. Gerald Brock, Director, 202-994-6351.

Illinois Institute of Technology, Graduate College, Armour College of Engineering and Sciences, Department of Computer Science, Chicago, IL 60616-3793. Offers computer science (MS, PhD); teaching (MST); telecommunications and software engineering (MTSE). Part-time and evening/weekend programs available. *Faculty:* 16 full-time (3 women), 15 part-time (3 women). *Students:* 161 full-time (34 women), 574 part-time (112 women); includes 230 minority (34 African Americans, 185 Asian Americans or Pacific Islanders, 11 Hispanic Americans), 239 international. Terminal master's awarded for partial completion of doctoral program. *Degree requirements:* For master's, computer language, thesis (for some programs), comprehensive exam required, foreign language not required; for doctorate, computer language, dissertation, comprehensive exam required, foreign language not required. *Entrance requirements:* For master's and doctorate, GRE (minimum score of 1200 required), TOEFL (minimum score of 550 required), undergraduate GPA of 3.0 required. *Application deadline:* For fall admission, 7/1; for spring admission, 11/1. Applications are processed on a rolling basis. Application fee: $30. Electronic applications accepted. *Unit head:* Dr. Bodgen Korel, Interim Chairman, 312-567-5150, Fax: 312-567-5067, E-mail: korel@charlie.cns.iit.edu. *Application contact:* Dr. S. Mohammad Shahidehpour, Dean of Graduate College, 312-567-3024, Fax: 312-567-7517, E-mail: grad@minna.cns.iit.edu.

Iona College, School of Arts and Science, Program in Telecommunications, New Rochelle, NY 10801-1890. Offers MS, Certificate. Part-time and evening/weekend programs available. *Faculty:* 4 full-time (0 women), 1 part-time (0 women). Average age 29. In 1998, 32 master's awarded. *Entrance requirements:* For master's, minimum GPA of 3.0. *Application deadline:* Applications are processed on a rolling basis. Application fee: $25. *Financial aid:* Unspecified assistantships available. *Unit head:* Dr. John Mallozzi, Chair, 914-633-2578. *Application contact:* Arlene Melillo, Director of Graduate Recruitment, 914-633-2328, Fax: 914-633-2023.

Michigan State University, Graduate School, College of Communication Arts and Sciences, Department of Telecommunication, East Lansing, MI 48824-1020. Offers telecommunication (MA); telecommunication-urban studies (MA). *Faculty:* 11 full-time (3 women). *Students:* 22 full-time (10 women), 58 part-time (23 women); includes 5 minority (2 African Americans, 2 Asian Americans or Pacific Islanders, 1 Hispanic American), 51 international. Average age 28. 104 applicants, 53% accepted. In 1998, 42 degrees awarded. *Degree requirements:* For master's, thesis required (for some programs), foreign language not required. *Entrance requirements:* For master's, GRE, TOEFL, minimum GPA of 3.25. *Application deadline:* Applications are processed on a rolling basis. Application fee: $30 ($40 for international students). *Financial aid:* In 1998–99, 2 research assistantships with tuition reimbursements (averaging $10,188 per year), 25 teaching assistantships with tuition reimbursements (averaging $9,944 per year) were awarded.; fellowships, career-related internships or fieldwork and institutionally-sponsored loans also available. Aid available to part-time students. Financial aid applicants required to submit FAFSA. *Faculty research:* New telecommunication technologies, cable, international communication, children and television, minority ownerships and programs. Total annual research expenditures: $85,000. *Unit head:* Dr. Thomas Muth, Interim Chair, 517-355-8372, Fax: 517-355-1292. *Application contact:* Dr. Charles Steinfield, Director of Graduate Studies, 517-355-4451.

New Jersey Institute of Technology, Office of Graduate Studies, Department of Electrical and Computer Engineering, Newark, NJ 07102-1982. Offers computer engineering (MS); electrical engineering (MS, PhD, Engineer), including biomedical systems (MS, PhD), communication and signal processing (MS, PhD), computer systems (MS, PhD), control systems (MS, PhD), microwave and lightwave engineering (MS, PhD), solid-state materials and devices (MS, PhD); power engineering (MS); telecommunications (MS). Part-time and evening/weekend programs available. Terminal master's awarded for partial completion of doctoral program. *Degree requirements:* For master's, thesis required (for some programs), foreign language not required; for doctorate, dissertation, residency required, foreign language not required. *Entrance requirements:* For master's, GRE General Test (minimum score of 450 on verbal section, 600 on quantitative, 550 on analytical required); for doctorate, GRE General Test (minimum score of 450 on verbal section, 600 on quantitative, 550 on analytical required), minimum graduate GPA of 3.5. Electronic applications accepted. *Faculty research:* Communications systems design, digital signal processing.

North Carolina State University, Graduate School, College of Management, Program in Management, Raleigh, NC 27695. Offers biotechnology (MS); computer science (MS); engineering (MS); forest resources management (MS); general business (MS); management information systems (MS); operations research (MS); statistics (MS); telecommunications systems engineering (MS); textile management (MS); total quality management (MS). Part-time programs available. *Faculty:* 40 full-time (9 women), 4 part-time (0 women). *Students:* 48 full-time (15 women), 156 part-time (43 women); includes 33 minority (16 African Americans, 15 Asian Americans or Pacific Islanders, 1 Hispanic American, 1 Native American), 4 international. *Degree requirements:* For master's, computer language required, foreign language and thesis not required. *Entrance requirements:* For master's, GRE or GMAT, TOEFL (minimum score of 550 required), minimum undergraduate GPA of 3.0. *Application deadline:* For fall admission, 6/25; for spring admission, 11/25. Applications are processed on a rolling basis. Application fee: $45. *Unit head:* Dr. Jack W. Wilson, Director of Graduate Programs, 919-515-4327, Fax: 919-515-6943, E-mail: jack_wilson@ncsu.edu. *Application contact:* Dr. Steven G. Allen, Director of Graduate Programs, 919-515-6941, Fax: 919-515-5073, E-mail: steve_allen@ncsu.edu.

Northwestern University, The Graduate School, Program in Telecommunications Science, Management, and Policy, Evanston, IL 60208. Offers MA, MEM, MS, Certificate. MS awarded through the Department of Electrical Engineering and Computer Science; MA awarded through the Departments of Comunication Studies and Radio/Television/Film; MEM awarded through the Department of Industrial Engineering and Management Sciences. Part-time programs available. *Faculty:* 22 full-time (2 women), 1 (woman) part-time. *Students:* 19 full-time (8 women); includes 7 minority (all Asian Americans or Pacific Islanders), 1 international. Average age 26. In 1998, 7 degrees awarded (14% entered university research/teaching, 86% found other work related to degree). *Entrance requirements:* For master's, GRE General Test, TOEFL. *Average time to degree:* Master's–2 years full-time. *Application deadline:* For fall admission, 1/15 (priority date). Applications are processed on a rolling basis. Application fee: $50 ($55 for international students). *Financial aid:* Fellowships, research assistantships, teaching assistantships, career-related internships or fieldwork, Federal Work-Study, and institutionally-sponsored loans available. Financial aid application deadline: 1/15; financial aid applicants required to submit FAFSA. *Faculty research:* Electronic media management. *Unit head:* Dr. Steven S. Wildman, Director, 847-491-3539, Fax: 847-467-1171, E-mail: s-wildman@nwu.edu. *Application contact:* Theomary Karamanis, Program Assistant, 847-491-3539, Fax: 847-467-1171, E-mail: telecomprog@nwu.edu.

See in-depth description on page 1511.

Pace University, School of Computer Science and Information Systems, New York, NY 10038. Offers computer communications and networks (Certificate); computer science (MS); computing studies (DPS); information systems (MS); New 9/1/1999 (DPS); object-oriented programming (Certificate); telecommunications (MS, Certificate). Part-time and evening/weekend programs available. *Faculty:* 36 full-time, 44 part-time. *Students:* 93 full-time (35 women), 468 part-time (170 women); includes 101 minority (40 African Americans, 41 Asian Americans or Pacific Islanders, 20 Hispanic Americans), 80 international. *Degree requirements:* For master's, computer language required, foreign language and thesis not required. *Entrance requirements:* For master's, GRE General Test. *Application deadline:* For fall admission, 7/31 (priority date); for spring admission, 11/30. Applications are processed on a rolling basis. Application fee: $60. Electronic applications accepted. *Unit head:* Dr. Susan Merritt, Dean,

914-422-4375. *Application contact:* Lois Rich, Associate Director, 914-422-4283, Fax: 914-422-4287, E-mail: gradwp@ny2.pace.edu.

Pennsylvania State University University Park Campus, Graduate School, College of Communications, Program in Telecommunications Studies, State College, University Park, PA 16802-1503. Offers MA. *Students:* 6 full-time (4 women), 1 part-time. *Entrance requirements:* For master's, GRE General Test. *Application deadline:* For fall admission, 3/30. Application fee: $50. *Unit head:* Dr. Richard Barton, Director of Graduate Studies, 814-865-3070.

Polytechnic University, Brooklyn Campus, Department of Electrical Engineering, Major in Telecommunication Networks, Brooklyn, NY 11201-2990. Offers MS. Part-time and evening/weekend programs available. *Students:* 10 full-time (0 women), 41 part-time (7 women); includes 13 minority (1 African American, 11 Asian Americans or Pacific Islanders, 1 Hispanic American), 8 international. Average age 33. 51 applicants, 45% accepted. In 1998, 14 degrees awarded. *Degree requirements:* For master's, thesis optional. *Entrance requirements:* For master's, BS in electrical engineering. *Application deadline:* Applications are processed on a rolling basis. Application fee: $45. Electronic applications accepted. *Financial aid:* Fellowships, research assistantships, teaching assistantships, institutionally-sponsored loans available. Aid available to part-time students. Financial aid applicants required to submit FAFSA. *Application contact:* John S. Kerge, Dean of Admissions, 718-260-3200, Fax: 718-260-3446, E-mail: admitme@poly.edu.

Polytechnic University, Farmingdale Campus, Graduate Programs, Department of Electrical Engineering, Major in Telecommunication Networks, Farmingdale, NY 11735-3995. Offers MS. Average age 33. 6 applicants, 67% accepted. *Degree requirements:* For master's, computer language, thesis required (for some programs). *Application deadline:* Applications are processed on a rolling basis. Application fee: $45. Electronic applications accepted. *Financial aid:* Institutionally-sponsored loans available. Aid available to part-time students. Financial aid applicants required to submit FAFSA. *Unit head:* John S. Kerge, Dean of Admissions, 718-260-3200, Fax: 718-260-3446, E-mail: admitme@poly.edu. *Application contact:* John S. Kerge, Dean of Admissions, 718-260-3200, Fax: 718-260-3446, E-mail: admitme@poly.edu.

Polytechnic University, Westchester Graduate Center, Graduate Programs, Department of Electrical Engineering, Major in Telecommunication Networks, Hawthorne, NY 10532-1507. Offers MS. *Students:* 1 full-time (0 women), 6 part-time (1 woman); includes 1 minority (Asian American or Pacific Islander), 1 international. Average age 33. 5 applicants, 40% accepted. In 1998, 2 degrees awarded. *Degree requirements:* For master's, computer language, thesis required (for some programs). *Application deadline:* Applications are processed on a rolling basis. Application fee: $45. *Application contact:* John S. Kerge, Dean of Admissions, 718-260-3200, Fax: 718-260-3446, E-mail: admitme@poly.edu.

Rochester Institute of Technology, Part-time and Graduate Admissions, College of Applied Science and Technology, Department of Computer Science and Information Technology, Program in Telecommunications Software Technology, Rochester, NY 14623-5604. Offers MS. 1 applicants, 0% accepted. In 1998, 2 degrees awarded. *Degree requirements:* For master's, computer language, thesis required. *Entrance requirements:* For master's, minimum GPA of 3.0. *Application deadline:* For fall admission, 3/1 (priority date). Applications are processed on a rolling basis. Application fee: $40. *Financial aid:* Unspecified assistantships available. *Unit head:* Dr. Rayno Niemi, Graduate Coordinator, 716-475-2202, E-mail: rdn@it.rit.edu.

Roosevelt University, Graduate Division, College of Arts and Sciences, School of Computer Science and Telecommunications, Program in Telecommunications, Chicago, IL 60605-1394. Offers MST. Part-time and evening/weekend programs available. *Degree requirements:* For master's, computer language required, foreign language and thesis not required. *Entrance requirements:* For master's, GRE. *Application deadline:* For fall admission, 6/1 (priority date). Applications are processed on a rolling basis. Application fee: $25 ($35 for international students). *Financial aid:* Application deadline: 2/15. *Faculty research:* Coding theory, mathematical models, network design, simulation models. Total annual research expenditures: $40,000. *Unit head:* Director of Graduate Admissions, 609-258-3034. *Application contact:* Joanne Canyon-Heller, Coordinator of Graduate Admissions, 312-341-3612, Fax: 312-341-3523, E-mail: applyru@roosevelt.edu.

Saint Mary's University of Minnesota, Graduate School, Program in Telecommunications, Winona, MN 55987-1399. Offers MS, MS/MA. MS/MA offered jointly with the Program in Management. Part-time and evening/weekend programs available. *Degree requirements:* For master's, thesis, colloquium required, foreign language not required. *Entrance requirements:* For master's, interview, minimum GPA of 2.75.

Southern Methodist University, School of Engineering and Applied Science, Department of Electrical Engineering, Dallas, TX 75275. Offers electrical engineering (MSEE, PhD); telecommunications (MS). Postbaccalaureate distance learning degree programs offered (no on-campus study). *Faculty:* 20 full-time (1 woman), 6 part-time (0 women). *Students:* 41 full-time (9 women), 355 part-time (55 women); includes 114 minority (19 African Americans, 73 Asian Americans or Pacific Islanders, 21 Hispanic Americans, 1 Native American), 63 international. *Degree requirements:* For master's, thesis optional, foreign language not required; for doctorate, dissertation, oral and written qualifying exams, oral final exam required. *Entrance requirements:* For master's, GRE General Test (minimum score of 650 on quantitative section required), TOEFL (minimum score of 550 required), minimum GPA of 3.0 in last 2 years; bachelor's degree in engineering, mathematics, or sciences; for doctorate, preliminary counseling exam, minimum GPA of 3.0, bachelor's degree in related field. *Application deadline:* For fall admission, 8/1 (priority date); for spring admission, 12/15. Applications are processed on a rolling basis. Application fee: $25. Tuition: Full-time $9,216; part-time $512 per credit hour. Required fees: $88 per credit hour. Part-time tuition and fees vary according to course load and campus/location. *Unit head:* Dr. Jerry D. Gibson, Chair, 214-768-3113, Fax: 214-768-3573, E-mail: gibson@seas.smu.edu. *Application contact:* Dr. Zeynep Celik-Butler, Assistant Dean for Graduate Studies and Research, 214-768-3979, Fax: 214-768-3845, E-mail: zcb@seas.smu.edu.

State University of New York Institute of Technology at Utica/Rome, School of Information Systems and Engineering Technology, Program in Telecommunications, Utica, NY 13504-3050. Offers MS. Part-time programs available. *Faculty:* 6 full-time (1 woman). *Students:* 6 full-time (1 woman), 20 part-time (4 women); includes 3 minority (2 African Americans, 1 Hispanic American) Average age 34. 15 applicants, 93% accepted. In 1998, 1 degree awarded (100% found work related to degree). *Degree requirements:* For master's, computer language, thesis required, foreign language not required. *Entrance requirements:* For master's, GRE General Test, TOEFL (minimum score of 550 required), minimum GPA of 3.0. *Average time to degree:* Master's–2 years full-time, 4 years part-time. *Application deadline:* For fall admission, 6/15 (priority date). Applications are processed on a rolling basis. Application fee: $50. *Financial aid:* In 1998–99, 19 students received aid, including 1 fellowship with full tuition reimbursement available (averaging $4,900 per year), 2 research assistantships with full tuition reimbursements available; Federal Work-Study also available. Aid available to part-time students. Financial aid applicants required to submit FAFSA. *Faculty research:* Wireless telecommunications, network design and simulation, broadband multimedia, telecommunications trade policy. *Unit head:* Dr. Eugene Newman, Chair, 315-792-7234, Fax: 315-792-7800, E-mail: fejn@sunyit.edu. *Application contact:* Marybeth Lyons, Director of Admissions, 315-792-7500, Fax: 315-792-7837, E-mail: smbl@sunyit.edu.

Syracuse University, Graduate School, School of Information Studies, Program in Telecommunications and Network Management, Syracuse, NY 13244-0003. Offers MS. Part-time and evening/weekend programs available. *Faculty:* 21. *Students:* 30 full-time (11 women), 67 part-time (21 women); includes 11 minority (5 African Americans, 3 Asian Americans or Pacific Islanders, 2 Hispanic Americans, 1 Native American), 50 international. Average age 31. 100 applicants, 81% accepted. In 1998, 27 degrees awarded. *Degree requirements:* For master's, internship or research project required, foreign language and thesis not required. *Entrance requirements:* For master's, GRE General Test (minimum combined score of 1000 required). *Application deadline:* Applications are processed on a rolling basis. Application fee: $40. Tuition: Full-time $13,992; part-time $583 per credit hour. *Financial aid:* Fellowships, research assistantships, teaching assistantships, career-related internships or fieldwork, Federal Work-Study, and tuition waivers (partial) available. Financial aid application deadline: 3/1. *Faculty research:* Multimedia, information resources management. *Unit head:* Tom Martin, Interim Director, 315-443-2911. *Application contact:* Barbara Settel, Assistant Dean, 315-443-2911.

See in-depth description on page 1513.

Université du Québec, Institut national de la recherche scientifique, Graduate Programs, Research Center—Telecommunications, Ste-Foy, PQ G1V 4C7, Canada. Offers information technology (Diploma); software engineering (M Sc); telecommunications (M Sc, PhD). Part-time programs available. *Degree requirements:* For master's and doctorate, thesis/dissertation required; for Diploma, thesis not required. *Entrance requirements:* For master's and Diploma, appropriate bachelor's degree, proficiency in French; for doctorate, appropriate master's degree, proficiency in French. *Faculty research:* Visual and verbal systems, communications networks.

University of Alberta, Faculty of Graduate Studies and Research, Department of Electrical and Computer Engineering, Edmonton, AB T6G 2E1, Canada. Offers computational optics (PhD); computer engineering (M Eng, M Sc, PhD); control systems (M Eng, M Sc, PhD); engineering management (M Eng); laser physics (M Sc, PhD); oil sands (M Eng, M Sc, PhD); plasma physics (M Sc, PhD); power engineering (M Eng, M Sc, PhD); telecommunications (M Eng, M Sc, PhD). Terminal master's awarded for partial completion of doctoral program. *Degree requirements:* For master's and doctorate, thesis/dissertation required, foreign language not required. *Entrance requirements:* For master's and doctorate, TOEFL (minimum score of 580 required; average 610). Electronic applications accepted. *Faculty research:* Controls, communications, microelectronics, electromagnetics.

University of California, San Diego, Graduate Studies and Research, Department of Electrical and Computer Engineering, La Jolla, CA 92093-5003. Offers applied ocean science (MS, PhD); applied physics (MS, PhD); communication theory and systems (MS, PhD); computer engineering (MS, PhD); electrical engineering (M Eng, MS, PhD); electronic circuits and systems (MS, PhD); intelligent systems, robotics and control (MS, PhD); photonics (MS, PhD); signal and image processing (MS, PhD). *Faculty:* 35. *Students:* 251 (24 women). *Entrance requirements:* For master's and doctorate, GRE General Test. Application fee: $40. *Unit head:* William Coles, Chair. *Application contact:* Graduate Coordinator, 619-534-6606.

University of Colorado at Boulder, Graduate School, College of Engineering and Applied Science, Interdisciplinary Telecommunications Program, Boulder, CO 80309. Offers ME, MS, MBA/MS. *Degree requirements:* For master's, comprehensive exam required. *Entrance requirements:* For master's, minimum undergraduate GPA of 3.0.

See in-depth description on page 1515.

University of Denver, University College, Denver, CO 80208. Offers applied communication (MSS); computer information systems (MCIS); environmental policy and management (MEPM); healthcare systems (MHS); liberal studies (MLS); library and information services (MLIS); public health (MPH); technology management (MoTM); telecommunications (MTEL). Part-time and evening/weekend programs available. Postbaccalaureate distance learning degree programs offered (no on-campus study). *Faculty:* 1 (woman) full-time, 553 part-time (181 women). *Students:* 1,529 (804 women); includes 189 minority (67 African Americans, 33 Asian Americans or Pacific Islanders, 75 Hispanic Americans, 14 Native Americans) 61 international. *Entrance requirements:* For master's, minimum undergraduate GPA of 3.0. *Application deadline:* For fall admission, 8/10; for spring admission, 2/22. Applications are processed on a rolling basis. Application fee: $25. *Unit head:* Peter Warren, Dean, 303-871-3268, Fax: 303-871-4047, E-mail: pwarren@du.edu. *Application contact:* Bryan Ehrlich, Admission Coordinator, 303-871-3969, Fax: 303-871-3303, E-mail: behrlich@du.edu.

University of Maryland, College Park, Graduate School, A. James Clark School of Engineering, Department of Electrical and Computer Engineering, Program in Telecommunications, College Park, MD 20742-5045. Offers MS. Part-time and evening/weekend programs available. *Students:* 14 full-time (4 women), 36 part-time (4 women); includes 17 minority (3 African Americans, 11 Asian Americans or Pacific Islanders, 2 Hispanic Americans, 1 Native American), 15 international. 57 applicants, 46% accepted. In 1998, 14 degrees awarded. *Degree requirements:* For master's, thesis or alternative required, foreign language not required. *Entrance requirements:* For master's, minimum GPA of 3.0. *Application deadline:* Applications are processed on a rolling basis. Application fee: $50 ($70 for international students). Tuition, state resident: part-time $272 per credit hour. Tuition, nonresident: part-time $475 per credit hour. Required fees: $632; $379 per year. *Financial aid:* Fellowships, research assistantships, teaching assistantships available. Financial aid applicants required to submit FAFSA. *Unit head:* Dr. Steven A. Tretter, Director, 301-405-3683. *Application contact:* Trudy Lindsey, Director, Graduate Admission and Records, 301-405-4198, Fax: 301-314-9305, E-mail: grschool@deans.umd.edu.

See in-depth description on page 1517.

University of Missouri–Kansas City, Program in Computer Science Telecommunications, Kansas City, MO 64110-2499. Offers computer networking (MS, PhD); software engineering (MS); telecommunications networking (MS, PhD). PhD offered through the School of Graduate Studies. Part-time programs available. *Faculty:* 15 full-time (1 woman). *Students:* 48 full-time (20 women), 74 part-time (21 women); includes 8 minority (2 African Americans, 6 Asian Americans or Pacific Islanders), 85 international. Average age 29. In 1998, 44 degrees awarded. *Degree requirements:* For master's, computer language required, foreign language not required; for doctorate, computer language, dissertation required, foreign language not required. *Entrance requirements:* For master's, GRE General Test (score in 75th percentile or higher on quantitative section required, 50th percentile or higher on verbal), minimum GPA of 3.0; for doctorate, GRE General Test (score in 85th percentile or higher on quantitative section required, 50th percentile or higher on verbal), minimum GPA of 3.5. *Application deadline:* For fall admission, 3/1 (priority date); for spring admission, 10/1. Applications are processed on a rolling basis. Application fee: $25. *Financial aid:* In 1998–99, 15 research assistantships, 15 teaching assistantships were awarded.; career-related internships or fieldwork, Federal Work-Study, institutionally-sponsored loans, and tuition waivers (partial) also available. Aid available to part-time students. Financial aid application deadline: 3/1. *Faculty research:* Multimedia networking, distributed systems/databases, data/network security. *Unit head:* Dr. Richard Hetherington, Director, 816-235-1193, Fax: 816-235-5159, E-mail: info@cstp.umkc.edu.

See in-depth description on page 1519.

University of New Haven, Graduate School, School of Business, Program in Business Administration, West Haven, CT 06516-1916. Offers accounting (MBA); business policy and strategy (MBA); computer and information science (MBA); finance (MBA); health care management (MBA); health care marketing (MBA); hotel and restaurant management (MBA); human resources management (MBA); international business logistics (MBA); management and organization (MBA); management science (MBA); marketing (MBA); operations research (MBA); public relations (MBA); telecommunications (MBA); travel and tourism administration (MBA). Part-time and evening/weekend programs available. *Students:* 65 full-time (18 women), 275 part-time (107 women); includes 26 minority (11 African Americans, 8 Asian Americans or Pacific Islanders, 7 Hispanic Americans), 67 international. *Degree requirements:* For master's, thesis or alternative required, foreign language not required. *Application deadline:* Applications are processed on a rolling basis. Application fee: $50. *Unit head:* Dr. Omid Nodoushani, Coordinator, 203-932-7123.

Telecommunications–Telecommunications Management

University of Pennsylvania, School of Engineering and Applied Science, Telecommunications and Networking Program, Philadelphia, PA 19104. Offers MSE. Part-time programs available. Electronic applications accepted.

See in-depth description on page 1521.

University of Pittsburgh, School of Information Sciences, Department of Information Science and Telecommunications, Program in Telecommunications, Pittsburgh, PA 15260. Offers MST, Certificate. *Faculty:* 6 full-time (1 woman), 2 part-time (0 women). *Students:* 45 full-time (7 women), 45 part-time (10 women); includes 10 minority (5 African Americans, 4 Asian Americans or Pacific Islanders, 1 Hispanic American), 33 international. 68 applicants, 94% accepted. In 1998, 38 degrees awarded. *Degree requirements:* For master's, computer language required, thesis optional, foreign language not required. *Entrance requirements:* For master's, GRE General Test (combined average 1645 on three sections), previous course work in computer programming and calculus. *Average time to degree:* Master's–2 years full-time, 4 years part-time. Application fee: $30 ($40 for international students). *Financial aid:* In 1998–99, 49 students received aid, including 2 fellowships (averaging $34,720 per year), 28 research assistantships (averaging $138,016 per year), 19 teaching assistantships (averaging $169,694 per year); career-related internships or fieldwork, grants, scholarships, and tuition waivers (full and partial) also available. Financial aid application deadline: 1/15. *Faculty research:* Telephony, photonic switching, telecommunications policy, network management, technical standards. Total annual research expenditures: $378,379. *Unit head:* Executive Engineering Program Office, 215-898-2897, Fax: 215-573-9673. *Application contact:* Ninette Kay, Admissions Coordinator, 412-624-5146, Fax: 412-624-5231, E-mail: nk@sis.pitt.edu.

See in-depth description on page 1523.

University of Southwestern Louisiana, Graduate School, College of Engineering, Department of Electrical and Computer Engineering, Program in Telecommunications, Lafayette, LA 70504. Offers MSTC. *Faculty:* 6 full-time (0 women). *Students:* 19 full-time (5 women), 12 part-time (1 woman); includes 3 minority (2 African Americans, 1 Hispanic American), 14 international. 39 applicants, 69% accepted. In 1998, 23 master's awarded. *Degree requirements:* Foreign language not required. *Entrance requirements:* For master's, GRE General Test, minimum GPA of 2.75. *Application deadline:* For fall admission, 5/15. Application fee: $5 ($15 for international students). *Financial aid:* In 1998–99, 2 fellowships with full tuition reimbursements (averaging $14,000 per year), 4 research assistantships with full tuition reimbursements (averaging $4,500 per year) were awarded. Financial aid application deadline: 5/1. *Unit head:* Dr. Kia Makki, Director, 318-482-5872.

The University of Texas at Dallas, Erik Jonsson School of Engineering and Computer Science, Programs in Electrical Engineering, Richardson, TX 75083-0688. Offers electrical engineering (MSEE, PhD); microelectronics (MSEE); telecommunications (MSEE). Part-time and evening/weekend programs available. *Students:* 90 full-time (23 women), 139 part-time (21 women); includes 53 minority (4 African Americans, 40 Asian Americans or Pacific Island-

ers, 9 Hispanic Americans), 94 international. *Degree requirements:* For master's, thesis (for some programs), minimum GPA of 3.0, thesis or major design project required, foreign language not required; for doctorate, dissertation, minimum GPA of 3.5 required, foreign language not required. *Entrance requirements:* For master's, GRE General Test (minimum score of 500 on verbal section, 700 on quantitative, 600 on analytical required), TOEFL (minimum score of 550 required), minimum GPA of 3.0 in related bachelor's degree; for doctorate, GRE General Test (minimum score of 500 on verbal section, 700 on quantitative, 600 on analytical required), TOEFL (minimum score of 550 required), minimum GPA of 3.5. *Application deadline:* For fall admission, 7/15; for spring admission, 11/15. Applications are processed on a rolling basis. Application fee: $25 ($75 for international students). *Unit head:* Dr. William Frensley, Head, 972-883-2412, Fax: 972-883-2710, E-mail: frensley@utdallas.edu. *Application contact:* Cynthia Stewart, Secretary, 972-883-2993, Fax: 972-883-2710, E-mail: ee-grad-info@utdallas.edu.

Western Illinois University, School of Graduate Studies, College of Education and Human Services, Department of Instructional Technology and Telecommunications, Macomb, IL 61455-1390. Offers MS. *Accreditation:* NCATE. *Faculty:* 9 full-time (3 women). *Students:* 13 full-time (3 women), 59 part-time (41 women), 5 international. Average age 40. 8 applicants, 50% accepted. In 1998, 30 degrees awarded. *Degree requirements:* For master's, thesis or alternative required, foreign language not required. *Entrance requirements:* For master's, GRE General Test. *Application deadline:* Applications are processed on a rolling basis. Application fee: $0 ($25 for international students). *Financial aid:* In 1998–99, 7 students received aid, including 7 research assistantships with full tuition reimbursements available (averaging $4,880 per year) Financial aid applicants required to submit FAFSA. *Faculty research:* Distance learning, educational technology. *Unit head:* Dr. Mo Hassan, Chairperson, 309-298-1952. *Application contact:* Barbara Baily, Director of Graduate Studies, 309-298-1806, Fax: 309-298-2245, E-mail: barb_baily@ccmail.wiu.edu.

Widener University, School of Engineering, Program in Electrical/Telecommunication Engineering, Chester, PA 19013-5792. Offers ME, ME/MBA. Part-time and evening/weekend programs available. *Faculty:* 4 part-time (1 woman). *Students:* 2 full-time (0 women), 7 part-time; includes 3 minority (1 African American, 2 Asian Americans or Pacific Islanders), 2 international. 5 applicants, 80% accepted. In 1998, 1 degree awarded. *Degree requirements:* For master's, thesis optional, foreign language not required. *Entrance requirements:* For master's, GMAT (ME/MBA). *Average time to degree:* Master's–2 years full-time, 4 years part-time. *Application deadline:* For fall admission, 8/1 (priority date); for spring admission, 12/1. Applications are processed on a rolling basis. Application fee: $25 ($300 for international students). *Financial aid:* In 1998–99, 1 teaching assistantship with full tuition reimbursement (averaging $7,500 per year) was awarded.; unspecified assistantships also available. Financial aid application deadline: 3/15. *Faculty research:* Signal and image processing, electromagnetics, telecommunications and computer network. *Unit head:* Dr. Alfred T. Johnson, Chairman, Department of Electrical/Telecommunication Engineering, 610-499-4053, Fax: 610-499-4059, E-mail: alfred.t.johnson@widener.edu.

Telecommunications Management

Alaska Pacific University, Graduate Programs, Business Administration Department, Program in Telecommunication Management, Anchorage, AK 99508-4672. Offers MBATM. Part-time and evening/weekend programs available. *Degree requirements:* For master's, thesis optional, foreign language not required. *Entrance requirements:* For master's, GMAT or GRE, minimum GPA of 3.0.

Capitol College, Graduate School, Laurel, MD 20708-9759. Offers electronic commerce management (MS); information and telecommunications systems management (MS); systems management (MS). Part-time and evening/weekend programs available. *Faculty:* 1 full-time (0 women), 24 part-time (4 women). *Students:* 6 full-time (3 women), 145 part-time (39 women); includes 23 minority (9 African Americans, 7 Asian Americans or Pacific Islanders, 6 Hispanic Americans, 1 Native American), 2 international. *Degree requirements:* For master's, foreign language and thesis not required. *Entrance requirements:* For master's, GRE General Test (minimum score of 500 on each section required), minimum GPA of 3.0. *Application deadline:* For fall admission, 7/1 (priority date); for winter admission, 12/1 (priority date); for spring admission, 3/1 (priority date). Applications are processed on a rolling basis. Application fee: $25 ($100 for international students). Electronic applications accepted. Tuition: Full-time $5,328; part-time $888 per course. Tuition and fees vary according to campus/location and program. *Unit head:* Dr. Joe Goldsmith, Dean of Graduate Studies, 301-369-2800, Fax: 301-953-3876. *Application contact:* Sandy Perriello, Coordinator of Graduate Administration, 703-998-5503, Fax: 703-379-8239, E-mail: gradschool@capitol-college.edu.

Carleton University, Faculty of Graduate Studies, Faculty of Engineering and Design, Ottawa-Carleton Institute for Electrical and Computer Engineering, Department of Systems and Computer Engineering, Program in Telecommunications Technology Management, Ottawa, ON K1S 5B6, Canada. Offers M Eng. *Students:* 5 full-time (2 women), 17 part-time (5 women). Average age 32. In 1998, 1 degree awarded. *Degree requirements:* For master's, thesis optional. *Entrance requirements:* For master's, TOEFL (minimum score of 550 required), honors degree. *Average time to degree:* Master's–1.1 years full-time. *Application deadline:* For fall admission, 3/1. Applications are processed on a rolling basis. Application fee: $35. *Financial aid:* Application deadline: 3/1. *Unit head:* Dr. David Peach, Director, Graduate Management Program, 805-756-7187, Fax: 805-756-0110, E-mail: dpeach@calpoly.edu. *Application contact:* Darlene Herbert, Administrator, 613-520-2600 Ext. 1511, Fax: 613-520-5727, E-mail: darlene@sce.carleton.ca.

Golden Gate University, School of Business, San Francisco, CA 94105-2968. Offers information systems (MBA); accounting (M Ac, MBA); business administration (EMBA, DBA); economics (MS); finance (MBA, MS); financial engineering (MS); financial planning (Certificate); human resource management (MS); human resources management (Certificate); international business (MBA); management (MBA); manufacturing management (MS, Certificate); marketing (MBA, MS); operations management (MBA); organizational behavior and development (MS); procurement and logistics management (MS, Certificate); professional export management (Certificate); project and systems management (MS, Certificate); public relations (MS, Certificate); telecommunications (MBA). Part-time and evening/weekend programs available. *Students:* 482 full-time (218 women), 841 part-time (409 women); includes 312 minority (55 African Americans, 193 Asian Americans or Pacific Islanders, 59 Hispanic Americans, 5 Native Americans), 350 international. *Degree requirements:* For master's and Certificate, foreign language and thesis not required; for doctorate, dissertation required, foreign language not required. *Entrance requirements:* For master's, GMAT (MBA), TOEFL (minimum score of 550 required), minimum GPA of 2.5 (MS). *Application deadline:* For fall admission, 7/1 (priority date). Applications are processed on a rolling basis. Application fee: $55 ($70 for international students). *Unit head:* Barbara Karlin, Dean, 415-442-7885, Fax: 415-543-2607. *Application contact:* Enrollment Services, 415-442-7800, Fax: 415-442-7807, E-mail: info@ggu.edu.

Golden Gate University, School of Technology and Industry, San Francisco, CA 94105-2968. Offers hospitality administration and tourism (MS); information systems (MS, Certificate); telecommunications management (MS, Certificate). Part-time and evening/weekend programs available. *Students:* 167 full-time (71 women), 398 part-time (144 women); includes 171 minority (43 African Americans, 105 Asian Americans or Pacific Islanders, 21 Hispanic Americans, 2

Native Americans), 161 international. *Degree requirements:* For master's and Certificate, computer language required, foreign language and thesis not required. *Entrance requirements:* For master's, GMAT (MBA), TOEFL (minimum score of 550 required), minimum GPA of 2.5. *Application deadline:* For fall admission, 7/1 (priority date). Applications are processed on a rolling basis. Application fee: $55 ($70 for international students). *Unit head:* James Koerlin, Dean, 415-442-6540, Fax: 415-442-7049. *Application contact:* Enrollment Services, 415-442-7800, Fax: 415-442-7807, E-mail: info@ggu.edu.

Keller Graduate School of Management, Graduate Program, Oakbrook Terrace, IL 60181. Offers accounting and financial management (MAFM); business administration (MBA); human resources management (MHRM); information systems management (MISM); project management (MPM); telecommunications management (MTM). Part-time and evening/weekend programs available. Postbaccalaureate distance learning degree programs offered (no on-campus study). *Faculty:* 15 full-time, 750 part-time. *Students:* 4,356. *Degree requirements:* For master's, business plan (MBA), Capstone Project (MHRM, MPM, MTM, MAFM) required, foreign language and thesis not required. *Entrance requirements:* For master's, GMAT, GRE General Test, or institutional assessment, interview. *Application deadline:* Applications are processed on a rolling basis. Application fee: $0. *Unit head:* Dr. Sherrill Hole, Director, Academic Affairs, 630-574-1894. *Application contact:* Michael J. Alexander, Director, Central Services, 630-574-1957, Fax: 630-574-1969, E-mail: malexander@keller.edu.

National University, Graduate Studies, School of Business and Technology, Department of Technology, La Jolla, CA 92037-1011. Offers e-commerce (MBA, MS); electronic engineering (MS); engineering management (MS); environmental management (MBA, MS); industrial engineering management (MS); software engineering (MS); technology management (MBA, MS); telecommunication systems management (MS). Part-time and evening/weekend programs available. Postbaccalaureate distance learning degree programs offered (minimal on-campus study). *Faculty:* 12 full-time, 125 part-time. *Students:* 305 (79 women); includes 122 minority (34 African Americans, 69 Asian Americans or Pacific Islanders, 17 Hispanic Americans, 2 Native Americans) 53 international. *Degree requirements:* For master's, foreign language and thesis not required. *Entrance requirements:* For master's, interview, minimum GPA of 2.5. *Application deadline:* Applications are processed on a rolling basis. Application fee: $60 ($100 for international students). Tuition: Full-time $7,830; part-time $870 per course. One-time fee: $60. Tuition and fees vary according to campus/location. *Unit head:* Dr. Leonid Preiser, Chair, 858-642-8425, Fax: 858-642-8716, E-mail: lpreiser@nu.edu. *Application contact:* Nancy Rohland, Director of Enrollment Management, 858-642-8180, Fax: 858-642-8709, E-mail: nrohland@nu.edu.

Northwestern University, The Graduate School, Program in Telecommunications Science, Management, and Policy, Evanston, IL 60208. Offers MA, MEM, MS, Certificate. MS awarded through the Department of Electrical Engineering and Computer Science; MA awarded through the Departments of Comunication Studies and Radio/Television/Film; MEM awarded through the Department of Industrial Engineering and Management Sciences. Part-time programs available. *Faculty:* 22 full-time (2 women), 1 (woman) part-time. *Students:* 19 full-time (8 women); includes 7 minority (all Asian Americans or Pacific Islanders), 1 international. Average age 26. In 1998, 7 degrees awarded (14% entered university research/teaching, 86% found other work related to degree). *Entrance requirements:* For master's, GRE General Test, TOEFL. *Average time to degree:* Master's–2 years full-time. *Application deadline:* For fall admission, 1/15 (priority date). Applications are processed on a rolling basis. Application fee: $50 ($55 for international students). *Financial aid:* Fellowships, research assistantships, teaching assistantships, career-related internships or fieldwork, Federal Work-Study, and institutionally-sponsored loans available. Financial aid application deadline: 1/15; financial aid applicants required to submit FAFSA. *Faculty research:* Electronic media management. *Unit head:* Dr. Steven S. Wildman, Director, 847-491-3539, Fax: 847-467-1171, E-mail: s-wildman@nwu.edu. *Application contact:* Theomary Karamanis, Program Assistant, 847-491-3539, Fax: 847-467-1171, E-mail: telecomprog@nwu.edu.

See in-depth description on page 1511.

Oklahoma State University, Graduate College, College of Business Administration, Program in Telecommunications Management, Stillwater, OK 74078. Offers MS. *Students:* 54 full-time

(16 women), 84 part-time (28 women); includes 13 minority (2 African Americans, 6 Asian Americans or Pacific Islanders, 2 Hispanic Americans, 3 Native Americans), 40 international. Average age 31. In 1998, 45 degrees awarded. *Degree requirements:* For master's, foreign language and thesis not required. *Entrance requirements:* For master's, GRE or GMAT, TOEFL. Application fee: $25. *Unit head:* Dr. Rick Wilson, Director, 405-744-9000.

Polytechnic University, Brooklyn Campus, Department of Management, Major in Telecommunications and Information Management, Brooklyn, NY 11201-2990. Offers MS. Average age 33. 3 applicants, 100% accepted. *Degree requirements:* For master's, thesis or alternative required. *Entrance requirements:* For master's, GMAT, minimum B average in undergraduate course work. *Application deadline:* Applications are processed on a rolling basis. Application fee: $45. Electronic applications accepted. *Application contact:* John S. Kerge, Dean of Admissions, 718-260-3200, Fax: 718-260-3446, E-mail: admitme@poly.edu.

Polytechnic University, Westchester Graduate Center, Graduate Programs, Division of Management, Major in Telecommunication and Computing Management, Hawthorne, NY 10532-1507. Offers MS. 37 applicants, 97% accepted. In 1998, 28 degrees awarded. *Degree requirements:* For master's, computer language required. *Application deadline:* Applications are processed on a rolling basis. Application fee: $45. Electronic applications accepted. *Unit head:* John S. Kerge, Dean of Admissions, 718-260-3200, Fax: 718-260-3446, E-mail: admitme@poly.edu. *Application contact:* John S. Kerge, Dean of Admissions, 718-260-3200, Fax: 718-260-3446, E-mail: admitme@poly.edu.

San Diego State University, Graduate and Research Affairs, College of Professional Studies and Fine Arts, School of Communication, Programs in Communication, San Diego, CA 92182. Offers advertising and public relations (MA); critical-cultural studies (MA); interaction studies (MA); intercultural and international studies (MA); new media studies (MA); news and information studies (MA); telecommunications and media management (MA). *Students:* 16 full-time (14 women), 45 part-time (29 women); includes 10 minority (2 African Americans, 3 Asian Americans or Pacific Islanders, 4 Hispanic Americans, 1 Native American), 3 international. *Degree requirements:* For master's, foreign language and thesis not required. *Entrance requirements:* For master's, GRE General Test (minimum combined score of 950 required), TOEFL (minimum score of 550 required). *Application deadline:* For fall admission, 3/1 (priority date); for spring admission, 11/1. Applications are processed on a rolling basis. Application fee: $55. *Unit head:* Dr. Michael Jensen, Associate Head, 518-276-6432, Fax: 518-276-6025, E-mail: burged@rpi.edu. *Application contact:* Brian Spitzberg, Graduate Coordinator, 619-594-7097, Fax: 619-594-6246, E-mail: spitz@mail.sdsu.edu.

Stevens Institute of Technology, Graduate School, Wesley J. Howe School of Technology Management, Program in Telecommunications Management, Hoboken, NJ 07030. Offers information management (MS, PhD); network planning and evaluation (MS, PhD); project management (MS, PhD); technology management marketing (MS, PhD); telecommunications management (Certificate). Offered in cooperation with the Program in Electrical Engineering. *Degree requirements:* For master's, computer language required, thesis optional, foreign language not required; for doctorate, computer language, dissertation required. *Entrance requirements:* For master's and doctorate, GMAT, GRE, TOEFL. Electronic applications accepted.

Syracuse University, Graduate School, School of Information Studies, Program in Telecommunications and Network Management, Syracuse, NY 13244-0003. Offers MS. Part-time and evening/weekend programs available. *Faculty:* 21. *Students:* 30 full-time (11 women), 67 part-time (21 women); includes 11 minority (5 African Americans, 3 Asian Americans or Pacific Islanders, 2 Hispanic Americans, 1 Native American), 50 international. Average age 31. 100 applicants, 81% accepted. In 1998, 27 degrees awarded. *Degree requirements:* For master's, internship or research project required, foreign language and thesis not required. *Entrance requirements:* For master's, GRE General Test (minimum combined score of 1000 required). *Application deadline:* Applications are processed on a rolling basis. Application fee: $40. Tuition: Full-time $13,992; part-time $583 per credit hour. *Financial aid:* Fellowships, research assistantships, teaching assistantships, career-related internships or fieldwork, Federal Work-Study, and tuition waivers (partial) available. Financial aid application deadline: 3/1. *Faculty research:* Multimedia, information resources management. *Unit head:* Tom Martin, Interim Director, 315-443-2911. *Application contact:* Barbara Settel, Assistant Dean, 315-443-2911.

See in-depth description on page 1513.

University of Colorado at Boulder, Graduate School, College of Engineering and Applied Science, Interdisciplinary Telecommunications Program, Boulder, CO 80309. Offers ME, MS, MBA/MS. *Degree requirements:* For master's, comprehensive exam required. *Entrance requirements:* For master's, minimum undergraduate GPA of 3.0.

See in-depth description on page 1515.

University of Dallas, Graduate School of Management, Program in Telecommunications Management, Irving, TX 75062-4736. Offers MBA, MM. Part-time programs available. *Students:* 151 (39 women). In 1998, 55 degrees awarded. *Degree requirements:* For master's, foreign language and thesis not required. *Entrance requirements:* For master's, GMAT (minimum score of 400 required), TOEFL (average 520), minimum GPA of 3.0. *Average time to degree:* Master's–1.3 years full-time, 2.5 years part-time. *Application deadline:* For fall admission, 8/6 (priority date); for spring admission, 12/8. Applications are processed on a rolling basis. Application fee: $25. *Financial aid:* Application deadline: 2/15. *Unit head:* Dr. Stan Kroder, Director, 972-721-5080, Fax: 972-721-5138. *Application contact:* Roxanne Del Rio, Director of Graduate Admissions, 972-721-5174, Fax: 972-721-4009, E-mail: admiss@gsm.udallas.edu.

University of Denver, Daniels College of Business, General Business Administration Program, Denver, CO 80208. Offers business administration (MBA); education management (MSM); health care management (MSM); management and communications (MSMC); management and general engineering (MSMGEN); management and telecommunications (MSMC); public health management (MSM); sports management (MSM). Part-time and evening/weekend programs available. *Faculty:* 76. *Students:* 306 (111 women); includes 16 minority (6 African Americans, 3 Asian Americans or Pacific Islanders, 6 Hispanic Americans, 1 Native American) 45 international. *Degree requirements:* For master's, foreign language and thesis not required. *Entrance requirements:* For master's, GMAT (average 545), LSAT (JD/MBA). *Application deadline:* For fall admission, 5/1 (priority date); for spring admission, 1/1. Applications are processed on a rolling basis. Application fee: $50. *Unit head:* Dr. Tom Howard, Director, 303-871-4402. *Application contact:* Jan Johnson, Executive Director, Student Services, 303-871-3416, Fax: 303-871-4466, E-mail: dcb@du.edu.

University of Maryland University College, Graduate School of Management and Technology, Program in Telecommunications Management, College Park, MD 20742-1600. Offers MS. Offered evenings and weekends only. Part-time and evening/weekend programs available. *Students:* 7 full-time (1 woman), 190 part-time (49 women); includes 97 minority (70 African Americans, 24 Asian Americans or Pacific Islanders, 3 Hispanic Americans), 8 international. 76 applicants, 97% accepted. In 1998, 34 degrees awarded. *Degree requirements:* For master's, thesis or alternative required, foreign language not required. *Entrance requirements:* For master's, previous course work in calculus and statistics. *Application deadline:* Applications are processed on a rolling basis. Application fee: $50. Electronic applications accepted. Tuition, state resident: full-time $5,058; part-time $281 per credit. Tuition, nonresident: full-time $6,876; part-time $382 per credit. Tuition and fees vary according to program. *Financial aid:* Federal Work-Study, grants, and scholarships available. Aid available to part-time students. Financial aid application deadline: 6/1; financial aid applicants required to submit FAFSA. *Unit head:* John Richardson, Director, 301-985-7200, Fax: 301-985-4611, E-mail: jrichard@ucsfs1.umd.edu. *Application contact:* Coordinator, Graduate Admissions, 301-985-7155, Fax: 301-985-7175, E-mail: gradinfo@nova.umuc.edu.

University of Miami, Graduate School, School of Business Administration, Department of Computer Information Systems/Telecommunications, Coral Gables, FL 33124. Offers computer information systems (MS); telecommunications management (Certificate). Certificate offered jointly with MBA or MS program, or as a 15 credit hour program for applicants with an undergraduate or master's degree. Part-time and evening/weekend programs available. *Faculty:* 7 full-time (1 woman), 6 part-time (2 women). *Students:* 33 full-time (12 women), 31 part-time (14 women). In 1998, 20 master's, 9 other advanced degrees awarded. *Degree requirements:* For master's and Certificate, foreign language and thesis not required. *Entrance requirements:* For master's, GMAT (minimum score of 400 required) or GRE, TOEFL (minimum score of 550 required; 213 for computer-based); for Certificate, TOEFL (minimum score of 550 required). *Application deadline:* For fall admission, 6/30 (priority date); for spring admission, 10/31. Applications are processed on a rolling basis. Application fee: $40. Electronic applications accepted. Tuition: Full-time $15,336; part-time $852 per credit. Required fees: $174. Tuition and fees vary according to program. *Financial aid:* In 1998–99, 5 research assistantships with partial tuition reimbursements were awarded.; career-related internships or fieldwork, Federal Work-Study, institutionally-sponsored loans, and scholarships also available. Financial aid application deadline: 3/1. *Faculty research:* Database management, expert systems, systems analysis and design, software engineering, information systems and strategic management. *Unit head:* Dr. Joel Stutz, Chairman, 305-284-6294, Fax: 305-284-5161, E-mail: jstutz@miami.edu. *Application contact:* Dierdre Lacativa, Director, Graduate Business Recruiting and Admissions, 305-284-4607, Fax: 305-284-1878, E-mail: gba@sba.miami.edu.

University of Pennsylvania, School of Engineering and Applied Science, Telecommunications and Networking Program, Philadelphia, PA 19104. Offers MSE. Part-time programs available. Electronic applications accepted.

See in-depth description on page 1521.

University of San Francisco, McLaren School of Business, Program in Business Administration, San Francisco, CA 94117-1080. Offers finance and banking (MBA); international business (MBA); management (MBA); marketing (MBA); telecommunications management and policy (MBA). *Faculty:* 34 full-time (6 women), 32 part-time (6 women). *Students:* 284 full-time (130 women), 260 part-time (117 women); includes 114 minority (10 African Americans, 86 Asian Americans or Pacific Islanders, 17 Hispanic Americans, 1 Native American), 150 international. *Degree requirements:* For master's, foreign language and thesis not required. *Entrance requirements:* For master's, GMAT (average 540), TOEFL (minimum score of 600 required), minimum undergraduate GPA of 3.2. *Application deadline:* For fall admission, 7/1 (priority date); for spring admission, 11/30. Applications are processed on a rolling basis. Application fee: $40 ($50 for international students). Tuition: Full-time $12,618; part-time $701 per unit. Tuition and fees vary according to course load, degree level, campus/location and program. *Unit head:* Cathy Fusco, Director, 415-422-6314, Fax: 415-422-2502, E-mail: mbausf@usfca.edu.

Webster University, School of Business and Technology, Department of Business, St. Louis, MO 63119-3194. Offers business (MA, MBA); computer resources and information management (MA, MBA); computer science/distributed systems (MS); finance (MA, MBA); health care management (MA); health services management (MA, MBA); human resources development (MA, MBA); human resources management (MA); international business (MA, MBA); management (MA, MBA); marketing (MA, MBA); procurement and acquisitions management (MA, MBA); public administration (MA); real estate management (MA, MBA); security management (MA, MBA); space systems management (MA, MBA, MS); telecommunications management (MA, MBA). *Faculty:* 5 full-time (1 woman). *Students:* 3,474 full-time (1,390 women), 1,592 part-time (603 women); includes 1,465 minority (1,021 African Americans, 169 Asian Americans or Pacific Islanders, 255 Hispanic Americans, 20 Native Americans), 397 international. *Degree requirements:* For master's, foreign language and thesis not required. *Application deadline:* Applications are processed on a rolling basis. Application fee: $25 ($50 for international students). *Unit head:* Lucille Berry, Chair, 314-968-7022, Fax: 314-968-7077, E-mail: berrylm@webster.edu. *Application contact:* Dr. Beth Russell, Director of Graduate Admissions, 314-968-7089, Fax: 314-968-7166, E-mail: russelmb@webster.edu.

Cross-Discipline Announcements

Carnegie Mellon University, Information Networking Institute, Pittsburgh, PA 15213-3891.

The MS in Information Networking is a cooperative endeavor of the Schools of Engineering, Computer Science, and Business, providing an alternative to the conventional one-year computer science or electrical engineering graduate program by integrating both and adding some features of an MBA program. Now in its eleventh year, the program carefully selects 35 people from the engineering and computer science disciplines to form each new class. The program provides technical electives aimed at several areas of specialization including (a) the telecommunications and computing industries, (b) wireless and mobile computing, (c) the financial services industry, and (d) systems integrating and consulting.

Monmouth University, Graduate School, West Long Branch, NJ 07764-1898.

The School of Science, Technology and Engineering offers graduate-level telecommunications programs designed to provide a broad perspective of and an interdisciplinary approach to telecommunications. The programs are interdisciplinary concentrations of guided electives that a student can include while pursuing a master's degree in either computer science, software engineering, or electronic engineering. The telecommunications programs include courses from all 3 disciplines.

DEPAUL UNIVERSITY

School of Computer Science, Telecommunications and Information Systems (CTI)
Master's Program in Telecommunications Systems

Programs of Study	The Master of Science (M.S.) in telecommunications systems program provides an integrated curriculum that covers all aspects of voice and data networking technologies as well as techniques for managing these technologies. Students graduate from this program with the skills necessary to efficiently use, design, and evaluate voice and data networks as network integrators, managers, or consultants. Classroom instruction comes from both full-time faculty members who are knowledgeable in network foundations and research as well as adjunct faculty members with significant practical experience as telecommunications specialists and consultants. Laboratory facilities provide an additional hands-on component.
	The program offers two concentrations, each of which can be modified in consultation with an academic adviser to suit a student's interests and goals. The standard concentration places more emphasis on the management and regulatory aspects of voice networks. The computer science concentration emphasizes data networking technologies.
	Courses are organized into three phases: prerequisite, core knowledge, and advanced. Most degree candidates are admitted conditionally into the prerequisite phase and become fully admitted to the graduate program by achieving a passing grade on the department's Graduate Assessment Examination or by proving equivalent competence in undergraduate course work or work experience. In order to obtain the M.S. degree, students must complete three core knowledge phase courses and ten advanced phase courses. A comprehensive examination covering material from core knowledge phase courses is required to enter the advanced phase.
Research Facilities	DePaul's Information Services Division (IS) houses a large network of computers and allows students access to a rich computing environment. The configuration includes several Sun SPARCcenters for student use. In addition, students have access to PC laboratories at the Loop and Lincoln Park campuses. There are numerous dial-up phone numbers available for off-campus work. DePaul's suburban campuses in the Lake County, Naperville, O'Hare, and South areas also offer excellent student laboratory facilities. Permanent student Internet-access accounts are available through a service called DePaul Online, as are dial-in connections.
	The Division of Telecommunications and Data Communications maintains three laboratories for exclusive use by students in this program. The Telecommunications Laboratory houses an AT Definity G3 PBX system with multiple analog, digital, and ISDN station sets; AUDIX voice mail; and additional adjunct equipment for CTI experimentation. The Local Area Networks Laboratory has several dozen workstations connected to Novell, Windows NT, OS/2 Warp, and UNIX file servers using Ethernet and Token Ring hubs and switches. A high-speed ATM switch and router provide backbone connectivity. These laboratories are completely managed by students and can be reserved for comprehensive experimentation that is not possible on production networks. The Multimedia Laboratory provides hardware and software for students to support high-bandwidth synchronized audio, video, and control applications among multiple high-performance Sun and NT workstations. Eight new laboratories open in September 1999: the computer-supported collaborative learning lab, the database teaching lab, the database research lab, the student project teamwork collaborative learning lab, the distributed systems lab, the human-computer interaction lab, the software engineering lab, and a second multimedia lab.
Financial Aid	The School provides a number of full and partial graduate assistantships carrying stipends. Application should be made directly to the School. There is one filing period for each academic year: February 2–April 1.
	The University's Financial Aid Office assists interested students in applying to the Federal Perkins Loan and Federal Stafford Student Loan programs. Students are encouraged to apply before May 1 to receive maximum consideration.
Cost of Study	Graduate tuition for the 1999–2000 academic year is $420 per quarter hour. Full-time graduate study generally consists of 8 to 12 hours per quarter.
Living and Housing Costs	There is no on-campus housing for graduate students, but the Housing Office has listings of apartments and residential hotels near the Lincoln Park campus. Apartments in the Chicago area are available at various rents.
Student Group	The School of Computer Science, Telecommunications and Information Systems (CTI) has approximately 1,200 students in the graduate program; nearly half attend full-time. This sizable enrollment allows the School to offer a wide variety of courses each quarter. Many students are already employed in a computer or computer-related profession and enroll in the program for career enhancement.
Student Outcomes	Graduates of this program have been hired as network designers, researchers, programmers, managers, and consultants by communications vendors, carriers, manufacturers, and consulting firms as well as a variety of end-user businesses.
Location	The city of Chicago offers DePaul students a wide range of cultural and recreational opportunities. The Loop campus is minutes from the Art Institute; Orchestra Hall; museums of art, natural history, and science; and the LaSalle Street business district. The 25-acre Lincoln Park campus is less than 1 mile from Lake Michigan beaches, the lakefront bicycle and running paths, the zoo, and other public recreational facilities. The stores, theaters, musical groups, and other attractions of the Lincoln Park community reflect the broad interests of people who live and work there. The Naperville, O'Hare, and South campuses serve students in the outlying suburbs.
The University	DePaul University was founded in 1898 by the Vincentian Fathers and is now one of the largest Catholic universities in the world. Urban in style, the University today still strives to maintain the heritage of St. Vincent DePaul.
	Numerous student organizations offer considerable opportunities for participation in both community and University activities. There are music performance groups, theater groups, student publications, sports, and honor and service societies. Athletic facilities include two gymnasiums, a swimming pool, racquetball courts, and extensive physical education equipment.
Applying	Master's students may begin their course work in any academic quarter. Ph.D. students are admitted twice per year. Application materials for the graduate programs in computer science, distributed systems, information systems, software engineering, telecommunication systems, or management information systems may be obtained by sending a request to the address given below. The application fee is $25.
Correspondence and Information	The CTI Web site (see address listed below) offers information and resources for current and prospective students. Students can order admission applications, view class schedules, visit faculty member home pages, and visit student home pages. Information can also be obtained by contacting:

School of Computer Science, Telecommunications and Information Systems
Graduate Programs
DePaul University
243 South Wabash
Chicago, Illinois 60604-2302
World Wide Web: http://www.cs.depaul.edu

DePaul University

THE FACULTY AND THEIR RESEARCH

L. Edward Allemand, Professor Emeritus; Ph.D., Louvain (Belgium), 1970. Information systems, human-computer interaction.

Ehab S. Al-Shaer, Assistant Professor; Ph.D., Old Dominion, 1998. Management of distributed systems, multimedia networks, multicast protocols.

Gary Andrus, Associate Professor; Ph.D., Wayne State, 1977. Formal language theory, compiler design.

Karen Bernstein, Assistant Professor; Ph.D., SUNY at Stony Brook, 1996. Programming environments, programming languages, software engineering, and concurrent systems.

André Berthiaume, Assistant Professor; Ph.D., Montreal, 1995. Quantum computation and quantum information processing.

Gregory Brewster, Assistant Professor; Ph.D., Wisconsin–Madison, 1994. Telecommunication systems, data communications, computer networks, performance analysis of communication systems.

Susy Chan, Associate Professor; Ph.D., Syracuse, 1979. IT management, planning, and strategies; systems analysis and design; electronic commerce.

I-Ping Chu, Associate Professor; Ph.D., SUNY at Stony Brook, 1981. Data communication, computer networks, combinatorial algorithms, database, distributed database.

Anthony Wai Man Chung, Associate Professor; Ph.D., Maryland, 1992. Communication networks, distributed systems, automated tools for software development, operating systems, programming languages.

Kamal Dahbur, Instructor; M.S., DePaul, 1993. Artificial intelligence, information systems.

Charles Earl, Visiting Assistant Professor; Ph.D., Chicago, 1998. Transformational planning and scheduling, mobile autonomous agents and machine learning for autonomous software agents, artificial intelligence, computational modeling.

Clark Elliott, Associate Professor; Ph.D., Northwestern, 1992. Artificial intelligence.

Helmut Epp, Associate Professor and Dean; Ph.D., Northwestern, 1966. Expert systems, artificial intelligence, computer security, hardware description languages.

Xiaowen Fang, Assistant Professor; Ph.D., Purdue, 1999. Human-computer interaction, information systems, Web applications.

Robert Fisher, Associate Professor; Ph.D., Harvard, 1975. Graphics, operating systems.

Jacob D. Furst, Assistant Professor; Ph.D., North Carolina, 1999. Image processing.

Gerald Gordon, Associate Professor; Ph.D., Berkeley, 1968. Computer vision.

Henry Harr, Associate Professor; Ph.D., IIT, 1988. Parallel processing, operating systems.

Alan Jeffrey, Associate Professor; Ph.D., Oxford, 1992. Semantics of programming languages.

Xiaoping Jia, Associate Professor; Ph.D., Northwestern, 1989. Software engineering, formal methods, object-oriented software development.

Richard Johnsonbaugh, Professor; Ph.D., Oregon, 1969. Combinatorial algorithms, pattern recognition.

Steve Jost, Associate Professor; Ph.D., Northwestern, 1985. Statistics, pattern recognition, image processing.

Martin Kalin, Professor; Ph.D., Northwestern, 1969. Programming languages, architecture, distributed systems.

George Knafl, Professor; Ph.D., Northwestern, 1978. Software quality, software reliability, data mining, statistical computing.

Linda V. Knight, Assistant Professor; Ph.D., DePaul, 1998. Information systems, artificial intelligence.

Vladimir Kulyukin, Assistant Professor; Ph.D., Chicago, 1998. Distributed information retrieval, artificial intelligence.

Glenn Lancaster, Associate Professor; Ph.D., California, Irvine, 1972. Compiler design.

Chengwen Liu, Associate Professor and Assistant Dean; Ph.D., Illinois, 1991. Database management systems, information retrieval, data compression.

King-Lup Liu, Visiting Assistant Professor; Ph.D., Illinois at Chicago, 1999. Database systems, information retrieval, multimedia retrieval.

Steve Lytinen, Associate Professor; Ph.D., Yale, 1984. Artificial intelligence, natural-language processing.

Will Marrero, Assistant Professor; Ph.D., Carnegie Mellon, 1999. Computer security, formal methods.

John McDonald, Assistant Professor; Ph.D., Northwestern, 1996. Computational geometry, mathematical modeling and simulation.

Craig S. Miller, Assistant Professor; Ph.D., Michigan, 1993. Human cognition, user modeling, machine learning.

David Miller, Associate Professor and Associate Dean; Ph.D., Chicago, 1981. Artificial intelligence, computation theory.

Daniel Mittleman, Assistant Professor; Ph.D., Arizona, 1995. Information systems, groupware, GSS, design of technology-supported physical environments.

Bamshad Mobasher, Assistant Professor; Ph.D., Iowa State, 1994. Data mining, autonomous software agents, knowledge-based systems.

Thomas Muscarello, Assistant Professor; Ph.D., Illinois, 1993. Hospital/medical informatics, police/fraud, information systems, software systems upgrades and maintenance, IT workforce issues.

Makoto Nakayama, Assistant Professor; Ph.D., UCLA, 1999. Strategic use of information systems, electronic commerce.

Corin Pitcher, Assistant Professor; Ph.D., Oxford, 1999. Semantics of programming languages, mobile code, security.

James W. Riley, Assistant Professor; Ph.D., North Carolina, 1999. Distributed computing, programming languages and computer design, software engineering, theoretical computer science.

John Rogers, Assistant Professor; Ph.D., Chicago, 1995. Computational complexity, mathematical logic, quantum computation.

LoriLee Sadler, Visiting Assistant Professor; Ph.D., Indiana, 1999. High-speed data networking, human-computer interaction, human cognition.

Marcus Schaefer, Assistant Professor; Ph.D., Chicago, 1999. Computational complexity theory, computability theory.

Eric J. Schwabe, Associate Professor; Ph.D., MIT, 1991. Algorithms, parallel computation.

Eric Sedgwick, Visiting Assistant Professor; Ph.D., Texas at Austin, 1997. Computational geometry and topology.

Amber Settle, Assistant Professor; Ph.D., Chicago, 1999. Distributed algorithms, cellular automata.

Paul A. Sisul, C.M., Instructor; M.Div., DeAndreis Institute of Theology, 1975. Language and applications.

Norma G. Sutcliffe, Visiting Assistant Professor; Ph.D., UCLA, 1997. Information systems, IT management, IT consulting.

Ian Sutherland, Assistant Professor; Ph.D., Northwestern, 1996. Fault tolerance, computer security, mathematical logic, semantics of real-number computation.

George K. Thiruvathukal, Assistant Professor; Ph.D., IIT, 1995. Concurrent, parallel, and distributed computing; object-oriented methods and programming; programming language design and implementation; software systems; computational science.

Noriko Tomuro, Instructor; Ph.D., DePaul, 1999. Artificial intelligence, natural-language processing.

Gary B. Weinstein, Instructor; M.S., Ohio, 1982. Database, operating systems.

Curt M. White, Visiting Associate Professor; Ph.D., Wayne State, 1986. Computer science education research, data communications, computer networks, genetic algorithms.

Nathaniel Whitmal, Assistant Professor; Ph.D., Northwestern, 1997. Signal and image processing, digital communications.

Rosalee Wolfe, Associate Professor; Ph.D., Indiana, 1987. Computer graphics, human-computer interaction.

NORTHWESTERN UNIVERSITY

Program in Telecommunications Science, Management, and Policy

Program of Study

Northwestern University offers an interdisciplinary master's program in telecommunications science, management, and policy that draws upon four distinct but related departments in the McCormick School of Engineering and Applied Science and the School of Speech. All students enrolled in the telecommunications program take a core curriculum covering the science, management, and policy of telecommunications. At the same time, students earn a master's degree in one of the four home departments involved in the program.

The Master's Program in Telecommunications Science, Management, and Policy is designed to give students a broad understanding of telecommunications and an area of specialization. Students who want to emphasize the science and technology of telecommunications can pursue a Master of Science degree in electrical and computer engineering. Those who want a concentration in electronic media management can earn a Master of Arts degree in radio/television/film. Those who want to specialize in information systems and engineering management can earn a Master of Engineering Management degree in industrial engineering and management sciences. Those who choose to emphasize telecommunications policy, economics, and the social impact of media and those interested in a social science approach to communication systems and technology can pursue a Master of Arts degree in communication studies.

Participating in the telecommunications science, management, and policy program adds nothing to the cost or duration of any of the departmental master's programs described above. Once the program of study is completed, the student receives a graduate certificate in telecommunications science, management, and policy and the degree conferred by his or her home department.

Research Facilities

Telecommunications students have access to the full resources of Northwestern University, including libraries in both the Chicago and the Evanston campuses, which hold more than 3 million volumes and 1,000 international and domestic newspapers. The School also conducts multidisciplinary research with its numerous research centers.

Financial Aid

Financial aid packages comprise a combination of loans, Federal Work-Study awards, and teaching and research assistantships. The latter two are only available to full-time students. The amount of the financial aid package is based on the student's current financial need and the availability of scholarship, Federal Work-Study, and loan funds.

Cost of Study

For the 1999–2000 academic year, tuition is $7266 per quarter (3 or 4 units of credit) for graduate students. Other quarterly costs (estimates) include books, $310; transportation, $527; and personal expenses, $1047.

Living and Housing Costs

The University assists students in finding suitable accommodations. Graduate housing is available at both the Evanston and Chicago campuses. Estimated costs for room and board are $12,840 for twelve months.

Student Group

Enrollment in the Program in Telecommunications Science, Management, and Policy was 19 students for the 1998–99 academic year. Ten were male and 9 were female; 8 were international students. All of them were enrolled full-time.

Location

All of the participating departments are located on the Evanston campus, situated along Lake Michigan, which affords a variety of scenic and recreational opportunities.

The University

Established May 31, 1850, Northwestern University is one of the nation's major private research universities. About 7,500 undergraduate and 5,200 graduate and professional school students study full-time on the Evanston and Chicago campuses. There are 1,600 full- and part-time faculty members.

Applying

Admission to the telecommunications program is a two-stage process. Students must first apply to one of the four home departments. At the time of application, they should note that telecommunications science, management, and policy is the desired area of specialization within the selected department. Once admitted to a home department, students are considered for admission to the telecommunications program. Ordinarily, students who meet the admission requirements of their home department are accepted into the telecommunications program. Entrance requirements include a bachelor's degree, GRE scores, and TOEFL scores for international students.

The application deadline is January 15 for financial aid consideration. For admission only, the application deadline is August 1 for United States citizens and June 3 for international applicants. The nonrefundable application fee is $50 for U.S. citizens and $55 for international applicants.

Correspondence and Information

Program in Telecommunications Science, Management, and Policy
Northwestern University
Harris Hall, Room 12
1881 Sheridan Road
Evanston, Illinois 60208-2222
Telephone: 847-491-3539
E-mail: telecomprog@nwu.edu
World Wide Web: http://nuinfo.nwu.edu/speech/departments/telecom

Northwestern University

THE FACULTY AND THEIR RESEARCH

Annette Barbier (Radio/Television/Film), M.F.A., Chicago. Teaches and does research in the areas of interactive multimedia, computer graphics, and animation; her video works have won numerous awards.

Ronald Braeutigam (Economics), Ph.D., Stanford. Teaches in the areas of regulatory and transportation economics; he has published on subjects including strategic uses of the administrative process.

Barbara Cherry (Communication Studies), Ph.D., Northwestern; J.D., Harvard. Teaches telecommunications policy and the legal, economic, and social implications of telecommunication technologies; her research focuses on the law and regulation of communication industries and technologies.

James Ettema (Communication Studies), Ph.D., Michigan. Teaches in the area of social and cultural aspects of media; he is author of many articles and monographs dealing with the social consequences of new information technologies.

Donald Frey (Industrial Engineering and Management Sciences), Ph.D., Michigan. Teaches information systems and technical entrepreneurship; he served for sixteen years as the chairman of the board and CEO of Bell & Howell and is a member of the Center for Information and Telecommunications Technology.

Abraham Haddad (Electrical and Computer Engineering), Ph.D., Princeton. Teaches stochastic systems, modeling, estimation, detection, and applications to communications and control; he is chair of Electrical Engineering and Computer Science.

Paul Hirsch (Management), Ph.D., Michigan. Allen Professor of Strategy and Organizational Behavior, he is an authority on organizational culture and career paths and has written extensively about the media.

Gerald Hoffman (Industrial Engineering and Management Sciences), Ph.D., Northwestern. Teaches in the area of applying information technology to business problems; he is president of a consulting firm and created the first strategic plan for information services at Standard Oil Company (now AMOCO).

Srikanta Kumar (Electrical and Computer Engineering), Ph.D., Yale. Teaches computer communications and networks; he has worked with General Electric and AT&T and has conducted research on stochastic scheduling, distributed communication, and adaptation and learning systems.

John Lavine (Journalism and Management), B.A., Carleton (Ottawa). A newspaper publisher and authority on media management, he is director of Northwestern's Newspaper Management Center.

C. C. Lee (Electrical and Computer Engineering), Ph.D., Princeton. Teaches statistical communication theory and computer networks; he has published many articles on topics such as signal processing, document coding and recognition, and computational complexity.

Lawrence Lichty (Radio/Television/Film), Ph.D., Ohio State. A nationally recognized expert on broadcast history and journalism, he has published widely on the television industry and the media's role in the war in Vietnam.

Peter Miller (Communication Studies), Ph.D., Michigan. Teaches in the areas of survey research and the social impact of mass media; he is an expert on television audience measurement.

Newton Minow (Communication Studies, Law, and Management), J.D., Northwestern. Teaches communication law and management; he is former chairman of the FCC and current director of the Annenberg Washington Program for Communication Policy Studies.

Rick Morris (Radio/Television/Film), J.D., Kansas; L.L.M., NYU. Teaches and researches policy, management, regulation, and new technologies.

John Panzar (Economics), Ph.D., Stanford. A nationally recognized expert in micro and regulatory economics whose research is widely cited; before coming to Northwestern, he was an economist at Bell Labs.

Roger Schank (Computer Science), Ph.D., Texas. One of the country's leading authorities on artificial intelligence, he is director of the Institute for the Learning Sciences.

James Schwoch (Radio/Television/Film), Ph.D., Northwestern. Teaches and researches media history, radio, and international communications.

Charles Thompson (Industrial Engineering and Management Science), Ph.D., Northwestern. Specializes in theories and methods for solving unstructured problems in the organization of management and technology; he is former director of Engineering Services at Admiral Corporation.

James Webster (Radio/Television/Film and Communication Studies), Ph.D., Indiana. Teaches telecommunications policy and audience analysis; he has published extensively on audience behavior and the impact of new technology and is an associate dean for the School of Speech.

Steven Wildman (Communication Studies), Ph.D., Stanford. Teaches and researches telecommunications policy, mass media economics, and the impact of communication technologies on organizations and markets; he is director of the Program in Telecommunications Science, Management, and Policy.

SYRACUSE UNIVERSITY

School of Information Studies
Master's Program in Telecommunications and Network Management

Programs of Study	The School of Information Studies offers a Master of Science in telecommunications and network management to prepare students for a growing number of careers in telecommunications dealing with the management, provision, and use of telecommunication technologies, including design and operation of voice and data networks. Graduates of the program work in a broad range of managerial, technical, and regulatory positions with telecommunication vendors, users, and policymakers.
	The School has an interdisciplinary faculty that teaches and conducts research in data administration and information engineering, expert systems, group-based decision support systems, information and communications policy, information resources management, management information systems, online retrieval systems, project management, strategic planning, and telecommunications management.
	The M.S. in telecommunications and network management requires the completion of 42 credit hours. Each student completes a 13-credit core, 6 credits of internship, and 23 credits of selected management and engineering courses. The program provides an integrated approach to the effective management, operation, and implementation of telecommunication systems within organizations. The School also has a cooperative program with the Ryerson Institute of Technology and offers courses in Toronto, Ontario, Canada. The program is also offered in a distance learning format, completed predominantly through home study with the use of the Internet but also requiring brief residencies in Syracuse.
Research Facilities	Access to computer facilities at Syracuse University and within the School is excellent. All students have access to data networks connecting the University's mainframe computer as well as client-server environments. Students have access to more than 150 terminals and 160 microcomputers in public clusters throughout the University. All students are provided with free computer accounts and unlimited access to the Internet. The School is located in the Center for Science and Technology, the University's most sophisticated facility for teaching and research in the areas of information science, computing, and information technology. Faculty members and students work with two research centers in the building, the Center for Advanced Technology in Computer Applications and Software Engineering (CASE) and the Northeast Parallel Architectures Center (NPAC).
Financial Aid	Fellowships, scholarships, and assistantships are available to full-time students. Most prestigious and competitive are Syracuse University graduate fellowships, which include a 30-credit scholarship and a stipend of $11,384 for the 1999–2000 academic year. University scholarships provide 24 hours of tuition, and graduate assistantships provided tuition and a stipend of $8465 per academic year. Tuition scholarships funded by the Gaylord Trust endowment and other small scholarships are available to part-time students.
	Loans are available through the Financial Aid Office (200 Archbold). For college work-study contracts, students work through the University Student Employment Office. These kinds of assistance are awarded according to federal financial-need guidelines.
Cost of Study	Tuition for 1999–2000 is $583 per graduate credit hour, or $1749 per 3-credit course. Fees are approximately $400 for one year of full-time study.
Living and Housing Costs	Academic-year living expenses are about $8000 for single students. The University has residence hall rooms and on-campus apartments for single and married graduate students. Many also live off campus.
Student Group	Syracuse University has about 15,000 students, including about 5,000 graduate students. Approximately 550 graduate students are enrolled in the School of Information Studies, many of them attending part-time. Ten percent are international students; the remainder come from all parts of the United States. Students have diverse backgrounds, with undergraduate majors in the liberal arts, business administration, computer science, and engineering.
Student Outcomes	Career opportunities for graduates of the telecommunications and network management program are excellent, and there is a high demand for experts in this field. Graduates find positions in this well-paying field through placements in three sectors of the industry: information systems positions within organizations requiring data and voice network management; telecommunications organizations involved in voice, data, or video transmissions; and communications businesses, such as phone companies and vendors of voice, data, and video systems. Recent graduates have obtained the following positions: manager of special services in the marketing department of LDDS/WorldCom; systems administrator in the client server technical services team; presale technical support specialist; and network and database manager.
Location	Syracuse, a city of 500,000, is set at the transportation crossroads of central New York State and is the commercial, industrial, medical, and cultural center for a wide area. Downtown Syracuse is only a 20-minute walk from the University, yet the campus is spacious and attractive. Winters are snowy and summers are pleasant. Lake Ontario, the Finger Lakes, and the Adirondack and Catskill Mountains are nearby. Boston, Toronto, New York, and Philadelphia are within a day's drive.
The School	The School of Information Studies is a leading center for innovative graduate programs in information management, and it stands out from other institutions that offer computer science, management, and related programs. The School focuses on information users and understanding user information needs. The interdisciplinary faculty combines expertise in information science, telecommunications, public administration, business management and management information systems, linguistics, computer science, library science, and communication. The School offers a unique undergraduate degree program in information management and technology as well as professional graduate programs in information resources management, telecommunications, and library science. A Ph.D. in information transfer is also offered. Students benefit from close interaction with the faculty.
Applying	Students are encouraged to apply for the fall semester, although admission is possible in either the fall, spring, or summer semester. Students applying for the master's-level program must have a bachelor's degree from an accredited undergraduate institution and an academic record satisfactory for admission to the graduate school; they must supply three letters of recommendation and an essay on their academic plans and professional goals. They must also earn a combined score of at least 1000 on the verbal and quantitative sections of the GRE General Test. Whenever possible, an interview is recommended. International students should plan to take the Test of English as a Foreign Language (TOEFL); a score of at least 550 is expected. Students interested in University fellowships must apply by January 10. Other financial aid applicants must submit all materials by March 15.
Correspondence and Information	School of Information Studies Syracuse University 4-206 Center for Science and Technology Syracuse, New York 13244-4100
	Telephone: 315-443-2911 E-mail: ist@syr.edu World Wide Web: http://istweb.syr.edu

Syracuse University

THE FACULTY AND THEIR RESEARCH

Robert Benjamin, Professor; B.S. (economics), Pennsylvania (Wharton), 1948. Strategic applications of information technology, managing information technology–enabled change.

Susan Bonzi, Associate Professor; Ph.D. (library and information science), Illinois, 1983. Image retrieval systems. Received the first Information Science Doctoral Dissertation award from American Society for Information Science (ASIS), 1982.

Kevin Crowston, Assistant Professor; Ph.D. (information technologies), MIT, 1991. Organizational implications of technology, coordination-intensive processes in human organizations.

Marta Dosa, Professor Emerita; Ph.D. (library science), Michigan, 1971. Environmental and health information, information planning in developing countries, international information policies. Funded research includes: Health Information Sharing Project (National Institutes of Health/National Library of Medicine), International Clearinghouse on Information Education (UNECLO), and International Federation for Documentation (FID). Received 1986 American Society for Information Science (ASIS) Outstanding Information Science Teacher Award.

Martha A. Garcia-Murillo, Assistant Professor; Ph.D. (international political economy and telecommunications), USC, 1998. Electronic commerce, business policy and strategy, international management, technology and innovation management.

Robert Heckman, Assistant Professor; Ph.D. (information systems), Pittsburgh, 1993. Vendor-provided information systems, user satisfaction, end-user computing.

Carol Hert, Assistant Professor; Ph.D. (information transfer), Syracuse, 1995. Information technology standards, library automation, information storage and retrieval theory.

Jeffrey Katzer, Professor; Ph.D. (communication), Michigan State, 1970. The information environment of managers; information behavior; organizational, economic, and social implications of the information age. Funded research has included: representation of overlaps in computerized information retrieval systems (National Science Foundation), impact of anaphoric resolution in retrieval performance (National Science Foundation). Author and coauthor of *Free Association Behavior and Human Language Processing, Evaluating Information.*

Barbara Kwasnik, Associate Professor; Ph.D. (library and information studies), Rutgers, 1989. Classification research, knowledge representation and organization, research methods. Dissertation received 1989 best dissertation awards from the American Society for Information Science (ASIS) and from the Association for Library and Information Science Education (ALISE). Fulbright Visiting Scholar grantee (Royal School of Librarianship), Copenhagen, Denmark, 1996.

R. David Lankes, Assistant Professor; Ph.D. (information transfer), Syracuse, 1998. Building and managing Internet services, designing multimedia with the Internet.

Antje B. Lemke, Professor Emerita; M.S.L.S., Syracuse, 1956. Study of the development of European libraries from the Age of Enlightenment to World War II; biography of Jacob and Wilhelm Grimm, with special emphasis on their contributions to librarianship and bibliography. Translator, *Out of My Life and Thought: An Autobiography of Albert Schweitzer.* Funded research: the Church and universities in Germany in the years of national socialism (Deutsche Forschungsgemeinschaft, Bonn). Awarded Syracuse University's Chancellor's Citation for Exceptional Academic Achievement, 1981.

Elizabeth Liddy, Professor; Ph.D. (information transfer), Syracuse, 1988. Indexing, data-mining, natural-language processing, information retrieval. Received ASIS Doctoral Dissertation Award and ALISE Doctoral Dissertation Award, 1988. Funded research: document retrieval using linguistic knowledge (DARPA) for development of DR-LINK.

Ian MacInnes, Assistant Professor; Ph.D. (political economy and public policy), USC, 1998. Electronic commerce, information technology and globalization, public policy, standardization and network economics.

Thomas H. Martin, Associate Professor; Ph.D. (communications), Stanford, 1974. Information policy, system design, human interaction with computers, human information processing, organizational communication and the foundations of information science. Assisted in the design of the Stanford Public Information Retrieval System (SPIRES).

Charles McClure, Distinguished Professor; Ph.D. (library and information studies), Rutgers, 1977. Management, planning, and evaluation of information services, federal information policy. Associate Editor, *Government Information Quarterly,* and coauthor, *Federal Information Policies in the 1990s.* Funded research: impacts and uses of national and statewide networks, electronic records management, and access to U.S. federal information.

David Molta, Assistant Professor; M.P.A., North Texas, 1986. Computer and operating systems, high-speed networks, client/server computing.

Milton L. Mueller, Associate Professor; Ph.D. (telecommunication), Pennsylvania, 1989. Telecommunication policy and deregulation, universal service.

Michael Nilan, Associate Professor; Ph.D. (communication research), Washington (Seattle), 1985. Employing user behaviors for the design of collaborative work environments in a global electronic network environment, user-based system design.

Jian Qin, Assistant Professor; Ph.D. (library and information science), Illinois at Urbana-Champaign, 1996. Web-based information architecture, bibliographic databases, communication in science and bibliometrics.

Steve Sawyer, Assistant Professor; D.B.A. (management information systems), Boston University, 1995. Work group performance and work group use of information technology, social and behavioral aspects of information technology, software development and software development management.

Barbara Settel, Associate Dean; M.L.S., Syracuse, 1976. Design and use of online retrieval systems, training end users, augmenting subject access to books in online catalogs.

Ruth V. Small, Associate Professor; Ph.D. (instructional design, development, and evaluation), Syracuse, 1986. Motivational aspects of information literacy, design and use of information and information technologies in education.

Zixiang (Alex) Tan, Assistant Professor; Ph.D. (telecommunications management and policy), Rutgers, 1996. Telecommunications policy and regulations, economic and social impacts of new technology, standardization policy, telecommunications in Asia.

Robert S. Taylor, Professor Emeritus; M.S. (library science), Columbia, 1950. Descriptions of organizational information environments and information-seeking behavior, definition of the information profession. Funded research: value-added processes in the information life cycle (National Science Foundation). Author, *Value Added Processes in Information Systems.*

Murali Venkatesh, Associate Professor; Ph.D. (management), Indiana, 1991. Group-based decision support systems, human-computer interaction, telecommunications.

Raymond F. von Dran, Professor and Dean; Ph.D. (library and information science), Wisconsin, 1976. Leadership and change in the management of communication and information technology; technology convergence and organizational change; competencies, curriculum, and organization structures in information education; effectiveness of modes of delivery in information education.

Rolf T. Wigand, Professor; Ph.D. (communication), Michigan State, 1975. Electronic commerce, information management, organizational communication, telecommunications policy, technology transfer. Author, *Organizations and Information Management,* Associate Editor, *The Information Society,* and Editor, *Communications.*

Ping Zhang, Assistant Professor; Ph. D. (information systems), Texas at Austin, 1995. Computer technology, information visualization for decision making, human-computer interaction.

Cooperating Adjunct Faculty

Bruce Derr, Senior Program Analyst, Academic Computing Services, Syracuse University. Microcomputers and information management.

Charles Ferguson, Account Executive, AT&T, Syracuse, New York. Telecommunications management.

William Gibbons, Former Associate Director of Marketing and Customer Network Design, NYNEX; current Vice President of the Central New York Communications Association; doctoral candidate in the information transfer program, Syracuse University.

Patricia Propert, Supervising Engineer, New York State Gas and Electric, Binghamton, New York. Telecommunications systems.

UNIVERSITY OF COLORADO AT BOULDER

Interdisciplinary Telecommunications Program

Programs of Study

For more than twenty-seven years, the University of Colorado's (CU) Interdisciplinary Telecommunications Program (ITP) has been a leader in telecommunications education. It offers a Master of Science (M.S.) and a Master of Engineering (M.E.) in telecommunications plus a joint M.B.A./M.S. in telecommunications (in conjunction with the CU College of Business Administration). The M.S. program requires 32 or 33 credit hours, and students are required to take courses in policy and regulation, economics, and technical aspects of telecommunications. Some courses on data, wireless communications, satellites, cable, optical fiber, voice and video communications systems, and network management are typically included in the technical portion of a student's program. Laboratory work on telephone switching systems and data communications are available, and additional material may be taken from both the electrical and computer engineering and the computer sciences departments. Courses on regulatory policy and standards at both the federal and international level are offered, along with courses on the economic aspects of the telecommunications industry. The M.S. program requires a thesis or a capstone project, and the M.E. program requires an extended project report. Graduates of the ITP typically work in the design, planning, and operation of communications systems. Representative employers include vendors, providers, user firms, governments, and consultants.

Research Facilities

The Interdisciplinary Telecommunications Program's Telecommunications Systems and Multi-Media Laboratory provides hands-on experience with the latest technologies in the field. It is available for research and training in digital and multimedia applications, data networking, video imaging, high-speed networking, and wireless systems. The ITP is also extensively supported by microcomputers, minicomputers, and mainframe computers. Terminals for student use are available throughout the Engineering Center where the program is housed.

The CU Engineering Center is a complex of classrooms, faculty areas, computing facilities, a library, and more than fifty research laboratories. Near the campus are government laboratories, including the National Institute of Standards and Technology, the National Oceanic and Atmospheric Administration, the National Center for Atmospheric Research, and the National Renewable Energy Lab. National laboratories and numerous industrial firms involve the program's students and faculty members in joint-research projects. Also, several major telecommunications companies are located in Colorado, making the ITP strategically situated for industry collaborations.

Financial Aid

Available financial aid includes graduate fellowships, research assistantships, teaching assistantships, and trainee positions. Some appointments cover tuition and fees and are comparable to those offered at other first-rate institutions. Loan funding is also available through outside sources.

Cost of Study

In the 1998–99 academic year, graduate in-state tuition for Colorado residents was $612 for up to 3 credit hours per semester. Nonresident graduate students paid $2487 for up to 3 credit hours. Student fees were approximately $350 per semester.

Living and Housing Costs

As of spring 1999, living expenses (with health insurance included) for a single student for twelve months are approximately $12,000. On-campus housing is $2644 per semester for a single dorm room with twenty-one meals a week. Information may be obtained from the Housing Office, Reservation Center, Campus Box 154, University of Colorado, Boulder, Colorado 80310.

Student Group

The Interdisciplinary Telecommunications Program, as of spring 1999, is composed of approximately 540 graduate students. About half are students on the Boulder campus and the other half are enrolled through the Center for Advanced Training in Engineering and Computer Science (CATECS), CU's distance learning program, and are studying via videotapes or satellite transmission in other parts of the U.S. and the world. About 30 percent of ITP students are international and 20 percent are women.

Location

The Boulder campus is located along the front range of the Rocky Mountains, where outdoor recreation opportunities include skiing, backpacking, fishing, mountain climbing, and cycling in a health and fitness–oriented environment. The area is rich with academic and cultural resources as well. Boulder is an affluent and educated community of 96,000 people located only 30 miles from Denver.

The University and The Program

Graduate study in engineering and applied science at the University of Colorado at Boulder is conducted within the framework of a large and diverse university with an international reputation. The mission of the University is to lead in discovery, communication, and the use of knowledge through instruction, research, and service to the public. As a comprehensive university, it provides each graduate student with an educational experience that is distinguished by the scope of its programs and course offerings, the outstanding quality of research facilities, the diversity of its student body, and the professionalism and dedication of its faculty.

Founded in 1971 with a grant from the National Science Foundation, the Interdisciplinary Telecommunications Program was developed in response to a need for people who have a comprehensive understanding of all the diverse elements of the profession. The program provides students with the opportunity to obtain the technical, business, and regulatory backgrounds necessary to be successful leaders in today's competitive global marketplace. The ITP offers challenging curriculum, extensive laboratories, and active relationships with telecommunications companies and personnel.

Applying

Admission and application information can be obtained by contacting the address below.

Correspondence and Information

Interdisciplinary Telecommunications Program
University of Colorado at Boulder
ECOT 317, Campus Box 530
Boulder, Colorado 80309-0530

Telephone: 303-492-8916
Fax: 303-492-1112
E-mail: itp@stripe.colorado.edu
World Wide Web: http://www.colorado.edu/ITP

University of Colorado at Boulder

THE FACULTY

James H. Alleman, Associate Professor; Ph.D. (economics), Colorado, 1976. Dr. Alleman was previously Director of the International Center for Telecommunications Management at the University of Nebraska, Director of Policy Research for GTE, and Economist for the International Telecommunications Union.

Gary L. Bardsley, Senior Instructor; M.S. (telecommunications), Colorado, 1991. Formerly Manager at US WEST, he has extensive experience in various phases of telecommunications.

Frank S. Barnes, Professor and Director of the ITP; Ph.D. (electrical engineering), Stanford, 1958; EE. Dr. Barnes conducts research in laser, fiber optics, microwave devices, and effects of electromagnetic fields on biological systems. He also teaches electronics and optical electronics and is a Fellow of the IEEE and AAAS. He was awarded the prestigious Distinguished Professor Award from CU in 1997.

Timothy X. Brown, Assistant Professor; Ph.D. (electrical engineering), Caltech, 1991; EE. Dr. Brown conducts research on mobile communication systems, switching, data networks, and neural networks and was formerly with the Jet Propulsion Laboratory and Bell Communications Research.

Michele H. Jackson, Assistant Professor; Ph.D. (speech communication), Minnesota, 1994. Dr. Jackson has numerous publications and teaches courses in the use of computer-based technologies in organizational communication.

Gerald A. Mitchell, Senior Instructor; M.S. (telecommunications), Colorado, 1991. Formerly on the technical staff with Bell Telephone Laboratories, a Manager at AT&T headquarters, and a Director at US WEST, he has extensive experience in most phases of telecommunications, with emphasis on traffic and systems engineering.

Jon R. Sauer, Professor; Ph.D. (physics), Tufts 1970. Formerly with AT&T, Bell Labs, Fermilab, and Indiana University, he also works on optical packet switching in the Optoelectronic Computing Systems Center. He has numerous publications and teaches courses in fiber optics and telecommunications computing.

Affiliated Faculty

Sharon K. Black, Lecturer; M.S. (telecommunications), Colorado, 1972; J.D., Denver, 1995. Previously Vice President of Telecommunications, COREMAR (subsidiary of NWNL Co.); Data Network Designer, Norwest Banks; and Analyst, Department of Commerce (NTIA), she has extensive experience in voice, data, satellite, microwave, and CATV systems. She currently practices international telecommunications law in Boulder.

Randall S. Bloomfield, Lecturer; M.E. (systems engineering), Virginia, 1979. An Electrical Engineer at the Institute for Telecommunications Service, NTIA, he has been a member of many U.S. delegations to standards-making bodies, especially the ITU and ANSI. He is a member of IEEE and AAAI.

Stanley E. Bush, Lecturer and Director of ITP's Laboratory; M.S. (electrical engineering), Rutgers, 1965; M.S. (business), Colorado State, 1973; EE. Formerly a technical supervisor at AT&T Bell Laboratories, he has extensive experience in the design and implementation of telecommunications networks. He has six patents and several published articles.

Harvey M. Gates, Associate Adjunct Professor; Ph.D. (electrical engineering), Denver, 1962; EE. Technical Vice President of BDM Corporation, Dr. Gates specializes in data communication networks and protocols. He has been a Manager for the U.S. Department of Commerce, National Telecommunications and Information Administration (NTIA).

Lewis L. House, Lecturer; Ph.D. (astrogeophysics), Colorado, 1962. President of Competitive Technology, Inc., and Competitive Intelligence Strategic Systems, he is formerly a Senior Director at US West and a Senior Scientist at the National Center for Atmospheric Research.

Richard B. Johnson, Lecturer; M.S. (telecommunications), 1982, J.D., 1975, Colorado. Senior Associate, HAI Consulting, he is formerly Director with US West Public Policy, Director of Network Planning with Sprint, Attorney with Teleport Communication Group and the federal government, and Telecommunications Engineer with GE. He has extensive experience in telecommunications policy development, regulatory and legislative activities and network planning, design, and operations.

Kenneth J. Klingenstein, Lecturer; Ph.D. (applied mathematics), Berkeley, 1975. He has been a leader in networking both nationally and internationally since 1985.

Robert A. Mercer, Adjunct Professor; Ph.D. (physics), Johns Hopkins, 1969. Dr. Mercer is the President of HAI Consulting, Inc., a telecommunications consulting firm. His work focuses on issues analysis and expert testimony related to the telecommunications infrastructure. Dr. Mercer has worked in the telecommunications field for more than twenty-five years and is a past member of the Board of Directors of the American National Standards Institute. He has taught numerous courses and seminars dealing with a variety of telecommunications topics.

Robbie Robertson, Lecturer; Ph.D. (mathematics), Colorado State, 1995. Senior computer scientist with Logicon Geodynamics, Inc., he conducts research in combinatorial optimization heuristics applied to problems that arise in sensor tracking systems and planning and scheduling systems. He has extensive practical experience in TCP/IP, UDP/IP, and World Wide Web applications development.

Timothy D. Schoechle, Lecturer; M.S. (telecommunications), Colorado, 1995. President of CyberLYNX Technologies, Inc., and Founder of BI Inc., he has more than twenty-five years of experience in the telecommunications, computer, semiconductor, and consumer electronics industries.

UNIVERSITY OF MARYLAND, COLLEGE PARK

Cross-Disciplinary M.S. in Telecommunications Program

Program of Study

The Cross-Disciplinary Master of Science in Telecommunications Program is designed to meet the increasing demands by the telecommunications industry for personnel who understand the technical aspects of complex telecommunication systems as well as the management, regulatory, and policy issues involved. The program is particularly well suited to students who have the goal of becoming technical managers in the telecommunications industry. The program has a technical core consisting of courses on telecommunications principles, systems, and networks that are taught by faculty members from the Department of Electrical Engineering and the Department of Computer Science and complemented by courses on telecommunications industry management and regulatory policy that are offered by the College of Business and Management and the School of Public Affairs. The curriculum is nearly equally divided between technical and business-oriented courses. Normally, students are required to have undergraduate degrees in engineering, computer science, math, or other technical areas because of the rigorous technical core.

Requirements for the degree include four technical courses, two courses on telecommunications policy and regulation, two courses on telecommunications management, two electives, 2 credits of a seminar, and a 3-credit project, for a total of 35 credit hours. The program can be pursued either full-time (completed in three semesters) or part-time. Classes are scheduled in the evening to accommodate students who work.

Research Facilities

Students can use the research facilities of the Department of Electrical Engineering, Department of Computer Science, College of Business and Management, and School of Public Affairs. These include extensive computer facilities with state-of-the-art personal computers and workstations that are located in several open laboratories; labs equipped with DSPs for signal processing applications, including speech coding; and network simulation facilities. The campus has an extensive library system for both technical and business research.

Financial Aid

Very limited financial aid in the form of reduced tuition is available based on excellence of academic performance and financial need.

Cost of Study

Tuition is currently $668 per credit.

Living and Housing Costs

Board and lodging are available in many private homes and apartments in College Park and the vicinity. Rooms in private homes range in cost from $250 to $350 a month, and one-bedroom apartments rent for an average of $600 per month. A list of accommodations, both University and private, is maintained by the University's's housing bureau.

Student Group

There were 50 students registered in the program during the spring 1999 semester, 22 of whom were full-time and 28 were part-time. There were 25 international students and 6 women. Fourteen students received financial aid in the form of partial tuition remission.

Student Outcomes

Graduates have been employed by companies and government agencies in the telecommunications industry, such as Bell Atlantic, BGE, GE Information Systems, Hughes Networks, LCC, MCIWorldCom, National Security Agency, Sprint, and Texas Instruments.

Location

The central campus of the University of Maryland is located in College Park, Maryland, a suburban area roughly between and within easy commuting distance of Washington and Baltimore. The campus is 12 miles from the White House. The museums, galleries, theaters, federal and special libraries, universities, concert halls, and abundant cultural activities of both cities offer students unlimited opportunities to participate in the cultural and social life of this thriving area. The immediate presence of many great national laboratories and technical facilities offers a particularly good opportunity for students.

The University and The Program

The University is one of the oldest and largest state universities in the country. The telecommunications program is located on the central campus, which has a total student population of approximately 33,000. The University offers many cultural and entertainment activities and has excellent athletic facilities.

Applying

Applicants should have a regionally accredited baccalaureate degree in engineering, computer science, or a related technical field with a B average or better. Application deadlines are August 1 (February 1 for international students) for the fall semester and December 1 (June 1 for international students) for the spring semester. Students should submit applications and an official college transcript to Graduate Admissions, 2107 Lee Building, University of Maryland, College Park, Maryland 20742. A copy of this material and three letters of recommendation must also be sent to the address below.

Correspondence and Information

Cross-Disciplinary M.S. in Telecommunications Program
Electrical Engineering Department
University of Maryland
2405 A. V. Williams Building
College Park, Maryland 20742

Telephone: 301-405-8189
Fax: 301-314-9281
E-mail: ashakuur@eng.umd.edu
World Wide Web: http://www.ee.umd.edu/ents

University of Maryland, College Park

THE FACULTY AND THEIR RESEARCH

Department of Electrical Engineering

William Destler, Professor and Dean, College of Engineering; Ph.D., Cornell. Microwave and millimeter-wave sources, accelerator technology.
Nariman Farvardin, Professor and Chair; Ph.D., Rennselaer. Communication systems, information theory, signal/image processing.
Evaggelos Geraniotis, Professor; Ph.D., Illinois. Communication networks, spread-spectrum systems, coding.
Armand Makowski, Professor; Ph.D., Kentucky. Stochastic control, queuing systems, applied stochastic processes.
Adrian Papamarcou, Associate Professor; Ph.D., Cornell. Statistical communications.
Mark Shayman, Professor; Ph.D., Harvard. Control theory, robotics, computer networks.
Steven A. Tretter, Associate Professor and Program Director; Ph.D., Princeton. Communication theory, coding, signal processing.

College of Business and Management

Michael Ball, Professor and Chair, Management Science and Statistics; Ph.D., Cornell. Applications of operations research to telecommunications, transportation systems, and manufacturing.
Curtis Grimm, Professor and Chair, Transportation, Business, and Public Policy; Ph.D., Berkeley. Strategic management in transportation industry, public policy issues in transportation and telecommunications.
Robert Krapfel, Associate Professor; Ph.D., Michigan State. Long-term buyer-seller relationships and associated risks.
Henry P. Sims Jr., Professor; Ph.D., Michigan State. Executive and managerial leadership, self-managing teams.
M. Susan Taylor, Professor and Chair, Management and Organization; Ph.D., Purdue. Career development, organizational recruitment, performance-feedback management.
Robert Windle, Associate Professor; Ph.D., Wisconsin. Transportation economics and policy issues.

School of Public Affairs

Steven Fetter, Associate Professor; Ph.D., Berkeley. International security and economic policy.

Department of Computer Science

Raymond Miller, Professor; Ph.D., Illinois. Parallel computation, distributed systems, computer networks, theory of computing.

UNIVERSITY OF MISSOURI–KANSAS CITY

Computer Science Telecommunications

Programs of Study

Computer Science Telecommunications (CST) at the University of Missouri–Kansas City (UMKC) offers courses leading to M.S.C.S. degrees in computer networking, software engineering, and telecommunications networking and Ph.D. degrees in computer networking, software architecture, and telecommunications networking through the University's interdisciplinary Ph.D. program. Computer Science Telecommunications holds an exciting and unique niche position, blending the telecommunications aspects of engineering, the networking aspects of computer science, and the software architecture aspects of information technology. The program offers in-depth education in the new technologies and skills most in demand in these areas. Graduate students have the opportunity to get a concentrated, state-of-the-art education in the most dynamic, challenging, and professionally significant specialty areas. The M.S. degrees are designed to prepare graduates for professional careers as project leaders and managers, to do research, and to go on to advanced studies. The M.S. degree requires 30 credit hours of course work in addition to the thesis or 36 credit hours with the nonthesis option. The Ph.D. degrees are designed to prepare graduates for research, teaching, and high-level management in the specialty areas of computer networking, software architecture, and telecommunications networking. The Ph.D. program entails course work beyond the master's degree and a dissertation based on original, supervised research in the specialty area. All of the graduate degrees address the theory, design, and application levels of communications software, hardware, and networks.

Research Facilities

CST is fully equipped with three instructional laboratories for teaching and student use. Equipment includes Alpha-based UNIX machines, a Windows NT Pentium-based network, fully networked Macintosh workstations, an Alpha-based VMS cluster, Sun computers, and several Alpha computers that run Linux. CST also houses state-of-the-art laboratories for graduate student and faculty research in each department. There is also a lab with graphic and character-based terminals. CST has a number of ATM switches and conducts research on a 38-gigahertz radio communications link. The location of UMKC in the telecommunications industry hub of Kansas City provides opportunities for students to do research and internships in cutting-edge industry laboratories by special arrangement. Library facilities include the UMKC Libraries, which maintain comprehensive journal subscriptions and offer a full range of reference services, and the private Linda Hall Library on campus, which is an internationally known science and technology resource.

The integration of communications technology into the academic enterprises of a comprehensive university poses a major challenge. A group composed of CST faculty members, staff members, and students is engaged in developing precommercial prototypes of software products derived from applied research in multimedia networking and distributed information and collaboration systems. As part of these initiatives, some UMKC courses are restructured so they may be offered asynchronously over the Internet, independent of time and place. Students at CST have the opportunity to participate in these and other exciting cutting-edge developments as research assistants, gaining valuable hands-on applied research experience in multimedia technologies.

Financial Aid

The Computer Science Telecommunications program has approximately forty graduate teaching and research assistantships each semester for quarter-time or half-time support. Research assistantships are also available for advanced students from faculty members with research grants and contracts. The Chancellor's Non-Resident Award, which covers the nonresident portion of UMKC tuition and fees for the first semester of study, is awarded to students who are nominated to the University Scholarship Office by the graduate faculty. Other forms of financial support include grants and loans from state and national programs.

Cost of Study

In 1999–2000, graduate tuition is $187.60 per credit hour for Missouri residents and $524.60 for nonresidents. For Computer Science Telecommunications courses, there is a fee of $17.50 per credit hour to support student-directed equipment purchases. Additional costs include application fees, books and course materials, and thesis processing.

Living and Housing Costs

The UMKC Office of Financial Aid suggests that the costs of housing, food, medical and dental insurance, and transportation average $10,200 per year for graduate students. On-campus housing is limited; however, off-campus housing is readily available.

Student Group

In fall 1998, the Computer Science Telecommunications program had 409 undergraduates, 131 master's students, and 31 Ph.D. students. Of the graduate students, approximately half were full-time. Seventy-three percent were men and 27 percent were women, and 79 percent were international students. Approximately half of the full-time students received financial aid. The CST graduate faculty seeks student applicants with a solid technical preparation and a strong interest in high-level applications and theoretical training in its specialized fields.

Student Outcomes

Computer Science Telecommunications graduates have gone on to technical, research, and managerial positions with industry leaders, including AT&T, Cisco Systems, Digital Equipment, GTE, Hewlett-Packard, IBM, Lucent, MCI, NorTel, Sprint, and Sprint PCS. Graduates also manage information systems and communications networks across a wide range of industries. The demand for graduates is high, and most students have a wide range of employment opportunities upon graduation.

Location

UMKC is located in the cultural heart of Kansas City, three blocks from the Nelson-Atkins Gallery of Art and the shopping and entertainment facilities of the Country Club Plaza. Major-league soccer, football, and baseball, in addition to numerous trendy shopping and historic sites, round out the recreational opportunities in this culturally rich, medium-sized (population 1.6 million) metropolitan area.

The University and The Program

UMKC is one of four campuses in the University of Missouri System. It was chartered in 1929 as the University of Kansas City, a private liberal arts college established by business and civic leaders, and in 1963 merged with the University of Missouri System to become UMKC. Computer Science Telecommunications was established in 1984 with support from the Sprint Corporation to meet the growing needs for research and highly trained employees in the telecommunications industry.

Applying

Application deadlines are March 1 for fall admission and financial aid consideration, May 1 for summer or fall, and October 1 for winter. Applicants should have a sound background in computer science and mathematics. A minimum GPA of 3.5 (for Ph.D.) or 3.0 (for M.S.) and a GRE quantitative score in at least the 85th (for Ph.D.) or 75th (for M.S.) percentile are required. International applicants also need a score of at least 550 on the TOEFL.

Correspondence and Information

For application and general information:

Computer Science Telecommunications
University of Missouri–Kansas City
5100 Rockhill Road, Room 207
Kansas City, Missouri 64110
Telephone: 816-235-1193
E-mail: info@cstp.umkc.edu
World Wide Web: http://www.cstp.umkc.edu

For specific questions about the program:

Dr. Richard G. Hetherington, Director and Professor
Computer Science Telecommunications
University of Missouri–Kansas City
5100 Rockhill Road, Room 207
Kansas City, Missouri 64110
Telephone: 816-235-1193
E-mail: hetherington@cstp.umkc.edu

University of Missouri–Kansas City

THE FACULTY AND THEIR RESEARCH

Jagan Agrawal, Professor; Ph.D., North Carolina State, 1972. Computer communications, telecommunications, digital communications, digital signal processing.

Kenneth Blundell, Associate Professor; Ph.D., Nottingham (England), 1977. Software engineering, artificial intelligence, neural networks.

Lein Harn, Professor; Ph.D., Minnesota, 1984. Digital signal processing, digital filter design, digital speech and image processing, cryptosystem design, data/video encryption.

Richard Hetherington, Professor and Director; Ph.D., Wisconsin–Madison, 1961. Computer networking, numerical analysis.

Mary Lou Hines, Assistant Professor; Ph.D., Kansas State, 1992. Object-oriented databases, software metrics, software engineering.

Vijay Kumar, Associate Professor; Ph.D., Southampton (England), 1983. Database management systems, distributed systems, main memory database management systems, concurrency control, database recovery, real-time systems.

Deep Medhi, Associate Professor; Ph.D., Wisconsin–Madison, 1987. Computer/communications network modeling, routing, and design; network survivability; large-scale optimization algorithms; network management.

E. K. Park, Professor; Ph.D., Northwestern, 1988. Software engineering, software architectures, object-oriented design/analysis, formal methods, programming methods, distributed systems, real-time processing.

Jerry Place, Associate Professor; Ph.D., Kansas, 1984. Distributed systems, parallel computer architecture, computer and network performance analysis, simulation and queuing theory, real-time systems.

Xiaojun Shen, Associate Professor; Ph.D., Illinois at Urbana-Champaign, 1989. Parallel processing, interconnection networks, algorithms.

Khosrow Sohraby, Professor; Ph.D., Toronto, 1985. Design and analysis of high-speed computer and communication networks, networking and design aspects of wireless and mobile communications, analysis of algorithms, parallel processing and large-scale computations.

Jerry Stach, Assistant Professor; Ph.D., Union (Ohio), 1995. Formalisms required for network software architecture, domain theory, concurrency algebras, computational models of network services.

Adrian Tang, Professor; Ph.D., Princeton, 1974. Computer networks, operating systems, programming languages and semantics, domain theory, protocol specifications.

Appie Van de Liefvoort, Associate Professor; Ph.D., Nebraska, 1982. Queuing theory and performance modeling, matrix-exponential distribution, performance modeling of computer and communication networks, algorithms and complexity.

UNIVERSITY OF PENNSYLVANIA

School of Engineering and Applied Science
Telecommunications and Networking Program

Program of Study

The Telecommunications and Networking Program (TCOM) is a unique professional master's program for both full- and part-time students. This program provides the most up-to-date knowledge—both technical and business—for the telecommunications and networking leaders of tomorrow. Multidisciplinary in nature, this cutting-edge program draws its faculty and courses from several School of Engineering and Applied Science departments, including Computer and Information Science, Electrical Engineering, and Systems Engineering, and from the Wharton School of Business. This multidisciplinary approach gives students the flexibility to tailor the curriculum to their specific interests, background, and career goals. The University of Pennsylvania is one of a small group of universities in the United States to grant a master's degree in telecommunications engineering.

The School of Engineering structured this program's innovative curriculum to respond to the constant change inherent in telecommunications and networking and to address the increasingly complex demands placed on current and future telecommunications managers in the private, public, and military sectors. Courses cover a broad range of telecommunications and networking issues while reinforcing a systems approach. Since the program takes a holistic rather than an individualistic approach, students learn everything about telecommunications systems, from hardware and software technologies to societal and management issues.

Many different perspectives, including computer and information science, electrical engineering, systems engineering, business, and policy, enrich the program. Each perspective is essential, but the combination is the key. These multiple perspectives allow for an exchange of ideas among a diverse group of faculty members and peers that extends beyond the classroom.

The degree requires ten course units, which can be completed in one year on a full-time basis. These are three basic telecommunications courses; one course on probability and stochastic processes; one course from a choice of business, policy, or societal issues; two additional telecommunications courses; and three electives. The three basic telecommunications courses are Networking—Technology, Protocols, and Practice; Networking—Theory and Fundamentals; and Broadband Networking. One of the electives, a telecommunications laboratory course, gives students hands-on experience with routers and Ethernet switches that are set up to emulate real-world networks. Business courses taught by Wharton faculty members in policy, marketing, and societal issues provide students with corporate survival skills.

Research Facilities

Most of the computers used for teaching are PCs with Windows NT or Windows 95. Some of the computers are SUN Workstations and servers with Sun's version of UNIX. Each computer is connected to an Ethernet network and supports the TCP/IP protocol. The Ethernets are gatewayed to the Penn Campus Network and the Internet. Students work with the vast array of utilities that Sun provides with its UNIX workstations as well as many other software packages and languages. Students can also use the wide variety of laboratories run by the program's contributing departments and schools. Laboratories include a Distributed Systems Laboratory, a Signal Processing Laboratory, and a Video Processing and Telecommunications Laboratory.

Financial Aid

Though financial aid is not available, students can apply for a variety of subsidized and unsubsidized loans with the assistance of the University Student Financial Services. Part-time students are also encouraged to check on their company's plan for tuition reimbursement. Many companies cover all or part of the cost of the program.

Cost of Study

Tuition for the academic year 1999–2000 is $23,670. There is a general fee of $1546 and a technology fee of $445 for full-time students. For part-time students, tuition is $2996 per course unit. The general fee is $188, and the technology fee is $60.

Living and Housing Costs

On-campus housing is available for both single and married students. Residences for single students cost $625 and up per month; for shared living, $510 and up per month; and for a private apartment, $815 and up per month. Housing costs for married couples range from $675 to $1065 per month. Many privately owned apartments exist in the immediate area.

Student Group

There are approximately 20,000 students at the University; more than 10,000 are enrolled in graduate and professional schools. Of these, 746 are in the graduate engineering program. The typical background of students in the Telecommunications and Networking Program consists of an undergraduate degree in engineering, physical science, or mathematics. Approximately 50 percent of the students in the program are part-time students. The student body is diverse and includes many international students.

Location

The University is located in West Philadelphia, close to the heart of the city. Philadelphia's renowned museums, concert halls, theaters, sports arenas, and parks provide students with rich cultural and recreational outlets. Equidistant from New York City and Washington D.C., Philadelphia is also conveniently close to the Jersey shore, Pennsylvania Dutch country, and the Poconos.

The University and The School

Since 1946, when the University of Pennsylvania dedicated ENIAC, the world's first large-scale, general-purpose computer, and offered its first graduate-level computer course, Penn has been at the forefront of computer technology. Continuing its commitment to technological advancement, in 1996, the School of Engineering and Applied Science created this new professional master's program in the ever-expanding field of telecommunications and networking.

Applying

Admission is competitive and is performed on a rolling basis. For the fall 2000 semester, applications are due by June 1, 2000. Candidates may apply by submitting an application to the address below or through Penn's online express application at http://sentry.isc.upenn.edu/ws/expressapp. Admission is based on the student's past record as well as letters of recommendation. The program is looking for students with an engineering, mathematics, or physical science degree from an accredited undergraduate program, although other degrees are considered. Students should have a grade point average of 3.0 on a 4.0 scale. Industrial experience is desirable but not required. All students whose native language is not English must take the Test of English as a Foreign Language (TOEFL); the minimum score accepted is 600.

Correspondence and Information

For information:
SEAS Professional Education Programs
Towne Building, Room 119
University of Pennsylvania
Philadelphia, Pennsylvania 19104-6391
Telephone: 215-898-0696
Fax: 215-573-9673
E-mail: tcom@seas.upenn.edu
WWW: http://www.seas.upenn.edu/com

For admissions:
Office of Graduate Admissions
School of Engineering and Applied Science
Towne Building, Room 111
University of Pennsylvania
Philadelphia, Pennsylvania 19104-6391
Telephone: 215-898-9246

University of Pennsylvania

THE FACULTY AND THEIR RESEARCH

Program Faculty

G. Anandalingam, Professor of Systems Engineering, Information Technology, and Operations and Information Management and Chair, Department of Systems Engineering; Ph.D., Harvard, 1981. Applied optimization, Bayesian analysis, game theory, and economic systems analysis; development of new implementations of simulated annealing and multilevel mathematical programs.

Magda El Zarki, Associate Professor of Electrical Engineering and Computer and Information Science and Director of the Telecommunications and Networking Program; Ph.D., Columbia, 1988. Packet video (audio/video synchronization, error concealment schemes, and quality of service measures) and wireless networks (channel allocation schemes, cell sizes, capacity versus performance versus receiver and base station complexity, media access schemes for multimedia services, decentralized control, and transmission of video services over wireless channels).

David Farber, Alfred Fitler Moore Professor of Telecommunication Systems and Director of the Distributed Systems Laboratory; M.S., Stevens, 1961. Design of high-speed networks, distributed operating systems capable of operating with these networks, and the integration of computers and communications to create superlarge geographical distributed clustered computer systems.

Michael Greenwald, Assistant Professor of Computer and Information Science; Ph.D., Stanford, 1998. Network protocol performance, congestion control, synchronization of distributed systems, structuring techniques for very-large-scale systems.

Roch Guerin, Alfred Fitler Moore Professor of Telecommunications Networks; Ph.D., Caltech, 1977. Networking and quality-of-service (traffic analysis, flow and congestion control, scheduling and buffer management algorithms, quality-of-service routing) and the intersection of the two.

Saleem Kassam, Professor of Electrical Engineering; Ph.D., Princeton, 1975. Nonlinear and adaptive filtering in signal processing and communications, video transmission and wireless channels, signal detection and estimation, robust signal processing, spatial array processing, imaging and image processing.

Insup Lee, Professor of Computer and Information Science; Ph.D., Wisconsin–Madison, 1983. Distributed systems, formal methods, mobile computing, operating systems, programming languages, real-time computing, software engineering.

Barry Silverman, Professor of Systems Engineering; Ph.D., Pennsylvania, 1977. Knowledge management (acquisition, modeling, inference, maintenance), intelligent software agents for human task performance support, medical informatics.

Jonathan Smith, Associate Professor of Computer and Information Science; Ph.D., Columbia, 1989. Distributed systems, operating systems, multimedia communications systems, applications of randomness, computer security and cryptology.

Santosh Venkatesh, Associate Professor of Electrical Engineering; Ph.D., Caltech, 1987. Neural networks, computational learning and pattern recognition.

Affiliated Faculty

Eric K. Clemons, Professor of Operations and Information Management and Management, Wharton School of Business; Ph.D., Cornell, 1976. Information technology and business strategy, information technology and financial markets, risk management of strategic information technology implementations, strategic implications of electronic commerce for channel power and profitability.

Susan B. Davidson, Professor of Computer and Information Science and Biotechnology Program; Director, Computational/Bioinformatics Track of the Biotechnology Program; and Co-Director, Center for Bioinformatics; Ph.D., Princeton, 1982. Distributed systems, database systems, real-time systems.

Jeshoshua Eliashberg, Sebastian S. Kresge Professor of Marketing and Professor of Operations and Information Management, Wharton School of Business; D.B.A., Indiana, 1978. Marketing research, marketing/operations research and development interface, international marketing, new product forecasting and planning models, marketing issues related to the media and entertainment industry.

Gerald Faulhaber, Professor of Public Policy and Management, Wharton School of Business; Ph.D., Princeton, 1975. Internet, technology, telecommunications, regulation, industrial organization, applied microeconomics and banking.

Carl Gunter, Associate Professor, Computer and Information Science, Ph.D., Wisconsin–Madison, 1985. Programming languages, semantics, logic, software specification.

Dwight L. Jaggard, Professor of Electrical Engineering and Associate Dean for Graduate Education and Research, School of Engineering and Applied Science; Ph.D., Caltech, 1976. Electromagnetic chirality, fractal electrodynamics, scattering from knots and imaging and inverse scattering.

Steven O. Kimbrough, Associate Professor of Operations and Information Management, Wharton School of Business; Ph.D., Wisconsin, 1982. Decision support systems, electronic commerce, artificial intelligence and computational rationality, logic modeling, evolutionary computation, including genetic algorithms and genetic programming.

Scott M. Nettles, Assistant Professor of Computer and Information Science; Ph.D., Carnegie Mellon, 1995. Design, implementation, and evaluation of programming languages, operating systems, and database systems; memory management and garbage collection; transaction and distributed computing support for high-level languages.

Lyle Ungar, Associate Professor of Computer and Information Sciences, secondary appointments in Chemical Engineering and Systems Engineering, and Director, SEAS Professional Education Programs; Ph.D., MIT, 1984. AI, machine-learning, and statistical methods for automatic construction of multivariable nonlinear models.

Yoram (Jerry) Wind, The Lauder Professor, Professor of Marketing, and Director, SEI Center for Advanced Studies in Management; Ph.D., Stanford, 1966. Marketing strategy, business portfolio analysis and strategy, new product and business development, global marketing strategy, expert systems for marketing decisions, marketing-driven corporate strategy and growth strategies.

UNIVERSITY OF PITTSBURGH

Department of Information Science and Telecommunications
Graduate Program in Telecommunications

Programs of Study

The telecommunications program offers a Master of Science in Telecommunications (M.S.T.) and a Certificate of Advanced Study (C.A.S.). The Ph.D. is offered in information science with a concentration in telecommunications. The telecommunications program is based in the Department of Information Science and Telecommunications and draws from resources in the Departments of Computer Science, Electrical Engineering, and Communication as well as the Katz Graduate School of Business. The M.S.T. is a 48-credit program of ten required courses and six electives. Areas of specialization include systems and technology, computer networking, and network administration, policy, and applications. At the end of the program, master's students are prepared to take jobs with telecommunications vendors, providers, user companies, consulting firms, and R&D labs as network designers and managers or telecommunications voice, data, or image specialists. The post-master's Certificate of Advanced Study allows students with a master's degree in a related field to explore a special area of interest or to update skills. The Ph.D. program provides research-oriented study and professional specialization in telecommunications. Requirements for the degree are 84 course or seminar credits beyond the bachelor's degree including the required courses for the M.S.T. degree at the University of Pittsburgh, an acceptable master's thesis or independent research paper, successful completion of the preliminary and comprehensive examinations, three terms of full-time academic study on campus, at least 18 dissertation credits, and submission and defense of a dissertation.

Research Facilities

Departmental computing and networking labs are housed in a modern 5,000-square-foot area. The school's computational facilities include a Sun Enterprise 4000 SMP ten-processor system with associated Sun ULTRAsparc 2 systems that collectively are arranged via ATM interconnects to supply both load-sharing and course-grain-distributed computing facilities in order to support research activities. The above cluster storage needs are addressed by two RAID arrays that facilitate large-scale data gathering and processing. A HYPERsparc upgraded Sun 670 serves as a general student access system. Workstations in the labs include Sun SPARC 5s, IPXs, and IPCs. Two labs are configured as classrooms with Pentium systems. The internal labs' network environment is being upgraded to employ a mixture of both ATM and fast Ethernet technology. Laser printing is provided throughout the labs. The telecommunications and network labs are built around a heterogeneous collection of UNIX workstations and microcomputer systems. Ethernet, token ring, and ATM test network environments are maintained for both research as well as to supplement classroom instruction. These labs also house training, diagnostic, and testing equipment for AT&T phone systems, T1 and T3 connections, and M13, FT3C, and TASI transmission equipment. All workstations and PCs in the labs are linked via a local area network to general-purpose University UNIX and VMS systems. Additional University computer facilities include an advanced technology and computer graphics lab and other labs located throughout the campus. All students have access to national and international electronic mail and assorted other network services.

Financial Aid

In 1998–99, 26 percent of full-time graduate students received financial aid from the University, most in the form of full or half graduate assistantships. Full graduate student assistants (GSAs) earned $4325 per term plus remission of tuition and assisted a faculty member for 20 hours per week. Half GSAs earn half the stipend and remission of half the tuition for 10 hours of work per week. Research assistantships and fellowships may be available, depending on external funding support. Financial aid awards are granted on the basis of academic achievement and financial need. Assistantships are normally awarded for the fall and spring terms; they may also be offered for the summer term if the budget permits.

Cost of Study

The tuition per term (four months) in 1998–99 for full-time study (9–15 credits) was $8680. Because Pitt is state-related, the University receives funding from the state, which enabled it to reduce tuition for Pennsylvania residents to $4054 per term. Tuition for part-time students was $720 per credit for out-of-state students and $348 per credit for residents of Pennsylvania.

Living and Housing Costs

Pittsburgh, ranked by Rand McNally among the most livable cities in the United States, is noted for its low cost of living. Average monthly rent is $400 for a one-bedroom apartment and $575 for a two-bedroom apartment. It is estimated that students require at least $2200 per term to cover living expenses exclusive of tuition. Comfortable and affordable housing in attractive residential neighborhoods is readily available within walking distance of the University.

Student Group

The current enrollment in telecommunications programs is 96 graduate students, of whom approximately 48 percent are full-time. Approximately 34 percent are international students and 20 percent are women. Members of minority groups represent about 12 percent of the U.S. students.

Location

The University is located in the heart of the city's educational center, with museums of art and natural history, music and lecture halls, Carnegie Mellon University and two smaller colleges, restaurants and shops, and a 450-acre park adjacent to the campus. The downtown corporate and cultural center is just a 10-minute bus ride away.

The University

The University of Pittsburgh, a privately organized state-related institution, enrolls approximately 32,000 students in twenty-two different schools. The telecommunications program is based in the School of Information Sciences, which has an enrollment of 733 students in one undergraduate and three graduate programs. This context provides collegial activity with the Department of Information Science and Telecommunications and the Department of Library and Information Science in such areas as distributed databases, digital libraries, parallel computing, human-computer interaction, and applications.

Applying

Applicants for all programs must submit a recent score from the GRE General Test. Requirements for admission to the M.S.T. degree program are a degree from an accredited college or university with a 3.0 GPA, working knowledge of a scientific programming language (preferably C), and 3-credit courses in calculus and probability. Ph.D. applicants must give evidence of superior scholarship, mastery of a specialized field of knowledge, and the ability to do significant and relevant research. They must have a master's degree from an accredited related program with a QPA of 3.3 or better and the same prerequisites as the M.S.T. degree program. Provisional acceptance into any of the degree programs may be granted to students lacking some of the prerequisites, with the condition that deficiencies be made up during the first two terms. A $30 application fee is required ($40 for international applicants). Deadlines for receipt of application materials are July 1 for September admission and November 1 for January admission. Applications for financial aid should be submitted by January 15 for fall term and October 1 for spring term. Candidates are usually notified of acceptance within six weeks of receipt of all application materials.

Correspondence and Information

Admissions Coordinator
505 SLIS Building
University of Pittsburgh
Pittsburgh, Pennsylvania 15260

Telephone: 412-624-5146
Fax: 412-624-2788
E-mail: arlene@tele.pitt.edu
World Wide Web: http://www.tele.pitt.edu

University of Pittsburgh

THE FACULTY AND THEIR RESEARCH

Toni Carbo, Professor and Dean, School of Information Sciences; Ph.D., Drexel. National and international information policies, measurement and use of scientific and technical information in the economy, education for the information professions.

Sujata Banerjee, Assistant Professor; Ph.D., USC. Design and analysis of high-speed networking protocols, traffic modeling, network reliability, concurrency control, failure recovery of distributed database systems.

Kenneth Sochats, Assistant Professor; M.S.E.E., M.B.A, Pittsburgh. Information networks, simulation, databases, artificial intelligence, management information systems (MIS), systems analysis and design, software engineering, network design, microcomputer applications, graphics.

Richard A. Thompson, Professor and Codirector, Telecommunications Program; Ph.D., Connecticut. Communications switching systems, especially photonic switching; intelligent networks; terminals, user services, and the human interface; fault tolerance and cellular automata; probabilistic formal languages.

David W. Tipper, Associate Professor; Ph.D., Arizona. Design and performance analysis of computer and telecommunication networks, control of communication networks, simulation methodology, queuing theory with emphasis on nonstationary/transient behavior, network survivability, application of control theory to communication networks and queuing systems.

Martin B. H. Weiss, Associate Professor and Codirector, Telecommunications Program; Ph.D., Carnegie Mellon. Telecommunications policy, technical standards, economic models, information system capacity management, network management and control.

James G. Williams, Professor; Ph.D., Pittsburgh. Information systems, networks, systems design, software engineering, simulation, system architecture, client server computing, database management.

Taieb Znati, Associate Professor; Ph.D., Michigan State. Real-time communication networks and protocols to support multimedia environments, multimedia synchronization and presentation, design and analysis of medium access control protocols to support distributed real-time systems, network performance.

Research Activities

Applications and services.
Computer networking protocols.
Distributed databases and distributed processing.
Economic models for telecommunications.
Gigabit networking.
Intelligent networks.
Network architecture and design.
Photonic switching.
Policy and regulation.
Switching network architectures and control.
TERN: Telecommunications Education and Research Network.

Research and Training Opportunities in Engineering and Applied Sciences

This part of Book 5 consists of one section covering research and training opportunities in engineering and applied sciences. The section contains a table of contents; a profile directory, which consists of brief profiles of the academic centers and institutes followed by 50-word and 100-word Announcements, if centers and institutes have chosen to submit such entries; and In-Depth Descriptions, which are more individualized statements, if centers and institutes have chosen to submit them.

Section 21
Research and Training Opportunities

Academic Centers and Institutes

The role and importance of academic centers and institutes in the graduate study experience has increased dramatically in recent years. In response to growing requests for information on such centers and institutes, the profiles in this section include the data on academic centers and institutes that were submitted in 1999 by each institution in response to Peterson's Supplemental Survey of Academic Centers and Institutes.

This section provides detailed information on university-owned and university-operated centers and institutes offering graduate students research or study opportunities. To qualify for inclusion in this section, a center or institute must be a formal and integral part of a graduate degree program. Such centers and institutes are separate from, but may sometimes maintain affiliations with, other special research facilities also located on the university's campus, such as laboratory or computer facilities.

Centers and institutes listed are academic in nature and do not include university business, administrative, or operational units or departments. Most have formally dedicated faculty and staff members associated with them and may provide training programs in the early part of a Ph.D. program. Many are interdisciplinary; however, some centers and institutes may focus on a single discipline, a major research project, or a specialized area of study. In some cases, graduate degrees may be awarded by the institute or research unit, although most do not.

Centers and institutes appear alphabetically by institution, followed by in-depth entries submitted by centers or institutes that chose to prepare detailed descriptions. The following items appear for each center or institute profile when available. Readers may contact centers and institutes directly for further information.

Name of Center or Institute. The name of the center or institute appears in boldface type.

Founding Year. The year the center or institute was established.

Academic Areas of Research and Training. Specific areas of graduate research or training listed (e.g., cancer research in molecular and cell biology, epidemiology, and psychology).

Degrees Offered. For those centers and institutes that do award degrees, these may include master's and/or doctoral degrees in specific academic fields or areas.

Graduate Students Served Last Academic Year. Figures are provided separately for the total number of students served in the last academic year and, for some institutions, how many students were served specifically at the master's, doctoral, and postgraduate levels, respectively.

Faculty. Figures are provided for the total number of faculty members associated with the center or institute and those associated with the center or institute but having their primary affiliation with another unit of the university.

Faculty Affiliations. The names of departments, programs, and other units with which faculty members are affiliated are listed here.

Annual Research Budget. Figures for the center or institute's annual research budget are listed.

Director. The name and title of the center or institute's director is provided, along with his/her address, telephone number, fax number, and e-mail address.

Information Contact. Provides the name and title of the person who should receive inquiries from interested students, with the address, telephone number, fax number, and e-mail address for this individual.

CONTENTS

Research and Training Opportunities

Alabama Agricultural and Mechanical University, Normal, AL 35762-1357.

The Howard J. Foster Center for Irradiation of Materials Founded in 1990. *Academic areas of research and training:* Provide research and services capabilities on materials processing and materials characterization; ion implantation, electron and ion microscopy; optical and electrical properties measurement; surface and interface processing. *Degrees offered:* None. *Graduate students served last academic year:* 12: 6 at the master's level; 6 at the doctoral level. *Faculty:* 24: 6 affiliated solely with the center. *Faculty affiliations:* Departments of Biology, Civil Engineering, Chemistry, Computer Science, Mathematics, Mechanical Engineering, Physics. *Annual research budget:* $1.2 million. *Director:* Dr. Daryush Ila, Director, 205-851-5866, Fax: 205-851-5868, E-mail: ila@cim.aamu.edu.

Research Institute Founded in 1999. *Academic areas of research and training:* Provide business type environment to conduct contracts for all agencies using the capabilities at the university. *Degrees offered:* None. *Graduate students served last academic year:* 6: 3 at the master's level; 1 at the doctoral level; 2 at the postgraduate level. *Faculty:* 68: 3 affiliated solely with the institute. *Faculty affiliations:* Departments of Biology, Chemistry, Civil Engineering, Computer Science, Environmental Science, Food Science, Mathematics, Mechanical Engineering, Physics, Urban Planning; School of Business. *Annual research budget:* $1 million. *Director:* Dr. Daryush Ila, Director, 205-851-5877, Fax: 205-851-5868, E-mail: ila@cim.aamu.edu.

Arizona State University, Tempe, AZ 85287.

Manufacturing Institute Founded in 1997. *Academic areas of research and training:* Virtual manufacturing, rapid prototyping, MEMS, laser fabrication. *Degrees offered:* None. *Graduate students served last academic year:* 25: 15 at the master's level; 5 at the doctoral level; 5 at the postgraduate level. *Faculty:* 23: 3 affiliated solely with the institute. *Faculty affiliations:* Departments of Computer Science and Engineering, Electrical Engineering, Industrial Engineering, Materials Engineering, Mechanical Engineering; College of Business. *Annual research budget:* $2 million. *Director:* Dr. Ampere Tseng, Co-Director, 480-965-8201, Fax: 480-965-2910, E-mail: ampere.tseng@asu.edu.

Auburn University, Auburn, Auburn University, AL 36849-0002.

Alabama Microelectronics Science and Technology Center (AMSTC) Founded in 1984. *Academic areas of research and training:* Device fabrication, device physics, microelectronics, packaging, stress sensors, low temperature electronics, high temperature electronics, electronics assembly, compound semiconductors, electronic cooling. *Degrees offered:* None. *Graduate students served last academic year:* 30: 20 at the master's level; 10 at the doctoral level. *Faculty:* 25. *Faculty affiliations:* Departments of Business, Chemistry, Electrical Engineering, Materials Engineering, Mechanical Engineering, Physics. *Annual research budget:* $1.5 million. *Director:* Dr. Richard C. Jaeger, Professor, 334-844-1871, Fax: 334-844-1888, E-mail: jaeger@eng.auburn.edu.

Materials Research and Education Center Founded in 1995. *Academic areas of research and training:* Materials engineering. *Degrees offered:* MS; PhD; M Mtl E. *Graduate students served last academic year:* 55: 12 at the master's level; 40 at the doctoral level; 3 at the postgraduate level. *Faculty:* 28: 8 affiliated solely with the center. *Faculty affiliations:* Departments of Chemical Engineering, Chemistry, Electrical Engineering, Mechanical Engineering, Physics; Interdepartmental Program in Textile Science. *Annual research budget:* $2 million. *Director:* Dr. Bryan A. Chin, Director, 334-844-3322, Fax: 334-844-3400, E-mail: bchin@eng.auburn.edu.

Pulp and Paper Research and Education Center Founded in 1986. *Academic areas of research and training:* Botany, cell science, chemical engineering, forestry, industrial engineering, materials engineering, mechanical engineering, microbiology, electrical engineering. *Degrees offered:* None. *Graduate students served last academic year:* 15: 6 at the master's level; 7 at the doctoral level; 2 at the postgraduate level. *Faculty:* 26: 1 affiliated solely with the center. *Faculty affiliations:* Departments of Botany and Microbiology, Cell Science, Chemical Engineering, Electrical Engineering, Forestry, Industrial Engineering, Materials Engineering, Mechanical Engineering. *Annual research budget:* $372,000. *Director:* Harry Cullinan, Director, 334-844-2016, Fax: 334-844-2045, E-mail: cullinan@eng.auburn.edu.

Thomas Walter Center for Technology Management Founded in 1989. *Academic areas of research and training:* Technology management, manufacturing, engineering practice. *Degrees offered:* None. *Graduate students served last academic year:* 9: 8 at the master's level; 1 at the doctoral level. *Faculty:* 42: 2 affiliated solely with the center. *Faculty affiliations:* Colleges of Business, Engineering. *Annual research budget:* $200,000. *Director:* Dr. M. Dayne Aldridge, Director, 334-844-4333, Fax: 334-844-1678, E-mail: aldridge@eng.auburn.edu.

Boston University, Boston, MA 02215.

BioMolecular Engineering Research Center Founded in 1991. *Academic areas of research and training:* Computational biology, bioinformatics. *Degrees offered:* None. *Graduate students served last academic year:* 6: 1 at the master's level; 3 at the doctoral level; 2 at the postgraduate level. *Faculty:* 9: 3 affiliated solely with the center. *Faculty affiliations:* Departments of Biology, Biomedical Engineering, Computer Science, Pharmacology and Experimental Therapeutics; Centers for Advanced Biotechnology, Human Genetics. *Director:* Dr. Temple F. Smith, Director, 617-353-7123, Fax: 617-353-7020. *Application contact:* Nancy Sands, Center Administrator, 617-353-7123, Fax: 617-353-7020, E-mail: sands@darwin.bu.edu.

Center for Advanced Biotechnology Founded in 1992. *Academic areas of research and training:* Biomaterials, bioanalytical procedures, genomics. *Degrees offered:* None. *Graduate students served last academic year:* 12: 2 at the master's level; 4 at the doctoral level; 6 at the postgraduate level. *Faculty:* 3: 3 affiliated solely with the center. *Faculty affiliations:* Departments of Biomedical Engineering, Chemistry. *Annual research budget:* $1 million. *Director:* Dr. Charles Cantor, Director, 617-353-8500, Fax: 617-353-8901, E-mail: ccantor@bu.edu. *Application contact:* Caitlin Ackley, Administrator, 617-353-8902, Fax: 617-353-8901, E-mail: cabstaff@darwin.bu.edu.

Center for Computational Science Founded in 1990. *Academic areas of research and training:* Interdisciplinary computational science. *Degrees offered:* None. *Graduate students served last academic year:* 5: 5 at the doctoral level. *Faculty:* 43: 3 affiliated solely with the center. *Faculty affiliations:* Departments of Aerospace and Mechanical Engineering, Chemistry, Cognitive and Neural Systems, Computer Science, Electrical and Computer Engineering, Geography, Manufacturing Engineering, Mathematics, Physics. *Annual research budget:* $1.6 million. *Director:* Dr. Claudio Rebbi, Director, 617-353-9058, Fax: 617-353-6062, E-mail: rebbi@bu.edu. *Application contact:* Dr. Ilona Lappo, Assistant Director, 617-353-5637, Fax: 617-353-6062, E-mail: lappo@bu.edu.

Bowling Green State University, Bowling Green, OH 43403.

Center for Materials Science Founded in 1995. *Academic areas of research and training:* Materials research. *Degrees offered:* None. *Graduate students served last academic year:* 13: 10 at the master's level; 3 at the doctoral level. *Faculty:* 11. *Faculty affiliations:* Departments of Biological Sciences, Chemistry, Family and Consumer Sciences, Geology, Physics and Astronomy, Technology Systems. *Annual research budget:* $200,000. *Director:* Dr. Robert I. Boughton, Director, 419-372-2421, Fax: 419-372-9938, E-mail: boughton@bgnet.bgsu.edu. *Application contact:* Jodi L. Sickler, Secretary, 419-372-7850, Fax: 419-372-9938, E-mail: matsci@newton.bgsu.edu.

Brigham Young University, Provo, UT 84602-1001.

Advanced Combustion Engineering Research Center Founded in 1986. *Academic areas of research and training:* Clean and efficient use of fossil fuels and waste materials, combustion, computer modeling. *Degrees offered:* None. *Graduate students served last academic year:* 34: 3 at the master's level; 14 at the doctoral level; 2 at the postgraduate level. *Faculty:* 20: 1 affiliated solely with the center. *Faculty affiliations:* Departments of Chemical Engineering, Chemistry and Biochemistry, Civil and Environmental Engineering, Fuels Engineering, Mechanical Engineering. *Annual research budget:* $6 million. *Director:* Dr. Thomas H. Fletcher, Director, 801-378-6236, Fax: 801-378-3831, E-mail: tom@harvey.et.byu.edu. *Application contact:* Koral Burt, Manager, 801-378-4126, Fax: 801-378-3831, E-mail: koral@et.byu.edu.

Brown University, Providence, RI 02912.

Lefschetz Center for Dynamical Systems Founded in 1964. *Academic areas of research and training:* Dynamical systems, nonlinear ordinary and partial differential equations and applications, stochastic systems theory. *Degrees offered:* PhD. *Graduate students served last academic year:* 13: 13 at the doctoral level. *Faculty:* 13. *Faculty affiliations:* Department of Mathematics; Divisions of Applied Mathematics, Engineering. *Annual research budget:* $1 million. *Director:* Prof. Christopher Jones, Director, 401-863-3696, Fax: 401-863-1355, E-mail: ckrtj@e151.cfm.brown.edu. *Application contact:* Winnie Isom, Secretary, 401-863-3724, Fax: 401-863-1355, E-mail: isom@151.cfm.brown.edu.

California Institute of Technology, Pasadena, CA 91125-0001.

Engineering Research Center for Neuromorphic Systems Engineering Founded in 1994. *Academic areas of research and training:* Neural networks, VLSI. *Degrees offered:* None. *Graduate students served last academic year:* 80: 3 at the master's level; 65 at the doctoral level; 12 at the postgraduate level. *Faculty:* 7. *Faculty affiliations:* Options in Computation and Neural Science, Computer Science, Electrical Engineering. *Director:* Dr. Pietro Perona, Director, 626-395-4867, Fax: 626-795-4765, E-mail: director @sunoptics.caltech.edu. *Application contact:* Janice Tucker, Administrator/Assistant to Director, 626-395-6254, Fax: 626-795-4765, E-mail: janice@sunoptics.caltech.edu.

California Polytechnic State University, San Luis Obispo, San Luis Obispo, CA 93407.

CAD Research Center Founded in 1986. *Academic areas of research and training:* Collaborative decision and support systems for complex problem situations, software design and development, artificial intelligence, computer-aided design, system integration. *Degrees offered:* None. *Graduate students served last academic year:* 75: 42 at the master's level; 4 at the postgraduate level. *Faculty:* 6: 2 affiliated solely with the center. *Faculty affiliations:* College of Business, Department of Computer Science. *Annual research budget:* $5 million. *Director:* Dr. Jens Pohl, Executive Director, 805-756-2841, Fax: 805-756-7567, E-mail: jpohl@calpoly.edu.

Irrigation Training and Research Center (ITRC) Founded in 1989. *Academic areas of research and training:* On-farm irrigation, irrigation districts, drip irrigation, irrigation management, automation. *Degrees offered:* None. *Graduate students served last academic year:* 6: 6 at the master's level. *Faculty:* 12: 3 affiliated solely with the center. *Faculty affiliations:* Department of Civil and Environmental Engineering; Programs in Bioresource and Agricultural Engineering, Crop Science, Soil Science. *Annual research budget:* $750,000. *Director:* Dr. Charles M. Burt, Director, 805-756-2379, Fax: 805-756-2433, E-mail: cburt@calpoly.edu.

California State University, Los Angeles, Los Angeles, CA 90032-8530.

Center for Information Resource Management *Academic areas of research and training:* Computer information systems, computer science, computer engineering. *Degrees offered:* None. *Graduate students served last academic year:* 15. *Faculty:* 2. *Faculty affiliations:* Program in Computer Information Systems. *Director:* Dr. Durward Jackson, Director, 213-343-2924, Fax: 213-343-5209, E-mail: djackson@calstatela.edu.

Carnegie Mellon University, Pittsburgh, PA 15213-3891.

Center for Building Performance and Diagnostics (CBPD) Founded in 1988. *Degrees offered:* MS; PhD. *Graduate students served last academic year:* 42: 13 at the master's level; 29 at the doctoral level. *Faculty:* 7. *Director:* Dr. Volker Hartkopf, Director, 412-268-2350, Fax: 412-268-6129.

Center for Electronic Design Automation Founded in 1982. *Academic areas of research and training:* Computer-aided design of VLSI circuits, systems and technology. *Degrees offered:* MS; PhD. *Graduate students served last academic year:* 77: 38 at the master's level; 35 at the doctoral level; 4 at the postgraduate level. *Faculty:* 14: 2 affiliated solely with the center. *Faculty affiliations:* Department of Electrical and Computer Engineering; School of Computer Science. *Annual research budget:* $6 million. *Director:* Donald E. Thomas, Director, 412-268-8889, Fax: 412-268-6662, E-mail: thomas@ece.cmu.edu. *Application contact:* Roxann Martin, Program Associate, 412-268-8889, Fax: 412-268-6662, E-mail: roxann@cedapo.ece.cmu.edu.

Center for Integrated Study of the Human Dimensions of Global Change Founded in 1991. *Academic areas of research and training:* Global environmental change, environment and energy policy, complex systems, decision analysis. *Degrees offered:* None. *Graduate students served last academic year:* 12: 12 at the doctoral level. *Faculty:* 38: 6 affiliated solely with the center. *Faculty affiliations:* Departments of Civil and Environmental Engineering, Electrical and Computer Engineering, Social and Decision Sciences; Graduate School of Industrial Administration; H. John Heinz III School of Public Policy and Management. *Annual research budget:* $2 million. *Director:* Dr. Hadi Dowlatabadi, Director, 412-268-3031, Fax: 412-268-3757, E-mail: hadi@cmu.edu.

Center for Light Microscope Imaging and Biotechnology Founded in 1991. *Academic areas of research and training:* Imaging technology, light microscopy and reagent chemistry, biomedical engineering, biology, chemistry, computer science, chemical engineering, physics. *Degrees offered:* None. *Graduate students served last academic year:* 17 at the doctoral level; 10 at the postgraduate level. *Faculty:* 20. *Faculty affiliations:* Departments of Biological Sciences, Chemistry, Chemical Engineering, Computer Science, Mechanical Engineering; Allegheny University of Health Sciences; University of Pittsburgh. *Annual research budget:* $1.8 million. *Director:* Dr. Alan Waggoner, Director, 412-268-3456, Fax: 412-268-6571. *Application contact:* Margie Zamborsky, Secretary to the Director, 412-268-3461, Fax: 412-268-6571, E-mail: mz0t@andrew.cmu.edu.

Computer-Aided Process Design Consortium (CAPD) Founded in 1985. *Academic areas of research and training:* Process synthesis, process optimization, process control, modeling and simulation, artificial intelligence, safety analysis, and scheduling and planning. *Degrees offered:* None. *Graduate students served last academic year:* 33: 2 at the master's level; 24 at the doctoral level; 7 at the postgraduate level. *Faculty:* 5. *Faculty affiliations:* Department of Chemical Engineering. *Annual research budget:* $1.5 million. *Director:* Dr. Lorenz T. Biegler, Co-Director, 412-268-2232, Fax: 412-268-7139, E-mail: lb01+@andrew.cmu.edu. *Application contact:* Laura Shaheen, 412-268-6344, Fax: 412-268-7137, E-mail: lr23@andrew.cmu.edu.

Data Storage Systems Center (DSSC) Founded in 1983. *Academic areas of research and training:* Developing advanced technology for data storage systems, magnetic recording, optical recording. *Degrees offered:* None. *Graduate students served last academic year:* 33 at the master's level; 72 at the doctoral level; 13 at the postgraduate level. *Faculty:* 26: 4 affiliated solely with the center. *Faculty affiliations:* Departments of Chemical Engineering, Computer

Science, Electrical and Computer Engineering, Materials Science and Engineering, Mechanical Engineering, Physics. *Director:* Dr. Robert M. White, Director, 412-268-6916, Fax: 412-268-3497, E-mail: white@ece.cmu.edu.

Green Design Initiative Founded in 1993. *Academic areas of research and training:* Environmentally conscious product, process, and policy design. *Degrees offered:* None. *Graduate students served last academic year:* 15: 5 at the master's level; 8 at the doctoral level; 2 at the postgraduate level. *Faculty:* 20: 5 affiliated solely with Green Design Initiative. *Faculty affiliations:* Departments of Civil Engineering, Engineering and Public Policy, Mechanical Engineering, Public Policy. *Annual research budget:* $1 million. *Director:* Dr. Lester Lave, Director, 412-268-8837, Fax: 412-268-6837. *Application contact:* Dr. Noellette Conway-Schempf, Executive Director, 412-268-2299, Fax: 412-268-6837, E-mail: nc0y@andrew.cmu.edu.

Case Western Reserve University, Cleveland, OH 44106.

Cardiac Bioelectricity Research and Training Center (CBRTC) Founded in 1994. *Academic areas of research and training:* Biomedical engineering, physiology, cardiology, cardiac electrophysiology, cardiac arrhythmias and sudden death. *Degrees offered:* None. *Graduate students served last academic year:* 30: 16 at the master's level; 8 at the doctoral level; 6 at the postgraduate level. *Faculty:* 33. *Faculty affiliations:* Departments of Anatomy, Biomedical Engineering, Physiology and Biophysics, Radiology, Surgery; School of Medicine. *Annual research budget:* $1.5 million. *Director:* Dr. Yoram Rudy, Director, 216-368-4051, Fax: 216-368-4969, E-mail: yxr@po.cwru.edu.

Center for Advanced Liquid Crystalline Optical Materials (ALCOM) *Academic areas of research and training:* Polymer physics. *Degrees offered:* None. *Graduate students served last academic year:* 9 at the doctoral level; 3 at the postgraduate level. *Faculty:* 12. *Faculty affiliations:* Departments of Macromolecular Science, Physics. *Annual research budget:* $800,000. *Director:* Dr. Jack L. Koenig, Director, 216-368-4176, Fax: 216-368-4171, E-mail: jlk6@po.cwru.edu.

Ernest B. Yeager Center for Electrochemical Sciences Founded in 1976. *Academic areas of research and training:* Batteries, biomedical sensors, corrosion, electrochemical engineering, electrochemical sensors, fuels cells, fundamental electrochemistry. *Degrees offered:* None. *Graduate students served last academic year:* 75: 10 at the master's level; 59 at the doctoral level; 6 at the postgraduate level. *Faculty:* 35. *Faculty affiliations:* Departments of Biomedical Engineering, Chemical Engineering, Chemistry, Electrical Engineering and Applied Physics, Macromolecular Science, Mechanical and Aerospace Engineering, Materials Science, and Physics; Edison Sensor Technology and Electronics Design Centers. *Annual research budget:* $4.5 million. *Director:* Dr. Robert F. Savinell, Director, 216-368-6525, Fax: 216-368-3016, E-mail: rfs2@po.cwru.edu. *Application contact:* Administrative Officer, 216-368-6525, Fax: 216-368-3016, E-mail: yces@cheme.cwru.edu.

Clarkson University, Potsdam, NY 13699.

Center for Advanced Materials Processing (CAMP) Founded in 1987. *Academic areas of research and training:* Colloid and surface science and engineering aspects of materials synthesis and processing with emphasis on fine particle technology, colloidal dispersions, and thin films. *Degrees offered:* None. *Graduate students served last academic year:* 65: 18 at the master's level; 31 at the doctoral level; 16 at the postgraduate level. *Faculty:* 18. *Faculty affiliations:* Departments of Chemical Engineering, Chemistry, Civil and Environmental Engineering, Mechanical and Aeronautical Engineering. *Annual research budget:* $3.2 million. *Director:* Dr. Raymond A. Mackay, Director, 315-268-2336, Fax: 315-268-7615, E-mail: mackay@clarkson.edu. *Application contact:* Edward P. McNamara, Deputy Director, E-mail: camp@clarkson.edu.

International Center for Gravity Materials Science and Applications Founded in 1991. *Academic areas of research and training:* Improvement of materials processing by microgravity or centrifugation, eutectic solidication, detached solidification, colloid coagulation, diamond film growth. *Degrees offered:* None. *Graduate students served last academic year:* 8: 4 at the master's level; 4 at the doctoral level. *Faculty:* 5: 1 affiliated solely with International Center for Gravity Materials Science and Applications. *Faculty affiliations:* Departments of Aeronautical Engineering, Chemical Engineering, Chemistry, Computer Engineering, Electrical Engineering, Mechanical Engineering. *Annual research budget:* $200,000. *Director:* Dr. Liya L. Regel, Director, 315-268-7672, Fax: 315-268-3833, E-mail: regel@clarkson.edu.

Clemson University, Clemson, SC 29634.

Bioengineering Alliance of South Carolina Founded in 1986. *Academic areas of research and training:* Bioengineering, biomaterials, biomechanics, biomedical engineering, bioinstrumentation, bioethics. *Degrees offered:* None. *Graduate students served last academic year:* 19: 16 at the master's level; 1 at the doctoral level; 2 at the postgraduate level. *Faculty:* 25: 15 affiliated solely with Bioengineering Alliance of South Carolina. *Faculty affiliations:* Medical University of South Carolina; University of South Carolina. *Annual research budget:* $500,000. *Director:* Dr. Subrata Saha, Director, 803-656-7603, Fax: 803-656-4466, E-mail: ssaha@clemson.edu.

Center for Advanced Engineering Fibers and Films Founded in 1988. *Academic areas of research and training:* Engineering fibers, films and composite materials, polymer processing, modeling and visualization, virtual reality. *Degrees offered:* None. *Graduate students served last academic year:* 50: 30 at the master's level; 20 at the doctoral level. *Faculty:* 30. *Faculty affiliations:* Departments of Bioengineering; Ceramics and Materials, Chemical, Mechanical Engineering; Chemistry; Computer Science; Mathematical Sciences; School of Textiles, Fiber, and Polymer Science. *Annual research budget:* $6.6 million. *Director:* Dr. Dan D. Edie, Director, 864-656-4535, Fax: 864-656-4557, E-mail: dan.edie@ces.clemson.edu. *Application contact:* Jane Jacobi, Administrative Director, 864-656-1050, Fax: 864-656-4466, E-mail: jane.jacobi@ces.clemson.edu.

Center for Advanced Manufacturing (CAM) Founded in 1982. *Academic areas of research and training:* CAD, manufacturing systems, materials and manufacturing processes, mechatronics and robotics, rapid phototyping/free-form fabrication, virtual reality. *Degrees offered:* None. *Graduate students served last academic year:* 30: 21 at the master's level; 8 at the doctoral level; 1 at the postgraduate level. *Faculty:* 20. *Faculty affiliations:* Departments of Bioengineering; Chemistry; Computer Science; Electrical and Computer, Industrial, Mechanical Engineering; Physics and Astronomy. *Annual research budget:* $1.4 million. *Director:* Dr. Frank W. Paul, Director, 803-656-3291, Fax: 803-656-4435, E-mail: frank.paul@ces.clemson.edu.

Electric Power Research Association (CUEPRA) Founded in 1985. *Academic areas of research and training:* Digital protection, distribution system automation, power quality, power system harmonics, power systems dynamics. *Degrees offered:* None. *Graduate students served last academic year:* 15: 7 at the master's level; 8 at the doctoral level. *Faculty:* 8: 4 affiliated solely with Electric Power Research Association (CUEPRA). *Faculty affiliations:* Department of Electrical and Computer Engineering. *Annual research budget:* $450,000. *Director:* Prof. Adly A. Girgis, Director, 864-656-5936, Fax: 864-656-1347, E-mail: adly.girgis@ces.clemson.edu.

South Carolina Institute for Energy Studies Founded in 1981. *Academic areas of research and training:* Fluidization, gas turbines, residential energy, waste-to-energy. *Degrees offered:* None. *Graduate students served last academic year:* 70: 40 at the master's level; 30 at the doctoral level. *Faculty:* 41: 6 affiliated solely with the center. *Faculty affiliations:* Departments of Chemical and Mechanical, Industrial Engineering. *Annual research budget:* $4.5 million. *Director:* Dr. Lawrence P. Golan, Director, 864-656-2267, Fax: 864-656-0142, E-mail: glawren@clemson.edu. *Application contact:* Glenda S. Black, Unit Administrator, 864-656-2267, Fax: 864-656-0142, E-mail: glenda@clemson.edu.

Colorado School of Mines, Golden, CO 80401-1887.

Advanced Steel Processing and Products Research Center (ASPPRC) Founded in 1984. *Academic areas of research and training:* Metallurgical engineering, materials and manufacturing. *Degrees offered:* MS; PhD. *Graduate students served last academic year:* 24: 16 at the master's level; 8 at the doctoral level. *Faculty:* 10: 8 affiliated solely with the center. *Faculty affiliations:* Departments of Engineering, Physics. *Annual research budget:* $1.2 million. *Director:* Dr. David K. Matlock, Director, 303-273-3775, Fax: 303-273-3016, E-mail: dmatlock@mines.edu. *Application contact:* Mimi M. Martin, Program Assistant, 303-273-3025, Fax: 303-273-3016, E-mail: mmartin@mines.edu.

Center for Research on Hydrates and Other Solids Founded in 1990. *Academic areas of research and training:* Chemical engineering, geology, petroleum engineering. *Degrees offered:* PhD. *Graduate students served last academic year:* 20: 5 at the master's level; 14 at the doctoral level; 1 at the postgraduate level. *Faculty:* 5: 1 affiliated solely with the center. *Faculty affiliations:* Departments of Geology, Petroleum Engineering. *Annual research budget:* $550,000. *Director:* Dr. E. Dendy Sloan, Director, 303-273-3723, Fax: 303-273-3730, E-mail: esloan@gashydrate.mines.colorado.edu.

Colorado Advanced Materials Institute (CAMI) Founded in 1983. *Academic areas of research and training:* Advanced materials. *Degrees offered:* None. *Faculty:* 1: 1 affiliated solely with the institute. *Faculty affiliations:* Departments of Chemical Engineering, Materials Science, Metallurgical and Materials Engineering, Physics. *Annual research budget:* $600,000. *Director:* Dr. Frederick J. Fraikor, Director, 303-273-3852, Fax: 303-273-3656, E-mail: ffraikor@mines.colorado.edu. *Application contact:* Roz Taylor, Program Administrator, 303-273-3852, Fax: 303-273-3656, E-mail: rtaylor@mines.edu.

Reservoir Characterization Project (RCP) Founded in 1985. *Academic areas of research and training:* Geology, geophysics, reservoir engineering. *Degrees offered:* None. *Graduate students served last academic year:* 12: 5 at the master's level; 7 at the doctoral level. *Faculty:* 6: 2 affiliated solely with the project. *Faculty affiliations:* Departments of Geology, Geophysics, Petroleum Engineering. *Annual research budget:* $1 million. *Director:* Dr. Thomas L. Davis, Director, 303-273-3938, Fax: 303-273-3478, E-mail: tdavis@mines.edu.

Colorado State University, Fort Collins, CO 80523-0015.

Flash Flood Laboratory Founded in 1997. *Academic areas of research and training:* End-to-end flash flood process including precipitation forecasting, hydrology, annecedent (soil) moisture, emergency warning, socio-economic impacts of flash floods. *Degrees offered:* None. *Graduate students served last academic year:* 2: 1 at the doctoral level; 1 at the postgraduate level. *Faculty:* 20. *Faculty affiliations:* Departments of Atmospheric Sciences, Civil Engineering, Economics, Sociology. *Annual research budget:* $45,000. *Director:* Dr. Thomas H. Vonder Haar, Director, Fax: 970-491-8241, E-mail: vonderhaar@cira.colostate.edu. *Application contact:* Kenneth E. Eis, Deputy Director, 970-491-8397, Fax: 970-491-8241, E-mail: eis@cira.colostate.edu.

Fluid Mechanics and Wind Engineering Program Founded in 1960. *Academic areas of research and training:* Wind engineering, hazard analysis, air pollution, windloads, pedestrian comfort, environmental impacts. *Degrees offered:* None. *Graduate students served last academic year:* 10: 5 at the master's level; 5 at the doctoral level. *Faculty:* 5: 3 affiliated solely with Fluid Mechanics and Wind Engineering Program. *Annual research budget:* $600,000. *Director:* Dr. Robert N. Meroney, Director, 970-491-8574, Fax: 970-491-8671, E-mail: meroney@engr.colostate.edu.

Geotechnical Engineering Laboratories Founded in 1975. *Academic areas of research and training:* Geotechnical and geoenvironmental engineering, natural hazards, earthquakes, expansive soils, dam safety, mine tailings, explosives. *Degrees offered:* None. *Graduate students served last academic year:* 19: 13 at the master's level; 5 at the doctoral level; 1 at the postgraduate level. *Faculty:* 12: 4 affiliated solely with the laboratories. *Faculty affiliations:* Departments of Chemical and Bioresource Engineering, Earth Resources, Groundwater and Environmental Hydrogeology; Program in Environmental Engineering. *Annual research budget:* $250,000. *Director:* Dr. Wayne A. Charlie, Director, 970-491-5048, Fax: 970-491-7727, E-mail: wcharlie@engr.colostate.edu.

Hydraulics and Hydromachinery Laboratories Founded in 1962. *Academic areas of research and training:* Hydraulics, river mechanics, erosion and sedimentation, dam safety, numerical modeling. *Degrees offered:* None. *Graduate students served last academic year:* 33: 20 at the master's level; 10 at the doctoral level; 3 at the postgraduate level. *Faculty:* 11: 11 affiliated solely with the laboratories. *Annual research budget:* $1.5 million. *Director:* Dr. Christopher I. Thornton, Director, 970-491-8493, Fax: 970-491-8462, E-mail: thornton@engr.colostate.edu.

Columbia University, New York, NY 10027.

Columbia Radiation Laboratory/Microelectronic Sciences Laboratories (CRL/MSL) Founded in 1942. *Academic areas of research and training:* Materials science, quantum electronics, opto-electronics, chemistry and physics of interfaces, environmental molecular science. *Degrees offered:* None. *Graduate students served last academic year:* 65: 50 at the doctoral level; 15 at the postgraduate level. *Faculty:* 24: 12 affiliated solely with the laboratory. *Faculty affiliations:* Departments of Applied Physics; Chemical Engineering, Materials Science, and Mining Engineering; Chemistry; Electrical Engineering. *Annual research budget:* $4.5 million. *Director:* Prof. George Flynn, Director, 212-854-4162, Fax: 212-932-1289, E-mail: flynn@chem.columbia.edu. *Application contact:* Linda Powers, Departmental Administrator, 212-854-3265, Fax: 212-854-1909, E-mail: lp215@columbia.edu.

Concordia University, Montréal, PQ H3G 1M8, Canada.

Centre for Building Studies (CBS) Founded in 1978. *Academic areas of research and training:* Building environment, building science, building structures, construction management, energy efficiency, building envelope. *Degrees offered:* None. *Graduate students served last academic year:* 91: 66 at the master's level; 25 at the doctoral level. *Faculty:* 14: 14 affiliated solely with the center. *Annual research budget:* $600,000. *Director:* Dr. T. Stathopoulos, Director, 514-848-3186, Fax: 514-848-7965, E-mail: statho@cbs-engr.concordia.ca. *Application contact:* Olga Soares, Administrative Assistant, 514-848-3195, Fax: 514-848-7965, E-mail: soares_o@manis.concordia.ca.

Concordia Centre for Composites (CONCOM) Founded in 1993. *Academic areas of research and training:* Composite structures and materials. *Degrees offered:* None. *Graduate students served last academic year:* 18: 6 at the master's level; 12 at the doctoral level. *Faculty:* 8: 5 affiliated solely with the center. *Faculty affiliations:* Departments of Chemistry, Electrical Engineering, Physics. *Annual research budget:* $300,000. *Director:* Dr. S. V. Hoa, Director, 514-848-3139, Fax: 514-848-3178, E-mail: hoasuon@vax2.concordia.ca. *Application contact:* Sophie Mérineau, Administrative Assistant, 514-848-3126, Fax: 514-848-3178, E-mail: sophmer@vax2.concordia.ca.

Concordia Computer-Aided Vehicle Engineering Research Centre (CONCAVE) Founded in 1986. *Academic areas of research and training:* Road and off-road vehicle dynamics, advanced vehicle suspension, vehicular vibration, intelligent vehicles, driver-vehicle interactions, driver vibration, hand-arm vibration, vehicular ergonomics, micromachining, microsensors, micromechatronics, freight vehicle dynamics, pavement loads. *Degrees offered:* None. *Graduate students served last academic year:* 22: 8 at the master's level; 12 at the doctoral level; 2 at the postgraduate level. *Faculty:* 10: 5 affiliated solely with the center. *Faculty affiliations:* Departments of Electrical and Computer Engineering, Mechanical Engineering; Programs in Exercise Science, Industrial Engineering. *Annual research budget:* $300,000. *Director:* Dr. A. K. W. Ahmed, Director, 514-848-7932, Fax: 514-848-8635, E-mail: waiz@vax2.concordia.ca. *Application contact:* Dr. Subhash Rakheja, Professor, 514-848-3162, Fax: 514-848-8635, E-mail: rakheja@vax2.concordia.ca.

Research and Training Opportunities

Cornell University, Ithaca, NY 14853-0001.

Center for Materials Research Founded in 1961. *Academic areas of research and training:* Advanced materials research, nano composites, thin films, nanostructures, energetic beams, magnetic materials, polymers. *Degrees offered:* None. *Graduate students served last academic year:* 45: 38 at the doctoral level; 7 at the postgraduate level. *Faculty:* 190: 95 affiliated solely with the center. *Faculty affiliations:* Departments of Applied Engineering and Physics, Chemical Engineering, Chemistry, Electrical Engineering, Materials Science and Engineering, Mechanical and Aerospace Engineering, Physics, Theoretical and Applied Mechanics. *Director:* Dr. Neil W. Ashcroft, Director, 607-255-4273, Fax: 607-255-3957, E-mail: director@ccmr.cornell.edu. *Application contact:* Dr. Helene R. Schember, Associate Director, 607-255-4274, Fax: 607-255-3957, E-mail: helene@ccmr.cornell.edu.

Cornell High Energy Synchrotron Source (CHESS) Founded in 1978. *Academic areas of research and training:* To operate a user-oriented national facility to provide state-of-the-art synchrotron radiation facilities for the scientific community. *Degrees offered:* None. *Graduate students served last academic year:* 358: 194 at the doctoral level; 164 at the postgraduate level. *Faculty:* 25: 2 affiliated solely with Cornell High Energy Synchrotron Source (CHESS). *Faculty affiliations:* Department of Biochemistry; Fields of Applied Physics, Chemical Engineering, Chemistry, Civil and Environmental Engineering, Material Science, Mechanical Engineering, Physics, Poultry Science, Lab of Ornithology. *Annual research budget:* $3 million. *Director:* Prof. Sol Gruner, Director, 607-255-3441, Fax: 607-255-9001, E-mail: smg26@cornell.edu.

Dalhousie University, Halifax, NS B3H 3J5, Canada.

Centre for Marine Vessel Development and Research Founded in 1989. *Academic areas of research and training:* Computational ship/marine hydrodynamics, computer simulation, computer-aided ship design, model test and analysis. *Degrees offered:* None. *Graduate students served last academic year:* 7: 2 at the master's level; 5 at the doctoral level. *Faculty:* 2: 2 affiliated solely with the center. *Faculty affiliations:* Program in Mechanical Engineering. *Annual research budget:* $200,000. *Director:* Dr. C. C. Hsiung, Director, 902-494-3819, Fax: 902-423-6711, E-mail: charles.hsiung@dal.ca.

École des Hautes Études Commerciales, Montréal, PQ H3T 2A7, Canada.

Groupe d'Études et de Recherche en Analyse des Décisions (GERAD) Founded in 1988. *Academic areas of research and training:* Operations research. *Degrees offered:* None. *Graduate students served last academic year:* 113: 64 at the master's level; 42 at the doctoral level; 7 at the postgraduate level. *Faculty:* 34. *Faculty affiliations:* Program in Quantitative Methods; École Polytechnique de Montréal Departments of Industrial Engineering, Mathematics; McGill University Faculty of Management; Université du Québec à Montréal Mathematics Program. *Annual research budget:* $3.6 million. *Director:* Dr. Pierre Hansen, Director, 514-340-5675, Fax: 514-340-5665, E-mail: gerad@crt.umontreal.ca.

École Polytechnique de Montréal, Montréal, PQ H3C 3A7, Canada.

Center for Research on Computation and its Applications (CERCA) Founded in 1992. *Academic areas of research and training:* Pharmaceutical chemistry, industrial mechanics, environmental forecasting, industrial geophysics, astrophysics, nanoelectronics, scientific visualization, high-performance computing. *Degrees offered:* None. *Graduate students served last academic year:* 39: 6 at the master's level; 17 at the doctoral level; 16 at the postgraduate level. *Faculty:* 22. *Faculty affiliations:* Departments of Atmospheric and Ocean Sciences, Chemistry, Computer Sciences, Mathematics and Statistics, Mechanical Engineering, and Physics. *Annual research budget:* $3.9 million. *Director:* Dr. Jean-Jacques Rousseau, Director, 514-369-5210, Fax: 514-369-3880, E-mail: rousseau@cerca.umontreal.ca. *Application contact:* Dr. Jean-Jacques Rousseau, Director, 514-369-5210, Fax: 514-369-3880, E-mail: rousseau@cerca.umontreal.ca.

Centre for Characterization and Microscopy of Materials Founded in 1989. *Academic areas of research and training:* Biological sciences, chemical engineering, materials and metallurgical engineering, mechanical engineering, physical engineering. *Degrees offered:* None. *Graduate students served last academic year:* 150: 60 at the master's level; 40 at the doctoral level; 50 at the postgraduate level. *Faculty:* 17: 9 affiliated solely with the center. *Faculty affiliations:* Departments of Chemical Engineering, Environment, Mechanical Engineering, Medicine, Occupational and Environmental Health, Pharmaceutics, Physical Engineering; Pulp and Paper Center. *Director:* Gilles L'Espérance, Director, 514-340-4532, Fax: 514-340-4468, E-mail: cm2@mail.polymtl.ca. *Application contact:* Elise Campeau, Administrative Technician, 514-340-4788, Fax: 514-340-4468.

Gas Technology Research Group (GREG) *Academic areas of research and training:* Applied mathematics, chemical engineering, mechanical engineering. *Degrees offered:* None. *Graduate students served last academic year:* 20: 10 at the master's level; 8 at the doctoral level; 2 at the postgraduate level. *Faculty:* 6: 1 affiliated solely with the group. *Faculty affiliations:* Departments of Chemical, Materials Science and Metallurgical, and Mechanical Engineering. *Annual research budget:* $60,000. *Director:* Prof. Danilo Klvana, Head, 514-340-4711, Fax: 514-340-4159, E-mail: dklvana@mailsrv.polymtl.ca.

Groupe d'Études et de Recherche en Analyse des Décisions (GERAD) Founded in 1988. *Academic areas of research and training:* Operations research. *Degrees offered:* None. *Graduate students served last academic year:* 113: 64 at the master's level; 42 at the doctoral level; 7 at the postgraduate level. *Faculty:* 32. *Faculty affiliations:* Departments of Industrial Engineering, Mathematics; École des Hautes Études Commerciales Program in Quantitative Methods; McGill University Faculty of Management; Université du Québec à Montr& ,eal Mathematics Program. *Annual research budget:* $3.6 million. *Director:* Dr. Pierre Hansen, Director, 514-340-5675, Fax: 514-340-5665, E-mail: gerad@crt.umontreal.ca.

Fisk University, Nashville, TN 37208-3051.

Center for Photonic Materials and Devices Founded in 1992. *Academic areas of research and training:* Chemistry, physics, materials science, photonics. *Degrees offered:* None. *Graduate students served last academic year:* 15: 15 at the master's level. *Faculty:* 17: 12 affiliated solely with the center. *Faculty affiliations:* Department of Physics. *Annual research budget:* $1.5 million. *Director:* Enrique Silberman, Director, 615-329-8620, Fax: 615-329-8634, E-mail: esilber@dubois.fisk.edu.

Florida Atlantic University, Boca Raton, FL 33431-0991.

Center for Applied Stochastics Research Founded in 1984. *Academic areas of research and training:* Applications of probability and statistics in various fields of engineering, predicting the response and safety of engineering structures in seismic events, analyzing possible motion instability of long span bridges in strong turbulent winds, reliability of marine and ocean structures due to random wave excitations, new measures for groundwater contamination. *Degrees offered:* None. *Graduate students served last academic year:* 3: 1 at the doctoral level; 1 at the postgraduate level. *Faculty:* 5: 1 affiliated solely with the center. *Faculty affiliations:* Departments of Mathematics, Mechanical Engineering, Ocean Engineering. *Annual research budget:* $190,655. *Director:* Dr. Y. K. Lin, Director, 561-297-3449, Fax: 561-297-2868, E-mail: linyk@casr.fau.edu.

Communications Technology Center Founded in 1990. *Academic areas of research and training:* High definition television, medical instrumentation. *Degrees offered:* MS; PhD. *Graduate students served last academic year:* 5: 4 at the master's level; 1 at the doctoral level. *Faculty:* 1: 1 affiliated solely with the center. *Annual research budget:* $1.5 million. *Director:* Dr. William E. Glenn, Director, 561-297-2343, Fax: 561-297-3418, E-mail: glenn@fau.edu.

Robotics Center Founded in 1986. *Academic areas of research and training:* Kinematic calibration, parallel manipulator, autonomous vehicle. *Degrees offered:* None. *Graduate students served last academic year:* 10: 3 at the master's level; 7 at the doctoral level. *Faculty:* 5. *Faculty affiliations:* Departments of Computer Science and Engineering, Electrical Engineering, Mechanical Engineering. *Annual research budget:* $100,000. *Director:* Dr. Oren Masory, Director, 407-367-2693, Fax: 407-367-2825, E-mail: masoryo@acc.fau.edu.

Florida Institute of Technology, Melbourne, FL 32901-6975.

Center for Airport Management and Development Founded in 1996. *Academic areas of research and training:* Enhance education and industry progress in all aspects of airport planning, design, development, operations, and management. *Degrees offered:* None. *Graduate students served last academic year:* 8: 8 at the master's level. *Faculty:* 14: 2 affiliated solely with the center. *Faculty affiliations:* Division of Engineering Science, School of Aeronautics. *Annual research budget:* $260,000. *Director:* Dr. Ballard M. Barker, Director, 407-674-7369, Fax: 407-674-7648, E-mail: fitcamd@fit.edu.

Center for Electronics Manufacturability Founded in 1990. *Academic areas of research and training:* Materials and manufacturing processes, multi-chip passivation. *Degrees offered:* None. *Graduate students served last academic year:* 10: 7 at the master's level; 3 at the doctoral level. *Faculty:* 11: 1 affiliated solely with the center. *Faculty affiliations:* Programs in Chemical Engineering, Computer Engineering, Computer Science, Electrical Engineering, Engineering Management, Physics. *Annual research budget:* $300,000. *Director:* Dr. Thomas J. Sanders, Director, 407-768-8000 Ext. 8769, Fax: 407-984-8461, E-mail: tsanders@ee.fit.edu.

Center for Remote Sensing Founded in 1995. *Academic areas of research and training:* Ocean remote sensing, atmospheric remote sensing, space physics and astronomy, environmental optics, electrical engineering, geographic information systems, hydrographic engineering, hyperspectral remote sensing of vegetation and water. *Degrees offered:* None. *Graduate students served last academic year:* 5: 4 at the master's level; 1 at the doctoral level. *Faculty:* 21. *Faculty affiliations:* Departments of Biological Sciences, Meteorology, Physics and Space Sciences; Programs in Electrical Engineering, Environmental Science, Ocean Engineering, Oceanography; School of Aeronautics. *Annual research budget:* $250,000. *Director:* Dr. Charles R. Bostater, Director, 407-768-8000 Ext. 7113, Fax: 407-773-0980, E-mail: bostater@probe.ocn.fit.edu.

Microelectronics Laboratory Founded in 1990. *Academic areas of research and training:* Materials and manufacturing processes. *Degrees offered:* None. *Graduate students served last academic year:* 5: 3 at the master's level; 2 at the doctoral level. *Faculty:* 6: 2 affiliated solely with the laboratory. *Faculty affiliations:* Programs in Computer Engineering, Electrical Engineering. *Annual research budget:* $100,000. *Director:* Dr. Thomas J. Sanders, Director, 407-768-8769, E-mail: tsanders@ee.fit.edu.

Wind and Hurricane Impact Research Laboratory (WHIRL) Founded in 1997. *Academic areas of research and training:* Wind hazard mitigation, wind engineering, wind-structure interaction, structural control, experimental fluids dynamics, meteorology. *Degrees offered:* Certificate. *Graduate students served last academic year:* 3: 3 at the master's level. *Faculty:* 15. *Faculty affiliations:* Departments of Aerospace Engineering, Mechanical Engineering; Programs in Civil Engineering, Environmental Science, Oceanography; Schools of Aeronautics, Business. *Director:* Dr. Jean-Paul Pinelli, Director, 407-674-8085, Fax: 407-674-7565, E-mail: pinelli@fit.edu.

Florida International University, Miami, FL 33199.

Hemispheric Center for Environmental Technology Founded in 1995. *Academic areas of research and training:* Research and development on the deactivation and decommissioning of nuclear facilities and the management and reduction of radioactive and hazardous wastes. *Degrees offered:* None. *Graduate students served last academic year:* 20: 17 at the master's level; 1 at the doctoral level; 2 at the postgraduate level. *Faculty:* 48: 37 affiliated solely with the center. *Faculty affiliations:* Departments of Civil Engineering, Computer Sciences, Geology, Mechanical Engineering; School of Business; Engineering Information Center; Latin American and Caribbean Center. *Annual research budget:* $7 million. *Director:* Dr. M. A. Ebadian, Director, 305-348-4238, Fax: 305-348-1697, E-mail: ebadian@eng.fiu.edu.

Florida State University, Tallahassee, FL 32306.

Center for Materials Research and Technology (MARTECH) Founded in 1985. *Academic areas of research and training:* Fabrication and characterization of low-dimensional materials, primarily magnetic, sensor technologies. *Degrees offered:* None. *Graduate students served last academic year:* 49: 5 at the master's level; 35 at the doctoral level; 9 at the postgraduate level. *Faculty:* 17: 2 affiliated solely with the center. *Faculty affiliations:* Departments of Chemistry, Physics; FAMU/FSU College of Engineering; National High Magnetic Field Laboratory; Supercomputer Computations Research Institute. *Director:* Dr. Stephan von Molnár, Director, 850-644-2246, Fax: 850-644-6504, E-mail: molnar@martech.fsu.edu. *Application contact:* Jessie Spencer, Coordinator of Administrative Services, 850-644-2830, Fax: 850-644-6504, E-mail: spencer@martech.fsu.edu.

George Mason University, Fairfax, VA 22030-4444.

Center for Secure Information Systems (CSIS) Founded in 1989. *Academic areas of research and training:* Information systems security. *Degrees offered:* Certificate. *Graduate students served last academic year:* 55: 30 at the master's level; 20 at the doctoral level; 5 at the postgraduate level. *Faculty:* 16: 8 affiliated solely with the center. *Faculty affiliations:* Departments of Computer Science, Information and Software Systems Engineering. *Annual research budget:* $750,000. *Director:* Dr. Sushil Jajodia, Director, 703-993-1653, Fax: 703-993-1638, E-mail: jajodia@gmu.edu.

The George Washington University, Washington, DC 20052.

Cyberspace Policy Institute Founded in 1994. *Academic areas of research and training:* Computer science, information policy, telemedicine security and privacy, Internet policy. *Degrees offered:* None. *Graduate students served last academic year:* 4: 3 at the master's level; 1 at the doctoral level. *Faculty:* 10. *Faculty affiliations:* Departments of Economics, Electrical Engineering and Computer Science, Management Science; Interdisciplinary Programs in Public Policy; National Law Center; Program in Telecommunications. *Annual research budget:* $30,000. *Director:* Dr. Lance J. Hoffman, Director, 202-994-4225, Fax: 202-994-5505, E-mail: hoffman@seas.gwu.edu.

Institute for Materials Science Founded in 1989. *Academic areas of research and training:* Materials science. *Degrees offered:* None. *Graduate students served last academic year:* 15: 15 at the doctoral level. *Faculty:* 10. *Faculty affiliations:* Departments of Chemistry; Civil, Mechanical, and Environmental Engineering; Electrical Engineering and Computer Science; Geology; Physics. *Annual research budget:* $850,000. *Director:* Dr. David E. Ramaker, Co-Director, 202-994-6934, Fax: 202-994-5873, E-mail: ramaker@gwu.edu.

Space Policy Institute Founded in 1987. *Academic areas of research and training:* National and international space policy issues and history of space activities. *Degrees offered:* None. *Graduate students served last academic year:* 20: 12 at the master's level; 8 at the doctoral level. *Faculty:* 2: 2 affiliated solely with the institute. *Annual research budget:* $750,000. *Director:* Dr. John M. Logsdon, Director, 202-994-7292, Fax: 202-994-1639, E-mail: logsdon@gwu.edu. *Application contact:* Michelle Treistman, Executive Assistant, 202-994-7292, Fax: 202-994-1639, E-mail: spi@gwu.edu.

Georgia Institute of Technology, Atlanta, GA 30332-0001.

The Center for Signal and Image Processing (CSIP) Founded in 1996. *Academic areas of research and training:* Digital signal processing, audio and video, radar, communications and networking, image processing, multimedia, neural networks, array processing, speech processing, statistical processing, ULSI. *Degrees offered:* None. *Graduate students served last academic year:* 100: 10 at the master's level; 90 at the doctoral level. *Faculty:* 15: 12 affiliated solely with the center. *Faculty affiliations:* College of Computing, Georgia Tech Research Institute. *Annual research budget:* $3 million. *Director:* Dr. Ronald Schafer, Chair, 404-894-2917, Fax: 404-894-8363, E-mail: ronald.schafer@ece.gatech.edu.

Composites Manufacturing Research Program Founded in 1990. *Academic areas of research and training:* Polymer and polymer matrix composites, materials and manufacturing processes. *Degrees offered:* None. *Graduate students served last academic year:* 22: 10 at the master's level; 10 at the doctoral level; 2 at the postgraduate level. *Faculty:* 2. *Faculty affiliations:* George W. Woodruff School of Mechanical Engineering, School of Chemical Engineering. *Director:* Dr. Jonathan S. Colton, Co-Director, 404-894-7407, Fax: 404-894-9342, E-mail: jonathan.colton@me.gatech.edu.

Fusion Research Center Founded in 1986. *Academic areas of research and training:* Fusion plasma physics and reactor design, plasma-materials interactions. *Degrees offered:* None. *Graduate students served last academic year:* 8: 2 at the master's level; 6 at the doctoral level. *Faculty:* 17: 5 affiliated solely with the center. *Faculty affiliations:* Schools of Mechanical Engineering, Physics. *Annual research budget:* $400,000. *Director:* Dr. Weston M. Stacey, Director, 404-894-3758, Fax: 404-894-3733, E-mail: weston.stacey@me.gatech.edu.

Multiuniversity Center for Integrated Diagnostics Founded in 1995. *Academic areas of research and training:* Basic research studies needed to address the elements essential to creating an integrated predictive diagnostics system, based on the concept of condition-based maintenance; thrust areas include mechanical system health monitoring, nondestructive evaluation, failure characterization and prediction methodology. *Degrees offered:* None. *Graduate students served last academic year:* 45: 13 at the master's level; 27 at the doctoral level; 5 at the postgraduate level. *Faculty:* 28: 1 affiliated solely with the center. *Faculty affiliations:* Schools of Civil and Environmental Engineering, Electrical and Computer Engineering, Material Science and Engineering, Mechanical Engineering; Northwestern University; University of Minnesota Departments of Electrical Engineering, Mechanical Engineering. *Annual research budget:* $2 million. *Director:* Ward O. Winer, Program Director, 404-894-3200, Fax: 404-894-8336, E-mail: ward.winer@me.gatech.edu. *Application contact:* Richard S. Cowan, Program Manager, 404-894-3270, Fax: 404-894-8336, E-mail: rick.cowan@me.gatech.edu.

Rapid Prototyping and Manufacturing Institute Founded in 1995. *Academic areas of research and training:* Development and deployment of rapid prototyping and manufacturing technologies through education, research, and service. *Degrees offered:* None. *Graduate students served last academic year:* 80: 50 at the master's level; 30 at the doctoral level. *Faculty:* 10: 1 affiliated solely with the institute. *Faculty affiliations:* College of Computing; George W. Woodruff School of Mechanical Engineering; Schools of Chemical Engineering, Materials Science and Engineering; Georgia Tech Research Institute. *Annual research budget:* $1.2 million. *Director:* Dr. David Rosen, Academic Director, 404-894-9668, Fax: 404-894-9342, E-mail: david.rosen@me.gatech.edu.

Systems Realization Laboratory (SRL) Founded in 1992. *Academic areas of research and training:* Design of open and sustainable systems, virtual and rapid protyping, environmentally conscious design and manufacture. *Degrees offered:* None. *Graduate students served last academic year:* 27: 15 at the master's level; 12 at the doctoral level. *Faculty:* 8: 4 affiliated solely with the laboratory. *Faculty affiliations:* Program in Mechanical Engineering. *Annual research budget:* $500,000. *Director:* Dr. Bert Bras, Director, 404-894-9667, Fax: 404-894-9342, E-mail: bert.bras@me.gatech.edu. *Application contact:* Dr. Farrokh Mistree, Professor, 404-894-8412, Fax: 404-894-9342, E-mail: farrokh.mistree@me.gatech.edu.

University Center of Excellence for Photovoltaics Research and Education (UCEP) Founded in 1992. *Degrees offered:* None. *Graduate students served last academic year:* 10: 10 at the doctoral level. *Faculty:* 16: 1 affiliated solely with the center. *Director:* Dr. Ajeet Rohatgi, Director, 404-894-7692, Fax: 404-894-5934, E-mail: arohatgi@ece.gatech.edu. *Application contact:* Rochelle Kraehe, Administrative Manager, 404-894-9269, Fax: 404-894-5934, E-mail: rkraehe@ece.gatech.edu.

Graduate School and University Center of the City University of New York, New York, NY 10036-8099.

Frank Stanton/Andrew Heiskell Center for Public Policy in Telecommunications and Information Systems Founded in 1988. *Academic areas of research and training:* Telecommunications policy, educational and economic impact of telecommunications on disadvantaged students and communities. *Degrees offered:* None. *Graduate students served last academic year:* 4: 3 at the doctoral level; 1 at the postgraduate level. *Faculty:* 4: 2 affiliated solely with the center. *Annual research budget:* $300,000. *Director:* Helen Birenbaum, Executive Director, 212-817-2105, Fax: 212-817-1583, E-mail: hbirenbaum@gc.cuny.edu.

Illinois Institute of Technology, Chicago, IL 60616-3793.

Advanced Building Materials and Systems Center Founded in 1988. *Academic areas of research and training:* Analysis and design of building materials, building structural systems, bridges. *Degrees offered:* None. *Graduate students served last academic year:* 5: 3 at the master's level; 2 at the doctoral level. *Faculty:* 4. *Faculty affiliations:* Department of Civil and Architectural Engineering, Physics Division. *Annual research budget:* $250,000. *Director:* Dr. Sidney A. Guralnick, Director, 312-567-3549, Fax: 312-567-3634, E-mail: ceguralnick@minna.iit.edu.

Center for Synchrotron Radiation Research and Instrumentation Founded in 1992. *Degrees offered:* None. *Graduate students served last academic year:* 5: 2 at the master's level; 2 at the doctoral level; 1 at the postgraduate level. *Faculty:* 11: 8 affiliated solely with the center. *Faculty affiliations:* Divisions of Mechanical and Aerospace Engineering, Physics. *Annual research budget:* $50,000. *Director:* Dr. Leroy D. Chapman, Director, 312-567-3575, Fax: 312-567-3576, E-mail: dean.chapman@iit.edu.

Center of Excellence for Polymer Science and Engineering Founded in 1990. *Academic areas of research and training:* Polymer processing, reaction engineering, recycling, characterization, and composites; polymer rheology; emulsion and suspension polymerization; non-linear optical polymers. *Degrees offered:* None. *Graduate students served last academic year:* 21: 12 at the master's level; 8 at the doctoral level; 1 at the postgraduate level. *Faculty:* 17. *Faculty affiliations:* Departments of Chemical and Environmental Engineering, Chemsitry, Computer Science and Applied Mathematics, Food Safety and Technology, Materials and Aerospace Engineering. *Annual research budget:* $400,000. *Director:* Dr. David C. Venerus, Acting Director, 312-567-5177, Fax: 312-567-8874, E-mail: chevenerus@iit.edu.

Announcement: The Center of Excellence for Polymer Science and Engineering of Illinois Institute of Technology began operation in 1990 with a $1.5-million challenge grant from the Amoco Foundation and has attracted matching corporate, foundation, and government grants. Has state-of-the-art facilities and equipment for polymer synthesis, processing, and characterization. Offers a concentration in polymer science and engineering to PhD, MS, and professional master's degree candidates in relevant engineering disciplines and in chemistry. Conducts extensive applied research for industry and basic research for government sponsors. Has developed talented plastics and rubber recycling technology based on solid-state shear extrusion principle.

Energy and Power Center Founded in 1989. *Academic areas of research and training:* Energy and energy-related environmental policy, energy economics, energy technology, sustainable energy systems, global climate change, energy, environment, economics. *Degrees offered:* None. *Graduate students served last academic year:* 100: 60 at the master's level; 40 at the doctoral level. *Faculty:* 24. *Faculty affiliations:* Chicago-Kent College of Law; Departments of Chemical and Environmental Engineering, Electrical and Computer Engineering, Mechanical Engineering, Materials and Aerospace Engineering; Stuart School of Business. *Annual research*

budget: $3 million. *Director:* Dr. Henry R. Linden, Director, 312-567-3095, Fax: 312-567-3967, E-mail: hlinden@alpha1.ais.iit.edu.

National Center for Food Safety and Technology Founded in 1988. *Academic areas of research and training:* Food safety. *Degrees offered:* MS. *Graduate students served last academic year:* 7: 7 at the master's level. *Faculty:* 15: 5 affiliated solely with the center. *Faculty affiliations:* Department of Chemical and Environmental Engineering, Program in Microbiology; Illinois Institute of Technology Research Institute. *Annual research budget:* $3.7 million. *Director:* Dr. Charles E. Sizer, Director, 708-563-1576, Fax: 708-563-1873.

Pritzker Institute of Medical Engineering Founded in 1981. *Academic areas of research and training:* Biomedical engineering, cardiovascular technology, implant electronics. *Degrees offered:* PhD. *Graduate students served last academic year:* 6: 5 at the master's level; 1 at the postgraduate level. *Faculty:* 26: 2 affiliated solely with the institute. *Faculty affiliations:* Departments of Biological, Chemical, and Physical Sciences; Chemical and Environmental Engineering; Computer Science; Electrical and Computer Engineering; Mechanical, Materials, and Aerospace Engineering; Institute of Psychology. *Annual research budget:* $1 million. *Director:* Dr. Robert C. Arzbaecher, Director, 312-567-5324, Fax: 312-567-5707, E-mail: arzbaecher@iit.edu.

Indiana University Bloomington, Bloomington, IN 47405.

Center for Research on Concepts and Cognition (CRCC) Founded in 1988. *Academic areas of research and training:* Emergent models of high-level perception and analogical thought in carefully-designed domains, mechanisms of creativity, human error-making, how discovery and creation work in mathematics and music, relationship between analogy and translation. *Degrees offered:* None. *Graduate students served last academic year:* 4: 4 at the doctoral level. *Faculty:* 1: 1 affiliated solely with the center. *Director:* Dr. Douglas R. Hofstadter, Director, 812-855-6965, Fax: 812-855-6966, E-mail: dughof@cogsci.indiana.edu. *Application contact:* Helga Keller, Administrative Assistant, 812-855-6965, Fax: 812-855-6966, E-mail: helga@cogsci.indiana.edu.

Environmental Systems Application Center (ESAC) Founded in 1972. *Academic areas of research and training:* Applied ecology, aquatic chemistry, groundwater modeling, hazardous waste management, lake and watershed management, risk assessment, wetlands ecology. *Degrees offered:* None. *Graduate students served last academic year:* 8: 8 at the master's level. *Faculty:* 7: 1 affiliated solely with the center. *Annual research budget:* $80,000. *Director:* William W. Jones, Director, 812-855-4556, Fax: 812-855-7802, E-mail: joneswi@indiana.edu.

Institute for Scientific Computing and Applied Mathematics Founded in 1986. *Academic areas of research and training:* Promotion of advanced research in applied mathematics with particular emphasis on scientific computing. *Degrees offered:* None. *Graduate students served last academic year:* 12: 10 at the doctoral level; 2 at the postgraduate level. *Faculty:* 18. *Faculty affiliations:* Departments of Astronomy, Chemistry, Computer Science, Economics, Geology, Mathematics. *Director:* Dr. Roger Temam, Director, 812-855-8521, Fax: 812-855-7850, E-mail: temam@indiana.edu. *Application contact:* Teresa Bunge, Administrative Secretary, 812-855-8521, Fax: 812-855-7850, E-mail: bunget@indiana.edu.

Iowa State University of Science and Technology, Ames, IA 50011.

Center for Sustainable Environmental Technologies (CSET) Founded in 1979. *Academic areas of research and training:* Develop and demonstrate sustainable environmental technologies, improving environmental performance of energy conversion processes and investigating the use of agricultural crops and residues as sustainable feedstocks for production of chemicals and energy. *Degrees offered:* None. *Graduate students served last academic year:* 16: 9 at the master's level; 5 at the doctoral level; 2 at the postgraduate level. *Faculty:* 30. *Faculty affiliations:* Departments of Aerospace Engineering and Engineering Mechanics; Agricultural and Biosystems Engineering; Agronomy; Animal Ecology; Animal Science; Biochemistry, Biophysics, and Molecular Biology; Botany; Chemical Engineering; Chemistry; Civil and Construction Engineering; Economics; Food Science and Human Nutrition; Forestry; Geological and Atmospheric Sciences; Materials Science and Engineering; Mechanical Engineering; Microbiology; Physics. *Annual research budget:* $700,000. *Director:* Dr. Robert C. Brown, Director, 515-294-7934, Fax: 515-294-3091, E-mail: rcbrown@iastate.edu. *Application contact:* Tonia M. McCarley, Program Assistant, 515-294-6555, Fax: 515-294-3091, E-mail: tmccarly@iastate.edu.

Computational Fluid Dynamics Center Founded in 1980. *Academic areas of research and training:* Computational fluid dynamics. *Degrees offered:* None. *Graduate students served last academic year:* 25: 12 at the master's level; 11 at the doctoral level; 2 at the postgraduate level. *Faculty:* 9. *Faculty affiliations:* Departments of Aerospace Engineering and Engineering Mechanics, Chemical Engineering, Civil and Construction Engineering, Mechanical Engineering. *Annual research budget:* $350,000. *Director:* Dr. John C. Tannehill, Manager, 515-294-5666, Fax: 515-294-3262, E-mail: johnt@iastate.edu.

Johns Hopkins University, Baltimore, MD 21218-2699.

Center for Nondestructive Evaluation (CNDE) Founded in 1984. *Academic areas of research and training:* Nondestructive characterization of materials. *Degrees offered:* None. *Graduate students served last academic year:* 25: 3 at the master's level; 23 at the doctoral level; 2 at the postgraduate level. *Faculty:* 46: 1 affiliated solely with the center. *Faculty affiliations:* Schools of Engineering, Medicine; Applied Physics Laboratory. *Annual research budget:* $2 million. *Director:* Dr. Robert E. Green, Director, 410-516-6115, Fax: 410-516-7249, E-mail: cnde@jhu.edu.

Kansas State University, Manhattan, KS 66506.

Center for Hazardous Substance Research Founded in 1985. *Academic areas of research and training:* Environmental science and engineering, hazardous substances, bioremediation, phytoremediation. *Degrees offered:* None. *Graduate students served last academic year:* 14: 6 at the master's level; 6 at the doctoral level; 2 at the postgraduate level. *Faculty:* 16: 1 affiliated solely with the center. *Faculty affiliations:* Departments of Agronomy, Biochemistry, Biological and Agricultural Engineering, Chemical Engineering, Chemistry, Civil Engineering, Geology. *Annual research budget:* $600,000. *Director:* Dr. Larry E. Erickson, Director, 785-532-2380, Fax: 785-532-5985, E-mail: lerick@ksu.edu. *Application contact:* Rita Shade, 785-532-6519, Fax: 785-532-5985, E-mail: ritam@ksu.edu.

Institute for Environmental Research Founded in 1963. *Academic areas of research and training:* Heating, ventilating, air conditioning system design, human thermal comfort. *Degrees offered:* None. *Graduate students served last academic year:* 8: 5 at the master's level; 3 at the doctoral level. *Faculty:* 7: 2 affiliated solely with the institute. *Faculty affiliations:* Departments of Chemical Engineering; Clothing, Textiles, and Interior Design; Industrial and Manufacturing Systems Engineering; Mechanical Engineering. *Annual research budget:* $500,000. *Director:* Dr. Mohammad H. Hosni, Director, 785-532-5620, Fax: 785-532-6642, E-mail: hosni@ksu.edu.

Lehigh University, Bethlehem, PA 18015-3094.

ATLSS Engineering Research Center Founded in 1986. *Academic areas of research and training:* Applied mechanics, computer science, experimentation, metallurgy and welding, sensor development, structural engineering. *Degrees offered:* None. *Graduate students served last academic year:* 33: 19 at the master's level; 11 at the doctoral level; 3 at the postgraduate level. *Faculty:* 12. *Faculty affiliations:* Departments of Civil Engineering, Manufacturing Systems, Materials Science, Mechanical Engineering, Structural Engineering, Corrosion Laboratory-Zettlemoyer Center for Surface Studies. *Annual research budget:* $4.5 million. *Director:* Dr. John W. Fisher, Director, 610-758-3535, Fax: 610-758-5553, E-mail: jwf2@lehigh.

Research and Training Opportunities

ATLSS Engineering Research Center (continued)
edu. *Application contact:* Dr. John E. Bower, Deputy Director, 610-758-3524, Fax: 610-758-5553, E-mail: jb0e@lehigh.edu.

Council on Tall Buildings and Urban Habitat Founded in 1969. *Academic areas of research and training:* Technology transfer, tall building statistics, architecture, engineering, urban planning, structures research. *Degrees offered:* None. *Graduate students served last academic year:* 1. *Faculty:* 1: 1 affiliated solely with Council on Tall Buildings and Urban Habitat. *Faculty affiliations:* Departments of Architecture, Civil and Environmental Engineering, Urban Planning. *Director:* Dr. Lynn S. Beedle, Director, 610-758-3515, Fax: 610-758-4522, E-mail: lsb0@lehigh.edu. *Application contact:* Dolores Rice, Publications Manager, 610-758-4602, Fax: 610-758-4522, E-mail: dbr1@lehigh.edu.

Zettlemoyer Center for Surface Studies (ZCSS) Founded in 1966. *Academic areas of research and training:* Environment, energy, safety, and health; high-technology communications; synthesis of chemicals and engineering materials; catalysis; coatings; printing inks. *Degrees offered:* None. *Graduate students served last academic year:* 30: 8 at the master's level; 12 at the doctoral level; 10 at the postgraduate level. *Faculty:* 17. *Faculty affiliations:* Departments of Chemical Engineering, Chemistry, Earth and Environmental Sciences, Electrical Engineering, Materials, Mechanical Engineering and Mechanics. *Annual research budget:* $2.5 million. *Director:* Dr. Richard G. Herman, Executive Director, 610-758-3486, Fax: 610-758-6555, E-mail: rgh1@lehigh.edu. *Application contact:* M. C. Sawyers, Administrative Coordinator, 610-758-3600, Fax: 610-758-6555, E-mail: mcs0@lehigh.edu.

Louisiana State University and Agricultural and Mechanical College, Baton Rouge, LA 70803.

Basin Research Institute (BRI) Founded in 1984. *Academic areas of research and training:* Oil and gas research, geology. *Degrees offered:* None. *Graduate students served last academic year:* 1: 1 at the doctoral level. *Faculty:* 4: 4 affiliated solely with the institute. *Annual research budget:* $450,000. *Director:* Dr. Chacko J. John, Director, 225-388-8328, Fax: 225-388-3662, E-mail: chacko@vartex.bri.lsu.edu.

Louisiana Forest Products Laboratory (LFPL) Founded in 1992. *Academic areas of research and training:* Wood products manufacturing, marketing, wood science and technology, environment and safety, wood products, international trade, value-added processing. *Degrees offered:* None. *Graduate students served last academic year:* 12: 4 at the master's level; 7 at the doctoral level; 1 at the postgraduate level. *Faculty:* 14: 6 affiliated solely with the laboratory. *Faculty affiliations:* College of Business Administration; Departments of Civil and Environmental Engineering, Industrial and Manufacturing Systems Engineering; School of Forestry, Wildlife, and Fisheries; Louisiana Tech University. *Annual research budget:* $650,000. *Director:* Dr. W. Ramsay Smith, Program Director, 225-388-4155, Fax: 225-388-4251, E-mail: wsmith@lsu.edu.

National Ports and Waterways Institute Founded in 1982. *Academic areas of research and training:* Maritime research, management, economics, finance, international trade, technology, government, operational analysis. *Degrees offered:* None. *Graduate students served last academic year:* 8: 4 at the master's level; 2 at the doctoral level; 2 at the postgraduate level. *Faculty:* 25: 10 affiliated solely with the institute. *Faculty affiliations:* Departments of Coastal Studies, Engineering, Management; The George Washington University Department of Engineering Management. *Annual research budget:* $1.5 million. *Director:* Dr. Anatoly Hochstein, Director, 703-276-7101, Fax: 703-276-7102, E-mail: npwi@seas.gwu.edu.

Remote Sensing and Image Processing Laboratory (RSIP) Founded in 1985. *Academic areas of research and training:* Electrical engineering and civil engineering: GIS/GPS real time processing, sensing and control, civil engineering infrastructure. *Degrees offered:* None. *Graduate students served last academic year:* 10: 5 at the master's level; 2 at the doctoral level; 3 at the postgraduate level. *Faculty:* 9: 2 affiliated solely with the laboratory. *Faculty affiliations:* Center for Coastal, Energy and Environmental Resources; Departments of Civil and Environmental Engineering, Electrical and Computer Engineering; Hazardous Substance Research Center; Louisiana Water Resources Research Institute. *Annual research budget:* $1.3 million. *Director:* Dr. W. David Constant, Director, 225-388-5267, Fax: 225-388-5263, E-mail: hscons@lsu.edu.

Louisiana Tech University, Ruston, LA 71272.

Trenchless Technology Center Founded in 1989. *Academic areas of research and training:* Trenchless technology, underground construction, infrastructure engineering, utility installation and repair. *Degrees offered:* None. *Graduate students served last academic year:* 14: 11 at the master's level; 3 at the doctoral level. *Faculty:* 26. *Faculty affiliations:* Departments of Chemical Engineering, Civil Engineering, Computer Science, Electrical Engineering, Geosciences, Management and Marketing, Mathematics and Statistics, Mechanical and Industrial Engineering. *Annual research budget:* $200,000. *Director:* Dr. Raymond Sterling, Director, 318-257-4072, Fax: 318-257-2777, E-mail: sterling@engr.latech.edu.

Marquette University, Milwaukee, WI 53201-1881.

Center for Industrial Processes and Productivity (CIPP) Founded in 1990. *Academic areas of research and training:* Deformation processing, surface mount technology, deburring and surface finishing, industrial ergonomics, rapid prototyping, engineering management. *Degrees offered:* None. *Graduate students served last academic year:* 101: 60 at the master's level; 40 at the doctoral level; 1 at the postgraduate level. *Faculty:* 13. *Faculty affiliations:* Departments of Civil and Environmental, Electrical and Computer, Mechanical and Industrial Engineering. *Director:* Dr. G. E. Otto Widera, Director, 414-288-7259, Fax: 414-288-1647, E-mail: geo.widera@marquette.edu. *Application contact:* Dr. Joseph P. Domblesky, Associate Director, 414-288-7832, Fax: 414-288-7082, E-mail: joe.domblesky@marquette.edu.

Signal Processing Research Center Founded in 1992. *Academic areas of research and training:* Signal processing, speech processing, speech enhancement, digital filter design, genetic algorithms. *Degrees offered:* None. *Graduate students served last academic year:* 9: 7 at the master's level; 2 at the doctoral level. *Faculty:* 5: 1 affiliated solely with the center. *Faculty affiliations:* Departments of Electrical and Computer Engineering, Biomedical Engineering. *Annual research budget:* $150,000. *Director:* Dr. James A. Heinen, Director, 414-288-3500, Fax: 414-288-5579, E-mail: james.heinen@marquette.edu. *Application contact:* Dr. James A. Heinen, Director, 414-288-3500, Fax: 414-288-5579, E-mail: james.heinen@marquette.edu.

Marshall University, Huntington, WV 25755-2020.

Center for Environmental, Geotechnical, and Applied Sciences Founded in 1993. *Academic areas of research and training:* Environmental engineering and science, technology management, information technology. *Degrees offered:* None. *Graduate students served last academic year:* 70: 70 at the master's level. *Faculty:* 12: 2 affiliated solely with the center. *Faculty affiliations:* Colleges of Business, Science; Graduate School of Engineering and Information Technology; School of Medicine. *Annual research budget:* $500,000. *Director:* Dr. James W. Hooper, Director, 304-696-5453, Fax: 304-696-5454, E-mail: hooper@marshall.edu. *Application contact:* Elizabeth E. Hanrahan, Program Coordinator, Senior, 304-696-5455, Fax: 304-696-5454, E-mail: hanrahan@marshall.edu.

Massachusetts Institute of Technology, Cambridge, MA 02139-4307.

Center for Educational Computing Initiatives (CECI) Founded in 1991. *Academic areas of research and training:* Development and use of computation and communications in education. *Degrees offered:* None. *Graduate students served last academic year:* 8: 5 at the master's level; 3 at the doctoral level. *Faculty:* 11: 1 affiliated solely with the center. *Faculty affiliations:* Departments of Civil and Environmental Engineering, Electrical Engineering and Computer Science; Programs in History, Literature. *Annual research budget:* $1 million. *Director:* Prof. Steven R. Lerman, Director, 617-253-4277, Fax: 617-253-8632, E-mail: lerman@mit.edu. *Application contact:* Pam Homsy, Assistant Director for Administration, 617-253-0113, Fax: 617-253-8632, E-mail: pam@ceci.mit.edu.

Center for Magnetic Resonance *Academic areas of research and training:* NMR technology, chemistry, DNP/EPR, spatial magnetic resonance. *Degrees offered:* None. *Graduate students served last academic year:* 31 at the doctoral level; 14 at the postgraduate level. *Faculty:* 2. *Faculty affiliations:* Departments of Chemistry, Nuclear Engineering. *Annual research budget:* $4 million. *Director:* Dr. Robert C. Griffin, Director, 617-253-5478, Fax: 617-253-5405, E-mail: griffin@ccnmr.mit.edu.

Center for Space Research Founded in 1966. *Academic areas of research and training:* Space life sciences, space plasma and gravitational physics, X-ray optical and planetary astronomy. *Degrees offered:* None. *Graduate students served last academic year:* 32: 32 at the doctoral level. *Faculty:* 19. *Faculty affiliations:* Departments of Aeronautics and Astronautics; Earth, Atmospheric, and Planetary Sciences; Physics. *Annual research budget:* $30 million. *Director:* Dr. Claude R. Canizares, Director, 617-253-7501, Fax: 617-253-3111, E-mail: crc@space.mit.edu.

Center for Technology, Policy, and Industrial Development Founded in 1985. *Academic areas of research and training:* Communications policy, hazardous substances management, international motor vehicles, materials systems, lean manufacturing, aerospace operations, aerospace labor, environmental policy, military logistics, lean sustainment practices, automobile production, industrial ecology, environmental law, mobility and transportation policy, Internet telephone. *Degrees offered:* None. *Graduate students served last academic year:* 120. *Faculty:* 60. *Faculty affiliations:* Sloan School of Management; School of Architecture and Planning; Departments of Aeronautics and Astronautics; Chemical Engineering; Chemistry; Civil and Environmental Engineering; Electrical Engineering; Engineering Systems; Materials Science; Mechanical Engineering; Political Science; Urban Studies. *Annual research budget:* $9 million. *Director:* Dr. Fred Moavenzaden, Director, 617-253-8973, Fax: 617-253-7140, E-mail: ctpidcom@mit.edu.

Francis Bitter Magnet Laboratory (FBML) Founded in 1967. *Academic areas of research and training:* Magnetic resonance, solid-state spectroscopy, condensed matter physics semiconductors, superconductivity, liquid crystals, magnet technology, imaging. *Degrees offered:* None. *Graduate students served last academic year:* 50: 30 at the doctoral level; 20 at the postgraduate level. *Faculty:* 15. *Faculty affiliations:* Departments of Biology, Chemistry, Nuclear Engineering, Physics. *Annual research budget:* $4 million. *Director:* Robert G. Griffin, Director, 617-253-5478, Fax: 617-253-5405, E-mail: rgg@mit.edu. *Application contact:* Elisabeth K.I. Shortsleeve, Secretary, 617-253-5478, Fax: 617-253-5405, E-mail: elka@mit.edu.

Materials Processing Center (MPC) Founded in 1980. *Academic areas of research and training:* Ceramic, electronic, metallic, and polymeric engineering materials; biomaterials, nanostructures, photonic and optoelectronic materials. *Degrees offered:* None. *Graduate students served last academic year:* 100: 60 at the master's level; 40 at the doctoral level. *Faculty:* 120. *Faculty affiliations:* Departments of Aeronautics and Astronautics, Biology, Chemical Engineering, Chemistry, Civil and Environmental Engineering, Electrical Engineering, Materials Science and Engineering, Mechanical Engineering, Ocean Engineering, Physics. *Annual research budget:* $6.5 million. *Director:* Dr. Lionel Kimerling, Director, 617-253-3217, Fax: 617-258-6900. *Application contact:* Frances M. Page, Administrative Staff, 617-253-5179, Fax: 617-258-6900, E-mail: fmpage@mit.edu.

Technology Laboratory for Advanced Composites (TELAC) Founded in 1978. *Academic areas of research and training:* Composite materials and their structures, including manufacture, design, behavior, failure, longevity, durability. *Degrees offered:* None. *Graduate students served last academic year:* 17: 13 at the master's level; 4 at the doctoral level. *Faculty:* 6. *Faculty affiliations:* Department of Aeronautics and Astronautics. *Director:* Dr. Paul A. Lagace, Co-Director, 617-253-3628, Fax: 617-253-0361, E-mail: pal@mit.edu.

McGill University, Montréal, PQ H3A 2T5, Canada.

Groupe d'Études et de Recherche en Analyse des Décisions (GERAD) Founded in 1988. *Academic areas of research and training:* Operations research. *Degrees offered:* None. *Graduate students served last academic year:* 68 at the master's level; 38 at the doctoral level; 4 at the postgraduate level. *Faculty:* 36. *Faculty affiliations:* Faculty of Management; École des Hautes Études Commerciales Program in Quantitative Methods; École Polytechnique de Montréal Departments of Industrial Engineering, Mathematics; Université du Québec à Montréal Mathematics Program. *Annual research budget:* $3.9 million. *Director:* Dr. Pierre Hansen, Director, 514-340-5675, Fax: 514-340-5665, E-mail: gerad@crt.umontreal.ca.

Michigan State University, East Lansing, MI 48824-1020.

Composite Materials and Structures Center (CMSC) Founded in 1985. *Academic areas of research and training:* Composite materials and processing (polymer, metal, ceramic and concrete matrix). *Degrees offered:* None. *Graduate students served last academic year:* 65: 23 at the master's level; 35 at the doctoral level; 7 at the postgraduate level. *Faculty:* 37: 2 affiliated solely with the center. *Faculty affiliations:* College of Engineering. *Annual research budget:* $5.4 million. *Director:* Dr. Lawrence T. Drzal, Director, 517-353-5466, Fax: 517-432-1634, E-mail: drzal@egr.msu.edu.

Michigan Technological University, Houghton, MI 49931-1295.

Institute of Materials Processing Founded in 1955. *Academic areas of research and training:* Applied research and development in: mineral processing, solid waste processing and utilization, metallic processing, mechanical alloying, material characterization. *Degrees offered:* None. *Graduate students served last academic year:* 2 at the master's level; 9 at the doctoral level; 2 at the postgraduate level. *Faculty:* 1. *Faculty affiliations:* Departments of Biological Sciences, Chemistry, Civil and Environmental Engineering, Geological Sciences, Mechanical Engineerin; Programs in Business Administration, Geological Engineering, Metallurgical and Materials Engineering. *Director:* Dr. Jim Hwang, Director, 906-487-2600, Fax: 906-487-2921, E-mail: jhwang@mtu.edu.

Keweenaw Research Center (KRC) Founded in 1965. *Academic areas of research and training:* Mechanical systems design modeling, measurement, and analysis; NVH; cold weather engineering. *Degrees offered:* None. *Graduate students served last academic year:* 1: 1 at the doctoral level. *Faculty:* 12. *Faculty affiliations:* Departments of Civil Engineering, Computer Science, Electrical Engineering, Mathematics, Mechanical Engineering, Metallurgy. *Annual research budget:* $3 million. *Director:* Jay Meldrum, Director, 906-487-3178, Fax: 906-487-2202, E-mail: jmeldrum@mtu.edu. *Application contact:* Gloria Strieter, Executive Secretary, 906-487-2750, Fax: 906-487-2202, E-mail: gjstriet@mtu.edu.

Mississippi State University, Mississippi State, MS 39762.

High Voltage Laboratory Founded in 1978. *Academic areas of research and training:* Evaluation of electrical insulation of transmission and distribution lines, aging of polymer insulators and power cables, lightning protection of power systems. *Degrees offered:* None. *Graduate students served last academic year:* 4: 3 at the master's level; 1 at the doctoral level. *Faculty:* 7: 2 affiliated solely with the laboratory. *Faculty affiliations:* Department of Electrical and Computer Engineering. *Director:* Dr. Stanislaw Grzybowski, Director, 662-325-2148, Fax: 662-325-2298, E-mail: stangrzy@ece.msstate.edu.

Montana State University–Bozeman, Bozeman, MT 59717.

EPICenter Project Founded in 1994. *Academic areas of research and training:* Research, development and demonstration of green building technologies, design practices, and construction. *Degrees offered:* None. *Graduate students served last academic year:* 4: 4 at the master's level. *Faculty:* 13: 1 affiliated solely with the center. *Faculty affiliations:* Departments of Architecture, Engineering, Health and Human Development, Nursing, Plant and Soil Sciences; Office of Rural Health. *Annual research budget:* $3 million. *Director:* Dr. Kath Williams, Director, 406-994-7713, Fax: 406-994-7980, E-mail: kathwms@montana.edu.

Optical Technology Center (OpTeC) Founded in 1995. *Academic areas of research and training:* Optical materials, lasers, and optoelectronic devices; sensors, micro-optical systems, holography, and coherent optics; chemistry, electrical engineering, physics. *Degrees offered:* None. *Graduate students served last academic year:* 39: 12 at the master's level; 15 at the doctoral level; 12 at the postgraduate level. *Faculty:* 11. *Faculty affiliations:* Departments of Chemistry, Electrical and Computer Engineering, Physics; Program in Biochemistry. *Annual research budget:* $2.5 million. *Director:* Dr. Lee Spangler, Director, 406-994-6279, Fax: 406-994-5407, E-mail: optec@physics.montana.edu. *Application contact:* Norma J. Hamilton, Administrative Assistant, 406-994-6279, Fax: 406-994-5407, E-mail: hamilton@montana.edu.

Reclamation Research Unit Founded in 1969. *Academic areas of research and training:* Mineland reclamation, restoration ecology, phyto-remediation. *Degrees offered:* None. *Graduate students served last academic year:* 12: 12 at the master's level. *Faculty:* 16: 4 affiliated solely with the unit. *Faculty affiliations:* Departments of Animal and Range Sciences, Biology, Civil Engineering, Land Resources and Environmental Sciences. *Annual research budget:* $400,000. *Director:* Dennis Neuman, Acting Director, 406-994-4821, Fax: 406-994-4876, E-mail: uasdn@montana.edu.

Montana Tech of The University of Montana, Butte, MT 59701-8997.

Center for Advanced Mineral and Metallurgical Processing (CAMP) Founded in 1989. *Academic areas of research and training:* Environmental engineering, metallurgical engineering, mineral economics, mineral processing, mining engineering, materials engineering. *Degrees offered:* None. *Graduate students served last academic year:* 7: 5 at the master's level; 2 at the postgraduate level. *Faculty:* 12: 1 affiliated solely with the center. *Faculty affiliations:* Programs in Chemistry and Geochemistry, Geological Engineering, Environmental Engineering, Metallurgy and Mineral Processing, Mineral Economics, and Mining Engineering. *Annual research budget:* $150,000. *Director:* Dr. Corby G. Anderson, Director, 406-496-4794, Fax: 406-496-4794, E-mail: canderson@mtech.edu. *Application contact:* Tami J. Patrick, Executive Assistant, 406-496-4652, Fax: 406-490-4794, E-mail: tpatrick@mtech.edu.

Morgan State University, Baltimore, MD 21251.

Transportation Institute Founded in 1992. *Academic areas of research and training:* Intermodal transportation, transportation and land use, social service transportation, ITS Education, transportation modeling. *Degrees offered:* MS. *Graduate students served last academic year:* 35: 30 at the master's level. *Faculty:* 14: 4 affiliated solely with the center. *Faculty affiliations:* Department of Transportation; Earl C. Graves School of Business and Management; Programs in City and Regional Planning, Civil Engineering, Physics, Psychology. *Annual research budget:* $2 million. *Director:* Dr. Z. Andrew Farkas, Interim Director, 443-885-3666, Fax: 443-319-3571, E-mail: zfarkas@moac.morgan.edu.

New Jersey Institute of Technology, Newark, NJ 07102-1982.

Center for Manufacturing Systems (CMS) Founded in 1989. *Academic areas of research and training:* Manufacturing, process improvement, product design, CAD/CAM, rapid prototyping. *Degrees offered:* None. *Graduate students served last academic year:* 100. *Faculty:* 21: 1 affiliated solely with the center. *Faculty affiliations:* Departments of Computer and Information Science, Electrical and Computer Engineering, Industrial and Manufacturing Engineering, Mechanical Engineering, Physics; Program in Chemical Engineering. *Annual research budget:* $2 million. *Director:* Dr. Donald Sebastian, Executive Director, 973-596-3615, Fax: 973-596-6056, E-mail: sebastian@njit.edu.

Hazardous Substance Management Research Center (HSMRC) Founded in 1984. *Academic areas of research and training:* Environmental science, environmental engineering, environmental policy, environmental management, environmental/health, pollution prevention, green/sustainable manufacturing. *Degrees offered:* None. *Graduate students served last academic year:* 55: 40 at the master's level; 10 at the doctoral level; 5 at the postgraduate level. *Faculty:* 42: 2 affiliated solely with the center. *Faculty affiliations:* Schools of Management, Medicine; Departments of Biology, Chemical Engineering, Chemistry, Civil Engineering, Environmental Science, Health, Public Policy. *Annual research budget:* $10 million. *Director:* Dr. Peter Lederman, Executive Director, 973-596-3233, Fax: 973-802-1946, E-mail: lederman@adm.njit.edu. *Application contact:* G. Margaret Griscavage, Assistant Director, Administration, 973-596-3233, Fax: 973-802-1946, E-mail: griscava@megahertz.njit.edu.

Institute for Transportation Founded in 1989. *Academic areas of research and training:* Transportation engineering, planning, management, operations; intelligent transportation systems; simulation studies; decision support systems. *Degrees offered:* MS; PhD. *Graduate students served last academic year:* 52: 37 at the master's level; 15 at the doctoral level. *Faculty:* 34: 5 affiliated solely with the institute. *Faculty affiliations:* Department of Civil and Environmental Engineering, Computer and Information Science, Electrical and Computer Engineering, Humanities and Social Sciences, Industrial and Manufacturing Engineering, Management, Mathematics, Mechanical Engineering. *Annual research budget:* $2.5 million. *Director:* Dr. Louis Pignataro, Executive Director.

Microelectronics Research Center Founded in 1989. *Academic areas of research and training:* Semiconductor microelectronics; microelectromechanical systems (MEMS); MEMS and CMOS circuit design; MEMS and CMOS microfabrication; anodic, thermal compression, and fusion water bonding. *Degrees offered:* None. *Graduate students served last academic year:* 27: 12 at the master's level; 12 at the doctoral level; 3 at the postgraduate level. *Faculty:* 18: 4 affiliated solely with the center. *Faculty affiliations:* Departments of Electrical and Computer Engineering, Mechanical Engineering, Physics; Programs in Chemical Engineering, Environmental Engineering; School of Industrial Management. *Annual research budget:* $1.1 million. *Director:* Dr. Kenneth Farmer, Director, 973-596-5714, Fax: 973-596-6495, E-mail: farmer@njit.edu.

New Mexico Institute of Mining and Technology, Socorro, NM 87801.

Energetic Materials Research and Testing Center (EMRTC) Founded in 1947. *Academic areas of research and training:* Energetic materials, explosives, shock physics, chemistry. *Degrees offered:* None. *Graduate students served last academic year:* 8: 7 at the master's level; 1 at the doctoral level. *Faculty:* 8: 4 affiliated solely with the center. *Faculty affiliations:* Departments of Chemistry, Material Engineering, Physics. *Annual research budget:* $8 million. *Director:* Dr. Jose Luis M. Cortez, Director, 505-835-5701, Fax: 505-835-5630, E-mail: jcortez@emrtc.nmt.edu.

North Carolina Agricultural and Technical State University, Greensboro, NC 27411.

Center for Aerospace Engineering (CAR) Founded in 1992. *Academic areas of research and training:* Aerospace research. *Degrees offered:* None. *Graduate students served last academic year:* 15: 12 at the master's level; 3 at the doctoral level. *Faculty:* 32: 16 affiliated solely with the center. *Faculty affiliations:* Programs in Electrical, Industrial, Mechanical Engineering. *Annual research budget:* $2 million. *Director:* Dr. Frederick Ferguson, Director, 336-334-7254, Fax: 336-334-7397, E-mail: fferguso@ucat.edu.

Center for Composite Materials Research Founded in 1988. *Academic areas of research and training:* Ceramic, metal, and polymeric composites; material characterization; testing; textile composites; fracture toughness; impact; structural integrity; polymer processing and fabrication. *Degrees offered:* None. *Graduate students served last academic year:* 28: 21 at the master's level; 7 at the doctoral level. *Faculty:* 12: 2 affiliated solely with the center. *Faculty affiliations:* Departments of Chemical Engineering, Mechanical Engineering. *Annual research budget:* $1.6 million. *Director:* Dr. V. Sarma Avva, Director, 336-334-7411, Fax: 336-334-7397, E-mail: avva@ncat.edu. *Application contact:* Cindia Hairston, Research Program Assistant, 336-334-7411, Fax: 336-334-7417, E-mail: cindia@ncat.edu.

North Carolina State University, Raleigh, NC 27695.

Center for Advanced Electronic Materials Processing (AEMP) Founded in 1989. *Academic areas of research and training:* Materials and manufacturing processes for microelectronics. *Degrees offered:* None. *Graduate students served last academic year:* 101: 31 at the master's level; 66 at the doctoral level; 4 at the postgraduate level. *Faculty:* 23: 3 affiliated solely with the center. *Faculty affiliations:* Departments of Chemical Engineering, Electrical and Computer Engineering, Materials Science and Engineering, Physics. *Annual research budget:* $5.5 million. *Director:* Dr. John R. Hauser, Director, 919-515-3001, Fax: 919-515-5055, E-mail: hauser@eos.ncsu.edu.

Center for Engineering Applications of Radioisotopes Founded in 1980. *Academic areas of research and training:* Industrial (including medical) radiation and radioisotope measurement applications, including the use of Monte Carlo simulation for design and interpretation, chemical nuclear engineering. *Degrees offered:* None. *Graduate students served last academic year:* 6: 5 at the doctoral level; 1 at the postgraduate level. *Faculty:* 5: 2 affiliated solely with the center. *Faculty affiliations:* Departments of Chemical Engineering, Nuclear Engineering. *Annual research budget:* $550,000. *Director:* Dr. Robin P. Gardner, Director, 919-515-3378, Fax: 919-515-5115, E-mail: gardner@ncsu.edu.

Electric Power Research Center (EPRC) Founded in 1985. *Academic areas of research and training:* Research involving electric power systems, nuclear power generation, and power quality. *Degrees offered:* None. *Graduate students served last academic year:* 12: 9 at the master's level; 3 at the doctoral level. *Faculty:* 4: 4 affiliated solely with the center. *Faculty affiliations:* Departments of Electrical and Computer Engineering, Nuclear Engineering. *Annual research budget:* $600,000. *Director:* Dr. Edward Vickery, Executive Director, 919-515-3517, Fax: 919-515-5108, E-mail: evickery@eds.ncsu.edu.

Furniture Manufacturing and Management Center Founded in 1991. *Academic areas of research and training:* Multidisciplinary applied research addressing problems of generic interest to furniture manufacturers: manufacturing processes, production scheduling, manufacturing system design. *Degrees offered:* None. *Graduate students served last academic year:* 10: 6 at the master's level; 4 at the doctoral level. *Faculty:* 8. *Faculty affiliations:* Department of Industrial Engineering. *Annual research budget:* $265,000. *Director:* Dr. C. Thomas Culbreth, Director, 919-515-3335, Fax: 919-515-1543, E-mail: culbreth@eos.ncsu.edu. *Application contact:* Jean Rollins, Administrative Assistant, 919-515-3335, Fax: 919-515-1543, E-mail: rollins@eos.ncsu.edu.

Hodges Wood Products Laboratory Founded in 1955. *Academic areas of research and training:* Wood science, manufacturing processes, machining and tooling research. *Degrees offered:* None. *Graduate students served last academic year:* 3 at the master's level; 1 at the doctoral level; 1 at the postgraduate level. *Faculty:* 4: 2 affiliated solely with the laboratory. *Faculty affiliations:* Department of Mechanical Engineering. *Annual research budget:* $300,000. *Director:* Dr. Myron Kelly, Director, 919-515-5808, Fax: 919-515-6302.

Integrated Manufacturing Systems Engineering Institute Founded in 1984. *Academic areas of research and training:* Integrated manufacturing systems. *Degrees offered:* MIMSE. *Graduate students served last academic year:* 12: 12 at the master's level. *Faculty:* 100: 50 affiliated solely with the laboratory. *Faculty affiliations:* Departments of Business and Economics, Civil Engineering, Computer Science, Electrical and Computer Engineering, Industrial Engineering, Materials Engineering, Mechanical Engineering, Textiles Engineering. *Director:* Dr. Larry M. Silverberg, Director, 919-515-5282, Fax: 919-515-1675, E-mail: silver@eos.ncsu.edu. *Application contact:* Nancy K. Evans, Program Assistant, 919-515-3808, Fax: 919-515-1675, E-mail: nancy_evans@imsei.ncsu.edu.

Materials Research Center Founded in 1984. *Academic areas of research and training:* Interdisciplinary program involved in the fundamental studies in the epitaxy of compound semiconductors. *Degrees offered:* None. *Graduate students served last academic year:* 45: 5 at the master's level; 30 at the doctoral level; 2 at the postgraduate level. *Faculty:* 17: 5 affiliated solely with the center. *Faculty affiliations:* Departments of Chemical Engineering, Chemistry, Electrical and Computer Engineering, Physics. *Annual research budget:* $3.8 million. *Director:* Dr. Robert F. Davis, Director, 919-515-2867, Fax: 919-515-3419, E-mail: robert_davis@ncsu.edu.

Precision Engineering Center (PEC) Founded in 1982. *Academic areas of research and training:* Precision metrology, manufacturing, and process control. *Degrees offered:* None. *Graduate students served last academic year:* 12: 6 at the master's level; 6 at the doctoral level. *Faculty:* 9. *Faculty affiliations:* Departments of Computer Science, Electrical and Computer Engineering, Materials Science, Mechanical and Aerospace Engineering, Physics. *Annual research budget:* $500,000. *Director:* Dr. Thomas A. Dow, Director, 919-515-3096, Fax: 919-515-3964, E-mail: thomas_dow@ncsu.edu. *Application contact:* Wendy Shearon, Administrative Assistant, 919-515-3096, Fax: 919-515-3964, E-mail: wendy_shearon@ncsu.edu.

Solid State Electronics Laboratory Founded in 1980. *Academic areas of research and training:* Semiconductor materials and device research. *Degrees offered:* None. *Graduate students served last academic year:* 45: 22 at the master's level; 23 at the doctoral level. *Faculty:* 12: 12 affiliated solely with the laboratory. *Annual research budget:* $2.5 million. *Director:* Dr. R. M. Kolbas, Director, 919-737-7350, E-mail: kolbas@eos.ncsu.edu.

Textile Protection and Comfort Center Founded in 1994. *Academic areas of research and training:* Improved performance in comfort for clothing and protective clothing systems in various end use scenarios. *Degrees offered:* None. *Graduate students served last academic year:* 5: 2 at the master's level; 3 at the doctoral level. *Faculty:* 6: 1 affiliated solely with the center. *Faculty affiliations:* Department of Chemistry, Program in Textile Engineering; Hohenstein Institute. *Annual research budget:* $800,000. *Director:* Dr. Roger L. Barker, Director, 919-515-6550, Fax: 919-515-2294, E-mail: roger_barker@ncsu.edu.

Workgroup for Intelligent Systems in Design and Manufacturing (WISDEM) Founded in 1995. *Academic areas of research and training:* Manufacturing systems, fuzzy logic, database applications, distributed fuzzy constraint systems, modeling imprecision in manufacturing and design. *Degrees offered:* None. *Graduate students served last academic year:* 8: 5 at the master's level; 3 at the doctoral level. *Faculty:* 4: 1 affiliated solely with the group. *Faculty affiliations:* Department of Industrial Engineering. *Annual research budget:* $100,000. *Director:* Dr. Robert E. Young, Director, 919-515-7201, Fax: 919-515-5281, E-mail: young@eos.ncsu.edu.

Northeastern University, Boston, MA 02115-5096.

Center for Electromagnetics Research (CER) Founded in 1984. *Academic areas of research and training:* Radar and microwave systems, ground penetrating radar, advanced signal processing, biological and medical optics, remote sensing, sensor/data fusion. *Degrees offered:* None. *Graduate students served last academic year:* 20: 14 at the master's level; 6 at the doctoral level. *Faculty:* 16. *Faculty affiliations:* Departments of Electrical and Computer Engineering; Mechanical, Industrial, and Manufacturing Engineering; Physics. *Annual research budget:* $1.6 million. *Director:* Michael B. Silevitch, Director, 617-373-5110, Fax: 617-373-8627.

Research and Training Opportunities

Northwestern University, Evanston, IL 60208.

Center for Catalysis and Surface Science Founded in 1983. *Academic areas of research and training:* Heterogeneous and homogeneous catalysis, surface science related to catalysis, energy production, environmental catalysis, chemicals production. *Degrees offered:* None. *Graduate students served last academic year:* 40: 30 at the doctoral level; 10 at the postgraduate level. *Faculty:* 19: 1 affiliated solely with the center. *Faculty affiliations:* Departments of Chemical Engineering, Chemistry, Civil Engineering, Materials Science and Engineering, Physics and Astronomy. *Annual research budget:* $2 million. *Director:* Dr. Peter Stair, Director, 847-491-5266, Fax: 847-467-1018, E-mail: pstair@nwu.edu. *Application contact:* Phyllis Long, Administrative Assistant, 847-491-4354, Fax: 847-467-1018, E-mail: p-long@nwu.edu.

The Ohio State University, Columbus, OH 43210.

Advanced Computing Center for the Arts and Design Founded in 1966. *Academic areas of research and training:* Computer graphics and animation within the arts and design, computer graphics software development, multimedia. *Degrees offered:* None. *Graduate students served last academic year:* 40: 30 at the master's level; 10 at the doctoral level. *Faculty:* 30: 5 affiliated solely with the center. *Faculty affiliations:* Austin E. Knowlton School of Architecture, College of the Arts, Department of Computer and Information Science. *Annual research budget:* $100,000. *Director:* Dr. Wayne E. Carlson, Director, 614-292-3416, Fax: 614-292-7776, E-mail: accad@cgrg.ohio-state.edu. *Application contact:* Elaine Smith, Assistant to the Director, 614-292-1053, Fax: 614-292-7776, E-mail: elaine@cgrg.ohio-state.edu.

Biomedical Engineering Center *Academic areas of research and training:* Cardiovascular, bioMEMS, orthopedics, corneal laser interactions. *Degrees offered:* MS; PhD. *Director:* Mauro Ferrari, Director, 614-292-4756, Fax: 614-292-7301, E-mail: ferrari.5@osu.edu. *Application contact:* Graduate Studies Assistant, 614-292-7152, Fax: 614-292-7301, E-mail: gradsec@chopin.bme.ohio-state.edu.

Center for Materials Research Founded in 1989. *Academic areas of research and training:* Interdisciplinary materials research, polymer research, high temperature materials, carbon/diamond research, semi-conductors, electronic materials, sensors, industrial partnerships, biomaterials, MEMS. *Degrees offered:* None. *Graduate students served last academic year:* 200. *Faculty:* 100. *Faculty affiliations:* Departments of Aerospace Engineering, Applied Mechanics and Aviation, Biomaterials Engineering, Chemical Engineering, Chemistry, Civil and Environmental Engineering and Geodetic Science, Consumer and Textile Sciences, Electrical Engineering, Industrial Welding and Systems Engineering, Materials Science and Engineering, Mechanical Engineering, Physics, Restorative and Prosthetic Dentistry, Statistics. *Annual research budget:* $150,000. *Director:* Dr. Arthur J. Epstein, Director, 614-292-5190, Fax: 614-292-3706, E-mail: cmr@mps.ohio-state.edu. *Application contact:* Fleda Crawford, Manager, 614-292-5190, Fax: 614-292-3706, E-mail: crawford.18@osu.edu.

The Engineering Research Center for Net Shape Manufacturing (ERC/NSM) Founded in 1986. *Academic areas of research and training:* Net shape manufacturing of parts, forging, stamping, tube hydroforming, high-speed machining. *Degrees offered:* None. *Graduate students served last academic year:* 30: 23 at the master's level; 5 at the doctoral level; 2 at the postgraduate level. *Faculty:* 6: 3 affiliated solely with the center. *Faculty affiliations:* Departments of Mechanical Engineering, Industrial Welding and Systems Engineering. *Annual research budget:* $900,000. *Director:* Dr. Taylan Altan, Director, 614-292-5063, Fax: 614-292-7219, E-mail: altan.1@osu.edu. *Application contact:* Dr. Taylan Altan, Director, 614-292-5063, Fax: 614-292-7219, E-mail: altan.1@osu.edu.

Water Resources Center Founded in 1965. *Academic areas of research and training:* Water quantity and quality, surface water and groundwater, water-related environmental science and engineering. *Degrees offered:* None. *Graduate students served last academic year:* 7: 5 at the master's level; 1 at the doctoral level; 1 at the postgraduate level. *Faculty:* 7. *Faculty affiliations:* Departments of Civil and Environmental Engineering and Geodetic Science; Food, Agricultural, and Biological Engineering; Geological Sciences; School of Natural Resources. *Annual research budget:* $300,000. *Director:* Dr. Earl Whitlatch, Director, 614-292-6108, Fax: 614-292-9448, E-mail: whitlatch.1@osu.edu. *Application contact:* Carol Moody, Administrative Assistant, 614-292-6108, Fax: 614-292-9448, E-mail: moody.5@osu.edu.

Ohio University, Athens, OH 45701-2979.

Avionics Engineering Research Center Founded in 1963. *Academic areas of research and training:* Avionics, Instrument Landing Systems, GPS, electromagnetic modeling, navigation and landing systems. *Degrees offered:* None. *Graduate students served last academic year:* 27: 12 at the master's level; 15 at the doctoral level. *Faculty:* 9: 5 affiliated solely with the center. *Faculty affiliations:* Departments of Computer Science, Electrical Engineering, Industrial Engineering, Mechanical Engineering. *Annual research budget:* $4.2 million. *Director:* Dr. James M. Rankin, Director, 740-593-1534, Fax: 740-593-1604, E-mail: rankinj@ohiou.edu.

Center for Advanced Materials Processing Founded in 1991. *Academic areas of research and training:* Materials processing, process design, modeling and control, thermal processing. *Degrees offered:* None. *Graduate students served last academic year:* 20: 12 at the master's level; 8 at the doctoral level. *Faculty:* 10. *Faculty affiliations:* Departments of Chemical Engineering, Chemistry, Industrial and Manufacturing Systems, Industrial Technology, Mechanical Engineering, Physics. *Annual research budget:* $400,000. *Director:* Dr. M. K. Alam, Director, 740-593-1558, Fax: 740-593-0476, E-mail: alam@bobcat.ent.ohiou.edu.

Center for Advanced Software Systems Integration Founded in 1995. *Academic areas of research and training:* Manufacturing software systems, computer integrated manufacturing, computer integrated enterprise. *Degrees offered:* None. *Graduate students served last academic year:* 8: 5 at the master's level; 3 at the doctoral level. *Faculty:* 16: 8 affiliated solely with the center. *Faculty affiliations:* Departments of Computer Science, Electrical and Computer Engineering, Industrial and Manufacturing Systems Engineering, Industrial Technology. *Annual research budget:* $210,000. *Director:* Dr. Charles M. Parks, Director, 740-593-1540, Fax: 740-593-0778, E-mail: cparks@bobcat.ent.ohiou.edu.

Institute for Telecommunication Studies Founded in 1983. *Academic areas of research and training:* International telecommunications, development communications, media training, audience research, new technology research, news flow. *Degrees offered:* None. *Graduate students served last academic year:* 12: 6 at the master's level; 6 at the doctoral level. *Faculty:* 20. *Faculty affiliations:* Center for International Studies; Colleges of Education, Engineering and Technology; Schools of Journalism, Telecommunications, Communications Systems Management. *Director:* Dr. Don Flournoy, Director, 740-593-4866, Fax: 740-593-9184, E-mail: don.flournoy@ohiou.edu.

Ohio Coal Research Center Founded in 1965. *Academic areas of research and training:* Coal combustion, coal cleaning, flue gas scrubbing. *Degrees offered:* None. *Graduate students served last academic year:* 4: 3 at the master's level; 1 at the doctoral level. *Faculty:* 5. *Faculty affiliations:* Departments of Chemical Engineering, Mechanical Engineering. *Annual research budget:* $250,000. *Director:* Dr. Michael Prudich, Director, 614-593-1501, Fax: 614-593-0873, E-mail: prudich@bobcat.ent.ohiou.edu.

Oklahoma State University, Stillwater, OK 74078.

Engineering Energy Laboratory Founded in 1975. *Academic areas of research and training:* Energy/power systems, renewable energy systems, power economics, power systems analysis. *Degrees offered:* None. *Graduate students served last academic year:* 7: 4 at the master's level; 3 at the doctoral level. *Faculty:* 2: 2 affiliated solely with the laboratory. *Annual research budget:* $30,000. *Director:* Dr. Rama Ramakumar, Director, 405-744-5157, Fax: 405-744-9198, E-mail: ramakum@master.ceat.okstate.edu.

Old Dominion University, Norfolk, VA 23529.

Center for Coastal Physical Oceanography Founded in 1991. *Academic areas of research and training:* Physical oceanography, virtual reality capabilities, ocean observing systems. *Degrees offered:* None. *Graduate students served last academic year:* 13: 2 at the master's level; 11 at the doctoral level. *Faculty:* 30: 15 affiliated solely with the center. *Faculty affiliations:* Departments of Computer Science; Mathematics; Ocean, Earth and Atmospheric Sciences. *Annual research budget:* $2.1 million. *Director:* Dr. Larry P. Atkinson, Director, 757-683-4945, Fax: 757-683-5550, E-mail: ccpoinfo@ccpo.odu.edu. *Application contact:* Carole E. Blett, Administrator, 757-683-4945, Fax: 757-683-5550, E-mail: carole@ccpo.odu.edu.

Center for Multidisciplinary Parallel-Vector Computation Founded in 1992. *Academic areas of research and training:* Academic and practical training in engineering, computer science, and applied mathematics in the areas of developing numerical algorithms and software for parallel-vector (finite element based) computational mechanics. *Degrees offered:* None. *Graduate students served last academic year:* 8: 4 at the master's level; 4 at the doctoral level. *Faculty:* 6. *Faculty affiliations:* Departments of Aerospace Engineering, Civil Engineering, Mechanical Engineering. *Annual research budget:* $100,000. *Director:* Dr. Duc T. Nguyen, Director, 757-683-3761, Fax: 757-683-5354, E-mail: nguyen@cee.odu.edu.

Center for Structural Acoustics and Fatigue Research Founded in 1989. *Academic areas of research and training:* Sonic fatigue, nonlinear random vibrations. *Degrees offered:* MS; PhD. *Graduate students served last academic year:* 3: 1 at the master's level; 2 at the doctoral level. *Faculty:* 2. *Faculty affiliations:* Departments of Aerospace Engineering, Mechanical Engineering. *Annual research budget:* $65,000. *Director:* Dr. Chuh Mei, Professor, 757-683-3733, Fax: 757-683-3200, E-mail: chmei@aero.odu.edu.

Institute for Computational and Applied Mechanics Founded in 1983. *Academic areas of research and training:* Computational fluid dynamics, computational mechanics, biomechanics, thermo-fluid interactions, combustion, aerodynamics, high-temperature gas dynamics. *Degrees offered:* None. *Graduate students served last academic year:* 16: 10 at the master's level; 6 at the doctoral level. *Faculty:* 52: 12 affiliated solely with the institute. *Faculty affiliations:* College of Engineering and Technology; Departments of Computer Sciences, Mathematics and Statistics. *Annual research budget:* $200,000. *Director:* Dr. Surendra N. Tiwari, Director, 757-683-6363, Fax: 757-683-5344, E-mail: suren@mem.odu.edu.

Oregon Graduate Institute of Science and Technology, Portland, OR 97291-1000.

Center for Coastal and Land-Margin Research Founded in 1991. *Academic areas of research and training:* Environmental science and engineering. *Degrees offered:* None. *Graduate students served last academic year:* 12: 2 at the master's level; 5 at the doctoral level; 5 at the postgraduate level. *Faculty:* 8. *Faculty affiliations:* Departments of Computer Science and Engineering, Environmental Science and Engineering. *Annual research budget:* $2 million. *Director:* Dr. Antonio Baptista, Director, 503-690-1147, Fax: 503-690-1273, E-mail: baptista@ccalmr.ogi.edu. *Application contact:* Daloris Flaming, Administrative Assistant, 503-690-1247, Fax: 503-690-1273, E-mail: daloris@ccalmr.ogi.edu.

Pacific Software Research Center *Academic areas of research and training:* Computer science and engineering, funcitonal programming, applied formal methods, domain specific languages, type systems, programming languages, software development. *Degrees offered:* None. *Graduate students served last academic year:* 15: 10 at the doctoral level; 5 at the postgraduate level. *Faculty:* 10. *Faculty affiliations:* Department of Computer Science and Engineering. *Annual research budget:* $1.6 million. *Director:* Dr. James Hook, Director, 503-748-1169, Fax: 503-748-1553, E-mail: hook@cse.ogi.edu. *Application contact:* Dr. James Hook, Director, 503-748-1169, Fax: 503-748-1553, E-mail: hook@cse.ogi.edu.

Pennsylvania State University Harrisburg Campus of the Capital College, Middletown, PA 17057-4898.

Environmental Technology Center Founded in 1993. *Academic areas of research and training:* Environmental pollution control, water, wastewater, pollution prevention, site remediation, microbiology, chemistry. *Degrees offered:* None. *Graduate students served last academic year:* 60: 60 at the master's level. *Faculty:* 12. *Faculty affiliations:* Schools of Business Administration; Public Affairs; Science, Engineering and Technology. *Annual research budget:* $100,000. *Director:* Dr. Charles Cole, Professor, 717-948-6133, Fax: 717-948-6401. *Application contact:* Dr. Charles Cole, Professor, 717-948-6133, Fax: 717-948-6401.

Pennsylvania State University Milton S. Hershey Medical Center, Hershey, PA 17033-2360.

Center for NMR Research Founded in 1990. *Academic areas of research and training:* MRI, NMR, imaging spectroscopy. *Degrees offered:* None. *Graduate students served last academic year:* 4: 1 at the master's level; 1 at the doctoral level; 2 at the postgraduate level. *Faculty:* 8: 4 affiliated solely with the center. *Faculty affiliations:* Departments of Medicine, Neurology, Pediatrics, Physics. *Annual research budget:* $800,000. *Director:* Dr. Michael Smith, Director, 717-531-6069, Fax: 717-531-8486, E-mail: mbsmith@psghs.edu. *Application contact:* Judith Perry, Administrative Assistant, 717-531-6069, Fax: 717-531-8486, E-mail: jperry@psghs.edu.

Pennsylvania State University University Park Campus, State College, University Park, PA 16802-1503.

Biomolecular Transport Dynamics Laboratory Founded in 1990. *Academic areas of research and training:* Bioengineering, cardiovascular studies, respiratory studies, drug delivery. *Degrees offered:* None. *Graduate students served last academic year:* 16: 4 at the master's level; 8 at the doctoral level; 4 at the postgraduate level. *Faculty:* 5. *Faculty affiliations:* Department of Chemical Engineering; Programs in Bioengineering, Physiology. *Annual research budget:* $1 million. *Director:* Dr. John Tarbell, Director, 814-863-4801, Fax: 814-865-7846, E-mail: jmt@psu.edu.

Biotechnology Institute-Life Sciences Consortium Founded in 1984. *Academic areas of research and training:* Biotechnology, electron microscopy, flow cytometry, computational biology, nucleic acid facility, hybridoma facility. *Degrees offered:* PhD. *Graduate students served last academic year:* 55: 1 at the master's level; 52 at the doctoral level; 2 at the postgraduate level. *Faculty:* 300. *Faculty affiliations:* Colleges of Agricultural Sciences, Liberal Arts; Departments of Biochemistry and Molecular Biology, Biology, Chemical Engineering, Chemistry, Food Technology, Horticulture, Nutrition, Plant Pathology, Veterinary Science; Milton S. Hershey Medical Center. *Annual research budget:* $2.1 million. *Director:* Dr. Nina Fedoroff, Director, 814-863-4576, Fax: 814-863-1357, E-mail: nvf1@psu.edu. *Application contact:* Judith Burns, Manager, Staff Services, 814-863-4576, Fax: 814-863-1357, E-mail: jeb2@psu.edu.

Center for Dielectric Studies Founded in 1982. *Academic areas of research and training:* Microelectronics, passive components, dielectrics, capacitors. *Degrees offered:* None. *Graduate students served last academic year:* 12: 5 at the master's level; 7 at the doctoral level. *Faculty:* 24. *Faculty affiliations:* Departments of Electrical Engineering, Materials Science and Engineering, Physics; Materials Research Laboratory. *Annual research budget:* $1.6 million. *Director:* Dr. Clive A. Randall, Director, 814-863-1328, Fax: 814-865-2326, E-mail: car4@psu.edu.

Center for Electronic Design, Communication and Computing Founded in 1993. *Academic areas of research and training:* Rapid system prototyping, VLSI design, computer architecture, embedded processing, digital signal processing. *Degrees offered:* None. *Graduate students served last academic year:* 5: 4 at the master's level; 1 at the doctoral level. *Faculty:* 5: 2 affiliated solely with the center. *Faculty affiliations:* Departments of Computer Science and

Engineering, Electrical Engineering. *Director:* Dr. Paul Hulina, Director, 814-865-1444, Fax: 814-863-7582, E-mail: pphecl@engr.psu.edu.

Center for Gas Turbines and Power   Founded in 1994. *Academic areas of research and training:* Aerospace/mechanical engineering, fluid dynamics, heat transfer, vibration, control, manufacturing, materials, inverse design related to gas turbines. *Degrees offered:* None. *Graduate students served last academic year:* 34: 11 at the master's level; 16 at the doctoral level; 7 at the postgraduate level. *Faculty:* 16: 8 affiliated solely with the center. *Faculty affiliations:* Departments of Aerospace Engineering, Engineering Science and Mechanics, Industrial and Manufacturing Engineering, Mechanical Engineering. *Annual research budget:* $1 million. *Director:* Dr. B. Lakshminarayana, Professor, 814-865-5551, Fax: 814-865-7092, E-mail: b1laer@engr.psu.edu.

Center for Medical Ultrasonic Transducer Engineering   Founded in 1994. *Academic areas of research and training:* Bioengineering, ultrasonics, ultrasonic transducers/arrays. *Degrees offered:* None. *Graduate students served last academic year:* 15: 4 at the master's level; 8 at the doctoral level; 3 at the postgraduate level. *Faculty:* 17: 9 affiliated solely with the center. *Faculty affiliations:* Departments of Acoustics, Bioengineering, Dermatolgy, Ophthalmology, Radiology. *Annual research budget:* $1 million. *Director:* Dr. K. Kirk Shung, Director, 814-865-1407, Fax: 814-863-0490, E-mail: kksbio@engr.psu.edu. *Application contact:* Tim Ritter, Manager, 814-863-6318, Fax: 814-863-0490, E-mail: tari45@psu.edu.

Combustion Laboratory   *Academic areas of research and training:* Diesel combustion, fuels, and emissions; applied catalysis for pollutant control; fuel additives; materials-combustion interactions. *Degrees offered:* None. *Graduate students served last academic year:* 12: 6 at the master's level; 4 at the doctoral level; 2 at the postgraduate level. *Faculty:* 8: 1 affiliated solely with the laboratory. *Faculty affiliations:* Departments of Chemical Engineering, Energy and Geo-Environmental Engineering, Materials Science and Engineering; Materials Characterization Laboratory; Materials Research Laboratory. *Annual research budget:* $500,000. *Director:* Dr. Andre L. Boehman, Director, 814-865-7839, Fax: 814-863-8892, E-mail: boehman@ems.psu.edu.

Consortium on Chemically Bonded Ceramics   Founded in 1991. *Academic areas of research and training:* Materials synthesis, new materials, materials characterization, low-temperature processing, cementitious materials, by-product utilization, zeolites, catalysts, clays, concrete for sustainability. *Degrees offered:* MS; PhD. *Graduate students served last academic year:* 14: 4 at the master's level; 5 at the doctoral level; 5 at the postgraduate level. *Faculty:* 12: 5 affiliated solely with the consortium. *Faculty affiliations:* Departments of Agronomy, Geoscience, Environmental Pollution Control, Materials Science and Engineering, Nuclear and Civil Engineering. *Annual research budget:* $600,000. *Director:* Dr. Della Roy, Director, 814-865-1196, Fax: 814-863-7040, E-mail: dellaroy@psu.edu.

Electro-Optics Laboratory   Founded in 1980. *Academic areas of research and training:* Optical computing, neural networks, photorefractive materials and devices, nonlinear liquid crystals, fiberoptic sensors, biooptic research, signal processing. *Degrees offered:* None. *Graduate students served last academic year:* 30. *Faculty:* 17: 7 affiliated solely with the laboratory. *Faculty affiliations:* Applied Research Laboratory, Material Research Laboratory. *Annual research budget:* $1 million. *Director:* Dr. Francis Yu, Director, 814-863-2989, Fax: 814-865-7065, E-mail: ftyece@engr.psu.edu. *Application contact:* Dr. Shizhuo Yin, Assistant Professor, 814-863-4256, Fax: 814-865-7065, E-mail: sxy105@psu.edu.

Materials Research Laboratory   Founded in 1962. *Academic areas of research and training:* Ceramic science, engineering science and mechanics, materials science and engineering, physics, electronic components, thin films, cement, particle technology, biomedical materials, organic piezoelectrics, ferroelectrics and electronic materials, diamond thin films, spectroscopic ellipsometry, actuators and transducer devices, microwave processing, capacitors, smart materials, powder mechanics, nanoscale materials, colloid science, phosphor technology, porous materials, piezoelectric single crystals, hydrothermal synthesis, textured materials, electrophoretic deposition, spray drying/pyrolsys, sintering and hot pressing. *Degrees offered:* None. *Graduate students served last academic year:* 102: 20 at the master's level; 60 at the doctoral level; 22 at the postgraduate level. *Faculty:* 43: 18 affiliated solely with the laboratory. *Faculty affiliations:* Departments of Agronomy, Ceramic Science and Engineering, Computer Science, Electrical Engineering, Electronic and Photonic Materials, Engineering Science and Mechanics, Materials Science, Mathematics, Physics; Intercollege Program in Materials. *Annual research budget:* $9 million. *Director:* Dr. Gary L. Messing, Director, 814-865-2262, Fax: 814-865-2326, E-mail: messing@mrl.psu.edu. *Application contact:* Wendy Bathgate, Staff Assistant, 814-865-1656, Fax: 814-865-2326, E-mail: web1@psu.edu.

Metal Casting Laboratory   Founded in 1991. *Academic areas of research and training:* Metal casting, foundry technology, cast metals and alloys, environmental studies. *Degrees offered:* None. *Graduate students served last academic year:* 12: 8 at the master's level; 4 at the doctoral level. *Faculty:* 6. *Faculty affiliations:* Departments of Civil and Environmental Engineering, Industrial and Manufacturing Engineering; Program in Metal Science and Engineering. *Annual research budget:* $600,000. *Director:* Dr. Robert Voigt, Director, 814-863-7290, Fax: 814-863-4745, E-mail: rcv2@psu.edu. *Application contact:* Robert J. Haefner, Metal Casting Center Manager, 814-863-2378, Fax: 814-863-4745, E-mail: rjh115@psu.edu.

Pennsylvania Mining and Mineral Resources Research Institute (PAMMRRI)   Founded in 1981. *Academic areas of research and training:* Mineral fields, mining. *Degrees offered:* None. *Graduate students served last academic year:* 2: 2 at the master's level. *Faculty:* 6: 1 affiliated solely with the institute. *Faculty affiliations:* Departments of Energy and GeoEnvironmental Engineering, Geosciences, Material Science and Engineering, Mineral Economics. *Director:* Dr. H. Reginald Hardy, Director, 814-863-1620, Fax: 814-865-3248, E-mail: h4h@psuvm.psu.edu.

Pennsylvania Transportation Institute (PTI)   Founded in 1968. *Academic areas of research and training:* Transportation research: pavements, materials and construction; transportation operations; transportation structures; vehicle systems and safety; specialties include architectural, civil, electrical, industrial, and mechanical engineering as well as agriculture, business logistics and management, economics, geography, psychology, and statistics. *Degrees offered:* None. *Graduate students served last academic year:* 57: 35 at the master's level; 22 at the doctoral level. *Faculty:* 51: 2 affiliated solely with the institute. *Faculty affiliations:* Departments of Architectural Engineering, Business Logistics, Civil Engineering, Economics, Industrial Engineering, Management and Organization, Mechanical Engineering, Psychology, Statistics. Affiliated centers and programs: Bus Testing and Research Center, Center for Intelligent Transportation Systems (CITranS), Crash Safety Research Center, LTAP—The Pennsylvania Local Roads Program, the Mid-Atlantic Universities Transportation Center (MAUTC), and the Northeast Center of Excellence for Pavement Technology (NECEPT). *Annual research budget:* $7 million. *Director:* Dr. Bohdan T. Kulakowski, Director, 814-865-1891, Fax: 814-865-3039, E-mail: btk1@psu.edu. *Application contact:* Dr. Bohdan T. Kulakowski, Director, 814-865-1891, Fax: 814-865-3039, E-mail: btk1@psu.edu.

P/M Lab   Founded in 1991. *Academic areas of research and training:* Engineering materials, especially those fabricated from powders using sintering technologies. *Degrees offered:* None. *Graduate students served last academic year:* 17: 7 at the master's level; 6 at the doctoral level; 4 at the postgraduate level. *Faculty:* 8: 5 affiliated solely with the laboratory. *Faculty affiliations:* Departments of Engineering Science and Mechanics, Industrial Engineering, Mechanical Engineering. *Annual research budget:* $2.3 million. *Director:* Dr. Randall M. German, Director, 814-863-8025, Fax: 814-863-8211, E-mail: rmg4@psu.edu.

Radiation Science and Engineering Center   Founded in 1955. *Academic areas of research and training:* Nuclear engineering, radiation, radiation science, reactor, nuclear reactor. *Degrees offered:* None. *Graduate students served last academic year:* 50: 35 at the master's level; 13 at the doctoral level; 2 at the postgraduate level. *Faculty:* 43: 3 affiliated solely with the center. *Faculty affiliations:* Colleges of Agriculture, Earth and Mineral Science, Engineering, Liberal Arts, Science. *Annual research budget:* $1.2 million. *Director:* Dr. C. Frederick Sears, Director, 814-865-6351, Fax: 814-863-4840, E-mail: cfsnuc@engr.psu.edu.

Research Center for the Engineering of Electronic and Acoustic Materials   Founded in 1985. *Academic areas of research and training:* EMI, RFI shielding materials; acoustic and radar absorbing materials; microwave processing of materials; smart materials and MEMS. *Degrees offered:* None. *Graduate students served last academic year:* 15: 8 at the master's level; 7 at the doctoral level. *Faculty:* 8: 5 affiliated solely with the center. *Faculty affiliations:* Departments of Electrical Engineering, Materials Science and Engineering, Mechanical Engineering, Polymer Science. *Annual research budget:* $1.6 million. *Director:* Dr. Vijay Varadan, Director, 814-863-4210, Fax: 814-865-3052, E-mail: vjvesm@engr.psu.edu.

Rock Mechanics Laboratory   Founded in 1965. *Academic areas of research and training:* Laboratory and field studies of geologic materials from an engineering point of view. *Degrees offered:* None. *Graduate students served last academic year:* 5: 2 at the master's level. *Faculty:* 10: 5 affiliated solely with the laboratory. *Faculty affiliations:* Programs in Geology, Geophysics, Mining Engineering, Petroleum and Natural Gas Engineering. *Director:* Dr. H. Reginald Hardy, Professor, 814-863-1620, Fax: 814-865-3248, E-mail: h4h@psuvm.psu.edu.

Polytechnic University, Brooklyn Campus, Brooklyn, NY 11201-2990.

Polytechnic Research Institute for Development and Enterprise (PRIDE)   Founded in 1995. *Academic areas of research and training:* Distanceless learning, life-long learning, telemedicine, multimedia technologies, information service and Internet technology. *Degrees offered:* None. *Graduate students served last academic year:* 12: 4 at the master's level; 6 at the doctoral level; 2 at the postgraduate level. *Faculty:* 14: 10 affiliated solely with the institute. *Faculty affiliations:* Centers for Advanced Technology in Telecommunications, Applied Large Scale Computing; Polymer Research Institute. *Annual research budget:* $1.5 million. *Director:* Dr. Ifay F. Chang, Executive Director, 718-260-3264, Fax: 914-323-2010, E-mail: ifay@pride-i2.poly.edu. *Application contact:* Qun Zhou, Assistant to the Director, 914-323-2037, Fax: 914-323-2010, E-mail: zhougun@pride-i2.poly.edu.

Princeton University, Princeton, NJ 08544-1019.

Center for Photonic and Optoelectronic Materials (POEM)   Founded in 1989. *Academic areas of research and training:* Photonic and optoelectronic materials, devices, and systems; ultra-fast communications, displays, photonic packaging, sensor and imaging technologies, nanostructures; micro and nanofabrication. *Degrees offered:* None. *Graduate students served last academic year:* 100: 7 at the master's level; 60 at the doctoral level; 33 at the postgraduate level. *Faculty:* 32. *Faculty affiliations:* Departments of Chemical Engineering, Chemistry, Electrical Engineering, Mechanical and Aerospace Engineering, Molecular Biology, Physics. *Annual research budget:* $8 million. *Director:* Dr. James C. Sturm, Director, 609-258-4454, Fax: 609-258-1954, E-mail: sturm@ee.princeton.edu. *Application contact:* Joseph Montemarano, Director for Industrial Liaison, 609-258-2267, Fax: 609-258-1954, E-mail: jmonte@princeton.edu.

Announcement: POEM was established in 1989 to expedite technology transfer from the laboratory to the marketplace through innovative collaborations supported by industry, government partners, and Princeton University. With more than 30 faculty members from 6 departments and more than 100 research scientists and staff members, POEM concentrates on advanced displays, sensor and imaging technologies, ultrafast laser techniques, photonic packaging, and multimedia research. State-of-the-art facilities address optoelectronic materials and devices, nanofabrication, and ultrafast lasers in applications ranging from computers and communications to medical diagnostics and the environment. POEM's close collaborations with industry provide unique training experiences and job opportunities for Princeton's brightest students. For more information, visit the Web site at http://www.poem.princeton.edu.

Princeton Materials Institute   Founded in 1990. *Academic areas of research and training:* Materials research on complex materials, polymers, ceramics, optoelectronic materials, biomaterials. *Degrees offered:* None. *Faculty:* 60. *Faculty affiliations:* Departments of Chemical Engineering, Chemistry, Civil and Environmental Engineering , Electrical Engineering, Mechanical and Aerospace Engineering, Molecular Biology, Physics. *Annual research budget:* $6 million. *Director:* Dr. William Russel, Director, 609-258-4580, Fax: 609-258-6878.

See in-depth description on page 1547.

The Spiro Lab   Founded in 1963. *Academic areas of research and training:* Bioinorganic chemistry, spectroscopy. *Degrees offered:* None. *Graduate students served last academic year:* 6 at the doctoral level; 4 at the postgraduate level. *Faculty:* 1: 1 affiliated solely with the laboratory. *Annual research budget:* $500,000. *Director:* Dr. Thomas G. Spiro, Professor, 609-258-3907, Fax: 609-258-3804, E-mail: spiro@princeton.edu.

Purdue University, West Lafayette, IN 47907.

Center for Agricultural Policy and Technology Assessment   Founded in 1989. *Academic areas of research and training:* Agricultural economics as applied to agricultural technologies. *Degrees offered:* None. *Graduate students served last academic year:* 3: 2 at the master's level; 1 at the doctoral level. *Faculty:* 6. *Faculty affiliations:* Departments of Agricultural and Biological Engineering, Agricultural Economics, Animal Sciences, Botany and Plant Pathology, Entomology, Foods and Nutrition. *Annual research budget:* $100,000. *Director:* Dr. Marshall A. Martin, Professor, 765-494-4268, Fax: 765-494-9176, E-mail: martin@agecon.purdue.edu.

Dauch Center for the Management of Manufacturing Enterprises (DCMME)   Founded in 1988. *Academic areas of research and training:* Manufacturing and technology management. *Degrees offered:* None. *Graduate students served last academic year:* 30: 28 at the master's level; 2 at the doctoral level. *Faculty:* 16: 1 affiliated solely with the center. *Faculty affiliations:* Schools of Engineering, Management, and Technology. *Annual research budget:* $75,000. *Director:* Dr. Herbert Moskowitz, Professor, 765-494-4421, Fax: 765-494-9658, E-mail: herbm@mgmt.purdue.edu. *Application contact:* Sarah Wassgren, Assistant Director, 765-494-4413, Fax: 765-494-9658, E-mail: swassgren@mgmt.purdue.edu.

Indiana Water Resources Research Center (IWRRC)   Founded in 1964. *Academic areas of research and training:* Ground and surface water contamination; transport of pollutants; agricultural needs and impacts; water infrastructure evaluation and rehabilitation; coastal and wetlands resources; water supply, reuse, and waste water treatment; hydrology and reservoir management; water management in the Great Lakes region; atmospheric and precipitation processes. *Degrees offered:* None. *Graduate students served last academic year:* 8: 4 at the master's level; 4 at the doctoral level. *Faculty:* 54: 1 affiliated solely with the center. *Faculty affiliations:* Departments of Agricultural Economics, Agronomy, Animal Sciences, Biological Sciences, Botany and Plant Pathology, Earth and Atmospheric Sciences, Entomology, Food Science, Forestry and Natural Resources, Pharmacology and Toxicology, Political Science, Sociology and Anthropology; Schools of Agricultural and Biological Engineering, Chemical Engineering, Civil Engineering, Management. *Annual research budget:* $62,298. *Director:* Dr. Jeff R. Wright, Professor, 765-494-2175, Fax: 765-494-2720, E-mail: wrightje@ecn.purdue.edu. *Application contact:* Kamie Redinbo, Assistant to the Director, 765-494-8041, Fax: 765-494-2720, E-mail: wrrc@ecn.purdue.edu.

Rensselaer Polytechnic Institute, Troy, NY 12180-3590.

Center for Automation Technologies (CAT)   Founded in 1988. *Academic areas of research and training:* Robotics, automation, manufacturing, process control, precision assembly, rapid prototyping, product data engineering, mechatronics, document process. *Degrees offered:* None. *Graduate students served last academic year:* 70: 50 at the master's level; 20 at the doctoral level. *Faculty:* 16. *Faculty affiliations:* Departments of Computer Science; Decision Science and Engineering Systems; Electrical, Computer, and Systems Engineering; Materials Science and Engineering; Mechanical and Aeronautical Engineering and Mechanics; School of Management. *Annual research budget:* $4 million. *Director:* Dr. Harry E. Stephanou, Director, 518-276-6156, Fax: 518-276-4897, E-mail: hes@cat.rpi.edu.

Research and Training Opportunities

Center for Image Processing Research Founded in 1986. *Academic areas of research and training:* Image and video processing, image and video compression and transmissions, multimedia systems, video conferencing systems, biomedical imaging, computer vision, pattern recognition, document processing. *Degrees offered:* None. *Graduate students served last academic year:* 43: 28 at the master's level; 12 at the doctoral level; 3 at the postgraduate level. *Faculty:* 27: 15 affiliated solely with the center. *Faculty affiliations:* Departments of Biomedical Engineering, Computer Science, Electrical, Computer and Systems Engineering, Mathematical Sciences. *Annual research budget:* $2 million. *Director:* Dr. James W. Modestino, Director, 518-276-6823, Fax: 518-276-6261, E-mail: modestino@ipl.rpi.edu.

Center for Infrastructure and Transportation Studies Founded in 1993. *Academic areas of research and training:* Developing new technologies and improving existing technologies that address infrastructure and transportation. *Degrees offered:* None. *Graduate students served last academic year:* 5: 4 at the master's level; 1 at the doctoral level. *Faculty:* 50: 1 affiliated solely with the center. *Faculty affiliations:* Departments of Aeronautical Engineering; Architecture; Chemical Engineering; Civil Engineering; Computer Science; Earth/Environmental Sciences; Electric Power Engineering; Electrical/Computer Systems Engineering; Environmental/Energy Engineering; Information Technology Services; Language, Literature and Communication; Materials Science and Engineering; Mathematics; Mechanical Engineering and Mechanics; Science and Technology Studies; Center for Integrated Electronics (CIE); Center for Automated Technologies; Electronics Manufacturing and Electronic Media. *Annual research budget:* $1 million. *Director:* Dr. George List, Director, 518-276-6362, Fax: 518-276-4833, E-mail: listq@rpi.edu.

Center for Integrated Electronics, Electronics Manufacturing, and Electronic Media Founded in 1982. *Academic areas of research and training:* Semiconductor technology, electronics, electronics manufacturing, research and education through electronic media. *Degrees offered:* None. *Graduate students served last academic year:* 70: 15 at the master's level; 55 at the doctoral level. *Faculty:* 80: 20 affiliated solely with the center. *Faculty affiliations:* Departments of Chemistry; Computer Science; Electrical, Computer, and Systems Engineering; Materials Science and Engineering; Mechanical Engineering; Aeronautical Engineering and Mechanics; Physics, App lied Physics, and Astronomy; Program in Chemical Engineering. *Annual research budget:* $9 million. *Director:* Dr. Don L. Millard, Director, 518-276-6724, Fax: 518-276-2990, E-mail: millard@rpi.edu. *Application contact:* Virginia F. Willigan, Manager of Operations, 518-276-2810, Fax: 518-276-8761, E-mail: williv@rpi.edu.

Center for Multiphase Research Founded in 1992. *Academic areas of research and training:* Modeling and computer simulation of multiphase flow and heat transfer, application in energy systems; nuclear systems; chemical processing; naval engineering; microgravity environment. *Degrees offered:* None. *Graduate students served last academic year:* 8: 2 at the master's level; 6 at the doctoral level. *Faculty:* 22. *Faculty affiliations:* Departments of Chemical Engineering, Civil Engineering, Computer Science, Environmental and Energy Engineering, Materials Science and Engineering, Mathematics, Mechanical Engineering, Physics. *Annual research budget:* $800,000. *Director:* Michael Z. Podowski, Director, 518-276-4000, Fax: 518-276-3055.

Center for Polymer Synthesis Founded in 1992. *Academic areas of research and training:* To provide a focus for the extensive effort in polymer synthesis and to integrate the multidisciplinary polymer efforts at Rensselaer; science programs deal with many aspects of research on the synthesis, properties, processing and applications of polymeric materials; special laboratories for the larger scale preparation of new monomers and polymers are also available. *Degrees offered:* None. *Graduate students served last academic year:* 65: 5 at the master's level; 50 at the doctoral level; 10 at the postgraduate level. *Faculty:* 14. *Faculty affiliations:* Departments of Chemical Engineering, Chemistry, Materials Science and Engineering, Physics. *Annual research budget:* $3 million. *Director:* Dr. Brian C. Benicewicz, 518-276-2534, Fax: 518-276-6434, E-mail: benice@rpi.edu. *Application contact:* Susan J. Mangione, Administrative Secretary, 518-276-6341, Fax: 518-276-6434, E-mail: mangis@rpi.edu.

Center for Services Research and Education Founded in 1990. *Academic areas of research and training:* Service sector, role of technology, information systems management, operation research, statistics. *Degrees offered:* None. *Graduate students served last academic year:* 25 at the master's level; 4 at the doctoral level. *Faculty:* 40: 20 affiliated solely with the center. *Faculty affiliations:* Departments of Civil Engineering, Decision Sciences and Engineering Systems; Lally School of Management and Technology. *Annual research budget:* $286,000. *Director:* Dr. Dan Berg, Director, 518-276-2895, Fax: 518-276-8227, E-mail: bergd@rpi.edu.

Electronics Agile Manufacturing Research Institute Founded in 1994. *Academic areas of research and training:* Information systems technologies to support product engineering in distributed manufacturing environments. *Degrees offered:* None. *Graduate students served last academic year:* 8: 3 at the master's level; 5 at the doctoral level. *Faculty:* 6. *Faculty affiliations:* Departments of Computer Science, Decision Sciences and Engineering Systems, Electrical Engineering; School of Management. *Annual research budget:* $2 million. *Director:* Dr. Robert Graves, Director, 518-276-6955, Fax: 518-276-2120, E-mail: graver@rpi.edu.

Flow Diagnostics Laboratory Founded in 1991. *Academic areas of research and training:* Aerodynamics, fluid mechanics, CFD, fluid dynamics, computation fluid dynamics, aircraft design. *Degrees offered:* None. *Graduate students served last academic year:* 17: 9 at the master's level; 7 at the doctoral level; 1 at the postgraduate level. *Faculty:* 8: 1 affiliated solely with the laboratory. *Faculty affiliations:* Departments of Aeronautical Engineering, Mechanical Engineering and Mechanics; Civil Engineering; Environmental and Energy Engineering. *Director:* Dr. Brian E. Thompson, Associate Professor, 518-276-6989, Fax: 518-276-6025, E-mail: thompson@rpi.edu.

Geotechnical Centrifuge Research Center Founded in 1989. *Academic areas of research and training:* Geotechnical engineering, soil dynamics, earthquake engineering, slope stability, soil remediation, geoenvironmental engineering. *Degrees offered:* None. *Graduate students served last academic year:* 10: 5 at the master's level; 5 at the doctoral level. *Faculty:* 5: 1 affiliated solely with the center. *Faculty affiliations:* Department of Civil Engineering. *Annual research budget:* $350,000. *Director:* Dr. Ricardo Dobry, Director, 518-276-6934, Fax: 518-276-4833, E-mail: dobryr@rpi.edu. *Application contact:* Dr. Tarek H. Abdoun, Manager, 518-276-6544, Fax: 518-276-4833, E-mail: abdout@rpi.edu.

Scientific Computation Research Center Founded in 1990. *Academic areas of research and training:* Development of high-performance computing methodologies with emphasis on parallel adaptive multiscale solution techniques; application areas include: CFD, nonlinear solid mechanics, materials processing, and structural acoustics. *Degrees offered:* MS; PhD. *Graduate students served last academic year:* 42: 8 at the master's level; 26 at the doctoral level; 8 at the postgraduate level. *Faculty:* 14. *Faculty affiliations:* Departments of Civil Engineering; Computer Science; Electrical, Computer, and Systems Engineering; Materials Science and Engineering; Programs in Aeronautical Engineering, Mathematics, Mechanical Engineering. *Annual research budget:* $2 million. *Director:* Dr. Mark S. Shephard, Director, 518-276-6795, Fax: 518-276-4886, E-mail: shephard@scorec.rpi.edu.

Rice University, Houston, TX 77251-1892.

The Center for Nanoscale Science and Technology (CNST) Founded in 1996. *Academic areas of research and training:* Single-walled carbon nanotubes, nanotechnology, geometrical molecules, buckyball, buckytubes, nanotubes, carbon atoms. *Degrees offered:* None. *Graduate students served last academic year:* 16: 6 at the doctoral level; 8 at the postgraduate level. *Faculty:* 15: 3 affiliated solely with the center. *Faculty affiliations:* Applied Physics Program; Departments of Chemistry, Electrical and Computer Engineering, Mechanical Engineering, Physics; Programs in Biology, Materials Science. *Annual research budget:* $1 million. *Director:* Dr. R. E. Smalley, Director, 713-527-4845, Fax: 713-285-5320, E-mail: res@cnst.rice.edu.

Computer and Information Technology Institute (CITI) Founded in 1988. *Academic areas of research and training:* Computation engineering, parallel and distributed computing, discrete optimization, digital signal processing, telecommunications and wireless networking, data analysis and modeling. *Degrees offered:* None. *Graduate students served last academic year:* 120: 20 at the master's level; 80 at the doctoral level; 20 at the postgraduate level. *Faculty:* 90. *Faculty affiliations:* Departments of Computer Science, Computational and Applied Mathematics, Electrical and Computer Engineering, Materials Science, Mechanical Engineering, Statistics. *Annual research budget:* $15 million. *Director:* Dr. Willy Zwaenepoel, Director, 713-285-5402, Fax: 713-285-5930, E-mail: willy@cs.rice.edu. *Application contact:* Anthony J. Elam, Executive Director, 713-527-4734, Fax: 713-285-5136, E-mail: elam@rice.edu.

The Institute of Biosciences and Bioengineering (IBB) Founded in 1986. *Academic areas of research and training:* Foster interdisciplinary interaction in biological, chemical, and engineering disciplines between research groups at Rice University and neighboring institutions. *Degrees offered:* None. *Graduate students served last academic year:* 146: 8 at the master's level; 94 at the doctoral level; 44 at the postgraduate level. *Faculty:* 60. *Faculty affiliations:* Departments of Biochemistry and Cell Biology, Bioengineering, Chemical Engineering, Chemistry, Electrical and Computer Engineering, Materials Science. *Annual research budget:* $8 million. *Director:* Dr. Larry V. McIntire, Chair, 713-527-6034, Fax: 713-285-5154, E-mail: ibb@rice.edu. *Application contact:* Diana L. Welch, Associate Director, 713-527-4671, Fax: 713-285-5154, E-mail: dwelch@rice.edu.

Rice Quantum Institute Founded in 1987. *Academic areas of research and training:* Facilitation and promotion of research in basic molecular physics, focusing on matter and electromagnetic energy at the scale where the quantum mechanical behavior dominates. *Degrees offered:* PhD. *Graduate students served last academic year:* 25: 25 at the doctoral level. *Faculty:* 40. *Faculty affiliations:* Divisions of Engineering, Natural Science. *Director:* Dr. Ken Smith, Director, 713-527-6028, Fax: 713-285-5935, E-mail: quantum@rice.edu. *Application contact:* Yvonne Creed, Executive Assistant, 713-527-6028, Fax: 713-285-5935.

Rochester Institute of Technology, Rochester, NY 14623-5604.

Center for Materials Science and Engineering Founded in 1981. *Academic areas of research and training:* Electronic materials, polymer science, complex fluids, thin films, magnetic materials, inorganic chemistry, microelectronic materials and processes. *Degrees offered:* MS. *Graduate students served last academic year:* 26: 26 at the master's level. *Faculty:* 28. *Faculty affiliations:* Departments of Chemistry, Mechanical Engineering, Microelectronic Engineering, Physics. *Director:* Robert Clark, Director, 716-475-2483, E-mail: racsse@rit.edu. *Application contact:* Dr. Peter Cardegna, Program Director, 716-475-2944, E-mail: pacsps@rit.edu.

Chester F. Carlson Center for Imaging Science Founded in 1985. *Academic areas of research and training:* Preparation for research or application of imaging modalities to engineering and science; research in astronomy, color, chemical (hard copy materials/processes) digital, electro-optical, medical, silver halide imaging; remote sensing, physics and chemistry of radiation sensitive materials/processes. *Degrees offered:* MS; PhD. *Graduate students served last academic year:* 87: 42 at the master's level; 41 at the doctoral level; 4 at the postgraduate level. *Faculty:* 21: 17 affiliated solely with the center. *Faculty affiliations:* Departments of Computer Science, Electrical Engineering, Microelectronic Engineering. *Annual research budget:* $1.7 million. *Director:* Dr. Ian Gatley, Director, 716-475-6220, Fax: 716-475-5988, E-mail: gatley@cis.rit.edu. *Application contact:* Marilyn Lockwood, Administrative Assistant, 716-475-5944, Fax: 716-475-5988, E-mail: lockwood@cis.rit.edu.

Rutgers, The State University of New Jersey, New Brunswick, New Brunswick, NJ 08903.

Center for Computer Aids for Industrial Productivity Founded in 1985. *Academic areas of research and training:* Application of high-speed scientific computing to the solution of industrial problems in areas such as machine vision, speech processing, computer-aided design, scientific visualization and quantification, multimedia information systems. *Degrees offered:* None. *Graduate students served last academic year:* 41: 20 at the master's level; 18 at the doctoral level; 3 at the postgraduate level. *Faculty:* 25: 5 affiliated solely with the center. *Faculty affiliations:* Departments of Biomedical Engineering, Computer Science, Electrical and Computer Engineering; Programs in Chemical and Biochemical Engineering, Industrial and Systems Engineering, Mathematics, Mechanical and Aerospace Engineering, Psychology. *Annual research budget:* $5.5 million. *Director:* Dr. James L. Flanagan, Director, 732-445-3443, Fax: 732-445-0547, E-mail: jlf@caip.rutgers.edu. *Application contact:* Dr. Edward J. Devinney, Senior Associate Director, 732-445-3443, Fax: 732-445-0547, E-mail: dasaro@caip.rutgers.edu.

Wireless Information Network Laboratory (WINLAB) Founded in 1989. *Academic areas of research and training:* Wireless and mobile communications. *Degrees offered:* None. *Graduate students served last academic year:* 11: 4 at the master's level; 5 at the doctoral level; 2 at the postgraduate level. *Faculty:* 10: 10 affiliated solely with the laboratory. *Faculty affiliations:* Department of Computer Science, Graduate School of Applied and Professional Psychology, RUTCOR–Rutgers University Center for Operations Research. *Annual research budget:* $1.5 million. *Director:* Dr. David Goodman, Director, 732-445-5261, Fax: 732-445-3693, E-mail: dgoodman@winlab.rutgers.edu. *Application contact:* Melissa Gelfman, Business Manager, 732-445-0283, Fax: 732-445-3693, E-mail: gelfman@winlab.rutgers.edu.

Stanford University, Stanford, CA 94305-9991.

Alliance for Innovative Manufacturing at Stanford (AIM@Stanford) Founded in 1984. *Academic areas of research and training:* Research and teaching in innovative manufacturing technologies. *Degrees offered:* None. *Graduate students served last academic year:* 80: 60 at the master's level; 20 at the doctoral level. *Faculty:* 40: 20 affiliated solely with Alliance for Innovative Manufacturing at Stanford (AIM@Stanford). *Faculty affiliations:* Graduate School of Business, School of Engineering. *Annual research budget:* $1.2 million. *Director:* Richard Reis, Director, 650-725-0919, Fax: 650-723-5034, E-mail: reis@stanford.edu. *Application contact:* Susan C. Hansen, Assistant Director, 650-723-9038, Fax: 650-723-5034, E-mail: susan.hansen@stanford.edu.

Center for Research on Information Storage Materials (CRISM) Founded in 1991. *Academic areas of research and training:* Materials or phenomena relevant to data storage, typically thin film magnetic materials; the science underpinning present technology. *Degrees offered:* None. *Graduate students served last academic year:* 25: 2 at the master's level; 23 at the doctoral level. *Faculty:* 8. *Faculty affiliations:* Departments of Applied Physics, Electrical Engineering, Materials Science and Engineering, Mechanical Engineering, Physics. *Annual research budget:* $2 million. *Director:* Dr. Robert L. White, Director, 650-723-4431, Fax: 650-723-4034, E-mail: white@ee.stanford.edu.

Center for the Study of Language and Information (CSLI) Founded in 1983. *Academic areas of research and training:* Communication, computer science, linguistics, mathematics, philosophy, psychology. *Degrees offered:* None. *Graduate students served last academic year:* 50: 50 at the doctoral level. *Faculty:* 50. *Faculty affiliations:* Departments of Communication, Computer Science, Linguistics, Mathematics, Philosophy, and Psychology. *Annual research budget:* $700,000. *Director:* John Perry, Director, 415-723-1224, Fax: 415-723-0758. *Application contact:* Michele King, Administrative Associate, 415-723-3084, E-mail: mking@csli.stanford.edu.

John A. Blume Earthquake Engineering Center Founded in 1975. *Academic areas of research and training:* Earthquake engineering, structural analysis and design, computational methods, structural control, structural health monitoring, seismic risk analysis, earthquake damage and loss estimation. *Degrees offered:* None. *Graduate students served last academic year:* 65: 30 at the master's level; 30 at the doctoral level; 5 at the postgraduate level. *Faculty:* 22: 7 affiliated solely with the center. *Faculty affiliations:* Departments of Computer Science, Electrical Engineering, Operations Research; School of Earth Sciences. *Director:* Dr. Anne S. Kiremidjian, Director, 650-723-4164, Fax: 650-725-9755, E-mail: ask@ce.stanford.edu. *Application contact:* Racquel Hagen, Administrative Assistant, 415-723-4150, Fax: 415-725-9755, E-mail: hagen@ce.stanford.edu.

Stanford Medical Informatics Founded in 1980. *Academic areas of research and training:* Use of information technology to solve problems in clinical medicine and biomedical sciences. *Degrees offered:* MS; PhD. *Graduate students served last academic year:* 26: 2 at the master's level; 24 at the doctoral level. *Faculty:* 47: 5 affiliated solely with Stanford Medical Informatics. *Faculty affiliations:* Departments of Computer Science, Economics, Electrical Engineering, Engineering-Economic Systems, Operations Research; Graduate School of Business; Schools of Education, Medicine. *Annual research budget:* $3.5 million. *Director:* Dr. Mark Musen, Head, 650-723-6979, Fax: 650-725-7944, E-mail: musen@stanford.edu. *Application contact:* Darlene Vian, Program Administrator, 650-725-3388, Fax: 650-498-4162, E-mail: vian@smi.stanford.edu.

Western Region Hazardous Substance Research Center Founded in 1989. *Academic areas of research and training:* Chemical engineering, chemistry, environmental engineering, environmental microbiology, hydrology, geology. *Degrees offered:* None. *Graduate students served last academic year:* 24: 3 at the master's level; 16 at the doctoral level; 5 at the postgraduate level. *Faculty:* 42: 21 affiliated solely with the center. *Faculty affiliations:* College of Veterinary Medicine; Departments of Civil and Environmental Engineering, Microbiology and Immunology, Petroleum Engineering; School of Earth Science; Oregon State University Departments of Botany and Plant Pathology, Chemistry, Civil Engineering. *Annual research budget:* $2 million. *Director:* Dr. Perry L. McCarty, Professor, 650-723-4131, Fax: 650-725-9474, E-mail: mccarty@ce.stanford.edu. *Application contact:* Sharon Parkinson, Administrative Assistant, 650-723-4123, Fax: 650-725-9474, E-mail: sharon@ce.stanford.edu.

State University of New York at Albany, Albany, NY 12222-0001.

Center for Technology in Government Founded in 1993. *Academic areas of research and training:* Information technology, government technology, public policy, computer science, technology applications, information science, business administration. *Degrees offered:* None. *Graduate students served last academic year:* 8: 8 at the doctoral level. *Faculty:* 6: 3 affiliated solely with the center. *Faculty affiliations:* Departments of Computer Science, Educational Administration, Geography and Planning, Management Science and Information Systems, Public Administration and Policy; School of Business, Information Science. *Annual research budget:* $1 million. *Director:* Dr. Sharon S. Dawes, Director, 518-442-3892, Fax: 518-442-3886, E-mail: info@ctg.albany.edu. *Application contact:* Stephanie Simon, Information Coordinator, 518-442-3895, Fax: 518-442-3886, E-mail: ssimon@ctg.albany.edu.

State University of New York at Buffalo, Buffalo, NY 14260.

Center for Biomedical Engineering Founded in 1988. *Academic areas of research and training:* Biomaterials, biomedical imaging, biomechanics, cellular and molecular biotechnology, rehabilitation. *Degrees offered:* None. *Graduate students served last academic year:* 100: 60 at the master's level; 40 at the doctoral level. *Faculty:* 24. *Faculty affiliations:* Departments of Chemical, Civil, Electrical and Computer, Industrial, Mechanical and Aerospace Engineering. *Annual research budget:* $2.4 million. *Director:* Dr. B. Barry Lieber, Director, 716-645-2593 Ext. 2313, Fax: 716-645-3875, E-mail: lieber@eng.buffalo.edu.

Industry/University Cooperative Research Center for Biosurfaces Founded in 1988. *Academic areas of research and training:* Biomaterials, biomedical engineering, biophysics, dental engineering, environmental quality control, medical devices, surface chemistry, cardiopulmonary research, bioaerosols. *Degrees offered:* None. *Graduate students served last academic year:* 21: 12 at the master's level; 5 at the doctoral level; 4 at the postgraduate level. *Faculty:* 56: 8 affiliated solely with the center. *Faculty affiliations:* Departments of Biomaterials, Biophysics, Chemistry, Mechanical and Chemical Engineering, Microbiology, Oral Diagnostic Sciences, Oral Surgery; Schools of Dental Medicine, Engineering and Applied Sciences; Alfred University New York State College of Ceramics; University of Memphis; University of Miami. *Annual research budget:* $1.2 million. *Director:* Dr. Robert E. Baier, Executive Director, 716-829-3560, Fax: 716-835-4872, E-mail: baier@acsu.buffalo.edu. *Application contact:* Dr. Anne E. Meyer, Buffalo Site Director, 716-829-3560, Fax: 716-835-4872, E-mail: aemeyer@acsu.buffalo.edu.

Stevens Institute of Technology, Hoboken, NJ 07030.

Highly Filled Materials Institute Founded in 1989. *Academic areas of research and training:* Filled materials processing, high degree of fill, processing, simulation, microstructural analysis. *Degrees offered:* None. *Graduate students served last academic year:* 11: 5 at the master's level; 6 at the doctoral level. *Faculty:* 5: 3 affiliated solely with the institute. *Faculty affiliations:* Program in Chemical Engineering. *Director:* Dr. Dilhan Kalyon, Director, 201-216-8225, Fax: 201-216-5601, E-mail: dkalyon@stevens-tech.edu. *Application contact:* Dr. Rahmi Yazici, Research Professor, 201-216-5030, Fax: 201-216-5601, E-mail: ryazici@stevens-tech.edu.

Syracuse University, Syracuse, NY 13244-0003.

Robert H. Brethen Operations Management Institute Founded in 1988. *Academic areas of research and training:* Manufacturing and service operations. *Degrees offered:* None. *Graduate students served last academic year:* 12: 10 at the master's level; 2 at the doctoral level. *Faculty:* 12. *Faculty affiliations:* Program in Quantitative Methods. *Director:* Prof. Fred F. Easton, Director, 315-443-3549, Fax: 315-443-5389, E-mail: ffeaston@mailbox.syr.edu.

Tennessee Technological University, Cookeville, TN 38505.

Center for Manufacturing Research and Technology Utilization Founded in 1984. *Academic areas of research and training:* Business, engineering, sciences. *Degrees offered:* None. *Graduate students served last academic year:* 70: 51 at the master's level; 14 at the doctoral level. *Faculty:* 35: 5 affiliated solely with the center. *Faculty affiliations:* Colleges of Business Administration, Education, Engineering. *Annual research budget:* $1.5 million. *Director:* Dr. Ted S. Lundy, Director, 931-372-3362, Fax: 931-372-6345, E-mail: tlundy@tntech.edu. *Application contact:* Dr. Ken R. Currie, Professor, 931-372-3847, Fax: 931-372-6345, E-mail: kcurrie@tntech.edu.

Texas A&M University, College Station, TX 77843.

Texas Engineering Experiment Station (TEES) Founded in 1914. *Academic areas of research and training:* Engineering and technology-oriented research for the enhancement of the educational system as well as the economic development of the state and the nation. *Degrees offered:* None. *Graduate students served last academic year:* 1,693: 980 at the master's level; 713 at the doctoral level. *Faculty:* 720: 360 affiliated solely with the station. *Faculty affiliations:* College of Engineering. *Annual research budget:* $66.3 million. *Director:* C. Roland Haden, Dean and Vice Chancellor for Engineering, 409-845-7203, Fax: 409-845-4925, E-mail: r-haden@teesmail.tamu.edu.

Texas Transportation Institute (TTI) Founded in 1950. *Academic areas of research and training:* Transportation materials, structures, and policy; transportation systems, economics, and planning. *Degrees offered:* None. *Graduate students served last academic year:* 168. *Faculty:* 40. *Faculty affiliations:* Departments of Architecture, Civil Engineering, Urban Planning. *Annual research budget:* $30 million. *Director:* Dr. Herbert H. Richardson, Director, 409-845-1713, Fax: 409-845-9356, E-mail: h-richardson@tamu.edu.

Texas Tech University, Lubbock, TX 79409.

Center for Agricultural Technology Transfer Founded in 1991. *Academic areas of research and training:* Computer and related equipment use. *Degrees offered:* None. *Graduate students served last academic year:* 10: 8 at the master's level; 2 at the doctoral level. *Faculty:* 6: 3 affiliated solely with the center. *Faculty affiliations:* Department of Agricultural Education and

Communications. *Director:* Dr. Paul Vaughn, Chairman, 806-742-2816, Fax: 806-742-2880, E-mail: zoprv@ttacs.ttu.edu.

Institute for Disaster Research (IDR) Founded in 1972. *Academic areas of research and training:* Civil engineering, fluid dynamics, mechanical engineering, structural engineering, wind damage mitigation, occupant safety. *Degrees offered:* None. *Graduate students served last academic year:* 8: 6 at the master's level; 2 at the doctoral level. *Faculty:* 6: 2 affiliated solely with the institute. *Faculty affiliations:* Wind Engineering Research Center. *Annual research budget:* $200,000. *Director:* Dr. James R. McDonald, Director, 806-742-3476, Fax: 806-742-3446, E-mail: jmcdonald@coe.ttu.edu.

Water Resources Center (WRC) Founded in 1965. *Academic areas of research and training:* Water resources engineering, groundwater, water resources management, ground water remediation and restoration, artificial recharge. *Degrees offered:* None. *Graduate students served last academic year:* 20: 16 at the master's level; 4 at the doctoral level. *Faculty:* 9: 1 affiliated solely with the center. *Faculty affiliations:* Departments of Biological Sciences, Chemical Engineering, Civil Engineering, Economics and Geography, Environmental Engineering, Plant and Soil Science. *Annual research budget:* $750,000. *Director:* Dr. Lloyd V. Urban, Director, 806-742-3597, Fax: 806-742-3449, E-mail: lurban@coettu.edu. *Application contact:* Jan E. Hudson, Assistant to the Director, 806-742-3597, Fax: 806-742-3449, E-mail: jhudson@coe.ttu.edu.

Université du Québec à Montréal, Montréal, PQ H3C 3P8, Canada.

Groupe d'Études et de Recherche en Analyse des Décisions (GERAD) Founded in 1988. *Academic areas of research and training:* Operations research. *Degrees offered:* None. *Graduate students served last academic year:* 113: 64 at the master's level; 42 at the doctoral level; 7 at the postgraduate level. *Faculty:* 34. *Faculty affiliations:* Mathematics Program; École des Hautes Études Commerciales Program in Quantitative Methods; École Polytechnique de Montréal Departments of Industrial Engineering, Mathematics; McGill University Faculty of Management. *Annual research budget:* $3.6 million. *Director:* Dr. Pierre Hansen, Director, 514-340-5675, Fax: 514-340-5665, E-mail: gerad@crt.umontreal.ca.

The University of Alabama at Birmingham, Birmingham, AL 35294.

Center for Telecommunications Education and Research Founded in 1984. *Academic areas of research and training:* Civil and environmental engineering, computer science, electrical and computer engineering, management, marketing, materials science and engineering, medicine, physics, public health. *Degrees offered:* None. *Graduate students served last academic year:* 11: 7 at the master's level; 3 at the doctoral level; 1 at the postgraduate level. *Faculty:* 61: 1 affiliated solely with the center. *Faculty affiliations:* Departments of Civil Engineering, Computer Science, Electrical Engineering, Materials Science and Engineering, Marketing, and Physics; Schools of Management, Medicine, and Public Health. *Director:* Dr. Gary J. Grimes, Executive Director, 205-934-3147, Fax: 205-934-8480, E-mail: ggrimes@eng.uab.edu.

Deep South Occupational Health and Safety Educational Resource Center Founded in 1979. *Academic areas of research and training:* Industrial hygiene, occupational and environmental medicine, occupational health nursing, occupational safety and ergonomics. *Degrees offered:* MS; PhD; MIE; MPH; MSN; MSPH; Dr PH; DSN. *Graduate students served last academic year:* 55: 39 at the master's level; 16 at the doctoral level. *Faculty:* 15. *Faculty affiliations:* Schools of Medicine, Nursing, Public Health; Auburn University College of Engineering. *Director:* Dr. R. Kent Destenstad, Associate Professor, 205-934-8488, Fax: 205-975-6341, E-mail: ehs@crl.soph.uab.edu. *Application contact:* Cherie Hunt, Program Coordinator, 205-934-8488, Fax: 205-975-6341, E-mail: ehs@crl.soph.uab.edu.

The Earth Center *Academic areas of research and training:* Environmental issues related to water and air pollution, utilization of by-products, environmental health programs. *Degrees offered:* None. *Graduate students served last academic year:* 50: 30 at the master's level; 15 at the doctoral level; 5 at the postgraduate level. *Faculty:* 100: 100 affiliated solely with the center. *Annual research budget:* $14 million. *Director:* Dr. Fouad H. Fouad, Executive Director, 205-934-8430, Fax: 205-934-9855, E-mail: ffouad@uab.edu. *Application contact:* Dr. Fouad H. Fouad, Executive Director, 205-934-8430, Fax: 205-934-9855, E-mail: ffouad@uab.edu.

Injury Control Research Center (ICRC) Founded in 1987. *Academic areas of research and training:* Biomechanics, biostatistics, epidemiology, injury control. *Degrees offered:* None. *Graduate students served last academic year:* 115: 70 at the master's level; 45 at the doctoral level. *Faculty:* 95: 5 affiliated solely with the center. *Faculty affiliations:* Schools of Engineering, Medicine, and Public Health. *Annual research budget:* $750,000. *Director:* Dr. Philip R. Fine, Director, 205-934-7845, Fax: 205-975-8143, E-mail: rfine@uab.edu. *Application contact:* Matthew Gunter, Assistant to the Director, 205-934-7845, Fax: 205-975-8143, E-mail: matt.gunter@ccc.uab.edu.

The University of Alabama in Huntsville, Huntsville, AL 35899.

Center for Automation and Robotics (CAR) *Degrees offered:* None. *Graduate students served last academic year:* 1: 1 at the master's level. *Director:* Dr. Bernard Schroer, Director, 256-890-6100, Fax: 256-890-6783, E-mail: schroerb@email.uah.edu.

Center for Space Plasma and Aeronomic Research (CSPAR) Founded in 1987. *Academic areas of research and training:* Space physics, astrophysics and space weather research. *Degrees offered:* None. *Graduate students served last academic year:* 22: 9 at the master's level; 12 at the doctoral level; 1 at the postgraduate level. *Faculty:* 21: 10 affiliated solely with the center. *Faculty affiliations:* Departments of Computer Science, Electrical and Computer Engineering, Mechanical and Aerospace Engineering, Physics. *Director:* Dr. S. T. Wu, Director.

Consortium for Materials Development in Space (CMDS) Founded in 1985. *Academic areas of research and training:* Materials and manufacturing processes. *Degrees offered:* None. *Graduate students served last academic year:* 10: 5 at the master's level; 5 at the doctoral level. *Faculty:* 8. *Faculty affiliations:* Departments of Biological Sciences, Chemical Engineering, Chemistry, Mechanical and Aerospace Engineering, Physics. *Annual research budget:* $3 million. *Director:* Dr. Bernard Schroer, Director, 205-890-6100, Fax: 205-890-6783.

Propulsion Research Center (PRC) Founded in 1991. *Academic areas of research and training:* Propulsion (rocket and air-breathing), turbomachinery bearings and seals, experimental uncertainty, combustion processes, aerothermochemistry. *Degrees offered:* None. *Graduate students served last academic year:* 20: 10 at the master's level; 10 at the doctoral level. *Faculty:* 8: 5 affiliated solely with the center. *Faculty affiliations:* Department of Mechanical and Aerospace Engineering. *Annual research budget:* $1.1 million. *Director:* Dr. Clark W. Hawk, Director, 205-890-7200, Fax: 205-890-7205, E-mail: hawkc@email.uah.edu.

The University of Arizona, Tucson, AZ 85721.

Center for Electronic Packaging Research (CEPR) Founded in 1991. *Academic areas of research and training:* Electrical and thermal performance of electronic packaging and interconnects, microelectronics and electromagnetics. *Degrees offered:* None. *Graduate students served last academic year:* 12: 8 at the master's level; 4 at the doctoral level. *Faculty:* 6. *Faculty affiliations:* Departments of Aerospace and Mechanical Engineering, Civil Engineering and Engineering Mechanics, Electrical and Computer Engineering. *Annual research budget:* $500,000. *Director:* Dr. John L. Prince, Director, 520-621-6189, Fax: 520-621-2999, E-mail: prince@ece.arizona.edu.

University of Arkansas, Fayetteville, AR 72701-1201.

Arkansas Biotechnology Center *Academic areas of research and training:* Biological science, food science, poultry science, water quality. *Degrees offered:* None. *Graduate students served last academic year:* 7: 3 at the master's level; 3 at the doctoral level; 1 at the

Research and Training Opportunities

Arkansas Biotechnology Center (continued)

postgraduate level. *Faculty:* 5. *Faculty affiliations:* Departments of Biological Science, Food Science, Geology, Poultry Science. *Annual research budget:* $1.5 million. *Director:* Collis Geren, Coordinator, 501-575-5900, Fax: 501-575-5908, E-mail: cgeren@comp.uark.edu.

Mack-Blackwell National Rural Transportation Study Center Founded in 1992. *Academic areas of research and training:* Highway engineering, modal and intermodal transportation management and engineering (waterways, trucking, railroads). *Degrees offered:* MS. *Graduate students served last academic year:* 25 at the master's level; 5 at the doctoral level. *Faculty:* 31: 1 affiliated solely with the center. *Faculty affiliations:* Departments of Civil Engineering, Computer Systems Engineering, Economics, Industrial Engineering, Marketing and Transportation, Political Science, Sociology. *Annual research budget:* $2 million. *Director:* Dr. Melissa S. Tooley, Director, 501-575-7957, Fax: 501-575-7168, E-mail: mst1@engr.uark.edu. *Application contact:* Dr. Robert P. Elliott, Chairman, 501-575-6028, Fax: 501-575-7168.

University of California, Berkeley, Berkeley, CA 94720-1500.

Electronics Research Laboratory Founded in 1951. *Academic areas of research and training:* Electrical engineering, computer sciences, bioelectronics, computer-aided design, electromagnetic radiation and microwave, integrated circuits, plasmas. *Degrees offered:* None. *Graduate students served last academic year:* 524: 74 at the master's level; 420 at the doctoral level; 30 at the postgraduate level. *Faculty:* 143: 116 affiliated solely with the laboratory. *Faculty affiliations:* Colleges of Chemistry, Environmental Design; Departments of Astronomy, Geology and Geophysics, Integrative Biology, Materials Science and Mineral Engineering, Mathematics, Mechanical Engineering, Physics. *Annual research budget:* $28.7 million. *Director:* Dr. Shankar Sastry, Director, 510-642-7200, Fax: 510-643-8426, E-mail: sastry@eecs.berkeley.edu. *Application contact:* Renate Valencia, Manager, 510-643-8424, Fax: 510-643-8426, E-mail: valencia@eecs.berkeley.edu.

Engineering Systems Research Center (ESRC) Founded in 1961. *Academic areas of research and training:* Manufacturing processes, management of technology, green manufacturing, mechatronics, robotics, bioengineering. *Degrees offered:* None. *Graduate students served last academic year:* 120: 18 at the master's level; 82 at the doctoral level; 20 at the postgraduate level. *Faculty:* 93: 75 affiliated solely with the center. *Faculty affiliations:* Departments of Electrical Engineering and ComputerSciences, Industrial Engineering and Operations Research, Nuclear Engineering; Haas School of Business; Schools of Optometry, Education, Public Health. *Annual research budget:* $1.5 million. *Director:* Dr. Masayoshi Tomizuka, Director, 510-642-4994, Fax: 510-643-0966, E-mail: esrc@euler.berkeley.edu. *Application contact:* Debra Richerson, Management Services Officer, 510-642-4994, Fax: 510-643-0966, E-mail: esrc@euler.berkeley.edu.

Institute of Transportation Studies (ITS) Founded in 1948. *Academic areas of research and training:* Transportation engineering, planning, policy, and infrastructure; intelligent transportation systems. *Degrees offered:* None. *Graduate students served last academic year:* 135: 45 at the master's level; 80 at the doctoral level; 10 at the postgraduate level. *Faculty:* 45. *Faculty affiliations:* Departments of City and Regional Planning, Civil and Environmental Engineering, Economics, Electrical Engineering and Computer Sciences, Industrial Engineering and Operations Research, Mechanical Engineering, Political Science; Haas School of Business; University of California, Davis; University of California, Irvine. *Annual research budget:* $18 million. *Director:* Dr. Martin Wachs, Director, 510-642-3585, Fax: 510-643-3955, E-mail: its@its.berkeley.edu.

Pacific Earthquake Engineering Research Center Founded in 1997. *Academic areas of research and training:* Earthquake engineering, engineering seismology, structural dynamics, geotechnical engineering, structural engineering. *Degrees offered:* None. *Graduate students served last academic year:* 60: 30 at the master's level; 30 at the doctoral level. *Faculty:* 40. *Faculty affiliations:* Departments of Architecture, Civil Engineering, Economics, Electrical Engineering, Public Policy, Seismology. *Annual research budget:* $7 million. *Director:* Jack P. Moehle, 510-231-9554, Fax: 510-231-9471, E-mail: moehle@euler.berkeley.edu.

Space Sciences Laboratory (SSL) Founded in 1959. *Academic areas of research and training:* Space research. *Degrees offered:* None. *Graduate students served last academic year:* 38. *Faculty:* 13. *Faculty affiliations:* Departments of Astronomy, Geology and Geophysics, Physics. *Annual research budget:* $21 million. *Director:* Robert P. Lin, Director, 510-642-1361, Fax: 510-643-7629, E-mail: mckee@ssl.berkeley.edu. *Application contact:* Judith Jones, Personnel Manager, 510-642-1528, E-mail: judy@ss1.berkeley.edu.

University of California Transportation Center (UCTC) Founded in 1989. *Academic areas of research and training:* Transportation policy, access to transportation. *Degrees offered:* None. *Graduate students served last academic year:* 208: 122 at the master's level; 86 at the doctoral level. *Faculty:* 200: 100 affiliated solely with the center. *Faculty affiliations:* Departments of City and Regional Planning, Civil and Environmental Engineering, Economics, Geography, Mechanical Engineering; Program in Ecology. *Annual research budget:* $800,000. *Director:* Prof. Elizabeth Deakin, Director, 510-642-4749, Fax: 510-643-5456, E-mail: edeakin@ix.netcom.com. *Application contact:* Briggs Nisbet, Assistant to the Director, 510-643-7378, Fax: 510-643-5456, E-mail: bnisbet@uclink.berkeley.edu.

University of California, Davis, Davis, CA 95616.

Center for Ecological Health Research Founded in 1992. *Academic areas of research and training:* Effects of multiple stresses on aquatic and terrestrial ecosystems. *Degrees offered:* None. *Graduate students served last academic year:* 19: 2 at the master's level; 16 at the doctoral level; 1 at the postgraduate level. *Faculty:* 36. *Faculty affiliations:* Colleges of Agricultural and Environmental Sciences, Engineering, Letters and Sciences; School of Veterinary Medicine. *Annual research budget:* $1 million. *Director:* Dr. Dennis E. Rolston, Director, 530-752-5028, Fax: 530-752-1552, E-mail: derolston@ucdavis.edu. *Application contact:* Cheryl Smith, Center Manager, 530-752-5028, Fax: 530-752-1552, E-mail: csmith@ucdavis.edu.

Institute of Transportation Studies Founded in 1991. *Academic areas of research and training:* Design and analysis of electric-drive vehicles and intelligent transportation system technology, transportation energy, environmental policy, travel behavior modeling and analysis. *Degrees offered:* MS; PhD. *Graduate students served last academic year:* 60: 40 at the master's level; 20 at the doctoral level. *Faculty:* 34: 4 affiliated solely with the institute. *Faculty affiliations:* Graduate School of Management; Programs in Chemical Engineering and Materials Science, Civil and Environmental Engineering, Economics, Environmental Sciences and Policy, Mechanical and Aeronautical Engineering. *Annual research budget:* $3.8 million. *Director:* Dr. Daniel Sperling, Director, 530-752-7434, Fax: 530-752-6572, E-mail: dsperling@ucdavis.edu. *Application contact:* Joan R. Byrne, Graduate Assistant, 530-752-0247, Fax: 530-752-6572, E-mail: itsgraduate@ucdavis.edu.

University of California, Irvine, Irvine, CA 92697.

Center for Research on Information Technology and Organizations (CRITO) Founded in 1991. *Academic areas of research and training:* Social and economic impact of information technology (IT), management of IT, globalization of IT production and use. *Degrees offered:* None. *Graduate students served last academic year:* 25: 15 at the master's level; 10 at the doctoral level. *Faculty:* 30: 15 affiliated solely with the center. *Faculty affiliations:* Departments of Education, Information and Computer Science; Graduate School of Management; School of Social Sciences. *Annual research budget:* $500,000. *Director:* Dr. Kenneth L. Kraemer, Director, 949-824-5246, Fax: 949-824-8091, E-mail: kkraemer@uci.edu. *Application contact:* Terri Pouliot, Assistant Director, 949-824-4507, Fax: 949-824-8091, E-mail: tpouliot@uci.edu.

Institute of Transportation Studies (ITS) Founded in 1973. *Academic areas of research and training:* Transportation research. *Degrees offered:* None. *Graduate students served last academic year:* 55: 17 at the master's level; 38 at the doctoral level. *Faculty:* 25. *Faculty*

affiliations: Departments of Civil and Environmental Engineering, Economics, Electrical and Computer Engineering, Information and Computer Sciences, Urban Planning; Graduate School of Management. *Annual research budget:* $3.5 million. *Director:* Dr. Wilfred W. Recker, Director, 949-824-5989, Fax: 949-824-8385, E-mail: wwrecker@uci.edu.

University of California, San Diego, La Jolla, CA 92093-5003.

Center for Magnetic Recording Research (CMRR) Founded in 1983. *Academic areas of research and training:* Magnetic recording, recording physics and micromagnetics, signal processing, materials research heads and media, micromagnetics and instrumentation, mechanics and tribology. *Degrees offered:* None. *Graduate students served last academic year:* 40: 31 at the doctoral level; 9 at the postgraduate level. *Faculty:* 17: 4 affiliated solely with the center. *Faculty affiliations:* Departments of Applied Mechanics and Engineering Sciences, Chemistry, Computer Science and Engineering, Electrical and Computer Engineering, Physics. *Annual research budget:* $2 million. *Director:* Dr. Sheldon Schultz, Director, 858-534-6210, Fax: 858-534-8059, E-mail: sschultz@ucsd.edu. *Application contact:* Cheryl Hacker, Management Services Officer, 858-534-6563, Fax: 858-534-8059, E-mail: chacker@ucsd.edu.

Institute for Pure and Applied Physical Science (IPAPS) *Academic areas of research and training:* Ferromagnetism, fluid mechanics, hydromagnetics, laser physics, numerical analysis, plasma physics, semiconductor heterostructures, solid surfaces, superconductivity, turbulence. *Degrees offered:* None. *Graduate students served last academic year:* 20: 10 at the master's level; 5 at the doctoral level; 5 at the postgraduate level. *Faculty:* 10: 2 affiliated solely with the institute. *Faculty affiliations:* Departments of Applied Mechanics and Engineering Sciences, Chemistry and Biochemistry, Electrical and Computer Engineering, Physics; Center for Magnetic Recording Research. *Annual research budget:* $2.5 million. *Director:* Dr. Brian Maple, Director, 858-534-3269, Fax: 858-534-1241, E-mail: mbmaple@ucsd.edu. *Application contact:* Donna Bott, Contact Person, 858-534-3560, Fax: 858-534-7649, E-mail: dbott@ucsd.edu.

Marine Physical Laboratory (MPL) Founded in 1946. *Academic areas of research and training:* Ocean science and technology, marine geophysics, ocean acoustics and optics, ocean chemistry. *Degrees offered:* None. *Graduate students served last academic year:* 25: 20 at the doctoral level; 5 at the postgraduate level. *Faculty:* 33: 30 affiliated solely with the laboratory. *Faculty affiliations:* Department of Physical Oceanography. *Annual research budget:* $13 million. *Director:* Dr. William Kuperman, Director, 858-534-1803, Fax: 858-822-0665, E-mail: wak@mpl.ucsd.edu. *Application contact:* Pat Jordan, Management Services Officer, 858-534-1802, Fax: 858-822-0665, E-mail: pjordan@vcsd.edu.

University of California, Santa Barbara, Santa Barbara, CA 93106.

Materials Research Laboratory (MRL) Founded in 1992. *Academic areas of research and training:* Biomineralization, biopolymers, catalysis, complex fluids, low-temperature synthesis, molecular sieves, polymers. *Degrees offered:* None. *Graduate students served last academic year:* 13 at the master's level; 49 at the doctoral level; 26 at the postgraduate level. *Faculty:* 34. *Faculty affiliations:* Departments of Biological Sciences, Chemical Engineering, Chemistry, Computer Science, Electrical and Computer Engineering, Geological Sciences, Materials, Mathematics, Mechanical and Environmental Engineering, Physics. *Annual research budget:* $3.6 million. *Director:* Anthony K. Cheetham, Director, 805-893-8767, Fax: 805-893-8797, E-mail: cheetham@iristew.ucsb.edu. *Application contact:* Maureen Evans, Management Services Officer, 805-893-8519, Fax: 805-893-8797, E-mail: maureen@mrl.ucsb.edu.

Optoelectronics Technology Center (OTC) Founded in 1990. *Academic areas of research and training:* To provide U.S. industry with the enabling material, device and circuit technologies for the next generation of advanced optoelectronic systems. *Degrees offered:* None. *Graduate students served last academic year:* 36: 30 at the doctoral level; 6 at the postgraduate level. *Faculty:* 16. *Faculty affiliations:* UCSB Department of Electrical and Computer Engineering; Cornell University Field of Electrical Engineering; University of California Los Angeles Departments of Electrical Engineering, Materials Science and Engineering; University of California San Diego Department of Electrical and Computer Engineering; University of Texas at Austin Department of Electrical and Computer Engineering; University of Southern California School of Engineering. *Annual research budget:* $3 million. *Director:* Dr. Larry A. Coldren, Director, 805-893-4486, Fax: 805-893-4500, E-mail: coldren@ece.ucsb.edu. *Application contact:* Giulia Brofferio, Program Manager, 805-893-7105, Fax: 805-893-4500, E-mail: giulia@ece.ucsb.edu.

University of Chicago, Chicago, IL 60637-1513.

Center for Advanced Radiation Sources (CARS) Founded in 1989. *Academic areas of research and training:* Applications of synchrotron radiation in geophysics, structural biology, chemistry, materials science, environmental science, soil science. *Degrees offered:* None. *Graduate students served last academic year:* 30: 30 at the doctoral level. *Faculty:* 5. *Faculty affiliations:* Departments of Biochemistry and Molecular Biology, Chemistry, Geophysical Sciences; James Franck Institute. *Annual research budget:* $5.5 million. *Director:* Dr. Keith Moffat, Director, 773-702-9951, Fax: 773-702-5454, E-mail: moffat@cars.uchicago.edu. *Application contact:* Joy Talsma, Executive Administrator, 773-702-9506, Fax: 773-702-5454, E-mail: talsma@cars.uchicago.edu.

University of Cincinnati, Cincinnati, OH 45221-0091.

Aerosol and Air Quality Research Laboratory Founded in 1985. *Academic areas of research and training:* Aerosol science and engineering, air quality and pollution control, environmental engineering, materials processing, environmentally benign processing. *Degrees offered:* MS; PhD. *Graduate students served last academic year:* 30: 10 at the master's level; 15 at the doctoral level; 5 at the postgraduate level. *Faculty:* 3: 3 affiliated solely with the laboratory. *Annual research budget:* $500,000. *Director:* Dr. Pratim Biswas, Director, 513-556-3697, Fax: 513-556-2599, E-mail: pratim.biswas@uc.edu.

Center for Cardiovascular Biomaterials Founded in 1993. *Academic areas of research and training:* Biomaterials, biocompatibility, biomimicry; cardiovascular studies. *Degrees offered:* None. *Graduate students served last academic year:* 10: 6 at the master's level; 4 at the doctoral level. *Faculty:* 8. *Faculty affiliations:* Departments of Engineering Mechanics, Internal Medicine (Cardiology), Materials Science and Engineering, Mechanical Engineering, Pharmacology, Radiology, Surgery. *Director:* Dr. Ronald W. Millard, Director, 513-558-2336, Fax: 513-558-1169, E-mail: ron.millard@uc.edu.

Center for Nanoscience and Technology Founded in 1988. *Academic areas of research and training:* Nanoelectronics, focused ion beam technology, wide bangap semiconductors, gallium nitride, silicon carbide, molecular beam epitaxy, chemical vapor deposition, photonic devices. *Degrees offered:* None. *Graduate students served last academic year:* 12: 1 at the master's level; 8 at the doctoral level; 3 at the postgraduate level. *Faculty:* 7: 1 affiliated solely with the center. *Faculty affiliations:* Departments of Computer Engineering, Computer Science; Program in Photonics. *Annual research budget:* $750,000. *Director:* Dr. Andrew J. Steckl, Director, Fax: 513-556-7326, E-mail: a.steckl@uc.edu.

University of Colorado at Boulder, Boulder, CO 80309.

Center for Membrane Applied Science and Technology Founded in 1990. *Academic areas of research and training:* Semipermeable membrane manufacture and use, polymeric and inorganic membrane science and technology. *Degrees offered:* None. *Graduate students served last academic year:* 14: 2 at the master's level; 10 at the doctoral level; 2 at the postgraduate level. *Faculty:* 35. *Faculty affiliations:* Departments of Chemical Engineering; Chemistry and Biochemistry; Civil, Environmental and Architectural Engineering; Mechanical Engineering; Physics; Colorado School of Mines; Colorado State University, University of Denver. *Annual research budget:* $622,000. *Director:* Dr. William Krantz, Co-Director, 303-492-

7517, Fax: 303-492-4637, E-mail: krantz@spot.colorado.edu. *Application contact:* Sandy Spahn, Center Coordinator, 303-492-7517, Fax: 303-492-4637, E-mail: spahn@spot.colorado.edu.

Colorado Center for Astrodynamics Research (CCAR) Founded in 1985. *Academic areas of research and training:* Astrodynamics, remote sensing, satellite oceanography, global positioning system research. *Degrees offered:* None. *Graduate students served last academic year:* 48: 15 at the master's level; 25 at the doctoral level; 8 at the postgraduate level. *Faculty:* 7: 7 affiliated solely with the center. *Annual research budget:* $2.5 million. *Director:* Dr. George H. Born, Director, 303-492-6677, Fax: 303-492-2825, E-mail: georgeb@colorado.edu.

Joint Center for Energy Management (JCEM) Founded in 1987. *Academic areas of research and training:* Energy systems for buildings, AI for buildings, renewable energy research, distributed electrical generation. *Degrees offered:* None. *Graduate students served last academic year:* 50: 35 at the master's level; 14 at the doctoral level; 1 at the postgraduate level. *Faculty:* 5: 5 affiliated solely with the center. *Annual research budget:* $500,000. *Director:* Dr. Jan F. Kreider, Founding Director, 303-492-7603, Fax: 303-492-7317, E-mail: kreider@bechtel.colorado.edu.

University of Colorado at Denver, Denver, CO 80217-3364.

Transportation Research Center Founded in 1986. *Academic areas of research and training:* Traffic engineering, transportation planning, travel demand forecasting, traffic safety, impact analysis. *Degrees offered:* None. *Graduate students served last academic year:* 32: 25 at the master's level; 5 at the doctoral level; 2 at the postgraduate level. *Faculty:* 3. *Faculty affiliations:* Department of Civil Engineering. *Annual research budget:* $200,000. *Director:* Dr. Bruce Janson, Director, 303-556-2831, Fax: 303-556-2368, E-mail: bjanson@carbon.cudenver.edu.

University of Connecticut, Storrs, CT 06269.

Connecticut Transportation Institute Founded in 1974. *Academic areas of research and training:* Transportation safety, transportation planning, highway engineering, materials. *Degrees offered:* None. *Graduate students served last academic year:* 14: 10 at the master's level; 4 at the doctoral level. *Faculty:* 18. *Faculty affiliations:* Department of Political Science; Fields of Civil Engineering, Economics, Environmental Engineering, Geography, Mathematics, Psychology; School of Business Administration. *Annual research budget:* $1.1 million. *Director:* Dr. Norman W. Garrick, Director, 860-486-2990, Fax: 860-486-2298, E-mail: norman.garrick@uconn.edu.

Environmental Research Institute Founded in 1987. *Academic areas of research and training:* Environmental research. *Degrees offered:* None. *Graduate students served last academic year:* 80: 6 at the postgraduate level. *Faculty:* 28. *Faculty affiliations:* Departments of Civil and Environmental Engineering, Chemical Engineering, Chemistry, Geology and Geophysics, Natural Resources Management and Engineering, Pathobiology, Plant Science. *Annual research budget:* $2.9 million. *Director:* Dr. George Hoag, Director, 860-486-4015, Fax: 860-486-5488, E-mail: www.eng2.uconn.edu/eri/.

Photonics Research Center Founded in 1992. *Academic areas of research and training:* Information gathering, information transport, energy conveyance, specifically photonic device (Fabry Perot and grating surface emitters), integrated optoelectronics fiber lasers, Bragg grating technology, embedded fiber optic sensor for Smart Structures Medical Imaging. *Degrees offered:* None. *Graduate students served last academic year:* 18: 9 at the master's level; 9 at the doctoral level. *Faculty:* 17: 2 affiliated solely with the center. *Faculty affiliations:* Department of Mechanical Engineering; Fields of Biomedical Science, Chemistry, Civil Engineering, Electrical and Systems Engineering, Environmental Engineering, Physics; Programs in Agriculture, MarineScience. *Annual research budget:* $2.5 million. *Director:* Dr. Chandra Roychoudhuri, Director, 860-486-2587, Fax: 860-486-1033, E-mail: chandra@engr.uconn.edu. *Application contact:* Lisa M. Mazzola, Technical Records Coordinator, 860-486-2886, Fax: 860-486-1033, E-mail: mazzola@engr.uconn.edu.

Taylor L. Booth Center for Computer Applications and Research Founded in 1981. *Academic areas of research and training:* Computing and computer research, computing applications in information systems, manufacturing, medical imaging, chemical processes, and psychology. *Degrees offered:* None. *Graduate students served last academic year:* 100: 50 at the master's level; 45 at the doctoral level; 5 at the postgraduate level. *Faculty:* 36: 1 affiliated solely with the center. *Faculty affiliations:* Departments of Electrical and Systems Engineering, Computer Science and Engineering, Civil and Environmental Engineering, Chemical Engineering, Mechanical Engineering, Psychology, Mathematics. *Annual research budget:* $1.4 million. *Director:* Dr. Peter B. Luh, Director, 860-486-5955, Fax: 860-486-1273, E-mail: luh@engr.uconn.edu.

University of Delaware, Newark, DE 19716.

Center for Catalytic Science and Technology Founded in 1978. *Academic areas of research and training:* Heterogeneous and homogeneous catalysis for chemical, petroleum, and environmental processes; surface chemistry and physics; novel materials for catalytic applications. *Degrees offered:* None. *Graduate students served last academic year:* 52: 2 at the master's level; 35 at the doctoral level; 12 at the postgraduate level. *Faculty:* 10. *Faculty affiliations:* Departments of Chemical Engineering, Chemistry and Biochemistry, Materials Science and Engineering. *Annual research budget:* $1.2 million. *Director:* Dr. Mark A. Barteau, Director, 302-831-8905, Fax: 302-831-2085, E-mail: barteau@che.udel.edu.

Center for Composite Materials Founded in 1974. *Academic areas of research and training:* Manufacture of affordable composites from renewable resources. *Degrees offered:* None. *Graduate students served last academic year:* 100. *Faculty:* 40: 20 affiliated solely with the center. *Faculty affiliations:* Colleges of Agriculture and Natural Resources, Business and Economics; Departments of Chemical Engineering, Mechanical Engineering, Plant Science. *Annual research budget:* $5 million. *Director:* Richard Wool, Professor, 302-831-3312, Fax: 302-831-8525, E-mail: richard_wool@ccm.udel.edu.

University of Detroit Mercy, Detroit, MI 48219-0900.

Center of Excellence in Environmental Engineering and Science Founded in 1992. *Academic areas of research and training:* Integration of the values and goals of the U.S. Environmental Protection Agency, biodegradable plastics, utilization of automotive shredder residue, recycling of thermosetting polymers, reuse of scrap tire rubber, recycling of mixed plastics and incinerator ashes. *Degrees offered:* None. *Graduate students served last academic year:* 6: 3 at the master's level; 2 at the doctoral level; 1 at the postgraduate level. *Faculty:* 25: 15 affiliated solely with the center. *Faculty affiliations:* Departments of Biology, Chemical Engineering, Chemistry, Civil and Environmental Engineering, Polymer Institute. *Annual research budget:* $1 million. *Director:* Dr. Daniel Klempner, Director, 313-993-3385, Fax: 313-993-1112, E-mail: klempndi@udmercy.edu.

University of Florida, Gainesville, FL 32611.

Engineering and Industrial Experiment Station (EIES) Founded in 1910. *Academic areas of research and training:* Aerospace, agricultural, chemical, civil, computer, electrical, environmental, industrial and systems, materials, and mechanical and biomedical engineering. *Degrees offered:* None. *Graduate students served last academic year:* 1,584: 810 at the master's level; 774 at the doctoral level. *Faculty:* 341: 322 affiliated solely with the station. *Faculty affiliations:* College of Education; Departments of Chemistry, Mathematics, Microbiology and Cell Science, Neuroscience, Dental Biomaterials, Physics, Psychology, Radiology, Statistics, Urban and Regional Planning. *Annual research budget:* $56.9 million. *Director:* Dr. Winfred M. Phillips, Dean, 352-392-6000, Fax: 352-392-9673, E-mail: wphil@eng.ufl.edu. *Application contact:* Dr. M. J. Ohanian, Associate Dean for Research and Administration, 352-392-0946, Fax: 352-392-9673, E-mail: johan@eng.ufl.edu.

University of Hawaii at Manoa, Honolulu, HI 96822.

Hawaii Natural Energy Institute (HNEI) Founded in 1974. *Academic areas of research and training:* Chemical engineering, mechanical engineering, ocean engineering, tropical agriculture, renewable energy, ocean resources development, marine biotechnology. *Degrees offered:* None. *Graduate students served last academic year:* 33: 16 at the master's level; 11 at the doctoral level; 6 at the postgraduate level. *Faculty:* 20: 9 affiliated solely with the institute. *Faculty affiliations:* Colleges of Engineering, Tropical Agriculture, Human Resources; Departments of Chemistry, Geology and Geophysics, Horticulture, Physics and Astronomy. *Annual research budget:* $2 million. *Director:* Dr. Patrick K. Takahashi, Director, 808-956-8890, Fax: 808-956-2336, E-mail: patrick@wiliki.eng.hawaii.edu.

University of Houston, Houston, TX 77004.

Bioengineering Research Center Founded in 1984. *Academic areas of research and training:* Bioengineering, biomedical engineering, biotechnology, molecular biology, biosignal analysis, medical imaging, mathematical modeling of physiological processes. *Degrees offered:* None. *Graduate students served last academic year:* 22: 13 at the master's level; 9 at the doctoral level. *Faculty:* 14. *Faculty affiliations:* Cullen College of Engineering. *Director:* Dr. Periklis Y. Ktonas, Director, 713-743-4429, Fax: 713-743-4444, E-mail: pktonas@uh.edu.

Center for Innovative Grouting Materials and Technology (CIGMAT) Founded in 1994. *Academic areas of research and training:* Civil engineering, environmental engineering, grouting materials, materials, coatings, pipes. *Degrees offered:* None. *Graduate students served last academic year:* 12: 8 at the master's level; 3 at the doctoral level; 1 at the postgraduate level. *Faculty:* 10. *Faculty affiliations:* College of Technology; Departments of Chemical Engineering, Chemistry, Civil and Environmental Engineering, Geosciences, Mechanical Enginineering. *Annual research budget:* $350,000. *Director:* Dr. C. Vipulanandan, Director, 713-743-4278, Fax: 713-743-4260, E-mail: cvipulanandan@uh.edu.

Space Vacuum Epitaxy Center (SVEC) Founded in 1986. *Academic areas of research and training:* Chemistry, electrical engineering, physics; development of thin film materials and devices for advanced electronics and space use. *Degrees offered:* None. *Graduate students served last academic year:* 24: 2 at the master's level; 18 at the doctoral level; 4 at the postgraduate level. *Faculty:* 23: 17 affiliated solely with the center. *Faculty affiliations:* Departments of Chemistry, Electrical and Computer Engineering, Physics; Program in Management; Law Center. *Annual research budget:* $3.5 million. *Director:* Dr. Alex Ignatiev, Director, 713-743-3621, Fax: 713-747-7724, E-mail: ignatiev@uh.edu. *Application contact:* Dr. Steven S. S. Pei, Associate Director, 713-743-3621, Fax: 713-747-7724, E-mail: spei@uh.edu.

Texas Center for Superconductivity (TCSUH) Founded in 1987. *Academic areas of research and training:* Multidisciplinary research and development center for high temperature superconductivity and related material sciences including: basic, applied, technology development and transfer, materials manufacturing. *Degrees offered:* None. *Graduate students served last academic year:* 88: 17 at the master's level; 52 at the doctoral level; 19 at the postgraduate level. *Faculty:* 73: 32 affiliated solely with the center. *Faculty affiliations:* Departments of Chemical Engineering, Chemistry, Electrical and Computer Engineering, Materials Engineering, Mechanical Engineering, Physics. *Annual research budget:* $6.2 million. *Director:* Dr. Paul C. W. Chu, Director, 713-743-8200, Fax: 713-743-8201, E-mail: cwchu@uh.edu. *Application contact:* Susan Warren Butler, Associate Director for Public Affairs, 713-743-8210, Fax: 713-743-8201, E-mail: sbutler@uh.edu.

University of Idaho, Moscow, ID 83844-4140.

Center for Applied Thermodynamic Studies (CATS) Founded in 1975. *Academic areas of research and training:* Binary and ternary mixtures, equation of state development, ground-coupled heat pumps, linear and nonlinear regression techniques, natural gas property formulations, transport property theory. *Degrees offered:* None. *Graduate students served last academic year:* 3: 3 at the postgraduate level. *Faculty:* 12: 6 affiliated solely with the center. *Faculty affiliations:* Departments of Chemical and Mechanical Engineering. *Annual research budget:* $47,000. *Director:* Dr. Richard T. Jacobsen, Director, 208-885-6479, Fax: 208-885-6645, E-mail: rtj@uidaho.edu. *Application contact:* Dr. Steven Penoncello, Professor and Chair, Mechanical Engineering, 208-885-6579, Fax: 208-885-9031, E-mail: stevep@uidaho.edu.

Environmental Biotechnology Institute Founded in 1984. *Academic areas of research and training:* Microbiology, biochemistry, bioprocess technology, chemistry, chemical engineering, forestry, genetic engineering, molecular biology. *Degrees offered:* None. *Graduate students served last academic year:* 18: 10 at the master's level; 6 at the doctoral level; 2 at the postgraduate level. *Faculty:* 70. *Faculty affiliations:* College of Forestry, Wildlife, and Range Sciences; Departments of Animal and Veterinary Medicine; Chemical Engineering; Chemistry; Microbiology, Molecular Biology and Biochemistry; Programs in Plant Science; Zoology. *Annual research budget:* $500,000. *Director:* Dr. Ronald L. Crawford, Director, 208-885-6580, Fax: 208-885-5741, E-mail: crawford@uidaho.edu.

Institute for Materials and Advanced Processes (IMAP) Founded in 1989. *Academic areas of research and training:* Innovative extraction techniques, plasma processing, synthesis of advanced materials by mechanical alloying. *Degrees offered:* None. *Graduate students served last academic year:* 18: 10 at the master's level; 5 at the doctoral level; 3 at the postgraduate level. *Faculty:* 25. *Faculty affiliations:* Departments of Chemical Engineering, Chemistry, Electrical Engineering, Forest Products, Mechanical Engineering, Metallurgy, Physics. *Annual research budget:* $850,000. *Director:* Dr. Francis H. Froes, Director, 208-885-7989, Fax: 208-885-5724, E-mail: imap.uidaho.edu.

National Institute for Advanced Transportation Technology (NCATT) Founded in 1991. *Academic areas of research and training:* Clean vehicle technology, traffic operations and control, infrastructure technology. *Degrees offered:* None. *Faculty:* 24. *Faculty affiliations:* Colleges of Engineering; Forestry, Wildlife, and Range Sciences; Departments of Geography, Psychology, Resource Recreation and Tourism; Program in Geological Engineering. *Annual research budget:* $1 million. *Director:* Dr. Michael Kyte, Director, 208-885-0576, Fax: 208-885-2877, E-mail: niatt@uidaho.edu.

University of Illinois at Chicago, Chicago, IL 60607-7128.

Center for Pharmaceutical Biotechnology Founded in 1993. *Academic areas of research and training:* Integration of biotechnology into pharmaceutical sciences. *Degrees offered:* PhD. *Graduate students served last academic year:* 20: 5 at the doctoral level; 15 at the postgraduate level. *Faculty:* 6: 6 affiliated solely with the center. *Annual research budget:* $1 million. *Director:* Dr. Michael E. Johnson, Director, 312-996-0796, Fax: 312-413-9303. *Application contact:* Aimee McConnell, Coordinator of Research Programs, 312-996-0796, Fax: 312-413-9303, E-mail: aimeemcc@uic.edu.

International Center for Software Engineering Founded in 1994. *Academic areas of research and training:* Distributed object computing, software engineering, software metrics, net-centric computing, simulation and modeling, telecommunication systems. *Degrees offered:* None. *Graduate students served last academic year:* 45: 30 at the master's level; 15 at the doctoral level. *Faculty:* 6: 3 affiliated solely with the center. *Faculty affiliations:* Concurrent Software System Lab, Distributed Real-Time Intelligent Systems Lab, Vision Interfaces and Systems Lab. *Annual research budget:* $500,000. *Director:* Dr. Carl K. Chang, Director, 312-996-4860, Fax: 312-413-0024, E-mail: ckchang@uic.edu.

University of Illinois at Urbana–Champaign, Urbana, IL 61801.

Center for Compound Semiconductor Microelectronics Founded in 1986. *Academic areas of research and training:* Compound semiconductors, III-V materials and devices, optoelectronics, microelectronics, wireless communications, microelectromechanical systems (MEMS). *Degrees offered:* None. *Graduate students served last academic year:* 48: 7 at the master's

Research and Training Opportunities

Center for Compound Semiconductor Microelectronics (continued)

level; 38 at the doctoral level; 3 at the postgraduate level. *Faculty:* 18. *Faculty affiliations:* Departments of Electrical and Computer Engineering, Materials Science and Engineering, Mechanical and Industrial Engineering. *Annual research budget:* $6.5 million. *Director:* Stephen G. Bishop, Director, 217-333-3097, Fax: 217-244-6375. *Application contact:* Stephen G. Bishop, Director, 217-333-3097, Fax: 217-244-6375.

Coordinated Science Laboratory Founded in 1951. *Academic areas of research and training:* Electronics, electronic materials, control, communication and computing systems, signal and image processing, VLSI circuits. *Degrees offered:* None. *Graduate students served last academic year:* 310: 100 at the master's level; 210 at the doctoral level. *Faculty:* 70. *Faculty affiliations:* Departments of Aeronautical and Astronautical Engineering, Computer Science, Electrical and Computer Engineering, General Engineering, Materials Science and Engineering; Graduate School of Library and Information Science. *Annual research budget:* $10 million. *Director:* W. Kenneth Jenkins, Director, 217-333-2511, Fax: 217-244-1764, E-mail: jenkins@uicsl.csl.uiuc.edu. *Application contact:* Virginia A. Winckler, Assistant Director, 217-333-2515, Fax: 217-244-1764, E-mail: vwinckle@uiuc.edu.

Frederick Seitz Materials Research Laboratory Founded in 1962. *Academic areas of research and training:* Materials science, condensed-matter physics, materials chemistry. *Degrees offered:* None. *Graduate students served last academic year:* 200: 170 at the doctoral level; 30 at the postgraduate level. *Faculty:* 90: 20 affiliated solely with the laboratory. *Faculty affiliations:* Departments of Chemical Engineering, Chemistry, Electrical Engineering, Materials Science, Mechanical Engineering, Physics, Theoretical and Applied Mechanics. *Annual research budget:* $16 million. *Director:* Dr. Joe Greene, Director, 217-333-1370, Fax: 217-244-2278, E-mail: greene@mrlxp2.mrl.uiuc.edu.

Mid-America Earthquake Center Founded in 1997. *Academic areas of research and training:* Earthquake engineering, structural engineering, geotechnical engineering, seismology, social science, economics. *Degrees offered:* None. *Graduate students served last academic year:* 81: 27 at the master's level; 54 at the doctoral level. *Faculty:* 54: 4 affiliated solely with the center. *Faculty affiliations:* College of Urban Planning and Public Affairs, Department of Civil and Materials Engineering, Georgia Institute of Technology Program in Geology; Massachusetts Institute of Technology, Saint Louis University, Texas A&M University, The University of Memphis, Washington University in St. Louis. *Annual research budget:* $4 million. *Director:* Dr. Daniel P. Abrams, Director, 217-333-0565, Fax: 217-333-3821, E-mail: d-abrams@staff.uiuc.edu. *Application contact:* Dr. Carolyn Sands, Assistant Director, 217-244-1795, Fax: 217-333-3821, E-mail: c-sands1@staff.uiuc.edu.

The University of Iowa, Iowa City, IA 52242-1316.

Center for Biocatalysis and Bioprocessing Founded in 1990. *Academic areas of research and training:* Biotransformations, new biocatalyst discovery, structure and function of biocatalysts, fermentation, bioremediation, biocatalysis in bio-active agent discovery, development. *Degrees offered:* None. *Graduate students served last academic year:* 50: 45 at the doctoral level; 5 at the postgraduate level. *Faculty:* 44: 1 affiliated solely with the center. *Faculty affiliations:* Departments of Biochemistry, Biological Sciences, Chemical and Biochemical Engineering, Chemistry, Civil and Environmental Engineering, Medicinal and Natural Products Chemistry, Microbiology. *Annual research budget:* $1.6 million. *Director:* John P. N. Rosazza, Director, 319-335-4900, Fax: 319-335-4901, E-mail: john-rosazza@uiowa.edu.

University of Kansas, Lawrence, KS 66045.

Information and Telecommunications Technology Center (ITTC) Founded in 1996. *Academic areas of research and training:* Information sciences, telecommunications. *Degrees offered:* None. *Graduate students served last academic year:* 98: 77 at the master's level; 18 at the doctoral level; 3 at the postgraduate level. *Faculty:* 25. *Faculty affiliations:* Departments of Chemical and Petroleum Engineering, Electrical Engineering and Computer Science, Mathematics. *Annual research budget:* $6.5 million. *Director:* Dr. Victor S. Frost, Acting Director, 913-864-4833, Fax: 913-864-0387, E-mail: frost@eecs.ukans.edu. *Application contact:* Judith C. Galas, Public Relations and Marketing Manager, 913-864-4776, Fax: 913-864-0387, E-mail: info@ittc.ukans.edu.

Transportation Center Founded in 1975. *Academic areas of research and training:* Civil engineering, computing, geography, urban planning. *Degrees offered:* None. *Graduate students served last academic year:* 11: 9 at the master's level; 2 at the doctoral level. *Faculty:* 9. *Faculty affiliations:* Departments of Civil Engineering, Urban Planning. *Annual research budget:* $800,000. *Director:* Dr. Joe Lee, Director, 785-864-3787, Fax: 785-864-3199.

University of Kentucky, Lexington, KY 40506-0032.

Center for Applied Energy Research (CAER) Founded in 1977. *Academic areas of research and training:* Energy research, fossil fuels, materials, environmental, chemical engineering, chemistry. *Degrees offered:* None. *Graduate students served last academic year:* 16: 8 at the master's level; 4 at the doctoral level; 4 at the postgraduate level. *Faculty:* 5. *Faculty affiliations:* Programs in Chemical and Materials Engineering, Chemistry, Mechanical Engineering, Physics. *Annual research budget:* $3.8 million. *Director:* Dr. Frank Derbyshire, Director, 606-257-0305, Fax: 606-257-0220, E-mail: derbyshire@caer.uky.edu. *Application contact:* Marybeth McAlister, Publications Manager, 606-257-0224, Fax: 606-257-0220, E-mail: mcalister@caer.uky.edu.

Center for Robotics and Manufacturing Systems Founded in 1986. *Academic areas of research and training:* Robotics, manufacturing systems, manufacturing processes, lean manufacturing, rapid prototyping. *Degrees offered:* None. *Graduate students served last academic year:* 32: 19 at the master's level; 10 at the doctoral level; 3 at the postgraduate level. *Faculty:* 12: 7 affiliated solely with the center. *Faculty affiliations:* Department of Mechanical Engineering; Programs in Chemical Engineering, Communications, Electrical Engineering, Materials Science. *Annual research budget:* $1.5 million. *Director:* Dr. Alan T. Male, Director, 606-257-6262 Ext. 205, Fax: 606-323-1035, E-mail: atmale@engr.uky.edu.

Kentucky Transportation Center Founded in 1979. *Academic areas of research and training:* Environmental analysis, geotechnology, impact analysis, materials testing, pavements, structures, intermodal, intelligent transportation systems, traffic/safety. *Degrees offered:* PhD; Certificate. *Graduate students served last academic year:* 42: 28 at the master's level; 14 at the doctoral level. *Faculty:* 15. *Faculty affiliations:* Programs in Business Administration, Civil Engineering, Economics, Geography, Geology. *Annual research budget:* $4.6 million. *Director:* Paul Toussaint, Director, 606-257-4513, Fax: 606-257-1815, E-mail: toussain@engr.uky.edu.

University of Maine, Orono, ME 04469.

The Laboratory for Surface Science and Technology (LASST) Founded in 1980. *Academic areas of research and training:* Surface science, interface science, thin film technology, microelectronics, sensor technology. *Degrees offered:* None. *Graduate students served last academic year:* 34: 8 at the master's level; 18 at the doctoral level; 8 at the postgraduate level. *Faculty:* 7: 2 affiliated solely with the laboratory. *Faculty affiliations:* Departments of Chemistry, Electrical and Computer Engineering, Physics and Astronomy. *Annual research budget:* $4.5 million. *Director:* Dr. Robert J. Lad, Director, 207-581-2257, Fax: 207-581-2255, E-mail: rjlad@maine.maine.edu.

University of Maryland, Baltimore, MD 21201-1627.

Medical Biotechnology Center Founded in 1987. *Academic areas of research and training:* Medical biotechnology, biophysics, molecular biology. *Degrees offered:* None. *Graduate students served last academic year:* 12: 2 at the doctoral level; 10 at the postgraduate level. *Faculty:* 30: 20 affiliated solely with the center. *Faculty affiliations:* Departments of Microbiology and Immunology, Pharmacology and Experimental Therapeutics, Physiology, Graduate Programs in Medicine, School of Pharmacy. *Annual research budget:* $12 million. *Director:* Dr. Jonathan C. Lederer, Director, 410-706-8181, Fax: 410-706-8184, E-mail: lederer@umbi.umd.edu. *Application contact:* Timothy P. Hughes, Assistant Director, 410-706-8181, Fax: 410-706-8184, E-mail: hughes@umbi.umd.edu.

University of Maryland, College Park, College Park, MD 20742-5045.

Center for Automation Research Founded in 1983. *Academic areas of research and training:* Computer vision, document processing, robotics. *Degrees offered:* None. *Graduate students served last academic year:* 40: 40 at the doctoral level. *Faculty:* 15: 10 affiliated solely with the center. *Faculty affiliations:* Departments of Computer Science, Electrical Engineering. *Annual research budget:* $3 million. *Director:* Dr. Azriel Rosenfeld, Director, 301-405-4526, Fax: 301-314-9115, E-mail: ar@cfar.umd.edu.

Center for Environmental Energy Engineering (CEEE) Founded in 1991. *Academic areas of research and training:* Energy conversion, energy and environment, enhanced heat transfer, vapor compression heat pumps, absorption heat pumps, refrigeration systems, natural gas building. *Degrees offered:* MS; PhD. *Graduate students served last academic year:* 30: 14 at the master's level; 14 at the doctoral level; 2 at the postgraduate level. *Faculty:* 7. *Faculty affiliations:* Departments of Chemical Engineering; Mechanical Engineering, Nuclear Engineering Program; Institute of Physical Science and Technology. *Annual research budget:* $1.5 million. *Director:* Dr. Reinhard Radermacher, Director, 301-405-5286, Fax: 301-405-2025, E-mail: rader@eng.umd.edu.

Center for Superconductivity Research Founded in 1989. *Academic areas of research and training:* Chemistry, electrical engineering, materials engineering, physics. *Degrees offered:* None. *Graduate students served last academic year:* 25: 20 at the doctoral level; 5 at the postgraduate level. *Faculty:* 15. *Faculty affiliations:* Departments of Chemistry, Electrical Engineering, Materials Engineering, Physics. *Annual research budget:* $3 million. *Director:* Dr. Richard L. Greene, Director, 301-405-6128, Fax: 301-405-3779, E-mail: rgreene@squid.umd.edu.

Computer-Aided Life-Cycle Engineering Electronic Products and Systems Consortium Founded in 1986. *Academic areas of research and training:* Electronic packaging, failure analysis, reliability. *Degrees offered:* None. *Graduate students served last academic year:* 50: 40 at the master's level; 10 at the doctoral level. *Faculty:* 15: 10 affiliated solely with the consortium. *Faculty affiliations:* Department of Mechanical Engineering. *Annual research budget:* $5 million. *Director:* Dr. Michael Pecht, Director, 301-405-5323, Fax: 301-314-9269, E-mail: pecht@calce-pc.umd.edu. *Application contact:* Joanyuan Lee, Program Analyst, 301-405-5323, Fax: 301-314-9269, E-mail: joanyuan@eng.umd.edu.

University of Massachusetts Amherst, Amherst, MA 01003.

Center for Intelligent Information Retrieval (CIIR) Founded in 1992. *Academic areas of research and training:* Computer science, information retrieval. *Degrees offered:* None. *Graduate students served last academic year:* 22: 20 at the doctoral level; 2 at the postgraduate level. *Faculty:* 7: 4 affiliated solely with the center. *Faculty affiliations:* Department of Computer Science. *Annual research budget:* $3.5 million. *Director:* Dr. W. Bruce Croft, Professor, 413-545-0463, Fax: 413-545-1789, E-mail: croft@cs.umass.edu. *Application contact:* Jean Ziemba, Administrative Director, 413-545-0463, Fax: 413-545-1789, E-mail: ziemba@cs.umass.edu.

Center for Real-Time and Intelligent Complex Computing Systems (CRICCS) Founded in 1992. *Academic areas of research and training:* Computer science, education, software engineering. *Degrees offered:* None. *Graduate students served last academic year:* 20: 10 at the master's level; 10 at the doctoral level. *Faculty:* 27: 10 affiliated solely with Center for Real-Time and Intelligent Complex Computing Systems (CRICCS). *Faculty affiliations:* Department of Computer Science. *Annual research budget:* $8 million. *Director:* Dr. W. Richards Adrion, Professor, 413-545-2475, Fax: 413-545-3729, E-mail: adrion@cs.umass.edu. *Application contact:* Wendy Cooper, Administrative Assistant, 413-545-2475, Fax: 413-545-3729, E-mail: cooper@cs.umass.edu.

University of Massachusetts Lowell, Lowell, MA 01854-2881.

Center for Advanced Materials (CAM) Founded in 1992. *Academic areas of research and training:* Chemical and electrical engineering, chemistry, materials science, physics, plastics engineering, polymer science. *Degrees offered:* None. *Graduate students served last academic year:* 5 at the master's level; 65 at the doctoral level; 10 at the postgraduate level. *Faculty:* 20: 6 affiliated solely with the center. *Faculty affiliations:* Departments of Chemical Engineering, Chemistry, Electrical Engineering, Physics and Applied Physics, and Plastics Engineering. *Annual research budget:* $500,000. *Director:* Dr. Sukant Tripathy, Director, 978-934-3687, Fax: 978-458-9571, E-mail: sukant_tripathy@uml.edu. *Application contact:* Dr. Daniel J. Sandman, Associate Director, 978-934-3835, Fax: 978-458-9571, E-mail: daniel_sandman@uml.edu.

Center for Atmospheric Research Founded in 1975. *Academic areas of research and training:* Electrical engineering, physics, atmosphere and space science. *Degrees offered:* None. *Graduate students served last academic year:* 5: 1 at the master's level; 4 at the doctoral level. *Faculty:* 6: 1 affiliated solely with the center. *Faculty affiliations:* College of Engineering, Departments of Atmospheric Science, Computer Science. *Annual research budget:* $2 million. *Director:* Dr. Bodo W. Reinisch, Director, 978-934-4903, Fax: 978-459-7915, E-mail: bodo_reinisch@uml.edu.

Center for Electromagnetic Materials and Optical Systems (CEMOS) Founded in 1988. *Academic areas of research and training:* Electrical engineering, electrooptics, optical engineering, physics, physical optics, electromagnetics, optical systems imaging. *Degrees offered:* MS; PhD; D Eng. *Graduate students served last academic year:* 18: 4 at the master's level; 8 at the doctoral level; 6 at the postgraduate level. *Faculty:* 9: 6 affiliated solely with the center. *Faculty affiliations:* Departments of Chemical Engineering, Electrical Engineering, Mathematics, Physics and Applied Physics. *Annual research budget:* $250,000. *Director:* Dr. M. A. Fiddy, Professor, 978-934-3306, Fax: 978-934-3027, E-mail: michael_fiddy@uml.edu.

Center for Environmental Engineering, Science and Technology (CEEST) Founded in 1993. *Academic areas of research and training:* Waste containment systems, contaminated site remediation research, geographic information systems, water quality control, air quality control, contaminant leachability, solid waste management, internat ional development. *Degrees offered:* None. *Graduate students served last academic year:* 25: 18 at the master's level; 5 at the doctoral level; 2 at the postgraduate level. *Faculty:* 21: 12 affiliated solely with the center. *Faculty affiliations:* Departments of Chemistry, Civil Engineering, Economics; Programs in Earth Science, Environmental Studies. *Annual research budget:* $500,000. *Director:* Dr. Hilary I. Inyang, Director, 978-934-2285, Fax: 978-934-3092, E-mail: ceest@caeds0.uml.edu.

University of Michigan, Ann Arbor, MI 48109.

The Artificial Intelligence Laboratory Founded in 1987. *Academic areas of research and training:* Artificial intelligence, robotics, computer science, human-computer interaction, computers and education, digital libraries. *Degrees offered:* None. *Graduate students served last academic year:* 47: 5 at the master's level; 40 at the doctoral level; 2 at the postgraduate level. *Faculty:* 15: 9 affiliated solely with the laboratory. *Faculty affiliations:* Departments of Atmospheric, Oceanic and Space Sciences; Linguistics; Mechanical Engineering; Nuclear Engineering and Radiological Sciences; Philosophy; Psychology; Mental Health Research Institute; School of Information. *Annual research budget:* $4 million. *Director:* Prof. Edmund H. Durfee, Director, 734-936-1563, Fax: 734-763-1260, E-mail: durfee@umich.edu.

Automotive Research Center (ARC) Founded in 1994. *Academic areas of research and training:* Modeling and simulation of ground vehicle systems. *Degrees offered:* None. *Gradu-*

Research and Training Opportunities

ate students served last academic year: 30: 12 at the master's level; 15 at the doctoral level; 3 at the postgraduate level. *Faculty:* 35. *Faculty affiliations:* College of Engineering in addition to varied departments related to ARC's subcontracting schools: Clemson University, Oakland University, University of Alaska-Fairbanks, University of Iowa, University of Tennessee, University of Wisconsin-Madison, and Wayne State University. *Annual research budget:* $3.7 million. *Director:* Dr. Panos Y. Papalambros, Director, 734-647-8401, Fax: 734-647-8403, E-mail: pyp@umich.edu. *Application contact:* Aloda Thomas, Administrator, 734-764-5263, Fax: 734-764-4256, E-mail: aloda@umich.edu.

The Center for Parallel Computing (CPC)   Founded in 1992. *Academic areas of research and training:* Parallel computing, supercomputing, data-intensive computing. *Degrees offered:* None. *Graduate students served last academic year:* 30: 15 at the master's level; 10 at the doctoral level; 5 at the postgraduate level. *Faculty:* 12. *Faculty affiliations:* Departments of Aerospace, Chemistry, Computer Science, Mathematics, Mechanical Engineering, Nuclear Engineering, Physics. *Annual research budget:* $2 million. *Director:* Dr. Quentin F. Stout, Director, 734-936-2310, Fax: 734-763-4540, E-mail: cpc-info@engin.umich.edu.

University of Minnesota, Twin Cities Campus, Minneapolis, MN 55455-0213.

Supercomputing Institute for Digital Simulation and Advanced Computation   Founded in 1983. *Academic areas of research and training:* Supercomputing research. *Degrees offered:* None. *Graduate students served last academic year:* 700. *Faculty:* 200. *Faculty affiliations:* Departments of Accounting, Astronomy, Cell Biology and Neuroanatomy, Chemistry, Economics, Electrical and Computer Engineering, Geography, Horticulture, Mathematics, Microbiology, Orthopaedic Surgery, Pharmacology, Physiology, Radiology, Wood and Paper Science; Bioprocess Technology Institute; College of Pharmacy; Institute for Math and Its Applications; St. Anthony Falls Lab; Theoretical Physics Institute. *Director:* Dr. Donald Truhlar, Director, 612-625-1818, Fax: 612-624-8861, E-mail: truhlar@umn.edu. *Application contact:* Michael J. Olesen, Assistant to the Director, 612-624-1356, Fax: 612-624-9565, E-mail: olesen@msi.umn.edu.

University of Mississippi, Oxford, University, MS 38677-9702.

Center for Computational Hydroscience and Engineering (CCHE)   Founded in 1983. *Academic areas of research and training:* Computational hydroscience and engineering. *Degrees offered:* MS; PhD. *Graduate students served last academic year:* 3 at the master's level; 5 at the doctoral level; 2 at the postgraduate level. *Faculty:* 9: 6 affiliated solely with Center for Computational Hydroscience and Engineering (CCHE). *Faculty affiliations:* Programs in Chemical, Civil, Mechanical Engineering. *Annual research budget:* $900,000. *Director:* Dr. Sam S. Y. Wang, Director, 601-232-7788, Fax: 601-232-7796, E-mail: wang@hydra.cche.olemiss.edu.

Composite Materials Research Group (CMRG)   Founded in 1990. *Academic areas of research and training:* Manufacture and testing of polymeric composites with emphasis on the pultrusion process. *Degrees offered:* None. *Graduate students served last academic year:* 11: 7 at the master's level; 4 at the doctoral level. *Faculty:* 6. *Faculty affiliations:* Departments of Civil Engineering, Material Science and Engineering, Mechanical Engineering. *Annual research budget:* $400,000. *Director:* Dr. James G. Vaughan, Director, 601-232-5378, Fax: 601-232-7219, E-mail: jgv@pultrusion.me.olemiss.edu.

The James Whitten National Center for Physical Acoustics (NCPA)   Founded in 1986. *Academic areas of research and training:* Physics of acoustics research and education, physics, mechanical engineering, electrical engineering, speech and hearing, biology. *Degrees offered:* None. *Graduate students served last academic year:* 35: 10 at the master's level; 20 at the doctoral level; 5 at the postgraduate level. *Faculty:* 13: 5 affiliated solely with the center. *Faculty affiliations:* Departments of Biology, Mechanical Engineering, Physics and Astronomy, Speech and Hearing. *Annual research budget:* $2.4 million. *Director:* Dr. Henry E. Bass, Director, 601-232-5840, Fax: 601-232-7494, E-mail: pabass@olemiss.edu.

University of Missouri–Columbia, Columbia, MO 65211.

Capsule Pipeline Research Center (CPRC)   Founded in 1991. *Academic areas of research and training:* Machine design, manufacturing coal logs, pipeline, hydraulic tests, freight transport. *Degrees offered:* None. *Graduate students served last academic year:* 26: 16 at the master's level; 4 at the doctoral level; 6 at the postgraduate level. *Faculty:* 11: 1 affiliated solely with the center. *Faculty affiliations:* Departments of Chemical, Civil, Industrial, Mechanical and Aerospace, and Mining Engineering; School of Law. *Annual research budget:* $700,000. *Director:* Dr. Henry Liu, Director, 573-882-2779, Fax: 573-884-4888, E-mail: liuh@missouri.edu. *Application contact:* Dr. Henry Liu, Director, 573-882-2779, Fax: 573-884-4888, E-mail: liuh@missouri.edu.

Power Electronics Research Center (PERC)   Founded in 1985. *Academic areas of research and training:* Power electronic materials and devices, power supplies, adjustable speed motor drives, harmonic compensation, soft computing control of power electronic converters. *Degrees offered:* None. *Graduate students served last academic year:* 10: 6 at the master's level; 4 at the doctoral level. *Faculty:* 6: 1 affiliated solely with Power Electronics Research Center (PERC). *Faculty affiliations:* Programs in Electronic Circuits, Electronic Materials and Devices, Power Systems. *Annual research budget:* $100,000. *Director:* Robert M. O'Connell, Director, 573-882-8373, Fax: 573-882-0397, E-mail: oconnellr@missouri.edu.

Research Reactor Center (MURR)   Founded in 1966. *Academic areas of research and training:* Radiation sciences and engineering, materials science, archaeology, nutrition, physics, anthropology, chemistry, nuclear engineering, life sciences, comparative oncology, radiopharmaceuticals. *Degrees offered:* None. *Faculty:* 31: 13 affiliated solely with the center. *Faculty affiliations:* Departments of Anthropology, Biochemistry, Chemical Engineering, Chemistry, Electrical and Computer Engineering, Food Science, Geological Sciences, Mechanical and Aerospace Engineering, Physics; Programs in Archaeology, Child Health, Nuclear Engineering, Nutrition. *Director:* Dr. Edward A. Deutsch, Director, 573-882-4211, Fax: 573-882-6360. *Application contact:* Christine M. Errante, Administrative Associate II, 573-882-5228, Fax: 573-882-6360, E-mail: errantec@missouri.edu.

University of Missouri–St. Louis, St. Louis, MO 63121-4499.

Center for Molecular Electronics   Founded in 1988. *Academic areas of research and training:* Electronic materials, microscopy, electronic devices, computer modeling, spectroscopy of solids. *Degrees offered:* None. *Graduate students served last academic year:* 16: 4 at the master's level; 8 at the doctoral level; 4 at the postgraduate level. *Faculty:* 12. *Faculty affiliations:* Departments of Chemistry, Physics. *Annual research budget:* $150,000. *Director:* Dr. Bernard Feldman, Director, 314-516-5019, Fax: 314-516-6152, E-mail: c4840@slvaxa.umsl.edu.

The University of Montana–Missoula, Missoula, MT 59812-0002.

Montana Biotechnology Center   Founded in 1996. *Academic areas of research and training:* Biotechnology, molecular and cell biology, HIV and infectious diseases. *Degrees offered:* None. *Graduate students served last academic year:* 5: 1 at the master's level; 2 at the doctoral level; 2 at the postgraduate level. *Faculty:* 1: 1 affiliated solely with the center. *Faculty affiliations:* Department of Chemistry, Division of Biological Sciences, Program in Wildlife Biology, Programs in Pharmaceutical Sciences, School of Forestry. *Annual research budget:* $400,000. *Director:* Dr. Jack Nunberg, Director, 406-243-6421, Fax: 406-243-6425, E-mail: nunberg@selway.umt.edu.

University of Nebraska–Lincoln, Lincoln, NE 68588.

Center for Biotechnology   Founded in 1986. *Academic areas of research and training:* Biotechnology. *Degrees offered:* None. *Graduate students served last academic year:* 130: 50

at the master's level; 50 at the doctoral level; 30 at the postgraduate level. *Faculty:* 166. *Annual research budget:* $2.7 million. *Director:* Dr. Anne K. Vidaver, Director, 402-472-2635, Fax: 402-472-3139.

Center for Ergonomics and Safety Research   Founded in 1991. *Academic areas of research and training:* Cumulative trauma disorders, sleep ergonomics, dental ergonomics, biomechanics, workplace design, construction ergonomics. *Degrees offered:* None. *Graduate students served last academic year:* 6: 5 at the master's level; 1 at the doctoral level. *Faculty:* 16: 10 affiliated solely with the center. *Faculty affiliations:* Departments of Civil Engineering, Computer Science and Engineering, Psychology; School of Dentistry. *Annual research budget:* $100,000. *Director:* Dr. Michael W. Riley, Director, 402-472-3495, Fax: 402-472-2410, E-mail: ieidrile@engunx.unl.edu. *Application contact:* Center Office, E-mail: ieidrile@engunx.unl.edu.

Center for Materials Research and Analysis (CMRA)   Founded in 1988. *Academic areas of research and training:* Materials preparation and characterization, nanostructured materials, magnetic materials. *Degrees offered:* None. *Graduate students served last academic year:* 122: 59 at the master's level; 63 at the doctoral level. *Faculty:* 67: 2 affiliated solely with the center. *Faculty affiliations:* Departments of Agronomy, Biological Systems Engineering, Chemical Engineering, Chemistry, Electrical Engineering, Engineering Mechanics, Geology, Mechanical Engineering, Physics and Astronomy. *Annual research budget:* $8.1 million. *Director:* Dr. David J. Sellmyer, Director, 402-472-7886, Fax: 402-472-2879, E-mail: cmra@unlinfo.unl.edu.

Center for Microelectronic and Optical Materials Research (CMOMR)   Founded in 1988. *Academic areas of research and training:* Thin films, bulk semiconductors, vacuum processing plasmas and plasma-solid interactions, ellipsometry optics. *Degrees offered:* None. *Graduate students served last academic year:* 10: 7 at the master's level; 2 at the doctoral level; 1 at the postgraduate level. *Faculty:* 8: 2 affiliated solely with the center. *Faculty affiliations:* Departments of Electrical Engineering, Mechanical Engineering, Physics and Astronomy. *Annual research budget:* $1.4 million. *Director:* Dr. John A. Woollam, Director, 402-472-1975, Fax: 402-472-7987, E-mail: jwoollam@unl.edu.

Mid-America Transportation Center (MATC)   Founded in 1995. *Academic areas of research and training:* Design and operation of surface transportation facilities and services. *Degrees offered:* None. *Graduate students served last academic year:* 72: 57 at the master's level; 15 at the doctoral level. *Faculty:* 68: 3 affiliated solely with the center. *Faculty affiliations:* Departments of Chemical Engineering, Civil Engineering, Community and Regional Planning, Computer Science and Engineering, Economics, Electrical Engineering, Industrial and Management Systems Engineering, Mechanical Engineering. *Annual research budget:* $2.5 million. *Director:* Dr. Patrick T. McCoy, Director, 402-472-5019, Fax: 402-472-0859, E-mail: pmccoy@unlinfo.unl.edu. *Application contact:* Dr. Patrick T. McCoy, Director, 402-472-5019, Fax: 402-472-0859, E-mail: pmccoy@unlinfo.unl.edu.

Nontraditional Manufacturing Research Center   Founded in 1989. *Academic areas of research and training:* Development of hardware and software for nontraditional manufacturing processes. *Degrees offered:* None. *Graduate students served last academic year:* 18: 17 at the master's level; 7 at the doctoral level. *Faculty:* 12: 5 affiliated solely with the center. *Faculty affiliations:* Departments of Chemical Engineering, Electrical Engineering, Engineering Mechanics, Mechanical Engineering. *Annual research budget:* $750,000. *Director:* K. P. Rajurkar, Director, 402-472-1385, Fax: 402-472-2410, E-mail: iermraju@engvms.unl.edu.

University of New Brunswick, Fredericton, NB E3B 5A3, Canada.

Institute of Biomedical Engineering   Founded in 1965. *Academic areas of research and training:* Biological signal processing, myo-electric control, upper limb prosthetics, ergonomics, biomechanics, motion analysis. *Degrees offered:* None. *Graduate students served last academic year:* 16: 10 at the master's level; 5 at the doctoral level. *Faculty:* 17: 3 affiliated solely with the institute. *Faculty affiliations:* Departments of Electrical Engineering, Mathematics and Statistics, Mechanical Engineering, Physics, Psychology; Faculties of Forestry, Kinesiology, Nursing. *Annual research budget:* $1.1 million. *Director:* Dr. E. N. Biden, Director, 506-453-4966, Fax: 506-453-4827, E-mail: biomed@unb.ca.

University of New Hampshire, Durham, NH 03824.

Environmental Research Group   Founded in 1987. *Academic areas of research and training:* Environmental engineering, fundamental and applied research, water treatment, waste utilization, in situ remediation, technology development and verification. *Degrees offered:* None. *Graduate students served last academic year:* 21: 18 at the master's level; 3 at the doctoral level. *Faculty:* 13. *Faculty affiliations:* Departments of Chemical Engineering, Civil Engineering, Microbiology. *Annual research budget:* $1.8 million. *Director:* Dr. Taylor Eighmy, Director, 603-862-2206, Fax: 603-862-2364, E-mail: taylor.eighmy@unh.edu.

University of New Mexico, Albuquerque, NM 87131-2039.

The Center for Autonomous Control Engineering   Founded in 1995. *Academic areas of research and training:* Autonomous control, space autonomy, satellites, mobile robotics, intelligent agents, image processing, software development, water treatment, space-born objects. *Degrees offered:* None. *Graduate students served last academic year:* 120: 15 at the master's level; 10 at the doctoral level. *Faculty:* 26: 8 affiliated solely with the center. *Faculty affiliations:* Departments of Aerospace Engineering, Biology, Chemistry, Civil Engineering, Computer Engineering, Computer Science, Earth Sciences, Electrical Engineering, Mathematics, Mechanical Engineering. *Annual research budget:* $2 million. *Director:* Dr. Mo Jamshidi, Director, 505-277-5538, Fax: 505-277-4681, E-mail: jamshidi@unm.edu. *Application contact:* Sandi Avrit, Administrative Assistant, 505-277-0319, Fax: 505-277-4681, E-mail: savrit@unm.edu.

The Center for High Technology Materials (CHTM)   Founded in 1983. *Academic areas of research and training:* Optoelectronics, microelectronics, optics, material science. *Degrees offered:* None. *Graduate students served last academic year:* 53: 53 at the doctoral level; 12 at the postgraduate level. *Faculty:* 25. *Faculty affiliations:* Departments of Chemical and Nuclear Engineering; Chemistry; Electrical and Computer Engineering; Physics and Astronomy. *Annual research budget:* $6 million. *Director:* Dr. Steven R. J. Brueck, Director, 505-272-7800, Fax: 505-272-7801, E-mail: bruck@chtm.unm.edu. *Application contact:* Dr. Robert B. Lauer, Associate Director, 505-272-7868, Fax: 505-272-7801, E-mail: blauer@chtm.unm.edu.

The Center for Micro-Engineered Materials (CMEM)   Founded in 1988. *Academic areas of research and training:* Materials science, nanocomposite materials chemistry, chemical engineering, polymer/ceramic composites. *Degrees offered:* None. *Graduate students served last academic year:* 18: 11 at the master's level; 4 at the doctoral level; 3 at the postgraduate level. *Faculty:* 14. *Faculty affiliations:* Departments of Chemical and Nuclear Engineering, Chemistry, Physics and Astronomy. *Annual research budget:* $2 million. *Director:* Dr. Abhaya K. Datye, Co-Director, 505-277-2833, Fax: 505-277-1024, E-mail: cmem@unm.edu.

The Microelectronics Research Center (MRC)   Founded in 1988. *Academic areas of research and training:* High performance electronics, VLSI, low power, radiation tolerant electronics, special purpose VLSI. *Degrees offered:* None. *Graduate students served last academic year:* 15: 11 at the master's level; 4 at the doctoral level. *Faculty:* 5: 3 affiliated solely with the center. *Faculty affiliations:* Departments of Computer Science, Electrical and Computer Engineering. *Annual research budget:* $3 million. *Director:* Dr. Gary Maki, Director, 505-272-7040, Fax: 505-272-7041, E-mail: maki@mrc.unm.edu.

University of New Orleans, New Orleans, LA 70148.

Energy Conversion and Conservation Center (ECCC)   Founded in 1995. *Academic areas of research and training:* Energy, conservation, thermal sciences, alternate energy sources. *Degrees offered:* None. *Graduate students served last academic year:* 2: 2 at the master's level. *Faculty:* 8: 2 affiliated solely with the center. *Faculty affiliations:* Concentration in Mechanical Engineering. *Annual research budget:* $2 million. *Director:* Dr. Edwin Russo, Chairman, 504-280-6632, Fax: 504-280-5539, E-mail: erusso@uno.edu.

Research and Training Opportunities

The University of North Carolina at Chapel Hill, Chapel Hill, NC 27599.

Center for Multiphase Research (CMR) Founded in 1992. *Academic areas of research and training:* Multiphase flow, contaminant transport, groundwater, geosciences, applied mathematics, scientific computing. *Degrees offered:* None. *Graduate students served last academic year:* 20: 5 at the master's level; 10 at the doctoral level; 5 at the postgraduate level. *Faculty:* 6. *Faculty affiliations:* Departments of Computer Sciences, Environmental Sciences and Engineering. *Annual research budget:* $1.3 million. *Director:* Dr. Cass T. Miller, Director, 919-966-2643, Fax: 919-966-7911, E-mail: casey_miller@unc.edu.

Microelectronic Systems Laboratory (MSL) Founded in 1981. *Academic areas of research and training:* Hardware systems, virtual/augmented environments (VR), imaging and IR optics, sensors and electro-optics. *Degrees offered:* None. *Graduate students served last academic year:* 20. *Faculty:* 8: 5 affiliated solely with the laboratory. *Faculty affiliations:* Department of Computer Science. *Director:* Leandra Vicci, Director, E-mail: vicci@cs.unc.edu.

University of Northern Iowa, Cedar Falls, IA 50614.

Recycling and Reuse Technology Transfer Center Founded in 1992. *Academic areas of research and training:* Industrial ecology, by-product reutilization studies, recycling policy analysis, recycling economic analysis. *Degrees offered:* None. *Graduate students served last academic year:* 15: 15 at the master's level. *Faculty:* 30. *Faculty affiliations:* Departments of Agronomy, Biology, Chemistry, Civil Engineering, Earth Science, Environmental Engineering, Public Policy; Program in Environmental Science/Technology. *Annual research budget:* $250,000. *Director:* Catherine L. Zeman, Director, 319-273-7090, Fax: 319-273-5815, E-mail: catherine.zeman@uni.edu.

University of North Texas, Denton, TX 76203.

Center for Network Neuroscience Founded in 1986. *Academic areas of research and training:* Neurobiology of nerve cell ensembles, network dynamics applications to neurotoxicology, drug development, biosensors. *Degrees offered:* None. *Graduate students served last academic year:* 6: 4 at the master's level; 2 at the doctoral level. *Faculty:* 7. *Faculty affiliations:* Departments of Biological Sciences, Computer Sciences, Physics, Speech and Hearing Sciences; Division of Biochemistry. *Annual research budget:* $150,000. *Director:* Dr. Guenter W. Gross, Director, 940-565-3615, Fax: 940-565-4136, E-mail: gross@nervous.cnns.unt.edu.

Information Systems Research Center Founded in 1986. *Academic areas of research and training:* Information technology. *Degrees offered:* None. *Graduate students served last academic year:* 20: 2 at the master's level; 18 at the doctoral level. *Faculty:* 26: 1 affiliated solely with the center. *Faculty affiliations:* Departments of Business Computer Information Systems, Computer Education, Computer Sciences, Information Sciences. *Annual research budget:* $130,000. *Director:* Dr. J. Wayne Spence, Director, 940-565-3128, Fax: 940-565-4317, E-mail: isrc@unt.edu.

Laboratory of Polymers and Composites (LAPOM) Founded in 1989. *Academic areas of research and training:* Service performance and reliability of polymer-based materials, ceramic and polymer composites, liquid crystals, friction, solution flow. *Degrees offered:* MS; PhD. *Graduate students served last academic year:* 12: 5 at the master's level; 5 at the doctoral level; 2 at the postgraduate level. *Faculty:* 8: 4 affiliated solely with the laboratory. *Faculty affiliations:* Departments of Engineering Technology, Materials Science. *Annual research budget:* $450,000. *Director:* Dr. Witold Brostow, Director, 940-565-4358, Fax: 940-565-4824, E-mail: brostow@unt.edu.

University of Oklahoma, Norman, OK 73019-0390.

Rock Mechanics Institute Founded in 1991. *Academic areas of research and training:* Oil and gas research, drilling and production, rock and rock mass properties. *Degrees offered:* None. *Graduate students served last academic year:* 21: 3 at the master's level; 10 at the doctoral level; 8 at the postgraduate level. *Faculty:* 6: 2 affiliated solely with the institute. *Faculty affiliations:* Schools of Business Administration, Civil Engineering and Environmental Science, Geology and Geodynamics, Petroleum and Geological Engineering. *Annual research budget:* $2 million. *Director:* Dr. Jean-Claude Roegiers, Director, 405-325-2900, Fax: 405-325-7511, E-mail: jc@rmg.ou.edu.

University of Oregon, Eugene, OR 97403.

Materials Science Institute Founded in 1985. *Academic areas of research and training:* Biotechnological materials, characterization of electronic materials/devices, experimental/theoretical geophysics, fundamental electrical/thermal transport properties of nanometer structures. *Degrees offered:* None. *Graduate students served last academic year:* 48: 1 at the master's level; 42 at the doctoral level; 5 at the postgraduate level. *Faculty:* 14. *Faculty affiliations:* Departments of Chemistry, Physics. *Annual research budget:* $1.5 million. *Director:* Dr. David Johnson, Professor, 541-346-4612, Fax: 541-346-3422, E-mail: davej@oregon.uoregon.edu. *Application contact:* Lucy Biggs, Office Coordinator, 541-346-4784, Fax: 541-346-3422, E-mail: lbiggs@oregon.uoregon.edu.

University of Pittsburgh, Pittsburgh, PA 15260.

Center for Biomedical Informatics Founded in 1996. *Academic areas of research and training:* Computer applications to health care, informatics, clinical systems, artificial intelligence. *Degrees offered:* None. *Graduate students served last academic year:* 35: 14 at the master's level; 16 at the doctoral level; 2 at the postgraduate level. *Faculty:* 39: 5 affiliated solely with the center. *Faculty affiliations:* Schools of Health and Rehabilitation Sciences, Information Sciences, Medicine. *Annual research budget:* $2.5 million. *Director:* Dr. Charles P. Friedman, Director, 412-647-7113, Fax: 412-647-7190, E-mail: cpf@cbmi.upmc.edu. *Application contact:* Joseph Cummings, Coordinator, 412-647-7113, Fax: 412-647-7190, E-mail: jcumm@cbmi.upmc.edu.

University of Puerto Rico, Mayagüez Campus, Mayagüez, PR 00681-5000.

Civil Infrastructure Research Center (CIRC) Founded in 1991. *Academic areas of research and training:* Civil infrastructure research in structures, soils, water resources, environmental engineering, transportation, construction management, technology transfer. *Degrees offered:* None. *Graduate students served last academic year:* 25: 20 at the master's level; 5 at the doctoral level. *Faculty:* 20. *Faculty affiliations:* Departments of Chemical Engineering, Civil and Environmental Engineering, Geology, Industrial Engineering; Programs in Architecture and Planning, Electrical and Computer Engineering, Social Sciences. *Annual research budget:* $1 million. *Director:* Dr. Antonio A. Gonzalez-Quevedo, Director, 787-265-3892, Fax: 787-265-3390, E-mail: antonio@rmce02.upr.clu.edu.

University of Regina, Regina, SK S4S 0A2, Canada.

Canadian Institute for Broadband and Information Network Technologies (CIBINT) Founded in 1989. *Academic areas of research and training:* Fiber optic network systems. *Degrees offered:* None. *Graduate students served last academic year:* 1: 1 at the master's level. *Faculty:* 4: 1 affiliated solely with the institute. *Faculty affiliations:* Faculties of Engineering, Graduate Studies and Research. *Annual research budget:* $200,000. *Director:* P. Van Vliet, President and CEO, 306-585-4381, Fax: 306-586-8202, E-mail: p.vanvliet@sk.sympatico.ca.

University of Rhode Island, Kingston, RI 02881.

Speech and Hearing Center Founded in 1969. *Academic areas of research and training:* Speech, language, and hearing services to children and adults of all age groups; specialty areas include child and adult language. *Degrees offered:* None. *Graduate students served last academic year:* 55: 55 at the master's level. *Faculty:* 8. *Faculty affiliations:* Department of Communicative Disorders. *Director:* Elizabeth C. Connors, Coordinator of Clinical Services, 401-874-5969, Fax: 401-874-4404, E-mail: billie@uri.edu.

University of Rochester, Rochester, NY 14627-0250.

Center for Superconducting Electronics Founded in 1992. *Academic areas of research and training:* Electrical engineering, digital circuits, optoelectronics, superconducting devices, quantum computing. *Degrees offered:* None. *Graduate students served last academic year:* 13: 11 at the doctoral level; 2 at the postgraduate level. *Faculty:* 5. *Faculty affiliations:* Department of Electrical and Computer Engineering. *Annual research budget:* $1 million. *Director:* Marc J. Feldman, Professor, 716-275-3799, Fax: 716-473-0486, E-mail: feldman@ece.rochester.edu.

The Laboratory for Laser Energetics *Academic areas of research and training:* Chemistry, electrical engineering, mechanical engineering, optics, physics. *Degrees offered:* None. *Graduate students served last academic year:* 124: 62 at the master's level; 60 at the doctoral level. *Faculty:* 31: 13 affiliated solely with the laboratory. *Faculty affiliations:* Departments of Chemistry, Electrical Engineering, Mechanical Engineering, Optics, Physics and Astronomy. *Annual research budget:* $38.6 million. *Director:* Prof. Robert L. McCrory, Director, 716-275-4973, Fax: 716-256-2586. *Application contact:* Jean M. Steve, Administrator, 716-275-5286, Fax: 716-256-2586, E-mail: jste@lle.rochester.edu.

Rochester Center for Biomedical Ultrasound Founded in 1986. *Academic areas of research and training:* Biomedical ultrasound, medical imaging, therapy, treatment. *Degrees offered:* None. *Graduate students served last academic year:* 20: 5 at the master's level; 10 at the doctoral level; 5 at the postgraduate level. *Faculty:* 70. *Faculty affiliations:* Departments of Anesthesiology, Biophysics, Biostatistics, Cardiology, Chemistry, Electrical Engineering, Environmental Health Sciences, Hematology, Mechanical Engineering, Obstetrics and Gynecology, Ophthalmology, Pediatrics, Radiology, Rehabilitation Medicine, Surgery, Urology. *Annual research budget:* $2 million. *Director:* Dr. Kevin J. Parker, Professor, 716-275-9542, Fax: 716-473-0486, E-mail: rcbu@ece.rochester.edu. *Application contact:* Pamela J. Clark, Business Manager, 716-275-9542, Fax: 716-473-0486, E-mail: pclark@ece.rochester.edu.

University of South Carolina, Columbia, SC 29208.

Center for Electrochemical Engineering (CEE) Founded in 1995. *Academic areas of research and training:* Electrochemical power sources such as fuel cells, batteries, capacitors. *Degrees offered:* None. *Graduate students served last academic year:* 25: 5 at the master's level; 18 at the doctoral level; 2 at the postgraduate level. *Faculty:* 16: 1 affiliated solely with the center. *Faculty affiliations:* Departments of Chemical Engineering, Chemistry and Biochemistry. *Annual research budget:* $2 million. *Director:* Dr. Ralph E. White, Director, 803-777-3270, Fax: 803-777-8265, E-mail: rew@sc.edu.

Center for Information Technology Founded in 1984. *Academic areas of research and training:* Information technology, enterprise information integration, distributed databases, multiagent systems. *Degrees offered:* None. *Graduate students served last academic year:* 12 at the master's level; 3 at the doctoral level; 1 at the postgraduate level. *Faculty:* 14: 4 affiliated solely with the center. *Faculty affiliations:* College of Business Administration, Department of Computer Science. *Annual research budget:* $300,000. *Director:* Dr. Michael N. Huhns, Director, 803-777-5921, Fax: 803-777-8045, E-mail: huhns@sc.edu.

Center for Mechanics of Materials and Nondestructive Evaluation Founded in 1982. *Academic areas of research and training:* Sensor development, non-contacting measurements, computer vision (2D and 3D), material characterization, fracture characterization, weldment modelling/experiments, weldmen and modelling, fiction stir welding, tank car modelling, damage tolerance assessment. *Degrees offered:* None. *Graduate students served last academic year:* 18: 5 at the master's level; 10 at the doctoral level; 3 at the postgraduate level. *Faculty:* 20: 10 affiliated solely with the center. *Faculty affiliations:* Departments of Chemical Engineering, Civil and Environmental Engineering, Mechanical Engineering. *Annual research budget:* $750,000. *Director:* Dr. Michael A. Sutton, Director, 803-777-7158, Fax: 803-777-0106, E-mail: sutton@sc.edu.

University of Southern California, Los Angeles, CA 90089.

Center for Composite Materials Founded in 1996. *Academic areas of research and training:* Composite materials, manufacturing processes, interface design, mechanical behavior. *Degrees offered:* None. *Graduate students served last academic year:* 15. *Faculty:* 10. *Faculty affiliations:* Departments of Aerospace Engineering, Chemical Engineering, Civil Engineering, Industrial and Systems Engineering, Materials Science and Engineering, Mechanical Engineering. *Annual research budget:* $1 million. *Director:* Dr. Steven R. Nutt, Director, 213-740-1634, Fax: 213-740-7744, E-mail: nutt@usc.edu.

Center for Electron Microscopy and Microanalysis (CEMMA) Founded in 1985. *Academic areas of research and training:* Materials analysis for physical and life sciences. *Degrees offered:* None. *Graduate students served last academic year:* 45: 10 at the master's level; 30 at the doctoral level; 5 at the postgraduate level. *Faculty:* 10. *Faculty affiliations:* Departments of Biological Sciences, Chemical Engineering, Chemistry, Electrical Engineering, Geology, Materials Science, Mechanical Engineering. *Annual research budget:* $100,000. *Director:* Dr. Steven R. Nutt, Director, 213-740-1634, Fax: 213-740-7744, E-mail: nutt@usc.edu. *Application contact:* Jack Worrall, Operations Director, 213-740-1990, E-mail: worrall@usc.edu.

Center for Telecommunications Management (CTM) Founded in 1985. *Academic areas of research and training:* Telecommunications policy, infrastructure of telecommunication industry, management and executive education in telecommunications. *Degrees offered:* None. *Graduate students served last academic year:* 15: 6 at the master's level; 7 at the doctoral level; 2 at the postgraduate level. *Faculty:* 23: 3 affiliated solely with the center. *Faculty affiliations:* Annenberg School for Communication, Graduate School of Business, School of Engineering. *Annual research budget:* $425,000. *Director:* Dr. Jack R. Borsting, Executive Director, 213-740-0980, Fax: 213-740-1602, E-mail: borsting@bus.usc.edu.

The University of Tampa, Tampa, FL 33606-1490.

Center for Quality Founded in 1991. *Academic areas of research and training:* General quality management, process re-engineering, quality measurement. *Degrees offered:* None. *Graduate students served last academic year:* 140: 100 at the master's level. *Faculty:* 1: 1 affiliated solely with the center. *Director:* Dr. Al C. Endres, Director, 813-258-7420, Fax: 813-258-7424, E-mail: acehm@aol.com.

University of Tennessee, Knoxville, Knoxville, TN 37996.

Measurement and Control Engineering Center Founded in 1985. *Academic areas of research and training:* Process control, analytical instruments, measurement. *Degrees offered:* None. *Graduate students served last academic year:* 26: 16 at the master's level; 8 at the doctoral level; 2 at the postgraduate level. *Faculty:* 11: 1 affiliated solely with the center. *Faculty affiliations:* Departments of Chemical Engineering, Chemistry, Electrical Engineering. *Annual research budget:* $350,000. *Director:* Dr. Arlene Garrison, Director, 423-974-2375, Fax: 423-974-4995, E-mail: agarrison@utk.edu.

Waste Management Research and Education Institute Founded in 1985. *Academic areas of research and training:* Waste management policy, clean technology, environmental biotechnology. *Degrees offered:* None. *Graduate students served last academic year:* 80. *Faculty:* 25. *Faculty affiliations:* Colleges of Education, Veterinary Medicine; Departments of Animal Science, Chemical Engineering, Civil and Environmental Engineering, Economics, Microbiology, Nuclear Engineering, Plant and Soil Science; Programs in Management Science, Rural Economy and Society; Center for Environmental Biotechnology; Energy, Environment, and Resources Center. *Annual research budget:* $3.5 million. *Director:* Dr. Gary Sayler, Director, 423-974-8080, Fax: 423-974-8027, E-mail: sayler@utk.edu. *Application contact:* Kimberly Davis, Assistant Director, 423-974-4251, Fax: 423-974-1838, E-mail: kdavis17@utk.edu.

University of Tennessee Space Institute, Tullahoma, TN 37388-9700.

Center for Laser Applications Founded in 1984. *Academic areas of research and training:* Materials and manufacturing processes. *Degrees offered:* None. *Graduate students served last academic year:* 26: 11 at the master's level; 15 at the doctoral level. *Faculty:* 18: 9 affiliated solely with the center. *Faculty affiliations:* Departments of Mechanical, Aerospace and Engineering Science; Physics. *Director:* Dr. Dennis R. Keefer, Chairman, 931-393-7475, Fax: 931-454-2271, E-mail: dkeefer@utsi.edu. *Application contact:* Carole A. Thomas, Administrative Assistant, 931-393-7485, Fax: 931-454-2271, E-mail: cthomas@utsi.edu.

The University of Texas at Arlington, Arlington, TX 76019.

Automation and Robotics Research Institute (ARRI) Founded in 1988. *Academic areas of research and training:* Manufacturing, new process development, automating advanced controls, reconfigurable systems, information systems, enterprise engineering. *Degrees offered:* None. *Graduate students served last academic year:* 60: 20 at the master's level; 6 at the doctoral level; 3 at the postgraduate level. *Faculty:* 26: 13 affiliated solely with the institute. *Faculty affiliations:* Departments of Computer Science and Engineering, Electrical Engineering; Programs in Industrial and Manufacturing Systems Engineering, Mechanical Engineering. *Annual research budget:* $6 million. *Director:* Dr. John J. Mills, Director, 817-272-5903, Fax: 817-272-5952, E-mail: jmills@arri.uta.edu.

Center for Advanced Polymer Research Founded in 1988. *Academic areas of research and training:* Polymer research, conductive polymers, electroluminescent polymers, plasma polymerization, anti-corrosion, light emission. *Degrees offered:* None. *Graduate students served last academic year:* 10: 8 at the doctoral level; 2 at the postgraduate level. *Faculty:* 6. *Faculty affiliations:* Departments of Chemistry and Biochemistry, Materials Science and Engineering. *Annual research budget:* $200,000. *Director:* Dr. Martin Pomerantz, Director, 817-272-3811, Fax: 817-273-3808, E-mail: pomerantz@uta.edu.

Human Performance Institute Founded in 1986. *Academic areas of research and training:* Modeling and measurement of human performance; sensory, motor, and integrated subsystems; prediction and determination of limiting performance resources; performance theory. *Degrees offered:* None. *Graduate students served last academic year:* 13: 9 at the master's level; 4 at the doctoral level. *Faculty:* 5. *Faculty affiliations:* Departments of Biology, Biomedical Engineering, Electrical Engineering; Departments of Music, Physical Therapy at affiliated institutions. *Annual research budget:* $250,000. *Director:* Dr. George V. Kondraske, Director, 817-273-2335, Fax: 817-273-2253, E-mail: kondraske@uta.edu.

Software Engineering Center for Telecommunications Founded in 1987. *Academic areas of research and training:* Software engineering, telecom applications, object-oriented testing, scenario-based prototyping, conceptual modeling, concurrent software engineering. *Degrees offered:* None. *Graduate students served last academic year:* 25: 20 at the master's level; 5 at the doctoral level. *Faculty:* 8: 4 affiliated solely with the center. *Faculty affiliations:* Department of Computer Science and Engineering. *Annual research budget:* $250,000. *Director:* Dr. Pei Hsia, Director, 817-272-3785, Fax: 817-272-3784, E-mail: hsia@cse.uta.edu.

The University of Texas at Austin, Austin, TX 78712-1111.

Artificial Intelligence Laboratory Founded in 1984. *Academic areas of research and training:* Automatic programming, automatic theorem proving, robotics, machine learning, neural networks, logical foundations, natural language processing. *Degrees offered:* None. *Graduate students served last academic year:* 30: 9 at the master's level; 20 at the doctoral level; 1 at the postgraduate level. *Faculty:* 9. *Faculty affiliations:* Department of Computer Sciences. *Annual research budget:* $200,000. *Director:* Dr. Gordon S. Novak, Director, 512-471-9569, Fax: 512-471-8885, E-mail: novak@cs.utexas.edu.

Bureau of Economic Geology Founded in 1909. *Academic areas of research and training:* Energy resources, oil and gas recovery, environmental geology, water resources, coastal processes. *Degrees offered:* None. *Graduate students served last academic year:* 50: 30 at the master's level; 15 at the doctoral level; 5 at the postgraduate level. *Faculty:* 5. *Faculty affiliations:* Departments of Geological Sciences, Petroleum and Geosystems Engineering. *Director:* Dr. William L. Fisher, Interim Director, 512-471-1534, Fax: 512-471-0140, E-mail: wfisher@mail.utexas.edu.

Center for Electromechanics Founded in 1974. *Academic areas of research and training:* Pulsed rotating generators for military and commercial applications, high rate industrial processes, electromagnetic accelerators (railguns) and their associated launch packages, flywheels, composite materials, power electronics, component technologies. *Degrees offered:* None. *Graduate students served last academic year:* 18: 17 at the master's level; 1 at the doctoral level. *Faculty:* 6: 3 affiliated solely with the center. *Faculty affiliations:* Departments of Electrical and Computer Engineering, Mechanical Engineering. *Director:* Dr. Robert H. Hebner, Director, 512-471-4496, Fax: 512-471-0781, E-mail: r.hebner@mail.utexas.edu. *Application contact:* Alan Walls, Deputy Director, 512-232-1643, Fax: 512-471-0781, E-mail: a.walls@mail.utexas.edu.

Center for Space Research Founded in 1981. *Academic areas of research and training:* Orbit determination, space geodesy, earth dynamics, environmental monitoring, planetary exploration, satellite remote sensing, laser ranging, altimetry and methods of distributed computing, earth systems science, GRACE, Icesat. *Degrees offered:* None. *Graduate students served last academic year:* 64: 10 at the master's level; 49 at the doctoral level; 5 at the postgraduate level. *Faculty:* 16: 6 affiliated solely with the center. *Faculty affiliations:* Departments of Aerospace Engineering and Engineering Mechanics, Computer Sciences, Geological Sciences, Mechanical Engineering. *Annual research budget:* $5.8 million. *Director:* Dr. Byron D. Tapley, Director, 512-471-5573, Fax: 512-232-2443, E-mail: tapley@csr.utexas.edu. *Application contact:* James M. Casey, Program Manager, 512-471-7370, Fax: 512-232-2443, E-mail: casey@csr.utexas.edu.

J. J. Pickle Research Campus Founded in 1963. *Academic areas of research and training:* Engineering science, social sciences. *Degrees offered:* None. *Graduate students served last academic year:* 550: 200 at the master's level; 300 at the doctoral level; 50 at the postgraduate level. *Faculty:* 200. *Faculty affiliations:* Colleges of Arts and Sciences, Business Administration, Engineering. *Director:* Dr. Juan M. Sanchez, Vice President for Research, 512-471-2877, Fax: 512-471-2827, E-mail: vp-research@mail.utexas.edu.

Offshore Technology Research Center Founded in 1988. *Academic areas of research and training:* Deepwater offshore structures, fluid structure interaction, composites in sea environment, characterization of ocean bottom. *Degrees offered:* None. *Graduate students served last academic year:* 23: 2 at the master's level; 20 at the doctoral level; 1 at the postgraduate level. *Faculty:* 22: 11 affiliated solely with the center. *Faculty affiliations:* Departments of Aerospace Engineering and Engineering Mechanics; Civil, Electrical and Computer, Mechanical Engineering. *Annual research budget:* $1.2 million. *Director:* Dr. John Tassoulas, Associate Director, 512-471-3753, Fax: 512-471-8477, E-mail: yannis@mail.utexas.edu. *Application contact:* Lorraine E. Sanchez, Senior Administrative Associate, 512-471-3753, Fax: 512-471-8477, E-mail: lorrain@mail.utexas.edu.

Texas Materials Institute Founded in 1998. *Academic areas of research and training:* Fracture, friction, and wear; high-temperature materials processing and mechanical behavior; micromechanical behavior of interfaces in electronic packaging systems; composite materials; catalysis; thin-film semiconductor materials; mesoscopic materials; picosecond phase transformations; structure-electronic property relationships of inorganic compounds; rheological properties of polymers and gels; engineering properties of polymers; magnetic properties of polymers and layered nanostructures; corrosion, stress corrosion, and corrosion fatigue of structural materials; computer modeling and prediction of phase equilibria; solid electrolytes and insertion compounds for batteries; fuel cells, sensors, and capacitors; solutions chemistry and sol gel. *Degrees offered:* None. *Graduate students served last academic year:* 59: 22 at the master's level; 31 at the doctoral level; 6 at the postgraduate level. *Faculty:* 40. *Faculty affiliations:* Departments of Aerospace Engineering and Engineering Mechanics; Chemical, Electrical and Computer, and Mechanical Engineering; Chemistry; Physics. *Annual research budget:* $919,636. *Director:* Dr. Donald R. Paul, Director, 512-475-8293, Fax: 512-475-8482, E-mail: tmi@mail.utexas.edu. *Application contact:* Lydia C. Griffith, Graduate Coordinator, 512-471-1504, Fax: 512-475-8482, E-mail: lcgriffith@mail.utexas.edu.

The University of Texas Health Science Center at San Antonio, San Antonio, TX 78284-6200.

The Center for Environmental Radiation Toxicology (CERT) Founded in 1993. *Academic areas of research and training:* Biological and health effects of ionizing and non-ionizing radiation, including microwaves, ultraviolet light, laser emissions, and ELF fields. *Degrees offered:* None. *Graduate students served last academic year:* 4: 3 at the master's level; 1 at the doctoral level. *Faculty:* 55. *Faculty affiliations:* Trinity University, University of Texas at San Antonio, U.S. Air Force Research Laboratory (Brooks Air Force Base, TX), Southwest Research Institute, Southwest Foundation for Biomedical Research. *Director:* Dr. Martin L. Meltz, Director, 210-567-5560, Fax: 210-567-3446, E-mail: meltz@uthscsa.edu.

University of Toledo, Toledo, OH 43606-3398.

The Center for Materials Science and Engineering (CMSE) Founded in 1997. *Academic areas of research and training:* Thin-film coatings, electronic materials, photonic materials, tribological materials, materials characterization, photovoltaics. *Degrees offered:* None. *Graduate students served last academic year:* 19: 10 at the master's level; 5 at the doctoral level; 4 at the postgraduate level. *Faculty:* 35. *Faculty affiliations:* Departments of Bioengineering; Chemical and Environmental Engineering; Chemistry; Electrical and Computer Engineering; Geology; Mechanical, Industrial, and Manufacturing Engineering; Physics and Astronomy. *Annual research budget:* $25,000. *Director:* Dr. Alvin Compaan, Director, 419-530-4787, Fax: 419-530-2723, E-mail: adc@physics.utoledo.edu.

University of Utah, Salt Lake City, UT 84112-1107.

Center for High Performance Computing *Academic areas of research and training:* Computer resources for high performance computing. *Degrees offered:* None. *Graduate students served last academic year:* 100: 60 at the doctoral level; 40 at the postgraduate level. *Faculty:* 31: 1 affiliated solely with the center. *Annual research budget:* $3 million. *Director:* Dr. Julio C. Facelli, Director, 801-581-5253, E-mail: facelli@chpc.utah.edu.

Nora Eccles Harrison Cardiovascular Research and Training Institute (CVRTI) Founded in 1967. *Academic areas of research and training:* Biochemistry, bioengineering, cardiology, physiology. *Degrees offered:* None. *Graduate students served last academic year:* 10: 2 at the master's level; 3 at the doctoral level; 5 at the postgraduate level. *Faculty:* 15: 15 affiliated solely with the institute. *Faculty affiliations:* Departments of Biochemistry, Bioengineering, Pharmacology, and Physiology; School of Medicine. *Annual research budget:* $2 million. *Director:* Director, 801-581-8183, Fax: 801-581-3128.

Rocky Mountain Center for Occupational and Environmental Health Founded in 1978. *Academic areas of research and training:* Ergonomics/industrial safety, industrial hygiene, occupational health nursing, occupational medicine. *Degrees offered:* None. *Graduate students served last academic year:* 24: 18 at the master's level; 6 at the doctoral level. *Faculty:* 39: 9 affiliated solely with the center. *Faculty affiliations:* Department of Family and Preventive Medicine. *Annual research budget:* $300,000. *Director:* Dr. Royce Moser, Director, 801-581-8719, Fax: 801-581-7224, E-mail: rmoser@rmcoeh.utah.edu.

Utah Engineering Experiment Station (UEES) Founded in 1909. *Academic areas of research and training:* Energy resources, mining, mineral processing, metallurgy, metals extraction, geology, geophysics, exploration-metal energy, coal cleaning, industrial construction management, industrial water treatment, acid neck drainage, wetlands and riparian reclamation, natural resources. *Degrees offered:* None. *Graduate students served last academic year:* 12: 4 at the master's level; 6 at the doctoral level; 2 at the postgraduate level. *Faculty:* 58: 2 affiliated solely with the station. *Faculty affiliations:* Colleges of Engineering, Mines and Earth Sciences. *Annual research budget:* $500,000. *Director:* Dr. Terrence D. Chatwin, Director, 801-581-6348, Fax: 801-581-5440, E-mail: terrence.chatwin@m.cc.utah.edu.

University of Virginia, Charlottesville, VA 22903.

Applied Electrophysics Laboratories (AEpL) Founded in 1967. *Academic areas of research and training:* Applied electromagnetics, circuits and fabrication technology, optics and quantum electronics, novel semiconductor and superconductor materials, devices and circuits, microelectronics education. *Degrees offered:* None. *Graduate students served last academic year:* 43: 18 at the master's level; 20 at the doctoral level; 5 at the postgraduate level. *Faculty:* 10. *Faculty affiliations:* Departments of Electrical Engineering, Materials Science and Engineering, Physics. *Annual research budget:* $2 million. *Director:* Dr. Thomas W. Crowe, Director, 804-924-7693, Fax: 804-924-8818, E-mail: twc8u@virginia.edu.

Center for Rotating Machinery and Controls (ROMAC) Founded in 1981. *Academic areas of research and training:* Analysis and design of turbomachinery, magnetic bearings, hydrodynamic bearings, machine vibrations. *Degrees offered:* None. *Graduate students served last academic year:* 30: 16 at the master's level; 14 at the doctoral level. *Faculty:* 12: 2 affiliated solely with the center. *Faculty affiliations:* Departments of Electrical Engineering, Mechanical and Aerospace Engineering. *Annual research budget:* $1.5 million. *Director:* Hossein Haj-Hariri, Director, 804-924-6290, Fax: 804-982-2037. *Application contact:* Tana Herndon, Office Manager/Conference Coordinator, 804-924-3292, Fax: 804-982-2246.

Center for Transportation Studies Founded in 1975. *Academic areas of research and training:* Civil engineering, planning, systems engineering, intelligent transportation technology, simulation and operations. *Degrees offered:* None. *Graduate students served last academic year:* 38: 9 at the master's level; 3 at the doctoral level; 17 at the postgraduate level. *Faculty:* 19: 4 affiliated solely with the center. *Faculty affiliations:* Departments of City and Environmental Planning; Electrical Engineering; Environmental Sciences; Mechanical, Aerospace, and Nuclear Engineering; Systems Engineering; Program in Commerce. *Annual research budget:* $1.5 million. *Director:* Dr. Michael J. Demetsky, Director, 804-924-6362, Fax: 804-982-2951, E-mail: mjd@virginia.edu. *Application contact:* Cindy Sites, Administrator, 804-924-4775, Fax: 804-982-2951, E-mail: css5b@virginia.edu.

Light Metals Center Founded in 1984. *Academic areas of research and training:* Development of light metal alloys, optimizing microstructures, interactions with industry, stimulate technical interaction between aerospace and automotive industries. *Degrees offered:* None. *Graduate students served last academic year:* 12: 2 at the master's level; 9 at the doctoral level; 1 at the postgraduate level. *Faculty:* 8: 4 affiliated solely with the center. *Faculty affiliations:* The Center for Electrochemical Science and Engineering. *Annual research budget:* $500,000. *Director:* Dr. Edgar A. Starke, Director, 804-924-6335, Fax: 804-924-1353, E-mail: eas1o@virginia.

University of Washington, Seattle, WA 98195.

Center for Videoendoscopic Surgery (CVES) Founded in 1993. *Academic areas of research and training:* Research and training in tools and techniques of videoendoscopic surgery. *Degrees offered:* None. *Graduate students served last academic year:* 5: 1 at the master's level; 4 at the postgraduate level. *Faculty:* 10. *Faculty affiliations:* Departments of Electrical

Research and Training Opportunities

Center for Videoendoscopic Surgery (CVES) (continued)
Engineering, Medical Education, Surgery; Human Interface Technology Lab. *Director:* Dr. Carlos A. Pellegrini, Co-Director, 206-543-3106, Fax: 206-543-8136.

University of West Florida, Pensacola, FL 32514-5750.

Center for Environmental Diagnostics and Bioremediation (CEDB) Founded in 1990. *Academic areas of research and training:* Bioremediation technologies for the safe degradation of hazardous and toxic wastes, bioindicators for assessing the health of the ecosystem and the degree of environmental pollution. *Degrees offered:* None. *Graduate students served last academic year:* 10: 9 at the master's level; 1 at the doctoral level. *Faculty:* 14: 10 affiliated solely with the center. *Faculty affiliations:* Department of Biology, Institute for Coastal and Estuarine Research (ICER). *Annual research budget:* $1.3 million. *Director:* Dr. K. Ranga Rao, Director, 850-474-2060, Fax: 850-474-3130, E-mail: rrao@uwf.edu. *Application contact:* Dr. Malcolm Shields, Assistant Director, 850-474-2060, Fax: 850-474-3130, E-mail: mshields@uwf.edu.

University of Wisconsin–Madison, Madison, WI 53706-1380.

Biotechnology Training Program Founded in 1989. *Degrees offered:* None. *Faculty affiliations:* Departments of Agronomy, Animal Health and Biomedical Sciences, Bacteriology, Biochemistry, Biomolecular Chemistry, Botany, Chemical Engineering, Chemistry, Computer Science, Dairy Science, Electrical Engineering, Food Science, Genetics, Horticulture, Mathematics, Medical Microbiology and Immunology, Oncology, Pathobiological Sciences, Pathology, Pediatrics, Pharmacology, Pharmacy, Plant Pathology, Soil Science, Zoology; Institute for Molecular Virology; Programs in Biophysics, Cell and Molecular Biology, Comparative Biosciences, Environmental Toxicology, Food Microbiology and Toxicology. *Director:* Dr. Malcolm Shields, Assistant Director, 850-474-2060, Fax: 850-474-3130, E-mail: mshields@uwf.edu. *Application contact:* Program Coordinator, 608-262-6753.

See in-depth description on page 1549.

Environmental Remote Sensing Center (ERSC) Founded in 1970. *Academic areas of research and training:* Environmental remote sensing, geographic information systems. *Degrees offered:* MS; PhD. *Graduate students served last academic year:* 37: 25 at the master's level; 10 at the doctoral level; 2 at the postgraduate level. *Faculty:* 13: 3 affiliated solely with the center. *Faculty affiliations:* Departments of Biological Systems Engineering, Civil and Environmental Engineering, Forestry, Geography, Soil Science; Institute for Environmental Studies; School of Business. *Annual research budget:* $500,000. *Director:* Dr. Thomas M. Lillesand, Director, 608-263-3251, Fax: 608-262-5964, E-mail: tmlilles@facstaff.wisc.edu.

Synchrotron Radiation Center (SRC) Founded in 1968. *Academic areas of research and training:* Electron storage ring provides high intensity photons from the X-ray region to the infrared for a variety of materials research problems and for microcircuit. *Degrees offered:* None. *Graduate students served last academic year:* 229: 125 at the master's level; 52 at the doctoral level; 52 at the postgraduate level. *Faculty:* 62. *Faculty affiliations:* Departments of Chemical Engineering, Chemistry, Electrical and Computer Engineering, Materials Science, Physics. *Annual research budget:* $3.8 million. *Director:* Dr. James W. Taylor, Executive Director, 608-877-2152, Fax: 608-877-2001, E-mail: jtaylor@src.wisc.edu. *Application contact:* Pamela D. Layton, Program Assistant 4, 608-877-2134, Fax: 608-877-2001, E-mail: playton@src.wisc.edu.

Wisconsin Center for Space Automation and Robotics (WCSAR) Founded in 1986. *Academic areas of research and training:* Genetic engineered plant material development, elite seed development; space flight payload design, system automation, and central system design. *Degrees offered:* MS; PhD. *Graduate students served last academic year:* 4: 4 at the master's level. *Faculty:* 10: 5 affiliated solely with the center. *Faculty affiliations:* Departments of Computer Science, Electrical and Computer Engineering, Genetics, Materials Engineering, Plant Physiology. *Annual research budget:* $2 million. *Director:* Dr. Weijia Zhou, Director, 608-262-5526, Fax: 608-262-9458, E-mail: wzhou@facstaff.wisc.edu.

University of Wisconsin–Milwaukee, Milwaukee, WI 53201-0413.

Center for By-Products Utilization (UWM-CBU) Founded in 1988. *Academic areas of research and training:* Recycling, construction materials, environmental issues. *Degrees offered:* None. *Graduate students served last academic year:* 10: 8 at the master's level; 2 at the doctoral level. *Faculty:* 21: 1 affiliated solely with the center. *Faculty affiliations:* Departments of Biological Sciences, Chemistry, Engineering, Geosciences, Physics; School of Business Administration. *Director:* Dr. Tarun R. Naik, Director, 414-229-6696, Fax: 414-229-6958, E-mail: tarun@uwm.edu. *Application contact:* Rudolph N. Kraus, Assistant Director, 414-229-4105, Fax: 414-229-6958, E-mail: rudik@uwm.edu.

University of Wyoming, Laramie, WY 82071.

Institute for Energy Research (IER) Founded in 1993. *Academic areas of research and training:* Geology, geophysics, mathematics, petroleum engineering. *Degrees offered:* None. *Graduate students served last academic year:* 14: 7 at the master's level; 7 at the doctoral level. *Faculty:* 2: 2 affiliated solely with the institute. *Faculty affiliations:* Departments of Chemical and Petroleum Engineering, Geology and Geophysics. *Annual research budget:* $3 million. *Director:* Dr. Ronald J. Steel, Interim Director, 307-766-4200, Fax: 307-766-2737, E-mail: rsteel@uwyo.edu. *Application contact:* Dr. Henry P. Heasler, Associate Director, 307-766-4200, Fax: 307-766-2737, E-mail: heasler@uwyo.edu.

Virginia Polytechnic Institute and State University, Blacksburg, VA 24061.

Antenna Group Founded in 1976. *Academic areas of research and training:* Design, construction, and test of antennas for wireless communications. *Degrees offered:* None. *Graduate students served last academic year:* 12: 4 at the master's level; 8 at the doctoral level. *Faculty:* 4: 4 affiliated solely with the group. *Annual research budget:* $500,000. *Director:* Prof. Warren L. Stutzman, Director, 540-231-6834, Fax: 540-231-3355, E-mail: stutzman@vt.edu.

Center for Intelligent Material Systems and Structures Founded in 1987. *Academic areas of research and training:* Vibration and control using smart materials and structures, development of new actuators. *Degrees offered:* None. *Graduate students served last academic year:* 26: 7 at the master's level; 15 at the doctoral level; 4 at the postgraduate level. *Faculty:* 14: 4 affiliated solely with the center. *Faculty affiliations:* Departments of Aerospace Engineering, Civil Engineering, Electrical Engineering, Engineering Science and Mechanics, Mechanical Engineering. *Annual research budget:* $1 million. *Director:* Prof. Daniel J. Inman, Director, 540-231-4704, Fax: 540-231-2903, E-mail: dinman@vt.edu. *Application contact:* Beth Howell, Program Manager, 540-231-2900, Fax: 540-231-2903, E-mail: bethrun@vt.edu.

Center for Power Electronics Systems (CPES) Founded in 1983. *Academic areas of research and training:* Energy efficiency, power conversion, electric propulsion, hybrid electric vehicles. *Degrees offered:* None. *Graduate students served last academic year:* 70: 40 at the master's level; 30 at the doctoral level. *Faculty:* 13: 7 affiliated solely with the center. *Faculty affiliations:* Departments of Electrical Engineering, Materials Science and Engineering, Mechanical Engineering. *Annual research budget:* $3 million. *Director:* Fred C. Lee, Director, 540-231-7716, Fax: 540-231-6390, E-mail: fclee@vt.edu.

Center for Wireless Telecommunications (CWT) Founded in 1993. *Academic areas of research and training:* Radio frequency (RF) systems and components, antennas, satellite communications, wireless networks, and business and regulatory issues affecting wireless telecommunications; spectrum allocation; embedded radio systems; advanced broadband wireless Internet access; radio spectrum auction and allocation strategies. *Degrees offered:* None. *Graduate students served last academic year:* 50. *Faculty:* 20: 3 affiliated solely with the center. *Faculty affiliations:* Departments of Computer Science; Economics; Electrical and Computer Engineering; Finance, Insurance, and Business Law; Management Science; Marketing; Geography. *Annual research budget:* $1 million. *Director:* Dr. Charles W. Bostian, Director, 540-231-5906, Fax: 540-231-3004, E-mail: bostian@vt.edu. *Application contact:* Melanie N. George, Manager of Technical Communications, 540-231-8037, Fax: 540-231-3004, E-mail: megeorge@vt.edu.

Mobile and Portable Radio Research Group (MPRG) Founded in 1991. *Academic areas of research and training:* Wireless communications. *Degrees offered:* None. *Graduate students served last academic year:* 8 at the master's level; 17 at the doctoral level; 3 at the postgraduate level. *Faculty:* 10: 5 affiliated solely with the group. *Faculty affiliations:* Program in Electrical and Computer Engineering. *Annual research budget:* $2 million. *Director:* Dr. Brian D. Woerner, Director, 540-231-2963, Fax: 540-231-2968, E-mail: woerner@vt.edu.

Virginia State University, Petersburg, VA 23806-0001.

Research Center on Magnetic Physics Founded in 1988. *Academic areas of research and training:* Physics, materials science, superconductivity, magnetism, nanostructured materials. *Degrees offered:* None. *Graduate students served last academic year:* 6: 6 at the master's level. *Faculty:* 3: 2 affiliated solely with the center. *Faculty affiliations:* Department of Physics. *Annual research budget:* $313,000. *Director:* Dr. Carey E. Stronach, Director, 804-524-5915, Fax: 804-524-5914, E-mail: cstronac@vsu.edu.

Washington State University, Pullman, WA 99164.

Electron Microscopy Center (EMC) Founded in 1964. *Academic areas of research and training:* Light microscopy; special projects in electron microscopy; special topics in electron microscopy, including SEM, TEM, confocal, x-ray, microanalysis, image analysis; research in biological sciences, food science, agricultural science, materials science, geology, biotechnology. *Degrees offered:* None. *Graduate students served last academic year:* 125: 50 at the master's level; 55 at the doctoral level; 20 at the postgraduate level. *Faculty:* 3: 1 affiliated solely with the center. *Faculty affiliations:* Departments of Botany, Zoology. *Director:* Dr. Vincent R. Franceschi, Director, 509-335-3025, Fax: 509-335-3517, E-mail: vfrances@mail.wsu.edu.

Wood Materials and Engineering Laboratory Founded in 1949. *Academic areas of research and training:* Integrated engineering approach to solving wood construction and industrial product problems; composite materials and nondestructive evaluation of wood materials and structures. *Degrees offered:* None. *Graduate students served last academic year:* 17: 10 at the master's level; 5 at the doctoral level; 2 at the postgraduate level. *Faculty:* 18: 5 affiliated solely with the laboratory. *Faculty affiliations:* Departments of Civil and Environmental Engineering, Natural Resource Sciences; Schools of Architecture, Mechanical and Materials Engineering. *Annual research budget:* $2 million. *Director:* Dr. Donald A. Bender, Director, 509-335-2829, Fax: 509-335-5077, E-mail: bender@wsu.edu.

Washington University in St. Louis, St. Louis, MO 63130-4899.

Applied Research Laboratory (ARL) Founded in 1989. *Academic areas of research and training:* Computer networking. *Degrees offered:* MS; PhD. *Graduate students served last academic year:* 15: 9 at the master's level; 6 at the doctoral level. *Faculty:* 12: 3 affiliated solely with the laboratory. *Faculty affiliations:* Departments of Computer Science, Electrical Engineering; Program in Biomedical Computing; Defense Advanced Research Projects Agency; NIF; NSF. *Annual research budget:* $2.3 million. *Director:* Dr. Jonathan S. Turner, Director, 314-935-8552, Fax: 314-935-7302, E-mail: jst@arl.wustl.edu. *Application contact:* Diana Ehrlich, Technical Assistant, 314-935-7534, Fax: 314-935-7302, E-mail: diana@arl.wustl.edu.

Center for Computational Mechanics Founded in 1973. *Academic areas of research and training:* Structural modeling and computation. *Degrees offered:* None. *Graduate students served last academic year:* 12: 7 at the master's level; 3 at the doctoral level; 2 at the postgraduate level. *Faculty:* 8: 2 affiliated solely with the center. *Faculty affiliations:* Departments of Mechanical Engineering, Systems Science and Mathematics. *Annual research budget:* $500,000. *Director:* David A. Peters, Director, 314-935-4337, Fax: 314-935-4014, E-mail: dap@mecf.wustl.edu. *Application contact:* Center Office, E-mail: dap@mecf.wustl.edu.

Center for Optimization and Semantic Control Founded in 1991. *Academic areas of research and training:* Optimization of large scale systems in transportation, health care, and business. *Degrees offered:* None. *Graduate students served last academic year:* 9: 9 at the doctoral level. *Faculty:* 13: 2 affiliated solely with the center. *Faculty affiliations:* Departments of Computer Science, Physics, Systems Science and Mathematics; John M. Olin School of Business; Program in Health Care. *Annual research budget:* $300,000. *Director:* Dr. Ervin Y. Rodin, Director, 314-935-6007, Fax: 314-935-6121, E-mail: rodin@rodin.wustl.edu.

Computer and Communications Research Center (CCRC) Founded in 1980. *Academic areas of research and training:* Telecommunications, computer architecture, parallel processing, systems performance evaluations. *Degrees offered:* None. *Graduate students served last academic year:* 14: 7 at the master's level; 7 at the doctoral level. *Faculty:* 6. *Faculty affiliations:* Departments of Computer Science, Electrical Engineering. *Annual research budget:* $600,000. *Director:* Dr. Mark Franklin, Director, 314-935-6107, Fax: 314-935-7302, E-mail: jbf@wuccrc.wustl.edu. *Application contact:* Dr. Mark Franklin, Director, 314-935-6107, Fax: 314-935-7302, E-mail: jbf@wuccrc.wustl.edu.

Computer Visualization Laboratory (CVL) Founded in 1990. *Academic areas of research and training:* Visualization, design methods, real-time graphics, computational geometry. *Degrees offered:* None. *Graduate students served last academic year:* 8: 7 at the doctoral level; 1 at the postgraduate level. *Faculty:* 2. *Faculty affiliations:* Department of Computer Science. *Director:* Dr. Catalin Roman, Director, 314-935-6132, Fax: 314-935-7302, E-mail: roman@cs.wustl.edu.

Electronic Signals and Systems Research Laboratory (ESSRL) Founded in 1986. *Academic areas of research and training:* Image processing, signal processing, computer vision, dynamic vision, automatic target recognition, information theory, ultrasound imaging, hyperspectral imaging. *Degrees offered:* None. *Graduate students served last academic year:* 22: 9 at the master's level; 13 at the doctoral level. *Faculty:* 12: 7 affiliated solely with the laboratory. *Faculty affiliations:* Departments of Electrical Engineering, Systems Science and Mathematics; Program in Radiology. *Annual research budget:* $750,000. *Director:* Dr. Joseph A. O'Sullivan, Director, 314-935-4173, Fax: 314-935-7500, E-mail: jao@essrl.wustl.edu.

Wayne State University, Detroit, MI 48202.

Center for Automotive Research Founded in 1980. *Academic areas of research and training:* Combustion engines, emission controls, diagnostics, friction and wear, mathematical modeling and computer simulation. *Degrees offered:* None. *Graduate students served last academic year:* 17: 5 at the master's level; 10 at the doctoral level; 2 at the postgraduate level. *Faculty:* 20: 10 affiliated solely with the center. *Faculty affiliations:* Departments of Chemistry, Mechanical Engineering, Physics and Astronomy; Programs in Chemical Engineering, Electrical Engineering. *Annual research budget:* $800,000. *Director:* Dr. Naeim A. Henein, Director, 313-577-3887, Fax: 313-577-8789, E-mail: henein@eng.wayne.edu.

Western Michigan University, Kalamazoo, MI 49008.

Digital Signal Processing Laboratory Founded in 1989. *Academic areas of research and training:* Adaptive differential pulse code modulation, alpha brain waves filter implementation, digital processing of EMG signals, DSP schemes for fiber optic rotation sensor scale factor linearization, digital filter design package. *Degrees offered:* None. *Graduate students served last academic year:* 3: 3 at the master's level. *Faculty:* 1: 1 affiliated solely with the laboratory. *Faculty affiliations:* Department of Electrical and Computer Engineering. *Annual research* *budget:* $100,000. *Director:* Dr. S. Hossein Mousavinezhad, Director, 616-387-4057, Fax: 616-387-4096, E-mail: h.mousavinezhad@wmich.edu.

Plastics Processing Capabilities Laboratory   Founded in 1989. *Academic areas of research and training:* Plastics processing, polymer testing. *Degrees offered:* None. *Graduate students served last academic year:* 2: 2 at the master's level. *Faculty:* 1. *Faculty affiliations:* Department of Industrial and Manufacturing Engineering. *Annual research budget:* $100,000.*Director:* Dr. Paul V. Engelmann, Director, 616-387-6527, Fax: 616-387-4075.

West Texas A&M University, Canyon, TX 79016-0001.

Alternative Energy Institute (AEI)   Founded in 1977. *Academic areas of research and training:* Solar energy, wind energy and turbines, wind turbine design. *Degrees offered:* None. *Graduate students served last academic year:* 4: 4 at the master's level. *Faculty:* 2. *Faculty affiliations:* Program in Engineering Technology. *Annual research budget:* $90,000. *Director:* Dr. Vaughn Nelson, Director, 806-651-2295, Fax: 806-651-2733, E-mail: aeimail@mail.wtamu.

edu. *Application contact:* Kenneth Starcher, Assistant Director, 806-651-2295, Fax: 806-651-2733, E-mail: aeimail@mail.wtamu.edu.

West Virginia University, Morgantown, WV 26506.

Constructed Facilities Center (CFC)   Founded in 1988. *Academic areas of research and training:* Advanced composite material technology and nondestructive evaluations into current construction practice; composite and hybridmaterials; nondestructive evaluation; construction of bridges, buildings, mineshafts, pipe lines. *Degrees offered:* None.   *Graduate students served last academic year:* 40: 30 at the master's level; 10 at the doctoral level. *Faculty:* 20: 10 affiliated solely with the center. *Faculty affiliations:* Departments of Chemical, Civil and Environmental, Electrical Engineering; Programs in Industrial Engineering, Mechanical Engineering. *Annual research budget:* $2.5 million. *Director:* Dr. Hota V. S. GangaRao, Director, 304-293-7608 Ext. 2634, Fax: 304-293-7459, E-mail: ghota@wvu.edu. *Application contact:* Dr. Robert Creese, Professor, 304-293-7608 Ext. 2711, Fax: 304-293-7459.

PRINCETON UNIVERSITY

Princeton Materials Institute

The Role of the Institute

Princeton University established the Princeton Materials Institute (PMI) in 1990 to promote multidisciplinary collaborations in research and education. It has 13 faculty members with shared appointments in other science and engineering departments: Chemical Engineering, Chemistry, Civil and Environmental Engineering, Electrical Engineering, Mechanical and Aerospace Engineering, Molecular Biology, and Physics. It also has more than 60 associated faculty members. The research program at PMI brings faculty members together to focus on areas such as complex, multifunctional, and bio-inspired materials and computational materials science, and collaborates closely with the Center for Photonic and Optoelectronic Materials (POEM). It is supported by several major research grants, such as the Princeton Center for Complex Materials (PCCM) supported by the NSF.

The Institute provides facilities and a physical environment that promote interdisciplinary interactions. There are many opportunities for thesis research, including polymers and complex fluids; bio-inspired and multifunctional materials; oxides having magnetic, photonic, electronic, and piezoelectric capabilities; composites and multi-layers; materials design through computational methods; and materials for the infrastructure.

Student Programs

Students are admitted to the materials program within one of the participating departments for study toward a Doctor of Philosophy degree. Information about applications may be obtained either through PMI or a department. The curriculum is based on a combination of offerings from PMI and the department. It is designed to suit the needs of each student. Thesis projects are chosen to take full advantage of the combined capabilities of the Institute and the department.

Facilities

The main facilities for PMI are housed in the nine-year-old William G. Bowen Hall, a 44,000-square-foot facility with office and laboratory space. Bowen Hall also houses an auditorium, a classroom, and a conference room and ample space for informal interaction between scientists and engineers, including two lounges and an atrium. Research benefits from PMI's central facilities in imaging, electron microscopy, characterization, computing, and testing. Other laboratories in Bowen Hall include Growth and Characterization of Optoelectronic Materials, Ceramic Materials, and Colloidal Crystals. Other facilities are available in the seven affiliated departments.

Funding

The Princeton Materials Institute derives most of its funding from individual investigator and multi-investigator grants from federal government sources. Two of the largest grants received by PMI include the Materials Research Science and Engineering Center (MRSEC) from the National Science Foundation and the Multidisciplinary University Research Initiative (MURI) from the Army Research Office. The MRSEC involves approximately thirty Princeton PIs from seven departments and is funded at approximately $3 million per year. Support for the MRSEC also includes funding for the administration of the center, shared facilities, outreach and education, and seed project support. The MURI award is $1 million per year and has involvement from five Princeton PIs and two additional universities (Drexel and Harvard). Total additional research funding from individual and multi-investigator grants total approximately $3 million in the current fiscal year. Located in the heart of the research corridor of New Jersey, PMI has also developed extensive relationships with industrial partners, including Exxon Research and Engineering, NEC Research, Hoechst-Celanese, Mobil, Lord Corporation, Dow Chemical, Dupont, Rhone-Poulenc, and others. PMI is also active within the Associated Institutions for Material Sciences (AIMS), which is a consortium of New Jersey's leading research institutions, including Rutgers University, Stevens Institute of Technology, the University of Medicine and Dentistry of New Jersey, David Sarnoff Research Center, and the New Jersey Institute of Technology.

Entrance

Interested students should either write directly to the Institute or to the department of their choice (chemical engineering, chemistry, civil and environmental engineering, electrical engineering, mechanical and aerospace engineering, molecular biology, or physics) to obtain the appropriate departmental information. The PMI Graduate committee chaired by Professor George Scherer helps match students with the most appropriate department and faculty members.

Student Support

Financial support is available through a variety of fellowships, teaching assistantships, and research assistantships funded both by federal agencies and industry.

Student Profile

Graduate students participate from the eight affiliated departments on a full-time basis. In 1998–99, there were approximately 55 students working on materials science projects supported by government, industry, and foundations. Faculty members look for the highest achieving students in their field and also for those who are interested in working in a multidisciplinary environment.

Correspondence and Information

Princeton Materials Institute
Bowen Hall
70 Prospect Avenue
Princeton University
Princeton, New Jersey 08540-5211

Telephone: 609-258-4580
Fax: 609-258-6878

Princeton University

THE FACULTY AND THEIR RESEARCH

Institute Faculty

Ilhan Aksay, Chemical Engineering. Biomimetics, processing of ceramics.
Roberto Car, Chemistry, Computational methods.
Robert Cava, Chemistry. New electronic and magnetic materials, solid-state chemistry.
Anthony Evans, Mechanical and Aerospace Engineering. Films, multilayers, multifunctional materials.
Stephen R. Forrest, Electrical Engineering and POEM. Photonic materials, molecular beam epitaxy.
Antoine Kahn, Electrical Engineering. Semiconductors, surfaces, interfaces.
Richard Register, Chemical Engineering. Block copolymers, polymer blends, interfaces.
William Russel, Chemical Engineering. Colloids, rheology.
George Scherer, Civil and Environmental Engineering. Sol-gel science, durability of cementitious materials.
Giacinto Scoles, Chemistry. Nanomaterials, clusters, organic thin films, and interfaces.
David Srolovitz, Mechanical and Aerospace Engineering. Modeling and simulation.
Zhigang Suo, Mechanical and Aerospace Engineering. Solid mechanics, small-scale materials.
Winston Soboyejo, Mechanical and Aerospace Materials. Mechanical performance, design.
Salvatore Torquato, Civil and Environmental Engineering. Statistical mechanics, structure and properties of heterogeneous materials.

Affiliated Faculty

Leland Allen, Chemistry.
Philip Anderson, Physics.
Robert Austin, Physics.
Jay Benzinger, Chemical Engineering.
Karen Bergman, Electrical Engineering.
Steven Bernasek, Chemistry.
Ravindra Bhatt, Electrical Engineering.
Andrew Bocarsly, Chemistry.
Jeff Carbeck, Chemical Engineering.
Jannette Carey, Chemistry.
Paul Chaikin, Physics.
Stephen Chou, Electrical Engineering.
Edward Cox, Molecular Biology.
Pablo Debenedetti, Chemical Engineering.
Charles Dismukes, Chemistry.
Fred Dryer, Mechanical and Aerospace Engineering.
Thomas Duffy, Geosciences.
John Groves, Chemistry.
Duncan Haldane, Physics.
Michael Hecht, Chemistry.
Peter Jaffe, Civil and Environmental Engineering.
George McLendon, Chemistry.
Nai-Phuan Ong, Physics.
Tullis Onstott, Geosciences.
Jean-Herve Prevost, Civil and Environmental Engineering.
Robert Prud'homme, Chemical Engineering.
Herschel Rabitz, Chemistry.
Barrie Royce, Mechanical and Aerospace Engineering.
Dudley Saville, Chemical Engineering.
Clarence Schutt, Chemistry.
Jeffrey Schwartz, Chemistry.
Mordechai Sergev, Electrical Engineering.
Mansour Shayegan, Electrical Engineering.
Lydia Sohn, Physics.
Zoltan Soos, Chemistry.
Thomas Spiro, Chemistry.
Malcolm Steinberg, Molecular Biology.
Jeffry Stock, Chemistry.
Sankaran Sundaresan, Chemical Engineering.
Sandra Troian, Chemical Engineering.
Daniel Tsui, Electrical Engineering.
Erik Vanmarcke, Civil and Environmental Engineering.
Warren S. Warren, Chemistry.

UNIVERSITY OF WISCONSIN–MADISON

Biotechnology Training Program

The Role of the Training Program

The University of Wisconsin–Madison offers a predoctoral training program in biotechnology. This program is taken as a minor course of study and does not grant a Ph.D. Instead, students earn a Ph.D. in a particular department with biotechnology as a minor. The objective of this Biotechnology Training Program is to educate a new cadre of scientists and engineers whose training and experience cross traditional boundaries. The program is designed to maintain the integrity of established Ph.D. programs. Thus, trainees receive Ph.D. degrees in their chosen major fields—for example, bacteriology/microbiology, biochemistry, chemical engineering, chemistry, pharmacy, molecular biology, genetics, computer science, biomedical engineering, civil/environmental engineering, or biophysics.

Student Programs

Each student completes a minor with a focus on cross-disciplinary training. A minor professor from a discipline outside the major field helps organize a minor degree course plan and oversees the training program. Each trainee participates in an industrial internship, normally during the summer after the first year of graduate school. This experience fosters an understanding and appreciation of the complex problems being addressed by the biotechnology industry. It also provides additional research opportunities and firsthand experience in interdisciplinary problem solving. Internship opportunities exist locally and with companies throughout the nation. In addition, students participate in a weekly student seminar that reinforces their cross-disciplinary approach. Students choose a major and minor professor from a list of more than 120 faculty members in forty different departments who are doing research related to biotechnology. Students are expected to complete their Ph.D. programs in four to six years. This training prepares scientists and engineers for careers in academia, industry, and government.

Facilities

The UW–Madison research laboratories are equipped with state-of-the-art instrumentation and have generous space for trainees. Equipment is available to support research on almost any aspect of modern biological, chemical, engineering, or computational research. Among the biotechnology-related resources on campus is the Biotechnology Center, which provides service facilities to UW faculty and students as well as industry. These include facilities for hybridoma production, information resources, plant biotechnology, DNA/protein synthesis and sequencing, and protein purification and a transgenic mouse facility. UW is also home to the Biochemistry Pilot Plant, which is well-known for studies in fermentation technology.

The UW College of Engineering sponsors thirty-four Engineering Centers, including Applied Microelectronics and Biomedical Engineering, and coordinates twelve industry consortia, including Bioremediation of Environmental Contaminants and Applied Water Pollution Control Research. Other major research facilities include forty libraries holding more than 5.4 million books, 50,000 serial titles, and 4 million forms; the National Nuclear Magnetic Resonance Facility; the Integrated Microscopy Resource for Biomedical Research, which has several electron microscopes; the Flow Cytometry Research Facility; the Regional Primate Research Center; the Synchroton Radiation Center; and the Biotron, which can simulate environmental conditions anywhere on earth. Many more research facilities exist on campus and are too numerous to list.

Funding

The major source of support for the Biotechnology Training Program (BTP) is a grant from the National Institutes of Health (NIH). The total amount of the NIH award for 1998–99 was $750,943, which included $387,684 for 33 stipends, $352,110 for tuition, and $9,900 for trainee travel. An individual trainee receives an annual stipend of $11,748 from NIH, as well as additional support from the UW Graduate School that increased the stipend to $14,600 per year in 1998–99. Tuition is paid by the BTP. Financial support is also provided by the enrolling department in most cases. The BTP is further supplemented by the Graduate School and UW Foundation. Usually, trainees will be supported as research assistants or teaching assistants after their period of support by the Biotechnology Training Program.

Entrance

Students interested in the Biotechnology Training Program should apply directly to the academic department of choice and indicate their interest in this program in their statement of purpose. To be considered for all types of financial support available, applications must be received before February 1.

Student Support

Financial support for students in the Biotechnology Training Program is currently available from a training grant from the National Institutes of Health, from faculty research grants, and from UW–Madison fellowships. NIH traineeships are generally awarded for a period of three years and pay for the student's tuition and fees as well as provide a monthly stipend for living expenses. Only U.S. citizens and permanent residents are eligible for NIH-funded traineeships, which are available on a competitive basis.

Student Profile

In 1998–99, the Biotechnology Training Program supported 33 trainees. Approximately half are from the biological sciences and half from majors in the physical sciences. Likewise, about half are men and half are women. All are full-time students. The NIH permits funding only for U.S. citizens and permanent residents in the program. International students who are not permanent residents are welcome to participate but cannot receive financial support from the program. Other students not needing BTP financial support are also welcome to participate in the program. Competitions for financial support typically receive 3 applicants for each available position. Students who are awarded BTP support, on average, have GPA's of 3.79 or above; have GRE scores of 610 verbal, 735 quantitative, and 725 analytic; and express a strong interest in cross-disciplinary training.

Correspondence and Information

Program Coordinator
Biotechnology Training Program
University of Wisconsin–Madison
1550 Linden Drive, Room 111
Madison, Wisconsin 53706

Telephone: 608-262-6753
E-mail: biotech@bact.wisc.edu
World Wide Web: http://www.bact.wisc.edu/biotech/btpindex

University of Wisconsin–Madison

AREAS OF RESEARCH

Agronomy. Physiology, biochemistry, molecular biology, genetics and breeding, and production of agronomic crop plants. World Wide Web: http://agronomy.wisc.edu.

Animal Health and Biomedical Sciences. Immunology and experimental hematology; molecular and cellular biology; ultrastructure; virology; microbiology; reproductive and cardiopulmonary physiology; in vitro fertilization and parasitology. World Wide Web: http://www.anatomy.wisc.edu.

Biochemistry. Prokaryotic and eukaryotic molecular biology; macromolecular structure and function; immunology; mechanistic enzymology; enzyme and metabolic regulation; plant biochemistry; microbial biochemistry; virology; molecular basis of vitamin function. World Wide Web: http://www.biochem.wisc.edu.

Biomedical Engineering. Application of engineering tools to biology and medicine. World Wide Web: http://www.engr.wisc.edu/bme.

Biomolecular Chemistry. Control of gene expression; characterization of tissue and plasma macromolecules; biochemistry of membrane transport; immunochemistry; developmental biology; endocrinology; neurochemistry; enzymology; biochemistry of viruses; control of intermediary metabolism. World Wide Web: http://www.biostat.wisc.edu/bmolchem/bmolchem.html.

Biophysics. Electrophysiology, ion channels, and sensory transduction; virus structure and function; X-ray crystallography and nuclear magnetic resonance spectroscopy of proteins and nucleic acids; electron paramagnetic resonance, optical, and vibrational spectroscopies; physical chemistry of biomolecules; enzyme mechanisms, catalysis, and kinetics; protein folding and dynamics, protein engineering, and biotechnology. World Wide Web: http://www.wisc.edu/grad/catalog/interdis/biophys.html.

Botany. Molecular biology; genetics, cellular, and developmental biology; structural botany; physiology; ecology; evolution; taxonomy; molecular systematics. Advanced instruction and opportunities for research are also available in phycology, lichenology, and mycology. World Wide Web: http://www.wisc.edu/botany.

Cell and Molecular Biology. This is a multidisciplinary program with 140 faculty members from thirty different departments enabling students to select a mentor and thesis research topic from a wide variety of fields. World Wide Web: http://www.bocklabs.wisc.edu.

Chemical Engineering. Bioreactor design; downstream processing; metabolic engineering and dynamics; enzyme kinetics; biomaterials; protein folding; protein interactions. World Wide Web: http://www.engr.wisc.edu/che

Chemistry. Analytical, bioanalytical, bioinorganic, bioorganic, biophysical, inorganic, organic, physical, and theoretical chemistry. World Wide Wrb: http://www.chem.wisc.edu.

Civil and Environmental Engineering. Bioremediation, water supply and waste-removal engineering. World Wide Web: http://www.engr.wisc.edu/bme

Comparative Biosciences. One example of research is prolactin-related growth factors expressed by the placenta: their structure, genes, regulation, receptor structure, and action. World Wide Web: http://www.vetmed.wisc.edu/cbs/cbs.html.

Computer Science. Machine learning (including neural network) techniques for analyzing biological sequences; computer-aided design of molecular biology experiments; string matching algorithms; advanced database technology to support scientific research. World Wide Web: http://www.cs.wisc.edu.

Dairy Science. Linkage of molecular and biochemical polymorphisms with quantitative trait loci in livestock; statistical models for hypothesis testing and schemes for implementation of marker-assisted selection. World Wide Web: http://www.wisc.edu/dysci/.

Electrical and Computer Engineering. Automatic control systems; biomedical engineering; communication and signal processing; computer engineering; electromagnetic fields and waves; energy and power systems; photonics; plasmas and controlled fusion; and solid state, quantum, and microelectronics. World Wide Web: http://www.eng.wisc.edu/ece.

Environmental Toxicology. Health-related chemicals in the environment. World Wide Web: http://www.wisc.edu/etc.

Food Microbiology and Toxicology. Physiology of toxin formation; safety of biotechnological processes; preservation of food by antibacterial enzymes; industrial fermentations for a carotenoid production; molecular characterization of virulence determinants of food-borne pathogens. World Wide Web: http://www.wisc.edu/fri.

Food Science. Microbial genetics; plant genetics; food safety; downstream processing of biotechnology products. World Wide Web: http://www.wisc.edu/foodsci.

Genetics. Plant genetics; population genetics; developmental genetics; molecular genetics; immunogenetics; neurogenetics; cytogenetics; viral genetics; bacterial genetics; mammalian genetics; behavioral genetics; medical genetics. World Wide Web: http://www.wisc.edu/genetics.

Horticulture. Identification, insertion, expression, and efficacy of genes important in vegetable, fruit, and tree crops, especially those (1) conferring resistance to chemical and biological stress, (2) controlling rates of protein turnover, and (3) determining membrane ATPases. World Wide Web: http://www.hort.wisc.edu.

Mathematics. Probability and stochastic processes (stochastic networks, population models, spatial processes); ordinary and partial differential equations (population models, reaction diffusion equations); algebra (combinatorics, graph theory, linear algebra). World Wide Web: http://www.math.wisc.edu.

Meat and Animal Science. Adaptation and development of molecular and cellular biology techniques for application in domestic species. World Wide Web: http://www.wisc.edu/animalsci.

Medical Microbiology and Immunology. Immunobiology and the molecular and cellular biology of host-parasite interactions involving bacteria, viruses, fungi, and protozoans. World Wide Web: http://www.biostat.wisc.edu/medmicro/home.html.

Microbiology. Prokaryotic and lower eukaryotic genetics; gene expression and its regulation; microbial physiology and diversity; molecular structure-function relationships; nucleic acid synthesis. World Wide Web: http://www.bact.wisc.edu.

Molecular Virology. Virus-host interactions; virus variation and evolution; gene expression and regulation in (+)strand RNA viruses; use of baculovirus-based expression vector systems; molecular biology of picornaviruses. World Wide Web: http://www.bocklabs.wisc.edu/welcome.html.

Oncology. Research opportunities span the disciplines central to cancer research, including biochemistry, chemical carcinogenesis, developmental genetics, and molecular virology of tumor viruses. E-mail: mcordle@oncology.wisc.edu.

Pathobiological Sciences. Molecular and cellular pathogenesis of bacterial, viral, and parasitic diseases; basic mechanisms of immunity to animal pathogens; transgenic animals as models to study pathogenesis of diseases, especially neurologic diseases. World Wide Web: http://www.vetmed.wisc.edu/pbs.

Pathology. Mechanisms of disease; cell and molecular biology of extracellular matrix and its receptors; growth factors; carcinogenesis; reactive oxygen metabolism; hematopoietic disorders; clinical chemistry; mechanisms/pathology of AIDS. World Wide Web: http://www.pathology.wisc.edu.

Pharmacology. Molecular aspects of receptor function; drug-receptor interaction. World Wide Web: http://www.wisc.edu/molpharm.

Pharmacy. Prokaryotic genetics, especially in relationship to antibiotic synthesis and production; stereoselective microbial synthesis of drugs and chemical intermediates. World Wide Web: http://www.pharmacy.wisc.edu.

Plant Pathology. Applied and fundamental aspects of interactions of bacteria, fungi, nematodes, and viruses with their plant hosts. World Wide Web: http://www.plantpath.wisc.edu.

Soil Science. Microbial degradation of organic pollutants, entailing investigations into underlying genetic mechanisms that direct the degradation of pollutants. World Wide Web: http://bob.soils.wisc.edu.

Zoology. Developmental biology; immunology; tumorigenesis; fertilization; cellular and intracellular motility; cytoskeletal structure and function; stimulus-secretion coupling; signal transduction pathways; neural structure-function organization. World Wide Web: http://www.wisc.edu/zoology.

Appendixes

This section contains two appendixes. The first, Institutional Changes Since the 1999 Edition, lists institutions that have closed, moved, merged, or changed their name or status since the last edition of the guides. The second, Abbreviations Used in the Guides, gives abbreviations of degree names, along with what those abbreviations stand for. These appendixes are identical in all six volumes of the Graduate Guides.

Institutional Changes
Since the 1999 Edition

Following is an alphabetical listing of institutions that have recently closed, moved, merged with other institutions, or changed their names or status. In the case of a name change, the former name appears first, followed by the new name.

Alfred Adler Institute of Minnesota (Hopkins, Minnesota): name changed to Alfred Adler Graduate School.

Allegheny University of the Health Sciences (Philadelphia, Pennsylvania): name changed to MCP Hahnemann University.

Brescia College (Owensboro, Kentucky): name changed to Brescia University.

California Baptist College (Riverside, California): name changed to California Baptist University.

Cornell University Medical College (New York, New York): name changed to Joan and Sanford I. Weill Medical College of Cornell University.

Denver Conservative Baptist Seminary (Denver, Colorado): name changed to Denver Seminary.

The Graduate School of America (Minneapolis, Minnesota): name changed to Capella University.

The Harid Conservatory (Boca Raton, Florida): merged with Lynn University (Boca Raton, Florida).

Huron International University (San Diego, California): closed.

International University (Englewood, Colorado): name changed to Jones International University.

Louisiana State University Medical Center (New Orleans and Shreveport, Louisiana): name changed to Louisiana State University Health Science Center.

Mankato State University (Mankato, Minnesota): name changed to Minnesota State University, Mankato.

Massachusetts College of Pharmacy and Allied Health Sciences (Boston, Massachusetts): name changed to Massachusetts College of Pharmacy and Health Sciences.

Mennonite College of Nursing (Bloomington, Illinois): merged with Illinois State University (Normal, Illinois).

Mercer University, Cecil B. Day Campus (Atlanta, Georgia): data profiled under Mercer University (Macon, Georgia).

Mount Sinai School of Medicine of the City University of New York (New York, New York): name changed to Mount Sinai School of Medicine of New York University.

Mount Vernon College (Washington, District of Columbia): merged with The George Washington University (Washington, District of Columbia).

The Naropa Institute (Boulder, Colorado): name changed to The Naropa University.

Northwest Nazarene College (Nampa, Idaho): name changed to Northwest Nazarene University.

Philadelphia College of Textiles and Science (Philadelphia, Pennsylvania): name changed to Philadelphia University.

Phillips University (Enid, Oklahoma): closed.

Regional Seminary of Saint Vincent de Paul in Florida, Inc. (Boynton Beach, Florida): name changed to Saint Vincent de Paul Regional Seminary.

Rockhurst College (Kansas City, Missouri): name changed to Rockhurst University.

Saint Leo College (Saint Leo, Florida): name changed to Saint Leo University.

Southern California College (Costa Mesa, California): name changed to Vanguard University of Southern California.

University of Massachusetts Medical Center at Worcester (Worcester, Massachusetts): name changed to University of Massachusetts Worcester.

Westminster College of Salt Lake City (Salt Lake City, Utah): name changed to Westminster College.

Abbreviations Used in the Guides

The following list includes abbreviations of degree names used in the profiles in the 2000 edition of the guides. Because some degrees (e.g., Doctor of Education) can be abbreviated in more than one way (e.g., D.Ed. or Ed.D.), and because the abbreviations used in the guides reflect the preferences of the individual colleges and universities, the list may include two or more abbreviations for a single degree.

Degrees

A Mus D	Doctor of Musical Arts
AC	Advanced Certificate
AD	Artist's Diploma
ADP	Artist's Diploma
Adv C	Advanced Certificate
Adv M	Advanced Master
Aerospace E	Aerospace Engineer
AGC	Advanced Graduate Certificate
ALM	Master of Liberal Arts
AM	Master of Arts
AMRS	Master of Arts in Religious Studies
APC	Advanced Professional Certificate
App Sc	Applied Scientist
Au D	Doctor of Audiology
B Th	Bachelor of Theology
C Phil	Certificate in Philosophy
CAES	Certificate of Advanced Educational Specialization
CAGS	Certificate of Advanced Graduate Studies
CAL	Certificate in Applied Linguistics
CAMS	Certificate of Advanced Management Studies
CAPS	Certificate of Advanced Professional Studies
CAS	Certificate of Advanced Studies
CASPA	Certificate of Advanced Study in Public Administration
CASR	Certificate in Advanced Social Research
CBHS	Certificate in Basic Health Sciences
CCJA	Certificate in Criminal Justice Administration
CE	Civil Engineer
CG	Certificate in Gerontology
CGS	Certificate of Graduate Studies
Ch E	Chemical Engineer
CHSS	Counseling and Human Services Specialist
CIF	Certificate in International Finance
CITS	Certificate of Individual Theological Studies
CLIS	Certificate of Library and Information Science
CMH	Certificate in Medical Humanities
CMS	Certificate in Ministerial Studies
	Certificate in Museum Studies
CNM	Certificate in Nonprofit Management
CP	Certificate in Performance
CPC	Certificate in Professional Counseling
	Certificate in Publication and Communication
CPH	Certificate in Public Health
CPM	Certificate in Public Management
CPS	Certificate of Professional Studies
CSD	Certificate in Spiritual Direction
CSE	Computer Systems Engineer
CSS	Certificate of Special Studies
CTS	Certificate of Theological Studies
CURP	Certificate in Urban and Regional Planning
D Arch	Doctor of Architecture
D Chem	Doctor of Chemistry
D Ed	Doctor of Education
D Eng	Doctor of Engineering
D Engr	Doctor of Engineering
D Env	Doctor of Environment
D Jur	Doctor of Jurisprudence
D Law	Doctor of Law
D Litt	Doctor of Letters
D Med Sc	Doctor of Medical Science
D Min	Doctor of Ministry
D Min PCC	Doctor of Ministry, Pastoral Care, and Counseling
D Miss	Doctor of Missiology
D Mus	Doctor of Music
D Mus A	Doctor of Musical Arts
D Mus Ed	Doctor of Music Education
D Phil	Doctor of Philosophy
D Ps	Doctor of Psychology
D Sc	Doctor of Science
D Sc D	Doctor of Science in Dentistry
D Th	Doctor of Theology
DA	Doctor of Arts
DA Ed	Doctor of Arts in Education
DAST	Diploma of Advanced Studies in Teaching
DBA	Doctor of Business Administration
DC	Doctor of Chiropractic
DCC	Doctor of Computer Science
DCD	Doctor of Communications Design
DCL	Doctor of Comparative Law
DCM	Doctor of Church Music
DCS	Doctor of Computer Science
DDN	Diplôme du Droit Notarial
DDS	Doctor of Dental Surgery
DE	Doctor of Engineering

DEM	Doctor of Educational Ministry		**Ed DCT**	Doctor of Education in College Teaching
DEPD	Diplôme Études Spécialisées		**Ed M**	Master of Education
DES	Doctor of Engineering Science		**Ed S**	Specialist in Education
DESS	Diplôme Études Supérieures Spécialisées		**EDM**	Executive Doctorate in Management
DFA	Doctor of Fine Arts		**EE**	Electrical Engineer
DFES	Doctor of Forestry and Environmental Studies			Environmental Engineer
DGP	Diploma in Graduate and Professional Studies		**EM**	Mining Engineer
DHA	Doctor of Health Administration		**EMBA**	Executive Master of Business Administration
DHCE	Doctor of Health Care Ethics		**EMCIS**	Executive Master of Computer Information Systems
DHL	Doctor of Hebrew Letters		**EMIB**	Executive Master of International Business
	Doctor of Hebrew Literature		**EMPA**	Executive Master of Public Affairs
DHS	Doctor of Human Services		**EMRA**	Executive Master of Rehabilitation Administration
DIBA	Doctor of International Business Administration		**EMS**	Executive Master of Science
Dip CS	Diploma in Christian Studies		**EMSF**	Executive Master of Science in Finance
DIT	Doctor of Industrial Technology		**EMSILR**	Executive Master of Science in Industrial and Labor Relations
DJ Ed	Doctor of Jewish Education			
DJS	Doctor of Jewish Studies		**Eng**	Engineer
DM	Doctor of Management		**Eng Sc D**	Doctor of Engineering Science
	Doctor of Music		**Engr**	Engineer
DMA	Doctor of Musical Arts		**Exec Ed D**	Executive Doctor of Education
DMD	Doctor of Dental Medicine		**Exec MBA**	Executive Master of Business Administration
DME	Doctor of Music Education		**Exec MIM**	Executive Master of International Management
DML	Doctor of Modern Languages		**Exec MPA**	Executive Master of Public Administration
DMM	Doctor of Music Ministry		**Exec MPH**	Executive Master of Public Health
DN Sc	Doctor of Nursing Science		**Exec MS**	Executive Master of Science
DNS	Doctor of Nursing Science		**GDPA**	Graduate Diploma in Public Administration
DO	Doctor of Osteopathy		**GDRE**	Graduate Diploma in Religious Education
DPA	Doctor of Public Administration		**Geol E**	Geological Engineer
DPC	Doctor of Pastoral Counseling		**GMBA**	Global Master of Business Administration
DPDS	Doctor of Planning and Development Studies		**GPD**	Graduate Performance Diploma
DPE	Doctor of Physical Education		**GPMBA**	Global Professional Master of Business Administration
DPH	Doctor of Public Health			
DPM	Doctor of Podiatric Medicine		**HS Dir**	Director of Health and Safety
DPS	Doctor of Professional Studies		**HSD**	Doctor of Health and Safety
DPT	Doctor of Physical Therapy		**IAMBA**	Information Age Master of Business Administration
Dr DES	Doctor of Design		**IEMBA**	International Executive Master of Business Administration
Dr OT	Doctor of Occupational Therapy			
Dr PH	Doctor of Public Health		**IMA**	Interdisciplinary Master of Arts
Dr Sc PT	Doctor of Science in Physical Therapy		**IMBA**	International Master of Business Administration
DS Sc	Doctor of Social Science		**IOE**	Industrial and Operations Engineer
DSM	Doctor of Sacred Music		**JCD**	Doctor of Canon Law
	Doctor of Sport Management		**JCL**	Licentiate in Canon Law
DSN	Doctor of Science in Nursing		**JD**	Juris Doctor
DSW	Doctor of Social Work		**JSD**	Doctor of Juridical Science
DV Sc	Doctor of Veterinary Science			Doctor of Jurisprudence
DVM	Doctor of Veterinary Medicine			Doctor of the Science of Law
EAA	Engineer in Aeronautics and Astronautics		**JSM**	Master of Science of Law
EAS	Education Administration Specialist		**L Th**	Licenciate in Theology
Ed D	Doctor of Education		**LL B**	Bachelor of Laws

LL D	Doctor of Laws
LL M	Master of Laws
LL M CL	Master of Laws in Comparative Law
LL M T	Master of Laws in Taxation
M Ac	Master of Accountancy
	Master of Accounting
	Master of Acupuncture
M Ac OM	Master of Acupuncture and Oriental Medicine
M Acc	Master of Accountancy
	Master of Accounting
M Acct	Master of Accountancy
	Master of Accounting
M Accy	Master of Accountancy
M Acy	Master of Accountancy
M Ad	Master of Administration
M Ad Ed	Master of Adult Education
M Adm	Master of Administration
M Adm Mgt	Master of Administrative Management
M Aero E	Master of Aerospace Engineering
M Ag	Master of Agriculture
M Ag Ed	Master of Agricultural Education
M Agr	Master of Agriculture
M Anesth Ed	Master of Anesthesiology Education
M App St	Master of Applied Statistics
M Appl Stat	Master of Applied Statistics
M Aq	Master of Aquaculture
M Arch	Master of Architecture
M Arch E	Master of Architectural Engineering
M Arch H	Master of Architectural History
M Arch UD	Master of Architecture in Urban Design
M Bio E	Master of Bioengineering
M Biomath	Master of Biomathematics
M Bus Ed	Master of Business Education
M Ch E	Master of Chemical Engineering
M Chem	Master of Chemistry
M Cl D	Master of Clinical Dentistry
M Cl Sc	Master of Clinical Science
M Co E	Master of Computer Engineering
M Comp E	Master of Computer Engineering
M Coun	Master of Counseling
M Cp E	Master of Computer Engineering
M Dec S	Master of Decision Sciences
M Dent Sc	Master of Dental Sciences
M Des	Master of Design
M Des S	Master of Design Studies
M Div	Master of Divinity
M Div CM	Master of Divinity in Church Music
M Ec	Master of Economics
M Econ	Master of Economics
M Ed	Master of Education
M Ed T	Master of Education in Teaching
M En	Master of Engineering
M En S	Master of Environmental Sciences
M Eng	Master of Engineering
M Eng Mgt	Master of Engineering Management
M Engr	Master of Engineering
M Env	Master of Environment
M Env Des	Master of Environmental Design
M Env E	Master of Environmental Engineering
M Env Sc	Master of Environmental Science
M Ext Ed	Master of Extension Education
M Fin	Master of Finance
M Fr	Master of French
M Gen E	Master of General Engineering
M Geo E	Master of Geological Engineering
M Geoenv E	Master of Geoenvironmental Engineering
M Hum	Master of Humanities
M Hum Svcs	Master of Human Services
M In Ed	Master of Industrial Education
M Kin	Master of Kinesiology
M Land Arch	Master of Landscape Architecture
M Lit M	Master of Liturgical Music
M Litt	Master of Letters
M Mat SE	Master of Material Science and Engineering
M Math	Master of Mathematics
M Med Sc	Master of Medical Science
M Mgmt	Master of Management
M Mgt	Master of Management
M Min	Master of Ministries
M Miss	Master of Missiology
M Mtl E	Master of Materials Engineering
M Mu	Master of Music
M Mu Ed	Master of Music Education
M Mus	Master of Music
M Mus Ed	Master of Music Education
M Nat Sci	Master of Natural Science
M Nurs	Master of Nursing
M Oc E	Master of Oceanographic Engineering
M Pharm	Master of Pharmacy
M Phil	Master of Philosophy
M Phil F	Master of Philosophical Foundations
M Pl	Master of Planning
M Pol	Master of Political Science
M Pr A	Master of Professional Accountancy
M Pr Met	Master of Professional Meteorology
M Prob S	Master of Probability and Statistics
M Prof Past	Master of Professional Pastoral

M Ps	Master of Psychology	**MAAE**	Master of Arts in Art Education
M Psych	Master of Psychology	**MAAT**	Master of Arts in Applied Theology
M Pub	Master of Publishing		Master of Arts in Art Therapy
M Rel	Master of Religion	**MAB**	Master of Agribusiness
M Rel Ed	Master of Religious Education	**MABC**	Master of Arts in Biblical Counseling
M Sc	Master of Science	**MABM**	Master of Agribusiness Management
M Sc A	Master of Science (Applied)	**MABS**	Master of Arts in Biblical Studies
M Sc BMC	Master of Science in Biomedical Communications	**MAC**	Master of Accounting
M Sc CS	Master of Science in Computer Science		Master of Analytical Chemistry
M Sc E	Master of Science in Engineering		Master of Art Conservation
M Sc Eng	Master of Science in Engineering		Master of Arts in Communication
M Sc Engr	Master of Science in Engineering		Master of Arts in Counseling
M Sc F	Master of Science in Forestry	**MACAT**	Master of Arts in Counseling Psychology: Art Therapy
M Sc FE	Master of Science in Forest Engineering	**MACCM**	Master of Arts in Church and Community Ministry
M Sc N	Master of Science in Nursing	**MACE**	Master of Arts in Christian Education
M Sc P	Master of Science in Planning		Master of Arts in Computer Education
M Sc Pl	Master of Science in Planning	**MACH**	Master of Arts in Church History
M Sc PT	Master of Science in Physical Therapy	**MACL**	Master of Arts in Classroom Psychology
M Sc T	Master of Science in Teaching	**MACM**	Master of Arts in Christian Ministries
M Soc	Master of Sociology		Master of Arts in Church Music
M Sp Ed	Master of Special Education		Master of Arts in Counseling Ministries
M Stat	Master of Statistics	**MACO**	Master of Arts in Counseling
M Sw E	Master of Software Engineering	**MACP**	Master of Arts in Counseling Psychology
M Sw En	Master of Software Engineering	**MACSE**	Master of Arts in Christian School Education
M Tax	Master of Taxation	**MACT**	Master of Arts in College Teaching
M Tech	Master of Technology	**MACTM**	Master of Applied Communication Theory and Methodology
M Th	Master of Theology	**MACY**	Master of Arts in Accountancy
M Th Past	Master of Pastoral Theology	**MAD**	Master of Applied Development
M Tox	Master of Toxicology	**MADH**	Master of Applied Development and Health
M Trans E	Master of Transportation Engineering	**MADR**	Master of Arts in Dispute Resolution
M Vet Sc	Master of Veterinary Science	**MAE**	Master of Aerospace Engineering
MA	Master of Arts		Master of Agricultural Economics
MA Comm	Master of Arts in Communication		Master of Applied Economics
MA Ed	Master of Arts in Education		Master of Architectural Engineering
MA Min	Master of Arts in Ministry		Master of Art Education
MA Missions	Master of Arts in Missions		Master of Arts in Education
MA Past St	Master of Arts in Pastoral Studies		Master of Arts in English
MA Ps	Master of Arts in Psychology		Master of Automotive Engineering
MA Psych	Master of Arts in Psychology	**MAES**	Master of Arts in Environmental Sciences
MA Sc	Master of Applied Science	**MAF**	Master of Arts in Finance
MA Th	Master of Arts in Theology	**MAFIS**	Master of Accountancy and Financial Information Systems
MA(R)	Master of Arts (Research)	**MAFLL**	Master of Arts in Foreign Language and Literature
MA(T)	Master of Arts in Teaching	**MAFM**	Master of Accounting and Financial Management
MAA	Master of Administrative Arts	**MAG**	Master of Applied Geography
	Master of Applied Anthropology	**MAGP**	Master of Arts in Gerontological Psychology
	Master of Arts in Administration	**MAGU**	Master of Urban Analysis and Management
		MAH	Master of Arts in Humanities
MAAA	Master of Arts in Arts Administration	**MAHA**	Master of Arts in Humanitarian Studies
MAABS	Master of Arts in Applied Behavioral Sciences	**MAHCD**	Master of Applied Human and Community Development

MAHL	Master of Arts in Hebrew Letters
MAHRM	Master of Arts in Human Resources Management
MAHS	Master of Arts in Human Services
MAICS	Master of Arts in Intercultural Studies
MAIDM	Master of Arts in Interior Design and Merchandising
MAIND	Master of Arts in Interior Design
MAIPE	Master of Arts in International Political Economy
MAIR	Master of Arts in Industrial Relations
MAIS	Master of Accounting and Information Systems
	Master of Arts in Intercultural Studies
	Master of Arts in Interdisciplinary Studies
	Master of Arts in International Studies
MAJ	Master of Arts in Journalism
MAJ Ed	Master of Arts in Jewish Education
MAJC	Master of Arts in Journalism and Communication
MAJCS	Master of Arts in Jewish Communal Service
MAJE	Master of Arts in Jewish Education
MAJS	Master of Arts in Jewish Studies
MALA	Master of Arts in Liberal Arts
	Master of Arts in Liturgical Arts
MALAS	Master of Arts in Latin American Studies
MALD	Master of Arts in Law and Diplomacy
MALER	Master of Arts in Labor and Employment Relations
MALIS	Master of Arts in Library and Information Science
MALL	Master of Arts in Liberal Learning
MALS	Master of Arts in Landscape Studies
	Master of Arts in Liberal Studies
MAM	Master of Agriculture and Management
	Master of Applied Mathematics
	Master of Applied Mechanics
	Master of Arts in Management
	Master of Arts Management
	Master of Avian Medicine
MAM Sc	Master of Applied Mathematical Science
MAMB	Master of Applied Molecular Biology
MAMC	Master of Arts in Mass Communication
MAME	Master of Arts in Missions/Evangelism
MAMFC	Master of Arts in Marriage and Family Counseling
MAMFCC	Master of Arts in Marriage, Family, and Child Counseling
MAMFT	Master of Arts in Marriage and Family Therapy
MAML	Master of Arts in School Media Librarianship
MAMM	Master of Arts in Ministry Management
	Master of Arts in Music Ministry
MAMS	Master of Applied Mathematical Sciences
	Master of Arts in Ministerial Studies
	Master of Associated Medical Sciences
MAMT	Master of Arts in Mathematics Teaching
MANM	Master of Arts in Nonprofit Management
MANT	Master of Arts in New Testament
MAO	Master of Arts in Organizational Psychology
MAOE	Master of Adult and Occupational Education

MAOM	Master of Arts in Organizational Management
MAOT	Master of Arts in Old Testament
MAP	Master of Applied Politics
	Master of Applied Psychology
	Master of Arts in Planning
	Master of Arts in Politics
	Master of Public Administration
MAP Min	Master of Arts in Pastoral Ministry
MAPA	Master of Arts in Public Administration
MAPC	Master of Arts in Pastoral Counseling
MAPEB	Master of Arts in Politics, Economics, and Business
MAPM	Master of Arts in Pastoral Ministry
MAPP	Master of Arts in Public Policy
MAPS	Master of Arts in Pastoral Studies
MAPW	Master of Arts in Professional Writing
MAR	Master of Arts in Religion
Mar Eng	Marine Engineer
MARC	Master of Arts in Religious Communication
MARE	Master of Arts in Religious Education
MARL	Master of Arts in Religious Leadership
MARS	Master of Arts in Religious Studies
MART	Master of Arts in Religion and Theology
MAS	Master of Accounting Science
	Master of Actuarial Science
	Master of Administrative Science
	Master of Aeronautical Science
	Master of American Studies
	Master of Applied Science
	Master of Applied Statistics
	Master of Archival Studies
MASA	Master of Advanced Studies in Architecture
MASAC	Master of Arts in Substance Abuse Counseling
MASD	Master of Arts in Spiritual Direction
MASF	Master of Arts in Spiritual Formation
MASLA	Master of Advanced Studies in Landscape Architecture
MASM	Master of Arts in Special Ministries
	Master of Arts in Specialized Ministries
MASP	Master of Arts in School Psychology
MASPAA	Master of Arts in Sports and Athletic Administration
MASS	Master of Applied Social Science
	Master of Arts in Social Science
MAT	Master of Arts in Teaching
	Master of Arts in Theology
Mat E	Materials Engineer
MATA	Master of Arts in Theology and the Arts
MATCM	Master of Acupuncture and Traditional Chinese Medicine
MATE	Master of Arts for the Teaching of English
MATESL	Master of Arts in Teaching English as a Second Language
MATESOL	Master of Arts in Teaching English to Speakers of Other Languages

MATEX	Master of Arts in Textiles		**MC Sc**	Master of Computer Science
MATFL	Master of Arts in Teaching Foreign Language		**MCA**	Master of Arts in Applied Criminology
MATH	Master of Arts in Therapy			Master of Commercial Aviation
MATI	Master of Administration of Information Technology		**MCC**	Master of Computer Science
MATL	Master of Arts in Teaching of Languages		**MCD**	Master of Communications Disorders
MATM	Master of Arts in Teaching of Mathematics		**MCE**	Master of Civil Engineering
MATS	Master of Arts in Theological Studies			Master of Computer Engineering
	Master of Arts in Transforming Spirituality			Master of Construction Engineering
MAUA	Master of Arts in Urban Affairs			Master of Continuing Education
MAUD	Master of Arts in Urban Design			Master of Control Engineering
MAUM	Master of Arts in Urban Ministry		**MCED**	Master of Community Economic Development
MAUPRD	Master of Arts in Urban Planning and Real Estate Development		**MCEM**	Master of Construction Engineering Management
MAURP	Master of Arts in Urban and Regional Planning		**MCG**	Master of Clinical Gerontology
MAW	Master of Arts in Writing		**MCH**	Master of Community Health
MAWB	Master of Arts in Wildlife Biology		**MCIS**	Master of Communication and Information Studies
MAWS	Master of Arts in Women's Studies			Master of Computer and Information Science
MBA	Master of Business Administration			Master of Computer Information Systems
MBA Arts	Master of Business Administration in Arts		**MCJ**	Master of Criminal Justice
MBA-EP	Master of Business Administration–Experienced Professionals		**MCJA**	Master of Criminal Justice Administration
MBA-PE	Master of Business Administration–Physician's Executive		**MCL**	Master of Canon Law
				Master of Christian Leadership
				Master of Civil Law
MBAA	Master of Business Administration in Aviation			Master of Comparative Law
MBAE	Master of Biological and Agricultural Engineering		**MCM**	Master of Church Management
	Master of Biosystems and Agricultural Engineering			Master of Church Ministry
MBAi	Master of Business Administration–International			Master of Church Music
MBAIB	Master of Business Administration in International Business			Master of City Management
				Master of Construction Management
MBAPA	Master of Business Administration–Physician Assistant		**MCMS**	Master of Clinical Medical Science
			MCP	Master of City Planning
MBATM	Master of Business in Telecommunication Management			Master of Community Planning
				Master of Community Psychology
MBC	Master of Building Construction			Master of Counseling Psychology
MBE	Master of Bilingual Education		**MCPD**	Master of Community Planning and Development
	Master of Biomedical Engineering		**MCRP**	Master of City and Regional Planning
	Master of Business Education		**MCS**	Master of Christian Studies
MBHCM	Master of Behavioral Health Care Management			Master of Combined Sciences
MBMSE	Master of Business Management and Software Engineering			Master of Communication Studies
				Master of Computer Science
MBOL	Master of Business and Organizational Leadership		**MCSM**	Master of Construction Science/Management
MBS	Master of Basic Science		**MD**	Doctor of Medicine
	Master of Behavioral Science		**MDA**	Master of Development Administration
	Master of Biblical Studies		**MDE**	Master of Developmental Economics
	Master of Biological Science			Master of Distance Education
	Master of Biomedical Sciences		**MDR**	Master of Dispute Resolution
	Master of Building Science		**MDS**	Master of Dental Surgery
	Master of Business Studies		**ME**	Master of Education
MBSI	Master of Business Information Science			Master of Engineering
MBT	Master of Business Taxation		**ME Sc**	Master of Engineering Science
MC	Master of Communication		**MEA**	Master of Engineering Administration
	Master of Counseling		**MEC**	Master of Electronic Commerce
			MECE	Master of Electrical and Computer Engineering
			Mech E	Mechanical Engineer
MC Ed	Master of Continuing Education		**MED**	Master of Education of the Deaf

MEDS	Master of Environmental Design Studies
MEE	Master of Electrical Engineering
	Master of Environmental Engineering
MEEM	Master of Environmental Engineering and Management
MEENE	Master of Engineering in Environmental Engineering
MEERM	Master of Earth and Environmental Resource Management
MEL	Master of Educational Leadership
MEM	Master of Ecosystem Management
	Master of Engineering Management
	Master of Environmental Management
	Master of Marketing
MEMS	Master of Emergency Medical Service
	Master of Engineering in Manufacturing Systems
MENVEGR	Master of Environmental Engineering
MEP	Master of Engineering Physics
	Master of Environmental Planning
MEPC	Master of Environmental Pollution Control
MEPD	Master of Education–Professional Development
MEPM	Master of Environmental Policy and Management
MES	Master of Engineering Science
	Master of Environmental Science
	Master of Environmental Studies
	Master of Special Education
MESM	Master of Environmental Science and Management
MESS	Master of Exercise and Sport Sciences
MET	Master of Education in Teaching
Met E	Metallurgical Engineer
METM	Master of Engineering and Technology Management
MEVE	Master of Environmental Engineering
MF	Master of Forestry
MFA	Master of Fine Arts
MFAS	Master of Fisheries and Aquatic Science
MFAW	Master of Fine Arts in Writing
MFC	Master of Forest Conservation
MFCC	Marriage and Family Counseling Certificate
	Marriage, Family, and Child Counseling
MFCS	Master of Family and Consumer Sciences
MFE	Master of Forest Engineering
MFR	Master of Forest Resources
MFRC	Master of Forest Resources and Conservation
MFS	Master of Family Studies
	Master of Food Science
	Master of Forensic Sciences
	Master of Forest Science
	Master of Forest Studies
	Master of French Studies
MFT	Master of Family Therapy
MGA	Master of Government Administration

MGCOD	Master of Group Counseling and Organizational Dynamics
MGD	Master of Graphic Design
MGE	Master of Geotechnical Engineering
MGH	Master of Geriatric Health
MGIS	Master of Geographic Information Science
MGP	Master of Gestion de Projet
MGPGP	Master of Group Process and Group Psychotherapy
MGS	Master of General Studies
	Master of Gerontological Studies
MH	Master of Humanities
MH Sc	Master of Health Sciences
MHA	Master of Health Administration
	Master of Healthcare Administration
	Master of Hospitality Administration
MHAMS	Master of Historical Administration and Museum Studies
MHCA	Master of Health Care Administration
MHCI	Master of Human-Computer Interaction
MHD	Master of Human Development
MHE	Master of Health Education
	Master of Home Economics
	Master of Human Ecology
MHE Ed	Master of Home Economics Education
MHHS	Master of Health and Human Services
MHK	Master of Human Kinetics
MHL	Master of Hebrew Literature
MHM	Master of Hospitality Management
	Master of Hotel Management
MHMS	Master of Health Management Systems
MHP	Master of Health Physics
	Master of Health Professions
	Master of Heritage Preservation
	Master of Historic Preservation
MHPA	Master of Heath Policy and Administration
MHPE	Master of Health Professions Education
	Master of Health Promotion and Education
MHR	Master of Human Resources
MHRD	Master in Human Resource Development
MHRDL	Master of Human Resource Development Leadership
MHRDOD	Master of Human Resource Development/ Organizational Development
MHRIM	Master of Hotel, Restaurant, and Institutional Management
MHRIR	Master of Human Resources and Industrial Relations
MHRLR	Master of Human Resources and Labor Relations
MHRM	Master of Human Resources Management
MHROD	Master of Human Resources and Organization Development

MHRTA	Master in Hotel, Restaurant, Tourism, and Administration	**MJ**	Master of Journalism Master of Jurisprudence
MHS	Master of Health Sciences Master of Healthcare Systems Master of Hispanic Studies Master of Human Services	**MJ Ed**	Master of Jewish Education
		MJA	Master of Justice Administration
		MJPM	Master of Justice Policy and Management
MHSA	Master of Health Services Administration Master of Human Services Administration	**MJS**	Master of Judaic Studies Master of Judicial Studies Master of Juridical Science
MHSE	Master of Health Science Education	**ML Arch**	Master of Landscape Architecture
MI	Master of Instruction	**MLA**	Master of Landscape Architecture Master of Liberal Arts
MI Arch	Master of Interior Architecture		
MI St	Master of Information Studies	**MLAS**	Master of Laboratory Animal Science
MIA	Master of Interior Architecture Master of International Affairs	**MLAUD**	Master of Landscape Architecture in Urban Development
MIAA	Master of International Affairs and Administration	**MLD**	Master of Leadership Studies
MIB	Master of International Business	**MLE**	Master of Applied Linguistics and Exegesis
MIBA	Master of International Business Administration	**MLERE**	Master of Land Economics and Real Estate
MIBS	Master of International Business Studies	**MLHR**	Master of Labor and Human Resources
MID	Master of Industrial Design Master of Interior Design	**MLI**	Master of Legal Institutions
		MLI Sc	Master of Library and Information Science
MIE	Master of Industrial Engineering	**MLIR**	Master of Labor and Industrial Relations
MIE Mgmt	Master of Industrial Engineering Management	**MLIS**	Master of Library and Information Science Master of Library and Information Services Master of Library and Information Studies
MIHE	Master of Integrated Humanities and Education		
MIHM	Master of International Health Management		
MIIM	Master of International and Intercultural Management	**MLLS**	Master of Leadership and Liberal Studies
		MLM	Master of Library Media
MIJ	Master of International Journalism	**MLRHR**	Master of Labor Relations and Human Resources
MILR	Master of Industrial and Labor Relations	**MLS**	Master of Legal Studies Master of Liberal Studies Master of Library Science Master of Life Sciences Master of Medical Laboratory Sciences
MIM	Master of Information Management Master of International Management		
MIMLA	Master of International Management for Latin America		
		MLSP	Master of Law and Social Policy
MIMOT	Master of International Management of Technology	**MM**	Master of Management Master of Ministry Master of Music
MIMS	Master of Information Management and Systems Master of Integrated Manufacturing Systems		
		MM Ed	Master of Music Education
MIP	Master of Infrastructure Planning Master of Intellectual Property	**MM Sc**	Master of Medical Science
		MM St	Master of Museum Studies
MIPP	Master of International Public Policy	**MMA**	Master of Management and Administration Master of Marine Affairs Master of Media Arts Master of Musical Arts
MIR	Master of Industrial Relations		
MIS	Master of Industrial Statistics Master of Information Science Master of Information Systems Master of Interdisciplinary Studies		
		MMAE	Master of Mechanical and Aerospace Engineering
		MMAS	Master of Military Art and Science
MISM	Master of Information Systems Management	**MMC**	Master of Mass Communications
MIT	Master in Teaching Master of Industrial Technology Master of Information Technology Master of Initial Teaching Master of International Trade	**MMCM**	Master of Music in Church Music
		MME	Master of Manufacturing Engineering Master of Mathematics for Educators Master of Mechanical Engineering Master of Medical Engineering Master of Mining Engineering Master of Music Education
MITA	Master of Information Technology Administration		
MITE	Master of Information Technology Education		
MITM	Master of International Technology Management	**MMF**	Master of Mathematical Finance

MMFT	Master of Marriage and Family Therapy
MMH	Master of Management in Hospitality Master of Medical History Master of Medical Humanities
MMIS	Master of Management Information Systems
MMM	Master of Manufacturing Management Master of Medical Management
MMME	Master of Metallurgical and Materials Engineering
MMP	Master of Marine Policy
MMPA	Master of Management and Professional Accounting
MMR	Master of Marketing Research
MMS	Master of Management Science Master of Management Studies Master of Manufacturing Systems Master of Marine Science Master of Marine Studies Master of Materials Science Master of Medical Science Master of Medieval Studies
MMSE	Master of Manufacturing Systems Engineering
MMT	Master of Music Therapy
MN	Master of Nursing
MN Sc	Master of Nursing Science
MNA	Master of Nonprofit Administration Master of Nurse Anesthesia Master of Nursing Administration
MNAS	Master of Natural and Applied Science
MNE	Master of Nuclear Engineering
MNM	Master of Nonprofit Management
MNO	Master of Nonprofit Organization
MNPL	Master of Not-for-Profit Leadership
MNR	Master of Natural Resources
MNRM	Master of Natural Resource Management
MNS	Master of Natural Science
MOA	Maître d'Orthophonie et d'Audiologie
MOB	Master of Organizational Behavior
MOD	Master of Organizational Development
MOH	Master of Occupational Health
MOL	Master of Organizational Leadership
MOM	Master of Manufacturing
MOR	Master of Operations Research
MOT	Master of Occupational Therapy
MoTM	Master of Technology Management
MP	Master of Planning
MP Ac	Master of Professional Accountancy
MP Acc	Master of Professional Accountancy Master of Professional Accounting
MP Acct	Master of Professional Accounting
MP Aff	Master of Public Affairs
MP Th	Master of Pastoral Theology

MPA	Master of Physician Assistant Master of Professional Accountancy Master of Professional Accounting Master of Public Administration Master of Public Affairs
MPA-URP	Master of Public Affairs and Urban and Regional Planning
MPAID	Master of Public Administration and International Development
MPAS	Master of Physical Activity Studies Master of Physician Assistant Studies Master of Public Art Studies
MPC	Master of Pastoral Counseling Master of Professional Communication Master of Professional Counseling
MPDS	Master of Planning and Development Studies
MPE	Master of Physical Education
MPEM	Master of Project Engineering and Management
MPH	Master of Public Health
MPHE	Master of Public Health Education
MPHTM	Master of Public Health and Tropical Medicine
MPIA	Master of Public and International Affairs
MPM	Master of Pest Management Master of Practical Ministries Master of Project Management Master of Public Management
MPP	Master of Public Policy
MPPA	Master of Public Policy Administration Master of Public Policy and Administration
MPPM	Master of Public and Private Management Master of Public Policy and Management
MPPPM	Master of Plant Protection and Pest Management
MPPUP	Master of Public Policy and Urban Planning
MPRTM	Master of Parks, Recreation, and Tourism Management
MPS	Master of Pastoral Studies Master of Policy Sciences Master of Political Science Master of Preservation Studies Master of Professional Studies Master of Public Service
MPSA	Master of Public Service Administration
MPSRE	Master of Professional Studies in Real Estate
MPT	Master of Physical Therapy
MPVM	Master of Preventive Veterinary Medicine
MPW	Master of Public Works
MQM	Master of Quality Management
MQS	Master of Quality Systems
MRC	Master of Rehabilitation Counseling
MRCP	Master of Regional and City Planning Master of Regional and Community Planning
MRE	Master of Religious Education
MRECM	Master of Real Estate and Construction Management

MRED	Master of Real Estate Development	**MS Pub P**	Master of Science in Public Policy
MRLS	Master of Resources Law Studies	**MS Sp Ed**	Master of Science in Special Education
MRM	Master of Rehabilitation Medicine	**MS Stat**	Master of Science in Statistics
	Master of Resources Management	**MS Text**	Master of Science in Textiles
MRP	Master of Regional Planning	**MS(R)**	Master of Science (Research)
MRRA	Master of Recreation Resources Administration	**MSA**	Master of School Administration
MRS	Master of Religious Studies		Master of Science Administration
MRTP	Master of Rural and Town Planning		Master of Science in Accountancy
MS	Master of Science		Master of Science in Accounting
MS Acct	Master of Science in Accounting		Master of Science in Administration
MS Accy	Master of Science in Accountancy		Master of Science in Agriculture
MS Admin	Master of Science in Administration		Master of Science in Anesthesia
MS Ag	Master of Science in Agriculture		Master of Science in Architecture
MS Arch	Master of Science in Architecture		Master of Science in Aviation
MS Arch St	Master of Science in Architectural Studies		Master of Sports Administration
MS Bio E	Master of Science in Biomedical Engineering	**MSA Phy**	Master of Science in Applied Physics
MS Biol	Master of Science in Biology	**MSAA**	Master of Science in Astronautics and Aeronautics
MS Bm E	Master of Science in Biomedical Engineering	**MSAAE**	Master of Science in Aeronautical and Astronautical Engineering
MS Ch E	Master of Science in Chemical Engineering	**MSABE**	Master of Science in Agricultural and Biological Engineering
MS Chem	Master of Science in Chemistry		
MS Civ E	Master of Science in Civil Engineering	**MSACC**	Master of Science in Accounting
MS Cp E	Master of Science in Computer Engineering	**MSAE**	Master of Science in Aeronautical Engineering
MS Eco	Master of Science in Economics		Master of Science in Aerospace Engineering
MS Econ	Master of Science in Economics		Master of Science in Agricultural Engineering
MS Ed	Master of Science in Education		Master of Science in Applied Economics
MS En E	Master of Science in Environmental Engineering		Master of Science in Architectural Engineering
MS Eng	Master of Science in Engineering		Master of Science in Art Education
MS Engr	Master of Science in Engineering	**MSAER**	Master of Science in Aerospace Engineering
MS Env E	Master of Science in Environmental Engineering	**MSAIS**	Master of Science in Accounting Information Systems
MS Int A	Master of Science in International Affairs		
MS Mat	Master of Science in Materials Engineering	**MSAM**	Master of Science in Advanced Management
MS Mat E	Master of Science in Materials Engineering		Master of Science in Applied Mathematics
MS Mat SE	Master of Science in Material Science and Engineering	**MSAP**	Master of Science in Applied Psychology
		MSAS	Master of Science in Architectural Studies
MS Math	Master of Science in Mathematics	**MSAT**	Master of Science in Advanced Technology
MS Met E	Master of Science in Metallurgical Engineering	**MSB**	Master of Science in Bible
MS Metr	Master of Science in Meteorology		Master of Science in Business
MS Mfg E	Master of Science in Manufacturing Engineering	**MSBA**	Master of Science in Business Administration
MS Mgt	Master of Science in Management	**MSBAE**	Master of Science in Biological and Agricultural Engineering
MS Min	Master of Science in Mining		Master of Science in Biosystems and Agricultural Engineering
MS Mt E	Master of Science in Materials Engineering		
MS Nsg	Master of Science in Nursing	**MSBE**	Master of Science in Biomedical Engineering
MS Pet E	Master of Science in Petroleum Engineering		Master of Science in Business Education
MS Phr	Master of Science in Pharmacy	**MSBENG**	Master of Science in Bioengineering
MS Phys	Master of Science in Physics	**MSBMS**	Master of Science in Basic Medical Science
MS Phys Op	Master of Science in Physiological Optics	**MSBS**	Master of Science in Biomedical Sciences
MS Poly	Master of Science in Polymers	**MSC**	Master of Science in Commerce
MS Psy	Master of Science in Psychology		Master of Science in Communication
			Master of Science in Computers
			Master of Science in Criminology
		MSCC	Master of Science in Christian Counseling
		MSCD	Master of Science in Communication Disorders
			Master of Science in Community Development

MSCE	Master of Science in Civil Engineering
	Master of Science in Clinical Epidemiology
	Master of Science in Computer Engineering
	Master of Science in Continuing Education
MSCEE	Master of Science in Civil and Environmental Engineering
MSCF	Master of Science in Computational Finance
MSCIS	Master of Science in Computer and Information Systems
	Master of Science in Computer Information Systems
MSCJ	Master of Science in Criminal Justice
MSCJA	Master of Science in Criminal Justice Administration
MSCLS	Master of Science in Clinical Laboratory Science
MSCNU	Master of Science in Clinical Nutrition
MSCP	Master of Science in Clinical Psychology
	Master of Science in Counseling Psychology
MSCRP	Master of Science in City and Regional Planning
	Master of Science in Community and Regional Planning
MSCS	Master of Science in Computer Science
MSCSD	Master of Science in Communication Sciences and Disorders
MSCSE	Master of Science in Computer and Systems Engineering
	Master of Science in Computer Science and Engineering
	Master of Science in Computer Systems Engineering
MSD	Master of Science in Dentistry
	Master of Science in Design
MSDD	Master of Software Design and Development
MSE	Master of Science Education
	Master of Science in Education
	Master of Science in Engineering
	Master of Software Engineering
	Master of Structural Engineering
MSE Mgt	Master of Science in Engineering Management
MSEAS	Master of Science in Earth and Atmospheric Sciences
MSEC	Master of Science in Economic Aspects of Chemistry
MSECE	Master of Science in Electrical and Computer Engineering
MSED	Master of Sustainable Economic Development
MSEE	Master of Science in Electrical Engineering
	Master of Science in Environmental Engineering
MSEH	Master of Science in Environmental Health
MSEL	Master of Science in Executive Leadership
	Master of Studies in Environmental Law
MSEM	Master of Science in Engineering Management
	Master of Science in Engineering Mechanics
	Master of Science in Engineering of Mines
	Master of Science in Environmental Management
MSENE	Master of Science in Environmental Engineering

MSES	Master of Science in Engineering Science
	Master of Science in Environmental Science
	Master of Science in Environmental Studies
MSESM	Master of Science in Engineering Science and Mechanics
MSESS	Master of Science in Exercise and Sport Studies
MSET	Master of Science in Engineering Technology
MSETM	Master of Science in Environmental Technology Management
MSEV	Master of Science in Environmental Engineering
MSF	Master of Science in Finance
	Master of Science in Forestry
MSFAM	Master of Science in Family Studies
MSFDE	Master of Science in Family Development Education
MSFM	Master of Financial Management
MSFOR	Master of Science in Forestry
MSFS	Master of Science in Financial Sciences
	Master of Science in Forensic Science
MSFT	Master of Science in Family Therapy
MSG	Master of Science in Gerontology
MSGC	Master of Science in Genetic Counseling
MSGFA	Master of Science in Global Financial Analysis
MSH	Master of Science in Health
	Master of Science in Hospice
MSH Ed	Master of Science in Health Education
MSHA	Master of Science in Health Administration
MSHCI	Master of Science in Human Computer Interaction
MSHCPM	Master of Science in Health Care Policy and Management
MSHCS	Master of Science in Human and Consumer Science
MSHES	Master of Science in Human Environmental Sciences
MSHFID	Master of Science in Human Factors in Information Design
MSHFS	Master of Science in Human Factors and Systems
MSHP	Master of Science in Health Professions
MSHR	Master of Science in Human Resources
MSHRM	Master of Science in Human Resource Management
MSHROD	Master of Science in Human Resources and Organizational Development
MSHS	Master of Science in Health Science
	Master of Science in Health Services
	Master of Science in Health Systems
MSHSA	Master of Science in Human Service Administration
MSHSE	Master of Science in Health Science Education
MSHT	Master of Science in History of Technology
MSI	Master of Science in Instruction
	Master of Science in Insurance
MSIA	Master of Science in Industrial Administration
MSIAM	Master of Science in Information Age Marketing

MSIB	Master of Science in International Business
MSIDM	Master of Science in Interior Design and Merchandising
MSIDT	Master of Science in Information Design and Technology
MSIE	Master of Science in Industrial Engineering Master of Science in International Economics
MSIL	Master of Science in International Logistics
MSIMC	Master of Science in Information Management and Communication Master of Science in Integrated Marketing Communications
MSIO	Master of Science in Industrial Optimization
MSIPC	Master of Science in Information Processing and Communications
MSIR	Master of Science in Industrial Relations
MSIS	Master of Science in Information Science Master of Science in Information Systems Master of Science in Interdisciplinary Studies
MSISM	Master of Science in Information Systems Management
MSIT	Master of Science in Industrial Technology Master of Science in Information Technology Master of Science in Instructional Technology
MSITM	Master of Science in Information Technology Management
MSJ	Master of Science in Journalism Master of Science in Jurisprudence
MSJBS	Master of Science in Japanese Business Studies
MSJJ	Master of Science in Juvenile Justice
MSJPS	Master of Science in Justice and Public Safety
MSJS	Master of Science in Jewish Studies
MSK	Master of Science in Kinesiology
MSL	Master of School Leadership Master of Science in Limnology Master of Studies in Law
MSLA	Master of Science in Legal Administration
MSLP	Master of Speech-Language Pathology
MSLS	Master of Science in Legal Studies Master of Science in Library Science Master of Science in Logistics Systems
MSLT	Master of Second Language Teaching
MSM	Master of Sacred Music Master of School Mathematics Master of Science in Management
MSMAE	Master of Science in Materials Engineering
MSMC	Master of Science in Management and Communications Master of Science in Mass Communications
MSMCS	Master of Science in Management and Computer Science
MSME	Master of Science in Mechanical Engineering
MSMFE	Master of Science in Manufacturing Engineering
MSMfSE	Master of Science in Manufacturing Systems Engineering

MSMGEN	Master of Science in Management and General Engineering
MSMI	Master of Science in Medical Illustration
MSMIS	Master of Science in Management Information Systems
MSMM	Master of Science in Manufacturing Management
MSMOT	Master of Science in Management of Technology
MSMS	Master of Science in Management Science
MSMSA	Master of Science in Management Systems Analysis
MSMSE	Master of Science in Manufacturing Systems Engineering Master of Science in Material Science and Engineering Master of Science in Mathematics and Science Education
MSMT	Master of Science in Medical Technology
MSN	Master of Science in Nursing
MSN(R)	Master of Science in Nursing (Research)
MSNA	Master of Science in Nurse Anesthesia
MSNE	Master of Science in Nuclear Engineering
MSNS	Master of Science in Natural Science
MSOB	Master of Science in Organizational Behavior
MSOD	Master of Science in Organizational Development
MSOL	Master of Science in Organizational Leadership
MSOM	Master of Science in Organization and Management Master of Science in Oriental Medicine
MSOR	Master of Science in Operations Research
MSOT	Master of Science in Occupational Technology Master of Science in Occupational Therapy
MSP	Master of Science in Pharmacy Master of Science in Planning Master of Speech Pathology
MSP Ex	Master of Science in Exercise Physiology
MSPA	Master of Science in Professional Accountancy Master of Science in Professional Accounting
MSPAS	Master of Science in Physician Assistant Studies
MSPC	Master of Science in Professional Communications
MSPE	Master of Science in Petroleum Engineering Master of Science in Physical Education
MSPFP	Master of Science in Personal Financial Planning
MSPG	Master of Science in Psychology
MSPH	Master of Science in Public Health
MSPHR	Master of Science in Pharmacy
MSPNGE	Master of Science in Petroleum and Natural Gas Engineering
MSPS	Master of Science in Pharmaceutical Science Master of Science in Psychological Services
MSPT	Master of Science in Physical Therapy
MSQSM	Master of Science in Quality Systems Management
MSR	Master of Science in Rehabilitation Sciences
MSRA	Master of Science in Recreation Administration

MSRC	Master of Science in Resource Conservation	**MTCM**	Master of Traditional Chinese Medicine
MSRE	Master of Science in Religious Education	**MTD**	Master of Training and Development
MSRMP	Master of Science in Radiological Medical Physics	**MTE**	Master of Teacher Education
MSRS	Master of Science in Recreational Studies	**MTEL**	Master of Telecommunications
MSRTM	Master of Science in Resort and Tourism Management	**MTESL**	Master in Teaching English as a Second Language
MSS	Master of Science in Sociology	**MTHM**	Master of Tourism and Hospitality Management
	Master of Science in Software	**MTI**	Master of Information Technology
	Master of Social Science	**MTLM**	Master of Transportation and Logistics Management
	Master of Social Services		
	Master of Special Studies	**MTM**	Master of Technology Management
	Master of Sports Science		Master of Telecommunications Management
MSSA	Master of Science in Social Administration		Master of the Teaching of Mathematics
MSSE	Master of Science in Software Engineering	**MTMH**	Master of Tropical Medicine and Hygiene
MSSI	Master of Science in Strategic Intelligence	**MTOM**	Master of Traditional Oriental Medicine
MSSL	Master of Science in Speech and Language	**MTP**	Master of Transpersonal Psychology
MSSM	Master of Science in Systems Management	**MTPW**	Master of Technical and Professional Writing
MSSPA	Master of Science in Student Personnel Administration	**MTS**	Master of Teaching Science
			Master of Theological Studies
MSSS	Master of Science in Systems Science	**MTSC**	Master of Technical and Scientific Communication
MSSW	Master of Science in Social Work	**MTSE**	Master of Telecommunications and Software Engineering
MST	Master of Science in Taxation		
	Master of Science in Teaching	**MTX**	Master of Taxation
	Master of Science in Technology	**MUA**	Master of Urban Affairs
	Master of Science in Telecommunications	**MUD**	Master of Urban Design
	Master of Science in Transportation	**MUP**	Master of Urban Planning
	Master of Science Teaching	**MUPDD**	Master of Urban Planning, Design, and Development
	Master of Science Technology		
	Master of Systems Technology	**MUPP**	Master of Urban Planning and Policy
MST Ch	Master of Science in Textile Chemistry	**MURP**	Master of Urban and Regional Planning
MSTA	Master of Science in Statistics		Master of Urban and Rural Planning
MSTC	Master of Science in Telecommunications	**MURPL**	Master of Urban and Regional Planning
MSTD	Master of Science in Training and Development	**MUS**	Master of Urban Studies
MSTE	Master of Science in Technical Education	**Mus AD**	Doctor of Musical Arts
	Master of Science in Textile Engineering	**Mus Doc**	Doctor of Music
	Master of Science in Transportation Engineering	**Mus M**	Master of Music
MSTM	Master of Science in Technical Management	**MVE**	Master of Vocational Education
	Master of Science in Technology Management	**MVT Ed**	Master of Vocational and Technical Education
MSUD	Master of Science in Urban Design	**MVTE**	Master of Vocational-Technical Education
MSUESM	Master of Science in Urban Environmental Systems Management	**MWC**	Master of Wildlife Conservation
		MWPS	Master of Wood and Paper Science
MSVE	Master of Science in Vocational Education	**MWR**	Master of Water Resources
MSW	Master of Social Work	**MWS**	Master of Women's Studies
MSWE	Master of Software Engineering	**MZS**	Master of Zoological Science
MSWREE	Master of Science in Water Resources and Environmental Engineering	**Nav Arch**	Naval Architecture
		Naval E	Naval Engineer
MT	Master of Taxation	**ND**	Doctor of Naturopathic Medicine
	Master of Teaching		Doctor of Nursing
	Master of Technology		
	Master of Textiles	**NE**	Nuclear Engineer
MTA	Master of Arts in Teaching	**NPMC**	Nonprofit Management Certificate
	Master of Tax Accounting	**Nuc E**	Nuclear Engineer
	Master of Teaching Arts		
	Master of Tourism Administration		
MTC	Master of Technical Communications		

Ocean E	Ocean Engineer		**SLPD**	Doctor of Speech-Language Pathology
OD	Doctor of Optometry		**SLS**	Specialist in Library Science
OTD	Doctor of Occupational Therapy		**SM**	Master of Science
PD	Professional Diploma		**SM Arch S**	Master of Science in Architectural Studies
PDD	Professional Development Degree		**SM Vis S**	Master of Science in Visual Studies
PE Dir	Director of Physical Education		**SMBT**	Master of Science in Building Technology
PED	Doctor of Physical Education		**SP**	Specialist Degree
PGC	Post-Graduate Certificate		**Sp C**	Specialist in Counseling
Ph L	Licentiate of Philosophy		**Sp Ed**	Specialist in Education
Pharm D	Doctor of Pharmacy		**Sp Ed S**	Special Education Specialist
PhD	Doctor of Philosophy		**SPS**	School Psychology Specialist
PMBA	Professional Master of Business Administration		**Spt**	Specialist Degree
PMC	Post Master's Certificate		**SSP**	Specialist in School Psychology
PMSA	Professional Master of Science in Accounting		**STB**	Bachelor of Sacred Theology
Psy D	Doctor of Psychology		**STD**	Doctor of Sacred Theology
Psy M	Master of Psychology		**STL**	Licentiate of Sacred Theology
Psy S	Specialist in Psychology		**STM**	Master of Sacred Theology
Re D	Doctor of Recreation		**Th D**	Doctor of Theology
Re Dir	Director of Recreation		**Th M**	Master of Theology
Rh D	Doctor of Rehabilitation		**TMBA**	Transnational Master of Business Administration
S Psy S	Specialist in Psychological Services		**V Ed S**	Vocational Education Specialist
SAS	School Administrator and Supervisor		**VMD**	Doctor of Veterinary Medicine
Sc D	Doctor of Science		**WEMBA**	Weekend Executive Master of Business Administration
Sc M	Master of Science		**XMA**	Executive Master of Arts
SCCT	Specialist in Community College Teaching		**XMBA**	Executive Master of Business Administration
SD	Doctor of Science			
SJD	Doctor of Juridical Science			

Indexes

There are three indexes in this section. The first, Index of In-Depth Descriptions and Announcements, gives page references for all programs that have chosen to place In-Depth Descriptions and Announcements in this volume. It is arranged alphabetically by institution; within institutions, the arrangement is alphabetical by subject area. It is not an index to all programs in the book's directories of profiles; readers must refer to the directories themselves for profile information on programs that have not submitted the additional, more individualized statements. The second index, Index of Directories and Subject Areas in Books 2–6, gives book references for the directories in Books 2–6, for example, "Industrial Design—Book 2," and also includes cross-references for subject area names not used in the directory structure, for example, "Computing Technology (*see* Computer Science)." The third index, Index of Directories and Subject Areas in This Book, gives page references for the directories in this volume and cross-references for subject area names not used in this volume's directory structure.

Index of In-Depth Descriptions and Announcements

Index of Directories and Subject Areas in Books 2–6

Following is an alphabetical listing of directories and subject areas in Books 2–6. Also listed are cross-references for subject area names not used in the directory structure of the guides, for example, "Arabic (*see* Near and Middle Eastern Languages)."

Accounting—Book 6

Acoustics—Book 4

Actuarial Science—Book 6

Acupuncture (*see* Oriental Medicine and Acupuncture)

Administration (*see* Arts Administration; Business Administration and Management; Educational Administration; Health Services Management and Hospital Administration; Industrial Administration; Public Policy and Administration; Sports Administration)

Adult Education—Book 6

Adult Nursing (*see* Medical/Surgical Nursing)

Advanced Practice Nursing—Book 6

Advertising and Public Relations—Book 6

Aeronautical Engineering (*see* Aerospace/Aeronautical Engineering)

Aerospace/Aeronautical Engineering—Book 5

Aerospace Studies (*see* Aerospace/Aeronautical Engineering)

African-American Studies—Book 2

African Languages and Literatures (*see* African Studies)

African Studies—Book 2

Agribusiness (*see* Agricultural Economics and Agribusiness)

Agricultural Economics and Agribusiness—Book 2

Agricultural Education—Book 6

Agricultural Engineering—Book 5

Agricultural Sciences—Book 4

Agronomy and Soil Sciences—Book 4

Alcohol Abuse Counseling (*see* Drug and Alcohol Abuse Counseling; Counselor Education)

Allied Health—Book 6

Allopathic Medicine—Book 6

American Indian Studies (*see* American Studies)

American Studies—Book 2

Analytical Chemistry—Book 4

Anatomy—Book 3

Animal Behavior—Book 3

Animal Sciences—Book 4

Anthropology—Book 2

Applied Arts and Design—Book 2

Applied History (*see* Public History)

Applied Mathematics—Book 4

Applied Mechanics (*see* Mechanics)

Applied Physics—Book 4

Applied Sciences (*see* Engineering and Applied Sciences)

Applied Statistics (*see* Statistics)

Aquaculture—Book 4

Arab Studies (*see* Near and Middle Eastern Studies)

Arabic (*see* Near and Middle Eastern Languages)

Archaeology—Book 2

Architectural Engineering—Book 5

Architectural History—Book 2

Architecture—Book 2

Archives Administration (*see* Public History)

Area and Cultural Studies (*see* African-American Studies; African Studies; American Studies; Asian Studies; Canadian Studies; East European and Russian Studies; Hispanic Studies; Jewish Studies; Latin American Studies; Near and Middle Eastern Studies; Northern Studies; Western European Studies; Women's Studies)

Art Education—Book 6

Art/Fine Arts—Book 2

Art History—Book 2

Arts Administration—Book 2

Art Therapy—Book 2

Artificial Intelligence/Robotics—Book 5

Asian-American Studies (*see* American Studies)

Asian Languages—Book 2

Asian Studies—Book 2

Astronautical Engineering (*see* Aerospace/Aeronautical Engineering)

Astronomy—Book 4

Astrophysical Sciences (*see* Astrophysics; Meteorology; Atmospheric Sciences; Planetary Sciences)

Astrophysics—Book 4

Athletics Administration (*see* Physical Education; Exercise and Sports Science; Kinesiology and Movement Studies)

Atmospheric Sciences—Book 4

Audiology (*see* Communication Disorders)

Bacteriology—Book 3

Banking (*see* Finance and Banking)

Behavioral Genetics (*see* Biopsychology)

Behavioral Sciences (*see* Biopsychology; Neuroscience; Psychology; Zoology)

Bible Studies (*see* Religion; Theology)

Bilingual and Bicultural Education (*see* Multilingual and Multicultural Education)

Biochemical Engineering—Book 5

Biochemistry—Book 3

Bioengineering—Book 5

Bioethics—Book 6

Biological and Biomedical Sciences—Book 3

Biological Chemistry (*see* Biochemistry)

Biological Engineering (*see* Bioengineering)

Biological Oceanography (*see* Marine Biology; Marine Sciences; Oceanography)

Biomathematics (*see* Biometrics)

Biomedical Engineering—Book 5

Biometrics—Book 4

Biophysics—Book 3

Biopsychology—Book 3

Biostatistics—Book 4

Biotechnology—Book 5

Black Studies (*see* African-American Studies)

Botany and Plant Sciences—Book 3

Breeding (*see* Animal Sciences; Botany and Plant Sciences; Genetics; Horticulture)

Broadcasting (*see* Communication; Media Studies)

Business Administration and Management—Book 6

Business Education—Book 6

Canadian Studies—Book 2

Index of Directories and
Subject Areas in This Book

NOTES

NOTES

NOTES

NOTES

NOTES

NOTES

NOTES

NOTES

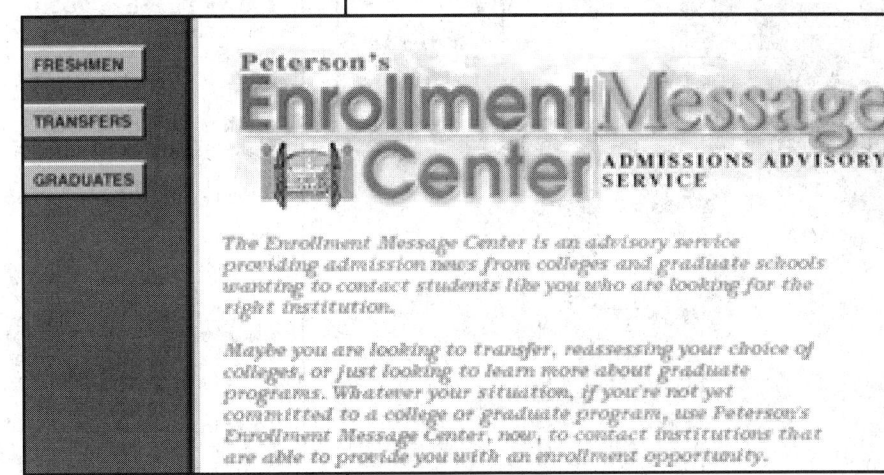